CHILTON'S

TRUCK and VAN REPAIR MANUAL 1979-86

Publisher Editor-In-Chief Kerry A. Freeman, S.A.E.

Managing Editors Peter M. Conti, Jr., W. Calvin Settle, Jr., S.A.E.

Assistant Managing Editor Nick D'Andrea

Senior Editors Debra Gaffney, Ken Grabowski, A.S.E., S.A.E.
Michael L. Grady, Richard J. Rivele, S.A.E.
Richard T. Smith, Jim Taylor
Ron Webb

Project Managers Benjamin E. Greisler, A.S.E., Martin J. Gunther, Jeffrey M. Hoffman
James Steele

Editorial Staff Peter A. Bilotta, A.S.E., Lawrence C. Braun, S.A.E., A.S.C.,
Thomas P. Browne III, Hugh J. Brulliea, Dean G. Callahan,
Michael M. Carroll, William C. Cottman, A.S.E., Robert B. Day,
Jr., Paul DeGuiseppi, A.S.E., Robert F. Dougherty, Jr., Robert E.
Doughten, Sam Fiorani, Andrew J. Folz, A.S.E., Edward J.
Giacomucci, A.S.E., Jacques Gordon, Neil Leonard, Kevin Maher,
Robert McAnally, Raymond K. Moore, Craig P. Nangle, A.S.E.,
Charles Ramsey, Roy Ripple, A.S.E., John H. Rutter, Don
Schnell, A.S.E., S.A.E., Larry E. Stiles, Anthony Tortorici, A.S.E.,
S.A.E., Thom Young

Assistant Production Manager Andrea M. Steiger

Production Assistants Marsha Park Herman, Monica Santa Maria, Margaret Stoner

Mechanical Artists Lisa Gressen, Kim Hayes

Director of Manufacturing Mike D'Imperio

OFFICERS

President, Chilton Enterprises David S. Loewith

Senior Vice President Ronald A. Hoxter

CHILTON BOOK COMPANY

ONE OF THE **DIVERSIFIED PUBLISHING COMPANIES,**
A PART OF **CAPITAL CITIES/ABC, INC.**

Manufactured in USA
© 1986 Chilton Book Company
Chilton Way, Radnor, PA 19089
ISBN 0-8019-7655-3
ISSN 0742-0315

0123456789 987654

HOW TO USE THIS MANUAL

This manual is arranged in two sections:

Truck Section

Truck sections are grouped by manufacturer (Chevrolet, Ford, etc.) and arranged in alphabetical order. The text and illustrations that comprise the service procedures in each Truck Section are arranged in the following order of systems and components: Tune-up, Engine Electrical, Engine Mechanical, Engine Lubrication, Engine Cooling, Emission Controls, Fuel System, Manual Transmission, Clutch, Automatic Transmission, Transaxle, Drive Axle, Rear Suspension, Front Suspension, Steering, Brakes, Heater, Radio, Windshield Wiper, Instrument Panel and Fuse Box.

Specification charts are always located at the front of each section. All illustrations are located as close as possible to the pertinent text. Procedures are for all models in the particular section unless specifically noted otherwise.

Unit Repair Section

The Unit Repair Section contains troubleshooting and overhaul procedures for the major components and systems of your truck. This portion of the book is intended to be used in conjunction with the Truck Sections.

For example: If your Truck's engine is misfiring and you do not know the cause, use the "Troubleshooting" portion of the Unit Repair Section to find the cause and its remedy. If the cause should prove to be defective piston rings which are allowing oil to foul the spark plugs, the remedy is to overhaul the engine. Then turn to the proper Truck Section to find the procedure for removing the engine from the Truck. After you have removed the engine, turn to the "Engine Rebuilding" chapter in the Unit Repair Section and follow the steps listed there to overhaul the engine.

Every major Unit Repair Section contains an Identification or Application chart to correlate the information contained in that section. The sections are usually arranged by brands, manufacturers or types of components rather than models of trucks.

All overhaul procedures in the Unit Repair Section begin with the component removed from the Truck. The reason for this division of material is an economic one. The steps involved in overhauling an engine are virtually the same for all engines. However, the operation of removing the engine from the truck varies greatly from model to model. By combining where possible, and separating where necessary, we are able to publish the maximum amount of information.

Locating Information

The Table of Contents, at the front of the book, lists the beginning of each Truck and Unit Repair Section in the manual.

The Index, also at the front of the book, is a comprehensive listing of all major mechanical sections and systems for every section in the book. The Index contains listings for Truck Sections as well as for corresponding Unit Repair Sections. Truck Section pages are prefixed with the letter "T". Unit Repair Sections are prefixed with the letter "U".

To find where a particular Truck Section is located in the book, you need only look in the Table of Contents. Once you have found the proper section, you may wish to find where specific procedures are located in that section. Turn to the Index and read across the top of the page until you reach the appropriate Truck Section. When the proper manufacturer's column has been found, read down the side column to the procedure or system for which you are looking. The intersection of the two columns will provide the page number(s) where the procedure is located.

Safety Notice

Proper service and repair procedures are vital to the safe, reliable operation of all motor vehicles, as well as the personal safety of those performing repairs. This manual outlines procedures for servicing and repairing vehicles using safe effective methods. The procedures contain many NOTES, CAUTIONS and WARNINGS which should be followed along with standard safety procedures to eliminate the possibility of personal injury or improper service which could damage the vehicle or compromise its safety.

It is important to note that repair procedures and techniques, tools and parts for servicing motor vehicles, as well as the skill and experience of the individual performing the work vary widely. It is not possible to anticipate all of the conceivable ways or conditions under which vehicles may be serviced, or to provide cautions as to all of the possible hazards that may result. Standard and accepted safety precautions and equipment should be used when handling toxic or flammable fluids, and safety goggles or other protection should be used during cutting, grinding, chiseling, prying, or any other process that can cause material removal or projectiles.

Some procedures require the use of tools specially designed for a specific purpose. Before substituting another tool or procedure, you must be completely satisfied that neither your personal safety, nor the performance of the vehicle will be endangered.

Part Numbers

Part numbers listed in this book are not recommendations by Chilton for any product by brand name. They are references that can be used with interchange manuals and aftermarket supplier catalogs to locate each brand supplier's discrete part number.

Although information in this manual is based on industry sources and is as complete as possible at the time of publication, the possibility exists that some truck manufacturers made later changes which could not be included here. While striving for total accuracy, Chilton Book Company cannot assume responsibility for any errors, changes, or omissions that may occur in the compilation of this data.

TRUCK SECTION

UNIT REPAIR SECTION

INDEX

R & I: Removal and Installation

Chevrolet/GMC	Datsun/Nissan	Dodge/Plymouth	Ford	International	Isuzu/LUV	Jeep	Mazda/Courier	Mitsubishi/D50/Arrow	Toyota	Volkswagen	
U5	U5	U5	U5	U5	U5	U5	U5	U5	U5	U5	Maintenance
T81	T125	T181	T303	T340	T372	T420	T477	T519	T596	T623	Manual Steering Gear
—	—	T162	—	—	—	T407	—	—	—	T618	Manual Transaxle R & I
—	—	U360	—	—	—	—	—	—	—	—	Manual Transaxle Overhaul
T57	T110	T162	T256	T333	T360	T407	T463	T504	T573	—	Manual Transmission R &
U352	U414	U362	U371	U352	U415	U354	U420	U431	U435	—	Manual Trans. Overhaul
T82	T126	T183	T306	T341	T374	T423	T480	T520	T600	T625	Master Cylinder R & I
U581	U581	U581	U581	U581	U581	U581	U581	U581	U581	U581	Master Cylinder Overhaul
T61	T113	T167	T276	T336	T362	T409	T467	T508	T578	T620	Neutral Safety Switch
T46	T105	T156	T240	T330	T356	T398	T453	T500	T558	T617	Oil Pan
T48	T105	T157	T245	T330	T356	T398	T454	T500	T558	T617	Oil Pump
T83	T128	T183	T306	T342	T374	T424	T481	T521	T601	T625	Parking Brake
T45	T104	T156	T239	T330	T356	T398	T452	T496	T551	T616	Piston and Connecting Rod
T81	T126	T181	T304	T340	T372	T420	T479	T520	T597	—	Power Steering Gear
T80	T126	T182	T303	T340	—	T420	T479	T520	T596	—	Power Steering Pump
T83	T128	T184	T313	T342	T375	T426	T482	T522	T603	T627	Radio
T73	T114	T175	T289	T339	T363	T416	T470	T509	T584	—	Rear Axle
T48	T105	T158	T246	T330	T356	T399	T455	T500	T559	T617	Rear Main Oil Seal
T77	T123	T179	T299	T339	T370	T422	T475	T518	T586	T621	Rear Suspension
T20	T96	T144	T212	T326	T349	T385	T437	T491	T536	T611	Regulator
T28	T100	T151	T224	T328	T352	T393	T441	T493	T543	T613	Rocker Arms
T74	T118	T180	T294	T341	T365	T422	T475	T513	T587	T621	Shock Absorbers
T1	T87	T131	T187	T321	T343	T377	T429	T485	T523	T605	Specifications
T77	T123	T179	T299	T336	T370	T422	T475	T518	T586	T621	Springs
T20	T97	T145	T212	T326	T350	T386	T437	T491	T536	T612	Starter
T78	T124	T180	T302	T340	T371	T419	T475	T518	T595	T622	Steering
T51	T107	T160	T249	T331	T357	T400	T458	T502	T563	T617	Thermostat
T39	T102	T152	T230	—	T354	T395	T447	T495	T548	T615	Timing Belt
T37	T101	T153	T230	T328	T354	T395	T444	T495	T548	T615	Timing Belt/Chain Cover
T39	T102	T153	T231	T328	T354	T395	T446	T495	T547	—	Timing Chain
T64	T113	T164	T277	T334	T361	T405	—	T504	T574	—	Transfer Cases
U749	U749	U749	U749	U749	U749	U749	U749	U749	U749	U749	Troubleshooting
T16	T92	T136	T206	T323	T346	T381	T433	T488	T529	T608	Tune-Up
—	—	—	—	—	—	T390	—	T495	T556	—	Turbocharger Service
T79	T124	T180	T302	T340	T371	T419	T475	T519	T595	T622	Turn Signal Switch
T76	T119	T177	—	T336	T366	T413	T473	T514	T593	—	Upper Control Arm
T28	T93	T138	T210	T324	T347	T381	T435	T488	T531	T609	Valve Adjustment
T51	T107	T159	T248	T331	T357	T400	T456	T502	T563	T618	Water Pump
U335	U335	U335	U335	U335	U335	U335	U335	U335	U335	U335	Wheel Alignment
T70	T126	T172	T280	T337	T367	T413	T474	T516	T594	T621	Wheel Bearings (Front)
T73	T114	T175	T289	T339	T363	T418	T469	T509	T584	T620	Wheel Bearings (Rear)
U585	T127	U585	U585	T341	T374	T423	U585	T521	U585	U585	Wheel Cylinders
T84	T129	T185	T314	T342	T376	T426	T482	T522	T603	T626	Windshield Wipers

Chevrolet/GMC

Pickups, Vans, Blazer, Jimmy, Suburban, S-Series

GENERAL ENGINE SPECIFICATIONS

Year	Engine No. Cyl Displacement (cu. in.)	Carburetor Type	Horsepower @ rpm ■	Torque @ rpm (ft lbs) ■	Bore and Stroke (in.)	Compression Ratio	Oil Pressure @ 2000 rpm
'79	6-250 LD	2 bbl	130 @ 3800	210 @ 2400	3.870 × 3.530	8.3:1	40–60
	6-250 Calif	2 bbl	125 @ 4000	205 @ 2000	3.870 × 3.530	8.3:1	40–60
	6-250 HD	2 bbl	130 @ 4000	205 @ 2000	3.870 × 3.530	8.3:1	40–60
	6-292	1 bbl	115 @ 3400	215 @ 1600	3.870 × 4.120	7.8:1	40–60
	8-305	2 bbl	140 @ 4000	240 @ 2000	3.740 × 3.480	8.4:1	45
	8-350 LD	4 bbl	165 @ 3600	270 @ 2000	4.000 × 3.480	8.2:1	45
	8-350 Hi Alt	4 bbl	155 @ 3600	260 @ 2000	4.000 × 3.480	8.2:1	45
	8-350 HD	4 bbl	165 @ 3800	255 @ 2800	4.000 × 3.480	8.3:1	45
	8-350	Diesel	120 @ 3600	220 @ 1600	4.057 × 3.385	22.5:1	35
	8-400 HD	4 bbl	180 @ 3600	310 @ 2400	4.125 × 3.750	8.2:1	40
	8-454 LD	4 bbl	205 @ 3600	335 @ 2800	4.250 × 4.000	8.0:1	40
	8-454 HD	4 bbl	210 @ 3800	340 @ 2800	4.250 × 4.000	7.9:1	40
'80	6-250	2 bbl	130 @ 4000	210 @ 2000	3.870 × 3.530	8.3:1	40–60
	6-250 Calif.	2 bbl	130 @ 4000	205 @ 2000	3.870 × 3.530	8.3:1	40–60
	6-292	1 bbl	115 @ 3400	215 @ 1600	3.870 × 4.120	7.8:1	40–60
	8-305	2 bbl	135 @ 4200	235 @ 2400	3.740 × 3.480	8.5:1	45
	8-350 LD	4 bbl	175 @ 4000	275 @ 2400	4.000 × 3.480	8.2:1	45
	8-350 LD Calif	4 bbl	170 @ 4000	275 @ 2000	4.000 × 3.480	8.2:1	45
	8-350 HD	4 bbl	165 @ 3800	255 @ 2800	4.000 × 3.480	8.3:1	45
	8-350	Diesel	125 @ 3600	225 @ 1600	4.057 × 3.385	22.5:1	35
	8-400 HD	4 bbl	180 @ 3600	310 @ 2400	4.125 × 3.750	8.3:1	40
	8-454 HD	4 bbl	210 @ 3800	340 @ 2800	4.250 × 4.000	7.9:1	40
'81	6-250	2 bbl	130 @ 4000	210 @ 2000	3.870 × 3.530	8.3:1	40–60
	6-250 Calif	2 bbl	130 @ 4000	205 @ 2000	3.870 × 3.530	8.3:1	40–60
	6-292	1 bbl	115 @ 3400	215 @ 1600	3.870 × 4.120	7.8:1	40–60
	8-305	2 bbl	135 @ 4200	235 @ 2400	3.740 × 3.480	8.5:1	45
	8-305	4 bbl	155 @ 4400	252 @ 2400	3.740 × 3.480	9.2:1	45
	8-350 LD	4 bbl	175 @ 4000	275 @ 2000	4.000 × 3.480	8.2:1	45
	8-350 HD	4 bbl	165 @ 3800	255 @ 2800	4.000 × 3.480	8.3:1	45
	8-350	Diesel	125 @ 3600	225 @ 1600	4.057 × 3.385	22.5:1	35
	8-454	4 bbl	210 @ 3800	340 @ 2800	4.250 × 4.000	7.9:1	40

GENERAL ENGINE SPECIFICATIONS

Year	Engine No. Cyl Displacement (cu. in.)	Carburetor Type	Horsepower @ rpm ■	Torque @ rpm (ft lbs) ■	Bore and Stroke (in.)	Compression Ratio	Oil Pressure @ 2000 rpm
'82	4-119	2 bbl	84 @ 4600	101 @ 3000	3.43 × 3.23	8.4:1	57
	4-121	2 bbl	83 @ 4600	108 @ 2400	3.50 × 3.15	9.3:1	45
	4-137	Diesel	58 @ 4300	93 @ 2200	3.46 × 3.62	21.0:1	55
	6-173	2 bbl	110 @ 4800	148 @ 2000	3.50 × 2.99	8.5:1	45
	6-250	2 bbl	130 @ 4000	210 @ 2000	3.870 × 3.530	8.3:1	40–60
	6-292	1 bbl	115 @ 3400	215 @ 1600	3.870 × 4.120	7.8:1	40–60
	8-305	4 bbl	140 @ 4200	240 @ 2400	3.740 × 3.480	8.5:1	45
	8-305	4 bbl	155 @ 4400	252 @ 2100	3.740 × 3.480	9.2:1	45
	8-350 LD	4 bbl	175 @ 4000	275 @ 2000	4.000 × 3.480	8.2:1	45
	8-350 HD	4 bbl	165 @ 3800	255 @ 2800	4.000 × 3.480	8.3:1	45
	8-379	Diesel	130 @ 3600	240 @ 2000	3.900 × 3.800	21.5:1	45
	8-454	4 bbl	210 @ 3800	340 @ 2800	4.250 × 4.000	7.9:1	40
'83	4-119	2 bbl	84 @ 4600	101 @ 3000	3.43 × 3.23	8.4:1	57
	4-121	2 bbl	83 @ 4600	108 @ 2400	3.50 × 3.15	9.3:1	45
	4-137	Diesel	58 @ 4300	93 @ 2200	3.46 × 3.62	21.0:1	55
	6-173	2 bbl	110 @ 4800	148 @ 2000	3.50 × 2.99	8.5:1	45
	6-250	2 bbl	130 @ 4000	210 @ 2000	3.870 × 3.530	8.3:1	40–60
	6-292	1 bbl	115 @ 3400	215 @ 1600	3.870 × 4.120	7.6:1	40–60
	8-305	4 bbl	140 @ 4200	240 @ 2400	3.740 × 3.480	8.5:1	45
	8-305	4 bbl	155 @ 4400	252 @ 2100	3.740 × 3.480	9.2:1	45
	8-350 LD	4 bbl	175 @ 4000	275 @ 2000	4.000 × 3.480	8.2:1	45
	8-350 HD	4 bbl	165 @ 3800	255 @ 2800	4.000 × 3.480	8.3:1	45
	8-379	Diesel	130 @ 3600	240 @ 2000	3.980 × 3.800	21.5:1	45
	8-454 HD	4 bbl	210 @ 3800	340 @ 2800	4.250 × 4.000	7.9:1	40
'84	4-119	2 bbl	84 @ 4600	101 @ 3000	3.43 × 3.23	8.4:1	57
	4-121	2 bbl	83 @ 4600	108 @ 2400	3.50 × 3.15	9.3:1	45
	4-137	Diesel	58 @ 4300	93 @ 2200	3.46 × 3.62	21.0:1	55
	6-173	2 bbl	110 @ 4800	148 @ 2000	3.50 × 2.99	8.5:1	45
	6-250	2 bbl	130 @ 4000	210 @ 2000	3.870 × 3.530	8.3:1	40–60
	6-292	1 bbl	115 @ 3400	215 @ 1600	3.870 × 4.120	7.8:1	40–60
	8-305	4 bbl	140 @ 4200	240 @ 2400	3.740 × 3.480	8.5:1	45
	8-305	4 bbl	155 @ 4400	252 @ 2100	3.740 × 3.480	9.2:1	45
	8-350 LD	4 bbl	175 @ 4000	275 @ 2000	4.000 × 3.480	8.2:1	45
	8-350 HD	4 bbl	165 @ 3800	255 @ 2800	4.000 × 3.480	8.3:1	45
	8-379	Diesel	130 @ 3600	240 @ 2000	3.980 × 3.800	21.5:1	45
	8-454 HD	4 bbl	210 @ 3800	340 @ 2800	4.250 × 4.000	7.9:1	40
'85	4-119	2bbl	82 @ 4600	101 @ 3000	3.43 × 3.23	8.4:1	57
	4-137	Diesel	62 @ 4300	96 @ 2200	3.46 × 3.26	21.0:1	55
	4-151	TBI	92 @ 4400	132 @ 2800	4.00 × 3.00	9.0:1	40–60
	6-173	2 bbl	110 @ 4800	145 @ 2100	3.50 × 2.99	8.5:1	45
	6-262	4 bbl	150 @ 4000	225 @ 2400	4.00 × 3.480	9.3:1	40–60
	8-305	4 bbl	155 @ 4000	245 @ 1600	3.740 × 3.480	8.6:1	45
	8-305	4 bbl	160 @ 4400	235 @ 2000	3.740 × 3.480	9.2:1	45
	8-350	4 bbl	165 @ 3800	275 @ 1600	4.00 × 3.480	8.2:1	45
	8-379	Diesel	130 @ 3600	240 @ 2000	3.980 × 3.800	21.5:1	45
	8-454	4 bbl	210 @ 3800	340 @ 2800	4.250 × 4.000	8.0:1	40

GENERAL ENGINE SPECIFICATIONS

Year	Engine No. Cyl Displacement (cu. in.)	Carburetor Type	Horsepower @ rpm ∎	Torque @ rpm (ft lbs) ∎	Bore and Stroke (in.)	Compression Ratio	Oil Pressure @ 2000 rpm
'86	4-119	2 bbl	82 @ 4600	101 @ 3000	3.43 × 3.23	8.4:1	57
	4-137	Diesel	62 @ 4300	96 @ 2200	3.46 × 3.62	21.0:1	55
	4-151	TBI	92 @ 4400	134 @ 2800	4.00 × 3.00	9.0:1	40–60
	6-173	2 bbl	110 @ 4800	145 @ 2100	3.50 × 2.99	8.5:1	45
	6-262	4 bbl	150 @ 4000	225 @ 2400	4.00 × 3.480	9.3:1	40–60
	8-305	4 bbl	155 @ 4400	245 @ 1600	4.00 × 3.480	8.6:1	45
	8-305	4 bbl	160 @ 4400	235 @ 2000	4.00 × 3.480	9.2:1	45
	8-350 LD	4 bbl	165 @ 3800	275 @ 1600	4.00 × 3.480	8.2:1	45
	8-350 HD	4 bbl	155 @ 4000	240 @ 2800	4.00 × 3.480	8.3:1	45
	8-379	Diesel	130 @ 3600	240 @ 2600	3.980 × 3.800	21.5:1	45
	8-454	4 bbl	230 @ 3800	360 @ 2800	4.250 × 4.000	8.0:1	40

∎ Horsepower and torque are SAE net figures. They are measured at the rear of the transmission with all accessories installed and operating. Since the figures vary when a given engine is installed in different models, some are representative rather than exact.
FI—Fuel Injection
TBI—Throttle body injection

CAPACITIES
Blazer and Jimmy

Year	Engine No. Cyl. Displacement (cu in.)	Engine Crankcase (qts) incl. Filter	Transmission Pts To Refill After Draining Manual 3-Speed	4-Speed	Automatic	Transfer Case (Pts)	Drive Axle (pts) Front/Rear	Gasoline Tank (gals) Std/Opt	Cooling System ∎ (qts) With Heater	A/C	With HD Cooling
'79	6-250	5	3.0	8.0	5.0	5.0 ④	5/3.5 ⑤	25/31	15	15.5	15.5
	8-305	5	3.0	8.0	5.0	5.0 ④	5/3.5 ⑤	25/31	17.5	17.5	17.5
	8-350	5	3.0	8.0	5.0	5.0 ④	5/3.5 ⑤	25/31	17.5	18	18
	8-400	5	—	—	5.0	5.0 ④	5/3.5 ⑤	25/31	18	19	19
'80–'81	6-250	5	3.0	8.0	5.0	5.0	5/3.5 ⑤	25/31	15	15.5	15.5
	8-305	5	3.0	8.0	5.0	5.0	5/3.5 ⑤	25/31	17.5	17.5	17.5
	8-350	5	3.0	8.0	5.0	5.0	5/3.5 ⑤	25/31	17.5	18	18
'82	6-250	5	2.0	8.0	6.0	5.0	5/ ⑦	25/31	15.0	15.5	—
	8-305	5	2.0	8.0	6.0	5.0	5/ ⑦	25/31	17.5	18.0	—
	8-350	5	2.0	8.0	6.0 ⑥	5.0	5/ ⑦	25/31	17.5	18.0	—
	8-379	7	—	8.0	6.0 ⑥	5.0	5/ ⑦	27/32	24.8	24.8	—
'83	6-250	5	3.0	8.0	6.0	5.0	5/ ⑦	25/31	15.0	15.5	—
	8-305	5	3.0	8.0	6.0	5.0	5/ ⑦	25/31	17.5	18.0	—
	8-350	5	3.0	8.0	6.0 ⑥	5.0	5/ ⑦	25/31	17.5	18.0	—
	8-379	7	—	8.0	6.0 ⑥	5.0	5/ ⑦	27/32	24.5	24.5	—
	8-454	7 ③	—	8.0	6.0 ⑥	5.0	5/ ⑦	25/31	23	24.5	—
'84	6-250	5	3.0	8.0	6.0	5.0	5/ ⑦	25/31	15.0	15.5	—
	8-305	5	3.0	8.0	6.0	5.0	5/ ⑦	25/31	17.5	18.0	—
	8-350	5	3.0	8.0	6.0 ⑥	5.0	5/ ⑦	25/31	17.5	18.0	—
	8-379	8	—	8.0	6.0 ⑥	5.0	5/ ⑦	27/32	24.5	24.5	—
	8-454	7 ③	—	8.0	6.0 ⑥	5.0	5/ ⑦	25/31	23	24.5	—

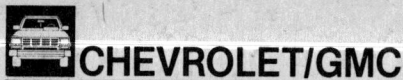

CAPACITIES
Blazer and Jimmy

Year	Engine No. Cyl. Displacement (cu in.)	Engine Crankcase (qts) incl. Filter	Transmission Pts To Refill After Draining — Manual 3-Speed	4-Speed	Automatic	Transfer Case (Pts)	Drive Axle (pts) Front/Rear	Gasoline Tank (gals) Std/Opt	Cooling System (qts) With Heater	A/C	With HD Cooling
'85–'86	8-305, 350	5	3.0	8.0	[1]	5.0	[2][3]	25/31	17.5	18	—
	8-379	7	3.0	8.0	[1]	5.0	[2][3]	27/32/41	23	24.5	—
	8-454	6	3.0	8.0	[1]	5.0	[2][3]	25/31/40	23	24.5	—

■Automatic transmission models have either the A/C or HD radiator; capacity may be increased on trucks with 4.11:1 axle ratios; when two figures are separated by a slash, the first is for 2 wheel drive

[1] Auto Trans capacities are:
TH350—6.3
TH400—9.0
700R4—10

[2] Front Axle Capacities are:
10–20 Series—4
30 Series—6

[3] Rear Axle Capacities are:
8½"—4¼

8⅞"—3½
9¾"—6.0
10½" (Chev)—6½
10½" (Dana)—7.2
12¼"—26.8

[4] 8¼ with full-time 4 wheel drive
[5] 8.5 ring gear—4.2
[6] with TH-M 400:7.0
with TH-M 700-R4: 10.0
[7] 8½" ring gear: 4.25
8⅞" ring gear: 3.5
9¾" ring gear (Dana): 6.0
10½" ring gear (Dana): 7.2
(Chev.): 6.5

CAPACITIES
Pick-Ups and Suburban

Year	Engine (No. Cyl.) Displacement (cu in.)	Engine Crankcase (qts) With Filter	Transmission (pts) Manual 3-spd	4-spd	Auto (Refill)	Drive Axle (pts) Front	Rear	Transfer Case (pts)	Fuel Tank (gals)	Cooling System (qts) w/o A/C	w/ A/C	HD
'79–'81	6-250	5	3.2[7]	8.0	[2]	5[8]	[6]	5[5]	20[10]	15.0	15.6	15.0
	6-292	6	3.2[7]	8.0	[2]	5[8]	[6]	5[5]	20[10]	14.8	15.4	14.8
	8-305	5	3.2[7]	8.0	[2]	5[8]	[6]	5[5]	20[10]	17.6	18.0	18.0
	8-350	5	3.2[7]	8.0	[2]	5[8]	[6]	5[5]	20[10]	17.6	18.0	18.0
	8-350 Diesel	7	—	—	5.0	—	[6]	—	20[10]	18.0	18.0	18.0
	8-400	5	3.2[7]	8.0	[2]	5[8]	[6]	5[5]	20[10]	20.4	20.4	20.4
	8-454	5[11]	3.2[7]	8.0	[2]	5[8]	[6]	5[5]	20[10]	24.4[9]	24.7	24.7
'82	6-250	5	3.0	8.0	6.0	5.0	[16]	5.0[13]	[14]	15.0	15.5	—
	6-292	6	3.0	8.0	6.0	5.0	[16]	5.0[13]	[14]	15.0	15.5	—
	8-305	5	3.0	8.0	6.0	5.0	[16]	5.0[13]	[14]	17.5	18.0	—
	8-350	5	3.0	8.0	[15]	5.0	[16]	5.0[13]	[14]	17.5	18.0	—
'83	6-250	5	3.0	8.0	6.0	5.0	[16]	10[1]	[14]	15	15.5	—
	8-305	5	3.0	8.0	6.0	5.0	[16]	10[1]	[14]	17.5	18	—
	8-350	5	3.0	8.0	6.0[6]	5.0	[16]	10[1]	[14]	17.5	18	—
	8-379	7	—	8.0	6.0[6]	5.0	[16]	10[1]	[14]	23	24.5	—
	8-454	5	—	8.0	6.0[6]	5.0	[16]	10[1]	[14]	23	24.5	—
'84	6-250	5	3.0	8.0	6.0	5.0	[16]	10[1]	[14]	15	15.5	—
	8-305	5	3.0	8.0	6.0	5.0	[16]	10[1]	[14]	17.5	18	—
	8-350	5	3.0	8.0	6.0[6]	5.0	[16]	10[1]	[14]	17.5	18	—
	8-379	7	—	8.0	6.0[6]	5.0	[16]	10[1]	[14]	23	24.5	—
	8-454	5	—	8.0	6.0[6]	5.0	[16]	10[1]	[14]	23	24.5	—

T4

CAPACITIES
Pick-Ups and Suburban

Year	Engine (No. Cyl.) Displacement (cu. in.)	Engine Crankcase (qts) With Filter	Transmission (pts) Manual 3-spd	Transmission (pts) Manual 4-spd	Transmission (pts) Auto (Refill)	Drive Axle (pts) Front	Drive Axle (pts) Rear	Transfer Case (pts)	Fuel Tank (gals)	Cooling System (qts) w/o A/C	Cooling System (qts) w/ A/C	Cooling System (qts) HD
'85–'86	6-262	5	3.0	8.0	6.3	⑰	⑯	4.0	⑭	10.9	10.9	—
	6-292	6	3.0	8.0	6.3	⑰	⑯	4.0	⑭	15.5	16	—
	8-305	5	3.0	8.0	6.3	⑰	⑯	4.0	⑭	17.5	18	—
	8-350	5	3.0	8.0	9.0	⑰	⑯	4.0	⑭	17.5	18	—
	8-379	7	3.0	8.0	9.0	⑰	⑯	6.0	⑭	23	24.5	—

① Heavy-duty 3-speed—3.5 pts
② Turbo Hydra-Matic 350—5.0 pts
   Turbo Hydra-Matic 400—7.5 pts
③ 3,300 and 3,500 lb Chevrolet axles—4.5 pts
   5,200 and 7,200 lb Chevrolet axles—6.5 pts
   5,500 lb Dana axles—6.0 pts
   11,000 lb Chevrolet axles—14.0 pts
④ 20 Series—21.0 gals
⑤ Full-time 4 wd—8.25 pts
⑥ 8½ in. ring gear—4.2 pts
   8⅞ in. ring gear (Chevrolet)—4.5 pts (3.5 pts '77–'78)
   10½ in. ring gear (Chevrolet)—5.4 pts
   10½ in. ring gear (Dana)—7.2 pts

12½ in. ring gear (Chevrolet)—26.8 pts
⑦ Tremec 3-spd—4.0 pts
   Muncie 3-spd—3.0 pts
⑧ 8½ in. ring gear—4.25 pts ('77–'81)
⑨ 22.8—'79–'81
⑩ 16.0 gal—short wheelbase models
⑪ 6 qts with filter, 5 qts without filter, '78–'81
⑫ OPT.: 20 gal
⑬ All 1-ton models: 10.0
⑭ Short bed w/single tank: 16 gal.
   Short bed w/dual tanks: 32 gal.
   Longbed w/single tank: 20 gal.
   Long bed gasoline models w/dual tanks, under 8600 lb GVWR: 32 gal

Long bed gasoline models with dual tanks over 8600 lb GVWR, and all diesel models w/dual tanks: 40 gal.
⑮ TH-M 350: 6.0
   TH-M 400: 7.0
   TH-M 700-R4:10.0
⑯ Ring gear   Capacity
   8½"   4.25
   8⅞"   3.5
   9¾" (Dana)   6.0
   10½" (Dana)   7.2
   10½" (Chev)   6.5
   12¼" (Dana)   26.8
⑰ Front Axle Capacities are:
   10–20 Series—4
   30 Series—6

CAPACITIES
S-Series, Blazer and Jimmy

Year	Engine (No. Cyl.) Displacement cu. in.	Crankcase (qts.)	Transmission (pts) 4 sp	Transmission (pts) 5 sp	Transmission (pts) Auto.	Transfer Case (pts)	Rear Drive Axle (pts)	Front Drive Axle (pts)	Gas Tank (gal)	Cooling System (qts) Manual	Cooling System (qts) Auto.
'82	4-119	4.0	2.7	2.7	7.0 ②	5.2	3.5	1.7	13.0 ①	9.6	9.6
	4-121	4.0	2.7	2.7	7.0 ②	5.2	3.5	1.7	13.0 ①	9.6	9.6
	4-137 Diesel	5.0	2.7	2.7	7.0 ②	5.2	3.5	1.7	13.0 ①	10.0	10.0
	6-173	4.5 ③	2.7	2.7	7.0 ②	5.2	3.5	1.7	13.0 ①	12.4	12.4
'83	4-119	4	2.7	2.7	7	10	3.5	3	13 ①	9.4	9.5
	4-121	4	2.7	2.7	7	10	3.5	3	13 ①	9.6	9.7
	4-137 Diesel	4	2.7	2.7	7	10	3.5	3	13 ①	10	10
	6-173	4.5 ③	2.7	2.7	7	10	3.5	3	13 ①	12	12
'84	4-119	4	2.7	2.7	7	10	3.5	3	13 ①	9.4	9.5
	4-121	4	2.7	2.7	7	10	3.5	3	13 ①	9.6	9.7
	4-137 Diesel	4	2.7	2.7	7	10	3.5	3	13 ①	10	10
	6-173	4.5 ③	2.7	2.7	7	10	3.5	3	13 ①	12	12
'85–'86	4-119	4	2.7	2.7	7	5.2	3.5	2.6	13 ①	9.5	—
	4-137	5.5	2.7	2.7	7	5.2	3.5	2.6	14 ①	11.5	12 ④
	4-151	3	2.7	2.7	7	5.2	3.5	2.6	13 ①	12	12
	6-173	4	2.7	2.7	7	5.2	3.5	2.6	13 ①	12	12

① Optional: 20.0 gal.
② Figure shown is for pan removal only. Total overhaul capacity is 19.0 pts.
③ 4 qts. without filter

④ Automatic and manual transmission figures are the same for heavy duty cooling or cooling with air conditioning

CAPACITIES
Vans (Except Astro)

Year	Model	Engine (No. Cyl.) Displacement (cu. in.)	Engine Crankcase (qts) With Filter	Transmission (pts) Manual 3-spd	Transmission (pts) Manual 4-spd	Transmission (pts) Automatic	Drive Axle (pts)	Gasoline Tank (gals)	Cooling System (qts) wo/AC	Cooling System (qts) w/AC
'79–'82	All	6-250	5	3.2	—	5 ③	3.5 ④	22/33	17	18.5
	All	8-305	5	3.2 ①	—	5 ③	3.5 ④	22/33	19.5	21.0
	10, 20, 1500, 2500	8-350	5	3.2 ①	—	5 ③	3.5 ④	22/33	20	21.5
	30, 3500	8-350	5	3.2 ①	—	5 ③	3.5 ④	22/33	20	21.5
	All	8-400	5	3.2 ①	—	5 ③	3.5 ④	22/33	20	21.5
'83	All	6-250	5	3	8	③	④	22/33	17	—
	All	8-305	5	3	8	③	④	22/33	19	20
	All	8-350	5	3	8	③	④	22/33	19	20
	All	8-379	7	3	8	③	④	22/33	24	24
'84	All	6-250	5	3	8	③	④	22/33	17	—
	All	8-305	5	3	8	③	④	22/33	19	20
	All	8-350	5	3	8	③	④	22/33	19	20
	All	8-379	7	3	8	③	④	22/33	24	24
'85–'86	All	6-252	5	3	8	6.3	④	22/23	11.1	11.1
	All	8-305	5	3	8	6.3	④	22/23	17	17
	All	8-350	5	3	8	⑤	④	22/23	17	17
	All	8-379	7	3	8	⑤	④	22/23	24 ⑥	24 ⑥

① 4 pts with top-cover Tremec three-speed and Saginaw three-speed
② 7 pts with 10,000 lb or higher GVW.
③ '82 TH-M 350: 6.0

TH-M 400: 7.0
TH-M 700-R4: 10.0
④ 9³/₄" ring gear: 6.0
10¹/₂" ring gear (Dana): 7.2
10¹/₂" ring gear (Chev): 6.5

12¹/₂" ring gear (Dana): 26.8
⑤ TH350—6.3
TH400—9.0
⑥ 25.6 w/HD cooling

CAPACITIES
Astro Vans

Year	Engine (No. Cyl.) Displacement (cu. in.)	Engine Crankcase (qts) With Filter	Transmission (pts) Manual 3-spd	Transmission (pts) Manual 4-spd	Transmission (pts) Automatic	Drive Axle (pts)	Gasoline Tank (gals)	Cooling System (qts) wo/AC	Cooling System (qts) w/AC
'85–'86	4-151	3	3.0	8.0	6.3	①	17/27	10	10
	6-262	5	3.0	8.0	6.3	①	17/27	13.5	13.5

① Capacities vary according to ring gear:
8¹/₂"—4.5
8⁷/₈"—3.5

9³/₄"—6.0
10¹/₂" (Chev)—6.5
10¹/₂" (Dana)—7.2

TUNE-UP SPECIFICATIONS
Pick-ups and Suburban

Year	Engine Displacement (cu in.)	Spark Plugs Type	Spark Plugs Gap (in.)	Distributor	Ignition Timing (deg) MT	Ignition Timing (deg) AT	Fuel Pump Pressure (psi)	Compression Pressure (psi) ●	Idle Speed (rpm)* MT	Idle Speed (rpm)* AT
'79	6-250 (LD Fed)	R46TS	0.035	Electronic	10B	10B	4.5–6.0	130	750	600
	6-250 ③	R46TS	0.035	Electronic	6B	8B	4.5–6.0	130	750	600
	6-292	R44T	0.035	Electronic	8B	8B	4.5–6.0	130	700	700
	8-305	R45TS	0.045	Electronic	6B	6B	7.5–9.0	150	600	500
	8-350 (LD)	R45TS	0.045	Electronic	8B	8B	7.5–9.0	150	700	500
	8-350 (HD)	R44T	0.045	Electronic	4B	4B	7.5–9.0	150	700	700(N)
	8-400	R45TS	0.045	Electronic	—	4B	7.5–9.0	150	—	500
	8-454 (LD)	R45TS	0.045	Electronic	8B	8B	7.5–9.0 ②	150	700	500
	8-454 (HD)	R44T	0.045	Electronic	—	4B	7.5–9.0 ②	150	—	700(N)

TUNE-UP SPECIFICATIONS
Pick-ups and Suburban

Year	Engine Displacement (cu in.)	Spark Plugs Type	Spark Plugs Gap (in.)	Distributor	Ignition Timing (deg) MT	Ignition Timing (deg) AT	Fuel Pump Pressure (psi)	Compression Pressure (psi) ●	Idle Speed (rpm)* MT	Idle Speed (rpm)* AT
'80	6-250 (LD Fed)	R46TS	0.035	Electronic	10B	10B	4.5–6.0	130	750	650
	6-250 (LD Calif)	R46TS	0.035	Electronic	10B	10B	4.5–6.0	130	750	600
	6-250 [3]	R46TS	0.035	Electronic	—	8B	4.5–6.0	130	—	600
	6-292	R44T	0.035	Electronic	8B	8B	4.5–6.0	130	700	700(N)
	8-305	R45TS	0.045	Electronic	8B	8B	7.5–9.0	150	600	500
	8-350 (LD)	R45TS	0.045	Electronic	8B	8B	7.5–9.0	150	700	500
	8-350 (HD Fed)	R44T	0.045	Electronic	4B	4B	7.5–9.0	150	700	700(N)
	8-350 (HD Calif)	R44T	0.045	Electronic	6B	6B	7.5–9.0	150	700	700(N)
	8-400 (HD Fed)	R44T	0.045	Electronic	—	4B	7.5–9.0	150	—	700(N)
	8-400 (HD Calif)	R44T	0.045	Electronic	—	6B	7.5–9.0	150	—	700(N)
	8-454	R44T	0.045	Electronic	4B	4B	7.5–9.0 [2]	150	700	700(N)
'81	6-250 (Fed)	R45TS	0.035	Electronic	10B	10B	4.5–6.0	130	750	650(D)
	6-250 (Calif)	R46TS	0.035	Electronic	10B	10B	4.5–6.0	130	750	650(D)
	6-292	R44T	0.035	Electronic	8B	8B	4.5–6.0	130	700	700(N)
	8-305 2 bbl	R45TS	0.045	Electronic	8B	8B	7.5–9.0	150	600	500(D)
	8-305 4 bbl	R45TS	0.045	Electronic	4B	4B [4]	7.5–9.0	150	700	500(D)
	8-350 (LD)	R45TS	0.045	Electronic	8B	8B [5]	7.5–9.0	150	700	500(D)
	8-350 (HD Fed)	R44T	0.045	Electronic	4B	4B	7.5–9.0	150	700	700(N)
	8-350 (HD Calif)	R44T	0.045	Electronic	6B	6B	7.5–9.0	150	700	700(N)
	8-350 Diesel	—	—	Electronic	—	8B [6]		450	—	575(D) [7]
	8-454	R44T	0.045	Electronic	4B	4B	7.5–9.0	150	700	700(N)
'82	6-250	R45TS	[8]	Electronic	[8]	[8]	4.5–6	—	[8]	[8]
	6-292	R44T	.035	Electronic	8	8	4–5	—	700	700
	8-305	R45TS	.045	Electronic	[8]	[8]	[8]	—	[8]	[8]
	8-350 LD	R45TS	.045	Electronic	[8]	[8]	[8]	—	[8]	[8]
	8-350 HD	R44T	.045	Electronic	[8]	[8]	[8]	—	[8]	[8]
	8-379	Diesel	—	—	[8]	[8]	[8]	—	[8]	[8]
	8-454	R44T	.045	Electronic	[8]	[8]	[8]	—	[8]	[8]
'83	6-250	R45TS	[8]	Electronic	[8]	[8]	4.5–6	—	[8]	[8]
	6-292	R44T	.035	Electronic	8	8	4–6	—	700	700
	8-305	R45TS	.045	Electronic	[8]	[8]	[8]	—	[8]	[8]
	8-350 LD	R45TS	.045	Electronic	[8]	[8]	[8]	—	[8]	[8]
	8-350 HD	R44T	.045	Electronic	[8]	[8]	[8]	—	[8]	[8]
	8-379	Diesel	—	—	[8]	[8]	[8]	—	[8]	[8]
	8-454	R44T	.045	Electronic	[8]	[8]	[8]	—	[8]	[8]
'84	6-250	R45TS	[8]	Electronic	[8]	[8]	4.5–6	—	[8]	[8]
	6-292	R44T	.035	Electronic	8	8	4.5–6	—	700	700
	8-305	R45TS	.045	Electronic	[8]	[8]	[8]	—	[8]	[8]
	8-350 LD	R45TS	.045	Electronic	[8]	[8]	[8]	—	[8]	[8]
	8-350 HD	R44T	.045	Electronic	[8]	[8]	[8]	—	[8]	[8]
	8-379	Diesel	—	—	[8]	[8]	[8]	—	[8]	[8]
	8-454	R44T	.045	Electronic	[8]	[8]	[8]	—	[8]	[8]
'85–'86	6-292	R43CTS	[8]	Electronic	[8]	[8]	4–6.5	—	[8]	[8]
	6-292	R44T	[8]	Electronic	[8]	[8]	4–6.5	—	[8]	[8]
	8-305	R45TS	[8]	Electronic	[8]	[8]	4–6.5	—	[8]	[8]
	8-350	R45TS	[8]	Electronic	[8]	[8]	4–6.5	—	[8]	[8]
	8-454	R44T	[8]	Electronic	[8]	[8]	4–6.5	—	[8]	[8]

NOTE: All engines use hydraulic valve lifters
NOTE: Part numbers in this chart are not recommendations by Chilton for any product by brand name.
NOTE: The underhood sticker often reflects tune-up changes made in production. Sticker figures must be used if they disagree with those in this chart.
● Maximum variation among cylinders—20 psi
B Before Top Dead Center
LD Light-duty
HD Heavy-duty

TUNE-UP SPECIFICATIONS
Pick-ups and Suburban

Year	Engine Displacement (cu in.)	Spark Plugs Type	Gap (in.)	Distributor	Ignition Timing (deg) MT	AT	Fuel Pump Pressure (psi)	Compression Pressure (psi) ●	Idle Speed (rpm)* MT	AT

Fed Federal (49 states)
Calif California only
MT Manual transmission
AT Automatic transmission
N Neutral
* Automatic transmission idle speed set in
   Drive unless otherwise indicated
NA Not available
— Not applicable
Hyd Hydraulic
① 700 rpm—California
② 5.5–7.0 with vapor return line
③ California C-20, C-2500 only
④ Calif.: 8B
   High Alt.: 2B
   Emission Label Code AAN: 6B
⑤ Calif.: 6B
   Calif. w/Emission Label Code
   AAD: 8B
⑥ Calif.: 5B
⑦ Calif.: 600(D)
⑧ See underhood sticker

TUNE-UP SPECIFICATIONS
Vans (Except Astro)

When analyzing compression results, look for uniformity among cylinders rather than specific pressures.

Year	Engine Cu In. Displacement	Spark Plugs Orig Type	Gap (in.)	Distributor	Ignition Timing ● (deg)▲ MT	AT	Fuel Pump Pressure (psi)	Curb Idle Speed (rpm) ● MT	AT
'79	6-250	R46TS	.035	Electronic	10B ⑥	10B ⑦	4.5–6	750	600
	8-305	R45TS	.045	Electronic	6B	6B	7.5–9	700	600
	8-350 ⑧	R45TS	.045	Electronic	8B	8B	7.5–9	700	500
	8-400 ⑧	R45TS	.045	Electronic	—	4B	7.5–9	—	500
'80–'81	6-250	R46TS	.035	Electronic	10B	8B ⑨	4–6	750	650(D)
	8-305 (2-bbl)	R45TS	.045	Electronic	8B	8B	7–9	700	600(D)
	8-305 (4-bbl)	R45TS	.045	Electronic	6B	4B	7–9	700	500(D)
	8-350	R45TS	.045	Electronic	8B ⑩	8B ⑪	7–9	700	500(D) ⑫
'82	6-250	R45TS	.045	Electronic	①	①	4–6	①	①
	8-305	R45TS	.045	Electronic	①	①	7–9	①	①
	8-350	R45TS	.045	Electronic	①	①	7–9	①	①
'83	6-250	R45TS	.045	Electronic	①	①	4–6	①	①
	8-305	R45TS	.045	Electronic	①	①	7–9	①	①
	8-350	R45TS	.045	Electronic	①	①	7–9	①	①
	8-379	Diesel	—	—	①	①	—	①	①

TUNE-UP SPECIFICATIONS
Vans (Except Astro)

When analyzing compression results, look for uniformity among cylinders rather than specific pressures.

Year	Engine Cu In. Displacement	Spark Plugs Orig Type	Gap (in.)	Distributor	Ignition Timing ● (deg)▲ MT	AT	Fuel Pump Pressure (psi)	Curb Idle Speed (rpm) ● MT	AT
'84	6-250	R45TS	.045	Electronic	①	①	4–6	①	①
	8-305	R45TS	.045	Electronic	①	①	7–9	①	①
	8-350	R45TS	.045	Electronic	①	①	7–9	①	①
	8-379	Diesel	—	—	①	①	—	①	①
'85–'86	6-252	R43CTS	①	Electronic	①	①	4–6.5	③	③
	8-305	R45TS	①	Electronic	①	①	4–6.5	③	③
	8-350	R45TS	②	Electronic	①	①	4–6.5	③	③
	8-379	—	—	Diesel	—	—	6.5–9	650	650 ④

NOTE: The underhood specifications sticker often reflects tune-up changes made in production. Sticker figures must be used if they disagree with those in this chart.

NOTE: Part numbers in this chart are not recommendations by Chilton for any product by brand name.

NOTE: All engines use hydraulic valve lifters.

● Figures in parentheses are for California, and are given only if they differ from the 49 state specification. Automatic transmission idle speeds are set in Drive, unless specified otherwise.

▲ At idle speed with vacuum advance hose disconnected and plugged, unless specified otherwise in the text.

N—Transmission in Neutral
D—Transmission in Drive
HD Heavy Duty
LD Light Duty

① See the underhood specifications sticker
② Vehicles w/HD emissions use R44T
③ If equipped w/ECM, no adjustment required
④ Adjust w/AT in Park
⑤ California only
⑥ G-20, G-30, 2500, 3500 series in Calif.—6B
⑦ G-20, G-30, 2500, 3500 series in Calif.—8B

⑧ Some G-30/3500 series vans differ. Check the underhood emission sticker.
⑨ High Alt.—10B
⑩ Fed 1 ton models—4B
   Calif ¾ and 1 ton models—6B
⑪ 1 ton models—6B
⑫ 1 ton models—700(N)
   Calif. ½ and ⅓ ton models—550 (D)

TUNE-UP SPECIFICATIONS
Astro Vans

When analyzing compression results, look for uniformity among cylinders rather than specific pressures.

Year	Engine Cu In. Displacement	Spark Plugs Orig Type	Gap (in.)	Distributor	Ignition Timing ● (deg)▲ MT	AT	Fuel Pump Pressure (psi)	Curb Idle Speed (rpm) ● MT	AT
'85–'86	4-151	R43TSX	①	Electronic	①	①	4–6.5	②	②
	6-262	R43CTS	①	Electronic	①	①	4–6.5	②	②

① Refer to underhood sticker
② Controlled by ECM and non-adjustable

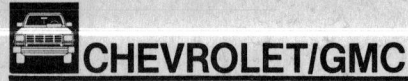

TUNE-UP SPECIFICATIONS
Blazer/Jimmy

When analyzing compression test results, look for uniformity among cylinders rather than specific pressures.

Year	Engine No. Cyl Displacement	Spark Plugs Orig Type	Spark Plugs Gap (in.)	Distributor	Ignition Timing (deg) Man Trans	Ignition Timing (deg) Auto Trans	Fuel Pump Pressure (psi)	Idle Speed (rpm) Man Trans	Idle Speed (rpm) Auto Trans ▲
'79	6-250	R46TS	.035	Electronic	10B	10B	4½–6	750	600
	8-305	R45TS	.045	Electronic	6B	6B	7–9	600	500
	8-350	R45TS	.045	Electronic	8B	8B	7–9	700	500
	8-400	R45TS	.060	Electronic	—	4B	7–9	—	500
'80–'81	6-250	R46TS	.035	Electronic	10B	10B	3.5–4.5	750	650(D)
	8-305	R45TS	.045	Electronic	4B	2B	7.0–8.5	700	500(D)
	8-350	R45TS	.045	Electronic	8B	8B	7.0–8.5	700	500(D)
'82	6-250	R45TS	.045	Electronic	①	①	4–6	①	①
	8-305	R45TS	.045	Electronic	①	①	7–9	①	①
	8-350	R45TS	.045	Electronic	①	①	7–9	①	①
	8-379	Diesel	—	—	①	①	—	①	①
'83	6-250	R45TS	.045	Electronic	①	①	4–6	①	①
	8-305	R45TS	.045	Electronic	①	①	7–9	①	①
	8-350	R45TS	.045	Electronic	①	①	7–9	①	①
	8-379	Diesel	—	—	①	①	—	①	①
	8-454	R44T	.045	Electronic	①	①	—	①	①
'84	6-250	R45TS	.045	Electronic	①	①	4–6	①	①
	8-305	R45TS	.045	Electronic	①	①	7–9	①	①
	8-350	R45TS	.045	Electronic	①	①	7–9	①	①
	8-379	Diesel	—	—	①	①	—	①	①
	8-454	R44T	.045	Electronic	①	①	—	①	①
'85–'86	8-305	R45TS	①	Electronic	①	①	4–6.5	①②	①②
	8-350 LD	R45TS	①	Electronic	①	①	4–6.5	①②	①②
	8-350 HD	R45TS	①	Electronic	①	①	4–6.5	①②	①②
	8-379	—	—	Diesel	—	—	6.5–9	①	①
	8-454	R44T	①	Electronic	①	①	4–6.5	①②	①②

NOTE: The underhood specifications sticker often reflects tuneup specification changes made in production. Sticker figures must be used if they disagree with those in this chart. Part numbers in this chart are not recommendations by Chilton for any product name.

NOTE: All engines use hydraulic valve lifters.

- Figures in parentheses are for California, and are given only when they differ from the 49 State models. When two idle speeds separated by a slash are given, the lower figure is with the solenoid disconnected.

▲ Automatic transmission idle speed set in Drive unless otherwise indicated

B Before Top Dead Center

N Neutral

TDC Top Dead Center

2WD Two wheel drive

4WD 4 wheel drive

① See under hood sticker

② Computer controlled on some models

TUNE-UP SPECIFICATIONS
S-Series Blazer/Jimmy

When analyzing compression results, look for uniformity among cylinders rather than specific pressures.

Year	Engine Cu In. Displacement	Spark Plugs Orig Type	Spark Plugs Gap (in.)	Distributor	Ignition Timing ● (deg)▲ MT	Ignition Timing ● (deg)▲ AT	Fuel Pump Pressure (psi)	Idle Speed (rpm) MT	Idle Speed (rpm) AT
'82	4–119	R-42XLS	.040	Electronic	6B	6B	3.0	800	900
	4–121	R-42CTS	.035	Electronic	12B	12B	5.0	750	700
	6-173	R-42TS	.040	Electronic	6B	10B	7.0	1000	750
'83	4–119	R-42XLS	.040	Electronic	①	①	3.0	①	①
	4–121	R-42CTS	.035	Electronic	①	①	5.0	①	①
	6-173	R-42TS	.040	Electronic	①	①	7.0	①	①
'84	4–119	R-42XLS	.040	Electronic	①	①	3.0	①	①
	4–121	R-42CTS	.035	Electronic	①	①	5.0	①	①
	6-173	R-42TS	.040	Electronic	①	①	7.0	①	①
'85–'86	4–119	R42XLS	①	Electronic	①	①	4–6.5	③	③
	4–137	—	—	Diesel	—	—	—	①	①
	4–151	R43TSX	①	Electronic	①	①	4–6.5	③	③
	4–173	R43CTS ②	①	Electronic	①	①	4–6.5	③	③

Note: The under hood specifications sticker often reflects tune-up specification changes made in production. Sticker figures must be used if they disagree with those in this chart.

① See under hood sticker
② Use R42TCS if vehicle is subjected to hard usage
③ Controlled by ECM, does not require adjustment

FIRING ORDERS

NOTE: Always replace spark plug wires one at a time.

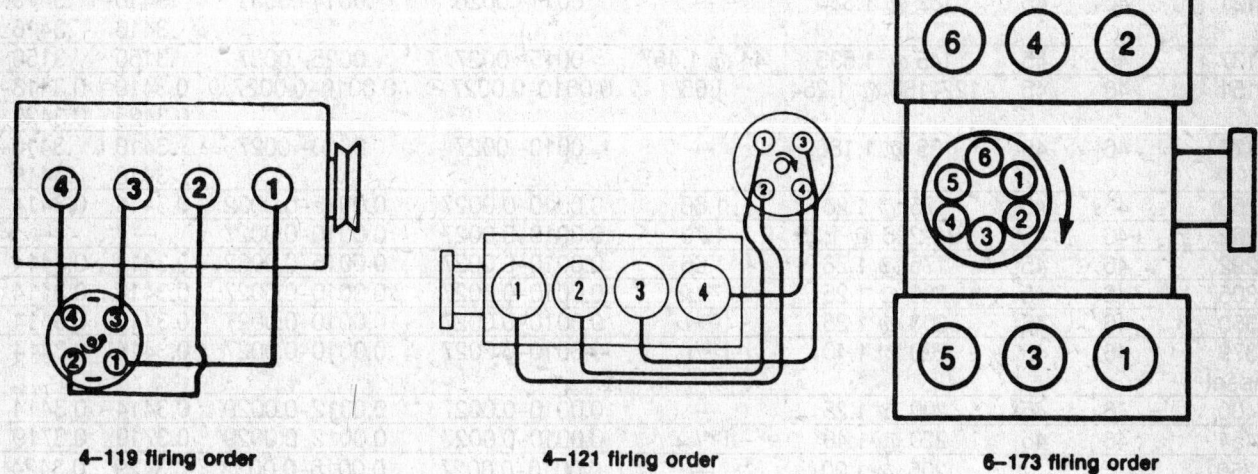

4–119 firing order   4–121 firing order   6–173 firing order

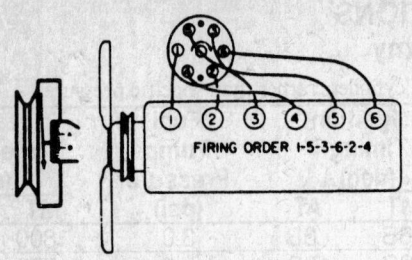

Six cylinder

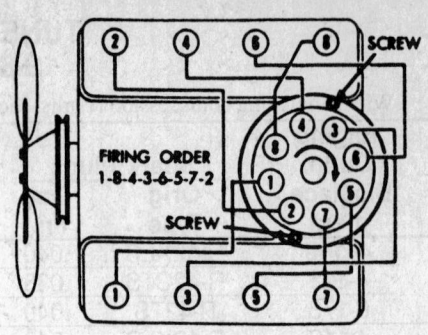

V8

2.5 L-4 firing order

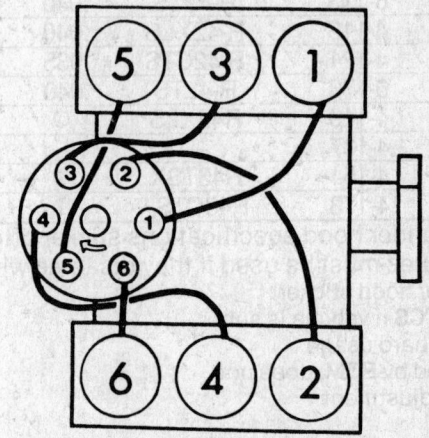

4.3 L V6 firing order

VALVE SPECIFICATIONS

Engine No. Cyl Displacement (cu in.)	Seat Angle (deg)	Face Angle (deg)	Spring Test Pressure (lbs @ in.)	Spring Installed Height (in.) [1]	Stem-to-Guide Clearance (in.)		Stem Diameter (in.)	
					Intake	Exhaust	Intake	Exhaust
4-119	45	45	35 @ 1.614	20 @ 1.515	.0009–.0022	.0015–.0031	.3102 min.	.3091 min.
4-121	46	45	182 @ 1.330	—	.0011–.0026	.0014–.0031	.3410– .3416	.3410– .3416
4-137	45	45	145 @ 1.535	44 @ 1.457	.0015–.0027	.0025–.0037	.3150	.3150
4-151	46	45	122–180 @ 1.254	1.66	0.0010–0.0027 [4]	0.0010–0.0027 [4]	0.3418– 0.3425	0.3418– 0.3425
6-173	46	45	195 @ 1.180	—	.0010–.0027	.0010–.0027	.3410– .3416	.3410– .3416
6-250	46	45	175 @ 1.26	1.66	0.0010–0.0027	0.0015–0.0032	0.3414	0.3414
6-262	46	45	194–206 @ 1.25	1.70	0.0010–0.0027	0.0010–0.0027	—	—
6-292	46	45	175 @ 1.26	1.66	0.0010–0.0027	0.0015–0.0032	0.3414	0.3414
8-305	46	45	200 @ 1.25	1 23/32	0.0010–0.0027	0.0010–0.0027	0.3414	0.3414
8-350	46	45	200 @ 1.25	1 23/32	0.0010–0.0027	0.0010–0.0027	0.3414	0.3414
8-379 Diesel	46	45	230 @ 1.40	1 13/15	0.0010–0.0027	0.0010–0.0027	0.3414	0.3414
8-400	46	45	200 @ 1.25	—	0.0010–0.0027	0.0012–0.0029	0.3414	0.3414
8-454	46	45	220 @ 1.40	1 51/64	0.0010–0.0027	0.0012–0.0029	0.3719	0.3719
8-350 Diesel	[2]	[3]	205 @ 1.00	—	0.0010–0.0027	0.0015–0.0032	0.3429	0.3424

[1] ± 1/32 in.

[2] Intake—45°
Exhaust—31°

[3] Intake—44°
Exhaust—30°

[4] As measured at top. At bottom, clearance is 0.0020–0.0037

[5] Face angle is 46° on 1984 and later models

CRANKSHAFT AND CONNECTING ROD SPECIFICATIONS

(All measurements are given in in.)

Year	Engine No. Cyl. Displacement (cu in.)	Crankshaft				Connecting Rod		
		Main Brg Journal Dia	Main Brg Oil Clearance	Shaft End-Play	Thrust on No.	Journal Diameter	Oil Clearance	Side Clearance
'82–'84	4-119	2.2050	0.008–0.0025	0.0117 max.	3	1.9290	0.0007–0.0030	0.0137 max.
	4-121	⑪	⑫	0.0020–0.0071	3	1.9990	0.0010–0.0031	0.0039–0.0240
	4-837	2.3590	0.0011–0.0033	0.0018	3	2.0837	0.0016–0.0047	0.0024
	6-173	2.4940	0.0017–0.0030	0.0020–0.0067	3	1.9980	0.0014–0.0032	0.0063–0.0173
'79–'81	6-250	2.2979–2.2994	Nos. 1–6 .0010–.0024 No. 7 .0016–.0035	.002–.006	7	1.999–2.000	.0010–.0026	.006–.017
	6-292	2.2979–2.2994	Nos. 1–6 .0010–.0024 No. 7 .0016–.0035	.002–.006	7	2.099–2.100	.0010–.0026	.006–.017
	8-305, 350, 400	⑤	.0008–.0020 ②	.002–.006	5	2.199–2.200 ⑫	.0013–.0035	.008–.014
	8-454	③	④	.006–.010	5	2.1985–2.1995	.0009–.0025	.013–.023
	8-350 Diesel	2.9993–3.0003	Nos. 1–4 .0005–.0021 No. 5 .0015–.0031	.0035–.0135	5	2.1238–2.1248	.0005–.0026	.006–.020
'82–'84	6-250	2.2979–2.2994	Nos. 1–6 0.0010–0.0024 No. 7 0.0016–0.0025 ⑩	0.002–0.006	7	1.999–2.000	0.0010–0.0026	0.006–0.017
	6-292	2.2979–2.2994	Nos. 1–6 0.0010–0.0024 No. 7 0.0016–0.0025 ⑩	0.002–0.006	7	2.099–2.100	0.0010–0.0026	0.006–0.017
	8-305	⑤	0.008–0.0020 ②	0.002–0.006	5	2.0988–2.0998	0.0013–0.0035	0.008–0.014
	8-350	⑤	0.008–0.0020 ②	0.002–0.006	5	2.0988–2.0998	0.0013–0.0035	0.008–0.014
	8-379 Diesel	⑧	⑨	0.002–0.007	5	2.3981–2.3991	0.0018–0.0039	0.007–0.025
'82–'86	8-454	⑦	④	0.006–0.010	5	2.2000–2.1990	0.0009–0.0025	0.013–0.023
'85–'86	4-151	2.300	0.0005–0.0022	0.0035–0.0085	5	2.000	0.0005–0.0026	0.005–0.022
	6-263	⑬	⑭	0.002–0.006	Rear	2.2497–2,2487	0.010–0.0032	0.007–0.015
	6-292	2.2979–2.2994	Nos. 1–6 0.0010–0.0024 No. 7 0.0016–0.0025	0.002–0.006	7	2.099–2.100	0.0010–0.0026	0.006–0.017
	8-305, 350	⑤	③	0.002–0.006	5	2.0988–2.0998	0.0013–0.0035	0.008–0.014
	8-379 Diesel	⑧	⑨	0.0039–0.0098	5	2.3981–2.3992	0.0018–0.0039	0.007–0.025

① No. 5—2.4479–2.4488
② Nos. 2–4—.0011–.0023
  No. 5—.0017–.0033
③ No. 1—2.7485–2.7494
  Nos. 2–4—2.7481–2.7490
  No. 5—2.7478–2.7488
④ Nos. 1–4—.0013–.0025
  No. 5—.0024–.0040
⑤ '79–'84 only: 305,
  350—No. 1—2.4484–2.4493
  Nos. 2–4—2.4481–2.4490

No. 5—2.4479–2.4488
  400—Nos. 1–4—2.6484–2.6493
  No. 5—2.6479–2.6488
⑥ '79–'81: 2.0988–2.0998
⑦ Nos. 1–4—2.7481–2.7490
  No. 5—2.7476–2,7486
⑧ Nos. 1–4—2.9495–2.9504
  No. 5—2.9492–2.9502
⑨ Nos. 1–4—0.0018–0.0033
  No. 5—0.0022–0.0037
⑩ '83–'84 .0016–.0035

⑪ Nos. 1–4—2.4940–2.4950
  No. 5—2.4930–2.4950
⑫ Nos. 1–4—.0006–.0019
  No. 5—.0014–.0027
⑬ Front—2.4484–2.4493
  Int.—2.4481–2.4990
  Rear—2.4479–2.4488
⑭ Front—.0008–.0020
  Int.—.0011–.0023
  Rear—.0017–.0032

PISTON AND RING SPECIFICATIONS

| Engine | Year | Piston to Bore Clearance | Ring Side Clearance | | | Ring Gap | | |
			Top Compression	Bottom Compression	Oil Control	Top Compression	Bottom Compression	Oil Control
4-119	'82-'84	.0018-.0026	.0059 Max	.0059 Max	.0059 Max	.012-.020	.008-.016	.008-.035
4-121	'82-'84	.0008-.0018	.0012-.0027	.0012-.0038	.0078 Max	.010-.020	.010-.020	.020-.055
4-137	'82-'84	.0062-.0070	.0018-.0028	.0012-.0021	.0008-.0021	.008-.016	.008-.016	.008-.016
4-151	'85-'86	0.0017-0.0041	0.0015-0.0030	0.0015-0.0030	Snug	0.010-0.022	0.010-0.020	0.015-0.055
6-173	'82-'84	.0017-.0027	.0012-.0028	.0016-.0037	.0078 Max	.010-.020	.010-.020	.020-.055
6-250	'77-'84	0.0005-0.0015 ②	0.0012-0.0027	0.0012-0.0032	.005 Max	0.010-0.020	0.010-0.020	0.015-0.055
6-263	'85-'86	0.0007-0.0017	0.0012-0.0032	0.0012-0.0032	0.002-0.007	0.010-0.020	0.010-0.025	0.015-0.055
6-292	'77-'84	0.0026-0.0036	0.0020-0.0040	0.0020-0.0040	.005 Max	0.010-0.020	0.010-0.020	0.015-0.055
6-292	'85-'86	0.0026-0.0036	0.0020-0.0040	0.0020-0.0040	0.005-0.0055	0.010-0.020	0.010-0.020	0.015-0.055
8-305	'79-'86	0.0007-0.0017	0.0012-0.0032	0.0012-0.0032	0.002-0.007	0.010-0.020	0.010-0.025	0.015-0.055
8-350	'79-'86	0.0007-0.0017	0.0012-0.0032	0.0012-0.0032	0.002-0.007	0.010-0.020	0.010-0.025	0.015-0.055
8-350 Diesel	'79-'81	0.0050-0.0060	0.0040-0.0060	0.0018-0.0038	0.001-0.005	0.015-0.025	0.015-0.025	0.015-0.055
8-379 Diesel	'82-'84	0.0040-0.0050	0.0030-0.0070	0.0015-0.0031	0.0016-0.0038	0.012-0.021	0.030-0.039	0.0098-0.020
8-379 Diesel	'85-'86	0.0035-0.0045 ⑤	0.0030-0.0070	0.0015-0.0031	0.0016-0.0037	0.0118-0.0216	0.0295-0.0393	0.0098-0.020
8-400	'79-'81	0.0014-0.0024	0.0012-0.0032	0.0012-0.0032	0.002-0.007	0.010-0.020	0.010-0.025	0.010-0.055
8-454	'79-'84	0.0014-0.0024 ③	0.0017-0.0032	0.0017-0.0032	0.002-0.007 ④	0.010-0.020	0.010-0.020	0.010-0.055
8-454	'85-'86	0.0030-0.0040	0.0017-0.0032	0.0017-0.0032	0.005-0.0065	0.010-0.020	0.010-0.020	0.015-0.055

① '78-'82: 0.0010-0.0020
② '83-'84: .0010-.0020
③ '83-'84: .0030-.0040
④ '83-'84: .005-.0065
⑤ Applies to Bohn pistons. Zollner—
0.0044-0.0054. Bore 7 & 8 must fit
.0005" looser

TORQUE SPECIFICATIONS

(ft. lb.)

| Engine | Cylinder Head Bolts | Rod Bearing Bolts | Main Bearing Bolts | Crankshaft Damper Bolt | Flywheel Bolts | Manifold | |
						Intake	Exhaust
119	72	43	75	87	76	17	16
121	70	37	70	75	50	23	25
137 Diesel	60	65	116-130	125-150	70	15	15
4-151	92	32	70	160	60-75	29	44
173	70	37	70	75	50	23	25
6-250	95	35	65	—	60	—	30 ①
6-263	65	45	70	60	55-75	30	20
6-292	95	35 ⑥	65	60	110 ⑧	35 ⑨	30
8-305	65	45	70	60	60	30	20
8-350	65	45	70	60	60	30	30 ①
8-350 Diesel	130 ②	42	120	200-310	60	40 ②	25
8-379 Diesel	88-103 ⑤	44-52	④	140-162	60	25-37	18-25
8-400	65	45	70	60	60	30	30
8-454	80	50 ③	110	65 ⑦	65	30	20

① End bolts only are torqued to 20 ft. lbs through 1984: on later models torque all bolts to 20 ft. lbs.
② Dip in oil
③ 7/16 in. bolts: 70
④ Inner: 105-117
Outer: 94-105

⑤ Applies to models through 1984: 1985 and later; torque in sequence to 20 ft. lbs, then again in sequence torque to 50 ft. lbs and finally to ¼ turn more—in sequence

⑥ 40: on 1985 and later models
⑦ 85: on 1985 and later models
⑧ 100: on 1985 and later models
⑨ 40: on 1985 and later models— exhaust-to-intake: 45 ft. lbs

BATTERY AND STARTER SPECIFICATIONS

	Battery			Starter [3]			
	Amp Hour		Ground			No Load Test	
						Amps	
Year	Capacity	Volts	Terminal	Identification	Volts	[1]	rpm
'79-'82	60	12	Neg	1108778 [2]	9	50–80	5500–10,500
	80	12	Neg	1187780 [2]	9	50–80	3500–6000
				1109056 [2]	9	50–80	5500–10,500
				1109052 [2]	9	65–95	7500–10,500
				1108776 [2]	9	65–95	7500–10,500
	125	12	Neg	1108776 [2]	9	65–95	7500–10,500
'83-'84	—	12	Neg	1109561	9	50–75	6000–11,900
	—	12	Neg	1109535	9	45–70	7000–11,900
	—	12	Neg	1998241	10	65–95	7500–10,500
	—	12	Neg	1998244	10	60–85	6800–10,500
	—	12	Neg	1998211	10	65–95	7500–10,500
	—	12	Neg	1998396	10	70–110	6500–10,700
	—	12	Neg	1998397	10	70–110	6500–10,700
	—	12	Neg	1109563	10	120–210	9000–13,400

[1] Solenoid included
[2] "R" terminal removed
[3] Brush spring tension is 35 oz. for all starters. Lock test is not recommended.

BATTERY AND STARTER SPECIFICATIONS

	Battery [1]			Starter			
		Test Load				No Load Test	
Year	Identification	(Amps)		Identification	Volts	Amps	rpm
'85-'86	1981103	200		1998427	10	50–75	6000–11,900
	1981110	190		1998437	10	60–90	6500–10,500
	1981200	230		1998438	10	70–110	6500–10,700
	1981109	150		1998439	10	70–110	6500–10,700
	1981102	170		1998441	10	70–110	6500–10,700
	1981104	250		1998443	10	70–110	6500–10,700
	1981108	370		1998444	10	—	—

[1] All batteries are 12 bolt, negative ground

ALTERNATOR AND REGULATOR SPECIFICATIONS

	Alternator			
	Part No. or	Field Current	Output	
Year	Manufacturer	@ 12 V	(amps)	Regulator
'79-'82	1102394	4.0–4.5	37	[1]
	1102491			
	1102889			
	1102485	4.0–4.5	42	[1]
	1102841, 87			
	1102480, 86	4.0–4.5	61	[1]
	1102886, 88			
	1101016, 28	4.0–4.5	80	[1]

ALTERNATOR AND REGULATOR SPECIFICATIONS
Alternator

Year	Part No. or Manufacturer	Field Current @ 12 V	Output (amps)	Regulator
'83-'84	1105185	4.0-4.5	37	①
	1100227	4.0-4.5	37	①
	1100204	4.0-4.5	37	①
	1100203	4.0-4.5	37	①
	1100207	4.0-4.5	66	①
	1100249	4.0-4.5	66	①
	1100275	4.0-4.5	66	①
	1100242	4.0-4.5	66	①
	1100208	4.0-4.5	66	①
	1100241	4.0-4.5	66	①
	1100209	4.0-4.5	78	①
	1100273	4.0-4.5	78	①
	1100276	4.0-4.5	78	①
	1100217	4.0-4.5	78	①
	1100259	4.0-4.5	78	①
'85-'86	1100203	4.0-4.5	37	①
	1100209	4.0-4.5	66	①
	1100217	4.0-4.5	78	①
	1100225	4.0-4.5	37	①
	1100241	4.0-4.5	66	①
	1100242	4.0-4.5	66	①
	1100207	4.0-4.5	66	①
	1100204	4.0-4.5	37	①
	1100208	4.0-4.5	66	①
	1101063	4.0-4.5	80	①
	1101064	4.0-4.5	80	①
	1100259	4.0-4.5	78	①
	1100229	4.0-4.5	42	①
	1100231	4.0-4.5	42	①
	1100293	4.0-4.5	85	①

—Not available
NA Not applicable
① All alternators use integral regulators

TUNE-UP

High Energy Ignition (HEI)

The General Motors HEI system is a pulse-triggered, transistor-controlled, inductive discharge ignition system. The entire HEI system is contained within the distributor cap.

The distributor, in addition to housing the mechanical and vacuum advance mechanisms, contains the ignition coil, the electronic control module, and the magnetic triggering device. The magnetic pick-up assembly contains a permanent magnet, a pole piece with internal "teeth," and a pick-up coil (not to be confused with the ignition coil).

In the HEI system, as in other electronic ignition systems, the breaker points have been replaced with an electronic switch, a transistor, which is located within the control module. This switching transistor performs the same function the points did in a conventional ignition system; it simply turns coil primary current on and off at the correct time. Essentially then, electronic and conventional ignition systems operate on the same principle.

The module which houses the switching transistor is controlled (turned on and off) by a magnetically generated impulse induced in the pick-up coil. When the teeth of the rotating timer align with the teeth of the pole piece, the induced voltage in the pick-up coil signals the electronic module to open the coil primary circuit. The primary current then decreases, and a high voltage is induced in the ignition coil secondary windings which is then directed through the rotor and high voltage leads (spark plug wires) to fire the spark plugs.

In essence then, the pick-up coil module system simply replaces the conventional breaker points and condenser. The condenser found within the distributor is for radio suppression purposes only and has nothing to do with the ignition process. The module automatically controls the dwell period, increasing it with increasing engine speed. Since dwell is automatically controlled, it cannot be adjusted. The module itself is non-adjustable and non-repairable and must be replaced if found defective.

HEI SYSTEM PRECAUTIONS

Timing Light Use

Inductive pick-up timing lights are the best kind to use if your truck is equipped with HEI. Timing lights which connect between the spark plug and the spark plug wire occasionally (not always) give false readings.

Spark Plug Wires

The plug wires used with HEI systems are of a different construction than conventional wires. When replacing them, make sure you get the correct wires, since conventional wires won't carry the voltage. Also, handle them carefully to avoid cracking or splitting them and never pierce them.

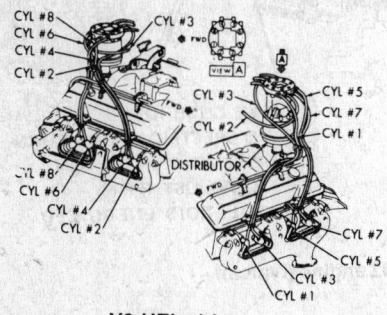

V8 HEI wiring

Tachometer Use

Not all tachometers will operate or indicate correctly when used on a HEI system. While some tachometers may give a reading, this does not necessarily mean the reading is correct. In addition, some tachometers hook up differently from others. If you can't figure out whether or not your tachometer will work on your truck, check with the tachometer manufacturer. Dwell readings, of course, have no significance at all.

HEI System Testers

Instruments designed specifically for testing HEI systems are available from several tool manufacturers. Some of these will even test the module itself. However, the tests given in the following section will require only an ohmmeter and a voltmeter.

Ignition Timing

Timing should be checked at each tuneup. It isn't likely to change much with HEI. The timing marks consist of a notch on the rim of the crankshaft pulley or vibration damper and a graduated scale attached to the engine front (timing) cover. A stroboscopic flash (dynamic) timing light must be used, as a static light is too inaccurate for emission controlled engines.

There are three basic types of timing light available. The first is a simple neon bulb with two wire connections. One wire connects to the spark plug terminal and the other plugs into the end of the spark plug wire for the No. 1 cylinder, thus connecting the light in series with the spark plug. This type of light is pretty dim and must be held very closely to the timing marks to be seen. Sometimes a dark corner has to be sought out to see the flash at all. This type of light is very inexpensive. The second type operates from the vehicle battery: two alli-

gator clips connect to the battery terminals, while an adapter enables a third clip to be connected between the No. 1 spark plug and wire. This type is a bit more expensive, but it provides a nice bright flash that you can see even in bright sunlight. It is the type most often seen in professional shops. The third type replaces the battery power source with 110 volt current.

1. To check and adjust the timing: Warm up the engine to normal operating temperature. Stop the engine and connect the timing light to the No. 1 (left front on V8, front on six and four cylinder) spark plug wire, either at the plug or at the distributor cap. Consult the engine compartment sticker for specific instructions for the particular vehicle/engine combination in question. On vehicles with Electronic Spark Timing, disconnect the four prong EST connector at the distributor so the engine will operate in the bypass timing mode. Otherwise, you will not be able to set the timing as the EST system will continuously attempt to compensate for changes in the position of the distributor.

─────── CAUTION ───────
Do not pierce the plug wire insulation with HEI; it will cause a miss. Use an inductive pickup timing light.

2. Clean off the timing marks and mark the pulley or damper notch and timing scale with white chalk. Disconnect and plug the vacuum line at the distributor. This is done to prevent any distributor vacuum advance. Check the underhood emission sticker for any other hoses or wires which may need to be disconnected.

3. Start the engine and adjust the idle speed to that specified in the "Tune-Up Specifications" chart. With automatic transmission, set the specified idle speed in Park. It will be too high, since it is normally (in most cases) adjusted in Drive. You can disconnect the idle solenoid, if any, to get the speed down. Otherwise, adjust the idle speed screw. This is done to prevent any centrifugal (mechanical) advance. The tachometer connects to the TACH terminal on the distributor and to a ground. Some tachometers must connect to the TACH terminal and to the positive battery terminal. Some tachometers won't work with HEI.

─────── CAUTION ───────
Never ground the HEI TACH terminal; serious system damage will result.

4. Aim the timing light at the pointer marks. Be careful not to touch the fan, because it may appear to be standing still. If the pulley or damper notch isn't aligned with the proper timing mark the timing will have to be adjusted.

5. Loosen the distributor base clamp locknut. You can buy trick wrenches which make this task a lot easier on V8s. Turn the distributor slowly to adjust the timing, holding it by the body and not the cap. Turn

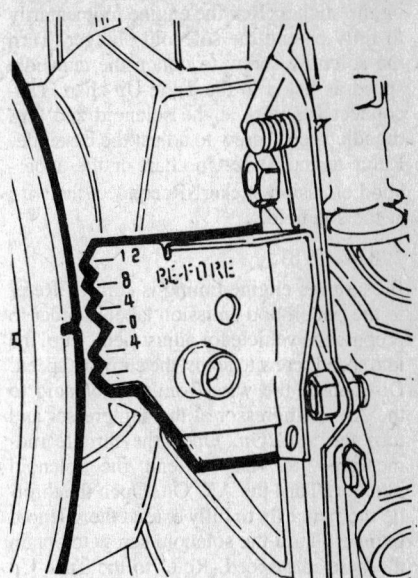

Typical timing marks

the distributor in the direction of rotor rotation to retard, and against the direction of rotation to advance.

6. Tighten the locknut. Check the timing again, in case the distributor moved slightly as you tightened it.

7. Replace the distributor vacuum line. Correct the idle speed.

8. Stop the engine and disconnect the timing light.

Carburetor

IDLE SPEED ADJUSTMENT

These procedures require the use of a tachometer. Tachometer hookup was explained earlier under Ignition Timing, Step 3. In some cases, the degree of accuracy required is greater than that available on a hand-held unit; a shop tachometer would be required to follow the instructions exactly.

─────── CAUTION ───────
When making idle speed adjustments: Block the wheels, set the parking brake, and don't stand in front of the truck.

1979–82
1 BBL
Start the engine and allow it to reach normal operating temperature. Be sure the choke is opened fully and linkage is off of the high cam step (fast idle). Turn the nut on the end of the solenoid to obtain the required idle rpm. Disconnect the lead wire to the solenoid and adjust base idle, reconnect wire when adjustment is complete.

2 BBL (SIX CYLINDER)
Be sure the ignition timing is correct. Refer to the underhood emission sticker and prepare the engine for adjustment as specified

on the sticker. Rev the engine momentarily to fully extend the solenoid plunger. Turn the solenoid screw to obtain the curb idle speed as listed in the Tune-Up chart. Disconnect the wire at the solenoid and turn the idle speed screw to adjust the base idle. Refer to the Tune-Up chart or the underhood emission sticker. Reconnect the wire at the solenoid.

2 BBL (V8)

Be sure the engine timing is correct. Refer to the underhood emission label in order to prepare the vehicle for adjustment. Turn the idle speed screw to adjust the curb idle speed. Disconnect the wire from the solenoid to the A/C compressor at the compressor and turn the A/C On. Open the throttle momentarily to fully extend the solenoid plunger. Turn the A/C On. Open the throttle momentarily to fully extend the solenoid plunger. Turn the solenoid screw to obtain the base idle speed. Refer to the Tune-Up chart or the underhood emission sticker for the correct speeds.

4 BBL

Check the ignition timing, adjust if necessary. On models not equipped with a solenoid: Be sure the idle speed screw is on the low step of the fast idle cam. Turn the idle adjusting screw to obtain the correct idle rpm. On models equipped with a solenoid: Turn the adjusting screw until the correct idle rpm is obtained with the A/C turned off. Disconnect the lead wire to the A/C compressor, turn the A/C switch on and adjust the solenoid plunger until correct idle rpm is reached. Check the engine decal to determine what gear is required when making the necessary adjustments. Reconnect the compressor wire after adjustments are completed.

1983 and LATER

Idle mixture adjustments are not performed except in case of major carburetor parts failure, or complete overhaul. On some models, idle speed is controlled by an electronic idle speed control which is not adjustable.

ASTRO 4.3L V6 (49 STATES) WITHOUT A/C

1. Check and/or adjust the ignition timing. Follow the instructions regarding preparing the engine for adjustments on the engine compartment sticker. Connect a tachometer as described in the manufacturer's instructions.

2. Refer to the engine compartment sticker for curb idle speed, and then adjust the idle speed screw (it passes through the body of the carburetor) to the specified rpm.

ASTRO 4.3L V6 (49 STATES) WITH A/C

1. Check and/or adjust the ignition timing as described above. Follow the instructions regarding preparing the engine for adjustments on the engine compartment sticker. Connect a tachometer as described in the manufacturer's instructions.

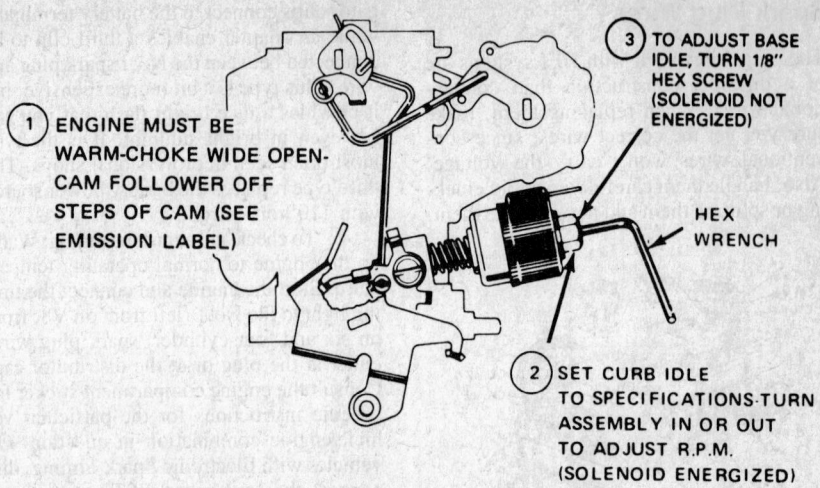

1. ENGINE MUST BE WARM-CHOKE WIDE OPEN-CAM FOLLOWER OFF STEPS OF CAM (SEE EMISSION LABEL)

3. TO ADJUST BASE IDLE, TURN 1/8" HEX SCREW (SOLENOID NOT ENERGIZED)

HEX WRENCH

2. SET CURB IDLE TO SPECIFICATIONS-TURN ASSEMBLY IN OR OUT TO ADJUST R.P.M. (SOLENOID ENERGIZED)

Idle speed adjustment, 292 engine (typical)

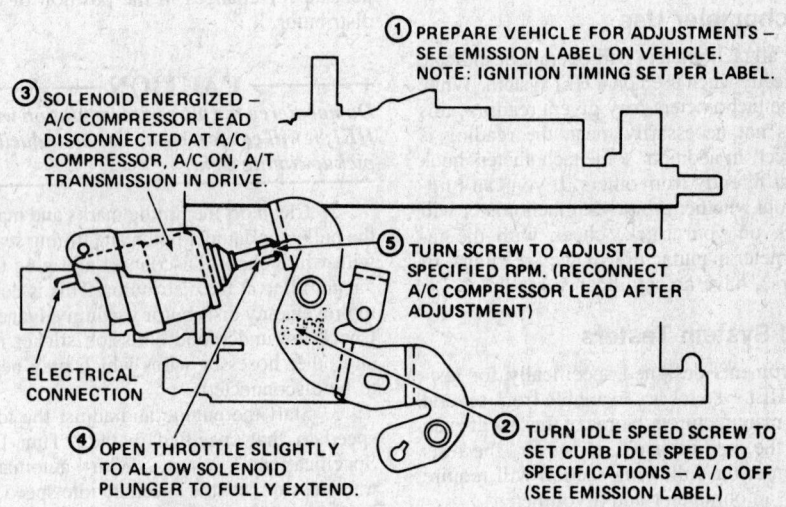

1. PREPARE VEHICLE FOR ADJUSTMENTS — SEE EMISSION LABEL ON VEHICLE. NOTE: IGNITION TIMING SET PER LABEL.

3. SOLENOID ENERGIZED — A/C COMPRESSOR LEAD DISCONNECTED AT A/C COMPRESSOR, A/C ON, A/T TRANSMISSION IN DRIVE.

5. TURN SCREW TO ADJUST TO SPECIFIED RPM. (RECONNECT A/C COMPRESSOR LEAD AFTER ADJUSTMENT)

ELECTRICAL CONNECTION

4. OPEN THROTTLE SLIGHTLY TO ALLOW SOLENOID PLUNGER TO FULLY EXTEND.

2. TURN IDLE SPEED SCREW TO SET CURB IDLE SPEED TO SPECIFICATIONS – A/C OFF (SEE EMISSION LABEL)

V8 2bbl idle speed adjustment with solenoid (typical)

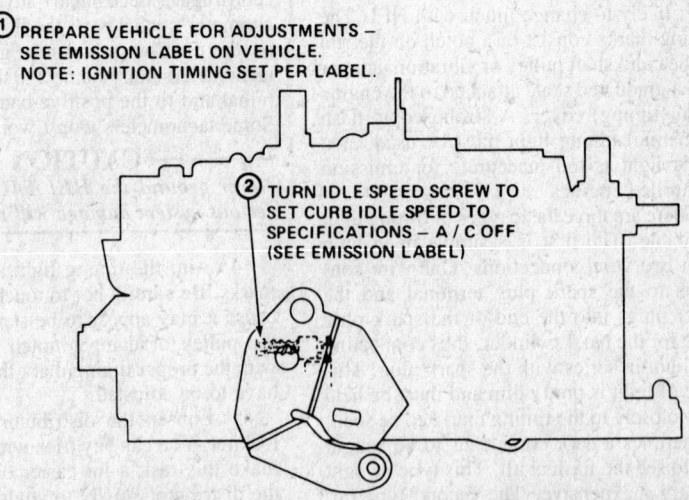

1. PREPARE VEHICLE FOR ADJUSTMENTS — SEE EMISSION LABEL ON VEHICLE. NOTE: IGNITION TIMING SET PER LABEL.

2. TURN IDLE SPEED SCREW TO SET CURB IDLE SPEED TO SPECIFICATIONS – A/C OFF (SEE EMISSION LABEL)

Typical 4bbl idle speed adjustment w/o solenoid

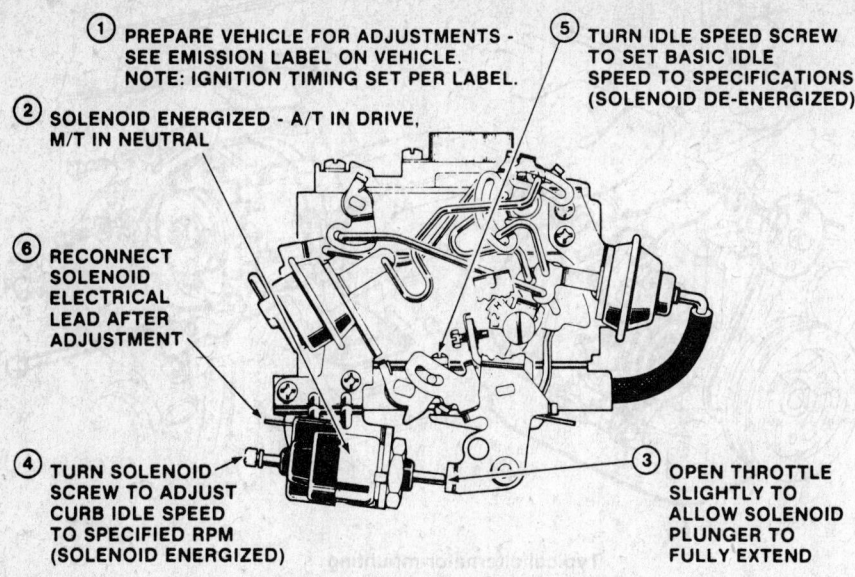

① PREPARE VEHICLE FOR ADJUSTMENTS - SEE EMISSION LABEL ON VEHICLE. NOTE: IGNITION TIMING SET PER LABEL.

② SOLENOID ENERGIZED - A/T IN DRIVE, M/T IN NEUTRAL

⑥ RECONNECT SOLENOID ELECTRICAL LEAD AFTER ADJUSTMENT

④ TURN SOLENOID SCREW TO ADJUST CURB IDLE SPEED TO SPECIFIED RPM (SOLENOID ENERGIZED)

⑤ TURN IDLE SPEED SCREW TO SET BASIC IDLE SPEED TO SPECIFICATIONS (SOLENOID DE-ENERGIZED)

③ OPEN THROTTLE SLIGHTLY TO ALLOW SOLENOID PLUNGER TO FULLY EXTEND

1979–82 250 six idle speed adjustment

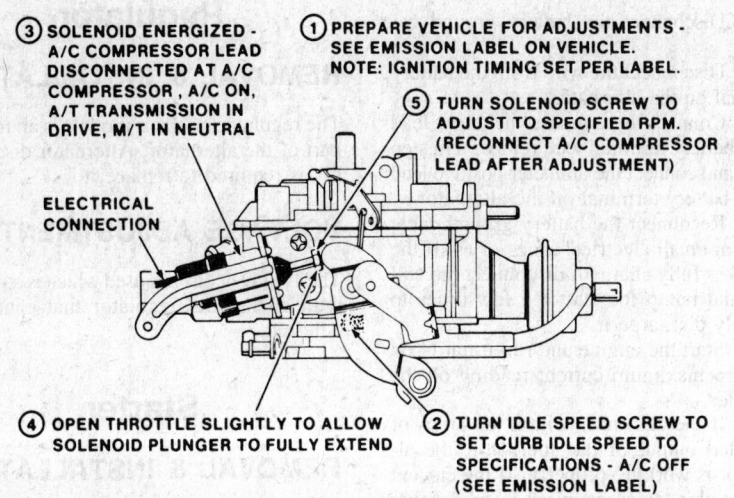

③ SOLENOID ENERGIZED - A/C COMPRESSOR LEAD DISCONNECTED AT A/C COMPRESSOR , A/C ON, A/T TRANSMISSION IN DRIVE, M/T IN NEUTRAL

ELECTRICAL CONNECTION

① PREPARE VEHICLE FOR ADJUSTMENTS - SEE EMISSION LABEL ON VEHICLE. NOTE: IGNITION TIMING SET PER LABEL.

⑤ TURN SOLENOID SCREW TO ADJUST TO SPECIFIED RPM. (RECONNECT A/C COMPRESSOR LEAD AFTER ADJUSTMENT)

④ OPEN THROTTLE SLIGHTLY TO ALLOW SOLENOID PLUNGER TO FULLY EXTEND

② TURN IDLE SPEED SCREW TO SET CURB IDLE SPEED TO SPECIFICATIONS - A/C OFF (SEE EMISSION LABEL)

Adjusting the Idle Speed Solenoid on Astro with the 4.3L V-6

2. Turn the air conditioning off. Refer to the engine compartment sticker for curb idle speed, and then adjust the idle speed screw (it passes through the body of the carburetor) to the specified rpm.

3. Disconnect the A/C compressor electrical connector. Turn the air conditioning on to energize the ISS (Idle Speed Solenoid). Open the throttle slightly to permit the ISS plunger to extend fully. If the vehicle has an automatic transmission, block the wheels and put the transmission in Drive.

4. Turn the solenoid screw to adjust the rpm to that specified for the ISS on the engine compartment sticker. When the adjustment is complete, reconnect the A/C compressor electrical connector.

ENGINE ELECTRICAL

Distributor

REMOVAL & INSTALLATION

1. Disconnect the negative battery cable. Disconnect the wiring harness connectors at the side of the distributor cap.

2. On the Astro:
   a. Remove the glove box.
   b. Remove the engine cover.

c. Remove the air cleaner attaching nut and then move the air cleaner aside.

d. Remove the distributor cap and lay it aside.

3. Disconnect the vacuum advance line.

4. Scribe a mark on the engine in line with the rotor and note the approximate position of the vacuum advance unit in relation to the engine.

5. Remove the distributor hold-down clamp and nut.

6. Lift the distributor from the engine.

7. To install the distributor with the engine undisturbed: (Go to Step 11 if the engine has been disturbed). Reinsert the distributor into its opening, aligning the previously made marks on the housing and the engine block.

8. The rotor may have to be turned either way a slight amount to align the rotor-to-housing marks.

9. Install the retaining clamp and bolt. Install the distributor cap, primary wire, and the vacuum hose. On the Astro, install the glove box, and air cleaner.

10. Start the engine and check the ignition timing. On the Astro, install the engine cover.

11. To install the distributor when the engine has been disturbed: Turn the engine so the No. 1 piston is at the top of its compression stroke. This may be determined by covering the No. 1 spark plug hole with your thumb and slowly turning the engine over. When the timing mark on the crankshaft pulley aligns with the 0 on the timing scale and your thumb is pushed out by compression, No. 1 piston is at top-dead-center (TDC).

12. Install the distributor to the engine block so that the vacuum advance unit points in the correct direction.

13. Turn the rotor so that it will point to the No. 1 terminal in the cap.

14. Install the distributor into the engine block. It may be necessary to turn the rotor a little in either direction in order to engage the gears.

15. Tap the starter a few times to ensure that the oil pump shaft is mated to the distributor shaft.

16. Bring the engine to No. 1 TDC again and check to see that the rotor is indeed pointing toward the No. 1 terminal of the cap.

17. After correct positioning is assured, turn the distributor housing so that it lines up as it did prior to distributor removal. Tighten the retaining clamp. Install the cap and hoses. On the Astro, install the glove box, and air cleaner. Check the timing. On the Astro, install the engine cover.

Alternator

ALTERNATOR PRECAUTIONS

1. When installing a battery, ensure that the ground polarity of the battery and the

ground polarity of the alternator and the regulator are the same.

2. When connecting a jumper battery, be certain that the correct terminals are connected.

3. When charging, connect the correct charger leads to the battery terminals.

4. Never operate the alternator on an open circuit. Be sure that all connections in the charging circuit are tight.

5. Do not short across or ground any of the terminals on the alternator or regulator.

6. Never polarize an AC system.

PRELIMINARY CHARGING SYSTEM TESTS

1. If you suspect a defect in your charging system, first perform these general checks before going on to more specific tests.

2. Check the condition of the alternator belt and tighten it if necessary.

3. Clean the battery cable connections at the battery. Make sure the connections between the battery wires and the battery clamps are good. Reconnect the negative terminal only and proceed to the next step.

4. With the key off, insert a test light between the positive terminal on the battery and the disconnected positive battery terminal clamp. If the test light comes on, there is a short in the electrical system of the van. The short must be repaired before proceeding. If the light does not come on, proceed to the next step.

NOTE: If the truck is equipped with an electric clock, the clock must be disconnected.

5. Check the charging system wiring for any obvious breaks or shorts.

6. Check the battery to make sure it is fully charged and in good condition.

CHARGING SYSTEM OPERATIONAL TEST

NOTE: You will need a current indicator to perform this test. If the current indicator is to give an accurate reading, the battery cables must be the same gauge and length as the original equipment.

1. With the engine running and all electrical systems turned off, place a current indicator over the positive battery cable.

2. If a charge of roughly five amps is recorded, the charging system is working. If a draw of about five amps is recorded, the system is not working. The needle moves toward the battery when a charge condition is indicated, and away from the battery when a draw condition is indicated.

3. If a draw is indicated, proceed with further testing. If an excessive charge (10–15 amps) is indicated, the regulator may be at fault.

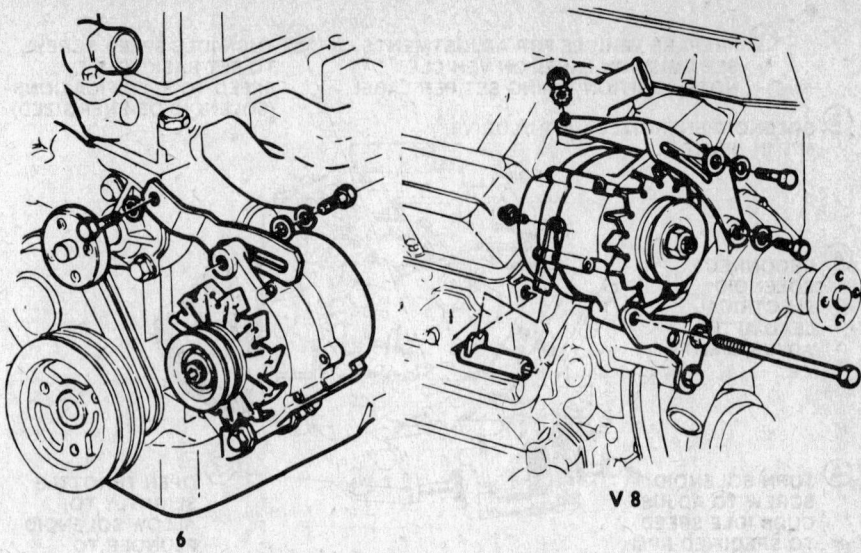

6

V 8

Typical alternator mounting

OUTPUT TEST

1. You will need an ammeter for this test.

2. Disconnect the battery ground cable.

3. Disconnect the wire from the battery terminal on the alternator.

4. Connect the ammeter negative lead to the battery terminal wire removed in step three, and connect the ammeter positive lead to the battery terminal on the alternator.

5. Reconnect the battery ground cable and turn on all electrical accessories. If the battery is fully charged, disconnect the coil wire and bump the starter a few times to partially discharge it.

6. Start the engine and run it until you obtain a maximum current reading on the ammeter.

7. If the current is within ten amps of the rated output of the alternator, the alternator is working properly. If the current is not within ten amps, insert a screwdriver in the test hole in the end frame of the alternator and ground the tab in the test hole against the side of the hole.

8. If the current is now within ten amps of the rated output, remove the alternator and have the voltage regulator replaced. If it is still below ten amps of rated output, have the alternator repaired.

REMOVAL & INSTALLATION

1. Disconnect the battery ground cable to prevent diode damage. On the Astro, remove the upper radiator fan shroud.

2. Disconnect and tag all wiring to the alternator.

3. Remove the alternator brace bolt.

4. Remove the drive belt.

5. Support the alternator and remove the mounting bolts. Remove the alternator.

6. Install the unit using the reverse procedure of removal. Adjust the belt to have

½ in. depression under thumb pressure on its longest run.

Regulator

REMOVAL & INSTALLATION

The regulator on these models is an integral part of the alternator. Alternator disassembly is required to replace it.

VOLTAGE ADJUSTMENT

The 10SI Delcotron is used which is equipped with an integral regulator that cannot be adjusted.

Starter

REMOVAL & INSTALLATION

The following is a general procedure for all trucks, and may vary slightly depending on model and series.

1. Disconnect the battery ground cable at the battery.

2. Raise and support the vehicle.

3. Disconnect and tag all wires at the solenoid terminal. On the Astro, it may be easier to unbolt and lower the starter, supporting it on a suitable block or jack and then detach the wiring.

4. Reinstall all nuts as soon as they are removed, since the thread sizes are different.

5. Remove the front bracket from the starter and the two mounting bolts. On engines with a solenoid heat shield, remove the front bracket upper bolt and detach the bracket from the starter. On the 2.5L engine in the Astro, remove the bolt attaching the rear heat shield at the engine mount and the nuts from the two starter through bolts and remove the shield.

6. Remove the front bracket bolt or nut. Lower the starter front end first, and then remove the unit from the truck. Make sure to retain any shims that are located between the starter on the flywheel end and the engine block.

7. Reverse the removal procedures to install the starter. Torque the two mounting bolts to 25–35 ft. lbs.

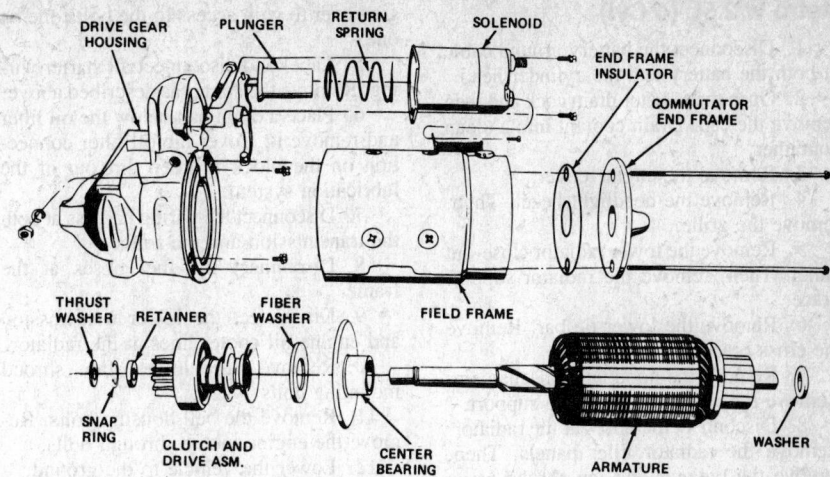

Exploded view of the 20MT starter used on the diesel.

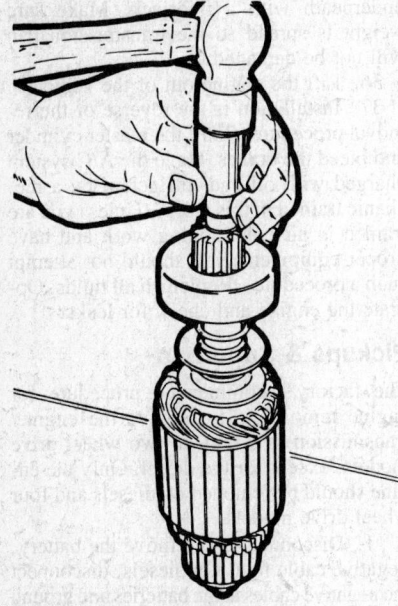

Use a piece of pipe to drive the retainer toward the snap-ring

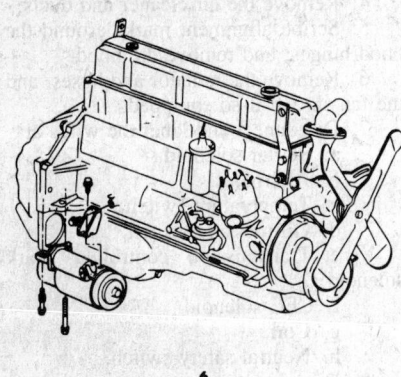

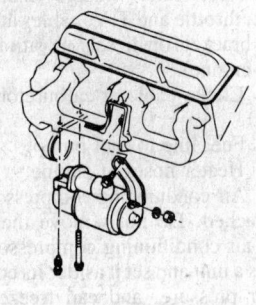

Typical starter mountings

ENGINE MECHANICAL

REMOVAL & INSTALLATION

Vans (Except Astro)

1. Disconnect the negative battery cable, then the positive battery cable, at the battery.
2. Drain the cooling system.
3. Remove the engine cover.
4. Remove the air cleaner. On V8s remove the air stove pipe.
5. Remove the grille. On the six cylinder, remove the grille cross brace. On V8s, remove the upper radiator support and the lower grille valance.
6. Disconnect the radiator hoses at the radiator.
7. On V8s, remove the radiator coolant reservoir bottle.
8. If the van is equipped with an automatic transmission, remove the fluid cooler lines from the radiator.

--- CAUTION ---
Discharging the air conditioning refrigerant should only be attempted by those who have the proper tools and training to do so, as serious personal injury may result. The refrigerant will instantly freeze any surface it comes in contact with, including your eyes.

9. Discharge the air conditioning system and remove the A/C vacuum reservoir. On the V8, remove the A/C condenser from in front of the radiator. On the six cylinder, remove the A/C compressor.
10. Remove the windshield washer jar and bracket.
11. Disconnect the accelerator linkage at the carburetor and remove the carburetor.
12. Remove the radiator support bracket and remove the radiator and the shroud.

13. On the six cylinder, remove the A/C compressor mounting bracket and position the compressor out of the way.
14. On V8s, disconnect the engine wiring harness from the firewall connection. On the six cylinder, disconnect the wiring at the alternator, distributor, oil pressure and temperature sending switches and the starter motor.
15. On V8s:
    a. Disconnect the heater hoses at the engine.
    b. Remove the thermostat housing.
    c. Remove the oil filler pipe.
    d. Remove the cruise control servo, servo bracket and transducer.
16. Raise the vehicle and drain the engine oil.
17. Remove the fuel line from the fuel tank at the fuel pump.
18. Disconnect the exhaust pipe at the manifold.
19. Remove the driveshaft and plug the end of the transmission.
20. Disconnect the transmission shift linkage and the speedometer cable.
21. Remove the transmission mounting bolts.
22. On the six cylinder with manual transmission, disconnect the clutch linkage and remove the clutch cross shaft.
23. On the V8, remove the engine mount bracket-to-frame bolts.
24. Remove the engine mount through bolts.
25. On the six cylinder:
    a. Lower the van and attach a lifting device to the engine.
    b. Raise the engine slightly and remove the right hand mount from the engine.
26. On the V8:
    a. Raise the engine slightly and remove the engine mounts. Support the engine with wood between the oil pan and the crossmember.
27. Remove the engine and transmission as one unit.
28. Reverse the removal procedure to install.

Astro w/2.5L (4 Cyl)

1. Disconnect the battery ground cable at both the battery and the cylinder head.

2. Open the radiator drain cock and then remove the cap. Drain coolant into a clean container.

3. Remove the engine cover.

4. Remove the headlight bezel. Then, remove the grille.

5. Remove the lower radiator close-out panel. Then, remove the radiator support brace.

6. Remove the lower tie bar. Remove the cross brace.

7. Remove the hood latch mechanism. Remove the radiator upper core support.

8. Disconnect the hoses at the radiator. Remove the radiator filler panels. Then, remove the radiator and fan shroud as an assembly.

9. Disconnect the engine harness at the bulkhead connector. Disconnect the harness at the ECM and pull it through the bulkhead.

10. Disconnect the heater hoses at the core.

11. Disconnect the accelerator cable. Disconnect the canister purge hose.

12. Remove the air cleaner and air cleaner adapter.

13. Remove the oil filler neck and the thermostat outlet.

14. Raise the vehicle and support it securely at approved jacking points.

15. Disconnect the exhaust pipe at the manifold.

16. Disconnect the wiring harness at both the transmission and the frame.

17. Label and then disconnect all starter wiring. Remove the starter as described above.

18. Remove the flywheel splash shield.

19. Disconnect the fuel hoses, being careful to collect any fuel that drains out in a metal container.

20. Remove the motor mount through bolts.

21. Remove the bell housing bolts.

22. Lower the vehicle.

23. Install a lifting crane to the engine lifting hooks. Support the transmission from underneath with a floorjack. Make sure weight is spread so the transmission pan will not be damaged.

24. Lift the engine out of the vehicle.

25. Installation is the reverse of the removal procedure. Replenish all fluids. Operate the engine and check for leaks.

Astro w/4.3L V6

1. Disconnect the battery ground cable. Drain the cooling system.

2. Raise the vehicle and support it securely at approved jacking points.

3. Disconnect both exhaust pipes at the manifolds.

4. Disconnect the strut rods at the flywheel inspection cover. Then, remove the inspection cover. Remove the torque converter-to-flywheel bolts by turning the engine over to gain access to the bolts one by one.

5. Label and disconnect all starter wiring. Remove the starter as described above.

6. Place a drain pan below the oil filter and remove it. Cover the oil filter connection on the block to keep dirt out of the lubrication system.

7. Disconnect the wiring harness at both the transmission and the frame.

8. Disconnect the fuel hoses at the frame.

9. Disconnect the lower transmission and engine oil cooler lines at the radiator.

10. Remove the lower fan shroud mounting bolts.

11. Remove the bell housing bolts. Remove the engine mount through bolts.

12. Lower the vehicle to the ground.

13. Remove the headlight bezels and the grille.

14. Remove the lower radiator close-out panel.

15. Remove the radiator support brace and core support cross brace.

16. Remove the lower tie bar.

17. Remove the hood latch mechanism.

18. Disconnect the master cylinder, plug all openings, and lay the master cylinder aside.

19. Remove the upper fan shroud. Remove the upper core support.

18. Disconnect the radiator hoses at the radiator. Disconnect the upper transmission cooler line and upper engine oil cooler line. Remove the radiator.

19. Have a mechanic who does air conditioning work discharge the refrigerant from the air conditioning (A/C) system, if the vehicle has one. Unless you are trained in air conditioning work and have proper equipment, you should not attempt such a procedure.

20. Remove the radiator filler panels.

21. Remove the engine cover and remove the A/C brace at the back of the compressor.

22. Disconnect the A/C compressor hoses leading to the accumulator and condenser and seal open ends.

23. Remove the air conditioning compressor and mounting bracket.

24. Remove the power steering pump as described later in this section.

25. Label and then disconnect all vacuum hoses leading from the engine to components mounted in the engine compartment.

26. Disconnect the engine wiring harness at the bulkhead.

27. Remove the right side kick panel. Then, disconnect the wiring harness at the ESC module and then push it through the bulkhead.

28. Disconnect the two refrigerant lines leading to the A/C accumulator and plug the openings. Disconnect the electrical connections for the pressure cycling switch. Then, remove the accumulator.

29. Disconnect the fuel line at the carburetor and collect any fuel that drains in a metal container.

30. Remove the diverter valve.

31. Remove the (automatic) transmission dipstick tube.

32. Disconnect the heater hoses at the heater core.

33. Disconnect and remove the horn.

34. Remove the Air Injection Reactor system check valves.

35. Install a lifting crane to the engine lifting hooks. Support the transmission from underneath with a floorjack. Make sure weight is spread so the transmission pan will not be damaged.

36. Lift the engine out of the vehicle.

37. Installation is the reverse of the removal procedure. Refill the master cylinder and bleed the brakes. Have the A/C system charged with oil and refrigerant by a mechanic trained in this work. (Unless you are trained in air conditioning work and have proper equipment, you should not attempt such a procedure). Replenish all fluids. Operate the engine and check for leaks.

Pickups & Suburban

The factory recommended procedure for engine removal is to remove the engine/transmission as a unit on two wheel drive models, except for the diesel. Only the engine should be removed on diesels and four wheel drive models.

1. Disconnect and remove the battery, negative cable first. On diesels, disconnect the negative cables at the batteries and ground wires at the inner fender panel.

2. Drain the cooling system.

3. Drain the engine oil.

4. Remove the air cleaner and ducts.

5. Scribe alignment marks around the hood hinges, and remove the hood.

6. Remove the radiator and hoses, and the fan shroud if so equipped.

7. Disconnect and label the wires at:

   a. Starter solenoid.

   b. Alternator.

   c. Temperature switch.

   d. Oil pressure switch.

   e. Transmission controlled spark solenoid.

   f. CEC solenoid.

   g. Coil.

   h. Neutral safety switch.

8. Disconnect:

   a. Accelerator linkage (hairpin at bellcrank, throttle and T.V. cables at intake manifold brackets on diesels. Position away from the engine).

   b. Choke cable at carburetor (if so equipped).

   c. Fuel line to fuel pump.

   d. Heater hoses at engine.

   e. Air conditioning compressor with hoses attached. Do not remove the hoses from the air conditioning compressor. Remove it as a unit and set it aside. Its contents are under pressure, and can freeze body tissue on contact.

   f. Transmission dipstick and tube on automatic transmission models, except for diesel. Plug the tube hole.

g. Oil dipstick and tube. Plug the hole.

h. Vacuum lines.

i. Oil pressure line to gauge, if so equipped.

j. Parking brake cable.

k. Power steering pump. This can be removed as a unit and set aside, without removing any of the hoses.

l. Engine ground straps.

m. Exhaust pipe (support if necessary).

9. Loosen and remove the fan belt, remove the fan blades and pulley. If you have the finned aluminum viscous drive fan clutch, keep it upright in its normal position. If the fluid leaks out, the unit will have to be replaced.

10. Remove the clutch cross-shaft.

11. Attach a chain or lifting device to the engine. If your engine doesn't have any lifting eyes, the usual locations are under the intake manifold bolts on V8s, or under the cylinder head bolts at either end on the sixes. You may have to remove the carburetor. Take the engine weight off the engine mounts, and unbolt the mounts. On all models except the gas engined C10, 1500, C20, and 2500, support and disconnect the transmission. With automatic transmission, remove the torque converter underpan and starter, unbolt the converter from the flywheel, detach the throttle linkage and vacuum modulator line, and unbolt the engine from the transmission. Be certain that the converter does not fall out. With manual transmission, unbolt the clutch housing from the engine.

12. On two wheel drive models, remove the driveshaft. Either drain the transmission or plug the driveshaft opening. Disconnect the speedometer cable at the transmission. Disconnect the shift linkage or lever, or the clutch linkage. Disconnect the transmission cooler lines, if so equipped. If you have an automatic or a four speed transmission, the rear crossmember must be removed. With the three speed, unbolt the transmission from the crossmember. Raise the engine/transmission assembly and pull it forward.

13. On diesels, remove the three bolts, transmission, right side; disconnect the wires to the starter and remove the starter.

14. On four wheel drive, raise and pull the engine forward until it is free of the transmission. On diesels, slightly raise the transmission, remove the three left transmission to engine bolts, and remove the engine.

15. On all trucks, lift the engine out slowly, making certain as you go that all lines between the engine and the truck have been disconnected.

16. Installation is as follows: On four wheel drive and diesels, lower the engine into place and align it with the transmission. Push the engine back gently and turn the crankshaft until the manual transmission shaft and clutch engage. Bolt the transmission to the engine. With automatic transmission, align the converter with the

flywheel, bolt the transmission to the engine, bolt the converter to the flywheel, replace the underpan and starter, and connect the throttle linkage and vacuum modulator line.

17. On two wheel drive, lower the engine/transmission unit into place. Replace the rear crossmember if removed. Bolt the three speed transmission back to the crossmember. Replace the driveshaft.

18. Install the engine mounts.

19. Replace all transmission connections and the clutch cross-shaft. Replace the fan, pulley, and belts.

20. Replace all the items removed from the engine earlier. Connect all the wires which were detached.

21. Replace the radiator and fan shroud, air cleaner, and battery or battery cables. Fill the cooling system and check the automatic transmission fuel level. Fill the crankcase with oil. Check for leaks.

Blazer & Jimmy

1. Disconnect the negative battery cable, then the positive battery cable.

2. Drain the cooling system.

3. Remove the air cleaner.

4. Scribe matchmarks on the hood hinges for reassembly and remove the hood.

5. Remove the radiator and fan shroud as outlined later in this chapter.

6. Disconnect and label (to avoid confusion) the wires at the following locations:

a. Starter solenoid.

b. Alternator.

c. Temperature sending switch.

d. Oil pressure sending switch.

e. Coil.

f. Vacuum advance solenoid and/or the CEC solenoid.

g. TCS solenoid (V8, if so equipped).

7. Disconnect the:

a. Accelerator linkage at the manifold.

b. Fuel line from the tank at the fuel pump.

c. Heater hoses at the engine block.

d. Oil pressure gauge and the vacuum lines at the engine.

e. Evaporative emission system lines at the carburetor and later the hose at the fuel vapor storage canister.

f. Power steering pump at the mounting bracket (lay the pump aside without disconnecting any of the lines).

g. Ground straps at the engine block.

h. Exhaust pipe at the manifold (hang the pipe from the frame with a wire).

i. TCS switch at the transmission (V8, if so equipped).

j. Vacuum line to the power brake unit at the manifold.

8. If equipped with air conditioning, unbolt the compressor at the bracket and lay it aside.

―――― **CAUTION** ――――
Do not disconnect any of the refrigerant lines. Evacuation of the air conditioning

system should only be performed by someone who has the proper skill and training to do so, as the refrigerant will instantly freeze anything it contacts, including your eyes.

9. Raise the vehicle and drain the engine oil.

10. Disconnect the exhaust pipe at the manifold.

11. Remove the flywheel splash shield or the converter housing cover, as applicable.

12. Remove the starter motor.

13. On automatic transmission models remove the converter-to-flywheel attaching bolts.

14. Remove the engine mount through bolts.

15. On four wheel drive models, remove the strut rods at the engine mount.

16. Remove the engine-to-bellhousing attaching bolts.

17. Lower the vehicle.

18. Using a floor jack, raise the transmission slightly.

19. Attach a lifting device to the engine and raise it slightly, taking the weight off the engine mounts.

20. Remove the engine mount-to-engine brackets.

21. Remove the engine.

22. Reverse the removal procedure to install.

S-Series

GAS ENGINES

1. Raise the hood and disconnect the battery cables.

2. Remove the skid plate and drain both the cooling system and the oil pan.

3. Remove the air cleaner assembly and vacuum hoses. Mark the vacuum hoses for reinstallation.

4. Disconnect all hoses, tubing and electrical leads from the engine and mark them for reinstallation.

5. Remove the radiator and fan blade assembly.

6. Disconnect the exhaust pipe from the exhaust manifold.

7. Raise the vehicle and, if equipped with a manual transmission, remove the clutch return spring and cable.

8. Remove the starter motor and fasten it to the frame rail with a piece of wire.

9. Remove the flywheel cover pan.

10. Remove the bell housing bolts and support the transmission.

11. Lift the engine slightly and remove the engine mount nuts.

12. Make certain that all lines, hoses, wires and cables have been disconnected from the engine and the frame.

13. Remove the engine from the vehicle with the front of the engine raised slightly.

14. Installation is the reverse of removal.

DIESEL ENGINE (2WD)

1. Raise engine hood.
2. Disconnect the battery ground cable.
3. Remove the hood.
4. Remove the battery assembly.
5. Remove under cover and drain the cooling system by opening the drain plugs on the radiator and on the cylinder block.
6. Remove the air cleaner assembly as follows:
   a. Remove the intake silencer.
   b. Remove the bolts fixing the air cleaner and loosen the clamp bolt.
   c. Lift the air cleaner slightly and disconnect the breather hose, then remove the air cleaner assembly.
7. Disconnect the upper water hose at the engine side.
8. Loosen the compressor drive belts by moving the power steering pump or idler (if so equipped).
9. Remove the cooling fan and fan shroud.
10. Disconnect the lower water hose at the engine side.
11. Remove the radiator grille.
12. Disconnect the radiator attaching bolts and remove the radiator.
13. Disconnect the accelerator control cable from the injection pump side.
14. Disconnect the air conditioner compressor control cable (if so equipped).
15. Disconnect the fuel hoses from the injection pump.
16. Disconnect the battery cable from the cylinder body.
17. Disconnect the transmission wiring.
18. Disconnect the vacuum hose from the fast idle actuator.
19. Disconnect the connector at fuel cut solenoid.
20. Disconnect the A/C compressor wiring.
21. Disconnect the heater hoses extending from the heater unit from the dash panel side.
22. Disconnect the hose for master-vac from the vacuum pump.
23. Disconnect vacuum hose from the vacuum pump.
24. Disconnect the generator wiring at the connector.
25. Disconnect the exhaust pipe from the exhaust manifold at the flange.
26. Remove the exhaust pipe mounting brake from the engine back plate.
27. Disconnect the starter motor wiring.
28. Disconnect the battery cable from starter motor.
29. Slide the gearshift lever boot upwards on the lever. Remove 2 gearshift lever attaching bolts and remove lever.
30. Place a pan under transmission to receive oil, disconnect speedometer cable at the transmission then disconnect the ground cable.
31. Disconnect the driveshaft at differential side.
32. Remove the driveshaft.

33. Remove return spring from clutch fork.
34. Disconnect clutch cable from hooked portion of clutch fork and pull it out forward through stiffener bracket.
35. Remove two bracket to transmission rear mount bolts and nuts.
36. Raise engine and transmission as required and remove (4) crossmember to frame bracket bolts.
37. Remove the rear mounting nuts from the transmission rear extension.
38. Disconnect electrical connectors at CRS switch and back-up lamp switch.
39. Remove the engine mounting bolt and nuts. Check that the engine is slightly lifted before removing the engine mounting bolt and nuts.
40. Check to make certain all the parts have been removed or disconnected from the engine that are fastened to the frame side. Remove the engine toward front of the vehicle by maneuvering the hoist, so that front part of the engine is lifted slightly. Install in the reverse order.

4WD

1. Raise engine hood.
2. Disconnect the battery ground cable.
3. Remove the engine hood.
4. Remove the battery assembly.
5. Remove under cover and drain the cooling system by opening the drain plugs on the radiator and on the cylinder block.
6. Remove the air cleaner as follows:
   a. Remove the intake silencer.
   b. Remove the bolts fixing the air cleaner and loosen the clamp bolt.
   c. Lift the air cleaner slightly and disconnect the breather hose, then remove the air cleaner assembly.
7. Disconnect the upper water hose at the engine side.
8. Loosen the compressor drive belts by moving the power steering oil pump or idler (if so equipped).
9. Remove the cooling fan and fan shroud.
10. Disconnect the lower water hose at the engine side.
11. Remove the radiator grille.
12. Disconnect the radiator attaching bolts and remove the radiator.
13. Disconnect the accelerator control cable from the injection pump side.
14. Disconnect the air conditioner compressor control cable (if so equipped).
15. Disconnect the fuel hoses from the injection pump.
16. Disconnect the battery cable from the cylinder body.
17. Disconnect the transmission wiring.
18. Disconnect the vacuum hose from the fast idle actuator.
19. Disconnect the connector at fuel cut solenoid.
20. Disconnect the A/C compressor wiring.
21. Disconnect the heater hoses extending from the heater unit from the dash panel side.

22. Disconnect the hose for master-vac from the vacuum pump.
23. Disconnect the vacuum hose from the vacuum pump.
24. Disconnect the generator wiring at the connector.
25. Disconnect the exhaust pipe from the exhaust manifold at the flange.
26. Remove the exhaust pipe mounting brake from the engine back plate.
27. Disconnect the starter motor wiring.
28. Disconnect the battery cable from starter motor.
29. Slide the transmission and transfer gearshift lever boot upwards on each lever, remove gearshift lever attaching bolts.
30. Remove return spring from transfer gear shift lever then remove levers.
31. Remove the transmission.
32. Remove the engine mounting bolts and nuts. Check that the engine is slightly lifted before removing the engine mounting bolts and nuts.
33. Check to make certain all the parts have been removed or disconnected from the engine that are fastened to the frame side. Remove the engine toward front of the vehicle by maneuvering the hoist, so that front part of the engine is lifted slightly. Install in the reverse order.

Cylinder Head

REMOVAL & INSTALLATION

4-119

1. Remove cam cover.
2. Remove EGR pipe clamp bolt at rear of cylinder head.
3. Raise vehicle and support the vehicle.
4. Disconnect exhaust pipe at exhaust manifold.
5. Lower the vehicle.
6. Drain cooling system.
7. Disconnect heater hoses at intake manifold and at front of cylinder head.
8. Remove A/C and/or P/S compressor or pump and lay them aside.
9. Disconnect accelerator linkage at carburetor, fuel line at carburetor, all necessary electrical connections, spark plug wires and necessary vacuum lines.
10. Rotate camshaft until No.4 cylinder is in firing position. Remove distributor cap and mark rotor to housing relationship. Remove the distributor.
11. Remove the fuel pump.
12. Lock the shoe on automatic adjuster in fully retracted position by depressing the adjuster lock lever with a screwdriver or equivalent in direction as indicated in the drawing.
13. Remove timing sprocket to camshaft bolt and remove the sprocket and the fuel pump drive cam from the camshaft. Keep the sprocket on the chain damper and tensioner—do not remove the sprocket from the chain.

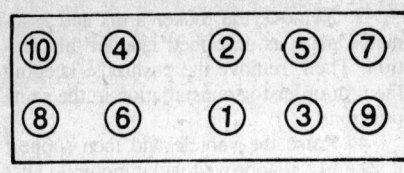

4-119 head bolt torque sequence

14. Disconnect AIR hose and check valve at air manifold.

15. Remove cylinder head to timing cover bolts.

16. Remove cylinder head bolts using Extension Bar Wrench J–24239–01; remove bolts in progressonal sequence, beginning with the outer bolts.

17. Remove the cylinder head, intake and exhaust manifold as an assembly.

18. Clean all gasket material from cylinder head and block surfaces.

NOTE: The gasket surfaces on both the head and block must be clean of any foreign matter and free of nicks or heavy scratches. Cylinder bolt threads in the block and threads on the bolts must be cleaned (dirt will affect bolt torque).

19. Place new gasket over dowel pins with "TOP" side of gasket up. Install the cylinder head in the reverse order. Tighten cylinder head bolts a little at a time in the proper sequence. Torque to 60 ft. lbs. and then retighten to specified torque of 72 ft. lb.

4-121

NOTE: The engine should be "overnight" cold before removing the cylinder head.

1. Disconnect the negative battery cable.

2. Drain the cooling system into a clean container; the coolant can be reused if it is still good.

3. Remove the air cleaner. Raise and support the front of the vehicle.

4. Remove the exhaust shield. Disconnect the exhaust pipe.

5. Remove the heater hose from the intake manifold and then lower the car.

6. Unscrew the mounting bolts and remove the engine lift bracket (includes air management).

7. Remove the distributor. Disconnect the vacuum manifold at the alternator bracket.

8. Tag and disconnect the remaining vacuum lines at the intake manifold and thermostat.

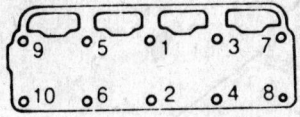

CYLINDER HEAD BOLT
TIGHTENING SEQUENCE

4-121 head bolt torque sequence

9. Remove the air management pipe at the exhaust check valve.

10. Disconnect the accelerator linkage at the carburetor and then remove the linkage bracket.

11. Tag and disconnect all necessary wires. Remove the upper radiator hose at the thermostat.

12. Remove the bolt attaching the dipstick tube and hot water bracket.

13. Remove the idler pulley. Remove the A.I.R. and power steering pump drive belts.

14. Remove the A.I.R. bracket-to-intake manifold bolt. If equipped with power steering, remove the air pump pulley, The A.I.R. thru-bolt and the power steering adjusting bracket.

15. Loosen the A.I.R. mounting bracket lower bolt so that the bracket will rotate.

16. Disconnect and plug the fuel line at the carburetor.

17. Remove the alternator. Remove the alternator brace from the head and then remove the upper mounting bracket.

18. Remove the cylinder head cover. Remove the rocker arms and push rods.

19. Remove the cylinder head bolts in the order given in the illustration. Remove the cylinder head with the carburetor, intake and exhaust manifolds still attached. To install, the gasket surfaces on both the head and the block must be clean of any foreign matter and free of any nicks or heavy scratches. Cylinder bolt threads in the block and the bolt must be clean.

20. Place a new cylinder head gasket in position over the dowel pins on the block. Carefully guide the cylinder head into position.

21. Coat the cylinder bolts with sealing compound and install them finger tight.

22. Using a torque wrench, gradually tighten the bolts in the proper sequence.

23. Installation of the remaining components is in the reverse order of removal.

4-137 Diesel

1. Remove the intake and exhaust manifolds.

2. Remove the intake and exhaust manifold gasket.

3. Drain the cooling system by opening the drain plugs on the radiator and on the cylinder block.

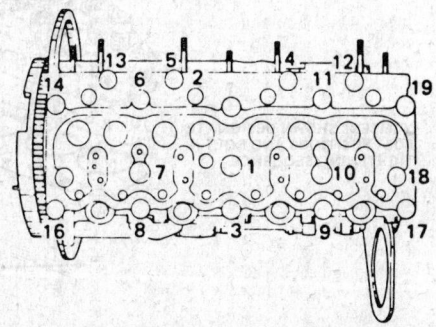

4-137 diesel head bolt torque sequence

4. Disconnect the upper water hose at the engine side.

5. Remove the cooling fan and fan shroud.

6. Remove the sleeve nuts and disconnect the injection pipes.

7. Remove the nozzle holder fixing nuts and remove the nozzle holder assembly.

8. Follow the rocker arm, bracket and shaft assembly removal steps.

9. Remove the pushrods.

10. Remove the joint bolt and disconnect the leak-off pipe.

11. Remove the 19 bolts fixing the cylinder head, then remove the cylinder head and gasket.

12. Install the cylinder head gasket with the TOP mark side up on the cylinder body by aligning the holes with the dowels.

13. Install the cylinder head. Tighten the mounting bolts in proper sequence.

14. Install the pushrod in position on the cylinder head.

15. Install the rocker arm assembly on the cylinder head. Tighten the bracket fixing bolts evenly in sequence commencing with the inner ones.

16. Follow the intake and exhaust manifold installation steps.

17. Install the cooling fan and fan shroud.

18. Connect the upper water hose to engine side.

19. Fill the engine cooling system.

6-173

LEFT SIDE

1. Raise and support the truck.

2. Drain the coolant from the block and lower the truck.

3. Remove the intake manifold.

4. Remove the crossover.

5. Remove the alternator and AIR pump brackets.

6. Remove the dipstick tube.

7. Loosen the rocker arm bolts and remove the pushrods. Keep the pushrods in the same order as removed.

8. Remove the cylinder head bolts in stages and in the reverse order of the tightening sequence.

9. Remove the cylinder head. Do not pry on the head to loosen it.

10. Installation is the reverse of removal. The words "This side Up" on the new cylinder head gasket should face upward. Coat the cylinder head bolts with sealer and torque to specifications in the sequence shown. Make sure the pushrods seat in the lifter seats and adjust the valves.

RIGHT SIDE

1. Raise and support the vehicle. Drain the coolant from the block.

2. Disconnect the exhaust pipe and lower the vehicle.

3. If equipped, remove the cruise control servo bracket.

4. Remove the air management valve and hose.5.

Remove the intake manifold. 6.

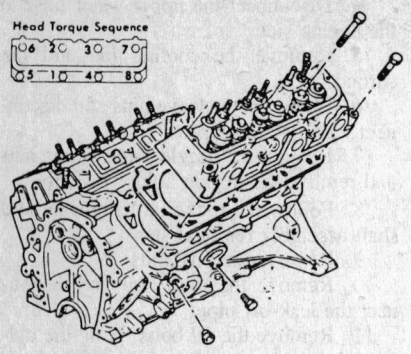

6–173 head bolt torque sequence

Remove the exhaust crossover.

7. Loosen the rocker arm nuts and remove the pushrods. Keep the pushrods in the order in which they were removed.

8. Remove the cylinder head bolts in stages and in the reverse order of the tightening sequence.

9. Remove the cylinder head, do not pry on the cylinder head to loosen it.

10. Installation is the reverse of removal. The words ''This Side Up'' on the new cylinder head gasket should face upwards. Coat the cylinder head bolts with sealer and tighten them to specifications in the sequence shown. Make sure the lower ends of the pushrods seat in the lifter seats and adjust the valves.

2.5L (4 Cyl)

1. Disconnect the negative battery cable at the battery and at the cylinder head and remove it. Remove the rocker cover as described later in this section.

2. Drain the cooling system. Disconnect the upper radiator hose at the intake manifold and thermostat housing, and remove the housing and thermostat. Disconnect or shift aside the heater hoses, if they might interfere with head removal.

3. Disconnect the accelerator cable.

4. Disconnect the alternator brace at the intake manifold. Then, remove the rear alternator bracket and move the alternator aside.

5. Remove the air conditioner compressor mounting bolts and brackets and place the compressor aside. Do not disconnect, disturb or stress the air conditioner hoses!

6. Carefully survey the cylinder head assembly and determine which vacuum lines will have to be disconnected. Then label and disconnect them.

7. Disconnect the fuel lines, collecting fuel that drains in a metal container.

8. Disconnect and remove the ignition coil. Disconnect the spark plug wires at the plugs.

9. Support the vehicle securely at approved jacking points. Disconnect the exhaust pipe at the manifold.

10. Disconnect the oxygen sensor.

11. Lower the vehicle to the ground.

12. Remove the rocker arms and associated parts as described later in this section. Then, remove the pushrods, keeping them in order for reinstallation in the same positions.

13. Loosen the head bolts in several stages, going from bolt to bolt, and turning each a fraction of a turn until tension is lost. Then, remove the bolts. Install a lifting crane and lift the head off the block. Rock the head, if necessary, to break the seal, don't pry it.

14. Carefully remove any dirt or gasket pieces that cling to the block deck or cylinder head gasket surface with a dull scraper. Clean the threads on the cylinder head bolts and those in the head must be clean and dry or the bolts will not torque properly.

15. Install a new gasket in position, right side up, over the dowel pins in the block deck. Note that the gasket will not align with the corners of the block if it is installed upside down.

16. Guide the head precisely into position.

17. Coat both the heads and the threads of the head bolts with sealer, and install them finger tight.

18. Torque the bolts in at least 3 equal stages, using the sequence shown in the illustration, to 92 ft. lbs. final torque.

19. Reverse the remaining removal procedures. Install the pushrods, rockers, balls, and nuts in original positions. Torque the nuts to 20 ft. lbs. Refill the cooling system, run the engine until it reaches operating temperature, and then recheck the coolant level and replenish coolant as necessary. Operate the engine and check for leaks.

4.3L V6

LEFT SIDE

1. Remove the intake manifold.

2. Disconnect the electrical harness at the rocker cover. Then remove the rocker cover.

3. Remove the rocker arms and associated parts as described later in this section. Then, remove the pushrods, keeping them in order for reinstallation in the same positions.

4. Raise the vehicle and then support it securely at approved jacking points. Disconnect the left side of the exhaust Y-pipe at the manifold.

5. Disconnect the air pump at the left side head.

6. Remove the exhaust manifold for this head as described later in this section.

7. Lower the vehicle to the ground. Remove the power steering pump as described later in this section.

8. Remove the A/C idler pulley. Remove the A/C compressor mounting bracket.

9. Note the routing and firing order and then disconnect the spark plug wires at the cylinder head. Remove the spark plugs.

10. Loosen the head bolts in several stages, going from bolt to bolt, and turning each a fraction of a turn until tension is lost. Then remove the bolts. Install a lifting crane and lift the head off the block. Rock the head, if necessary, to break the seal, don't pry it.

11. Carefully remove any dirt or gasket pieces that cling to the block deck or cylinder head gasket surface with a dull scraper. Clean the threads on the cylinder head bolts and those in the head. Threads must be clean and dry or the bolts will not torque properly.

12. If the gasket is made of steel, coat both sides of the new gasket with sealer, making sure the coating is THIN and regular. You can use a paint roller very effectively to do this. If the gasket is steel/asbestos, USE NO SEALER.

13. Position the gasket on the block deck with the bead up and so that is is located by the dowel pins.

14. Guide the cylinder head into position over the dowel pins and position it

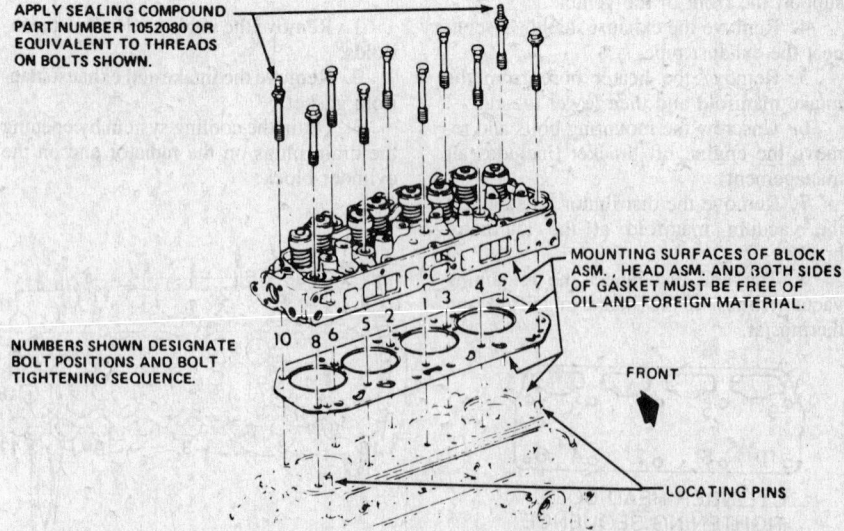

APPLY SEALING COMPOUND PART NUMBER 1052080 OR EQUIVALENT TO THREADS ON BOLTS SHOWN.

MOUNTING SURFACES OF BLOCK ASM., HEAD ASM. AND BOTH SIDES OF GASKET MUST BE FREE OF OIL AND FOREIGN MATERIAL.

NUMBERS SHOWN DESIGNATE BOLT POSITIONS AND BOLT TIGHTENING SEQUENCE.

FRONT

LOCATING PINS

Torque the head bolts in the sequence shown in several stages (2.5 L engine)

squarely over them and the gasket. Install the bolts finger tight.

15. Torque the bolts in at least 3 equal stages, using the sequence shown in the illustration, to 67 ft. lbs. final torque.

16. Install the remaining parts in reverse of the installation procedure. Note that the pushrods must be installed in the same positions as before. Adjust the valves as described below. Refill the cooling system, run the engine until it reaches operating temperature, and then recheck the coolant level and replenish coolant as necessary. Operate the engine and check for leaks.

RIGHT SIDE

1. Remove the intake manifold as described later in this section.

2. Raise and support the vehicle at approved jacking points. Disconnect the exhaust Y-pipe at the exhaust manifold. Then, lower the vehicle.

3. Remove the exhaust manifold for this head as described later in this section.

4. Label and then disconnect the plug wires.

5. Remove the PCV hose and the oil filler tube.

6. Disconnect the Air Injection Reactor pipe and the wiring harness at the back of the right head. Remove the engine ground wire also located there.

7. Remove the rocker cover.

8. Remove the spark plugs.

9. Remove the alternator lower mounting bolt; then, remove the alternator as described earlier in this section and set it aside.

10. Remove the rocker arms and associated parts as described later in this section. Then, remove the pushrods, keeping them in order for reinstallation in the same positions.

11. Loosen the head bolts in several stages, going from bolt to bolt, and turning each a fraction of a turn until tension is lost. Then remove the bolts. Install a lifting crane and lift the head off the block. Rock the head, if necessary, to break the seal, don't pry it.

12. Carefully remove any dirt or gasket pieces that cling to the block deck or cylinder head gasket surface with a dull scraper. Clean the threads on the cylinder head bolts and those in the head must be clean and dry or the bolts will not torque properly.

13. Follow Steps 12–16 of the precedure for the left side cylinder head directly above.

Inline Six Cylinder

1. Drain the cooling system and remove the air cleaner. Disconnect the PCV hose. If equipped, disconnect the air injection hose.

2. Disconnect the accelerator pedal rod at the bellcrank on the manifold, and the fuel and vacuum lines at the carburetor.

3. Disconnect the exhaust pipe at the manifold flange, then remove the manifold bolts and clamps and remove the manifolds and carburetor as an assembly.

4. Remove the fuel and vacuum line

retaining clip from the water outlet. Then disconnect the wire harness from the heat sending unit and coil, leaving the harness clear of clips on the rocker arm cover.

5. Disconnect the radiator hose at the water outlet housing and the battery ground strap at the cylinder head.

6. Disconnect the wires and remove the spark plugs. Disconnect the coil-to-distributor primary wire lead at the coil and remove the coil on models without HEI.

7. Remove the rocker arm cover. Back off the rocker arm nuts, pivot the rocker arms to clear the pushrods and remove the pushrods.

8. Remove the cylinder head bolts, cylinder head and gasket.

9. To install: Place a new cylinder head gasket over the dowel pins in the cylinder block with the head up. Do not use sealer on composition steel/asbestos gaskets.

10. Guide and lower the cylinder head into place over the dowels and gasket.

11. Use sealant on the cylinder head bolts, install and tighten them down snugly.

12. Tighten the cylinder head bolts a little at aa time with a torque wrench in the correct sequence. Final torque should be as specified.

13. Install the valve pushrods down through the cylinder head openings and seat them in their lifter sockets.

14. Install the rocker arms, balls and nuts and tighten the rocker arm nuts until all pushrod play is taken up.

15. Install the thermostat, the thermostat housing and the water outlet using new gaskets. Then connect the radiator hose.

16. Install the temperature sending switch.

17. Install the spark plugs.

18. Use new plug gaskets (if required) and torque to specifications.

19. Install the coil, then connect the heat sending unit and coil primary wires, and the battery ground cable at the cylinder head.

20. Clean the surfaces and install a new gasket over the manifold studs. Install the manifold. Install the bolts and clamps and torque as specified.

21. Connect the throttle linkage.

22. Connect the PCV fuel and vacuum lines and secure the lines in the clip at the water outlet. Connect the air injection line.

23. Fill the cooling system and check for leaks.

24. Adjust the valve lash as explained later.

25. Install the rocker arm cover and position the wiring harness in the clips.

26. Clean and install the air cleaner.

V8 Gas Engines

1. Remove the intake manifold.

2. Remove the exhaust manifolds.

3. If the truck is equipped with air conditioning, remove the A/C compressor and the forward mounting bracket and lay the compressor aside. Do not disconnect any of the refrigerant lines.

4. Remove the rocker covers. Back off the rocker arm nuts and pivot the rocker arms out of the way so that the pushrods can be removed. Identify the pushrods so that they can be installed in their original positions.

5. Remove the cylinder head bolts and remove the heads.

6. Install the cylinder heads using new gaskets. Coat a steel gasket on both sides with sealer. If a composition gasket is used, do not use sealer.

7. Clean the bolts, apply sealer to the threads, and install them hand tight.

8. Tighten the head bolts a little at a time in the proper sequence. Head bolt torque is listed in the Torque Specifications chart.

9. Install the intake and exhaust manifolds and components in the reverse order of removal. Adjust the rocker arms aand check ignition timing.

Diesel

1. Remove the intake manifold.

2. Remove the rocker arm cover(s), after removing any accessory brackets which interfere with cover removal.

3. Disconnect and label the glow plug wiring.

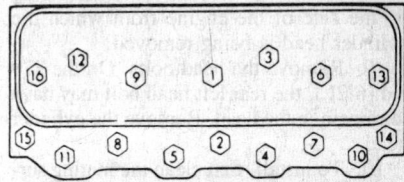

8-454 head bolt torque sequence

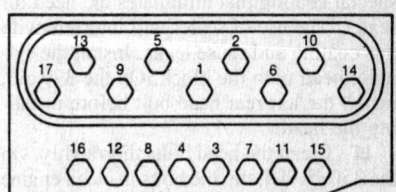

Small block V8 head bolt tightening sequence

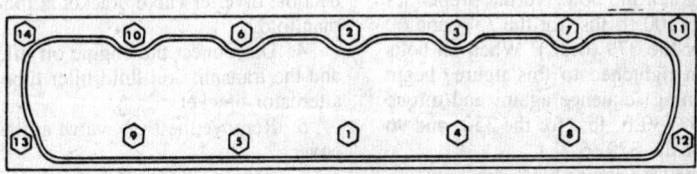

6-250, 292 head bolt torque sequence

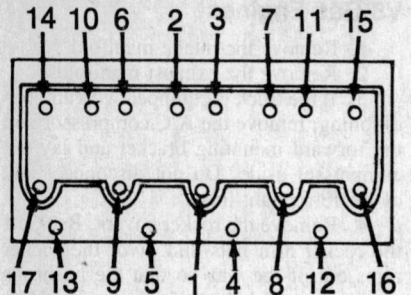

8-379 (6.2L) diesel head bolt torque sequence

8-350 (5.7L) diesel head bolt torque sequence

4. On the right cylinder head, remove the ground strap from the head.

5. Remove the rocker arm bolts, the bridge pivots, the rocker arms, and the pushrods, keeping all the parts in order so that they can be returned to their original positions. It is a good practice to number or mark the parts to avoid interchanging them.

6. Remove the fuel return lines from the nozzles.

7. Remove the exhaust manifold(s).

8. Remove the engine block drain plug on the side of the engine from which the cylinder head is being removed.

9. Remove the head bolts. On the 379 cid (6.2L), the rear left head bolt may have to remain in the head. Remove the cylinder head.

10. To install, first clean the mating surfaces thoroughly. Install new head gaskets on the engine block. Do NOT coat the gaskets with any sealer. The gaskets have a special coating that eliminates the need for sealer. The use of sealer will interfere with this coating and cause leaks. Install the cylinder head onto the block. On the 379 cid, install the left rear head bolt before installing the head.

11. Clean the head bolts thoroughly. On the 350 (5.7L), dip the bolts in clean engine oil and install into the cylinder block until the heads of the bolts lightly contact the cylinder head. On the 379 cid, coat the bolt threads with G sealer.

12. Tighten the bolts, in the proper sequence, to 100 ft. lbs. for the 350, and 60 ft. lbs. for the 379 (6.2L). When all bolts have been tightened to this figure, begin the tightening sequence again, and torque all bolts to 130 ft. lbs. for the 350, and 96 ft. lbs. for the 379 (6.2L).

13. Install the engine block drain plug(s), the exhaust manifold(s), the fuel return lines,

the glow plug wiring, and the ground strap for the right cylinder head.

14. After disassembling, cleaning, and reassembling the valve lifters, bleed them down and install them into the engine. Install the pushrods, rocker arms, and pivots into their original locations.

15. Install the intake manifold.

16. Install the rocker cover(s). The covers do not use gaskets, but are sealed with a bead of RTV (room temperature vulcanizing) silicone sealer instead. Apply a $\frac{3}{32}$ inch bead of RTV sealer, G.M. #1052289 or the equivalent, to the clean and dry mating surface of the rocker arm cover. Run the bead of sealer to the inside of the bolt holes. Install the cover to the head within 10 minutes (while the sealer is still wet).

Rocker Arm Cover

REMOVAL & INSTALLATION

Except Astro 4.3L

1. Disconnect the negative battery cable.

2. Remove the air cleaner.

3. Disconnect the crankcase ventilation hose at the rocker arm cover.

4. Disconnect the wiring from the rocker cover clips. If so equipped, disconnect the Air Injection Reactor hoses at the diverter valve. Then, disconnect the diverter valve bracket at the intake manifold.

5. Remove the carburetor heat stove pipe, on models so equipped. On diesels, remove the injection lines.

6. If the truck is equipped with A/C, remove the compressor rear brace. Do not disconnect any of the refrigerant lines.

7. Remove the rocker arm attaching bolts and remove the cover. If the cover is difficult to remove, gently tap the front of the cover rearward with your hand or a rubber mallet. If this still does not work, CAREFULLY pry the cover off. Be very careful not to distort the sealing surface.

8. On installation, apply a $\frac{3}{16}$ in. bead of sealer to the mating surface after removing all the old loose sealer.

9. Reverse the removal procedure to install.

Astro 4.3L V6

RIGHT SIDE

1. Disconnect the battery negative cable and remove the engine cover.

2. Remove the air cleaner.

3. Disconnect the Air Injection Reactor hoses at the diverter valve. Then, disconnect the diverter valve bracket at the intake manifold.

4. Disconnect the engine oil filler tube and the transmission fluid filler tube at the alternator bracket.

5. Remove the PCV valve at the valve cover.

6. Remove the bolts retaining the Air Injection Reactor pipes at the back of the

cylinder head and move the hose out of the way.

7. Remove the distributor cap and wires. Then, remove the rocker cover bolts and remove the rocker cover.

8. Scrape all remaining pieces of gasket off both sealing surfaces with a dull scraper. Coat both sides of a new gasket with sealer, and install it and the cover with all boltholes aligned.

9. Install the bolts and torque alternately in several stages to 4 ft. lbs.

10. Reverse the remaining removal procedures to complete the installation.

LEFT SIDE

1. Disconnect the battery negative cable and remove the engine cover.

2. Remove the air cleaner.

3. Disconnect the vacuum pipe at the carburetor.

4. Disconnect the electrical harness at the rocker cover.

5. Disconnect the detent and accelerator cables at the carburetor. Then, remove the bracket for the detent and accelerator cables at the intake manifold.

6. Remove the rocker arm cover bolts. Remove the rocker cover.

7. Scrape all remaining pieces of gasket off both sealing surfaces with a dull scraper. Coat both sides of a new gasket with sealer, and install it and the cover with all boltholes aligned.

8. Install the bolts and torque alternately in several stages to 4 ft. lbs.

9. Reverse the remaining removal procedures to complete the installation. Adjust the detent and accelerator cables.

Valve System

VALVE LASH ADJUSTMENT

4–119 & 4–137 Diesel

NOTE: The valves are adjusted with the engine cold.

1. Make sure that the cylinder head and camshaft retaining bolts are tightened to the proper torque.

2. Remove the camshaft carrier side cover.

3. Turn the crankshaft with a wrench on the front pulley attaching bolt or by bumping the engine with the starter or remote starter button until the No.1 piston is at TDC of the compression stroke. You can tell when the piston is coming up on the compression stroke by removing the spark plug and placing your thumb over the hole and you will feel air being forced out of the spark plug hole past your thumb. Stop turning the crankshaft when the TDC timing mark on the crankshaft pulley is directly aligned with the timing pointer.

4. With the No.1 piston at TDC of the compression stroke, check the clearance between the rocker arm an the camshaft with the proper thickness feeler gauge on

Nos. 1 and 2 intake valves and Nos. 1 and 3 exhaust valves.

5. Adjust the clearance by loosening the locknut with an open-end wrench, turning the adjuster screw with a phillips head screwdriver and retightening the locknut. The proper thickness feeler gauge should pass between the camshaft and the rocker with a slight drag when the clearance is correct.

6. Turn the crankshaft one full turn to position the No. 4 piston at TDC of its compression stroke. Adjust the remaining valves: Nos. 2 and 4 exhaust and Nos. 3 and 4 intake in the same manner as outlined in Step 5.

7. Install the camshaft carrier side-cover.

Gas Engines Except 4.3L

1. Remove the rocker covers and gaskets.

2. Adjust the valves on inline six cylinder engines as follows:

a. Mark the distributor housing with a piece of chalk at Nos. 1 and 6 plug wire positions. Remove the distributor cap with the plug wires attached.

b. Crank the engine until the distributor rotor points to No. 1 cylinder (piston on compression stroke at TDC). At this point, adjust the following valves: No. 1 Exhaust and Intake; No. 2 Intake; No. 3 Exhaust; No. 4 Intake; No. 5 Exhaust

c. Back out the adjusting nut until lash is felt at the pushrod, then turn the adjusting nut in until all lash is removed. This can be determined by checking pushrod end-play while turning the adjusting nut. When all play has been removed, turn the adjusting nut in 1 full turn.

d. Crank the engine until the distributor rotor points to No. 6 cylinder (on compression stroke at TDC). The following valves can be adjusted: No. 2 Exhaust; No. 3 Intake; No. 4 Exhaust; No. 5 Intake; No. 6 Intake and Exhaust.

3. Adjust the valves on V8 engines as follows:

a. Crank the engine until the mark on the damper aligns with the TDC or 0° mark on the timing tab and the engine is in No. 1 firing position. This can be determined by placing the fingers on the No. 1 cylinder valves as the marks align. If the valves do not move, it is in No. 1 firing position. If the valves move, it is in No. 6 firing position and the crankshaft should be rotated one more revolution to the No. 1 firing position.

b. The adjustment is made in the same manner as 6 cylinder engines.

c. With the engine in No. 1 firing position, the following valves can be adjusted: Exhaust Nos. 1,3,4,8; Intake Nos. 1,2,5,7

d. Crank the engine 1 full revolution until the marks are again in alignment. This is No. 6 firing position. The following valves can now be adjusted: Exhaust Nos. 2,5,6,7; Intake Nos. 3,4,6,8

4. Reinstall the rocker arm covers using new gaskets.

5. Install the distributor cap and wire assembly.

4.3L V6

1. Remove the valve covers. If you have just completed reassembly of the valve train after parts have been replaced, it is wise to check that all the pushrods are properly seated in the lifter sockets. Crank the engine until it reaches the center or ''0'' mark on the timing tab on the front cover. Put your fingers in contact with both No. 1 cylinder rocker arms and feel for motion while cranking the engine. If, as the crankshaft comes up on the ''0'' mark there is no motion in the rockers, the engine is in proper position (at TDC of the compression stroke). If there is motion, turn the engine another 360°, following the same procedure, to get No. 1 cylinder to firing position.

2. In this position adjust: Exhaust valves Nos. 1, 5 and 6; Intake valves Nos. 1, 2, 3. Back out the adjusting nut until lash is felt at the pushrod. Then tighten the nut down very gradually until lash just disappears. A precise way to do this is to rotate the pushrod with your fingers. When valve lash is lost, the effort required to turn the pushrod abruptly increases a great deal. Note the position of the adjusting nut precisely at this point. Then turn it downward (clockwise) exactly one more turn (360°). Repeat the adjustment procedure for each of the valves listed with the engine in this position.3.

Turn the engine one full turn (360°) until the timing mark again reaches the ''0'' mark. The engine will now be in No. 4 firing position. Adjust the remaining valves: Exhaust: Nos. 2, 3 and 4; Intake: Nos. 4, 5, 6. 4.

Reinstall the valve covers with new gaskets as described above.

VALVE GUIDES

Valve guides are integral with the cylinder head on all engines. Valve guide bores may be reamed to accommodate oversize valves. If wear permits, valve guides can be knurled to allow the retention of standard valves. Maximum allowable valve stem-to-guide bore clearances are listed on the Valve Specifications Chart.

Rocker Arms

REMOVAL & INSTALLATION

4-119

1. Remove the camshaft carrier as outlined under Cylinder Head Removal.

2. Remove the rocker spring from the pivot and lift the rocker from the cylinder head. Be careful not to lose the rocker guide

resting on the top of each of the valves.

3. Install in the reverse order of removal.

4-137 Diesel

1. Remove the rocker cover.

2. Remove the 8 bolts fixing the rocker arm brackets in sequence commencing with the outer ones.

3. Remove the rocker arm, bracket and shaft assembly.

4. To install, follow the removal procedure in reverse order.

5. Tighten the bracket fixing bolts evenly in sequence commencing with the inner ones to 15 ft. lbs.

4-121

1. Remove the air cleaner. Remove the cylinder head cover.

2. Remove the rocker arm nut and ball. Lift the rocker arm off the stud. Always keep the rocker arm assemblies together and install them on the same stud. Remove the push rods.

3. To install: Coat the bearing surfaces of the rocker arms and the rocker arm balls with ''Molykote'' or its equivalent.

4. Install the push rods making sure that they seat properly in the lifter.

5. Install the rocker arms, balls and nuts. Tighten the rocker arm nuts until all lash is eliminated.

6. Adjust the valves when the lifter is on the base circle of a camshaft lobe: Crank the engine until the mark on the crankshaft pulley lines up with the '0' mark on the timing tab. Make sure that the engine is in the No. 1 firing position. Place your fingers on the No. 1 rocker arms as the mark on the crank pulley comes near the '0' mark. If the valves are not moving, the engine is in the No. 1 firing position. If the valves move,

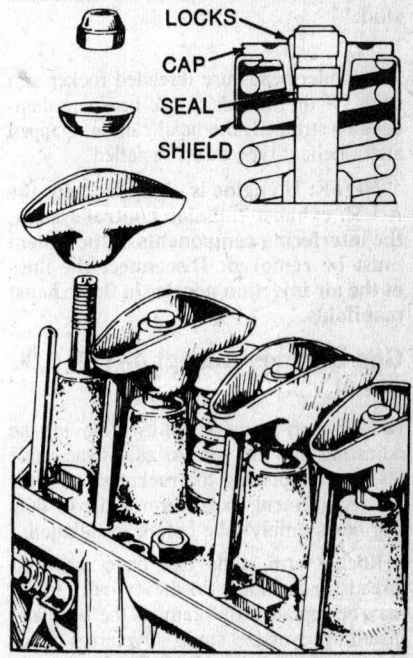

Rocker arm components

the engine is in the No.4 firing position; rotate the engine one complete revolution and it will be in the No.1 position.

7. When the engine is in the No.1 firing position, adjust the following valves: Exhaust Nos.1, 3; Intake Nos.1, 2.

8. Back the adjusting nut out until lash can be felt at the push rod, then turn the nut until all lash is removed (this can be determined by rotating the push rod while turning the adjusting nut). When all lash has been removed, turn the nut in 1½ additional turns, this will center the lifter plunger.

9. Crank the engine one complete revolution until the timing tab and the '0' mark are again in alignment. Now the engine is in the No.4 firing position. Adjust the following valves: Exhaust Nos.2, 4; Intake Nos.3, 4.

10. Installation of the remaining components is in the reverse order of removal.

6-173

NOTE: Some engines are assembled using RTV (Room Temperature Vulcanizing silicone sealant in place of rocker arm cover gasket. If the engine was assembled using RTV, never use a gasket when reassembling. Conversely, if the engine was assembled using a rocker arm cover gasket, never replace it with RTV. When using RTV, an ⅛ in. bead is sufficient. Always run the bead on the inside of the bolt holes.

Rocker arms are removed by removing the adjusting nut. Be sure to adjust valve lash after replacing rocker arms.

NOTE: When replacing an exhaust rocker, move an old intake rocker arm to the exhaust rocker arm stud and install the new rocker arm on the intake stud.

Cylinder heads use threaded rocker arm studs. If the threads in the head are damaged or stripped, the head can be retapped and a helical type insert installed.

NOTE: If engine is equipped with the A.I.R. exhaust emission control system, the interfering components of the system must be removed. Disconnect the lines at the air injection nozzles in the exhaust manifolds.

Gas Engines Except 4-119, 2.5L & 4.3L

Rocker arms are removed by removing the adjusting nut. Be sure to adjust the valve lash after replacing the rocker arms. Coat the replacement rocker arm and ball with engine assembly lube before installation.

Rocker arm studs that have damaged threads or are loose in the cylinder heads may be replaced by reaming the bore and installing oversize studs. Oversizes available are .003 and .013 in. The bores may

also be tapped and screw-in studs installed. Several aftermarket companies produce complete rocker arm stud kits with installation tools.

2.5L (4 cyl)

1. Remove the rocker cover as described above.

2. Remove the rocker bolt, ball, and then the rocker arm. Keep them in order for installation in the same positions if they might be reused. Note that if you are replacing the pushrod only, you can loosen the nut until it is nearly to the top of the sutd and then swing the rocker out of the way without removing it. The pushrod may now be removed and replaced.

3. If replacing the pushrod, install it and make sure it seats in the lifter. Then, coat the wear surfaces of any new rocker, ball, or pushrod parts with a lubricant designed for engine rebuilding purposes (an example is Molykote™).

4. Install the parts in proper order, making sure the top of the pushrod engages with the end of the rocker where it is recessed. Install the bolt and torque it to 20 ft. lbs. The bolt must not be overtorqued!

5. Install the rocker cover.

4.3L V6

1. Remove the rocker covers.

2. Remove the rocker arm nuts, and then the rocker arm balls, rocker arms, and pushrods (if necessary). Keep them in order for installation in the same positions if they might be reused.

3. If replacing the pushrod, install it and make sure it seats in the lifter. Then, coat the wear surfaces of any new rocker, ball, or pushrod parts with a lubricant designed for engine rebuilding purposes (an example is Molykote™).

4. Install the parts in proper order, making sure the top of the pushrod engages with the end of the rocker where it is recessed. Just start the nut onto the stud, its final position will be reached in the next step.

5. Adjust the valves (see the valve adjustment procedure).

6. Install the rocker cover as described above.

Diesel Engines Except 4-137

1. Remove the air cleaner, high pressure fuel lines to the injectors, and the rocker arm cover.

2. Remove the arm pivot bolts and the pivot(s). Remove the rocker arms. The use of bridged pivots require that the rocker arms be removed in pairs.

3. To install, position the set of rocker arms in the original locations.

4. Lubricate the pivot contact surfaces and install the pivot(s).

5. Install the pivot bolts. Tighten the bolts alternately and evenly to 25 ft. lbs., following the bleed down procedure outlined previously under Valve Lash Adjustment for diesel engines.

6. Install the rocker arm cover as outlined in Step 16 of the diesel engine cylinder head removal and installation procedure. Install the fuel lines and the air cleaner.

VALVE LIFTER SERVICE

Diesel Engines except 4-137

Whenever the rocker arms and the intake manifold have been removed, the valve lifters must be removed, disassembled, assembled while submerged in diesel fuel or kerosene and bled down using a specially weighted press. The lifters also must be disassembled, reassembled while submerged and bled down on the press whenever they are removed. Note that if the rocker arms have been removed but the intake manifold has not been disturbed, the lifters can be bled down as outlined in the "Valve Lash Adjustment" procedure for diesel engines. The following procedure is to be used for lifter removal and installation, or whenever both the rocker arms and the intake manifold have been disturbed.

1. Remove the intake manifold.

2. Remove the rocker covers, the rocker arms, and the pushrods. Keep all the parts in order so that they may be installed in their original locations.

3. Remove the valve lifters.

4. To disassemble the lifters: Remove the retainer ring with a small screwdriver.

5. Remove the pushrod seat. Remove the oil metering valve. Remove the plunger and plunger spring. Remove the check valve retainer from the plunger, and remove the valve and its spring.

6. Clean all parts in a safe solvent. Check for burrs, nicks, scoring, or excessive wear, and replace as necessary.

7. Check the lifter foot for excessive wear: Place a straightedge across the lifter foot. Hold the lifter at eye level. Check for light appearing between the lifter foot and the straightedge. If light is visible, indicating a concave surface, the lifter should be replaced and the camshaft inspected. If the cam lobe is worn across the full width of the cam base circle (opposite the high lobe of the cam), the camshaft should be replaced. Wear at the center of the cam base circle is normal. Wear across the full width of the nose of the lobe is also normal.

8. After the lifter parts have been cleaned, assemble the valve disc spring and retainer into the plunger. Be sure the retainer flange is pressed tightly against the bottom of the recess in the plunger.

9. Install the plunger spring over the check retainer.

10. Hold the plunger with the spring up. Insert into the lifter body. Hold the plunger vertically while doing this to avoid cocking the spring.

11. Fill the reservoir of G.M. tool J-5790 with kerosene to within ½ inch of the top

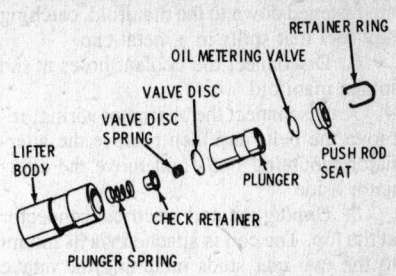

Exploded view of the 5.7L diesel valve lifter

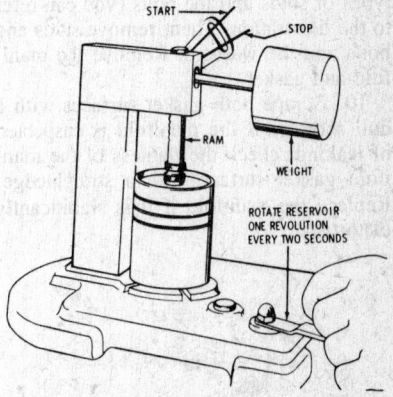

A specially weighted hand press must be used to assemble the diesel valve lifters

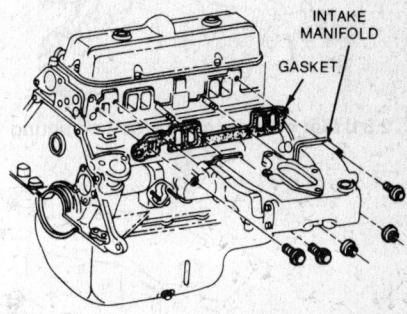

4-119 intake manifold

of the reservoir. This tool is a specially weighted press with provision for reservoir rotation.

12. Place the valve lifter assembly into the reservoir. Position the oil control valve and the pushrod seat onto the plunger.

13. Install a ¼ inch steel ball onto the pushrod seat. Lower the tester ram until it contacts the steel ball. Do not press on the ram. Allow the ram to move downward by its own weight, until the air bubbles expelled from the lifter assembly disappear.

14. Raise the ram, then allow it to lower by its own weight. Repeat this operation until all air is expelled from the lifter. Do not attempt to hasten the process by pumping the ram up and down.

15. After all air has been expelled, allow the ram to descend, bleeding the lifter, until the retaining ring groove is exposed. Install the retaining ring.

16. Adjust the ram screw so that it contacts the steel ball in the pushrod seat at the same time as the pointer is at the start line.

17. Raise the arm of the tester, then start the bleed down test by resting the ram on the steel ball and starting a timer. Rotate the reservoir one revolution every two seconds, and time the indicator from the start line to the stop line. Acceptable leak down time is 6 seconds minimum for used lifters, and from 9–60 seconds for a new lifter.

18. If the lifter leak down rate falls within the specified limit, the lifter may be reused. If not, new lifters should be installed in the engine.

19. If new lifters are to be installed, they must first be filled with kerosene or diesel fuel. Install the lifter in the tester. Fill the reservoir to within ½ inch of the top, and fill the lifter as outlined in Steps 13–15.

20. To install the lifters: Coat the foot of the lifter with G.M. lubricant #562458 or the equivalent.

21. Install the lifters into their original positions. Install the pushrods into their original positions.

22. Install the intake manifold.

23. Install the rocker arms and pivots.

24. Bleed down the lifters as outlined under Valve Lash Adjustment for the diesel engine.

25. Install the rocker covers.

Intake Manifold

REMOVAL & INSTALLATION

4-119

1. Disconnect the battery ground cable and remove the air cleaner assembly.

2. Remove the EGR pipe clamp bolt at the rear of the cylinder head.

3. Raise the vehicle and remove the EGR pipe from the intake and exhaust manifolds.

4. Remove the EGR valve and bracket assembly from the intake manifold.

5. Lower the vehicle and drain the cooling system.

6. Remove the upper coolant hoses from the manifold.

7. Disconnect the accelerator linkage, vacuum lines, electrical wiring and fuel line from the intake manifold.

8. Remove the retaining nuts and remove the manifold from the cylinder head.

9. Remove the lower heater hose while holding the manifold away from the engine. Remove the manifold from the vehicle.

10. Installation is the reverse of removal.

4-137 Diesel

1. Raise engine hood.

2. Remove the bolts fixing the air cleaner and loosen the clamp bolt.

3. Lift the air cleaner slightly and disconnect the breather hose, then remove the air cleaner assembly.

4. Remove the 2 bolts and 4 nuts mounting the intake manifold.

5. Remove the intake manifold.

6. Installation is the reverse of removal. Torque the bolts to 15 ft. lbs.

4-121

1. Disconnect the negative battery cable.

2. Remove the air cleaner. Drain the cooling system.

3. Tag and disconnect all necessary vacuum lines and wires. Remove the idler pulley.

4. Remove the A.I.R. drive belt. If equipped with power steering, remove the drive belt and then remove the pump with the lines attached. Position the pump out of the way.

5. Remove the A.I.R. bracket-to-intake manifold bolt. Remove the air pump pulley.

6. If equipped with power steering, remove the A.I.R. thru-bolt and then the power steering adjusting bracket.

7. Loosen the lower bolt on the air pump mounting bracket so that the bracket will rotate.

8. Disconnect the fuel line at the carburetor. Disconnect the carburetor linkage and then remove the carburetor.

9. Lift off the Early Fuel Evaporation (EFE) heater grid.

10. Remove the distributor.

11. Remove the mounting bolts and nuts and remove the intake manifold. Make sure to disconnect the heater hose and condenser from the bottom of the intake manifold before you lift it all the way out.

12. Using a new gasket, replace the manifold, tightening the nuts and bolts to specification.

13. Installation of the remaining components is in the reverse order of removal. Adjust all necessary drive belts and check the ignition timing.

2.5L (4 cyl)

1. Disconnect the negative battery cable. On the Astro van, remove the glovebox assembly and the engine cover.

2. Remove the air cleaner assembly. Drain the cooling system.

3. Disconnect the rail that mounts the vacuum pipes at both the thermostat housing and the exhaust manifold.

4. Go around the engine and label (if necessary) and then disconnect all vacuum and electrical connectors that would interfere with manifold removal. Where these lines are routed so as to be mounted on or near the manifold, disconnect mounts or harnesses also.

5. Disconnect the accelerator cable at the TBI throttle body. Disconnect fuel lines at the TBI throttle body and anywhere they

are fastened down to the manifold, catching any fuel that spills in a metal cup.

6. Disconnect the coolant hoses at the intake manifold.

7. Disconnect the alternator wiring, remove the belt, and then remove the alternator mounting bolts and move the alternator aside.

8. Unplug the coil electrical connector at the top. The coil is attached via its mount to the two rear studs mounting the intake manifold. Remove the ignition coil attacting nuts at the intake manifold, and remove the coil assembly.

9. Note locations of the two different types of studs and the bolts (you can refer to the illustration. Then, remove studs and bolts and the washers. Remove the manifold and gasket.

10. Scrape both gasket surfaces with a dull scraper. If the manifold is suspected of leaking, check the flatness of the manifold gasket surface with a straightedge. Replace the manifold if it is significantly distorted.

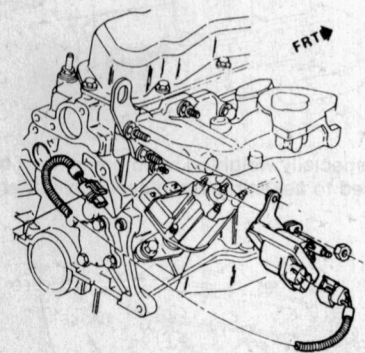

2.5 L four cylinder engine coil mounting

11. Install a new gasket with all holes lined up and then position the manifold against it. Install all the studs and then install finger tight.

12. First torque the studs numbered 1 (in the illustration) to 25 ft. lbs.; then torque those numbered 2 to 37 ft. lbs.; finally torque those numbered 3 to 28 ft. lbs. Reverse the remaining removal procedures to complete the installation.

4.3L V6

1. Disconnect the negative battery cable. Remove the engine cover.

2. Remove the air cleaner. Drain the cooling system.

3. Mark and then remove the distributor cap and plug wires. Disconnect the Electronic Spark Control connector. Then, remove the distributor (you can refer to the removal procedure located earlier in this section, if necessary).

4. Disconnect the transmission and accelerator cables at the throttle linkage.

5. Remove the air conditioning compressor rear brace, leaving the compressor and lines in position.

6. Disconnect both the transmission and engine oil filler tubes at the alternator brace.

7. Remove the air conditioner compressor idler pulley at the alternator brace. Then, remove the alternator brace.

8. Disconnect the fuel line at the carburetor, catching any fuel that spills in a metal cup.

9. Label and then remove any vacuum hoses and electrical connectors at the carburetor which will interfere with carburetor and manifold removal.

10. Remove Air Injection Reactor hoses and their brackets.

11. Disconnect the heater hose at the manifold.

12. Remove the manifold bolts and remove the manifold.

14. Clean all RTV sealant and any other foreign material from the gasket and seal surfaces on the manifold, block and heads with a degreaser.

15. Install the gaskets on the heads and then run a $\frac{3}{16}$ in. bead of RTV sealer #1052917 or equivalent on the front and rear ridges of the cylinder case. The bead must also be extended $\frac{1}{2}$ in. up each cylinder head to seal the manifold side gaskets and retain them during manifold installation. Apply an appropriate sealer at the water passages as well.

16. Carefully put the manifold in position precisely, with all thread holes and passages precisely lined up. Install the studs and bolts finger tight. Torque them to 30 ft. lbs. first in the order shown at the top of the illustration. Then, repeat the torquing operation at 30 ft. lbs. in the second sequence.

17. Perform the remaining steps in the reverse of the removal procedure. Operate the engine and check for leaks.

6–173

1. Remove the rocker covers.

2. Drain the cooling system.

3. If equipped, remove the AIR pump and bracket.

4. Remove the distributor cap. Mark the position of the ignition rotor in relation to the distributor body, and remove the distributor. Do not crank the engine with the distributor removed.

5. Remove the heater and radiator hoses from the intake manifold.

6. Remove the power brake vacuum hose.

7. Disconnect and label the vacuum hoses. Remove the EFE pipe from the rear of the manifold.

8. Remove the carburetor linkage. Disconnect and plug the fuel line.

9. Remove the manifold retaining bolts and nuts.

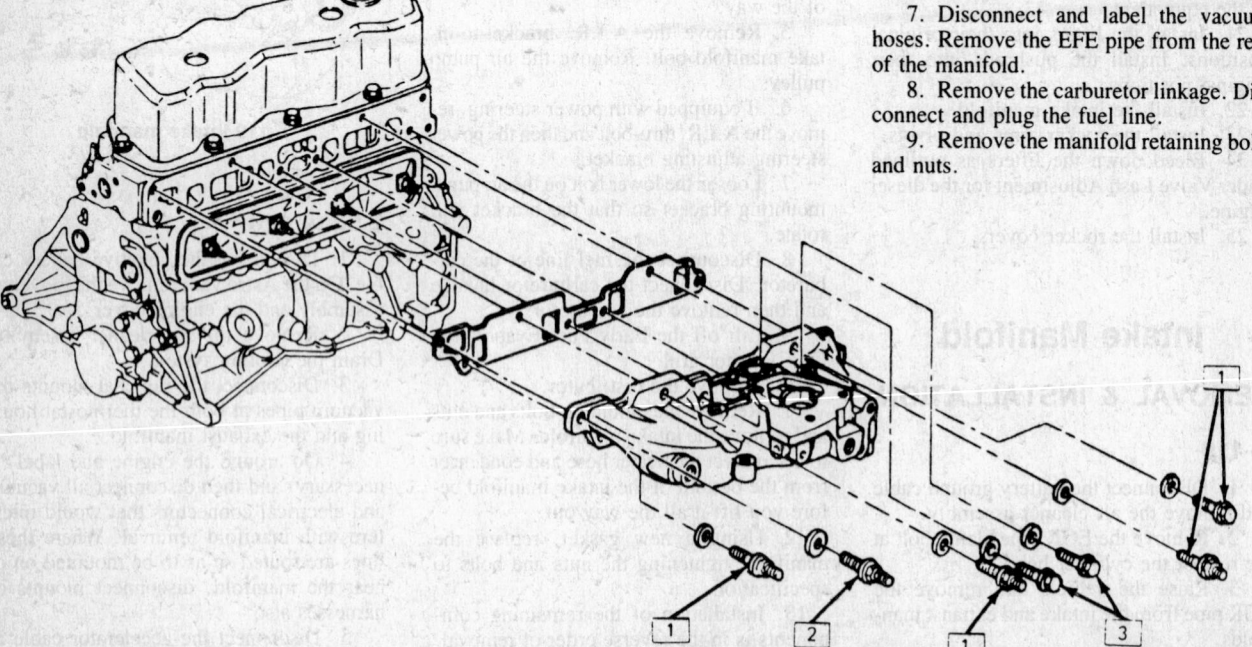

2.5 L four Cylinder engine intake manifold mounting. Bolts 1 are torqued to 25 ft.-lbs.; 2 to 37 fts.-lbs.; and 3 to 28 ft.-lbs.

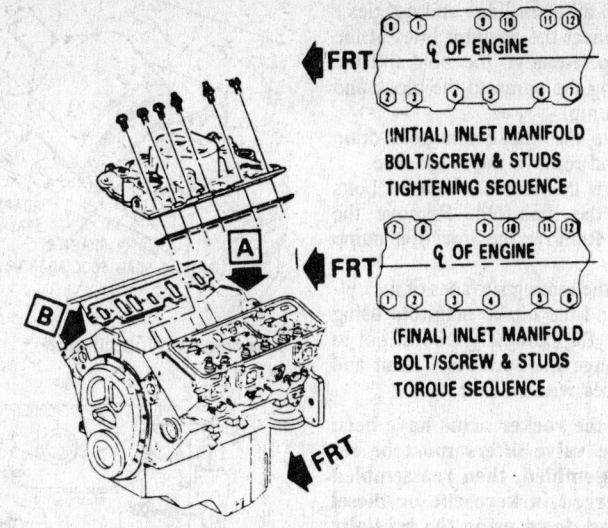

(INITIAL) INLET MANIFOLD
BOLT/SCREW & STUDS
TIGHTENING SEQUENCE

(FINAL) INLET MANIFOLD
BOLT/SCREW & STUDS
TORQUE SEQUENCE

Torque the manifold bolts in the sequence shown on the 4.3 L V-6

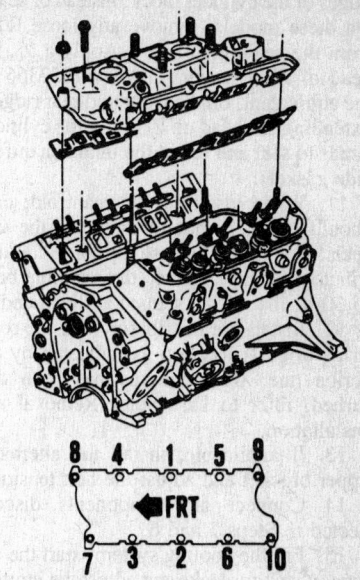

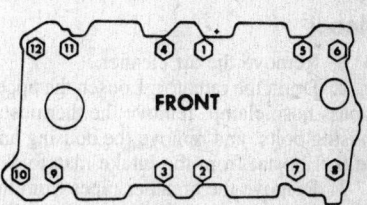

6–173 intake manifold torque sequence

Small block V8 intake manifold torque sequence

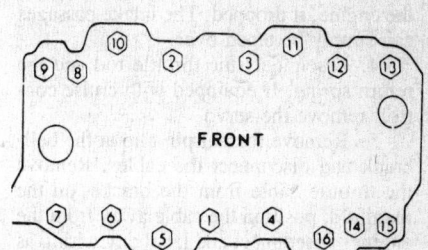

8-454 intake manifold torque sequence

10. Remove the intake manifold. Remove and discard the gaskets, and scrape off the old silicone seal from the front and rear ridges.

11. To install: The gaskets are marked for right and left side installation; do not interchange them. Clean the sealing surface of the engine block, and apply a $3/16$ in. bead of silicone sealer to each ridge.

12. Install the new gaskets onto the heads. The gaskets will have to be cut slightly to fit past the center pushrods. Do not cut any more material than necessary. Hold the gaskets in place by extending the ridge bead of sealer $1/4$ in. onto the gasket ends.

13. Install the intake manifold. The area between the ridges and the manifold should be completely sealed.

14. Install the retaining bolts and nuts, and tighten in sequence to 23 ft. lbs. Do not overtighten; the manifold is made from aluminum, and can be warped or cracked with excessive force.

15. The rest of installation is the reverse of removal. Adjust the ignition timing after installation, and check the coolant level after the engine has warmed up.

Six Cylinder with Combination Manifold

1. The intake and exhaust manifolds are removed as an assembly. Remove the air cleaner.

2. Disconnect the throttle rods at the bellcrank and remove the throttle return spring.

3. Disconnect the fuel and vacuum lines at the carburetor. Plug the fuel line. Disconnect the choke cable at the carburetor.

4. Disconnect the crankcase ventilation hose at the carburetor. Disconnect the vapor hose at the evaporativ canister, if so equipped.

5. Disconnect the exhaust pipe at the manifold flange and discard the packing.

6. Disconnect the EGR valve hose (if equipped).

7. Remove the manifold attaching bolts and clamps and remove the manifold assembly. Discard the gaskets.

8. The manifold assembly can be separated by removing 1 bolt and 2 nuts at the center. Don't tighten these down all the way until the manifold assembly is installed on the engine.

9. Check the manifold for straightness along the exhaust port faces. If it is distorted more than 0.015 in. it should be replaced. Clean all mounting faces.

10. Installation is the reverse of removal. Use all new gaskets. Bolt torques are given in the Torque Specifications.

V8 Except Diesel

1. Remove the air cleaner.
2. Drain the radiator.
3. Disconnect:
   a. Battery cables at the battery.
   b. Upper radiator and heater hoses at the manifolds.
   c. Crankcase ventilation hoses as required.
   d. Fuel line and choke cable at the carburetor.
   e. Accelerator linkage at the carburetor.
   f. Vacuum hose at the distributor.
   g. Power brake hose at the carburetor base or manifold, if applicable.
   h. Temperature sending switch wires.
   i. Water pump by-pass at the water pump (Mark IV only).
4. Remove the distributor cap and scribe the rotor position relative to the distributor body.
5. Remove the distributor.
6. As required, remove the oil filler bracket, air cleaner bracket, air compressor and bracket and accelerator bellcrank.
7. Remove the manifold-to-head at-taching bolts then remove the manifold and carburetor as an assembly.

8. If the manifold is to be replaced, transfer the carburetor (and mounting studs), water outlet and thermostat (use a new gasket) heater hose adapter, EGR valve (use new gasket) and, if applicable, TVS switch and the choke coil. All engines use a carburetor heat choke tube which must be transferred to a new manifold.

9. Before installing the manifold, thoroughly clean the gasket and seal surfaces of the cylinder heads and manifold.

10. Install the manifold end seals, folding the tabs if applicable, and the manifold/head gaskets, using a sealing compound around the water passages, 1978 and later models use RTV (Room Temperature Vulcanizing) silicone seal at the front and rear

ridges of the cylinder block, instead of seals. On these models, remove any loose RTV from the sealing surfaces. Apply a $\frac{3}{16}$ in. bead of RTV sealer, G.M. #1052366 or the equivalent, on the front and rear ridges, extending the bead up $\frac{1}{2}$ in. on the cylinder heads to seal and retain the intake manifold side gaskets.

11. When installing the manifold, care should be taken not to dislocate the end seals. It is helpful to use a pilot in the distributor opening. Tighten the manifold bolts to 30 ft. lbs. in the sequence illustrated.

12. Install the distributor with the rotor in its original location as indicated by the scribe line. If the engine has been disturbed, refer to Distributor Removal and Installation.

13. If applicable, install the alternator upper bracket and adjust the belt tension.

14. Connect all components disconnected in Steps 3 and 6.

15. Fill the cooling system, start the engine, check for leaks and adjust the ignition timing and carburetor idle speed and mixture.

Diesel

1. Remove the air cleaner.

2. Drain the radiator. Loosen the upper bypass hose clamp, remove the thermostat housing bolts, and remove the housing and the thermostat from the intake manifold.

3. Remove the breather pipes from the rocker covers and the air crossover. Remove the air crossover. It is a good idea to cover the air intakes in the manifold to prevent nuts and bolts from falling down into the engine, if dropped. The intake passages can simply be taped over.

4. Disconnect the throttle rod and the return spring. If equipped with cruise control, remove the servo.

5. Remove the hairpin clip at the bellcrank and disconnect the cables. Remove the throttle cable from the bracket on the manifold; position the cable away from the engine. Disconnect and label any wiring as necessary.

6. Remove the alternator bracket as necessary. If the truck is equipped with air conditioning, remove the compressor mounting bolts and move the compressor aside, without disconnecting any of the hoses. Remove the compressor mounting bracket from the intake manifold.

7. Disconnect the fuel line from the pump and the fuel filter. Remove the fuel filter and bracket.

8. Disconnect the fuel return line from the injection pump. Using two wrenches to prevent the lines from being twisted, disconnect the injection pump lines at the nozzles.

CAUTION
Do not bend the injection pump lines!

9. Remove the three nuts retaining the injection pump, using G.M. special tool No. J26987 or the equivalent. Remove the

pump and cap all open lines and nozzles.

10. Disconnect the vacuum lines at the vacuum pump. Remove the bolt and the bracket holding the pump to the block and remove the pump.

11. Remove the intake manifold drain tube clamp and remove the drain tube.

12. Remove the intake manifold bolts and remove the manifold. Remove the adapter seal. Remove the injection pump adapter.

13. Clean the mating surfaces of the cylinder heads and the intake manifold using a putty knife. Be extremely careful not to scratch or gouge the surfaces. Clean and dry the surfaces with solvent.

NOTE: If the rocker arms have been removed, the valve lifters must be removed, disassembled, then reassembled while submerged in kerosene or diesel fuel, then bled down using the specially weighted press (see the procedure earlier in this section). Do not install the manifold until the affected lifters have been serviced.

14. To install the manifold: Coat both sides of the gasket surface that seal the intake manifold to the cylinder heads with G.M. sealer 1050805 or the equivalent. Position the intake manifold gaskets on the cylinder heads. Install the end seals, making sure that the ends are positioned under the cylinder heads.

15. Carefully lower the intake manifold into place on the engine.

16. Clean the intake manifold bolts thoroughly, then dip them in clean engine oil. Install the bolts and tighten to 15 ft. lbs. in the sequence shown. Next, tighten all the bolts to 30 ft. lbs., in sequence, and finally tighten to 40 ft. lbs. in sequence.

17. Install the intake manifold drain tube and clamp.

18. File the mark from the injection pump adapter.

CAUTION
Do not file the mark from the injection pump.

19. Place the engine on TDC for the No.1 cylinder. The mark on the harmonic balancer on the crankshaft will be aligned with the zero mark on the timing tab, and both valves for No.1 cylinder will be closed. The index mark on the injection pump driven gear should be offset to the right when No.1 is at TDC. Check that all these conditions are met before continuing.

20. Apply chassis grease to the seal area on the adapter, the tapered edge and the seal area on the intake manifold. Install the adapter but leave the bolts loose.

21. Apply chassis grease to the inside and outside diameters of the adapter seal, and to the seal installing tool, G.M. J28425. Install the seal onto the tool.

22. Push the seal onto the injection pump adapter, using the tool (J–28425 or the equivalent). Remove the tool and inspect the seal to see if it is properly positioned.

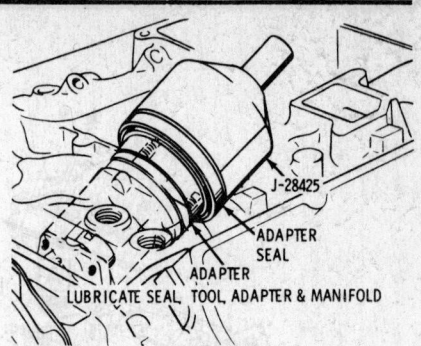

Adapter seal installation

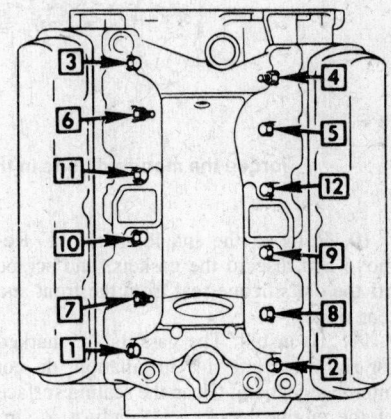

8-350 diesel intake manifold torque sequence

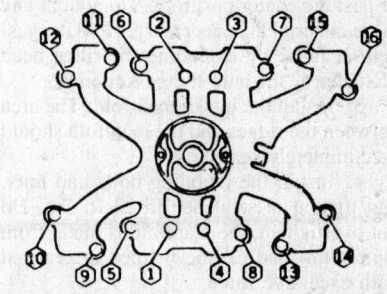

8-379 diesel intake manifold torque sequence

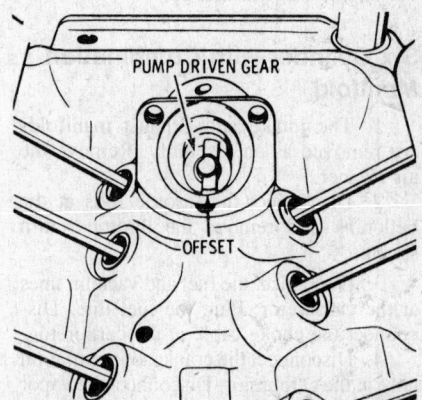

The index mark on the injection pump driven gear will be offset to the right when the no. 1 cylinder is at TDC

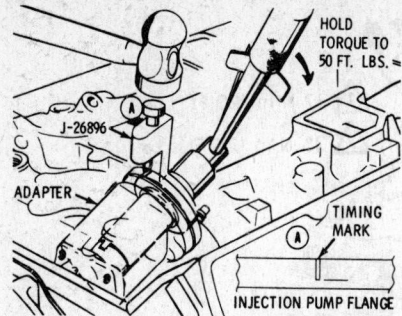

Injection pump timing mark application

23. Tighten the adapter bolts to 25 ft. lbs.

24. Install a timing tool, G.M. J26896 or the equivalent, into the injection pump adapter. Tighten the tool toward No.1 cylinder to 50 ft. lbs. While holding the tool and adapter at this torque, mark the injection pump adapter by striking the marking pin with a hammer. Remove the tool.

25. Remove the protective caps from the lines. Line up the offset tang on the injection pump driveshaft with the pump driven gear. Install the pump.

26. Install the three retaining nuts and lockwashers for the injection pump but do not tighten the nuts. Connect the injection pump lines to the nozzles. Use two wrenches to tighten the lines (25 ft. lbs.).

——————— CAUTION ———————
Do not bend or twist the injection pump lines.
—————————————————————

27. Connect the fuel return lines to the pump.

28. Align the injection pump mark with the adapter mark and tighten the nuts. Use a ¾ in. open end wrench on the boss at the front of the injection pump to aid in rotating the pump to align the marks. Tighten the nuts to 18 ft. lbs.

29. Adjust the throttle rod and return spring.

30. Install the fuel filter and bracket and install the fuel line to the pump and the filter.

31. Install the vacuum pump and the vacuum lines. Do not operate the engine without the vacuum pump installed, it is the drive for the engine oil pump.

32. Connect the wiring.

33. Install the alternator and air conditioning compressor brackets.

34. Install the cable in the bracket and bellcrank, then install the bellcrank.

35. Connect the throttle rod and the return spring.

36. Remove the tape from the air intakes and install the air crossover. Install the breather tubes and the flow control valve at the air crossover. Connect the upper radiator hose and the heater hose, install the thermostat and thermostat housing, fill the cooling system, start the engine and check for leaks.

Exhaust Manifold

REMOVAL & INSTALLATION

4–119

1. Disconnect the battery ground cable and remove the air cleaner assembly.

2. Remove the EGR pipe clamp bolt at the rear of the cylinder head.

3. Raise the vehicle and remove the EGR pipe from the intake and exhaust manifolds.

4. Separate the exhaust pipe from the manifold.

5. Remove the manifold shield and remove the heat stove.

6. Remove the manifold retaining nuts and remove the manifold from the engine.

7. Installation is the reverse of removal.

4–137 Diesel

1. Raise engine hood.

2. Remove the bolts fixing the air cleaner and loosen the clamp bolt.

3. Lift the air cleaner slightly and disconnect the breather hose, then remove the air cleaner assembly.

4. Disconnect the exhaust pipe from the exhaust manifold at the flange.

5. Remove the 3 nuts fixing the exhaust manifold, then remove the engine hanger and exhaust manifold.

6. Installation is the reverse of removal. Torque the bolts to 15 ft. lbs.

4–121

1. Disconnect the negative battery cable.

2. Remove the air cleaner. Remove the exhaust manifold shield. Raise and support the front of the vehicle.

3. Disconnect the exhaust pipe at the manifold and then lower the vehicle.

4. Disconnect the air management-to-check valve hose and remove the bracket. Disconnect the oxygen sensor lead wire.

5. Remove the alternator belt. Remove the alternator adjusting bolts, loosen the pivot bolt and pivot the alternator upward.

6. Remove the alternator brace and the A.I.R. pipes bracket bolt.

7. Unscrew the mounting bolts and remove the exhaust manifold. The manifold should be removed with the A.I.R. plumbing as an assembly. If the manifold is to be replaced, transfer the plumbing to the new one.

8. Clean the mating surface on the manifold and the head, position the manifold and tighten the bolts to the proper specifications.

9. Installation of the remaining components is in the reverse order of removal.

2.5L (4 cyl)

1. Disconnect the negative battery cable. On the Astro van, remove the glove box and the engine cover.

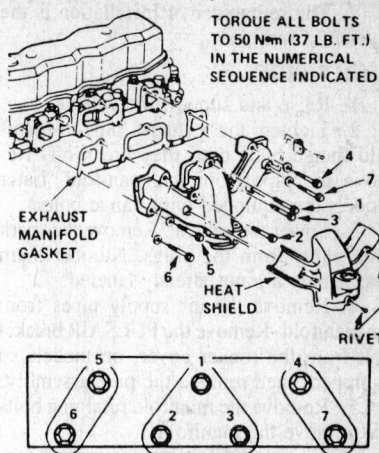

TORQUE ALL BOLTS TO 50 N•m (37 LB. FT.) IN THE NUMERICAL SEQUENCE INDICATED

BOLT LOCATIONS

Torque sequence for torquing the exhaust manifold bolts on the 2.5 L four cylinder engine

2. Remove the exhaust heat stove pipe at the manifold. Disconnect the oxygen sensor wire at the oxygen sensor.

3. Raise the vehicle and support it securely at approved locations. Disconnect the exhaust pipe at the manifold.

4. Lower the vehicle again and remove the rear air conditioner compressor bracket, leaving the front compressor bracket and compressor in place.

5. Remove the bolts and remove the manifold.

6. Use a dull-edged scraper and scrape any carbon or gasket pieces from the gasket surfaces. If you suspect the manifold may have been leaking due to distortion, check the flatness of its gasket surface with a straightedge; replace it if a significant distortion is found.

7. Install a new gasket as shown in the illustration. Install the manifold and all bolts, just finger tight.

8. Torque the bolts to 44 ft. lbs. in the sequence shown in the illustration.

9. Reverse the remaining removal procedures to complete the installation. Operate the engine and check carefully for leaks, repairing any problems as necessary.

6–173

LEFT SIDE

1. Remove the air cleaner. Remove the carburetor heat stove pipe.

2. Remove the air supply plumbing from the exhaust manifold.

3. Raise and support the car. Unbolt and remove the exhaust pipe at the manifold.

4. Unbolt and remove the manifold.

5. To install: Clean the mating surfaces of the cylinder head and manifold. Install the manifold onto the head, and install the retaining bolts finger tight.

6. Tighten the manifold bolts in a circular pattern, working from the center to the ends, to 25 ft. lbs. in two stages.

7. Connect the exhaust pipe to the manifold.

8. The remainder of installation is the reverse of removal.

RIGHT SIDE

1. Raise and support the vehicle.

2. Tighten the exhaust pipe-to-manifold flange bolts until they break off. Remove the pipe from the manifold. Later models are equipped with flange bolts.

3. Lower the vehicle. Remove the spark plug wires from the plugs. Number them first if they are not already labeled.

4. Remove the air supply pipes from the manifold. Remove the PULSAIR bracket bolt from the rocker cover, on models so equipped, then remove the pipe assembly.

5. Remove the manifold retaining bolts and remove the manifold.

6. To install: Clean the mating surfaces of the cylinder head and manifold. Position the manifold against the head and install the retaining bolts finger tight.

7. Tighten the bolts in a circular pattern, working from the center to the ends, to 25 ft. lbs. in two stages.

8. Install the air supply system.

9. Install the spark plug wires. Raise and support the car. Connect the exhaust pipe to the manifold and install new flange bolts.

Six Cylinder With Integral Head

1. Remove the air cleaner. Disconnect negative battery terminal.

2. Remove the power steering pump and, if equipped, the AIR pump.

3. Remove the EFE (early fuel evaporation) valve bracket.

4. Disconnect the throttle controls and the throttle return spring.

5. Disconnect the exhaust pipe at the manifold flange. Disconnect the converter bracket at the transmission mount, if so equipped. If equipped with manifold converter, disconnect the exhaust pipe from the converter, and remove the converter.

6. Remove the manifold attaching bolts and remove the manifold. Discard the gasket.

7. Check for cracks in the manifold before it is replaced.

8. Install a new gasket on the exhaust manifold.

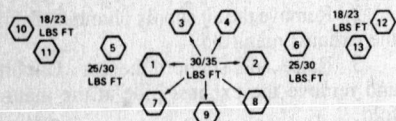

Six cylinder integral manifold torque sequence

9. Clean and oil the bolts and install the bolts, torquing them to specifications.

10. Connect the exhaust pipe, throttle controls, and return spring. Install the air cleaner, start the engine, and check for leaks.

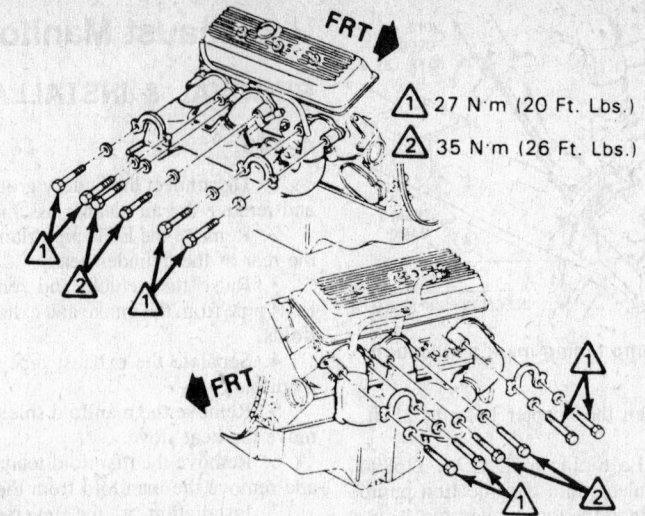

Torque sequence for torquing the exhaust manifold bolts on the 4.3 L V-6

4.3L V6

RIGHT SIDE

1. Disconnect the negative battery cable. Raise the vehicle and support it at approved locations.

2. Disconnect the exhaust pipe at the manifold. Then, lower the vehicle back to the floor.

3. Remove the engine cover. Disconnect the Air Injection Reactor hose at the check valve.

4. Remove the exhaust manifold bolts. Then, disconnect the AIR pipe bracket at the head.

5. Remove the manifold.

6. Using a dull-edged scraper, scrape any carbon or gasket pieces from the gasket surfaces. If you suspect the manifold may have been leaking due to distortion, check the flatness of its gasket surface with a straightedge; replace it if a significant distortion is found.

7. Install a new gasket in the proper position and then install the manifold in position over it. Install the bolts finger tight.

8. Refer to the applicable part of the illustration and torque the bolts to the specification given for each bolt according to its numbered location. Complete the remaining steps of the installation procedure in reverse of removal.

LEFT SIDE

1. Disconnect the negative battery cable. Raise the vehicle and support it securely at approved positions.

2. Disconnect the exhaust pipe at the manifold.

3. Disconnect the Air Injection Reactor pipe at the head.

4. Remove the manifold bolts and remove the manifold.

5. Using a dull-edged scraper, scrape any carbon or gasket pieces from the gasket surfaces. If you suspect the manifold may

have been leaking due to distortion, check the flatness of its gasket surface with a straightedge; replace it if a significant distortion is found.

6. Install a new gasket in the proper position and then install the manifold in position over it. Install the bolts finger tight.

7. Refer to the applicable part of the illustration and torque the bolts to the specification given for each bolt according to its numbered location. Complete the remaining steps of the installation procedure in reverse of removal.

V8 Except Diesel

1. If equipped with AIR, remove the air injector assembly. The ¼ in. pipe threads in the manifold are straight cut threads. Do not use a ¼ in. tapered pipe tap to clean the threads.

2. Disconnect the battery.

3. If equipped, remove the carburetor heat stove pipe.

4. Remove the spark plug wire heat shields. On Mark IV, remove spark plugs.

5. On the left exhaust manifold, disconnect and remove the alternator.

6. Disconnect the exhaust pipe from the manifold and hang it from the frame out of the way.

7. Bend the locktabs and remove the end bolts, then the center bolts. Remove the manifold.

NOTE: A ⁹⁄₁₆ in. thin wall 6-point socket, sharpened at the leading edge and tapped onto the head of the bolt, simplifies bending the locktabs.

8. Installation is the reverse of removal. Clean all mating surfaces and use new gaskets. Torque all bolts to specifications from the inside working out.

Diesel

LEFT SIDE

1. Remove the air cleaner.

2. Remove the lower alternator bracket.

3. Raise the truck and remove the exhaust pipe from the manifold flange.

4. Lower the truck. Bend the locktabs away from the manifold mounting bolts. Remove the bolts and remove the manifold from above. Do not lose the locktabs, and do not use the washers for the bolts, which go under the locktabs.

5. Installation is the reverse. Tighten the manifold bolts to 25 ft. lbs. in two stages, working in a circular pattern from the center to the ends. Then tighten the front bolt to 30 ft. lbs.

RIGHT SIDE

1. Raise and support the truck. Remove the bolts retaining the exhaust pipe to the manifold flange.

2. Bend the locktabs away from the manifold mounting bolts. Remove the bolts and remove the manifold. Do not lose the locktabs and the washers for the bolts, which go under the locktabs.

3. Installation is the reverse. Tighten the bolts to 25 ft. lbs. in two progressive steps, working in a circular pattern from the center to the ends.

Timing Cover

REMOVAL & INSTALLATION

4–119

1. Remove the cylinder head.
2. Remove the oil pan.
3. Remove the oil pickup tube from the oil pump.
4. Remove the harmonic balancer.
5. Remove the AIR pump drive belt.
6. On air conditioned vehicles: Remove the compressor, with lines still connected, and lay it to one side. Remove the compressor mounting brackets. If equipped with power steering, remove the pump, with hoses attached, and bracket and lay aside.
7. Remove the distributor cap and then remove the distributor.
8. Remove the front cover attaching bolts and remove the front cover.
9. Remove and discard the front cover to block gasket.
10. Install a new gasket onto cylinder block.
11. Align the oil pump drive gear punch mark with the oil filter side of cover; then align the center of dowel pin with alignment mark on oil pump case.
12. Rotate the crankcase until Nos. 1 and 4 cylinders are at top dead center.
13. Install the front cover by engaging the pinion gear with the oil pump drive gear on the crankshaft.
14. Check that the punch mark on the oil pump drive gear is turned to the rear side as viewed through clearance between front cover and cylinder block.
15. Check that the slit at the end of oil pump shaft is parallel with front face of cylinder block and that it is offset forward.

16. With all parts correctly installed, install and tighten front cover bolts.
17. Reverse Steps 1–7 of Removal procedure.
18. Check engine timing.
19. Check for leaks.

4–121

NOTE: The following procedure requires the use of a special tool.

1. Remove the engine drive belts.
2. Although not absolutely necessary, removal of the right front inner fender splash shield will facilitate access to the front cover.
3. Unscrew the center bolt from the crankshaft pulley and slide the pulley and hub from the crankshaft.
4. Remove the alternator lower bracket.
5. Remove the oil pan-to-front cover bolts.
6. Remove the front cover-to-block bolts and then remove the front cover. If the front cover is difficult to remove, use a plastic mallet.
7. The surfaces of the block and front cover must be clean and free of oil. Apply a $\frac{1}{8}$ in. bead of RTV sealant to the cover. The sealant must be wet to the touch when the bolts are torqued down.

NOTE: When applying RTV sealant to the front cover, be sure to keep it out of the bolt holes.

8. Position the front cover on the block using a centering tool (J23042) and tighten the screws.
9. Installation of the remaining components is in the reverse order of removal. The oil seal can be replaced with the cover either on or off the engine. If the cover is on the engine, remove the crankshaft pulley and hub first. Pry out the seal using a large screwdriver, being careful not to distort the seal mating surface. Install the new seal so that the open side or helical side is towards the engine. Press it into place with a seal driver made for the purpose. Install the hub if removed.

4–137 Diesel

1. Remove the radiator.
2. Remove the compressor drive belt by moving the power steering oil or idler (if so equipped).
3. Loosen the alternator adjusting plate bolt and fixed bolt, then remove the fan belt.
4. Remove the 4 bolts mounting the crankshaft pulley and remove the crankshaft pulley.
5. Remove the bolts mounting the timing pulley housing covers, then remove the covers.
6. Installation is the reverse of removal.

2.5L (4 cyl)

1. Disconnect the negative battery cable.

2. Disconnect the power steering pump reservoir fan shroud at the timing cover. Remove the upper radiator fan shroud.

3. Loosen the bolts on the fan and pulley. Loosen the drive belts. Then remove the fan and pulley.

4. Remove the alternator as described earlier in this section. Then, remove the alternator brace and front and rear brackets.

5. Remove the front crankshaft pulley. Then, remove the hub bolt.

6. Remove the crankshaft hub with a puller.

7. Drain the cooling system. Disconnect the lower radiator hose at the water pump.

8. Remove the front cover bolts (including the two that also attach the oil pan) and remove the front cover.

9. Clean the gasket surfaces on the block, cover and oil pan. Then, apply a continuous bead of RTV sealer on the block and pan side of the cover. Keep the sealer out of the boltholes.

NOTE: The cover must be installed using a centering tool J34995 or equivalent; otherwise the installation of the hub after the cover is back in place will damage the seal, or at least result in seal leakage due to imperfect alignment.

10. Install the centering tool into the front seal. Then, install the cover as you install the centering tool over the front of the crankshaft. Install the two cover-to-oil pan bolts finger tight. Then, install the remaining cover bolts finger tight.

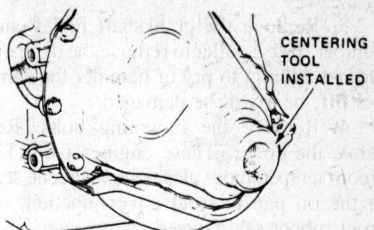

Installing the centering tool for the timing gear cover on the 2.5 L four

11. Torque all the bolts to 90 inch lbs.
12. Remove the centering tool.
13. Coat the front cover oil seal contact area of the hub with engine oil. Then, position the hub on the crankshaft with the keyway aligned with the key on the shaft and slide it into position until it bottoms on the crankshaft. Install the center bolt and torque it to 160 ft. lbs.
14. Complete the installation in reverse of the removal procedure.

6–173

1. Remove the water pump.
2. Remove the compressor without disconnecting any A/C lines and lay it aside.
3. Remove harmonic balancer, using a puller.

NOTE: The outer ring (weight) of the harmonic balancer is bonded to the hub

with rubber. The balancer must be removed with a puller which acts on the inner hub only. Pulling on the outer portion of the balancer will break the rubber bond or destroy the tuning of the torsional damper.

4. Disconnect the lower radiator hose and heater hose.
5. Remove timing gear cover attaching screws, and cover and gasket.
6. Clean all the gasket mounting surfaces on the front cover and block. Apply a continuous 3/32 in. bead of sealer (1052357 or equivalent) to front cover sealing surface and around coolant passage ports and central bolt holes.
7. Apply a bead of silicone sealer to the oil pan-to-cylinder block joint.
8. Install a centering tool in the crankcase snout hole in the front cover and install the cover.
9. Install the front cover bolts finger tight, remove the centering tool and tighten the cover bolts. Install the harmonic balancer, pulley, water pump, belts, radiator, and all other parts.

Inline Six Cylinder

1. Remove the radiator after draining it.
2. Remove the fan, pulley, and belt. Remove any power steering and/or AIR pump drive belts. Remove any braces for the above pumps which will interfere with cover removal and position the pumps out of the way.
3. Remove the crankshaft pulley and damper. Use a puller to remove the damper. Do not attempt to pry or hammer the damper off, or it will be damaged.
4. Remove the mounting bolts. Remove the cover. These engines use RTV (room temperature vulcanizing) silicone seal at the oil pan to front cover junction; no front rubber seal is used.
5. Apply a 3/16 in. bead of RTV silicone seal on the cover sealing surface.
6. Coat the front cover gasket with sealer and use a 1/8 in. bead of silicone sealer at the oil pan to cylinder block joint. Replace the damper before tightening the cover bolts down, so that the cover seal will align. The damper must be drawn into place. Hammering it will destroy it.
7. Replace the oil pan if it was removed, and fill the crankcase with oil.

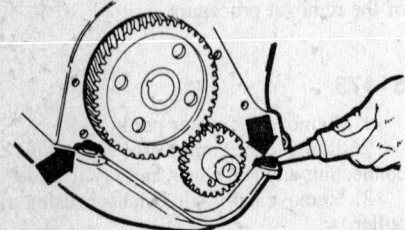

Apply sealer to the six cylinder timing cover at the areas shown

4.3L V6

1. Disconnect the negative battery cable.
2. Remove the drive belts and front pulley.
3. Raise and support the vehicle at approved jacking points for access. Then install a threaded puller such as J23523-1 or equivalent to the front hub. Turn the puller screw to remove the hub from the crankshaft.
4. Drain the cooling system. Remove the water pump.
5. Remove the front cover attaching screws and remove the front cover. Remove the gasket and discard it. Scrape the gasket surfaces with a dull scraper to remove any remaining gasket material. Cut any remaining gasket material protruding from the joint between the oil pan gasket and the block with a sharp knife.
6. Apply a 1/8 in. bead of RTV sealer such as 1052917 or equivalent to the joint formed between the oil pan and block, keeping the sealer out of bolt holes. Coat a new gasket with gasket sealer and apply it to the cover with all holes lined up.
7. Install the timing cover-to-oil pan seal. Coat the bottom of this seal with clean engine oil and then position the timing cover over the end of the crankshaft. Loosely install the upper cover attaching screws to hold the cover in place.
8. Press downward on the cover so the dowels on the block enter the holes in the cover without binding and without distortion in the cover. Tighten the attaching screws alternately and evenly to hold the cover in this position. Then, install the remaining cover screws just finger tight.
9. Torque all the front cover attaching screws to 7 ft. lbs.
10. Coat the front cover seal contact area on the front hub with clean engine oil. Place the hub in position over the crankshaft and key.

NOTE: The front hub and damper must be installed with a tool designed especially for this purpose. This is because the intertial weight section of the damper is attached to the hub with a rubber material that cannot tolerate any end thrust.

11. Use a damper installing tool such as J23523 or equivalent to pull the hub into position. When installing the threaded end of the tool into the center of the crankshaft, make sure that at least 1/2 in. of thread engagement is obtained. Then, install the plate, thrust bearing and nut and turn the nut to force the hub onto the crankshaft. When the hub bottoms out, remove the tool and install the hub retaining bolt into the center of the crankshaft. Torque it to 60 ft. lbs.
12. Perform the remaining steps of the installation procedure in reverse order.

Small Block V8

1. Drain the oil.
2. Drain and remove the radiator.

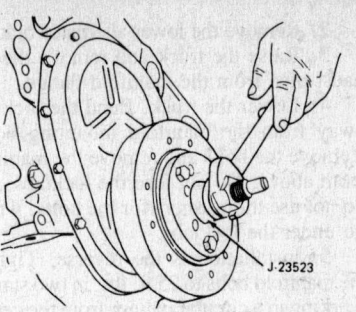

Installing the torsional damper on the 4.3 L V-6 after the timing cover is in place

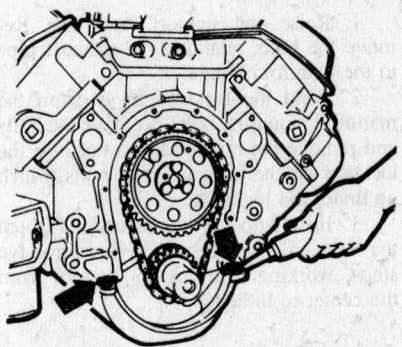

Apply sealer to the V8 cover pads at the areas shown

3. Remove the fan, pulley and belt. Remove any power steering and/or AIR pump drive belts. Remove any braces for these pumps which will interfere with cover removal and position the pumps out of the way.
4. Remove the water pump.
5. Remove the crankshaft pulley and damper. Use a puller on the damper. Do not attempt to pry or hammer the damper off.
6. Rotate the retaining bolts, and remove the timing cover.
7. Clean the gasket surfaces on the block and the front cover.
8. Use a sharp knife to trim any excess oil pan gasket material which protrudes from the oil pan-to-engine block junction.
9. Apply a 1/8 in. bead of RTV silicone sealer, G.M. 1052366 or the equivalent, to the joint of the oil pan and cylinder block.
10. Coat the front cover gasket with sealer and install the gasket onto the cover.
11. Install the front cover-to-oil pan seal. Lightly coat the bottom of the seal with clean engine oil and position over the crankshaft end.
12. Loosely install the front cover upper attaching bolts.
13. Press downward on the cover so that the dowels in the block are aligned with the holes in the cover. While holding the cover in position, tighten the upper attaching bolts alternately and evenly.
14. Install the remaining bolts and tighten all the bolts to specification.
15. Install the torsional damper and the water pump.

Mark IV V8

1. Remove the torsional damper and water pump.
2. Remove the two oil pan-to-front cover attaching screws.
3. Remove the front cover-to-block attaching screws.
4. Pull the cover slightly forward to permit cutting the oil pan front seal.
5. Using a sharp knife, cut the oil pan front seal flush with the cylinder block at both sides of the cover.
6. Remove the front cover and the portion of the oil pan front seal. Remove the front cover gasket.
7. Clean the gasket mating surfaces.
8. Cut the tabs from a new oil pan front seal, using a sharp knife to get a clean cut.
9. Install the seal on the front cover pressing the tips into the holes in the front cover.
10. Coat a new gasket with gasket sealer and install the gasket on the cover.
11. Apply a ⅛ in. bead of RTV sealant to the joint formed at the junction of the oil pan and cylinder block.
12. Place the front cover in position.
13. Align the cover over the dowel pins in the block.
14. Further installation is the reverse of removal.

Diesel

1. Drain cooling system. Disconnect radiator hoses and bypass hose.
2. Remove all belts, fan and fan pulley, crankshaft pulley and harmonic balancer, and accessory brackets. The harmonic balancer must be removed with a puller which pulls from the rear center of the balancer. Any other type of puller, such as a universal claw type which pulls on the outside of the hub, can destroy the balancer. The outside ring of the balancer is bonded in rubber to the hub; by pulling on the outside, it is possible to break that bond.
3. Remove cover-to-block attaching bolts and remove the cover, timing indicator and water pump assembly.
4. Remove the front cover and dowel pins. It may be necessary to grind a flat on the pins to get a rough surface for gripping.
5. To install: Grind a chamfer on one end of each dowel pin.
6. Cut excess material from front end of oil pan gasket on each side of engine block.
7. Clean block, oil pan, and front cover mating surfaces with solvent.
8. Trim about ⅛ in. from each end of a new front pan seal, using a sharp knife to insure a straight cut.
9. Install the new front cover gasket on engine block and new front seal on front cover. Apply sealer to gasket around coolant holes and place on block.
10. Apply silicone sealer at junction of block, pan, and front cover.
11. Place the cover on block and press downward to compress the seal. Rotate cover

left and right and guide pan seal into cavity using a small screwdriver.
12. Apply engine oil to bolts (threads and heads). Install two bolts finger tight to hold cover in place.
13. Install two dowel pins chamfered end first.
14. Install timing indicator and water pump assembly. Torque bolts evenly to 13 ft. lbs. for the water pump bolts, and 35 ft. lbs. for cover bolts.
15. Apply lubricant to balancer seal surface. Install balancer, and balancer bolt. Torque to approximately 250 ft. lbs.
16. Install brackets. Connect bypass hose and radiator hoses. Install crankshaft pulley and four attaching bolts. Torque to 20 ft. lbs.
17. Install fan pulley, fan, and four attaching bolts. Torque to 20 ft. lbs. Install belts and adjust tension. Fill radiator. Road test and check for leaks.

Timing Gear Cover Oil Seal

REPLACEMENT

All Engines

The seal may be replaced with the cover either on or off the engine.
1. With the cover removed: Pry the old seal from the cover using a wooden or plastic pick to prevent damage to the sealing lip. A plastic knitting needle makes a good removal tool.
2. Oil the lip of the new seal. Place a support under the cover so that it is not damaged when the seal is installed.
3. Install the seal so that the open side of the seal is toward the inside of the cover. Drive the seal into place with a tool made for the purpose (G.M. tool J23042 or the equivalent).
4. The seal can also be replaced with the cover in place on the engine: Remove the torsional damper. Pry the old seal from the cover, as outlined in Step 1.
5. Oil the lip of the new seal and place it into position, with the open side of the seal toward the engine. Drive the seal into position with a tool designed for the purpose (G.M. tool J23042 or the equivalent).
6. Install the damper.

Timing Chain, Gear or Belt

REMOVAL & INSTALLATION

4-119

1. Remove front cover assembly.
2. Lock the shoe on automatic adjuster in fully retracted position by depressing the adjuster lock lever in direction as shown.
3. Remove timing chain from crankshaft sprocket.

4. Check timing sprockets for wear or damage. If crankshaft sprocket must be replaced, remove sprocket and pinion gear from crankshaft using puller J25031.
5. Check timing chain for wear or damage; replace as necessary. Measure distance (L) with chain stretched with a pull of approximately 22 lbs (98 N). Standard (L) value is 15 in, (381mm); replace chain if (L) is greater than 15.1 in. (385mm).
6. Remove attaching bolt and remove automatic chain adjuster.
7. Check that the shoe becomes locked when shoe is pushed in with the lock lever released.
8. Check that lock is released when the shoe is pushed in. The adjuster assembly must be replaced if rack teeth are found to be worn excessively.
9. Remove "E" clip and remove chain tensioner. Check tensioner for wear or damage; replace as necessary.
10. Inspect tensioner pin for wear or damage. If replacement is necessary, remove pin from cylinder block using locking pliers. Lubricate NEW pin tensioner with clean engine oil. Start new pin in block, place tensioner over appropriate pin. Place "E" clip on pin and then tap pin into block, using a hammer, until clip just clear tensioner. Check tensioner and adjuster for freedom of rotation on pins.
11. Inspect guide for wear or damage and plugged lower oil jet. If replacement or cleaning is necessary, remove guide bolts, guide and oil jet. Install new guide and upper attaching bolt. Install lower oil jet and bolt so that oil port is pointed toward crankshaft as shown.
12. Install timing sprocket and pinion gear (groove side toward front cover). Align key grooves with key on crankshaft, then drive into position using Installating Tool J26587.
13. Turn crankshaft so that key is turned toward cylinder head side (Nos. 1 and 4 pistons at top dead center).
14. Install the timing chain by aligning mark plate on chain with mark on crankshaft timing sprocket. The side of the chain with the mark plate is on the front side and the side of chain with the most links between mark plates is on the chain guide side. Keep the timing chain engaged with the camshaft timing sprocket until the camshaft timing sprocket is installed on camshaft.
15. Install the camshaft timing sprocket so that marked side of sprocket faces forward and so that the triangular mark aligns with the chain mark plate.
16. Install the automatic chain adjuster.
17. Release lock by depressing the shoe on adjuster by hand, and check to make certain the chain is properly tensioned when lock is released.
18. Install front cover assembly.

4-121

1. Remove the front cover as previously detailed.

Installing the front cover oil seal (cover off)

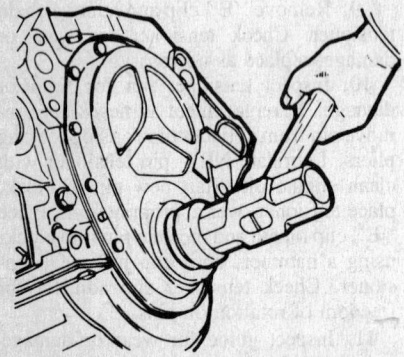

Installing the front cover oil seal (cover installed)

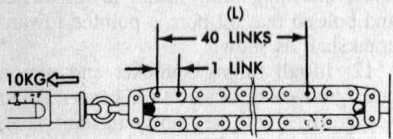

4-119 checking timing chain for wear

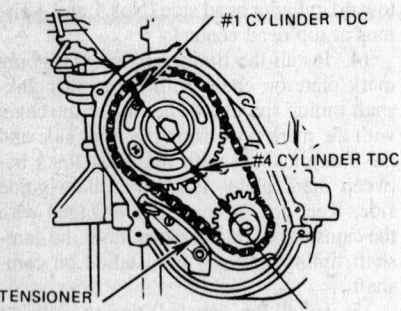

4-121 timing chain alignment

2. Place the No.1 piston at TDC of the compression stroke so that the marks on the camshaft and crankshaft sprockets are in alignment (see illustration).

3. Loosen the timing chain tensioner nut as far as possible without actually removing it.

4. Remove the camshaft sprocket bolts and remove the sprocket and chain together. If the sprocket does not slide from the camshaft easily, a light blow with a soft mallet at the lower edge of the sprocket will dislodge it.

5. Use a gear puller (J2288-8-20) and remove the crankshaft sprocket.

6. Press the crankshaft sprocket back onto the crankshaft.

7. Install the timing chain over the camshaft sprocket and then around the crankshaft sprocket. Make sure that the marks on the two sprockets are in alignment (see illustration). Lubricate the thrust surface with Molykote or its equivalent.

8. Align the dowel in the camshaft with the dowel hole in the sprocket and then install the sprocket onto the camshaft. Use the mounting bolts to draw the sprocket onto the camshaft and then tighten them to 27-33 ft. lb.

9. Lubricate the timing chain with clean engine oil. Tighten the chain tensioner.

10. Installation of the remaining components is in the reverse order of removal.

4-137 Diesel

1. Follow the timing pulley housing cover removal steps.

2. Remove the bolts fixing the injection pump timing pulley flange, then remove the flange.

3. When removing tension spring, avoid using excess force, or distortion of spring will result.

4. Remove the fixing nut of the tension pulley, then remove the tension pulley and tension center.

5. Remove the timing belt. Avoid twisting or kinking the belt and keep it free from water, oil, dust and other foreign matter. No attempt should be made to readjust belt tension. If the belt has been loosened through service of the timing system, it should be replaced with a new one.

6. Check that the setting marks on the crank pulley, injection pump timing pulley, and camshaft timing pulley are in alignment, then install the timing belt in sequence of crankshaft timing pulley, camshaft timing pulley, and injection pump timing pulley. Make an adjustment, so that slackness of the belt is taken up by the tension pulley. When installing the belt, care should be taken so as not to damage the belt.

7. Install the tension center and tension pulley, making certain the end of the tension center is in proper contact with two pins on the timing pulley housing.

8. Hand-tighten the nut, so that tension pulley can slide freely.

9. Install the tension spring correctly and semitighten the tension pulley fixing nut.

10. Turn the crankshaft 2 turns in normal direction of rotation to permit seating of the belt. Further rotate the crankshaft 90 degrees beyond top dead center to settle the injection pump. Never attempt to turn the crankshaft in reverse direction.

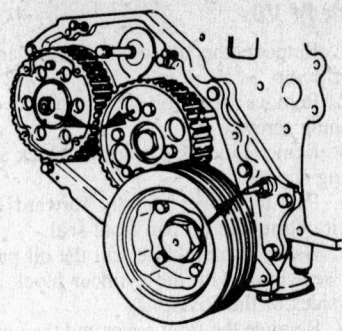

Diesel timing mark alignment

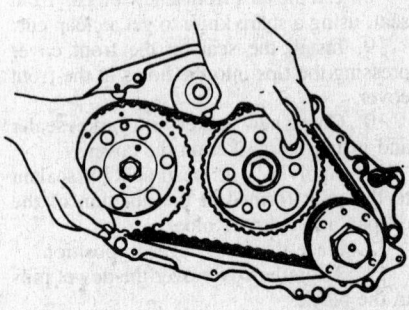

Installing the diesel timing belt

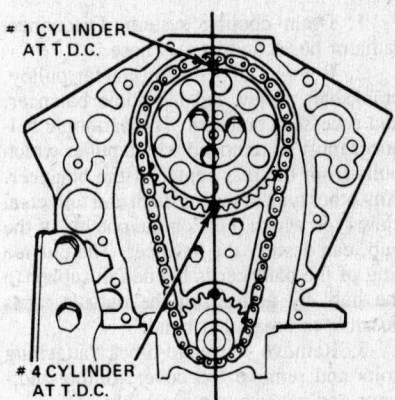

6-173 timing chain alignment

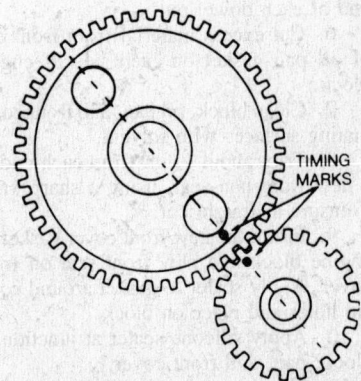

Six cylinder timing gear alignment

11. Loosen the tension pulley fixing nut completely, allowing the pulley to take up looseness of the belt. Then, tighten the nut to 78–95 ft. lbs.

12. Install the flange on the injection pump pulley. The hole in the outer circumference of the flange should be aligned with the timing mark "△" on the injection pump pulley.

13. Turn the crankshaft 2 turns in normal direction of rotation to bring the piston in No.1 cylinder to top dead center on compression stroke and check that the mark "△" on the timing pulley is in alignment with the hole in the flange.

14. The belt tension should be checked at a point between the injection pump pulley and crankshaft pulley using tool J29771, to a pull of 33–55 lbs.

15. Adjust valve cleaances.

16. Install remaining parts in the reverse order of removal.

2.5L (4 cyl)

NOTE: To perform this procedure, you'll need a press, and a press plate adapter J971 or equivalent.

1. Remove the timing cover. Remove the camshaft.

2. The crankshaft gear can be removed by simply sliding it off the keyed end of the crankshaft. Use a puller if the gear will not slide off easily.

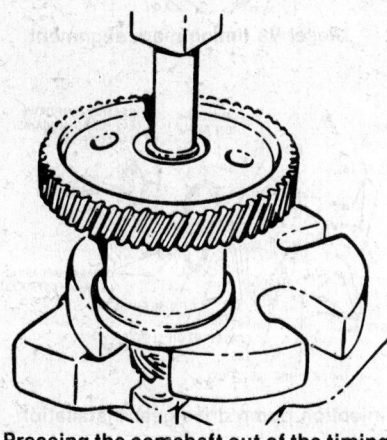

Pressing the camshaft out of the timing gear (2.5 L four cylinder engine)

3. To remove the camshaft timing gear, support the camshaft/gear assembly gear upward in the groove in the press plate adapter (J971 or equivalent) in such a manner that the back of the gear is firmly supported. The remainder of the crankshaft will hang down below. Use a socket or similar cylindrical object that is just slightly smaller than the inside diameter in the camshaft gear to force the camshaft downward and out of the gear. Position the adapter so the Woodruff key will be forced downward and out of the gear without damage; it must be located over the opening in the plate. Apply pressure with the press and force the camshaft out of the center of the gear.

4. To install the timing gear, slide the camshaft into a slit in the adapter that will permit it to be supported securely by the rear of the front bearing journal. Then, place the gear spacer ring and thrust plate over the end of the shaft, and install the woodruff key in the shaft keyway.

5. Put the camshaft gear into position and then use J21474–13, J21795–1 or equivalent to spread the load over the top of the gear. Press the gear onto the shaft until it bottoms out against the spacer ring. Then, check the end clearance of the thrust plate. It must be 0.0015–0.0050 in. If the clearance is less than .015 in., the spacer ring must be replaced.

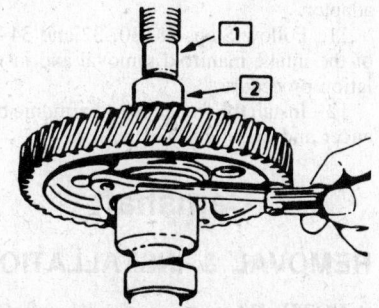

Checking the end clearance of the thrust plate (2.5 L four cylinder engine). (1) is the arbor press and (2) is the adapter designed to spread the load over the top surface of the gear.

6. Install the crankshaft sprocket onto the crankshaft over the woodruff key. Install the camshaft. Install the timing cover.

6–173

To replace the chain, remove the crankcase front cover. This will allow access to the timing chain. Crank the engine until the marks punched on both sprockets are closest to one another and in line between the shaft centers. Take out the three bolts that hold the camshaft sprocket to the camshaft. This sprocket is a light press fit on the camshaft and will come off readily. It is located by a dowel. The chain comes off with the camshaft sprocket. A gear puller will be required to remove the camshaft sprocket.

Without disturbing the position of the engine, mount the new crank sprocket on the shaft, then mount the chain over the camshaft sprocket. Arrange the camshaft sprocket in such a way that the timing marks will line up between the shaft centers and the camshaft locating dowel will enter the dowel hole in the cam sprocket.

Place the cam sprocket, with its chain mounted over it, in position on the front of the camshaft and pull up with the three bolts that hold it to the camshaft.

After the sprockets are in place, turn the engine two full revolutions to make certain that the timing marks are in correct alignment between the shaft centers.

Inline Six Cylinder

The six cylinder camshaft is gear driven.

To remove the camshaft gear, remove the camshaft and press the gear off.

——— CAUTION ———

The thrust plate must be positioned so that the Woodruff key in the shaft does not damage it when the shaft is pressed out of the gear. Support the hub of the gear or the gear will be seriously damaged.

The crankshaft gear may be removed with a gear puller while in place in the block.

V6 and V8 Except Diesel

These models are equipped with a timing chain. To replace the chain, remove the radiator core, water pump harmonic balancer, and the crankcase front cover. This will allow access to the timing chain. Crank the engine until the zero marks punched on both sprockets are closest to one another and in line between the shaft centers. Then, take out the three bolts that hold the camshaft gear to the camshaft. This gear is a light press fit on the camshaft and should come off easily. If it does not come off readily, tap the lower edge of the sprocket lightly with a plastic mallet to dislodge it. It is located by a dowel.

The chain comes off with the camshaft gear. A gear puller will be required to remove the crankshaft gear.

Without disturbing the position of the engine, mount the new crank gear on the shaft, then mount the chain over the camshaft gear. Arrange the camshaft gear in such a way that the timing marks will line up between the shaft centers and the camshaft locating dowel will enter the dowel hole in the cam sprocket.

Place the cam sprocket, with its chain mounted over it, in position on the front of the camshaft and pull up with the three bolts that hold it to the camshaft. On the V6, torque the bolts to 20 ft. lbs.

After the gears are in place, turn the engine two full revolutions to make certain that the timing marks are in correct alignment between the shaft centers.

Diesel

1. Remove the crankshaft pulley, harmonic balancer and front cover. Be certain to use the correct tool for balancer removal, to avoid damaging the part.

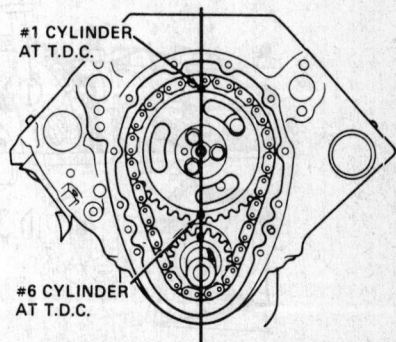

1979–82 V8 timing sprocket alignment

T41

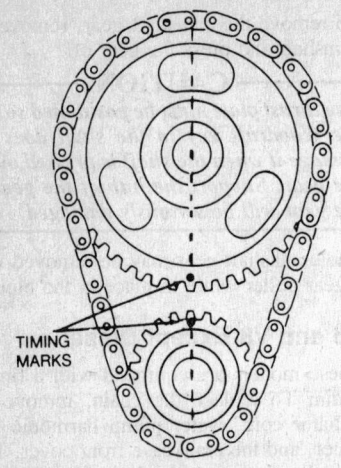

V8 timing mark alignment

2. Align the timing marks on the crankshaft and camshaft sprockets by rotating the crankshaft in the direction of normal engine rotation. The marks should be closest to one another and in line between the shaft centers, as shown in the accompanying diagram. Do not disturb the position of the engine.

3. Remove the oil slinger from the crankshaft. Remove the camshaft sprocket retaining nut.

4. Remove the crankshaft sprocket. The sprocket-to-crankshaft fit is such that a puller may be necessary. If possible, the crankshaft key should be removed before using the puller. If this is not possible, align the puller so that the fingers of the tool do not overlap the end of the key when the sprocket is removed. The keyway is machined only partway in the crankshaft sprocket, and breakage can occur if the sprocket is improperly removed.

5. Remove the timing chain and camshaft sprocket.

6. The fuel pump eccentric is behind the crankshaft sprocket, and may be removed if necessary.

7. Install the key in the crankshaft, if removed. Install the fuel pump eccentric, if removed.

8. Install the camshaft sprocket, crankshaft sprocket, and the timing chain together, with the timing marks aligned.

Tighten the camshaft sprocket retaining bolt to 65 ft. lbs.

NOTE: When the two timing marks are in alignment and closest together, the No.6 cylinder is at TDC. To obtain TDC for No.1 cylinder, slowly rotate the crankshaft one full revolution. This will move the camshaft sprocket timing mark to the top. No.1 cylinder will then be at TDC.

9. Install the oil slinger.

10. The injection pump must be retimed. Follow Steps 1–5 and 7–9 of the intake manifold removal and installation procedure. Remove the injection pump adapter and the seal from the injection pump adapter.

11. Follow Steps 18–30, 32 and 34–36 of the intake manifold removal and installation procedure.

12. Install the front cover, harmonic balancer and the crankshaft pulley.

Camshaft

REMOVAL & INSTALLATION

NOTE: Whenever a new camshaft is installed, it is recommended that all valve lifters be replaced, to insure the durability of the camshaft lobes and the valve lifter feet.

4–119

1. Remove cam cover.

2. Rotate camshaft until No.4 cylinder is in firing position. Remove distributor cap and mark rotor to housing position. Remove distributor.

3. Remove the fuel pump.

4. Lock the shoe on automatic adjuster in fully retracted position by depressing the adjuster lock lever with a screwdriver or equivalent in direction as indicated. After locking the automatic adjuster, check that the chain is in free state.

5. Remove the timing sprocket to camshaft bolt and remove the sprocket and fuel pump drive cam from the camshaft. Keep the timing sprocket on the chain damper

and tensioner without removing the chain from the sprocket.

6. Remove rocker arm, shaft and bracket assembly.

7. Remove the camshaft assembly.

8. Apply a generous amount of clean engine oil to the camshaft and journals of cylinder head.

9. Install the camshaft assembly.

10. Install the rocker arm, shaft and bracket assembly as outlined previously.

11. Check that the mark on the No.1 rocker arm shaft bracket is in alignment with the mark on the camshaft and that the crankshaft pulley groove is aligned with the TDC mark ("0" mark) on the front cover.

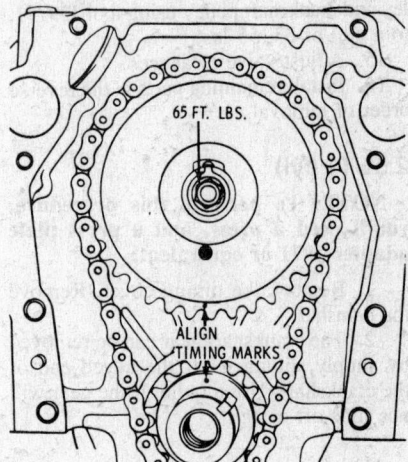

Diesel V8 timing mark alignment

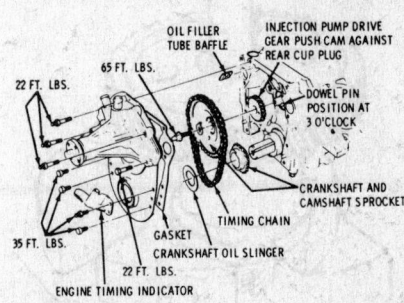

Injection pump drive gear installation

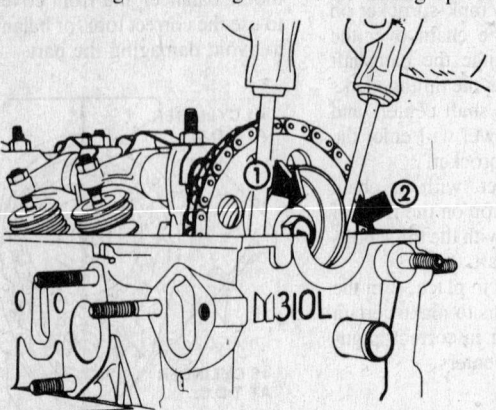

Depressing the adjusting lock lever

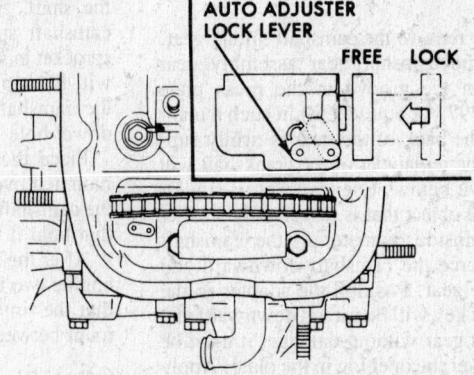

Locking the timing chain adapter

12. Assemble the timing sprocket to the camshaft by aligning it with the pin on the camshaft; use care not to remove the chain from the sprocket.

13. Install the fuel pump drive cam and install the sprocket retaining bolt and washer. Remove the half moon seal in front end of head; then install torque wrench and torque bolt to 60 ft. lbs. replace halfmoon seal in cylinder head.

14. Install the distributor.

15. Release lock by depressing the shoe on adjuster with a screwdriver or equivalent, check timing chain for proper tension.

16. Check valve timing; rotor and mark on distributor housing should be in alignment when No. 4 piston is in firing position. Timing mark on crank pulley should align with TDC mark (''O'' mark) on front cover.

17. Reinstall distributor cap.

18. Reinstall cam cover.

4-137 Diesel

1. Remove the camshaft carrier.

2. Remove the two bolts retaining the thrust plate in position on the front of the camshaft carrier.

3. Remove the thrust plate and carefully slide the camshaft out through the front of the carrier.

4. Install the camshaft in the carrier in the reverse order of removal, coating it liberally with engine oil before sliding it into position. Exercise care not to damage the camshaft bearing journals during the installation.

4-121

1. Remove the engine.

2. Remove the intake manifold.

3. Remove the cylinder head cover, pivot the rocker arms to the sides, and remove the pushrods, keeping them in order. Remove the valve lifters, keeping them in order. There are special tools which make lifter removal easier.

4. Remove the front cover.

5. Remove the distributor.

6. Remove the fuel pump and its pushrod.

7. Remove the timing chain and sprocket as described earlier in this chapter.

8. Carefully pull the camshaft from the block, being sure that the camshaft lobes do not contact the bearings.

9. To install, lubricate the camshaft journals with clean engine oil. Lubricate the lobes with Molykote or the equivalent. Install the camshaft into the engine, being extremely careful not to contact the bearings with the cam lobes.

10. Install the timing chain and sprocket. Install the fuel pump and pushrod. Install the timing cover. Install the distributor.

11. Install the valve lifters. If a new camshaft has been installed, new lifters should be used to ensure durability of the cam lobe.

12. Install the pushrods and rocker arms and the intake manifold. Adjust the valve

lash after installing the engine. Install the cylinder head cover.

2.5L (4 Cyl)

1. Disconnect the negative battery cable. Drain the cooling system.

2. Remove the alternator and bracket as described earlier in this section.

3. Remove the brace connecting the intake manifold to the block. Disconnect the lower radiator and heater hoses.

4. Remove the oil pressure sending unit.

5. Remove the wiring harness brackets from the side cover. Then, remove the side cover mounting nuts and remove the side cover.

6. Remove the power steering reservoir at the fan shroud.

7. Loosen the fan and water pump drive pulley bolts and then remove the power steering and air conditioning compressor belts.

8. Remove the fan and pulley.

9. Remove the crankshaft hub mounting bolt and remove the hub with a puller.

10. Remove the front cover bolts and remove the front cover.

11. Remove the distributor as described earlier in this section.

12. Remove the oil pump driveshaft cover, bearing and driveshaft.

6-173

Follow the 6-173 engine removal procedure then remove the camshaft as follows:

1. Remove the intake manifold, valve lifters and timing chain cover. If the vehicle is equipped with air conditioning, unbolt the condenser and move it aside without disconnecting any lines.

2. Remove fuel pump and pump pushrod.

3. Remove camshaft sprocket bolts, sprocket and timing chain. A light blow to the lower edge of a tight sprocket should free it (use a plastic mallet).

4. Install two bolts in cam bolt holes and pull cam from block.

5. To install, reverse removal procedure aligning the sprocket timing marks.

Inline Six Cylinder

1. In addition to removing the timing gear cover, remove the grille and radiator. If the truck has air conditioning, the condenser must be moved to provide room for camshaft removal. It may be possible to unbolt the condenser and move it aside, far enough for clearance. Otherwise, the air conditioning system must be discharged and the condenser removed. Do not disconnect any of the air conditioning lines unless you are thoroughly familiar with A/C systems and the hazards involved. It is recommended that you have the system discharged by a professional.

─── **CAUTION** ───

Compressed refrigerant expands (boils) into the atmosphere at a temperature of − 21°F

or less. It will freeze any surface it contacts, including your skin or eyes.

2. Remove the valve cover and gasket, loosen all the valve rocker arm nuts and pivot the arms clear of the pushrods.

3. Remove the distributor and fuel pump.

4. Remove the side covers and gaskets. Remove the pushrods and valve lifters.

5. Align the timing gear marks and remove the two camshaft thrust plate retaining screws by working through the holes in the camshaft gear.

6. Remove the camshaft and gear assembly by pulling it out through the front of the block. Be careful not to dislodge the camshaft bearings.

NOTE: If renewing either camshaft or camshaft gear, the gear must be pressed off the camshaft. Install press plates under the camshaft gear and press the camshaft from the gear. The thrust plate must be positioned so that the Woodruff key in the camshaft does not damage the shaft when the camshaft is pressed from the gear. Support the hub of the gear to prevent damage to the part. The replacement parts must be assembled in the same manner (under pressure). In placing the gear on the camshaft, press the gear onto the shaft until it bottoms against the gear spacer ring. The end clearance of the thrust plate should be 0.001–0.005 in.

7. Install the camshaft assembly in the engine. Pre-lube the cam lobes with E.O.S. or SAE 90 gear lubricant. Do not dislodge the cam bearings when inserting the camshaft.

8. Turn the crankshaft and camshaft to align and bring the timing marks together. Push the camshaft into this aligned position.

9. Runout on either crankshaft or camshaft gear should not exceed 0.003 in.

10. Backlash between the two gears should be between 0.004–0.006 in.

11. Install the timing gear cover and gasket.

12. Install the oil pan and gaskets.

13. Install the harmonic balancer.

14. Line up the keyway in the balancer with the key on the crankshaft and drive the balancer onto the shaft until it bottoms against the crankshaft gear.

15. Install the valve lifters and pushrods. Install the side covers with new gaskets. Later models use RTV sealer instead of gaskets. Attach the coil wires; install the fuel pump.

16. Install the distributor and set the timing as described under the Distributor Removal and Installation section at the beginning of the chapter.

17. Pivot the rocker arms over the pushrods and adjust the valves.

18. Add oil to the engine. Install and adjust the fan belt.

19. Install the radiator or shroud.

20. Install the grille assembly.

21. Fill the cooling system, start the engine and check for leaks.

22. Check and adjust the timing.

4.3L V6

1. Disconnect the battery ground cable. Remove the engine cover.

2. Remove the air cleaner. Drain the cooling system.

3. Remove the distributor and carburetor as described earlier in this section.

4. Remove the rear air conditioner brace.

5. Disconnect the transmission and engine oil filler tubes at the alternator bracket.

6. Disconnect the A/C idler pulley.

7. Remove the alternator adjusting bracket and set it aside.

8. Remove the diverter valve bracket. Then, disconnect the Air Injection Reactor hoses at the diverter valve.

9. Remove the intake manifold, referring to the procedure above as necessary.

10. Remove the upper fan shroud. Remove the power steering pump as described later in this section.

11. Remove the air pump and its mounting bracket.

12. Remove the fan and pulley. Then, remove the water pump as described later in this section.

13. Remove the engine balancer hub and the timing cover as described earlier in this section.

14. Turn the engine to align the timing marks with the camshaft mark at the top rather than at the bottom. This will put No. 1 cyl. at TDC.

15. Remove the timing chain and cam sprocket. Remove the fuel pump.

16. Disconnect the radiator hoses and transmission cooler lines (plug the transmission cooler lines).

17. Remove the lower fan shroud. Disconnect the brake master cylinder lines at the master cylinder and plug all openings.

18. Remove the valve covers. Remove the rocker arms and pushrods, keeping all parts in order.

19. Remove the lifters, keeping them in order for replacement in the same positions.

20. Install two $\frac{5}{16}$-18 x 4'' bolts into the bolt holes in the end of the camshaft. Then, remove the camshaft carefully. Since all

bearing journals are the same diameter, you must be careful to pull the camshaft straight out, moving it slowly to avoid nicking the edges of the journals.

21. Inspect the camshaft journals for out-of-round with a michrometer. The minimum diameter must not be more than 0.001 in. less than the maximum diameter. Otherwise, replace the camshaft.

22. Coat the bearing journals of the camshaft with a lubricant designed to lubricate new parts during engine rebuilding (''Molykote'' or equivalent).

23. If a new camshaft is being installed, transfer the bolts used as guides during removal to the new shaft. Install the camshaft, being careful, again, not to damage the bearing journals.

24. Remove the two bolts installed in the end of the camshaft to guide it.

25. Refer to the procedure, located earlier in this section, on installation of the timing chain and sprockets. Make sure the chain is installed on the two sprockets so the timing marks will be in proper position (No.1 cyl. at TDC). Align the dowel in the camshaft with the dowel hole in the camshaft sprocket and then install the sprocket. Install the bolts and tightn them alternately and evenly to draw the sprocket onto the camshaft. Torque the bolts to 20 ft. lbs.

26. Install the remaining parts in reverse order. You should use new lifters if the camshaft has been replaced, in order to guarantee maximum life of the lifters and camshaft. Coat the lifter feet with ''Molykote'' or equivalent. also. Adjust the valves. Refill and thoroughly bleed the brake system. Operate the engine and check for leaks.

V8 Except Diesel

1. Remove the intake manifold, valve lifters and timing chain cover.

2. Remove the grille and radiator. If the truck has air conditioning, the condenser must be moved to provide room for camshaft removal. It may be possible to

unbolt the condenser and move it aside, far enough for clearance. Otherwise, the air conditioning system must be discharged and the condenser removed. Do not disconnect any of the air conditioning lines unless you are thoroughly familiar with A/C systems and the hazards involved.

3. Remove the fuel pump and pump pushrod.

4. Remove the camshaft sprocket bolts, sprocket and timing chain. A light blow to the lower edge of a tight sprocket should free it (use a plastic mallet).

5. Install two $\frac{7}{16}$-18 × 14 in. bolts in the cam bolt holes and pull the cam from the block.

6. To install, reverse the removal procedure aligning the timing marks.

NOTE: Pre-lube the cam lobes with E.O.S. or SAE 90 gear lubricant. Do not dislodge the cam bearings when installing the camshaft.

350 (5.7L) Diesel

Removal of the camshaft also requires removal of the injection pump drive and driven gears, removal of the intake manifold, disassembly of the valve lifters, and re-timing of the injection pump.

1. Disconnect the negative battery cables. Drain the coolant. Remove the radiator.

2. Remove the intake manifold and gasket and the front and rear intake manifold seals. Refer to the intake manifold removal and installation procedure.

3. Remove the balancer pulley and the balancer. Remove the engine front cover.

4. Remove the valve covers. Remove the rocker arms, pushrods and valve lifters. Be sure to keep the parts in order so that they may be returned to their original positions.

5. If the truck has air conditioning, the condenser must be moved to provide room for camshaft removal. It may be possible to unbolt the condenser and move it aside far enough for clearance. Otherwise, the air conditioning system must be discharged and the condenser removed. Do not disconnect any of the air conditioning lines unless you are thoroughly familiar with A/C systems and the hazards involved.

6. Remove the camshaft sprocket retaining bolt, and remove the timing chain and sprockets.

7. Position the camshaft dowel pin at the 3 o'clock position.

8. Push the camshaft rearward and hold it there, being careful not to dislodge the oil gallery plug at the rear of the engine. Remove the pump drive gear by sliding it from the camshaft while rocking the pump driven gear.

9. To remove the pump driven gear, remove the injection pump adapter, remove the snap-ring, and remove the selective washer. Remove the driven gear and spring.

10. Remove the camshaft by sliding it out the front of the engine. Be extremely

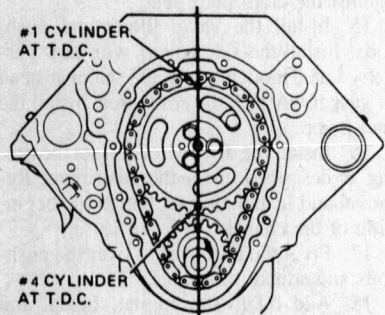

Alignment of camshaft and crankshaft sprocket timing marks on the 4.3 L V6

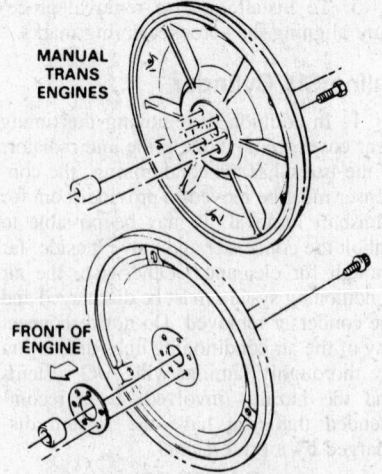

Installing the 2.5 L four flywheel

careful not to allow the cam lobes to contact any of the bearings, or the journals to dislodge the bearings during camshaft removal. Do not force the camshaft, or bearing damage will result.

11. Coat the camshaft and the cam bearings with G.M. lubricant #562458 or the equivalent.

12. Carefully slide the camshaft into position in the engine.

13. Install the timing chain and sprockets in proper alignment.

14. Check the injection pump driven gear bushing and replace as necessary.

15. Install the injection pump driven gear, spring, shim, and snap-ring. Check the gear end-play. If the end-play is not within 0.002–0.005 in., replace the shim to obtain the specified clearance. Shims are available in 0.003 in. increments, from 0.080 to 0.115 in.

16. Position the camshaft dowel pin at the 3 o'clock position. Align the zero marks on the pump drive gear and pump driven gear. Hold the camshaft in the rearward position and slide the pump drive gear onto the camshaft.

17. Disassemble, assemble, and bleed the valve lifters as outlined earlier. Install the lifters, pushrods, rocker arms, and pivots. Install the injection pump adapter and injection pump, re-timing the engine as outlined in the intake manifold removal and installation procedure.

18. Install the intake manifold, as outlined earlier. Install the rocker covers.

19. Install the engine front cover, balancer and pulley.

20. Install the radiator. Install the air conditioning condenser, if equipped. Fill the cooling system. Check the automatic transmission fluid level. Connect the negative battery cables.

379 (6.2L) Diesel

1. Disconnect the battery.
2. Raise the vehicle.
3. Drain the radiator and block.
4. Disconnect the exhaust pipe at the exhaust manifolds.
5. Remove the fan shroud lower attaching bolts.
6. Lower the vehicle.
7. Remove the fan shroud upper attaching bolts.
8. Remove the radiator.
9. Remove the fan.
10. Remove the vacuum pump.
11. Remove the intake manifold as previously outlined.
12. Remove the injection pump lines at the pump and nozzles. Cap the injection nozzles to prevent dirt from entering fuel (tag injection lines for reinstallation).
13. Remove the water pump.
14. Remove the injection pump gear.
15. Scribe a mark on the front cover aligning the line on the injection pump flange to the front cover.

16. Remove the power steering pump and the generator and lay aside.

18. If equipped with A/C remove the compressor and lay aside.

19. Remove the rocker arm covers as previously outlined.

20. Remove the rocker arm shaft assembly and pushrods. Place the parts in a rack so they may be reinstalled in the same location.

21. Remove the thermostat housing/crossover from the cylinder heads.

22. Remove the cylinder head as previously outlined with the exhaust manifolds attached.

23. Remove the valve lifter clamps, guide plates and valve lifters. Place the parts in a rack so they may be reinstalled in the same location.

24. Remove the front cover.
25. Remove the timing chain.
26. Remove the fuel pump.
27. Remove the cam retainer plate.
28. If equipped with A/C remove the A/C condensor mounting bolts, and with the aid of an assistant, lift the condensor.
29. Remove the camshaft.

NOTE: Whenever a new camshaft is installed, replacement of all valve lifters, oil filter, and new oil is recommended to insure durability of the camshaft lobes and lifters. Whenever a new camshaft is installed coat the camshaft lobes with Molykote or its equivalent.

30. Lubricate the camshaft journals with engine oil and install the camshaft.

31. Install the retainer plate (20 ft. lbs.).
32. Install the fuel pump.
33. Install the timing chain.
34. Install the front cover.
35. Install the valve lifters, guide plates and clamps, rotate the crankshaft to insure the valve lifters are free to travel.
36. Install the cylinder head.
37. Install the rocker arm shaft assembly and pushrods. Care must be taken to insure the pushrods are installed properly.
38. Install the rocker arm covers.
39. Install the injection pump to the front cover, making sure the lines on the pump and the scribe line on the front cover are aligned.
40. Install the injection pump driven gear, making sure the gears are aligned. Anytime the timing chain, gears, or sprockets are replaced, it will be necessary to retime the engine.
41. Install the water pump.
42. Install the injection lines.
43. Install the generator, power steering and A/C.
44. Install the crank pulley.
45. Install the fan.
46. Install the drive belts and adjust as necessary.
47. Install the fan shroud.
48. Install the radiator, and fill with coolant.
49. Connect the necessary wires and hoses.

50. Raise the vehicle.
51. Connect the exhaust pipes to the exhaust manifold.
52. Lower the vehicle.
53. Install the vacuum pump.
54. Connect the secondary fuel filter lines (with adapter).
55. Install the cylinder head covers J29664–1.
56. Connect the battery.
57. Start the engine and check for leaks.
58. Stop the engine.
59. Remove the protective covers.
60. Loosen the vacuum pump hold-down and disconnect the secondary filter (with adapter) from the fuel lines.
61. Install the intake manifold.

Piston and Connecting Rods

Refer to the Engine Rebuilding section in Unit Repair for details.

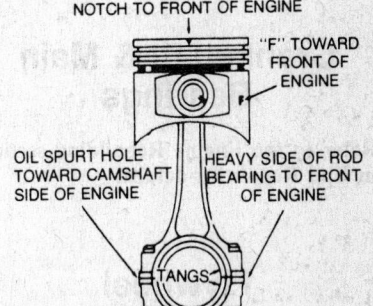

6-250 and 292 piston/rod assembly

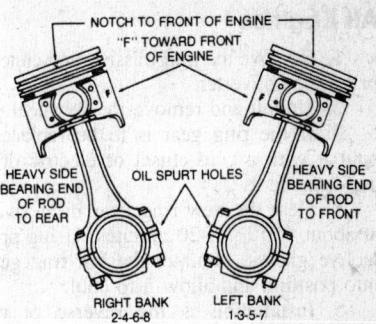

Small block V8 piston/rod assembly

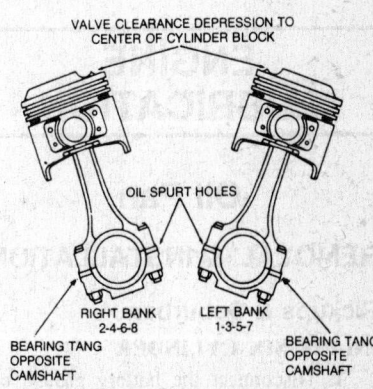

8-454 piston/rod assembly

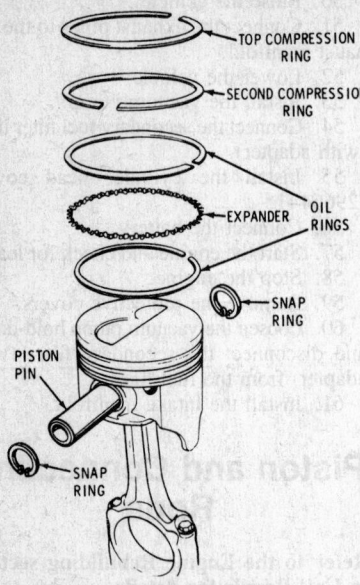

Diesel piston assembly

Crankshaft & Main Bearings

Refer to the Engine Rebuilding section in Unit Repair for details.

Flywheel

REMOVAL & INSTALLATION

All Engines

1. Remove the transmission and clutch, or torque converter.
2. Unbolt and remove the flywheel.
3. If the ring gear is to be replaced, split it with a cold chisel or electric drill and remove it.
4. Heat the new ring gear in an oven at about 450°F for 20 minutes. Using protective gloves, hammer the hot ring gear into position and allow it to cool.
5. Installation is the reverse of removal.

ENGINE LUBRICATION

Oil Pan

REMOVAL & INSTALLATION

Pickups & Suburban

INLINE SIX CYLINDER

1. Disconnect the battery ground cable.

2. Raise and support the vehicle. Disconnect the starter leaving the wires attached and swing it out of the way.
3. If there is not enough clearance, remove the bolts securing the engine mounts to the crossmember and raise the engine high enough to insert a 2 in. × 4 in. piece of wood between the engine mounts and the crossmember brackets.
4. Drain the engine oil.
5. Remove the flywheel or converter cover.
6. Remove the oil pan.
7. Clean all gasket surfaces and install a new seal in the rear main bearing groove and a new seal in the crankcase front cover. Installation is the reverse of removal. Install new side gaskets on the block, but do not use sealer. Fill the engine with oil and run the engine, checking for leaks.

V8 EXCEPT DIESEL

1. Drain the engine oil.
2. Remove the oil dipstick and tube.
3. If necessary, remove the exhaust crossover. On 454s, remove the air cleaner, fan shroud, and distributor cap.
4. Remove the flywheel or converter cover. Remove the starter. On 454s, remove the oil pressure line from the block. On four wheel drive models with automatic transmission, remove the strut rods at the motor mounts.
5. On 454s only, remove the engine mount through bolts and raise the engine.
6. Remove the oil pan and discard the gaskets.
7. Installation is the reverse of removal. Clean all gasket surfaces and use new gaskets to assemble. Use gasket sealer to retain the side gaskets to the cylinder block. Install a new oil pan rear seal in the rear main bearing cap slot with the ends butting the side gaskets. Install a new front seal in the crankcase front cover with the ends butting the side gaskets. Fill the engine with oil and check for leaks.

DIESEL

1. Remove the drive and vacuum pump.
2. Disconnect the battery cables.
3. Remove the fan shroud attaching screws and pull the shroud up from the clips.
4. Block the rear wheels and jack up the front of the truck. Drain the oil.
5. Remove the flywheel cover.
6. Remove the starter and solenoid.
7. Remove both of the engine mount through bolts and raise the engine. Loosen the right hand mount and remove the left hand mount.
8. Unbolt and remove the oil pan.

NOTE: If extended work is to be done, the mounts should be reinstalled and the engine lowered to the frame brackets.

9. To install: After cleaning the mounting surfaces thoroughly, apply sealer to both sides of the pan gaskets and install the gaskets on the block.
10. Install the front and rear rubber seals.

11. Apply a thin coat of all purpose grease on the seals, and install the oil pan. Torque the bolts to 10 ft. lbs. in a circular sequence, starting in the middle and working out.
12. Further installation is the reverse of removal. Fill the engine with oil and check for leaks.

Vans except Astro

INLINE SIX CYLINDER

1. Disconnect the negative battery cable and remove the engine cover.
2. Remove the air cleaner and studs.
3. Remove the fan finger guard.
4. Remove the radiator upper supporting brackets.
5. Raise and safely support the van.
6. On vans with manual transmissions: Disconnect the clutch cross shaft from the left front mounting bracket. Remove the transmission-to-bellhousing upper bolt. Remove the transmission rear mounting bolts and install two $7/16 \times 3$ in. bolts. Raise the transmission and place a small piece of 2 × 4 wood in between the mount and the crossmember.
7. Remove the starter motor.
8. Drain the engine oil.
9. Remove the engine mount through bolts.
10. Raise the engine slightly and place small 2 × 4 wooden blocks in between the mount and the block.
11. Remove the flywheel splash shield or the converter cover, as applicable.
12. Remove the oil pan attaching bolts and remove the oil pan.
13. Clean the gasket surface thoroughly and use a new gasket on installation.
14. Reverse to install.

V8

1. Drain the engine oil.
2. Remove the oil dipstick and tube.
3. If necessary, remove the exhaust pipe crossover.
4. If equipped with automatic transmission, remove the converter housing pan.
5. Remove the starter brace and bolt and swing the starter aside.
6. Remove the oil pan and discard the gaskets.
7. Installation is the reverse of removal. Clean all gasket surfaces and use new gaskets to assemble. Use gasket sealer to retain the side gaskets to the cylinder block. Install a new oil pan rear seal in the rear main bearing cap slot with the ends butting the side gaskets. Install a new front seal in the crankcase front cover with the ends butting the side gaskets. Fill the engine with oil and check for leaks.

Blazer & Jimmy

INLINE SIX CYLINDER

1. Disconnect the battery ground cable.
2. Raise and support the vehicle. Disconnect the starter leaving the wires attached and swing it out of the way.

3. Remove the flywheel or converter cover.

4. Drain the engine oil.

5. On some models, there may be enough clearance to remove the oil pan without raising the engine. If there isn't, remove the through bolts securing the engine mounts to the brackets and raise the engine high enough to insert a 2 in. × 4 in. piece of wood between the engine mounts and brackets.

6. Lower the engine so it rests on the blocks, in a slightly raised position. This should provide enough clearance.

7. Remove the oil pan attaching screws.

8. Remove the oil pan.

9. Clean all gasket surfaces and install a new seal in the rear main bearing groove and a new seal in the crankcase front cover. Installation is the reverse of removal. Install new side gaskets on the block, but do not use sealer. Fill the engine with oil and run the engine, checking for leaks.

V8 EXCEPT DIESEL

NOTE: For diesel engine procedures, follow the previous Pickup Truck instructions.

1. Drain the engine oil.

2. If necessary, remove the exhaust pipe crossover.

3. If equipped with automatic transmission, remove the converter housing pan.

4. On four wheel drive models equipped with automatic transmission, remove the strut rods at the engine mounts.

5. Remove the oil pan and discard the gaskets.

6. Installation is the reverse of removal. Clean all gasket surfaces and use new gaskets to assemble. Use gasket sealer to retain the side gaskets to the cylinder block. Install a new oil pan rear seal in the rear main bearing cap slot with the ends butting the side gaskets. Install a new front seal in the crankcase front cover with the ends butting the side gaskets. Reassemble, fill the engine with oil and check for leaks.

S–Series

4–119

NOTE: On 4-wheel drive the engine must be removed before removing the oil pan.

1. Disconnect the negative battery terminal.

2. Jack up your vehicle and support it with jackstands.

3. Drain the oil.

4. Remove the front splash shield.

5. Remove the front crossmember, if necessary.

6. Disconnect the relay rod at the idler arm and lower the relay rod.

7. Remove the left side bellhousing bracket.

8. Remove the vacuum line at the oil pan.

9. Remove the oil pan bolts and the pan.

NOTE: It may be necessary to remove the motor mounts and jack up the engine in order to remove the oil pan.

10. Installation is the reverse of removal. Tighten the retaining bolts to 43 inch lbs.

4–121

1. Disconnect the negative battery cable.

2. Drain the crankcase. Raise and support the front of the vehicle.

3. Remove the A/C brace if so equipped.

4. Remove the exhaust shield and disconnect the exhaust pipe at the manifold.

5. Remove the starter motor and position it out of the way.

6. Remove the flywheel cover. Remove the oil pan.

NOTE: Prior to oil pan installation, check that the sealing surfaces on the pan, cylinder block and front cover are clean and free of oil. If installing the old pan, be sure that all old RTV has been removed.

7. Apply a $\frac{1}{8}$ in. bead of RTV sealant to the oil pan sealing surface. Use a new oil pan rear seal and install the pan in place. Tighten the bolts to 9–13 ft. lbs.

8. Installation of the remaining components is in the reverse order of removal.

4–137 Diesel

The engine must be removed from the truck.

1. With the engine on a work stand, unbolt and remove the oil pan from the crankcase.

2. Discard the gasket and clean the gasket surfaces.

3. At this time, the crankcase may also be removed from the block. Discard the gasket and seals and clean the gasket surfaces.

4. Install the oil pan and/or crankcase using new gaskets coated with sealer. Torque the oil pan bolts to 5–9 ft. lb.; the crankcase bolts to 15 ft. lb.

6–173

The engine must be removed from the truck.

1. With the engine on a work stand, unbolt and remove the pan.

2. Discard the gasket and clean the gasket surfaces.

3. The oil pan does not use a preformed gasket. Rather it is sealed with RTV gasket material. Make sure that the sealing surfaces are free of oil and old RTV material.

4. Run a $\frac{1}{8}$ in. bead of sealer along the entire sealing surface of the pan.

5. Place the pan on the engine and finger tighten the bolts. Torque the smaller bolts to 6–9 ft. lb.; the larger bolts to 15–22 ft. lb.

Astro

2.5L (4 Cyl)

1. Disconnect the negative battery cable. Raise the vehicle and support it at approved jacking points. Drain the engine oil.

2. Label and then disconnect the starter wiring. Remove the flywheel splash shield. Unbolt and remove the starter.

3. Disconnect the exhaust pipe at the manifold. Disconnect all exhaust hangers.

4. Remove the oil pan bolts and remove the oil pan.

5. Remove all RTV sealant and oil from the sealing surfaces on the pan, cylinder case and front cover. Make sure RTV sealant is removed from blind attaching holes.

6. Squeeze a $\frac{1}{8}$ in. bead of RTV sealer over the entire sealing flange of the oil pan. Install the pan square in position, and then install all bolts finger tight. Torque alternately and evenly to 75 inch lbs.

7. Reverse the remaining removal procedures to install. Refill the oil pan, operate the engine and check for leaks.

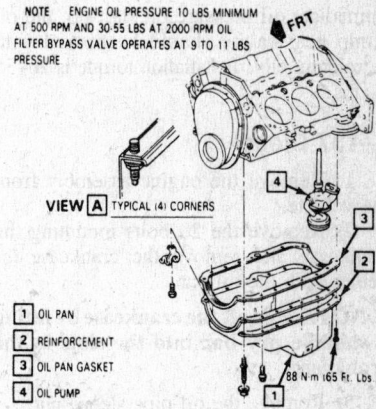

NOTE: ENGINE OIL PRESSURE 10 LBS MINIMUM AT 500 RPM AND 30-55 LBS AT 2000 RPM OIL FILTER BYPASS VALVE OPERATES AT 9 TO 11 LBS PRESSURE

VIEW A TYPICAL (4) CORNERS

1 OIL PAN
2 REINFORCEMENT
3 OIL PAN GASKET
4 OIL PUMP
88 N·m (65 Ft. Lbs.)

Mounting the oil pan on the 2.5 L V6

4.3L V6

1. Disconnect the negative battery cable. Raise the vehicle and support it at approved jacking points. Drain the engine oil.

2. Disconnect the exhaust pipes at the manifolds.

3. Remove the engine strut rods at the inspection cover and then remove the inspection cover.

4. Label and then disconnect the starter wiring. Then, remove the starter mounting bolts, and remove the starter.

5. Install a suitable jack under the engine and support it. Then, remove the bolts attaching both engine mounts to the block. Raise the engine until there is adequate clearance to remove the oil pan.

6. Remove the oil pan mounting bolts and remove the pan.

7. Clean all gasket surfaces thoroughly. Install the pan in reverse order, using all new gaskets and seals. Torque the $\frac{5}{16}$-18 mounting bolts to 14 ft. lbs. and the $\frac{1}{2}$-20 mounting bolts to 7 ft. lbs.

8. Complete the installation in reverse order, torquing the engine mount-to-block bolts to 35 ft. lbs. Make sure the crankcase

is refilled to the proper level with clean oil. Operate the engine and check for leaks.

Oil Pump

REMOVAL & INSTALLATION

4-119

1. Drain and remove the oil pan.
2. Disconnect the oil feed pipe.
3. Remove the two bolts securing the oil pump to the cylinder block and remove the oil pump.
4. Install in the reverse order of removal.

4-121

1. Remove the engine oil pan.
2. Remove the pump attaching bolts and carefully lower the pump.
3. Install in reverse order. To ensure immediate oil pressure on start-up, the oil pump gear cavity should be packed with petroleum jelly. Installation torque is 26–35 ft. lbs.

4-137 Diesel

1. Remove the engine assembly from the vehicle.
2. Remove the 20 bolts mounting the crankcase and remove the crankcase together with the oil pan.

NOTE: Pry off the crankcase by fitting a suitable pry bar into the slots in the crankcase.

3. Remove the oil pipe sleeve nut.
4. Remove the 2 bolts fixing the oil pump and remove the oil pump with oil pipe.
5. Install the oil pipe and leave the joints semi-tight.
6. Fully tighten the oil pump fixing screws, then tighten the oil pipe joints.
7. Reverse the removal procedure for the remaining parts.

6-173

1. Remove the oil pan.
2. Unbolt and remove the oil pump and pickup.
3. Installation is the reverse of removal. Torque the pump bolts to 26–35 ft. lb. Before installing an oil pump, fill it with clean oil.

2.5L (4 Cyl)

1. Remove the oil pan.
2. Remove the bolt that retains the pickup to the block. Then, remove the bolts retaining the oil pump-to-block positioning flange to the block, and remove that flange and the oil pump.
3. If the pump has been disassembled, is being replaced, or for any reason oil has been removed from it, it must be primed. It can either be filled with oil before in-

stalling the cover plate (and oil kept within the pump during handling), or the entire pump cavity can be filled with petroleum jelly. IF THE PUMP IS NOT PRIMED, THE ENGINE COULD BE DAMAGED BEFORE IT RECEIVES ADEQUATE LUBRICATION WHEN YOU START IT.
4. To install the pump, first align the slot in the oil pump shaft with the tang on the oil pump driveshaft in the block. Then, slide the pump into place (it should slide easily); the pump positioning flange will fit over the lower driveshaft bushing (note that no gasket is required). Install the mounting bolts and torque to 22 ft. lbs.
5. Install the oil pan.

1	PUMP BODY	6	SPRING RETAINER
2	PICKUP TUBE	7	COVER SCREWS
3	PICKUP SCREW ASSEMBLY	8	COVER
4	PRESSURE REGULATOR VALVE	9	IDLER GEAR
5	PRESSURE REGULATOR SPRING	10	DRIVE GEAR AND SHAFT

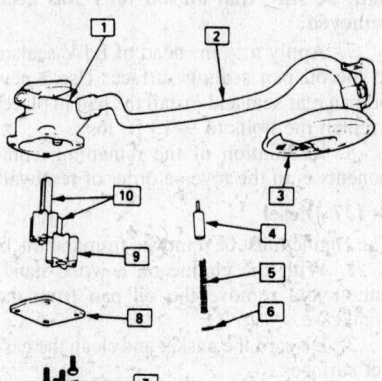

Exploded view of the oil pump for the 2.5 L four

Inline Six Cylinder

1. Drain the oil and remove the oil pan.
2. Remove the 2 flanged mounting bolts and remove the pickup pipe bolt.
3. Remove the pump and screen as an assembly.
4. To install, align the oil pump driveshafts with the distributor tang and install the oil pump. Position the flange over the distributor lower bushing, using no gasket. The oil pump should slide easily into place. If not, remove it and reposition the slot to align with the distributor tang.
5. Reinstall the oil pan and fill the engine with oil.

4.3L V6

1. Remove the oil pan.
2. Remove the bolt attaching the pump to the rear main bearing cap. Remove the pump and the extension shaft, which will come out behind it.
3. If the pump has been disassembled, is being replaced, or for any reason oil has been removed from it, it must be primed.

It can either be filled with oil before installing the cover plate (and oil kept within the pump during handling), or the entire pump cavity can be filled with petroleum jelly. IF THE PUMP IS NOT PRIMED, THE ENGINE COULD BE DAMAGED BEFORE IT RECEIVES ADEQUATE LUBRICATION WHEN YOU START IT.
4. Engage the extension shaft with the oil pump shaft. Align the slot on the top of the extension shaft with the drive tang on the lower end of the distributor drive shaft, and then position the pump at the rear maing bearing cap so the mounting bolt can be installed. Install the bolt, torquing to 65 ft. lbs.
5. Install the oil pan.

V8 and Diesel

1. Drain the oil and remove the oil pan.
2. Remove the bolt (two bolts on diesels) holding the pump to the rear main bearing cap. Remove the pump and extension shaft.
3. To install, assemble the pump and extension shaft to the rear main bearing cap aligning the slot on the top of the extension shaft with the drive tang on the distributor driveshaft. The installed position of the oil pump screen is with the bottom edge parallel to the oil pan rails. Further installation is the reverse of removal.

Oil Cooler

REMOVAL & INSTALLATION

4-137 Diesel

1. Place a suitable size tray under the oil filter to receive oil and water flowing out from the filter.
2. Drain the cooling system by opening the drain plugs on the radiator and on the cylinder block.
3. Remove the oil cooler water drain plug and drain the water.
4. Disconnect the oil cooler hoses at the cooler side.
5. Remove the oil filter cartridge using filter wrench.
6. Remove the nut fixing the oil cooler, then remove the oil cooler assembly.
7. Install the cooler using a new O-ring. Torque to 55–60 ft. lbs.

Rear Main Oil Seal

REPLACEMENT

All Engines Except Below

Both halves of the rear main oil seal can be replaced without removing the crankshaft. Always replace the upper and lower seal together. The lip should face the front of the engine. Be very careful that you do not break the sealing bead in the channel on the outside portion of the seal while installing it. An installation tool can be fabricated to protect the seal bead.

1. Remove the oil pan, oil pump and rear main bearing cap.

2. Remove the oil seal from the bearing cap by prying it out with a suitable tool.

3. Remove the upper half of the seal with a small punch. Drive it around far enough to be gripped with pliers.

4. Clean the crankshaft and bearing cap.

5. Coat the lips and bead of the seal with light engine oil, keeping oil from the ends of the seal.

6. Position the fabricated tool between the crankshaft and seal seat.

7. Position the seal between the crankshaft and tip of the tool so that the seal bead contacts the tip of the tool. The oil seal lip should face forward.

8. Roll the seal around the crankshaft using the tool to protect the seal bead from the sharp corners of the crankcase.

9. The installation tool should be left installed until the seal is properly positioned with both ends flush with the block.

10. Remove the tool.

11. Install the other half of the seal in the bearing cap using the tool in the same manner as before. Light thumb pressure should install the seal.

12. Install the bearing cap with sealant applied to the mating areas of the cap and block. Keep sealant from the ends of the seal.

13. Torque the main bearing cap retaining bolts to 10–12 ft. lbs. Tap the end of the crankshaft first rearward, then forward with a lead hammer. This will line up the rear main bearing and the crankshaft thrust surfaces. Tighten the main bearing cap to specification.

14. Further installation is the reverse of removal.

Diesel

It is not necessary to remove the crankshaft to correct seal leaks at the rear main bearing.

1. Drain oil and remove oil pan. Remove rear main bearing cap.

2. Insert a packing tool such as a screwdriver or a punch against one end of the seal in the cylinder block and drive the old seal gently into the groove until it is packed tight. This varies from $1/4$–$3/4$ in., depending on the pack required. Be careful not to nick the main bearing when packing the seal.

3. Repeat this procedure on the other end of the cylinder block seal.

4. Measure the amount the seal was driven up on one side. Add $1/16$ in., then use a razor blade to cut this length from the old seal removed from the bearing cap. Repeat for the other side.

5. Place a drop of sealer on each end of the cut pieces of seal.

6. Work these two pieces of seal into the cylinder block with two small screwdrivers. Pack them into the block firmly.

Trim the ends of the seal flush with the block.

NOTE: Place a piece of shim stock or strip of metal between the seal and the crankshaft to protect the bearing surface before trimming.

7. Clean the bearing cap and seal grooves.

8. Install a new seal into the bearing cap, packing by hand.

9. Using a seal installer, pack the seal firmly into the groove. These tools are generally available in automotive parts stores.

10. Cut the seal flush with the mating surface of the bearing cap. Pack the seal end fibers away from the edges, toward the center with a screwdriver.

11. Clean the bearing insert and install in the bearing cap.

12. Clean the crankshaft bearing journal, and the mating surface of the bearing cap. Place a dab of sealer, such as Loctite® 496, on the mating surface of the cap. Lay a piece of Plastigage® on the bearing surface.

13. Install the bearing cap, lubricate the bolt threads with engine oil, and install. Torque the bolts to 120 ft. lbs. on the 350 and 70 ft. lbs. on the 379. Check the bearing clearance. If clearance is excessive, check for frayed seal edges. When the clearance is correct, retorque the cap.

14. Install the oil pan.

4–119

1. Disconnect the negative battery terminal.

2. Remove the oil pan.

3. Remove the transmission.

NOTE: On manual transmissions, remove the clutch assembly.

4. Unbolt the starter and tie it out of the way.

5. Remove the flywheel.

6. Remove the rear main seal retainer.

7. Remove the oil seal and discard it.

8. Install the new oil seal.

9. Installation is the reverse of removal. Fill the space between the seal lips with grease and lubricate the seal lips with engine oil.

4–121

1. Remove the oil pan and pump.

2. Remove the rear main bearing cap.

3. Gently pack the upper seal into the groove approximately $1/4$ in. on each side.

4. Measure the amount the seal was driven in on one side and add $1/16$ in. Cut this length from the old lower cap seal. Be sure to get a sharp cut. Repeat for the other side.

5. Place the piece of cut seal into the groove and pack the seal into the block. Do this for each side.

6. Install a piece of Plastigage or the equivalent on the bearing journal. Install the rear cap and tighten to 75 ft. lbs. Re-

move the cap and check the gauge for bearing clearance. If out of specification, the ends of the seal may be frayed or not flush, preventing the cap from proper seating. Correct as required.

7. Clean the journal, and apply a thin film of sealer to the mating surfaces of the cap and tighten to 70 ft. lbs. Install the pan and pump.

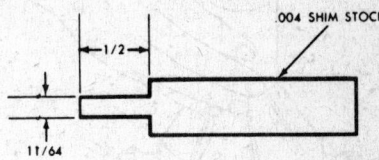

Fabricated oil seal installation tool

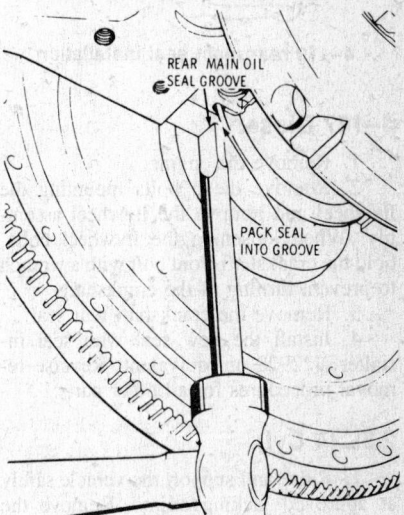

Pack the old seal into the groove

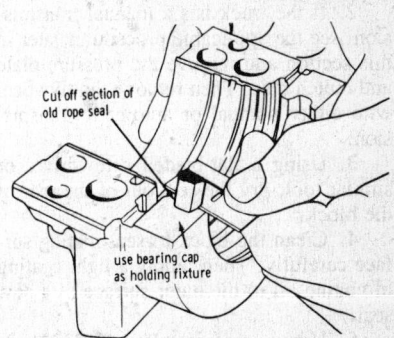

Use the bearing cap as a holding fixture for cutting the old rope seal

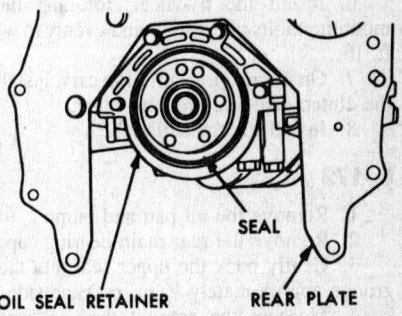

4–119 rear main seal

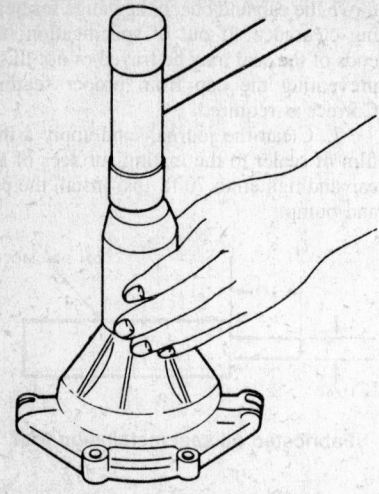

4-119 rear main seal installation

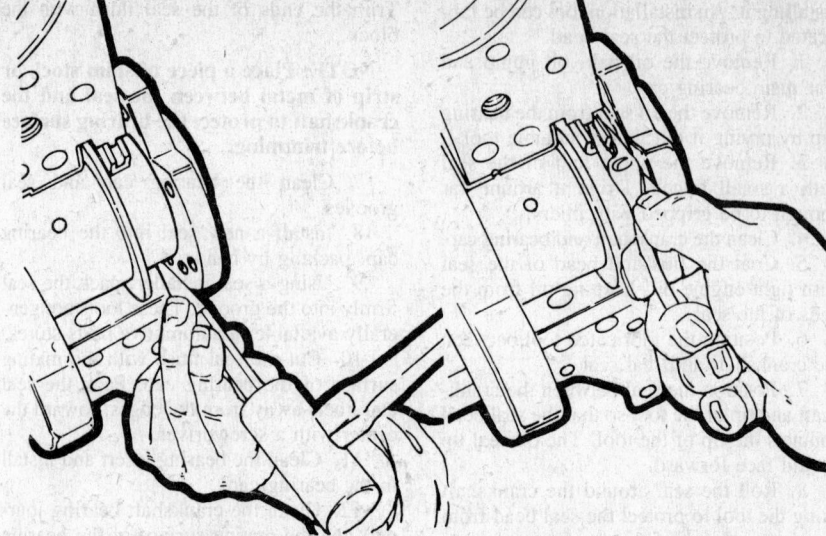

Removing the upper half of the oil seal

4-137 Diesel

1. Remove the engine.
2. Remove the 6 bolts mounting the flywheel and remove the flywheel assembly. When loosening the flywheel bolts, hold the crankshaft front bolt with a wrench to prevent turning of the crankshaft.
3. Remove the crankshaft rear seal.
4. Install the new seal with seal installer, J22928 or equivalent. Reverse removal procedures for all other parts.

2.5L (4 Cyl)

1. Raise and support the vehicle safely at approved jacking points. Remove the transmission as described later in this section.
2. If the truck has a manual transmission, see the applicable procedures later in this section and remove the pressure plate and clutch disc. Then remove tha flywheel with either manual or automatic transmssion.
3. Using a flat-bladed screwdriver or similar tool, pry the seal out of the rear of the block.
4. Clean the block-to-seal mating surface carefully. Then, apply a light coating of engine oil to the outer surface of a new seal.
5. Use a seal installer (J-34924 or equivalent) to tap a new seal into position. Make sure the seal seats squarely.
6. Install the flywheel, torquing the mounting bolts alternately and evenly to 44 ft. lbs.
7. On manual transmission cars, install the clutch disc and pressure plate.
8. Install the transmission.

6-173

1. Remove the oil pan and pump.
2. Remove the rear main bearing cap.
3. Gently pack the upper seal into the groove approximately ¼ in. on each side.
4. Measure the amount the seal was driven in on one side and add ¹/₁₆ in. Cut

this length from the old lower cap seal. Be sure to get a sharp cut. Repeat for the other side.
5. Place the piece of cut seal into the groove and pack the seal into the block. Do this for each side.

NOTE: G.M. makes a guide tool (J-29114-1) which bolts to the block via on oil pan bolt hole, and a packing tool (J-29114-2) which are machined to provide a built-in stop for the installation of the short cut pieces. Using the packing tool, work the short pieces of seal onto the guide tool, then pack them into the block with the packing tool.

6. Install a new lower seal in the rear main cap.
7. Install a piece of Plastigage or the equivalent on the bearing journal. Install the rear cap and tighten to 70 ft. lbs. Remove the cap and check the gauge for bearing clearance. If out of specification, the ends of the seal may be frayed or not flush, preventing the cap from proper sealing. Correct as required.
8. Clean the journal, and apply a thin film of sealer to the mating surfaces of the cap and block. Do not allow any sealer to get onto the journal or bearing. Install the bearing cap and tighten to 70 ft. lbs. Install the pan and pump.

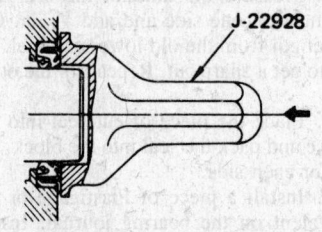

4-137 diesel rear main seal installation

ENGINE COOLING

Radiator

REMOVAL & INSTALLATION

All Except Astro

1. Drain the coolant.
2. Disconnect the hoses and automatic transmission cooler lines (if equipped). Plug the cooler lines. Diesels have transmission cooler and oil cooler lines.
3. Disconnect the coolant recovery system hose.
4. If the vehicle is equipped with a fan shroud, detach the shroud and hang it over the fan to provide clearance.
5. On six cylinder engines, remove the finger guard.
6. Remove the mounting panel from the radiator support and remove the upper mounting pads.
7. Lift the radiator up and out of the truck. Lift the shroud out if necessary.
8. Installation is the reverse of removal. Fill the cooling system and check the automatic transmission fluid level, and run the engine, checking for leaks.

Astro

1. Disconnect the negative battery cable. Loosen the radiator drain cock and drain the cooling system into a clean container.
2. Disconnect the master cylinder, unbolt it, and lay it aside. Plug all openings.
3. Remove the upper fan shroud. Disconnect the radiator hoses.
4. Disconnect the upper transmission cooler lines and engine oil cooler lines.
5. Raise the vehicle and support it securely at approved jacking points. Disconnect the lower transmission cooler and engine oil cooler lines. Remove the lower fan shroud.

6. Disconnect the overflow tube.

7. Lift the radiator upward above the lower fan shroud and remove it.

8. Install the radiator in reverse order, making sure the upper shroud and insulator fit properly over the top of the radiator. After refilling the brake system, bleed the system thoroughly. Refill the cooling system with a 50/50 antifreeze and water mix and then start and run the engine until the thermostat opens. Refill the radiator as necessary, install the cap, and run the engine, checking for leaks.

Water Pump

REMOVAL & INSTALLATION

Except Below

1. Drain the radiator and loosen the fan pulley bolts.

2. Disconnect the heater hose and radiator. Disconnect the lower radiator hose at the water pump.

3. Loosen the alternator swivel bolt and remove the fan belt. Remove the fan bolts, fan and pulley.

4. Remove the water pump attaching bolts and remove the pump and gasket from the engine. On inline engines, remove the water pump straight out of the block to avoid damaging the impeller.

Do not store viscous drive (thermostatic) fan clutches in any other position than the normal installed position. They should be supported so that the clutch disc remains vertical; otherwise, silicone fluid may leak out.

5. Installation is the reverse of removal. Clean the gasket surfaces and install new gaskets. Coat the gasket with sealer. On the 2.5L, coat the water pump bolts with a sealer such as GM part 1052080 or the equivalent. A $5/16$ in. $\times$ 24 $\times$ 1 in. guide stud installed in one hole of the fan will make installing the fan onto the hub easier. It can be removed after the other 3 bolts are started. Fill the cooling system and adjust the fan belt tension.

4-119 and 4-137 Diesel

1. Disconnect the battery ground. Drain the cooling system.

2. Remove the front cover. Disconnect the hoses at the pump.

3. On models without air conditioning, remove the fan.

4. On models with air conditioning, remove the fan belt, fan pulley, fan, air pump pulley and fan set plate.

5. Unbolt and remove the pump.

6. Clean the gasket surfaces thoroughly.

7. Installation is the reverse of removal. Always use a new gasket.

4-121

1. Disconnect the battery ground.

2. Drain the cooling system.

3. Remove all drive belts. Disconnect the hoses at the pump.

4. Remove the alternator.

5. Unbolt and remove the pump.

6. Thoroughly clean the gasket surfaces. Discard the old gasket.

7. Using a new gasket, install the pump and assemble all other components in reverse order of removal.

6-173

1. Disconnect the battery ground.

2. Drain the cooling system.

3. Disconnect the hoses at the pump.

4. Unbolt and remove the pump.

5. Thoroughly clean the sealing surfaces of old gasket material. This engine uses RTV sealant in place of a gasket.

6. Place a $3/32$ in. bead of sealer on the water pump mating surface. Coat the bolt threads with pipe compound and mount the pump on the engine.

7. Assemble all other components in reverse order of removal.

Thermostat

The factory installed thermostat is a 195°F unit.

NOTE: Poor heater output and slow warmup is often caused by a thermostat stuck in the open position; occasionally one sticks shut causing immediate overheating. Do not attempt to correct a chronic overheating condition by permanently removing the thermostat. Thermostat flow restriction is designed into the system; without it, localized overheating due to turbulence may occur.

REMOVAL & INSTALLATION

1. Drain approximately $1/3$ the coolant. This will reduce the coolant level to below the level of the thermostat housing.

2. It is not necessary to remove the upper radiator hose from the thermostat housing. Remove the 2 retaining bolts from the thermostat housing (located on the front top of V8 intake manifolds and directly in front of the valve cover on six cylinder engines) and remove the thermostat.

3. To test the thermostat, place it in hot water or a solution of 33% glycol, 25° above the temperature stamped on the valve. Submerge the valve and agitate the solution. The valve should open fully. Remove the thermostat and place it in the same solution 10° below the temperature stamped on the valve. The valve should close completely.

4. Installation is the reverse of removal. Use a new gasket and sealer. Refill the cooling system and run the engine, checking for leaks.

EMISSION CONTROLS

Refer to Emissions in the Unit Repair section.

GASOLINE FUEL SYSTEM

Fuel Pump

TESTING THE FUEL PUMP

Fuel pumps should always be tested on the vehicle. The larger line between the pump and tank is the suction side of the system and the smaller line, between the pump and carburetor, is the pressure side. A leak in the pressure side would be apparent because of dripping fuel. A leak in the suction side is usually only apparent because of a reduced volume of fuel delivered to the pressure side.

1. Tighten any loose line connections and look for any kinks or restrictions.

2. Disconnect the fuel line at the carburetor. Disconnect the distributor-to-coil primary wire. Place a container at the end of the fuel line and crank the engine a few revolutions. If little or no gasoline flows from the line, either the fuel pump is not working or the line is plugged. Disconnect the line at the pump and the tank; blow through the line with compressed air and try again. Reconnect the line. If the problem is traced to the tank, the tank and gauge unit must be removed to check the condition of the inlet filter screen.

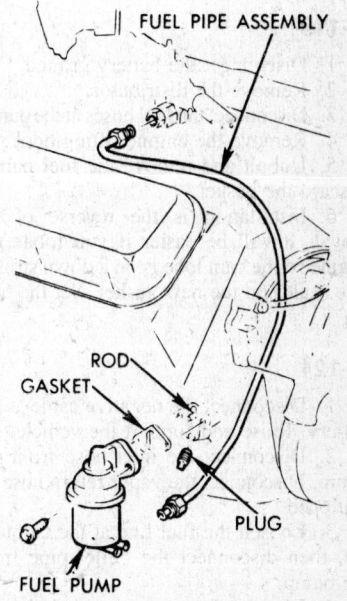

Typical gasoline fuel pump

3. If fuel flows in good volume, check the fuel pump pressure to be sure.

4. Attach a pressure gauge to the pressure side of the fuel line.

5. Run the engine and note the reading on the gauge. Stop the engine and compare the reading with the specifications listed in the Tune-Up Specifications. If the pump is operating properly, the pressure will be as specified and will be constant at idle speed. If pressure varies or is too high or low, the pump should be replaced.

6. Remove the pressure gauge.

REMOVAL & INSTALLATION

Except Below

NOTE: When you connect the fuel pump outlet fitting, always use 2 wrenches to avoid damaging the pump.

1. Disconnect the fuel intake and outlet lines at the pump and plug the pump intake line.

2. On V8 engines, you can remove the upper bolt from the right front engine mounting boss (on the front of the block) and insert a long bolt $\frac{3}{8}$-16 × 2 in.) to hold the fuel pump pushrod.

3. Remove the two pump mounting bolts and lockwashers; remove the pump and its gasket.

4. If the rocker arm pushrod is to be removed from V8s, remove the two adapter bolts and lockwashers and remove the adapter and its gasket.

5. Install the fuel pump with a new gasket reversing the removal procedure. Heavy grease can be used to hold the fuel pump pushrod up while installing the pump, if you didn't install the long bolt in Step 2. Coat the mating surfaces with sealer.

6. Connect the fuel lines and check for leaks.

4–119

1. Disconnect the battery ground.
2. Remove the distributor.
3. Disconnect the fuel hoses at the pump.
4. Remove the engine lifting hook.
5. Unbolt and remove the fuel pump. Discard the gasket.
6. Installation is the reverse of removal. It will be easier if you rotate the engine so the cam lobe is on a down stroke. Use sealer on the new gasket. Set the timing.

4–121

1. Disconnect the negative cable at the battery. Raise and support the vehicle.
2. Disconnect the inlet hose from the pump. Disconnect the vapor return hose, if equipped.
3. Loosen the fuel line at the carburetor, then disconnect the outlet pipe from the pump.
4. Remove the two mounting bolts and remove the pump from the engine.

5. To install, place a new gasket on the pump and install the pump on the engine. Tighten the two mounting bolts alternately and evenly.

6. Install the pump outlet pipe. This is easier if the pipe is disconnected from the carburetor. Tighten the fitting while backing up the pump nut with another wrench. Install the pipe at the carburetor.

7. Install the inlet and vapor hoses. Lower the car, connect the negative battery cable, start the engine, and check for leaks.

6–173

1. Disconnect the battery ground.
2. Disconnect the fuel hoses at the pump.
3. Unbolt and remove the pump. Discard the gasket.
4. Installation is the reverse of removal. Use sealer on the new gasket. It will be easier if you rotate the engine so that the cam lobe is on a down stroke.

Carburetor

REMOVAL & INSTALLATION

NOTE: This procedure is a general one that covers all models. Remember that each carburetor application may differ slightly.

1. Remove the air cleaner and its gasket.
2. Disconnect the fuel and vacuum lines from the carburetor.
3. Disconnect the choke coil rod or heated air line tube.
4. Disconnect the throttle linkage.
5. On automatic transmission trucks, disconnect the throttle valve linkage.
6. Remove the CEC valve vacuum hose and electrical connector.
7. Remove the idle stop electrical wiring from the idle stop solenoid, if so equipped.
8. Remove the carburetor attaching nuts and/or bolts, gasket or insulator, and remove the carburetor.
9. Install the carburetor using a reverse of the removal procedure. Use a new gasket and fill the float bowl with gasoline to ease starting the engine.

For carburetor specifications and adjustments, refer to the Unit Repair section.

Fuel Tank

REMOVAL & INSTALLATION

1. Drain the tank.
2. Raise and support the truck on jackstands.
3. Disconnect the wiring and ground strap at the tank.
4. Disconnect the filler neck hose and vent hose at the tank.

5. Disconnect the fuel feed line and vapor line at the tank.
6. Place a floor jack under the tank to take up its weight.
7. Remove the fuel tank support bolts and lower the tank.
8. Installation is the reverse of removal.

DIESEL FUEL SYSTEM

Fuel Supply Pump

REMOVAL & INSTALLATION

The fuel supply pump is serviced in the same manner as the fuel pump on the gasoline engine.

Fuel Filter

REMOVAL & INSTALLATION

The fuel filter is a square assembly located at the back of the engine above the intake manifold. Disconnect the fuel lines and remove the filter. Install the lines to the new filter. Start the engine and check for leaks.

Fuel Injection Pump and Lines

REMOVAL & INSTALLATION

Except 4–137

NOTE: This procedure contains throttle rod and transmission cable adjustments.

1. Remove the air cleaner.
2. Remove the filters and pipes from the valve covers and air crossover.
3. Remove the air crossover and cap the intake manifold with screened covers (tool J–26996–1) or tape.
4. Disconnect the throttle rod and return spring.
5. Remove the ballcrank.
6. Remove the throttle and transmission cables from the intake manifold brackets.
7. Disconnect the fuel lines from the filter and remove the filter.
8. Disconnect the fuel inlet line at the pump.
9. Remove the rear A/C compressor brace and remove the fuel line.
10. Disconnect the fuel return line from the injection pump.
11. Remove the clamps and pull the fuel return lines from each injection nozzle.
12. Using two wrenches, disconnect the high pressure lines at the nozzles.

13. Remove the three injection pump retaining nuts with tool J–26987 or its equivalent.

14. Remove the pump and cap all lines and nozzles.

15. To install: Remove the protective caps from all lines and nozzles. Place the engine on TDC for the No. 1 cylinder. The mark on the harmonic balancer on the crankshaft will be aligned with the zero mark on the timing tab, and both valves for No.1 cylinder will be closed. The index mark on the injection pump driven gear should be offset to the right when No. 1 is at TDC. Check that all of these conditions are met before continuing.

16. Line up the offset tang on the pump driveshaft with the pump driven gear and install the pump.

17. Install, but do not tighten the pump retaining nuts.

18. Connect the high pressure lines at the nozzles.

19. Using two wrenches, torque the high pressure line nuts to 25 ft. lbs.

20. Connect the fuel return lines to the nozzles and pump.

21. Align the timing mark on the injection pump with the line on the timing mark adapter and torque the mounting nuts to 35 ft. lbs. A ¾ in. open end wrench on the boss at the front of the injection pump will aid in rotating the pump to align the marks.

22. To adjust the throttle rod: Remove the clip from the cruise control rod and remove the rod from the bellcrank. Loosen the locknut on the throttle rod a few turns, then shorten the rod several turns. Rotate the bellcrank to the full throttle stop, then lengthen the throttle rod until the injection pump lever contacts the injection pump full throttle stop, then release the bellcrank. Tighten the throttle rod locknut.

23. Install the fuel inlet line between the transfer pump and the filter.

24. Install the rear A/C compressor brace.

25. Install the bellcrank and clip.

26. Connect the throttle rod and return spring.

27. To adjust the transmission cable: Push the snap-lock to the disengaged position. Rotate the injection pump lever to the full throttle stop and hold it there. Push in the snap-lock until it is flush. Release the injection pump lever.

28. Start the engine and check for fuel leaks.

29. Remove the screened covers or tape and install the air crossover.

30. Install the tubes in the air flow control valve in the air crossover and install the ventilation filters in the valve covers.

31. Install the air cleaner.

32. Start the engine and allow it to run for two minutes. Stop the engine, let it stand for two minutes, then restart. This permits the air to bleed off within the pump.

4–137

1. Raise engine hood.

2. Disconnect the battery ground cable.

3. Remove the battery.

4. Remove the under cover.

5. Drain the cooling system by opening the drain plugs on the radiator and on the cylinder block.

6. Disconnect the upper water hose at the engine side.

7. Loosen the compressor drive belt by moving the power steering oil pump or idler. (If so equipped.)

8. Remove the cooling fan and fan shroud.

9. Disconnect the lower water hose at the engine side.

10. Remove the air conditioner compressor (If so equipped.)

11. Remove the fan belt.

12. Remove the crankshaft pulley.

13. Remove the timing pulley housing covers.

14. Remove the tension spring and fixing bolt, then remove the tension center and pulley.

15. Remove the timing belt.

16. Remove the engine control cable and wiring harness of the fuel cut solenoid.

17. Remove the fuel hoses and injection pipes. Use a wrench to hold the delivery holder when loosening the sleeve nuts on the injection pump side.

18. Install a 6mm bolt (with pitch of 1.25) into threaded hole in the timing pulley housing through the hole in pulley to prevent turning of the pulley. Remove the bolts fixing the injection pump timing pulley, then remove the pulley using pulley puller.

19. Remove injection pump flange fixing nuts and rear bracket bolts, then remove the injection pump.

20. Install the injection pump by aligning notched line on the flange with the line on the front bracket.

21. Install the injection pump timing pulley by aligning it with the key groove. Torque to 42–52 ft. lbs.

22. Bring the piston in No. 1 cylinder to top dead center on compression stroke and align marks on the timing pulleys.

23. Follow the timing belt installation steps.

24. Check the injection timing.

25. To install remaining parts, follow the removal steps in reverse order.

SLOW IDLE SPEED ADJUSTMENT

Except 4–137 and 8–379

1. Run the engine to normal operating temperature.

2. Insert the probe of a magnetic pickup tachometer into the timing indicator hole.

3. Set the parking brake and block the drive wheels.

4. Place the transmission in Drive and turn the A/C Off.

5. Turn the slow idle screw on the injection pump to obtain the idle specification on the emission control label.

4–137

1. Set parking brake and block drive wheels.

2. Place transmission in neutral.

3. Start and normalize the engine. Engine coolant temperature: above 80°C (176°F).

4. Set the engine tachometer.

5. If the idle speed deviates from the specified range of 700–800 rpm, loosen the idle speed adjusting screw lock nut.

6. Turn the adjusting screw in or out until the idle speed is in the correct range. After tightening the lock nut lock it in place.

Idle Speed Adjustment

8–379

1. All idle speeds are to be set within 25 rpm of specified value.

2. Set parking brake and block drive wheels.

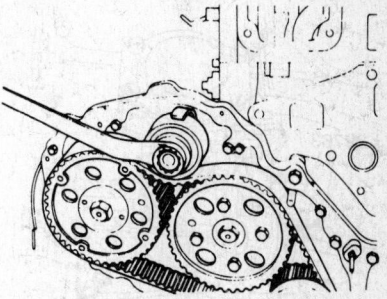

Loosening tension pulley

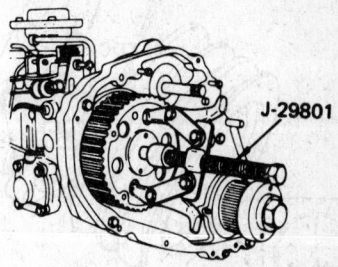

J-29801

Removing timing pulley

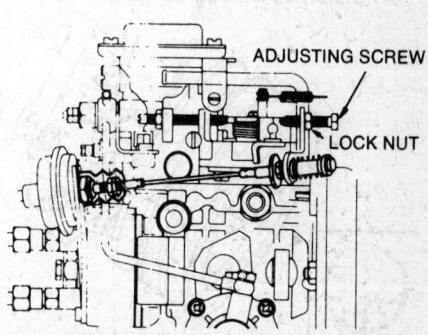

ADJUSTING SCREW

LOCK NUT

4–137 diesel idle speed adjustment points

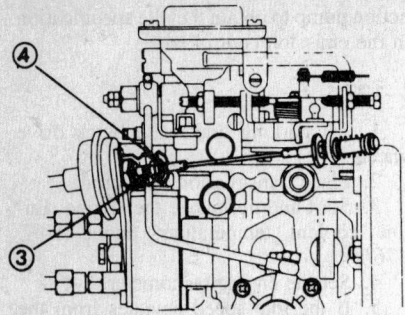

4-137 diesel fast idle speed adjustment points

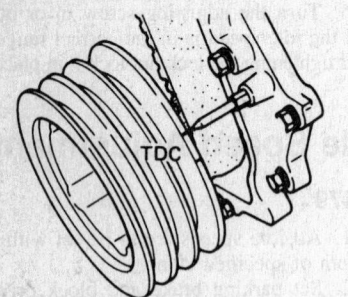

#1 piston at TDC

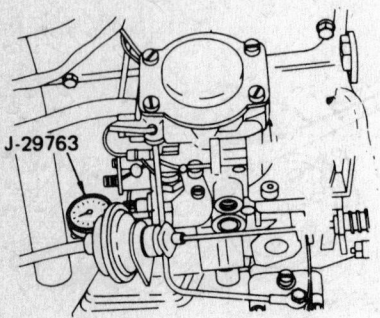

J-29763

Static timing gauge installed

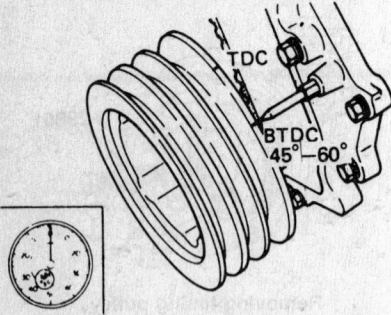

#1 piston 45-60 degrees before TDC

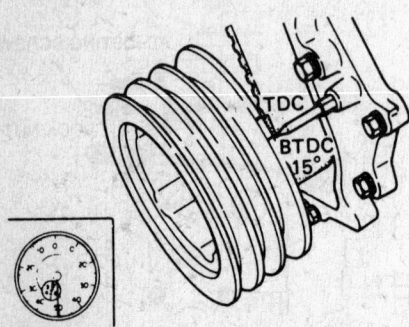

#1 piston 15 degrees before TDC

3. Engine must be at normal operating temperature. Air cleaner should be on and all accessories should be turned off.

4. Install, tool J–26925, diesel tachometer or equivalent per manufacturers instructions.

5. Adjust low idle speed screw on fuel injection pump to an engine speed of 650 RPM in neutral or park for automatic transmission and for manual transmissions.

6. Adjust fast idle speed as follows: Remove connector from fast idle solenoid. Use an insulated jumper wire from battery positive terminal to solenoid terminal to energize solenoid. Open throttle momentarily to ensure that the fast idle solenoid plunger is energized and fully extended. Adjust the extended plunger by turning the hex head to an engine fast idle speed of 800 RPM in neutral. Remove jumper wire and reinstall connector to fast idle solenoid.

7. Remove the tachometer.

FAST IDLE SOLENOID ADJUSTMENT

1979 Except 4–137

1. Set the parking brake and block the drive wheels.

2. Run the engine to normal operating temperature.

3. Place the transmission in Drive and disconnect the compressor clutch wire. Turn the A/C On. On trucks without A/C, disconnect the solenoid wire, and connect jumper wires to the solenoid terminals. Ground one of the wires and connect the other to a 12 volt battery to activate the solenoid.

4. Adjust the fast idle solenoid plunger to obtain 650 rpm.

1980 and Later, Except 4–137 and 8–379

1. With the ignition off, disconnect the single green wire from the fast idle relay located on the front of the firewall.

2. Set the parking brake and block the drive wheels.

3. Start the engine and adjust the solenoid (energized) to the specifications on the underhood emission control label.

4. Turn off the engine and reconnect the green wire.

4–137

1. Start and normalize the engine. Engine coolant temperature: above 80°C (176°F).

2. Set the engine tachometer.

3. Disconnect the hoses from the vacuum switch valve, then connect a pipe (4 mm dia.) in position between the hoses.

4. Loosen adjust nut and adjust engine idle speed by moving the nut. Fast idle should be 900–950 rpm.

5. Tighten the nut.

6. Remove engine tachometer.

CRUISE CONTROL SERVO RELAY ROD ADJUSTMENT

1. Turn the engine Off.

2. Adjust the rod to minimum slack then put the clip in the first free hole closest to the bellcrank, but within the servo ball.

INJECTION TIMING ADJUSTMENT

Except 4–137 and 8–379

For the engine to be properly timed, the lines on the top of the injection pump adapter and the flange of the injection pump must be aligned.

1. The engine must be off for resetting the timing.

2. Loosen the three pump retaining nuts with J–26987, an injection pump intake manifold wrench, or its equivalent.

3. Align the timing marks and torque the pump retaining nuts to 35 ft. lbs.

NOTE:The use of a ¾ in. open end wrench on the boss at the front of the pump will aid in rotating the pump to align the marks.

4. Adjust the throttle rod. (See Fuel Injection Pump Removal and Installation, Step 22.)

4–137

1. Check that notched line on the injection pump flange is in alignment with notched line on the injection pump front bracket.

2. Bring the piston in No. 1 cylinder to top dead center on compression stroke by turning the crankshaft as necessary.

3. With the timing pulley housing cover removed, check that timing belt is properly tensioned, and that timing marks are aligned.

4. Disconnect the injection pipe from the injection pump and remove the distributor head screw, then install static timing gauge. Set the lift approximately 1mm (0.04 in.) from the plunger.

5. Use a wrench to hold the delivery holder when loosening the sleeve nuts on the injection pump side.

6. Bring the piston in No. 1 cylinder to a point 45–60 degrees before top dead center by turning the crankshaft, then calibrate the dial indicator to zero.

7. Turn the crankshaft pulley slightly in both directions and check that gauge indication is stable.

8. Turn the crankshaft in normal direction of rotation, and take the reading of the dial indicator when the timing mark (15 degrees) on the crankshaft pulley is in alignment with the pointer. Reading should be 0.020 in.

9. If the reading of dial indicator deviates from the specified range, hold crankshaft in position 15 degrees before top dead center and loosen two nuts on injection pump flange.

10. Move the injection pump to a point where dial indicator gives reading of 0.020 in., then tighten the pump flange nuts.

8-379

For the engine to be properly timed, the marks on the top of the engine front cover and the injection pump flange must be aligned. The engine must be off when the timing is reset. On Federal models, align scribe marks. On California models, align half circles. If the marks are not aligned, adjustment is necessary.

1. Loosen the three pump retaining nuts.
2. Align mark on injection pump with mark on front cover and tighten nuts to 40 Nm (30 ft. lbs.).
3. Adjust throttle rod.
4. Set engine to TDC No. 1 cylinder (firing).
5. Install timing fixing J–33042 or equivalent in F.I. pump location. Do not use gasket.
6. Slot of F.I. pump gear to be in vertical 6 o'clock position. (If not, remove fixture and rotate engine crankshaft 360°). The timing marks on gears will be aligned.
7. Fasten gear to fixture, and tighten.
8. Install on 10mm nut to housing upper stud to hold fixture flange nut to be "finger" tight.
9. Torque large bolt (18mm head) counterclockwise (toward left bank) to 50 ft. lbs. Tighten 10mm nut.
10. Insure crankshaft has not rotated (and fixture did not bind on 10mm nut).
11. Strike scriber with mallet to mark "TDC" on front housing.
12. Remove timing fixture.
13. Install fuel injection pump with gasket.
14. Install one 8mm bolt to attach gear to pump hub and tighten to specification.
15. Align timing marks on F.I. pump to front housing mark. Tighten to specification (3) 10mm attachment nuts.
16. Rotate engine and install remaining (2) pump gear attaching bolts and tighten to specification.

Injection Nozzle

REMOVAL & INSTALLATION

1979 8-350 Diesel

1. Remove the fuel return line from the nozzle.
2. Remove the nozzle hold-down clamp and spacer using tool J–26952.
3. Cap the high pressure line and nozzle tip. The nozzle tip is highly susceptible to damage and must be protected at all times.
4. If an old nozzle is to be reinstalled, a new compression seal and carbon stop seal must be installed after removal of the used seals.
5. Remove the caps and install the nozzle, spacer and clamp. Torque to 25 ft. lbs.

6. Replace return line, start the engine and check for leaks.

1980 and Later 8-350 Diesel

The injection nozzles on these engines are simply unbolted from the cylinder head, after the fuel lines are removed, in similar fashion to a spark plug. Be careful not to damage the nozzle end and make sure you remove the copper nozzle gasket from the cylinder head if it does not come off with the nozzle.

Clean the carbon off the tip of the nozzle with a soft brass wire brush and install the nozzles, with gaskets.

NOTE: 1981 and later models use two type of injectors, CAV Lucas and Diesel Equipment. When installing the inlet fittings, torque the Diesel Equipment injector fitting to 45 ft. lbs. and the CAV Lucas to 25 ft. lbs.

8-379

NOTE: Nozzles used in Pickups and Blazers are different than those used in Vans and are not interchangeable.

1. Disconnect batteries.
2. Disconnect fuel line clip.
3. Remove fuel return hose.
4. Remove fuel injection line as previously outlined.
5. Remove injection nozzle using tool J–29873 or equivalent whenever possible.

NOTE: When removing an injection nozzle, use tool J–29873 or equivalent. Be sure to remove the nozzle using the 30mm hex. Failure to do so will result in damage to the injection nozzle. Always cap the nozzle and lines to prevent damage and contamination.

Injection Pump Adapter, Adapter Seal, and New Adapter Timing Mark

REMOVAL & INSTALLATION

NOTE: Skip Steps 4 and 9 if a new adapter is not being installed.

1. Remove injection pump and lines.
2. Remove the injection pump adapter.
3. Remove the seal from the adapter.
4. File the timing mark from the adapter. Do not file the mark off the pump.
5. Position the engine at TDC of No.1 cylinder. Align the mark on the balancer with the zero mark on the indicator. The index is offset to the right when No.1 is at TDC.
6. Apply chassis lube to the seal areas. Install, but do not tighten the injection pump.
7. Install the new seal on the adapter using tool J–28425, or its equivalent.
8. Torque the adapter bolts to 25 ft. lbs.

9. Install timing tool J–26896 into the injection pump adapter. Torque the tool, toward No.1 cylinder, to 50 ft. lbs. Mark the injection pump adapter. Remove the tool and install the injection pump.

Injection Pump

REMOVAL & INSTALLATION

8-379 Pickup & Blazer

1. Disconnect batteries.
2. Remove fan.
3. Remove fan shroud.
4. Remove intake manifold.
5. Remove fuel lines.
6. Disconnect accelerator cable at injection pump, and detent cable where applicable.
7. Disconnect necessary wires and hoses at injection pump.
8. Disconnect fuel return line at top of injection pump.

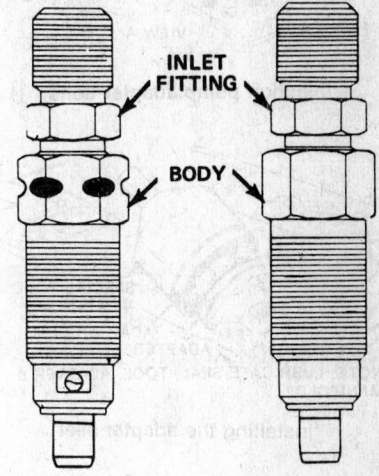

DIESEL EQUIPMENT   **C.A.V. LUCAS**

Diesel injector identification—1980 and later

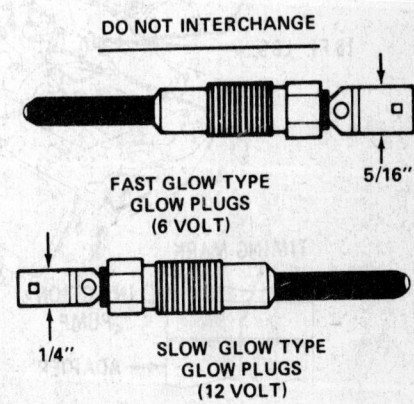

Glow plug identification

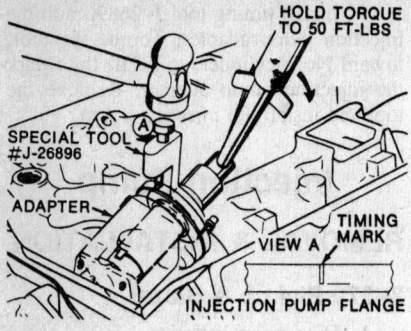

Marking the injection pump adapter

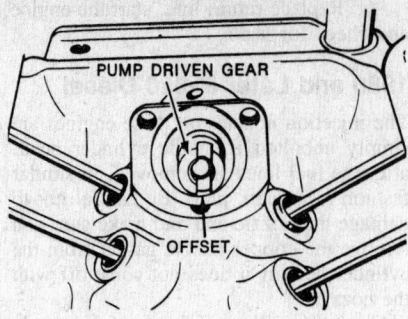

Offset on the diesel fuel pump driven gear

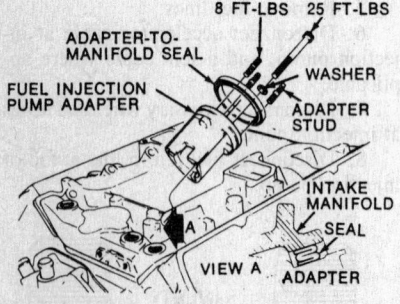

Injection pump adapter bolts

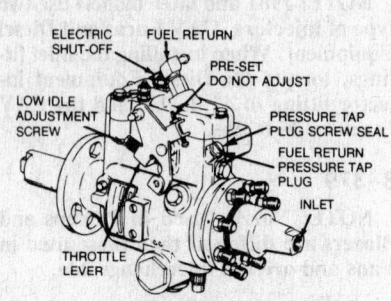

Injection pump

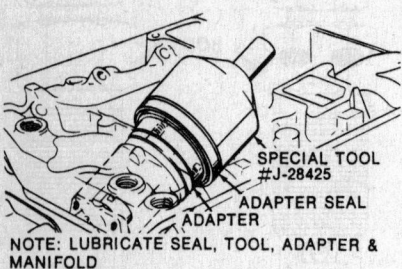

Installing the adapter seal

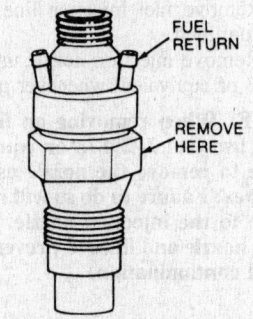

8–379 injection nozzle

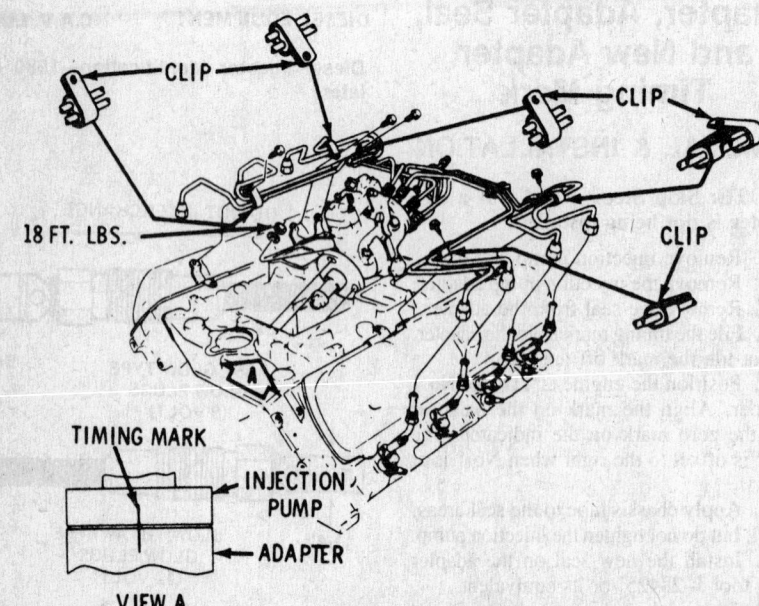

Diesel injection pump timing marks

9. Disconnect fuel feed line at injection pump.

10. Remove A/C hose retainer bracket if equipped with A/C.

11. Remove oil fill tube, includes C.D.R.V. vent hose assembly.

12. Remove grommet.

13. Scribe or paint a mark on front cover and injection pump flange.

14. It will be necessary to rotate engine in order to gain access to driven gear to injection pump retaining bolts through the oil filler neck hole.

15. Remove injection pump to front cover attaching nuts.

16. Remove pump and cap all open lines and nozzles.

17. Install pump per the following: Replace gasket and align locating pin on pump hub with slot in injection pump driven gear. At the same time, align timing marks.

18. Attach injection pump to front cover, torque nuts to 30 ft. lbs. Align timing marks before fully torquing nuts.

19. Install drive gear to injection pump bolts, torque bolts to 20 ft. lbs.

20. Install oil fill tube, includes C.D.R.V. vent hose assembly.

21. Install grommet.

22. Install A/C hose retainer bracket if equipped.

23. Install fuel feed line at injection pump, torque to 20 ft. lbs..

24. Install fuel return line at top of injection pump.

25. Connect necessary wires and hoses.

26. Connect accelerator cable.

27. Connect injection lines.

28. Install intake manifold.

29. Install fan shroud and fan.

30. Connect batteries. Start engine and check for leaks.

Vans

1. Remove intake manifold.

2. Remove air cleaner inlet hose (rotate snorkel up).

3. Remove hood latch, disconnect cable and move aside.

4. Remove windshield washer bottle.

5. Remove fan shroud bolts.

6. Remove upper shroud.

7. Disconnect rubber hose from oil fill tube.

8. Disconnect oil fill tube attaching nuts and remove oil fill tube.

9. Remove oil fill tube grommet.

10. Rotate engine as necessary and remove drive gear to pump bolts.

11. Remove fuel filter and bracket—includes line to injection pump.

12. Disconnect wire looms from injection lines.

13. Disconnect injection lines at brackets. Also disconnect oil pan dipstick tube at left cylinder head.

14. Disconnect electrical connections at injection pump.

15. If equipped with automatic transmission, disconnect T.V. cable.

16. Disconnect accelerator cable.
17. Disconnect injection lines at nozzles Numbers 2, 4, 5, 6, 7, 8.
18. Raise vehicle.
19. Disconnect Number 1 and 3 injection lines at nozzles.
20. Cover nozzles Numbers 1, 3, 5, 7.
21. Lower vehicle.
22. Cover nozzles Number 2, 4, 6, 8.
23. Disconnect injection lines at pump and remove lines. Tag lines for reinstallation.
24. Cap all lines.
25. Disconnect fuel return line.
26. Scribe a mark on front cover and pump flange for reinstallation.
27. Remove pump to front cover attaching nuts.
28. Remove injection pump and cap all open discharge fittings.
29. Install using the following sequence; Replace gasket. Align locating pin on pump hub with slot in injection pump gear. At the same time, align timing marks.
30. Attach injection pump to front cover, torque nuts to 30 ft. lbs.. Align timing marks before fully torquing nuts.
31. Attach pump to drive gear, torque bolts to 20 ft. lbs..
32.. For the remainder of installation procedures, reverse removal steps.

Fuel Injection Line

REMOVAL & INSTALLATION

Pickup & Blazer

1. Disconnect batteries.
2. Disconnect air cleaner bracket at valve cover.
3. Remove crankcase ventilator bracket and move aside.
4. Disconnect secondary filter lines.
5. Remove secondary filter adapter.
6. Loosen vacuum pump hold-down clamp and rotate pump in order to gain access to intake manifold bolt.
7. Remove intake manifold bolts. Injection line clips are retained by the same bolts.
8. Remove intake manifold.
9. Install protective covers J–29664–1 or equivalent.
10. Remove injection line clips at loom brackets.
11. Remove injection lines at nozzles and cover nozzles with protective caps.
12. Remove injection lines at pump and tag lines for reinstallation.
13. Remove fuel line from injection pump.
14. Installation is the reverse of removal.

Vans

1. Disconnect batteries.
2. Remove engine cover.

3. Remove intake manifold.
4. Install protective covers J–29664–1 or equivalent.
5. Remove injection line clips at loom brackets.
6. Raise vehicle (left bank only).
7. Remove injection lines at nozzles and cover nozzles with protective caps.
8. Lower vehicle (left bank only).
9. Remove injection lines at pump and tag lines for reinstallation.
10. Installation is the reverse of removal.

GLOW PLUGS

There are two types of glow plugs used on General Motors diesels: the "fast glow" type and the "slow glow" type. The fast glow type use pulsing current applied to 6 volt glow plugs while the slow glow type use continuous current applied to 12 volt glow plugs.

An easy way to tell the plugs apart is that the fast glow (6 volt) plugs have a $5/16$ in. wide electrical connector plug while the slow glow (12 volt) connector plug is $1/4$ in. wide. Do not attempt to interchange any parts of these two glow plug systems.

MANUAL TRANSMISSION

2WD except Vans

REMOVAL & INSTALLATION

1. Raise the vehicle and support on jackstands.
2. Drain the transmission.
3. Disconnect the speedometer cable, TCS switch and back-up lamp wire at the transmission.
4. Disconnect the shift control levers or shift control from the transmission. On 4 speeds, remove the gearshift lever by pressing down firmly on the slotted collar plate with a pair of channel lock pliers and rotating counterclockwise. Plug the opening to keep out dirt.
5. Disconnect the parking brake lever and controls (if used).
6. Remove the driveshaft after marking the position of shaft to flange.
7. Position a jack under the transmission to support the weight of the transmission.
8. Remove the crossmember. Visually inspect to see if other equipment, brackets or lines, must be removed to permit removal of transmission.

NOTE: Mark the position of the crossmember when removing to prevent incorrect installation. The tapered surface should face the rear.

9. Remove the flywheel housing underpan.
10. Remove the top two transmission-to-housing bolts and insert two guide pins. The use of guide pins will not only support the transmission but will prevent damage to the clutch disc. Guide pins can be made by taking two bolts, the same as those just removed, only longer, and cutting off the heads. Slot them for a screwdriver. Be sure to support the clutch release bearing and support assembly during removal of the transmission. This will prevent the release bearing from falling out of the flywheel housing.
11. Remove the two remaining bolts and slide transmission straight back from engine. Use care to keep the transmission drive gear straight in line with clutch disc hub. Be sure to support release bearing when removing transmission to avoid having bearing fall into flywheel housing.
12. When the transmission is free from the engine, move from under the vehicle.
13. To install the transmission: Place the transmission on guide pins, slide forward starting the main drive gear into the clutch disc's splines. Place the transmission in gear and rotate transmission flange or output yoke to aid the entry of the main drive gear into the disc's splines. Make sure the clutch release bearing is in position.

— CAUTION —
Avoid springing the clutch when the transmission is being installed on the engine. Do not force the transmission into the clutch disc hub. Do not let the transmission hang unsupported in the splined portion of the clutch disc.

14. Install the two lower transmission mounting bolts, and flywheel lower pan (if equipped).
15. Remove the guide pins and install upper mounting bolts. Torque to 75 ft. lbs.
16. Install the driveshaft, watch alignment marks. Install the crossmember according to the alignment marks.
17. Connect the parking brake, back-up lamp and TCS switch (if used).
18. Connect the shift levers, or install the shifter, and adjust if needed.
19. Connect the speedometer cable and refill transmission.
20. Lower the vehicle and road test.

Vans—Except Astro

REMOVAL & INSTALLATION

1. Raise and support the van.
2. Drain the transmission. The Tremec top-cover transmission is drained by removing the lower case to extension housing bolt.
3. Disconnect the speedometer cable, back-up light and TCS switch.
4. Remove the shift controls from the transmission.

5. Disconnect the driveshaft and remove it from the vehicle.

6. Support the transmission with a floor jack.

7. Inspect the transmission to be sure that all necessary components have been removed or disconnected.

8. Mark the front of the crossmember to be sure that it is installed correctly.

9. Support the clutch release bearing to prevent it from falling out of the flywheel housing when the transmission is removed.

10. Remove the flywheel housing under pan and transmission mounting bolts.

11. Move the transmission slowly away from the engine, keeping the mainshaft in alignment with the clutch disc hub. Be sure that the transmission is supported.

12. Remove the transmission from under the vehicle.

13. Installation is the reverse of removal. Lightly coat the mainshaft with high temperature grease. Do not use much grease, since, under normal operation, the grease will be thrown onto the clutch, causing it to fail.

14. Tighten the transmission to flywheel housing bolts to 75 ft. lbs. Fill the transmission with lubricant. Road test the vehicle.

Astro Vans

1. Raise and support the vehicle and drain the transmission fluid. Disconnect and remove the driveshaft as described later in this section. Plug the opening at the rear of the transmission with a clean rag.

2. Disconnect the speedometer cable. Mark and then disconnect all electrical connectors at the transmission.

3. Disconnect the shift linkage at the shifter. Then, remove the shifter support attaching bolts the transmission.

4. Remove the transmission mount attaching bolts.

5. Support the transmission in a secure manner from below. Then, remove the crossmember attaching bolts. Remove the crossmember from the vehicle.

6. Finally remove the transmisson attaching bolts and remove the transmission from the vehicle.

7. To install, reverse the removal procedure, noting the following points:

a. Apply a light coating of high temperature grease fo both the main drive gear bearing retainer and the splined portion of the transmission drive shaft. This is important to ensure free sliding of clutch and transmission parts during assembly.

b. Observe torque specifications: Transmission-to-clutch housing bolts 48 ft. lbs.; Crossmember to body bolts 35 ft. lbs.; Mount-to-crossmember bolts 25 ft. lbs.; Mount-to-transmission bolts 45 ft. lbs. Make sure to adjust the shift linkage and refill the transmission to the proper level with approved fluid.

4WD Models

REMOVAL & INSTALLATION

3 Speed

1. Jack up the vehicle and support it safely on stands. Remove the skid plate, if any.

2. Drain the transmission and transfer case. Remove the speedometer cable and the TCS switch from the side of the transmission.

3. Disconnect the driveshafts and secure them out of the way.

4. Remove the shifter lever by removing the pivot bolt to the adapter assembly. You can then push the shifter up out of the way.

5. Remove the bolts attaching the strut to the right side of the transfer case and to the rear of the engine, and remove the strut.

6. While supporting the transfer case securely, remove the attaching bolts to the adapter.

7. Remove the transfer case securing bolts from the frame and lower and remove the transfer case. (The case is attached to the right side of the frame.)8. Disconnect the shift rods from the transmission. 9. While holding the rear of the engine with a jack, remove the adapter mounting bolts.

10. Remove the upper transmission bolts and insert two guide pins to keep the assembly aligned. See the two wheel drive procedure (Step 10) for details on making these.

11. Remove the flywheel pan and the lower transmission bolts.

12. Pull the transmission and the adapter straight back on the guide pins until the input shaft is free of the clutch disc.

13. The transmission and the adapter are removed as an assembly. The adapter can be separated once the assembly is out.

14. Installation is the reverse of removal. Place the transmission in gear and turn the output shaft to align the clutch splines. Transmission bolt torque is 75 ft. lbs. See Transfer Case Removal and Installation for adapter bolt torques.

4 Speed

1. Remove the shifter boots and the floor mat or carpeting from the front passenger compartment.

2. Remove the transmission shift lever. See Step 4 of the two wheel drive procedure for details on removing the lever. It may be necessary to remove the center floor outlet from the heater to complete the next step. Remove the center console, if so equipped.

3. Remove the transmission cover after releasing the attaching screws. It will be necessary to rotate the cover 90° to clear the transfer case shift lever.

4. Disconnect the transfer case shift le-

ver link assembly and the lever from the adapter. Remove the skid plate, if any.

5. Remove the back-up light, the TCS switch, and the speedometer cable from the side of the transmission.

6. Raise and support the truck. Support the engine. Drain the transmission and the transfer case. Detach the transmission and the transfer case. Detach both driveshafts and secure them out of the way.

7. Remove the transmission-to-frame bolts. To do this, it will be necessary to open the locking tabs. Remove the transfer case-to-frame bracket bolts.

8. While supporting the transmission and transfer case, remove the crossmember bolts and the crossmember. It will be necessary to rotate the crossmember to remove it from the frame.

9. Remove the lower clutch housing cover.

NOTE: On V8 engines it is necessary to remove the exhaust crossover pipe.

10. Remove the transmission-to-clutch housing bolts. Remove the upper bolts first and install guide pins. See Step 10 of the two wheel drive procedure for details.

11. Slide the transmission back until the main drive gear clears the clutch assembly and then lower the unit.

12. Install the transfer case on the transmission as an assembly. Attach the assembly to the clutch housing. Put the transmission in gear and turn the output shaft to align the clutch splines. Torque the transmission bolts to 75 ft. lbs.

13. Install the clutch housing cover and, on V8 models, the exhaust pipe.

14. Install the frame crossmember, the retaining adapter, and the transfer case.

15. The front and rear transfer case yoke locknuts must be torqued to 150 ft. lbs.

16. Install the front and rear driveshafts.

17. Connect the speedometer cable, backup lights, and TCS switches.

18. Fill the transmission and the transfer case to the proper level.

19. Position the transfer case shift lever and the shift lever link on the shift rail bar.

20. Install the transmission floor cover and the center heating duct.

21. Install the center console, if so equipped.

22. Install the transmission shift lever.

S Series Trucks

REMOVAL & INSTALLATION

4–Speed 77.5mm

NOTE: On 4WD models, see Transfer Case Removal and Installation

1. Disconnect the battery ground.

2. Remove the upper starter retaining nut.

3. Remove the shift lever boot attaching screws and slide the boot up the shift lever.

4. Disconnect the shift lever at the transmission.

5. Disconnect the electrical connector at the transmission.

6. Raise and support the truck on jackstands.

7. Remove the driveshaft.

8. Disconnect the exhaust pipe at the manifold.

9. Disconnect the exhaust pipe at the manifold.

10. Disconnect the clutch cable at the transmission.

11. Place a floor jack under the transmission to take up its weight.

12. Remove the transmission mount bolts.

13. Remove the catalytic converter hanger.

14. Remove the crossmember.

15. Remove the lower dust cover bolts.

16. Remove the lower starter bolt.

17. Unbolt the transmission from the engine and lower it on the jack.

18. Installation is the reverse of removal. Torque the transmission-to-engine bolts to 25 ft. lbs. on the 4 cylinder and 55 ft. lbs. on the 6 cylinder. Torque the transmission mount-to-transmission bolts to 35 ft. lbs.; the crossmember-to-frame bolts to 25 ft. lbs.; the dust cover bolts to 7 ft. lbs.

4–Speed and 5–Speed

NOTE: On 4WD models, see Transfer Case Removal and Installation

1. Disconnect the battery ground.

2. Remove the shift lever boot screws and slide the boot up the lever.

3. Shift the transmission into neutral and remove the shift lever.

4. Raise and support the truck on jackstands.

5. Remove the driveshaft.

6. Disconnect the speedometer cable and wiring at the transmission.

7. Disconnect the clutch cable at the transmission.

8. Place a floor jack under the transmission and take up its weight. Remove the transmission mount bolts.

9. Remove the catalytic converter hanger.

10. Remove the crossmember.

11. Remove the dust cover bolts.

12. Unbolt the transmission from the bell housing and lower the jack. It will be necessary to pull the transmission back a ways to clear the clutch. Installation is the reverse of removal. Lightly grease the input shaft splines with chassis lube. Torque the transmission mount bolts to 35 ft. lbs.; the mount-to-crossmember bolts to 25 ft. lbs.; the crossmember-to-frame bolts to 25 ft. lbs; the transmission-to-bell housing bolts to 55 ft. lbs.

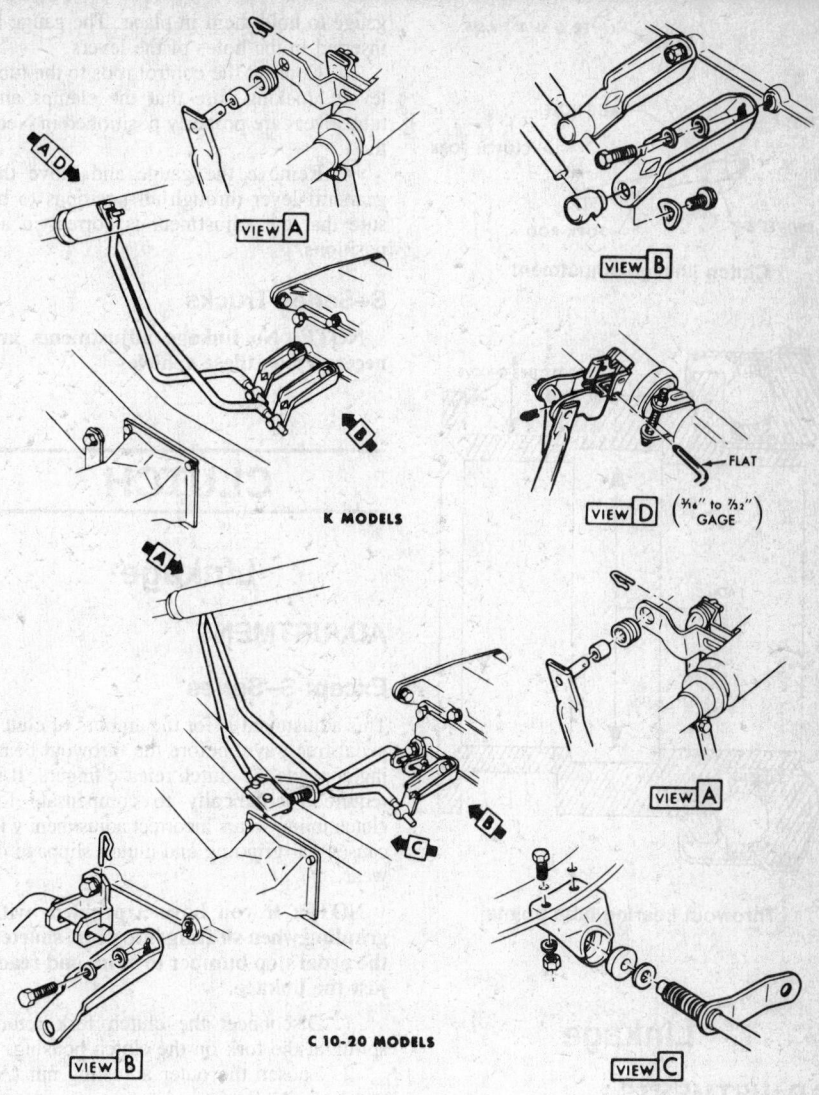

Three speed column shift controls, except vans

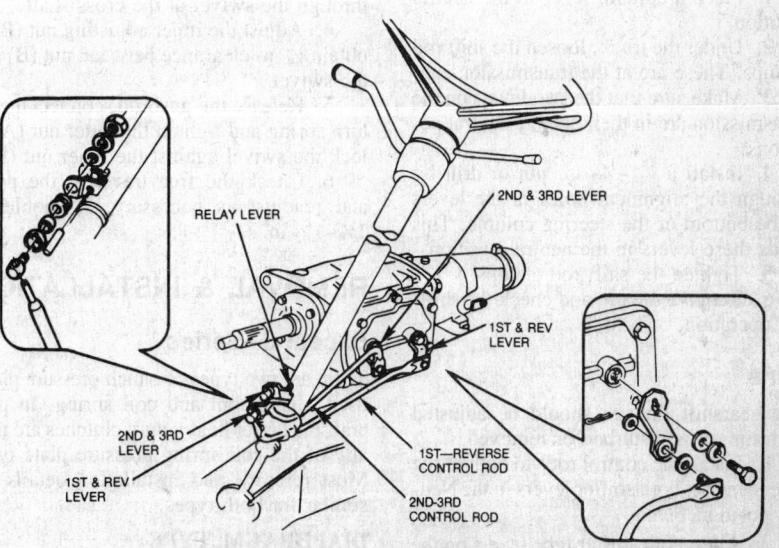

Van 3-speed column shift details

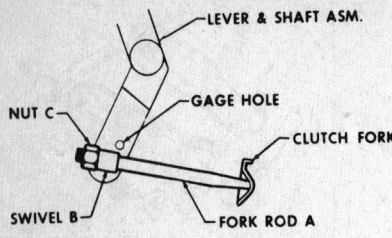

Clutch linkage adjustment

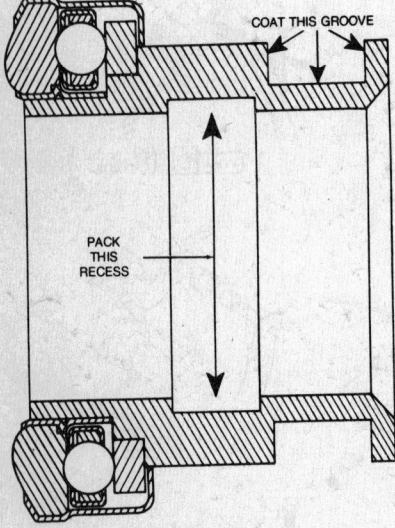

Throwout bearing lube points

Linkage

ADJUSTMENTS

Except Vans

3 SPEED COLUMN SHIFT

1. Place a column lever in the neutral position.
2. Under the truck, loosen the shift rod clamps. These are at the transmission end.
3. Make sure that the two levers on the transmission are in their center, neutral positions.
4. Install a $\frac{3}{16}$–$\frac{7}{32}$ in. pin or drill bit through the alignment holes in the levers at the bottom of the steering column. This holds these levers in the neutral position.
5. Tighten the shift rod clamps.
6. Remove the pin and check the shifting operation.

Vans

The gearshift linkage should be adjusted each time it is disturbed or removed.

1. Install the control rods to both of the levers and set both shifter levers in the Neutral position.
2. Align both shifter tube levers on the mast jacket in Neutral. Install a $\frac{3}{16}$–$\frac{7}{16}$ in.

gauge to hold them in place. The gauge is inserted in the holes of the levers.

3. Connect the control rods to the tube levers, making sure that the clamps and tube levers are properly positioned in Neutral.
4. Remove the gauge and move the gearshift lever through all positions to be sure that the adjustment is correct in all positions.

S–Series Trucks

NOTE: No linkage adjustments are necessary on these vehicles

CLUTCH

Linkage

ADJUSTMENT

Except S–Series

This adjustment is for the amount of clutch pedal free travel before the throwout bearing contacts the clutch release fingers. It is required periodically to compensate for clutch lining wear. Incorrect adjustment will cause gear grinding and clutch slippage or wear.

NOTE: If you have a problem with grinding when shifting into gear, shorten the pedal stop bumper to $\frac{3}{8}$ in. and readjust the linkage.

1. Disconnect the clutch fork return spring at the fork on the clutch housing.
2. Loosen the outer adjusting nut (A) and back it off approximately $\frac{1}{2}$ in. from the swivel.
3. Hold the clutch fork pushrod against the fork to move the throwout bearing against the clutch fingers. The pushrod will slide through the swivel at the cross-shaft.
4. Adjust the inner adjusting nut (B) to obtain $\frac{1}{4}$ in. clearance between nut (B) and the swivel.
5. Release the pushrod, connect the return spring and tighten the outer nut (A) to lock the swivel against the inner nut (B).
6. Check the free travel at the pedal and readjust as necessary. It should be $1\frac{1}{4}$–$1\frac{1}{2}$ in.

REMOVAL & INSTALLATION

Except S–Series

There are two types of clutch pressure plates used, diaphragm and coil spring. In general, the larger heavy duty clutches are usually of the coil spring pressure plate type. Most removal and installation details are similar for both types.

DIAPHRAGM TYPE

1. Remove the transmission.

2. Disconnect the fork pushrod and remove the flywheel housing. Remove the clutch throwout bearing from the fork.
3. Remove the clutch fork by pressing it away from the ball mounting with a screwdriver until the fork snaps loose from the ball or remove the ball stud from the clutch housing.
4. Install a pilot tool (an old mainshaft makes a good pilot tool) to hold the clutch while you are removing it.

NOTE: Before removing the clutch from the flywheel, matchmark the flywheel, the clutch cover and one of the pressure plate lugs. These parts must be reassembled in their original positions as they are a balanced assembly.

5. Loosen the clutch attaching bolts one turn at a time to prevent distortion of the clutch cover until the tension is released.
6. Remove the clutch pilot tool and the clutch from the vehicle. Inspect the flywheel and pressure plate for discoloration, scoring or wear marks. The flywheel can be refaced if necessary, otherwise replace the parts. Also inspect the clutch fork and throwout bearing for looseness or wear. Replace if either is evident.
7. Install the pressure plate in the cover assembly, aligning the notch in the pressure plate with the notch in the cover flange.
8. Install the pressure plate retracting springs, lockwashers and the drive strap to the pressure plate bolts. Torque to 11 ft. lbs.
9. Turn the flywheel until the X mark is at the bottom.
10. Install the clutch disc, pressure plate and cover using an old mainshaft as an aligning tool.
11. Turn the clutch until the X mark on the clutch cover aligns with the X mark on the flywheel.
12. Install the attaching bolts and tighten them a little at a time in a crisscross pattern until the spring pressure is taken up.
13. Remove the aligning tool.
14. Pack the clutch ball fork seat with a small amount of high temperature grease. Too much grease will cause slippage. Install a new retainer in the groove of the clutch fork, if necessary. Install the retainer with the high side up and the open end on the horizontal.
15. If the clutch fork ball was removed, reinstall it in the clutch housing and snap the clutch fork onto the ball.
16. Lubricate the inside of the throwout bearing collar and the throwout fork groove with a small amount of graphite grease.
17. Install the throwout bearing.
18. Install the flywheel housing and transmission.
19. Further installation is the reverse of removal. Adjust the clutch linkage.

COIL SPRING TYPE

Basically, the same procedures apply to diaphragm clutch removal as to coil spring clutch removal. When loosening the clutch holding bolts, loosen them only a turn or two at a time in order to avoid bending the rim of the cover. It will be helpful to place wood or metal spacers, about ³⁄₈ in. thick, between the clutch levers and the cover to hold the levers down as the holding bolts are being removed or when the clutch is being removed from the engine.

S–Series Truck

ADJUSTMENT

1. Lift up on the pedal to allow the self adjuster to adjust the cable length.
2. Depress the pedal several times to set the pawl into mesh with the detent teeth.
3. Check the linkage for lost motion caused by loose or worn swivels, mounting brackets or a damaged cable.

REMOVAL & INSTALLATION

1. Remove the transmission.
2. Remove the flywheel housing.
3. Remove the clutch fork.
4. Insert a clutch alignment tool in the clutch hub and into the crankshaft pilot bearing.
5. Check for an X or other painted mark on the pressure plate and flywheel. If there isn't any mark, mark the assembly for installation purposes.
6. Loosen the pressure plate bolts, evenly and alternately, a little at a time until spring tension is released. Remove the pressure plate and drive plate.
7. Check the flywheel for cracks, wear, scoring or other damage. Check the pilot bearing for wear. Replace it by yanking it out with a slide hammer and driving in a new one with a wood or plastic hammer.
8. Installation is the reverse of removal. Use the alignment tool to aid installation. The raised hub of the driven plate faces the transmission. Align the mating marks and tighten the bolts evenly and alternately to 20 ft. lbs.

AUTOMATIC TRANSMISSION

ADJUSTMENTS

Shift Linkage

EXCEPT VANS

1. The shift tube and lever assembly must be free in the mast jacket.
2. Lift the selector lever toward the steering wheel and allow the selector lever to be positioned in Dive by the detent. Do not use the selector lever pointer as a reference.
3. Release the selector lever. The lever should not be able to go into Low unless the lever is lifted. A properly adjusted linkage will prevent the lever from moving beyond both the Neutral and Drive detents unless the lever is lifted.
6. If adjustment is required, remove the screw (A) and spring washer from the swivel clamp (B).
7. Set the transmission lever (C) in Neutral by moving it to L_1 and then three detents clockwise.
8. Put the transmission selector lever in Neutral as determined by the mechanical stop in the steering column. Do not use the indicator pointer as a reference. The pointer is the last thing to be adjusted.
9. Assemble the swivel spring and washer to the lever (D) and tighten.
10. Readjust the Neutral safety switch if necessary.
11. To adjust the shift position indicator, remove the column cover at the bottom of the instrument panel and loosen the screw to move the pointer.
12. Check the operation. With the switch in RUN, and the transmission in Reverse, be sure that the key cannot be removed and that the steering wheel is locked. With the key in LOCK and the shift lever in PARK, be sure that the key can be removed, the steering wheel is locked, and that the transmission remains in PARK when the steering column is locked.

VANS

CAUTION

Any inaccuracies in this procedure may lead to premature transmission failure due to operation without the controls in the full detent position. Such operation will result in reduced oil pressure, and therefore only partial engagement of the drive clutches. Partial engagement of the clutches with sufficient pressure to cause apparent normal operation will result in transmission failure after only a few miles of operation.

1. Remove the nut (F) and slide off the washers, grommet, bushings, and clamp (E). Remove the swivel (D).
2. Remove the retainer, grommets and the transmission lever (C) from the shaft assembly.
3. Set the transmission lever (C) in the Neutral position either by moving the lever (C) counterclockwise to the L_1 position, then clockwise three steps to the Neutral position, or by moving the lever (C) clockwise to the Park position, then counterclockwise two steps to the Neutral position.
4. Set the column shift lever in the Neutral position by rotating the shift lever until it locks into the stop in the column. Do not use the gear select pointer as a reference to position the column shift lever.
5. Attach rod (A) to the shaft assembly (B) as shown.

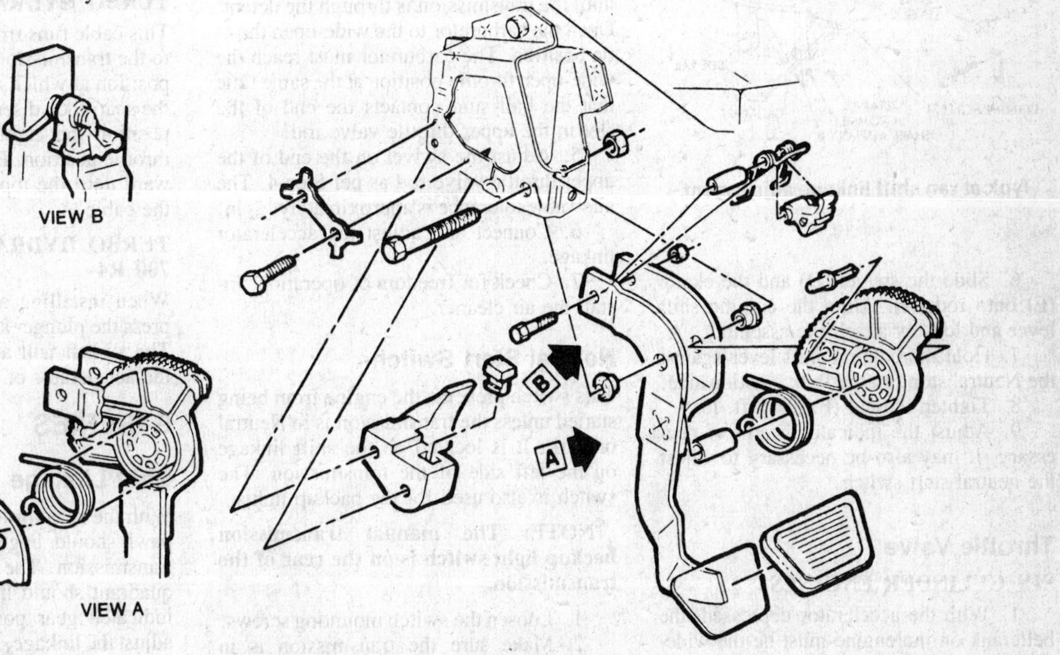

S–Series self–adjusting clutch mechanism

CHEVROLET/GMC

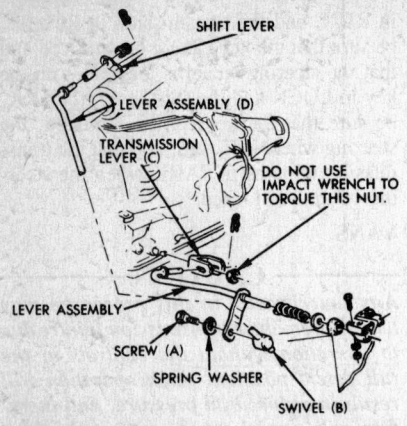

Automatic transmission shift linkage adjustment—except vans

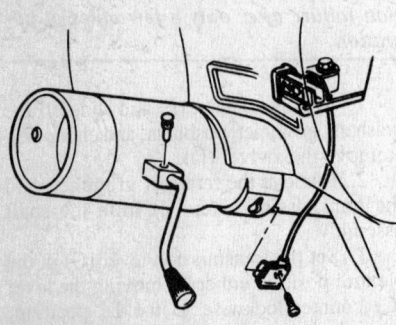

The adjustment point for the transmission shift position indicator is accessible after removing the lower column cover

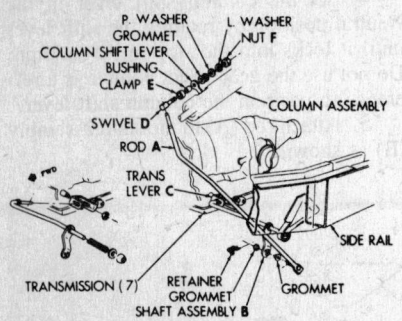

Typical van shift linkage adjustment

6. Slide the swivel (D) and the clamp (E) onto rod (A). Align the column shift lever and loosely attach the assembly.

7. Hold the column shift lever against the Neutral stop, on the Park position side.

8. Tighten the nut (F) to 18 ft. lbs.

9. Adjust the indicator needle if necessary. It may also be necessary to adjust the neutral start switch.

Throttle Valve
SIX CYLINDER ENGINES

1. With the accelerator depressed, the bellcrank on the engine must be the wide-open throttle position.

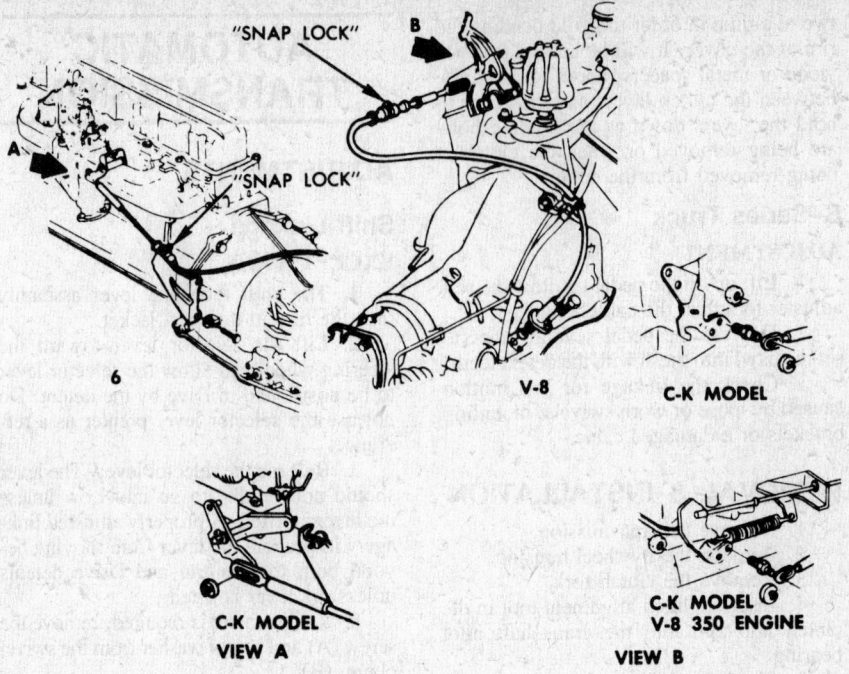

Detent cable adjustment—except diesel

2. The dash lever must be 1/64–1/16 in. off the lever stop and the transmission lever must be against the transmission internal stop.

V8 ENGINES

1. Remove the air cleaner.
2. Disconnect the accelerator linkage at the carburetor.
3. Disconnect the accelerator return spring and the throttle valve rod return springs.
4. Pull the throttle valve rod forward until the transmission is through the detent. Open the carburetor to the wide-open throttle position. The carburetor must reach the wide-open throttle position at the same time that the ball stud contacts the end of the slot in the upper throttle valve rod.
5. Adjust the swivel on the end of the upper throttle valve rod as per Step 4. The allowable tolerance is approximately 1/32 in.
6. Connect and adjust the accelerator linkage.
7. Check for freedom of operation. Install the air cleaner.

Neutral Start Switch

This switch prevents the engine from being started unless the transmission is in Neutral or Park. It is located on the shift linkage on the left side of the transmission. The switch is also used for the backup lights.

NOTE: The manual transmission backup light switch is on the rear of the transmission.

1. Loosen the switch mounting screws.
2. Make sure the transmission is in Neutral.

3. Insert a pin through the hole in the switch actuating arm into the switch body to hold the switch in the Neutral position. Adjust as necessary to make the pin fit.
4. Tighten the adjustment. Remove the pin.
5. Check that the engine can be started only in Park and Neutral and that the backup lights go on only in Reverse. Adjust as necessary.

Downshift
TURBO HYDRA-MATIC 350

This cable runs from the carburetor linkage to the transmission. It regulates the throttle position at which a downshift occurs. With the snap-lock disengaged from the bracket, position the carburetor at the wide open throttle position. Push the snap-lock downward until the top is flush with the rest of the cable.

TURBO HYDRA-MATIC 400 and 700 R4

When installing a new downshift switch, press the plunger as far forward as possible. The switch will adjust itself the first time the accelerator oi floorboarded.

S-SERIES
Shift Linkage

With the selector lever in Park, the parking pawl should engage and immobilize the transmission. The pointer on the indicator quadrant should line up properly with the indicated gear position in all ranges. To adjust the linkage, raise an support the truck on jackstands. Place the selector in Park.

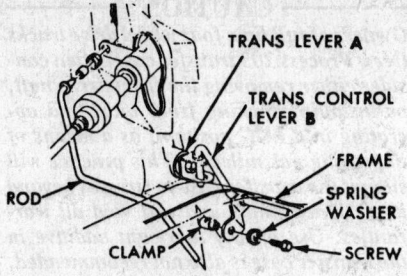

Transmission linkage adjustment

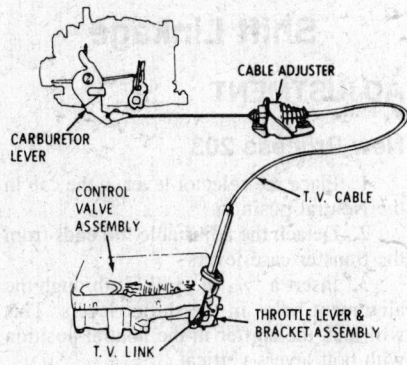

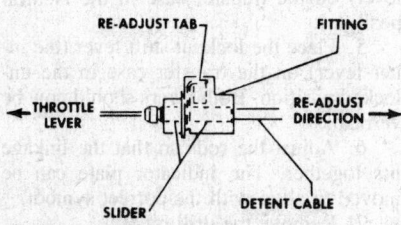

Throttle valve cable adjustment point

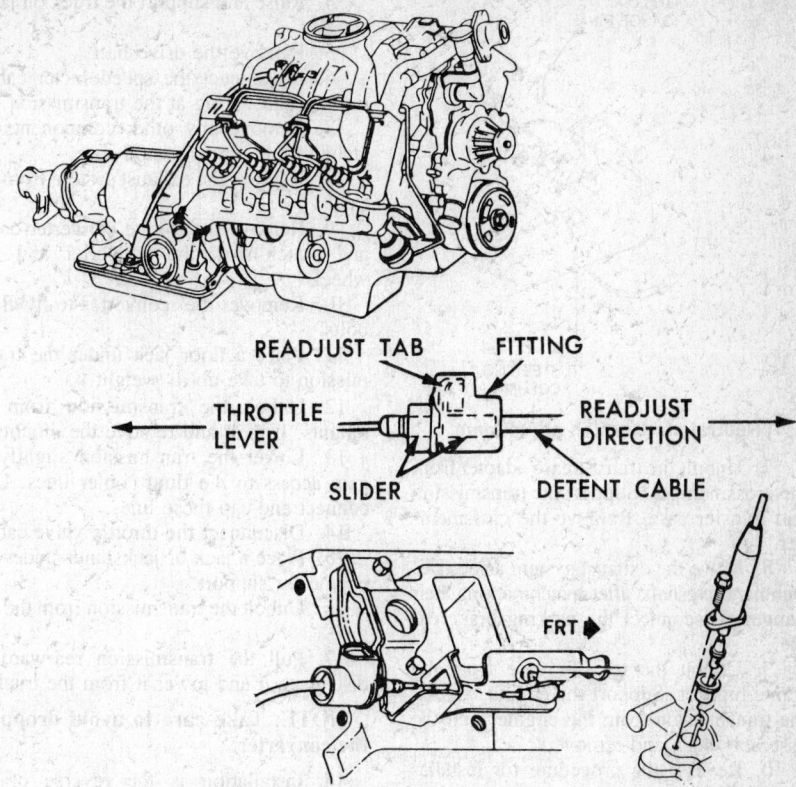

Throttle valve/detent cable adjustment—diesel

Loosen the locknut on the linkage arm at the transmission and make sure that the transmission lever is fully in the Park position. Tighten the locknut.

Throttle Valve

1. With the engine off, depress the readjusting tab.
2. Move the slider back through the fitting in the direction away from the throttle body until the slider stops against the fitting. Release the adjusting tab.
3. Open the carburetor throttle plate to the wide open position. This will automatically adjust the cable. Release the throttle plate.

Transmission

REMOVAL & INSTALLATION

NOTE: It would be best to drain the transmission before starting. It may be necessary to disconnect and remove the exhaust crossover pipe on V8s, and to disconnect the catalytic converter and remove its support bracket, on models so equipped.

2WD

1. Disconnect the battery ground cable. Disconnect the detent cable at the carburetor.
2. Raise and support the truck.
3. Remove the driveshaft, after matchmarking its flanges.
4. Disconnect the speedometer cable, downshift cable, vacuum modulator line, shift linkage, and fluid cooler lines at the transmission. Remove the filler tube. On the Astro only, remove the transmission support brace bolts at the converter cover; and disconnect the exhaust crossover pipe at the exhaust manifolds.
5. Support the transmission and unbolt the rear mount from the crossmember. On the Astro, raise the transmission very slightly. Remove the crossmember. On the Astro, slide the crossmember rearward and then remove it.
6. Remove the torque converter underpan, matchmark the flywheel and converter, and remove the converter bolts.
7. Support the engine and lower the transmission slightly for access to the upper transmission to engine bolts.
8. Remove the transmission to engine bolts and pull the transmission back. Rig up a strap or keep the front of the transmission up so the converter doesn't fall out.
9. Reverse the procedure for installation. Bolt the transmission to the engine first. Torque to 30 ft. lbs. except on the Astro. On the Astro, torque these bolts to

35 ft. lbs on the V6 and 60 ft. lbs on the inline four., then the converter to the flywheel (35 ft. lbs.). Make sure that the converter attaching lugs are flush and that the converter can turn freely before installing the bolts. Tighten the bolts finger tight, then torque to specification, to ensure proper converter alignment. If the oil filler tube has been removed, install a new seal.

NOTE: Lubricate the internal yoke splines at the transmission end of the driveshaft with lithium base grease. The grease should seep out through the vent hole.

4WD

1. Disconnect the battery ground cable and remove the transmission dipstick. Detach the downshift cable at the carburetor. Remove the transfer case shift lever knob and boot.
2. Raise an support the truck.
3. Remove the skid plate, if any. Remove the flywheel cover.
4. Matchmark the flywheel and torque converter, remove the bolts, and secure the converter so it doesn't fall out of the transmission.
5. Detach the shift linkage, speedometer cable, vacuum modulator line downshift cable, and cooler lines at the transmission. Remove the filler tube.
6. Remove the exhaust crossover pipe to manifold bolts.

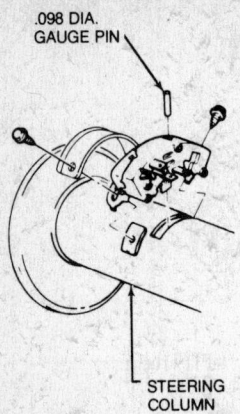

.098 DIA.
GAUGE PIN

STEERING
COLUMN

Neutral start switch adjustment

7. Unbolt the transfercase adapter from the crossmember. Support the transmission and transfer case. Remove the crossmember.

8. Move the exhaust system aside. Detach the driveshafts after matchmarking their flanges. Disconnect the parking brake cable.

9. Unbolt the transfer case from the frame bracket. Support the engine. Unbolt the transmission from the engine, pull the assembly back and remove.

10. Reverse the procedure for installation. both the transmission to the engine first (30 ft.lbs), then the converter to the flywheel (34 ft. lbs.). Make sure that the converter attaching lugs are flush and that the converter can turn freely before installing the bolts. See Transfer Case Removal and Installation for adapter bolt torques.

Pan & Filter

REMOVAL & INSTALLATION

Except S–Series

1. Jack up your truck and support it with jackstands.

2. Loosen the pan bolts on the transmission.

3. Remove all pan bolts except for 2 or 3 at one corner.

4. Gently tap the pan using a rubber mallet and drain the oil into an appropriate container.

5. Remove the pan and remove the filter. Clean the pan with a lint free rag.

6. Install a new filter and pan gasket.

7. Add new transmission fluid to the proper level. Start your truck and check for leaks.

S–Series Trucks

NOTE: On 4WD models, see Transfer Case Removal and Installation

1. Remove the air cleaner assembly.

2. Disconnect the throttle valve cable at the carburetor.

3. On the 4–119 engine, remove the upper starter bolt.

4. Raise and support the truck on jackstands.

5. Remove the driveshaft.

6. Disconnect the speedometer cable, linkage and wiring at the transmission.

7. Remove any other components attached to the transmission case.

8. Remove the exhaust system from the truck.

9. Remove the torque converter cover and match-mark the converter and flywheel.

10. Remove the converter-to-flywheel bolts.

11. Place a floor jack under the transmission to take up its weight.

12. Unbolt the transmission from its mounts. Unbolt and remove the mounts.

13. Lower the transmission slightly to gain access to the fluid cooler lines. Disconnect and cap these lines.

14. Disconnect the throttle valve cable.

15. Place a jack or jackstands under the engine for support.

16. Unbolt the transmission from the engine.

17. Pull the transmission rearward to disengage it and lower it from the truck.

NOTE: Take care to avoid dropping the converter.

18. Installation is the reverse of removal. Match up the mating marks on the converter. Torque the transmission-to-engine bolts to 25 ft. lbs. on 4 cylinder models and 5 ft. lbs. on the 6 cylinder; torque the converter-to-flywheel bolts to 35 ft. lbs.; torque the transmission-to-mount bolts to 35 ft. lbs.; the mount-to-frame bolts to 25 ft. lbs.

TRANSFER CASE

There are three transfer cases used. The New Process 205 is used in part time systems with all transmissions in 1980, and with manual transmissions only through 1979. It has a large New Process emblem on the back of the case. The full time New Process 203 is used with automatics through 1979. It can be identified by the H LOC and L LOC positions on the shifter.

The aluminum case New Process 208 was introduced in 1981 on K10 and 20 models. S–Series trucks use the New Process 207. This unit is an aluminum case model with chain drive. Proper fluid for this unit is Dexron II automatic transmission fluid.

NOTE: Models with a New Process 203 full time four wheel drive transfer case, especially with manual transmissions, may give a front wheel "chatter" or vibration on sharp turns. This is a normal characteristic of this drivetrain combination. If it occurs shortly after shifting out of a LOC position, the transfer case is probably still locked up. This should correct itself after about a mile of driving, or can be alleviated by backing up for a short distance.

Shift Linkage

ADJUSTMENT

New Process 203

1. Place the selector lever in the cab in the Neutral position.

2. Detach the adjustable rod ends from the transfer case levers.

3. Insert a $^{11}/_{64}$ in. drill bit through the alignment holes in the shifter levers. This will lock the shifter in the neutral position with bolt levers vertical.

4. Place the range shift lever (the outer lever) on the transfer case in the Neutral position.

5. Place the lockout shift lever (the inner lever) on the transfer case in the unlocked position. Both levers should now be vertical.

6. Adjust the rods so that the linkage fits together. The indicator plate can be moved to align with the correct symbol.

7. Remove the drill bit.

New Process 208

1. Put transfer case lever in 4HI detent.

2. Push lower shifter lever forward to 4HI stop.

3. Install rod swivel in shift lever hole.

4. Hang 0.200 thick gauge cover rod behind swivel.

5. Run rear rod nut A against gauge with shifter against 4HI stop.

6. Remove gauge and push swivel rearward against nut A.

7. Run front rod nut B against swivel and tighten.

New Process 207

1. Loosen the transfer case switch bolt and the case shift lever pivot bolts.

2. Shift the transfer case to the 4H position.

3. Remove the console and slide the boot up the lever.

4. Install a $^{5}/_{16}$ in. drill bit through the shifter and into the switch bracket.

5. Install a bolt at the case lever to lock it in position.

6. Tighten the switch bracket bolt to 30 ft. lbs. and the shifter pivot bolt to 100 ft. lbs.

7. Remove the bolt you installed to lock the lever.

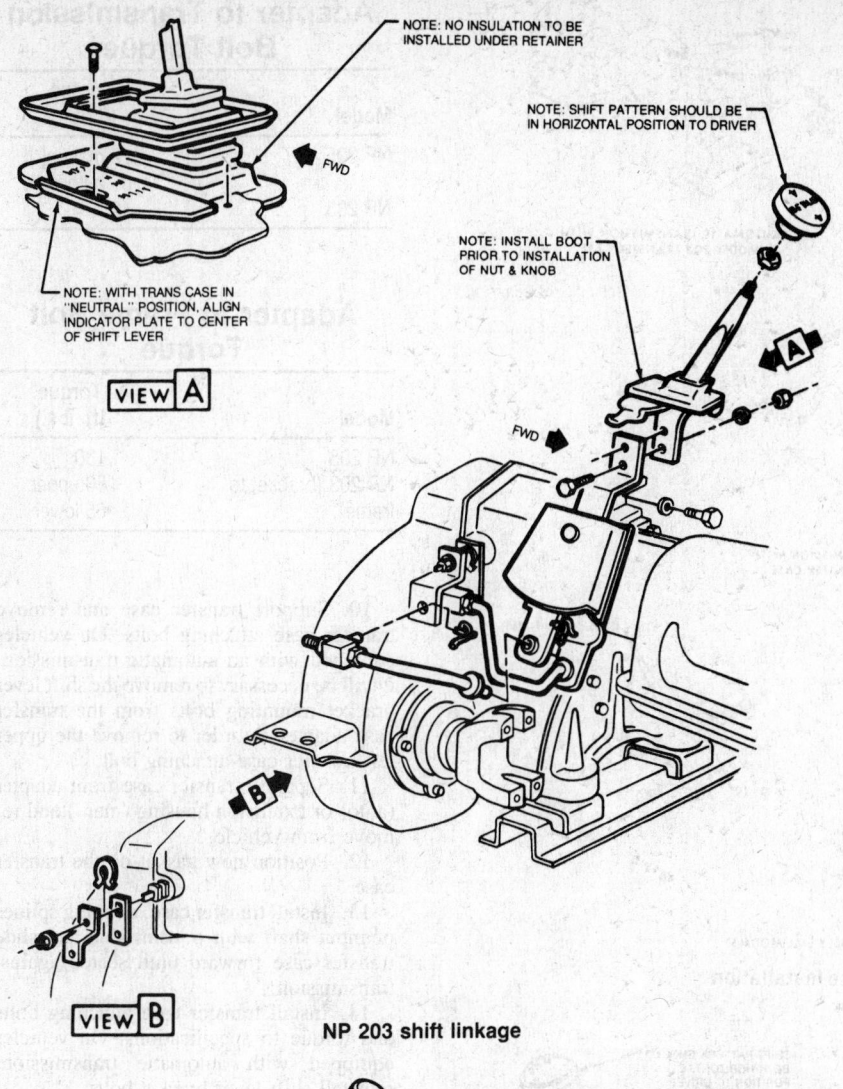

NOTE: NO INSULATION TO BE INSTALLED UNDER RETAINER

NOTE: SHIFT PATTERN SHOULD BE IN HORIZONTAL POSITION TO DRIVER

NOTE: INSTALL BOOT PRIOR TO INSTALLATION OF NUT & KNOB

FWD

NOTE: WITH TRANS CASE IN "NEUTRAL" POSITION, ALIGN INDICATOR PLATE TO CENTER OF SHIFT LEVER

VIEW A

FWD

VIEW B

NP 203 shift linkage

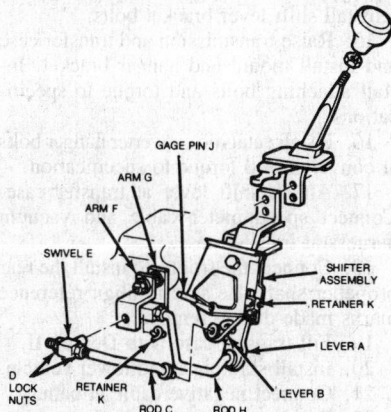

GAGE PIN J

ARM G

ARM F

SWIVEL E

D LOCK NUTS

RETAINER K

ROD C

SHIFTER ASSEMBLY

RETAINER K

LEVER A

LEVER B

ROD H

NP 203 linkage adjustment

8. Remove the drill bit. Check the drill bit. Check the linkage action.

REMOVAL & INSTALLATION

NP 203 and 205

1. Raise and support the truck.
2. Drain the transfer case.

3. Disconnect the speedometer cable, back-up light switch, and the TCS switch.
4. If necessary, remove the skid plate and crossmember support.
5. Disconnect the front and rear driveshafts and support them out of the way. On New Process 205 models, disconnect the shift lever rod from the shift rail link. On New Process 203 models, disconnect the shift levers at the transfer case.
6. Remove the transfer case-to-frame mounting bolts.
7. Support the transfer case and remove the bolts attaching the transfer case to transmission adapter.
8. Move the transfer case to the rear until the input shaft clears the adaptor and lower the transfer case from the truck.
9. To install the transfer case: Lifting the transfer case on a transmission jack, attach the case to the adapter using through bolts. Torque to specification.
10. Remove the transmission jack and install the transfer case-to-frame rail bolts. Make certain to bend the locking tabs after installation.
11. Connect the shift linkage.
12. Connect the front driveshaft to the

Adapter to Transfer Case Bolt Torque

Model	Year	Torque (ft. lbs.)
NP 205	'75–'82	25
NP 203	'75–'79	38

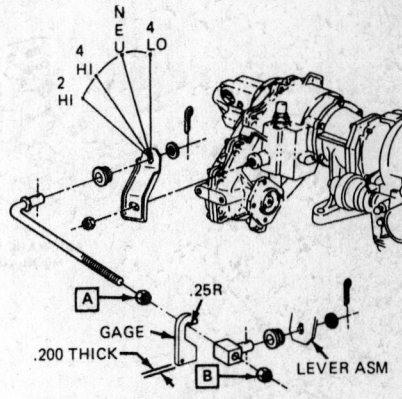

NEU

4 HI

4 LO

4 HI

2 HI

A

.25R

GAGE

.200 THICK

B

LEVER ASM

NP 208 linkage adjustment

front transfer case output shaft and the rear drive shaft to the rear output shaft.
13. Install the crossmember and skid plate, if equipped.
14. Connect the speedometer cable, back-up light, and TCS switches.
15. Fill the transfer case to the proper level with lubricant.
16. Lower the vehicle.

NOTE: Recheck all bolt torques. When attaching the driveshafts, make sure that the flange locknuts are torqued to specifications.

NP 207

1. Shift transfer case into 4 Hi.
2. Disconnect negative cable at battery.
3. Raise vehicle and remove skid plate.
4. Drain lubricant from output shaft yoke and propeller shaft for assembly reference. Disconnect front propeller shaft from transfer case.
5. Mark transfer case front output shaft yoke and propeller shaft for assembly reference. Disconnect the front propeller shaft from transfer case.
6. Mark rear axle yoke and propeller shaft for assembly reference. Remove rear propeller shaft.
7. Disconnect speedometer cable and vacuum harness at transfer case. Remove shift lever from transfer case.
8. Remove catalytic converter hanger bolts at converter.
9. Raise transmission and transfer case and remove transmission mount attaching bolts. Remove mount and catalytic converter hanger and lower transmission and transfer case.

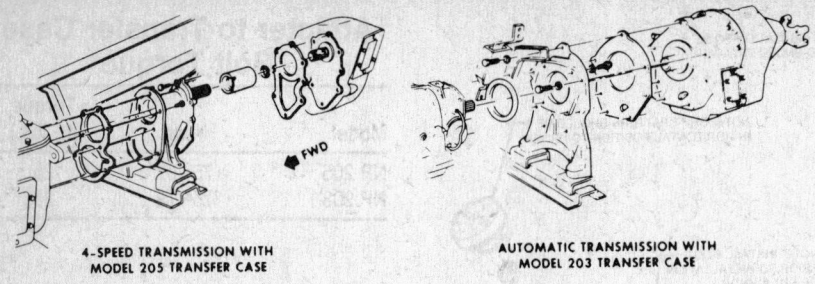

4-SPEED TRANSMISSION WITH
MODEL 205 TRANSFER CASE

AUTOMATIC TRANSMISSION WITH
MODEL 203 TRANSFER CASE

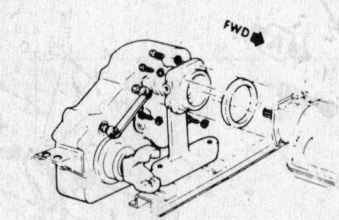

AUTOMATIC TRANSMISSION WITH
MODEL 205 TRANSFER CASE

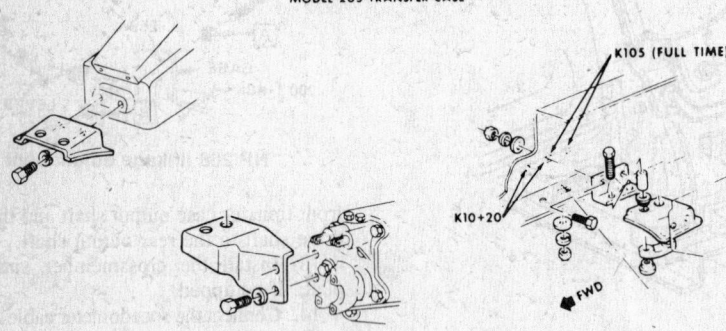

K105 (FULL TIME)

K10+20

SUPPORT AND BRACKET ASSEMBLY (ALL MODELS)

Typical transfer case installation

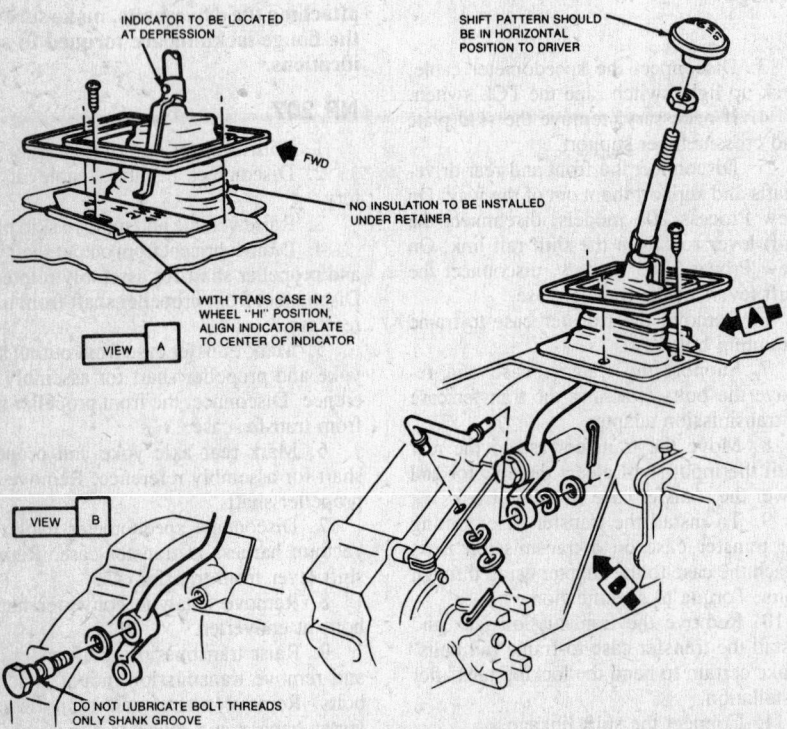

INDICATOR TO BE LOCATED
AT DEPRESSION

SHIFT PATTERN SHOULD
BE IN HORIZONTAL
POSITION TO DRIVER

NO INSULATION TO BE INSTALLED
UNDER RETAINER

WITH TRANS CASE IN 2
WHEEL "HI" POSITION,
ALIGN INDICATOR PLATE
TO CENTER OF INDICATOR

VIEW A

VIEW B

DO NOT LUBRICATE BOLT THREADS
ONLY SHANK GROOVE

NP 205 linkage

Adapter to Transmission Bolt Torque

Model	Torque (ft. lbs.)
NP 205	22 manual
	35 automatic
NP 203	40

Adapter to Frame Bolt Torque

Model	Torque (ft. lbs.)
NP 205	130
NP 203 (bracket to frame)	50 upper
	65 lower

10. Support transfer case and remove transfer case attaching bolts. On vehicles equipped with an automatic transmission, it will be necessary to remove the shift lever bracket mounting bolts from the transfer case adapter in order to remove the upper left transfer case attaching bolt.

11. Separate transfer case from adapter (auto) or extension housing (man.) and remove from vehicle.

12. Position new gasket on the transfer case.

13. Install transfer case, aligning splines of input shaft with transmission and slide transfer case forward until seated against transmission.

14. Install transfer case attaching bolts and torque to specifications. On vehicles equipped with automatic transmission, reinstall shift lever bracket bolts.

15. Raise transmission and transfer case and install mount and hanger bracket. Install attaching bolts and torque to specifications.

16. Install catalytic converter hanger bolts at converter and torque to specification.

17. Attach shift lever at transfer case. Connect speedometer cable and vacuum harness at transfer case.

18. Connect the front and install the rear propeller shaft. Be sure to align reference marks made during removal.

19. Fill transfer case with Dexron II.

20. Install skid plate and lower vehicle.

21. Connect negative cable at battery.

22. Road test vehicle, check to make sure vehicle shifts and operates into all ranges.

NP 208

1. Place the transfer case in 4H.

2. Raise the vehicle.

3. Drain the lubricant from the transfer case.

4. Remove the cotter pin from the shift lever swivel.

5. Mark the transfer case front and rear output shaft yokes and propeller shafts for assembly alignment reference.

6. Disconnect the speedometer cable and indicator switch wires.

7. Disconnect the front propeller shaft at the transfer case yoke.

8. Disconnect the parking brake cable guide from the pivot located on right from rail, if necessary.

9. Remove the engine strut rod from the transfer case on automatic transmission models.

10. Place a support under the transfer case and remove the transfer case-to-transmission adapter bolts.

11. Move the transfer case assembly rearward until free of the transmission output shaft and remove the assembly.

12. Remove all gasket material from the rear of the transmission adapter housing.

13. Install the transmission-to-transfer case gasket on the transmission.

14. Shift the transfer case to 4H position if not done previously.

15. Rotate the transfer case output shaft (by turning yoke) until the transmission output shaft gear engages the transfer case input shaft. Move the transfer case forward until the case seats against the transmission. Be sure the transfer case is flush against transmission. Severe damage to the transfer case will result if the attaching bolts are tightened while the transfer case is cocked or in a bind.

16. Install the transfer case attaching bolts. Tighten the bolts to 30 ft. lbs.

17. Connect the speedometer driven gear to the transfer case.

18. Connect the front and rear propeller shafts to the transfer case. Be sure to align the shafts-to-yokes using the reference marks made during removal. Tighten the shaft-to-yoke clamp strap nuts to 15 ft. lbs.

19. Remove the support stand from under the transfer case.

20. Connect the parking brake cable if disconnected.

21. Attach the cotter pin to the shift lever swivel.

22. Connect the engine strut to the transfer case on automatic models.

23. Fill the transfer case with Dexron® II.

24. Lower the vehicle.

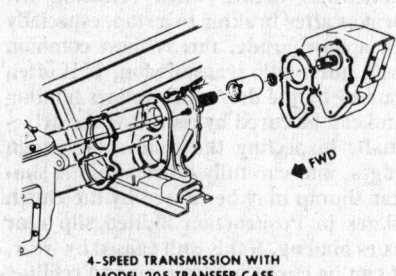

4-SPEED TRANSMISSION WITH MODEL 205 TRANSFER CASE

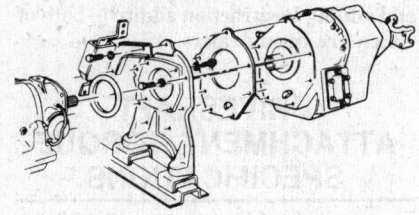

AUTOMATIC TRANSMISSION WITH MODEL 203 TRANSFER CASE

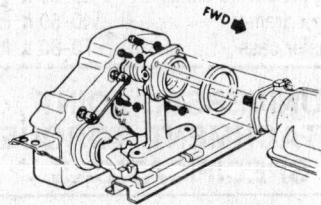

AUTOMATIC TRANSMISSION WITH MODEL 205 TRANSFER CASE

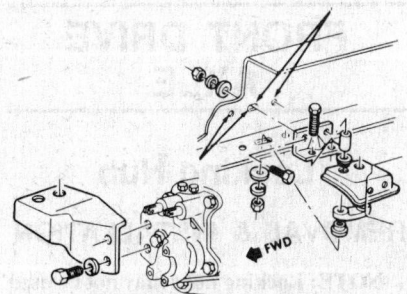

SUPPORT AND BRACKET ASSEMBLY (ALL MODELS)

Transfer case adapters

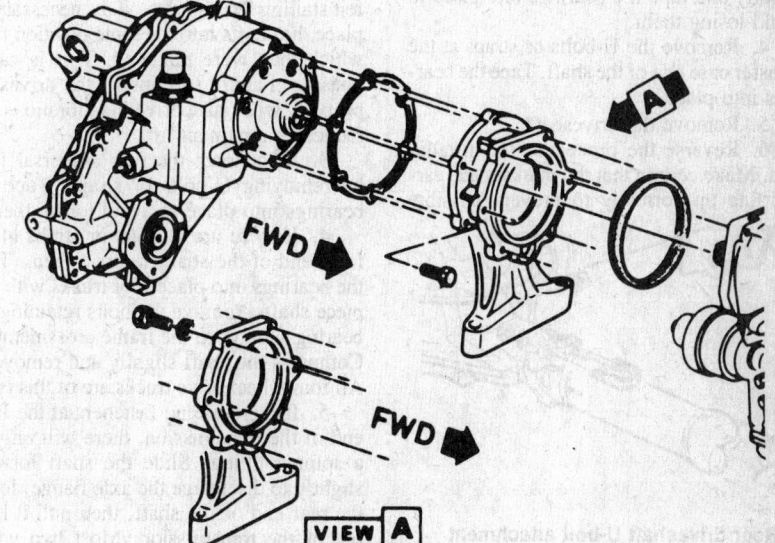

WITH AUTOMATIC TRANSMISSION

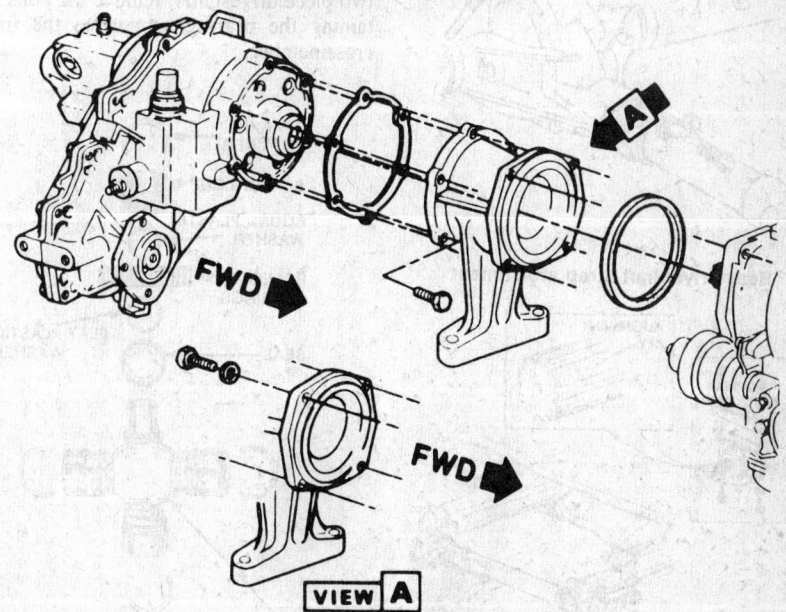

WITH MANUAL TRANSMISSION

NP 208 installation

T67

DRIVELINE

Front Driveshaft (4WD Only)

REMOVAL & INSTALLATION

1. Jack the front of the vehicle so that the front wheels are off the ground. Block the rear wheels and safely support the truck on stands.

2. Scribe aligning marks on the driveshaft and the pinion flange to aid in reassembly.

3. Remove the U-bolts or straps at the axle end of the shaft. Compress the shaft slightly and tape the bearings into place to avoid losing them.

4. Remove the U-bolts or straps at the transfer case end of the shaft. Tape the bearings into place.

5. Remove the driveshaft.

6. Reverse the procedure for installation. Make certain that the marks made earlier line up correctly to prevent possible

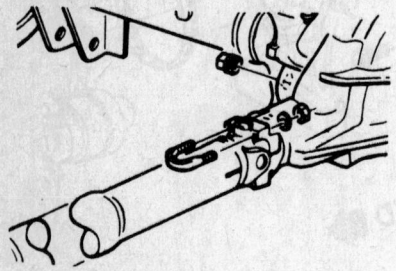

Rear driveshaft U-bolt attachment

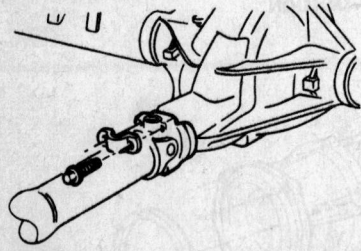

Rear driveshaft strap attachment

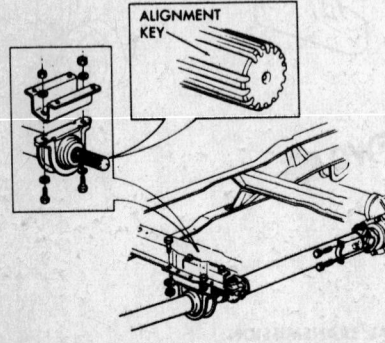

32 splined shaft U-joint alignment keyway

imbalances. Be sure that the constant velocity joint (the big double one) is at the transfer case end.

Rear Driveshaft

REMOVAL & INSTALLATION

1. Raise and safely support the rear of the truck as necessary. There is less chance of lubricant leakage from the rear of the transmission on two wheel drive models if the rear is raised. block the front wheels.

2. Scribe alignment marks on the driveshaft and flange of the rear axle., and transfer case or transmission. If the truck is equipped with a two piece driveshaft, be certain to also scribe marks at the center joint near the splined connection. When reinstalling driveshafts, it is necessary to place the shafts into the same position from which they were removed. This is called phasing. Failure to reinstall the driveshaft properly will cause driveline vibrations and reduced component lift.

3. Disconnect the rear universal joint by removing U-bolts or straps. Tape the bearings into place to avoid losing them.

4. If there are U-bolts or straps at the front end of the shaft, remove them. Tape the bearings into place. For trucks with two piece shafts, remove the bolts retaining the bearing support to the frame crossmember. Compress the shaft slightly and remove it. All four wheel drive trucks are of this type.

5. If there are no fasteners at the front end of the transmission, there will only be a splined fitting. Slide the shaft forward slightly to disengage the axle flange, lower the rear end of the shaft, then pull it back out of the transmission. Most two wheel drive trucks are of this type. For trucks with two piece driveshafts, remove the bolts retaining the bearing support to the from crossmember.

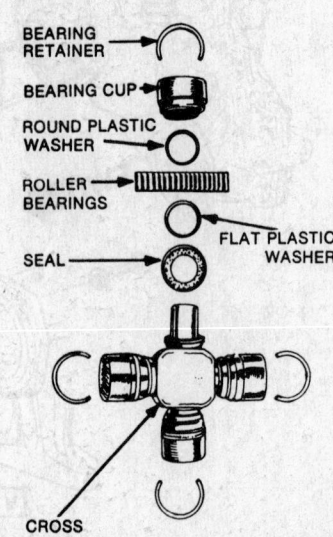

BEARING RETAINER

BEARING CUP

ROUND PLASTIC WASHER

ROLLER BEARINGS

SEAL

FLAT PLASTIC WASHER

CROSS

Injection molded retainer U-joint repair kit

6. Reverse the procedure for installation. It may be tricky to get the scribed alignment marks to match up on trucks with two piece driveshaft. For those models only, the following instructions may be of some help. First, slide the grease cap and gasket onto the rear splines. On 4-wheel drive models with 16 splines, after installing the front shaft to the transmission and bolting the support to the crossmember, arrange the front trunnion vertically and the second trunion horizontally. Most models with 32 splines have an alignment key. The driveshaft cannot be replaced incorrectly. Simply match up the key with the keyway.

7. On 2WD automatic transmission models, lubricate the internal yoke splines at the transmission end of the shaft with lithium base grease. The grease should seep out through the vent hole.

NOTE: A thump in the rear driveshaft sometimes occurs when releasing the brakes after braking to a stop, especially on a downgrade. this is most common with automatic transmission. It is often caused by the driveshaft splines binding and can be cured by removing the driveshaft, inspecting the splines for rough edges, and carefully lubricating. A similar thump may be caused by the clutch plates in Positraction limited slip rear axles binding. If this isn't caused by wear, it can be cured by draining and refilling the rear axle with the special lubricant and adding Positraction additive, both of which are from dealers.

DRIVESHAFT ATTACHMENT TORQUE SPECIFICATIONS

To rear axle (strap)	12–17 ft. lbs.
To rear axle (U-bolt)	18–22 ft. lbs.
Bearing support to hanger	20–30 ft. lbs.
Hanger to frame	40–50 ft. lbs.
To transfer case	70–80 ft. lbs.

UNIVERSAL JOINT ATTACHMENT TORQUE SPECIFICATIONS

Strap attachments	15 ft. lbs.
U-bolt attachments	20 ft. lbs.

FRONT DRIVE AXLE

Locking Hub

REMOVAL & INSTALLATION

NOTE: Locking hubs may not be used with full time four wheel drive. Locking hubs should be run in the lock position for at least 10 miles each month to assure proper differential lubrication. This procedure requires snap-ring pliers and a special hub nut wrench. It isn't very easy without them. You will have to modify this procedure if you have non-factory installed hubs.

1. Set the hub in the Lock position.
2. Remove the outer retaining plate allen head bolts and take off the plate, O-ring, and knob.
3. Take out the large snap-ring inside the hub and remove the outer clutch retaining ring and actuating cam body.
4. Relieve pressure on the axle shaft snap-ring and remove it.
5. Take out the axle shaft sleeve and clutch ring assembly and the inner clutch ring and bushing assembly. Remove the spring and retainer plate.
6. Clean all the hub components in a safe solvent and dry them. Lubricate everything with a high temperature grease.
7. Install the spring retainer plate with the flange side to the bearing and seat it against the outer bearing cup.
8. Install the spring with the large end against the retainer plate.

NOTE: When the spring is properly installed and seated it will extend past the spindle nuts about $7/8$ in.

9. Place the inner clutch ring and bushing assembly into the axle shaft sleeve and clutch ring assembly. Install these components, push in, and install the axle shaft snap-ring. If there are two axle shaft snap ring grooves, use the inner one.

NOTE: You can install a $7/16$ in. bolt in the axle and pull outward on it to aid in seating the snap-ring.

10. Install the actuating cam body with the cams out. Replace the outer clutch retaining ring and then the internal snap-ring.
11. Install a new O-ring, then install the actuating knob and retaining plate in the lock position. The grooves in the knob must fit into the actuator cam body. Install the cover bolts and seals.

Axle Shaft

REMOVAL & INSTALLATION

NOTE: This procedure requires snapring pliers and a special hub nut wrench. It is very difficult without them.

1. Remove the wheel and tire
.2. For K10 or K1500 models and K20 or K2500 models with locking front hubs: Lock the hubs. Remove the outer retaining plate Allen head bolts and take off the plate, O-ring, and knob. Take out the large snapring inside the hub and remove the outer clutch retaining ring and actuating cam body. This is a lot easier with snap-ring pliers. Relieve pressure on the axle shaft snap-ring and remove it. Take out the axle shaft sleeve and clutch ring assembly and the inner clutch

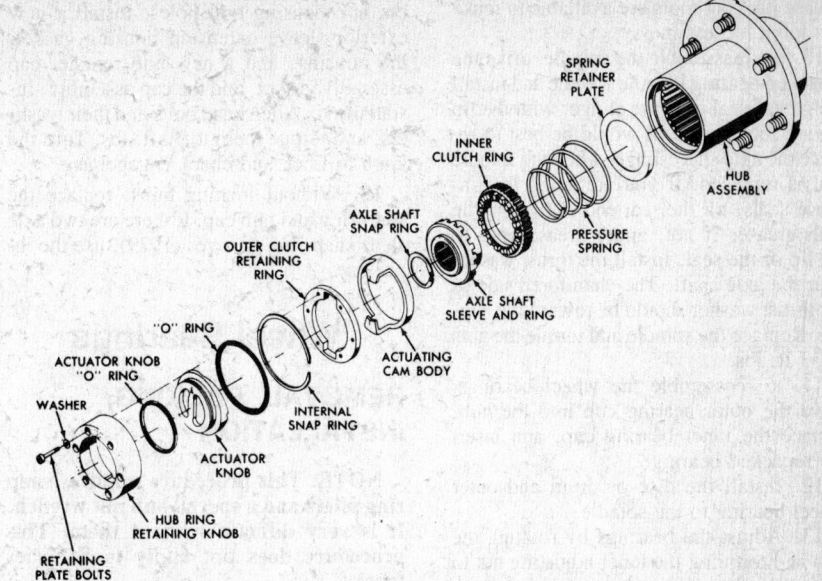

Details of the locking hubs used on all K-10 and 1500 models and 1977 and later K-20 and 2500 models

ring and bushing assembly. Remove the spring and retainer plate.

NOTE: You will have to modify this procedure for either of the models mentioned above if you have non-factory installed locking hubs.

3. If you don't have locking front hubs, remove the hub cap and snap-ring. Next, remove the drive gear and pressure spring. To prevent the spring from popping out, place a hand over the drive gear and use a screwdriver to pry the gear out. Remove the spring.
4. Remove the wheel bearing outer lock nut, lock ring, and wheel bearing inner adjusting nut. A special wrench is required.
5. Remove the brake disc assembly and outer wheel bearing. Remove the spring retainer plate if you don't have locking hubs. Pull out the axle shaft and universal assembly. When installing the shaft, turn it slowly to mesh with splines.
6. Remove the oil seal and inner bearing cone from the hub using a brass drift and hammer. Discard the oil seal. Use the drift to remove the inner and outer bearing cups.
7. Check the condition of the spindle bearing. If you have drum brakes, remove the grease retainer, gasket, and backing plate after removing the bolts. Unbolt the spindle and tap it with a soft hammer to break it loose. remove the spindle and check the condition of the thrust washer, replacing it if worn. Now you can remove the oil seal and spindle roller bearing.

NOTE: The spindle bearings must be greased each time the wheel bearings are serviced.

8. Clean all parts in solvent, dry, and check for wear or damage.

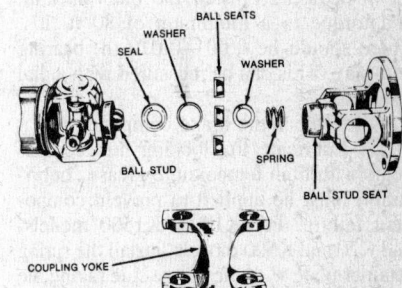

Exploded view of a constant velocity joint

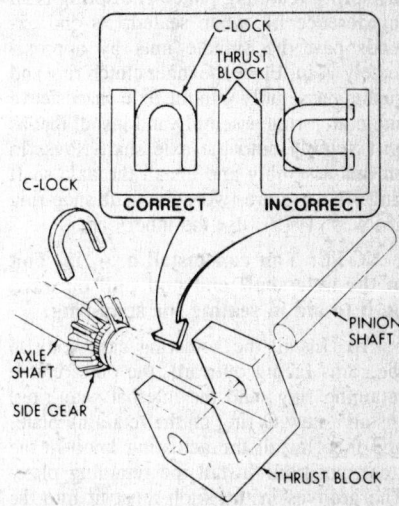

Correct C-lock positioning on locking differentials

9. Pack both wheel bearings (and the spindle bearing) using wheel bearing grease. Place a healthy glob of grease in the palm of one hand and force the edge of the bearing into it so that grease fills the bearing. Do this until the whole bearing is packed.

Grease packing tools are available to make this job a lot less messy.

10. To reassemble the spindle: drive the repacked bearing into the spindle and install the grease seal onto the slinger with the lip toward the spindle. It would be best to replace the axle shaft slinger when the spindle seal is replaced. If you are using the improve seals, fill the seal end of the spindle with grease. If not, apply grease only to the lip of the seal. Install the thrust washer over the axle shaft. The chamfered side of the thrust washer should be toward the slinger. Replace the spindle and torque the nuts to 33 ft. lbs.

11. to reassemble the wheel bearings: drive the outer bearing cup into the hub, replace the inner bearing cup, and insert the repacked bearing.

12. Install the disc or drum and outer wheel bearing to the spindle.

13. Adjust the bearings by rotating the hub and torquing the inner adjusting nut to 50 ft. lbs. then loosening it and retorquing to 35 ft. lbs. Next back the nut off $3/8$ turn or less. Turn the nut to the nearest hole in the lockwasher. Install the outer locknut and torque to a minumum of 80 ft. lbs. There should be 0.001–0.010 in. bearing end-play. This can be measured with a dial indictor.

14. Replace the brake components.

15. Lubricate the locking hub components with high temperature grease. Lubrication must be applied to prevent component failure. For K10 or K1500 models, and K20 and K500 models, install the spring retainer plate with the flange side facing the bearing over the spindle nuts and seat it against the bearing outer cup. Install the pressure spring with the large end against the spring retaining plate. The spring is an interference fit; when seated, its end extends past the spindle nuts by approximately $7/8$ in. Place the inner clutch ring and bushing assembly into the axle shaft sleeve and clutch ring assembly and install that as an assembly onto the axle shaft. Press in on this assembly and install the axle shaft ring. If there are two axle shaft snap-ring grooves (1979), use the inner one.

NOTE: You can install a $7/16$ in. bolt in the axle shaft end and pull outward on it to aid in seating the snap-ring.

16. Install the actuating cam body in the cams facing outward, the outer clutch retaining ring, and the internal snap-ring. Install a new O-ring on the retaining plate, and then install the actuating knob in the Lock position. Install the retaining plate. The grooves in the knob must fit into the actuator cam body. Install the seals and six cover bolts and torque them to 30 ft. lbs. Turn the knob to the Free position and check for proper operation.

NOTE: Remove the head from a 5 in. long $3/4$ in. bolt and use this to align the hub assembly.

17. Install the headless bolt into one of the hub housing bolt holes. Install a new exterior sleeve extension housing gasket, the housing, and a new hub retainer cap assembly gasket, and the cap assembly. Install the six Allen head bolts and their washers, and torque them to 30 ft. lbs. Turn the knob to Lock and check engagement.

18. Without locking hubs, replace the snap-ring and hub cap. If there are two axle shaft snap-ring grooves (1979), use the inner one.

Wheel Bearings

REMOVAL, PACKING, INSTALLATION

NOTE: This procedure requires snap ring pliers and a special hub nut wrench. It is very difficult without them. This procedure does not apply to S–Series trucks.

1. Remove the wheel and tire.
2. For $1/2$ and $3/4$ ton trucks with locking front hubs, lock the hubs. Remove the outer retaining plate Allen head bolts and take off the plate, O-ring, and knob. Take out the large snap ring inside the hub and remove the outer clutch retaining ring and actuating am body. This is a lot easier with snap ring pliers. Relieve pressure on the axle shaft snap ring and remove it. Take out the axle shaft sleeve and clutch ring assembly and the inner clutch ring and bushing assembly and the inner clutch ring and bushing assembly. Remove the spring and retainer plate.
3. If you don't have locking front hubs, remove the hub cap and snap ring. Next remove the drive gear and pressure spring. To prevent the spring from popping out, place a hand over the drive gear and pry the gear out. Remove the spring.
4. Remove the wheel bearing outer lock nut, lock ring, and wheel bearing inner adjusting nut. A special wrench is required.
5. Remove the brake disc assembly and outer wheel bearing. Remove the spring retainer plate if you don't have locking hubs.
6. Remove the oil seal and inner bearing cone from the hub using a brass drift and hammer. Discard the oil seal. Use the drift to remove the inner and outer bearing cups.
7. Check the condition of the spindle bearing. Unbolt the spindle and tap it with a soft hammer to break it loose. Remove the spindle and check the condition of the thrust washer, replacing it if worn. Now you can remove the oil seal and spindle roller bearing.

NOTE: The spindle bearings must be greased each time the wheel bearings are serviced.

8. Clean all parts in solvent, dry and check for wear or damage.
9. Pack both wheel bearings (and the spindle bearings) using wheel bearing grease.

Place a healthy glob of grease in the palm of one hand and force the edge of the bearing into it so that grease fills the bearing. Do this until the wheel bearing is packed. Grease packing tools are available to make this job easier.

10. To reassemble the spindle: drive the repacked bearing into the spindle and install the grease seal onto the slinger with the lip toward he spindle. It would be best to replace the axle shaft slinger when the spindle seal is replaced.

NOTE: If you are using the improved seals, fill the seal end of the spindle with grease. If not, apply grease only to the lip of the seal. Install the thrust washer over the axle shaft. On late 1982 and later models, the chamfered side of the thrust washer should be toward the slinger. Replace the spindle and torque the nuts to, 33 ft. lbs. for 1979–80: 65 ft. lbs. for 1981 and later models.

11. To reassemble the wheel bearings: drive the outer bearing cup into the hub, replace the inner bearing cup, and insert the repacked bearing.

12. Install the disc or drum and outer wheel bearing to the spindle.

13. Adjust the bearings by rotating the hub and torquing the inner adjusting nut to 50 ft. lbs. Next, back the nut off $3/8$ turn or less. Turn the nut to the nearest hole in the lock-washer. Install the outer locknut and torque to minimum of 80 ft. lbs. for 1979–80: 160–205 ft. lbs. for 1981 and later $1/2$ and $3/4$ ton and 65 ft. lbs on 1 ton vehicles. There should be 0.001–0.010 in. bearing end play. This can be measured with a dial indicator.

14. Replace the brake components.

15. Lubricate the locking hub components with high temperature grease. Lubrication must be applied to prevent component failure. Install the spring retainer plate with the flange side facing the bearing over the spindle nuts and seat it against the bearing outer cup. Install the pressure spring with the large end against the spring retaining plate. The spring is an interference fit; when seated, its end extends past the spindle nuts by approximately $7/8$ in. Place the inner clutch ring and bushing assembly and install that as an assembly onto the axle shaft. Press in on this assembly and install that as an assembly onto the axle shaft. Press in on this assembly and install the axle shaft ring. If there are two axle shafts snap ring grooves (1979), use the inner one.

S–Series Trucks

REMOVAL & INSTALLATION

Tube/Shaft Assembly

1. Disconnect negative battery cable.
2. Disconnect shift cable from vacuum actuator by disengaging locking spring. then push actuator diaphragm in to release cable.

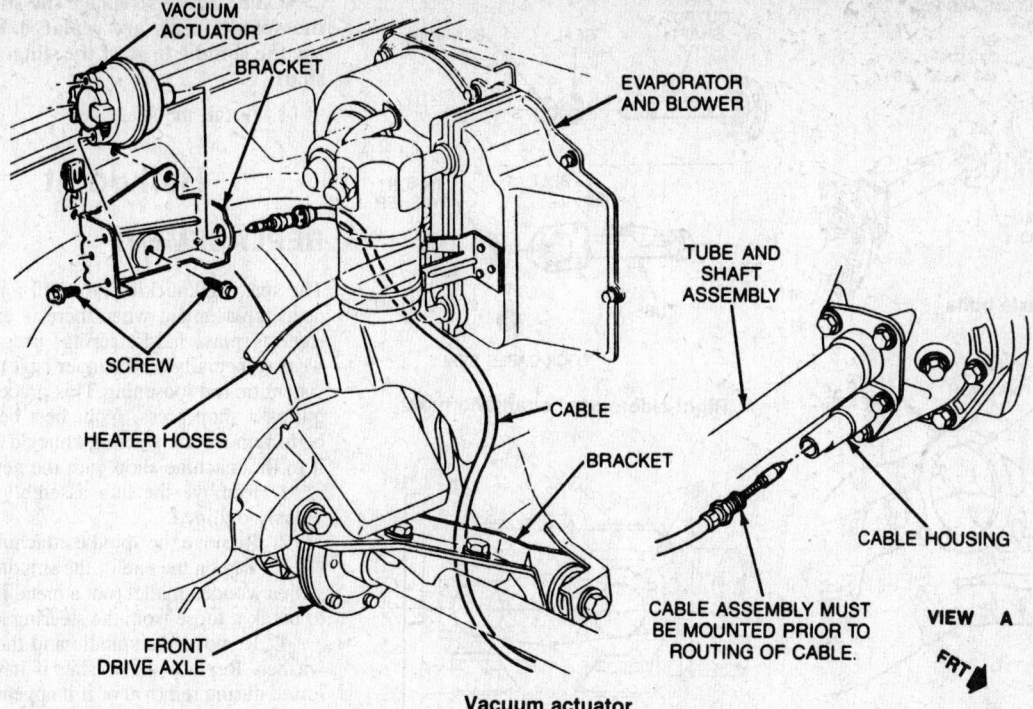

Vacuum actuator

3. Unlock steering wheel at steering column so linkage is free to move.

4. Raise vehicle and place jackstands under the frame.

5. Remove front wheels.

6. Remove engine drive belt shield.

7. Remove front axle skid plate (if equipped).

8. Place support under right hand lower control arm and disconnect right hand upper, ball joint, then remove support so control arm will hang free.

9. Disconnect right hand drive axle shaft from tube assembly by removing six bolts. Keep axle from turning by inserting a drift through opening in top of brake caliper into corresponding vane of brake rotor.

10. Disconnect four wheel drive indicator light electrical connection from switch.

11. Remove three bolts securing cable and switch housing to carrier and pull housing away to gain access to cable locking spring. do not unscrew cable coupling nut unless cable is being replaced.

12. Disconnect cable from shift fork shaft by lifting spring over slot in shift fork.

13. Remove two bolts securing tube bracket to frame.

14. Remove remaining two upper bolts securing tube assembly to carrier.

15. Remove tube assembly by working around drive axle. Be careful not to allow sleeve, thrust washers, connector, and output shaft to fall out of carrier or be damaged when removing tube.

16. Install sleeve, thrust washers, connector and output shaft in carrier. Apply sealer 1052357, Loctite® 514 or equivalent on tube to carrier surface. Be sure to install thrust washer. Apply grease to washer to hold it in place during assembly.

17. Install tube and shaft assembly to carrier and install bolt at one o'clock position but do not torque. Pull assembly down and install cable and switch housing, and remaining four bolts. Torque all bolts to 45–60 ft. lbs.

18. Install two bolts securing tube to frame and torque.

19. Check operation of four wheel drive mechanism using Tool J–33799. Insert tool into shift fork and check for rotation of axle shaft.

20. Remove tool and install shift cable switch housing by pushing cable through into fork shaft hole. Cable will automatically snap in place.

21. Connect four wheel drive indicator light electrical connection to switch.

22. Install support under right hand lower control arm to raise arm and connect upper ball joint.

23. Install right-hand drive axle to axle tube by installing one bolt first, then, rotate axle to install remaining five bolts. Hold axle from turning by inserting a drift through opening in top of brake caliper into corresponding vane of brake rotor. Tighten bolts to 53–63 ft. lbs.

24. Install front axle skid plate, if equipped.

25. Install engine drive belt shield.

26. Install front wheels.

27. Lower vehicle.

28. Connect shift cable to vacuum actuator by pushing cable end into vacuum actuator shaft hole. Cable will snap in place automatically.

29. Connect negative battery cable.

Shift Cable

1. Disengage shift cable from vacuum actuator by disengaging locking spring, then push actuator diaphragm in to release cable. Squeeze the two locking fingers of the cable with pliers, then pull cable out of bracket hole.

2. Raise vehicle and remove three bolts securing cable and switch housing to carrier and pull housing away to gain access to cable locking spring. Disconnect cable from shaft fork shaft by lifting spring over slot in shift fork.

3. Unscrew cable from housing.

4. Remove cable from vehicle.

5. Install cable observing proper routing.

6. Install cable and switch housing to carrier using three attaching bolts. Torque mounting bolts to 30–40 ft. lbs.

7. Guide cable through switch housing into fork shaft hole and push cable in. Cable will automatically snap in place. Start turning coupling nut by hand, to avoid cross threading, then torque nut to 71–106 inch lbs. Do not over torque nut as this will cause thread damage to plastic housing.

8. Lower vehicle.

9. Connect shift cable to vacuum actuator by pressing cable into bracket hole. Cable and housing will snap in place automatically.

10. Check cable operation.

Differential Carrier Right Half Output Shaft and Tube

1. Remove right-hand output shaft from tube by striking inside of flange with a soft face hammer while holding tube.

2. Remove output shaft tub seal by prying out of tube.

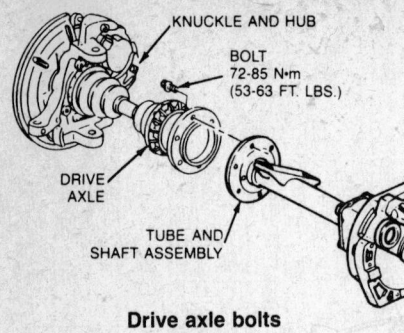

Drive axle bolts

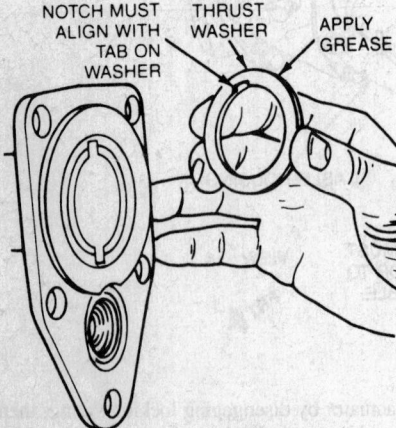

NOTCH MUST ALIGN WITH TAB ON WASHER — THRUST WASHER — APPLY GREASE

Thrust washer

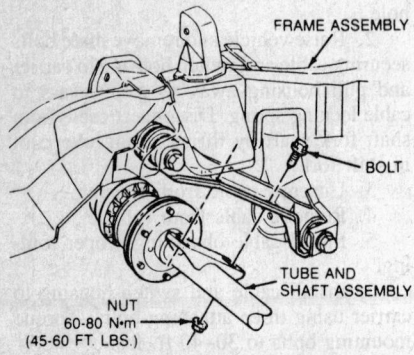

FRAME ASSEMBLY — BOLT — TUBE AND SHAFT ASSEMBLY — NUT 60-80 N·m (45-60 FT. LBS.)

Tube–to–frame attachment

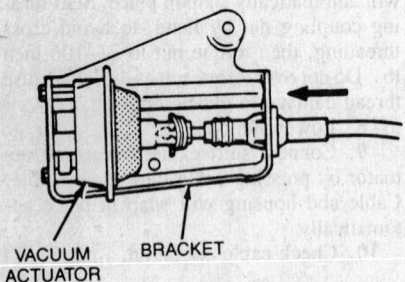

VACUUM ACTUATOR — BRACKET

Cable–to–vacuum actuator attachment

3. Remove output shaft tube bearing using J–29369–2.

4. Remove differential shift cable housing seal by driving out with a punch or similar tool.

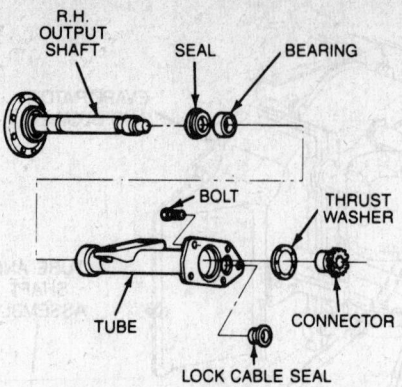

R.H. OUTPUT SHAFT — SEAL — BEARING — BOLT — THRUST WASHER — TUBE — CONNECTOR — LOCK CABLE SEAL

Right side output shaft and tube

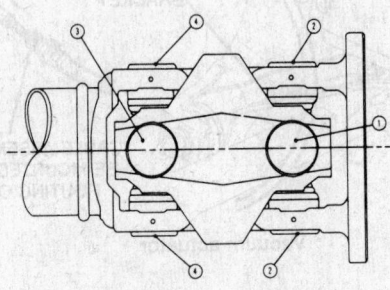

CV joint disassembly sequence

5. Install output shaft tube bearing using tool J–33844. Tool must be flush with tube when bearing is correctly installed.

6. Install output shaft tube seal using tool J–33893. Flange of seal must be flush with tube outer surface when seal is installed.

7. Install output shaft into tube and seat by striking flange with a soft face hammer.

8. Install differential shift cable housing seal using J–33799.

Axle Shaft U-Joint

1. Remove the axle shaft.

2. Squeeze the ends of the trunnion bearings in a vise to relieve the load on the snap-rings. Remove the snap-rings.

3. Support the yoke in a vise and drive on one end of the trunnion bearing with a brass drift enough to drive the opposite bearing from the yoke.

4. Support the other side of the yoke and drive the other bearing out.

5. Remove the trunnion.

6. Clean and check all parts. You can buy U-joint repair kits to replace all the wearing parts.

7. Lubricate the bearings with wheel bearing grease.

8. Replace the trunnion and press the bearings into the yoke and over the trunnion hubs far enough to install the lock rings.

9. Hold the trunnion in one hand and tap the yoke lightly to seal the bearings against the lock rings.

10. The axle slingers can be pressed off the shafts.

NOTE: Always replace the slingers if the spindle seals are replaced. You can use the spindle to start the slinger on the shaft.

11. Install the shaft.

Ball Joint

REPLACEMENT

The steering knuckle pivot ball joints may need replacement when there is excessive steering play, hard steering, irregular tire wear (especially on the inner edge), or persistent tie rod loosening. This procedure requires a shop press. Your best bet would be to remove the steering knuckle and take it to the machine shop with the new parts.

1. Remove the hub assembly as previously outlined.

2. Remove the spindle attaching bolts.

3. Tap on the end of the spindle lightly with a wooden mallet (not a metal hammer) to break it loose from the steering knuckle.

4. Remove the spindle and the bronze washer. Replace the washer if it was distorted during removal or if it appears worn.

5. Remove the cotter pin from the tie rod nut.

6. Loosen the tie rod nut and tap on the nut with a wooden mallet in order to break the studs loose from the knuckle arm.

7. Remove the nuts and disconnect the tie rod.

8. Remove the steering arm attaching nuts. Use new, self-locking nuts on installation.

9. Remove the cotter pin from the upper ball joint socket nut.

10. Remove the retaining nuts from the upper and lower ball joint sockets.

11. Remove the knuckle by forcing a wedge between the lower ball stud and the yoke, then between the upper ball stud and the yoke.

NOTE: If you have to loosen the upper ball stud adjusting sleeve to remove the knuckle, don't loosen it more than two turns. The soft threads on the yoke are easily damaged.

12. Remove the lower ball joint snap ring.

13. Remove the lower ball joint as illustrated using special tool J–9519–10 (or a similar C-clamp), J–23454–1 (a solid metal punch), and J–6383–3 (or a piece of 2½ in. outer diameter steel pipe with a ³⁄₁₆ in. wall thickness cut to a length of 2½ in.). The lower ball joint must be removed before the upper ball joint can be serviced.

14. Press the new lower ball joint into the knuckle and install the snap-ring. The lower joint doesn't have a cotter pin hole.

15. Press the upper ball joint into the knuckle.

16. Position the knuckle to the yoke. Install new stud nuts finger tight.

17. Push up on the knuckle and tighten the lower nut to 70 ft. lbs.

18. Using a spanner wrench, install and torque the upper ball stud adjusting sleeve to 50 ft. lbs. Torque the upper stud nut to 100 ft. lbs. and install the cotter pin. Don't loosen the nut, but make it tighter to line up the cotter pin hole.

19. Replace the steering arm, using new nuts and torquing to 90 ft. lbs.

20. Attach the tie rod to the steering arm. Tighten the nuts to 45 ft. lbs.

21. Check the knuckle turning torque with a spring scale hooked to the tie rod hole in the steering arm. With the knuckle straight ahead, measure the right angle pull to keep the knuckle turning after initial breakaway, in both directions. The pull should be 25 lbs. or less.

22. Replace the axle shaft and other components. Tighten the steering linkage nuts to 45 ft. lbs.

Pinion Seal

REPLACEMENT

1. Mark the drive shaft and pinion flange so they can be reassembled in the same position.

2. Disconnect the driveshaft from the pinion flange and support shaft up in body tunnel by wiring drive shaft to the exhaust pipe. If joint bearings are not retained by a retainer strap, use a piece of tape to hold bearings on their journals.

3. Mark the position of the pinion flange, pinion shaft and nut so the proper pinion bearing pre-load can be maintained.

4. Remove pinion flange nut and washer.

5. With suitable container in place to hold any fluid that may drain from rear axle, remove pinion flange.

6. Remove oil seal by driving it out of carrier with a blunt chisel. Do not damage carrier.

7. Examine seal surface of pinion flange for tool marks, nicks, or damage, such as a groove worn by the seal. If damaged, replace the flange.

8. Examine carrier bore and remove any burrs that might cause leaks around the O.D. of the seal.

9. Installation is the reverse of removal. Apply Special Seal Lubricant No. 1050169 or equivalent to the O.D. of the pinion flange and sealing lip of new seal.

REAR AXLE

Some models are equipped with a locking differential rear axle. If you're not sure which one is in your truck, block the front wheels and jack up the rear of the truck. With the transmission in Neutral, spin one of the rear wheels in a forward motion with your hands. If the other wheel travels in the same direction, it is a locking differential.

Axle Shaft, Bearing and Seal

REMOVAL & INSTALLATION

All Axles Except Floating and Locking Differentials

1. Support the axle on jackstands.

2. Remove the wheels and brake drums.

3. Clean off the differential cover area, loosen the cover to drain the lubricant, and remove the cover.

4. Turn the differential until you can reach the differential pinion shaft lockscrew. Remove the lockscrew and the pinion shaft.

5. Push in on the axle end. Remove the C-lock from the inner (button) end of the shaft.

6. Remove the shaft, being careful of the oil seal.

7. You can pry the oil seal out of the housing by placing the inner end of the axle shaft behind the steel case of the seal, then prying it out carefully.

8. A puller or a slide hammer is required to remove the bearing from the housing.

9. Pack the new or reused bearing with wheel bearing grease and lubricate the cavity between the seal lips with the same grease.

10. The bearing has to be driven into the housing. Don't use a drift, you might cock the bearing in its bore. Use a piece of pipe or a large socket instead. Drive only on the outer bearing race. In a similar manner, drive the seal in flush with the end of the tube.

11. Slide the shaft into place, turning it slowly until the splines are engaged with the differential. Be careful of the oil seal.

12. Install the C-lock on the inner axle end. Pull the shaft out so that the C-lock seats in the counterbore of the differential side gear.

13. Position the differential pinion shaft through the case and the pinion gears, aligning the lockscrew hole. Install the lockscrew.

14. Install the cover with a new gasket and tighten the bolts evenly in a criss-cross pattern.

15. Fill the axle with lubricant.

16. Replace the brake drum and wheels.

Locking Differential Axles

This axle uses a thrust block on the differential pinion shaft.

1. Follow Steps 1–3 of the preceding procedure.

2. Rotate the differential case so that you can remove the lockscrew and support the pinion shaft so it can't fall into the housing. Remove the differential pinion shaft lockscrew.

3. Carefully pull the pinion shaft partway out and rotate the differential case until the shaft touches the housing at the top.

4. Use a screwdriver to position the C-lock with its open end directly inward. You

can't push in the axle shaft until you do this. Do not force the axle shaft in.

5. Push the axle shaft in and remove the C-lock. Remove the axle shaft and repeat Steps 4–5 for the other shaft.

6. Follow Steps 7–11 of the preceding procedure.

7. Keep the pinion shaft partway out of the differential case while installing the C-lock on the axle shaft. Put the C-lock on the axle shaft and carefully pull out on the axle shaft until the C-lock is clear of the thrust block.

8. Follow Steps 13–16 of the previous procedure.

Floating Differentials

Some 20 and 2500 models and all 30 and 3500 models use axles of full floating design. The procedures are the same for locking the non-locking axles. The best way to remove the bearings from the wheel hub is with an arbor press. Use of a press reduces the chances of damaging the bearing races, cocking the bearing in its bore, or scoring the hub walls. A local machine shop is probably equipped with the tools to remove and install bearings and seals. However, if one is not available, the hammer and drift method outlined can be used.

1. Support the axle on jackstands.

2. Remove the wheels.

3. Remove the bolts and lock washers that attach the axle shaft flange to the hub.

4. Rap on the flange with a soft faced hammer to loosen the shaft. Grip the rib on the end of the flange with a pair of locking pliers, and twist to start shaft removal. Remove the shaft from the axle tube.

5. The hub and drum assembly must be removed to remove the bearings and oil seals. You will need a large socket to remove and later adjust the bearing adjustment nut. There are also tools available which resemble the four wheel drive front wheel bearing adjusting tool.

6. Disengage the tang of the locknut retainer from the slot or flat of the locknut, then remove the locknut from the housing tube, using the earlier mentioned tool.

7. Disengage the tang of the retainer from the slot or flat of the adjusting nut and remove the retainer from the housing tube.

8. Remove the adjusting nut from the housing tube with the tool mentioned earlier.

9. Remove the thrust washer from the housing tube.

10. Pull the hub and drum straight off the axle housing.

11. Remove the oil seal and discard.

12. Use a hammer and a long drift to knock the inner bearing cup, and oil seal from the hub assembly.

13. Remove the outer bearing snap-ring with a pair of pliers. It may be necessary to tap the bearing outer race away from the retaining ring slightly by tapping on the ring to remove the ring.

14. Drive the outer bearing from the hub with a hammer and drift.

15. To reinstall the bearings, place the outer bearing into the hub. The larger outside diameter of the bearing should face the outer end of the hub. Drive the bearing into the hub using a washer that will cover both the inner and outer races of the bearing. Place a socket on the top of this washer, then drive the bearing into place with a series of light taps. If available, an arbor press should be used for this job.

16. Drive the bearing past the snap-ring groove, and install the snap-ring. Then turning the hub assembly over, drive the bearing back against the snap-ring. Again, protect the bearing by placing a washer on top of it. You can use the thrust washer that fits between the bearing and the adjusting nut for this job.

17. Place the inner bearing into the hub. The thick edge should be toward the shoulder in the hub. Press the bearing into the hub until it seats against the shoulder, using a washer and socket as outlined earlier. Make certain that the bearing is not cocked and that it is fully seated on the shoulder.

18. Pack the cavity between the oil seal lips with front wheel bearing grease, and position it in the hub bore. Carefully press it into place on top of the inner bearing.

19. Pack the wheel bearings with the grease, and lightly coat the inside diameter of the hub bearing contact surface and the outside diameter of the axle housing tube.

20. Make sure that the inner bearing, oil seal, axle housing oil deflector, and outer bearing are properly positioned. Install the hub and drum assembly on the axle housing, excercising care so as not to damage the oil seal or dislocate other internal components.

21. Install the thrust washer so that the tang on the inside diameter of the washer is in the keyway on the axle housing.

22. Install the adjusting nut. Tighten to 50 ft. lbs., at the same time rotating the hub to make sure that all the bearing surfaces are in contact. Back off the nut and retighten to 35 ft. lbs., then back off ¼ of a turn.

23. Install the tanged retainer against the inner adjusting nut. Align the adjusting nut so that the short tang of the retainer will engage the nearest slot on the adjusting nut.

24. Install the outer locknut and tighten to 65 ft. lbs. Bend the long tang of the retainer into the slot of the outer nut. This method of adjustment should provide 0.001–0.010 in. of end play.

25. Place a new gasket over the axle shaft and position the axle shaft in the housing so that the shaft splines enter the differential side gear. Position the gasket so that the holes are in alignment, and install the flange-to-hub attaching bolts. Torque to 155 ft. lbs.

NOTE: To prevent lubricant from leaking through the flange holes, apply a non-hardening sealer to the bolt threads. Use the sealer sparingly.

26. Install the wheels.

FRONT SUSPENSION

Two wheel drive models use coil spring independent front suspension. A stabilizer (sway) bar is optional to minimize body lean and sway in curves. Four wheel drive models have a non-independent leaf spring front suspension. A stabilizer bar is standard. A steering linkage damper is standard on late models. Heavy duty shock absorbers, springs, and stabilizer bars have been optional for most models. S–Series trucks (2WD) front suspension is similar to standard size pickups. Four wheel drive models use a torsion bar-type front suspension.

Springs

REMOVAL & INSTALLATION

NOTE: Springs, particularly coil springs, are under considerable tension. Be very careful when removing and installing them; they can exert enough force to cause very serious injuries.

2WD Models

1. Raise the vehicle and support it under the frame so that the control arms will hang free.

2. Remove the lower shock absorber mounting bolt. Detach the stabilizer bar from the lower control arm.

3. Place a floor jack under the lower control arm crosshaft.

NOTE: As a safety precaution, install a chain through the spring and lower control arm.

4. Raise the jack. This will remove the tension of the lower control arm so that the two U-bolts which secure the cross-shaft can be removed.

5. Lower the control arm slowly by releasing the floor jack to the point where the spring can be removed.

6. Remove the spring.

7. Place the spring on the control arm and then, using a jack, slowly raise the control arm. Use the safety chain as described in Step 3.

8. Place the control arm cross-shaft onto the crossmember and then install the U-bolts and the attaching nuts. Make certain that the indexing hole in the cross-shaft is lined up with the crossmember stud.

9. Torque the U-bolt nuts to 85 ft. lbs. Remove the safety chain.

10. Install the lower part of the shock absorber and the stabilizer bar.

11. Lower the vehicle.

4WD Models

1. Raise and support the truck under the front axle and frame so that the tension on the springs is relieved.

2. Remove the shackle upper retaining bolt and the front spring eye bolt.

3. Remove the spring-to-axle U-bolt nuts. Pull off the spring, the lowerplate, and the spring pads.

4. Remove the shackle-to-spring bolts, bushings and shackle. To replace the bushing, place the spring onto a press or vise and press out the bushing. Press in the new bushing. The new bushing should protrude evenly on both sides of the spring.

5. Install the spring shackle bushings into the spring and then attach the shackle. Do not tighten the bolt.

6. Place the upper spring cushion onto the spring.

7. Place the front of the spring into the frame and install the bolt but do not tighten it.

8. Position the shackle bushing into the frame and attach the rear shackle but do not tighten it.

9. Install the lower spring pad and the spring retainer plate. Tighten the U-bolts to 150 ft. lbs.

10. Torque the rear spring shackle bolts an the rear eye bolts to 50 ft. lbs. and the front eye bolts to 90 ft. lbs.

11. Lower the vehicle.

Torsion Bar (S–Series 4WD)

REMOVAL & INSTALLATION

1. Raise and safely support the vehicle.

2. Remove the torsion bar adjusting screw.

3. Remove support retainer attaching nuts and bolts.

4. Slide torsion bar forward in lower control arm until torsion bar clears support. Pull down on bar and remove from control arm.

5. Count the number of turns when removing the torsion bar for easy reinstallation. Apply lubricant to top of adjusting arm and adjusting bolt for easy reinstallation. Also apply lubricant to hex ends of torsion bar. Installation is the reverse of removal.

Shock Absorbers

The usual procedure for testing shock absorbers is to stand on the bumper at the end nearest the shock being tested and start the vehicle bouncing up and down. Step off; the vehicle should come to rest within one bounce cycle. Another good test is to drive the vehicle over a bumpy road. Bouncing over bumps is normal, but the shock absorbers should stop the bouncing, after the pump is passed, within one or two cycles.

REMOVAL & INSTALLATION

This usual procedure is to replace shock

absorbers in axle pairs, to provide equal damping. Heavy duty replacements are available for firmer control.

1. Raise and support the front axle as necessary.

2. Remove the bolt and nut from the lower shock end.

3. On two wheel drive original equipment shocks, remove the upper stud nut from inside the frame. On four wheel drive shocks, remove the upper bolt and nut.

4. Purge the new shock of air by extending it in its normal position and compressing it while inverted. Do this several times. It is normal for there to be more resistance to extension than to compression.

5. Install the shock absorber. Tighten the two wheel drive upper stud nut (inside the frame) to 8 ft. lbs. and the four wheel drive upper bolt to 65 ft. lbs. Tighten the 2WD lower bolt to 60 ft. lbs. Tighten the 4WD lower bolt to 65 ft. lbs.

Steering Knuckle S–Series

REMOVAL & INSTALLATION

1. Raise front of vehicle and support with floor stands under front lift points. Remove the wheel. Spring tension is needed to assist in breaking ball joint studs loose from steering knuckle. Do not place stands under lower control arm.

2. Remove caliper.

3. Remove hub and rotor assembly.

4. Remove the three bolts attaching shield to knuckle.

5. Remove tie-rod end from knuckle using Tool J–6627 or equivalent.

6. Carefully remove knuckle seal if knuckle is to be replaced.

7. Remove ball studs from steering knuckle using tool J–23742 or equivalent.

--- **CAUTION** ---

Floor jack must remain under control arm spring seat during removal and installation to retain spring and control arm in position.

8. Position a floor jack under lower control arm near spring seat and raise jack until it supports lower control arm.

9. Raise upper control arm to disengage ball joint stud from knuckle.

10. Raise knuckle from lower ball joint stud and remove knuckle. Inspect the tapered hole in the steering knuckle. Remove any dirt. If out-of-roundness, deformation, or damage is noted, the knuckle MUST be replaced.

11. Insert upper and lower ball joint studs into knuckle and install nuts.

12. Install shield to knuckle seal and splash shield. Torque attaching bolts to 10 ft. lbs.

13. Install tie rod end into knuckle. Install tool J–29193 or equivalent and torque

to 15 ft. lbs. Remove tool and install nut to 40 ft. lbs.

14. Replace wheel bearings. Install hub and disc assembly.

15. Adjust wheel bearings. Install caliper.

16. Install remaining parts in reverse order of removal.

Steering Knuckle

2WD Model

REMOVAL & INSTALLATION

1. Raise vehicle and support lower control arms.

2. Remove wheel.

3. Remove caliper.

4. Remove disc splash shield bolts securing the shield to the steering knuckle. Remove shield. Disconnect the tie-rod ends.

5. Remove upper and lower ball stud cotter pins and loosen ball stud nuts. Free steering knuckle from ball studs by installing Special Tool J–23742 or equivalent. Remove ball stud nuts and withdraw steering knuckle.

6. Place knuckle in position and insert upper and lower ball studs into knuckle bosses. The steering knuckle hole, ball stud and nut should be free of dirt and grease before tightening nut.

7. Install ball stud nuts and tighten nut to 80–100 ft. lbs. If necessary, tighten one more notch to insert cotter pins. Do not loosen nut to insert cotter pin.

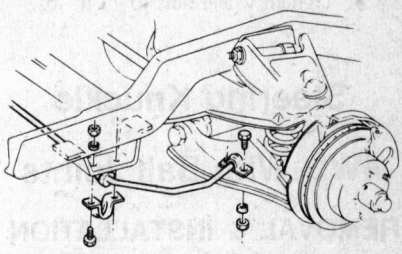

Two wheel drive coil spring and stabilizer

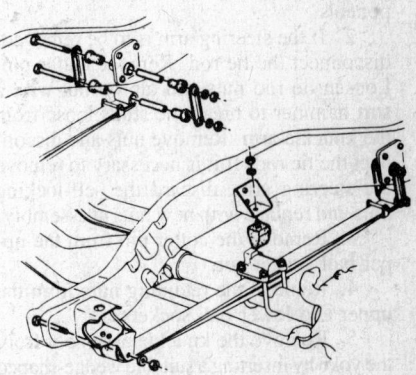

Four wheel drive front leaf spring

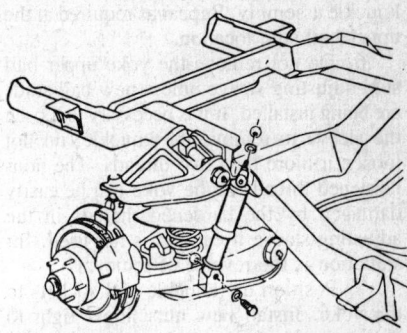

Two wheel drive front shock absorber

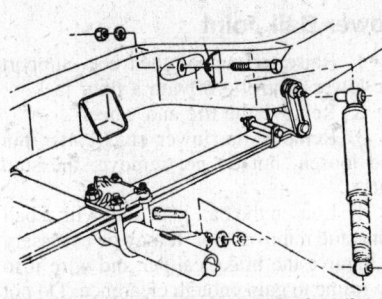

Four wheel drive front shock absorber

DO NOT BACK OFF NUT TO INSTALL NEW COTTER PIN.

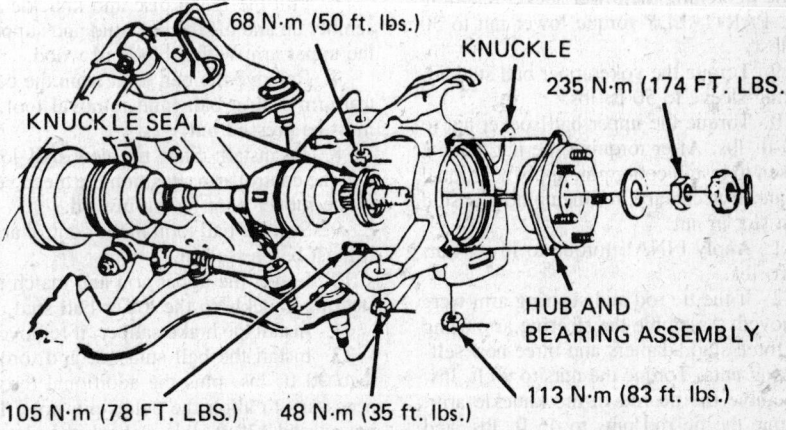

Removal and installation of hub, bearing assembly, knuckle and seal

8. Reverse remaining removal procedure, and tighten splash shield mounting bolt. Tighten two caliper assembly mounting bolts to 35 ft. lbs. torque.
4. Adjust wheel bearings.
5. Tighten wheel nuts to 75 ft. lbs.

Steering Knuckle

4WD With Ball Joints

REMOVAL & INSTALLATION

1. Remove the automatic locking hub, hub-and-disc assembly, and spindle components.
2. If the steering arm is to be removed, disconnect the tie rod. Remove cotter pin. Loosen tie rod nuts and tap on nut with a soft hammer to break the studs loose from the knuckle arm. Remove nuts and disconnect the tie rod. If it is necessary to remove the steering arm, discard the self-locking nuts and replace with new nuts at assembly.
3. Remove the cotter pin from the upper ball socket nut.
4. Remove the retaining nuts from the upper and lower ball sockets.
5. Remove the knuckle assembly from the yoke by inserting a suitable wedge-shaped tool between the lower ball stud and the yoke and tapping on the tool to release the knuckle assembly. Repeat as required at the upper ball stud location.
6. Do not remove the yoke upper ball stud adjusting sleeve unless new ball studs are being installed, If it is necessary to loosen the sleeve to remove the knuckle, do not loosen it more than two threads. The non-hardened threads in the yoke can be easily damaged by the hardened threads in the adjusting sleeve if caution is not used. Installation is the reverse of removal.
7. Position the knuckle and sockets to the yoke. Install new nuts finger tight to the upper (the nut with the cotter pin slot) and lower ball socket studs.
8. Push up on the knuckle (to keep the ball socket from turning in the knuckle) while tightening the lower socket retaining nut. PARTIALLY torque lower nut to 30 ft. lbs.
9. Torque the yoke upper ball stud adjusting sleeve to 50 ft. lbs.
10. Torque the upper ball socket nut to 100 ft. lbs. After torquing the nut, do not loosen to install cotter pin, apply additional torque, if necessary, to line up hole in stud with slot in nut.
11. Apply FINAL torque to lower nut, 70 ft. lbs.
12. If the tie rod and steering arm were removed: Assemble the steering arm using the three stud adapters and three new self-locking nuts. Torque the nuts to 90 ft. lbs. Assemble the tie rod to the knuckle arm. Torque the tie rod nuts to 45 ft. lbs. and install cotter pin.

Ball Joints

Service procedures for 4WD models are found under Steering Knuckle.

INSPECTION

Excessive ball joint wear will usually show up as wear on the inside of the front tires. Don't jump to conclusions; front end misalignment can give the same symptom.

Upper Ball Joint

1. Raise and support the truck; let the control arms hang freely.
2. Measure the distance between the tip of the ball joint stud and the tip of the grease fitting below the ball joint.
3. Move the support to the control arm and allow the wheel and tire to hang free. Measure the distance again. If the vibration between the two measurements exceed $\frac{3}{32}$ in. the ball joint should be replaced.

NOTE: This is the manufacturer's recommended wear limit. Your state inspection regulations may disagree. Follow the state regulations if they are more strict.

REMOVAL & INSTALLATION

NOTE: Observe the Caution under Front Spring Removal and Installation when working with ball joints.

Lower Ball Joint

1. Raise and support the truck. Support the lower control arm with a floor jack.
2. Remove the tire and wheel.
3. Remove the lower stud cotter pin and loosen, but do not remove, the stud nut.
4. Loosen the ball joint stud with a ball joint stud removal tool. It may be necessary to remove the brake caliper and wire it to the frame to gain enough clearance. Do not let the caliper hang by the hose.
5. When the stud is loose, remove the tool and ball stud nut.
6. Pull the brake disc and knuckle assembly up and off the ball stud and support the upper arm with a block of wood.
7. Remove the ball joint from the control arm with a ball joint removal tool. It must be pressed out.
8. To install: Start the new ball joint into the control arm. Position the bleed vent in the rubber boot facing inward.
9. Seat the ball joint in the control arm. It must be pressed in.
10. Lower the upper arm and match the steering knuckle to the lower ball stud.
11. Install the brake caliper, if removed.
12. Install the ball stud nut and torque it to 90 ft. lbs. plus the additional torque necessary to align the cotter pin hole. Do not exceed 130 ft. lbs. or back the nut off to align the holes with the pin.

13. Install a new lube fitting and lubricate the new joint.
14. Install the tire and wheel.
15. Lower the truck.

Upper Ball Joint

1. Raise and support the truck.
2. Support the lower control arm with a floor jack.
3. Remove the cotter pin from the upper ball stud and loosen, but do not remove the stud nut.
4. Using a ball joint stud removal tool, loosen the ball stud in the steering knuckle. When the stud is loose, remove the tool and the stud nut. It will be necessary to remove the brake caliper and wire it to the frame to gain clearance. Do not allow the caliper to hang by the hose.
5. Drill out the rivets. Remove the ball joint assembly.
6. Install the service ball joint, using the nuts supplied. Tighten the nuts to 45 ft. lbs.
7. Torque the ball stud nut to 50 ft. lbs. plus the additional torque required to align the cotter pin. Do not exceed 90 ft. lbs. and never back the nut off to align the pin.
8. Install a new cotter pin.
9. Install a new lube fitting and lubricate the new joint.

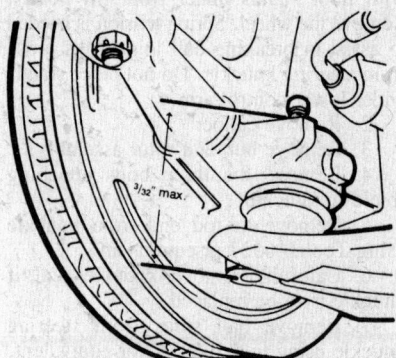

Lower ball joint inspection

10. If removed, install the brake caliper.
11. Install the wheel and tire and lower the truck.

Upper Control Arm Except S–Series

REMOVAL & INSTALLATION

1. Raise and support the truck on jackstands.
2. Support the lower control arm with a floor jack.
3. Remove the wheel and tire.
4. Remove the cotter pin from the upper control arm ball stud and loosen the stud nut one turn.
5. Loosen the upper control arm ball stud in the steering knuckle using a ball joint stud removal tool. Remove the nut

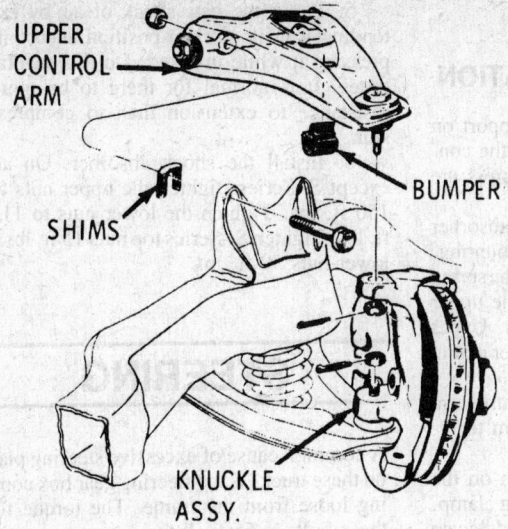

Upper control arm

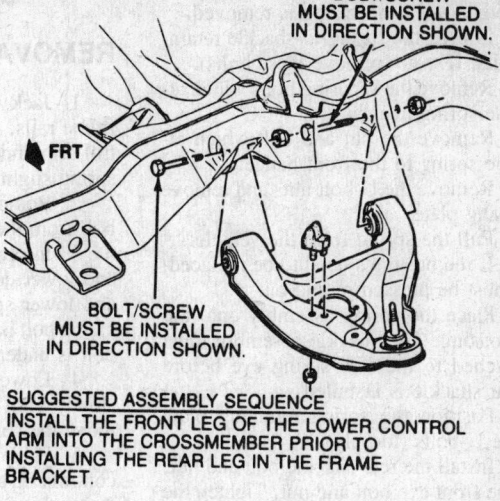

BOLT/SCREW
MUST BE INSTALLED
IN DIRECTION SHOWN.

FRT

BOLT/SCREW
MUST BE INSTALLED
IN DIRECTION SHOWN.

SUGGESTED ASSEMBLY SEQUENCE
INSTALL THE FRONT LEG OF THE LOWER CONTROL
ARM INTO THE CROSSMEMBER PRIOR TO
INSTALLING THE REAR LEG IN THE FRAME
BRACKET.

Lower control arm attachment

from the ball stud and raise the upper arm to clear the steering knuckle. It will be necessary to remove the brake caliper and wire it to the frame to gain clearance.

6. Remove the nuts securing the control arm shaft to the frame and remove the control arm. Tape the shims and spacers together and tag for proper reassembly.

7. Installation is the reverse of removal. Place the control arm in position and install the nuts. Before tightening the nuts to 70 ft. lbs., insert the caster and camber shims in the same order as when removed. Have the front end alignment checked, and as necessary, adjusted.

Lower Control Arm Except S–Series

REMOVAL & INSTALLATION

1. Raise and support the truck on jackstands.
2. Remove the spring. See Spring Removal and Installation.
3. Support the inboard end of the control arm after spring removal.
4. Remove the cotter pin from the lower ball stud and loosen the nut one turn.
5. Loosen th lower ball stud in the steering knuckle using a ball joint stud removal tool. When the stud is loose, remove the nut from the stud. It will be necessary to remove the brake caliper and wire it to the frame to gain clearance.
6. Remove the lower control arm.
7. Installation is the reverse of removal.

Upper Control Arm S–Series

REMOVAL & INSTALLATION

1. Note the location of the shims.

Alignment shims are to be installed in the same position from which they were removed. Remove nuts and shims. Raise front of vehicle and support lower control arm with floor stands. The floor jack must remain under control arm spring seat during removal and installation to retain spring and control arm in position. Since the weight of the vehicle is used to relieve spring tension on the upper control arm, the floor stands must be positioned between the spring seats and ball joints of the lower control arms for maximum leverage.

2. Remove wheel, then loosen the upper ball joint from the steering knuckle.
3. Support hub assembly to prevent weight from damaging brake hose.
4. It is necessary to remove the upper control arm attaching bolts to allow clearance to remove upper control arm assembly.
5. Remove upper control arm.
6. Position upper control arm attaching bolts loosely in the frame and install pivot shaft on the attaching bolts. Note that the inner pivot bolts must be installed with the bolt heads to the front (on the front bushing) and to the rear (on the rear bushing).
7. Install alignment shims in their original position between the pivot shaft and frame on their respective bolts. Torque nuts to 45 ft. lbs.
8. Remove the temporary support from the hub assembly, then connect ball joint to steering knuckle.
9. Install wheel, then check wheel alignment, and adjust if necessary.

Lower Control Arm S–Series

REMOVAL & INSTALLATION

1. Remove coil spring.
2. Remove lower ball joint and stud.
3. After stud breaks loose, hold up on

lower control arm. Remove control arm.

4. Guide lower control arm out of opening in splash shield with a putty knife or similar tool.
5. Install lower ball joint stud into knuckle and tighten nut.
6. Install spring.
7. Check front alignment. Reset as required.

Stabilizer Bar S–Series

REMOVAL & INSTALLATION

1. Raise and support the vehicle.
2. Disconnect each side of stabilizer linkage by removing nut from link bolt, pull bolt from linkage and remove retainers, grommets and spacer.
3. Remove bracket to frame or body bolts and remove stabilizer shaft, rubber bushings and brackets.
4. To replace, reverse sequence of operations, being sure to install with the identification forming on the right side of the vehicle. The rubber bushings should be positioned squarely in the brackets with the slit in the bushings facing the front car. Torque stabilizer link nut to 13 ft. lbs. and the bracket bolts to 24 ft. lbs.

REAR SUSPENSION

Leaf Springs—Except S–Series

REMOVAL & INSTALLATION

1. Raise the vehicle and support it on the frame so that the springs will hang.

Position a floor jack under the rear to hold the weight when the spring is removed.

2. Loosen the spring-to-shackle retaining bolts. (Do not remove these bolts).

3. Remove the securing bolts which attach the spring hanger.

4. Remove the nut and bolt which attach the spring to the front hanger.

5. Remove the U-bolt nuts and remove the spring plate.

6. Pull the spring from the vehicle.

7. If the bushings need to be replaced, they must be pressed in and out.

8. Place the spring assembly onto the axle housing. The shackle assembly must be attached to the rear spring eye before the rear shackle is installed.

9. Position the spring retaining plate and the U-bolts (loosely).

10. Install the rear shackle bolt and nut, then the front eye bolt and nut. Tighten the bolts to 110 ft. lbs.

S–Series Trucks

1. Raise vehicle and support the body chassis and axle separately, so that the load on the spring is relieved.

2. Loosen, but do not remove, spring-to-shackle retaining nut.

3. Remove the U-bolt retaining nuts, withdraw the U-bolts, and rotate the spring anchor plate, on the shock absorber, to clear the spring.

4. Remove the nut and bolt securing the shackle to the frame. Be careful to restrain the spring, since the spring is now free to rotate about the front hanger bolt.

5. Remove the nut and bolt at the front hanger, then remove the spring from the vehicle.

6. Clean axle spring pad.

7. Attach spring to vehicle at the front hanger by installing the nut and bolt. Do not apply final torque at this time.

8. Install the shackle-to-frame attaching nut and bolt, but do not apply final torque at this time. Be sure that: The shackle is loosely attached to the rear spring eye before attaching the shackle to the frame, and the shackle must be positioned with its open end toward the front of the vehicle and the axle must be in position above the spring before attaching the shackle to the frame.

9. Position axle spring pad onto the spring so that the center bolt head is seated in the pilot hole of the spring pad seat.

10. Rotate the anchor plate, on shock absorber, underneath the spring assembly and install the U-bolts. Align the anchor plate, install the retaining nuts evenly (handtight), then tighten diagonally opposite nuts to 34 ft. lbs.

11. Lower the vehicle so that the weight of the vehicle is supported by the suspension components. Torque the U-bolt nuts, spring eye bolt nuts (front and rear), and shackle-to-frame bolt nuts to 85 ft. lbs.

12. Lower the vehicle.

Coil Springs

REMOVAL & INSTALLATION

1. Jack up the vehicle and support on frame rails. Position a jack under the control arm and raise enough to compress the spring slightly.

2. Remove the lower shock absorber bolt from the lower control arm mounting.

3. Insert a safety chain through the spring and lower control arm. Remove the upper and lower spring retaining clamps. Upper clamp bolt is inside the spring, lower clamp bolt is under the control arm.

4. Lower the jack under the control arm slowly until there is suffcient room to remove the spring.

5. Place the spring in position on the control arm and install the mounting clamp. Raise the control arm slightly and locate the upper clamp. Install the spring in the reverse order of removal.

Shock Absorbers

REMOVAL & INSTALLATION

The usual procedure is to replace shock absorbers in axle pairs, to provide equal damping. Heavy duty replacements are available for firmer control. Air adjustable shock absorbers can be used to maintain a level ride with heavy loads or when towing.

1. Raise and support the rear axle as necessary.

2. If air shocks are installed, bleed the air and detach the lines.

3. Remove the nut and washer at the top.

4. Remove the nut, washer, and bolt at the bottom.

5. Purge the new shock of air by extending it in its normal position and compressing it while inverted. Do this several times. It is normal for there to be more resistance to extension than to compression.

6. Install the shock absorber. On all except S–Series, tighten the upper nuts to 150 ft. lbs. Tighten the lower nuts to 114 ft. lbs. Tighten S–Series top nuts 15 ft. lbs., lower nuts 50 ft. lbs.

STEERING

A common cause of excessive steering play on these trucks is the steering gear box coming loose from the frame. The torque for these bolts is 65 ft. lbs.

Steering Wheel

REMOVAL & INSTALLATION

Except S–Series

1. Disconnect the battery ground cable.

2. Remove the horn button. Remove the receiving cup, bellville washer, and bushing (if equipped).

3. Mark the steering wheel-to-steering shaft relationship.

4. Remove the snap-ring from the steering shaft.

5. Remove the nut and washer from the steering shaft.

6. Remove the steering wheel with a puller.

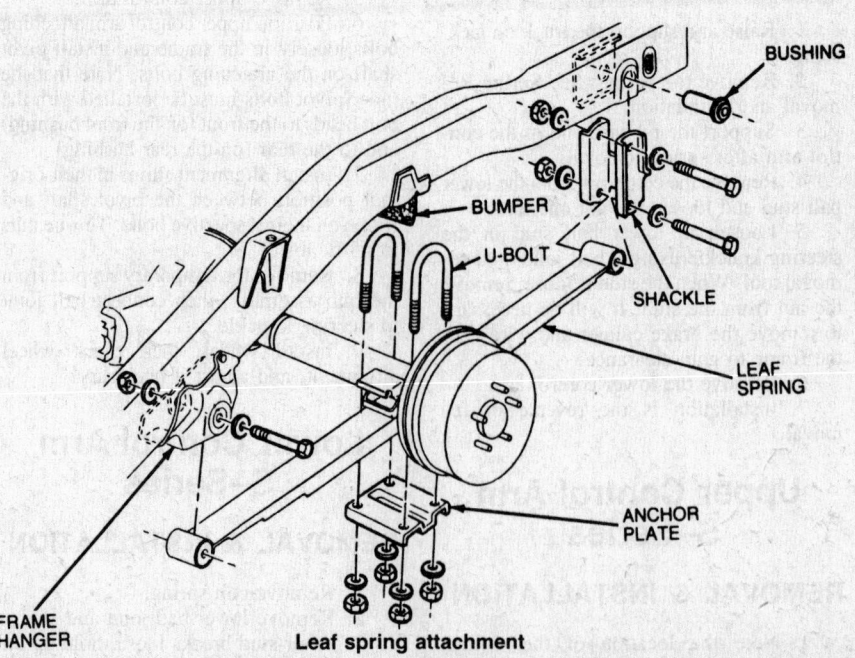

BUSHING

BUMPER

U-BOLT

SHACKLE

LEAF SPRING

ANCHOR PLATE

FRAME HANGER

Leaf spring attachment

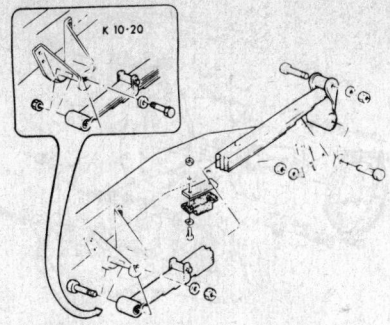

Typical leaf spring installation

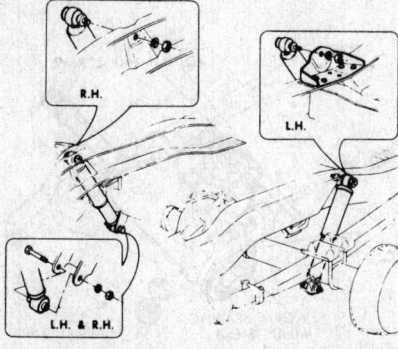

Staggered rear shock absorber details

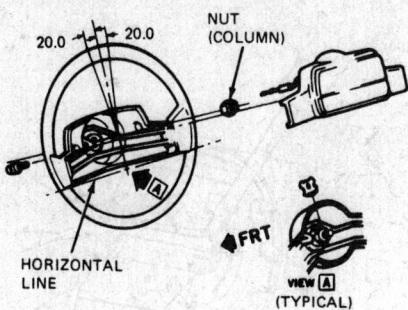

Steering wheel—typical

---CAUTION---
Don't hammer on the steering shaft.

7. Installation is the reverse of removal. The turn signal control assembly must be in the neutral position to prevent damaging the cancelling cam and control assembly. Tighten the nut to 30 ft. lbs.

S–Series

1. Disconnect battery ground cable.
2. Remove steering wheel shroud screws on underside of steering wheel.
3. Lift steering wheel shroud and horn contact lead assembly from the steering wheel.
4. Remove snap ring.
5. Remove steering wheel nut.
6. Using Tool J–2927, thread puller anchor screws into threaded holes provided in steering wheel. Turn center bolt of tool clockwise (butting against the steering shaft) to remove the steering wheel.

NOTE: Do not hammer on puller while turning. The tool centering adapters need not be installed.

7. Installation is the reverse of removal. Torque steering wheel nut 30 ft. lbs. Do not over-torque shaft nut or steering wheel rub may result.

Turn Signal Switch

REPLACEMENT

Except S–Series

1. Remove the steering wheel
2. Remove the column to instrument panel trim cover. Loosen the three cover screws and lift the cover off the shaft or place a screwdriver in the cover slot and pry out to free the cover, depending on year and model.
3. The round lockplate must be pushed down to remove the wire snap-ring from the shaft. A special tool is available to do this. The tool is an inverted U-shape with a hole for the shaft. The shaft nut is used to force it down. Pry the wire snap-ring out of the shaft groove.
4. Remove the tool and lift the lockplate off the shaft.
5. Slip the canceling cam, upper bearing preload spring, and thrust washer off the shaft.
6. Remove the turn signal lever screw and the lever. Push the flasher knob in and unscrew it.
7. Pull the switch connector out of the mast jacket and tape the upper part to facilitate switch removal. On tilt wheels, place the turn signal and shifter housing in the low position and remove the harness cover.
8. Remove the three switch mounting screws. Remove the switch by pulling it straight up while guiding the wiring harness cover through the column.
9. Install the replacement switch by working the connector and cover down through the housing and under the bracket. On tilt models, the connector is worked down through the housing, under the bracket, and then cover is installed on the harness.
10. Install the switch mounting screw and the connector on the mast jacket bracket. Install the column to instrument panel trim plate.
11. Install the flasher knob and the turn signal lever.
12. With the turn signal lever in neutral and the flasher knob out, slide the thrust washer, upper bearing pre-load spring, and canceling cam onto the shaft.
13. Position the lockplate on the shaft and press it down until a new snap-ring can be inserted in the shaft groove.
14. Install the cover and the steering wheel.

S–Series

1. Remove the steering wheel.

2. Pry out the steering shaft lock cover.
3. Remove the retaining ring and shaft lock.
4. Remove the canceling cam and spring.
5. Remove the switch actuator arm, unscrew and remove the switch and unplug the wire connector.
6. Installation is the reverse of removal.

Ignition Switch and Lock Cylinder

REMOVAL & INSTALLATION

Except S–Series

1. Remove the steering wheel and the turn signal switch as previously outlined. It is not necessary to completely remove the turn signal switch. Just pull the switch out far enough so it can hang out of the steering column shift. Do not disconnect the wiring harness.
2. With the lock cylinder in the Run position, remove the lock cylinder attaching screw and the cylinder.
3. To install, align the cylinder key with the keyway in the housing and rotate the key all the way clockwise while holding the cylinder body.
4. Insert the cylinder into the housing and install the attaching screw.
5. Install the turn signal switch and the steering wheel as previously outlined.

S–Series Trucks

1. Disconnect the battery ground.
2. Turn the lock to the RUN position.
3. Remove the turn signal switch.
4. Remove the cylinder retaining screw.
5. Pull out the cylinder.
6. Installation is the reverse of removal. Turn the cylinder to the STOP position while installing.

Ignition Switch

REMOVAL & INSTALLATION

Except S–Series

1. The switch is on the steering column, behind the instrument panel. Lower the steering column, making sure that it is supported. Extreme care is necessary to prevent damage to the collapsible column.
2. Make sure the switch is in the Lock position. If the lock cylinder is out, pull the switch rod up to the stop, then go down one detent.
3. Remove the two screws and the switch.
4. Before installation, make sure the switch is in the lock position. The switch can be moved to the Lock position using a screwdriver inserted into the locking rod slot.

5. Install the switch using the original screws. Use of screws that are too long could prevent the column from collapsing on impact.

6. Reinstall the column.

S–Series

1. Remove the steering wheel, lock cylinder, turn signal switch, shift lever, shift lever bowl, shift bowl shroud and bowl lower bearing.

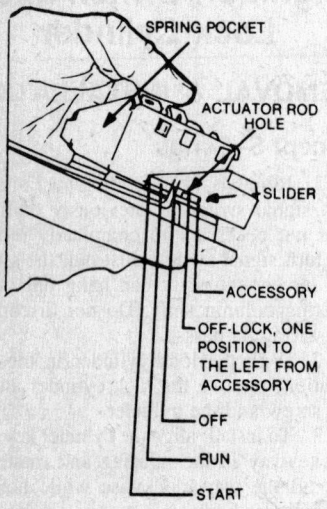

Ignition switch details

2. Unbolt and remove the ignition switch and dimmer switch from the column.

3. For installation, on all columns except key release type, move the switch slider to the extreme left position, then two detents right. this is the OFF-LOCK position. On key release columns, leave the slider in the extreme left position. This is ACCESSORIES. To adjust the dimmer switch, depress the switch slightly to allow insertion of a $3/32$ in. drill bit into the hole above the actuator rod. Force the switch upward to take up any lash, then tighten the screw. On tilt columns, the ACC position on the ignition switch is the extreme right position. On these columns, move the slider two detents left to the OFF-LOCK position.

Power Steering Pump

REMOVAL & INSTALLATION

1. Disconnect the hoses at the pump. When the hoses are disconnected, secure the ends in a raised position to prevent leakage. Cap the ends of the hoses to prevent the entrance of dirt.

2. Cap the pump fittings.

3. Loosen the bracket-to-pump mounting nuts.

4. Remove the pump drive belt.

5. Remove the bracket-to-pump bolts and remove the pump from the truck.

6. Installation is the reverse of removal. Fill the reservoir and bleed the pump by turning the pulley counterclockwise (as viewed from the front) until bubble stop forming. Bleed the system. Adjust the belt tension.

System Bleeding

1. Fill the reservoir to the proper level and let the fluid remain undisturbed for at least 2 minutes.

2. Start the engine and run it for only about 2 seconds.

3. Add fluid as necessary.

4. Repeat Steps 1–3 until the level remains constant.

5. Raise the front of the vehicle so that the front wheels are off the ground. Set the parking brake and block both rear wheels front and rear. Manual transmission should be in Neutral; automatic transmission should be in Park.

6. Start the engine and run it at approximately 1,500 rpm.

7. Turn the wheels (off the ground) to the right and left, lightly contacting the stops.

8. Add fluid as necessary.

9. Lower the vehicle and turn the wheels right and left on the ground.

10. Check the level and refill as necessary.

11. If the fluid is extremely foamy, let the truck stand for a few minutes with the engine off and repeat the above procedure. Check the belt tension and check for a bent or loose pulley. The pulley should not wobble with the engine running.

12. Check that no hoses are contacting any parts of the truck, particularly sheet metal.

13. Check the level and refill as necessary. This step and the next are very important. When filling, follow Steps 1–10.

14. Check for air in the fluid. Aerated fluid appears milky. If air is present, repeat the above operations. If it is obvious that the pump will not respond to bleeding after several attempts, a pressure test may be required.

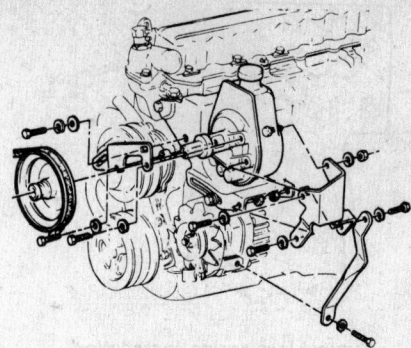

6-250 power steering pump mounting

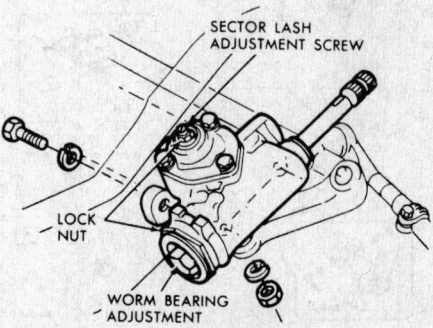

Steering gear adjustment points

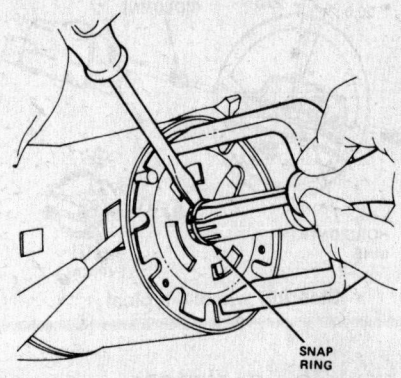

Removing the lockplate retaining ring

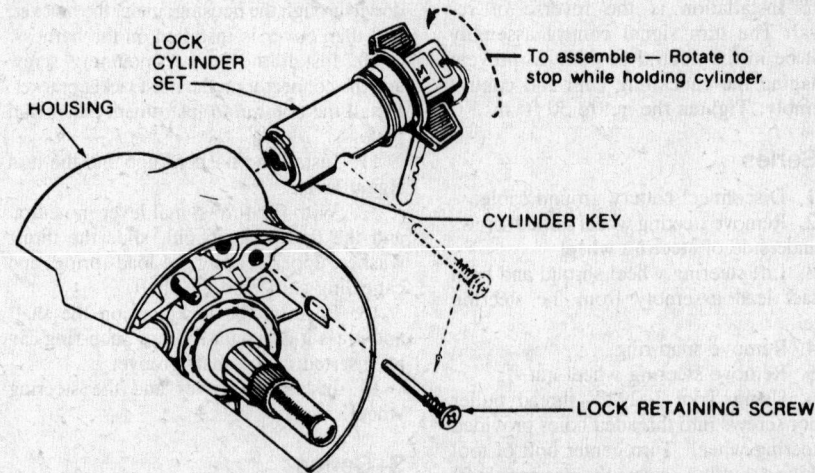

Typical lock cylinder removal

...

CLAMPING INSTRUCTIONS FOR
ALL LIGHT TRUCKS.

A. All bolts must be installed in direction shown.
B. Rotate both inner and outer tie rod sockets rearward to limit of ball stud travel.
C. Position clamps within angles shown.
D. Tighten clamps.
E. With this same rearward rotation, all bolt centerlines must be between angles shown.

CAUTION: CLAMPS MUST BE BETWEEN AND CLEAR OF DIMPLES BEFORE TORQUING NUT.

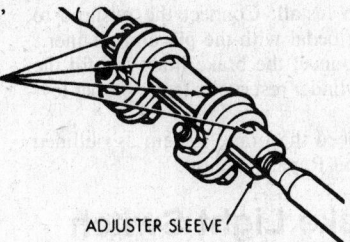

ADJUSTER SLEEVE

FWD

VERTICAL LINE
ADJUSTER SLEEVE
CLAMP
45°

CAUTION: CENTERLINE OF SLOT IN CLAMP MUST BE IN THIS RANGE OF ADJUSTMENT.

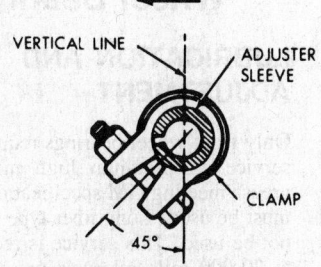

NOTE: IMPORTANT - SLOT IN ADJUSTER SLEEVE MUST NOT BE WITHIN OPEN AREA OF CLAMP JAWS OR CLOSER THAN .10 TO THE EDGE OF CLAMP JAW OPENING. ROTATE CLAMP TO MEET REQUIREMENTS WITHIN PROPER POSITION AS SHOWN.

Tie rod sleeve clamp installation

Power Steering Gear

ADJUSTMENTS

Refer to the Unit Repair Section.

REMOVAL & INSTALLATION

1. Disconnect hoses at gear. When hoses are disconnected, secure ends in raised position to prevent drainage of oil Cap or tape the ends of the hoses to prevent entrance of dirt.
2. Install two plugs in gear fittings to prevent entrance of dirt.
3. Remove the flexible coupling to steering shaft flange bolts. Mark the relationship of the universal yoke to the stub shaft.
4. Mark the relationship of the Pitman arm to the Pitman shaft. Remove the Pitman shaft nut or Pitman arm from the Pitman shaft using puller J-6632.
5. Remove the steering gear to frame bolts and remove the gear assembly.
6. On G, C, and K models, remove the flexible coupling pinch bolt and remove the coupling from the steering gear stub shaft.
7. Install the flexible coupling onto the steering gear stub shaft, aligning the flat in the coupling with the flat on the shaft. Push the coupling onto the shaft until the stub shaft bottoms on the coupling reinforcement. Install the pinch bolt.

NOTE: **The coupling bolt must pass through the shaft undercut, or damage to the components could occur.**

8. Place the steering gear in position, guiding the coupling bolt into the steering shaft flange.
9. Install the steering gear to frame bolts.
10. If flexible coupling alignment pin plastic spacers were used, make sure they are buttoned on the pins, tighten the flange bolt nuts and then remove the plastic spacers.
11. If flexible coupling alignment pin plastic spacers were not used, center the pins in the slots in the steering shaft flange and then install and torque the flange bolt nuts.
12. Install the Pitman arm onto the Pitman shaft, lining up the marks made at removal. Install the Pitman shaft nut.
13. Remove the plugs and caps from the steering gear and hoses and connect the hoses to the gear. Tighten the hose fittings.

Manual Steering Gear

ADJUSTMENTS

Refer to the Unit Repair Section.

REMOVAL & INSTALLATION

1. Set the front wheels in straight ahead position by driving vehicle a short distance on a flat surface.

2. Remove the flexible coupling to steering shaft flange bolts. Mark the relationship of the universal yoke to the wormshaft.
3. Mark the relationship of the Pitman arm to the Pitman shaft. Remove the Pitman shaft nut or Pitman arm pinch bolt and then remove the Pitman arm from the Pitman shaft, using puller J-6632.
4. Remove the steering gear to frame bolts and remove the gear assembly.
5. Remove the flexible coupling pinch bolt and remove the coupling from the steering gear wormshaft.
6. Install the flexible coupling onto the steering gear wormshaft, aligning the flat in the coupling with the flat on the shaft. Push the coupling onto the shaft until the wormshaft bottoms on the coupling reinforcement. Install the pinch bolt and torque to 24 ft. lbs. The coupling bolt must pass through the shaft undercut.
7. Place the steering gear in position, guiding the coupling bolt into the steering shaft flange.
8. Install the steering gear to frame bolts and torque to 70 ft. lbs. or 55 ft. lbs. for S-Series trucks.
9. If flexible coupling alignment pin plastic spacers were used, make sure they are bottomed on the pins, torque the flange bolt nuts to 25 ft. lbs., and then remove the plastic spacers.
10. If flexible coupling alignment pin plastic spacers were not used, center the pins in the slots in the steering shaft flange and then install and torque the flange bolt nuts to 25 ft. lbs.
11. Install the Pitman arm onto the Pitman shaft, lining up the marks made at removal. Install the Pitman shaft nut torque to 185 ft. lbs.

Tie Rod Ends

REMOVAL & INSTALLATION

1. Raise the front of the truck and support it safely on jackstands.
2. Remove the tie rod end stud cotter pin and nut.
3. Use a tie rod end ball joint removal to loosen the stud.
4. Remove the inner stud in the same way.
5. Loosen the tie rod adjuster sleeve clamp nuts.
6. Unscrew the tie rod end from the threaded sleeve. The threads may be left or right hand threads. Count the number of turns required to remove it.
7. To install, grease the threads and turn the new tie rod end in as many turns as were needed to remove it. This will give approximately correct toe-in. tighten the clamp bolts.
8. Tighten the stud nuts to 45 ft. lbs. and install new cotter pins. You may tighten the nut to align the cotter pin, but don't loosen it.
9. Adjust the toe-in.

Relay Rod

REMOVAL & INSTALLATION

1. Raise and safely support the vehicle.
2. Disconnect the inner ends of tie rods from relay rod.
3. Remove the nut from relay and rod ball stud attachment at Pitman arm.
4. Detach relay rod from pitman arm by using tool such as J–24319–01. Shift steering linkage as required to free pitman arm from relay rod.
5. Remove nut from idler arm and remove relay rod from idler arm.
6. Installation is the reverse of removal. Torque the nuts to 40 ft. lbs. Adjust toe-in if necessary.

Idler Arm

REMOVAL & INSTALLATION

1. Raise and safely support the vehicle.
2. Remove idler arm to frame nuts, washers, and bolts.
3. Remove nut from idler arm to relay rod ball stud.
4. Disconnect relay rod from idler arm by using J–24319–01 or similar puller.
5. Remove idler arm.
6. Installation is the reverse of removal. Torque nuts to 30 ft. lbs.

BRAKE SYSTEM

Refer to the Unit Repair Section.

ADJUSTMENT

Rear Drum Brakes

These brakes equipped with self-adjusters and no manual adjustment should be necessary, except when brake linings are replaced.

Front Disc Brakes

These brakes are inherently self-adjusting and no adjustment is ever necessary or possible.

Master Cylinder

REMOVAL & INSTALLATION

NOTE: Clean any master cylinder parts in alcohol or brake fluid. Never use mineral based cleaning solvents such as gasoline, kerosene, carbon-tetrachloride, acetone, or paint thinner as these will destroy rubber parts.

1. Using a clean cloth, wipe the master cylinder an its lines to remove excess dirt and then place cloths under the unit to absorb spilled fluid.

2. Remove the hydraulic lines from the master cylinder and plug the outlets to prevent the entrance of foreign material.
3. Disconnect the brake pushrod from the brake pedal on non-powered brakes.
4. Remove the attaching bolts and remove the master cylinder from the firewall or the brake booster.
5. To install: Connect the pushrod to the brake pedal with the pin and retainer.
6. Connect the brake lines and fill the master cylinder reservoirs to the proper levels.
7. Bleed the brake system as outlined in the Unit Repair Section.

Brake Light Switch

ADJUSTMENT

With pedal in fully released position, the

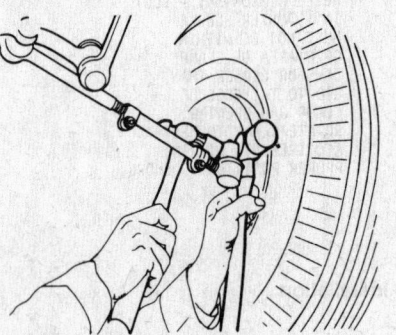

Freeing the tie rod end

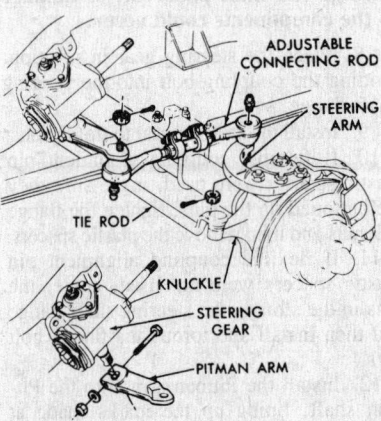

Four wheel drive steering linkage

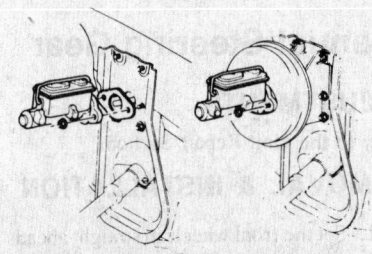

Typical master cylinder installations

stop light switch plunger should be fully depressed against the pedal shank. Adjust switch by moving in or out as necessary.

1. Make certain that the tubular clip is in brake pedal mounting bracket.
2. With brake pedal depressed, insert switch into tubular clip until switch body seats on clip. Audible clicks can be heard as the threaded portion of the switch is pushed through the clip toward the brake pedal.
3. Pull brake pedal fully rearward against pedal stop until audible clicking sounds can no longer be heard. Switch will be moved in tubular clip providing adjustment.
4. Release brake pedal and then repeat Step 3 to assure that no audible clicking sounds remain.

Wheel Bearings

LUBRICATION AND ADJUSTMENT

Only front wheel bearings require periodic service. A premium high melting point grease meeting GM specifications 6031–M must be used. Long fiber type grease must not be used. This service is recommended at 20,000 mile intervals or whenever the truck has been driven in water up to the hubs.

1. Remove the wheel and tire assembly, and the brake drum or brake caliper.
2. Remove the hub and disc as an assembly. Remove the caliper mounting bolts and insert a block between the brake pads as the caliper is removed. Remove the caliper and wire it out of the way. Do not allow the caliper to hang by the brake hose.
3. Pry out the grease cap, cotter pin, spindle nut, and washer, then remove the hub. Do not drop the wheel bearings.
4. Remove the outer roller bearing assembly from the hub. The inner bearing assembly will remain in the hub and may be removed after prying out the inner seal. Discard the seal.
5. Clean all parts in solvent (air dry) and check for excessive wear or damage.
6. Using a hammer and drift, remove the bearing caps from the hub. When installing new cups, make sure that they are not cocked and that they are fully seated against the hub shoulder.
7. Pack both wheel bearings using high melting point wheel bearing grease made fro disc brakes. Ordinary grease will melt and ooze out, ruining the pads. Place a healthy glob of grease in the palm of one hand and force the edge of the bearing into it so that the grease fills the bearing. Do this until the whole bearing is packed. Grease packing tools are available to make this job a lot less messy. There are also tools which make it possible to grease the inner bearing without removing it or the disc from the spindle.
8. Place the inner bearing in the hub and install a new inner seal, making sure that the seal flange faces the bearing cup.

9. Carefully install the wheel hub over the spindle.

10. Using your hands, firmly press the outer bearing into the hub. Install the spindle washer and nut.

11. To adjust the bearings, spin the wheel hub by hand and tighten the nut till it is just snug (12 ft. lbs.). Back off the nut till it is loose, then tighten it finger tight. Loosen the nut until either hole in the spindle lines up with a slot in the nut and insert a new cotter pin. There should be 0.001–0.005 in., end-play. this can be measured with a dial indicator, if you wish.

12. Replace the dust cap, wheel and tire.

Parking Brakes

ADJUSTMENT

Before attempting parking brake adjustment, make sure that the rear brakes are fully adjusted by making several stops in reverse.

Except S–Series

1. Raise and support the rear axle. Release the parking brake.
2. Apply the pedal four clicks.
3. Adjust the cable equalizer nut under the truck until a moderate drag can be felt when the rear wheels are turned forward.
4. Release the parking brake and check that there is no drag when the wheels are turned forward.

NOTE: If the parking brake cable is replaced, prestretch it by applying the parking brake hard about three times before attempting adjustment.

S–Series

Adjustment of parking brake system is necessary whenever the parking brake cables have been disconnected.

1. Set parking brake pedal at specified number of clicks: 2WD pickups—8 clicks. 4WD pickups—10 clicks. All Blazers—10 clicks
2. Raise and suitably support vehicle.
3. Place a properly calibrated cable tension gauge on the designated rear cable as close to the equalizer as practical. LH cable on 2WD pickups. RH cable on 4WD pickups. RH cable on all Blazers
4. Drive the adjusting nut until specified tension is indicated on gauge. 200–220 lbs. for 2WD pickups. 140–150 lbs. for 4WD pickups and 140–150 lbs. on Blazers
5. Cables are not to be adjusted too tightly, as this will result in brake drag.
6. Remove support and lower vehicle.

CABLE REPLACEMENT

Except S–Series
FRONT CABLE

1. Raise and safely support the vehicle.

2. Remove adusting nut from equalizer.
3. Remove retainer clip from rear portion of front cable at frame and from lever arm.
4. Disconnect front brake cable from parking brake pedal or lever assemblies. Remove front brake cable. On some models, it may assist installation of new cable if a heavy cord is tied to the other end of cable in order to guide new cable through proper routing.
5. Install cable by reversing removal procedure.
6. Adjust parking brake.

CENTER CABLE

1. Raise and safely support the vehicle.
2. Remove adjusting nut from equalizer.
3. Unbolt connector at each end and disengage hooks and guides.
4. Install new cable by reversing removal procedure.
5. Adjust parking brake.
6. Apply parking brake 3 times with heavy pressure and repeat adjustment.

REAR CABLE

1. Raise and safely support the vehicle.
2. Remove rear wheel and brake drum.
3. Loosen adjusting nut at equalizer.
4. Disengage rear cable at connector.
5. Bend retainer fingers.
6. Disengage cable at brake shoe operating lever.
7. Install new cable by reversing removal procedure.

S–Series
FRONT CABLE

1. Raise and suitably support the vehicle.
2. Loosen adjuster nut and disconnect front cable from connector. Compress retainer fingers and loosen at frame.
3. Remove support and lower vehicle.
4. Remove windshield washer bottle.
5. Disconnect cable from parking brake pedal assembly, compress retainer fingers and remove cable.
6. Install cable by reversing procedure.
7. Adjust parking brake.

CENTER CABLE

1. Raise and suitably support the vehicle.
2. Remove adjuster nut at equalizer and pull cable from equalizer.
3. Disconnect cable at retainers.
4. Install cable by reversing procedure and adjust parking brake.

LEFT/RIGHT REAR

1. Raise and suitably support the vehicle.
2. Mark relationship of tire and wheel assembly to axle flange and remove.
3. Remove brake drum.
4. Loosen equalizer and disconnect cable at center retainer.

5. Compress plastic retainer fingers and remove retainer from frame bracket.
6. Remove rear brake shoe and disconnect cable.
7. Remove rear brake shoe and disconnect cable.
8. Install cable by reversing procedure. Make sure cable is routed properly and securely retained.
9. Adjust parking brake cable and lower vehicle.

CHASSIS ELECTRICAL

Blower—Except Vans and S–Series

REMOVAL & INSTALLATION

1. Disconnect the battery leads, open the hood, and securely support it.
2. Mark the position of the blower motor in relation to its case. Remove the electrical connection at the motor.
3. Remove the blower attaching screws and remove the assembly. Pry gently on the flange if the sealer sticks.
4. The blower wheel can be removed from the motor shaft by removing the nut at the center.
5. Assemble the blower wheel to the motor with the open end of the wheel away from the motor and install the unit into the blower case. Connect the ground strap and the electrical connection.

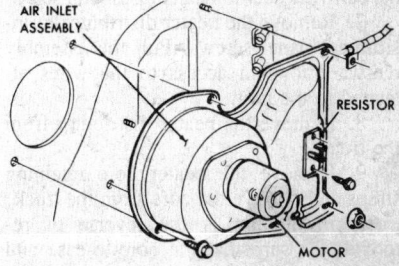

Heater blower assembly—except vans

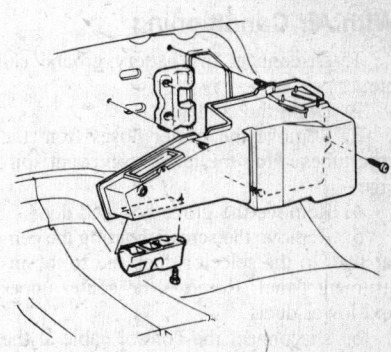

Heater distributor—except vans

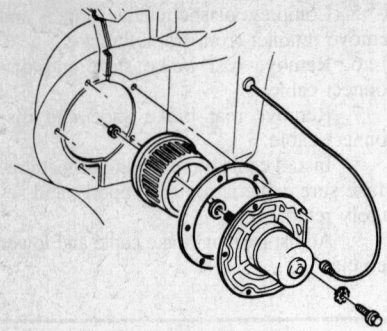

Van blower motor

6. Position the hood hinge using the scribe marks and check the hood alignment.

7. Connect the battery.

Core—Except Vans & S–Series

REMOVAL & INSTALLATION

Without Air Conditioning

1. Disconnect the battery ground cable.

2. Disconnect the heater hoses at the core tubes and drain the engine coolant. Plug the core tubes to prevent spillage.

3. Remove the nuts from the distributor air ducts in the engine compartment.

4. Remove the glove compartment and door.

5. Disconnect the ''Air-Defrost'' and ''Temperature'' door cables.

6. Remove the floor outlet and remove the defroster duct-to-heater distributor screw.

7. Remove the heater distributor-to-instrument panel screws. Pull the assembly rearward to gain access to the wires attached to the unit.

8. Remove the heater distributor from the truck.

9. Remove the heater core retaining straps and remove the core from the truck.

10. Installation is the reverse of removal. Be sure that the core-to-case and case-to-dash panel sealer is intact. Fill the cooling system and check for leaks.

With Air Conditioning

1. Disconnect the battery ground cable.

2. Drain the coolant.

3. Remove the heater hoses from the core tubes. Plug the tubes to prevent spillage.

4. Remove the glove box and door.

5. Remove the screws holding the center duct to the selector duct and to the instrument panel. Remove the center upper and lower ducts.

6. Disconnect the control cable at the temperature door.

7. Remove the three stud nuts from the firewall. Remove the selector duct to firewall screw inside the truck.

8. Pull the selector duct assembly back until the core tubes clear the firewall, then lower it to disconnect the vacuum and electrical connections.

9. Remove the selector duct assembly and remove the core from the truck.

10. Reverse the procedure for installation.

Blower—Vans

REMOVAL & INSTALLATION

1. Disconnect the battery cables. Remove the battery.

2. Unclip the blower motor lead wire.

3. Remove the blower attaching screws.

4. Remove the blower assembly. It may be necessary to pry gently on the blower flange. Sometimes the sealer acts as an adhesive.

5. If the motor is being replaced, remove the nut attaching the blower wheel to the blower motor shaft and separate the two.

REMOVAL & INSTALLATION

1. Disconnect the battery ground.

2. Disconnect all wires at the blower.

3. Remove the vacuum tank if so equipped.

4. Remove the blower motor mounting screws and lift out the blower.

5. Installation is the reverse of removal.

Heater Core—S–Series

REMOVAL & INSTALLATION

1. Disconnect the battery ground.

2. Drain the cooling system.

3. Remove the heater hoses at the core tubes.

4. Remove the core over retaining screws.

5. Remove the retainers at each end of the core.

6. Lift out the core.

7. Installation is the reverse of removal.

Radio

REMOVAL & INSTALLATION

Except Vans & S–Series

1. Remove the negative battery cable and the control knobs and the bezels from the radio control shafts.

2. On AM radios, remove the support bracket stud nut and its lockwasher.

3. On AM/FM radios, remove the support bracket-to-instrument panel screws.

4. Lifting the rear edge of the radio, push the radio forward until the control shafts clear the instrument panel. Then lower the radio far enough so that the electrical connections can be disconnected.

5. Remove the power lead, and antenna wires and then pull out the unit.

6. To install the radio, reverse the above procedure.

Vans

1. Disconnect the ground cable from the battery.

2. Remove the engine cover.

3. Remove the air cleaner from the carburetor.

4. Remove the stud in the carburetor which holds the air cleaner.

5. Cover the carburetor with a clean rag.

6. Remove the knobs, washers and nuts from the front of the radio.

7. Remove the rear bracket screw and bracket from the radio.

8. Remove the radio through the engine access area. Lower the radio far enough to detach the wiring.

9. Remove the radio.

10. Installation is the reverse of removal.

S–Series Trucks

1. Disconnect the battery ground.

2. Remove the instrument panel center bezel.

3. Remove the four screws from the radio bracket an pull the radio forward.

4. Disconnect the antenna and electrical wires. Pull the radio out.

5. Installation is the reverse of removal.

Windshield Wiper Motor

REMOVAL & INSTALLATION

Except Vans & S–Series

1. Make sure the wipers are in the park position.

2. Disconnect the battery ground cable.

3. Disconnect the wiring and hoses at the windshield washer pump.

4. Remove the plastic air intake cover screen and reach in to loosen the wiper drive rod attaching screws. There is a small access hole provided. Remove the drive rod from the motor crank arm.

5. Unbolt and remove the motor.

6. Reverse the procedure for installation, making sure to lubricate the motor crank arm pivot.

NOTE: Failure of the washers to operate or to shut off is often caused by grease or dirt on the electromagnetic contacts. Simply unplug the wire and pull off the plastic cover for access. Likewise, failure of the wipers to park is often caused

by grease or dirt on the park switch contacts. The park switch is under the cove behind the pump.

Vans

1. Be sure that the wiper motor is in PARK position. The wiper arms should be in their normal OFF position.
2. Open the hood and disconnect the battery ground cable.
3. Remove the exposed cowl cover screws with the hood up.
4. Remove the wiper arms. This can be done by pulling the wiper arms away from the glass to release the clip underneath. The wiper arms are splined to the shafts and can be pulled off.
5. Remove the remaining screws securing the cowl panel and remove it.
6. Loosen the nuts holding the transmission linkage to the wiper motor crank arm.
7. Disconnect the power feed to the wiper arm at the connector next to the radio.
8. Remove the flex hose from the left defroster outlet to gain access to the wiper motor screws.
9. Remove the one screw holding the left-hand heater duct to the engine shroud and move the heater duct down and out.
10. Remove the windshield washer hoses from the pump.
11. Remove the 3 screws holding the wiper motor to the cowl and lift the wiper motor out from under the dash.
12. Installation is the reverse of removal. Install the wiper motor in the PARK position.

S—Series

1. Disconnect the battery ground.
2. Remove the wiper arms.
3. Remove the cowl, vent and grille.
4. Loosen, but do not remove, the nuts which hold the drive link to the motor crank arm.
5. Detach the drive link from the crank arm.
6. Disconnect the wiring at the motor.
7. Remove the motor mounting screws. Turn the motor upward and outward and remove it.
8. Installation is the reverse of removal. Torque the attaching screws to 50–75 in. lbs.

Wiper Linkage

REMOVAL & INSTALLATION

S—Series

1. Disconnect the battery ground.
2. Remove the wiper arms.
3. Remove the cowl vent and grille.
4. Remove the nut securing the linkage to the motor.
5. Remove the screw securing the linkage to the motor.

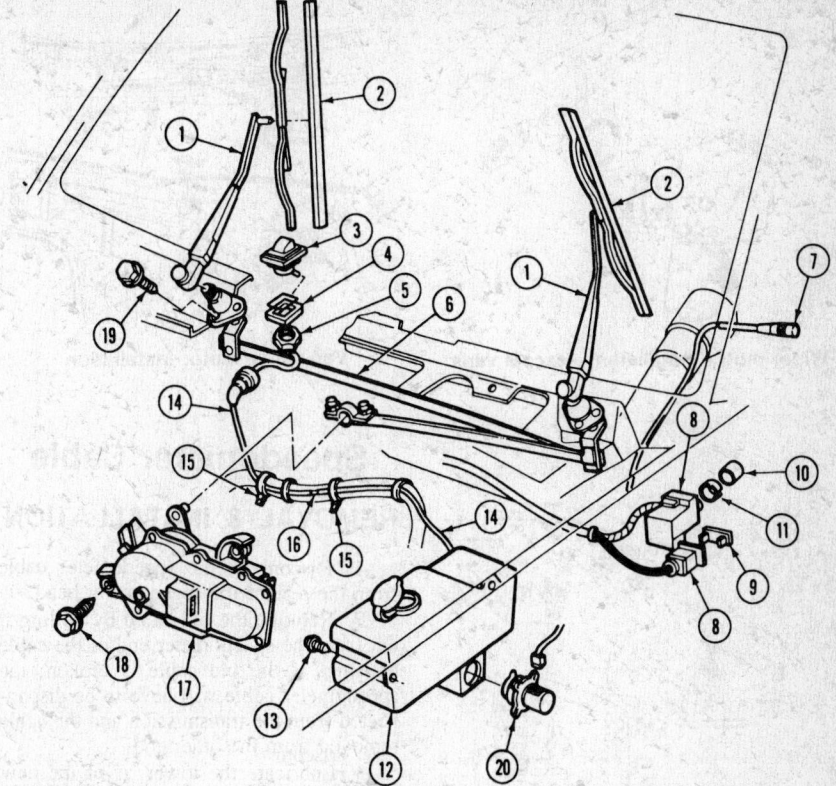

1. Arm, windshield wiper
2. Blade
   insert
3. Nozzle
4. Spacer, nozzle
5. Nut, type R stamped (M16)
6. Transmission, left hand
   transmission, right hand
7. Lever
8. Module
9. Lens, pulse switch
10. Knob, pulse switch
11. Nut, pulse module retaining
12. Reserovir
13. Bolt, (M6 × 1 × 25)
14. Hose, (5/32" ID)
15. Strap
16. Connector
17. Motor assembly
18. Bolt (M5 × .8 × 28)
19. Screw, (M6.3 × 1.69 × 20)
20. Pump

S—Series windshield wiper system

6. Installation is the reverse of removal. Torque the screws and nuts to 50–80 inch lbs.

Instrument Cluster

REMOVAL & INSTALLATION

Except Vans & S—Series

1. Disconnect battery ground cable.
2. Remove headlamp switch control knob and radio control knobs.
3. Remove eight screws and remove instrument bezel.
4. Reach up under instrument cluster and disconnect speedometer by first depressing the tang on the rear of the speedometer head, then pulling cable free from head as tang is depressed.
5. Disconnect oil pressure gauge line at fitting in engine compartment.
6. Pull instrument cluster out just far enough to disconnect line from oil pressure gauge.
7. Remove cluster.
8. Install cluster in reverse order of removal.

Vans

1. Open the hood and disconnect the battery ground cable.
2. Reach up under the dash and disconnect the speedometer cable by first depressing the tang on the rear of the speedometer head and detaching the cable as the tang is depressed.
3. Unplug the instrument panel harness connector.
4. Disconnect and plug the oil pressure gauge line (if equipped).
5. Remove the two nuts from the instrument cluster studs.
6. Pull the top of the cluster away from instrument panel and lift out the bottom of the cluster.
7. Remove the cluster.
8. Installation is the reverse of removal. The clips at the top of the cluster slip into the openings in the instrument panel after the bottom of the cluster in installed.

S—Series

1. Disconnect the battery ground.
2. Remove the five screws retaining, the cluster trim plate and carefully lift off the plate.

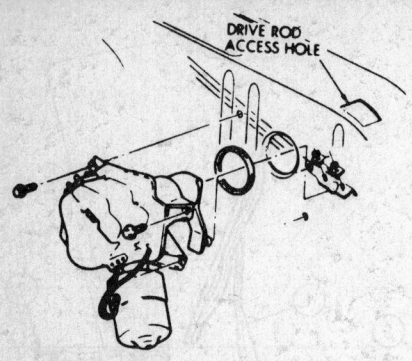

Wiper motor installation—except vans

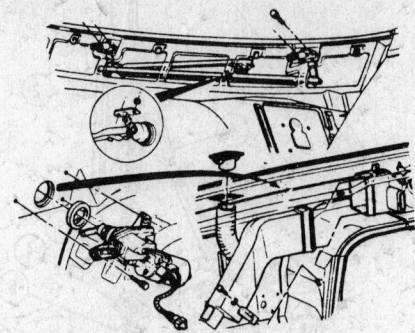

Van wiper motor installation

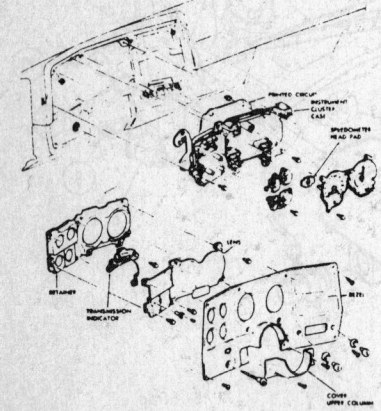

Instrument cluster—except vans

3. Remove the instrument panel face plate.

4. Remove the cluster lens.

5. Disconnect the speedometer cable.

6. Disconnect the cluster electrical connector.

7. Lift out the cluster.

8. Installation is the reverse of removal.

Speedometer Cable

REMOVAL & INSTALLATION

1. Disconnect the speedometer cable from the rear of the speedometer head.

2. Remove the old cable by pulling it out from the speedometer end of the cable housing. If the old cable is broken, the speedometer cable will have to be disconnected from the transmission and the cable removed from the other end.

3. Lubricate the lower $^3/_4$ of the new cable with speedometer cable lubricant and feed the cable into the cable housing.

Headlight

REMOVAL & INSTALLATION

1. Remove the headlight bezel by releasing the attaching screws.

2. Remove the spring (if any) from the retaining ring and turn the unit to disengage it from the headlamp adjusting screws.

3. Disconnect the wiring harness connector.

NOTE: Do not disturb the adjusting screws.

4. Remove the retaining ring and the headlamp.

5. Position the new sealed beam unit in the retaining ring.

6. Attach the wiring connector.

7. Install the headlamp assembly, twisting the ring slightly to engage the adjusting screws.

8. Install the retaining ring spring and check the operation of the unit. Install the bezel.

Fusible Links

Fusible links are sections of wire, with special insulation, designed to melt under electrical overload. Replacements are simply spliced into the wire in most cases.

Circuit Breakers

A circuit breaker is an electrical switch which breaks the circuit in case of an overload. All models have a circuit breaker in the headlight switch to protect the headlight and parking light systems. An overload may cause the lights to flash on and off.

Fuses and Flashers

The fuse block is mounted to the firewall, inside the truck, to the left of the steering column. The turn signal flasher and hazard warning flasher plug into the fuse block. Each fuse receptacle is marked as to the circuits it protects and the correct amperage. In-line fuses are also used to protect some circuits.

Datsun/Nissan

GENERAL ENGINE SPECIFICATIONS

Year	Engine Displacement cc (cu. in.)	Carb Type	Advertised Horsepower @ rpm	Advertised Torque @ rpm (ft. lbs.)	Bore × Stroke (in.)	Advertised Compression Ratio	Oil Pressure (psi/idle)
'79–'80	1952 (119)	2 bbl	97@5600	102@3200	3.35 × 3.39	8.5:1	50–57
'81	2164 (132)	Diesel	61@4000	102@1800	3.27 × 3.94	21.6:1	60
	2187 (133.5)	2 bbl	98@4000	117@1800	3.43 × 3.62	8.5:1	60
'82–'83	2187 (133.5)	2 bbl	98@4000	117@1800	3.43 × 3.62	8.5:1	60
	2164 (132)	Diesel	61@4000	102@1800	3.27 × 3.94	21.6:1	60
'84–'86	1952 (119)	2 bbl	97@5600	102@3200	3.35 × 3.39	9.4:1	60
	2389 (146)	2 bbl	103@4800	134@2800	3.50 × 3.78	8.3:1	60
	2488 (152)	Diesel	70@4000	115@2000	3.50 × 3.94	21.4:1	60

① Calif.—21.9:1

GASOLINE ENGINE TUNE-UP SPECIFICATIONS

Year	Engine Displacement cu. in. (cc)	Spark Plug Type	Spark Plug Gap (in.)	Distributor Point Dwell (deg)	Distributor Point Gap (in.)	Ignition Timing (deg) MT	Ignition Timing (deg) AT	Intake Valve Opens (deg)	Fuel Pump Pressure (psi)	Compression Pressure (psi)▲	Idle Speed (rpm) MT	Idle Speed (rpm) AT	Valve Clearance (in.)● In	Valve Clearance (in.)● Ex
'79	4-119 (1952)	BP6ES-11	0.041	—	④	12B	12B	16B	3.0–3.9 ③	171 ②	650	630 ①	0.010	0.012
'80	4-119 (1952)	BP6ES-11	0.041	—	④	12B ⑤	12B ⑤	16B	3.0–3.9 ③	171 ②	600	600	0.010	0.012
'81	4-133.5 (2187)	BP6ES ⑥	0.033	—	④	5B	5B	16B	3.0–3.9	171 ②	650 ⑦	650	0.012	0.012
'82–'83	4-133.5 (2187)	⑧	0.033	—	④	3B	3B	16B	3.0–3.9	171 ②	650 ⑦	650	0.012	0.012

GASOLINE ENGINE TUNE-UP SPECIFICATIONS

Year	Engine Displacement cu. in. (cc)	Spark Plug Type	Gap (in.)	Distributor Point Dwell (deg)	Point Gap (in.)	Ignition Timing (deg) MT	AT	Intake Valve Opens (deg)	Fuel Pump Pressure (psi)	Compression Pressure (psi)▲	Idle Speed (rpm) MT	AT	Valve Clearance (in.)● In	Ex
'84–'86	4-119 (1952)	⑧	0.033	—	④	5B	—	16B	2.7–3.4	171 ②	600	—	0.012	0.012
	4-146 (2389)	⑧	0.033	—	④	3B	3B	16B	2.7–3.4	171 ②	650 ⑦	650	0.012	0.012

NOTE: Part number in this chart are not recommendations by Chilton for any product by brand name.

● Measured with engine hot
▲ Lowest reading must be at least 80% of the highest
B Before top dead center
① Transmission in Drive
② 128 psi minimum
③ 4.6 or less with electric fuel pump (air conditioned models)
④ 0.012–0.020, Transistor ignition air gap
⑤ Calif. Heavy Duty models: 10B
⑥ Canada: BPR6ES
⑦ 4-WD: 800
⑧ Intake side: BPR6ES
   Exhaust side: BPR5ES

DIESEL ENGINE TUNE-UP SPECIFICATIONS

Year	Injector Opening Pressure (psi)	Low Idle (rpm)	Dashpot Speed (rpm)	Valve Clearance (in.) Intake	Exhaust	Intake Valve Opens (deg.)	Injection Timing rpm	Firing Order
'81–'83	1422.5	550–700	1280–1350	.014	.014	28	20 BTDC	1-3-4-2
'84–'86	1422.5	650–800	1280–1350	.014	.014	28	18 BTDC	1-3-4-2

FIRING ORDERS

NOTE: Always remove spark plug wires one at a time.

L20B firing order: 1-3-4-2

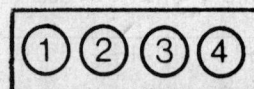

SD22, SD25 firing order: 1-3-4-2

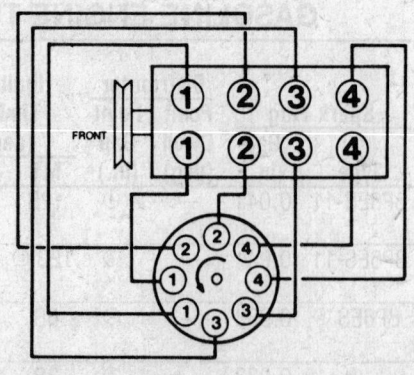

Z20, Z22, Z24 firing order: 1-3-4-2

CAPACITIES

Year	Engine Displacement cc (cu. in.)	Crankcase qts w/filter	wo/filter	Transmission pts 4sp	5sp	Auto	Transfer Case pts	Rear Drive Axle pts	Front Drive Axle pts	Gas Tank gal	Cooling System qts Manual	Auto
'79	1952 (119.1)	4.5	4.0	3.6	4.2	11.7	—	2.1	—	13.2 ①	9.3	9.2
'80	1952 (119.1)	4.5	4.0	3.6	4.2	11.7	3.0	2.6	2.1	13.2 ①	9.3	9.2
'81 2-WD	2187 (133.5)	4.6	4.12	3.6	4.2	11.7	—	2.6	—	13.2 ①	10.75	10.6
4-WD	2187 (133.5)	4.5	3.8	3.6	4.2	11.7	3.0	2.6	2.1	15.8 ②	10.75	10.6
Diesel	2164 (132)	5.8	—	—	4.2	—		2.6	—	15.8 ②	—	—
'82–'83 2-WD	2187 (133.5)	4.6	4.12	3.6	4.2	11.7		2.6	—	13.25 ①	10.75	10.6
4-WD	2187 (133.5)	4.5	3.8	3.6	4.2	11.7	3.0	2.6	2.1	15.8 ②	10.75	10.6
Diesel	2164 (132)	6.4 ③	—	—	4.2	—		2.6	—	13.25 ①	10.5	—
'84–'86 2-WD	1952(119) 2389(146)	4.3	3.9	—	4.2	11.7		2.6 ④⑤	—	13.25 ①	10.75	10
4-WD	1952(119) 2389(146)	4.5	4.0	—	4.2	—	3.0	2.6 ⑤	2.1	15.8 ②	10.75	—
Diesel	2488(152)	5.3	4.8	—	4.2	—		2.6	—	13.25 ①	11.1	—

① Long bed: 64(16.8)
② Longbed w/4-WD:75(19.8)
③ '83:5.2(5.5)
④ With dual rear wheels: 1.30(2.7)
⑤ '85–'86 Canada—1.50(3.1)
—Not applicable

CRANKSHAFT AND CONNECTING ROD SPECIFICATIONS

(all measurements given in in.)

Year	Engine Displacement cc (cu. in.)	Crankshaft Main Brg. Journal Dia	Main Brg. Oil Clearance	Shaft End-Play	Thrust on No.	Connecting Rod Journal Dia	Oil Clearance	Side Clearance
'79–'80	1952 (119)	2.3599–2.3604	0.0008–0.0026	0.0020–0.0071	3	1.9670–1.9675	0.0009–0.0026	0.008–0.012
'81	2164 (132)	2.7918–2.7988	0.0013–0.0038	0.0024–0.0094	3	2.0840–2.0906	0.0013–0.0038	0.0039–0.0079
	2187 (133.5)	2.1631–2.1636	0.0008–0.0024	0.0021–0.0070	3	1.9670–1.9675	0.0010–0.0022	0.008–0.012
'82–'83	2164 (132)	2.7916–2.7921	0.0014–0.0037	0.0024–0.0055	3	2.0832–2.0837	0.0014–0.0034	0.004–0.008
	2187 (133.5)	2.1631–2.1636	0.0008–0.0024	0.0021–0.0071	3	1.9670–1.9675	0.0010–0.0022	0.008–0.012
'84–'86	1952 (119)	2.1631–2.1636	0.0008–0.0024	0.0020–0.0071	3	1.9670–1.9675	0.0005–0.0021	0.008–0.012
	2389 (146)	2.1631–2.1636	0.0008–0.0024	0.0020–0.0071	3	1.9670–1.9675	0.0005–0.0021	0.008–0.012
	2488 (152)	2.7916–2.7921	0.0014–0.0034	0.0024–0.0055	3	2.0832–2.0837	0.0014–0.0032	0.004–0.008

VALVE SPECIFICATIONS

Year	Engine Displacement cc (cu in.)	Seat Angle (deg)	Face Angle (deg)	Spring Test Pressure (lbs. @ in.) Outer	Inner	Free Length (in.) Outer	Inner	Stem-to-Guide Clearance (in.) Intake	Exhaust	Stem Diameter (in.) Intake	Exhaust
'79–'80	1952 (119)	45	45 ①	47@ 1.58	27@1.38	1.97	1.77	0.0008– 0.0021	0.0016– 0.0029	0.3139	0.3131
'81	2164 (132)	45.5	44.5	33@ 1.634	—	1.929	—	0.0006– 0.0018	0.0016– 0.0028	0.3137– 0.3143	0.3137– 0.3143
	2187 (133.5)	45.5	44.5	51@ 1.575	24@1.378	1.959	1.736	0.0008– 0.0021	0.0016– 0.0029	0.3136– 0.3142	0.3128– 0.3134
'82–'83	2164 (132)	45.5	44.5	134.7@ 1.197	—	1.9764	—	0.0006– 0.0018	0.0016– 0.0028	0.3138– 0.3144	0.3128– 0.3134
	2187 (133.5)	45.5	44.5	115.3@ 1.180	57@0.98	1.9594	1.7362	0.0008– 0.0021	0.0016– 0.0029	0.3136– 0.3142	0.3128– 0.3134
'84–'86	1952 (119)	45.5	44.5	115.3@ 1.180	57@0.98	1.959	1.7362	0.0008– 0.0021	0.0016– 0.0029	0.3136– 0.3142	0.3128– 0.3134
	2389 (146)	45.5	44.5	115.3@ 1.180	57@0.98	1.959	1.7362	0.0008– 0.0021	0.0016– 0.0029	0.3136– 0.3142	0.3128– 0.3134
	2488 (152)	45.5	44.5	148@ 1.224	—	1.982		0.0006– 0.0018	0.0016– 0.0028	0.3138– 0.3144	0.3128– 0.3134

① 45°30'–'79

PISTON AND RING SPECIFICATIONS

All measurements are given in inches.

Year	Engine Disp. (cu in.)	Piston-to-Bore Clearance	Ring Gap Top Compression	Bottom Compression	Oil Control	Ring Side Clearance Top Compression	Bottom Compression	Oil Control
'79–'80	1952 (119)	0.0010–0.0018	0.0098–0.0157	0.0118–0.0197	0.0118–0.0354	0.0016–0.0029	0.0012–0.0028	Snug
'81	2187 (133.5)	0.0010–0.0018	0.0098–0.0157	0.0059–0.0118	0.0118–0.0354	0.0016–0.0029	0.0012–0.0025	Snug
	2164 (132)	0.0047–0.0075	0.0118–0.0197	0.0118–0.0197	0.0018–0.0197	0.0024–0.0039	0.0016–0.0031 ①	0.0008–0.0024
'82–'83	2187 (133.5)	0.0010–0.0018	0.0098–0.0157	0.0059–0.0118	0.0118–0.0354	0.0016–0.0029	0.0012–0.0025	Snug
	2164 (132)	0.0016–0.0043	0.0118–0.0177	0.0079–0.0138	0.0059–0.0118	0.0024–0.0039	0.0016–0.0031 ①	0.0008–0.0024
'84–'86	1952 (119)	0.0010–0.0018	0.0098–0.0157	0.0059–0.0118	0.0118–0.0354	0.0016–0.0029	0.0012–0.0025	Snug
	2389 (146)	0.0010–0.0018	0.0098–0.0157	0.0059–0.0118	0.0118–0.0354	0.0016–0.0029	0.0012–0.0025	Snug
	2488 (152)	0.0031–0.0041	0.0118–0.0177	0.0079–0.0138	0.0059–0.0118	0.0024–0.0039	0.0016–0.0031 ①	0.0008–0.0024

① Diesel: 2nd and 3rd rings

TORQUE SPECIFICATIONS

(All readings in ft. lbs. unless noted)

Year	Engine Displacement cc (cu in.)	Cylinder Head Bolts	Rod Bearing Bolts	Main Bearing Bolts	Crankshaft Pulley Bolts	Flywheel to Crankshaft Bolts	Manifolds Intake	Exhaust
'79–'80	1952 (119)	61	37	37	102	109	11	11
'81–'83	2187 (133.5)	51–58	33–40	33–40	87–116	101–116	12–15	12–16
	2164 (132)	94 large 40 small	36–40	109–116 ①	217–239	33–36	11–13	11–13
'84–'86	1952 (119)	②	33–40	33–40	87–116	101–116	12–15	12–15
	2389 (146)	②	33–40	33–40	87–116	101–116	12–15	12–15
	2488 (152)	94 large 40 small	49–52	123–127	217–239	108–123	11–13	11–13

① 1982–'83: 123–127
② Five step torque procedure: Step 1; 22 ft. lbs., Step 2; 58 ft. lbs., Step 3; loosen all bolts completely, Step 4; 22 ft. lbs., Step 5; 54 to 61 ft.lbs.

BATTERY AND STARTER SPECIFICATIONS

Year	Engine Displacement cc (cu in.)	Battery Amp Hour Capacity	Battery Volts	Battery Ground	Starter Lock Test Amps	Starter Lock Test Volts	Starter Lock Test Torque (ft. lbs.)	Starter No Load Test Amps	Starter No Load Test Volts	Starter No Load Test rpm	Brush Spring Tension (oz)
'79–'80	1952 (119)	①	12	Neg	Not Recommended			60—	12	7000 + ②	56
'81–'83	2187 (133.5)	①	12	Neg	Not Recommended			60 ③	11.5 ④	6000–7000 ⑤	56
	2164 (132)	①	12	Neg	800	5.0	21.0	150	12	3500	123.2
'84–'86	1952 (119)	①	12	Neg	Not Recommended			60 ③	11.5	6000–7000 ⑤	56
	2389 (146)	①	12	Neg	Not Recommended			60 ③	11.5	6000–7000 ⑤	56
	2488 (152)	①	12	Neg	Not Recommended			150	12	3500	123.2

① 60 and 80 amp hour batteries were available
② 6000 + if equipped with automatic transmission
③ Canada: 100
④ Canada: 11.0
⑤ Canada: 3900

ALTERNATOR AND REGULATOR SPECIFICATIONS

Year	Alternator Manufacturer and/or Part Number	Alternator Output @ Alternator rpm	Charge Indicator Relay Back Gap (in.)	Charge Indicator Relay Air Gap (in.)	Charge Indicator Relay Point Gap (in.)	Voltage Regulator Back Gap (in.)	Voltage Regulator Air Gap (in.)	Voltage Regulator Point Gap (in.)	Regulated Voltage
'79–'80	Hitachi LT135-44	27.5 @ 2500	—Transistorized Non-Adjustable Relay—						14.4–15.0
	LR138-01 ①	30.0 @ 2500							
'81	Hitachi LR150-98	50 @ 5000	—Transistorized Non-Adjustable Relay—						14.4–15.0
	LR160-78	60 @ 5000							
	LR150-52	50 @ 5000							
	LR160-78	60 @ 5000							
'82	Hitachi LR150-98B	40 @ 2500	—Transistorized Non-Adjustable Relay—						14.4–15.0
	LR160-78 & 78B	50 @ 2500							
	LR160-97B	52 @ 2500							
'83	Hitachi LR150-98B	40 @ 2500	—Transistorized Non-Adjustable Relay—						14.4–15.0
	LR160-78 & 78B	50 @ 2500							
	LR160-97B	42 @ 2500							
	LR150-133E	42 @ 2500							
'84	Hitachi LR150-98B	40 @ 2500	—Transistorized Non-Adjustable Relay—						14.4–15.0
	LR160-78B	50 @ 2500							
	LR150-177	50 @ 2500							
	LR160-120	50 @ 2500							
	LR150-403	42 @ 2500							
	LR155-401	50 @ 2500							
'85–'86	Hitachi LR150-98B	40 @ 2500	—Transistorized Non-Adjustable Relay—						14.4–15.0
	LR150-197B	40 @ 2500							
	LR160-78B	50 @ 2500							
	LR160-140B	50 @ 2500							
	LR150-177	40 @ 2500							
	LR150-194B	40 @ 2500							
	LR160-120	50 @ 2500							
	LR150-403	42 @ 2500							
	LR155-401	50 @ 2500							

① With air conditioning   —Not applicable

BRAKE SPECIFICATIONS

(All measurements are given in in.)

| Year | Master Cylinder Bore | Wheel Cylinder or Caliper Bore | | Piston-to-Bore Clearance | Brake Drum or Rotor Diameter | | Minimum Lining Thickness | Brake Disc | |
		Front	Rear		Front	Rear		Minimum Thickness	Maximum Run-out
'79	0.813	2.125	0.625	0.006	10.67	10.00	0.08 (disc) 0.06 (drum)	0.413	0.0059
'80–'83	0.875	2.125	0.625	0.006	10.67	10.00	0.08 (disc) 0.06 (drum)	0.413	0.0059
'84–'86	0.945	2.386	0.687 ①	0.006	9.84 ②	10.00 ③	0.08 (disc) 0.06 (drum)	0.787	0.0028

① 4WD: 0.630
② 4WD: 10.50
③ With dual rear wheels: 8.66

Years	Engine Designation	Displacement (cc)
1979–80	L20B	1952
1981–83	Z22	2187
1981–83	SD22	2164
1984–86	Z20	1952
1984–86	Z24	2389
1984–86	SD25	2488

TUNE-UP

Electronic Ignition

The electronic ignition differs from its conventional counterpart only in the distributor component area. The secondary side of the ignition system is the same as a conventional breaker points system.

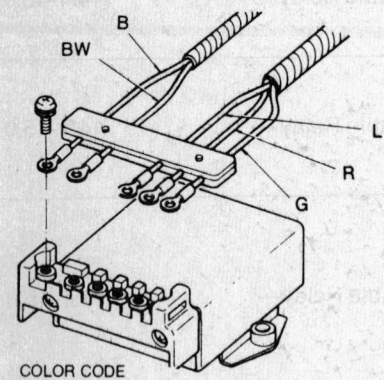

COLOR CODE

B : BLACK
BW : BLACK WITH WHITE STRIPE
R : RED
G : GREEN
L : BLUE

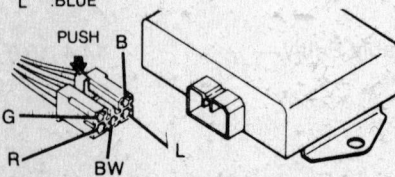

Electronic control unit connection

Located in the distributor, in addition to the normal ignition rotor, is a four spoke rotor (reluctor) which rests on the distributor shaft where the breaker points cam is found on earlier systems. A pick-up coil, consisting of a magnet, a ring type pickup surrounding the reluctor and wiring, rests on the "breaker plate" next to the reluctor. An integrated circuit (IC) ignition unit is mounted on the side of the distributor.

When a reluctor spoke is not aligned with the pick-up coil, it generates large lines of flux between itself, the magnet, and the pick-up coil. This large flux variation results in a high generated voltage in the pick-up coil, preventing current from flowing to the pick-up coil. When a reluctor spoke lines up with the pick-up coil, the flux variation is low and zero voltage is generated, allowing current to flow to the pick-up coil. Ignition primary current is then cut off by the electronic unit, allowing the field in the ignition coil to collapse, inducing high secondary voltage in the conventional manner. The high voltage then flows through the distributor to the spark plug, as usual.

Because no points or condenser are used, and because dwell is determined by the electronic unit, no adjustments are necessary. Ignition timing is checked in the usual way, but unless the distributor is disturbed it is not likely to ever change very much. Service consists of inspection of the distributor cap, rotor, and ignition wires, replacing when necessary. These parts can be expected to last for at least 40,000 miles. In addition, the reluctor air gap should be checked periodically.

1. The distributor cap is held on by two clips. Release them with a screwdriver and lift the cap straight up and off, with the wires attached. Inspect the cap for cracks, carbon tracks, or a worn center contact. Replace it if necessary, transferring the wires one at a time from the old cap to the new.

2. Pull the ignition rotor (not the spoked reluctor) straight up to remove. Replace it if its contacts are worn, burned, or pitted. Do not file the contacts. To replace, press it firmly onto the shaft. It only goes on one way, so be sure it is fully seated.

3. Before replacing the ignition rotor, check the reluctor air gap. Use a non-magnetic feeler gauge. Rotate the engine until a reluctor spoke is aligned with the pick-up coil (either bump the engine around with the starter, or turn it with a wrench on the crankshaft pulley bolt). The gap should measure 0.012–0.020 in. Adjustment, if necessary, is made by loosening the pick-up coil mounting screws and shifting its position to center the pick-up coil (ring) around the reluctor. Tighten the screws and recheck the gap.

4. Inspect the wires for cracks or brittleness. Replace them one at a time to prevent crosswiring, carefully pressing the replacement wires into place. The cores of electronic wires are more susceptible to breakage than those of standard wires, so treat them gently.

For electronic ignition system troubleshooting, refer to the Electrical section in Unit Repair.

Ignition Timing

NOTE: Ignition timing is the measurement, in degrees of crankshaft rotation, of the point at which the spark plugs fire in each of the cylinders. It is measured in degrees before or after Top Dead Center (TDC) of the compression stroke.

Because it takes a fraction of a second for the spark plug to ignite the mixture in the cylinder, the spark plug must fire a little before the piston reaches TDC. Otherwise, the mixture will not be completely ignited as the piston passes TDC and the full power of the explosion will not be used by the engine.

The timing measurement is given in degrees of crankshaft rotation before the piston reaches TDC (BTDC). If the setting for the ignition timing is 5° BTDC, the spark plug must fire 5° before each piston reaches TDC. This only holds true, however, when the engine is at idle speed.

As the engine speed increases, the pistons go faster. The spark plugs have to ignite the fuel even sooner if it is to be completely ignited when the piston reaches TDC. To do this, the distributor has a means to advance the timing of the spark as the engine speed increases. This is accomplished by centrifugal weights within the distributor and a vacuum diaphragm, mounted on the side of the distributor. On Datsun pick-ups, it is not necessary to disconnect the vacuum line from the diaphragm when the ignition timing is being set.

If the ignition is set too far advanced (BTDC), the ignition and expansion of the fuel in the cylinder will occur too soon and tend to force the piston down while it is still traveling up. This causes engine ping. If the ignition spark is set too far retarded, after TDC (ATDC), the piston will have already passed TDC and started on its way down when the fuel is ignited. This will cause the piston to be forced down for only a portion of its travel. This will result in poor engine performance and lack of power.

Timing marks consist of a notch on the rim of the crankshaft pulley and a scale of degrees attached to the front of the engine. The notch corresponds to the position of the piston in the No. 1 cylinder. A stroboscopic (dynamic) timing light is used, which is hooked into the circuit of the No. 1 cylinder spark plug. Every time the spark plug fires, the timing light flashes. By aiming the timing light at the timing marks, the exact position of the piston within the cylinder can be read, since the stroboscopic flash makes the mark on the pulley appear to be standing still. Proper timing is indicated when the notch is aligned with the correct number on the scale. There are three basic types of timing light available. The first is a simple neon bulb with two wire connections (one for the spark plug and one for the plug wire, connecting the light in a

series). This type of light is quite dim, and must be held closely to the marks to be seen, but it is quite inexpensive. The second type of light operates from the truck battery. Two alligator clips connect to the battery terminals, while a third wire connects to the spark plug with an adapter. This type of light is more expensive, but the xenon bulb provides a nice bright flash which can even be seen in sunlight. The third type replaces the battery source with 110 volt house current. Some timing lights have other functions built into them, such as dwell meters, tachometers, or remote starting switches. These are convenient, in that they reduce the tangle of wires under the hood, but may duplicate the functions of tools you already have.

Your Datsun has electronic ignition, you should use a timing light with an inductive pickup. This pickup simply clamps onto the No. 1 plug wire, eliminating the adapter. It is not susceptible to crossfiring or false triggering, which may occur with a conventional light, due to the greater voltages produced by electronic ignition.

ADJUSTMENT

1. Locate the timing marks on the crankshaft pulley and the front of the engine.

2. Clean off the timing marks, so that you can see them.

3. Use chalk or white paint to color the mark on the crankshaft pulley and the mark on the scale which will indicate the correct timing when aligned with the notch on the crankshaft pulley.

4. Attach a tachometer to the engine.

5. Attach a timing light to the engine, according to the manufacturer's instructions.

6. Refer to the underhood sticker to determine if the vacuum hose is to be left connected to the distributor. If no instructions differ, leave the vacuum line connected to the distributor vacuum diaphragm.

7. Check to make sure that all of the wires clear the fan and then start the engine. Allow the engine to reach normal operating temperature.

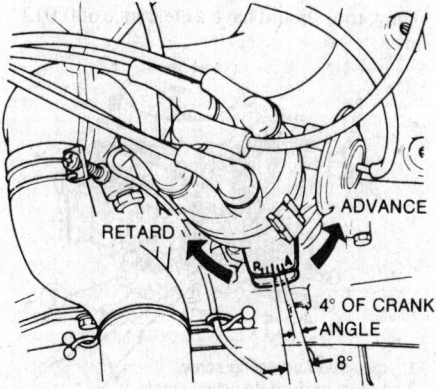

Ignition timing marks on the distributor

9. Adjust the idle to the correct setting.

10. Aim the timing light at the timing marks. If the marks which you put on the pulley and the engine are aligned when the light flashes, the timing is correct. Turn off the engine and remove the tachometer and the timing light. If the marks are not in alignment, proceed with the following steps.

11. Turn off the engine.

12. Loosen the distributor lockbolt just enough so that the distributor can be turned with a little effort.

13. Start the engine. Keep the wires of the timing light clear of the fan.

14. With the timing light aimed at the pulley and the marks on the engine, turn the distributor in the direction of rotor rotation to retard the spark, and in the opposite direction of rotor rotation to advance the spark. Align the marks on the pulley and the engine with the flashes of the timing light.

15. Tighten the distributor lockbolt and recheck the timing.

Valve Lash

Valve adjustment determines how far the valves enter the cylinder and how long they stay open and closed.

If the valve clearance is too large, part of the lift of the camshaft will be used in removing the excessive clearance. Consequently, the valve will not be opening for as long as it should. This condition has two effects; the valve train components will emit a tapping sound as they take up the excessive clearance and the engine will perform poorly because the valves don't open fully and allow the proper amount of gases to flow into and out of the engine.

If the valve clearance is too small, the intake valves and the exhaust valves will open too far and they will not fully seat on the cylinder head when they close. When a valve seats itself on the cylinder head, it does two things: it seals the combustion chamber so that none of the gases in the cylinder escape and it cools itself by transferring some of the heat it absorbs from the combustion in the cylinder to the cylinder head and to the engine's cooling system. If the valve clearance is too small, the engine will run poorly because of the gases escaping from the combustion chamber. The valves will also become overheated and will warp, since they cannot transfer heat unless they are touching the valve seat in the cylinder head.

NOTE: While all valve adjustments must be made as accurately as possible, it is better to have the valve adjustment slightly loose than slightly tight, as a burned valve may result from overly tight adjustments.

ADJUSTMENT

1979–80 Gasoline Engine

1. The valves are adjusted with the engine at normal operating temperature. Oil temperature, and the resultant parts expansion, is much more important than water temperature. Run the engine for at least fifteen minutes to ensure that all the parts have reached their full expansion. After the engine is warmed up, shut it off.

2. Purchase either a new gasket or some silicone gasket seal before removing the camshaft cover. Note the location of any wires and hoses which may interfere with cam cover removal, disconnect them and move them aside. Then remove the bolts which hold the cam cover in place and remove the cam cover.

3. Place a wrench on the crankshaft pulley bolt and turn the engine over until the valves for No. 1 cylinder are closed. If you have not done this before, it is a good idea to turn the engine over slowly several times and watch the valve action until you have a clear idea of just when the valve is closed.

NOTE: On all engine except the Z–series, the valves are closed when both cam lobes are pointing up. On the Z–series engines the valves are closed when the cam lobe points straight down.

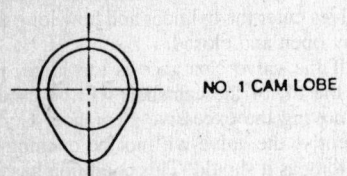

NO. 1 CAM LOBE

Cam lobe pointing straight down—Z series engines

4. Check the clearance of the intake and exhaust valves. You can differentiate between them by lining them up with the tubes of the intake and exhaust manifolds. The correct size feeler gauge should pass between the base circle of the cam and the rocker arm with just a slight drag. Be sure the feeler gauge is inserted straight and not on an angle.

5. If the valves need adjustment, loosen the locking nut and then adjust the clearance with the adjusting screw. You will probably find it necessary to hold the locking nut while you turn the adjuster. After you have the correct clearance, tighten the locking nut and recheck the clearance. Remember, it's better to have them too loose than too tight, especially exhaust valves.

6. Repeat this procedure until you have checked and/or adjusted all the valves. Keep in mind that all that is necessary is to have the valves closed and the camshaft lobes pointing up. It is not particularly important what stroke the engine is on.

7. Install the cam cover gasket, the cam cover, and any wires and hoses which were removed.

1981 Z22 Engine

1. The valves must be adjusted with the engine warm, so start the car and run the engine until the needle on the temperature gauge reaches the middle of the gauge. After the engine is warm, shut it off.

2. Purchase either a new gasket or some silicone gasket sealer before removing the camshaft cover. Counting on the old gasket to be in good shape is a losing proposition; always use new gaskets. Note the location of any wires and hoses which may interfere with cam cover removal, disconnect them and move them to one side. Remove the bolts holding the cover in place and remove the cover. Remember, the engine will be hot, so be careful.

3. Place a wrench on the crankshaft pulley bolt and turn the engine over until the first cam lobe behind the camshaft timing chain sprocket is pointing straight down.

NOTE: If you decide to turn the engine by "bumping" it with the starter, be sure to disconnect the high tension wire from the coil(s) to prevent the engine from accidentally starting and spewing oil all over the engine compartment.

── **CAUTION** ──
Never attempt to turn the engine by using a wrench on the camshaft sprocket bolt; there is a one to two turning ratio between the camshaft and the crankshaft which will put a tremendous strain on the timing chain.

4. See the illustration marked "Primary adjustment" and adjust valves (1), (4), (6), and (7) to 0.012 in. using a flat-bladed feeler gauge. The feeler gauge should pass between the valve stem and the rocker arm screw with a very slight drag. Insert the feeler gauge straight, not at an angle.

5. If the clearance is not within specified value, loosen the rocker arm lock nut and turn the rocker arm screw to obtain the proper clearance. After correct clearance is obtained, tighten the lock nut.

6. Turn the engine over so that the first cam lobe behind the camshaft timing chain sprocket is pointing straight up and adjust the valves marked (2), (3), (5), and (8) in the "Secondary adjustment" illustration. They, too, should have a clearance of 0.012 in.

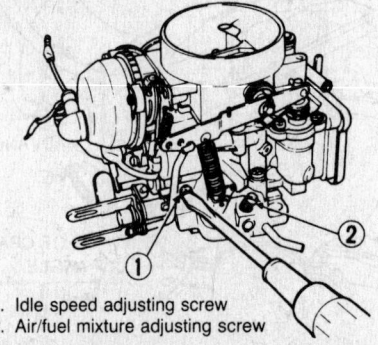

1. Idle speed adjusting screw
2. Air/fuel mixture adjusting screw
Idle speed and mixture adjustments

7. Install the cam cover gasket, the cam cover and any wires and hoses which were removed.

1983–86 Z20, Z22, Z24 and 1981–86 Diesel Engines

NOTE: Adjustment should be made with the engine hot. Valve clearance is 0.012 in.

1. Run the engine to normal operating temperatures.

2. Shut off the engine and remove the valve cover.

3. Rotate the crankshaft until the timing marks indicate that No. 1 cylinder is at TDC of the compression stroke. If you're not sure of which stroke you're on, remember the No. 1 spark plug and hold your thumb over the hole. Pressure will be felt as the piston starts up on the compression stroke.

4. With No. 1 at TDC you can set valves 1, 2, 4 and 6 on the gasoline engines and 1, 2, 3 and 5 on the diesels, counting from the front.

5. In a similar manner, put No.4 piston at TDC compression. You may now set valves 3, 5, 7, 8 on the gasoline engines and 4, 6, 7 and 8 on the diesels.

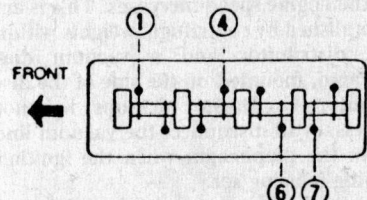

Primary adjustment on Z-series engines through 1982

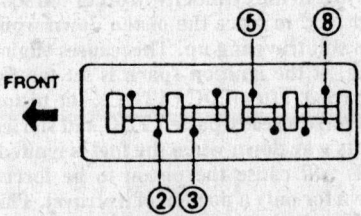

Secondary adjustment on Z-series engines through 1982

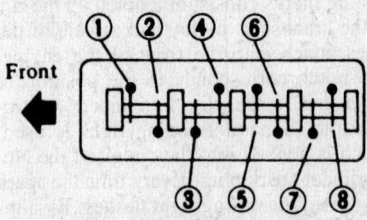

1983-86 Z-series valve arrangement

Carburetor

For adjustment and specification information, refer to the Carburetor section in Unit Repair.

When the engine in your Datsun is running, the air-fuel mixture from the car-

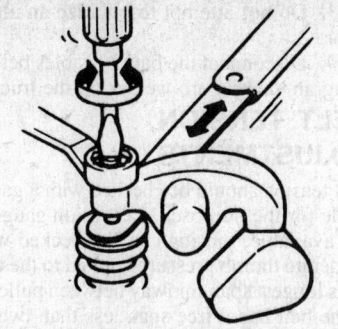

1983-86 Z-series valve arrangement

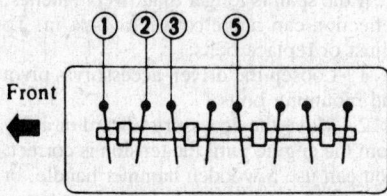

1983-86 Z-series and all diesels, valves adjusted with No. 1 piston at TDC

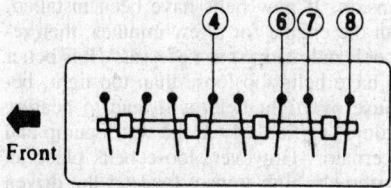

1983-86 Z-series and all diesels, valves adjusted with No. 4 piston at TDC

buretor is being drawn into the engine by a partial vacuum which is created by the movement of the pistons downward on the intake stroke. The amount of air-fuel mixture that enters into the engine is controlled by the throttle plates in the bottom of the carburetor. When the engine is not running the throttle plates are closed, completely blocking off the bottom of the carburetor from the inside of the engine. The throttle plates are connected by the throttle linkage to the accelerator pedal in the passenger compartment of the Datsun. When you depress the pedal, you open the throttle plates in the carburetor to admit more air-fuel mixture to the engine.

When the engine is not running, the throttle plates are closed. When the engine is idling, it is necessary to have the throttle plates open slightly. To prevent having to hold your foot on the pedal when the engine is idling, an idle speed adjusting screw is added to the carburetor linkage.

The idle adjusting screw contacts a lever (throttle lever) on the outside of the carburetor. When the screw is turned, it either opens or closes the throttle plates of the carburetor, raising or lowering the idle speed of the engine. This screw is called the curb idle adjusting screw.

A special mixture circuit is incorporated into the carburetor to enable the engine to run smoothly at idle. This circuit is controlled by the mixture screw, which determines the amount of fuel admitted at idle.

Idle Speed & Mixture

ADJUSTMENT

1979–81

1. Start the engine and run it until it reaches operating temperature.

2. Allow the engine idle speed to stabilize by running the engine at idle for at least one minute.

3. If it hasn't already been done, check and adjust the ignition timing to the proper setting.

4. Turn off the engine and connect a tachometer to the engine.

5. Disconnect and plug the air hose between the three way connector and the check valve, if equipped. Start the engine. With the transmission in Neutral, check the idle speed on the tachometer. If the reading on the tachometer is correct, turn the idle adjusting screw clockwise with a screwdriver to increase the idle speed and counterclockwise to decrease it.

6. With an automatic transmission in Drive (wheels chocked and parking brake applied) or a manual transmission in Neutral, turn the mixture screw out until the engine rpm starts to drop due to an overly rich mixture.

7. Turn the screw in past the starting point until the engine rpm start to drop because of a too lean mixture. The rpm drop should be 45–55 rpm with manual transmission, or 25–35 rpm with automatic transmission (in Drive). If the mixture limiter cap will not allow this adjustment, remove it, make the adjustment, and reinstall it.

8. Install the air hose. If the engine speed increases, reduce it with the idle speed screw.

1982 and Later

1. Connect a tachometer according to the manufacturer's instructions.

2. Turn all the accessories and lights OFF. Make sure that the wheels are straight ahead on models with power steering.

3. Block the wheels.

4. Run the engine at 2,000 rpm for 2 minutes with the transmission in Park or Neutral.

5. Run the engine at normal idle speed for 1 minute in Park or Neutral.

6. Check the idle speed using the figures provided on your underhood sticker. If the indicated idle speed does not agree with the specified speed, adjust the idle by turning the throttle adjusting screw.

NOTE: Idle limiter caps are installed on the mixture adjusting screws so that an incorrect adjustment cannot be made. If a satisfactory idle cannot be obtained within the range of the limiter caps or if the limiter caps prevent access to the mixture screws, remove them and make the adjustment as outlined above. Reinstall the limiter caps so that the cap can

be turned only $\frac{1}{8}$ of a turn counterclockwise before it reaches the stop. Have the engine checked with a CO meter after making the adjustment.

ENGINE ELECTRICAL

Distributor

REMOVAL & INSTALLATION

1. Unlatch the two distributor cap retaining clips and remove the distributor cap with the plug wires attached.

2. Note the position of the rotor in relation to the base. Scribe a mark on the base of the distributor and on the engine block to facilitate reinstallation. Align the marks with the direction the metal tip of the rotor is pointing.

3. Disconnect the vacuum advance hose from the distributor. Remove the bolt which holds the distributor to the engine.

4. Lift the distributor assembly from the engine.

5. Insert the distributor shaft and assembly into the engine. Line up the mark on the distributor and the one on the engine with the metal tip of the rotor. Make sure that the vacuum advance diaphragm is pointed in the same direction as it was pointed originally. This will be done automatically if the marks on the engine and the distributor are lined up with the rotor.

6. Install the distributor hold-down bolt and clamp. Leave the screw loose enough so that you can move the distributor with heavy hand pressure.

7. Connect the distributor wiring harness. Install the distributor cap on the distributor housing. Secure the distributor cap with the spring clips.

8. Adjust the ignition timing as necessary.

NOTE: If the crankshaft has been turned or the engine disturbed in any manner (i.e., disassembled and rebuilt) while the distributor was removed, or if the marks were not drawn, it will be necessary to initially time the engine. Follow the procedure given below.

9. It is necessary to place the No. 1 cylinder in the firing position to correctly install the distributor. To locate this position, the ignition timing marks on the crankshaft front pulley are used.

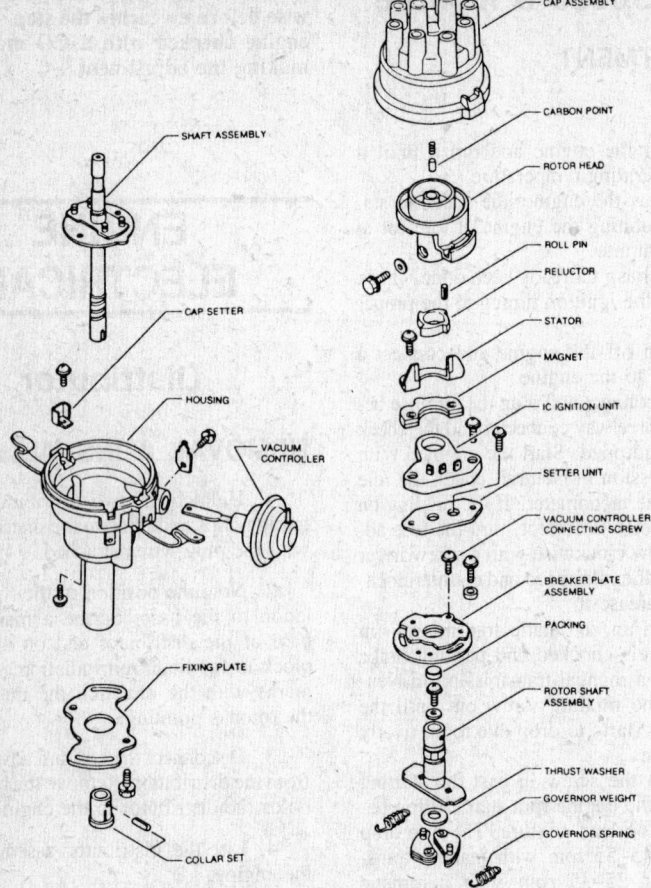

Z-series engine distributor—typical

10. Remove the No. 1 cylinder spark plug. Turn the crankshaft until the piston in the No. 1 cylinder is moving up on the compression stroke. This can be determined by placing your thumb over the spark plug hole and feeling the air being forced out of the cylinder. Stop turning the crankshaft when the timing marks that are used to time the engine are aligned.

11. Oil the distributor housing lightly where the distributor bears on the cylinder block.

12. Install the distributor so that the rotor, which is mounted on the shaft, points toward the No. 1 spark plug terminal tower position when the cap is installed. Of course you won't be able to see the direction in which the rotor is pointing if the cap is on the distributor. Lay the cap on the top of the distributor and make a mark on the side of the distributor housing just below the No. 1 spark plug terminal. Make sure that the rotor points toward that mark when you install the distributor.

13. When the distributor shaft has reached the bottom of the hole, move the rotor back and forth slightly until the driving lug on the end of the shaft enters the slots cut in the end of the oil pump shaft and the distributor assembly slides down into place.

14. Install the distributor hold-down bolt. Readjust the timing as necessary.

Alternator

PRECAUTIONS

To prevent damage to the alternator and regulator, the following precautionary measures must be taken when working with the electrical system.

1. Never reverse battery connections. Always check the battery polarity visually. This is to be done before any connections are made to be sure that all of the connections correspond to the battery ground polarity of the truck.

2. Booster batteries for starting must be connected properly. Make sure that the positive cable of the booster battery is connected to the positive terminal of the battery which is getting the boost.

3. Disconnect the battery cables before using a fast charger; the charger has a tendency to force current through the diodes in the opposite direction for which they were designed. This burns out the diodes.

4. Never use a fast charger as a booster for starting the vehicle.

5. Never disconnect the voltage regulator while the engine is running.

6. Do not ground the alternator output terminal.

7. Do not operate the alternator on an open circuit with the field energized.

8. Do not attempt to polarize an alternator.

9. Disconnect the battery cables before using an electric arc welder on the truck.

BELT TENSION ADJUSTMENTS

Belt tension should be checked with a gauge made for the purpose. If a tension gauge is not available, tension can be checked with moderate thumb pressure applied to the belt at its longest span midway between pulleys. If the belt has a free span less than twelve inches, it should deflect approximately $\frac{1}{8}-\frac{1}{4}$ in. If the span is longer than twelve inches, deflection can range between $\frac{1}{8}-\frac{3}{8}$ in. To adjust or replace belts:

1. Loosen the driven accessory's pivot and mounting bolts.

2. Move the accessory toward or away from the engine until the tension is correct. You can use a wooden hammer handle, or broomstick, as a lever, but do not use anything metallic, such as a prybar.

3. Tighten the bolts and recheck the tension. If new belts have been installed, run the engine for a few minutes, then recheck and readjust as necessary. It is better to have belts too loose than too tight, because overtight belts will lead to bearing failure, particularly in the water pump and alternator. However, loose belts place an extremely high impact load on the driven component due to the whipping action of the belt.

REMOVAL & INSTALLATION

1. Disconnect the negative battery terminal.

2. Disconnect the two lead wires and connector from the alternator.

3. Loosen the drive belt adjusting bolt and remove the belt.

4. Unscrew the alternator attaching bolts and remove the alternator from the vehicle.

5. Install the alternator in the reverse order of removal.

For further servicing information refer to the Electrical section in Unit Repair.

Regulator

REMOVAL & INSTALLATION

The transistorized regulator is soldered to the brush assembly inside the alternator. It is non-adjustable, and must be replaced together with the brush assembly if faulty.

1. Remove the alternator.

2. Remove the through bolts and separate the front cover from the stator housing.

3. Unsolder the wire connecting the diode plate to the brush at the brush terminal.

4. Remove the bolt retaining the diode plate to the rear cover.

5. Remove the nut securing the battery terminal bolt.

6. Lift the stator slightly, together with the diode plate, to gain access to the diode plate screw. Remove the screw.

NOTE: On some models the regulator and brush assembly is riveted to the diode assembly. The solder and rivets must be removed, then reinstalled and stake new ones.

7. Separate the stator and diode, and remove the brush and regulator assembly.

8. Assembly is the reverse. Apply soldering heat sparingly, carrying out the operation as quickly as possible, to avoid heat damage to the transistors and diodes. Before assembling the alternator halves, bend a piece of wire in an "L" and slip it through the rear cover next to the brushes. Use the wire to hold the brushes in a retracted position until the case halves are assembled. Remove the wire carefully, to prevent damage to the slip rings.

Starter
REMOVAL & INSTALLATION

1. Disconnect the negative battery cable from the battery.

2. Disconnect the starter wiring at the starter, taking note of the positions for correct reinstallation.

3. Remove the bolts attaching the starter to the engine and remove the starter from the vehicle.

4. Install the starter in the reverse order of removal.

For further starter motor servicing, refer to the Electrical section of Unit Repair.

ENGINE MECHANICAL

Engine

REMOVAL & INSTALLATION

It is much easier to remove the engine and the transmission together as an assembly than to remove only the engine from the engine compartment. After the engine and transmission are removed from the vehicle, the two can be separated.

1. Disconnect the battery ground cable. Remove the battery.

2. Mark the location of the hood hinges on the body in order to facilitate installation and remove the hood.

3. Remove the air cleaner after disconnecting the PCV hose from the rocker cover.

4. Drain the radiator of coolant and the engine crankcase of oil.

5. Disconnect the upper and lower radiator hoses from the engine. Disconnect and plug the automatic transmission cooler lines at the radiator, if so equipped. Use a flare nut wrench if one is available.

6. Remove the bolts or brackets securing the radiator and remove the radiator from the vehicle.

7. Disconnect the engine ground cable at the cylinder head.

8. Disconnect the electrical leads at the starter, alternator, distributor, the high-tension ignition coil cable, and the oil pressure and temperature sending units' wires.

9. Disconnect the fuel pump (or filter on electric pump models), the heater hose at the engine side, and the choke wire and accelerator cable at the carburetor. Disconnect the emission hoses or wires to the carbon canister, air pump, B.C.D.D. solenoid, and fuel cut solenoid; the vacuum hose to the brake booster (on models so equipped), and any other wires or hoses running to the engine. Tag all wires as they are disconnected for assembly.

10. Remove the transmission control linkage from the transmission; in the case of an automatic transmission, remove the cross-shaft assembly from the transmission. Remove the selector rod from the selector lever on the automatic transmission. On manual transmissions, lift the rubber boot and remove the nut or C-clip from the shift lever and detach the shift lever from the transmission.

11. Remove the two bolts securing the clutch slave cylinder. Disconnect the clutch slave cylinder and flexible tubing as an assembly.

12. Disconnect the speedometer cable and the back-up light wiring (and neutral switch, if equipped) from the rear section of the transmission.

13. Disconnect the exhaust pipe from the exhaust manifold.

14. Disconnect the driveshaft center bearing bracket from the third crossmember of the frame. Disconnect the driveshaft at the differential housing. Remove the driveshaft assembly from the vehicle and plug the rear end of the transmission extension housing to prevent loss of transmission lubricant.

15. Attach a suitable lifting device to the engine and lift the engine slightly.

16. Remove the front engine mount bolts on both sides of the engine.

17. Place a jack under the transmission and lift it slightly.

18. Loosen the two combination engine rear mounting/transmission mounting bolts. On models with a catalytic converter, loosen the two exhaust pipe hanger bolts.

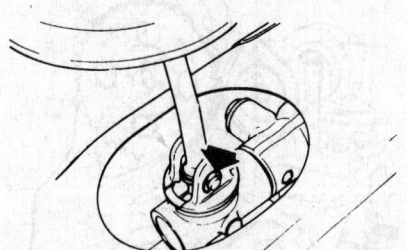

Remove the C-clip and pin on later models for shift lever removal

19. Remove the bolts securing the idler arm to the frame, and push down the tie-rod.

20. Pull the engine toward the front as far as possible and carefully raise the engine with the transmission up and out of the vehicle.

21. Install the engine in the reverse order of removal. Do not connect any parts to the engine or transmission until the engine and transmission are in place on the engine/transmission mounts and secured by the mounting bolts. Secure the rear support first and then the front engine mounts, using the upper bolt hole as a guide.

Cylinder Head

REMOVAL & INSTALLATION
L20B Engines

1. Crank the engine until the No. 1 piston is at TDC of the compression stroke and disconnect the negative battery cable, drain the cooling system, and remove the air cleaner and attending hoses.

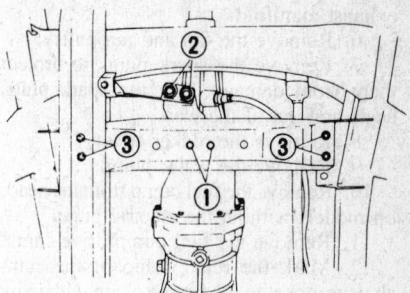

1. Engine mount bolts
2. Exhaust pipe bolts
3. Crossmember bolts

Engine and transmission cross member removal

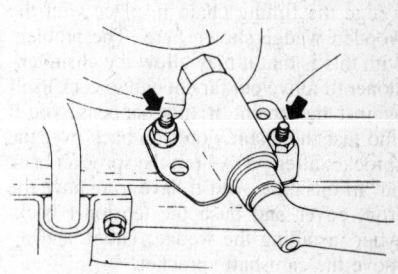

Idler arm removal

2. Remove the alternator.

3. If equipped with air conditioning, unbolt the compressor and move it aside onto the fender. Do not detach any of the compressor lines; the escaping refrigerant will freeze any surface it contacts, including your skin.

4. Disconnect the carburetor throttle linkage, the fuel line and any other vacuum lines or electrical leads, and remove the carburetor.

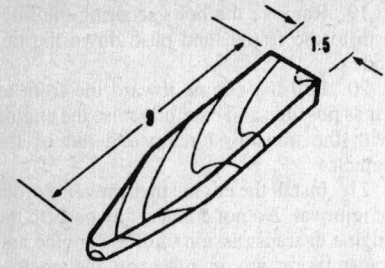

Dimensions for fabricating the wooden wedge used to support the timing chain

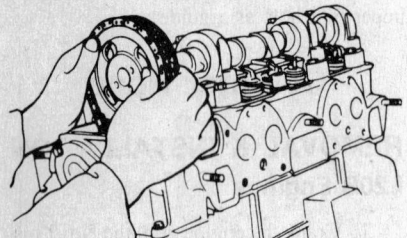

Removing the camshaft sprocket and chain

5. Disconnect the exhaust pipe from the exhaust manifold.

6. Remove the fan and fan pulley.

7. Remove the spark plugs to protect them from damage. Lay the spark plugs aside and out of the way.

8. Remove the rocker cover.

9. Remove the water pump.

10. Remove the fuel pump from the head, on models without the electric pump.

11. Remove the fuel pump drive cam.

12. Mark the relationship of the camshaft sprocket to the timing chain with paint or chalk. If this is done, it will not be necessary to locate the factory timing marks. Before removing the camshaft sprocket, it will be necessary to wedge the chain in place so that it will not fall down into the front cover. The factory procedure is to wedge the timing chain in place with the wooden wedge shown here. The problem with this is that it may allow the chain tensioner to move out far enough to cock itself against the chain. If this happens, you'll find that the chain won't go back over the sprocket after you've put the sprocket back on. In this case, you'll have to remove the front cover and push the tensioner back. After installing the wedge, unbolt and remove the camshaft sprocket.

13. Loosen and remove the cylinder head bolts. You will need a 10 mm Allen wrench to remove the head bolts. Keep the bolts in order, because they are different sizes. Lift the cylinder head assembly from the engine. Remove the intake and exhaust manifolds as necessary.

14. Thoroughly clean the cylinder block and head mating surfaces. Check the block and head for flatness before installing the head. Install a new cylinder head gasket. Do not use sealer on the cylinder head gasket.

15. With the crankshaft turned so that the No. 1 piston is at TDC of the compression stroke (if not already done so as mentioned in Step 1), make sure that the camshaft sprocket timing mark and the oblong groove in the camshaft retaining plate are aligned.

16. Place the cylinder head in position on the cylinder block, being careful not to allow any of the valves to come in contact with any of the pistons. Do not rotate the crankshaft or camshaft separately because of possible damage which might occur to the valves.

17. Temporarily tighten the two center right and left cylinder head bolts to 14.5 ft. lbs.

18. Install the camshaft sprocket together with the timing chain to the camshaft. Make sure that the marks you made earlier line up. If the chain will not stretch over the sprocket, the problem lies in the tensioner. See "Timing Chain Removal and Installation" for timing procedure, if necessary.

19. Install the cylinder head bolts. Note that there are two sizes of bolts used; the longer bolts are installed on the driver side of the engine with a smaller bolt in the center position. The remaining small bolts are installed on the opposite side of the cylinder head.

20. Tighten the cylinder head bolts in three stages: first to 29 ft. lbs, second to 43 ft. lbs, and lastly to 47–62 ft. lbs. Tighten the cylinder head bolts on all models in the proper sequence.

21. Install and assemble the remaining components of the engine in the reverse order of removal.

Z20, Z22 and Z24 Engine

1. Complete Steps 1–5 under L20 Overhead Camshaft Engine.

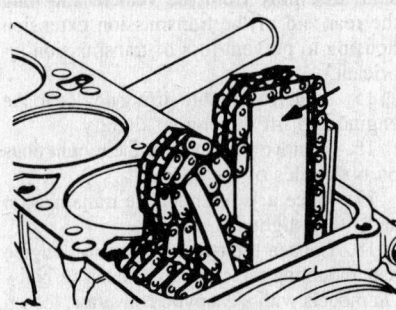

Support the timing chain with a wedge

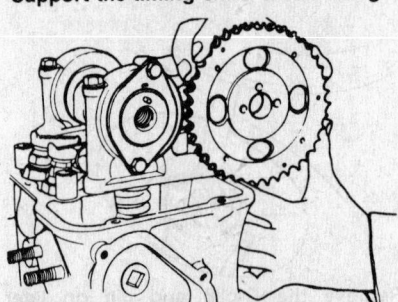

Installing the camshaft sprocket

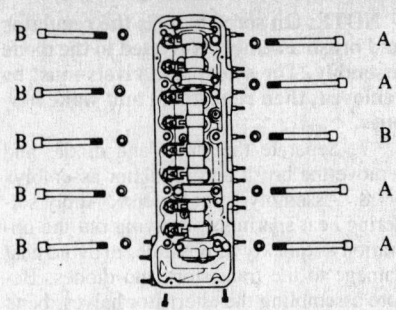

Different size cylinder head bolts

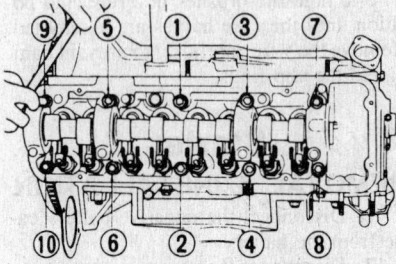

L20B head bolt tightening sequence

2. Disconnect the throttle linkage, the air cleaner or its intake hose assembly (fuel injection). Disconnect the fuel line, the return fuel line and any other vacuum lines or electrical leads. Label the lines, wires and hoses for reinstallation location. Remove the carburetor to avoid damaging it while removing the head.

3. Remove the EGR tube from around the rear of the engine.

4. Remove the exhaust air induction tubes from around the front of the engine.

5. Unbolt the exhaust manifold from the exhaust pipe. Remove the fuel pump.

6. Remove the PCV valve from around the rear of the engine if necessary.

7. Remove the spark plugs to protect them from damage. Remove the valve cover.

8. Mark the relationship of the camshaft sprocket to the timing chain with paint or chalk. If this is done, it will not be necessary to locate the factory timing marks. Before removing the camshaft sprocket, it will be necessary to wedge the chain in place so that it will not fall down into the front cover. The factory procedure is to wedge the timing chain in place with the wooden wedge shown here. The problem with this procedure is that it may allow the chain tensioner to move out far enough to cock itself against the chain. If this happens, you'll find that the chain won't go back on. In this case, you'll have to remove the front cover and push the tensioner back. After you've wedged the chain, unbolt the camshaft sprocket and remove it.

9. Working from both ends in, loosen the cylinder head bolts and remove them. Remove the bolts securing the cylinder head to the front cover assembly.

10. Lift the cylinder head off the engine block. It may be necessary to tap the head lightly with a copper or brass mallet to loosen it.

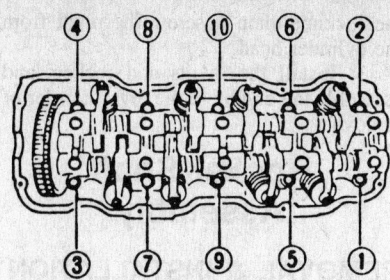

Z series engine head bolt loosening sequence

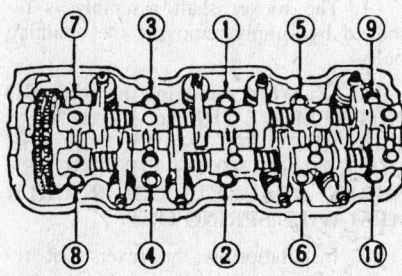

Z series engine head bolt tightening sequence

11. To install the cylinder head: Thoroughly clean the cylinder block and head surfaces and check both for warpage.

12. Fit the new head gasket. Don't use sealant. Make sure that no open valves are in the way of raised pistons, and do not rotate the crankshaft or camshaft separately because of possible damage which might occur to the valves.

13. Temporarily tighten the two center right and left cylinder head bolts to 14 ft. lbs.

14. Install the camshaft sprocket together with the timing chain to the camshaft. Make sure the marks you made earlier line up with each other. Refer to "Timing Chain Removal and Installation".

15. Install the cylinder head bolts and torque them to 20 ft. lbs., then 40 ft. lbs., then 58 ft. lbs. in the order shown in the illustration.

16. Assemble the rest of the components in the reverse order of disassembly. It is always wise to drain the crankcase oil after the cylinder head has been installed to avoid coolant contamination.

SD22 and SD25 Diesel

1. Remove the air cleaner.

2. Remove the crankcase vent hose and remove the intake and exhaust manifolds. These are bolted together.

3. Remove the alternator, bracket and belts.

4. Disconnect the coolant hose between the head and the oil cooler.

5. Remove the fuel filter assembly.

6. Disconnect the injection lines from the pump and the injectors. Cap all openings at once.

7. Remove the bypass hoses between the coolant pump and the thermostat housing.

8. Remove the fan.

9. Remove the rocker arm cover.

10. Remove the rocker arm shaft assembly.

11. Remove the pushrods and keep them in order.

12. Remove the fuel return lines.

13. Remove the nozzles from the head.

14. Remove the cylinder head bolts in the sequence shown.

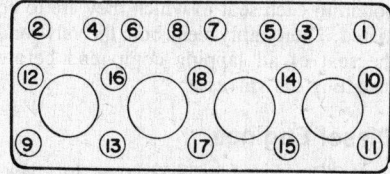

SD22, SD25 head bolt loosening sequence

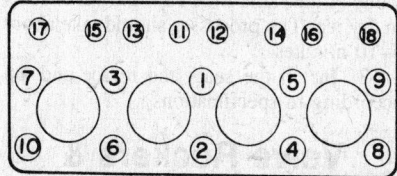

SD22, SD25 head bolt tightening sequence

15. Attach a hoist to the head and lift it clear of the block. On occasion, the precombustion chambers may fall out, especially if the head is bumped or handled roughly. Take care that they are returned to their original positions if this occurs.

16. Remove the head gasket and O-rings.

17. Clean and inspect all parts.

18. Check the head with a straight-edge. Maximum warpage is .0079. Do not remove more than .011 from the head.

19. Place a new cylinder head gasket on the block with the stainless steel inset side facing up.

20. Install the O-rings around the water and oil passages.

21. Position the head on the block.

22. Coat the head bolts with clean engine oil and torque them in sequence, in stages as follows:
Large: 43, 94; Small: 21, 36.

23. Install the pushrods, pressing down and turning them to be sure of proper seating.

24. Install the rocker arm shaft assembly, torquing the bolts to 18 ft. lbs. in sequence from the center to each end.

25. Install the injection nozzles.

26. Install all other parts in reverse order of removal.

Valve Guides

REMOVAL & INSTALLATION

NOTE: The valve guides on diesel engines are not replaceable.

1. With the cylinder head removed from the engine, and the valves removed from

the head, use a drift and a hammer or press. Drive the valve guides out from the combustion chamber side toward the rocker cover side. A heated cylinder head will facilitate the operation.

2. Ream the cylinder head side guide hole at room temperature. The guide hole should be 0.4719–0.4723 in. for standard valves and 0.4797–0.4802 in. for 0.0079 in. oversize valves which are available for service.

3. After heating the cylinder head to 302–392°F, press the new valve guide carefully into the cylinder head. The top of the valve guide should protrude out the top of the guide hole 0.4173 in.

4. Ream the bore of the valve guide with the valve guide pressed into the cylinder head. The standard valve guide bore size is 0.3150–0.3157 in.

5. Assemble the cylinder head and install it on the engine in the reverse order of removal.

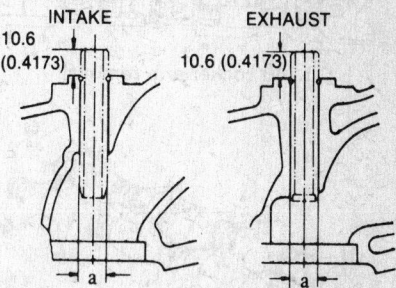

Gasoline engine valve guide installation

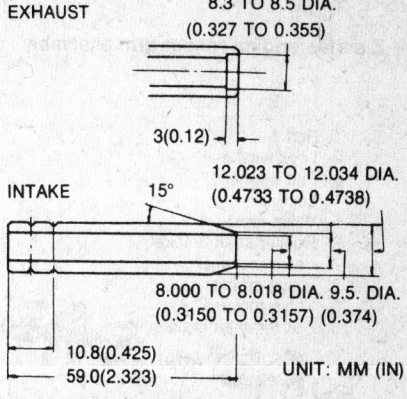

Valve guide dimensions

Valve Seat
REMOVAL & INSTALLATION
Gasoline Engines

1. With the cylinder head removed from the engine and the valves removed from the cylinder head, old valve seat inserts can be removed by boring them out until they collapse. Be careful that the boring doesn't continue beyond the bottom face of the insert recess in the cylinder head.

2. Select the suitable valve seat insert and check its outside diameter.

3. Machine the cylinder head recess using the center of the valve guide as the center of the valve seat insert so that the insert will have the correct fit.

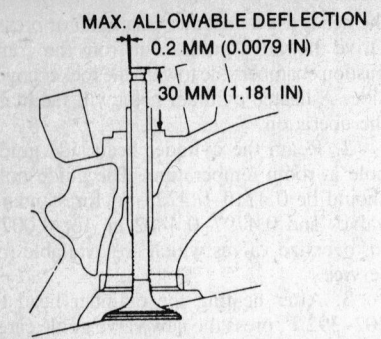

MAX. ALLOWABLE DEFLECTION
0.2 MM (0.0079 IN)
30 MM (1.181 IN)

Measuring the valve stem-to-guide clearance

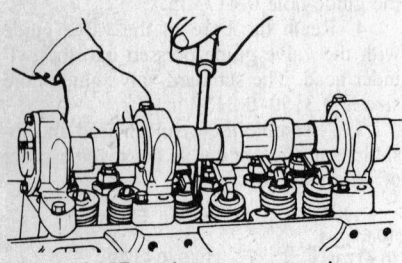

L20B rocker arm removal

Z series engine rocker arm assembly

4. Ream the cylinder head recess at room temperature.

5. Heat the cylinder head to 302–392°F.

6. Fit the insert, making sure that it seats fully in the recess in the cylinder head. Peen the insert with a punch in at least four places equally spaced around its circumference.

7. Grind the valve seats to the proper angle.

8. Lap the valves with lapping compound to each seat to which they are to be mated. Thoroughly clean both the valve and the seat of all lapping compound before installing the valves.

Diesel Engines

1. The seats may be removed by cracking with a cold chisel.

2. Immerse the head in water at 175°F while at the same time cool the valve seats in dry ice. The processes should take about 5–10 minutes.

3. Install the seats and reface and lap according to specifications.

Valve Rockers & Rocker Pivots

REMOVAL & INSTALLATION

L20B

1. Loosen the rocker pivot locknut, lower the pivot by screwing it down into the cylinder head, and remove the rocker arm by pressing down on the valve spring.

2. To remove the rocker pivots, loosen the locknut, then unscrew the pivot from the cylinder head.

3. Install the pivots and rockers and assemble the engine in the reverse order of removal.

Rocker Shaft Assembly

REMOVAL & INSTALLATION

Z20, Z22 and Z24

1. The rocker shaft assembly is removed by simply removing the retaining bolts.

NOTE: When removing the bolts, DO NOT REMOVE THE No. 1 AND NO. 5 BRACKET BOLTS FROM THE ROCKER STAND OR THE ROCKER SHAFT BRACKET AND ROCKER ARM WILL SPRING OUT.

2. Installation is the reverse of removal. Torque the bolts evenly from the ends toward the center to 11–18 ft. lbs.

SD22 and SD25 Diesel

1. Remove the shaft retaining bolts evenly, from the center towards the ends.

2. Lift the shaft assembly off the head.

3. If you are disassembling the shaft and rocker arms, it may be necessary to immerse the assembly in water heated to 160°F for a few minutes to free the rocker arms. NEVER HAMMER THEM OFF!

4. Installation is the reverse of removal. Torque the retaining bolts evenly from the ends toward the center to 14–18 ft. lbs.

Intake Manifold

REMOVAL & INSTALLATION

All Engines

1. Remove the air cleaner assembly together with all of the attending hoses. Remove the EGR tube.

NOTE: It is important to replace the gasket whenever the intake manifold is removed. Because the intake and exhaust manifolds share a common gasket, whenever the intake manifold is removed, the exhaust manifold must also be removed, so that the gasket can be replaced.

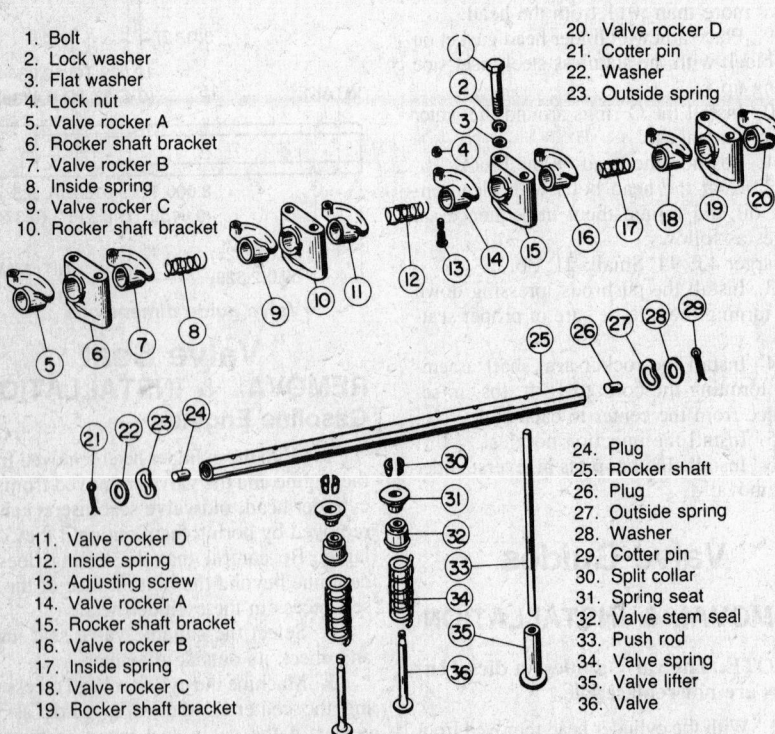

1. Bolt
2. Lock washer
3. Flat washer
4. Lock nut
5. Valve rocker A
6. Rocker shaft bracket
7. Valve rocker B
8. Inside spring
9. Valve rocker C
10. Rocker shaft bracket
11. Valve rocker D
12. Inside spring
13. Adjusting screw
14. Valve rocker A
15. Rocker shaft bracket
16. Valve rocker B
17. Inside spring
18. Valve rocker C
19. Rocker shaft bracket
20. Valve rocker D
21. Cotter pin
22. Washer
23. Outside spring
24. Plug
25. Rocker shaft
26. Plug
27. Outside spring
28. Washer
29. Cotter pin
30. Split collar
31. Spring seat
32. Valve stem seal
33. Push rod
34. Valve spring
35. Valve lifter
36. Valve

SD22, SD25 rocker arm assembly

Intake manifold and gasket

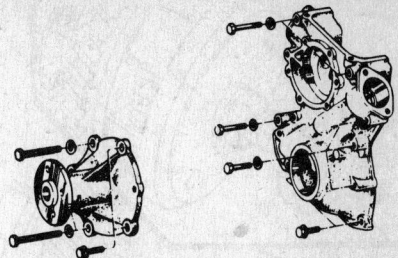

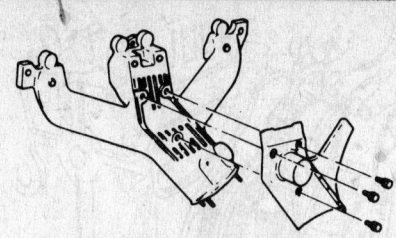

Exhaust manifold and heat stove

2. Disconnect the throttle linkage, fuel, and vacuum lines from the carburetor. Label all wires and hoses as they are removed to simplify installation.

3. The carburetor can be removed from the manifold at this point or can be removed as an assembly with the intake manifold.

4. Loosen the intake manifold attaching nuts, working from the two ends toward the center, and then remove them.

5. Remove the intake manifold from the engine.

6. Install the intake manifold in the reverse order of removal. Always use a new gasket when installing the manifold; air leaks will cause burnt valves. Tighten the manifold bolts from the center outwards, in two progressive steps, to 9–12 ft. lbs.

Exhaust Manifold

REMOVAL & INSTALLATION

All Engines

1. Remove the air cleaner assembly.
2. Disconnect the exhaust pipe from the exhaust manifold.

It is not absolutely necessary to replace the gasket when only the exhaust manifold is removed, unless the gasket is damaged, or leaks develop.

3. Loosen and remove the exhaust manifold attaching nuts and remove the manifold from the engine.

4. Install the exhaust manifold in the reverse order of removal. Use new gaskets at the cylinder head (if necessary) and exhaust pipe. Tighten the mounting bolts in a circular pattern, working from the center to the ends, in two progressive steps to the figures in the torque chart.

Timing Gear Cover

REMOVAL & INSTALLATION

Gasoline Engines

1. Disconnect the negative battery cable from the battery, drain the cooling system, and remove the radiator together with the upper and lower radiator hoses.

2. Loosen the alternator drive belt adjusting screw and remove the drive belt. Remove the bolts which attach the alternator bracket to the engine and set the alternator aside out of the way.

3. Remove the distributor.

4. Remove the oil pump attaching screws, and take out the pump and its drive spindle.

5. Remove the cooling fan and the fan pulley together with the drive belt.

6. Remove the water pump.

7. Remove the crankshaft pulley bolt and remove the crankshaft pulley.

8. Remove the bolts holding the front cover to the front of the cylinder block, the four bolts which retain the front of the oil pan to the bottom of the front cover and the two bolts which are screwed down through the front of the cylinder head and into the top of the front cover.

9. Carefully pry the front cover off the front of the engine.

10. Cut the exposed front section of the oil pan gasket away from the oil pan. Do the same to the gasket at the top of the front cover. Remove the two side gaskets and clean all of the mating surfaces.

11. Cut the portions needed from a new oil pan gasket and top front cover gasket.

12. Apply sealer to all of the gaskets and position them on the engine in their proper places.

13. Apply a light coating of oil to the crankshaft oil seal and carefully mount the front cover to the front of the engine and install all of the mounting bolts. Tighten the 8mm bolts to 7–12 ft. lbs. and the 6mm bolts to 3–6 ft. lbs. Tighten the oil pan attaching bolts to 4–7 ft. lbs.

14. Before installing the oil pump, place the gasket over the shaft and make sure that the mark on the drive spindle faces (aligned with) the oil pump hole. On L20B engines, install the oil pump so that the projection on the top of the shaft is located in the exact position as when it was removed or is in the 11:25 o'clock position with the piston in the No. 1 cylinder is placed at TDC on the compression stroke, if the engine was disturbed since disassembly. Tighten the oil pump attaching screws to 8–10 ft. lbs. See Oil Pump Removal and Installation procedures for "Z" engines.

APPLY SEALANT AT THESE POINTS

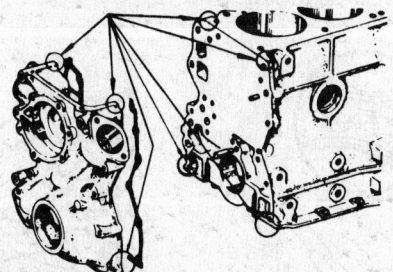

Gasoline engine front cover installation

Diesel Engines

NOTE: A 41mm (1.614 in.) socket is needed for this procedure.

1. Remove the fan and pulley.
2. Remove the water pump bypass hose and allow the cooling system to drain below the level of the water pump.

Gasoline engine front cover bolts

Crankshaft pulley nut removal

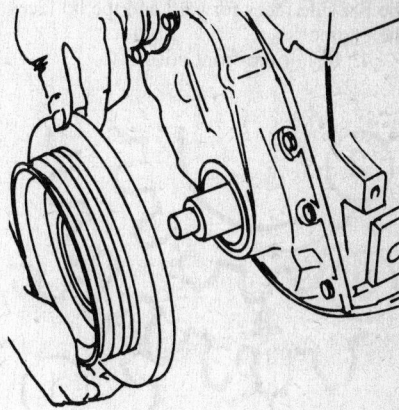

Lifting the pulley assembly

3. Remove the three bolts and lift the water pump and gasket off the block. Discard the gasket.

4. Remove the crankshaft pulley nut with a 41mm socket.

5. Drive the pulley from the crankshaft with a wooden or plastic mallet.

6. Remove the five bolts and lift the timing gear cover from the case.

7. Installation is the reverse of removal. Always replace the cover oil seal and use a new cover gasket. Torque the cover bolts to 8 ft. lbs., the crankshaft pulley nut to 238 ft. lbs., and the water pump bolts to 8 ft. lbs. for the 8mm bolt and 16 ft. lbs. for the 10mm bolts.

NOTE: Do not tighten the water pump bolts until the belt adjuster is installed when installing the alternator.

Timing Gear Cover Oil Seal

REMOVAL & INSTALLATION

1. Remove the front cover.

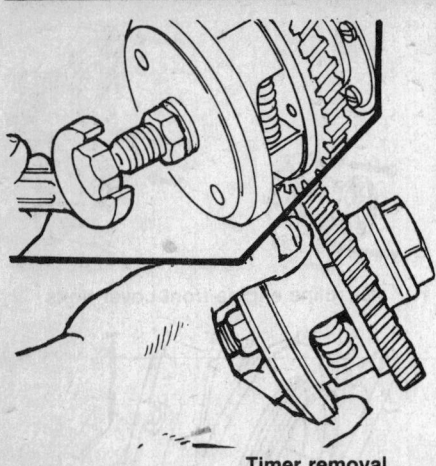

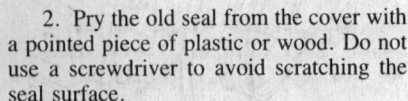

Timer removal

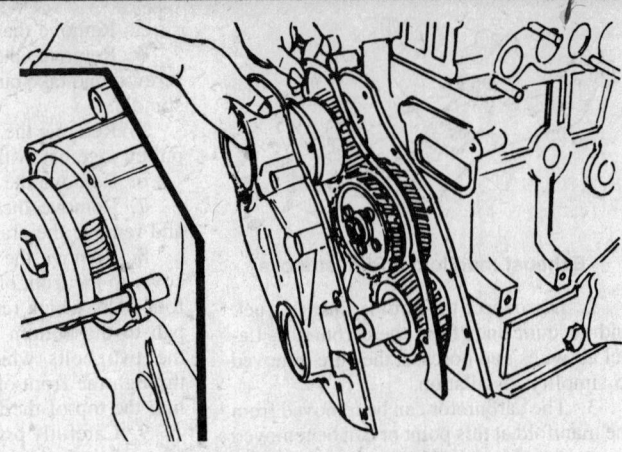

Timing gear case removal

2. Pry the old seal from the cover with a pointed piece of plastic or wood. Do not use a screwdriver to avoid scratching the seal surface.

3. Oil the lip of the new seal. Do not use grease. Press it into place, making sure the flat side faces forward and the lip faces the engine.

4. Install the front cover.

Timing Gears and Case

NOTE: The following requires use of special tools.

REMOVAL & INSTALLATION

SD22 and SD25 Diesel

1. Remove the timing gear cover.
2. Remove the timing gear round nut.
3. Using timer extractor 57926–581, thread the tool into the timer weight holder. Remove the timing gear assembly by threading in the extractor tool bolt.
4. Unbolt and remove the camshaft gear set.
5. Remove the oil slinger. Unbolt the crankshaft gear and remove it with a gear puller.
6. Install the camshaft gear.
7. Install the crankshaft gear and oil slinger while carefully aligning the timing marks as shown. Measure the gear backlash. Backlash should be 0.0028–0.0079 in.
8. With the No. 1 piston at TDC, mesh the timing gear and idler gear at the "Y" marks. After aligning the gear with the keyway, secure the timer assembly with a lock washer and the round nut. Torque the nut to 50–58 ft. lbs.

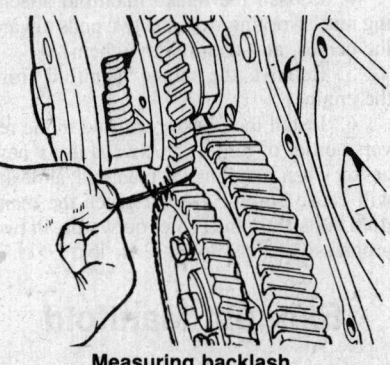

Measuring backlash

9. Install the cover.

NOTE: If the gear case oil jet was removed, install in the relationship as shown.

Timing Chain and Tensioner

REMOVAL & INSTALLATION

Gasoline Engines

1. Before beginning any disassembly procedures, position the No. 1 piston at TDC on the compression stroke.

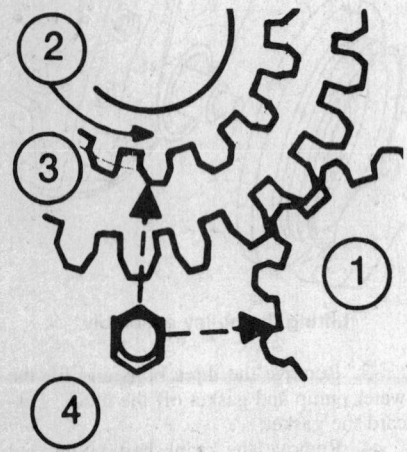

1. Crankshaft gear
2. Idler
3. Camshaft gear
4. Oil jet

Oil jet orientation

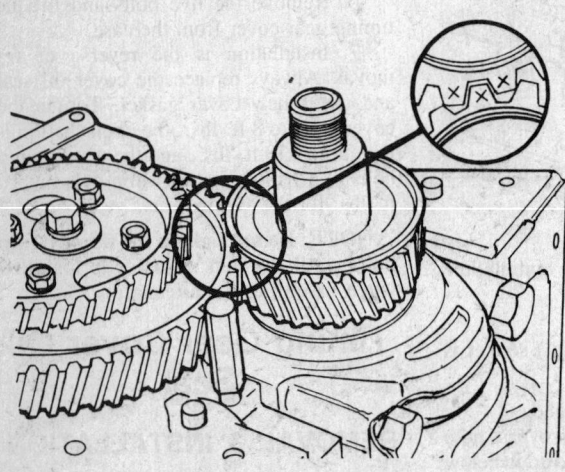

Timing mark alignment

Oil slinger installation

2. Remove the front cover. Remove the camshaft cover.

3. With the No. 1 piston at TDC, the timing marks on the camshaft sprocket and the timing chain should be visible. Mark both of them with paint. Also mark the relationship of the camshaft sprocket to the camshaft. At this point you will see that there are three sets of timing marks and locating holes in the sprocket. They are for making adjustments to compensate for timing chain stretch. See the "Timing Chain Adjustment" section following for details.

4. With the timing marks on the cam sprocket clearly marked, locate and mark the timing marks on the crankshaft sprocket. Also mark the chain timing mark. Of course, if the chain is not to be reused, marking it is useless.

5. Unbolt the camshaft sprocket and remove the sprocket along with the chain. As you remove the chain, hold it where the chain tensioner contacts it. When the chain is removed, the tensioner is going to come apart. Hold on to it and you won't lose any of the parts. The crankshaft sprocket can be removed with a puller, if necessary. There is no need to remove the chain guide unless it is being replaced.

6. Install the timing chain and the camshaft sprocket together after first positioning the chain over the crankshaft sprocket. Position the sprocket so that the marks made earlier line up. This is assuming that the engine has not been disturbed. The camshaft and the crankshaft keys should both be pointing upward. If a new chain and/or gear is being installed, position the sprocket so that the timing marks on the chain align with the marks on the sprocket (with both keys pointing up). The marks are on the right-hand side of the sprockets as you face the engine. The L20B engine has 44 pins between the mating marks of the chain and

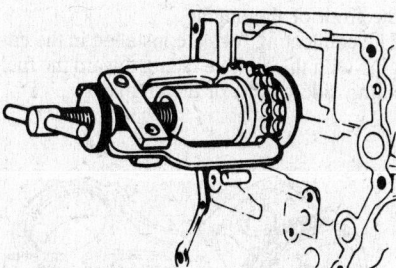

Gasoline engine crankshaft sprocket removal

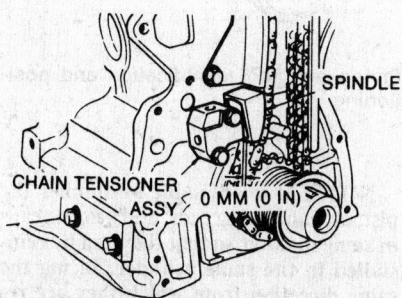

Installing the timing chain tensioner

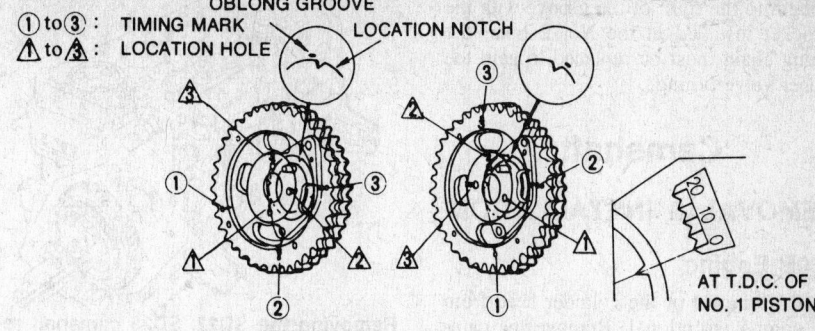

① to ③ : TIMING MARK
⚠ to ⚠ : LOCATION HOLE
OBLONG GROOVE
LOCATION NOTCH
AT T.D.C. OF NO. 1 PISTON

AFTER ADJUSTMENT   BEFORE ADJUSTMENT

Adjusting the camshaft sprocket to obtain proper valve timing due to a worn timing chain

sprockets when the chain is installed correctly. Count the pins. There are two pins per chain link. This is an important step. If you do not get the exact number of pins between the timing marks, valve timing will be incorrect, and the engine will either not run at all or run very badly. Z20 and Z22 engines do not use the pin counting method for finding correct valve timing. Instead,

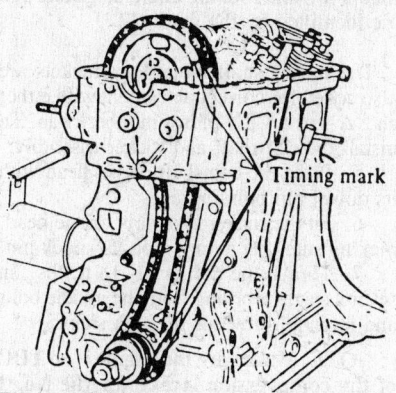

Timing mark

Timing chain and sprocket alignments—Z20, Z22 and Z24 engines

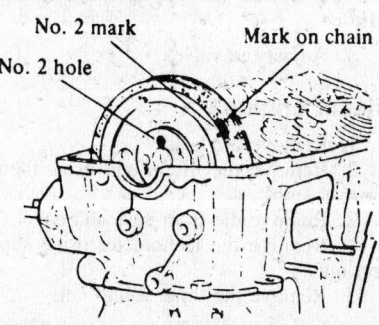

No. 2 mark
No. 2 hole
Mark on chain

Use the No. 2 mark and the hole to align the camshaft—Z20, Z22 and Z24 engines

position the key in the crankshaft sprocket so that it is pointing upward and install the camshaft sprocket on the camshaft with its dowel pin at the top using the No.2 (No. 1 for L24 and L28) mounting hole and timing mark. The painted links of the chain should be on the right hand side of the sprockets as you face the engine. See the illustration.

7. Install the chain tensioner. Adjust the protrusion of the chain tensioner spindle to zero clearance.

8. With a new seal installed in the front cover and a light coat of oil applied to the seal, assemble the remaining components of the engine in the reverse order of disassembly.

TIMING CHAIN ADJUSTMENT

When the timing chain stretches excessively, the valve timing will be adversely affected. If the stretch of the chain roller links is excessive, adjust the camshaft sprocket location by transferring the camshaft set position of the camshaft sprocket from the factory position of No. 1 to No. 2 hole as follows:

1. Turn the crankshaft until the No. 1 piston is at TDC on its compression stroke. Examine whether the camshaft sprocket location notch is to the left of the oblong groove on the camshaft retaining plate. If the notch in the sprocket is to the left of the groove in the retaining plate, then the chain is stretched and needs adjusting.

2. Remove the camshaft sprocket together with the chain and reinstall the sprocket and chain with the locating dowel on the camshaft inserted into the No. 2 hole of the sprocket and the timing mark on the timing chain aligned with the No. 2 mark on the sprocket. The amount of modification is 4° of the crankshaft rotation.

3. Recheck the valve timing as outlined in Step 1. The notch in the sprocket should be to the right of the groove in the camshaft retaining plate.

4. If and when the notch cannot be

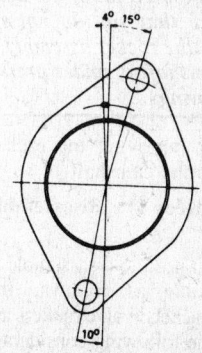

The camshaft retaining plate

brought to the right of the groove with the sprocket installed in the No. 2 hole, the timing chain must be replaced to gain the proper valve timing.

Camshaft

REMOVAL & INSTALLATION

L20B Engine

1. Removal of the cylinder head from the engine is optional. Remove the camshaft sprocket from the camshaft together with the timing chain.

2. Loosen the valve rocker pivot locknut and remove the rocker arm by pressing down on the valve spring. Remove all of the rocker arms in this manner.

3. Remove the two retaining nuts on the camshaft retainer plate at the front of the cylinder head and carefully slide the camshaft out of the camshaft carrier.

4. Lightly coat the camshaft bearings with clean motor oil and carefully slide the camshaft in place in the camshaft carrier.

5. Install the camshaft retainer plate with the oblong groove in the face of the plate facing toward the front of the engine.

6. Check the valve timing as outlined under "Timing Chain Removal and Installation" and install the timing sprocket on the camshaft, tightening the bolt together with the fuel pump cam to 86–116 ft. lbs.

7. Install the rocker arms by pressing down the valve springs with a screwdriver and install the valve rocker springs.

8. Install the cylinder head, if it was removed, and assemble the rest of the engine in the reverse order of removal.

Z20 and Z22 Engines

1. Removal of the cylinder head from the engine is optional. Remove the camshaft sprocket from the camshaft together with the timing chain, after setting the No. 1 piston at TDC on its compression stroke.

2. Loosen the bolts holding the rocker shaft assembly in place and remove the six center bolts. Do not pull the four end bolts out of the rocker assembly because they hold the unit together.

— **CAUTION** —

When loosening the bolts, work from the ends in and loosen all of the bolts a little at a time so that you do not strain the camshaft or the rocker assembly. Remember, the camshaft is under pressure from the valve springs.

3. After removing the rocker assembly, remove the camshaft.

NOTE: Keep the disassembled parts in order.

4. If you need to disassemble the rocker unit, assemble as follows. Install the mounting brackets, valve rockers and springs observing the following considerations. The rocker shafts are different. Both have punch

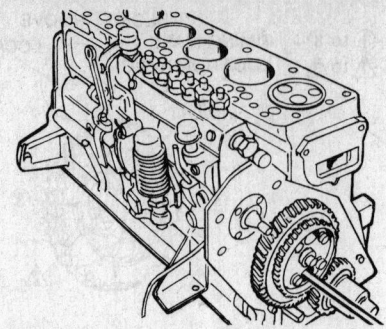

Removing the SD22, SD25 camshaft retaining bolts

marks in the ends that face the front of the engine. The rocker shaft that goes on the side of the intake manifold has two slits in its end just below the punch mark. The exhaust side rocker shaft does not have slits.

The rocker arm for the intake and exhaust valves are interchangeable between cylinders one and three and are identified by the mark "1". Similarly, the rockers for cylinders two and four are interchangeable and are identified by the mark "2".

The rocker shaft mounting brackets are also coded for correct placement with either an "A" or a "Z" plus a number code. To install the camshaft and rocker assembly:

5. Place the camshaft on the head with its dowel pin pointing up.

6. Fit the rocker assembly on the head, making sure you mount it on its knock pin.

7. Torque the bolt to 11–18 ft. lbs., in several stages working from the middle bolts and moving outwards on both sides.

NOTE: Make sure the engine is on TDC of the compression stroke for the No. 1 piston or you may damage some valves. See the section on timing chain installation.

8. Adjust the valves.

Diesel Engines

1. Remove the head.
2. Remove the lifters and mark them for reassembly.
3. Remove the front case and cover.
4. Remove the tachometer drive support nuts.
5. Remove the timer round nut.

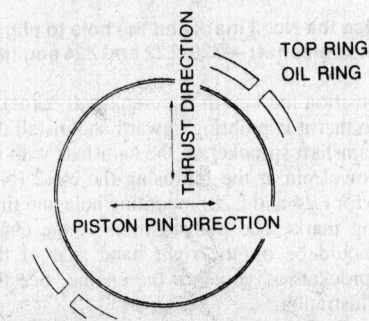

Arrangement of the piston ring gaps around the piston

6. Thread the timer extractor, ST 57926–581, into the timer weight holder. Remove the timer assembly by tightening the extractor bolt.

7. Remove the oil pump drive spindle.

8. Remove the camshaft locating plate bolts and carefully slide the camshaft from the engine.

9. Coat the camshaft with clean engine oil and carefully slide it into the block. Install the locating plate.

10. Install the oil pump drive spindle by aligning the oil pump drive shaft groove and the camshaft oil pump drive gear with the spindle.

11. Install all other parts in reverse order.

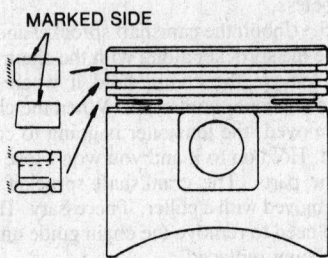

Piston ring installation

Pistons and Connecting Rods

PISTON AND CONNECTING ROD IDENTIFICATION AND POSITIONING

The pistons are marked with a notch in the piston head. When installed in the engine, the notch markings are to be facing toward the front of the engine.
The connecting rods are installed in the engine with the oil hole facing toward the fuel pump side (right) of the engine.

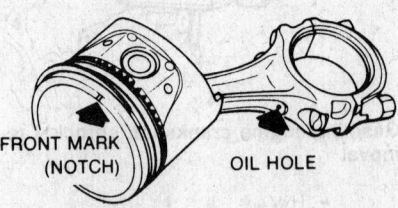

Piston and rod identification and positioning

NOTE: It is advisable to number the pistons, connecting rods, and bearing caps in some manner so that they can be reinstalled in the same cylinder, facing the same direction from which they are removed.

ENGINE LUBRICATION

Oil Pan

REMOVAL & INSTALLATION

To remove the oil pan it will be necessary to unbolt the motor mounts and jack the engine to gain clearance. Block the engine in position and drain the oil. Remove the attaching screws and remove the oil pan and gasket. Install the oil pan in the reverse order with a new gasket. Apply a thin bead of silicone seal to the engine block at the junction of the block and front cover, and the junction of the block and main bearing cap. Then apply a thin coat of silicone seal to the new oil pan gasket, install the gasket to the block, and install the pan. Tighten the pan bolts in a circular pattern from the center to the ends, to 4–7 ft. lbs. Overtightening will distort the pan lip, causing leakage.

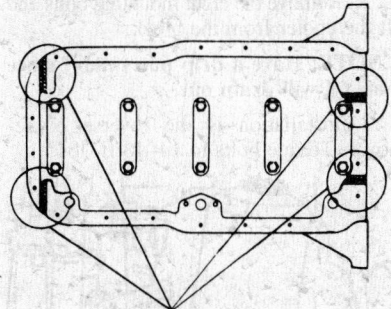

APPLY SEALANT AT THESE POINTS

Apply a thin bead of silicone sealer to these areas before installation

Rear Main Oil Seal

REPLACEMENT

In order to replace the rear main oil seal, the rear main bearing cap must be removed. Removal of the rear main bearing cap requires the use of a special rear main bearing cap puller. Also, the oil seal is installed with a special crankshaft rear oil seal drift.

1. Remove the engine and transmission assembly from the vehicle.
2. Remove the transmission from the engine.
3. Remove the clutch from the flywheel.
4. Remove the flywheel from the crankshaft.
5. Remove the rear main bearing cap together with the bearing cap side seals.
6. Remove the rear main oil seal from around the crankshaft.
7. Apply oil to the sealing lip of the oil seal and install the seal around the crankshaft using a suitable tool.

8. Apply sealer to the rear main bearing cap as indicated and install the rear main bearing cap and tighten the cap bolts to 33–40 ft. lbs.
9. Apply sealant to the rear main bearing cap side seals and install the side seals, driving the seals into place with a suitable drift.
10. Assemble the engine and install it in the vehicle in the reverse order of removal.

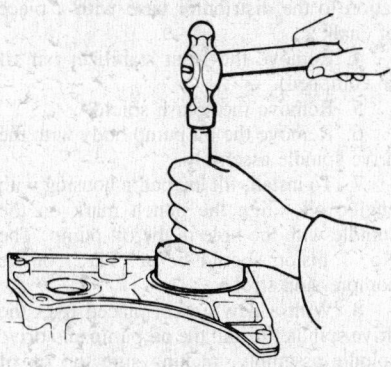

Installing the rear main seal

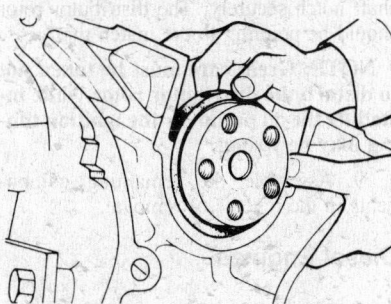

Removing the rear main seal

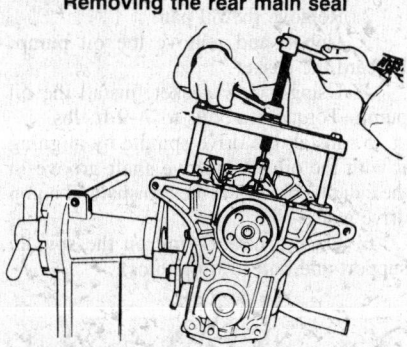

Removing the rear main bearing cap with a puller

Removing the gasoline engine oil pump

Oil Pump

REMOVAL & INSTALLATION

1979 Gas Engine

The oil pump is mounted externally on the engine, eliminating the need to remove the oil pan in order to remove the oil pump.
1. Remove the distributor.
2. Drain the engine oil.
3. Remove the front stabilizer.
4. Remove the splash shield board.
5. Remove the oil pump body with the drive spindle assembly.
6. Before installing the oil pump in the engine, turn the crankshaft so that the No. 1 piston is at TDC of the compression stroke.

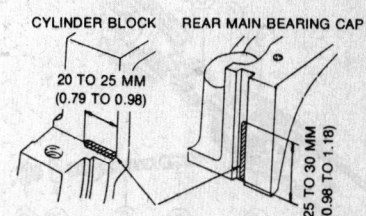

CYLINDER BLOCK   REAR MAIN BEARING CAP

20 TO 25 MM (0.79 TO 0.98)

25 TO 30 MM (0.98 TO 1.18)

Application of sealer to the rear main bearing cap

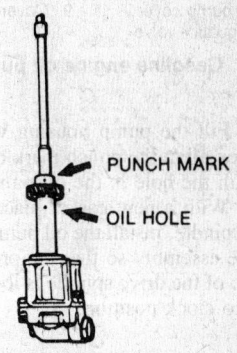

PUNCH MARK

OIL HOLE

Aligning the punch mark on the spindle with the hole in the oil pump on gasoline engines

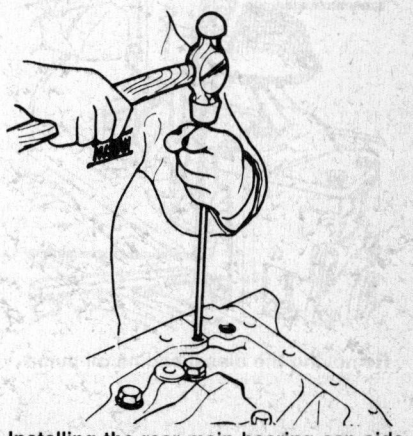

Installing the rear main bearing cap side seals

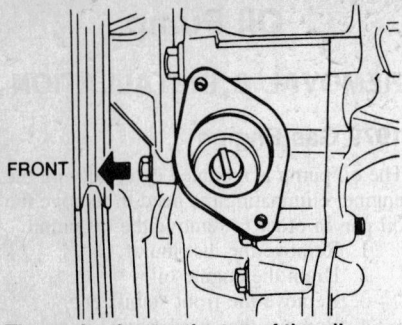

The projection on the top of the oil pump drive spindle located at the 11:25 o'clock position. The smaller crescent formed by the notch faces forward.

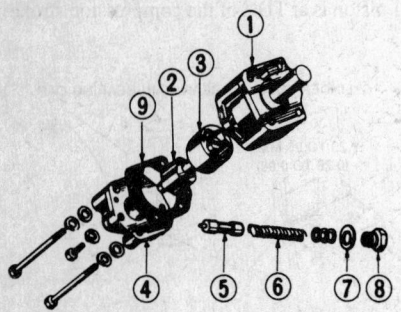

1. Oil pump body
2. Inner rotor and shaft
3. Outer rotor
4. Oil pump cover
5. Regulator valve
6. Regulator spring
7. Washer
8. Regulator cap
9. Cover gasket

Gasoline engine oil pump

7. Fill the pump housing with engine oil, then align the punch mark on the spindle with the hole in the oil pump.

8. With a new gasket placed over the drive spindle, install the oil pump and drive spindle assembly so that the projection on the top of the drive spindle is located in the 11:25 o'clock position.

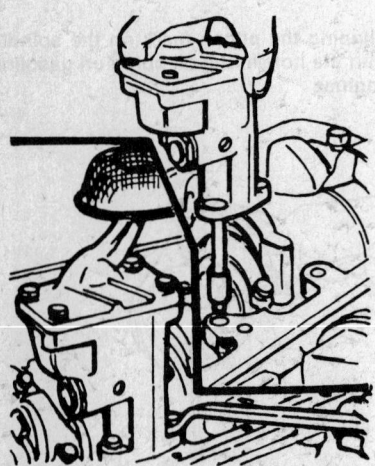

Removing the diesel engine oil pump

9. Install the distributor with the metal tip of the rotor pointing toward the No. 1 spark plug tower of the distributor.

1980 and Later Gas Engines

1. Drain the crankcase.
2. Turn the crankshaft so that the No. 1 piston is at TDC on its compression stroke.
3. Remove the distributor cap and mark the position of the distributor rotor in relation to the distributor base with a piece of chalk.
4. Remove the front stabilizer bar (if so equipped).
5. Remove the splash shield.
6. Remove the oil pump body with the drive spindle assembly.
7. To install, fill the pump housing with engine oil, align the punch mark on the spindle with the hole in the oil pump. The No. 1 piston should be at (TDC) on its compression stroke.
8. With a new gasket placed over the drive spindle, install the oil pump and drive spindle assembly, making sure the tip of the drive spindle fits into the distributor shaft notch securely. The distributor rotor should be pointing to the match mark.

NOTE: Great care must be taken not to disturb the distributor rotor while installing the oil pump, or the ignition timing may be wrong.

9. Assemble the remaining components in the reverse of removal.

Diesel Engines

1. Remove the oil pump drive spindle.
2. Remove the oil pan.
3. Unbolt and remove the oil pump. Discard the gasket.
4. Using a new gasket, install the oil pump. Torque the bolts to 7–9 ft. lbs.
5. Install the drive spindle by aligning it with the oil pump drive shaft groove in the cylinder block and the camshaft oil pump drive gear.
6. Place a new O-ring on the spindle support and bolt it to the block.

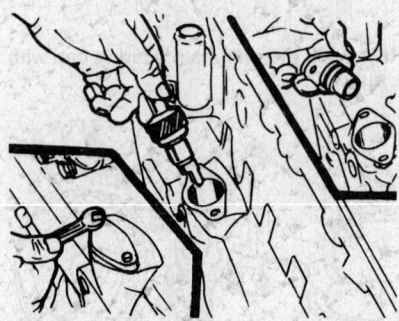

Diesel engine oil pump drive spindle removal

Oil Filter Canister Assembly

REMOVAL & INSTALLATION

SD22 and SD25 Diesel

1. Remove the bolts at the oil filter end of the oil inlet and outlet lines.
2. Remove the four filter assembly mounting bolts and separate the filter from the block.

NOTE: Have a drip pan ready, since some oil will drain out.

3. Installation is the reverse of removal. Torque the bolts to 14–18 ft. lbs.

Oil Cooler

REMOVAL & INSTALLATION

SD22 and SD25 Diesel

1. Remove the water hose from the cooler.
2. Remove the eight mounting bolts and lift the cooler from the block.

NOTE: Have a drip pan ready, since some oil will drain out.

3. Installation is the reverse of removal. Torque bolts to 14–18 ft. lbs

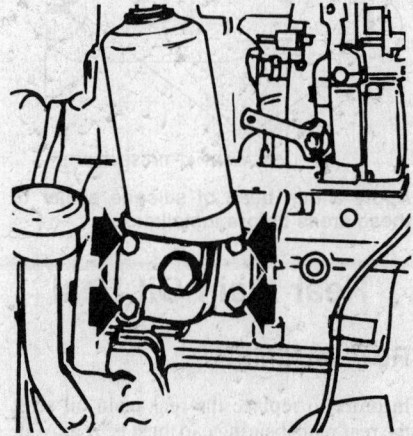

Oil filter showing the four mounting bolts

Oil cooler removal

ENGINE COOLING

Radiator

REMOVAL & INSTALLATION

1. Drain the engine coolant into a clean container.
2. Remove the front grille.
3. Disconnect the upper and lower radiator hoses. On a truck with an automatic transmission, disconnect the fluid cooler inlet and outlet lines from the radiator. Plug the lines to prevent the loss of transmission fluid and the entrance of dirt. Remove the fan shroud, if equipped.
4. Remove the bolts retaining the radiator from the radiator side supports and remove the radiator upward.
5. Install the radiator in the reverse order of removal.

Water Pump

REMOVAL & INSTALLATION

1. Drain the engine coolant into a clean container. On the diesel engine, remove the bypass hose from the pump.
2. Loosen the four bolts retaining the fan shroud to the radiator and remove the shroud.
3. Loosen the belt, then remove the fan and pulley from the water pump hub.
4. Remove the bolts (3 on the diesel engine and 5 on the gasoline engine) retaining the pump and remove the pump together with the gasket from the front cover.

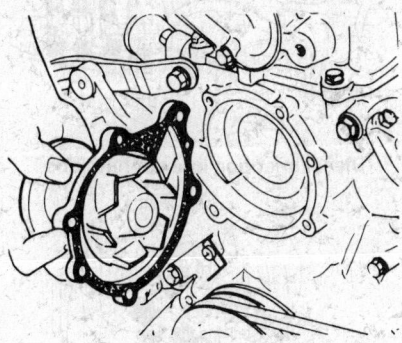

Removing the water pump

5. Remove all traces of gasket material and install the water pump in the reverse order with a new gasket and sealer. Tighten the bolts uniformly.

Thermostat

The factory-installed thermostat opening temperature is 180°F for trucks sold in the U.S., 190°F for trucks sold in Canada.

REMOVAL & INSTALLATION

1. Drain the engine coolant into a clean container so that the level is below the thermostat housing.
2. Disconnect the upper radiator hose at the water outlet.

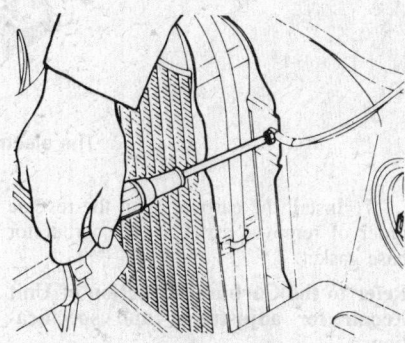

Removing the radiator securing bolts

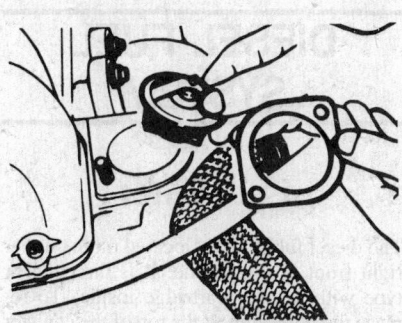

Removing the thermostat

3. Loosen the two securing nuts and remove the water outlet, gasket, and the thermostat from the thermostat housing.
4. Install the thermostat in the reverse order of removal, using a new gasket with sealer and with the thermostat spring toward the inside of the engine.

GASOLINE FUEL SYSTEM

If the fuel pump is suspected as being faulty, tests for both pressure and volume should be performed. Never replace the pump without performing these simple tests first. Always check all hoses for leaks or clogs before testing the pump.

Fuel Filter

The gasoline fuel filter is located on the right inner fender in the engine compartment. It is a disposable cartridge type, and should be replaced every 24,000 miles.

To replace the filter, loosen the clamps on the fuel lines and slide the clamps down the hoses past the point to which the filter pipes extend. Gently twist and pull on the hoses to remove them from the filter. Be careful, because some gas will spill from the bottom fuel line. Remove the old filter from the clip, install a new filter, and replace the hoses and clamps. Be certain that the line from the fuel tank connects to the fuel inlet, and that the carburetor line connects to the outlet. Start the engine and check for leaks.

Mechanical Fuel Pump

The fuel pump is a mechanically-operated, diaphragm-type driven by the fuel pump eccentric cam on the front of the camshaft. Design of the fuel pump permits disassembly, cleaning, and repair or replacement of defective parts.

TESTING

1. Disconnect the line between the carburetor and the pump at the carburetor.
2. Connect a fuel pump pressure gauge into the line.
3. Start the engine. The pressure should be between 3.0–3.9 psi. There is usually enough gas in the float bowl to perform this test.
4. If the pressure is ok, perform a capacity test. Remove the gauge from the line. Use a graduated container to catch the gas from the fuel line. Fill the carburetor float bowl with gas. Run the engine for one minute at about 1,000 rpm. The pump should deliver 1,000cc in one minute or less.

REMOVAL & INSTALLATION

1. Disconnect the two fuel lines from the fuel pump. Be sure to keep the line leading from the fuel tank up high to prevent the excessive loss of fuel.
2. Remove the two fuel pump mounting nuts and remove the fuel pump assembly from the side of the engine.
3. Install the fuel pump in the reverse order of removal, using a new gasket and sealer on the mating surface.

Electric Fuel Pump

The pump is usually mounted on a bracket located on the right frame rail next to the fuel tank. There is a filter mounted in the body of the pump, which does not normally require service. The pump can be disassembled, if necessary, but all electronic parts within the body (one transistor, two diodes, and three resistors) must be replaced as an assembly.

TESTING

1. Disconnect the hose from the pump outlet at the pump.

2. Connect a length of hose to the outlet. The hose should have an inside diameter of ¼ in. (6mm.). The diameter of the hose is important for accurate measurements.

3. Raise the end of the hose above the level of the pump. Turn the ignition switch on and catch the gasoline in a graduated container. Pump output should be 1,400cc in one minute or less. Fuel pump pressure should be between 3.1–3.8 psi.

REMOVAL & INSTALLATION

1. Remove the inlet and outlet hoses, catching the fuel that drains in a metal container.

2. Disconnect the wiring at the connector.

3. Remove the two bolts securing the pump to the bracket and remove the pump.

4. Installation is the reverse. Replace the hose clamps if their condition warrants.

Carburetor

REMOVAL & INSTALLATION

1. Remove the air cleaner.

2. Disconnect and label the fuel and vacuum lines from the carburetor.

3. Remove the throttle linkage.

4. Remove the four nuts and washers retaining the carburetor to the manifold.

5. Lift the carburetor from the manifold.

6. Remove and discard the gasket used between the carburetor and the manifold.

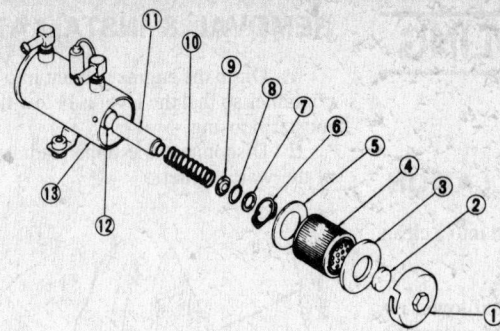

The electric fuel pump

1. End cover
2. Magnet
3. Gasket
4. Filter
5. Gasket
6. Retainer
7. Washer
8. O-ring
9. Inlet valve
10. Return spring
11. Plunger
12. Plunger cylinder
13. Body

7. Install the carburetor in the reverse order of removal using a new carburetor base gasket.

Refer to the Carburetor section of Unit Repair for adjustment and specifications.

DIESEL FUEL SYSTEM

Fuel Filter

The diesel fuel filter is located on the upper right front of the engine. It is the canister type with a paper cartridge inside. To replace the filter, unbolt the top of the canister and lift out the old filter. Insert the new filter using new gaskets and O-rings. Re-place the top. The filter should be replaced every 6,000 miles.

Injection Pump

REMOVAL & INSTALLATION

NOTE: The following procedure requires the use of special tools. In some applications, this procedure is best done with the engine removed from the vehicle.

1. Remove the inlet and outlet lines from the oil cooler.

2. Remove the bolts (4) and separate the oil filter and lines from the cooler.

3. Remove the coolant hose between the oil cooler and the head.

4. Remove the bolts (10) and separate the cooler from the block.

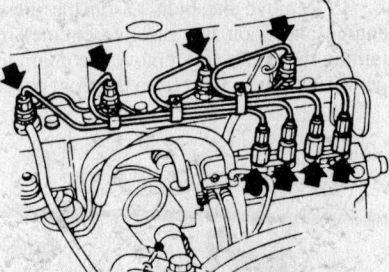

Diesel injection line connections

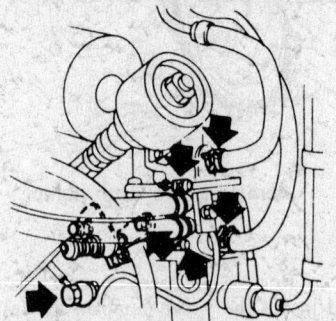

Hose connection points

5. Disconnect the fuel lines and remove the fuel filter from the bracket.

6. Remove the injection lines from the nozzles and pump. Cover all openings immediately.

7. Remove the fan, spacer and pulley from the water pump.

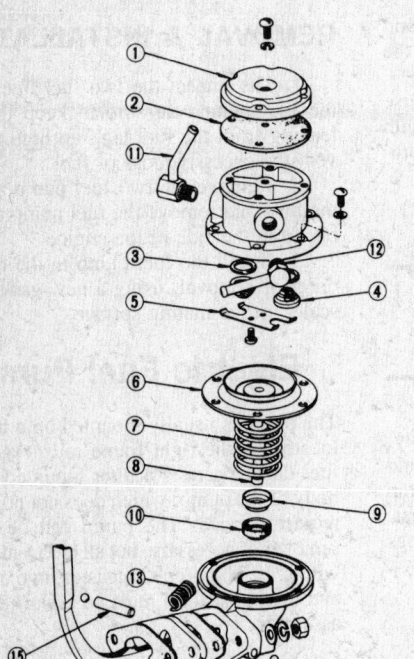

1. Fuel pump cap
2. Cap gasket
3. Valve packing
4. Fuel pump valve
5. Valve retainer
6. Diaphragm
7. Diaphragm spring
8. Pull rod
9. Lower body seal washer
10. Lower body seal
11. Inlet connector
12. Outlet connector
13. Rocker arm spring
14. Rocker arm
15. Rocker arm side pin
16. Fuel pump packing
17. Spacer—fuel pump to cylinder block

The mechanical fuel pump

8. Remove the bypass hose from the pump and thermostat housing.

9. Remove the three bolts and lift off the water pump and gasket.

10. Remove the inspection cover and pointer from the flywheel housing and lock the flywheel in place with a locking tool.

11. Flatten the lockwasher and remove the crankshaft pulley nut.

12. Tap evenly around the edge of the pulley using a brass drift, until the cone protrudes from the pulley. Remove the cone.

13. Drive the pulley and damper from the crankshaft with a soft mallet.

14. Remove the inner cover from the timing gear case.

15. Pry out the oil seal.

16. Remove the mounting bolts and tap the case loose with a soft mallet.

17. Remove the tachometer drive support nuts.

18. Remove the timer round nut.

19. Thread the timer extractor, special tool 57926–581 into the timer weight holder. Remove the timer assembly by tightening the extractor bolt.

20. Unbolt and separate the injection pump from the front end plate.

21. Temporarily install the injection pump and gasket on the front plate.

22. Check the timing marks and bring No. 1 piston to TDC.

23. Mesh the injection pump drive gear and idler gear at the timing marks.

24. After aligning the injection pump keyway, install the lockwasher and round nut and torque to 50–58 ft. lbs.

25. Install the tachometer drive coupling.

26. Check the backlash between the pump drive gear and the idler gear. Backlash should be 0.0028–0.0079 in. Adjust if necessary.

27. Remove the No. 1 cylinder holder clamp, loosen the delivery valve and pull out the delivery spring. Tighten the valve holder to 22–25 ft. lbs.

28. Connect the fuel supply lines.

29. Bring the No. 1 piston to 20° BTDC. This can be done by aligning the first mark, in normal rotation, on the crankshaft pulley with the raised line on the gear case.

30. Hand prime the pump. Push the pump in all the way toward the block. Move the pump slowly away from the block until the fuel just stops flowing from the valve holder. Lock the pump in place.

31. Remove the delivery holder and assemble the spring. Torque the holder to 22–25 ft. lbs.

32. Install remaining parts in reverse order of removal. Oil filter bolt torque is 15–18 ft. lbs.

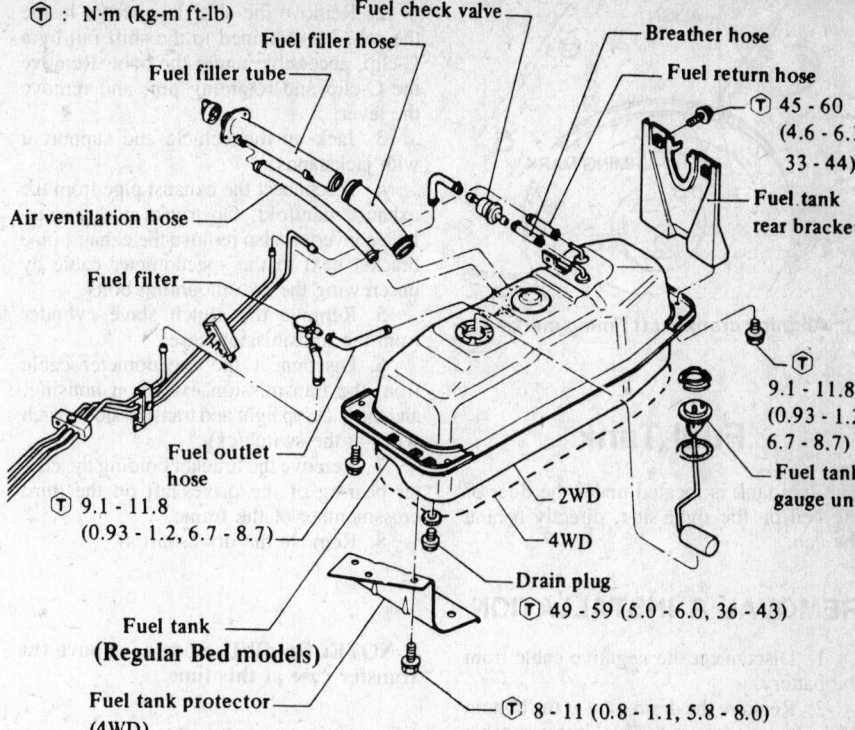

ⓣ : N·m (kg-m ft-lb)

Fuel check valve

Fuel filler hose

Breather hose

Fuel filler tube

Fuel return hose

ⓣ 45 - 60 (4.6 - 6.1, 33 - 44)

Air ventilation hose

Fuel tank rear bracket

Fuel filter

9.1 - 11.8 (0.93 - 1.2, 6.7 - 8.7)

Fuel tank gauge unit

Fuel outlet hose

ⓣ 9.1 - 11.8 (0.93 - 1.2, 6.7 - 8.7)

2WD

4WD

Drain plug

ⓣ 49 - 59 (5.0 - 6.0, 36 - 43)

Fuel tank
(Regular Bed models)

Fuel tank protector (4WD)

ⓣ 8 - 11 (0.8 - 1.1, 5.8 - 8.0)

Fuel tank and lines—typical all models, 1980 and later

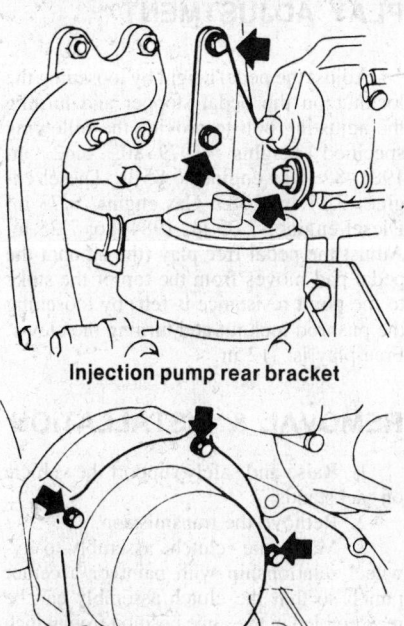

Injection pump rear bracket

Timer cover

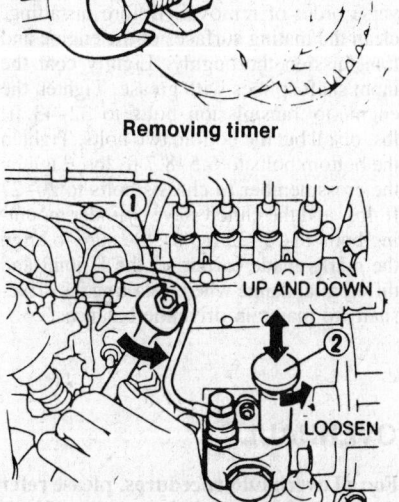

ST19530000

Removing timer

UP AND DOWN

LOOSEN

Bleeding the system

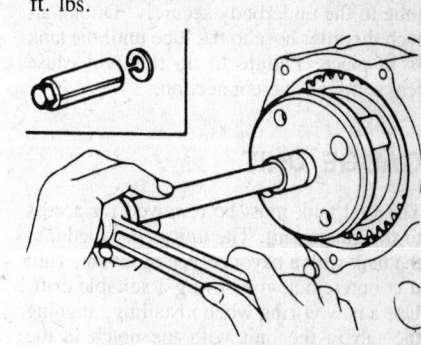

Round nut installation

Injection pump timing mark alignment

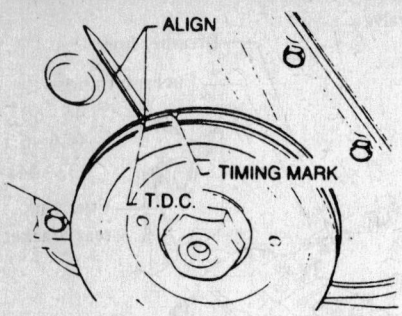

Aligning crankshaft timing marks

Fuel Tank

The fuel tank is located under the floor of the bed on the right side, directly behind the cab.

REMOVAL & INSTALLATION

1. Disconnect the negative cable from the battery.
2. Remove the drain plug at the bottom of the tank and drain the fuel into a suitable container.
3. Disconnect the filler tube from the filler hose.
4. Disconnect the ventilation hoses, the fuel return hose and fuel outlet hose from the tank. Disconnect the gauge unit wires at the electrical connector.
5. Remove the mounting bolts and remove the fuel tank.
6. Installation is the reverse. Install the clamps securely, but do not crimp any of the lines. Install the clips holding the fuel tube to the underbody securely. Do not attach the filler hose to the tube until the tank is in place. Failure to do this will cause leaks around the connection.

GAUGE UNIT

The fuel tank must be removed for access to the gauge unit. The unit is installed into the tank with a bayonet-type of mount. Turn it counterclockwise, using a suitable drift. Use a new O-ring when installing, aligning the tab in the unit with the notch in the tank.

MANUAL TRANSMISSION

REMOVAL & INSTALLATION

1. Disconnect the battery ground cable from the battery.

2. Remove the shift lever from inside the cab. It is retained to the shift rail by a C-clip, accessible under the boot. Remove the C-clip and retaining pin, and remove the lever.
3. Jack up the vehicle and support it with jackstands.
4. Disconnect the exhaust pipe from the exhaust manifold. On trucks with a catalytic converter, also remove the exhaust pipe bracket next to the speedometer cable by unscrewing the two mounting bolts.
5. Remove the clutch slave cylinder from the transmission case.
6. Disconnect the speedometer cable from the transmission extension housing, and the back-up light and transmission switch wires at the switch(es).
7. Remove the bracket holding the center bearing of the driveshaft on the third crossmember of the frame.
8. Remove the driveshaft(s).

NOTE: On 4WD models, remove the transfer case at this time.

9. Support the engine with a jack located under the oil pan. Place a block of wood between the jack and the oil pan to prevent damage to the oil pan. Support the transmission with a jack.
10. Remove the rear engine mount securing bolts and the crossmember mounting bolts. Only the rear engine/transmission extension housing mount is removable from the crossmember.
11. Remove the starter motor.
12. Remove the bolts securing the transmission to the engine, pull the transmission toward the rear until the transmission mainshaft is free of the back of the engine. Separate the transmission from the engine, then lower the transmission out from under the truck.
14. Install the transmission in the reverse order of removal. Before installing, clean the mating surfaces of the engine and transmission thoroughly. Lightly coat the input shaft splines with grease. Tighten the engine-to-transmission bolts to 32–43 ft. lbs. on all but the bottom two bolts. Tighten the bottom bolts to 6.5–8.7 ft. lbs. Tighten the crossmember to chassis bolts to 20–27 ft. lbs. and the clutch slave cylinder mounting bolts to 18–22 ft. lbs. Be sure to align the marks made earlier on the U-joint and differential flange when installing the driveshaft, to maintain driveline balance.

OVERHAUL

For all overhaul procedures, please refer the Manual Transmission section of Unit Repair.

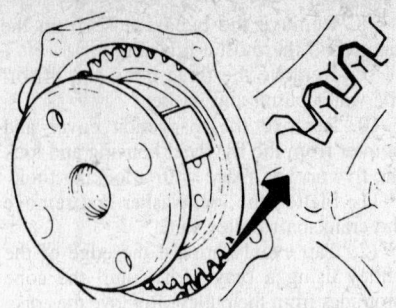

Aligning timer Y-marks

CLUTCH

The clutch is a hydraulically-operated single-plate, dry friction disc, diaphragm spring type.

The clutch is operated by a clutch pedal which is mechanically connected to a clutch master cylinder. When the pedal is depressed, the piston in the master cylinder is moved in the master cylinder bore. This movement compresses the fluid in the master cylinder causing hydraulic pressure which is transferred through a tube to the slave cylinder. The slave cylinder is mounted to the clutch housing with its piston connected to the clutch release lever. The hydraulic pressure in the slave cylinder forces the slave cylinder piston to travel out the cylinder bore and move the clutch release lever, disengaging the clutch.

PEDAL HEIGHT AND FREE-PLAY ADJUSTMENT

Adjust the pedal height by loosening the locknut on the pedal stopper and turning the adjusting bolt to provide the following specified heights. 1979–80: 6.42 in. 1981–82: Gas engine, 6.85 in. Diesel engine, 7.17 in. 1983: Gas engine, 6.73 in. Diesel engine, 7.05 in. 1984–86: 7.25 in. Adjust the pedal free play (the amount the pedal pad moves from the top of the stoke to the point resistance is felt) by loosening the pushrod locknut and turning the clevis. Free play is: .12 in.

REMOVAL & INSTALLATION

1. Raise and safely support the vehicle on jackstands.
2. Remove the transmission.
3. Mark the clutch assembly-to-flywheel relationship with paint or a center punch so that the clutch assembly can be reassembled in the same position from which it is removed. Insert a clutch aligning tool (dummy shaft) into the hub. This tool is available from your dealer or an auto parts store. It is important to support the weight

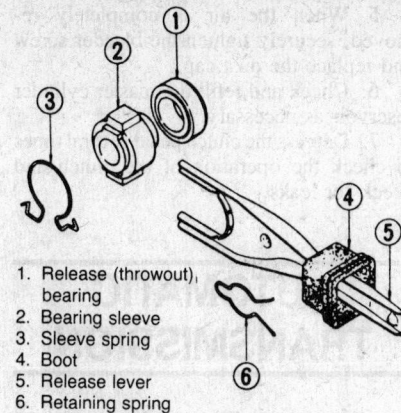

1. Release (throwout) bearing
2. Bearing sleeve
3. Sleeve spring
4. Boot
5. Release lever
6. Retaining spring

Clutch release mechanism

of the clutch while the retaining bolts are being removed.

4. Loosen the six clutch cover-to-flywheel attaching bolts, one turn at a time in an alternating sequence, until the spring tension is relieved to avoid distorting or bending the clutch cover. Remove the clutch assembly.

5. Inspect the flywheel for scoring, roughness, or signs of overheating. Light scoring may be cleaned up with emery cloth, but any deep grooves or scoring warrant replacement or refacing (if possible) of the flywheel. If the clutch facings or flywheel are oily, inspect the transmission front cover oil seal, the pilot bushing, and engine rear seals, etc. for leakage, and correct before replacing the clutch. If the pilot bushing in the crankshaft is worn, replace it. Install it using a soft hammer. The factory-supplied part does not have to be oiled, but check the procedure if you are using an aftermarket part. Inspect the clutch cover for wear or scoring, and replace as necessary. The pressure plate and spring cannot be disassembled; you must replace the clutch cover as an assembly.

6. Inspect the clutch release bearing. If it is rough or noisy, it should be replaced. The bearing can be removed from the sleeve with a puller; this requires a press to install the new bearing. After installation, coat the groove in the sleeve, the contact surfaces of the release lever, pivot pin and sleeve, and the release bearing contact surfaces on the transmission front cover with a light coat of grease. Be careful not to use too much grease, which will run at high temperatures and get onto the clutch facings. Reinstall the release bearing on the lever.

7. Apply a thin coat of grease to the pressure plate wire ring, diaphragm spring, clutch cover grooves and the drive bosses on the pressure plate.

8. Apply a thin coat of Lubriplate® to the splines in the driven plate. Slide the clutch disc onto the splines, and move it back and forth several times. Remove the disc and wipe off the excess lubricant. Be very careful not to get any grease on the clutch facings.

9. Assemble the clutch cover and the clutch plate on the clutch alignment arbor.

10. Align the marks made on the clutch cover and the flywheel (if the old cover is being used) and install the six clutch cover-to-flywheel attaching bolts. Three dowels are used to locate the clutch cover on the flywheel properly. Tighten the bolts in an alternating sequence one turn at a time to 12–15 ft. lbs. Remove the aligning arbor.

11. Install the transmission.

Clutch Master Cylinder

REMOVAL & INSTALLATION

1. Disconnect the clutch pedal arm from the pushrod clevis. Remove the dust cover (boot) from the master cylinder body and pushrod. It will not go through the firewall without tearing.

2. Disconnect the clutch hydraulic line from the master cylinder.

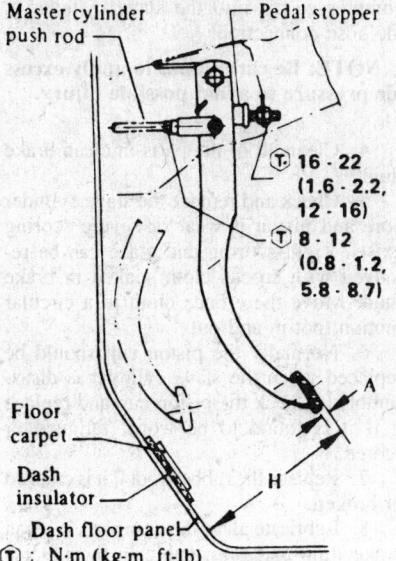

Master cylinder push rod — Pedal stopper

Ⓣ 16 - 22
(1.6 - 2.2, 12 - 16)

Ⓣ 8 - 12
(0.8 - 1.2, 5.8 - 8.7)

Floor carpet
Dash insulator
Dash floor panel
A
H
Ⓣ : N·m (kg-m, ft-lb)

Clutch pedal adjustment: "A" is pedal free play, "H" is pedal height

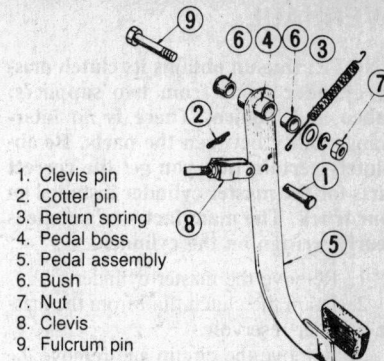

1. Clevis pin
2. Cotter pin
3. Return spring
4. Pedal boss
5. Pedal assembly
6. Bush
7. Nut
8. Clevis
9. Fulcrum pin

Clutch pedal assembly

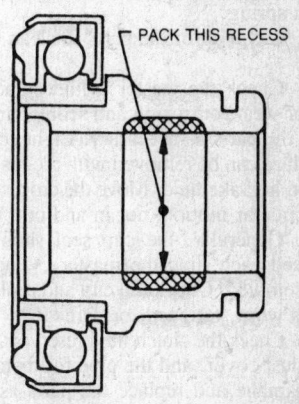

PACK THIS RECESS

Coat the area indicated in the bearing sleeve with grease

NOTE: Take precautions to keep brake fluid from coming in contact with any painted surfaces.

3. Remove the nuts attaching the master cylinder and remove the master cylinder and pushrod toward the engine compartment side.

4. Install the master cylinder in the reverse order of removal and bleed the clutch hydraulic system.

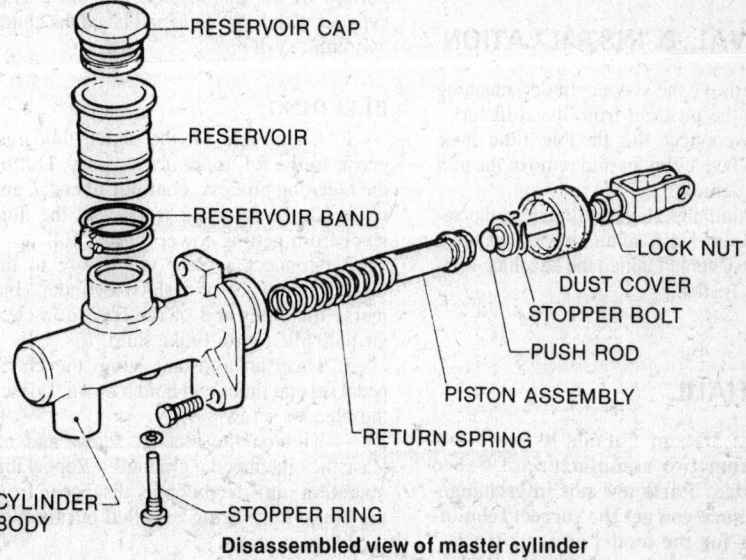

RESERVOIR CAP
RESERVOIR
RESERVOIR BAND
LOCK NUT
DUST COVER
STOPPER BOLT
PUSH ROD
PISTON ASSEMBLY
RETURN SPRING
CYLINDER BODY
STOPPER RING

Disassembled view of master cylinder

OVERHAUL

NOTE: Datsun obtains its clutch master cylinder parts from two suppliers: Nabco and Tokico. There is no interchangeability between the parts. Be absolutely certain that you get the correct parts for the master cylinder installed on your truck. The manufacturer's name is clearly written on the cylinder.

1. Remove the master cylinder.
2. Drain the clutch fluid from the master cylinder reservoir.
3. Remove the circlip and remove the pushrod.
4. Remove the stopper, piston, cup, and return spring.
5. Clean all of the parts in clean brake fluid.
6. Check the master cylinder and piston for wear, corrosion and scores, and replace the parts as necessary. Light scoring and glaze can be removed with crocus cloth soaked in brake fluid. Move the crocus cloth in a circular motion; not in and out.
7. Generally, the cup seal should be replaced each time the master cylinder is disassembled. Check the cup and replace it if it is worn, fatigued, or damaged.
8. Check the clutch fluid reservoir, filler cap, dust cover, and the pipe for distortion and damage and replace the parts as necessary.
9. Lubricate all new parts with clean brake fluid.
10. Reassemble the master cylinder parts in the reverse order of disassembly, taking note of the following: Reinstall the cup seal carefully to prevent damaging the lipped portions; Adjust the height of the clutch pedal after installing the master cylinder in position on the vehicle; Fill the master cylinder and clutch fluid reservoir and then bleed the clutch hydraulic system.

Clutch Slave Cylinder

REMOVAL & INSTALLATION

1. Remove the slave cylinder attaching bolts and the pushrod from the shift fork.
2. Disconnect the flexible fluid hose from the slave cylinder and remove the unit from the vehicle.
3. Install the slave cylinder in the reverse order of removal and bleed the clutch hydraulic system. Tighten the attaching bolts to 18–25 ft. lbs.

OVERHAUL

NOTE: Datsun obtains its slave cylinders from two manufacturers: Nabco and Tokico. Parts are not interchangeable. Be sure you get the correct rebuilding parts for the model on your truck.

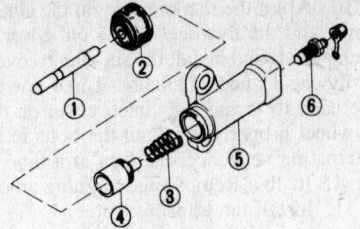

1. Push rod
2. Dust cover
3. Piston spring
4. Piston
5. Operating cylinder
6. Bleeder screw

Clutch slave cylinder

1. Remove the slave cylinder from the vehicle.
2. Remove the pushrod and boot.
3. Force out the piston by blowing compressed air into the slave cylinder at the hose connection.

NOTE: Be careful not to apply excess air pressure to avoid possible injury.

4. Clean all of the parts in clean brake fluid.
5. Check and replace the slave cylinder bore and piston if wear or severe scoring exists. Light scoring and glaze can be removed with crocus cloth soaked in brake fluid. Move the crocus cloth in a circular motion; not in and out.
6. Normally the piston cup should be replaced when the slave cylinder is disassembled. Check the piston cup and replace it if it is found to be worn, fatigued or scored.
7. Replace the rubber boot if it is cracked or broken.
8. Lubricate all of the new parts in clean brake fluid and reassemble in the reverse order of disassembly, taking note of the following: Use care when reassembling the piston cup to prevent damaging the lipped portion of the piston cup; Fill the master cylinder with brake fluid and bleed the clutch hydraulic system.

BLEEDING

1. Check and fill the clutch fluid reservoir to the full mark if necessary. During the bleeding process, continue to check and replenish the reservoir to prevent the fluid level from getting lower than ½ full.
2. Connect a clear vinyl hose to the bleeder screw on the slave cylinder. Immerse the other end of the hose in a clear jar half filled with brake fluid.
3. Have an assistant pump the clutch pedal several times and hold it down. Loosen the bleeder screw slowly.
4. Tighten the bleeder screw and release the clutch pedal gradually. Repeat this operation until air bubbles disappear from the brake fluid being expelled out through the bleeder screw.

5. When the air is completely removed, securely tighten the bleeder screw and replace the dust cap.
6. Check and refill the master cylinder reservoir as necessary.
7. Depress the clutch pedal several times to check the operation of the clutch and check for leaks.

AUTOMATIC TRANSMISSION

Pan

REMOVAL & INSTALLATION

Loosen the automatic transmission pan attaching bolts more at one corner than the other three corners. Allow the fluid to drain out the one corner. Remove all of the pan attaching bolts and remove the pan. Install the pan in the reverse order of removal. Always use a new gasket.

ADJUSTMENTS

Shift Linkage

1. Loosen the control lever-to-linkage rod locknut.
2. Place the shift lever in D and the control lever on the transmission in the D detent position.
3. Tighten the locknut and move the shift lever through all positions making sure that the detent is felt in each position and the transmission responds properly to each gear selection. Make double certain that Park engages properly and holds the truck. If proper adjustment cannot be made in each detent, replace all shift lever and linkage grommets.

Kickdown Switch & Downshift Solenoid

With the ignition switch in the ON position and the engine off, when the accelerator

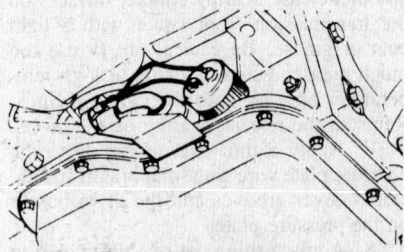

Downshift solenoid

pedal is depressed fully, the kick-down switch contacts should be closed and the downshift solenoid activated, emitting a clicking sound. If the components fail to operate in this manner, check for continuity first at the switch and then at the solenoid if the switch checks out as being satisfac-

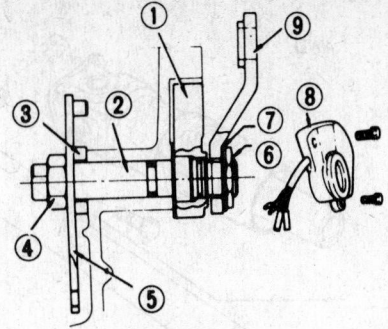

tory. Replace either of the components as necessary.

Neutral Safety Switch

The neutral safety switch is located on the transmission range selector lever. The switch operates the back-up lights and controls the operation of the starter. The starter should only operate when the transmission is in Park or Neutral. To adjust the neutral safety switch, unscrew the securing nut of the range selector lever and the two bolts securing the switch body. Remove the machine screw under the switch body. Adjust the shift selector to the Neutral position (in vertical position and detent clicks). Move the switch slightly aside so that the screw hole will be aligned with the pin hole of the internal rotor combined with the manual shaft and check their alignment by inserting a 0.080 in. (2.0mm) diameter pin into the holes. A #47 drill bit will work for this. Fasten the switch body with the bolts, pull out the pin, and tighten the screw into the hole. Connect the selector lever. If the neutral safety switch does not perform satisfactorily after adjustment, replace it with a new one.

Transmission

REMOVAL & INSTALLATION

1. Disconnect the negative battery cable.

2. Disconnect the shaft from the accelerator linkage.

3. Raise and support the truck.

4. Matchmark the U-joint and differential flange and disconnect them. Remove the center bearing mounting bolts and remove the driveshaft. Plug the transmission extension housing.

5. Disconnect the exhaust pipe from the manifold and discard the gasket. Use a new gasket upon assembly. On trucks with a catalytic converter, disconnect the exhaust pipe bracket.

6. Disconnect the shift linkage at the transmission.

7. Disconnect the neutral switch wires. Disconnect the vacuum hose from the diaphragm, and the wire from the downshift solenoid. Disconnect the speedometer cable from the extension housing.

8. Remove the fluid filler tube.

9. Disconnect the fluid cooler lines at the transmission. Use a flare nut wrench if one is available.

10. Support the engine with a jack under the oil pan, placing a wooden block between the pan and the jack as a buffer. Also support the transmission with a jack.

11. Remove the torque converter cover. Matchmark the converter and the drive plate for reassembly; they were balanced as a unit at the factory. Remove the bolts attaching the converter to the drive plate (flywheel). You will have to rotate the engine to do

1. Inhibitor switch
2. Manual shaft
3. Washer
4. Nut
5. Manual plate
6. Washer
7. Nut
8. Inhibitor switch
9. Selector lever

Neutral safety switch

this, using a wrench on the crankshaft pulley bolt.

12. Remove the bolts for the rear engine mount and the crossmember. Remove the crossmember.

13. Remove the starter.

14. Remove the transmission-to-engine bolts. Lower the transmission back and down, out from under the truck.

15. Before installing the transmission, check the drive plate runout with a dial indicator. Turn the crankshaft one full turn. Maximum allowable runout is 0.012 in. (0.3mm). Replace the drive plate if runout exceeds 0.020 in. (0.5mm); otherwise, reface it.

16. When installing the torque converter, be sure to line up the notch in the converter with the projection on the oil pump. Align the marks made during removal and bolt the converter to the drive plate, tightening the bolts to 29–36 ft. lbs. Then rotate the engine a few turns to make sure the transmission rotates freely without binding. The engine-to-transmission bolt torque is 29–36 ft. lbs. Adjust the shift linkage and neutral switch after installation.

TRANSFER CASE

REMOVAL & INSTALLATION

1. Disconnect the battery ground.

2. Raise the vehicle and support it on stands.

3. Remove the transfer case shield.

4. Remove the primary driveshaft securing nuts.

5. Remove the front and rear driveshafts.

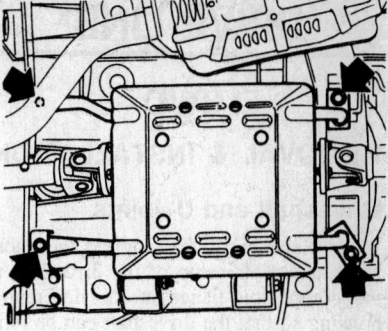

Transfer case shield bolts

6. Disconnect the 4WD switch wire.

7. Disconnect the speedometer cable.

8. Remove the exhaust pipe.

9. Support the transfer case with a jack.

10. Temporarily loosen the transfer case insulator bolts.

11. Remove the shift lever rubber boot from the floor.

12. Unbolt and remove the transfer case.

13. Installation is the reverse of removal. Torque all mounting bolts to 20–26 ft. lbs.

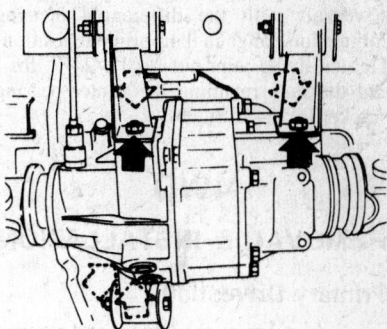

Transfer case mounting bolts

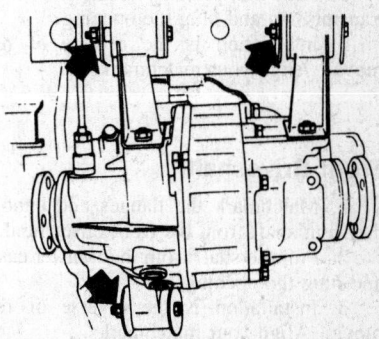

Transfer case insulator bolts

OVERHAUL

Refer to the **Transfer Case** section of Unit Repair for servicing procedures.

DRIVE LINE

2WD

REMOVAL & INSTALLATION

Driveshaft and U-Joints

1. Raise and safely support the truck. Mark the relationship of the driveshaft to the companion flange at the differential housing so that the driveshaft can be reinstalled in the same position.

2. Remove the bolts retaining the center bearing bracket.

3. Remove the bolts connecting the driveshaft to the companion flange at the differential housing (and at the front flange on early models).

4. Move the driveshaft assembly toward the rear of the truck, passing it under the rear axle, removing the sleeve yoke from the transmission. Watch for oil leaking out the end of the transmission. Plug if necessary.

5. Install the driveshaft in the reverse order of removal. Be careful not to bang the sleeve yoke into the seal inside the transmission extension housing, which will damage the seal, causing leakage. Align the driveshaft with the differential housing companion flange in their original position. Tighten the U-joint nuts to 17–24 ft. lbs., and the nuts retaining the center bearing bracket to 12–16 ft. lbs.

4WD

REMOVAL & INSTALLATION

Primary Driveshaft

1. Matchmark the flanges and separate the primary driveshaft from the transfer case.

2. Remove the transfer case.

3. Pull the primary shaft from the transmission and plug the opening.

4. Installation is the reverse of removal. Align your matchmarks.

Front Driveshaft

1. Matchmark the flanges and unbolt the front shaft from the front differential.

2. Pull the shaft from the transfer case and plug the opening.

3. Installation is the reverse of removal. Align your matchmarks.

Rear Driveshaft

1. Matchmark the flanges and unbolt the shaft from the rear differential.

2. Remove the center bearing bracket.

3. Pull the shaft from the transmission and plug the opening.

4. Installation is the reverse of removal. Align your matchmarks.

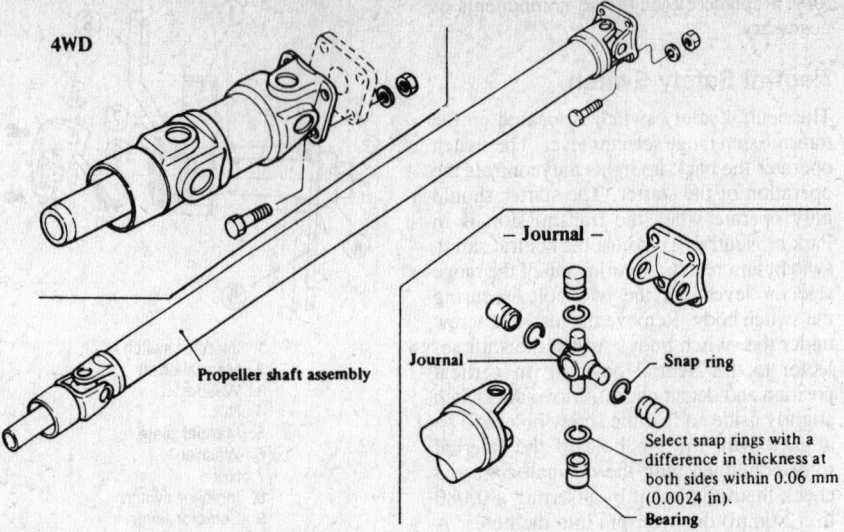

Front axle propellor shaft: the left insert shows the primary driveshaft which connects the transfer case with the transmission

U-Joint Overhaul

Refer to the U-Joint/CV-Joint section of Unit Repair for procedures.

Center Bearing

REPLACEMENT

The center bearing is a sealed unit which must be replaced as an assembly if defective.

1. Remove the driveshaft.

2. Matchmark the flange yoke and the companion flange which connect the front half of the driveshaft to the rear. Also matchmark the companion flange and the front driveshaft. Remove the bolts and separate the shafts.

3. You must devise a way to hold the driveshaft while unbolting the companion flange from the front driveshaft. Do not place the front driveshaft tube in a vise, because the chances are it will get crushed. The best way is to grip the flange somehow while loosening the nut. It is going to require some strength to remove.

4. Slide the companion flange off the front driveshaft and remove the center bearing from its mount.

5. The new bearing is already lubricated. Install it into the mount, making sure that the seals and so on are facing the same way as when removed.

6. Slide the companion flange onto the front driveshaft, aligning the marks made during removal. Install the washer and locknut. Tighten the nut to 145–175 ft. lbs. Check that the bearing rotates freely around the driveshaft.

7. Connect the companion flange to the flange yoke, aligning the marks made during disassembly. Torque to 18–23 ft. lbs.

8. Install the driveshaft, aligning the marks made at the axle flange during removal.

REAR AXLE

Axle Shaft, Bearing and Seal

REMOVAL & INSTALLATION

Single Rear Wheels

1. Raise the rear of the vehicle and support it. Remove the rear wheel and tire.

2. Disconnect the rear parking brake cable by removing the adjusting nut and clamps.

3. Disconnect the brake tube at the rear brake backing plate. Plug the end of the brake tube to prevent loss of brake fluid.

4. Remove the brake drum.

5. Remove the nuts securing the wheel bearing retainer to the brake backing plate.

6. Pull out the axle shaft assembly together with the brake backing plate using a slide hammer.

7. Remove the oil seal in the axle housing if necessary. It can be pried out with a screwdriver. Oil the lips of the new seal and install it carefully to avoid damage to the lip.

8. To replace the bearing, unbend and

Axle shaft end play is adjusted by the addition or subtraction of shims behind the brake backing plate

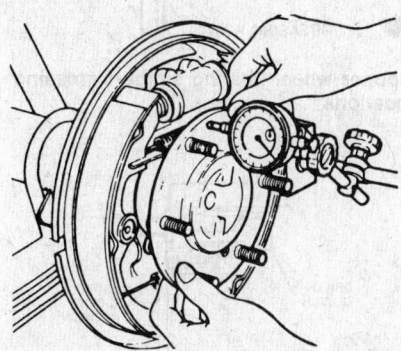

Measure the axial end-play with a dial indicator

discard the lockwasher. Remove the locknut with a soft drift and a hammer.

9. Press the old bearing and cage off the shaft.

10. Remove the oil seal in the cage. Use a brass drift to remove the bearing cup after the seal has been removed.

11. Install the new cup with a brass drift. Install a new oil seal over the bearing cup. Lubricate the area between the seal lips with grease after installation.

12. Place the bearing cage and spacer on the axle shaft, then fit the bearing, tapping it into place with a soft drift and light hammer blows.

13. Place the flat bearing lockwasher over the bearing, then the new nut lockwasher. Install the locknut, tightening to 108 ft. lbs. Continue to tighten after that until the grooves line up with the lockwasher tabs. The nut can be tightened up to 145 ft. lbs. Bend the lockwasher tabs into place.

14. Lubricate the bearing and the recess in the axle housing with wheel bearing grease. Coat the axle splines with gear oil. Coat the seal surface of the shaft with grease.

15. Install the axle shaft in the reverse order of removal. The axle end-play should be 0.012–0.035 in. The end-play is adjusted by adding or removing shims behind the brake backing plate. Tighten the backing plate attaching nuts to 39–46 ft. lbs.

Dual Rear Wheels

1. Follow Steps 1–6 of the procedure for single rear wheels.

2. Remove the attaching screws and detach the lockwasher from the rear wheel bearing nut.

3. Remove the rear wheel bearing nut.

4. Remove the bearings and seal and drive out the races with a brass drift.

5. Coat the axles shaft splines with 90W gear oil and coat the seal lip with chassis lube.

6. Install new bearing races with an installing drive and pack the hub with chassis lube.

7. Park each bearing and the O-ring with chassis lube.

8. Install the bearings and axle shaft. Be careful not to damage the seal with the shaft. Always use new seals. Make sure that the axle shaft end play is 0.0031 in. Observe the following torques: Wheel bearing locknut, 123–145 ft. lbs. Backing plate nut, 62–80 ft. lbs. Wheel lugs, 159–188 ft. lbs.

FRONT AXLE

Free Running Hub

REMOVAL & INSTALLATION

Through 1983

1. Raise and support the front axle on stands.

2. Set the hub in the lock position.

3. Remove the driven clutch by turning it clockwise.

NOTE: A pin is located inside the hub case to lock the driven clutch. Pull and turn the clutch while attracting the pin with a magnet.

4. Remove the lock pin.

5. Set the hub at the free position.

6. Screw the driven clutch into place by turning it counterclockwise until it bottoms.

7. Turn it clockwise until it aligns with the bolt hole.

8. Install the lock pin.

1984 and Later

1. Raise and support the front axle on jackstands.

2. Remove the wheels.

3. Remove the locking hub cover using Torx® wrench.

4. Pull out the locking hub.

5. Remove the snapring and pull out the drive clutch.

6. Remove the snapring and remove the wheel bearing locknut using tool KV40104300, or its equivalent.

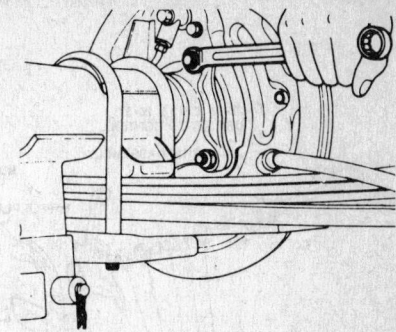

Disconnecting the brake backing plate from the axle housing

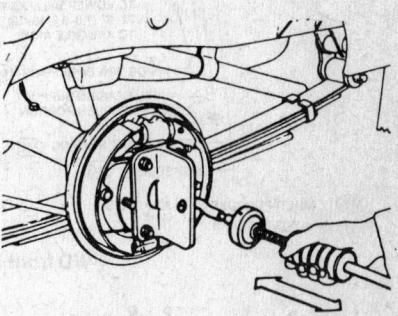

Removing the axle shaft with a slide hammer

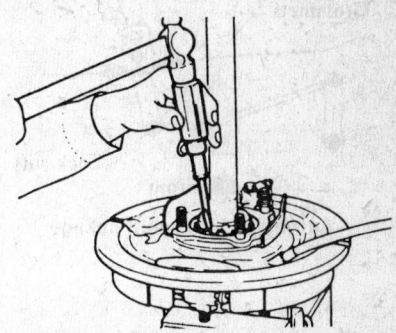

Unbend the lockwasher to remove the bearing locknut; use a new nut at installation

Axle Shaft

REMOVAL & INSTALLATION

Through 1983

1. Remove the free running hub.

2. Remove the snap-ring and remove the driven clutch.

3. Remove the rebound bumper.

4. Disconnect the stabilizer bar at the lower link.

5. Remove the bolts attaching the axle shaft to the carrier. DO NOT REMOVE THE RUBBER BOOTS!

6. Pull the axle shaft from the suspension.

7. Installation is the reverse of re-

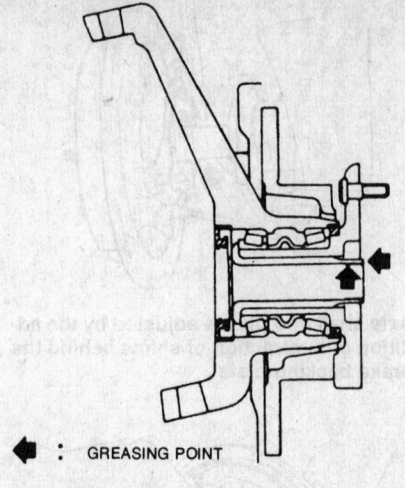

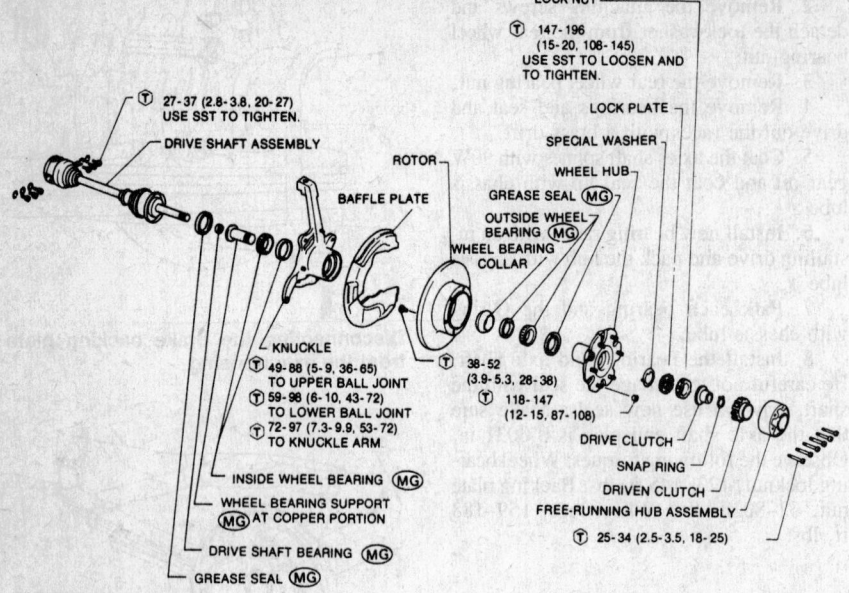

LOCK NUT
147-196
(15-20, 108-145)
USE SST TO LOOSEN AND
TO TIGHTEN.

LOCK PLATE

SPECIAL WASHER

WHEEL HUB

GREASE SEAL (MG)

OUTSIDE WHEEL
BEARING (MG)

WHEEL BEARING
COLLAR

27-37 (2.8-3.8, 20-27)
USE SST TO TIGHTEN.

DRIVE SHAFT ASSEMBLY

ROTOR

BAFFLE PLATE

KNUCKLE

49-88 (5-9, 36-65)
TO UPPER BALL JOINT
59-98 (6-10, 43-72)
TO LOWER BALL JOINT
72-97 (7.3-9.9, 53-72)
TO KNUCKLE ARM

38-52
(3.9-5.3, 28-38)

118-147
(12-15, 87-108)

DRIVE CLUTCH

SNAP RING

DRIVEN CLUTCH

FREE-RUNNING HUB ASSEMBLY

25-34 (2.5-3.5, 18-25)

INSIDE WHEEL BEARING (MG)

WHEEL BEARING SUPPORT
(MG) AT COPPER PORTION

DRIVE SHAFT BEARING (MG)

GREASE SEAL (MG)

(MG) : MULTI-PURPOSE GREASE POINTS
(T) : N·m (KG-M, FT-LB)

4-WD front axle through 1983

⬅ : GREASING POINT

Copper wheel bearing support greasing locations

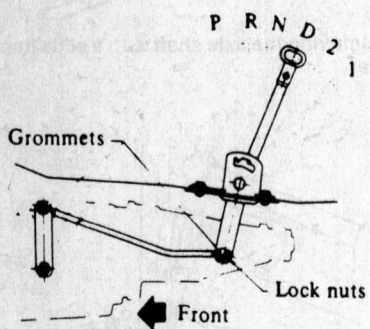

P R N D 2 1

Grommets

Lock nuts

⬅ Front

Automatic trans shift linkage

moval. Observe the following points: Apply multi-purpose wheel bearing grease to the copper portion of the wheel bearing support. Adjust the axle shaft axial end play by installing the proper thickness of snaprings on the end of the shaft. Torque the axle shaft flange bolts to 20–27 ft. lbs.; the free running hub bolts to 18–25 ft. lbs.; the stabilizer bar-to-frame bolts to 12–16 ft. lbs.; the stabilizer bar-to-lower link bolts to 12–16 ft. lbs. and the wheel nuts to 87–108 ft. lbs.

1984 and Later

1. Remove the locking hub.
2. Remove the snapring and drive clutch.
3. Disconnect the lower ball joint.
4. Disconnect the lower end of the shock absorber.
5. Disconnect the driveshaft from the front axle.
6. Pull the axle shaft from the housing. It helps to turn the steering wheel to the right when pulling the right shaft and left when pulling the left shaft.
7. Installation is the reverse of removal. Note the following: Apply chassis lube to all bearing surfaces. Before installing the shaft make sure that the spacer is in place. Adjust the axle shaft end play by using various thicknesses of snaprings. The end play should be 0.004–0.012 in. Observe the following torques: Driveshaft-to-axle, 20–27 ft. lb. Locking hug, 18–25 ft. lb. Ball joint-to-spindle, 87–123 ft. lb.

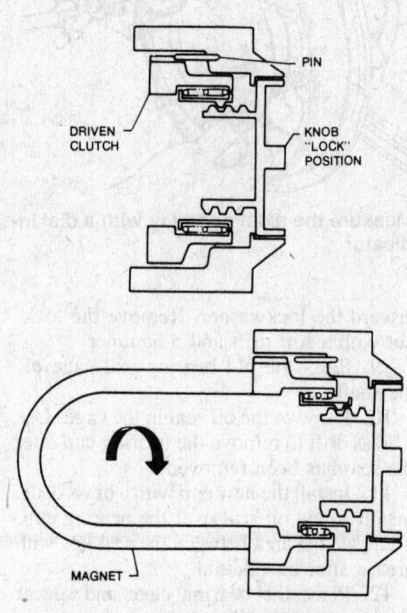

PIN

DRIVEN CLUTCH

KNOB "LOCK" POSITION

MAGNET

Hub removal sequence

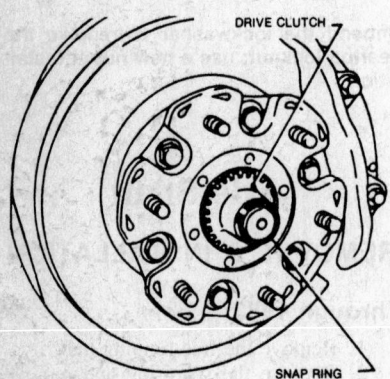

DRIVE CLUTCH

SNAP RING

Free-running hub snap-ring and driven clutch location

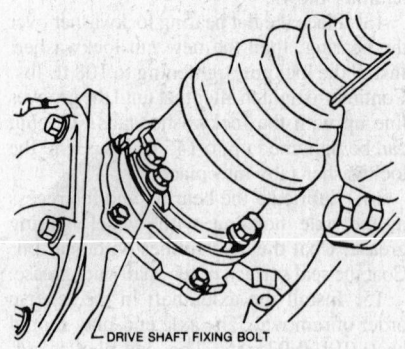

DRIVE SHAFT FIXING BOLT

Axle shaft-to-carrier bolts

FRONT SUSPENSION

2WD Models

The front suspension on Datsun pickups is of an independent, unequal length control arm type with torsion bar springs. The control arms are attached to a bracket which is welded to the frame at their inner pivot points and to the steering knuckle/spindle support at their outer points. The torsion bar has splines on each end; the front end is installed to the torque arm which is attached to the lower control arm and the rear end is installed to the torsion bar anchor which is secured to the chassis frame. Fore-and-aft movement of the front suspension is controlled by strut bars connected to the lower control arms at one end and mounted to the chassis frame at the forward end. An optional torsion bar type stabilizer is connected to the lower control arms through vertical stabilizer links. The steering knuckle/spindle is attached to the steering knuckle/spindle support by ball joints.

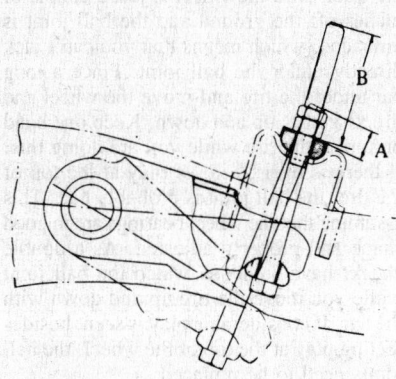

Installing the torsion bar anchor end on 2-wheel drive models

Torsion Bars

REMOVAL & INSTALLATION

Through 1983

1. Jack up the front of the vehicle and support it with jackstands. Remove the wheel. On trucks with a catalytic converter, the converter must be removed if the left torsion bar is being removed.
2. Loosen the ride height adjusting nuts at the anchor (rear) end of the torsion bar, allowing the anchor arm to hang down.
3. Remove the dust cover at the rear end of the torsion bar and remove the snap-ring.
4. Pull the anchor arm rearward and off the torsion bar. Withdraw the torsion bar

from the lower control arm and remove it from under the vehicle.

5. Before installing the torsion bar, apply a light coat of grease to the splines. Install the torsion bar to the lower control arm.

NOTE: The torsion bars are marked on the end with an "L" (left) or an "R" (right). The torsion bars must be installed on the same side from which they are removed.

6. Install the anchor arm on the rear end of the torsion bar. Dimension "A" must be: 0.28–0.67 in. Note that there are two different methods used for measuring this distance. Be sure you are using the correct illustration for your truck.
7. Install a new retaining snap-ring and dust boot to the anchor arm end of the torsion bar.

NOTE: Always use a new snap-ring. Never reinstall the old one.

8. Tighten the adjusting nut until the link protrudes above the support bracket 2.36–2.76 in. on all models (dimension "B").
9. Install the wheel and lower the vehicle.
10. Adjust the vehicle ride height with the truck at curb weight (full tank of gas and no passengers). Refer to "Ride Height Adjustment." Tighten the locknut to 23–30 ft. lbs.

1984 and Later

1. Block the rear wheels and raise and support the front end with jackstands under the frame rails.
2. Remove the torsion bar spring anchor bolt.
3. Remove the dust cover and remove the snap ring from the anchor arm.
4. Pull the anchor arm off toward the rear.
5. Pull the torsion bar out toward the rear.
6. Remove the torsion bar spring torque arm.
7. Check the torsion bars for wear, cracks or other damage. Replace them if they are suspect.

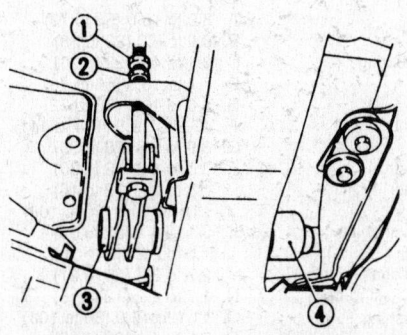

1. Lock nut
2. Adjusting nut
3. Anchor arm
4. Dust cover

2-wheel drive torsion bar anchor

8. Install the torque arm on the lower link. Torque the outer side to 20–27 ft. lb.; the inner side to 26–33 ft. lb.
9. Install the snap ring and dust cover on the torsion bar.
10. Coat the splines on the torsion bar with chassis lube and install it in the torque arm. The torsion bars are marked L and R and are not interchangeable.
11. Place the lower link in position so that clearance between it and the rebound bumper is 0.
12. Install the anchor arm so that dimension G in the illustration is: Gas engine; Left side: 4.33–4.72 in. Right side: 5.12–5.51 in. Diesel engine; Left side: 4.53–4.92 in. Right side: 5.31–5.71 in.

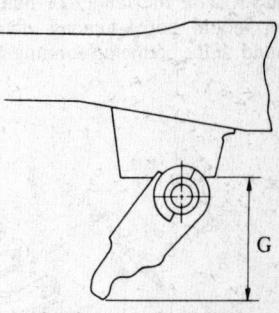

Anchor arm installation

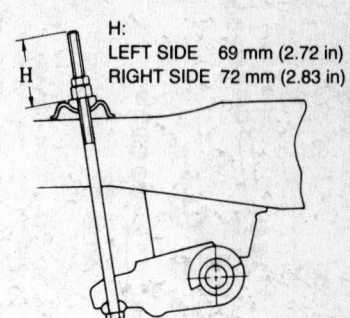

H:
LEFT SIDE 69 mm (2.72 in)
RIGHT SIDE 72 mm (2.83 in)

Anchor arm preliminary adjustment

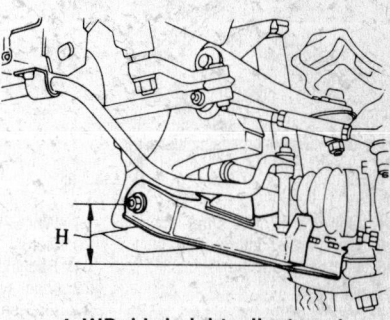

4-WD ride height adjustment

13. Temporarily tighten the anchor arm bolt so that dimension H in the illustration is 2.72 in. on the left side and 2.83 in. on the right side.
14. Install the snap ring on the anchor arm, turning it to make sure that it is completely in the groove.
15. Install the dust cover.
16. Lower the truck so that it is resting on its wheels. Turn the anchor bolt adjusting nut so that dimension H in the illustra-

tion is 4.65–4.80 in. on regular cab models and 4.45–4.61 in. on King Cab models. Torque the adjusting nut to 22–30 ft. lb.

Shock Absorbers

TESTING

Visually inspect the shock absorber. If there is evidence of leakage and the shock absorber is covered with oil, the shock is defective and must be replaced. If there is no sign of excessive leakage (a small amount of weeping is natural) bounce the truck at one corner by pressing up and down on the fender or bumper. When you have the truck bouncing as much as you can, stop bouncing it, and release the fender or bumper. The truck should stop bouncing after the first rebound. If the bouncing continues past

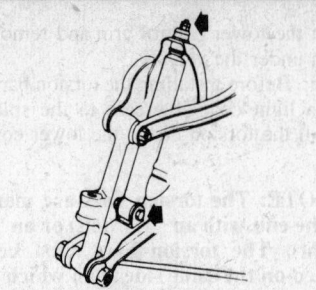

Front shock absorber attaching bolts

the center point of the bounce more than once, the shock absorbers are worn and should be replaced.

REMOVAL & INSTALLATION

1. Jack up the vehicle and support it. Remove the wheel.

2. Hold the upper stem of the shock absorber and remove the nuts, washer, and rubber bushing.

3. Remove the bolt from the lower end of the shock absorber and remove the shock absorber from the vehicle.

4. Install the shock absorber in the reverse order of removal. Replace all of the rubber bushings with new ones if a new shock absorber is being installed. Install the lower retaining bolt from the front of the truck. Tighten the upper attaching nut to 12–16 ft. lbs. and the lower nut to 23–30 ft. lbs.

Ball Joints

INSPECTION

The ball joint should be replaced when play becomes excessive. Datsun does not publish specifications on just what constitutes excessive play, relying instead on a method of determining the force (in inch pounds) required to keep the ball joint turning. This method is not very helpful to the backyard mechanic since it involves removing the ball joint, which is what we are trying to avoid in the first place. An effective way to determine ball joint play is to jack up the truck until the wheel is just a couple of inches off the ground and the ball joint is unloaded, which means that you can't jack directly under the ball joint. Place a long bar under the tire and move the wheel and tire assembly up and down. Keep one hand on top of the tire while you are doing this. If there is over ¼ in. of play at the top of the tire, the ball joint is probably bad. This assuming that the wheel bearings are in good shape and properly adjusted. As a double check, have someone watch the ball joint while you move the tire up and down with the bar. If considerable play is seen, besides feeling play at the top of the wheel, the ball joints need to be replaced.

REMOVAL & INSTALLATION

Upper

THROUGH 1983

1. Raise and support the truck on stands placed on the frame rails.

2. Remove the wheels.

3. Loosen the torsion bar anchor lock and adjusting nuts to relieve spring tension.

4. Remove and discard the cotter pin from the ball joint stud and remove the nut. Separate the stud from the knuckle spindle with a ball joint removal tool.

5. Loosen the bolts retaining the ball joint to the control arm, and remove the joint.

6. Install the new ball joint into the control arm, tightening the bolts to 12–16 ft. lbs. Install the ball joint stud into the knuckle spindle and install the nut. Tighten the nut to 60 ft. lbs., then continue to tighten until the holes align (limit: 72 ft. lbs.). In-

1. Upper arm pivot shaft	19. Wheel hub	lbs.)
2. Camber adjusting shim	20. Rotor	A. 8.0 to 10.0 (58 to 72)
3. Rebound bumper	21. Backing plate	B. 3.9 to 5.3 (28 to 38)
4. Bushing	22. Inner wheel bearing	C. 4.1 to 4.1 (22 to 30)
5. Upper arm	23. Grease seal	D. 7.7 to 10.5 (56 to 76)
6. Upper ball joint	24. Spacer	E. 8.0 to 10.0 (58 to 72)
7. Knuckle spindle	25. Lower ball joint	F. 17.2 to 19.5 (124 to 141)
8. Torsion bar	26. Lower arm	G. 3.9 to 5.3 (28 to 38)
9. Boot	27. Tension rod (strut)	H. 3.0 to 4.2 (22 to 30)
10. Anchor arm	28. Shock absorber	I. 1.6 to 2.2 (12 to 16)
11. Anchor arm adjusting bolt	29. Torque arm	J. 11.1 to 15.0 (90 to 108)
12. Adjusting nut	30. Lower arm pivot shaft	K. 3.1 to 4.1 (22 to 30)
13. Cotter pin	31. Bumper stop	L. 1.6 to 2.2 (12 to 16)
14. O-ring	32. Stabilizer (optional)	M. 2.7 to 3.7 (20 to 27)
15. Hub cap	33. Stabilizer connecting bolt	N. 1.7 to 2.2 (12 to 16)
16. Spindle nut	34. Lower arm bushing	O. 11.1 to 15.0 (80 to 108)
17. Washer	35. Stabilizer collar	P. 3.6 to 4.6 (26 to 33)
18. Outer wheel bearing	Tightening torque kg-m (ft.	Q. 3.9 to 5.3 (28 to 38)

1979 and later front suspension typical

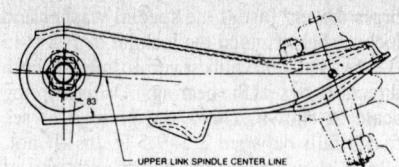

1979–83 upper arm pivot shaft installation before tightening shaft nuts

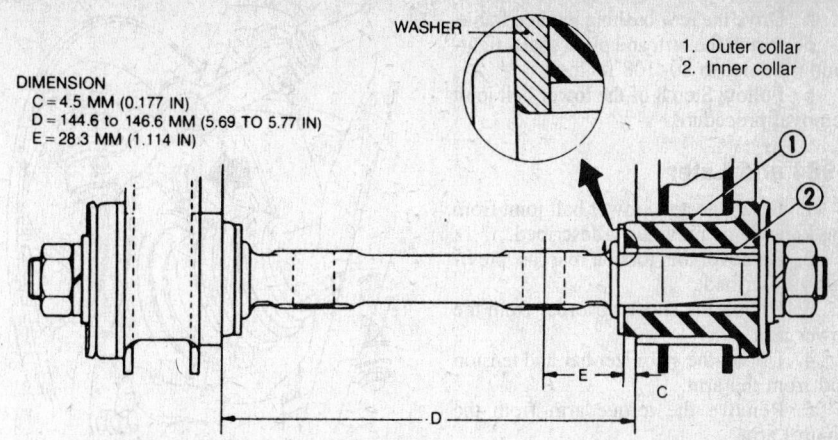

DIMENSION
C = 4.5 MM (0.177 IN)
D = 144.6 to 146.6 MM (5.69 TO 5.77 IN)
E = 28.3 MM (1.114 IN)

WASHER

1. Outer collar
2. Inner collar

Upper pivot shaft and bushing dimension for models through 1983

stall a new cotter pin. Install the wheel, lower the truck, and adjust the ride height. Have the alignment checked.

1984 AND LATER

1. Raise and support the front end with jackstands under the frame rails.
2. Remove the front wheels.
3. Support the lower control arm with a floor jack and remove the upper ball joint-to-knuckle nut.
4. Using a ball joint separator, such as tool ST29020001, remove the ball joint from the knuckle.
5. Unbolt the ball joint from the upper arm.
6. Installation is the reverse of removal. Have the front end alignment checked. Note the following torques: Ball joint-to-support arm: 12–15 ft. lb. Ball joint-to-knuckle: 58–72 ft. lb.

Lower

THROUGH 1983

1. Perform Steps 1–2 of the upper ball joint removal procedure. Remove the lower shock absorber mounting bolt.
2. Loosen the torsion bar spring anchor lock and adjusting nuts, and remove the anchor arm bolt from the anchor arm.
3. Remove the snap-ring, then move the anchor arm and torsion bar fully rearward.
4. Disconnect the stabilizer bar from the lower arm, if equipped.
5. Disconnect the strut (tension rod) from the lower arm.
6. Remove and discard the cotter pin from the ball joint stud, and remove the nut. Separate the ball joint from the knuckle spindle with a ball joint removal tool.
7. Remove the attaching bolts, and remove the ball joint from the lower arm.
8. Install the new ball joint in the arm, tightening the bolts to 28–38 ft. lbs. Install the ball joint stud into the knuckle spindle and tighten the nut to 127 ft. lbs. Continue to tighten until the holes align, then install the new cotter pin (limit: 141 ft. lbs.). The torsion bar ride height must be adjusted after assembly.

1984 AND LATER

1. Raise and support the front end on jackstands under the frame rails.
2. Remove the front wheels.
3. Remove the torsion bar as previously described.
4. Unbolt the shock absorber from the lower arm.
5. Remove the ball joint nut.

6. Using a ball joint separator such as tool ST29020001, separate the ball joint from the knuckle.
7. Unbolt the ball joint from the lower arm.
8. Installation is the reverse of removal. Observe the following torques: Ball joint-to-lower arm: 28–38 ft. lb. Ball joint-to-knuckle: 87–123 ft. lb.

Upper Control Arm

REMOVAL & INSTALLATION

Through 1983

1. Perform Steps 1–4 of the upper ball joint removal procedure.
2. Remove the bolts retaining the upper arm pivot shaft and remove the shaft, arm, and camber adjusting shims from the body. Note the location of the shims so that they may be installed in their original positions during assembly.
3. To remove the upper arm shaft and bushings from the arm, remove the nuts and washers from the shaft. Use a press, first on one end of the shaft and then the other, to press out the bushings. Remove the shaft.
4. To install, coat the bushing with soapy water and press it into the upper arm. Install the washers onto the shaft and install the shaft into the arm. Be sure the chamfered side of the washer is against the shaft flange. Measure the distance between the outer collar on the bushing and the washer on the shaft (dimension "C" in the illustration); it should exceed 4.5mm (0.177 in.).
5. Press the other bushing into the arm, and check dimension "C." Also check the distance between the outer collars of both bushings (dimension "D") and the distance between the end of the bushing and the centerline of the shaft mounting bolt bore (dimension "E"). D should be 144.6–146.6mm (5.69–5.77 in.); E should be 28.3mm (1.114 in.).
6. Rotate the pivot shaft in the arm until the angle is as specified in the illustration. Install the nuts and washers on the shaft. Tighten the bushings to 56–76 ft. lbs.

7. Install the upper arm and shaft assembly to the body, replacing the camber shims in their original locations. Tighten the bolts to 80–108 ft. lbs.
8. Follow Step 6 of the upper ball joint removal procedure.

1984 and Later

1. Separate the ball joint from the knuckle as previously described.
2. Unbolt the control arm spindle. Lift out the control arm.
3. The bushings may now be pressed out from both sides of the control arm.
4. Apply a soapy solution to new bushings and press them into position in one end of the arm, so that the flange on the bushing firmly contacts the end surface of the upper link collar.
5. Install the spindle and press in the remaining bushings.

NOTE: The inner washers are installed with the rounded edges facing inwards.

6. Temporarily tighten the spindle end nuts.
7. Install the upper ball joint. Tighten the bolts to 12–15 ft. lb.
8. Bolt the control arm to the frame. Tighten the bolts to 80–100 ft. lb.
9. Tighten the spindle end nut with camber adjusting shims. Torque the nuts to 56–76 ft. lb. Check the dimensions A and B shown in the illustration. Install the ball joint to the knuckle and check front end alignment.

Lower Control Arm
REMOVAL & INSTALLATION

Through 1983

1. Perform Steps 1–6 of the lower ball joint removal procedure.
2. Remove the lower arm pivot shaft bolt and washer. Tap the pivot shaft out of the bushing. Push down on the torsion bar and remove the lower arm.
3. Use a bushing driver to tap the lower arm bushing from the frame.

4. Drive the new bushing into the frame.

5. Install the arm and pivot shaft, tightening the nut to 80–108 ft. lbs.

6. Follow Step 8 of the lower ball joint removal procedure.

1984 and Later

1. Disconnect the lower ball joint from the knuckle as previously described.

2. Remove the torsion bar as previously described.

3. Unbolt the shock absorber from the lower arm.

4. Unbolt the stabilizer bar and tension rod from the arm.

5. Remove the torque arm from the control arm.

6. Unbolt the control arm from the frame.

7. Unbolt the ball joint.

8. Using a driver, remove the lower arm spindle busings from the frame.

9. Obtain new bushings.

10. Install the ball joint on the arm. tighten the bolts to 28–38 ft. lb.

11. Using a driver, install the new bushings, coated with a soapy solution, in the frame.

12. Install the torque arm on the torsion bar.

13. Install the control arm on the frame.

14. Install the torque arm on the control arm. Torque the inner side bolt to 26–33 ft. lb. and the outer side to 20–27 ft. lb.

15. Install the ball joint on the knuckle.

16. Connect the shock absorber.

17. Install the tension rod and stabilizer bar.

18. Lower the truck so that it is resting normally on its wheels.

19. Turn the anchor bolt adjusting nut so that dimension H, in the illustration is 4.45–4.61 in. for King Cab models and 4.65–4.80 for regular models.

20. Check the front end alignment.

4WD Models

REMOVAL & INSTALLATION

Hub & Knuckle

THROUGH 1983

1. Block the rear wheels, raise and support the front of the vehicle on jackstands.

2. Remove the wheels.

3. Remove the brake caliper and suspend it out of the way. DO NOT DISCONNECT THE BRAKE HOSE!

4. Remove the axle shaft.

5. Remove the bolt securing the knuckle arm to the knuckle.

6. Loosen, but do not remove the upper and lower ball joint tightening nuts.

7. Separate the ball joints from the knuckle with a ball joint removing tool.

— CAUTION —
NEVER REMOVE THE BALLJOINT NUT IN THE PREVIOUS STEP!

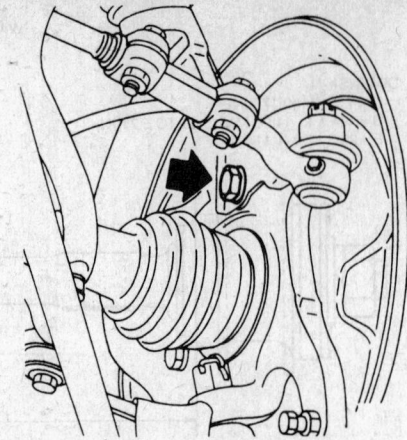

Knuckle arm attaching bolt

8. Jack up the lower link and remove the ball joint tightening nut.

9. Remove the knuckle.

10. Unbend the lockwasher with a screwdriver and remove the front hub locknut.

11. Remove the lockwasher and special washer.

12. Push the wheel bearing support out of the hub.

13. Separate the knuckle from the hub with a puller.

14. Remove the bearing collar.

15. Remove the inside bearing and seal. Drive the race out with a brass driver.

16. Separate the hub from the rotor.

17. Knock the hub on a wood block to move the outer bearing away from the hub surface, then pull it the rest of the way with a bearing puller. Remove the grease seal.

18. Remove the axle shaft bearing from the bearing support with a brass driver.

19. Clean and thoroughly repack the bearings.

20. Assembly is the reverse of removal. Note the following points: Install the bearing outer race into each side of the knuckle with a brass driver. Install the outer grease seal and bearing with a brass driver. Pack the seal lip with wheel bearing grease. Be sure that the seal faces the right direction. The wheel bearing collar thickness determines end-play.

21. Determine what thickness to use as follows: Install the collar which was removed. Install the inside bearing with a

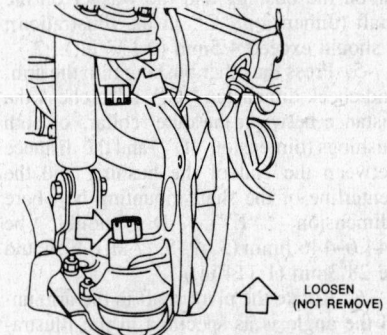

Upper and lower ball joint nuts
LOOSEN (NOT REMOVE)

brass driver. Install the special washer and lockwasher. Tighten the locknut to 108–145 ft. lbs. Turn the hub several times in both directions to seat the bearings. Using a spring scale as shown, check the preload to see that it falls between 2.2–9.5 ft. lbs. If not, adjust by replacing the collar with one of a different thickness, as shown by the number stamped on the collar. The larger the number, the thicker the collar. When preload has been correctly set, secure the nut by bending the lockwasher tip.

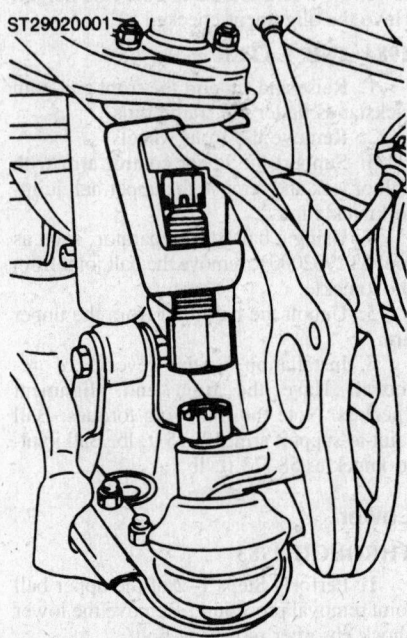

ST29020001

Removing the ball joint using a ball joint tool

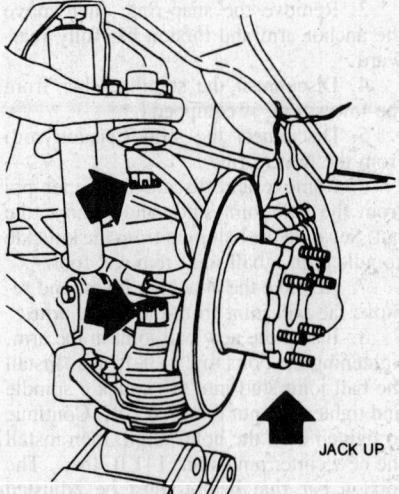

JACK UP

Removing the ball joint tightening nuts

1984 AND LATER

1. Raise and support the front end with jackstands under the frame rails.

2. Remove the front wheels.

3. Remove the calipers and suspend them out of the way.

4. Remove the locking hub assembly.

5. Remove the tie rod using tool HT72520000, or equivalent.

6. Unbolt the knuckle arm from the knuckle.

7. Support the lower control arm with floor jack and remove the upper and lower ball joints-to-knuckle nuts.

8. Separate the ball joints from the knuckle.

9. Remove the snap ring and lock washer from the hub.

10. Remove the hub locknut using tool KV40104300 or equivalent.

11. Separate the hub and rotor from the knuckle.

12. Assembly and installation are the reverse of removal and disassembly.

Shock Absorbers

See the 2WD section.

Torsion Bars

THROUGH 1983

1. Raise and support the front end on jackstands.

2. Remove the torsion bar spring anchor bolt.

3. Pull the anchor arm out rearward.

4. Pull the torsion bar spring rearward.

5. Remove the torsion bar.

6. Install the torsion arm to the lower link. Torque the outer bolt to 20–27 ft. lbs. and the inner to 26–33 ft. lbs.

7. Place a coating of grease on the torsion bar spring serrated end and install it in the torsion arm.

8. Install the anchor arm to the serrated end of the torsion bar spring and install the anchor arm adjusting bolt. Turn the bolt until the bottom of the nut is about ½ in. from the end of the bolt.

9. Install the dust cover.

10. Adjust the anchor arm position until the distance between the end of the dust cover and the bottom of the nut is about 2½ in.

11. Lower the vehicle.

12. Turn the anchor bolt adjusting nut until the center line of the lower link spindle

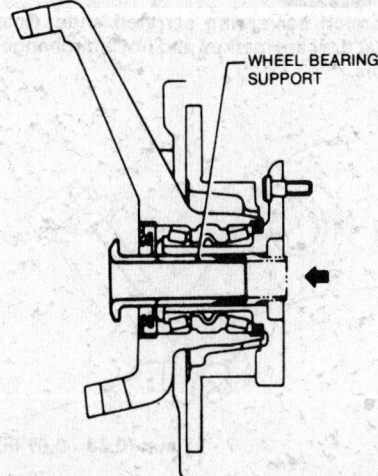

Wheel bearing support greasing location

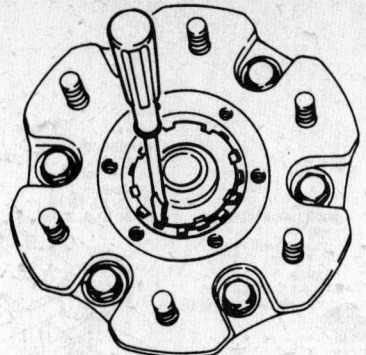

Unbending the lockwasher

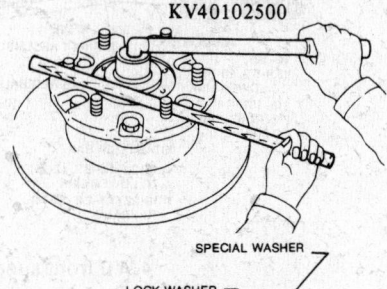

Removing the special washer and lockwasher

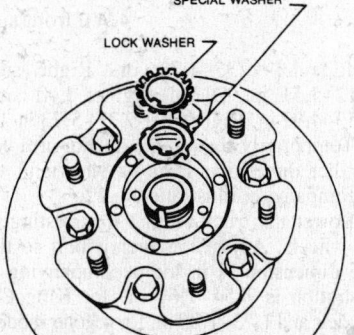

Using a puller to separate the hub from the knuckle

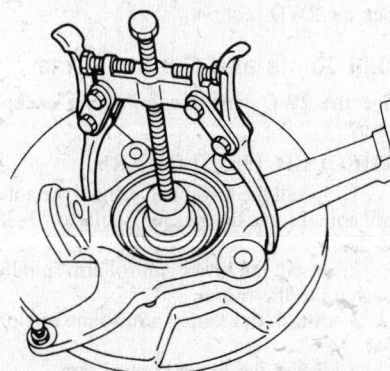

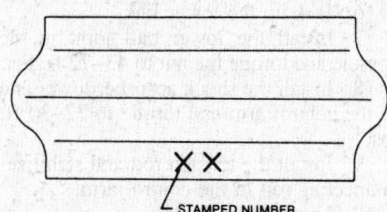

Bearing collars are stamped with a number to indicate thickness

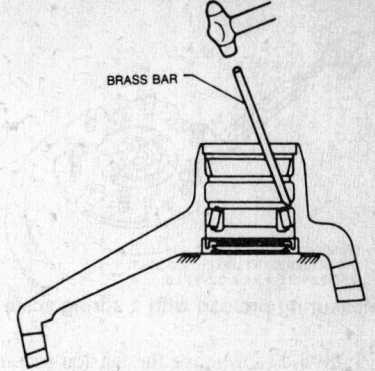

Driving out the inner bearing and seal

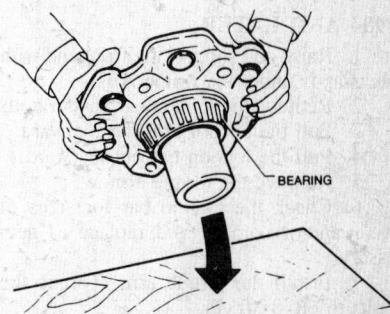

Hit the hub on a block of wood to break the outer bearing loose from the hub face

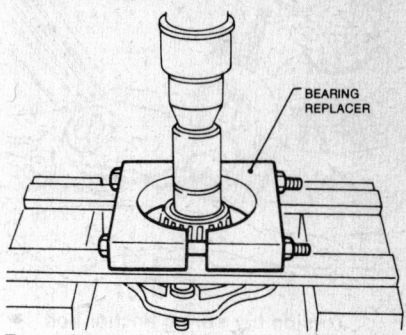

Removing the outer bearing with a puller

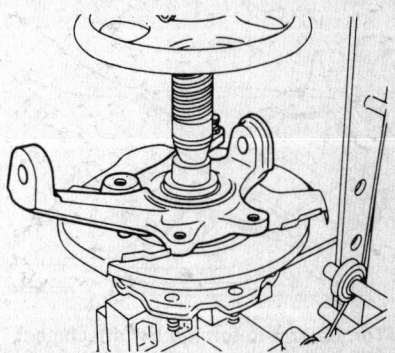

Installing the outer bearing into the knuckle

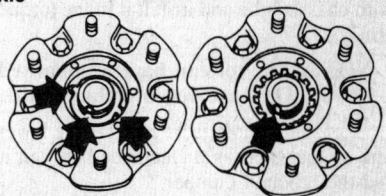

Proper positioning of the lockwasher and special washer

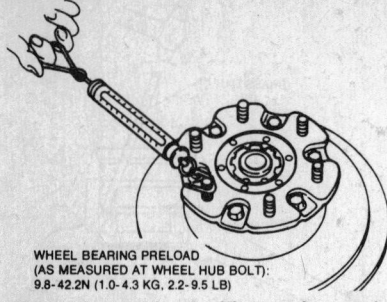

WHEEL BEARING PRELOAD
(AS MEASURED AT WHEEL HUB BOLT):
9.8–42.2N (1.0–4.3 KG, 2.2–9.5 LB)

Measuring preload with a spring scale

is 5.28–5.47 in. above the tension rod attaching bolts. This is dimension H in the accompanying figure.

1984 AND LATER

1. Raise and support the front end with jackstands under the frame rails.
2. Remove the torsion bar anchor bolt.
3. Pull the anchor arm out rearward.
4. Pull the torsion bar out rearward.
5. Remove the torque arm.
6. Check the torsion bar for signs of wear and/or damage and replace as necessary.
7. Install the torque arm. Torque the bolts to 66–87 ft. lb.

Torsion bar spring anchor bolt

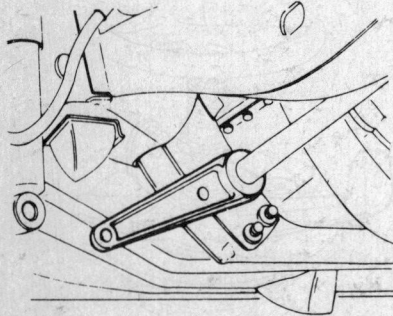

Torque arm-to-torsion bar attachment

8. Coat the splines on the torsion bar with chassis lube and install it in the torque arm.

   NOTE: The torsion bars are marked L and R. They are not interchangeable.

9. Using a floor jack, position the lower link so that there is 0 clearance between it and the rebound bumper.
10. Install the anchor arm so that dimension G in the illustration is: Gas engine;

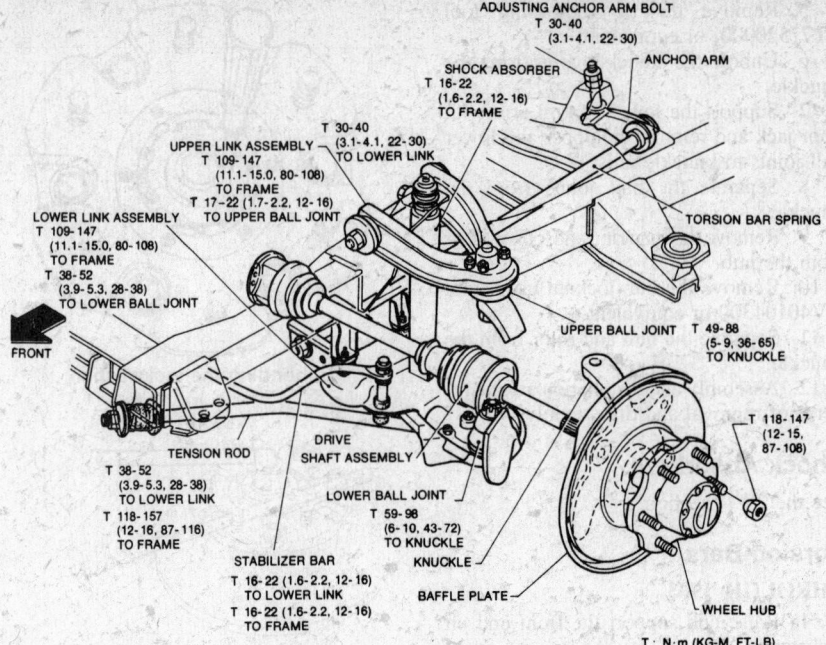

ADJUSTING ANCHOR ARM BOLT
T 30–40
(3.1–4.1, 22–30)

ANCHOR ARM

SHOCK ABSORBER
T 16–22
(1.6–2.2, 12–16)
TO FRAME

T 30–40
(3.1–4.1, 22–30)
TO LOWER LINK

UPPER LINK ASSEMBLY
T 109–147
(11.1–15.0, 80–108)
TO FRAME
T 17–22 (1.7–2.2, 12–16)
TO UPPER BALL JOINT

TORSION BAR SPRING

LOWER LINK ASSEMBLY
T 109–147
(11.1–15.0, 80–108)
TO FRAME
T 38–52
(3.9–5.3, 28–38)
TO LOWER BALL JOINT

UPPER BALL JOINT   T 49–88
(5–9, 36–65)
TO KNUCKLE

FRONT

T 118–147
(12–15,
87–108)

TENSION ROD

DRIVE
SHAFT ASSEMBLY

T 38–52
(3.9–5.3, 28–38)
TO LOWER LINK
T 118–157
(12–16, 87–116)
TO FRAME

LOWER BALL JOINT
T 59–98
(6–10, 43–72)
TO KNUCKLE

STABILIZER BAR
T 16–22 (1.6–2.2, 12–16)
TO LOWER LINK
T 16–22 (1.6–2.2, 12–16)
TO FRAME

KNUCKLE

BAFFLE PLATE

WHEEL HUB

T : N·m (KG-M, FT-LB)

4-WD front suspension through 1983

Left side: 4.33–4.72 in. Right side: 5.12–5.51 in. Diesel engine; Left side: 4.53–4.92 in. Right side: 5.31–5.73 in.11.

Temporarily tighten the anchor arm bolt so that dimension H is as shown in the accompanying illustrations. 12.

Lower the truck so that it is resting on its wheels. Adjust the anchor bolt so that the dimension H in the accompanying illustration is 1.54–1.69 in. for King Cab models and 1.73–1.89 in. for regular models.

Tension Bar and Stabilizer Link

See the 2WD section.

Ball Joints and Control Arms

See the 2WD with the following exceptions:

THROUGH 1983 (LOWER)

1. Install the lower ball joint to the control arm. Torque the attaching bolts to 28–38 ft. lbs.
2. Install the lower control arm spindle bushing to the frame.
3. Attach the torque arm to the torsion bar.
4. Install the lower control arm.
5. Install the torque arm on the lower control arm. Torque the inner side to 26–33 ft. lbs. and the outer side to 20–27 ft. lbs.
6. Jack up the lower link.
7. Install the lower ball joint in the spindle and torque the nut to 43–72 ft. lbs.
8. Install the shock absorber lower end to the control arm and torque to 22–30 ft. lbs.
9. Install the tension rod and stabilizer connecting rod to the control arm.
10. Lower the vehicle.
11. Turn the anchor bolt adjusting nut to obtain the dimension specified in Step

12 of Torsion Bar Removal and Installation.

12. After installation, check wheel alignment.

1984 AND LATER (LOWER)

1. Remove the torsion bar as described above.

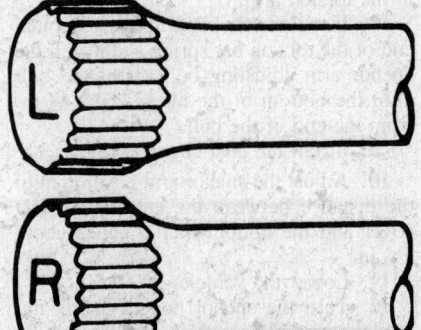

Torsion bar spring serrated ends. Note that they are marked and not interchangeable

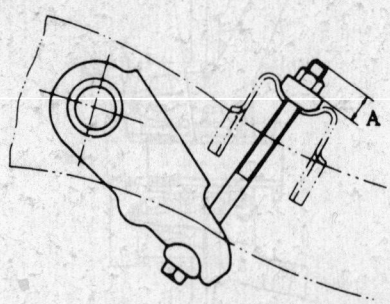

A: 7 - 17 mm (0.28 - 0.67 in)

Anchor arm adjusting bolt measurement

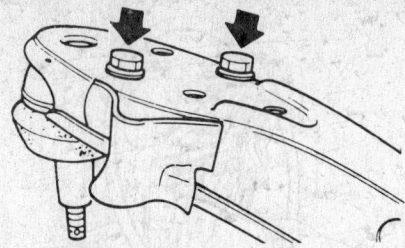

Ball joint-to-control arm bolts

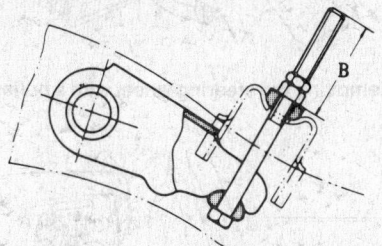

B: 60 - 70 mm (2.36 - 2.76 in)

Anchor arm position adjustment

2. Remove the stabilizer bar as described above.

3. Support the lower control arm with floor jack.

4. Remove the ball joint-to-knuckle nut. Separate the ball joint from the knuckle as described above.

5. Unbolt the control arm from the frame, removing the rear bolts first.

6. Using a suitable driver, remove the bushing from the frame.

7. Replace the bushings if at all worn or damaged.

8. Installation is the reverse of removal. Torque the control arm mounting bolts to 66–87 ft. lb.

REAR SUSPENSION

Springs

REMOVAL & INSTALLATION

———— CAUTION ————

The leaf springs are under a considerable amount of tension. Be very careful when removing or installing them; they can exert enough force to cause serious injuries.

1. Jack up the rear of the truck and support it with jackstands placed under the frame.

2. Disconnect the shock absorbers at their lower end.

3. Remove the nuts securing the U-bolts around the axle housing.

4. Place a jack under the rear axle housing and raise the housing to remove the weight off the spring.

5. Remove the nuts from the spring

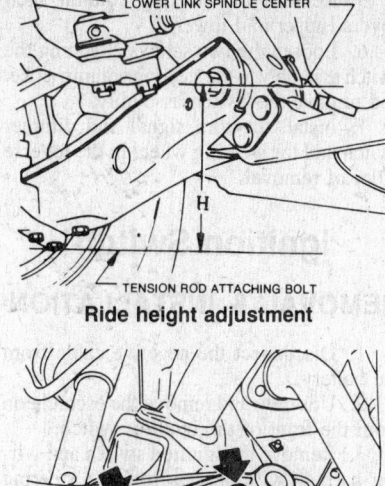

Ride height adjustment

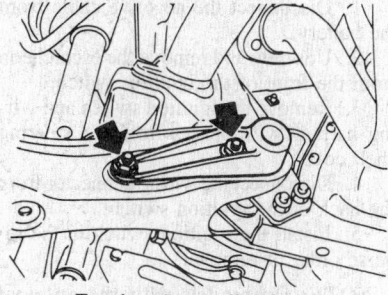

Torsion arm-to-link bolts

shackles, drive out the shackle pins and remove the spring from the vehicle.

6. Install the spring in the reverse order of removal. The weight of the truck must be on the rear wheels before tightening the front pin, shackle, and shock absorber attaching nuts. Tighten the front pin and shackle nuts to 83–94 ft. lbs. (37–50 ft. lbs. for the spring shackle nuts, through 1982), the U-bolt nuts to 53–72 ft. lbs., and the shock absorber lower end nut to 12–16 ft. lbs. for 2WD models. 22–30 ft. lbs. for 4WD models.

Shock Absorbers
INSPECTION AND TESTING

Inspect and test the rear shock absorbers in the same manner as outlined for the front shock absorbers.

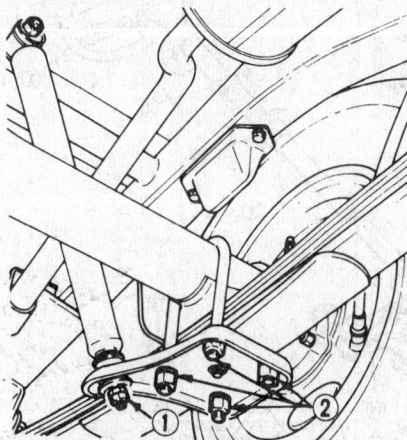

1. Shock absorber lower attaching nut
2. U-bolt attaching nut

Shock absorber lower end and U-bolt attaching nuts

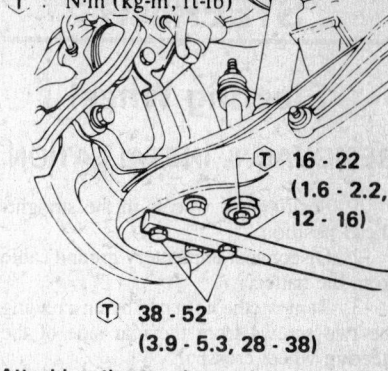

T : N·m (kg-m, ft-lb)

T 16 - 22
(1.6 - 2.2, 12 - 16)

T 38 - 52
(3.9 - 5.3, 28 - 38)

Attaching the tension rod and stabilizer bar to the lower control arm

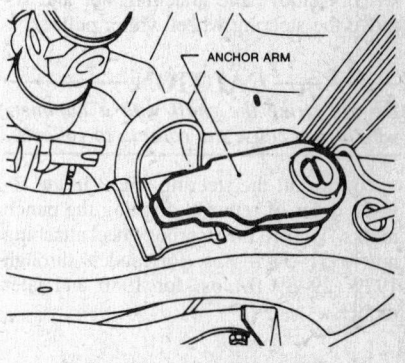

Torsion bar anchor arm

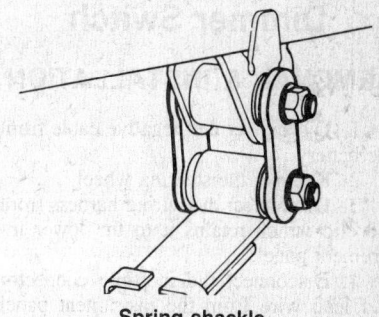

Spring shackle

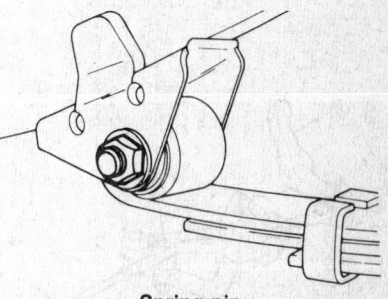

Spring pin

REMOVAL & INSTALLATION

The rear shock absorbers are removed simply by removing the upper and lower attaching nuts, and removing the component from the vehicle. They are installed in the reverse order. The weight of the vehicle must be on the rear wheels before tightening the shock absorber attaching nuts to; Upper 22–30 ft. lbs. Lower, 2WD; 12–16 ft. lbs. 4WD; 22–30 ft. lbs.

STEERING

Steering Wheel

REMOVAL & INSTALLATION

1. Position the wheels in the straight-ahead position.
2. Disconnect the battery ground cable from the battery.
3. Remove the horn pad by unscrewing the two screws from the rear side of the steering wheel crossbar.
4. Punch mark the top of the steering column shaft and the steering wheel flange.
5. Remove the attaching nut and remove the steering wheel with a puller.

—————— CAUTION ——————
Do not strike the shaft with a hammer, which may cause the column to collapse.

6. Install the steering wheel in the reverse order of removal aligning the punch marks. Tighten the steering wheel attaching nut to 51–54 ft. lbs. for models through 1979. 29–39 ft. lbs. for 1980 and later models.

Turn Signal and Dimmer Switch

REMOVAL & INSTALLATION

1. Disconnect the negative cable from the battery.
2. Remove the steering wheel.
3. Disconnect the wiring harness from the clip which retains it to the lower instrument panel.
4. Disconnect the multiple connector and lead wire from the instrument panel wiring harness.

5. Remove the steering column shell covers (upper and lower).
6. Loosen the two screws attaching the switch assembly to the steering column jacket and remove the switch assembly.
7. Install the turn signal and dimmer switch and the steering wheel in the reverse order of removal.

Ignition Switch

REMOVAL & INSTALLATION

1. Disconnect the negative cable from the battery.
2. Unscrew and remove the escutcheon from the front of the ignition switch.
3. Remove the ignition switch and wiring harness with spacer from the steering shell cover.
4. Disconnect the wiring connector from the back of the ignition switch.
5. Install the ignition switch in the reverse order of removal.

NOTE: On models with the optional steering lock cylinder, remove the switch by removing the two retaining screws from the back of the steering lock cylinder.

Steering Lock

REMOVAL & INSTALLATION

1. Remove the ignition switch.
2. Drill out the two shear screws.
3. Remove the two other normal type screws and dismount the steering lock from the steering jacket tube.
4. Install a new steering lock in the reverse order, being sure to tighten the two shear screws until they shear.

Removing the steering wheel with a puller

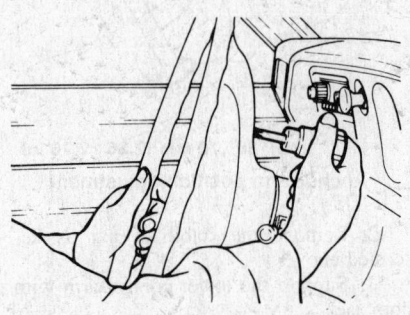

Removing the horn pad from the steering wheel

Steering Linkage

REMOVAL & INSTALLATION

1979

1. Jack up the front of the truck and support it with jackstands placed under the frame.
2. Remove the cotter pins and nuts securing the side rod ball studs to the steering knuckle/spindle arms.
3. Use a puller to disconnect the side rod ball studs from the steering knuckle arms. If a puller is not available, strike the side of the steering knuckle arm boss with a hammer, backing it up with a heavy hammer on the opposite side, and at the same time having an assistant pull the ball stud out of the steering knuckle arm.

NOTE: Do not strike the ball stud head, the ball socket on the side rod, or the side rod with the hammer.

4. Remove the nut securing the steering gear arm on the sector shaft and remove the gear arm with a puller. If a puller is not available, and the steering gear arm need not be removed, disconnect the side arm and tie-rod ball studs from the steering gear arm in the same manner as outlined in Step 3.
5. Remove the idler arm assembly from the frame by unscrewing the two attaching nuts.
6. Install the steering linkage in the reverse order of removal. Tighten the ball stud nuts to 40–55 ft. lbs. idler arm assem-

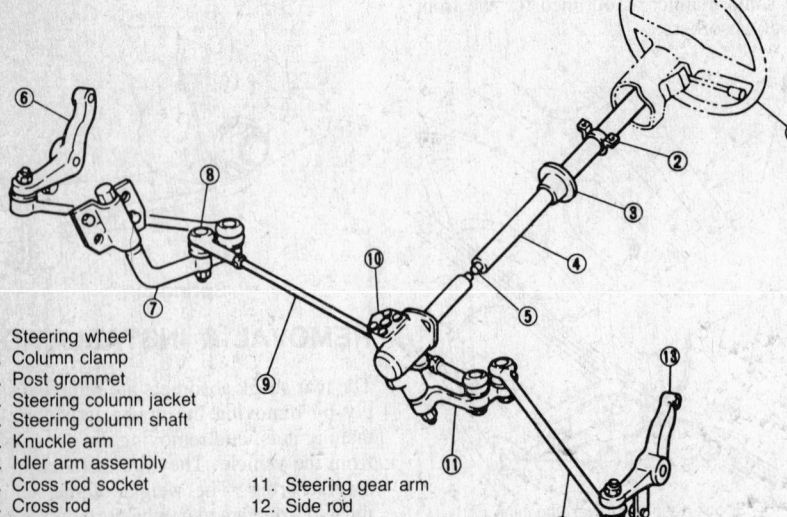

1. Steering wheel
2. Column clamp
3. Post grommet
4. Steering column jacket
5. Steering column shaft
6. Knuckle arm
7. Idler arm assembly
8. Cross rod socket
9. Cross rod
10. Steering gear assembly
11. Steering gear arm
12. Side rod
13. Knuckle arm

1979 steering linkage system

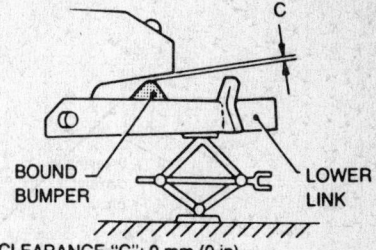

bly attaching nuts to 23–27 ft. lbs., and the tie-rod adjustment locknuts to 58–72 ft. lbs. Adjust the toe-in and steering angle.

1980 and Later

1. Raise and support the front end with jackstands under the frame rails.

2. Mark the pitman arm and shaft for installation purposes. Remove the pitman arm with a puller designed for the purpose such as special tool 290200001.

3. Remove the idler arm.

4. Remove the side rods from the knuckles using a special tool such as HT72520000.

5. Installation is the reverse of removal. Observe the following torques: Pitman arm-to-shaft, manual steering: 94–108

ft. lb. Power steering: 101–130 ft. lb. Idler arm-to-frame: 36–51 ft. lb. Side rods-to-knuckle: 40–72 ft. lb.

6. Check wheel alignment.

Steering Gear

REMOVAL & INSTALLATION

Manual Steering

1. Raise and support the truck on jackstands.

2. Unbolt the wormshaft at the rubber coupling.

3. Matchmark the idler arm and sector shaft, and with the wheels in a straight ahead position, remove the idler arm-to-sector shaft nut and remove the idler arm with a puller.

4. Unbolt and remove the gear from the frame.

5. Installation is the reverse of removal. Torque the wormshaft coupling bolt to 29–36 ft. lbs.; the idler arm nut to 94–108

CLEARANCE "C": 0 mm (0 in)

Positioning lower control arm on 1984 and later models

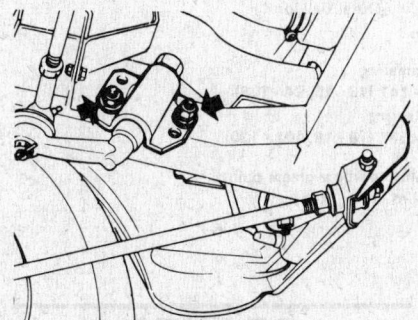

The idler arm attaching nuts

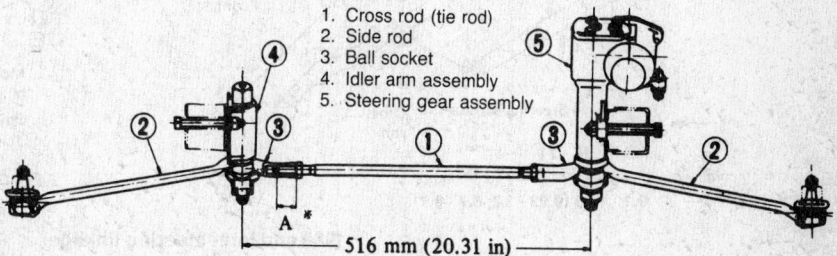

1. Cross rod (tie rod)
2. Side rod
3. Ball socket
4. Idler arm assembly
5. Steering gear assembly

*After adjustment of toe-in, be sure that dimension "A" at both ends of cross rod is not less than 20 mm (0.79 in).

Steering linkage installation dimensions, all models

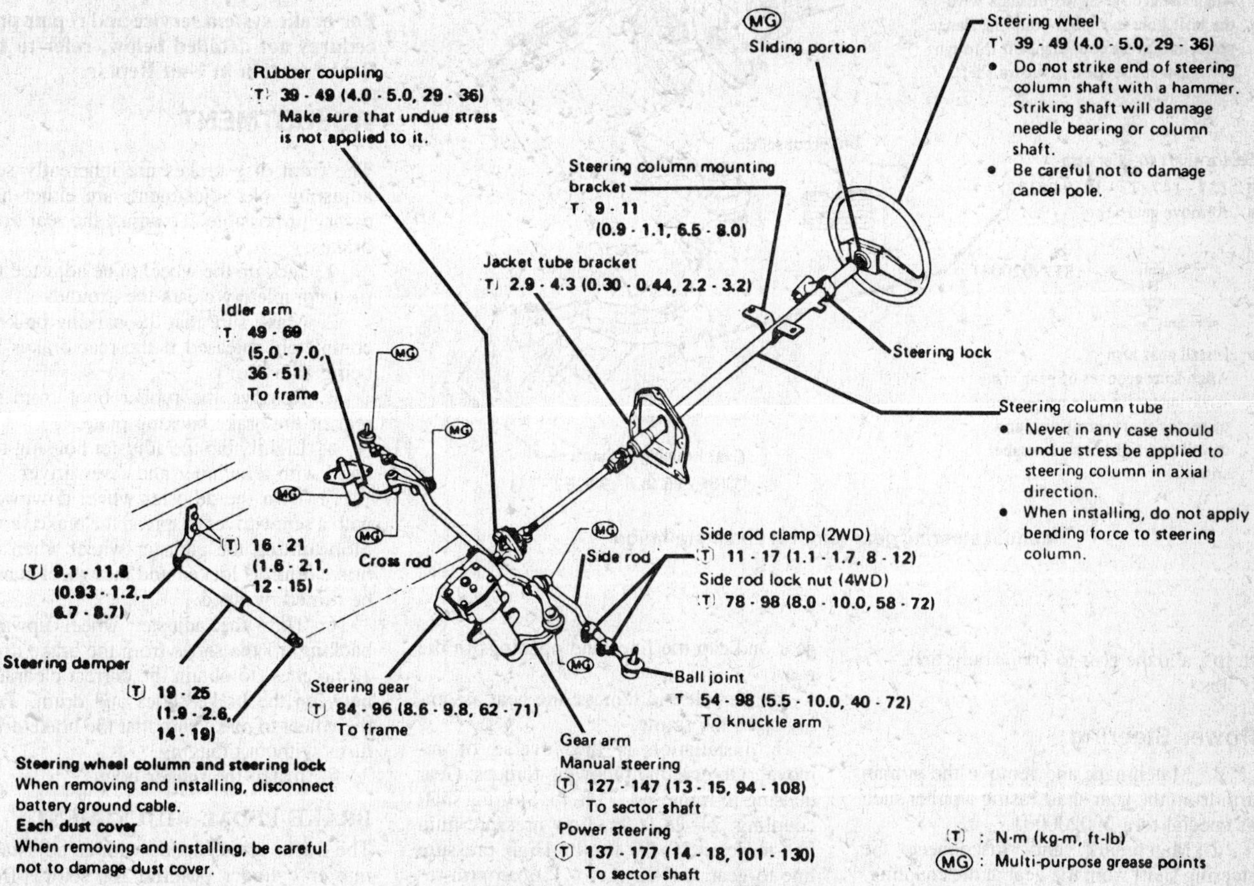

Rubber coupling
Ⓣ 39 - 49 (4.0 - 5.0, 29 - 36)
Make sure that undue stress is not applied to it.

Steering column mounting bracket
Ⓣ 9 - 11
(0.9 - 1.1, 6.5 - 8.0)

Jacket tube bracket
Ⓣ 2.9 - 4.3 (0.30 - 0.44, 2.2 - 3.2)

Idler arm
Ⓣ 49 - 69
(5.0 - 7.0, 36 - 51)
To frame

Ⓣ 16 - 21
(1.6 - 2.1, 12 - 15)

Ⓣ 9.1 - 11.8
(0.93 - 1.2, 6.7 - 8.7)

Cross rod

Steering damper
Ⓣ 19 - 25
(1.9 - 2.6, 14 - 19)

Steering gear
Ⓣ 84 - 96 (8.6 - 9.8, 62 - 71)
To frame

Steering wheel column and steering lock
When removing and installing, disconnect battery ground cable.
Each dust cover
When removing and installing, be careful not to damage dust cover.

Sliding portion

Steering wheel
Ⓣ 39 - 49 (4.0 - 5.0, 29 - 36)
● Do not strike end of steering column shaft with a hammer. Striking shaft will damage needle bearing or column shaft.
● Be careful not to damage cancel pole.

Steering lock

Steering column tube
● Never in any case should undue stress be applied to steering column in axial direction.
● When installing, do not apply bending force to steering column.

Side rod clamp (2WD)
Ⓣ 11 - 17 (1.1 - 1.7, 8 - 12)
Side rod lock nut (4WD)
Ⓣ 78 - 98 (8.0 - 10.0, 58 - 72)

Side rod

Ball joint
Ⓣ 54 - 98 (5.5 - 10.0, 40 - 72)
To knuckle arm

Gear arm
Manual steering
Ⓣ 127 - 147 (13 - 15, 94 - 108)
To sector shaft
Power steering
Ⓣ 137 - 177 (14 - 18, 101 - 130)
To sector shaft

Ⓣ : N·m (kg-m, ft-lb)
Ⓜ🄶 : Multi-purpose grease points

1980 and later steering system

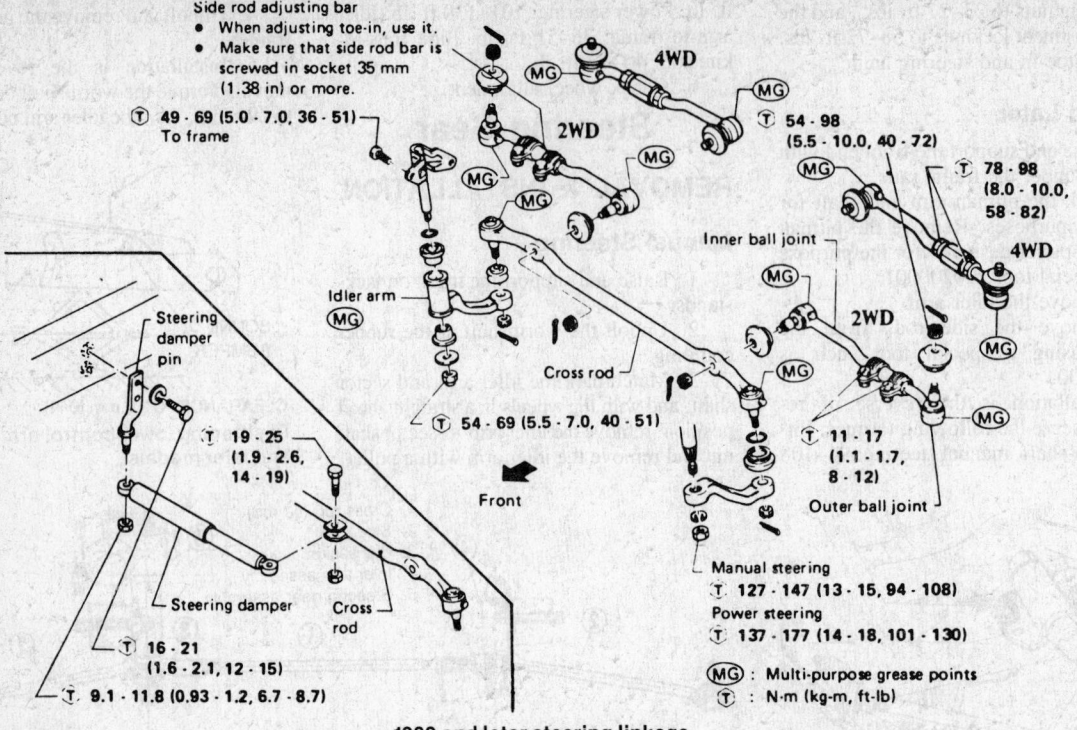

Side rod adjusting bar
● When adjusting toe-in, use it.
● Make sure that side rod bar is screwed in socket 35 mm (1.38 in) or more.
Ⓣ 49 - 69 (5.0 - 7.0, 36 - 51)
To frame

4WD
2WD

Ⓣ 54 - 98 (5.5 - 10.0, 40 - 72)

Ⓣ 78 - 98 (8.0 - 10.0, 58 - 82)
4WD

Inner ball joint

2WD

Idler arm

Cross rod

Ⓣ 54 - 69 (5.5 - 7.0, 40 - 51)

Front

Ⓣ 11 - 17 (1.1 - 1.7, 8 - 12)

Outer ball joint

Steering damper pin

Ⓣ 19 - 25 (1.9 - 2.6, 14 - 19)

Steering damper

Cross rod

Ⓣ 16 - 21 (1.6 - 2.1, 12 - 15)

Ⓣ 9.1 - 11.8 (0.93 - 1.2, 6.7 - 8.7)

Manual steering
Ⓣ 127 - 147 (13 - 15, 94 - 108)
Power steering
Ⓣ 137 - 177 (14 - 18, 101 - 130)

(MG) : Multi-purpose grease points
Ⓣ : N·m (kg-m, ft-lb)

1980 and later steering linkage

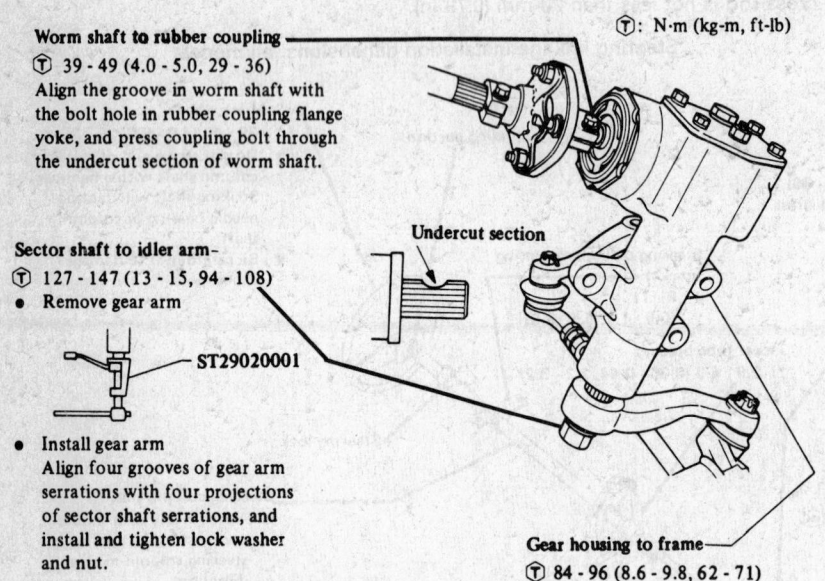

Ⓣ: N·m (kg-m, ft-lb)

Worm shaft to rubber coupling
Ⓣ 39 - 49 (4.0 - 5.0, 29 - 36)
Align the groove in worm shaft with the bolt hole in rubber coupling flange yoke, and press coupling bolt through the undercut section of worm shaft.

Undercut section

Sector shaft to idler arm
Ⓣ 127 - 147 (13 - 15, 94 - 108)
● Remove gear arm

ST29020001

● Install gear arm
Align four grooves of gear arm serrations with four projections of sector shaft serrations, and install and tighten lock washer and nut.

Gear housing to frame
Ⓣ 84 - 96 (8.6 - 9.8, 62 - 71)

Manual steering gear removal and installation

ft. lbs. and the gear-to-frame bolts to 62–71 ft. lbs.

Power Steering

1. Matchmark and remove the pitman arm from the gear shaft, using a puller such as special tool 290200001.

2. Matchmark and disconnect the steering shaft from the gear at the coupling.

3. Disconnect the fluid lines from the gear and cap the lines and openings in the gear.

4. Unbolt and remove the gear assembly from the frame.

5. Installation is the reverse of removal. Observe the following torques: Gear housing-to-frame: 62–71 ft. lb. Steering shaft coupling: 24–28 ft. lb. Low pressure fluid line-to-gear: 29–36 ft. lb. High pressure line-to-gear: 14–22 ft. lb. Pitman arm-to-sector shaft: 101–130 ft. lb.

BRAKES

For brake system service and repair procedures not detailed below, refer to the Brake section in Unit Repair.

ADJUSTMENT

The front disc brakes are inherently self-adjusting. No adjustments are either necessary or possible. To adjust the rear drum brakes:

1. Jack up the wheel to be adjusted until it completely clears the ground.

2. Make sure that the parking brake is completely released if the rear brakes are being adjusted.

3. Remove the rubber boot from the rear of the brake backing plate.

4. Lightly tap the adjuster housing forward with a hammer and screwdriver.

5. Turn the adjuster wheel downward with a screwdriver to spread the brake shoes. Stop turning the adjuster wheel when the brake drum is locked and the wheel cannot be turned by hand.

6. Turn the adjuster wheel upward, backing off the shoes from the brake drum 12 notches, to obtain the correct clearance between the brake shoes and drum. Turn the wheel to make sure that the brake drum turns without dragging.

7. Install the rubber boot.

BRAKE PEDAL ADJUSTMENT

The trucks are equipped with an adjustable master cylinder pushrod for setting free-play.

1. Adjust the height of the stop light switch so that the top surface of the brake pedal is off the surface of the floor board (without rugs): 6.06 in., 1979 and 6.7 in. for 1980 and later. Tighten the stop light switch locknut.

2. Adjust the length of the master cylinder pushrod (booster pushrod on vehicles so equipped) clevis so that the specified free-play exists between the brake pedal and the pushrod. Free-play should be 0.04–0.20 in.

3. Operate the brake pedal to make sure that it operates freely with no noise or interference.

Wheel Bearings

The following is for 2WD models only. For 4WD vehicles, refer to the Front Drive Axle section.

Only the front wheel bearings require periodic service. The lubricant to use is high temperature disc brake wheel bearing grease meeting NLGI No.2 specifications. You will not need any special tools for this job, although the use of a torque wrench is strongly recommended for accurate measurement of bearing preload. The most important thing to remember when working with the wheel bearings is that although they are basically durable, in some ways they are remarkably fragile. Mishandling, grit, misalignment, scratches, improper preload, etc. will quickly destroy any roller bearing, no matter how well hardened during manufacture.

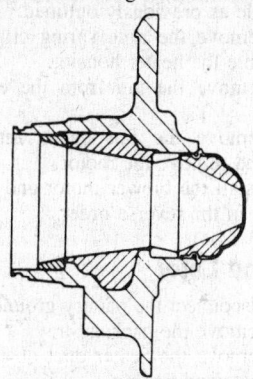

Fill the shaded portion of the hub and grease cap with wheel bearing grease. Also coat the cups with grease

REMOVAL & INSTALLATION

1. Loosen the wheel nuts, raise the truck, and remove the wheel and tire. Remove the brake drum or brake caliper.

2. It is not necessary to remove the drum or disc from the hub. The outer wheel bearing will come off with the hub. Simply pull the hub and disc or drum assembly toward you off the spindle. Be sure to catch the bearing before it falls to the ground.

3. From the inner side of the hub, remove the inner grease seal, and lift the inner bearing from the hub. Discard the grease seal.

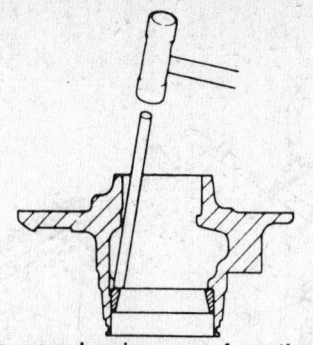

Drive worn bearing cups from the hub with a soft drift and hammer

4. Clean the bearings in solvent and allow them to air dry. You risk leaving bits of lint in the races if you dry them with a rag. Clean the grease cap, nuts, spindle, and the races in the hub thoroughly, and allow the parts to dry.

5. Inspect the bearing carefully. If they are worn, cracked, pitted, burned, scored, etc., they should be replaced, along with the bearing cups in which they run in the hub. Do not mix old and new parts.

6. If the cups are worn, remove them from the hub, by using a brass rod as drift.

7. To install: If the old cups were removed, install the new inner and outer cups into the hub, using either a tool made for the purpose, or a socket or piece of pipe of a large enough diameter to press on the outside rim of the cup only.

--- **CAUTION** ---

Use care not to cock the bearing cups in the hub. If they are not fully seated, the bearings will be impossible to adjust properly.

8. Pack the inside area of the hub and cups with grease, according to the illustration given. Pack the inside of the grease cap while you're at it, but do not install the cap into the hub.

9. Pack the inner bearing with grease. Place a large glob of grease into the palm of one hand and push the inner bearing through it with a sliding motion. The grease must be forced through the side of the bearing and in between each roller. Continue until the grease begins to ooze out the other side through the gaps between the rollers; the bearing must be completely packed with

grease. Install the inner bearing into its cup in the hub, then press a new grease seal into place over it.

10. Install the hub and rotor or drum assembly onto the spindle. Pack the outer bearing with grease in the same manner as the inner bearing, then install the outer bearing into place in the hub.

11. Apply a thin coat of grease to the washer and the threaded portion of the spindle, then loosely install the washer and adjusting nut. Go on to the bearing preload adjustment following.

BEARING PRELOAD ADJUSTMENT

1. While turning the hub forward, tighten the adjusting nut to 22–29 ft. lbs.

2. Rotate the hub a few more times to snug down the bearings.

3. Retighten the nut to 22–29 ft. lbs. Unscrew the adjusting nut $\frac{1}{8}$ of a turn (45 degrees). Install the lock nut (castellated nut) and snug it down against the adjusting nut until one of its grooves lines up with the hole in the spindle. It is okay to tighten the adjusting nut up to 15 degrees to allow the lock nut holes to align. Install a new cotter pin, bending its ends around the lock nut.

4. Install the wheel and a couple of lug nuts. Check the axial play of the wheel by shaking it back and forth. The bearing free play should feel close to zero, but the wheel should spin freely.

5. If the bearing play is correct, remove the wheel, replace the caliper, then install the wheel and grease cap.

Rear Wheel Cylinders

REMOVAL & INSTALLATION

1. Jack up the vehicle, remove the wheel, brake drum, and brake shoes.

2. Disconnect the brake tube from the rear of the wheel cylinder.

3. Remove the wheel cylinder retaining nuts and remove the wheel cylinder from the backing plate.

4. Install the wheel cylinder and assemble the brake in the reverse order of removal and disassembly.

5. Bleed the brake hydraulic system.

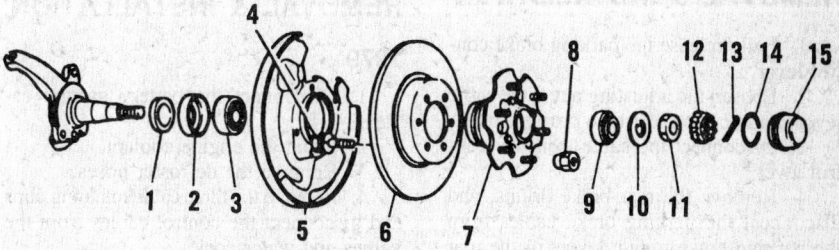

1. Spacer	4. Hub bolt	7. Hub	10. Washer	13. Cotter pin
2. Grease seal	5. Backing plate	8. Lug nut	11. Adjusting nut	14. O-ring
3. Inner bearing	6. Disc (rotor)	9. Outer bearing	12. Lock (castle) nut	15. Grease cap

2-wheel drive hub and bearings with disc brakes. Drum brakes are similar

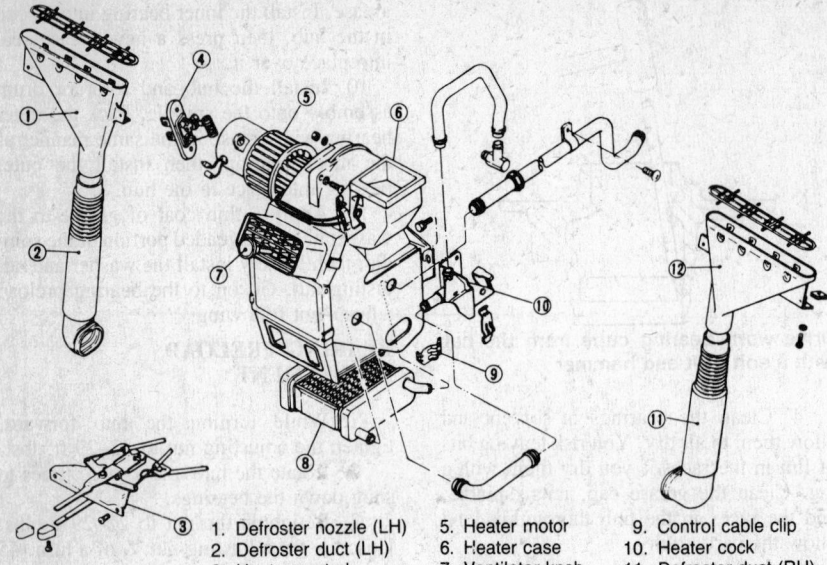

1. Defroster nozzle (LH)
2. Defroster duct (LH)
3. Heater control
4. Resistor
5. Heater motor
6. Heater case
7. Ventilator knob
8. Heater core
9. Control cable clip
10. Heater cock
11. Defroster duct (RH)
12. Defroster nozzle (RH)

1979 heater assembly

Parking Brake Cable

ADJUSTMENT

1. Jack up the rear of the vehicle until the rear wheels clear the ground.
2. Adjust the rear brakes.
3. Loosen the locknut at the parking cable lever assembly mounted on the driveshaft center bearing crossmember.
4. Turn the adjusting nut until the parking brake control lever operating stroke is between 3–4 in.
5. Release the parking brake and make sure that the rear wheels turn freely with no drag.
6. Lower the vehicle.

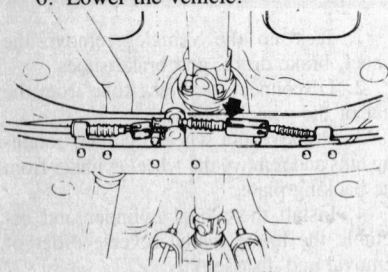

Parking brake adjusting nut

REMOVAL & INSTALLATION

1. Fully release the parking brake control lever.
2. Loosen the adjusting nut at the cable lever mounted to the frame crossmember.
3. Disconnect the cable from the control lever.
4. Remove the rear brake drums, and disconnect the parking brake cables from the parking brake toggle levers of the rear service brake assemblies.
5. Remove lockplate, spring and clip, and pull the parking brake cable out toward the cable lever.

6. Remove the cotter pin at the cable lever and disconnect the cable.
7. Install the cables in the reverse order of removal. Apply a light coat of grease to the cables to make sure that they slide properly. Adjust the parking brake cables.

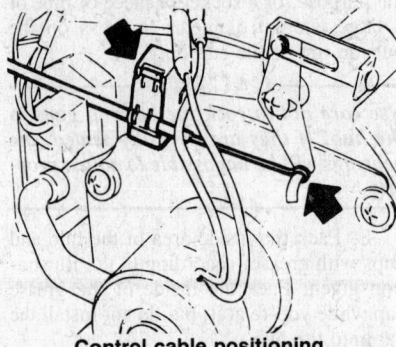

Control cable positioning

CHASSIS ELECTRICAL

Heater Assembly
REMOVAL & INSTALLATION

1979

1. Disconnect the battery ground cable.
2. Drain the engine coolant.
3. Remove the defroster hoses.
4. Remove the three cable retaining clips and disconnect the control cables from the valves and water cock.
5. Disconnect the two fan motor leads from each connector.
6. Disconnect the two resistor lead wires from each connector.

7. Disconnect the water hoses from the heater core and water cock.
8. Remove the three heater housing mounting bolts and remove the heater assembly from the vehicle.
9. Install the heater assembly in the reverse order of removal.

1980 and Later

1. Disconnect the battery ground cable.
2. Drain the cooling system.
3. On models with air conditioning, disconnect the heater hose from the engine.
4. On models without air conditioning, remove the heater duct and disconnect the heater hose at the heater.
5. Remove the console box and instrument panel assembly.
6. Disconnect the air intake control cable from the blower.
7. On models equipped with A/C, remove the blower. Remove the evaporator unit nuts and bolts, but do not remove the evaporator unit.
8. Remove the heater assembly.
9. Installation is the reverse of removal. Adjust the control cable for proper operation.

Blower
REMOVAL & INSTALLATION

1979

1. Remove the heater assembly from the vehicle as previously outlined.
2. Remove the nine spring clips and disassemble the heater housing.
3. Remove the fan from the electric motor.
4. Remove the fan motor retaining screws and remove the motor.
5. Install the blower motor and heater assembly in the reverse order.

1980 and Later

1. Disconnect the battery ground.
2. Remove the package tray.
3. Remove the heater duct on models without air conditioning.
4. Remove the resistor connector and disconnect the control cable.
5. Remove the blower.
6. Installation is the reverse of removal. Adjust the control cable for proper operation.

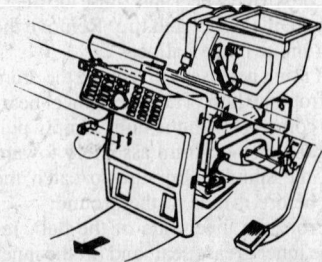

Removing the heater assembly front cover

Heater Core

REMOVAL & INSTALLATION

1979

1. Drain the engine coolant.
2. Remove the defroster hoses.
3. Disconnect the water hoses from the inlet and outlet pipes of the heater core.
4. Remove the four clips and front cover.
5. Remove the heater core from the heater housing.
6. Install the heater core in the reverse order of removal.

1980 and Later

See Heater Assembly procedures.

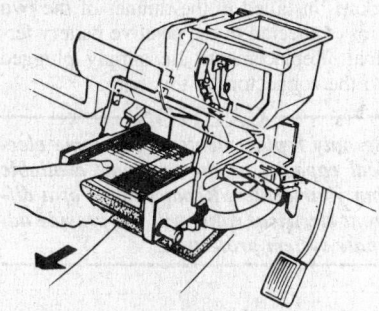

Removing the heater core

Radio

Observe the following cautions when working on the radio:

1. Always observe the proper polarity of the connections (positive to positive and negative to negative).
2. Never operate the radio without a speaker, to prevent damage to the output transistors. If a new speaker is installed, make sure it has the correct impedance (ohms) for the radio. If a new antenna or antenna cable is used, or if poor AM reception is noted, the antenna trimmer can be adjusted. Tune the radio to a weak station around 1400kc. Adjust the trimmer screw until best reception and maximum volume are obtained. The trimmer screw for the factory-installed radio is located above the left knob on the front of the radio in most cases. For best FM reception, raise the antenna to 31 inches. For best AM reception, raise the antenna to its full height.

REMOVAL & INSTALLATION

1979

1. Pull the knobs off the radio control shafts.
2. Remove the radio retaining nuts and washer from the radio control shafts.
3. Remove the bezel plate from the front of the radio.
4. Disconnect the antenna cable and the power and speaker wires from under the instrument panel.
5. Remove the radio from the instrument panel.

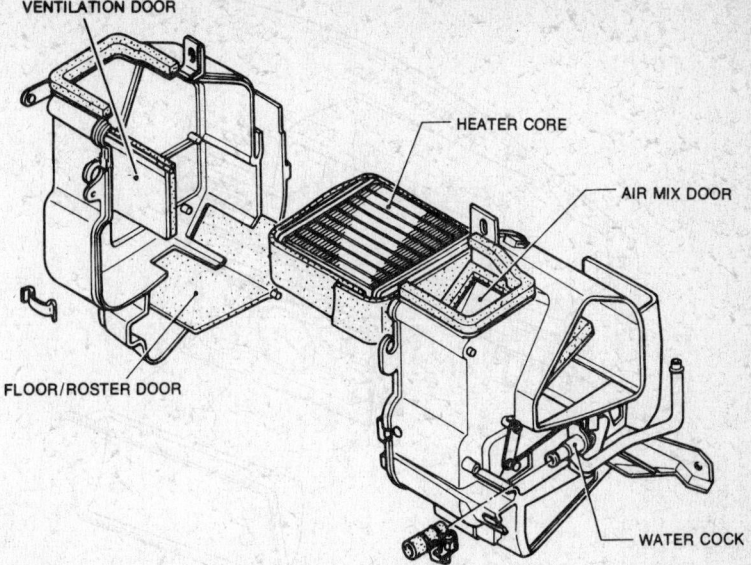

1980 and later heater assembly

6. Install the radio in the reverse order of removal.

1980 and Later

1. Disconnect the battery ground.
2. Remove the ash tray and heater/air conditioner control panel.
3. Disconnect the wiring plug at the back of the radio.
4. Remove the plug covering the mounting screws, remove the screws and pull the radio from the dash.
5. Disconnect the wiring harness and the antenna.
6. Installation is the reverse of removal.

Windshield Wiper Motor

REMOVAL & INSTALLATION

1. Remove the wiper blades and arms as an assembly from the pivots. The arms are retained to the pivots by nuts. Remove the nuts and pull the arms straight off.
2. Remove the cowl top grille. It is retained by four screws at its front edge. Remove the screws and pull the grille forward to disengage the tabs at the rear.
3. Remove the stop ring which connects the wiper motor arm to the connecting rod.
4. Disconnect the wiper motor harness at the connector on the wiper motor body from under the instrument panel.
5. Remove the three retaining screws and pull the wiper motor outward and remove the motor from the vehicle.
6. Install the wiper motor in the reverse order of removal. The wiper arms should be installed so that the blades are 0.98 in. (25mm) above, and parallel to, the windshield molding. If the motor has been run,

be sure the motor and linkage is in its parked position before installing the wiper arms. To do this, turn the ignition switch on, and cycle the motor three or four times. Shut off the motor with the wiper switch (not the ignition switch), and allow the motor to return to the park position.

Wiper Linkage

REMOVAL & INSTALLATION

1. Remove the wiper blade and arm from the pivot.
2. Remove the cowl top grille.
3. Remove the two flange nuts retaining the wiper linkage pivot to the cowl top.
4. Remove the stop ring which retains the connecting rod to the wiper motor arm.
5. Remove the wiper motor linkage assembly from the truck.
6. Install the linkage in the reverse order of removal.

Instrument Cluster

REMOVAL & INSTALLATION

1979

1. Disconnect the battery ground (negative) cable.
2. Working through the openings of the instrument cluster cover, remove the three screws retaining the cluster cover to the instrument panel and remove cover.
3. From underneath the instrument panel, remove the one screw retaining the cluster assembly to the lower instrument panel.
4. Withdraw the cluster lid slightly. Press the windshield wiper control knob in, turn it counterclockwise and pull it off the switch. Remove the headlight/parking light switch knob in the same manner.

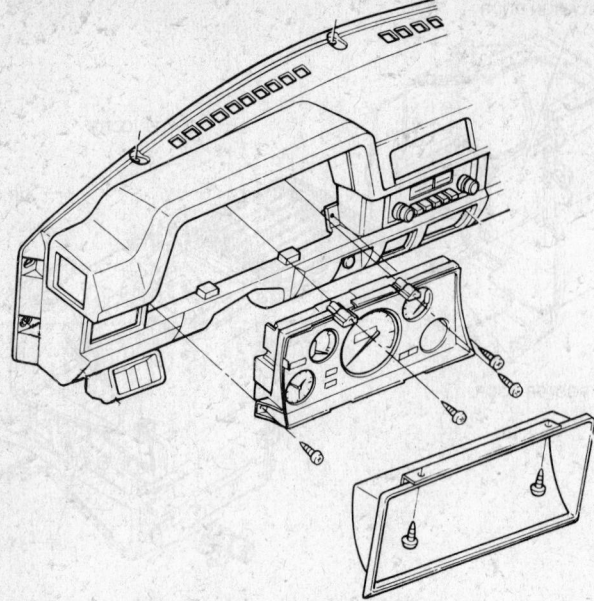

1984 and later cluster removal

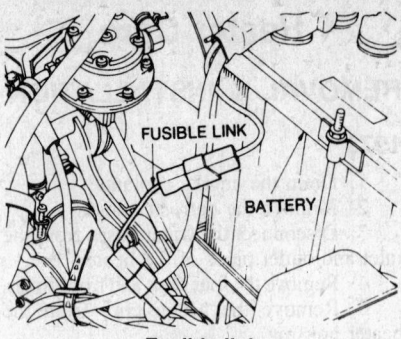

Fusible link

5. From behind the instrument cluster, disconnect the speedometer cable at the speedometer head and the multiple connector from the printed circuit.

6. On vehicles with a clock, disconnect the wires at each connection on the instrument panel printed circuit.

7. Remove the four screws retaining the cluster assembly to the cluster lid.

8. Remove the instrument cluster assembly from under the instrument panel.

9. Install the instrument cluster in the reverse order of removal.

1980 and Later

1. Disconnect the battery ground.
2. Remove the cluster lid.
3. Remove the cluster assembly.
4. Remove the gauges from the cluster, individually.
5. Installation is the reverse of removal.

Speedometer Cable

REPLACEMENT

1. Reach up under the instrument panel and disconnect the cable housing from the back of the speedometer. It is attached by a knurled knob which simply unscrews.

2. Pull the cable from the cable housing. If the cable is broken, the other half of the cable will have to be removed from the transmission end. Unscrew the retaining knob and remove the cable from the transmission extension housing.

3. Lubricate the cable with graphite powder (sold as speedometer cable lubricant, curiously enough) and feed the cable into the housing. It is best to start at the speedometer end and feed the cable down towards the transmission. It is also usually necessary to unscrew the transmission connection and install the cable end to the gear, then reconnect the housing to the transmission. Slip the cable end into the speedometer, and reconnect the cable housing.

Headlights

REMOVAL & INSTALLATION

1. Remove the radiator grille retaining screws and remove the radiator grille.

2. Loosen and remove, if necessary, the three retaining ring screws. Do not disturb the aiming adjusting screws.

3. Remove the retaining ring by rotating it clockwise.

4. Remove the headlight from the mounting ring and disconnect the electrical connector from behind the light.

5. Change the headlight and connect the wiring connector to the new light.

6. Place the headlight in position so that the three locating tabs behind the light fit in the three holes on the mounting ring.

7. Install the headlight retaining ring and tighten the retaining screws.

8. Install the radiator grille.

Fusible Links

A fusible link is a protective device used in an electrical circuit. When current increases beyond a certain amperage, the fusible metal wire of the link melts, thus breaking the electrical circuit and preventing further damage to other components and wiring. Whenever a fusible link is melted because of a short circuit, correct the cause before installing a new fusible link.

There is only one fusible link in Datsun pickups installed in the thinner of the two wires connected to the positive battery terminal. Replacements are simply plugged into the connectors in this wire.

CAUTION

Use only replacements of the same electrical capacity as the original, available from your dealer. Replacements of a different electrical value will not provide adequate system protection.

Fuses

Fuses protect all the major electrical systems in the truck. In case of an electrical overload, the fuse melts, breaking the circuit and stopping the flow of electricity.

If a fuse blows, the cause should be investigated and corrected before the installation of a new fuse. This, however, is easier to say than to do. Because each fuse protects a limited number of components, your job is narrowed down somewhat. Begin your investigation by looking for obvious fraying, loose connections, breaks in insulation, etc. Electrical problems are almost always a real headache to solve, but patience and persistence, coupled with logic, usually provide a solution.

The amperage of each fuse and the circuit it protects are marked on the cover of the fusebox, which is located under the instrument panel next to the steering column.

Flashers and Relays

The turn signal and four-way hazard flashers are located under the instrument panel on opposite sides of the steering column. The turn signal flasher is the larger of the two. Replacement is made by unplugging the old flasher and plugging in the new one.

Relays are used for the horn, headlights, wiper, heater, choke heater, catalyst floor sensor, air conditioner compressor, and transmission switches, although obviously not all relays are used on all models. All relays used are grouped together, and mounted on the right fender in the engine compartment.

Dodge/Plymouth
Pickups, Vans, Ramcharger, Trail Duster, Rampage

GENERAL ENGINE SPECIFICATIONS

Year	Engine Displacement (cu in.)	Horsepower (@ rpm)	Torque @ rpm (ft lbs)	Bore × Stroke (in.)	Compression Ratio	Oil Pressure (psi @ rpm)
'79	6-225	90 @ 3600	160 @ 1600	3.40 × 4.12	8.4 : 1	30-70 @ 2000
	6-243 Diesel	100 @ 3700	163 @ 2200	3.62 × 3.94	20 : 1	43 @ 1000
	8-318	140 @ 4000	245 @ 1600	3.91 × 3.31	8.6 : 1	30-80 @ 2000
	8-360	155 @ 3600	270 @ 2400	4.00 × 3.58	8.5 : 1	30-80 @ 2000
'80-'81 ④	6-225	90 @ 3600	160 @ 1600	3.40 × 4.12	8.4 : 1	35-65 @ 2000
	8-318 ②	120 @ 3600	245 @ 1600	3.91 × 3.31	8.5 : 1	35-65 @ 2000
	8-318 ③	155 @ 4000	240 @ 2000	3.91 × 3.31	8.5 : 1	35-65 @ 2000
	8-360 ②	130 @ 3200	255 @ 2000	4.00 × 3.58	8.4 : 1	35-65 @ 2000
	8-360 ③	185 @ 4000	275 @ 2000	4.00 × 3.58	8.0 : 1	35-65 @ 2000
'82-'84 ④	4-135	84 @ 4800	111 @ 2400	3.44 × 3.62	8.5 : 1	50 @ 2000
	4-155.9	92 @ 4500	131 @ 2500	3.59 × 3.86	8.2 : 1	56 @ 2000
	6-225	90 @ 3600	160 @ 1600	3.40 × 4.12	8.4 : 1	35-65 @ 2000
	8-318 ②	120 @ 3600	245 @ 1600	3.91 × 3.31	8.5 : 1	35-65 @ 2000
	8-318 ③	155 @ 4000	240 @ 2000	3.91 × 3.31	8.5 : 1	35-65 @ 2000
	8-360 ②	130 @ 3200	255 @ 2000	4.00 × 3.58	8.4 : 1	35-65 @ 2000
	8-360 ③	185 @ 4000	275 @ 2000	4.00 × 3.58	8.0 : 1	35-65 @ 2000
'85-'86	4-135	96 @ 5200	119 @ 3200	3.44 × 3.62	9.5 : 1	25-80 @ 3000
	4-153.9	104 @ 4800	142 @ 2800	3.59 × 3.86	8.7 : 1	45-90 @ 3000
	6-225	90 @ 3600	160 @ 1600	3.40 × 4.12	8.4 : 1	25-70 @ 3000
	— 8-318 ②	104 @ 3600	140 @ 3600	3.91 × 3.31	9.0 : 1	30-80 @ 3000
	8-318 ③	155 @ 4000	240 @ 2000	3.91 × 3.31	9.0 : 1	30-80 @ 3000
	8-360 ②	130 @ 3200	255 @ 2000	4.00 × 3.58	—	30-80 @ 3000
	8-360 ③	185 @ 4000	275 @ 2000	4.00 × 3.58	—	30-80 @ 3000

① 10 more hp with two barrel carburetor
② 2 barrel carb
③ 4 barrel carb
④ Horsepower and torque are SAE net figures. They are measured at the rear of the transmission with all accessories installed and operating. Since the figures vary when a given engine is installed in different models, some ratings are representative rather than exact

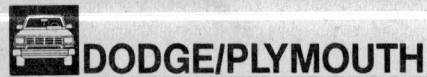

TUNE-UP SPECIFICATIONS

Year	Engine No. Cyl. Displacement (cu in.)	Spark Plugs Original Type	Gap (in.)	Distributor	Ignition Timing (±2°) (deg)▲ MT	AT	Intake Valve Opens (deg)	Fuel Pump Pressure (psi)	Idle Speed (rpm)• MT	AT	Valve Clearance (in.) In	Ex
'79	6-225 LD	P-560PR	.035	Electronic	12B	12B	16B	3.5-5.0	675	675	.010	.020
	6-225 LD Calif.	P-560PR	.035	Electronic	8B	8B	16B	3.5-5.0	800	800	.010	.020
	6-225 HD	P-560PR	.035	Electronic	12B	12B	16B	3.5-5.0	675	675	.010	.020
	6-243 Diesel	—	—	—	—	—	18B	32B	—	650	.012	.012
	V8-318 LD	P-64PR	.035	Electronic	12B	12B	10B	5.0-7.0	680	680	Hyd.	Hyd.
	V8-318 LD Calif.	P-64PR	.035	Electronic	6B	6B[1]	10B	5.0-7.0	750	750	Hyd.	Hyd.
	V8-318 HD	P-64PR	.035	Electronic	2A	2A[2]	10B	5.0-7.0	750[3]	750[3]	Hyd.	Hyd.
	V8-360 LD	P-65PR	.035	Electronic	10B[3]	10B[3]	18B	5.0-7.0	750[3]	750[3]	Hyd.	Hyd.
	V8-360 LD Calif.	P-65PR	.035	Electronic	10B	10B	18B	5.0-7.0	750	750	Hyd.	Hyd.
	V8-360 HD	P-65PR	.035	Electronic	4B	4B	18B	5.0-7.0	750	750	Hyd.	Hyd.
	V8-360 HD Calif.	P-65PR	.035	Electronic	—	4B	18B	5.0-7.0	750	750	Hyd.	Hyd.
'80	6-225 LD	560PR	.035	Electronic	12B	12B	16B	3.5-5.0	600[23]	600[23]	.010	.020
	6-225 MD Calif.	560PR	.035	Electronic	12B	12B	16B	3.5-5.0	800	800	.010	.020
	6-225 HD Canada	560PR	.035	Electronic	12B	12B	16B	3.5-5.0	675	675	.010	.020
	V8-318 LD [24]	64PR	.035	Electronic	12B	12B	10B	5.0-7.0	600	600	Hyd.	Hyd.
	V8-318 LD [25]	64PR	.035	Electronic	10B	10B	10B	5.0-7.0	750	750	Hyd.	Hyd.
	V8-318 MD Calif.	64PR	.035	Electronic	10B[4]	10B[4]	10B	5.0-7.0	750	750	Hyd.	Hyd.
	V8-318 HD	64PR	.035	Electronic	8B[4]	8B[4]	10B	5.0-7.0	750	750	Hyd.	Hyd.
	V8-360 Canada	65PR	.035	Electronic	—	4B[5]	18B	5.0-7.0	—	750	Hyd.	Hyd.
	V8-360 Canada	65PR	.035	Electronic	12B	12B[6]	18B	5.0-7.0	650	650	Hyd.	Hyd.
	V8-360 LD	65PR	.035	Electronic	12B	12B	18B	5.0-7.0	650	650	Hyd.	Hyd.
	V8-360 LD	65PR	.035	Electronic	—	10B	18B	5.0-7.0	—	750	Hyd.	Hyd.
	V8-360 MD	65PR	.035	Electronic	10B	10B	18B	5.0-7.0	750	750	Hyd.	Hyd.
	V8-360 HD	65PR	.035	Electronic	10B	10B	18B	5.0-7.0	750	750	Hyd.	Hyd.
	V8-360 HD	65PR	.035	Electronic	4B	4B	18B	5.0-7.0	700	700	Hyd.	Hyd.
'81	6-225 LD	560PR	.035	Electronic	12B	16B	6B	3.5-5.0	[8]	[8]	Hyd.	Hyd.
	6-225 HD	560PR	.035	Electronic	12B	16B	6B	3.5-5.0	725	750	Hyd.	Hyd.
	V8-318 LD [5]	64PR	.035	Electronic	10B	16B	10B	5.0-7.0	650	650	Hyd.	Hyd.
	V8-318 HD Canada [5]	64PR	.035	Electronic	2A	2A	10B	5.0-7.0	750	750	Hyd.	Hyd.
	V8-318 LD [7]	64PR	.035	Electronic	12B	16B	10B	5.0-7.0	750	750	Hyd.	Hyd.
	V8-318 HD [7]	64PR	.035	Electronic	12B	12B	10B	5.0-7.0	750	750	Hyd.	Hyd.
	V8-360-1 LD [7]	64PR	.035	Electronic	12B	16B	18B	5.0-7.0	600	625	Hyd.	Hyd.
	V8-360-1 HD [5]	64PR	.035	Electronic	—	4B	18B	5.0-7.0	—	750	Hyd.	Hyd.
	V8-360-1 HD [7]	64PR	.035	Electronic	—	4B	18B	5.0-7.0	—	700	Hyd.	Hyd.
	V8-360-3 HD	73SR	.035	Electronic	—	4B	18B	5.0-7.0	—	700	Hyd.	Hyd.
	V8-360-3 HD	73SR	.035	Electronic	—	10B	18B	5.0-7.0	—	750	Hyd.	Hyd.
'82	4-135	65PR	.035	Electronic	12B	12B	14B	4.5-6.0	850	900	Hyd.	Hyd.
	6-225-1	560PR	.035	Electronic	12B	16B	6B	3.5-5.0	600	600	Hyd.	Hyd.
	V8-318 [5]	64PR	.035	Distributor	12B	12B	10B	4.75-6.25	750	750	Hyd.	Hyd.
	V8-318 [9]	64PR	.035	Electronic	12B	16B	10B	4.75-6.25	750	750	Hyd.	Hyd.
	V8-360 [5]	65PR	.035	Electronic	4B	4B	18B	5.0-7.0	—	700	Hyd.	Hyd.
	V8-360 [7]	65PR	.035	Electronic	—	4B	18B	5.0-7.0	—	700	Hyd.	Hyd.
'83-'84	4-135	65PR	.035	Electronic	12B	12B	14B	4.5-6.0	800	800	Hyd.	Hyd.
	6-225	560PR	.035	Electronic	12B	16B	6B	3.5-5.0	[9]	[9]	Hyd.	Hyd.
	8-318 [5]	64PR	.035	Electronic	12B	12B	10B	4.75-6.25	[10]	[10]	Hyd.	Hyd.
	8-318 [7]	64PR	.035	Electronic	12B	12B	10B	4.75-6.25	750	750	Hyd.	Hyd.
	8-360 [5]	65PR	.035	Electronic	—	4B	18B	5.0-7.0	—	750	Hyd.	Hyd.
	8-360 [7]	RF10	.035	Electronic	4B	4B	18B	5.0-7.0	700	700	Hyd.	Hyd.
'85-'86	4-135	RN12Y	.035	Electronic	6B	6B	16B	4.5-6.0	850	900	Hyd.	Hyd.
	4-155.9	RN12Y	.035-.040	Electronic	7B	7B	25B	—	800	800	Hyd.	Hyd.
	4-155.9HA	RN12Y	.035-.040	Electronic	12B	12B	25B	—	850	850	Hyd.	Hyd.
	6-225	RBL16Y	.035	Electronic	12B	16B	6B	4.0-5.5	725	750	Hyd.	Hyd.
	6-225 Calif.	RBL16Y	.035	Electronic	12B	16B	6B	4.0-5.5	775	775	Hyd.	Hyd.
	8-318	RN12YC	.035	Electronic	12B	12B	10B	5.75-7.25	700	750	Hyd.	Hyd.
	8-318 Calif.	RN12YC	.035	Electronic	8B	8B	10B	5.75-7.25	650	650	Hyd.	Hyd.
	8-318HA	RN12YC	.035	Electronic	8B	8B	10B	5.75-7.25	650	650	Hyd.	Hyd.

TUNE-UP SPECIFICATIONS

Year	Engine No. Cyl. Displacement (cu in.)	Spark Plugs Original Type	Gap (in.)	Distributor	Ignition Timing (±2°) (deg)▲ MT	AT	Intake Valve Opens (deg)	Fuel Pump Pressure (psi)	Idle Speed (rpm) • MT	AT	Valve Clearance (in.) In	Ex
'85–'86	8-360	RN12YC	.035	Electronic	16B	16B	18B	5.75–7.25	800	800	Hyd.	Hyd.
	8-360HA	RN12YC	.035	Electronic	16B	16B	18B	5.75–7.25	710	710	Hyd.	Hyd.
	8-360HD	RN12YC	.035	Electronic	10B	10B	18B	5.75–7.25	710	710	Hyd.	Hyd.

NOTE: The underhood specifications sticker often reflects tune-up specifications changes made in production. Sticker figures must be used if they disagree with those in this chart.
NOTE: All Canadian specifications are the same as Federal specifications unless noted otherwise.
• Manual in neutral, automatic in park.
— Not Applicable
▲ With vacuum advance disconnected and plugged.
  B Before Top Dead Center
  A After Top Dead Center
  TDC Top Dead Center
  CAP Cleaner Air Package
  Fed. All states except California
  (Min.) Minimum
  Hyd. Hydraulic valve lifters; no service adjustment
  LD Light Duty emissions
  MD Medium Duty emissions
  HD Heavy Duty emissions

① For GVWR under 6000 lbs.—8B
② Canadian models w/4bbl—8B
③ Canadian models w/4bbl—4B @ 700 rpm
④ Canadian HD engine w/2bbl—2A
⑤ 2bbl
⑥ Canadian w/4bbl and distributor 4111487—10B
⑦ 4bbl
⑧ Federal—600: Calif.—800: Canada—725
⑨ Fed 1bbl—600MT, 650 AT: Calif.—750 All: Canada 1bbl—725 MT, 750 AT: All 2bbl—700
⑩ Fed & Canada—750: HA & Calif.—700

FIRING ORDERS

NOTE: To avoid confusion, always replace spark plug wires one at a time.

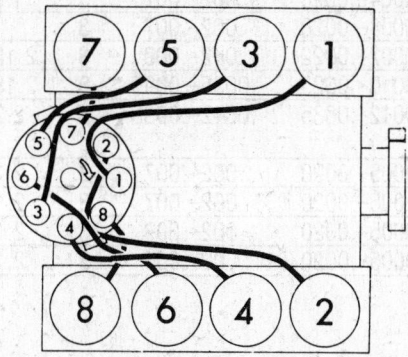

318, 360 V8 engines

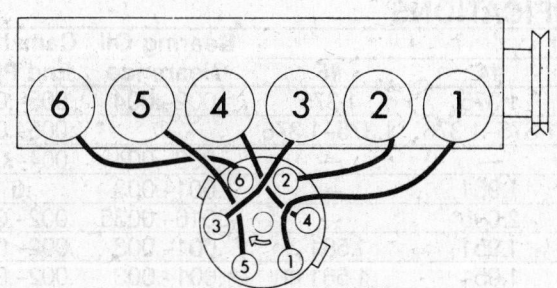

6 cylinder engines

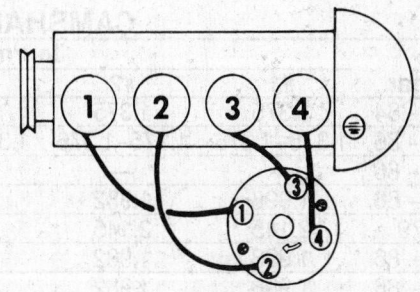

4 cylinder engines

VALVE SPECIFICATIONS

Year	Engine Displacement (cu in.)	Seat Angle (deg)	Face Angle (deg)	Spring Test Pressure (lbs @ in.)	Spring Installed Height (in.)	Stem to Guide Clearance (in.)		Stem Diameter (in.)	
						Intake	Exhaust	Intake	Exhaust
'82–'86	4-135	45	45	175 @ 1.22	1.65	.001–.003	.002–.004	.312–.313	.311–.312
'84–'86	4-155.9	45	45	16 @ 1.59	1.59	.0012–.0024	.0020–.0035	.315	.315
'79–'86	6-225	45	①	137–150 @ 1⁵/₁₆	1¹¹/₁₆	.001–.003	.002–.004	.372–.373	.371–.372
'79	6-243 Diesel	45	45	NA	1⁴⁹/₆₄	.002–.003	.003–.004	.314	.314
'79–'81	8-318	45	45	②	③	.001–.003	.002–.004	.372–.373	.371–.372
'82–'86	8-318	45	45	④	④	.001–.003	.002–.004	.372–.373	.371–.372
'79–'81	8-360	45	45	②	③	.001–.003	.002–.004	.372–.373	.371–.372
'82–'86	8-360	45	45	②	②	.001–.003	.002–.004	.372–.373	.371–.372

① Intake; 45°
 Exhaust; 43°
② Intake; 170–184 @ 1⁵/₁₆
 Exhaust; 181–197 @ 1¹/₁₆
③ Intake; 1¹¹/₁₆
 Exhaust; 1³³/₆₄
④ Intake; 170–184 @ 1⁵/₁₆
 Exhaust; 180–194 @ 1¹/₁₆

CRANKSHAFT AND CONNECTING ROD SPECIFICATIONS

(All measurements given in in.)

Year	Engine Displacement (cu in.)	Crankshaft				Connecting Rod		
		Main Brg Journal Dia	Main Brg Oil Clearance	Shaft End Play	Thrust on No.	Journal Dia	Oil Clearance	Side Clearance
'82–'86	4-135	2.362–2.363	.0004–.0026 ①	.002–.007	3	1.968–1.969	.0004–.0026 ②	.005–.015 ③
'84–'86	4-155.9	2.3622	.0008–.0028	.002–.007	3	2.0866	.0008–.0028	.004–.010
'79–'80	6-225	2.7495–2.7505	.0002–.0022	.002–.009	3	2.1865–2.1875	.0002–.0022	.006–.025
'81–'84	6-225	2.7495–2.7505	.0010–.0025	.0035–.0095	3	2.1865–2.1875	.0010–.0022	.007–.013
'79	6-243 Diesel	2.754–2.755	.0012–.0035	.0012–.0035	7	2.281–2.282	.0015–.0044	.006–.018
'79–'80	8-318	2.4995–2.5005	.0005–.0020	.002–.007	3	2.124–2.125	.0005–.0025	.006–.014
'81–'86	8-318	2.4995–2.5005	.0005–.0020 ④	.002–.007	3	2.124–2.125	.0005–.0022	.006–.014
'79–'80	8-360	2.8095–2.8105	.0005–.0020	.002–.009	3	2.124–2.125	.0005–.0025	.006–.014
'81–'86	8-360	2.8095–2.8105	.0005–.0020 ④	.002–.009	3	2.124–2.125	.0005–.0022	.006–.014

① '85–'86: .0003–.0031
② '85–'86: .0008–.0034
③ '85–'86: .005–.013
④ '85–'86: #1 Brg .0005–.0015 max.

CAMSHAFT SPECIFICATIONS

Engine	Year	Journal Diameter					Bearing Oil Clearance	Camshaft End Play
		#1	#2	#3	#4	#5		
4-135	'82–'84	1.375	1.375	1.375	1.375	1.375	.002–.004	.005–.013
4-135	'85–'86	1.375–1.376	1.375–1.376	1.375–1.376	1.375–1.376	1.375–1.376	①	.005–.013
4-155.9	'84–'86	—	—	—	—	—	.002–.004	.004–.008
6-225	'79–'86	1.998	1.982	1.967	1.951	—	.001–.003	.0
6-243	'79	2.145	2.145	2.125	2.086	—	.0016–.0035	.002–.008
8-318	'79–'86	1.998	1.982	1.967	1.951	1.561 ②	.001–.003	.002–.010
8-360	'79–'86	1.998	1.982	1.967	1.951	1.561 ②	.001–.003	.002–.010

① Inspect the camshaft journals for scoring and replace
 if scored or if lobe wear exceeds .010 in.
② 1985–86 models—1.5605

TORQUE SPECIFICATIONS

Year	Engine Displacement (cu in.)	Cylinder Head Bolts	Rod Bearing Bolts	Main Bearing Bolts	Crankshaft Bolt	Flywheel-to-Crankshaft Bolts	Manifold Intake	Manifold Exhaust
'82–'86	4-135	①⑥	②	③	50	55 ④	15 ⑤	15 ⑤
'84–'86	4-155.9	⑥⑦	34	58	87	—	⑧	⑧
'77–'86	6-225	70	45	85	Press fit	55	20 ⑨	10
'79	Diesel	90	65	69 ⑩	289	80	—	—
'79–'86	8-318 8-360	105	45	85	100	55	40	20 ⑪

① '82–'85 models: Tighten in 4 steps—30, 45, 45, plus ¼ turn
'86 models: Head bolt size has been increased to 11 mm: Tighten in 4 steps—45, 65, 65, plus ¼ turn
② 40 plus ¼ turn more
③ 30 plus ¼ turn more
④ '86—70
⑤ '86—16.6
⑥ Camcap bolts—135 engine 165 inch lbs. 155.9 engine 160 inch lbs.
⑦ Front-head to cover bolts—160 inch lbs. Headbolts 69 cold
⑧ 180 inch lbs.
⑨ Stud—30 Nut—20
⑩ 80 ft. lbs. on bolts with "H" mark
⑪ 15 ft. lbs. on nuts

PISTON AND RING SPECIFICATIONS

Engine	Year	Piston-to-Bore Clearance	Ring Gap Top Compression	Ring Gap Bottom Compression	Ring Gap Oil Control	Ring Side Clearance Top Compression	Ring Side Clearance Bottom Compression	Ring Side Clearance Oil Control
4-135	'83–'86	.0005–.0240	.011–.021	.011–.021	.015–.055	.0015–.0031	.0015–.0037	.00024–.00075
4-155.9	'83–'86	.0008–.0016	.010–.018	.010–.018	.0078–.035	.0024–.0039	.0008–.0024	.00020–.00035
6-225	'79–'86	.0005–.0015	.010–.020	.010–.020	.015–.055	.0015–.0030	.0015–.0030	.0002–.0050
6-243	'79	.0008–.0020	.012–.020	.010–.020	.010–.020	.0010–.0020	.0010–.0020	.0010–.0020
8-318	'79–'86	.0005–.0015	.010–.020	.010–.020	.015–.055	.0015–.0030	.0015–.0030	.0002–.0050
8-360	'79–'86	.0005–.0015	.010–.020	.010–.020	.015–.055	.0015–.0030	.0015–.0030	.0002–.0050

CAPACITIES

Year	Engine Displacement Cu In. No. of Cyls	Engine Crankcase (qts) With Filter	Engine Crankcase (qts) Without Filter	Transmission (pts) Manual	Transmission (pts) Automatic	Drive Axle (pts)	Transfer Case (pts)	Cooling System (qts)
'79–'80	6-225	6	5	①	16⅔	③	9 ④	12 ⑤
	V8-318	6	5	①	16⅔	③	9 ④	16 ⑤
	V8-360	6	5	①	16⅔	③	9 ④	14½ ⑤
'81	6-225	6	5	①	⑥	③	④	12 ②
	V8-318	6	5	①	⑥	③	④	16 ②
	V8-360	6	5	①	⑥	③	④	14½ ②
'82	4-135	4	3.5	4	15 ⑦	⑦	④	7
	6-225	6	5	①	⑥	③	④	12 ②
	V8-318	6	5	①	⑥	③	④	16 ②
	V8-360	6	5	①	⑥	③	④	14½ ②
'83–'84	4-135	4	3.5	⑧	⑨	⑩	④	9
	4-155.9	5	4.5	⑧	⑨	⑩	④	9
'85–'86	4-135	4	3.5	4.6	17.8	⑩	④	8.5
	4-155.9	5	4.5	4.6	17.8	⑩	④	9.5

CAPACITIES

Year	Engine Displacement Cu In. No. of Cyls	Engine Crankcase (qts) With Filter	Without Filter	Transmission (pts) Manual	Automatic	Drive Axle (pts)	Transfer Case (pts)	Cooling System (qts)
'83–'86	6-225	6	5	7 ⑪	⑥	③	④	12 ②
	V8-318	6	5	7 ⑪	⑥	③	④	16 ②
	V8-360	6	5	7 ⑪	⑥	③	④	14.5 ②

① 4.25 w/A230
  3.50 w/A390
  7.50 w/4 overdrive
  7.50 w/NP435
② Add 1 at. w/aux. rear heater or A/C
③ See Rear Axle Chart
④ '79—9 pts
  '80—NP208; 6 pts, NP205; 4.5 pts
⑤ Add 2 qts. for HD cooling or A/C
⑥ A904T/A999—17.1 (dry)
  A727—7.7 w/o converter drain
⑦ Auto Transaxle differential—2.4 ('82 only)
⑧ 4 speed—4
  5 speed—4.6
⑨ 8.9 qts. w/ converter drain. 4 qts w/o converter drain
⑩ Models from '83 share common sump between trans & diff.
⑪ A833 overdrive—7.5

REAR AXLE IDENTIFICATION CHART

Identification	Ring Gear Size (in.)	Maker and Model	Approximately Capacity (pts) ①
10 bolt cover, front filler plug	8³⁄₈	Chrysler	4.4
12 bolt cover, filler plug in cover	9¹⁄₄	Chrysler	4.5
10 bolt cover, filler plug in cover	9³⁄₄	Spicer 60	6.0
10 bolt cover, filler plug in cover	10¹⁄₂	Spicer 70	6.5

①These are design capacities and may differ slightly from those given in the Capacities Chart.

TUNE UP

Electronic Ignition

Refer to the Electrical section of Unit Repair for electronic ignition testing.

An electronic ignition system is installed as standard equipment. The distributor housing, cap, rotor and advance unit are the same used on point type systems. A magnetic pickup and control (reluctor) have replaced the points and distributor cam. A condenser is no longer necessary. The only maintenance required with electronic ignition is inspection of the wiring, cap, rotor and periodic cleaning and changing of the spark plugs. In later model years, depending on the model, the distributor might have two pick-ups. One controls the start mode; the other, the run mode. If the distributor needs parts replacement or servicing, maintain the proper air gap between the pickup and reluctor is necessary. The 2.2 liter four cylinder engine uses a Hall Effect type of pickup assembly in which a spinning rotor

mounted onto the distributor shaft effects a stationary pickup unit. On this type of distributor, no gap adjustment is possible. The electronic ignition system eliminates the contact points and the need to set dwell. Hence, the dwell is non-adjustable and cannot be altered in any way.

Air Gap

Except 4 Cylinder Engines

The air gap adjustment is not a regular maintenance item. However, if the gap is too wide a no start condition can exist. Air gap is not adjustable on either of the four cylinder engines.

NOTE: When checking the pickup air gap use a non-magnetic feeler gauge. If a non-magnetic feeler-gauge is not available, use brass shim stock of the proper thickness. The reluctor teeth may appear ragged at the edges, but no attempt should be made to clean them. A sharp edge is necessary to quickly decrease the magnetic field and induce negative voltage in

the pickup coil. If the teeth are rounded, the voltage signal to the control unit may be erratic.

——— CAUTION ———

Do not touch the round transistor mounted on the control unit when the ignition switch is "On", it can produce a nasty electrical shock.

ADJUSTMENT

Models with a single pickup require 0.006 inch adjustment. Dual pickup equipped distributors require; Start pickup (identified by a two male prong distributor connector) .006 inch; Run pickup (identified by a male and female connector) 0.012 inch up to 1984 and 0.008 in. for later models.

Single Pickup Distributor

1. Align a reluctor tooth with the pickup coil tooth. Loosen the pickup coil hold-down screw.

2. Insert a non-magnetic feeler gauge of the proper size between the reluctor tooth and the pickup coil tooth. Adjust the air

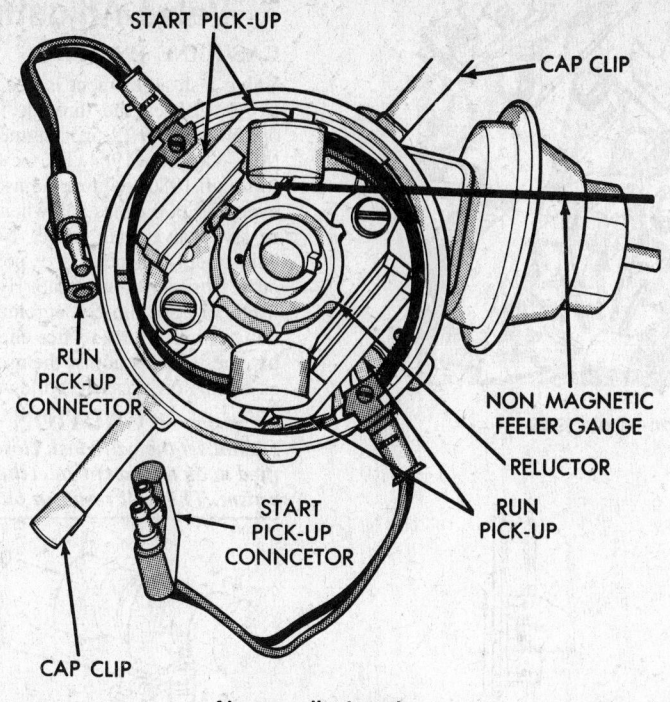

Air gap adjustment

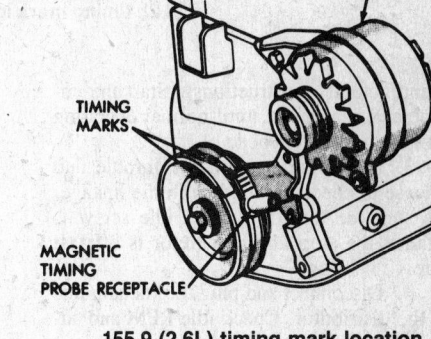

135 (2.2L) timing mark location—through 1984

155.9 (2.6L) timing mark location

gap so that contact is made between the reluctor tooth, feeler gauge and pickup coil tooth. Tighten the holddown screw.

3. Remove the feeler gauge. No force should be required to remove the feeler gauge.

4. Check the air gap with a larger size (about 0.002 inch larger) feeler gauge. It should not fit into the air gap so do not try to force it.

Dual Pickup Distributor

The procedure is the same as for the single pickup distributor. Adjust the start pickup first and then the run pickup. Test each adjustment with a larger feeler gauge as mentioned in the single pickup distributor air gap adjustment.

Ignition Timing

NOTE: On engines with electronic ignition, your timing light may or may not work, depending on the construction of the light. Consult the manufacturer of the light if in doubt.

ADJUSTMENT

135 (2.2L) 4 Cylinder

1. Connect a timing light according to the manufacturer's instructions.

2. Run the engine to normal operating temperature.

3. Make sure the idle speed is correct.

4. Loosen the distributor holddown screw just enough so that the distributor can be rotated.

5. Ground the carburetor switch (if equipped). Disconnect and plug the vacuum line(s) to the distributor control. If the engine is equipped with an E.S.C. (Electronic Spark Control) computer, disconnect the vacuum advance line at the computer.

6. Remove the timing hole access cover and aim the timing light at the hole in the clutch housing. Carefully rotate the distributor until the timing marks are aligned.

7. Tighten the distributor and recheck the timing.

8. Check, and if necessary adjust, the idle speed.

155.9 (2.6L) 4 Cylinder

1. Locate, clean and mark the timing scale and pointer on the crank pulley and front cover.

2. Connect a timing light and tachometer to the engine (follow the equipment

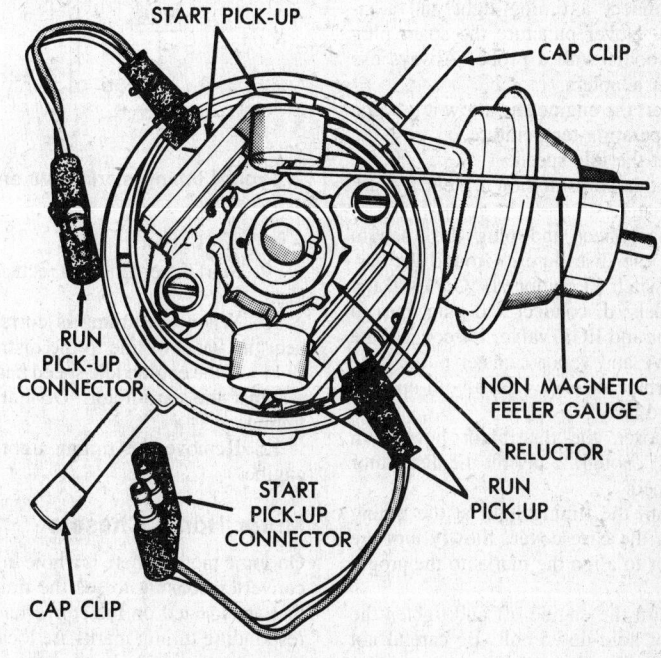

Dual pickup distributor

2.2L timing mark location—from 1985

manufacturer's instructions). Start the engine and run at idle until normal operating temperature is reached.

3. Momentarily open the throttle and release to check for binding in the linkage. Be sure the throttle linkage idle screw is against the stop and carburetor is off fast idle.

4. Disconnect and plug the vacuum line at the distributor. Check idle RPM and adjust if necessary.

5. Aim the timing light at the crankshaft pulley and check the timing. Loosen the distributor locknut and rotate the distributor in the direction necessary to align the timing marks. Tighten the locknut and recheck timing. Check idle RPM, readjust idle and timing as necessary.

6. Reconnect the distributor vacuum line. Shut off engine and disconnect the tachometer and timing light.

6 & 8 Cylinder Engines

1. Connect a timing light and a tachometer. Never puncture the spark plug wires or boots with a probe. Always use the proper adapters.

2. Start the engine and allow it to reach normal operating temperature.

3. Set the idle speed.

4. Put the transmission in Neutral; Park for automatic.

5. Disconnect and plug the vacuum line(s) at the distributor. Ground the carburetor switch (if equipped). On 1981 and later models; disconnect and plug lines to distributor and EGR valve. Disconnect the PCV valve and vapor canister purge hose at the carburetor end. Leave both open to underhood air.

6. Loosen the distributor hold-down screw just enough to permit the distributor to be turned.

7. Aim the timing light at the timing marks on the case cover. Slowly turn the distributor to align the marks to the proper setting.

8. Turn the engine off and tighten the distributor hold-down bolt. Be careful not to move the distributor while you are tightening.

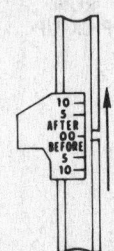

Typical ignition timing marks—6 cylinder engine

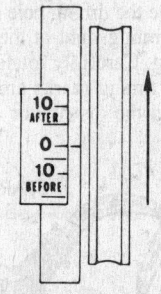

Typical timing marks—V8 engine

9. Start the engine and recheck the timing.

10. When the timing is correct, reconnect the vacuum line to the distributor.

11. If the engine idle speed has changed, readjust the carburetor. Do not reset the timing.

12. Remove the timing light from the engine.

Motor Home Chassis

On some models there is a hole in the torque converter housing to see the timing marks that are located on the converter. The corresponding timing marks are located on the housing. Timing light hook up is the same as on crankshaft pulley timing.

Valve Adjustment

GASOLINE ENGINES

Valve lash adjustment is necessary on the 6 cyl. 225 engine through 1980 and the optional 155.9 (2.6L) engine used in the Mini-Van from 1984. The 6 cyl. 225 engine (through 1980) requires adjustment at least every 20,000 miles, or when there is excessive valve train noise. No valve lash adjustment is necessary or possible on any other Chrysler-built engine. Hydraulic valve lifters or lash adjusters automatically maintain zero clearance. After engine reassembly these lifters adjust themselves as soon as engine oil pressure builds up.

—————— CAUTION ——————

Do not set the valve lash closer than specified in an attempt to quiet the valve mechanism. This will result in burned valves.

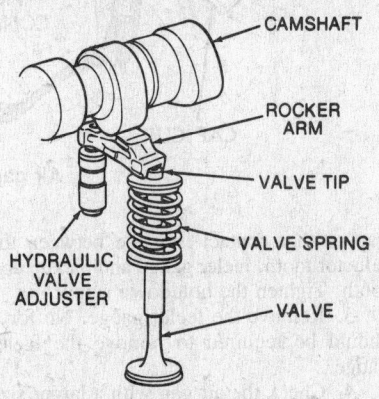

Hydraulic valve adjuster used on (2.2) 4 cylinder engine

155.9 (2.6L) 4 Cylinder

NOTE: A jet valve is added on USA models. The jet valve adjuster is located on the intake valve rocker arm and must be adjusted before the intake valve.

1. Start the engine and allow it to reach normal operating temperature.

NOTE: Do not run the engine with the rocker arm cover removed, oil will be sprayed on to the hot exhaust manifold.

2. Shut off engine and remove the rocker arm cover.

3. Watch the valve operation on No. 1 cylinder (No. 1 cylinder on transverse mounted engines is on the driver's side) while turning the crankshaft to close the exhaust valve and have the intake valve just begin to open. This places the No. 4 cylinder on TDC of its firing stroke and permits the adjustment of the valves.

4. Jet valves must be adjusted before the intake valve. To adjust the jet valves: Loosen the intake valve lock nut and back off the adjustment screw two or more turns. Loosen the lock nut on the jet valve adjusting screw. Turn the jet valve adjusting screw counterclockwise and insert a 0.006 in. feeler gauge between the valve stem and

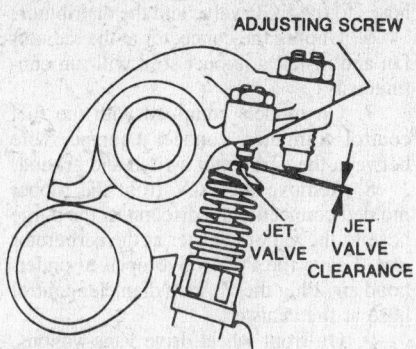

Adjusting the jet valve on 155.9 (2.6L) engines

the adjusting screw. Tighten the adjusting screw until it touches the feeler gauge.

NOTE: The jet valve spring is weak, be careful not to force the jet valve in.

5. After adjustment is made, hold the adjusting screw with a screwdriver and tighten the lock nut. Proceed to adjust the intake and the exhaust valves on the same cylinder as the jet valve you finished adjusting. Adjust by loosening the locknut and passing a feeler gauge of the correct thickness between the bottom of the rocker arm and top of the valve stem. If the clearance is too great or too small, turn the adjusting screw until the gauge will pass through with a slight drag. Tighten the locknut and proceed to the next valve. Refer to the chart in Step 3 for the adjusting sequence.

6–225 through 1980

1. The engine must be at normal operating temperature. Mark the crankshaft pulley into three equal 120° segments, starting at the TDC mark.

2. Remove the valve (rocker) cover and the distributor cap.

3. Set the engine at TDC on the No. 1 cylinder by aligning the mark on the crankshaft pulley with the "0" mark on the timing cover pointer. The distributor rotor should point at the position of the No. 1 spark plug wire in the distributor cap. Both rocker arms on the No. 1 cylinder should be free to move slightly.

4. The cylinders are numbered from the front to rear.

5. The lash is measured between the rocker arm and the end of the valve. To check the lash, insert the correct size feeler gauge between the rocker arm and the valve. Press down lightly on the other end of the rocker arm. If the gauge cannot be inserted, loosen the self-locking adjustment nut on the top of the rocker arm. Tighten the nut until the gauge can just be inserted and withdrawn without buckling.

6. After both valves for the No. 1 cylinder are adjusted, turn the engine so that the pulley turns 120° in the normal direction of rotation (clockwise). The distributor ro-

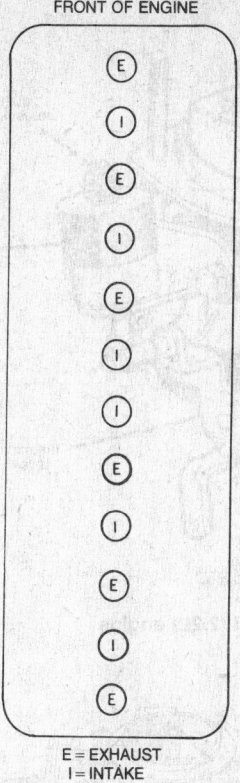

FRONT OF ENGINE

E = EXHAUST
I = INTAKE

Valve location—6 cylinder engine

tor will turn 60°, since it turns at half engine speed.

7. Check that the rocker arms are free and adjust the valves for the next cylinder in the firing order, No.5. The firing order is 1–5–3–6–2–4.

8. Turn the engine 120° to adjust each of the remaining cylinders in the firing order. When you are done the crankshaft will have made two complete revolutions (720°) and the distributor rotor one complete revolution (360°).

9. Replace the rocker cover with a new gasket. Replace the distributor cap. Start the engine and check for leaks.

DIESEL ENGINE

6–243

Valve adjustment is required on the diesel engine every 36,000 miles.

──────── CAUTION ────────
Do not set the valve lash closer than specified in an attempt to quiet the lifters. This will only result in burned valves.

1. The engine must be cold for this adjustment. Mark the crankshaft pulley into three equal 120° sections, starting at the TDC or "0" mark.

2. Remove the valve cover.

3. Disconnect the fuel line from the injector on cylinder No. 1. Be very careful not to kink the line.

4. Using a socket and a breaker bar on

the crankshaft bolt, turn the engine in the same direction as it normally runs until the "0" mark on the crankshaft damper rear face is aligned with the pointer on the bottom of the timing gear case. As you are turning, keep an eye on the fuel line that was disconnected. Fuel will squirt out of the line a few degrees before you reach the "0" mark. If it doesn't squirt out, you've got the wrong cylinder at TDC. Turn the engine another 360° (one revolution) and you will have No. 1 at TDC.

5. The cylinders are numbered from front to rear.

6. The lash is measured between the rocker arm and the end of the valve. To check the lash, insert a 0.012 in. feeler gauge between the rocker arm and the valve. Adjust the proper clearance by loosening the locknut on the rocker arm and turning the adjusting screw. After the adjustment is made, tighten the locknut and recheck the clearance to be sure it did not change as the locknut was being tightened.

7. After both valves for the No. 1 cylinder are adjusted, turn the engine so that the pulley turns 120° in the normal direction of rotation (clockwise). The distributor rotor will turn 60°, since it turns at half engine speed.

8. Check that the rocker arms are free and adjust the valves for the next cylinder in the firing order, No.5. The firing order is 1–5–3–6–2–4.

9. Turn the engine 120° to adjust each of the remaining cylinders in the firing order. When you are done the crankshaft will have made two complete revolutions (720°).

10. Replace the rocker cover with a new gasket. Attach the fuel line to the injector. Start the engine and check for leaks.

Carburetor

ADJUSTMENT

Idle Speed & Mixture

NOTE: Various calibrations are used, refer to the emissions decal under the hood for procedures when adjusting idle or mixture.

Before suspecting the carburetor as the cause of poor performance or rough idle, check the ignition system thoroughly, including the distributor, timing, spark plugs and wires. Also check the air cleaner, evaporative emission system, PCV system, EGR valve and engine compression. Check the intake manifold, vacuum hoses and other vacuum connections for leaks and cracks. Mixture, on today's carburetors, is preset at the factory and tampering with the mixture screw will produce little change. Some carburetors have hidden mixture screws which makes adjustment impossible unless special equipment is used. When performing a tune-up, adjust the idle speed to the specified rpm (tachometer connected), using the idle speed screw or the solenoid adjustment.

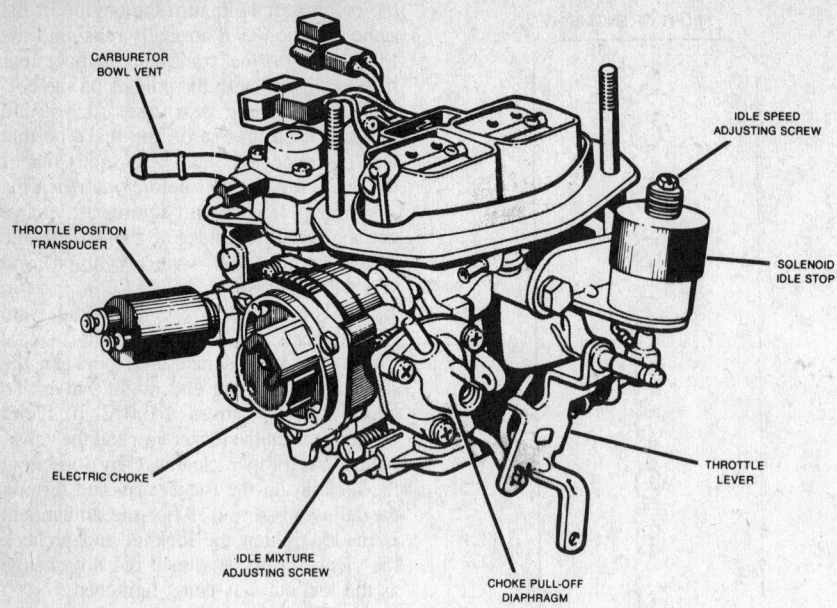

Idle speed adjusting location—135 (2.2L) engine

6 & 8 CYLINDER

NOTE: Adjust with the air cleaner installed.

1. Run the engine at fast idle to stabilize engine temperature.
2. Make sure that the choke plate is fully opened.
3. Attach a tachometer to the engine. With electronic ignition, connect the meter to the negative primary coil terminal and to a ground.

NOTE: Not all tachometers or dwell/tachometers will work with electronic ignition; some may be damaged. Check the manufacturer's instructions carefully.

4. Connect an exhaust analyzer to the engine and insert the probe as far into the tailpipe as possible. On vehicles with dual exhaust, insert the probe into the left tailpipe as this is the side without the heat riser valve.
5. Check ignition timing and adjust it as required.
6. If the truck has air conditioning, turn the air conditioner off. On six-cylinder engines, turn the headlights on high beam.
7. Place the manual transmission in neutral; put the automatic in Park. Make sure the hot idle compensator valve (if any) on the carburetor is fully seated in the closed position.
8. Turn the engine idle speed adjustment screw in or out to adjust idle speed to specification. If the carburetor has an electric solenoid, turn the solenoid adjusting screw in or out to obtain the specified rpm. Then, adjust the curb idle speed screw until it just touches the stop on the carburetor body. Now, back the curb idle speed adjusting screw out one full turn.
9. Turn each idle mixture adjustment screw $\frac{1}{16}$ turn richer (counterclockwise).

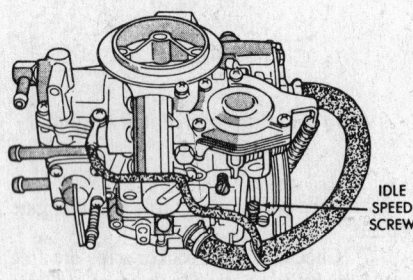

Idle speed adjusting—155.9 (2.6L) engines

Wait 30 seconds and observe the reading on the exhaust gas analyzer. Continue this procedure until the meter indicates a definite increase in the richness of the mixture.

NOTE: This step is very important. A carburetor that is set too lean will cause the exhaust gas analyzer to give a false reading indicating a rich mixture. Because of this, the carburetor must be known to have a rich mixture to verify the reading on the exhaust gas analyzer.

10. After verifying the reading obtained on the meter, adjust the mixture screws to get the air/fuel ratio and percentage of CO indicated on the engine compartment sticker. Turn the mixture screws clockwise (leaner) to raise the meter reading or counterclockwise (richer) to lower the meter reading.

135 (2.2L) 4 CYLINDER

1. Set the parking brake and place the transmission in neutral.
2. Turn off all the lights and accessories.
3. Connect a tachometer to the engine following the manufacturer's instructions.
4. Start the engine and allow it to reach normal operating temperature.

5. Disconnect and plug the vacuum hoses to the EGR valve and the distributor.
6. Unplug the connector at the radiator fan and install a jumper so it will run continuously.
7. On models equipped with the fuel control computer, connect a jumper wire between the carburetor switch and ground.
8. Remove the PCV from the rubber molded connector and disconnect the purge hose to the vapor canister at the carburetor end. Leave the PCV valve open to underhood air. Plug the $\frac{3}{16}$ inch diameter control hose at the canister.
9. On front wheel drive vans/wagons, unplug the electrical connector at the electric fan motor and jumper the connections so the fan will run continuously.
10. If the vehicle has a 6520 carburetor (with oxygen sensor) disconnect the oxygen sensor at the test connector located on the left fender shield.
11. Read the RPM indicated on the tachometer. If the RPM is not the same as the idle set rpm specified on the emissions label, turn the idle speed screw (on top of the solenoid) to correct it.
12. Unplug and reconnect all hoses and reconnect all wires.

155.9 (2.6L) 4 CYLINDER

1. Set the parking brake and place the transmission in neutral.
2. Turn off all the lights and accessories and disconnect the cooling fan.
3. Connect a tachometer to the engine following the manufacturer's instructions.
4. Start the engine and allow it to reach normal operating temperature. Then, run the engine at 2500 rpm for 10 seconds and allow it to return to idle. Make sure two minutes elapses after this before you check idle speed.
5. Check the timing and adjust if necessary.
6. Remove the timing light and read the rpm indicated on the tachometer. It is not the same as the curb idle specified on the emission label adjust the idle speed adjusting screw. The screw is accessible through the hole in the choke cover plate using a long narrow screwdriver at a 45° angle inwards.
7. After adjusting the curb idle speed, press the A/C button on. With the compressor running, set the engine speed to 900 rpm by turning the idle up adjusting screw. The idle up adjusting screw is accessible through a hole in the choke cover plate using a long narrow shaft screwdriver at a 45° angle downwards.
8. Turn the engine off, disconnect the tachometer and reconnect the cooling fan.

IDLE SPEED SOLENOID

This solenoid is energized whenever the ignition circuit is on. Its function is to allow the throttle plates to close farther when the ignition is switched off, thereby preventing engine over-running.

1. Bring the engine to operating temperature and attach a tachometer.

2. With the engine running, adjust the solenoid screw to the proper rpm.

3. Adjust the slow curb idle screw until the screw end just contacts the stop on the carburetor body. Back the screw off one full turn.

4. Test the above procedure by disconnecting the solenoid wire at the connector. Be sure not to let the lead short to the engine. The solenoid should de-energize and idle speed should drop down below normal. Now reconnect the wire. After you reconnect the solenoid, move the throttle linkage by hand since the solenoid isn't strong enough to move it.

IDLE MIXTURE PLUGS

NOTE: Tampering with the carburetor is a violation of Federal Law. Adjustment of the idle air fuel mixture can only be done under certain circumstances. The idle mixture adjusting screws are covered with tamper-proof plugs. If the plugs are removed and the mixture adjusted, then new plugs and roll pins must be installed.

The mixture should be adjusted only if an idle defect remains after a complete diagnosis reveals that no other fault exists, such as a bad spark plug wire or a vacuum leak. Also, the computer controlled fuel system must be operating properly. Adjustment to the idle mixture should be performed after a major carburetor overhaul.

1. On 6 & 8 cylinder models: Remove the carburetor from the engine.

2. Hold the carburetor in a suitable fixture for removing the roll pin and concealment plug.

3. Drill a small pilot hole in the casting into the base gasket surface and at a 45° angle toward the concealment plug. Redrill the hole to a larger size.

4. Insert a blunt punch into the hole and drive out the plug.

5. Insert a sharp punch through the idle mixture adjusting hole and slide out the roll pin.

6. Install the carburetor on the engine and perform the idle mixture adjustments using the propane enrichment method.

7. Remove the carburetor and install the roll pin and concealment plug in the carburetor.

8. Install the carburetor.

PROPANE ENRICHMENT
6 & 8 Cylinder (Except As Listed)

1. On 6 & 8 cylinder models: Place the transmission in neutral position and set the parking brake. Turn all lights and accessories off. Connect a tachometer and a timing light to the engine. Start the engine and allow it to warm up on the second stop of the fast idle cam. Do this until normal operation temperature is reached, then return the engine to idle.

2. Disconnect and plug the EGR vacuum hose and the distributor vacuum hose. Check the engine timing and adjust if nec-

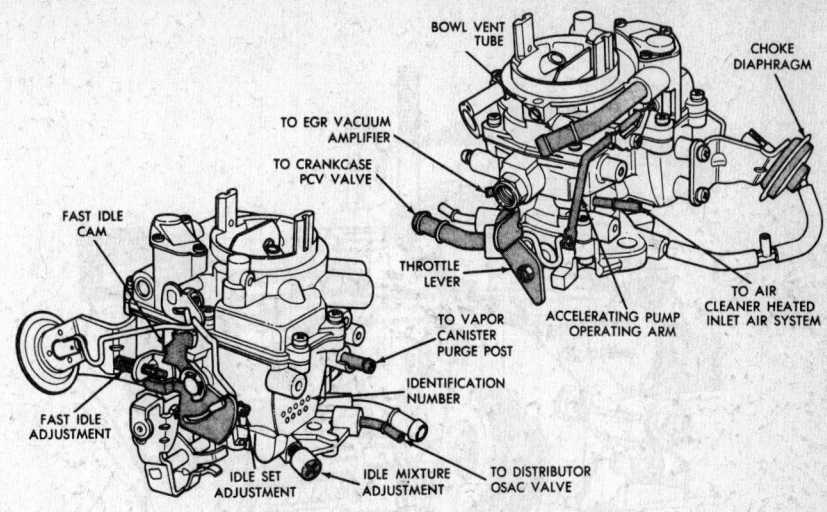

Typical Holley 2 barrel carburetor

essary. Disconnect the heated air door vacuum hose at the carburetor nipple and in it's place, install the propane supply hose. On 440 Cu. In. engines without the heated air system, insert the propane supply hose 12 in. into the air cleaner snorkel. Make sure that the propane bottle is in an upright and safe position. Remove the PCV valve from the cylinder head cover and disconnect the purge hose to the vapor canister at the carburetor end. Leave both open to underhood air.

3. Open the propane main flow valve. With the air cleaner in place, slowly open the propane metering valve until the maximum engine rpm is reached. When too much propane is added, the engine rpm will decrease. Fine tune the metering valve to obtain the highest engine rpm.

4. With the propane still flowing, adjust the idle speed screw to attain the propane rpm specified on the emissions label. If there has been a change in the maximum rpm, readjust the idle speed screw to the specified propane rpm.

5. Turn off the propane main valve and allow the engine speed to stabilize. With the air cleaner in place, slowly adjust the idle air mixture screws to achieve the smoothest idle at the specified idle set rpm. Pause between adjustments to allow the engine speed to stabilize. If it appears necessary to remove the limiter caps to reach the idle set rpm, first check for engine malfunctions and vacuum leaks. If idle limiter caps are removed, service caps must be installed with the tang against the maximum rich stop.

6. Turn the propane main valve on. If the maximum speed is more than 25 rpm different than the specified propane rpm, repeat Steps 3 through 6.

7. Turn both the propane main valve and the metering valve off. Remove the tachometer. Remove the propane supply hose and reinstall the heated air door vacuum hose (except models without heated air).

Unplug and reinstall the vacuum hose to the EGR valve and to the distributor.

8. Replace the PCV valve. Reconnect the canister purge hose to its proper place. A variation in the engine RPM may occur, but do not readjust.

2.2L Four Cylinder Engine

1. Remove the concealment plug. Set the parking brake. Make sure the transaxle is in neutral. Turn off all lights and accessories.

2. Connect a tachometer to the engine. Start the engine and warm it up while on the second step of the fast idle cam. When the engine reaches normal operating temperature, open the throttle and bring the engine down to normal idle speed.

3. Unplug the connector at the radiator fan (front wheel drive vehicles only) and jumper it so the fan runs continuously. Remove the PCV valve from the crankcase vent and allow it to draw underhood air. If the vehicle has an oxygen sensor, unplug the oxygen feedback system test connector located on the left fender shield.

4. Disconnect the vacuum harness from the CVSCC valve and plug both hoses. Disconnect the wiring from the kicker solenoid on the left fender shield.

5. Check to make sure the two propane valves are both fully closed and that the bottle is upright and in a safe location. Disconnect the vacuum hose leading to the heated air sensor at the three-way connector. Connect the propane hose from the propane bottle in place of this hose.

6. Open the main propane valve. Then, very gradually and slowly open the propane metering valve while watching the tachometer reading until maximum rpm is reached. Note that the air cleaner must remain in place as you do this. Note that opening the propane metering valve rapidly will usually cause the carburetor setting to jump to a rich condition and slow the engine down.

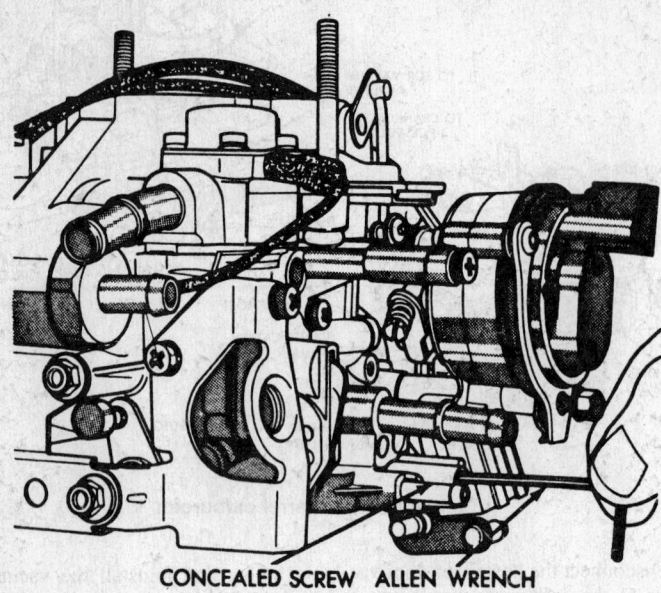

CONCEALED SCREW   ALLEN WRENCH

Propane assisted idle mixture adjustment—2.2L engine

You'll find a point where turning the metering valve open farther will just begin slowing the rpm back down. Optimize the valve opening very carefully, allowing time for the engine to adjust, so the engine runs at the highest rpm possible.

7. Referring to the engine compartment sticker, adjust the idle speed screw (located on top of the carburetor solenoid) to get the rpm specified for propane enrichment. Increase the engine rpm to 2,500 for 15 seconds. Then, allow the engine to return to idle. Repeat the last part of Step 6 in order to optimize the rpm. If this raises rpm, repeat the idle speed adjustment. Go back and forth in this way until the engine runs at the specified rpm with the propane adjustment at the optimum level. Keep clearing out the engine by running it at 2,500 rpm as described above.

8. Now, turn off the main propane valve and allow the engine rpm to stabilize. Then, adjust the idle mixture screw to get the specified idle set rpm. Pause after each change in the setting of the screw in order for rpm to stabilize. Increase the engine rpm to 2,500 for 15 seconds, allow it to return to idle, and recheck the rpm. Adjust if necessary.

9. Now, recheck the accuracy of the adjustment by again using propane enrichment to optimize the rpm as described in Step 6. If the reading is more than 25 rpm different from the propane enrichment rpm, repeat Steps 6–8, as you have not produced optimum mixture at the specified enrichment rpm.

10. Turn off both propane valves. Disconnect the propane tank and reconnect the vacuum hose. Replace the concealment plug. Reverse all remaining portions of Steps 1–4.

225 Engine w/Holley 6145 Electronic Feedback Carburetor

Note: To perform this procedure you will need propane enrichment equipment and a precisely regulated vacuum supply.

1. Remove the concealment plug which gives access to the mixture screw. Connect a tachometer according to manufacturer's instructions. With the parking brake on and transmission in Neutral, start the engine and operate it on the second step of the fast idle cam until it is hot. Then, open the throttle to bring the engine to normal idle speed.

2. Disconnect and plug the EGR valve vacuum line at the EGR valve. Jumper the carburetor switch to a good ground. Leave the air cleaner in place.

3. Trace the vacuum hose leading to the choke diaphragm back to the Tee and disconnect this hose only; then connect the propane supply hose in its place.

4. Make sure the propane bottle is securely upright and in a safe location. Then, pull the PCV valve from the valve cover and allow it to draw underhood air. Disconnect the control hose (which is $\frac{3}{16}$ in. in diameter) from the charcoal canister and plug it.

5. Being careful not to touch the hot exhaust manifold, and pulling directly on the bullet connector only, disconnect the oxygen sensor harness lead from the oxygen sensor. Then, jumper the harness lead to ground.

───── **CAUTION** ─────
Make sure you don't put any stress on the wire to the oxygen sensor as you do this.

6. Wait two minutes to allow the effect of disconnecting the oxygen sensor to take full effect. As you wait, disconnect the vacuum line at the vacuum transducer on the Spark Control Computer. Then, connect an auxiliary vacuum supply to the vacuum transducer and set it at 61 in. of vacuum. When the two minutes have elapsed, proceed to the next step.

7. Open the main propane valve and then open the metering valve very slowly, as adding too much propane will cause the engine rpm to suddenly decrease by running too rich. You'll find a point where turning the metering valve open farther will just begin slowing the rpm back down. Optimize the valve opening very carefully, allowing time for the engine to adjust, so the engine runs at the highest rpm possible.

8. Referring to the engine compartment sticker, adjust the idle speed screw (located on top of the carburetor solenoid) to get the rpm specified for propane enrichment. Repeat the last part of Step 7 in order to optimize the rpm. If this raises rpm, repeat the idle speed adjustment. Go back and forth in this way until the engine runs at the specified rpm with the propane adjustment at the optimum level.

9. Turn off the propane metering valve and main valve. Allow the engine to run for one minute in order to allow the rpm to stabilize. Then, adjust the mixture screw very slowly, pausing after each change to allow the engine to stabilize, until you achieve the specified idle rpm and, at the same time, the smoothest possible idle.

10. Again, open the main propane valve and then adjust the metering valve carefully to optimize engine rpm without changing the throttle setting. Measure the rpm, it should be no more than 25 rpm more than in Step 9. If it is, repeat Steps 7–9, as you have failed to get the optimum propane mixture level at the specified propane rpm. Retest as necessary.

11. Turn off both propane valves. Disconnect the propane line and restore all vacuum and electrical connections changed at the beginning of the procedure. Install a new concealment plug over the mixture adjusting screw.

ENGINE ELECTRICAL

Distributor

REMOVAL & INSTALLATION

1. Remove splash shield (if equipped). Disconnect the vacuum line(s) at the distributor.

2. Disconnect the pickup lead wire connector or connectors from the wiring harness.

3. On the distributor used on the 2.6 liter engine, you need not remove the cap but may merely remove the wires from the cap. On other engines, unfasten the clips or screws that retain the distributor cap and lift off the cap.

NOTE: The spark plug and coil wires used on the 2.2L engine are equipped with forked type distributor cap connectors. The cap must be removed from the distributor and the forked clip compressed before the wire can be removed from the cap.

4. Bump the engine around until the rotor is pointing at No. 1 cylinder firing position and the timing marks on the front case and crank pulley are aligned. Disconnect the negative battery cable from the battery.

5. Mark the distributor body and the engine block to indicate the position of the distributor in the block. Mark the distributor body to indicate the rotor position. These marks are used as guides when installing the distributor.

6. Remove the distributor holddown bolt and bracket. Carefully lift the distributor from the engine. The shaft may rotate slightly as the distributor is removed. Make a note of where the movement stops. That point is where the rotor must point when the distributor is reinstalled into the block.

7. If the crankshaft has not been rotated while the distributor was removed from the engine, installation is the reverse of the removal procedure. Use the reference marks made before removal to correctly position the distributor in the block. The shaft may have to be rotated slightly to engage the cam gear (4 and 6 cyl.) or intermediate shaft gear (V8).

8. If the crankshaft was rotated or otherwise distributed (e.g., during engine rebuilding) after the distributor was removed, proceed as follows.

9. On 4 or 6 cylinders engines: Remove No. 1 spark plug and, with your thumb plugging the hole, rotate the engine until No. 1 piston is up on compression at top dead center. You'll feel the pressure of the compression stroke with your thumb and the "0" mark on the crankshaft pulley hub will be aligned with the timing pointer.

10. Turn the rotor to a position just ahead of the No. 1 distributor cap terminal.

11. Lower the distributor into the opening, engaging the distributor gear with the drive gear on camshaft. With the distributor fully seated in the engine, the rotor should be under the cap No. 1 tower.

12. Install the cap, tighten the hold-down bracket bolt. Connect the wiring and the vacuum hose. Check the timing with a timing light. Adjust if necessary.

13. On V8 engines: Rotate the crankshaft until No. 1 cylinder is at top dead center (TDC) of the compression stroke. To do this, remove the spark plug from cylinder No. 1 and place your thumb over the hole. Slowly turn the engine by hand

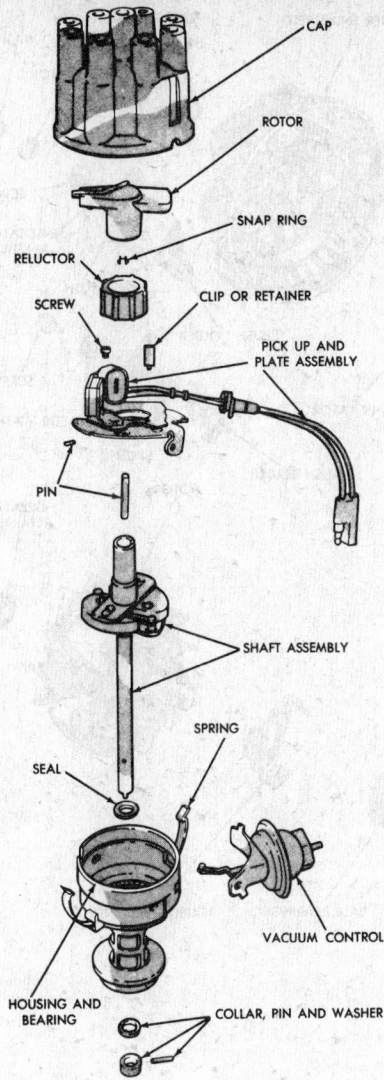

Typical single pickup distributor

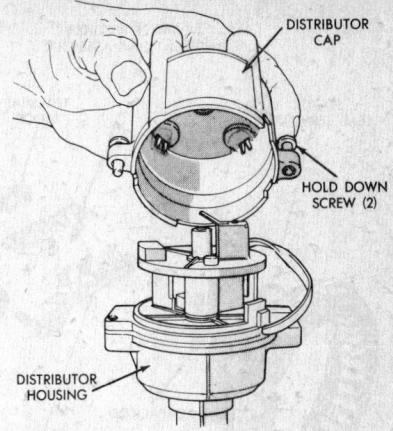

Typical distributor—4 cylinder engine

in the normal direction of rotation until compression is felt at the hole. The "0" mark on the crankshaft pulley should be aligned with the pointer on the timing case cover.

14. Hold the distributor over the mounting pad on the cylinder block so that the distributor body flange coincides with the mounting pad and the rotor points to the No. 1 cylinder firing position.

15. Install the distributor while holding the rotor in position, allowing it to move only enough to engage the slot in the drive gear.

16. Install the cap, tighten the hold-down bracket bolt. Connect the wiring and vacuum hose. Check the timing with a timing light. Adjust if necessary.

Alternator

PRECAUTIONS

The following are a few precautions to observe when servicing the alternator.

1. Never switch battery polarity.

2. When installing a battery, always connect the non-grounded (positive) terminal first.

3. Never disconnect the battery while the engine is running.

4. If the molded connector is disconnected from the alternator, do not ground the hot wire.

5. Never run the alternator with the main output cable disconnected.

6. Never electric weld around the truck without disconnecting the alternator.

7. Never apply any voltage in excess of battery voltage during testing.

8. Never "jump" a battery for starting purposes with more than 12 volts.

BELT TENSION ADJUSTMENT

NOTE: On some models it may be necessary to remove the lower splash shield to gain clearance when installing a new drive belt.

Belt tension should be checked with a gauge made for the purpose. If a gauge is not available, tension can be checked with moderate thumb pressure applied to the belt at its longest span midway between pulleys. If the belt has a free span less than twelve inches, it should deflect approximately $\frac{1}{4}$ inch. If the span is longer than twelve inches, deflection can range between $\frac{1}{4}$ and $\frac{3}{8}$ inches.

1. Loosen the driven accessory's pivot and mounting bolts.

2. Move the accessory toward or away from the engine until the tension is correct. You can use a wooden hammer handle or broomstick as a lever, but do not use anything metallic.

3. Tighten the bolts and recheck the tension. If new belts have been installed, run the engine for a few minutes, then recheck and readjust as necessary.

NOTE: If the driven component has two drive belts, the belts should be replaced in pairs to maintain proper ten-

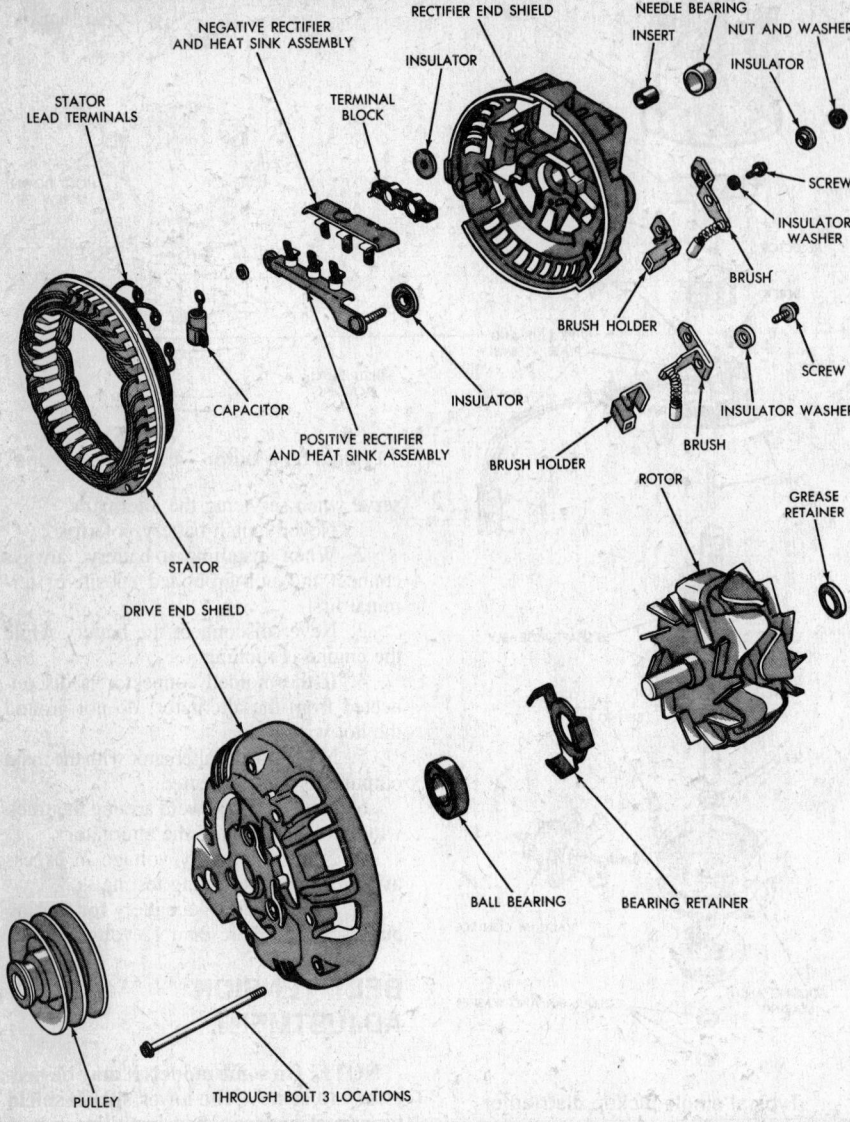

Exploded view of conventional alternator

sion. It is better to have belts too loose than too tight, because overtight belts will lead to bearing failure, particularly in the water pump and alternator. However, loose belts place an extremely high impact load on the driven component due to the whipping action of the belt.

REMOVAL & INSTALLATION

1. Turn the ignition switch off. Disconnect the battery ground cable at the battery.

2. Disconnect and label the alternator output (BATT) lead. Then, disconnect the field (FLD) leads ("R" and "L" on the Mitsubishi alternator). Finally, disconnect the ground wire. If the wiring is retained by a retainer, remove its mounting nut and remove the retainer.

3. Loosen the alternator adjusting bolt and swing the alternator in toward the engine. Disengage the alternator drive belt.

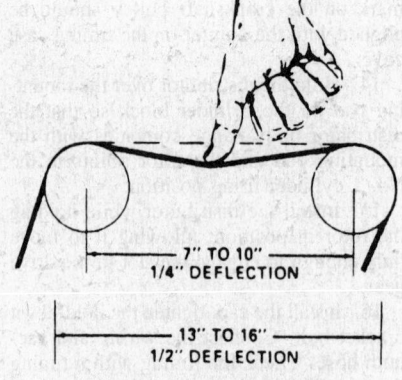

Checking drive belt deflection

4. Remove the alternator mounting bolts and remove the alternator from the vehicle.

5. Installation is the reverse of removal. Be sure to connect all ground wires and leads securely.

6. Adjust the belt tension.

Regulator

ADJUSTMENTS

The electronic voltage regulator has no moving parts and requires no adjustment after it leaves the factory. Any repairs are accomplished by replacement with a new regulator. Some 2.6 engines may use a Chrysler built external regulator with the Chrysler built alternator. If so, refer to the above procedures for regulator removal. Most 2.6 engines are equipped with a Mitsubishi-built alternator that contains a built-in regulator. See the alternator replacement procedures for removal and installation.

REMOVAL & INSTALLATION

1. Release the spring clips and pull off the regulator wiring plug.
2. Unbolt and remove the regulator.
3. Installation is the reverse of removal. Be sure that the spring clips engage the wiring plug.

Procedures for servicing Alternators and Starter Motors are found in the Electrical section of Unit Repair.

Starter

REMOVAL & INSTALLATION

1. Disconnect the ground cable at the battery.
2. Remove the cable from the starter. Remove the heat shield clamp and heat shield if so equipped.
3. Disconnect the solenoid leads at their solenoid terminals.
4. Remove the starter attachment bolts and withdraw the starter from the engine flywheel housing. On some models with automatic transmissions, the oil cooler tube bracket will interfere with the starter removal. In this case, remove the starter attachment bolts, slide the cooler tube bracket off the stud, and then withdraw the starter.
5. Installation is the reverse of the above. Be sure that the starter and flywheel housing mating surfaces are free of dirt and oil to make a good electrical contact.

ENGINE MECHANICAL

Engine

REMOVAL & INSTALLATION

4 Cylinder W/Manual Transaxle

1. Scribe the hood hinge outlines on the hood and remove the hood.

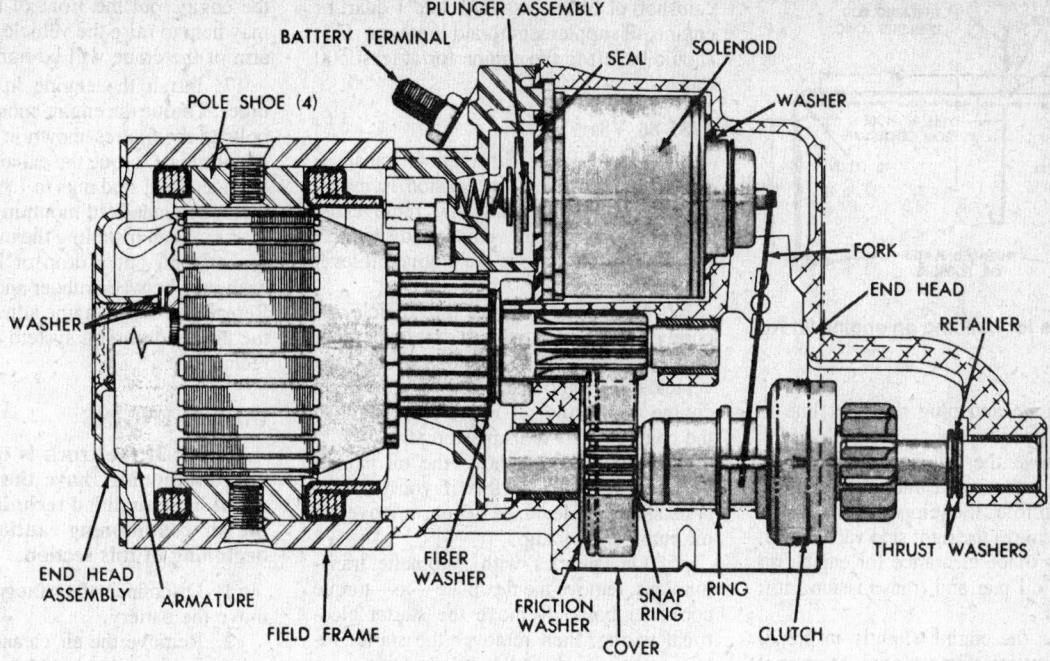

PLUNGER ASSEMBLY
BATTERY TERMINAL
SEAL
SOLENOID
WASHER
POLE SHOE (4)
FORK
END HEAD
RETAINER
WASHER
THRUST WASHERS
END HEAD
ASSEMBLY
ARMATURE
FIBER
WASHER
SNAP
RING
RING
CLUTCH
FIELD FRAME
FRICTION
WASHER
COVER

Conventional gear reduction starter

2. Drain the cooling system. Disconnect the battery cables and remove the battery.

3. Remove all water hoses from the radiator and engine. Remove the radiator and cooling fan assembly.

4. Remove the air cleaner, duct and hose assembly. Label vacuum hoses for reinstallation identification.

5. Remove the A/C compressor and mounting brackets. Leave the hoses attached and position the compressor out of the way.

6. Remove the power steering pump and brackets. Leave the hoses attached and position the pump out of the way.

7. Position a suitable container under the oil filter to catch any spilled oil and remove the oil filter.

8. Disconnect and identify all electrical connectors that will interfere with engine removal. Disconnect the fuel lines, accelerator linkage, heater hoses and air pump hoses.

9. Remove the alternator. Disconnect the clutch cable from the throwout bearing arm.

10. Remove the lower transaxle case lower cover. Disconnect the exhaust pipe from the exhaust manifold.

11. Remove the starter motor. Install a suitable jack under the transaxle assembly.

12. Install an engine lifting sling and attach to a chain hoist. Take slack out of the lifting chain. Raise transmission jack until contact is made with the transaxle.

13. Remove the inner right side splash shield. Remove the engine ground to chassis bonding strap.

14. Remove the transaxle to engine mounting bolts.

15. Remove the "through" nut(s) and bolt(s) from the front engine mount and anti-roll struts. Remove the left mount through bolt or insulator mounting bolts. Raise or lower the engine slightly with the chain hoist to relieve pressure on the through bolts.

16. Raise engine, separate from the transaxle and remove from vehicle.

17. To install the engine; lower into position with chain hoist and engage transaxle in drive.

18. Position engine against the transaxle and install mounting "through bolts". Do not tighten until all mount bolts have been installed. Tighten to 40 ft. lbs.

19. Install the transaxle to engine mounting bolts and tighten to 70 ft. lbs.

20. Remove the engine lifting sling and transaxle support jack.

21. Install remaining components in the reverse order of removal.

22. Fill the cooling system, add oil (if necessary), connect the battery. Start the engine and allow to reach normal operating temperature. Check for leaks. Check and reset timing and idle speed if necessary.

4 Cylinder W/Auto Transaxle

1. Follow Steps 1–10 of the Manual Transaxle procedure.

2. Remove the lower transaxle case lower cover, mark converter and flywheel for installation reference. Remove the converter to flywheel mounting bolts. Attach a "C" clamp on the front of the converter housing to prevent the torque converter from falling out when the engine is removed.

3. Proceed with the following steps of manual transaxle procedure. When install-

ing the engine, converter and flywheel mounting holes and reference marks must line up.

Rear Wheel Drive Vans & Wagons

EXCEPT DIESEL, 4 CYLINDER & 1985–86 V8s

NOTE: Engine removal is a complicated operation. A floor jack is a necessity and you will probably have to fabricate several stands and attaching apparatus. On vehicles equipped with air conditioning, before removing the engine, have an air conditioning expert discharge the system. Then, disconnect the compressor suction and discharge lines and effectively seal the openings.

1. Disconnect the battery and drain the coolant from the radiator and engine block. Drain the engine oil. On V8s, remove the oil filter.

2. Remove the engine cover, air cleaner, and starter.

3. Remove the front bumper, grille, and support brace. Disconnect both radiator hoses and remove the radiator and support brace as a unit.

4. Remove the power steering and air pumps with the hoses attached and lay them aside.

5. Disconnect the throttle linkage, heater and vacuum hoses and all electrical connections to the ignition, alternator, and all other electrical connections.

6. Remove the alternator, fan, pulley, and drive belts.

7. Remove the heater blower motor.

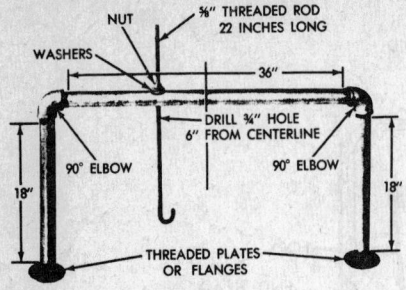

Dimensions for building an engine lift for vans

8. Remove and plug the inlet line to the fuel pump.

9. Remove the oil dipstick tube. On V8s, remove the intake manifold and left exhaust manifold. If equipped with air conditioning, remove the right side valve cover.

10. To provide clearance for engine removal, the oil pan and transmission must be removed.

11. Raise the engine slightly in preparation for transmission removal. Support it with an engine lifting fixture. This tool can be fabricated from galvanized pipe fittings obtained locally. Use only galvanized parts with an inside diameter of 1½ in. or larger. Be sure they are firmly threaded together to assure maximum strength.

12. Raise the vehicle and support it on jackstands. Remove the starter and distributor.

13. Remove the driveshaft and engine rear support. Remove the rear support by removing the rear mount through-bolt and the U-shaped bracket from the crossmember. Remove the insulator from the bottom face of the transmission housing.

14. If equipped with an automatic transmission, remove the transmission intact with the filler tube and the torque converter separated from the drive plate.

15. Raise the rear of the engine approximately 2 in. and remove the clutch or drive plate and the flywheel. On V8s, position the cut-out in the crankshaft flange at 3 o'clock. Remove the oil pan screws and lower the oil pan far enough to reach inside and turn the oil pump pick-up tube slightly to the right to clear the pan. Remove the oil pan.

16. Lower the vehicle.

17. Using a boom hoist attached to the engine with the shortest hook-up possible, take up all tension and support the engine. The boom hoist is the ideal tool to use. If one is not available, it may be possible to support the engine on a stationary hoist and roll the vehicle out from under the engine.

18. Remove the engine front mounts and insulators.

19. Carefully remove the engine from the vehicle.

20. Installation is the reverse of removal. Check all fluid levels and perform all tune-up adjustments if the engine was rebuilt. If the engine was rebuilt or new

camshaft or lifters installed, add 1 quart of engine oil supplement to aid break-in. This should be left in the engine for at least 500 miles.

1985-86 V8s

1. On vehicles with manual transmissions, refer to the Transmission Removal & Installation procedure and remove the transmission. On all vehicles, disconnect the battery and remove the engine oil level dipstick.

2. Raise and support the vehicle securely. Remove the exhaust crossover pipe.

3. Remove the inspection cover, if the vehicle has an automatic transmission. Drain engine oil. While oil is draining, remove the engine—to—transmission strut.

4. Unbolt and remove the oil pan; it may be easiest to do this if you turn the crankshaft for clearance. Then, remove the oil pump and pickup.

5. On vehicles with automatic transmissions, remove the flex plate—to—torque converter bolts. Remove the starter electrical wiring; then remover the starter retaining bolts and set the starter aside.

6. On vehicles with automatic transmissions, remove the lower transmission bell housing to engine bolts.

7. Remove the lower nuts only from the engine mount insulators.

8. Lower the vehicle and drain the cooling system. Remove the engine cover, air cleaner, and carburetor. If the vehicle is air conditioner equipped, have someone who is trained in air conditioning work and has the proper tools discharge the system. Then, disconnect the compressor and condenser lines and tightly seal all openings.

9. Remove the front bumper, grille, and support brace. Disconnect both radiator hoses, and then remove the radiator (and if A/C equipped), condenser, and support as an assembly. Unbolt the A/C compressor and set it aside, if the vehicle is A/C equipped.

10. Unbolt the power steering pump and set it aside, leaving hoses connected and being careful not to put stress on them. Remove the air pump.

11. Disconnect the throttle linkage, heater and vacuum hoses, and all electrical connections to the ignition coil, alternator, and other engine accessories. Remove the alternator.

12. Remove the fan blade or fan clutch and blade unit and the V-belts.

13. Disconnect the flexible fuel line to the fuel pump and then securely cap the openings to prevent leaks.

14. Remove the left side exhaust manifold and heat shield. Mark and then remove the spark plug wires and distributor cap.

15. Attach a portable hydraulic crane type lift to the intake manifold. Place a floor jack under the automatic transmission and support it securely so as not to damage the pan.

16. Remove the upper bell housing bolts (on cars with automatics). Carefully guide

the engine out the front of the vehicle. It may help to raise the vehicle slightly so the arm of the crane will be horizontal.

17. Install the engine in exact reverse order. Torque the engine mounting nuts and bolts to the figures shown in the applicable illustration. Torque the exhaust manifold to cylinder head stud nuts to 15 ft. lbs. Torque the intake manifold mounting screws to 45 ft. lbs. When installing the oil pan, use new gaskets and put a drop of RTV sealer at each joint between rubber and cork gaskets. Remake all basic engine adjustments. Have the air conditioning system evacuated and recharged.

DIESEL ENGINE

NOTE: If the truck is equipped with air conditioning, have the system evacuated by a qualified technician. Refer to the air conditioning caution note at the beginning of this section.

1. Disconnect the battery cables and remove the battery.

2. Remove the air cleaner.

3. Scribe matchmarks on the hood hinges and remove the hood.

4. Drain the cooling system.

5. Remove the upper and lower radiator hoses.

6. Remove the coolant reserve tank.

7. Raise the truck and safely support on jackstands. Disconnect and remove the transmission oil cooler lines from the radiator. Remove the lower radiator and fan shroud mounting screws.

8. Lower the truck and remove the upper radiator and fan shroud mounting screws and remove the radiator and the fan shroud.

9. Disconnect the heater hoses from the engine and push them aside.

10. Disconnect the speedometer cable housing from the engine.

11. Disconnect the electrical connections at the alternator, temperature sending unit, starter relay-to-solenoid wires, the oil gauge sending unit and the injection pump control motor. Set the wiring harness aside.

12. Disconnect and plug the fuel line at the transfer pump inlet. Disconnect and cap return line at the injector lines bleed-back connection.

13. Disconnect and remove the injection pump linkage. Disconnect and remove the accelerator and throttle cable linkage.

14. Disconnect the starter motor wire from the solenoid. Remove the starter motor.

15. Remove the battery ground cable from the engine block.

16. Disconnect and plug the power steering hoses at the power steering gear.

17. Raise the truck and disconnect the exhaust pipe from the exhaust manifold.

18. Drain the engine oil and remove the dipstick tube from the oil pan. Remove the transmission cooler line and road draft tube bracket from the oil pan.

19. Remove the oil pan bolts from the oil pan.

TIGHTENING TORQUE

Ⓐ	65 FT. LBS. (88 N•m)
Ⓑ	75 FT. LBS. (102 N•m)

FLANGE NUT
SCREW Ⓐ
AIR INJECTION TUBE BRACKET
FLANGE NUT
SCREW Ⓐ
BRACKET (RIGHT)
FLANGE NUT
INSULATOR
NUT Ⓑ
WASHER
SCREW Ⓐ
BRACKET (LEFT)
WASHER
NUT Ⓑ
INSULATOR
POSITION OF LOCATING PIN
FRONT
FRONT
POSITION OF LOCATING PIN
NUT Ⓑ
RIGHT SIDE
LEFT SIDE

Engine mount torque—V8 from 1985

20. Remove the transmission inspection plate. Remove the oil pan on vans to gain clearance if necessary. Use a new gasket on installation.

21. Remove the four flex plate-to-torque converter cover bolts.

22. Remove the exhaust pipe bracket and the lower bell housing bolts.

23. Remove any other brackets that can interfere with removal.

24. Support the transmission with a floor jack.

25. Remove the cylinder head (valve) cover and the gasket.

26. Attach a boom hoist to the engine, wrapping the chain as tight and close as possible.

27. Remove the four bolts and six nuts from the engine mounts.

28. Remove the two upper bell housing bolts.

29. Roll the boom hoist back, removing the engine from the van or truck.

30. Installation is the reverse of removal.

Pickups, Ramcharger & Trailduster

1. Drain the coolant from the radiator and cylinder block.

2. Disconnect the battery ground cable. Remove the battery on V8 models.

3. Scribe the outline of the hood hinges and remove the hood.

4. If equipped with air conditioning, remove the compressor with lines attached and lay it aside.

--- **CAUTION** ---

Do not disconnect any refrigerant lines. Bodily injury could result.

5. Disconnect the electrical connec-tions at the alternator, ignition coil, tem-perature and oil pressure sending units, starter-to-solenoid, and engine/body ground.

6. Remove the air cleaner. Disconnect throttle and transmission linkage at the carburetor.

7. Remove the distributor cap, wires and rotor.

8. Disconnect and plug the fuel pump line.

9. Disconnect the radiator and heater hoses. Disconnect and plug the oil cooler lines.

10. Remove the fan, spacer, fluid drive, and radiator. Do not store the fan drive unit with the shaft pointing downward. Fluid will leak out.

11. Raise the truck and support the rear of the engine.

12. Disconnect the exhaust pipes at the manifolds.

13. Remove the starter on V8 models.

14. Remove the automatic transmission dust cover and attach a C-clamp to the front bottom of the torque converter housing to prevent the converter from falling out. Remove the drive plate bolts from the torque converter. On manual transmission models, remove the rear crossmember, transmis-sion, transfer case and adapter. You can leave the transfer case in place on six-cyl-inder models.

15. Support the transmission and re-move the transmission attaching bolts.

16. Lower the truck and attach a lifting sling and hoist to the engine.

17. Remove the front motor mount bolt stud nuts and washers.

18. Carefully remove the engine.

19. Installation is the reverse of re-moval. Fill the engine with coolant and fresh oil. Adjust the transmission linkage, car-buretor, and ignition timing.

Intake Manifold

REMOVAL & INSTALLATION

155.9 (2.6L) 4 Cylinder

1. Disconnect the negative battery ca-ble.

2. Drain the cooling system and dis-connect the hoses from the water pump to the intake manifold.

3. Disconnect the carburetor air horn adapter and move to one side.

4. Disconnect the vacuum hoses and throttle linkage from the carburetor.

5. Disconnect the fuel inlet line at the fuel filter.

6. Remove the fuel filter and fuel pump and move to one side.

7. Remove the intake manifold retain-ing nuts and washers and remove the man-ifold.

8. Installation is the reverse of re-moval. Tighten the retaining nuts to 150 in. lbs.

243 (3.9L) Diesel

1. Disconnect the negative battery ca-ble.

2. Remove the air cleaner assembly and mounting brackets from the intake mani-fold.

3. Disconnect or remove any linkage, controls or hoses that will interfere with manifold removal.

4. Disconnect the hoses from the fuel filter at the transfer pump and injection pump. Drain the fuel filter and remove (with mounting brackets) from the intake mani-fold.

5. Disconnect the injector lines for cyl-inders 3 and 6 from the injector pump if necessary for clearance. Disconnect the clamps that attach the fuel lines to the man-ifold and push the lines up and out of the way.

6. Remove the manifold to cylinder head attaching fasteners and remove the mani-fold. Clean all gasket mounting surfaces.

7. Install a new intake manifold gasket and the manifold in the reverse order of removal.

V8 Engines

1. Drain cooling system and discon-

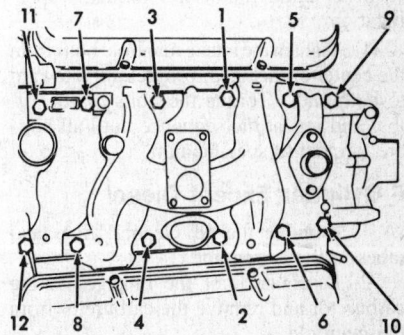

V8 intake manifold tightening sequence

nect battery, negative cable first.

2. Remove alternator, carburetor air cleaner, and fuel line.

3. Disconnect accelerator linkage.

4. Remove vacuum control between carburetor and distributor.

5. Remove the distributor cap and wires.

6. Disconnect coil wires, temperature sending unit wire, heater hoses and bypass hose.

7. Remove intake manifold, ignition coil and carburetor as an assembly.

8. Installation is the reverse of the above procedure. Tighten the intake manifold to head bolts in the sequence illustrated, from center alternating out.

Combination Manifold

REMOVAL & INSTALLATION

135 (2.2L) 4 Cylinder

1. Disconnect the battery and drain the cooling system.

2. Remove the air cleaner and disconnect all vacuum lines, electrical wiring and fuel lines from the carburetor. Remove the throttle linkage.

3. Loosen the power steering pump and remove the drive belt. Remove the power brake vacuum hose from the intake manifold.

4. On vehicles equipped with A.I.R., remove the coupling hose from the diverter valve to exhaust manifold air injection tube assembly.

5. Remove the water hoses from the water crossover and raise the vehicle. Remove the exhaust pipe from the manifold.

6. Remove the intake manifold support bracket and the EGR tube.

7. On vehicles equipped with A.I.R., remove the four air injection tube bolts and remove the air injection tube assembly from the exhaust manifold.

8. Remove the intake manifold and then remove the exhaust manifold.

9. If necessary remove the water crossover cover from the intake manifold.

10. Installation is the reverse of removal. Use new manifold gaskets. Tighten the exhaust manifold from the center and progress outward to both ends. Torque the manifold nuts to 16.7 ft. lbs. and repeat the sequence until all bolts are torqued to specification.

11. Tighten the intake manifold bolts from the center of the head, progressing outward to both ends. Torque the bolts to 16.7 ft. lbs. and repeat the sequence until all bolts are torqued to specification.

6 Cylinder Except Diesel

1. Remove the air cleaner, lines and tubes to the carburetor.

2. Disconnect all the linkages to the carburetor and remove the carburetor from the manifold.

3. Disconnect the exhaust pipe from the

manifold, remove the manifold attaching washers and retaining nuts, and remove the manifold from the cylinder head.

4. Separate the exhaust manifold from the intake manifold, if necessary, and install a new gasket between the two upon reassembly.

NOTE: Do not tighten the three securing bolts until the manifold assembly has been installed on the cylinder head.

5. Position the manifold on the cylinder head using a new gasket, and install the conical and triangular washers, the retaining nuts, and torque the retaining nuts and the three securing bolts to the specified torque.

6. Attach the exhaust pipe to the exhaust manifold flange.

7. Install the carburetor and attach all the lines, tubes, and linkages. Install the air cleaner assembly.

Exhaust Manifold

REMOVAL & INSTALLATION

155.9 (2.6L) 4 Cylinder

1. Disconnect the battery.

2. Remove the air cleaner.

3. Remove the belt from the power steering pump.

4. Raise the vehicle and make sure it is supported safely.

5. Remove the exhaust pipe from the manifold or converter.

6. Disconnect the air injection tube assembly from the exhaust manifold and lower the vehicle.

7. Remove the power steering pump assembly and move to one side.

8. Remove the heat cowl from the exhaust manifold.

9. Remove the exhaust manifold retaining nuts and remove the assembly from the vehicle.

10. Remove the carburetor air heater from the manifold.

11. Separate the exhaust manifold from the catalytic converter by removing the retaining screws.

12. Installation is the reverse of removal. Use a new gasket between the exhaust manifold and the front catalytic converter and torque the mounting screws to 24 ft. lbs. Use a new manifold gasket and coat the cylinder head side lightly with sealer. Torque the manifold center mounting nuts to 150 in. lbs. then torque the outer mounting nuts to 150 in. lbs.

243 (3.9L) Diesel

NOTE: Refer to "Cylinder Head" section for procedures.

V8 Engines

1. Disconnect the exhaust manifold at the flange where it mates to the exhaust pipe.

2. If the vehicle is equipped with air injection and/or a carburetor-heated air stove, remove them.

3. Remove the exhaust manifold by removing the securing bolts and washers. To reach these bolts, it may be necessary to jack the engine slightly off its front mounts. When the exhaust manifold is removed, sometimes the securing studs will screw out with the nuts. If this occurs, the studs must be replaced with the aid of sealing compound on the coarse thread ends. If this is not done, water leaks may develop at the studs.

4. To install, reverse the removal procedures. On the center branch of the 318 and the 360 exhaust manifold, no conical washers are used.

Cylinder Head

REMOVAL & INSTALLATION

135 (2.2L) 4 Cylinder

———— CAUTION ————

Do not perform this operation on a warm engine. Remove the head bolts in reverse of installation sequence shown. Loosen evenly in several steps. Do not attempt to slide the cylinder head off of the block. Lift the head straight up and off of the engine block.

1. Disconnect the negative battery cable. Drain the cooling system.

2. Remove the air cleaner assembly. Mark the various hoses for installation identification.

3. Disconnect all lines, hoses, wiring harnesses, etc. from the manifold, carburetor and cylinder head.

4. Disconnect the accelerator linkage. Remove the carburetor. Disconnect the converter and exhaust pipe. Remove the intake and exhaust manifolds.

5. Remove the upper part of the timing case (front cover).

135 (2.2L) cylinder head torque sequence. Models using 11 mm bolts require higher torque, refer to specifications chart

6. Turn the engine by hand until all gear timing marks line up (engine at TDC, No. 1 piston).

7. Loosen the drive belt tensioner and slip the timing belt off of the camshaft gear.

8. If the car is equipped with air conditioning, remove the compressor and mounting brackets, place out of the way. Do not disconnect any of the compressor lines unless the system is safely bled of freon.

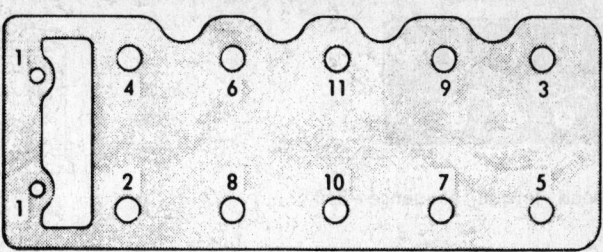
9. Remove the valve cover, gaskets and seals. Remove the head bolts in the reverse order of the tightening sequence.

10. Lift off the cylinder head, clean all gasket surfaces.

11. Installation is in the reverse order of removal. Refer to the timing belt replacement section to check camshaft timing. Make sure all gasket surfaces are cleaned and free of deep nicks or scratches. Always install new gaskets and seals. Tighten bolts in the order shown in illustrations. Make sure all timing marks are aligned before installing the drive belt. The drive belt is correctly tensioned when possible to twist 90° with the thumb and index finger midway between the cam and intermediate shafts.

155.9 (2.6L) 4 Cylinder

1. Disconnect the battery and drain the cooling system. Disconnect the upper radiator hose.

2. Remove the breather hoses and purge hose.

3. Remove the air cleaner and fuel line.

4. Remove the vacuum hose at the distributor and purge control valve. Remove alternator belt. Remove power steering pump and A/C compressor with brackets and lines attached. DO NOT DISCONNECT COMPRESSOR LINES.

5. Disconnect the spark plug wires after marking them for reinstallation.

6. Turn engine to No. 1 piston on TDC (top dead center) on the compression stroke. Remove the distributor cap, and distributor by removing the retainer nut and pulling the unit out.

7. Disconnect the heater hose at the intake manifold.

8. Disconnect the water temperature gauge unit wire.

9. Disconnect the fuel hoses and plug the line leading to the gas tank to prevent fuel leakage.

10. Remove the fuel pump mounting nuts or bolts and remove the pump assembly. Remove the insulator and gaskets.

11. Disconnect the exhaust pipe at the exhaust manifold flange.

12. Remove the rocker cover.

13. Remove its breather and semi-circular seal.

14. After slightly loosening the camshaft sprocket bolt confirm that No. 1 piston is at TDC on compression stroke (both valves closed).

NOTE: Never turn the engine over using the camshaft bolt; it puts undue strain on the chain and other components.

15. Mark the timing chain and sprocket with white paint for alignment reference. Remove the camshaft sprocket bolt and distributor drive gear. Remove the camshaft sprocket and allow it to rest with the chain attached on the holder below.

16. Remove the cylinder head bolts in the sequence shown. Head bolts should be loosened in two or three stages to prevent head warpage.

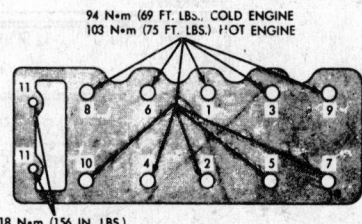

Cylinder head loosening sequence

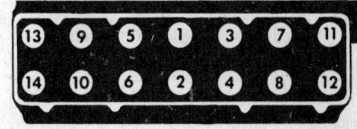

155.9 (2.6L) cylinder head torque sequence

17. Remove the cylinder head and cylinder head gasket.

18. Install the cylinder head and start the bolts. Then, tighten the bolts numbered 1–10 only in the sequence shown to 34 ft. lbs.

19. Now torque the same bolts (Numbers 1–10 only) in the same sequence to 69 ft. lbs. Tighten the bolts numbered 11 just slightly.

20. Then, complete the basic installation procedures in reverse order, including refilling the cooling system. Run the engine until it is hot. Finally, again remove the cam cover, and torque bolts 1–10 in sequence to 75 ft. lbs. Finally, torque bolts 11 to 156 in. lbs.

6 Cylinder Except Diesel

1. Drain the cooling system and disconnect the battery.

2. Remove the air cleaner and the fuel line from the carburetor.

3. Disconnect the accelerator linkage.

4. Remove the vacuum advance line from between the carburetor and the distributor.

5. Disconnect the cables from the spark plugs.

6. Disconnect the heater hose and the clamp which secures the by-pass hose.

7. Disconnect the water temperature sending unit.

8. Disconnect the exhaust pipe at the exhaust manifold flange. Disconnect the diverter valve line (if equipped) from the intake manifold and remove the air tube assembly from the cylinder head.

9. Remove the intake and exhaust manifolds and the carburetor as an assembly.

10. Remove the closed ventilation system, the evaporative control system (if so equipped), and the valve cover.

11. Remove the rocker arm and shaft assembly.

12. Remove the pushrods and keep them

Cylinder head torque sequence—6 cylinder

in order to ensure installation in their original locations.

13. Remove the head bolts and remove the cylinder head.

14. Clean all of the gasket surfaces of the engine block and the cylinder head, and install the spark plugs.

15. Inspect all surfaces with a straightedge. If warpage is indicated, measure the amount. This amount must not exceed 0.00075 times the span length in any direction. For example, if a 12 in. span is 0.004 warped, the maximum allowable is $12 \times 0.00075 = 0.009$ in. In this case, the head is within limits. If warpage exceeds the specified limits, either replace the head or lightly machine the head gasket surface.

16. Coat a new cylinder head gasket with sealer, install the gasket, and install the cylinder head.

17. Install the cylinder head bolts. Torque the cylinder head bolts to 50 ft lbs. in the sequence indicated in the illustration. Repeat this sequence to retorque all the head bolts to specifications.

18. Reverse the removal procedure. Steps 1–12, to complete the installation. When installing the intake and exhaust manifold assembly, loosen the 3 bolts which secure the intake manifold to the exhaust manifold to maintain proper alignment. After installation, torque the 3 bolts in this sequence: inner bolts, then outer bolts. Refer to the manifold section, later in this chapter, for the proper tightening sequence. Check the valve adjustment.

V8

1. Drain the cooling system and disconnect the battery ground cable.

2. Remove the alternator, air cleaner, and fuel line.

3. Disconnect the accelerator linkage.

4. Remove the vacuum advance line running between the carburetor and the distributor.

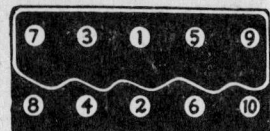

Cylinder head torque sequence—318, 360-V8

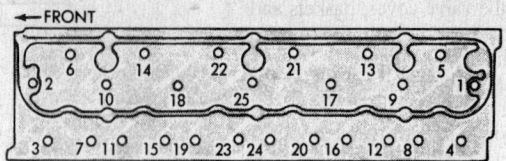

Diesel cylinder head bolt loosening sequence

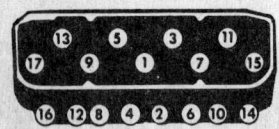

Cylinder head torque sequence—400, 440-V8

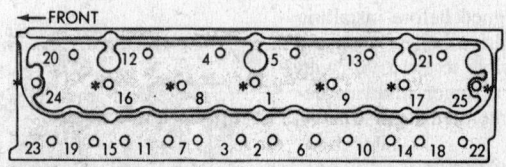

*BOLTS TO BE TIGHTENED TOGETHER WITH THE ROCKER SHAFT BRACKETS.

Diesel cylinder head bolt tightening sequence

5. Remove the distributor cap and wires as an assembly.

6. Disconnect the coil wires, water temperature sending unit, heater hoses, and by-pass hose.

7. Remove the closed ventilation system, the evaporative control system (if so equipped), and the valve covers.

8. Remove the intake manifold, ignition coil, and the carburetor as an assembly.

9. Remove the exhaust manifolds.

10. Remove the rocker and shaft assemblies.

11. Remove the pushrods and keep them in order to ensure installation in their original locations.

12. Remove the head bolts from each cylinder head and remove the cylinder heads.

13. Clean all the gasket surfaces of the engine block and the cylinder heads. Install the spark plugs.

14. Inspect all surfaces with a straightedge. If warpage is indicated, measure the amount. This amount must not exceed 0.00075 times the span length in any direction. For example, if a 12 in. span is 0.004 in. warped, the maximum allowable difference is 12 × 0.00075 = 0.009 in. In this case, the head is within limits. If the warpage exceeds the specified limits, either replace the head or lightly machine the head gasket surface.

15. Coat new cylinder head gaskets with sealer, install the gaskets, and install the cylinder heads.

16. Install the cylinder head bolts. Torque the cylinder head bolts to 50 ft. lbs. in the sequence indicated. Repeat this sequence to retorque all the cylinder head bolts to specifications.

17. Reverse the removal procedure Steps 1–12 to complete the installation.

Diesel Engine

1. Drain the cooling system.

2. Disconnect the negative battery cable.

3. Remove the air cleaner.

4. Disconnect the hoses from the fuel filter at the transfer pump and the injection pump. Drain the filter and remove it from the manifold.

5. Remove the manifold nuts and air cleaner mounting bracket attaching nuts.

6. Disconnect the injection lines for cylinders 3 and 6 from the injection pump.

7. Remove the intake manifold and gaskets from the head.

8. Push the exhaust manifold shield to one side.

9. Remove the heater hose and the by-pass hose.

10. Remove the thermostat housing and the upper radiator hose from the water manifold. Remove the spray gasket.

11. Disconnect the temperature sending unit wire.

12. Disconnect the fuel line mounting brackets from the cylinder head and push the fuel lines aside.

13. Remove the three exhaust manifold bridges.

14. Remove the water manifold and gasket from the cylinder head.

15. Raise the truck and support with jackstands.

16. Disconnect the exhaust pipe from the exhaust manifold.

17. Lower the truck. Remove the exhaust manifold, heat shield and gasket.

18. Disconnect and remove the wire from the glow plug buss bar.

19. Remove the injection lines from the injection pump.

20. Disconnect the fuel injection line from the head. Remove the bracket and the ground strap.

21. Disconnect the alternator bracket and the engine lifting fixture. Push them aside.

22. Remove the cylinder head cover and the gasket.

23. Loosen and remove the cylinder head bolts in the sequence illustrated.

24. Lift out the rocker arm and shaft assembly.

25. Remove the push rods, keeping them in order. The push rods MUST be installed in their original location.

26. Remove the injector tubes, injector holders and the injectors.

27. Disconnect and remove the glow plug buss bar.

28. Remove the six glow plugs from the cylinder head.

29. Remove the cylinder head. Check the head for cracks, damage or evidence of water leaks. Clean all the oil, grease, scale, sealant and carbon from the head. Thoroughly clean the gasket surfaces. Also check each combustion chamber jet for cracks or melting. If a jet is cracked or melted, remove it with a push rod inserted through a glow plug bore. Inspect all cylinder head surfaces with a straightedge. Out-of-flatness must not exceed 0.010 in. If it does, a surface grinder must be used to bring the head to an out-of-flatness of less than 0.006 in.

30. Install the glow plugs in the head. Tighten them firmly.

31. Install the glow plug buss bar. Be sure the connections are good.

32. Install the injectors, injector tubes and the injector holders in the head. Tighten the nozzle holder attaching nuts to 37 ft. lbs.

33. Coat the new gasket lightly with sealer. Place the gasket on the block and place the cylinder head over the dowels.

34. Install the cylinder head bolts and tighten them in the sequence illustrated to 90.4 ft lbs. Do not install the head bolts which retain the rocker shaft assembly.

35. Install the push rods in their original locations.

36. Install the rocker arm and shaft assembly. Tighten the mounting bolts, the same as the cylinder head bolts, to 90.4 ft. lbs.

37. Adjust the valve clearance to 0.012 in. at top dead center of each compression stroke.

38. Install the cylinder head cover and gasket.

39. Install the alternator bracket and the engine lifting fixture.

40. Connect the fuel lines to the injection pump (except Nos. 3 and 6). Install the bracket and the ground strap.

41. Install the fuel line to the transfer pump.

42. Install the exhaust manifold and the heat shield assembly, using a new gasket.

43. Raise the truck and support it safely. Attach the exhaust pipe to the exhaust manifold.

44. Lower the truck and install the water manifold on the head using a new gasket.

45. Install the three exhaust manifold bridges.

46. Install the fuel lines in the bracket.

47. Connect the temperature sending unit wire.

48. Using a new gasket, install the thermostat housing. Attach the upper radiator hose to the thermostat housing.

49. Install the bypass and heater hoses.

50. Install the exhaust manifold heat shield and the exhaust manifold.

51. Using a new gasket and spray shield, install the air intake manifold.

52. Connect the injection lines from cylinders 3 and 6 to the injection pump.

53. Install the fuel filter to the back of the manifold.

54. Connect the fuel hoses.

55. Install the air cleaner bracket and install the air cleaner.

56. Fill the cooling system and connect the battery cables.

Rocker Arms and Shafts

REMOVAL & INSTALLATION

135 (2.2L) 4 Cylinder

————— CAUTION —————

When depressing the valve spring with Chrysler tool 4682, or the equivalent, the valve locks can become dislocated. Check and make sure both locks are fully seated in the valve grooves and retainer.

1. Remove the valve cover.

2. Rotate the camshaft until the lobe base is on the rocker arm that is to be removed.

3. Slightly depress the valve spring using Chrysler tool 4682 or equivalent. Slide the rocker off the lash adjuster and valve tip and remove. Label the rocker arms for position identification. Proceed to next rocker arm and repeat Steps 2 and 3.

4. Install in reserve order. Check the valve keys, be sure they are not dislocated.

155.9 (2.6L) 4 Cylinder

1. Turn the engine until No. 1 piston

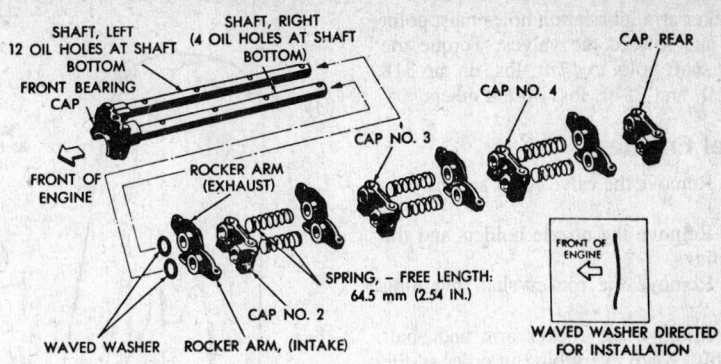

155.9 (2.6L) engine—rocker arm and shaft assembly

is at TDC (top dead center) of the compression stroke. Remove the distributor cap and confirm that the rotor tip is pointed at the No. 1 plug wire location. Disconnect the negative battery cable. Remove the distributor.

2. Remove the water pump cover (upper shield) and valve cover.

3. Confirm that No. 1 piston is at TDC of the compression stroke. Take white paint and mark the timing chain in line with the camshaft sprocket timing mark.

3. Remove the camshaft sprocket bolt, distributor drive gear and sprocket with chain meshed. Secure sprocket and chain in holder.

4. Loosen the camshaft bearing bracket and rocker arm assembly mounting bolts. Start at each end and work toward the center. Do not remove the retaining bolts. When all retaining bolts are loose, lift assembly from the cylinder head.

5. Remove the bolts from the camshaft bearing caps and remove the rocker shafts and arms. Keep all parts in order. Note the way the rocker shafts, rocker arms, springs and wave washers are mounted, the left shaft has 12 oil holes which face down. The right shaft has 4 oil holes that face down.

6. Lubricate all parts and assemble in reverse order. Secure the assembled shafts in position with retaining bolts through the cam bearing caps and install on head.

6 Cylinder (Except Diesel)

The rocker arm shaft has 12 straight steel rocker arms arranged on it with hardened steel spacers fitted between each pair of rocker arms. The shaft is secured by bolts and steel retainers which are attached to the 7 cylinder head brackets. To remove the rocker arm and shaft:

1. Remove the closed ventilation system.

2. Remove the evaporative control system (if so equipped).

3. Remove the valve cover and its gasket.

4. Remove the rocker shaft bolts and retainers.

5. Remove the rocker arm and shaft assembly.

6. Reverse the above for installation.

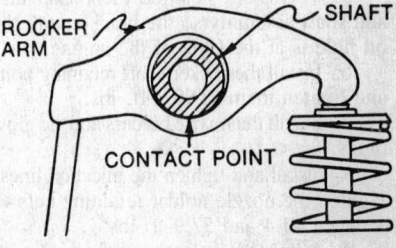

Inspect the rocker-arm-to-rocker shaft contact area

The oil hole on the end of the shaft must be on top and point toward the front of the engine to provide proper lubrication to the rocker arms. The special bolt goes to the rear.

7. Torque all bolts to 25 ft. lbs.

8. Temporarily set the intake valve tappet at 0.015 in. and the exhaust valve at 0.025 in.

9. Run the engine at 550 rpm until it is fully warmed up and adjust the valves.

V8

The stamped steel rocker arms are arranged on one rocker arm shaft per cylinder head. Because the angle of the pushrods tends to force the rocker arm pairs to absorb the side thrust of each rocker arm. The shaft is secured by bolts and steel retainers attached to the brackets on the cylinder head. To remove the arm and shaft from each cylinder head:

1. Disconnect the spark plug wires.

2. Disconnect the closed ventilation system and evaporative control system (if so equipped) from the valve cover.

3. Remove each valve cover and gasket.

4. Remove the rocker shaft bolts and retainer.

5. Remove each rocker arm and shaft assembly. Keep everything in order for installation in the original position.

6. Reverse the above for installation. The notch on the end of both 318 and 360 rocker shafts should point to the engine centerline and toward the front of the engine on the left cylinder head and toward the rear on the right side. On the 400 and 440,

the rocker arm lubrication holes must point down and toward the valves. Torque the rocker shaft bolts to 17 ft. lbs. on the 318 and 360, and 25 ft. lbs. on the others.

Diesel Engine

1. Remove the valve cover and the gasket.

2. Remove the nozzle holders and the glow plugs.

3. Remove the rocker shaft retaining bolts.

4. Remove the rocker arm and shaft assembly. Keep everything in order so the rocker assembly can be installed in the original position.

5. To install, position the rocker arm and shaft assembly so the bracket with the oil hole is at the front of the engine.

6. Install the rocker shaft retaining bolts and tighten them to 90.4 ft. lbs.

7. Install the nozzle holders and the glow plugs.

8. Install and tighten the injection lines. Tighten the nozzle holder retaining nuts to between 43.4 and 57.9 ft. lbs.

9. Adjust the valves.

Valve Stem Oil Seal

REPLACEMENT

If valve stem oil seals are found to be the cause of excessive oil consumption, they may be replaced without removing the cylinder heads.

1. Remove the air cleaner.

2. Remove the rocker arm covers and spark plugs.

3. Detach the coil wire from the distributor.

4. Turn the engine so that No. 1 cylinder is at Top Dead Center on the compression stroke. Both valves for No. 1 cylinder should be fully closed and the crankshaft damper timing mark at TDC. The distributor rotor will point at the No. 1 spark plug wire location in the cap.

5. Remove the rocker shaft and install a dummy shaft for spring compressor use.

6. Apply 90–100 psi air pressure to No. 1 cylinder, using a spark plug hole air hose adaptor.

7. Use a valve spring compressor to compress each No. 1 cylinder valve spring and remove the retainer locks and the spring. Remove the old seals.

8. Install a cup shield on the exhaust valve stem. Position it down against the valve guide.

9. Push the intake valve stem seal firmly and squarely over the valve guide.

10. Compress the valve spring only enough to install the lock.

11. Repeat the operation on each successive cylinder in the firing order, making sure that the crankshaft is exactly on TDC for each cylinder.

12. Replace the rocker arms, covers, spark plugs, and coil wire.

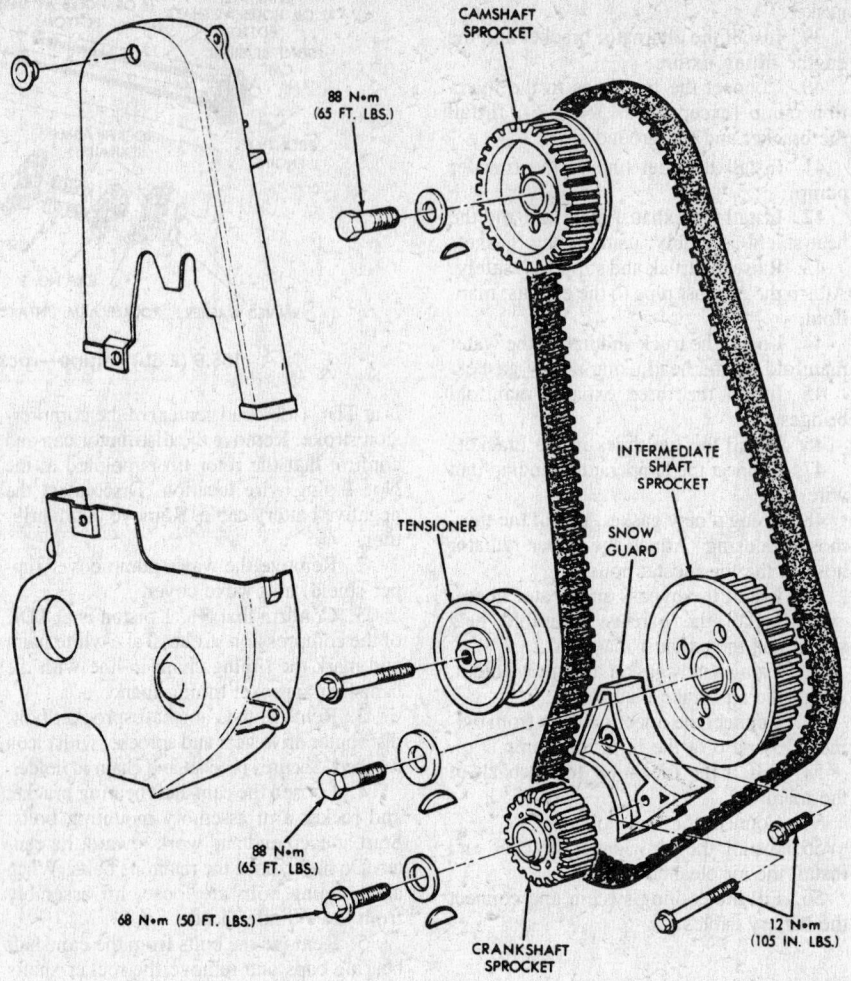

Timing belt and components—135 (2.2L) engine

Timing Cover, Belt and Sprockets

REMOVAL & INSTALLATION

135 (2.2L) 4 Cylinder

TIMING COVER

1. Loosen the alternator lock screw and adjusting screw. Remove the drive belt.

2. Remove the power steering pump lock screw. Remove the pivot bolt and nut. Remove the drive belt. Remove the power steering pump and mounting bracket. The hoses need not be disconnected, locate the pump out of the way.

3. Loosen and remove the water pump pulley mounting screws and remove the pulley.

4. From under the vehicle remove the right inner splash shield.

5. Remove the crankshaft pulley.

6. The upper part of the timing cover is retained by nuts, the lower part is retained

with screws. Remove the fasteners and the two halves of the timing cover.

7. Reverse the removal order for installation.

TIMING BELT

1. Follow Steps 1–6 of the Timing Cover Removal and Installation section.

2. Place a jack under the engine with a piece of wood separating it from the jacking point.

3. Remove the right engine mounting bolt and raise the engine slightly. Be sure the engine is supported securely.

4. Loosen the belt tensioner and remove the timing belt.

5. Turn the crankshaft until the dot mark on the sprocket is at about two o'clock. Turn the intermediate shaft sprocket until the dot mark is at about eight o'clock. Line up the crankshaft and intermediate sprocket marks.

6. Turn the camshaft until the arrows on the mounting hub are in line with the front (No. 1) camshaft retaining cap flat

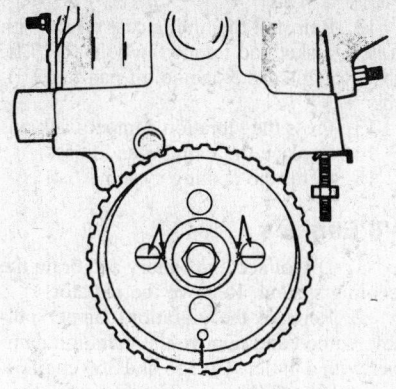

Camshaft timing—135 (2.2L) engine

spots. The small hole in the camshaft sprocket must be at the top and be in a vertical center line with the engine.

7. Install the timing belt. Adjust and tighten the belt tensioner.

8. Adjust the tensioner by turning the large tensioner hex to the right. Tension should be correct when the belt can be twisted 90° with the thumb and the forefinger, midway between the camshaft and the intermediate sprocket.

9. Complete the belt installation by reversing the removal steps.

NOTE: After applying the belt tensioner, rotate the engine two complete revolutions and recheck the timing marks for alignment.

TIMING SPROCKETS

The camshaft, intermediate shaft, and crankshaft sprockets are located by keys on their respective shafts and each is retained by a bolt. To remove any or all of the pulleys, first remove the timing belt cover and belt and then use the following procedure.

NOTE: When removing the crankshaft pulley, don't remove the four socket head bolts which retain the outer belt pulley to the timing belt pulley.

1. Remove the center bolt.

2. Gently pry the pulley off the shaft.

3. If the pulley is stubborn in coming off, use a gear puller. Don't hammer on the pulley.

4. Remove the pulley and key.

5. Install the pulley in the reverse order of removal.

6. Tighten the center bolt to 58 ft. lbs.

7. Install the timing belt, check valve timing, tension belt, and install the cover.

Timing Chain, Cover, "Silent Shafts" and Tensioner

REMOVAL & INSTALLATION

155.9 (2.6L) 4 Cylinder

1. Bring the engine to No. 1 piston at

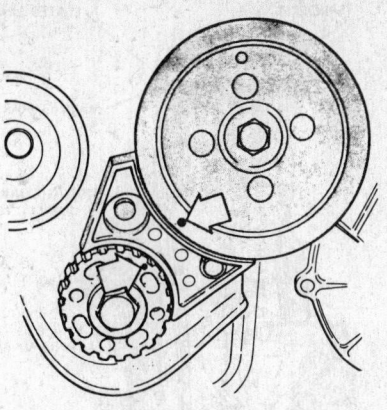

Crankshaft and intermediate shaft timing alignment—135 (2.2L) engine

TDC (top dead center) of the compression stroke.

2. Disconnect the negative battery cable. Remove the air cleaner assembly.

3. Remove the alternator drive belt. Disconnect the spark plug wires from the plugs, free wires from supports. Remove the distributor with cap and wires attached.

4. Remove the air conditioner compressor drive belt. Remove the compressor and mounting brackets, with lines attached; and position out of the way.

5. Remove the power steering drive belt. Remove the power steering pump and brackets, with lines attached, and position out of the way.

6. Raise and support the front of the vehicle on jackstands.

7. Remove the right front inner splash shield. Drain the engine oil. Remove the crankshaft pulley.

8. Place a floor jack under the engine with a piece of wood mounted between jack and lifting point.

9. Raise the jack until contact is made with the engine. Relieve pressure by jacking slightly and remove the center bolt from the right engine mount.

10. Remove the engine oil dipstick. Disconnect all vacuum hoses that run across the valve cover. Remove the valve cover.

11. Remove the two front cylinder head to timing case cover bolts (bolts in front of the cam sprocket). DO NOT LOOSEN ANY OTHER CYLINDER HEAD BOLTS.

12. Remove the oil pan retaining bolts and lower the oil pan. Remove the timing indicator and engine mounting plate from the timing chain case cover. Remove the remaining bolts retaining the chain cover and remove the cover.

13. Remove the three "silent shaft" drive chain guides. Remove the left side silent shaft and right side oil pump drive sprocket retaining bolts.

14. Remove the silent shaft drive chain, crankshaft sprocket and silent shaft sprockets.

15. Remove the camshaft sprocket retaining bolt. Remove the distributor drive gear. Remove the sprocket holder bracket and right and left timing chain guides.

16. Depress the timing chain tensioner and remove the timing chain, camshaft sprocket and crankshaft sprocket. Remove tensioner, spring and washer from the oil pump.

17. If the silent shafts require service, remove the thrust plate or oil pump retaining screws, remove plate or pump and shaft.

18. Clean all parts, especially gasket mounting surfaces. Inspect all parts for cracks, damage or wear. Replace worn parts.

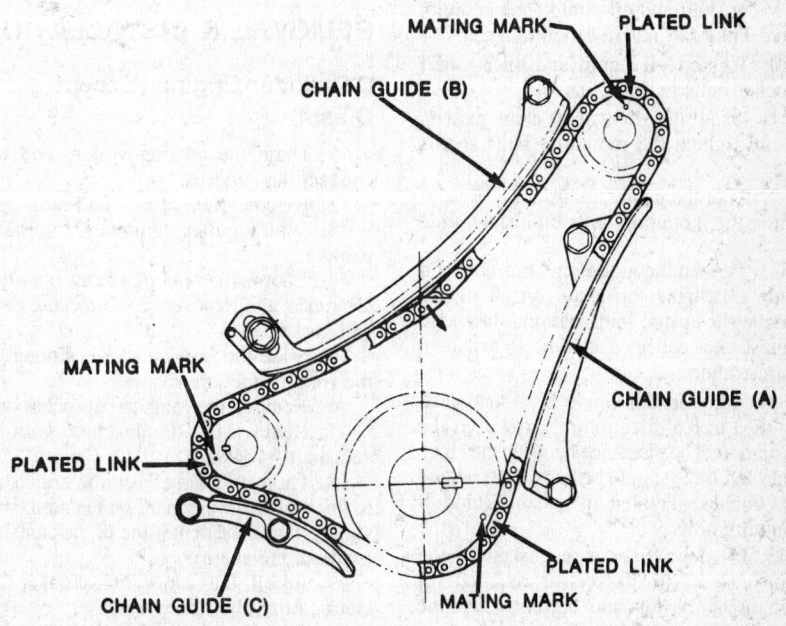

"Silent Shaft" balance system timing mark alignment—155.9 (2.6L) engine

19. Install the left side silent shaft and thrust plate with a new O-ring. Tighten the retaining bolts to 71 in. lbs.

20. Install the right side silent shaft, prime the oil pump with fresh oil and install. Tighten the retaining bolts to 71 in. lbs.

21. Verify that No. 1 piston is at TDC on the compression stroke (keyway at approx. 3 o'clock). Make sure the dowel pin hole on the front of the camshaft is in the vertical position at 12 o'clock.

22. Install the cam sprocket holder. Install the right and left chain guides. Install the tensioner spring, washer and shoe on the oil pump body.

23. Position the crankshaft and camshaft sprockets on the timing chain with timing marks aligned. The crank and camshaft sprockets have a punch mark on one gear tooth. The timing chain is equipped with two plated links. The marked tooth on each gear should be installed in the plated link.

24. Using both hands, lift the gears and chain with marks aligned, slide the gears onto their respective shafts with dowel hole and keyway in proper position. Verify gear marks and plated links are aligned.

25. Install the dowel pin, distributor drive gear and sprocket bolt on the camshaft. Torque bolt to 40 ft. lbs.

26. Install the silent shaft chain drive sprocket on the crankshaft.

27. Install the oil pump and silent shaft drive sprockets in the silent drive chain with the punch marked tooth on each sprocket inserted into the plated links on the timing chain.

28. Hold the sprockets and chain with both hands, lift and align the remaining plated link with the punch marked tooth on the crankshaft sprocket.

29. Install plated link over the punch marked tooth on the crank sprocket. Install the silent shaft and oil pump sprocket with plated links and marked tooth aligned.

30. Tighten oil pump and silent shaft sprocket bolts to 25 ft. lbs.

31. Install the silent shaft chain guides. Do not tighten the mounting bolts at this time.

32. After the guides are installed loosely, tighten the mounting bolts on chain Guide A.

33. Tighten the mounting bolts on chain Guide C. Shake the chain on all of the sprockets to ensure snug seating. Make sure chain slack is collected at point (P) as shown in illustration.

34. Adjust chain Guide B so that slack is pulled in the direction of arrow F in the illustration. The clearance between the chain guide and links should be between 0.04 and 0.14 inches. Tighten the chain Guide B mounting bolts.

35. Fit new cover case gaskets to the chain case. Trim gaskets as required for snug fit at the top and bottom. Coat the gaskets with sealer and install case cover.

36. Installation from this point is in the reverse order of removal.

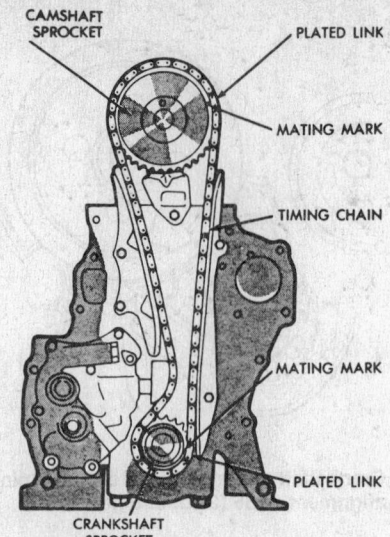

Timing gears and chain alignment—155.9 (2.6L) engine

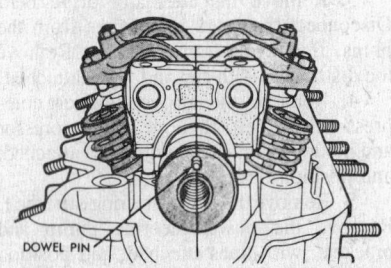

Camshaft alignment—155.9 (2.6L) engine

Timing Cover and Chain

REMOVAL & INSTALLATION

6 Cylinder Engine (Except Diesel)

1. Drain the cooling system and disconnect the battery.

2. Remove the radiator and fan.

3. With a puller, remove the vibration damper.

4. Loosen the oil pan bolts to allow clearance, and remove the timing case cover and gasket.

5. Slide the crankshaft oil slinger off the front of the crankshaft.

6. Remove the camshaft sprocket bolt.

7. Remove the timing chain with the camshaft sprocket.

8. On installation: Turn the crankshaft to line up the timing mark on the crankshaft sprocket with the centerline of the camshaft (without the chain).

9. Install the camshaft sprocket and chain. Align the timing marks.

10. Torque the camshaft sprocket bolt to 35 ft. lbs.

11. Replace the oil slinger.

12. Reinstall the timing case cover with a new gasket and torque the bolts to 17 ft. lbs. Retighten the engine oil pan to 17 ft. lbs.

13. Press the vibration damper back on.

14. Replace the radiator and hoses.

15. Refill the cooling system.

V8 Engines

1. Disconnect the battery and drain the cooling system. Remove the radiator.

2. Remove the vibration damper pulley. Unbolt and remove the vibration damper with a puller. On 318 and 360 engines, remove the fuel lines and fuel pump, then loosen the oil pan bolts and remove the front bolt on each side.

3. Remove the timing gear cover and the crankshaft oil slinger.

4. On 318 and 360 engines, remove the camshaft sprocket lockbolt, securing cup washer, and fuel pump eccentric. Remove the timing chain with both sprockets.

5. To begin the installation procedure, place the camshaft and crankshaft sprockets on a flat surface with the timing indicators on an imaginary centerline through both sprocket bores. Place the timing chain around both sprockets. Be sure that the timing marks are in alignment.

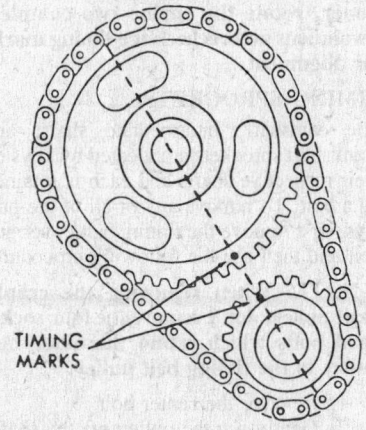

Timing mark alignment—6 cylinder engine

CAUTION
When installing the timing chain, have an assistant support the camshaft with a suitable tool to prevent it from contacting the plug in the rear of the engine block. Remove the distributor and the oil pump/distributor drive gear. Position the suitable tool against the rear side of the cam gear and be careful not to damage the cam lobes.

---

6. Turn the crankshaft and camshaft to align them with the keyway location in the crankshaft sprocket and the keyway or dowel hole in the camshaft sprocket.

7. Lift the sprockets and timing chain while keeping the sprockets tight against the chain in the correct position. Slide both

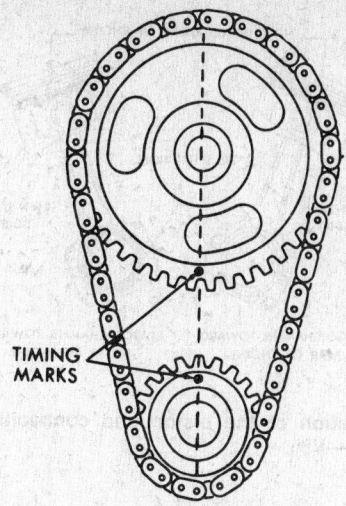

Timing mark alignment—V8 engine

sprockets evenly onto their respective shafts.

8. Use a straightedge to measure the alignment of the sprocket timing marks. They must be perfectly aligned.

9. On 318 and 360 engines, install the fuel pump eccentric, cup washer, and camshaft sprocket lockbolt and torque to 35 ft. lbs. If camshaft end play exceeds 0.010 in., install a new thrust plate. It should be 0.002–0.006 in. with the new plate.

CHECKING TIMING CHAIN SLACK

1. Position a scale (ruler or straightedge) next to the timing chain to detect any movement in the chain.

2. Place a torque wrench and socket on the camshaft sprocket attaching bolt. Apply either 30 ft. lbs. (if the cylinder heads are installed on the engine) or 15 ft. lbs. (cylinder heads removed) of force to the bolt and rotate the bolt in the direction of crankshaft rotation in order to remove all slack from the chain.

3. While applying torque to the camshaft sprocket bolt, the crankshaft should not be allowed to rotate. It may be necessary to block the crankshaft to prevent rotation.

4. Position the scale over the edge of a timing chain link and apply an equal amount of torque in the opposite direction. If the movement of the chain exceeds $\frac{1}{8}$ in., replace the chain.

Timing Cover Seal

REPLACEMENT

NOTE: A seal remover and installer tool is required to prevent seal damage.

1. Using a seal puller, separate the seal from the retainer.

2. Pull the seal from the case.

3. To install the seal place it face down in the case with the seal lips downward.

4. Seat the seal tightly against the cover face. There should be a maximum clearance of 0.0014 in. between the seal and the cover. Be careful not to over-compress the seal.

Timing Gear (Diesel)

REMOVAL & INSTALLATION

1. Remove the timing gear cover, the gasket and the front oil seal. Remove the idler pulley bracket.

2. Align the timing marks.

3. Using a puller, remove the camshaft drive gear.

4. Turn the injection pump to allow the notch in the drive gear to pass by the idler gear teeth.

5. Loosen the idler gear mounting bolt. Remove the thrust plate and remove the idler gear.

6. Be sure that the crankshaft is set with No. 1 cylinder at TDC.

7. Install the idler gear on the shaft so the marks on the camshaft drive gear match up with the marks on the idler gear.

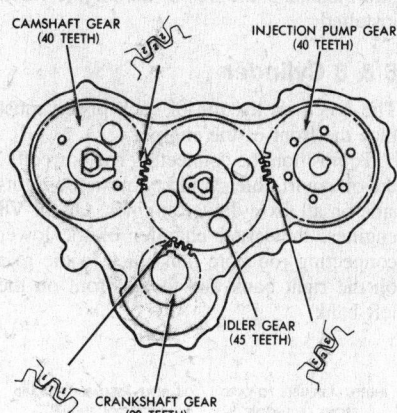

Timing gear alignment—diesel engine

8. Install the thrust plate and the hold down bolt.

9. Install the camshaft gear and the thrust plate on the camshaft. Be sure all the marks line up with the marks on the idler gear. Tighten the hold down bolt.

10. Put the injection pump in position and mesh the pump drive gear with the idler gear so the marks on the drive gear match up properly with the marks on the idler gear. Be sure the pump mounting flange scale is set at the proper injection point.

11. Install the mounting nuts to the timing gear case and tighten them.

12. Connect the fuel feed line and the filter hoses to the pump.

13. Bleed the air from the fuel system.

14. Connect the injector pipes.

15. Check the idler gear for end play using a feeler gauge between the gear and the thrust plate. It should be between 0.002 and 0.006 in. If it exceeds 0.014 in., replace the thrust plate.

16. Install a new front oil seal (using a new gasket), the timing gear cover and the crankshaft drive pulley.

Camshaft

REMOVAL & INSTALLATION

135 (2.2L) 4 Cylinder

1. Remove the timing belt cover.

2. Remove the timing belt.

3. Remove the air cleaner assembly.

4. Remove the valve cover.

5. Remove the Nos. 1, 3, and 5 camshaft bearing caps.

6. Loosen caps 2 and 4 diagonally and in increments.

7. Lift the camshaft out.

8. Lubricate the camshaft journals and lobes with engine assembly lubricant and position it in the head.

9. Install new oil seal.

10. Install the Nos. 1, 3, 5 bearing caps and torque the nuts to 14 ft. lbs.

11. Install the Nos. 2 and 4 caps and diagonally torque the nuts to 14 ft. lbs.

— **CAUTION** —

All bearing caps are slightly offset. They should be installed so the numbers on the cap read right side up from the driver's seat.

12. Position a dial indicator so that the feeler touches the front end of the camshaft. Check for end play. Play should not exceed 0.006 in.

13. Place a new seal on the No. 1 bearing cap. If necessary, replace the end plug in the head.

14. Follow the procedures under Timing Belt Removal and Installation for belt installation and timing.

15. Check the valve clearance and ignition timing.

155.9 (2.6L) 4 Cylinder

1. Remove the breather hoses and purge hose.

2. Remove the air cleaner and fuel line.

3. Remove the fuel pump. Remove the distributor.

4. Disconnect the spark plug cables.

5. Remove the rocker cover.

6. Remove the breather and semi-circular seal.

7. After slightly loosening the camshaft sprocket bolt, turn the crankshaft until No. 1 piston is at Top Dead Center on compression stroke (both valves closed).

8. Remove the camshaft sprocket bolt and distributor drive gear.

9. Remove the camshaft sprocket with chain and allow it to rest on the camshaft sprocket holder.

10. Remove the camshaft bearing cap tightening bolts. Do not remove the front and rear bearing cap bolts altogether, but keep them inserted in the bearing caps so

that the rocker assembly can be removed as a unit.

11. Remove the rocker arms, rocker shafts and bearing caps as an assembly.

12. Remove the camshaft.

13. Installation is the reverse of removal. Lubricate the camshaft lobes and bearings and fit camshaft into head. Install the assembled rocker arm shaft assembly.

6 Cylinder Engine

1. Remove the cylinder head, timing gear cover, camshaft sprocket, and timing chain.

2. Remove the valve tappets, keeping them in order to ensure installation in their original locations.

3. Remove the crankshaft sprocket.

4. Remove the distributor and oil pump.

5. Remove the fuel pump.

6. Install a long bolt into the front of the camshaft to facilitate its removal.

7. Remove the camshaft, being careful not to damage the cam bearings with the cam lobes.

8. Prior to installation, lubricate the camshaft lobes and bearing journals. It is recommended that 1 pt. of crankcase conditioner be added to the initial crankcase oil fill.

9. Install the camshaft in the engine block. From this point, reverse the removal procedure.

V8 Engines

1. Remove the intake manifold, cylinder head covers, rocker arm assemblies, push rods, and valve tappets, keeping them in order to insure the installation in their original locations.

2. Remove the timing gear cover, the camshaft and crankshaft sprockets, and the timing chain.

3. Remove the distributor and lift out the oil pump and distributor driveshaft. On 400 and 440 cu. in. engines, remove the fuel pump to allow the push rod to drop away from the cam eccentric.

4. Remove the camshaft thrust plate (on 318 and 360).

5. Install a long bolt into the front of the camshaft and remove the camshaft, being careful not to damage the cam bearings with the cam lobes.

6. Prior to installation, lubricate the camshaft lobes and bearing journals. It is recommended that 1 pt. of Crankcase Conditioner be added to the initial crankcase oil fill. Insert the camshaft into the engine block within 2 in. of its final position in the block.

7. Have an assistant support the camshaft with a suitable tool to prevent the camshaft from contacting the plug in the rear of the engine block. Position the suitable tool against the rear side of the cam gear and be careful not to damage the cam lobes.

8. Replace the camshaft thrust plate. If camshaft end play exceeds 0.010 in., install a new thrust plate. It should be 0.002–0.006 in. with the new plate.

9. Install the timing chain and sprockets, timing gear cover, and pulley.

10. Install the tappets, pushrods, rocker arms, and cylinder head covers. Install fuel pump, if removed.

11. Install the distributor and oil pump driveshaft. Install the distributor.

12. After starting the engine, adjust the ignition timing.

Pistons and Connecting Rods

Refer to the Engine Rebuilding section in Unit Repair for procedures.

4 Cylinder

The piston crown is marked with an arrow which must point toward the drive belt end of the engine when installed. The connecting rod and cap are marked with rectangular forge marks which must be mated when assembled and which must be on the intermediate shaft side of the engine when installed.

6 & 8 Cylinder

The notch on the top of each piston must face the front of the engine.

To position the connecting rod correctly, the oil squirt hole should point to the right-side on all six-cylinder engines. On all V8 engines, the larger chamfer of the lower connecting rod bore must face to the rear on the right bank and to the front on the left bank.

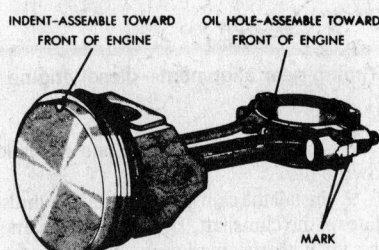

135 (2.2L) engine piston installation

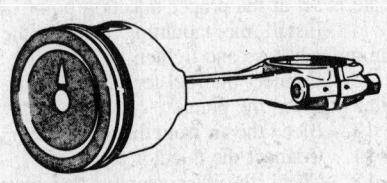

155.9 (2.6L) engine piston installation (mark faces front)

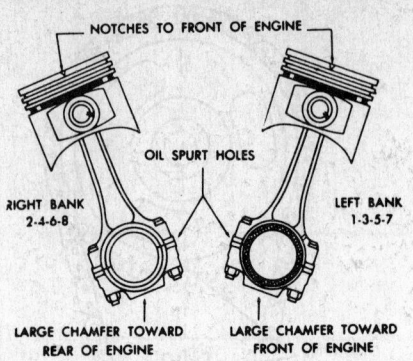

Relation of the piston and connecting rod—V8

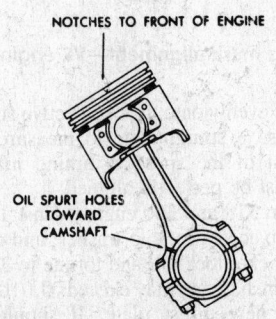

Relation of the piston and connecting rod—6 cylinder

Crankshaft & Main Bearings

Refer to the Engine Rebuilding section in Unit Repair for procedures.

ENGINE LUBRICATION

Oil Pan

REMOVAL & INSTALLATION

4 Cylinder

1. Raise and safely support the truck on jackstands. Drain the oil pan.

2. Support the pan and remove the attaching bolts.

3. Lower the pan and discard the gaskets.

4. Clean all gasket surfaces thoroughly and install the pan using gasket sealer and a new gasket.

NOTE: 1985 and later models employ end gaskets and formed-in-place side gaskets formed with RTV sealer. Observe these procedures/precautions when installing oil pans on these models: Scrape all gasket surfaces clean with a wire brush or scraper. Make sure gasket rails are flat, and if they are not, flatten them with a hammer and flat plate. Remove all oil and dirt. Make sure old RTV is removed from blind attaching holes. Apply the gasket material sparingly——so the bead is only about 0.12 in. in diameter. Be certain beads surround each mounting hole. Use new end gaskets. Place the pan squarely against the cylinder block so sealer is not smeared. Torque the pan in place while the sealer is still wet——within 10 minutes.

5. Torque the pan bolts to 7 ft. lbs.
6. Refill the pan, start the engine, and check for leaks.

6 Cylinder Vans & Pickups

1. Disconnect the battery and remove the dipstick.
2. Remove the engine cover and remove the starter and air cleaner.
3. Raise the van and support on jackstands and drain the crankcase oil.
4. Install an engine support as described under "Engine Removal."
5. Disconnect and tie out of the way: driveshaft, transmission linkage, and exhaust pipe at the manifold.
6. Remove the clutch torque shaft (if equipped) and the oil cooler lines (if equipped).
7. Disconnect the speedometer cable and electrical connections to the transmission.
8. Remove the support bracket, inspection plate, and drive plate-to-converter attaching screws if equipped.
9. Remove the bolts which attach the transmission to the clutch housing. Carefully work the transmission and converter rearward off the engine dowels and disengage the converter hub from the end of the crankshaft. Remove the transmission.
10. Support the rear of the engine and raise it two inches.
11. Remove the oil pan attaching bolts. Positioning the crankshaft so that the counterweights will clear the pan, rotate the pan to the steering gear side and remove it. You may have to turn the pump pickup tube for clearance.
12. Installation is the reverse of removal. Make sure that the pickup screen contacts the bottom of the pan. Fill the engine with oil and check for leaks.
13. Remove the oil pan attaching screws and position the crankshaft so that the pan will clear the counterweights. Remove the pan.
14. Installation is the reverse of removal. Check all fluid levels and be sure that there are no leaks.

318 & 360 V8 Vans & Pickups

1. Disconnect the battery ground cable. Remove the dipstick and tube, engine cover, and air cleaner.
2. Disconnect the throttle linkage at the rear of the engine and the clutch or automatic transmission linkage.
3. Raise the engine slightly and support it with the device described under "Engine Removal."
4. Raise and support the vehicle and drain the oil. Remove the starter.
5. Remove the driveshaft and engine rear support.
6. Remove the transmission from the van. Remove the automatic transmission with the filler tube installed and the torque converter separated from the drive plate.
7. Remove the clutch assembly and flywheel (or driveplate) from the crankshaft.
8. Raise the engine about 2 in.
9. Rotate the crankshaft so that the counterweights will clear the oil pan. Maximum clearance is with the notch in the crankshaft flange at the 3 o'clock position. Remove the oil pan. It will be necessary to reach inside the oil pan and turn the oil pick-up tube and strainer slightly to the right to clear the pan.
10. Installation is the reverse of removal. On '85 and later models, put a drop of RTV sealer on the joints between rubber and cork gaskets. Be sure to check all fluid levels and be sure that there are no leaks.

2WD Ramcharger and Trail Duster

1. Disconnect the battery cable and remove the dipstick.
2. Raise and support the truck.
3. Drain the oil.
4. Remove the torque converter or clutch housing brace.
5. If necessary, remove the exhaust pipe.
6. Remove the oil pan bolts and remove the pan.

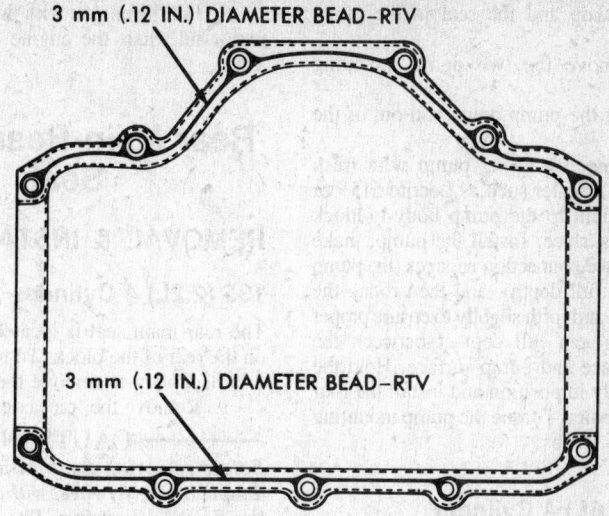

Applying RTV sealer to the 2.2L engine oil pan

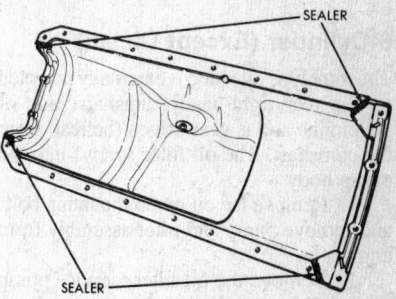

Apply ⅛ inch beads of silicone sealer to the corners of the oil pan—6 cylinder shown

7. Installation is the reverse of removal.

4WD Ramcharger and Trail Duster

1. Raise vehicle and support on jackstands.
2. Remove the two front engine mounting bolts.
3. Remove the left-side support, connecting the converter housing and cylinder block.
4. Raise the engine approximately 2 in.
5. Drain oil.
6. Remove the oil pan bolts, lower pan down and to the rear. (Do not turn oil pickup out of position).

Oil Pump

REMOVAL & INSTALLATION

135 (2.2L) 4 Cylinder

1. Raise and safely support the truck. Drain oil. Remove the oil pan.
2. Remove the two pickup mounting bolts (one on the No. 3 main bearing cap bolt and one on the oil pump) and remove

the oil pickup and the seal in the pump intake.

3. Remove the two pump mounting bolts.

4. Pull the pump down and out of the engine.

5. Prime, by filling pump with fresh oil. Apply a sealer such as Loctite 515® or the equivalent to the pump body-to-block machined surface. Install the pump, make sure the drive connection engages (the pump is inserted full depth), and then rotate the pump back and forth slightly to ensure proper positioning and full contact between the block surface and pump surface. Hold the pump firmly in position and install the four mounting bolts. Torque the pump mounting bolts to 16 ft. lbs.

155.9 (2.6L) 4 Cylinder

NOTE: Refer to the "Timing Chain, Cover; Silent Shaft" section for oil pump removal and installation procedures.

6 Cylinder (Except Diesel)

The rotor type oil pump is externally mounted on the rear right-hand (camshaft) side of the engine and is gear driven (helical) from the camshaft. The oil filter screws into the pump body.

1. Remove the oil pump mounting bolts and remove pump and filter assembly from engine.

2. Remove the oil filter, service pump or replace as required.

3. Prime, by filling pump with fresh oil. Install pump and new mounting gasket in the reverse order of removal after cleaning all gasket surfaces.

Diesel Engine

1. Raise and safely support the vehicle. Drain engine oil. Remove the oil pan, oil pickup tube, strainer and all old gaskets.

2. Remove the oil pump joint bolt, filter assembly to pump tube and the oil pump.

3. Clean all gasket surfaces. Prime by filling pump with fresh oil.

4. Installation is in the reverse order of removal. Always use new gaskets.

318, 360 V8 Engines

NOTE: It is necessary to remove the oil pan, and to remove the oil pump from the rear main bearing cap to service the oil pump.

1. Raise and safely support the vehicle. Drain the engine oil and remove the oil pan.

2. Remove the oil pump mounting bolts and remove the oil pump from the rear main bearing cap. Service or replace pump as necessary.

3. Prime the oil pump before installation by filling the rotor cavity with engine oil. Install the oil pump on the engine and tighten attaching bolts to 30 ft. lbs.

4. Continue the installation in the reverse order of the removal.

5. Fill the engine with the proper grade motor oil. Start the engine and check for leaks.

Rear Main Bearing Oil Seal

REMOVAL & INSTALLATION

135 (2.2L) 4 Cylinder

The rear main seal is located in a housing on the rear of the block. To replace the seal it is necessary to remove the engine.

1. Remove the transaxle and flywheel.

———————— CAUTION ————————

Before removing the transaxle, align the dimple on the flywheel with the pointer on the flywheel housing. The transaxle will not mate with the engine during installation unless this alignment is observed.

2. Very carefully, pry the old seal out of the support ring.

3. Coat the new seal with clean engine oil and press it into place with a flat piece of metal. Take great care not to scratch the seal or crankshaft.

4. Install the flywheel and transaxle.

155.9 (2.6L) 4 Cylinder

The rear main oil seal is located in a housing on the rear of the block. To replace the seal, remove the transaxle and do the work from underneath the vehicle or remove the engine and do the work on the bench.

1. Remove the housing from the block.

2. Remove the separator from the housing.

3. Pry out the old seal.

4. Lightly oil the replacement seal. The oil seal should be installed so that the seal plate fits into the inner contact surface of the seal case. Install the separator with the oil holes facing down.

6 and 8 Cylinder Engines (Except Diesel)

Service replacement seals are of the split rubber type composition. This type of seal makes it possible to replace the upper rear seal without removing the crankshaft. The seal must be used as an upper and lower set and cannot be used with the rope type seal.

NOTE: Rope type seals are included in overhaul gasket sets, for use when the crankshaft has been removed, on all engines, except the 360 V8, which uses only the composition seal.

The following procedure is for removing the rope type rear main seal and replacing it with the rubber type seal.

1. Remove the oil pan, and both the rear seal retainer and the rear main bearing cap, if separate.

2. Remove the lower rope seal from the cap or retainer by prying the seal out of the groove.

3. With the use of suitable tools, either pull or push the seal from its seat, while rotating the crankshaft, being careful not to damage the surface of the journal. If necessary, loosen all the main bearing caps slightly, to lower the crankshaft, which will aid in the removal and replacement of the seal.

4. Clean and lubricate the crankshaft journal. Hold the seal tight against the crankshaft with the painted stripe to the rear, and install the seal into the block groove.

5. Rotate the crankshaft while pushing the seal into the groove. Be careful that the sharp edges of the block groove, do not cut or nick the rear of the seal.

6. Install the lower half of the seal into the lower seal retainer or the main bearing cap, if separate, with the paint stripe facing to the rear.

7. Install the lower seal retainer and/or the rear main bearing cap. Torque all main bearing caps to specifications.

8. Install the oil pan, add oil and check for oil leaks.

Diesel Engine

The rear main oil seal is a one piece design mounted with a sleeve. Two side seals are also used in grooves located on No.7 main bearing cap.

1. Oil pan and flywheel are removed.

2. Rear seal retainer bolts, retainer and No.7 main bearing cap are removed.

3. The two side seals are installed after No.7 main bearing cap has been mounted and torqued to specs. Apply gasket sealer to the seals. Slide the seals into place, do not "drive" them in. When properly installed, the seal should conform to the corners of the mounting grooves.

4. The rear main seal is retained by a sleeve and plate. Remove the old seal from the retainer, clean the retainer, mount a new seal, lubricate the lips of the seal and install seal sleeve and retainer. Torque the mounting bolts to 2.2 ft. lbs. Install the flywheel, oil pan etc.

ENGINE COOLING

Radiator

REMOVAL & INSTALLATION

4 Cylinder

1. Disconnect negative battery cable. Open the radiator drain cock.

2. Remove the upper and lower hoses and coolant recovery tank connecting tube.

3. Disconnect wiring harness from fan

motor. Remove the upper shroud mounting bolts.

4. Remove the shroud and fan assembly.

5. Disconnect and plug automatic transmission fluid cooling lines, if equipped. Remove the top radiator attaching bolts.

6. Remove the bottom radiator attaching bolts, if equipped.

7. Lift radiator from engine compartment.

8. Installation is the reverse of removal. During installation, first seat the radiator in lower extruded holes, unless the bottom is bolted.

6 & 8 Cylinder Engines

1. Drain the cooling system.

2. Disconnect the battery ground cable.

3. Detach the upper hose from the radiator.

4. Remove the shroud mounting nuts and position it out of the way.

5. Remove the radiator top mounting screws. If equipped with air conditioning, remove the condenser attaching screws, accessible through the grille. Do not disconnect any air conditioning lines.

6. Raise the vehicle and support it. Disconnect and plug the automatic transmission cooler lines and cap the openings in the cooler.

7. Hold the radiator in place and remove the lower mounting screws. Carefully lower the radiator out of the van or lift it up and out if a pickup truck.

8. Installation is in the reverse of removal. Check all fluid levels and run the engine, making sure that there are no leaks.

Water Pump

REMOVAL & INSTALLATION

135 (2.2L) 4 Cylinder

1. Drain the cooling system.

2. Remove the upper radiator hose.

3. Without discharging the system, remove the air conditioning compressor from the engine brackets and set to one side.

4. Remove the alternator and move to one side.

5. Disconnect the lower radiator hose and the bypass hose and remove the water pump by removing the pump to engine retaining screws.

6. Installation is the reverse of removal. Tighten the top three retaining screws to 250 in. lbs. and the lower screw to 50 ft. lbs.

155.9 (2.6L) 4 Cylinder

1. Drain the cooling system.

2. Remove the radiator hose, by-pass hose and heater hose from the water pump.

3. Remove the drive pulley shield.

4. Remove the locking screw and pivot screws.

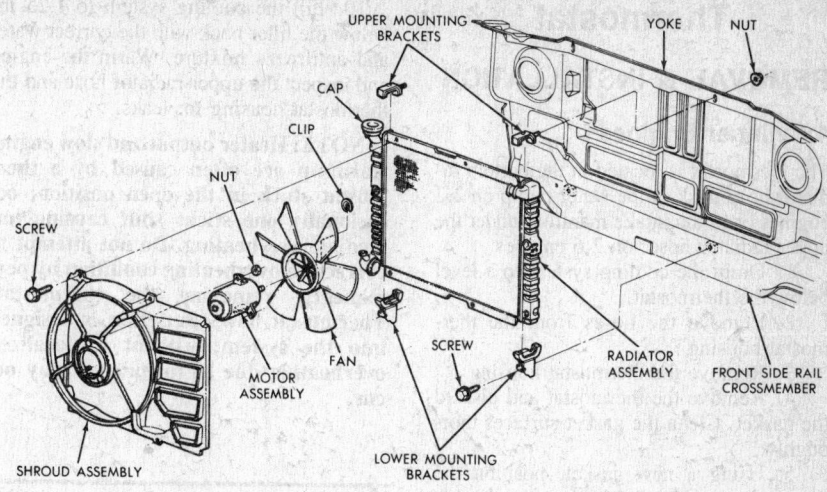

Typical 4 cylinder engine cooling system components

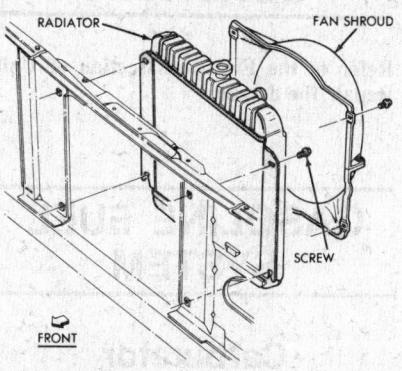

Typical radiator and fan shroud mounting—6 and V8 engines

5. Remove the drive belt and water pump from the engine.

6. Install new mounting gasket and pump. Installation is the reverse of removal. Tighten mounting bolts to 80 in. lbs. After adjusting the belt tension tighten the locking screw and pivot screws to 204 in. lbs. Tighten the drive pulley shield to 105 in. lbs.

6 Cylinder (Except Diesel)

NOTE: This job can sometimes be done without removing the radiator on models without air-conditioning, if there is enough room to get at the water pump bolts.

1. Remove the radiator.

2. Remove all drive belts.

3. Remove the fan, spacer, pulley, and bolts as an assembly.

4. If equipped with an air pump, remove the pump brackets with the hoses attached and tie it out of the way.

5. Disconnect the heater hose and all other hoses from the pump.

6. Remove the pump from the block.

7. Installation is the reverse of removal. Fill the cooling system and adjust the tension of the drive belts.

Diesel Engine

1. Drain the cooling system and remove the heater hoses and the bypass hose.

2. Loosen the alternator mounting bolts and remove the belt.

3. Remove the cooling fan, the spacer and the drive pulley.

4. Remove the water pump mounting bolts and remove the pump.

5. Reverse to install.

318 & 360 V8

1. Remove the radiator. Note that on 1985–86 models without air conditioning, it may be left in place.

2. Loosen all accessories that are belt driven and remove all the drive belts.

3. On engines without air conditioning, remove the alternator bracket attaching bolts and tie the alternator and bracket out of the way.

4. On engines with air conditioning, remove the idler pulley assembly, alternator, and adjusting bracket.

5. Remove the fan blade, spacer (or fluid unit), pulley, and bolts as an assembly.

NOTE: To prevent silicone fluid from draining into the drive bearing and ruining the lubricant, do not place the thermostatic fan drive unit with the shaft pointing downward.

6. Disconnect all hoses from the water pump.

7. Remove the air conditioning compressor front mounting bolts.

8. Remove the water pump-to-compressor front bracket bolts and the bracket.

NOTE: Do not disconnect any refrigerant lines from the compressor.

9. Remove the water pump.

10. Installation is the reverse of removal. Fill with coolant and check for leaks.

Thermostat

REMOVAL & INSTALLATION

4 Cylinder Engines

The thermostat is located in the bottom radiator hose neck in the water pump on 2.2 engines or in the intake manifold under the upper radiator hose, on 2.6 engines.

1. Drain the cooling system to a level below the thermostat.
2. Remove the hoses from the thermostat housing.
3. Remove the thermostat housing.
4. Remove the thermostat and discard the gasket. Clean the gasket surfaces thoroughly.
5. Using a new gasket, position the thermostat and install the housing and bolts. Make sure that the thermostat is seated properly.
6. Refill the cooling system.

6 & 8 Cylinder Engines

1. Drain the cooling system to below the level of the thermostat.
2. Remove the upper radiator hose from the thermostat housing. Note the positioning of the thermostat. It is important that the thermostat is correctly installed.
3. Withdraw the housing bolts and remove the housing and the thermostat.
4. Check to make sure that the thermostat valve closes tightly. If the valve does not close completely due to foreign material, carefully clean the sealing edge of the valve while being careful not to damage the sealing edge. If the valve does not close tightly after it has been cleaned, a new thermostat must be installed.
5. Immerse the thermostat in a container of warm water so that its pellet is completely covered and does not touch the bottom or sides of the container.
6. Heat the water and, while stirring the water continuously (to ensure uniform temperature), check the water temperature with a thermometer at the point when a 0.001 in. feeler gauge can be inserted in the valve opening at a water temperature with ± 5 degrees of the standard thermostat opening temperature. If the thermostat does not open within the temperature range, replace it with a new thermostat.
7. Continue heating the water to a temperature of approximately 20 degrees higher than the standard thermostat opening temperature. At this point, the thermostat should be fully open. If it is not, install a new thermostat.
8. To install, use a new gasket and position the thermostat so that its pellet end (the part with the spring) is toward the engine block. On the six, the vent hole must be up. Gaskets should be first soaked in water. It is also permissible to use a bead of rubber sealer. Refit the thermostat housing and tighten its securing bolts.
9. Refit the upper radiator hose.

10. Fill the cooling system to 1.25 in. below the filler neck with the correct water and antifreeze mixture. Warm the engine and inspect the upper radiator hose and the thermostat housing for leaks.

NOTE: Heater output and slow engine warm-up are often caused by a thermostat stuck in the open position; occasionally one sticks shut causing immediate overheating. Do not attempt to correct an overheating condition by permanently removing the thermostat. Thermostat flow restriction is designed into the system; without it, localized overheating due to turbulence may occur.

EMISSION CONTROLS

Refer to the Emissions section of Unit Repair for details.

GASOLINE FUEL SYSTEM

Carburetor

REMOVAL & INSTALLATION

The following is a general removal procedure for all carburetors.
1. Disconnect the battery ground cable.
2. Remove the air cleaner. On the 2.6L engine carburetor, remove the air intake housing from the air horn.
3. Remove the fuel tank pressure-vacuum filler cap. The tank could be under a small amount of pressure. On the 2.6L engine carburetor, drain the radiator until the coolant level is below the carburetor, and then disconnect the coolant hoses at the carburetor.
4. Disconnect and plug the fuel lines. Use two wrenches to avoid twisting the fuel line. A metal container is also useful to catch any fuel which spills from the lines.
5. Disconnect the throttle and choke linkage.
6. Disconnect any vacuum lines.
7. Remove the mounting bolts (and, on the 2.6L engine carb, the nut).
8. Carefully remove the carburetor from the engine and carry it in a level position to a clean work place.
9. Installation is the reverse of removal. Adjust the curb idle speed. Make sure to refill the cooling system, if necessary.

Refer to the Carburetor section in Unit Repair for adjustments and specifications.

Fuel Filter

REMOVAL & INSTALLATION

4 Cylinder Engines

Two filters are used. One is part of the fuel pickup in the fuel tank. The other is a sealed paper unit located in the carburetor inlet (2.2 engines) or an in-line filter (2.6 engines). The tank unit does not usually need replacing, but can be replaced if necessary. The carburetor filter should be replaced periodically. To replace the inlet filter, place a rag or container under the inlet and disconnect the fuel line. Unscrew the inlet fitting. The filter has a spring behind it so take care not to lose it. Replacement is the reverse of removal.

The inline filter can be removed by placing a rag under the filter and removing the clamps. Pull the filter out of the lines and discard it. Install a new filter, noting the direction of fuel flow, usually indicated by an arrow.

6 & 8 Cylinder Engines

Locate the filter in the fuel line between the fuel pump and the carburetor. Using hose-clamp pliers, remove the attaching clamps and pull the filter off. Reverse this procedure for installation. Be sure that the arrow on the filter is pointing toward the carburetor (direction of fuel flow). Replace the filter every 30,000 miles.

NOTE: Some filters have a third line, the purpose of which is to prevent vapor lock by allowing fuel vapors to return to the tank.

Mechanical Fuel Pump

REMOVAL & INSTALLATION

1. Disconnect the fuel lines from the inlet, output a return (if equipped) sides of the fuel pump.
2. Plug these lines to prevent gasoline from leaking out.
3. Unbolt the retaining bolts from the fuel pump and remove the fuel pump from the engine.
4. Remove the old gasket from the engine and/or fuel pump. Note that on the pump used on the 2.2L engine, there is a spacer with a gasket on either side.
5. Clean all mounting surfaces.
6. Using a new gasket or, on the 2.2L engine, two new gaskets, install the fuel pump. Installation is the reverse of removal.

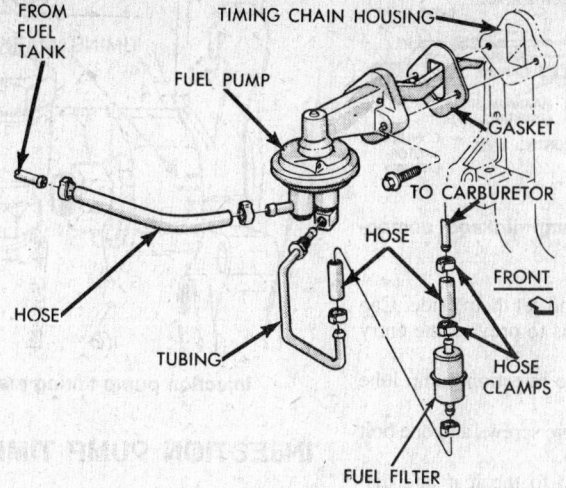

318 and 360 V8 engine fuel pump details

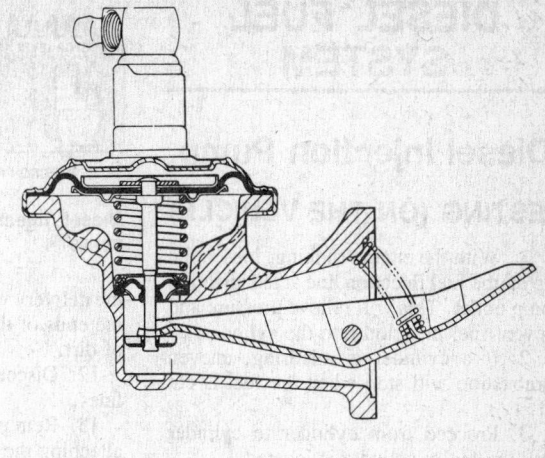

Fuel pump—6 cylinder

Fuel Tank

REMOVAL & INSTALLATION

Rear Drive Vans & Pickups

1. Disconnect the battery ground cable.

2. Remove the fuel tank filler cap.

3. Pump all fuel from the tank into an approved holding tank. Raise the vehicle on a lift.

4. Disconnect the fuel line and wire lead to the gauge unit. Remove the ground strap.

5. Remove the vent hose shield and the hose clamps from the hoses running to the vapor vent tube.

6. Remove the filler tube hose clamps and disconnect the hose from the tank.

7. Place a transmission jack under the center of the tank and apply sufficient pressure to support the tank.

8. Disconnect the two J-bolts and remove the retaining straps at the rear of the tank. Lower the tank from the vehicle. Feed the two vent tube hoses and filler tube vent hose through the grommets in the frame as the tank is being lowered. Remove the tank gauge unit.

9. Inspect the fuel filter, and if it is clogged or damaged, replace it.

10. Insert a new gasket in the recess of the fuel gauge opening and slide the gauge into the tank. Align the positioning tangs on the gauge with those on the tank. Install the lock ring, and tighten securely.

11. Position the tank on a jack and lift it into place, feeding the vent hoses through the grommets on the way up.

12. Connect the J-bolts and retaining straps, and tighten. Remove the jack.

13. Connect the filler tube and all vent hoses.

14. Connect the fuel supply line, ground strap, and gauge unit wire lead.

15. Refill the tank and inspect it for leaks. Connect the battery ground cable.

Front Wheel Drive Vans

1. Disconnect the battery ground cable.

2. Remove the fuel tank filler cap.

3. Disconnect the fuel supply hose and then pump all fuel from the tank into an approved holding tank via the metal supply tube at the right front shock tower.

4. Remove the screws attaching the filler tube to the inner and outer quarter panel. Raise the vehicle on a lift.

5. Disconnect the lines and wiring at the tank. Then, support the tank in a secure manner from underneath with a floor jack. Remove the bolts from the tank mounting straps.

6. Lower the tank slightly for access, and then work the filler tube out of the tank. Lower the tank farther and disconnect the vapor separator rollover valve hose. Remove the tank and insulator pad from the vehicle.

7. To install, first position the tank on the jack; then connect the vapor separator rollover valve hose and position the insulator pad on the tank.

───── **CAUTION** ─────

Check to make sure the vapor vent hose is securely attached to the tank. Then, watch as you install the tank to make sure the hose is not pinched between the tank and floorpan.

8. Raise the tank almost to its installed position and work the filler tube into the tank. Put the tank in installed position, install the strap bolts, and then torque them to 40 ft. lbs. Check to make sure the straps are not twisted before and after the bolts are installed. Remove the jack.

9. Connect the fuel lines and wiring.

10. If the van had a gasket between the filler tube and quarter panel, make sure to reinstall it. Then, install the mounting screws holding the tube to the quarter panel, and torque them to 17 in. lbs.

11. Reconnect the battery ground cable, fill the tank and install the filler cap. Check for leaks.

Ramcharger and Trail Duster

1. If there is a tank skid plate, remove it.

2. Disconnect the battery ground cable.

3. Remove the tank filler cap.

4. Pump or siphon the contents of the tank into a safe container.

5. Raise the vehicle on a hoist and disconnect the fuel line and tank sending unit wire. Remove the ground strap or wire.

6. Remove the hose clamps from the vent dome hose.

7. Remove the filler tube hose clamps. Detach the hoses from the tank.

8. Support the tank with a padded transmission jack.

9. Disconnect the two J-bolts and remove the straps at the rear of the tank.

10. Remove the tank gauge sending unit.

11. Use a new tank gauge sending unit gasket. Check the filter on the end of the fuel suction tube.

12. Use a new or undamaged tank to frame insulator. Raise the tank into position.

13. Connect the J-bolts and retaining straps and tighten.

14. Connect the filler tube and all hoses. Tighten the clamps.

15. Connect the fuel line, ground strap or wire, and tank sending unit wire. Make sure that all fuel line heat shields are in place.

16. Reconnect the battery ground cable and replace the skid plate.

DIESEL FUEL SYSTEM

Diesel Injection Pump

TESTING (ON THE VEHICLE)

1. With the engine running, loosen the cap on the fuel injection line at the injection pump outlet. This will relieve pressure and prevent fuel injection into the cylinder.
2. If a cylinder is misfiring, uneven combustion will stop when the fuel is cut off.
3. Proceed from cylinder to cylinder until the faulty cylinder is located.
4. Perform a compression test on the cylinder in question. If the cylinder in question meets the compression specifications, replace the injection pump.

REMOVAL & INSTALLATION

1. Disconnect the batteries.
2. Disconnect the fuel shutoff rod at the stop lever.
3. Remove the power steering pump and the mounting bracket from the engine and set it aside.
4. Thoroughly clean the area around the hose fittings and the injection pipes.
5. Drain the engine oil and remove the dipstick and the dipstick tube.
6. Disconnect the throttle cable and the linkage from the injection pump control lever.
7. Remove the throttle control bracket assembly from the crankcase, injection pump and the control bracket and set it aside.
8. Disconnect the fuel supply line to the fuel feed pump and set it aside, loosening the anchoring clamps as necessary.
9. Disconnect the fuel hoses leading to the filter from the feed pump and the injection pump. Replace the screws and seals to prevent dirt from getting into the pump.
10. Rotate the engine so piston No. 1 is approximately 7° before top dead center on the compression stroke.
11. Disconnect the injection pipes from

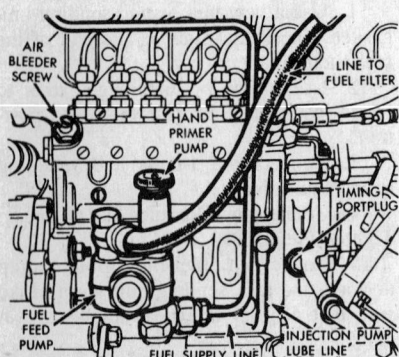

Diesel injection pump details

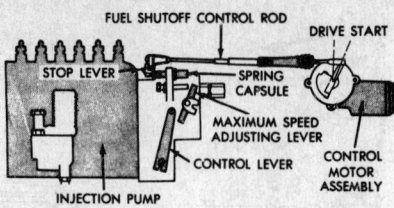

Diesel injection pump linkage components

the delivery valves and set them aside. Cap the ends of the valves to prevent the entry of dirt.

12. Disconnect the injection pump lube lines.
13. Remove the five screws and one bolt attaching the pump.
14. Pull the pump to the rear and disengage it from the front plate and timing gear case. Rotate the pump toward the crankcase and continue pulling it to the rear until the automatic timer is freed.
15. Loosen the four nuts attaching the pump to the mounting flange plate and align the center timing mark on the pump flange with the pointer on the plate. Tighten the four nuts.
16. Be sure that the O-ring is in place on the forward face of the pump mounting flange.
17. Remove the threaded timing port plug on the governor housing behind the control lever to expose the camshaft bushing timing mark. Rotate the pump drive gear to align the timing mark on the camshaft bushing with the pointer on the governor. The guide plate notch will be at approximately the 8 o'clock position as viewed from the front. Be sure the engine is positioned as described in Step 10 of the removal procedure.
18. Insert the automatic timer into the timing gear case and with the injection pump rotated against the crankshaft, rotate the pump driver gear to mesh the drive and idler gears. Do not force the pump into position.
19. Push the pump forward into the case. Rotate it away from the crankcase to align the attachment holes.
20. Attach the pump to the timing gear case.
21. Rotate the engine crankshaft in the opposite direction of normal operation until the crankshaft reaches the 18° before TDC mark on the crankshaft pulley. The governor pointer and the injection pump camshaft bushing timing marks should now be aligned. If they are not aligned, the pump has been installed incorrectly and must be removed and reinstalled.
22. Install the governor housing timing port plug and proceed with the pump installation by reversing the remainder of the removal procedure. Do not connect the No. 1 injecting pipe, fuel control rod or the batteries.
23. Bleed the air from the fuel filter and the injection pump by removing the air bleeder screws. Time the injection pump.

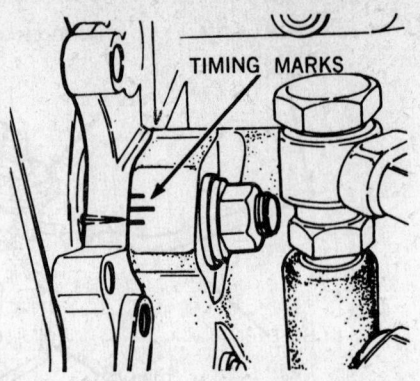

Injection pump timing marks

INJECTION PUMP TIMING

1. Disconnect the batteries and the fuel shut off rod at the stop lever.
2. Rotate the crankshaft in the direction of normal operation until No. 1 cylinder reaches top dead center of the compression stroke. This is done by aligning the lines on the crankshaft pulley rear face with the pointer on the bottom of the case.
3. Remove the forward oil filler cap on the rocker cover and check the No. 1 cylinder valves for looseness. If they are loose, you are at TDC.
4. Rotate the crankshaft in the normal direction of engine operation 1¾ turns.
5. Disconnect No. 1 injection pipe from the delivery valve holder.
6. Turn the crankshaft in the normal direction of engine operation in small steps. Stop when fuel begins to flow from the delivery valve holder. Injection begins at this point. The control lever must be in the idle position.
7. Read the injection timing point from the scale on the back of the crankshaft damper. If the timing is correct, the timing mark should be at the value shown on the Vehicle Emission Control Information label on the rocker cover minus 2 degrees.
8. If the timing point determined differs from the standard value, minus 2 degrees, loosen the four pump-to-flange plate nuts and rotate the pump (toward the crankcase to advance the timing, away from the crankcase to retard it) to correct the difference. Each mark on the injection pump timing scale represents 6 degrees. Tighten the flange plate nuts and repeat the timing procedure to be sure the timing is correct.

MANUAL TRANSMISSION/ TRANSAXLE

Refer to the Transmission or Transfer Case section of the Unit Repair for service procedures.

Manual Transaxle

1. Disconnect the battery cable and install a lifting eye on number 4 cylinder exhaust manifold bolt. Install a locally fabricated engine support fixture across the shock towers in the engine compartment.

2. Disconnect the gearshift linkage and clutch cable from the transaxle.

3. Remove the left front splash shield.

4. Remove the right and left drive shafts from the transaxle and support them. Do not let the drive axles hang free.

5. Remove the anti-rotational link from the transaxle and support the transmission with a suitable jack. Install a chain around the jack and transmission for safety.

6. Remove the speedometer adaptor and pinion from the transaxle.

7. Remove the engine mount from the front crossmember and the front mount insulator through bolt. Remove the upper bell housing bolts.

8. Remove the left engine mount and remove the lower bell housing bolts.

9. Pry the transaxle away from the engine and lower the assembly from the vehicle.

NOTE: When installing the transaxle use two bolts with the heads removed to act as locating guides. Place guides in the top two transaxle to engine block mounting location. When transaxle is in place, remove guides and install mounting bolts.

10. Installation is the reverse of removal. Refill the transaxle with Dexron® II or equivalent automatic transmission fluid. Adjust the clutch free play and transmission linkage.

11. Torque the engine mounting bolts to 40 ft. lbs. Torque the transmission to cylinder block bolts to 70 ft. lbs. Torque the hub nut to 180 ft. lbs. and torque the lug nuts to 80 ft. lbs.

Manual Transmission

3 Speed 2WD Models

1. Raise and safely support. Drain lubricant.

2. Disconnect and match-mark the driveshaft. On the sliding spline type, disconnect driveshaft at the rear universal joint, then carefully pull the shaft yoke out of the transmission extension housing. Do not nick or scratch the splines.

3. Disconnect gearshift control rods and speedometer cable.

4. Remove backup light switch if so equipped.

5. Support engine.

6. Remove crossmember and rubber insulator on models with A–390 transmission. On all other models, unbolt the insulator or mount from the crossmember. Support the transmission with a jack.

7. Remove transmission to clutch housing bolts.

8. Slide transmission rearward until pinion shaft clears clutch completely, then lower transmission from vehicle.

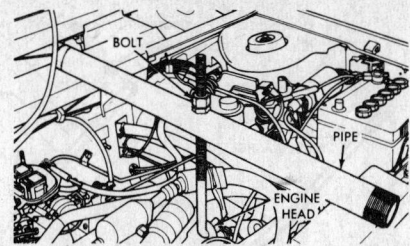

Fabricated engine support fixture

9. Installation is the reverse order of the above procedure. Before inserting transmission drive shaft into clutch, make sure clutch housing bore, disc and face are aligned. Tighten clutch housing to transmission bolts to 50 ft. lbs. torque.

10. Fill with lubricant.

11. Adjust shift linkage.

12. Road test.

4 Speed 2WD Models

1. Raise and support the truck. Shift transmission into any gear. Drain lubricant.

2. Disconnect universal joint and loosen yoke retaining nut.

3. Disconnect parking brake (if attached to transmission) and speedometer cables at transmission.

4. Remove lever retainer by pressing down, rotating retainer counterclockwise slightly, then releasing.

5. Remove lever and its springs and washers.

6. Support the rear of the engine and remove the crossmember. Remove transmission to clutch bell housing retaining bolts and pull transmission rearward until drive pinion clears clutch, then remove transmission.

7. To install, place ½ teaspoon of short fibre grease in pinion shaft pilot bushing, taking care not to get any grease on flywheel face.

8. Align clutch disc and backing plate with a spare drive pinion shaft or clutch aligning tool, then carefully install transmission.

9. Install transmission to bell housing bolts, tightening to 50 ft. lbs. torque. Replace the crossmember.

10. Install gear shift lever, shift into any gear and tighten yoke nut to 95–105 ft. lbs. torque.

11. Install universal joint, speedometer cable and brake cable.

12. Adjust clutch.

13. Install transmission drain plug and fill transmission with lubricant.

14. Road test.

3 & 4 Speed 4WD Models

1. Raise and support the truck. Drain lubricant.

2. Remove the skid plate, if any.

3. Disconnect the speedometer cable.

4. Disconnect and match-mark the front and rear driveshafts. Suspend each shaft from a convenient place; do not allow them to hang free.

5. Disconnect the shift rods at the transfer case. On 4-speed transmissions, remove the shift lever retainer by pressing down and turning it counterclockwise. Remove the shift lever springs and washers.

6. Remove the rear driveshaft. Matchmark the driveshaft and rear U-joints before removing the driveshaft.

7. Support the transfer case.

8. Remove the extension-to-transfer case mounting bolts.

9. Move the transfer case rearward to disengage the front input shaft spline.

10. Lower and remove the transfer case.

11. Disconnect the back-up light switch.

12. Support the engine.

13. Support the transmission.

14. Remove the transmission crossmember.

15. Remove the transmission-to-clutch housing bolts.

16. Slide the transmission rearward until the mainshaft clears the clutch disc.

17. Lower and remove the transmission.

18. Installation is the reverse of removal. The transmission pilot bushing in the end of the crankshaft requires high-temperature grease. Multipurpose grease should be used. Do not lubricate the end of the mainshaft, clutch splines, or clutch release levers. Adjust the gearshift linkage on 3-speed transmissions.

Overdrive 4 Four Speed

1. Disconnect the negative battery cable. Raise and support truck. Drain lubricant.

2. Remove the retaining screws from the floor pan boot and slide the boot up and off the shift lever.

3. Remove the shift lever, retaining clips, washers, and control rods from the shift unit.

4. Remove the two bolts and washers which secure the shift unit to mounting plate on the extension housing and remove the unit.

5. Drain the fluid from the transmission.

6. Disconnect the propeller shaft at the rear universal joint, marking the parts for re-installation in the same position. Carefully pull the shaft yoke out of the transmission extension housing.

7. Disconnect the speedometer cable and backup light switch leads.

8. Install engine support fixture C–3487–A or equivalent. Be sure that the support points are tight against the oil pan flange.

9. Raise engine slightly with the support fixture. Disconnect the extension housing from the center crossmember.

10. Support the transmission with a suitable jack and remove the center crossmember.

11. Remove the transmission to clutch housing bolts. Slide the transmission toward the rear until drive pinion shaft clears

the clutch disc, before lowering the transmission. Remove the transmission.

12. Installation is the reverse of removal. Use high temperature multi-purpose grease on the pilot bushing in the end of the crankshaft, around the inner end of the pinion shaft pilot bushing in the flywheel and on the pinion bearing retainer release bearing area. Do not lubricate the end of the pinion shaft, clutch disc splines or clutch release levers.

13. Torque the clutch housing bolts to 50 ft. lbs. Tighten the crossmember bolts to 30 ft. lbs. Torque engine and transmission mounts to 50 ft. lbs.

14. Fill the transmission with 7.5 pints of Dexron II® (or equivalent) automatic transmission fluid.

15. Road test the vehicle.

Transfer Case

REMOVAL & INSTALLATION

1. Raise and support the truck.
2. Remove the skid plate, if any.
3. Drain the transfer case by removing the bottom bolt from the front output rear cover.
4. Disconnect the speedometer cable.
5. Disconnect the front and rear output shafts. Suspend these from a convenient location; do not allow them to hang free.
6. Disconnect the shift rods at the transfer case.
7. Support the transfer case.
8. Remove the adaptor-to-transfer case mounting bolts and move the transfer case rearward to disengage the front input splines.
9. Lower and remove the transfer case.
10. Installation is the reverse of removal. Adjust the linkage.

Shift Linkage

ADJUSTMENT

Vans & Pickups

A230, A250

1. Adjust the length of the 2–3 shift rod so the position of the shift lever on the steering column will be correct.
2. Assemble the 1st-reverse and 2–3 shift rods, and place each in its normal position, secured with a clip. Loosen both swivel clamp bolts.
3. Move the 2–3 shift lever into 3rd position (this means moving the forward lever forward). Move the steering column lever until it is about five degrees above the horizontal. Tighten the shift rod swivel clamp bolt.
4. Shift the transmission to neutral. Place a suitable tool between the crossover blade and the 2–3 lever at the steering column so that both lever pins are engaged by the crossover blade.
5. Set the 1st-reverse lever in neutral.

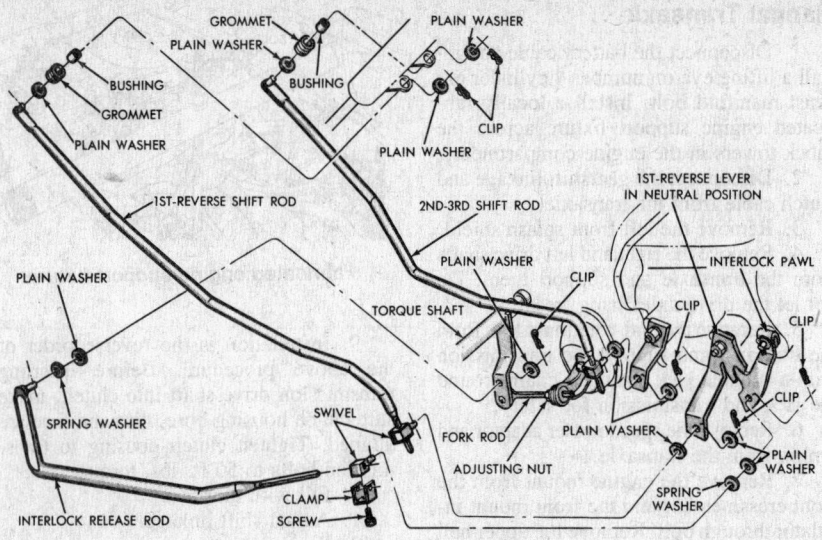

Shift linkage A250 transmission

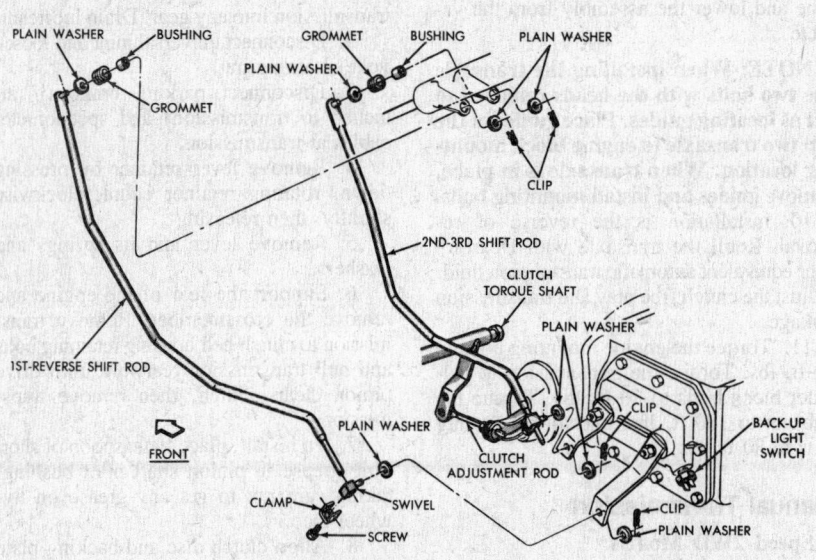

Shift linkage A230 transmission

Tighten the swivel clamp bolt.

6. Remove the tool from the cross over blade, and check all shifts for smoothness.

Ramcharger & Trail Duster

A230

1. Remove both shift rod swivels from the transmission shift levers. Make sure that the transmission shift levers are in the neutral position (middle detent).
2. Move the shift lever to line up the locating slots in the bottom of the steering column shift housing and bearing housing.
3. Place a suitable tool between the crossover blade and the 2nd and 3rd lever at the steering column, so that both lever pins are engaged by the crossover blade.
4. Set the 1st-reverse lever on the

transmission to the reverse position (rotate clockwise).

5. Adjust the 1st-reverse rod swivel by loosening the clamp bolt and sliding the swivel along the rod so it will enter the 1st-reverse lever at the transmission. Install the washers and the clip. Tighten the swivel bolt.

6. Remove the gearshift housing locating tool, and shift the transmission into the neutral position.

7. Adjust the 2nd–3rd rod swivel by loosening the clamp bolt and sliding the swivel along the rod so it will enter the 2nd–3rd lever at the transmission. Install the washers and the clip. Tighten the swivel bolt.

8. Remove the tool from the cross-over blade at the steering column and shift the

transmission through all the gears to check the adjustment and the crossover smoothness.

All Models

A–390

1. Loosen both shift rod swivels. Make sure that the transmission shift levers are in the neutral position (middle detent).

2. Move the shift lever to line up the locating slots in the bottom of the steering column shift housing and bearing housing. Install a suitable tool in the slot.

3. Place a suitable tool between the crossover blade and the 2nd–3rd lever at the steering column so that both lever pins are engaged by the crossover blade.

4. Tighten both rod swivel bolts. Remove the gearshift housing locating tool.

5. Remove the tool from the crossover blade at the steering column and shift the transmission through all gears to check adjustment and crossover smoothness.

6. Check for proper operation of the steering column lock in reverse. With the proper linkage adjustment, the ignition should lock in reverse only, with hands off the gearshift lever.

Overdrive 4 Speed

1. Place the floorshift lever in Neutral. Insert a ¼ in. drill bit through the bottom of the shifter to hold the levers in place.

2. Detach the shift rods. Make sure that the three transmission levers are in their Neutral detents.

3. Adjust the shift rods to make the length exactly right to fit into the transmission levers. Start with the 1st–2nd shift rod. It may be necessary to remove the clip at the shifter end of the rod to rotate this rod.

4. Replace the washers and the clips.

5. Remove the drill bit and check the shifting action.

Transaxle

A double ended pin that is used to lock the linkage in place prior to adjustment.

1. Remove the screw from the top and reinsert the other end, locking the linkage in place.

2. The linkage is locked in the Neutral detent between 1st and 2nd gears.

3. Align the marks on the linkage.

4. Remove the pin and replace it in its original location. Check the operation of the shift linkage.

Clutch Interlock

ADJUSTMENT

A–250 3 Speed

This adjustment is required only on the A–250 3 speed transmission. This is a top cover unit used only as base equipment on

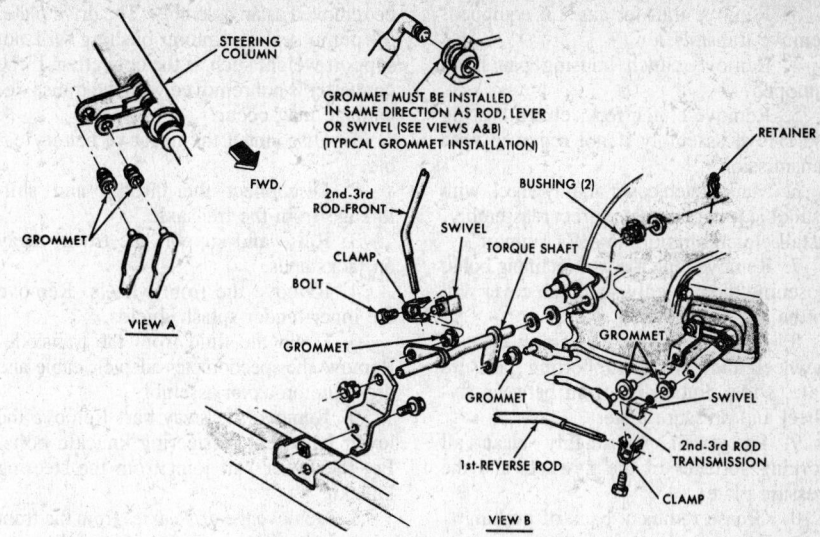

Shift linkage A390 transmission

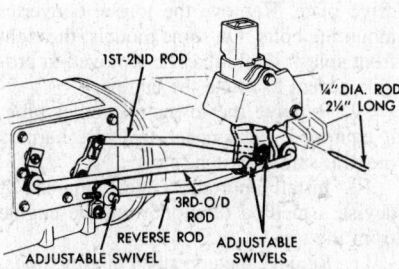

Shift linkage Overdrive 4 adjustment

light duty six cylinder models. It has synchromesh only on second and third gears.

1. Disconnect the clutch rod swivel from the interlock pawl. Adjust the clutch pedal free play.

2. Shift the transmission to neutral. Loosen swivel clamp bolt and slide the swivel onto the rod until the pawl is positioned fully within the slot in the first-reverse lever. Install the washers and clip.

3. Hold the interlock pawl forward and tighten the swivel clamp bolt. The clutch pedal must be in fully returned position during the adjustment.

NOTE: Do not pull the clutch rod rearward to engage the swivel in the pawl.

4. Shift the transmission into first and reverse and release the clutch pedal while in either gear to check for normal clutch action. Then, shift halfway between neutral and either gear and release clutch. The interlock should hold it to within one or two inches of the floor.

CLUTCH

ADJUSTMENT

4 Cylinder Engines

Models with the A460, A465 and A525

transaxles are equipped with self-adjusting clutches. The clutch cable cannot be adjusted.

6 & 8 Cylinder Engines

The only adjustment required is pedal free-play. Adjust the clutch actuating fork rod by turning the self-locking adjusting nut to provide ⅛ in. (³⁄₃₂ in. on vans and wagons) free movement at the end of the fork. This will provide the recommended 1½ in. (1 in. on Vans, Wagons, Ramcharger and Trailduster) freeplay at the pedal.

REMOVAL & INSTALLATION

4 Cylinder Engines

1. Remove the transaxle.

2. Mark the clutch cover and flywheel for reinstallation position if reused.

3. Install a clutch aligning tool for support purposes.

4. Loosen the attaching bolts one or two turns at a time in succession to avoid bending the cover flange.

5. Remove the pressure plate and clutch disc.

6. Service the throwout bearing and fork if necessary.

7. Inspect for oil leakage from the rear main oil seal. Service if necessary. Examine the flywheel surface, replace the flywheel if badly scored.

8. Position the clutch disc and pressure plate assembly on the flywheel. Install clutch aligning tool and install the clutch in the reverse order of removal. Tighten the mounting bolts in stages. Torque bolts to 250 in. lbs.

6 & 8 Cylinder Engines

1. Raise and support truck. Support the engine on a suitable jack.

2. Remove crossmember.

3. Remove transfer case, if equipped. Remove transmission.

4. Remove clutch housing pan if so equipped.

5. Remove clutch fork, clutch bearing and sleeve assembly if not removed with transmission.

6. Mark clutch cover and flywheel, with a suitable tool to assure correct reassembly. Install clutch aligning tool for support.

7. Remove clutch cover retaining bolts, loosening them evenly so clutch cover will not be distorted.

8. Pull pressure plate assembly clear of flywheel and, while supporting pressure plate, slide clutch disc from between flywheel and pressure plate.

9. To install, thoroughly clean all working surfaces of the flywheel and the pressure plate.

10. Grease radius at back of bushing.

11. Rotate clutch cover and pressure plate assembly for maximum clearance between flywheel and frame crossmember if crossmember was not removed during clutch removal.

12. Tilt top edge of clutch cover and pressure plate assembly back and move it up into the clutch housing. Support clutch cover and pressure plate assembly and slide clutch disc into position.

13. Position clutch disc and plate against flywheel and insert spare transmission main drive gear shaft or clutch installing tool through clutch disc hub and into main drive pilot bearing.

14. Rotate clutch cover until the punch marks on cover and flywheel line up.

15. Bolt cover loosely to flywheel. Tighten cover bolts a few turns at a time, in progression, until tight. Then tighten bolts to 20 ft. lbs. torque.

16. Install transmission.

17. Install frame crossmembers and insulator, tighten all bolts.

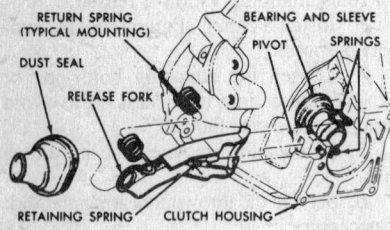

Clutch release fork, bearing and sleeve

AUTOMATIC TRANSAXLE

REMOVAL & INSTALLATION

The automatic transaxle can be removed by itself with the engine still mounted in the truck, but the transaxle and converter must be removed as an assembly. The drive plate, oil pump seal and pump bushing will not support weight such as the converter. If the converter is not removed with the transaxle, damage may occur.

1. Disconnect the negative battery cable.

2. Disconnect the throttle and shift linkage from the transaxle.

3. Raise and support the truck safely on jackstands.

4. Remove the front wheels. Remove the inner fender splash shields.

5. Drain the fluid from the transaxle. Remove the speedometer adapter, cable and drive pinion as an assembly.

6. Remove the sway bar. Remove the lower ball joint to steering knuckle bolts. Pry the lower ball joint from the steering knuckle.

7. Remove the drive axles from the front wheel hubs. See shaft removal in following section.

8. Matchmark the torque converter and drive plate. Remove the torque converter mounting bolts. On some models, the right front splash shield may be removed to provide access to rotate the engine.

9. Remove the lower oil cooler tube, if equipped. Disconnect any wire harness (neutral safety switch, etc.)

10. Install a portable engine support, or devise a method of supporting the engine from the top.

11. Remove the upper bell housing bolts. Remove the engine mount bracket from the front crossmember. Support the transmission.

12. Remove the front mount insulator ''through'' bolts and the bell housing mount.

13. Remove the long ''through'' bolt from the left hand engine mount.

14. Raise the transaxle, remove any remaining mounting bolts and separate from the engine. Slowly lower the transaxle to the ground.

15. Installation is in the reverse order of removal. Fill the transaxle with the correct amount of Dexron® II. Check for leaks.

NOTE: Pan gaskets are formed with RTV sealant, follow the directions on the tube when applying.

Shift Linkage

ADJUSTMENT

NOTE: When it is necessary to disconnect the linkage cable from the lever, which uses plastic grommets as retainers, the grommets should be replaced.

1. Make sure that the adjustable swivel block is free to slide on the shift cable.

2. Place the shift lever in Park.

3. With the linkage assembled, and the swivel lock bolt loose, move the shift on the transaxle all the way to the rear detent.

4. Tighten the adjuster swivel lock bolt to 8 ft. lb.

5. Check the linkage action.

NOTE: The automatic transmission gear selector release button may pop up in the knob when shifting from PARK to DRIVE. This is caused by inadequate retention of the selector release knob retaining tab. The release button will always work but the loose button can be annoying. A sleeve and washers are available to cure this condition. If these are unavailable, do the following:

1. Remove the release button.

2. Cut and fold a standard paper match stem as shown.

3. Using tweezers, insert the folded match as far as possible into the clearance slot as shown. The match should be below the knob surface.

4. Insert the button, taking care not to break the button stem.

Throttle Cable

ADJUSTMENT

1. Adjust the idle speed as previously described.

2. Run the engine to normal operating temperature.

3. Loosen the adjustment bracket lock screw.

4. Make sure the adjustment bracket is free to slide in its slot.

5. Hold the transmission lever firmly rearward against its internal stop and tighten the adjustment bracket lock screw to 105 in. lbs.

6. Test the cable operation.

Band

ADJUSTMENTS

Front (Kickdown) Band

Chrysler recommends that the band be adjusted at each fluid change. The adjustment screw is located on the left side of the case.

1. Loosen the locknut and back off the nut about five full turns.

2. Tighten the band adjusting screw to 72 in. lbs.

3. Back off the adjusting screw exactly 2.5 turns.

4. Hold the adjusting screw and tighten the locknut to 35 ft. lbs.

Rear (Low and Reverse) Band

1. Drain fluid and remove transaxle oil pan.

2. Loosen and back off lock nut approximately 5 turns. Using an inch-pound torque wrench, tighten adjusting screw to 41 in. lbs.

3. Back off adjusting screw 3.5 turns.

4. Tighten lock nut to 10 ft. lbs.

5. Reinstall oil pan and fill with proper fluid.

Neutral Start Switch

The neutral start circuit is the center contact of the three-terminal switch located in the transmission case.

1. Remove the wiring connector and test for continuity between the center pin and the case. Continuity should exist only in Park and Neutral.

2. Remove the switch and check that the operating lever fingers are centered in the switch opening.

3. Install the switch and a new seal and tighten to 24 ft. lbs. Retest with a lamp.

4. Replace the lost transmission fluid.

5. If shift linkage adjustment is correct and the switch still malfunctions, replace the switch.

Pan, Fluid and Filter

REMOVAL & INSTALLATION

NOTE: RTV silicone sealer is used in place of a pan gasket.

Chrysler recommends no fluid or filter changes during the normal service life. Severe usage requires a fluid and filter change every 15,000 miles. Severe usage is defined as:

1. More than 50% heavy city traffic during 90°F weather.

2. Commercial operation or trailer towing.

NOTE: When changing the fluid, only Dexron® II fluid should be used. A filter change should be performed at every fluid change.

1. Raise the vehicle and support it on jackstands.

2. Place a large container under the pan, loosen the pan bolts and tap at one corner to break it loose. Drain the fluid.

3. When the fluid is drained remove the pan bolts.

4. Remove the retaining screws and replace the filter. Tighten the screws to 35 in. lbs.

5. Clean the fluid pan, peel off the old RTV silicone sealer and install the pan, using a ⅛ inch bead of new RTV sealer. Always run the sealer bead inside the bolt holes. Tighten the pan bolts to 165 in. lbs.

6. Pour four quarts of Dexron® II fluid through the filler tube.

7. Start the engine and idle it for at least 2 minutes. Set the parking brake and move the selector through each position, ending in Park.

8. Add sufficient fluid to bring the level to the FULL mark on the dipstick. The level should be checked in Park, with the engine idling at normal operating temperatures.

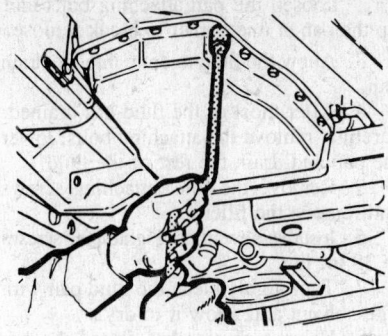
Removing the transmission oil pan bolts

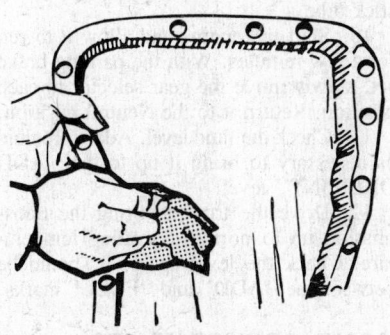
Clean pan with safe solvent and a rag

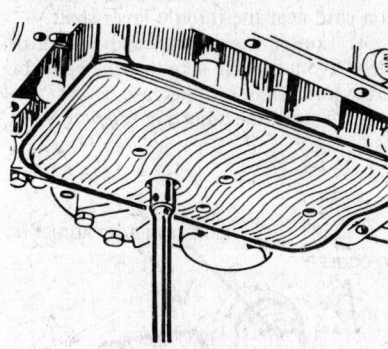
Remove filter attaching screws

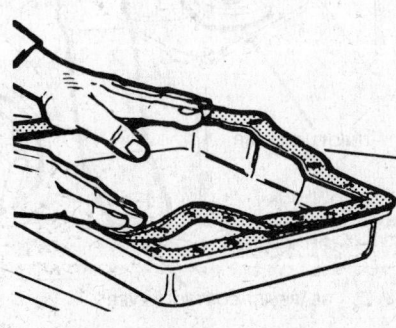
Install a new gasket

AUTOMATIC TRANSMISSION

REMOVAL & INSTALLATION

1. Remove the transmission and converter as an assembly; otherwise the converter drive plate pump bushing, and oil seal will be damaged. The drive plate will not support a load. Therefore, none of the weight of transmission should be allowed to rest on the plate during removal. Remove the transfer case, as necessary.

2. Attach a remote control starter switch to the starter solenoid so the engine can be rotated from under the vehicle.

3. Disconnect high tension cable from the ignition coil.

4. Remove splash shields.

5. Drain the transmission fluid.

6. Mark converter and drive plate to aid in reassembly.

7. Rotate the engine with the remote control switch to locate two converter to drive plate bolts at 5 and 7 o'clock positions. Remove the two bolts, rotate engine again and remove the other two bolts.

— **CAUTION** —

Do not rotate converter on drive plate by prying with a screwdriver or similar tool as drive plate might become distorted. Also the starter should never be engaged if drive plate is not attached to converter with at least one bolt or if transmission case to engine block bolts have been loosened.

8. Disconnect battery ground cable. Remove engine to transmission struts, if necessary. You may have to drop the exhaust system on some models.

9. Remove the starter.

10. Remove wire from the neutral starting switch.

11. Remove gearshift cable or rod from the transmission and the lever.

12. Disconnect the throttle rod from left side of transmission.

13. Disconnect the oil cooler lines at transmission and remove the oil filler tube. Disconnect the speedometer cable.

14. Disconnect the driveshaft.

15. Install engine support fixture to hold up the rear of the engine.

16. Raise transmission slightly with jack to relieve load and remove support bracket or crossmember. Remove all bell housing bolts and carefully work transmission and converter rearward off engine dowels and disengage converter hub from end of crankshaft.

— **CAUTION** —

Attach a small C-clamp to edge of bell housing to hold converter in place during transmission removal; otherwise the front pump bushing might be damaged.

NOTE: Install transmission and converter as an assembly. The drive plate will not support a load. Do not allow weight of transmission to rest on the plate during installation.

17. Rotate pump rotors until the rotor lugs are vertical.

18. Carefully slide converter assembly over input shaft and reaction shaft. Make sure converter impeller shaft slots are also vertical and fully engage front pump inner rotor lugs.

19. Use a "C" clamp on edge of converter housing to hold converter in place during transmission installation.

20. Converter drive plate should be free of distortion and drive plate to crankshaft bolts tightened to 55 ft. lbs. torque.

21. Using a jack, position transmission and converter assembly in alignment with engine.

22. Rotate converter so mark on converter (made during removal) will align with mark on drive plate. The offset holes in plate are located next to the ⅛ hole in inner circle of the plate. A stamped "V" mark identifies the offset hole in converter front cover. Carefully work transmission assembly forward over engine block dowels with converter hub entering the crankshaft opening.

23. Install converter housing to engine bolts and tighten to 28 ft. lbs.

24. Install the two lower drive plate to converter bolts and tighten to 270 in. lbs. torque.

25. Install engine to transmission struts, if required. Install starting motor and connect battery ground cable.

26. Rotate engine and install two remaining drive plate to converter bolts.

27. Install crossmember and tighten attaching bolts to 90 ft. lbs. torque. Lower transmission so that extension housing is aligned and rests on the rear mount. Install bolts and tighten to 40 ft. lbs. torque.

28. Remove transmission jack and engine support fixture, then install tie-bars under the transmission.

29. Replace the driveshaft.

30. Connect oil cooler lines, install oil filler tube and connect the speedometer cable.

31. Connect gearshift cable or rod and torqueshaft assembly to the transmission case and to the lever.

32. Connect throttle rod to the lever at left side of transmission bell housing.

33. Connect wire to back-up and neutral starting switch.

34. Install cover plate in front of the converter assembly.

35. Refill transmission with fluid.

36. Adjust throttle and shift linkage.

PAN AND FILTER SERVICE

1. Raise the front of the truck and support it on jackstands. Place a large drain pan under the transmission.

2. Loosen the pan attaching bolts and tap the pan at one corner to break it loose.

3. Allow the fluid to drain into the drain pan.

4. After most of the fluid has drained, carefully remove the attaching bolts, lower the pan and drain the rest of the fluid.

5. Remove the filter attaching screws and remove the filter.

6. Install a new filter. Tighten the screws to 35 in. lbs.

7. Thoroughly clean the fluid pan with safe solvent and allow it to dry.

8. Using a new gasket, install the pan to the transmission. Tighten the attaching bolts to 150 in. lbs.

9. Pour four quarts of Dexron® automatic transmission fluid in through the dipstick tube.

10. Start the engine and allow it to run for a few minutes. With the parking brake set, slowly move the gear selector to each position. Return it to the Neutral position.

11. Check the fluid level. Add more fluid as necessary to bring it up to the "ADD ONE PINT" level.

12. Drive the truck to bring the transmission up to normal operating temperature. Check the level again. It should be between the "ADD" and "FULL" marks.

BAND ADJUSTMENTS

Kickdown Band

The kickdown band adjusting screw is located on the left-hand side of the transmission case near the throttle lever shaft.

1. Loosen the locknut and back it off about five turns. Be sure that the adjusting screw turns freely in the case.

2. Torque the adjusting screw to 72 in. lbs.

3. Back off the adjusting screw as follows: V8: 2½ turns, 6 cyl: 2 turns, Diesel: 2 turns. Tighten the locknut to 30 ft. lb.

Low and Reverse Band

The pan must be removed from the transmission to gain access to the low and reverse band adjusting screw.

1. Remove the skid plate, if any. Drain the transmission fluid and remove the pan.

2. Loosen the band adjusting screw locknut and back if off about five turns. Be sure that the adjusting screw turns freely in the lever.

3. Torque the adjusting screw to 72 in. lbs.

4. Back off the adjusting screw: All LoadFlite through 1980; 2 turns. 1981 and later A904T and A999 (wide ratio); 4 turns. 1981 and later A727; 2 turns. Keep the adjusting screw from turning, tighten and torque the locknut to 30 ft. lbs.

5. Use a new gasket and install the transmission pan. Torque the pan bolts to 150 in. lbs. Refill the transmission with Dexron® II fluid.

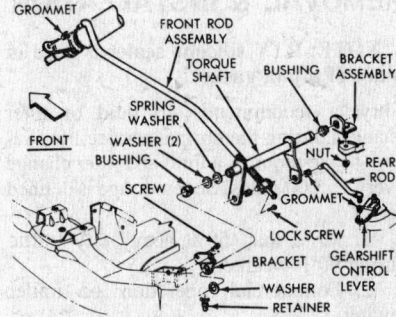

1979 and later gearshift linkage—typical

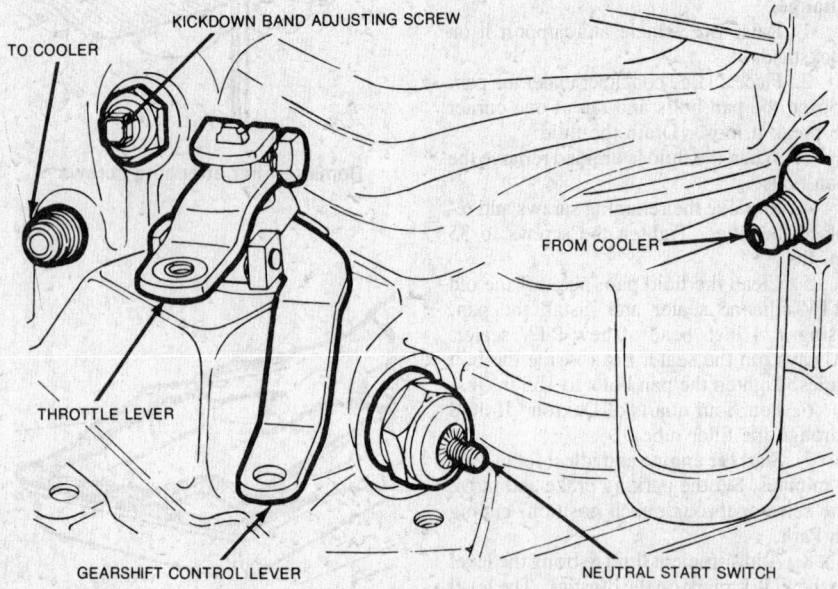

Loadflite adjusting points

SHIFT LINKAGE ADJUSTMENT

NOTE: To insure proper adjustment, it is suggested that new linkage grommets be installed.

1. Place the gearshift lever in the Park position.

2. Move the shift control lever on the transmission all the way to the rear (in the Park detent).

3. Set the adjustable rod to the proper length and install it with no load in either direction. Tighten the swivel bolt.

4. The shift linkage must be free of binding and be positive in all positions. Make sure that the engine can start only when the gearshift lever is in the Park or Neutral position. Be sure that the gearshift lever will not jump into an unwanted gear.

THROTTLE KICKDOWN ROD ADJUSTMENT

1. Warm the engine up to normal operating temperature. Turn the engine Off.

2. Block the choke plate fully open.

3. Raise and safely support the vehicle.

4. Loosen the adjusting swivel lock screw.

5. To insure the proper alignment, the swivel must be free to slide along the flat end of the throttle rod so the preload spring action is not restricted. Clean the parts if necessary.

6. Hold the transmission lever firmly forward against the internal stop and tighten the swivel lock screw to 100 in. lbs. Linkage backlash will automatically be taken up by the preload spring.

NEUTRAL START SWITCH ADJUSTMENT

The neutral safety switch is thread mounted into the transmission case. When the gearshift lever is placed in either the Park or Neutral position, a cam, which is attached to the transmission throttle lever inside the transmission, contacts the neutral safety switch and provides a ground to complete the starter solenoid circuit.

The back-up light switch is incorporated into the neutral safety switch. The center terminal is for the neutral safety switch and the two outer terminals are for the backup lamps.

There is no adjustment for the switch. If a malfunction occurs, the switch must be removed and replaced.

To remove the switch, disconnect the electrical leads and unscrew the switch. Use a drain pan to catch the transmission fluid. Using a new seal, install the new switch and torque it to 24 ft. lbs. Refill the transmission.

DRIVESHAFT & U-JOINTS

REMOVAL & INSTALLATION

Rear Wheel Drive (Single Section Type)

This driveshaft has a universal joint at either end and no external supports.

1. Raise and support the truck with the rear higher.

2. Matchmark the shaft and pinion flange to assure proper balance at installation.

3. Remove both rear U-joint roller and bushing clamps from the rear axle pinion flange. Do not disturb the retaining strap which holds the bushing assemblies on the U-joint cross.

NOTE: Do not allow the driveshaft to hang during removal. Suspend it from the frame with a piece of wire. Before removing the driveshaft, raise the rear end of the truck to prevent loss of transmission fluid.

4. Slide the driveshaft, with the front sliding yoke, off the transmission output shaft.

5. Installation is the reverse of removal. Align the matchmarks made during removal.

Two-Section Type

This driveshaft has a universal joint at either end, with a third universal joint and a support bearing at the center.

1. Matchmark the shaft and the rear axle pinion hub yoke. Matchmark the center bearing spline and slip yoke.

NOTE: Do not allow the driveshaft to hang down during removal. Suspend it from the frame. Raise the rear of the truck to prevent loss of transmission fluid.

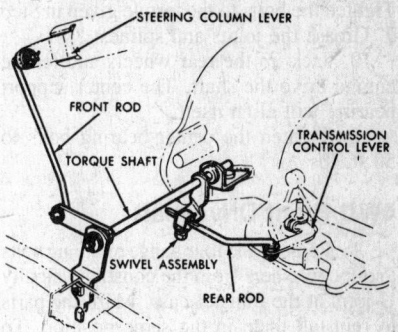

Typical gearshift linkage—through 1978

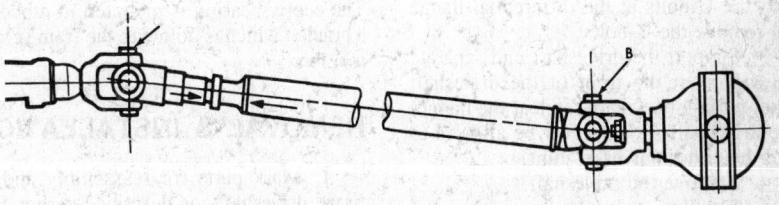

Single shaft-yoke alignment

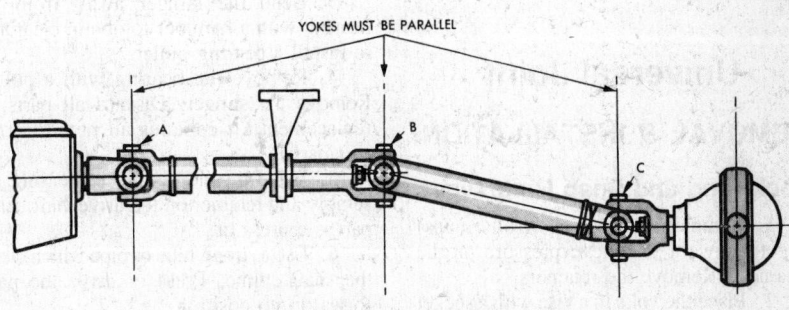

Two-piece shaft-yoke vertical alignment

LOW AND REVERSE BAND ADJUSTING SCREW

FILTER

BACK-UP LIGHT AND NEUTRAL START SWITCH

Low and reverse band adjusting screw location

2. Remove both rear U-joint roller and bushing assembly clamps from the rear axle pinion yoke. Do not disturb the retaining strap used to hold the bushing assemblies on the U-joint cross.

3. Slide the rear half of the shaft off the front shaft splines at the center bearing. Remove the rear half.

4. At the transmission end of the front half, remove the bushing retaining bolts and clamps, after matchmarking. If there is a driveshaft brake, there will be flange nuts.

5. Unbolt the center bearing mounting nuts and bolts and remove the front half of the shaft.

6. On installation, align the matchmarks at the transmission and start all the bolts and nuts at the front U-joint and the center support bearing.

7. Tighten $\frac{1}{4}$ in. clamp bolts to 170 in. lbs, and $\frac{5}{16}$ in. bolts to 300 in. lbs. Tighten driveshaft brake flange nuts to 35 ft. lbs. Leave the center bearing bolts just snug.

8. Align the rear shaft matchmarks and slide the yoke onto the front shaft splines.

9. Align the rear U-joint matchmarks and install the bushing clamps and bolts. Tighten the bolts to the torque given in Step 7. Grease the joints and splines.

10. Jack up the rear wheels and let the engine drive the shaft. The center support bearing will align itself.

11. Tighten the center bearing bolts to 50 ft. lbs.

4WD Front Driveshaft

1. Remove the four flange retaining bolts and lockwashers from the constant velocity U-joint at the transfer case. Mark the parts to reinstall them in the same position. To prevent the constant velocity joint from turning while removing the nuts, use a press bar.

2. Remove the nuts and lockwashers from the U-bolts at the differential flange and remove the U-bolts.

3. Support the driveshaft and separate the U-joint at the front of the driveshaft yoke, pulling backward to clear the flange. The driveshaft should never be allowed to hang by either universal joint.

4. Remove the driveshaft.

5. Installation is the reverse of removal.

Universal Joint

REMOVAL & INSTALLATION

Lock Ring and Snap Ring Types

1. Hammer the bushings (roller cups) slightly inward to relieve pressure on the retainers. Remove the retainers.

2. Place the yoke in a vise with a socket bigger than the bushing on one side and one smaller than the bushing on the other side.

3. Apply pressure, forcing one bushing out into the larger socket.

4. Reverse the vise and socket arrangement to remove the other bushing and the cross.

5. On installation, press the new bushings in just far enough to install the retainers.

Strap Clamp Type (Rear Axle Yoke)

Unbolt strap bolts and remove straps, bushings, seals and washer retainers. Install new components as required. When assembling, grease bearings. Install with grease fitting parallel to other fittings in drive train. Tighten strap bolts to 20 ft. lbs. torque.

Constant Velocity U-Joint

This is the double universal joint used in the front driveshaft on four-wheel drive models. These are disassembled in the same way as the snap-ring type U-Joint. Original equipment U-joints are held together by plastic retainers which shear when pressed out. The bearing cups in the center part of the joint should be pressed out before those in the yoke. Original equipment constant velocity joints cannot be reassembled. Replacement part kits have bearing cups with grooves for retaining rings.

Slip Joints

When reassembling slip joints make sure that arrows stamped on each side are matched. This will assure proper universal joint alignment.

Center Bearing

When two or more driveshafts are used in tandem, a rubber mounted center bearing supports the center portion of the drive line. The center bearing is mounted in rubber in a bracket which is bolted to the frame crossmember.

REMOVAL & INSTALLATION

1. Mark parts for reassembly and remove driveshafts as described earlier.

2. Place the front shaft in a vise and pull the bearing support and insulator away from the bearing.

3. Bend the slinger away from the bearing with a hammer to obtain clearance to install a bearing puller.

4. Remove the bearing with a puller. Remove the slinger. Discard all parts. A replacement kit contains all necessary repair parts.

5. Place the new slinger, bearing assembly and retainer on the driveshaft. Each part is a press fit.

6. Use a strong tube or pipe which clears the shaft spline. Press or drive the parts forward into position.

7. Connect the two piece driveshaft and install in the reverse order of removal.

4WD FRONT DRIVE AXLE

Warn Front Locking Hubs

REMOVAL & INSTALLATION

1. Straighten the lock tabs and remove the six hub mounting bolts.

2. Tap the hub gently with a mallet to remove.

3. Separate the clutch assembly from the body assembly.

4. Remove the snap ring from the rear of the body assembly, using snap-ring pliers. Slip the axle shaft hub out of the body from the front.

5. Remove the Allen screw from the inner side of the clutch, and remove the bronze dial assembly from the front side of the clutch housing assembly.

6. Remove the clutch assembly from the rear of the housing, complete with the twelve roller pins.

7. Coat the moving parts with a waterproof grease.

8. Slide the axle shaft hub into the body from the front, and replace the snap-ring.

9. Replace the bronze dial assembly and the inner clutch. Tighten the Allen screw and stake the edge of the screw with a center punch to prevent loosening.

10. With the dial in the free position, rotate the outer clutch body into the inner assembly until it bottoms in the housing. Back it up to the nearest hole and install the roller pins.

11. Position the hub and clutch assembly together with a new gasket in between.

12. Position the hub assembly over the end of the axle and replace the six hub mounting bolts and lock tabs.

13. Torque the bolts to 35 ft. lbs. and bend the lock tabs to anchor the bolts.

14. Verify the operation by road testing.

Dana and "Dualmatic" Front Locking Hubs

REMOVAL & INSTALLATION

1. Place hub in lock position. Remove Allen head mounting bolts and washers.

2. Carefully remove retainer, O-ring seal and knob. Separate knob from retainer.

3. Remove large internal snap-ring. Slide retainer ring and cam from hub.

4. While pressing against sleeve and ring assembly, remove axle shaft snap-ring. Relieve pressure and remove sleeve and ring, ring and bushing, spring and plate.

5. Inspect all parts for wear, nicks and burrs. Replace all parts which appear questionable.

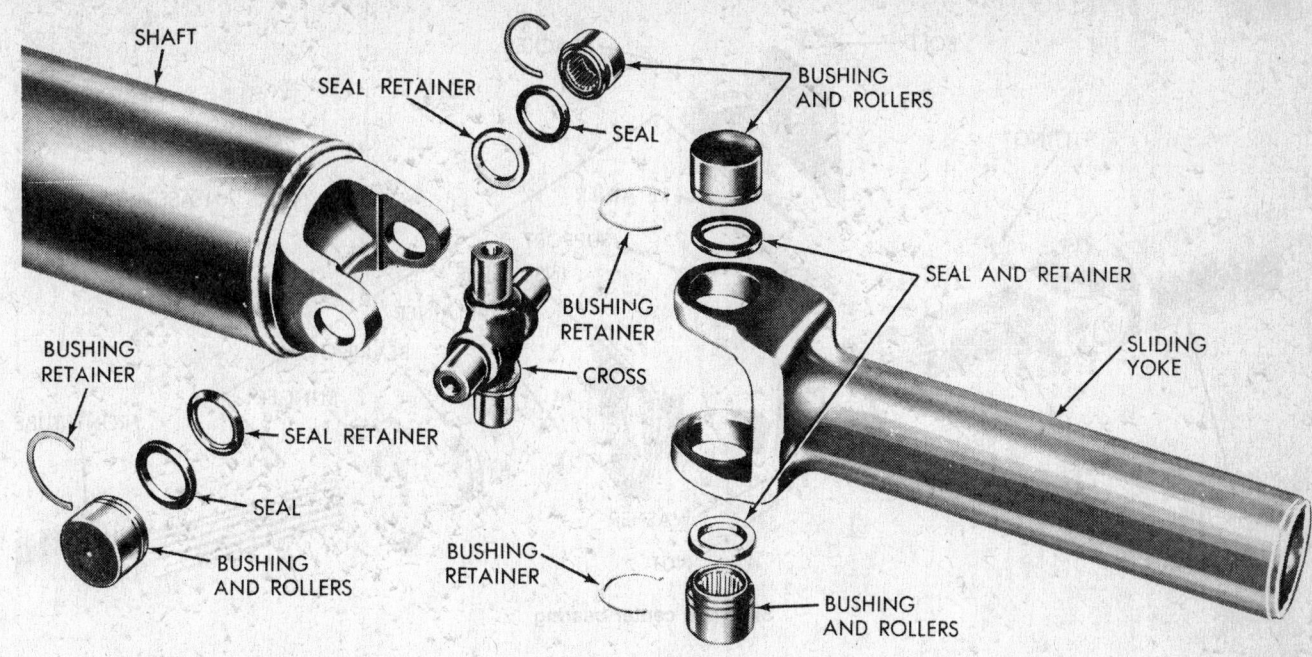

Sliding yoke type universal joint

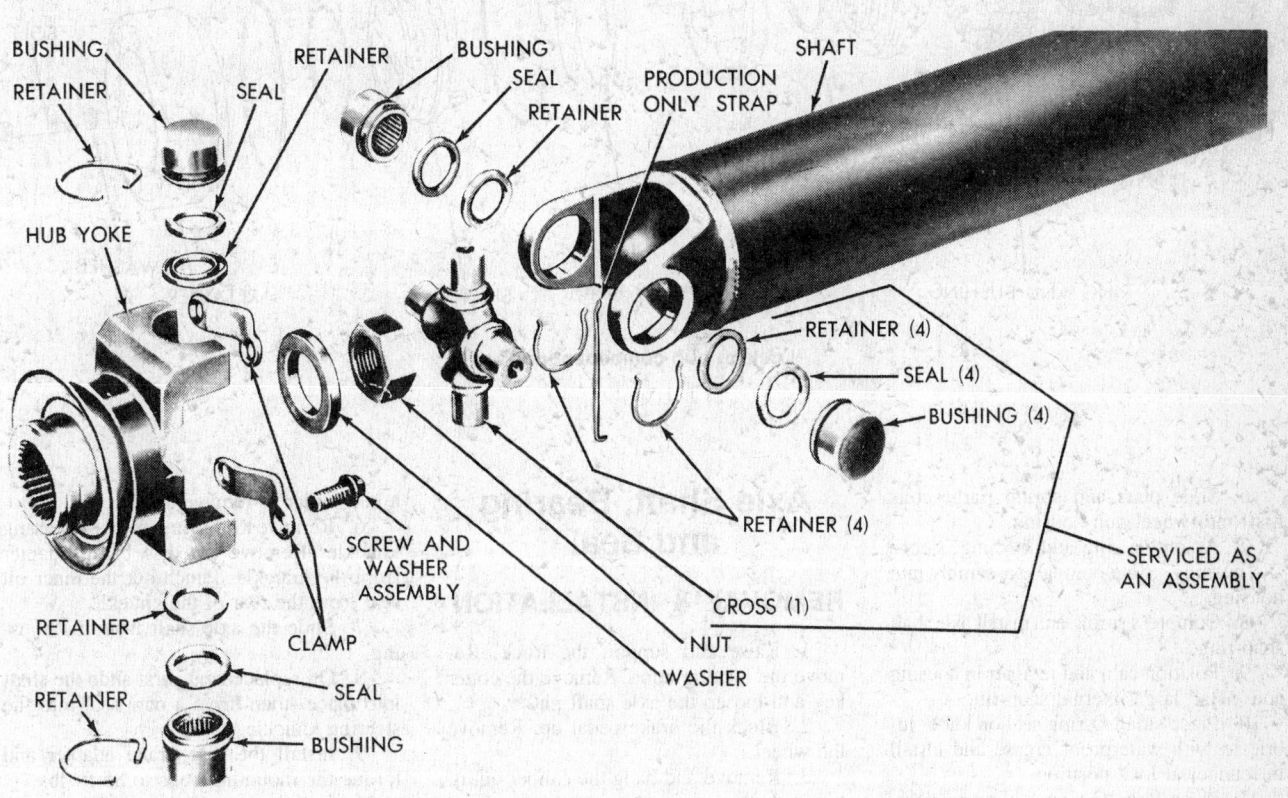

Hub yoke type universal joint

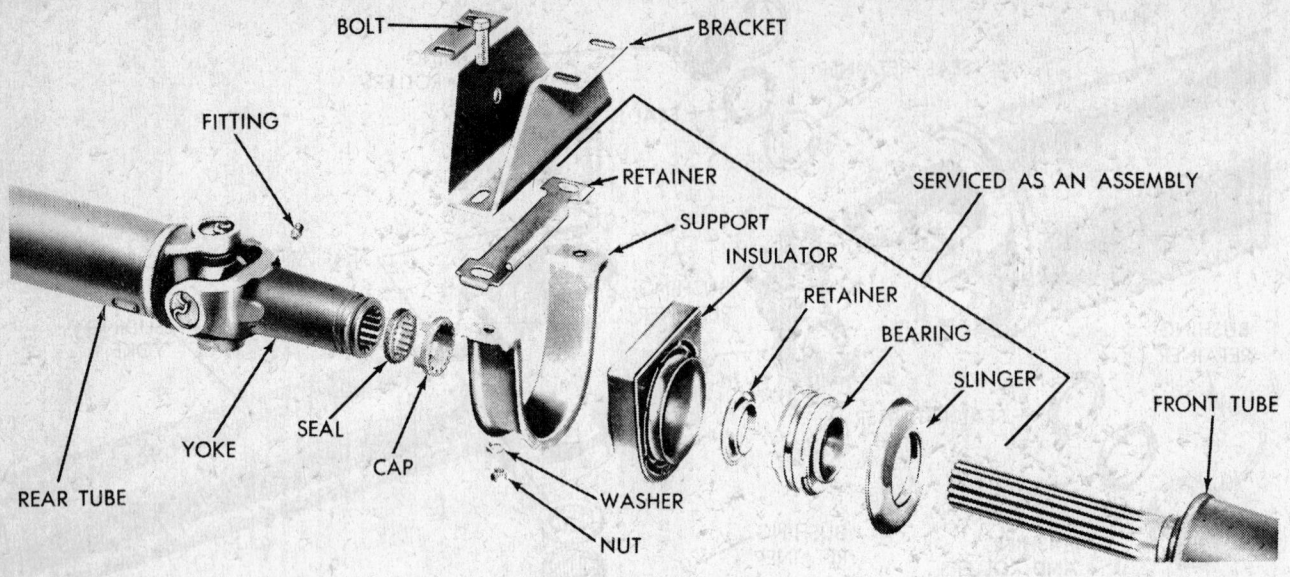

Slip yoke center bearing

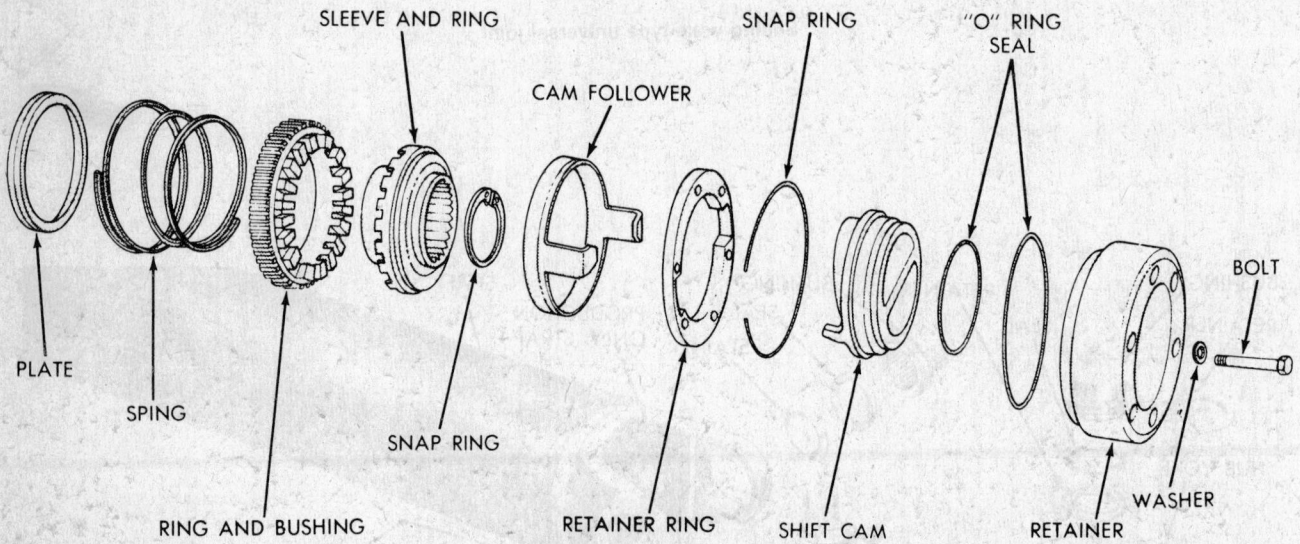

Locking hub component parts

6. Slide plate and spring (large coils first) into wheel hub housing.

7. Assemble ring and bushing, sleeve and bushing. Slide complete assembly into housing.

8. Compress spring and install axle shaft snap-ring.

9. Position cam and retainer in housing and install large internal snap-ring.

10. Place small O-ring seal on knob, lubricate with waterproof grease and install in retainer at lock position.

11. Place large O-ring seal on retainer. Align retainer and retainer ring and install washers and Allen head mounting screws.

12. Check operation.

Axle Shaft, Bearing and Seal

REMOVAL & INSTALLATION

1. Raise and support the truck. Remove the locking hubs. Remove the cotter key and loosen the axle shaft nut.

2. Block the brake pedal up. Remove the wheel.

3. Remove and hang the caliper out of the way. Remove the inner pad.

4. Working through the hole in the disc hub, remove the socket head capscrews.

5. Remove the axle shaft nut and use a hub puller to remove the disc and hub.

6. Remove the O-ring from the steering knuckle. Remove the disc brake adapter from the knuckle. Punch out the inner oil seal from the rear of the knuckle.

7. Slide the axle shaft from the housing.

8. On replacement, first slide the shaft into place, then drive a new seal into the steering knuckle.

9. Install the disc brake adapter and torque the mounting bolts to 85 ft. lbs.

10. Install a new O-ring on the steering knuckle.

11. Slide the disc and hub, retainer and bearing assembly over the shaft and start it

into the housing. Install the axle shaft nut.

12. Install the capscrews holding the retainer to the steering knuckle flange. Tighten them to 30 ft. lbs. in a criss-cross pattern.

13. Torque the axle shaft nut to 100 ft. lbs. and tighten further until the cotter key can be installed.

14. Locate the inner pad on the adapter with the shoe flanges in the adapter ways. Slide the caliper into position, being careful not to pull the dust boot from its grooves.

15. Install the anti-rattle springs, making sure that the inner one is on top of the retainer spring plate. Tighten the retaining clips to 200 in. lbs.

16. Install the wheel and lower the vehicle.

4WD Steering Knuckle Service

REMOVAL & INSTALLATION

The ball joints should be replaced if there is any looseness or end-play. The steering knuckle and ball joint must be removed to replace the ball joint.

1. Refer to the Front Drive Axle Removal and Replacement, on Exposed U-Joint, Ball Joint Type, Steps 1 through 6.

2. Remove and discard the O-ring from the steering knuckle.

3. Remove the capscrews from the brake splash shield and remove the splash shield. Remove the brake disc adapter from the steering knuckle.

4. Disconnect the tie-rod from the steering knuckle. On the left side, disconnect the drag link from the steering knuckle arm.

5. Using a punch and hammer, remove the inner oil seal from the rear of the steering knuckle.

6. Carefully, slide the outer and inner axle shaft complete with U-joint from the axle housing.

7. On the left-side, remove the steering knuckle arm by tapping it to loosen the tapered dowels.

8. Remove the cotter pin from the upper ball joint nut. Remove the upper and lower ball joint nuts and discard the lower nut.

9. Separate the steering knuckle from the axle housing yoke with a brass drift and hammer. Remove and discard the sleeve from the upper ball joint yoke on the axle housing.

10. Position the steering knuckle upside down in a vise with soft jaws and remove the snap-ring from the lower ball joint.

11. Press the lower and upper ball joints from the steering knuckle individually.

12. Position the steering knuckle right side up in a vise with soft jaws. Press the lower ball joint into position and install the snap-ring.

13. Press the upper ball joint into position. Install new boots on both ball joints.

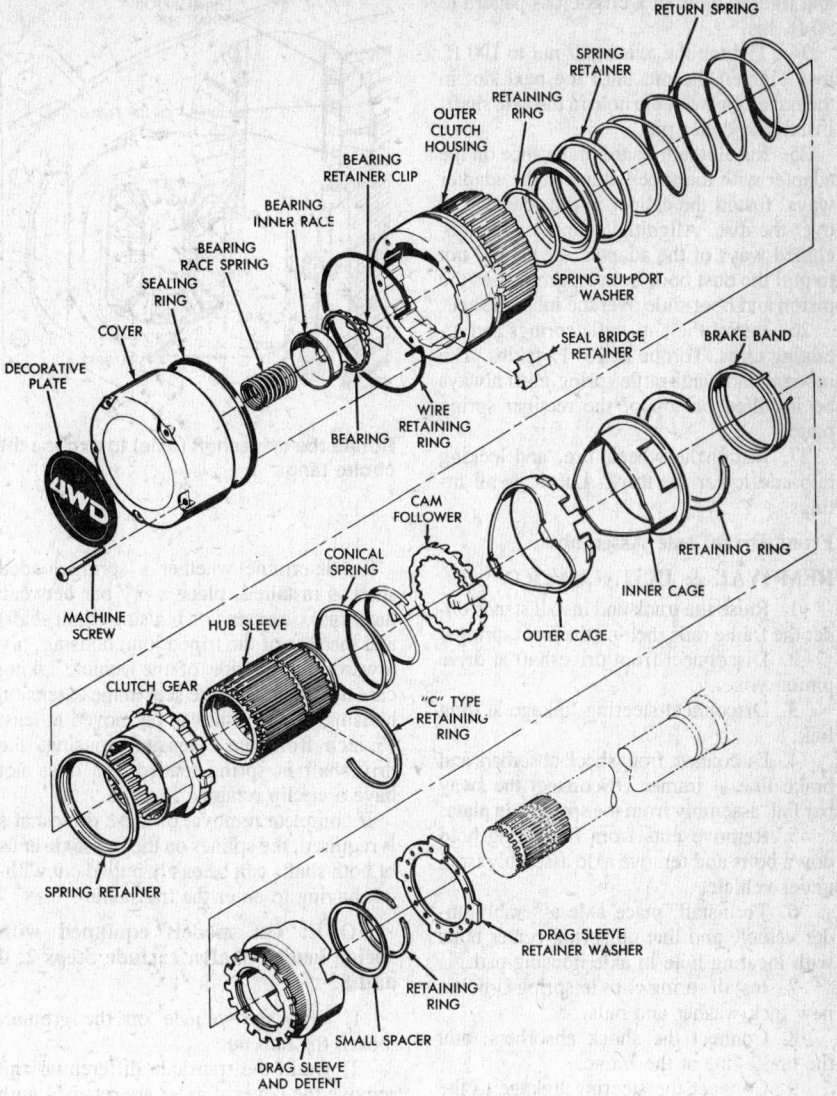

Automatic locking hub—disassembled

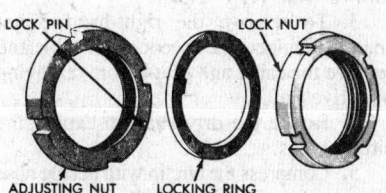

Late model wheel bearing adjusting nut, lock nut and ring—4WD

14. Screw a new sleeve into the upper ball joint yoke on the axle housing, leaving about two threads showing at the top.

15. Install the steering knuckle on the axle housing yoke and install a new lower ball joint nut, tightening it to 80 ft. lbs.

16. Tighten the sleeve in the upper ball joint yoke to 40 ft. lbs. Install the upper ball joint nut and tighten it to 100 ft. lbs. Align the cotter key hole in the stud with the slot in the castellated nut and install the cotter pin. Do not loosen the nut to align the holes.

17. On the left-side, position the steering knuckle arm over the studs on the steering knuckle. Install the tapered dowels and nuts. Tighten the nuts to 90 ft. lbs. Install the drag link on the steering arm. Install the nut and tighten it to 60 ft. lbs. Install the cotter pin.

18. Install the tie-rod end on the steering knuckle. Tighten the nut to 45 ft. lbs. and install the cotter pin.

19. Install the axle shaft. Install the brake splash shield and tighten the screws to 13 ft. lbs. Install the brake disc adapter and tighten the bolts to 85 ft. lbs.

20. Install a new O-ring in the steering knuckle.

21. Clean any rust from the axle shaft splines.

22. Carefully, slide the hub, rotor and retainer, and bearing onto the axle shaft and start it into the housing. Install the axle shaft nut.

23. Align the retainer with the steering knuckle flange. Install the retainer screws

and tighten them in a criss-cross pattern to 30 ft. lbs.

24. Tighten the axle shaft nut to 100 ft. lbs. Tighten the nut until the next slot in the nut aligns with the hole in the axle shaft. Install the cotter pin.

25. Install the inboard brake shoe on the adapter with the shoe flanges in the adapter ways. Install the caliper in the adapter and over the disc. Align the caliper on the machined ways of the adapter. Be careful not to pull the dust boot from its grooves as the piston and boot slide over the inboard shoe.

26. Install the anti-rattle springs and retaining clips. Torque to 16–17 ft. lbs. The inboard shoe anti-rattle spring must always be installed on top of the retainer spring plate.

27. Install the wheel, tire, and locking hub and lower the truck. Lubricate all fittings.

Front Drive Axle Assembly

REMOVAL & INSTALLATION

1. Raise the truck and install stands under the frame rails, behind the front springs.
2. Disconnect front driveshaft at drive pinion yoke.
3. Disconnect steering linkage at drag link.
4. Disconnect front shock absorbers and brake line at frame. Disconnect the sway bar link assembly from the spring clip plate.
5. Remove nuts from the spring hold down bolts and remove axle assembly from under vehicle.
6. To install, place axle assembly under vehicle and line up spring center bolts with locating hole in axle housing pad.
7. Install spring clips or spring U-bolts, new lock washer and nuts.
8. Connect the shock absorbers, and the brake line at the frame.
9. Connect the steering linkage to the drag link, and the driveshaft to the pinion yoke. Check lubricants and bleed the brakes.
10. Lower the vehicle and test the operation.

FRONT WHEEL DRIVE

Refer to the Transmission section of Unit Repair for service procedures.

Drive Axles

REMOVAL & INSTALLATION

The inboard CV joints on early production models may be retained by circlips in the differential side gears. On later production models, the shafts incorporate the use of a spring within the right and left inboard tripod joint assemblies and do not use a circlip.

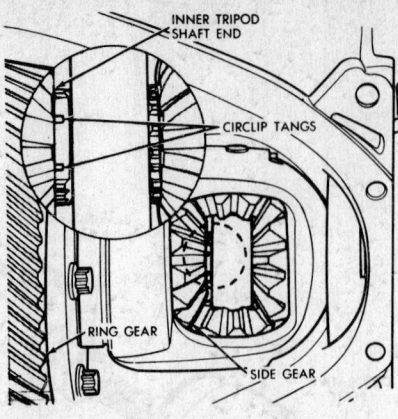

Rotate the driveshaft (axle) to expose the circlip tangs

To determine whether a spring loaded shaft is installed, place a pry bar between the transaxle extension housing (right shaft) and the face of the tripod joint housing, pry towards the outside of the vehicle taking care not to damage the seal in the extension housing. If the joint can be moved at least ½ inch from the extension housing, the driveshaft is spring loaded and does not have a circlip retainer.

If complete removal of these driveshafts is required, the splines on the transaxle ends of both shafts can be easily pulled out without having to enter the transaxle.

NOTE: On models equipped with spring loaded shafts, exclude Steps 2, 4 and 5.

1. With the vehicle on the ground, loosen the hub nut.
2. Drain the transaxle differential and remove the cover if axles are retained with circlips. Any time the transaxle differential cover is removed, a new gasket should be formed from RTV sealant.
3. To remove the right-hand driveshaft, disconnect the speedometer cable and remove the cable and gear before removing the driveshaft.
4. Rotate the driveshaft to expose the circlip tangs.
5. Compress the circlip with needle nose pliers and push the shaft into the side gear cavity.
6. Remove the clamp bolt from the ball stud and steering knuckle.
7. Separate the ball joint stud from the steering knuckle, by prying against the knuckle leg and control arm.
8. Separate the outer CV (constant velocity) joint splined shaft from the hub by holding the CV housing and moving the hub away. Do not pry on the slinger or outer CV joint.
9. Support the shaft at the CV joints, remove the six allen head screws (if equipped) from the transaxle drive flange and remove the shaft. Do not pull on the shaft.

NOTE: Removal of the left shaft may be made easier by inserting the blade of a thin prybar between the differential pinion shaft and prying against the end face of the shaft.

10. Installation is the reverse of removal. Be sure the circlip tangs are positioned against the flattened end of the shaft before installing the shaft. A quick thrust will lock the circlip in the groove. Tighten the hub nut with the wheels on the ground to 180–200 ft. lbs.

Steering Knuckle (FWD)

REMOVAL & INSTALLATION

Service or repair to the bearing, hub, brake dust shield or the steering knuckle itself will require removal of the knuckle. Before attempting this operation, be aware that to reassemble the components it is necessary to torque the front hub nut to at least 200 ft. lbs. You will need a large torque wrench to read that high and a great deal of strength to attain that much torque on the nut.

1. Remove the cotter pin and nut-lock.
2. Loosen the hub nut while the car is resting on the wheels with the brakes applied.

NOTE: The hub and driveshaft are splined together through the knuckle and retained by the hub nut.

3. Raise and support the vehicle.
4. Remove the wheel and tire.
5. Remove the hub nut. Be sure the splined driveshaft is free to separate from the spline in hub when the knuckle is removed.
6. Disconnect the tie rod end from the steering arm.
7. Disconnect the brake hose retainer from the strut.
8. Remove the clamp bolt holding the ball joint stud in the steering knuckle.
9. Remove the brake caliper adapter screw and washers.
10. Support the caliper on a wire hook.
11. Remove the brake disc.
12. Matchmark the camber adjusting cams and loosen both bolts.
13. Support the steering knuckle and remove the cam adjusting and through-bolts. Remove the upper knuckle leg out of the strut bracket and lift the knuckle from the ball joint stud. Do not allow the driveshaft to hang during this procedure.
14. Service procedures requiring hub removal also require that a new bearing be installed.
15. Installation is the reverse of removal. A new hub nut is required. When the car is resting on the wheels, with the brakes applied, tighten the hub nut to 180–200 ft. lbs.

REAR AXLE IDENTIFICATION CHART

Identification	Ring Gear Size (in.)	Maker and Model	Approximately Capacity (pts)①
10 bolt cover, front filler plug	8⅜	Chrysler	4.4
12 bolt cover, filler plug in cover	9¼	Chrysler	4.5
10 bolt cover, filler plug in cover	9¾	Spicer 60	6.0
10 bolt cover, filler plug in cover	10½	Spicer 70	7.0 thru 1976, 6.5 for 1977 and later

① These are design capacities and may differ slightly from those given in the Capacities Chart.

REAR AXLE

Refer to the Drive Axle section in Unit Repair for service procedures.

Axle Shaft and Bearing

REMOVAL & INSTALLATION

8⅜ and 9¼ Inch Rear

NOTE: There is no provision for axle shaft end-play adjustment on these axles.

1. Jack up the vehicle and remove the rear wheels.
2. Clean all dirt from the housing cover and remove the housing cover to drain the lubricant.
3. Remove the brake drum.
4. Rotate the differential case until the differential pinion shaft lockscrew can be removed. Remove the lockscrew and pinion shaft.
5. Push the axle shafts toward the center of the vehicle and remove the C-locks from the grooves on the axle shafts.
6. Pull the axle shafts from the housing, being careful not to damage the bearing which remains in the housing.
7. Inspect the axle shaft bearings and replace any doubtful parts. Whenever the axle shaft is replaced, the bearings should also be replaced.
8. Remove the axle shaft seal from the bore in the housing, using the button end of the axle shaft.
9. Remove the axle shaft bearing from the housing. Do not reuse the bearing or the seal.
10. Check the bearing shoulder in the axle housing for imperfections. These should be corrected with a file.
11. Clean the axle shaft bearing cavity.
12. Grease and install the axle shaft bearing in the cavity. Be sure that the bearing is not cocked and that it is seated firmly against the shoulder.
13. Install the axle shaft bearing seal. It should be seated beyond the end of the flange face.
14. Insert the axle shaft, making sure that the splines do not damage the seal. Be sure that the splines are properly engaged with the differential side gear splines.

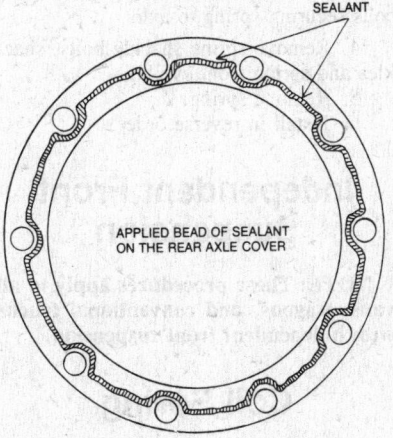

Rear axle cover sealant application

15. Install the C-locks in the grooves on the axles shafts. Pull the shafts outward so that the C-locks seat in the counterbore of the differential side gears.
16. Install the differential pinion shaft through the case and pinions. Install the lockscrew and secure it in position.
17. Clean the housing and gasket surfaces. Install the cover and a new gasket. Refill the axle with the specified lubricant.

NOTE: Replacement differential cover gaskets may not be available. The use of gel type nonsticking sealant is recommended.

18. Install the brake drum and wheel.

9¾ and 10½ Inch Rear

1. Remove the axle shaft flange nuts and washers.
2. Rap the axle shafts sharply in the center of the flange with a hammer to free the dowels.
3. Remove the tapered dowels and axle shafts. Some models are equipped with bolts rather than dowels.
4. Clean the gasket area with solvent and install a new flange gasket.
5. Install the axle shaft into the housing.
6. If the axle has an outer wheel bearing seal, install new gaskets on each side of the seal mounting flange.
7. Install the tapered dowels, lockwashers, and nuts. Torque the nuts to 40–70

ft. lbs. with ⁷⁄₁₆ in. studs and 65–105 with ½ in. Some axles have bolts instead of studs and tapered dowels. Bolt torque is 45–75 ft. lbs.

BEARING ADJUSTMENT

1. Raise and support the rear axle.
2. Remove the axle shaft, outer nut, and lockring.
3. Rotate the wheel and tire and tighten the adjusting nut until a slight binding is felt. Back the nut off ⅙ turn.
4. Install the lockring and jamnut. Don't overtighten the jamnut or you will change the adjustment.
5. Install a new gasket and the axle shaft. Lower the vehicle.

Rear Wheel Bearings (FWD Models)

The rear wheel bearings, on Front Wheel Drive models, should be inspected and re-lubricated whenever the rear brakes are serviced or at least every 30,000 miles. Repack the bearings with high temperature multi-purpose grease.

Check the lubricant to see if it is contaminated. If it contains dirt or has a milky appearance indicating the presence of water, the bearings should be cleaned and repacked.

Clean the bearings in kerosene, mineral spirits or other suitable cleaning fluid. Do not dry them by spinning the bearings. Allow them to air dry.

1. Raise and support the truck with the rear wheels off the floor.
2. Remove the wheel grease cap, cotter pin, nut-lock and bearing adjusting nut.
3. Remove the thrust washer and bearing.
4. Remove the drum from the spindle.
5. Thoroughly clean the old lubricant from the bearings and hub cavity. Inspect the bearing rollers for pitting or other signs of wear. Light discoloration is normal.
6. Repack the bearings with high temperature multi-purpose EP grease and add a small amount of new grease to the hub cavity. Be sure to force the lubricant between all rollers in the bearing.
7. Install the drum on the spindle after coating the polished spindle surfaces with wheel bearing lubricant.
8. Install the outer bearing cone, thrust washer and adjusting nut.

9. Tighten the adjusting nut to 20–25 ft. lbs. while rotating the wheel.

10. Back off the adjusting nut to completely release the preload from the bearing.

11. Tighten the adjusting nut finger tight.

12. Position the nut-lock with one pair of slots in line with the cotter pin hole. Install the cotter pin.

13. Clean and install the grease cap and wheel.

14. Lower the truck.

FRONT SUSPENSION

Rear Wheel Drive Models

King Pins (I–Beam Axle)

REMOVAL & INSTALLATION

1. Raise the front of the vehicle and safely support it on stands. Remove the wheels, calipers, and rotors.

2. Remove the brake support plate attaching bolts, and remove the support plate from the steering knuckle. Secure the plate to the frame so that it does not hang by the brake hose. If equipped with air brakes, disconnect the push rod and remove the air chamber.

3. Remove the steering arm from the steering knuckle.

4. Remove the pivot pin locking screw or pin from the knuckle. Some models may have two locking screws.

5. Remove the upper pivot pin oil seal plug from the knuckle and drive the pivot pin down, forcing the lower oil seal plug from its seat.

NOTE: On some models, a lock ring is used to hold the oil seal plug in place. Others have caps with hold-down screws.

6. Remove the knuckle from the axle and if equipped with bronze bushings, press the old ones out and press the new ones in and line ream them to fit the new pin. If equipped with Delrin or Zytel type bushings, bronze bushings must be used as service replacements. Upon installation of the bushings, align the grease hole in the bushing to that of the grease fitting hole in the knuckle.

7. Install the knuckle on the axle. Position the thrust bearing, and install the pivot pin through the knuckle and axle, securing it with the locking pins or screws.

8. Install the oil seal plugs and secure them by staking in four locations.

9. Lubricate to assure grease channels

are open. Complete the assembly by reversing Step 1 through 3.

Leaf Spring

REMOVAL & INSTALLATION

1. Raise truck until weight is removed from springs.

2. Install stands under side frame members as a safety precaution.

3. Disconnect the sway bar at the spring plate. Remove nuts, lockwashers and U-bolts securing spring to axle.

4. Remove spring shackle bolts, shackles and spring front eye bolt.

5. Remove spring.

6. Install in reverse order.

Independent Front Suspension

NOTE: These procedures apply to all vans, wagons, and conventional trucks with independent front suspension.

Coil Spring

REMOVAL & INSTALLATION

1. Raise the vehicle and support it with jackstands under the front ends of the frame rails.

2. Remove the wheel.

3. Remove the shock absorber and upper shock absorber bushing and sleeve.

4. If equipped, remove the sway bar.

5. Remove the strut.

6. Install a spring compressor and tighten finger-tight.

7. Remove the cotter pins and ball joint nuts.

8. Install a ball joint breaker tool and turn the threaded portion of the tool to lock it against the lower stud.

9. Spread the tool to place the lower stud under pressure, then strike the steering knuckle sharply with a hammer to free the stud. Do not attempt to force the stud out of the steering knuckle with the tool.

10. Remove the tool. Slowly release the spring compressor until all tension is relieved from the spring.

11. Remove the spring compressor and spring.

12. Installation is the reverse of removal. Compress the spring until the ball joint can be properly positioned in the steering knuckle.

Shock Absorber

REMOVAL & INSTALLATION

1. Raise and support the vehicle with jackstands positioned at the extreme front ends of the frame rails.

2. Remove the wheel.

3. Remove the upper nut and retainer.

4. Remove the two lower mounting bolts.

5. Remove the shock absorber.

6. Installation is the reverse of removal.

Upper Control Arm

REMOVAL & INSTALLATION

NOTE: Any time the control arm is removed, it is necessary to align the front end.

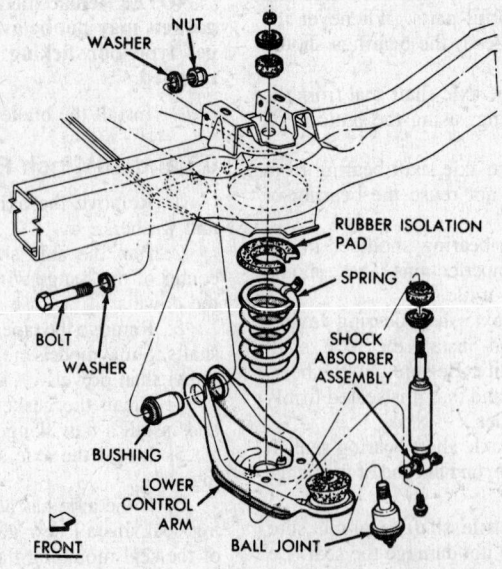

1979 and later coil spring and control arm components

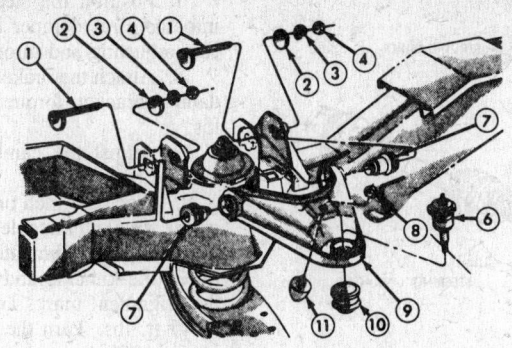

1. Cam and bolt assembly
2. Cam
3. Lock washer
4. Nut
6. Ball joint assembly
7. Bushing assembly
8. Lock nut
9. Upper control arm
10. Upper ball joint assembly
11. Bumper assembly

Typical upper control arm—through 1978

1. Raise and support the vehicle with jackstands under the frame rails.
2. Remove the wheel.
3. Remove the shock absorber and shock absorber upper bushing and sleeve.
4. Install a spring compressor and tighten it finger-tight.
5. Remove the cotter pins and ball joint nuts.
6. Install a ball joint breaker and turn the threaded portion of the tool, locking it securely against the upper stud. Spread the tool enough to place the upper ball joint under pressure and strike the steering knuckle sharply to loosen the stud. Do not attempt to remove the stud from the steering knuckle with the tool.
7. Remove the tool.
8. Remove the eccentric pivot bolts, after making their relative positions in the control arm.
9. Remove the upper control arm.
10. Installation is the reverse of removal. Tighten the ball joint nuts to 135 ft. lbs. Tighten the eccentric pivot bolts to 70 ft. lbs.
11. Adjust the caster and camber.

Lower Control Arm

REMOVAL & INSTALLATION

1. Follow the procedure outlined under ''Coil Spring Removal and Installation.''
2. Remove the mounting bolt from the crossmember.
3. Remove the lower control arm from the vehicle.
4. Installation is the reverse of removal. After the vehicle has been lowered to the ground, tighten the mounting bolt to 210 ft. lbs.

Lower Ball Joint

REMOVAL & INSTALLATION

1. Remove the lower control arm.
2. Remove the ball joint seal.
3. Using an arbor press and a sleeve, press the ball joint from the control arm.
4. Installation is the reverse of removal. Be sure that the ball joint is fully seated. Install a new ball joint seal.
5. Install the lower control arm. Be sure to install the ball joint cotter pins.

Upper Ball Joint

REMOVAL & INSTALLATION

1. Install a jack under the outer end of the lower control arm and raise the vehicle.
2. Remove the wheel.
3. Remove the ball joint nuts. Using a ball joint breaker, loosen the upper ball joint.
4. Unscrew the ball joint from the control arm.
5. Screw a new ball joint into the control arm and tighten 125 ft. lbs.
6. Install the new ball joint seal, using a 2 in. socket. Be sure that the seal is seated on the ball joint housing.
7. Insert the ball joint into the steering knuckle and install the ball joint nuts. Tighten the nuts to 135 ft. lbs. and install the cotter pins.
8. Install the wheel and lower the truck to the ground.

Wheel Bearings

It is recommended that the front wheel bearings be cleaned, inspected and repacked periodically and as soon as possible after the front hubs have been submerged in water.

NOTE: Sodium based grease is not compatible with lithium based grease. Be careful not to mix the two types. The best way to prevent this is to completely clean all of the old grease from the hub assembly before installing any new grease.

Before handling the bearings there are a few things that you should remember to do and try to avoid.
DO the following:
1. Remove all outside dirt from the housing before exposing the bearing.
2. Treat a used bearing as gently as you would a new one.
3. Work with clean tools in clean surroundings.
4. Use clean, dry canvas gloves, or at least clean, dry hands.
5. Clean solvents and flushing fluids are a must.
6. Use clean paper when laying out the bearings to dry.
7. Protect disassembled bearings from rust and dirt. Cover them up.
8. Use clean rags to wipe bearings.
9. Keep the bearings in oil-proof paper when they are to be stored or are not in use.
10. Clean the inside of the housing before replacing the bearing.

Do NOT do the following:
1. Don't work in dirty surroundings.
2. Don't use dirty, chipped, or damaged tools.
3. Try not to work on wooden work benches or use wooden mallets.
4. Don't handle bearings with dirty or moist hands.
5. Do not use gasoline for cleaning; use a safe solvent.
6. Do not spin-dry bearings with compressed air. They will be damaged.
7. Do not spin unclean bearings.
8. Avoid using cotton waste or dirty cloths to wipe bearings.
9. Try not to scratch or nick bearing surfaces.
10. Do not allow the bearing to come in contact with dirt or rust at any time.

REPACKING

NOTE: Sodium based grease is not compatible with lithium based grease. Be careful not to mix the two types. The best way to prevent this is to completely clean all of the old grease from the hub assembly before installing any new grease.

1. Raise the front of the vehicle and place jackstands under the vehicle. Remove the wheel.
2. Remove the front hub grease cap and driving hub snap-ring.
3. Remove the splined driving hub and the pressure spring. This may require slight prying with a pry bar.

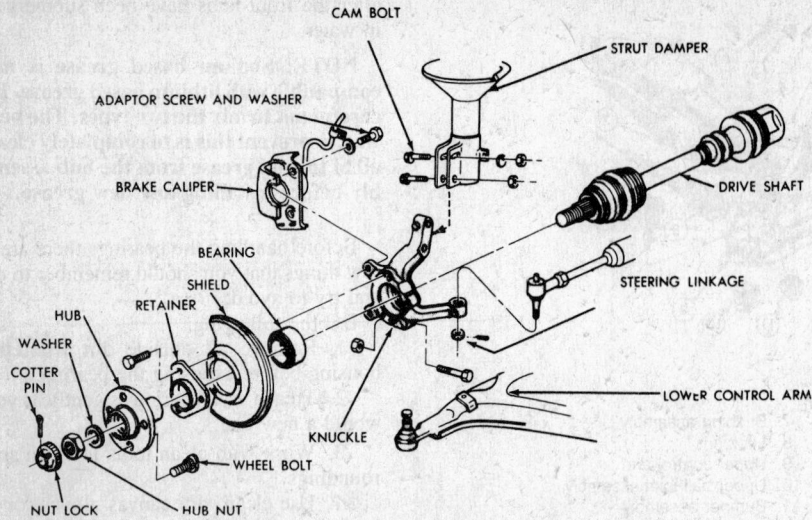

CAM BOLT

STRUT DAMPER

ADAPTOR SCREW AND WASHER

DRIVE SHAFT

BRAKE CALIPER

BEARING

STEERING LINKAGE

SHIELD

RETAINER

HUB

WASHER

COTTER PIN

LOWER CONTROL ARM

KNUCKLE

WHEEL BOLT

NUT LOCK

HUB NUT

Exploded view front suspension—Rampage

4. Remove the wheel bearing locknut, lockring, and adjusting nut.

5. Removal brake caliper suspend on wire out of the way.

6. Carefully drive out the inner bearing cone and grease seal from the hub.

7. Inspect the bearing cups (races) for cracks and pits. If the cups are excessively worn or there are pits or cracks visible, replace them along with the cones. The cups are removed from the hub by driving them out with a drift pin. They are installed in the same manner.

8. If it is determined that the cups are in satisfactory condition and are to remain in the hub, clean and inspect the cones (bearings). Refer to the bearing diagnosis chart. Replace the bearings if necessary. If it is necessary to replace either the cone or the cup, both parts should be replaced as a unit.

9. Thoroughly clean all components in a suitable solvent and blow them dry with compressed air or allow them to dry while resting on clean paper. Do not spin the bearings with compressed air while drying them.

10. Cover the spindle with a cloth and brush all loose dust and dirt from the brake assembly. Remove the cloth and thoroughly clean the inside of the hub and the spindle.

11. Pack the inside of the hub with wheel bearing grease. Add grease to the hub until the grease is flush with the inside diameter of the bearing cup.

12. Pack the bearing cone and roller assemblies with wheel bearing grease. A bearing packer is desirable for this operation. If a packer is not available, place a large portion of grease into the palm of your hand and sliding the edge of the roller cage through the grease with your other hand, work as much grease in between the rollers as possible.

13. Position the inner bearing into the inner bearing cup and install the new grease seal.

14. Carefully position the hub assembly onto the spindle. Be careful not to damage the new seal. Install the rotor or caliper.

15. Place the outer bearing into position on the spindle and into the bearing cup.

16. Install the bearing adjusting nut and tighten it while rotating the hub back and forth to seat the bearings.

17. Back off the adjusting nut about ¼ turn.

18. Assemble the lockring by turning the nut to the nearest notch where the dowel pin will enter. Install the outer locknut.

19. Install the pressure spring retainer, spring, the driving hub and driving hub snapring. This is for vehicles without freerunning hubs.

20. Install the grease cap and adjust the brakes, if they were backed off to remove the hub assembly. Remove the jackstands and lower the vehicle.

Front Wheel Drive Models

Strut Assembly

REMOVAL & INSTALLATION

1. Loosen the wheel lug nuts.

2. Raise the vehicle and remove the wheel assembly.

3. Mark the camber cam to damper bracket position before removing the cam adjusting bolt.

4. Remove the cam adjusting bolt, the through bolt and the brake hose to damper bracket retaining screw.

5. Remove the strut damper to fender shield (strut tower) mounting nut and washer assembly.

6. Install strut assembly into fender (strut tower) and install nut and washer assemblies. Torque the retaining nuts to 20 ft. lbs.

7. Position the steering knuckle neck into the strut damper bracket, install the cam adjusting and through bolts.

8. Attach the brake hose retainer to the damper bracket, torque the screws to 10 ft. lbs.

9. Adjust the camber to the original marked position.

10. Place a 4 inch or larger "C" clamp on the strut and knuckle. Tighten the clamp just enough to eliminate any looseness between the knuckle and the strut. Recheck the alignment marks and tighten the bolts to 45 ft. lbs. Turn the bolts an additional ¼ turn (90°) beyond the specified torque.

11. Install the wheel and tire assembly. Torque the wheel nuts to 80 ft. lbs.

STRUT TYPE COIL SPRING

1. Use a coil spring compressor (strut type) to compress the coil spring.

2. Hold the strut rod while removing the strut rod.

3. Remove the strut damper mount assembly and the coil spring.

If both strut springs are to be removed and reinstalled, mark them so they can be returned to their original position.

――――――― CAUTION ―――――――
If Chrysler special tool spring compressor tool L–4514 is used, do not open the tool jaws beyond 9¼ inches.
―――――――――――――――――――

4. Reinstall the strut rod nut and torque it to 60 ft. lbs. before releasing the spring compressor.

Lower Control Arm

REMOVAL & INSTALLATION

1. Raise and safely support the vehicle.

2. Remove the front inner pivot through bolt, rear stub strut nut, retainer and bushing.

3. Remove the ball joint to steering knuckle clamp bolt.

4. Separate the ball joint stud from steering knuckle.

――――――― CAUTION ―――――――
Do not pull the steering knuckle away from the vehicle, while the ball joint is disconnected, it can separate the inner C/V joint.
―――――――――――――――――――

5. Remove the sway bar to control arm end bushing retainer nuts and rotate the control arm over the sway bar. Remove rear stub strut bushing, sleeve and retainer.

6. Install retainer, bushing and sleeve on stub strut.

7. Position control arm over sway bar and install rear stub strut and front pivot into crossmember.

8. Install front pivot bolt and loosely assemble the nut.

9. Install the stub strut bushing and retainer and loosely assemble the nut.

10. Install ball joint stud into steering knuckle and install the clamp bolt. Tighten the clamp bolt to 50 ft. lbs.

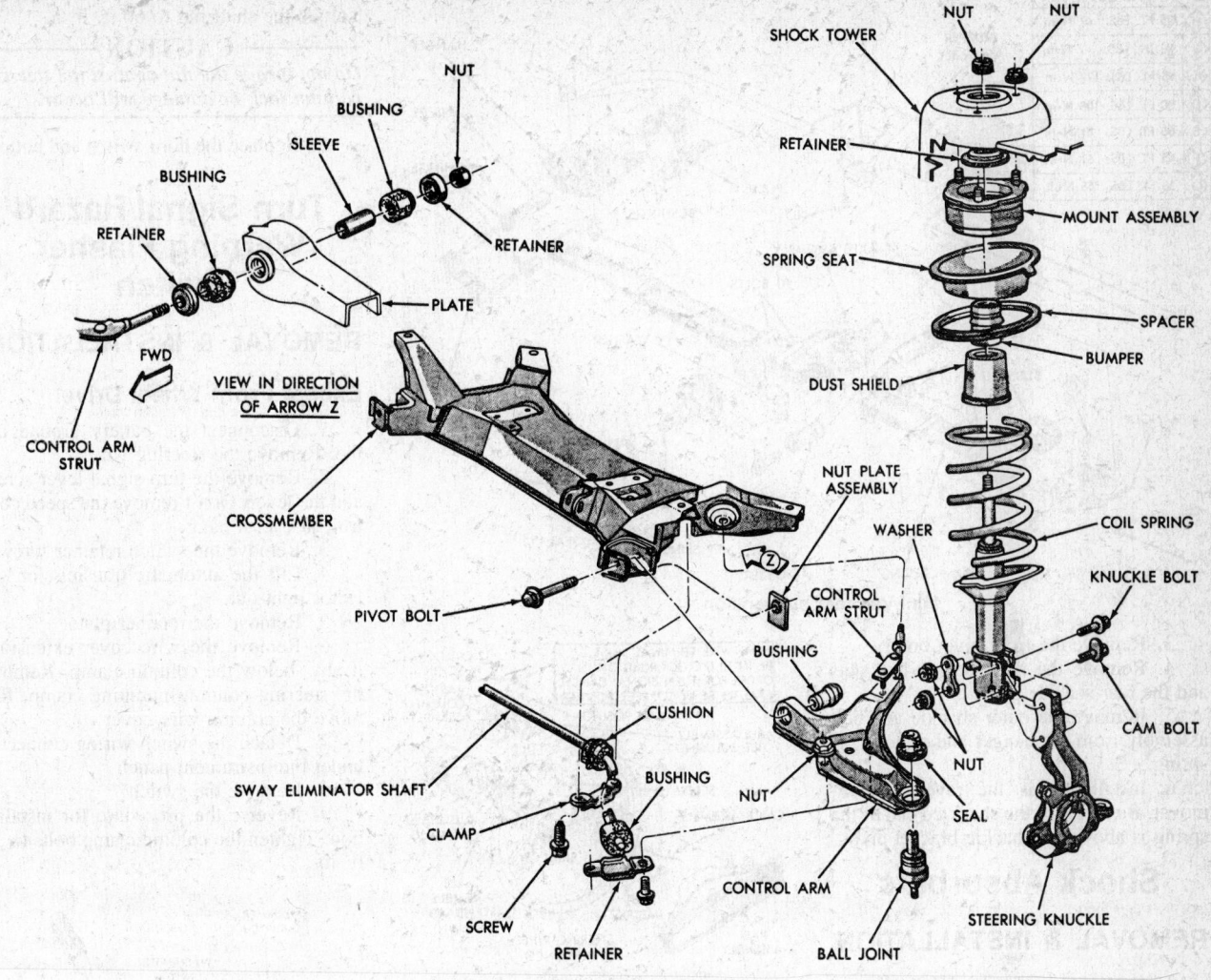

Mini-Van front suspension

11. Position sway bar end bushing retainer to control arm. Install retainer bolts and tighten nuts to 22 ft. lbs.

12. Lower the vehicle and support the control arm to design height. Tighten the front pivot bolt to 105 ft. lbs. and the stub strut nut to 70 ft. lbs.

Lower Ball Joint

INSPECTION

1. With the weight of the vehicle resting on the road wheels, grasp the grease fitting and attempt to move it.

2. No mechanical force is necessary. If the ball joint is worn the grease fitting will move easily. Replace worn ball joints.

REMOVAL & INSTALLATION

The lower front ball joints are pressed into the lower control arm and an arbor press will be needed to remove and install them.

1. Pry the seal from the ball joint.

2. Position special tool C–4699–2 or equivalent to support the lower control arm while receiving ball joint assembly.

3. Use a 1 1/16 inch deep well socket in the arbor press to remove the ball joint.

4. Position the ball joint into the control arm cavity.

5. Position the control arm assembly in the press with special tool C–4699–1 or equivalent supporting the control arm.

6. Align the assembly and press until the ball joint housing ledge stops against control arm cavity down flange.

7. Support the ball joint housing with special tool C–4699–2 or equivalent and position a new seal over the ball joint stud.

8. Using a 1 1/2 inch socket in an arbor press force the seal to seat against the control arm.

WHEEL BEARING ADJUSTMENT

1. Remove the wheel and tire assembly.

2. Remove the hub lock nut cotter pin and lock nut.

3. Loosen the hub nut and apply the brakes.

4. With brakes applied, tighten hub nut to 180–200 ft. lbs.

5. Install nut lock and new cotter pin. Wrap cotter pin tightly against nut lock.

6. Install wheel and tire assembly and tighten wheel nuts to 80 ft. lbs.

REAR SUSPENSION

Leaf Springs

REMOVAL & INSTALLATION

1. Raise the truck and support the rear with jackstands under the bumper brackets. Be sure that the front wheels are chocked and that the parking brake is set. Support the axle with a jack.

2. Remove the U-bolts and the U-bolt plate that holds the axle to the springs.

LET	TORQUE	
A	35 FT. LBS.	47 N•m
B	70 IN. LBS.	7 N•m
C	95 FT. LBS.	129 N•m
D	80 FT. LBS.	108 N•m
E	60 FT. LBS.	81 N•m
F	45 FT. LBS.	61 N•m
G	50 FT. LBS.	68 N•m

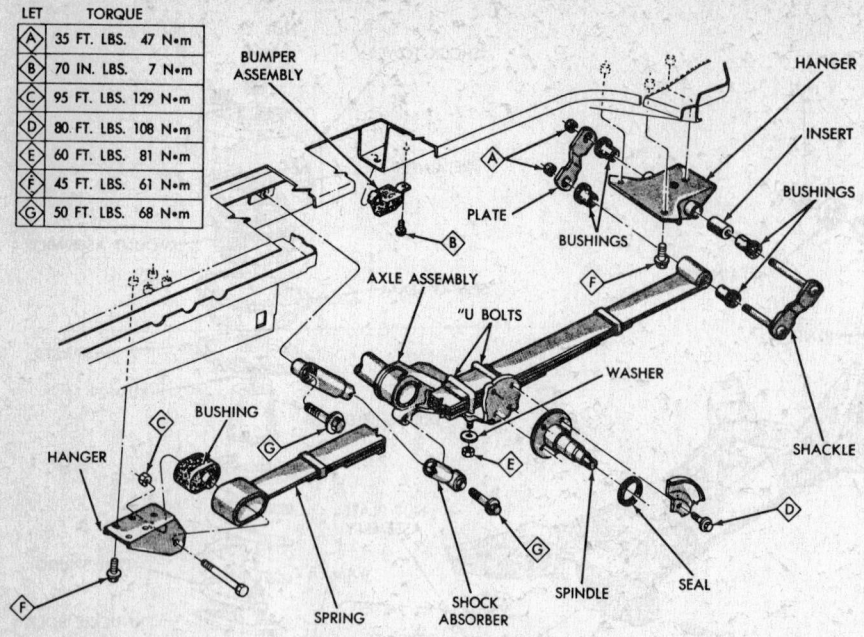

Mini-Van rear suspension

3. Remove the front pivot bolt.

4. Remove the rear shackle bolt nuts and the rear shackle plate.

5. Remove the outer shackle and bolt assembly from the hanger and remove the spring.

6. Installation is the reverse of removal. Be sure that the shackled end of the spring is above the shackle bracket pivot.

Shock Absorbers

REMOVAL & INSTALLATION

1. Jack and support the truck. Remove the wheel.

2. Remove the nut from the stud or bolt at the upper end. Remove the stud or bolt from the upper end.

3. Remove the lower nut at the bushing end.

4. Pivot the shock absorber and washers from the lower stud.

5. Remove the shock absorber and washers from the lower stud.

6. Installation is the reverse of removal. Purge the new shock of air by extending it in its normal position and compressing it while inverted. Do this several times. It is normal for there to be more resistance to extension than to compression.

STEERING

Steering Wheel

REMOVAL & INSTALLATION
Except Front Wheel Drive

1. Remove the horn button from the

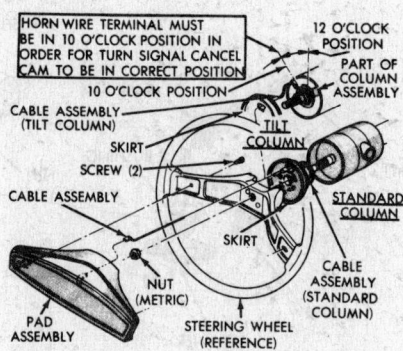

1978 and later steering wheel—typical

retainer by rotating it to the left, or remove the horn pad from the retainer by removing the two screws from underneath.

2. Disconnect the horn wire from the horn switch terminal.

3. Remove three screws and lift out the horn switch and button or pad retainer assembly.

4. Back the steering wheel retaining nut off to the top of the shaft.

5. Install a steering wheel puller and draw the steering wheel from the steering shaft splines.

6. Remove the steering wheel nut and pull the steering wheel off of the steering shaft.

7. Installation is the reverse of removal. Tighten the nut to 60 ft. lbs.

Front Wheel Drive

1. Remove the horn button or pad and horn switch. See Step 1 in previous section.

2. Remove the steering wheel nut.

3. Using a steering wheel puller, remove the steering wheel.

4. Align the master serration in the wheel

hub with the missing tooth on the shaft. Torque the shaft nut to 60 ft. lbs.

— **CAUTION** —
Do not torque the nut against the steering column lock or damage will occur.

5. Replace the horn switch and button.

Turn Signal/Hazard Warning Flasher Switch

REMOVAL & INSTALLATION
Except Front Wheel Drive

1. Disconnect the battery ground cable. Remove the steering wheel.

2. Remove the turn signal lever screw and the lever. Don't remove the speed control, just let it hang.

3. Remove the switch retainer screws.

4. Lift the automatic transmission selector light out.

5. Remove the retainer plate.

6. Remove the wire cover extension, if any, below the column clamp. Remove the steering column mounting clamp. Remove the column wire cover.

7. Detach the switch wiring connector under the instrument panel.

8. Remove the switch.

9. Reverse the procedure for installation. Tighten the column clamp bolts to 30 ft. lbs.

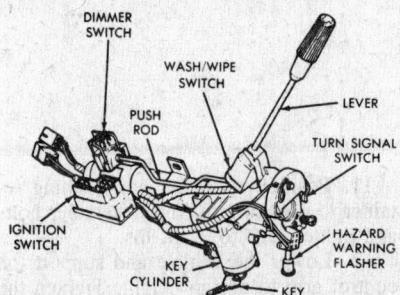

Steering column switch components—Rampage

Front Wheel Drive
WITHOUT TILT WHEEL

1. Disconnect the negative battery cable.

2. Remove the steering wheel as described earlier.

3. On vehicles equipped with intermittent wipe or intermittent wipe with speed control, remove the two screws that attach the turn signal lever cover to the lock housing and remove the turn signal lever cover.

4. Remove the wash/wipe switch assembly.

5. Pull the hider up the control stalk and remove the two screws that attach the control stalk sleeve to the wash/wipe switch.

6. Rotate the control stalk shaft to the full clockwise position and remove the shaft from the switch by pulling straight out of the switch.

7. Remove the turn signal switch and upper bearing retainer screws. Remove the retainer and lift the switch up and out.

8. Installation is the reverse of removal.

WITH TILT WHEEL

1. Disconnect the negative battery cable.

2. Remove the steering wheel as previously described.

3. Remove the tilt lever and push the hazard warning knob in and unscrew it to remove it.

4. Remove the ignition key lamp assembly.

5. Pull the knob off the wash/wipe switch assembly.

6. Pull the hider up the stalk and remove the two screws that attach the sleeve to the wash/wipe switch and remove the sleeve.

7. Rotate the shaft in the wiper switch to the full clockwise position and remove the shaft by pulling straight out of the wash/wipe switch.

8. Remove the plastic cover from the lock plate. Depress the lock plate with tool C–4156, or suitable substitute. Pry the retaining ring out of the groove. Remove the lock plate, canceling cam and upper bearing spring.

9. Remove the switch actuator screw and arm.

10. Remove the three turn signal switch attaching screws and place the shift bowl in low position. Wrap a piece of tape around the connector and wires to prevent snagging then remove the switch and wires.

11. Installation is the reverse of removal.

Manual Steering Gear

REMOVAL & INSTALLATION

Except Front Wheel Drive
RAMCHARGER AND TRAIL DUSTER

1. Remove the two bolts from the wormshaft coupling.

2. Remove the steering arm from the steering gear using a suitable tool.

3. Remove the steering gear-to-frame bolts and remove the unit from the vehicle.

4. To install, position the steering gear to the frame and install the mounting bolts.

5. Install the steering arm and place the front wheels in the straight ahead position.

6. Place the steering wheel in the straight ahead position.

7. Install the wormshaft-to-column coupling bolts.

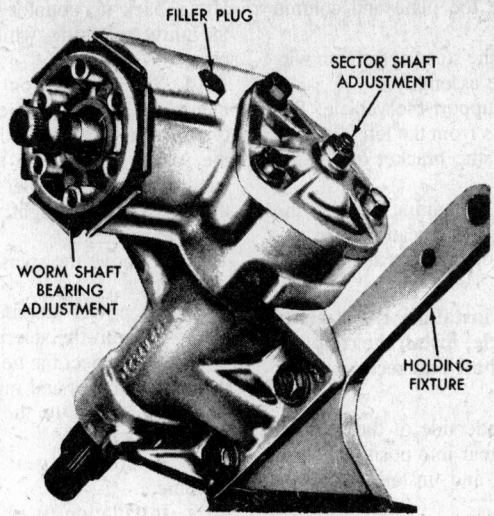

Manual steering gear adjustment points—typical

VANS AND PICKUPS

1. Disconnect the battery.

2. Raise and support the vehicle. Disconnect the "rubber and fabric" coupling (leaving the lower half of the coupling on the wormshaft).

3. Disconnect the shift linkage at the steering column.

4. Remove the steering arm retaining nut and washer. With a suitable tool remove the steering arm from the sector shaft.

5. Remove the three gear mounting bolts. Lower the vehicle from the hoist and remove the toe plate and column support bolts.

6. Disconnect the wiring and remove the column assembly.

7. Raise and support the vehicle. Remove the steering gear through the opening on the inboard side of the frame. (It may be helpful to remove the three bolts from the left idle arm bracket and move the bracket out of the way to provide additional clearance.)

NOTE: If the lower half of the steering coupling was removed from the gear, reinstall it on the wormshaft and secure it with a roll pin before installing the gear into the vehicle.

8. From underneath the vehicle, place the steering gear in position and install the three mounting bolts.

9. Reinstall the idler arm bracket, if it was removed.

10. Install the steering arm on the sector shaft. Install the washer and retaining nut.

11. Install the steering column assembly. Connect the steering column wiring. Install the shift linkage at the steering column.

12. Connect the steering shaft coupling at the wormshaft.

13. Connect the battery.

Power Steering Gear

REMOVAL & INSTALLATION

Except Front Wheel Drive

1. Raise the hood and remove the battery.

2. Disconnect the wires from the windshield washer pump. Remove the windshield washer reservoir mounting screws and position the reservoir out of the way.

3. Disconnect the power steering hoses at the steering gear. Cap the fittings at the steering gear and tie the hoses above the fluid level in the pump reservoir to prevent oil leakage.

4. Raise and safely support the vehicle. Disconnect the "rubber and fabric" coupling at the steering gear (leaving the lower half of the coupling on the wormshaft).

5. Disconnect the shift linkage at the steering column.

6. Remove the steering arm shield if so equipped. Remove the nut and washer, then with a suitable tool, remove the steering arm from the sector shaft.

7. Remove the mounting bolt on the left side of the gear.

8. Lower the vehicle and remove one of the two remaining steering gear mounting bolts.

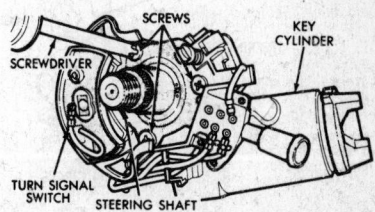

Turn signal switch—typical of Rampage

9. Remove the toe plate and column support bolts.

10. Disconnect the steering column wiring and remove the assembly.

11. Raise and support the vehicle. Remove the three bolts from the left idler arm bracket and swing the bracket out of the way.

12. Remove the remaining bolt and the steering gear from underneath of the vehicle, through the opening on the inboard side of the frame.

NOTE: Before installing the steering gear into the vehicle, install the coupling half of the wormshaft and secure it with the roll pin.

13. From the underside of the vehicle, place the steering gear into position on the mounting bracket and install the three mounting bolts.

14. Continue the installation in the reverse order of the removal.

15. Lower the vehicle, start the engine and turn the steering wheel several times from stop to stop to bleed the system of air.

16. Stop the engine and check the fluid level, correct if necessary. Inspect for leaks.

Front Wheel Drive

REMOVAL & INSTALLATION
Manual or Power Steering Gear

1. Raise and support truck. Remove the front wheels.

2. Remove the tie rod ends, using a suitable puller.

3. Drive out the lower roll pin attaching the pinion shaft to the lower universal joint.

Use a back up counter weight, to protect the universal joint, while driving the roll pin.

4. Support the front suspension crossmember with a hydraulic jack. Remove the two rear nuts attaching the crossmember to the frame. Loosen the two front bolts attaching the crossmember to frame and lower the crossmember slightly for access to the boot seal shields.

5. Remove the splash shields and boot seal shields.

6. On power steering models remove the hoses from the steering gear box.

7. Disconnect the tie rod ends from the steering knuckles and remove the bolts attaching the gear to the front suspension crossmember.

8. Remove the gear from driver's side of vehicle.

9. Installation is the reverse of removal.

10. Torque the steering gear to crossmember bolts to 20.8 ft. lbs.

Power Steering Pump

REMOVAL & INSTALLATION

Except Front Wheel Drive

1. Loosen the pump mounting and locking bolts and remove the drive belt.

2. Disconnect and plug both hoses.

3. Remove the mounting and locking bolts and remove the pump.

4. Install the pump on the engine and install the mounting and locking bolts.

5. Install and adjust the drive belt. Tighten the mounting bolts to 30 ft. lbs.

6. Connect the pressure and return hoses. Route the hoses in the same manner as they were prior to removal. They should be routed smoothly with no sharp bends. Tighten the pump end hose fitting to 25 ft. lbs. The hoses should remain at least 1 in. away from pulleys, battery case, and brake lines and at least 2 in. away from exhaust manifolds. If equipped, the protective sponge sleeves should be used to protect the hoses from contact with other parts.

7. Fill the pump with the specified power steering fluid or the equivalent.

8. Start the engine and turn the steering wheel lock-to-lock several times to bleed the system. Check for leaks and recheck the fluid level.

Front Wheel Drive

1. Disconnect the power steering hoses from the pump.

2. Remove the adjusting bolt and slip off the belt.

3. Support the pump, remove the mounting bolts and lift out the pump.

4. Installation is the reverse of removal. Adjust the belt to specifications.

Tie Rod Ends

REMOVAL & INSTALLATION

Except Front Wheel Drive

1. Raise the front of the truck and support it safely on jackstands.

2. Remove the cotter pin and the nut from the end of the tie rod.

3. Install a puller and apply sufficient pressure to release the tie rod end from the knuckle. Measure or count the number of exposed threads.

4. Loosen the tie rod sleeve clamping bolt and unscrew the tie rod end.

5. Install the new new tie rod end into the sleeve, to the number of turns counted when removing the old end.

2. Connect the tie rod end to the knuckle and tighten the nut as follows: 1/2 in. nut to 45 ft. lbs.: 9/16 in. nut to 55 ft. lbs.: 5/8 in. nut to 75 ft. lbs. Install the cotter pin.

3. Lower the truck and adjust the toe-in.

4. Tighten the clamping bolts.

Front Wheel Drive

1. Raise the front of the truck and safely support it on jackstands.

2. Remove the cotter pin and the nut from the end of the tie rod.

3. Install a puller and apply sufficient pressure to release the tie rod from the knuckle arm.

4. Loosen the jam nut and unscrew the tie rod end.

5. Installation is in the reverse order of removal. Have the toe adjustment checked.

	TORQUE	
LET	NEWTON METERS	POUNDS
Ⓐ	28	250 INCH
Ⓑ	47	35 FOOT
Ⓒ	75	55 FOOT

Labels: TIE ROD ADJUSTING NUT (2) Ⓒ, BOLT AND WASHER ASSEMBLY (2) Ⓐ, BOLT AND WASHER ASSEMBLY (2) Ⓐ, STEERING KNUCKLE (2), TIE ROD END (2), COTTER PIN (2), NUT (2) Ⓑ, MOUNTING BRACKET, BUSHING, STEERING GEAR, BRACKET, FRONT CROSSMEMBER, VIEW IN CIRCLE Z

Typical steering assembly mounting—Mini-Van

BRAKES

Refer to the Brakes section of Unit Repair for service procedures.

CHASSIS ELECTRICAL

Heater Core

REMOVAL & INSTALLATION

Without Air Conditioning (Except Front Wheel Drive)

1. Disconnect the battery ground cable.
2. Drain the radiator.

NOTE: Remove the radiator and grille for clearance, if necessary.

3. Cover the alternator with a waterproof cover.
4. Disconnect the blower motor resistor and ground wires from the heater.
5. Disconnect and plug the heater core hoses.
6. Disconnect the control cables and underdash braces.
7. Remove the retaining screws from the water valve. Do not disconnect the hoses from the water valve; place the water valve with hoses attached to one side.
8. Remove the blower motor cooler tube.
9. Remove the nuts holding the housing to the mounting studs and tip the complete unit for removal.
10. To remove the heater core: Remove the retaining nuts and lift the blower assembly out of the housing.
11. Remove the cover retaining nuts and lift the cover off the housing.
12. Remove the core retaining screws and lift the core out of the housing.
13. Installation is the reverse of removal. Fill the cooling system.
14. Let the engine warm up with the heater on, then check the coolant level.

With A/C Vans and Pickups (Except Front Wheel Drive)

— CAUTION —

The air conditioning system must be discharged to remove the heater core. Do not attempt this if you are not familiar with air conditioning service; have someone trained in this work discharge the system.

1. Disconnect the battery ground cable and drain the coolant.
2. Remove the grille, condenser, and radiator, if necessary.

3. Place a waterproof cover over the alternator.
4. Disconnect the heater hoses at the water valve and remove the valve and bracket. Disconnect and cap the refrigerant lines.
5. Remove the glovebox, spot cooler bezel, and appearance shield.
6. Working through the glovebox opening, remove the evaporator housing to firewall screws and nuts.
7. Remove the wiper motor. Detach all evaporator housing vacuum and electrical connections. Detach the blower motor cooling hose and the drain hoses.
8. Remove the two 2¼ in. bolts from the crossbar and the four screws from the sealplate on the front of the housing. Separate the evaporator and blower motor housings, remove the evaporator housing.
9. Remove the receiver drier and cap all the openings. Carefully pry the heater core out, leaving the air seal at the front intact.
10. On installation, connect the hoses to the core. Position the evaporator housing on top of the blower housing. Install the mounting screws and nuts.
11. Position the crossbar under the lip on the blower housing opening and install the two 2¼ in. bolts. Install the four seal plate screws at the front of the housing.
12. Replace the wiper motor and connect the vacuum and electrical lines. Connect the blower motor cooler hose and the drain hoses.
13. Connect the heater hoses to the water valve.
14. Install the receiver drier and connect the refrigerant lines.
15. Install the radiator, condenser, and grille.
16. Replace the glove box, spot cooler bezel, and appearance shield.
17. Install the battery ground cable and fill the cooling system. Let the engine warm up with the *heater on*, then check the coolant. Have the air conditioner charged.

Front Wheel Drive

Heater Assembly and Core

Front Wheel Drive with A/C

Through 1984

NOTE: Removal of the Heater-Evaporator Unit is required for core removal. Two people will be required to perform the operation. Discharge, evacuation and recharge and leak testing of the refrigerant system is necessary.

— CAUTION —

This work should be performed only by a trained technician who is equipped with the proper tools. Have the system discharged (or do so yourself if you are properly trained and equipped) before attempting removal.

1. Discharge system before opening any lines and drain engine coolant. Disconnect negative battery cable.
2. Disconnect heater hoses at heater core. Plug heater core tube openings to prevent coolant from spilling out when assembly is removed.
3. Disconnect vacuum lines at brake booster and at water valve.
4. Remove "H" valve.
5. Remove condensate drain tube.
6. Remove evaporator heater assembly to dash retaining nuts.
7. Remove wire connector from resistor block, push out dash grommet, feed wire through grommet hole into passenger compartment.
8. Remove the steering wheel and lower the steering column to the drivers seat.
9. Remove lower instrument panel. The lower instrument panel must be disconnected to the point that the right side can be moved rearward and rested on the passenger seat. The left side of the instrument panel can remain electrically connected. This will disconnect the blower motor and resistor block wiring, the temperature control cable and vacuum harness from the instrument panel control.
10. Remove the evaporator heater unit hangar strap and swing out of the way.
11. Pull the unit rearward and out of vehicle.
12. Place evaporator heater assembly on work bench in position as would be viewed by front seat passenger.
13. Remove one (1) screw retaining the vacuum harness. Feed harness through hole in cover.
14. Remove thirteen (13) screws from cover, remove cover. Temperature control door will come out with the cover. Remove the nut and lever from the door shaft to remove the temperature door from the cover.
15. Remove screw from heater core tube retaining bracket and lift core out of unit.
16. Blower assembly, sound helmet, motor and wheel. Remove five (5) screws from the sound helmet.
17. To remove blower wheel remove retaining clamp from blower wheel hub and slide blower wheel from blower motor shaft.
18. Remove motor by taking out three (3) screws in helmet. Remove motor with wires and grommet.
19. Place heater core into unit and fasten with screws.
20. Reinstall blower wheel onto blower motor shaft and secure with retaining clamp.
21. Feed blower motor wires through hole in sound helmet. Lower blower assembly (rubber seal in place) into helmet. Pull wiring grommet into place and install three (3) mounting screws. Entire blower assembly and helmet can be installed into fan scroll with five (5) screws.
22. Install cover: Line up to housing using pilot pin, temperature door and screw holes for alignment. Drive thirteen (13) screws.
23. Vacuum Harness—Route through

holes in cover. Drive one (1) screw to retain vacuum harness. Make all vacuum hose attachments to actuators.

— **CAUTION** —

Care must be taken that the vacuum lines to the engine compartment do not hang up on the accelerator, or become trapped between the assembly and dash. Such a situation will result in kinked lines and the necessity of a second unit removal to free them. Proper routing of these lines as the unit is positioned to the dash may require two people.

24. Place unit on the floor as far forward under the instrument panel as possible.

25. Raise unit taking care not to catch studs in dashliner. Route vacuum lines through opening.

26. Position assembly in place and attach hanger strap.

27. Reinstall assembly to dash retaining nuts and tighten.

28. Install condensate drain tube.

29. Remove plugs from heater core tube openings and install heater hoses.

30. Install vacuum lines (Black line to brake booster and grey line to water valve).

31. Install ''H'' valve. See ''H'' valve installation.

32. Make sure that the evaporator heater assembly seal is properly aligned with top of center distribution duct opening.

33. Reinstall lower instrument panel.

34. Route resistor block wires through dash and install remaining parts in reverse order of removal. Recharge system.

1985–86

NOTE: Removal of the Heater-Evaporator Unit is required for core removal. Two people will be required to perform the operation. Discharge, evacuation and recharge and leak testing of the refrigerant system is necessary.

— **CAUTION** —

This work should be performed only by a trained technician who is equipped with the proper tools. Have the system discharged (or do so yourself if you are properly trained and equipped) before attempting removal.

1. Discharge system before opening any lines and drain engine coolant. Disconnect negative battery cable.

2. Remove the lower instrument panel on the passengers' side. Remove the seven attaching screws and remove the steering column cover from underneath the steering column.

3. Remove the lower column reinforcement mounting screws (3) and remove the lower column reinforcement (located under the column).

4. Remove the right side cowl and sill trim. Remove the bolt fastening the right side of the instrument panel to the right cowl. Then, loosen the two brackets holding the lower edge to the A/C/Heater or Heater housing.

5. Remove the instrument panel trim moulding–covering reinforcement. Remove the screws fastening the right side of the instrument panel to the steering column.

6. Disconnect the source vacuum line at the brake booster (in the engine compartment). If the vehicle has air conditioning, also disconnect the vacuum line at the water valve.

7. Clamp off the heater core hoses to prevent leakage, Then, remove the hoses from the core. Cap off the openings in the core to retain coolant.

8. Disconnect the air conditioning lines at the ''H'' valve. *Cap off these openings in a secure manner to keep moisture out of the A/C system.*

9. Remove the four nuts from the studs where the heater or heater/A/C assembly is mounted. Then, pull the right side of the lower instrument panel rearward until it reaches the passenger seat. With the panel in this position, disconnect the blower motor, resistor wiring, and temperature control cable.

10. Disconnect the hanger strap from the unit and bend it rearward for clearance. Then, pull the assembly rearward from the dashboard and out of the vehicle.

11. Place evaporator heater assembly on work bench in position as would be viewed by front seat passenger.

12. Remove one (1) screw retaining the vacuum harness. Feed harness through hole in cover.

13. Remove thirteen (13) screws from cover, remove cover. Temperature control door will come out with the cover. Remove the nut and lever from the door shaft to remove the temperature door from the cover.

14. Remove screw from heater core tube retaining bracket and lift core out of unit.

15. Install the core and then the unit in reverse order. Make sure to properly refill the cooling system and recharge the air conditioning system (the air conditioning work to be done only by someone who is qualified and has the proper tools). Run the engine and check for coolant leaks.

Auxiliary Heater, Core and Blower

The auxiliary heater, in some vans, is a rear-mounted recirculating heater, temperature controlled by a Bowden cable.

REMOVAL & INSTALLATION

1. Drain the radiator.

2. Disconnect the negative battery cable.

3. From under the vehicle, disconnect the intake and outlet hoses from the heater. Remove the water valve with hoses attached.

4. Remove the screws that mount the heater to the floor pan and disconnect the wiring to the heater.

5. Remove the heater cover attaching screws. The heater core is attached to the heater cover and will come out with the cover.

6. Remove the core from the cover.

7. To remove the auxiliary heater blower: Remove the screws attaching the blower motor to the heater and remove the blower motor.

8. Installation is the reverse of removal. Be sure to fill the cooling system and check for leaks. Check the operation of the auxiliary heater.

Radio

REMOVAL & INSTALLATION

1979 and Later (Except Front Wheel Drive)

1. Disconnect the negative battery cable.

2. Remove the seven instrument panel and bezel attaching screws. Pull the bezel off the retaining clip.

3. Remove the five instrument cluster screws.

4. Pull the cluster out far enough to gain access to the speedometer cable. Push the cable spring clip toward the cluster and disconnect the cable.

5. Remove the right and left printed circuit board multiple connectors.

6. Remove the instrument cluster.

7. Remove the radio mounting screws.

8. Remove the ground strap screw.

9. Pull the radio out of the instrument panel and disconnect the wiring.

10. Reverse the removal procedure to install.

Front Wheel Drive

1. Remove the seven bezel attaching screws and open the glove compartment.

2. Remove the bezel, guiding the right end around the glove compartment and away from the panel.

3. Disconnect the radio ground strap and remove the two radio mounting screws.

4. Pull the radio from the panel and disconnect the wiring and antenna lead.

5. Installation is the reverse of removal.

TRIMMING THE ANTENNA

All radios are trimmed at the factory and should require no further trimmer adjustment unless the radio is being installed after repair, or if trimmer adjustment is desired because of poor performance.

1. Extend the antenna to full length or to 31–33 in. for best FM reception.

2. Tune the radio to a weak signal between 1400 and 1600 kilocycles on the AM band.

3. Increase the radio volume and set the tone control to maximum treble.

4. The trimmer screw on most radios is located lower rear right-hand corner of the radio and can be reached by inserting a screwdriver into the recess hole.

5. Adjust the trimmer by turning it back and forth until the peak response in volume is obtained.

Wiper Motor

REMOVAL & INSTALLATION

Except Front Wheel Drive

The wiper motor is removed from under the hood. It is not necessary to remove the cowl grille.

1. Unplug the electrical wiring at the motor.

2. Remove the 3 mounting bolts from the wiper motor flange.

3. Lower the motor down far enough to gain access to the crank arm to drive link bushing. Pry the bushing from the crank arm.

4. Remove the motor.

5. Hold the drive crank with a wrench while removing the crank nut. Remove the drive crank from the motor.

6. Installation is the reverse of removal. Check and adjust (if necessary) the wiper arm park position.

Front Wheel Drive

1. Place the wipers in the Park position.

2. Remove wiper arms and blades, disconnect hoses from connectors.

3. Open the hood assembly.

4. Remove cowl top plenum grill and disconnect washer hose from connector.

5. Remove cowl plenum chamber plastic screen.

6. Remove the wiper pivot screws.

7. Disengage the pivots from the cowl top mounting positions.

8. Push pivots down into plenum chamber.

9. Disconnect wiper motor wiring harness from suppressor.

10. Remove three (3) mounting nuts from wiper motor.

11. Remove wiper motor assembly and linkage.

12. Clamp motor crank in a vise and remove nut from end of motor shaft. Do not rotate motor output shaft from park position. Remove crank from motor.

13. Do the necessary service needed on the motor, pivots, links and cranks.

DRIVE LINK

1. Remove the wiper arms.

2. Remove the cowl grille cover.

3. Reach through the access hole and remove the drive link from the connecting arm and the crank arm. Remove the connecting link pins by prying.

4. Remove the drive link through the access hole.

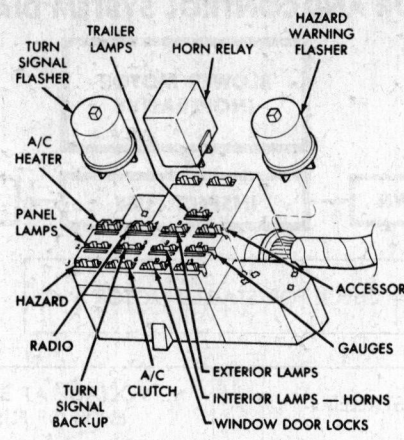

Typical fuse block–1981 shown

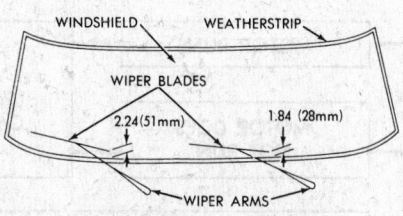

ADJUST WIPER ARM SO HEEL OF BLADE IS ABOVE THE WEATHERSTRIP IN PARK POSITION AS SHOWN ±.25

Adjusting wiper arms

5. Reverse to install. Use slip joint pliers to snap the links and pins together.

CONNECTING LINK
Except Front Wheel Drive

1. Remove the cowl grille cover.

2. Reach through the access hole and remove the connecting link from the drive link and the pivot pins by prying the bushings apart.

3. Remove the connecting link through the access hole.

4. Reverse to install. Use slip joint pliers to snap the links and pins together.

WIPER ARM ADJUSTMENT
Except Front Wheel Drive

To determine if an adjustment is required, apply a downward force of 25 oz parallel to the windshield glass at the tip of the wiper arm (where the blade is attached to the arm). With this force applied, pull the wiper blade away from the glass several times to prevent glass friction from affecting downward movement. The clearance between the tip of the wiper blade and the windshield molding should be as specified in the illustration. If the clearance is not as specified, reposition the wiper arm.

To remove the arm, lift it and look for a spring retainer at the bottom. Hold the retainer out of the way, wiggle the arm, and pull it off. To replace it, just push it

on, making sure that the latch is out of the way.

Instrument Cluster

REMOVAL & INSTALLATION

Except Front Wheel Drive

For instrument cluster removal and installation procedures, refer to Steps 1–6 of the 1979 and Later Radio Removal and Installation procedure.

Front Wheel Drive

1. Open the ash tray, remove tray and disconnect lighter.

2. From underneath panel, remove the mounting screw that attach upper and lower panels.

3. Push housing forward and lower rear to remove.

4. Reverse procedure for installation.

Headlights

REMOVAL & INSTALLATION

1. Remove the headlight bezel attaching screws and remove the bezel. On some models it may not be necessary to remove the bezel, depending on whether the headlight retaining ring is accessible.

2. Remove the headlight ring attaching screws and remove the ring. Do not touch the headlight aiming screws.

3. Pull the sealed beam out slightly and unplug the connector.

4. Reverse to install. As long as the adjusting screws were not moved, the headlight will be held in proper adjustment by the retaining ring.

Fuses and Fusible Links

The fuse panel is usually located on the left side of the steering column, either on the kick panel, steering colum or firewall.

Always replace a blown fuse with one of the same rating. If the same fuse continues to blow out, check the circuit for overload or short circuit.

When a fusible link blows it is very important to find out why it blew. They are placed in vehicles electrical system for protection against dead shorts to ground which can be caused by electrical component failure or various wiring failure. Do not just replace a fusible link to correct a problem, find out what caused the problem.

HEATER BLOWER MOTOR AND CONTROL SYSTEM DIAGNOSIS (ELECTRICAL)

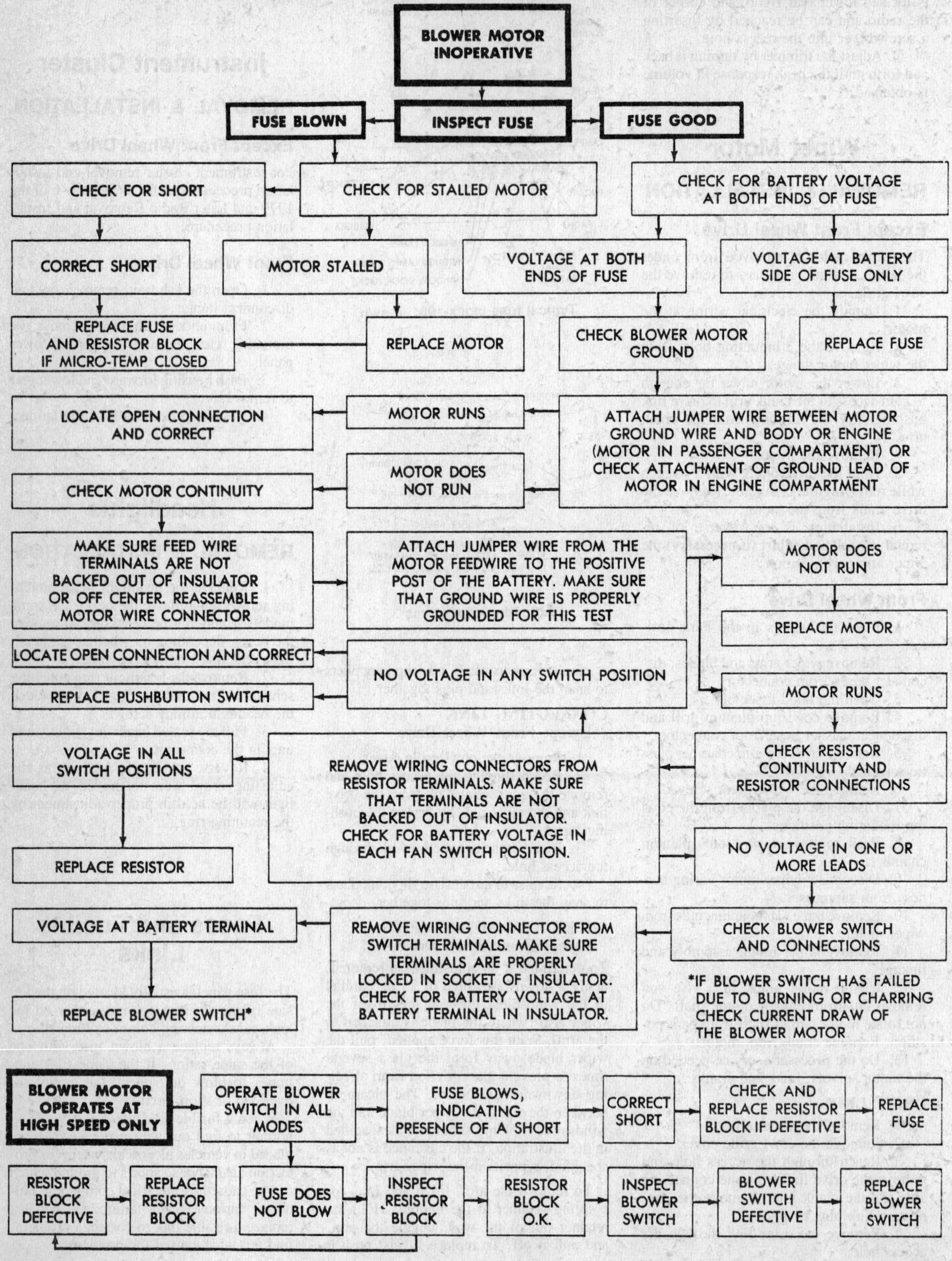

BLOWER MOTOR INOPERATIVE

FUSE BLOWN ← **INSPECT FUSE** → **FUSE GOOD**

CHECK FOR SHORT ← **CHECK FOR STALLED MOTOR**

CHECK FOR BATTERY VOLTAGE AT BOTH ENDS OF FUSE

CORRECT SHORT

MOTOR STALLED

VOLTAGE AT BOTH ENDS OF FUSE

VOLTAGE AT BATTERY SIDE OF FUSE ONLY

REPLACE FUSE AND RESISTOR BLOCK IF MICRO-TEMP CLOSED ← **REPLACE MOTOR**

CHECK BLOWER MOTOR GROUND

REPLACE FUSE

LOCATE OPEN CONNECTION AND CORRECT ← **MOTOR RUNS** ← **ATTACH JUMPER WIRE BETWEEN MOTOR GROUND WIRE AND BODY OR ENGINE (MOTOR IN PASSENGER COMPARTMENT) OR CHECK ATTACHMENT OF GROUND LEAD OF MOTOR IN ENGINE COMPARTMENT**

CHECK MOTOR CONTINUITY ← **MOTOR DOES NOT RUN** ←

MAKE SURE FEED WIRE TERMINALS ARE NOT BACKED OUT OF INSULATOR OR OFF CENTER. REASSEMBLE MOTOR WIRE CONNECTOR

ATTACH JUMPER WIRE FROM THE MOTOR FEEDWIRE TO THE POSITIVE POST OF THE BATTERY. MAKE SURE THAT GROUND WIRE IS PROPERLY GROUNDED FOR THIS TEST

MOTOR DOES NOT RUN

REPLACE MOTOR

LOCATE OPEN CONNECTION AND CORRECT ←

NO VOLTAGE IN ANY SWITCH POSITION

MOTOR RUNS

REPLACE PUSHBUTTON SWITCH ←

VOLTAGE IN ALL SWITCH POSITIONS ← **REMOVE WIRING CONNECTORS FROM RESISTOR TERMINALS. MAKE SURE THAT TERMINALS ARE NOT BACKED OUT OF INSULATOR. CHECK FOR BATTERY VOLTAGE IN EACH FAN SWITCH POSITION.**

CHECK RESISTOR CONTINUITY AND RESISTOR CONNECTIONS

REPLACE RESISTOR

NO VOLTAGE IN ONE OR MORE LEADS

VOLTAGE AT BATTERY TERMINAL ← **REMOVE WIRING CONNECTOR FROM SWITCH TERMINALS. MAKE SURE TERMINALS ARE PROPERLY LOCKED IN SOCKET OF INSULATOR. CHECK FOR BATTERY VOLTAGE AT BATTERY TERMINAL IN INSULATOR.**

CHECK BLOWER SWITCH AND CONNECTIONS

REPLACE BLOWER SWITCH*

*IF BLOWER SWITCH HAS FAILED DUE TO BURNING OR CHARRING CHECK CURRENT DRAW OF THE BLOWER MOTOR

BLOWER MOTOR OPERATES AT HIGH SPEED ONLY → **OPERATE BLOWER SWITCH IN ALL MODES** → **FUSE BLOWS, INDICATING PRESENCE OF A SHORT** → **CORRECT SHORT** → **CHECK AND REPLACE RESISTOR BLOCK IF DEFECTIVE** → **REPLACE FUSE**

RESISTOR BLOCK DEFECTIVE → **REPLACE RESISTOR BLOCK**

FUSE DOES NOT BLOW → **INSPECT RESISTOR BLOCK** → **RESISTOR BLOCK O.K.** → **INSPECT BLOWER SWITCH** → **BLOWER SWITCH DEFECTIVE** → **REPLACE BLOWER SWITCH**

Ford
Pickups, Vans, Bronco, Bronco II, Ranger

ENGINE IDENTIFICATION

C.I.D.	Liter	C.I.D.	Liter	C.I.D.	Liter
122	2.0	232	3.8	302	5.0
140	2.3	255	4.1	351	5.8
173	2.8	300	4.9	400	6.6
179	2.9			460	7.5

ENGINE CODE SPECIFICATIONS
Gasoline Engine Codes

(The Engine Identification Code letter is the fourth character in the vehicle identification number for Ford vehicles. See the chart below for code letter information.)

	Model Year and Engine Code							
Engines	'79	'80	'81	'82	'83	'84	'85	'86
122-4	—	—	—	—	K	K	K	K
140-4	—	—	—	—	Z	Z	Z	Z
173-V6	—	—	—	—	S	S	S	—
179-V6	—	—	—	—	—	—	—	④
300—6 Cylinder 1V	B	E	E	E ①	Y	Y	Y	Y
232-V6	—	—	—	3	3	3	—	—
255—V8	—	—	D	D	—	—	—	—
302—V8	G	F	F	F	F	F	F	F
351—V8	H	G	G ②	W	G	G	G ③	G
400—V8	S	Z	Z	Z	—	—		
460—V8 Econoline	A	L	L	L	L	L	L	L
Light Truck	J	L	L	L	L	L	L	L

① 9—Gasoline engine for LP
conversion
② W = E-100 Windsor only, E-150
less leaded fuel
G = E-150, with leaded fuel E-250,
E-350
③ 4 bbl.
④ Not available at time of
publication

FORD

GENERAL ENGINE SPECIFICATIONS

Cu. In. Displace-ment	Year	Bore × Stroke	Firing Order	Developed Horse Power @ rpm	Developed Torque @ rpm	Com-pression Ratio	Carbu-retor	Valve Lifter Type	Normal Oil Pressure (p.s.i.)
FOUR CYLINDER									
122	'83–'86	3.52 × 3.13	1.3.4.2	73 @ 4000	107 @ 2406	9.0:1	1V	Hyd.	40–60
140	'83–'86	3.78 × 3.13	1-3-4-2	79 @ 3800 ⑦	124 @ 2400 ⑧	9.0:1	1V	Hyd.	40–60
SIX CYLINDER									
173-V6	'83–'85	3.66 × 2.70	1-4-2-5-3-6	115 @ 4600	150 @ 2600	8.7:1	2V	Mech.	40–60
179-V6	'86	3.66 × 2.83	1-4-2-5-3-6	140 @ 4600	170 @ 2600	9.3:1	FI	Hyd.	40–60
300	'79–'86	4.00 × 3.98	1-5-3-6-2-4	120 @ 3400 ①	229 @ 1400 ②	8.0:1 ③	1V	Hyd.	40–60
232-V6	'82–'84	3.81 × 3.39	1-4-2-5-3-6	112 @ 4000	175 @ 2600	8.6:1	2V	Hyd.	54–59
EIGHT CYLINDER									
255	'81–'82	3.68 × 3.00	1-5-4-2-6-3-7-8	111 @ 3400	194 @ 1600	8.2:1	2V	Hyd.	40–60
302	'78–'84	4.00 × 3.00	1-5-4-2-6-3-7-8	④	⑤	8.0:1 ⑥	2V	Hyd.	40–60
351W	'79–'86	4.00 × 3.50	1-3-7-2-6-5-4-8	N.A.	N.A.	N.A.	2V	Hyd.	40–65
351M	'79	4.00 × 3.50	1-3-7-2-6-5-4-8	N.A.	N.A.	N.A.	2V	Hyd.	50–75
351	'83–'85	4.00 × 3.50	1-3-7-2-6-5-4-8	210 @ 4000	305 @ 2800	8.3:1	4V	Hyd.	40–65
400	'79–'82	4.00 × 4.00	1-3-7-2-6-5-4-8	N.A.	N.A.	N.A.	2V	Hyd.	50–75
460	'79–'86	4.36 × 3.85	1-5-4-2-6-3-7-8	N.A.	N.A.	N.A.	4V	Hyd.	40–65

① '79: E-100 exc Calif.; all E-150: 117 @ 3000
  E-250 Manual trans.:
  114 @ 3000
  E-250 Automatic trans.:
  116 @ 3200
  E-350: 114 @ 3000
  '80–'81: Bronco, F-100, 250 (49 States)— 119 @ 3200
  E-100, 250 (49 States)— 115 @ 3200
  All Caif. 116 @ 3200
  '82: N.A.
② '79: E-100, 150: 243 @ 1600
  E-250 Manual trans.:
  234 @ 1600
  E-250 Automatic trans.:
  247 @ 1000
  E-350: 247 @ 1000
  '80–'81: Bronco, F-100, 250 (49 States)— 243 @ 1200
  E-100, 250 (49 States)— 241 @ 1200
  All Calif.—244 @ 1200
  '82: N.A.
③ '79–'80: E-100, 150, 250—8.9:1
  E-350—8.0:1
  '81–'82: 8.9:1
④ '79: E-100, 150 exc Calif.:
  135 @ 3400
  E-100 Calif.: 129 @ 3200
  E-150 Calif.: 137 @ 3400
  E-250: 136 @ 3400

'80–'81: Bronco, F-100, 250 (49 States)—137 @ 3600
  E-100, 250 (49 States)— 138 @ 3600
  Bronco, F-150 (4 × 4), 250, E-100-250 Calif.: 136 @ 3600
  F-100, 150 (4 × 2) Calif.: 133 @ 3400
  '82: N.A.
⑤ '79: E-100, 150 exc Calif.:
  243 @ 2000
  E-100 Calif.: 238 @ 2400
  E-150 Calif.: 245 @ 2000
  E-250: 235 @ 2400
  '80–'81: Bronco, F-100, 250 (49 States)—239 @ 1800
  E-100, 250 (49 States)— 242 @ 1800
  Bronco, F-150 (4 × 4) Calif.—235 @ 2000
  '82: N.A.
⑥ '79: E-100, 150, 250: 8.9:1
  E-350: 8.0:1
  '80–'82: 8.4:1
⑦ Automatic Trans; 82 @ 4200
⑧ Automatic Trans; 126 @ 2200

TUNE-UP SPECIFICATIONS
1979–80

(For 1979 and 1980 Tune-Up Specifications consult the Vehicle Emissions Control Label, which is located on the engine of the vehicle. This decal will contain a calibration number which when used in conjunction with the chart below will yield the required tune-up information. If the information given in this chart disagrees with the information on the decal, use the information on the decal.)

Calibration	Spark Plug Gap	Ignition Timing	Fast Idle RPM High Cam	Kick Down	Curb Idle rpm A/C ① Off/On	Non A/C	Tsp Off rpm A/C	Non A/C
9-51G-RO	.042–.046	6°BTDC	—	1600	700	700	500	500
9-51J-RO	.042–.046	6°BTDC	—	1600	700	700	500	500
9-51K-RO	.042–.046	6°BTDC	—	1600	700	700	500	500
9-51L-RO	.042–.046	6°BTDC	—	1600	700	700	500	500
9-51M-RO	.042–.046	6°BTDC	—	1600	700	700	500	500
9-51S-RO	.042–.046	6°BTDC	—	1600	700	700	500	500
9-51T-RO	.042–.046	6°BTDC	—	1600	700	700	500	500
9-52G-RO	.042–.046	10°BTDC	—	1600	550	550	500	500
9-52J-RO	.042–.046	10°BTDC	—	1600	550	550	500	500
9-52L-RO	.042–.046	10°BTDC	—	1600	550	550	500	500
9-52M-RO	.042–.046	10°BTDC	—	1600	550	550	500	500
9-53G-RO	.042–.046	6°BTDC	2000	—	700	700	—	—
9-53H-RO	.042–.046	4°BTDC	2000	—	700	700	—	—
9-54G-RO	.042–.046	8°BTDC	2000	—	600	600	550	550
9-54H-RO	.042–.046	6°BTDC	2000	—	600	600	550	550
9-54J-RO	.042–.046	6°BTDC	2000	—	600	600	550	550
9-54R-RO	.042–.046	6°BTDC	2000	—	600	600	550	550
9-54S-RO	.042–.046	8°BTDC	2000	—	650	650	550	550
9-54T-RO	.042–.046	6°BTDC	2000	—	600	600	550	550
9-54U-RO	.042–.046	6°BTDC	2000	—	650	650	550	550
9-59H-RO	.042–.046	10°BTDC	2000	—	650	650	—	—
9-59J-RO	.042–.046	10°BTDC	2000	—	650	650	—	—
9-59K-RO	.042–.046	10°BTDC	2000	—	650	650	—	—
9-59S-RO	.042–.046	8°BTDC	2000	—	650	650	—	—
9-59T-RO	.042–.046	10°BTDC	2000	—	650	650	—	—
9-60G-RO	.042–.046	6°BTDC	2000	—	550	550	—	—
9-60H-RO	.042–.046	6°BTDC	2000	—	550	550	—	—
9-60J-RO	.042–.046	6°BTDC	2000	—	550	550	—	—
9-60L-RO	.042–.046	6°BTDC	2000	—	550	550	—	—
9-60M-RO	.042–.046	6°BTDC	2000	—	550	550	—	—
9-60S-RO	.042–.046	10°BTDC	2100	—	550	550	—	—
9-61G-RO	.042–.046	10°BTDC	2000	—	650	650	—	—
9-61H-RO	.042–.046	10°BTDC	2000	—	650	650	—	—
9-62J-RO	.042–.046	6°BTDC	1900	—	550	550	—	—
9-62M-RO	.042–.046	6°BTDC	1900	—	550	550	—	—
9-63H-RO	.042–.046	4°BTDC	—	1500	800	800	500	500
9-64G-RO	.042–.046	10°BTDC	2200	—	600	600	500	500
9-64H-RO	.042–.046	12°BTDC	2200	—	600	600	500	500
9-64S-RO	.042–.046	8°BTDC	2200	—	600	600	500	500
9-66G-RO	.042–.046	14°BTDC	—	1600	650	650	800 ②	800
7-76J-R11	.042–.046	6°BTDC	—	1700	650	650	525 ②	525
7-93J-RO	.042–.046	10°BTDC	2500	—	600	600	—	—
7-95J-RO	.042–.046	10°BTDC	2500	—	600	600	—	—
9-71J-RO	.042–.046	6°BTDC	1750	—	600	600	—	—
9-72J-RO	.042–.046	12°BTDC	2000	—	600	600	500	500
9-73-RO	.042–.046	3°BTDC	1750	—	600	600	—	—
9-74J-RO	.042–.046	3°BTDC	2000	—	600	600	500	500
9-77J-RO	.042–.046	12°BTDC	—	1600	700	700	500	500
9-77M-RO	.042–.046	12°BTDC	2550	—	—	700	—	500
9-78J-RO	.042–.046	12°BTDC	—	1600	550	550	500	500
9-83G-RO	.042–.046	6°BTDC	2200	—	—	600	—	—
9-83H-RO	.042–.046	2°BTDC	2500	—	—	600	—	—

TUNE-UP SPECIFICATIONS
1979–80

(For 1979 and 1980 Tune-Up Specifications consult the Vehicle Emissions Control Label, which is located on the engine of the vehicle. This decal will contain a calibration number which when used in conjunction with the chart below will yield the required tune-up information. If the information given in this chart disagrees with the information on the decal, use the information on the decal.)

Calibration	Spark Plug Gap	Ignition Timing	Fast Idle RPM High Cam	Fast Idle RPM Kick Down	Curb Idle rpm A/C ① Off/On	Curb Idle rpm Non A/C	Tsp Off rpm A/C	Tsp Off rpm Non A/C
9-87G-RO	.042–.046	8°BTDC	2700	—	—	600	—	—
9-97J-RO	.042–.046	8°BTDC	—	1600	650	—	—	—
9-97J-R11	.042–.046	8°BTDC	—	1600	650	—	—	—

① Only for A/C-TSP equipped, A/C compressor electromagnetic clutch deenergized.

TUNE-UP SPECIFICATIONS
1981

(For 1981 Tune-Up Specifications consult the Vehicle Emissions Control Label, which is located on the engine of the vehicle. This decal will contain a calibration number which when used in conjunction with the chart below will yield the required tune-up information. If the information given in this chart disagrees with the information on the decal, use the information on the decal.)

Calibration Number	Engine	Spark Plug Gap	Ignition Timing °BTDC	Timing RPM	Fast Idle rpm High CAM	Fast Idle rpm Kick Down	Curb Idle rpm A/C On	Curb Idle rpm A/C off	Curb Idle rpm Non A/C
1-57G-R1	4.1L	.042-.046	4	800	2200	—	—	—	750
1-57G-R10	4.1L	.042-.046	4	800	2050	—	—	—	700
1-58-R0	4.1L	.042-.046	10	800	2000	—	—	—	575
1-51D-R0	4.9L	.042-.046	6	800	—	1400	700	600	600
1-51D-R10	4.9L	.042-.046	6	800	—	1250	650	550	550
1-51D-R12	4.9L	.042-.046	6	800	—	1250	650	550	550
1-51E-R0	4.9L	.042-.046	6	800	—	1400	700	600	600
1-51F-R0	4.9L	.042-.046	6	800	—	1250	650	550	550
1-51G-R0	4.9L	.042-.046	6	800	—	1250	650	550	550
1-51H-R0	4.9L	.042-.046	6	800	—	1250	650	550	550
1-51K-R0	4.9L	.042-.046	6	800	—	1250	650	550	550
1-51L-R0	4.9L	.042-.046	6	800	—	1250	650	550	550
1-51E-R10	4.9L	.042-.046	6	800	—	1400	700	600	600
1-51F-R10	4.9L	.042-.046	6	800	—	1250	650	550	550
1-51G-R10	4.9L	.042-.046	6	800	—	1250	650	550	550
1-51H-R10	4.9L	.042-.046	6	800	—	1250	650	550	550
1-51K-R10	4.9L	.042-.046	6	800	—	1250	650	550	550
1-51L-R10	4.9L	.042-.046	6	800	—	1250	650	550	550
1-51S-R0	4.9L	.042-.046	6	800	—	1400	700	600	600
1-51S-R10	4.9L	.042-.046	6	800	—	1250	650	550	550
1-51T-R0	4.9L	.042-.046	6	800	—	120	650	550	550
1-52G-R0	4.9L	.042-.046	10	800	—	1400	—	—	550
1-52H-R0	4.9L	.042-.046	10	800	—	1250	—	—	500
1-52K-R0	4.9L	.042-.046	10	800	—	1250	—	—	500
1-52L-R0	4.9L	.042-.046	10	800	—	1250	—	—	500
1-52G-R10	4.9L	.042-.046	10	800	—	1400	—	—	550
1-52H-R10	4.9L	.042-.046	10	800	—	1250	—	—	500
1-52K-R10	4.9L	.042-.046	10	800	—	1250	—	—	500
1-52L-R10	4.9L	.042-.046	10	800	—	1250	—	—	500
1-52S-R0	4.9L	.042-.046	10	800	—	1400	—	—	550
1-52T-R0	4.9L	.042-.046	10	800	—	1250	—	—	550
5-77-R1	4.9L	.042-.046	10	800	—	1500	—	—	600(A)
5-78-R1	4.9L	.042-.046	10	800	—	1500	—	—	700(M)
9-77J-R12	4.9L	.042-.046	12	800	1600	—	—	—	700
9-77S-R10	4.9L	.042-.046	10	800	1600	—	—	—	700
9-78J-R0	4.9L	.042-.046	12	800	—	1500	—	—	550

TUNE-UP SPECIFICATIONS
1981

(For 1981 Tune-Up Specifications consult the Vehicle Emissions Control Label, which is located on the engine of the vehicle. This decal will contain a calibration number which when used in conjunction with the chart below will yield the required tune-up information. If the information given in this chart disagrees with the information on the decal, use the information on the decal.)

Calibration Number	Engine	Spark Plug Gap	Ignition Timing °BTDC	Timing RPM	Fast Idle rpm High CAM	Fast Idle rpm Kick Down	Curb Idle rpm A/C On	Curb Idle rpm A/C off	Curb Idle rpm Non A/C
9-78J-R11	4.9L	.042-.046	12	800	1600	—	—	—	550
1-53D-R0	5.0L	.042-.046	8	800	2200	—	—	—	700
1-53F-R0	5.0L	.042-.046	8	800	2050	—	—	—	650
1-53G-R0	5.0L	.042-.046	8	800	2050	—	—	—	650
1-53H-R0	5.0L	.042-.046	8	800	2050	—	—	—	650
1-53K-R0	5.0L	.042-.046	8	800	2050	—	—	—	650
1-53D-R10	5.0L	.042-.046	8	800	2200	—	—	—	700
1-53G-R10	5.0L	.042-.046	8	800	2050	—	—	—	650
1-53K-R10	5.0L	.042-.046	8	800	2050	—	—	—	650
1-53D-R12	5.0L	.042-.046	8	800	2050	—	—	—	650
1-53F-R11	5.0L	.042-.046	8	800	2050	—	—	—	650
1-53G-R12	5.0L	.042-.046	8	800	2050	—	—	—	650
1-53H-R11	5.0L	.042-.046	8	800	2050	—	—	—	650
1-53K-R13	5.0L	.042-.046	8	800	2050	—	—	—	650
1-54D-R1	5.0L	.042-.046	8	800	2000	—	—	—	575
1-54K-R0	5.0L	.042-.046	8	800	1850	—	—	—	525
1-54F-R0	5.0L	.042-.046	8	800	2000	—	—	—	575
1-54G-R0	5.0L	.042-.046	8	800	2000	—	—	—	575
1-54H-R0	5.0L	.042-.046	8	800	2000	—	—	—	575
1-54L-R2	5.0L	.042-.046	8	800	2000	—	—	—	575
1-54L-R10	5.0L	.042-.046	8	800	1850	—	—	—	525
1-54P-R0	5.0L	.042-.046	—	—	1350	—	—	—	—
1-54R-R0	5.0L	.042-.046	—	—	1350	—	—	—	—
1-54P-R10	5.0L	.042-.046	—	—	1350	—	—	—	—
1-54R-R10	5.0L	.042-.046	—	—	1200	—	—	—	—
7-79-R1	5.0L	.042-.046	6	800	—	1250	—	—	750
7-80-R0	5.0L	.042-.046	6	800	—	1500	—	—	650
1-59A-R0	5.8L	.042-.046	10	800	2000	—	650	—	650
1-59B-R0	5.8L	.042-.046	10	800	2000	—	650	—	650
1-59G-R0	5.8L	.042-.046	10	800	2000	—	650	—	650
1-59H-R0	5.8L	.042-.046	10	800	2000	—	650	—	650
1-59K-R0	5.8L	.042-.046	10	800	2000	—	650	—	650
1-59A-R10	5.8L	.042-.046	10	800	2000	—	650	—	650
1-59B-R10	5.8L	.042-.046	10	800	1850	—	600	—	600
1-59G-R10	5.8L	.042-.046	10	800	1850	—	600	—	600
1-59H-R10	5.8L	.042-.046	10	800	1850	—	600	—	600
1-59K-R10	5.8L	.042-.046	10	800	1850	—	600	—	600
1-60A-R0	5.8L	.042-.046	6	800	2200	—	—	—	—
1-60B-R0	5.8L	.042-.046	6	800	2200	—	—	—	—
1-60H-R1	5.8L	.042-.046	6	800	2000	—	625	550	550
1-60J-R0	5.8L	.042-.046	6	800	2000	—	625	550	550
1-60K-R0	5.8L	.042-.046	6	800	2000	—	625	550	550
1-60A-R10	5.8L	.042-.046	6	800	2000	—	625	550	550
1-60B-R10	5.8L	.042-.046	6	800	1850	—	575	500	500
1-60H-R10	5.8L	.042-.046	6	800	1850	—	575	500	500
1-60J-R10	5.8L	.042-.046	6	800	1850	—	575	500	500
1-60K-R10	5.8L	.042-.046	6	800	1850	—	575	500	500
1-63T-R0	5.8L	.042-.046	—	—	1700	—	—	—	—
1-64A-R0	5.8L	.042-.046	8	800	2000	—	625	550	550
1-64G-R1	5.8L	.042-.046	10	600	2000	—	625	550	550
1-64H-R2	5.8L	.042-.046	10	600	2000	—	625	550	550
1-64R-R1	5.8L	.042-.046	—	—	1650	—	—	—	—

FORD

TUNE-UP SPECIFICATIONS
1981

(For 1981 Tune-Up Specifications consult the Vehicle Emissions Control Label, which is located on the engine of the vehicle. This decal will contain a calibration number which when used in conjunction with the chart below will yield the required tune-up information. If the information given in this chart disagrees with the information on the decal, use the information on the decal.)

Calibration Number	Engine	Spark Plug Gap	Ignition Timing °BTDC	Timing RPM	Fast Idle rpm High CAM	Fast Idle rpm Kick Down	Curb Idle rpm A/C On	Curb Idle rpm A/C off	Curb Idle rpm Non A/C
1-64S-R0	5.8L	.042-.046	—	—	1500	—	—	—	—
1-64T-R0	5.8L	.042-.046	—	—	1500	—	—	—	—
7-76J-R11	5.8L	.042-.046	6	800	1700	—	—	—	600
9-71J-R10	5.8L	.042-.046	10	800	1750	—	—	—	600
9-71J-R11	5.8L	.042-.046	10	800	1750	—	—	—	600
9-72J-R11	5.8L	.042-.046	10	800	2000	—	—	—	600
9-72J-R12	5.8L	.042-.046	10	800	2000	—	—	—	600
9-83G-R12	6.1L	.042-.046	6	800	2200	—	—	—	600
9-83H-R11	6.1L	.042-.046	6	800	2500	—	—	—	600
9-83H-R14	6.1L	.042-.046	2	800	2500	—	—	—	600
9-73J-R11	6.6L	.042-.046	6	800	1750	—	—	—	600
9-73J-R12	6.6L	.042-.046	6	800	1750	—	—	—	600
9-74J-R11	6.6L	.042-.046	3	800	2000	—	—	—	600
9-74J-R12	6.6L	.042-.046	6	800	2000	—	—	—	600
9-87G-R11	7.0L	.042-.046	6	800	2700	—	—	—	600
9-97J-R0	7.5L	.042-.046	8	800	1600	—	—	—	650
7-93J-R0	7.8L	.038-.042	10	800	2500	—	—	—	600
7-95J-R0	8.8L	.038-.042	10	800	2500	—	—	—	600

TUNE-UP SPECIFICATIONS
1982

(For 1982 Tune-Up Specifications consult the Vehicle Emissions Control Label, which is located on the engine of the vehicle. This decal will contain a calibration number which when used in conjunction with the chart below will yield the required tune-up information. If the information given in this chart disagrees with the information on the decal, use the information on the decal.)

Calibration	Engine	Spark Plug Gap	Ignition Timing	Fast Idle rpm	Curb Idle rpm
2-54R-R0	5.0L	.042-.046		1350	—
2-54X-R1	5.0L	.042-.046	12° BTDC	2100	650
1-63T-R0	5.8L	.042-.046	—	1700	—
1-63T-R10B	5.8L	.042-.046	—	1700	—
1-64H-R2	5.8L	.042-.046	10° BTDC	2000	625
1-64R-R1	5.8L	.042-.046	—	1650	—
1-64S-R0	5.8L	.042-.046	—	1650	—
1-64T-R0	5.8L	.042-.046	—	1650	—
1-64T-R10	5.8L	.042-.046	—	1650	—
2-63Y-R10B	5.8L	.042-.046	—	1700	—
2-64X-R0	5.8L	.042-.046	14° BTDC	2000	625
2-64Y-R10B	5.8L	.042-.046	—	1650	—
9-77J-R12	4.9L	.042-.046	12° BTDC	1600	—
9-77G-R10	4.9L	.042-.046	10° BTDC	1600	—
9-78J-R0	4.9L	.042-.046	12° BTDC	1600	—
9-78J-R11	4.9L	.042-.046	12° BTDC	1600	—
2-75J-R17	5.8L	.042-.046	5° BTDC	1500	700 ④
2-76J-R17	5.8L	.042-.046	5° BTDC	1500	—
7-75J-R14	5.8L	.042-.046	6° BTDC	1500	700 ④
7-76J-R11	5.8L	.042-.046	6° BTDC	1700	—
7-76J-R13	5.8L	.042-.046	12° BTDC	1600	500
7-76J-R14	5.8L	.042-.046	6° BTDC	1700	—
7-76J-R15	5.8L	.042-.046	6° BTDC	1700	—
9-83G-R12	6.1L	.042-.046	6° BTDC	2200	600

T192

TUNE-UP SPECIFICATIONS
1982

(For 1982 Tune-Up Specifications consult the Vehicle Emissions Control Label, which is located on the engine of the vehicle. This decal will contain a calibration number which when used in conjunction with the chart below will yield the required tune-up information. If the information given in this chart disagrees with the information on the decal, use the information on the decal.)

Calibration	Engine	Spark Plug Gap	Ignition Timing	Fast Idle rpm	Curb Idle rpm
9-83H-R11	6.1L	.042-.046	6° BTDC	2500	600
9-83H-R14	6.1L	.042-.046	2° BTDC	2500	600
9-73J-R11	6.6L	.042-.046	6° BTDC	1750	600 ④
9-73J-R12	6.6L	.042-.046	6° BTDC	1750	600 ④
9-73J-R13	6.6L	.042-.046	6° BTDC	1750	600 ④
9-73J-R14	6.6L	.042-.046	6° BTDC	1750	600 ④
9-74J-R11	6.6L	.042-.046	6° BTDC	2000	—
9-74J-R12	6.6L	.042-.046	6° BTDC	2000	—
9-74J-R13	6.6L	.042-.046	3° BTDC	2000	—
9-74J-R14	6.6L	.042-.046	6° BTDC	2000	—
9-87G-R11	7.0L	.042-.046	6° BTDC	—	600
9-97J-R12	7.5L	.042-.046	8° BTDC	—	650

① A/C on—50 RPM Less if NON A/C
② A/C on—600 NON A/C
③ 100 RPM Less for NON A/C or A/C off
④ NON A/C.

TUNE-UP SPECIFICATIONS
1983

(For 1983 Tune-Up Specifications consult the Vehicle Emissions Control Label, which is located on the engine of the vehicle. This decal will contain a calibration number which when used in conjunction with the chart below will yield the required tune-up information. If the information given in this chart disagrees with the information on the decal, use the information on the decal.)

Calibration	Engine	Spark Plug Gap	Ignition Timing	Fast Idle rpm	Curb Idle rpm
3-41D-R01	2.0L	.032-.036	6° BTDC	2000	800
3-41D-R10	2.0L	.032-.036	6° BTDC	2000	800
3-41P-R02	2.0L	.032-.036	6° BTDC	2000	800
3-41P-R11	2.0L	.032-.036	6° BTDC	2000	800
3-41P-R12	2.0L	.032-.036	6° BTDC	2000	800
3-49S-R01	2.3L	.032-.036	6° BTDC	2000	850 ①
3-49S-R10	2.3L	.032-.036	6° BTDC	2000	850 ①
3-49S-R11	2.3L	.032-.036	6° BTDC	2000	850 ①
3-49X-R01	2.3L	.032-.036	10° BTDC	2000	850
3-49X-R11	2.3L	.032-.036	10° BTDC	2000	850
3-50S-R01	2.3L	.032-.036	8° BTDC	2000	750
3-50S-R01	2.3L	.032-.036	8° BTDC	2000	800
3-50X-R10	2.3L	.032-.036	8° BTDC	2000	800
3-50X-R11	2.3L	.032-.036	10° BTDC	2000	800
3-49G-R20	2.3L	.042-.046	6° BTDC	2000	850 ①
3-49H-R17	2.3L	.042-.046	6° BTDC	2000	850 ①
3-49S-R16	2.3L	.042-.046	6° BTDC	2000	850 ①
3-49T-R20	2.3L	.042-.046	6° BTDC	2000	850 ①
3-49T-R20	2.3L	.042-.046	6° BTDC	2000	850 ①
3-49Y-R19	2.3L	.042-.046	10° BTDC	2000	850 ①
3-50S-R18	2.3L	.042-.046	6° BTDC	2000	800
3-50Y-R18	2.3L	.042-.046	10° BTDC	2000	800
4-61F-R00	2.8L	.042-.046	10° BTDC	3000	850-900 ②
4-62D-R00	2.8L	.042-.046	10° BTDC	3000	850-900 ②
4-62S-R01	2.8L	.042-.046	10° BTDC	3000	850-900 ②
4-62S-R10	2.8L	.042-.046	10° BTDC	3000	800-900 ③
3-55D-R00	3.8L	.042-.046	2° BTDC	1300	550
3-56D-R00	3.8L	.042-.046	10° BTDC	2200	550
3-51D-R00	4.9L	.042-.046	6° BTDC	1600	700 ④
3-51E-R01	4.9L	.042-.046	6° BTDC	1600	700 ④

 **FORD**

TUNE-UP SPECIFICATIONS
1983

(For 1983 Tune-Up Specifications consult the Vehicle Emissions Control Label, which is located on the engine of the vehicle. This decal will contain a calibration number which when used in conjunction with the chart below will yield the required tune-up information. If the information given in this chart disagrees with the information on the decal, use the information on the decal.)

Calibration	Engine	Spark Plug Gap	Ignition Timing	Fast Idle rpm	Curb Idle rpm
3-51F-R00	4.9L	.042-.046	6° BTDC	1600	700 ④
3-51G-R00	4.9L	.042-.046	6° BTDC	1600	700 ④
3-51H-R00	4.9L	.042-.046	6° BTDC	1600	700 ④
3-51K-R00	4.9L	.042-.046	6° BTDC	1600	700 ④
3-51L-R00	4.9L	.042-.046	6° BTDC	1600	700 ④
3-51P-R00	4.9L	.042-.046	10° BTDC	1600	500
3-51R-R00	4.9L	.042-.046	6° BTDC	1600	700 ④
3-51R-R10	4.9L	.042-.046	6° BTDC	1600	700 ④
3-51S-R00	4.9L	.042-.046	6° BTDC	1600	700 ④
3-51S-R10	4.9L	.042-.046	6° BTDC	1600	700 ④
3-51T-R00	4.9L	.042-.046	6° BTDC	1600	700 ④
3-51T-R10	4.9L	.042-.046	6° BTDC	1600	700 ④
3-51V-R00	4.9L	.042-.046	10° BTDC	1600	700 ④
3-51X-R00	4.9L	.042-.046	10° BTDC	1600	700 ④
3-51Z-R00	4.9L	.042-.046	10° BTDC	1600	700 ④
3-52E-R00	4.9L	.042-.046	10° BTDC	1600	600 ⑤
3-52F-R00	4.9L	.042-.046	10° BTDC	1600	600 ⑤
3-52G-R00	4.9L	.042-.046	10° BTDC	1600	600 ⑤
3-52K-R00	4.9L	.042-.046	10° BTDC	1600	600 ⑤
3-52R-R00	4.9L	.042-.046	10° BTDC	1600	600 ⑤
3-52R-R10	4.9L	.042-.046	10° BTDC	1600	600 ⑤
3-52S-R00	4.9L	.042-.046	10° BTDC	1600	600 ⑤
3-52S-R10	4.9L	.042-.046	10° BTDC	1600	600 ⑤
3-52T-R00	4.9L	.042-.046	10° BTDC	1600	600 ⑤
3-52T-R10	4.9L	.042-.046	10° BTDC	1600	600 ⑤
3-52V-R00	4.9L	.042-.046	14° BTDC	1600	600 ⑤
3-52Y-R00	4.9L	.042-.046	14° BTDC	1600	600 ⑤
3-52Z-R00	4.9L	.042-.046	14° BTDC	1600	600 ⑤
3-53F-R00	5.0L	.042-.046	8° BTDC	2100	700 ⑥
3-53G-R00	5.0L	.042-.046	8° BTDC	2100	700 ⑥
3-531L-R00	5.0L	.042-.046	8° BTDC	2100	700 ⑥
3-53L-R00	5.0L	.042-.046	8° BTDC	2100	700 ⑥
3-53W-R00	5.0L	.042-.046	12° BTDC	2100	700 ⑥
3-53Y-R00	5.0L	.042-.046	12° BTDC	2100	700 ⑥
3-53Z-R00	5.0L	.042-.046	12° BTDC	2100	700 ⑥
3-54E-R00	5.0L	.042-.046	8° BTDC	2250	675 ⑦
3-54F-R00	5.0L	.042-.046	8° BTDC	2250	675 ⑦
3-54J-R00	5.0L	.042-.046	8° BTDC	2250	675 ⑦
3-54L-R00	5.0L	.042-.046	8° BTDC	2250	675 ⑦
3-54P-R00	5.0L	.042-.046	8° BTDC	2250	575
3-54R-R00	5.0L	.042-.046	8° BTDC	2250	575
3-54T-R00	5.0L	.042-.046	8° BTDC	2250	575
3-54W-R00	5.0L	.042-.046	12° BTDC	2250	600 ⑧
3-54Y-R00	5.0L	.042-.046	12° BTDC	2250	675 ⑦
3-54Z-R00	5.0L	.042-.046	12° BTDC	2250	675 ⑦
1-63T-R15B	5.8L	.042-.046	—	2000	1400 ⑨
1-63Y-R14B	5.8L	.042-.046	—	2000	1400 ⑨
2-64Y-R14B	5.8L	.042-.046	—	2000	1400 ⑨
1-63T-R12	5.8L	.042-.046	—	1700	750
1-63T-R13	5.8L	.042-.046	—	1700	750
1-64H-R02	5.8L	.042-.046	10° BTDC	2000	625 ⑩
1-64T-R12	5.8L	.042-.046	—	1650	600
1-64T-R13	5.8L	.042-.046	—	1650	600
2-63Y-R11	5.8L	.042-.046	—	1700	750

TUNE-UP SPECIFICATIONS
1983

(For 1983 Tune-Up Specifications consult the Vehicle Emissions Control Label, which is located on the engine of the vehicle. This decal will contain a calibration number which when used in conjunction with the chart below will yield the required tune-up information. If the information given in this chart disagrees with the information on the decal, use the information on the decal.)

Calibration	Engine	Spark Plug Gap	Ignition Timing	Fast Idle rpm	Curb Idle rpm
2-63Y-R12	5.8L	.042-.046	—	1700	750
2-64X-R00	5.8L	.042-.046	14° BTDC	2000	625 ⑩
2-64Y-R11	5.8L	.042-.046	—	1650	600
2-64Y-R12	5.8L	.042-.046	—	1650	600
5-77-R01	4.9L	.042-.046	10° BTDC	1500	600 ⑪
5-78-R01	4.9L	.042-.046	10° BTDC	1500	600 ⑪
9-77J-R12	4.9L	.042-.046	12° BTDC	1600	700
9-77S-R10	4.9L	.042-.046	10° BTDC	1600	700
9-78J-R00	4.9L	.042-.046	12° BTDC	1600	550
9-78J-R11	4.9L	.042-.046	12° BTDC	1600	550
7-79-R01	5.0L	.042-.046	6° BTDC	1250	750
7-80-R00	5.0L	.042-.046	6° BTDC	1600	650
2-75A-R10	5.8L	.042-.046	8° BTDC	1500	650
2-75J-R20	5.8L	.042-.046	8° BTDC	1500	650
2-76A-R10	5.8L	.042-.046	8° BTDC	1500	650
2-76J-R20	5.8L	.042-.046	8° BTDC	1500	650
9-83G-R12	6.1L	.042-.046	6° BTDC	2200	600
9-83G-R14	6.1L	.042-.046	6° BTDC	1600	600
9-83H-R11	6.1L	.042-.046	6° BTDC	2500	600
9-83H-R14	6.1L	.042-.046	2° BTDC	2500	600
9-87G-R11	7.0L	.042-.046	6° BTDC	2700	600
9-97J-R13	7.5L	.042-.046	8° BTDC	1600	600
9-98S-R00	7.5L	.042-.046	6° BTDC	1500	600

① 800—w/o Power Steering
② 750-800—Auto. Trans
③ 700-800—Auto. Trans
④ 600—Non A/c or A/c off
⑤ 550—Non A/c or A/c off
⑥ 800—A/c on
⑦ 600—Non A/c or A/c off
⑧ 675—A/c on
⑨ 900—Throttle solenoid on, 600 throttle solenoid off
⑩ 550—A/c off and Non A/c
⑪ 700—Manual Trans.

TUNE-UP SPECIFICATIONS
1984

(For 1984 Tune-Up Specifications consult the Vehicle Emissions Control Label, which is located on the engine of the vehicle. This decal will contain a calibration number which when used in conjunction with the chart below will yield the required tune-up information. If the information given in this chart disagrees with the information on the decal, use the information on the decal.)

Calibration	Engine	Spark Plug Gap	Ignition Timing	Fast Idle rpm	Curb Idle rpm
3-41P-R15	2.0L	.032-.036	8° BTDC	2000	800
3-41S-R18	2.0L	.032-.036	9° BTDC	2000	800
3-49G-R20	2.3L	.042-.046	6° BTDC	2000	850 ①
3-49G-R20	2.3L	.042-.046	6° BTDC	2200	850 ①
3-49G-R17	2.3L	.042-.046	6° BTDC	2000	850 ①
3-49S-R16	2.3L	.042-.046	6° BTDC	2000	850 ①
3-49T-R20	2.3L	.042-.046	6° BTDC	2000	850 ①
3-49Y-R20	2.3L	.042-.046	10° BTDC	2000	850 ①
3-50H-R18	2.3L	.042-.046	6° BTDC	2000	800
3-50S-R18	2.3L	.042-.046	6° BTDC	2000	800
3-50Y-R18	2.3L	.042-.046	10° BTDC	2000	800
4-61F-R00	2.8L	.042-.046	10° BTDC	3000	800-900 ②
4-61F-R10	2.8L	.042-.046	10° BTDC	3000	800-900 ②
4-61G-R00	2.8L	.042-.046	10° BTDC	3000	700-800

TUNE-UP SPECIFICATIONS
1984

(For 1984 Tune-Up Specifications consult the Vehicle Emissions Control Label, which is located on the engine of the vehicle. This decal will contain a calibration number which when used in conjunction with the chart below will yield the required tune-up information. If the information given in this chart disagrees with the information on the decal, use the information on the decal.)

Calibration	Engine	Spark Plug Gap	Ignition Timing	Fast Idle rpm	Curb Idle rpm
4-61K-R01	2.8L	.042-.046	10° BTDC	3000	850-950 [2]
4-61S-R00	2.8L	.042-.046	10° BTDC	3000	800-900 [2]
4-62D-R00	2.8L	.042-.046	10° BTDC	3000	800-900 [2]
4-62D-R10	2.8L	.042-.046	10° BTDC	3000	800-900 [2]
4-62S-R01	2.8L	.042-.046	10° BTDC	3000	800-900 [2]
4-62S-R10	2.8L	.042-.046	10° BTDC	3000	800-900 [2]
4-51D-R01	4.9L	.042-.046	10° BTDC	1600	600-700 [3]
4-51E-R00	4.9L	.042-.046	10° BTDC	1600	600-700 [3]
4-51K-R00	4.9L	.042-.046	10° BTDC	1600	600-700 [3]
4-51L-R00	4.9L	.042-.046	10° BTDC	1600	600-700 [3]
4-51R-R00	4.9L	.042-.046	10° BTDC	1600	600-700 [3]
4-51S-R00	4.9L	.042-.046	10° BTDC	1600	600-700 [3]
4-51S-R01	4.9L	.042-.046	10° BTDC	1600	600-700 [3]
4-51S-R02	4.9L	.042-.046	10° BTDC	1600	600-700 [3]
4-51T-R00	4.9L	.042-.046	10° BTDC	1600	600-700 [3]
4-51Z-R00	4.9L	.042-.046	10° BTDC	1600	600-700 [3]
4-52L-R00	4.9L	.042-.046	10° BTDC	1600	600-700 [3]
4-52R-R00	4.9L	.042-.046	10° BTDC	1600	600-700 [3]
4-52S-R00	4.9L	.042-.046	10° BTDC	1600	600-700 [3]
4-52T-R00	4.9L	.042-.046	10° BTDC	1600	600-700 [3]
4-52W-R00	4.9L	.042-.046	10° BTDC	1600	600-700 [3]
4-53F-R00	5.0L	.042-.046	8° BTDC	2100	800 [4]
4-53F-R10	5.0L	.042-.046	8° BTDC	2100	800 [4]
4-53G-R00	5.0L	.042-.046	8° BTDC	2100	800 [4]
4-53G-R10	5.0L	.042-.046	8° BTDC	2100	800 [4]
4-53K-R00	5.0L	.042-.046	8° BTDC	2100	800 [4]
4-53K-R10	5.0L	.042-.046	8° BTDC	2100	800 [4]
4-53Z-R00	5.0L	.042-.046	8° BTDC	2100	800 [4]
4-53Z-R10	5.0L	.042-.046	8° BTDC	2100	800 [4]
4-54E-R00	5.0L	.042-.046	8° BTDC	2100	800 [4]
4-54E-R10	5.0L	.042-.046	8° BTDC	2100	800 [4]
4-54J-R00	5.0L	.042-.046	8° BTDC	2100	800 [4]
4-54J-R10	5.0L	.042-.046	8° BTDC	2100	800 [4]
4-54L-R00	5.0L	.042-.046	8° BTDC	2100	800 [4]
4-54L-R10	5.0L	.042-.046	8° BTDC	2100	800 [4]
4-54R-R00	5.0L	.042-.046	10° BTDC	2000	575
4-54R-R10	5.0L	.042-.046	10° BTDC	2000	575
4-54T-R00	5.0L	.042-.046	10° BTDC	2000	575
4-54T-R10	5.0L	.042-.046	10° BTDC	2000	575
4-54W-R00	5.0L	.042-.046	12° BTDC	2100	675 [5]
4-54W-R10	5.0L	.042-.046	12° BTDC	2100	675 [5]
4-63H-R00	5.8L	.042-.046	10° BTDC	2000	750
4-64H-R00	5.8L	.042-.046	10° BTDC	2000	600
4-64H-R00	5.8L	.042-.046	10° BTDC	2000	600
4-64T-R00	5.8L	.042-.046	10° BTDC	2000	600
4-64T-R00	5.8L	.042-.046	10° BTDC	2000	600
4-64Y-R00	5.8L	.042-.046	10° BTDC	2000	600
5-77-R01	4.9L	.042-.046	10° BTDC	1500	600 [6]
5-78-R01	4.9L	.042-.046	10° BTDC	1500	600 [6]
9-77J-R12	4.9L	.042-.046	12° BTDC	1600	700 [7]
9-78J-R00	4.9L	.042-.046	12° BTDC	1600	550
9-78J-R11	4.9L	.042-.046	12° BTDC	1600	550
7-79-R01	5.0L	.042-.046	6° BTDC	1500	750 [8]

TUNE-UP SPECIFICATIONS
1984

(For 1984 Tune-Up Specifications consult the Vehicle Emissions Control Label, which is located on the engine of the vehicle. This decal will contain a calibration number which when used in conjunction with the chart below will yield the required tune-up information. If the information given in this chart disagrees with the information on the decal, use the information on the decal.)

Calibration	Engine	Spark Plug Gap	Ignition Timing	Fast Idle rpm	Curb Idle rpm
7-80-R00	5.0L	.042-.046	6° BTDC	1500	750 ⑧
2-75A-R10	5.8L	.042-.046	8° BTDC	1500	800 ⑧
2-75J-R20	5.8L	.042-.046	8° BTDC	1500	700 ⑧
2-76A-R10	5.8L	.042-.046	8° BTDC	1500	800 ⑧
2-76J-R20	5.8L	.042-.046	8° BTDC	1500	700 ⑧
9-83G-R12	6.1L	.042-.046	6° BTDC	2200	600
9-83G-R14	6.1L	.042-.046	6° BTDC	1600	600
9-83H-R11	6.1L	.042-.046	6° BTDC	2500	600
9-87G-R11	7.0L	.042-.046	6° BTDC	2700	600
9-97J-R10	7.5L	.042-.046	8° BTDC	1600	800 ⑨
3-98S-R10	7.5L	.042-.046	8° BTDC	1600	800 ⑨
4-98S-R00	7.5L	.042-.046	8° BTDC	1600	800 ⑨
4-37A-R00	2.0L	—	—	1450	725-775
4-37B-R00	2.0L	—	—	1450	725-775
3-47D-R00	2.2L	—	—	1450	780-830
4-68J-R00	6.9L	—	—	—	650-700
4-68X-R00	6.9L	—	—	—	650-700

① 800 without P/S
② 700-800 RPM—Auto. Trans. in DRIVE
③ 550-650 RPM—Auto. Trans. in DRIVE
④ 700—Non a/c or A/C off
⑤ 700—Non a/c or A/C off
⑥ 700—Manual Trans.
⑦ 600—TSP off
⑧ 650—"D" auto trans; 525—TSP off
⑨ 650—DRIVE

TUNE-UP SPECIFICATIONS
1985-86

(For 1985–86 Tune-Up Specifications consult the Vehicle Emissions Control Label, which is located on the engine of the vehicle. This decal will contain a calibration number which when used in conjunction with the chart below will yield the required tune-up information. If the information given in this chart disagrees with the information on the decal, use the information on the decal.)

Calibration	Engine	Spark Plug Gap	Ignition Timing	Fast Idle rpm	Curb Idle rpm
5-41D-R00	2.0L	.042-.046	6°BTDC	1700	775-825
5-41D-R10	2.0L	.042-.046	6°BTDC	1700	775-825
5-49F-R01	2.3L	.042-.046	10°BTDC	—	575-725
5-49S-R01	2.3L	.042-.046	10°BTDC	—	575-725
5-50H-R02	2.3L	.042-.046	10°BTDC	—	625-775
5-50S-R02	2.3L	.042-.046	10°BTDC	—	625-775
5-61F-R01	2.8L	.042-.046	10°BTDC	3000	800-900 ①
5-61S-R01	2.8L	.042-.046	14°BTDC	3200	800-900 ①
5-62E-R01	2.8L	.042-.046	10°BTDC	3000	800-900 ①
5-62R-R01	2.8L	.042-.046	10°BTDC	3000	800-900 ①
4-51R-R00	4.9L	.042-.046	10°BTDC	1600	600-700 ②
4-51S-R02	4.9L	.042-.046	10°BTDC	1600	600-700 ②
4-51T-R00	4.9L	.042-.046	10°BTDC	1600	600-700 ②
4-52G-R00	4.9L	.042-.046	10°BTDC	1600	600-700 ②
4-52G-R10	4.9L	.042-.046	10°BTDC	1600	600-700 ②
4-52L-R00	4.9L	.042-.046	10°BTDC	1600	600-700 ②
4-52L-R10	4.9L	.042-.046	10°BTDC	1600	600-700 ②
4-52R-R00	4.9L	.042-.046	10°BTDC	1600	600-700 ②
4-52S-R00	4.9L	.042-.046	10°BTDC	1600	600-700 ②
4-52S-R10	4.9L	.042-.046	10°BTDC	1600	600-700 ②
4-52T-R00	4.9L	.042-.046	10°BTDC	1600	600-700 ②

TUNE-UP SPECIFICATIONS
1985–86

(For 1985–86 Tune-Up Specifications consult the Vehicle Emissions Control Label, which is located on the engine of the vehicle. This decal will contain a calibration number which when used in conjunction with the chart below will yield the required tune-up information. If the information given in this chart disagrees with the information on the decal, use the information on the decal.)

Calibration	Engine	Spark Plug Gap	Ignition Timing	Fast Idle rpm	Curb Idle rpm
5-51D-R00	4.9L	.042–.046	10°BTDC	1600	600–700 ②
5-51E-R00	4.9L	.042–.046	10°BTDC	1600	600–700 ②
5-51F-R00	4.9L	.042–.046	10°BTDC	1600	600–700 ②
5-51H-R00	4.9L	.042–.046	10°BTDC	1600	600–700 ②
5-51K-R00	4.9L	.042–.046	10°BTDC	1600	600–700 ②
5-51L-R00	4.9L	.042–.046	10°BTDC	1600	600–700 ②
5-51V-R00	4.9L	.042–.046	10°BTDC	1600	600–700 ②
5-51Z-R00	4.9L	.042–.046	10°BTDC	1600	600–700 ②
5-52E-R00	4.9L	.042–.046	10°BTDC	1600	600–700 ②
5-52K-R00	4.9L	.042–.046	10°BTDC	1600	600–700 ②
5-52W-R00	4.9L	.042–.046	10°BTDC	1600	600–700 ②
5-52Y-R00	4.9L	.042–.046	10°BTDC	1600	600–700 ②
4-54R-R12	5.0L	.042–.046	10°BTDC	2000	575D
4-54R-R13	5.0L	.042–.046	10°BTDC	2000	575D
4-54R-R14	5.0L	.042–.046	10°BTDC	2000	575D
5-53D-R00	5.0L	.042–.046	10°BTDC	—	—
5-53D-R01	5.0L	.042–.046	8°BTDC	—	—
5-53F-R00	5.0L	.042–.046	10°BTDC	—	—
5-53F-R01	5.0L	.042–.046	8°BTDC	—	—
5-53H-R00	5.0L	.042–.046	10°BTDC	—	—
5-53H-R01	5.0L	.042–.046	8°BTDC	—	—
5-54Q-R00	5.0L	.042–.046	10°BTDC	—	—
5-54Q-R01	5.0L	.042–.046	10°BTDC	—	—
5-54S-R00	5.0L	.042–.046	10°BTDC	—	—
5-54S-R01	5.0L	.042–.046	10°BTDC	—	—
5-54W-R00	5.0L	.042–.046	10°BTDC	—	—
5-54X-R00	5.0L	.042–.046	10°BTDC	—	—
4-64G-R00	5.8L(F)	.042–.046	10°BTDC	1900	650D
4-64G-R02	5.8L(F)	.042–.046	10°BTDC	1900	650D
4-64G-R02	5.8L(E)	.042–.046	10°BTDC	1900	650D
4-64T-R00	5.8L(F)	.042–.046	10°BTDC	2000	600D
4-64T-R00	5.8L(E)	.042–.046	10°BTDC	2000	600D
4-64Z-R10	5.8L(F)	.042–.046	14°BTDC	1900	650D
4-64Z-R10	5.8L(E)	.042–.046	14°BTDC	1900	650D
5-63H-R00	5.8L	.042–.046	10°BTDC	2000	700
5-63Y-R00	5.8L	.042–.046	10°BTDC	2000	700

① 700–800 in Drive A/T
② 550–650 in Drive A/T

FIRING ORDER

NOTE: To avoid confusion, always replace spark plug wires one at a time.

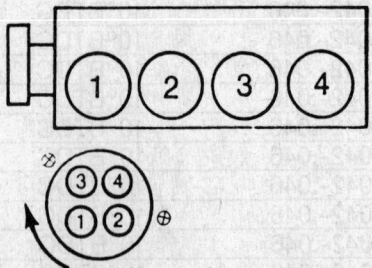

122 (2.0L), 140 (2.3L) 4 cylinder

FIRING ORDER

NOTE: To avoid confusion, always replace spark plug wires one at a time.

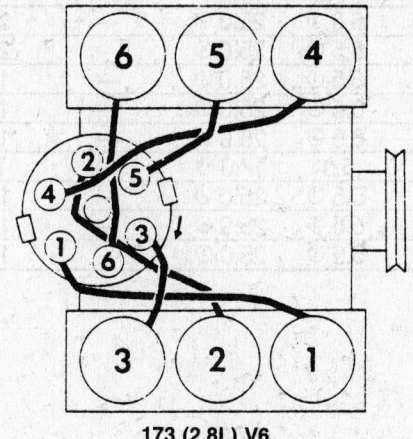

173 (2.8L) V6

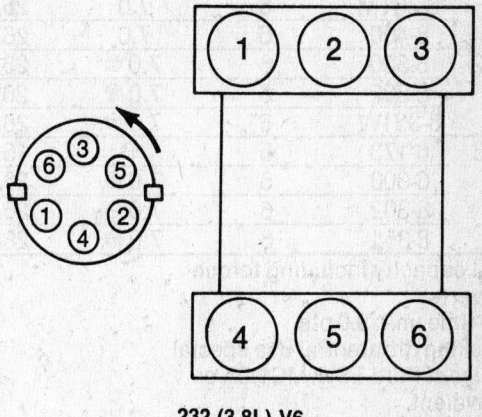

232 (3.8L) V6

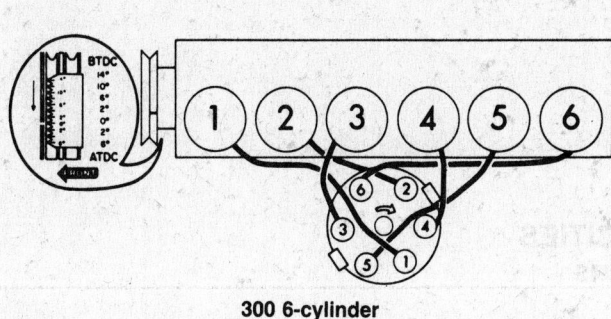

300 6-cylinder

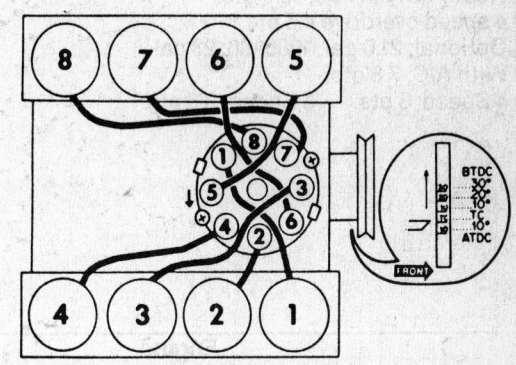

255, 302, 460 V8

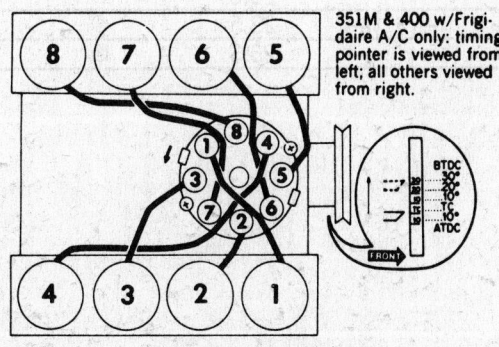

351M & 400 w/Frigidaire A/C only: timing pointer is viewed from left; all others viewed from right.

351W, 351M, 400 V8

CAPACITIES
Bronco and Bronco II

Year	Engine	Crankcase With Filter (qts)	Transmission (pts) Manual	Automatic ①	Transfer Case (pts)	Axle (pts) Front	Rear	Gasoline Tank (gals) Main	Auxilliary	Cooling System (qts)
'79	8-351M	6	7.0	26.4	4.0②	5.8	6.5③	25.0④	—	20.0⑤
	8-400	6	7.0	26.4	4.0②	5.8	6.5③	25.0④	—	22.0⑥
'80–'82	6-300	6	7.0⑩	26.8	6.5	4.0	6.5③	25.0④	—	13.0⑦
	8-302	6	7.0⑩	26.8	6.5	4.0	6.5③	25.0④	—	13.0⑧
	8-351W	6	7.0⑩	26.8	6.5	4.0	6.5③	25.0④	—	15.0⑨
'83–'86	6-173	5	⑬	15.8	3.0	1.0	5.5	17.0⑪	—	7.2⑫
	6-300	6	7.0⑩	26.8	7.0	4.0	5.5③	25.0④	—	13.0⑦
	8-302	6	7.0⑩	26.8	7.0	4.0	5.5③	25.0④	—	13.0⑧
	8-351	6	7.0⑩	26.8	7.0	4.0	5.5③	25.0④	—	15.0⑨

① Total capacity including torque converter
② Full-time unit: 9.0 pts.
③ If locking differential use special lubricant Ford ESW-MC119A or equivalent
④ Optional tank; 32 gal.
⑤ Heavy duty; 22.0 qts.
⑥ Heavy duty; 24.0 qts.
⑦ A/C or heavy duty; 14.0 qts.
⑧ Heavy duty; 14.0 qts.
⑨ Heavy duty or A/C; 16.0 qts.
⑩ 4 speed overdrive 4.5 pts.
⑪ Optional; 21.0 gal. 1985–86: 23 gal.
⑫ With A/C; 7.8 qts.
⑬ 4 Speed: 3 pts.   5 Speed; 3.6 pts.

CAPACITIES
Vans

Year	Engine Displacement Cu In.	Engine Crankcase (qts) With Filter	Without Filter	Transmission (pts) Manual 3-spd	4-spd	Automatic ■	Drive Axle ▲ (pts)	Gasoline Tank (gals)	Cooling System (qts) W/AC	W/O	W/Extra Cooling
'79–'86	300	6.0	5.0	3.5	5.0	20.0①	②	③	20.0	15.0	—
	302	6.0	5.0	3.5	5.0	20.0①	②	③	17.5	15.0④	17.5
	351W	6.0	5.0	—	—	23.5	②	③	20.0	20.0	21.0
	420 diesel	10.0	9.0	—	—	23.5	②	③	31.0	31.0	—
	460	6.0	5.0	—	—	23.5	②	③	28.0	28.0	—

① C-6:24.5
② Ford Axles: 6.5
  Dana model 61-1:5.0
  Dana model 70:5.5
③ E-100, 150 with 124 inch wheel base: 18.0
  All others: 22.1
  Optional auxiliary: 18.0 except
  E-350 cutaway chassis: 40.0
④ With auto trans.: 17.5

▲ For limited slip units, add 1 oz of limited slip additive
■ Includes torque converter

CRANKCASE AND COOLING SYSTEM (QTS) CAPACITIES
Pickup

Year	Engine and Model	Crankcase Oil and Filter Change	Cooling System					
			Standard System		Extra Cooling		With A/C	Super Cooling
			Man Trans	Auto Trans	Man Trans	Auto Trans		
'79–'86	122(2.0L)	5.0	6.5	6.5	—	—	7.2	—
	140(2.3L)	6.0 ③	6.5	6.5	—	—	7.2	—
	173 eng (2.8L)	5.0	13.0	13.0	—	—	14.0	—
	300 eng., F-100/250/350	6.0	13.0	14.0	13.0	14.0	17.0	—
	232, 255, 302 eng., F-100/150	6.0	15.0 ①	15.0 ①	15.0 ①	18.0 ①	18.0 ①	—
	351 eng., F-100/150/350	6.0	17.0	—	17.0	—	—	24.0
	400 eng., F/250/350	6.0	18.0	—	18.0	—	18.0	24.0
	420 diesel All	10.0	29.0	29.0	29.0	29.0	29.0	29.0
	460 eng., 150/250/350	6.0	24.0 ②	24.0 ②	24.0 ②	24.0 ②	24.0 ②	24.0 ②

SRW—Single rear wheels
DRW—Dual rear wheels
① 13 qts—1983 and later
   15 qts—1983 and later
② 1985 and later: 16.5 Man Trans
              17.5 All others
③ If stamped steel pan: 6 qts.

TRANSMISSION AND TRANSFER CASE CAPACITIES
Pickup

Year	Model	Type	Capacity (pints)
'79–'86	F-100/250	Ford 3.03, 3-sp	3.5
	F-150/250/350	Warner T-18 4-sp	7.0
	F-100/150/250/350	New Process 435 4-sp	
		with extension	7.0
		without extension	6.5
	All F Series	Clark 4-sp Overdrive	5.0
	F-150/350	Warner T19B 4-sp	7.0
	F-100/150 F-150/350	C-4 Automatic	20.0
	F-250 4 × 4	C-6 Automatic	
		351, 400, 600 engs.	26.75
		All others	23.5
	F-150/250 4 × 4	New Process 205, 2-sp Transfer Case	4.5
	F-150/250/350 4 × 4	New Process 203 Full Time Transfer Case	9.0
	F-150/250	New Process 208 Transfer Case	7.0
	F-250/350	Borg Warner 1345 Transfer Case	6.5
	F-100/150 F-250/350	C-5 Automatic	22.0
	All Models, exc Ranger	Automatic Overdrive (AOD)	24.0
	Ranger	4 and 5 speed	3.0
		C-3	15.4
		C-5 (4 × 2)	15.0
		C-5 (4 × 4)	15.6
		Warner 1350	3.0

DRIVE AXLES CAPACITIES
Pickup

Year	Model	Type	Capacity (pints)
'79–'86	F-100/150	Ford	5.5
	F-250, 4 × 2, F-250/350 4 × 4	Dana 60-2	7.0
	F-250	Dana 61-2	6.0
	F-350	Dana 70	7.0
	F-150/250 4 × 4	Dana 44-9F Front axle	3.5*
	F-250/350 4 × 4	Dana 60-7F Front axle	6.0
	F-250/350 4 × 4 HD	Dana 50-IFS	4.1
	F-250/350 SRW	Dana 61-1 (Rear)	6.0
	F-350 DRW	Dana 70 HD (Rear)	7.4
	F-250 4 × 2 F-250 4 × 4	Dana 44 IFS (Front Axle)	3.8
	Ranger 4 × 4	Dana 28 (Front Axle)	1.0
		6-¾ RG (Rear Axle)	3.0
		7.5 RG (Rear Axle)	4.0

*4.0—with New Process 205 2-sp Transfer Case   SEW-Single Rear Wheels   DRW—Dual Rear Wheels

FUEL TANK CAPACITIES
Pickup

Year	Model	Standard	Optional
'79–'82	F-100/250/350 4 × 2 (Reg Cab)	19.2	20.2
	F-250 4 × 4 (Reg Cab)	19.2	—
	F-250/350 4 × 2, (Crew Cab)	20.2	—
	Super Cab	19.2	19.5
	F-350 4 × 2 (Reg Cab)	20.2	19.0
	F-250/350 4 × 4, (Reg Cab)	26.0	19.2
'83–'86	F-100 4 × 2 (Reg Cab)	16.5	19.0 Aft./Axle
	F-150 4 × 2 (Crew Cab)	16.0	19.0 Aft./Axle
	F-150 4 × 4 (Reg Cab)	16.5	19.0 Aft./Axle
	F-250 4 × 2 (Crew Cab)	16.5	19.0 Aft./Axle
	F-150 4 × 2 (Reg Cab)	19.0	19.0 Aft./Axle
	F-150 4 × 4 (Reg Cab)	19.0	19.0 Aft./Axle
	F-150 4 × 2 (Crew Cab)	19.0	19.0 Aft./Axle
	F-150 4 × 4 (Crew Cab)	19.0	19.0 Aft./Axle
	F-250 4 × 2 (Reg Cab)	19.0	19.0 Aft./Axle
	F-250 4 × 2 (Crew Cab)	19.0	19.0 Aft./Axle
	F-250 4 × 4 (All)	19.0	19.0 Aft./Axle
	F-350 4 × 2 (Reg Cab)	19.0	19.0 Aft./Axle
	F-350 4 × 2 (Super Cab)	19.0	19.0 Aft./Axle
	F-350 4 × 2 (Cab Chassis)	19.0 ①	19.0 ① Aft./Axle
	Ranger 15.2 (SWB) 17.0 (LWB)	—	13.0 Aft./Axle

Aft./Axle—After Rear Axle   ▲ Variations are possible depending on optional equipment
① 22.1

VALVE SPECIFICATIONS

Cu. In. Displace-ment	Year	Lash (Hot) Inches Int.	Lash (Hot) Inches Exh.	Angle Degree Face	Angle Degree Seat	Stem Dia. Inches Int.	Stem Dia. Inches Exh.	Stem Clearance Intake	Stem Clearance Exhaust	Valve Life Inches	Valve Spring Lbs.@Inches Open	Valve Spring Lbs.@Inches Closed	Spring Free Length Inch
							FOUR CYLINDER						
122	'83–'86	Zero		44	45	.3416	.3411	.0010–.0027	.0015–.0032	NA	①	NA	NA
140	'83–'86	Zero		44	45	.3416	.3411	.0010–.0027	.0015–.0032	NA	②	NA	NA

VALVE SPECIFICATIONS

Cu. In. Displace-ment	Year	Lash (Hot) Inches		Angle Degree		Stem Dia. Inches		Stem Clearance		Valve Life Inches	Valve Spring Lbs.@Inches		Spring Free Length Inch
		Int.	Exh.	Face	Seat	Int.	Exh.	Intake	Exhaust		Open	Closed	
SIX CYLINDER													
173V6	'83–'86	③		44	45	.3159	.3149	.0008–.0025	.0008–.0035	NA	152@1.22	NA	NA
300	'79–'86	Zero		44	45	.3420	.3420	.0010–.0027	.0010–.0027	.249	192@1.18	80@1.58 ④	1.99 ⑤
232V6	'82–'84	Zero		44	45	.3422	.3415	.0010–.0025	.0015–.0032	⑥	⑦	⑧	NA
EIGHT CYLINDER													
255	'81–'82	Zero		46	45	.3420	.3415	.0010–.0027	.0010–.0027	⑨	⑩	⑪	NA
302	'79–'86	Zero		44	45	.3420	.3415	.0010–.0027	.0015–.0032	.2375 ⑫	200@1.36	78@1.78 ⑬	2.04 ⑭
351 W	'79–'86	Zero		44	45	.3420	.3415	.0010–.0027	.0015–.0032	.2600	200@1.36	78@1.78	2.04 ⑭
351 M	'79–'80	Zero		44	45	.3420	.3415	.0010–.0027	.0015–.0032	.2350 ⑭	226@1.39	80@1.82 ⑬	2.06
400	'79–'82	Zero		44	45	.3420	.3415	.0010–.0027	.0015–.0032	.2480	225@1.39	80@1.82 ⑮	2.06
460	'79–'86	Zero		44	45	.3420	.3420	.0010–.0027	.0010–.0027	.2530	252@1.33	80@1.81	2.07

① Intake: 71–79 @ 1.52
  Exhaust: 142–156 @ 1.12
② Intake: 71–79 @ 1.56
  Exhaust: 159–175 @ 1.16
③ Cold Adjustment
  Intake: .014
  Exhaust: .016
④ Open: 1.95 @ 1.30:
  Closed: 80 @ 1.70
⑤ Exhaust: 1.87
⑥ Intake: 0.415 in.
  Exhaust: 0.417 in.
⑦ Loaded: 215 lbs @ 1.39 in.
  75 lbs @ 1.70 in.

⑧ Assembled height 1.70–1.78 in.
⑨ Intake: .2375 in.
  Exhaust: .2474 in.
⑩ Intake: 74–82 lbs @ 1.78 in.
    190–212 lbs @ 1.36 in.
  Exhaust: 76–84 lbs @ 1.60 in.
    190–210 lbs @ 1.20 in.
⑪ Intake: 2.04 in.
  Exhaust: 1.85 in.
⑫ Exhaust: .2474 in.
⑬ Exhaust: Opened: 200 @ 120
    Closed: 80 @ 160
⑭ Exhaust 1.85 in.
⑮ '79–'80: Exhaust closed 84 @ 1.68 in.

CRANKSHAFT BEARING JOURNAL SPECIFICATIONS

(Inches)

Cu. In. Displace-ment	Year	Main Bearing Journals				Thrust On No.	Connecting Rod Bearing Journals		
		Journal Diameter	Oil Clearance	Shaft End-play			Journal Diameter	Oil Clearance (max)	Side Clearance
FOUR CYLINDER									
122	'83–'86	2.3982–2.3990	.0008–.0015	.004–.008		3	2.0472	.0008–.0015	.0035–.0105
140	'83–'86	2.3982–2.3990	.0008–.0015	.004–.008		3	2.0472	.0008–.0015	.0035–.0105
SIX CYLINDER									
173V6	'83–'85	2.2441	.0008–.0015	.012		3	2.1260	.0006–.0016	.004–.011
300		2.3982–2.3990	.0009–.0028	.004–.008		5	2.1228–2.1236	.0009–.0027	.006–.013
232-V6	'82–'84	2.5190	.0009–.0027	.004–.008		3	2.3107	.0009–.0027	.004–.011
EIGHT CYLINDER									
255	'81–'82	2.2490	.0005–.0024 ①	.004–.008		3	2.1232	.0008–.0025	.010–.020
302	'79–'86	2.2482–2.2490	.0005–.0015 ④	.004–.008		3	2.1228–2.1236	.0010–.0015	.010–.020
351W	'79–'86	2.9994–3.0002	②	.004–.008		3	2.3103–2.3111	.0008–.0026	.010–.020
351M	'79–'80	2.9994–3.0002	.0008–.0025 ③	.004–.008		3	2.3103–2.3111	.0008–.0025	.010–.020
400	'79–'82	2.9994–3.0002	.0008–.0025 ③	.004–.008		3	2.3103–2.3111	.0008–.0025	.010–.020
460	'79–'86	2.9994–3.0002	.0008–.0015 ③	.004–.008		3	2.4992–2.5000	.008–.0015	.010–.020

① No. 1 .0001–.0020.
② No. 1—.0005–.0015;
  All others .0008–.0015.
③ Maximum
④ No. 1—.0001–.0015

PISTON AND RING SPECIFICATIONS

All measurements given in inches

Year	Engine	Piston to Bore Clearance	Ring Side Clearance			Ring Gap		
			Top Compression	Bottom Compression	Oil Control	Top Compression	Bottom Compression	Oil Control ①
'83–'86	4-122	.0014–.0022	.0020–.0040	.0020–.0040	snug	.0100–.0200	.0100–.0200	.015–.055
'83–'86	4-134	.0021–.0031	.0020–.0035	.0016–.0031	.0012–.0028	.0157–.0217	.0118–.0157	.0138–.0217
'83–'86	4-140	.0014–.0022	.0020–.0040	.0020–.0040	snug	.0100–.0200	.0100–.0200	.015–.055
'83–'86	6-173	.0011–.0019	.0020–.0033	.0020–.0033	snug	.0150–.0230	.0150–.0230	.015–.055
'82–'84	6-232	.0014–.0022	.0020–.0040	.0020–.0040	snug	.0100–.0200	.0100–.0200	.015–.055
'79–'86	6-300	.0014–.0022	.0019–.0036	.0020–.0040	snug	.0100–.0200	.0100–.0200	.010–.035
'82	8-255	.0014–.0024	.0020–.0040	.0020–.0040	snug	.0100–.0200	.0100–.0200	.015–.055
'79–'86	8-302	.0018–.0026	.0020–.0040	.0020–.0040	snug	.010–.020	.0100–.0200	.015–.035
'79–'80	8-351M	.0014–.0022	.0019–.0036	.0020–.0040	snug	.010–.020	.0100–.0200	.010–.035
'79–'86	8-351W	.0022–.0030	.0019–.0036	.0020–.0040	snug	.010–.020	.0100–.0200	.015–.035
'79–'82	8-400	.0014–.0022	.0019–.0036	.0020–.0040	snug	.0100–.0200	.0100–.0200	.015–.035
'79–'86	8-460	.0022–.0300	.0019–.0036	.0020–.0040	snug	.0100–.0200	.0100–.0200	.010–.035
'83–'86	8-420	.0055–.0075	.0020–.0040	.0020–.0040	.0010–.0030	.0140–.0240	.0100–.0240	.0600–.0700

① Steel rails

TORQUE SPECIFICATIONS

(All readings in ft. lbs.)

Year	Engine	Cylinder Head Bolts	Rod Bearing Bolts	Main Bearing Bolts	Crankshaft Pulley Bolt	Flywheel-to-Crankshaft Bolts	Manifold	
							Intake	Exhaust
'83–'86	4-122	①	②	①	100–120	56–64	14–21 ③	16–23 ③
'83–'86	4-134	80–85	50–54	80–85	253–289	95–137	12–17 ③	17–20 ③
'83–'86	4-140	①	②	①	100–120	56–64	14–21 ③	16–23 ③
'83–'86	6-173	④	19–24	65–75	85–96	47–52	15–18 ③	20–30 ③
'82–'84	6-232	⑤	30–36	62–81	85–100	75–85	18	15–22
'79–'86	6-300	70–75	40–45	60–70	130–150	75–85	22–32	28–33
'82	8-255	65–72	19–24	60–70	70–90	75–85	18–20	18–24
'79–'86	8-302	65–70	19–24	60–70	70–90	75–85	23–25	12–16 ⑥
'79–'80	8-351M	95–105	40–45	95–105	70–90	75–85	⑦	18–24
'79–'86	8-351W	105–112	40–45	95–105	70–90	75–85	23–25	18–24
'79–'80	8-400	95–105	40–45	95–105	70–90	75–85	⑦	18–24
'83–'86	420 (diesel)	⑧	46–51 ⑨	95 ⑩	90	44–50 ⑪	24	30
'79–'86	8-460	130–140	40–45	95–105	70–90	75–85	25–30	28–33

① Torque bolts in two steps:
   Step 1: 50–60 ft. lbs.   Step 2: 80–90 ft. lbs.
② Torque nuts in two steps:
   Step 1: 25–30 ft. lbs.   Step 2: 30–36 ft. lbs.
③ Torque in stages, recheck after engine is warmed
④ Torque bolts in three steps:
   Step 1: 29–40 ft. lbs.   Step 2: 40–51 ft. lbs.
   Step 3: 70–85 ft. lbs.
⑤ Tighten in four steps:
   Step 1: 47 ft. lbs.   Step 2: 55 ft. lbs.
   Step 3: 63 ft. lbs.   Step 4: 74 ft. lbs.
   Back-off all bolts 2–3 turns and retorque in four steps.
⑥ '80 and later: 18–24
⑦ 3/8": 22–32
   5/16": 17–25
⑧ Refer to text for sequence procedure
⑨ Two steps: 1st—38 ft. lbs. then 46–51 ft lbs.
⑩ Two steps: 1st—75 ft. lbs. then 95 ft. lbs.
⑪ Apply locking sealer to threads

ALTERNATOR SPECIFICATIONS

Year	Color Code	Output Amps	Output Watts	Field Current Amps	Cut-In rpm	Brush Length Inches New	Brush Length Inches Limit
'79–'81	Orange	40	600	2.9	400	1/2	5/16
	Green	60	900	2.9	400	1/2	5/16
	Green ①	60	900	4.0	400	1/2	3/16
'82–'86	Orange	40	600	2.9	900	1/2	5/16
	Green	60	900	6.0	1025	1/2	5/16
	Black	70	1050	6.0	780	1/2	1/4
	Red	100	1500	6.0	930	1/2	1/4

① Blue ink on pulley face

BATTERY AND STARTER SPECIFICATIONS

Year	Engine	Battery Ampere/Hour Capacity	Battery Volts	Battery Ground	Starter Lock Test Amps	Starter Lock Test Volts	Starter Lock Test Torque (ft lbs)	Starter No Load Test Amps	Starter No Load Test Volts	Starter No Load Test rpm	Brush Spring Tension (oz)
'79	All	41	12	Neg	460	5	9.0	70	12	9500	40
		53	12	Neg	670	5	15.5	80	12	9500	80
		68	12	Neg							
'80–'86	All	36	12	Neg	460	5	9.0	70	12	9500	40
		45	12	Neg	670	5	15.5	80	12	9500	80
		63	12	Neg							
		81	12	Neg							

① Prestolite model with 400 cid engine

BRAKE SPECIFICATIONS
Pickups and Vans

Year	Model	Master Cylinder Bore	Caliper Bore	Wheel Cylinder Bore Front	Wheel Cylinder Bore Rear	Rotor Diameter	Rotor Minimum Thickness	Rotor Maximum Run-out	Brake Drum Diameter Front	Brake Drum Diameter Rear	Machined Oversize Front	Machined Oversize Rear
'79–'80	F&E-100, 150	1.00	2.875	—	.9375	11.54	1.180	.003	—	11.03	—	11.09
	F&E-250	1.062	2.180	—	1.000	12.50	1.120	.003	—	12.0	—	12.06
	F&E-350	1.062	2.180	—	1.062	12.50	1.120	.003	—	12.0	—	12.06
'81	F-100 ①	1.00	2.599	—	.9375	10.97	1.120 ⑧	.003	—	10.0	—	10.09
'82–'86	Ranger 4×2	.9375	2.597	—	.750	10.28	.81	.002	—	9.0	—	9.06
	Ranger 4×4	.9375	2.597	—	.750	10.86	.81	.002	—	9.0	—	9.06
	F-100 ②, F-150 4×2, E-100, E-150 ③	1.00 ④	2.875	—	.9375	11.65	1.120	.003	—	11.03	—	11.09
	F-150 4×4, E-150 ⑤	1.00 ④	2.875	—	.9375	11.65	1.120	.003	—	11.03	—	11.09
	F-250 4×2 ⑥	1.00	2.875	—	.9375	12.59	1.120	.003	—	12.0	—	12.06
	F-250 Super cab, F-200 ⑦ 4×2, E-250, E-350	1.062	2.180	—	1.062 ⑨	12.56	1.180	.003	—	12.0	—	12.06
	F-350 4×4, F-350 4×4	1.062	2.180	—	1.062 ⑨	12.48	1.180	.003	—	12.0	—	12.06

① w/std payload package
② exc. std payload package
③ w/GVWR under 6350 lb.
④ E-150: 1.062
⑤ w/GVWR over 6350 lb.
⑥ Regular cab w/payload under 8500 lb.
⑦ Regular cab w/payload over 8500 lb.
⑧ 1982: .810 or LD models
⑨ '83–'84 1.00

BRAKE SPECIFICATIONS
Bronco and Bronco II
(All measurements in inches)

Year	Master Cylinder Bore	Caliper Bore	Wheel Cylinder Bore		Rotor Diameter	Rotor Minimum Thickness	Rotor Maximum Run-Out	Brake Drum Diameter		Machined Oversize	
			Front	Rear				Front	Rear	Front	Rear
'79–'81	1.00	2.875	—	.9375	11.72	1.120	.003	—	11.03	—	11.09
'82–'86	1.00	2.875	—	.9375	11.65	1.120	.003	—	11.03	—	11.09
'83–'86 Bronco II (4×2)	.9375	2.597	—	.750	10.28	.81	.002	—	9.0	—	9.06
Bronco II (4×4)	.9375	2.597	—	.750	10.86	.81	.002	—	9.0	—	9.06

TUNE UP

Electronic Ignition

Ford truck engines are equipped with one of the following systems; Dura Spark II, Dura Spark III or Thick Film Integrated (TFI) Universal Distributor (Hall Effect) which is controlled by the EECIV (electronic engine control) system.

Procedures for diagnosis and repair of the ignition system are found in the Electrical Chapter of the Unit Repair Section.

On Dura Spark distributors, the contact breaker points and condenser in the distributor are replaced by a permanent magnet low-voltage generator. The generator consists of an armature with four, six or eight teeth mounted on the top of the distributor shaft, and a permanent magnet inside a small coil. The coil is riveted in place to provide a preset air gap with the armature. The distributor base, cap, rotor and vacuum and centrifugal spark advance are about the same as the conventional system. The distributor is wired to a solid state module in the engine compartment. Inside the module is an electronic circuit board which consists of inner connecting resistors, capacitors, transistors and diodes. The module senses a signal from the magnetic generator to perform the switching function of conventional points and it senses and controls dwell. Unless a malfunction occurs, or the distributor is moved or replaced, the initial ignition timing remains constant. Because the low voltage coil in the distributor is riveted in position, gap adjustment is not possible.

On some engines, depending on calibration, the Dura Spark II system may use a standard module or a "Universal Ignition Module". The Universal Ignition Module (UIM) is capable of providing spark timing retard in response to barometric or engine sensors, or MCU signal.

Trucks equipped with EECIV use a universal distributor design which incorporates an integrally mounted TFI–IV module. The distributor uses a "Hall Effect" vane switch

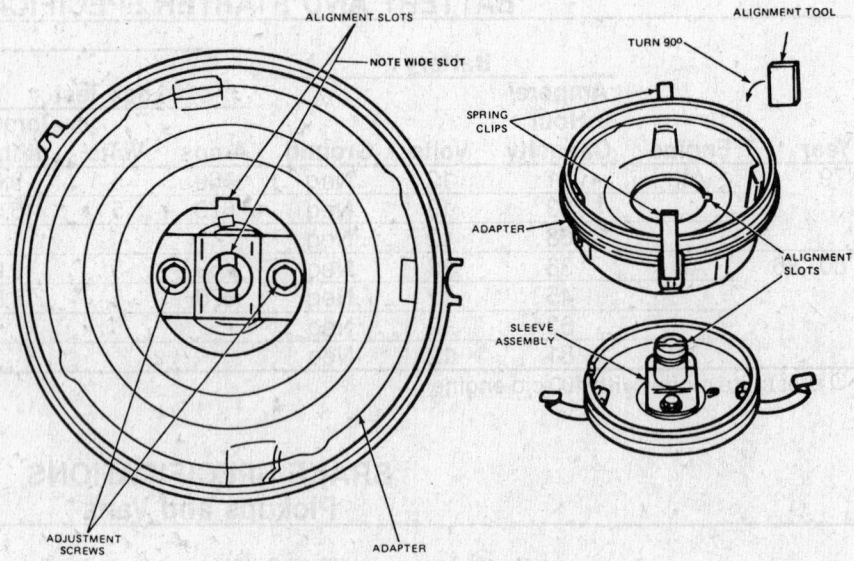

DuraSpark III rotor alignment

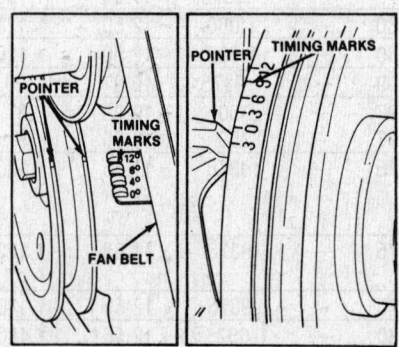

Typical timing marks: left, block mounted; right, pulley mounted

stator assembly and has provision for fixed octane adjustment. A new cap, adapter and rotor are designed for use on the Universal Distributor. The Thick Film Integrated (TFI) module is contained in molded thermo-plastic and is mounted on the distributor base. The TFI–IV features a "push start" mode which allows push starting of the vehicle, if necessary. The TFI–IV system uses an "E-Core" ignition coil, which replaces the oil-filled coil found on other systems.

EEC BI-LEVEL ROTOR & DISTRIBUTOR CAP

Conventional distributors are restricted to approximately 20 degrees of centrifugal distributor advance. In order to permit a flexibility in choosing engine calibrations, based totally on engine need, the EEC system was designed to allow up to 50 degrees distributor advance.

The distributor rotor and cap electrodes have been re-designed to handle the additional spark advance capability. The technique involves using two separate levels of secondary voltage distribution. Both the rotor and cap have an upper and lower electrode levels.

As the rotor turns, one of the high voltage electrode pick-up arms is aligned with one arm of the distributor cap center electrode plate, allowing high voltage to be transmitted from the plate through the rotor, distributor cap, and plug wire to the appropriate spark plug.

The numbers in the top of the distributor cap are spark plug wire/cylinder identification numbers only.

In a conventional distributor, the firing order follows the circular path of the rotor. In an EEC distributor, however, the upper and lower level electrodes fire alternately in a pattern that jumps from one side of the cap to the other.

NOTE: Refer to Distributor Removal and Installation for rotor alignment instructions.

Ignition Timing

Ignition timing is the measurement, in degrees of crankshaft rotation, of the point at which the spark plugs fire in each of the cylinders. It is measured in degrees before or after Top Dead Center (TDC) of the compression stroke. Ignition timing is controlled by turning the distributor body in the engine.

Ideally, the air/fuel mixture in the cylinder will be ignited by the spark plug just as the piston passes TDC of the compression stroke. If this happens, the piston will be beginning the power stroke just as the compressed and ignited air/fuel mixture starts to expand. The expansion of the air/fuel mixture then forces the piston down on the power stroke and turns the crankshaft.

Because it takes a fraction of a second for the spark plug to ignite the mixture in the cylinder, the spark plug must fire a little before the piston reaches TDC. Otherwise, the mixture will not be completely ignited as the piston passes TDC and the full power of the explosion will not be used by the engine.

The timing measurement is given in degrees of crankshaft rotation before the piston reaches TDC (BTDC). If the setting for the ignition timing is 5° BTDC, each spark plug must fire 5° before each piston reaches TDC. This only holds true, however, when the engine is at idle speed.

As the engine speed increases, the pistons go faster. The spark plugs have to ignite the fuel even sooner if it is to be completely ignited when the piston reaches TDC. To do this, the distributor has a means to advance the timing of the spark as the engine speed increases. This is accomplished by centrifugal weights within the distributor and a vacuum diaphragm mounted on the side of the distributor. It is usually necessary to disconnect the vacuum lines from the diaphragm when the ignition timing is being set, however this is not true in all cases. Always refer to the emission decal in the engine compartment to determine if the lines should be disconnected and plugged.

If the ignition is set too far advanced (BTDC), the ignition and expansion of the fuel in the cylinder will occur too soon and tend to force the piston down while it is still traveling up. This causes engine ping.

If the ignition spark is set too far retarded after TDC (ATDC), the piston will have already passed TDC and started on its way down when the fuel is ignited. This will cause the piston to be forced down for only a portion of its travel. This will result in poor engine performance and lack of power.

The timing is best checked with an inductive type timing light. This device is connected in series with the No. 1 spark plug. The current that fires the spark plug also causes the timing light to flash.

Timing marks are located at the front of the engine and consist of a pointer or scale on the timing case or belt cover and reference mark or marks on the crankshaft pulley. A scale of degrees is either on the pointer plate or pulley, depending on year, model and engine. When the engine is running, and curb idle adjusted, the timing light is aimed at the mark on the crankshaft pulley and the scale.

TACHOMETER HOOKUP

Models equipped with a "conventional" type coil have an adapter on the top of the coil that provides a clip marked "Tach Test". On models (TFI) equipped with an "E" type coil, the tach connection is made at the back of the wire harness connector. A cut-out is provided and the tachometer lead wire alligator clip can be connected to the dark green/yellow dotted wire of the electrical harness plug.

ADJUSTMENT

NOTE: On models equipped with EEC III, all ignition timing is controlled by the EEC III module. Initial ignition timing is not adjustable and no attempt at adjustment should be made. For a description of EEC systems, refer to the Unit Repair sections on "Electronic Ignition Systems" and on "Emission Controls". Models equipped with the TFI–IV ignition system do not require ignition timing as routine maintenance, since the system adjusts itself. An initial check can be made. Refer to the following two procedure paragraphs.

1. Locate the timing marks on the crankshaft pulley and the front of the engine.

2. Clean the timing marks so that you can see them.

3. Mark the timing marks with a piece of chalk or with paint. Color the mark on the scale that will indicate the correct timing when it is aligned with the mark on the pulley or the pointer. It is also helpful to mark the notch in the pulley or the tip of the pointer with a small dab of color.

4. Attach a tachometer to the engine. (See tachometer hook-up instructions).

5. Attach a timing light according to the manufacturer's instructions. If the timing light has three wires, one is attached to the No. 1 spark plug with an adapter. The

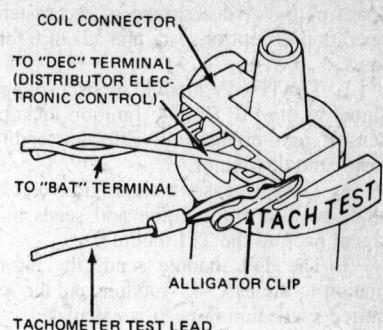

Attaching a tachometer lead to the coil connector

other wires are connected to the battery. The red wire goes to the positive side of the battery and the black wire is connected to the negative terminal of the battery.

NOTE: Refer to emissions decal in the engine compartment to determine if the vacuum line(s) should be disconnected.

6. Disconnect the vacuum line to the distributor at the distributor and plug the vacuum line. A golf tee does a fine job.

7. Check to make sure that all of the wires clear the fan and then start the engine.

8. Adjust the idle to the correct setting.

9. Aim the timing light at the timing marks. If the marks that you put on the flywheel or pulley and the engine are aligned when the light flashes, the timing is correct. Turn off the engine and remove the tachometer and the timing light. If the marks are not in alignment, proceed with the following steps.

10. Turn off the engine.

11. Loosen the distributor lockbolt just enough so that the distributor can be turned with a little effort.

12. Start the engine. Keep the wires of the timing light clear of the fan.

13. With the timing light aimed at the pulley and the marks on the engine, turn the distributor in the direction of rotor rotation to retard the spark, and in the opposite direction of rotor rotation to advance the spark. Align the marks on the pulley and the engine with the flashes of the timing light.

14. When the marks are aligned, tighten the distributor lockbolt and recheck the timing with the timing light to make sure that the distributor did not move when you tightened the lockbolt.

15. Turn off the engine and remove the timing light.

MODELS EQUIPPED WITH EEC–IV AND TFI

The connection between the EEC–IV microprocessor and the Thick Film Integrated (TFI) ignition module is called the SPOUT circuit. SPOUT simply means "spark out" since a signal carried to the TFI shuts off the coil and produces a spark for firing the

spark plugs. A description of the system operation to control spark and advance follows:

1. The TFI–IV module sends a voltage signal to the PIP (Profile Ignition Pickup) sensor, part of the Hall Effect assembly inside the distributor.

2. The PIP sensor then provides crankshaft position information and sends this signal back to the TFI module.

3. The TFI module sends the information to the EEC–IV module and the required spark timing need is calculated.

4. The required timing information goes back to the TFI module through electrical circuitry (the SPOUT) and the coil turns off the primary circuit to fire the spark plugs at the precise time.

5. The TFI module also determines dwell and limits primary circuit to a safe value. If there is an open in the SPOUT signal wire, the TFI module will use the PIP sensor to provide spark; but the engine will only run at basic timing setting.

6. To check basic timing the SPOUT wire must be disconnected. An inline connector is provided on the "yellow with green dot" or "black" single SPOUT wire. The wire is located between the distributor and engine harness. With the wire disconnected the TFI module is locked into no advance and basic timing may be checked and adjusted if necessary. See proceeding paragraph.

Idle Mixture Adjustment

CARBURETORS

NOTE: For this procedure, Ford recommends a propane enrichment procedure. This requires special equipment not available to the general public. In lieu of this equipment the following procedure may be followed to obtain a satisfactory idle mixture.

1. Block the wheels, set the parking brake and run the engine to bring it to normal operating temperature.

2. Disconnect the hose between the emission canister and the air cleaner.

3. On engines equipped with the Thermactor air injection system, the routing of the vacuum lines connected to the dump valve will have to be temporarily changed. Mark them for reconnection before switching them.

4. For valves with one or two vacuum lines at the side, disconnect and plug the lines.

5. For valves with one vacuum line at the top, check the line to see if it is connected to the intake manifold or an intake manifold source such as the carburetor or distributor vacuum line. If not, remove and plug the line at the dump valve and connect a temporary length of vacuum hose from the dump valve fitting to a source of intake manifold vacuum.

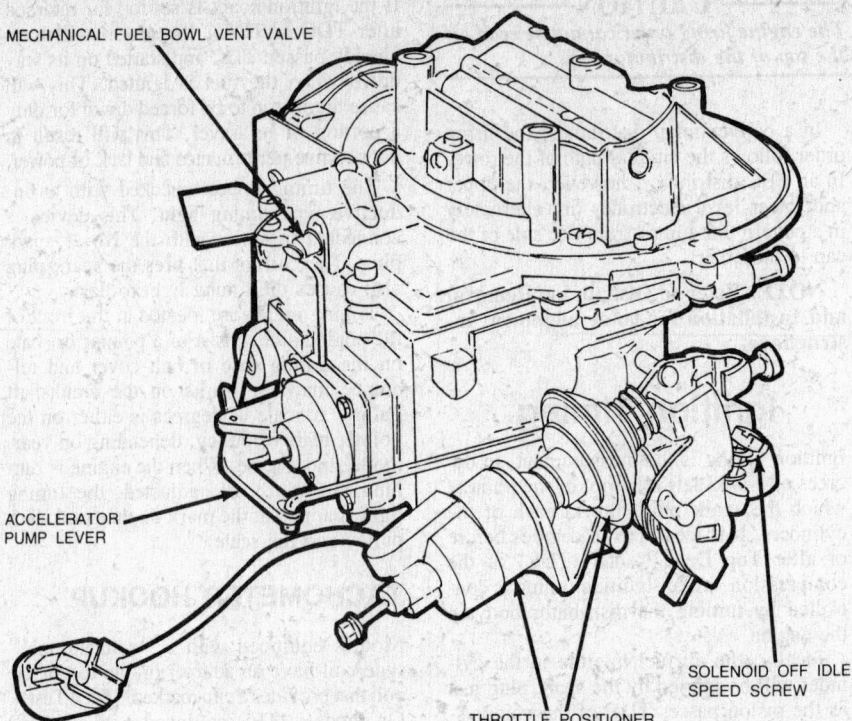

MECHANICAL FUEL BOWL VENT VALVE

ACCELERATOR PUMP LEVER

SOLENOID OFF IDLE SPEED SCREW

THROTTLE POSITIONER

Motorcraft 2150 carburetor with solenoid

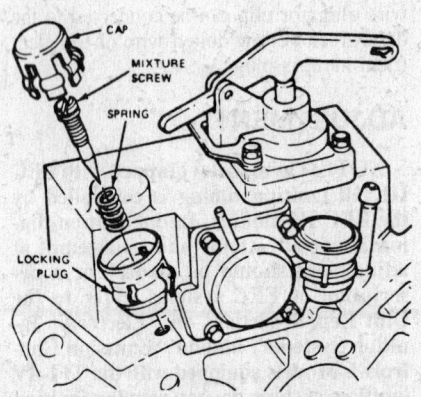

CAP

MIXTURE SCREW

SPRING

LOCKING PLUG

Some 1980 and later models have 2-piece metal plugs in place of the plastic limiter caps on the idle mixture adjusting screws. These caps should be carefully removed before attempting any adjustments.

6. Remove the limiter caps from the mixture screws by CAREFULLY cutting them with a sharp knife.

7. Place the transmission in neutral and run the engine at 2500 rpm for 15 seconds.

8. Place the automatic transmission in Drive; the manual in neutral.

9. Adjust the idle speed to the higher of the two figures given on the underhood sticker.

10. Turn the idle mixture screw(s) to obtain the highest possible rpm, leaving the screw(s) in the leanest position that will maintain this rpm.

11. Repeat Steps 7 thru 10 until further

adjustment of the mixture screw(s) does not increase the rpm.

12. Turn the screw(s) in until the lower of the two idle speed figures is reached. Turn the screw(s) in 1/4 turn increments each to insure a balance.

13. Turn the engine off and remove the tachometer. Reinstall all equipment.

NOTE: Rough idle, that cannot be corrected by normal service procedures may be caused by leakage between the EGR valve body and diaphragm. To determine if this is the cause: Tighten the EGR bolts to 15 ft. lbs. Connect a vacuum gauge to the intake manifold. Lift to exert a sideways pressure on the diaphragm housing. If the idle changes or the reading on the vacuum gauge varies, replace the EGR valve.

Idle Speed Adjustment

CARBURETORS

Through 1982

1. Remove the air cleaner and disconnect and plug the vacuum lines.

2. Block the wheels, apply the parking brake, turn off all accessories, start the engine and run it to normalize underhood temperatures.

3. Check that the choke plate is fully open and connect a tachometer according to the manufacturer's instructions.

4. Check the throttle stop positioner (TSP) off speed as follows: Collapse the

plunger by forcing the throttle lever against it. Place the transmission in neutral and check the engine speed. If necessary, adjust to specified TSP Off speed with the throttle adjusting screw. See the underhood sticker.

5. Place the manual transmission in neutral; the automatic in Drive and make certain the TSP plunger is extended.

6. Turn the TSP until the specified idle speed is obtained.

7. Install the air cleaner and connect the vacuum lines. Check the idle speed. Adjust, if necessary, with the air cleaner on.

1983 and Later

122 (2.0L) AND 140 (2.3L) YFA–IV & YFA IV–FB (FEEDBACK)

1. Block the wheels and apply parking brake. Place the transmission in Neutral or Park.

2. Bring engine to normal operating temperature.

3. Place A/C selector in the Off position.

4. Place transmission in specified position as referred to on emissions decal.

5. Check/adjust curb idle RPM. If adjustment is required, turn the hex head adjustment at the rear of the TSP (throttle solenoid positioner) housing.

6. Place the transmission in Neutral or Park. Rev the engine momentarily. Place transmission in specified position and recheck curb idle RPM. Readjust if required.

7. Turn the ignition key to the Off position.

8. If a curb idle RPM adjustment was required and the carburetor is equipped with a dashpot, adjust the dashpot clearance to specification as follows: Turn key to On position. Open throttle to allow TSP solenoid plunger to extend to the curb idle position. Collapse dashpot plunger to maximum extent. Measure clearance between tip of plunger and extension pad on throttle vent lever. If required, adjust to specification. Tighten dashpot locknut. Recheck clearance. Turn key to Off position.

9. If curb idle adjustment was required, check/adjust the bowl vent setting as follows: Turn ignition key to the On position to activate the TSP (engine not running). Open throttle to allow the TSP solenoid plunger to extend to the curb idle position. Secure the choke plate in the wide open position. Open throttle so that the throttle vent lever does not touch the bowl vent rod. Close the throttle to the idle set position and measure the travel of the fuel bowl vent rod from the open throttle position. Travel of the bowl vent rod should be within specification (.100 to .150 inches). If out of specification, bend the throttle vent lever at notch to obtain required travel.

10. Remove all test equipment and reinstall air cleaner assembly. Tighten the holddown bolt to specification.

300 (4.9L) YFA–IV & YFA–IV–FB

1. Block the wheels and apply the

parking brake. Place the transmission in Neutral or Park.

2. Bring engine to normal operating temperature.

3. Place A/C Heat Selector to Off position.

4. Place transmission in specified gear.

5. Check/adjust curb idle RPM as follows: TSP dashpot. Insure that TSP is activated using a ⅜ inch open end wrench, adjust curb idle RPM by rotating the nut directly behind the dashpot housing. Adjust curb idle RPM by turning the idle RPM speed screw. Front mounted TSP (same as A/C kicker on all other calibrations) insure that TSP is activated. After loosening lock nut, adjust curb idle RPM by rotating TSP solenoid until specified RPM is obtained. Tighten locknut.

6. Check/adjust anti-diesel (TSP Off). Manually collapse the TSP by rotating the carb throttle shaft lever until the TSP Off adjusting screw contacts the carburetor body. If adjustment is required, turn the TSP Off adjusting screw while holding the lever adjustment screw against the stop.

7. Place the transmission in Neutral or Park. Rev the engine momentarily. Place the transmission in specified position and recheck curb idle rpm. Readjust if required.

8. Check/adjust dashpot clearance to .120 ± .030.

9. If a final curb idle speed adjustment is required, the bowl vent setting must be checked as follows: Stop the engine and turn the ignition key to the On position, so that the TSP dashpot or TSP is activated but the engine is not running (where applicable). Secure the choke plate in the wide-open position. Open the throttle, so that the throttle vent lever does not touch the fuel bowl vent rod. Close the throttle, and measure the travel of the fuel bowl vent rod from the open throttle position. Travel of the fuel bowl vent rod should be within .100 to .150 in. If out of specification, bend the

throttle vent lever to obtain the required travel. Remove all test equipment, and tighten the air cleaner holddown bolt to specification.

10. Whenever it is required to adjust engine idle speed by more than 50 rpm, the adjustment screw on the AOD linkage lever at the carburetor should also be readjusted.

173 (2.8L), 232 (3.8L) & 302 (5.0L) 2150–2V FB (FEEDBACK)

1. Set parking brake and block wheels.

2. Place the transmission in Park.

3. Bring the engine to normal operating temperature.

4. Disconnect the electric connector on the EVAP purge solenoid.

5. Disconnect and plug the vacuum hose to the VOTM kicker.

6. Place the transmission in Drive position.

7. Check/adjust curb idle rpm, if adjustment is required: Adjust with the the curb idle speed screw or the saddle bracket adjusting screw, depending on how equipped.

8. Place the transmission in Neutral or Park. Rev the engine momentarily. Place the transmission in Drive position and recheck curb idle rpm. Readjust if required.

9. Remove the plug from the vacuum hose to the VOTM kicker and reconnect.

10. Reconnect the electrical connector on the EVAP purge solenoid.

302 (5.0L) 2150–2V (NON-FEEDBACK)

1. Set parking brake and block wheels.

2. Place the transmission in Neutral or Park.

3. Bring engine to normal operating temperature.

4. Place A/C Heat selector to Off position.

5. Disconnect and plug vacuum hose to thermactor air bypass valve.

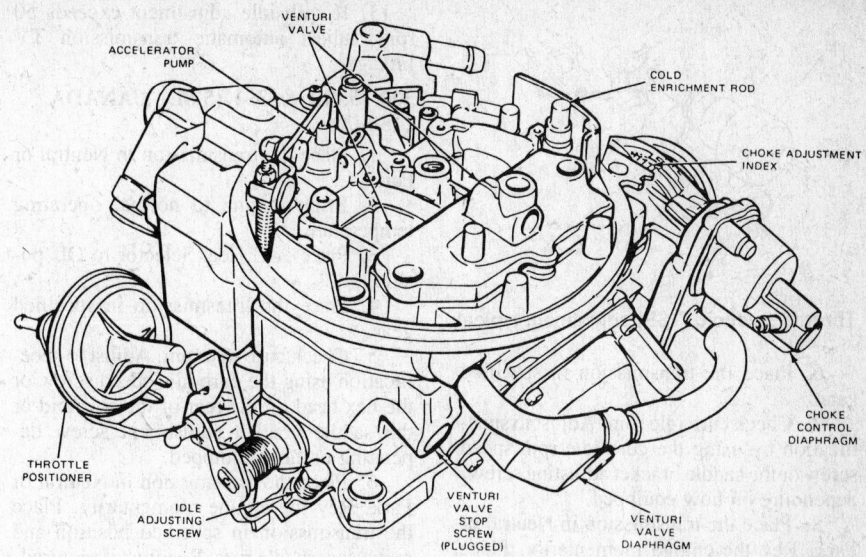

Motocraft 7200VV carburetor

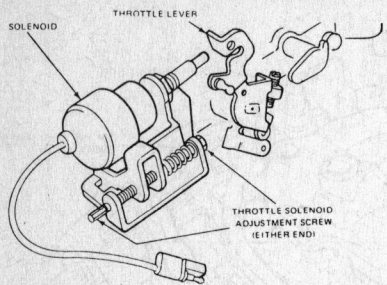

Throttle solenoid (TSP) adjustment-typical

6. Place the transmission in specified gear.

7. Check curb idle rpm. Adjust to specification by using the curb idle rpm speed screw or the saddle bracket adjusting screw, depending on how equipped.

8. Place the transmission in Neutral or Park. Rev the engine momentarily. Place the transmission in specified position, and recheck curb idle rpm. Readjust if required.

9. Remove plug from vacuum hose to thermactor air bypass valve and reconnect.

10. Whenever it is required to adjust engine idle speed by more than 50 rpm, the adjustment screw on the AOD linkage lever at the carburetor should also be readjusted.

351 (5.8L) 2150–2V OR 7200 VV

1. Block the wheels and apply parking brake. Place the transmission in Neutral or Park.

2. Bring the engine to normal operating temperature.

3. Disconnect purge hose on canister side of evaporator purge solenoid. Check to ensure that purge vacuum is present (solenoid has opened and will require 3 to 5 minute wait after starting engine followed by a short time at part-throttle). Reconnect purge hose.

4. Disconnect and plug the vacuum hose to the VOTM kicker.

5. Place the transmission in specified position.

6. Check/adjust curb idle rpm. If adjustment is required, adjust with the curb idle speed screw or the saddle bracket adjusting screw (ensure curb idle speed screw is not touching throttle shaft lever).

7. Place the transmission in Neutral or Park. Rev the engine momentarily. Place the transmission in specified position and recheck curb idle rpm. Readjust if required.

8. Check/adjust throttle position sensor (TPS).

9. Remove the plug from the vacuum hose to the VOTM kicker and reconnect.

10. Apply a slight pressure on top of the nylon nut located on the accelerator pump to take up the linkage clearance.

11. Turn the nylon nut on the accelerator pump rod clockwise until a .010 ± .005 clearance is obtained between the top of the accelerator pump and the pump lever.

12. Turn the accelerator pump rod nut one turn counterclockwise to set the lever lash preload.

13. If curb idle adjustment exceeds 50 rpm, adjust automatic transmission TV linkage.

302 (5.0L) & 351 (5.8L) CANADA 2150–2V

1. Place the transmission in Neutral or Park.

2. Bring engine to normal operating temperature.

3. Place A/C Heat Selector to Off position.

4. Place the transmission in specified gear.

5. Check curb idle rpm. Adjust to specification using the curb idle speed screw or the hex head on the rear of the solenoid or the saddle bracket adjustment screw depending on how equipped.

6. Place the transmission in Neutral or Park. Rev the engine momentarily. Place the transmission in specified position and recheck curb idle rpm. Readjust if required.

7. TSP Off: With transmission in specified gear, collapse the solenoid plunger, and set specified TSP Off speed on the speed screw.

8. Disconnect vacuum hose to decel throttle control modulator and plug (if so equipped).

9. Connect a slave vacuum from manifold vacuum to the decel throttle control modulator (if so equipped).

10. Check/adjust decel throttle control rpm. Adjust if necessary.

11. Remove slave vacuum hose.

12. Remove plug from decel throttle control modulator hose and reconnect.

460 (7.5L)

1. Block the wheels and apply parking brake.

2. Run engine until normal operating temperature is reached.

3. Place the vehicle in Park or Neutral, A/C in Off position, and set parking brake.

4. Remove air cleaner.

5. Disconnect and plug decel throttle control kicker diaphragm vacuum hose.

6. Connect a slave vacuum hose from an engine manifold vacuum source to the decel throttle control kicker.

7. Run engine at approximately 2500 rpm for 15 seconds, then release the throttle.

8. If decel throttle control rpm is not within ± 50 rpm of specification, adjust the kicker.

9. Disconnect the slave vacuum hose and allow engine to return to curb idle.

10. Adjust curb idle, if necessary, using the curb idle adjusting screw.

11. Rev the engine momentarily, recheck curb idle and adjust if necessary.

12. Reconnect the decel throttle control vacuum hose to the diaphragm.

13. Reinstall the air cleaner.

DIESEL

NOTE: A special tachometer is required to check engine RPM on a diesel engine.

134 (2.2L)

1. Block the wheels and apply the parking brake.

2. Start and run engine until the normal operating temperature is reached. Shut off engine.

3. Connect diesel engine tachometer.

4. Start engine and check RPM. Refer to emissions decal for latest specifications. RPM is usually adjusted in Neutral for manual transmissions and Drive for automatic models.

5. The adjustment bolt is located on the bell crank at the top of the injector pump. The upper bolt is for curb idle, the lower for max speed.

6. Loosen the locknut. Turn the adjustment screw clockwise to increase RPM, counter-clockwise to lower the RPM.

7. Tighten the locknut. Increase engine speed several times and recheck idle. Readjust if necessary.

420 (6.9L)

1. Block the wheels and apply the parking brake.

2. Bring the engine to normal operating temperature. Shut off engine.

3. Connect diesel engine tachometer.

4. Start the engine and check RPM. Refer to the emissions decal for latest specifications. RPM is usually adjusted in Neutral for manual transmissions and Drive for automatic models.

5. Turn the idle speed adjusting screw in the required direction to increase or decrease RPM. The adjusting screw is located on the top of the injector pump above the cold start valve.

6. Place the gear selector in neutral, if automatic, and speed up engine several times. Recheck idle RPM, readjust if necessary.

Valve Adjustment

GASOLINE ENGINES

NOTE: Refer to the "Rocker Arm/Rocker Shaft and Stud Section" for valve adjustment procedures for the 173 (2.8L) V6 engine.

All gasoline engines used, except the 2.8L V6, are equipped with hydraulic valve lifters or lash adjusters. Hydraulic valve lifters or lash adjusters operate with zero clearance in the valve train, and because of this the rocker arms are non-adjustable. The only means by which valve system clearances can be altered is by installing over or undersize pushrods except on OHC (overhead cam) models; but, because of the hydraulic lifter's natural ability to compensate for slack in the valve train, all components of all the valve system should be checked for wear if there is excessive play in the system. Refer to Rocker Arm section.

DIESEL ENGINES

134 (2.2L) Diesel

1. Warm the engine until normal operating temperature is reached.

2. Remove the valve cover. Check the head bolt torque in sequence.

3. Turn the engine to bring the No. 1 piston to TDC (top dead center) of the compression stroke.

4. Adjust the following valves: No. 1 Intake and Exhaust. No. 2 Intake. No. 3 Exhaust.

5. Rotate the crankshaft 360° and bring No. 4 piston to TDC of the compression stroke.

6. Adjust the following valves: No. 2 Exhaust. No. 3 Intake. No. 4 Intake and Exhaust.

7. To adjust the valves, loosen the locknut on the rocker arm. Rotate the adjusting screw clockwise to reduce clearance, counter-clockwise to increase clearance. Clearance is checked with a flat feeler gauge that is passed between the rocker arm and valve stem.

WHEN NO. 1 CYLINDER IS AT TOP DEAD ENTER

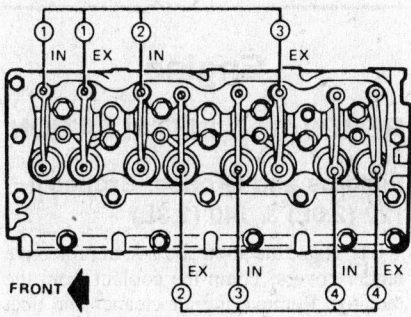

WHEN NO. 4 CYLINDER IS AT TOP DEAD ENTER

Valve adjusting sequence 134 (2.2L) diesel

8. After adjustments are made, be sure the locknuts are tight. Be sure mounting surfaces are clean. Install the valve cover and new valve cover gasket.

420 (6.9L)

The 6.9L diesel engine is equipped with hydraulic lifters that minimize engine noise and maintain zero lash or tappet clearance. This eliminates the need for periodic adjustment. The hydraulic tappets also incorporate camshaft roller followers for improved camshaft wear characteristics.

IGNITION SYSTEM

Complete service and troubleshooting information on Dura Spark and TFI systems can be found in the "Electronic Ignition" section of Unit Repair.

Distributor

REMOVAL & INSTALLATION

On certain models equipped with EEC, the distributor is locked into position and all timing control is handled by the EEC module. No timing adjustment is required or possible. Rotor alignment is critical on bi-level systems, and any required servicing should be done by someone familiar with the system. See the Emissions Section in Unit Repair for rotor alignment.

1. Remove the air cleaner on V6 and V8 engines. On 4 and 6 cylinder in-line engines, removal of a thermactor (air) pump mounting bolt and drive belt will allow the pump to be moved to the side and permit access to the distributor. If necessary, disconnect the thermactor air filter and lines as well.

2. Remove the distributor cap and position the cap and ignition wires to the side.

3. Disconnect the wire harness plug from the distributor connector. Disconnect and plug the vacuum hoses from the vacuum diaphragm assembly (if equipped).

4. Rotate the engine (in normal direction of rotation) until No. 1 piston is on TDC (Top Dead Center) of the compression stroke. The TDC mark on the crankshaft pulley and the pointer should align. Rotor tip pointing at No. 1 spark plug wire position on distributor cap.

5. On DuraSpark I or II, turn the engine a slight bit more (if required) to align the stator (pick-up coil) assembly pole with an (the closest) armature pole. On DuraSpark III, the distributor sleeve groove (when looking down from the top) and the cap adaptor alignment slot should align. On models equipped with EECIV (1984 and later), remove the rotor (2 screws) and note the position of the "polarizing square" and shaft plate for reinstallation reference.

6. Scribe a mark on the distributor body and engine block to indicate the position of the rotor tip and position of the distributor in the engine. DuraSpark III and some EECIV system distributors are equipped with a notched base and will only locate at one position on the engine.

7. Remove the holddown bolt and clamp located at the base of the distributor. (Some DuraSpark III and EECIV system distributors are equipped with a special holddown bolt that requires a Torx head wrench for removal). Remove the distributor from the engine. Pay attention to the direction the rotor tip points if it moves from the No. 1 position the drive gear disengages. For reinstallation purposes, the rotor should be at this point to insure proper gear mesh and timing.

8. Avoid turning the engine, if possible, while the distributor is removed. If the engine is turned from TDC position, TDC timing marks will have to be reset before the distributor is installed; Steps 4 and 5.

9. Position the distributor in the engine

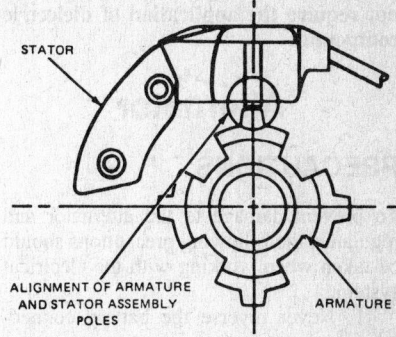

TOOTH MUST BE PERFECTLY ALIGNED WITH TIMING MARKS

Align armature and stator (4 cylinder shown)

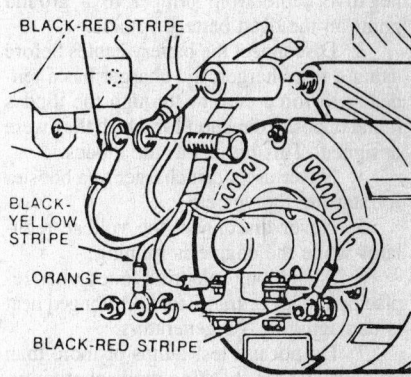

Typical alternator connections and mounting (color codes may differ)

with the rotor aligned to the marks made on the distributor, or at the position the rotor pointed when the distributor was removed. The stator and armature or "polarizing square" and shaft plate should also be aligned. Engage the oil pump intermediate shaft and insert the distributor until fully seated on the engine, if the distributor does not fully seat, turn the engine slightly to fully engage the intermediate shaft.

10. Follow the above procedures on models equipped with an indexed distributor base. Make sure when positioning the distributor that the slot in the distributor base will engage the block tab and the sleeve/adaptor slots are aligned.

11. After the distributor has been fully seated onto he block, recheck the timing mark and rotor alignment. Install the holddown bracket and bolt. On models equipped with an indexed base, tighten the mounting bolt. On other models, snug the mounting bolt so the distributor can be turned for ignition timing purposes.

12. The rest of the installation is in the reverse order of removal. Check and reset the ignition timing.

NOTE: A silicone compound is used on rotor tips, distributor cap contacts and on the inside of the connectors on the spark plugs cable and module couplers. Always apply Silicone Dielectric

Compound after servicing any component of the ignition system. Various models use a multi-point rotor which does not require the application of dielectric compound.

Alternator

PRECAUTIONS

To prevent damage to the alternator and regulator, the following precautions should be taken when working with the electrical system.

1. Never reverse the battery connections.

2. Booster batteries for starting must be connected properly: positive-to-positive and negative cable from jumper to a ground point on the dead battery vehicle.

3. Disconnect the battery cables before using a fast charger; the charger has a tendency to force current through the diodes in the opposite direction for which they were designed. This burns out the diodes.

4. Never use a fast charger as a booster for starting the vehicle.

5. Never disconnect the voltage regulator while the engine is running.

6. Avoid long soldering times when replacing diodes or transistors. Prolonged heat is damaging to AC generators.

7. Do not use test lamps of more than 12 volts (V) for checking diode continuity.

8. Do not short across or ground any of the terminals on the AC generator.

9. The polarity of the battery, generator, and regulator must be matched and considered before making any electrical connections within the system.

10. Never operate the alternator on an open circuit. Make sure that all connections within the circuit are clean and tight.

11. Disconnect the battery terminals when performing any service on the electrical system. This will eliminate the possibility of accidental reversal of polarity.

12. Disconnect the battery ground cable if arc welding is to be done on any part of the car.

REMOVAL & INSTALLATION

1. Open the hood and disconnect the battery ground cable.

2. Remove the adjusting arm bolt.

3. Remove the alternator through-bolt. Remove the drive belt from the alternator pulley and lower the alternator.

4. Label all of the leads to the alternator so that you can install them correctly and disconnect the leads from the alternator.

5. Remove the alternator from the vehicle.

6. To install, reverse the above procedure.

Belt Tension Adjustment

The fan belt drives the alternator and water pump. If the belt is too loose, it will slip and the alternator will not be able to produce its rated current. Also, the water pump will not operate efficiently and the engine could overheat. Check the tension of the belt by pushing your thumb down on the longest span of the belt, midway between the pulleys. Belt deflection should be approximately ½ in. To adjust belt tension, proceed as follows:

1. Loosen the alternator mounting bolt and the adjusting arm bolts.

2. Apply pressure on the alternator front housing only, moving the alternator away from the engine to tighten the belt. Do not apply pressure to the rear of the cast aluminum housing of an alternator; damage to the housing could result.

3. Tighten the alternator mounting bolt and the adjusting arm bolts when the correct tension is reached.

Regulator

The alternator regular has been designed to control the charging system's rate of charge. The electromechanical regulator is calibrated at the factory and is not adjustable.

REMOVAL & INSTALLATION

1. Disconnect the negative terminal of the battery. On some models it may be necessary to move the battery.

2. Disconnect all of the electrical leads (harness connectors) at the regulator.

3. Remove all of the hold-down screws, then remove the unit from the vehicle.

4. Install the new voltage regulator using the hold-down screws from the old one, or new ones if they are provided with the replacement regulator. Tighten the hold-down screws.

5. Connect all the harness connectors to the new regulator.

Refer to the Electrical section in Unit Repair for alternator and starter motor servicing.

Starter Motor

REMOVAL & INSTALLATION

Positive Engagement Type

1. Disconnect the positive battery terminal.

2. Raise the vehicle and disconnect the starter cable at the starter terminal.

3. Remove all of the starter attaching bolts that attach the starter to the bellhousing.

4. Remove the starter from the engine.

5. Install the starter in the reverse order of removal.

Solenoid Actuated Type

1. Disconnect the battery ground cable.

2. Raise the vehicle and disconnect the cables and wires at the starter solenoid.

3. Turn the front wheels to the right and remove the two bolts attaching the steering idler arm to the frame.

4. Remove the starter mounting bolts and remove the starter.

5. Install in the reverse order of removal.

STARTER RELAY

REPLACEMENT

The starter relay is mounted on the inside of the right wheel well. To replace it, disconnect the negative battery cable from the battery, disconnect all of the electrical leads from the relay and remove the relay from the fender wall. Replace in the reverse order of removal.

ENGINE MECHANICAL

Engine

REMOVAL & INSTALLATION

Pickups and Bronco, Bronco II 122 (2.0L) & 140 (2.3L)

1. Raise the hood and install protective fender covers. Drain the coolant from the radiator. Remove the air cleaner and duct assembly.

2. Disconnect the battery ground cable at the engine and disconnect the battery positive cable at the battery and set aside.

3. Mark the location of the hood hinges and remove the hood.

4. Disconnect the upper and lower radiator hoses from the engine. Remove the radiator shroud screws. Remove the radiator upper supports.

5. Remove engine fan and shroud assembly. Then remove the radiator. Remove the oil fill cap.

6. Disconnect the coil primary wire at the coil. Disconnect the oil pressure and the water temperature sending unit wires from the sending units.

7. Disconnect the alternator wire from the alternator, the starter cable from the starter and the accelerator cable from the carburetor. If so equipped, disconnect the transmission kickdown rod.

8. If so equipped, remove the A/C compressor from the mounting bracket and position it out of the way, leaving the refrigerant lines attached.

9. Disconnect the power brake vacuum hose. Disconnect the chassis fuel line from the fuel pump. Disconnect the heater hoses from the engine.

10. Remove the engine mount nuts. Raise the vehicle and safely support on jackstands.

11. Drain engine oil from the crankcase. Remove the starter motor.

12. Disconnect the muffler exhaust inlet pipe at the exhaust manifold.

13. Remove the dust cover (manual transmission) or converter inspection plate (automatic transmission).

14. On vehicles with a manual transmission, remove the flywheel housing cover lower attaching bolts. On vehicles with automatic transmissions, remove the converter-to-flywheel bolts, then remove the converter housing lower attaching bolts.

15. Remove clutch slave cylinder (manual transmission). Lower the vehicle.

16. Support the transmission and flywheel or converter housing with a jack.

17. Remove the flywheel housing or converter housing upper attaching bolts.

18. Attach the engine lifting hooks to the existing lifting brackets. Carefully, so as not to damage any components, lift the engine out of the vehicle.

19. To install the engine: If clutch was removed, reinstall. Carefully lower the engine into the engine compartment. On a vehicle with automatic transmission, start the converter pilot into the crankshaft. On a vehicle with a manual transmission, start the transmission main drive gear into the clutch disc. It may be necessary to adjust the position of the transmission in relation to the engine if the input shaft will not enter the clutch disc. If the engine hangs up after the shaft enters, turn the crankshaft in the clockwise direction slowly (transmission in gear), until the shaft splines mesh with the clutch disc splines.

20. Install the flywheel or converter housing upper attaching bolts. Remove the engine lifting hooks from the lifting brackets.

21. Remove the jack from under the transmission. Raise the vehicle and safely support on jackstands.

22. On a vehicle with a manual transmission, install the flywheel lower housing bolts and tighten to specifications. On a vehicle with an automatic transmission, attach the converter to the flywheel bolts and tighten to specifications. Install the converter housing-to-engine bolts and tighten to specifications.

23. Install clutch slave cylinder.

24. Install the dust cover (manual transmission) or converter inspection plate (automatic transmission). Connect the exhaust inlet pipe to the exhaust manifold.

25. Install the starter motor and connect the starter cables.

26. Lower the vehicle. Install the engine mount nuts and tighten to 65–85 ft. lbs.

27. Connect the heater hoses to the engine. Connect the chassis fuel line to the fuel pump. Connect the power brake vacuum hose.

28. Connect the alternator wire to the alternator, connect the accelerator cable to the carburetor. If so equipped, connect the transmission kickdown rod. If so equipped, install the A/C compressor to the mounting bracket.

29. Connect the coil primary wire at the coil. Connect the oil pressure and water temperature sending unit wires. Install oil fill cap.

30. Install the radiator and secure with upper support brackets. Install the fan and shroud assembly. Connect upper and lower radiator hoses.

31. Install the hood and align.

32. Install the air cleaner assembly. Fill and bleed the cooling system.

33. Fill the crankcase with specified oil. Connect battery ground cable to engine and battery positive cable to battery.

34. Start the engine and check for leaks.

134 (2.2L) Diesel

1. Open hood and install protective fender covers. Mark location of hood hinges and remove hood.

2. Disconnect battery ground cables from both batteries. Disconnect battery ground cables at engine.

3. Drain coolant from radiator.

4. Disconnect air intake hose from air cleaner and intake manifold.

5. Disconnect upper and lower radiator hoses from engine. Remove engine cooling fan. Remove radiator shroud screws. Remove radiator upper supports and remove radiator and shroud.

6. Disconnect radio ground strap, if so equipped.

7. Remove No.2 glow plug relay from firewall, with harness attached, and lay on engine.

8. Disconnect engine wiring harness at main connector located on left fender apron. Disconnect starter cable from starter.

9. Disconnect accelerator cable and speed control cable, if so equipped, from injection pump.

10. Remove cold start cable from injection pump.

CAUTION

Do not disconnect air conditioning lines or discharge the system unless the proper equipment is on hand and you are familiar with the procedure. Have the system discharged by a qualified mechanic prior to start of engine removal.

11. Discharge A/C system and remove A/C refrigerant lines and position out of the way.

12. Remove pressure and return hoses from power steering pump, if so equipped.

13. Disconnect vacuum fitting from vacuum pump and position fitting and vacuum hoses out of the way.

14. Disconnect and cap fuel inlet line at fuel line heater and fuel return line at injection pump.

15. Disconnect heater hoses from engine.

16. Loosen engine insulator nuts. Raise vehicle and safely support on jackstands.

17. Drain engine oil from oil pan and remove primary oil filter.

18. Disconnect oil pressure sender hose from oil filter mounting adapter.

19. Disconnect muffler inlet pipe at exhaust manifold.

20. Remove bottom engine insulator nuts. Remove transmission bolts. Lower vehicle. Attach engine lifting sling and chain hoist.

21. Carefully lift engine out of vehicle to avoid damage to components.

22. Install engine on work stand, if necessary.

23. When installing the engine: Carefully lower engine into engine compartment to avoid damage to components.

24. Install two top transmission-to-engine attaching bolts. Remove engine lifting sling.

25. Raise vehicle and safely support on jackstands.

26. Install engine insulator nuts and tighten to specification.

27. Install remaining transmission-to-engine attaching bolts and tighten all bolts to specification.

28. Connect muffler inlet pipe to exhaust manifold and tighten to specification.

29. Install oil pressure sender hose and install new oil filter as described in this Section.

30. Lower vehicle.

31. Tighten upper engine insulator nuts to specification.

32. Connect heater hoses to engine. Connect fuel inlet line to fuel line heater and fuel return line to injection pump. Connect vacuum fitting and hoses to vacuum pump. Connect pressure and return hoses to power steering pump, if so equipped. Check and add power steering fluid.

33. Install A/C refrigerant lines and charge system, if so equipped.

NOTE: System can be charged after engine installation is completed.

34. Connect cold start cable to injection pump. Connect accelerator cable and speed control cable, if so equipped, to injection pump.

35. Connect engine wiring harness to main wiring harness at left fender apron. Connect radio ground strap, if so equipped.

36. Position radiator in vehicle, install radiator upper support brackets and tighten to specification. Install radiator fan shroud and tighten to specification. Install radiator fan and tighten to specification.

37. Connect upper and lower radiator hoses to engine and tighten clamps to specification. Connect air intake hose to air cleaner and intake manifold.

38. Fill and bleed cooling system.

39. Fill crankcase with specified quantity and quality of oil.

40. Connect battery ground cables to engine. Connect battery ground cables to both batteries.

41. Run engine and check for oil, fuel and coolant leaks. Close hood.

173 (2.8L) V6 and 179 (2.9L) V6

1. Disconnect the battery ground cable and drain the cooling system.

2. Remove the hood after scribing hinge positions. Remove the air cleaner and intake duct assembly.

3. Remove or disconnect thermactor system parts that will interfere with removal or installation of the engine.

4. Disconnect the radiator upper and lower hoses at the radiator. Remove the fan shroud attaching bolts and position the shroud over the fan. Remove the radiator and shroud.

5. Remove the alternator and bracket. Position the alternator out of the way. Disconnect the alternator ground wire from the cylinder block.

6. Remove A/C compressor and power steering and position out of way, if so equipped.

------ CAUTION ------

Do not disconnect air conditioning lines or discharge the system unless the proper equipment is on hand and you are familiar with the procedure. Have the system discharged by a qualified mechanic prior to start of engine removal.

7. Disconnect the heater hoses at the block and water pump.

8. Remove the ground wires from the cylinder block.

9. Disconnect the fuel tank to fuel pump fuel line at the fuel pump. Plug the fuel tank line.

10. Disconnect the throttle cable linkage at the carburetor and intake manifold.

11. Disconnect the primary wires from the ignition coil. Disconnect the brake booster vacuum hose. Disconnect the wiring from the oil pressure and engine coolant temperature senders.

12. Raise the vehicle and secure with safety stands. Disconnect the muffler inlet pipes at the exhaust manifolds.

13. Disconnect the starter cable and remove the starter.

14. Remove the engine front support to crossmember attaching nuts or through bolts.

15. If equipped with automatic transmission, remove the converter inspection cover and disconnect the flywheel from the converter.

16. Remove the kickdown rod. Remove the converter housing-to-cylinder block bolts and the adapter plate-to-converter housing bolt.

17. On vehicles equipped with a manual transmission, remove the clutch linkage and mounting bolts. Lower the vehicle.

18. Attach engine lifting sling and hoist to lifting brackets at exhaust manifolds.

19. Position a jack under the transmission. Raise the engine slightly and carefully pull it from the transmission. Carefully lift the engine out of the engine compartment

so that the rear cover plate is not bent or components damaged.

20. When installing the engine: Attach engine lifting sling and hoist to lifting brackets at exhaust manifolds.

21. Lower the engine carefully into the engine compartment. Make sure the exhaust manifolds are properly aligned with the muffler inlet pipes.

22. On a vehicle with a manual transmission, start the transmission main shaft into the clutch disc. It may be necessary to adjust the position of the transmission in relation to the engine if the input shaft will not enter the clutch disc. If the engine hangs up after the shaft enters, turn the crankshaft slowly (transmission in gear) until the shaft splines mesh with the clutch disc splines. On a vehicle with an automatic transmission, start the converter pilot into the crankshaft. Install the clutch housing or converter housing upper bolts, making sure that the dowels in the cylinder block engage the flywheel housing. Remove the jack from under the transmission. Remove the lifting sling.

23. On a vehicle with an automatic transmission, position the kickdown rod on the transmission and engine. Raise the vehicle and secure with safety stands. On a vehicle with an automatic transmission, position the transmission linkage bracket and install the remaining converter housing bolts. Install the adapter plate-to-converter housing bolt. Install the converter-to-flywheel nuts and install the inspection cover. Connect the kickdown rod on the transmission.

24. Install the starter and connect the cable.

25. Connect the muffler inlet pipes at the exhaust manifolds.

26. Install the engine front support nuts and washer attaching it to the crossmember or through bolts. Lower the vehicle.

27. Install the battery ground cable. Connect the ignition coil primary wires, then connect the coolant temperature sending unit and oil pressure sending unit. Connect brake booster vacuum hose. Install the throttle linkage.

28. Connect the fuel tank line at the fuel pump. Connect the ground cable at the cylinder block. Connect the heater hoses to the water pump and cylinder block.

29. Install the alternator and bracket. Connect the alternator ground wire to the cylinder block. Install the drive belt and adjust the belt tension to specifications.

30. Install A/C compressor and power steering pump, if so equipped.

31. Position the fan shroud over the fan. Install the radiator and connect the radiator upper and lower hoses. Install the fan shroud attaching bolts. Fill and bleed the cooling system. Fill the crankcase with the proper grade and quantity of oil. Install thermactor parts removed or disconnected. Reconnect battery ground cable.

32. Charge A/C system if so equipped.

NOTE: System can be charged after engine installation is completed.

33. Operate the engine at fast idle until it reaches normal operating temperature and check all gaskets and hose connections for leaks. Adjust ignition timing and idle speed. Install the air cleaner and intake duct. Install and adjust the hood.

300 (4.9L)

1. Drain the cooling system and the crankcase. Remove the hood and the air cleaner. Disconnect the negative battery cable.

2. Disconnect the heater hose from the water pump and coolant outlet housing. Disconnect the flexible fuel line from the fuel pump.

3. Remove the radiator.

4. Remove the fan, water pump pulley, and fan belt.

5. Disconnect the accelerator cable at the carburetor. Remove the throttle return spring. On trucks equipped with power brakes, disconnect the brake booster vacuum hose at the intake manifold. On trucks with automatic transmissions, disconnect the transmission kickdown rod at the bellcrank assembly.

6. Disconnect the exhaust pipe from the exhaust manifold.

7. Disconnect the body ground strap and the battery ground cable from the engine.

8. Disconnect the engine wiring harness at the ignition coil, the coolant temperature sending unit, and the oil pressure sending unit. Position the wiring harness out of the way.

9. Remove the alternator mounting bolts and position the alternator out of the way.

10. On a truck equipped with power steering, remove the power steering pump from the mounting brackets and move it to one side, leaving the lines attached.

11. Raise and safely support the truck. Remove the starter and automatic transmission filler tube bracket, if so equipped. Also, remove the rear engine plate upper right bolt.

12. On manual transmission equipped trucks, remove the flywheel housing lower attaching bolts and disconnect the clutch return spring.

13. On automatic transmission equipped trucks, remove the converter housing access cover assembly and remove the flywheel-to-converter attaching nuts. Secure the converter in the housing. Remove the transmission oil cooler lines from the retaining clip at the engine. Remove the lower converter housing-to-engine attaching bolts.

14. Remove the nut from each of the two front engine mounts.

15. Lower the vehicle and position a jack under the transmission and support it. Remove the remaining bellhousing-to-engine attaching bolts.

16. Attach the engine lifting device and raise the engine slightly and carefully pull it from the transmission. Lift the engine out of the vehicle.

17. To install the engine: Place a new gasket on the muffler inlet pipe.

18. Carefully lower the engine into the truck. Make sure that the dowels in the engine block engage the holes in the bellhousing.

19. On manual transmission equipped trucks, start the transmission input shaft into the clutch disc. It may be necessary to adjust the position of the engine or transmission in order for the input shaft to enter the clutch disc. If necessary, turn the crankshaft until the input shaft splines mesh with the clutch disc splines.

20. On automatic transmission equipped trucks, start the converter pilot into the crankshaft. Unsecure the converter in the housing.

21. Install the bellhousing upper attaching bolts. Remove the jack supporting the transmission.

22. Lower the engine until it rests on the engine mounts. Remove the lifting device.

23. Install the engine mount nuts and tighten them to 45–55 ft. lbs.

24. Install the automatic transmission coil cooler lines bracket, if so equipped.

25. Install the remaining bellhousing attaching bolts.

26. Connect the clutch return spring, if so equipped.

27. Install the starter and connect the starter cable. Attach the automatic transmission fluid filler tube bracket, if so equipped.

28. On trucks with automatic transmissions, install the transmission oil cooler lines in the bracket at the cylinder block.

29. Connect the exhaust pipe to the exhaust manifold. Tighten the nuts to 25–35 ft. lbs.

30. Connect the engine ground strap and negative battery cable.

31. On a truck with an automatic transmission, connect the kick-down rod to the bellcrank assembly on the intake manifold.

32. Connect the accelerator linkage to the carburetor and install the return spring.

33. On a truck with power brakes, connect the brake booster vacuum line to the intake manifold.

34. Connect the coil primary wire, oil pressure and coolant temperature sending unit wires, fuel line, heater hoses, and the battery positive cable.

35. Install the alternator to its mounting bracket. Install the power steering pump to its bracket, if so equipped.

36. Install the water pump pulley, spacer, fan, and fan belt. Adjust the belt tension.

37. Install the radiator and connect the upper and lower radiator hoses to the radiator and engine. Connect the automatic transmission oil cooler lines, if so equipped.

38. Install and adjust the hood.

39. Fill the cooling system. Fill the crankcase.

40. Start the engine and check for leaks. Bleed the cooling system. Adjust the clutch pedal free-play or the automatic transmission control linkage. Install the air cleaner.

232 (3.8L) V6 and V8   Exc 420 (6.9L) Diesel and 460 (7.5L) V8

1. Drain the cooling system and crankcase. Remove the hood.

2. Disconnect the battery, negative cable first, and alternator cables from the cylinder block.

3. Remove the air cleaner and intake duct assembly, plus the crankcase ventilation hose.

4. Disconnect the upper and lower radiator hoses, and, if so equipped, the automatic transmission oil cooler lines.

5. Remove the fan shroud and lay it over the fan. Remove the radiator and fan, shroud, fan, spacer, pulley, and belt.

6. Disconnect the alternator leads and the alternator adjusting bolts. Allow the alternator to swing down out of the way.

7. Disconnect the oil pressure sending unit lead from the sending unit.

8. Disconnect the fuel tank-to-pump fuel line at the fuel pump and plug the line.

9. Disconnect the accelerator linkage at the carburetor. Disconnect the automatic transmission kick-down rod and remove the return spring, if so equipped.

10. Disconnect the heater hoses from the water pump and intake manifold. Disconnect the temperature sending unit wire from the sending unit.

11. Remove the upper bellhousing-to-engine attaching bolts.

12. Disconnect the primary wire from the coil. Remove the wiring harness from the left rocker arm cover and position the wires out of the way. Disconnect the ground strap from the cylinder block.

13. Raise the front of the truck and disconnect the starter cable from the starter. Remove the starter.

14. Disconnect the exhaust pipe from the exhaust manifolds.

15. Disconnect the engine mounts from the brackets on the frame.

16. On trucks with automatic transmissions, remove the converter inspection plate and remove the torque converter-to-flywheel attaching bolts.

17. Remove the remaining bellhousing-to-engine attaching bolts.

18. Lower the vehicle and support the transmission with a jack.

19. Install an engine lifting device.

NOTE: On the V6, the intake manifold is aluminum. If a lifting device is attached to the manifold, all manifold bolts must be installed.

20. Raise the engine slightly and carefully pull it out from the transmission. Lift the engine out of the engine compartment.

21. Install the engine in the reverse order of removal. Make sure that the dowels in the engine block engage the holes in the bellhousing through the rear cover plate. If the engine hangs up after the transmission input shaft enters the clutch disc (manual transmission only), turn the crankshaft with the transmission in gear until the input shaft splines mesh with the clutch disc splines.

22. Tighten the exhaust pipe-to-exhaust manifold nuts to 25–35 ft. lbs., and all others as follows: $\frac{1}{4}$ in. 20; 6–9 ft. lbs. $\frac{5}{16}$ in. 18; 12–18 ft. lbs. $\frac{3}{8}$ in. 16; 22–32 ft. lbs. $\frac{7}{16}$ in. 14; 45–57 ft. lbs. $\frac{1}{2}$ in. 13; 55–80 ft. lbs. $\frac{9}{16}$ in. 85–120 ft. lbs.

460 (7.5L) V8

1. Remove the hood.

2. Drain the cooling system, the radiator and the cylinder block.

3. Disconnect the negative battery cable and remove the air cleaner assembly.

4. Disconnect the upper and lower radiator hoses and the transmission oil cooler lines from the radiator.

5. Remove the fan shroud from the radiator and remove the fan from the water pump. Remove the fan and shroud from the engine compartment.

6. Remove the upper support and remove the radiator.

7. If the truck is equipped with air conditioning, remove the compressor from the engine and position it out of the way. If the compressor must be removed completely, loosen the air conditioning service valves (disconnect) carefully to discharge the air conditioning system. Remove the compressor.

8. Remove the power steering pump from the engine, if so equipped, and position it to one side. Do not disconnect the fluid lines.

9. Disconnect the fuel pump inlet line from the pump and plug the line.

10. Remove the alternator drive belts and disconnect the alternator from the engine, positioning it aside.

11. Disconnect the ground cable from the right front corner of the engine.

12. Disconnect the heater hoses.

13. Remove the transmission fluid filler tube attaching bolt from the right-side valve cover and position the tube out of the way.

14. Disconnect all vacuum lines at the rear of the intake manifold.

15. Disconnect the speed control cable at the carburetor, if so equipped. Disconnect the accelerator rod and the transmission kickdown rod and secure them out of the way.

16. Disconnect the engine wiring harness at the connector on the fire wall.

17. Raise the vehicle and disconnect the exhaust pipes at the exhaust manifolds.

18. Disconnect the starter cable and remove the starter. Bring the starter forward and rotate the solenoid outward to remove the assembly.

19. Remove the access cover from the converter housing and remove the flywheel-to-converter attaching nuts. Remove the lower converter housing-to-engine attaching bolts.

20. Remove the engine mount through-bolts attaching the rubber insulators to the frame brackets.

21. Lower the vehicle and place a jack under the transmission to support it.

22. Remove the converter housing-to-engine block attaching bolts (left-side).

23. Disconnect the coil wire and remove the coil and bracket assembly from the intake manifold.

24. Attach the engine lifting device and carefully lift the engine from the engine compartment.

25. Install the engine in the reverse order of removal. Tighten the alternator pivot bolt to 45–57 ft. lbs. and all the rest of the nuts and bolts as is outlined in Step 21 of the preceding ''V8 except 460 Removal and Installation'' procedure.

420 (6.9L) Diesel

1. Open the hood. Disconnect the battery ground cables from both batteries.

2. Scribe alignment marks at the hood hinges and remove the hood.

3. Drain the cooling system.

4. Remove the air cleaner and intake duct assembly. Install an intake manifold cover over the air intake opening.

5. Remove the radiator fan shroud halves.

6. Remove the fan and clutch assembly. The retaining nut is equipped with left handed threads, remove by turning clockwise.

7. Disconnect the upper and lower hoses from the radiator.

8. Disconnect the automatic transmission oil cooler lines at the radiator, if so equipped.

9. Remove the radiator.

10. Loosen A/C compressor, if so equipped, and remove the drive belt.

11. Remove the A/C compressor, if so equipped, and position it on the radiator upper support.

─────── **CAUTION** ───────

If compressor cannot be secured with lines connected, do not disconnect the lines unless you are familiar with discharging the system and have the proper tools. Have the system discharged prior to the start of engine removal.

─────────────────────────

12. Loosen the power steering pump and remove the drive belt. Remove the power steering pump and position it out of the way on left side of engine compartment.

13. Disconnect the fuel supply line heater and alternator wires at the alternator. Disconnect the oil pressure sending unit wire at the sending unit. Remove the oil pressure sender from the dash panel and lay it on the engine.

14. Disconnect the accelerator cable from the injection pump. Disconnect the speed control cable from the injection pump, if so equipped. Remove the accelerator cable bracket with cables attached, from the intake manifold and position out of the way.

15. Disconnect the transmission kickdown rod from the injection pump, if so equipped. Disconnect the main wiring harness connector from the right side of engine. Disconnect the engine ground strap from the rear of engine. Disconnect the fuel return hose from left rear of engine.

16. Remove the two upper transmission-to-engine attaching bolts.

17. Disconnect the heater hoses from the water pump and the right cylinder head. Disconnect the water temperature sender wire from the sender on left front of engine block. Disconnect the water temperature overheat light switch wire from the switch on top front of left cylinder head. Position wires out of the way.

18. Raise vehicle and safely support on jackstands.

19. Disconnect both battery ground cables from the lower front of engine.

20. Disconnect and cap the fuel inlet line at fuel supply pump.

21. Disconnect the starter cables at the starter motor.

22. Disconnect the muffler inlet pipes at the exhaust manifolds.

23. Disconnect the engine insulators from No. 1 crossmember. Remove the flywheel inspection plate. Remove the four converter-to-flywheel attaching nuts, if so equipped. Lower vehicle.

24. Support the transmission with a floorjack. Remove the four lower transmission to engine attaching bolts.

25. Attach an engine lifting sling and chain hoist. Raise the engine high enough to clear number one crossmember and pull forward.

26. Rotate the front of the engine approximately 45 degrees to the left and lift it out of the engine compartment.

27. When installing the engine; lower engine into engine compartment. Use care not to damage windshield wiper motor when installing engine in vehicle.

28. Start the transmission main shaft into the clutch disc. It may be necessary to adjust position of transmission in relation to engine if mainshaft binds or will not enter clutch disc. If the engine hangs up after main shaft enters clutch disc, rotate crankshaft slowly (transmission in gear) until mainshaft splines mesh with clutch disc splines. Align converter to flywheel studs, if so equipped.

29. Lower into engine insulator brackets on number one crossmember.

30. Install the four lower transmission to engine attaching bolts and tighten. Remove engine lifting sling. Raise the vehicle and safely support on jackstands.

31. Install the four convertor to flywheel attaching nuts, if so equipped. Install the flywheel inspection plate.

32. Install the engine insulator support to crossmember bracket attaching nuts and washers. Connect the muffler inlet pipes to exhaust manifolds. Connect both battery ground cables to the lower front of the engine. Connect starter cables to starter. Install the fuel pump inlet line on fuel pump. Lower vehicle.

33. Connect the water temperature sender wire to sender on left front of engine block. Connect the wire to water temperature ov-

erheat light switch on top of left cylinder head. Install the heater hoses on right cylinder head and water pump and tighten clamps.

34. Connect the engine ground strap at rear of engine. Connect the fuel return hose at left rear of engine. Connect the transmission kickdown rod, if so equipped.

35. Install the accelerator cable bracket on the intake manifold. Connect the accelerator cable to the injection pump. Connect the speed control cable, if so equipped, to injection pump.

36. Install the oil pressure sender on dash panel. Connect the oil pressure gauge sender wire to oil pressure sender.

37. Connect the fuel supply line heater and alternator wires to alternator.

38. Install the power steering pump and drive belt. Do not adjust belt at this time.

39. Install A/C compressor and drive belt. Adjust A/C compressor and power steering pump drive belts.

NOTE: The A/C system can be recharged after engine installation is completed.

40. Install the radiator. Connect the automatic transmission oil cooler lines at the radiator, if so equipped. Connect the upper and lower radiator hoses to the radiator and tighten hose clamps. Fill and bleed the cooling system.

41. Install the fan and clutch assembly. Remember, left hand thread. Turn counterclockwise to tighten.

42. Install the radiator fan shroud halves.

42. Remove intake manifold cover, and install the air cleaner. Install the intake duct assembly.

43. Install hood using scribe marks drawn on hood at removal.

44. Connect the battery ground cables at both batteries. Check the engine oil level and fill as needed with the specified type and grade of oil. Run the engine and check for fuel, oil and coolant leaks.

VANS

300 (4.9L)

1. Take off the engine cover, drain the coolant, remove the air cleaner, and disconnect the battery.

2. Remove the bumper, grille, and gravel deflector.

3. Detach the upper radiator hose at the engine. Remove the alternator splash shield and detach the lower hose at the radiator. Remove the radiator and shroud, if any.

4. Disconnect the engine heater hoses and the alternator wires. Remove the power steering pump and support.

5. Disconnect and plug the fuel line at the pump.

6. Detach from the engine: distributor and gauge sending unit wires, brake booster hose, accelerator cable and bracket.

7. Disconnect the automatic transmission kickdown linkage at the bellcrank.

8. Remove the exhaust manifold heat deflector and unbolt the pipe from the manifold.

9. Disconnect the automatic transmission vacuum line from the intake manifold and from the junction. Remove the transmission dipstick tube support bolt at the intake manifold.

10. Remove the upper engine-to-transmission bolts.

11. Remove the starter. Remove the flywheel inspection cover. Remove the four automatic transmission torque converter nuts, then remove to front engine support nuts. Take off the oil filter.

12. Remove the rest of the transmission-to-engine fasteners, then lift the engine out from the engine compartment with a floor crane.

13. To replace the engine, lower it into place and start the mounting bolts. Install the upper transmission bolts, the converter nuts, and the lower transmission bolts. Tighten the mounting bolts. Install all the items removed in the previous steps.

V8 Engines

NOTE: Refer to the proceeding pickup truck section for unit disconnection details on the 6.9L diesel engine.

1. Take off the engine cover, drain the coolant, remove the air cleaner, and disconnect the battery. Remove the bumper, grille, and gravel deflector. Remove the upper grille support bracket, hood lock support, and air conditioning condenser upper mounting brackets.

2. With air conditioning, the system must be discharged to remove the condenser. Do not attempt to do this yourself, unless you are trained in air conditioning. Disconnect the lines at the compressor.

3. Remove the accelerator cable bracket and the heater hoses. Detach the radiator hoses and the automatic transmission cooler lines, if any. Remove the fan shroud, fan, and radiator.

4. Pivot the alternator in and detach the wires.

5. Remove the air cleaner, duct and valve, exhaust manifold shroud, and flex tube.

6. Disconnect the automatic transmission shift rod.

7. Disconnect the fuel and choke lines, detach the vacuum lines, and remove the carburetor and spacer.

8. Remove the oil filter. Detach the exhaust pipe from the manifold. Unbolt the automatic transmission tube bracket from the cylinder head. Remove the starter.

9. Remove the engine mount bolts. With automatic, remove the converter inspection cover and unbolt the converter from the flex plate.

10. Unbolt the engine ground cable and support the transmission.

11. Remove the power steering front bracket. Detach only one vacuum line at the rear of the intake manifold. Disconnect

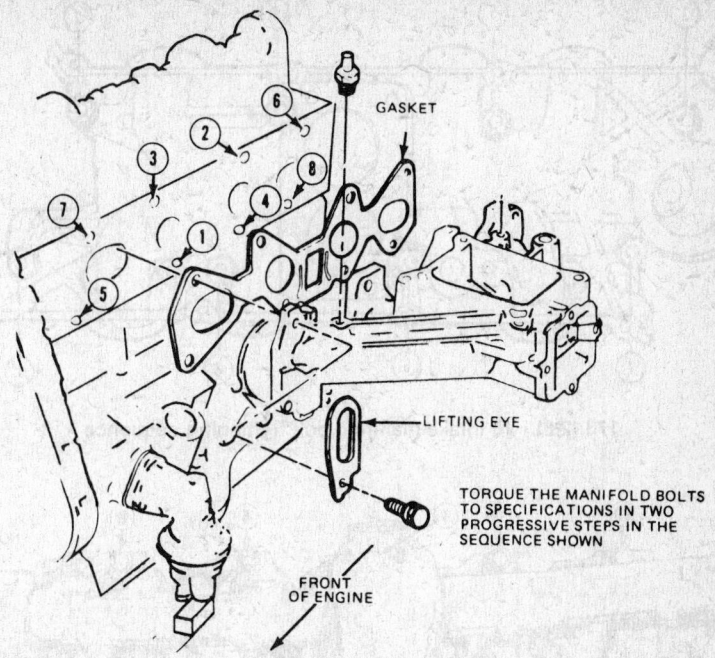

122 (2.0L), 140 (2.3L) intake manifold bolt tightening sequence

the engine wiring loom. Remove the speed control servo from the manifold. Detach the compressor clutch wire.

12. Install a lifting bracket and attach a floor crane. Remove the transmission-to-engine bolts, making sure the transmission is supported. Remove the engine.

13. To install the engine, align the converter to the flex plate and the engine dowels to the transmission. With manual transmission, start the transmission shaft into the clutch disc. You may have to turn the crankshaft slowly with the transmission in gear. Install the transmission bolts, then the mounting bolts. Install all the items removed in the previous steps.

Intake Manifold

REMOVAL & INSTALLATION

122 (2.0L) and 140 (2.3L)

1. Drain the cooling system. Remove the air cleaner and duct assembly. Disconnect the negative battery cable.

2. Disconnect the accelerator cable, vacuum hoses (as required) and the hot water hose at the manifold fitting. Be sure to identify all vacuum hoses for proper reinstallation.

3. Remove the engine oil dipstick. Disconnect the heat tube at the EGR (exhaust gas recirculation) valve. Disconnect the fuel line at the carburetor fuel fitting.

4. Remove the dipstick retaining bolt from the intake manifold.

5. Disconnect and remove the PCV at the engine and intake manifold.

6. Remove the distributor cap and position the cap and wires out of the way,

after removing the plastic plug connector from the valve cover.

7. Remove the intake manifold retaining bolts. Remove the manifold from the engine.

8. Clean all gasket mounting surfaces.

9. Install a new mounting gasket and intake manifold on the engine. Torque the bolts in proper sequence. The rest of the installation is in the reverse order of removal.

134 (2.2L) Diesel

1. Disconnect the battery ground cables from both batteries.

2. Disconnect the air inlet hose from the air cleaner and intake manifold. Disconnect and remove the fuel injection lines from the nozzles and injection pump. Cap all lines and fittings to prevent dirt pickup.

3. Remove the nut that attaching the lower fuel return line brace to the intake manifold.

4. Disconnect and remove the lower fuel line from the injector pump and upper fuel return line.

5. Remove the air conditioner compressor with the lines attached and position out of the way. Remove the power steering pump and rear support with the lines still attached and position out of the way.

6. Remove the air inlet adapter, dropping resistor (electrical measuring device) and the gaskets.

7. Disconnect the fuel filter inlet line, remove the fuel filter mounting bracket from the cylinder head and position the filter assembly out of the way.

8. Remove the mounting nuts for the fuel line heater assembly to intake manifold and position the heater out of the way.

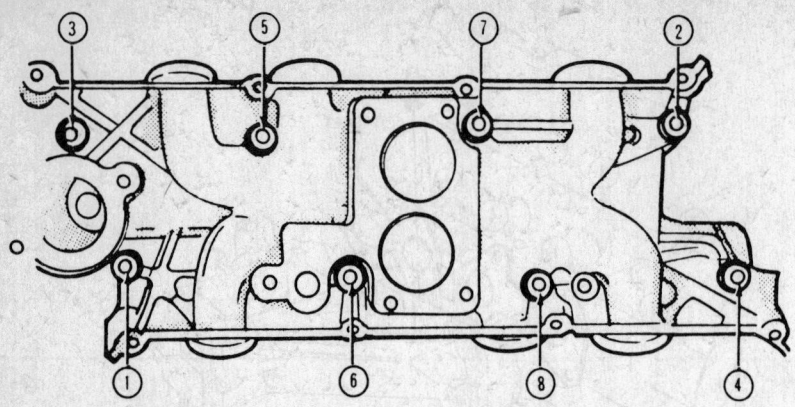

173 (2.8L) V6 intake manifold bolt tightening sequence

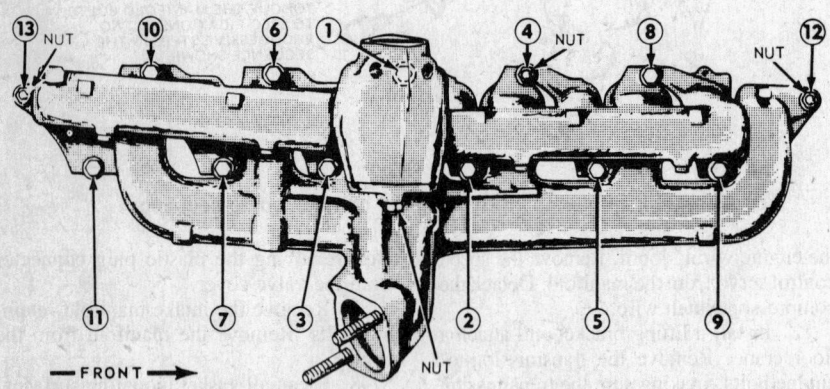

←FRONT→

6-300 manifold bolt tightening sequence

9. Remove the nuts that attach the intake manifold to the cylinder head. Remove the intake manifold and gasket.

10. Clean all gasket mounting surfaces.

11. Use a new intake manifold mounting gasket. Install the intake manifold in the reverse order of removal.

12. Do not tighten the mounting nuts until No.3 lower nut that holds the fuel return line bracket is installed. After installation of the No.3 nut, tighten all of the mounting nuts.

173 (2.8L) V6

1. Drain the cooling system. Remove the air cleaner and duct assembly.

2. Disconnect the negative battery cable. Disconnect the accelerator cable from the carburetor linkage.

3. Disconnect and remove the upper radiator hose. Disconnect and remove the bypass hose from the intake manifold and thermostat housing.

4. Remove the distributor cap and spark plug wires as an assembly. Turn the engine till No. 1 piston is at TDC (top dead center) on the compression stroke. Remove the distributor.

5. Remove any vacuum lines and controls that will interfere with the intake manifold removal. Label all hoses for identification.

6. Remove both valve covers. Remove the manifold mounting nuts and bolts. Remove the manifold. Tap the manifold lightly with a plastic mallet (if necessary) to break the gasket seal.

7. Remove all old gasket material and sealing compound from the mounting surfaces.

8. Apply sealing compound to the joining surfaces. Place the intake mounting gasket into position. Make sure that the tab on the right bank head gasket fits into the cutout of the manifold gasket. Apply sealing compound to the intake manifold bolt bosses and install the intake manifold. Tighten the mounting nuts and bolts in the proper torque sequence.

9. Install the distributor and the rest of the removed components in reverse order.

10. Refill the cooling system, start the engine and check for coolant or oil leaks.

11. Check idle RPM and ignition timing. Adjust if necessary.

300 (4.9L)

The intake and exhaust manifolds on these engines are known as combination manifolds and are serviced as a unit.

1. Remove the air cleaner. Disconnect the choke cable at the carburetor. Disconnect the accelerator cable or rod at the carburetor. Remove the accelerator retracting spring.

2. On a vehicle with automatic transmission, remove the kick-down rod-retracting spring. Remove the accelerator rod bellcrank assembly.

3. Disconnect the fuel inlet line and the distributor vacuum line from the carburetor.

4. Disconnect the muffler inlet pipe from the exhaust manifold.

5. Disconnect the power brake vacuum line, if so equipped.

6. Remove the bolts and nuts attaching the manifolds to the cylinder head. Lift the manifold assemblies from the engine. Remove and discard the gaskets.

7. To separate the manifolds, remove the nuts joining the intake and exhaust manifolds.

8. Clean the mating surfaces of the cylinder head and the manifolds.

9. If the intake and exhaust manifolds have been separated, coat the mating surfaces lightly with graphite grease and place the exhaust manifold over the studs on the intake manifold. Install the lockwashers and nuts. Tighten them finger tight.

10. Install a new intake manifold gasket.

11. Coat the mating surfaces lightly with graphite grease. Place the manifold assemblies in position against the cylinder head. Make sure that the gaskets have not become dislodged. Install the attaching washers, bolts and nuts. Tighten the attaching nuts and bolts in the proper sequence to 26 ft. lbs. If the intake and exhaust manifolds were separated, tighten the nuts joining them.

12. Position a new gasket on the muffler inlet pipe and connect the inlet pipe to the exhaust manifold.

13. Connect the crankcase vent hose to the intake manifold inlet tube and position the hose clamp.

14. Connect the fuel inlet line and the distributor vacuum line to the carburetor.

15. Connect the accelerator cable to the carburetor and install the retracting spring. Connect the choke cable to the carburetor.

16. On a vehicle with an automatic transmission, install the bellcrank assembly and the kick-down rod retracting spring. Adjust the transmission control linkage.

17. Install the air cleaner.

232 (3.8L) V6 and V8's

1. Drain the cooling system, remove the air cleaner and the intake duct assembly.

2. Disconnect the accelerator rod from the carburetor and remove the accelerator retracting spring. Disconnect the automatic transmission kick-down rod at the carburetor, if so equipped.

3. Disconnect the high-tension lead and all other wires from the ignition coil.

NOTE: Distributor removal is not necessary on 3.8L V6 engines, disregard steps pretaining to its removal.

4. Disconnect the spark plug wires from the spark plugs by grasping the rubber boots

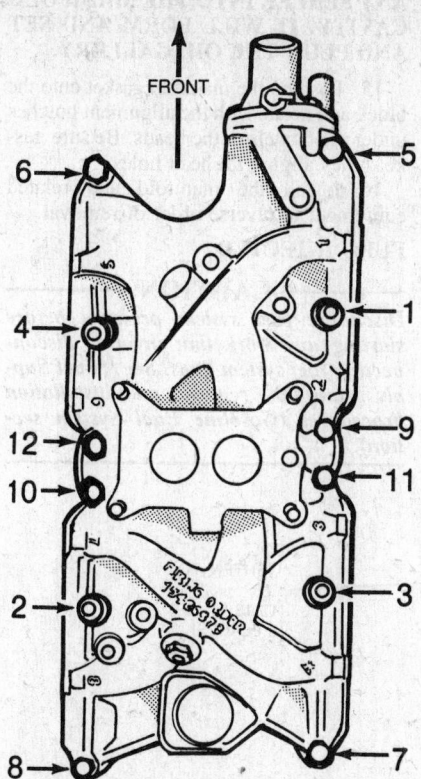

8-255, 302, 351W intake manifold bolt tightening sequence

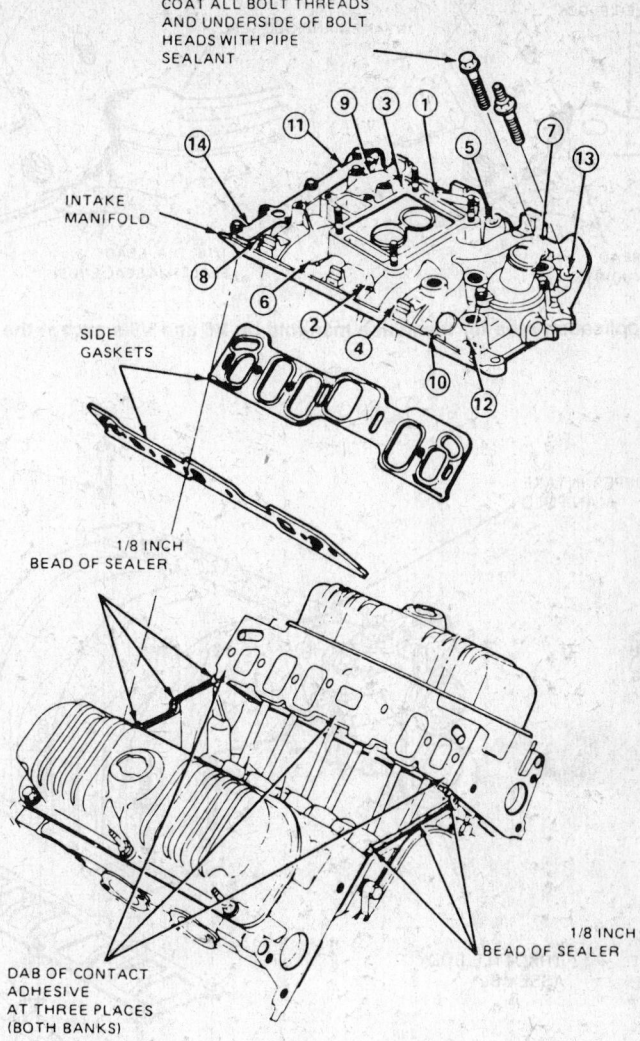

232 (3.8L) V6 intake manifold bolt tightening sequence

and twisting and pulling at the same time. Remove the wires from the brackets on the rocker covers. Remove the distributor cap and spark plug wire assembly.

5. Remove the carburetor fuel inlet line and the distributor vacuum line from the carburetor. (See Note above Step 4).

6. Remove the distributor lockbolt and remove the distributor and vacuum line. See "Distributor Removal and Installation."

7. Disconnect the upper radiator hose from the coolant outlet housing and the water temperature sending unit wire at the sending unit. Remove the heater hose from the intake manifold.

8. Loosen the clamp on the water pump bypass hose at the coolant outlet housing and slide the hose off the outlet housing.

9. Disconnect the PCV hose at the rocker cover.

10. If the engine is equipped with the Thermactor exhaust emission control system, remove the air pump to cylinder head air hose at the air pump and position it out of the way. Also remove the air hose at the backfire suppressor valve. Remove the air hose bracket from the valve rocker arm cover and position the air hose out of the way. Remove EGR valve tube on V6 models.

11. Remove the intake manifold and carburetor as an assembly. It may be necessary to pry the intake manifold from the cylinder head. Remove all traces of the intake manifold-to-cylinder head gaskets and the two end seals from both the manifold

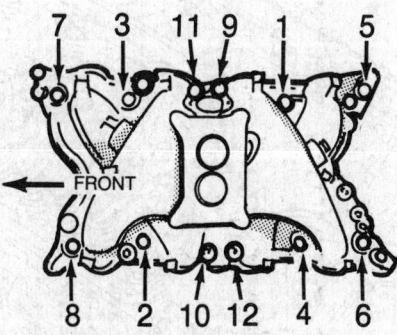

8-351M, 400 intake manifold bolt tightening sequence

and the other mating surfaces of the engine. Installation is as follows:

12. Clean the mating surfaces of the intake manifold, cylinder heads and block with laquer thinner or similar solvent. On V8 engines: Apply a 1/8 in. bead of silicone-rubber RTV sealant at the points shown in the accompanying diagram.

NOTE: The 3.8L V6 engine does not use end seals. RTV sealant is used. Apply 1/8 inch bead of sealant to each end of the engine block at the points where the intake manifold rests. Assembly must occur within 15 minutes of sealant application.

—— **CAUTION** ——

Do not apply sealer to the waffle portions of the seals as the sealer will rupture the end seal material.

13. On V8 engines: Position new seals on the block and press the seal locating extensions into the holes in the mating surfaces.

14. Apply a 1/16 in. bead of sealer to the outer end of each manifold seal for the full length of the seal (4 places). As before, do not apply sealer to the waffle portion of the end seals.

NOTE: RTV sealer sets in about 15 minutes, depending on brand, so work quickly but carefully. DO NOT DROP

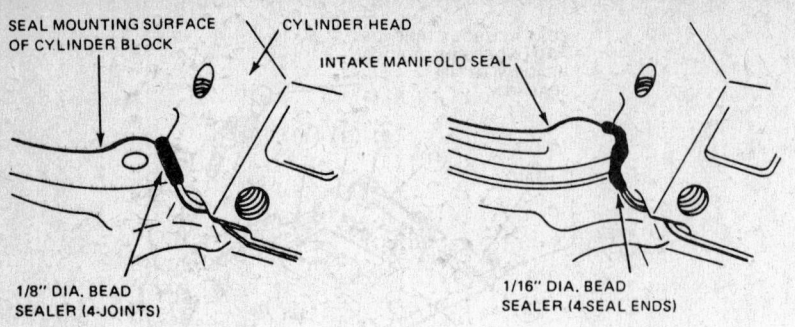

1/8" DIA. BEAD
SEALER (4-JOINTS)

1/16" DIA. BEAD
SEALER (4-SEAL ENDS)

SEAL MOUNTING SURFACE
OF CYLINDER BLOCK

CYLINDER HEAD

INTAKE MANIFOLD SEAL

Sealer application area for the intake manifold on V6 and V8's except the 460

ANY SEALER INTO THE MANIFOLD CAVITY. IT WILL FORM AND SET AND PLUG THE OIL GALLERY.

15. Position the manifold gasket onto the block and heads with the alignment notches under the dowels in the heads. Be sure gasket holes align with head holes.

16. Install the manifold and related equipment in reverse order of removal.

FUEL INJECTED

─────── **CAUTION** ───────

Discharge fuel system pressure before starting any work that involves disconnecting fuel system lines. See "Fuel Supply Manifold" removal and installation procedures (Gasoline Fuel System section).

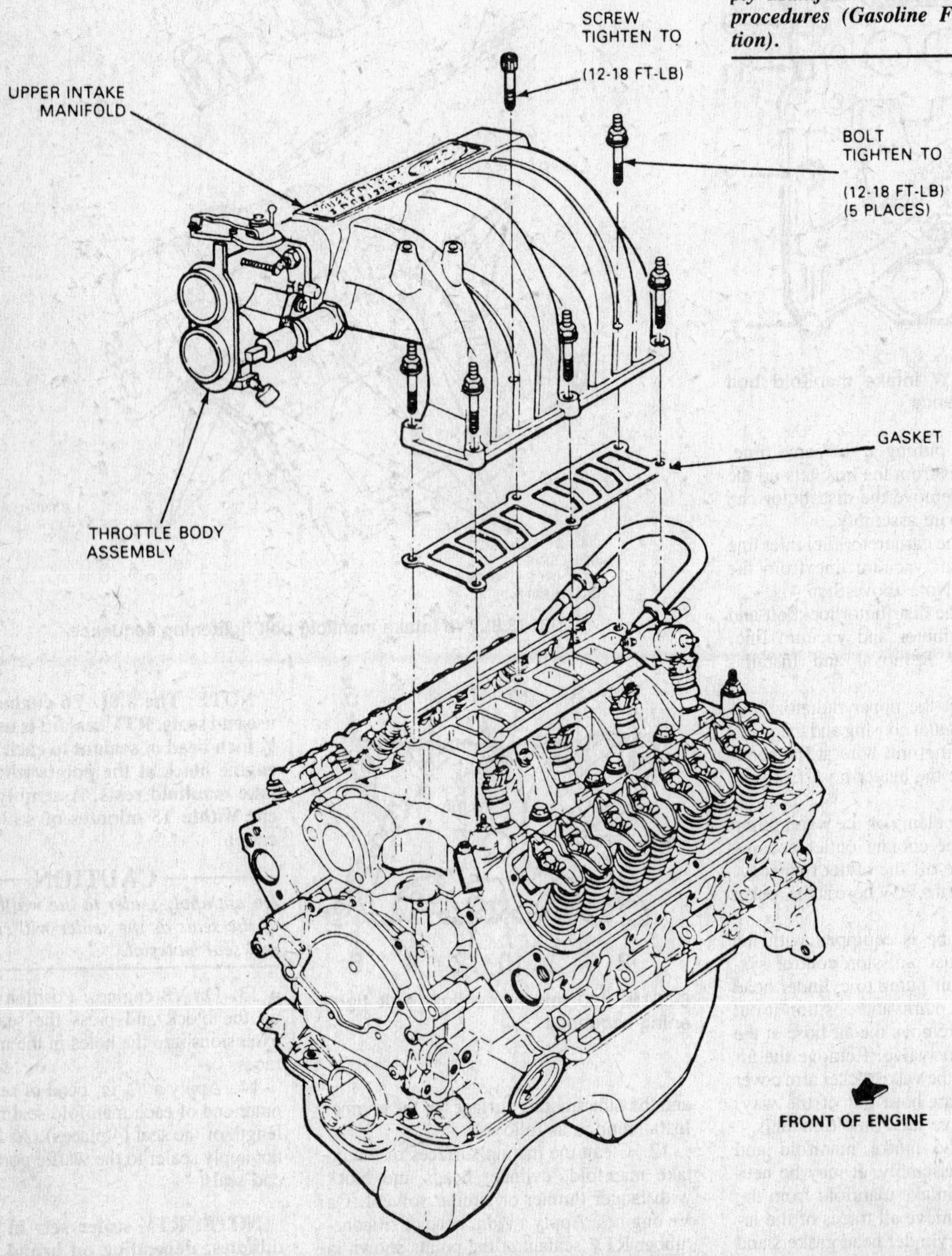

SCREW
TIGHTEN TO

(12-18 FT-LB)

BOLT
TIGHTEN TO

(12-18 FT-LB)
(5 PLACES)

UPPER INTAKE
MANIFOLD

THROTTLE BODY
ASSEMBLY

GASKET

FRONT OF ENGINE

Upper manifold installation—fuel injection

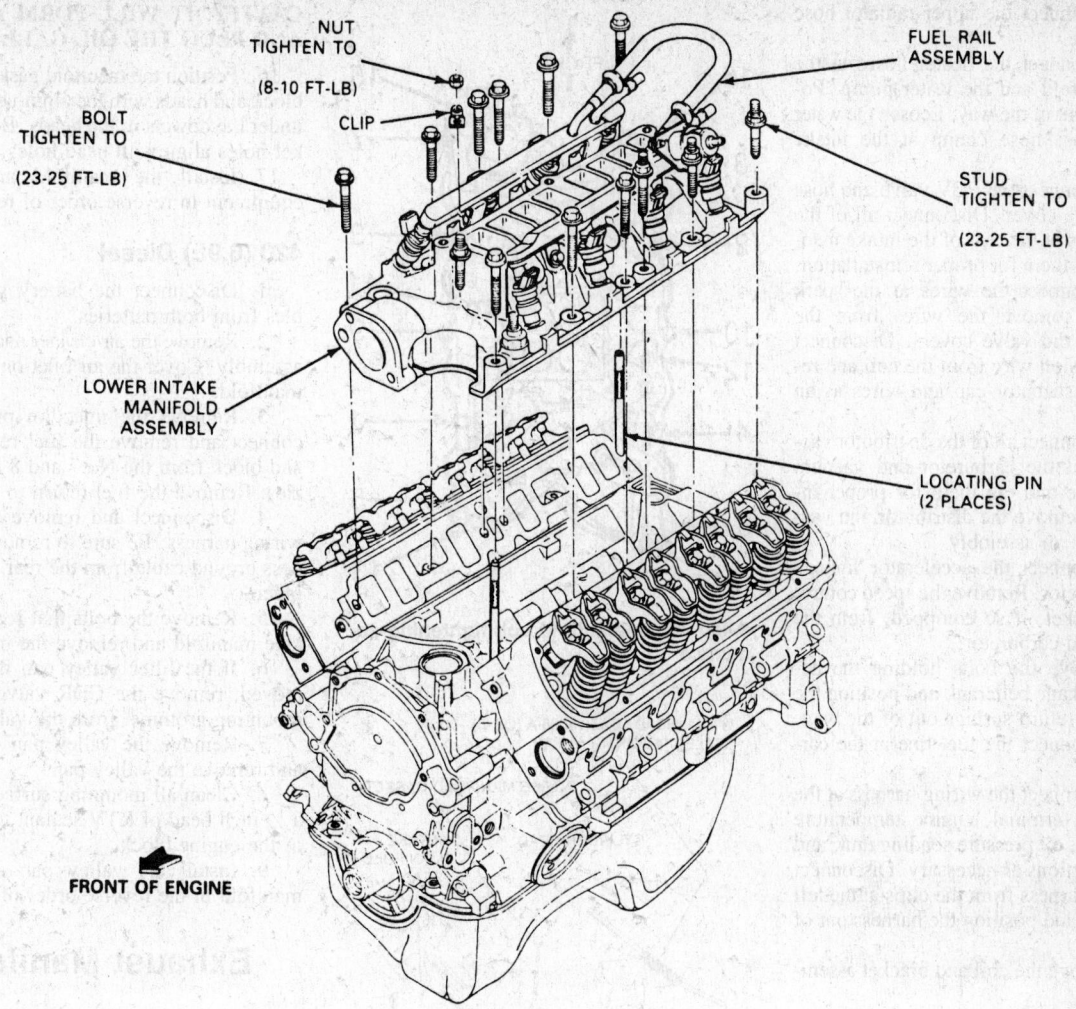

NUT
TIGHTEN TO
(8-10 FT-LB)

BOLT
TIGHTEN TO
(23-25 FT-LB)

CLIP

FUEL RAIL
ASSEMBLY

STUD
TIGHTEN TO
(23-25 FT-LB)

LOWER INTAKE
MANIFOLD
ASSEMBLY

LOCATING PIN
(2 PLACES)

FRONT OF ENGINE

Lower manifold installation—fuel injection

1. To remove the upper manifold: Remove the air cleaner. Disconnect the electrical connectors at the air bypass valve, throttle position sensor and EGR position sensor.

2. Disconnect the throttle linkage at the throttle ball and the AOD transmission linkage from the throttle body. Remove the bolts that secure the bracket to the intake and position the bracket and cables out of the way.

3. Disconnect the upper manifold vacuum fitting connections by removing all the vacuum lines at the vacuum tree (label lines for position identification). Remove the vacuum lines to the EGR valve and fuel pressure regulator.

4. Disconnect the PCV system by disconnecting the hose from the fitting at the rear of the upper manifold.

5. Remove the two canister purge lines from the fittings at the throttle body.

6. Disconnect the EGR tube from the EGR valve by loosening the flange nut.

7. Remove the bolt from the upper intake support bracket to upper manifold. Remove the upper manifold retaining

bolts and remove the upper intake manifold and throttle body as an assembly.

8. Clean and inspect all mounting surfaces of the upper and lower intake manifolds.

9. Position a new mounting gasket on the lower intake manifold and install the upper manifold in the reverse order of removal. Mounting bolts are torqued to 12–18 ft. lbs.

10. To remove the lower intake manifold: Upper manifold and throttle body must be removed first.

11. Drain the cooling system. Remove the distributor assembly, cap and wires.

12. Disconnect the electrical connectors at the engine coolant temperature sensor and sending unit, at the air charge temperature sensor and at the knock sensor.

13. Disconnect the injector wiring harness from the main harness assembly. Remove the ground wire from the intake manifold stud. The ground wire must be installed at the same position it was removed from.

14. Disconnect the fuel supply and return lines from the fuel rails.

15. Remove the upper radiator hose from

the thermostat housing. Remove the bypass hose. Remove the heater outlet hose at the intake manifold.

16. Remove the air cleaner mounting bracket. Remove the intake manifold mounting bolts and studs. Pay attention to the location of the bolts and studs for reinstallation. Remove the lower intake manifold assembly.

17. Clean and inspect the mounting surfaces of the heads and manifold.

18. Apply a $\frac{1}{16}$ inch bead of RTV sealer to the ends of the manifold seals (at the junction point of the seals and gaskets). Install the end seals and intake gaskets on the cylinder heads. The gaskets must interlock with the seal tabs.

19. Install locator bolts at opposite ends of each head and carefully lower the intake manifold into position. Install and tighten the mounting bolts and studs to 23–25 ft. lbs. Install the remaining components in the reverse order of removal.

460 (7.5L)

1. Drain the cooling system and remove the air cleaner assembly.

2. Disconnect the upper radiator hose at the engine.

3. Disconnect the heater hoses at the intake manifold and the water pump. Position them out of the way. Loosen the water pump by-pass hose clamp at the intake manifold.

4. Disconnect the PCV valve and hose at right valve cover. Disconnect all of the vacuum lines at the rear of the intake manifold and tag them for proper reinstallation.

5. Disconnect the wires at the spark plugs, and remove the wires from the brackets on the valve covers. Disconnect the high-tension wire from the coil and remove the distributor cap and wires as an assembly.

6. Disconnect all of the distributor vacuum lines at the carburetor and vacuum control valve and tag them for proper installation. Remove the distributor and vacuum lines as an assembly.

7. Disconnect the accelerator linkage at the carburetor. Remove the speed control linkage bracket, if so equipped, from the manifold and carburetor.

8. Remove the bolts holding the accelerator linkage bellcrank and position the linkage and return springs out of the way.

9. Disconnect the fuel line at the carburetor.

10. Disconnect the wiring harness at the coil battery terminal, engine temperature sending unit, oil pressure sending unit, and other connections as necessary. Disconnect the wiring harness from the clips at the left valve cover and position the harness out of the way.

11. Remove the coil and bracket assembly.

12. Remove the intake manifold attaching bolts and lift the manifold and carburetor from the engine as an assembly. It may be necessary to pry the manifold away from the cylinder heads. Do not damage the gasket sealing surfaces. Installation is as follows:

13. Clean the mating surfaces of the intake manifold, cylinder heads and block with laquer thinner or similar solvent. Apply a 1/8 in. bead of silicone-rubber RTV sealant at the points shown in the accompanying diagram.

— CAUTION —
Do not apply sealer to the waffle portions of the seals as the sealer will rupture the end seal material.

14. Position new seals on the block and press the seal locating extensions into the holes in the mating surfaces.

15. Apply a 1/16 in. bead of sealer to the outer end of each manifold seal for the full length of the seal (4 places). As before, do not apply sealer to the waffle portion of the end seals.

NOTE: RTV sealer sets in about 15 minutes, depending on brand, so work quickly but carefully. **DO NOT DROP ANY SEALER INTO THE MANIFOLD**

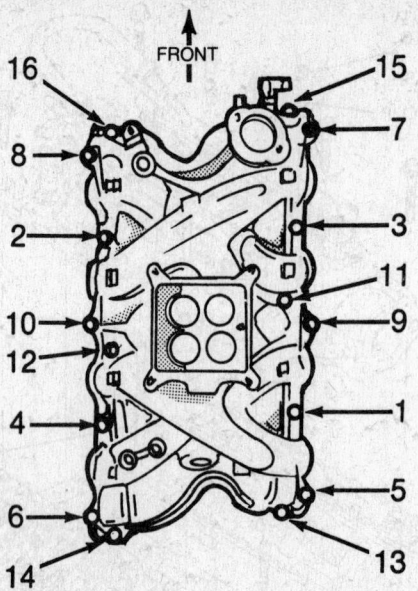
8-460 intake manifold bolt tightening sequence

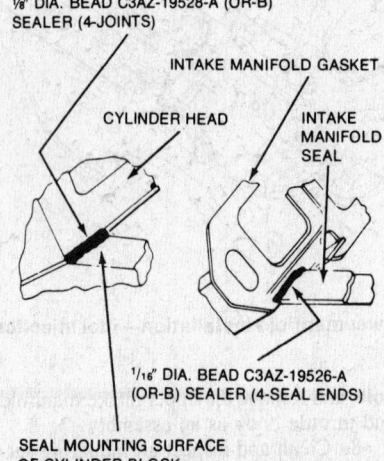
RTV sealer application area for the intake manifold on the 8-460

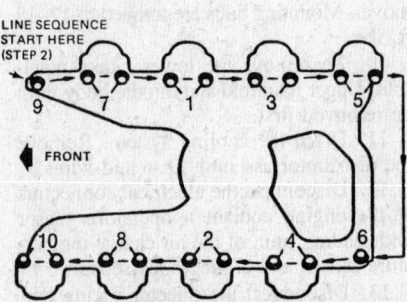
420 (6.9L) diesel intake manifold bolt tightening sequence. Torque in 2 steps;
1. Torque bolts to 24 ft. lbs. in numbered sequence.
2. Torque bolts to 24 ft. lbs. in line sequence.

CAVITY. IT WILL FORM AND SET AND PLUG THE OIL GALLERY.

16. Position the manifold gasket onto the block and heads with the alignment notches under the dowels in the heads. Be sure gasket holes align with head holes.

17. Install the manifold and related equipment in reverse order of removal.

420 (6.9L) Diesel

1. Disconnect the battery ground cables from both batteries.

2. Remove the air cleaner and duct hose assembly. Cover the air inlet on the intake manifold.

3. Remove the injection pump. Disconnect and remove the fuel return hoses and block from the No.7 and 8 (rear) nozzles. Remove the fuel return to tank line.

4. Disconnect and remove the engine wiring harness. Be sure to remove the harness ground cable from the rear of the cylinder.

5. Remove the bolts that retain the intake manifold and remove the manifold.

6. If the lifter valley pan is to be removed; remove the CDR valve tube and mounting grommet from the valley pan.

7. Remove the valley pan drain plug and remove the valley pan.

8. Clean all mounting surfaces. Apply a 1/8 inch bead of RTV sealant to each end of the engine block.

9. Install the valley pan and intake manifold in the reverse order of removal.

Exhaust Manifold

REMOVAL & INSTALLATION

122 (2.0L) and 140 (2.3L)

1. Remove the air cleaner and duct assembly. Disconnect the negative battery cable.

2. Remove the EGR line at the exhaust manifold. Loosen the EGR tube. Remove the check valve at the exhaust manifold and disconnect the hose at the end of the air by-pass valve.

3. Remove the bracket attaching the heater hoses to the valve cover. Disconnect the exhaust pipe from the exhaust manifold.

4. Remove the exhaust manifold mounting bolts/nuts and remove the manifold.

5. Install the exhaust manifold in the reverse order.

134 (2.2L) Diesel

1. Disconnect the ground cables from both batteries.

2. Disconnect the exhaust pipe from the manifold.

3. Remove the heater hose bracket from the valve cover and exhaust manifold studs.

4. Remove the vacuum pump support brace and bracket. Remove the bolt that attaches the engine oil dipstick tube support bracket to the exhaust manifold.

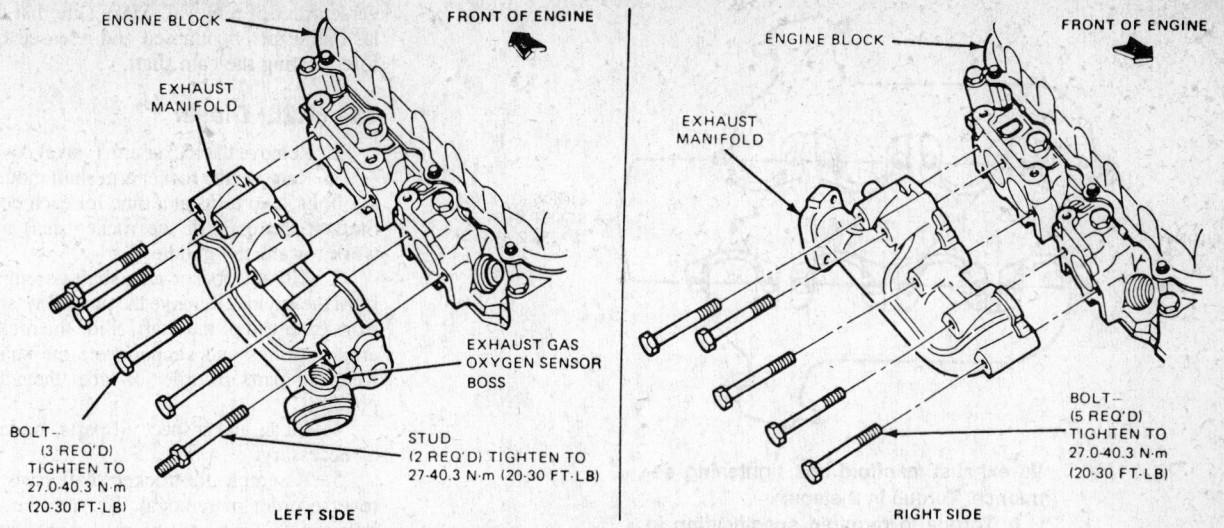

173 (2.8L) V6 exhaust manifold bolt tightening sequence

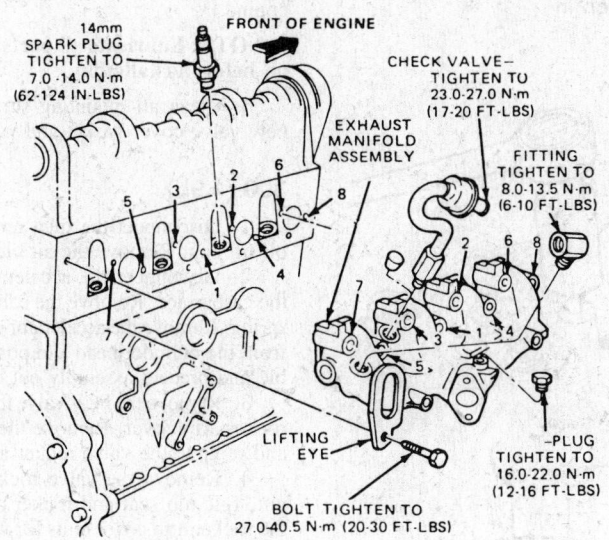

122 (2.0L), 140 (2.3L) exhaust manifold bolt tightening sequence

5. Remove the nuts that attach the exhaust manifold to the engine and remove the manifold.

6. Clean all gasket mounting surfaces. Install a new mounting gasket and install the exhaust manifold and components in the reverse order of removal.

173 (2.8L) V6 and 179 (2.9L) V6

1. Disconnect the negative battery cable. Remove the air cleaner and duct assembly.

2. Remove the left side heat shroud from the exhaust manifold. Remove any thermactor system parts that will interfere with manifold removal. Disconnect the choke heat tube at the carburetor.

3. Disconnect the exhaust pipes from the exhaust manifolds. Remove the mounting nuts from exhaust manifold studs. Remove the exhaust manifolds.

4. Install in the reverse order using new exhaust pipe to manifold gaskets.

300 (4.9L)

The intake and exhaust manifold on these engines are known as combination manifolds and are serviced as a unit. See "Intake Manifold Removal and Installation."

232 (3.8L) V6 and V8 Except 420 (6.9L) Diesel

1. Remove the air cleaner if the manifold being removed has the carburetor heat stove attached to it. On 351 and 400 remove the oil filter.

2. Remove the dipstick tube bracket bolt/nut on the 302 V8. On 351 and 400 V8 vehicles with a column mounted automatic transmission lever, disconnect the selector lever cross-shaft for clearance. On 1981 and later models, disconnect the EGO sensor, if equipped.

3. Remove any thermactor parts that will interfere with manifold removal.

4. Disconnect the exhaust pipe or catalytic converter from the exhaust manifold. Remove and discard the donut gasket.

5. Disconnect the EGR downtube. Remove the exhaust manifold attaching screws and remove the manifold from the cylinder head.

6. Install the exhaust manifold in the reverse order of removal. Apply a light coat of graphite grease to the mating surface of the manifold. Install and tighten the at-

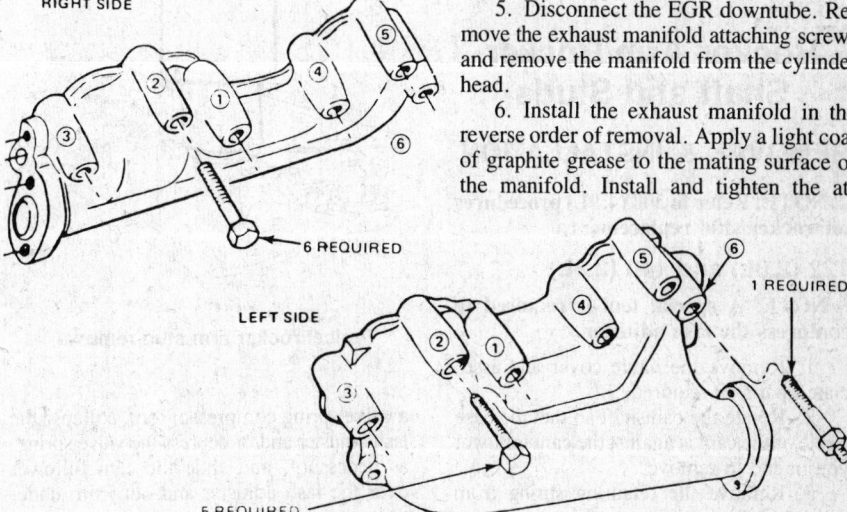

232 (3.8L) V6 exhaust manifold bolt tightening sequence

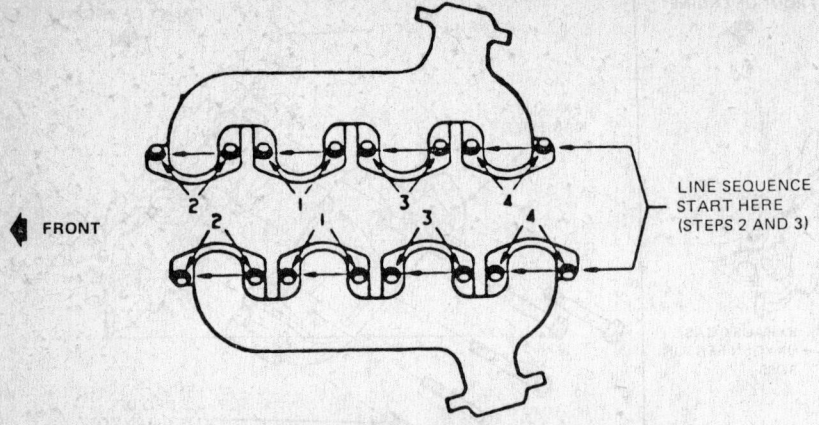

FRONT

LINE SEQUENCE
START HERE
(STEPS 2 AND 3)

V8 exhaust manifold bolt tightening sequence. Torque in 2 steps;
1. **Torque to required specification in numbered sequence.**
2. **Torque to required specification in line sequence.**

taching bolts, starting from the center and working to both ends alternately. Tighten to the proper specifications.

420 (6.9L) Diesel

1. Disconnect the ground cables from both batteries.

2. Raise and safely support the front of the vehicle on jackstands.

3. Disconnect the exhaust pipes from the manifolds.

4. If the right side manifold is to be removed, lower the vehicle. Remove the left side manifold from underneath while vehicle is raised.

5. The manifold attaching bolts are retained by lock tabs. Use a suitable tool to bend the tabs away from the bolt heads.

6. Clean all mounting surfaces. Apply anti-seize compound to the mounting bolts. Use new gaskets. Reverse the removal procedure for installation.

Rocker Arm/Rocker Shaft and Studs

REMOVAL & INSTALLATION

NOTE: Refer to 300 (4.9L) procedures for rocker stud replacement.

122 (2.0L) and 140 (2.3L)

NOTE: A special tool is required to compress the lash adjuster.

1. Remove the valve cover and associated parts as required.

2. Rotate the camshaft so that the base circle of the cam is against the cam follower you intend to remove.

3. Remove the retaining spring from the cam follower, if so equipped.

4. Using special tool T74P-6565-B or

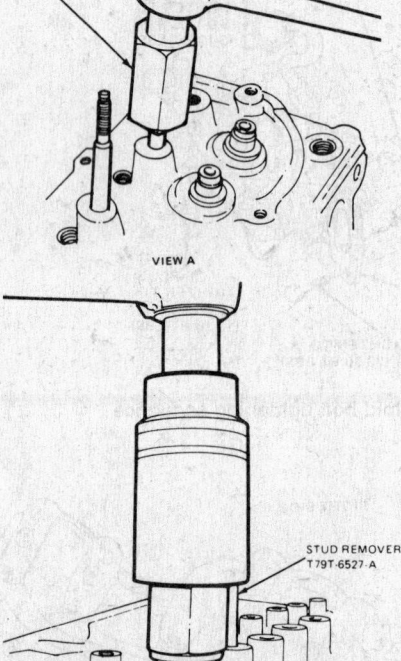

T79T-6527-A

VIEW A

STUD REMOVER
T79T-6527-A

VIEW B

Typical rocker arm stud removal

a valve spring compressor tool, collapse the lash adjuster and/or depress the valve spring, as necessary, and slide the cam follower over the lash adjuster and out from under the camshaft.

5. Install the cam follower in the re-

verse order of removal. Make sure that the lash adjuster is collapsed and released before rotating the cam shaft.

134 (2.2L) Diesel

1. Remove the rocker arm (valve) cover.

2. Remove the rocker arm shaft mounting bolts, two turns at a time for each bolt. Start at the ends of the rocker shaft and work toward the middle.

3. Lift the rocker arm shaft assembly from the engine. Remove the pin and washer from each end of the shaft. Slide the rocker arms, springs and supports off the shaft. Keep all parts in order or label them for position.

4. Clean and inspect all parts, replace as necessary.

5. Assemble the rocker shaft parts in reverse order of removal. Be sure the oil holes in the shaft are pointed downward. Reinstall the rocker shaft assembly on the engine.

NOTE: Lubricate all parts with motor oil before installation.

6. Clean all mounting surfaces, use a new valve cover gasket and valve cover.

300 (4.9L)

1. Disconnect the inlet air hose at the oil fill cap. Remove the air cleaner.

2. Disconnect the accelerator cable at the carburetor. Remove the cable retracting spring. Remove the accelerator cable bracket from the cylinder head and position the cable and bracket assembly out of the way.

3. Remove the PCV valve from the valve rocker arm cover. Remove the cover bolts and remove the valve rocker arm cover.

4. Remove the valve rocker arm stud nut, fulcrum seat and rocker arm. Inspect the rocker arm cover bolts for worn or damaged seals under the bolt heads and replace as necessary. If it is necessary to remove a rocker arm stud, Tool T79T-6527-A is available. A 0.006 oversize reamer T62F-6527-B3 or equivalent and a 0.015 inch oversize reamer T62F-6527-B5 or equivalent are available. For 0.010 inch oversize studs, use reamer T66P-6527-B or equivalent. To press in replacement studs, use stud replacer T79T-6527-B or equivalent for 6-300. Rocker arm studs that are broken or have damaged threads may be replaced with standard studs. Loose studs in the head may be replaced with 0.006, 0.010 or 0.015 inch oversize studs which area available for service. When going from a standard size rocker arm stud to a 0.010 or 0.015 inch oversize stud, always use the 0.006 inch oversize reamer before finish reaming with the 0.010 or 0.015 inch oversize reamer.

5. Position the sleeve of the rocker arm stud remover over the stud with the bearing end down. Thread the puller into the sleeve and over the stud until it is fully bottomed. Hold the sleeve with a wrench; then, rotate the puller clockwise to remove the stud. If

the rocker arm stud was broken off flush with the stud boss, use an easy-out to remove the broken stud following the instructions of the tool manufacturer.

6. If a loose rocker arm stud is being replaced, ream the stud bore using the proper reamer (or reamers in sequence) for the selected oversize stud. Make sure the metal particles do not enter the valve area.

7. Coat the end of the stud with Lubriplate® or it's equivalent. Align the stud with the stud bore; then, tap the sliding driver until it bottoms. When the driver contacts the stud boss, the stud is installed to its correct height.

8. Apply Lubriplate® or equivalent to the top of the valve stem and at the push rod guide in the cylinder head.

9. Apply Lubriplate® or equivalent to the rocker arm fulcrum seat and the fulcrum seat socket in the rocker arm. Install the valve rocker arm, fulcrum seat and stud nut.

10. Clean the valve rocker arm cover and the cylinder head gasket surface. Place the new gasket in the cover making sure that the tabs of the gasket engage in the notches provided in the cover.

11. Install the cover on the cylinder head. Make sure the gasket seats evenly all around the head. Partially tighten the cover bolts in sequence, starting at the middle bolts. Then tighten the bolts to 3–5 ft. lb.

12. Install the PCV valve in the rocker arm cover. Install the accelerator cable bracket on the cylinder head and connect the cable to the carburetor.

13. Connect the the inlet air hose to the oil fill cap.

14. Install air cleaner.

VALVE ADJUSTMENT

173 (2.8L) V6

1. Remove the air cleaner assembly and disconnect the negative battery cable.

2. Remove the Thermactor air bypass valve and its mounting bracket.

3. Remove the two engine lifting eyes; remove the alternator drive belt. loosen the alternator mounting bolts and swing the alternator outward toward the fender.

4. Remove the plug wires and remove the rocker covers.

5. When removing the rocker covers, remove or reposition any wires or hoses which block the removal of the rocker covers.

6. Torque the rocker arm support bolts to 46 ft. lbs.

7. Reconnect the battery cable, place the transmission in Neutral (manual) or Park (automatic), and apply the parking brake.

8. Place a finger on the adjusting screw of the intake valve rocker arm for cylinder No.5. Cylinder numbering is shown under Firing Order at the start of the section. Valve arrangement, from front to rear, on the left bank is I–E–I–E–I; on the right it is

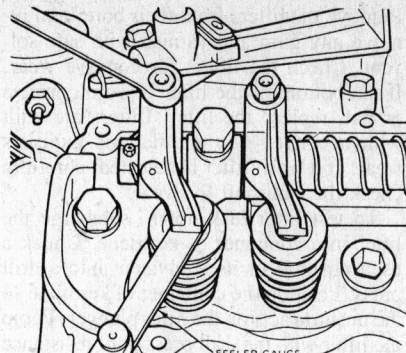

Adjusting the valves-173 (2.8L) V6 engine

I–E–I–E–E–I. You will be able to feel the rocker arm begin to move.

9. Use a remote starter switch or manual means to turn the engine until you can just feel the valve begin to open. Now the engine is in position to adjust the intake and exhaust valves on the No. 1 cylinder.

10. Adjust the No. 1 cylinder intake valve so that a 0.014 in. feeler gauge has a slight drag while a 0.015 in. feeler gauge is a tight fit. To decrease lash, turn the adjusting screw clockwise; to increase lash, turn the adjusting screw counterclockwise. There are no locknuts to tighten; the adjusting screws are self-locking.

--- **CAUTION** ---

Do not use a step-type, "go-no go" feeler gauge. When checking lash, insert the feeler gauge and move it parallel with the crankshaft. Do not move it in and out perpendicular with the crankshaft; this will give an erroneous feel which will result in overtightened valves.

11. Adjust the exhaust valve the same way so that a 0.016 in. feeler gauge has a slight drag, while a 0.017 in. gauge is a tight fit.

12. The rest of the valves are adjusted in the same way, in their firing order (1–4–2–5–3–6), by positioning the engine according to the following chart:

Intake Valve Just Opening	Adjust both Valves for this Cylinder (Intake—0.014 in.; Exhaust—0.016 in.)
5	1
3	4
6	2
1	5
4	3
2	6

13. Remove all the old gasket material from the cylinder heads and rocker cover gasket surfaces, and disconnect the negative cable from the battery.

14. Remove the spark plug wires and reinstall the rocker arm covers.

15. Reinstall any hoses and wires which were removed.

16. Reinstall the spark plug wires, the alternator drive belt, and the Thermactor air bypass valve and its mounting bracket.

17. Reconnect the battery cable, replace the air cleaner assembly, start the engine, and check for leaks.

232 (3.8L) V6 and V8s Except 420 (6.9L) Diesel

These engines are equipped with individually mounted rocker arms. Use the fol-

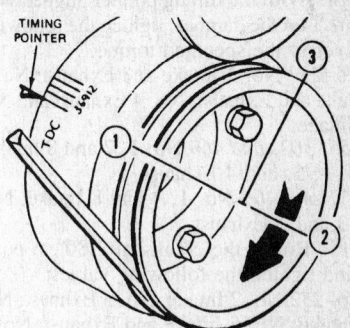

POSITION 1 – No. 1 at TDC at end of compression stroke.
POSITION 2 – Rotate the crankshaft 180 degrees (one half revolution) clockwise from POSITION 1.
POSITION 3 – Rotate the crankshaft 270 degrees (three quarter revolution) clockwise from POSITION 2.

Crankshaft pulley marking instructions for rocker arm installation on V8 engines

lowing procedure to remove the rocker arms:

1. Disconnect the choke heat chamber air hose, the air cleaner and inlet duct assembly, the choke heat tube, PCV valve and hose, and the EGR hoses (if so equipped).

2. On models so equipped, disconnect the Thermactor by-pass valve and air supply hoses.

3. Label and disconnect the spark plug wires at the plugs. Remove the plug wires from the looms.

4. Remove the valve cover attaching bolts and remove the cover(s).

5. Remove the valve rocker arm stud nut.

ADJUSTMENT

This adjustment is actually part of the installation procedure for the individually mounted rocker arms found on the V6 and V8 engines, and is necessary to achieve an accurate torque value for each rocker arm nut.

By its nature, an hydraulic valve lifter will expand when it is not under load. Thus, when the rocker arms are removed and the pressure via the pushrod is taken off the lifter, the lifter expands to its maximum. If the lifter happens to be at the top of the camshaft lobe when the rocker arm is being reinstalled, a large amount of torque would be necessary when tightening the rocker arm nut just to overcome the pressure of the expanded lifter. This makes it very difficult to get an accurate torque setting with

individually mounted rocker arms. For this reason, the rocker arms are installed in a certain sequence which corresponds to the low points of the camshaft lobes.

1. Crank the engine until No.1 cylinder is at TDC of the compression stroke and the timing pointer is aligned with the mark on the crankshaft damper.

2. Scribe a mark on the damper at this point.

3. Scribe two additional marks on the damper (see illustration).

4. With the timing pointer aligned with mark 1 on the damper, tighten the following valves to the specified torque:

● V6–232 No. 1 Intake and Exhaust; No. 3 Intake and Exhaust; No. 4 Exhaust and No. 6 Intake.

● 255, 302, and 460 No. 1, 7 and 8 Intake; No. 1, 5, and 4 Exhaust

● 351 and 400 No. 1, 4 and 8 Intake; No. 1, 3 and 7 Exhaust.

5. Rotate the crankshaft 180° to point 2 and tighten the following valves:

● V6–232 No. 2 Intake; No. 3 Exhaust; No. 4 Intake; No. 5 Intake and Exhaust; No. 6 Exhaust.

● 255, 302, and 460 No. 5 and 4 Intake; No. 2 and 6 Exhaust

● 351 and 400 No. 3 and 7 Intake; No. 2 and 6 Exhaust

6. Rotate the crankshaft 270° to point 3 and tighten the following valves:

● 302 and 460 No. 2, 3, and 6 Intake; No. 7, 3 and 8 Exhaust

● 351 and 400 No. 2, 5 and 6 Intake; No. 4, 5 and 8 Exhaust

7. On 232, 255, 302 and 351W engines, tighten nut until it contacts the rocker shoulder, then torque to 18–20 ft. lbs.; 351C and 400 engines, tighten bolt to 18–25 ft. lbs.; 460 engine, tighten nut until it contacts rocker shoulder, then torque to 18–22 ft. lbs.

420 (6.9L) Diesel

1. Disconnect the ground cables from both batteries.

2. Remove the valve cover retaining bolts and the valve covers.

3. Remove the rocker arm retaining bolts and posts. Keep the rockers and posts in order and identify for reinstallation to original positions.

4. Turn the engine until the timing mark is at the 11 o'clock position as viewed from the front of the engine.

5. Install all rocker arms, posts and retaining bolts and tighten.

6. Clean all gasket mounting surfaces, install new valve cover gaskets and reinstall the valve covers to the engine. Reconnect the battery cables, start the engine and check for oil leaks.

HYDRAULIC VALVE LIFTER INSPECTION

NOTE: The lifters used on diesel engines require a special test fluid, kerosene is not satisfactory.

Remove the lifters from their bores and remove any gum and varnish with safe solvent. Check the lifters for concave wear. If the bottom of the lifter is worn concave or flat, replace the lifter. Lifters are built with a convex bottom, flatness indicates wear. If a worn lifter is detected, carefully check the camshaft for wear.

To test lifter leak down, submerge the lifter in a container of kerosene. Chuck a used pushrod or its equivalent into a drill press. Position the container of kerosene so the pushrod acts on the lifter plunger. Pump the lifter with the drill press until resistance increases. Pump several more times to bleed any air from the lifter. Apply very firm, constant pressure to the lifter and observe the rate which fluid bleeds out of the lifter. If the lifter bleeds down very quickly (less than 15 seconds), the lifter should be replaced. If the time exceeds 60 seconds, the lifter is sticking and should be cleaned or replaced. If the lifter is operating properly (leak down time 15–60 seconds) and not worn, lubricate and reinstall in engine.

NOTE: Always inspect the valve pushrods for wear, straightness and oil blockage. Damaged pushrods will cause erratic valve operation.

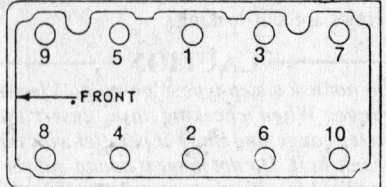

122 (2.0L), 140 (2.3L) 4 cylinder head bolt tightening sequence

Cylinder Head

REMOVAL & INSTALLATION

122 (2.0L) and 140 (2.3L)

1. Drain the cooling system. Disconnect the negative battery cable.

2. Remove the air cleaner.

3. Remove the valve cover.

NOTE: On models with air conditioning, remove the mounting bolts and the drive belt, and position the compressor, with hoses attached, out of the way. Remove the compressor upper mounting bracket from the cylinder head.

——————— CAUTION ———————
If the compressor refrigerant lines do not have enough slack to permit repositioning of the compressor without first disconnecting the refrigerant lines, the air conditioning system will have to be evacuated by a trained air conditioning serviceman. Under no circumstances should an untrained person attempt to disconnect the air conditioning refrigerant lines.

4. Remove the intake and exhaust manifolds from the head.

5. Remove the camshaft drive belt cover. Note the location of the belt cover attaching screws that have rubber grommets.

6. Loosen the drive belt tensioner and remove the bolt.

7. Remove the water outlet elbow from the cylinder head with the hose attached.

8. Remove the cylinder head attaching bolts.

9. Remove the cylinder head from the engine.

10. Clean all gasket material and carbon from the top of the cylinder block and pistons and from the bottom of the cylinder head.

11. Position a new cylinder head gasket on the engine and place the head on the engine.

NOTE: If you encounter difficulty in positioning the cylinder head on the engine block, it may be necessary to install guide studs in the block to correctly align the head and the block. To fabricate guide studs, obtain two new cylinder head bolts and cut their heads off with a hack saw. Install the bolts in the holes in the engine block which correspond with cylinder head bolt holes Nos. 3 and 4, as identified in the cylinder head bolt tightening sequence illustration. Then, install the head gasket and head over the bolts. Install the cylinder head attaching bolts, replacing the studs with the original head bolts.

12. Using a torque wrench, tighten the head bolts in proper sequence.

13. Install the camshaft drive belt.

14. Install the camshaft drive belt cover and its attaching bolts. Make sure the rubber grommets are installed on the bolts. Tighten the bolts to 6–13 ft. lbs.

15. Install the water outlet elbow and a new gasket on the engine and tighten the attaching bolts to 12–15 ft. lbs.

16. Install the intake and exhaust manifolds.

17. Assemble the rest of the components in reverse order of removal.

134 (2.2L) Diesel

1. Disconnect the ground cables from both batteries.

2. Mark the hood hinges for realignment on installation and remove the hood. Drain the cooling system.

3. Disconnect the breather hose from the valve cover and remove the intake hose and breather hose from the air cleaner and intake manifold.

4. Remove the heater hose bracket from the valve cover and exhaust manifold. Disconnect the heater hoses from the water pump and thermostat housing and position tube assembly out of the way.

5. Remove the vacuum pump support brace from the pump bracket and cylinder head.

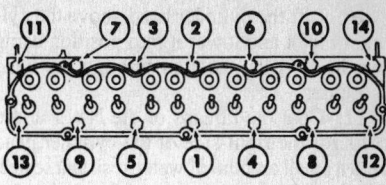

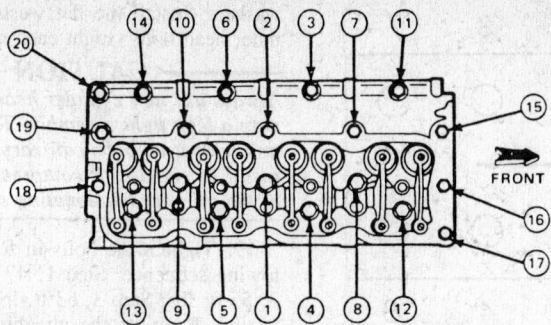

134 (2.2L) diesel head bolt tightening sequence

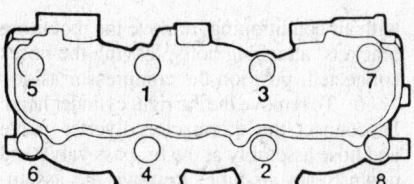

6-300 cylinder head bolt torque sequence

6. Loosen and remove the alternator and vacuum pump drive belts. Loosen and remove the A/C compressor and/or power steering drive belt.

7. Disconnect the brake booster vacuum hose and remove the vacuum pump.

8. Disconnect the exhaust pipe from the exhaust manifold. Disconnect the coolant thermoswitch and coolant temperature sender wiring harness.

9. Disconnect and remove the fuel injection lines from the injector nozzles and pump. Cap all lines and fittings to prevent dirt from entering the system.

10. Disconnect the engine wire harness from the alternator, the glow plug harness and dropping resistor and position the harness out of the way.

11. Disconnect the fuel lines from both sides of the fuel heater. Remove the fuel filter assembly from the mounting bracket and position out of the way with the fuel line attached.

12. Loosen the lower No. 3 intake port nut and the bolt on the injection pump; disconnect the lower fuel return line from the intake manifold stud and the upper fuel return line.

13. If equipped with power steering, remove the bolt that attaches the pump rear support bracket to the cylinder head.

14. Remove the upper radiator hose. Disconnect the by-pass hose from the thermostat housing.

15. Remove the A/C compressor and position out of the way with the lines still attached.

—————— CAUTION ——————

Do not disconnect the compressor lines unless the proper tools are on hand to discharge the system and you are familiar with the procedure.

16. Remove the valve cover, rocker arm shaft assembly and the pushrods. Identify the pushrods and keep them in order for return to their original position.

17. Remove the cylinder head attaching bolts, starting at the ends of the head, working alternately toward the center. Remove the cylinder head from the truck.

18. Clean all gasket mounting surfaces. Install the cylinder head in the reverse order of removal. Torque the cylinder head bolts in the proper sequence.

173 (2.8L) V6 head bolt tightening sequence

173 (2.8L) V6 and 179 (2.9L) V6

1. Remove the air cleaner assembly and disconnect the negative battery cable, and accelerator linkage. Drain the cooling system.

2. Remove the distributor cap with the spark plug wires attached. Remove the distributor vacuum line and distributor. Remove the hose from the water pump to the water outlet which is on the carburetor.

3. Remove the valve covers, fuel line and filter, carburetor, and the intake manifold.

4. Remove the rocker arm shaft and oil baffles. Remove the pushrods, keeping them in the proper sequence for installation.

5. Remove the exhaust manifold.

6. Remove the cylinder head retaining bolts and remove the cylinder heads and gaskets.

7. Remove all gasket material and carbon from the engine block and cylinder heads.

8. Place the head gaskets on the engine block. Pay attention, the left and right gaskets are not interchangeable.

9. Install guide studs in the engine block. Install the cylinder head assembles on the engine block one at a time. Tighten the cylinder head bolts in sequence.

10. Install the intake and exhaust manifolds.

11. Install the pushrods in the proper sequence. Install the oil baffles and the rocker arm shaft assemblies. Adjust the valve clearances.

12. Install the valve covers with new gaskets.

13. Install the distributor and set the ignition timing.

14. Install the carburetor and the distributor cap with the spark plug wires.

15. Connect the accelerator linkage, fuel line, with fuel filter installed, and distributor vacuum line to the carburetor. Fill the cooling system.

300 (4.9L)

1. Drain the cooling system. Remove the air cleaner. Remove the oil filler tube. Disconnect the negative battery cable.

2. Disconnect the muffler inlet pipe at the exhaust manifold. Pull the muffler inlet pipe down. Remove the gasket.

3. Disconnect the accelerator rod or cable retracting spring. Disconnect the choke control cable if applicable and the accelerator rod at the carburetor.

4. Disconnect the transmission kickdown rod. Disconnect the accelerator linkage at the bellcrank assembly.

5. Disconnect the fuel inlet line at the fuel filter hose, and the distributor vacuum line at the carburetor. Disconnect other vacuum lines as necessary for accessibility and identify them for proper connection.

6. Remove the radiator upper hose at the coolant outlet housing.

7. Disconnect the distributor vacuum line at the distributor. Disconnect the carburetor fuel inlet line at the fuel pump. Remove the lines as an assembly.

8. Disconnect the spark plug wires at the spark plugs and the temperature sending unit wire at the sending unit.

9. Grasp the PCV vent hose near the PCV valve and pull the valve out of the grommet in the valve rocker arm cover. Disconnect the PCV vent hose at the hose fitting in the intake manifold spacer and remove the vent hose and PCV valve.

10. Disconnect the carburetor air vent tube and remove the valve rocker arm cover.

11. Remove the valve rocker arm shaft assembly. Remove the pushrods in sequence so that they can be identified and reinstalled in their original positions.

12. Remove the cylinder head bolts and remove the cylinder head. Do not pry between the cylinder head and the block as the gasket surfaces may be damaged.

13. To install the cylinder head: Clean the head and block gasket surfaces. If the cylinder head was removed for a gasket change, check the flatness of the cylinder head and block.

14. Apply sealer to both sides of the new cylinder head gasket, depending on maker of gasket, refer to gasket manufacturers instructions. Position the gasket on the cylinder block.

15. Install a new gasket on the flange of the muffler inlet pipe.

16. Lift the cylinder head above the cylinder block and lower it into position using two head bolts installed through the head as guides.

17. Coat the threads of the No. 1 and 6 bolts for the right-side of the cylinder head with a small amount of water-resistant sealer. Oil the threads of the remaining bolts. Install, but do not tighten, two bolts at the opposite ends of the head to hold the head and gasket in position.

18. The cylinder head bolts are tightened in 3 progressive steps. Torque them (in the proper sequence) to 55 ft. lbs., then 65 ft. lbs., and finally to 75 ft. lbs.

19. Apply Lubriplate® to both ends of the pushrods and install them in their original positions.

20. Install the valve rocker arm shaft assembly.

21. Adjust the valves, as necessary.

22. Install the muffler inlet pipe lockwashers and attaching nuts.

23. Connect the radiator upper hose at the coolant outlet housing.

24. Position the distributor vacuum line and the carburetor fuel inlet line on the engine. Connect the fuel line at the fuel filter hose and install a new clamp. Install the distributor vacuum line at the carburetor. Connect the accelerator linkage at the bellcrank assembly. Connect the transmission kickdown rod.

25. Connect the accelerator rod retracting spring. Connect the choke control cable (if applicable) and the accelerator rod at the carburetor.

26. Connect the distributor vacuum line at the distributor. Connect the carburetor fuel inlet line at the fuel pump. Connect all the vacuum lines using their previous identification for proper connection.

27. Connect the temperature sending unit wire at the sending unit. Connect the spark plug wires. Connect the battery cable at the cylinder head.

28. Fill the cooling system.

29. Install the valve rocker cover. Connect the carburetor air vent tube.

30. Connect the PCV vent hose at the carburetor spacer fitting. Insert the PCV valve with the vent hose attached, into the valve rocker arm cover grommet. Install the air cleaner, start the engine and check for leaks.

232 (3.8L) V6

1. Drain the cooling system.

2. Disconnect the cable from the battery negative terminal.

3. Remove the air cleaner assembly including air intake duct and heat tube.

4. Loosen the accessory drive belt idler. Remove the drive belt.

5. To remove the left cylinder head: If equipped with power steering, remove the pump mounting brackets' attaching bolts, leaving the hoses connected, place the pump/bracket assembly aside in a position to prevent the fluid from leaking out. If equipped

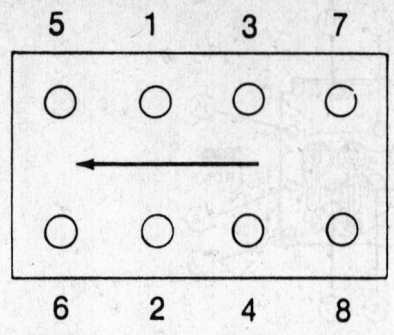

232 (3.8L) V6 head bolt tightening sequence

with air conditioning, remove the mounting brackets' attaching bolts, leaving the hoses connected, position the compressor aside.

6. To remove the the right cylinder head: Disconnect the thermactor diverter valve and hose assembly at the by-pass valve and downstream air tube. Remove the assembly. Remove the accessory drive idler. Remove the alternator. Remove the thermactor pump pulley. Remove the thermactor pump. Remove the alternator bracket. Remove the PCV valve.

7. Remove the intake manifold.

8. Remove the valve rocker arm cover attaching screws. Loosen the silicone rubber gasketing material by inserting a putty knife under the cover flange. Work the cover loose and remove.

CAUTION

The plastic rocker arm covers will break if excessive prying is applied.

---

9. Remove the exhaust manifold(s).

10. Loosen the rocker arm fulcrum attaching bolts enough to allow the rocker arm to be lifted off the pushrod and rotated to one side.

11. Remove the pushrods. The position of each rod should be installed in the original position during assembly.

12. Remove the cylinder head attaching bolts. Remove the cylinder head(s).

13. Remove and discard the old cylinder head gasket(s). Discard the cylinder head bolts.

14. Lightly oil all bolt and stud bolt threads before installation except those specifying special sealant.

15. Clean the cylinder head, intake manifold, valve rocker arm cover and cylinder head gasket surfaces. If the cylinder head was removed for a cylinder head gasket replacement, check the flatness of the cylinder head and block gasket surfaces.

16. Position new head gasket(s) on the cylinder block using the dowels for alignment.

17. Position the cylinder heads to the block.

18. Apply a thin coating of pipe sealant or equivalent to the threads of the short cylinder head bolts (nearest to the exhaust manifold). Do not apply sealant to the long bolts. Lightly oil the cylinder head bolt flat

washers. Install the flat washers and cylinder head bolts (Eight each side).

CAUTION

Always use new cylinder head bolts to assure a leak tight assembly. Torque retention with used bolts can vary, which may result in coolant or compression leakage at the cylinder head mating surface area.

---

19. Tighten the bolts in four steps following sequence: Step 1, 47 ft. lbs: Step 2, 55 ft. lbs: Step 3, 63 ft. lbs: Step 4, 74 ft. lbs. Back-off the attaching bolts 2–3 turns. Repeat tightening sequence.

NOTE: When the cylinder head attaching bolts have been tightened using the above sequential procedure, it is not necessary to retighten the bolts after extended engine operation. However, the bolts can be checked for tightness if desired.

20. Dip each pushrod end in heavy engine oil. Install the pushrods in their original position. For each valve rotate the crankshaft until the tappet rests on the heel (base circle) of the camshaft lobe.

21. Position the rocker arms over the pushrods, install the fulcrums, and tighten the fulcrum attaching bolts to 61–132 in. lbs.

CAUTION

Fulcrums must be fully seated in cylinder head and pushrods must be seated in rocker arm sockets prior to final tightening.

---

22. Lubricate all rocker arm assemblies with heavy engine oil. Finally tighten the fulcrum bolts to 19–25 ft. lbs. For final tightening, the camshaft may be in any position.

If the original valve train components are being installed, a valve clearance check is not required. If a component has been replaced, perform a valve clearance check.

23. Install the exhaust manifold(s).

24. Apply a 1/8–3/16 inch bead of RTV silicone sealant to the rocker arm cover flange. Make sure the sealer fills the channel in the cover flange. The rocker arm cover must be installed within 15 minutes after the silicone sealer application. After this time, the sealer may start to set-up, and its sealing effectiveness may be reduced.

25. Position the cover on the cylinder head and install the attaching bolts. Note the location of the wiring harness routing clips and spark plug wire routing clip stud bolts. Tighten the attaching bolts to 36–60 in. lbs. torque.

26. Install the intake manifold.

27. Install the spark plugs.

28. Connect the secondary wires to the spark plugs.

29. Install the oil fill cap. If equipped with air conditioning, install the compressor mounting and support brackets. Install the remaining parts in the reverse order of removal.

30. Fill the cooling system with the specified coolant.

— CAUTION —

This engine has an aluminum cylinder head and requires a special unique corrosion inhibited coolant formulation to avoid radiator damage.

31. Start the engine and check for coolant, fuel, and oil leaks.

32. Check and, if necessary, adjust the curb idle speed.

33. Install the air cleaner assembly including the air intake duct and heat tube.

V8 Except 420 (6.9L) Diesel & 460 (7.5L)

1. Remove the intake manifolds and the carburetor as an assembly.

2. Remove the rocker arm cover(s).

3. If the right cylinder head is to be removed, loosen the alternator adjusting arm bolt and remove the alternator mounting bracket bolt and spacer. Swing the alternator down and out of the way. Remove the air cleaner inlet duct from the right cylinder head assembly. On 351 and 400 remove the ground strap at the rear of the head. If the left cylinder head is being removed, remove the bolts fastening the accelerator shaft assembly at the front of the cylinder head. On vehicles equipped with air conditioning, the system must be discharged and the compressor removed. The procedure is best left to an air conditioning specialist. Persons not familiar with A/C systems can be easily injured when working on the systems.

4. Disconnect the exhaust manifold(s) from the muffler inlet pipe(s).

5. Loosen the rocker arm stud nuts so that the rocker arms can be rotated to the side. Remove the pushrods and identify them so that they can be reinstalled in their original positions.

6. Remove the cylinder head bolts and lift the cylinder head from the block.

7. Clean the cylinder head, intake manifold, and the valve cover and head gasket surfaces.

8. A specially treated composition head gasket is used. Do not apply sealer to a composition gasket. Position the new gasket over the locating dowels on the cylinder block. Then, position the cylinder head on the block and install the attaching bolts.

9. The cylinder head bolts are tightened in 3 progressive steps. Tighten all the bolts in the proper sequence to 50 ft. lbs., 60 ft. lbs., and finally to 70 ft. lbs. of torque on the 255, 302 and 351W. On 351M and 400 V8s, tighten to 70, 80 then 95–105 ft. lbs.

10. Clean the pushrods. Blow out the oil passage in the rods with compressed air. Check the pushrods for straightness. Never try to straighten a pushrod; always replace it.

11. Apply Lubriplate® to the ends of the

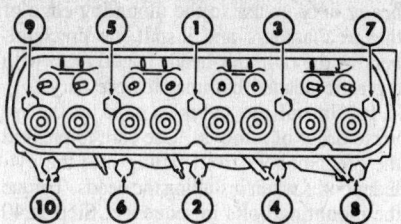

Cylinder head bolt torque sequence, all V8's

pushrods and install them in their original positions.

12. Apply Lubriplate® to the rocker arms and their fulcrum seats and install the rocker arms. Adjust the valves.

13. Position a new gasket(s) on the muffler inlet pipe(s) as necessary. Connect the exhaust manifold(s) at the muffler inlet pipe(s).

14. If the right cylinder head was removed, install the alternator, ignition coil and air cleaner duct on the right cylinder head. Adjust the drive belt. If the left cylinder head was removed, install the accelerator shaft assembly at the front of the cylinder head.

15. Clean the valve rocker arm cover and the cylinder head gasket surfaces. Place the new gaskets in the covers, making sure that the tabs of the gasket engage the notches provided in the cover. Install the compressor. If the system has be bled, recharging can be done after cylinder head operations are completed.

16. Install the intake manifold and all remaining parts.

460 (7.5L) V8

1. Disconnect the negative battery cable. Remove the intake manifold and carburetor as an assembly.

2. Disconnect the exhaust pipe from the exhaust manifold.

3. Loosen the air conditioning compressor drive belt, if so equipped.

4. Loosen the alternator attaching bolts and remove the bolt attaching the alternator bracket to the right cylinder head.

5. Disconnect the air conditioning compressor from the engine and move it aside, out of the way.

— CAUTION —

Do not disconnect the compressor lines unless the proper tools are on hand to discharge the system and you are familiar with the procedure.

6. Remove the bolts securing the power steering reservoir bracket to the left cylinder head. Position the reservoir and bracket out of the way.

7. Remove the valve rocker arm covers. Remove the rocker arm bolts, rocker arms, oil deflectors, fulcrums and pushrods in sequence so that they can be reinstalled in their original positions.

8. Remove the cylinder head bolts and lift the head and exhaust manifold off the

engine. If necessary, pry at the forward corners of the cylinder head against the casting bosses provided on the cylinder block. Do not damage the gasket mating surfaces of the cylinder head and block by prying against them.

9. Remove all gasket material from the cylinder head and block. Clean all gasket material from the mating surfaces of the intake manifold. If the exhaust manifold was removed, clean the mating surfaces of the cylinder head and exhaust manifold. Apply a thin coat of graphite grease to the cylinder head exhaust port areas and install the exhaust manifold.

10. Position two long cylinder head bolts in the two rear lower bolt holes of the left cylinder head. Place a long cylinder head bolt in the rear lower bolt hole of the right cylinder head. Use rubber bands to keep the bolts in position until the cylinder heads are installed on the cylinder block.

11. Position new cylinder head gaskets on the cylinder block dowels. Do not apply sealer to the gaskets, heads, or block.

12. Place the cylinder heads on the block, guiding the exhaust manifold studs into the exhaust pipe connections. Install the remaining cylinder head bolts. The longer bolts go in the lower row of holes.

13. Tighten all the cylinder head attaching bolts in the proper sequence in three stages: 75 ft. lbs., 105 ft. lbs., and finally, to 135 ft. lbs. When this procedure is used, it is not necessary to retorque the heads after extended use.

14. Make sure that the oil holes in the pushrods are open and install the pushrods in their original positions. Place a dab of Lubriplate® to the ends of the pushrods before installing them.

15. Lubricate and install the valve rockers. Make sure that the pushrods remain seated in their lifters.

16. Connect the exhaust pipes to the exhaust manifolds.

17. Install the intake manifold and carburetor assembly. Tighten the intake manifold attaching bolts in the proper sequence to 25–30 ft. lbs.

18. Install the air conditioning compressor to the engine.

19. Install the power steering reservoir to the engine.

20. Apply oil-resistant sealer to one side of the new valve cover gaskets and lay the cemented side in place in the valve covers. Install the covers.

21. Install the alternator on the right cylinder head and adjust the alternator drive belt tension.

22. Adjust the air conditioning compressor drive belt tension.

23. Fill the radiator with coolant.

24. Start the engine and check for leaks.

420 (6.9L) V8 Diesel

1. Disconnect the ground cables from both batteries. Drain the cooling system.

2. Remove the radiator shroud halves.

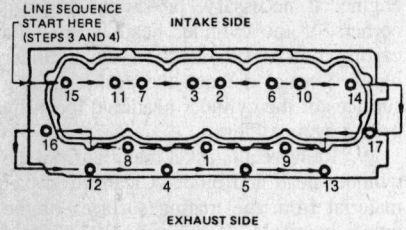

LINE SEQUENCE
START HERE
(STEPS 3 AND 4)
INTAKE SIDE

15 11 7   3   2   6   10 14
16   17
        8   1   9
    12   4   5   13

EXHAUST SIDE

420 (6.9L) diesel head bolt tightening sequence. Torque in 4 steps;

1. **Torque bolts to 40 ft. lbs. in numbered sequence.**
2. **Torque bolts to 65 ft. lbs. in numbered sequence.**
3. **Torque bolts to 75 ft. lbs. in line sequence.**
4. **Repeat step 3.**

Remove the fan and clutch assembly using Ford special tool T83T6312A and B or the equivalent. The attaching nut is equipped with a left hand thread, remove by turning clockwise.

3. Disconnect the alternator and fuel heater wiring harness from the alternator. Remove the alternator. Remove the vacuum pump.

4. Remove the fuel filter, cap all lines and fittings to prevent dirt from entering the system. Remove the alternator, vacuum pump and fuel filter mounting brackets, leave the fuel filter attached.

5. Remove the heater hose from the cylinder head. Remove the injector pump, cap all lines and fittings.

6. Remove the intake manifold and valley cover.

7. Raise and safely support the front of the vehicle with jackstands.

8. Disconnect the exhaust pipes from the exhaust manifolds.

9. Remove the clamp holding the engine oil dipstick tube in position on the right side cylinder head. Remove the bolt securing the transmission fluid dipstick tube to the rear of the cylinder head.

10. Lower the vehicle. Remove the engine oil dipstick and tube from the right side.

11. Remove the valve covers, rocker arms and pushrods. Identify all parts and keep in order for installation in the original positions.

12. Remove the injector nozzles and glow plugs.

13. Remove the cylinder head attaching bolts. Remove the cylinder heads from the engine.

—————— **CAUTION** ——————
The prechambers may fall out of the cylinder head on removal. Prevent damage to the prechambers.

14. Prechambers can be removed using a brass drift and suitable hammer.

15. Clean all gasket mounting surfaces. Clean and inspect the prechambers and ports for cracks. Apply a light coating of extra heavy duty grease to the mounting edge of the prechambers and install the prechambers in the cylinder head. Lightly tap with a plastic headed hammer if necessary.

16. Install the cylinder head in the reverse order of removal. Use care to prevent the prechambers from falling into the cylinder bores when installing the heads. Torque the mounting bolts in sequence. Step 1; 40 ft. lbs: Step 2; 65 ft. lbs: Step 3; 75 ft. lbs: Step 4; Repeat Step 3.

17. If necessary, purge the high pressure fuel lines of air by loosening the connector one half to one turn. Crank the engine until a solid flow of air free fuel comes out.

Timing Cover and Belt

REMOVAL & INSTALLATION

122 (2.0L) and 140 (2.3L)

The correct installation and adjustment of the camshaft drive belt is mandatory if the engine is to run properly. The camshaft controls the opening of the engine valves through coordination of the movement of the camshaft and the crankshaft. When any given piston is on the intake stroke the cor-

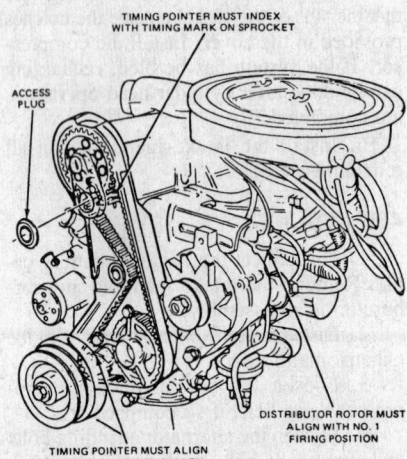

TIMING POINTER MUST INDEX
WITH TIMING MARK ON SPROCKET

ACCESS
PLUG

DISTRIBUTOR ROTOR MUST
ALIGN WITH NO. 1
FIRING POSITION

TIMING POINTER MUST ALIGN
WITH TDC MARK ON PULLEY

122 (2.0L), 140 (2.3L) timing mark alignment

responding intake valve must be open to admit air/fuel mixture into the cylinder. When the same piston is on the compression and power strokes, both valves in that cylinder must be closed. When the piston is on the exhaust stroke, the exhaust valve for that cylinder must be open. If the opening and closing of the valves is not coordinated with the movements of the pistons, the engine will run very poorly, if at all.

The camshaft drive belt also turns the engine auxiliary shaft. The distributor is driven by the engine auxiliary shaft. Since the distributor controls ignition timing, the auxiliary shaft must be coordinated with the camshaft and the crankshaft, because both valves in any given cylinder must be closed and the piston in that cylinder near the top of the compression stroke when the spark plug fires.

Due to this complex interrelationship between the camshaft, the crankshaft and the auxiliary shaft, the cogged pulleys on each component must be aligned when the camshaft drive belt is installed.

Should the camshaft drive belt jump timing by a tooth or two, the engine could still run; but very poorly. To visually check for correct timing of the crankshaft, auxiliary shaft, and the camshaft:

There is an access plug provided in the cam drive belt cover so that the camshaft timing can be checked without removing the drive belt cover.

1. Remove the access plug.
2. Turn the crankshaft until the timing marks on the crankshaft indicate TDC.
3. Make sure that the timing mark on the camshaft drive sprocket is aligned with the pointer on the inner belt cover. Also, the rotor of the distributor must align with the No. 1 cylinder firing position.

—————— **CAUTION** ——————
Never turn the crankshaft of any of the overhead cam engines in the opposite direction of normal rotation. Backward rotation of the crankshaft may cause the timing belt to slip and alter the timing.

4. To replace the timing belt, set the engine to TDC with No. 1 piston on the compression stroke. The crankshaft and camshaft timing marks should align with their respective pointers and the distributor rotor should point to the No. 1 plug tower.

5. Loosen the adjustment bolts on the alternator and accessories and remove the drive belts. To provide clearance for removing the camshaft belt, remove the fan and pulley.

6. Remove the belt outer cover.

7. Remove the distributor cap from the distributor and position it out of the way.

8. Loosen the belt tensioner adjustment and pivot bolts. Lever the tensioner away from the belt and retighten the adjustment bolt to hold it away.

9. Remove the crankshaft bolt and pulley. Remove the belt guide behind the pulley.

10. Remove the camshaft drive belt.

11. Install the new belt over the crankshaft pulley first, then counterclockwise over the auxiliary shaft sprocket and the camshaft sprocket. Adjust the belt fore and aft so that it is centered on the sprockets.

12. Loosen the tensioner adjustment bolt, allowing it to spring back against the belt.

13. Remove the spark plugs and rotate the crankshaft two complete turns in the normal rotation direction to remove any belt slack. Turn the crankshaft until the timing check marks are lined up. If the timing has slipped, remove the belt and repeat the procedure.

14. Tighten the tensioner adjustment bolt to 14–21 ft. lbs., and the pivot bolt to 28–40 ft. lbs.

15. Replace the belt guide and crank-shaft pulley, distributor cap, belt outer cover, fan and pulley, drive belts and accessories. Adjust the accessory drive belt tension. Start the engine and check the ignition timing.

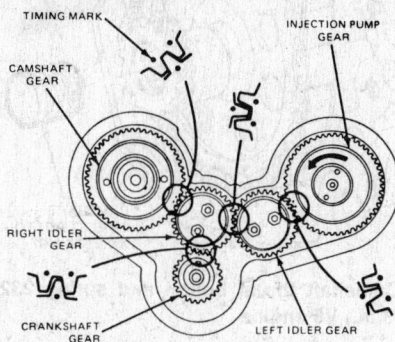

134 (2.2L) diesel timing mark alignment

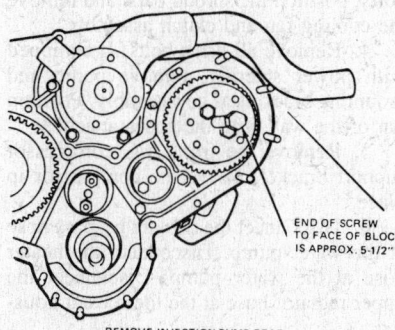

134 (2.2L) diesel injection pump gear removal

Timing Cover/Seal and Timing Chain/Gears

REMOVAL & INSTALLATION

134 (2.2L) Diesel

1. Bring the engine to No. 1 piston at TDC on the compression stroke.
2. Disconnect the ground cables from the batteries. Drain the cooling system.
3. Remove the radiator fan shroud and cooling fan. Drain the engine oil from the crankcase.
4. Loosen the idler pulley and remove the A/C compressor belt. Remove the power steering belt. Remove the power steering pump and mounting bracket, position out of the way with the hoses attached.
5. Loosen and remove the alternator and vacuum pump drive belts.
6. Remove the water pump. Using a suitable puller, remove the crankshaft pulley.
7. Remove the nuts and bolts retaining the timing case cover to the engine block. Remove the timing case cover.
8. Remove the engine oil pan.
9. Verify that all timing marks are aligned. Rotate the engine, if necessary, to align marks.
10. Remove the bolt attaching the camshaft gear and remove the washer and friction gear.
11. Remove the bolt attaching the injection pump gear and remove the washer and friction gear.
12. Install Ford tool T83T6306A or equivalent on to the camshaft drive gear and remove the gear. Attach the puller to the injection pump drive gear and remove the gear.
13. Remove the nuts attaching the idler gears after marking reference points on the idler gears for reinstallation position. Remove the idler gear assemblies.
14. Remove the nuts attaching the injection pump to the timing gear case. Support the injection pump in position.
15. Remove the bolts that attaching the timing gear case to the engine block and remove the case if necessary.
16. Clean all gasket mounting surfaces. Clean all parts, replace as necessary.
17. Remove the old oil seal from the front cover and replace.
18. Position the timing gear cover case with a new mounting gasket and install.
19. Install the timing gears as follows: Verify that the crankshaft and the right idler pulley timing marks align and install the right idler gear assembly. Install the camshaft gear so that the timing marks align with the timing mark on the right idler gear. Install the left idler gear assembly so that the timing marks align with the timing mark on the right idler gear. Install the injection pump gear so that the timing marks align with the timing mark on the left idler gear. Install all friction gears, washers, nuts and bolts on the gears.
20. Install the timing case covers using a new mounting gasket.
21. Install the remaining components in the reverse order of removal.

173 (2.8L) V6

FRONT COVER

1. Disconnect negative battery cable. Remove the oil pan.
2. Drain the coolant. Remove the radiator and any other parts to provide necessary clearance.
3. If equipped with air conditioning, unbolt the compressor and bracket and move them aside; do not disconnect the A/C lines.
4. Remove alternator, thermactor, and drive belt(s).
5. Remove the fan.
6. Remove the drive pulley from the crankshaft.
7. Remove the front cover retaining bolts. If necessary, tap cover lightly with a plastic hammer to break gasket seal. Remove front cover. If front cover plate gasket needs replacement, remove two screws and remove plate. If necessary, remove guide sleeves from cylinder block.
8. Clean the mating surfaces of gasket

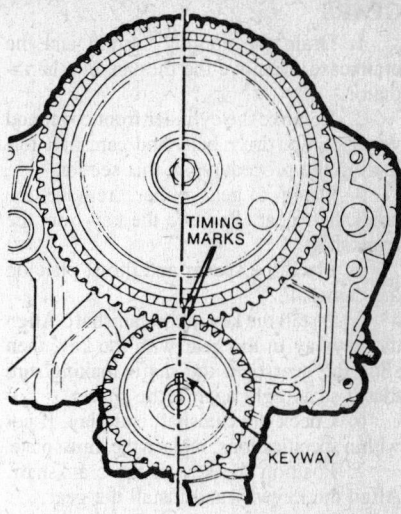

173 (2.3L) V6 timing mark alignment

material. Apply sealing compound to the gasket surfaces on the cylinder block and back side of the front cover plate. Position the gasket and front cover plate on cylinder block. Temporarily install four front cover screws to position the gasket and cover plate in place. Install and tighten two cover plate attaching bolts, then remove four screws that were temporarily installed.

9. If removed, fit new seal rings to the guide sleeves and, with no sealer used, insert the sleeves in the cylinder block with the chamfered side of the sleeve toward the front cover.
10. Apply sealing compound to front cover gasket surface. Place the gasket in position on front cover.
11. Place the front cover on the engine and start all retaining screws two or three turns. Center the cover by inserting tool in the oil seal.
12. Tighten the front cover attaching screws to specifications.
13. Install the belt drive pulley and tighten the attaching bolt to specifications.
14. Install the oil pan.
15. Install the water pump, water hose, A/C compressor, alternator and drive belt(s). Adjust the drive belt tension to specifications.
16. Fill the cooling system to the proper level with the specified coolant.
17. Operate the engine at fast idle speed and check for coolant and oil leaks.

If the guide sleeves were removed, install them with new seal rings but do not use sealing compound.

FRONT COVER SEAL

1. Support the front cover to prevent damage while driving out the seal.
2. Drive out the seal from the front cover.
3. Support the front cover to prevent damage while installing the seal.
4. Coat the new front cover oil seal with a light grease. Install the new seal in the front cover.

FORD

GEARS

1. Drain the cooling system and the crankcase. Remove the oil pan and the radiator.

2. Remove the cylinder front cover and water pump, drive belt, and camshaft following the procedures in this section.

3. Using a gear puller, remove the crankshaft gear. Remove the key from the crankshaft.

4. Place the spacer and thrust plate on the camshaft.

5. Install the key in the camshaft. Align the keyway in the gear with the key, then slide the gear onto the shaft, making sure that it seats tight against the spacer.

6. Check the camshaft end play. If not within specifications, replace the thrust plate.

7. Position the key in the crankshaft. Align the keyway and install the gear.

8. Install the cylinder front cover following the procedures in this section. Replace the oil pan and the radiator.

9. Fill the cooling system and crankcase.

10. Start the engine and adjust the ignition timing. Operate the engine at fast idle and check all hose connections and gaskets for leaks.

300 (4.9L)

1. Bring the engine to No. 1 piston at TDC (top dead center) on the compression stroke. Drain the cooling system. Disconnect negative battery cable.

2. Remove the radiator and shroud.

3. Remove the alternator adjusting arm bolt, loosen the drive belt and swing the alternator arm aside. Remove the fan, drive belts and pulleys.

4. Remove the screw and washer from the end of the crankshaft, remove the crankshaft damper.

5. Remove the front oil pan and front cover attaching screws.

Be careful not to get foreign material in the crankcase during service work, or the crankcase oil will have to be changed.

6. Remove the cylinder front cover and discard the gasket. It is a good idea to replace the crankshaft oil seal when the cylinder front cover is removed.

7. Drive out the crankshaft oil seal with a pin punch. Clean the seal bore in the cover.

8. Remove the camshaft and crankshaft gears using a suitable puller. Install the new gears, camshaft first using Ford tool T65L6306A or the equivalent. Do not hammer on the gears. Install the crankshaft gear over the drive key and install with tool. Verify that the timing marks on both gears are aligned. Install the crankshaft oil slinger.

9. Coat a new crankshaft oil seal with grease and install the seal in the cover. Drive the seal in until it is fully seated in the seal bore.

10. Cut the old front oil pan seal flush

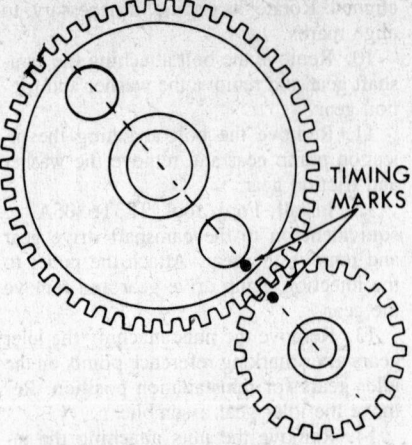

6-300 valve timing mark alignment

at the cylinder block/pan junction and remove the old seal material.

11. Clean all gasket surfaces.

12. Cut and fit a new pan seal flush to the cylinder block pan junction. Use the old seal as a pattern.

13. Coat the gasket surfaces of the block and cover with a resistant sealer. Position a new front cover gasket on the cylinder block.

14. Align the pan seal locating tabs with the pan holes. Pull the seal tabs through until the seal is completely seated. Apply a silicone sealer to the block/pan junction.

15. Position the front cover assembly over the end of the crankshaft and against the cylinder block. Start the cover and pan attaching screw. Slide the cover alignment tool over the crank stub and into the seal bore of the cover. Install the alternator adjusting arm, tighten all attaching screws to specification.

Tighten the oil pan screws first (compressing the pan seal) to obtain the proper alignment of the cover.

16. Lubricate the crank stub, damper hub I.D. and the seal rubbing surface with Lubriplate. Align the damper keyway with the key on the crankshaft and install the damper.

17. Install the washer and capscrew into the damper and tighten specification.

18. Install the pulleys, drive belts, and fan. Adjust all drive belts to correct tension.

19. Install the radiator and shroud. Connect all cooling system hoses.

20. Fill and bleed the cooling system. If no foreign material has entered the crankcase during service work, it is not necessary to change the engine oil.

21. Operate the engine at fast idle and check for coolant and oil leaks.

232 (3.8L) V6

1. Bring No. 1 piston to T.D.C. (top dead center) of the compression stroke. Drain the cooling system. Disconnect the negative battery cable.

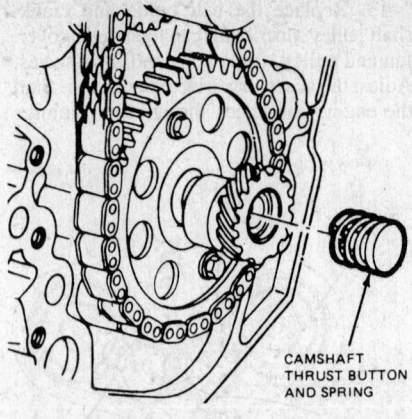

Camshaft thrust button and spring 232 (3.8L) V6 engine

2. Remove the air cleaner and duct assembly.

3. Remove the fan shroud attaching bolts, position the shroud back and remove the cooling fan and clutch assembly.

4. Remove all drive belts. If equipped with power steering, remove pump and mounting brackets as an assembly. Position out of the way with the hoses attached.

5. Remove the front A/C compressor support bracket. Leave the compressor in place.

6. Disconnect the coolant by-pass hose at the water pump. Disconnect the heater hose at the water pump. Disconnect the upper radiator hose at the thermostat housing.

7. Disconnect the coil wire from the distributor cap and remove the distributor cap with plug wires attached. Remove the distributor.

8. Raise and safely support on jackstands.

9. Remove the crankshaft belt drive and damper using the suitable puller.

10. Remove the fuel pump shield (if equipped). Disconnect and plug all lines to and from the fuel pump. Remove the fuel pump.

11. Remove the oil filter and disconnect the lower radiator hose at the water pump.

12. Remove the oil pan after draining the engine oil. The front cover cannot be removed without removing the oil pan.

13. Lower the car. Remove the front cover attaching bolts (the water pump need not be removed).

— **CAUTION** —

A front cover attaching bolt is located behind the oil filter mounting adapter, be sure the bolt is removed before attempting cover removal.

14. Remove the timing indicator and remove the front cover and water pump as an assembly.

15. Remove the camshaft thrust button and spring from the end of the camshaft.

16. Remove the camshaft gear retaining bolts.

17. Remove the camshaft gear, crank-

shaft gear and timing chain. If the crankshaft gear is difficult to remove, evenly pry on the sides of the gear with two small pry bars.

18. Clean all gasket mounting surfaces on the front cover, engine block and fuel pump mountings. Install a new crankshaft oil seal in the front cover.

19. Verify that the No. 1 piston is at TDC on the compression stroke. The keyway on the crankshaft should be at 12 o'-clock.

20. Lubricate the timing chain with engine oil. Install the two gears and chain as an assembly.

21. Make sure the timing marks on the gears are positioned across from each other. Install the camshaft retaining bolts.

22. Install the camshaft thrust button and spring. Lubricate the thrust button with Polyethylene Grease before installation. The thrust button and spring must be bottomed out in the camshaft seat and must not be allowed to fall out during cover installation.

23. Apply sealer to the gasket mounting surfaces, install a new front cover mounting gasket and the front cover using the two dowels for alignment. Position the timing indicator in place.

24. Apply pipe sealer to the cover mounting bolts before installation. Install bolts and tighten.

25. Install the remaining components in the reverse order of removal.

NOTE: When installing the fuel pump, turn the engine until the least resistance is encountered on the pump lever. Evenly tighten the mounting bolts, the cover is aluminum and the treads can be damaged.

255 (4.2L), 302 (5.0L) and 351 (5.8L) V8 Except Econoline

1. Bring the engine to No. 1 cylinder at TDC (top dead center) on the compression stroke. Disconnect the negative battery cable at the battery. Drain the cooling system.

2. Remove the fan shroud to radiator attaching bolts. Position shroud over the fan.

3. Disconnect the radiator lower hose, heater hose and by-pass hose at the water pump. Remove the drive belts, fan, fan spacer, and pulley.

4. Remove the fan shroud.

5. Loosen the alternator pivot bolt and bolt attaching the alternator adjusting arm to the water pump.

6. Remove the crankshaft pulley from the crankshaft vibration damper. Remove the damper attaching bolt and washer. Install a pulley, the vibration damper and remove the damper.

7. Disconnect the fuel pump outlet line from the fuel pump. Remove the fuel pump to one side with the flexible fuel line still attached.

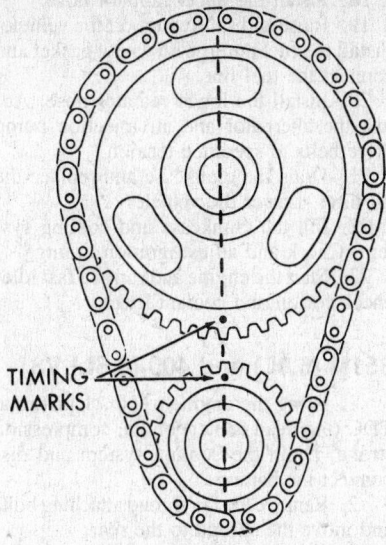

TIMING MARKS

V6 and V8 valve timing mark alignment

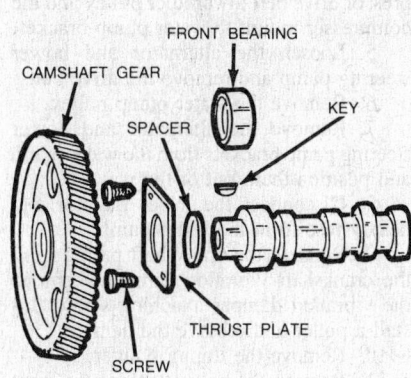

FRONT BEARING
CAMSHAFT GEAR
SPACER
KEY
THRUST PLATE
SCREW

Front end of the camshaft and related components

8. Remove the oil dipstick and the bolt attaching the dipstick to the exhaust manifold.

9. Remove the oil pan to cylinder front cover attaching bolts. Use a knife with a thin blade to cut the oil pan gasket flush with the cylinder block face prior to separating the cover from the cylinder block. Remove the cylinder front cover and water pump as an assembly.

10. Discard the cylinder front cover gasket. Remove the crankshaft front oil slinger.

11. Check the timing chain deflection. The method for checking timing chain deflection is outlined at the end of this section. If the deflection exceeds specification, replace the chain and sprockets as follows: Crank the engine until the timing marks on the sprockets are correctly aligned. Remove the camshaft sprocket capscrew, washers, and fuel pump eccentric. Slide both sprockets and the timing chain forward and remove the chain and sprockets as an assembly. Position the sprockets and timing chain on the camshaft. Be sure that the timing

marks are properly aligned. Install the fuel pump, eccentric, washers, and camshaft sprocket capscrew. Tighten the capscrew to specification.

12. Install the crankshaft front oil slinger.

13. Clean the cylinder front cover, oil pan and block gasket surfaces. Clean the oil pan gasket surface where the oil pan and front cover fasten.

14. Install a new crankshaft front oil seal.

15. Lubricate the timing chain and fuel pump eccentric with a heavy engine oil.

16. Coat the gasket surface of the oil pan with sealer, then cut and position the required sections of a new gasket on the oil pan and apply sealer at the corners. Install the pan seal as required. Coat the gasket surfaces of the block and cover with sealer, and position a new gasket on the block.

17. Position the cylinder front cover on the cylinder block. Use care when installing the cover to avoid seal damage or possible gasket dislocation.

18. Install the cylinder front cover to seal alignment tool.

19. It may be necessary to force the cover downward to slightly compress the pan gasket. This operation can be facilitated by using a suitable tool at the front cover attaching hole locations.

20. Coat the threads of the attaching bolts with a oil-resistant sealer and install the bolts. While pushing in on the alignment tool, tighten the oil pan to cover attaching bolts to specification. Tighten the cover to block attaching bolts to specification. Remove the alignment tool.

21. Apply Lubriplate or equivalent to the oil seal rubbing surface of the vibration damper inner hub to prevent damage to the seal. Apply a white lead and oil mixture to the front of the crankshaft for damper installation.

22. Line up the crankshaft vibration damper keyway with the key in the crankshaft. Install the vibration damper on the crankshaft. Install the capscrew and washer and tighten to specification. Install the crankshaft pulley.

23. Lubricate the fuel pump lever with heavy engine oil and install the pump using a new gasket. Connect the fuel pump outlet pipe.

24. Install the alternator pivot bolt and bolt attaching the alternator adjusting arm to the water pump.

25. Position the fan shroud over the water pump. Install the pulley, spacer and fan. Install and adjust the drive belts and adjust to specified tension. Connect the radiator, heater, and by-pass hoses. Position the fan shroud on the radiator and install the attaching bolts.

26. Fill and bleed the cooling system.

27. Run the engine at fast idle and check for coolant and oil leaks. Check the coolant level. Check and adjust the ignition timing.

28. Install the air cleaner and intake duct assembly including the crankcase ventilation hose.

FORD

302 (5.0L) and 351W (5.8L) V8 Econoline

1. Bring engine to No. 1 cylinder at TDC (top dead center) on the compression stroke. Disconnect the negative battery cable at the battery. Drain the radiator.

2. Remove the air conditioning idler pulley, bracket and drive belt if equipped.

3. Remove the upper radiator hose. Remove the fan and shroud as an assembly. Raise and safely support the vehicle.

4. Loosen the thermactor and alternator drive belts.

5. Disconnect the lower radiator hose at the water pump. Disconnect the fuel line at the fuel pump and remove the pump. Lower the vehicle.

6. Remove the by-pass hose. Remove the power steering pump drive belt if equipped. Remove the water pump pulley and disconnect the heater hose at the water pump.

7. Remove the air condition compressor upper bracket and the power steering pump mount.

8. Remove the crankshaft pulley. Remove the oil pan to front cover bolts. Remove the front cover.

9. Check timing chain deflection, as outlined at the end of this section. If the deflection exceeds specification, replace the chain and sprockets as follows: Crank the engine until the timing marks on the sprockets are correctly aligned. Remove the camshaft sprocket capscrew, washers, and fuel pump eccentric. Slide both sprockets and the timing chain forward and remove the chain and sprockets as an assembly. Position the sprockets and timing chain on the camshaft. Be sure that the timing marks are properly aligned. Install the fuel pump, eccentric, washers, and camshaft sprocket capscrew. Tighten the capscrew to specification.

10. Clean the front cover, fuel pump, and damper. Lubricate the crankshaft front seal. Clean the gasket surface at the pan and trim the gasket. Clean the front cover gasket surface at the block.

11. Replace the oil seal in the front cover. Position the gasket on the front cylinder cover. Apply a silicone sealer to the oil pan and cylinder block junction. Cut the pan gasket and position on pan and front cover.

12. Install the front cover, fuel pump, and crankshaft pulley.

13. Install the power steering pump and water pump by-pass hose. Connect the heater hose at the water pump.

14. Install the air conditioning compressor upper bracket, water pump pulley and power steering drive belt.

15. Install the alternator belt, thermactor belt, and fan/shroud assembly.

16. Adjust the power steering pump drive belt tension to specification.

17. Install the air conditioning drive belt idler pulley and bracket. Install the air conditioning drive belt and tighten to specification.

18. Install the upper radiator hose.

19. Raise and safely support the vehicle. Install the fuel pump with a new gasket and connect the fuel line.

20. Install the lower radiator hose. Adjust the alternator and air injection pump drive belts to specified tension.

21. Drain the crankcase and replace the oil filter. Lower the vehicle.

22. Fill the crankcase and cooling system. Check and adjust ignition timing.

23. Start the engine and run at a fast idle, check for oil and coolant leaks.

351M (5.8L) and 400 (6.6L) V8

1. Bring the engine to No. 1 piston at TDC (top dead center) on the compression stroke. Drain the cooling system and disconnect the battery.

2. Remove the fan shroud attaching bolts and move the shroud to the rear.

3. Remove the fan and spacer from the water pump shaft.

4. Remove the air conditioner compressor drive belt lower idler pulley and the compressor mount to water pump bracket.

5. Loosen the alternator and power steering pump and remove the drive belts.

6. Remove the water pump pulley.

7. Remove the alternator and power steering pump brackets from the water pump and position them out of the way.

8. Disconnect the lower radiator and heater hose from the water pump.

9. Remove the crankshaft pulley from the crankshaft vibration damper. Remove the vibration damper attaching screw. Install a puller and remove the damper.

10. Remove the timing pointer.

11. Remove the bolts attaching the front cylinder cover to the cylinder block. Remove the front cover and water pump assembly.

12. Disconnect the fuel pump outlet line from the pump. Remove the fuel pump attaching bolts and lay the pump to one side with the flexible line still attached.

13. Discard the cylinder front cover gasket and oil pan seal.

14. Check the timing chain deflection, as outlined at the end of this section.

15. If the timing chain deflection exceeds specification, proceed as follows: Crank the engine until the timing marks on the sprockets are aligned. Remove the camshaft sprocket capscrew, washer, and two piece fuel pump eccentric. Slide both sprockets and the timing chain forward, and remove them as an assembly. Position the sprockets and timing chain on the camshaft and crankshaft. Be certain that the timing marks on the sprockets are correctly aligned. Install the two piece fuel pump eccentric, washers, and camshaft sprocket capscrew. Tighten the camshaft capscrew to specification. Make sure that the outer fuel pump eccentric sleeve rotates freely.

16. Coat a new fuel pump gasket with oil resistant sealer and position the fuel pump and gasket on the cylinder block with the fuel pump arm resting on the eccentric outer sleeve. Install the pump attaching bolt and nut and tighten to specification. Connect the fuel pump outlet line.

17. Remove the front crankshaft seal from the front cover. Clean the cylinder front cover and the engine block gasket surfaces.

18. Coat the gasket surfaces of the block and cover with sealer, and position a new gasket on the cylinder block alignment dowels.

19. Position the cylinder front cover and water pump assembly on the cylinder block alignment dowels.

20. Coat the threads of the attaching bolts with an oil resistant sealer and install the timing pointer and attaching bolts. Tighten the bolts to specifications.

21. Install the front cover oil seal into the cylinder front cover.

22. Apply Lubriplate® or its equivalent to the oil seal rubbing surface of the vibration damper inner hub to prevent damage to the seal. Apply a white lead and oil mixture to the front of the crankshaft for damper installation.

23. Line up the crankshaft vibration damper keyway with the key on the crankshaft. Install the vibration damper on the crankshaft by pressing on with appropriate tool. Install the capscrew and washer, tighten to specification. Install the crankshaft pulley.

24. Connect the heater hose and the lower radiator hose to the water pump.

25. Install the air conditioner compressor to water pump bracket and lower idler pulley.

26. Position the alternator bracket and power steering pump bracket on the water pump and install the bolts.

27. Position the water pump pulley on the water pump shaft and install the drive belts.

28. Place the fan shroud over the pulley, and install the fan and spacer.

29. Position the fan shroud over the radiator and install the attaching bolts.

30. Adjust the drive belts to specification.

31. Raise the vehicle and remove the oil pan and install new gasket and seals as described in "Oil Pan Removal and Installation."

32. Lower the vehicle. Fill the crankcase. Fill and bleed the cooling system. Connect the battery cable.

33. Operate the engine until normal operating temperature has been reached and check for oil or coolant leaks.

460 (7.5L) V8

1. Bring the engine to No. 1 piston at TDC (top dead center) on the compression stroke. Drain the cooling system and crankcase.

2. Remove the radiator shroud and fan.

3. Disconnect the upper and lower radiator hoses, and the automatic transmission oil cooler lines from the radiator.

4. Remove the radiator upper support and remove the radiator.

5. Loosen the alternator attaching bolts and air conditioning compressor idler pulley and remove the drive belts with the water pump pulley. Remove the bolts attaching the compressor support to the water pump and remove the bracket (support), if so equipped.

6. Remove the crankshaft pulley from the vibration damper. Remove the bolt and washer attaching the crankshaft damper and remove the damper with a puller. Remove the Woodruff key from the crankshaft.

7. Loosen the by-pass hose at the water pump, and disconnect the heater return tube at the water pump.

8. Disconnect and plug the fuel inlet and outlet lines at the fuel pump, and remove the fuel pump.

9. Remove the bolts attaching the front cover to the cylinder block. Cut the oil pan seal flush with the cylinder block face with a thin knife blade prior to separating the cover from the cylinder block. Remove the cover and water pump as an assembly. Discard the front cover gasket and oil pan seal.

10. Transfer the water pump if a new cover is going to be installed. Clean all of the gasket sealing surfaces on both the front cover and the cylinder block.

11. Check the timing chain deflection, as outlined at the end of this section. If timing chain deflection exceeds specification, proceed as follows: Crank the engine until the timing marks on the sprockets are aligned. Remove the camshaft sprocket capscrew, washer, and two piece fuel pump eccentric. Slide both sprockets and the timing chain forward, and remove them as an assembly. Position the sprockets and timing chain on the camshaft and crankshaft. Be certain that the timing marks on the sprockets are correctly aligned. Install the two piece fuel pump eccentric, washers, and camshaft sprocket capscrew. Tighten the camshaft capscrew to specification.

12. Coat the gasket surface of the oil pan with sealer. Cut and position the required sections of a new seal on the oil pan. Apply sealer to the corners.

13. Coat the gasket surfaces of the cylinder block and cover with sealer and position the new gasket on the block.

14. Position the front cover on the cylinder block. Use care not to damage the seal and gasket or mislocate them.

15. Coat the front cover attaching screws with sealer and install them.

It may be necessary to force the front cover downward to compress the oil pan seal in order to install the front cover attaching bolts. Use a drift to engage the cover screw holes through the cover and pry downward.

16. Assemble and install the remaining components in the reverse order of removal. Tighten the front cover bolts to 15–20 ft. lbs., the water pump attaching screws to 12–15 ft. lbs., the crankshaft damper to

70–90 ft. lbs., the crankshaft pulley to 35–50 ft. lbs., fuel pump to 19–27 ft. lbs., the oil pan bolts to 9–11 ft. lbs. for the $5/16$ inch screws to 7–9 ft. lbs. for the $1/4$ inch screws, and the alternator pivot bolt to 45–50 ft. lbs.

420 (6.9L) Diesel

FRONT COVER AND OIL SEAL

1. Disconnect the battery ground cables from both batteries. Drain cooling system. Remove the air cleaner and install intake air opening.

2. Remove the radiator fan shroud halves. Remove the fan and clutch assembly using Tool T83T–6312–A and B. Left hand thread. Remove by turning nut clockwise.

3. Remove the injection pump and adapter. Remove the water pump. Remove power steering pump and bracket, place out of the way with hoses attached. Remove A/C compressor front mounting bracket. Remove alternator, if necessary.

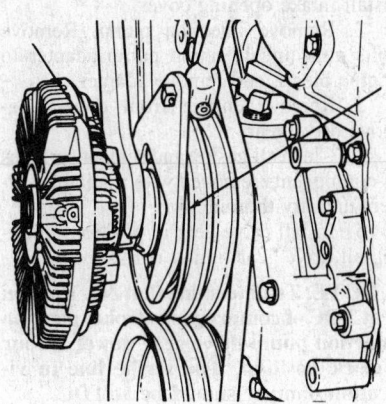

Fan clutch removal

NOTE: If A/C compressor is in the way, remove with hoses attached and position out of the way.

4. Raise vehicle and safely support on jackstands.

5. Remove crankshaft pulley and vibration damper. Remove ground cables at front of engine.

6. Remove five bolts attaching front cover to engine block and oil pan.

7. Lower vehicle.

8. Remove the bolts attaching engine front cover to engine block, and remove front cover.

9. Support engine front cover, and using an arbor press and suitable driver, drive crankshaft seal out of front cover.

10. Remove old gasket material and clean engine block, engine front cover, and oil pan sealing surfaces with a suitable solvent and dry throughly.

11. Clean water pump sealing surface.

12. Coat new front crankshaft oil seal with Polyethylene grease. Install new oil seal using front crankshaft seal replacer, a suitable spacer, and an arbor press.

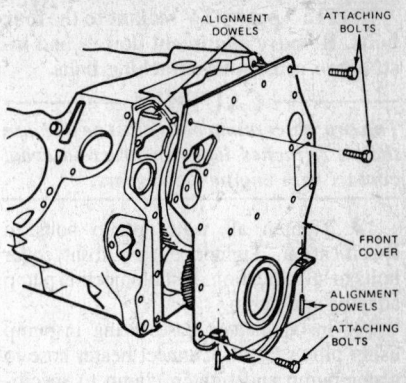

Front cover installation on the 420 (6.9L) diesel

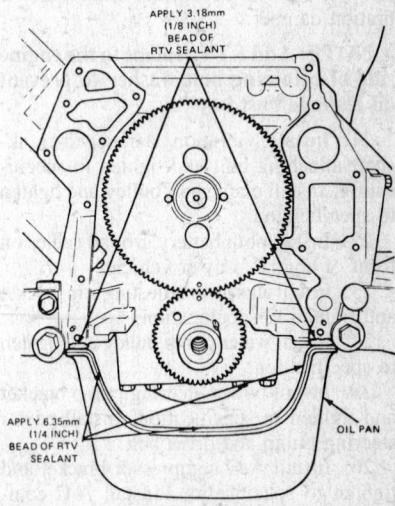

Front cover sealant application 420 (6.9L) diesel

——— CAUTION ———
Support engine front cover.

13. Bottom out tool on front cover surface. Seal is automatically installed at proper depth.

14. Install fabricated alignment dowels, on engine block to align front cover and gaskets. Apply gasket sealer on engine block sealing surfaces. Install gaskets on engine block.

NOTE: RTV Sealant should be applied immediately prior to front cover installation.

15. Apply a $1/8$ inch bead of RTV sealant on front of engine block.

16. Apply a $1/4$ inch bead of RTV sealant on oil pan mounting surface. Install engine front cover in position and install three attaching bolts. Remove alignment dowels from engine and oil pan and install and hand tighten remaining front cover bolts.

17. Install fabricated alignment dowels in engine block, if necessary. Install water pump gasket on engine front cover alignment dowels. Install water pump and hand tighten bolts.

NOTE: Apply RTV sealant to the four bolts. Remove alignment dowels and install two remaining attaching bolts.

— CAUTION —

Top two water pump bolts must be no more than 1¼ inches long, in order to avoid contact with engine drive gears.

18. Tighten all water pump bolts to specification. Tighten engine front cover bolts to specification. Install injection pump adaptor and pump.

19. Install heater hose fitting in pump using pipe sealant. Connect heater hose to water pump and tighten clamp to specification.

20. Raise vehicle and safely support on jackstands. Lubricate damper seal nose with clean engine oil and install crankshaft vibration damper.

NOTE: Add RTV sealant to the engine side of retaining bolt washer to prevent oil leakage past keyway.

21. Install vibration damper-to-crankshaft attaching bolt and tighten to specification. Install crankshaft pulley and tighten to specification.

22. Install both battery ground cables on front of engine. Lower vehicle.

23. Install alternator adjusting arm bracket and tighten to specification.

24. Install water pump pulley and tighten to specification.

25. Install power steering pump bracket and tighten to specification. Install power steering pump and drive belt.

26. Install A/C compressor bracket and tighten to specification. Install A/C compressor and drive belt.

27. Install alternator adjusting arm and install alternator and vacuum pump drive belts.

28. Adjust alternator, vacuum pump, power steering pump and A/C compressor drive belts.

29. Refill and bleed cooling system.

30. Connect battery ground cables to both batteries.

31. Remove intake manifold cover, install air cleaner and tighten to specification.

32. Run engine and check for coolant and oil leaks.

33. Install fan and clutch assembly. Install radiator fan shroud halves.

CRANKSHAFT DRIVE GEAR

1. Complete front cover removal procedures.

2. Install crankshaft drive gear remover Tool T83T–6316–A, and using a breaker bar to prevent crankshaft rotation, or flywheel holding Tool T74R–6375–A, remove crankshaft gear.

3. Install crankshaft gear using Tool T83T–6316–B aligning crankshaft drive gear timing mark with crankshaft drive gear timing mark.

NOTE: Gear may be heated to 300–350°F for ease of installation. Heat in oven. Do not use torch.

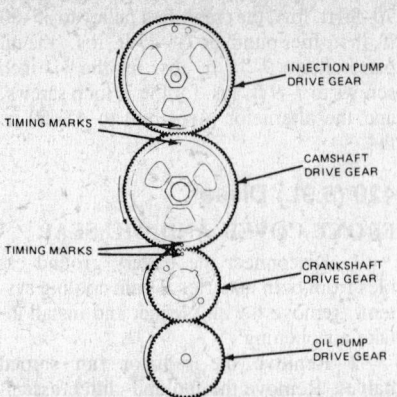

Timing mark alignment 420 (6.9L) diesel

4. Complete front cover installation procedures.

INJECTION PUMP DRIVE GEAR AND ADAPTER

1. Disconnect battery ground cables from both batteries. Remove air cleaner and install intake opening cover.

2. Remove injection pump. Remove bolts attaching injection pump adapter to engine block, and remove adapter.

3. Remove engine front cover. Remove drive gear.

4. Clean all gasket and sealant surfaces of components removed with a suitable solvent and dry thoroughly.

5. Install drive gear in position, aligning all drive gear timing marks.

NOTE: To determine that No. 1 piston is at TDC of compression stroke, position injection pump drive gear dowel at four o'clock position. The scribe line in vibration damper should be at TDC.

— CAUTION —

Use extreme care to avoid disturbing injection pump drive gear, once it is in position.

6. Install engine front cover. Apply a ⅛ inch bead of RTV Sealant along bottom surface or injection pump adapter.

NOTE: RTV should be applied immediately prior to adapter installation.

7. Install injection pump adaptor. Apply sealer to bolt threads before assembly.

NOTE: With injection pump adapter installed the injection pump drive gear cannot "jump" timing.

8. Install all removed components. Run engine and check for leaks.

NOTE: If necessary, purge the high pressure fuel lines of air by loosening the connector one half to one turn and crank the engine until a solid flow of fuel, free of air bubbles flow from the connection.

CAMSHAFT DRIVE GEAR, FUEL PUMP CAM, SPACER AND THRUST PLATE

1. Complete front cover removal procedures.

2. Remove camshaft allen screw.

3. Install gear puller, Tool T83T–6316–A and remove gear. Remove fuel supply pump, if necessary.

4. Install gear puller, Tool T77E–4220–B and shaft protector T83T–6316–A and remove fuel pump cam and spacer, if necessary.

5. Remove bolts attaching thrust plate, and remove thrust plate, if necessary.

6. Install new thrust plate, if removed.

7. Install spacer and fuel pump cam against camshaft thrust flange, using installation sleeve and replacer Tool T83T–6316–B, if removed.

8. Install camshaft drive gear against fuel pump cam, aligning timing mark with timing mark on crankshaft drive gear, using installation sleeve and replacer Tool T83T–6316–B.

9. Install camshaft allen screw and tighten to specification.

10. Install fuel pump, if removed.

11. Install front cover, following previous procedure.

CHECKING TIMING CHAIN DEFLECTION EXCEPT 232 (3.8L) V6 ENGINE

To measure timing chain deflection, rotate crankshaft clockwise to take up slack on the left side of chain. Choose a reference point and measure distance from this point and the chain. Rotate crankshaft in the opposite direction to take up slack on the right side of the chain. Force the left (slack) side of the chain out and measure the distance to the reference point chosen earlier. The difference between the two measurements is the deflection.

Timing chain should be replaced if deflection measurement exceed specified limit. The deflection measurement should not exceed ½ inch.

TIMING CHAIN DEFLECTION 232 (3.8L) V6 ENGINE

1. Remove the right valve rocker arm cover.

2. Loosen the No.3 exhaust valve rocker arm and rotate it to one side.

3. Install a dial indicator on the end of the push rod, using proper adapter tools.

4. Turn the crankshaft clockwise until the No. 1 piston is at TDC. The damper mark should point to TDC on the timing degree indicator. This will also take up the slack on the right side of the chain.

5. Zero the dial indicator needle.

6. Turn the crankshaft slowly counterclockwise until the slightest movement is seen on the dial indicator. Stop and observe the damper timing mark for the number of degrees of travel from TDC.

7. If the reading on the timing degree indicator exceeds 6°, replace the timing chain and sprockets.

CAMSHAFT ENDPLAY MEASUREMENT

The camshaft gears used on some engines are easily damaged if pried upon while the valve train load is on the camshaft. Loosen rocker arm nuts or rocker arm shaft support bolts before checking camshaft endplay.

Push camshaft toward rear of engine, install and zero a dial indicator, then pry between camshaft gear and block to pull the camshaft forward. If endplay is excessive, check for correct installation of spacer. If spacer is installed correctly, then replace thrust plate.

MEASURING TIMING GEAR BACKLASH

Use a dial indicator installed on block to measure timing gear backlash. Hold gear firmly against the block while making measurement. If excessive backlash exists, replace both gears.

Camshaft

REMOVAL & INSTALLATION

NOTE: When installing the camshaft refer to the proceeding section for gear alignment.

122 (2.0L) and 140 (2.3L)

NOTE: The following procedure covers camshaft removal and installation with the cylinder head on or off the engine. If the cylinder head has been removed start at Step 9.

1. Drain the cooling system. Remove the air cleaner assembly and disconnect the negative battery cable.
2. Remove the spark plug wires from the plugs, disconnect the retainer from the valve cover and position the wires out of the way. Disconnect rubber vacuum lines as necessary.
3. Remove all drive belts. Remove the alternator mounting bracket-to-cylinder head mounting bolts, position bracket and alternator out of the way.
4. Disconnect and remove the upper radiator hose. Disconnect the radiator shroud.
5. Remove the fan blades and water pump pulley and fan shroud. Remove cam belt and valve covers.
6. Align engine timing marks at TDC for No. 1 cylinder. Remove cam drive belt.
7. Jack up the front of the vehicle and support on jackstands. Remove the front motor mount bolts. Disconnect the lower radiator hose from the radiator. Disconnect and plug the automatic transmission cooler lines.
8. Position a piece of wood on a floor jack and raise the engine carefully as far as it will go. Place blocks of wood between the engine mounts and crossmember pedestals.

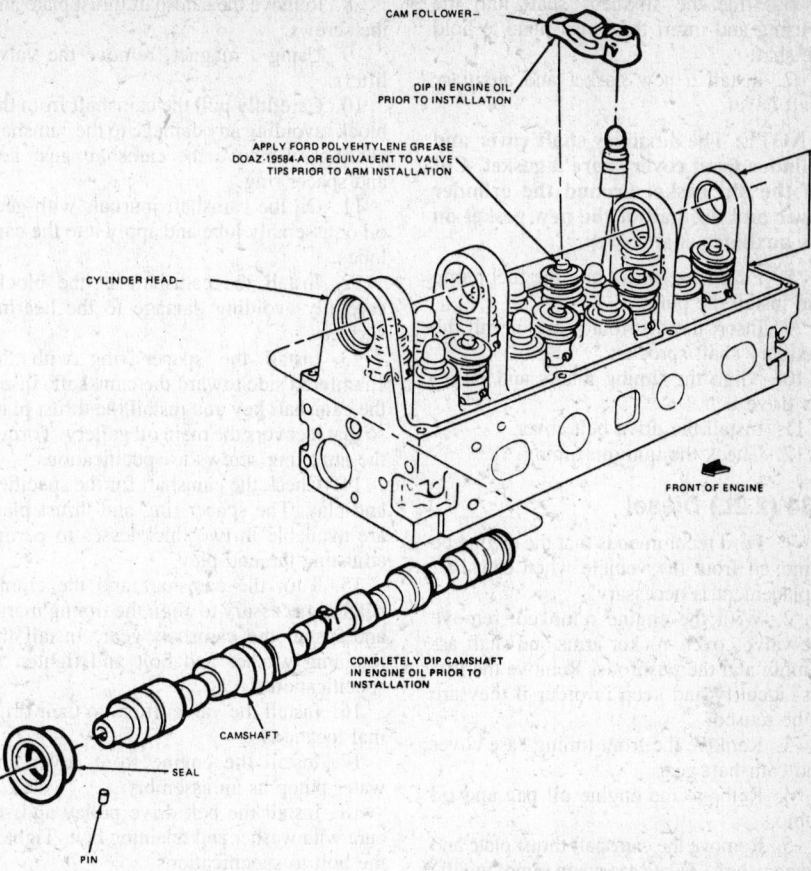

Camshaft removal/installation 122 (2.0L), 140 (2.3L) 4 cylinder

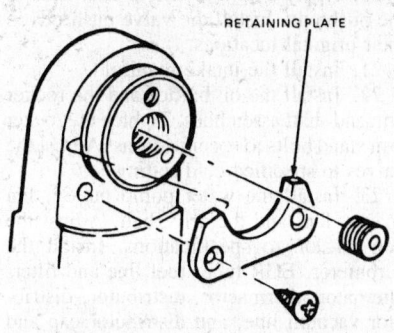

Camshaft retainer plate-122 (2.0L), 140 (2.3L) 4 cylinder

9. Remove the rocker arms.
10. Remove the camshaft drive gear and belt guide using a suitable puller. Remove the front oil seal with a sheet metal screw and slide hammer.
11. Remove the camshaft retainer located on the rear mounting stand by unbolting the two bolts.
12. Remove the camshaft by carefully withdrawing toward the front of the engine. Caution should be used to prevent damage to cam bearings, lobes and journals.
13. Check the camshaft journals and lobes for wear. Inspect the cam bearings, if worn (unless the proper bearing installing tool is on hand), the cylinder head must be removed for new bearings to be installed by a machine shop.

14. Cam installation is in the reverse order of removal. See following notes.

NOTE: Coat the camshaft with heavy SF oil before sliding it into the cylinder head. Install a new front seal. Apply a coat of sealer or teflon tape to the cam drive gear bolt before installation.

— CAUTION —

After any procedure requiring removal of the rocker arms, each lash adjuster must be fully collapsed after assembly, then released. This must be done before the camshaft is turned.

Auxiliary Shaft

1. Remove the camshaft drive belt cover.
2. Remove the drive belt. Remove the auxiliary shaft sprocket. A puller may be necessary to remove the sprocket.
3. Remove the distributor and fuel pump.
4. Remove the auxiliary shaft cover and thrust plate.
5. Withdraw the auxiliary shaft from the block.

NOTE: The distributor drive gear and the fuel pump eccentric on the auxiliary shaft must not be allowed to touch the auxiliary shaft bearings during removal and installation. Completely coat the shaft with oil before sliding it into place.

6. Slide the auxiliary shaft into the housing and insert the thrust plate to hold the shaft.

7. Install a new gasket and auxiliary shaft cover.

NOTE: The auxiliary shaft cover and cylinder front cover share a gasket. Cut off the old gasket around the cylinder cover and use half of the new gasket on the auxiliary shaft cover.

8. Fit a new gasket into the fuel pump and install the pump.

9. Insert the distributor and install the auxiliary shaft sprocket.

10. Align the timing marks and install the drive belt.

11. Install the drive belt cover.

12. Check the ignition timing.

134 (2.2L) Diesel

1. Ford recommends that the engine be removed from the vehicle when camshaft replacement is necessary.

2. With the engine removed; remove the valve cover, rocker arms and shaft assembly and the pushrods. Remove the lifters, identify and keep in order if they are to be reused.

3. Remove the front timing case cover and camshaft gear.

4. Remove the engine oil pan and oil pump.

5. Remove the camshaft thrust plate and the camshaft. Take care when removing the camshaft not to damage lobes or bearings.

6. Apply oil to the camshaft bearings and bearing journals. Apply Polyethylene grease to the camshaft lobes and install the camshaft into the engine.

7. Reinstall components in the reverse order of removal.

173 (2.8L) V6

1. Disconnect the negative battery cable from the battery. Bring engine to No. 1 cylinder at TDC (top dead center) on the compression stroke. Drain the coolant and remove the radiator, fan, spacer, water pump pulley and the drive belt.

2. Remove the distributor cap with spark plug wires as an assembly. Remove the distributor vacuum line, distributor, alternator, thermactor, rocker arm covers, fuel line and filter, carburetor, EGR tube, and intake manifold. Remove the spark plug wire boots.

3. Drain the crankcase. Remove the rocker arm and the shaft assemblies. Lift out the pushrods and place in a marked rack so they can be reinstalled in the same location.

4. Remove the oil pan.

5. Remove the drive sprocket attaching bolt and slide the sprocket off the end of the shaft.

6. Remove the engine front cover and water pump as an assembly.

7. Remove the camshaft gear retaining bolt and slide the gear off the camshaft.

8. Remove the camshaft thrust plate and the screws.

9. Using a magnet, remove the valve lifters.

10. Carefully pull the camshaft from the block, avoiding any damage to the camshaft bearings. Remove the camshaft gear key and spacer ring.

11. Oil the camshaft journals with gear oil or assembly lube and apply it to the cam lobes.

12. Install the camshaft in the block, carefully avoiding damage to the bearing surfaces.

13. Install the spacer ring with the chamfered side toward the camshaft. Insert the camshaft key and install the thrust plate so that it covers the main oil gallery. Torque the attaching screws to specifications.

14. Check the camshaft for the specified end-play. The spacer ring and thrust plate are available in two thicknesses to permit adjusting the end-play.

15. Turn the camshaft and the crankshaft as necessary to align the timing marks and install the camshaft gear. Install the retaining washer and bolt and tighten to specifications.

16. Install the valve lifters to their original locations.

17. Install the engine front cover and water pump as an assembly.

18. Install the belt drive pulley and secure with washer and retaining bolt. Tighten the bolt to specifications.

19. Install the oil pan.

20. Apply a light grease to both ends of the pushrods. Install the valve pushrods in their original locations.

21. Install the intake manifold.

22. Install the oil baffles and the rocker arm and shaft assemblies. Tighten the rocker arm stand bolts to specifications. Adjust the valves to specified cold setting.

23. Install the water pump pulley, fan spacer, fan, and the drive belt. Adjust the belt tension to specifications. Install the carburetor, EGR tube, fuel line and filter, alternator, thermactor, distributor, distributor vacuum line, and distributor cap and wires. Install the radiator. Fill the cooling system to the proper level with the specified coolant. Adjust the ignition timing.

24. Install the rocker arm covers.

25. Run the engine and check the idle speed.

26. Run the engine at fast idle speed and check for coolant and oil leaks.

300 (4.9L)

1. Drain the cooling system. Disconnect the negative battery cable. Remove the radiator shroud and radiator. On some models it may be necessary to remove the grille and radiator support for necessary clearance.

2. Remove the front cover following procedure described in Front Cover Removal and Installation.

3. Remove air cleaner and crankcase vent tube at the rocker cover.

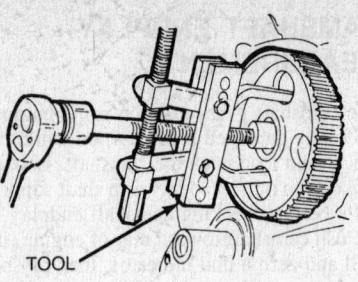

Camshaft gear removal 6-300 engine

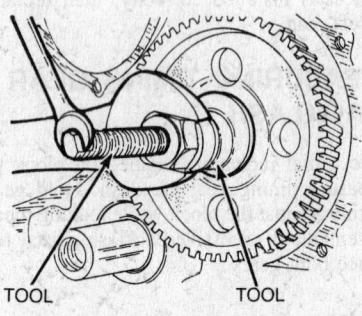

6-300 camshaft sprocket installation

4. Disconnect accelerator cable, choke cable and hand throttle cable (if so equipped). Remove accelerator cable retracting spring.

5. If applicable, remove air compressor and power steering belts.

6. Disconnect oil filler hose from rocker cover.

7. Remove distributor cap and wiring as an assembly, then disconnect vacuum line and primary wire and remove distributor.

8. Remove fuel pump.

9. Remove valve rocker cover, loosen rocker arm stud nuts and move rocker arms to one side. Remove push rods, identifying each so that they may be installed in their original locations.

10. Remove push rod cover and valve lifters, identifying the position of each.

11. Turn crankshaft to align timing marks, remove camshaft thrust plate bolts and carefully pull camshaft and gear from block. Metal camshaft gear (300 HD) is bolted onto camshaft and fiber gear (300 LD) is pressed on and must be removed with an arbor press.

12. To install camshaft, oil journals and apply Lubriplate to lobes, then carefully install camshaft, spacer, thrustplate and gear as an assembly, making sure timing marks are aligned, then tightening thrustplate bolts to 19–20 ft. lbs. Do not rotate crankshaft until distributor is installed.

13. Install front cover, referring to Front Cover Removal and Installation for correct procedure.

14. Install valve lifters, then the pushrods in their original locations. Apply heavy engine oil to the lifters and Lubriplate to the pushrods.

15. Install in order the following com-

ponents, referring to appropriate sections by topic for detailed instructions if necessary and using new gaskets with sealer: pushrod cover, valve rocker cover (adjust valve lash first), distributor (rotor in No. 1 cylinder firing position), fuel pump, distributor cap and wiring assembly, crankcase ventilation valve (in rocker cover), oil filler hose, accelerator cable and retracting spring, choke cable, hand throttle cable, front cylinder cover, water pump pulley, fan, belt, air compressor and power steering belts, radiator, hood latch, grill and air cleaner.

16. Fill crankcase if drained.

17. Install grille, support, radiator, hoses, etc. Fill and bleed cooling system, checking for leaks.

18. Set the ignition timing, then connect distributor vacuum line.

19. Adjust carburetor idle speed and idle fuel mixture.

232 (3.8L) V6 and V8's Except 420 (6.9L) Diesel

1. Disconnect the negative battery cable. Drain cooling system. Bring engine to No. 1 piston at TDC. Remove the intake manifold and valley pan, if so equipped.

2. Remove the rocker covers, loosen the rockers on their pivots and remove the pushrod. The pushrods must be reinstalled in their original positions.

3. Remove the valve lifters in sequence with a magnet. They must be replaced in their original positions.

4. Remove the timing gear cover and timing chain and sprockets.

5. In addition to the radiator and air conditioning condenser, if so equipped, it may be necessary to remove the front grille assembly and the hood lock assembly to gain the necessary clearance to slide the camshaft out the front of the engine.

6. Remove the camshaft thrust plate attaching screws and carefully slide the camshaft out of its bearing bores. Use extra caution not to scratch the bearing journals with the camshaft lobes.

7. Install the camshaft in the reverse order of removal. Coat the camshaft with engine oil liberally before installing it. Slide the camshaft into the engine very carefully so as not to scratch the bearing bores with the camshaft lobes. Install the camshaft thrust plate and tighten the attaching screws to 9–12 ft. lbs. Measure the camshaft end-play. If the end-play is more than 0.009 in., replace the thrust plate. Assemble the remaining components in the reverse order of removal.

420 (6.9L) Diesel

1. Remove engine from vehicle. Attach engine to stand or suitable device. Remove injection pump and adapter, intake manifold and tappets, engine front cover and fuel supply pump.

2. Remove camshaft drive gear, fuel

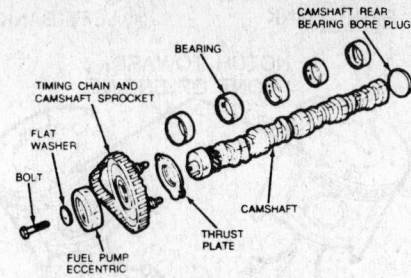

V8 camshaft and related parts

supply pump cam, spacer and thrust plate from the camshaft.

3. Carefully remove camshaft with Ford Tool T65L–6250–A and adapter, 14–0314, by pulling toward front of engine. Use care to avoid damaging camshaft bearings.

4. Camshaft lobes are to be coated with Polyethylene Grease and journals lubricated with recommended quality engine oil before installation.

5. Oil camshaft journal and apply Polyethylene Grease to lobes. Carefully slide camshaft through bearings. Install new camshaft thrust plate onto cylinder block and tighten to specification.

6. Install spacer and fuel pump cam against camshaft thrust flange using installation sleeve and replacer Ford Tool T83T–6316–B.

7. Install camshaft drive gear against fuel pump cam, aligning timing mark with timing mark on crankshaft drivegear using installation sleeve and replacer Ford Tool T83T–6316–B.

8. Install camshaft allen screw and tighten.

9. Install fuel supply pump.

10. Install new crankshaft oil seal in engine front cover. Install engine front cover.

11. Install water pump. Install injection pump adapter.

12. Lubricate tappets and bores with recommended quality engine oil and install tappets in their original positions. Install tappet guides. Install tappet guide retainer.

13. Position pushrods, copper colored ends toward rocker arms, into their respective tappets making sure they are seated fully in pushrod seats. Install rocker arms and valve covers with new gaskets.

14. Install intake manifold.

15. Install injection pump.

16. Install engine in vehicle.

Pistons and Connecting Rods

REMOVAL & INSTALLATION

Refer to the Engine Rebuilding Chapter in the Unit Repair Section for instructions.

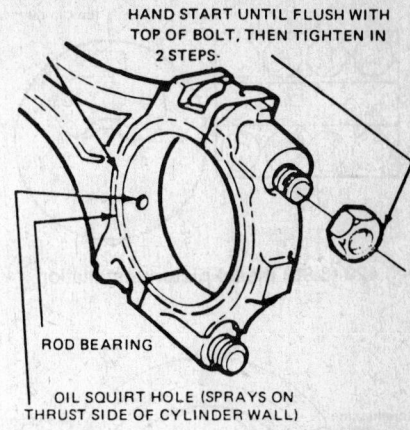

122 (2.0L), 140 (2.3L) engines-the piston is installed with the notches toward the front of the engine and the oil squirt hole in the connecting rod as shown

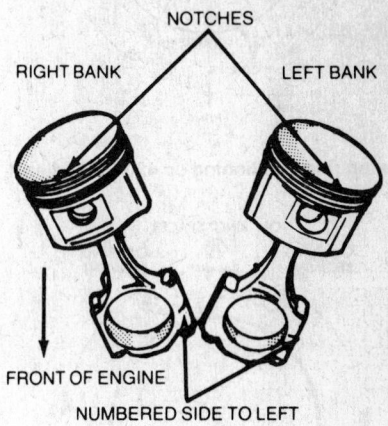

173 (2.8L) V6 piston installation

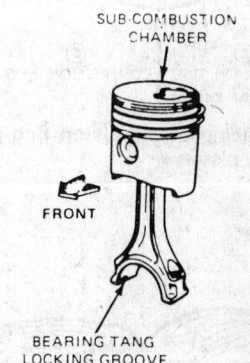

134 (2.2L) diesel piston installation

RING POSITIONING

An illustration is provided for proper ring end gap location. Position the rings as shown prior to piston installation. If the instruction sheet that generally comes with a set of new piston rings differs, install as per instruction sheet.

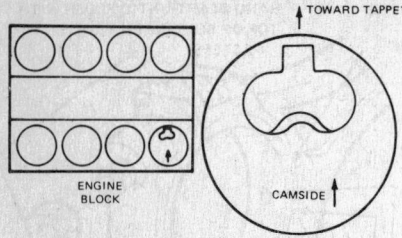

420 (6.9L) diesel piston installation

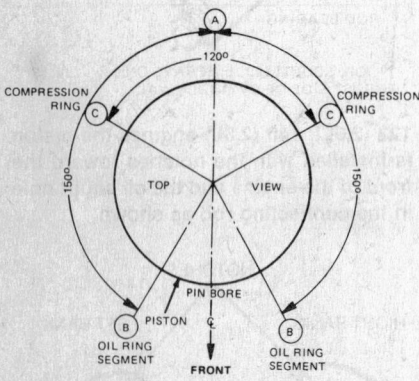

Piston ring positioning on 420 (6.9L) diesel

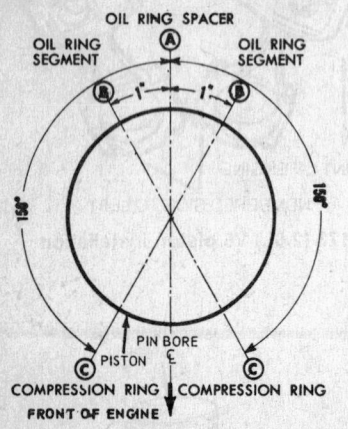

Proper spacing of the piston ring gaps around the piston

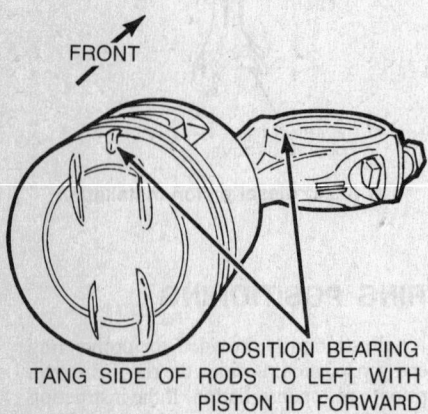

POSITION BEARING TANG SIDE OF RODS TO LEFT WITH PISTON ID FORWARD

Connecting rod and piston positioning on the 6-300

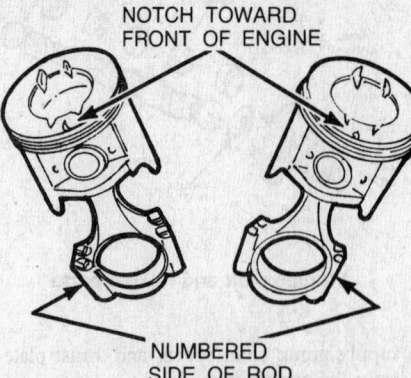

Connecting rod and piston position for 232 V6, 255, 302, 351, 400, 460 V8 engines

Crankshaft and Main Bearings

REMOVAL & INSTALLATION

Refer to the Engine Rebuilding Chapter in the Unit Repair Section for instructions.

Flywheel

REMOVAL & INSTALLATION

1. If the vehicle is equipped with a manual transmission, remove the transmission and flywheel housing, clutch pressure and disc.

2. If the vehicle is equipped with an automatic transmission, remove the transmission and converter housing with converter mounted. Remove the converter to flywheel mounting bolts first, make sure the converter does not come off of the transmission shaft when removing the transmission.

3. Remove the flywheel mounting bolts and the flywheel. Inspect and replace the pilot bearing if necessary.

4. Installation is in the reverse order of removal.

ENGINE LUBRICATION

Oil Pan

REMOVAL & INSTALLATION

122 (2.0L) and 140 (2.3L)

1. Disconnect the negative battery cable.

2. Remove air cleaner assembly. Remove oil dipstick. Remove engine mount retaining nuts.

3. Remove oil cooler lines at the radiator, if so equipped. Remove (2) bolts retaining the fan shroud to the radiator and remove shroud.

4. Remove radiator retaining bolts (automatic only). Position radiator upward and wire to the hood (automatic only).

5. Raise the vehicle and safely support on jackstands.

6. Drain oil from crankcase.

7. Remove starter cable from starter and remove starter.

8. Disconnect the exhaust manifold tube to the inlet pipe bracket at the thermactor check valve.

9. Remove transmission mount retaining nuts to the crossmember.

10. Remove bellcrank from converter housing (automatic only).

11. Remove oil cooler lines from retainer at the block (automatic only).

12. Remove front crossmember (automatic only).

13. Disconnect right front lower shock absorber mount (manual only).

14. Position jack under engine, raise and block with a piece of wood approximately 2½ inches high. Remove jack.

15. Position jack under the transmission and raise slightly (automatic only).

16. Remove oil pan retaining bolts, lower pan to the chassis. Remove oil pump drive and pick-up tube assembly.

17. Remove oil pan (out the front, automatic only) (out the rear, manual only).

18. Clean oil pan and inspect for damage. Clean oil pan gasket surface at the cylinder block. Clean oil pump exterior and oil pump pick-up tube screen.

19. Position oil pan gasket and end seals to the cylinder block (use contact cement to retain).

20. Position oil pan to the crossmember.

21. Install oil pump and pick-up tube assembly. Install oil pan to cylinder block with retaining bolts.

22. Lower jack under transmission (automatic only).

23. Position jack under engine, raise slightly, and remove wood spacer block.

24. Replace oil filter.

25. Connect the exhaust manifold tube to the inlet pipe bracket at the thermactor check valve.

26. Install transmission mount to the crossmember.

27. Install oil cooler lines to the retainer at the block (automatic only).

28. Install bellcrank to converter housing (automatic only.)

29. Install right front lower shock absorber mount (manual only). Install front crossmember (automatic only).

30. Install starter and connect cable. Lower vehicle.

31. Install engine mount bolts.

32. Locate the radiator to the supports and install the (2) retaining bracket bolts

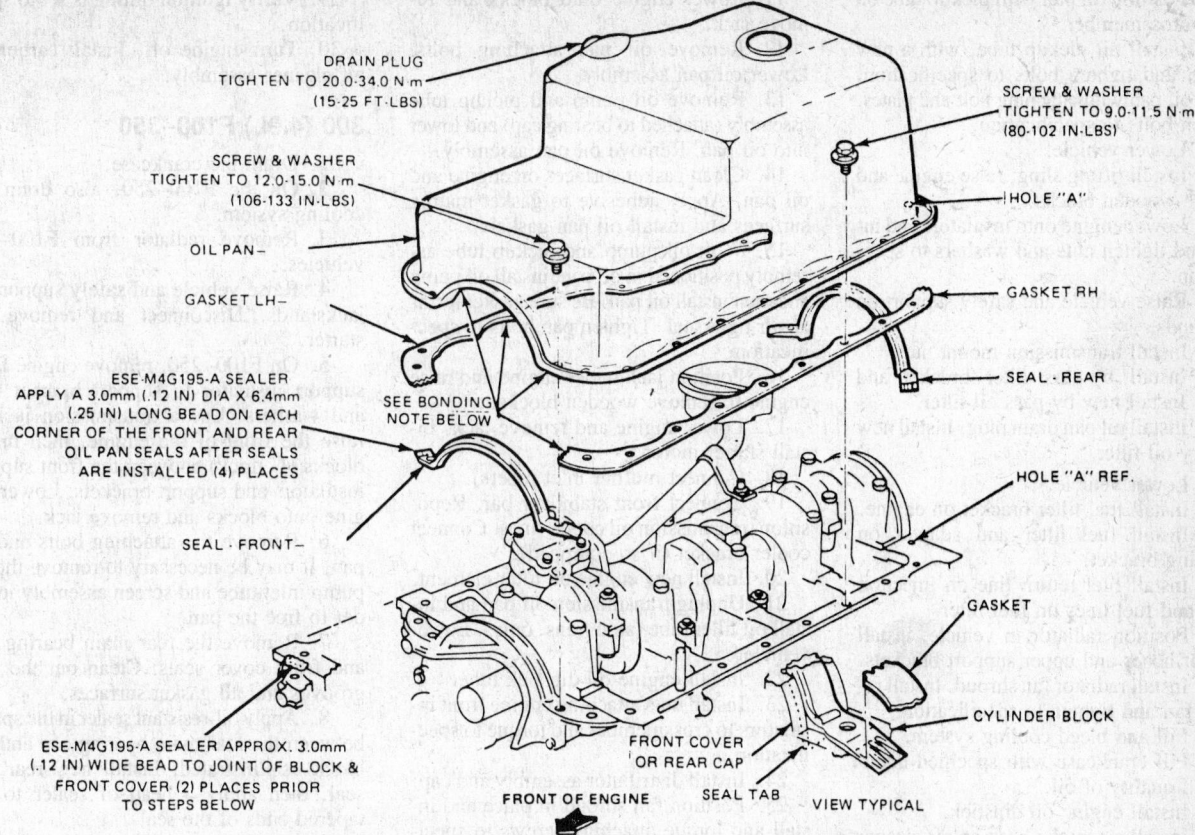

DRAIN PLUG
TIGHTEN TO 20 0-34.0 N·m
(15-25 FT·LBS)

SCREW & WASHER
TIGHTEN TO 9.0-11.5 N·m
(80-102 IN·LBS)

SCREW & WASHER
TIGHTEN TO 12.0-15.0 N·m
(106-133 IN·LBS)

HOLE "A"

OIL PAN

GASKET LH

GASKET RH

ESE·M4G195·A SEALER
APPLY A 3.0mm (.12 IN) DIA X 6.4mm
(.25 IN) LONG BEAD ON EACH
CORNER OF THE FRONT AND REAR
OIL PAN SEALS AFTER SEALS
ARE INSTALLED (4) PLACES

SEE BONDING
NOTE BELOW

SEAL · REAR

HOLE "A" REF.

SEAL · FRONT

GASKET

ESE·M4G195·A SEALER APPROX. 3.0mm
(.12 IN) WIDE BEAD TO JOINT OF BLOCK &
FRONT COVER (2) PLACES PRIOR
TO STEPS BELOW

FRONT COVER
OR REAR CAP

CYLINDER BLOCK

FRONT OF ENGINE   SEAL TAB   VIEW TYPICAL

THERMAL BONDING INSTRUCTIONS · OIL PAN GASKETS TO BE BONDED SECURELY TO OIL PAN
USING A THERMAL PROCESS MEETING THE REQUIREMENTS OF THE (ES·DOAE·6584·A OR EQUIVALENT)
ADHESIVE COATING SPECIFICATION · IF NECESSARY IN PLACE OF THERMAL BONDING USE ADHESIVE
(ESE·M2G52·A OR B OR EQUIVALENT) APPLY EVENLY TO OIL PAN FLANGE & TO PAN SIDE OF GASKETS
ALLOW ADHESIVE TO DRY PAST "WET" STAGE THEN INSTALL GASKETS TO OIL PAN.

1. APPLY SEALER AS NOTED ABOVE
2. INSTALL SEALS TO FRONT COVER & REAR BEARING CAP · PRESS SEAL TABS FIRMLY INTO BLOCK
3. INSTALL (2) GUIDE PINS
4. INSTALL OIL PAN OVER GUIDE PINS & SECURE WITH (4) BOLTS
5. INSTALL (18) BOLTS
6. TORQUE ALL BOLTS IN SEQUENCE CLOCKWISE FROM HOLE "A" AS NOTED ABOVE

Oil pan and gasket installation 122 (2.0L), 140 (2.3L) engines

(automatic only). Install fan shroud on the radiator.

33. Connect oil cooler lines to the radiator (automatic only).

34. Install air cleaner assembly.

35. Install oil dipstick. Fill crankcase with oil.

36. Start engine and check for leaks.

134 (2.2L) Diesel

1. Disconnect the ground cables from both batteries.

2. Remove engine oil dipstick. Disconnect air intake hose from air cleaner and intake manifold.

3. Drain coolant. Remove engine fan. Remove engine fan shroud.

4. Disconnect radiator hoses. Remove radiator upper support brackets and remove radiator and fan shroud.

5. Disconnect and cap fuel inlet and outlet lines at fuel filter and return line at injection pump.

6. Remove fuel filter assembly from mounting bracket. Remove the fuel filter mounting bracket from cylinder head.

7. Remove nuts and washers attaching engine brackets to insulators.

8. Raise vehicle and safely support on jackstands.

9. Loosen transmission insulator bolts at rear of transmission. Remove bottom engine insulator.

10. Drain the engine oil from crankcase. Remove primary oil filter from left side of engine.

11. Remove by-pass filter mounting bracket and hoses.

12. Lower vehicle.

13. Attach engine lifting sling and hoist. Raise engine until insulator studs clear insulators. Slide engine forward, then raise engine approximately 3 inches.

14. Install a wooden block 3 inches high between left mount and bracket. Install a wooden block 4¼ inches high between right mount and bracket. Lower engine.

15. Remove lifting sling and raise vehicle.

16. Remove oil pan attaching bolts, and lower oil pan onto cross member.

17. Disconnect oil pickup from oil pump and bearing cap, and lay in oil pan.

18. Move oil pan forward and up between front of engine and front body sheet metal.

NOTE: If additional clearance is needed, move A/C condensor forward.

19. Clean gasket mating surfaces of oil pan and engine block with a suitable solvent and dry thoroughly. Apply ⅛ inch bead of Silicone Sealer on split line between engine block and engine front cover and along side rails.

20. Locate oil pan gaskets in position with Gasket Cement and make sure that gasket tabs are seated in seal cap grooves.

21. Press front and rear oil pan seals in seal cap grooves with both ends of seals contacting oil pan gaskets.

22. Apply ⅛ inch bead of sealer at ends of oil pan seals where they meet oil pan gaskets.

23. Position oil pan with pickup tube on No. 1 crossmember.

24. Install oil pickup tube, with a new gasket, and tighten bolts to specification. Install oil pan with attaching bolt and plates. Tighten bolts to specification.

25. Lower vehicle.

26. Install lifting sling, raise engine and remove wooden blocks.

27. Lower engine onto insulators and install and tighten nuts and washers to specification.

28. Raise vehicle and safely support on jackstands.

29. Install transmission mount nuts.

30. Install by-pass filter bracket and hoses. Install new by-pass oil filter.

31. Install oil pan drain plug. Install new primary oil filter.

32. Lower vehicle.

33. Install fuel filter bracket on engine.

34. Install fuel filter and adaptor on mounting bracket.

35. Install fuel return line on injection pump and fuel lines on fuel filter.

36. Position radiator in vehicle, install radiator hoses and upper support brackets.

37. Install radiator fan shroud. Install radiator fan and tighten to specification.

38. Fill and bleed cooling system.

39. Fill crankcase with specified quantity and quality of oil.

40. Install engine oil dipstick.

41. Install air intake hose on air cleaner and intake manifold.

42. Connect battery ground cables to both batteries.

43. Run engine and check for oil, fuel and coolant leaks.

173 (2.8L) V6 and 179 (2.9L) V6

1. Disconnect negative battery cable. Remove carburetor air cleaner assembly.

2. Remove fan shroud and position over fan.

3. Remove distributor cap, position forward of dash panel. Remove distributor and cover bore opening.

4. Remove nuts attaching engine front insulators to cross member. Remove engine oil dipstick tube.

5. Raise vehicle and safely support on jackstands.

6. Drain engine crankcase. Remove transmission fluid filler tube and plug pan hole (auto trans. only).

7. Remove engine oil filter element. Disconnect muffler inlet pipe(s).

8. Disconnect oil cooler bracket and lower (if so equipped). Remove starter motor.

9. Position out of way, transmission oil cooler lines (if so equipped). Disconnect front stabilizer bar and position forward.

10. Position jack under engine and raise engine maximum height (until it touches dash panel) and install wooden blocks between front insulator mounts and No.2 crossmember.

11. Lower engine onto blocks and remove jack.

12. Remove oil pan attaching bolts. Lower oil pan assembly.

13. Remove oil pump and pickup tube assembly (attached to bearing cap) and lower into oil pan. Remove oil pan assembly.

14. Clean gasket surfaces on engine and oil pan. Apply adhesive to gasket mating surfaces and install oil pan gaskets.

15. With oil pump and pickup tube assembly positioned in oil pan, install oil pump and then install oil pan. Be sure gasket forms an air tight seal. Tighten pan bolts to specification.

16. Position jack under engine and raise engine to remove wooden blocks.

17. Lower engine and remove jack. Install starter motor.

18. Connect muffler inlet pipe(s).

19. Connect front stabilizer bar. Reposition transmission oil cooler lines. Connect cooler bracket (if so equipped).

20. Install new engine oil filter element.

21. Unplug transmission oil pan and install oil filler tube (auto trans. only). Lower vehicle.

22. Install engine oil dipstick tube.

23. Install nuts attaching engine front insulators to crossmember and torque to specification.

24. Install distributor assembly and cap.

25. Position fan shroud in place and install and torque attaching screws to specification.

26. Fill engine crankcase with specified amount of oil and transmission with specified amount of fluid.

27. Connect battery negative cable.

28. Start engine, allow to run until normal operating temperature and check for oil or fluid leaks.

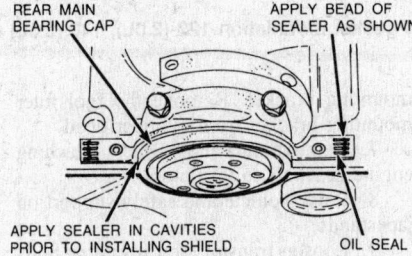

6-300 oil pan rear seal installation

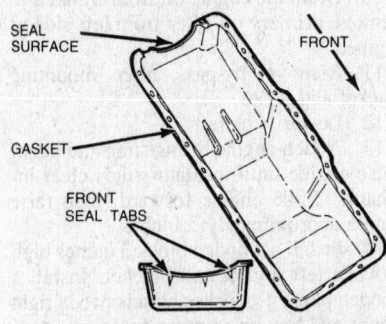

6-300 oil pan front seal installation

29. Verify ignition timing is set to specification.

30. Turn engine off. Install carburetor air cleaner assembly.

300 (4.9L) F100–350

1. Drain the crankcase.

2. On the F100–250, also drain the cooling system.

3. Remove radiator from F100–250 vehicles.

4. Raise vehicle and safely support on jackstands. Disconnect and remove the starter.

5. On F100–250, remove engine front support insulator to support bracket nuts and washers. Use a transmission jack to raise the front of the engine, then install blocks (1' thick) between the front support insulators and support brackets. Lower engine onto blocks and remove jack.

6. Remove the attaching bolts and oil pan. It may be necessary to remove the oil pump inlet tube and screen assembly in order to free the pan.

7. Remove the rear main bearing cap and front cover seals. Clean out the seal grooves and all gasket surfaces.

8. Apply oil-resistant sealer in the spaces between the rear main bearing cap and the block as illustrated. Install new rear cap seal, then apply a bead of sealer to the tapered ends of the seal.

9. Install new oil pan side gaskets with sealer and position the front cover seal.

10. Clean oil pump pick-up assembly and place it in the pan.

11. Position pan under the engine and install pick-up assembly.

12. Install pan and attaching bolts, tightening to 10—12 ft. lbs.

13. Raise engine enough with a jack and remove wood blocks. Lower engine and install washers and nuts on the support insulator studs, tightening to 40–60 ft. lbs.

14. Install starter and starter cable on F100—250 trucks.

15. Lower vehicle and install radiator if it was removed.

16. Fill crankcase and cooling system and start engine to check for leaks.

300 (4.9L) E100–350

1. Remove the engine cover. Remove the air cleaner and the carburetor.

2. If equipped with air conditioning, discharge the system and remove the compressor.

3. If the vehicle is an E350, disconnect the thermactor check valve inlet hose and remove the check valve. Remove the EGR valve.

4. Remove the radiator hoses. Unbolt the fan shroud and position on the fan. If equipped with automatic transmission, disconnect the cooler lines and remove the oil filler tube.

5. Remove exhaust inlet pipe to manifold nuts. Raise the vehicle on a hoist and disconnect and plug fuel pump inlet line.

Remove the starter. Remove alternator splash shield and front engine support nuts.

6. Remove the power steering return line clip which is located in front of the No. 1 crossmember.

7. Raise the engine and place 3 in. blocks under the engine mounts. Remove the oil pan dipstick tube.

8. Remove the oil pan bolts and remove the oil pan. Remove the pickup tube and screen from the oil pump.

9. Clean the oil pan, tube and screen assembly and the gasket surfaces of the block and oil pan.

10. Install the oil pump and screen assembly, if removed. Cement a new oil pan gasket on the oil pan. Position a new oil pan to cylinder front cover seal on the oil pan. Position the rear seal to the rear bearing cap and apply sealer. Install the oil pan.

11. Install the dipstick tube and lower the engine. Install the support nuts, starter, and connect the fuel line.

12. Install the lower radiator hose. Connect the transmission cooler lines and the transmission fill tube, if equipped.

13. Install the power steering return line clip and position the line.

14. Install the alternator splash shield and lower the hoist. Install the EGR valve and the carburetor. Connect the exhaust.

15. On E350 models, install the thermactor check valve and connect the inlet hose.

16. Install the fan shroud and the upper radiator hose. Fill the cooling system.

17. Install the air conditioning compressor and charge the system.

18. Replace the oil filter and fill the crankcase. Start the engine and check for leaks. Adjust the carburetor curb idle speed. Install the air cleaner.

232 (3.8L) V6

1. Disconnect the cable from the battery negative cable.

2. Remove the air cleaner and duct assembly.

3. Remove the bolts attaching the fan shroud to the radiator and position the shroud over the fan.

4. Remove the engine oil dipstick.

5. Raise the vehicle and safely support on jackstands. Drain the engine oil and replace the drain plug.

6. Remove the oil filter.

7. Disconnect the muffler inlet pipes from the exhaust manifolds. Remove the clamp attaching inlet pipe to converter pipe and remove inlet pipe from vehicle.

8. Disconnect the transmission shift linkage at the transmission.

9. Disconnect the transmission cooler lines at the radiator if so equipped.

10. Remove the nuts attaching the engine supports to the chassis brackets.

11. Raise the engine as high as possible, and place wood blocks between the engine supports and the chassis brackets. Remove the jack.

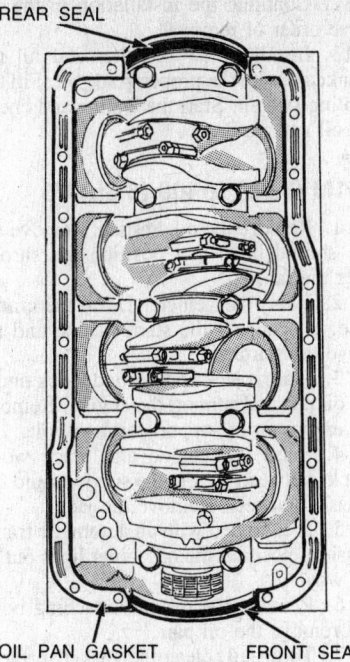

REAR SEAL

OIL PAN GASKET   FRONT SEAL

8-302 oil pan gasket and seals

12. Remove the oil pan attaching bolts and drop the oil pan. Remove the oil pickup and tube assembly and let them lay in the oil pan. Remove the oil pan from the vehicle.

13. Clean oil pan and sealing surfaces. Inspect the gasket sealing surfaces for damages and distortion due to overtightening of the bolts. Repair and straighten as required.

14. Trial fit the oil pan to the cylinder block. Make sure enough clearance has been provided to allow the oil pan to be installed without the sealant being scraped off when the pan is positioned under the engine.

15. Remove the oil pan.

16. Lay the oil pick-up and tube assembly and place them in the oil pan.

17. Place the oil pan in position on No. 1 crossmember.

18. Install the oil pick-up and tube assembly using a new gasket. Make sure the support bracket engages the stud on the No.2 main bearing cap attaching bolt.

19. Tighten the attaching bolts to 15–22 ft. lbs. and the attaching nut to 15–22 ft. lbs.

20. Install a new oil pan rear seal. Using a small screwdriver, work the tabs on each end of the seal into the gap between the rear main cap and cylinder block. With the tabs positioned work the edge of the seal into the seal groove in the rear main cap.

21. Install the oil pan as follows: Apply a ⅛ inch bead of RTV Sealer to the seam where the front cover and cylinder block join. Apply a ⅛ inch bead of RTV Sealer to each end of the rear seal where the rear main cap and cylinder block join. Apply a ⅛ inch bead of RTV Sealer along the oil pan rails on the cylinder block. As the bead crosses the front cover increase the bead

width to ¼ inch. Position oil pan to bottom of engine and secure with bolts. Tighten bolts to 7–8 ft. lbs.

22. Raise the engine and remove the wood blocks. Lower the engine and install the insulator-to-chassis bracket nuts and washers. Tighten the nuts to 50–70 ft. lbs.

23. Position the inlet pipe to the converter pipe and secure with attaching clamp. Connect inlet pipe to exhaust manifolds and secure with attaching nuts. Tighten to 25–38 ft. lbs.

24. Connect the transmission shift linkage at the transmission.

25. Connect the transmission cooler lines at the radiator if so equipped.

26. Install a new oil filter.

27. Lower the vehicle. Install the air cleaner assembly. Position fan shroud to radiator brace and secure with bolts.

28. Connect the battery negative cable to the battery.

29. Install the engine oil dipstick.

30. Fill crankcase with oil. Run the engine and check for possible leaks.

302 (5.0L) & 351W (5.8L) V8

BRONCO

1. Remove the air cleaner and duct assembly. Remove the oil dipstick tube. Drain the engine oil.

2. Remove the oil pan bolts and remove the oil pan.

3. To install, clean the oil pan and the cylinder block of all old gasket material. Position a new oil pan gasket and end seals to the cylinder block.

4. Clean and install the oil pump pickup tube and screen assembly, if removed.

5. Install the oil pan to the cylinder block. Install the oil dipstick tube, air cleaner and duct assembly.

6. Fill the crankcase with the proper oil. Start the engine and check for leaks.

F SERIES

1. Remove the oil dipstick. Remove the bolts attaching the fan shroud to the radiator and position the shroud over the fan.

2. Remove the nuts and lockwashers attaching the engine support insulators to the chassis bracket.

3. Disconnect the oil cooler line at the left side of the radiator, if equipped with automatic transmission.

4. Raise the engine and place wood blocks under the engine supports. Drain the crankcase.

5. Remove the oil pan bolts and lower the oil pan onto the crossmember.

6. Remove the oil pump pick-up tube and screen. Lower this assembly, into the oil pan. Remove the oil pan.

7. To install, clean the oil pan, inlet tube and gasket surfaces. Position a new oil pan gasket and seals to the cylinder block.

8. Install the oil pick-up tube and screen to the oil pump, and install the lower attaching bolt and gasket loosely. Place the oil pan on the crossmember. Install the up-

per pick-up tube bolt. Tighten both pick-up tube bolts.

9. Install the oil pan. Remove the wood blocks and lower the engine.

10. Install the insulator-to-chassis bracket nuts and washers.

11. Connect the automatic transmission cooler line, if equipped. Install the fan shroud attaching bolts.

12. Fill the crankcase with oil. Install the oil dipstick. Start the engine and check for leaks.

ECONOLINE

1. Disconnect the battery and remove engine cover. Remove the air cleaner. Drain the cooling system.

2. If equipped with power steering remove the pump and position it out of the way. If so equipped, remove the air conditioning compressor retainer and position the compressor out of the way.

3. Disconnect the radiator hoses. Remove the fan shroud bolts and oil filler tube. Remove the oil dipstick bolt. Raise the vehicle on a hoist.

4. Remove the alternator splash shield. If equipped, disconnect the automatic transmission cooler lines at the radiator.

5. Disconnect and plug the fuel line at the fuel pump. Remove the engine mount nuts. Drain the engine oil. Remove the dipstick tube. Disconnect the muffler inlet pipe from the exhaust manifolds.

6. If equipped, remove the automatic transmission dipstick and tube. Disconnect the manual linkage at the transmission. Remove the center driveshaft support and remove the driveshaft from the transmission.

7. Place a transmission jack under the oil pan and insert a wooden block between the pan and jack.

––––––––––– CAUTION –––––––––––

The engine and transmission assembly will pivot around the rear engine mount. The engine assembly must be raised four inches (measured from the front motor mounts). The engine must remain centered in the engine compartment to obtain this much lift.

8. Raise the engine and transmission assembly. Insert wooden blocks to support the engine in its uppermost position.

9. Remove the oil pan bolts and lower the oil pan. Unbolt the oil pump and the oil pick-up tube and lay them in the oil pan. Remove the oil pan from the vehicle.

NOTE: The oil pump must be removed along with the removal of the oil pan. When installing the oil pump refer to the procedure for Oil Pump Removal and Installation.

10. To install, clean the oil pan, oil pick-up tube, oil pump and gasket surfaces. Position new gasket and seals to the engine block.

11. Position the oil pan with the oil pump to vehicle and install the oil pump. Install the oil pan.

12. Continue the installation in the reverse order of removal.

13. Install a new oil filter and fill the crankcase with the proper grade oil. Fill the cooling system. Start the engine and check for oil and water.

351M (5.8L) & 400 (6.6L) V8

1. Remove the oil dipstick. Remove the fan shroud bolts and position the shroud over the fan.

2. Raise the vehicle. Drain the crankcase. Disconnect the starter cable and remove the starter.

3. Place a jack and a wood block under the oil pan and support the engine. Remove the engine front support through bolts.

4. Raise the engine and place wood blocks between the engine supports and the chassis brackets. Remove the jack.

5. If equipped with an automatic transmission, position the oil cooler lines out of the way.

6. Remove the oil pan attaching bolts and remove the oil pan.

7. To install, clean the gasket surfaces of the block, oil pan, oil pick-up tube, and screen. Coat the block surface and the oil pan gasket with sealer. Position the oil pan gasket to the cylinder block.

8. Position the oil pan front seal on the cylinder front cover plate. Position the oil pan rear seal on the rear main bearing cap. Be sure that the tabs on both the front and rear seals are over the oil pan gasket.

9. Position and install the oil pan. Continue the installation in the reverse order of removal. Fill the crankcase. Start the engine and check for oil leaks.

460 (7.5L) V8

EXCEPT ECONOLINE

1. Disconnect the battery ground cable. Disconnect the radiator shroud and position it over the fan.

2. Raise the vehicle on a hoist and drain the crankcase. Remove the oil filter.

3. Remove the through bolt from each engine support. Place a floor jack under the front edge of the oil pan, with a block of wood between the jack and the oil pan. Raise the engine just high enough to insert 1¼' blocks of wood between the insulators and the brackets. Remove the floor jack.

4. Remove the oil pan bolts and remove the oil pan. It may be necessary to rotate the crankshaft to provide clearance between the pan and the crankshaft counterweights.

5. To install, clean the gasket surfaces of the block and the oil pan. Coat both surfaces with sealer. Position the oil pan gasket on the cylinder block. Position the oil pan front seal on the cylinder front cover. Position the oil pan rear seal on the rear main bearing cap. Be sure that the tabs on both the front and rear seals are over the oil pan gasket.

6. Position and install the oil pan. Continue the installation in the reverse order of removal.

7. Replace the oil filter and fill the crankcase. Start the engine and check for oil leaks.

ECONOLINE

1. Remove the engine cover, disconnect the battery and drain the cooling system.

2. Remove the air cleaner assembly. Disconnect the throttle and transmission linkage at the carburetor. Disconnect the power brake vacuum lines.

3. Disconnect the fuel line, choke lines and remove the carburetor air cleaner adaptor from the carburetor.

4. Disconnect the radiator hoses. If equipped, disconnect the oil cooler lines. Remove the fan assembly and remove the radiator. If equipped, remove the power steering pump and position it aside.

5. Remove the front engine mount attaching bolts. Remove the engine oil dipstick tube from the exhaust manifold. Remove the oil filler tube and bracket.

6. If so equipped, rotate the air conditioning lines (at the rear of the compressor) down to clear the dash (or remove them).

7. Raise the vehicle and safely support on jackstands, drain the crankcase and remove the oil filter.

8. Remove the muffler inlet pipe assembly. Disconnect the manual and kickdown linkage from the transmission. Remove the driveshaft and coupling shaft assembly. Remove the transmission tube assembly.

9. Remove the dipstick and tube from the oil pan. Place a transmission jack under the engine oil pan. Insert a wood block between the jack surface and the oil pan. Jack the engine upward, pivoting on the rear mount until the transmission contacts the floor. Block the engine in position.

NOTE: The engine must remain centralized to obtain the maximum height. The engine must be raised four inches at the mounts to remove the oil pan.

10. Remove the oil pan bolts and lower the oil pan. Remove the oil pump and pick-up tube attachments and drop them into the oil pan. Remove the oil pan rearward from the vehicle.

NOTE: The oil pump must be removed when removing the oil pan. When installing refer to the procedure for Oil Pump Removal and Installation.

11. To install, clean the oil pan gasket surface at the cylinder block, the oil pan assembly, the oil pump pick-up tube, and the screen.

12. Position the oil pan gaskets and end seals to the cylinder block using sealer. Position the oil pan with the oil pump and pick-up tube assembly to the chassis and install the oil pump assembly. Position and install the oil pan. Continue the installation in the reverse order of the removal.

13. Fill the cooling system, replace the

oil filter, fill the crankcase and connect the battery. Start the engine and check for oil and water leaks.

420 (6.9L) Diesel

1. Disconnect battery ground cable from both batteries.
2. Remove engine oil level dipstick. Remove transmission oil level dipstick, if so equipped.
3. Remove air cleaner and install intake opening cover.
4. Remove fan and clutch assembly using Ford Tool T83T–6312–A and B or equivalent. The attachment has a left hand thread. Remove by turning nut counterclockwise.
5. Drain cooling system. Disconnect lower radiator hose.
6. Disconnect power steering return hose from pump. Plug hose and pump to prevent contamination of the system.
7. Disconnect alternator wiring harness and fuel line heater connector from alternator.
8. Raise vehicle and safely support on jackstands.
9. Disconnect and plug transmission oil cooler lines from radiator, if so equipped.
10. Disconnect and plug fuel pump inlet fuel line.
11. Drain crankcase and remove oil filter. Remove bolt attaching transmission oil filler tube to engine block and remove tube.
12. Disconnect muffler inlet pipe from exhaust manifolds.
13. Disconnect muffler inlet pipe at muffler flange and remove inlet pipe.
14. Remove upper inlet pipe mounting stud from right exhaust manifold.
15. Remove nuts and washers attaching engine insulators to No. 1 crossmember. Lower vehicle.
16. Install lifting sling and raise engine until transmission housing contacts body.
17. Install wood blocks (2¾ inch LH side, 2 inch RH side) between engine insulators and crossmember.
18. Lower engine so that blocks support engine.
19. Raise vehicle and safely support.
20. Remove flywheel inspection plate.
21. Position fuel pump inlet line at rear of No. 1 crossmember and transmission oil cooler lines, if so equipped, out of the way.
22. Remove oil pan attaching bolts.
23. Remove oil pump and pick-up tube from engine and lay in oil pan.
24. Remove oil pan by pulling down and toward rear of vehicle.

NOTE: Crankshaft may have to be turned to reposition counterweights to aid in removal of oil pan.

25. Remove oil pump pickup tube from oil pump, if required.
26. Remove old gasket material and clean mating surfaces of oil pan, engine block and front and rear covers with a suitable solvent and dry thoroughly.

27. Clean mating surfaces of oil pickup tube. Inspect for cracks, and assemble to oil pump with new gasket, if removed. Tighten nuts to specification.
28. Prime oil pump with recommended engine oil. Rotate pump drive gear to distribute oil within pump body. Place oil pump and pick-up tube in oil pan.
29. Place oil pan in position on No. 1 crossmember.
30. Install oil pump and pick-up tube and tighten to specifications.
31. Apply ⅛ inch bead of RTV Sealant on side rails of engine block oil pan mating surface, and a ¼ inch bead of RTV Sealant on ends of engine oil pan mating surface on front and rear covers, and in mating corners.
32. Install locally fabricated oil pan installation dowels in position.
33. Position oil pan on engine and install attaching bolts. Remove oil pan locating dowels and install two remaining oil pan bolts. Tighten all oil pan bolts.
34. Install flywheel inspection plate and tighten to specifications. Lower vehicle.
35. Raise engine and remove wooden engine support blocks.
36. Lower engine onto No. 1 crossmember and remove lifting sling. Raise vehicle and safely support.
37. Install nuts and washers attaching engine insulators to No. 1 crossmember and tighten.
38. Install upper muffler inlet pipe mounting stud on right exhaust manifold.
39. Position muffler inlet pipe in vehicle and connect muffler inlet pipe to muffler flange, using a new gasket and tighten. Connect muffler inlet pipe to exhaust manifolds and tighten.
40. Install transmission oil filler tube, using a new O-ring and tighten attaching bolt. Install oil pan drain plug and new oil filter and tighten.
41. Connect fuel pump inlet line to fuel pump and tighten to specification.

NOTE: Make sure fuel line clip is reinstalled in No. 1 crossmember.

42. Connect transmission oil cooler lines and tighten to specification, if so equipped. Lower vehicle.
43. Connect alternator wiring harness and fuel line heater connector to alternator.
44. Connect power steering return hose to power steering pump.
45. Connect lower radiator hose clamp and tighten. Install radiator fan and clutch assembly using suitable tools. Left hand thread. Install by turning nut counterclockwise.
46. Remove intake manifold cover, and install air cleaner and tighten.
47. Install engine oil and transmission oil dipsticks.
48. Refill cooling system.
49. Fill crankcase with specified quantity, quality, and viscosity of engine oil.
50. Connect battery ground cables to both batteries.

51. Run engine and check for oil, fuel and coolant leaks.
52. Check power steering fluid and add, if necessary.

Oil Pump

REMOVAL & INSTALLATION

Except 232 (3.8L) V6

1. Remove the oil pan.
2. Remove the oil pump inlet tube and screen assembly.

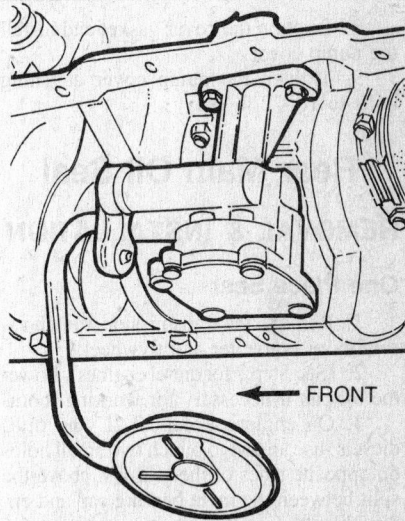

Typical oil pump installation

3. Remove the oil pump attaching bolts and remove the oil pump gasket and intermediate driveshaft.
4. Before installing the oil pump, prime it by filling the inlet and outlet port with engine oil and rotating the shaft of the pump to distribute it.
5. Position the intermediate driveshaft into the distributor socket.
6. Position the new gasket on the pump body and insert the intermediate driveshaft into the pump body.
7. Install the pump and intermediate driveshaft as an assembly. Do not force the pump if it does not seal readily. The driveshaft may be misaligned with the distributor shaft. To align it, rotate the intermediate driveshaft into a new position.
8. Install the oil pump attaching bolts and torque them to 12–15 ft. lbs. on the 6 cylinder engines and to 20–25 ft. lbs. on the V8 engines.
9. Install the oil pan.

232 (3.8L) V6

1. If necessary remove the oil filter.
2. Remove the oil pump cover attaching bolts and remove the cover.
3. Lift the pump gears of the pocket in the front cover.

4. Remove the cover gasket. Discard the gasket.

5. If necessary, remove the pump gears from the cover.

6. Pack the gear pocket with petroleum jelly. DO NOT USE CHASSIS LUBRICANTS.

7. Install the gears in the cover pocket making sure the petroleum jelly fills all voids between the gears and the pocket.

— CAUTION —

Failure to properly pack the oil pump gears with petroleum jelly may result in failure of the pump to prime when the engine is started.

8. Position the cover gasket and install the pump cover.

9. Tighten the pump cover attaching bolts to 18–22 ft. lbs.

Rear Main Oil Seal

REMOVAL & INSTALLATION

One Piece Seal

1. Remove the transmission, clutch assembly or converter and flywheel.

2. (See Step 7 for diesel engines). Lower the oil pan if necessary for working room.

3. On engines except 2.2L and 6.9L diesels, use an awl to punch two small holes on opposite sides of the seal just above the split between the main bearing cap and engine block. Install a sheet metal screw in each hole. Use two small pry bars and pry evenly on both screws using two small blocks of wood as a fulcrum point for the pry bars. Use caution throughout to avoid scratching or damage to the oil seal mounting surfaces.

4. When the seal has been removed, clean the mounting recess.

5. Coat the seal and block mounting

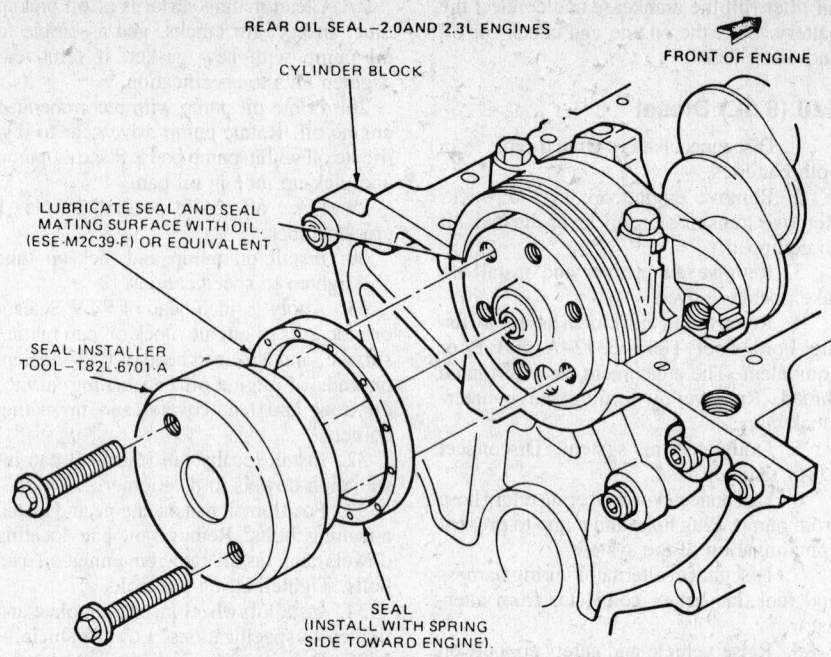

REAR OIL SEAL – 2.0 AND 2.3L ENGINES

CYLINDER BLOCK

FRONT OF ENGINE

LUBRICATE SEAL AND SEAL MATING SURFACE WITH OIL. (ESE-M2C39-F) OR EQUIVALENT

SEAL INSTALLER TOOL–T82L-6701-A

SEAL (INSTALL WITH SPRING SIDE TOWARD ENGINE)

NOTE: REAR FACE OF SEAL MUST BE WITHIN 0.127mm (0.005-INCH) OF THE REAR FACE OF THE BLOCK.

Rear main oil seal installation, typical of models using a one piece seal

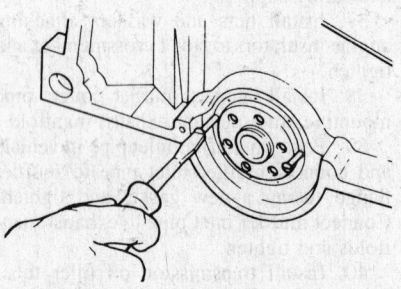

Install sheet metal screws to help in removal of a one piece rear main oil seal

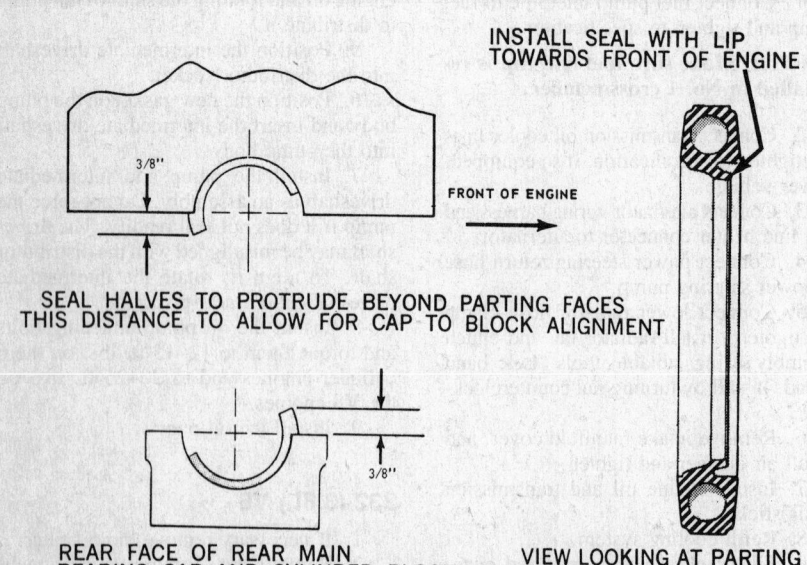

3/8"

3/8"

FRONT OF ENGINE

SEAL HALVES TO PROTRUDE BEYOND PARTING FACES THIS DISTANCE TO ALLOW FOR CAP TO BLOCK ALIGNMENT

INSTALL SEAL WITH LIP TOWARDS FRONT OF ENGINE

REAR FACE OF REAR MAIN BEARING CAP AND CYLINDER BLOCK

VIEW LOOKING AT PARTING FACE OF SPLIT, LIP TYPE CRANKSHAFT SEAL

V8 rear main oil seal positioning

surfaces with oil. Apply white lube to the contact surface of the seal and crankshaft. Start the seal into the mounting recess and install with seal mounting tool Ford number T82L–6701–A or equivalent.

6. Install the remaining components in the reverse.

7. On the 2.2L and 6.9L diesel engines; the oil seal is one piece but mounted on a retaining plate. Remove the mounting plate from the rear of the engine and replace the seal. Reinstall in reverse order of removal.

Split Seal

Remove the oil pan. In some cases it may be necessary to remove the oil pump pickup and screen or the whole pump assembly.

1. Loosen all main bearing caps, lowering the crankshaft slightly, but not more than 1/32'.

2. Remove the rear main bearing cap.

3. Remove the seal halves from cap and block. Use a seal removing tool on the block half or install a small metal screw in one end so that the seal may be pulled out.

— CAUTION —

Do not damage or scratch the crankshaft seal surfaces.

4. If so equipped, remove the oil seal retaining pin from the bearing cap.

5. Thoroughly clean seal grooves in block and cap with brush and solvent.

6. Dip seal halves in engine oil.

7. Carefully install upper half of seal with the lip facing toward the front of the engine until 3/8 inch is left protruding below parting surface. Be careful not to scrape seal.

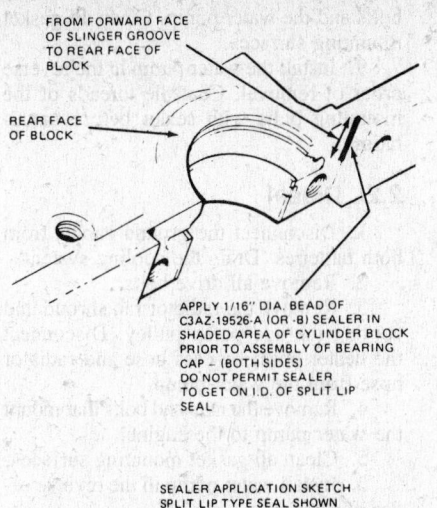

FROM FORWARD FACE OF SLINGER GROOVE TO REAR FACE OF BLOCK

REAR FACE OF BLOCK

APPLY 1/16" DIA. BEAD OF C3AZ-19526-A (OR -B) SEALER IN SHADED AREA OF CYLINDER BLOCK PRIOR TO ASSEMBLY OF BEARING CAP – (BOTH SIDES) DO NOT PERMIT SEALER TO GET ON I.D. OF SPLIT LIP SEAL

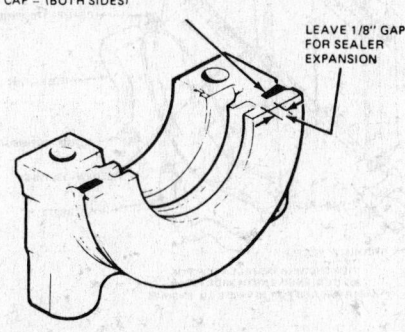

APPLY 1/16" DIA. BEAD OF C3AZ-19526-A (OR -B) SEALER AS INDICATED ON BEARING CAP – (BOTH SIDES)

LEAVE 1/8" GAP FOR SEALER EXPANSION

SEALER APPLICATION SKETCH SPLIT LIP TYPE SEAL SHOWN BASIC APPLICATION AREAS FOR OTHER SEAL INSTALLATIONS ARE THE SAME.

Applying RTV sealant to the main bearing cap on all V8's

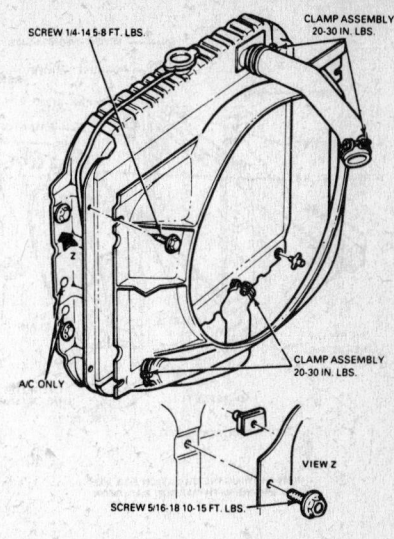

CLAMP ASSEMBLY 20-30 IN. LBS.

SCREW 1/4-14 5-8 FT. LBS.

CLAMP ASSEMBLY 20-30 IN. LBS.

A/C ONLY

VIEW Z

SCREW 5/16-18 10-15 FT. LBS.

Typical radiator installation, Bronco w/V8 400

8. Tighten all but the rear main bearing caps to specified torque.

9. Install lower seal half in the rear main bearing cap with the lip facing toward the front of the engine. Apply a light coat of oil-resistant sealer to the rear of the top mating surface of the cap. Do not apply sealer to the area forward of the side seal groove.

10. Install rear main bearing cap and tighten bolts to specified torque.

11. Install oil pump and oil pan.

12. Fill crankcase and operate engine to check for leaks.

SCREW-N606676-S43B
BRACKET
BOLT
PAD-8124
NUT 383375-S2
RADIATOR-8005
SNAP HOSE INTO SHROUD
SHROUD
SCREW-383152-S2
HOSE-8260
OVERFLOW HOSE-381440
PAD-8125
FAN GUARD-8A611
HOSE-8B273
CLIP-383375-S2
PAD-8B370
CLAMP-379995-S8
HOSE-8260
SCREW 383152-S2
RECOVERY BOTTLE
CLAMP-379995-S8
HOSE-8B273
CLAMP 379995-S8

THIS VIEW FOR A/C ONLY
STANDARD & EXTRA COOLING SAME AS A/C EXCEPT AS SHOWN

Typical 6-300 radiator installation

COOLING SYSTEM

The satisfactory performance of any engine is controlled to a great extent by the proper operation of the cooling system. The engine block is fully waterjacketed to prevent distortion of the cylinder walls. Directed cooling and water holes in the cylinder head causes water to flow past the valve seats, which are one of the hottest parts of any engine, to carry heat away from the valves and seats.

The minimum temperature of the coolant is controlled by the thermostat, mounted in the coolant outlet passage of the engine. When the coolant temperature is below the temperature rating of the thermostat, the thermostat remains closed and the coolant is directed through the radiator by-pass hose to the water pump and back into the engine. When the coolant temperature reaches the temperature rating of the thermostat, the thermostat opens and allows coolant to flow past it and into the top of the radiator. The radiator dissipates the excess engine heat before the coolant is recirculated through the engine.

The cooling system is pressurized and operating pressure is regulated by the rating of the radiator cap which contains a relief valve. The reason for a pressurized cooling system is to allow for higher engine operating temperatures with a higher coolant boiling point.

Radiator

REMOVAL & INSTALLATION

1. Drain the cooling system.
2. Disconnect the transmission cooling lines from the bottom of the radiator, if so equipped.
3. Remove the bolts mounting the shroud or shroud halves if so equipped, and position the shroud over the fan, clear the radiator.
4. Disconnect the upper and lower hoses from the radiator.

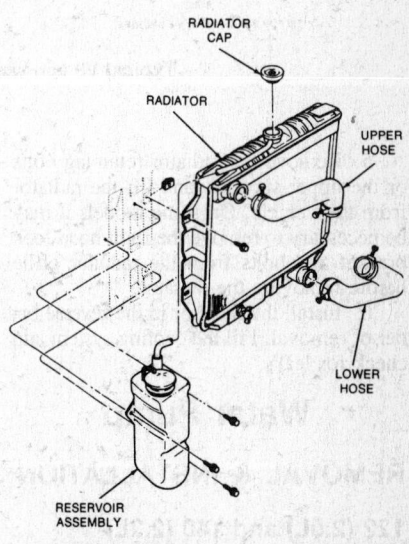

RADIATOR CAP
RADIATOR
UPPER HOSE
LOWER HOSE
RESERVOIR ASSEMBLY

Typical coolant recovery system

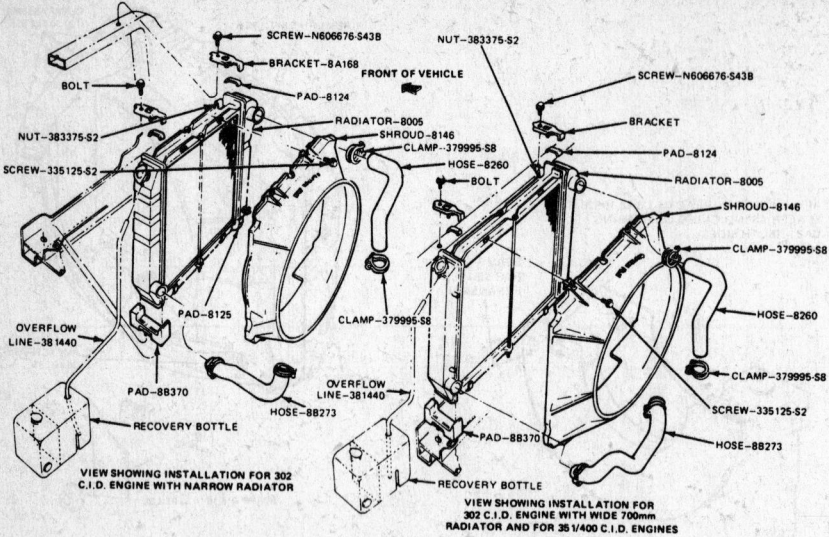

V8 radiator installation typical of Bronco and pick-ups

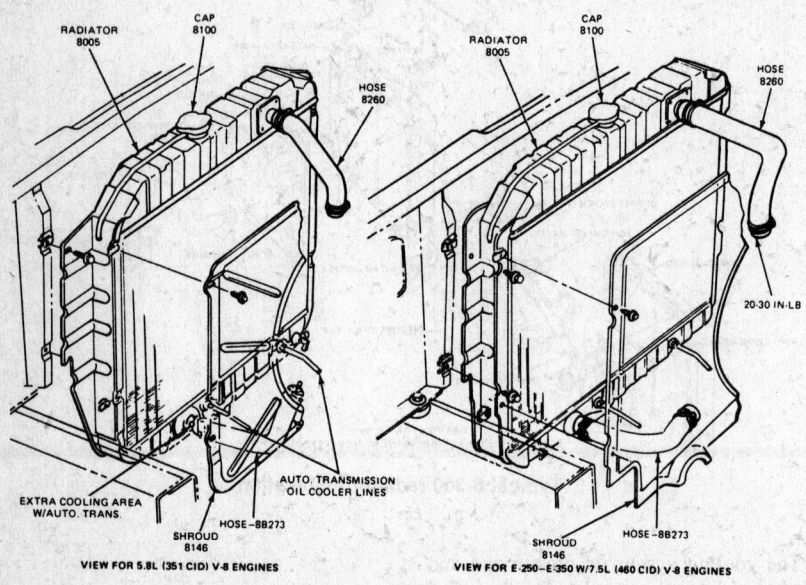

Typical V8 van radiator installation

5. Remove the radiator retaining bolts or the upper supports and lift the radiator from the vehicle. On some models it may be necessary to remove the right hood lock bracket and bolts from the radiator grille before removing the radiator.

6. Install the radiator in the reverse order of removal. Fill the cooling system and check for leaks.

Water Pump

REMOVAL & INSTALLATION

122 (2.0L) and 140 (2.3L)

1. Disconnect the negative battery ca-

ble. Drain the cooling system. Loosen and remove the drive belt.

2. Remove the two bolts that retain the fan shroud and position the shroud back over the fan.

3. Remove the four bolts that retain the cooling fan. Remove the fan and shroud.

4. Loosen and remove the power steering and A/C compressor drive belts.

5. Remove the water pump pulley and the vent hose to the emissions canister.

6. Remove the heater hose at the water pump.

7. Remove the cam belt cover. Remove the lower radiator hose from the water pump.

8. Remove the water pump mounting

bolts and the water pump. Clean all gasket mounting surfaces.

9. Install the water pump in the reverse order of removal. Coat the threads of the mounting bolts with sealer before installation.

2.2L Diesel

1. Disconnect the ground cables from both batteries. Drain the cooling system.

2. Remove all drive belts.

3. Remove the radiator fan shroud and cooling fan and pump pulley. Disconnect the heater hose, by-pass hose and radiator hose from the water pump.

4. Remove the nuts and bolts that mount the water pump to the engine.

5. Clean all gasket mounting surfaces.

6. Install water pump in the reverse order of removal.

173 (2.8L) V6

1. Disconnect the negative battery cable. Drain the cooling system.

2. Loosen and remove drive belts. Remove pump pulley. Disconnect all the water hoses from the water pump and thermostat housing.

3. Remove the radiator shroud (if necessary) and cooling fan and clutch assembly. The fan clutch assembly mounting nut is equipped with a left hand thread, remove by turning clockwise.

4. Remove the mounting bolts and water pump, water inlet and thermostat housing as an assembly.

5. Clean all gasket mounting surfaces. Transfer parts to the new pump.

6. Install the water pump in the reverse order of removal.

232 (3.8L) V6

1. Disconnect the negative battery cable. Drain the cooling system.

2. Remove the air cleaner and duct assembly.

3. Remove drive belts and pump pulley. Remove the fan shroud and cooling fan/clutch assembly.

4. Remove the power steering pump with mounting brackets and hoses attached. Position out of the way.

5. Remove the A/C compressor front mounting bracket. Leave the compressor in place.

6. Disconnect the by-pass hose from the water pump. Disconnect the heater and radiator hose at the water pump.

7. Remove the mounting bolts and the water pump.

8. Clean all gasket mounting surfaces. Installation is in the reverse order of removal.

300 (4.9L)

1. Disconnect the negative battery cable. Drain the cooling system.

2. Disconnect the lower radiator hose from the water pump.

3. Remove the drive belt, fan and water pump pulley. Remove the alternator and air pump belts.

4. Disconnect the heater hose at the water pump.

5. Remove the water pump.

6. Before installing the old water pump, clean the gasket mounting surfaces on the pump and on the cylinder block. If a new water pump is being installed, remove the heater hose fitting from the old pump and install it on the new one. Coat the new gaskets with sealer on both sides and install the water pump in the reverse order of removal.

255 (4.2L), 302 (5.0L), 351W (5.8L) V8

1. Drain the cooling system.

2. Remove the bolts securing the fan shroud to the radiator, if so equipped, and position the shroud over the fan.

3. Disconnect the lower radiator hose, heater hose and by-pass hose at the water pump. Remove the drive belts, fan, fan spacer and pulley. Remove the fan shroud, if so equipped.

4. Loosen the alternator pivot bolt and the bolt attaching the alternator adjusting arm to the water pump.

5. Remove the bolts securing the water pump to the timing chain cover and remove the water pump.

6. Install the water pump in the reverse order of removal, using a new gasket.

351M (5.8L), 400 (6.6L), 460 (7.5L) V8

1. Drain the cooling system and remove the fan shroud attaching bolts.

2. Remove the fan assembly attaching screws and remove the shroud and fan.

3. Loosen the power steering pump attaching bolts.

4. If the truck is equipped with air conditioning, loosen the compressor attaching bolts, and remove the air conditioning compressor and power steering pump drive belts.

5. Loosen the alternator pivot bolt. Remove the two attaching bolts and spacer. Remove the drive belt, then rotate the bracket out of the way.

6. Remove the three air conditioning compressor attaching bolts and secure the compressor out of the way.

7. Remove the power steering pump attaching bolts and position the pump to one side.

8. Remove the air conditioner bracket attaching bolts and remove the bracket.

9. Disconnect the lower radiator hose and heater hose from the water pump.

10. Loosen the by-pass hose clamp at the water pump.

11. Remove the remaining water pump attaching bolts and remove the pump from the front cover. Remove the separator plate

from the pump. Discard the gaskets.

12. Remove all gasket material from all of the mating surfaces.

13. Install the water pump in the reverse order of removal, using a new gasket and waterproof sealer. When the water pump is first positioned to the front cover of the engine, install only those bolts not used to secure the air conditioner and alternator brackets.

420 (6.9L) Diesel

1. Disconnect battery ground cables from both batteries. Drain cooling system.

2. Remove radiator fan shroud halves.

3. Remove fan and clutch assembly using suitable tools. Attached by left hand thread: Remove by turning nut clockwise.

4. Loosen power steering pump and A/C compressor and remove drive belts.

5. Loosen vacuum pump and remove drive belt. Loosen alternator and remove drive belt.

6. Remove water pump pulley.

7. Disconnect heater hose from water pump. Remove heater hose fitting from water pump.

8. Remove alternator adjusting arm and adjusting arm bracket. Remove A/C compressor with hoses attached, and position out of the way. Remove A/C compressor brackets.

9. Remove power steering pump and bracket and position out of the way.

10. Remove bolts attaching water pump to front cover and remove pump.

11. Clean water pump and engine front cover mating surfaces with solvent.

12. Install fabricated dowel pins for water pump alignment.

13. Install water pump with new gasket and tighten to specifications.

NOTE: Coat two bolts and two bottom bolts with RTV Sealer before installation.

14. Install alternator adjusting arm bracket.

15. Install water pump pulley.

16. Coat heater hose fitting with pipe sealant and install in water pump.

17. Connect heater hose to water pump and tighten clamp to specifications.

18. Install power steering pump bracket. Install power steering pump and drive belt. Install A/C compressor bracket. Install A/C compressor and drive belt.

19. Install alternator adjusting arm and alternator drive belt. Install vacuum pump drive belt.

20. Adjust accessory drive belts.

21. Install fan and clutch assembly. Left hand thread. Turn nut counterclockwise to tighten.

22. Install radiator fan shroud halves. Fill and bleed cooling system.

23. Connect battery ground cables to both batteries. Run engine and check for coolant leaks.

Thermostat

REMOVAL & INSTALLATION

1. Drain the cooling system to a level below the coolant outlet housing. Use the petcock valve at the bottom of the radiator to drain the system, or disconnect the lower hose at the radiator.

2. Disconnect the by-pass hose, if equipped. Remove the coolant outlet housing retaining bolts and slide the housing with the hose attached to one side.

3. Turn the thermostat counterclockwise to unlock it from the outlet.

4. Remove the gasket from the engine block and clean both mating surfaces.

5. To install the thermostat, coat a new gasket with water-resistant sealer and position it on the outlet of the engine. The gasket must be in place before the thermostat is installed.

6. Install the thermostat with the bridge (opposite end from the spring) inside the elbow connection and turn it clockwise to lock it in position with the bridge against the flats cast into the elbow connection.

7. Position the elbow connection onto the mounting surface of the outlet so that the thermostat flange is resting on the gasket and install the retaining bolts.

8. Fill the radiator and operate the engine until it reaches operating temperature. Check the coolant level and adjust as necessary.

NOTE: It is a good practice to check the operation of a new thermostat before it is installed in an engine. Place the thermostat in a pan of boiling water. If it does not open more than $^{1}/_{4}$ in., do not install it in the engine.

EMISSION CONTROLS

Refer to Emission Controls in the Unit Repair section for a description and service procedures for the various systems.

GASOLINE FUEL SYSTEM

Carburetor

REMOVAL & INSTALLATION

1. Disconnect the negative battery cable. Remove the air cleaner and duct assembly.

2. Remove the throttle cable or rod from

the throttle lever. Disconnect the distributor vacuum line EGR vacuum line, if so equipped, the inline fuel filter, and the choke heat tube at the carburetor.

3. Disconnect the choke clean air tube from the air horn. Disconnect the choke actuating cable, if so equipped.

4. Remove the carburetor retaining nuts then remove the carburetor. Remove the carburetor mounting gasket, spacer (if so equipped), and the lower gasket from the intake manifold.

5. Before installing the carburetor, clean the gasket mounting surfaces of the spacer and carburetor. Place the spacer between two new gaskets and position the spacer and gaskets on the intake manifold. Position the carburetor body flange, snug the nuts, then alternately tighten each nut in a criss-cross pattern.

6. Connect the inline fuel filter, throttle cable, choke heat tube, distributor vacuum line, EGR vacuum line, and choke cable.

7. Connect the choke clean air line to the air horn.

8. Adjust the engine idle speed, the idle fuel mixture and anti-stall dashpot (if so equipped). Install the air cleaner.

Refer to the Carburetor Section in Unit Repair for adjustments and specifications.

Fuel Filter

REMOVAL & INSTALLATION

In-line/Screw-in Fuel Filter

1. Remove the air cleaner and duct assembly.

2. If the connection between the filter and inlet is rubber, release the two clamps and slide the line away from the filter.

3. Unscrew the fuel filter from the carburetor inlet fitting using the proper size openend wrench, usually $^{11}/_{16}$ inch.

4. Screw in the new filter and install the gas line using the new hose and clamps contained in the filter kit. Start the engine and and check for leaks. Install the air cleaner and duct assembly.

5. If the filter is connected with a steel line and fitting, use the proper size wrench on the filter hex fitting to hold the filter, and the proper size flare fitting or openend wrench to loosen the gas line. Disconnect the gas line and remove the fuel filter.

6. Install the new filter into the carburetor inlet fitting. Install gas line into the filter while holding the filter with a wrench. Start the engine and check for fuel leaks. Install the air cleaner and duct assembly.

Behind Inlet Fitting

1. Remove the air cleaner and duct assembly.

2. Hold the carburetor inlet fitting with the proper size openend wrench and disconnect the gas line.

3. Unscrew the inlet fitting from the carburetor. Position a rag underneath the fitting to absorb any spill. Dispose of the rag safely after the fitting has been removed.

4. Unscrew the fitting and remove with gasket, filter and spring.

5. Install the spring, new filter and gasket into the fuel inlet of the carburetor.

6. Hand start the inlet fitting into the carburetor and tighten with a wrench.

7. Hand start the fuel line fitting, hold the filter with the proper size wrench and tighten the fuel line.

8. Start the engine and check for fuel leaks.

9. Install the remainder of the parts.

Air Intake Throttle Body Fuel Injection

REMOVAL & INSTALLATION

1. Disconnect the air intake hose.

2. Disconnect the throttle position sensor and air by-pass valve connectors.

3. Remove the four throttle body mounting nuts and carefully separate the air throttle body from the upper intake manifold.

4. Remove and discard the mounting gasket. Clean all mounting surfaces using care not to damage the gasket surfaces of the throttle body and manifold. Do not allow any material to drop into the intake manifold.

5. Install the throttle body in the reverse order of removal. The mounting nuts are tightened to 12–15 ft. lbs.

Fuel Supply Manifold

REMOVAL & INSTALLATION

1. Remove the gas tank fill cap. Relieve fuel system pressure by locating and disconnecting the electrical connection to either the fuel pump relay, the inertia switch or the in-line high pressure fuel pump. Crank the engine for about ten seconds. If the engine starts, crank for an additional five seconds after the engine stalls. Reconnect the connector. Disconnect the negative battery cable. Remove the upper intake manifold assembly.

NOTE: Special tool T81P–19623–G or equivalent is necessary to release the garter springs that secure the fuel line/hose connections.

2. Disconnect the fuel crossover hose from the fuel supply manifold. Disconnect the fuel supply and return line connections at the fuel supply manifold.

3. Remove the two fuel supply manifold retaining bolts. Carefully disengage the manifold from the fuel injectors and remove the manifold.

4. When installing: Make sure the injector caps are clean and free of contamination. Place the fuel supply manifold over each injector and seat the injectors into the manifold. Make sure the caps are seated firmly.

5. Torque the fuel supply manifold retaining bolts to 15–22 ft. lbs. Install the remaining components in the reverse order of removal.

NOTE: Fuel injectors may be serviced after the fuel supply manifold is removed. Grasp the injector and pull up on it while gently rocking injector from side to side. Inspect the mounting O-rings and replace any that show deterioration.

Mechanical Fuel Pump

REMOVAL & INSTALLATION

1. Loosen the threaded fittings to the fuel pump (use the proper size flare wrench), do not remove the lines at this time.

2. Loosen the fuel pump mounting bolts one or two turns. Loosen the pump and gasket from the engine or front cover. Rotate the engine, in the proper direction, while checking the tension on the fuel pump. When the cam or eccentric lobe is near the low point pressure on the fuel pump arm will be greatly reduced. This is especially important on engines using an aluminum front cover to help prevent thread stripping.

3. Have a rag handy to catch fuel spill and disconnect all lines from the fuel pump. Dispose of the rag safely.

4. Remove the fuel pump mounting bolts, the fuel pump and mounting gasket.

5. Clean all mounting surfaces. Apply oil resistant sealer to mounting surfaces. Install the fuel pump and new gasket in the reverse order of removal. Start the engine and check for leaks.

TESTING

Incorrect fuel pump pressure and low volume (flow rate) are the two most likely fuel pump troubles that will affect engine performance. Low pressure will cause a lean mixture and fuel starvation at high speeds and excessive pressure will cause high fuel consumption and carburetor flooding.

To determine that the fuel pump is in satisfactory operating condition, tests for both fuel pump pressure and volume should be performed.

The tests are performed with the fuel pump installed on the engine and the engine at normal operating temperature and at idle speed.

Before the test, make sure that the replaceable fuel filter has been changed at the proper mileage interval. If in doubt, install a new filter.

Pressure Test

1. Remove the air cleaner assembly.

Disconnect the fuel inlet line of the fuel filter at the carburetor. Use care to prevent fire, due to fuel spillage. Place an absorbent cloth under the connection before removing the line to catch any fuel that might flow out of the line.

2. Connect a pressure gauge, a restrictor and a flexible hose between the fuel filter and the carburetor.

3. Position the flexible hose and the restrictor so that the fuel can be discharged into a suitable, graduated container.

4. Before taking a pressure reading, operate the engine at the specified idle rpm and vent the system into the container by opening the hose restrictor momentarily.

5. Close the hose restrictor, allow the pressure to stabilize and note the reading. The pressure should be 5 psi. If the pump pressure is not within 4–6 psi and the fuel lines and filter are in satisfactory condition, the pump is defective and should be replaced. If the pump pressure is within the proper range, perform the test for fuel volume.

Volume Test

1. Operate the engine at the specified idle rpm.

2. Open the hose restrictor and catch the fuel in the container while observing the time it takes to pump 1 pint. 1 pint should be pumped in 20 seconds. If the pump does not pump to specifications, check for proper fuel tank venting or a restriction in the fuel line leading from the fuel tank to the carburetor before replacing the fuel pump.

Electric Fuel Pump

Two electric pumps are used on injected models; a low pressure boost pump mounted in the gas tank and a high pressure pump mounted on the vehicle frame. Models equipped with the 7.5L engine use a single low pressure pump mounted in the gas tank,

On injected models the low pressure pump is used to provide pressurized fuel to the inlet of the high pressure pump and helps prevent noise and heating problems. The externally mounted high pressure pump is capable of supplying 15.9 gallons of fuel an hour. System pressure is controlled by a pressure regulator mounted on the engine.

On internal fuel tank mounted pumps tank removal is required. Frame mounted models can be accessed from under the vehicle. Prior to servicing release system pressure (see proceeding Fuel Supply Manifold details). Disconnect the negative battery cable prior to pump removal.

REMOVAL & INSTALLATION

In-Tank Pump

1. Disconnect the negative battery cable.

2. Depressurize the system and drain

as much gas from the tank by pumping out through the filler neck.

3. Raise the back of the vehicle and safely support on jackstands.

4. Disconnect the fuel supply, return and vent lines at the right and left side of the frame.

5. Disconnect the wiring harness to the fuel pump.

6. Support the gas tank, loosen and remove the mounting straps. Remove the gas tank.

7. Disconnect the lines and harness at the pump flange.

8. Clean the outside of the mounting flange and retaining ring. Turn the fuel pump lock ring counterclockwise and remove.

9. Remove the fuel pump.

10. Clean the mounting surfaces. Put a light coat of grease on the mounting sufaces and on the new sealing ring. Install the new fuel pump.

11. Installation is in the reverse order of removal. Fill the tank with at least 10 gals. of gas. Turn the ignition key ON for three seconds. Repeat 6 or 7 times until the fuel system is pressurized. Check for any fitting leaks. Start the engine and check for leaks.

External Pump

1. Disconnect the negative battery cable.

2. Depressurize the fuel system.

3. Raise and support the rear of the vehicle on jackstands.

4. Disconnect the inlet and outlet fuel lines.

5. Remove the pump from the mounting bracket.

7. Install in reverse order, make sure the pump is indexed correctly in the mounting bracket insulator.

"Quick-Connect" Line Fittings

REMOVAL & INSTALLATION

NOTE: "Quick-Connect" (push) type fittings must be disconnected using proper procedures or the fitting may be damaged. Two types of retainers are used on the push connect fittings. Line sizes of ⅜ in. and 5/16 in. use a "hairpin" clip retainer. ¼ line connectors use a "duck bill" clip retainer.

Hairpin Clip

1. Clean all dirt and/or grease from the fitting. Spread the two clip legs about an ⅛ inch each to disengage from the fitting and pull the clip outward from the fitting. Use finger pressure only, do not use any tools.

2. Grasp the fitting and hose assembly and pull away from the steel line. Twist the fitting and hose assembly slightly while pulling, if necessary, when a sticking condition exists.

3. Inspect the hairpin clip for damage, replace the clip if necessary. Reinstall the clip in position on the fitting.

4. Inspect the fitting and inside of the connector to insure freedom of dirt or obstruction. Install fitting into the connector and push together. A click will be heard when the hairpin snaps into proper connection. Pull on the line to insure full engagement.

Duck Bill Clip

1. A special tool is available from Ford for removing the retaining clips (Ford Tool No. T82L–9500–AH). If the tool is not on hand see Step 2. Align the slot on the push connector disconnect tool with either tab on the retaining clip. Pull the line from the connector.

2. If the special clip tool is not available, use a pair of narrow 6 in. channel lock pliers with a jaw width of 0.2 in. or less. Align the jaws of the pliers with the openings of the fitting case and compress the part of the retaining clip that engages the case. Compressing the retaining clip will release the fitting which may be pulled from the connector. Both sides of the clip must be compressed at the same time to disengage.

3. Inspect the retaining clip, fitting end and connector. Replace the clip if any damage is apparent.

4. Push the line into the steel connector until a click is heard, indicting the clip is in place. Pull on the line to check engagement.

Fuel Tank

REMOVAL & INSTALLATION

In-Cab Fuel Tank

1. Siphon the fuel from the tank into a suitable container through the filler neck.

2. Move the seat to the full forward position and tilt the seat forward.

3. Disconnect the fuel gauge sending unit wire and fuel line from the tank. Disconnect the vapor vent chamber and vapor lines of the fuel evaporative emission control system.

4. Loosen the filler neck hose clamp at the tank end of the hose, and pull the filler neck away from the tank.

5. Remove the fuel tank retaining nuts and bolts and lift the tank out of the cab. If the tank is being replaced, remove the fuel gauge sending unit and install it in the new tank.

6. Install the fuel tank in the reverse order of removal.

In-Frame Fuel Tank

1. Drain the fuel from the tank into a suitable container by either removing the drain plug, if so equipped, or siphoning through the filler cap opening.

2. Disconnect the fuel gauge sending unit wire and fuel outlet line.

3. Disconnect the air relief tube from the filler neck and fuel tank.

4. Loosen the filler neck hose clamp at the fuel tank and pull the filler neck away from the tank.

5. Remove the retaining strap mounting nuts and bolts and lower the tank to the floor.

6. If a new tank is being installed, change over the fuel gauge sending unit to the new tank.

7. Install the fuel tank in the reverse order of removal.

Behind-The-Axle Fuel Tank

1. Raise the rear of the truck.

2. Disconnect the negative battery cable.

3. Disconnect the fuel gauge sending unit wire at the fuel tank.

4. Remove the fuel drain plug or siphon the fuel from the tank into a suitable container.

5. Loosen the fuel line hose clamps, slide the clamps forward and disconnect the fuel line at the fuel gauge sending unit.

6. If the sending unit is to be removed, turn the unit retaining ring counterclockwise and remove the sending unit, retaining ring and gasket. Discard the gasket.

7. Loosen the clamps on the fuel filler pipe and vent hose as necessary and disconnect the filler pipe hose and vent hose from the tank.

8. If the tank is the metal type, support the tank and remove the bolts attaching the tank supports to the frame. Carefully lower the tank and disconnect the vent tube from the vapor emission control valve in the top of the tank. Finish removing the filler pipe and filler pipe vent hose if not possible previously. Remove the tank from under the vehicle.

9. If the tank is the plastic type, support the tank and remove the bolts attaching the combination skid plate and tank support to the frame. Carefully lower the tank and disconnect the vent tube from the vapor emission control valve in the top of the tank. Finish removing the filler pipe and filler pipe vent hose if it was not possible previously. Remove the skid plate and tank from under the vehicle. Remove the skid plate from the tank.

10. Install the tank in the reverse order of removal.

DIESEL FUEL SYSTEM

Adjustments

INJECTOR TIMING

134 (2.2L)

NOTE: Special Tools Ford 14-0303,

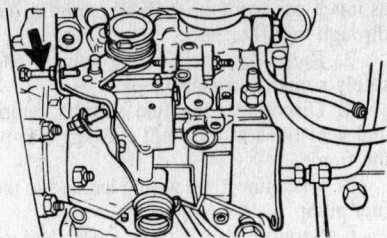

Idle speed adjustment 134 (2.2L) diesel

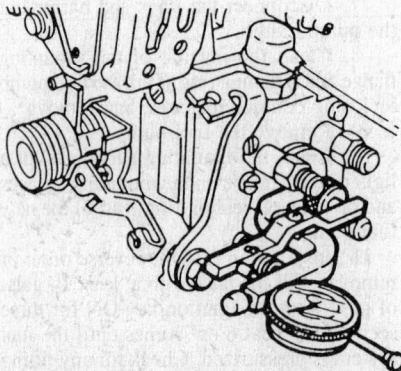

Injection timing gauge installation on the 134 (2.2L) diesel

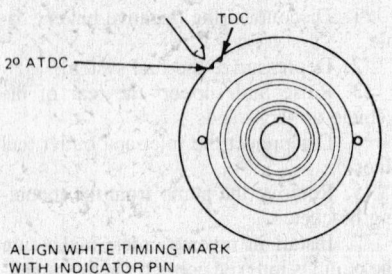

ALIGN WHITE TIMING MARK WITH INDICATOR PIN

Aligning the 134 (2.2L) diesel timing marks

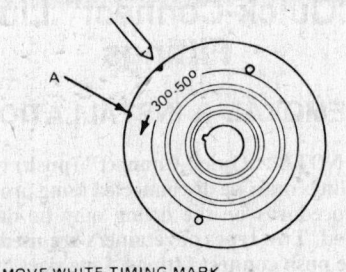

MOVE WHITE TIMING MARK FROM 2° ATDC TO POSITION A

Moving the crankshaft pulley timing mark on the 134 (2.2L) diesel

Static Timing Gauge Adapter and D82L4201A, Metric Dial Indicator, or the equivalents are necessary to set or check the injector timing.

1. Disconnect both battery ground cables. Remove the air inlet hose from the air cleaner and intake manifold.

2. Remove the distributor head plug bolt and washer from the injection pump.

3. Install the Timing Gauge Adapter and Metric Dial Indicator so that the indi-

cator pointer is in contact with the injector pump plunger and gauge reads approximately (0.08 inch).

4. Align the 2° ATDC (after top dead center) on the crankshaft pulley with the indicator on the timing case cover.

5. Slowly turn the engine counterclockwise until the dial indicator pointer stops moving (approximately 30°–50°).

6. Adjust the dial indicator to 0 (Zero). Confirm that the dial indicator does not move from Zero, by rotating the crankshaft slightly right and left.

7. Turn the crankshaft clockwise until the timing mark aligns with the cover indicator. The dial indicator should read 1, plus or minus 0.0008 inch). If the reading is not within specifications, adjust the timing as follows: Loosen the injection pump mounting nuts and bolts. Rotate the injection pump counterclockwise (reverse direction of engine rotation) past the correct timing position, then clockwise until the timing is correct. This procedure will eliminate gear backlash. Repeat Steps 5, 6, and 7 to check that the timing is properly adjusted.

8. Remove the dial indicator and adapter. Install the injector head gasket and plug. Install all removed parts.

9. Run engine, check and adjust idle RPM. Check for fuel leaks.

420 (6.9L)

NOTE: Special equipment, a Dynamic Timing Meter, Ford D83T–6002–A or the equivalent is necessary to set or check the "dynamic" injection timing. Both static and dynamic methods follow.

STATIC TIMING

1. Loosen the injection pump to mounting nuts.

2. Rotate the injection pump to bring the mark on the pump into alignment with the mark on pump mounting adapter.

3. Visually recheck the alignment of the timing marks and tighten injection pump mounting nuts.

DYNAMIC TIMING

1. Bring the engine up to normal operating temperature.

2. Stop the engine and install a dynamic timing meter, Rotunda; 78–0100 or equivalent, by placing the magnetic probe pick-up into the probe hole.

3. Remove the No. 1 glow plug wire and remove the glow plug, install luminosity probe and tighten to 12 ft. lbs. Install the photocell over the probe.

4. Connect a dynamic timing meter to the battery and adjust the offset of the meter.

5. Set the transmission in neutral and raise the rear wheels off the ground. Using Rotunda; 14–0302, throttle control, set the engine speed to 1400 rpm with no accessory load. Observe the injection timing on the dynamic timing meter.

6. If dynamic timing is not within ±

2° of specification, then injection pump timing will require adjustment.

7. Turn the engine off. Note the timing mark alignment. Loosen the injection pump-to-adapter nuts.

8. Rotate the injection pump clockwise (when viewed from the front of engine) to retard and counter-clockwise to advance timing. Two degrees of dynamic timing is approximately .030 inch of timing mark movement.

9. Start the engine and recheck the timing. If the timing is not within ± 1° of specification, repeat Steps 7 through 9.

10. Turn off the engine. Remove the dynamic timing equipment. Lightly coat the glow plug threads with anti-seize compound, install the glow plug and tighten to 12 ft. lbs. Connect the glow plug wires.

Fuel Supply Pump
REMOVAL & INSTALLATION
134 (2.2L)

1. Disconnect both ground cables from the batteries.

2. Disconnect and cap the fuel inlet and outlet lines. Remove the nuts and bolts mounting the pump to mounting bracket and remove the pump.

3. Install in the reverse order of removal.

420 (6.9L)

The pump used on the 6.9L Diesel Engine is mounted on the front cover. Refer to the Gasoline Engines fuel pump section for procedures.

Fuel Filter
REMOVAL & INSTALLATION

1. Remove the spin-on filter by turning counterclockwise as viewed from the bottom of the filter.

2. Clean the filter mounting flange. Coat the new filter sealing lip with diesel fuel.

3. Install and tighten filter until the gasket touches the mounting flange. Tighten an additional one-half turn. Refer to the instructions with the new filter, if they call for more than one-half turn tightening-follow the instructions.

Water Separator
SERVICING
134 (2.2L)

Water should be drained from the fuel sedimenter whenever the warning light comes on or every 5,000 miles. More frequent drain intervals may be required depending on fuel quality and vehicle usage.

The instrument panel warning lamp (WATER IN FUEL) will glow when approximately 0.53 quarts of water has accumulated in the sedimenter. When the warning lamp glows, shut off the engine as soon as safely possible. A suitable drain pan or container should be placed under the sedimenter, which is mounted inside the frame rail, underneath the driver's side of the cab. To drain the fuel sedimenter, pull up on the T-handle (located on the cab floor behind the driver's seat) until resistance is felt. Turn the ignition switch to the On position, so the warning lamp glows and hold T-handle up for approximately 45 seconds after lamp goes out.

To stop draining fuel, release T-handle and inspect sedimenter to verify that draining has stopped. Discard drained fluid suitably.

420 (6.9L)

The 6.9L diesel engine is equipped with a fuel/water separator in the fuel supply line. A "Water in Fuel" indicator light is provided on the instrument panel to alert the operator. The light should glow when the ignition switch is in the START position to indicate proper light and water sensor function. If the light glows continuously while the engine is running, the water must be drained from the separator as soon as practical to prevent damage to the fuel injection system.

F SERIES

1. Stop the vehicle and shut off the engine.

NOTE: Failure to shut off engine prior to draining separator will cause air to enter the fuel system.

2. Unscrew the vent 2½ to 3 turns. The vent is located on the top center of the fuel/water separator unit.

3. Unscrew the water drain located on the bottom of the fuel/water separator 1½ to 2 turns and drain water. Use an appropriate container.

4. After water is completely drained, close the water drain fingertight.

5. Tighten the vent until snug, then turn it an additional ¼ turn.

6. Restart the engine and check "Water in Fuel" indicator light. The light should not glow. If it continues to glow, have fuel system checked and repaired.

E SERIES

1. Stop the vehicle and shut Off the engine.

2. Locate the water/fuel separator drain cable knob, attached to the upper cowl flange on the left side of the vehicle, under the hood.

3. Place an approved container under the separator; which is accessible behind the left front wheel.

4. Pull the knob out and hold for 45 seconds.

5. Release knob and remove container.

6. Restart the engine and check "Water in Fuel" indicator lamp. The lamp should not be lit. If it continues to stay lit, have fuel system checked and repaired. The electrical sensor is the only replaceable item on the water/fuel separator. This assembly is threaded into the top of the separator. The remainder of the water/fuel separator is serviced as a complete unit.

CAUTION

When draining the water/fuel separator, the water must be drained into an approved container.

Fuel Injector Lines
REMOVAL & INSTALLATION
420 (6.9L)

NOTE: Before removing any fuel lines, clean exterior with clean fuel oil, or solvent to prevent entry of dirt into fuel system when fuel lines are removed. Blow dry with compressed air.

1. Disconnect battery ground cables from both batteries.

2. Remove air cleaner and cap intake manifold opening with Ford Tool T83P-9424-A or equivalent.

3. Disconnect accelerator cable and speed control cable, if so equipped, from injection pump.

4. Remove accelerator cable bracket from intake manifold and position out of the way with cable(s) attached.

NOTE: To prevent fuel system contamination, cap all fuel lines and fittings with protective cap set.

5. Disconnect fuel line from fuel filter to injection pump and cap all fittings.

6. Disconnect and cap nozzle fuel lines at nozzles.

7. Remove the fuel line clamps from fuel lines to be removed.

8. Remove and cap injection pump inlet elbow.

9. Remove and cap inlet fitting adapter.

10. Remove injection nozzle lines, one at a time, from injection pump.

NOTE: Fuel lines must be removed following this sequence; 5-6-4-8-3-1-7-2. Install caps on each end of each fuel line and pump fitting as it is removed and identify each fuel line accordingly.

11. Install fuel lines on injection pump, one at a time and tighten to 22 ft. lbs.

NOTE: Fuel lines must be installed in the sequence; 2-7-1-3-8-4-6-5.

12. Clean old sealant from injection pump elbow, using clean solvent, and dry thoroughly.

13. Apply a light coating of pipe sealant on elbow threads.

14. Install elbow in injection pump

adapter and tighten to a minimum of 6 ft. lbs. then tighten further, if necessary, to align elbow with injection pump fuel inlet lines, but do not exceed 360° of rotation or 10 ft. lbs.

15. Remove caps from fuel lines and connect lines to nozzles and tighten to 22 ft. lbs.

16. Uncap and connect fuel line from fuel filler to injection pump and tighten.

17. Install fuel line retaining clamps and tighten.

18. Install accelerator cable bracket on intake manifold.

19. Connect accelerator and speed control cable, if so equipped, to injection pump throttle lever.

20. Remove intake manifold cover, and install air cleaner.

21. Connect battery ground cables to both batteries.

22. Run engine and check for fuel leaks.

23. If necessary, purge high pressure fuel lines of air by loosening connector one half to one turn and cranking engine until solid fuel, free from bubbles, flows from connection.

— CAUTION —

Keep eyes and hands away from nozzle spray. Fuel spraying from the nozzle under high pressure can penetrate the skin.

Injection Pump

REMOVAL & INSTALLATION

134 (2.2L)

1. Disconnect battery ground cables from both batteries.

2. Remove radiator fan and shroud. Loosen and remove A/C compressor/power steering pump drive belt and idler pulley, if so equipped. Remove injection pump drive gear cover and gasket.

3. Rotate engine until injection pump drive gear keyway is at TDC.

4. Remove large nut and washer attaching drive gear to injection pump.

NOTE: Care should be taken not to drop washer into timing gear case.

5. Disconnect intake hose from air cleaner and intake manifold.

6. Disconnect throttle cable and speed control cable, if so equipped.

7. Disconnect and cap fuel inlet line at injection pump.

8. Disconnect fuel shut-off solenoid lead at injection pump.

9. Disconnect and remove fuel injection lines from nozzles and injection pump. Cap all fuel lines and fittings.

10. Disconnect lower fuel return line from injection pump and fuel hoses. Loosen lower No.3 intake port nut and remove fuel return line.

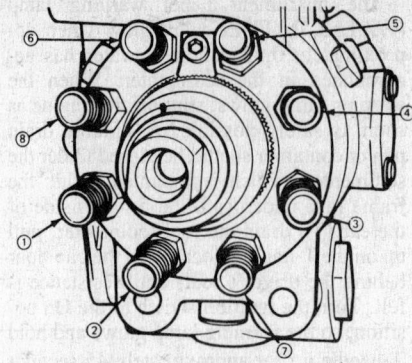

Injector pump cylinder numbering on the 420 (6.9L) diesel

11. Remove two nuts attaching injection pump to front timing gear cover and one bolt attaching pump to rear support bracket.

12. Install Gear and Hub Remover, Tool T83T–6306–A or equivalent, in drive gear cover and attach to injection pump drive gear. Rotate screw clockwise until injection pump disengages from drive gear. Remove the injection pump.

NOTE: Carefully remove injection pump to avoid dropping key into timing gear case. Disconnect cold start cable before removing injection pump from vehicle. Connect cold start cable to pump before positioning injection pump in timing gear case.

13. Install injection pump in position in timing gear case aligning key with keyway in drive gear in TDC position.

NOTE: Use care to avoid dropping key in timing gear case.

14. Install nuts and washers attaching injection pump to timing gear case and tighten to draw injection pump into position.

NOTE: Do not tighten to specification at this time.

15. Install bolt attaching injection pump to rear support. Install washer and nut attaching injection drive gear to injection pump and tighten.

16. Install injection pump drive gear cover, with new gasket, on timing gear case cover and tighten.

17. Adjust injection timing at this time.

18. Install lower fuel return line to injection pump and intake manifold stud. Tighten Banjo bolt on injection pump and nut on intake manifold. Install connecting fuel hoses and clamps. Install fuel injection lines to injection pump and nozzles and tighten.

19. Connect lead to fuel shut-off solenoid on injection pump. Connect fuel inlet line to injection pump and install hose clamp.

20. Install throttle cable and speed control cable, if so equipped.

21. Air bleed fuel system.

22. Install intake hose on air cleaner and intake manifold.

23. Install A/C compressor/power steering pump drive belt and idler pulley, if so equipped and tighten.

24. Install radiator shroud and radiator fan.

25. Connect battery ground cables to both batteries.

26. Run engine and check for oil and fuel leaks.

420 (6.9L)

NOTE: Before removing any fuel lines, clean exterior with clean fuel oil or solvent to prevent entry of dirt into engine when fuel lines are removed.

— CAUTION —

Do not wash or steam clean engine while engine is running. Serious damage to injection pump could occur.

1. Disconnect battery ground cables from both batteries.

2. Remove engine oil filter neck.

3. Remove bolts attaching injection pump to drive gear.

4. Disconnect electrical connectors to injection pump.

5. Disconnect accelerator cable and speed control cable from throttle lever, if so equipped.

6. Remove air cleaner and install intake opening cover.

7. Remove accelerator cable bracket, with cables attached, from intake manifold and position out of the way.

NOTE: All fuel lines and fittings must be capped, to prevent fuel contamination.

8. Remove fuel filter-to-injection pump fuel line and cap fittings.

9. Remove and cap injection pump inlet elbow.

10. Remove and cap injection pump fitting adapter.

11. Remove fuel return line on injection pump, rotate out of the way, and cap all fittings.

NOTE: It is not necessary to remove injection lines from injection pump to remove injection pump. If lines are to be removed, loosen injection line fittings at injection pump before removing it from engine.

12. Remove fuel injection lines from nozzles and cap lines and nozzles.

13. Remove three nuts attaching injection pump to injection pump adapter.

14. If injection pump is to be replaced, loosen injection line retaining clips and injection nozzle fuel lines and cap all fittings. Do not install injection nozzle fuel lines until new pump is installed in engine.

15. Lift injection pump, with nozzle lines attached, up and out of engine compartment.

16. Install new O-ring on drive gear end of injection pump.

17. Move injection pump down and into position.

18. Position alignment dowel on injection pump into alignment hole on drive gear.

19. Install bolts attaching injection pump to drive gear and tighten.

20. Install nuts attaching injection pump to adapter. Align scribe lines on injection pump flange and injection pump adapter and tighten to 14 ft. lbs.

21. If injection nozzle fuel lines were removed from injection pump install at this time.

22. Remove caps from nozzles and fuel lines and install fuel line nuts on nozzles and tighten to 22 ft. lbs.

23. Connect fuel return line to injection pump.

24. Install injection pump fitting adapter with a new O-ring.

25. Clean old sealant from injection pump elbow threads, using clean solvent, and dry thoroughly. Apply a light coating of pipe sealant on elbow threads.

26. Install elbow in injection pump adapter and tighten to a minimum of 6 ft. lbs. Then tighten further, if necessary, to align elbow with injection pump fuel inlet line, but do not exceed 360° of rotation or 10 ft. lbs.

27. Remove caps and connect fuel filter-to-injection pump fuel line.

28. Install accelerator cable bracket to intake manifold.

29. Remove intake manifold cover and install air cleaner.

30. Connect accelerator and speed control cable, if so equipped, to throttle lever.

31. Install electrical connectors on injection pump.

32. Clean injection pump adapter and oil filler neck sealing surfaces.

33. Apply a ⅛ inch bead of RTV Sealant on adapter housing.

34. Install oil filter neck and tighten.

35. Connect battery ground cables to both batteries. Run engine and check for fuel leaks.

36. If necessary, purge high pressure fuel lines of air by loosening connector one half to one turn and cranking engine until solid fuel, free from bubbles flows from connection.

37. Check and adjust injection pump timing.

Injectors

REMOVAL & INSTALLATION

134 (2.2L)

1. Disconnect battery ground cables from both batteries.

2. Disconnect and remove injection lines from nozzles and injection pump. Cap all lines and fittings.

3. Remove fuel return line and gaskets.

4. Remove bolts attaching fuel line heater clamp to cylinder head and position heater out of the way.

5. Remove nozzles, using a 27mm deepwell socket.

6. Remove nozzle washer (copper) and nozzle gasket (steel), using Tool T71P-19703–C or equivalent.

7. Clean nozzle assemblies with Nozzle Cleaning Kit, Rotunda 14–0301 or equivalent, and a suitable solvent, and dry thoroughly. Clean nozzle seats in cylinder head with Nozzle Seat Cleaner, T83T-9527–B or equivalant.

8. Position new nozzle washers and gaskets in nozzle seats, install nozzles and tighten.

NOTE: Install nozzle gaskets with blue side face up (toward nozzle).

9. Position fuel line heater clamps, install attaching bolts, and tighten to specification.

10. Install fuel return line with new gaskets on nozzles.

11. Install injection lines on nozzles and injection pump and tighten line nuts.

12. Connect battery ground cables to both batteries. Run engine and check for fuel leaks.

420 (6.9L)

NOTE: Before removing nozzle assemblies, clean exterior of each nozzle assembly and the surrounding area with clean fuel oil or solvent to prevent entry of dirt into engine when nozzle assemblies are removed. Also, clean fuel inlet and fuel leak-off piping connections. Blow dry with compressed air.

1. Remove fuel line retaining clamp(s) from effected nozzle line(s).

2. Disconnect nozzle fuel inlet (high pressure) and fuel leak-off tees from each nozzle assembly and position out of the way. Cover open ends of fuel inlet lines and nozzles to prevent entry of dirt.

3. Remove injection nozzles by turning counterclockwise. Pull nozzle assembly with copper washer from engine. Be careful not to strike nozzle tip against any hard surface during removal. Cover nozzle assembly fuel inlet opening and nozzle tip with plastic cap to prevent entry of dirt.

NOTE: Remove copper injector nozzle gasket from nozzle bore with Tool T71P-19703–C, or equivalent, if not attached to nozzle tip.

4. Place nozzle assemblies in a fabricated holder as they are removed from the heads. The holder should be marked with numbers corresponding to the cylinder numbering of the engine. Use of this holder permits replacing nozzles in their respective ports in the cylinder heads.

5. Thoroughly clean nozzle bore in cylinder head before reinserting nozzle assembly with nozzle seat cleaner, Tool T83T-9527–A or equivalent. Pay particular attention to seating surface, in order that no small particles of metal or carbon will cause assembly to be cocked or permit blow-by of combustion gases. Blow out particles with compressed air.

6. Remove protective cap and install a new copper gasket on nozzle assembly, with a small dab of grease.

NOTE: Anti-Seize Compound or equivalent should be used on nozzle threads to aid installation and future removal.

7. Install nozzle assembly into cylinder head nozzle bore. Be careful that nozzle tip does not strike against recess wall.

8. Tighten nozzle assembly.

9. Remove protective caps from nozzle assemblies and fuel lines. Install leak-off tees to nozzle assembly.

NOTE: Install two new O-ring seals for each fuel return tee.

10. Connect high pressure fuel line and tighten using a fuel line flare wrench.

11. Install fuel line retainer clamp(s), and tighten.

12. Start engine.

13. If necessary, purge high pressure fuel lines of air by loosening connector one half to one turn and cranking engine until solid fuel, free from bubbles flows from connection.

14. Check for fuel leakage at high pressure connections.

TESTING

Where ideal conditions of good combustion, specified engine temperature control and absolutely clean fuel prevail, nozzles require little attention. Nozzle trouble is usually indicated by one or more of the following symptoms:

1. Smoky exhaust (black)
2. Loss of power
3. Misfiring
4. Increased fuel consumption
5. Combustion knock
6. Engine overheating

Where faulty nozzle operation is sus-

pected on an engine that is misfiring or puffing black smoke, a simple test can be made to determine which cylinder is causing the difficulty.

With the engine running at a speed that makes the problem most pronounced, momentarily loosen the high pressure fuel inlet line connection on one nozzle assembly sufficiently to "cut-out" the cylinder (one half to one turn) to leak off the fuel charge to the cylinder. Then tighten to specifications.

Check each cylinder in the same manner. If one is found where loosening makes no difference in the irregular operature or causes puffing black smoke to stop, the injection nozzle for the cylinder should be serviced or replaced.

—————— CAUTION ——————
Keep eyes and hands away from nozzle spray. Fuel spraying from the nozzle under high pressure can penetrate the skin and cause infection. Medical attention should be provided immediately in the event of skin penetration.

Fuel Control

On-off fuel control is provided by an electric solenoid located in the diesel injection pump housing cover. Current is supplied to the solenoid when the ignition switch is turned on. If no fuel is supplied with the ignition switch in the on position, check for current at the solenoid terminal before condemning the solenoid.

Fuel Cut-Off Solenoid

REMOVAL & INSTALLATION

1. Disconnect battery ground cables from both batteries.
2. Remove connector from fuel cut-off solenoid.
3. Remove fuel cut-off solenoid assembly.
4. Install fuel cut-off solenoid, with new O-ring, and tighten.
5. Install connector on fuel cut-off solenoid.
6. Connect battery ground cables to both batteries. Run engine and check for fuel leaks.

Glow Plug System

The "quick start; afterglow" system is used to enable the engine to start more quickly when the engine is cold. It consists of the flour glow plugs, the control module, two relays, a glow plug resistor assembly, coolant temperature switch, clutch and neutral switches and connecting wiring. Relay power and feedback circuits are protected by fuse links in the wiring harness. The control module is protected by a separate 10A fuse in the fuse panel.

When the ignition switch is turned to the ON position, a Wait-to-Start signal appears near the cold-start knob on the panel. When the signal appears, relay No. 1 also closes and full system voltage is applied to the glow plugs. If engine coolant temperature is below 30°C (86°F), relay No.2 also closes at this time. After three seconds, the control module turns off the Wait-to-Start light indicating that the engine is ready for starting. If the ignition switch is left in the ON position about three seconds more without cranking, the control opens relay No. 1 and current to the plugs stops to prevent overheating. However, if coolant temperature is below 30°C (86°F) when relay No. 1 opens, relay No.2 remains closed to apply reduced voltage to the plugs through the glow plug resistor until the ignition switch is turned off.

When the engine is cranked, the control module cycles relay No. 1 intermittently. Thus, glow plug voltage will alternate between 12 and four volts, during cranking, with relay No.2 closed, or between 12 and zero volts with relay No.2 open. After the engine starts, alternator output signals the control module to stop the No. 1 relay cycling and the afterglow function takes over.

If the engine coolant temperature is below 30°C (86°F), the No.2 relay remains closed. This applies reduced (4.2 to 5.3) voltage to the glow plugs through the glow plug resistor. When the vehicle is under way (clutch and neutral switches closed), or coolant temperature is above 30°C (86°F), the control module opens relay No.2, cutting off all current to the glow plugs.

TESTING THE GLOW PLUGS

1. Disconnect the leads from each glow plug. Connect one lead of the ohmmeter to the glow plug terminal and the other lead to a good ground. Set the ohmmeter on the X1 scale. Test each glow plug in the like manner.
2. If the meter indicates less than one ohm, the problem is not with the glow plug.
3. If the ohmmeter indicates one or more ohms, replace the glow plug and retest.

REMOVAL & INSTALLATION

1. Disconnect battery ground cables from both batteries.
2. Disconnect glow plug harness from glow plugs.
3. Using a 12mm deepwell socket, remove glow plugs.
4. Install glow plugs, using a 12mm deepwell socket, and tighten.
5. Install glow plug harness on glow plugs and tighten.
6. Connect battery ground cables to both batteries.

420 (6.9L)

The 6.9L diesel engine utilizes an electric

glow plug system to aid in the start of the engine. The function of this system is to pre-heat the combustion chamber to aid ignition of the fuel.

The system consists of eight glow plugs (one for each cylinder), control switch, power relay, after glow relay, wait lamp latching relay, wait lamp and the eight fusible links located between the harness and the glow plug terminal.

On initial start with cold engine, the glow plug system operates as follows: The glow plug control switch energizes the power relay (which is a magnetic switch) and the power relay contacts close. Battery current energizes the glow plugs. Current to the glow plugs and a wait lamp will be shut off when the glow plugs are hot enough. This takes from 2 to 10 seconds after the key is first turned on. When the wait lamp goes off, the engine is ready to start. After the engine is started the glow plugs begin an on-off cycle for about 40 to 90 seconds. This cycle helps to clear start-up smoke. The control switch (the brain of the operation) is threaded into the left cylinder head coolant jacket. The control unit senses engine coolant temperature. Since the control unit senses temperature and glow plug operation the glow plug system will not be activated unless needed. On a restart (warm engine) the glow plug system will not be activated unless the coolant temperature drops below 165°F (91°C).

The fast start system utilizes 6 volt glow plugs in a 12 volt system to achieve rapid heating of the glow plug, a cycling device is required in the circuit.

—————— CAUTION ——————
Never bypass the power relay of the glow plug system. Constant battery current (12 volts) to glow plugs will cause them to overheat and fail.

TESTING THE GLOW PLUG

1. Disconnect the leads from the glow plug. Connect one lead of an ohmmeter to the glow plug terminal and the other lead to the metal case of the glow plug. Set the ohmmeter to the X1 scale. Test each glow plug.
2. If the meter indicates less than 2 ohms the problem is not with the glow plug.
3. If the meter indicates 2 ohms or more replace the glow plug and retest.

MANUAL TRANSMISSION

For overhaul procedures refer to the Unit Repair Section.

Linkage

ADJUSTMENT
EXCEPT VANS

Ford 3.03, Warner T85N, T87, T89

1. Place the shifter in the Neutral position and insert a gauge pin ($\frac{3}{16}$ in. diameter) through the steering column shift levers and the locating hole in the spacer.

2. If the shift rods at the transmission are equipped with threaded sleeves, adjust the sleeves so that they enter the shift levers on the transmission easily with the shift levers in the Neutral position. Now lengthen the rods seven turns of the sleeves and insert them into the shift levers.

3. If the shift rods are slotted, loosen the attaching nut, make sure that the transmission shift levers are in the Neutral position, then retighten the attaching nuts.

4. Remove the gauge pin and check the operation of the shift linkage.

Four-Speed Overdrive w/External Linkage

1. Attach the shift rods to the levers.

2. Rotate the output shaft to determine that the transmission is in neutral.

3. Insert an alignment pin into the shift control assembly alignment hole.

4. Attach the slotted end of the shift rods over the flats of the studs in the shift control assembly.

5. Install the locknuts and remove the alignment pin.

ADJUSTMENT

VANS

3–Speed

1. Place the gearshift lever in the Neutral position.

2. Loosen the adjustment nuts on the transmission shift levers sufficiently to allow the shift rods to slide freely on the transmission shift levers.

3. Insert a $\frac{3}{16}$ in. rod through the pilot hole in the shift tube mounting bracket until it enters the adjustment hole of both the upper and lower shift lever.

4. Place the transmission shift levers in the Neutral position and tighten the adjustment nuts on the transmission shift levers.

5. Remove the $\frac{1}{4}$ in. rod from the pilot hole, and check the operation of the gearshift lever in all gear positions.

4–Speed Overdrive w/External Linkage

1. Disconnect the 3 shift rods from the shifter assembly.

2. Insert a .25′ diameter pin through

the alignment hole in the shifter assembly. Make sure the levers are in the neutral position.

3. Align the 3 transmission levers as follows: forward lever (3rd–4th lever) in the mid-position (neutral), rearward lever (1st–2nd lever) in the mid-position (neutral), and middle lever (reverse lever) rotate counterclockwise to the neutral position.

4. Rotate the output shaft to assure that the transmission is in neutral.

5. Attach the slotted end of the shift rods over the slots of the studs in the shifter assembly. Install and tighten the locknuts to 15–20 ft. lbs.

6. Remove the alignment pin. Check for proper operation.

Pickup Trucks

REMOVAL & INSTALLATION

Ford 3.03, Warner T85N, T87, T89

1. Raise the vehicle and support it with jackstands. Support the engine with a jack and a block of wood placed under the oil pan.

2. Drain the lubricant out of the transmission by removing the drain plug if so equipped, or removing the lower extension housing-to-transmission bolt.

3. Position a transmission jack under the transmission.

4. Disconnect the gearshift linkage at the transmission.

5. On the Warner T–85N model, disconnect the solenoid and governor wires at the connectors near the solenoid. Remove the overdrive wiring harness from its clip on the transmission. Disconnect the overdrive cable.

6. If the vehicle has 4WD, remove the transfer case shift lever bracket from the transmission.

7. Disconnect the speedometer cable.

8. Disconnect the driveshaft from the differential and transmission and remove it from the vehicle.

9. Raise the transmission if necessary and remove the rear support.

10. Remove the transmission-to-flywheel housing attaching bolts.

11. Move the transmission to the rear until the input shaft clears the flywheel housing and lower the transmission out and from under the vehicle.

NOTE: Do not depress the clutch pedal while the transmission is removed.

12. Before installing the transmission, apply a light film of grease to the release bearing inner hub surface, the release lever fulcrum and fork, and the front bearing retainer of the transmission. Do not apply excessive grease because it will fly off due to centrifugal force and contaminate the clutch disc.

13. Install the transmission in the re-

verse order of removal. It may be necessary to turn the output shaft with the transmission in gear to align the input shaft splines with the splines in the clutch disc. Fill the transmission with lubricant and adjust the shift linkage.

Warner T–18 4 Speed

1. Disconnect the back-up light switch at the rear of the gearshift housing cover.

2. Remove the rubber boot, floor mat, and the body floor pan cover. Remove the gearshift lever. Remove the weatherpad.

3. Raise the vehicle and support it with jackstands. Position a transmission jack under the transmission and disconnect the speedometer cable.

4. Disconnect the driveshaft from the transmission and wire it up to one side.

5. Remove the rear transmission support.

6. Remove the transmission attaching bolts.

7. Move the transmission to the rear until the input shaft clears the flywheel housing and lower the transmission.

8. Before installing the transmission, apply a light film of grease to the inner hub surface of the clutch release bearing, the release lever fulcrum and the front bearing retainer of the transmission. Do not apply excessive grease because it will fly off onto the clutch disc.

9. Install the transmission in the reverse order of removal. It may be necessary to turn the output shaft with the transmission in gear to align the input shaft splines with the splines in the clutch disc. Fill the transmission with lubricant if it was drained.

Warner T19B 4 Speed

4 × 2

1. Remove the floor mat, and the body floor pan cover, and remove the gearshift lever shift ball and boot as an assembly. Remove the weather pad.

2. Raise the vehicle and position safety stands. Position a suitable jack under the transmission, and disconnect the speedometer cable.

3. Disconnect the back-up lamp switch located at the rear of the gear shift housing cover.

4. Disconnect the drive shaft or coupling shaft and clutch linkage from the transmission and wire it to one side.

5. Remove the transmission rear insulator and lower retainer. Remove the crossmember. Remove the transmission attaching bolts.

6. Move the transmission to the rear until the in shaft clears the clutch housing. Lower transmission.

7. Place the transmission on a suitable jack install guide studs in the clutch housing and raise the transmission until the input shaft splines are aligned with the clutch disc splines. The clutch release bearing and hub must be properly positioned in the release lever fork.

8. Slide the transmission forward on the guide stud until it is in position on the clutch housing. Install the attaching bolts and tighten them to 45–50 ft. lbs. Remove the guide studs and install the lower attaching bolts.

9. Install the crossmember. Position the insulator and retainer between the transmission and crossmember. Install bolts and tighten to 45–60 ft. lbs. Install the nut retaining the insulator and retainer to crossmember. Tighten to 50–70 ft. lbs. Remove the transmission jack.

10. Connect the speedometer cable and driven gear and clutch linkage.

11. Install the bolts attaching the front U-joint of the coupling shaft to the transmission output shaft flange. Install the transmission rear support and upper and lower absorbers. Connect the back-up lamp switch.

12. Install the shift lever, boot and shift ball as an assembly and lubricate the spherical ball seat with Multi-Purpose Long-Life Lubricant C1AZ–1959(ESA–M1C75–B) or equivalent.

13. Install the weather pad. Install the floor pan cover and floor mat.

4 × 4

1. Open door and cover seat. Remove the screws holding the floor mat.

2. Remove the screws holding the access cover to the floor pan. Place the shift lever in the reverse position and remove the cover.

3. Remove the insulator and dust cover.

4. Remove the transfer case shift lever, shift ball and boot as an assembly.

5. Remove transmission shift lever, shift ball and boot as an assembly.

6. Raise and safely support the vehicle.

7. Remove the drain plug and drain the transmission. Disconnect the rear driveshaft from the transfer case and wire it out of the way.

8. Disconnect the front driveshaft from the transfer case and wire it out of the way.

9. Remove the retainer ring that holds the shift link in place and remove the shift link from transfer case.

10. Remove the speedometer cable from the transfer case.

11. Position a suitable jack under the transfer case. Remove the six bolts holding the transfer case to the transmission and lower the transfer case from the vehicle.

12. Remove the eight bolts that hold the rear support bracket to the transmission.

13. Position a suitable jack under the transmission and remove the rear support bracket and brace.

14. Remove the four bolts that hold the transmission to the bell housing. Remove the transmission from the vehicle.

15. Place the transmission on a transmission jack and install it in the vehicle installing two guide studs in the bell housing top holes, to guide the transmission into position.

16. Install the two lower bolts. Remove

the guide studs and install the upper bolts.

17. Place the rear support bracket in position and install the eight retaining bolts.

18. Install the two bolts at the rear support insulator bracket. Remove the transmission jack.

19. Position the transfer case on a suitable jack and install the six retaining bolts and gasket. Position the transfer case on the transmission and tighten the bolts.

20. Install the transfer case shift link and retaining ring. Position and install the speedometer cable. Remove wire and connect front driveshaft. Remove wire and connect rear driveshaft.

21. Fill transfer case with Dexron II, automatic transmission fluid. The manual transmission with Standard Transmission Lubricant (SAE 80W) lubricant. Lower vehicle.

22. Remove fabricated dirt shield and prepare gasket area.

23. Position gasket and shift cover.

24. Install two pilot bolts, then install remaining shift cover retaining bolts.

25. Install transfer case shift lever, shift ball and boot as an assembly and transmission shift lever, shift ball and boot as an assembly.

26. Install dust cover and insulator. Install access cover to floor pan screws. Install the floor mat screws. Install the boot area screws.

New Process 435 4 Speed

1. Remove the rubber boot and floor mat.

2. Remove the floor pan, transmission cover plate, and weather pad. It may be necessary to remove the seat assembly.

3. Disconnect the back-up light switch located in the rear of the gearshift housing cover.

4. Raise the vehicle and place jackstands under the frame to support it. Place a transmission jack under the transmission and disconnect the speedometer cable.

5. Disconnect the parking brake lever from its linkage and remove the gearshift housing.

6. Disconnect the driveshaft.

7. Remove the transmission rear support.

8. Remove the transmission-to-flywheel housing attaching bolts, slide the transmission rearward until the input shaft clears the flywheel housing and lower it out from under the truck.

9. Before installing the transmission, apply a light film of grease to the inner hub surface of the clutch release bearing, release lever fulcrum and fork, and the front bearing retainer of the transmission. Do not apply excessive grease because it will fly off and contaminate the clutch disc.

10. Install the transmission in the reverse order of removal. It may be necessary to turn the output shaft with the transmission in gear to align the input shaft splines with the splines in the clutch disc. The front

bearing retainer is installed through the clutch release bearing.

Ford Four Speed Overdrive

1. Raise the truck and support it on jackstands.

2. Mark the driveshaft so that it can be installed in the same position.

3. Disconnect the driveshaft at the rear U-joint and slide it off the transmission output shaft.

4. Disconnect the speedometer cable from the transmission.

5. Remove the shift rods from the levers and the shift control from the extension housing.

6. Support the engine and remove the extension housing-to-crossmember bolts.

7. Support the transmission on a jack and unbolt it from the engine.

8. Move the transmission and jack rearward until clear. If necessary, lower the engine enough for clearance.

9. Installation is the reverse of removal. It is a good idea to install and snug down the upper transmission-to-engine bolts first, then the lower. For linkage adjustment, see the beginning of this chapter. Check the fluid level.

Single Rail Four Speed Overdrive

1. Raise and safely support the vehicle. Drain the lubricant.

2. Mark the driveshaft so that it can be installed in the same position then disconnect the driveshaft from the rear U-joint flange. Slide the driveshaft off the transmission output shaft.

3. Disconnect the speedometer cable from the extension housing.

4. Remove the three screws securing the shift tower to the turret assembly.

5. Remove the shift tower from the turret assembly.

6. Support the engine with a transmission jack and remove the bolts which attach the rear extension housing to the engine.

7. Raise the rear of the engine high enough to remove the weight from the crossmember.

8. Remove the bolts retaining the crossmember to the frame side supports and remove the crossmember.

9. Support the transmission on a jack and remove the bolts that attach the transmission to the flywheel housing.

10. Move the transmission and jack rearward until the transmission input shaft clears the flywheel housing.

NOTE: If necessary, lower the engine enough to obtain clearance for transmission removal. Do not depress the clutch pedal while the transmission is removed.

11. Before installing the transmission install two guide pins in the flywheel housing lower mounting bolt holes.

12. Move the transmission forward on

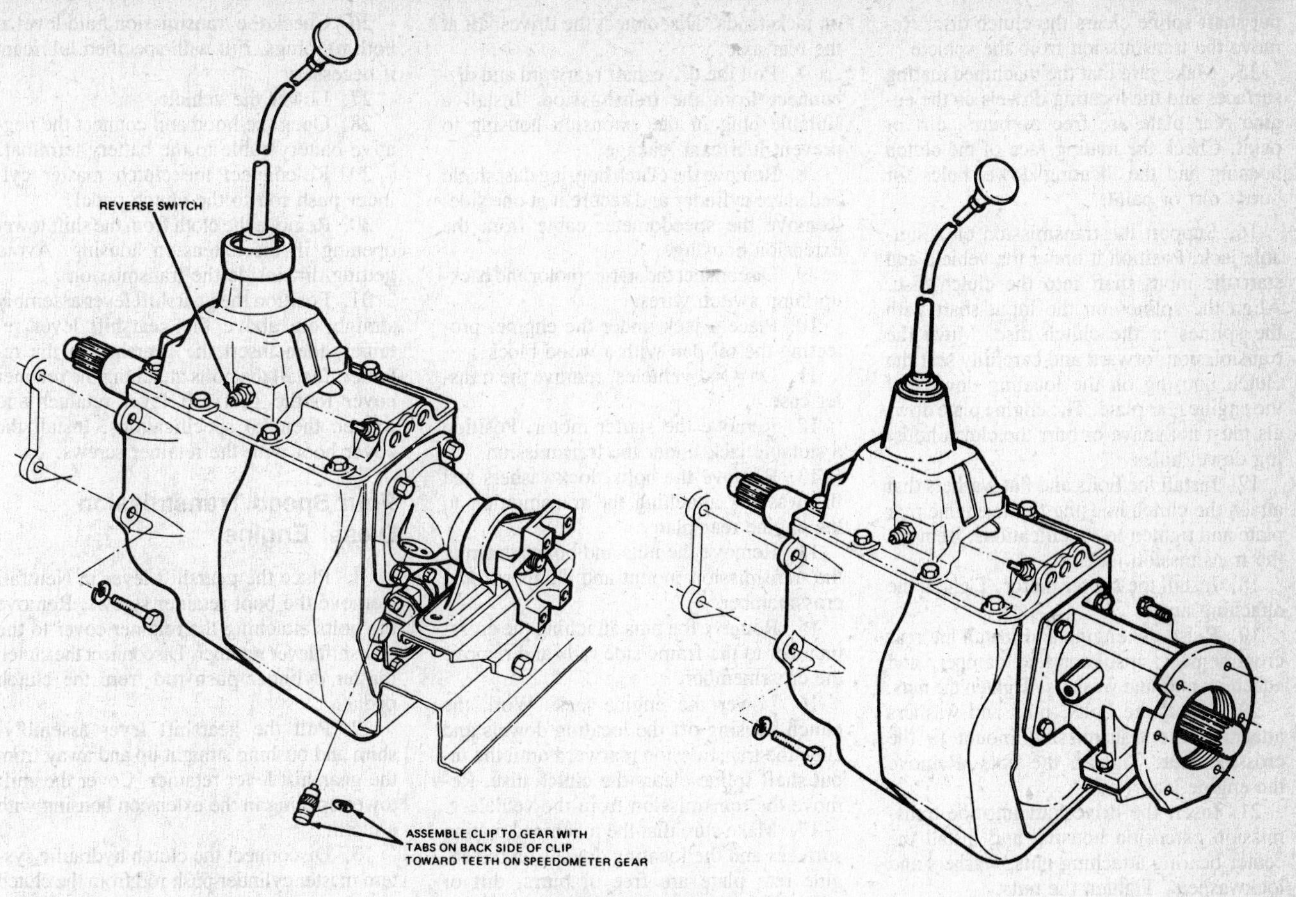

REVERSE SWITCH

ASSEMBLE CLIP TO GEAR WITH
TABS ON BACK SIDE OF CLIP
TOWARD TEETH ON SPEEDOMETER GEAR

NP435 installation

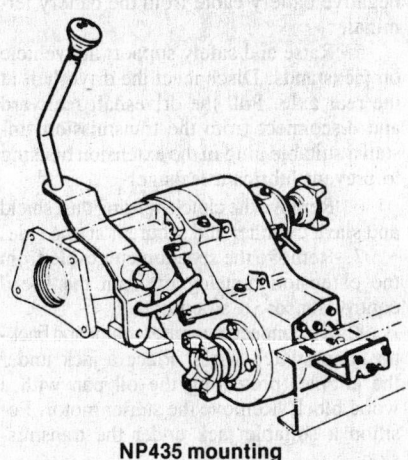

NP435 mounting

the guide pins until the input shaft splines enter the clutch hub splines and the case is positioned against the flywheel housing.

13. Install the two upper transmission to flywheel housing mounting bolts snug, and then remove the two guide pins. Install the two lower mounting bolts.

14. Raise the rear of the engine and install the crossmember and attaching bolts then lower the engine.

15. With the transmission extension housing resting on the engine rear support, install the transmission extension housing attaching bolts.

16. Position the shift tower to the extension housing and secure with the three screws.

17. Reconnect the speedometer cable to the extension housing.

18. Slide the forward end of the driveshaft over the transmission output shaft. Connect the driveshaft to the rear U-joint flange.

19. Fill with transmission lubricant and lower the vehicle.

Ranger and Bronco II

REMOVAL & INSTALLATION

Five Speed Overdrive
Diesel Engine

1. Place the gearshift lever in neutral. Remove the boot retainer screws. Remove the bolts attaching the retainer cover to the gearshift lever retainer. Disconnect the clutch master cylinder push rod from the clutch pedal.

2. Pull the gearshift lever assembly, shim and bushing straight up and away from the gearshift lever retainer. Cover the shift tower opening in the extension housing with a cloth.

3. Disconnect the clutch hydraulic sys-

tem master cylinder push rod from the clutch pedal.

4. Open the hood and disconnect the negative battery cable from the battery terminal.

5. Raise and safely support the vehicle on jackstands. Disconnect the driveshaft at the rear. Pull the driveshaft rearward and disconnect from the transmission. Install a suitable plug in the extension housing to prevent lubricant leakage.

6. Remove the clutch housing dust shield and slave cylinder and secure it at one side.

7. Remove the speedometer cable from the extension housing.

8. Disconnect the starter motor and back-up lamp switch wires.

9. Place a jack under the engine, protecting the oil pan with a wood block.

10. Remove the starter motor. Position a suitable jack under the transmission.

11. Remove the bolts, lockwashers and flat washers attaching the transmission to the engine rear plate.

12. Remove the nuts and bolts attaching the transmission mount and damper to the crossmember.

13. Remove the nuts attaching the crossmember to the frame side rails and remove the crossmember.

14. Lower the engine jack. Work the clutch housing off the locating dowels and slide the transmission rearward until the in-

put shaft spline clears the clutch disc. Remove the transmission from the vehicle.

15. Make sure that the machined mating surfaces and the locating dowels on the engine rear plate are free of burrs, dirt or paint. Check the mating face of the clutch housing and the locating dowel holes for burrs, dirt or paint.

16. Support the transmission on a suitable jack. Position it under the vehicle and start the input shaft into the clutch disc. Align the splines on the input shaft with the splines in the clutch disc. Move the transmission forward and carefully seat the clutch housing on the locating dowels of the engine rear plate. The engine plate dowels must not shave or burr the clutch housing dowel holes.

17. Install the bolts and flat washers that attach the clutch housing to the engine rear plate and tighten to specifications. Remove the transmission jack.

18. Install the starter motor. Tighten the attaching nuts.

19. Raise the engine and install the rear crossmember, insulator and damper, and attaching nuts and washers. Tighten the nuts.

20. Install the bolts, nuts and washers attaching the transmission mount to the crossmember. Tighten the nuts. Remove the engine jack.

21. Insert the driveshaft into the transmission extension housing and install the center bearing attaching nuts, washers and lockwashers. Tighten the nuts.

22. Connect the driveshaft to the rear axle drive flange. Tighten the attaching nuts.

23. Connect the starter and back-up lamp switch wires.

24. Install the clutch slave cylinder and dust shield on the clutch housing. Install the speedometer cable.

25. Check the transmission fluid level at both fill plugs. Fill with lubricant if necessary.

26. Lower the vehicle.

27. Open the hood and connect the negative battery cable to the battery terminal.

Five Speed Transmission Gas Engine

1. Place the gearshift lever in neutral. Remove the boot retainer screws. Remove the bolts attaching the retainer cover to the gearshift lever retainer. Disconnect the clutch master cylinder push rod from the clutch pedal.

2. Pull the gearshift lever assembly, shim and bushing straight up and away from the gearshift lever retainer.

3. Cover the shift tower opening in the extension housing with a cloth.

4. Disconnect the clutch hydraulic system master cylinder push rod from the clutch pedal.

5. Open the hood and disconnect the negative battery cable from the battery terminal.

6. Raise and safely support the vehicle

on jackstands. Disconnect the driveshaft at the rear axle.

7. Pull the driveshaft rearward and disconnect from the transmission. Install a suitable plug in the extension housing to prevent lubricant leakage.

8. Remove the clutch housing dust shield and slave cylinder and secure it at one side. Remove the speedometer cable from the extension housing.

9. Disconnect the starter motor and back-up lamp switch wires.

10. Place a jack under the engine, protecting the oil pan with a wood block.

11. On 4 × 4 vehicles, remove the transfer case.

12. Remove the starter motor. Position a suitable jack under the transmission.

13. Remove the bolts, lockwashers and flat washers attaching the transmission to the engine rear plate.

14. Remove the nuts and bolts attaching the transmission mount and damper to the crossmember.

15. Remove the nuts attaching the crossmember to the frame side rails and remove the crossmember.

16. Lower the engine jack. Work the clutch housing off the locating dowels and slide the transmission rearward until the input shaft spline clears the clutch disc. Remove the transmission from the vehicle.

17. Make sure that the machined mating surfaces and the locating dowels on the engine rear plate are free of burrs, dirt or paint. Check the mating face of the clutch housing and the locating dowel holes for burrs, dirt or paint.

18. Mount the transmission on a suitable jack. Position it under the vehicle and start the input shaft into the clutch disc. Align the splines on the input shaft with the splines in the clutch disc. Move the transmission forward and carefully seat the clutch housing on the locating dowels of the engine rear plate. The engine plate dowels must not shave or burr the clutch housing dowel holes. Install the bolts and flat washers that attach the clutch housing to the engine rear plate and tighten to specifications. Remove the transmission jack.

19. Install the starter motor. Tighten the attaching nuts.

20. Raise the engine and install the rear crossmember, insulator and damper, and attaching nuts and washers.

21. Install the bolts, nuts and washers attaching the transmission mount to the crossmember. Remove the engine jack.

22. On 4 × 4 vehicles, install the transfer case.

23. Insert the driveshaft into the transmission extension housing and install the center bearing attaching nuts, washers and lockwashers.

24. Connect the driveshaft to the rear axle drive flange.

25. Connect the starter and back-up lamp switch wires. Install the clutch slave cylinder and dust shield on the clutch housing. Install the speedometer cable.

26. Check the transmission fluid level at both fill plugs. Fill with specified lubricant if necessary.

27. Lower the vehicle.

28. Open the hood and connect the negative battery cable to the battery terminal.

29. Re-connect the clutch master cylinder push rod to the clutch pedal.

30. Remove the cloth from the shift tower opening in the extension housing. Avoid getting dirt inside the transmission.

31. Position the gearshift lever assembly straight up above the gearshift lever retainer, then insert the gearshift in the retainer. Install the bolts attaching the retainer cover to the gearshift lever retainer and tighten them to specifications. Install the cover boot with the retainer screws.

Four Speed Transmission Diesel Engine

1. Place the gearshift lever in Neutral. Remove the boot retainer screws. Remove the bolts attaching the retainer cover to the gearshift lever retainer. Disconnect the clutch master cylinder push rod from the clutch pedal.

2. Pull the gearshift lever assembly, shim and bushing straight up and away from the gearshift lever retainer. Cover the shift tower opening in the extension housing with a cloth.

3. Disconnect the clutch hydraulic system master cylinder push rod from the clutch pedal.

4. Open the hood and disconnect the negative battery cable from the battery terminal.

5. Raise and safely support the vehicle on jackstands. Disconnect the driveshaft at the rear axle. Pull the driveshaft rearward and disconnect from the transmission. Install a suitable plug in the extension housing to prevent lubricant leakage.

6. Remove the clutch housing dust shield and slave cylinder and secure it at one side.

7. Remove the speedometer cable from the extension housing or from the speed control sensor, if so equipped.

8. Disconnect the starter motor and back-up lamp switch wires. Place a jack under the engine, protecting the oil pan with a wood block. Remove the starter motor. Position a suitable jack under the transmission.

9. Remove the bolts attaching the transmission to the engine rear plate. Remove the nuts and bolts attaching the transmission mount and damper to the crossmember.

10. Remove the nuts attaching the crossmember to the frame side rails and remove the crossmember.

11. Lower the engine jack. Work the clutch housing off the locating dowels and slide the transmission rearward until the input shaft spline clears the clutch disc. Remove the transmission from the vehicle.

13. Make sure that the machined mating surfaces and the locating dowels on the en-

gine rear plate are free of burrs, dirt or paint. Check the mating face of the clutch housing and the locating dowel holes for burrs, dirt or paint. Mount the transmission on a suitable jack. Position it under the vehicle and start the input shaft into the clutch disc. Align the splines on the input shaft with the splines in the clutch disc. Move the transmission forward and carefully seat the clutch housing on the locating dowels of the engine rear plate. The engine plate dowels must not shave or burr the clutch housing dowel holes.

14. Install the bolts that attach the clutch housing to the engine rear plate. Remove the transmission jack.

15. Install the starter motor.

16. Raise the engine and install the rear crossmember and attaching nuts and washers.

17. Install the bolts, nuts and washers attaching the transmission mount and damper to the crossmember. Remove the engine jack.

18. Insert the driveshaft into the transmission extension housing and install the center bearing attaching nuts, washers and lockwashers.

19. Connect the driveshaft to the rear axle drive flange.

20. Connect the starter and back-up lamp switch wires. Install the clutch slave cylinder and dust shield on the clutch housing. Install the speedometer cable.

21. Check the transmission fluid level at the fill plug. Fill with specified lubricant if necessary. Lower the vehicle.

22. Open the hood and connect the negative battery cable to the battery terminal.

23. Reconnect the clutch master cylinder push rod to the clutch pedal.

24. Remove the cloth from the shift tower opening in the extension housing. Avoid getting dirt inside the transmission.

25. Position the gearshift lever assembly straight up above the gearshift lever retainer, then insert the gearshift in the retainer. Install the bolts attaching the retainer cover to the gearshift lever retainer.

26. Install the cover boot with the retainer screws.

Transfer Case

REMOVAL & INSTALLATION

1. Raise and safely support the vehicle. Remove the skid plate from frame.

2. Place a drain pan under transfer case, remove the drain plug and drain fluid from the transfer case.

3. Disconnect the four-wheel drive indicator switch wire connector at the transfer case.

4. Disconnect the front driveshaft from the axle input yoke.

5. Loosen the clamp retaining the front driveshaft boot to the transfer case, and pull the driveshaft and front boot assembly out of the transfer case front output shaft.

6. Disconnect the rear driveshaft from the transfer case output shaft yoke.

7. Disconnect the speedometer driven gear from the transfer case rear cover. Disconnect the vent hose from the control lever.

8. Loosen or remove the large bolt and the small bolt retaining the shifter to the extension housing. Pull on the control lever until the bushing slides off the transfer case shift lever pin. If necessary, unscrew the shift lever from the control lever. Remove the heat shield from the transfer case.

CAUTION

The catalytic converter is located beside the heat shield. Be careful when working around the catalytic converter because of the extremely high temperatures generated by the converter.

9. Support the transfer case with a suitable jack. Remove the five bolts retaining the transfer case to the transmission and the extension housing.

10. Slide the transfer case rearward off the transmission output shaft and lower the transfer case from the vehicle. Remove the gasket from between the transfer case and extension housing.

11. Place a new gasket between the transfer case and the extension housing.

12. Raise the transfer case with a suitable jack so that the transmission output shaft aligns with the splined transfer case input shaft. Slide the transfer case forward onto the transmission output shafts and onto the dowel pin. Install the five bolts retaining the transfer case to the extension housing. Tighten bolts to 25–35 ft. lbs.

13. Remove the transmission jack from the transfer case.

14. Install the heat shield on the transfer case. Tighten the bolts to 27–37 ft. lbs.

15. Move the control lever until the bushing is in position over the transfer case shift lever pin. Install and hand start the attaching bolts. First, tighten the large bolt retaining the shifter to the extension housing to 70–90 ft. lbs., then the small bolt to 31–42 ft. lbs.

NOTE: Always tighten the large bolt retaining the shifter to the extension housing before tightening the small bolt.

16. Install the vent assembly so the white marking on the hose is in position in the notch in the shifter. The upper end of the vent hose should be two inches above the top of the shifter and positioned inside of the shift lever boot.

17. Connect the speedometer driven gear to the transfer case rear cover. Tighten the screw to 20–25 in. lbs.

18. Connect the rear driveshaft to the transfer case output shaft yoke. Tighten the bolts to 12–15 ft. lbs.

19. Clean the transfer case front output shaft female splines. Apply 5–8 grams of Multi-Purpose Long-Life Lubricant, C1AZ-19590-B(ESA-M1C175-B) or

equivalent to the splines. Insert the front driveshaft male spline.

20. Connect the front driveshaft to the axle input yoke. Tighten the bolts to 12–15 ft. lbs.

21. Push the driveshaft boot to engage the external groove on the transfer case front output shaft. Secure with a clamp.

22. Connect the four-wheel drive indicator switch wire connector at the transfer case. Install the drain plug and tighten to 14–22 ft. lbs. Remove the fill plug and install 3 pints of Dexron II, automatic transmission fluid. Install fill plug and tighten to 14–22 ft. lbs.

23. Install the skid plate to frame. Tighten nuts and bolts to 22–30 ft. lbs. Lower the vehicle.

Bronco

3.03 3 SPEED

Removal & Installation

1. Shift the transfer case into Neutral.

2. Remove the bolts attaching the fan shroud to the radiator support, if so equipped.

3. Raise and safely support the vehicle.

4. Support the transfer case shield with a jack and remove the bolts that attach the shield to the frame side rails. Remove the shield.

5. Drain the transmission and transfer case lubricant. To drain the transmission lubricant, remove the lower extension housing-to-transmission bolt.

6. Disconnect the front and rear driveshafts at the transfer case.

7. Disconnect the speedometer cable at the transfer case.

8. Disconnect the T.R.S. switch, if so equipped.

9. Disconnect the shift rods from the transmission shift levers. Place the First-Reverse gear shift lever into the First gear position and insert the fabricated tool. The tool consists of a length of rod, the same diameter as the holes in the shift levers, which is bent in such a way to fit in the holes in the two shift levers and hold them in the position stated above. More important, this tool will prevent the input shaft roller bearings from dropping into the transmission and output shaft. THIS TOOL IS A MUST.

10. Support the engine with a jack.

11. Remove the two cotter pins, bolts, washers, plate and insulators that secure the crossmember to the transfer case adapter.

12. Remove the crossmember-to-frame side support attaching bolts.

13. Position a transmission jack under the transfer case and remove the upper insulators from the crossmember. Remove the crossmember.

14. Roll back the boot enclosing the transfer case shift linkage. Remove the threaded cap holding the shift lever assem-

bly to the shift bracket. Remove the shift lever assembly.

15. Remove the two lower bolts attaching the transmission to the flywheel housing.

16. Reposition the transmission jack under the transmission and secure it with a chain.

17. Remove the two upper bolts securing the transmission to the flywheel housing. Move the transmission and transfer case rearward and downward out of the vehicle.

18. Move the assembly to a bench and remove the transfer case-to-transmission attaching bolts.

19. Slide the transmission assembly off the transfer case. To install the transmission:

20. Position the transfer case to the transmission. Apply an oil-resistant sealer to the bolt threads and install the attaching bolts. Tighten to 42–50 ft. lbs.

21. Position the transmission and transfer case on a transmission jack and secure them with a chain.

22. Raise the transmission and transfer case assembly into position and install the transmission case to the flywheel housing.

23. Install the two upper and two lower transmission attaching bolts and torque them to 37–42 ft. lbs.

24. Position the transfer case shift lever and install the threaded cap to the shift bracket. Reposition the rubber boot.

25. Raise the transmission and transfer case high enough to provide clearance for installing the crossmember. Position the upper insulators to the crossmember and install the crossmember-to-frame side support attaching bolts.

26. Align the bolt holes in the transfer case adapter with those in the crossmember, then lower the transmission and remove the jack.

27. Install the crossmember-to-transfer case adapter bolts, nuts, insulators, plates and washers. Tighten the nuts and secure them with cotter pins.

28. Remove the engine jack.

29. Remove the fabricated tool and connect each shift rod to its respective lever on the transmission. Adjust the linkage.

30. Connect the speedometer cable.

31. Connect the T.R.S. switch, if so equipped.

32. Install the front and rear driveshafts to the transfer case.

33. Fill the transmission and transfer case to the bottom of the filler hole with the recommended lubricant.

34. Position the transfer case shield to the frame side rails and install the attaching bolts.

35. Lower the vehicle.

36. Install the fan shroud, if so equipped.

37. Check the operation of the transfer case and the transmission shift linkage.

NP435

1. Remove the rubber boot and floor mat.

2. Remove the weather pad. It may be necessary first to remove the seat assembly.

3. Disconnect the back-up light switch located in the rear of the gearshift housing cover.

4. Raise the vehicle and position safety stands. Position a transmission jack under the transmission, and disconnect the speedometer cable.

5. Disconnect the parking brake lever from its linkage, and remove the gearshift housing.

6. Disconnect the driveshaft or coupling shaft. Remove the bolts that attach the coupling shaft center support to the crossmember and wire the coupling shaft and driveshaft to one side. Remove the transfer case.

7. Remove the transmission attaching bolts at the clutch housing, and remove the transmission.

8. Before installing the transmission apply a light film of chassis lubricant to the release lever fulcrum and fork. Do not apply a thick coat of grease to these parts, as it will work out and contaminate the clutch disc.

9. Place the transmission on a transmission jack, and raise the transmission until the input shaft splines are aligned with the clutch disc splines. The clutch release bearing and hub must be properly positioned in the release lever fork.

10. Install guide studs in the clutch housing and slide the transmission forward on the guide studs until it is in position on the clutch housing. Install the attaching bolts or nuts, and tighten them to the following torques: $7/16$–14: 40–50 ft. lbs. $5/8$–11: 120–150 ft. lbs. $9/16$–12: 90–115 ft. lbs. $5/8$–18C: 120–150 ft. lbs. $9/16$–18C: 90–115 ft. lbs.

11. Remove the guide studs and install the two lower attaching bolts. Install the bolts attaching the coupling shaft center support to the crossmember. Tighten the bolts to 40–50 ft. lbs.

12. Connect the driveshaft or coupling shaft and the speedometer cable. Tighten the U-joint nuts.

13. Connect the back-up light switch wire.

14. Install the transmission cover plate. Install the seat assembly if it was removed.

15. Install weather pad, pad retainer, floor mat, and rubber boot.

T–18 through 1979

1. Open door cover seat.

2. Remove shift knobs.

3. Remove the four screws attaching the transmission shift lever boot assembly.

4. Remove the four screws holding the floor mat.

5. Remove the eleven screws holding the access cover to the floor pan. Place the shift lever in the reverse position and remove the cover.

6. Remove the insulator and dust cover.

7. Remove the transfer case shift lever.

8. Remove the eight bolts holding the shift cover and gasket.

9. Use cardboard or heavy paper to fabricate a suitable cover for the shift cover opening to protect the transmission from dirt during removal.

10. Raise and safely support the vehicle on jackstands.

11. Remove the drain plug and drain the transmission.

12. Disconnect the rear driveshaft from the transfer case and wire it out of the way.

13. Disconnect the front driveshaft from the transfer case and wire it out of the way.

14. Remove the cotter key that holds the shift link in place and remove the shift link.

15. Remove the speedometer cable from the transfer case.

16. Position a transmission jack under the transfer case. Remove the six bolts holding the transfer case to the transmission and lower the transfer case from the vehicle.

17. Remove the eight bolts that hold the rear support bracket to the transmission.

18. Position a transmission jack under the transmission and remove the rear support bracket and brace.

19. Remove the four bolts that hold the transmission to the bell housing.

20. Remove the transmission from the vehicle.

21. Place the transmission on a transmission jack and install it in the vehicle installing two guide studs in the bell housing top holes, to guide the transmission into position.

22. Install the two lower bolts. Remove the guide studs and install the upper bolts.

23. Place the rear support bracket in position and install the eight retaining bolts.

24. Install the two bolts at the rear support insulator bracket. Remove the transmission jack.

25. Position the transfer case on the transmission jack and install the six retaining bolts and gasket. Position the transfer case on the transmission and tighten the bolts to 50–60 ft. lbs.

26. Install the transfer case shift link and cotter pin.

27. Position and install the speedometer cable.

28. Remove wire and connect front driveshaft.

29. Remove wire and connect rear driveshaft.

30. Fill transfer case and manual transmission with lubricant.

31. Lower the vehicle.

32. Remove fabricated dirt shield and prepare gasket area.

33. Position gasket and shift cover.

34. Install two pilot bolts, then install remaining shift cover retaining bolts.

35. Install transfer case shift handle.

36. Install dust cover and insulator.

37. Install access cover to floor pan screws.

38. Install the four floor mat screws.

39. Install the four boot area screws.
40. Install the shift knobs.

T–18 from 1980

1. Open door cover seat.
2. Remove shift knobs.
3. Remove the four screws attaching the transmission shift lever boot assembly.
4. Remove the four screws holding the floor mat.
5. Remove the eleven screws holding the access cover to the floor pan. Place the shift lever in the reverse position and remove the cover.
6. Remove the insulator and dust cover.
7. Remove the transfer case shift lever.
8. Remove transmission shift lever.
9. Raise and safely support the vehicle.
10. Remove the drain plug and drain the transmission.
11. Disconnect the rear driveshaft from the transfer case and wire it out of the way.
12. Disconnect the front driveshaft from the transfer case and wire it out of the way.
13. Remove the retainer ring that holds the shift link in place and remove the shift link from transfer case.
14. Remove the speedometer cable from the transfer case.
15. Position a transmission jack under the transfer case. Remove the six bolts holding the transfer case to the transmission and lower the transfer case from the vehicle.
16. Remove the eight bolts that hold the rear support bracket to the transmission.
17. Position a transmission jack under the transmission and remove the rear support bracket and brace.
18. Remove the four bolts that hold the transmission to the bell housing.
19. Remove the transmission from the vehicle.
21. Place the transmission on a transmission jack and install it in the vehicle installing two guide studs in the bell housing top holes, to guide the transmission into position.
22. Install the two lower bolts. Remove the guide studs and install the upper bolts.
23. Place the rear support bracket in position and install the eight retaining bolts. Torque to 35–50 ft. lbs.
24. Install the two bolts at the rear support insulator bracket. Remove the transmission jack.
25. Position the transfer case on the transmission jack and install the six retaining bolts and gasket. Position the transfer case on the transmission and tighten the bolts to 28–33 ft. lbs.
26. Install the transfer case shift link and retainer ring.
27. Position and install the speedometer cable.
28. Remove wire and connect front driveshaft.
29. Remove wire and connect rear driveshaft.
30. Fill transfer case and transmission.

31. Lower the vehicle.
32. Remove fabricated dirt shield and prepare gasket area.
33. Position gasket and shift cover.
34. Install two pilot bolts, then install remaining shift cover retaining bolts.
35. Install transfer case shift handle and transmission shift lever.
36. Install dust cover and insulator.
37. Install access cover to floor pan screws.
38. Install the four floor mat screws.
39. Install the four boot area screws.
40. Install the shift knobs.

Single Rail Overdrive

1. Raise the vehicle and support it on jackstands.
2. Mark the driveshaft so that it may be installed in the same relative position. Disconnect the driveshaft from the rear U-joint flange. Slide the driveshaft off the transmission output shaft and install an extension housing seal installation tool, or rags into the extension housing to prevent lubricant leakage.
3. Disconnect the speedometer cable from the extension housing.
4. Remove three screws securing shift lever to turret assembly.
5. Remove shift lever from turret assembly.
6. Support the engine with a transmission jack and remove the extension housing-to-engine rear support attaching bolts.
7. Raise the rear of the engine high enough to remove the weight from the crossmember. Remove the bolts retaining the crossmember to the frame side supports and remove the crossmember.
8. Support the transmission on a jack and remove the bolts that attach the transmission to the flywheel housing.
9. Move the transmission and jack rearward until the transmission input shaft clears the flywheel housing. If necessary, lower the engine enough to obtain clearance for transmission removal.

————— CAUTION —————
Do not depress the clutch pedal while the transmission is removed.

10. Make sure that the mounting surface of the transmission and the flywheel housing are free of dirt, paint, and burrs. Install two guide pins in the flywheel housing lower mounting bolt holes. Move the transmission forward on the guide pins until the input shaft splines enter the clutch hub splines and the case is positioned against the flywheel housing.
11. Install the two upper transmission to flywheel housing mounting bolts snug, and then remove the two guide pins. Install the two lower mounting bolts. Torque all mounting bolts to specifications.
12. Raise the rear of the engine and install the crossmember. Install and torque the crossmember attaching bolts to specifications, then lower the engine.

13. With the transmission extension housing resting on the engine rear support, install the transmission extension housing attaching bolts. Torque the bolts to specifications.
14. Position shift tower to extension housing and secure with three screws.
15. Connect the speedometer cable to the extension housing.
16. Remove the extension housing installation tool and slide the forward end of the driveshaft over the transmission output shaft. Connect the driveshaft to the rear U-joint flange.
17. Fill the transmission to the proper level with the specified lubricant.
18. Lower the truck. Check the shift and crossover motion for full shift engagement and smooth crossover operation.

Vans

REMOVAL & INSTALLATION

3 Speed

1. Raise the vehicle and support on jackstands. Drain the lubricant from the transmission by removing the lower extension housing-to-transmission bolt.
2. Disconnect the driveshaft from the flange at the transmission. Secure the front end of the driveshaft out of the way by tying it up with a length of wire.
3. Disconnect the speedometer cable from the extension housing and disconnect the gearshift rods from the transmission. Disconnect the transmission regulated spark switch, if so equipped.
4. Position a transmission jack under the transmission. Chain the transmission to the jack.
5. Raise the transmission slightly and remove the 4 bolts which retain the transmission support crossmember to the frame side rails. Remove the bolt which retains the transmission extension housing to the crossmember.
6. Remove the 4 transmission-to-flywheel housing bolts.
7. Position a bar under the rear of the engine to support it.
8. Remove the transmission from the vehicle by lowering the jack. To install the transmission:
9. Make sure that the machined surfaces of the transmission case and the flywheel housing are free of dirt, paint, and burrs.
10. Install a guide pin in each lower mounting bolt hole.
11. Start the input shaft through the release bearing. Align the splines on the input shaft with the splines in the clutch disc. Move the transmission forward on the guide pins until the input shaft pilot enters the bearing or bushing in the crankshaft. If the transmission front bearing retainer binds up on the clutch release bearing hub, work the release bearing lever until the hub slides

onto the transmission front bearing retainer. Install the two upper mounting bolts and lockwashers which attach the flywheel housing to the transmission. Remove the two guide pins and install the lower mounting bolts and lockwashers.

12. Raise the jack slightly and remove the engine support bar. Position the support crossmember on the frame side rails and install the retaining bolts. Install the extension housing-to-crossmember retaining bolt.

13. Connect the gearshift rods and the speedometer cable. Connect the transmission regulated spark switch lead, if so equipped.

14. Install the driveshaft.

15. Fill the transmission to the bottom of the filler hole with the proper lubricant.

16. Adjust the clutch pedal free-play and the shift linkage as required.

4 Speed Overdrive

1. Raise the vehicle and support on jackstands.

2. Mark the driveshaft so that it may be installed in the same relative position. Disconnect the driveshaft from the rear U-joint flange. Slide the driveshaft off the transmission output shaft and install the extension housing seal installation tool into the extension housing to prevent lubricant leakage.

3. Disconnect the speedometer cable from the extension housing.

4. Remove the retaining clips, flat washers, and spring washers that secure the shift rods to the shift levers. Remove the bolts connecting the shift control to the transmission extension housing. Remove the nut connecting the shift control to the transmission case.

NOTE: A '6' and '8' is stamped on transmission extension housing by the shift control plate bolt holes. The '6' and '8' refer to either a 6 or 8 cylinder engine application. The shift control plate bolts must be placed in the right holes for proper plate positioning dependent upon engine application.

5. Remove the rear transmission support connecting bolts attaching the support on the crossmember to the transmission extension housing.

6. Support the engine with a transmission jack and remove the extension housing-to-engine rear support attaching bolts.

7. Raise the rear of the engine high enough to remove the weight from the crossmember. Remove the bolts retaining the crossmember to the frame side supports and remove the crossmember.

8. Support the transmission on a jack and remove the bolts that attach the transmission to the flywheel housing.

9. Move the transmission and jack rearward until the transmission input shaft clears the flywheel housing. If necessary, lower the engine enough to obtain clearance for transmission removal. Do not depress the clutch pedal while the transmission is removed.

10. Make sure that the mounting surfaces of the transmission and the flywheel housing are free of dirt, paint, and burrs. Install two guide pins in the flywheel housing lower mounting bolt holes. Move the transmission forward on the guide pins until the input shaft splines enter the clutch hub splines and the case is positioned against the flywheel housing.

11. Install the two upper transmission to flywheel housing mounting bolts snug, and then remove the two guide pins. Install the two lower mounting bolts. Tighten all mounting bolts to 40–45 ft. lbs.

12. Raise the rear of the engine and install the crossmember. Install and torque the crossmember attaching bolts to 20–30 ft. lbs. then lower the engine.

13. With the transmission extension housing resting on the engine rear support, install the transmission extension housing attaching bolts. Tighten the bolts to 42–50 ft. lbs.

14. Install the transmission support bolts and tighten to 40–50 ft. lbs.

15. Position the shift control bracket on the stud on the transmission case and on the bolt attaching holes (holes marked either '6' or '8' dependent upon 6 or 8 cylinder engine application) on the transmission extension housing. Install and hand tighten connecting bolts. The bracket must be placed in the proper position for correct shift control operation. Tighten the nut connecting the bracket to the transmission case to 22–30 ft. lbs. Tighten the bolts to 22–30 ft. lbs.

16. Secure each shift rod to its respective lever with the spring washer, flat washer, and retaining pin.

17. Connect the speedometer cable to the extension housing.

18. Remove the extension housing installation tool and slide the forward end of the driveshaft over the transmission output shaft. Connect the driveshaft to the rear U-joint flange. Adjust the linkage.

19. Fill the transmission to the proper level with the specified lubricant.

20. Lower the vehicle. Check the shift and crossover motion for full shift engagement and smooth crossover operation.

CLUTCH

Pickup

NOTE: Refer to the Ranger/Bronco II Section for instructions on those vehicles.

FREEPLAY ADJUSTMENT MANUAL LINKAGE

1. Measure the clutch pedal free-play

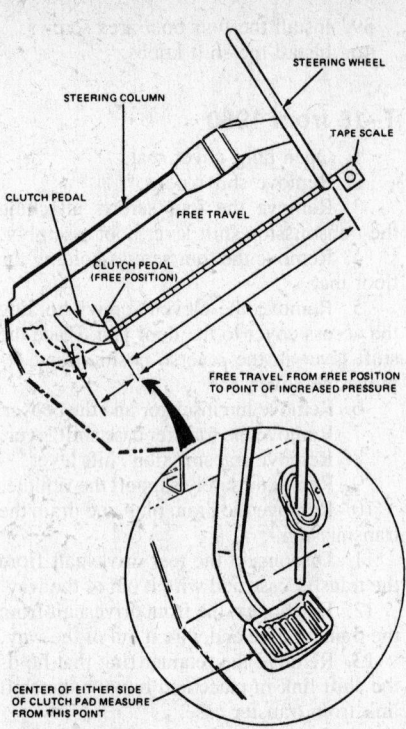

Clutch freeplay measurement—Bronco and pickup

by depressing the pedal slowly until the free-play between the release bearing assembly and the pressure plate is removed. Note this measurement. The difference between this measurement and when the pedal is not depressed is the free-play measurement.

2. If the free-play measurement is less than ½–¾ inch, the clutch linkage must be adjusted.

3. Loosen the two jam nuts on the release rod under the truck and back off both nuts several turns.

4. Loosen or tighten the first jam nut (nearest the release lever) against the bullet (rod extension) until a free-play measurement of ¾–1½ in. is obtained. A free-play measurement closer to 1½ in. is more desirable.

5. When the correct free-play measurement is obtained, hold the first jam nut in position and securely tighten the other nut against the first.

6. Recheck the free-play adjustment. Total pedal travel is fixed and is not adjustable.

HYDRAULIC CLUTCH

The Hydraulic clutch system consists of a combination clutch and master cylinder assembly, a slave cylinder and connecting tubing. The slave cylinder is mounted on the bell housing. The hydraulic clutch system provides automatic clutch adjustment. No adjustment of the clutch linkage or pedal position is required.

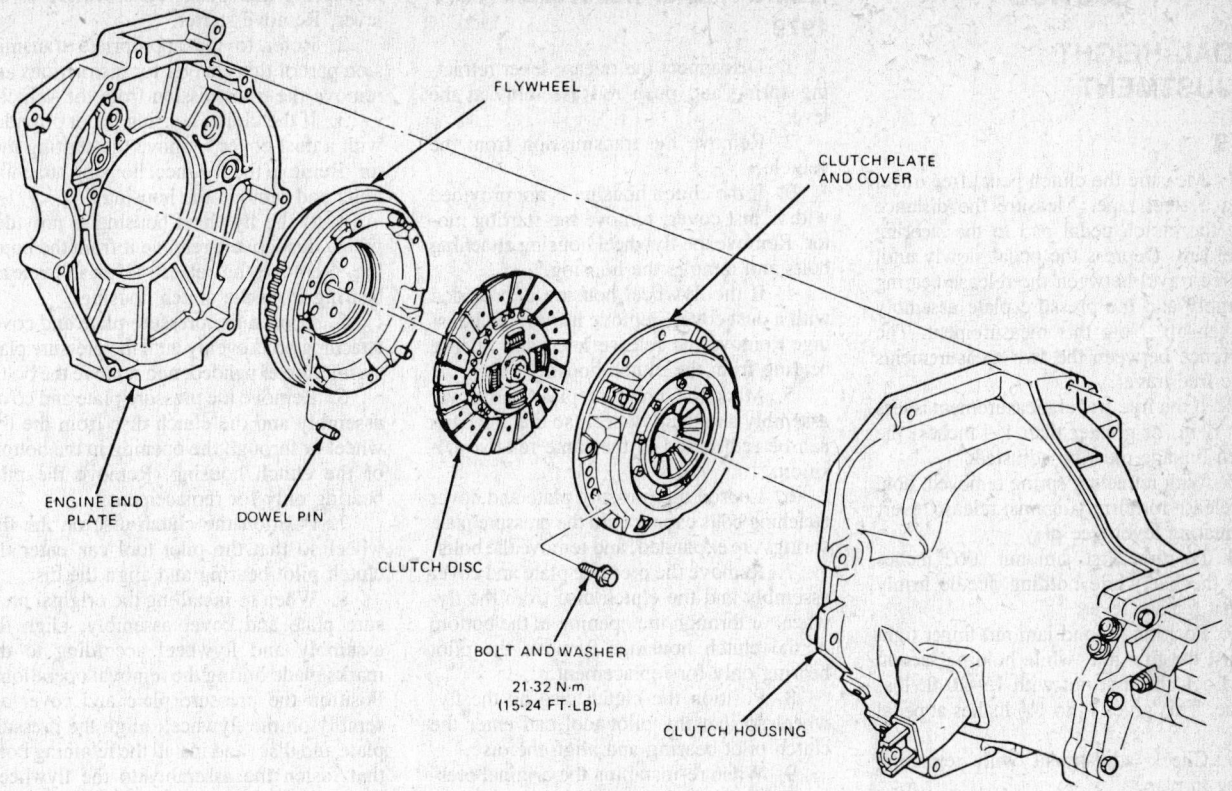

FLYWHEEL

CLUTCH PLATE
AND COVER

ENGINE END
PLATE

DOWEL PIN

CLUTCH DISC

BOLT AND WASHER

21-32 N·m
(15-24 FT-LB)

CLUTCH HOUSING

Clutch installation-134 (2.2L) diesel

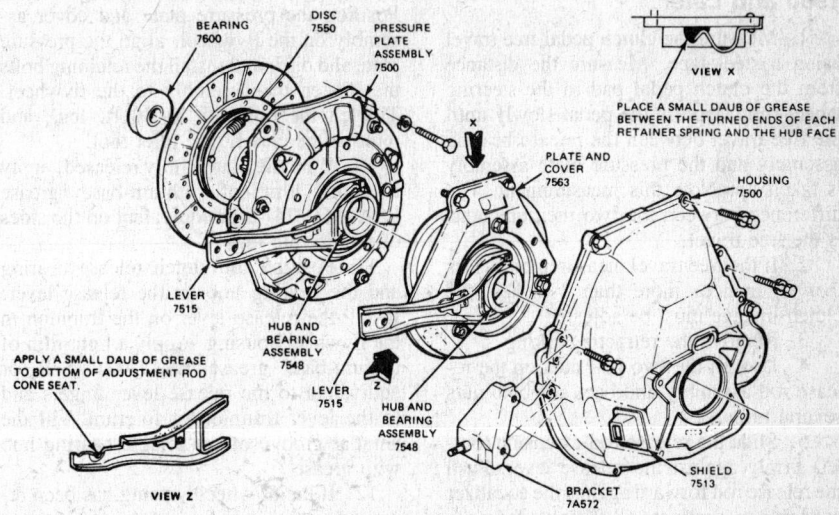

BEARING
7600

DISC
7550

PRESSURE
PLATE
ASSEMBLY
7500

VIEW X

PLACE A SMALL DAUB OF GREASE
BETWEEN THE TURNED ENDS OF EACH
RETAINER SPRING AND THE HUB FACE

PLATE
AND
COVER
7563

HOUSING
7500

LEVER
7515

HUB AND
BEARING
ASSEMBLY
7548

APPLY A SMALL DAUB OF GREASE
TO BOTTOM OF ADJUSTMENT ROD
CONE SEAT.

LEVER
7515

HUB AND
BEARING
ASSEMBLY
7548

SHIELD
7513

VIEW Z

BRACKET
7A572

8-351, 400 clutch installation

REMOVAL & INSTALLATION

1. Disconnect the release lever retracting spring and pushrod assembly. If equipped with an hydraulic clutch, remove the slave cylinder with line connected.

2. Remove the transmission.

3. If the clutch housing does not have a dust cover, remove the starter. Remove the flywheel housing attaching bolts and remove the housing.

4. If the flywheel housing does have a dust cover, remove the cover and then remove the release lever and bearing from the clutch housing.

5. Mark the pressure plate and cover assembly and the flywheel so that they can be reinstalled in the same relative position.

6. Loosen the pressure plate and cover attaching bolts evenly in a staggered sequence a turn at a time until the pressure plate springs are relieved of their tension. Remove the attaching bolts.

7. Remove the pressure plate and cover assembly and the clutch disc from the flywheel.

8. Position the clutch disc on the flywheel so that an aligning tool or spare transmission mainshaft can enter the clutch pilot bearing and align the disc.

9. When reinstalling the original pressure plate and cover assembly, align the assembly and flywheel according to the marks made during removal. Position the pressure plate and cover assembly on the

flywheel, align the pressure plate and disc, and install the retaining bolts. Tighten the bolts in an alternating sequence a few turns at a time until 15–20 ft. lbs. is reached.

10. Remove the tool used to align the clutch disc.

11. With the clutch fully released, apply a light coat of grease on the sides of the driving lugs.

12. Position the clutch release bearing and the bearing hub on the release lever. Install the release lever on the fulcrum in the flywheel housing. Apply a light coating

of grease to the release lever fingers and the fulcrum. Fill the groove of the release bearing hub with grease.

13. If the flywheel housing has been removed, position it against the rear engine cover plate and install the attaching bolts and tighten them to 40–50 ft. lbs.

14. Install the starter motor.

15. Install the transmission.

16. Connect the release lever retracting spring and install the dust cover, if so equipped.

17. Adjust the clutch linkage.

Bronco

PEDAL HEIGHT ADJUSTMENT

1979

1. Measure the clutch pedal free travel using a steel tape. Measure the distance from the clutch pedal pad to the steering wheel rim. Depress the pedal slowly until the free travel between the release bearing assembly and the pressure plate assembly is taken up. Note this measurement. The difference between the two measurements is the free travel.

2. If the free-travel measurement is less than ¾ in. or greater than 1½ inches, the clutch linkage must be adjusted.

3. With retracting spring removed, hold the release rod firmly against release lever, eliminating lever free play.

4. Position first jam nut .062 inches from the bar while holding needle firmly against release lever.

5. Position second jam nut finger tight against the first nut, while holding second nut. Lock the first nut with 15–20 ft. lbs. torque. This gives ¾ to 1½ inches at pedal pad.

6. Check adjustment with retracting spring in place.

1980 and Later

1. Measure the clutch pedal free travel using a steel tape. Measure the distance from the clutch pedal pad to the steering wheel rim. Depress the pedal slowly until the free travel between the release bearing assembly and the pressure plate assembly is taken up. Note this measurement. The difference between the two measurements is the free travel.

2. If the free travel measurement is less than ½ inch or more than 2 inches, the clutch linkage must be adjusted.

3. Remove the retracting spring.

4. Loosen the two jam nuts on the release rod assembly and back off both nuts several turns.

5. Slide the release rod extension (bullet) firmly against the release lever. Push the release rod forward against the equalizer bar lever to eliminate all freeplay from the linkage system.

6. Insert 0.135 inch thick gauge between the jam nut and bullet. Tighten the first jam nut finger tight against the gauge with all freeplay eliminated.

7. Tighten the second jam nut finger tight against the first jam nut. Hold the first nut and tighten the second jam nut to 15–20 ft. lbs. Freeplay should measure ¾–1½ inches at the pedal.

8. With the recommended free travel obtained, and holding the first jam nut, position and securely tighten the second jam nut against the first jam nut.

9. Re-check the pedal free travel.

REMOVAL & INSTALLATION

1979

1. Disconnect the release lever retracting spring and push rod assembly at the lever.

2. Remove the transmission from the vehicle.

3. If the clutch housing is not provided with a dust cover, remove the starting motor. Remove the flywheel housing attaching bolts and remove the housing.

4. If the flywheel housing is provided with a dust cover, remove it from the housing. Remove the release lever and release bearing from the clutch housing.

5. Mark the pressure plate and cover assembly and the flywheel, so that the parts can be reinstalled in the same relative position.

6. Loosen the pressure plate and cover attaching bolts evenly until the pressure plate springs are expanded, and remove the bolts.

7. Remove the pressure plate and cover assembly and the clutch disc from the flywheel or through the opening in the bottom of the clutch housing. Remove the pilot bearing only for replacement.

8. Position the clutch disc on the flywheel so that the pilot tool can enter the clutch pilot bearing and align the disc.

9. When re-installing the original pressure plate and cover assembly, align the assembly and flywheel according to the marks made during the removal operations. Position the pressure plate and cover assembly on the flywheel, align the pressure plate and disc, and install the retaining bolts that fasten the assembly to the flywheel. Tighten the bolts to 20–30 ft. lbs., and remove the clutch disc pilot tool.

10. With the clutch fully released, apply a light film of lithium-base grease ESA–M1C75–B or equivalent on the sides of the driving lugs.

11. Position the clutch release bearing and the bearing hub on the release lever. Install the release lever on the trunnion in the flywheel housing. Apply a light film of lithium-base grease ESA–M1C75–B or equivalent to the release lever fingers and to the lever trunnion or fulcrum. Fill the annular groove of the release bearing hub with grease.

12. If the flywheel housing has been removed, position it against the engine rear cover plate and install the attaching bolts. Tighten the bolts to 40–50 ft. lbs.

13. Install the starter motor. Install the transmission assembly on the clutch housing. Tighten the bolts to 50–60 ft. lbs.

14. Install the slave cylinder on vehicles so equipped, and tighten the bolts.

15. Adjust the release lever push rod assembly. Connect the release lever retracting spring.

16. Install the clutch housing dust cover if so equipped.

1980 and Later

1. Disconnect the release lever retracting spring and push rod assembly at the lever. Remove starter.

2. Refer to the appropriate transmission part of this chapter for instructions and remove the transmission from the vehicle.

3. If the clutch housing is not provided with a dust cover, remove the starting motor. Remove the flywheel housing attaching bolts and remove the housing.

4. If the flywheel housing is provided with a dust cover, remove it from the housing. Remove the release lever and release bearing from the clutch housing.

5. Loosen the pressure plate and cover attaching bolts evenly until the pressure plate springs are expanded, and remove the bolts.

6. Remove the pressure plate and cover assembly and the clutch disc from the flywheel or through the opening in the bottom of the clutch housing. Remove the pilot bearing only for replacement.

7. Position the clutch disc on the flywheel so that the pilot tool can enter the clutch pilot bearing and align the disc.

8. When re-installing the original pressure plate and cover assembly, align the assembly and flywheel according to the marks made during the removal operations. Position the pressure plate and cover assembly on the flywheel, align the pressure plate and disc, and install the retaining bolts that fasten the assembly to the flywheel. Tighten the bolts to 20–30 ft. lbs., and remove the clutch disc pilot tool.

9. Position the clutch release bearing and the bearing hub on the release lever. Install the release lever on the pivot bar pedestal in the flywheel housing. Apply a light film of lithium-base grease ESA–M1C75–B or equivalent to the release lever fingers and to the lever pivot ball. Fill the annular groove of the release bearing hub with grease.

10. If the flywheel housing has been removed, position it against the engine rear cover plate and install the attaching bolts. Tighten the bolts to 40–50 ft. lbs.

11. Install the starter motor. Install the transmission assembly on the clutch housing. Tighten the bolts.

12. Adjust the release lever push rod assembly. Connect the release lever retracting spring.

13. Install the clutch housing dust cover if so equipped.

Ranger and Bronco II

REMOVAL & INSTALLATION

1. Disconnect the clutch hydraulic system master cylinder from the clutch pedal.

2. Raise the vehicle and safely support on jack stands. Remove the dust shield from the clutch housing.

3. Disconnect the hydraulic clutch

linkage from the housing and release lever. Remove the starter.

4. Remove the bolts attaching the clutch housing to the engine block. Note the direction in which the bolts are installed.

5. Index the driveshaft to the companion flange and remove the driveshaft.

6. Remove the nuts attaching the transmission and insulator to the No.2 crossmember support. Raise the transmission with a suitable jack. Remove the No.2 crossmember support. Lower the transmission and clutch housing.

7. Remove the release lever, and hub and bearing. Mark the assembled position of the pressure plate and cover to the flywheel (for re-assembly).

8. Loosen the pressure plate and cover attaching bolts evenly until the pressure plate springs are expanded, and remove the bolts.

9. Remove the pressure plate and cover assembly and the clutch disc from the flywheel. These parts can be removed through the opening in the bottom of the clutch housing on models where the housing is fitted with a dust cover. Remove the pilot bearing only for replacement.

10. Position the clutch disc on the flywheel so that the Clutch Alignment Shaft D79T–7550–A or equivalent can enter the clutch pilot bearing and align the disc.

11. When re-installing the original pressure plate and cover assembly, align the assembly and flywheel according to the marks made during the removal operations. Position the pressure plate and cover assembly on the flywheel, align the pressure plate and disc, and install the retaining bolts that fasten the assembly to the flywheel. Tighten the bolts to 15–24 ft. lbs., and remove the clutch disc pilot tool.

12. Position the clutch release bearing and the bearing hub on the release lever. Install the release lever on the release lever seat in the flywheel housing. Apply a light film of lithium-base grease C1AZ–19590–B (ESA–M1C75–B) or equivalent to the release lever fingers and to the lever pivot ball. Fill the annular groove of the release bearing hub with grease.

13. Raise the transmission and clutch housing into position. Install the No.2 crossmember support to the frame. Install connecting nuts, bolts and washers.

14. Lower the transmission and insulator into the support. Install and tighten nuts. Remove the transmission jack.

15. Install the driveshaft, making sure the index marks on the driveshaft are aligned with the marks on the companion flange.

16. Install the bolts attaching the housing to the engine block in the correct position as removed. Tighten to 28–38 ft. lbs.

17. Install the hydraulic clutch linkage on the housing in position with the release lever. Install dust shield. Install starter.

18. Lower the vehicle and connect the clutch hydraulic system master cylinder to the clutch pedal. Check clutch for proper operation.

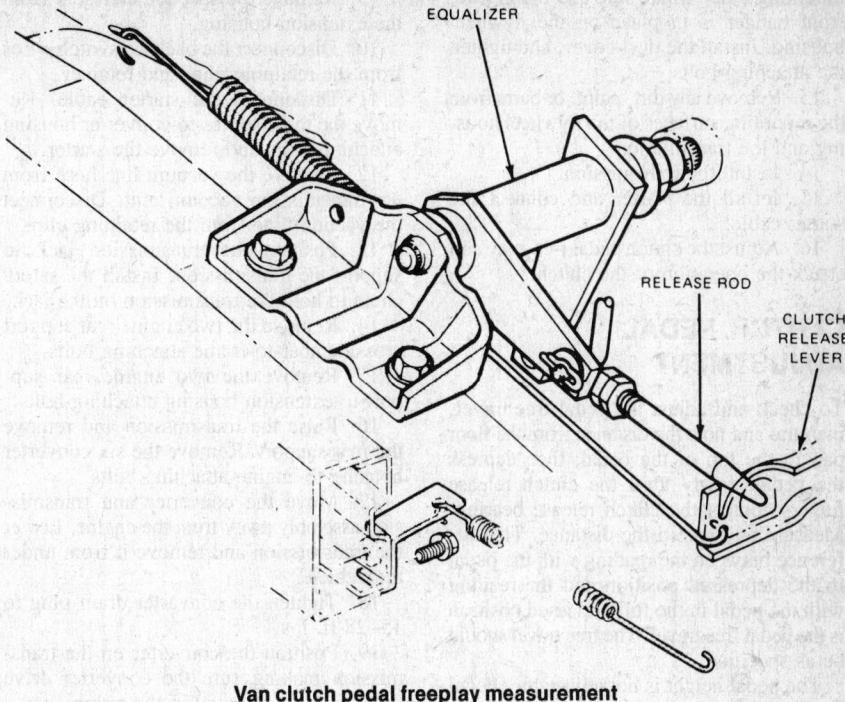

Van clutch pedal freeplay measurement

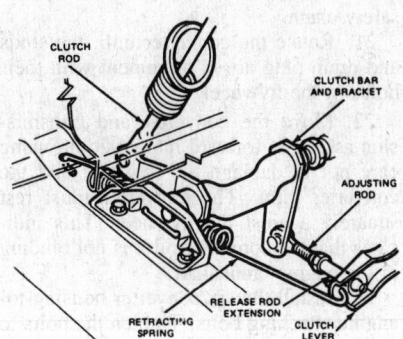

1980 and later van clutch pedal free travel adjustment

Vans

FREEPLAY ADJUSTMENT

To check and adjust the pedal free travel measure and note the distance from the floor pan to the top of the pedal: then depress the pedal slowly until the clutch release fingers contact the clutch release bearing. Measure and record the distance. The difference between the reading with the pedal in the depressed position and the reading with the pedal in the fully released position is the pedal free travel. The free travel should be as specified.

The pedal height is not adjustable. Pedal height is adjusted by loosening the nut securing the clutch pedal eccentric bumper and rotating the bumper until the clutch pedal height is within $1\frac{1}{4}$–$1\frac{1}{2}$ inches.

REMOVAL & INSTALLATION

1. Disconnect the cable from the starter and remove the starter.

2. Remove the transmission.

3. Disconnect the release lever retracting spring and release the rod.

4. Remove the hub and release bearing assembly.

5. Remove the flywheel housing-to-engine bolts, and lower the flywheel housing.

6. Remove the pressure plate and the disc from the flywheel. Unscrew the attaching bolts a few turns at a time, in a staggered sequence to prevent distortion of the pressure plate.

7. Wash the flywheel surface with alcohol. Do not use an oil-base cleaner, carbon tetrachloride or gasoline. To install the clutch:

8. Place the clutch disc and the pressure plate and cover assembly in position on the flywheel. Start the retaining bolts until finger-tight.

9. Align the clutch disc with a clutch arbor (an old mainshaft works well) and then evenly torque the bolts to 23–28 ft. lbs.

10. Do not grease the release lever pivot assembly. Crimp the dust seal tabs flush against the flywheel housing. Attach the springs of the release bearing hub to the ends of the release fork. Be careful not to distort the springs.

11. Fill the groove in the clutch release bearing hub with lithium-base grease. Wipe the excess grease from the hub.

12. Position the flywheel housing and release lever assembly, and install the

mounting bolts. Make sure that the muffler front hanger is in place on the flywheel housing. Install the dust cover, and tighten the attaching bolts.

13. Remove any dirt, paint, or burrs from the mounting surfaces of the flywheel housing and the transmission.

14. Install the transmission.

15. Install the starter and connect the starter cable.

16. Adjust the clutch pedal free-play and check the operation of the clutch.

CLUTCH PEDAL ADJUSTMENT

To check and adjust the pedal free travel, measure and note the distance from the floor pan to the top of the pedal; then depress the pedal slowly until the clutch release fingers contact the clutch release bearing. Measure and record the distance. The difference between the reading with the pedal in the depressed position and the reading with the pedal in the fully released position is the pedal free travel. The free travel should be as specified.

The pedal height is not adjustable. Pedal height is adjusted by loosening the nut securing the clutch pedal eccentric bumper and rotating the bumper until the clutch pedal height is within $1\frac{1}{4}$–$1\frac{1}{2}$ inches.

AUTOMATIC TRANSMISSION

REMOVAL & INSTALLATION

C4, C5

F100–250

1. Raise the vehicle and support on jackstands.

2. Place the drain pan under the transmission fluid pan. Remove the fluid filler tube from the pan and drain the transmission fluid.

3. Remove the converter drain plug access cover from the lower end of the converter housing.

4. Remove the converter-to-flywheel attaching nuts. Place a wrench on the crankshaft pulley attaching bolt to turn the converter to gain access to the nuts.

5. With the wrench on the crankshaft pulley attaching bolt, turn the converter to gain access to the converter drain plug and remove the plug. Place a drain pan under the converter to catch the fluid. With fluid drained, reinstall the plug.

6. Remove the driveshaft.

7. Disconnect the oil cooler lines from the transmission.

8. Disconnect the manual and downshift linkage rods from the transmission control levers.

9. Remove the speedometer gear from the extension housing.

10. Disconnect the back-up switch wires from the retaining clips and retainer.

11. Disconnect the starter cable. Remove the three starter-to-converter housing attaching bolts and remove the starter.

12. Remove the vacuum line hose from the transmission vacuum unit. Disconnect the vacuum line from the retaining clip.

13. Position the transmission jack to support the transmission. Install the safety chain to hold the transmission on the jack.

14. Remove the two engine rear support crossmember-to-frame attaching bolts.

15. Remove the two engine rear support-to-extension housing attaching bolts.

16. Raise the transmission and remove the rear support. Remove the six converter housing-to-engine attaching bolts.

17. Move the converter and transmission assembly away from the engine. Lower the transmission and remove it from under the vehicle.

18. Tighten the converter drain plug to 15–28 ft. lbs.

19. Position the converter on the transmission making sure the converter drive flats are fully engaged in the pump gear.

20. With the converter properly installed, place the transmission on the jack. Secure the transmission to the jack with the safety chain.

21. Rotate the converter until the studs and drain plug are in alignment with their holes in the flywheel.

22. Move the converter and transmission assembly forward into position, using care not to damage the flywheel and the converter pilot. The converter must rest squarely against the flywheel. This indicates that the converter pilot is not binding in the engine crankshaft.

23. Install the six converter housing-to-engine attaching bolts. Tighten the bolts to 40–50 ft. lbs.

24. Install the converter-to-flywheel attaching nuts. Tighten the nuts to 20–34 ft. lbs. Remove the safety chain from the transmission.

25. Install the rear support. Install the rear support-to-extension housing attaching bolts. Tighten the bolts.

26. Position the starter into the converter housing and install the three attaching bolts. Tighten the bolts to 20–30 ft. lbs. Install the starter cable.

26. Remove the transmission jack.

27. Connect the transmission filler tube to the transmission pan. Connect the oil cooler lines to the transmission.

28. Attach the back-up switch wires to the connector clip.

29. Install the speedometer driven gear in the extension housing. Tighten the attaching bolt.

30. Connect the transmission linkage rods to the transmission control levers. When making transmission control attachments, new retaining rings and grommets should always be used. Attach the shift rod to the

steering column shift lever. Align the flats of the adjusting stud with the flats of the rod slot and insert the stud through the rod. Assemble the adjusting stud nut and washer to a loose fit. Perform a linkage adjustment.

32. Install the driveshaft.

33. Install the vacuum line in the retaining clip. Connect the vacuum line to the diaphragm unit.

34. At the front lower area of the converter housing, install the lower cover and the control lever dust shield. Install the attaching bolts. Tighten the bolts.

35. Secure the fluid filler tube to the pan. Tighten the fitting to 32–42 ft. lbs.

36. Lower the vehicle.

37. Fill the transmission to the proper level.

38. Raise the vehicle and check for transmission fluid leakage. Lower the vehicle and adjust the throttle and manual linkage.

C6 and FMX

BRONCO, F150–350

1. Remove the two upper converter housing-to-engine bolts.

2. Remove the bolt securing the fluid filler tube to the engine cylinder head.

3. Raise the vehicle and support on jackstands.

4. Place the drain pan under the transmission fluid pan. Starting at the rear of the pan and working toward the front, loosen the attaching bolts and allow the fluid to drain.

5. Remove all of the pan attaching bolts except two at the front, to allow the fluid to further drain. With fluid drained, install two bolts on the rear side of the pan to temporarily hold it in place.

6. Remove the converter drain plug access cover from the lower end of the converter housing.

7. Remove the converter-to-flywheel attaching nuts. Place a wrench on the crankshaft pulley attaching bolt to turn the converter to gain access to the nuts.

8. With the wrench on the crankshaft pulley attaching bolt, turn the converter to gain access to the converter drain plug. Place a drain pan under the converter to catch the fluid and remove the plug. After the fluid has been drained, re-install the plug.

9. Disconnect the driveshaft from the rear axle and slide shaft rearward from the transmission. Install a seal installation tool in the extension housing to prevent fluid leakage.

10. Disconnect the speedometer cable from the extension housing.

11. Disconnect the downshift and manual linkage rods from the levers at the transmission.

12. Disconnect the oil cooler lines from the transmission.

13. Remove the vacuum hose from the vacuum diaphragm unit. Remove the vacuum line retaining clip.

14. Disconnect the cable from the ter-

minal on the starter motor. Remove the three attaching bolts and remove the starter motor.

15. On F 150–350 (4 × 4) and Bronco vehicles, remove the transfer case.

16. Remove the two engine rear support and insulator assembly-to-attaching bolts.

17. Remove the two engine rear support and insulator assembly-to-extension housing attaching bolts.

18. Remove the six bolts securing the No.2 crossmember to the frame side rails.

19. Raise the transmission with a transmission jack and remove both crossmembers.

20. Secure the transmission to the jack with the safety chain.

21. Remove the remaining converter housing-to-engine attaching bolts.

22. Move the transmission away from the engine. Lower the jack and remove the converter and transmission assembly from under the vehicle.

23. Tighten the converter drain plug.

24. Position the converter on the transmission making sure the converter drive flats are fully engaged in the pump gear.

25. With the converter properly installed, place the transmission on the jack. Secure the transmission to the jack with the chain.

26. Rotate the converter until the studs and drain plug are in alignment with their holes in the flywheel.

27. Move the converter and transmission assembly forward into position, using care not to damage the flywheel and the converter pilot. The converter must rest squarely against the flywheel. This indicates that the converter pilot is not binding in the engine crankshaft.

28. Install and tighten the converter housing-to-engine attaching bolts.

29. Remove the transmission jack safety chain from around the transmission.

30. Position the No.2 crossmember to the frame side rails. Install and tighten the attaching bolts.

31. Install transfer case on F 150–350 (4 × 4) and Bronco.

32. Positon the engine rear support and insulator assembly above the crossmember. Install the rear suport and insulator assembly-to-extension housing mounting bolts and tighten the bolts.

33. Lower the transmission and remove the jack.

34. Secure the engine rear support and insulator assembly to the crossmember with the attaching bolts and tighten them.

35. Connect the vacuum line to the vacuum diaphragm making sure that the line is in the retaining clip.

36. Connect the oil cooler lines to the transmission.

37. Connect the downshift and manual linkage rods to their respective levers on the transmission.

38. Connect the speedometer cable to the extension housing.

39. Secure the starter motor in place with the attaching bolts. Connect the cable to the terminal on the starter.

40. Install a new O-ring on the lower end of the transmission filler tube and insert the tube in the case.

41. Secure the converter-to-flywheel attaching nuts and tighten them.

42. Install the converter housing access cover and secure it with the attaching bolts.

43. Connect the driveshaft.

44. Adjust the shift linkage as required.

45. Lower the vehicle. Then install the two upper converter housing-to-engine bolts and tighten them.

46. Position the transmission fluid filler tube to the cylinder head and secure with the attaching bolt.

47. Make sure the drain pan is securely attached, and fill the transmission to the correct level with the specified fluid.

E100–350

1. Working from inside the vehicle, remove the engine compartment cover.

2. Disconnect the neutral start switch wires at the plug connector.

3. If the vehicle is equipped with a V-8 engine, remove the flexhose from the air cleaner heat tube.

4. Remove the upper converter housing-to-engine attaching bolts (three bolts on 6-cylinder engines; four bolts on V8 engines).

5. Raise the vehicle and support on jackstands.

6. Place the drain pan under the transmission fluid pan. Starting at the rear of the pan and working toward the front, loosen the attaching bolts and allow the fluid to drain. Finally remove all of the pan attaching bolts except two at the front, to allow the fluid to further drain. With fluid drained, install two bolts on the rear side of the pan to temporarily hold it in place.

7. Remove the converter drain plug access cover from the lower end of the converter housing.

8. Remove the converter-to-flywheel attaching nuts. Place a wrench on the crankshaft pulley attaching bolt to turn the converter to gain access to the nuts.

9. With the wrench on the crankshaft pulley attaching bolt, turn the converter to gain access to the converter drain plug. Place a drain pan under the converter to catch the fluid. Then, remove the plug. With fluid drained, re-install the plug.

10. Disconnect the drive shaft.

11. Remove fluid filler tube.

12. Disconnect the starter cable at the starter. Remove the starter-to-converter housing attaching bolts and remove the starter.

13. Position the engine support bar (Tool T65E–6000–JO) to the frame and engine oil pan flanges.

14. Disconnect the cooler lines from the transmission. Disconnect the vacuum line from the vacuum diaphragm unit. Remove the vacuum line from the retaining clip at the transmission.

15. Remove the speedometer driven gear from the extension housing.

16. Disconnect the manual and downshift linkage rods from the transmission control levers.

17. Position a transmission jack to support the transmission. Install the safety chain to hold the transmission.

18. Remove the bolts and nuts securing the rear support and insulator assembly to the crossmember. Remove the six bolts retaining the crossmember to the side rails and remove the two support gussets. Raise the transmission with the jack and remove the crossmember.

19. Remove the bolt that retains the transmission filler tube to the cylinder block. Lift the filler tube and dipstick from the transmission.

20. Remove the remaining converter housing-to-engine attaching bolts. Lower the jack and remove the converter and transmission assembly from under the vehicle.

21. Remove the converter and mount the transmission in a holding fixture.

22. Tighten the converter drain plug.

23. Position the converter on the transmission making sure the converter drive flats are fully engaged in the pump gear.

24. With the converter properly installed, place the transmission on the jack. Secure the transmission to the jack with the safety chain.

25. Rotate the converter until the studs and drain plug are in alignment with their holes in the flywheel.

26. Move the converter and transmission assembly forward into position, using care not to damage the flywheel and the converter pilot. The converter must rest squarely against the flywheel. This indicates that the converter pilot is not binding in the engine crankshaft.

27. Install the lower converter housing-to-engine attaching bolts. Tighten the bolts. Install the converter-to-flywheel attaching nuts. Tighten the nuts.

28. Install the crossmember. Install the rear support and insulator assembly-to-crossmember attaching bolts and nuts. Tighten the bolts.

29. Remove the safety chain and remove the jack from under the vehicle. Remove the engine support bar.

30. Install a new O-ring on the lower end of the transmission filler tube and insert the tube and dipstick in the case.

31. Connect the vacuum line to the vacuum diaphragm making sure the line is secured in the retaining clip.

32. Connect the cooler lines to the transmission.

33. Install the speedometer driven gear into the extension housing. Tighten the attaching bolt.

34. Connect the transmission linkage rods to the transmission control levers. When making transmission control attachments new retaining ring and grommet should always be used. Attach the shift rod to the

steering column shift lever. Align the flats of the adjusting stud with the flats of the rod slot and insert the stud through the rod. Assemble the adjusting stud nut and washer to a loose fit. Perform a linkage adjustment.

35. Install the converter housing access cover and tighten the attaching bolts.

36. Position the starter into the converter housing and install the attaching bolts. Tighten the bolts. Install the starter cable.

37. Install the driveshaft.

38. Lower the vehicle.

39. Install the upper converter housing-to-engine attaching bolts. Tighten the bolts.

40. On V8 engines, install the flex hose to the air cleaner heat tube. Install the bolt that retains the filler tube to the cylinder block.

41. Connect the neutral start switch wires at the plug connector.

42. Make sure the transmission fluid pan is securely attached, and fill the transmission to the proper level with the specified fluid.

43. Raise the vehicle and check for transmission fluid leakage. Lower the vehicle and adjust the downshift and manual linkage.

44. Install the engine compartment cover.

Automatic Overdrive

1. Raise the vehicle and support on jackstands.

2. Place the drain pan under the transmission fluid pan. Starting at the rear of the pan and working toward the front, loosen the attaching bolts and allow the fluid to drain. Finally remove all of the pan attaching bolts except two at the front, to allow the fluid to further drain. With fluid drained, install two bolts on the rear side of the pan to temporarily hold it in place.

3. Remove the converter drain plug access cover from the lower end of the converter housing.

4. Remove the converter-to-flywheel attaching nuts. Place a wrench on the crankshaft pulley attaching bolt to turn the converter to gain access to the nuts.

5. Place a drain pan under the converter to catch the fluid. With the wrench on the crankshaft pulley attaching bolt, turn the converter to gain access to the converter drain plug and remove the plug. After the fluid has been drained, reinstall the plug.

6. Disconnect the driveshaft from the rear axle and slide shaft rearward from the transmission. Install a seal installation tool in the extension housing to prevent fluid leakage.

7. Disconnect the cable from the terminal on the starter motor. Remove the three attaching bolts and remove the starter motor. Disconnect the neutral start switch wires at the plug connector.

8. Remove the rear mount-to-crossmember attaching bolts and the two crossmember-to-frame attaching bolts.

9. Remove the two engine rear support-to-extension housing attaching bolts.

10. Disconnect the TV linkage rod from the transmission TV lever. Disconnect the manual rod from the transmission manual lever at the transmission.

11. Remove the two bolts securing the bellcrank bracket to the converter housing.

12. Raise the transmission with a transmission jack to provide clearance to remove the crossmember. Remove the rear mount from the crossmember and remove the crossmember from the side supports.

13. Lower the transmission to gain access to the oil cooler lines.

14. Disconnect each oil line from the fittings on the transmission.

15. Disconnect the speedometer cable from the extension housing.

16. Remove the bolt that secures the transmission fluid filler tube to the cylinder block. Lift the filler tube and the dipstick from the transmission.

17. Secure the transmission to the jack with the chain.

18. Remove the converter housing-to-cylinder block attaching bolts.

19. Carefully move the transmission and converter assembly away from the engine and, at the same time, lower the jack to clear the underside of the vehicle.

20. Remove the converter and mount the transmission in a holding fixture.

21. Tighten the converter drain plug.

22. Position the converter on the transmission, making sure the converter drive flats are fully engaged in the pump gear by rotating the converter.

23. With the converter properly installed, place the transmission on the jack. Secure the transmission to the jack with a chain.

24. Rotate the converter until the studs and drain plug are in alignment with the holes in the flywheel.

25. Move the converter and transmission assembly forward into position, using care not to damage the flywheel and the converter pilot. The converter must rest squarely against the flywheel. This indicates that the converter pilot is not binding in the engine crankshaft.

26. Install and tighten the converter housing-to-engine attaching bolts to 40–50 ft. lbs.

27. Remove the safety chain from around the transmission.

28. Install a new O-ring on the lower end of the transmission filler tube. Insert the tube in the transmission case and secure the tube to the engine with the attaching bolt.

29. Connect the speedometer cable to the extension housing.

30. Connect the oil cooler lines to the right side of transmission case.

31. Position the crossmember on the side supports. Position the rear mount on the crossmember and install the attaching bolt and nut.

32. Secure the engine rear support to the extension housing and tighten the bolts to 16–20 ft. lbs.

33. Lower the transmission and remove the jack.

C3
RANGER/BRONCO II

1. Raise the vehicle and safely support on jackstands. Place a drain pan under the transmission fluid pan. Starting at the rear of the pan and working toward the front, loosen the attaching bolts and allow the fluid to drain. Then remove all of the pan attaching bolts except two at the front, to allow the fluid to further drain. After all the fluid has drained, install two bolts on the rear side of the pan to temporarily hold it in place.

2. Remove the converter drain plug access cover and adapter plate bolts from the lower end of the converter housing.

3. Remove the four flywheel to converter attaching nuts. Crank the engine to turn the converter to gain access to the nuts, using a wrench on the crankshaft pulley attaching bolt. On belt driven overhead camshaft engines, never turn the engine backwards.

4. Crank the engine until the converter drain plug is accessible and remove the plug. Place a drain pan under the converter to catch the fluid. After all the fluid has been drained from the converter, reinstall the plug and tighten to 20–30 ft. lbs. Remove the driveshaft. Install cover, plastic bag etc. over end of extension housing.

5. Remove the speedometer cable from the extension housing. Disconnect the shift rod at the transmission manual lever. Disconnect the downshift rod at the transmission downshift lever.

6. Remove the starter-to-converter housing attaching bolts and position the starter out of the way.

7. Disconnect the neutral start switch wires from the switch. Remove the vacuum line from the transmission vacuum modulator.

8. Position a suitable jack under the transmission and raise it slightly.

9. Remove the engine rear support-to-crossmember bolts. Remove the crossmember-to-frame side support attaching bolts and remove the crossmember insulator and support and damper.

10. Lower the jack under the transmission and allow the transmission to hang.

11. Position a jack to the front of the engine and raise the engine to gain access to the two upper converter housing-to-engine attaching bolts.

12. Disconnect the oil cooler lines at the transmission. Plug all openings to keep out dirt.

13. Remove the lower converter housing-to-engine attaching bolts. Remove the transmission filler tube.

14. Secure the transmission to the jack with a safety chain.

15. Remove the two upper converter housing-to-engine attaching bolts. Move the

transmission to the rear and down to remove it from under the vehicle.

16. Position the converter to the transmission making sure the converter hub is fully engaged in the pump. With the converter properly installed, place the transmission on the jack and secure with safety chain.

17. Rotate the converter so the drive studs and drain plug are in alignment with their holes in the flywheel. With the transmission mounted on a transmission jack, move the converter and transmission assembly forward into position being careful not to damage the flywheel and the converter pilot.

—————— CAUTION ——————
During this move, to avoid damage, do not allow the transmission to get into a nosed down position as this will cause the converter to move forward and disengage from the pump gear. The converter must rest squarely against the flywheel. This indicates that the converter pilot is not binding in the engine crankshaft.

18. Install the two upper converter housing-to-engine attaching bolts and tighten to 28–38 ft. lbs.

19. Remove the safety chain from the transmission. Insert the filler tube in the stub tube and secure it to the cylinder block with the attaching bolt. Tighten the bolt to 28–38 ft. lbs. If the stub tube is loosened or dislodged, it should be replaced. Install the oil cooler lines in the retaining clip at the cylinder block. Connect the lines to the transmission case.

20. Remove the jack supporting the front of the engine. Raise the transmission. Position the crossmember, insulator and support and damper to the frame side supports and install the attaching bolts. Tighten the bolts to 20–30 ft. lbs.

21. Lower the transmission and install the rear engine support-to-crossmember nut. Tighten the bolt to 60–80 ft. lbs.

22. Remove the transmission jack. Install the vacuum hose on the transmission vacuum unit. Install the vacuum line into the retaining clip.

23. Connect the neutral start switch plug to the switch. Install the starter and tighten the attaching bolts to 15–20 ft. lbs.

24. Install the four flywheel-to-converter attaching nuts. When assembling the flywheel to the converter, first install the attaching nuts and tighten to 20–34 ft. lbs.

25. Install the converter drain plug access cover and adaptor plate bolts. Tighten the bolts to 12–16 ft. lbs.

26. Connect the muffler inlet pipe to the exhaust manifold.

27. Connect the transmission shift rod to the manual lever. Connect the downshift rod to the downshift lever.

28. Connect the speedometer cable to the extension housing. Install the driveshaft. Tighten the companion flange U-bolt attaching nuts to 70–95 ft. lbs.

29. Adjust the manual and downshift linkage as required.

30. Lower the vehicle. Fill the transmission to the proper level with the specified fluid. Pour in five quarts of fluid; then run the engine and add fluid as required. Check the transmission, converter assembly and oil cooler lines for leaks.

C5 (4×2)

1. Raise the vehicle and safely support on jackstands. Place the drain pan under the transmission fluid pan. Starting at the rear of the pan and working toward the front, loosen the attaching bolts and allow the fluid to drain. Finally remove all of the pan attaching bolts except two at the front, to allow the fluid to further drain. Finally remove all of the pan attaching bolts except two at the front, to allow the fluid to further drain. With fluid drained, install two bolts on the rear side of the pan to temporarily hold it in place.

2. Remove the converter drain plug access cover from the lower end of the converter housing.

3. Remove the converter-to-flywheel attaching nuts. Place a wrench on the crankshaft pulley attaching bolt to turn the converter to gain access to the nuts.

4. Place a drain pan under the converter to catch the fluid. With the wrench on the crankshaft pulley attaching bolt, turn the converter to gain access to the converter drain plug and remove the plug. After the fluid has been drained, reinstall the plug.

5. Disconnect the driveshaft from the rear axle and slide shaft rearward from the transmission. Install a suitable cover in the extension housing to prevent fluid leakage. Mark the rear driveshaft yoke and axle flange so they can be installed in their original position.

6. Disconnect the cable from the terminal on the starter motor. Remove the three attaching bolts and remove the starter motor. Disconnect the neutral start switch wires at the plug connector.

7. Remove the rear mount-to-crossmember attaching nuts and the two crossmember-to-frame attaching bolts. Remove the right and left gusset.

8. Remove the two engine rear insulator-to-extension housing attaching bolts.

9. Disconnect the TV linkage rod from the transmission TV lever. Disconnect the manual rod from the transmission manual lever at the transmission.

10. Remove the two bolts securing the bellcrank bracket to the converter housing.

11. Raise the transmission with a suitable jack to provide clearance to remove the crossmember. Remove the rear mount from the crossmember and remove the crossmember from the side supports. Lower the transmission to gain access to the oil cooler lines. Disconnect each oil line from the fittings on the transmission.

12. Disconnect the speedometer cable from the extension housing.

13. Remove the bolt that secures the transmission fluid filler tube to the cylinder block. Lift the filler tube and the dipstick from the transmission.

14. Secure the transmission to the jack with the chain. Remove the converter housing-to-cylinder block attaching bolts.

15. Carefully move the transmission and converter assembly away from the engine and, at the same time, lower the jack to clear the underside of the vehicle.

16. Tighten the converter drain plug to specifications. Position the converter on the transmission, making sure the converter drive flats are fully engaged in the pump gear by rotating the converter.

17. With the converter properly installed, place the transmission on the jack. Secure the transmission to the jack with a chain.

18. Rotate the converter until the studs and drain plug are in alignment with the holes in the flywheel. Move the converter and transmission assembly forward into position, using care not to damage the flywheel and the converter pilot. The converter must rest squarely against the flywheel. This indicates that the converter pilot is not binding in the engine crankshaft.

19. Install and tighten the converter housing-to-engine attaching bolts to specification.

20. Remove the safety chain from around the transmission.

21. Install the new O-ring on the lower end of the transmission filler tube. Insert the tube in the transmission case and secure the tube to the engine with the attaching bolt.

22. Connect the speedometer cable to the extension housing.

23. Connect the oil cooler lines to the right side of transmission case.

24. Secure the engine rear support to the extension housing and tighten the bolts to specification.

25. Position the crossmember on the side supports. Lower the transmission and remove the jack. Secure the crossmember to the side supports with the attaching bolts.

26. Position the damper assembly over the engine rear support studs. (The painted face of the damper is facing forward when installed in the vehicle.) Secure the rear engine support to the crossmember.

27. Position the bellcrank to the converter housing and install the two attaching bolts.

28. Connect the TV linkage rod to the transmission TV lever. Connect the manual linkage rod to the manual lever at the transmission.

29. Secure the converter-to-flywheel attaching nuts and tighten them to specification.

30. Install the converter housing access cover and secure it with the attaching bolts.

31. Secure the starter motor in place with the attaching bolts. Connect the cable to the terminal on the starter. Connect the neutral start switch wires at the plug connector.

32. Connect the driveshaft to the rear axle so the index marks on the companion flange and the rear yoke are aligned. Lubricate the slip yoke with grease. Adjust the shift linkage as required.

33. Adjust throttle linkage.

34. Lower the vehicle. Fill the transmission to the correct level with the specified fluid. Start the engine and shift transmission to all ranges, then recheck fluid level.

C5 (4 × 4)

1. Remove the bolt securing the fluid filler tube to the engine valve cover bracket.

2. Place a drain pan under the transmission fluid pan. Starting at the rear of the pan and working towards the front, loosen the attaching bolts and allow the fluid to drain. Finally, remove all of the pan attaching bolts except two at the front, to allow the fluid to drain further. With fluid drained, install two bolts on the rear side of the pan to temporarily hold it in place.

3. Remove the converter drain plug access cover from the lower end of the converter housing. Remove the converter-to-flywheel attaching nuts. Place a wrench on the crankshaft pulley attaching bolt to turn the converter to gain access to the nuts.

4. Place a drain pan under the converter to catch the fluid. With the wrench on the crankshaft pulley attaching bolt, turn the converter to gain access to the converter drain plug and remove the plug.

5. After the fluid has been drained, reinstall the cable from the terminal at the starter motor. Remove the three attaching bolts and remove the starter motor. Disconnect the neutral start switch wires at the plug connector.

6. Remove the rear mount-to-crossmember attaching nuts and the two crossmember-to-frame attaching bolts. Remove the right and left gusset.

7. Remove the two engine rear insulator-to-extension housing attaching bolts.

8. Disconnect the TV linkage rod from the transmission TV lever. Disconnect the manual rod from the transmission manual lever at the transmission. Disconnect the downshift and manual linkage rods from the levers on the transmission.

9. Remove the vacuum hose from the vacuum diaphragm unit. Remove the vacuum line from the retaining clip.

10. Remove the two bolts securing the bellcrank bracket to the converter housing.

11. Remove the transfer case. Refer to the Transfer Case section behind Manual Transmissions.

12. Raise the transmission with a transmission jack to provide clearance to remove the crossmember. Remove the rear mount from the crossmember and remove the crossmember from the side supports.

13. Lower the transmission to gain access to the oil cooler lines.

14. Disconnect each oil line from the fittings on the transmission.

15. Disconnect the speedometer cable from the extension housing.

16. Secure the transmission to the jack with the chain. Remove the converter housing-to-cylinder block attaching bolts.

17. Carefully move the transmission and converter assembly away from the engine and, at the same time, lower the jack to clear the underside of the vehicle.

18. Position the converter on the transmission, making sure the converter drive flats are fully engaged in the pump gear by rotating the converter.

19. With the converter properly installed, place the transmission on the jack. Secure the transmission to the jack with a chain.

20. Rotate the converter until the studs and drain plug are in alignment with the holes in the flywheel.

21. Move the converter and transmission assembly forward into position, using care not to damage the flywheel and the converter pilot. The converter must rest squarely against the flywheel. This indicates that the converter pilot is not binding in the engine crankshaft.

22. Install and tighten the converter housing-to-engine attaching bolts.

23. Remove the safety chain from around the transmission.

24. Install a new O-ring on the lower end of the transmission filler tube. Insert the tube in the transmission case.

25. Connect the speedometer cable to the extension housing.

26. Connect the oil cooler lines to the right of the transmission case.

27. Position the crossmember on the side supports. Position the rear mount insulator on the crossmember and install the attaching bolts and nuts.

28. Install the transfer case.

29. Secure the engine rear support to the extension housing. Lower the transmission and remove the jack.

30. Secure the crossmember to the side supports with the attaching bolts and tighten to specification.

31. Position the bellcrank to the converter housing and install the two attaching bolts.

32. Connect the downshift and manual linkage rods to their respective levers on the transmission.

33. Connect the vacuum line to the vacuum diaphragm making sure that the line is in the retaining clip.

34. Secure the converter-to-flywheel attaching nuts. Install the converter housing access cover and secure it with the attaching bolts.

35. Secure the starter motor in place with the attaching bolts. Connect the cable to the terminal on the starter. Connect the neutral start switch wires at the plug connector.

36. Adjust the shift linkage as required. Lower the vehicle.

37. Position the transmission fluid filler tube to the valve cover bracket and secure with the attaching bolt. Fill the transmission

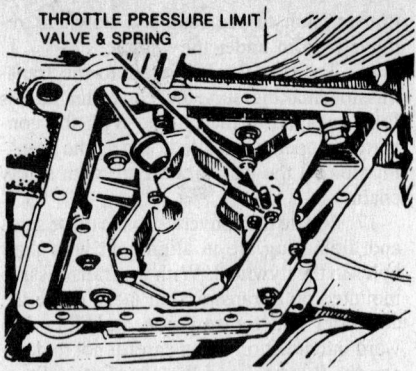

C4 throttle pressure limiter valve which is held in place in the valve body by the filter. The valve is installed with the larger end toward the valve body. The spring fits over the valve stem.

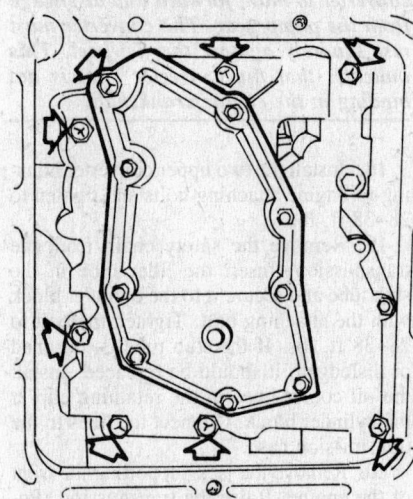

View showing the filter screen which is attached to the lower valve body. Bolts indicated by arrows hold the valve body. If the screen only is being serviced, do not remove these bolts

to the correct level. Start the engine and shift the transmission to all ranges, then recheck the fluid level.

Pan

REMOVAL & INSTALLATION

1. Raise and support the vehicle on jackstands.

2. Place a drain pan under the transmission.

3. Loosen the pan attaching bolts and drain the fluid from the transmission.

4. When the fluid has drained to the level of the pan flange, remove the remaining pan bolts working from the rear and both sides of the pan to allow it to drop and drain slowly.

5. When all of the fluid has drained, remove the pan and clean it thoroughly. Discard the pan gasket.

6. Clean the transmission oil pan and transmission mating surfaces.

7. Install the transmission oil pan in the reverse order of removal, torquing the attaching bolts to 12–16 ft. lbs. and using a new gasket. Fill the transmission with 3 qts. of the correct type fluid, check the operation of the transmission and check for leakage.

NOTE: When starting the engine after the transmission fluid has been drained, do not race the engine. Move the gearshift selector through all of the ranges before moving the vehicle.

FILTER SERVICE

1. Remove the transmission oil pan and gasket.

2. Remove the screws holding the fine mesh screen to the lower valve body.

NOTE: Be careful not to lose the throttle pressure limit valve and spring when separating the oil screen from the valve body on a C4.

3. Install the new filter screen and transmission oil pan gasket in the reverse order of removal.

FRONT BAND ADJUSTMENT

FMX

1. Remove the transmission oil pan.

2. Loosen the front servo adjusting screw locknut.

3. Pull back on the actuating rod, and insert a ¼ in. spacer between the adjusting screw and the servo piston stem.

4. Tighten the adjusting screw to 10 in. lbs.

5. Remove the spacer and tighten the adjusting screw an additional ¾ turn.

6. Hold the adjusting screw stationary and tighten the locknut. Tighten the locknut to 20–25 ft. lbs.

7. Install the oil pan and a new gasket in the reverse order of removal.

C3

1. Remove the downshift rod from the transmission downshift lever. Clean all of the dirt away from the bank adjusting nut and screw area. Remove and discard the locknut.

2. Tighten the adjusting screw to 10 ft. lbs. Back off the adjusting screw exactly two turns.

3. Install a new locknut, hold the adjusting screw in position and tighten the locknut to 35–45 ft. lbs. Install the downshift rod.

INTERMEDIATE BAND ADJUSTMENT

C4, C5 and C6

1. Raise and support the vehicle on jackstands.

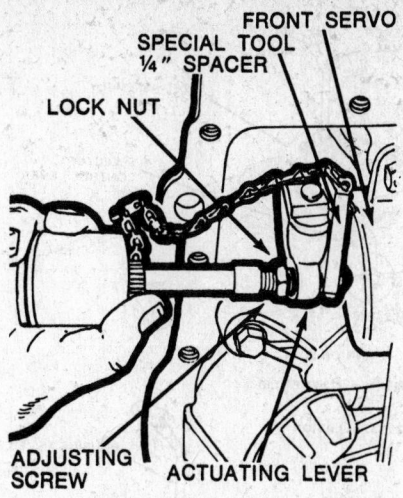

FMX front band adjustment

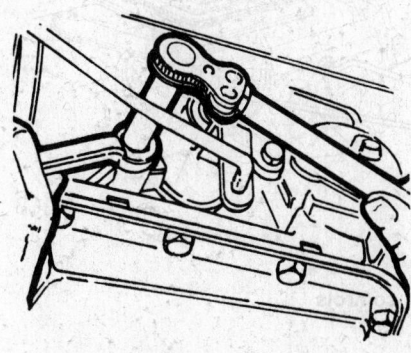

C4 intermediate band adjustment

2. Clean all dirt away from the band adjusting screw. Remove and discard the locknut.

3. Install a new locknut and tighten the adjusting screw to 10 ft. lbs.

4. On the C4 transmission; back off the adjusting screw exactly 1¾ turns. On the C5; back off the adjusting screw exactly 4¼ turns. On the C6 transmission; back off the adjusting screw exactly 1½ turns.

5. Hold the adjusting screw from turning and tighten the locknut to 35–45 ft. lbs.

6. Remove the jackstands and lower the vehicle.

LOW-REVERSE BAND ADJUSTMENT

C4, C5

1. Clean all dirt from around the band adjusting screw and remove and discard the locknut.

2. Install a new locknut on the adjusting screw. Using a torque wrench, tighten the adjusting screw to 10 ft. lbs.

3. Back off the adjusting screw exactly 3 full turns.

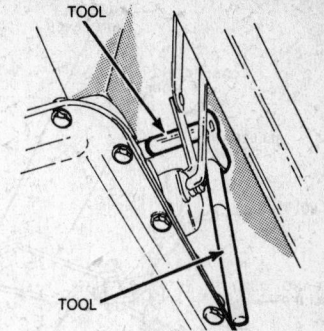

C6 intermediate band adjustment

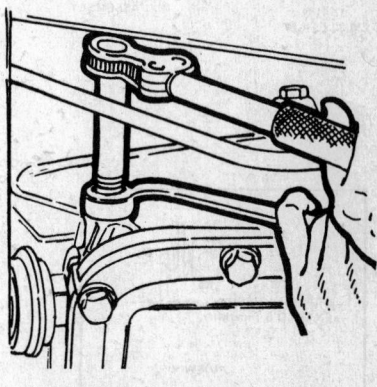

Low-reverse band adjustment

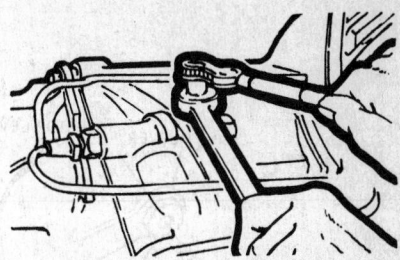

FMX rear band adjustment

4. Hold the adjusting screw steady and tighten the locknut to 35–45 ft. lbs.

REAR BAND ADJUSTMENT

FMX

1. Remove all dirt away from the adjusting screw threads then oil the threads.

2. Loosen the rear band adjusting screw locknut.

3. Tighten the adjusting screw to 10 ft. lbs.

4. Back off the adjusting screw exactly 1½ turns.

5. Hold the adjusting screw stationary and tighten the adjusting screw locknut to 35–40 ft. lbs.

6. Reinstall the oil pan and a new gasket.

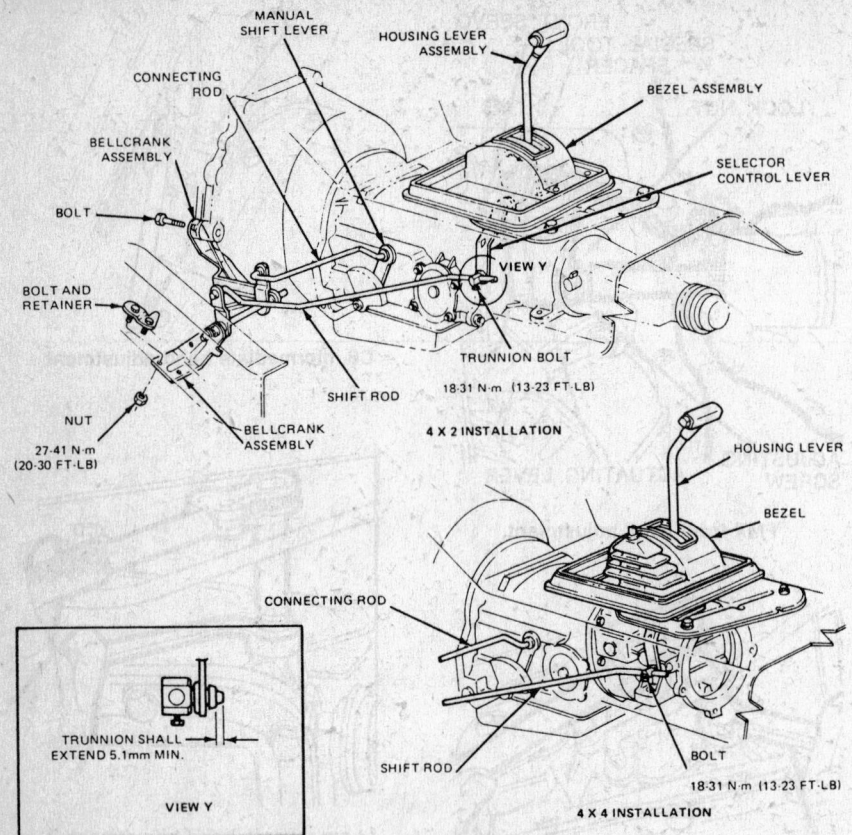

MANUAL SHIFT LEVER

HOUSING LEVER ASSEMBLY

CONNECTING ROD

BEZEL ASSEMBLY

BELLCRANK ASSEMBLY

SELECTOR CONTROL LEVER

BOLT

BOLT AND RETAINER

VIEW Y

NUT
27-41 N·m
(20-30 FT·LB)

BELLCRANK ASSEMBLY

SHIFT ROD

TRUNNION BOLT
18-31 N·m (13-23 FT-LB)

4 X 2 INSTALLATION

HOUSING LEVER

BEZEL

CONNECTING ROD

TRUNNION SHALL EXTEND 5.1mm MIN.

VIEW Y

SHIFT ROD

BOLT
18-31 N·m (13-23 FT-LB)

4 X 4 INSTALLATION

C5 floor shift controls

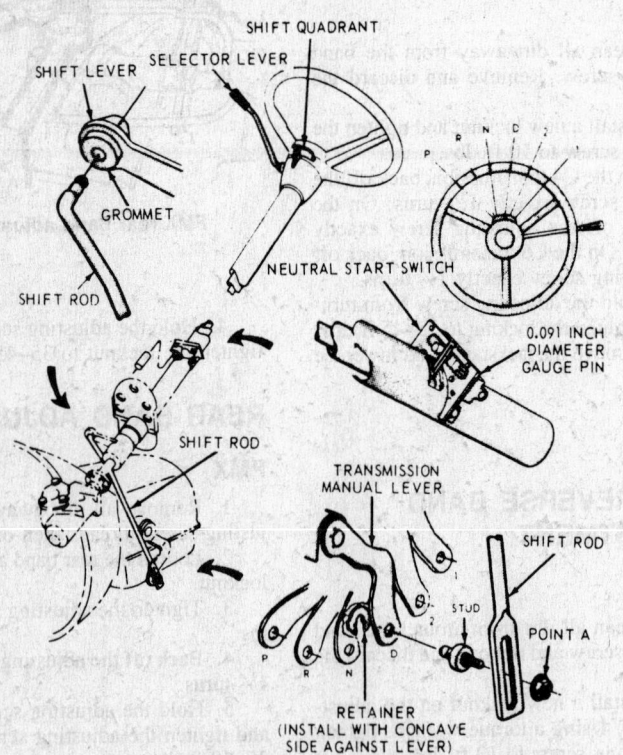

SHIFT QUADRANT

SHIFT LEVER

SELECTOR LEVER

GROMMET

NEUTRAL START SWITCH

P R N D

SHIFT ROD

0.091 INCH DIAMETER GAUGE PIN

SHIFT ROD

TRANSMISSION MANUAL LEVER

SHIFT ROD

STUD

POINT A

P R N D

RETAINER
(INSTALL WITH CONCAVE SIDE AGAINST LEVER)

C6 shift linkage adjustment

SHIFT LINKAGE ADJUSTMENT

1. With the engine stopped, place the transmission selector lever at the steering column in the D position against the D stop.

2. Loosen the shift rod adjusting nut at the transmission lever.

3. Shift the manual lever at the transmission to the D position, two detents from the rear. On an F100 with 4WD, move the bellcrank lever.

4. With the selector lever and transmission manual lever in the D position, tighten the adjusting nut to 12–18 ft. lbs. Do not allow the rod or shift lever to move while tightening the nut.

5. Check the operation of the shift linkage.

THROTTLE VALVE LINKAGE ADJUSTMENT

Automatic Overdrive

ADJUSTMENT AT THE CARBURETOR w/ROD

The TV control linkage may be adjusted at the carburetor using the following procedure:

1. Check that engine idle speed is set at specification.

2. De-cam the fast idle cam on the carburetor so that the throttle lever is at its idle stop. Place shift lever in N (neutral), set park brake (engine off).

3. Backout linkage lever adjusting screw all the way (screw end is flush with lever face).

4. Turn in adjusting screw until a thin shim (.005 inch max.) or piece of writing paper fits snugly between end of screw and Throttle Lever. To eliminate effect of friction, push linkage lever forward (tending to close gap) and release before checking clearance between end of screw and throttle lever. Do not apply any load on levers with tools or hands while checking gap.

5. Turn in adjusting screw an additional four turns. (Four turns are preferred. Two turns minimum is permissible if screw travel is limited).

6. If it is not possible to turn in adjusting screw at least two additional turns or if there was insufficient screw adjusting capacity to obtain an initial gap in Step 2 above, refer to Linkage Adjustment at Transmission. Whenever it is required to adjust idle speed by more than 50 rpm, the adjustment screw on the linkage lever at the carburetor should also be readjusted. After making any idle speed adjustments, always verify the linkage lever and throttle lever are in contact with the throttle lever at its idle stop and the shift lever is in N (neutral).

Idle Speed Change	Turns on Linkage Lever Adjustment Screw
Less than 50 rpm	No change required
50 to 100 rpm increase	1½ turns out
50 to 100 rpm decrease	1½ turns in
100 to 150 rpm increase	2½ turns out
100 to 150 rpm decrease	2½ turns in

ADJUSTMENT AT TRANSMISSION

The linkage lever adjustment screw has limited adjustment capability. If it is not possible to adjust the TV linkage using this screw, the length of the TV control rod assembly must be readjusted using the following procedure. This procedure must also be followed whenever a new TV control rod asssembly is installed. This procedure requires placing the vehicle on jackstands to give access to the linkage components at the transmission TV control lever.

1. Set the engine curb idle speed to specification.
2. With engine off, de-cam the fast idle cam on the carburetor so that the throttle lever is against the idle stop. Place shift lever in Neutral and set park brake (engine off).
3. Set the linkage lever adjustment screw at its approximately mid-range.
4. If a new TV control rod assembly is being installed, connect the rod to the linkage lever at the carburetor.

— CAUTION —
The following steps involve working in proximity to the exhaust system. Allow the exhaust system to cool before proceeding.

5. Raise the vehicle and support on jackstands.
6. Using a 13 mm box end wrench, loosen the bolt on the sliding trunnion block on the TV control rod assembly. Remove any corrosion from the control rod and free-up the trunnion block so that it slides freely on the control rod. Insert pin into transmission lever grommet.
7. Push up on the lower end of the control rod to insure that the linkage lever at carburetor is firmly against the throttle lever. Release force on rod. Rod must stay up.
8. Push the TV control lever on the transmission up against its internal stop with a firm force (approximately 5 pounds) and tighten the bolt on the trunnion block. Do not relax force on lever until nut is tightened.
9. Lower the vehicle and verify that the throttle lever is still against the idle stop. If not, repeat Steps 2 through 9.

ADJUSTMENT w/CABLE

Whenever it is required to adjust the idle speed by more than 150 RPM, the TV control cable should be readjusted. Failure to do so may result in the symptoms due to a "too short" cable if the idle speed was increased or a "too long" cable if the idle speed was reduced.

1. Check and set, if necessary, engine idle speed to specification with and without TSP activated.
2. Shut engine off. Remove air cleaner. Set parking brake block wheels, and put selector in "N". (Do not put selector in "P").
3. Verify that the cable routing is free of sharp bends or pressure points and that the cable operates freely. Lubricate the TV lever ball stud. Check for damage to cable or rubber boot.
4. Unlock the locking tab at the carburetor end by pushing up from below, and prying up the rest of the way to free the cable.
5. A retention spring must be installed

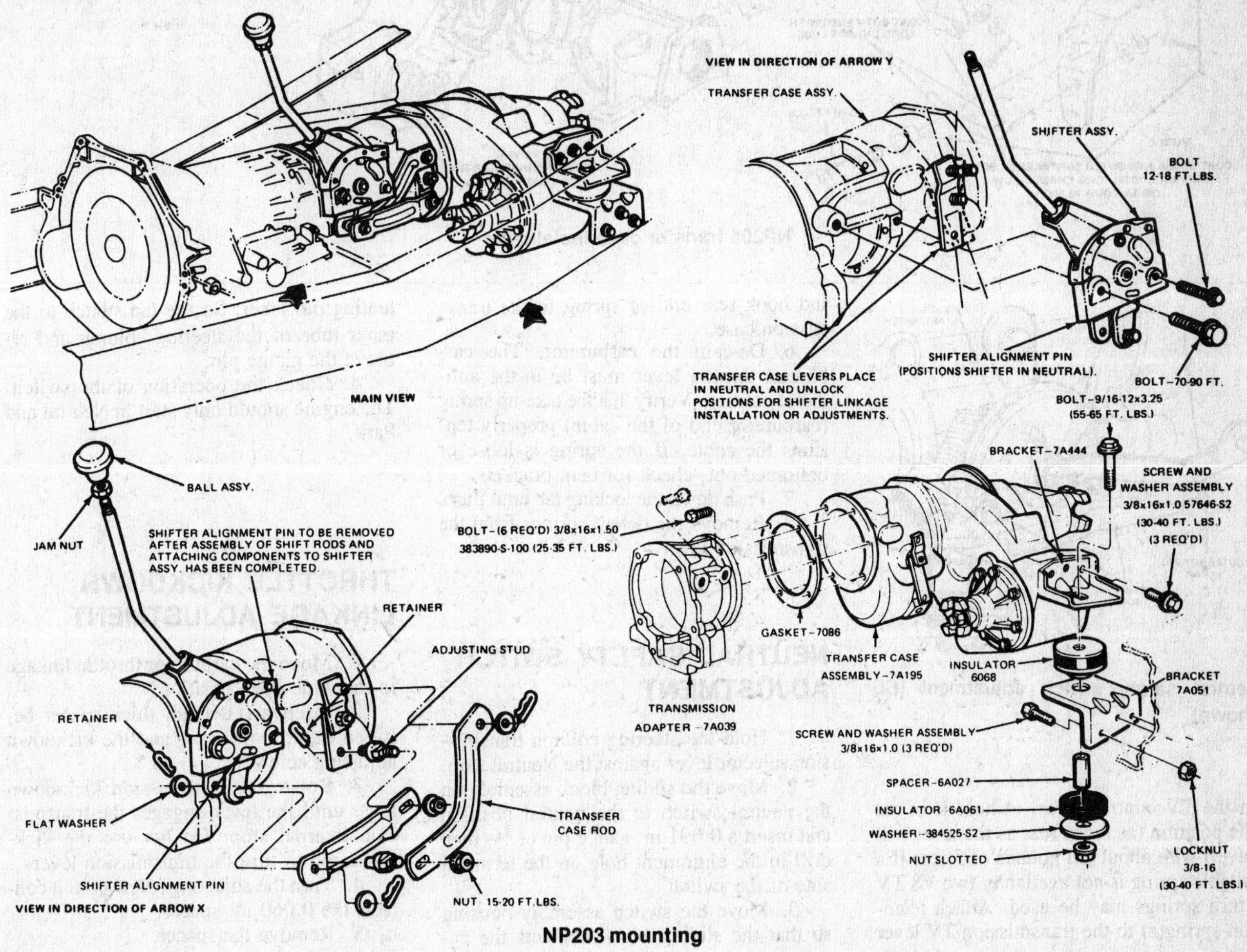

NP203 mounting

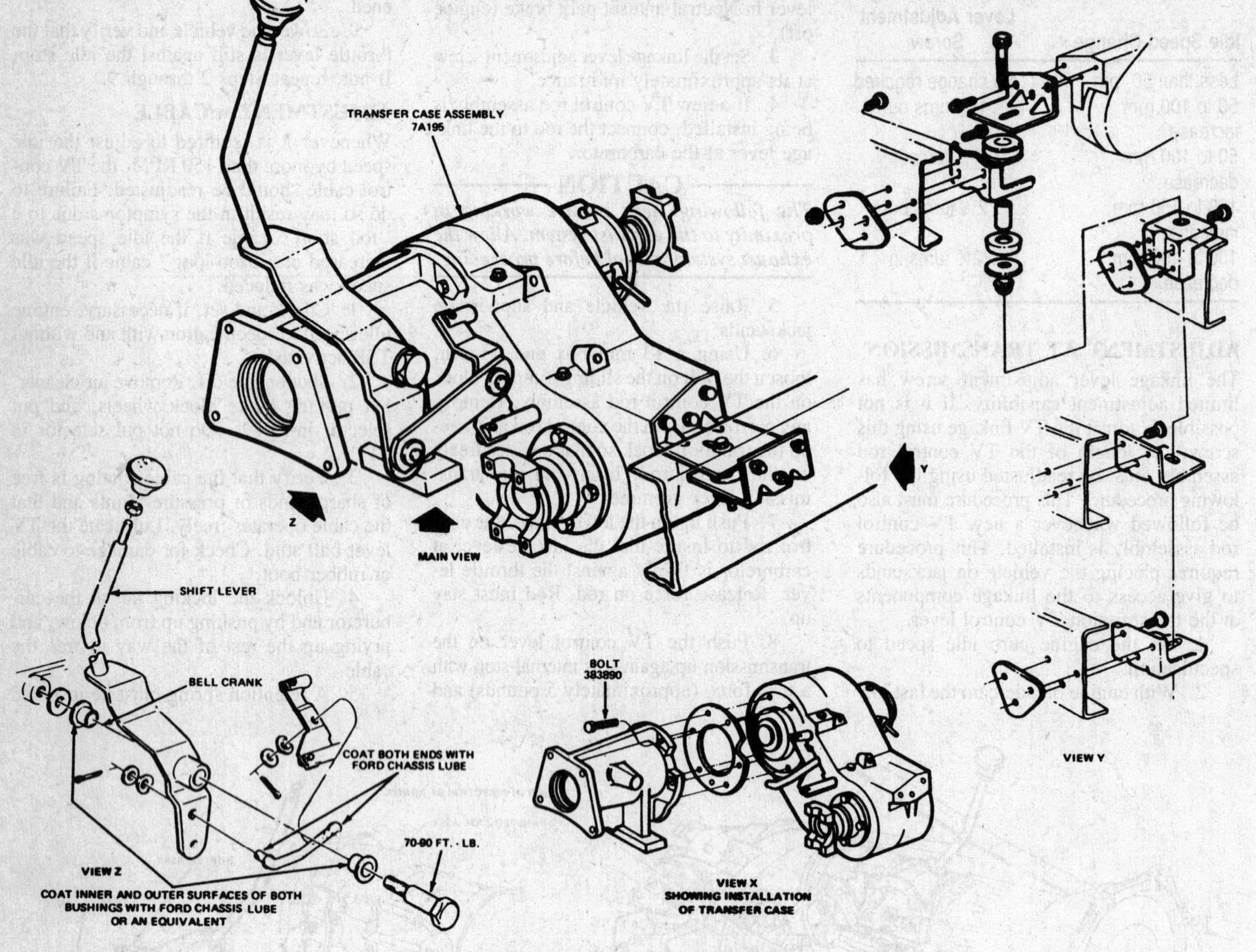

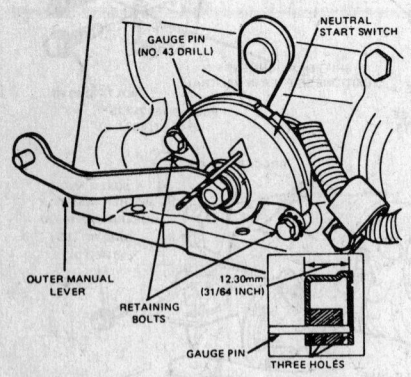

TRANSFER CASE ASSEMBLY
7A195

MAIN VIEW

SHIFT LEVER

Z

BELL CRANK

COAT BOTH ENDS WITH
FORD CHASSIS LUBE

VIEW Z
COAT INNER AND OUTER SURFACES OF BOTH
BUSHINGS WITH FORD CHASSIS LUBE
OR AN EQUIVALENT

70-90 FT. - LB.

BOLT
383890

VIEW X
SHOWING INSTALLATION
OF TRANSFER CASE

VIEW Y

NP205 transfer case installation

NEUTRAL
START SWITCH

GAUGE PIN
(NO. 43 DRILL)

OUTER MANUAL
LEVER

RETAINING
BOLTS

GAUGE PIN

12.30mm
(31/64 INCH)

THREE HOLES

Neutral safety switch adjustment (C5 shown)

on the TV control lever, to hold it in the idle position (as far to rear as the lever will travel) with about ten pounds of force. If a suitable spring is not available, two V8 TV return springs may be used. Attach retention spring(s) to the transmission TV lever

and hook rear end of spring to the transmission case.

6. De-cam the carburetor. The carburetor throttle lever must be in the anti-diesel position. Verify that the take-up spring (carburetor end of the cable) properly tensions the cable. If the spring is loose or bottomed out, check for bent brackets.

7. Push down the locking tab until flush.

8. Remove the detent springs from the transmission lever.

NEUTRAL SAFETY SWITCH ADJUSTMENT

1. Hold the steering column transmission selector lever against the Neutral stop.

2. Move the sliding block assembly on the neutral switch to the neutral position and insert a 0.091 in. gauge pin or 3/32 inch drill in the alignment hole on the terminal side of the switch.

3. Move the switch assembly housing so that the sliding block contacts the ac-

tuating pin lever. Secure the switch to the outer tube of the steering column and remove the gauge pin.

4. Check the operation of the switch. The engine should only start in Neutral and Park.

THROTTLE KICKDOWN LINKAGE ADJUSTMENT

1. Move the carburetor throttle linkage to the wide open position.

2. Insert a 0.060 in. thick spacer between the throttle lever and the kickdown adjusting screw.

3. Rotate the transmission kick-down lever until the lever engages the transmission internal stop. Do not use the kickdown rod to turn the transmission lever.

4. Turn the adjusting screw until it contacts the 0.060 in. spacer.

5. Remove the spacer.

TRANSFER CASE

Pickup

NOTE: For Bronco II and Ranger refer to the previous Manual Transmission section.

REMOVAL & INSTALLATION

Dana Model 21

1. Raise the vehicle and support on jackstands.
2. Disconnect the front and rear driveshafts at the transfer case.
3. Disconnect the shift rod from the transfer case shift lever arm.
4. Remove the bolts and nuts which attach the transfer case to the transmission extension housing and remove the transfer case from under the truck.
5. Install the transfer case in the reverse order of removal, using a new gasket between the transfer case and the transmission extension housing.

New Process Model 205

1. Drain the transfer case and disconnect the rear axle driveshaft from the flange at the transfer case.
2. Disconnect the front wheel driveshaft and the transmission output shaft at the transfer case if on an F250. On an F100, disconnect the front wheel driveshaft and remove the bolts attaching the transmission adapter to the transfer case.
3. Disconnect the shift selector rod and the speedometer cable at the transfer case.
4. Place a transmission jack under the transfer case and secure it with a chain.
5. Remove the transfer case mounting bolts and remove the unit from under the vehicle.
6. Install the transfer case in the reverse order of removal.

New Process Model 203

1. Drain the transfer case by removing the power take-off lower bolts and the front output rear cover lower bolts.
2. Disconnect the front axle drive shaft from the flange of the transfer case.
3. Disconnect the shift rods from the transfer case.
4. Disconnect the speedometer cable and lockout lamp switch wire from the transfer case rear output shaft housing.
5. Remove the bolts which attach the transfer case to the transmission adapter. Disconnect the rear axle drive shaft at the transfer case flange.
6. Position a transmission jack under the transfer case and secure it to the jack.
7. Remove the transfer case mounting bolts and remove the transfer case.
8. Install the transfer case in the reverse order of removal.

New Process Model 208

1. Raise the vehicle and support on jackstands and drain the fluid from the transfer case.
2. Disconnect the four wheel drive indicator switch wire connector at the transfer case.
3. Disconnect the speedometer driven gear from the transfer case rear bearing retainer.
4. Remove the nut retaining the transmission shift lever assembly to the transfer case.
5. Remove the skid plate from the frame, if so equipped.
6. Remove the heat shield from the frame.
7. Support the transfer case with a transmission jack or equivalent.
8. Disconnect the front driveshaft from the front output shaft yoke.
9. Disconnect the rear driveshaft from the rear output shaft yoke.
10. Remove the bolts retaining the transfer case to the transmission adapter.
11. Lower the transfer case from the vehicle.
12. When installing place a new gasket between the transfer case and the adapter.
13. Raise the transfer case with a transmission jack so the transmission output shaft aligns with the splined transfer case input shaft.
14. Install the bolts retaining the case to the adapter.
15. Connect the rear driveshaft to the rear output shaft yoke.
16. Connect the front driveshaft to the front output yoke.
17. Remove the transmission jack from the transfer case.
18. Position the heat shield to the frame crossmember and mounting lug to the transfer case and install and tighten the bolts and screw.
19. Install the skid plate to the frame.
20. Install the shift lever to the transfer case and tighten the retaining nut.
21. Install the speedometer driven gear to the transfer case.
22. Connect the four wheel drive indicator switch wire to the transfer case.
23. Install the drain plug. Remove the filler plug and install six pints of Dexron® II type transmission fluid.
24. Lower the vehicle.

Borg Warner Model 1345

1. Raise the vehicle and support on jackstands.
2. Drain the fluid from the transfer case.
3. Disconnect the four wheel drive indicator switch wire connector at the transfer case.
4. Remove the skid plate from the frame, if so equipped.
5. Disconnect the front driveshaft from the front output yoke.
6. Disconnect the rear driveshaft from the rear output shaft yoke.

7. Disconnect the speedometer driven gear from the transfer case rear bearing retainer.
8. Remove the retaining rings and shift rod from the transfer case shift lever.
9. Disconnect the vent hose from the transfer case.
10. Remove the heat shield from the frame.
11. Support the transfer case with a transmission jack.
12. Remove the bolts retaining the transfer case to the transmission adapter.
13. Lower the transfer case from the vehicle.
14. When installing place a new gasket between the transfer case and the adapter.
15. Raise the transfer case with the transmission jack so that the transmission output shaft aligns with the splined transfer case input shaft. Install the bolts retaining the transfer case to the adapter.
16. Remove the transmission jack from the transfer case.
17. Connect the rear driveshaft to the rear output shaft yoke.
18. Install the shift lever to the transfer case and install the retaining nut.
19. Connect the speedometer driven gear to the transfer case.
20. Connect the four wheel drive indicator switch wire connector at the transfer case.
21. Connect the front driveshaft to the front output yoke.
22. Position the heat shield to the frame crossmember and the mounting lug on the transfer case. Install and tighten the retaining bolts.
23. Install the skid plate to the frame.
24. Install the drain plug. Remove the filler plug and install six pints of Dexron® II type transmission fluid or equivalent.
25. Lower the vehicle.

TRANSFER CASE SHIFT LINKAGE ADJUSTMENT

New Process Model 205
MANUAL TRANSMISSION

Adjust the length of the shift rod between the transfer case and the shift lever with the lever in 4WD-Low so that the distance between the rear face of the transmission and the shift lever-to-rod clevis pin is 3.94 to 3.82 in.

AUTOMATIC TRANSMISSION

Adjust the length of the shift rod between the transfer case and the shift lever with the lever in 4WD-Low so that the distance between the upper surface of the automatic transmission extension housing and the upper horizontal edge of the shift lever is 0.640 to 0.600 in.

New Process 203 Full-Time 4WD

1. Place the shift in the neutral position.

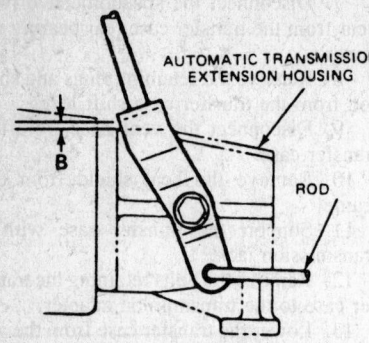

VIEW IN DIRECTION OF ARROW

ADJUSTMENT (AUTOMATIC TRANSMISSION)

TRANSFER CASE MUST BE IN 4 WHEEL LOW
(CROSS LINK ASSEMBLY ALL THE WAY IN)

DIMENSION "B" MUST BE $\frac{.64}{.60}$

ADJUST ROD (7B051) TO CORRECT LENGTH
AND TIGHTEN LOCK NUT AT CLEVIS

NP205 transfer case installation showing the adjustment of the shift linkage with automatic transmission

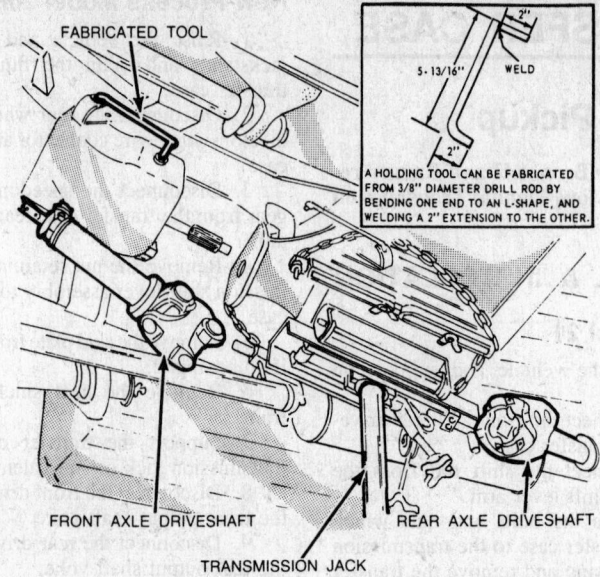

Removal or installation of the Dana 20 transfer case and the dimensions of the holding tool which must be used

2. Remove the two adjusting stud nuts.

3. Install a 0.025 inch diameter alignment pin (1.25 inches long through the shifter assembly.

4. Align the bottom transfer case lever (Lock lever) by rotating clockwise to the forward position.

5. Align the top transfer case lever (range lever) by placing in the middle or neutral position.

6. Re-position the two shift rods and tighten the adjusting stud nuts to 15–20 ft. lbs.

7. Remove the alignment pin from the shifter assembly.

Bronco

NOTE: For Bronco II, refer to the Manual Transmission Section

REMOVAL & INSTALLATION

Dana 20

1. Shift the transfer case into Neutral.

2. Remove the bolts attaching the fan shroud to the radiator support.

3. Raise and safely support the vehicle.

4. Support the transfer case shield with a jack and remove the bolts that attach the shield to the frame side rails. Remove the shield.

5. Drain the transmission and transfer case lubricant.

6. Disconnect the front and rear driveshafts at the transfer case.

7. Disconnect the speedometer cable at the transfer case.

8. If equipped with a manual transmission, disconnect the shift rods from the transmission shift levers. Then, place the First-Reverse gear shift lever in to the First

gear position and insert the fabricated tool. (See transmission removal and installation). This tool will prevent the input shaft roller bearings from dropping into the transmission case when separating the transfer case from the transmission and output shaft.

9. Support the engine with a jack.

10. Remove the two cotter pins, bolts, washers, plates and insulators that secure the crossmember to the transfer case adapter.

11. Remove the crossmember to frame side support attaching bolts.

12. Raise the transmission and remove the upper insulators from the crossmember. Remove the crossmember.

13. Remove the carpet from around the shift levers. Remove the bolts holding the shifter to the transmission adapter. Remove the lower spring from the shifter. Remove the boot from the bottom of the shifter. Bend up the left side of the floor opening as required to remove the shifter assembly.

14. Secure the transfer case to a transmission jack and remove the transfer case adapter-to-transmission attaching bolts.

15. Move the transfer case and jack rearward until it clears the transmission output shaft. Lower the transfer case.

16. Installation is the reverse of the removal procedure.

New Process 205

1. Drain the transfer case. Disconnect the rear axle driveshaft and front driveshaft from the flange at the transfer case.

2. Disconnect the shift selector rod steady rest and the speedometer cable at the transfer case.

3. Secure the transfer case to a transmission jack, and remove the mounting bolts.

4. Remove transfer case, and place it on a floor stand or workbench.

5. Remove the transfer case from the

floor stand or bench and place it on the transmission jack.

6. Raise the transfer case into position and attach the mounting bolts. Tighten the bolts and nuts to 20–40 ft. lbs.

7. Connect the shift selector rod, the speedometer cable, and steady rest.

8. Connect the front and rear axle drive shafts and tighten the universal joint U-bolt nuts.

9. Fill the transfer case to filler plug level with SAE 80W/90 oil. Tighten drain plug to 25–35 ft. lbs.

New Process 203

1. Drain the transfer case by removing the power take-off lower bolts and the front output rear cover lower bolts.

2. Disconnect the front axle drive shaft from the flange at the transfer case.

3. Disconnect the shift rods from the transfer case.

4. Disconnect the speedometer cable and lockout lamp switch wire from the transfer case rear output shaft housing.

5. Remove the transfer case-to-transmission adapter attaching bolts. Disconnect the rear axle driveshaft at the transfer case flange.

6. Position a transmission jack under the transfer case and secure it to the jack.

7. Remove the transfer case mounting bracket support to frame crossmember nuts, bolts, spacers and upper absorbers and remove the transfer case.

8. Using a chain fall, place the transfer case on a suitable work bench.

9. Using a chain fall, secure the transfer case on a transmission jack.

10. Position the transfer case in the truck, aligning the mounting bracket supports with the lower absorbers. Align the transfer case-to-transmission attaching bolts.

11. Install the transfer case mounting bracket support to frame crossmember bolts with upper absorbers and spacers. Tighten all mounting bolts to 40–50 ft. lbs.

12. Remove the transmission jack.

13. Connect the speedometer cable and the lockout switch wire to the transfer case.

14. Connect the rear axle driveshaft to the transfer case rear output flange.

15. Install the shift rods on the transfer case and adjust the shift linkage.

16. Connect the front axle driveshaft to the transfer case flange.

17. Fill the transfer case with SAE 80W/90 oil. Tighten the filler plug to 25–35 ft. lbs.

New Process 208

1. Raise the vehicle and support on jackstands.

2. Place a drain pan under transfer case, remove drain plug and drain fluid from transfer case.

3. Disconnect four wheel drive indicator switch wire connector at transfer case.

4. Disconnect speedometer driven gear from transfer case rear bearing retainer.

5. Remove nut retaining transmission shift lever assembly to transfer case.

6. If so equipped, remove skid plate from frame.

7. Remove heat shield from frame.

--- CAUTION ---

Catalytic converter is located beside the heat shield. Be careful when working around catalytic converter because of the extremely high temperatures generated by the converter.

8. Support transfer case with transmission jack.

9. Disconnect front driveshaft from front output shaft yoke.

10. Disconnect rear driveshaft from rear output shaft yoke.

11. Remove the bolts retaining transfer case to transmission adapter. Remove gasket between transfer case and adapter.

12. Lower transfer case from vehicle.

13. Place a new gasket between transfer case and adapter.

14. Raise transfer case with transmission jack so transmission output shaft aligns with splined transfer case input shaft. Install bolts retaining transfer case to adapter. Tighten bolts to 30–40 ft. lbs.

15. Connect rear driveshaft to rear output shaft yoke.

16. Connect front driveshaft to front output yoke.

17. Remove transmission jack from transfer case.

18. Position heat shield to frame crossmember and mounting lug on transfer case. Install and tighten bolts and screw to 11–16 ft. lbs.

19. Install skid plate to frame. Tighten nuts and bolts.

20. Install shift lever to transfer case. Install retaining nut.

21. Connect speedometer driven gear to transfer case.

22. Connect four-wheel drive indicator switch wire connector at transfer case.

23. Install drain plug. Remove filler plug and install 2.8 liters (six pints) of automatic transmission fluid Dexron® II. Install filler plug.

24. Lower vehicle.

Shift Lever Adjustment

NOTE: The NP 205 and 208 do not require adjustment.

NP 203

1. Place shift lever in neutral position.

2. Remove two adjusting stud nuts.

3. Install 0.25 inch diameter alignment pin (1.25 inches long) through shifter assembly.

4. Align the two transfer case levers as follows: Bottom lever, (lock lever): Rotate clockwise to the forward position. Top lever, (range lever): Place in the Mid-position or Neutral position.

5. Re-position the two shift rods and tighten new adjusting stud nuts to 15–20 ft. lbs.

6. Remove alignment pin from shifter assembly.

DRIVELINE

Driveshaft

REMOVAL & INSTALLATION

4WD

1. Mark shaft and flange for installation in same position. To remove the rear driveshaft, disconnect the double Cardan joint from the flange at the transfer case and the single U-joint from the flange at the rear axle. Remove the driveshaft.

2. To remove the front driveshaft, disconnect the double Cardan joint from the flange at the transfer case and the single U-joint from the front axle. Remove the driveshaft.

3. Installation is the reverse of removal. Torque driveshaft-to-transfer case bolts to 20–25 ft. lbs.; driveshaft to axle bolts to 8–15 ft. lbs.

2WD

1. Mark shaft and flange for installation in same position. Unscrew the nuts attaching the U-bolts to the flange at the

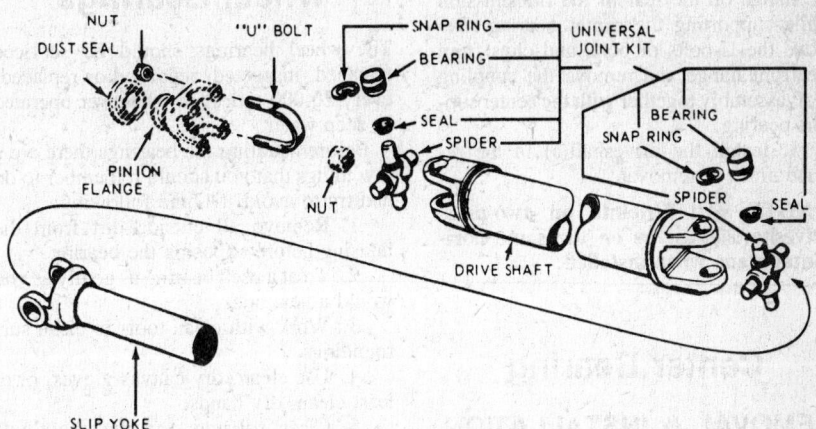

One-piece driveshaft with a slip yoke

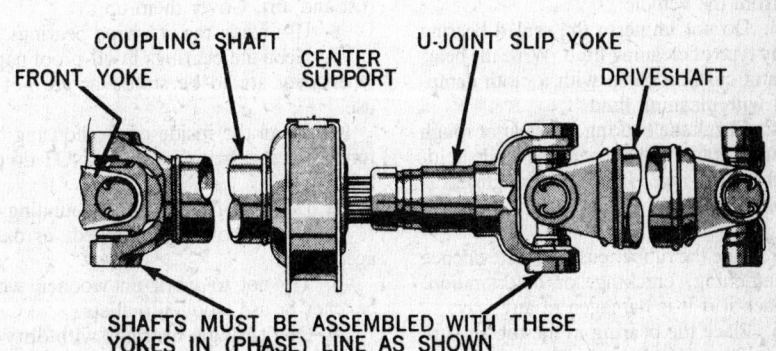

SHAFTS MUST BE ASSEMBLED WITH THESE YOKES IN (PHASE) LINE AS SHOWN

Two-piece driveshaft with a slip yoke at the transmission end

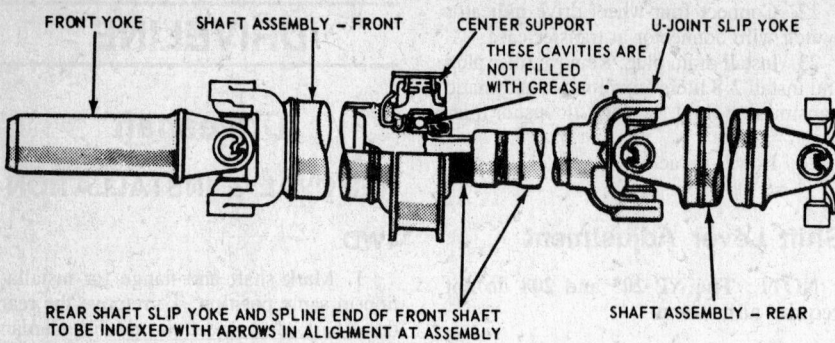

FRONT YOKE   SHAFT ASSEMBLY – FRONT   CENTER SUPPORT   U-JOINT SLIP YOKE

THESE CAVITIES ARE NOT FILLED WITH GREASE

REAR SHAFT SLIP YOKE AND SPLINE END OF FRONT SHAFT TO BE INDEXED WITH ARROWS IN ALIGMENT AT ASSEMBLY

SHAFT ASSEMBLY – REAR

Two-piece driveshaft with a fixed yoke at the transmission end

rear axle, or remove the bolts and clips. Remove the U-bolts or bolts and clips and allow the rear of the driveshaft to drop down. Slide the front of the driveshaft out of the rear of the transmission, transfer case, or the center support bearing. Remove the driveshaft from the vehicle.

2. On those vehicles equipped with two driveshafts and a center support bearing, unscrew the attaching bolts holding the center support bearing to the frame. If equipped with a sliding yoke at the transmission, slide the coupling shaft out of the rear of the extension housing. Otherwise, remove the nuts from the U-bolts or bolts and clips holding the front of the coupling shaft to the flange on the rear of the transmission while supporting the center bearing. Remove the U-bolts or bolts and clips from the front flange and remove the coupling shaft assembly together with the center support bearing.

3. Install the driveshaft(s) in the reverse order of removal.

NOTE: All U-joints on two-piece driveshafts must be on the same horizontal plane when installed.

Center Bearing

REMOVAL & INSTALLATION

1. Remove the driveshafts.

2. Remove the two center support bearing attaching bolts and remove the assembly from the vehicle.

3. Do not immerse the sealed bearing in any type of cleaning fluid. Wipe the bearing and cushion clean with a cloth dampened with cleaning fluid.

4. Check the bearing for wear or rough action by rotating the inner race while holding the outer race. If wear or roughness is evident, replace the bearing.

Examine the rubber cushion for evidence of hardening, cracking, or deterioration. Replace it if it is damaged in any way.

5. Place the bearing in the rubber support and the rubber support in the U-shaped support and install the bearing in the reverse order of removal.

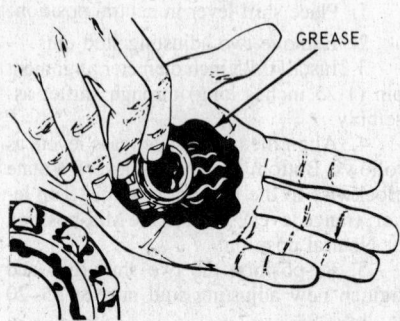

GREASE

Hand-packing the wheel bearings

Wheel Bearings

The wheel bearings should be serviced (cleaned, inspected, repacked or replaced) every 20,000 miles, or whenever operated in deep water.

Before handling the bearings there are a few things that you should remember to do and try to avoid. DO the following:

1. Remove all outside dirt from the housing before exposing the bearing.

2. Treat a used bearing as gently as you would a new one.

3. Work with clean tools in clean surroundings.

4. Use clean, dry canvas gloves, or at least clean, dry hands.

5. Clean solvents and flushing fluids are a must.

6. Use clean paper when laying out the bearings to dry.

7. Protect disassembled bearings from rust and dirt. Cover them up.

8. Use clean rags to wipe bearings.

9. Keep the bearings in oil-proof paper when they are to be stored or are not in use.

10. Clean the inside of the housing before replacing the bearing. Do NOT do the following:

1. Don't work in dirty surroundings.

2. Don't use dirty, chipped, or damaged tools.

3. Try not to work on wooden work benches or use wooden mallets.

4. Don't handle bearings with dirty or moist hands.

5. Do not use gasoline for cleaning; use a safe solvent.

6. Do not spin-dry bearings with compressed air. They will be damaged.

7. Do not spin unclean bearings.

8. Avoid using cotton waste or dirty cloths to wipe bearings.

9. Try not to scratch or nick bearing surfaces.

10. Do not allow the bearing to come in contact with dirt or rust at any time.

SERVICE

2WD Front

1. Jack the truck up until the wheel to be serviced is off the ground and can spin freely. It is easier to check all the bearings at the same time. If the equipment needed is available, raise the front end of the truck so that both front wheels are off the ground. Use jackstands to support the vehicle. Make sure that the truck is completely stable before proceeding any further.

2. Remove the lug nuts and remove the wheel/tire assembly from the hub. It is necessary to remove the caliper assembly from the rotor and caliper support. Do not disconnect the brake line from the caliper. Simply hang the caliper with a length of heavy wire above the hub. Be careful not to strain the flexible brake tube.

3. Remove the grease cap with a screwdriver or pliers.

4. Remove the cotter pin and discard it. Cotter pins should never be reused.

5. Remove the nut lock, adjusting nut, and washer from the spindle.

6. Wiggle the hub so that the outer wheel bearing comes loose and can be removed. Remove the outer bearing.

7. Remove the hub from the spindle and place it on a work surface, supported by two blocks of wood under the hub.

8. Place a block of wood or drift pin through the spindle hole and tap out the inner grease seal. Tap lightly so not to damage the bearing. When the seal falls out, so will the inner bearing. Discard the seal.9.

Place all of the bearings, nuts, nut locks, washers and grease caps in a container of solvent. Use a light soft brush to thoroughly clean each part. Make sure that every bit of dirt and grease is rinsed off, then place each cleaned part on an absorbent cloth or paper and allow them to dry completely.

10. Clean the inside of the hub, including the bearing races, and the spindle. Remove all traces of old lubricant from these components.

11. Inspect the bearings for pitting, flat spots, rust, and rough areas. Check the races in the hub and the spindle for the same defects and rub them clean with a cloth that has been soaked in solvent. If the races show hair line cracks or worn shiny areas, they must be replaced. The races are installed in the hub with a press fit and are removed by driving them out with a suitable punch or drift. Place the new races squarely onto the hub and place a block of wood over them. Drive the race into place with

a hammer, striking the block of wood. Never hit the race with any metal object. Replacement seals, bearings, and other required parts can be bought at an auto parts store. The old parts should be taken along to be compared with the replacement parts to ensure a perfect match.

12. Pack the wheel bearings with grease. There are special devices made for the specific purpose of greasing bearings, but if one is not available, pack the wheel bearings by hand. Put a large dab of grease in the palm of your hand and push the bearing through it with a sliding motion. The grease must be forced through the side of the bearing and in between each roller. Continue until the grease begins to ooze out the other side and through the gaps between the rollers; the bearing must be completely packed with grease.

NOTE: Sodium based grease is not compatible with lithium based grease. Be careful not to mix the two types. The best way to prevent this is to completely clean all of the old grease from the hub and spindle before installing any new grease.

13. Turn the hub assembly over so that the inner side faces up, making sure that the race and inner area are clean, and drop the inner wheel bearing into place. Using a hammer and a block of wood, tap the new grease seal in place. Never hit the seal with the hammer directly. Move the block of wood around the circumference until it is properly seated.

14. Slide the hub assembly onto the spindle and push it as far as it will go, making sure that it has completely covered the brake shoes. Keep the hub centered on the spindle to prevent damage to the grease seal and the spindle threads.

15. Place the outer wheel bearing in place over the spindle. Press it in until it is snug. Place the washer on the spindle after the bearing. Screw on the spindle nut and turn it down until a slight binding is felt.

16. With a torque wrench, tighten the nut to 17–25 ft. lbs. to seat the bearings. Install the nut lock over the nut so that the cotter pin hole in the spindle is aligned with a slot in the nut lock. Back off the adjusting nut and the nut lock two slots of the nut lock and install the cotter pin.

17. Bend the longer of the two ends opposite the looped end out and over the end of the spindle. Trim both ends of the cotter pin just enough so that the grease cap will fit, leaving the bent end shaped over the end of the spindle.

18. Install the grease cap, brake caliper if so equipped, and the wheel/tire assembly. The wheel should rotate freely with no noise or noticeable end-play.

4WD Front Without Free-Running Hubs

EXCEPT RANGER & BRONCO II

NOTE: Sodium based grease is not compatible with lithium based grease. Be

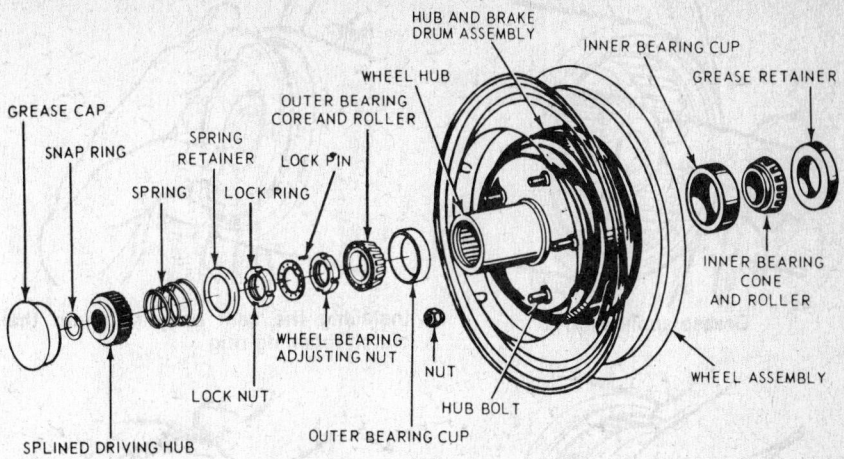

Typical front hub without free-wheeling (locking hubs)

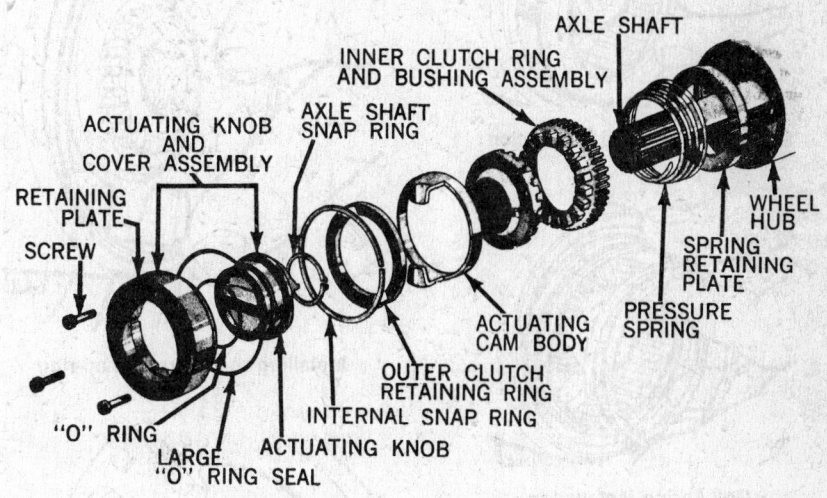

Typical internal locking hub through 1979

careful not to mix the two types. The best way to prevent this is to completely clean all of the old grease from the hub assembly before installing any new grease.

1. Raise the front of the vehicle and place jackstands under the vehicle. Remove the wheel.

2. Remove the front hub grease cap and driving hub snap-ring. On models equipped with free-running hubs, remove the retainer knob hub ring, actuator knob, snap-ring, outer clutch retaining ring, and actuating cam body.

3. Remove the splined driving hub and the pressure spring. This may require slight prying with an appropriate tool.

4. Remove the wheel bearing locknut, lockring, and adjusting nut.

5. Remove the brake caliper. Suspend the caliper out of the way and remove the hub and rotor assembly.

6. Carefully drive out the inner bearing cone and grease seal from the hub.

7. Inspect the bearing cups (races) for cracks and pits. If the cups are excessively worn or there are pits or cracks visible, replace them along with the cones. The cups

are removed from the hub by driving them out with a drift pin. They are installed in the same manner.

8. If it is determined that the cups are in satisfactory condition and are to remain in the hub, clean and inspect the cones (bearings). Refer to the bearing diagnosis chart. Replace the bearings if necessary. If it is necessary to replace either the cone or the cup, both parts should be replaced as a unit.

9. Thoroughly clean all components in a suitable solvent and blow them dry with compressed air or allow them to dry while resting on clean paper.

CAUTION
Do not spin the bearings with compressed air while drying them.

10. Cover the spindle with a cloth and brush all loose dust and dirt from the brake assembly. Remove the cloth and thoroughly clean the inside of the hub and the spindle.

11. Pack the inside of the hub with wheel bearing grease. Add grease to the hub until

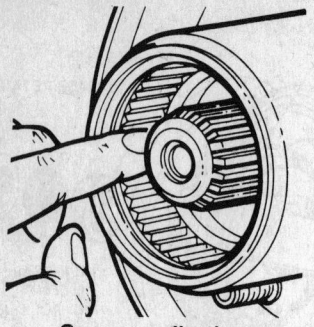

Grease application

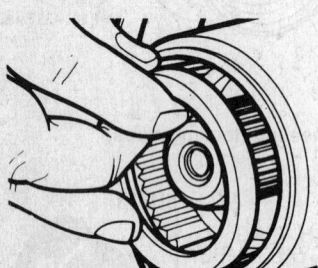

Spring retainer ring installation

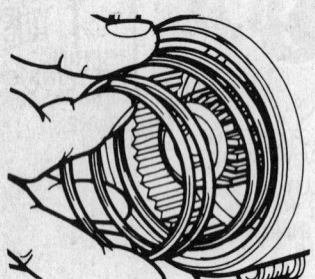

Coil spring installation

Axle shaft sleeve and ring, and inner clutch ring installation

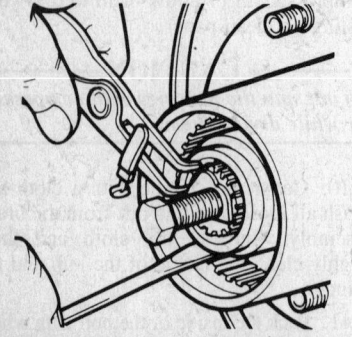

Installing the axle shaft snap-ring

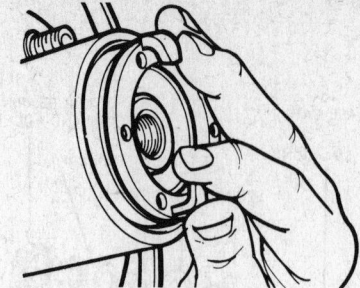

Installing the cam body ring into the clutch retaining ring

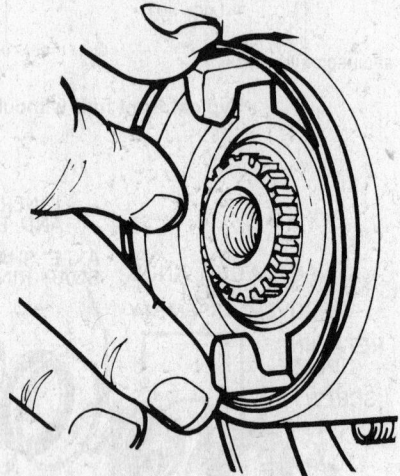

Installing the internal snap-ring

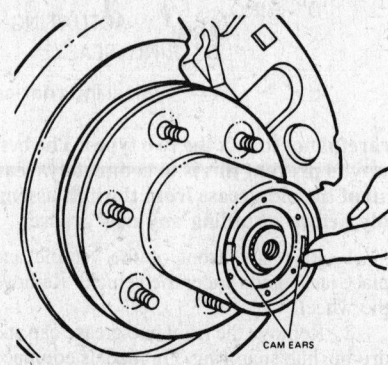

Applying a small amount of grease on the ears of the cam

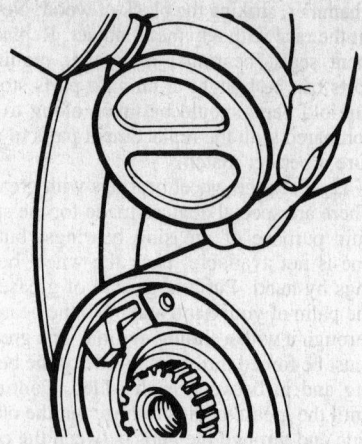

Lubricating the selector knob

14. Carefully position the hub assembly onto the spindle. Be careful not to damage the new seal. Install the caliper.

15. Place the outer bearing into position on the spindle and into the bearing cup.

16. Install the bearing adjusting nut and tighten it to 50 ft. lbs. while rotating the hub back and forth to seat the bearings.

17. Back off the adjusting nut about 90°.

18. Assemble the lockring by turning the nut to the nearest notch where the dowel pin will enter.

19. Install the outer locknut and torque to 80–100 ft. lbs. The final end-play of the wheel on the spindle should be 0.001–0.010 in.

20. Install the pressure spring retainer, spring, the driving hub and driving hub snap-ring.

21. Install the grease cap and adjust the brakes, if they were backed off to remove the hub assembly. Remove the jackstands and lower the vehicle.

4WD Front With Free-Running Hub

1979
EXCEPT RANGER & BRONCO II

1. Remove the free-running hub bolts and washers.

2. Remove the hub ring and the knob. Wipe the parts clean.

3. Remove the internal snap-ring from the groove in the hub.

4. Remove the cam body ring and clutch retainer (as an assembly) from the hub. Disassemble the parts.

5. Remove the axle shaft snap-ring. For easier snap-ring removal, push inward on the axle shaft sleeve ring and, at the same time, pull out on the axle with a bolt.

6. Remove the axle shaft sleeve ring and inner clutch ring. A slight rocking of the hub may make them slide out easier.

7. Remove the pressure spring.

8. Remove the spring retainer ring.

9. Grease the hub inner spline with Moly grease or equivalent.

10. Install the spring retainer ring, po-

the grease is flush with the inside diameter of the bearing cup.

12. Pack the bearing cone and roller assemblies with wheel bearing grease. A bearing packer is desirable for this operation. If a packer is not available, place a large portion of grease into the palm of your hand and sliding the edge of the roller cage through the grease with your other hand, work as much grease in between the rollers as possible.

13. Position the inner bearing into the inner bearing cup and install the new grease seal.

sitioned as shown with recessed undercut area going in first. Be sure ring seats against the bearing.

11. Install the coil spring with large end entering first.

12. Grease with Moly grease or equivalent and install the axle shaft sleeve and ring and the inner clutch ring. Be sure that the teeth are meshed together in a locked position for easy assembly. It may be necessary to rock the hub back and forth for spline alignment. Keep the two gears in locked position.

13. Install the axle shaft snap-ring. Push inward on gear and, if necessary, pull out axle shaft with bolt to allow groove clearance on shaft for the snap-ring. Be sure snap-ring is fully seated in the snap-ring groove on the shaft.

14. Install the actuating cam body ring into the outer clutch retaining ring. Assemble into hub.

15. Install the internal snap-ring. Be sure snap-ring is fully seated in the snap-ring groove of the hub.

16. Apply a small amount of Moly grease or equivalent on the ears of the cam.

17. Apply a small amount of Parker O-ring lube or an equivalent lube in groove of actuating knob before assembling outer O-ring.

18. Assemble knob in hub ring and assemble to axle with knob in locked position. Assemble screws and washers alternately and evenly, making sure the retainer ring is not cocked in the hub.

19. Tighten the six lock-out hub bolts to 30–35 in. lbs. Be sure the washers are under each retaining screw. Each free-running hub will fit either wheel. Do not drive vehicle until you are sure that both free-running hubs are functioning properly.

FROM 1980

Manual Hubs
Except Ranger & Bronco II

1. To remove hub, first separate cap assembly from body assembly by removing the six (6) socket head capscrews from the cap assembly and slip apart.

2. Remove snap-ring (retainer ring) from the end of the axle shaft.

3. Remove the lock ring seated in the groove of the wheel hub. The body assembly will now slide out of the wheel hub. If necessary, use an appropriate puller to remove the body assembly.

4. Install hub in reverse order of removal. Torque socket head capscrews to 30–35 in. lbs.

Automatic Locking Hubs
Except Ranger & Bronco II

1. Remove capscrews and remove hub cap assembly from spindle.

2. Remove capscrew from end of axle shaft.

3. Remove lock ring seated in the groove

of the wheel hub with a knife blade or with a small sharp awl with the tip bent in a hook.

4. Remove body assembly from spindle. If body assembly does not slide out easily, use an appropriate puller.

5. Unscrew all three sets in the spindle locknut until the heads are flush with the edge of the locknut. Remove outer spindle locknut with tool T80T–4000–V, automatic hub lock nut wrench.

6. Reinstall in reverse order of removal. Tighten the outer spindle locknut to 15–20 ft. lbs. with special tool T80T–4000–V, automatic hub lock nut wrench. Tighten down all three set screws. Firmly push in body assembly until the friction shoes are on top of the spindle outer locknut.

7. Install capscrew into the axle shaft and tighten to 35–50 ft. lbs.

8. Place cap on spindle and install capscrews. Tighten to 35–50 in. lbs. Turn dial firmly from stop to stop, causing the dialing mechanism to engage the body spline.

NOTE: Be sure both hub dials are in the same position; "AUTO" or "LOCK."

BEARING REPLACEMENT OR REPACKING

1. Raise the vehicle and support on safety stands.

2. If equipped with free-running hubs refer to Free-Running Hub Removal and Installation.

3. Remove the front hub grease cap and driving hub snap-ring.

4. Remove the splined driving hub and the pressure spring. This may require a slight prying assist.

5. Remove the wheel bearing lock nut, lock ring, and adjusting nut using tool T59T–1197–B, or equivalent.

6. Remove the hub and disc assembly. The outer wheel bearing and spring retainer will slide out as the hub is removed.

7. Remove the spindle retaining nuts, then carefully remove the spindle from the knuckle studs and axle shaft.

8. Clean all old grease from the needle bearings and wipe clean the spindle face that mates with the spindle bore seal.

9. Remove the spindle bore seal, V-seal, and thrust washer from the outer axle shaft. Clean any old grease or dirt from these parts and replace those that show signs of excessive wear.

10. Using Multi-Purpose Lubricant, Ford Specification ESA–M1C75–B or equivalent, thoroughly lubricate the needle bearing and pack the spindle face that mates with the spindle bore seal.

11. Assemble the V-seal in the spindle bore next to the needle bearing. Assemble the spindle bore seal on the axle shaft.

12. Assemble the spindle with the axle shaft on the knuckle studs. Adjust the retaining nuts to 50–60 ft. lbs.

13. Carefully drive the inner bearing cone

and grease seal out of the hub using Tool T69L–1102–A.

14. Inspect the bearing cups for pits or cracks. If necessary, remove them with a drift. If new cups are installed, install new bearings.

15. Lubricate the bearings with Multi-Purpose Lubricant Ford Specification, ESA–M1C7–B or equivalent. Clean all old grease from the hub. Pack the cones and rollers. If a bearing packer is not available, work as much lubricant as possible between the rollers and the cages.

16. Position the inner bearing cone and roller in the inner cup and install the grease retainer.

17. Carefully position the hub and disc assembly on the spindle.

18. Install the outer bearing cone and roller, and the adjusting nut.

19. Using a torque wrench, tighten the bearing adjusting nut to 50 ft. lbs., while rotating the wheel back and forth to seat the bearings.

20. Back off the adjusting nut approximately 90 degrees.

21. Assemble the lock ring by turning the nut to the nearest hole and inserting the dowel pin. Note: The dowel pin must seat in a lock ring hole for proper bearing adjustment and wheel retention.

22. Install the outer lock nut and tighten to 50–80 ft. lbs. Final end play of the wheel on the spindle should be 0.001 to 0.010 inch.

23. Install the pressure spring and driving hub snap-ring.

24. Apply non-hardening sealer to the seating edge of the grease cap, and install the grease cap.

25. Adjust the brake if it was backed off.

26. Remove the safety stands and lower the vehicle.

Bronco II and Ranger

MANUAL HUB

1. Raise the vehicle and install safety stands.

2. Remove the wheel lug nuts and remove the wheel and tire assembly.

3. Remove the retainer washers from the lug nut studs and remove the manual locking hub assembly. To remove the internal hub lock assembly from the outer body assembly, remove the outer lock ring seated in the hub body groove. The internal assembly, spring and clutch gear will now slide out of the hub body.

— **CAUTION** —
Do not remove the screw from the plastic dial.

4. Rebuild the hub assembly in the reverse order of disassembly.

5. Install the manual locking hub assembly over the spindle and place the retainer washers on the lug nut studs.

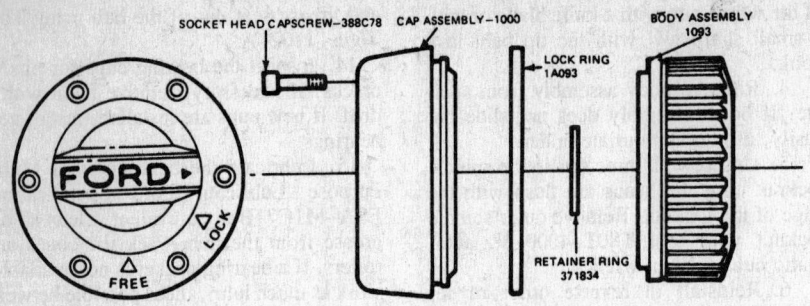

SOCKET HEAD CAPSCREW-388C78 CAP ASSEMBLY-1000 BODY ASSEMBLY 1093 LOCK RING 1A093 RETAINER RING 371834

1980 and later manual locking hubs

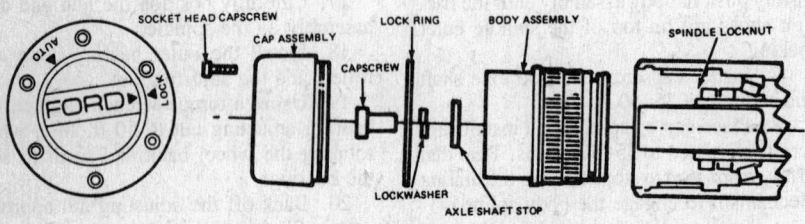

SOCKET HEAD CAPSCREW CAP ASSEMBLY CAPSCREW LOCK RING BODY ASSEMBLY SPINDLE LOCKNUT LOCKWASHER AXLE SHAFT STOP

1980 and later automatic locking hubs

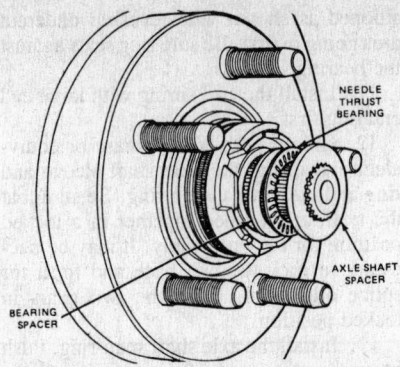

NEEDLE THRUST BEARING AXLE SHAFT SPACER BEARING SPACER

Axle shaft spacer, thrust bearing and bearing spacer on Ranger/Bronco II with automatic locking hubs- 4x4

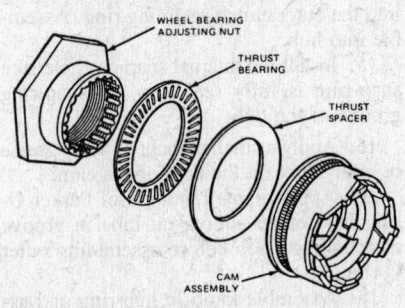

WHEEL BEARING ADJUSTING NUT THRUST BEARING THRUST SPACER CAM ASSEMBLY

Wheel bearing adjusting nut assembly-Ranger/Bronco II 4x4

6. Install the wheel and tire assembly. Install the lug nuts and tighten to 85–115 ft. lbs.

ADJUSTMENT

1. Raise the vehicle and support on safety stands. Remove the wheel lug nuts and remove the wheel and tire assembly.

2. Remove the retainer washers from the lug nut studs and remove the manual locking hub assembly from the spindle.

3. Remove the snap ring from the end of the spindle shaft.

4. Remove the axle shaft spacer, needle thrust bearing and the bearing spacer.

5. Remove the outer wheel bearing locknut from the spindle, using a four-prong spindle nut spanner wrench. Make sure the tabs on the tool engage the slots in the lock-nut. Remove the locknut washer from the spindle.

6. Loosen the inner wheel bearing locknut using Four Prong Spindle Nut Spanner Wrench or equivalent. Make sure that the tabs on the tool engage the slots in the locknut and that the slot in the tool is over the pin on the locknut.

7. Tighten the inner locknut to 35 ft. lbs. to seat the bearings. Spin the rotor and back off the inner locknut ¼ turn. Install the lockwasher on the spindle. It may be necessary to turn the inner locknut slightly so that the pin on the locknut aligns with the closest hole in the lockwasher.

8. Install the outer wheel bearing lock-nut using Four-Prong Spindle Nut Spanner Wrench or equivalent.

9. Tighten locknut to 150 ft. lbs.

10. Install the bearing thrust spacer, needle thrust bearing and axle shaft spacer.

11. Clip the snap ring onto the end of the spindle.

12. Install the manual hub assembly over the spindle. Install the retainer washers.

13. Install the wheel and tire assembly. Install and tighten lugnuts to 85–115 ft. lbs.

14. Check the end play of the wheel and tire assembly on the spindle. End play should be 0.001–0.003 inch.

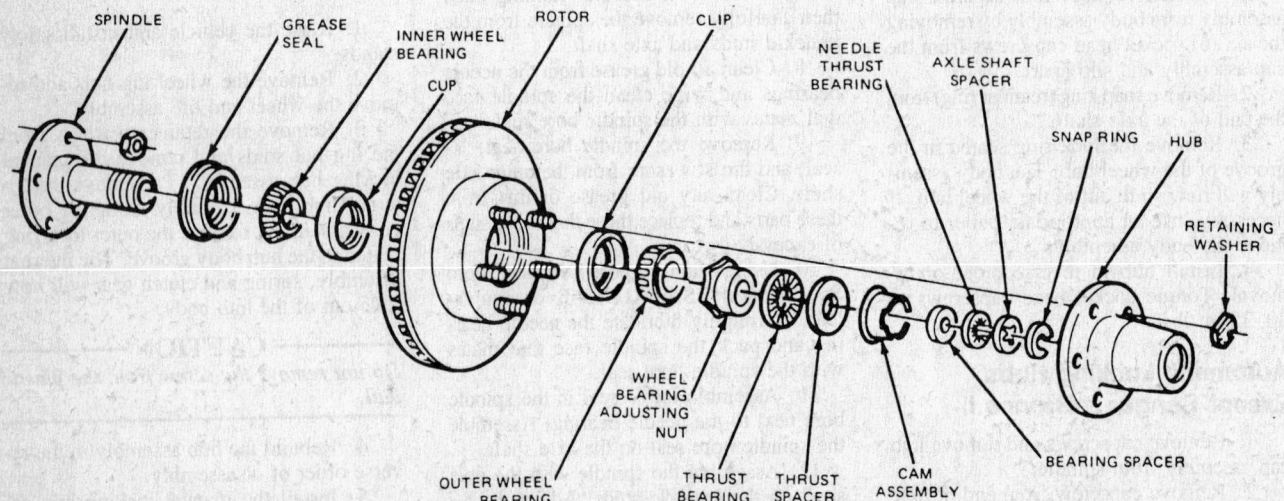

SPINDLE GREASE SEAL INNER WHEEL BEARING CUP ROTOR CLIP NEEDLE THRUST BEARING AXLE SHAFT SPACER SNAP RING HUB RETAINING WASHER OUTER WHEEL BEARING WHEEL BEARING ADJUSTING NUT THRUST BEARING THRUST SPACER CAM ASSEMBLY BEARING SPACER

Automatic locking hubs used on the Ranger/Bronco II 4x4

AUTOMATIC LOCKING HUBS

1. Raise the vehicle and support on safety stands. Remove the wheel lug nuts and remove the wheel and tire assembly.

2. Remove the retainer washers from the lug nut studs and remove the automatic locking hub assembly from the spindle.

3. Remove the snap ring from the end of the spindle shaft.

4. Remove the axle shaft spacer, needle thrust bearing and the bearing spacer. Being careful not to damage the plastic moving cam, pull the cam assembly off the wheel bearing adjusting nut and remove the thrust washer and needle thrust bearing from the adjusting nut.

5. Loosen the wheel bearing adjusting nut from the spindle using a 2¾ inch hex socket tool.

6. While rotating the hub and rotor assembly, tighten the wheel bearing adjusting nut to 35 ft. lbs. to seat the bearings, then back off the nut ¼ turn (90°).

7. Retighten the adjusting nut to 16 in. lb. using a torque wrench. Align the closest hole in the wheel bearing adjusting nut with the center of the spindle keyway slot. Advance the nut to the next hole if required.

8. Install the locknut needle bearing and thrust washer in the order of removal and push or press the cam assembly onto the locknut by lining up the key in the fixed cam with the spindle keyway.

9. Install the bearing thrust washer, needle thrust bearing and axle shaft spacer. Clip the snap ring onto the end of the spindle.

10. Install the automatic locking hub assembly over the spindle by lining up the three legs in the hub assembly with three pockets in the cam assembly. Install the retainer washers.

11. Install the wheel and tire assembly. Install and tighten lugnuts to 85–115 ft. lbs.

12. Final end play of the wheel on the spindle should be 0.001–0.003 inch).

FRONT DRIVE AXLE

Axle Shaft

REMOVAL & INSTALLATION

1979 F100–150

1. Raise and support the vehicle on jackstands.

2. Remove the grease cap. Remove the driving hub retaining snap-ring, then slide splined driving hub from between axle shaft and wheel hub.

3. Remove driving hub spring.

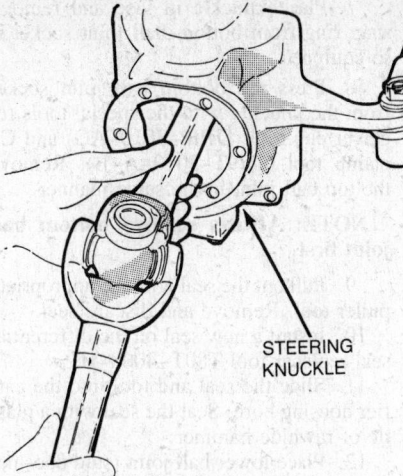

Typical removal/installation of axle shaft through 1979

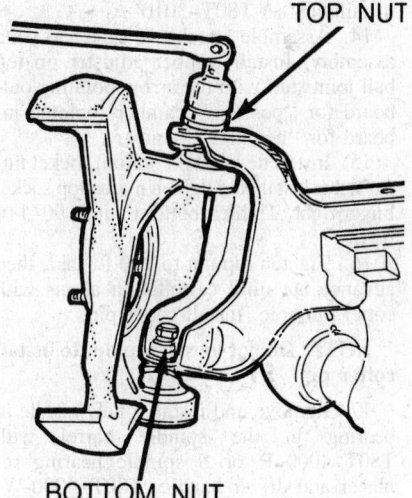

Removing the steering knuckle nut on the Dana 44-IFS

NOTE: If equipped with free-wheeling hubs, see hub removal.

4. Remove lock nut, washer, and wheel bearing adjusting nut from spindle. Remove wheel, hub and drum as an assembly. The wheel outer bearing will be forced off the spindle at the same time. Remove wheel inner bearing cone.

5. Remove capscrews that attach brake backing plate and spindle to steering knuckle. Remove brake backing plate and secure it to one side. Carefully remove the spindle.

6. Pull shaft assembly from axle housing, working universal joint through bore in steering knuckle.

7. To install reverse the removal procedure.

1979 F250–350

1. Raise and support the front of the truck.

2. Remove the wheel.

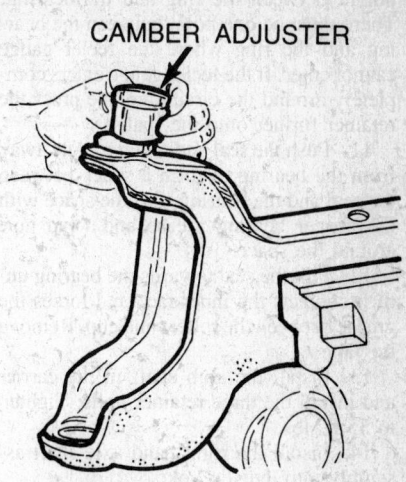

Removing the camber bushing on the Dana 44-IFS

3. Remove the caliper from the rotor and suspend it from the frame.

4. Remove the dust cap, cotter pin, nut, washer and outer bearing and remove the rotor from the spindle.

5. Remove the inner bearing cone and seal.

6. Remove the axle shaft, working the U-joint through the bore of the steering knuckle. Be careful not to damage the seal.

7. Installation is the reverse of removal.

From 1980 F150–350

1. Remove the axle shaft assembly. Remove the axle bearing as follows:

2. Remove the stub assembly by removing the three bolts attaching the retainer plate to the carrier housing.

3. Place the axle shaft in a vise and drill a ½' hole in the outside of the bearing retaining ring to a depth of ¾ of the thickness of the ring.

4. Place a chisel across the hole and strike sharply with a hammer to remove the retaining ring. Replace the bearing retaining ring upon assembly.

5. Press the bearing from the axle shaft using special tools axle bearing remover T80T–4000M and sleeve T80T–4000L.

6. Remove the steal and retainer plate from the stub shaft.

7. To install place the new seal and retainer plate on the shaft.

8. Place the bearing on the shaft with the large radius on the inner race facing the yoke end of the shaft.

9. Press the bearing onto the shaft using an axle bearing replacer T80T–4000–N and a pinion bearing cone remover T71P–4621–B. A 0.0015 inch feeler gauge should not fit between the bearing seat and the bearing.

10. Using the same special tools in Step 9 press the bearing retainer ring onto the stub shaft. A 0.0015 inch feeler gauge should

not fit between the ring and the bearing. There must be one point between the bearing and the ring where the feeler gauge cannot enter. If the feeler gauge enters completely around the circumference press the retainer further onto the shaft.

11. Push the seal and retainer plate away from the bearing to form a space between the seal and the bearing. Fill the space with the proper bearing grease and wrap tape around the space.

12. Pull the seal towards the bearing until it contacts the inner race and forces the grease between the rollers and cup. Remove the tape.

13. Install the stub shaft in the carrier and install the three retainer bolts. Tighten to 35 ft. lbs.

14. Install the right hand axle shaft assembly into the slip yoke.

15. Install the spindle.

1979 Bronco

1. Raise the vehicle and support it on jackstands.

2. Remove the front wheels.

3. On drum brake models, remove the hub, brake drum, backing plate and spindle. Tie the backing plate up to avoid damage to the hose.

4. On disc brake models, remove the hubs, calipers and rotors. Tie up the calipers to avoid damage to the hoses. Remove the nuts that attach the brake support bracket, dust shield and spindle.

5. Pull the axle shaft from the housing, carefully working the U-joint through the steering knuckle.

6. Install the shaft in reverse order of removal.

From 1980 Bronco

NOTE: This procedure requires the use of special tools.

1. Remove spindle nuts and remove spindle. It may be necessary to tap the spindle with a rawhide or plastic hammer to break the spindle loose. Remove spindle, splash shield and axle shaft assembly.

2. Place the spindle in a vise with a shop towel around the spindle to protect the spindle from damage. Using a slide hammer T50T–100–A and seal remover, tool 1175–AC remove the axle shaft seal and then the needle bearing from the spindle bar.

3. If the tie rod has not been removed, then remove cotter key from the tie rod nut and then remove nut. Tap on the tie rod stud to free it from the steering arm.

4. Remove the cotter pin from the top ball joint stud. Loosen the nut on the top stud and the bottom nut inside the knuckle. Remove the top nut.

5. Sharply hit the top stud with a plastic or rawhide hammer to free the knuckle from the tube yoke. Remove and discard bottom nut. Use new nut upon assembly.

6. Remove camber adjuster with Pitman arm puller T64P–3590–F.

7. Place knuckle in vise and remove snap ring from bottom ball joint socket if so equipped.

8. Press the bottom ball joint socket from the knuckle with the special tools receiver cup tool (D79P–3010–AG) and C1 clamp tool (D79T–3010–A–B). Remove the top ball joint in the same manner.

NOTE: Always remove bottom ball joint first.

9. Pull out the seal with the appropriate puller tool. Remove and discard seal.

10. Install a new seal on the differential seal replacer tool T80T–4000–H.

11. Slide the seal and tool into the carrier housing bore. Seat the seal with a plastic or rawhide hammer.

12. Place lower ball joint (stud does not have a cotter key hole in stud) in knuckle and press into position using ball joint installation set T80T–3010–A.

13. Install the upper ball joint (stud has cotter key hole) in knuckle with ball joint installation set T80T–3010–A.

14. Assemble knuckle to tube and yoke assembly. Install camber adjuster on top ball joint stud with the arrow pointing outboard for "positive" camber, pointed inboard for "negative" camber.

15. Install new nut on bottom socket finger tight. Install and tighten nut on top socket finger tight. Tighten bottom nut to 90–110 ft. lbs.

16. Tighten top nut to 100 ft. lbs., then advance nut until castellation aligns with cotter pin hole. Install cotter pin.

NOTE: Do not loosen top nut to install cotter pin.

17. Remove and install a new needle to bearing in the spindle barrel with T80T–4000–R or S spindle bearing replacer and driver handle, T80T–4000–W. Install a new seal with tool T80T–4000–W. Install a new seal with tool T80T–400–T or U, seal replacer and T80T–4000–W driven handle.

18. Install the axle shaft assembly into the housing. Install the splash shield and spindle. Install and tighten the spindle attaching nuts.

Ranger and Bronco II

DANA 28

1. Raise the vehicle and install safety stands. Remove the wheel and tire assembly. Remove the caliper.

2. Remove hub locks, wheel bearings, and lock nuts.

3. Remove the hub and rotor. Remove the outer wheel bearing cone. Remove the grease seal from the rotor. Remove the inner wheel bearing.

4. Remove the inner and outer bearing cups from the rotor. Remove the nuts retaining the spindle to the steering knuckle. Tap the spindle with a plastic hammer to jar the spindle from the knuckle. Remove the splash shield.

5. Remove the shaft and joint assembly by pulling the assembly out of the carrier.

6. On the right side of the carrier, remove and discard the keystone clamp from the shaft and joint assembly and the stub shaft. Slide the rubber boot onto the stub shaft and pull the shaft and joint assembly from the splines of the stub shaft.

7. Place the spindle in a vise on the second step of the spindle. Wrap a shop towel around the spindle or use a brass-jawed vise to protect the spindle. Remove the oil seal and needle bearing from the spindle with slide hammer and seal remover. If required, remove the seal from the shaft, by driving off with a hammer.

8. Clean all dirt and grease from the spindle bearing bore. Bearing bores must be free from nicks and burrs. Place the bearing in the bore with the manufacturer's identification facing outward. Drive the bearing into the bore using Spindle Bearing Replacer, T83T–3123–A and Driver Handle T80T–4000–W or equivalents.

9. Install the grease seal in the bearing bore with the lip side of the seal facing towards the tool. Drive the seal in the bore with Spindle Bearing Replacer, T83T–3123–A and Driver Handle T80–4000–W or equivalents. Coat the bearing seal lip with Multi-Purpose Long Life Lubricant, C1AZ–19590–B (ESA–M1C75–B) or equivalent.

10. If removed, install a new shaft seal. Place the shaft in a press, and install the seal with Spindle/Axle Seal Installer, T83T–3132–A, or equivalent.

11. On the right side of the carrier, install the rubber boot and new keystone clamps on the stub shaft slip yoke. Since the splines on the shaft are phased, there is only one way to assemble the right shaft and joint assembly into the slip yoke. Align the missing spline in the slip yoke barrel with the gapless male spline on the shaft and joint assembly. Slide the right shaft and joint assembly into the slip yoke making sure the splines are fully engaged. Slide the boot over the assembly and crimp the keystone clamp using Clamp Pliers.

12. On the left side of the carrier slide the shaft and joint assembly through the knuckle and engage the splines on the shaft in the carrier.

13. Install the splash shield and spindle onto the steering knuckle. Install and tighten the spindle nuts to 35–45 ft. lbs.

14. Drive the bearing cups into the rotor using bearing cup replacer T73T–4222–B and Driver Handle, T80T–4000–W or equivalents.

15. Pack the inner and outer wheel bearings and the lip of the oil seal with Multi-Purpose Long-Life Lubricant, C1AZ–19590–B (ESA–M1C75–B) or equivalent.

16. Place the inner wheel bearing in the inner cup. Drive the grease seal into the bore with Hub Seal Replacer, T83T–1175–B and Driver Handle, T80T–4000–W or equivalents. Coat the bearing seal lip with

multipurpose long life lubricant, C1AZ–19590–B (ESA–M1C75–B) or equivalent.

17. Install the rotor on the spindle. Install the outer wheel bearing into cup.

NOTE: Verify that the grease seal lip totally encircles the spindle.

18. Install the wheel bearing, locknut, thrust bearing, snap ring, and locking hubs.

RIGHT SIDE STUB AXLE AND CARRIER

1. Remove the nuts and U-bolts connecting the driveshaft to the yoke. Disconnect the driveshaft from the yoke. Wire the driveshaft out of the way, so it will not interfere in the carrier removal process.

2. Remove both spindles and the Left and Right Shaft and U-joint assemblies as described previously.

3. Support the carrier with a suitable jack and remove the bolts retaining the carrier to the support arm. Separate the carrier from the support arm and drain the lubricant from the carrier. Remove the carrier from the vehicle.

4. Place the carrier in a holding fixture.

5. Rotate the slip yoke and shaft assembly so the open side of the snap ring is exposed.

6. Remove the snap ring from the shaft. Remove the slip yoke and shaft assembly from the carrier.

7. Remove the oil seal and caged needle bearings at the same time, using Slide Hammer, T50T–100–A and Collet, D80L–100–A or equivalents. Discard the seal and needle bearing.

8. Make sure the bearing bore is free from nicks and burrs. Install a new caged needle bearing on Needle Bearing Replacer, T83T–1244–A or equivalent, with the manufacturer name and part number facing outward towards the tool. Drive the needle bearing until it is seated in the bore.

9. Coat the seal with Long-Life Multi-Purpose Lubricant, C1AZ–19590–B (ESA–M1C75–B) or equivalent. Drive the seal into the carrier using Needle Bearing Replacer T83T–1244–A or equivalent.

10. Install the slip yoke and shaft assembly into the carrier so the groove in the shaft is visible in the differential case.

11. Install the snap ring in the groove in the shaft. Force the snap ring into position with a suitable tool. Remove the carrier from the holding fixture.

NOTE: Do not tap on the center of the snap ring. This may damage the snap ring.

12. Clean all traces of gasket RTV sealant from the surfaces of the carrier and support arm and make sure the surfaces are free from dirt and oil. Apply a bead of RTV sealant, D6AZ–19562–A (clear) or B (black) (ESB–M4G92–A and ESE–M4G195–A) or equivalent, in a bead between ¼–⅜ inches wide. The bead should be continuous and should not pass through or outside the holes.

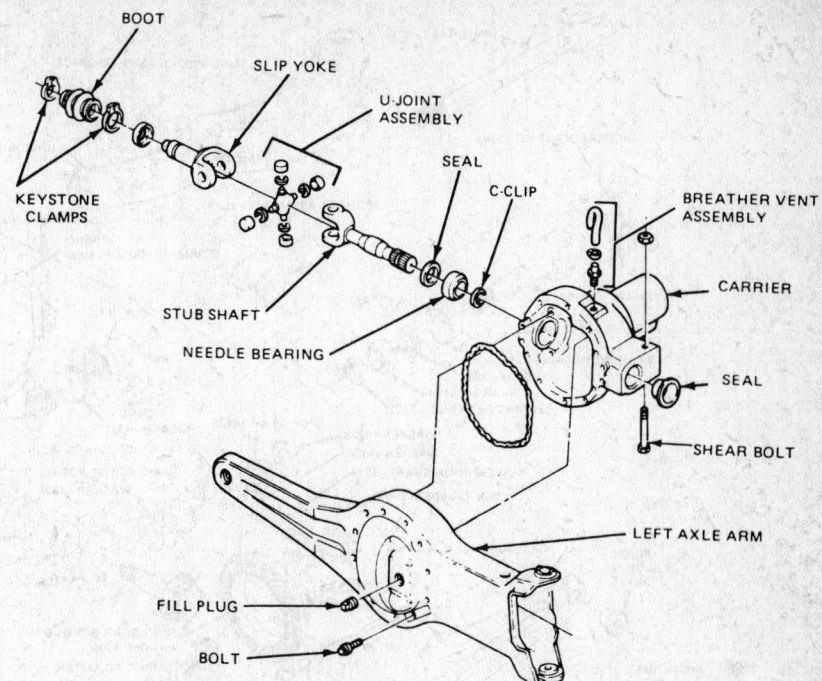

Carrier, slip yoke and stub axle- Ranger/Bronco II 4x4

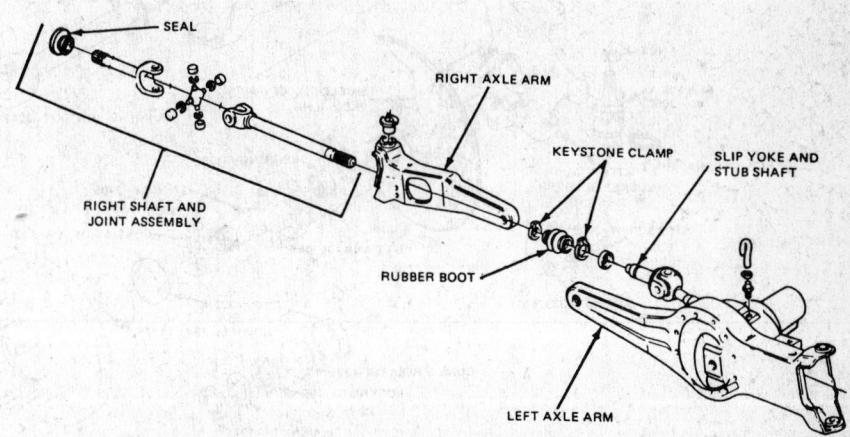

Right axle and joint assembly Ranger/Bronco II 4x4

NOTE: The carrier must be installed on the support arm within five minutes after applying the RTV sealant.

13. Position the carrier on a suitable jack and install it in position on the support arm using guide pins to align. Install the attaching bolts and hand tighten. Tighten the bolts in a clockwise or counter-clockwise pattern to 40–50 ft. lbs.

14. Install the shear bolt retaining the carrier to the axle arm and tighten to 75–95 ft. lbs.

15. Install both spindles and the left and right shaft and joint assemblies as described in the removal and installation portion of this section.Connect the driveshaft to the yoke. Install the nuts and U-bolts and tighten to 8–15 ft. lbs.

Axle Shaft Bearing

1980 and LATER MODELS WITH DANA 44IFS, 44IFSHD OR 50IFS AXLES

This procedure requires the use of special tools.

REMOVAL & INSTALLATION

1. Remove the axle shaft assembly as described in this part under Axle Shaft and Steering Knuckle.

2. Remove the stub assembly by removing 3 bolts attaching retainer plate to carrier housing.

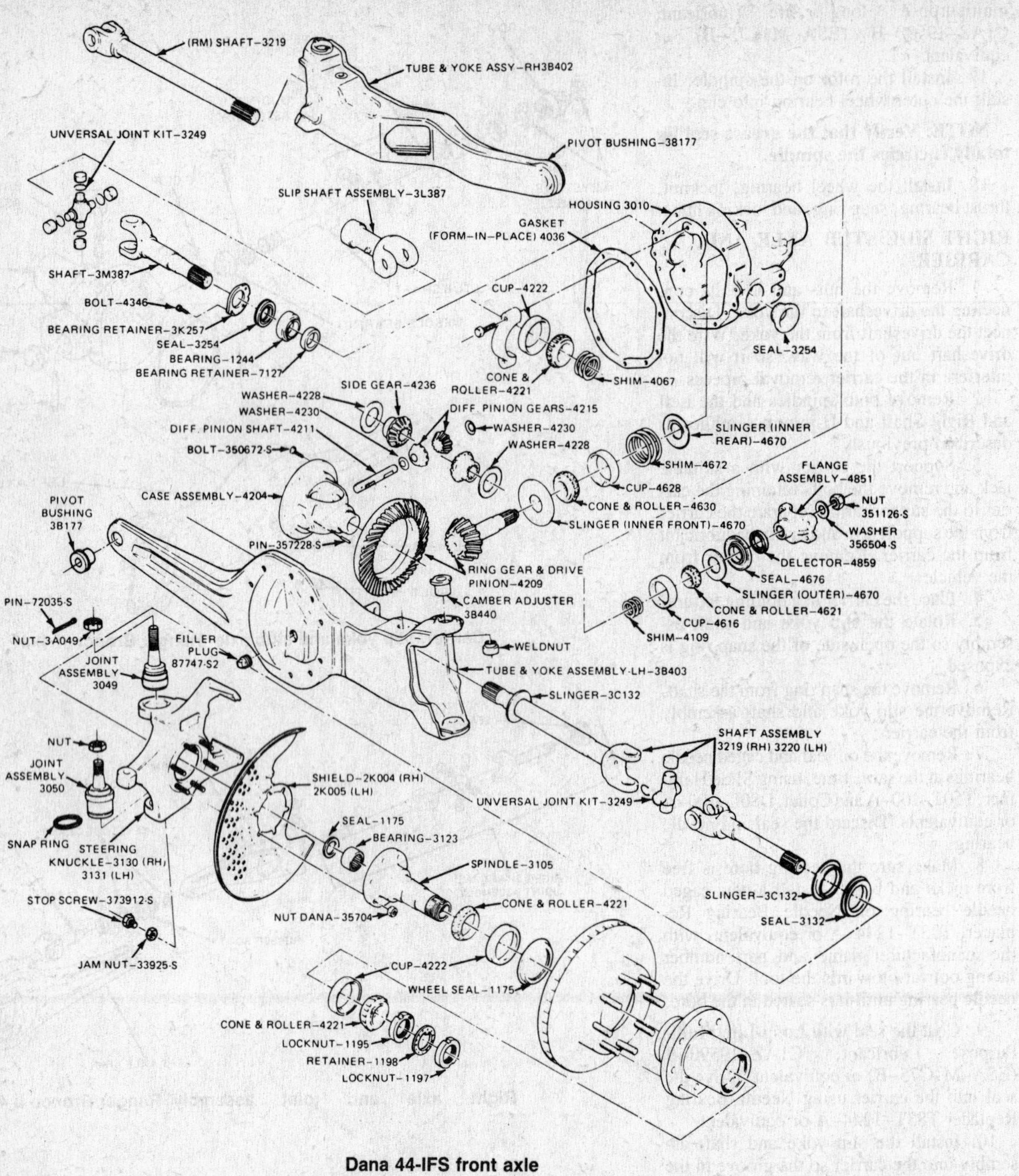

Dana 44-IFS front axle

3. Place the axle shaft in a vise and drill a 6.35mm (¼ inch) hole in the outside of the bearing retaining ring to a depth ¾ the thickness of the ring.

NOTE: Do not drill through the ring because this will damage the axle shaft.

4. With a chisel placed across the hole, strike sharply with a hammer to remove the retaining ring. Replace bearing retaining ring upon assembly.

5. Press the bearing from the axle shaft with the special tools T80 axle bearing remover T80T-4000-M and sleeve T80T-4000-L.

NOTE: Do not use a torch to aid in bearing removal or the stub shaft will be damaged.

6. Remove the seal and retainer plate from the stub shaft. Discard seal and replace with new seal upon assembly.

7. Inspect the retainer plate and stub shaft for distortion, nicks or burns. Replace if necessary.

8. Install retainer plate and new seal on shaft. Coat oil seal with grease.

9. Place the bearing on the shaft. The large radius on the inner race must face the yoke end of the shaft.

10. Press the bearing onto the shaft until completely seated. A 0.0015 inch feeler gauge should not fit between the bearing seat and bearing.

11. Use axle bearing replacer T80T-4000-N and pinion bearing cone remover T71P-4621-B to press the bearing retainer ring onto the stub shaft. Press the bearing retainer ring until completely seated. A 0.038mm 0.0015 inch feeler gauge should not fit between the ring and bearing. There must be one point between the bearing and ring where the feeler gauge cannot enter. If feeler gauge enters completely around the

circumference press the retainer further onto the shaft.

12. Push the seal and retainer plate away from the bearing to form a space between the seal and bearing. Fill the space with wheel bearing grease meeting Ford specification ESA–M1C75B or equivalent.

13. With the space filled with grease, wrap tape around the space.

14. Pull the seal towards the bearing until it contacts the inner race. This will force grease between the rollers and cup. Remove tape.

NOTE: If grease is not visible on the small end of the rollers, repeat Steps 6 through 8 until grease is visible. Install the slip yoke and U-joint to stub shaft.

15. Install the stub shaft in the carrier and install 3 retainer bolts. Torque to 30–40 ft. lbs. Install right hand axle shaft assembly into slip yoke.

16. Install splash shield and spindle.

Axle Shaft U-Joint Overhaul

Follow the procedures outlined under Axle Shaft Removal and Installation to gain access to the U-joints. Overhaul them as described under U-joints.

Pinion Seal

REMOVAL & INSTALLATION

NOTE: A torque wrench capable of at least 225 ft. lbs. is required for pinion seal installation.

——— CAUTION ———
Some models use a collapsible spacer to set pinion depth and preload. When replacing the pinion seal always install a new spacer. Never tighten the pinion nut more than 225 ft. lbs. or the spacer will be compressed too far.

1. Raise and safely support the vehicle with jackstands under the frame rails. Allow the axle to drop to rebound position for working clearance.

2. Mark the companion flanges and U-joints for correct reinstallation position.

3. Remove the drive shaft. Use a suitable tool to hold the companion flange. Remove the pinion nut and companion flange.

4. Use a slide hammer and hook or sheet metal screw to remove the oil seal.

5. If the vehicle uses a collapsible spacer, install new spacer. Install a new pinion seal after lubricating the sealing surfaces. Use a suitable seal driver. Install the companion flange and pinion nut. On models using a spacer, tighten the nut to 225 ft. lbs. On other models, pinion nut torque is 200–220 ft. lbs.

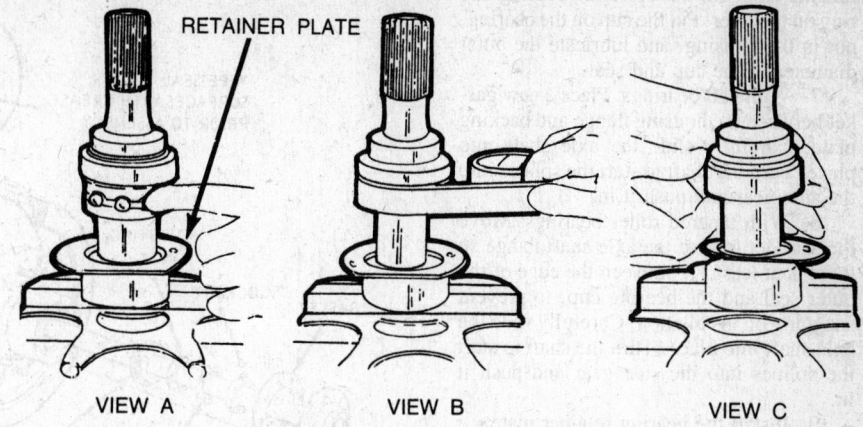

VIEW A   VIEW B   VIEW C

Lubricating the axle bearing on the Dana 44-IFS

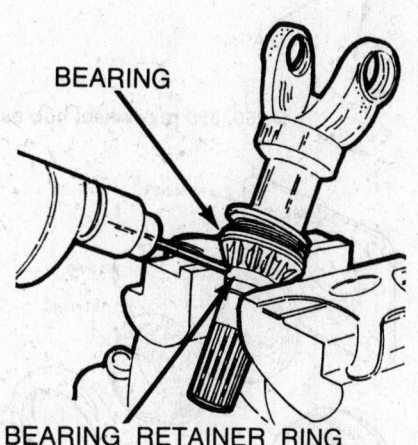

Drilling the stub shaft bearing retaining ring on the Dana 44-IFS

___

REAR AXLE

Refer to Rear Axles in the Unit Repair Section for overhaul procedures.

Axle Shaft and Bearing

REMOVAL & INSTALLATION

Removable Carrier Type

NOTE: The following procedure requires the use of special tools, including a shop press.

1. Raise and support the vehicle. Remove the wheel/tire assembly from the brake drum.

2. Remove the clips which secure the brake drum to the axle flange, then remove the drum from the flange.

3. Working through the hole provided in each axle shaft flange, remove the nuts which secure the wheel bearing retainer plate.

4. Pull the axle shaft assembly out of the axle housing. You may need a slide hammer.

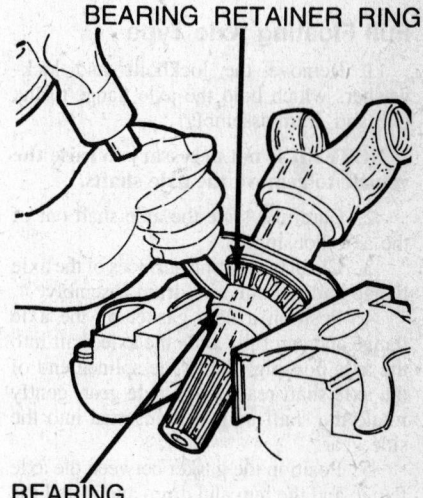

Removing the bearing retainer ring on the Dana 44-IFS

NOTE: The brake backing plate must not be dislodged. Install one nut to hold the plate in place after the axle shaft is removed.

5. If the axle has ball bearings: Loosen the bearing retainer ring by nicking it in several places with a cold chisel, then slide it off the axle shaft. On models equipped with a thick retaining ring drill a 1/4–1/2 in. hole part way through the ring, then break it with a cold chisel. A hydraulic press is needed to press the bearing off and to press the new one on. Press the new bearing and the new retainer ring on separately. Use a slide hammer to pull the old seal out of the axle housing. Carefully drive the new seal evenly into the axle housing, preferably with a seal drive tool.

6. If the axle has tapered roller bearings. Use a slide hammer to remove the bearing cup from the axle housing. Drill a 1/4–1/2 1/2 in. hole part way through the bearing retainer ring, then break it with a cold chisel. A hydraulic press is needed to press the bearing off and remove the seal. Press on the new seal and bearing, then the new

retainer ring. Do not press the bearing and ring on together. Put the cup on the bearing, not in the housing, and lubricate the outer diameter of the cup and seal.

7. With ball bearings: Place a new gasket between the housing flange and backing plate. Carefully slide the axle shaft into place. Turn the shaft to start the splines into the side gear and push it in.

8. With tapered roller bearings: Move the seal out toward the axle shaft flange so there is at least $3/32$ between the edge of the outer seal and the bearing cup, to prevent snagging on installation. Carefully slide the axle shaft into place. Turn the shaft to start the splines into the side gear and push it in.

9. Install the bearing retainer plate.

10. Replace the brake drum and the wheel and tire.

Full Floating Axle Type

1. Remove the lockbolts and lockwashers which hold the axle flange to the hub and drum assembly.

NOTE: It is not necessary to raise the vehicle to remove the axle shafts.

2. Carefully slide the axle shaft out of the axle housing.

3. Clean the mating surfaces of the axle flange and the hub and drum assembly.

4. Position a new gasket on the axle flange and carefully slide the axle shaft into the axle housing. When the splined end of the axle shaft reaches the side gear, gently rotate the shaft until it is inserted into the side gear.

5. Position the gasket between the axle flange and the hub and drum and install the lockbolts and lockwashers.

REAR WHEEL BEARINGS

F250–350, E250–350

The wheel bearings on the full floating rear axle are packed with wheel bearing grease. Axle lubricant can also flow into the wheel hubs and bearings, however, wheel bearing grease is the primary lubricant. The wheel bearing grease provides lubrication until the axle lubricant reaches the bearings during normal operation.

1. Set the parking brake and loosen the axle shaft bolts.

2. Raise the rear wheels off the floor and place jackstands under the rear axle housing so that the axle is parallel with the floor.

3. Remove the axle shaft bolts.

4. Remove the axle shaft and gaskets.

5. With the axle shaft removed, remove the gasket from the axle shaft flange studs.

6. Bend the lockwasher tab away from the locknut, and then remove the locknut, lockwasher, and the adjusting nut.

7. Remove the outer bearing cone and pull the wheel straight off the axle.

8. With a piece of hardwood which will

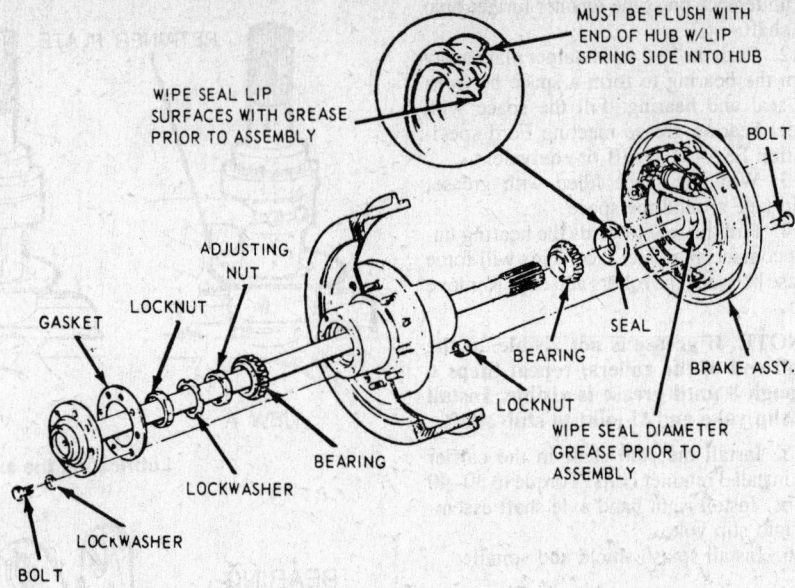

F-250, 350 rear wheel hub assembly with full-floating axles

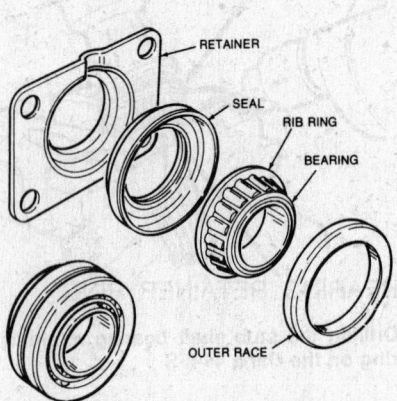

Tapered roller bearings used on some E-100, 150 and 200 axle shafts

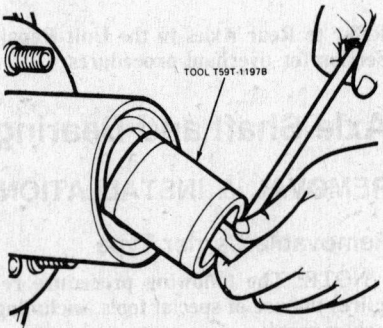

Lock nut, lock ring and adjusting nut removal

just clear the outer bearing cup, drive the inner bearing cone and inner seal out of the wheel hub.

9. Wash all the old grease or axle lubricant out of the wheel hub, using a suitable solvent.

10. Wash the bearing cups and rollers and inspect them for pitting, galling, and uneven wear patterns. Inspect the roller for end wear.

11. If the bearing cups are to be replaced, drive them out with a drift. Install the new cups with a block of wood and hammer or press them in.

12. If the bearing cups are properly seated, a 0.0015 in. feeler gauge will not fit between the cup and the wheel hub.

13. Pack each bearing cone and roller with a bearing packer or in the manner previously outlined for the front wheel bearings on 2WD trucks.

14. Place the inner bearing cone and roller assembly in the wheel hub. Install a new inner seal in the hub.

15. Install the wheel.

16. Install and tighten the bearing adjusting nut to 50–80 ft. lbs. while rotating the wheel.

17. Back off (loosen) the adjusting nut $3/8$ of a turn.

18. Apply axle lube to a new lockwasher and install it with the smooth side out.

19. Install the locknut and tighten it to 90–110 ft. lbs. The wheel must rotate freely after the locknut is tightened. The wheel end-play should be within 0.001–0.010 in.

20. Bend two lockwasher tabs inward over an adjusting nut flat and two lockwasher tabs outward over the locknut flat.

21. Install the axle shaft, gasket, lockbolts, and washers. Tighten the bolts to 40–50 ft. lbs.

22. Adjust the brakes, if necessary.

Ranger & Bronco II

A 6¾ inch, integral Carrier is used on some Ranger and Bronco II models. Refer to the Removable Carrier section for procedures.

Integral Carrier "C"-Lock Type

1. Raise and safely support the vehicle on jackstands.

2. Remove the wheels and tires from the brake drums.

3. Place a drain pan under the housing and drain the lubricant by loosening the housing cover.

4. Remove the locks securing the brake drums to the axle shaft flanges and remove the drums.

5. Remove the housing cover and gasket.

6. Remove the side gear pinion shaft lockbolt and the side gear pinion shaft.

7.. Push the axle shafts inward and remove the C-locks from the inner end of the axle shafts. Temporarily replace the shaft and lockbolt to retain the differential gears in position.

8. Remove the axle shafts with a slide hammer. Be sure the seal is not damaged by the splines on the axle shaft

.9. Remove the bearing and oil seal from the housing. Both the seal and bearing can be removed with a slide hammer

10. Two types of bearings are used on some axles, one requiring a press fit and the other a loose fit. A loose fitting bearing does not necessarily indicate excessive wear.

11. Inspect the axle shaft housing and axle shafts for burrs or other irregularities. Replace any work or damaged parts. A light yellow color on the bearing journal of the axle shaft is normal, and does not require replacement of the axle shaft. Slight pitting and wear is also normal.

12. Lightly coat the wheel bearing rollers with axle lubricant. Install the bearings in the axle housing until the bearing seats firmly against the shoulder.

13. Wipe all lubricant from the oil seal bore, before installing the seal.

14.. Inspect the original seals for wear. If necessary, these may be replaced with new seals, which are prepacked with lubricant and do not require soaking.

15. Install the oil seal.

16. Remove the lockbolt and pinion shaft. Carefully slide the axle shafts into place. Be careful that you do not damage the seal with the splined end of the axle shaft. Engage the splined end of the shaft with the differential side gears.

17. Install the axle shaft C-locks on the inner end of the axle shafts and seat the C-locks in the counterbore of the differential side gears.

18. Rotate the differential pinion gears until the differential pinion shaft can be installed. Install the differential pinion shaft lockbolt. Tighten to 15–22 ft. lbs.

19. Install the brake drum on the axle shaft flange.

20. Install the wheel and tire on the brake drum and tighten the attaching nuts.

21. Clean the gasket surface of the rear housing and install a new cover gasket and the housing cover. Some covers do not use a gasket. On these models, apply a bead of silicone sealer on the gasket surface. The bead should run inside of the bolt holes.

22. Raise the rear axle so that it is in the running position. Add the amount of specified lubricant to bring the lubricant level to ½ in. below the filler hole.

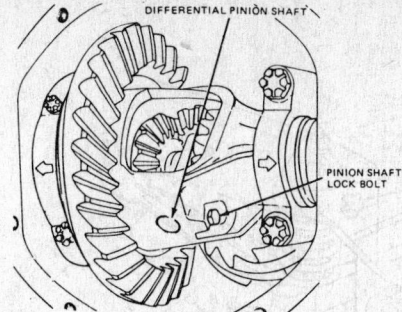

Integral rear with "C-locks"-pinion shaft and lock bolt location

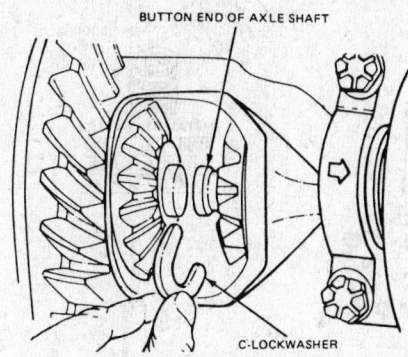

"C-lock" removal/installation

FRONT SUSPENSION

Springs

REMOVAL & INSTALLATION

2WD

1. Raise the front of the vehicle and place jackstands under the frame and a jack under the axle. Remove wheel and tire assemblies. Remove the brake calipers and suspend with wire so that there is no tension on the brake hose.

2. Disconnect the shock absorber from the lower bracket.

3. Remove the rebound bracket on Ranger and Bronco II, remove the lower spring retainer.

4. On models except Ranger and Bronco II: Remove the two spring upper retainer attaching bolts from the top of the spring upper seat and remove the retainer.

5. Remove the nut attaching the spring lower retainer to the lower seat and axle and remove the retainer.

6. Lower the axle and remove the spring. Some downward pressure using a prybar may be required.

7. Place the spring in position and raise the front axle.

8. Position the spring lower retainer over

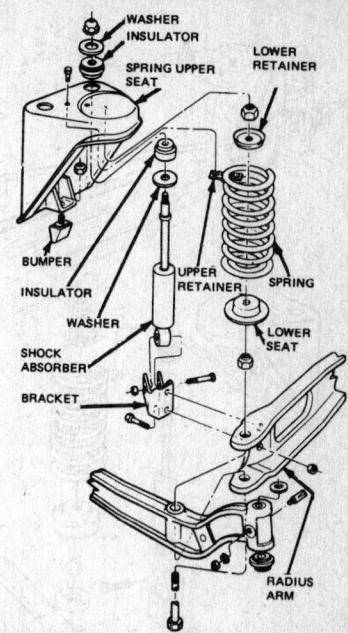

Exploded view of the front spring assembly and shock absorber on 2WD pick-ups

the stud and lower seat, and install the two attaching bolts.

9. Position the upper retainer over the spring coil and against the spring upper seat, and install the two attaching bolts.

10. Tighten the upper and lower retainer attaching nuts and bolts to 15–25 ft. lbs.

11. Connect the shock absorber to the lower bracket and install the rebound bracket.

12. Remove the jack and safety stands.

4WD F100–150 1979

1. Raise the vehicle until the tires are a few inches off the ground and place jackstands under the frame side rails. Position a hydraulic floor jack under the center of the front axle housing.

2. Remove the shock absorber-to-lower bracket attaching bolt and nut.

3. Remove the spring lower retainer attaching bolts from the inside of the spring coil.

4. Lower the axle enough to relieve tension from the spring.

5. Remove the spring upper retainer attaching bolts and nuts and remove the upper retainer.

6. Remove the spring, lower retainer and the lower seat from the front spring:

7. Position the upper retainer over the spring coil and loosely install the attaching bolts and nuts.

8. Position the spring lower seat, and the lower retainer to the frame spring pocket and the radius arm.

9. Raise the axle up into position and install the two lower retainer attaching bolts and tighten them.

10. Tighten the upper retainer attaching bolts.

11. Position the shock absorber to the

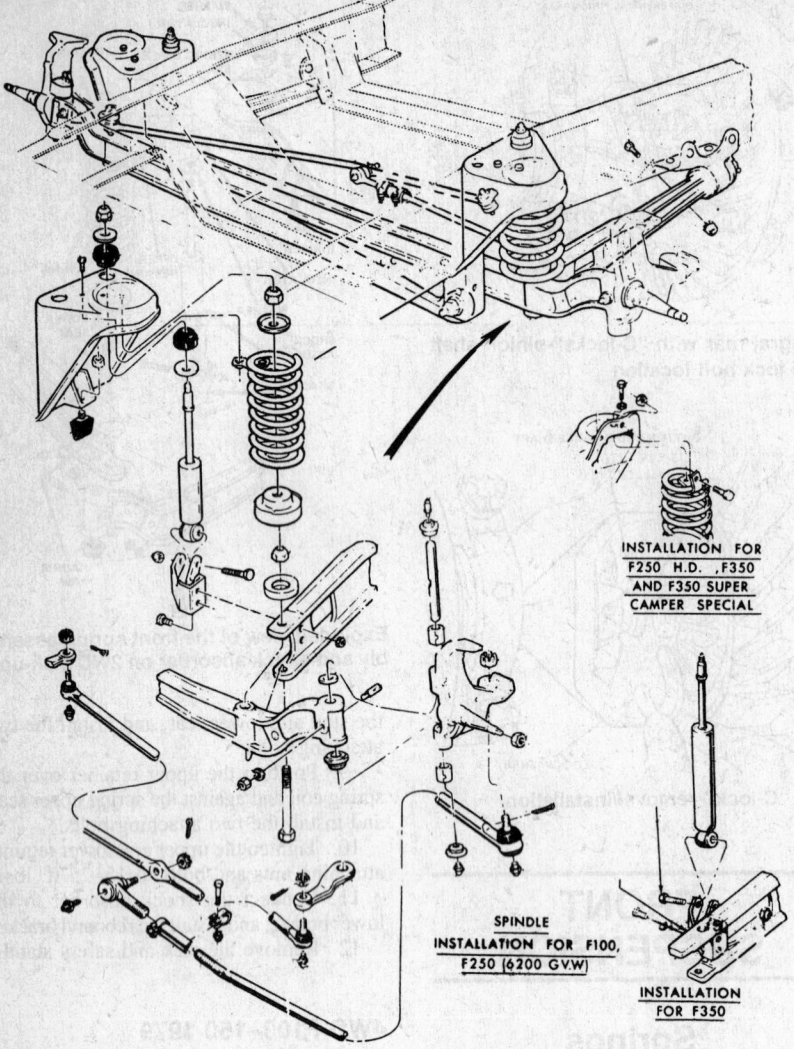

INSTALLATION FOR
F250 H.D. ,F350
AND F350 SUPER
CAMPER SPECIAL

SPINDLE
INSTALLATION FOR F100,
F250 (6200 GVW)

INSTALLATION
FOR F350

2WD twin I-beam front suspension

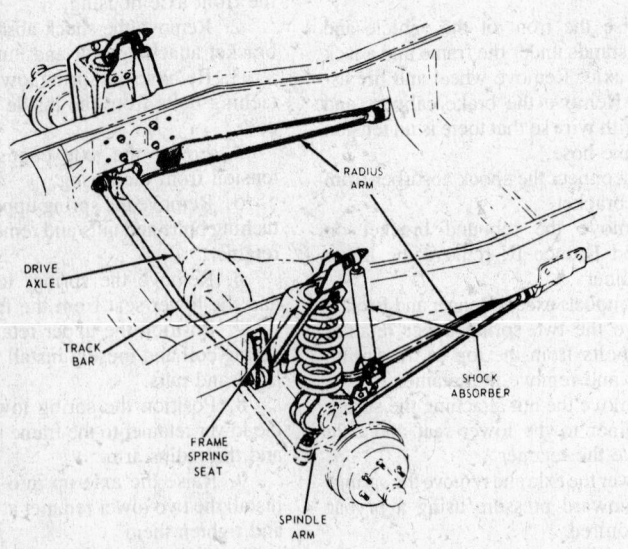

RADIUS
ARM

DRIVE
AXLE

TRACK
BAR

SHOCK
ABSORBER

FRAME
SPRING
SEAT

SPINDLE
ARM

F-100, 150 4WD front suspension

lower bracket and install the attaching bolt and nut.

12. Remove the jackstands and lower the vehicle.

1979 Bronco

1. Raise the vehicle and support on jackstands. Remove wheels and calipers. Support the calipers with wire, do not allow tension on the brake hose. Remove the shock absorber-to-lower bracket attaching bolt and nut.

2. Remove two spring lower retainer attaching bolts from inside of the spring coil.

3. Remove two spring upper retainer attaching bolts and nuts and remove the upper retainer.

4. Position safety stands under the frame side rails and lower the axle enough to relieve tension from the spring. Remove the spring, lower retainer, and lower the spring from the vehicle.

5. Position the spring, spring lower seat, and lower retainer to the frame spring pocket and the radius arm. Position the spring seat and the lower retainer.

6. Position the upper retainer over the spring coil and loosely install the two attaching bolts and nuts.

7. Install the two lower retainer attaching bolts and tighten to 80–120 ft. lbs.

8. Tighten the upper retainer attaching bolts to 20–30 ft. lbs.

9. Position the shock absorber to the lower bracket and install the attaching bolt and nut. Tighten the bolt and nut to 40–60 ft. lbs. Remove safety stands and lower the vehicle.

4WD F250–350 1979

1. Raise the vehicle until the weight is off the front spring with the wheels still touching the floor.

2. Disconnect the lower end of the shock absorber from the axle. Remove the U-bolts and spacer.

3. Remove the nut from the shackle bolt retaining spring at the front and drive the shackle bolt out with a drift.

4. Remove the nuts and shackle bar at the rear shackle bracket. Drive out the two shackle bolts and remove the spring.

5. Position the spring on the spring seat cap and install the shackle bolts through the shackle bracket and spring.

6. Install the shackle bar and nuts on the shackle bolts and tighten them to 90–130 ft. lbs.

7. Position the front of the spring to allow the front shackle bolts to be installed. Tighten the nut to 90–130 ft. lbs.

8. Position the U-bolt spacer and place the U-bolts in position through the holes in the spring seat cap. Install the U-bolt nuts, but do not tighten them yet.

9. Connect the lower end of the shock absorber to the axle.

10. Lower the vehicle and tighten the U-bolt nuts to 100–135 ft. lbs.

4WD Bronco, F150 From 1980

1. Raise the vehicle and remove the shock absorber lower attaching bolt and nut.

2. Remove the spring lower retainer nuts from inside of the spring coil.

3. Remove the upper spring retainer by removing the attaching screw.

4. Position safety stands under the frame side rails and lower the axle enough to relieve tension from the spring.

NOTE: The axle must be supported on the jack throughout spring removal, and must not be permitted to hang from the brake hose. If the length of the brake hose does not provide sufficient clearance it may be necessary to remove and support the brake caliper.

5. Remove the spring lower retainer and lower the spring from the vehicle.

6. To install place the spring in position and slowly raise the front axle. Make sure the springs are positioned correctly in the upper spring seats.

7. Install the lower spring retainer and torque the nut to 50 ft. lbs.

8. Position the upper retainer over the spring coil and install the attaching screws.

9. Position the shock absorber to the lower bracket and torque the attaching bolt and nut to 53 ft. lbs.

10. Remove the safety stands and lower the vehicle.

4WD F250–350 From 1980

1. Raise the vehicle frame until the weight is off the front spring with the wheels still touching the floor. Support the axle to prevent rotation.

2. Disconnect the lower end of the shock absorber from the U-bolt spacer. Remove the U-bolts. U-bolt cap and spacer.

3. Remove the nut from the hanger bolt retaining the spring at the rear and drive out the hanger bolt.

4. Remove the nut connecting the front shackle and spring eye and drive out the shackle bolt and remove the spring.

5. To install position the spring on the spring seat. Install the shackle bolt through the shackle and spring. Torque the nuts to 135 ft. lbs.

6. Position the rear of the spring and install the hanger bolt. Torque the nut to 175 ft. lbs.

7. Position the U-bolt spacer and place the U-bolts in position through the holes in the spring seat cap. Install but do not tighten the U-bolt nuts.

8. Connect the lower end of the shock absorber to the U-bolt spacer.

9. Lower the vehicle and tighten the U-bolt nuts to 100 ft. lbs.

4WD Ranger and Bronco II

1. Raise the vehicle and install safety stands. Position a jack beneath the spring under the axle. Raise the jack and compress the spring.

2. Remove the bolt and nut retaining the shock absorber to the radius arm. Slide the shock out from the bracket.

3. Remove the nut that retains the spring to the axle and radius arm. Remove the retainer.

4. Slowly lower the axle until all spring tension is released and adequate clearance

exists to remove the spring. Remove the spring by rotating upper coil out of tabs in upper spring seat. Remove the spacer and seat.

NOTE: The axle must be supported on the jack throughout spring removal and installation, and must not be permitted to hang by the brake hose. If the length of the brake hose is not sufficient to provide adequate clearance for removal and installation of the spring, the disc brake caliper must be removed from the spindle. After removal, the caliper must be placed on the frame or otherwise supported to prevent suspending the caliper from the caliper hose. These precautions are absolutely necessary to prevent serious damage to the tube portion of the caliper hose assembly.

5. If required, remove the stud from the axle assembly.

6. If removed, install the stud in the axle. Tighten to 160–220 ft. lbs.

7. Install the lower seat and spacer over the stud/bolt. Position upper end of spring so end of coil fits into spring stop in upper spring seat and top coil fits over upper spring retainer.

8. Rotate spring into position. Slowly raise axle until lower end of spring is in position on the lower insulator.

9. Install the retainer and nut on the stud. Tighten nut to 70–100 ft. lbs.

10. Position the shock in the lower bracket. Install nut and bolt and tighten to 42–72 ft. lbs.

11. Remove the jack.

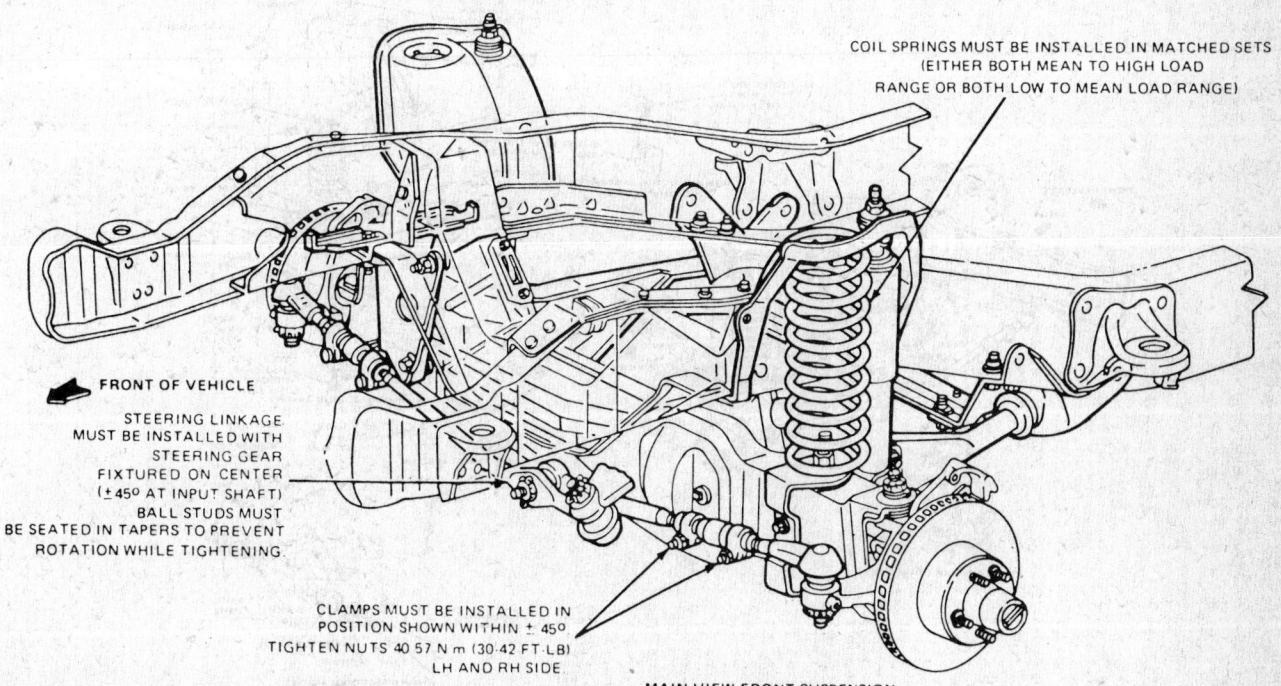

COIL SPRINGS MUST BE INSTALLED IN MATCHED SETS (EITHER BOTH MEAN TO HIGH LOAD RANGE OR BOTH LOW TO MEAN LOAD RANGE)

FRONT OF VEHICLE

STEERING LINKAGE MUST BE INSTALLED WITH STEERING GEAR FIXTURED ON CENTER (±45° AT INPUT SHAFT) BALL STUDS MUST BE SEATED IN TAPERS TO PREVENT ROTATION WHILE TIGHTENING.

CLAMPS MUST BE INSTALLED IN POSITION SHOWN WITHIN ±45° TIGHTEN NUTS 40 57 N m (30-42 FT-LB) LH AND RH SIDE

MAIN VIEW FRONT SUSPENSION

Ranger/Bronco II 4x4 front suspension

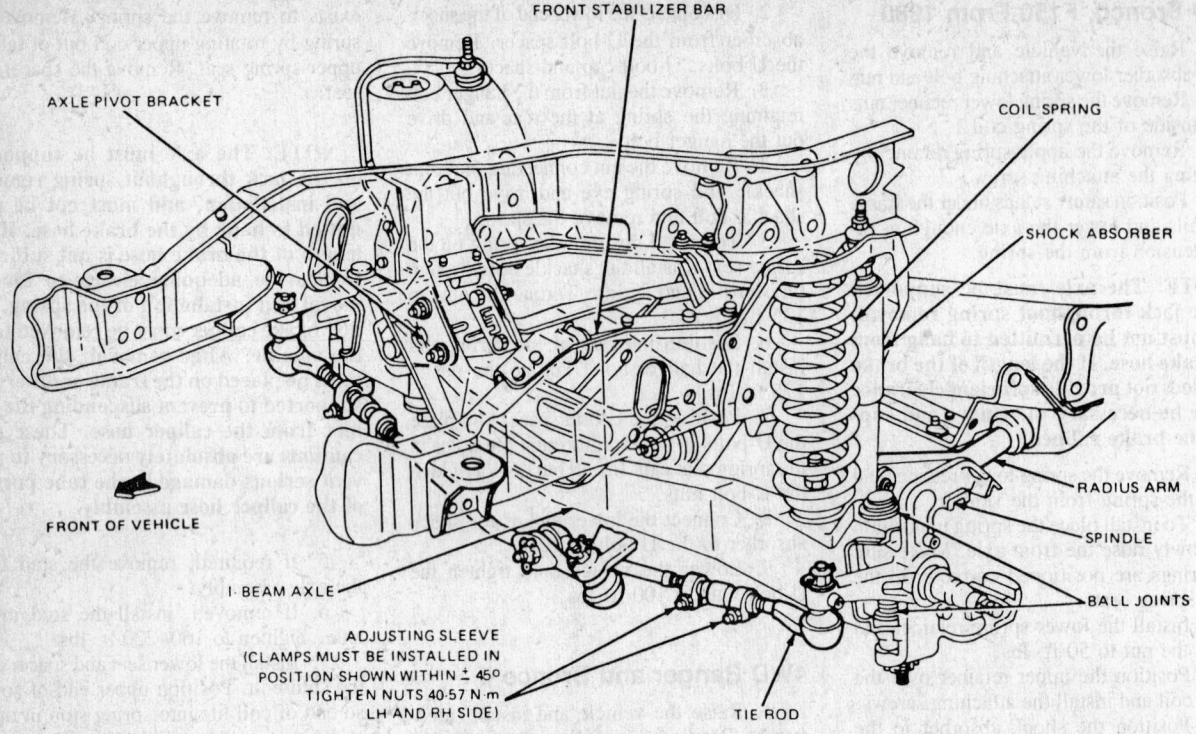

AXLE PIVOT BRACKET

FRONT STABILIZER BAR

COIL SPRING

SHOCK ABSORBER

RADIUS ARM

SPINDLE

BALL JOINTS

FRONT OF VEHICLE

I-BEAM AXLE

ADJUSTING SLEEVE
(CLAMPS MUST BE INSTALLED IN
POSITION SHOWN WITHIN ± 45°
TIGHTEN NUTS 40-57 N·m
LH AND RH SIDE)

TIE ROD

Ranger/Bronco II 2x4 front suspension

Shock Absorbers

REMOVAL & INSTALLATION
2WD

1. Insert a wrench from the rear side of the upper spring seat to hold the upper

shock retaining nut. Loosen the stud by using another wrench on the hex on the shaft.

2. Remove the bolt and nut at the lower end.

3. On installation, make sure to get the washers and insulators in the right place. Tighten the upper nut by turning the hex

on the shaft Replace the lower bolt. It is recommended that new rubber insulators be used.

4WD

1. Remove the bolt and nut attaching

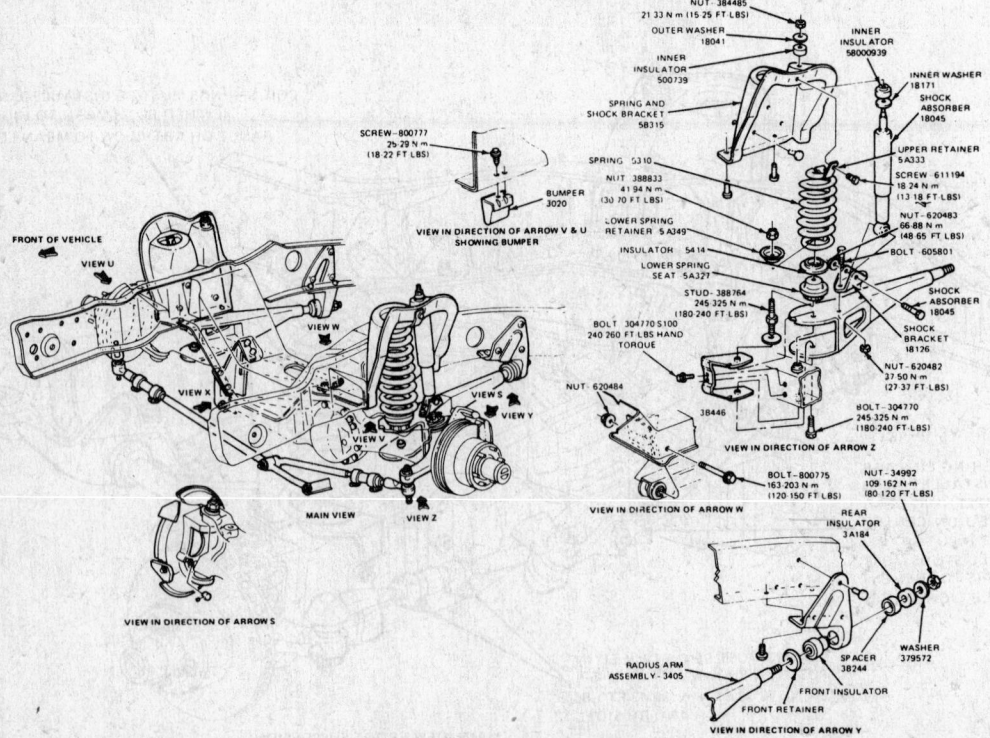

1980 F-150 4WD front suspension

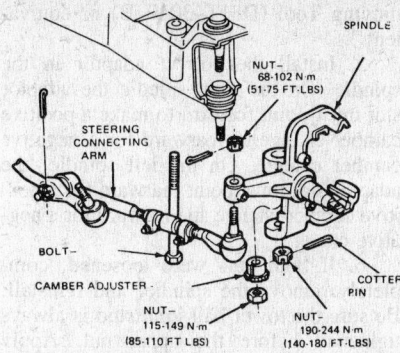

Spindle, camber adaptor and ball joints-Ranger/Bronco II

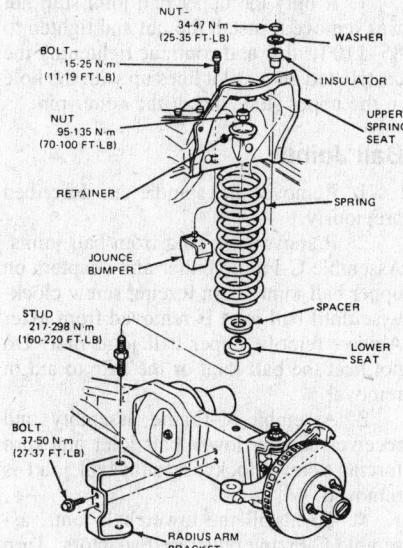

Coil spring removal/installation Ranger/Bronco II 4x4

the shock absorber to the lower bracket on the radius arm.

2. Remove the nut, washer and insulator from the shock absorber at the frame bracket and remove the shock absorber.

3. Position the washer and insulator on the shock absorber rod and position the shock absorber to the frame bracket.

4. Position the insulator and washer on the shock absorber rod and install the attaching nut loosely.

5. Position the shock absorber to the lower bracket and install the attaching bolt and nut loosely.

6. Tighten the lower attaching bolts to 40–70 ft. lbs., and the upper attaching bolts to 25–30 ft. lbs.

Front Wheel Spindle and King Pin 2WD

REMOVAL & INSTALLATION

1. Raise the front of the truck until the front wheel clears the ground and place jackstands under the frame.

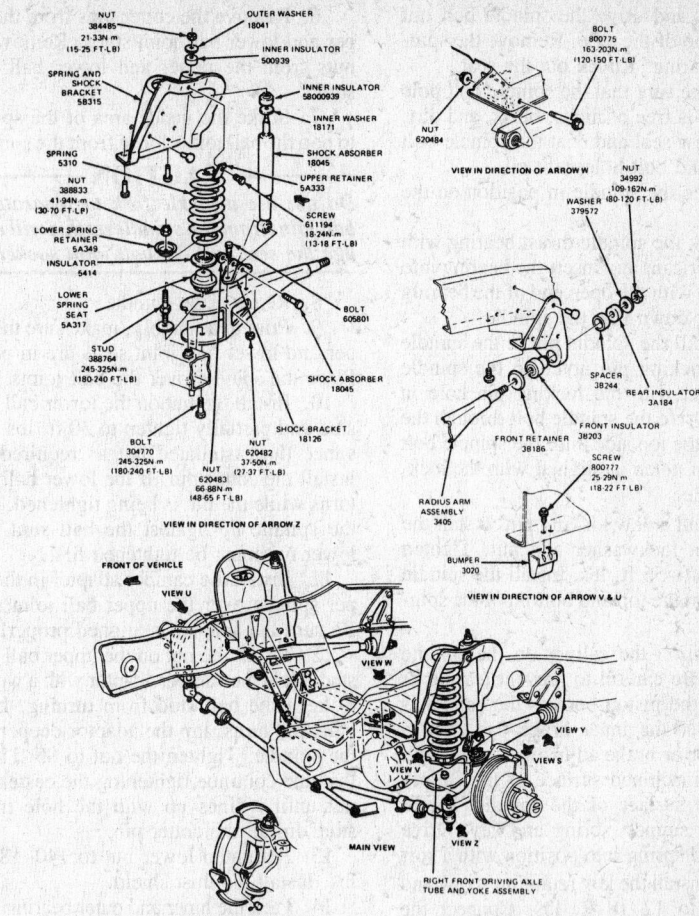

1981–82 4WD pick-up and Bronco front suspension

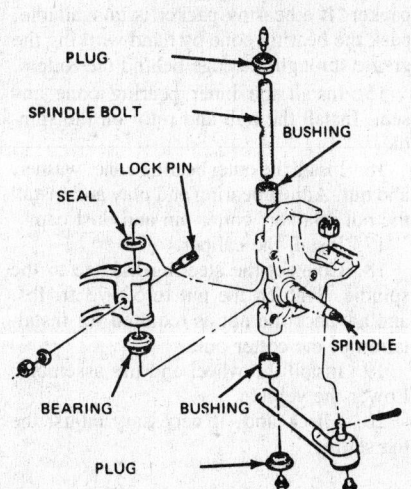

Spindle installation on light duty 2WD pick-ups

2. Remove the wheel and tire.

3. Remove the caliper key retaining screw. Drive out the caliper support key and spring with brass drift and hammer. Remove the caliper from the spindle by pushing the caliper downward against the spindle assembly and rotating the upper end of the caliper upward and out of the spindle assembly. It is not necessary to disconnect

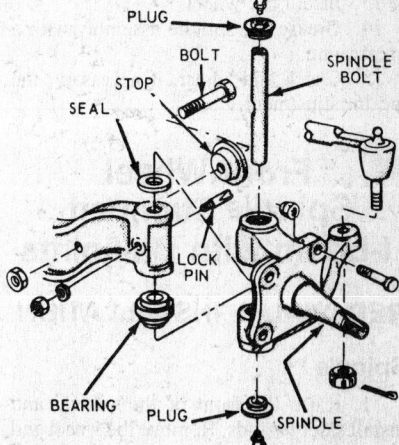

Spindle installation on heavy duty 2WD pick-ups

the brake fluid hose. Wire the caliper to a suspension part to remove the weight of the caliper from the hose. Disconnect the steering linkage from the spindle arm.

4. Disconnect the steering linkage from the integral spindle and spindle arm.

5. Remove the nut and lockwasher from the locking pin, and remove the locking pin.

6. Remove the upper and lower spindle

bolt plugs, and drive the spindle bolt out from the top of the axle. Remove the spindle and bearing. Knock out the seal.

7. Make sure that the spindle bolt hole in the axle is free of nicks, burrs, and dirt. Install a new seal and coat the spindle bolt bushings and bolt hole with oil.

8. Place the spindle in position on the axle.

9. Pack the spindle thrust bearing with chassis lubricant and insert the bearing into the spindle with the open end of the bearing seal facing down into the spindle.

10. Install the spindle pin in the spindle with the locking pin notch in the spindle bolt aligned with the locking pin hole in the axle. Drive the spindle bolt through the axle from the top side until the spindle bolt locking pin notch is aligned with the locking pin hole.

11. Install a new locking pin. Install the locking pin lockwasher and nut. Tighten the nut to 40–55 ft. lbs. Install the spindle bolt plugs at the top and bottom of the spindle bolt.

12. Position the caliper on the spindle assembly. Be careful to prevent tearing or cutting of the piston boot as the caliper is slipped over the inner brake pad. Use a screwdriver or brake adjusting tool to hold the upper machined surface of the caliper against the surface of the spindle. Install the caliper support spring and key. Drive the key and spring into position with a soft hammer. Install the key retaining screw and tighten it to 12–18 ft. lbs. Connect the steering linkage to the spindle and tighten the nut to 50–70 ft. lbs. advancing the nut as necessary to install the cotter pin.

13. Install the wheel.

14. Grease the spindle assembly with a grease gun.

15. Check and adjust, if necessary, the toe-in adjustment.

Front Wheel Spindle Stamped I-Beam with Balljoints

REMOVAL & INSTALLATION

Spindle

1. Raise the front of the vehicle and install safety stands. Remove the wheel and tire assembly.

2. Remove the caliper assembly from the rotor and hold it out of the way with wire.

3. Remove the dust cap, cotter pin, nut retainer, nut, washer, and outer bearing, and remove the rotor from the spindle.

4. Remove inner bearing cone and seal. Discard the seal. Remove brake dust shield.

5. Disconnect the steering linkage from the integral spindle and spindle arm by removing the cotter pin and nut and then removing the tie rod end from the spindle arm.

6. Remove the cotter pins from the upper and lower ball joint studs. Remove the nuts from the upper and lower ball joint stud.

7. Strike the inside area of the spindle to pop the ball joints loose from the spindle.

--- CAUTION ---

Do not use a pickle fork to separate the ball joint from the spindle as this will damage the seal and the ball joint socket.

8. Remove the spindle.

9. Prior to assembly, make sure the upper and lower ball joint seals are in place. Place the spindle over the ball joints.

10. Install the nut on the lower ball joint stud and partially tighten to 30 ft. lbs. Advance the castellated nut as required and install the cotter pin. If the lower ball stud turns while the nut is being tightened, push the spindle up against the ball stud. The lower nut must be tightened first.

11. Install the camber adapter in the upper spindle over the upper ball joint stud. Be sure the adapter is aligned properly.

12. Install the nut on the upper ball joint stud. Hold the camber adaptor with a wrench to keep the ball stud from turning. If the ball stud turns, tap the adaptor deeper into the spindle. Tighten the nut to 85–110 ft. lbs. and continue tightening the castellated nut until it lines up with the hole in the stud. Install the cotter pin.

13. Retighten lower nut to 140–180 ft. lbs. Install the dust shield.

14. Pack the inner and outer bearing cone with C1AZ–19590–B (ESA–M1C75–B) or equivalent bearing grease. Use a bearing packer. If a bearing packer is unavailable, pack the bearing cone by hand working the grease through the cage behind the rollers.

15. Install the inner bearing cone and seal. Install the hub and rotor on the spindle.

16. Install the outer bearing cone, washer, and nut. Adjust bearing end play and install the nut retainer, cotter pin and dust cap.

17. Install the caliper.

18. Connect the steering linkage to the spindle. Tighten the nut to 52–73 ft. lbs. and advance the nut as required for installation of the cotter pin.

19. Install the wheel and tire assembly. Lower the vehicle.

20. Check and, if necessary adjust the toe setting.

Camber Adjuster

1. Remove the cotter pin and nut from the upper ball joint stud.

2. Strike the inside of the spindle to pop the upper ball joint from the spindle.

3. If the upper ball joint does not pop loose, remove the cotter pin and back the lower ball joint nut about half way down the lower ball joint stud, and strike the side of the lower spindle.

4. Remove the camber adapter (camber adjustment sleeve) using Ball Joint Re-

moving Tool (D81T–3010–B) or equivalent.

5. Install the correct adaptor in the spindle. On the right spindle the adaptor slot must point forward to make a positive camber change or rearward for a negative camber change. On the left spindle, the adaptor slot must point rearward for a positive camber change and forward for a negative change.

6. If both nuts were loosened, completely remove the spindle, and reinstall. Be sure the lower ball joint stud is always tightened before the upper nut. Apply Locktite No. 242 (D5AZ–19554–A) or equivalent to stud threads before installing nut.

7. If only the upper ball joint stud nut was removed, install the nut and tighten to 85–110 ft. lbs. and continue tightening the castellated nut until it lines up with the hole in the upper stud. Install the cotter pin.

Ball Joints

1. Remove the spindle as described previously.

2. Remove snap ring from ball joints. Assemble C-Frame puller and adapters on upper ball joint. Turn forcing screw clockwise until ball joint is removed from axle. Always remove upper ball joint first. Do not heat the ball joint or the axle to aid in removal.

3. Assemble C-Frame assembly and receiver cup on lower ball joint and turn forcing screw clockwise until ball joint is removed.

4. To install the lower ball joint, assemble C-Frame puller and adaptors. Turn forcing screw clockwise until ball joint is seated. Lower ball joint must be installed first. Do not heat the ball joint or axle to aid in installation.

5. Install the snap ring onto the ball joint. To install the upper ball joint, assemble the C-frame and repeat Steps 1 and 2.

6. Install the spindle.

Front Wheel Spindle 4WD

REMOVAL & INSTALLATION

NOTE: This procedure also includes bearing and seal replacement and repacking.

1. Raise the vehicle and install safety stands.

2. If equipped with manual locking hubs remove the hubs as follows: Remove the six socket heat bolts from the cap assembly and separate the cap assembly from the body. Remove the snap-ring from the end of the axle shaft. Remove the lock ring seated in the groove of the wheel hub and slide the body assembly out of the wheel hub.

3. If equipped with automatic locking hubs remove the hubs as follows: Remove the bolts and remove the hub cap assembly

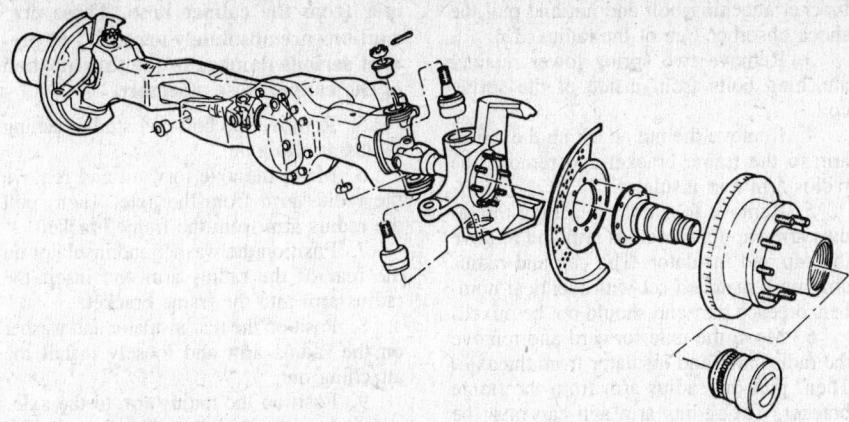

1980 and later knuckle and spindle assembly on 4WD vehicles

from the spindle. Remove the bolt from the end of the shaft. Remove the lock ring seated in the groove of the wheel hub. Remove the body assembly from the spindle. Use a puller if necessary. Unscrew all three set screws in the spindle locknut until the heads are flush with the edge of the locknut. Remove the outer spindle locknut with tool T80T–4000–V, automatic hub locknut wrench.

4. Remove the front hub grease cap and driving hub snap-ring.

5. Remove the splined driving hub and the pressure spring. Slightly pry off if necessary.

6. Remove the wheel bearing locknut, lock ring and adjusting nut using special tool T59T–1197–B or equivalent.

7. Remove the hub and disc assembly. The outer wheel bearing and spring retainer will slide out as the hub is removed.

8. Remove the spindle nuts and remove the spindle, splash shield and axle shaft assembly.

NOTE: It may be necessary to break the spindle loose with a plastic hammer.

9. Clean all old grease from the needle bearings and wipe clean the spindle face that mates with the spindle bore seal.

10. Remove the spindle bore seal, V-seal, and thrust washer from the outer axle shaft. Clean and replace if necessary.

11. Using Multi-Purpose Lubricant Ford ESA–M1C75–B or equivalent, thoroughly lubricate the needle bearing and pack the spindle face that mates with the spindle bore seal.

12. Position the V-seal in the spindle bore next to the needle bearing. Assemble the spindle bore seal on the axle shaft.

13. Assemble the spindle with the axle shaft on the knuckle studs and tighten the retaining nuts to 75 ft. lbs.

14. Carefully drive the inner bearing cone and grease seal out of the hub using tool T77F–1102–A or equivalent.

15. Inspect the inner bearing cups and if necessary remove with a drift.

NOTE: If new cups are installed, install new bearings.

16. Lubricate the bearings with the lubricant specified earlier and clean all old grease from the hub. Pack the cones and rollers with lubricant. Try to pack as much as possible between the rollers and the cages.

17. Position the inner bearing cone and roller in the inner cup and install the grease retainer.

18. Install the hub and disc assembly on the spindle.

19. Install the outer bearing cone and roller, and the adjusting nut.

20. Using tool T59T–1197–B or equivalent and a torque wrench tighten the bearing adjusting nut to 50 ft. lbs. while rotating the wheel back and forth. Back off the adjusting nut no more than 90 degrees.

21. Assemble the lock ring by turning the nut to the nearest hole and inserting the dowel pin.

NOTE: The dowel pin must seat in the lock ring hole for proper bearing adjustment and wheel retention.

22. Install the outer lock nut and tighten to 65 ft. lbs. Final end play on the wheel and spindle should be 0.001–0.006 inch.

23. Adjust the brake if necessary and lower the vehicle.

Knuckle and Ball Joint (4WD)

REPLACEMENT

NOTE: A combination ball joint puller/press and a special spanner wrench are needed for this job. If these aren't available the job should not be attempted.

1. Follow the procedures under Axle Shaft Removal.

2. Disconnect the connecting rod end from the knuckle.

3. Remove the cotter pin from the upper ball socket and loosen the upper and lower ball socket nuts. Discard the nut from the lower ball socket after the knuckle breaks loose from the yoke.

4. Remove the knuckle from the yoke.

If the upper socket remains in the yoke, remove it by hitting the top of the stud with a soft-faced hammer. Discard the socket and adjusting sleeve.

5. Remove the bottom socket with a ball joint puller (available at most auto parts stores) after first removing the snap ring.

6. For installation: Place the knuckle in a vise and assemble the bottom socket. Place the new socket into the knuckle making sure it isn't cocked, place the driver over the socket, place the forcing screw into the socket and force the socket into the knuckle.

7. Make sure that the socket shoulder is seated against the knuckle. Use a .0015 in. feeler gauge between the socket seat and the knuckle.

8. The gauge should not enter the area of minimum contact. Install the snap ring.

9. Assemble the top socket into the knuckle. Assemble the holding plate onto the backing plate screw. Tighten the nuts snugly. Place a new socket into the knuckle. Be sure it is not cocked. Place a driver over the socket and force the socket assembly into the knuckle. Using a .0015 in. gauge, check the fit at the shoulder. The gauge should not enter the area of minimum contact.

10. Install a new adjusting sleeve into the top of the yoke leaving about two threads exposed.

11. Assemble the knuckle and yoke. Install a new nut on the bottom socket and make it finger tight.

12. Place a wrench and step plate over the adjusting sleeve and install the puller so that it grasps the step plate. Tighten the forcing screw to pull the knuckle assembly into the yoke. With torque still applied, tighten the nut to 70–90 ft. lbs. If the bottom stud should turn with the nut, add more torque to the puller forcing screw. Remove the puller, step plate and holding plate.

13. Tighten the adjusting sleeve to 40 ft. lbs. and remove the wrench.

14. Install the top socket nut and torque it to 100 ft. lbs. Line up the cotter pin hole by tightening, not loosening, the nut. Install the cotter pin and test the steering effort with a spring scale attached to the knuckle. Pull should not exceed 26 ft. lbs. If it does, the ball joints will have to be replaced.

15. Connect the steering linkage to the knuckle. Torque it to 40 ft. lbs.

16. Install the axle shaft as described in the Axle Shaft Removal and Installation procedure.

Radius Arm

REMOVAL & INSTALLATION

All Except Bronco, 1980 and Later F150 with 4WD, Ranger and Bronco II

1. Raise the front of the vehicle and

place safety stands under the frame and a jack under the wheel or axle.

2. Disconnect the shock absorber from the radius arm bracket.

3. Remove the two spring upper retainer attaching bolts from the top of the spring upper seat and remove the retainer.

4. Remove the nut which attaches the spring lower retainer to the lower seat and axle and remove the retainer.

5. Lower the axle and remove the spring.

6. Disconnect the steering rod from the spindle arm.

7. Remove the spring lower seat and shim from the radius arm. Then, remove the bolt and nut which attach the radius arm to the axle.

8. Remove the cotter pin, nut and washer from the radius arm rear attachment.

9. Remove the bushing from the radius arm and remove the radius arm from the vehicle.

10. Remove the inner bushing from the radius arm.

11. Position the radius arm to the axle and install the bolt and nut finger-tight.

12. Install the inner bushing on the radius arm and position the arm to the frame bracket.

13. Install the bushing, washer, and attaching nut. Tighten the nut and install the cotter pin.

14. Connect the steering rod to the spindle arm and install the attaching nut. Tighten the radius arm-to-axle attaching bolt and nut.

F150 4WD From 1980

1. Raise the vehicle and position safety stands under the frame side rails.

2. Remove the shock absorber lower attaching bolt and nut and pull the shock absorber free of the radius arm.

3. Remove the lower spring retaining bolt from the inside of the spring coil.

4. Remove the nut attaching the radius arm to the frame bracket and remove the radius arm rear insulator. Lower the axle and allow the axle to move forward.

NOTE: The axle must be supported on the jack throughout spring removal, and must not be permitted to hang from the brake hose. If the length of the brake hose does not provide sufficient clearance it may be necessary to remove and support the brake caliper.

5. Remove the bolt and stud attaching the radius arm to the axle.

6. Move the axle forward and remove the radius arm from the axle. Then, pull the radius arm from the frame bracket.

7. Installation is the reverse of removal. Install new bolts and the stud type bolt which attach the radius arm to the axle and tighten to 210 ft. lbs. Tighten the radius arm rear attaching nut to 100 ft. lbs.

1979 Bronco

1. Raise the vehicle and position safety stands under the frame side rails.

2. Remove the shock absorber-to-lower bracket attaching bolt and nut and pull the shock absorber free of the radius arm.

3. Remove two spring lower retainer attaching bolts from inside of the spring coil.

4. Remove the nut attaching the radius arm to the frame bracket and remove the radius arm rear insulator.

5. Remove four bolts attaching the radius arm cap to the radius arm and remove the cap and insulator. The cap and radius arm are a matched set with identical numbers on each part and should not be mixed.

6. Move the axle forward and remove the radius arm and insulator from the axle. Then, pull the radius arm from the frame bracket. The radius arm and cap must be identified by a T on each piece in addition to a number (1 through 100).7.

Position the washer and insulator on the rear of the radius arm and insert the radius arm and insulator into the frame bracket. 8.

Position the rear insulator and washer on the radius arm and loosely install the attaching nut. 9.

Position the insulator on the axle and position the radius arm to the insulator and axle.

10. Position the front insulator to the axle and install the radius arm cap with the numbers on the radius arm and cap together. Tighten the attaching bolts diagonally in pairs to 90–110 ft. lbs.

11. Position the spring lower seat and retainer to the spring and axle. Install the two attaching bolts. Tighten the bolts to 45–55 ft. lbs.

12. Tighten the radius rod rear attaching nut to 80–120 ft. lbs.

13. Position the shock absorber to the lower bracket and install the attaching bolt and nut. Tighten the nut to 40–60 ft. lbs. Remove safety stands and lower the vehicle.

1980 and Later Bronco

1. Raise the vehicle and position safety stands under the frame side rails.

2. Remove the shock absorber-to-lower bracket attaching bolt and nut and pull the shock absorber free of the radius arm.

3. Remove spring lower retainer attaching bolt from inside of the spring coil.

4. Remove the nut attaching the radius arm to the frame bracket and remove the radius arm rear insulator. Lower the axle and allow axle to move forward.

NOTE: The axle must be supported on the jack throughout spring removal and installation, and must not be permitted to hang by the brake hose. If the length of the brake hose is not sufficient to provide adequate clearance for removal and installation of the spring, the disc brake caliper must be removed from the spindle. After removal, the caliper must be placed on the frame or otherwise

supported to prevent suspending the caliper from the caliper hose. These precautions are absolutely necessary to prevent serious damage to the tube portion of the caliper hose assembly.

5. Remove the bolt and stud attaching radius arm to axle.

6. Move the axle forward and remove the radius arm from the axle. Then, pull the radius arm from the frame bracket.

7. Position the washer and insulator on the rear of the radius arm and insert the radius arm into the frame bracket.

8. Position the rear insulator and washer on the radius arm and loosely install the attaching nut.

9. Position the radius arm to the axle.

10. Install new bolts and study-type bolt attaching radius arm to axle. Tighten to 180–240 ft. lbs.

11. Position the spring lower seat, spring insulator and retainer to the spring and axle. Install the two attaching bolts. Tighten the nuts to 30–70 ft. lbs.

12. Tighten the radius rod rear attaching nut to 80–120 ft. lbs.

13. Position the shock absorber to the lower bracket and install the attaching bolt and nut. Tighten the nut to 40–60 ft. lbs. Remove safety stands and lower the vehicle.

Ranger and Bronco II

1. Raise the front of the vehicle and place safety stands under the frame. Place a jack under the axle.

NOTE: The axle must be supported on the jack throughout spring removal and installation, and must not be permitted to hang by the brake hose. If the length of the brake hose is not sufficient to provide adequate clearance for removal and installation of the spring, the disc brake caliper must be removed from the spindle. After removal, the caliper must be placed on the frame or otherwise supported to prevent suspending the caliper from the caliper hose. These precautions are absolutely necessary to prevent serious damage to the tube portion of the caliper hose asssembly.

2. Disconnect the lower end of the shock absorber from the shock lower bracket (bolt and nut).

3. Remove the front spring. Loosen the axle pivot bolt.

4. Remove the spring lower seat and stud from the radius arm, and then remove the bolts that attach the radius arm to the axle and front bracket.

5. Remove the nut, rear washer and insulator from the rear side of the radius arm rear bracket.

6. Remove the radius arm from the vehicle, and remove the inner insulator and retainer from the radius arm stud.

7. Install in reverse order.

Stabilizer Bar

REMOVAL & INSTALLATION

Bronco and 4WD Pickups

1979

1. Remove locknut, washers, and insulator to remove link assemblies from stabilizer bar. Remove nuts, bolts, and washers connecting link assemblies to frame.

2. Remove nuts on U-bolts to remove stabilizer bar from retainers. Remove stabilizer bar. Remove U-bolts, brackets and retainers.

3. Place bracket assemblies on axle aligning holes in brackets with alignment pins on axles.

4. Install U-bolts through bracket assembly. Position stabilizer bar on brackets. Install retainer and tighten nuts to 35–55 ft. lbs.

5. Install link assemblies on frame. Connect link assemblies to stabilizer bar. Tighten link to stabilizer bar nuts to 18–25 ft. lbs. Tighten link to frame nuts to 40–60 ft. lbs.

FROM 1980

1. Remove nuts, bolts and washers connecting the stabilizer bar to connecting links. Remove nuts and bolts of the stabilizer bar retainer.

2. Remove stabilizer bar insulator assembly.

3. To remove the stabilizer bar mounting bracket, the coil spring must be removed as described above under spring removal. Remove the lower spring seat. The bracket attaching stud and bracket can now be removed.

4. To install the stabilizer bar mounting brackets, locate the brackets so that the locating tang is positioned in the radius arm notch (or quad shock bracket notch if vehicle has quad shocks). Install a new stud. Torque to 180–220 ft. lbs. A new stud is required because of the adhesive on the threads. Reposition the spring lower seat and reinstall the spring and retainers.

5. To reinstall the stabilizer bar insulator assembly, assemble all nuts, bolts and washers to the bar, brackets, retainers and links loosely. With the bar positioned correctly, torque retainer nuts to 32–35 ft. lbs. with retainer around the insulator. Then torque all remaining nuts at the link assemblies to 41–50 ft. lbs.

RANGER AND BRONCO II

1. Remove the nuts and U-bolts retaining the lower shock bracket/stabilizer bar bushing to radius arm.

2. Remove retainers and remove the stabilizer bar and bushing.

3. Place stabilizer bar in position on the radius arm and bracket.

4. Install retainers and U-bolts. Tighten retainer bolts to 35–50 ft. lbs. Tighten U-bolt nuts to 48–64 ft. lbs.

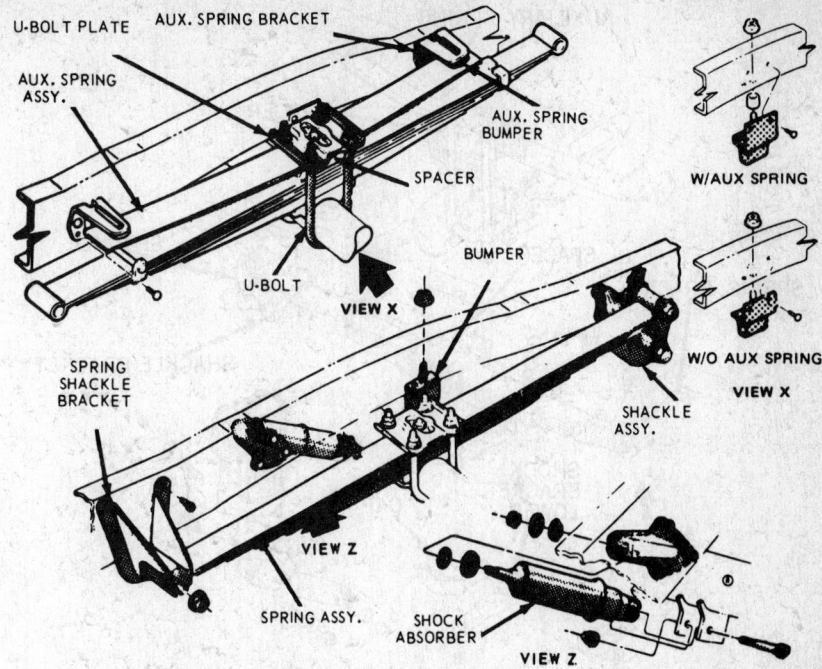

Rear spring installation, F-100, 150 2WD and 4WD, and F-250 2WD

REAR SUSPENSION

Springs

REMOVAL & INSTALLATION

2WD Pickups and Ranger/Bronco II

1. Raise the vehicle by the frame until the weight is off the rear spring with the tires still on the floor.

2. Remove the nuts from the spring U-bolts and drive the U-bolts from the U-bolt plate. Remove the auxiliary spring and spacer, if so equipped.

3. Remove the spring-to-bracket nut and bolt at the front of the spring.

4. Remove the upper and lower shackle nuts and bolts at the rear of the spring and remove the spring and shackle assembly from the rear shackle bracket.

5. Remove the bushings in the spring or shackle, if they are worn or damaged, and install new ones.

6. Position the spring in the shackle and install the upper shackle-to-spring nut and bolt with the bolt head facing outward.

7. Position the front end of the spring in the bracket and install the nut and bolt.

8. Position the shackle in the rear bracket and install the nut and bolt.

9. Position the spring on top of the axle with the spring center bolt centered in the hole provided in the seat. Install the auxiliary spring and spacer, if so equipped.

10. Install the spring U-bolts, plate, and nuts.

11. Lower the vehicle and tighten the attaching hardware as follows: U-bolt nuts, $\frac{1}{2}$ in. 45–70 ft. lbs: $\frac{9}{16}$ in. 85–115 ft. lbs: Front spring hanger, $\frac{9}{16}$ in. 75–105 ft. lbs: $\frac{5}{8}$ in. 150–190 ft. lbs: Rear spring hanger 75–105 ft. lbs.

4WD Pickups

1. Raise the truck by the frame until the weight is off the rear springs and the wheels are still touching the ground.

2. Remove the nuts from the spring U-bolts and drive the U-bolts out of the spring seat cap. Remove the spring cap. Remove the auxiliary spring and spacer, if so equipped.

3. Remove the shackle pin lockbolts from each end of the spring. Insert a drift in the hole provided in the frame from the inner side and drive the shackle pin out of each spring bracket.

4. Remove the spring and shackle from the truck. Remove the spring-to-axle spacer.

5. Drive the remaining shackle pin out of the rear spring eye and remove the shackle from the spring.

6. After checking and replacing worn or damaged bushings, nuts and bolts, and broken or weak springs, position the shackle to the rear spring eye.

7. Install the shackle pin through the shackle and spring eye with the lubrication fitting on the shackle pin facing outboard.

8. Align the shackle pin lockbolt groove with the lockbolt hole in the shackle, and install the lockbolt, washer, and nut.

9. Position the spring on the axle with the spring center bolt in the hole provided

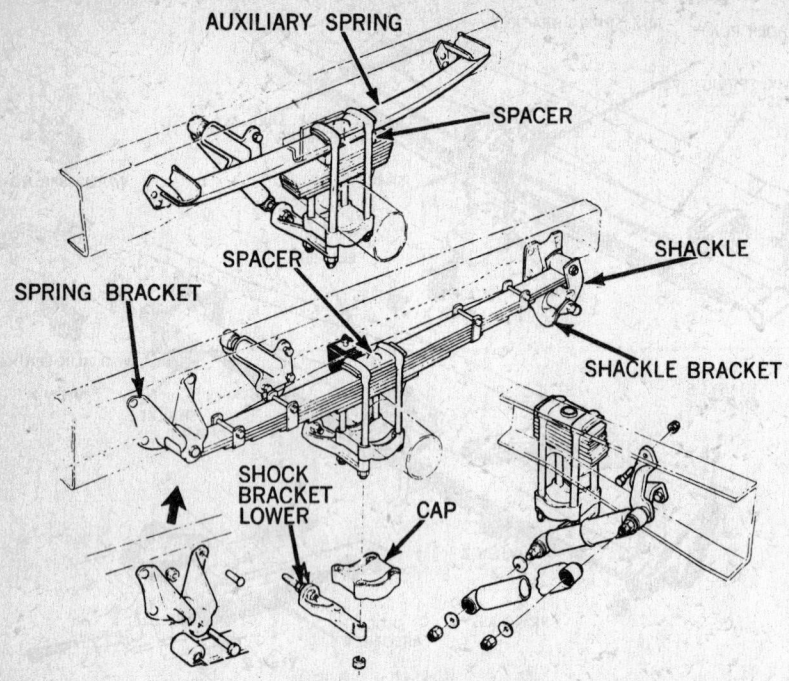

Rear spring installation, F-350

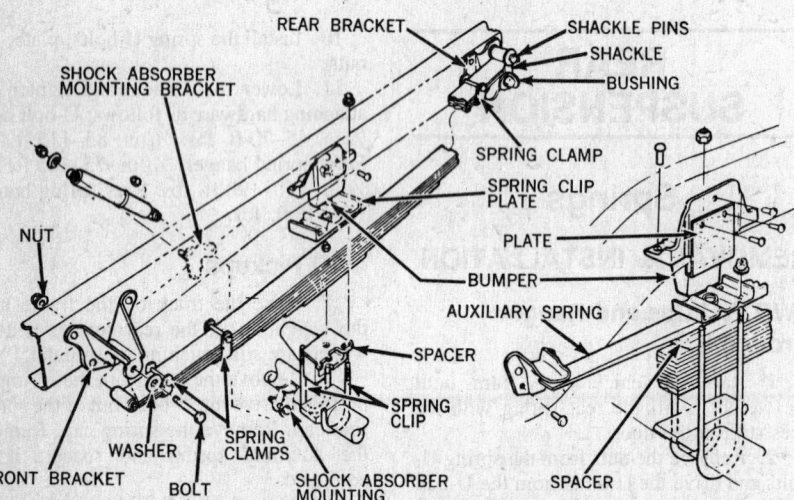

Rear spring installation, F-250 4WD

in the axle spring seat or spacer. Install the spacer between the spring seat and the spring. Make sure that the spacer dowel is positioned in the pilot hole of the axle spring seat.

10. Install the shackle pin through the shackle and rear bracket. The lubrication fitting on the shackle pin faces outboard. Align the pin groove with the lock bolt hole in the bracket and install the lockbolt, washer, and nut.

11. Install the shackle pin at the front bracket and spring eye in the same manner as above.

12. Install the auxiliary spring and spacer, if so equipped. Place the spring cap on top of the spring at the center bolt and place the spring U-bolts over the spring assembly and axle.

13. Position the spring seat cap, and install the nuts on the spring U-bolts.

14. Lower the truck to the ground and tighten the attaching hardware.

Bronco

1. Raise the vehicle and install jackstands under the frame. The vehicle must be supported in such a way that the rear axle hangs free with the tire a few inches off the ground. Place a hydraulic floor jack under the center of the axle housing.

2. Disconnect the shock absorber from the axle.

3. Remove the U-bolt attaching nuts and remove the two U-bolts and the spring clip plate.

4. Lower the axle to relieve the spring

tension and remove the nut from the spring front attaching bolt.

5. Remove the spring front attaching bolt from the spring and hanger with a drift.

6. Remove the nut from the shackle-to-hanger attaching bolt and drive the bolt from the shackle and hanger with a drift and remove the spring from the vehicle.

7. Remove the nut from the spring rear attaching bolt. Drive the bolt out of the spring and shackle with a drift.

8. To install the rear spring: Position the shackle (closed section facing toward the front of the vehicle) to the spring rear eye and install the bolt and nut.

9. Position the spring front eye and bushing to the spring front hanger, and install the attaching bolt and nut.

10. Position the spring rear eye and bushing to the shackle, and install the attaching bolt and nut.

11. Raise the axle to the spring and install the U-bolts and spring clip plate.

12. Torque the U-bolt nuts and spring front and rear attaching bolt nuts to 45–60 ft. lbs.

13. Remove the jackstands and lower the vehicle.

NOTE: Squeaky rear springs can be corrected by tightening the front and rear eye bolts to 150–204 ft. lbs., then raising and supporting the rear of the vehicle so that the rear springs hang, spreading the leaves. Apply a silicone based lubricant for a distance of three inches in from each leaf tip.

E100–200

1. Raise the rear of the vehicle and support the chassis with jackstands. Support the rear axle with a floor jack or hoist.

2. Disconnect the lower end of the shock absorber from the bracket on the axle housing.

3. Remove the two U-bolts and plate.

4. Lower the axle and remove the upper and lower rear shackle bolts.

5. Pull the rear shackle assembly and rubber bushings from the bracket and spring.

6. Remove the nut and mounting bolt which secure the front end of the spring. Remove the spring assembly from the front shackle bracket.

7. Install new rubber bushings in the rear shackle bracket and in the rear eye of the replacement spring.

8. Assemble the front eye of the spring to the front shackle bracket with the front mounting bolt and nut. Do not tighten the nut.

9. Mount the rear end of the spring with the upper bolt of the rear shackle assembly passing through the eye of the spring. Insert the lower bolt through the rear spring hanger.

10. Assemble the spring center bolt in the pilot hole in the axle and install the plate. Install the U-bolts through the plate. Do not tighten the attaching nuts at this time.

11. Raise the axle with a floor jack or

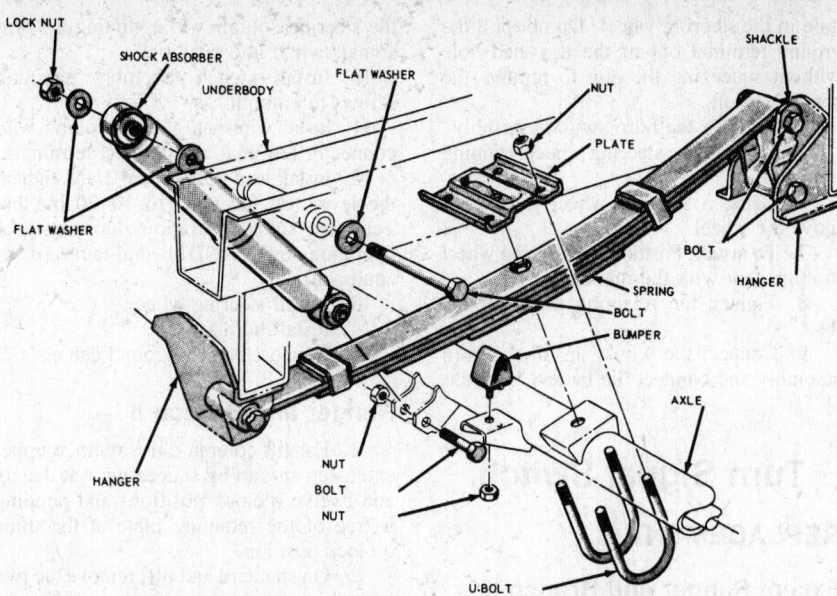

Rear spring installation, E-100, 150 and 200

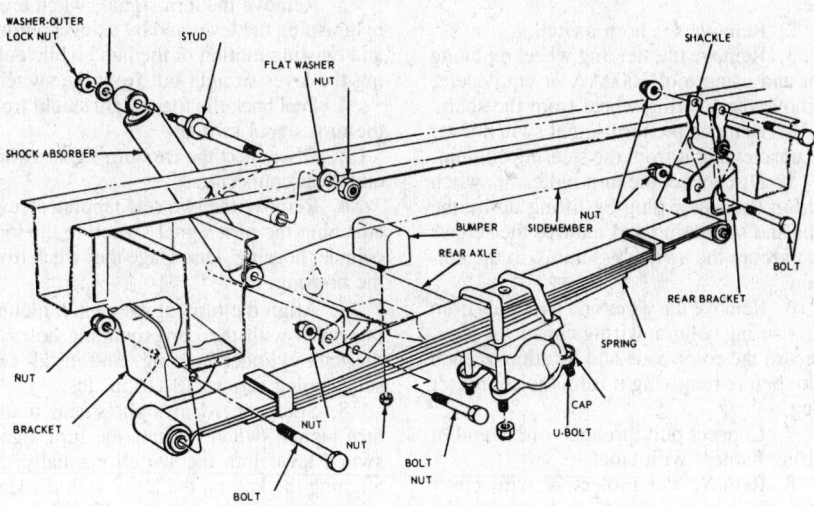

Rear spring installation, E-250, 300 and 350

Shock Absorbers

REMOVAL & INSTALLATION

1. Raise the vehicle and place jackstands under the frame.

2. Remove the shock absorber-to-upper bracket attaching nut and washers, and bushing from the shock absorber rod or if mounted with nut and bolt, remove as lower.

3. Remove the shock absorber-to-axle attaching bolt. Drive the bolts from the axle bracket and shock absorber with a brass drift and remove the shock absorber.

4. Position the washers and bushing on the shock absorber rod and position the shock absorber at the upper bracket.

5. Position the bushing and washers on the shock absorber rod and install the attaching nut loosely.

6. Position the shock absorber at the axle housing bracket and install the attaching bolt and nut. Tighten the lower nut to 40–60 ft. lbs. and the upper nut to 15–25 ft. lbs. If attached at top with similar mounting, torque is same.

FRONT END ALIGNMENT

CASTER

The caster angles are designed into the front axle and cannot be adjusted.

CAMBER

The camber angles are designed into the front axle and cannot be adjusted except on 1983 and later 2WD vehicles.

1983 and Later (2WD)

Camber is adjusted by replacing the camber adapter on the upper ball joint stud. Adapters are available in 0°, ½, 1° and 1½° increments.

If camber needs adjustment, replace the camber adapter as described under camber adapter. On the right spindle, the slot in the camber adapter must point forward for a positive camber change and rearward for a negative camber change. On the left spindle, the slot in the adapter must point rearward for a positive camber change and forward for a negative camber change.

TOE-IN ADJUSTMENT

All Models

Toe-in can be measured by either a front end alignment machine or by the following method:

With the front wheels in the straightahead position, measure the distance between the extreme front and the extreme

hoist until the vehicle is free of the jackstands. Connect the lower end of the shock absorber to the bracket on the axle housing.

12. Tighten the spring front mounting bolt and nut, the rear shackle nuts and the U-bolt nuts.

13. Remove the jackstands and lower the vehicle.

E250–350

1. Raise the rear of the vehicle and support the chassis with jackstands. Support the rear axle with a floor jack or hoist.

2. Disconnect the lower end of the shock absorber from the bracket on the axle housing.

3. Remove the two spring U-bolts and the spring cap.

4. Lower the axle and remove the spring front bolt from the hanger.

5. Remove the two attaching bolts from

the rear of the spring. Remove the spring and shackle.

6. Assemble the upper end of the shackle to the spring with the attaching bolt.

7. Connect the front of the spring to the front bracket with the attaching bolt.

8. Assemble the spring and shackle to the rear bracket with the attaching bolt.

9. Place the spring plate over the head of the center bolt.

10. Raise the axle with a jack. Install the center bolt through the pilot hole in the pad on the axle housing.

11. Install the spring U-bolts, cap and attaching nuts. Tighten the nuts snugly.

12. Connect the lower end of the shock absorber to the lower bracket.

13. Tighten the spring front mounting bolt and nut, the rear shackle nuts and spring U-bolt nuts.

14. Remove the jackstands and lower the vehicle.

rear of the front wheels. In other words, measure the distance across the undercarriage of the vehicle between the two front edges and the two rear edges of the two front wheels. Both of these measurements (front and rear of the two wheels) must be taken at an equal distance from the floor and at the approximate centerline of the spindle. The difference between these two distances is the amount that the wheels toe-in or toe-out. The wheels should always be adjusted to toe-in according to specifications.

1. Loosen the clamp bolts at each end of the left tie-rod, seen from the front of the vehicle. Rotate the connecting rod tube until the correct toe-in is obtained, then tighten the clamp bolts.

2. Recheck the toe-in to make sure that no changes occurred when the bolts were tightened.

NOTE: The clamps should be positioned ³⁄₁₆ in. from the end of the rod with the clamp bolts in a vertical position in front of the tube, with the nut down.

STEERING

Steering Wheel

REMOVAL & INSTALLATION

Through 1982

1. Disconnect the battery ground and mark the steering wheel-to-column alignment.

2. Remove one screw from the underside of each spoke and lift the horn assembly from the wheel. On vehicles with a sport wheel option, pry or unscrew (try unscrewing first) the button cover.

3. Disconnect the horn switch wires by pulling the spade terminal from the blade connector. Squeeze or pinch the J-clip ground wire terminal fully and pull it out of the

hole in the steering wheel. Do not pull the ground terminal out of the threaded hole without squeezing the clip to remove the spring tension.

4. Remove the horn switch assembly.

5. Remove the steering wheel retaining nut.

6. Using a steering wheel puller, remove the wheel.

7. To install: Position the steering wheel in alignment with the marks.

8. Tighten the retaining nut to 50 ft. lbs.

9. Connect the wires, install the horn assembly and connect the battery ground.

Turn Signal Switch

REPLACEMENT

Except Ranger and Bronco II

1. Disconnect the battery ground cable.

2. Remove the horn switch.

3. Remove the steering wheel retaining nut and using tool 3600AA or equivalent, remove the steering wheel from the shaft.

4. Remove the turn signal switch lever by unscrewing it from the steering column.

5. Disconnect the turn indicator switch wiring connector plug by lifting up on the tabs and separating and remove the screws that secure the switch assembly to the column.

6. Remove the wires and terminals from the steering column wiring connector plug. Record the color code and location of each wire before removing it from the connector plug.

7. Connect pull through wire to end of wiring harness with tape.

8. Remove the protective wire cover from the wiring harness and remove the switch and wires through the top of the column.

9. Tape the loose ends of the new turn signal switch wires to the pull-through wire or cord. Carefully pull the wires through

the steering column while guiding the turn signal switch into position.

10. Install switch assembly retaining screws to column.

11. Install wires into steering column wire connector terminal and connect terminals.

12. Install turn signal lever. Hand-tighten the lever (on flat side) to 10–20 in. lbs. Test turn signal operation, hazard signal operation and PRND21 dial-lamp (if so equipped).

13. Install steering wheel.

14. Install horn switch.

15. Connect battery ground cable.

Ranger and Bronco II

1. For tilt column only, remove upper extension shroud by squeezing it at the six and twelve o'clock positions and popping it free of the retaining plate at the three o'clock position.

2. On standard and tilt, remove the two trim shroud halves by removing the two attaching screws.

3. Remove the turn signal switch lever by grasping the lever and by using a pulling and twisting motion of the hand while pulling the lever straight out from the switch.

4. Peel back the foam sight shield from the turn signal switch.

5. Disconnect the two turn signal switch electrical connectors.

6. Remove the two self-tapping screws attaching the turn signal switch to the lock cylinder housing. Disengage the switch from the housing.

7. Align the turn signal switch mounting holes with the corresponding holes in the lock cylinder housing, and install two self-tapping screws 18–26 in. lbs.

8. Stick the foam sight shield to the turn signal switch. Install the turn signal switch lever into the switch manually, by aligning the key on the lever with the keyway in the switch and by pushing the lever toward the switch to full engagement.

9. Install the two turn signal switch electrical connectors to full engagement.

10. Install the steering column trim shrouds.

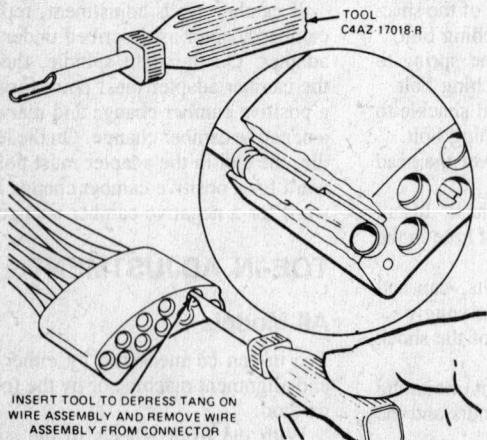

Typical turn signal switch connector

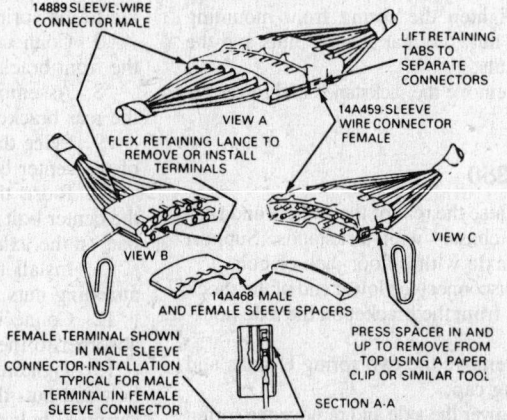

Typical turn signal switch terminal removal

Power Steering Pump

REMOVAL & INSTALLATION

Except CII Pump

1. Disconnect the pressure and return lines from the pump and plug them to prevent loss of fluid or entrance of dirt into the system.
2. Loosen the belt tension adjusting bolt all the way.
3. Remove the bolts attaching the pump mounting bracket to the air conditioning bracket (if equipped).
4. Remove the pump, mounting bracket and pulley assembly.
5. Install the pump, bracket and pulley assembly and loosely attach the bolts that secure the pump mounting bracket to the air conditioning bracket (if equipped).
6. Install the drive belts on the pulley.
7. Loosely install the belt tension adjusting nut.
8. Pry between the pump adjustment bracket and the engine block until correct tension is achieved. While still holding this tension, tighten the adjusting bolt.
9. Tighten all attaching bolts.
10. Connect the pressure and return lines to the pump.
11. Fill the reservoir with power steering fluid. Bleed the air from the system by turning the steering wheel from left to right several times. Inspect for leaks.

Quick Connect Power Steering Fitting Service

The quick connect power steering fitting, under certain conditions may leak and/or result in improper engagement. The leak can be caused by a cut O-ring, imperfections in the outlet fitting inside diameter, or improperly machined O-ring groove. Improper engagement can be caused by an improperly machined tube end, tube nut, snap ring, outlet fitting, or gear port.

If a leak occurs, the O-ring should be replaced with quick connect O-rings ($\frac{3}{8}$ tube end: 388749S; $\frac{5}{16}$ tube end: 388748S). The O-rings that are used on the tube-O power steering fitting should not be used on the quick connect fitting because of dimensional and material changes. If O-ring replacement does not solve the leak problem, outlet fitting replacement and, lastly, hose replacement should be made.

If improper engagement occurs due to a missing or bent snap ring, or improperly machined tube nut, it may be repaired with a service snap ring kit (kit includes a new tube nut). The system should then be properly filled, the engine started, and the steering wheel cycled from lock-to-lock to test for positive engagement. If the hose assembly still does not engage, replace the entire hose assembly.

Quick connect hose assemblies for service have tube nuts, snap rings, and O-rings already attached.

When the quick connect tube nut is tightened or loosened, a tube nut wrench, not an open end wrench, is recommended. An open end wrench may result in tube nut deformation under excessive torque conditions. Care must be taken not to overtighten the tube nut. Tighten to 10–15 ft. lbs. Swivel and/or end play of the quick connect fittings is normal, and does not indicate an undertightened fitting.

CII Pump

NOTE: The CII pump is equipped with a fiberglass reservoir and can be identified by the reservoir. Never pry against the fiberglass, as damage will occur. The 3.8L V6 with a serpentine belt driving the power steering pump uses a separate idler pulley on a slider-type bracket for belt tension adjustment. To adjust or remove the belt tension, loosen the bolts in the slider slots and tighten the adjusting belt as required to obtain the correct belt tension.

1. To remove the power steering fluid from the pump reservoir, disconnect the fluid return hose at the reservoir and drain the fluid into a container. Remove the pressure hose from the pump.
2. Remove the bolts from the pump adjustment bracket. Loosen the pump sufficiently to remove the belt off the pulley. Remove the pump (still attached to the adjustment bracket) from the support bracket.
3. Remove the pulley from the pump if required.
4. Remove the bolts attaching the adjustment bracket to the pump and remove the pump.
5. Place the adjustment bracket on the pump. Install and tighten the bolts to specification listed at the end of this Section.
6. Install the pulley on the pump if removed.
7. Place the pump with adjustment bracket and pulley on the support bracket.

Install the bolts connecting the support bracket to the adjustment bracket.
8. Place the belt on the pulley and adjust belt tension. Tighten bolts on adjustment bracket.
9. Install the pressure hose to the pump fitting.
10. Connect the return hose to the pump, and tighten the clamp.
11. Fill the reservoir with specified power steering fluid, start the engine and turn the steering wheel from stop to stop to remove air from the system.
11. Check for leaks and recheck the fluid level. Add fluid if necessary.

Manual Steering Gear

REMOVAL & INSTALLATION

1979

1. Raise and safely support the vehicle. Remove the Pitman arm attaching nut and lockwasher.
2. Remove the Pitman arm from the sector shaft using tool T64P–3590–F.
3. Remove bolts, nuts and flat washers

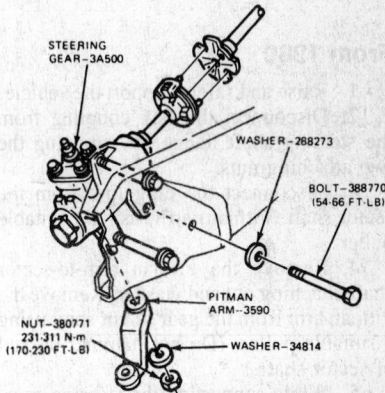

Typical steering gear installation-from 1980

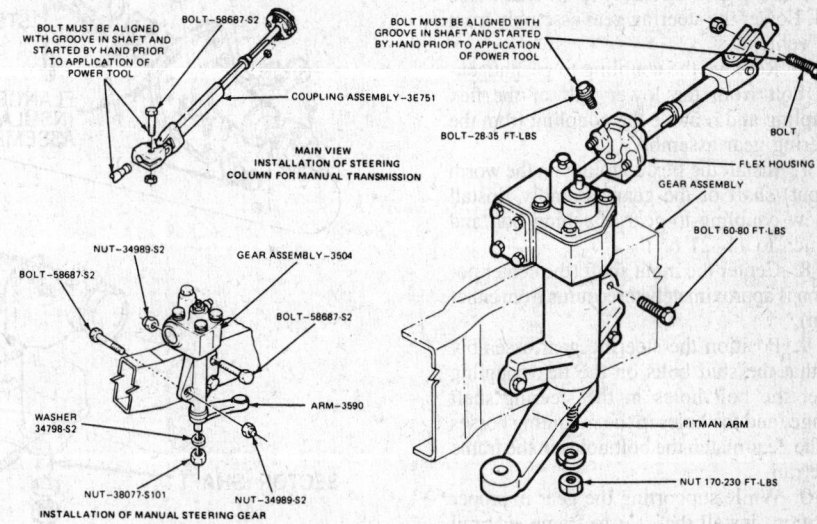

Typical steering gear installation-thru 1979

that attach the steering gear to the frame side rail, and lower the vehicle.

4. Remove the flex coupling bolt and nut from the coupling clamp. Loosen the clamp from the coupling at the end of the steering column and separate the coupling from the steering gear input shaft by pushing the steering shaft toward the steering column. Discard the clamp, bolt, and nut.

5. Remove and discard the flex coupling clamp from the steering gear input shaft, and remove the steering gear from the frame side rail.

6. Place the steering gear on the frame side rail and install the attaching bolts, nuts, and flat washers.

7. Place the flex coupling and a new clamp on the steering gear input shaft and install the clamp bolt and nut. Tighten the bolt and nut to 28–35 ft. lbs.

8. Install a new steering shaft clamp at the end of the steering column and tighten the bolt and nut to 20–30 ft. lbs.

9. Raise the vehicle and tighten the steering gear attaching bolts and nuts to 60–80 ft. lbs.

10. Place the Pitman arm on the sector shaft and install the washer and attaching nut. Tighten the nut to 170–230 ft. lbs. Lower the vehicle, and fill the gear with lubricant SAE–90EP oil.

From 1980

1. Raise and safely support the vehicle.

2. Disconnect the flex coupling from the steering shaft flange by removing the two attaching nuts.

3. Disconnect the drag link from the sector shaft (Pitman) arm, using a suitable puller.

4. Remove the Pitman arm-to-sector shaft attaching nut and washer. Remove the Pitman arm from the gear sector shaft using a suitable puller. (Do not hammer on end of sector shaft.)

5. While supporting the steering gear, remove the bolts and washers that attach the steering gear assembly to the frame side rail. Lower the steering gear assembly from the vehicle.

6. Remove the coupling to gear attaching bolt from the lower half of the flex coupling and remove the coupling from the steering gear assembly.

7. Install the flex coupling on the worm (input) shaft of the gear assembly. Install a new coupling-to-gear attaching bolt and tighten to 11–21 ft. lbs.

8. Center the input shaft (the center position is approximately three turns from either stop).

9. Position the steering gear assembly so that the stud bolts on the flex coupling enter the bolt holes in the steering shaft flange, and the holes in the mounting bosses of the gear match the bolt holes in the frame side rail.

10. While supporting the gear in proper position, install the gear-to-frame side rail attaching bolts and washers and tighten to

70 ft. lbs. If new gear-to-frame bolts and washers are required, use only Grade 9 bolts.

11. Connect the drag link to the Pitman arm, install the drag link ball stud nut, and tighten to 50–75 ft. lbs. Then install the cotter pin.

12. Assemble the Pitman arm on the sector shaft pointing downward. Install the attaching nut and washer, and tighten to 170–230 ft. lbs.

13. Secure the flex coupling to the steering shaft flange with the two attaching nuts and tighten to 28–35 ft. lbs.

Power Steering Gear

REMOVAL & INSTALLATION

1. Disconnect the pressure and return lines from the steering gear. Plug the lines and the ports in the gear to prevent entry of dirt. Disconnect brake lines from the steering gear bracket.

2. Remove the bolts that secure the flex coupling to the steering gear and to the column steering shaft assembly.

3. Raise the vehicle and remove the Pitman arm attaching nut, and washer.

4. Remove the Pitman arm from the sector shaft using tool T64P–3590–F. Remove the tool from the Pitman arm. Do not damage the seals.

5. On vehicles with standard transmission remove the clutch release lever retracting spring to provide clearance for removing the steering gear.

6. Support the steering gear, and remove the steering gear attaching bolts.

7. Work the steering gear free of the

flex coupling. Remove the steering gear from the vehicle.

8. Slide the flex coupling into place on the steering shaft assembly. Turn the steering wheel so the spokes are in the horizontal position.

9. Center the steering gear input shaft.

10. Slide the steering gear input shaft into the flex coupling and into place on the frame side rail. Install the attaching bolts and tighten to 60–80 ft. lbs.

11. Be sure the wheels are in the straight ahead position, then install the Pitman arm on the sector shaft. Install the Pitman arm attaching washer and nut. Tighten nut to 170–230 ft. lbs.

12. Connect and tighten the pressure and the return lines to the steering gear. Reinstall the brake lines on the steering gear bracket.

13. Disconnect the coil wire. Fill the reservoir. Turn on the ignition and turn the steering wheel from left to right to distribute the fluid.

14. Re-check fluid level and add fluid, if necessary. Connect the coil wire, start the engine and turn the steering wheel from side to side. Inspect for fluid leaks.

Steering Linkage Connecting Rods

Replace the drag link if a ball stud is excessively loose or if the drag link is bent. Do not attempt to straighten a drag link. Replace the connecting rod if the ball stud is excessively loose, if the connecting rod is bent or if the threads are stripped. Do not attempt to straighten connecting rod. Always check to insure that the adjustment sleeve and clamp stops are correctly installed on the Bronco.

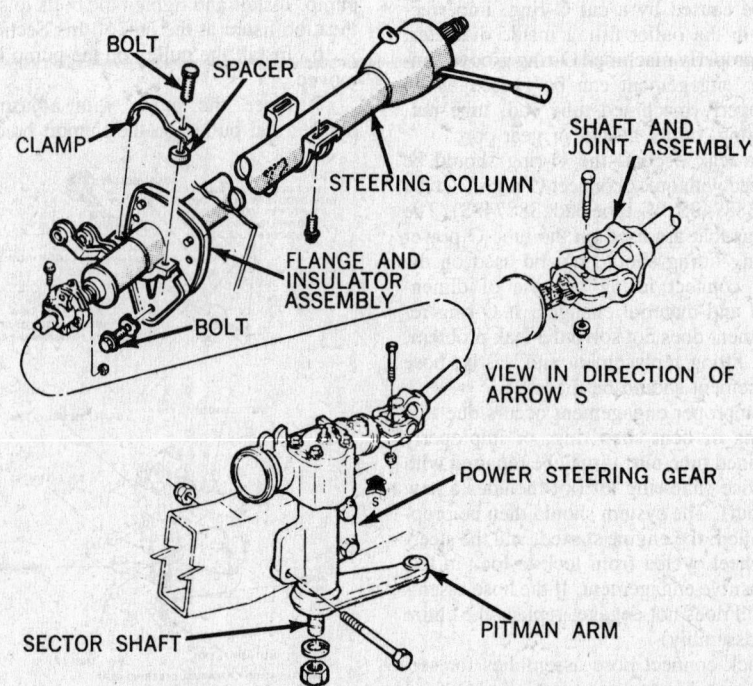

Power steering gear installation-thru 1979

REMOVAL & INSTALLATION

Vans & 2WD Pickups

Replace the drag link if a ball stud is excessively loose or if the drag link is bent. Do not attempt to straighten a drag link. Replace the connecting rod if the ball stud is excessively loose, if the connecting rod is bent or if the threads are stripped. Do not attempt to straighten connecting rod. After installing a connecting rod or adjusting toe-in check to insure that the adjustment sleeve clamps are correctly positioned on the F and E100 and F and E150 and to insure that the clamp stop is correctly installed on the F and E250 and F and E350.

1. Remove the cotter pins and nuts from the drag link, ball studs and from the right connecting rod ball stud.

2. Remove the right connecting rod ball stud from the drag link.

3. Remove the drag link ball studs from the spindle and the Pitman arm.

4. Position the new drag link, ball studs in the spindle, and Pitman arm and install nuts.

5. Position the right connecting rod ball stud in the drag link and install nut.

6. Tighten the nuts to 50–75 ft. lbs. and install the cotter pins.

7. Remove the cotter pin and nut from the connecting rod.

8. Remove the ball stud from the mating part.

9. Loosen the clamp bolt and turn the rod out of the adjustment sleeve. Count the number of turns for approximate position when installing.

10. Lubricate the threads of the new connecting rod, and turn it into the adjustment sleeve to about the same distance the old rods were installed. This will provide an approximate toe-in setting. Position the connecting rod ball studs in the spindle arms.

11. Install the nuts on to the connecting rod ball studs, tighten the nut to 50–75 ft. lbs. and install the cotter pin.

12. Check the toe-in and adjust, if necessary. After checking or adjusting toe-in, center the adjustment sleeve clamps between the locating nibs, position the clamps and tighten the nuts to 29–41 ft. lbs.

4WD Pickups

1. Raise the vehicle and support on jackstands. Disconnect the drag link from the spindle connecting rod end.

2. Disconnect the right spindle connecting rod end from the right spindle arm.

3. Disconnect the left spindle connecting rod ends from the left spindle arm and remove the spindle connecting rod ends from the truck.

4. Place the connecting rod ends in a vise and loosen the connecting rod tube clamps.

5. Remove the short (right) rod end from the connecting rod tube and remove the

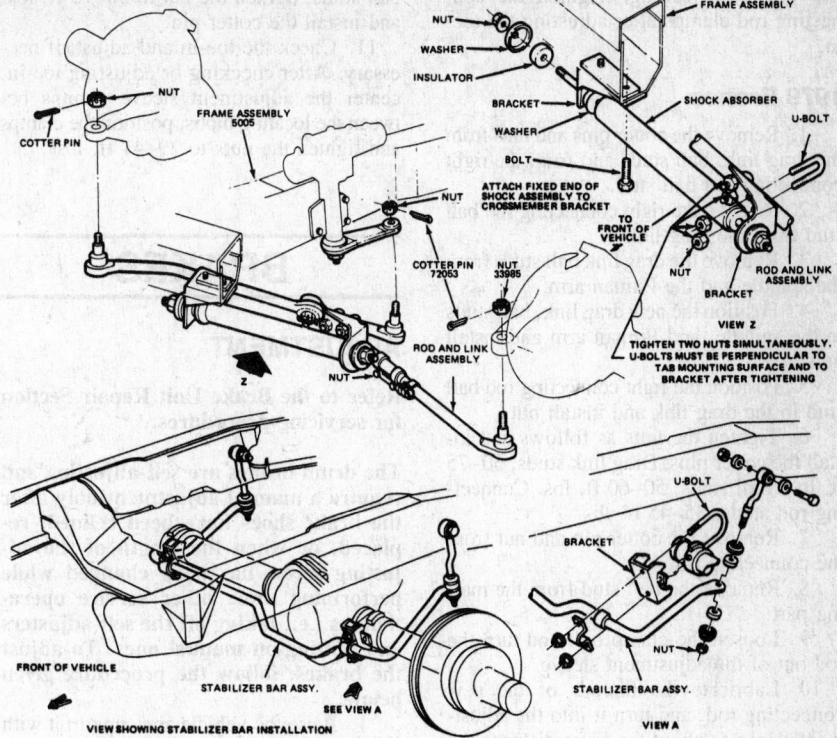

1979 Bronco steering linkage

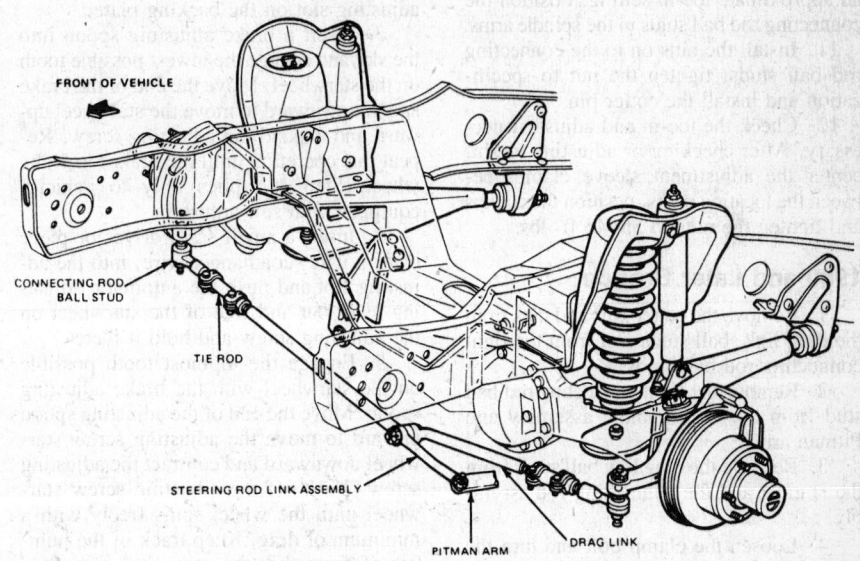

1980 and later Bronco steering linkage

tube from the long (left) connecting rod end.

6. Clean and oil the threads on all the parts to be reused.

7. Install the connecting rod tube and clamps on the left spindle connecting rod end. Don't tighten the clamps yet.

8. Install the right connecting rod end in the tube and remove the assembly from the vise.

9. Install new dust seals on the left spindle connecting rod end and position the end on the left spindle arm.

10. Install the connecting rod end at-

taching nut, tighten it, and install the cotter pin.

11. Install new dust seals on the right spindle connecting rod end and position the end on the right spindle arm. Install the attaching nut, tighten it, and install the cotter pin.

12. Install new seals on the drag link ball stud and position the drag link on the spindle connecting rod end. Install the attaching nut, tighten it, and install the cotter pin.

13. Lubricate the spindle connecting rod ends and drag link.

14. Lower the vehicle and check and ad-

just the toe-in setting. Tighten the connecting rod clamps after adjusting the toe-in.

1979 Bronco

1. Remove the cotter pins and nuts from the drag link, ball studs and from the right connecting rod ball stud.

2. Remove the right connecting rod ball stud from the drag link.

3. Remove the drag link ball studs from the spindle and the Pitman arm.

4. Position the new drag link, ball studs in the spindle, and Pitman arm and install nuts.

5. Position the right connecting rod ball stud in the drag link and install nut.

6. Tighten the nuts as follows and install the cotter pins: Drag link studs, 50–75 ft. lbs. Ball studs, 50–60 ft. lbs. Connecting rod studs, 35–45 ft. lbs.

7. Remove the cotter pin and nut from the connecting rod.

8. Remove the ball stud from the mating part.

9. Loosen the clamp bolt and turn the rod out of the adjustment sleeve.

10. Lubricate the threads of the new connecting rod, and turn it into the adjustment sleeve to about the same distance the old rods were installed. This will provide an approximate toe-in setting. Position the connecting rod ball studs in the spindle arms.

11. Install the nuts on to the connecting rod ball studs, tighten the nut to specification and install the cotter pin.

12. Check the toe-in and adjust, if necessary. After checking or adjusting toe-in, center the adjustment sleeve clamps between the locating nibbs, position the clamps and tighten the nuts to 35–45 ft. lbs.

1980 and Later Bronco

1. Remove the cotter pins and nuts from the drag link, ball studs and from the right connecting rod ball studs.

2. Remove the right connecting rod ball stud from the right spindle assembly and Pitman arm.

3. Remove the drag link ball studs from the spindle and the connecting rod assembly.

4. Loosen the clamp bolt and turn the rod out of the adjustment sleeve.

5. Lubricate the threads of the new connecting rod, and turn it into the adjustment sleeve to about the same distance the old rods were installed. This will provide an approximate toe-in setting. Position the connecting rod ball studs in the spindle arms.

6. Position the new drag link, ball studs in the spindle, and connecting rod assembly and install nuts.

7. Position the right connecting rod ball stud in the drag link and install nut.

8. Tighten all the nuts to 50–75 ft. lbs. and install the cotter pins.

9. Remove the cotter pin and nut from the left connecting rod.

10. Install the nuts on the connecting rod

ball studs, tighten the nut to 50–75 ft. lbs. and install the cotter pin.

11. Check the toe-in and adjust, if necessary. After checking or adjusting toe-in, center the adjustment sleeve clamps between the locating nibbs, position the clamps and tighten the nuts to 29–41 ft. lbs.

BRAKES

ADJUSTMENT

Refer to the Brake Unit Repair Section for servicing procedures.

The drum brakes are self-adjusting and require a manual adjustment only after the brake shoes have been relined, replaced, or when the length of the adjusting screw has been changed while performing some other service operation, as i.e., taking off the self-adjusters and putting on manual ones. To adjust the brakes, follow the procedure given below:

1. Raise the vehicle and support it with safety stands.

2. Remove the rubber plug from the adjusting slot on the backing plate.

3. Insert a brake adjusting spoon into the slot and engage the lowest possible tooth on the starwheel. Move the end of the brake spoon downward to move the starwheel upward and expand the adjusting screw. Repeat this operation until the brakes lock the wheel. Step 4 applies only to vehicles equipped with self-adjusters.

4. Insert a small screwdriver or piece of firm wire (coathanger wire) into the adjusting slot and push the automatic adjusting lever out and free of the starwheel on the adjusting screw and hold it there.

5. Engage the topmost tooth possible on the starwheel with the brake adjusting spoon. Move the end of the adjusting spoon upward to move the adjusting screw starwheel downward and contract the adjusting screw. Back off the adjusting screw starwheel until the wheel spins freely with a minimum of drag. Keep track of the number of turns that the starwheel is backed off, or the number of strokes taken with the brake adjusting spoon.

6. Repeat this operation for the other side. When backing off the brakes on the other side, the starwheel adjuster must be backed off the same number of turns to prevent side-to-side brake pull.

7. Repeat this operation on the other set of brakes (front or rear).

8. When all 4 brakes are adjusted, on vehicles equipped with self-adjusting brakes, make several stops while backing the vehicle, to equalize the brakes at all of the wheels. On vehicles not equipped with self-adjusters, make a few low-speed stops while going forward to check for brake pull. If

the front end of the car has a tendency to pull to one side when the brakes are applied, back off the adjustment of the brake assembly on the side the vehicle pulls to.

9. Remove the safety stands and lower the vehicle. Road test the vehicle.

Master Cylinder

REMOVAL & INSTALLATION

1. With the engine turned off, push the brake pedal down to expel vacuum from the brake booster system.

2. Disconnect the hydraulic lines from the brake master cylinder.

3. Remove the brake booster-to-master cylinder retaining nuts and lock washers. Remove the master cylinder from the brake booster.

4. Before installing the master cylinder, check the distance from the outer end of the booster assembly push rod to the front face of the brake booster assembly. Turn the push rod adjusting screw in or out as required to obtain .931–.946 in. for Bronco, E100–250 and F100–250; .980–.995 in. for E & F350.

5. Position the master cylinder assembly over the booster push rod and onto the two studs on the booster assembly. Install the attaching nuts and lockwashers and tighten to 20–30 ft. lbs.

6. Loosely connect the hydraulic brake system lines to the master cylinder.

7. Bleed the hydraulic brake system. Centralize the differential valve. Then, fill the dual master cylinder reservoirs with DOT 3 brake fluid to within $\frac{1}{4}$ inch of the top. Install the gasket and reservoir cover.

Parking Brake

ADJUSTMENT

1979

PRE-TENSION PROCEDURE

NOTE: These procedures require a special tool available at most good auto supply dealers or your Ford dealer.

1. Depress the parking brake pedal until the parking brake control is in the second tooth (two notches or clicks).

2. Attach a Burroughs gauge, service tool No. BT–33–75 W2–25, or equivalent, to the LH rear cable and adjust the cable tension, registered on the gauge to 250 pounds, by tightening the equalizer nut. Hold for 5 minutes and release pedal.

3. Back off the equalizer nut until zero pounds of tension is registered on the gauge.

FINAL ADJUSTMENT

1. Position the parking brake pedal as outlined under pre-tension procedure.

2. Adjust the final tension to the mean 70 lbs. tension (50–90 lbs. specs) as reg-

istered on the Burroughs gauge by tightening the equalizer nut.

3. Remove the gauge and release the parking brake.

4. Check the clearance between the parking brake lever and the cam plate. The clearance should be 0.015 inch with the brakes fully released. If the clearance is not within specifications, readjust the parking brake cable.

5. Place the parking brake pedal in the fully released position, then check the slack in the parking brake two rear cables. The cables should be tight enough to provide full application of the rear brake shoes, when the parking brake lever or foot pedal is placed in the fully applied position, yet loose enough to ensure complete release of the brake shoes when the lever is in the released position.

1980 and Later

1. Make sure the brake drums are cold for correct adjustment.

2. Depress the parking brake pedal until the parking brake control is in the second tooth (two notches or two clicks).

3. Attach a Rotunda cable tension gauge (model 210018) or equivalent behind the equalizer assembly (either toward the right or left rear drum assembly).

4. Turn the equalizer adjusting nut until the tension reads 250 ft. lbs. as read on the cable tension gauge.

5. Back off the equalizer adjusting nut until the tension reads 50 ft. lbs. on the cable tension gauge.

6. For the final adjustment, retighten the equalizer adjusting nut until the tension reads between 60–100 ft. lbs. as read on the cable tension gauge.

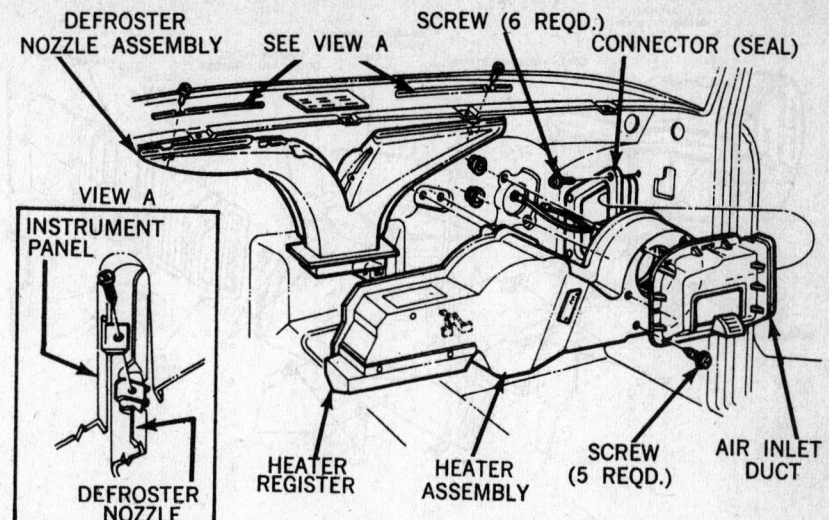

Heater installation–thru 1979

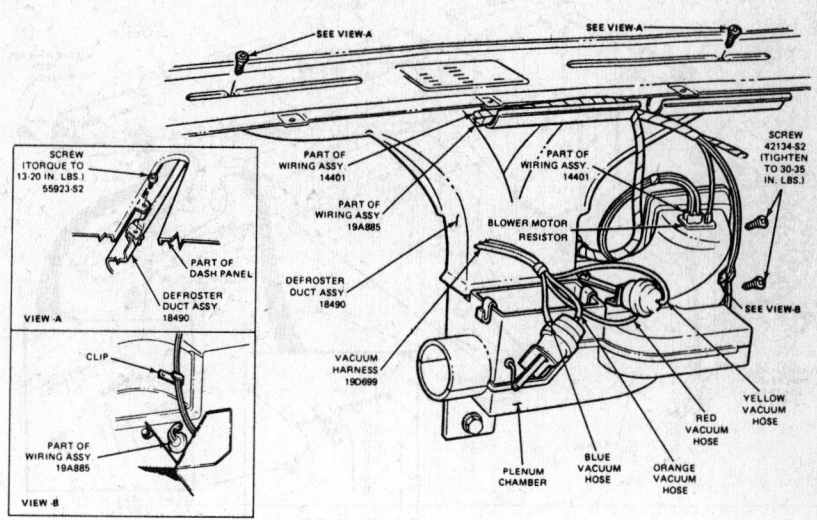

1979 plenum chamber and nozzle

CHASSIS ELECTRICAL

NOTE: Always disconnect the battery ground cable before working on electrical equipment.

Blower Motor

REMOVAL & INSTALLATION

Vans, w/o Air Conditioning

1. Disconnect the orange (orange with black strips) motor lead wire. Remove the ground wire screw from the firewall.

2. Disconnect the blower motor cooling tube.

3. Remove the four mounting plate screws and the motor assembly.

4. Reverse the procedure for installation.

Vans, w/Air Conditioning

1. Disconnect the resistor electrical leads on the front of the blower cover inside the truck.

2. Remove the blower cover.

3. Push the wiring grommet forward out of the housing hole.

4. Remove the blower motor mounting plate. Remove the blower motor.

5. Reverse the procedure for installation.

1979 Bronco and Pickups; Standard Heater, Without Air Conditioning

1. Disconnect the temperature and function control Bowden cables from the heater housing. This must be done to prevent damage to the cables.

2. Disconnect the wires from the blower resistor.

3. Remove five screws attaching the air inlet (vent) duct to the heater housing.

4. Disconnect the blower wires.

5. Drain the radiator and remove the heater hoses from the heater core.

6. Remove three heater stud retaining nuts and remove heater.

7. Remove gasket between the heater hose ends and the dash panel at core tubes.

8. Remove two screws and two nuts attaching blower to heater.

9. Remove blower fan from motor shaft, and remove motor from mounting plate.

10. Install new motor on mounting plate and install blower fan on motor shaft.

11. Install blower and motor in heater.

12. Position heater assembly in vehicle and install three stud retaining nuts.

13. Connect heater hoses to heater core and fill radiator.

14. Connect blower motor wires.

15. Place defroster nozzle on heater so that the defroster and heater openings are

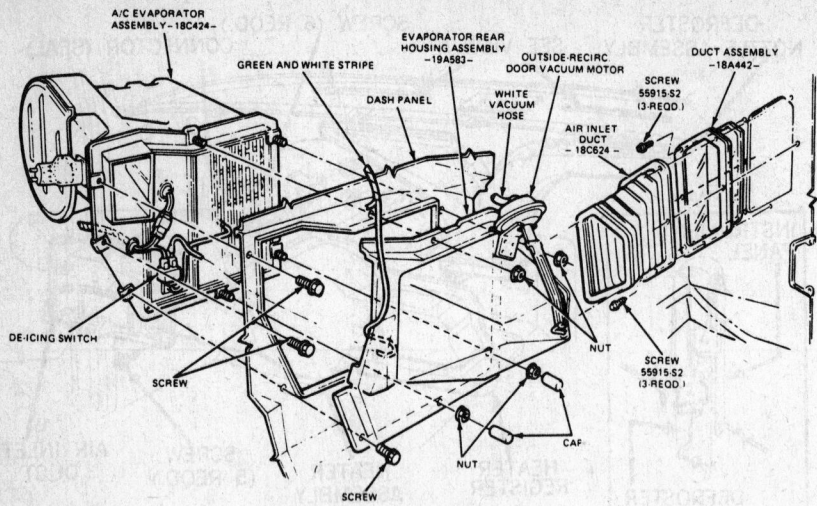

1979 evaporator rear housing

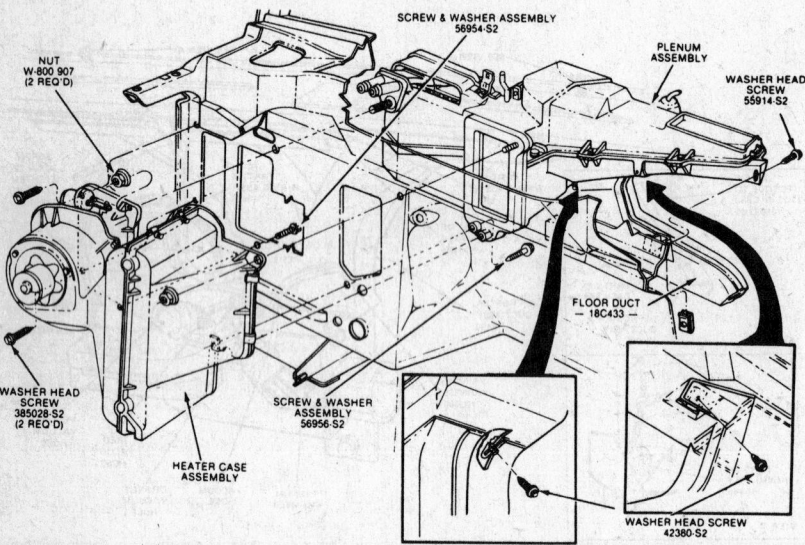

Typical heater case and plenum-from 1980

in the up position and there is no air leak around the seal.

16. Install air inlet (vent) duct to heater. Push duct firmly against seal on side cowl and tighten five attaching screws.

17. Connect wires to blower motor resistor.

18. Connect temperature and function control cables to heater, and adjust the cables.

19. Re-install gasket between the heater hose ends and the dash panel at core ends.

20. Fill cooling system.

1979 Bronco and Pickups; Deluxe Hi-Lo Heater, Without Air Conditioning

1. Disconnect the battery cable, remove the carburetor air cleaner and partially drain the coolant system.

2. Remove the heater hoses from the heater core.

3. Remove the glove box liner and remove the register duct by pulling from the instrument panel register and releasing the clip at the plenum.

4. Disconnect the right cowl outside air inlet vacuum hose from the outside-recirculating door vacuum motor.

5. Remove the rear housing from under the instrument panel. Remove the outside air inlet duct from the rear housing (4 nuts and 1 bolt) and install one upper nut to retain heater housing-to-dash after rear housing is removed.

6. Remove two screws retaining plenum-to-dash (above transmission tunnel) and two screws to heater housing and remove the plenum.

7. Install a piece of protective tape on "A" pillar inner cowl panel, at lower right corner of instrument panel.

8. Remove the lower right instrument panel-to-"A" pillar bolt and lower the center instrument panel brace, bolt and nut.

9. Position the instrument panel rear-

ward and install the "A" pillar bolt to hold the panel in the rearward position.

10. Remove the heater core (3 screws retaining 2 plates).

11. Remove the temperature blend door (snaps off).

12. Remove the temperature blend door arm support (2 screws) and pivot arm retainer (1 screw).

13. Remove blower motor (2 screws) and remove blower wheel.

14. Transfer blower to blower motor and panel assembly.

15. Install door arm pivot retainer (1 screw) and door arm support (2 screws).

16. Install the temperature blend door (snaps on).

17. Install heater core.

18. Install the plenum (4 screws).

19. Connect blower wires.

20. Remove heater housing upper retaining nut and install the heater outlet (4 nuts and 1 bolt). Position the air inlet duct.

21. Connect the white vacuum hose to the outside-recirculating door vacuum motor.

22. Reposition the instrument panel, install the retaining bolts and remove the protective tape at the "A" pillar inner cowl panel, lower right corner of instrument panel.

23. Install the right register duct assembly and install the glove box liner.

24. Connect heater hoses to the heater core assembly.

25. Fill cooling system, install the air cleaner and connect the battery cable to the battery.

26. Check blower motor operation.

1979 Bronco and Pickups With Air Conditioning

REMOVAL (WITHOUT DISCHARGING THE A/C SYSTEM)

1. Disconnect the battery cable, remove the carburetor air cleaner and partially drain the coolant system.

2. Remove the heater hoses from the heater core.

3. From under the hood, remove A/C hose support bracket from the cowl (one screw).

4. Remove the insulation tape from the expansion valve and sensing bulb. Then remove the cover plate and seal from the evaporator housing at the expansion valve (two screws).

5. Remove the glove box liner and remove the A/C duct by pulling from the instrument panel register and releasing the clip at the plenum.

6. Disconnect the right cowl fresh air inlet vacuum hose from the fresh air door vacuum motor.

7. Remove the evaporator rear housing from under the instrument panel. Then, remove the fresh air inlet tube from the evaporator rear housing (4 nuts and 1 bolt) and install one upper nut to retain evaporator housing-to-dash after rear housing is removed.

8. Disconnect wires from the de-icing switch and pull capillary tube out of evaporator core. Remove the de-icing switch mounting plate (four screws).

9. Remove two screws retaining plenum-to-dash (above transmission tunnel) and two screws to evaporator case and remove the plenum.

10. Install a piece of protective tape on "A" pillar inner cowl panel, at lower right corner of instrument panel.

11. Then, remove the lower right instrument panel-to-"A" pillar bolt and lower the center instrument panel brace, bolt and nut.

12. Position the instrument panel rearward and install the "A" pillar bolt to hold the panel in the rearward position.

13. Remove four evaporator retaining screws.

14. Position the evaporator away from the case and secure it rearward and upward. Remove evaporator sealing grommet.

15. Remove heater core (3 screws retaining 2 plates).

16. Remove A/C-heat door (snaps off).

17. Remove A/C-heat door arm support (2 screws) and pivot arm retainer (1 screw).

18. Remove blower motor (2 screws) and remove blower wheel.

INSTALLATION

1. Transfer blower wheel to blower motor and panel assembly.

2. Install door arm pivot retainer (1 screw) and door arm support (2 screws).

3. Install A/C-heat door (snaps on).

4. Install heater core.

5. Remove the retainer that held the evaporator away from the case, install evaporator and tube sealing grommet.

6. Install the plenum (4 screws).

7. Install the de-icing switch mounting plate, install de-icing switch capillary tube back into evaporator core and position blower wire grommet.

8. Connect blower and de-icing switch wires.

9. Remove upper evaporator case retaining nut and install the evaporator outlet (4 nuts and 1 bolt). Then, position the air inlet bellows.

10. Connect the right cowl fresh air inlet vacuum hose to the fresh air door vacuum motor.

11. Reposition the instrument panel, install the retaining bolts and remove the protective tape at the "A" pillar inner cowl panel, lower right corner of instrument panel.

12. Install the right A/C duct assembly and install the glove box liner.

13. Install seal and cover plate to the evaporator case at the expansion valve.

14. Install insulation tape over the expansion valve and sensing bulb.

15. Install the A/C hose support bracket-to-cowl.

16. Connect heater hoses to the heater core assembly.

17. Fill cooling system, install the carburetor air cleaner and connect the battery cable to the battery.

18. Check blower motor operation.

1980 and Later, Bronco and Pickups, Comfort Vent Heaters, Without Air Conditioning

1. Disconnect the motor wires at the hard shell connectors.

2. Disconnect the blower motor air cooling tube from the motor.

3. Remove four blower motor mounting plate attaching screws and remove the motor and wheel assembly from the blower housing.

4. Remove the hub clamp spring from the blower wheel hub and the retainer from the motor shaft. Then, remove the blower wheel from the motor shaft.

5. Position the blower wheel on the blower motor shaft. Then, install a new hub clamp spring on the blower hub as shown. The hub clamp spring is included with a new blower wheel but not with the blower motor.

6. Install a new flange gasket on the blower motor flange.

7. Position the blower motor and wheel assembly in the blower housing and install the four attaching screws. The wire clamp should be installed under the screw closest to the resistor assembly.

8. Cement the blower motor air tube on the nipple of the blower housing with RTV silicone adhesive.

9. Connect the blower motor wires at the hard shell connectors.

10. Check the blower motor for proper operation.

1980 and Later Bronco and Pickups Standard & High Output Heaters, Without Air Conditioning

1. Disconnect the motor wire at the hard shell connector and the ground wire at the ground screw.

2. Remove four screws attaching the blower motor and wheel to the heater case.

3. Remove the blower motor and wheel from the heater case.

4. Remove the blower wheel hub clamp spring and the tab lock washer from the motor shaft. Then, pull the blower wheel from the motor shaft.

5. Install the blower wheel on the blower motor shaft.

6. Install the hub clamp spring on the blower hub.

7. Position the blower motor and wheel to the heater case, and install the four attaching screws.

8. Connect the blower motor wires and check the blower motor for proper operation.

1980 and Later Bronco and Pickups With Air Conditioning

REMOVAL (WITHOUT DISCHARGING THE A/C SYSTEM)

1. Disconnect the motor wires at the hard shell connectors.

2. Disconnect the blower motor air cooling tube from the motor.

3. Remove four blower motor mounting plate attaching screws and remove the motor and wheel assembly from the blower housing.

4. Remove the hub clamp spring from the blower wheel hub and the retainer from the motor shaft. Then, remove the blower wheel from the motor shaft.

INSTALLATION

1. Position the blower wheel on the blower motor shaft to the dimension shown. Then, install a new hub clamp spring on the blower hub as shown. The hub clamp spring is included with a new blower wheel but not with the blower motor.

2. Install a new flange gasket on the blower motor flange.

3. Position the blower motor and wheel assembly in the blower housing and install the four attaching screws. The wire clamp should be installed under the screw closest to the resistor assembly.

4. Cement the blower motor air tube on the nipple of the blower housing with RTV silicone adhesive.

5. Connect the blower motor wires at the hard shell connectors.

6. Check the blower motor for proper operation.

Ranger and Bronco II w/o A/C

1. Remove the nut from the bottom of the plenum assembly just to the right of the heater core access cover.

2. Open the hood and disconnect the electrical harness from the blower motor by pushing on the connector tab and pulling off the connector.

3. Remove the mounting bolt from the upper right side of firewall. Remove the mounting nuts from the heater case assembly.

4. Pull the blower motor assembly from the firewall case.

5. Install in the reverse order.

Ranger and Bronco II with A/C

1. Open the hood and disconnect the heater blower wiring harness from the blower by pushing down on the locking tab and pulling the connector from the motor.

2. Remove the emission control, if equipped, from the front of the blower case. Three bolts hold the "box" in position.

3. Disconnect the blower motor cooling tube. Remove the blower plate mounting screws and remove the blower motor from the case on the firewall.

4. Install in the reverse order.

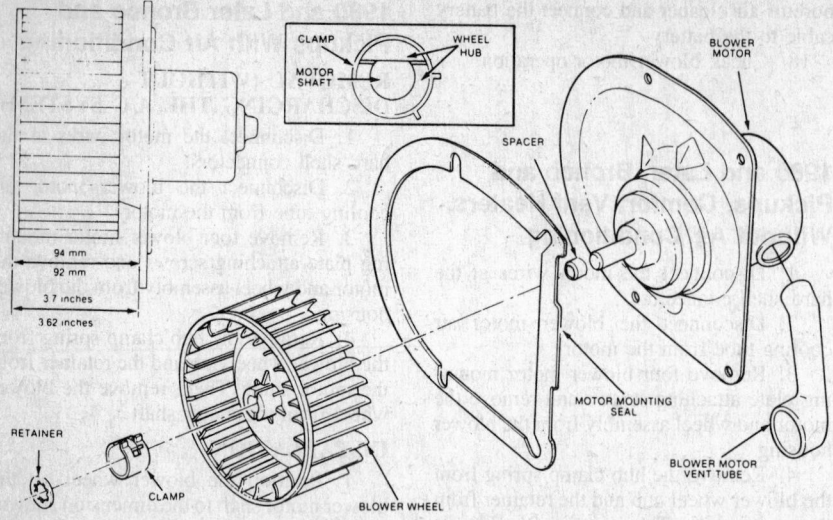

Typical blower motor and wheel with A/C-from 1980

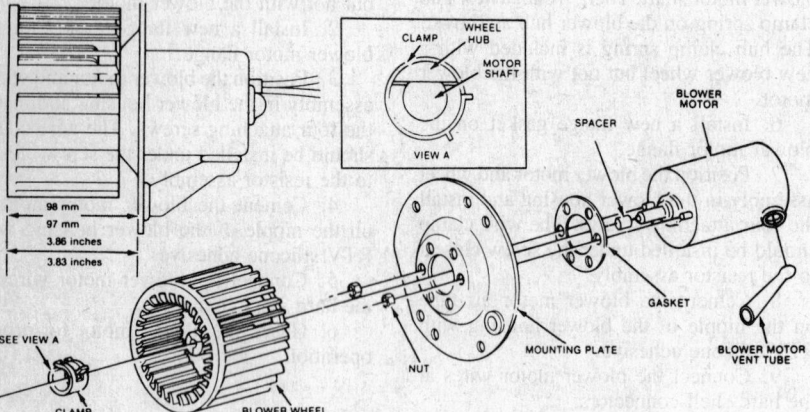

Typical blower motor and wheel without A/C-from 1980

Heater Core

REMOVAL & INSTALLATION

Vans Without Air Conditioning

1. Drain the coolant; remove the battery.
2. Disconnect the resistor wiring harness and the orange blower motor lead. Remove the ground wire screw from the firewall.
3. Detach the heater hoses and the plastic hose retaining strap.
4. Remove the five mounting screws inside the truck.
5. Remove the heater assembly.
6. Cut the seal at the top and bottom edge of the core retainer. Remove the two screws and the retainer. Slide the core and seal out of the case.
7. Reverse the procedure for installation.

Vans With Air Conditioning

1. Disconnect the resistor electrical leads

on the front of the blower cover inside the truck. Detach the vacuum line from the vacuum motor. Remove the blower cover.
2. Remove the nut and push washer from the air door shaft. Remove the control cable from the bracket and the air door shaft.
3. Remove the blower motor housing and the air door housing.
4. Drain the coolant and detach the heater hoses.
5. Remove the heater core retaining brackets. Remove the core and seal assembly.
6. Reverse the procedure for installation.

1979 Bronco and Pickups Without Air Conditioning

1. Disconnect the temperature and function control Bowden cables from the heater housing. This must be done to prevent damage to the cables.
2. Disconnect the wires from the blower resistor.
3. Remove five screws attaching the air inlet (vent) duct to the heater housing.

4. Disconnect the blower wires.
5. Drain the radiator and remove the heater hoses from the heater core.
6. Remove three heater stud retaining nuts and remove heater.
7. Remove gasket between the heater hose ends and the dash panel at core tubes.
8. Remove heater core cover and gasket (four screws).
9. Pull heater core and lower support from heater.
10. Install foam gaskets on heater core and install in heater assembly.
11. Install the core seal and cover plate.
12. Position heater assembly in vehicle and install three stud retaining nuts.
13. Connect heater hoses to heater core and fill radiator.
14. Connect blower motor wires.
15. Place defroster nozzle on heater so that the defroster and heater openings are in the up position and there is no air leak around the seal.
16. Install air inlet (vent) duct to heater. Push duct firmly against seal on side cowl and tighten five attaching screws.
17. Connect wires to blower motor resistor.
18. Connect temperature and function control cables to heater, and adjust the cables.
19. Re-install gasket between the heater hose ends and the dash panel at core ends.
20. Fill the cooling system.

1979 Bronco and Pickups, Deluxe Hi-Lo Heater, Without Air Conditioning

1. Disconnect the battery cable, remove the carburetor air cleaner and partially drain the coolant system.
2. Remove the heater hoses from the heater core.
3. Remove the glove box liner and remove the register duct by pulling from the instrument panel register and releasing the clip at the plenum.
4. Disconnect the right cowl outside air inlet vacuum hose from the outside-recirculating door vacuum motor.
5. Remove the rear housing from under the instrument panel. Remove the outside air inlet duct from the rear housing (4 nuts and 1 bolt) and install one upper nut to retain heater housing-to-dash after rear housing is removed.
6. Remove two screws retaining plenum-to-dash (above transmission tunnel) and two screws to heater housing and remove the plenum.
7. Install a piece of protective tape on "A" pillar inner cowl panel, at lower right corner of instrument panel.
8. Remove the lower right instrument panel-to-"A" pillar bolt and lower the center instrument panel brace, bolt and nut.
9. Position the instrument panel rearward and install the "A" pillar bolt to hold the panel in the rearward position.

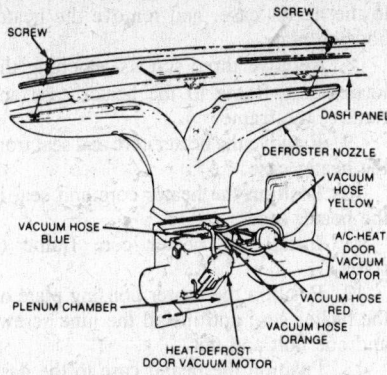

1979 A/C-heater control connections

10. Remove heater core (3 screws retaining 2 plates).

11. Remove the temperature blend door (snaps off).

12. Remove the temperature blend door arm support (2 screws) and pivot arm retainer (1 screw).

13. Remove blower motor (2 screws) and remove blower wheel.

14. Transfer blower wheel to blower motor and panel assembly.

15. Install door arm pivot retainer (1 screw) and door arm support (2 screws).

16. Install the temperature blend door (snaps on).

17. Install heater core.

18. Install the plenum (4 screws).

19. Connect blower wires.

20. Remove heater housing upper retaining nut and install the heater outlet (4 nuts and 1 bolt). Position the air inlet duct.

21. Connect the white vacuum hose to the outside-recirculating door vacuum motor.

22. Reposition the instrument panel, install the retaining bolts and remove the protective tape at the "A" pillar inner cowl panel, lower right corner of instrument panel.

23. Install the right register duct assembly and install the glove box liner.

24. Connect heater hoses to the heater core assembly.

25. Fill cooling system, install the air cleaner and connect the battery cable to the battery.

26. Check bower motor operation.

1979 Bronco and Pickups With Air Conditioning

REMOVAL (WITHOUT DISCHARGING A/C SYSTEM)

1. Disconnect the battery cable, remove the carburetor air cleaner and partially drain the coolant system.

2. Remove the heater hoses from the heater core.

3. From under the hood, remove A/C hose support bracket from the cowl (one screw).

4. Remove the insulation tape from the expansion valve and sensing bulb. Then remove the cover plate and seal from the

evaporator housing at the expansion valve (two screws).

5. Remove the glove box liner and remove the A/C duct by pulling from the instrument panel register and releasing the clip at the plenum.

6. Disconnect the right cowl fresh air inlet vacuum hose from the fresh air door vacuum motor.

7. Remove the evaporator rear housing from under the instrument panel. Then, remove the fresh air inlet tube from the evaporator rear housing (4 nuts and 1 bolt) and install one upper nut to retain evaporator housing-to-dash after rear housing is removed.

8. Disconnect wires from the de-icing switch and pull capillary tube out of evaporator core. Remove the de-icing switch mounting plate (four screws).

9. Remove two screws retaining plenum-to-dash (above transmission tunnel) and two screws to evaporator case and remove the plenum.

10. Install a piece of protective tape on "A" pillar inner cowl panel, at lower right corner of instrument panel.

11. Then, remove the lower right instrument panel-to-"A" pillar bolt and lower the center instrument panel brace, bolt and nut.

12. Position the instrument panel and rearward and install the "A" pillar bolt to hold the panel in the rearward position.

13. Remove four evaporator retaining screws.

14. Position the evaporator away from the case and secure it rearward and upward. Remove evaporator sealing grommet.

15. Remove heater core (3 screws retaining 2 plates).

16. Remove A/C-heat door (snaps off).

17. Remove A/C-heat door arm support (2 screws) and pivot arm retainer (1 screw).

INSTALLATION

1. Install door arm pivot retainer (1 screw) and door arm support (2 screws).

2. Install A/C-heat door (snaps on).

3. Install heater core.

4. Remove the retainer that held the evaporator away from the case, install evaporator and tube sealing grommet.

5. Install the plenum (4 screws).

6. Install the de-icing switch mounting plate, install de-icing switch capillary tube back into evaporator core and position blower wire grommet.

7. Connect blower and de-icing switch wires.

8. Remove upper evaporator case retaining nut and install the evaporator outlet (4 nuts and 1 bolt). Then, position the air inlet bellows.

9. Connect the right cowl fresh air inlet vacuum hose to the fresh air door vacuum motor.

10. Reposition the instrument panel, install the retaining bolts and remove the protective tape at the "A" pillar inner cowl panel, lower right corner of instrument panel.

11. Install the right A/C duct assembly and install the glove box liner.

12. Install seal and cover plate to the evaporator case at the expansion valve.

13. Install insulation tape over the expansion valve and sensing bulb.

14. Install the A/C hose support bracket-to-cowl.

15. Connect heater hoses to the heater core assembly.

16. Fill cooling system, install the carburetor air cleaner and connect the battery cable to the battery.

17. Check blower motor operation.

1980 and Later Bronco and Pickups, Comfort Vent Heaters, Without Air Conditioning

1. Disconnect the heater hoses from the heater core tubes and plug the hoses with suitable ⅝ inch plugs.

2. Remove the glove compartment liner.

3. Remove two spring clips attaching the heater core cover to the plenum along the top edge of the heater core cover.

4. Remove eight screws attaching the heater core cover to the plenum and remove the cover.

5. Remove the heater core from the plenum taking care not to spill coolant from the core.

6. Install the heater core in the plenum.

7. Install the heater core cover (eight (8) screws and two spring clips along the top edge of the cover).

8. Install the glove compartment liner.

9. Connect the heater hoses to the heater core. Tighten the hose clamps.

10. Add coolant to raise the coolant level to specification.

11. Check the system for proper operation and for coolant leaks.

1980 and Later Bronco and Pickups, Standard & High Output Heaters, Without Air Conditioning

1. Disconnect the temperature cable from the temperature blend door and the mounting bracket on top of the heater case.

2. Disconnect the wires from the blower motor resistor and the blower motor.

3. Disconnect the heater hoses from the heater core and plug the hoses with suitable ⅝ inch plugs.

4. Working under the instrument panel, remove two nuts retaining the left end of the heater case and the right end of the plenum to the dash panel.

5. In the engine compartment, remove one screw attaching the top center of the heater case to the dash panel.

6. Remove two screws attaching the right end of the heater case to the dash panel, and remove the heater case from the vehicle.

7. Remove nine screws and one (1) bolt and nut attaching the heater housing plate

FORD

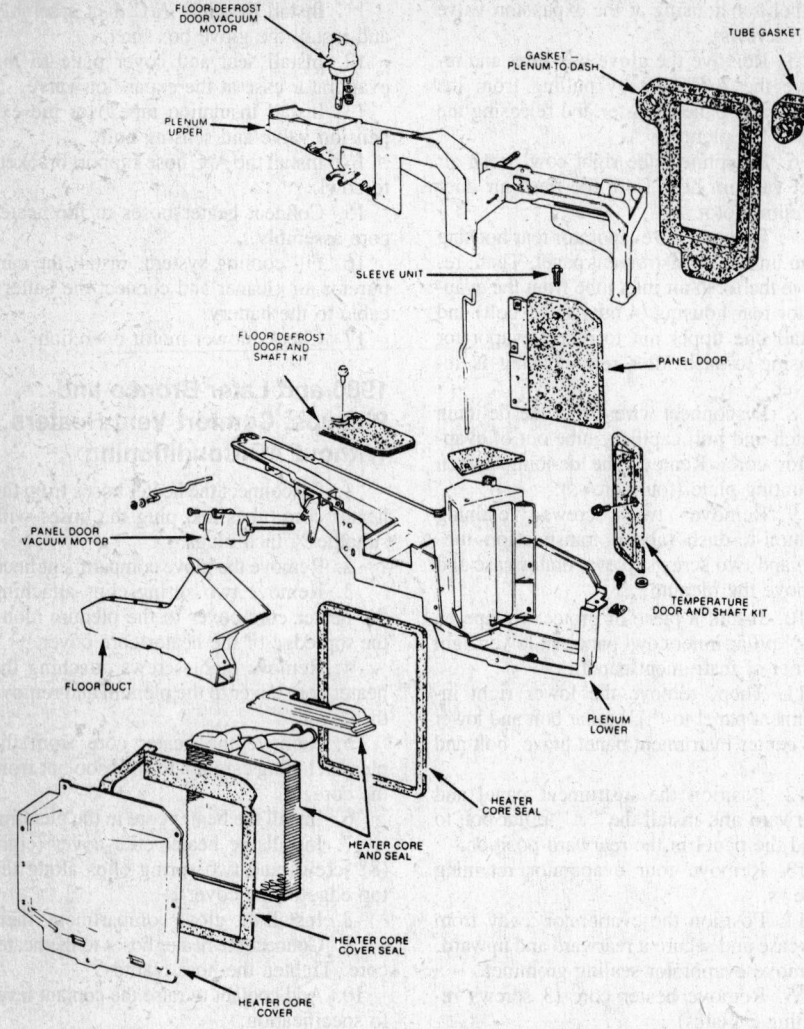

Typical comfort heater plenum-from 1980

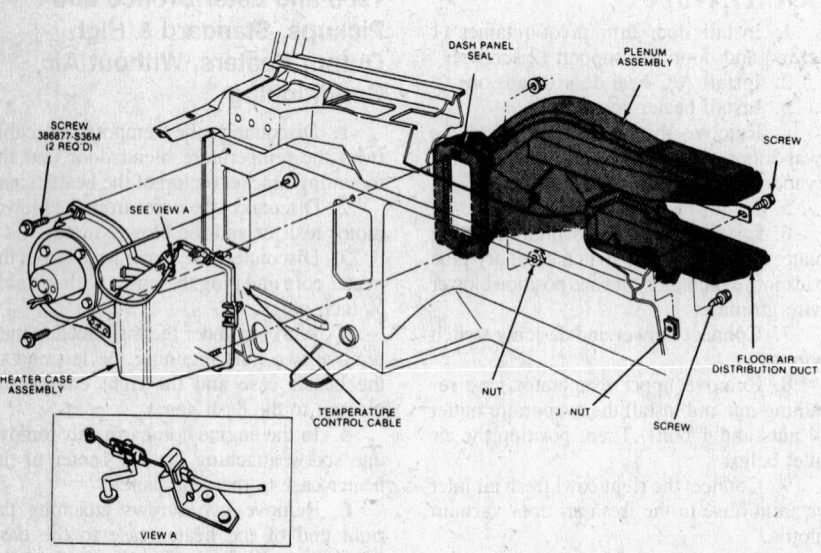

Heater case and plenum assemblies on 1980 and later standard and high output units without A/C

T312

to the heater case, and remove the heater housing plate.

8. Remove three screws attaching the heater core frame to the heater case and remove the frame.

9. Remove the heater core and seal from the heater case.

10. Position the heater core and seal in the heater case.

11. Install the heater core frame (3 screws).

12. Position the heater housing plate on the heater case and install the nine screws and one bolt and nut.

13. Position the heater case to the dash panel and install the three attaching screws.

14. Working in the passenger compartment, install two nuts to retain the heater case and plenum right end to the dash panel.

15. Connect the heater hoses to the heater core. Tighten the hose clamps.

16. Connect the wires to the blower motor resistor assembly.

17. Connect the blower motor wires.

18. Position (slide) the self-adjusting clip on the temperature cable to a position approximately one inch from the cable end loop.

19. Snap the temperature cable on the cable mounting bracket of the heater case. Then, position the self-adjusting clip on the door crank arm.

20. Adjust the temperature cable.

21. Check the system for proper operation.

1980 and Later Bronco and Pickups With Air Conditioning

REMOVAL (WITHOUT DISCHARGING THE A/C SYSTEM)

1. Disconnect the heater hoses from the heater core tubes and plug the hoses with suitable $5/8$ inch plugs.

2. Remove the glove compartment liner.

3. Remove eight screws attaching the heater core cover to the plenum and remove the cover.

4. Remove the heater core from the plenum taking care not to spill coolant from the core.

INSTALLATION

1. Install the heater core in the plenum.

2. Install the heater core cover (eight screws).

3. Install the glove compartment liner.

4. Connect the heater hoses to the heater core. Tighten the hose clamps.

5. Add coolant to raise the coolant level to specification.

Ranger and Bronco II w/o A/C

1. Allow the engine to cool down completely. Drain the cooling system to a point that is below the heater hoses.

2. Disconnect the heater hoses from the heater core tubes. Plug the core tubes.

3. From under the dash, remove the screws that attach the access cover to the plenum assembly. Remove the access cover.

4. Pull the core down and out of the plenum assembly.

5. Install in the reverse order. Fill cooling system, start the engine and check for leaks.

Ranger and Bronco II with A/C

1. Refer to the without A/C section. Procedure is the same.

Auxiliary Heater Case (With or Without A/C)

VANS
Removal & Installation

1. Remove the first bench seat (if so equipped).

2. Remove the auxiliary heater and/or air conditioning cover assembly attaching screws and remove the cover.

3. Position the cover assembly to the body side panel and install the attaching screws.

4. Install the bench seat (if removed) and tighten the retaining bolts 25–45 ft. lbs.

Auxiliary Heater Core and Seal Assembly

VANS
Removal & Installation

1. Remove the first bench seat (if so equipped).

2. Remove auxiliary heater and/or air conditioning cover attaching screws (15) and remove the cover.

3. Partially drain the engine coolant from the coolant system.

4. Remove the heater hoses from the auxiliary heater core assembly (2 clamps).

5. Pull the wiring assembly away from the heater core seal.

6. Slide the heater core and seal assembly out of the housing slot.

7. Slide the heater core and seal assembly into the housing slot (position the wiring to one side).

8. Install the heater hoses to the heater core assembly (2 clamps).

9. Fill the cooling system to specification.

10. Position the cover assembly to the body side panel and install the attaching screws (15).

11. Install the bench seat (if removed) and tighten the retaining bolts 25–45 ft. lbs.

RADIO

REMOVAL & INSTALLATION
Vans

1. Detach the battery ground cable.

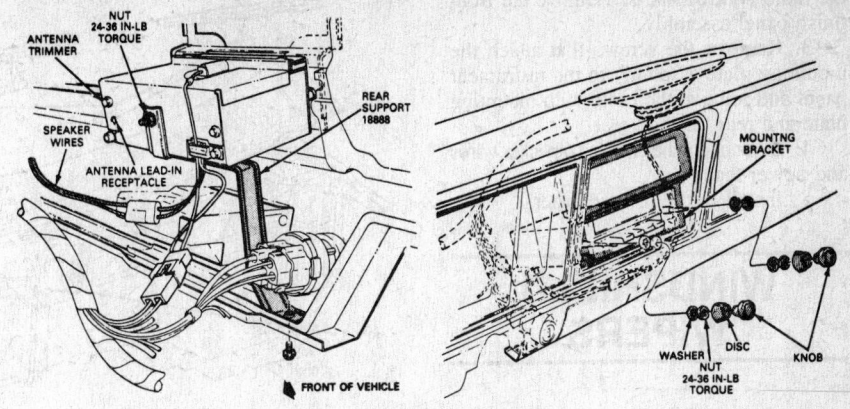

Typical van radio installation

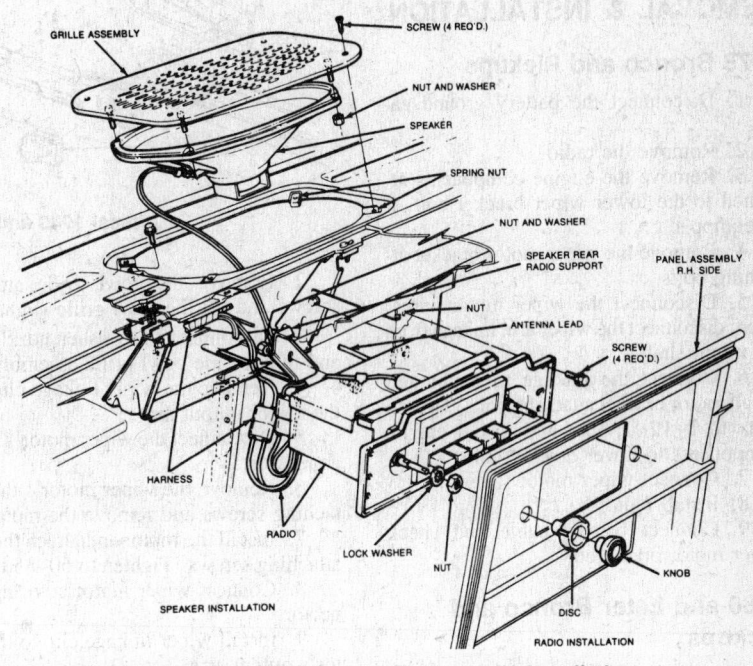

Typical pick-up, 1978 and later Bronco radio installation

2. Remove the heater and A/C control knobs. Remove the lighter.

3. Remove the radio knobs and discs.

4. If the truck has a lighter, snap out the name plate at the right side to remove the panel attaching screw.

5. Remove the five finish panel screws.

6. Very carefully pry out the cluster panel in two places.

7. Detach the antenna lead and speaker wires.

8. Remove the two nuts and washers and the mounting plate.

9. Remove the four front radio attaching screws. Remove the rear support nut and washer, and remove the radio.

10. Reverse the procedure for installation.

Bronco and Pickups

1. Disconnect the battery ground cable.

2. On 1979 models, remove the ash tray and bracket.

3. Disconnect the antenna, speakers and radio lead.

4. Remove the bolt attaching the radio rear support to the lower edge of the instrument panel.

5. Remove the knobs and discs from the radio control shafts.

6. Remove the retaining nuts from the control shafts and remove the bezel.

7. Remove the nuts and washers from the control shafts and remove the radio from the panel.

8. Installation is the reverse of removal.

Ranger and Bronco II

1. Disconnect the negative battery cable.

2. Remove the knobs and discs from

the radio control shafts. Remove the front finish panel assembly.

3. Remove the screws that attach the mounting plate assembly to the instrument panel and remove the radio with mounting plate and rear bracket.

4. Disconnect the antenna, speaker wires and power lead.

5. Install in the reverse order.

WINDSHIELD WIPERS

Motor

REMOVAL & INSTALLATION

1979 Bronco and Pickups

1. Disconnect the battery ground cable.

2. Remove the radio.

3. Remove the engine components attached to the lower wiper bracket bolt, if so equipped.

4. Remove the wiper motor bracket attaching bolts.

5. Disconnect the wiper motor wires. Then, disconnect the wiper arm linkage from the motor shaft.

6. Connect the linkage to motor and install motor bracket attaching bolts. Tighten bolts to 8–12 ft. lbs. and install engine components to lower bracket bolts.

7. Connect wiper motor wires.

8. Install radio.

9. Connect battery cable and check wiper motor operation.

1980 and Later Bronco and Pickups

1. Disconnect the battery ground cable.

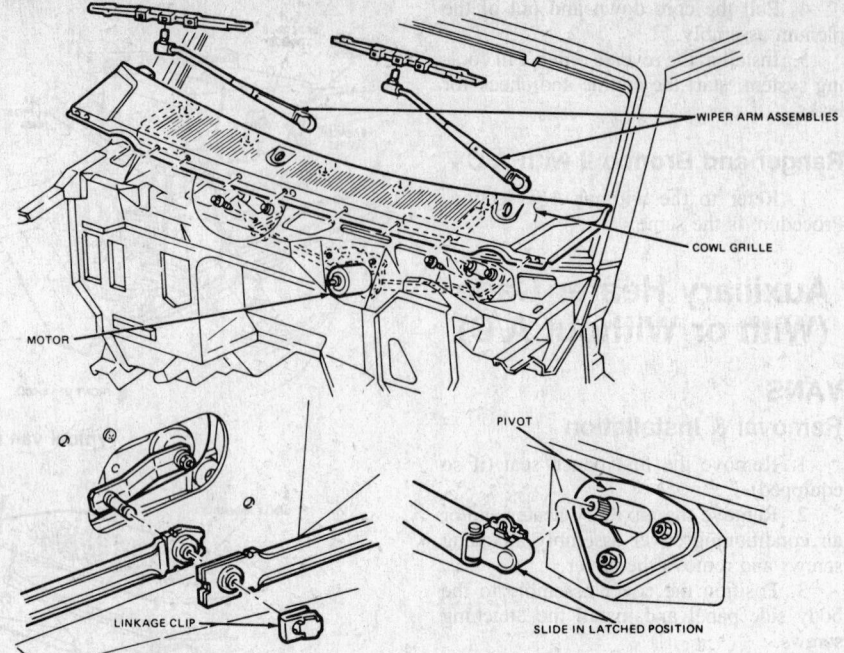

Typical 1980 and later wiper motor installation

2. Remove the cowl grille attaching screws and lift the cowl grille slightly.

3. Disconnect the washer nozzle hose and remove the cowl grille assembly.

4. Remove the wiper linkage clip from the motor output arm.

5. Disconnect the wiper motor's wiring connector.

6. Remove the wiper motor's three attaching screws and remove the motor.

7. Install the motor and attach the three attaching screws. Tighten to 60–85 in. lbs.

8. Connect wiper motor's wiring connector.

9. Install wiper linkage clip to the motor's output arm.

10. Connect the washer nozzle hose and install the cowl assembly and attaching screws.

11. Install both wiper arm assemblies.

12. Connect battery ground cable.

Vans

1. Disconnect the battery ground cable. Remove the fuse panel and bracket.

2. Disconnect the motor wires.

3. Remove the arms and blades.

4. Remove the outer air inlet cowl. Take off the motor linkage clip.

5. Unbolt and remove the motor.

6. Reverse the procedure for installation.

Ranger and Bronco II

1. Turn on the ignition and wiper switches. When the wiper blades are straight up and down on the windshield, turn off the ignition switch so that the wiper arms and blades remain in position.

2. Disconnect the negative battery cable.

3. Remove the wiper arm and blade from the right side. Remove the pivot nut from the wiper arm drive and allow the pivot to drop into the cowl.

4. Remove the access cover from the right side of the firewall. Reach in and disconnect the linkage from the wiper motor drive arm.

5. Disconnect the wiring at the wiper motor. Remove the wiper motor mounting bolts and remove the motor.

6. Install in the reverse order. Make sure the linkage lock clips are snapped in position.

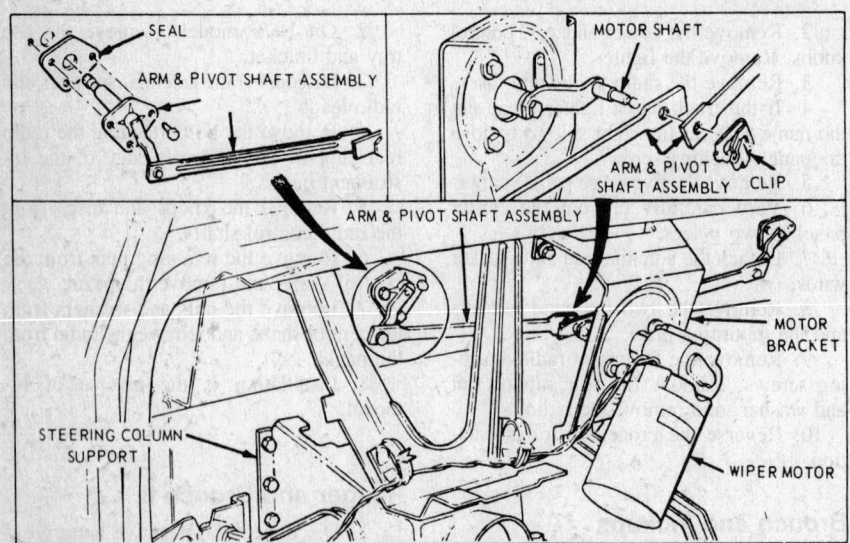

Typical wiper motor installation, except Bronco—through 1979

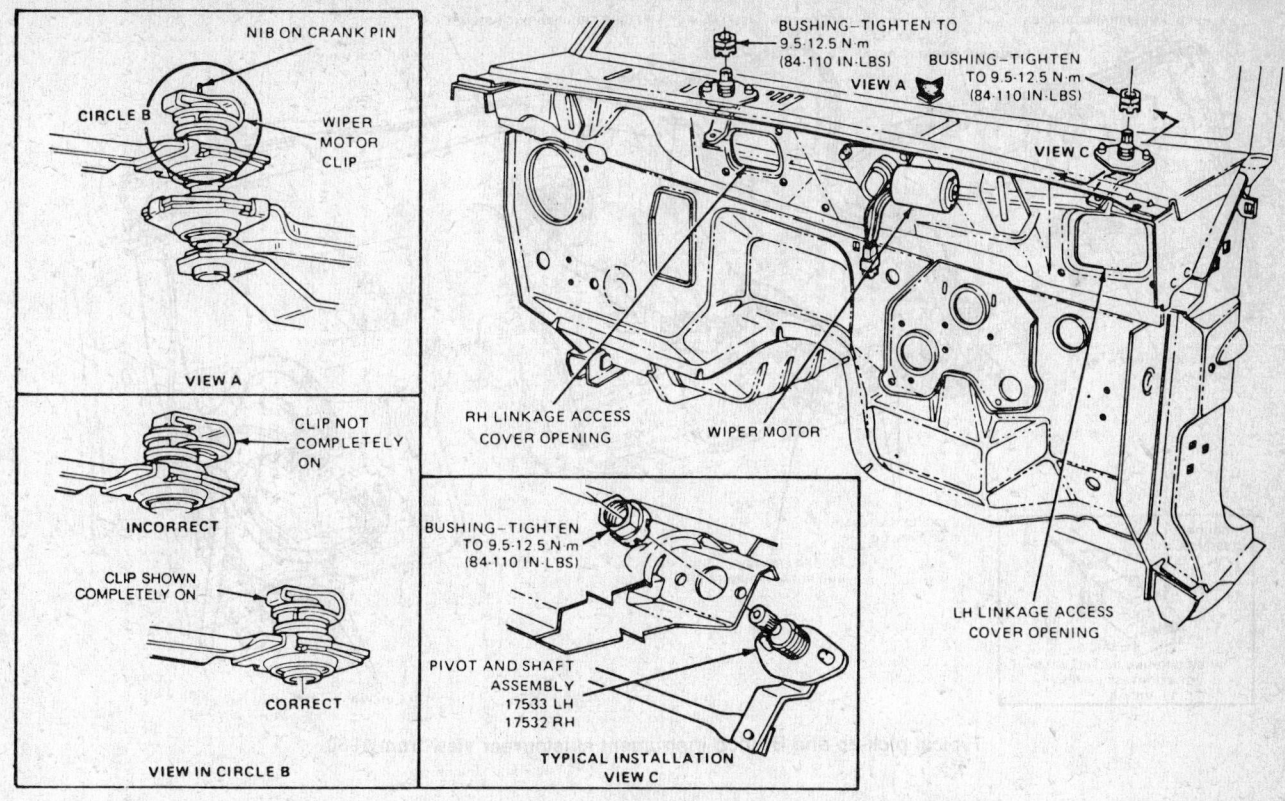

Wiper linkage and motor removal/installation- Ranger/Bronco II

Linkage

REMOVAL & INSTALLATION

Vans

1. Disconnect the battery ground cable.

2. Remove the wiper blades and arms. Detach the washer hoses.

3. Remove the cowl grille.

4. Remove the linkage clips. Remove the pivot to cowl screws and remove the assembly.

1979 Bronco and Pickups

1. Open the hood and disconnect the battery ground cable. Remove the arm and blade assemblies from the pivot shafts.

2. Reach under the instrument panel and disconnect the speedometer cable from the rear of the instrument cluster.

3. Remove the instrument cluster bezel.

4. Loosen the three bolts retaining the wiper motor bracket to the cowl. This will allow access between the cowl panel and the link assembly.

5. Remove the clip retaining the motor drive arm to the link assemblies.

6. Through the cluster bezel opening, remove the retaining bolts from the left pivot assembly. Remove the left pivot and link assembly from under the instrument panel.

7. Remove the glove box assembly.

8. Remove the three bolts retaining the

right pivot and link assembly to the cowl panel.

9. Disconnect the right link assembly from the drive arm and remove the right pivot and link assembly.

10. Place gaskets on the pivot shafts and position the shafts to the cowl panel and install the retaining bolts.

11. Install the glove box assembly.

12. Position the link assemblies to the motor drive arm and install the retaining clip.

13. Tighten the bolts retaining the motor bracket to the cowl and then re-install engine components to lower bracket bolt.

14. Install the wiper arm and blade assemblies.

15. Position and install the instrument cluster bezel.

16. Connect the speedometer cable.

17. Connect the battery ground cable and close the hood and check the operation of the wipers.

1980 and Later Bronco, Pickups

1. Disconnect the battery ground cable.

2. Remove both wiper arm assemblies.

3. Remove the cowl grille attaching screws and lift the cowl grille slightly.

4. Disconnect the washer nozzle hose and remove the cowl grille assembly.

5. Remove the wiper linkage clip from the motor output arm and pull the linkage from the output arm.

6. Remove the pivot body to cowl

screws and remove the linkage and pivot shaft assembly (three screws on each side). The left and right pivots and linkage are independent and can be serviced separately.

7. Attach the linkage and pivot shaft assembly to cowl with attaching screws.

8. Replace the linkage to the output arm and attach the linkage clip.

9. Connect the washer nozzle hose and cowl grills assembly.

10. Attach cowl grille attaching screws.

11. Replace both wiper arm assemblies.

12. Connect battery ground cable.

Ranger and Bronco II

1. Refer to the wiper motor removal section and follow procedure until the linkage has been disconnected from the wiper motor.

2. Slide the right side linkage and pivot out through the access hole.

3. Remove the left side wiper arm and blade assembly.

4. Remove the left side access cover. Remove the pivot nut and slide the linkage and pivot through the access hole.

5. Install in the reverse order.

Wiper Arm Assembly

REPLACEMENT

1979

Bend the arm backwards at the joint next to the pivot. Now, pull the arm straight off

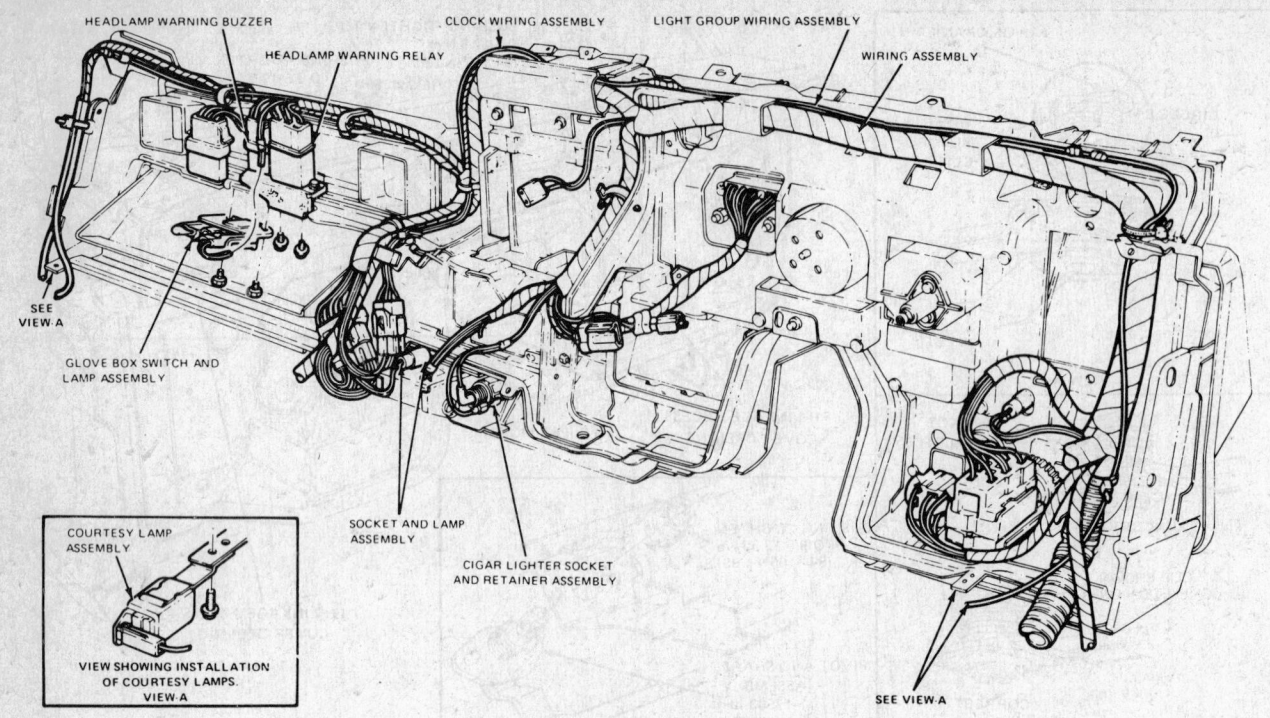

Typical pick-up and Bronco instrument cluster rear view-from 1980

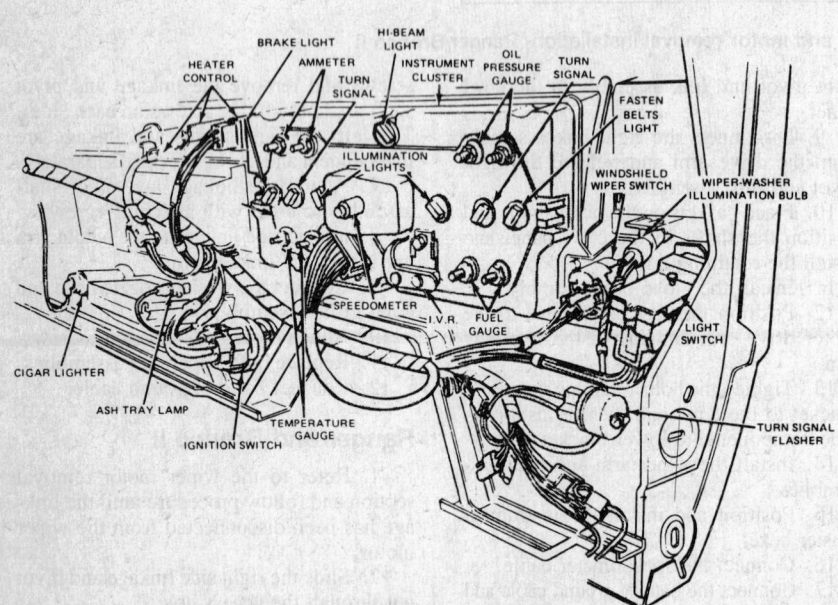

Bronco and pickup instrument panel—rear view

the splined pivot shaft. To replace the arm, hold it in the bent position and slide it on the pivot.

1980 and Later

Raise the blade end of the arm off of the windshield and move the slide latch away from the pivot shaft. This will unlock the wiper arm from the pivot shaft and hold the blade end of the arm off of the glass at the same time. The wiper arm can now be pulled off of the pivot shaft without the aid of any tools.

Blade Assembly to Wiper Arm

REPLACEMENT

1979

Wiper blades are used from two different manufacturers. Trico and Anco blades come in two types. With a bayonet type, the blade saddle slides over the end of the arm and is engaged by a locking stud. With the bot-

tom type, a screw and nut is used to retain the blade on the arm.

BAYONET TYPE

To remove a Trico type blade, press down on the arm to unlatch the top stud. Depress the tab on the saddle and pull the blade from the arm. To remove an Anco type blade, press inward on the tab and pull the blade from the arm. To install a new blade assembly, slide the blade saddle over the end of the wiper arm so that the locking stud snaps into place.

SIDE SADDLE PIN TYPE

To remove a pin type Trico-type blade, insert an appropriate tool into the spring release opening of the blade saddle, depress the spring clip and pull the blade from the arm. To install, push the blade saddle on to the pin, so that the spring clip engages the pin. Be sure the blade is securely attached to the arm.

From 1980

1. Cycle arm and blade assembly to a position on the windshield where removal of blade assembly can be performed without difficulty. Turn ignition key off at desired position.

2. With blade assembly resting on windshield, grasp either end of the wiper blade frame and pull away from windshield, then pull blade assembly from pin.

NOTE: Rubber element extends past frame. To prevent damage to the blade element, be sure to grasp blade frame and not the end of the blade element.

3. To install, push blade assembly onto pin until fully seated. Be sure blade is securely attached to the wiper arm.

INSTRUMENT CLUSTER

REMOVAL & INSTALLATION

1979 Bronco and Pickups

1. Disconnect the battery ground cable.

2. Remove the radio knobs from the radio shafts (if so equipped).

3. Remove the fuel gauge switch knob (if so equipped), heater control knobs and wiper-washer knob. Use a hook-shaped tool to release each knob lock tab.

4. Remove the knob and shaft from the light switch.

5. Remove one nut and washer from each radio control shaft, and remove the radio bezel.

6. Remove the cluster trim cover. The attaching screws are located as follows: four screws along top of bezel; one screw between the lights and wiper-washer switch, and two screws below the radio. Then, disconnect the A/C duct (if so equipped), and illumination light from the bezel. The illumination light is located between the lights and wiper-washer switches. Remove four cluster attaching screws, disconnect the speedometer cable and wire connector from the printed circuit, and remove the cluster.

7. Position cluster to opening and connect the multiple connector and the speedometer cable. Connect the A/C duct and A/C illumination light (if so equipped) and install the four cluster retaining screws.

8. Install the trim cover.

9. Install the radio bezel (if so equipped).

10. Install the light switch knob and shaft.

11. Install the heater control knobs and the wiper-washer control knobs.

12. Install the radio knobs, (if so equipped).

13. Connect the battery cable, and check the operation of all gauges, lights and signals.

1980 and Later Bronco, Pickups

1. Disconnect the battery ground cable.

2. Remove the fuel gauge switch knob (if so equipped), and wiper-washer knob. Use a hook tool to release each knob lock tab.

3. Remove the knob from the headlamp and windshield wiper switch. Remove the fog lamp switch knob, if so equipped.

4. Remove steering column shroud. Care must be taken not to damage transmission control selector indicator (PRNDL)

cable on vehicles equipped with automatic transmission.

5. On vehicles equipped with automatic transmission, remove loop on indicator cable assembly from retainer pin. Remove bracket screw from cable bracket and slide bracket out of slot in tube.

6. Remove the cluster trim cover. Remove four cluster attaching screws, disconnect the speedometer cable, wire connector from the printed circuit, 4 × 4 indicator light and remove the cluster.

7. Position cluster to opening and connect the multiple connector, the speedometer cable and 4 × 4 indicator light. Install the four cluster retaining screws.

8. If so equipped, place loop on transmission indicator cable assembly over retainer on column.

9. Position the tab on steering column bracket into slot on column. Align and attach screw.

10. Place transmission selector lever on steering column into "D" position.

11. Adjust slotted bracket so the pin is within the letter band.

12. Install the trim cover.

13. Install the headlamp switch knob. If so equipped, install the fog lamp switch.

14. Install the wiper-washer control knobs.

15. Connect the battery cable, and check the operation of all gauges, lights and signals.

Vans

1. Disconnect the battery ground cable.

2. Remove two steering column shroud to panel retaining screws and remove shroud.

3. Loosen bolts which attach the column to the B and C Support to provide sufficient clearance for cluster removal. (Required for tilt steering column vehicles only).

4. Remove seven instrument cluster to panel retaining screws.

5. Position cluster part away from the panel for access to the back of the cluster to disconnect the speedometer. If there is not sufficient access to disengage the speedometer cable from the speedometer, it may be necessary to remove the speedometer cable at the transmission and pull cable through cowl, to allow room to reach the speedometer quick disconnect.

6. Disconnect the harness connector plug from the printed circuit board and remove the cluster assembly from the instrument panel.

7. Apply approximately 3/16 inch diameter ball of silicone lubricant or equivalent in the drive hole of the speedometer head.

8. Position the cluster near its opening in the instrument panel.

9. Connect the harness connector plug to the printed circuit board.

10. Connect the speedometer cable (quick disconnect) to the speedometer head. Con-

nect the speedometer cable and housing assembly to the transmission (if removed).

11. Install the seven instrument cluster-to-panel retaining screws and connect the battery ground cable.

12. Check operation of all gauges, lights, and signals.

13. Reinstall the steering column.

14. Position steering column shroud to instrument panel and install two screws.

Ranger and Bronco II

1. Disconnect the negative battery cable.

2. Remove the two steering column shroud-to-panel retaining screws and remove the shroud.

3. Remove the lower instrument panel trim. Detach the trim cover from the panel by removing the retaining screws.

4. Remove the four instrument cluster to panel retaining screws.

5. Pull the cluster slightly away from the panel and disconnect the speedometer cable. If there is not enough room to disconnect the cable, disconnect from transmission and gently pull through the firewall until enough slack is gained.

6. Disconnect the wiring harness connector from the back of the cluster. Disconnect any bulb and socket assemblies from the cluster. Remove the cluster.

7. Install in the reverse order of removal.

Speedometer Cable Core

REMOVAL & INSTALLATION

1. Reach up behind the cluster and disconnect the cable by depressing the quick disconnect tab and pulling the cable away.

2. Remove the cable from the casing. If the cable is broken, raise the vehicle on a hoist and disconnect the cable from the transmission.

3. Remove the cable from the casing.

4. To remove the casing from the vehicle, pull it through the floor pan.

5. To replace the cable, slide the new cable into the casing and connect it at the transmission.

6. Route the cable through the floor pan and position the grommet in its groove in the floor.

7. Push the cable onto the speedometer head.

Ignition Switch

NOTE: Refer to the steering section for Ranger/Bronco II procedures.

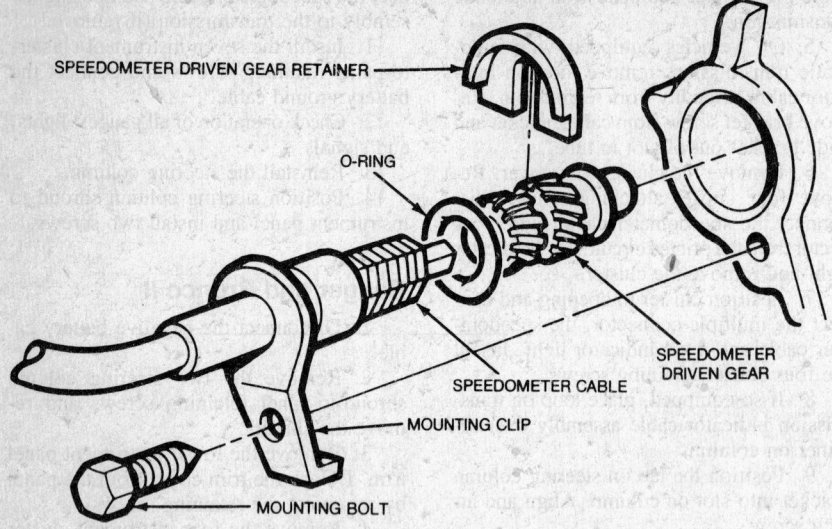

SPEEDOMETER DRIVEN GEAR RETAINER

O-RING

SPEEDOMETER CABLE

SPEEDOMETER DRIVEN GEAR

MOUNTING CLIP

MOUNTING BOLT

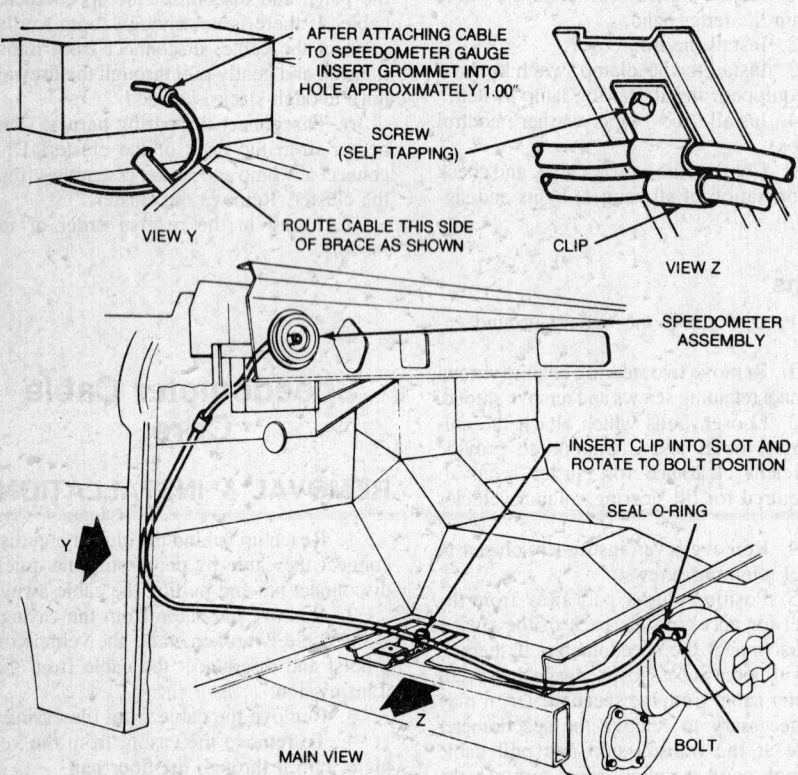

AFTER ATTACHING CABLE TO SPEEDOMETER GAUGE INSERT GROMMET INTO HOLE APPROXIMATELY 1.00"

SCREW (SELF TAPPING)

ROUTE CABLE THIS SIDE OF BRACE AS SHOWN

VIEW Y

CLIP

VIEW Z

SPEEDOMETER ASSEMBLY

INSERT CLIP INTO SLOT AND ROTATE TO BOLT POSITION

SEAL O-RING

Y

Z

MAIN VIEW

BOLT

Speedometer cable installation, typical

REMOVAL & INSTALLATION

1979

1. Disconnect the battery ground cable.

2. Turn the ignition key to Accessories and slightly depress the release pin in the face of the lock cylinder.

3. Turn the key counterclockwise and pull the key and lock assembly out of the switch.

4. From under the instrument panel, press in on the rear of the switch 1/8 turn counterclockwise.

5. Remove the bezel and switch. Remove the retainer and spring.

6. Remove the nut from the back of the switch.

7. Remove the accessory and gauge feed wires from the accessory terminal of the switch. Pull the insulated plug from the rear of the switch.

8. To install: insert a screwdriver into the lock opening of the switch and turn the slot in the switch to the full counterclockwise position.

9. Connect the insulated plug and wires to the back of the switch. Connect the accessory and gauge wires to the switch and install the retaining nut.

10. Place the bezel and switch in the switch opening, press the switch toward the instrument panel and rotate it 1/8 turn to lock it.

11. Position the spring and retainer on the switch with the open face of the retainer away from the switch. Place the switch in the opening.

12. Press the switch toward the instrument panel and install the bezel.

13. Place the key in the cylinder and turn the key to the accessory position. Place the lock and key in the switch, depress the release pin slightly, and turn the key counterclockwise. Push the new lock cylinder into the switch. Turn the key to check the operation.

14. Connect the battery.

From 1980

1. Disconnect the battery ground cable.

2. Remove steering column shroud and lower the steering column.

3. Disconnect the switch wiring at the multiple plug.

4. Remove the two nuts that retain the switch to the steering column.

5. Lift the switch vertically upward to disengage the actuator rod from the switch and remove switch.

6. When installing the ignition switch, both the locking mechanism at the top of the column and the switch itself must be in LOCK position for correct adjustment. To hold the mechanical parts of the column in LOCK position, move the shift lever into PARK (with automatic transmissions) or REVERSE (with manual transmissions), turn the key to LOCK position, and remove the key. New replacement switches, when received, are already pinned in LOCK position by a metal shipping pin inserted in a locking hole on the side of the switch.

7. Engage the actuator rod in the switch.

8. Position the switch on the column and install the retaining nuts, but do not tighten them.

9. Move the switch up and down along the column to locate the mid-position of rod lash, and then tighten the retaining nuts.

10. Remove the locking pin, connect the battery cable, and check for proper start in PARK or NEUTRAL. Also check to make certain that the start circuit cannot be actuated in the DRIVE and REVERSE position.

11. Raise the steering column into position at instrument panel. Install steering column shroud.

Headlights

REMOVAL & INSTALLATION

Vans

1. Remove the screws retaining the headlight trim ring and remove the trim ring.

2. Loosen the headlight retaining ring screws, rotate the ring counterclockwise and remove it. Do not disturb the adjusting screw settings.

3. Pull the headlight bulb forward and disconnect the wiring assembly plug from the hub.

4. Connect the wiring assembly plug to the new bulb. Place the bulb in position, making sure that the locating tabs of the bulb are fitted in the positioning slots.

5. Install the headlight retaining ring, slipping the ring tabs over the screws and rotating the ring clockwise as far as possible. Tighten the screws.

6. Place the headlight trim ring into position, and install the retaining screws.

7. Check the operation of the headlight.

1979 and Later Pickup, Bronco, Ranger and Bronco II Models

1. Remove the attaching screws and remove the headlamp door attaching screws and remove the headlamp door.

2. Remove the headlight retaining ring screws, and remove the retaining ring. Do not disturb the adjusting screw settings.

3. Pull the headlight bulb forward and disconnect the wiring assembly plug from the bulb.

4. Connect the wiring assembly plug to the new bulb. Place the bulb in position, making sure that the locating tabs of the bulb are fitted in the positioning slots.

5. Install the headlight retaining ring.

6. Place the headlight trim ring or door into position, and install the retaining screws.

Headlight Switch

REMOVAL & INSTALLATION

1. Disconnect the battery ground cable.

2. Depending on the year and model remove the wiper-washer and fog lamp switch knob if they will interfere with the headlight switch knob removal. Check the switch body (behind dash, see Step 3) for a release button. Press in on the button and remove the knob and shaft assembly. If not equipped with a release button, a hook tool may be necessary for knob removal.

3. Remove the steering column shrouds and cluster panel finish panel if they interfere with the required clearance for working behind the dash.

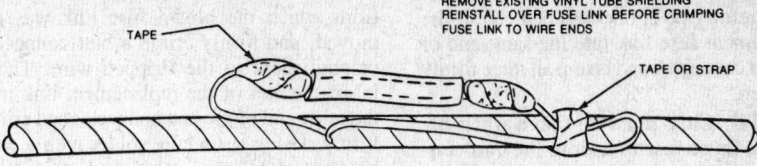

TYPICAL REPAIR USING THE SPECIAL #17 GA. (9.00" LONG-YELLOW) FUSE LINK REQUIRED FOR THE AIR/COND. CIRCUITS (2) #687E and #261A LOCATED IN THE ENGINE COMPARTMENT

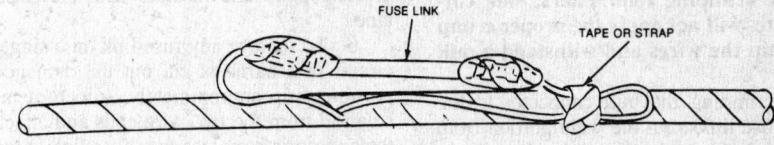

TYPICAL REPAIR FOR ANY IN-LINE FUSE LINK USING THE SPECIFIED GAUGE FUSE LINK FOR THE SPECIFIC CIRCUIT

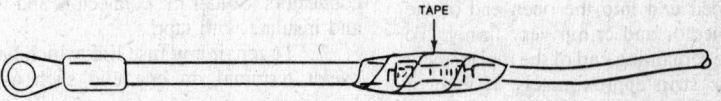

TYPICAL REPAIR USING THE EYELET TERMINAL FUSE LINK OF THE SPECIFIED GAUGE FOR ATTACHMENT TO A CIRCUIT WIRE END

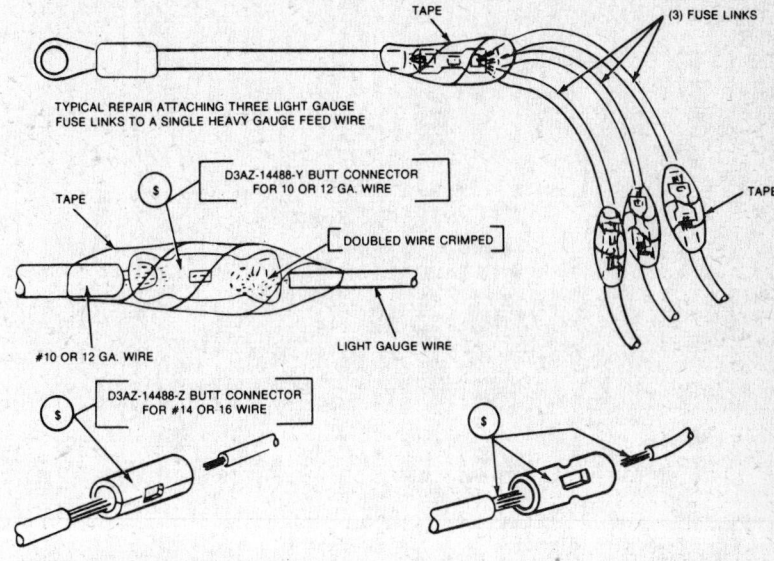

FUSIBLE LINK REPAIR PROCEDURE

General fuse link repair procedure

4. Unscrew the switch mounting nut from the front of the dash. Remove the switch from the back of the dash and disconnect the wiring harness.

5. Install in reverse order.

FUSE LINK

The fuse link is a short length of special, Hypalon (high temperature) insulated wire, integral with the engine compartment wiring harness and should not be confused with standard wire. It is several wire gauges smaller than the circuit which it protects. Under no circumstances should a fuse link replacement repair be made using a length of standard wire cut from bulk stock or from another wiring harness. To repair any blown fuse link use the following procedure:

1. Determine which circuit is dam-aged, its location and the cause of the open fuse link. If the damaged fuse link is one of three fed by a common No. 10 or 12 gauge feed wire, determine the specific affected circuit.

2. Disconnect the negative battery cable.

3. Cut the damaged fuse link from the wiring harness and discard it. If the fuse link is one of three circuits fed by a single feed wire, cut it out of the harness at each splice end and discard it.

4. Identify and procure the proper fuse link and butt connectors for attaching the fuse link to the harness.

5. To repair any fuse link in a 3-link group with one feed: After cutting the open link out of the harness, cut each of the remaining undamaged fuse links close to the feed wire weld. Strip approximately ½ inch of insulation from the detached ends of the two good fuse links. Then insert two

wire ends into one end of a butt connector and carefully push one stripped end of the replacement fuse link into the same end of the butt connector and crimp all three firmly together.

NOTE: Care must be taken when fitting the three fuse links into the butt connector as the internal diameter is a snug fit for three wires. Make sure to use a proper crimping tool. Pliers, side cutters, etc. will not apply the proper crimp to retain the wires and withstand a pull test.

After crimping the butt connector to the three fuse links, cut the weld portion from the feed wire and strip approximately ½ inch of insulation from the cut end. Insert the stripped end into the open end of the butt connector and crimp very firmly. To attach the remaining end of the replacement fuse link, strip approximately ½ inch of insulation from the wire end of the circuit from which the blown fuse link was removed, and firmly crimp a butt connector or equivalent to the stripped wire. Then, insert the end of the replacement link into the other end of the butt connector and crimp firmly. Using rosin core solder with a consistency of 60 percent tin and 40 percent lead, solder the connectors and the wires at the repairs and insulate with electrical tape.

6. To replace any fuse link on a single circuit in a harness, cut out the damaged portion, strip approximately ½ inch of insulation from the two wire ends and attach the appropriate replacement fuse link to the stripped wire ends with two proper size butt connectors. Solder the connectors and wires and insulate with tape.

7. To repair any fuse link which has an eyelet terminal on one end such as the charging circuit, cut off the open fuse link behind the weld, strip approximately ½ inch of insulation from the cut end and attach the appropriate new eyelet fuse link to the cut stripped wire with an appropriate size butt connector. Solder the connectors and wires at the repair and insulate with tape.

8. Connect the negative battery cable to the battery and test the system for proper operation.

NOTE: Do not mistake a resistor wire for a fuse link. The resistor wire is generally longer and has print stating, "Resistor-don't cut or splice". When attaching a single No. 16, 17, 18 or 20 gauge fuse link to a heavy gauge wire, always double the stripped wire end of the fuse link before inserting and crimping it into the butt connector for positive wire retention.

International
Scout, Scout II

GENERAL ENGINE SPECIFICATIONS

Year	Engine No. Cyl Displacement (cu. in.)	Carburetor Type	Horsepower @ rpm	Torque @ rpm (ft lbs)	Bore × Stroke (in.)	Compression Ratio	Oil Pressure @ 2000 rpm (psi)
'79–'80	4-196	1V	111 @ 4400	180 @ 2000	4.125 × 3.656	8.1:1	50
'80	6-198 (Turbo)	Diesel	101 @ 3800	175 @ 2200	3.27 × 3.94	20.8:1	—
'79	6-198	Diesel	73 @ 3200	133 @ 1600	3.27 × 3.94	22.0:1	—
'79–'80	8-304	2V	147 @ 3900	240 @ 2400	3.875 × 3.218	8.2:1	45
'79–'80	8-345	2V	157 @ 3800	266 @ 2400	3.875 × 3.656	8.1:1	45

VALVE SPECIFICATIONS

Year	Engine No. Cyl Displacement (cu in.)	Seat Angle (deg)	Face Angle (deg)	Spring Test Pressure (lbs @ in.)	Spring Free Height	Stem-to-Guide Clearance (in.) Intake	Stem-to-Guide Clearance (in.) Exhaust	Stem Diameter (in.) Intake	Stem Diameter (in.) Exhaust
'79–'80	4-196	①	①	188 @ 1.428	2.065	.001–.0035	.0015–.004	.372	.415
'80	6-198	45	45	57 @ 1.57	1.89	.0006–.0018	.0016–.0028	.314	.314
'79	6-198	45	45	33 @ 1.634	1.929	.0006–.0018	.0016–.0028	.314	.314
'79–'80	8-304	45	45	180–195 @ 1.429	2.065	.001–.0035	.0015–.004	.372	.371
'79–'80	8-345	45	45	180–195 @ 1.429	2.065	.001–.0035	.0015–.004	.372	.371

① Intake 30°; Exhaust 45°

TUNE-UP SPECIFICATIONS

Year	Engine No. Cyl Displacement (cu. in.)	Spark Plugs Gap (in.)	Distributor	Ignition Timing (deg) Man	Ignition Timing (deg) Auto	Fuel Pump Pressure (psi)	Idle Speed (rpm) Man	Idle Speed (rpm) Auto
'79–'80	4-196	.035	Electronic	TDC	5B	5	575	600
'79–'80	8-304	.030	Electronic	TDC	TDC	5	575	600
'79–'80	8-345	.035	Electronic	TDC	5B	5	650	700

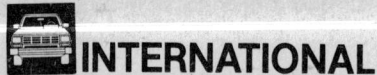

FIRING ORDER

NOTE: To avoid confusion, always replace spark plug wires one at a time.

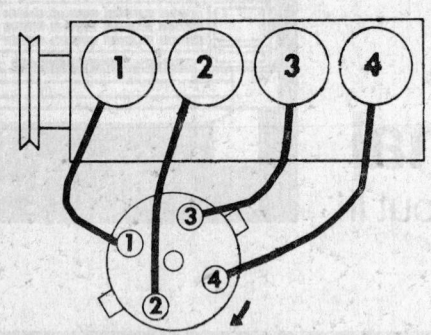

4 cylinder: 1-3-4-2

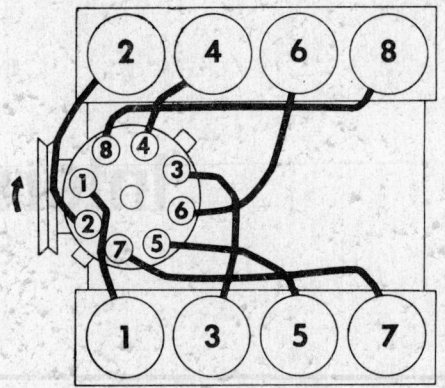

8 cylinder: 1-8-4-3-6-5-7-2

CRANKSHAFT AND CONNECTING ROD SPECIFICATIONS

(All measurements given in inches)

Year	Engine No Cyl Displacement (cu in.)	Crankshaft				Connecting Rod		
		Main Brg Journal Dia	Main Brg Oil Clearance	Shaft End-play	Thrust on No	Journal Diameter	Oil Clearance	Side Clearance
'79–'80	4-196	2.7484–2.7494	.001–.004	.003–.008	3	2.373–2.374	.0011–.0036	.004–.011
'80	6-198	2.795	.0016–.0036	.0024–.0055	3	2.087	.0012–.0036	.0039–.0079
'79	6-198	2.7918–2.7988	.0013–.0038	.0024–.0094	3	2.0840–2.0906	.0013–.0038	.0039–.0079
'79–'80	8-304	2.7484–2.7494	.001–.004	.003–.008	3	2.373–2.374	.0011–.0036	.008–.0016
'79–'80	8-345	2.7484–2.7494	.001–.004	.003–.008	3	2.373–2.374	.0011–.0036	.008–.0016

CAMSHAFT SPECIFICATIONS

(All measurements in inches)

Engine	Journal Diameter					Bearing Clearance	Valve Lift		Camshaft End Play
	1	2	3	4	5		Intake	Exhaust	
4-196	2.099–2.100	2.089–2.090	2.079–2.080	2.069–2.070	2.059–2.060	.0015–.0035	.440	.395	.006–.014
6-198 ①	2.024	2.016	2.008	2.000	—	.0063–.0079	.358	.358	.0032–.0111
6-198 ②	1.774–1.789	1.714–1.729	1.714–1.729	1.608–1.629	—	.0059–.0079 ③	.248	.248	.0032–.0102
8-304	2.099–2.100	2.089–2.090	2.079–2.080	2.069–2.070	2.059–2.060	.0015–.0035	.440	.395	.006–.014
8-345	2.099–2.100	2.089–2.090	2.079–2.080	2.069–2.070	2.059–2.060	.0015–.0035	.440	.395	.006–.014

① 1980
② 1979
③ No. 4 bearing

PISTON AND RING SPECIFICATIONS

(All measurements given in inches)

Year	Engine.	Piston to Bore Clearance	Ring Side Clearance			Ring Gap		
			Top Compression	Bottom Compression	Oil Control	Top Compression	Bottom Compression	Oil Control
'79–80	4-196	.0035	.0015–.003	.0015–.003	.002–.0035	.013–.023	.013–.023	.013–.028
'80	6-198	.0047–.0075	.0024–.0039	.0016–.0032 ①	.0008–.0024	.0059–.0197	.0059–.0197 ①	.0059–.0197
'79	6-198	.0047–.0067	.0024–.0039	.0016–.0032 ①	.0008–.0024	.0118–.0197	.0118–.0197 ①	.0118–.0197
'79–'80	8-304	.0035	.0015–.003	.0015–.003	.000–.0084	.010–.020	.010–.020	.015–.055
'79–'80	8-345	.0035	.0015–.003	.0015–.003	.000–.0084	.010–.020	.010–.020	.015–.055

① 2nd and 3rd compression rings

TORQUE SPECIFICATIONS

(All readings in ft. lbs.)

Year	Engine	Cylinder Head Bolts	Rod Bearing Bolts	Main Bearing Bolts	In.-Ex. Manifold Bolts	Crankshaft Damper Bolt	Flywheel	Injector Nozzle Holder	Injection Pump
'80	6-198	94 large 36 small	37–41	108–116	11–13	217–239	33–36	50–65	15–18
'79	6-198	94 large 36 small	36–40	109–116	11–13	217–239	33–36	50–65	15–18

TORQUE SPECIFICATIONS

(All readings in ft. lbs.)

Year	Engine	Cylinder Head Bolts	Rod Bearing Bolts	Main Bearing Bolts	Crankshaft Pulley Bolt	Flywheel-to-Crankshaft Bolts	Manifold	
							Intake	Exhaust
'79–'80	4-196	90–100	40–45	75–80	100–110	45–55	40–45	40–45
'79–'80	8-304	90–100	45–55	75–85	100–110	45–55	40–45	40–45
	8-345	90–100	45–55	75–85	100–110	45–55	40–45	40–45

TUNE-UP

Electronic Ignition System

The electronic control unit is built into the distributor body instead of being a separate unit as on earlier models. The red wire from the distributor connects to the coil positive terminal. The brown wire from the distributor connects to the coil negative terminal and the third wire from the distributor, white in color, connects to the deceleration throttle modulator when used. Because primary voltage (low voltage) current is regulated within the electronic control unit, a ballast resistor or resistance wire is not required in the primary circuit.

ADJUSTING AIR GAP

The air gap should be adjusted with the use of a brass (non-metallic) feeler gauge, placed between the center line of the sensor and a aligned tooth of the trigger wheel. Loosen the sensor hold down screw and move the sensor until the gauge will pass through with a slight drag. Tighten the hold down screw. Dwell angle is determined by the angle between the adjacent teeth of the trigger wheel and by the air gap between the ends of the wheel teeth and the center line of the sensor. Since no wearing surfaces exist on the trigger wheel and sensor, dwell remains constant and no adjustment is required after the initial sensor air gap is made. Gap should be 0.008.

For electronic ignition troubleshooting, refer to the Electrical section in Unit Repair.

Ignition Timing

NOTE; The timing light is connected to the No. 1 spark plug wire on 4 cylinder engines: No. 8 plug wire on V8 engines (304 and 345).

Ignition timing is the measurement in degrees of crankshaft rotation at the instant

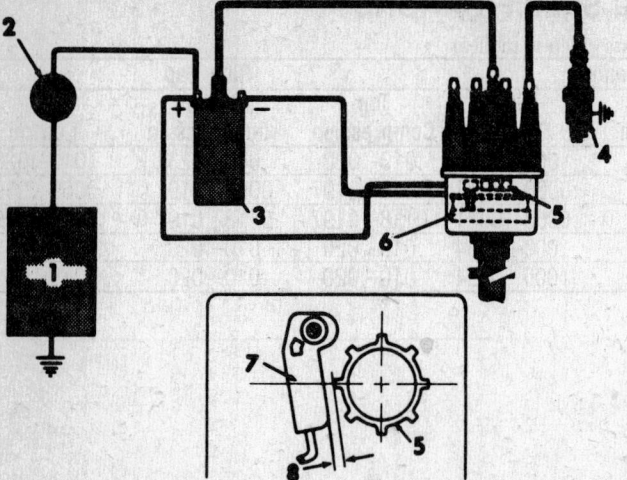

1. Battery
2. Ignition switch
3. Ignition coil
4. Spark plug
5. Trigger wheel
6. Electronic control
7. Sensor
8. Air gap

Electronic ignition with built-in control unit

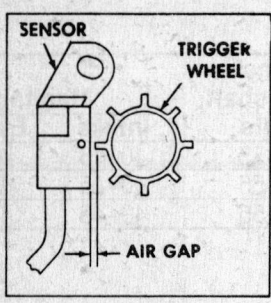

Air gap—4 cylinder

Air gap—V8

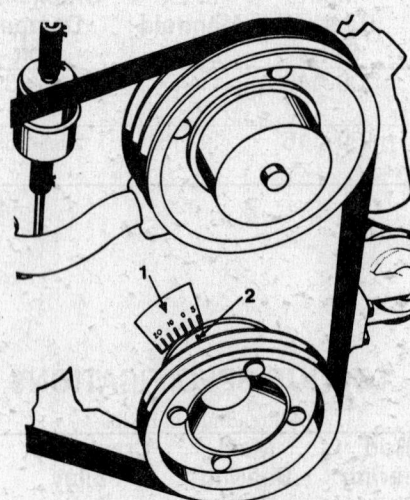

Typical timing marks

the spark plugs in the cylinders fire, in relation to the location of the piston, while the piston is on the compression stroke.

1. Remove and plug the vacuum line(s) from the distributor advance/retard mechanism. Be sure there is no vacuum leak at the line(s).

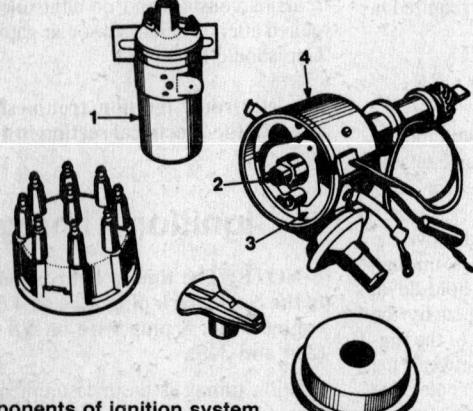

1. Ignition coil
2. Trigger wheel
3. Electronic control
4. Distributor

Components of ignition system

2. Connect the timing light.

3. Mark the crank pulley and timing quadrant lines with chalk or paint to make them easier to see.

4. Start the engine and aim the timing light at the timing quadrant.

5. The timing is adjusted by loosening the distributor hold-down nut and turning the distributor either clockwise or counterclockwise as required until the mark on the pulley aligns with the correct degree marking on the quadrant.

6. Tighten the hold-down nut and recheck the timing, readjust if necessary.

7. Stop the engine, remove the timing light and reconnect the vacuum line(s).

Valve Lash

GAS ENGINES

The valve lash is not adjustable since hydraulic valve lifters are employed. If a lifter should become noisy; dirt, grit or metal chips may have becomed lodged in the lifter body. This problem may be alleviated by frequently changing the engine oil and filter. Remove the lifter, clean or replace it if the noise persists. Check the other valve train components for wear, replace as necessary.

DIESEL ENGINES

1. Remove the valve cover.

2. Turn the engine until No. 1 piston is at TDC (top dead center) on the compression stroke.

3. Adjust the following valves (by cylinder number shown) to 0.014 in.: No. 1 Exhaust and Intake valve. No. 2 Exhaust only. No. 3 Intake only. No. 4 Exhaust only. No. 5 Intake only.

4. Turn the engine until No. 6 piston is at TDC (top dead center) on the compression stroke.

5. Adjust the following valves (by cylinder number shown) to 0.014 in.: No. 2 Intake only. No. 3 Exhaust only. No. 4 Intake only. No. 5 Exhaust only. No. 6 Intake and Exhaust.

6. Adjustment is made by loosening the rocker arm screw locknut and turning the screw as necessary to allow the proper size feeler gauge blade to pass, with a slight drag, between the bottom of the adjustment screw and the top of the valve stem. Tighten the locknut while holding the screw stationary. Install the valve cover.

Idle Speed & Mixture

GAS ENGINES

To comply with the mandated emission control requirements, certain procedures must be followed when adjusting the air/fuel mixture and speed. The engine must be at normal operating temperature, choke open, air cleaner installed, ignition timing correct and the parking brake applied. The following procedures apply to all carburetors, with minor deviations possible, depending upon the carburetor used.

Observe the following precautions when adjusting the idle mixture and speed.

1. Do not idle the engine for longer than three minutes at a time.

2. After each three minute interval, increase the engine speed to 2,000 rpm for one minute.

3. Continue with the idle adjustment and repeat Step 2 as necessary.

PRELIMINARY IDLE SETTING (AFTER CARBURETOR OVERHAUL)

1. Connect a tachometer to the engine.

2. Connect a test vacuum gauge to the engine intake manifold.

3. Operate the engine at a fast idle speed to bring the operating temperature to normal.

4. Adjust the carburetor to the specified idle speed.

5. Adjust the idle mixture screw(s) and idle speed screw to obtain ''Lean best idle'' at the specified speed.

NOTE: ''Lean best idle'' is the point at which intake manifold vacuum starts to drop as shown on the gauge.

6. Install the colored (service) plastic cap(s) with the tab fully turned counterclockwise against the stop.

7. Adjust the idle speed to specifications.

8. Make final idle adjustments to obtain the recommended idle setting.

IDLE ADJUSTMENT (LEAN DROP METHOD)

1. Connect a tachometer to the engine.

2. Rotate the idle adjusting screw(s) counterclockwise against the stops.

3. Adjust the idle speed to give an engine speed 25 rpm higher than the specified idle speed.

4. Rotate the idle mixture screw(s) clockwise slowly and equally (if two) until the specified speed is obtained.

5. If the engine is rough or the specified idle speed cannot be attained, remove the limiter cap(s) and continue the adjustment as outlined in Step 4, until the specified rpm is attained and the engine is smooth.

6. Install new plastic limiter cap(s) with the tab fully counterclockwise against the stop.

7. Readjust as necessary to maintain the specified rpm.

IDLE ADJUSTMENT (EXHAUST ANALYZER METHOD)

When exhaust analyzer equipment is used,

the following procedure is recommended to be used to adjust the idle mixture and speed.

1. Connect a tachometer to the engine and insert the exhaust analyzer into the exhaust pipe.

NOTE: Refer to the manufacturers instructions for complete connection procedures.

2. Operate the engine for fifteen minutes at fast idle speed (approximately 1000–1200 rpm), to bring engine to normal operating temperature and to stabilize the temperature of the exhaust analyzer.

3. Calibrate the test equipment as per the manufacturers instructions.

NOTE: If the combustion analyzer does not respond to changes in the mixture quality, check for leaks or restrictions in the sample lines. The thermal conductivity instruments used in the analyzer are both temperature and pressure sensitive, and require a definite sample flow. Refer to the manufacturer's instructions as necessary.

4. Adjust the idle mixture screw(s) counterclockwise against the tab stop.

5. Adjust the idle speed screw to obtain the specified idle speed.

6. Observe the analyzer dial and adjust the idle mixture screw(s) clockwise by $\frac{1}{16}$ turn increments to obtain the specified idle mixture setting and readjust the idle speed as necessary.

7. If the idle speed and mixture cannot be obtained, remove the idle limiter cap(s).

NOTE: To prevent damage to the mixture screw(s) or seat, file or grind the side of the plastic cap. Do not pry cap off.

8. With the engine operating, adjust the mixture screw(s) to obtain the ''lean best idle'' at the specified idle speed.

NOTE: ''Lean best idle'' is the point at which maximum manifold vacuum begins to drop as shown on the gauge

9. Install new plastic limiter cap(s) with the tab fully counterclockwise against the stop.

10. Readjust the idle mixture screw(s) to obtain the recommended CO setting.

NOTE: After completing the idle adjustment procedure, if unsatisfactory idle operation still exists, a recheck of the ignition system, crankcase ventilation system, timing advance system, air induction system, exhaust gas recirculation system, or hot idle compensation system should be made.

Idle Speed

DIESEL ENGINE

1. Adjust the low idle speed at the low idle speed adjustment screw at the rear of the injection pump.

2. Connect a diesel tachometer to the engine. Start the engine and operate until coolant reaches normal operating temperature.

3. Allow the engine to idle and note the idle speed shown on the tachometer. If the low idle speed is not within specifications (700–750 rpm) adjust as required.

4. Loosen the buffer locknut. Back off the buffer screw adjustment. Loosen the locknut on the idle speed adjustment screw. Turn the idle screw to obtain the correct idle speed. Tighten the locknut.

5. Turn the buffer screw in until it touches or increases idle. Increase the idle speed by 10–25 rpm. Back off the buffer screw one full turn. Tighten locknut.

ENGINE ELECTRICAL

Distributor

REMOVAL

1. Remove the distributor wire harness connected to the coil.

2. Remove the distributor cap by either unsnapping the retaining clips or twisting the retaining clamps with a screwdriver. Bump the engine until the top dead center marks on the timing case and pointer are aligned on the compression stroke.

3. Using chalk or paint, carefully mark the position of the distributor rotor in relation to the distributor housing, and the position of the distributor housing in relation to the engine block. When this is done, you should have a line on the distributor housing directly in line with the tip of the rotor and another line on the engine block directly in line with the mark on the distributor housing. These markings are important because the distributor must be installed in the exact same position to avoid having to retime the engine.

4. Remove the vacuum line(s) from the advance diaphragm.

5. Remove the bolt and bracket that attaches the distributor to the engine block and carefully pull the unit straight up and out of the engine.

INSTALLATION (ENGINE NOT DISTURBED)

1. Before installing the distributor, replace the distributor mounting gasket in the engine counterbore.

2. Turn the rotor so the tip is aligned with the mark made on the housing during removal. Then turn the rotor about $\frac{1}{8}$ turn in the opposite direction of distributor rotation past the mark.

3. Lower the distributor into the mounting hole and align the mark on the distributor housing with the one on the engine block.

NOTE: It may be necessary to move the rotor slightly to start the gear into mesh with the camshaft gear, but the rotor should align with the mark on the distributor housing when the distributor is down in place.

4. Install the distributor hold-down bolt and lockwasher but do not fully tighten.

5. Connect the distributor wire to the negative pole of the coil.

6. Install the distributor cap making certain that all the leads are all the way into the cap. Secure the cap.

NOTE: If the distributor cap is misaligned on its locating slots, the rotor and the cap will be damaged.

7. Start the engine and set the timing.

INSTALLATION WHEN THE ENGINE HAS BEEN DISTURBED

If the engine was turned over with the distributor removed or if you have installed the distributor incorrectly and the engine will not start, it will be necessary to time the engine from scratch.

1. Remove No.1 spark plug. Have an assistant rotate the engine while you hold your finger over the plug port. Continue rotating slowly until you feel pressure and the timing mark on the crankshaft pulley lines up with the mark on the timing tab.

2. Install a new distributor mounting gasket in the counterbore of the engine.

3. Locate the No.1 spark plug wire tower on the distributor cap.

4. Scribe a locating mark on the body of the distributor directly below the No.1 spark plug wire tower with the cap installed.

5. Remove the cap, install the distributor, and align the rotor with the mark on the housing.

6. It may be necessary to turn the rotor slightly to get the distributor and the camshaft gears to mesh but the rotor should still align when the distributor is bottomed.

7. Install the distributor hold-down bolt and hand tighten.

8. Install the No.1 spark plug, the distributor lead to the coil, and any plug wires which were removed.

9. Install the distributor cap making certain that the tang on the distributor body aligns with the slot in the distributor cap.

10. Start the engine and check the timing.

Alternator

PRECAUTIONS

Rectifiers and regulators in alternator systems are easily damaged by incorrect polarity. Observe the following precautions when wiring and testing circuits:

1. Always be certain of battery polarity.

2. Always connect booster battery negative to a grounded point of the vehicle requiring jumped after connecting positive to positive.

3. Never ground the alternator output terminal.

4. When adjusting the voltage regulator, be careful not to short with adjusting tool.

5. Before making any tests, turn off the ignition switch and disconnect the battery ground.

6. Never use a fast charger with the battery connected unless the charging unit is equipped with a special alternator protector.

7. Never try to polarize the alternator regulator, this will cause severe damage to the regulator and alternator.

REMOVAL & INSTALLATION

1. Disconnect the negative battery cable.

2. Remove the wire terminals from the rear of the alternator.

3. Loosen the adjusting strap and pivot bolts. Push inward on the alternator to loosen the belt and slip it off the pulley.

4. Remove the adjusting strap and pivot bolts, and remove the alternator from the engine.

5. Installation is in the reverse of the removal. Adjust the belt to have no more than $\frac{1}{2}$ in. deflection on the longest span of the belt.

Voltage Regulators

Two types of voltage regulators are used to control the output of the alternators. One type is the internal unit, mounted with-in the alternator, and the second is an external type, normally mounted on the inner fender panel or the firewall.

REMOVAL & INSTALLATION

External Type

1. Disconnect clamp lead at the negative terminal of battery.

2. Disconnect the wiring harness connector at regulator terminals.

3. Remove mounting screws and regulator unit from vehicle.

4. To install, reverse the above procedure.

5. Reconnect cable clamp to battery terminal, checking polarity first.

Internal Type

1. Remove the alternator as outlined.

2. Mark and separate the front housing from the rear housing.

3. Remove the diode trio screws and nuts, and remove the trio assembly.

4. Remove the two remaining screws in the regulator, and remove the brush holder and the regulator from the rear housing.

5. Installation is the reverse order of the removal, assuring that the insulated sleeves are installed on the proper screws, during installation.

NOTE: Refer to the Unit Repair Section for more repair information.

Starter

For servicing and overhauling starter motors, see General Repair Section.

REMOVAL & INSTALLATION

1. Disconnect cable clamp from negative terminal of battery.

2. Disconnect cable and wire leads from terminals of solenoid assembly, identifying leads with tags. If the solenoid is not mounted directly on the starter motor, disconnect the cable from the solenoid to the motor at the motor terminal.

3. Remove the starter motor mounting bolts or stud nuts.

4. Pull the starter assembly forward to clear housing and remove the starter.

5. To install, reverse the above procedure, installing new tang lockwashers where removed.

ENGINE MECHANICAL

Engine

REMOVAL & INSTALLATION

The following is an outline of general engine removal. Removal procedure will vary due to the variety of body models and accessory equipment. Before lifting out engine be certain that everything has been disconnected. Remove anything that might be in the way of the actual lifting.

1. Drain the coolant from radiator and engine block.

2. Drain the crankcase oil.

3. Disconnect the negative battery ground cable and remove cable clamp from hot terminal of battery.

4. Remove all water hoses to the radiator and heater.

5. Remove the fan blades and fan shroud.

6. Remove any radiator cross-brace rods or brackets.

7. Remove radiator mounting bolts and lift out radiator.

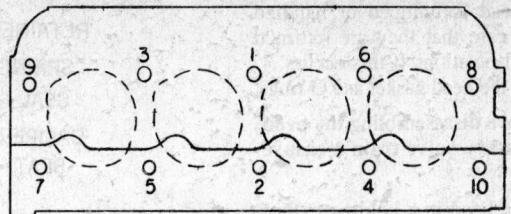

8. Remove hood hinge bracket mounting bolts and remove hood assembly.

9. Disconnect and remove the air cleaner assembly from the engine. Remove breather hose from air cleaner, if applicable.

10. Disconnect the fuel pump inlet line.

11. Remove the vacuum lines from manifold and all other components, and lines from the air compressor, and air pump, if applicable.

12. Disconnect the throttle linkage, choke control wire and hand throttle control wire, if applicable. On V8 engines remove the carburetor, if necessary, for the fitting of the lifting fixture.

13. If so equipped, disconnect wire from heater control valve.

14. Disconnect all wiring from the engine as follows: Water temperature gauge sender. Oil pressure gauge sender. Alternator wires. Primary ignition wire harness to the resistor or distributor. Starter solenoid wires and battery cable.

15. If so equipped, disconnect the tachometer drive. Disconnect the exhaust pipes at manifolds.

---------- **CAUTION** ----------

If there is not enough slack in the air conditioner compressor lines to leave them attached to the compressor when it is dismounted and positioned out of the way, have the air conditioning system "bled" by a professional using the proper tools and safe procedures.

---

17. If so equipped, remove the automatic transmission filler tube, remove the A//C compressor with lines attached and position out of the way and disconnect the power steering pump lines from the pump.

18. Install lifting fixtures and suitable sling.

19. Connect hoisting equipment to lifting fixture and hoist enough to support engine.

20. Remove bell housing mounting bolts. On V8 engines the flywheel housing front cover is removed before the flywheel housing is removed from crankcase.

21. Disconnect clutch linkage.

22. Remove front engine mounting bolts. On some models it is easier to unbolt the mount from the frame crossmember.

23. Remove side engine mount bolts.

24. Slow raise the engine. In hoisting out engine, first pull engine forward to clear clutch assembly from transmission, then tilt front up and carefully raise out of the chassis.

25. Installation of the engine is in general the reverse of the above described procedure. Be careful when installing that wires are not pinched between engine and frame. Lower the engine until transmission main drive gear spline can be aligned with the clutch driven disc. The weight of the engine must remain supported until the bell housing is secured to flywheel housing. After engine has been secured to chassis, remove hoisting equipment and lifting fixtures.

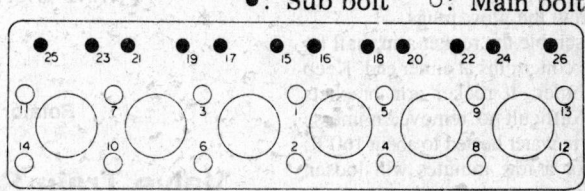

Gas engine—head tightening sequence

●: Sub bolt   ○: Main bolt

Diesel head bolt torque sequence

Cylinder Head

REMOVAL & INSTALLATION

Gas Engines

1. Disconnect the negative battery cable. Remove spark plug wires and spark plugs. Remove intake and exhaust manifolds as described in the following section. On V8's, this may entail removal of the air compressor and air compressor mounting bracket.

2. Remove valve covers and gaskets.

3. Loosen rocker arm shaft bracket bolts and remove the rocker arm assembly.

NOTE: Be sure to remove and keep track of the two dowel sleeves on the end brackets of the rocker arm assembly.

4. Remove the pushrods, marking them so that they may be installed in their same locations.

5. Remove cylinder head bolts.

6. When lifting off cylinder, do not lose the two locating dowel sleeves.

7. Installation is basically the reverse of the above procedure, with the exception of the following additional steps.

8. Be sure to use a new head gasket and to reinstall dowel sleeves when positioning the head and mounting the rocker assembly. Reinstall pushrods in their original locations.

9. On all engines, turn engine crankshaft until leading edge of balance weight on crankshaft pulley is aligned with the zero degree mark on the timing indicator before installing rocker arm assembly.

10. On 4–196 engines, be sure to install rocker assembly so that the oil feed shaft bracket is third from the rear.

11. On V8 engines, install rocker arm assembly so that the notches at the end of the shaft are facing upward. Oil feed brackets are third from the rear on the right (even numbers) bank and third from the front on the left (odd numbers) bank.

---------- **CAUTION** ----------

Do not use a power wrench on heads of engines with hydraulic lifters. Torque head bolts slowly so that the leakdown of the lifters may relieve strain from the valve train.

---

12. On all engines, tighten the head bolts in the sequence illustrated to 90–100 ft. lbs.

13. Retorque head bolts to the specified torque after 1000 miles of operation.

14. Install rocker covers and any other equipment removed for head work. Replace rocker cover gasket if necessary.

Diesel Engine

1. Remove the air cleaner.

2. Remove the crankcase vent hose and remove the intake and exhaust manifolds.

3. Remove the alternator, bracket and belts.

4. Disconnect the coolant hose between the head and the oil cooler.

5. Remove the fuel filter assembly.

6. Disconnect the injection lines from the pump and the injectors. Cap all openings at once.

7. Remove the bypass hoses between the coolant pump and the thermostat housing.

8. Remove the fan.

9. Remove the rocker arm cover.

10. Remove the rocker arm shaft assembly.

11. Remove the pushrods and keep them in order.

12. Remove the fuel return lines.

13. Remove the nozzles from the head.

14. Remove the cylinder head bolts in the sequence shown.

15. Attach a hoist to the head and lift it clear of the block. On occasion, the precombustion chambers may fall out, espe-

cially if the head is bumped or handled roughly. Take care that they are returned to their original positions if this occurs.

16. Remove the head gasket and O-rings.

NOTE: Before disassembling the head, make all necessary valve train measurements.

17. Place the head in a holding fixture.

18. Disassemble the valves. Mark all parts for assembly in their original positions.

19. Remove the retaining wire and lift off the valve stem seals.

20. Unscrew the glow plugs.

21. Disassemble the rocker arm shaft by removing the cotter pins at either end. Keep all parts in order. If rocker arm brackets prove to be difficult to remove, immerse the assembly in water heated to about 160°F. Immersion for a few minutes will loosen the parts.

22. Clean and inspect all parts.

23. Check the head with a straightedge. Maximum warpage is 0.0079 in. Do not remove more than 0.011 in. from the head.

24. Check the valve springs for free length and tilt. Free length must not be less than 1.850 in. and tilt must not exceed 0.03937 in. (1mm).

25. Valve seats may be removed by cracking with a cold chisel or with a valve seat remover. New valve seats should be put in dry ice for five minutes prior to installation, at the same time the head should be immersed in 175°F water.

26. Assemble the head in reverse order of disassembly.

27. Place a new cylinder head gasket on the block with the stainless steel inset side facing up.

28. Install the O-rings around the water and oil passages.

29. Position the head on the block.

30. Coat the head bolts with clean engine oil and torque in sequence, in stages as follows: Large bolts: First stage 43 ft. lbs. Second stage 94 ft. lbs. Small bolts: First stage 21 ft. lbs. Second stage 36 ft. lbs.

31. Install the pushrods, pressing down and turning them to be sure of proper seating.

32. Install the rocker arm shaft assembly, torquing the bolts to 18 ft. lbs. in sequence from the center to each end.

33. Install the injection nozzles.

34. Install all other parts in reverse order of removal.

VALVE ROTATORS

Gas Engines

On the 4–196, V–304 and V–345 engines, rotators are used between the valve spring and the cylinder head on the exhaust valve only.

NOTE: Keep the valves and their related parts together so they may be reinstalled in their respective positions.

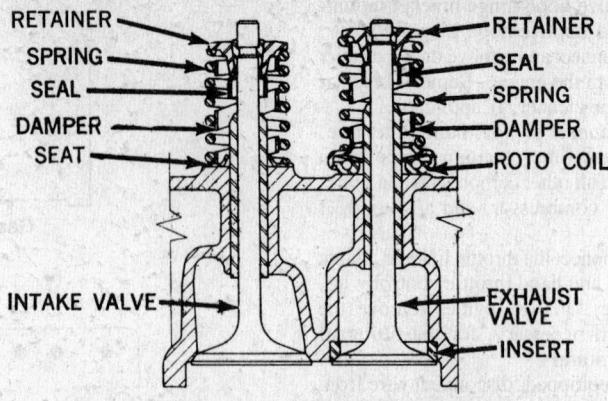

Rotator used under the exhaust valve spring

Valve Train Service

The 4–196 and V8 engines utilized hydraulic lifters for which there is no lash adjustment. Excess noise in the valve train of these engines indicates that service is required. Instructions for servicing hydraulic lifters may be found in the Unit Repair Section. Valve removal, service, and installation procedures may be found in the Engine Rebuilding section. See specifications table at the beginning of this section for valve spring and valve seat angle specifications.

ROCKER ARM

Removal & Installation

1. Remove rocker cover and gasket.

2. Remove rocker arm assembly mounting bolts and flat washers.

3. Remove rocker assembly.

4. If applicable, remove clip-ring and retainer to disassemble rocker components. Be sure to keep all parts in order so that they may be replaced in their original positions.

5. Clean all parts thoroughly, making sure that oil passages are clear. If necessary to remove plugs from ends of shaft, drill a hole in one plug, knock out the other with a steel rod, then knock out the drilled plug.

6. Inspect shaft for wear and warpage. Replace bent or worn shaft.

Intake Manifold

REMOVAL & INSTALLATION

Gas Engines

1. If engine is in vehicle, remove air cleaner and, if applicable, governor vacuum line.

2. Disconnect throttle linkage, choke cable and fuel line.

3. Remove carburetor.

4. On V8 304 and 345 engines, disconnect hose from thermostat housing and bracket for spark plug wires.

5. On 4 cylinder models, remove coil, coil mounting bracket and ignition resistor from intake manifold.

6. Remove positive crankcase ventilation pipe and vacuum line.

7. Remove mounting bolts, manifold and gasket.

8. Installation is the reverse of the above procedure. Install new gaskets and tighten the mounting bolts from the center out, torquing to 40–45 ft. lbs.

Exhaust Manifold

REMOVAL & INSTALLATION

Gas Engines

1. Disconnect exhaust pipe from manifold.

2. Unbolt exhaust manifold from head.

3. Remove manifold.

4. Installation is the reverse of the above procedure. Install new manifold-to-head gasket and new manifold-to-pipe gasket.

5. Torque manifold-to-head bolts to 25–30 ft. lbs.

Timing Case & Gears

REMOVAL & INSTALLATION

Crankshaft Pulley (Gas Engines)

Accessibility of the crankcase pulley and front (timing) cover will vary according to the model. On some vehicles the timing case will be accessible only if the engine is completely removed. The following instructions are general and apply to most front cover repairs and service.

1. Drain cooling system.

2. Disconnect radiator hoses and remove radiator. In some cases the radiator shroud and truck hood must be removed.

3. Loosen front engine mounts and jack up engine enough to provide access to the crankshaft pulley with a puller.

4. Loosen and remove fan belts and remove fan blades.

5. Remove crankshaft pulley retaining bolt. The vibration damper behind the pulley must be removed with a puller.

6. Using a suitable puller, remove the pulley from the crankshaft. On some models the pulley is in two pieces and the pulley must be unbolted from its hub before the hub is removed with a puller.

Front Oil Seal

1. Remove crankshaft pulley as described in the preceding procedure, Steps 1–6.

2. Remove seal. It is preferable to use an appropriate seal puller. Use a new gasket when installing front cover and be sure to align cover before tightening.

3. Install a new seal using a suitable seal installing tool if possible. Lubricate first and be careful not to damage seal or seating surface of cover.

4. Install crankshaft pulley, fan belt and fan blades.

5. Lower engine and tighten mounting bolts.

6. Install radiator, shroud, hoses and whatever else was removed.

7. Fill cooling system.

Timing Gears

Timing gears can be removed without disassembling the engine. In some cases, however, the engine must be removed.

1. Remove crankcase pulley as described in Steps 1–6 of Crankshaft Pulley Removal.

2. Remove engine front cover.

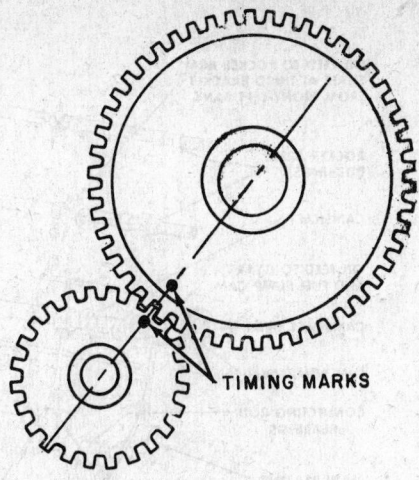

Timing gear alignment

3. Rotate engine to align timing marks on crankshaft gear and camshaft gear.

4. To remove either gear, remove bolt and washer. Use a suitable puller.

NOTE: Replace both the cam gear and crankshaft gear, due to being serviced in matched sets.

5. Use a suitable installing tool to install gears. Lubricate with engine oil and insert key in shaft to align gear. Align timing marks as illustrated. Be careful not to damage threads on shaft. Install and tighten retaining bolt.

6. Rotate engine to check that gears are not binding.

7. Check gear backlash with a dial indicator. It should be within 0.0005–0.0045 in. on OHV4 and V8 engines.

8. Use a new gasket when installing front cover and be sure to align cover before tightening. On some models there is an oil slinger on the crankshaft.

9. Install crankshaft pulley, belt and fan blades. Tighten retaining bolt.

10. Lower engine and tighten engine mounting bolts.

11. Install radiator and hoses.

12. Fill cooling system.

Camshaft

REMOVAL & INSTALLATION

On most models, it is possible to remove the camshaft with the engine remaining in the vehicle. However, the body grille work, radiator, A/C condenser (if equipped), hood, bumper, and braces must be removed to allow clearance for the camshaft to be withdrawn from the engine block. In some cases, it would be more advantageous to remove the engine from the vehicle to replace the camshaft. The decision would depend upon the individual and his shop facilities.

1. Remove the intake manifolds on the V8 engines, and the rocker covers on all engines.

2. Remove the rocker arms or assemblies, pushrods and tappets.

3. Remove the distributor and mechanical fuel pump.

4. Remove the oil pan and oil pump, if necessary.

5. Remove the crankshaft pulley as previously described.

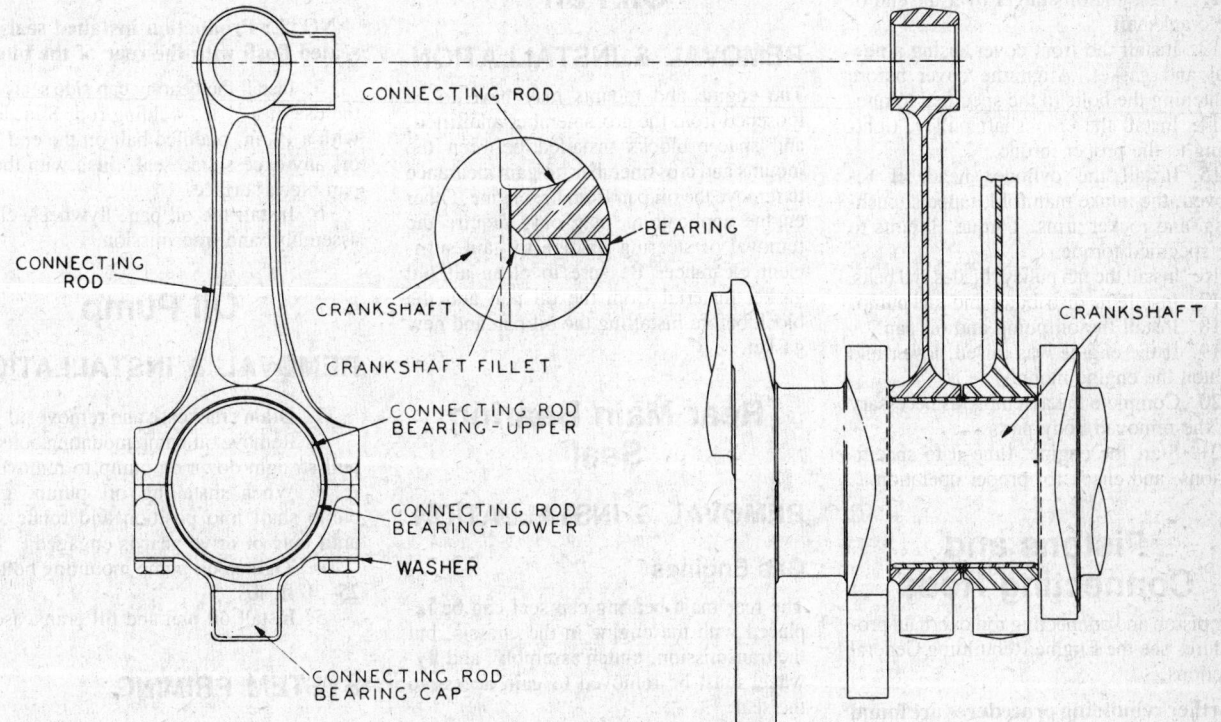

Installation of connecting rod to crankshaft

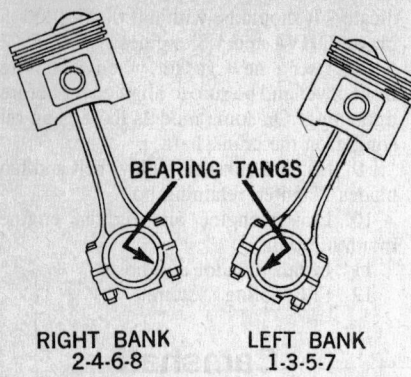

BEARING TANGS

RIGHT BANK	LEFT BANK
2-4-6-8	1-3-5-7

V8 engine—piston and rod assembly

6. Remove the front timing cover, gasket and seal.

7. Remove the two screws securing the camshaft thrust flange to the block.

8. Remove the camshaft and gear. To prevent nicking and damaging the camshaft or bearings, use a camshaft removal and installation tool, which is an extension on the front of the camshaft to act as a handle.

9. When installing the camshaft and gear, coat the bearing surfaces and lobes with lubricant and use the installing tool if possible, to aid in the installation of the camshaft. Make sure the gear timing marks align properly.

10. Working through the two large holes in the camshaft gear, install the two thrust flange screws and tighten to proper torque specifications.

11. Check timing gear backlash. If the end play exceeds the allowable limits, replace the thrust flange.

12. Place the oil slinger over the end of the crankshaft.

13. Install the front cover, using a new seal and gasket. Align the cover before tightening the bolts to the specified torque.

14. Install the crankshaft pulley, tightening to the proper torque.

15. Install the cylinder head, if removed, the intake manifold, tappets, pushrods, and rocker arms. Torque all bolts to the specified torque.

16. Install the fan pulley, blades and belts.

17. Install the distributor and fuel pump.

18. Install the oil pump and oil pan.

19. If the engine was raised, lower and tighten the engine mounts.

20. Complete the assembly as necessary for the removed body parts.

21. Start the engine, time it to specifications, and check for proper operation.

Pistons and Connecting Rods

For piston and connecting rod overhaul procedures see the Engine Rebuilding General Sections.

Further rebuilding procedures are found in the Engine Rebuilding Unit Repair section.

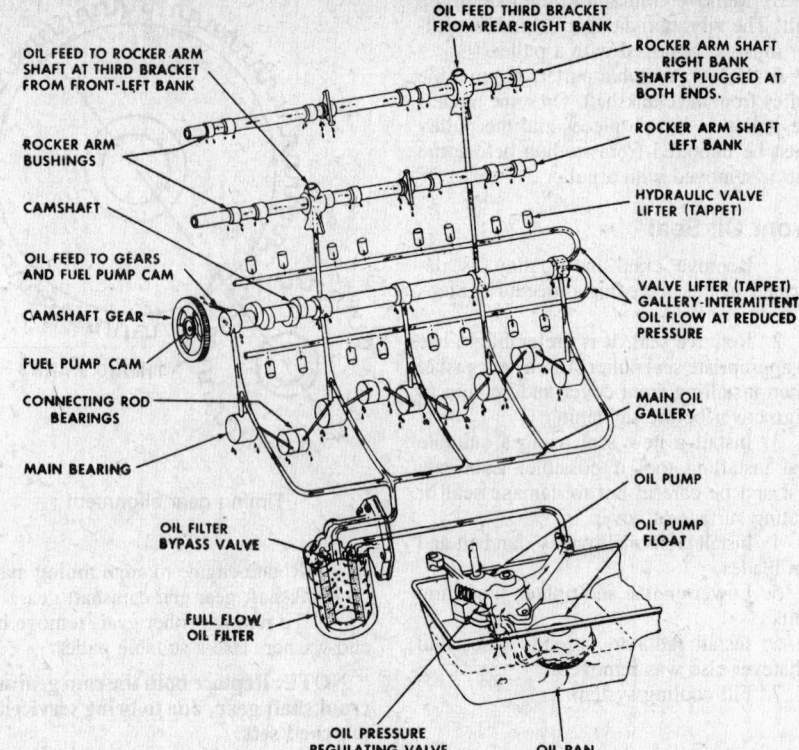

Lubrication system—V8 engine

ENGINE LUBRICATION

Oil Pan

REMOVAL & INSTALLATION

The engine and mounts may have to be loosened from the crossmember and lifted, and spacer blocks installed between the mounts and crossmember, to gain clearance to remove the oil pan from the engine. Other engine applications may only require the removal of steering linkage to gain sufficient clearance. Be sure to clean all old gasket material from the oil pan and the block before installing the oil pan and new gasket.

Rear Main Bearing Seal

REMOVAL & INSTALLATION

Gas Engines

The rear main bearing cap seal can be replaced with the engine in the chassis, but the transmission, clutch assembly, and flywheel must be removed to gain access to the seal.

1. Remove the transmission, clutch assembly, and the flywheel.

2. Remove the engine oil pan.

3. With a slide hammer with a screw end adapter pierce the seal and remove it from the recess in the cap and block.

4. Lubricate the new seal, seat it squarely with a seal installer tool .085 inch from the rear face of the block.

NOTE: Production installed seals are seated flush with the rear of the block.

5. Install the bearing cap side seals with the use of a ⅛ in. welding rod, 8 in. long, with a 5/32 in. puddled ball on the end. Cut off any excess side seal, flush with the oil pan block surface.

6. Install the oil pan, flywheel, clutch assembly, and transmission.

Oil Pump

REMOVAL & INSTALLATION

1. Drain crankcase and remove oil pan.

2. Remove oil pump mounting bolts and pull straight down on pump to remove.

3. When installing oil pump, guide pump shaft into position and rotate shaft until tang of drive gear is engaged.

4. Tighten oil pump mounting bolts to: 25–30 ft. lbs.

5. Install oil pan and fill crankcase.

SYSTEM PRIMING

The recommended procedure to prime the internal parts and the oil pump is to attach a bearing leak detector or similar tool to a suitable fitting on the oil gallery, located on the left side of the engine block. Inject enough oil into the engine to fill the oil filter and the various passage ways for the lubrication system. Disconnect the primary coil wire and turn the engine over, while the priming operation is in process. Do not overfill the crankcase when this method is used. This type of priming will minimize the possibility of scuffing or heat build-up in the areas of friction, which could cause premature engine failure.

ENGINE COOLING

The cooling system is a closed type, utilizing a two valve pressure cap. One valve is used to relieve excessive pressure from the system, and the second valve is used to allow atmospheric air to enter the system during the cooling down period. The engine temperature is controlled by a thermostat, located on the front of the engine block or cylinder head. The coolant is forced through the engine and radiator by the water pump, located on the front of the engine, which is belt driven by the crankshaft pulley.

CAUTION

To avoid personal injury, remove the pressure cap from the radiator in two steps. Loosen the cap to its first notch and allow the pressure to escape through the overflow pipe. After the pressure has been released, press on the cap and continue to turn until the prongs on the cap disengage from the radiator neck.

Water Pump

REMOVAL & INSTALLATION

1. Drain cooling system.
2. If radiator shrouds hinder access they must be removed before proceeding.
3. Loosen alternator pivot bolts and adjusting bolt on bracket to relieve tension on the fan belt and remove belt from water pump pulley.
4. Remove all pipes and hoses connected to the water pump.
5. Remove mounting bolts or stud nuts and water pump.
6. Installation is the reverse of the above procedure. Be sure to install new gaskets and, if applicable, new O-rings on pipe end fittings.

EMISSION CONTROLS

Refer to the Emission Control Unit Repair Section.

1. Lever, cam
2. Spring, cam lever return
3. Plug
4. Pin
5. Diaphragm assembly
6. Screw and lockwasher
7. Housing, valve assembly
8. Screw and lockwasher
9. Diaphragm, air dome
10. Air dome and filter assembly
11. Gasket, filter bowl
12. Filter
13. Elbow
14. Spring
15. Filter bowl
16. Retainer
17. Washer
18. Bolt
19. Lockwasher
20. Pump body
21. Pin, cam lever
22. Gasket, mounting

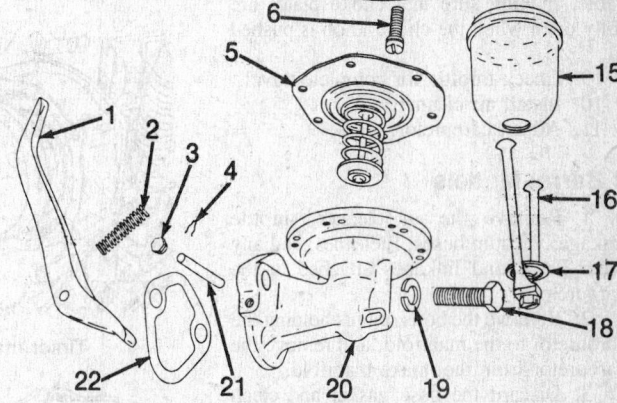

Exploded view of a typical fuel pump

GASOLINE FUEL SYSTEM

Fuel Pump

REMOVAL & INSTALLATION

1. Remove the fuel inlet pipe or hose and the outlet fuel pipe to the carburetor from the fuel pump fittings.
2. Remove the attaching bolts from the fuel pump housing to engine block and remove the fuel pump.
3. Clean the gasket surfaces of all gasket particles.
4. Install new gasket on the fuel pump mounting flange and install the fuel pump operating arm into the hole in the block, and into contact with the eccentric lobe on the camshaft.
5. Install the attaching bolts and tighten the pump to the block securely.
6. Install the inlet hose or pipe, and the outlet pipe to the fuel pump and tighten securely to avoid air leaks.

PRESSURE TEST

1. Disconnect fuel line at carburetor inlet and attach pressure gauge between the inlet and disconnected line.
2. Start engine and take reading. Consult Tune-up Specifications at the beginning of this section for correct pump pressure.

3. When engine is stopped, the pressure should remain constant or very slowly return to zero.

CAPACITY TEST

1. Disconnect fuel line from the fuel pump.
2. Connect a piece of hose to the line so that fuel can be directed into a measuring container.
3. Start engine and note time it takes to fill a pint container. Pump should fill one pint within 20–30 seconds.

Carburetor

REMOVAL & INSTALLATION

Holley Model 1940

1. Remove the air cleaner, fuel lines, vacuum lines and any other lines or linkage attached to the carburetor.
2. Remove the attaching bolts from the base of the carburetor and remove the carburetor from the manifold. Remove and discard the old gasket from under the carburetor.
3. To install reverse the removal procedure making sure to install a new gasket under the carburetor base.

Two Barrel Models

1. Remove air cleaner, throttle linkage and choke cable.

2. Disconnect fuel line and distributor vacuum lines.

3. Remove bolts from mounting studs and lift off carburetor.

4. To install, clean manifold mating surface and install a new flange gasket.

5. Install carburetor but do not tighten down stud nuts.

6. Connect fuel line and vacuum lines.

7. Tighten nuts on mounting studs in an alternating fashion so that flange gasket compresses evenly for a good seal.

8. Connect throttle linkage and choke cable, making sure that choke plates are fully open when the choke knob is pushed in.

9. Check throttle for complete travel.

10. Install air cleaner.

11. Adjust carburetor.

4 Barrel Models

1. Remove the air cleaner, throttle linkage, vacuum hoses, fuel lines, and any other hoses and linkages attached to the carburetor.

2. Remove the bolts or nuts holding the carburetor to the manifold, and remove the carburetor from the intake manifold.

3. Discard the base gasket and clean the base and manifold surface of gasket particles.

4. To install the carburetor, reverse the removal procedure, using a new base gasket.

5. Adjust the idle speed and air mixture.

Refer to the Unit Repair Section for adjustments and specifications..

DIESEL FUEL SYSTEM

NOTE: This section contains procedures requiring many special tools. For reference purposes, International Harvester factory numbers are included in the text. These tools may be available from a number of different sources. We encourage you to shop competitively for price and quality and to cross-reference these numbers as necessary in purchasing your tools.

Injection Pump

REMOVAL & INSTALLATION

NOTE: In some applications, this procedure is best done with the engine removed from the vehicle.

1. Remove the inlet and outlet lines from the oil cooler.

2. Remove the bolts (4) and separate the oil filter and lines from the cooler.

Fuel filter removal

Timer installation

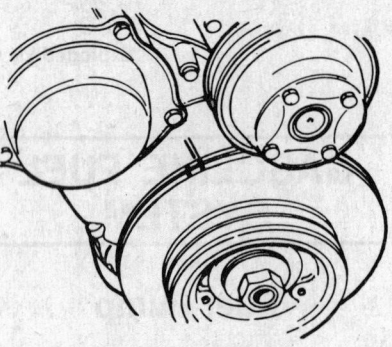

#1 piston at 20° BTDC

3. Remove the coolant hose between the oil cooler and the head.

4. Remove the bolts (10) and separate the cooler from the block.

5. Disconnect the fuel lines and remove the fuel filter from the bracket.

6. Remove the injection lines from the nozzles and pump. Cover all openings immediately.

7. Remove the fan, spacer and pulley from the water pump.

8. Remove the bypass hose from the pump and thermostat housing.

9. Remove the three bolts and lift off the water pump and gasket.

10. Remove the inspection cover and pointer from the flywheel housing and lock the flywheel in place with a locking tool.

11. Flatten the lockwasher and remove the crankshaft pulley nut.

12. Tap evenly around the edge of the pulley using a brass drift, until the cone protrudes from the pulley. Remove the cone.

13. Drive the pulley and damper from the crankshaft with a soft mallet.

14. Remove the inner cover from the timing gear case.

15. Pry out the oil seal.

16. Remove the mounting bolts and tap the case loose with a soft mallet.

17. Remove the tachometer drive support nuts.

18. Remove the timer round nut.

19. Thread the timer extractor, special tool #57926–581 into the timer weight holder. Remove the timer assembly by tightening the extractor bolt.

20. Unbolt and separate the injection pump from the front end plate.

21. Temporarily install the injection pump and gasket on the front plate.

22. Check the timing marks and bring the #1 piston to TDC.

23. Mesh the injection pump drive gear and idler gear at the timing marks.

24. After aligning the injection pump keyway, install the lockwasher and round nut and torque to 50–58 ft. lbs.

25. Install the tachometer drive coupling.

26. Check the backlash between the pump drive gear and the idler gear. Backlash should be 0.0028–0.0079 in. Adjust if necessary.

27. Remove the No. 1 cylinder holder clamp, loosen the delivery valve and pull out the delivery spring. Tighten the valve holder to 22–25 ft. lbs.

28. Connect the fuel supply lines.

29. Bring the No. 1 piston to 20° BTDC. This can be done by aligning the first mark, in normal rotation, on the crankshaft pulley with the raised line on the gear case.

30. Hand prime the pump. Push the pump in all the way toward the block. Move the pump slowly away from the block until the fuel just stops flowing from the valve holder. Lock the pump in place.

31. Remove the delivery holder and assemble the spring. Torque the holder to 22–25 ft. lbs.

32. Install remaining parts in reverse order of removal.

NOTE: Oil filter bolt torque is 15–18 ft. lbs.

Injection Nozzle

REMOVAL, OVERHAUL & INSTALLATION

1. Loosen the injection lines at the pump

and nozzles and remove the lines. Cap the openings immediately.

2. Unscrew the injector and holder from the head.

3. Secure the nozzle holder in a vise and remove the lock nut.

4. Remove the nipple.

5. Remove the nozzle holder body from the nozzle nut.

6. Remove the spacer collar and push-rod.

7. Remove the nozzle holder body from the vise and remove the nozzle spring and adjusting shims.

NOTE: The adjusting shims may be removed with a piece of wire, but great care must be taken to avoid damage to the nozzle tip.

8. Clean fuel oil may be used to clean all parts. Inspect all parts for damage and good fit.

9. Assemble the nozzle in reverse order of disassembly.

10. Install the nozzle in a tester.

11. Operate the tester lever at 1 stroke per second and read the pressure at injection. The pointer will oscillate slightly during injection.

12. Increase or decrease the thickness of the nozzle spring adjusting shims until opening pressure is 1,422.3 psi. A total of 31 different shims are available. A shim thickness of .05mm equals a difference of 85.338 psi.

13. Install the nozzles and lines. Torque the nozzles to 50–65 ft. lbs.

CLUTCH

Clutch Assembly

REMOVAL & INSTALLATION

1. Remove transmission. Extreme care should be taken to support the transmission until it is completely removed so that the main shaft splines will clear the driven member. For transmission removal procedures see Transmission Removal and Installation.

2. Remove flywheel housing cover.

3. Disconnect clevis yoke from clutch release lever.

4. Compress clutch assembly. On 9 spring clutches, the pressure plate is drilled and tapped so that three retaining cap screws and flat washers may be installed. Tighten the cap screws until flat washers and cap screw heads are seated on the back plate. On the 11, 12 and 10 (6 spring, open back plate type) clutches, three retaining spacers are used to hold the clutch assembly compressed during removal. Slightly loosen the back plate to flywheel mounting screws to wedge the retaining spacers into place. On

the 10 six spring (full back plate type) clutch, three $\frac{5}{8} \times 3 \times \frac{1}{4}$ hardwood blocks are used to compress the clutch during removal. Loosen back plate to flywheel retaining bolts enough to wedge the blocks between the back plate inner flange and release fingers.

5. Remove back plate to flywheel bolts and remove back plate assembly and driven disc.

6. When removing the clutch assembly, observe that the balance mark (spot of white paint) on the back plate flange is located as near as possible to the balance mark (''L'') stamped on the flywheel face. These balance marks should be located in the same relative position at clutch installation. If there are no marks, scribe a line to indicate correct position.

7. To install clutch, position the clutch driven member so that the long portion of the hub is toward the rear (all except the 10 6 spring open back plate type, which may be fitted either way). Clutch must be compressed for correct installation.

8. Place clutch assembly over the driven member on the flywheel so that the balance mark (spot of white paint) is as near as possible to the flywheel balance mark (''L''). Loosely install two or three back plate to flywheel mounting screws.

9. Using a clutch aligning arbor or transmission main drive gear shaft to hold the driven member in place, complete installation of the remaining back plate to flywheel mounting screws and lockwashers. Tighten capscrews alternately and evenly.

blocks or retaining spacers which were used to hold the clutch compressed.

11. Install transmission as described in Transmission Removal and Installation.

12. Connect linkage to clutch release lever.

13. Install flywheel housing cover. Adjust linkage or cable as described.

Clutch Adjustment

1. Measure and correct the clutch pedal height to approximately 9 inches.

NOTE: On some models it may be necessary to increase the clutch pedal height setting slightly over the amount specified, in order to obtain complete clutch release.

2. Disconnect the return spring on release fork.

3. Loosen the nut on the cable or linkage rod.

4. Hold the pedal assembly against the pedal stop and lengthen or shorten the rod or cable to obtain zero clearance at the release bearing face and the pressure plate fingers.

5. After obtaining zero clearance, lengthen or shorten cable or linkage to obtain $\frac{3}{32}$ in. between the bearing face and the fingers of the pressure plate.

6. Tighten nut on the cable or linkage rod.

7. Reconnect the return spring.

MANUAL TRANSMISSION

For all overhaul procedures, refer to "Manual Transmission Overhaul" in the Unit Repair section.

REMOVAL & INSTALLATION

Removal and installation of manual transmissions will vary in detail, depending on which vehicle is being serviced. The following general procedure includes the basic steps common to all models.

1. Access to the transmission may be improved by removing cab floor panels if vehicle is equipped.

2. Raise vehicle on a hoist or jack up and support with jack stands.

3. Drain the transmission lubricant.

4. Disconnect drive shaft at the transmission. If the vehicle is equipped with a transfer case which is not mounted directly to the transmission, disconnect the shaft between the transfer case and transmission at the yoke. If the vehicle is equipped with a transfer case which is mounted directly to the transmission, it must be removed with the transmission as a unit and the forward and rear drive shafts must be disconnected. Secure shaft out of the way with wire.

5. Disconnect shift linkage from transmission shift levers. If the vehicle is equipped with a transfer case which is mounted directly to the transmission, disconnect the shift linkage from the transfer case shift levers.

6. If the vehicle is equipped with a transmission mounted handbrake, disconnect the handbrake cable at the relay lever.

7. Disconnect speedometer cable from the transmission.

8. On some models it may be necessary to remove the starter motor.

9. Support the rear of engine by means of a hydraulic jack.

10. Remove the transmission mounting bolts and insulators at the engine rear crossmember. If possible, remove the rear engine crossmember. Remove gear shift lever and housing from top of transmission if applicable.

11. Attach suitable hoisting equipment or jack to transmission and raise enough to support the transmission assembly.

12. Remove top transmission to clutch housing bolts and install transmission guide pins.

13. Remove remaining transmission to clutch housing bolts.

14. Carefully pull transmission rear-

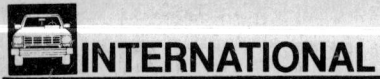

ward, keeping it in line until the main drive gear shaft is clear of the clutch.

CAUTION

Extreme care must be exercised to insure that the weight of the transmission does not rest on the hub of the clutch driven disc.

15. Depending on vehicle model, either lift the transmission up through the floorboard and out the right door or lower it with a jack.

16. Installation is the reverse of the above procedure.

17. Fill transmission with fluid.

Shift Linkage

Different types of transmissions are used which may require the shift linkage to be either mounted in the transmission and controlled by a shift lever, or to have a slight lever mounted remotely with linkage rods connecting the lever to the transmission. No adjustment is provided when the linkage is mounted in the transmission. When the shift lever is remotely mounted, the connecting rods have adjustment provisions. The adjustments are made with the shift control and the transmission arms in the neutral position, and the control rods adjusted to enter either the transmission arms or the shift lever arms with a free fit. Normally the control rods are threaded and trunnions and jam nuts are used to position the rods.

LINKAGE AND CABLE

Shifter rods connect the shift arms of the transfer case to the shift lever arms. Nonadjustable and adjustable links are used on the various models of vehicles. To insure the proper alignment of the rods to the arms, use the following procedure.

Linkage Adjustment

1. Place the shift lever in the neutral position.

2. Remove the shift control rod at the transfer case.

3. Assure that the shift arm of the transfer case is in the center or neutral position.

4. If the control rod is adjustable, position the trunnion or clevis to align with the hole in the shift arm of the transfer case.

5. If the control rod is non-adjustable and the rod does not line up with the hole in the shift arm of the transfer case, replacement or bending will be necessary for the control rod.

6. Reconnect the control rod to the transfer case shift arm and check for proper operation.

Cable Adjustment

A pull on the cable will engage the gears. To disengage, merely push the control cable in. To adjust, follow this procedure.

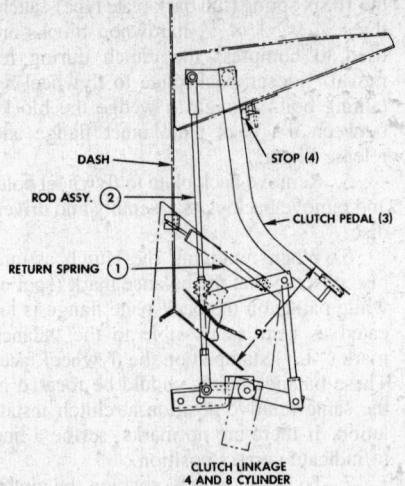

Clutch linkage

1. Pull the control cable knob out approximately two inches and block in this position.

2. Loosen the cable mounting housing jam nut.

3. Remove the two cable mounting housing bolts.

4. Unscrew the cable mounting housing away from the transfer case housing.

5. Confirm the inner clevis is positioned in the engaged position.

6. Turn the cable housing down the cable jacket to a snug fit against the gasket on the transfer case mounting boss. Install the two retaining screws.

7. Turn the jam nut down the cable jacket and secure against the cable mounting housing.

8. Remove the control cable knob block and operate the cable assembly to check the shifter operation.

CABLE

Removal & Installation

1. Leave the control knob pushed in.

2. Loosen the cable jam nut at the cable mounting housing on the transfer case and turn it back to the end of the threads.

3. Remove the two bolts holding the cable mounting housing to the transfer case.

4. Unscrew the housing all the way to the jam nut.

5. Pull the cable mounting housing forward until the inner cable jam nut is clear.

6. Loosen the inner cable jam nut at the shift clevis and unhook the cable end pin from the clevis.

7. Position the shift cable to obtain working clearance.

8. Turn the jam nut and the cable mounting housing to the bottom of the thread end.

9. With the transfer case shifter assembly in the fully engaged position, install the cable mounting housing gasket to the case.

NOTE: The clevis is pulled out to engage.

10. Connect the cable end pin to the shifter assembly and secure with the jam nut.

NOTE: Confirm that the pin is installed flush with the cable end.

11. Block the control knob out approximately two inches.

12. Turn the cable mounting housing down the cable jacket to a snug fit against the gasket on the transfer case mounting boss.

13. Secure the cable mounting housing with the two mounting bolts to the transfer case.

14. Turn the jam nut down the cable jacket and lock against the cable mounting housing.

5. Remove the block from the shift control cable knob and operate the cable to check the shifter operation.

TRANSFER CASE

REMOVAL & INSTALLATION

Frame Mounted

1. Drain the transfer case and disconnect the rear axle drive shaft at the transfer case.

2. Disconnect the front drive shaft at the transfer case.

3. Disconnect the speedometer cable, and indicator light switch wire, if equipped.

4. Disconnect the shift linkage. If equipped with a shift cable, refer to the cable removal and installation outlined previously.

5. Place a transmission jack under the transfer case and remove the mounting bolts from the frame to case.

6. Remove the transfer case from the vehicle.

7. The installation of the transfer case is the reverse of removal.

Transmission Mounted

1. Disconnect the rear driveshaft at the transfer case and drain the case assembly.

2. Disconnect the front driveshaft at the transfer case.

3. Disconnect the speedometer cable and the indicator light switch wire, if equipped.

4. Disconnect the shift linkage or cable.

5. Place a transmission jack under the transfer case and remove the flange bolts holding the transfer case to the transmission.

6. Pull the transfer case rearward to disengage the transmission output shaft from the coupler.

7. Lower the transfer case and remove from the vehicle.

8. The installation of the transfer case is the reverse of removal.

AUTOMATIC TRANSMISSION

Model T-407

REMOVAL & INSTALLATION

NOTE: The transmission and converter must be removed as a unit assembly. Damage can result to the converter drive plate, pump bushing, or to the pump seal, if the converter is allowed to remain on the converter drive plate.

1. Connect a remote switch to the starter solenoid so that the engine can be rotated from under the vehicle.

2. Disconnect the coil high tension cable.

3. Raise the vehicle and support safely.

4. Remove the engine rear crossmember on 4 × 4 vehicles, if necessary.

5. Remove the cover plate from the front of the converter housing to provide access to the converter drain plug and mounting bolts.

6. Rotate the engine to bring the drain plug to the six o'clock position. Drain the converter and loosen the pan bolts to drain the transmission.

7. Mark the converter and drive plate to aid in the assembly. Rotate the engine to locate the converter-to-drive plate bolts and remove the bolts.

8. Disconnect the negative battery cable and remove the starter motor assembly.

9. Disconnect the wires from the back-up light and neutral start switch.

10. Disconnect the gearshift cable or rod and bellcrank from the transmission.

11. Disconnect the throttle rod from the left side of the transmission.

12. Disconnect the cooler lines at the transmission and remove the filler tube.

13. Disconnect the speedometer cable, and move cable away from the transmission.

14. Disconnect the front universal joint and secure the shaft out of the way.

15. On vehicles equipped with parking brake mounted on the rear extension, remove the parking brake cable.

16. On vehicles equipped with dual exhaust, the left exhaust system may have to be removed.

17. Install an engine support fixture to hold the rear of the engine.

18. Raise the transmission slightly, and remove the support crossmember holding the rear mount assembly.

19. Remove all bell housing bolts.

20. Carefully move the transmission assembly rearward off the block dowels and disengage the converter hub from the end of the crankshaft. Place a converter holding tool on the bell housing to hold the converter in place.

21. Lower the transmission assembly and remove the transmission from the vehicle.

22. To remove the converter assembly from the transmission, remove the holding tool and carefully slide the converter out of the transmission.

23. Rotate the pump rotors with tool SE-2402 or its equivalent, so that the lugs on the pump inner rotor are vertical.

24. Position the converter so that the impeller shaft slots are vertical and carefully slide the converter assembly over the input shaft and reaction shaft. Make sure that the converter slots fully engage the pump inner rotor lugs.

NOTE: The surface of the converter front cover lug should be at least ½ inch to the rear of a straightedge, placed on the face of the bell housing, when the converter is pushed all the way into the transmission.

25. Install the converter holding tool to hold the converter in place.

26. Position the transmission on a jack assembly and move the unit under the vehicle.

27. Rotate the converter to align the previously made marks on the drive plate and converter.

28. Raise the transmission and align with the engine. Install a pilot stud to aid in the alignment of the converter to the drive plate. Carefully work the transmission assembly forward over the engine block dowels with the converter hub entering the crankshaft opening.

29. Install the converter housing bolts and tighten to specified torque.

30. Install the crossmember and mount at the rear of the transmission. Remove the engine support fixture.

31. Install the oil filler tube and speedometer cable.

32. Connect the throttle rod and the gear shift rod to the transmission levers.

33. Connect the wires to the neutral start and back-up light switch.

34. Install the drive shaft and front universal joint.

35. Install the starter motor assembly.

36. Remove the pilot stud from the converter and install the bolts to the converter-drive plate assembly.

37. Install the cooler lines to the transmission.

38. Install the converter access plate on the front of the converter housing.

39. If the left exhaust system was removed, replace the pipes and brackets.

40. Install the parking brake cable and adjust, if equipped with the extension housing parking brake assembly.

41. Adjust the shift and throttle linkage.

42. Fill the transmission and connect the negative battery cable, if not done, and start the engine. Recheck the fluid level and refill as necessary.

TRANSMISSION FLUID DRAIN AND REFILL

1. Raise the vehicle on a jack or hoist. Support safely.

2. Place a large drain container under the transmission oil pan.

3. Loosen the pan bolts and tap one corner of the pan to break it loose, allowing the fluid to drain.

4. Remove the access plate from the front of the converter housing. Remove the converter drain plug and allow the fluid to drain.

5. Remove and clean the pan, remove the fluid filter and discard.

6. Install a new filter assembly on the valve body and tighten the screws securely.

7. Using a new pan gasket, install the pan and tighten the bolts securely.

8. Install and tighten the converter drain plug.

9. Install the converter housing access plate.

10. Install six quarts of transmission fluid into the transmission. Start the engine and allow to run for two minutes. Check the fluid level and add enough oil to bring the level to the "ADD ONE PINT" mark.

11. Recheck the level after moving the selector lever through all the gear positions and after the transmission has reached normal operating temperature. The level should be between the "FULL" mark and the "ADD ONE PINT" mark.

KICKDOWN BAND ADJUSTMENT

NOTE: The kickdown band is located on the left side of the transmission case near the throttle lever shaft.

1. Loosen the locknut and back off approximately five turns.

2. Tighten the adjusting screw to 10 ft. lbs.

3. Back off the adjusting screw 2¼ turns with the 6 and 8 cylinder engines. Hold the adjusting screw in position and tighten the lock nut to 29 ft. lbs.

LOW AND REVERSE BAND ADJUSTMENT

1. Raise the vehicle, support safely, drain the transmission fluid, and remove the pan.

2. Loosen the lock nut on the adjusting screw.

3. Tighten the adjusting screw to 10 ft. lbs.

4. Back off adjusting screw 2¼ turns. Tighten the lock nut to 30 ft. lbs.

5. Install the pan using a new pan gasket.

6. Fill the transmission with fluid, start the engine and recheck the level. Add as necessary.

BACK-UP LIGHT AND NEUTRAL START SWITCH

No provisions are made for any adjustments of the back-up light and neutral start switch. The neutral start circuit is controlled by the inner terminal and the back-up light circuits are controlled by the two outside terminals.

The replacement of the switch is accomplished by unscrewing the switch from the transmission case, and screwing a new switch into the case. Since fluid leakage will occur when removing the switch, fluid must be added after the new switch is installed.

SHIFT LINKAGE

Adjustable Cable Control

1. Install cable conduit anchor clamps at both ends.
2. Install swivel on the control lever so that a distance of .55 inch exists from the end of the cable to the opposite side of the trunnion. Tighten the jam nut securely.
3. With the control in PARK position and transmission lever in the full rearward position (PARK detent), adjust the yoke so that the rod end pin installs freely and secure the yoke nut and install the cotter pin.

Column Shift

1. Assemble all linkage parts, but leave the upper control rod bolt loose.
2. Place the selector lever in DRIVE position.
3. Move the shift control lever on the transmission to the DRIVE position.
4. Tighten the upper bolt on the control rod to 14–16 ft. lbs.
5. Check the adjustment as follows: Shift effort must be free and detents feel crisp. All gate stops must be positive. Key start must only occur in the PARK or NEUTRAL positions. Detent positions must be in proper relationship to the transmission lever positions.

THROTTLE VALVE LINKAGE ADJUSTMENT

1. With the engine off and an assistant holding the accelerator pedal to the floor, check for full carburetor throttle plate opening.
2. If necessary, adjust the throttle cable and pedal floor stop to obtain wide open throttle.
3. If necessary, adjust the idle speed of the engine with the use of a tachometer and with the engine at normal operating temperature and the carburetor off the fast idle cam. Adjust the curb idle speed, (throttle stop solenoid activated) with the transmission in neutral and the air conditioning in the OFF position.

NOTE: Be sure that carburetor is not being held open by a deceleration valve dashpot, solenoid valve, or a vacuum throttle modulator valve.

─────── CAUTION ───────
All components in the throttle control and transmission linkage system must operate freely with absolutely no sticking, excessive friction, or interference from other chassis components.

DRIVETRAIN

Front Driveshaft

The front driveshaft connects from the transfer case to the front axle companion flange. The U-joints are attached to the differential by bearing flanges on older models and U-bolts on newer ones.

REMOVAL & INSTALLATION

1. Raise the front of the vehicle and place it on jackstands.
2. Place the transfer case in gear.
3. Remove the attaching bolts from the flange on the transfer case. Lower the driveshaft slowly to the ground.
4. Remove the attaching bolts from the flange on the front axle and pull the shaft from the vehicle.
5. To install. Attach the front and rear universal joints.
6. Lower the vehicle.

Rear Driveshaft

REMOVAL & INSTALLATION

2WD

1. Raise the vehicle and support it securely.
2. Remove the attaching bolts from the companion flange and lower the end of the driveshaft to the floor.
3. Partially drain the transmission lubricant.
4. Pull the yoke from the rear of the transmission.
5. To install. Insert the yoke onto the transmission output shaft. Be careful not to damage the rear transmission seal.
6. Attach the rear U-joint to the companion flange.
7. Fill the transmission to the proper level.
8. Lower the vehicle.

4WD

1. Raise and securely support the vehicle.

2. Disconnect the rear driveshaft bolts from the companion flange.
3. Remove the driveshaft-to-transfer case flange attaching bolts.
4. Remove the driveshaft.
5. To install: Attach the front U-joint to the transfer case flange.
6. Connect the rear end of the driveshaft to the companion flange of the rear axle.
7. Lower the vehicle.

U-Joints (Driveshaft)

Refer to the Unit Repair Section for overhaul procedures.

FRONT DRIVE AXLE

Leaf Spring

REMOVAL & INSTALLATION

1. Raise the vehicle and support on the frame rails behind the front springs with floor stands.
2. Remove the shock absorber from the spring.
3. Remove the U-bolts, spring bumpers and retainer, or the U-bolt seat.
4. Remove the lubricators, if used.
5. Remove the nuts from the shackles and bracket pins.
6. Slide the spring off the bracket and shackle pins.
7. Remove the spring from the vehicle.
8. Installation is the reverse of removal. Tighten all nuts and bolts securely.

Front Drive Axle Removal

1. Jack up truck until load is removed from springs and block up frame to safely hold weight.
2. Drain lubricant from main housing and, if applicable, from wheel end housings.
3. Disconnect brakes.
4. Disconnect drag link from ball stud bracket.
5. Disconnect drive shaft from pinion shaft yoke.
6. Supporting axle with a portable floor jack, remove spring U-bolts.
7. Roll axle assembly out from under truck.
8. To install, reverse the above procedure.

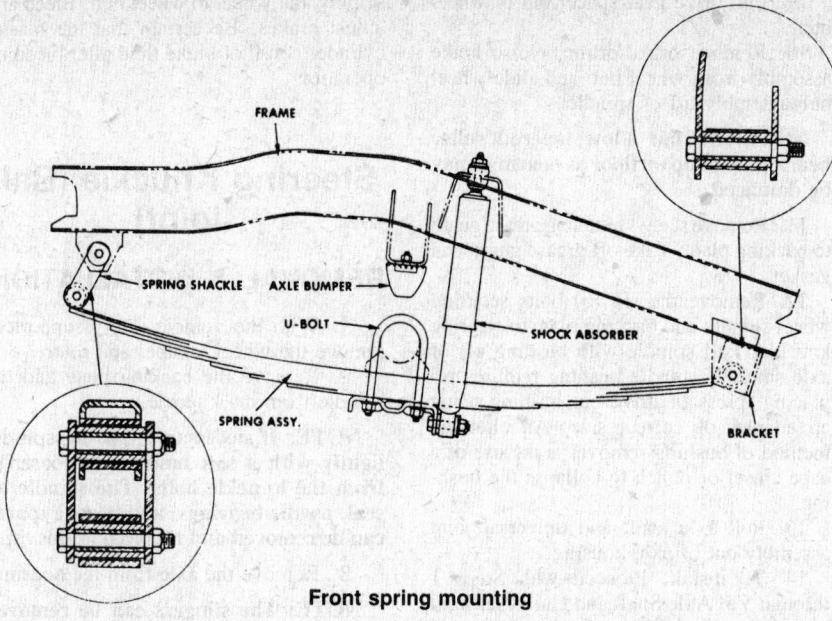

Front spring mounting

Front Drive Axle Adjustments

Preload on the knuckle bearings of these front axles must be maintained at all times. Check for looseness each time knuckle is lubricated.

1. Jack up front end of truck until off-center weight of the wheel is relieved (wheel just barely touching ground).
2. Remove wheel and wheel adapter from hub.
3. Disconnect tie rod and drag link.
4. Remove axle shaft.
5. To remove play (check for play by pushing and pulling on top and bottom of knuckle) and increase preload drag, turn adjusting bolt into back of knuckle. Preload should read (spring scale hooked into end of steering arm) 12 lbs.

Front Wheel Bearing Adjustment

1. While continuously rotating the front wheel, gently tighten the wheel bearing adjusting nut until it places tension on the bearing or thrust washer. If you notice possible brake drag, adjust the brakes to make sure brake shoe drag does not interfere with the adjustment.
2. Continue rotating the wheel while torquing the adjusting nut to 30 ft. lbs.
3. Then, back off the adjusting nut 1/4–1/3 of a turn. Install the wheel bearing nut lockwasher and jam nut. Torque the jam nut to 125–150 ft. lbs.
4. Bend the lockwasher over the jam nut and/or locknut.

NOTE: For overhaul procedures, see Unit Repair Section.

Front Axle Shaft

REMOVAL & INSTALLATION

Axles Having Drive Flange

1. Raise vehicle, support with floor stands and remove wheel from vehicle.
2. Remove grease cap and snap-ring from end of axle shaft.
3. Remove drive flange cap screws, lock-washer, flange and gasket. If equipped with locking hubs, bend up locking tab, take out capscrews and remove clutch body.

NOTE: Lift off clutch body holding it erect so as not to let drive pins fall out of body. If they do fall out, be certain to install them during reassembly.

4. Remove hub body. Loosen setscrew

and unscrew drag shoe from spindle. Remove brake drum countersunk setscrews, where applicable and remove drum.

5. Bend the lip on the wheel bearing lockwasher away from the outer wheel bearing nut and remove the nut and lockwasher. Remove wheel bearing adjusting nut (inner) and bearing lockwasher.
6. Remove the wheel hub with wheel bearing.
7. Remove backing plate and wheel spindle retaining bolts and lockwashers. Support backing plate to prevent damage to brake hose if hose is not disconnected.
8. Remove wheel bearing spindle with bushing. If spindle bushing requires replacing, press out bushing using an adapter of correct size. An alternate method of bushing removal is the use of a cape chisel or punch to collapse the bushing.
9. Pull axle shaft and universal joint assembly out of axle housing.
10. Insert axle shaft and universal joint assembly into axle housing. Position splined end of axle shaft into differential pinion gear and push into place.
11. If wheel bearing spindle bushing was removed, press new bushing into spindle using an installer tool or adapter of proper size. Lubricate ID of bushing with chassis lube when installed to provide initial lubrication. Bushing should be pressed in until bushing flange is seated against shoulder in spindle. Assemble wheel spindle and backing plate to steering knuckle. Secure with six (6) bolts and lockwashers and tighten to specifications. Connect hydraulic brake fluid line if disconnected.
12. Pack wheel bearings using a pressure lubricator or by carefully working lubricant into bearing cones by hand. Slide lubricated inner wheel bearing on spindle until it stops against spindle shoulder.
13. Apply thin coating of lubricant specified for wheel bearings to seal lip and install seal into wheel hub using an adapter of correct diameter. Lip of seal should ex-

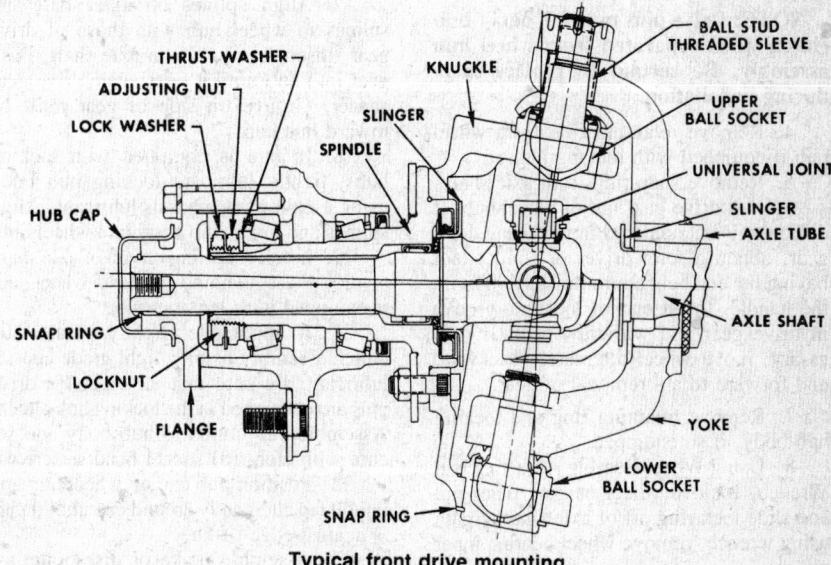

Typical front drive mounting

tend towards wheel (away from backing plate assembly).

14. Assemble wheel hub on spindle. Install lubricated outer wheel bearing cone on spindle. Push cone on spindle until it rests against bearing cup.

15. Install wheel bearing lockwasher and adjusting (inner) nut. Tighten adjusting nut until there is a slight drag on the bearings when the hub is turned; then back-off approximately one-sixth turn.

16. Install tang-type lockwasher and lock nut (outer). Tighten nut and bend lockwasher tang over lock nut. If axle is equipped with locking hubs, install drag shoe on spindle and tighten setscrew.

17. Align splines of drive flange with those of axle shaft and secure drive flange and new gasket to wheel hub with capscrews and lockwashers. Tighten capscrews securely. If equipped with locking hubs, lightly lubricate hub body and clutch using a light grade chassis lubricant an install new gasket, hub body, snap ring and hub clutch. Be certain that all drive pins are positioned in locking hub clutch when clutch is installed. Secure hub clutch to wheel hub with capscrews and lock. Tighten to specifications and bend tang over head of capscrew.

18. Install snap-ring and grease cup if not equipped with locking hubs.

19. Assemble brake drum and wheels to wheel hub. Bleed and adjust brakes. Be certain that master cylinder is full of brake fluid after completing bleeding operation.

Axles Having Drive Gear

1. Raise and support vehicle with floor stands placed under frame rails. Remove wheel from vehicle.

2. Lightly tap alternately around edge of hub cap with hammer and screwdriver or similar tool until hub cap is removed.

3. If axle is equipped with locking hubs, remove the eight (8) socket-head setscrews securing hub clutch assembly to wheel hub assembly.

NOTE: Drive pins may fall out of hub clutch when separated from wheel hub assembly. Be certain to replace them during installation.

4. Remove retaining ring from wheel hub if equipped with locking hubs.

5. Remove snap-ring from axle shaft.

6. Pull drive gear out of wheel hub. If difficulty is encountered in removing drive gear, obtain a screwdriver or similar tool having the end bent approximately 90° with the handle. Insert end of tool into groove in drive gear and withdraw gear. If necessary, move wheel alternately backward and forward to aid removal of gear.

7. Remove retaining ring and locking hub body, if so equipped.

8. Using Wheel Bearing Adjusting Nut Wrench, remove wheel bearing outer nut and slide lock ring off of axle shaft. Again using wrench, remove wheel bearing inner nut.

9. Pull drive gear spacer out of wheel hub.

10. Remove brake drum or disc brake assembly from wheel hub and slide wheel hub assembly off of spindle.

NOTE: Do not allow tapered roller bearings to drop on floor as bearings may be damaged.

11. Remove screws retaining grease guard to backing plate. Take off grease guard and gasket.

12. Remove the six (6) bolts securing wheel spindle and backing plate to steering knuckle. Pull spindle with bushing off of axle shaft. If spindle bushing requires replacing, press or drive out bushing using an adapter of correct size. An alternate method of bushing removal is the use of a cape chisel or punch to collapse the bushing.

13. Pull axle shaft and universal joint assembly out of axle housing.

14. To install. Proceed with Steps 1 through 5 of Axle Shaft and Universal Joint Installation (Axles having drive flange).

15. Insert drive gear spacer over spindle and against outer wheel bearing cup.

16. Position wheel bearing inner adjusting nut Wheel Bearing Adjusting Nut wrench with pin in nut extending toward handlle end of wrench. Install nut on spindle and tighten until it is snug against outer wheel bearing; then loosen adjusting nut $\frac{1}{4}$ turn. Align tang on adjusting nut lock ring with groove in wheel spindle. Slide ring on spindle and index pin on adjusting nut with hole in lock ring. If pin will not index with hole in lock ring, turn adjusting nut to the left (Loosen) until it will index.

NOTE: When attempting to index pin with hole in lock ring, turn nut very slightly since adjusting nut should be locked with first hole in lock ring past $\frac{1}{4}$ turn lose. Position wheel bearing outer nut in adjusting nut wrench and install on spindle. Tighten nut securely.

17. Align splines on axle shaft and splines in wheel hub with those of drive gear. Insert drive gear on axle shaft. Push gear into hub until it rests again drive gear spacer. Groove on side of gear must be toward hub cap.

18. If axle is equipped with locking hubs, lightly lubricate locking hub body using a light grade chassis lubricant. Align splines and insert hub body into wheel hub.

19. Install snap-ring on end of axle shaft.

20. Place retaining groove in wheel hub, if equipped with locking hub.

21. If applicable, lightly grease hub clutch assembly using a light grade chassis lubricant. Be sure that all eight (8) drive pins are positioned in the locking hub clutch. Assemble hub clutch to hub body and secure with eight (8) socket head setscrews.

22. Position hub cap on wheel hub and lightly tap alternately around cap until flange is against edge of hub.

23. Assemble brake or disc brake as-

sembly and wheel to wheel hub. Bleed and adjust brakes. Be certain that the master cylinder is full of brake fluid after bleeding operation.

Steering Knuckle (Ball Joint)

REMOVAL & INSTALLATION

1. With the vehicle safely supported, remove the wheel, caliper and rotor.

2. Remove the backing plate and the spindle from the knuckle.

NOTE: If necessary, tap the spindle lightly with a soft hammer to loosen it from the knuckle bolts. The spindle oil seal, needle bearings, and bronze spacer can be removed and replaced at this time.

3. Remove the axle from the housing.

NOTE: The slingers can be removed from the axle by using pullers or tapping the axle through the slingers.

4. Disconnect and remove the tie rod from the steering arm.

5. Remove the cotter pin from the upper ball socket stud and remove the nut.

6. Remove the nut from the lowerball socket stud and discard.

NOTE: This nut is of a special torque design and should only be used one time.

7. Remove the lower ball socket snapring (used on 4×4 applications only), and unseat the upper and lower ball socket studs with a lead hammer or with a puller tool arrangement, to separate the knuckle from the yoke.

NOTE: If the upper ball socket stud remains in the yoke flange, remove it by striking it on the stud with a soft hammer.

8. With the aid of puller tools or a press and ram, remove the bottom ball socket.

9. Reverse the knuckle and remove the upper ball socket.

10. With the aid of a special socket, remove the threaded sleeve in the top flange of the yoke.

11. Assemble the lower ball socket into the knuckle with a press and ram or a puller tool arrangements, making sure that the ball socket is firmly seated against the knuckle. Install the snap-ring on the 4×4 application.

12. Assemble the upper ball socket into the knuckle with a press and ram or a puller type tool arrangement, making sure that the ball socket is firmly seated against the knuckle.

NOTE: Use a 0.0015 in. feeler gauge blade between the socket and knuckle. The blade should not enter at the minimum area of contact.

13. Install new threaded sleeve into the

top flange of the yoke, leaving approximately two threads exposed.

14. Install the knuckle assembly to the yoke, using a new nut on the lower ball socket stud. Torque the lower nut to 80 ft. lbs.

15. With the use of a special socket, torque the threaded sleeve to 50 ft. lbs. in the upper yoke flange.

16. Install the top ball socket stud nut and torque to 100 ft. lbs. Align the cotterpin holes between the stud and the castellated nut. Do not loosen nut to align the holes. Install the cotter pin.

17. Assemble the tie rod to the steering arm.

18. Assure that slingers are properly installed on the axle shaft and install the shaft into the housing.

19. Position the spindle over the axle end with the bronze bushing in place.

20. Install the backing plate.

21. Install the hub rotor, caliper and wheel assembly, and lower the vehicle.

CHECKING BALL JOINT PLAY

To check the ball sockets for excessive looseness, raise the vehicle and attach a dial indicator to the lower yoke or axle tube and set the indicator against the knuckle or lower ball socket, with a loaded pressure so as to read in both directions. Grasp the wheel at the top and bottom and move the wheel inward and outward. If the total indicator reading exceeds 0.020 in., both the upper and lower ball sockets should be replaced.

Front Drive Locking Hubs

Three types of locking hubs are used: Manual, Lock-O-Matic and Warn. Manual locking hubs are either engaged or disengaged, depending on how they are set. Lock-O-Matic hubs, when in "free" position,

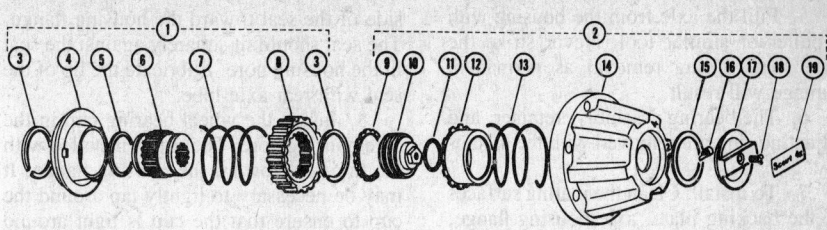

1. Clutch and bearing assembly
2. Cap assembly
3. Retaining ring
4. Bearing hub
5. Washer
6. Hub
7. Compression spring
8. Clutch ring
9. Clutch nut
10. Dial screw
11. O-ring seal
12. Clutch cup
13. Compression spring
14. Hub cap
15. Washer
16. Detent dial
17. Control dial
18. Screw
19. Label

Components of a manual locking hub

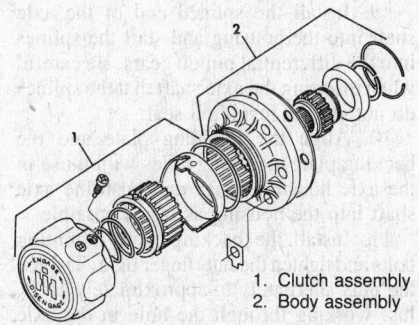

1. Clutch assembly
2. Body assembly

Components of a Dualmatic locking hub

automatically engage axle and wheel when forward torque is applied by the axle shaft. Thus, whenever front wheel drive is disengaged at the transmission, the wheels freewheel. "Lock" position is required only when engine braking control on the front wheel is desired.

FRONT LOCKING HUBS

1. Bend up tabs on mounting bolt lock washers.

2. Remove mounting bolts using a thin-walled socket or appropriate hex wrench (externally splined type).

3. When clutch body is lifted off, im-

mediately tilt it up so that the drive pins do not fall out.

4. Remove lock ring holding hub body onto axle shaft and pull off hub body.

5. Remove drag shoe (Lock-O-Matic only) from axle spindle by loosening hex-head set screw and unscrew drag shoe.

6. To install, reverse the above procedure.

REAR AXLE

Bearing and Seal

REPLACEMENT

Adjustable Bearing Ring Type

1. Raise and support the vehicle securely using jackstands.

2. Remove the rear wheels.

3. Remove the metal retaining clips from the wheel lugs and pull off the brake drum. It may be necessary to back off the brake shoe adjuster to prevent binding between the shoes and the drum.

4. Remove the backing plate retaining bolts. These bolts also run through the axle retaining plate.

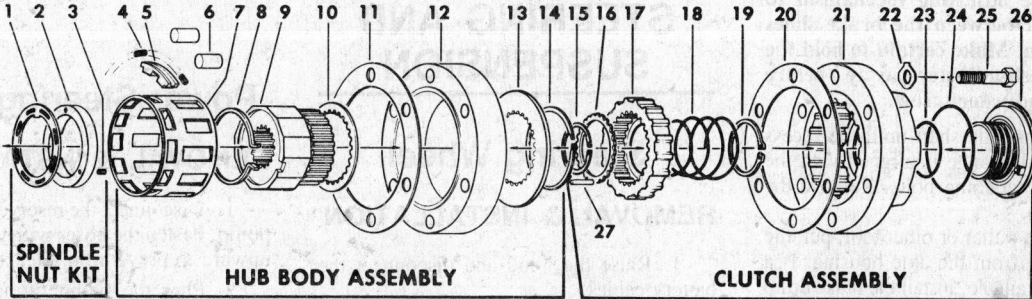

SPINDLE NUT KIT	HUB BODY ASSEMBLY	CLUTCH ASSEMBLY

1. Lockwasher
2. Spindle nut
3. Set screw
4. Garter spring
5. Friction shoe
6. Roller
7. Cage
8. Centering spring
9. Shaft axle hub
10. Centering spring
11. Gasket
12. Hub body
13. Washer, bearing
14. Retaining ring
15. Retaining ring
16. Clutch nut
17. Clutch, ring and cup
18. Compression ring
19. Retaining ring
20. Gasket
21. Cap
22. Lockwasher tab
23. O-ring
24. U-ring
25. Cap screw
26. Control dial
27. Groove pin

Components of an automatic locking hub

5. Pull the axle from the housing with a puller or similar tool. Never strike the axle shaft during removal as permanent damage will result.

6. The bearing assembly, retainer, and adjusting ring are removed with the assembly.

7. To install: Clean the mating surfaces of the backing plate, axle housing flange, and the wheel bearing retainer.

8. Install a new gasket between the backing plate and the axle housing flange.

9. Place a new gasket on the front face of the backing plate so that when the axle is installed the gasket will be between the backing plate and the wheel bearing retainer.

10. Turn the adjusting ring into the bearing retainer approximately two turns. This is to prevent excessive pressure on the wheel bearings or movement of the bearing cone or the clinch ring on the axle shaft.

NOTE: Do not thread the adjusting ring completely into the bearing retainer until the retainer and the backing plate have been bolted to the axle housing flange and tightened to the correct torque.

11. Install the axle shaft assembly. Do not damage the axle seal when inserting the shaft.

12. Install the adjuster ring lock and secure the lock.

13. Secure the bearing retainer and the backing plate to the axle housing flange and torque to 40–45 ft. lbs.

14. Coat the threads on the adjuster with a good waterproof sealer and install the adjusting ring into the wheel bearing retainer.

15. Adjust the axle shaft end-play.

16. Install the brake drums and retaining clips.

Unit Bearing Type

1. Raise the rear of the vehicle and support securely.

2. Remove the tires and wheels.

3. Remove the brake drums.

NOTE: It might be necessary to back off the brake adjusting mechanism to avoid contact between the brake shoes and the drum. Make certain to hold the automatic self-adjuster away before turning the adjuster screw.

4. Turn the axle shaft until the access hole in the axle flange is aligned with the backing plate retaining bolts. Remove the bolts and nuts.

5. Using a puller or other tool, pull the axle assembly from the axle housing. If a puller is not available, install the brake drum backwards and tighten two lug nuts. Exert outward pressure to remove the outside seal from the housing bore.

6. Remove the wheel bearing cup and the inside oil seal from the axle housing. The bearing cup is a loose fit in the housing but the clearance may not be sufficient to permit removal with your fingers.

7. Install the new oil seal with the closed side of the seal toward the housing flange. The seal should fit squarely against the seat in the housing bore. Lubricate the lip of the seal with rear axle lube.

8. Install the wheel bearing cup in the axle housing bore. The cup is installed with the thin side toward the housing flange. It may be necessary to lightly tap around the cup to ensure that the cup is tight around its seat.

NOTE: If a new wheel bearing is used, the bearing cup will be bonded to the rib ring and the cup will be installed with the axle shaft. Do not break the bond or install the cup separately.

9. Install the splined end of the axle shaft into the housing and start the splines into the differential pinion gears. Be careful when inserting the axle shaft that the splines do not cut the inside oil seal.

10. Align the retaining plate and the backing plate mounting holes with those in the axle housing flange and push the axle shaft into the housing as far as possible.

11. Install the backing plate mounting bolts and tighten the nuts finger tight. Tighten the nuts alternately to approximately 15 ft. lbs. working through the hole in the axle shaft flange. The nuts should be tightened so that the seal and the wheel bearing are drawn tight against their seats.

12. Torque the backing plate mounting bolts to 50–60 ft. lbs.

13. Adjust the brakes until the brake drum will just slip over the shoes. Install the drum.

14. Install the rear wheels and tighten the lug nuts. Back the vehicle up and apply the brakes. This will activate the automatic brake adjusters.

Locking Differentials

For overhaul procedures of differentials with ''NoSPIN'' and ''PowrLok'' locking units, see Rear Axle in the General Repair Section.

STEERING AND SUSPENSION

Steering Wheel

REMOVAL & INSTALLATION

1. Raise the hood and disconnect the battery cables.

2. Remove the horn cap, spring, and horn button baseplate.

3. Remove the retaining bolt and washer from the center shaft. Note the position of the long screw.

4. Using a puller, remove the wheel from the steering column.

5. To install, reverse the removal procedure.

Turn Signal Switch

REPLACEMENT

1. Disconnect the battery cables.

2. Remove the horn button and spring.

3. Remove the three screws which hold the horn button retaining plate.

4. Remove the steering wheel.

5. Remove the horn button contact ring located at the upper end of the column assembly.

6. Remove the turn signal switch retaining screws and pull out the switch assembly.

7. Installation is in the reverse of removal.

Steering Gear

For manual steering gear overhaul, see the Unit Repair Section.

STEERING GEAR REMOVAL

1. Loosen collar clamp at bottom of steering wheel column. Disconnect any wiring.

2. Remove nut or loosen clamp bolt which secures steering arm to lever shaft, removing steering arm from lever shaft using a suitable puller if necessary.

3. Remove mounting bolts and steering gear assembly.

4. To install, reverse the above procedure, taking special care not to bind steering column if there is no universal joint.

DRAG LINK ADJUSTMENT

To adjust the drag link, remove the cotter pin and turn the adjusting plug in the desired direction. If excess play is present in the link, turn the adjusting plug inward until it is tight and then back off to the first cotter pin hole. Install a new cotter pin of the correct size.

Power Steering Pump

REMOVAL & INSTALLATION

1. Disconnect the reservoir hoses at the pump. Fasten the hoses with the ends raised upward to prevent fluid leakage.

2. Plug the pump fittings to prevent leakage of oil from the pump.

3. Loosen the pump-to-bracket mounting bolts, lean the pump to one side, and remove the drive belt.

4. Remove the bolts which hold the pump to the mounting bracket and remove the pump assembly.

5. Installation is the reverse of removal.

Tie Rod End

REMOVAL & INSTALLATION

1. Remove the cotter pins and retaining nuts at both ends of the tie rod and from the end of the connecting rod where it attaches to the tie rod.

2. Remove the nut attaching the steering damper push rod to the tie rod bracket and move the damper aside.

3. Remove the tie rod ends from the steering arms and connecting rod with a puller.

4. Count the number of threads showing on the tie rod before removing the ends, as a guide to installation.

5. Loosen the adjusting tube clamp bolts and unthread the ends.

6. Installation is the reverse of removal. Adjust toe-in, if necessary.

Shock Absorbers

REMOVAL & INSTALLATION

NOTE: Before installing new shocks, they should be purged of air. To do this, hold the shock upright and fully extend it, then invert and compress it. Do this several times.

1. Remove the locknuts and washers.

2. Pull the shock absorber eyes and rubber bushings from the mounting pins.

3. Install the shocks in the reverse order of the removal procedure.

NOTE: Squeaking usually occurs when movement takes place between the rubber bushings and the metal parts. The squeaking may be eliminated by placing the bushings under greater pressure. This is accomplished either by adding additional washers or by tightening the locknuts. Do not use mineral lubricant to stop the squeaking as it will deteriorate the rubber.

Front End Alignment

Refer to the Unit Repair Section for specifications.

BRAKES

Late model trucks are equipped with a dual hydraulic brake system in which there are separate hydraulic systems for the front and rear brakes. In this dual system a warning light switch operates a warning light on the dashboard when there is a pressure failure in either the front or rear system. A power system may be employed to reduce the effort applied to the brake pedal.

For all brake system service procedures not detailed below, refer to "Brakes" in the Unit Repair section.

Master Cylinder

REMOVAL & INSTALLATION

1. Disconnect hydraulic lines from master cylinder.

2. Disconnect master cylinder pushrod at brake pedal and remove nuts securing cylinder to dash panel.

3. If master cylinder is mounted on power unit, remove nuts securing master cylinder to power unit and remove cylinder from vehicle.

4. Installation is the reverse of the above procedure.

5. Bleed the brake system.

RESETTING THE WARNING LIGHT SWITCH

Once a difference of 85–150 psi pressure between the front and rear systems has activated the warning light switch, it will not go off by itself and must be manually reset.

1. Clean switch and disconnect wire from terminal.

2. Unscrew and completely remove switch from body. This will allow the pistons to center and hold the switch in "off" position.

3. Screw switch back into body and reconnect wire to terminal.

4. If fluid is in the switch cavity, press brake pedal to see if pistol O-ring seals are leaking. If there is leakage, the O-rings must be replaced.

5. Warning light switch should be checked periodically for proper function and the presence of foreign matter and dirt.

Wheel Cylinder (Rear Wheel)

REMOVAL & INSTALLATION

1. Raise the rear of the vehicle and support it safely.

2. Remove the drum.

3. Remove the brake lining from the brake support plate.

4. Remove the hydraulic line from the wheel cylinder.

5. Remove the wheel cylinder attaching bolts and remove the cylinder.

6. Install the wheel cylinder on the brake support plate.

7. Install the hydraulic line to the wheel cylinder.

8. Install the brake shoes on the brake support plate.

9. Install the drum.

5. Install the wheel assembly.

6. Bleed the system and refill the master cylinder.

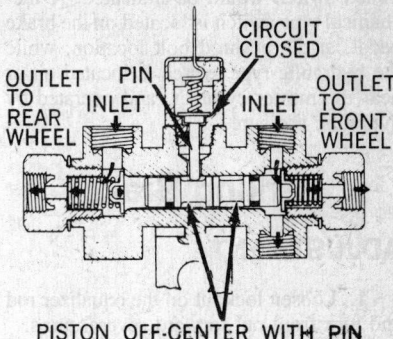

PISTON OFF-CENTER WITH PIN UNEQUAL FLUID PRESSURE AT OUTLETS

Warning light switch circuit closed

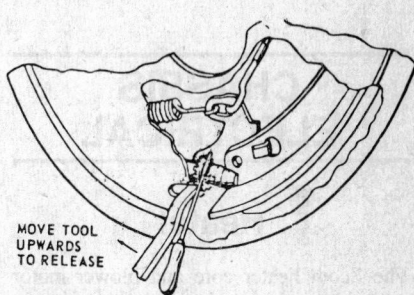

Backing off adjustment—self-adjusting brakes

Disc Brakes

NOTE: Refer to the Unit Repair Section.

Brake Pedal Adjustment

There are no provisions available for the adjustment of the brake pedal height. However, it should be checked to determine if sufficient height exists. Corrections can only be made by replacement of parts, alignment, or straightening of the affected parts. To determine if sufficient pedal height exists, open a wheel cylinder bleed valve to simulate a failed system, and depress the brake pedal. The pedal should not contact the floor board during this test.

NOTE: Close the bleeder valve before releasing the brake pedal. The brake warning light switch will have to be reset after the test is completed.

Stoplight Switch Adjustment

No stoplight switch adjustments are provided. If the stop lamps are inoperative, a defective switch, defective bulbs, loose or broken connections, or an improper positioned switch would be indicated. A mechanical type switch is located on the brake pedal, at the pushrod bolt location, while the hydraulic type switch is located on or near the master cylinder, and operated by hydraulic pressure.

Parking Brake

ADJUSTMENT

1. Loosen locknut on the equalizer rod and turn front nut forward several turns.
2. Turn the locknut (rear) forward just enough to remove any slack but not so much that the brake shoes lift of their anchors.
3. Tighten both nuts against the equalizer.

CHASSIS ELECTRICAL

Heater

The Scout heater core and blower motor assembly are located in the right rear corner of the engine compartment.

Heater Core

REMOVAL & INSTALLATION

1. Drain the cooling system and remove the negative battery cable.
2. Remove the heater hoses from the heater core outlets.
3. Remove the windshield washer bottle from the firewall.
4. Remove the cover plate from the heater box and remove the heater core from the housing.
5. Remove the core end cover. Do not damage the core fins during the removal and installation procedure.

6. Installation is the reverse of removal. Fill the cooling system and check the operation of the heater system.

Heater Blower

REMOVAL & INSTALLATION

1. Disconnect the negative battery cable.
2. Disconnect the wire harness to the blower motor.
3. Remove the six sheet metal screws securing motor to the blower housing. Remove the motor with the blower.
4. Loosen and remove the bolt from the end of the motor shaft and detach the squirrel cage.
5. To install reverse the removal procedure.

Radio

REMOVAL & INSTALLATION

1. Disconnect the negative battery cable.
2. Remove the bolts which hold the radio to the radio support.
3. Remove the attaching screws which hold the radio cover bezel to the instrument panel.
4. Pull the radio from the instrument panel and disconnect the radio lead wires.
5. Remove the radio from the car.
6. Installation is in reverse order of removal.

Wiper Motor and Linkage

REMOVAL & INSTALLATION

1. Raise the hood and disconnect the windshield washer hose from the cowl.

2. Remove the attaching screws and pull off the cowl.
3. Remove the bolts from the wiper mounting bracket.
4. Release the attaching link from the motor to the wiper arm linkage.
5. Disconnect the motor wiring.
6. Remove the motor with the link assembly from the vehicle.
7. To install. Position the wiper motor on the mounting bracket and connect the wiring.
8. Connect the wiper linkage to the motor.
9. Install the cowl and the windshield washer hose.

Instrument Cluster

REMOVAL & INSTALLATION

Scouts use eight securing screws to mount the dash. Remove the screws and pull the panel out. It will be necessary to remove the speedometer cable and the electrical connections from the gauges if the cluster is to be totally removed. If an oil pressure gauge is used, plug the capillary tube to prevent leakage. To install, reverse the removal procedure.

Headlights

REMOVAL & INSTALLATION

1. Remove the screws retaining the headlight door and remove the door.
2. Remove the screws retaining the retaining ring and remove the ring.
3. Pull the headlight out, disconnect the wire harness and remove the headlight from the vehicle.
4. Install the headlight in the reverse order of removal.

Isuzu/LUV

GENERAL ENGINE SPECIFICATIONS

Engine Displacement cc (cu. in.)	Carb Type	Horsepower @ rpm	Torque @ rpm (ft. lbs.)	Bore × Stroke (in.)	Compression Ratio	Oil Pressure psi @ 1400 rpm
1800 (110.8)	2-bbl	80 @ 4800	95 @ 3000	3.31 × 3.23	8.5:1	57
1949 (119.0)	2-bbl	82 @ 4600	101 @ 3000	3.42 × 3.23	8.4:1	56
2254 (137.0)	2-bbl	96 @ 4600	123 @ 3000	3.52 × 3.54	8.3:1	57
2238 (136.6)	Turbo Diesel	80 @ 4000	128 @ 2200	3.46 × 3.62	21.0:1	55
2238 (136.6)	Diesel	58 @ 4300	93 @ 2200	3.46 × 3.62	21.0:1	55

GASOLINE ENGINE TUNE-UP SPECIFICATIONS

Engine Displacement cc (cu in.)	Spark Plug Type	Gap (in.)	Distributor Point Dwell (deg)	Point Gap (in.)	Ignition Timing (deg) MT	AT	Intake Valve Opens (deg)	Fuel Pump Pressure (psi)	Idle Speed (rpm) MT	AT	Valve Clearance (in.) In.	Ex.
1800 (110.8)	BPR-6511 ①	0.040 ①	Electronic ②		6B	6B	21B	3.0	800	900	0.006	0.010
1949 (119.0)	BPR-6ES11	0.040	Electronic		6B	6B	21B	3.0	800	900	0.006	0.010
2254 (137.0)	NA	NA	Electronic		NA	NA	NA	NA	NA	NA	NA	NA

NA—not available at time of publication
① Models with point ignition:
    Plugs: BPR-6ES
    Gap: 0.030
② Models with point ignition:
    Gap: 0.016–0.020
    Dwell: 47–57

ISUZU/LUV

DIESEL ENGINE TUNE-UP SPECIFICATIONS

Injector Opening Pressure (psi)	Low Idle (rpm)	Valve Clearance (in.)		Intake Valve Opens (deg)	Injection Timing (deg)	Firing Order
		Intake	Exhaust			
1493 ①	750 ②	0.016	0.016	16B	15B ④	1-3-4-2

① LUV-2133
② MT-AT: 850 rpm
③ Adjustment cold
④ Federal—California: BB

CAPACITIES

Engine Displacement cc (cu in.)	Crankcase l (qts)	Transmission l (pts)			Transfer Case l (pts)	Rear Drive Axle l (pts)	Front Drive Axle l (pts)	Gas Tank l (gal)	Cooling System l (qts)	
		4 sp.	5 sp.	Auto					Manual	Auto
1800 (110.8)	5.72 ①	2.7	2.7	14	5.2	2.7	1.7	13	6.7	6.4
1949 (119.0)	4.1	2.7	3.3	12.8	5.2	3.2	2.1	13.2 ②	8.5	8.2
2254 (137.0)	NA	2.7	3.3	12.8	5.2	3.2	2.1	13.2 ②	8.5	8.2
2238 (136.6)	6.0	2.7	3.3	12.8	5.2	3.2	2.1	13.2 ②	9.5	11.2

NA—not available at time of publication
① 1981 and 82: 4.2 qts
② Optional 19.1 gal (Long WB)

FIRING ORDERS

NOTE: To avoid confusion, always replace spark plug wires one at a time.

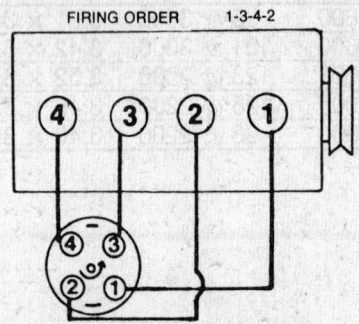

Gasoline engine firing order

CRANKSHAFT AND CONNECTING ROD SPECIFICATIONS

(All measurements given in in.)

Engine Displacement cc (cu in.)	Crankshaft				Connecting Rod		
	Main Brg. Journal Dia.	Main Brg Oil Clearance	Shaft End-Play	Thrust on No.	Journal Dia.	Oil Clearance	Side Clearance
1800 (110.8)	2.2050	0.0008-0.0025	0.0117	3	1.9290	0.0007-0.0030	0.0137
1949 (119.0)	2.2050	0.0008-0.0025	0.0117	3	1.9290	0.0007-0.0030	0.0137
2254 (137.0)	NA	NA	NA	NA	NA	NA	NA
2238 (136.6)	2.3593	0.0011-0.0033	0.0018	3	2.0837	0.0016-0.0047	0.0024

NA—not available at time of publication

VALVE SPECIFICATIONS

Engine Displacement cc (cu in.)	Seat Angle (deg)	Face Angle (deg)	Spring Test Pressure (lbs. @ in.)		Stem-to-Guide Clearance (in.)		Stem Diameter (in.)	
			Outer	Inner	Intake	Exhaust	Intake	Exhaust
1800 (110.8)	45	45	35 @ 1.614	20 @ 1.516	0.0009–0.0022	0.0015–0.0031	0.3102	0.3091
1949 (119.0)	45	45	35 @ 1.614	20 @ 1.516	0.0009–0.0022	0.0015–0.0031	0.3102	0.3091
2254 (137.0)	45	45	NA	NA	NA	NA	NA	NA
2238 (136.6)	45	45	145 @ 1.535	44 @ 1.457	0.0015–0.0027	0.0025–0.0037	0.3150	0.3150

NA—not available at time of publication

PISTON AND RING SPECIFICATIONS

(All measurements in inches)

Engine Displacement cu. in. (cc)	Piston Clearance	Ring Gap			Ring Side Clearance		
		Top Compression	Bottom Compression	Oil Control	Top Compression	Bottom Compression	Oil Control
110.8 (1817)	0.0018–0.0026	0.008–0.016	0.008–0.016	0.008–0.035	0.0059	0.0059	0.0059
119.0 (1949)	0.0018–0.0026	0.014–0.020	0.014–0.020	0.008–0.035	0.0059	0.0059	0.0059
137.0 (2254)	NA	NA	NA	NA	NA	NA	NA
136.6 (2238)	0.0062–0.0070	0.0079–0.0158	0.0079–0.0158	0.0079–0.0158	0.0018–0.0028	0.0012–0.0021	0.0008–0.0021

NA—not available at time of publication

TORQUE SPECIFICATIONS

(All readings in ft. lbs. unless noted)

Engine Displacement cc (cu in.)	Cylinder Head Bolts	Rod Bearing Bolts	Main Bearing Bolts	Crankshaft Pulley Bolts	Flywheel to Crankshaft Bolts	Manifolds	
						Intake	Exhaust
1800 (110.8)	72 ①	43	72	87	69	13	15
1949 (119.0)	72 ①	43	72	87	69	13	15
2254 (137.0)	NA	NA	NA	NA	NA	NA	NA
2238 (136.6)	65 ②	61	123	136	68	15	15

NA—not available at time of publication
① Tighten to 61 ft. lbs. first. Retighten to 72 ft. lbs.
② Tighten in 2 steps: 44 ft. lbs. first, then 65 ft. lbs. final torque.

ALTERNATOR AND REGULATOR SPECIFICATIONS

Alternator		Voltage Regulator					
Manufacturer and/or Part Number	Output Amps @ Generator rpm	Charge Indicator Relay		Voltage Regulator			
		Back Gap (in.)	Point Gap (in.)	Back Gap (in.)	Air Gap (in.)	Point Gap (in.)	Regulated Voltage
LT140-126	40	0.036	0.020	0.015	0.014	0.012	13.8–14.8
LT150-144	50	0.036	0.020	0.015	0.014	0.012	13.8–14.8
LT150-131B	50	0.036	0.020	0.015	0.014	0.012	13.8–14.8

BATTERY AND STARTER SPECIFICATIONS

Engine Displacement cc (cu in.)	Battery			Starter							Brush Spring Tension (oz)
				Lock Test			No Load Test				
	Amp Hour Capacity	Volts	Ground	Amps	Volts	Torque (ft. lbs.)	Amps	Volts	RPM		
1800-1949 (110.8) (119.0)	50	12	N	Not Recommended							56
2238 (136.6)	80	12	N	Not Recommended							56

BRAKE SPECIFICATIONS

(All measurements are given in in.)

Master Cylinder Bore	Wheel Cylinder or Caliper Bore		Piston-to-Bore Clearance	Brake rotor or Drum Diameter		Minimum Lining Thickness	Brake Disc	
	Front	Rear		Front	Rear		Minimum Thickness	Maximum Run-out
0.875	—	1.00	0.006	10.00	10.00	0.039	0.453	0.005 ①

① Rate of change must not exceed 0.001 in. in 30°

TUNE-UP

Breaker Points and Condenser

When replacing the points, always install a new condenser.

Remember that a change in the point gap or dwell also changes the ignition timing. Therefore, if the points are adjusted, you must also correct the ignition timing.

INSPECTION

1. Disconnect the high-tension wire from the top of the distributor.
2. Remove the distributor cap by prying off the spring clips on the sides of the cap.
3. Remove the rotor from the distributor shaft by pulling it straight up. Examine the condition of the rotor. If it is cracked or the brass tip is excessively worn or burned, the rotor should be replaced. If not excessively worn, clean the brass tip with fine emery paper until shiny.
4. Pry open the contacts of the points with a screwdriver and check the condition of the contacts. If they are excessively worn, burned or pitted, they should be replaced.
5. If the points are in good condition, adjust them and replace the rotor and the distributor cap. If the points need to be replaced, follow the replacement procedure given below.

REMOVAL & INSTALLATION

1. Remove the coil high-tension wire from top of the distributor cap. Remove the distributor cap from the distributor and place it out of the way. Remove the rotor from the distributor shaft.
2. Loosen the screw that holds the condenser lead to the body of the breaker points and remove the condenser lead from the points.
3. Remove the screw that holds and grounds the condenser to the distributor body. Remove the condenser from the distributor and discard it.
4. Remove the points assembly attaching screws and adjustment lockscrews. A screwdriver with a holding mechanism will come in handy here, so that you don't drop a screw into the distributor and have to remove the entire distributor to retrieve it.
5. Remove the points by lifting them straight up and off the locating dowel on the plate. Wipe off the cam and apply new cam lubricant. Discard the old set of points.
6. Slip the new set of points onto the locating dowel and install the screws that hold the assembly onto the plate. Do not tighten them all the way.
7. Attach the new condenser to the plate with the ground screw.
8. Attach the condenser lead to the points at the proper place.
9. Apply a small amount of cam lubricant to the shaft where the rubbing block of the points touches.

ADJUSTMENT WITH A FEELER GAUGE

1. If the contact points of the assembly are not parallel, bend the stationary contact so that they make contact across the entire surface of the contacts. Bend only the stationary bracket part of the point assembly; not the moveable contact.
2. With the vehicle in neutral and parking brake on, turn the engine until the rubbing block of the points is on the crest of one of the high points of the distributor cam. You can do this by either having an assistant turn the ignition switch "on" and then releasing it quickly ("bumping" the engine) or by turning the crankshaft pulley bolt with a wrench or socket wrench.
3. Place the correct size feeler gauge between the contacts. Make sure it is parallel with the contact surfaces.
4. With your free hand, insert a screwdriver into the notch provided for adjustment or into the eccentric adjusting screw, then twist the screwdriver to either increase or decrease the gap to the proper setting.
5. Tighten the adjustment lockscrew, and turn the engine until the rubbing block of the points rests on the next crest of the distributor cam. Recheck the contact gap to make sure that it didn't change when the lockscrew was tightened.
6. Replace the rotor and distributor cap, and the high-tension wire that connects the top of the distributor and the coil. Make sure that the rotor is firmly seated all the way onto the distributor shaft and that the tab of the rotor is aligned with the notch in the shaft. Align the tab in the base of the distributor cap with the notch in the distributor body. Make sure that the cap is firmly seated on the distributor and that the retainer springs are in place. Make sure that the end of the high-tension wire is firmly placed in the top of the distributor and the coil.

ADJUSTMENT WITH A DWELL METER

1. Adjust the points with a feeler gauge as previously described.

2. Connect the dwell meter to the ignition circuit according to the manufacturer's instructions. One lead of the meter is connected to a ground and the other lead is connected to the distributor post on the coil. An adapter is usually provided for this purpose.

3. If the dwell meter has a set line on it, adjust the meter to zero the indicator.

4. Start the engine.

NOTE: Be careful when working on any vehicle while the engine is running. Make sure that the transmission is in Neutral and that the parking brake is applied. Keep hands, clothing, tools and the wires of the test instruments clear of the rotating fan blades.

5. Observe the reading on the dwell meter. If the reading is within the specified range, turn off the engine and remove the dwell meter.

NOTE: If the meter does not have a scale for 4 cylinder engines, multiply the 8 cylinder reading by two.

6. If the reading is above the specified range, the breaker point gap is too small. If the reading is below the specified range, the gap is too large. In either case, the engine must be stopped and the gap adjusted in the manner previously covered. After making the adjustment, start the engine and check the reading on the dwell meter. When the correct reading is obtained, disconnect the dwell meter.

7. Check the adjustment of the ignition timing.

Electronic Ignition

AIR GAP SETTING

All 1981 and later models have electronic ignition. The only adjustment possible on this ignition is setting of the air gap.

1. Remove the distributor cap and O-ring.

2. Remove the rotor.

3. Use a feeler gauge to measure the air gap at the pick up coil projection. The gap should be 0.008–0.016 in. Adjust if necessary.

4. Loosen the screws and move the signal generator until the gap is correct. Tighten the screws and recheck the gap.

NOTE: The electrical parts in this system are not repairable. If found to be defective they must be replaced.

You can also check the signal generator by using an ohmmeter to determine its resistance. It should be 140–180 ohms. If the resistance is not correct, it must be replaced.

Ignition Timing

NOTE: For diesel injection timing, see Diesel Fuel System section.

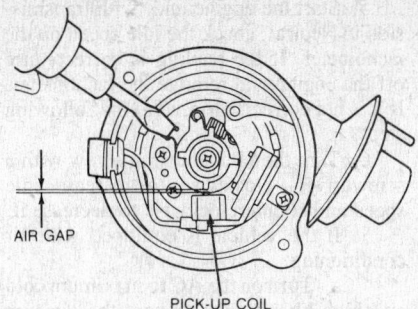

Adjusting the air gap

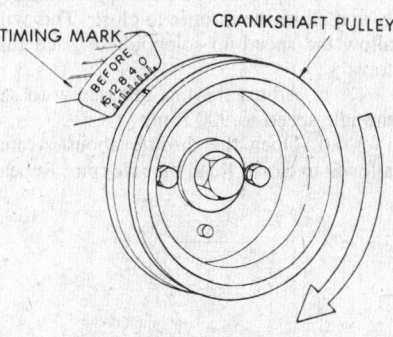

Ignition timing marks

The timing marks are located near the front crankshaft pulley and consist of a pointer with graduations attached to the engine block and mark on the crankshaft pulley.

1. Set the dwell angle to the proper specification.

2. Locate the timing marks on the crankshaft pulley and the front of the engine.

3. Clean off the timing marks, so that you can see them.

4. Use chalk or white paint to color the mark on the pointer that will indicate the correct timing, when aligned with the crankshaft pulley. It is also helpful to mark the notch on the crankshaft pulley with a small dab of color.

5. Attach a tachometer to the engine.

6. Attach a timing light to the engine according to the manufacturer's instructions. If the timing light has three wires, one, usually green or blue, is attached to the No. 1 spark plug with an adapter. The other wires are connected to the battery. The red wire goes to the positive side of the battery and the black wire is connected to the negative terminal of the battery.

7. Disconnect the vacuum line to the distributor at the distributor and plug the vacuum line. A golf tee does a good job.

8. Make sure that all wires from the timing light and tachometer are clear of the fan and belts. Start the engine.

9. Adjust the idle to the correct setting and rpm.

10. Aim the timing light at the timing marks. If the marks that you put on the pulley and the engine are aligned when the light flashes, the timing is correct. Turn off

the engine and remove the tachometer and the timing light. If the marks are not in alignment, proceed with the following steps.

11. Turn off the engine.

12. Loosen the distributor lockbolt just enough so that the distributor can be turned with a little effort.

13. Start the engine. Keep the wires of the timing light clear of the fan.

14. With the timing light aimed at the pulley and the marks on the engine, turn the distributor in the direction of rotor rotation to retard the spark, and in the opposite direction of the rotor rotation to advance the spark. Align the marks on the pulley and the engine with the flashes of the timing light.

15. Tighten distributor lockbolt, unplug vacuum line and test instruments, and reconnect.

Valve Lash

NOTE: The valves are adjusted with the engine COLD. It is best to allow an engine to sit overnight before beginning a valve adjustment. While all valve adjustments must be made as accurately as possible, it is better to have the valve adjustment slightly loose than slightly tight, as a burned valve may result from overly tight adjustments.

1. Make sure that both the cylinder head and camshaft retaining bolts are tightened to the proper torque.

2. Remove the camshaft carrier side-cover, and discard the gasket if it is torn, cracked, or worn in any way.

Cylinder Valve	1	2	3	4
Intake	○	○	●	●
Exhaust	○	●	○	●

Note: ○ When piston in No. 1 cylinder is at TDC on compression stroke.

● When piston in No. 4 cylinder is at TDC on compression stroke.

Valve adjusting sequence

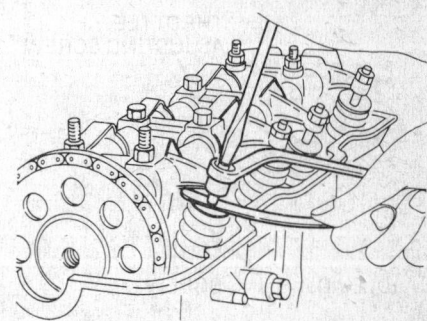

Valve clearance adjustment

3. Turn the crankshaft with a wrench on the front pulley attaching bolt or by "bumping" the engine with the starter until the No. 1 piston is at TDC of the compression stroke. You can tell when the piston is coming up on the compression stroke by removing the spark plug and placing your thumb over the hole; you will feel air being forced out of the spark plug hole past your thumb. Both valves on No. 1 cylinder will be closed. Stop turning the crankshaft when the TDC timing mark on the crankshaft pulley is directly aligned with the timing mark pointer.

4. With the No. 1 piston at TDC of the compression stroke, check the clearance between the rocker arm and valve stem with the proper thickness feeler gauge on Nos. 1 and 2, intake valves and Nos. 1 and 3 exhaust valves.

5. Adjust the clearance by loosening the locknut with an open-end wrench, turning the adjusting screw with a phillips head screwdriver and retightening the locknut. The proper thickness feeler gauge should pass between the camshaft or valve stem and the rocker with a slight drag when the clearance is corrected.

6. Turn the crankshaft one full turn to position the No. 4 piston at TDC of its compression stroke. Adjust the remaining valves: Nos. 2 and 4 exhaust and Nos. 3 and 4 intake in the same manner as outlined in Step 5.

7. Install the camshaft carrier side-cover with a new gasket and sealer.

Carburetor

For adjustment and specification information, refer to the Carburetor section in Unit Repair.

IDLE SPEED & MIXTURE ADJUSTMENT

Through 1980

1. Start the engine and run it until it reaches operating temperature.

2. If it hasn't already been done, check and adjust the ignition timing. After you have set the timing, turn off the engine.

3. Attach tachometer to the engine.

4. Remove the air cleaner.

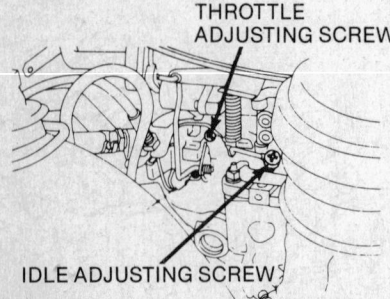

THROTTLE ADJUSTING SCREW

IDLE ADJUSTING SCREW

Idle speed and mixture adjustment screws

5. Start the engine and, with transmission in Neutral, check the idle speed on the tachometer. If the reading is correct, turn off the engine and remove the tachometer. If it is not correct, proceed to the following steps.

6. Turn the idle adjusting screw with a screwdriver—clockwise to increase idle speed and counterclockwise to decrease it.

7. If the vehicle is equipped with air conditioning:

    a. Turn on the AC to maximum cold and high blower. Disconnect the vacuum line to the air cleaner housing air compensator and plug the inlet manifold;

    b. Open the throttle approximately ⅓ and allow the throttle to close. This will allow the speed-up solenoid to reach full travel;

    c. Adjust the fast-idle screw to set the idle speed to 900 rpm;

    d. Open the throttle about ⅓ and allow it to close. Read the idle rpm. Repeat

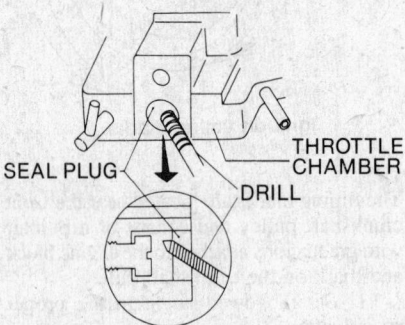

SEAL PLUG   THROTTLE CHAMBER   DRILL

Removing the idle mixture plug

Step C until the correct reading is obtained. Shut off the engine.

8. Turn the mixture adjusting screw all the way. Seat the needle tip lightly to avoid damaging the tip. Back the screw out 3½ turns.

9. Start the engine. Turn the mixture screw until the maximum engine rpm is achieved.

10. Reset the engine idle speed.

11. Turn the idle mixture screw clockwise (lean) until the engine speed drops 50 rpm.

12. Reset the idle mixture screw ½ turn counter-clockwise (rich) from Step 11 position.

13. Rest the throttle adjusting screw to required idle speed.

14. Unplug and reconnect any vacuum lines that may have been disconnected.

1981 and Later Models

The idle mixture adjustment is the same as previously described with the following exceptions:

1. Make the idle speed adjustment with the engine at normal operating temperature. Be sure the choke is fully opened, air conditioning off and the air cleaner installed.

2. Disconnect and plug the distributor vacuum, the canister purge and EGR vacuum lines. Shut off the vacuum to the idle compensator by bending the rubber hose.

3. Adjust to required idle speed with throttle adjusting screw.

4. If equipped with air conditioning, turn A/C to max cold and high blower.

5. Open throttle to approx. ⅓ and allow throttle to close, the speed up solenoid should activate. Adjust speed up solenoid screw until 900 rpm is reached. In order to adjust the idle mixture you must first remove the plug that covers the mixture screw.

6. To remove the plug, first remove the carburetor and turn it upside down.

7. Remove the plug carefully with a hammer and screwdriver (see illustration).

8. Reinstall the carburetor.

9. Turn the mixture screw all the way in, and then back it out 2 turns (Federal) and 1 turn (California). Readjust the idle if necessary.

ENGINE ELECTRICAL

Distributor

REMOVAL & INSTALLATION

1. Remove the high-tension wires from the distributor cap terminal towers, noting their positions to assure correct reassembly.

2. Remove the primary lead from the coil terminal.

3. Disconnect the vacuum line.

4. Unlatch the two distributor cap retaining clips and remove the distributor cap.

5. Note the position of the rotor in relation to the base. Scribe a mark on the base of the distributor and on the engine block to facilitate reinstallation. Align the marks with the direction the metal tip of the rotor is pointing.

6. Remove the bolt which holds the distributor to the engine.

7. Lift the distributor assembly from the engine.

8. Insert the distributor into the engine. Line up the mark on the engine with the metal tip of the rotor. Make sure that the vacuum advance diaphragm is pointed in the same direction as it was pointed originally. This will be done automatically if the marks on the engine and the distributor are lined up with the rotor.

9. Install the distributor hold-down bolt and clamp. Leave the screw loose enough so that you can move the distributor with heavy hand pressure.

10. Connect the primary wire to the coil. Install the distributor cap on the distributor

housing. Secure the distributor cap with the spring clips.

11. Install the spark plug wires, checking the wire placement with the firing order diagram in the front of this section. Make sure the wires are pressed all the way into the top of the distributor cap and firmly onto the spark plugs.

12. Adjust the point dwell and set the ignition timing.

NOTE: If the crankshaft has been turned or the engine disturbed in any manner while the distributor was removed, or if the marks were not drawn, it will be necessary to initially time the engine. Follow the procedure given below:

1. It is necessary to place the No. 1 cylinder in the firing position to correctly install the distributor. To locate this position, the ignition timing marks on the crankshaft front pulley are used.

2. Remove the No. 1 cylinder spark plug. Turn the crankshaft until the piston in the No. 1 cylinder is moving up on the compression stroke. This can be determined by placing your thumb over the spark plug hole and feeling the air being forced out of the cylinder. Stop turning the crankshaft when the timing marks that are used to time the engine are aligned.

3. Oil the distributor housing lightly where the distributor bears on the cylinder block.

4. Install the distributor so that the rotor, which is mounted on the shaft, points toward the No. 1 spark plug terminal tower position when the cap is installed. Of course you won't be able to see the direction in which the rotor is pointing if the cap is on the distributor. Lay the cap on the top of the distributor and make a mark on the side of the distributor housing just below the No. 1 spark plug terminal. Make sure that the rotor points toward the mark when you install the distributor.

5. When the distributor shaft has reached the bottom of the hole, move the rotor back and forth slightly until the driving lug on the end of the shaft enters the slots cut in the end of the oil pump shaft and the distributor assembly slides down into place.

6. When the distributor is correctly installed, the breaker points should be in such a position that they are just ready to break contact with each other. This is accomplished by rotating the distributor body after it has been installed in the engine. Once again, line up the marks that you made before the distributor was removed from the engine.

7. Install the distributor hold-down bolt.

8. Install the spark plug into the No. 1 spark plug hole and continue from Step 9 of the distributor installation procedure.

Alternator
PRECAUTIONS

To prevent damage to the alternator and

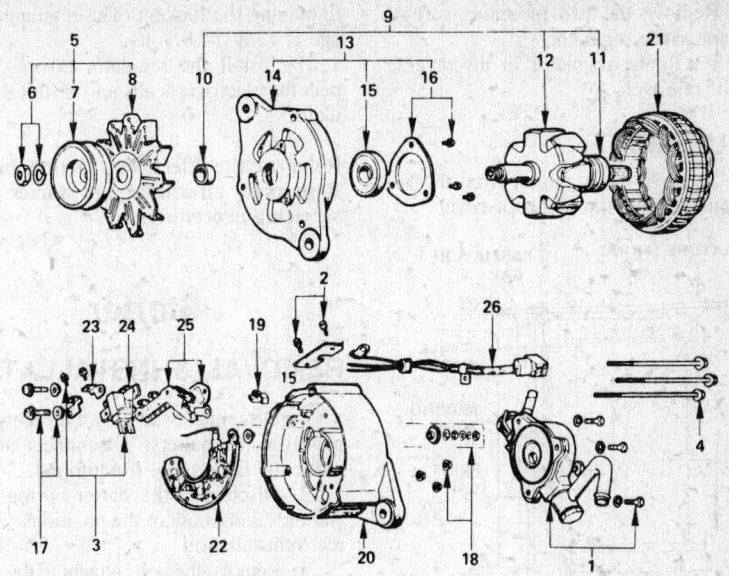

1. Vacuum pump
2. Cover
3. Brush
4. Through bolt
5. Pulley assembly
6. Pulley nut
7. Pulley
8. Fan
9. Rotor assembly
10. Spacer
11. Ball bearing
12. Rotor
13. Front cover assembly
14. Front cover
15. Ball bearing
16. Bearing retainer
17. Screw
18. Terminal bolt and nut
19. Lead wire
20. Rear cover
21. Stator
22. Diode
23. Holder plate
24. Brush holder
25. IC regulator assembly
26. Lead wire

Exploded view of the alternator

regulator, the following precautionary measures must be taken when working with the electrical system.

● Never reverse battery connections. Always check the battery polarity visually. This is to be done before any connections are made to be sure that all of the connections correspond to the battery ground polarity. Booster batteries for starting must be connected properly. Make sure that the positive cable of the booster battery is connected to the positive terminal of the battery that is getting the boost. Connect the negative boost cable from booster battery to a good chassis ground on the vehicle with the dead battery. Disconnect the battery cables before using a fast charger; the charger has a tendency to force current through the diodes in the opposite direction for which they were designed. This burns out the diodes. Never use a fast charger as a booster for starting the vehicle. Never disconnect the voltage regulator while the engine is running. Do not ground the alternator output terminal.

● Do not operate the alternator on an open circuit with the field energized. Do not attempt to polarize an alternator.

REMOVAL & INSTALLATION

1. Remove the air pump.

2. Disconnect the battery ground cable before disconnecting the cable from the alternator "A" terminal. This is a hot cable connected directly to the battery.

3. Disconnect the alternator circuit at

the connector and disconnect the cable from the "A" terminal.

4. Remove the mounting bolts on the lower part of the alternator and the fan belt adjusting bolt. Pull belt clear of alternator pulley and remove alternator.

5. Install the alternator in the reverse order of removal and tighten the fan belt and air pump belt tension.

BELT TENSION ADJUSTMENT

Any engine V-belt is correctly tensioned when the longest span of belt between pulleys can be depressed about ¼ in. in the middle by moderate thumb pressure. To adjust, loosen the accessory's slotted adjusting bracket bolt. If the hinge bolt is very tight, it may be necessary to loosen it slightly to move the item.

NOTE: Be careful not to overtighten belts, as this will damage the bearings, particularly in air or water pumps and alternators.

Regulator
REMOVAL & INSTALLATION

1. Remove the negative battery cable from the battery.

2. Disconnect the electrical leads at the regulator, taking note to the positions in order to facilitate correct reconnection.

3. Remove the two mounting screws and remove the regulator.

4. Install the regulator in the reverse order of removal.

ADJUSTMENT

1. Remove the regulator from the vehicle and remove the regulator cover.

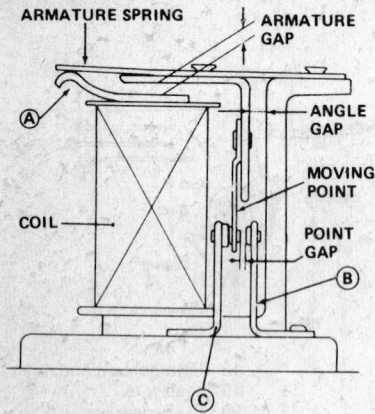

Voltage relay adjustment

2. If the contact points are rough, carefully dress them with fine emery paper (a 600-grit autobody paper works fine).

3. Check and adjust the core gap first, and then the point gap. Adjustment of the yoke gap is unnecessary.

4. Adjust the core gap by loosening the screws attaching the contact set to the yoke. Move the contact set up or down as required. The standard core gap is 0.024–0.039 in. Tighten the attaching screw.

5. Adjust the point gap by loosening the screw attaching the upper contact. Move the upper contact up or down as required. The standard point gap is 0.012–0.016 in.

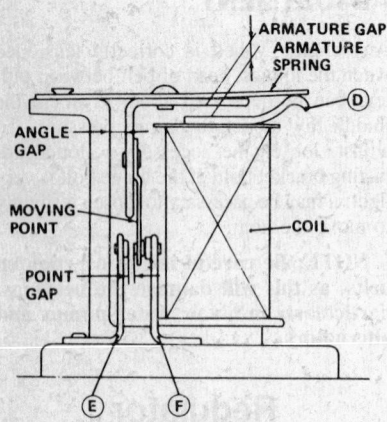

Voltage regulator adjustment

6. Adjust the regulated voltage by means of the adjusting screw. Turn the adjusting screw in to increase voltage and out to reduce voltage. When the correct adjustment is obtained, secure the adjusting screw by

tightening the locknut. The regulated voltage is 13.8–14.8 volts.

7. Install the regulator cover, reconnect the electrical leads and install the regulator.

Refer to the Electrical section in Unit Repairs for alternator and starter motor servicing procedures.

Starter

REMOVAL & INSTALLATION

1. Disconnect the negative battery cable from the battery. Disconnect and remove the EGR pipe, if equipped.

2. Disconnect the starter wiring at the starter, taking note of the positions for correct reinstallation.

3. Remove the bolts attaching the starter to the engine and remove the starter from the vehicle.

4. Install the starter in the reverse order of removal.

ENGINE MECHANICAL

Engine

REMOVAL & INSTALLATION

Gasoline Engine

NOTE: On pickup models with 2WD, the engine and transmission may be removed as an assembly. 4WD models require transmission separation prior to engine removal.

1. Raise and remove hood and disconnect the battery cables.

2. Remove the skid plate and drain both the cooling system and the oil pan.

3. Remove the air cleaner assembly and the many vacuum hoses. Mark all vacuum hoses and the places they connect for installation.

4. Disconnect all hoses, tubing and electrical leads from the engine and mark them where they connect for reinstallation.

5. Remove the radiator and fan blade assembly.

6. Using penetrating oil where necessary, disconnect the exhaust pipe from the exhaust manifold.

7. Raise the vehicle and remove the clutch return spring and the clutch cable, if equipped with a manual transmission.

8. Remove the starter motor. On 2WD, remove driveshaft. Remove transmission mount bolts. Remove gearshift lever assembly.

9. On 4WD; remove the flywheel cover pan.

10. On 4WD; remove the bell housing bolts and support the transmission with a floor jack or suitable stand.

11. Lift the engine slightly and remove the engine mount nuts.

12. Make certain that all lines, hoses, cables and wires have been disconnected from the engine and frame.

13. Lift the engine from the vehicle with the front of the engine raised slightly to clear the transmission input shaft.

14. Installation is the reverse of removal paying special attention to the following: Fill the cooling system. Fill the crankcase with engine oil. Check and adjust the clutch pedal free play. Start the engine, run at idle and check for leakage.

15. Adjust the following: Fan belt tension. Valve clearances. Ignition timing. Engine idle speed.

Diesel Engine

2WD

1. Raise engine hood.

2. Disconnect the battery ground cable.

3. Remove the engine hood.

4. Remove the battery assembly.

5. Remove under cover and drain the cooling system by opening the drain plugs on the radiator and on the cylinder block.

6. Remove the air cleaner assembly as follows: Remove the intake silencer. Remove the bolts fixing the air cleaner and loosen the clamp bolt. Lift the air cleaner slightly and disconnect the breather hose, then remove the air cleaner assembly.

7. Disconnect the upper water hose at the engine side.

8. Loosen the compressor drive belts by moving the power steering oil pump or idler if so equipped.

9. Remove the cooling fan and fan shroud.

10. Disconnect the lower water hose at the engine side.

11. Remove the radiator grille.

12. Remove the radiator attaching bolts and remove the radiator.

13. Disconnect the accelerator control cable from the injection pump side.

14. Disconnect the air conditioner compressor control cable. (If so equipped).

15. Disconnect the fuel hoses from the injection pump.

16. Disconnect the battery cable from the cylinder body.

17. Disconnect the transmission wiring.

18. Disconnect the vacuum hose from the fast idle actuator.

19. Disconnect the connector at fuel cut solenoid.

20. Disconnect the A/C compressor wiring, sensing resistor and thermoswitch connectors.

21. Disconnect the heater hoses extending from the heater unit from the dash panel side.

22. Disconnect the hose for master-vac from the vacuum pump.

23. Disconnect vacuum hose from the vacuum pump.

24. Disconnect the generator wiring at the connector.

25. Disconnect the exhaust pipe from the exhaust manifold at the flange.

26. Remove the exhaust pipe mounting brake from the engine back plate.

27. Disconnect the starter motor wiring.

28. Disconnect the battery cable from starter motor.

29. Slide the gearshift lever boot upwards on the lever. Remove 2 gearshift lever attaching bolts and remove lever.

30. Place a pan under transmission to receive oil, disconnect speedometer cable at the transmission then disconnect the ground cable.

31. Disconnect the propeller shaft at differential side.

32. Remove the propeller shaft.

33. Remove return spring from clutch fork.

34. Disconnect clutch cable from hooked portion of clutch fork and pull it out forward through stiffener bracket.

35. Remove two bracket to transmission rear mount bolts and nuts.

36. Raise engine and transmission as required and remove (4) crossmember to frame bracket bolts.

37. Remove the rear mounting nuts from the transmission rear extension.

38. Disconnect electrical connectors at CRS switch and back-up lamp switch.

39. Remove the engine mounting bolt and nuts. Check that the engine is slightly lifted before removing the engine mounting bolt and nuts.

40. Engine removal: Check to make certain all the parts have been removed or disconnected from the engine that are fastened to the frame side. Remove the engine toward front of the vehicle by adjusting the hoist, so that front part of the engine is lifted slightly above the level.

41. Installation is the reverse of removal paying special attention to the following: Fill the cooling system. Fill the crankcase with engine oil. Check and adjust the clutch pedal free play. Start the engine, run at idle and check for leakage.

42. Adjust the following: Fan belt tension. Valve clearances. Ignition timing. Engine idle speed.

4WD

1. Raise engine hood.

2. Disconnect the battery ground cable.

3. Remove the engine hood.

4. Remove the battery assembly.

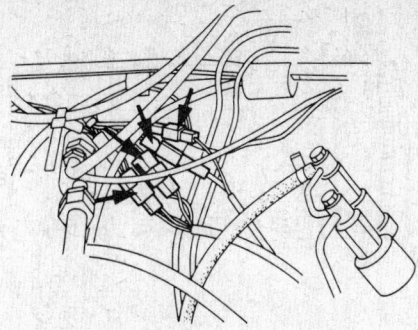

Disconnect points for the air compressor switch, sensing resistor, and thermoswitch connectors

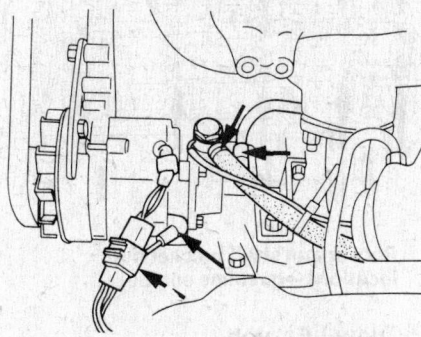

Vacuum hose and generator wiring removal points

5. Remove under cover and drain the cooling system by opening the drain plugs on the radiator and on the cylinder block.

6. Remove the air cleaner as follows: Remove the intake silencer. Remove the bolts fixing the air cleaner and loosen the clamp bolt. Lift the air cleaner slightly and disconnect the breather hose, then remove the air cleaner assembly.

7. Disconnect the upper water hose at the engine side.

8. Loosen the compressor drive belts by moving the power steering oil pump or idler. (If so equipped).

9. Remove the cooling fan and fan shroud.

10. Disconnect the lower water hose at the engine side.

11. Remove the radiator grille.

12. Remove the radiator attaching bolts and remove the radiator.

13. Disconnect the accelerator control cable from the injection pump side.

14. Disconnect the air conditioner compressor control cable. (If so equipped).

15. Disconnect the fuel hose from the injection pump.

16. Disconnect the battery cable from the cylinder body.

17. Disconnect the transmission wiring.

18. Disconnect the vacuum hose from the fast idle actuator.

19. Disconnect the connector at fuel cut solenoid.

20. Disconnect the A/C compressor switch wiring, sensing resistor, and the thermoswitch connectors.

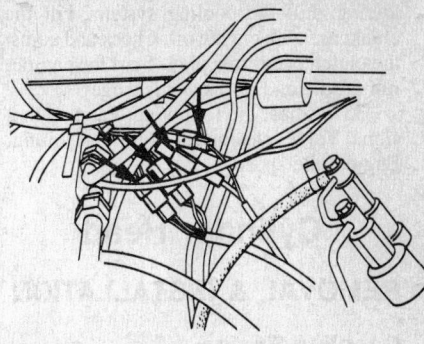

Disconnect points for the air compressor switch, sensing resistor, and thermoswitch connectors

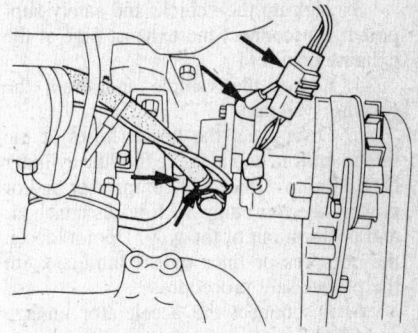

Vacuum hose and generator wiring removal points

21. Disconnect the heater hoses extending from the heater unit from the dash panel side.

22. Disconnect the hose for master-vac from the vacuum pump.

23. Disconnect the vacuum hose from the vacuum pump.

24. Disconnect the generator wiring at the connector.

25. Disconnect the exhaust pipe from the exhaust manifold at the flange.

26. Remove the exhaust pipe mounting brake from the engine back plate.

27. Disconnect the starter motor wiring.

28. Disconnect the battery cable from starter motor.

29. Slide the transmission and transfer gearshift lever boot upwards on each lever, remove gearshift lever attaching bolts.

30. Remove return spring from transfer gear shift lever then remove levers.

31. Remove the transmission.

32. Remove the engine mounting bolts and nuts. Check that the engine is slightly lifted before removing the engine mounting bolts and nuts.

33. Engine removal: Check to make certain all the parts have been removed or disconnected frame the engine that are fastened to the frame side. Remove the engine toward front of the vehicle by maneuvering the hoist, so that front part of the engine is lifted slightly above the level.

34. Installation is the reverse of removal paying special attention to the fol-

lowing: Fill the cooling system. Fill the crankcase with engine oil. Check and adjust the clutch pedal free play. Start the engine, run at idle and check for leakage.

35. Adjust the following: Fan belt tension. Valve clearances. Ignition timing. Engine idle speed.

Cylinder Head

REMOVAL & INSTALLATION

Gasoline Engine

1. Remove the cam cover.
2. Remove the EGR pipe clamp bolt at the rear of the cylinder head.
3. Jack up the vehicle and safely support it. Disconnect the exhaust pipe at the exhaust manifold.
4. Lower the vehicle and drain the cooling system.
5. Disconnect the heater hoses at the inlet manifold and at the rear of the cylinder head. Remove the A/C compressor and/or power steering pump with hoses attached, and position out of the way. Do not loosen the compressor lines unless familiar with the proper safe procedure.
6. Disconnect the accelerator linkage and fuel line at the carburetor, all necessary electrical connections, spark plug wires and vacuum lines.
7. Rotate the camshaft until the No.4 cylinder is in the firing position. Remove the distributor cap and mark the rotor to housing relationship. Remove distributor and the fuel pump.
8. Lock the timing chain adjuster by depressing and turning the automatic adjuster side pin 90° clockwise.
9. Remove the timing sprocket to camshaft bolt and remove the sprocket from the camshaft.

NOTE: Keep the sprocket on the chain damper and chain.

10. Disconnect the AIR hose and the check valve at the exhaust manifold.
11. Remove the cylinder head to timing cover bolts.
12. Remove the cylinder head bolts in a progressional sequence, starting with the outer bolts.
13. Remove the cylinder head, intake and exhaust manifold as a unit.

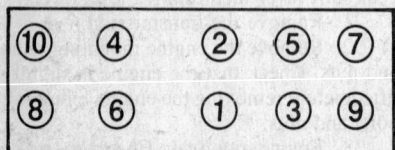

FRONT ──────────────▶

Gasoline engine head bolt torque sequence

14. To install, reverse the removal procedure and tighten the bolts in the sequence and torque shown in the specifications table in the front of the section.

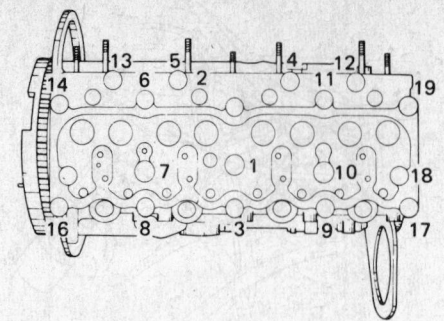

Diesel engine head bolt torque sequence

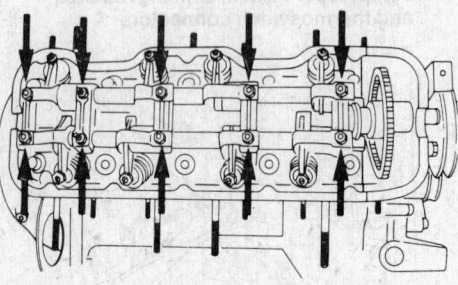

Rocker arm shaft bracket nut locations—gasoline engine

Diesel Engine

1. Follow the intake and exhaust manifold removal steps.
2. Remove the intake and exhaust manifold gasket.
3. Drain the cooling system by opening the drain plugs on the radiator and on the cylinder block.
4. Disconnect the upper water hose at the engine side.
5. Remove the cooling fan and fan shroud.
6. Remove the sleeve nuts and disconnect the injection pipes.
7. Remove the nozzle holder fixing nuts and remove the nozzle holder assembly.
8. Follow the rocker arm, bracket and shaft assembly removal steps.
9. Remove the pushrods.
10. Remove the joint bolt and disconnect the leak-off pipe.
11. Remove the 19 bolts fixing the cylinder head then remove the cylinder head and gasket.
12. Install the cylinder head gasket with the TOP mark side up on the cylinder body by aligning the holes with the dowels.
13. Install the cylinder head.
14. Install the pushrod in position on the cylinder head.
15. Install the rocker arm assembly on the cylinder head. Tighten the bracket fixing bolts evenly in sequence commencing with the inner ones.
16. Follow the intake and exhaust manifold installation steps.
17. Install the cooling fan and fan shroud.
18. Connect the upper water hose to engine side.
19. Fill the engine cooling system.

VALVE GUIDE REPLACEMENT

1. With the cylinder head removed from the vehicle and the valves removed from the head, drive the guides out toward the upper face of the cylinder head with a suitable driver. The valve guides cannot be driven out downward because they are secured in place with a snap-ring.
2. Lubricate the outside of the new valve guide with oil. Press it all the way into position, from the upper face of the cylinder head, until it is brought in contact with the snap-ring. Allowable interference between the cylinder head and the valve guide is 0.0016 in.

Valve Rockers

REMOVAL & INSTALLATION

Gasoline Engine

1. Remove the cam cover.
2. Loosen the rocker arm shaft bracket nuts a little at a time, in sequence, starting with the outer brackets.
3. Remove the nuts from the rocker arm shaft brackets. Remove shaft assembly.
4. To disassembly rocker and shafts: Remove the spring from the rocker arm shaft and remove the rocker brackets and arms. Keep parts in order for reassembly.
5. Before installing apply a generous amount of clean engine oil to the rocker arm shaft, rocker arms and valve stems.
6. Install the longer shaft on the exhaust valve side and the shorter shaft on the intake side so that the aligning marks on the shafts are turned on the front (timing) side of the engine.
7. Assemble the rocker arm shaft brackets and rocker arms to the shafts so that the cylinder number that is on the upper face of the brackets is pointed toward the front of the engine.
8. Align the mark on the No. 1 rocker arm shaft bracket with the mark on the intake and exhaust valve side rocker arm shaft.
9. Make certain the amount of projection of the rocker arm shaft beyond the face of the No. 1 rocker arm shaft bracket is longer on the exhaust side shaft than on the intake shaft when the rocker arm shaft stud holes are aligned with the rocker arm shaft bracket stud holes.
10. Place the rocker arm shaft springs in position between the shaft bracket and rocker arm.
11. Check that the punch mark on the rocker arm shaft is turned upward, then install the rocker arm shaft bracket assembly onto the cylinder head studs. Align the mark on the camshaft with the mark on the No. 1 rocker arm shaft bracket.
12. Tighten the rocker arm shaft brackets stud nuts to 16 ft. lbs.

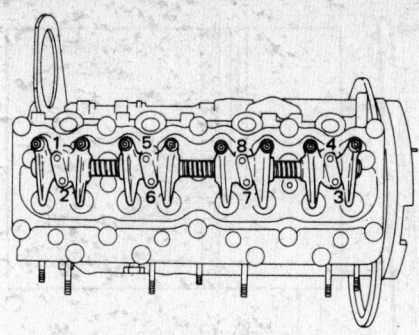

Rocker arm shaft bracket nut locations—diesel engine

NOTE: Hold the rocker arm springs with an adjustable wrench while torquing nuts to prevent damage to the spring. Start with the center nut and work outward.

13. Adjust the valves and install the camshaft cover, with a new gasket and sealer.

Diesel Engine

1. Remove the rocker cover.
2. Remove the 8 bolts fixing the rocker arm brackets in sequence commencing with the outer ones.
3. Remove the rocker arm, bracket and shaft assembly.
4. To install, follow the removal procedure in reverse order.
5. Tighten the bracket fixing bolts evenly in sequence commencing with the inner ones to 15 ft. lbs.

Intake Manifold

REMOVAL & INSTALLATION

Gasoline Engine

1. Disconnect the battery ground cable and remove the air cleaner assembly.
2. Remove the EGR pipe clamp bolt at the rear of the cylinder head.
3. Raise the vehicle and remove the EGR pipe from the intake and exhaust manifolds.
4. Remove the EGR valve and bracket assembly from the intake manifold.
5. Lower the vehicle and drain the cooling system.
6. Remove the upper coolant hoses from the manifold.
7. Disconnect the accelerator linkage, vacuum lines, electrical wiring and fuel line from the intake manifold.
8. Remove the retaining nuts and remove the manifold from the cylinder head.
9. Remove the lower heater hose while holding the manifold away from the engine. Remove the manifold from the vehicle.
10. Installation is reverse of removal.

Diesel Engine

1. Raise engine hood.

2. Remove the bolts fixing the air cleaner and loosen the clamp bolt.
3. Lift the air cleaner slightly and disconnect the breather hose, then remove the air cleaner assembly.
4. Remove the 2 bolts and 4 nuts fixing the intake manifold.
5. Remove the intake manifold.
6. Installation is the reverse of removal. Torque the bolts to 15 ft. lbs.

Exhaust Manifold

REMOVAL & INSTALLATION

Gasoline Engine

1. Disconnect the battery ground cable and remove the air cleaner assembly.
2. Remove the EGR pipe clamp bolt at the rear of the cylinder head.
3. Raise the vehicle and remove the EGR pipe from the intake and exhaust manifolds.
4. Separate the exhaust pipe from the manifold.
5. Remove the manifold shield and remove the heat stove.
6. Remove the manifold retaining nuts and remove the manifold from the engine.
7. Installation is the reverse of removal.

Diesel Engine

1. Raise engine hood.
2. Remove the bolts fixing the air cleaner and loosen the clamp bolt.
3. Lift the air cleaner slightly and disconnect the breather hose, then remove the air cleaner assembly.

4. Disconnect the exhaust pipe from the exhaust manifold at the flange.
5. Remove the 3 nuts fixing the exhaust manifold, then remove the engine hanger and exhaust manifold.
6. Installation is the reverse of removal. Torque the bolts to 15 ft. lbs.

Timing Gear Cover

REMOVAL & INSTALLATION

Gasoline Engine

1. Remove the cylinder head.
2. Remove the oil pan.
3. Remove the oil pickup tube from the oil pump.
4. Remove the harmonic balancer. (See Timing Cover Seal—Removal and Installation).
5. Remove the air pump drive belt.
6. If equipped with air conditioning, remove the compressor and lay it to one side. Then remove the mounting brackets.
7. Remove the distributor cap and the distributor.
8. Remove the front cover attaching bolts then the cover.
9. Install a new gasket onto the cylinder block.
10. Align the oil pump drive gear punch mark with the oil filter side of the cover; then align the center of the dowel pin with the alignment mark on the oil pump case.
11. Rotate the crankshaft until the No. 1 and the No. 4 cylinders are at TDC.
12. Install the front cover by engaging the pinion gear with the oil pump drive gear on the crankshaft.

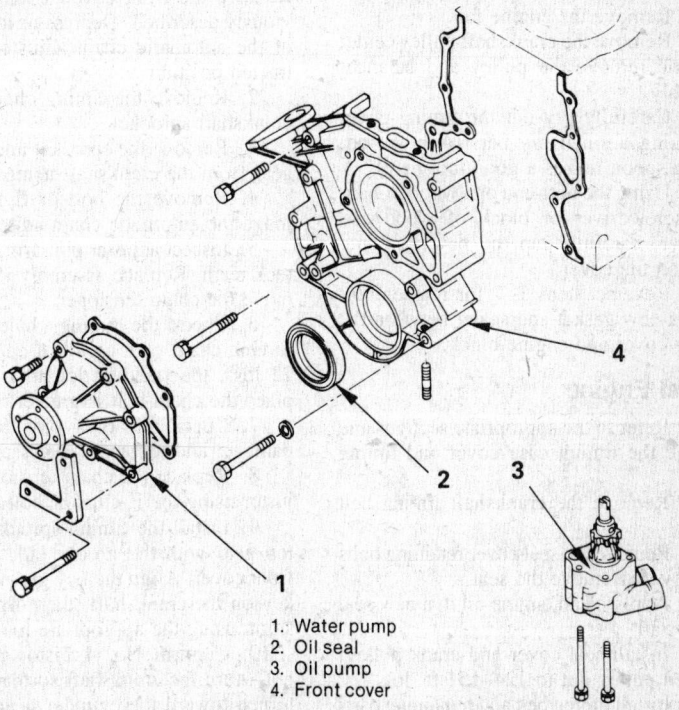

1. Water pump
2. Oil seal
3. Oil pump
4. Front cover

Front timing cover and water pump assemblies

13. Check that the punch mark on the oil pump drive gear is turned to the rear side as viewed through the clearance between the front cover and the cylinder block.

14. Check that the slit at the end of the oil pump shaft is parallel with the front face of the cylinder block and is offset forward.

15. Reverse Steps 1–7 for assembly, using new gaskets and sealer on the oil pan, and a new head gasket.

Diesel Engine

1. Remove the radiator.

2. Remove the compressor drive belt by moving the power steering oil pump or idler. (If so equipped.)

3. Loosen the generator adjust plate bolt and fixing bolt, then remove the fan belt.

4. Remove the 4 bolts fixing the crankshaft pulley and remove the crankshaft pulley.

5. Remove the bolts fixing the timing pulley housing covers, then remove the covers.

6. Installation is the reverse of removal.

Timing Cover Seal
REMOVAL & INSTALLATION

Gasoline Engine

1. Disconnect the negative battery cable.

2. Drain the cooling system.

3. Disconnect the radiator inlet and outlet hoses.

4. Remove the radiator assembly.

5. Remove the alternator and compressor drive belts.

6. Remove the engine fan.

7. Remove the crankshaft pulley center bolt and remove the pulley and balancer assembly.

8. Carefully pry out the timing cover seal using a small pry bar. (A brake adjusting spoon makes a good tool here).

9. Using the butt-end of a wooden-handled screwdriver or block of wood and hammer, carefully tap the new seal into place on the cover.

10. Reverse Steps 1–7 for reassembly, using a new gasket and sealer between the timing cover and engine block.

Diesel Engine

1. Refer to the appropriate sections and remove the timing case cover and timing belt.

2. Remove the crankshaft timing belt pulley.

3. Remove the seal cover retaining bolts and cover. Remove the seal.

4. Apply clean engine oil to a new seal and install.

5. Install seal cover and crank pulley. Tighten pulley nut to 124–151 ft. lbs.

6. Install timing belt and remaining parts in reverse order of removal.

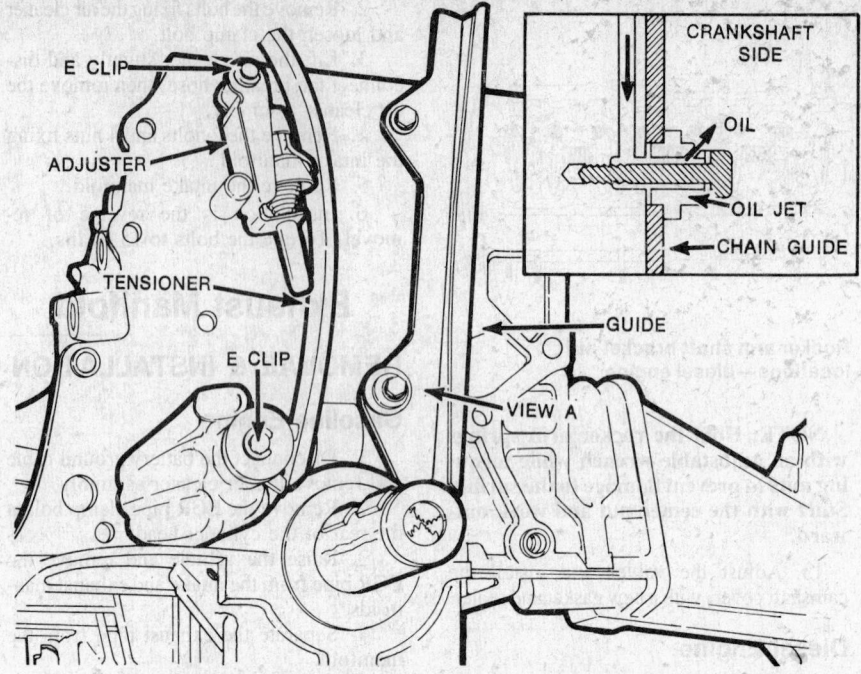

Timing chain adjuster

Timing Chain, Sprockets and Tensioner

REMOVAL & INSTALLATION

Gasoline Engine

1. Bring engine to No. 1 piston at TDC (top dead center) on the compression stroke. Remove the front cover assembly as previously described. Depress or lock the shoe of the automatic chain adjuster in the retracted position.

2. Remove the timing chain from the crankshaft sprocket.

3. Remove the sprocket and the pinion gear from the crankshaft using a puller.

4. Remove the bolt or E-clip and remove the automatic chain adjuster.

5. Inspect adjuster pin, arm, wedge and rack teeth. Replace assembly if worn. Remove the chain tensioner.

6. Check the timing chain for wear. Stretch chain with a pull of approximately 22 lbs., the standard length is 15.00; replace the chain if it is greater than 15.16.

7. Check tensioner pins for wear or damage, and replace if necessary.

8. Replace the chain tensioner and adjuster using the E-clips or bolt.

9. Install the timing sprocket and pinion gear with the groove side toward the front cover. Align the key grooves with the key on the crankshaft, then drive into position using the appropriate tool.

10. Confirm No. 1 piston at TDC, if not—turn the crankshaft so that the key is turned toward the cylinder head side (No. 1 and No. 4 pistons at top dead center).

11. Install the timing chain by aligning the mark plate on the chain with the mark on the crankshaft timing sprocket. The side of the chain with the mark plate is on the front side and the side of the chain with the most links between the mark plates is on the chain guide side.

12. Install the camshaft timing sprocket so that the mark side of the sprocket faces forward and so that the triangular mark aligns with the chain mark plate.

NOTE: Keep the timing chain engaged with the camshaft timing sprocket until the sprocket is installed on the camshaft.

13. Install the front cover assembly, using a new gasket and sealer.

Timing Belt
REMOVAL & INSTALLATION

Diesel Engine

1. Follow the timing cover removal steps.

2. Remove the bolts fixing the injection pump timing pulley flange, then remove the flange.

3. When removing tension spring, avoid using excess force, or distortion of spring will result.

4. Remove the fixing nut of the tension pulley, then remove the tension pulley and tension center.

5. Remove the timing belt. Avoid twisting or kinking the belt and keep it free from water, oil, dust and other foreign matter.

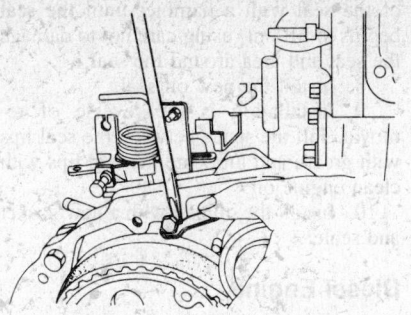

Removing the tension spring

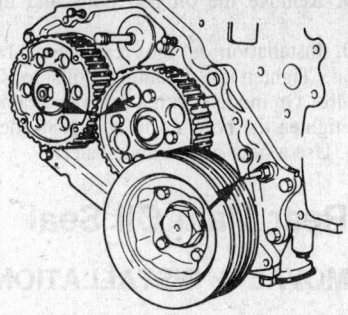

Timing marks aligned on the diesel

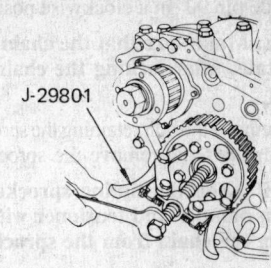

J-29801

Removing camshaft pulley

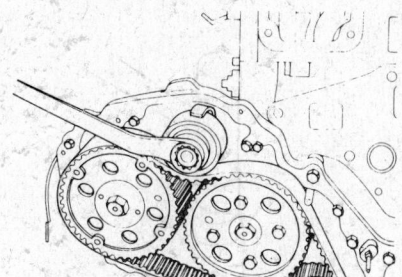

Removing the tension pulley

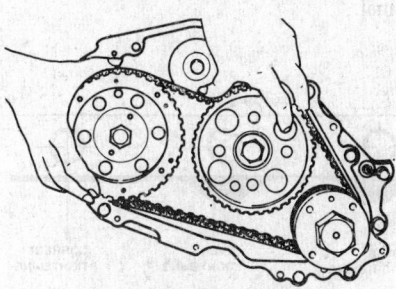

Installing belt

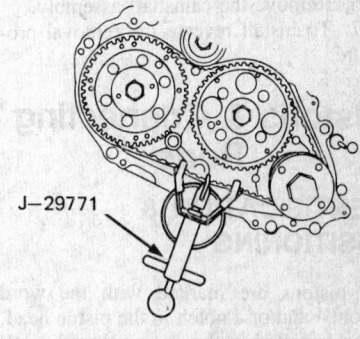

J—29771

Checking belt tension

NOTE: No attempt should be made to readjust belt tension. If the belt has been loosened through service of the timing system, it should be replaced with a new one.

6. Check that the setting marks on the crank pulley, injection pump timing pulley, and camshaft timing pulley are in alignment, then install the timing belt in sequence of crankshaft timing pulley, camshaft timing pulley, and injection pump timing pulley. Make an adjustment so that slackness of the belt is taken up by the tension pulley. When installing the belt, care should be taken so as not to damage the belt.

7. Install the tension center and tension pulley, making certain the end of the tension center is in proper contact with two pins on the timing pulley housing.

8. Hand-tighten the nut so that tension pulley can slide freely.

9. Install the tension spring correctly and semitighten the tension pulley fixing nut. Tighten to 22–36 ft. lbs.

10. Turn the crankshaft 2 turns in normal direction of rotation to permit seating of the belt. Further rotate the crankshaft 90 degrees beyond top dead center to settle the injection pump. Never attempt to turn the crankshaft in reverse direction.

11. Loosen the tension pulley fixing nut completely, allowing the pulley to take up looseness of the belt. Then, tighten the nut to 78–95 ft. lbs.

12. Install the flange on the injection pump pulley. The hole in the outer circumference of the flange should be aligned with the timing mark "△" on the injection pump pulley.

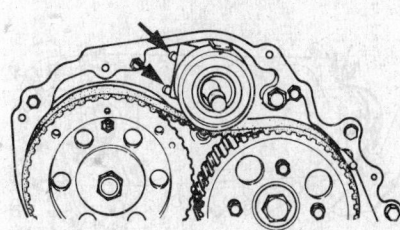

The tension pulley making proper contact with the two pins on the housing

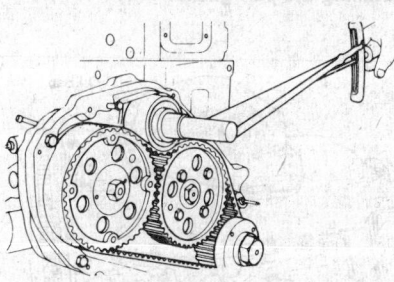

Torquing the tension pulley

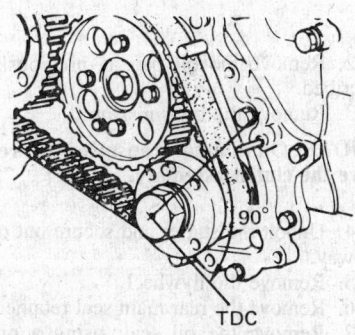

TDC

Bringing the #1 cylinder to TDC

13. Turn the crankshaft 2 turns in normal direction of rotation to bring the piston in No. 1 cylinder to top dead center on compression stroke and check that the mark "△" on the timing pulley is in alignment with the hole in the flange.

14. The belt tension should be checked at a point between the injection pump pulley and crankshaft pulley using tool J29771, to a pull of 33–55 lbs.

15. Adjust valve clearances.

16. Install remaining parts in the reverse order of removal.

Camshaft

REMOVAL & INSTALLATION

Gasoline Engine

1. Remove the cam cover.
2. Rotate the engine until the No. 4 cylinder is at TDC (top dead center) on the compression stroke. Remove the distributor cap and mark the rotor to housing position.
3. Lock the timing chain adjuster by depressing and turning the automatic ad-

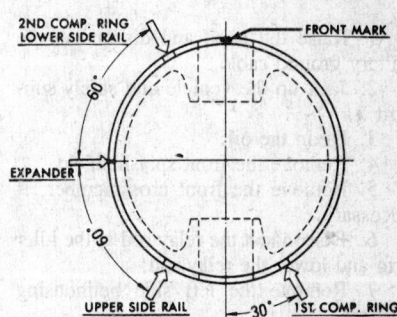

Piston ring positioning

juster slide pin 90° in a clockwise position.

NOTE: Make sure that the chain is in a free state, after locking the chain adjuster.

4. Remove the bolt retaining the sprocket to the camshaft and remove the sprocket.

NOTE: Keep the timing sprocket on the chain damper and tensioner without removing the chain from the sprocket.

5. Remove the rocker arm, shaft and bracket assembly.
6. Remove the camshaft assembly.
7. To install reverse the removal procedure.

Pistons & Connecting Rods

IDENTIFICATION & POSITIONING

The pistons are marked with the word "Front" and/or a notch in the piston head. When installed in the engine the "Front" and notch markings are to be facing the front of the engine. The connecting rods are numbered corresponding to the cylinders in which they are to be installed. Install the connecting rods in their correct cylinders with the marking to the right of the notch in the piston (looking from the rear of the engine).

ENGINE LUBRICATION

Oil Pan

REMOVAL & INSTALLATION

All models have a one-piece stamped steel oil pan attached to the crankcase.

NOTE: On 4WD and diesel models, the engine must be removed before removing the oil pan.

1. Raise the hood and disconnect the battery ground cable.
2. Jack up the vehicle and safely support it.
3. Drain the oil.
4. Remove the front splash shield.
5. Remove the front crossmember, if necessary.
6. Disconnect the relay rod at the idler arm and lower the relay rod.
7. Remove the left side bellhousing bracket.
8. Disconnect the vacuum line at the oil pan.

9. Remove the oil pan bolts and the pan.
10. Installation is the reverse of removal. Tighten the retaining bolts to 43 inch lbs. On models with a separate crankcase tighten the bolts 15 ft. lbs. (180 inch lbs.). Use a new gasket and sealer.

Rear Main Oil Seal

REMOVAL & INSTALLATION

Gasoline Engine

1. Disconnect the negative battery terminal.

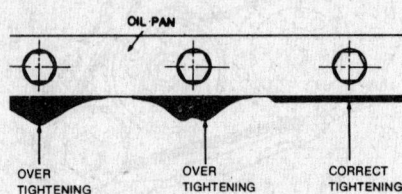

Correct tightening of the oil pan bolts

Use sealer at the points indicated when installing the pan gasket

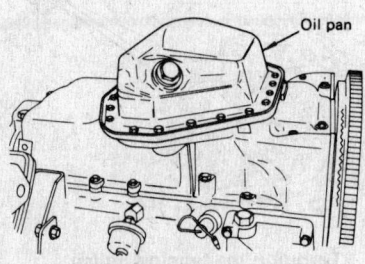

Diesel oil pan (early models)

2. Remove the oil pan as previously described.
3. Remove the transmission.

NOTE: On manual transmissions, remove the clutch assembly.

4. Unbolt the starter and secure out of the way.
5. Remove the flywheel.
6. Remove the rear main seal retainer.
7. Remove the oil seal, using a pin punch. Work the punch around the diameter

of the seal with a hammer until the seal begins to lift out, using care not to damage the seat and area around the seal.

8. Install the new oil seal.
9. Installation is the reverse of removal. Fill the space between the seal lips with grease and lubricate the seal lips with clean engine oil.
10. Install the oil pan with a new gasket and sealer.

Diesel Engine

1. Follow the engine assembly removal steps.

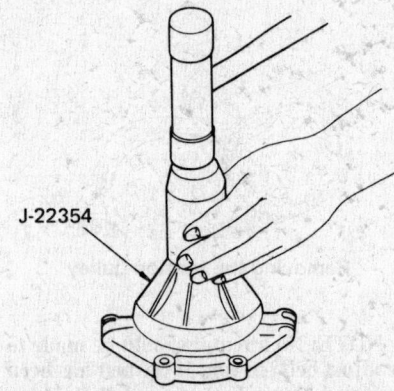

Installing the crankshaft rear seal on gasoline engines

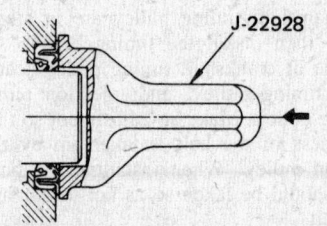

Installing the rear main seal on diesel engines

2. Remove the 6 bolts mounting the flywheel and remove the flywheel assembly. When loosening the flywheel bolts, hold the crankshaft front bolt with a wrench to prevent turning of the crankshaft.
3. Remove the crankshaft rear seal.
4. Install the new seal with seal installer J22928 or equivalent. Reverse removal procedures for all other parts.

Oil Pump

REMOVAL & INSTALLATION

Gasoline Engine

1. Drain and remove the oil pan.
2. Disconnect the oil feed pipe.
3. Remove the two bolts securing the

oil pump to the cylinder block and remove the oil pump.

4. Install in the reverse order of removal.

Diesel Engine

1. Follow the engine assembly removal steps.

2. Remove the 20 bolts fixing the crankcase and remove the crankcase together with the oil pan.

NOTE: Pry off the crankcase by fitting a screwdriver into the slots in the crankcase.

3. Remove the oil pipe sleeve nut.

4. Remove the 3 bolts fixing the oil pump and remove the oil pump with oil pipe.

5. Install the oil pipe and leave the joints semi-tight.

6. Fully tighten the oil pump fixing screws, then tighten the oil pipe joints.

7. Reverse the removal procedure for the remaining parts.

Oil Cooler

REMOVAL & INSTALLATION

Diesel Engine

1. Place a suitable size tray under the oil filter to receive oil and water flowing out from the filter.

2. Drain the cooling system by opening the drain plugs on the radiator and on the cylinder block.

3. Remove the oil cooler water drain plug and drain the water.

4. Disconnect the oil cooler hoses at the cooler side.

5. Remove the oil filter cartridge using filter wrench.

6. Remove the nut fixing the oil cooler, then remove the oil cooler assembly.

7. Install the cooler using a new O-ring. Torque to 55–60 ft. lbs.

ENGINE COOLING

Radiator

REMOVAL & INSTALLATION

1. Drain the radiator by opening the drain cock on the lower part of the radiator.

2. Disconnect the radiator upper, lower and expansion tank hoses.

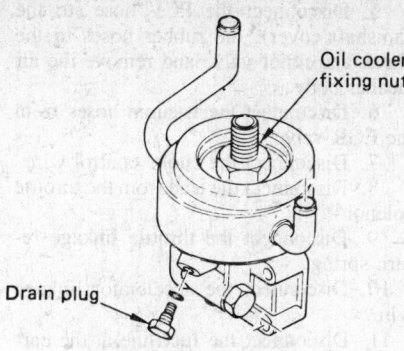

Oil cooler fixing nut

Drain plug

Diesel oil cooler

3. Remove the surge tank (if so equipped) and the radiator fan shroud.

4. Remove the four bolts retaining the radiator and remove the radiator assembly.

5. Install the radiator in the reverse order of removal.

Water Pump

REMOVAL & INSTALLATION

1. Disconnect the battery ground cable.

2. Remove the lower cover and drain the cooling system. Remove the coolant hoses from the pump body.

3. On non-air conditioned gasoline models, remove the fan blades and pulleys from the hub.

4. On air conditioned gasoline models, remove the air pump and alternator belts, fan blades and pulleys.

5. On diesel models, remove drive belts and crankshaft drive pulley. Remove fan pulley.

6. Remove the water pump assembly.

7. Installation is the reverse of removal.

Thermostat

REMOVAL & INSTALLATION

1. Drain the radiator by opening the petcock on the bottom of the radiator.

2. Remove the air cleaner assembly. Disconnect the upper and/or lower radiator hoses for necessary clearance.

3. Disconnect the water outlet from the engine.

4. Remove the thermostat.

5. Replace the thermostat in the reverse order of removal, using a new gasket under the new outlet housing and making sure that the thermostat is placed so that the spring end is inside the engine.

GASOLINE FUEL SYSTEM

Fuel Filter

REMOVAL & INSTALLATION

A fuel filter is located in the fuel line leading from the fuel tank to the fuel pump. On the pickup models the filter is located at the rear near the fuel tank and on the Trooper II models it is located near the air filter in the engine compartment. If the fuel line is suspected of being clogged (rough running, hesitation, uneven throttle response), check the fuel filter. The fuel filter element should be checked once a year and replaced if excessively dirty.

NOTE: 1979–80 models require filter element replacement every 15,000 miles, 1981 and later models at 30,000 mile intervals.

Mechanical Fuel Pump

REMOVAL & INSTALLATION

The fuel pump is a mechanically-operated, diaphragm-type driven by the fuel pump eccentric cam on the jackshaft.

1. Remove the distributor assembly and the high tension cables.

2. Disconnect the fuel lines at the fuel pump. Be careful not to lose the joint bolt gaskets when removing the joint bolt (if so equipped).

3. Remove the two fuel pump mounting nuts and remove the fuel pump assembly from the side of the engine.

4. Install the fuel pump in the reverse order of removal, using a new gasket and sealer on the mating surface.

Electric Fuel Pump

The fuel pump is of the electro-magnetic type and is installed on the inner face of the third crossmember at the left hand side. This fuel pump is a totally enclosed type and cannot be disassembled.

REMOVAL & INSTALLATION

1. Disconnect the negative battery cable and wiring connection to the fuel pump. Disconnect the hoses at the fuel pump.

2. Remove the two bolts and one nut mounting the fuel pump and remove the fuel pump assembly.

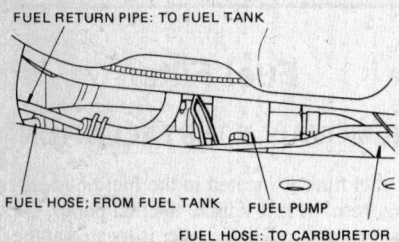

FUEL RETURN PIPE: TO FUEL TANK

FUEL HOSE; FROM FUEL TANK   FUEL PUMP

FUEL HOSE: TO CARBURETOR

Fuel pump

Carburetor

REMOVAL & INSTALLATION

1. Remove the air cleaner wing nut and disconnect the rubber hoses from the clips on the air cleaner cover and the vacuum hose from the vacuum motor.

2. Remove the bracket bolts at the air cleaner and remove the air cleaner cover and filter element.

3. Disconnect the hot air hose (to the hot air duct), the air hose to the air pump at the air cleaner, and the vacuum hose at the joint nipple side of the intake manifold.

4. Loosen the bolt clamping the air cleaner to the carburetor. Separate the air cleaner body from the carburetor but do not remove it completely as the hoses remain connected.

5. Disconnect the PCV hose (to the camshaft cover), the rubber hoses to the check and relief valve and remove the air cleaner body.

6. Disconnect the vacuum hoses from the EGR valve.

7. Disconnect the choke control wire.

8. Disconnect the lead from the throttle solenoid.

9. Disconnect the throttle linkage return spring.

10. Disconnect the accelerator linkage wire.

11. Disconnect the fuel line at the carburetor.

12. Remove the four retaining nuts and lockwashers securing the carburetor to the manifold and remove the carburetor.

13. Install the carburetor in the reverse order of removal.

Refer to the Carburetor section in Unit Repair for adjustments and specifications.

THROTTLE LINKAGE ADJUSTMENT

When the primary throttle valve is opened to an angle of 50° from its closed position, the adjust plate which is interlocked with the primary throttle valve, is brought into contact with the return plate. When the primary throttle valve is opened farther, the return plate is pulled apart from the stopper allowing the secondary throtttle valve to open.

1. Measure the clearance between the primary throttle valve and the wall of the throttle chamber at the center of the throttle valve when the adjust plate is brought into contact with the return plate. Standard clearance is 0.26–0.32 in.

2. If necessary, make the adjustment by bending the tab of the return plate.

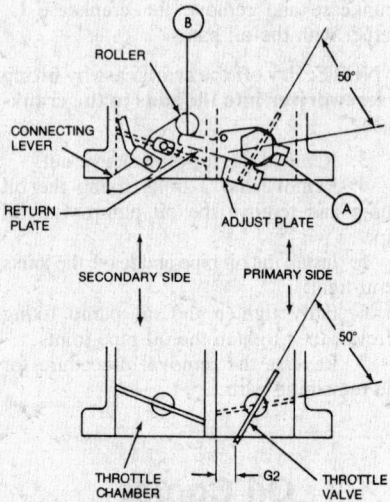

Throttle linkage adjustment

DIESEL FUEL SYSTEM

Fuel Filter

REMOVAL & INSTALLATION

1. Disconnect water separator sensor wiring at the connector and water drain hose.

2. Remove the filter using a filter wrench. Use care so as not to spill fuel within the cartridge.

3. Drain the cartridge into a suitable pan, then remove the sensor from the filter cartridge.

4. Install the sensor on a new filter cartridge, applying clean diesel fuel to the O-ring before installation.

5. Turn in the filter until sealing face is brought into contact with the O-ring. Further tighten ⅔ of a turn. Connect the connector to the sensor and install the water drain hose.6.

Fill the cartridge with fuel by operating the priming pump handle 30 to 40 times.

NOTE: The force needed to operate the priming pump increases when the filter becomes filled.

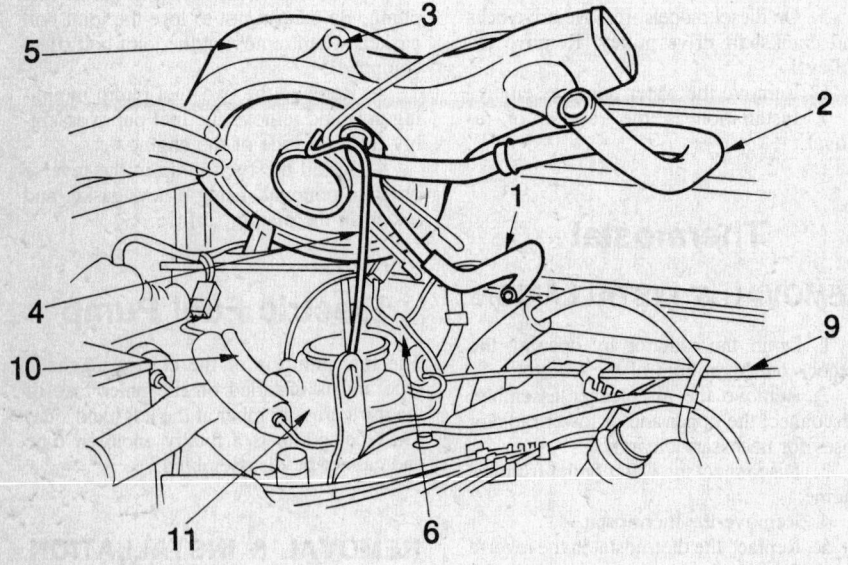

1. P.C.V. hose
2. A.I.R hose
3. Bolts; attaching air cleaner
4. T.C.A. vacuum hose and air duct
5. Air cleaner
6. Throttle return spring
9. Accel control cable
10. Fuel pipes
11. Carburetor assembly

Removing the carburetor

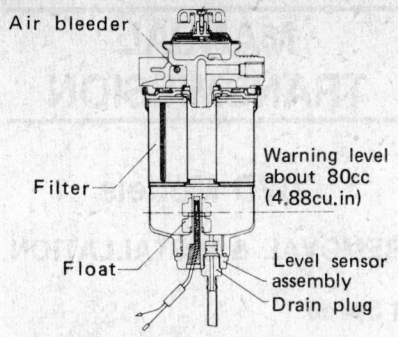

Diesel engine fuel filter/water separator

7. Start the engine and check for fuel leaks.

WATER SEPARATOR DRAINING PROCEDURE

1. Place a container (approx. ½ gal. capacity) at the end of the vinyl hose beneath the drain plug of the separator.
2. Loosen the drain plug approximately four turns.
3. Continue to operate the priming pump ten times by hand.
4. Tighten the drain plug, and continue to operate the priming pump again several strokes.
5. Start the engine and check for fuel leakage from around the sealing portions. Also check to see that the "FILTER" indicator light has turned off.

Injection Timing

1. Check that notched line on the injection pump flange is in alignment with notched line on the injection pump front bracket.

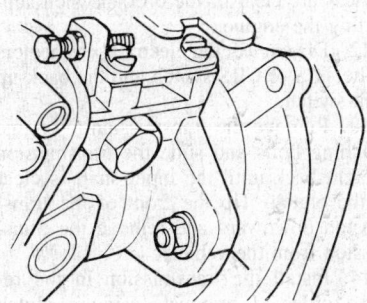

Injection pump and flange alignment

2. Bring the piston in No. 1 cylinder to top dead center on compression stroke by turning the crankshaft as necessary.
3. With the timing pulley housing cover removed, check that timing belt is properly tensioned and that timing marks are aligned.
4. Disconnect the injection pipe from the injection pump and remove the distrib-

utor head screw, then install static timing gauge. Set the lift approximately 1mm (0.04 in.) from the plunger.
5. Use a wrench to hold the delivery holder when loosening the sleeve nuts on the injection pump side.
6. Bring the piston in No. 1 cylinder to a point 45–60 degrees before top dead center by turning the crankshaft, then calibrate the dial indicator to zero.

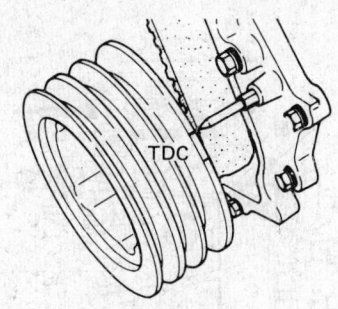

#1 piston at TDC

7. Turn the crankshaft pulley slightly in both directions and check that gauge indication is stable.
8. Turn the crankshaft in normal direction of rotation, and take the reading of the dial indicator when the timing mark (15 degrees) on the crankshaft pulley is in alignment with the pointer. Reading should be 0.020 in.
9. If the reading of dial indicator deviates from the specified range, hold crankshaft in position 15 degrees before to dead center and loosen two nuts on injection pump flange.
10. Move the injection pump to a point where dial indicator gives reading of 0.020 in., then tighten pump flange nuts.

IDLE SPEED ADJUSTMENT

1. Set parking brake and block drive wheels.
2. Place transmission in neutral.
3. Start and normalize the engine. Engine coolant temperature: above 80°C (176°F).
4. Set the engine tachometer.
5. If the idle speed deviates from the specified range of 700–800 rpm, loosen the idle speed adjusting screw lock nut.
6. Turn the adjusting screw in or out until the idle speed is in the correct range. After tightening the lock nut, lock it in place

FAST IDLE SPEED

1. Start and normalize the engine. Engine coolant temperature above 80°C (176°F).
2. Set the engine tachometer.
3. Disconnect the hoses from the vac-

Timing marks aligned

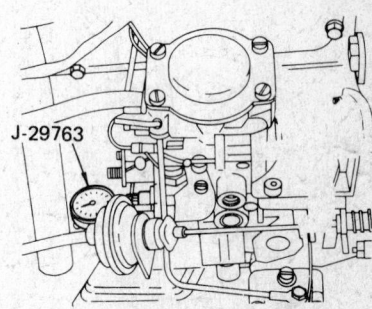

Static timing gauge installed

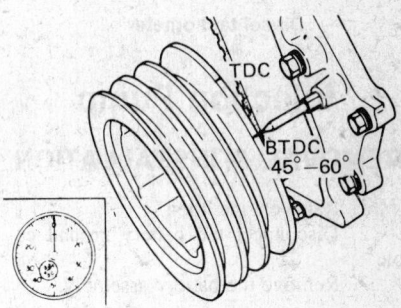

#1 piston 45–60 degrees before TDC

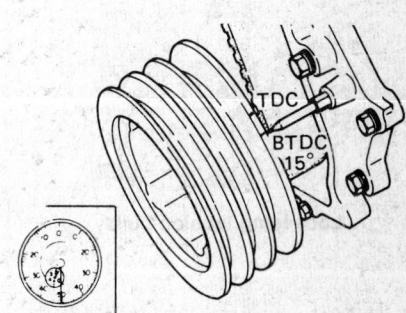

#1 piston 15 degrees before TDC

uum switch valve, then connect a pipe (4mm dia.) in position between the hoses.
4. Loosen adjust nut and adjust engine idle speed by moving the nut. Fast idle should be 900–950 rpm.
5. Tighten the nut.
6. Remove engine tachometer.

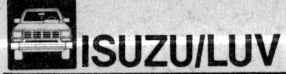

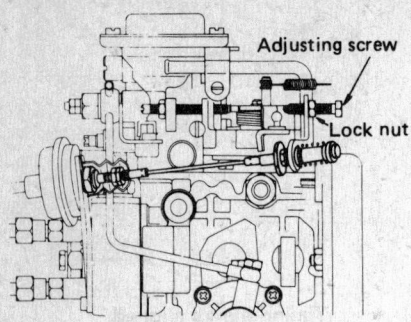

Idle speed adjustment points

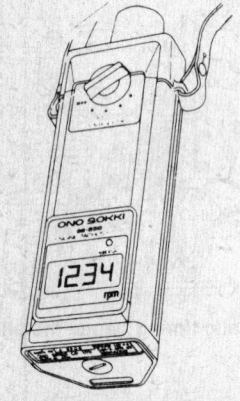

Diesel tachometer

Injection Pump

REMOVAL & INSTALLATION

1. Raise engine hood.
2. Disconnect the battery ground cable.
3. Remove the battery assembly.

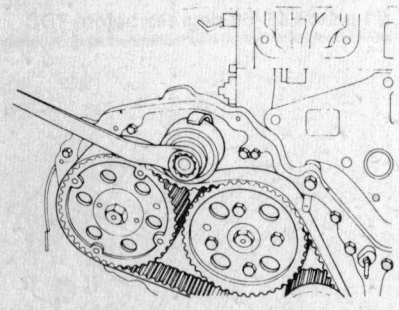

Loosening tension pulley

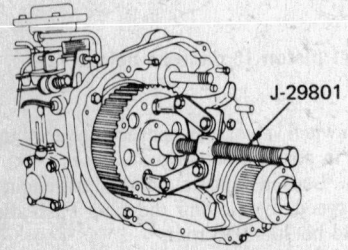

Removing timing pulley

---

4. Remove the under cover.
5. Drain the cooling system by opening the drain plugs on the radiator and on the cylinder block.
6. Disconnect the upper water hose at the engine side.
7. Loosen the compressor drive belt by moving the power-steering oil pump or idler. (If so equipped.)
8. Remove the cooling fan and fan shroud.

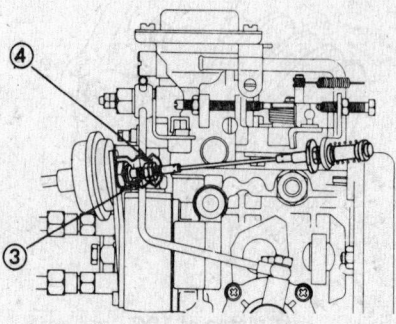

Fast idle adjusting points

9. Disconnect the lower water hose at the engine side.
10. Remove the air conditioner compressor. (If so equipped.)
11. Remove the fan belt.
12. Remove the crankshaft pulley.
13. Remove the timing pulley housing covers.
14. Remove the tension spring and fixing bolt, then remove the tension center and pulley.
15. Remove the timing belt.
16. Remove the engine control cable and wiring harness of the fuel cut solenoid.
17. Remove the fuel hoses and injection pipes. Use a wrench to hold the delivery holder when loosening the sleeve nuts on the injection pump side.
18. Install a 6mm bolt (with pitch of 1.25) into threaded hole in the timing pulley housing through the hole in pulley to prevent turning of the pulley. Remove the bolts fixing the injection pump timing pulley, then remove the pulley using pulley puller.
19. Remove injection pump flange fixing nuts and rear bracket bolts, then remove the injection pump.
20. Install the injection pump by aligning notched line on the flange with the line on the front bracket.
21. Install the injection pump timing pulley by aligning it with the key groove. Torque to 42–52 ft. lbs.
22. Bring the piston in No. 1 cylinder to top dead center on compression stroke and align marks on the timing pulleys.
23. Follow the timing belt installation steps.
24. Check the injection timing.
25. To install remaining parts, follow the removal steps in reverse order.

MANUAL TRANSMISSION

2WD Models

REMOVAL & INSTALLATION

4 Speed

1. Disconnect the negative battery cable.
2. Remove the air cleaner assembly, and disconnect the accelerator linkage at the carburetor throttle lever.
3. Slide the gearshift lever boot upward on the lever, remove the two gearshift lever attaching bolts and remove the lever.
4. Remove the starter attaching bolts and lay the starter assembly aside.
5. Jack up the vehicle and safely support it with jackstands. Disconnect the exhaust pipe at the flange and disconnect the exhaust pipe hanger at the transmission.
6. Disconnect the speedometer cable at the transmission and disconnect the driveshaft at the differential. Either drain the transmission oil or have a rag ready to plug the output shaft opening (at rear of trans) to prevent spillage. Remove the driveshaft.
7. Disconnect the clutch cable from the bell housing and clutch fork.
8. Remove the bolts attaching the stiffeners, then remove the skid plate (all models).
9. Remove the three frame bracket-to-transmission rear mounting bolts.
10. Raise the engine and transmission as required and remove the four crossmember-to-frame bracket bolts.
11. Lower the engine and transmission assembly and support the rear of the engine. Make sure all jacks, jackstands and support devices are clear of the oil pan when supporting the engine.
12. Disconnect the electrical connectors at the TCS or CRS switch and the back-up light switch.
13. Remove the transmission-to-engine attaching bolts and slide the transmission straight back until the input shaft is clear of the clutch. Tip the front of the transmission downward and remove the transmission from the vehicle.
14. Install the transmission in the reverse order of removal, using a clutch aligning arbor or discarded transmission input shaft to align the clutch disc and pilot bearing, if the clutch was also removed. Coat the input shaft lightly with Lubriplate® or similar grease before installation.

5 Speed

1. Disconnect battery ground cable.
2. Slide the gearshift lever boot upwards on the lever, remove two gearshift lever attaching bolts and remove lever.

3. Remove starter attaching bolts and lay starter assembly aside.

4. Raise vehicle on hoist and disconnect exhaust pipe hanger at transmission.

5. Place pan under transmission to catch the oil and drain transmission. Disconnect the speedometer cable at the transmission, disconnect ground cable and disconnect propeller shaft at differential. Remove propeller shaft.

6. Remove return spring from clutch fork and remove the clutch cable.

7. Take out two bolts (lower bolts) mounting the flywheel stone guard.

8. Remove two frame bracket to transmission rear mount bolts and nuts.

9. Raise engine and transmission as required and remove four crossmember to frame bracket bolts.

10. Remove the rear mounting (2 nuts) from the transmission rear extension.

11. Lower engine and transmission assembly and support rear of engine.

12. Disconnect electrical connector at back-up lamp switch.

13. Remove transmission to engine attaching bolts.

14. Pull transmission straight back until disengaged from clutch. Tip front of transmission downward and remove.

15. Position transmission in vehicle and slide forward guiding clutch gear into pilot bearing.

16. Install transmission to engine attaching bolts.

17. Raise and lower engine and transmission as required and install crossmember frame bracket and rear mount.

18. Install the two bolts (lower bolts) mounting the flywheel stone guard.

19. Remove plug (if used) from rear extension and install propeller shaft.

20. Connect speedometer cable, ground cable and exhaust pipe hanger.

21. Connect clutch cable and adjust shift fork as outlined in Clutch Section.

22. Connect electrical connector at back-up lamp switch. Fill transmission with lubricant.

23. Lower vehicle and install starter assembly.

24. Connect battery negative cable.

25. Install gearshift lever and adjust clutch pedal height as outlined in Clutch Section.

12. Check transmission operation.

4WD Models

REMOVAL & INSTALLATION

Transmission & Transfer Case

1. Disconnect the negative battery terminal.

2. Drain the transmission oil.

3. Slide the shift lever boots upward and unbolt each lever.

4. Remove the return spring from the transfer case shift lever and remove both levers.

5. Remove the starter attaching bolts and remove the starter assembly.

6. Jack up your vehicle and support it with jackstands. Disconnect the exhaust pipe from the manifold and disconnect the pipe support from the transmission.

7. Disconnect the speedometer cable at the transmission. Disconnect the rear driveshaft at the differential. Disconnect the ground strap.

8. Remove the rear driveshaft from the transfer case. Remove the front driveshaft.

9. Disconnect the clutch return spring.

10. Disconnect the clutch cable at the fork.

11. Remove the flywheel stoneguard.

12. Remove the transmission rear crossmember bolts.

13. Raise the engine and transmission, and remove the rear crossmember-to-frame bolts.

14. Remove the rear mounting bolts from the transfer case.

15. Unbolt and remove the side case from the transmission.

16. Remove the stud bolt from the transfer case.

17. Lower the engine and transmission assembly and support the rear of the engine.

18. Disconnect the CRS switch and backup light switch.

19. Remove the shifter cover and gasket from the transfer case.

20. Remove the transmission-to-engine bolts. When removing the transmission, turn the side case fitting face of the case downward and pull the case straight back until free from the clutch. Tip the front of the transmission downward and remove it.

21. Installation is the reverse of removal. Torque shifter cover bolts to 14 ft. lbs.

Refer to the Transmission or Transfer Case section in Unit Repair for servicing procedures.

CLUTCH

REMOVAL & INSTALLATION

1. Jack up the vehicle and safely support it using jackstands.

2. Remove the transmission.

3. Mark the clutch assembly-to-flywheel relationship with a center punch or scribe so that the clutch assembly can be reassembled in the same position from which it is removed.

4. Loosen the six clutch cover-to-flywheel attaching bolts, one turn at a time in an alternating sequence, until the spring tension is relieved to avoid distorting or bending the clutch cover.

5. Support the clutch pressure plate and cover assembly with a clutch aligning arbor, then remove the bolts and the clutch assembly.

6. Apply a thin coat of grease to the pressure plate wire ring, diaphragm spring,

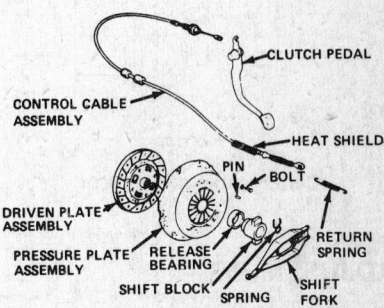

Clutch components

clutch cover grooves and the drive bosses on the pressure plate.

7. Apply a thin coat of Lubriplate® to the splines in the driven plate.

8. Assemble the clutch cover and pressure plate and the driven plate on a clutch alignment arbor.

9. Align the marks made on the clutch cover and flywheel and install the six clutch cover-to-flywheel attaching bolts. Tighten the bolts to 50 in. lbs. Remove the aligning arbor.

PEDAL HEIGHT ADJUSTMENT

Adjust the pedal stop so that the clutch and brake pedals are the same height. Make sure that the clutch switch is in contact with the clutch pedal bracket.

Clutch Cable

REMOVAL & INSTALLATION

1. Loosen the clutch cable lock and adjusting nuts. Remove the clutch cable clip

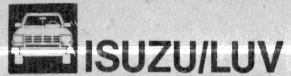
at the engine compartment location.

2. Raise the vehicle and remove the spring from the shift fork end.

3. Disconnect the cable end from the shift fork and pull the cable assembly through the bracket.

4. Lower the vehicle enough to disengage the hooked part of the clutch pedal from the cable eye. Pull the cable assembly towards the engine compartment and remove the cable from the vehicle.

5. Installation is the reverse of removal.

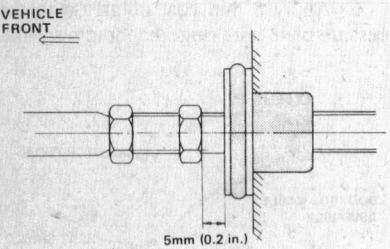

VEHICLE FRONT

5mm (0.2 in.)

Clutch cable adjustment

ADJUSTMENT

1. Pull the outer cable forward and turn the adjusting nut inward until the rubber lip on the washer damper touches the firewall.

2. Depress and release the clutch pedal a few times.

3. Pull the outer cable forward again and fully tighten the adjusting nut. Back the nut off to provide a 0.20 in. clearance.

4. Release the outer cable and tighten the nut.

AUTOMATIC TRANSMISSION

REMOVAL & INSTALLATION

1. Disconnect the negative battery cable and remove the throttle valve cable from the carburetor.

2. Remove the transmission dipstick assembly.

3. Raise the vehicle and remove the pan from the converter housing. Drain the transmission fluid.

4. Remove the starter assembly.

5. Have a clean rag ready to use as a plug for the end of the transmission. Disconnect the driveshaft and remove it from the vehicle. Plug the rear of the transmission to avoid oil leakage.

6. Disconnect the shift lever control rod from the transmission shift lever.

7. Remove the exhaust pipe bracket. Remove the speedometer cable from the transmission.

8. Remove the oil cooler lines and position them along the vehicle frame to prevent damage.

9. Remove the bolts and nuts coupling the drive plate to the converter.

10. Remove the bolts holding the frame bracket to the transmission rear mount.

11. Raise the engine and transmission assembly and remove the frame bracket from the cross member. Remove the rear mount from the transmission.

12. Remove the bell housing bolts and move the transmission and converter assembly.

─────── CAUTION ───────
Do not allow the torque converter to drop from the transmission during removal.

13. Install in reverse order.

Shift Linkage

ADJUSTMENT

1. Loosen the control rod lock nuts so that trunnion will slide on the control rod.

2. Turn the manual shaft of the transmission counterclockwise, viewed from the left side of the transmission, as far as it will go.

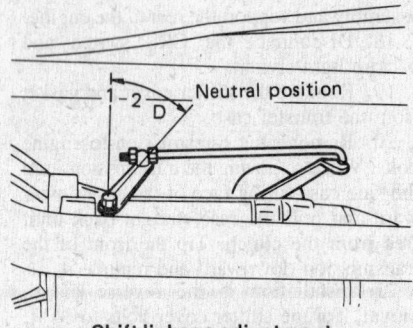

Neutral position

1–2 D

Shift linkage adjustment

3. Back off the manual shaft three stops to the neutral position.

4. Holding the shaft in this position, move the shift lever to the neutral position and push the shift control lower lever rearward to remove play. Tighten the lock nuts.

5. Check for proper operation of the transmission in all transmission ranges.

Throttle Valve Control Cable

REMOVAL & INSTALLATION

1. Loosen the throttle valve control cable adjusting nuts and disconnect the cable from the carburetor throttle lever by removing the pin.

2. Remove the throttle valve cable clip from the right side of the cylinder body.

3. Remove the bolt holding the throttle cable to the transmission and pull the cable upward. Disconnect the end of the inner cable from the throttle lever link on the transmission side.

4. Remove the cable assembly from the vehicle.

5. Installation is the reverse of removal.

ADJUSTMENT

1. Loosen the throttle valve control cable adjusting nuts.

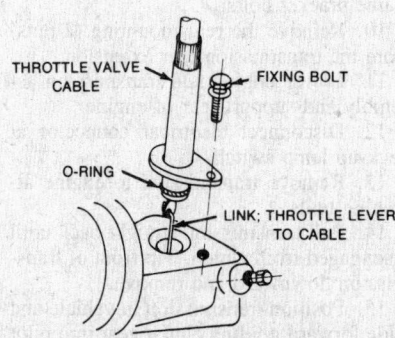

THROTTLE VALVE CABLE — FIXING BOLT

O-RING

LINK; THROTTLE LEVER TO CABLE

Throttle valve cable removal

2. Open the carburetor throttle lever to the wide open position and adjust the inner cable by turning the adjustment nut (lower) on the outer cable by hand so that the inner cable has a free play of approximately 0.040 in.

3. Tighten the lock nut (upper) securely.

4. Make sure that the stroke of the inner cable from the wide open position to the closed position is within the range of 1.37–1.41 in.

Inhibitor Switch

ADJUSTMENT

1. Loosen the screws holding the switch. Move the switch body so that the center of the moveable part of the switch, aligns with the neutral position indicator line on the steel case when the shift lever is in the neutral position.

2. Tighten the holding screws, and make sure that the engine does not start in gear.

Intermediate Band

The bands are not adjustable. A selective band apply pin is installed at time of assembly or overhaul that compensates for normal band wear.

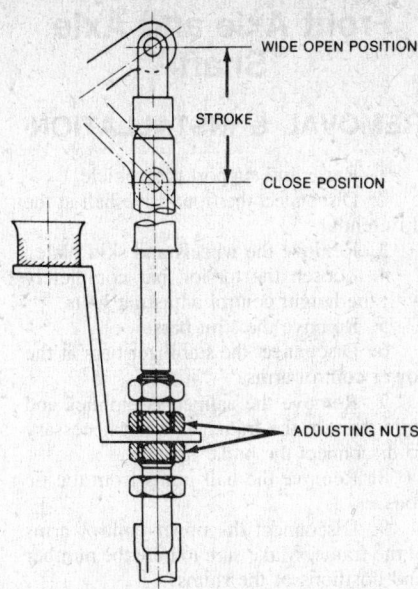

Throttle valve cable adjustment

Transmission Fluid

DRAIN & REFILL

1. Jack up the vehicle and support it with jackstands. Support the transmission at the vibration damper, and remove the oil pan retaining bolts from the front and sides of the fluid pan.

2. Loosen the rear fluid pan bolts approximately four (4) turns.

———— CAUTION ————

If the vehicle has been running, the transmission fluid will be HOT, exceeding 350°F.

3. Carefully pry the front of the fluid pan loose from the transmission case using a small pry bar or brake adjusting spoon. Allow the fluid to drain into a waste container.

4. Remove and clean the pan.

5. Remove the two retaining screen-to-valve body bolts, screen and gasket and clean the screen thoroughly.

6. Install screen and new gasket in place on valve body. Tighten retaining bolts to 6–10 ft. lbs.

7. Install the fluid pan with a new gasket and torque the pan retaining bolts to 10–13 ft. lbs.

8. Lower the vehicle and install six pints of Dexron® II in the transmission. Start the engine and move the selector lever through each gear position.

9. Recheck the fluid level and fill to the following levels. Fluid at room temperature: Level should be 1/8–3/8 in. below the add mark on the dip stick. Fluid at normal operating temperature should be at the full mark on the dip stick.

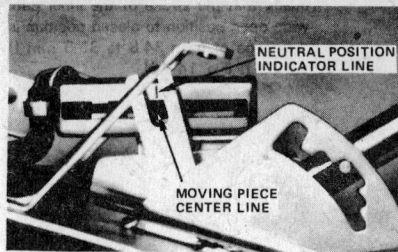

Inhibitor switch adjustment

NOTE: Normal operating temperature is reached after approximately 15 miles of highway type driving or equivalent.

DRIVE LINE

Rear Driveshaft & U-Joints

REMOVAL & INSTALLATION

Short Wheel Base Pickup Models

NOTE: Match-mark all parts for installation.

1. Disconnect the driveshaft rear flange from the differential pinion flange.

2. Pull the one piece driveshaft from the end of the transmission housing cover.

3. Plug or cover the transmission housing cover end to prevent lubricant leakage.

4. Installation is the reverse of removal.

Long Wheel Base Pickup & Trooper II Models

NOTE: Match-mark all parts for installation.

1. Disconnect the flanged yokes between the front and rear driveshafts.

2. Disconnect the rear driveshaft flange from the differential pinion flange and remove the rear shaft.

3. Remove the center bearing support bracket from the fourth crossmember and pull the front driveshaft from the rear of the transmission housing cover.

4. Install a plug or cover the transmission housing cover end to prevent lubricant loss.

5. Installation is the reverse of removal.

Front Driveshaft

REMOVAL & INSTALLATION

1. Jack up the vehicle and support it with jackstands.

NOTE: You must raise the vehicle enough for the front wheels to turn freely.

2. Place the transmission and transfer case in the neutral position.

3. Match-mark each end of the driveshaft and the flanges on the rear and transfer case.

4. Remove the U-bolts from the rear and the transfer case flanges.

5. Remove the driveshaft.

6. Installation is the reverse of removal.

Refer to the U–Joint/CV–Joint section in Unit Repair for servicing procedures.

Rear Axle Shaft

REMOVAL & INSTALLATION

1. Raise and safely support the truck.

2. Remove the rear wheel cover and the wheel and tire.

3. Remove the brake drum, brake shoes and disconnect the parking brake inner cable.

4. Disconnect the brake line at the wheel cylinder and plug the end of the line.

5. Remove the four nuts from the bearing holder through-bolts from the inside of the brake backing plate.

6. Using an axle puller, pull out the axle shaft assembly. Never strike the brake backing plate with a hammer in an attempt to remove the axle shaft.

7. Install the axle shaft in the reverse order of removal, tightening the bearing holding plate attaching nuts to 55 ft. lbs., bleeding the brake hydraulic system after installing the brakes and adjusting the parking brake cable as necessary.

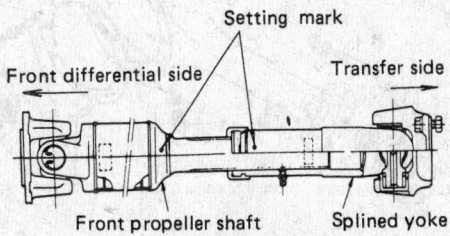

Front driveshaft assembly

1.1a Both; differential side
2. Bolt; flange
2a Propeller shaft assembly
3. Propeller shaft assembly; 2nd
4. Bolt; center bearing bracket
5. Propeller shaft assembly; 1st

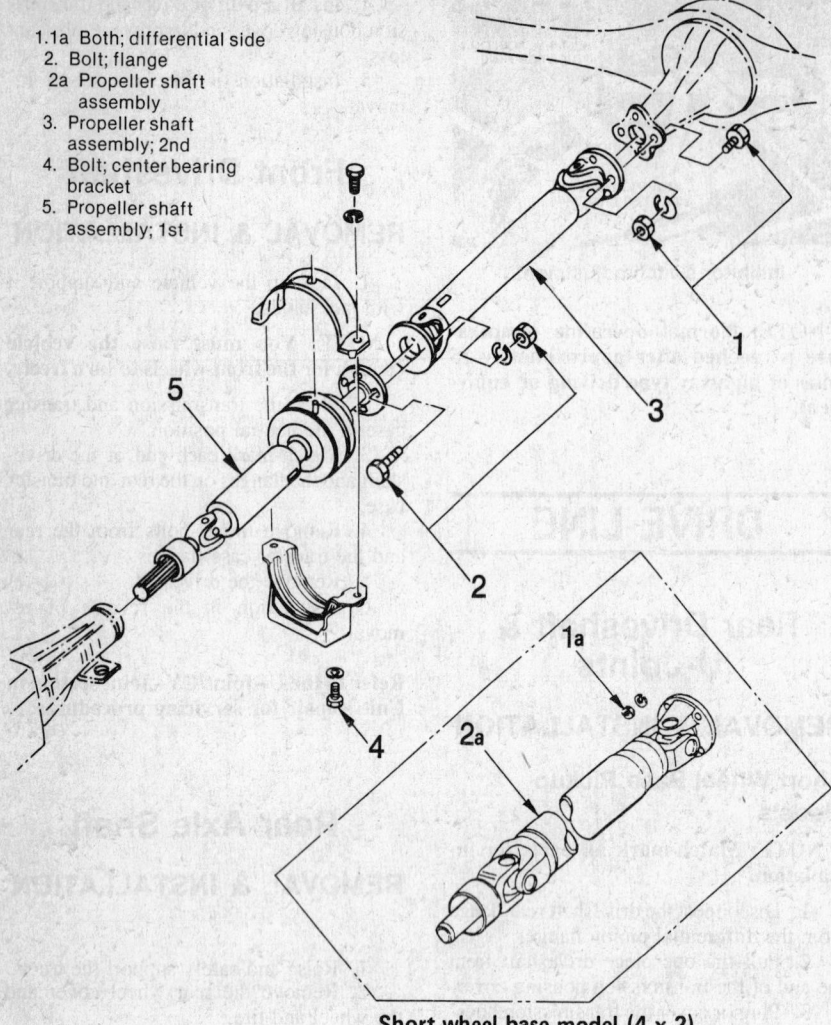

Short wheel base model (4 x 2)

Rear driveshaft assembly

1. Differetial carrier and case assembly
2. Mounting bolt
3. Gasket
4. Drain plug
5. Filler plug
6. Vent
7. Through-bolt
8. Oil seal
9. Shims
10. Locknut
11. Lockwasher
12. Axle shaft bearing
13. Bearing holder
14. Grease seal
15. Axle shaft
16. Wheel stud
17. Brake drum
18. Wheel nut
19. Drum-to-flange screw

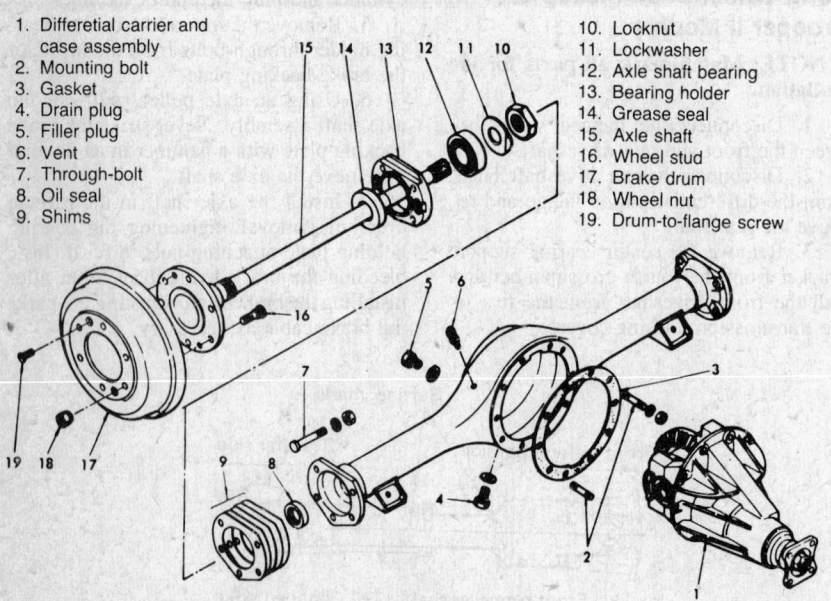

Axle shaft and housing

Front Axle and Axle Shaft

REMOVAL & INSTALLATION

1. Raise and support the vehicle.
2. Disconnect the front driveshaft at the differential.
3. Remove the wheels and skid plate.
4. Loosen the torsion bar completely with the height control adjusting bolts.
5. Remove the strut bars.
6. Disconnect the stabilizer bars at the lower control arms.
7. Remove the caliper assemblies and wire them to the frame. It is not necessary to disconnect the brake lines.
8. Remove the ball joints from the tie rods.
9. Disconnect the upper control arms at the frame. Make sure to note the number and positions of the shims.
10. Remove the steering link ends from the lower control arms.
11. Disconnect the shock absorbers from the lower control arms.
12. Disconnect the lower control arms from the frame.
13. Remove the free wheeling hub. See Front Bearing Removal.
14. Remove the rotors and upper links.
15. Remove the pitman arm and idler arm along with the steering linkage assembly.
16. Support the differential housing with a jack, lower it clear of the vehicle and roll it out. Take care to avoid damaging the birfield joints.
17. Drain the differential case and remove the four bolts attaching the axle mounting bracket to the case.
18. Pull the shaft assemblies from the case on both sides.
19. Installation is the reverse of removal.

Item	Ft. Lbs.
Axle shaft-to-case	43
Differential case-to-frame	15
Pitman arm-to-sector shaft	160
Idler arm-to-pivot shaft	87
Lower control arm-to-frame	94
Ball joint castellated nut	100

FRONT SUSPENSION— 2 WD

Torsion Bars

REMOVAL & INSTALLATION

1. Jack up the front of the vehicle and support it with jackstands.

2. Remove the adjusting bolt from the height control arm.

3. Mark the location and remove the height control arm from the torsion bar and the third crossmember.

4. Mark the location and withdraw the torsion bar from the lower control arm.

5. For installation, apply a generous amount of grease to the serrated ends of the torsion bar.

6. Hold the rubber bumpers in contact with the lower control arm. Jack the vehicle up under the lower control arm to accomplish this.

7. Insert the front end of the torsion bar into the control arm.

8. Install the height control arm in position so that its end is reaching the adjusting bolt. Be sure to lubricate the part

of the height control arm that fits into the chassis with grease.

9. Install a new cotter pin in the control arm.

10. Turn the adjusting bolt to the location marked before removal.

11. Lower the vehicle and check the vehicle height.

Shock Absorbers

REMOVAL & INSTALLATION

1. Raise the vehicle and support it with jackstands.

1. Torsion bar
2. Height control arm
3. Upper pivot nut
4. Lower pivot nut
5. Height control bolt
6. Stopper plate
7. Bolt
8. Strut rod
9. Strut rod bushing
10. Strut rod washer
11. Tube
12. Nut
13. Bolt
14. Bolt
15. Shock absorber
16. Shock absorber bushing
17. Bushing retainer
18. Nut
19. Bolt
20. Nut
21. Dust cover
22. Screw
23. Lower control arm bumper
24. Upper control arm bumper
25. Stabilizer bar
26. Bolt
27. Stabilizer bar bushing
28. Stabilizer bar upper clamp
29. Stabilizer bar lower clamp
30. Bolt
31. Nut
32. Bolt
33. Nut
34. Bracket, stabilizer bar to frame
35. Bolt
36. Bushing
37. Washer
38. Nut
39. Nut
40. Nut

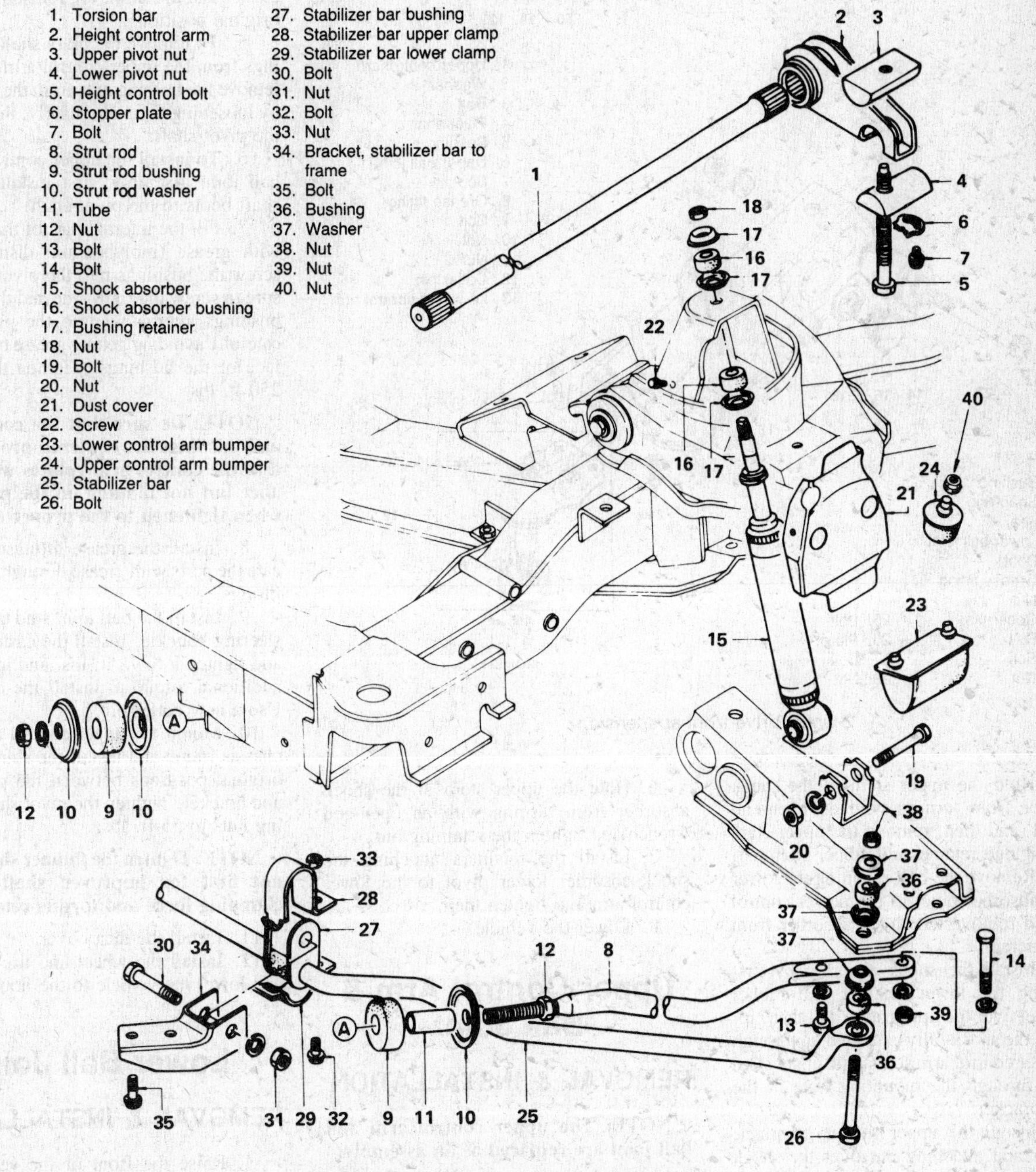

2-wheel drive torsion, strut and stabilizer bars

T365

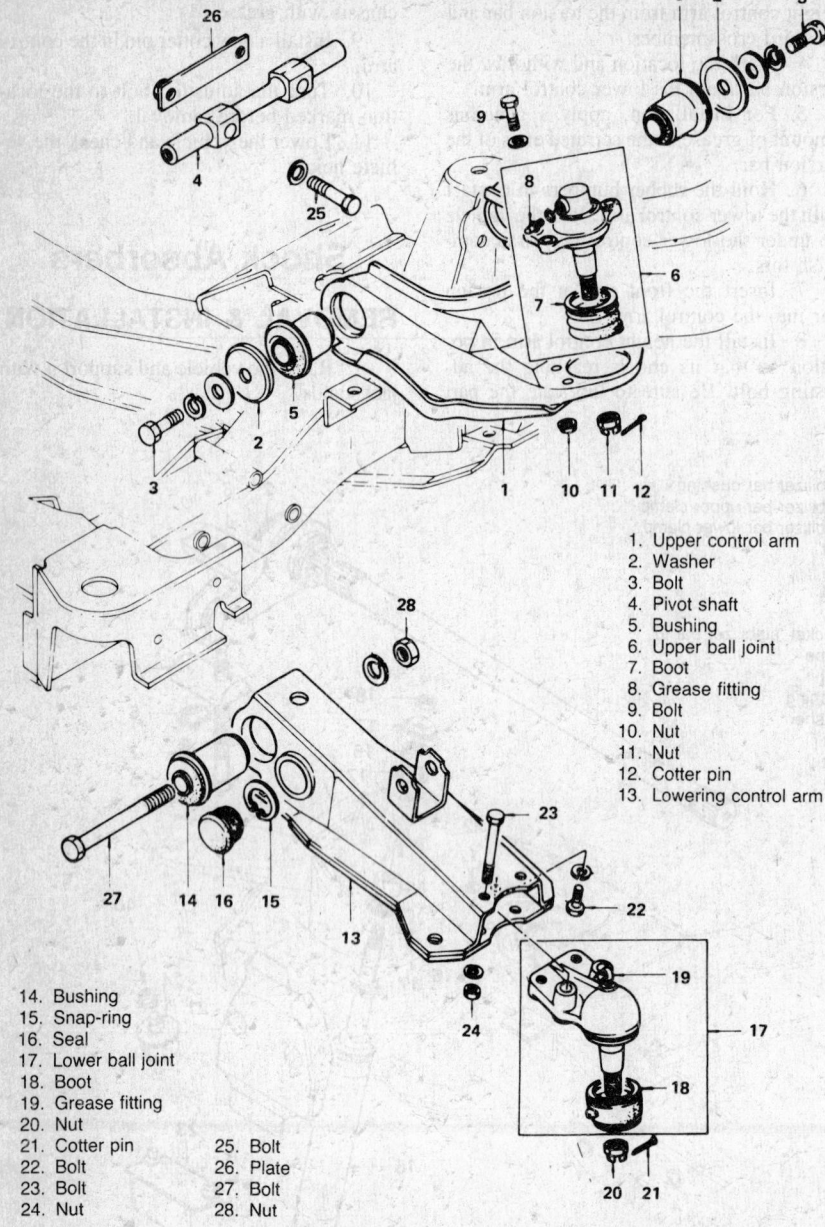

1. Upper control arm
2. Washer
3. Bolt
4. Pivot shaft
5. Bushing
6. Upper ball joint
7. Boot
8. Grease fitting
9. Bolt
10. Nut
11. Nut
12. Cotter pin
13. Lowering control arm

14. Bushing
15. Snap-ring
16. Seal
17. Lower ball joint
18. Boot
19. Grease fitting
20. Nut
21. Cotter pin   25. Bolt
22. Bolt   26. Plate
23. Bolt   27. Bolt
24. Nut   28. Nut

2-wheel drive front suspension

2. Hold the upper stem of the shock absorber from turning with an open-end wrench, and then, remove the upper stem retaining nut, retainer and rubber grommet.

3. Remove the bolt retaining the lower shock absorber pivot to the power control arm and remove the shock absorber from the vehicle.

4. Install the shock absorber by first installing the lower retainer and rubber grommet over the upper stem and then, installing the shock fully extended up through the upper control arm so that the upper stem passes through the mounting hole in the frame bracket.

5. Install the upper rubber grommet, retainer and attaching nut over the shock absorber upper stem.

6. Hold the upper stem of the shock absorber from turning with an open-end wrench and tighten the retaining nut.

7. Install the retainers attaching the shock absorber lower pivot to the lower control arm and tighten them.

8. Lower the vehicle.

Upper Control Arm & Ball Joint

REMOVAL & INSTALLATION

NOTE: The upper control arm and ball joint are replaced as an assembly.

1. Raise the vehicle and support it on jackstands placed under the lower control arms.

2. Remove the wheel and tire assembly.

3. Remove the cotter pin nut fastening the upper control arm and upper ball joint assembly and disconnect the upper control arm from the steering knuckle.

NOTE: Do not allow the steering knuckle to hang by the flexible brake line. Wire the steering knuckle up to the frame temporarily.

4. Remove the two bolts from the upper pivot shaft and remove the upper control arm from the bracket. Be sure to note the position and number of shims used for adjusting the camber and caster angles when removing the upper control arm. This is to ensure that the shims are reinstalled in their original positions.

5. To remove the pivot shaft and bushings from the upper control arm assembly, remove the bushing nuts from the pivot shaft by loosening them alternately, then remove the pivot shaft.

6. To install the upper control arm and ball joint assembly, first install the pivot shaft boots to the pivot shaft.

7. Fill the internal part of the bushings with grease (molybdenum disulfide) and screw the bushings into the pivot shaft. Be sure to screw the right-side and the left-side bushings alternately into the pivot shafts carefully avoiding getting grease on the outer face of the bushings. Tighten the nuts to 250 ft. lbs.

NOTE: Be sure that the control arm and bushings are centered properly and that the control arm rotates with resistance but not binding on the pivot shaft when tightened to the proper torque.

8. Install the grease fittings and lubricate the parts with grease through the grease fittings.

9. Install the ball joint stud through the steering knuckle. Install the castellated nut and tighten it to 75 ft. lbs. and just enough additional torque to install the cotter pin. Use a new cotter pin.

10. Mount the upper control arm to the chassis frame and install the shims in their original positions between the pivot shaft and bracket. Tighten the pivot shaft attaching nuts to 55 ft. lbs.

NOTE: Tighten the thinner shim pack's nut first for improved shaft-to-frame clamping force and torque retention.

11. Install the dust cover.

12. Install the wheel and tire assembly and lower the vehicle to the floor.

Lower Ball Joint

REMOVAL & INSTALLATION

1. Raise the front of the vehicle and support it with jackstands.

2. Remove the wheel and tire assembly.

3. Remove the cotter pin and castellated nut which retains the ball joint to the steering knuckle.

4. Remove the two bolts retaining the lower ball joint and strut rod.

5. Remove the remaining two bolts.

6. Remove the ball joint.

7. Install the lower ball joint by mounting the joint to the lower control arm and tightening the four bolts to 45 ft. lbs.

8. Install the ball joint stud into the steering knuckle and install the castellated nut and torque it to 75 ft. lbs. and just enough additional torque to align the cotter pin hole with one of the castellations on the nut. Install a new cotter pin.

9. Lubricate the lower ball joint through the grease fitting.

10. Install the wheel and tire assembly and lower the vehicle to the ground.

Lower Control Arm

REMOVAL & INSTALLATION

1. Jack up the vehicle and support it with jackstands.

2. Remove the wheel and tire.

3. Remove the strut bar by removing the frame side bracket and the double nuts, washer and the rubber bushing from the front side of the strut bar. Next, remove the two bolts fastening the strut bar to the lower control arm and remove the bar.

4. Disconnect the stabilizer bar from the lower control arm.

5. Remove the torsion bar.

6. Disconnect the shock absorber from the lower control arm.

7. If you so desire, remove the lower ball joint from the lower control arm joint at this time.

8. Remove the retaining nut and drive out the bolt holding the lower control arm to the chassis with a soft metal drift. Remove the lower control arm from the vehicle.

9. To install the lower control arm, first, install the lower ball joint to the lower control arm. Tighten the retaining nuts to 45 ft. lbs.

10. Mount the lower control arm to the frame. Drive the bolt into position carefully with a soft metal drift. Use care not to damage the serrated portions. Tighten the nut on the end of the pivot bolt to 135 ft. lbs.

11. Install the stabilizer bar to the lower control arm.

12. Place the washers and bushings on the strut rod and install it through the frame bracket. Install the second set of washers and bushings on the strut rod together with the lockwashers and nut. Leave the nut loose temporarily.

13. Install the strut rod to the lower control arm and tighten the bolts to 45 ft. lbs.

14. Assemble the lower ball joint to the steering knuckle.

15. Assemble the wheel and tire and lower the vehicle.

16. Tighten the first strut bar-to-chassis frame attaching nut to 175 ft. lbs., and the second locknut to 55 ft. lbs.

Front Wheel Bearings

REMOVAL & INSTALLATION

1. Remove the hub assembly.

2. Remove the outer roller bearing assembly from the hub. Pry out the inner bearing lip seal and remove the inner bearing assembly.

3. Wash all parts in a cleaning solvent and dry with compressed air.

4. Check the bearings for pitting or scoring. Also check for smooth rotation and lack of noise.

5. Thoroughly lubricate the bearings with new wheel bearing lubricant.

6. Apply a light coat of lubricant to the spindle and inside surface of the hub.

7. Place the inner bearing in the race of the hub and install a new grease seal.

8. Install the hub assembly on the spindle.

9. Install the outer wheel bearing, washer and adjust nut.

10. Adjust the wheel bearings as outlined below.

11. Install the dust cap on the hub.

12. Install the brake caliper and support assembly.

13. Install the wheel and tighten the nuts.

ADJUSTMENT

1. With the wheel raised, remove the hub cap and dust cap and then remove the cotter pin and nut retainer from the end of the spindle.

2. While rotating the wheel, tighten the spindle nut to 22 ft. lbs.

3. Turn the hub 2–3 turns and loosen the nut just enough so that it can be turned with your fingers.

4. Turn the nut all the way in with your fingers and check to be sure the hub has no free play.

5. Measure the starting torque by pulling one of the wheel hub studs with a pull scale. Tighten the spindle nut so that the pull scale reads 1.1–2.6 lbs. when the hub begins to rotate.

NOTE: Make sure that the brake pads are not in contact with the drum when measuring rotating torque.

6. Install the nut retainer, new cotter pin, dust cap and hub cap.

Locking Hub

REMOVAL & INSTALLATION

1. Jack up your vehicle and support it with jackstands.

2. Place the transfer case in the 2H position.

3. Set the hubs in the free position.

4. Remove the hub cover bolts and remove the hub cover.

Cover Assembly

1. While pushing the follower toward the knob, turn the clutch assembly clockwise, and then remove the clutch assembly from the knob.

2. Remove the snap-ring and remove the knob from the cover.

NOTE: Do not lose the detent ball.

3. Remove the ball and spring from the knob.

4. Remove the X-ring from the knob by pressing it off with your fingers.

NOTE: Do not use a screwdriver to remove this ring because it may scratch the ring.

5. Remove the compression spring, retaining spring, and the follower from the clutch assembly.

6. Remove the retaining spring from the clutch assembly by turning it counterclockwise.

Body Assembly

1. Remove the snap-ring and then remove the inner assembly from the body.

2. Separate the ring, inner, and spacer by removing the snap-ring.

3. Installation is the reverse of removal with the following suggestions. Apply grease to the X-ring, the inner cover and the outside cicumference of the knob.

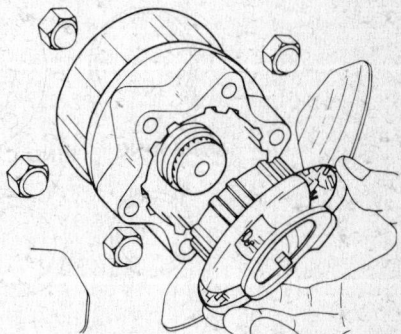

Front hub cover removal

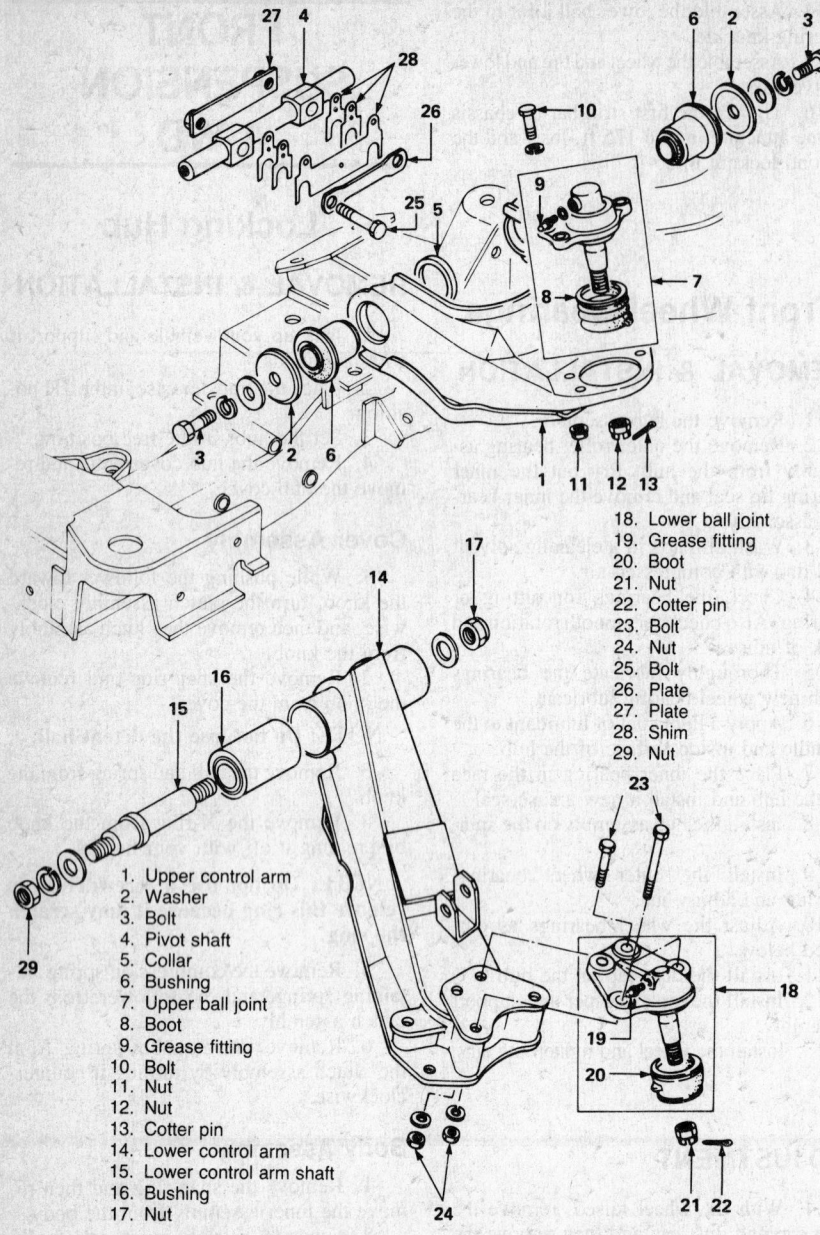

18. Lower ball joint
19. Grease fitting
20. Boot
21. Nut
22. Cotter pin
23. Bolt
24. Nut
25. Bolt
26. Plate
27. Plate
28. Shim
29. Nut

1. Upper control arm
2. Washer
3. Bolt
4. Pivot shaft
5. Collar
6. Bushing
7. Upper ball joint
8. Boot
9. Grease fitting
10. Bolt
11. Nut
12. Nut
13. Cotter pin
14. Lower control arm
15. Lower control arm shaft
16. Bushing
17. Nut

4-wheel drive front suspension

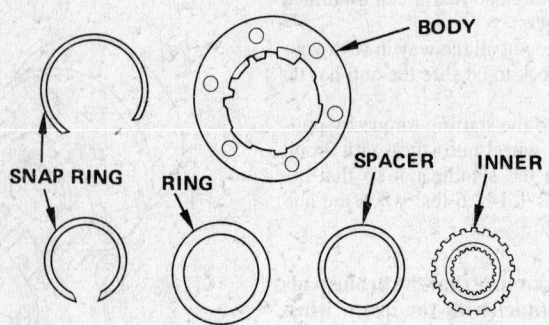

Hub body components

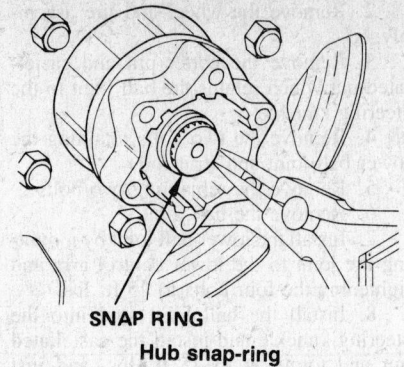

SNAP RING
Hub snap-ring

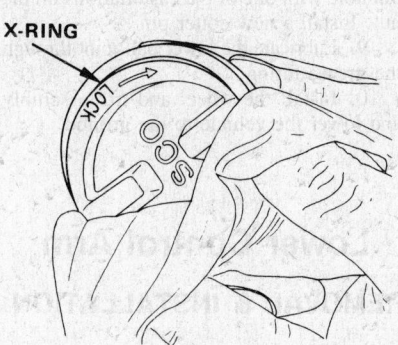

Removing X-ring

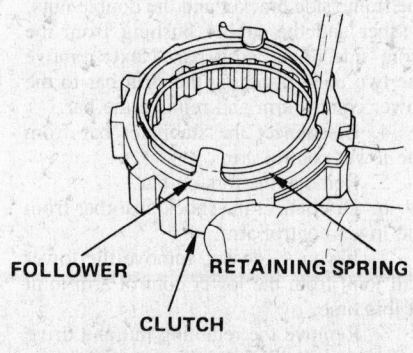

FOLLOWER   RETAINING SPRING
CLUTCH
Removing clutch follower

Front Wheel Hub

REMOVAL & INSTALLATION

1. Jack up your vehicle and support it with jackstands.

2. Remove the front wheel.

3. Remove the free wheeling hub as previously outlined.

4. Remove the brake caliper and tie it out of the way.

5. Remove the lockwasher and hub nut.

6. Remove the hub and rotor assembly.

NOTE: Do not drop any of the wheel bearings as damage could result.

7. Installation is the reverse of removal.

Front End Alignment

RIDE HEIGHT ADJUSTMENT

NOTE: The ride height should be measured with a full tank of gas, spare tire, jack, no passengers, and with the tires inflated to the correct pressure.

1. Place the vehicle on a smooth level floor and bounce the front end several times. Raise the vehicle and then allow it to settle to a normal height.

2. Measure the distance between the bottom of the lower ball joint stud which fits through the steering knuckle and the ground and the distance between the frame crossmember that the lower control arm attaches to and the ground. The difference between these two measurements should be 2.52 in. (1.54 in. with the vehicle loaded to GVW).

3. Adjust the vehicle height by first loosening the nuts on the front end of the strut bar and then turning the vehicle height adjusting bolt. Turn the bolt clockwise to raise the vehicle. As an additional check, measure the clearance between the rubber bumper and the lower control arm. The clearance should be ⅞ in.

4. Check the ride height at the front of the vehicle as outlined in Step 2 above and the ride height at the rear axle by measuring the clearance between the top of the axle and the bottom of the frame where the frame rises to clear the axle. The clearance between the frame and axle at this point should be 7.90 in. (6.26 in. with the vehicle loaded to GVW).

5. After obtaining the correct clearances, securely tighten the strut bar attaching nuts to the proper torque.

Front Wheel Bearings

REMOVAL & INSTALLATION

1. Raise and support the front end. Place the hub in 2H.

2. Remove the free wheeling hub cover assembly.

3. Remove the snap-ring and shims from the spindle.

4. Remove the free wheeling hub body and lock washer.

5. Remove the outer roller bearing assembly from the hub with your fingers.

6. Using a brass or wood drift, drive out the inner bearing assembly along with the oil seal. Replace the seal.

7. Wash all parts in a non-flammable solvent.

8. Check all parts for cracks or wear. Thoroughly lubricate all bearing parts with a high-temperature (molybdenum-disulfide) wheel bearing grease. Remove any excess. Apply about 2 ounces of the grease to the hub.

9. Lightly coat the spindle with the same grease.

1. Knuckle
2. Oil seal
3. Washer
4. Bearing
5. Adaptor
6. Shield
7. Retainer ring
8. Oil seal
9. Hub bearing inner
10. Dust shield
11. Bolt
12. Wheel pin
13. Bolt
14. Hub and disc
15. Wheel bearing outer
16. Hub nut
17. Lock washer
18. Shim
19. Snap-ring
20. Ring
21. Spacer
22. Inner
23. Body
24. Clutch
25. Retaining spring
26. Follower
27. Compression spring
28. Snap-ring
29. Detent ball and spring
30. Knob
31. X–Ring
32. Cover
33. Bolt

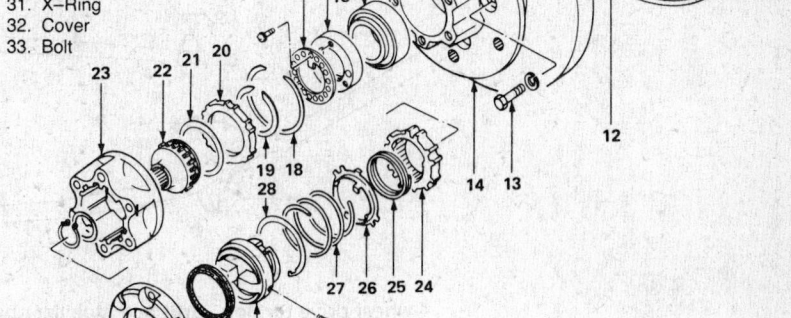

4-wheel drive front hub

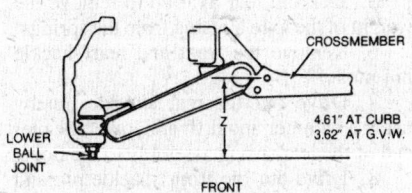

Ride height measurement, front

FREE POSITION

Front hub knob

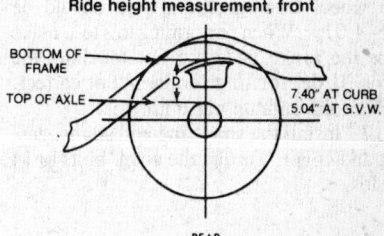

Ride height measurement, rear

10. Place the inner bearing into the hub race and install a new seal and retaining ring.

11. Carefully install the hub on the spindle and install the outer bearing.

12. Install the spindle nut.

13. While rotating the hub, tighten the hub so that the wheel can just be turned by hand.

14. Turn the hub 2–3 turns and back off the nut just enough so that it can be loosened with the fingers.

1. Torsion bar
2. Rubber seat
3. Height control arm
4. Pivot Nut
5. Height control seat and bolt
6. Strut bar
7. Strut bar bushing
8. Strut bar washer
9. Nut
10. Bolt
11. Bolt
12. Nut
13. Shock absorber
14. Bushing
15. Retainer
16. Nut
17. Bolt
18. Nut
19. Lower control arm bumper
20. Bolt
21. Upper control arm bumper
22. Stabilizer bar
23. Stabilizer bar bushing
24. Stabilizer bar support
25. Stabilizer bar bracket
26. Bolt
27. Bushing
28. Washer
29. Nut
30. Bolt
31. Bushing
32. Washer
33. Nut

4-wheel drive torsion, strut and stabilizer bars

15. Finger-tighten the nut so that all play is taken up at the bearing.

16. Attach a pull scale to one of the lugs and check the amount of pull needed to start the wheel turning. Initial pull should be 2.6–4.0 lbs. When performing this test, make sure the brake pads are not touching the rotor. If the rotating torque is not correct, tighten the spindle nut until it is.

17. Install the snap-ring and shims, gasket and cover. Torque the cover bolts to 14 ft. lbs.

REAR SUSPENSION

Springs

REMOVAL & INSTALLATION

1. Jack up the rear of the vehicle and place jackstands under the frame near the rear end of the rear spring brackets.

2. Remove the rear shock absorbers.

3. Remove the parking brake cable clips.

4. Remove the nuts from the U-bolts holding the springs to the axle housing.

5. Jack the rear axle up to remove the weight of the axle housing from the springs.

6. Remove the front and rear shackle pin nuts.

7. Drive out the rear shackle pin by using a hammer and drift and lower the rear end of the leaf spring assembly to the floor.

8. Drive out the front shackle pin and remove the leaf spring assembly rearward.

9. Remove the shackle pin from the rear spring bracket and remove the shackle.

10. Check the leaf springs for cracks, wear and broken leaves. Replace any leaves found to be cracked, broken, fatigued or seriously worn.

11. Check the shackles for bending and the pins for wear.

12. Check the U-bolts for distortion or other damage.

13. Mount the shackle to the bracket.

14. Align the front end of the leaf spring assembly with the front bracket and install the shackle pin.

15. Align the rear end of the leaf spring assembly with the shackle and install the shackle pin.

16. Loosely install the shackle pin nuts and install the U-bolts. Tighten the U-bolt nuts to 40 ft. lbs.

17. Install the shock absorbers.

18. Clip the parking brake cable to the bracket.

19. Remove the jackstands and lower the vehicle so that the vehicle weight is on the leaf springs.

20. Tighten the shackle pin nuts to 130 ft. lbs.

Shock Absorbers

REMOVAL & INSTALLATION

Remove the rear shock absorbers by loosening and removing the upper and lower attaching nuts and pulling the shock absorber ends off the mounting studs, together with the washers and rubber bushings. Install the shock absorbers in the reverse order of removal, making sure that you use new rubber bushings and that they are installed correctly in the bevel shaped mounting holes in the end of the shock absorbers.

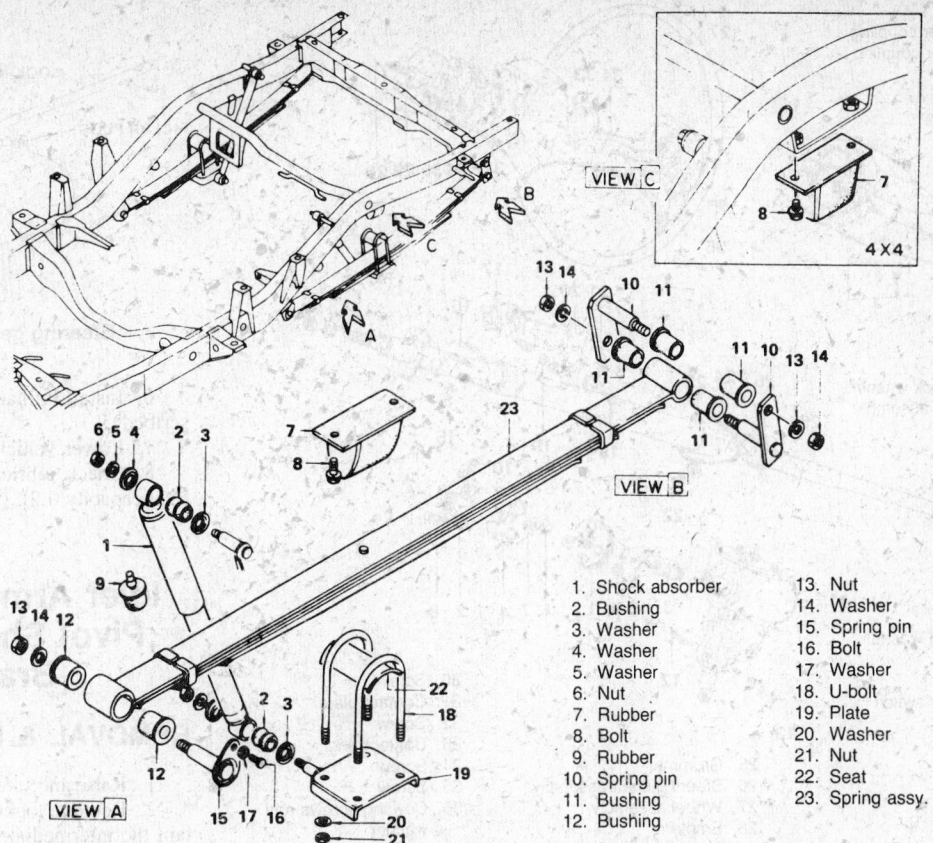

1. Shock absorber	13. Nut
2. Bushing	14. Washer
3. Washer	15. Spring pin
4. Washer	16. Bolt
5. Washer	17. Washer
6. Nut	18. U-bolt
7. Rubber	19. Plate
8. Bolt	20. Washer
9. Rubber	21. Nut
10. Spring pin	22. Seat
11. Bushing	23. Spring assy.
12. Bushing	

Rear suspension

STEERING

Steering Wheel

REMOVAL & INSTALLATION

1. Disconnect the battery ground cable.

2. Remove the horn shroud and spring by pushing and turning it counterclockwise. Remove the horn contact ring and wire.

3. Remove the steering wheel-to-steering shaft retaining nut, washer and lockwasher.

4. Mark the relative position of the steering wheel and shaft to each other.

5. Remove the steering column cowling by removing the four attaching screws and washers.

6. Remove the steering wheel from the shaft with a puller.

NOTE: Under no circumstances is the steering shaft to be hammered upon, jarred, or leaned upon. The steering column is a collapsible, energy-absorbing type and can be easily damaged through mistreatment.

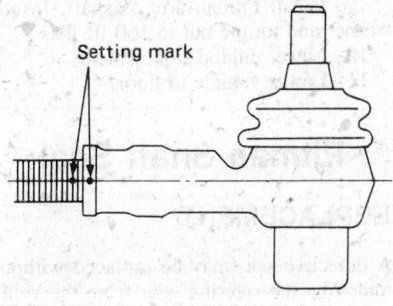

Match-mark the tie rod ends before removal

7. Install the steering wheel in the reverse order of removal, aligning the marks made on the steering wheel and the shaft. Draw the steering wheel onto the shaft with the attaching nut.

Turn Signal & Dimmer Switch

REMOVAL & INSTALLATION

1. Disconnect the battery ground cable.

2. Remove the five screws retaining the steering column cowling and remove the cowling.

3. Remove the wire connectors from the switch.

4. Remove the switch by removing the two screws which retain the switch clamp to the steering column mast jacket.

5. Replace the switch in the reverse order of removal.

Tie Rod Ends

REMOVAL & INSTALLATION

1. Jack up your vehicle and support it with jackstands.

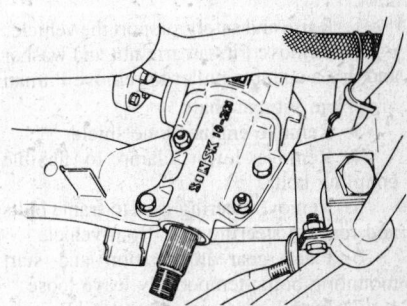

Steering gear mounting

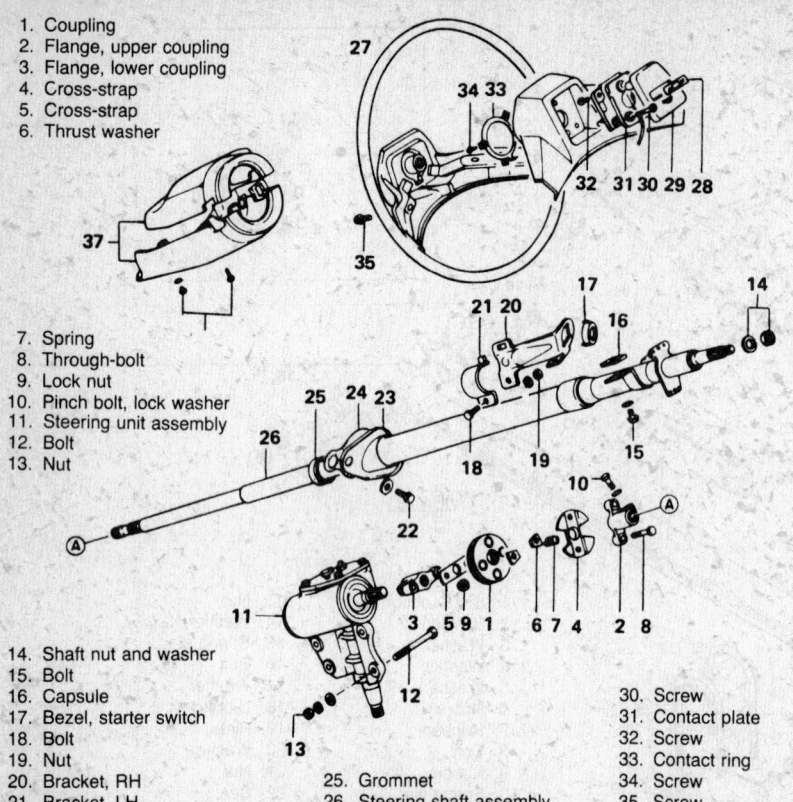

1. Coupling
2. Flange, upper coupling
3. Flange, lower coupling
4. Cross-strap
5. Cross-strap
6. Thrust washer
7. Spring
8. Through-bolt
9. Lock nut
10. Pinch bolt, lock washer
11. Steering unit assembly
12. Bolt
13. Nut
14. Shaft nut and washer
15. Bolt
16. Capsule
17. Bezel, starter switch
18. Bolt
19. Nut
20. Bracket, RH
21. Bracket, LH
22. Bolt
23. Bracket
24. Gasket
25. Grommet
26. Steering shaft assembly
27. Wheel assembly
28. Emblem
29. Button
30. Screw
31. Contact plate
32. Screw
33. Contact ring
34. Screw
35. Screw
36. Cowling screws and washer
37. Column cowling

Steering column

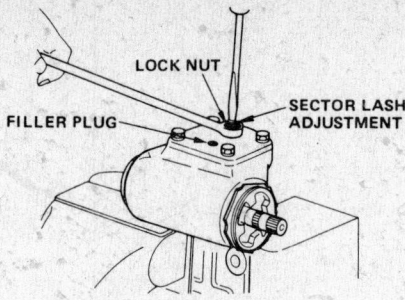

Steering gear adjustment

2. Match-mark the tie rod ends and the sleeves for installation.

3. Remove the cotter pin and nut from the tie rod end, and loosen the clamping bolts on the sleeve.

4. Use a puller to remove the tie rod from the steering knuckle.

5. Unscrew the tie rod and remove it.

6. Check it for damage and replace it if necessary.

7. Installation is the reverse of removal.

NOTE: Make sure to use a new cotter pin when installing the tie rod.

Steering Gear

REMOVAL & INSTALLATION

1. Raise and safely support the vehicle.

2. Remove Pitman arm nut and washer and use a suitable puller to remove Pitman arm from Pitman shaft.

3. Remove engine stone shield.

4. Remove lower clamp to flexible coupling bolts.

5. Remove steering gear to frame bolts and remove steering gear from vehicle.

6. Place gear in position and start mounting bolts; temporarily leave loose.

7. Install steering gear flexible coupling bolts and torque to 22 ft. lbs.

8. Torque steering column mounting bolts to 13 ft. lbs.

9. Install Pitman arm to shaft. Install washer and torque nut to 160 ft. lbs.

10. Lower engine stone shield.

11. Lower vehicle to floor.

Pitman Shaft Seal

REPLACEMENT

A defective seal may be replaced without removing the steering gear from the vehicle. If it has been diagnosed that the seal should be replaced proceed as follows:

1. Raise and safely support the vehicle.

2. Remove Pitman arm as previously described.

3. Wipe clean the area around the seal.

4. Using an awl, screwdriver or similar tool, pry out the old seal being careful not to damage the housing bore.

NOTE: Inspect the lubricant in the gear for contamination. If the lubricant is contaminated in any way, the gear must be removed from the vehicle and completely overhauled.

5. Coat the new Pitman shaft seal with steering gear lubricant (or equivalent). Position the seal in the Pitman shaft bore and tap into position using a suitable size socket.

6. Install Pitman arm as previously described.

7. Lower vehicle to floor.

8. Check lubricant level in gear box; total capacity 0.2L (7 oz.). Do not overfill.

Idler Arm, Idler Arm Pivot Shaft and/or Bracket

REMOVAL & INSTALLATION

1. Raise and safely support the vehicle.

2. Remove lockwasher and nut that retain the intermediate rod to the idle arm.

3. The ball studs for the above may be removed using tool J21687–02 or equivalent.

4. Remove the four bolts, nuts and washers retaining the bracket to the frame, and remove the idler arm pivot shaft and bracket with idler arm.

5. If the idler arm is being replaced: Remove idler arm to idler arm pivot shaft nut and lockwashers. Remove the idler arm from the pivot shaft.

6. Install bracket and shaft assembly to frame and torque to 29 ft. lbs.

7. Install the idler arm to the shaft and torque nut to 87 ft. lbs.

8. Install the ball studs and intermediate rod to the idler arm. Torque the castellated nuts to 50 ft. lbs. and just enough additional to line up cotter pin holes. Install new cotter pins.

9. Lubricate idler arm pivot shaft.

10. Lower vehicle to floor.

Intermediate Rod and Tie Rods

REMOVAL & INSTALLATION

1. Raise and safely support the vehicle.

2. Remove cotter pin from the ball studs connecting tie rods to intermediate rod and steering damper. Remove the castellated nuts and disconnect the parts using special tool J21687–02 or equivalent.

3. Remove the nut and lockwasher on ball stud connecting the intermediate rod to

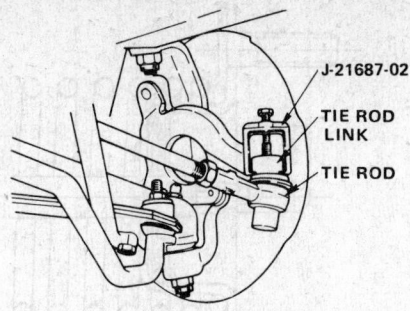

Tie rod end removal

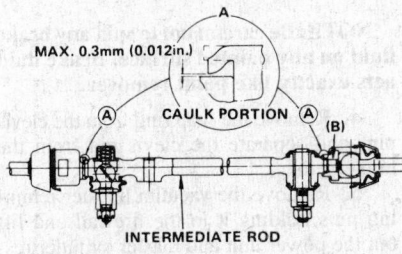

Installing tie rod to intermediate rod

idler arm. Disconnect the parts using special tool J21687–02 or equivalent.

4. Remove the intermediate rod together with tie rods.

5. If the tie rod is replaced, disconnect the intermediate rod from tie rod.

6. If the tie rod has been removed: Apply liquid gasket (Felcobond 401) to portion B. Install and tighten tie rod end to specified torque of 65 ft. lbs. Caulk two portions (upper and lower portions) of A.

7. Make sure threads on ball studs and nuts are clean and smooth.

8. Install intermediate rod to idler arm, install lockwasher, nut, and torque nut to 50 ft. lbs.

9. Raise end of rod and install on Pitman arm. Torque nut to 44 ft. lbs., then advance nut just enough to insert cotter pin and install new cotter pin.

10. Install intermediate rod to steering damper end. Torque nut to 87 ft. lbs., then advance nut just enough to insert cotter pin and install new cotter pin.

11. Install the tie rods to adapter, torque nut to 44 ft. lbs., then advance nut just enough to insert cotter pin and install new cotter pin, and lubricate tie rod ball studs.

Steering Damper

REPLACEMENT

1. Raise vehicle and safely support the vehicle.

2. Remove cotter pin and castellated nut on stud connecting the damper end with intermediate rod.

3. Remove the cotter pin and castellated nut on the stud connecting the damper end to second crossmember (on 2WD) or to idler arm bracket (on 4WD).

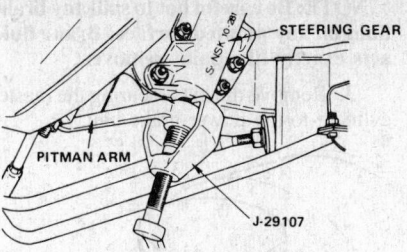

Pitman arm removal

4. Remove the steering damper.
5. To install, reverse Steps 1–4.

Pitman Arm

REMOVAL & INSTALLATION

1. Raise and safely support the vehicle.

2. Remove nut and lockwasher retaining Pitman arm to Pitman shaft.

3. Using tool J29107 or equivalent remove Pitman arm from pitman shaft.

4. Remove cotter pins and castellated nuts retaining the Pitman arm to the tie rod and intermediate rod.

5. The ball studs may be removed from the Pitman arm using tool J21687–02.

6. Remove the Pitman arm.

7. Install Pitman arm to intermediate rod and tie rod ball studs. Install castellated nuts (do not torque yet).

8. Install Pitman arm to Pitman shaft.

9. Install Pitman shaft lockwashers and nut, and torque to 160 ft. lbs.

10. Torque ball stud nuts to 44 ft. lbs. and just enough additional to align cotter pin hole. Install new cotter pin.

11. Lower vehicle to floor.

Steering Column and Shaft Assembly

REMOVAL & INSTALLATION

1. Disconnect the battery ground cable.

2. Remove upper coupling clamp pinch bolt from the steering shaft flexible coupling in the engine compartment.

NOTE: Apply a setting mark across the steering shaft and coupling clamp.

3. Disconnect the combination switch and ignition switch wiring at the harness connector.

4. Remove the two column to instrument panel bolts.

5. Remove the steering column toward the cab being careful not to damage or jar the shaft as it is an energy absorbing unit.

6. Remove the five screws that retain the steering column cowling.

7. Remove the steering wheel as outlined previously.

8. Remove two screws that retain the combination switch to column on flange.

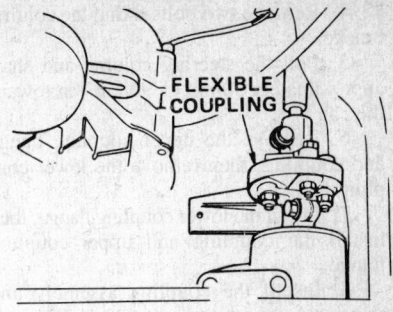

Steering column flexible coupling

9. Remove the combination switch assembly.

10. Remove bolt that retains the ignition switch bracket.

11. Install the combination switch on the column flange using 2 screws.

12. Install the ignition switch to the column as previously described.

13. Install the steering wheel lockwasher and nut and torque the nut to 29 ft. lbs.

14. Install the horn shroud.

15. Install the steering column cowling.

16. Place column through cowl and move into position.

17. Install end of steering shaft into flexible coupling clamp. Install coupling to worm gear shaft. Torque pinch bolts to 22 ft. lbs.

NOTE: Align setting marks applied at disassembly when connecting steering shaft end to coupling clamp.

18. Install column to instrument panel unless vehicle is setting on its wheels or suspension.

19. Tighten the column to instrument panel bolts. Tightening torque 11 ft. lbs.

20. Make connections of combination switch and ignition switch wiring at the connector.

21. Check operation of horn and combination switch.

22. Check for interference between steering column cowling and steering wheel.

23. Check alignment between front wheels and steering wheel.

24. Check steering wheel free-play when the steering unit assembly is installed in position on the vehicle. The standard steering wheel play is about 10 mm (0.4 in.) when measured at the outside diameter of the steering wheel.

Steering Shaft Flexible Coupling

REMOVAL & INSTALLATION

1. Jack up front of vehicle until wheels are just off the ground.

2. Remove coupling through-bolts and lock nut. Only two bolts can be removed.

3. Remove the pinch bolts on the upper and lower flanges of the coupling.

4. Remove two bolts fixing the column bracket.

5. Pull the steering column and shaft approximately 50mm (1.968 in.) in toward the cab.

6. Remove the upper coupling flange and coupling, then remove the lower coupling flange.

7. Install the lower coupling flange, then install the coupling and upper coupling flange.

8. Install the coupling assembly and torque the through bolts to 22 ft. lbs.

9. Install pinch bolts and torque to 22 ft. lbs.

10. Lower vehicle to floor.

11. Install and torque the column bracket bolts to 13 ft. lbs.

BRAKES

ADJUSTMENTS

All brakes are self-adjusting. The front brakes self-adjust when the vehicle is moved slowly back and forth while applying the service brakes. The rear brakes are self-adjust when the parking brake is set and released.

Master Cylinder

REMOVAL & INSTALLATION

1. Set the parking brake and chock the wheels to prevent the car from rolling.

2. Open the hood and disconnect the front and rear brake brake lines from the master cylinder. Place rags under the master cylinder to catch any leaking fluid.

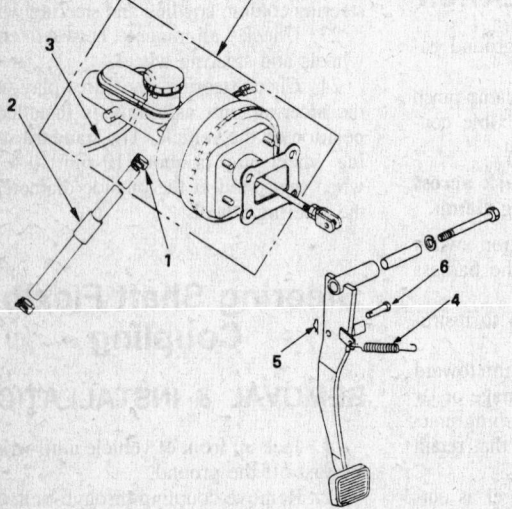

Power booster (Master Vac) mounting

NOTE: Be careful not to spill any brake fluid on any painted surface. Brake fluid acts exactly like paint remover.

3. Remove the nuts securing the master cylinder to the power brake unit.

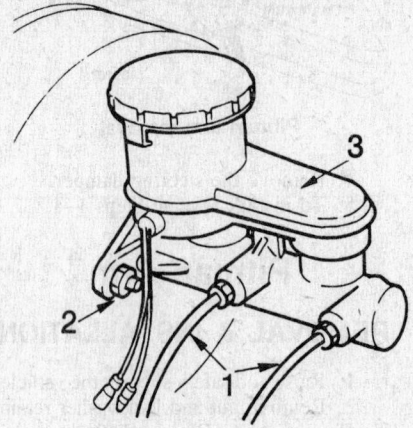

1. Brake pipe
2. Nut; master cylinder to vacuum servo
3. Master cylinder assembly

Master cylinder removal points

Power Brake (Master Vac) Booster

REMOVAL & INSTALLATION

1. Set the parking brake and chock the wheels to prevent the car from rolling.

2. Open the hood and loosen the clamp attaching the vacuum hose to the vacuum booster and remove the hose.

3. Disconnect the front and rear brake brake lines from the master cylinder. Place rags under the master cylinder to catch any leaking fluid.

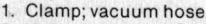

1. Clamp; vacuum hose
2. Vacuum hose
3. Brake pipe
4. Return spring; brake pedal
5. Snap ring
6. Pin; push rod to brake pedal
7. Master vac with master cylinder
8. Master cylinder assembly
9. Master vac assembly

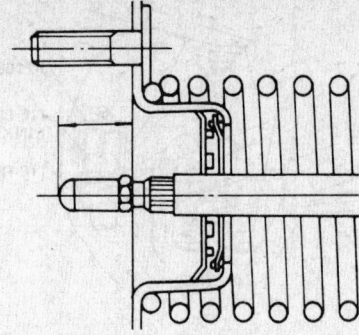

Master Vac pushrod measurement

NOTE: Be careful not to spill any brake fluid on any painted surface. Brake fluid acts exactly like paint remover.

4. Remove the snap ring from the clevis pin, and separate the clevis pin from the brake pedal.

5. Remove the vacuum booster retaining nuts holding it to the firewall and lift out the power unit and master cylinder/reservoir as an assembly.

6. Installation is the reverse of removal. Check the distance from the flange face of the vacuum booster to the end of the push-rod before installation of the master cylinder. The distance should be 0.709–0.717 in. If the measurement deviates from the specified range, make an adjustment with the lock nut at the end of the push-rod.

Wheel Cylinder
REMOVAL & INSTALLATION

1. Remove the brake shoes.

2. Disconnect the hydraulic brake line at the wheel cylinder.

3. Remove the wheel cylinder attaching bolts from the backing plate.

4. Cap the openings of the brake line and the wheel cylinder.

5. Installation is the reverse of removal. Bleed the brake system.

Parking Brake

ADJUSTMENT

NOTE: Adjustment of the parking brake is necessary every time the rear brake cables are disconnected or after overhauling the rear brake assembly.

1. Fully release the parking brake lever and check the cable for free movement.

2. Firmly grab the second relay lever rod. Rotate the adjusting nut until all the slack disappears from the cable. Set the adjusting nut.

3. Pull the parking brake up to its fully set position three or four times.

4. If the parking brake is properly adjusted, the traveling range should be between 12–14 notches. If the travel is incorrect, readjust to specifications.

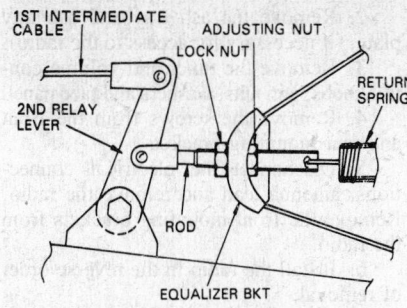

Parking brake adjustment

REMOVAL & INSTALLATION

1. Raise and support the vehicle on jackstands.

2. Loosen the adjusting nut and remove the lever return spring, and then the adjusting nut.

3. Remove the cotter pin from the retaining pin on the second lever assembly and remove the front cable.

4. Remove the two cotter pins from the retaing pins on the intermediate cable and remove the cable.

5. Remove the retaining clips from the rear fixing brackets and lower the rear brake cables.

6. Remove the rear brake drums, then remove the rear brake shoes and disconnect the rear brake cables from the lever in the rear brake shoes.

7. Installation is the reverse of the removal procedure.

CHASSIS ELECTRICAL

Blower

REMOVAL & INSTALLATION

1. Disconnect the battery ground cable.

2. Disconnect the blower motor electrical leads.

3. Remove the blower-to-heater core screws and remove the blower motor assembly.

4. Install in the reverse order.

Heater Core

REMOVAL & INSTALLATION

1. Disconnect the battery ground cable.

2. Place a drain pan under the heater hoses at the heater and remove the heater hoses from the core tubes, securing the heater hoses in a raised position to prevent further loss of coolant. Plug or tape the heater core tubes to prevent spillage of coolant in the passenger compartment when removing.

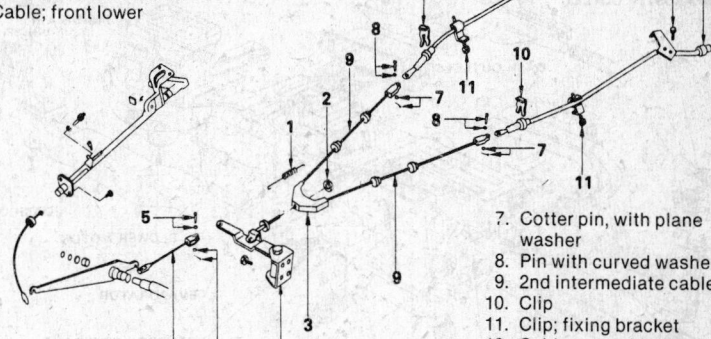

1. Return spring
2. Nut
3. Equalizer bracket
4. Cotter pin, with plane washer
5. Pin with curved washer
6. Cable; front lower

7. Cotter pin, with plane washer
8. Pin with curved washer
9. 2nd intermediate cable
10. Clip
11. Clip; fixing bracket
12. Cable assembly; rear
13. 2nd relay lever assembly

Parking brake cable assembly

3. Remove the five parcel shelf attaching screws and remove the shelf.

4. Loosen the air diverter and defroster door bowden cable clamps at the heater case and disconnect the cables from the doors.

5. Disconnect the blower resistor leads.

6. Remove the control assembly-to-instrument panel screws and swing the control to the left and lay it on the floor. Be careful not to kink the water valve bowden cable.

7. Remove the four heater-to-firewall screws. Pull the heater rearward until the core tubes clear the dash opening, then remove the heater by moving it to the right and down.

8. Remove the core tube clamp screw and remove the clamp.

9. Remove the seven screws and separate the heater case halves.

10. Remove the core from the case.

11. Install and assemble the heater core and heater case in the reverse order of removal, using new seals around the heater core.

Ignition Switch

REMOVAL & INSTALLATION

1. Disconnect the negative battery terminal.

2. Remove the multi-connector from the rear of the ignition switch.

3. Remove the lock nut from the front of the switch. Remove the switch from the rear of the instrument panel.

4. Installation is the reverse of removal.

NOTE: Align the locating lug on the switch body with the slot in the instrument panel mounting hole during installation.

Radio

REMOVAL & INSTALLATION

1. Disconnect the battery ground cable.

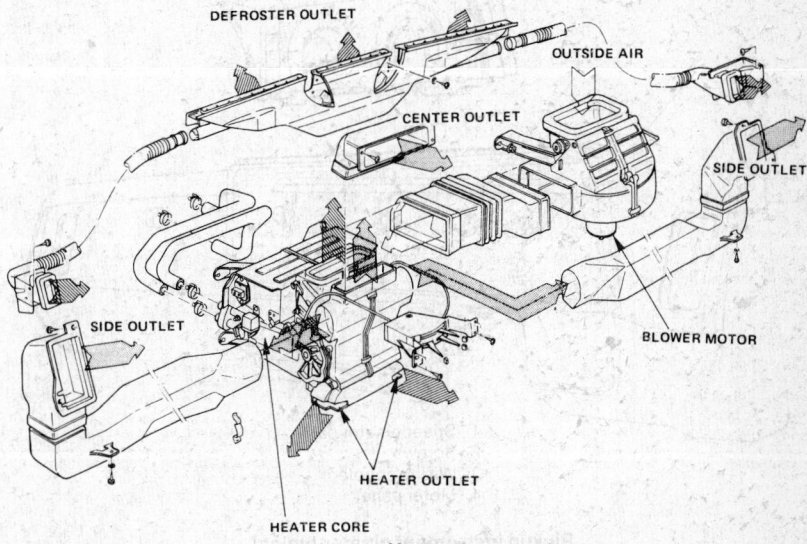

Heater

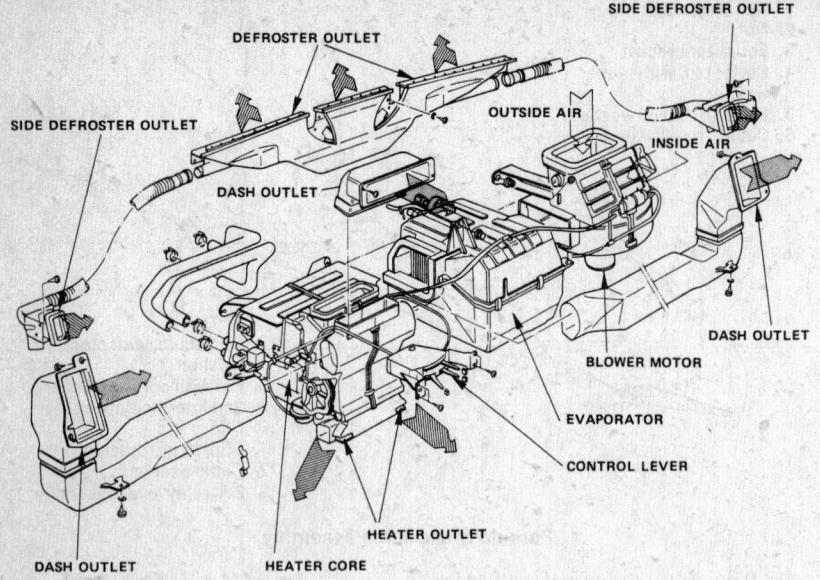

SIDE DEFROSTER OUTLET

DEFROSTER OUTLET

SIDE DEFROSTER OUTLET

OUTSIDE AIR

INSIDE AIR

DASH OUTLET

DASH OUTLET

BLOWER MOTOR

EVAPORATOR

DASH OUTLET

CONTROL LEVER

HEATER OUTLET

DASH OUTLET

HEATER CORE

Heater-air conditioner

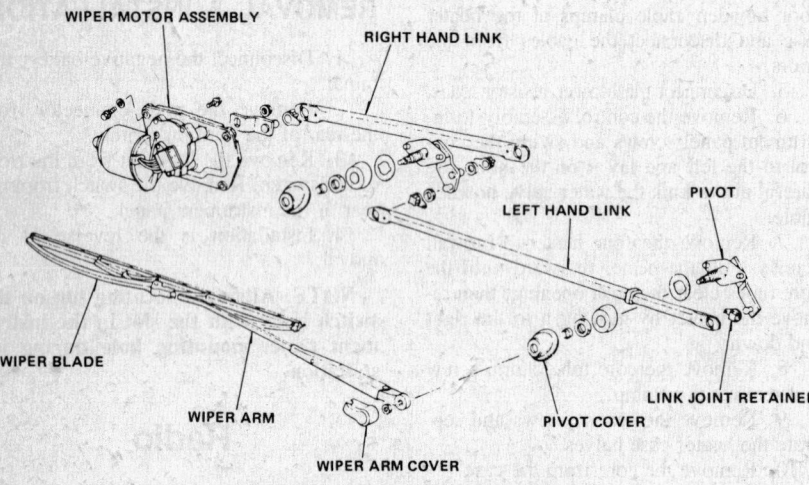

WIPER MOTOR ASSEMBLY

RIGHT HAND LINK

LEFT HAND LINK

PIVOT

WIPER BLADE

WIPER ARM

LINK JOINT RETAINER

PIVOT COVER

WIPER ARM COVER

Windshield wiper motor and linkage

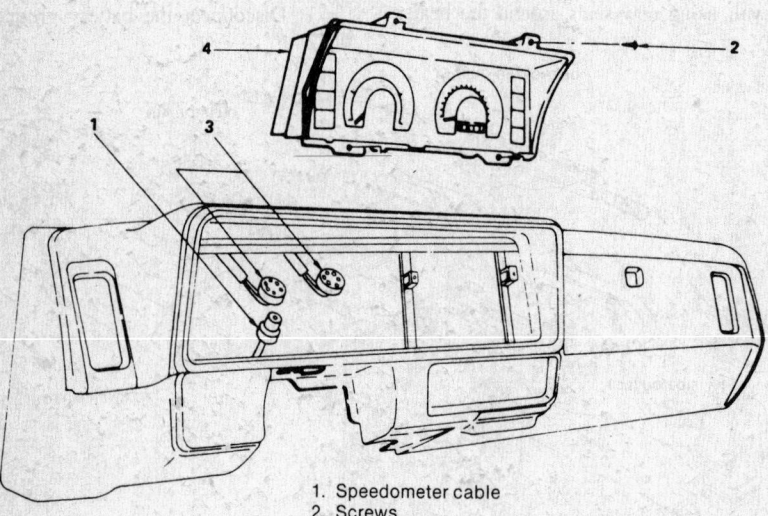

4

2

1

3

1. Speedometer cable
2. Screws
3. Wiring coupler
4. Meter panel

Pickup instrument cluster-typical

2. Remove the ash tray and ash tray plate, (if necessary for access to the radio).

3. Remove the tuner and volume control knobs, jam nuts, washers and face panel.

4. Remove the screws from the front and rear mounting brackets.

5. Disconnect the electrical connections, antenna lead and remove the radio. Remove the front mounting brackets from the radio.

6. Install the radio in the reverse order of removal.

Windshield Wiper Motor and Linkage
REMOVAL & INSTALLATION

1. Remove the wiper blades and arms.
2. Remove the two bolts attaching the pivot.
3. Remove the four wiper motor mounting bolts and remove the wiper motor and linkage.
4. To remove the motor independently, take out the motor shaft nut and three bolts and then pull off the connector and disconnect the ground cable.
5. Install and assemble in the reverse order of removal. Make sure to install the wiper motor linkage so that it is not twisted or touching any adjacent parts.

Instrument Cluster

REMOVAL & INSTALLATION

Pickup Models

1. Disconnect the negative battery cable.
2. Disconnect the speedometer cable.
3. Remove the screws and/or nuts securing the instrument cluster and pull the assembly part way out.
4. Disconnect the wiring harness at the connector and remove the instrument panel.
5. Install the instrument panel in the reverse order of removal.

Trooper II Models

1. Disconnect the negative battery cable.
2. Remove the steering wheel.
3. Remove the screws and/or nuts securing the instrument cluster and pull the assembly part way out.
4. Disconnect the wiring harness at the connector and remove the instrument panel.
5. Install the instrument panel in the reverse order of removal.

Fuse Box Location

The fuse box is located on the left (drivers) side of the valance panel inside the engine compartment. A fusible link is also used to protect the wiring harness, which is located near the positive battery cable.

Jeep

GENERAL ENGINE SPECIFICATIONS

Year	Engine Cu In. Displacement	Carburetor Type	Horsepower @ rpm ①	Torque @ rpm (ft lbs) ①	Bore and Stroke (in.)	Advertised Compression Ratio	Oil Pressure @ 30 mph (psi)
'79	6-258	2 bbl	110 @ 3200	210 @ 1800	3.750 × 3.895	8.0:1	37
	8-304	2 bbl	125 @ 3200	220 @ 2400	3.750 × 3.440	8.40:1	37
	8-360	2 bbl	175 @ 4000	285 @ 2900	4.080 × 3.440	8.25:1	37
'80	6-258	2 bbl	110 @ 3200	210 @ 1800	3.750 × 3.895	8.0:1	37
	8-304	2 bbl	125 @ 3200	220 @ 2400	3.750 × 3.440	8.40:1	37
	8-360	2 bbl	175 @ 4000	285 @ 2900	4.080 × 3.440	8.25:1	37
'81	4-151	2 bbl	87 @ 4400	128 @ 2400	4.00 × 3.00	8.3:1	36-41
	6-258	2 bbl	110 @ 3200	210 @ 1800	3.750 × 3.895	8.0:1	37
	8-304	2 bbl	125 @ 3200	220 @ 2400	3.750 × 3.440	8.40:1	37
	8-360	2 bbl	175 @ 4000	285 @ 2900	4.080 × 3.440	8.25:1	37
'82	4-151	2 bbl	87 @ 4400	128 @ 2400	4.00 × 3.00	8.3:1	36-41
	6-258	2 bbl	110 @ 3200	210 @ 1800	3.750 × 3.895	8.0:1	37
	8-360	2 bbl	175 @ 4000	285 @ 2900	4.080 × 3.440	8.25:1	37
'83	4-150	1 bbl	83 @ 4200	116 @ 2600	3.876 × 3.188	9.2:1	40
	4-151	2 bbl	87 @ 4400	128 @ 2400	4.00 × 3.00	8.3:1	36-41
	6-258	2 bbl	110 @ 3200	210 @ 1800	3.750 × 3.895	8.0:1	37
	8-360	2 bbl	175 @ 4000	285 @ 2900	4.080 × 3.440	8.25:1	37
'84	4-150	1 bbl	83 @ 4200	116 @ 2600	3.876 × 3.188	9.2:1	40
	6-173	2 bbl	110 @ 4800	148 @ 2000	3.50 × 2.99	8.5:1	45
	6-258	2 bbl	110 @ 3200	210 @ 1800	3.750 × 3.895	8.0:1	37
	8-360	2 bbl	175 @ 4000	285 @ 2900	4.080 × 3.440	8.25:1	37
'85-'86	4-126	DIESEL	N/A	N/A	3.358 × 3.503	21.5:1	11.6 ③
	4-150	1 bbl ②	83 @ 4200	116 @ 2600	3.876 × 3.188	9.2:1	40
	6-173	2 bbl	110 @ 4800	148 @ 2000	3.50 × 2.99	8.5:1	45
	6-258	2 bbl	110 @ 3200	210 @ 1800	3.750 × 3.895	8.0:1	37
	8-360	2 bbl	175 @ 4000	285 @ 2900	4.080 × 3.440	8.25:1	37

① Horsepower and torque are SAE net figures. They are measured at the rear of the transmission with all accessories installed and operating. Since the figures vary when a given engine is installed in different models, some are representative rather than exact.

② Some vehicles are equipped with fuel injection

③ 11.6 at Idle
45.5 at 3000

 **JEEP**

TUNE-UP SPECIFICATIONS

Year	Engine No. Cyl Displacement (cu. in.)	hp	Spark Plugs Type	Gap (in.)	Distributor Point Dwell (deg)	Ignition Timing (deg) [2]	Valves Intake Opens [3] (deg)	Fuel Pump Pressure (psi)	Idle Speed Man Trans	Idle Speed Auto Trans
'79	6—258	110	N-13L	.035	Electronic	8B	14½	4–5	700	600
	8—304	125	N-12Y	.035	Electronic	8B [1]	14¾	5–6½	700(750)	600
	8—360	175	N-12Y	.035	Electronic	8B	14¾	5–6½	800	600
'80	6—258	110	N-13L	.035	Electronic	8B	14½	4–5	700	600
	8—304	125	N-12Y	.035	Electronic	8B [1]	14¾	5–6½	700(500)	600
	8—360	175	N-12Y	.035	Electronic	8B	14¾	5–6½	800	600
'81	4—151	87	RBL13Y6	.060	Electronic	14 [4]	33	4–5¾	500	650
	6—258	110	N-13L	.035	Electronic	8B	14½	4–5	700	600
	8—304	125	N-12Y	.035	Electronic	8B [1]	14¾	5–6½	700	600
	8—360	175	N-12Y	.035	Electronic	8B	14¾	5–6½	800	600
'82	4—151	87	RBL13Y6	.060	Electronic	14 [4]	33	4–5¾	500	650
	6—258	110	N-13L	.035	Electronic	8B	14½	4–5	700	600
	8—360	175	N-12Y	.035	Electronic	8B	14¾	5–6½	800	600
'83	4—150	—	RFN14LY	.035	Electronic	12B	27	6½–8	750	750
	4—151	87	R44T5X	.060	Electronic	10B [5]	33	6½–8	900	700
	6—173	110	RUT2YC	.041	Electronic	10B	21	5–7	750	750
	6—258	110	RFN14LY	.035	Electronic	8B [2]	14½	4–5	700	600
	8—360	175	RN12Y	.035	Electronic	8B	14¾	5–6	800	600
'84	4—150	—	RFN14LY	.035	Electronic	12B	27	6½–8	750	750
	6—173	110	RUT2YC	.041	Electronic	10B	21	5–7	750	750
	6—258	110	RFN14LY	.035	Electronic	8B [6]	14½	4–5	700	600
	8—360	175	RN12Y	.035	Electronic	8B	14¾	5–6	800	600
'85–'86	4—126	—	NONE		NONE	—	—	N/A	750-850	750-850
	4—150	—	RFN14LY	.035	Electronic	12B	27	6½–8	750	750
	6—173	110	RUT2YC	.041	Electronic	10B	21	5–7	750	750
	6—258	110	RFN14LY	.035	Electronic	8B [6]	14½	4–5	700	600
	8—360	175	RN12Y	.035	Electronic	8B	14¾	5–6	800	600

NOTE: If the information given in this chart disagrees with the information on the engine tune-up decal, use the specifications on the decal.

NOTE: Figures in parentheses are for California engines

B—Before top dead center (BTDC)

[1] w/manual trans.—5B, CJ model only

[2] With vacuum advance disconnected

[3] All figures before TDC

[4] Both manual and automatic transmissions

[5] w/auto. trans; 12B

[6] Calif. CJ w/man. trans; 6B

FIRING ORDERS

NOTE: To avoid confusion, always replace spark plug wires one at a time.

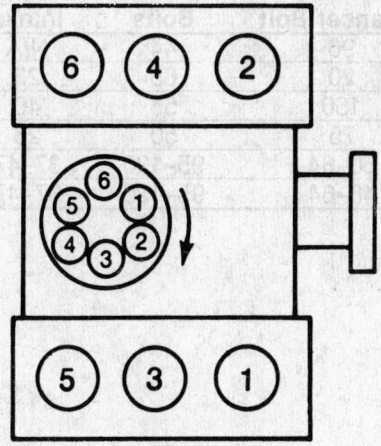

V6–173 firing order

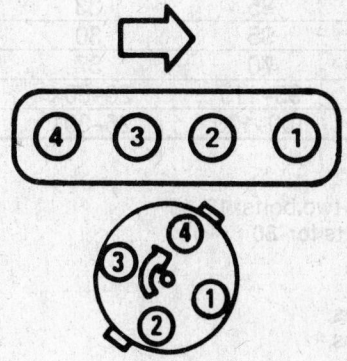

4–150 firing order

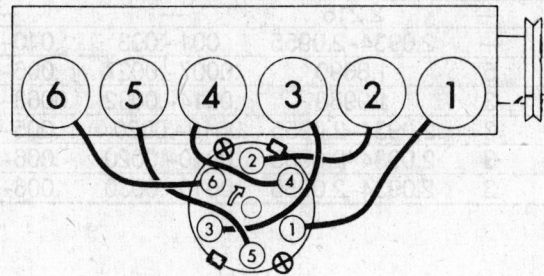

6 cylinder firing order: 1-5-3-6-2-4

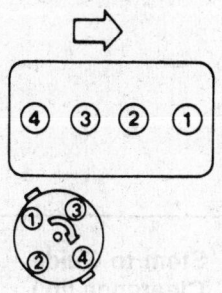

4 cylinder firing order: 1-3-4-2

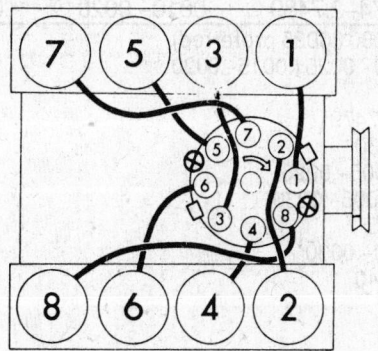

V8 firing order: 1-8-4-3-6-5-7-2

TORQUE SPECIFICATIONS

(All readings in ft. lbs)

Engine No. cyl Displacement (cu in.)	Cylinder Head Bolts	Rod Bearing Bolts	Main Bearing Bolts	Crankshaft Balancer Bolt	Flywheel to Crankshaft Bolts	Manifold Intake	Manifold Exhaust
4-126	④	48	69	96	44	N/A	N/A
4-150	85	33	80	20	65	23	23
4-151	95	30	65	160	55	40	40
6-173	70	37	70	75	50	23	25
6-258	95–115	26–30 ①	75–85	50–64	95–120	37–47	20–30 ②
8-304, 360	100–120	26–30 ①	90–105	48–64	95–120	37–47	20–30 ③

① 30–35 for '80
② 18–20 for '80
③ 20–30, center two bolts; 12–18, outer four bolts for '80
1st—22 ft. lbs.
2nd—37 ft. lbs.
3rd—70–77 ft. lbs.
4th—70–77 ft. lbs.

CRANKSHAFT AND CONNECTING ROD SPECIFICATIONS

(All measurements given in inches)

Engine	Crankshaft Main Bearing Journal Dia	Crankshaft Main Bearing Oil Clearance	Crankshaft Shaft End Play	Crankshaft Thrust on No.	Connecting Rod Journal Dia	Connecting Rod Oil Clearance	Connecting Rod Side Clearance
4-126	2.475	—	—	—	2.216	—	—
4-150	2.4996–2.5001	.0001–.0025	.0015–.0065	—	2.0934–2.0955	.001–.003	.010–.019
4-151	2.2988	.0005–.0022	.0035–.0085	5	1.8690	.0007–.0027	.006–.022
6-173	2.4940	.0017–.0030	.0020–.0067	3	1.9980	.0014–.0032	.0063–.0173
6-258	2.4986–2.5001	.0010–.0020 ①	.0015–.0065 ⑤	3	2.0934–2.0955	.0010–.0020 ②	.005–.014
V8-304	2.7474–2.7489 ③	.0010–.0020 ④	.003–.008 ⑥	3	2.0934–2.0955	.0010–.0020	.006–.018
V8-360	2.7474–2.7489 ③	.0010–.0020 ④	.003–.008	3	2.0934–2.0955	.0010–.0030	.006–.018

① '79: .0010–.0030 (.0025 preferred)
② '80–'82: .0010–.0025 (.0015–.0020 preferred)
③ #5: 2.7464–2.7479
④ #5: .0020–.0030
⑤ '81–'82: #1: .005–.0026
   2, 3, 4, 5, 6: .0005–.0030
   7: .0011–.0035
⑥ '81–'82: .0010–.0030
   #5: .0020–.0040

VALVE SPECIFICATIONS

Engine No. Cyl. Displacement (cu in.)	Seat Angle ① (deg)	Face Angle ② (deg)	Spring Test Pressure (lbs @ in.)	Spring Installed Height (in.)	Stem to Guide Clearance (in.) Intake	Stem to Guide Clearance (in.) Exhaust	Stem Diameter (in.) Intake	Stem Diameter (in.) Exhaust
4-126		45	51 @ 1.54	1.779	—	—	.3140	.3140
4-150	—	44	212 @ 1.20	1.82	.0010–.0030	.0010–.0030	.3412	.3412
4-151	46	45	82 @ 1.66 ④	1²¹⁄₃₂ ③④	.0010–.0027	.0010–.0027	.3422	.3422
6-173	46 ⑤	45 ⑤	195 @ 1.18	1.57	.0010–.0027	.0010–.0027	.3416	.3416
6-258	44¹⁄₂	44	100 @ 1¹³⁄₁₆ ④	2¹⁵⁄₆₄ ③④	.0010–.0030	.0010–.0030	.3720	.3720
8-304	44¹⁄₂	44	84 @ 1¹³⁄₁₆ ④	2¹⁵⁄₆₄ ③④	.0010–.0030	.0010–.0030	.3720	.3720
8-360	44¹⁄₂	44	84 @ 1¹³⁄₁₆ ④	2¹⁵⁄₆₄ ③④	.0010–.0030	.0010–.0030	.3720	.3720

① Exhaust valve seat angle given; all intake valve seat angles are 30° unless otherwise noted
② Exhaust valve face angle given; all intake valve angles are 29° unless otherwise noted
③ 80–2"
④ Without rotaters
⑤ Intake or Exhaust V6 only

PISTON RING SPECIFICATIONS

All measurements given in inches

Engine	Ring Gap			Ring Side Clearance			Piston to Bore Clearance
	Top Compression	Bottom Compression	Oil Control	Top Compression	Bottom Compression	Oil Control	
4-126	—	—					
4-150	.010–.020	.010–.020	.010–.025	.0017–.0032	.0017–.0032	.001–.008	.0009–.0017
4-151	.010–.020	.010–.020	.010–.020	.0025–.0033	.0025–.0033	.0025–.0033	.0025–.0033
6-173	.010–.020	.010–.020	.020–.055	.0012–.0028	.0016–.0037	.0078 max.	.0017–.0027
6-258	.010–.020	.010–.020	.010–.025	.0015–.003	.0015–.003	.001–.008	.0009–.0017
8-304	.010–.020	.010–.020	.010–.025	.0015–.0035	.0015–.003	.0011–.008	.0010–.0018
8-360	.010–.020	.010–.020	.015–.045	.0015–.0035	.0015–.0035	.000–.007	.0012–.0020

TUNE-UP

Electronic Ignition

Refer to the Electrical section for component troubleshooting on electronic ignition systems.

All models are equipped with electronic ignition. The HEI (high energy ignition) system used on the V6 (173 cu. in.) engines uses an externally mounted coil. There are two different methods for controlling spark timing. 49 state vehicles use a conventional type distributor with vacuum advance. California vehicles do not use a vacuum advance. Advance is controlled by the ECM (Electronic Control Module).

Ignition Timing

1. Locate the timing marks on the crankshaft pulley and the front of the timing case cover.
2. Clean off the timing marks, so that you can see them.
3. Use chalk or white paint to color the mark on the scale that will indicate the correct timing, when aligned with the mark on the pulley or the pointer. It is also helpful to mark the notch in the pulley or the tip of the pointer with a small dab of color.
4. Attach a tachometer to the engine.
5. Attach a timing light to the engine.
6. Disconnect the vacuum lines to the distributor at the distributor and plug the vacuum lines. Loosen the distributor lockbolt just enough so that the distributor can be turned with a little resistance.

7. Check to make sure that all of the wires clear the fan and then start the engine.
8. Adjust the idle to the correct specification.
9. With the timing light aimed at the pulley and the marks on the engine, turn the distributor in the direction of rotor rotation to retard the spark, and in the opposite direction of rotor rotation to advance the spark. Align the marks on the pulley and the engine with the flushes of the timing light.
10. When the marks are aligned, tighten the distributor locknut and recheck the timing with the timing light to make sure that the distributor did not move when you tightened the locknut.
11. Turn off the engine and disconnect the test equipment.

MAGNETIC TIMING PROBE EXCEPT DIESEL

A bracket and hole are cast into the timing case cover for the use of a magnetic timing probe, connected to a special electronic timing meter for precise ignition timing. The probe is inserted into the hole of the bracket until the vibration damper is touched. When the engine is started, the probe is automatically spaced away from the damper by the damper's eccentricity, or being slightly out of center. The probe senses a

milled slot on the damper and compensating for the bracket's 9.5° ATDC position, registers the reading on the timing meter. Any necessary corrections can then be made to the ignition timing.

NOTE: Do not use the probe bracket and hole the check the ignition timing, using a conventional timing light.

Valve Lash Adjustment

Manual valve lash adjustment is not necessary nor possible since hydraulic valve lifters are used on all models except on vehicles equipped with the diesel engine.

ADJUSTMENT

Diesel Engine

1. Be sure that the engine is cold before adjusting the valves. Remove the valve cover.
2. Set No. 1 cylinder to TDC on the compression stroke and check the valve clearance of number one and number two intake and number one and number three exhaust valves. Adjust as required.

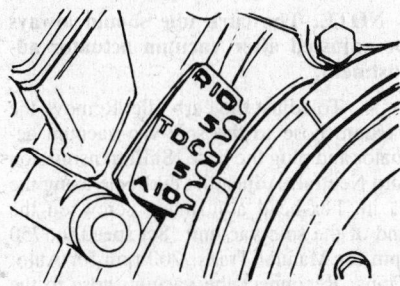

V8 engine timing marks

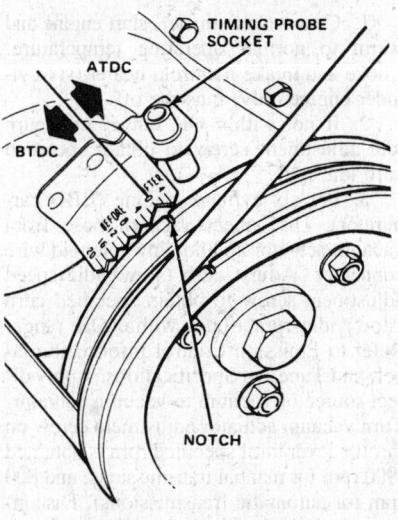

4–151 timing marks

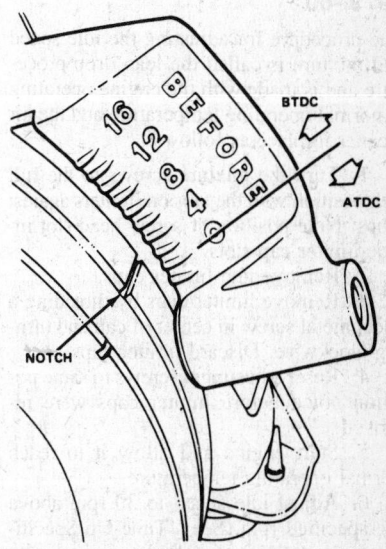

6 cylinder timing marks

NOTE: The No. 1 cylinder is located at the flywheel end of the engine.

3. Rotate the crankshaft 360 degrees and check the clearance of the number three and number four intake and number two and number four exhaust valves. Adjust as required.

4. To adjust, loosen locknut and turn adjustment screw as necessary. As each adjustment screw is tightened, be sure that the bottom of the screw is aligned with the valve stem. If the adjustment screw is not aligned with the stem when tightened, the stem could bend. Tighten locknut.

4. The exhaust valve adjustment specification is .010 in. The intake valve adjustment specification is .008 in.

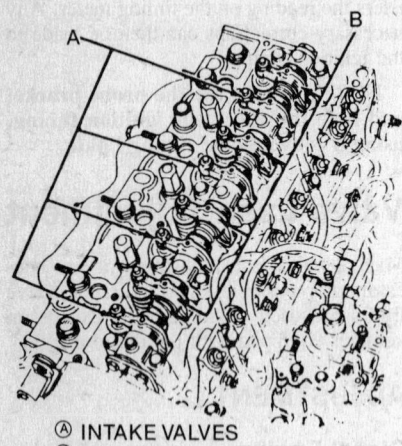

Ⓐ INTAKE VALVES
Ⓑ EXHAUST VALVES

Diesel engine valve adjustment sequence

Idle Speed

ADJUSTMENT

1979–80

The procedure for adjusting the idle speed and mixture is called the lean drop procedure and is made with the engine operating at normal operating temperature and the air cleaner in place as follows:

1. Turn the mixture screws to the full rich position with the tabs on limiters against stops. Note position of screw head slot inside limiter cap slots.

2. Remove idle limiter caps.

3. Remove limiter caps by threading a sheet metal screw in center of cap and turning clockwise. Discard limiter caps.

4. Reset adjustment screws to same position noted before limiter caps were removed.

5. Start engine and allow it to reach normal operating temperature.

6. Adjust idle speed to 30 rpm above the specified rpm. See "Tune-Up Specifications" chart. On 6 cylinder engines with throttle stop solenoid, turn solenoid in or

out to obtain specified rpm. On V8 engines with throttle stop solenoid, turn hex screw on throttle stop solenoid carriage to obtain specified rpm. This is done with solenoid wire connected. Tighten solenoid locknut, if so equipped. Disconnect solenoid wire and adjust curb idle speed screw to obtain idle speed of 500 rpm. Reconnect the solenoid wire.

7. Starting from full rich stop position, as was determined before limiter caps were removed, turn mixture adjusting screws clockwise (leaner) until a loss of engine speed is noticed.

8. Turn screws counterclockwise (richer) until the highest rpm reading is obtained at lean best idle setting. The lean best idle setting is on the lean side of the highest rpm setting without changing rpm.

9. If the idle speed changed more than 30 rpm during the mixture adjustment procedure, reset the idle speed to 30 rpm above the specified rpm with idle speed adjusting screw or the throttle stop solenoid and repeat the mixture adjustment.

10. Install new limiter caps over mixture adjusting screws with tabs positioned against full rich stops. Be careful not to disturb idle mixture setting while installing caps.

1981–82

Idle mixture screws on these carburetors are sealed with plugs or dowel pins. A mixture adjustment must be undertaken ONLY when the carburetor is overhauled, the throttle body replaced, or the engine does not meet required emission standards. Since expensive testing equipment is needed to properly set the mixture, only the idle speed adjusting procedure is given below.

NOTE: The adjustment is made with the manual transmission in neutral and the automatic in drive. Therefore, make certain that the vehicles parking brake is set firmly, and that the wheels are blocked. It may be a good idea to have someone in the vehicle with their foot on the brake.

1. Connect tachometer, start engine and warm to normal operating temperature. Choke and intake manifold heater (six-cylinder engine only) must be off.

2. If not within OK range, turn curb idle adjustment screw to obtain specified curb idle rpm.

3. For six cylinder engine (BBD carburetor): Disconnect vacuum hose from vacuum actuator and holding solenoid wire connector. Adjust curb (slow) idle speed adjustment screw to obtain specified curb (slow) idle rpm if not within OK range. Refer to Emission Control Information label, and Tune-Up Specifications. Apply direct source of vacuum to vacuum actuator. Turn vacuum actuator adjustment screw on throttle lever until specified rpm is obtained (900 rpm for manual transmissions, and 800 rpm for automatic transmissions). Disconnect manifold vacuum source from vacuum actuator. With jumper wire apply battery

voltage (12V) to energize holding solenoid. Turn A/C on, if equipped.

NOTE: Throttle must be opened manually to allow Sol-Vac throttle positioner to be extended.

With Sol-Vac throttle positioner extended, idle speed should be 650 rpm for automatic transmission equipped vehicles and 750 rpm for manual transmission equipped vehicles. If idle speed is not within tolerance, adjust Sol-Vac (hex-head adjustment screw) to obtain specified rpm. Remove jumper wire from Sol-Vac holding solenoid wire connector. Connect Sol-Vac holding solenoid wire connector. Connect original hose to vacuum actuator.

4. For four and eight cylinder engines (2SE, E2SE or 2150 carburetor, turn nut on solenoid plunger or hex screw on solenoid carriage to obtain specified idle rpm: Tighten locknut, if equipped. Disconnect solenoid wire connector and adjust curb idle screw to obtain 500 rpm idle speed. Connect solenoid wire connector. If model 2150 carburetor (eight-cylinder engine, is equipped with dashpot. With throttle at curb idle position, fully depress dashpot stem and measure clearance between stem and throttle lever. Clearance should be 0.032 inch. Adjust by loosening locknut and turning dashpot.

1983–84

4 CYLINDER

1. Fully warm-up the engine.

2. Check the choke fast idle adjustment: Disconnect and plug the EGR valve vacuum hose. Position the fast idle adjustment screw on the second step of the fast idle cam with the transmission in neutral. Adjust fast idle speed to: 2000 rpm for Manual Trans. and 2300 rpm For Auto. Trans. Allow the throttle to return to normal curb idle and reconnect the EGR vacuum hose.

3. To adjust the Sole-Vac Vacuum Actuator: Remove the vacuum hose to the vacuum actuator and plug the hose. Connect an external vacuum source to the actuator and apply 10–15 inches Hg. of vacuum to the actuator. Shift transmission to Neutral. Adjust idle speed to the following rpm using the vacuum actuator adjustment screw on the throttle lever: 850 rpm for Auto. Trans. 950 rpm for Manual Trans. The adjustment is made with all accessories turned off.

NOTE: The curb idle should always be adjusted after vacuum actuator adjustment.

4. To adjust the curb idle: Remove the vacuum hose to the sole-vac vacuum actuator and plug the hose. Shift transmission into Neutral. Adjust the curb idle using the ¼ in. hex-head adjustment screw on the end of the sole-vac unit. Set speed to: 750 rpm for Manual Trans. 700 rpm for Auto. Trans. Reconnect the vacuum hose to the vacuum actuator.

NOTE: Engine speed will vary 10–30 rpm during this mode due to closed loop fuel control.

5. To adjust the TRC (Anti-Diesel): The TRC screw is preset at the factory and should not require adjustment. However, to check adjustment, the screw should be ¾ turn from closed throttle position.

SIX CYLINDER

1. Sole-Vac Vacuum Actuator Adjustment: Disconnect and plug the vacuum hose to the sole-vac vacuum actuator. Disconnect the sole-vac electrical connector. Connect an external vacuum source to the vacuum actuator and apply 10–15 inches (Hg.) of vacuum. Open throttle for at least 3.0 seconds (1200 rpm); then close throttle. Set the speed using the vacuum actuator adjustment screw on the throttle lever to obtain specified rpm. Disconnect the external vacuum source. Reconnect the sole-vac vacuum hose and electrical connector.

2. Sole-Vac Holding Solenoid Adjustment

NOTE: The sole-vac vacuum actuator adjustment should always precede the sole-vac solenoid adjustment.

Disconnect and plug the vacuum hose to the sole-vac vacuum actuator. Disconnect the sole-vac electrical connector. Energize the sole-vac holding solenoid with either of the two following methods: Apply battery voltage (+ 12V) to the solenoid, or reconnect the sole-vac electrical connector and turn on the rear window defogger or turn on the air conditioner with the compressor disconnected. Open throttle for at least 3.0 seconds (1200 rpm) to allow the sole-vac holding solenoid to fully extend. Set the speed using the ¼ hex-head adjustment screw on the end of the sole-vac unit to obtain the specified rpm. Reopen the throttle above 1200 rpm to insure the correct holding position and reset speed if necessary. Reconnect the vacuum hose to the sole-vac actuator. Reconnect the sole-vac electrical connector if disconnected.

1985–86

4 CYLINDER W/YFA CARBURETOR

1. The TRC (anti-diesel) adjustment screw is statically set at ¾ of turn from the throttle valve closed position during the factory assembly and does not normally require readjustment. Should this adjustment be required, turn the adjustment screw counterclockwise to the throttle plate closed position and then turn the screw clockwise ¾ turn.

2. To adjust the Sole-Vac vacuum actuator: Connect a tachometer to the ignition coil TACH wire connector.

3. Place the transmission in the NEUTRAL position and lock the parking brake.

4. Start the engine and allow it to reach normal operating temperature.

5. Connect an external vacuum source

to the SOLE-VAC vacuum actuator and apply 10–15 in. Hg. of vacuum. Plug the engine vacuum hose.

6. Adjust the vacuum actuator until an engine speed of approximately 1000 rpm is achieved.

NOTE: Refer to the Vehicle Emission Control Information Label for the latest specifications for the particular engine being adjusted.

7. Remove the vacuum source from the vacuum actuator and retain the plug in the vacuum hose from the engine.8.

Turn the hex-head curb idle speed adjustment screw until the speed of 500 rpm is obtained.

NOTE: Refer to the Vehicle Emission Control Label for the latest specifications for the particular engine being adjusted.

8. Stop the engine and connect the engine vacuum hose to the vacuum actuator.

9. Remove the tachometer from the engine.

6–258 W/BBD CARBURETOR AND CEC SYSTEM

NOTE: The carburetor choke and intake manifold heater must be off. This occurs when the engine coolant heats to approximately 160 degrees F.

1. Have the engine at normal operating temperature. Connect a tachometer to the ignition coil negative (TACH) terminal.

2. Remove the vacuum hose to the SOLE-VAC vacuum actuator unit. Plug the vacuum hose. Disconnect the holding solenoid wire connector.

3. Adjust the curb (slow) idle speed screw to obtain the correct curb idle speed. Refer to the specifications under Idle Speed or refer to the Emission Information label, under the hood, for the correct curb idle engine rpm.

4. Apply a direct source of vacuum to the vacuum actuator, using a hand vacuum pump or its equivalent. When the SOLE-VAC throttle positioner is fully extended, turn the vacuum actuator adjustment screw on the throttler lever until the specified engine rpm is obtained. Disconnect the vacuum source from the vacuum actuator.

5. With a jumper wire, apply battery voltage (12 volts) to energize the holding solenoid.

NOTE: The holding wire connector can be installed and either the rear window defroster or the air conditioner (with the compressor clutch wire disconnected) can be turned on to energize the holding solenoid.

6. Hold the throttle open manually to allow the throttle positioner to fully extend.

NOTE: Without the vacuum actuator, the throttle must be opened manually to allow the SOLE-VAC throttle positioner to fully extend.

7. If the holding solenoid idle speed is

not within specifications, adjust the idle using the ¼ in. hex-headed adjustment screw on the end of the SOLE-VAC unit. Adjust to specifications.

8. Disconnect the jumper wire from the SOLE-VAC holding solenoid wire connector, if used. Connect the wire connector to the SOLE-VAC unit, if not connected. Install the original vacuum hose to the vacuum actuator.

9. Remove the tachometer and if disconnected, connect the compressor clutch wire. Install any other component that was previously removed.

V8–360 CID ENGINE WITH MODEL 2150 CARBURETOR

NOTE: If the vehicle is equipped with automatic transmission, lock the parking brake, chock the wheels and place the selector lever in DRIVE position before adjusting the idle speed.

1. Connect a tachometer to the ignition coil negative terminal.

2. Start the engine and allow it to reach normal operating temperature.

3. Turn the hex-head adjustment screw on the solenoid carriage to obtain the correct engine speed of 600 rpm. (Verify from Emission Control label, located under the engine hood).

4. Disconnect the solenoid wire connector and adjust the curb idle speed screw to obtain 500 rpm. (Verify from Emission Control label, located under the engine hood).

5. Re-connect the solenoid wire connector and stop the engine.

6. If equipped with a dashpot, position the throttle at the curb idle position and depress the dashpot stem.

7. Measure the clearance between the stem and the throttle lever. A clearance of 0.032 in. should exist.

8. Adjust the clearance as required by loosening the locknut and turning the dashpot until the correct clearance is obtained. Tighten the dashpot locknut.

V6–173 CID ENGINE

NOTE: Some California vehicles using the V6 engine, are equipped with a 2200 hour engine timer. The timer activates a solenoid to control operation of the carburetor secondary vacuum brake after 2200 hours of vehicle operation. The timer is not a serviceable component and must not be disassembled. In the event of a timer malfunction, the complete engine wiring harness must be replaced.

MODEL 2SE CARBURETOR (49 STATES AND CANADA)

1. Connect a tachometer to the ignition coil negative terminal or to the pigtail wire connector above the heater blower motor.

2. Disconnect the plug the vacuum hose at the distributor vacuum advance.

3. If necessary, adjust the ignition timing with the engine speed at or below specifications.

4. Reconnect the vacuum hose to the distributor vacuum advance unit.

5. Disconnect the deceleration valve hose and canister purge hose. Plug the hose and remove the air cleaner assembly.

6. If equipped with air conditioning, turn the control switch to the ON position and open the throttle momentarily to insure the solenoid armature is fully extended. Adjust the solenoid idle speed adjusting screw to obtain the specified engine curb idle speed rpm. Turn the A/C control switch to the OFF position.

7. If not equipped with A/C, adjust the engine idle speed rpm with the solenoid idle speed adjusting screw. Disconnect the solenoid wire and adjust the curb idle.

8. Install the air cleaner assembly. Connect all hoses and other connections.

MODEL E2SE CARBURETOR (CALIFORNIA)

1. Connect a tachometer to the ignition system. Start the engine and operate to normal operating temperature.

2. Turn off all accessories including the A/C system.

3. Put the manual transmission equipped vehicles in NEUTRAL and the A/T equipped vehicles in DRIVE with the parking brake locked and the wheels chocked.

4. Adjust the curb idle speed adjusting screw to obtain the specified rpm of 700 for both manual and automatic transmission equipped vehicles.

5. Disconnect the vacuum hose from the idle kick actuator and connect an outside vacuum source to the actuator. Apply 15 in. Hg. of vacuum to the actuator.

6. Adjust the actuator hex-head adjustment screw for the specified rpm of 1200 with both types of transmissions in the NEUTRAL position.

7. Stop the engine, remove the tachometer and vacuum pump. Install the vacuum hose to the actuator.

MODEL 4–126 CID DIESEL ENGINE WITH TURBOCHARGER

1. The idle speed is adjusted on the injection pump linkage.

2. Loosen the screw locknut, adjust the idle speed to 800 ± 50 rpm with the adjusting screw and tighten the locknut.

ENGINE ELECTRICAL

Distributor

REMOVAL & INSTALLATION

1. Remove the air cleaner assembly and any other component that will interfere with distributor removal.

2. Disconnect the distributor wiring connector at the harness plug.

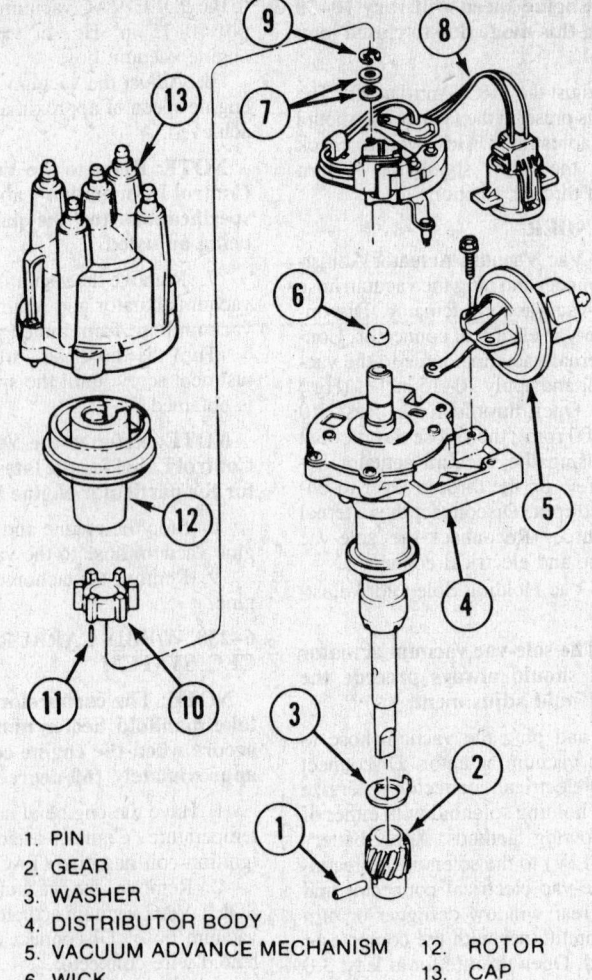

1. PIN
2. GEAR
3. WASHER
4. DISTRIBUTOR BODY
5. VACUUM ADVANCE MECHANISM
6. WICK
7. WASHERS
8. PICK-UP COIL
9. RETAINER
10. TRIGGER WHEEL
11. PIN
12. ROTOR
13. CAP

4–150 distributor

NOTE: The wire connector will contain special conductive grease. Do not remove the grease. The same grease will also be found on the metal parts of the rotor. Do not remove it even if it looks charred.

3. If the distributor is equipped with a vacuum diaphragm, disconnect the vacuum hose(s).

4. Release the distributor cap retainers, remove the cap and position it out of the way. If it is necessary to disconnect any plug wires, be sure to mark them for proper location.

5. Use chalk or paint and carefully mark the distributor body that the center of the rotor tip points to. Mark the engine block in relation to the mark on the distributor body. When this is done, the tip of the rotor, the line on the distributor and the engine block mark should match. When the

distributor is reinstalled in the engine the distributor must be in the same position if correct ignition timing is to be maintained.

6. Remove the distributor hold-down bolt(s), lockwasher and clamp. Lift the distributor from the engine.

NOTE: The shaft and rotor will move slightly (about an ⅛ of a turn) away from the mark on the distributor body when the distributor is removed. When reinstalling the distributor, the rotor must be keyed to the same position, to insure gear mesh.

7. Align the rotor with the mark on the distributor body. Turn the rotor about an ⅛ turn counterclockwise and install the distributor into the engine. The rotor should turn back to align with the mark if the distributor body has not rotated and is fully seated. Install the hold-down clamp, lockwasher and bolt(s) but do not tighten fully.

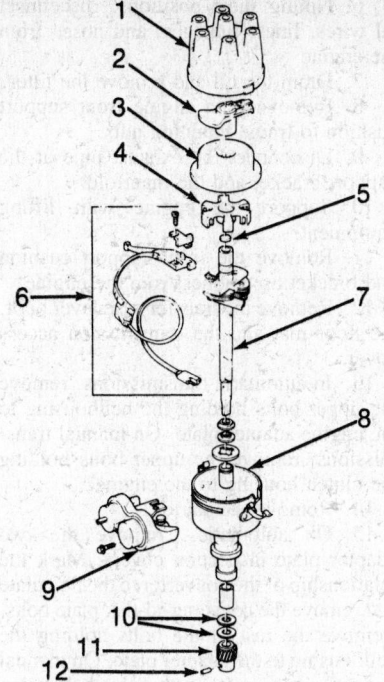

1. Distributor cap
2. Rotor
3. Dust shield
4. Trigger wheel
5. Felt wick
6. Sensor assembly
7. Shaft assembly
8. Housing
9. Vacuum control
10. Shim
11. Drive gear
12. Pin

Exploded view of typical 6 cylinder distributor

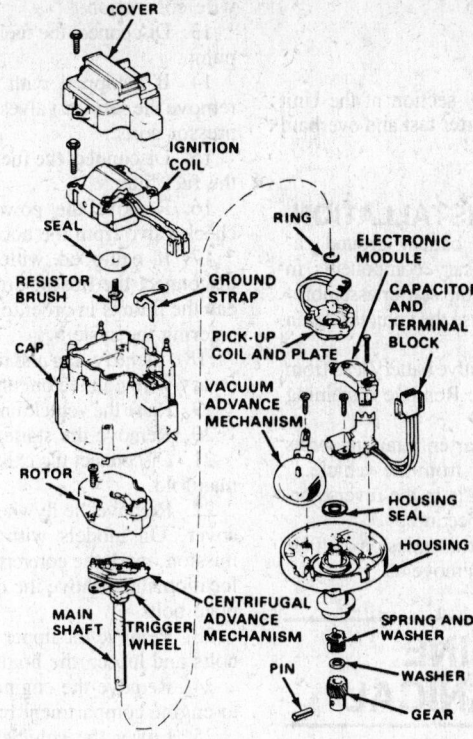

4–151 distributor

8. Connect the wiring harness and install the distributor cap. If any plug wires were removed from the cap reinstall them.

9. Start the engine, allow it to reach normal operating temperature and check/reset the ignition timing.

NOTE: If the engine has been turned with the distributor removed, or if marks were not drawn, it will be necessary to initially time the engine. Follow the next procedure.

10. If the engine was turned while the distributor was removed, it will be necessary to place the No.1 cylinder in the firing position to correctly install the distributor.

11. Remove the No. 1 cylinder spark plug. Turn the engine until the piston in No. 1 cylinder is moving up on the compression stroke. This can be determined by placing your thumb over the spark plug hole and feeling the air being forced out of the cylinder. Stop turning the engine when the marks on the crankshaft pulley and the timing gear cover are in alignment at TDC (''O'' mark).

12. Install the rotor and align the rotor tip with No. 1 spark plug terminal tower position when the cap is installed. To locate No. 1 position, temporarily install the distributor cap and mark the side of the distributor body just below the cap terminal tower.

13. Remove the cap. Align the rotor with the No.1 mark on the distributor body.

14. Refer to proceeding distributor installation procedures.

Alternator

Refer to the Electrical section of the Unit Repair for detailed alternator test and overhaul procedures.

1. The polarity of the battery, generator and regulator must be matched and considered before making any electrical connections in the system.

2. When connecting a booster battery, be sure to connect the negative battery terminal to ground and the positive battery terminals together.

3. When connecting a charger to the battery, connect the charger positive lead to the battery positive terminal. Connect the charger negative lead to the battery negative terminal.

4. Never operate the AC generator on open circuit. Be sure that all connections in the circuit are clean and tight.

5. Do not short across or ground any of the terminals on the AC generator.

6. Do not attempt to polarize the AC generator.

7. Do not use test lamps of more than 12V for checking diode continuity.

8. Avoid long soldering times when replacing diodes or transistors. Prolonged heat is damaging to these units.

9. Disconnect the battery ground terminal when servicing any AC system. This will prevent the possibility of accidentally reversing polarity.

REMOVAL & INSTALLATION

1. Remove the negative battery cable from the battery. Remove all necessary components in order to gain access to the alternator assembly.

2. Remove the wire terminals attached to the rear of the alternator.

3. Loosen the bolt holding the adjusting bar and the pivot bolt at the opposite side of the alternator.

4. Move the alternator inward to relieve the belt tension and remove the belt.

5. Remove the adjusting bar and pivot bolts and remove the alternator from the engine.

6. Installation is the reverse of the removal procedure.

7. When installing the belt, adjust to allow ½ inch play on the longest run between the pulleys.

Voltage Regulator

The voltage regulator is an integral part of the alternator. Refer to the Unit Repair Section for the integral voltage regulator removal and installation procedures.

Starter

REFERENCE

Refer to the Electrical section of the Unit Repair for detailed starter test and overhaul procedures.

REMOVAL & INSTALLATION

1. Disconnect the battery ground cable. Remove all necessary components in order to gain access to the starter assembly.

2. If necessary, raise the vehicle to gain working clearance.

3. Remove the positive battery lead from the starter or solenoid. Remove remaining wires as necessary.

4. Remove the starter retaining bolts and remove the starter from the vehicle.

5. The installation is in the reverse order of the removal procedure.

On some vehicles, the transmission oil filler tube may have to be removed.

ENGINE MECHANICAL

Engine

REMOVAL & INSTALLATION

150 FOUR CYLINDER

1. Disconnect the battery cables, negative first.

2. Remove the air cleaner.

3. Remove the hood. Discharge the A/C system using the proper precautions.

4. Drain the radiator.

5. Remove the lower radiator hose.

6. Remove the upper radiator hose and coolant recovery hose.

7. Remove the fan shroud. If equipped, disconnect the transmission fluid cooler lines.

8. Remove the radiator. If equipped, remove the A/C condenser.

9. Remove the fan assembly and install a $\frac{5}{16} \times \frac{1}{2}$ inch SAE capscrew through the fan pulley into the water pump flange to maintain the pulley and water pump in alignment when the crankshaft is rotated.

10. Disconnect the heater hoses.

11. Disconnect the throttle linkages, cruise control cable (if equipped) and throttle valve rod.

12. Disconnect the wires from the starter motor solenoid and disconnect CEC System wire harness connector. If equipped with fuel injection, release fuel pressure and disconnect all fuel injection harness connectors. Disconnect the fuel line from the fuel pump.

13. If equipped with fuel injection, disconnect the quick connect fuel lines at the inner fender panel by squeezing the two retaining tabs against the fuel tube. Pull the fuel tube and the retainer from the quick

connect fitting. Disconnect the TDC sensor wire connection.

13. Disconnect the fuel line from the fuel pump.

14. If equipped with air conditioning, remove the service valves and cap the compressor ports.

15. Disconnect the fuel return hose from the fuel filter.

16. Remove the power brake vacuum check valve from the booster, if equipped.

17. If equipped with power steering, disconnect the hoses, drain the pump and cap the fittings in order to prevent dirt from entering the system.

:18. Identify, tag and disconnect all necessary wire connectors and vacuum hoses.

19. Raise the vehicle and support it safely.

20. Remove the starter.

21. Disconnect the exhaust pipe from the manifold.

22. Remove the flywheel housing access cover. On models with automatic transmission, mark the converter and drive plate location and remove the converter-to-drive plate bolts.

23. Remove the upper flywheel housing bolts and loosen the bottom bolts.

24. Remove the engine mount cushion-to-engine compartment bracket bolts.

25. Lower the vehicle. Attach a lifting device to the engine.

26. Raise the engine off the front supports.

27. Place a support stand under the converter (or flywheel) housing.

28. Remove the remaining converter (or flywheel) housing bolts.

29. Lift the engine out of the engine compartment.

30. Installation is the reverse of removal.

Item	Ft. lbs.
Exhaust pipe to manifold	17
Torque converter drive plate to crankshaft bolts	60
Clutch housing spacer–to–block	27
Clutch housing–to–block	43
Engine mounts	35

151 FOUR CYLINDER

1. Disconnect the negative battery cable. Remove the hood. Remove the air cleaner.

2. Drain the coolant. Disconnect the radiator hoses. Disconnect the automatic transmission lines from the radiator. If there is a radiator shroud, remove it, then remove the radiator.

3. Remove the fan and spacer.

4. Remove and set aside the power steering pump and belt. Do not disconnect any of the hydraulic lines.

5. Bleed the compressor refrigerant charge, observe all safety precautions. Remove the condenser and receiver assembly.

6. Noting their positions, disconnect all wires, lines, linkages, and hoses from the engine.

7. Drain the oil and remove the filter.

8. Remove both engine front support cushion to frame retaining nuts.

9. Disconnect the exhaust pipe at the support bracket and the manifold.

10. Support the engine with lifting equipment.

11. Remove the front support cushion and bracket assemblies from the engine.

12. Remove the transfer case lever boot, the floor mat and the transmission access cover.

13. In automatic transmissions, remove the upper bolts holding the bellhousing to the engine adapter plate. On manual transmissions, remove the upper bolts holding the clutch housing to the engine.

14. Remove the starter.

15. On automatics, remove the two adapter plate inspection covers. Mark the relationship of the converter to the flex plate and remove the converter-to-flex plate bolts. Remove the rest of the bolts holding the bellhousing to the adapter plate. On manual transmissions, remove the clutch housing lower cover and the rest of the bolts holding the clutch housing to the engine.

16. Support the transmission with a floor jack and remove the engine by pulling it forward and upward.

17. Installation is the reverse of the removal procedure.

Item	Ft. Lbs.
Clutch/converter housing-to-engine	35
Clutch slave cylinder	18
Engine mount nuts	34
Starter mounting bolts	27
Starter bracket nut	40 in. lbs.
Exhaust pipe-to-manifold nuts	35

173 SIX CYLINDER

1. Disconnect the battery cable.

2. Remove the air cleaner.

3. Remove the hood. Discharge the A/C system using the proper safety precautions.

4. Drain the radiator.

5. Remove the lower radiator hose.

6. Remove the upper radiator hose and coolant recovery hose.

7. Remove the fan shroud.

8. If equipped, disconnect the transmission fluid cooler lines.

9. Remove the radiator. If equipped with A/C remove the condenser.

10. Remove the fan assembly.

11. Remove the heater hoses.

12. Disconnect the throttle linkage, including the cruise control cable and automatic transmission throttle valve cable.

13. Remove the power brake booster hose.

14. Identify, mark and disconnect the necessary wire connectors and vacuum hoses.

15. Remove the power steering pump assembly and lay aside.

16. Disconnect the fuel line at the fuel pump.

17. Disconnect the hoses from the A/C compressor.

18. Raise and support the vehicle.

19. Remove the exhaust pipes from the exhaust manifold.

20. Disconnect the exhaust pipe at the catalytic converter flange and allow the exhaust pipe to drop out of the way.

21. Remove the flywheel/converter housing access cover.

22. Remove the torque converter bolts.

23. Disconnect the wires at the starter.

24. Remove the flywheel/converter housing bolts.

25. Lower the vehicle.

26. Place a support under the transmission.

27. Remove the air pump and hose from the bracket.

28. Attach an engine lifting device to the engine.

29. Remove the engine mount through bolts.

30. Disconnect the ground strap at the rear of the left cylinder head.

31. Remove the engine from the engine compartment.

32. Installation is the reverse of removal.

Item	Ft. lb.
Engine mounting bracket	70–92
Engine torque strut	30–40
Drive plate-to-torque converter	25–35
Transmission-to-block	48–63

258 SIX CYLINDER

1. Remove the hood after marking the hinge locations.

2. Disconnect the negative battery cable. Remove the air cleaner.

3. Drain the coolant. Disconnect the radiator hoses. Disconnect automatic transmission cooler lines from the radiator. If there is a radiator shroud, remove it, then remove the radiator.

4. Remove the fan assembly and install a $5/16 \times 1/2$ inch SAE capscrew through the fan pulley into the water pump flange to maintain the pulley and water pump in alignment when the crankshaft is rotated.

5. Remove and set aside the power steering pump and belt. Do not disconnect the hydraulic lines. If equipped, remove the power brake vacuum check valve from the booster.

6. Bleed the compressor refrigerant charge. Be sure to observe all safety precautions. Remove the condenser and receiver assembly.

7. Disconnect all wires, lines, linkage, and hoses from the engine.

8. Raise and support the vehicle safely. Drain the oil and remove the filter. Remove the starter. Remove the engine ground strap.

9. Remove both engine front support cushion-to-frame retaining nuts.

10. Disconnect the exhaust pipe at the support bracket and the manifold.

11. Support the engine with the lifting equipment.

12. Remove the front support cushion and bracket assemblies from the engine.

13. Remove the flywheel housing access cover. On models with automatic transmission, mark the converter and drive plate location and remove the converter-to-drive plate bolts.

14. Remove the upper flywheel housing bolts and loosen the bottom bolts.

15. Remove the engine mount cushion-to-engine compartment bracket bolts.

16. Lower the vehicle. Attach a lifting device to the engine.

17. Raise the engine off the front supports.

18. Place a support stand under the converter (or flywheel) housing.

19. Remove the remaining converter (or flywheel) housing bolts.

20. Lift the engine out of the engine compartment.

21. Installation is the reverse of removal.

DIESEL

1. Disconnect the battery cables and remove the battery. Remove the hood.

2. If equipped, remove the skid plate.

3. Drain the radiator. Remove the air cleaner assembly.

4. If equipped, discharge the A/C compressor. Be sure to observe all safety precautions.

5. Disconnect the radiator hoses and remove the "E" clip from the bottom of the radiator.

6. Raise and support the vehicle safely. If the vehicle is equipped with automatic transmission disconnect the oil cooler lines at the radiator.

7. Remoce the splash shield from the oil pan. Lower the vehicle.

8. Loosen the radiator shroud and remove the radiator fan assembly. Remove the shroud and the splash shield.

9. Remove the radiator and the condenser assembly from the vehicle. Remove the inner cooler.

10. Remove the exhaust shield from the manifold. Disconnect the hoses at the remote oil filter. Remove the oil filter.

11. Tag and disconnect all vacuum hoses and electrical connections. Disconnect and plug the fuel inlet and outlet lines at the fuel pump.

11. If equipped with automatic transmission, remove the left motor mount through bolt retaining nut.

12. Remove the motor mount retaining bolts. Disconnect the accelerator cable. Raise

and support the vehicle safely.

13. Disconnect and drain the power steering hoses at the power steering pump.

14. Disconnect the exhaust pipe at the exhaust manifold. Remove the motor mount retaining nuts.

15. Support the engine. Remove the left motor mount bolts, automatic transmission equipped vehicles, remove the left motor mount.

16. Remove the starter.

17. If the vehicle is equipped with automatic transmission, mark and remove the converter to drive plate bolts through the starter opening. Install the left motor mount and retaining bolts finger tight. Install the motor mount cushion through bolt. Remove the engine support.

18. Remove the transmission to engine retaining bolts.

19. Lower the vehicle. Remove the remaining engine to transmission retaining bolts.

20. Remove the power steering pump from the engine. Remove the oil separator and disconnect the hoses. Disconnect the heater hoses.

21. Remove the remaining engine to transmission retaining bolt.

22. Remove the reference pressure regulator from the dash panel. Install the engine lifting device and position a jack under the transmission.

23. Remove the engine from the vehicle.

24. Installation is the reverse of the removal procedure.

V8

1. Disconnect the negative battery cable. Remove the hood.

2. Remove the air cleaner assembly.

3. Drain the radiator. Remove the upper and lower radiator hoses. If equipped with automatic transmission, disconnect the cooler lines.

4. Remove the radiator from the vehicle.

5. Remove the fan assembly.

6. If equipped with power steering remove the pump and lay it aside. Do not disconnect the hoses from the pump.

7. If the vehicle is equipped with A/C discharge the system . Observe all the required safety precautions. Remove the condenser and the receiver.

8. Remove the battery and the battery tray from the vehicle.

9. Remove the heater core housing and charcoal canister from the firewall as required.

10. If equipped, remove cruise command vacuum servo bellows and mounting bracket as a complete assembly.

11. On some CJ models it may be necessary to remove the left front support cushion and bracket from cylinder block.

12. Disconnect all wires, lines linkage, and hoses which are connected to the engine.

13. If equipped with automatic trans-

mission, disconnect the transmission filler tube bracket from the right cylinder head. Do not remove the filler tube from the transmission.

14. Remove both engine front support cushion-to-frame retaining nuts.

15. Support the weight of the engine with a lifting device.

16. On some vehicles it will be necessary to remove the transfer case shift lever boot, floor (if so equipped), and transmission access cover.

17. Remove the upper bolts which secure the transmission bellhousing to the engine adapter plate on vehicles equipped with automatic transmission. If equipped with manual transmission, remove the upper bolts which secure the clutch housing to the engine.

18. Disconnect the exhaust pipes at the exhaust manifolds and support bracket.

19. Remove the starter motor.

20. Support the transmission with a floor jack.

21. If equipped with automatic transmission, remove the two engine adapter plate inspection covers. Mark the assembled position of the converter and flex plate and remove the converter-to-flex plate cap screws. Remove the remaining bolts which secure the transmission bellhousing to the engine adapter plate.

22. If equipped with manual transmission, remove the clutch housing lower cover and the remaining bolts which secure the clutch housing to the engine.

23. Remove the engine by pulling upward and forward.

NOTE: If equipped with power brakes, care must be taken to avoid damaging the power unit while removing the engine.

24. Installation is the reverse of the removal procedure.

Intake Manifold

REMOVAL & INSTALLATION

150 Four Cylinder

NOTE: It may be necessary to remove the carburetor or the throttle body from the intake manifold before the manifold is removed.

1. Disconnect the negative battery cable. Drain the radiator.

2. Remove the air cleaner. Disconnect the fuel pipe. Remove the carburetor or the throttle body, as required.

3. Disconnect the coolant hoses from the intake manifold.

4. Disconnect the throttle cable from the bellcrank.

5. Disconnect the PCV valve vacuum hose from the intake manifold.

6. If equipped, remove the vacuum advance CTO valve vacuum hoses.

7. Disconnect the system coolant temperature sender wire connector (located on the intake manifold). Disconnect the air temperature sensor wire, if equipped.

8. Disconnect the vacuum hose from the EGR valve.

9. On vehicles equipped with power steering remove the power steering pump and its mounting bracket. Do not detach the power steering pump hoses.

10. Disconnect the intake manifold electric heater wire connector, as required.

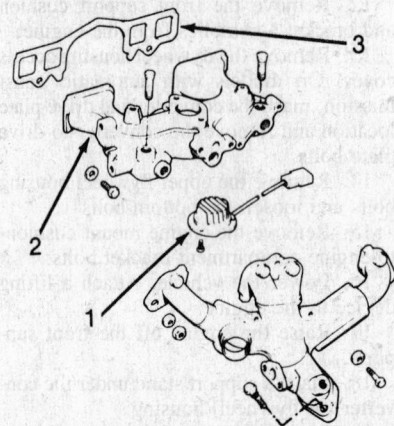

4–150 intake manifold

11. Disconnect the throttle valve linkage, if equipped with automatic transmission.

12. Disconnect the EGR valve tube from the intake manifold.

13. Remove the intake manifold attaching screws, nuts and clamps. Remove the intake manifold. Discard the gasket.

14. Clean the mating surfaces of the manifold and cylinder head.

NOTE: If the manifold is being replaced, ensure all fittings, etc. are transferred to the replacement manifold.

15. Installation is the reverse of removal. Torque manifold bolts to 23 ft. lbs.

151 Four Cylinder

1. Disconnect the negative battery cable.

2. Remove air cleaner and PCV valve hose.

3. Drain cooling system.

4. Tag and remove vacuum hoses (ensure distributor vacuum advance hose is removed).

5. Disconnect fuel pipe and electrical wire connections from carburetor.

6. Disconnect carburetor throttle linkage. Remove carburetor and carburetor spacer.

7. Remove bellcrank and throttle linkage brackets and move to one side for clearance.

8. Remove heater hose at intake manifold.

9. Remove alternator. Note position of spacers for installation.

10. Remove manifold-to-cylinder head bolts and remove manifold.

11. Position replacement gasket and install replacement manifold on cylinder head. Start all bolts.

12. Tighten manifold-to-cylinder head bolts using sequence. Tighten all bolts with 37 ft. lbs. torque.

13. Connect heater hose to intake manifold.

14. Install bellcrank and throttle linkage brackets.

15. Connect carburetor throttle linkage to brackets and bellcrank.

16. Install carburetor spacer and tighten bolts with 15 ft. lbs. torque.

17. Install carburetor and gasket. Tighten nuts with 15 ft. lbs. torque.

18. Install fuel pipe and electrical wire connections. Install vacuum hoses.

19. Connect the negative battery cable.

20. Refill cooling system. Start engine and inspect for leaks.

21. Install air cleaner and PCV valve hose.

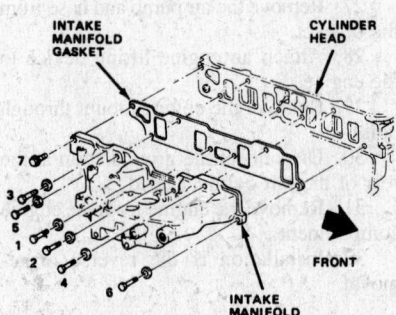

4–151 intake manifold bolt torque sequence

173 Six Cylinder

NOTE: It may be necessary to remove the carburetor from the intake manifold before the manifold is removed.

1. Disconnect the negative battery cable. Remove the rocker covers. Drain the radiator.

2. If equipped with A/C disconnect the compressor and move it to on side. Disconnect the spark plugs wires at the spark plugs. Disconnect the wires at the ignition coil.

3. If equipped, remove the air pump and bracket.

4. Remove the distributor cap. Mark the position of the ignition rotor in relation to the distributor body, and remove the distributor. Do not crank the engine with the distributor removed.

5. Remove the EGR valve. Remove the air hose. Disconnect the charcoal canister hoses. Remove the pipe bracket from the left cylinder head, if equipped.

6. Remove the diverter valve. Remove the power brake vacuum hose. Remove the heater and radiator hoses from the intake manifold.

7. Disconnect and label the vacuum hoses. If equipped, remove the EFE pipe from the rear of the manifold. Disconnect the coolant temperatyure switches.

8. Remove the carburetor linkage. Disconnect and plug the fuel line.

9. Remove the manifold retaining bolts and nuts.

10. Remove the intake manifold. Remove and discard the gaskets, and scrape off the old silicone seal from the front and rear ridges.

11. The gaskets are marked for right and left side installation; do not interchange them. Clean the sealing surface of the engine block, and apply a $\frac{3}{16}$ in. bead of silicone sealer to each ridge.

12. Install the new gaskets onto the heads. The gaskets will have to be cut slightly to fit past the center pushrods. Do not cut any more material than necessary. Hold the gaskets in place by extending the ridge bead of sealer $\frac{1}{4}$ in. onto the gasket ends.

13. Install the intake manifold. The area between the ridges and the manifold should be completely sealed.

14. Install the retaining bolts and nuts, and tighten in sequence to 23 ft. lbs. Do not overtighten; the manifold is made from aluminum, and can be warped or cracked with excessive force.

15. The rest of installation is the reverse of removal. Adjust the ignition timing after installation, and check the coolant level after the engine has warmed up.

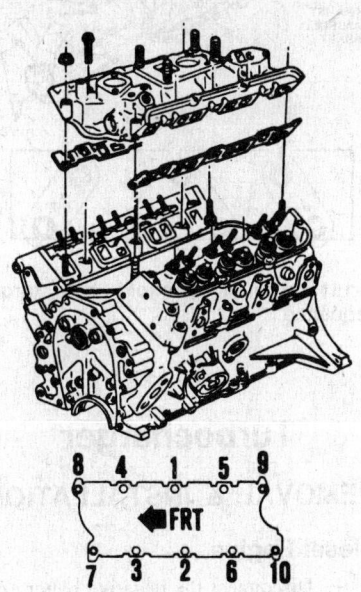

6-173 intake manifold

258 Six Cylinder

NOTE: The intake and exhaust manifold are mounted externally on the left side of the engine and are attached to the cylinder head. They are removed as a unit.

1. Disconnect the negative battery cable. Remove the air cleaner and carburetor.

2. Disconnect the accelerator cable from the accelerator bellcrank.

3. Disconnect the PCV vacuum hose from the intake manifold. Remove the vacuum advance CTO valve vacuum hoses as required.

4. Disconnect the distributor vacuum hose and electrical wires at the TCS solenoid vacuum valves. If equipped, disconnect the CEC system coolant temperature sender wire connector on the intake manifold.

5. Remove the TCS solenoid vacuum valve and bracket from the intake manifold. In some cases it might not be necessary to remove the TCS unit.

6. Disconnect the EGR valve and back pressure sensor hoses. If equipped, disconnect the electric heater wire connector. Remove trhe carburetor from the vehicle.

7. Remove the power steering mounting bracket and pump and set it aside without disconnecting the hoses. Remove air pump, if equipped.

8. Remove the EGR valve and back-pressure sensor. If equipped, remove air conditioning drive belt idler assembly from cylinder head.

9. Disconnect the exhaust pipe from the manifold flange. If equipped with automatic transmission disconnect the throttle valve linkage.

10. Remove the manifold attaching bolts, nuts and clamps.

11. Separate the intake manifold and exhaust manifold from the engine as an assembly, and discard the gasket.

12. If either manifold is to be replaced, they should be separated at the heat riser area.

13. Clean the mating surface of the manifolds and the cylinder head before replacing the manifolds. Replace them in reverse order of the above procedure with new gasket. Tighten the bolts and nuts to the specified torque in the proper sequence.

Diesel

1. Disconnect the negative battery cable. Disconnect the air inlet hose at the intake manifold.

2. Remove all necessary components in order to gain access to the intake manifold retaining bolts.

3. Tag and remove all vacuum hoses and electrical connections that are attached to the intake manifold.

4. Remove the intake manifold retaining bolts. Remove the assembly from the vehicle. Discard the intake manifold gaskets.

5. Installation is the reverse of the removal procedure.

V8

1. Disconnect the negative battery cable. Drain the coolant from the radiator.

2. Remove the air cleaner assembly.

3. Disconnect the spark plug wires.

4. Disconnect the upper radiator hose

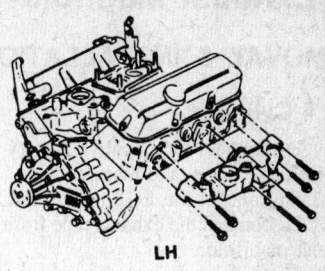

LH

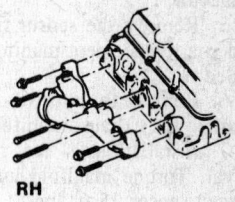

RH

6-173 exhaust manifold

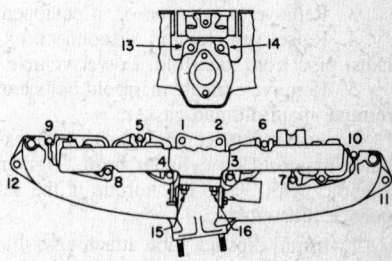

6 cylinder engine intake manifold bolt tightening sequence

and the by-pass hose from the intake manifold. Disconnect the heater hose from the rear of the manifold.

5. Disconnect the ignition coil bracket and lay the coil aside.

6. If equipped, Disconnect the TCS solenoid vacuum valve from the right side valve cover.

7. Disconnect all lines, hoses, linkages and wires from the carburetor and intake manifold and TCS components as required. Remove carburetor.

8. Disconnect the air delivery hoses at the air distribution manifolds.

9. Disconnect the air pump diverter valve and lay the valve and the bracket assembly, including the hoses, forward of the engine.

10. Remove the intake manifold after removing the cap bolts that hold it in place. Remove and discard the side gaskets and the end seals.

11. Clean the mating surfaces of the intake manifold and the cylinder head before replacing the intake manifold. Use new gaskets and tighten the cap bolts to the correct torque. Install in reverse order of the above procedure.

Exhaust Manifold

REMOVAL & INSTALLATION

150 4 Cylinder

1. Disconnect the negative battery cable. Remove the intake manifold.
2. Disconnect the EGR valve tube.
3. Disconnect the exhaust pipe from the exhaust manifold.
4. Disconnect the oxygen sensor wire connector.
5. Remove the sensor from the manifold if a replacement manifold is to be installed.
6. Remove the nuts from the end studs. Remove the exhaust manifold.
7. Installation is the reverse of removal. Torque manifold nuts 23 ft. lbs., oxygen sensor 35 ft. lbs.

151 Four Cylinder

1. Disconnect the negative battery cable. Remove air cleaner and heated air tube.
2. Remove engine oil dipstick tube attaching bolt.
3. Remove oxygen sensor, if equipped.
4. Raise vehicle and disconnect exhaust pipe from manifold. Lower vehicle.
5. Remove exhaust manifold bolts and remove manifold and gasket.
6. Install replacement gasket and exhaust manifold on cylinder head. Tighten all bolts with 39 ft. lbs. torque in the sequence illustrated.
7. Install dipstick tube attaching bolt.
8. Install heated air tube and air cleaner.
9. Install oxygen sensor, if removed.
10. Raise vehicle and connect exhaust pipe to manifold. Tighten bolts with 35 ft. lbs. torque. Lower vehicle.

173 Six Cylinder

LEFT SIDE

1. Disconnect the negative battery cable. Remove the air cleaner. Remove the carburetor heat stove pipe.
2. Remove the air supply plumbing from the exhaust manifold.
3. Raise and support the vehicle safely. Unbolt and remove the exhaust pipe at the manifold.
4. Unbolt and remove the manifold.
5. Clean the mating surfaces of the cylinder head and manifold. Install the manifold onto the head, and install the retaining bolts finger tight.
2. Tighten the manifold bolts in a circular pattern, working from the center to the ends, to 25 ft. lbs. in two stages.
3. Connect the exhaust pipe to the manifold.
4. The remainder of installation is the reverse of removal.

RIGHT SIDE

1. Disconnect the negative battery cable. Raise and support the vehicle safely.

Drain the engine oil. Remove the oil pan. Drain the engine oil and remove the oil pan.
2. Remove the oil pump retaining screws, oil pump, and gasket from the engine block.
3. Remove the cover retaining screws, cover, and gasket from the pump body.
4. Measure the gear end clearance between the gears and the face of the oil pump body.
5. Measure the gear lobe clearance to the pump body sides.
6. Remove the gears and shaft from the body.
7. Remove the cotter pin, spring retainer, spring, and oil pressure relief valve from the pump body.
8. Repair or replace defective components as required.
9. Installation is the reverse of the removal procedure. Be sure to use new gaskets and seals as required.

258 Six Cylinder

NOTE: The intake and exhaust manifolds must be removed together. See the procedure for removing and installing the intake manifold.

Diesel

1. Disconnect the negative battery. Remove the intake manifold.
2. Disconnect the exhaust pipe from the adapter.
3. Remove the oil supply pipe and the oil return hose from the turbocharger assembly.
4. Disconnect the turbocharger air inlet and outlet hoses.
5. Remove the turbocharger retaining bolts. Remove the turbocharger from the vehicle.
6. Remove the exhaust manifold retaining bolts. Remove the exhaust manifold and gasket. Discard the gasket.
7. Installation is the reverse of the removal procedure.

V8

1. Disconnect the negative battery cable. Disconnect the spark plug wires.
2. Disconnect the air delivery hose at the distribution manifold.
3. Remove the air distribution manifold and the injection tubes.
4. Disconnect the exhaust pipe at the manifold.
5. Remove the exhaust manifold attaching bolts and washers along with the spark plug shields.
6. Separate the exhaust manifold from the cylinder head.
7. Install in reverse order of the above procedure. Clean the matting surfaces and tighten the attaching bolts to the correct torque.

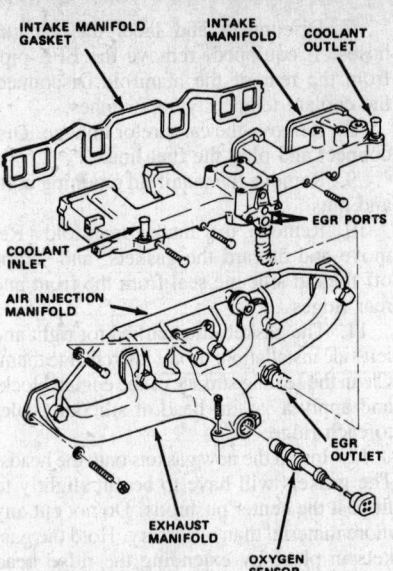

Intake and exhaust manifolds on 6 cylinder engine with oxygen sensor

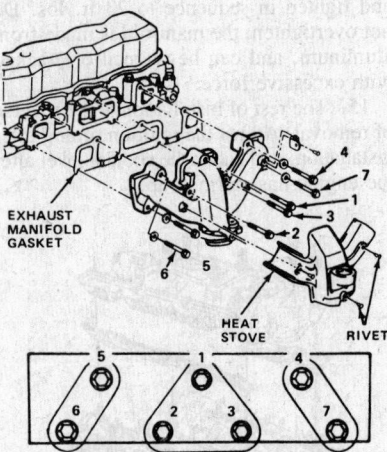

4–151 exhaust manifold bolt torque sequence

Turbocharger

REMOVAL & INSTALLATION

Diesel Engine

1. Disconnect the negative battery cable.
2. Remove all the necessary components in order to gain access to the turbocharger retaining bolts.
3. Disconnect the exhaust pipe flange. Remove the oil supply pipe. Remove the oil return hose.
4. Remove the turbocharger retaining bolts. Remove the turbocharger from the vehicle.
5. Installation is the reverse of the removal procedure.

Cylinder Head

REMOVAL & INSTALLATION

150 Four Cylinder

1. Disconnect the battery cable.
2. Drain the coolant and disconnect the hoses at the thermostat housing.
3. Remove the air cleaner.
4. Remove the valve cover.
5. Remove the rocker arms, bridge and pivot assemblies. Remove the push rods.

NOTE: Retain the push rods, bridge, pivot and rocker arms in the same order as removed to facilitate installation in the original positions.

6. Disconnect the power steering pump bracket. Set the pump and bracket aside. Do not disconnect the hoses.
7. Remove the intake and exhaust manifolds from the cylinder head.
8. If equipped with air conditioning, perform the following: Remove the air conditioner compressor drive belt. Loosen the alternator drive belt. Remove the A/C compressor/alternator bracket-to-cylinder head mounting screw.

NOTE: The serpentine drive belt tension is released by loosening the alternator.

9. Remove the bolts from the A/C compressor (if equipped) and alternator mounting bracket and set the compressor aside.
10. Disconnect the ignition wires and remove the spark plugs. Disconnect the temperature sending unit wire connector.
11. Remove the cylinder head bolts, cylinder head and gasket.
12. Thoroughly clean the machined surfaces of the cylinder head and block. Remove all gasket material and cement.
13. Installation is the reverse of removal, with the following recommendations.
14. Apply an even coat of sealing compound, or equivalent, to both sides of the replacement cylinder head gasket and position the gasket on the cylinder block with the word TOP facing upward.
15. Coat the threads of the stud bolt in the number eight sequence position with Permatex sealant or equivalent. Torque to 75 ft. lbs.

NOTE: Do not apply sealing compound to the cylinder head and block machined surfaces. Do not allow the sealing compound to enter the cylinder bores.

16. Torque the head bolts to 85 ft. lbs. in the proper sequence.

151 Four Cylinder

1. Disconnect the negative battery cable. Drain the cooling system and disconnect the hoses at the thermostat housing.
2. Remove the cylinder head cover

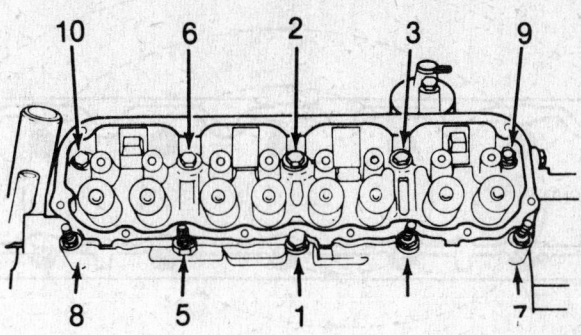

4–150 cylinder head torque sequence

(valve cover), the gasket, the rocker arm assembly, and the pushrods.

NOTE: The pushrods and rockers must be replaced in their original positions.

3. Remove the intake and exhaust manifold from the cylinder head.
4. Disconnect the spark plug wires and remove the spark plugs to avoid damaging them.
5. Remove air conditioning drive belt idler bracket from cylinder head. Loosen alternator belt and remove bracket-to-head mounting screw. Remove compressor mounting bracket and set the unit aside.
6. Disconnect the temperature sending unit wire, ignition coil and bracket assembly and battery ground cable from the engine.
7. Remove the cylinder head bolts, the cylinder head and gasket from the block.
8. To install, reverse the above procedure. Tighten the cylinder head bolts to the specified torque, in the proper sequence.

173 Six Cylinder

LEFT SIDE

1. Disconnect the negative battery cable. Raise and support the vehicle safely. Disconnect the exhaust pipe from the exhaust manifold.
2. Drain the coolant from the block and lower the vehicle.
3. Remove the intake manifold.
4. Remove the exhaust manifold.
5. If equipped, remove the powewr steering pump and bracket.
6. Remove the dipstick tube.
7. Loosen the rocker arm bolts and remove the pushrods. Keep the pushrods in the same order as removed.
8. Remove the cylinder head bolts in stages and in the reverse order of the tightening sequence.
9. Remove the cylinder head. Do not pry on the head to loosen it.
10. Installation is the reverse of removal.

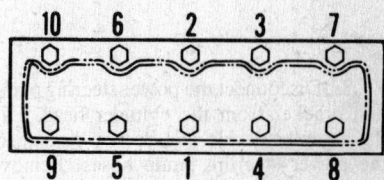

4–151 cylinder head bolt torque sequence

NOTE: The words "This Side Up" on the new cylinder head gasket should face upward. Coat the cylinder head bolts with sealer and torque to specifications.

RIGHT SIDE

1. Disconnect the negative battery cable. Raise and support the vehicle safely. Drain the coolant from the block.
2. Disconnect the exhaust pipe and lower the vehicle.
3. If equipped, remove the cruise control servo bracket.
4. Remove the alternator and air pump bracket assembly.
5. Remove the intake manifold.
6. Loosen the rocker arm nuts and remove the pushrods. Keep the pushrods in the order in which they were removed.
7. Remove the cylinder head bolts in stages and in the reverse order of the tightening sequence.
8. Remove the cylinder head. Do not pry on the cylinder head to loosen it.
9. Installation is the reverse of removal. The words "This Side Up" on the new cylinder head gasket should face upwards. Coat the cylinder head bolts with sealer and tighten them to specification.

258 Six Cylinder

1. Disconnect the negative battery cable. Drain the cooling system and disconnect the hoses at the thermostat housing. Remove the air cleaner.
2. Remove the valve cover, the gasket, the rocker arm assembly, and the pushrods.

NOTE: The pushrods and rockers must be replaced in their original positions.

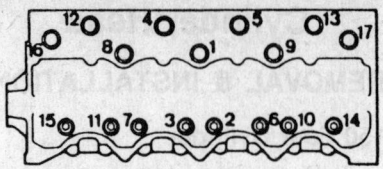

Diesel engine head bolt torque sequence

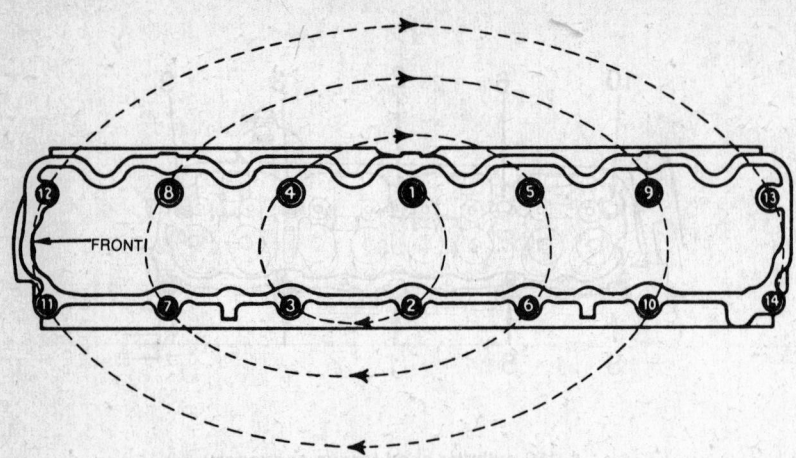

6 cylinder engine cylinder head bolt tightening sequence

3. Disconnect the power steering pump and bracket from the cylinder head. Lay the assembly aside and do not disconnect the power steering pump hoses. Remove the intake and exhaust manifold from the cylinder head.

4. Disconnect the spark plug wires and remove the spark plugs to avoid damaging them.

5. Remove air conditioning drive belt idler bracket from cylinder head. Loosen alternator belt and remove bracket-to-head mounting screw. Remove compressor mounting bracket and set the unit aside.

NOTE: On vehicles so equipped, the serpentine drive belt tension is released by loosening the alternator.

6. Disconnect the temperature sending unit wire, ignition coil and bracket assembly and battery ground cable from the engine.

7. Remove the cylinder head bolts, the cylinder head and gasket from the block.

8. To install, reverse the above procedure. Tighten the cylinder head bolts to the specified torque, in the proper sequence.

Diesel

1. Disconnect the negative battery cable.

2. Remove the intake manifold. Remove the exhaust manifold.

3. Remove the valve cover. Drain the engine coolant. Remove the timing belt cover.

4. Install sprocket holding tool MOT–854 or equivalent and remove the camshaft sprocket retaining bolt. Remove the special tool.

5. Loosen the bolts and move the tensioner away from the timing belt. Retighten the tensioner bolts.

6. Remove the timing belt from the sprockets.

NOTE: If it is necessary to remove the fuel injection pump sprocket use special tool BVI–28–01 or BVI–859 to accomplish this procedure.

7. Disconnect the fuel pipe fittings from the injectors. Plug them in order to prevent dirt from entering the system.

8. Disconnect the fuel pipe fittings from the fuel injection pump. Plug them in order to prevent dirt from entering the system.

9. Remove the fuel pipes from there mountings on the engine. Remove all hoses and connectors from the fuel injection pump.

10. Remove the injection pump retaining bolts. Remove the fuel injection pump and its mounting brackets, as an assembly, from the vehicle.

11. Remove the the retaining bolts and nuts from the cylinder head. Loosen pivot bolt but do not remove it. Remove the remaining cylinder head bolts.

12. Place a block of wood against the cylinder head and tap it with a hammer in order to loosen the cylinder head gasket. The pivot movement will be minimal due to the small clearance between the studs and the cylinder head. Remove the pivot bolt from the cylinder head.

13. Remove the retaining bolts and the rocker arm shaft assembly from the cylinder head.

NOTE: Do not lift the cylinder head from the cylinder block until the gasket is completely loosened from the cylinder liners. Otherwise, the inner seals could be broken.

14. Remove the cylinder head and the gasket from the engine block.

15. Installation is the reverse of the removal procedure. Be sure that the new cylinder head gasket is positioned properly on the cylinder head and that it is the correct thickness for piston protrusion.

16. Torque the cylinder head retaining bolts to 22 ft. lbs., then to 37 ft. lbs, then

to 70–77 ft. lbs. Once all the bolts are tightened recheck the torque.

NOTE: The cylinder head bolts must be retightened after the cylinder head is installed in the vehicle. Operate the engine for a minimum of twenty minutes. Allow the engine to cool for a minimum of two and one half hours. Loosen each cylinder head bolt in sequence about one-half turn. Then retighten in the proper sequence and torque to 70–77 ft. lbs. For the final tightening, tighten the bolts again in sequence without loosening them to 70–77 ft. lbs.

V8

1. Disconnect the negative battery cable. Drain the cooling system and the cylinder block.

2. When removing the right cylinder head, it may be necessary to remove the heater core housing from the firewall.

3. Remove the valve cover(s) and gasket(s).

4. Remove the rocker arm assemblies and push rods.

NOTE: The valve train components must be replaced in their original positions.

5. Remove the spark plugs to avoid damaging them.

6. Remove the intake manifold with the carburetor still attached.

7. Remove the exhaust pipes at the flange of the exhaust manifold. When replacing the exhaust pipes, it is advisable to install new gaskets at the flange.

8. Loosen all of the drive belts.

9. Disconnect negative battery cable at cylinder head. Remove air conditioning compressor mount bracket and alternator support brace from cylinder head.

10. Disconnect the air pump and power steering pump brackets from the cylinder head.

11. Remove the cylinder head bolts and lift the head(s) from the cylinder block.

12. Remove the cylinder head gasket(s) from the head(s) or the block.

13. To install, reverse the above procedure.

NOTE: Apply an even coat of sealing compound to both sides of the new head gasket only. Wire brush the cylinder head bolts, then lightly oil them prior to installation. First, tighten all bolts to 80 ft. lbs., then tighten them to the specified torque. Follow the correct tightening sequence.

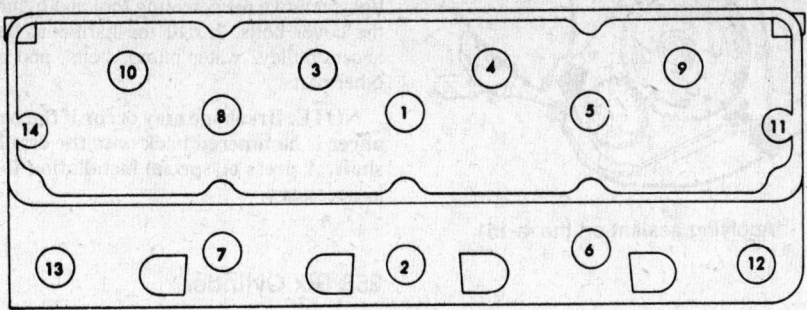

V8 engine cylinder head bolt tightening sequence

Rocker Arm Assemblies

REMOVAL & INSTALLATION

150 Four Cyl & 258 Six Cyl

1. Disconnect the negative battery cable. Remove the valve cover.

2. Remove the two capscrews at each bridge and pivot assembly.

3. Remove the bridges, pivots and rocker arms. Keep them in order as they must be installed in the same position as they were removed.

4. Installation is the reverse of the removal procedure. Torque the capscrews to 19 ft. lbs. Be sure to use new gaskets as required.

151 Four Cyl & 173 Six Cyl

1. Disconnect the negative battery cable. Remove the valve cover and gasket.

2. Remove the rocker arm nut and ball.

3. Lift the rocker arm off the rocker arm stud, always keep the rocker arm, nut and ball together and always assemble them on the same stud.

4. Remove the pushrod from its bore. Make sure the pushrods are always returned to the same bore with the same end in the block.

5. Reverse the above procedure for installation. Tighten the rocker arm nut to 20 ft. lbs.

NOTE: On the six cylinder engine tighten the rocker arm nut until it just touches the valve stem. Rotate the engine until number one piston is at TDC of the compression stroke. The "0" on the timing scale should be aligned with the timing pointer and the rotor should be at the number one spark plug tower of the distributor cap. The folowing valves can now be adjusted. Exhaust valves:1–2–3, Intake valves:1–5–6. Turn the adjusting nut until it backs off the stem slightly, then tighteit until it just touches the stem. Then turn the nut one and one-half turns more to center the tappet plunger. Rotate the engine one complete revolution

more. This will bring number four piston to TDC on the compression stroke. At this point the following valves should be adjusted. Exhaust valves:4–5–6, Intake valves:2–3–4.

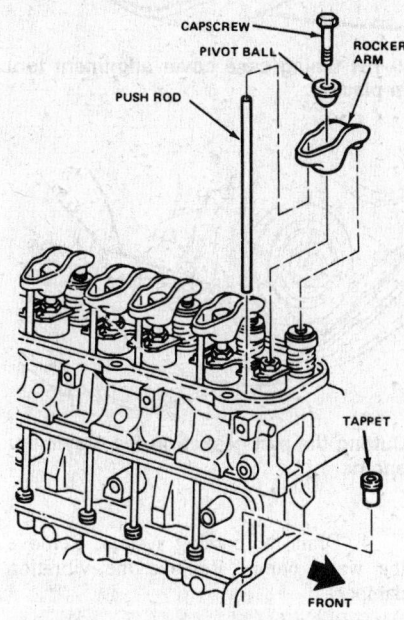

4–151 rocker arm and pushrod removal

Diesel

1. Disconnect the negative battery cable. Remove the cylinder head cover and gasket.

2. Remove the valve cover.

3. Remove the rocker shaft retaining bolts. Remove the rocker arm shaft assembly from the vehicle.

4. Installation is the reverse of the removal procedure. Be sure to use new gaskets and adjust the valves as required.

5. Be sure that the engine is cold before adjusting the valves.

6. Set number one cylinder to TDC on the compression stroke and check the valve clearance of number one and number two intake and number one and number three exhaust valves. Adjust as required.

7. Rotate the crankshaft 360 degrees and check the clearance of the number three and number four intake and number two and number four exhaust valves. Adjust as required.

NOTE: The number one cylinder is located at the flywheel end of the engine.

8. To adjust, loosen locknut and turn adjustment screw as necessary. As each adjustment screw is tightened, be sure that the bottom of the screw is aligned with the valve stem. If the adjustment screw is not aligned with the stem when tightened, the stem could bend. Tighten locknut.

9. The exhaust valve adjustment specification is 0.010 in. The intake valve adjustment specification is 0.008 in.

V8

1. Remove the cylinder head cover and gasket.

2. Loosen the bridged pivot capscrews a turn at a time, so as not to break the bridge.

3. Remove the rocker arm and bridge assembly from the cylinder head.

4. Install the rocker arms and bridge assembly on the cylinder head, and align the pushrods.

5. Install the capscrews and tighten each one a turn at a time to avoid breaking the bridge. Tighten the capscrews to 19 ft. lbs. torque.

6. Install the cylinder head cover with a new gasket and torque the cover bolts to 50 inch lbs.

Crankshaft Pulley (Vibration Damper)

REMOVAL & INSTALLATION

1. Remove the fan shroud, as required. Remove drive belts from pulley.

2. Remove the retaining bolts and separate the pulley from the vibration damper.

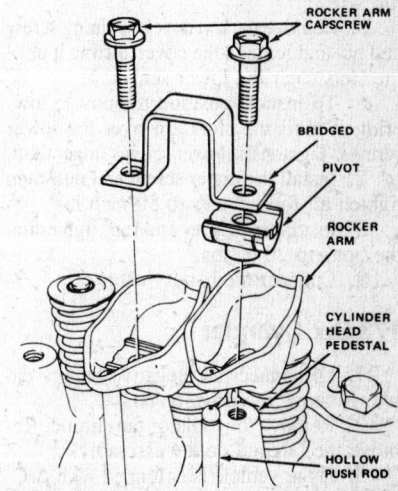

Bridged pivot type rocker arms found on 6 and V8 engines

3. Remove the vibration damper retaining bolt from the crankshaft end.

4. Using a vibration damper puller, remove the damper from the crankshaft.

5. Upon installation, align the key slot of the pulley hub to the crankshaft key. Complete the assembly in the reverse order of removal. Torque the retaining bolts to specifications.

Timing Gear Cover and Oil Seal

REMOVAL & INSTALLATION

150 Four Cylinder

1. Disconnect the negative battery cable. Remove the radiator fan shroud, if equipped.

2. Remove the vibration pulley and damper assembly. Remove the fan and hub assembly.

3. If equipped, remove the A/C compressor. Remove the alternator bracket assembly from the cylinder head and position it aside.

4. Remove the oil pan to timing case cover retaining bolts, and the cover to cylinder block bolts.

5. Remove the timing case cover, front seal and gasket from the engine block.

6. Upon installation, cut off the oil pan side gasket end tabs and the oil pan front seal tabs flush with the front face of the cylinder block and remove the gasket tabs.

7. Using new gaskets as required complete the installation in the reverse order of the removal procedure.

151 Four Cylinder

1. Raise the hood.

2. Disconnect the negative battery cable.

3. Remove the fan and spacer.

4. Loosen the two lower cover retaining screws.

5. Remove the top cover retaining screw and nut and remove the cover, lifting it until the slots clear the lower screws.

6. To install, position the cover, lowering it until the slots are over the lower screws. Tighten the lower screws finger tight.

7. Install the upper screw and nut, then tighten all four screws to 50 inch lbs.

8. Install the spacer and fan, tightening the bolts to 20 ft. lbs.

9. Connect the battery cable.

173 Six Cylinder

1. Disconnect the negative battery cable. Remove the drive belts.

2. Remove the radiator fan shroud. Remove the fan and pulley assembly.

3. If the vehicle is equipped with A/C, remove the compressor from the mounting bracket. Remove the mounting bracket.

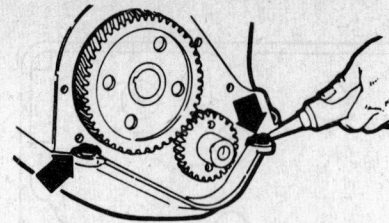

Applying sealant on the 4–151

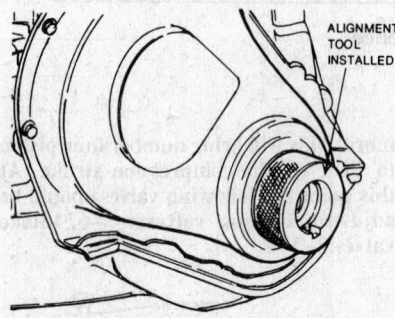

4–151 timing case cover alignment tool in place

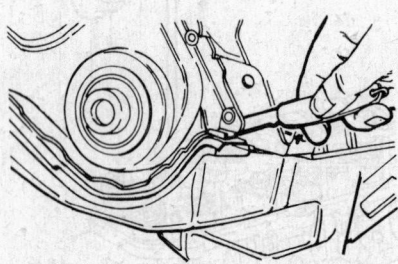

Cutting the pan gasket on the 4 cylinder engine

4. Drain the cooling system. Remove the water pump. Remove the vibration damper.

NOTE: On some vehicles the outer ring (weight) of the harmonic balancer is bonded to the hub with rubber. The balancer must be removed with a puller which acts on the inner hub only. Pulling on the outer portion of the balancer will break the rubber bond or destroy the tuning of the torsional damper.

5. Disconnect the lower radiator hose and heater hose.

6. Remove timing gear cover attaching screws, and cover and gasket.

7. Clean all the gasket mounting surfaces on the front cover and block. Apply a continuous $3/32$ in. bead of sealer to front cover sealing surface and around coolant passage ports and central bolt holes.

8. Apply a bead of silicone sealer to the oil pan-to-cylinder block joint.

9. Install a centering tool in the crankcase snout hole in the front cover and install the cover.

10. Install the front cover bolts finger tight, remove the centering tool and tighten the cover bolts. Install the harmonic balancer, pulley, water pump, belts, and all other parts.

NOTE: Breakage may occur if the balancer is hammered back onto the crankshaft. A press or special installation tool is necessary.

258 Six Cylinder

1. Disconnect the negative battery cable. Remove the drive belts, engine fan and hub assembly, the accessory pulley and vibration damper. Remove the A/C compressor and alternator bracket assembly.

2. Remove the oil pan to timing chain cover screws and the screws that attach the cover to the block.

3. Raise the timing chain cover just high enough to detach the retaining nibs of the oil pan neoprene seal from the bottom side of the cover. This must be done to prevent pulling the seal end tabs away from the tongues of the oil pan gaskets which would cause a leak.

4. Remove the timing chain cover and gasket from the engine.

5. Using the proper tool cut off the oil pan seal end tabs flush with the front face of the cylinder block and remove the seal. Clean the timing chain cover, oil pan, and cylinder block surfaces.

6. Remove the crankshaft oil seal from the timing chain cover.

7. Install in reverse order of the above procedure. It will be necessary to cut the same amount from the end tabs of a new oil pan seal as was cut from the original seal, before installing the new gasket. Be sure to use gasket sealer on both sides of the timing cover gasket.

Diesel

1. Disconnect the negative battery cable.

2. Remove all necesary components in order to gain access to the timing belt cover bolts.

3. Remove the timing belt cover retaining bolts. Remove the timing belt cover from the engine.

4. Installation is the reverse of the removal procedure.

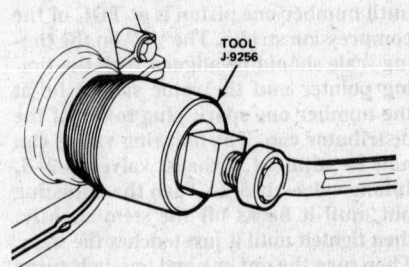

6 cylinder engine timing case cover seal removal

V8

1. Remove the negative battery cable.
2. Drain the cooling system and disconnect the radiator hoses and by-pass hose.
3. Remove all of the drive belts and the fan and spacer assembly. Remove air conditioning compressor and bracket assembly from the engine, if equipped. Do not disconnect air conditioning hoses.
4. Remove the alternator and the front portion of the alternator bracket as an assembly.
5. Disconnect the heater hose.
6. Remove the power steering pump, and/or the air pump, and the mounting bracket as an assembly. Do not disconnect the power steering hoses.
7. Remove the distributor cap and note the position of the rotor. Remove the distributor.
8. Remove the fuel pump.
9. Remove the vibration damper and pulley.
10. Remove the two front oil pan bolts and the bolts which secure the timing chain cover to the engine block.

NOTE: The timing gear cover retaining bolts vary in length and must be installed in the same locations from which they were removed.

11. Remove the cover by pulling forward until it is free of the locating dowel pins. Remove the oil slinger.
12. Clean the gasket surface of the cover and the engine block.
13. Pry out the original seal from inside the timing chain cover and clean the seal bore.
14. Drive the new seal into place from the inside with a block of wood until it contacts the outer flange of the cover.
15. Apply a light film of motor oil to the lips of the new seal.
16. Before reinstalling the timing gear cover, remove the lower locating dowel pin from the engine block. The pin is required for correct alignment of the cover and must either be reused or a replacement dowel pin installed after the cover is in position.
17. Cut both sides of the oil pan gasket flush with the engine block with a razor blade.
18. Trim a new gasket to correspond to the amount cut off at the oil pan.
19. Apply seal to both sides of the new gasket and install the gasket on the timing case cover.
20. Install the new front oil pan seal.
21. Align the tongues of the new oil pan gasket pieces with the oil pan seal and cement them into place on the cover.
22. Apply a bead of sealer to the cutoff edges of the original oil pan gaskets.
23. Place the timing case cover into position and install the front oil pan bolts. Tighten the bolts slowly and evenly until the cover aligns with the upper locating dowel.
24. Install the lower dowel through the

cover and drive it into the corresponding hole in the engine block.
25. Install the cover retaining bolts in the same locations from which they were removed, tightened to 25 ft. lbs.
26. Assemble the remaining components in the reverse order of removal.

Trim the timing gear cover gasket as indicated before installing on V8 engine

Timing Chain, Gears or Belt

REMOVAL & INSTALLATION

150 Four Cylinder

1. Disconnect the negative battery cable. Remove the timing case cover.
2. Rotate the crankshaft until the zero timing mark on the crankshaft sprocket is closest to and on center line with the mark on the cam sprocket.
3. Remove the oil slinger from the crankshaft.
4. Remove the camshaft retaining bolt and remove the sprockets and chain as an assembly.
5. Installation is the reverse of removal, with the following recommendations: Turn the tensioner lever to the unlock (down) position. Pull the tensioner block toward the tensioner lever to compress the spring. Hold the block and turn the tensioner lever to the lock (up) position. Torque the camshaft sprocket retaining bolt to 50 ft. lbs.

151 Four Cylinder

1. Raise the hood and install a bolt in the hood hold open link.
2. Disconnect the negative battery cable.
3. Remove the air conditioning and alternator belts.
4. Remove the crankshaft pulley and the four pulley-to-sprocket bolts. Remove the pulley and damper or washer as applicable.

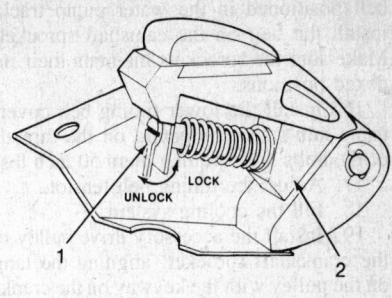

4–150 timing chain tensioner

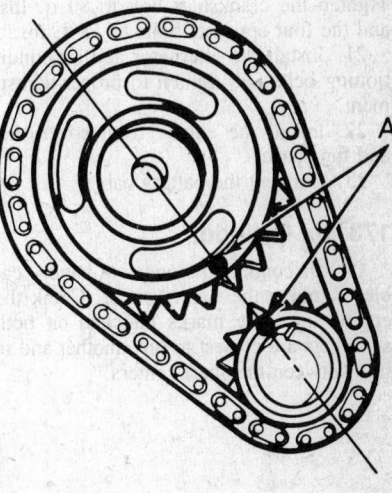

4–150 timing mark alignment

NOTE: It is not necessary to remove the pulley if only the camshaft sprocket is being removed.

5. Drain the engine coolant and loosen the water pump bolts to relieve tension on the timing belt.
6. Remove the timing belt lower cover.
7. Remove the timing belt.
8. Align one of the holes in the camshaft sprocket with the bolt behind the sprocket. Using a socket on the bolt to keep the sprocket from rotating, remove the sprocket retaining bolt and washer.
9. Remove the camshaft sprocket.
10. The crankshaft sprocket may be removed with a puller.
11. Press the crankshaft sprocket back on. Make sure that the timing mark is facing out and that the key is installed.
12. To install the camshaft sprocket, align the dowel in the camshaft with the locating hole in the end of the camshaft sprocket.
13. Install the sprocket retaining bolt, tightening to 80 ft. lbs.
14. Align the timing mark on the camshaft sprocket with the notch on the timing belt upper cover and the crankshaft timing mark with the cast rib on the oil pump cover.
15. Install the timing belt on the crankshaft sprocket, then with the back of the

belt positioned in the water pump track, install the belt on the camshaft sprocket. Make sure the sprockets maintain their indexed positions.

16. Install the lower timing belt cover, using anti-sieze compound on the threads of the bolts and torquing them 50 inch lbs.

17. Adjust the timing belt tension.

18. Fill the cooling system.

19. Install the accessory drive pulley to the crankshaft sprocket, aligning the tang on the pulley with the keyway on the crankshaft. Install the damper locating dowel in the locating hole of the sprocket.

20. Loosely install the four sprocket bolts, then install the crankshaft (center) bolt. Tighten the crankshaft bolt to 80 ft. lbs. and the four sprocket bolts to 15 ft. lbs.

21. Install the alternator and air conditioning belts and tighten to proper adjustment.

22. Install the engine front cover, fan and fan spacer.

23. Connect the battery cable.

173 Six Cylinder

1. Disconnect the negative battery cable. Remove the timing cover. Crank the engine until the marks punched on both sprockets are closest to one another and in line between the shaft centers.

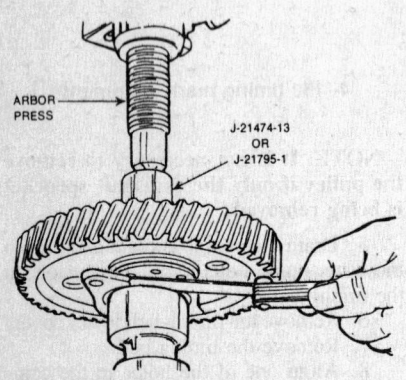

Installing 4-151 camshaft timing gear and measuring thrust plate end clearance

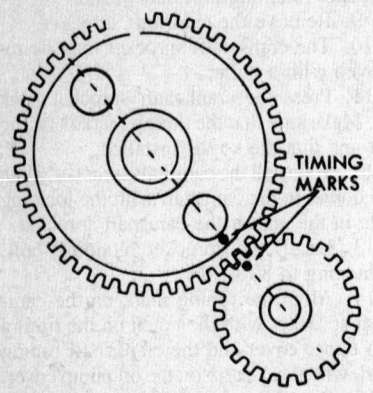

Timing gear alignment—4-151 engine

2. Remove the three bolts that hold the camshaft sprocket to the camshaft. This sprocket is a light press fit on the camshaft and is located by a dowel. The chain comes off with the camshaft sprocket. A gear puller may be required to remove the camshaft sprocket.

3. Without disturbing the position of the engine, mount the new crank sprocket on the shaft, then mount the chain over the camshaft sprocket. Arrange the camshaft sprocket in such a way that the timing marks will line up between the shaft centers and the camshaft locating dowel will enter the dowel hole in the cam sprocket.

4. Place the cam sprocket, with its chain mounted over it, in position on the front of the camshaft and pull up with the three bolts that hold it to the camshaft.

5. After the sprockets are in place, turn the engine two full revolutions to make certain that the timing marks are in correct alignment between the shaft centers.

6. Continue the installation in the reverse order of the removal procedure.

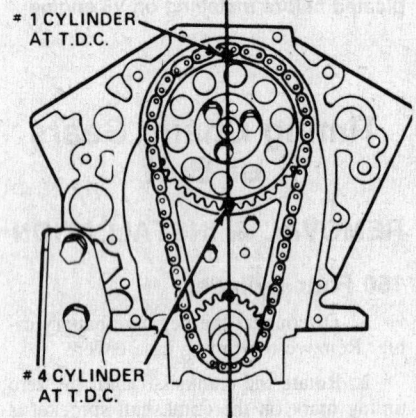

6–173 timing mark alignment

258 Six Cylinder

1. Disconnect the negative battery cable. Remove the drive belts, engine fan and hub assembly, accessory pulley, vibration damper and timing chain cover.

2. Remove the oil seal from the timing chain cover.

3. Remove the camshaft sprocket retaining bolt and washer.

4. Rotate the crankshaft until the timing mark on the crankshaft sprocket is closest to and in a center line with the timing pointer of the camshaft sprocket.

5. Remove the crankshaft sprocket, camshaft sprocket and timing chain as an assembly. Disassemble the chain and sprockets.

6. Assemble the timing chain, crankshaft sprocket and camshaft sprocket with the timing marks aligned.

7. Install the assembly to the crankshaft and the camshaft.

8. Install the camshaft sprocket retaining bolt and washer and tighten to 45–55 ft. lbs.

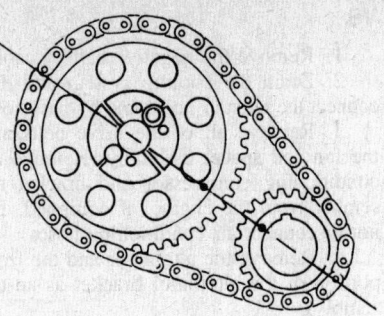

6-258 cylinder timing gear alignment.

4. Install the timing chain cover and a new oil seal.

5. Install the vibration damper, torque the retaining bolt to 80 ft. lbs., accessory pulley, engine fan and hub assembly and drive belts. Tighten the belts to the proper tension.

Diesel

1. Disconnect the negative battery cable.

2. Remove the timing belt cover.

3. Install sprocket holding tool MOT–854 or equivalent and remove the camshaft sprocket retaining bolt. Remove the special tool.

4. Loosen the bolts and move the chain tensioner away from the timing belt. Tighten the tensioner bolts.

5. Remove the timing belt from the sprockets.

6. If it is necessary to remove the fuel injection pump sprocket, use tools BVI–28–01 and BVI–859, or equivalent.

7. Installation is the reverse of the removal procedure.

V8

1. Disconnect the negative battery cable. Remove the timing chain cover and gasket.

2. Remove the crankshaft oil slinger.

3. Remove the camshaft sprocket retaining bolt and washer, distributor drive gear and fuel pump eccentric.

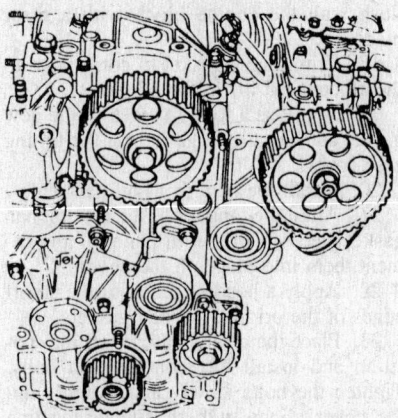

Alignment of timing belt sprockets— diesel engine

4. Rotate the crankshaft until the timing mark on the crankshaft sprocket is adjacent to, and on a center line with, the timing mark on the camshaft sprocket.

5. Remove the crankshaft sprocket, camshaft sprocket and timing chain as an assembly. Disassemble the chain and sprockets.

6. Assemble the timing chain, crankshaft sprocket and camshaft sprocket with the timing marks on both sprockets aligned.

7. Install the assembly to the crankshaft and the camshaft.

8. Install the fuel pump eccentric, distributor drive gear, washer and retaining bolt. Tighten the bolt to 25–35 ft. lbs.

NOTE: The fuel pump eccentric must be installed with the stamped word "REAR" facing the camshaft sprocket.

4. Install the crankshaft oil slinger.

5. Install the timing chain cover using a new gasket and oil seal.

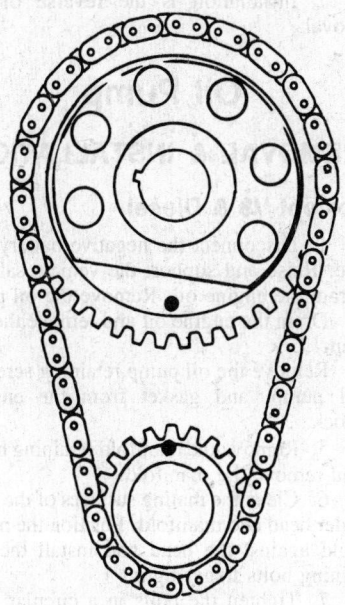

Timing gear alignment—V8 engines

Camshaft

NOTE: Caution must be taken when performing this procedure. Camshaft bearings are coated with babbit material, which can be damaged by scraping the cam lobes across the bearing.

REMOVAL & INSTALLATION

150 Four Cylinder

1. Disconnect the negative battery cable. Drain the cooling system.

2. Remove the shroud and the radiator. If equipped with A/C remove the condenser.

3. Remove the fan and water pump pulley.

4. Remove the rocker cover, rocker arms, and pushrods.

5. Remove the distributor, carefully noting its position for installation.

6. Remove the spark plugs and fuel pump. Remove the valve cover.

7. Remove the rocker arms, bridges and pivots, and pushrods. Be sure to keep these components in order for reinstallation.

8. Remove the valve lifters from the cylinder head using tool J21884 or equivalent.

9. Remove the timing case cover. Remove the timing chain and sprockets.

NOTE: If the cam appears to have been rubbing against the timing case cover, examine the oil pressure relief holes in the rear cam journal to be sure that they are free of debris.

10. Carefully remove the camshaft from the engine.

11. Installation is the reverse of the removal procedure.

151 Four Cylinder

1. Disconnect the negative battery cable. Drain the cooling system.

2. Remove the shroud and the radiator. If equipped with A/C it may be necessary to remove the condenser.

3. Remove the fan and water pump pulley.

4. Remove the rocker cover, rocker arms, and pushrods.

5. Remove the distributor, carefully noting its position for installation.

6. Remove the spark plugs and fuel pump.

7. Remove the pushrod cover and gasket, then remove the lifters.

8. Remove the crankshaft hub and timing gear cover.

9. Remove the two camshaft thrust plate screws by working through the holes in the gear.

10. Remove the camshaft and gear assembly by pulling it through the front of the block.

11. Install in reverse order, make sure camshaft surfaces are dust free and lubed with oil. Torque the thrust plate screws to 75 inch lbs.

173 Six Cylinder

1. Disconnect the negative battery cable. Drain the cooling system.

2. Remove the radiator. If equipped with A/C remove the condenser. Remove the intake manifold and valve lifters. Remove fuel pump and pump pushrod.

3. Remove the timing cover. Remove camshaft sprocket bolts, sprocket and timing chain. A light blow to the lower edge of a tight sprocket should free it (use a plastic mallet).

4. Install two bolts in cam bolt holes and pull cam from block.

5. Installation is the reverse of the removal procedure.

258 Six Cylinder

1. Disconnect the negative battery cable. Drain and remove radiator.

2. If equipped, remove air conditioning condenser and receiver assembly.

3. Remove fuel pump, distributor and ignition wires.

4. Remove cylinder head cover and gasket.

5. Remove rocker arms, bridged pivot assemblies and pushrods. Be sure to replace these parts in the same order as removed.

6. Remove cylinder head and gasket. Remove the valve lifters.

7. Remove timing case cover.

8. Remove timing chain and sprockets.

9. Remove the front bumper or grill as required.

10. Carefully remove the camshaft from the engine.

11. Installation is the reverse of removal.

Diesel

1. Disconnect the negative battery cable.

2. Drain the cooling system. Remove the valve cover. Remove the timing belt cover.

3. Remove the cylinder head. Remove the rocker arm shaft. Remove the camshaft gear. Remove the oil seal from the cylinder head by prying it out using a suitable tool. Remove the camshaft from the cylinder head.

4. Installation is the reverse of the removal procedure.

V8

1. Disconnect the negative battery cable. Drain and remove radiator.

2. If equipped, remove air conditioning condenser and receiver assembly as a charged unit.

3. Remove fuel pump, distributor and ignition wires.

4. Remove cylinder head cover and gasket.

5. Remove drive belts, fan, and hub assembly.

6. Remove intake manifold.

7. Remove rocker arms, bridged pivot assemblies and pushrods. Be sure to replace these parts in the same order as removed.

8. Remove the valve lifters.

9. Remove timing case cover.

10. Remove distributor drive gear and fuel pump eccentric from the camshaft.

NOTE: The fuel pump eccentric must be installed with the word "REAR" facing the camshaft sprocket.

11. Remove timing chain and sprockets.

12. Remove the front bumper or grill, and hood latch support bracket as required.

13. Carefully remove the camshaft from the engine.

14. Installation is the reverse of removal.

Pistons and Connecting Rods

IDENTIFICATION

NOTE: If there is a notch on the top of the piston or an "F" mark anywhere on the piston, it must face the front of the engine.

4–151

The letter "F" or the notches in the edge of the piston, goes toward the front of the engine. On the 151 four cylinder engine the notch on the connecting rod should be opposite the notch on the piston.

150, 258 & V8

The connecting rod caps are stamped with the number of the cylinder to which they belong. Replace them in there original positions. The numbered sides and squirt hole must face the camshaft when assembled in the six cylinder engine. The numbered sides must face out on the V8 engines when assembled.

V6

There is a machined hole or a cast notch "E" in the top of all pistons to indicate proper installation. The piston assemblies should aleays be installed with the hole or notch toward the front of the engine.

Diesel

Mark the connecting rods and rod bearing caps on the intermediate shaft side of the cylinder block with the number of the corresponding cylinder. Number one cylinder is located at the flywheel/drive plate end of the engine block. When removing, remove each connecting rod, cylinder liner and piston as a complete assembly. Each piston and cylinder liner are matched as a set be sure that they are marked properly for installation. Install the assembly according to the marks made during the removal stage of the overhaul.

ENGINE LUBRICATION

Oil Pan

REMOVAL & INSTALLATION

150 Four Cylinder

1. Disconnect the negative battery cable. Raise and support the vehicle safely. Drain the engine oil.
2. Disconnect the exhaust pipe at the exhaust manifold. Disconnect the exhaust hanger at the catalytic converter. Lower the pipe.
3. Remove the starter. Remove the torque converter housing access cover.
4. Remove the oil pan retaining bolts. Remove the oil pan from the vehicle.
5. Installation is the reverse of the removal procedure.

151 Four Cylinder

1. Disconnect the negative battery cable. Raise and support the vehicle safely. Drain engine oil, and remove starter.
2. On CJ models: place jack under transmission bellhousing. Disconnect right engine support cushion bracket from block and raise engine to allow clearance for the oil pan.
3. Remove oil pan, also remove the front and rear neoprene seals and side gaskets.
4. Installation is the reverse of removal.

173 Six Cylinder

1. Disconnect the battery ground.
2. Disconnect the right exhaust pipe at the manifold.
3. Raise and support the vehicle safely.
4. Disconnect the left exhaust pipe at the manifold.
5. Remove the starter.
6. Remove the flywheel access cover.
7. Disconnect the exhaust pipe at the converter and lower it so that the Y portion rests on the upper control arms.
8. Unbolt and remove the oil pan.
9. Clean all RTV gasket material from the mating surfaces.
10. Installation is the reverse of removal. Apply RTV in a ⅛ in. bead on the pan lip.

258 Six Cylinder

1. Disconnect the negative battery cable. Raise and support the vehicle safdely. Drain engine oil, and remove starter. Remove the flywheel cover access housing.
2. On some vehicles it may be necessary to place a jack under the transmission bellhousing. Disconnect right engine support cushion bracket from block and raise engine to allow clearance for the oil pan.
3. Remove oil pan, also remove the front and rear neoprene seals and side gaskets.
4. Installation is the reverse of removal.

Diesel

1. Disconnect the negative battery cable. Raise and support the vehicle safely. Remove the converter housing shield, as required.
2. Drain the engine oil . This engine has two oil drain plugs, both must be opened.
3. Remove all the necessary components in order to gain access to the oil pan retaining bolts.
4. Remove the oil pan retaining bolts. Remove the oil pan from the engine.
5. Installation is the reverse of the removal procedure.

V8

1. Disconnect the negative batttery cable. Raise and support the vehicle safely. Drain engine oil and remove starter. Remove the converter housing access cover.
2. On some vehicles it may be necessary to remove the frame cross bar and automatic transmission lines.
3. If required, cut the corner of engine mount on right side to provide clearance for pan removal.
4. If equipped with manual transmission, bend tabs down on dust shield.
5. Remove oil pan bolts and pan.
6. Remove oil pan front and rear neoprene oil seals.
7. Installation is the reverse of removal.

Oil Pump

REMOVAL & INSTALLATION

Except V8 & Diesel

1. Disconnect the negative battery cable. Raise and support the vehicle safely. Drain the engine oil. Remove the oil pan.
   Drain the engine oil and remove the oil pan. 2.
   Remove the oil pump retaining screws, oil pump, and gasket from the engine block. 3.
5. Remove the manifold retaining bolts and remove the manifold.
6. Clean the mating surfaces of the cylinder head and manifold. Position the manifold against the head and install the retaining bolts finger tight.
7. Tighten the bolts in a circular pattern, working from the center to the ends, to 25 ft. lbs. in two stages.
8. Install the air supply system. Install the spark plug wires. If equipped install the cruise control servo.
9. Raise and support the vehicle safely. Connect the exhaust pipe to the manifold.

NOTE: Fill the pump gear cavity with petroleum jelly prior to the installation of the pump cover, to insure self priming.

V8

1. Disconnect the negative battery cable. Drain the engine oil. Remove the engine oil filter.
2. Remove the engine oil pump cover from the timing chain cover. Remove the oil pump gears and shaft.
3. Remove the oil pressure relief valve from the body.

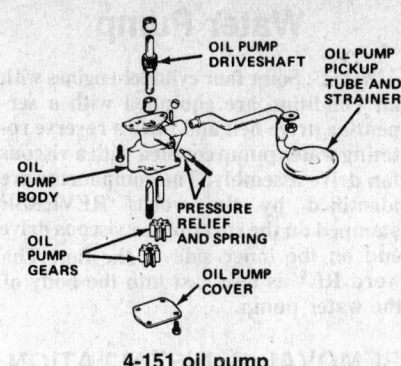

4-151 oil pump

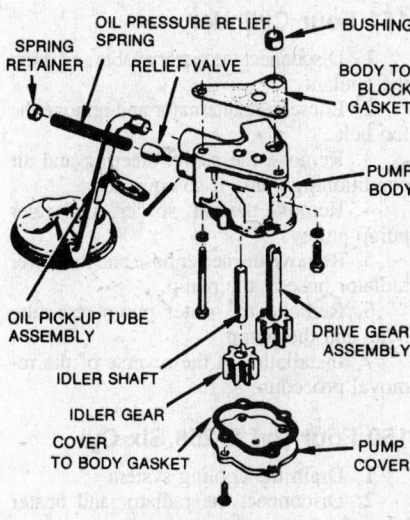

Inline six oil pump

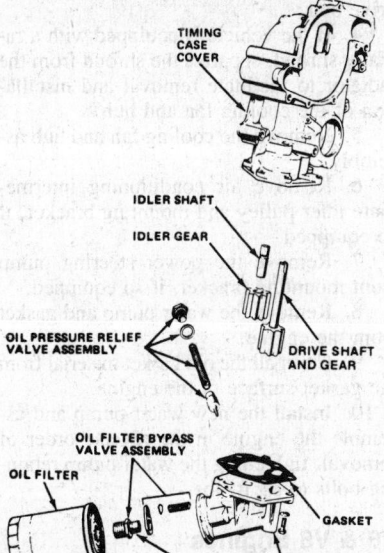

Typical V8 oil pump

4. Inspect the gears for abnormal wear, chips, looseness on the shafts, galling, and scoring.

5. Inspect the cover and cavity for breaks, cracks, distortion, and abnormal wear.

6. Install the gears into the pump cavity, and with the use of a straight edge and feeler gauge, check the gear to housing clearance.

7. If the clearances measure out of the allowable span, the timing chain cover and gears should be replaced.

8. Installation is the reverse of the removal procedure. be sure to use new gaskets and seals as required.

NOTE: Fill the pump gear cavity with petroleum jelly prior to the installation of the pump cover, to insure self priming.

Diesel

1. Disconnect the negative battery cable.

2. Remove the vacuum pump along with the oil pump drive gear.

3. Remove the timing belt cover. Loosen the intermediate shaft drive sprocket using tool MOT–855 or equivalent.

4. Remove the intermediate shaft bolt, sprocket, cover, clamp plate and intermediate shaft.

5. Raise and support the vehicle safely. Drain the engine oil. Remove the oil pan.

6. Remove the piston skirt cooling oil jet assembly to oil pump pipe.

7. Remove the oil pump retaining bolts. Remove the oil pump.

8. Be sure that the oil pump locating dowels are in place on the pump.

9. Inspect the gears for abnormal wear, chips, looseness on the shafts, galling, and scoring.

10. Inspect the cover and cavity for breaks, cracks, distortion, and abnormal wear.

11. Install the gears into the pump cavity, and with the use of a straight edge and feeler gauge, check the gear to housing clearance.

12. Repair or replace defective components as required.

13. Installation is the reverse of the removal procedure. Be sure to use new gaskets and seals as required.

Rear Main Bearing Oil Seal

REMOVAL & INSTALLATION

1. Disconnect the negative battery cable. Raise the vehicle and support it safely.

2. Drain the engine oil. Remove the oil pan.

3. Remove the oil pan gaskets and neoprene seals. Clean all sealing surfaces.

4. Remove the rear main bearing cap and discard the bottom oil seal.

5. Loosen the remaining main bearing caps to allow the crankshaft to drop slightly.

6. Using a brass drift and hammer, tap the upper oil seal until enough seal is exposed on the opposite side of the crankshaft

to permit pulling the seal from the engine block.

7. Coat the block contacting surface of the seal with soap, and the lip of the seal with engine oil and install the seal into the engine block.

NOTE: The lip of the seal must face the front of the engine.

8. Coat the lower seal in the same manner as the upper and install in the rear main bearing cap. Install RTV silicone sealer or equivalent to the lower seal end tabs before installation.

9. Install sealer on both chamfered edges of the rear main bearing cap and install to the block.

NOTE: Do not apply sealer to the mating surface of the bearing cap and the engine block. The main bearing oil clearance could be changed.

10. Tighten all main bearing caps to the proper torque.

11. Install the oil pan using new gaskets and seals. Add the necessary oil to the oil pan, lower the vehicle, start the engine and inspect for oil leaks.

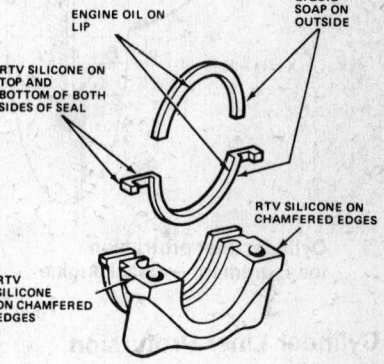

Rear main bearing oil seal installation—6 and V8 engines

Wet Cylinder Liners

REMOVAL & INSTALLATION

Diesel

1. Remove the engine from the vehicle.

2. Remove the cylinder head.

3. Remove the oil pan.

4. Remove the cylinder liner along with the piston and connecting rod assembly.

5. Separate the piston assembly from the cylinder liner. Remove the O-ring seal and the plastic ring from the cylinder liner.

6. If the liner is going to be replaced, be sure to also replace the piston assembly, as they are a matched set.

7. Upon installation, install the piston assembly into the cylinder liner with tool MOT–851 or equivalent.

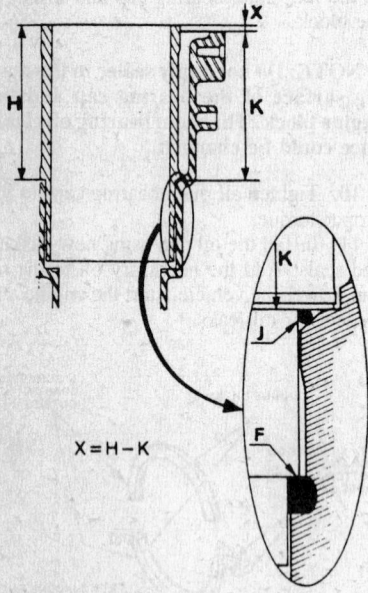

NOTE: The cylinder liners require a rubber O-ring seal and a plastic ring to provide a seal between the liner and the cylinder block, as each liner is supported by the cylinder block. The correct liner protrusion "X" which is above the cylinder block is achieved by close matching tolerances when the cylinder liner and block are manufactured. If replacement liners are required, the liner protrusions above the cylinder block must be measured and all the cylinder liners rearranged according to the results of the measurements.

Cylinder liner protrusion measurement—diesel engine

Cylinder Liner Protrusion Measurement

1. Insert each reusable cylinder liner in its original position in the cylinder block. If applicable, insert the replacement liner in the cylinder block.

2. Install tool MOT–LM and MOT–251–01 or equivalent, on the engine block and tighten the screw clamp. Position tool MOT–252–01 or equivalent, across each cylinder liner, in turn, and secure it with tool MOT–853 or equivalent. Tighten the tool retaining bolts gradually and torque them to 37 ft. lbs. This will assure that each cylinder liner will be firmly in contact with the cylinder block

3. Measure the protrusion "X" of each cylinder liner above the cylinder block using the dial indicator and block gauge. The correct specification is 0.0019–0.0047 in.

4. If an out of specification cylinder liner protrusion is measured, install a replacement liner. Measure the protrusion to determine if the cylinder block or the cylinder liner is defective.

5. With all cylinder liner protrusions within specification arrange them so that

the difference in protrusion between any two adjacent liners does not exceed 0.0016 in.

6. The protrusions are stepped down from the number one cylinder to the number four cylinder or from the number four cylinder to the number one cylinder.

7. When the correct cylinder liner protrusion arrangement has been determined, match each piston and connecting rod assembly with its original liner and remark each according to the new position in the cylinder block.

ENGINE COOLING

Radiator

REMOVAL & INSTALLATION

NOTE: On 1984 and later Wagoneer and Cherokee and 1985 and later Comanchee the A/C condenser must be removed with the radiator. This involves purging the refrigerant. Be sure to use the proper precautions as freon can be dangerous if proper safety conditions are not taken.

1. Drain the radiator by opening the drain cock and removing the radiator pressure cap.

2. Remove the upper and lower hose clamps and hoses at the radiator.

3. Disconnect the automatic transmission oil cooler lines at the radiator, if so equipped. Remove the radiator shroud from the radiator, if so equipped.

4. Remove all attaching screws that secure the radiator to the radiator body support.

5. Remove the radiator.

6. Replace in reverse order of the above procedure.

Thermostat

REMOVAL & INSTALLATION

The thermostat is located in the water outlet housing at the front or on top of the engine. On the V6 and V8 engines the water outlet housing is located in front of the intake manifold.

To remove the thermostats from all of these engines, first drain the cooling system. Remove the two attaching screws and lift the housing from the engine. Remove the thermostat and the gasket. To install, place the thermostat in the housing with the spring inside the engine. Install a new gasket with a small amount of sealing compound applied to both sides. Install the water outlet and tighten the attaching bolts to 30 ft. lbs. Refill the cooling system.

Water Pump

NOTE: Some four cylinder engines with air condition are equipped with a serpentine drive belt and have a reverse rotating water pump coupled with a viscous fan drive assembly. The components are identified by the words REVERSE stamped on the cover of the viscous drive and on the inner side of the fan. The word REV is also cast into the body of the water pump.

REMOVAL & INSTALLATION

151 Four Cylinder

1. Disconnect the battery cable and drain the coolant.

2. Loosen the alternator and remove the fan belt.

3. Remove the power steering and air conditioning belts, if so equipped.

4. Remove the fan, spacer, and water pump pulley.

5. Remove the heater hose and the lower radiator hose at the pump.

6. Remove the water pump retaining bolts and the pump.

7. Installation is the reverse of the removal procedure.

150 Four Cyl & 258 Six Cyl

1. Drain the cooling system.

2. Disconnect the radiator and heater hoses from the water pump.

3. Loosen the alternator adjustment strap screw, upper pivot bolt, and remove the drive belt.

4. If the vehicle is equipped with a radiator shroud, separate the shroud from the radiator to facilitate removal and installation of the cooling fan and hub.

5. Remove the cooling fan and hub assembly.

6. Remove air conditioning intermediate idler pulley and mounting bracket, if so equipped.

7. Remove the power steering pump front mounting bracket, if so equipped.

8. Remove the water pump and gasket from the engine.

9. Clean all the old gasket material from the gasket surface of the engine.

10. Install the new water pump and assemble the engine in the reverse order of removal, tightening the water pump retaining bolts to 13 ft. lbs.

V6 & V8 Engines

1. Disconnect the negative battery cable.

2. Drain the radiator and disconnect the upper radiator hose at the radiator.

3. Loosen all the drive belts.

4. Remove the fan and hub assembly.

5. Separate the radiator shroud from the radiator, if so equipped.

6. If the vehicle is equipped with a viscous fan, remove the fan assembly and

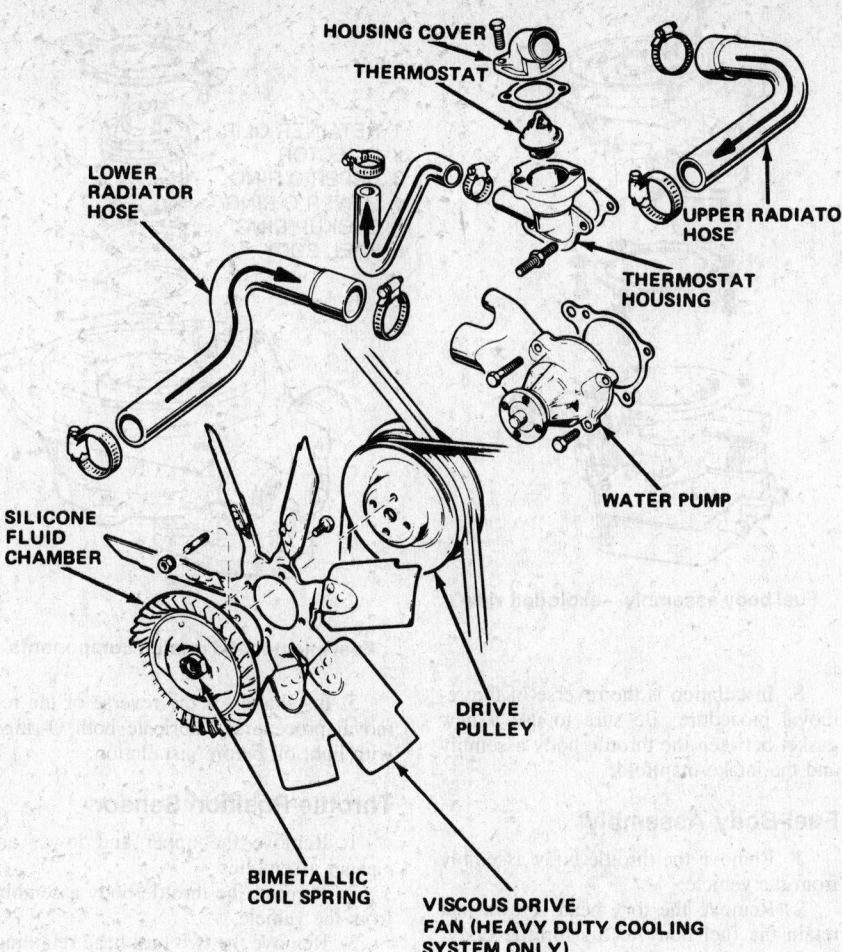

LOWER
RADIATOR
HOSE

HOUSING COVER
THERMOSTAT

UPPER RADIATOR
HOSE

THERMOSTAT
HOUSING

WATER PUMP

SILICONE
FLUID
CHAMBER

DRIVE
PULLEY

BIMETALLIC
COIL SPRING

VISCOUS DRIVE
FAN (HEAVY DUTY COOLING
SYSTEM ONLY)

4-151 water pump and related parts

shroud all at the same time. Do not unbolt the fan blades.

NOTE: The studs in the water pump may back out of the water pump while removing the nuts, preventing the fan assembly from clearing the water pump. If this happens, install a double nut on the stud(s) and remove the studs.

7. On some vehicles equipped with air conditioning, install a double nut on the air conditioning compressor bracket to water pump stud and remove the stud. Removal of this stud eliminates removing the compressor mounting bracket.

8. Remove the alternator and mounting bracket assembly and place it aside. Do not disconnect the alternator wires.

9. Remove the two nuts attaching the power steering pump to the rear half of the pump mounting bracket, if so equipped.

10. Remove the two bolts attaching the front half of the bracket to the rear half.

11. Remove the remaining upper bolt from the inner air pump support brace, loosen the lower bolt and drop the brace away from the power steering front bracket.

12. Remove the front half of the power steering bracket from the water pump mounting stud.

13. Disconnect the heater hose, by-pass hose, and lower radiator hose at the water pump.

14. Remove the water pump and gasket from the timing chain cover and clean all old gasket material from the gasket surface of the timing chain cover.

15. Install the new water pump and assemble the remaining components in the reverse order of removal, tightening the water pump-to-engine block screws to 25 ft. lbs. and the water pump-to-timing case cover screws to 4 ft. lbs. (48 inch lbs.)

Diesel

1. Disconnect the negative battery cable. Drain the engine coolant.

2. Remove all necessary components in order to gain access to the water pump assembly.

3. Remove the coolant hose from the water pump. Remove the drive belts. Remove the fan and hub assembly.

4. It is not necessary to remove the timing belt tensioner. Use a long strap and clip in order to retain the timing belt tensioner plunger in place.

5. Remove the water pump retaining bolts. Remove the water pump assembly from the vehicle.

6. Installation is the reverse of the removal procedure. Be sure to use a new water pump gasket.

EMISSION CONTROLS

Refer to the Emission section in Unit Repair for information.

GASOLINE FUEL SYSTEM

Fuel Pump

REMOVAL & INSTALLATION

Mechanical

1. Remove all the necessary components in order to gain access to the fuel pump. Disconnect the fuel lines.

2. Remove the two attaching bolts that hold the fuel pump to the engine and lift the fuel pump off of the engine.

3. Before installing the fuel pump, make sure that all of the mating surfaces are clean.

4. Cement a new gasket to the mating surface of the fuel pump.

5. Position the fuel pump on the cylinder block so that the cam lever of the pump rests on the camshaft.

6. Secure the pump to the engine with the two bolts and lock washers.

7. Connect the fuel lines to the fuel pump.

Electric

1. Disconnect the negative battery cable. Remove all necessary components in order to gain access to the fuel tank sending unit.

2. Drain the fuel from the fuel tank. Raise and support the vehicle safely.

3. Remove the fuel inlet and outlet hoses from the sending unit. Remove the sending unit wires.

4. Remove the sending unit retaining lock ring. Remove the sending unit, which incorporates the electric fuel pump, along with the O-ring seal from the fuel tank.

4. Installation is the reverse of the removal procedure. Be sure to use a new O-ring seal.

TESTING

Volume Check

Disconnect the fuel line from the carburetor. Place the open end in a suitable container. Start the engine and operate it at

normal idle speed. The pump should deliver at least one pin in 30 seconds.

Pressure Check

Disconnect the fuel line at the carburetor. Disconnect the fuel return line from the fuel filter if so equipped, and plug the nipple on the filter. Install a T-fitting on the open end of the fuel line and refit the line to the carburetor. Plug a pressure gauge into the remaining opening of the T-fitting. The hose leading to the pressure gauge should not be any longer than 6 inches. Start the engine and let it run at idle speed. Bleed any air out of the hose between the gauge and the T-fitting. Pressure readings are given in the Tune-Up Specifications Chart.

Carburetors

NOTE: For detailed rebuilding and settings refer to the Unit Repair Section on carburetors.

REMOVAL & INSTALLATION

To remove the carburetor from any engine, first remove the air cleaner from the top of the carburetor. Remove all lines and hoses, noting their positions to facilitate installation. Remove all throttle and choke linkage at the carburetor. Remove the carburetor attaching nuts which hold it to the intake manifold. Lift the carburetor from the engine along with the carburetor base gasket. Discard the gasket. Install the carburetor in the reverse order of removal, using a new base gasket.

Fuel Injection Components

REMOVAL & INSTALLATION

Throttle Body

1. Disconnect the negative battery cable. Remove the upper air cleaner assembly.
2. Remove the lower air cleaner assembly retaining bolts. Remove the lower air cleaner assembly.
3. Remove the throttle cable and the return spring. Disconnect the wire harness connector from the injector.
4. Disconnect the wire harness connector from the wide open throttle switch. Disconnect the wire harness connector from the ISC motor.
5. Disconnect the fuel supply pipe from the throttle body. Disconnect the fuel return pipe from the throttle body.
6. Disconnect the vacuum hoses from the throttle body assembly. Disconnect the potentiometer wire connector.
7. Remove the throttle body to manifold retaining bolts. Remove the throttle body assembly from the intake manifold.

Fuel body assembly—exploded view

8. Installation is the reverse of the removal procedure. Be sure to use a new gasket between the throttle body assembly and the intake manifold.

Fuel Body Assembly

1. Remove the throttle body assembly from the vehicle.
2. Remove the torx head screws that retain the fuel body to the throttle body. Remove and discard the gasket.
3. Installation is the reverse of the removal procedure. Be sure to use a new gasket.

Fuel Pressure Regulator

1. Remove the throttle body assembly from the vehicle.
2. Remove the three retaining screws that hold the pressure regulator to the fuel body.
3. Remove the pressure regulator assembly. Note the location of the components for reassembly.. Discard the gasket.
4. Installation is the reverse of the removal procedure. Be sure to use a new gasket.

Fuel Injector

1. Remove the air cleaner and hose assembly.
2. Remove the fuel injector wire. Remove the fuel injector retainer clip screws. Remove the fuel injector retainer clip.
3. Using a small pair of pliers, gently grasp the center collar of the injector, between the electrical terminals, and carefully remove the injector using a lifting-twisting motion.
4. Discard the upper and lower O-rings. Note that the back up ring fits over the upper O-ring.

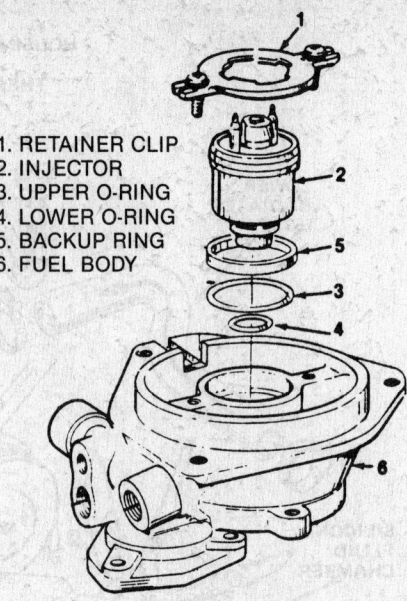

1. RETAINER CLIP
2. INJECTOR
3. UPPER O-RING
4. LOWER O-RING
5. BACKUP RING
6. FUEL BODY

Fuel injector and related components

5. Installation is the reverse of the removal procedure. Lubricate both O-rings with light oil before installation.

Throttle Position Sensor

1. Remove the upper and lower air cleaner assemblies.
2. Remove the throttle body assembly from the vehicle.
3. Remove the two torx head retaining screws holding the TPS assembly to the throttle body.
4. Remove the throttle position sensor from the throttle shaft lever.
5. Installation is the reverse of the removal procedure.

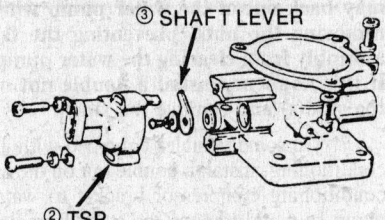

③ SHAFT LEVER

② TSP

Throttle position sensor assembly

Idle Speed Actuator Motor

NOTE: The closed throttle switch is integral with the motor.

1. Disconnect the throttle return spring. Disconnect the wire harness connector from the motor.
2. Remove the motor to bracket retaining nuts. Be usre to use a back up wrench as not to remove the motor studs which hold the motor together.
3. Remove the motor from the bracket.
4. Installation is the reverse of the removal procedure.

DIESEL FUEL SYSTEM

Injection Pump

REMOVAL & INSTALLATION

1. Disconnect the negative battery cable. Remove the timing belt cover.

2. Install sprocket holding tool MOT–854. Remove the camshaft sprocket retaining bolt. Remove the special tool.

3. Loosen the tensioner bolts and move the timing belt tensioner away from the timing belt. Tighten the tensioner bolts.

4. Remove the timing belt from the injection pump sprocket.

5. Disconnect all the fuel pipe fittings from the injectors. Plug the injectors to prevent dirt from entering the fuel system.

6. Disconnect the fuel pipe fittings from the fuel injection pump. Plug the fittings to prevent dirt from entering the system.

7. Remove the fuel line fittings from the vehicle. Remove all hoses and connectors from the injection pump assembly.

8. Remove the fuel injection pump retaining bolts. Remove the fuel injection pump and bracket from the engine as an assembly.

9. To install, attach the pump mounting brackets and the pump to the cylinder head using the retaining bolts.

10. Position the camshaft sprocket timing mark at the 12 o'clock position. Position the fuel injection pump sprocket at the 12 o'clock position

11. Install the timing belt. Continue the installation in the reverse order of the removal procedure. Adjust the injection pump timing.

Injectors

REMOVAL & INSTALLATION

1. Disconnect the negative battery cable. Remove all necessary components in order to gain access to the fuel injectors.

2. Carefully remove the fuel injector lines. Remove the fuel injectors from there mountings in the cylinder head.

3. When reinstalling the fuel injectors be sure to use new copper washers and heat shields.

DIESEL INJECTION PUMP STATIC TIMING ADJUSTMENT

1. Loosen the injection pump adjustment screw. Turn the control cable clevis pin about one-quarter of a turn in order to disengage the cold start system control.

2. Rotate the crankshaft two revolutions counterclockwise and align the camshaft and the injection pump sprocket timing marks with the marks on the timing mark cover.

3. Install special tool MOT–861 in the crankshaft counterweight top dead center slot.

4. Remove the screw plug which is located between the four high pressure fuel outlets at the rear of the fuel injection pump.

5. Remove the cooper washer from the screw plug.

6. Install dial indicator support, tool MOT–856 or equivalent in place of the screw plug. Insert the stem of the dial indicator into the support tool.

7. Remove the top dead center rod, tool MOT–861, from the crankshaft counterweight TDC slot.

8. Slowly turn the crankshaft counterclockwise until the dial indicator pointer stops moving. Zero the dial indicator pointer.

9. Slowly turn the crankshaft clockwise until the top dead center rod, tool MOT–861, can be inserted into the TDC slot in the crankshaft counterweight.

10. At TDC the dial indicator pointer should indicate a piston lift travel distance of 0.031–0.033 in.

11. If the pump piston travel lift distance is not within specification, the pump position must be adjusted.

12. Loosen the pump adjusting bolts. Observe the dial indicator gauge. To increase the piston lift, rotate the pump toward the engine and then away from the engine. To decrease the piston lift, rotate the pump away from the engine.

NOTE: Adjust the piston lift by rotating the pump away from the engine. This is the normal direction of pump rotation. When increasing the piston lift, rotate the pump toward the engine until the piston lift is greater than the specification, then rotate the pump away from the engine until the correct piston lift is indicated on the dial indicator gauge.

13. Tighten the injection pump adjustment bolts. Remove the top dead center rod, tool MOT–861, from the crankshaft counterweight.

14. Observe the dial indicator gauge. Rotate the crankshaft two complete revolutions until the top dead center rod can be inserted into the crankshaft counterweight TDC slot.

15. The dial indicator pointer should return to the correct injection pump timing specification, this will indicate that the injection pump timing is correct.

16. Remove the tool and reinstall all components that were removed.

Fuel Tank

REMOVAL & INSTALLATION

Exc 1984 and Later Wagoneer, Cherokee and Comanche

The fuel tank is attached to the frame by brackets and bolts. The brackets are attached to the tank at the seam flange.

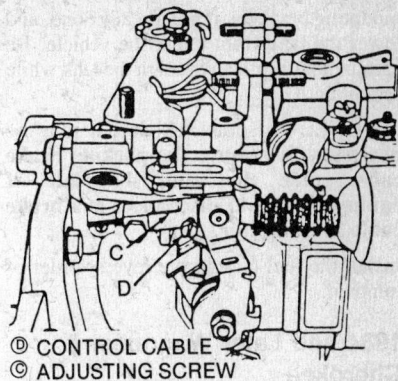

Ⓓ CONTROL CABLE
Ⓒ ADJUSTING SCREW

Diesel engine injection pump adjusting screw location

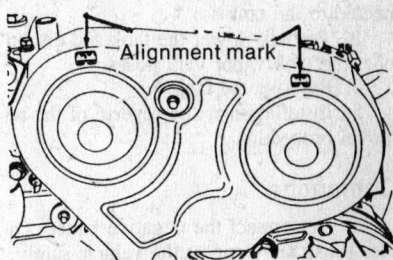

Diesel engine injection pump timing cover locating marks

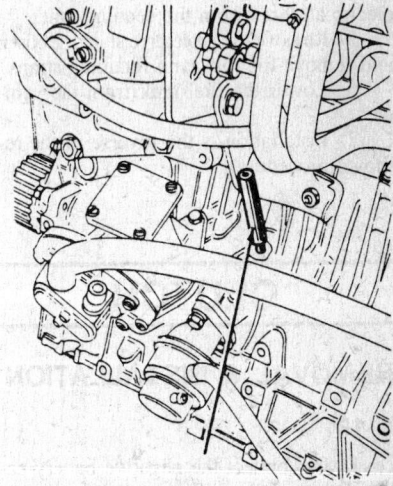

Diesel engine injection pump—top dead center rod positioning

NOTE: Before removing the fuel tank, make sure the level of the fuel inside the tank is at least below any of the various hoses connected. It is best to either drain or siphon the majority of fuel out of the tank to make it easier to handle while removing it.

To remove the tank, loosen all of the clamps retaining the hoses to the tank and disconnect the hoses from the tank. It may be necessary to remove the fuel tank-to-mounting bracket screws and lower the tank slightly to gain access to some of the connecting hoses. Disconnect the tank from the

mounting brackets, if not already done, and lower the tank from under the vehicle. Be careful not to spill any fuel in the tank while removing it.

NOTE: On some vehicles, it may be necessary to remove the parking brake cable guide clips and skid-plate, if equipped, and to disconnect one brake cable at connector.

Install the fuel tank in the reverse order of removal.

1984 and Later Wagoneer & Cherokee

1. Drain the fuel tank. Place a jack under the skid plate and remove the strap nuts.
2. Disconnect all hoses and wires connected to the tank.
3. Partially lower the tank and disconnect the tank vapor vent hoses.
4. Remove the tank.
5. Installation is the reverse of the removal procedure.

Comanche

1. Disconnect the negative battery cable. Raise and support the vehicle safely.
2. Remove the rear propeller shaft.
3. If equipped, remove the skid plate.
4. Disconnect the fuel inlet and outlet lines at the sending unit. Disconnect the electrical wires from the sending unit.
5. Remove the protective shield. Loosen and remove the fuel tank retaining straps.
6. Lower the fuel tank from the vehicle.
7. Installation is the reverse of the removal procedure.

CLUTCH

REMOVAL & INSTALLATION

4–151

1. Disconnect the negative battery cable. Remove the shift lever boot.
2. Remove the shift lever assembly.
3. Raise the vehicle and support it on jack stands.
4. Remove the transmission and transfer case.
5. Remove the slave cylinder-to-clutch housing bolts.
6. Disengage the slave cylinder pushrod from the throwout lever and move the cylinder out of the way.
7. Remove the starter.
8. Remove the throwout bearing.
9. Unbolt and remove the clutch housing.
10. Mark the position of the clutch pressure plate and remove the pressure plate bolts evenly, a little at a time in rotation.
11. Remove the pilot bushing lubricat-

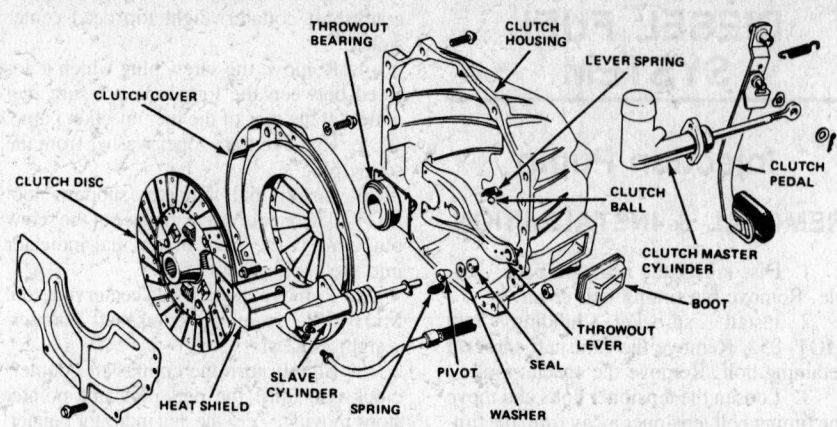

4-151, exploded view clutch assembly

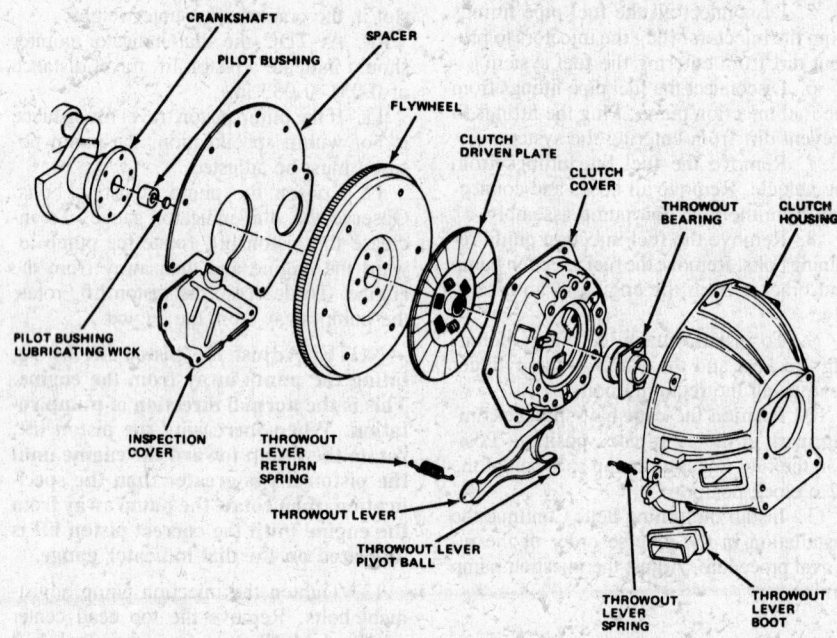

Typical 6 and V8 clutch assembly

ing wick from its bore in the crankshaft and soak the wick in clean engine oil.

12. Installation is the reverse of removal. Torque the pressure plate bolts to 23 ft. lbs., tightening them evenly, a little at a time in rotation. Torque the clutch housing to 54 ft. lbs.; the transmission-to-clutch housing bolts to 54 ft. lbs.; the transfer case-to-transmission bolts to 30 ft. lbs.

4–150, V6–173 & Diesel

1. Disconnect the negative battery cable. Remove the transmission/transfer case assembly.
2. Mark the position of the clutch pressure plate in relation to the flywheel for installation.
3. Loosen the pressure plate bolts evenly, a little at a time each! Failure to loosen the bolts evenly, in rotation, will cause warping of the pressure plate.

4. When all spring tension is relieved from the pressure plate, back out the bolts and remove the pressure plate and driven plate.
5. If the pilot busing is equipped with a lubricating wick, remove it and soak it in clean engine oil.
6. Installation is the reverse of removal. Tighten the pressure plate bolts evenly, in rotation in three or four different steps, to: 4–150: 23 ft. lb. 6–173: 16 ft. lb.

6–258 & V8

1. Disconnect the negative battery cable. Remove the transmission.
2. Remove the starter.
3. Remove the throwout bearing and sleeve assembly.
4. Remove the bell housing.
5. Mark the clutch cover, pressure plate

and the flywheel with a center punch so that these parts can be later installed in the same position.

6. Remove the clutch cover-to-flywheel attaching bolts. When removing these bolts, loosen them in rotation, one or two turns at a time, until the spring tension is released. The clutch cover is a steel stamping which could be warped by improper removal procedures, resulting in clutch chatter when reused.

7. Remove the clutch assembly from the flywheel.

8. Installation is the reverse of the removal procedure.

CLUTCH PEDAL FREE-PLAY ADJUSTMENT

NOTE: Some vehicles are equipped with a hydraulic clutch which is non–adjustable.

1. Lift the pedal up against the stop.
2. Raise and support the vehicle safely. Loosen and release the rod adjuster jam nut.
3. Turn the release rod adjuster in or out to obtain the proper adjustment.
4. Adjust the pedal free-play to about one inch.
5. Tighten the jam nut.
6. Lower the vehicle.

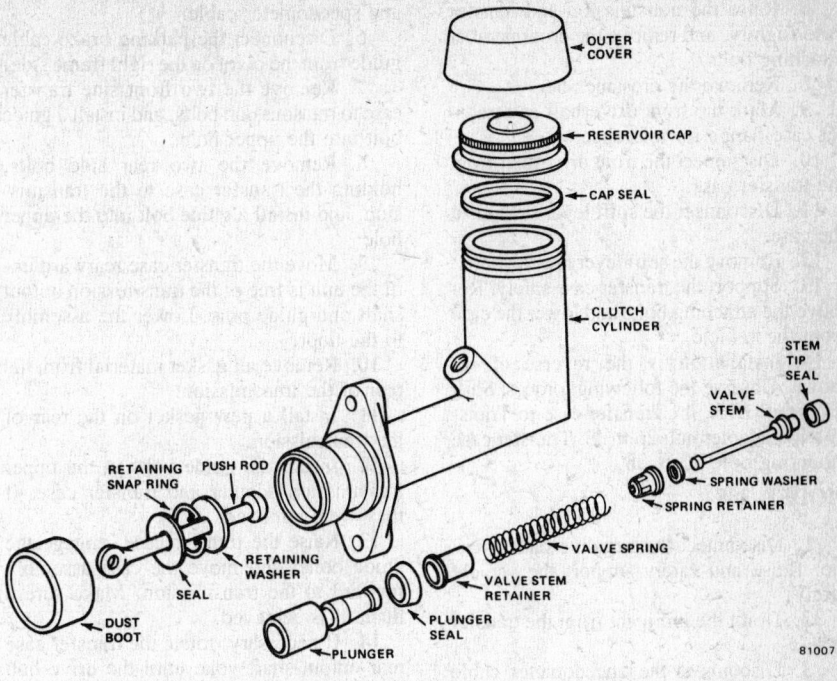

Clutch master cylinder

Clutch Master Cylinder

REMOVAL & INSTALLATION

1. Disconnect the negative battery cable. Disconnect the hydraulic line at the cylinder.
2. Cap the hydraulic line to prevent dirt from entering.
3. Remove the cotter pin and washer securing the clutch pushrod to the pedal arm.
4. Unbolt the master cylinder from the firewall.
5. Installation is the reverse of removal. Torque the mounting nuts to 11 ft. lb. on 4–151 models and 19 ft. lb on all other models. Fill the reservoir with clean brake fluid.

Clutch Slave Cylinder

REMOVAL & INSTALLATION

1. Disconnect the negative battery cable. Disconnect the hydraulic line at the cylinder.
2. Remove the bolts attaching the cylinder to the clutch cover housing. On 4–151, remove the throwout lever-to-pushrod retaining spring.
3. Remove the cylinder and cap the hydraulic line.
4. Installation is the reverse of removal. Torque the attaching bolts to 16 ft. lb.

TRANSFER CASE

Refer to the Transfer Case General Repair Section for troubleshooting and overhaul procedures.

REMOVAL & INSTALLATION
MODEL 20

1. Disconnect the negative battery cable. Remove shift lever knob, boot, and shift lever.
2. Remove floor covering and remove transmission access cover from floorpan.
3. Drain lubricant from transfer case. On CJ models drain the transmission also.
4. If equipped, disconnect torque reaction bracket from crossmember. Disconnect speedometer at transfer case.
5. On CJ models, place support stand under clutch housing to support engine and transmission, and remove rear crossmember.
6. Disconnect front and rear driveshafts at transfer case, making sure to mark shaft yokes for assembly.
7. On Cherokee and Truck models, disconnect parking brake cable at equalizer and exhaust pipe support bracket at transfer case.
8. Remove bolts attaching transfer case to transmission and remove transfer case.

NOTE: One transfer case attaching bolt must be removed from front end of the case. This bolt is located at the bottom right corner of the transmission.

9. Install a new gasket on the transmission.
10. Shift the transfer case into 4WD low and install the case assembly on the guide bolts.
11. Rotate the transfer case output shaft until the transmission main shaft gear engages the rear output shaft gear of the transfer case.
12. Slide the transfer case forward until the two units mate flush.
13. Install one upper bolt, remove the dowel guide bolts and install the remaining bolts. Torque to 30 ft. lbs.
14. Connect the speedometer cable and parking brake cable.
15. Install the propeller shafts after aligning the indexing marks.
16. Fill the unit with gear lube, and lower the vehicle.
17. Install the transfer case shift lever, boot, and knob.

MODEL 207

1. Shift the unit into 4–High. Disconnect the negative battery cable.
2. Raise and support the vehicle safely.
3. Drain the lubricant from the case.
4. Mark the rear axle yoke and drive shaft for reference.
5. Remove the rear driveshaft.
6. Disconnect the speedometer cable, vacuum hoses and vent hose from the case.

7. Raise the transmission and transfer case slightly, and remove the crossmember attaching bolts.

8. Remove the crossmember.

9. Mark the front driveshaft and transfer case flange for reference.

10. Disconnect the front driveshaft from the transfer case.

11. Disconnect the shift lever linkage at the case.

12. Remove the shift lever bracket bolts.

13. Support the transfer case safely. Remove the attaching bolts and lower the case from the vehicle.

14. Installation is the reverse of removal. Observe the following torques: Shift lever nut: 18 ft. lb. Transfer case-to-Transmission adapter nut: 26 ft. lb. Transfer case mounting bolt: 22 ft. lb.

MODEL 208

1. Disconnect the negative battery cable. Raise and safely support the vehicle safely.

2. Drain the lubricant from the transfer case.

3. Disconnect the speedometer cable and indicator switch wires. Disconnect the transfer case shift lever link at the operating lever.

4. Support the rear of the transmission and remove the rear crossmember.

5. Mark the transfer case front and rear output shaft yokes and driveshafts for assembly alignment reference.

6. Disconnect the front and rear driveshafts at the transfer case yokes. Secure the shafts to the frame rails with wires to keep them out of the way.

7. Disconnect the parking brake cable guide from the pivot located on the right frame rail, if necessary.

8. Remove the bolts that attach the exhaust pipe support bracket to transfer case, if necessary.

9. Remove the bolts that attach the transfer case to the transmission.

10. Move the transfer case assembly rearward until it is free of the transmission output shaft. Remove and lower the transfer case.

11. Installation is in the reverse order of removal. Torque the transfer case mounting bolts to 40 ft. lbs.

MODEL 219

1. Disconnect the negative battery cable. Raise and support the vehicle safely.

2. Remove reduction unit on Cherokee, Wagoneer, and Truck models, if equipped.

3. Index the marks on the front and rear yokes and propeller shafts for proper alignment during assembly.

4. Disconnect both the front and rear propeller shafts. On CJ7 models, place support stand under transmission and remove crossmember.

5. Mark the vacuum diaphragm control for identification during the assembly, and then disconnect the vacuum hoses, wiring, and speedometer cable.

6. Disconnect the parking brake cable guide from the pivot on the right frame side.

7. Remove the two front side transfer case to transmission bolts, and install a guide bolt into the upper hole.

8. Remove the two rear side bolts, holding the transfer case to the transmission, and install a guide bolt into the upper hole.

9. Move the transfer case rearward until the unit is free of the transmission output shaft and guide pins. Lower the assembly to the floor.

10. Remove all gasket material from the rear of the transmission.

11. Install a new gasket on the rear of the transmission.

12. Install the guide bolts in the upper transmission adapter and transfer case, if they were removed.

13. Raise the transfer case, engage the guide bolts, and move the case assembly forward to the transmission. Make sure a flush fit is achieved.

14. If necessary, rotate the transfer case rear output shaft yoke until the drive hub splines align with the transmission output shaft.

15. Install front and rear attaching bolts, and remove the guide bolts during this operation.

16. Attach the exhaust pipe bracket support, if removed.

17. Align the propeller shaft and indexing marks on the yokes and attach the propeller shafts.

18. Connect the speedometer cable, wiring, and vacuum hoses.

19. Connect the parking brake cable guide to the pivot bracket on the right frame side.

20. Install the specified lubricant, and lower the vehicle.

MODEL 228 AND 229

1. Disconnect the negative battery cable. Raise and support the vehicle safely.

2. Drain the lubricant from the transfer case.

3. Disconnect the speedometer cable and vent hose. Disconnect the transfer case shift lever link at the operating lever.

4. Support the weight of the transmission safely. Remove the rear crossmember.

5. Mark the relation of the front and rear driveshafts with their yokes for reference.

6. Disconnect the front and rear driveshafts at the yokes.

7. Disconnect the shift motor vacuum hoses.

8. Disconnect the transfer case shift linkage.

9. Support the transfer case with a jack and remove the transfer case-to-transmission bolts.

10. Move the transfer case rearward and lower it from the vehicle.

11. Installation is the reverse of removal. Don't install any mounting bolts until all parts are aligned. Make sure that all splined shafts mesh properly before tightening bolts. Torque the transfer case-to-transmission bolts to 40 ft. lb. and the driveshaft yoke nuts to 10 ft. lb. (120 inch lb.)

MODEL 300

1. Disconnect the negative battery cable.

2. On models with a manual transmission; remove the shift lever knob, trim ring and boot from the transmission and transfer case shift levers.

3. Remove the floor covering and remove the transmission access cover from the floorpan.

4. Raise and safely support the vehicle. Drain the lubricant from the transfer case.

5. Support the engine and transmission under the clutch bell housing and remove the rear crossmember.

6. Mark the transfer case front and rear yokes and driveshafts for assembly alignment reference.

7. Disconnect the front and rear driveshafts at the transfer case. Secure the shafts out of the way.

8. Disconnect the speedometer cable at the transfer case. Disconnect the parking brake cable at the equalizer and the exhaust pipe support bracket at the transfer case to gain any needed clearance.

9. Remove the bolts mounting the transfer case to the transmission. Remove the transfer case.

10. Installation is in the reverse order of removal. The transfer case should be shifted into the 4L position before installation. Rotate the output shaft yoke until the transmission output shaft gear engages the transfer case input shaft. Torque the mounting bolts to 30 ft. lbs.

Linkage Adjustments

The Shifter rails of the transfer case lever assembly connect to the shifter rails of the transfer case either directly or through non-adjustable links on vehicles with manual transmissions. An adjustable trunnion is provided on the lower shift rod to provide desired adjustment on automatic transmission equipped models. The linkage should be lubricated periodically.

QUADRA-TRAC®

Since the Quadra-Trac® system is a "full time 4WD" system, and is constantly engaged in 4WD, there is no "shift linkage" as such. There are two features which can be operated manually concerning the transfer case: the "Lock-Out" feature and the engagement of the optional "Low Range Reduction Unit."

Since the "Lock-Out" feature is a vacuum actuated unit, there are no external adjustments that can be made other than making sure that all vacuum lines are in place, connected and not damaged in any way.

The reduction unit is actuated by a shift cable and can be adjusted in the following manner:

1. Loosen the nut which clamps the cable to the shift lever pivot. Be sure that the cable can move freely in the pivot.

2. Move the reduction shift lever to the most rearward detent position (Hi-Range position).

3. Push the Low Range lever inward until it stops. Pull the Low Range lever out slightly, no more than $1/16$ in.

4. Tighten the cable clamp nut at the reduction unit shift lever.

NOTE: This procedure only applies to the Quadra-Trac transfer case equipped with a reduction unit.

MODEL 207 AND 229 RANGE CONTROL LINKAGE ADJUSTMENT

1. Place the range control lever in the 2WD position (high range position for model 229).

2. Insert a $1/8$ inch spacer between the shift gate and the lever.

3. Hold the lever in this position and place the transfer case lever in 2WD (high range position for model 229).

4. Adjust the link at the block, A (207) or B (229), to provide a free pin at the case outer lever.

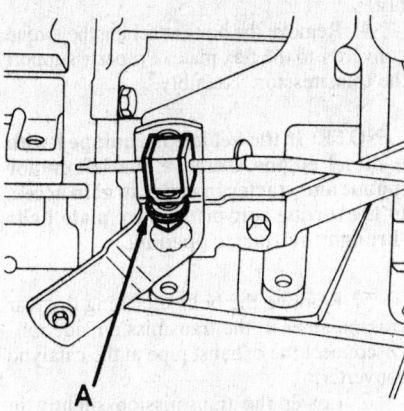

A is the adjusting point on the 207

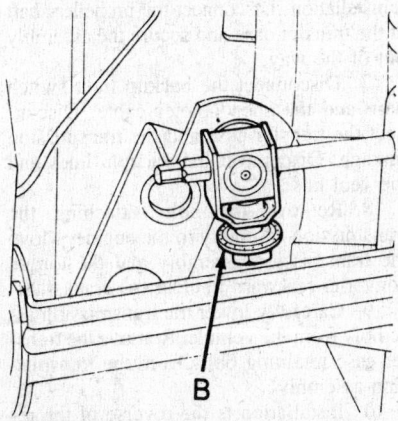

B is the adjusting point on the 229

MANUAL TRANSMISSION

Refer to the Manual Transmission General Repair Section for application, troubleshooting and overhaul.

REMOVAL & INSTALLATION

Except Cherokee, Wagoneer & Comanche

1. Disconnect the negative battery cable. Remove the shift level knobs, trim rings, and bolts.

2. If the vehicle is equipped with the T4 or T5 transmission, Remove the bolts attaching the transmission shift lever housing to the transmission. Remove the lever and the housing.

3. If the vehicle is equipped with the T176 transmission press and turn the transmission shift lever retainer counterclockwise to release the lever. Remove the lever, boot, spring and seat as an assembly.

4. Raise the vehicle and support it safely.

5. Mark the rear propeller shaft and the transfer case yoke for reassembly. Disconnect the rear propeller shaft at the transfer case yoke. Move the shaft aside and secure it out of the way.

6. Support the rear of the engine by placing a support under the clutch housing.

7. Remove the nuts and bolts attaching the rear crossmember to the frame rails and rear support cushion. Remove the rear crossmember from the vehicle.

8. Disconnect the speedometer cable, back-up light switch, four wheel drive indicator switch wire and the transfer case vent hose at the transfer case assembly.

9. Mark the front propeller shaft and the transfer case yoke for reassembly. Disconnect the front propeller shaft from the transfer case yoke. Move the shaft to the side and position it out of the way.

10. On T4, T5 and T176 transmissions except the T176 transmission in the Grand Wagoneer and truck remove the transfer case shift lever as follows. Remove the shift lever retaining nut. Remove the cotter pins that retain the shift control link pins in the shift rods and remove the pins. Remove the shifter shaft and disengage the shift lever from the shift control links. Slide the lever upward in the boot in order to move the lever out of the way.

11. If the vehicle is a Grand Wagoneer or truck equipped with the T176 transmission, remove the cotter pin and washers that connect the link to the shift lever and disconnect the link from the shift lever.

12. Support the transmission and transfer case assembly.

13. Remove the bolts attaching to the clutch housing. Remove the transmission and transfer case assembly. Remove the bolts attaching the transfer case to the transmission and remove the transmission.

14. Installation is the reverse of the removal procedure.

Cherokee, Wagoneer & Comanche

1. Disconnect the negative battery cable. Raise the shift lever boot and remove the upper part of the console.

2. Remove the lower part of the console.

3. Remove the inner boot.

4. Remove the shift lever using tool J34635.

5. Raise and support the vehicle safely. Drain the transmission and transfer case, if equipped. Mark the relation between the rear driveshaft and the transmission output yoke. Remove the driveshaft.

6. If the vehicle is equipped with a transfer case, position a support stand under the transfer case in order to support the weight of both the transmission and the transfer case assembly.

7. If the vehicle is not equipped with a transfer case, position a support under the transmission assemblt in order to support the weight of the transmission.

8. Remove the nuts and bolts attaching the rear crossmember to the frame rails and rear support cushion. Remove the rear crossmember.

8. Disconnect the speedometer cable, backup light switch. If equipped, disconnect the transfer case vent hose.

9. Disconnect the transfer case vacuum hoses and linkage. Remove the clutch slave cylinder.

10. If equipped, mark the front propeller shaft and transfer case yoke for reinstallation. Move the shaft aside and secure it out of the way.

11. Properly support the transmission and the transfer case as required.

12. Remove the transmission-to-engine mounting bolts, move the assembly rearward and lower it from the vehicle.

13. If the vehicle is equipped with a transfer case remove the transfer case to transmission retaining bolts.

14. Installation is the reverse of removal. Make sure that the shaft splines mesh before tightening the mounting bolts. Observe the following torques: Crossmember-to-body: 30 ft. lb. Crossmember-to-transmission: 33 ft. lb. Transmission-to-engine: 28 ft. lb. Transfer case-to-transmission adapter: 26 ft. lb. U-joints-to-yokes: 170 inch lb.

LINKAGE ADJUSTMENTS

The shift lever is connected to the transfer case shift rails through rods and nonadjustable links, therefore external adjustments are not possible.

AUTOMATIC TRANSMISSION

REMOVAL & INSTALLATION

1979

1. Disconnect the negative battery cable. Remove transmission dipstick.

2. If vehicle is equipped with radiator shroud, remove bolts attaching shroud to core support. Raise vehicle and support safety.

3. Mark front and rear universal joints and axle yokes.

4. On Cherokee, Wagoneer, and Truck models, remove parking brake cable jamnut and adjuster nut, remove clip attaching parking brake cable to crossmember and pull out of crossmember.

5. On CJ, and Truck models, with low range reduction unit, disconnect shift rod at reduction unit shift lever and remove reduction unit. On all other models, remove reduction unit shift lever from shift shaft and remove reduction unit.

6. Disconnect speedometer cable. Disconnect and mark emergency drive control vacuum lines and indicator lamp wire. Remove bolt attaching vacuum line routing bracket to rear of transfer case.

7. Disconnect detent solenoid wire at transmission case connector. Remove starter. Remove converter housing inspection cover and mark the torque converter. Remove converter-to-drive plate attaching bolts.

8. Support the transmission and remove the rear crossmember. Disconnect exhaust system components where necessary.

9. Remove spring clip and flat washer attaching transmission gearshift rod trunion to outer range selector lever. Do not loosen trunnion locknut. Disengage gearshift rod and trunnion from outer range selector lever. Remove spring clip and spring attaching outer range selector lever to transmission selector lever. Remove bolts attaching outer range selector lever bracket and bushing to frame and remove bracket, lever, and bushing as an assembly.

10. Disconnect front propeller shaft at transfer case yoke and secure shaft. Disconnect transmission oil cooler lines. Disconnect engine-to-modulator vacuum hose, and remove transmission filler tube.

11. Position a support stand under the engine. Remove the transmission filler tube.

12. Remove the converter housing to engine attaching bolts and move the transmission assembly rearward with a supporting jack until it clears the crankshaft.

13. Hold the converter in place and lower the transmission from the vehicle.

14. The installation of the transmission is in the reverse of the removal procedure. Be sure the converter is properly aligned to the drive plate during installation.

1980 and Later (exc 1984 and Later Cherokee, Wagoneer & Comanche)

1. Disconnect the negative battery cable.

2. Disconnect the fan shroud, if equipped.

3. Disconnect the transmission oil fill tube top bracket.

4. Raise the vehicle and support it safely.

5. Remove the inspection cover from the lower part of the converter housing.

6. Remove the oil filler tube and dipstick.

7. Remove the starter.

8. Mark the driveshafts and yokes for position. Disconnect the driveshafts from the transfer case yokes. Secure the shafts to the frame with wire so they are out of the way.

9. On eight cylinder models, disconnect the exhaust pipes at the exhaust manifolds and remove, if necessary, to gain clearance.

10. Drain the transfer case lubricant and transmission fluid.

11. Disconnect the speedometer cable, gearshift linkage, throttle linkage and the wires to the neutral safety switch.

12. Mark the converter drive plate and converter for location reference.

13. Remove the bolts that attach the converter to the drive plate.

14. Support the transmission-transfer case assembly on a suitable jack. Be sure the transmission assembly is firmly chained or secured for removal.

15. Remove the rear crossmember. Lower the transmission slightly and disconnect the oil cooler lines.

16. Remove the bolts that mount the transmission to the engine.

17. Move the transmission and converter back and away from the engine. Make sure the converter breaks loose from the drive plate and is firmly mounted on the transmission.

18. Hold the converter in position and lower the transmission assembly until the converter housing clears the engine.

19. Remove the transmission and transfer case assembly from under the vehicle.

20. Remove the transfer case.

21. Installation is in the reverse order of removal. Be sure to line up the marks on the converter and drive plate when reinstalling. Transmission mounting bolts are torqued to 28 ft.

1984 and Later Cherokee, Wagoneer & Comanche

2WD

1. Disconnect the negative battery cable. Raise and support the vehicle safely.

2. Mark the rear propeller shaft and yoke for reassembly. Disconnect and remove the rear propeller shaft.

3. Remove the torque converter inspection cover. Mark the converter drive plate and converter assembly for reassembly.

4. Remove the bolts attaching the torque converter to the flex plate. Properly support the transmission assembly.

5. Remove the bolts attaching the rear crossmember to the transmission side rail. Disconnect the exhaust pipe at the catalytic converter.

6. Lower the transmission slightly in order to disconnect the fluid cooler lines.

7. Disconnect the backup light switch wire and the speedometer cable. Disconnect the transmission linkage.

8. Remove the bolts attaching the transmission assembly to the engine. Move the transmission assembly and the torque converter rearward to clear the crankshaft.

9. Carefully lower the transmission assembly from the vehicle.

10. Installation is the reverse of the removal procedure.

4WD

1. Disconnect the negative battery cable. Raise and support the vehicle safely.

2. Mark the rear propeller shaft and yoke for reassembly. Disconnect and remove the rear propeller shaft.

3. Remove the torque converter inspection cover. Mark the converter drive plate and converter assembly for reassembly.

4. Remove the bolts attaching the torque converter to the flex plate. Properly support the transmission assembly.

NOTE: If the vehicle is equipped with a diesel engine, remove the left motor mount and starter in order to gain access to the torque converter drive plate bolts through the starter opening.

5. Remove the bolts attaching the rear crossmember to the transmission side rail. Disconnect the exhaust pipe at the catalytic converter.

6. Lower the transmission slightly in order to disconnect the fluid cooler lines. Mark the front propeller shaft assembly for reinstallation. Disconnect the propeller shaft at the transfer case and secure the assembly out of the way.

7. Disconnect the backup light switch wire and the speedometer cable. Disconnect the transfer case and the transmission linkage. Disconnect the vacuum lines and the vent hose.

8. Remove the bolts attaching the transmission assembly to the engine. Move the transmission assembly and the torque converter rearward to clear the crankshaft.

9. Carefully lower the transmission assembly from the vehicle. Remove the transfer case retaining bolts from the transmission assembly.

10. Installation is the reverse of the removal procedure.

Linkage

ADJUSTMENT

1979

1. Place the steering column gearshift lever in the Neutral position.
2. Raise the vehicle on a hoist.
3. Loosen the locknut on the gearshift rod trunnion just enough to permit movement of the gearshift rod in the trunnion.
4. Place the outer range selector lever at the transmission, fully into the Neutral detent position and tighten the locknut at the trunnion to 9 ft. lbs.
5. Lower the car and operate the steering column gearshift in all ranges. The car should start in Park and Neutral only and the column gearshift lever engage properly in all detent positions.

1980 and Later

1. Raise and safely support the vehicle.
2. Loosen the shift rod trunnion jamnuts.
3. Remove the lockpin that retains the shift rod trunnion to the bell crank. Disengage the trunnion and shift rod at the bell crank.
4. Place the gear shift lever in the Park position and lock the steering column.
5. Move the transmission lever rearward into the Park detent. Be sure the lever is as far rearward as it will go.
6. Check the engagement of the Park detent by trying to rotate the driveshaft (rear wheels must be off of the ground). The shaft will not rotate if the Park detent is engaged.
7. Adjust the trunnion until it will fit in the bell crank arm freely. Tighten the jamnuts. Install the lock pin.
8. Check engine starting in Park and Neutral, be sure it will not start in any other gear.

Band

ADJUSTMENT

1979

No provisions are made for the external band adjustments of this transmission. Only during the assembly, can a different sized pin be installed in the rear band apply system, to compensate for lining wear.

Front Band

1980 AND LATER

The front band adjusting screw is located on the left side of the transmission case just above the manual valve and throttle control levers.

1. Raise and safely support the vehicle.
2. Loosen the adjusting screw locknut and back if off five turns.
3. Check the adjusting screw to make sure it turns freely, lubricate it if necessary.

---

4. Tighten the adjusting screw to 36 inch lbs.
5. Back of the adjusting screw two turns. Tighten the locknut. Do not allow the adjusting screw to turn when tightening the locknut.

Rear Band

1980 AND LATER

NOTE: The transmission oil pan must be removed to gain access to the adjusting screw.

1. Raise and safely support the vehicle.
2. Drain the transmission fluid and remove the oil pan.
3. The adjusting screw is located on the right rear side above the rear side edge of the filter.
4. Loosen the locknut. Tighten the adjusting screw to 41 inch lbs. Back off the adjusting screw four turns on models 904 through 1983 and 7 turns on 904 in 1984 and later vehicles, four turns on 999, and two turns on model 727. Hold the adjusting screw so that it will not turn and tighten the locknut.
5. Install the oil pan and new gasket.
6. Lower the vehicle and fill the transmission to the correct level.

Neutral Switch Adjustment

1979

1. Apply the parking brake.
2. Check and adjust the manual linkage, if necessary.
3. Remove the Neutral switch from the steering column.
4. Place the selector lever in Park and lock the steering column.
5. Move the switch actuating lever until it is aligned with the letter "P" stamped on the back of the switch.
6. Insert a $\frac{3}{32}$ in. drill in the hole located below the letter "N" stamped on the back of the switch.
7. Move the switch actuating lever until it stops against the drill.
8. Position the switch on the steering column, install the attaching screws and remove the drill.
9. Check the operation of the switch. The engine should start in Park and Neutral only. The backup light should glow only in the Reverse position.

1980 AND LATER

The neutral safety switch is located on the side of the transmission by the manual linkage. It is an electrical switch that is thread mounted. The neutral starting section of the switch is contained in the center terminal of the three terminal switch. The other terminals control the backup lights.

1. Raise and support the vehicle safely. Remove the wiring connector from the switch. Test for continuity between the center terminal pin and the transmission case. Continuity should exist only when the transmission control is in Park or Neutral.

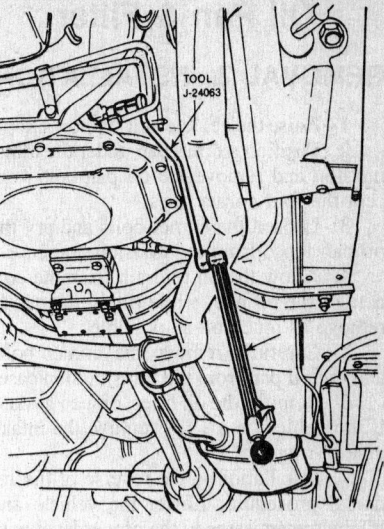

Front band adjustment-TorqueFlite

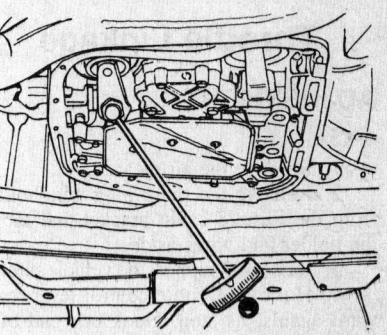

Rear band adjustment-TorqueFlite

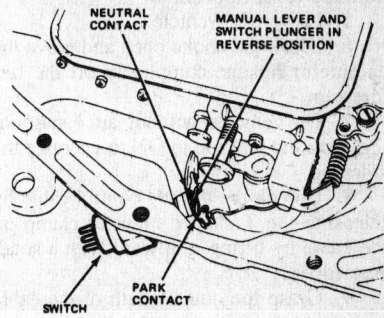

Torqueflite neutral start and backup light switch

2. If test shows that the switch is defective, check the gearshift linkage adjustment before replacing the switch.
3. Remove the switch from the transmission. A certain amount of fluid will leak out when the switch is removed, have a container ready to catch the fluid.
4. Move the gearshift lever to Park and neutral positions. Inspect the switch operating lever fingers and manual lever and shaft for proper alignment with the switch opening in the transmission case.
5. Install a new switch and seal into the transmission case. Tighten to 24 ft. lbs. Test for continuity.
6. Lower the vehicle and correct the transmission fluid level.

Oil Pan & Filter

REMOVAL & INSTALLATION

1. Raise the vehicle and support safely.
2. Position a drain pan under the transmission and remove the oil pan bolts, except the four corner ones.
3. Loosen the corner bolts and pry the oil pan loose from the transmission case.
4. Allow the oil to drain from the corners of the oil pan, while tilting the pan to remove as much oil as possible.
5. Carefully remove the corner bolts and the oil pan from the transmission case.
6. Remove the oil filter, oil pan gasket. If the vehicle is a 1979 remove the intake tube O-ring seal.
7. Installation is the reverse of the removal procedure. Lower the vehicle and fill the transmission to the proper level with automatic transmission fluid.

Throttle Linkage

ADJUSTMENT

4–151

1. Remove the air cleaner.
2. Remove the spark plug wire holder from the throttle cable bracket and move the holder and wires aside.
3. Raise and support the vehicle safely.
4. Hold the throttle control lever rearward against its stop. Hook one end of a spare spring to the lever and hook the oposite end to any convenient point. This will hold the lever in position.
5. Lower the vehicle.
6. Block the choke open and move the carburetor linkage completely off the fast idle cam.
7. On vehicles without air conditioning, turn the ignition to ON to energize the solenoid.
8. Unlock the throttle control cable by releasing the T shaped adjuster clamp on the cable by lifting it upward with a small screwdriver.
9. Grasp the outer sheath of the cable and move the cable and sheath forward to remove any load on the cable bell crank.
10. Adjust the cable by removing the cable and sheath rearward until there is no play at all between the plastic cable and the bell crank ball.
11. When play has been eliminated, lock the cable by pressing the T shaped clamp downward until it snaps into place.
12. Turn the ignition off. Install all parts and remove the spare spring.

4–150 w/Carburetor

1. Disconnect the throttle control rod spring at the carburetor.
2. Raise and support the vehicle safely.
3. Use the throttle control rod spring to hold the transmission throttle control lever forward against its stop.

4. Hook one end of a spring on the throttle control lever and the other end of the spring on the throttle linkage bellcrank bracket, which is attached to the torque converter housing.
5. Lower the vehicle. Block the choke in the open position. Set the carburetor throttle off the fast idle cam.
6. Turn the ignition key to the "ON" position in order to energize the solenoid.
7. Open the throttle halfway to allow the solenoid to lock. Return the carburetor to the idle position.
8. Loosen the retaining bolt on the throttle control adjusting link. Do not remove the spring clip and nylon washer.
9. Pull the end of the link to eliminate lash. Tighten the retaining bolt. Turn the ignition to the "OFF" position.
10. Raise and support the vehicle safely. Remove the throttle control rod spring from the linkage. Lower the vehicle. Install the spring on the throttle control rod.

4–150 w/Fuel Injection

NOTE: An idle speed assembly exerciser box is required in order to make this adjustment. The purpose of this special tool is to by-pass the idle speed motor.

1. Be sure that the vehicle ignition key is in the "OFF" position. Raise and support the vehicle safely.
2. Hook one end of a spring on the throttle control lever and the other end of the spring on the throttle linkage bellcrank bracket, which is attached to the torque converter housing. Lower the vehicle.
3. Disconnect the idle speed actuator motor wire harness and connect the idle speed assembly exerciser box. Upon connection, the adjustment light should turn off and the ready light should turn on.
4. Loosen the retaining bolt on the throttle control adjusting link. Pull the end of the link in order to eliminate lash and tighten the link retaining bolt.
5. Press the extend button on the idle speed assembly exerciser box until the idle speed actuator motor ratchets.
6. Disconnect the idle speed assembly exerciser box and reconnect the idle speed actuator motor wiring harness.
7. Raise and safely support the vehicle. Remove the spring from the linkage. Lower the vehicle.

Diesel

NOTE: Special tool J35514, throttle valve lever and cable adjusting tool, will be required in order to complete this procedure.

1. Disconnect the throttle valve cable from the pin on the throttle valve lever.
2. Disconnect the cable from the transmission throttle lever. Remove and discard the cable.
3. Set the injection pump automatic advance lever at the curb idle position. Loosen

the screw and turn the cable clevis one-quater of a turn. The lever should be seated against the stop in the curb idle position.
4. Install the throttle valve lever and cable adjusting tool. Loosen the tool thumbscrew and move the sliding legs rearward. Position the notched leg on the cable bracket. Rest the legs on the lever. Move the legs forward until the rear leg lightly touches the cable attaching pin. Retighten the thumbscrew.
5. Check the throttle valve lever travel by moving the lever to the wide open throttle position. The cable attaching pin should now lightly touch the forward leg of the sliding legs on the special tool.
6. If the pin does not touch the forward leg of the tool, or if the pin tends to move the tool too much the throttle lever travel must be adjusted.
7. To adjust the throttle lever travel, hold the lever in the wide open throttle position. Loosen the lever screws and move the lever in or out in order to adjust the travel. The pin should lightly touch the forward leg of the tool when the lever travel is adjusted properly at wide open throttle. Tighten the screws to 66 inch lbs. Return the lever to the curb idle position and verify that the pin is again lightly touching the forward leg of the tool.

NOTE: If the pin does not lightly touch the rear leg at the curb idle position, loosen the thumbscrew. Re set the rear sliding leg of the tool so that it touches the pin. Repeat the above steps until correct lever travel is obtained.

8. Install a new throttle valve cable.

6–173

1. Remove the air cleaner assembly. Raise and support the vehicle safely.
2. Hold the throttle control lever rearward against its stop. Use a spring to hold the lever. Hook one end of the spring to the lever and the other end to a convenient mounting point.
3. Lower the vehicle. Block the choke open and set the carburetor linkage off of the fast idle cam.
4. Unlock the throttle control cable by releasing the T-shaped cable adjuster clamp. Release the clamp by lifting it upward using a suitable tool.
5. Grasp the cable outer sheath and move the cable and sheath forward, this will remove any cable load on the throttle cable bellcrank.
6. Adjust the cable by moving the cable and sheath rearward until there is zero lash between the plastic cable end and the bell-crank ball.
7. When this has been accomplished, lock the cable by pressing the T-shaped adjuster clamp downward.
8. Install the air cleaner assembly. Raise and support the vehicle safely. Remove the spring from its mounting. Lower the vehicle.

6-258

1. Disconnect the throttle control rod spring at the carburetor.

2. Raise and support the vehicle safely.

3. Use the throttle control rod spring to hold the throttle control lever forward against its stop, by hooking one end of the spring on the throttle control lever and the other end on the throttle linkage bell crank bracket which is attached to the transmission housing.

4. Block the choke plate open and move the throttle linkage off the fast idle cam.

5. On carburetors equipped with a throttle operated solenoid valve, turn the ignition ON to energize the solenoid, then open the throttle halfway to allow the solenoid to lock and return the carburetor to the idle position.

6. Loosen the retaining bolt on the throttle control adjusting link. Do not remover the spring clip and nylon washer.

7. Pull on the end of the link to eliminate play and tighten the retaining bolt.

8. Remove the throttle control rod spring and install it on the control rod from where it came.

9. Lower the vehicle.

V8

1. Disconnect the throttle control rod spring at the carburetor.

2. Raise and support the vehicle safely.

3. Use the throttle control rod spring to hold the transmission throttle valve control lever against its stop.

4. Block the choke plate open and make sure the throttle linkage is off the fast idle cam.

NOTE: On carburetors equipped with a throttle operated solenoid valve, turn the ignition to ON to energize the solenoid. Then turn the throttle halfway to allow the solenoid to lock and return the carburetor to idle.

5. Loosen the retaining bolt on the throttle control rod adjuster link. Remove the spring clip and move the nylon washer to the rear of the link.

6. Push on the end of the link to eliminate play and tighten the link retaining bolt.

7. Install the nylon washer and spring clip.

8. Remove the throttle control rod spring and install it in its intended position.

9. Lower the vehicle.

FRONT DRIVE AXLE

Front Locking Hubs

Front drive hubs are serviced as either a complete assembly or a sub assembly, such as the hub body or hub clutch assembly.

Do not attempt to disassemble these units. If the entire hub assembly or subassembly has to be replaced it must be replaced as a complete unit.

REMOVAL & INSTALLATION

CJ Models

1. Remove the bolts and the tabbed lockwashers that attach the hub body to the axle hub. Save the bolts and the washer.

2. Remove the hub body and gasket. Discard the gasket.

3. Do not turn the hub control dial once the hub body has been removed.

4. Remove the retaining ring from the axle shaft. Remove the hub clutch and bearing assembly.

5. Clean and inspect the components for wear and damage. Replace defective components as required.

6. Installation is the reverse of the removal procedure.

7. Turn the control hub dials to the 4 X 2 position and rotate the wheels. They should rotate freely, if not check the hub installation. Be sure that the controls are in the fully engaged position.

Cherokee, Grand Wagoneer & Truck

1. Remove the socket head screws from the hub body assembly.

2. Remove the large retaining ring from the axle hub. Remove the small retaining ring from the axle shaft.

3. Remove the hub and clutch assembly.

4. Clean and inspect the components for wear and damage. Replace defective components as required.

5. Installation is the reverse of the removal procedure.

6. Turn both controls to the FREE position and rotate the wheels. They must rotate freely, if they drag check the hub installation. Be sure that the control dials are in the fully engaged position.

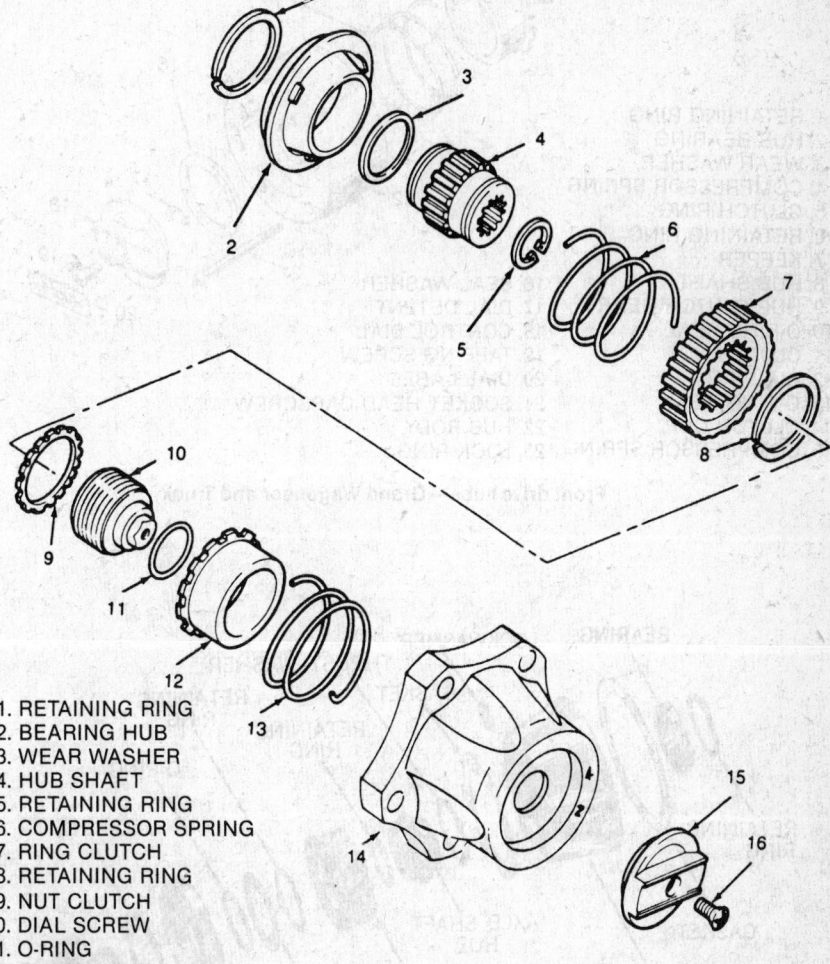

1. RETAINING RING
2. BEARING HUB
3. WEAR WASHER
4. HUB SHAFT
5. RETAINING RING
6. COMPRESSOR SPRING
7. RING CLUTCH
8. RETAINING RING
9. NUT CLUTCH
10. DIAL SCREW
11. O-RING
12. CLUTCH CUP
13. COMPRESSOR SPRING
14. HUB
15. CONTROL DIAL
16. SCREW

Front drive hubs—CJ and Scrambler

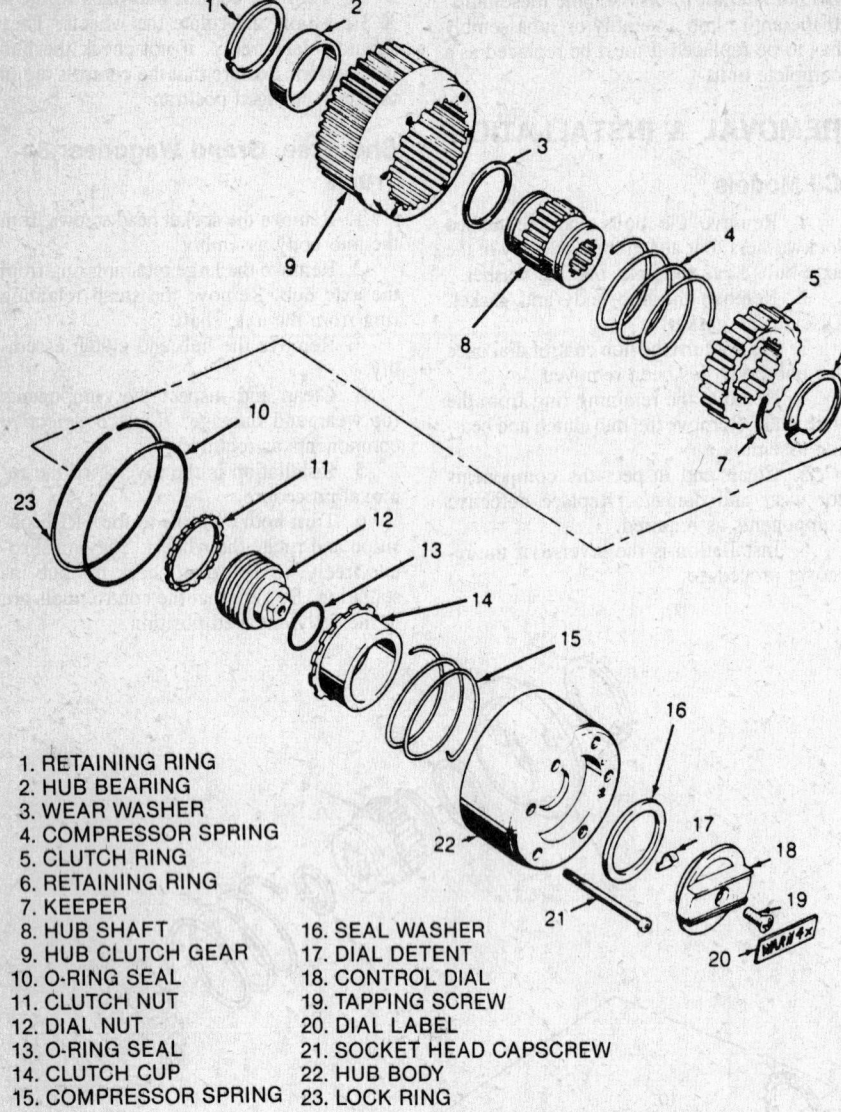

1. RETAINING RING
2. HUB BEARING
3. WEAR WASHER
4. COMPRESSOR SPRING
5. CLUTCH RING
6. RETAINING RING
7. KEEPER
8. HUB SHAFT
9. HUB CLUTCH GEAR
10. O-RING SEAL
11. CLUTCH NUT
12. DIAL NUT
13. O-RING SEAL
14. CLUTCH CUP
15. COMPRESSOR SPRING
16. SEAL WASHER
17. DIAL DETENT
18. CONTROL DIAL
19. TAPPING SCREW
20. DIAL LABEL
21. SOCKET HEAD CAPSCREW
22. HUB BODY
23. LOCK RING

Front drive hubs—Grand Wagoneer and Truck

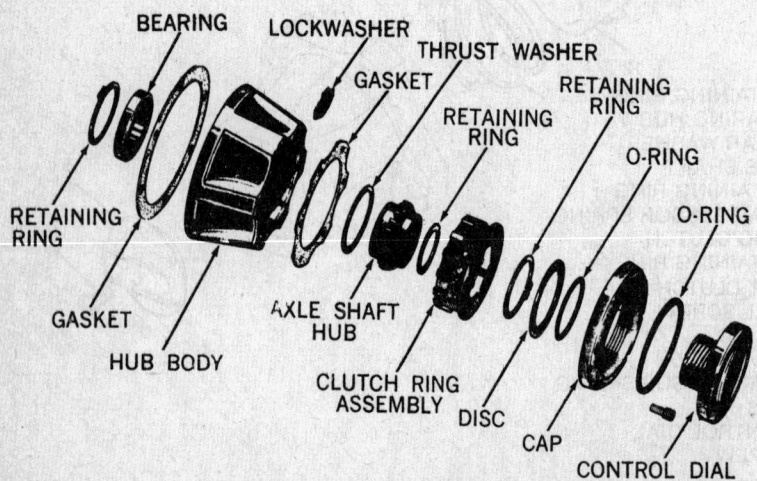

Exploded view of a manual front hub

Axle Shaft

REMOVAL & INSTALLATION

CJ and Scrambler

1. Raise and support the vehicle safely. Remove the tire and wheel. Remove the disc brake caliper.

2. Remove the bolts attaching the front hub to the axle and remove the hub body and gasket.

3. Remove the retaining ring from the axle shaft. Remove the hub clutch and bearing assembly from the axle.

4. Straighten the lip of the lock washer. Remove the outer lock nut, lock washer, inner locknut and tabbed washer. Use tool J25103, or equivalent, in order to remove the locknut.

5. Remove the outer bearing and remove the disc brake rotor. Remove the disc brake caliper adapter and splash shield. Remove the axle spindle.

6. Remove the axle shaft and universal joint assembly.

7. Installation is the reverse of the removal procedure.

Cherokee, Grand Wagoneer & Truck

1. Raise and support the vehicle safely. Remove the disc brake caliper.

2. On vehicles without front hubs, remove the rotor hub cap. Remove the axle shaft snap ring, drive gear, pressure spring and spring retainer.

3. On models with front hubs, remove the socket head screws from the hub body. Remove the hub body and the large retaining ring. Remove the small retaining ring from the axle shaft. Remove the hub clutch assembly from the axle.

4. Remove the outer locknut, washer and inner locknut using tool J6893-03 or equivalent.

5. Remove the rotor. The spring retainer and the outer bearing are removed with the rotor.

6. Remove the nuts and bolts attaching the spindle and support shield. Remove the spindle and the support shield.

7. Remove the axle shaft.

8. Installation is the reverse of the removal procedure.

1984 and Later Wagoneer, Cherokee & Comanche

1. Raise and support the vehicle safely.

2. Remove the wheels, calipers and rotors.

3. Remove the cotter pin, locknut and axle hub nut.

4. Remove the hub-to-knuckle attaching bolts.

5. Remove the hub and splash shield from the steering knuckle.

To remove the left shaft:

6. Remove the axle shaft from the housing.

To remove the right shaft:

7. Disconnect the vacuum harness from the shift motor.

8. Remove the shift motor from the housing.

9. Remove the axle shaft from the housing.

10. To install the right axle shaft first be sure that the shift collar is in position on the intermediate shaft and that the axle shaft is fully engaged in the intermediate shaft end.

11. Install the shift motor, making sure that the fork engages with the collar. Tighten the bolts to 8 ft. lb.

12. On the left side, install the axle shaft in the housing.

13. Partially fill the hub cavity of the knuckle with chassis lube and install the hub and splash shield.

14. Tighten the hub bolts to 75 ft. lb.

15. Install the hub washer and nut. Torque the nut to 175 ft. lb. Install the locknut. Install a new cotter pin.

16. Install the rotor, caliper and wheel.

Wheel Bearing

ADJUSTMENT

CJ and Scrambler

1. Raise and support the vehicle safely.

2. Remove the bolts attaching the front hub to the hub rotor. Remove the hub body and gasket.

3. Remove the snap ring from the axle shaft and remove the hub clutch assembly.

4. Straighten the lip of the outer lock nut tabbed washer. Remove the outer locknut and tabbed washer.

5. Loosen and then tighten the inner locknut to 50 ft. lbs. Rotate the wheel while tightening the nut to seat the bearing properly.

6. Back off the inner locknut about one-sixth of a turn as you are turning the wheel. The wheel must rotate freely.

7. Install the tabbed washer and the outer locknut.

8. Torque the outer locknut to 50 ft. lbs.

9. Recheck the bearing adjustment. The wheel must rotate freely. Correct as required.

10. Install components as they were removed.

Cherokee, Wagoneer & Truck

1. Raise and support the vehicle safely.

2. If the vehicle is not equipped with front hubs, Remove the wheel cover and hubcap. Remove the drive gear snap ring. Remove the drive gear, pressure spring and spring cup.

3. If the vehicle is equipped with front hubs, Remove the socket head screws from the hub body and remove the body from the hub clutch assembly. Remove the large retaining ring from the hub. Remove the small retaining ring from the axle shaft. Remove the hub and clutch assembly. Remove the outer locknut and lock washer.

4. Seat the bearings by loosening and then tightening them to 50 ft. lbs. Back off the inner locknut about one-sixth of a turn while rotating the wheel.

5. Install the lock washer. Align one of the lock washer holes with the peg on the inner locknut and install the washer on the nut.

6. Install and torque the outer locknut to 50 ft. lbs. Recheck the bearing adjustment. Correct as required.

7. On vehicles without front hubs, The spring cup must be installed so the recessed side faces the bearing and the flat side faces the pressure spring. The pressure spring should contact the flat side of the cup only.

8. Install the spring cup and the pressure spring. Install the drive gear and the drive gear snap ring.

9. If the vehicle is equipped with front hubs, Install the clutch assembly. Install the small retaining ring on the axle shaft. Install the large retaining ring on the hub.

10. Install the hub body on the hub clutch. Install the socket head screws in the hub and torque tham to 30 inch lbs.

11. Lower the vehicle.

1984 and Later Cherokee, Wagoneer & Comanche

TYPE ONE

1. Raise and support the vehicle safely. Remove the wheel. Remove the disc brake caliper as required. Remove the cotter pin, locknut and axle hub nut.

2. Tighten the hub bolts to 75 ft. lbs.

3. Install the hub washer and nut and tighten the hub nut to 175 ft. lbs. Install the locknut and new cotter pins.

4. Install the caliper, if removed. Install the wheel. Lower the vehicle.

TYPE TWO

1. Raise and support the vehicle safely. Remove the wheel. Remove the caliper, as required.

2. Remove the grease cap, cotter pin and nut retainer.

3. Tighten the spindle nut to 25 ft. lbs. while rotating the rotor.

4. Loosen the spindle nut about one half turn while rotating the wheel. Torque the nut to 19 ft. lbs.

5. Install removed components.

Steering Knuckle

REMOVAL & INSTALLATION

EXCEPT 1984 AND LATER CHEROKEE, WAGONEER & COMANCHE

1. Remove the axle assembly from the vehicle.

2. Disconnect the tie rod end at the steering knuckle arm.

3. Remove and discard the lower ball stud jamnut.

4. Remove the cotter pin from the upper ball stud. Loosen the stud nut until the top edge of the nut is flush with the top of the stud.

5. Unseat the upper and lower ball studs using a hammer. Remove the upper ball stud nut and the steering knuckle.

6. Remove the upper ball stud split ring seat using tool J23447 or tool J25158.

7. Installation is the reverse of the removal procedure.

1984 AND LATER WAGONEER, CHEROKEE & COMANCHE

1. Remove the outer axle shaft.

2. Remove the caliper anchor plate from the knuckle.

3. Remove the knuckle-to-ball joint cotter pins and nuts.

4. Drive the knuckle out with a brass hammer.

NOTE: A split ring seat (3) is located in the bottom of the knuckle. During installation, this ring seat must be set to a depth of 5.23mm (.206). Measure the depth to the top of the ring seat (4).

5. Installation is the reverse of removal. Tighten the knckle retaining nuts to 75 ft. lb. and the caliper anchor bolts to 77 ft. lb.

Typical front hub and rotor assembly

INNER BEARING
BEARING CUP
HUB AND ROTOR ASSEMBLY
OUTER BEARING
BEARING CUP
PRESSURE SPRING
SPRING CUP
DRIVE GEAR
OIL SEAL
STUD
SNAP RING
INNER LOCKNUT
LOCK WASHER
OUTER LOCKNUT
HUB CAP

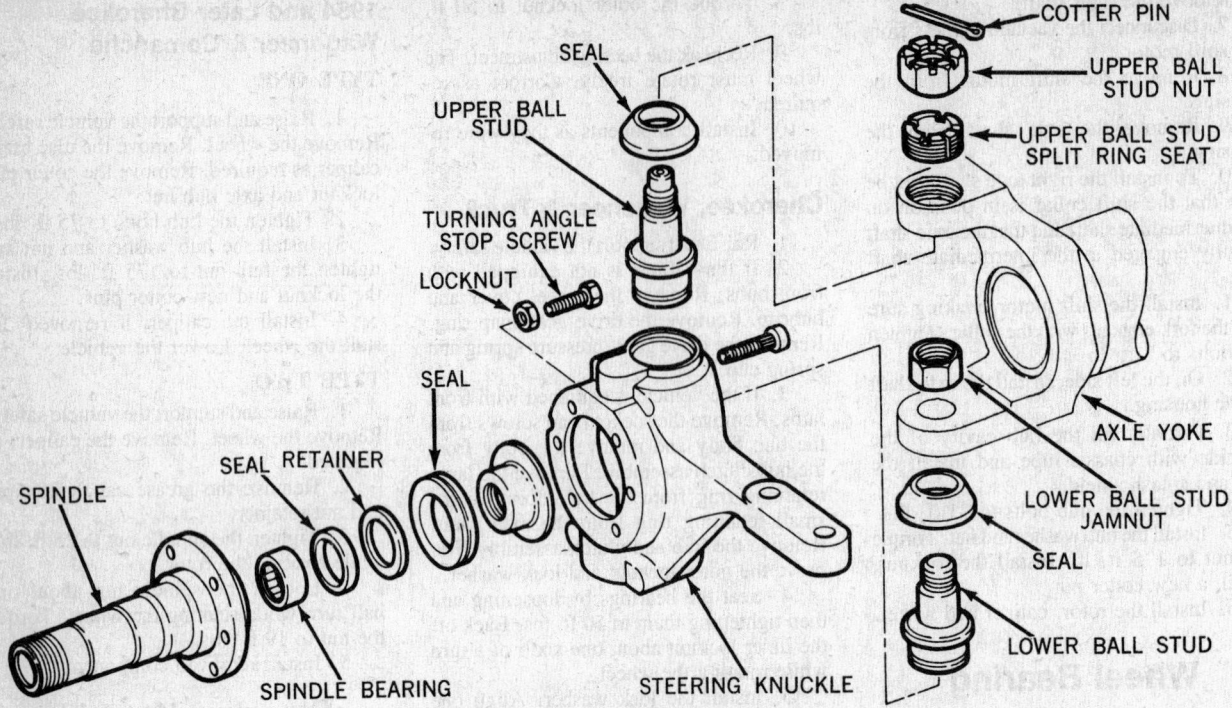

CUTTER PIN

UPPER BALL STUD NUT

UPPER BALL STUD SPLIT RING SEAT

SEAL

UPPER BALL STUD

TURNING ANGLE STOP SCREW

LOCKNUT

SEAL

SEAL RETAINER

SPINDLE

SPINDLE BEARING

STEERING KNUCKLE

AXLE YOKE

LOWER BALL STUD JAMNUT

SEAL

LOWER BALL STUD

Steering knuckle components-typical

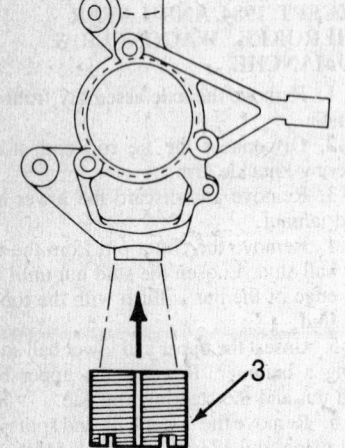

Split retaining ring

Ball Joints

REMOVAL & INSTALLATION

CJ, Scrambler, Cherokee, Grand Wagoneer & Truck

LOWER BALL STUD

1. Remove the steering knuckle from

the vehicle. Position the assembly in a suitable holding fixture with the upper ball stud pointing downward.

2. Attach tool J2511–1, or equivalent to the spindle mating surface of the knuckle assembly. Position tool J25211–3 on the lower ball stud.

3. Assemble and install the puller on the steering knuckle. Hook one arm of the puller in the plate of tool J25211–1 and the opposite arm of the tool in the steering knuckle.

4. Tighten the puller screw to press the lower stud out of the knuckle.

5. Remove the tools from the knuckle.

6. Installation is the reversde of the removal procedure.

CJ & Scrambler

UPPER BALL STUD

1. Remove the steering knuckle from the vehicle. Position the assembly in a suitable holding fixture with the upper ball stud pointing downward.

2. Remove both arms from tool J25215. Place button J25211–3 on the upper ball stud.

3. Install adapter tool J25211–4 on the nut end of the puller screw so that the adapter shoulder faces the nut end of the screw.

4. Insert the nut end of the puller screw through the upper ball stud hole in the knuckle. Hold the adapter and the frame against the knuckle.

5. Remove the lower ball stud from the knuckle.

6. Installation is the reverse of the removal procedure.

Cherokee, Grand Wagoneer & Truck

UPPER BALL STUD

1. Remove the steering knuckle from the vehicle. Position the assembly in a suitable holding fixture with the upper ball stud pointing downward.

2. Remove both arms from tool J25215. Place button J25211–3 on the upper ball stud.

3. Thread the puller frame halfway onto the puller screw. Insert the nut end of the screw through the lower ball stud hole in the steering knuckle. Position the puller frame against the knuckle and the puller screw against tool J25211–3.

4. Tighten the puller screw and press the upper ball stud out of the steering knuckle.

5. Installation is the reverse of the removal procedure.

1984 and Later Wagoneer, Cherokee & Comanche

UPPER BALL JOINT

1. Remove the steering knuckle from the vehicle.

2. Position the receiver tool (1) over the top of the ball joint. Set the adapter tool (2) in a C-clamp (3) and position the clamp so that tightening the clamp screw will remove the ball joint.

3. To install, use the same C-clamp, with adapter tool as illustrated.

LOWER BALL JOINT

1. With the knuckle removed from the vehicle, use a C-clamp, with receiver tool (6) and adapter tool (7) as illustrated, to force out the ball joint.

2. To install, use the same C-clamp, with adapter tools (8 and 9), as illustrated.

Upper Ball Joint

ADJUSTMENT

Except 1984 and Later Wagoneer, Cherokee & Comanche

Adjustment of the upper ball joint is necessary only when there is excessive play in the steering, persistent loosening of the steering linkage, or abnormal wear of the tires.

1. Raise and support the vehicle safely. Remove the front tires.

2. If the vehicle is equipped with a steering damper, disconnect it at the tie rod and move it aside.

3. Unlock the steering column. Disconnect the steering connecting rod. Disconnect the connecting rod at the right side of the tie rod.

4. Remove the cotter pin and the retaining nut attaching the tie rod to the right side steering knuckle.

5. Rotate both steering knuckles through a complete arc several times. Work from the right side of the vehicle when rotating the knuckles.

6. Install a torque wrench on the tie rod retaining nut and check the torque. The torque wrench must be positioned at a ninety degree angle to the steering knuckle arm in order to obtain a correct reading.

7. Rotate the steering knuckles slowly through a complete arc and measure the torque required to rotate the knuckles.

8. If the reading is less than 25 ft. lbs. turning effort is within specification. If not then procede as follows.

9. Disconnect the tie rod from both steering knuckles. Install a one half by one inch bolt, flat washer and nut in the tie rod stud mounting hole in one of the steering knuckles. Tighten the bolt and nut.

10. Install the torque wrench according to Step 6.

11 Rotate the steering knuckles slowly through a complete arc and measure the torque required to rotate the knuckles.

12 If the reading is less than 10 ft. lbs. turning effort is within specification and the defect is not related to the knuckle ball studs.

13. Install components that were removed to accomplish this procedure.14.

Lower the vehicle.

Front Axle Assembly

REMOVAL & INSTALLATION

Except 1984 and Later Wagoneer, Cherokee & Comanche

1. Raise and support the vehicle safely. Remove the wheel covers and wheels.

2. Index the propeller shaft to the differential yoke for the proper alignment upon installation. Disconnect the propeller shaft at the axle yoke and secure the shaft to the frame rail.

3. Disconnect the steering linkage from the steering knuckles. Disconnect the shock absorbers at the axle housing.

4. If the vehicle is equipped with a stabilizer bar, remove the nuts attaching the stabilizer bar connecting links to the spring tie plates.

5. On vehicles equipped with sway bar, remove nuts attaching sway bar connecting links to spring tie plates.

6. Disconnect the breather tube from the axle housing. Disconnect the stabilizer bar link bolts at the spring clips.

7. Remove the brake calipers, hub and rotor, and the brake shield.

8. Remove the U-bolts and the tie plates.

9. Support the assembly on a jack and loosen the nuts securing the rear shackles, but do not remove the bolts.

10. Remove the front spring shackle bolts. Lower the springs to the floor.

11. Pull the jack and axle housing from underneath the vehicle.

12. Installation is the reverse of the removal procedure. Check the front end alignment, as required.

1984 and Later Wagoneer, Cherokee & Comanche

1. Raise and support the vehicle safely.

2. Remove the wheels, calipers and rotors.

3. Disconnect all vacuum hoses at the axle.

4. Mark the relation between the front driveshaft and yoke.

5. Disconnect the stabilizer bar, rod and center link, front driveshaft, shock absorbers, steering damper, track bar.

6. Place a floor jack under the axle to take up the weight.

7. Disconnect the upper and lower control arms at the axle and lower the axle from the truck.

8. Installation is the reverse of the removal procedure.

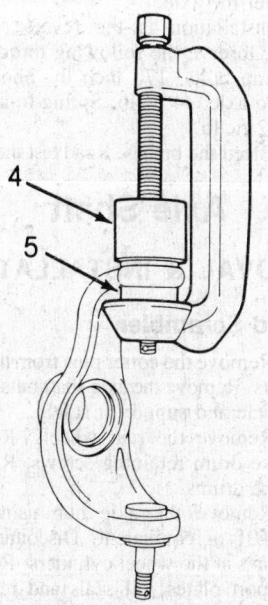

Upper ball joint removal

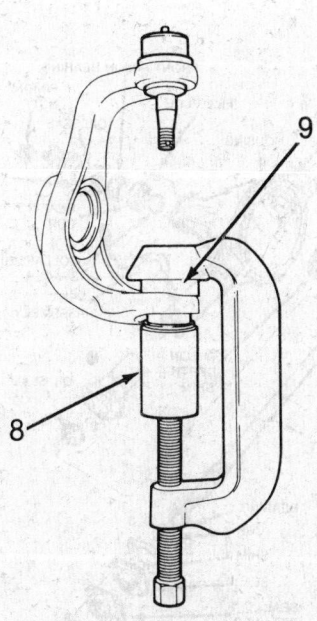

Lower ball joint removal

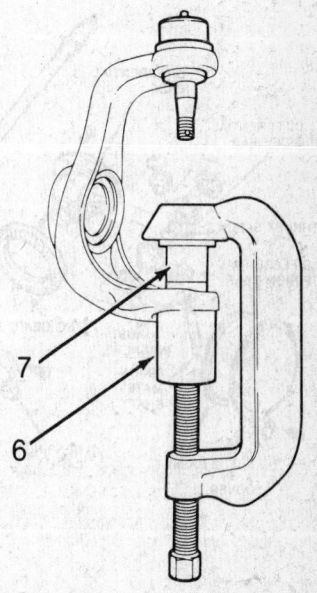

Lower ball joint installation

NOTE: Discard the U-joint straps new replacement straps must be used whenever the straps are removed.

Driveshaft

REMOVAL & INSTALLATION

In order to remove the front and rear driveshafts, first match-mark the driveshaft and yoke for reference, unscrew the attaching nuts from the universal joint's U-bolts, remove the U-bolts and slide the shaft forward or backward toward the slip-joint. The shaft can then be removed from the end yokes and removed from under the vehicle. Install the driveshaft in the reverse order.

NOTE: Some driveshafts are marked at the slip-joints with arrows on the spline and sleeve yoke. When installing the driveshaft, align the arrows to have the yokes at the front and rear of the shaft in the same parallel plane.

Refer to the U–Joint/CV–Joint section of Unit Repair for servicing procedures.

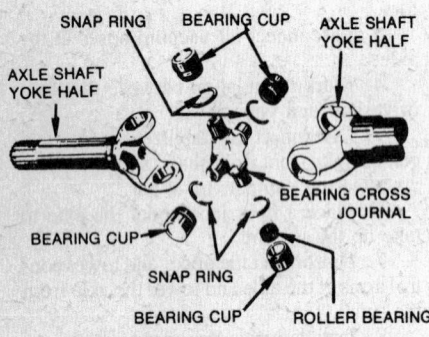

Exploded view of typical U-joint

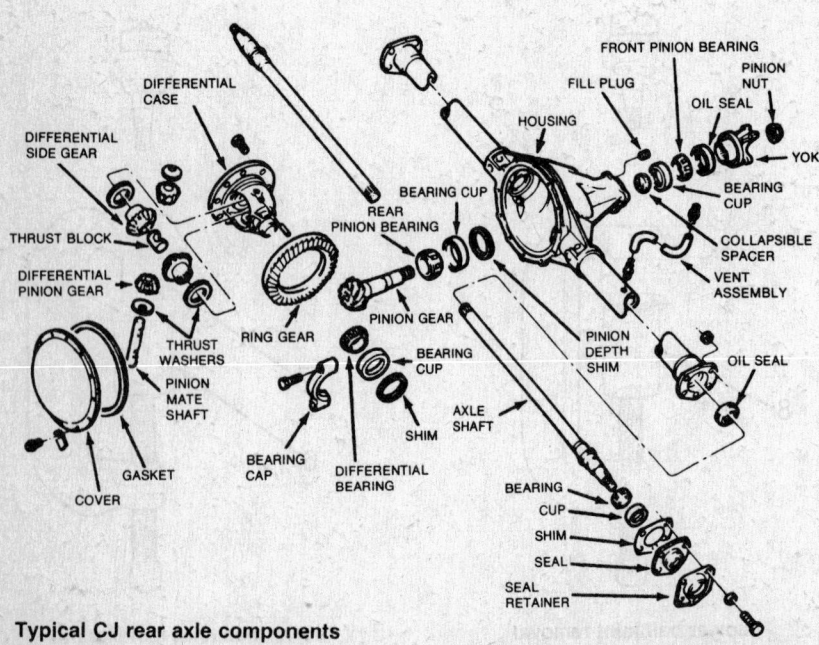

Typical CJ rear axle components

REAR AXLE

Refer to the Drive Axle General Repair Section for application, troubleshooting, and overhaul procedures.

Rear Axle Assembly

REMOVAL & INSTALLATION

CJ & Scrambler

1. Remove the cotter pins from the axle shaft nuts. Remove the axle shaft nuts. Raise the vehicle and support it safely.
2. Remove the rear wheels. Remove the brake drum retaining screws. Remove the brake drums.
3. Remove the axle hub using tool J25109–01 or equivalent. Disconnect the brake lines at the wheel cylinders. Remove the support plates, oil seals and retainers and the end play shims.
4. Axle shaft end play shims are installed on the left side of the axle only.
5. Remove the axle shafts using tool J2498 or equivalent. Drain the lubricant and remove the axle housing cover.
6. Disconnect the parking brake cables at the equalizer. Mark the propeller shaft for reinstallation. Disconnect the shaft at the axle yoke.
7. Disconnect the flexible brake hose at the body floorpan bracket. Disconnect the vent hose at the axle tube.
8. Support the rear axle assembly. Remove the spring U-bolts, spring plates and spring clip plate if the vehicle is equipped with a stabilizer bar.
9. Remove the rear axle from the vehicle.

10. Installation is the reverse of the removal procedure.

Cherokee, Grand Wagoneer & Truck

1. Raise the vehicle and support it safely.
2. Remove the rear wheels.
3. Place an indexing mark on the rear yoke and propeller shaft, and disconnect the shaft.
4. Disconnect the shock absorbers from the axle tubes.
5. Disconnect the brake hose from the tee fitting on the axle housing.
6. Disconnect the parking brake cable at the frame mounting.
7. Remove U-bolts. On vehicles with spring mounted above axle, disconnect spring at rear shackle.
8. Support the axle on a jack, remove the spring clips, and remove the axle assembly from under the vehicle.
9. Installation is the reverse of removal.

NOTE: Bleed and adjust brakes accordingly.

1984 and Later Cherokee, Wagoneer & Comanche

1. Raise and support the vehicle safely.
2. Remove the wheels and brake drums.
3. Disconnect the shock absorbers.
4. Disconnect the brake hose at the frame rail.
5. Disconnect the parking brake cables at the equalizer.
6. Mark the relation between the driveshaft and yoke, and disconnect the driveshaft.
7. Place a floor jack under the axle to take up the weight.
8. Remove the axle-to-spring U-bolts and lower the axle.
9. Installation is the reverse of removal. Observe the following torques: U-joint strap bolts: 170 inch lb. Shock absorber-to-axle: 44 ft. lb. Spring-to-axle U-bolts: 52 ft. lb.
10. Bleed the brakes. Road test the truck.

Axle Shaft

REMOVAL & INSTALLATION

CJ and Scrambler

1. Remove the cotter pins from the axle shaft nuts. Remove the axle shaft nuts. Raise the vehicle and support it safely.
2. Remove the rear wheels. Remove the brake drum retaining screws. Remove the brake drums.
3. Remove the axle hub using tool J25109–01 or equivalent. Disconnect the brake lines at the wheel cylinders. Remove the support plates, oil seals and retainers and the end play shims.
4. Axle shaft end play shims are installed on the left side of the axle only.

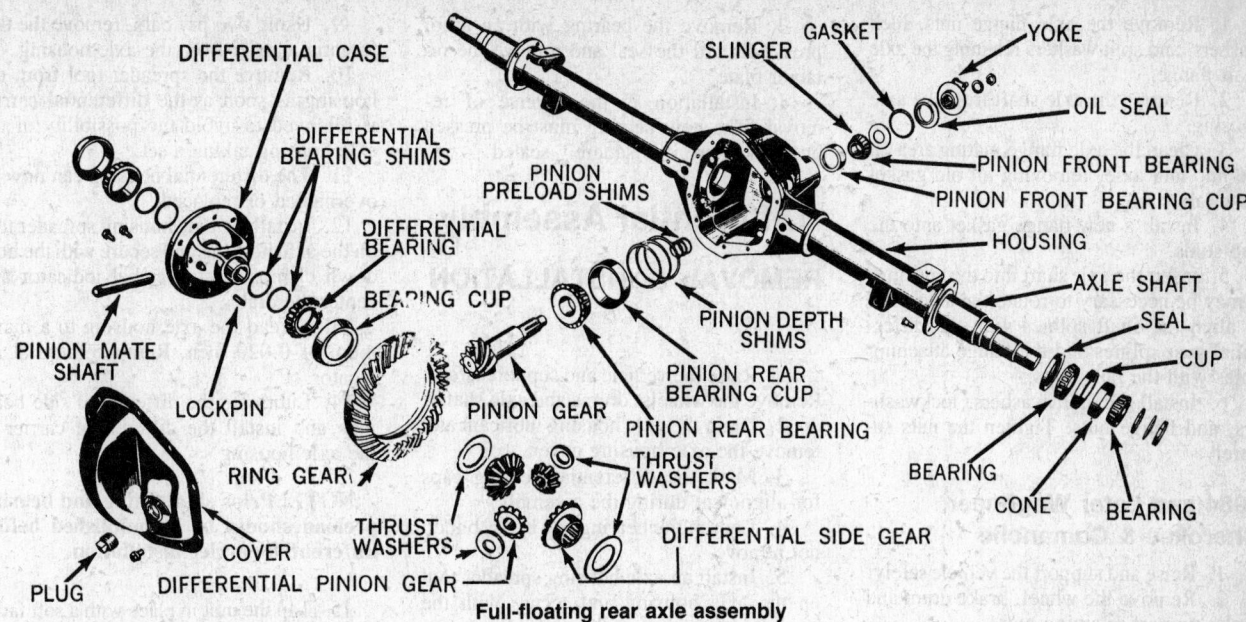

DIFFERENTIAL CASE

DIFFERENTIAL BEARING SHIMS

DIFFERENTIAL BEARING

BEARING CUP

PINION MATE SHAFT

LOCKPIN

RING GEAR

COVER

PLUG

PINION PRELOAD SHIMS

PINION GEAR

THRUST WASHERS

DIFFERENTIAL PINION GEAR

PINION DEPTH SHIMS

PINION REAR BEARING CUP

PINION REAR BEARING

THRUST WASHERS

DIFFERENTIAL SIDE GEAR

SLINGER GASKET YOKE

OIL SEAL

PINION FRONT BEARING

PINION FRONT BEARING CUP

HOUSING

AXLE SHAFT

SEAL

CUP

BEARING

CONE BEARING

Full-floating rear axle assembly

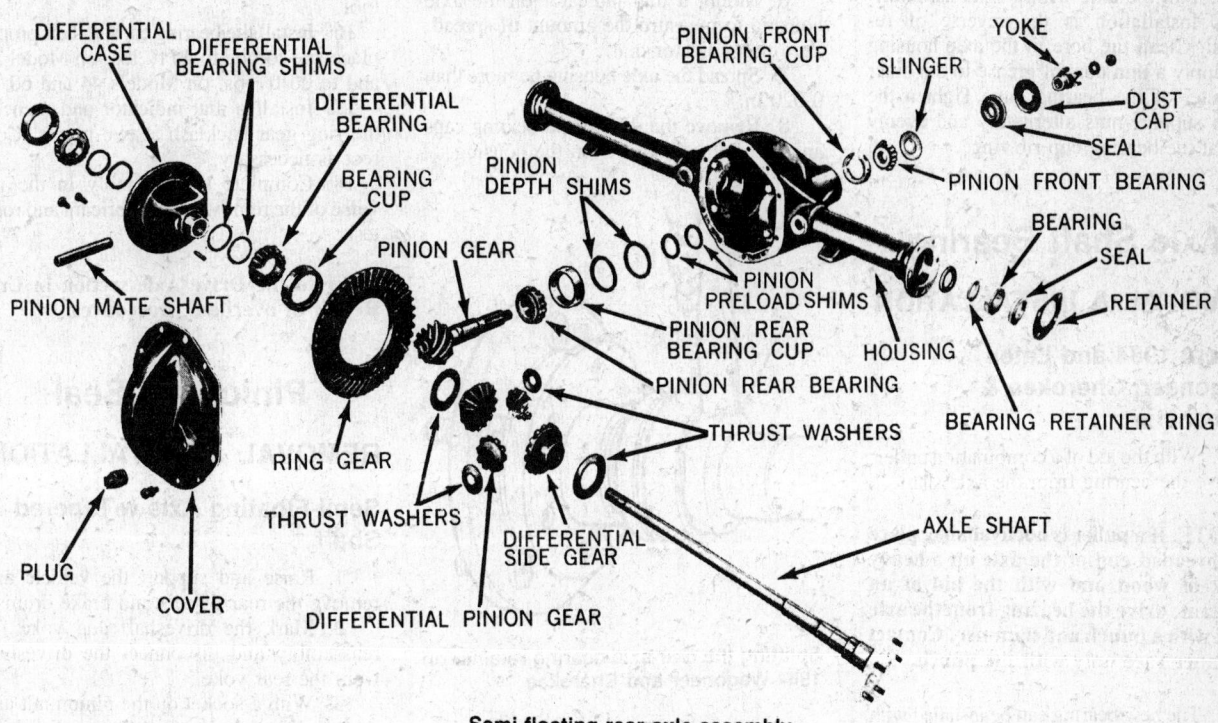

DIFFERENTIAL CASE

DIFFERENTIAL BEARING SHIMS

DIFFERENTIAL BEARING

BEARING CUP

PINION MATE SHAFT

PINION DEPTH SHIMS

PINION GEAR

RING GEAR

THRUST WASHERS

PLUG

COVER

DIFFERENTIAL PINION GEAR

DIFFERENTIAL SIDE GEAR

PINION FRONT BEARING CUP

SLINGER

PINION PRELOAD SHIMS

PINION REAR BEARING CUP

PINION REAR BEARING

THRUST WASHERS

HOUSING

AXLE SHAFT

YOKE

DUST CAP

SEAL

PINION FRONT BEARING

BEARING

SEAL

RETAINER

BEARING RETAINER RING

Semi-floating rear axle assembly

5. Remove the axle shafts using tool J2498 or equivalent.

6. Installation is the reverse of the removal procedure.

NOTE: On vehicles equipped with a Trac Loc differential, do not rotate the differential gears unless both axles are in position. If one shaft is removed and the other shaft rotated, the side gear splines will become misaligned and prevent installation of the replacement shaft.

Cherokee, Grand Wagoneer & Truck

1. Raise and support the vehicle safely. Remove the wheels. Remove the brake drum.

2. Remove the nuts attaching the support plate and retainer to the axle tube flange using the access hole in the axle shaft flange.

3. Position adapter tool J21579, or equivalent and a slide hammer on the axle shaft flange. Remove the axle shaft from the rear axle assembly.

4. If the cup portion of the wheel bearing assembly remains in the axle assembly, remove it using tool J2619–01 or J26941.

5. Installation is the reverse of the removal procedure.

Full Floating Axle Shaft

It is not necessary to raise the rear wheels in order to remove the rear axle shaft on full-floating rear axles.

1. Remove the axle flange nuts, lock washers, and split washers retaining the axle shaft flange.

2. Remove the axle shaft from the axle housing.

3. Clean the axle flange mating area on the hub and axle, removing all old gasket material.

4. Install a new flange gasket onto the hub studs.

5. Insert the axle shaft into the housing. It may be necessary to rotate the axle shaft to align the shaft splines with the differential gear splines and the flange attaching holes with the hub studs.

6. Install the split washers, lockwashers, and flange nuts. Tighten the nuts securely.

1984 and Later Wagoneer, Cherokee & Comanche

1. Raise and support the vehicle safely.

2. Remove the wheel, brake drum and brake support retaining nuts.

3. Pull the axle with a slide hammer.

4. Installation is the reverse of removal. Clean the bore in the axle housing and apply a thin coat of grease to the outer diameter of the bearing cup. Tighten the brake support nuts alternately and evenly to seat the bearing cup rib ring.

Axle Shaft Bearing

REMOVAL & INSTALLATION

Except 1984 and Later Wagoneer, Cherokee & Comanche

1. With the aid of a combination puller, remove the bearing from the axle shaft.

NOTE: If a puller is not available, place the threaded end of the axle on a heavy block of wood and with the aid of an assistant, drive the bearing from the axle shaft with a punch and hammer. Contact the inner race only with the punch.

2. The new bearing can be installed with the use of a combination puller, or with the use of a length of pipe, fitted to the diameter of the inner bearing race, and slipped over the axle end to contact and drive the bearing to its seat on the axle shaft.

3. Lubricate the bearing with wheel bearing grease, making sure the grease fills the cavities between the bearing rollers.

1984 and Later Wagoneer, Cherokee & Comanche

1. Position the axle shaft in a vise.

2. Remove the retaining ring by drilling a ¼ inch hole about ¾ of the way through the ring, then using a cold chisel over the hole, split the ring.

3. Remove the bearing with an arbor press, discard the seal and remove the retainer plate.

4. Installation is the reverse of removal. The new bearing must be pressed on. Make sure it is squarely seated.

Differential Assembly

REMOVAL & INSTALLATION

1. Raise the vehicle and support safely. Remove the wheels, drums and axle shafts.

2. Drain the axle housing lubricant and remove the axle housing cover.

3. Mark the differential bearing caps for alignment during the assembly.

4. Loosen the bearing cap bolts, but do not remove.

5. Install an axle housing spreader tool on the axle housing and secure with the hold-down clamps.

6. Mount a dial indicator on the axle housing to measure the amount of spread. Zero the indicator dial.

7. Spread the axle housing no more than 0.020 in.

8. Remove the differential bearing caps and the dial indicator from the housing.

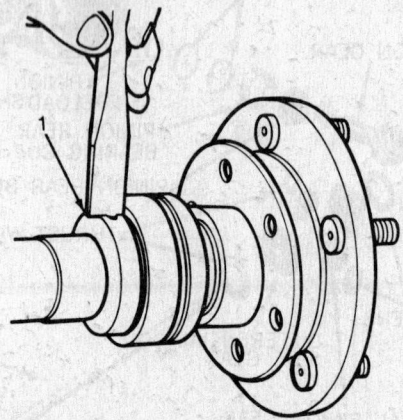

Splitting the rear axle bearing retainer on 1984 Wagoneer and Cherokee

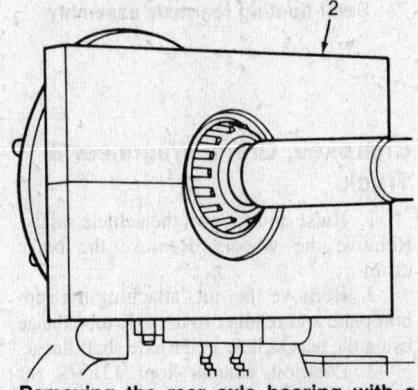

Removing the rear axle bearing with a press on 1984 Wagoneer and Cherokee

9. Using two pry bars, remove the differential carrier from the axle housing.

10. Remove the spreader tool from the housing as soon as the differential carrier is removed to avoid the possibility of the axle housing taking a set.

11. The differential housing can now be overhauled or replaced.

12. Install the axle housing spreader tool on the axle housing and secure with the hold down clamps. Install a dial indicator and center the dial.

13. Spread the axle housing to a maximum of 0.020 inch. Remove the dial indicator.

14. Lubricate the differential side bearings and install the differential carrier in the axle housing.

NOTE: Prior shim fitting and bearing preload should be accomplished before differential carrier installation.

15. Tap the unit in place with a soft faced hammer. Remove the axle housing spreader tool.

16. Install the bearing caps in their proper place and torque to 40 ft. lbs. on Model 30 and to 80 ft. lbs. on Models 44 and 60.

17. Install a dial indicator and recheck the ring gear backlash at two points. Correct as necessary.

18. Complete the assembly in the reverse of the removal, add lubricant and road test.

Refer to the Drive Axle section in Unit Repair of overhaul procedures.

Pinion Oil Seal

REMOVAL & INSTALLATION

Semi-Floating Axle w/Tapered Shaft

1. Raise and support the vehicle and remove the rear wheels and brake drums.

2. Mark the driveshaft and yoke for reassembly and disconnect the driveshaft from the rear yoke.

3. With a socket on the pinion nut and an inch lb. torque wrench, rotate the drive pinion several revolutions. Check and record the torque required to turn the drive pinion.

4. Remove the pinion nut. Use a flange holding tool to hold the flange while removing the pinion nut. Discard the pinion nut.

5. Mark the yoke and the drive pinion shaft for reassembly reference.

6. Remove the rear yoke with a puller.

7. Inspect the seal surface of the yoke and replace it with a new one if the seal surface is pitted, grooved, or otherwise damaged.

8. Remove the pinion oil seal.

9. Before installing the new seal, coat

the lip of the seal with rear axle lubricant.

10. Install the seal, driving it into place with the proper driving tool.

11. Install the yoke on the pinion shaft. Align the marks made on the pinion shaft and yoke during disassembly.

12. Install a new pinion nut. Tighten nut until end play is removed from the pinion bearing. Do not overtighten.

13. Check the torque required to turn the drive pinion. The pinion must be turned several revolutions to obtain an accurate reading.

14. Tighten the pinion nut to obtain the torque reading observed during disassembly (Step 3) plus 5 inch lbs. Tighten the nut minutely each time, to avoid overtightening. Do not loosen and then retighten the nut.

NOTE: If the desired torque is exceeded a new collapsible pinion spacer sleeve must be installed and the pinion gear preload reset.

15. Install the driveshaft, aligning the index marks made during disassembly. Install the rear brake drums and wheels.

Semi-Floating & Full-Floating Axles w/Flange Shaft

1. Raise and support the vehicle.

2. Mark the driveshaft and yoke for reference during assembly and disconnect the driveshaft at the yoke.

3. Remove the pinion shaft nut and washer.

4. Remove the yoke from the pinion shaft, using a puller.

5. Remove the pinion shaft oil seal.

6. Install the new seal with a suitable driver.

7. Install the pinion shaft washer and nut. Tighten the nut to 210 ft. lbs. on the semi-floating axles and 260 ft. lbs. on the full-floating axles.

8. Align the index marks on the driveshaft and yoke and install the driveshaft. Tighten the attaching bolts or nuts to 16 ft. lbs.

9. Remove the supports and lower the vehicle.

Bearing Adjustment w/Floating Axle

1. Raise the vehicle so that the wheel can be rotated. Support the vehicle safely.

2. Remove the axle shaft.

3. Straighten the lip of the lock washer and remove the lock washer and lock nut.

4. Tighten the adjusting nut and rotate the wheel until binding exists. Back off the adjusting nut 1/16 turn until the wheel rotates freely without any lateral shake.

5. Replace the lock washer and tighten the lock nut, bending the lip of the lock washer over the lock nut.

6. Install the axle with a new gasket and tighten the axle nuts securely.

STEERING

Steering Wheel

REMOVAL & INSTALLATION

1. Disconnect the negative battery cable.

2. Place the front wheels in the straight-ahead position.

3. Remove the horn button from the steering wheel. Turn the botton until the locktab on the button align with the notches in the contact cup and pull upward to remove it. With the sport wheel, just pull button up.

4. Remove the steering wheel nut and washer.

5. If the Jeep is equipped with a sport style steering wheel, remove the horn button, nut and washer, bottom retaining ring, and horn contact ring.

6. Remove the plastic horn contact cup retainer and remove the cup and contact plate from the steering wheel.

7. Remove the horn contact pin and bushing from the steering wheel.

8. Paint or scribe alignment marks on the steering wheel and shaft for reference during assembly.

9. Remove the steering wheel using a puller.

10. Install the steering wheel in the reverse order, tightening the nut to 30 ft. lbs. for 1979–83, and 25 ft. lb. for 1984 and later.

Turn Signal Switch

REPLACEMENT

1979

1. Disconnect the negative battery cable.

2. Remove the steering wheel.

3. Depress the lockplate and pry the round wire snap-ring from the steering shaft groove. A lockplate compressor tool is available for compressing the lockplate.

4. Remove the lockplate, directional signal canceling cam, upper bearing preload spring, and thrust washer from the steering shaft.

5. Move the directional signal actuating lever to the right turn position and remove the lever.

6. Depress the hazard warning light switch and remove the button by turning it counterclockwise.

7. Remove the directional signal wiring harness connector block from its mounting bracket on the right-side of the lower column.

8. On vehicles equipped with an automatic transmission, use a stiff wire, such as a paper clip, to depress the lock tab which retains the shift quadrant light wire in the connector block.

9. Remove the directional signal switch retaining screws and pull the switch and wiring harness from the steering column.

10. Guide the wiring harness of the new switch into position and carefully align the switch assembly. Make sure that the actuating lever pivot is correctly aligned and seated in the upper housing pivot boss prior to installing the retaining screws.

11. Install the directional signal lever and actuate the directional signal switch to assure correct operation.

12. Place the thrust washer, spring, and directional signal canceling cam on the upper end of the steering shaft.

13. Align the lockplate splines with the steering shaft splines and place the lockplate in position with the directional signal canceling cam shaft protruding through the dogleg opening in the lockplate.

14. Install the snap-ring.

15. Install the anti-theft cover.

16. Install the steering wheel and connect the negative battery cable.

17. Check the operation of the turn signal switch.

1980 and Later

1. Disconnect the battery ground.

2. Cover the painted areas of the column.

3. Remove the column-to-dash bezel.

4. Loosen the toe plate screws.

5. With tilt columns, place the column in the non-tilt position.

6. Remove the steering wheel.

7. Remove the lockplate cover.

8. Compress the lockplate and unseat the steering shaft snap-ring as follows: Check the steering shaft nut threads. Metric threads have an identifying groove in the steering wheel splines. SAE threads do not. With SAE threads use a compressor tool such as tool J23653 to compress the lockplate and remove the snap-ring. If the shaft has metric threads, replace the forcing screw in the compressor with metric forcing screw J23653–4 before using.

9. Remove the compressor and snap-ring.

10. Remove the lockplate, canceling cam and upper bearing preload spring.

11. Place the turn signal lever in the right turn position and remove the lever.

12. Remove the hazard warning knob. Press the knob inward and turn counterclockwise to remove it.

13. Remove the wiring harness protectors.

14. Disconnect the wiring harness connectors.

15. Remove the turn signal switch attaching screws and lift out the switch.

Ignition Switch

REPLACEMENT

The ignition switch is on top of the lower

part of the steering column, inside the vehicle. Some vehicle applications require the removal of lower trim panel in order to gain access to the switch assembly.

1. Put the key in the lock and turn to the Off-unlocked position.

2. Disconnect the battery ground cable.

3. Detach the wire connectors at the switch.

4. Remove the switch screws.

5. Disconnect the actuating rod from the switch and remove the switch.

6. Move the switch slider all the way down the column. Move it back toward the steering wheel two clicks to the center Off-unlocked position.

7. Engage the column actuating rod in the switch slider and fasten the switch down.

8. Connect the wire connectors, then the battery ground cable.

Power Steering Pump

REMOVAL & INSTALLATION

If the power steering pump has to be removed to service another component, it is not necessary to remove the hoses from the pump. Just disconnect the mounting fixtures and lift the pump away from the engine and lay it out of the way. The only time the power steering hoses have to be removed from the pump is when the pump has to be removed from the vehicle for service or replacement.

1. Disconnect the negative battery cable. Remove all the necessary components in order to gain access to the power steering retaining bolts.

2. Remove the pump drive belt tension adjusting bolt. Disconnect the belt from the pump.

3. Disconnect the return and pressure hoses from the pump. Cover the hose connector and union on the pump and open ends of the hoses to avoid the entrance of dirt.

4. On some V8 engines it will be necessary to, remove the front bracket from the engine. Remove the two nuts which secure the rear of the pump to the bracket, and the two bolts which secure the front of the pump to the bracket and remove the pump.

5. To install, position the pump in the bracket and install the rear attaching screws. Install the front bracket, as required.

6. Connect the hydraulic hoses. Adjust the drive belt tension.

7. Fill the pump reservoir to the correct level.

8. Start the engine and wait for at least three minutes before turning the steering wheel. Check the level frequently during this time.

9. Slowly turn the steering wheel through its entire range a few times with the engine running. Recheck the level and inspect for possible leaks.

NOTE: If air becomes trapped in the fluid, the pump may become noisy until all of the air is out. This may take some time since trapped air does not bleed out rapidly.

Manual Steering Gear

REMOVAL & INSTALLATION

Except 1984 and Later Wagoneer, Cherokee & Comanche

1. Remove the intermediate shaft-to-wormshaft coupling clamp bolt and disconnect the intermediate shaft.

2. Remove the pitman arm nut and lockwasher.

3. Using a puller, remove the pitman arm from the shaft.

4. On some vehicles you will have to raise the left side of the vehicle slightly to relieve tension on the left front spring and rest the frame on a jackstand.

5. Remove the steering gear lower bracket-to-frame bolts.

6. Remove the bolts attaching the steering gear upper bracket to the crossmember. One of these bolts is a Torx® head bolt. This bolt, and some others may be removed with the aid of a 9 inch extension. Remove the gear.

NOTE: Loctite 271® or similar material must be applied to all attaching bolt threads prior to installation.

7. Position the tie plate upper and lower mounting brackets on the gear and install the bolts. Torque the bracket-to-gear bolts to 70 ft. lbs. and the bracket-to-tie plate bolt to 55 ft. lbs.

8. Align and engage the intermediate shaft coupling with the steering gear wormshaft splines.

9. Position the steering gear on the frame and install the mounting bolts. Torque the bolts to 55 ft. lbs. Install the pitman arm and torque the nut to 185 ft. lbs.

NOTE: The steering gear may produce a slight roughness, this can be eliminated by turning the steering wheel full left and right 10–15 times.

1984 and Later Wagoneer, Cherokee & Comanche

1. Disconnnect the steering shaft from the gear.

2. Raise and support the vehicle safely.

3. Disconnect the center link from the pitman arm.

4. Remove the front stabilizer bar.

5. Remove the pitman arm nut, mark the relation between the arm and shaft, and using a puller, pull the pitman arm from the shaft.

6. Remove the gear attaching bolts and remove the gear.

7. Installation is the reverse of re-

moval. Observe the following torques: Center link-to-pitman arm: 35 ft. lb. Stabilizer bar-to-link: 27 ft. lb. Pitman arm-to-gear: 185 ft. lb. Steering gear-to-frame: 65 ft. lb.

Power Steering Gear

REMOVAL & INSTALLATION

Except 1984 and Later Wagoneer, Cherokee & Comanche

1. Disconnect the hoses at the gear and raise them above the pump to prevent fluid loss.

2. Remove the clamp bolt and nut attaching the intermediate shaft coupling to the steering gear stub shaft and disconnect the intermediate shaft.

3. Mark the pitman shaft and arm for alignment. Remove the pitman nut and lockwasher and remove the pitman arm with a puller.

4. On some vehicles it will be necessary to raise the left side of the vehicle slightly to relieve tension from the spring. Support with a jack stand under the frame.

5. Remove the three lower steering gear mounting bracket-to-frame bolts.

6. Remove the two steering gear-to-crossmember upper bolts. Remove the gear and brackets as an assembly.

7. Remove the brackets from the gear.

NOTE: Prior to installation, all bolts must be coated with Loctite 271® or its equivalent.

8. Position the mounting brackets on the gear and torque the bolts to 70 ft. lbs.

9. Align and connect the intermediate shaft coupling to the steering gear stub shaft.

10. Position the steering gear on the frame and crossmember. Install and tighten the bolts to 55 ft. lbs.

11. Lower the vehicle.

12. Install the intermediate shaft coupling-to-steering gear stub shaft clamp bolt and nut. Tighten the nut to 45 ft. lbs.

13. Align and install the Pitman arm, nut and lockwasher. Torque the nut to 185 ft. lbs. Stake the nut in two places.

14. Connect the hoses. Torque the hose connections to 25 ft. lbs.

1984 and Later Wagoneer, Cherokee & Comanche

1. Place the wheels in a straight ahead position.

2. Position a drain pan under the gear.

3. Disconnect the hoses at the gear and raise and secure the hose ends to prevent draining. Cap the hose ends.

4. Disconnect the intermediate shaft from the stub shaft.

5. Raise and support the front of the vehicle safely. Mark the relation between the pitman arm and shaft.

6. Disconnect the center link from the pitman arm and remove the stabilizer bar.

7. Using a puller, remove the pitman arm.

8. Unbolt and remove the gear.

9. Installation is the reverse of removal. Observe the following torques: Steering gear-to-frame: 65 ft. lb. Pitman arm-to-shaft: 185 ft. lb. Stabilizer bar-to-frame: 55 ft. lb. Stabilizer bar-to-link: 27 ft. lb. Center link-to-pitman arm: 35 ft. lb.

Refer to the Steering section in Unit Repair for adjustment procedures.

Tie Rod End

REMOVAL & INSTALLATION

1. Remove the cotter pins and retaining nuts at both ends of the tie rod and from the end of the connecting rod where it attaches to the tie rod.

2. Remove the nut attaching the steering damper push rod to the tie rod bracket and move the damper aside.

3. Remove the tie rod ends from the steering arms and connecting rod. It may be necessary to use a puller on some vehicles.

4. Count the number of threads showing on the tie rod before removing the ends, as a guide to installation.

5. Loosen the adjusting tube clamp bolts and unthread the ends.

6. Installation is the reverse of removal. Torque the connecting rod-to-tie rod nut to 70 ft. lbs.

7. Adjust toe-in, if necessary.

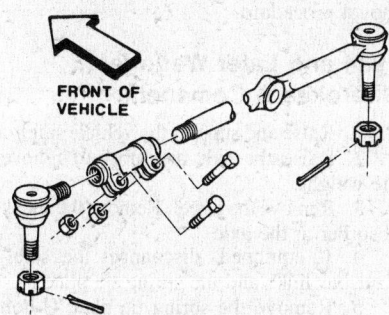

Typical tie-rod assembly

FRONT SUSPENSION

Front Leaf Spring

REMOVAL & INSTALLATION

1. Raise the vehicle and support it safely.

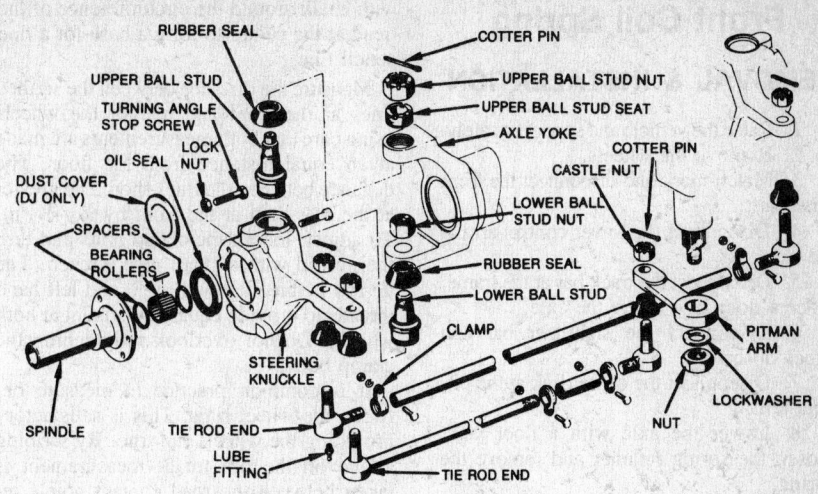

Typical steering linkage

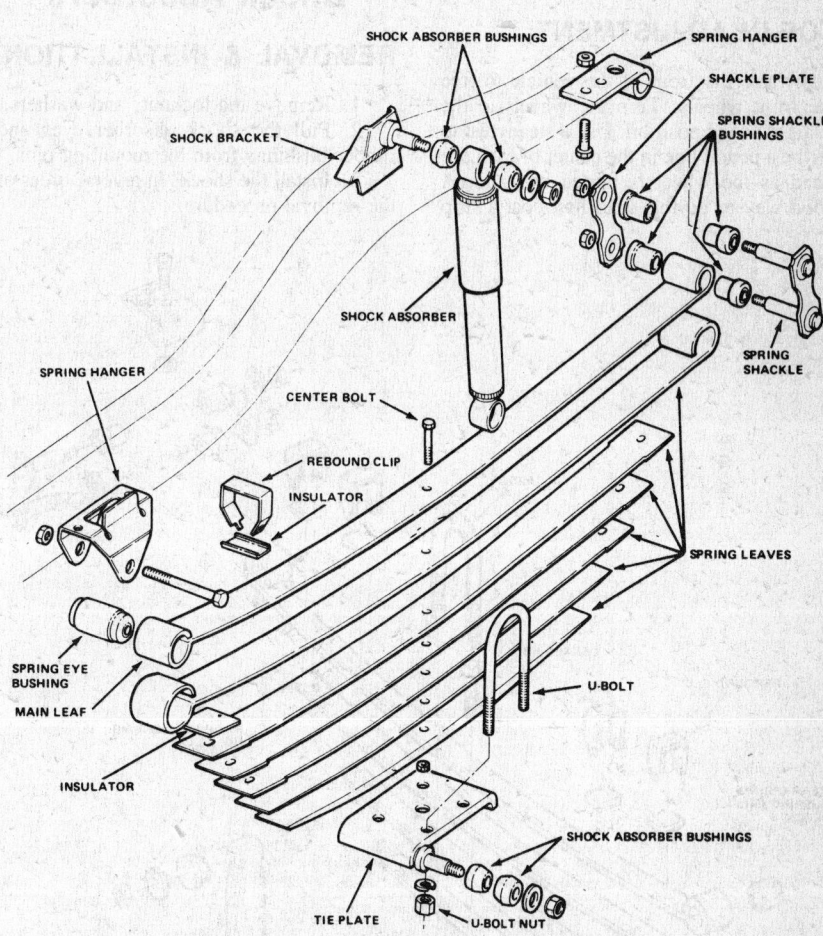

Typical front spring and shock absorber

2. Position a jack under the axle. Raise the axle to relieve the springs of the axle weight.

3. If equipped, disconnect the stabilizer bar. Remove the spring U-bolts and tie plates.

4. Remove the bolt attaching the spring front eye to the shackle.

5. Remove the bolt attaching the spring rear eye to the shackle.

6. Remove the spring from its mounting.

7. Installation is the reverse of the removal procedure.

Front Coil Spring

REMOVAL & INSTALLATION

1. Raise the vehicle and support it safely.
2. Remove the wheels.
3. Match-mark and disconnect the front driveshaft.
4. Disconnect the lower control arm at the axle.
5. Disconnect the track bar at the frame. Place a floor jack under the axle.
6. Disconnect the stabilizer bar and shock absorbers.
7. Disconnect the center link at the pitman arm.
8. Lower the axle with a floor jack, loosen the spring retainer and remove the spring.
9. Installation is the reverse of removal.

TOE-IN ADJUSTMENT

First raise the front of the vehicle to free the front wheels. Turn the wheels to the straight ahead position. Use a steadyrest to scribe a pencil line in the center of each tire tread as the wheel is turned by hand. A good way to do this is to first coat a strip with chalk around the circumference of the tread at the center to form a base for a fine pencil line.

Measure the distance between the scribed lines at the front and rear of the wheels using care that both measurements are made at an equal distance from the floor. The distance between the lines should be greater at the rear than at the front by $\frac{3}{64}$–$\frac{3}{32}$ in. To adjust, loosen the clamp bolts and turn the tie rod with a small pipe wrench. The tie rod is threaded with right and left hand threads to provide equal adjustment at both wheels. Do not overlook retightening the clamp bolts.

It is common practice to measure between the wheel rims. This is satisfactory providing the wheels run true. By scribing a line on the tire tread, measurement is taken between the road contact points reducing error caused by wheel run-out.

Shock Absorbers

REMOVAL & INSTALLATION

1. Remove the locknuts and washers.
2. Pull the shock absorber eyes and rubber bushings from the mounting pins.
3. Install the shocks in reverse order of the removal procedure.

REAR SUSPENSION

Leaf Spring

REMOVAL & INSTALLATION

Mounted Below the Axle

1. Raise the vehicle and support the axle.
2. Disconnect the shock absorber and stabilizer bar, if so equipped.
3. Remove the U-bolts and tie plates.
4. Disconnect the front and rear ends of the spring and remove the spring.
5. The spring can be disassembled by removing the spring rebound clips and the center bolt.
6. Installation is the reverse of the removal procedure.

Mounted Above the Axle

1. Raise the vehicle and support the frame ahead of the axle.
2. If the left side spring is being removed, remove the fuel tank skid plate.
3. Remove the wheel.
4. Disconnect the shock absorber.
5. Remove the tie plate U-bolts and the tie plate. Remove the bolt attaching the spring rear eye to the spring shackle.
6. Remove the bolt attaching the spring front eye to the spring hanger on the frame rail.
7. Remove the spring from the vehicle.
8. Installation is the reverse of the removal procedure.

1984 and Later Wagoneer, Cherokee & Comanche

1. Raise and support the vehicle safely.
2. Raise the axle assembly to relieve the weight.
3. Remove the wheel. Remove the shock absorber at the axle.
4. If equipped, disconnect the stabilizer bar links and the spring tie plate.
5. Remove the spring tie plate U-bolt and the spring tie plate. Remove the rear eye to spring shackle bolt and the front eye to bracket bolt.
6. Lower the spring from the vehicle.
7. Installation is the reverse of the removal procedure.

Shock Absorber

REMOVAL & INSTALLATION

1. Raise the vehicle for working clearance and support safely.
2. Place a jack under the axle assembly and raise to relieve the springs of axle weight and to place the shock absorber in its mid stroke.

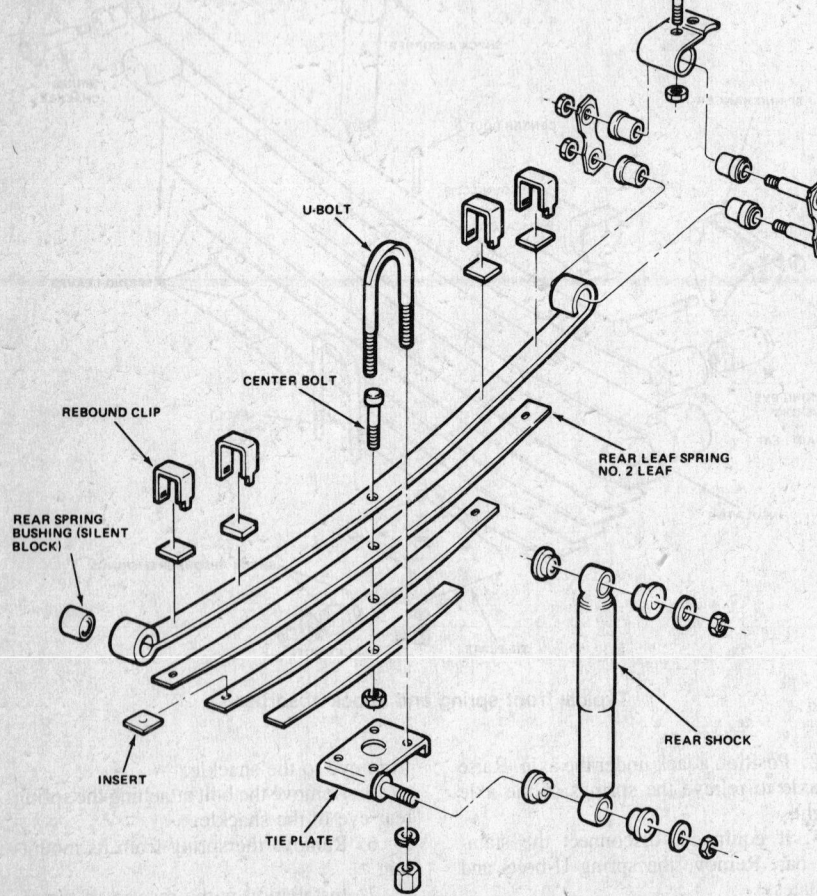

U-BOLT

CENTER BOLT

REBOUND CLIP

REAR SPRING BUSHING (SILENT BLOCK)

REAR LEAF SPRING NO. 2 LEAF

REAR SHOCK

INSERT

TIE PLATE

Typical rear spring and shock absorber assembly

3. Remove the retaining nuts or bolts and remove the shock absorber from the vehicle.

4. Install the new shock absorber and tighten the attaching nuts or bolts. (upper bolts: 15 ft. lb. lower nut: 44 ft. lb.)

5. Lower the vehicle to the ground.

BRAKES

Refer to the Brakes General Repair Section for detail troubleshooting and brake hydraulic system repair procedures.

Master Cylinder

REMOVAL & INSTALLATION

1. Disconnect and plug the brake lines at the master cylinder.

2. Disconnect the master cylinder push rod at the brake pedal on vehicles with manual brakes.

4. Remove all attaching bolts and nuts and lift the master cylinder from the vehicle.

5. Install the master cylinder in the reverse order of removal. Torque the mounting nuts to 10–12 ft. lb. Bleed the hydraulic system.

Power Unit

REMOVAL & INSTALLATION

1. Disconnect brake pedal pushrod rod at brake pedal.

2. Disconnect vacuum hose from booster check valve.

3. Remove attaching nuts and separate master cylinder from brake booster. Do not disconnect brake lines at master cylinder.

4. On some vehicles it may be necessary to remove the bolts holding the power unit bellcrank to the dash panel then remove the power unit and bellcrank as one assembly. Remove the bellcrank from the original power unit and lubricate the pivot pins with chassis lubricant before installing it on the replacement unit.

5. Remove the bolts attaching the power unit to the dash panel and remove the unit.

6. Installation is the reverse of removal. Torque the booster-to-firewall nuts to 30 ft. lb.; the master cylinder-to-booster nuts to 10–12 ft. lb.

NOTE: When replacing the power brake unit, use the push rod that is supplied with the new unit, as it has been correctly gauged and preset to the new unit. Torque the booster-to-firewall nuts to 30 ft. lb.; the master cylinder-to-booster nuts to 10–12 ft. lb.

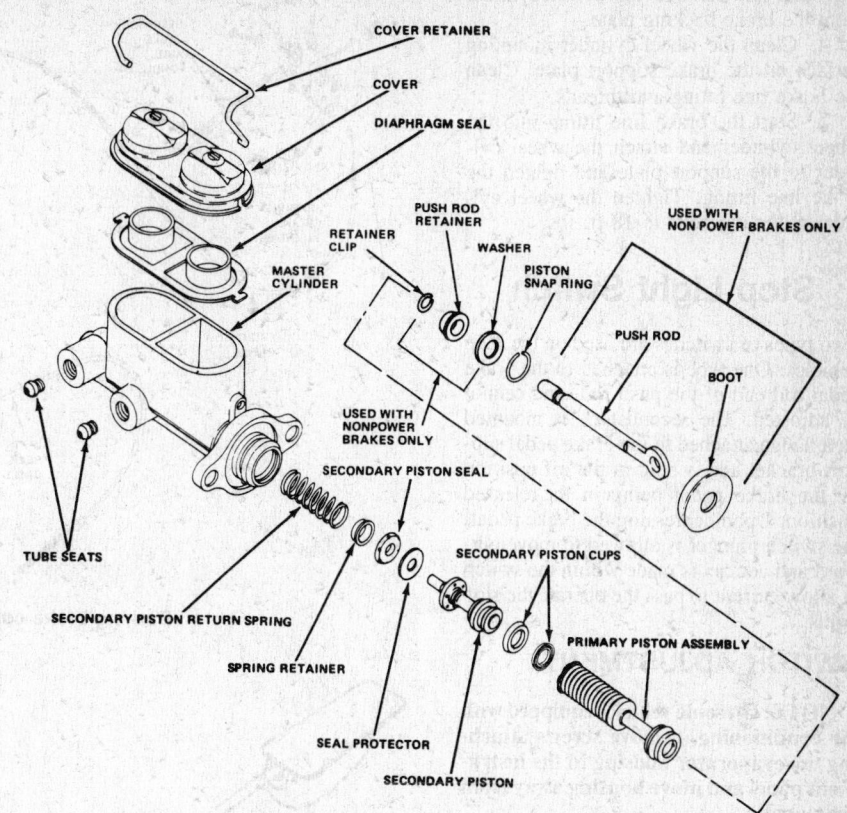

Exploded view of a typical master cylinder

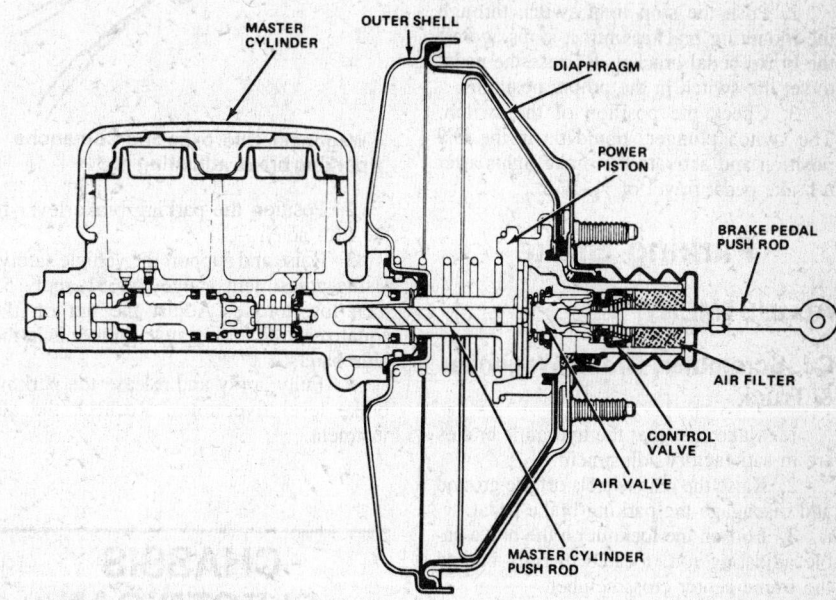

Power brake booster—typical

Wheel Cylinder

REMOVAL & INSTALLATION

1. Raise and support the vehicle and remove the brake drums and brake shoes.

2. Disconnect the brake line. Do not bend the line away from the wheel cylinder. When the cylinder is removed from the support plate, the line will separate from the wheel cylinder easily.

3. Remove the wheel cylinder mount-

ing bolts and remove the wheel cylinder from the brake backing plate.

4. Clean the wheel cylinder mounting surface on the brake support plate. Clean the brake line fitting and threads.

5. Start the brake line fitting into the wheel cylinder and attach the wheel cylinder to the support plate and tighten the brake line fitting. Tighten the wheel cylinder mounting bolts to 18 ft. lbs.

Stop Light Switch

Two types of switches are used on the Jeep vehicles. One type is attached to the brake pedal rod end of the push rod, and cannot be adjusted. The second type is mounted on a flange attached to the brake pedal support bracket and is held in the off position by the brake pedal being in its released position. Upon depressing the brake pedal, the switch plunger is allowed to move outward and contact is made within the switch to allow current to pass the operate the stop lights.

SWITCH ADJUSTMENT

NOTE: On some vehicles equipped with air conditioning, remove screws attaching the evaporator housing to the instrument panel and move housing away from the panel.

1. Hold the brake pedal in the applied position.

2. Push the stop light switch through the mounting bracket until it stops against the brake pedal bracket. Release the pedal to set the switch in the proper position.

3. Check the position of the switch. The switch plunger should be in the ON position and activate the brake lights after a brake pedal travel of $3/8$–$5/8$ in.

Parking Brake

ADJUSTMENT

CJ, Scrambler, Grand Wagoneer & Truck

1. Make sure that the hydraulic brakes are in satisfactory adjustment.

2. Raise the rear wheels off the ground and disengage the parking brake pedal.

3. Loosen the locknut on the brake cable adjusting rod, located directly behind the frame center crossmember.

4. Spin the wheels and tighten the adjustment until the rear wheels drag slightly. Loosen the adjustment until there is no drag and the wheels spin freely.

5. Tighten the locknut to lock the adjusting nut.

Cherokee, Wagoneer & Comanche

1. Make sure that the hydraulic brakes are is satisfactory adjustment.

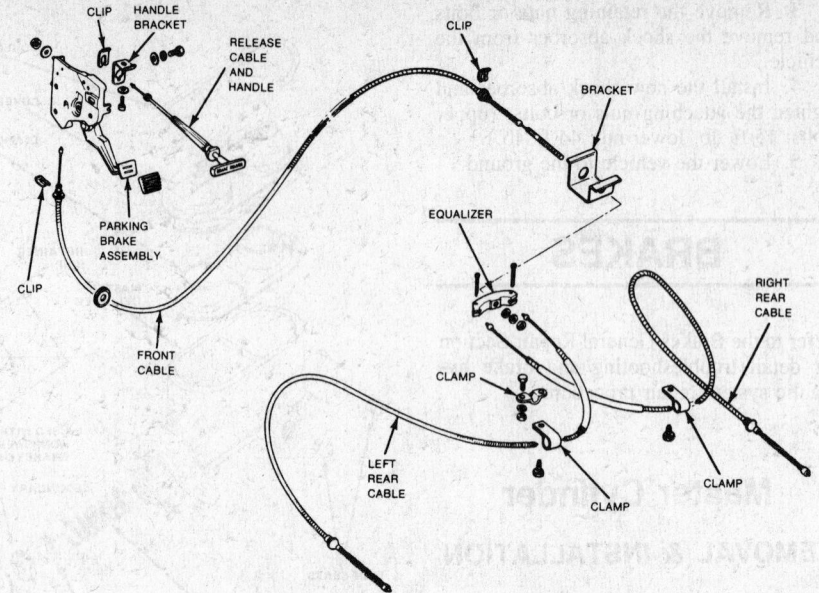

Parking brake components—typical

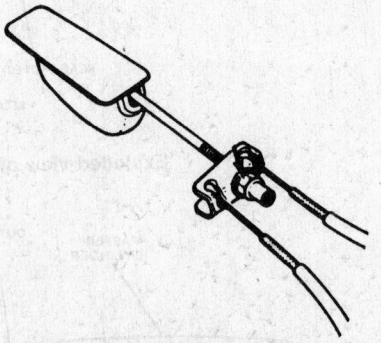

Wagoneer, Cherokee and Comanche parking brake adjusting nut

2. Position the parking brake lever in the fifth notch.

3. Raise and support the vehicle safely. Using adjustment gauge J34651 apply 55 inch lbs. torque. Adjust the nut on the equalizer so that the gauge pointer is in the blue band.

4. Fully apply and release the parking brake lever five times. Recheck the adjustment.

CHASSIS ELECTRICAL

Heater Core

REMOVAL & INSTALLATION

1980 and Later CJ & Scrambler

1. Remove the battery, drain the cooling system, and disconnect the heater hoses.

2. Disconnect the damper door control cable.

3. Disconnect the blower motor wire harness and ground wire at the switch and instrument panel.

4. Remove the defroster duct hose. On some vehicles it may be necessary to first remove the glove box.

5. If equipped, disconnect the heater to air deflector duct at the heater housing.

6. Remove the nuts from the heater housing studs, protruding into the engine compartment.

7. Remove the heater housing assembly from the vehicle and remove the core from the housing.

8. Installation is the reverse of the removal procedure.

1980 and Later Cherokee, Grand Wagoneer & Truck

1. Remove the negative battery cable and drain the cooling system.

2. Disconnect the temperature control cable from the blend air door.

3. Remove the heater hoses and blower motor resistor wires.

4. Remove the heater core housing to dash panel attaching screws or nuts, projecting into the engine compartment.

5. Remove the heater housing assembly from the vehicle.

6. Separate the halves of the housing, after scribing a mark on the two halves. Remove the core retaining screws and remove the heater core.

7. Installation is the reverse of the removal procedure.

1984 and Later Cherokee, Wagoneer & Comanche

1. Disconnect the negative battery cable. Drain the cooling system.

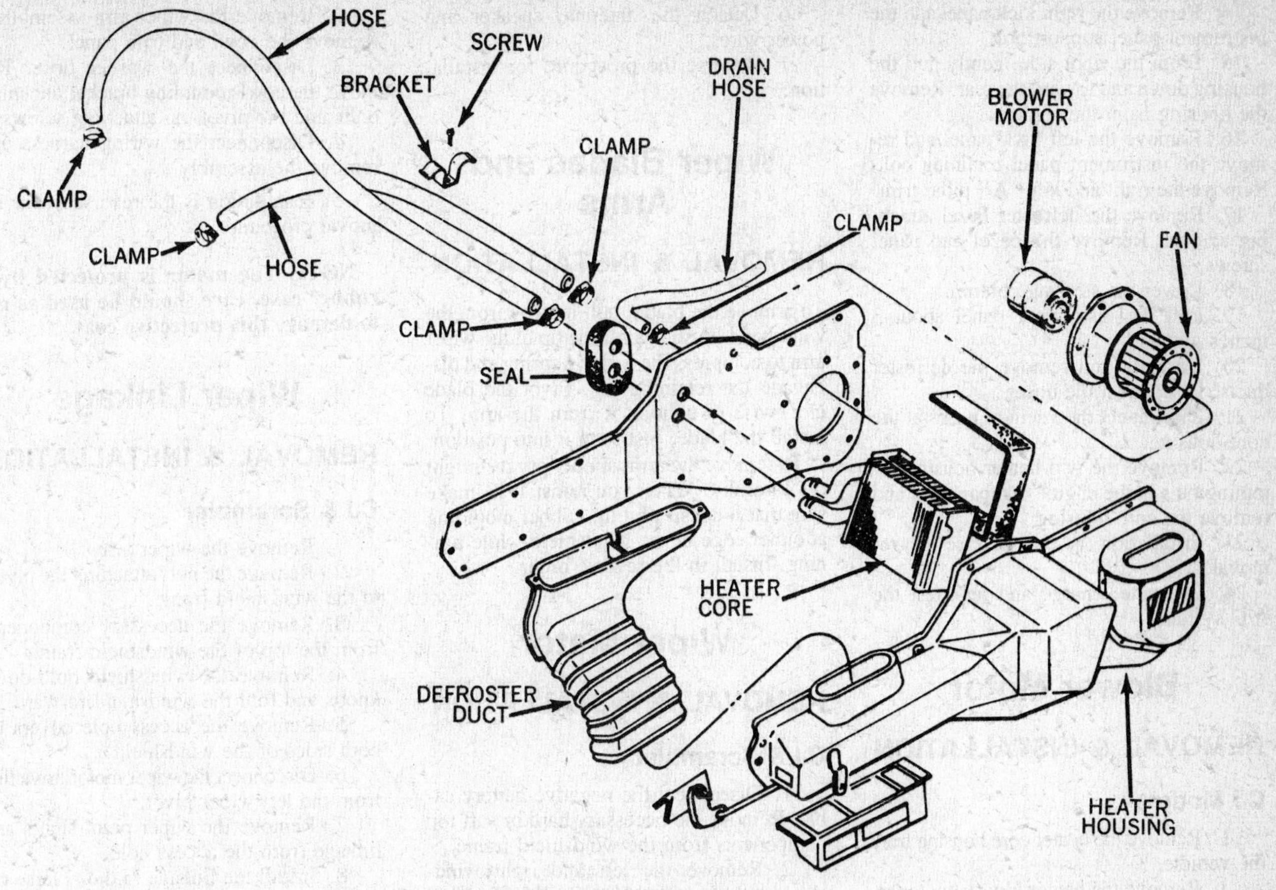

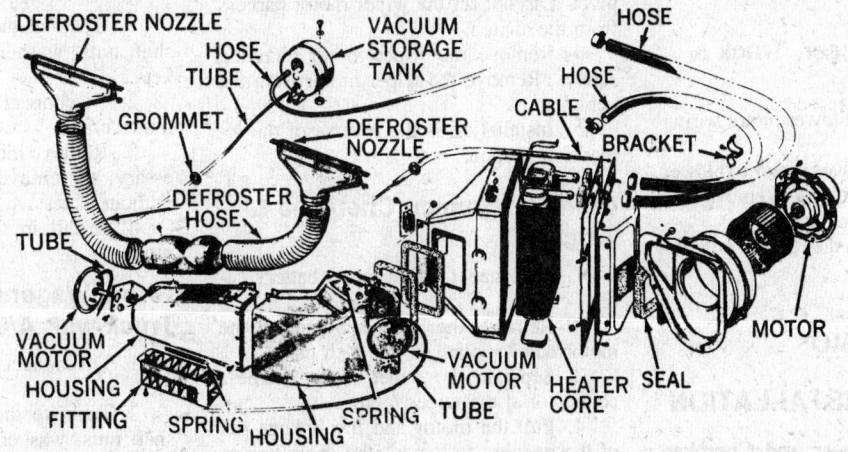

Typical CJ heater/defroster assembly

Grand Wagoneer, Cherokee and Truck heating system and related components

2. Disconnect the hoses at the core tubes.

3. Disconnect the battery ground cable.

4. Discharge the refrigerant from the A/C system.

5. Disconnect the refrigerant lines at the expansion valve.

6. Disconnect the blower motor wires and vent tube.

7. If equipped, Remove the console.

8. Remove the lower instrument panel section.

9. Disconnect all wires and hoses at the case.

10. Cut the plastic retaining strap which holds the evaporator/blower housing to the heater core housing.

11. Disconnect the heater control cables.

12. Remove the clip at the rear of the blower housing flange and remove the retaining screws.

13. Remove the housing attaching nuts from the studs on the engine compartment side of the firewall. Remove the evaporator drain tube.

14. Remove the right kick panel and the instrument panel support bolt.

15. From the right side, gently pull the housing down and toward the rear. Remove the housing from the vehicle.

16. Remove the left kick panel and remove the instrument panel retaining bolt. Remove the right and left "A" pillar trim.

17. Remove the defroster bezel attaching screws. Remove the bezel and panel screws.

18. Lower the steering column.

19. Pull the instrument panel about 3 inches outward.

20. Unscrew and remove the defroster ducts. Disconnect the hoses.

21. Disconnect the vacuum hoses at the core housing.

22. Remove the two heater housing retaining nuts in the engine compartment and remove the core housing.

23. Installation is the reverse of removal.

24. Evacuate, charge and leak test the A/C system.

Blower Motor

REMOVAL & INSTALLATION

CJ Models

1. Remove the heater core housing from the vehicle.

2. Remove the heater blower motor retaining screws. Remove the blower motor from the heater core housing.

3. Installation is the reverse of the removal procedure.

Cherokee, Wagoneer, Truck & Comanche

1. Disconnect the blower motor wiring connector.

2. Remove the blower motor to blower housing mounting screws. Remove the blower motor and fan assembly.

3. Installation is in the reverse order of removal.

Radio

REMOVAL & INSTALLATION

NOTE: On Wagoneer and Cherokee models the glove box door, liner and lock striker must be removed. The radio is removed through the glove box.

1. Disconnect the battery ground cable.

2. Remove the control knobs, nuts, and bezel.

3. With air conditioning, remove the screws and lower the assembly.

4. Disconnect the radio bracket from the instrument panel.

5. Tilt the radio down and remove it toward the steering wheel.

6. Detach the antenna, speaker and power wires.

7. Reverse the procedure for installation.

Wiper Blades and Arms

REMOVAL & INSTALLATION

To remove the blade, pull it away from the windshield. Push against the tip of the wiper arm to compress the locking spring and disengage the retaining pin. Pivot the blade clockwise to unhook it from the arm. To install the blade, just snap it into position.

To remove the arm, simply pry it straight off carefully. When you reinstall it, make sure that it doesn't hit the rubber moulding at either edge of the windshield while running. Install in the reverse order.

Wiper Motor

REMOVAL & INSTALLATION

CJ & Scrambler

1. Disconnect the negative battery cable. Remove the necessary hard or soft top components from the windshield frame.

2. Remove the left and right windshield holddown knobs and fold the shield forward.

3. Remove the left access hole cover. Disconnect the drive link from the left wiper pivot. Disconnect the wiper motor harness from the switch.

4. Remove the wiper motor retaining screws. Remove the wiper motor from the vehicle.

5. Installation is the reverse of the removal procedure.

Grand Wagoneer, Cherokee & Truck

1. Disconnect the negative battery cable.

2. Remove the screws attaching the motor adapter plate to the dash panel.

3. Separate the wiper wiring harness connector at the wiper motor.

4. Pull the motor and the linkage out of the opening to expose the drive link to crank stud retaining clip.

5. Raise up the lock tab of the clip and slide the clip off of the stud.

6. Remove the wiper motor from the vehicle.

7. Installation is the reverse of the removal procedure.

1984 and Later Wagoneer, Cherokee & Comanche

1. Disconnect the negative battery cable.

2. Remove the wiper arm assemblies. Remove the cowl and trim panel.

3. Disconnect the washer hose. Remove the cowl mounting bracket attaching bolts and the pivot pin attaching screws.

4. Disconnect the wiring harness and remove the assembly.

5. Installation is the reverse of the removal procedure.

NOTE The motor is protected by a rubber case, care should be used as not to damage this protective coat.

Wiper Linkage

REMOVAL & INSTALLATION

CJ & Scrambler

1. Remove the wiper arms.

2. Remove the nuts attaching the pivots to the windshield frame.

3. Remove the necessary components from the top of the windshield frame.

4. Remove the windshield hold-down knobs and fold the windshield forward.

5. Remove the access hole covers on both sides of the windshield.

6. Disconnect the wiper motor drive link from the left wiper pivot.

7. Remove the wiper pivot shafts and linkage from the access hole.

8. Install the linkage in the reverse order.

Grand Wagoneer, Cherokee & Truck without A/C

1. Remove the wiper arms and pivot shaft nuts, washers, escutcheons and gaskets.

2. Disconnect the drive arm from the motor crank.

3. Remove individual links where necessary, to remove the pivot shaft bodies without excessive interference.

4. Install in the reverse order of removal.

Grand Wagoneer, Cherokee & Truck with A/C

1. Disconnect the negative battery cable.

2. Remove the wiper arms and pivot shaft nuts, washers, escutcheons and gaskets.

3. Remove the instrument cluster. Remove the left defroster duct.

4. Disconnect the drive arm from the motor crank.

5. Lower the glove box in order to gain access to the right linkage clip. Remove the linkage clip.

6. Remove the screws attaching the left pivot shaft body. Remove the left pivot shaft body and linkage assembly through the instrument cluster opening.

7. Installation is the reverse of the removal procedure.

Instrument Cluster

REMOVAL & INSTALLATION

CJ & Scrambler

1. Disconnect the negative battery cable.

2. Disconnect the speedometer cable from the back of the speedometer.

3. Remove the instrument cluster attaching screws/nuts and remove the cluster.

4. Disconnect the instrument cluster electrical connectors and remove the cluster from the vehicle.

5. Install in the reverse order.

Wagoneer, Cherokee, Grand Wagoneer & Truck

1. Disconnect the battery ground cable. Remove the cluster retaining screws.

2. Disconnect the speedometer cable.

3. Disconnect the cluster terminal pin plug. Disconnect the four terminal connector.

4. Mark the electrical connectors and hoses, disconnect them and the blend door air cable.

6. Remove the heater control panel lamps. Disconnect the heater temperature control wire from the lever. Remove the cluster.

7. Installation is the reverse of removal.

1984 and Later Wagoneer, Cherokee & Comanche

1. Disconnect the battery ground.

2. Remove the four instrument panel bezel screws and lift off the bezel. The bezel unsnaps.

3. Remove the cigar lighter housing screws.

4. Remove the rocker switch housing screws.

5. Remove the cluster screws.

6. Disconnect the speedometer cable, pull the cluster out slowly and disconnect two multi-connectors at the cluster back. Remove the cluster.

7. Installation is the reverse of removal.

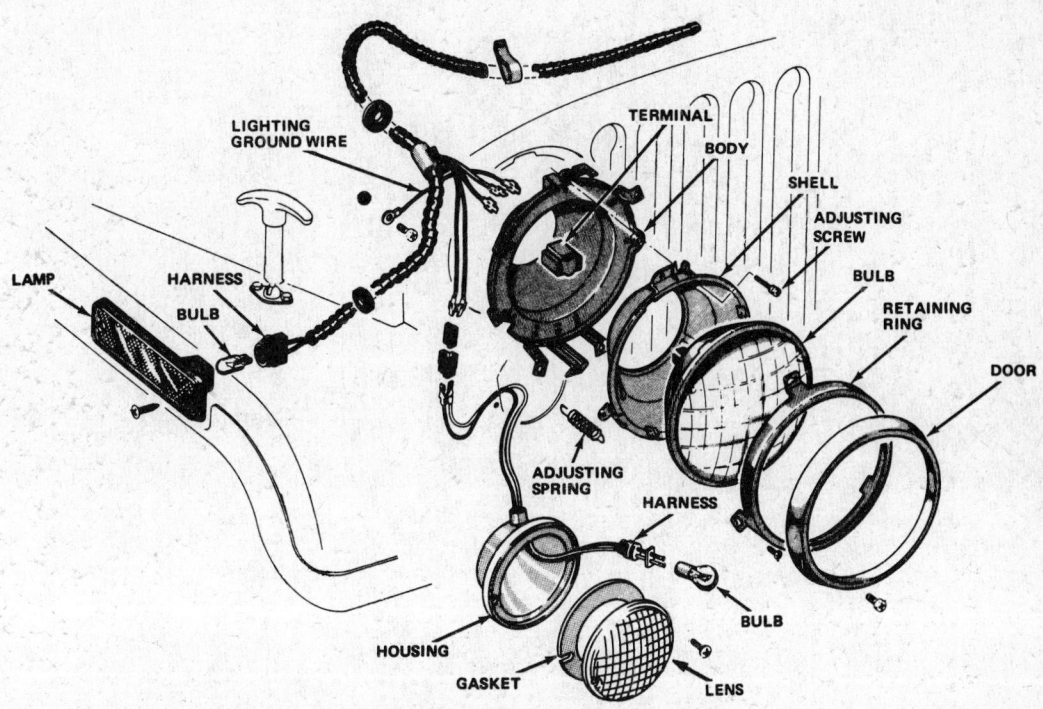

Typical front end lighting arrangement

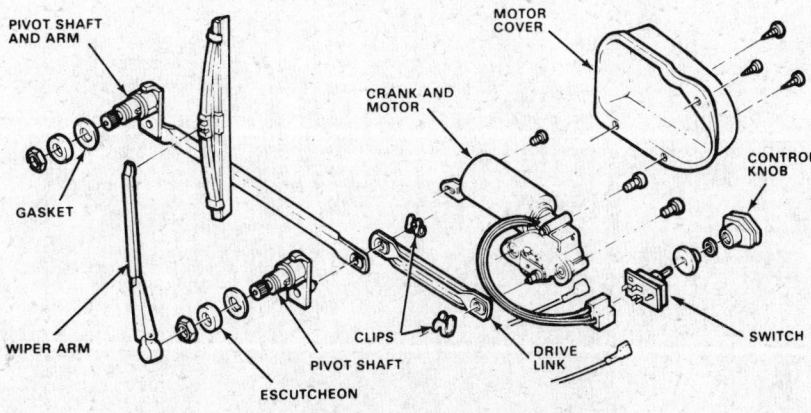

Typical windshield wiper components

Speedometer Cable

REPLACEMENT

1. Reach up behind the center of the speedometer head. The cable is connected by a threaded ring. Unscrew the ring and pull the cable sheath from the head.

2. The cable core can be pulled from the sheath.

3. If the core is broken, detach the other end of the sheath from the transmission. Pull out the broken end.

4. When installing the cable, apply a very small amount of speedometer cable graphite lubricant.

4. Pull the headlamp out and disconnect the wire harness.

5. Install in reverse order of the above procedure. Check for proper seating of the lamp in its mounting ring and check for proper alignment.

Fusible Links

Fusible links are sections of wire, with special insulation, designed to melt under electrical overload. There is usually one in the main wire from the battery, and near the alternator output side. If one melts, it must be replaced with a new link of the correct amperage rating. Never replace a melted link with ordinary wire; you run the risk of melting your entire wiring harness.

Headlight

REMOVAL & INSTALLATION

1. Remove the attaching screws from the headlight trim ring. Pull out slightly at the bottom and push up to disengage the upper retaining tab.

2. Remove the trim ring.

3. Remove the three retaining screws from the retaining ring.

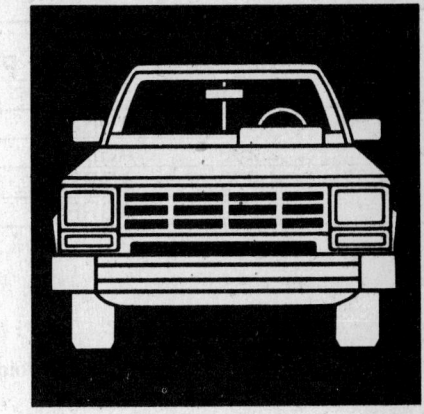

Mazda/Courier

ENGINE IDENTIFICATION CHART

No. Cyl.	Displacement			Fuel System	Application	Manufacturer
	cu. in.	cc	Liters			
4	120.2	1,970	2.0	2-bbl	1979–82 Courier 1979–84 Mazda	Mazda
4	121.9	1,998	2.0	2-bbl	1986 Mazda	Mazda
4	134.7	2,209	2.2	Diesel	1982 Courier 1982–84 Mazda	Mazda
4	140.2	2,299	2.3	2-bbl	1979–82 Courier	Ford

NOTE: Mazda did not market a 1985 model truck. The 1986 models were introduced in mid-year 1985. For that reason, there will be no reference to 1985 trucks in this section.

GENERAL ENGINE SPECIFICATIONS

Years	Engine (cc)	Fuel System Type	Horsepower @ rpm	Torque ft. lb. @ rpm	Bore × Stroke	Comp. Ratio	Oil Press. (psi.) @ 2000 rpm
'79-'84	1,970	2-bbl	77 @ 4300	109 @ 2400	3.150 × 3.860	8.6:1	50–64
	2,299	2-bbl	92 @ 5000	121 @ 3000	3.780 × 3.126	9.0:1	40–60
'82-'84	2,209	Diesel	58 @ 4000	88 @ 2500	3.500 × 3.500	21.0:1	55–60
'86	1,998	2-bbl	80 @ 4500	110 @ 2500	3.390 × 3.390	8.6:1	55–60

TUNE-UP SPECIFICATIONS
Mazda

Years	Engine (cc)	Spark Plugs		Ignition Timing (deg.)		Valve* Clearance		Idle Speed (N)	
		Type	Gap	Man. Trans.	Auto. Trans.	In.	Exh.	Man. Trans.	Auto. Trans.
'79-'84	1,970	BPR-6ES	0.031	8B	8B	0.012	0.012	650	650
'82-'84	2,209	Diesel		2A**	—	0.012	0.012	700	—
'86	1,998	BPR-5ES	0.031	6B	—	0.012	0.012	850	—

*Valve side; engine hot
**Diesel injection static timing
B: Before top dead center
A: After top dead center
N: Manual transmission in neutral;
   automatic transmission in drive

TUNE-UP SPECIFICATIONS
Courier

Years	Engine (cc)	Spark Plugs Type	Spark Plugs Gap	Timing (deg.) Man. Trans.	Timing (deg.) Auto. Trans.	Valve Clearance In.	Valve Clearance Exh.	Idle Speed Man. Trans.	Idle Speed Auto. Trans.
'79-'80	1,970	AGR32	0.031	8B	8B	0.012	0.012	650	650
	2,299	AGRF52	0.043	6B	6B	Hyd.	Hyd.	800	700
'81-'82	1,970	AGR32	0.031	8B	8B	0.012	0.012	650	650
	2,299	AGRF52	0.034	6B	6B	Hyd.	Hyd.	850	700

*Valve side; engine hot

FIRING ORDER

NOTE: To avoid confusion, always replace spark plug wires one at a time.

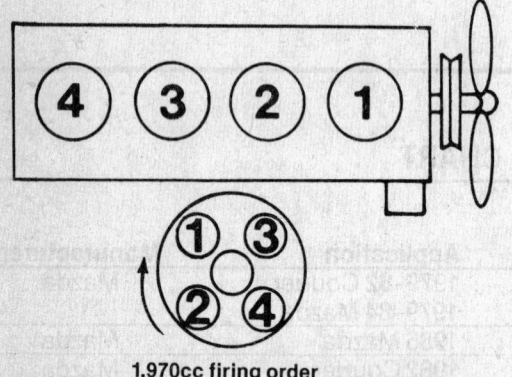

1,970cc firing order

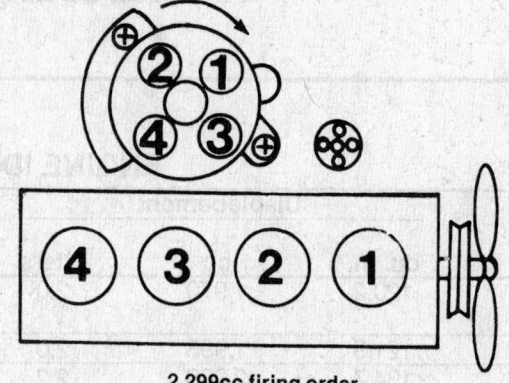

2,299cc firing order

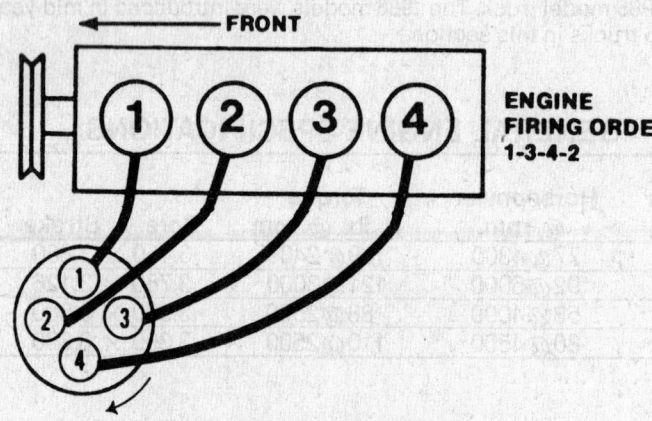

ENGINE FIRING ORDER 1-3-4-2

1986 1,998cc engine firing order

CAPACITIES CHART

Years	Engine (cc)	Crankcase Includes Filter (qt)	Transmission (pts) 4-sp	Transmission (pts) 5-sp	Transmission (pts) Auto.	Drive Axle (pts)	Fuel Tank (gal)	Cooling System (qt) w/AC	Cooling System (qt) wo/AC
'79-'84	1,970	5.0	3.2	3.6	13.2	2.8	①	7.5	7.5
'79-'82	2,299	5.0	3.2	3.6	13.2	2.8	①	8.8	8.8
'82-'84	2,209	5.3	—	3.6	—	2.8	①	—	11.1
'86	1,998	4.5	3.0	2.6	—	2.8	②	7.9	7.9

① Short bed: 15.0
Long bed: 17.5

② Short bed: 14.6
Long bed: 15.6

VALVE SPECIFICATIONS

Engine (cc)	Seat Angle (deg)	Face Angle (deg)	Spring Test Pressure (1 lbs. @ in.)	Spring Installed Height (in.)	Stem to Guide Clearance (in.)		Stem Diameter (in.)	
					Intake	Exhaust	Intake	Exhaust
1,970	45	45	①	②	0.0007-0.0021	0.0007-0.0023	0.3150	0.3150
1,998	45	45	③	④	0.0010-0.0024	0.0010-0.0024	0.3177-0.3185	0.3159-0.3165
2,209	⑤	45	⑥	⑦	0.0015-0.0046	0.0020-0.0051	0.3150	0.3150
2,299	45	44	75 @ 1.56	1.702	0.0010-0.0027	0.0015-0.0032	0.3416-0.3423	0.3411-0.3418

① Inner: 20.9 @ 1.26
   Outer: 31.4 @ 1.34
② 1979-81 Inner: 1.260
   Outer: 1.339
  1982-84 Inner: 1.306
   Outer: 1.385

③ Outer: 96.3 @ 2.007
   Inner: 95.8 @ 1.722
④ Outer: 2.007
   Inner: 1.722
⑤ Intake: 45
   Exhaust: 30

⑥ Inner: 28 @ 1.488
   Outer: 39.6 @ 1.587
⑦ Inner: 1.488
   Outer: 1.587

CAMSHAFT SPECIFICATIONS

Engine (cc)	Journal Diameter					Bearing Clearance	Elevation		End Play
	1	2	3	4	5		Int.	Exh.	
1,970	1.7717	1.7717	1.7717	—	—	①	1.7731	1.7718	0.004
1,998	1.2575-1.2584	1.2563-1.2573	1.2563-1.2573	1.2563-1.2573	1.2575-1.2584	②	1.5040	1.5040	0.004
2,209	2.0473	2.0374	2.0177	—	—	0.0024-0.0047	1.6767	1.6767	0.004
2,299	1.7713-1.7720	1.7713-1.7720	1.7713-1.7720	1.7713-1.7720	—	0.0010-0.0030	0.2437	0.2437	0.004

① #1 & 3: 0.0017
  #2: 0.0021
② #1 & 5: 0.0014-0.0033
  #2, 3, 4: 0.0026-0.0045

CRANKSHAFT AND CONNECTING ROD SPECIFICATIONS

Engine (cc)	Crankshaft				Connecting Rod		
	Main Bearing Journal Dia.	Main Bearing Oil Clearance	Shaft End Play	Thrust on No.	Journal Dia.	Oil Clearance	Side Clearance
1,970	2.4804	0.0012-0.0020	0.0030-0.0090	3	2.0866	0.0011-0.0030	0.0040-0.0080
1,998	2.3597-2.3604	0.0012-0.0019	0.0030-0.0070	3	2.0050-2.0060	0.0010-0.0026	0.0040-0.0100
2,209	2.5591	0.0016-0.0036	0.0055-0.0154	3	2.0866	0.0014-0.0030	0.0094-0.0134
2,299	2.3390-2.3982	0.0008-0.0015	0.0040-0.0080	4	2.0464-2.0472	0.0011-0.0030	0.0035-0.0105

PISTON AND RING SPECIFICATIONS

Engine (cc)	Ring Gap			Ring Side Clearance			Piston Clearance
	#1 Compr.	#2 Compr.	Oil Control	#1 Compr.	#2 Compr.	Oil Control	
1,970	0.0080-0.0160	0.0080-0.0160	0.0120-0.0350	0.0012-0.0028	0.0012-0.0025	snug	①
1,998	0.0080-0.0120	0.0060-0.0120	0.0120-0.0350	0.0012-0.0028	0.0012-0.0028	snug	0.0014-0.0030
2,209	0.0157-0.0217	0.0118-0.0157	0.0138-0.0217	0.0020-0.0035	0.0016-0.0031	0.0012-0.0028	0.0021-0.0031

PISTON AND RING SPECIFICATIONS

Engine (cc)	Ring Gap			Ring Side Clearance			Piston Clearance
	#1 Compr.	#2 Compr.	Oil Control	#1 Compr.	#2 Compr.	Oil Control	
2,299	0.0100–0.0200	0.0100–0.0200	0.0150–0.0550	0.0020–0.0040	0.0020–0.0040	snug	0.0014–0.0022

① 1979–81: 0.0014–0.0030
   1982–84: 0.0019–0.0025

TORQUE SPECIFICATIONS

Engine	Cyl. Head	Conn. Rod	Main Bearing	Crankshaft Damper	Flywheel	Manifold Intake	Manifold Exhaust
1,970	①	30–33	61–65	101–108	112–118	14–19	16–21
1,998	59–64	37–41	61–65	9–12	71–76	14–19	16–21
2,209	80–85	50–54	80–85	145–181	95–137	11–17	11–17
2,299	80–90	30–36	80–90	100–120	54–64	14–21	27–28 ②

① 1979–81 Torque cold to 59–64 ft. lb.; then retorque hot to 69–72 ft. lb.
   1982–84 Torque cold to 65–69 ft. lb.; then retorque hot to 69–72 ft. lb.
② Torque in two steps: 5–7; 14–21

BRAKE SPECIFICATIONS

Year & Model	Master Cyl. Bore	Brake Disc			Brake Drum			Wheel Cyl. or Caliper Bore	
		Original Thickness	Minimum Thickness	Maximum Run-Out	Orig. Inside Dia.	Max. Wear Limit	Maximum Machine O/S	Front	Rear
'79–'84 B2000 Courier	0.875	0.4724	0.4331	0.0039	10.24	10.28	—	2.125	0.875
'82–'84 Diesel	0.813	0.7874	0.7480	0.0039	10.24	10.28	—	2.125	0.875
'86 B2000	0.875	0.7874	0.7086	0.0016	10.24	10.31	—	2.125	0.750

ALTERNATOR AND REGULATOR SPECIFICATIONS

Engine (cc)	Year	Alternator			Regulator		
		Field Current @ 12 v (amps)	Output (amps)	Regulated Volts @ 75° F	Air Gap (in.)	Point Gap (in.)	Point Gap (in.)
1,970	'79–'81	30	45	14–15	0.04	0.015	0.04
	'82–'84	23	45	13.5	—	—	—
1,998	'86	21	55	14.7	—	—	—
2,209	'82–'84	28	80	13.5	—	—	—
2,299	'79–'82	33	45	13.5	—	—	—

STARTER SPECIFICATIONS

Engine (cc)	Year	Lock Test			No-Load Test			Brush* Spring Tension (oz.)
		Amps	Volts	Torque (ft. lb.)	Amps	Volts	rpm	
1,970	'79–'84	310	5.0	5.4	53	11.5	6800	49–63
1,998	'86	430	5.0	7.2	60	11.5	6500	50–60
2,209	'82–'84	1050	2.0	21.6	180	11.0	3800	49–63
2,299	'79–'82	not recommended			50	11.0	5000	35–40

TUNE-UP

Electronic Ignition

COURIER

The 1,970cc and 2,299cc engines use an electronic ignition system which eliminates the points and condenser. Located in the distributor, in addition to the normal ignition rotor, is a four spoke signal rotor which turns on the shaft in the same manner as the points cam found in conventional systems. A magnetic pickup coil is the only other component in the distributor. The system also includes an ignition module, mounted on the left fender apron on 2,299cc engines or right fender apron on 1979 1,970cc engines; the module is mounted on the distributor on the 1980 and later 1,970cc engines. A conventional ignition coil is also used.

When a rotor spoke is not lined up with the pickup coil, it generates large lines of flux between itself, the magnet, and the pickup coil. This large flux variation results in a high generated voltage in the pickup coil, preventing battery current from flowing to the pickup coil. When a rotor spoke lines up with the coil, the flux variation is low. Thus, zero voltage is generated in the pickup coil, allowing current to flow to it. Ignition primary current is then cut off by the ignition module, causing high voltage to be induced in the ignition coil secondary windings. The high voltage flows through the distributor to the spark plug.

Because no points or condenser are used, and because dwell is determined by the module, no adjustments are necessary. Ignition timing is checked in the usual way, but unless the distributor is disturbed it is not likely to change.

Service for the electronic ignition consists of inspection of the distributor cap, rotor, and ignition wires, replacing when necessary. These parts can be expected to last for at least 40,000 miles.

1. The cap is held on by two screws. After unscrewing them, lift the cap straight up and off, with the wires attached. Inspect the cap for cracks, carbon paths, or a worn center contact. Replace it if necessary, transferring the wires one at a time from the old cap to the new.

2. On all models except 1980 and later 1,970cc engines, pull the ignition rotor (not the spoked timing rotor) straight up to remove. The ignition rotor on 1980 and later 1,970cc engines is retained by two screws. Use a magnetic screwdriver to remove the screws, and remove the rotor from the distributor shaft. Inspect the rotor, and replace it if the contacts are worn, burned or pitted. Do not file the contacts. Replace the rotor: On 2,299cc and 1979 1,970cc engines, press the rotor onto the shaft, and check to see that it is fully seated; on 1980 and later 1,970cc models, replace the retaining screws and washers.

3. Inspect the wires for cracks or brittleness. Replace them one at a time to prevent crosswiring, carefully pressing the replacement wires into place. The cores of electronic ignition wires are more susceptible to breakage than those of standard wires, so handle them carefully.

MAZDA

1. The distributor cap is held on by two screws. Unscrew them and lift the cap straight up with the wires still attached. Inspect the cap for cracks, carbon tracking and worn contacts. Replace it, if necessary, transferring the wires one at a time to avoid miswiring.

2. On 1979 models, pull the rotor straight up to remove it. On 1980 and later models, the rotor is held in place by two screws. Use a magnetic screwdriver to remove them and lift off the rotor. Replace the rotor if it appears worn, burned or pitted.

3. Inspect the wires for cracks or brittleness. Replace them, one at a time, if they appear at all suspect. Avoid bending the wires sharply, or kinking them, as the carbon cores are subject to such damage.

For electronic ignition component troubleshooting, refer to the Electrical section in Unit Repair.

Ignition Timing

ADJUSTMENT

Courier

Timing should be checked at each tune-up and any time the points are adjusted or replaced. It is not likely to change very much on engines with electronic ignition.

If the ignition is set too far advanced (BTDC), the ignition and expansion of the fuel in the cylinder will occur too soon and tend to force the piston down while it is still traveling up. This causes engine ping. If the ignition spark is set too far retarded, after TDC (ATDC), the piston will have already passed TDC and started on its way down when the fuel is ignited. This will cause the piston to be forced down for only a portion of its travel, resulting in poor engine performance and lack of power. To check and adjust the timing:

1. Warm the engine to normal operating temperature. Stop the engine and clean off the notches in the crankshaft pulley. Use some paint or chalk to make the marks more visible, if necessary. You will probably have to "bump" the engine around with the starter to get the notches into an accessible position.

2. There are two notches cut into the crankshaft pulley. Looking down at the marks, the Top Dead Center (TDC) mark is on the exhaust system side of the engine. The one on the carburetor side indicates the proper number of degrees before TDC (BTDC) at which the spark plug is to fire. TDC is not colored at all, and the BTDC timing mark is colored white.

3. Disconnect and plug the vacuum hose at the distributor. The line must be plugged to prevent a vacuum leak.

4. Connect the timing light according to the manufacturer's instructions to the No. 1 spark plug (at the front of the engine). Connect a tachometer according to the manufacturer's instructions.

5. Set the parking brake, block the front wheels, put the transmission in Neutral, and start the engine.

CAUTION

Keep your hands, hair, clothes, and the various wires clear of the fan, belts and pulleys.

---

6. Reduce the idle speed by means of the idle speed screw. Adjust the idle to 700–750 rpm (800–850 rpm with the 2,299cc and manual transmission). This prevents the centrifugal advance mechanism from affecting the timing. Aim the timing light at the timing marks. If the BTDC mark on the pulley and the pointer are lined up, the timing is okay.

7. If the timing marks are not aligned, adjust by loosening the distributor hold-down clamp bolt at the base of the distributor shaft and turning the distributor slightly in one direction or the other until the marks coincide. Hold the distributor body, not the cap. When the marks are aligned, tighten the bolt.

8. Check the timing again to be sure it didn't change during the tightening process.

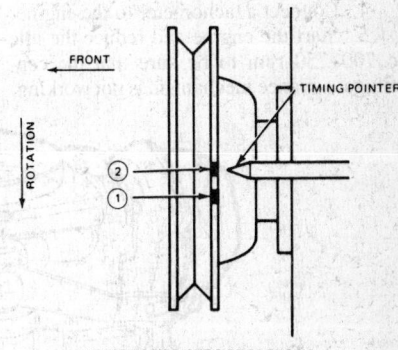

VIEWED FROM RIGHT SIDE OF ENGINE

1. Denotes 5° BTDC on 1975–77 and 1980 models; denotes 8° BTDC on 1978–79 and 1981–82 models.
2. Denotes TDC—all models

Timing mark identification. The ATDC mark is to the right of top dead center (above it in the illustration).

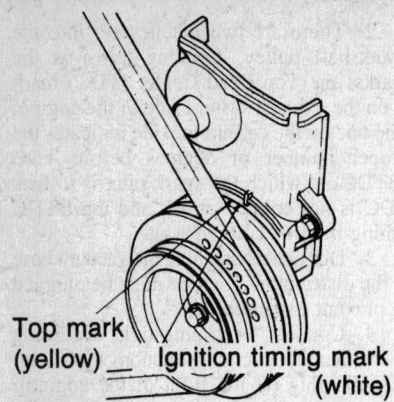

Top mark (yellow) — Ignition timing mark (white)

1986 1,998cc engine ignition timing marks

9. Check the centrifugal advance mechanism by accelerating the engine to 2000 rpm. The timing should advance (the marks should move away from the pointer, towards the carburetor side of the engine). Note the speed at which the advance begins.

10. Unplug and reattach the vacuum hose. Accelerate the engine to 2000 rpm. The engine speed at which advance begins should be sooner (and the advance should be greater) than with only the centrifugal advance operating. If this does not occur, the vacuum advance unit should be inspected for free operation, and the hose checked for vacuum. Reset the idle to specification.

11. Shut off the engine, and disconnect the timing light and tachometer.

Mazda

1. Raise the hood and clean and mark the timing marks. Chalk or fluorescent paint makes a good, visible mark.

2. Disconnect the vacuum line at the distributor and plug the disconnected line.

3. Connect a timing light to the front (No. 1) cylinder, a power source and ground. Follow the manufacturer's instructions.

4. Connect a tachometer to the engine.

5. Start the engine and reduce the idle to 700–750 rpm to be sure that the centrifugal advance mechanism is not working.

6. With the engine running, shine the timing light at the timing pointer and observe the position of the pointer in relation to the timing marks on the crankshaft pulley. All engines have two timing marks. Looking straight down on the marks, the one on the left is TDC, the one on the right is BTDC.

6. With the engine running, shine the timing light at the timing pointer and observe the position of the pointer in relation to the timing mark on the crankshaft pulley.

7. If the timing is not as specified, adjust the timing by loosening the distributor holddown bolt and rotating the distributor in the proper direction. When the proper ignition timing is obtained, tighten the holddown bolt on the distributor.

8. Check the centrifugal advance mechanism by accelerating the engine to about 2,000 rpm. If the ignition timing advances, the mechanism is working properly.

9. Stop the engine and remove the timing light.

10. Reset the idle to specifications.

11. Remove the tachometer.

Idle Speed & Mixture Adjustments

COURIER

1. Put the transmission in Neutral.

2. Connect a tachometer to the engine.

3. Start the engine and allow it to reach normal operating temperature. The choke should be fully open and the air cleaner on.

4. Run the engine for 3 minutes at 2000 rpm. Disconnect the canister purge hose between the canister and the air cleaner case.

5. Set the curb idle speed to specifications, using the curb idle speed adjusting screw. The correct idle speed can be found in the "Tune-Up Specifications" chart located at the beginning of this section, or on the emission control information label in the engine compartment.

6. The carburetor is probably equipped with idle limiter caps, which are small caps designed to provide a limited range of mixture adjustment.

7. The mixture can be checked only with the truck hooked up to an exhuast emission HC/CO analyzer. Your dealer, state inspection center, or service station may have one of these. The mixture is adjusted after the idle speed. The engine should be warm, choke open, and air cleaner on.

8. Disconnect the air hose between the air pump and the check valve, and plug the port of the check valve.

9. Turn the idle mixture adjusting screw in or out to obtain the specified idle/CO setting. The proper figure is given on the emission control information label in the engine compartment.

10. Unplug the check valve port and reconnect the hose. Connect the canister purge hose.

Supplemental Checks

If a satisfactory idle speed cannot be obtained after the normal idle adjustments, check the following:

1. Vacuum leaks.
2. Ignition system wiring.
3. Spark plugs.
4. Dwell angle.
5. Distributor point condition, and
6. Ignition timing.

If the idle condition does not improve, check these items, and correct, if necessary:

    a. Fuel level
    b. Crankcase ventilation system
    c. Valve clearance
    d. Engine compression

If a satisfactory idle still cannot be obtained, have someone with an exhaust gas analyzer look into the problem for a possible too lean mixture or over-rich mixture.

MAZDA

Gasoline Engines

1979–80

1. Connect a tachometer to the engine. Warm the engine to normal operating temperature.

2. Run the engine (in neutral) to 2,000 rpm for a minute or two.

3. Allow the engine to return to idle. Adjust the idle to specifications by means of the idle speed screw.

4. The mixture can only be adjusted with the air of an exhaust gas analyzer. Connect the CO meter to the exhaust and note the reading.

5. On models sold in California, disconnect the air hose between the air pump and the check valve, and plug the port of the check valve. On 49 States vehicles, disconnect the air hose between the air cleaner and reed valve and plug the port of the reed valve.

6. Adjust the mixture by means of the mixture screw until the CO concentration is 2–4%.

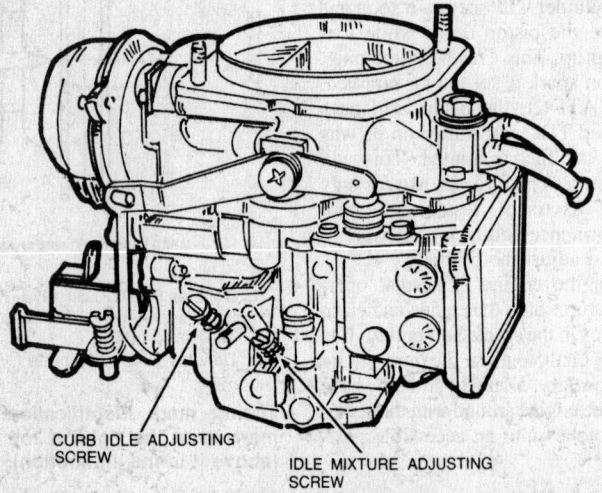

CURB IDLE ADJUSTING SCREW

IDLE MIXTURE ADJUSTING SCREW

Carburetor adjusting screws

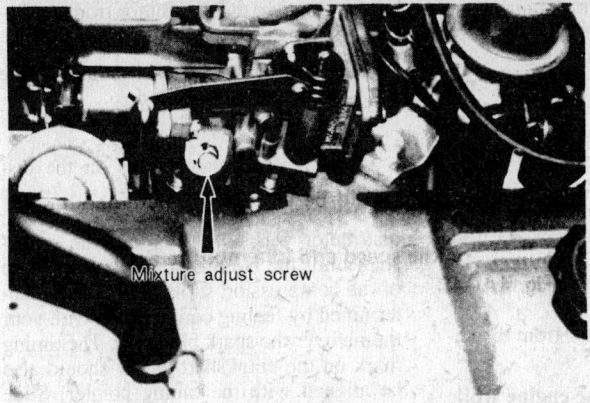

Mixture adjusting screw location on 81-84 engines

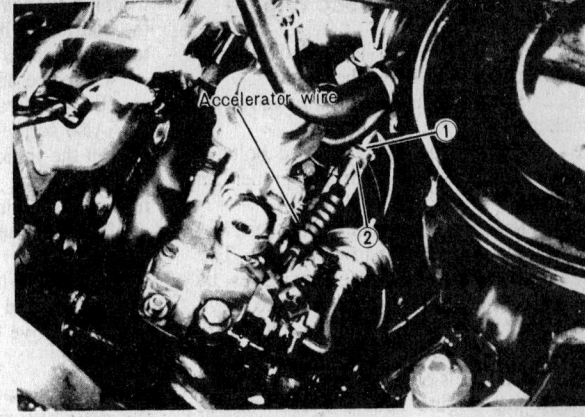

Diesel throttle linkage adjustment

7. If the limiter cap was removed from the mixture screw, reinstall it.

1981–86

1. Remove the mixture screw by pulling out the entire shell while turning it counterclockwise. Install a new mixture screw.

2. Turn the new screw very slowly until it seats *very lightly*. Then, reverse rotation and turn three turns.

3. Connect an exhaust gas analyzer to the vehicle.

4. On California vehicles, disconnect the air hose between the air bypass valve and check valve and plug the check valve port. On 49 States vehicles, disconnect the air hose connecting the air cleaner and reed valve and plug the port in the reed valve.

5. Turn the throttle adjusting screw until idle speed (engine hot) is 620 rpm in Neutral for California vehicles and 670 rpm for 49 States vehicles. Then, turn the mix-

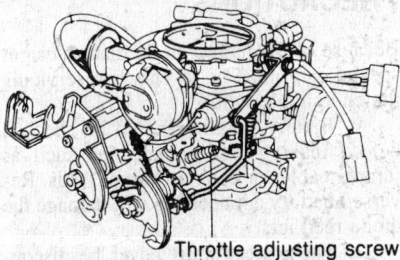

1986 1,998cc throttle adjusting screw location

Diesel idle speed adjustment

ture screw to get the highest idle rpm. Repeat these two procedures until the idle speed meets specs at the mixture giving the highest rpm.

6. Turn the mixture screw clockwise until rpm is 600 rpm for California vehicles and 650 rpm for 49 States vehicles.

7. Check CO concentration. If it is less than 1%, turn the mixture screw counterclockwise $1/4$ turn.

8. Reconnect the disconnected air hose (with plugs you installed removed) and then recheck and, if necessary, readjust idle speed.

9. Install a blind plug into the mixture adjusting screw shell.

Diesel Engine

1. Run the engine until it reaches normal operating temperature. Check idle speed with a special diesel tachometer at the injection pump. If idle speed is not 700 rpm, check the play in the accelerator cable (next step) and then adjust the idle speed adjusting bolt.

2. Check the play in the accelerator linkage. It should be 0.04–0.12 in. If the play has to be adjusted, loosen the locknut (1), and then turn the adjusting nut (2) clockwise to increase the play or counterclockwise to decrease it. Tighten the locknut.

3. Then, loosen the locknut (3) for the idle adjusting bolt, and turn the adjusting bolt clockwise to increase the idle speed or counterclockwise to decrease it. Tighten the locknut.

4. Open the throttle slightly and then release it to check that the throttle linkage permits the injection pump to return to idle. Adjust the throttle linkage to provide free operation, if necessary.

Valve Adjustment

NOTE: While all valve adjustments must be made as accurately as possible, it is better to have the valve adjustment slightly loose than slightly tight, as burned valves may result from overly tight adjustments.

2,299cc Engine

Valve clearance in the 2,299cc engine are automatically adjusted by hydraulic valve lash adjusters, which resemble hydraulic valve lifters both in appearance and in function. Adjustments are not necessary.

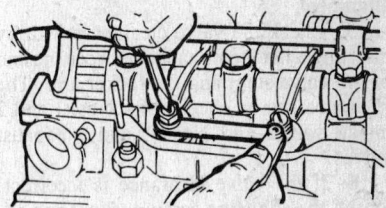

Checking the valve clearances at the valve

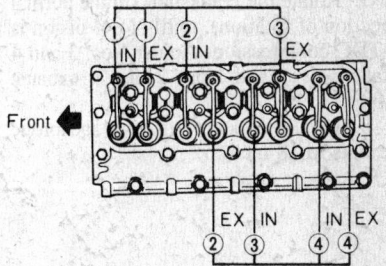

Diesel valve adjustment sequence

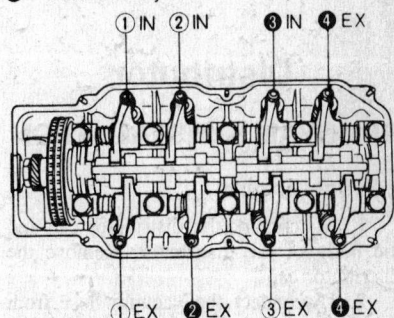

1,970cc and 1,998cc engine valve adjustment sequence

All except the 2,299cc Engine

1. Run the engine until normal operating temperature is reached.

2. Shut off the engine and remove the rocker cover.

3. Torque the cylinder head bolts to their specified torque, in the proper sequence.

4. Rotate the crankshaft so that the No. 1 cylinder (front) is in the firing position. This can be determined, on gasoline engines, by removing the spark plug from the No. 1 cylinder and putting your thumb over the spark plug port. When compression is felt, the No. 1 cylinder is on the compression stroke. Rotate the engine with a wrench on the crankshaft pulley and stop it at TDC of the compression stroke on the No. 1 cylinder, as confirmed by the alignment of the TDC mark in the crankshaft pulley and the timing pointer. On the diesel, No. 1 TDC can be determined by loosening the No. 1 injection pipe at the injection pump. Whe the engine is slowly cranked, fuel will squirt from the loosened fitting as No. 1 piston approaches TDC. Confirm TDC by the timing mark alignment.

5. Check the valve clearances with by inserting a flat feeler gauge between the end of the valve stem and the rocker arm. The clearance can be checked for Nos. 1 and 2 intake valves and Nos. 1 and 3 exhaust valves.

6. If the valve clearance is incorrect, loosen the adjusting screw locknut and adjust the clearance by turning the adjusting screw with the feeler blade inserted. Hold the adjusting screw in the correct position and tighten the locknut.

7. Rotate the crankshaft (in the normal direction of rotation), until No. 4 piston is at TDC compression. Adjust Nos. 3 and 4 intake valves and Nos. 2 and 4 exhaust valves.

8. Install the rocker arm cover and torque the nuts to 18 inch lb.

ENGINE ELECTRICAL

Distributor

REMOVAL & INSTALLATION

2,299cc Engine

1. Matchmark the distributor cap and the body of the distributor. Remove the distributor cap.

2. Disconnect the vacuum hose from the diaphragm.

3. Remove the rubber plug from the timing belt cover. Scribe matchmarks on the distributor body and the cylinder block to indicate their relative positions.

4. Scribe another mark on the distributor body indicating the position of the rotor. Also scribe a mark on the cam pulley and on the indicator inside the timing belt cover to mark the position of the pulley relative to the indicator.

5. Disconnect the primary wires from the distributor.

6. Remove the distributor holddown nut, lockwasher and flat washer.

7. Remove the distributor from the engine.

NOTE: Do not crank the engine while the distributor is removed. All distributors have a helical drive gear, which will cause the rotor to turn slightly when the distributor is removed. If the amount of movement is noted during removal, the rotor can be adjusted slightly before installation to compensate. Lubricate the drive gear and distributor shaft with engine oil before installation.

8. If the engine was cranked while the distributor was removed, turn the crankshaft until No. 1 cylinder is at the top of the compression stroke. This can be determined by feeling compression with your thumb through the spark plug port. The TDC mark on the crankshaft pulley should also be aligned with the timing pointer. Match the mark on the cam pulley and the TDC mark on the crankshaft pulley. Slide the distributor into the engine with the rotor pointing to No. 1 firing position.

9. If the engine has not been cranked while the distributor was removed, slide the distributor (with the O-ring) into the engine, aligning the matchmarks made during removal.

10. Install the flat washer, lockwasher and hold-down nut, but do not tighten the nut.

11. Install the distributor cap and connect the primary wires.

12. Set the ignition timing as previously outlined and tighten the holddown nut.

13. Connect the vacuum line.

1,970cc and 1,998cc Engines

1. Matchmark the distributor cap and the body of the distributor. Remove the distributor cap.

2. Disconnect the vacuum hose from the diaphragm. Disconnect the electrical wire at the distributor, if so equipped.

3. Scribe matchmarks on the distributor body and the cylinder block to indicate the relative positions.

4. Scribe another mark on the distributor body indicating the position of the rotor.

5. Disconnect the primary wires from the distributor.

6. Remove the distributor holddown nut, lockwasher and flat washer.

7. Remove the distributor from the engine.

NOTE: Do not crank the engine while the distributor is removed.

8. Align the matchmarks on the distributor gear and body.

9. If the engine was cranked while the distributor was removed, turn the crankshaft until the No. 1 cylinder is at the top of the compression stroke. This can be determined by feeling compression with your thumb over the spark plug port. The timing mark on the crankshaft pulley should also be aligned with the timing pointer. Slide the distributor into the engine with the rotor pointing to the No. 1 cylinder firing position (see Firing Order).

10. If the engine has not been cranked while the distributor was removed, slide the distributor (with the O-ring) into the engine, aligning the matchmarks made during removal.

11. Install the flat washer, lockwasher and holddown nut, but do not tighten the nut.

12. Install the distributor cap and connect the primary wires.

13. Set the ignition timing, and tighten the holddown nut.

14. Connect the vacuum line wire, if so equipped.

Alternator

PRECAUTIONS

Because of the nature of alternator design, special care must be taken when servicing the charging system.

1. Battery polarity should be checked before making any connections such as jumper cables or battery charger leads. Reversed battery connections will damage the diode rectifiers.

2. The battery must never be disconnected while the alternator is running because the regulator will be ruined.

3. Always disconnect the battery ground cable before replacing the alternator.

4. Do not attempt to polarize an alternator.

5. Do not short across or ground any alternator terminals.

6. Always disconnect the battery ground cable before removing the alternator output cable.

7. If electric arc welding equipment is to be used on the car, first disconnect the battery and alternator cables. Never operate the car with the electric arc welding equipment attached.

8. If the battery is to be ''quick charged'', disconnect the negative cable from the battery.

REMOVAL & INSTALLATION

1. Disconnect the negative battery terminal.
2. Label and disconnect all wiring.
3. Remove the alternator adjusting bolt.
4. Remove the V-belt.
5. On the diesel, remove the vacuum and oil hoses.
6. Remove the remaining attaching bolts and remove the alternator.
7. Install in exact reverse order, adjusting belt tension as described below.

BELT TENSION ADJUSTMENT

1. Check the alternator drive belt tension by applying thumb pressure to the belt midway between the fan pulley and the alternator. It should deflect $3/8$–$1/2$ in. with a new belt or $1/2$–$5/8$ in. with a used belt.
2. If it does not have the correct tension, loosen the alternator mounting bolts slightly and loosen the adjusting arm bolt.
3. Push the alternator in the direction required to obtain proper belt deflection.

——— **CAUTION** ———
Do not pry or pound on the alternator housing.

4. Tighten the adjusting arm bolt to 20 ft. lb. Tighten the alternator mounting bolts securely.
5. Recheck the tension.

Regulator

REMOVAL & INSTALLATION

NOTE: 1982 and later trucks use an integral regulator, built into the alternator. No adjustments are possible.

1. Raise the hood and disconnect the negative battery cable.
2. Disconnect the regulator wires at the multiple connector.
3. Remove the two regulator attaching screws and remove the regulator from the splash shield.
4. Position the regulator on the fender splash shield and install the two attaching screws.
5. Connect the regulator wires at the multiple connector.
6. Connect the negative battery cable.
7. Start the engine and be sure that the charging system indicator light goes out.

Starter

REMOVAL & INSTALLATION

1. Raise the hood and disconnect the battery ground cable.

2. Remove the carburetor air cleaner and air intake tube.
3. Disconnect the battery cable from the starter solenoid battery terminal.
4. Pull the ignition switch wire from the solenoid terminal.
5. Raise and support the truck on jackstands.
6. Working under the truck, remove the two starter attaching bolts, washers and nuts.
7. Tilt the drive end of the starter and remove the starter by working it out below the emission system hoses.
8. Install the starter and two bolts, washers and nuts.
9. Connect the ignition switch wire to the solenoid terminal.
10. Connect the battery cable to the solenoid battery terminal.
11. Install the carburetor air cleaner and air intake tube.
12. Connect the ground cable to the battery.
13. Lower the truck to the ground and check the operation of the starter.

For starter motor servicing, refer to the Electrical section of Unit Repair.

ENGINE MECHANICAL

Engine

REMOVAL & INSTALLATION

2,299cc Engine

The engine is removed through the engine compartment leaving the transmission in place. When installing nuts or bolts, lubricate the threads with light engine oil, but do not oil threads which require oil-resistant or water-resistant sealer.

NOTE: Always label all disconnected hoses as they are removed, as an aid to correct installation.

1. Scribe the locations of the hood hinges and remove the hood.
2. Drain the coolant.
3. Remove the air cleaner and the heat stove.
4. Disconnect the radiator hoses at the radiator. If you have an automatic transmission, also disconnect and plug the two fluid cooler lines at the radiator.
5. Unbolt and remove the radiator. You will first have to loosen the shroud mounting bolts and slide the shroud rearward.
6. Disconnect the accelerator linkage at the carburetor.

7. Disconnect and plug the fuel line at the carburetor.
8. Remove the linkage attaching nuts at the intake manifold and remove the linkage.
9. Disconnect the cable at the air bypass valve (if equipped).
10. Disconnect the choke cable.
11. Disconnect the battery cables, negative cable first.
12. Disconnect the coil high tension wire at the distributor. Disconnect the coil lead wire.
13. Remove the fan.
14. Loosen the alternator retaining bolts and remove the alternator drive belt.
15. If equipped, remove the Thermactor air pump drive belt.
16. Remove the alternator bracket and adjusting arm bolts. Position the alternator out of the way. If your Courier has air conditioning, unbolt the compressor and set it aside without detaching any lines.

——— **CAUTION** ———
Never open any air conditioning lines. Escaping compressed refrigerant can freeze any body surface, including the eyes, that it contacts. It also decomposes into a poisonous gas in the presence of flame.

17. If equipped, disconnect the Thermactor hoses at the pump. Remove the Thermactor bracket and adjusting arm bolt and lay the Thermactor pump aside. Disconnect the brake vacuum booster hose at the engine, and the vacuum lines at the vacuum amplifier.
18. Remove the heater hoses.
19. If equipped, disconnect the Thermactor air filter hose at the air by-pass valve.
20. Disconnect the oil pressure gauge lead wire and boot from the sending unit.
21. Disconnect the battery ground cable from the block.
22. Disconnect the wires from the starter solenoid.
23. Raise and support the vehicle and drain the oil from the engine.
24. Remove the engine front lower skid plate.
25. Disconnect the exhaust pipe at the exhaust manifold. Also unbolt the exhaust pipe hanger from the transmission.
26. Remove the lower transmission-to-engine bolts at this time. With automatic transmission, first disconnect the vacuum hose from the diaphragm. Unbolt the access cover from the lower end of the torque converter housing. Matchmark the driveplate (flywheel) and the torque converter for later alignment, then remove the four bolts connecting the driveplate to the torque converter. Disconnect the throttle linkage from the transmission.
27. Remove the lower exhaust pipe bolt and let the pipe hang on the 1,970cc. Remove the engine support bolts and remove the right engine mount bracket.

28. Lower the truck.

29. Remove the starter upper bolts and remove the starter.

30. Support the transmission with a jack. Remove the remaining transmission-to-engine bolts.

31. Install a lifting sling on the engine at the engine hanger brackets. Remove the clutch slave cylinder from the transmission.

32. Attach a sling to the hoist and pull the engine forward until it clears the transmission shaft. Be sure that the transmission is not dislodged from the jack. Unbolt the engine from the mounts.

33. Lift the engine from the truck. If your truck has an automatic transmission, be careful not to let the torque converter fall out of its housing.

34. Lower the engine into the chassis. Loosely install several clutch housing bolts.

35. Install the engine mount bolts. Install the right engine mount and the clutch slave cylinder.

36. Remove the lifting sling and transmission jack. Be sure that the mainshaft is engaged with the engine.

37. Install the starter and upper attaching bolt.

38. Raise and support the truck.

39. Install the flywheel housing bolts with manual transmission. With automatic, line up the marks made earlier on the driveplate and the torque converter, and install the four bolts. Torque them to 25–36 ft. lb. Install the torque converter housing-to-engine bolts, and torque to 23–34 ft. lb. Replace the access cover on the torque converter housing.

40. Connect the exhaust pipe to the exhaust manifold. Install the bracket bolt. Install the exhaust pipe hanger to the transmission.

41. Install the lower starter bolts.

42. Install the engine front skid plate.

43. Connect the starter wires and lower the vehicle.

44. Install the accelerator linkage. Connect the fuel line and choke cable. With automatics, also connect the throttle linkage and the vacuum hose.

45. Connect the heater hoses. If equipped, connect the Thermactor air filter hose. Connect the brake vacuum booster hose and the vacuum line at the amplifier.

46. Install the battery ground cable to the block. Install the coil wires.

47. Install the alternator on the bracket and install the bolts. Install the air conditioning compressor, if removed.

48. If equipped, install the Thermactor air pump.

49. Install the fan.

50. Install the drive belts and adjust the tension.

51. Install the radiator. Connect the upper and lower hoses. Connect the automatic transmission cooler lines, if equipped.

52. Connect the oil pressure sending unit.

53. Fill the cooling system and crankcase with the specified type and amount of fluid. Install the air cleaner and heat stove.

54. Install the hood. Connect the battery cables. Start the engine and check for leaks and proper operation.

1,970cc Engine

1. Scribe the locations of the hood hinges and remove the hood.

2. Remove the engine splash shield.

3. Drain the coolant.

4. Drain the engine oil.

5. Disconnect the battery cables and remove the battery.

6. Disconnect the primary wire and coil wire from the distributor.

7. Disconnect the wires at the alternator and disconnect the plug from the rear of the alternator. Remove the drive belts from the alternator, thermactor and air conditioning compressor. Remove each of these items from their mounting brackets and position them out of the way. It is not necessary to disconnect and refrigerant lines.

8. Disconnect the wire from the oil pressure switch.

9. Disconnect the engine ground wire.

10. Remove the air cleaner and heat insulator.

11. Disconnect the breather hose from the rocker cover.

12. Disconnect the water temperature gauge wire and solenoid valve wire.

13. Disconnect the starter wires.

14. Remove the upper and lower radiator hoses. On models with automatic transmission, disconnect the cooling lines at the radiator tank.

15. Remove the bolts attaching the radiator cowling. The cowling can only be removed after the radiator has been removed.

16. Unbolt and remove the radiator and cowling.

17. Disconnect the heater hoses from the intake manifold.

18. Disconnect the throttle cable from the carburetor and remove the throttle linkage from the rocker cover attaching point or the intake manifold clamps, on later models.

19. Disconnect the choke cable from the carburetor.

20. Disconnect the fuel ventilation hose from the oil separator.

21. Disconnect the fuel line at the carburetor and plug the fuel line.

22. Raise and support the truck on jackstands. Remove the starter.

23. Disconnect the exhaust pipe from the manifold.

24. Remove the engine skid plate. Remove the clutch cover plate. On trucks with automatic transmission, disconnect the vacuum line at the modulator. Remove the torque converter access cover, matchmark the converter and drive plate and remove the four converter-to-drive plate bolts.

25. Support the transmission with a jack and remove the bolts attaching the engine to the transmission.

26. Unbolt the right and left engine mounts.

27. Attach a lifting sling to the engine and pull the engine forward until it clears the clutch shaft.

28. Lift the engine from the truck.

29. Installation is the reverse of removal. Be sure to check all fluid levels. Observe the following torques:

● Converter-to-drive plate: 25–35 ft. lb.
● Transmission-to-engine: 25–35 ft. lb.

1,998cc Engine

1. Scribe the locations of the hood hinges and remove the hood.

2. Remove the engine splash shield.

3. Drain the coolant.

4. Drain the engine oil.

5. Disconnect the battery cables and remove the battery.

6. Remove the air cleaner and the oil dipstick.

7. Remove the radiator shroud and the engine fan. Place the fan in an upright position to avoid fluid loss from the fan clutch.

8. Disconnect and tag all wires, hoses, cables, pipes and linkage from the engine.

9. Remove the 3-way solenoid valves, but don't disconnect the vacuum tubes.

10. Remove the duty solenoid valves, but don't disconnect the vacuum tubes.

11. Remove the emissions canister.

12. Remove the radiator.

13. Disconnect the exhaust pipe at the manifold and remove the exhaust manifold.

14. Dismount the air conditioning compressor and position it out of the way. Don't disconnect any refrigerant lines.

15. Dismount the power steering pump and position it out of the way without disconnecting any hoses.

16. Raise and support the truck on jackstands.

17. Remove the starter.

18. Attach a lifting sling and shop crane to the engine lifting eyes and take up the weight of the engine.

19. Support the transmission with a floor jack and remove the transmission-to-engine bolts.

20. Remove the engine support plates and mounting nuts, push the engine forward to clear the transmission and lift it out of the truck.

21. Installation is the reverse of removal. Torque the exhaust manifold-to-engine nuts to 16–21 ft. lb.

Diesel

1. Scribe the locations of the hood hinges and remove the hood.

2. Remove the engine splash shield.

3. Drain the coolant.

4. Drain the engine oil.

5. Disconnect the battery cables and remove the battery.

6. Remove the air cleaner and the oil dipstick.

7. Remove the radiator shroud and the engine fan. Place the fan in an upright position to avoid fluid loss from the fan clutch.

8. Disconnect and tag all wires, hoses, cables, pipes and linkage from the engine.

9. Remove the clutch release cylinder.

10. Remove the oil cooler.

11. Remove the radiator.

12. Disconnect the exhaust pipe at the manifold and remove the exhaust manifold.

13. Dismount the air conditioning compressor and position it out of the way. Don't disconnect any refrigerant lines.

14. Dismount the power steering pump and position it out of the way without disconnecting any hoses.

15. Raise and support the truck on jackstands.

16. Attach a lifting sling and shop crane to the engine lifting eyes and take up the weight of the engine.

17. Support the transmission with a floor jack and remove the transmission-to-engine bolts.

19. Remove the engine support plates and mounting nuts, push the engine forward to clear the transmission and lift it out of the truck.

20. Installation is the reverse of removal.

Remove arrowed engine mount bolts

Cylinder Head
REMOVAL & INSTALLATION

1,970cc Engine

NOTE: The engine must be cold before proceeding.

1. Drain the cooling system.

2. Scribe alignment marks around the hood hinges and remove the hood.

3. Remove the air cleaner.

4. Disconnect the coil wire and vacuum line from the distributor.

5. Rotate the crankshaft to put the No. 1 cylinder at TDC on the compression stroke.

6. Remove the plug wires and distributor cap as a unit.

7. Remove the distributor.

8. Remove the rocker arm cover.

9. Raise and support the truck. Disconnect the exhaust pipe from the manifold.

10. Remove the accelerator linkage. Disconnect and tag all wiring, cable and hoses from the engine.

11. Remove the water pump. Remove the nut, washer and distributor gear from the camshaft.

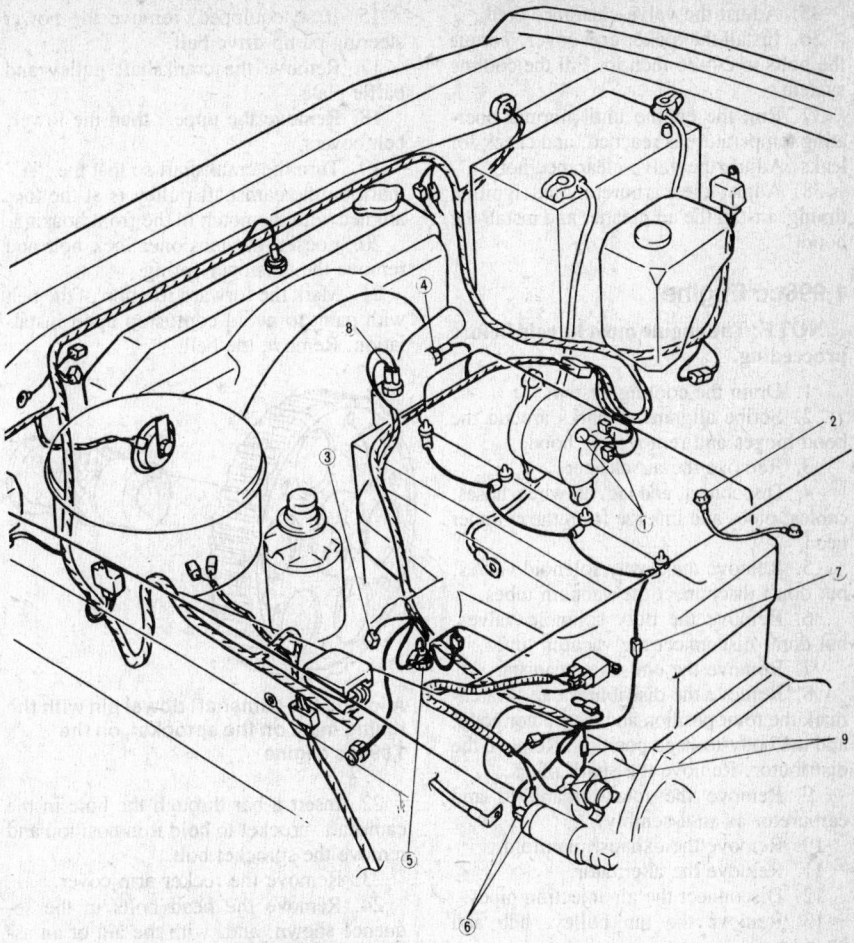

Disconnect the wiring in the order shown

12. Remove the nut, washer and camshaft gear. Support the timing chain from falling into the timing chain case. Do not remove the cam gear from the timing chain. The relationship between the chain and gear teeth should not be disturbed.

13. Remove the cylinder head bolts, and cylinder head-to-front cover bolt.

14. Remove the rocker arm assembly.

15. Remove the camshaft and camshaft gear.

16. Lift off the cylinder head.

17. Remove all tension from the timing chain.

18. Clean the rocker cover gasket surface at the head and the cover. Clean the head gasket surface at the head and the block. Clean the water pump gasket surface at the head and the block. Clean the water pump gasket surface at the head gasket surface and the front cover.

19. Check the cylinder head flatness with a straightedge and feeler blades. It should not exceed 0.003 in. in any six in. span or 0.006 in. overall. If necessary, the cylinder head can be milled, not to exceed 0.008 in.

20. Clean the cylinder head bolt holes of oil and dirt.

21. Position a new head gasket on the cylinder block.

22. Install the cylinder head on the block using the guides at either end of the block.

23. Install the camshaft on the head and camshaft gear.

24. Install the rocker arm assembly.

25. Install the head bolts. Torque the bolts to specification.

26. Install the camshaft gear washer and nut.

27. Install the distributor gear, washer and nut.

28. Time the engine. Follow the instructions under Timing Chain Tensioner Adjustment.

29. Adjust the timing chain tension. See Timing Chain Tensioner Adjustment.

30. Connect the exhaust pipe to the exhaust manifold. Lower the truck.

31. Install the distributor, distributor cap and plug wires.

32. Install the lower intake bracket bolt.

33. Install the accelerator linkage.

34. Connect the vacuum line and coil wire.

1,970cc engine cylinder head bolt torque sequence

35. Adjust the valve clearance cold.

36. Install the rocker arm cover. Torque the bolts to 24–36 inch lb. Fill the cooling system.

37. Run the engine until normal operating temperature is reached, and check for leaks. Adjust the valve clearance hot.

38. Adjust the carburetor and ignition timing. Install the air cleaner and install the hood.

1,998cc Engine

NOTE: The engine must be cold before proceeding.

1. Drain the cooling system.

2. Scribe alignment marks around the hood hinges and remove the hood.

3. Remove the air cleaner.

4. Disconnect, and tag, all wires, hoses, cables, pipes and linkage from the cylinder head.

5. Remove the 3-way solenoid valves, but don't disconnect the vacuum tubes.

6. Remove the duty solenoid valves, but don't disconnect the vacuum tubes.

7. Remove the emissions canister.

8. Remove the distributor cap. Match-mark the rotor position and distributor body, and the body-to-head position. Remove the distributor. Remove the spark plugs.

9. Remove the intake manifold and carburetor as an assembly.

10. Remove the exhaust manifold.

11. Remove the alternator.

12. Disconnect the air injection pipes.

13. Remove the fan pulley, hub and bracket.

14. If so equipped, remove the air conditioning compressor drive belt.

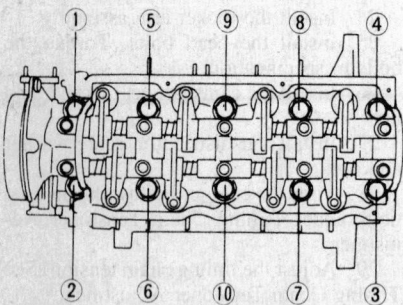

1,998 cc engine head bolt loosening sequence

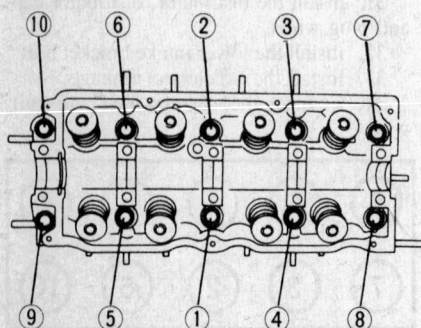

1,998cc engine head bolt tightening sequence

15. If so equipped, remove the power steering pump drive belt.

17. Remove the crankshaft pulley and baffle plate.

18. Remove the upper, then the lower, belt covers.

19. Turn the crankshaft so that the "A" mark on the camshaft pulley is at the top, aligned with the notch in the front housing.

20. Loosen the tensioner lock bolt and remove the tensioner spring.

21. Mark the forward rotation of the belt with paint to avoid confusion upon installation. Remove the belt.

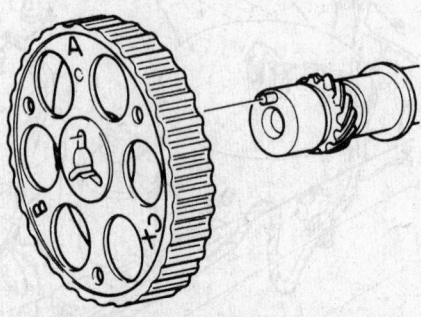

Aligning the camshaft dowel pin with the timing mark on the sprocket, on the 1,998cc engine

22. Insert a bar through the hole in the camshaft sprocket to hold it in position and remove the sprocket bolt.

23. Remove the rocker arm cover.

24. Remove the head bolts in the sequence shown, and, with the aid of an assistant or a lifting device, lift off the head.

25. Clean the head and block mating surfaces thoroughly and install a new head gasket on the block.

26. Install the head and tighten the head bolts, in the order shown, to 60–64 ft. lb. If new head bolts are being used, make sure you use the new, surface treated plain washers.

27. Install the camshaft pulley with the dowel pin on the camshaft engaging the pulley slot just below the A mark on the pulley. Tighten the bolt to 40–48 ft. lb. The timing mark on the front housing and the A mark must be aligned.

28. Lubricate the distributor O-ring with clean engine oil, align all the matchmarks and install the distributor.

29. Replace the belt if it has been contaminated by oil or grease, or shows any sign of damage, wear, cracks or peeling.

30. Make sure that the timing mark on the camshaft is aligned as described above, and that the timing mark (notch) on the crankshaft sprocket is aligned with the triangular shaped mark on the front housing.

31. Install the tensioner and spring, positioning the tensioner all the way to the intake manifold side and temporarily secure it there with the lock bolt.

32. Install the belt onto the sprockets from YOUR right side. If you are reusing the original belt, make sure you follow the directional mark previously made.

33. Loosen the lock bolt so that the tensioner applies tension to the belt.

34. Turn the crankshaft two full revolutions in the direction of normal rotation. This will apply equal tension to all points of the belt.

35. Make sure that the timing marks are still aligned. If not, repeat the belt installation procedure.

36. Tighten the tensioner lock bolt to 30–35 ft. lb.

37. Measure the timing belt tension by pressing on the belt at the midpoint of the longest straight run. Belt deflection should be 11–13mm. If not, repeat the belt adjustment procedure, above.

38. Installation of all other parts is the reverse of removal. Observe the following torques:

Belt covers: 80 inch lb.
Fan bracket: 40 ft. lb.
Exhaust manifold: 16–21 ft. lb.
Intake manifold: 14–19 ft. lb.
Spark plugs: 11–17 ft. lb.
Rocker arm cover: 24–36 inch lb.

When installing the drive belts on the various accessories, check the belt deflection.

Diesel

NOTE: The engine must be cold before proceeding.

1. Drain the cooling system.

2. Scribe alignment marks around the hood hinges and remove the hood.

3. Remove the air cleaner.

4. Disconnect, and tag, all wires, hoses, cables, pipes and linkage from the cylinder head.

5. Remove the injection lines and injectors.

6. Remove the intake and exhaust manifolds.

7. Dismount the alternator and move it out of the way.

8. Remove the rocker arm cover. Remove the rocker arm assembly and lift out the pushrods, keeping them in order for proper installation.

9. Remove the cylinder head bolts and, with the aid of an assistant, lift off the head.

10. Installation is the reverse of removal. Make sure that the mating surfaces of the head and block are absolutely clean. Use a new head gasket. When inserting the pushrods, make certain that they bottom in the depressed part of the tappet. Tighten the head bolts, in the order shown, to 80–85 ft. lb., in three equal steps. Torque the rocker cover bolts to 24–36 inch lb.

2,299cc Engine

It is easier to remove the cylinder head if the intake and exhaust manifolds are removed first. It is not necessary to remove the camshaft. The engine should be cold to reduce the chance of warpage when the cylinder head bolts are removed.

1. Drain the cooling system.

2. Remove the air cleaner. Remove the

spark plug wires from the plugs. Remove the spark plugs.

3. Remove the valve cover. It is held on by eight bolts.

4. Remove the intake and exhaust manifolds from the head. See the appropriate procedures for each.

5. Remove the camshaft drive belt cover. There are tubular spacers underneath the two bolts directly above the crankshaft pulley.

6. Loosen the drive belt tensioner and slip the belt off of the cam sprocket and tensioner. See the Belt Tensioner Adjustment section. It is not necessary to remove the tensioner from the head.

7. Remove the coolant outlet elbow, with the hose attached, from the cylinder head.

8. Loosen the cylinder head bolts a little at a time in the reverse order of the torque sequence shown for tightening.

9. Remove the cylinder head.

10. Clean all the gasket material and sealer from the cylinder head, engine block, valve cover, and water outlet. Be careful not to scratch the surfaces. Vacuum out any particles which fall into the cylinders or passages, being careful not to nick the cylinder walls.

11. Check the cylinder head for flatness. It should not exceed 0.003 in. in any 6 in. span, or 0.006 in. overall. If the head must be machined, do not remove more than 0.010 in. from the original surface. Remove any burrs or scratches with an oil stone.

12. Clean the cylinder head and block bolt holes of any oil or dirt.

13. Place a new gasket on the cylinder block.

14. Position the cam with the pin in the position shown in the illustration.

15. Lower the head carefully onto the block and gasket.

16. Lightly oil the threads of the cylinder head bolts before installation. Torque the bolts to specification in at least three passes, increasing the amount of torque used each time. Torque the bolts in the sequence shown.

17. Slip the camshaft drive belt back over the cam sprocket and tensioner, then time the camshaft. See the "Camshaft Timing Belt" removal and installation procedure for instructions.

18. Install the camshaft drive belt cover and its attaching bolts. Make sure the two

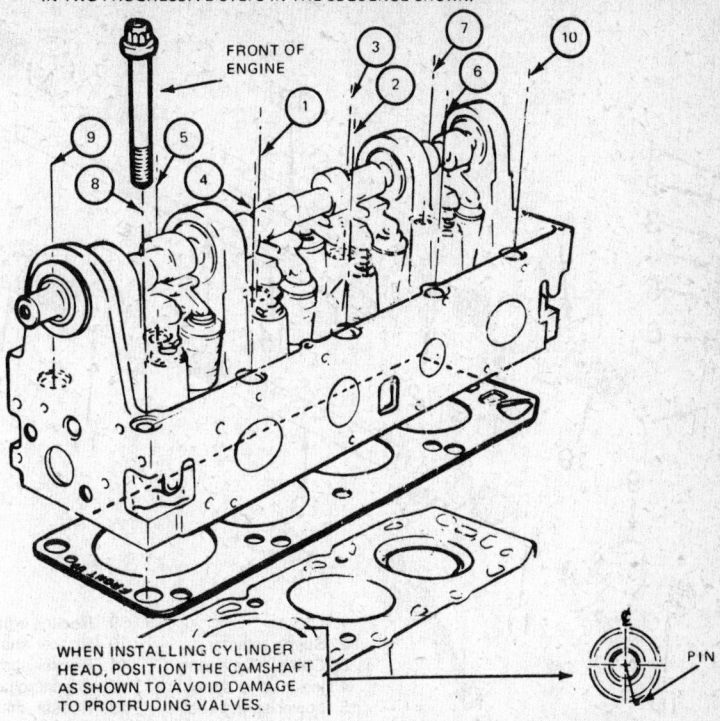

TORQUE THE CYLINDER HEAD BOLTS TO SPECIFICATIONS IN TWO PROGRESSIVE STEPS IN THE SEQUENCE SHOWN.

WHEN INSTALLING CYLINDER HEAD, POSITION THE CAMSHAFT AS SHOWN TO AVOID DAMAGE TO PROTRUDING VALVES.

Cylinder head removal and installation for the 2,299cc engine

spacers are installed correctly. Tighten the bolts to 6–13 ft. lb.

19. Install the water outlet elbow and a new gasket onto the head. Tighten the bolts to 12–15 ft. lb.

20. Install the intake and exhaust manifolds. See the appropriate section for each.

21. Install the valve cover, using a new gasket. Use an oil resistant sealer, such as silicone, on both sides of the gasket. Install the air cleaner. Install the spark plugs and wires.

22. Fill the cooling system.

Rocker Arms and Shafts
REMOVAL & INSTALLATION

1,970cc Engine

This operation should only be performed on a cold engine; the bolts which hold the rocker shafts in place also hold the cylinder head to the block.

1. Raise the hood and cover the fenders.

2. Disconnect the choke cable, if so equipped.

3. If equipped, disconnect the air bypass valve cable.

4. Disconnect the spark plug wires.

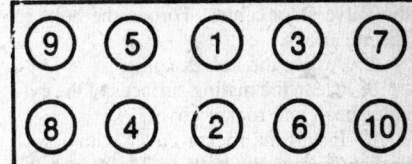

2,299cc engine cylinder head torque sequence

Remove the wires from the spark plug wire clips on the rocker covers and position them out of the way.

5. Remove the rocker cover and discard the gasket.

6. Remove the rocker arm shaft attaching bolts evenly and remove the rocker arm shafts.

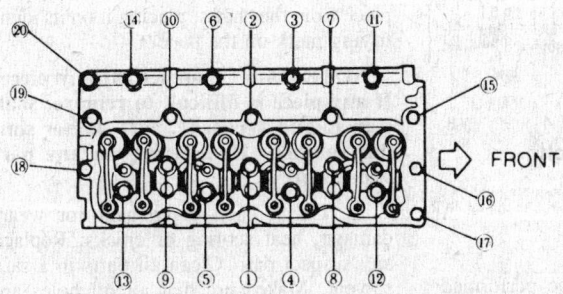

Diesel engine head bolt torque sequence

FRONT

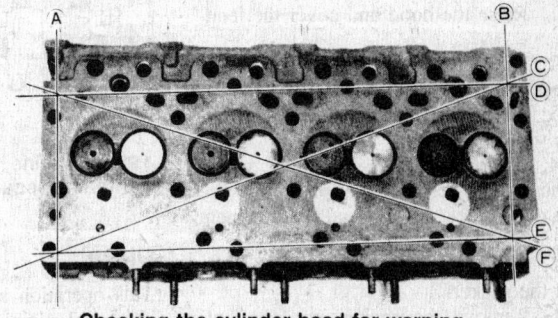

Checking the cylinder head for warping

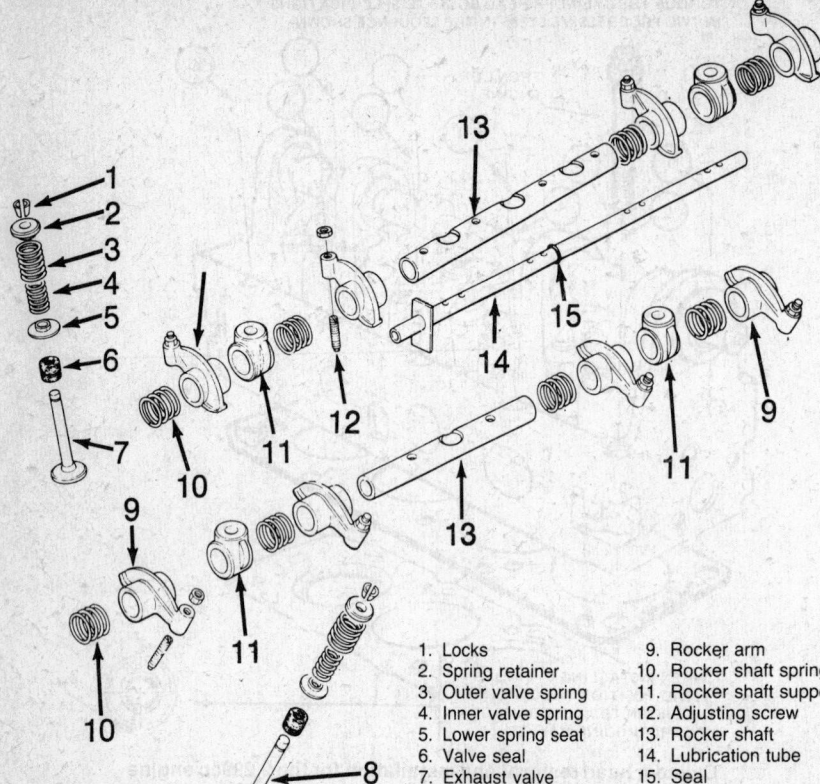

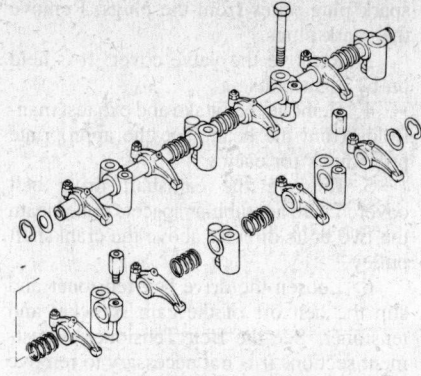

Rocker arms and shafts

1. Locks
2. Spring retainer
3. Outer valve spring
4. Inner valve spring
5. Lower spring seat
6. Valve seal
7. Exhaust valve
8. Intake valve
9. Rocker arm
10. Rocker shaft spring
11. Rocker shaft support
12. Adjusting screw
13. Rocker shaft
14. Lubrication tube
15. Seal

Rocker arm shaft assembly for the 1,970cc engine

7. Install the rocker arm assemblies on the cylinder head. Install the balls on each rocker arm as shown. Temporarily tighten the cylinder head bolts to specifications and offset each rocker arm support 0.04 in. from the valve stem center. Torque the bolts to specifications.

8. Adjust the valves cold.

9. Clean the mating surfaces of the cylinder head and rocker cover.

10. Install the rocker cover with a new gasket. Torque the bolts to 24–36 inch lb.

11. Install the spark plug wire on the plugs. Place the wires in the clips on the rocker cover. Connect the choke and air by-pass valve cable.

12. Start the engine and check for leaks.

13. Allow the engine to reach operating temperature, torque the head bolts to specifications and adjust the valves hot.

1,998cc Engine

1. Raise the hood and cover the fenders.

2. Disconnect the accelerator cable, if necessary.

3. If equipped, disconnect the air by-pass valve cable.

4. Disconnect the spark plug wires. Remove the wires from the spark plug wire clips on the rocker covers and position them out of the way.

5. Remove the rocker cover and discard the gasket.

6. Remove the rocker arm shaft at-taching bolts evenly, in the order shown, and remove the rocker arm shafts.

7. Install the rocker arm assemblies on the cylinder head. Torque the bolts, in the order shown, to 13–20 ft. lb.

8. Check the valve adjustment and reset, if necessary.

9. Clean the mating surfaces of the cylinder head and rocker cover.

10. Install the rocker cover with a new gasket. Torque the bolts to 24–36 inch lb.

11. Install the spark plug wire on the plugs. Place the wires in the clips on the rocker cover. Connect the choke and air by-pass valve cable.

12. Start the engine and check for leaks.

13. Allow the engine to reach operating temperature, torque the head bolts to specifications and adjust the valves hot.

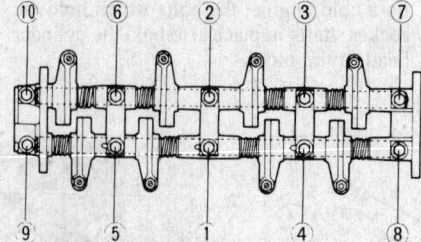

1,998cc engine rocker arm bolt tightening sequence

Diesel

This operation should only be performed on a cold engine; the bolts which hold the rocker shafts in place also hold the cylinder head to the block.

1. Raise the hood and cover the fenders.

2. Disconnect the accelerator cable, if necessary.

3. Remove the rocker cover and discard the gasket.

4. Remove the rocker arm shaft at-taching bolts evenly and remove the rocker arm shafts. Lift out the pushrods, keeping them in order for proper installation. When installing the pushrods, make sure they are seated in the depressed bottom section of the tappets.

5. Install the rocker arm assemblies on the cylinder head. Torque the bolts to specifications.

6. Adjust the valves cold.

7. Clean the mating surfaces of the cylinder head and rocker cover.

9. Install the rocker cover with a new gasket. Torque the bolts to 24–36 inch lb.

10. Install the spark plug wire on the plugs. Place the wires in the clips on the rocker cover. Connect the choke and air by-pass valve cable.

11. Start the engine and check for leaks.

12. Allow the engine to reach operating temperature, torque the head bolts to specifications and adjust the valves hot.

DISASSEMBLY

All Except the Diesel

NOTE: Don't mix up the parts! Keep them identified!

1. Lay out a clean piece of heavy paper marked with a location for each component.

2. Remove the end caps and slide each piece from the shafts, placing it on its identifying mark on the paper.

NOTE: Don't hammer off any piece. If any piece is difficult to remove, soak it in Liquid Wrench®, WD–40® or similar solution. Hammering on any part will distort it!

3. Check each component for wear, damage, heat scoring or cracks. Replace any suspect part. Clean all parts in a safe solvent. Make sure that all oil holes are clear.

4. Assembly is the reverse of disassembly.

Diesel Engine

NOTE: Don't mix up the parts! Keep them identified!

1. Lay out a clean piece of heavy paper marked with a location for each component.

2. Remove the snap rings and washers from the ends of the shaft. Slide each piece from the shaft, placing it on its identifying mark on the paper.

NOTE: Don't hammer off any piece. If any piece is difficult to remove, soak it in Liquid Wrench®, WD–40® or similar solution. Hammering on any part will distort it!

3. Check each component for wear, damage, heat scoring or cracks. Replace any suspect part. Clean all parts in a safe solvent. Make sure that all oil holes are clear.

4. Assembly is the reverse of disassembly.

Intake Manifold

REMOVAL & INSTALLATION

1,970cc and 1,998cc Engines

1. Drain the cooling system.
2. Remove the air cleaner.
3. Remove the accelerator linkage.
4. Disconnect the choke cable and fuel line. Plug the fuel line.
5. Disconnect the PCV valve hose.
6. Disconnect the heater return hose and by-pass hose.
7. Remove the intake manifold-to-cylinder head attaching nuts.
8. Remove the manifold and carburetor as an assembly.
9. Clean the gasket mating surfaces.
10. Install a new gasket and the manifold on the studs. Torque the attaching nuts to specification, working from the center outward.
11. Connect the PCV valve hose to the manifold.
12. Connect the by-pass and heater return hoses.
13. Install the accelerator linkage.

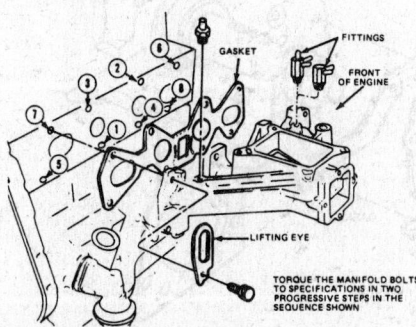

2,299cc engine intake manifold torque sequence

14. Connect the fuel line and choke cable.
15. Replace the air cleaner.
16. Fill the cooling system. Run the engine and check for leaks.

Diesel Engine

Fig. 55

1. Remove the air inlet tube.
2. Remove the vacuum sensing line.
3. Bleed the fuel system and remove the injection lines. Cap the lines to prevent the entrance of dirt.
4. Disconnect the accelerator linkage and any other hose or wire connected to the manifold.
5. Remove the fuel return line.
6. Unbolt and remove the manifold.
7. Discard the gasket and thoroughly clean the gasket surfaces of the head and manifold.
8. Installation is the reverse of removal. Always use a new gasket. Torque the manifold bolts to 11–17 ft. lb.

2,299cc Engine

1. Drain the cooling system and remove the air cleaner assembly.
2. Disconnect the accelerator cable at the carburetor.
3. Disconnect all of the vacuum hoses from the carburetor and the intake manifold. Tag the hoses so that they may be reinstalled correctly.
4. Disconnect the heat tube at the EGR valve.
5. Remove the engine oil dipstick and tube assembly.
6. Disconnect the fuel lines at the carburetor.
7. Remove the PCV valve.
8. Remove the two distributor cap screws and move the distributor cap aside.

9. Remove the intake manifold retaining bolts and remove the manifold and carburetor as an assembly.
10. Clean the gasket mating surfaces.
11. Installation is the reverse of the previous steps. Replenish the cooling system with coolant and check the system for leaks after the engine is started.

NOTE: Torque the manifold fasteners according to the accompanying illustration.

Exhaust Manifold

REMOVAL & INSTALLATION

2,299cc Engine

1. Remove the hot air duct which runs to the air cleaner case, if equipped. Remove the air injection nozzles or air pipe assembly, if equipped, on the 1,970cc. If the engine has a hot air duct, remove the upper and lower heat insulators. If the engine has an EGR valve, remove it.
2. Raise and support the truck.
3. Remove the two attaching nuts from the exhaust pipe at the manifold. Discard the gold gasket.
4. Disconnect the air pump check valve hose.
5. Remove the manifold.
6. Apply a light film of graphite grease to the exhaust manifold mating surfaces before installation.
7. Install the manifold on the studs and install the attaching nuts. Final torque is 27–38 ft. lb. Follow the pattern illustrated.
8. Install a new exhaust pipe gasket. Connect the exhaust pipe and torque the nuts to 15 ft. lbs.
9. Connect the hose to the air pump check valve. Install the EGR valve, heat

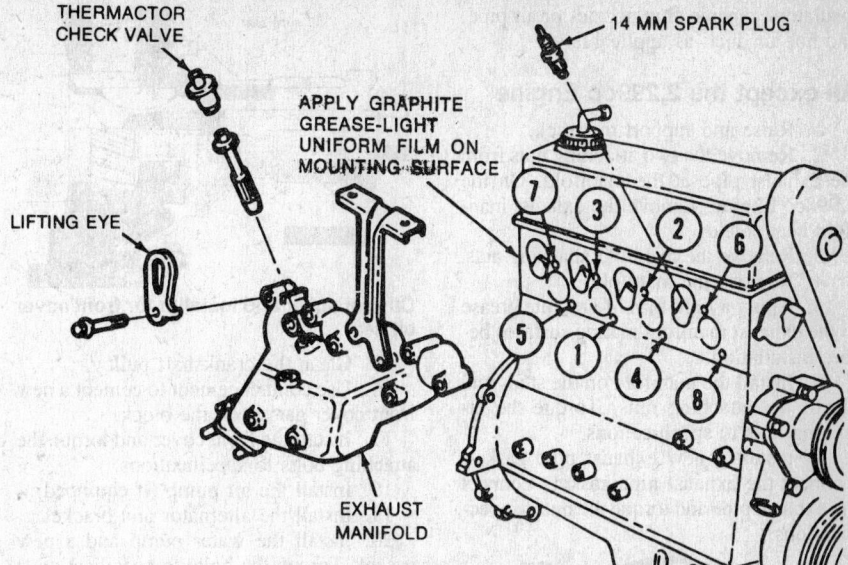

TORQUE THE MANIFOLD BOLTS TO SPECIFICATIONS IN TWO PROGRESSIVE STEPS IN THE SEQUENCE SHOWN

2,299cc engine exhaust manifold torque sequence

insulators, air injection nozzles or air pipe, and hot air duct, as applicable.

All except the 2,299cc Engine

1. Raise and support the truck.
2. Remove the two attaching nuts from the exhaust pipe at the manifold. On the 1,998cc Engine, remove the exhaust manifold heat shield.
3. Remove the manifold attaching nuts.
4. Remove the manifold.
5. Apply a light film of graphite grease to the exhaust manifold mating surfaces before installation.
6. Install the manifold on the studs and install the attaching nuts. Torque the attaching nuts to specifications.
7. Install a new exhaust pipe gasket. Connect the exhaust pipe gasket. Connect the exhaust pipe and torque the nuts to specifications.

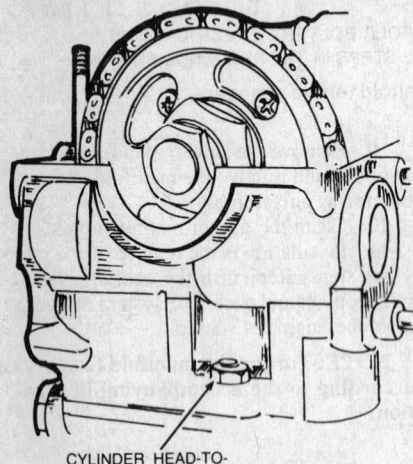

CYLINDER HEAD-TO-FRONT COVER BOLT

Cylinder head-to-front cover bolt location on the 1,970cc engine

Front Cover

REMOVAL & INSTALLATION

1,970cc Engine

1. Scribe alignment marks on the hood hinges and remove the hood.
2. Drain the cooling system.
3. Disconnect the upper and lower radiator hose. Remove the radiator.
4. Remove the accessory drive belts.
5. Remove the crankshaft pulley and the water pump.
6. Remove the cylinder head-to-front cover bolt.
7. Raise and support the truck.
8. Remove the engine skid plate.
9. Disconnect the emission line from the oil pan. Drain the oil from the engine.
10. Remove the oil pan.
11. Remove the alternator and bracket and lay the alternator aside.
12. Remove the steel tube from the front of the engine.
13. Unbolt and remove the front cover.
14. Clean all the gasket mating surfaces.

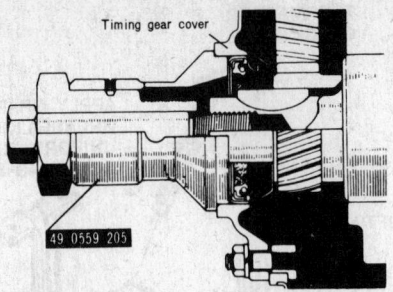

Timing gear cover

49 0559 205

Oil seal puller and installer for front cover oil seal

15. Clean the crankshaft pulley.
16. Use contact cement to cement a new front cover gasket on the block.
17. Install the front cover and torque the attaching bolts to specifications.
18. Install the air pump (if equipped).
19. Install the alternator and bracket.
20. Install the water pump and a new gasket. Torque the bolts to specifications.
21. Connect the by-pass hose and heater hose to the water pump.
22. Install the crankshaft pulley and attaching bolt. Torque the bolt to specifications.
23. Install the alternator belts, and the water pump pulley.
24. Install the fan. Adjust the tension of the belt(s).
25. Install the radiator and the upper and lower hoses.
26. Install the air cleaner.
27. Install the oil pan the the emission line.
28. Install the engine skid plate.
29. Lower the truck to the ground.
30. Fill the engine with oil and fill the cooling system. Run the engine and check for leaks.
31. Install the hood.

Diesel

1. Disconnect the battery ground.
2. Drain the cooling system.
3. Remove the fan and fan shroud.
4. Remove all drive belts from the engine.

5. Remove the power steering pump and its bracket, and position it out of the way, without disconnecting the hoses.
6. Remove the water pump.
7. Remove the crankshaft pulley bolt and, using a puller, remove the pulley.

Unbolt and remove the timing gear cover. Discard the gasket. The front cover seal may be replaced at this time. Coat the outer circumference of the new seal with sealer, and the sealing surface with clean engine oil. 8.

Installation is the reverse of removal. Make sure that the gasket mating surfaces are clean. Use a new gasket coated with sealer. Tighten the timing gear cover bolts to 12–17 ft. lb.; the crankshaft pulley bolts to 150–180 ft. lb.

Timing Belt Covers

REMOVAL & INSTALLATION

1,998cc Engine

1. Disconnect the battery ground.
2. Remove the distributor.
3. Remove the fan and radiator shroud.
4. Remove the alternator.
5. Disconnect the air injection pipes.
6. Remove the fan pulley, hub and bracket.
7. If so equipped, remove the air conditioning compressor drive belt.

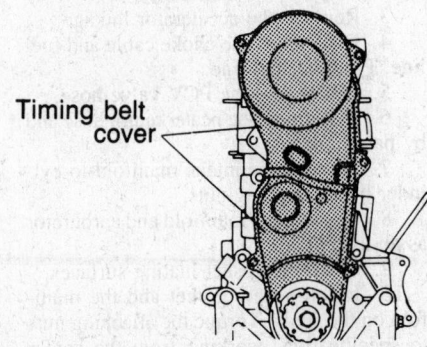

Timing belt cover

1,998cc engine timing belt covers

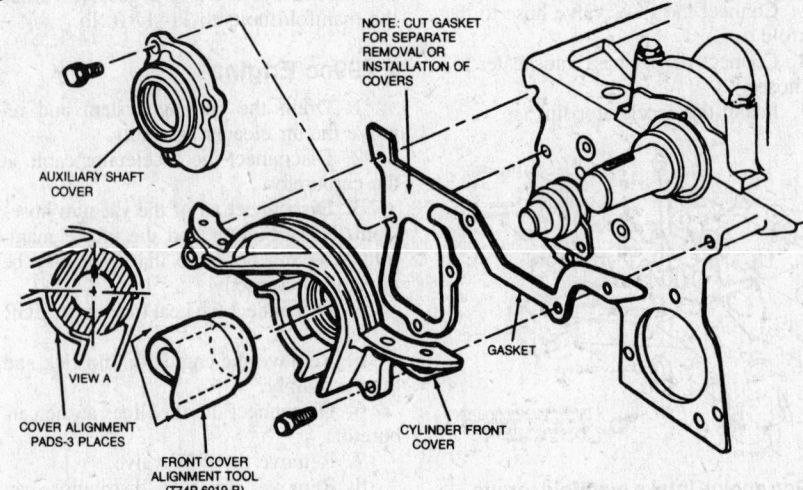

NOTE: CUT GASKET FOR SEPARATE REMOVAL OR INSTALLATION OF COVERS

AUXILIARY SHAFT COVER

GASKET

VIEW A

COVER ALIGNMENT PADS-3 PLACES

FRONT COVER ALIGNMENT TOOL (T74P-6019-B)

CYLINDER FRONT COVER

Front cover installation details for the 2,299cc engine

8. If so equipped, remove the power steering pump drive belt.

9. Remove the crankshaft pulley and baffle plate.

10. Remove the upper, then the lower, belt covers.

11. Installation is the reverse of removal.

2,299cc Engine

There are actually two front covers on the 2,299cc engine, which share a common gasket. You will need either the special Ford tool or a stepped socket (or pipe) to align the crankshaft cover upon reinstallation, and a thin-fingered gear puller to remove the crankshaft sprocket.

1. Drain the cooling system. Remove the upper and lower radiator hoses at the radiator. If your Courier has an automatic transmission, remove and plug the fluid cooler lines from the bottom of the radiator. Remove the radiator.

2. Remove the drive belts to the alternator, air pump and air conditioning compressor, if so equipped, and move the components out of the way. Do NOT disconnect any air conditioning lines. Unbolt and remove the fan and its pulley from the water pump shaft.

3. Remove the camshaft belt outer cover. There are tubular spacers underneath the two bolts directly above the crankshaft pulley.

4. Remove the crankshaft pulley. It is held on with a single center bolt. Remove the belt guide underneath the pulley. It is pressed onto the crankshaft over the locating key.

5. Loosen the cam belt tensioner and remove the belt. See the Timing Belt and Tensioner procedure section.

6. Remove the belt inner cover.

7. Use a puller to remove the crankshaft cam belt sprocket. Be sure that the puller does not cock the sprocket on the shaft.

8. Drain the engine oil and remove the oil pan.

9. Unbolt and remove the cylinder front cover and the auxiliary shaft cover, and the common gasket. If only one cover is to be removed, cut the gasket around the remaining cover, then use the necessary half of a new gasket when the cover is replaced.

10. Before reinstalling the cover(s), clean all the gasket surfaces thoroughly. Pry out the old seals from the covers, but do not install new shaft seals until the covers are in place. Position a new gasket on the front of the engine, install the covers and bolts, but do not tighten them. Using either the Ford tool (illustrated) or a stepped pipe, align the cylinder cover and the crankshaft, so that the timing belt will not interfere with the front cover. Torque the bolts to 6–9 ft. lb. with the tool in place.

11. Install new shaft seals after the covers have been installed. Oil the lips of the seals before installation. The rest of the installation process is the reverse of re-

moval. Be sure to use sealer on the oil pan gasket, front cover and rear main bearing cap seals. No special tool is necessary to replace the crankshaft sprocket; just align it with the key and press it into place. After assembly, fill the crankcase with oil, the cooling system with coolant, and adjust the belt tension. Start the engine, check for leaks, and adjust the initial ignition and engine (belt) timing.

Front Cover Oil Seal

REMOVAL & INSTALLATION

1979–84 1,970cc Engines

The front cover oil seal can be removed and a new one installed without removing the front cover.

1. Scribe alignment marks on the hood hinges and remove the hood.

2. Drain the cooling system.

3. Disconnect the upper and lower radiator hoses and remove the radiator.

4. Remove the drive belt(s).

5. Remove the crankshaft pulley.

6. Pry the front oil seal from the front cover.

7. Clean the pulley and seal area.

8. Press a new front seal into position (flush).

9. Install the crankshaft pulley and torque the bolt to specifications.

10. Install the drive belt(s) and adjust the tension.

11. Install the radiator and connect the upper and lower hoses. Fill the cooling system.

12. Start the engine and check for leaks.

13. Install the hood.

B2200 Diesel

See the previous front cover removal procedures.

2,299cc Engine

The cylinder and auxiliary shaft front cover seals can be replaced without removing the covers. Follow Steps 1–5 and 7 of the 2,299cc Engine Front Cover procedure. Ford recommends the use of a puller to remove the old seal; you can also use a sharp pointed piece of plastic or wood to pry it out. Do not use a screwdriver, though, because you may damage the seal seat. Coat the lips of the new seal with engine oil and press it into place. Ford has a threaded arbor press available for this job. You should also be able to press it home using a socket with a diameter slightly smaller than that of the seal (the socket should just clear the seal seat in the front cover). Place a block of wood on the socket, and tap the seal into place with light hammer blows on the wood.

After installing the new seals, replace the crankshaft sprocket, cam belt, belt guide, crankshaft pulley, belt outer cover, fan drive belts, and radiator. Replace the coolant, adjust the tension of the various belts, and check the initial ignition timing and the engine (belt) timing.

Front Housing and Camshaft Oil Seal

REMOVAL & INSTALLATION

1,998cc Engine

1. Disconnect the battery ground.
2. Drain the cooling system.

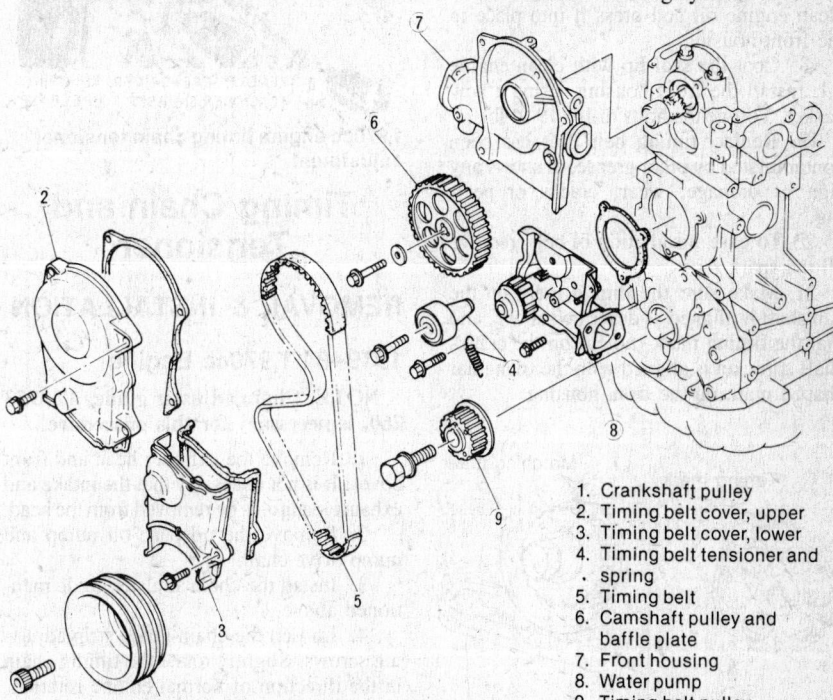

1. Crankshaft pulley
2. Timing belt cover, upper
3. Timing belt cover, lower
4. Timing belt tensioner and spring
5. Timing belt
6. Camshaft pulley and baffle plate
7. Front housing
8. Water pump
9. Timing belt pulley

Timing belt and front housing components for the 1,998cc engine

3. Remove the distributor.

4. Remove the fan shroud and fan.

5. Remove the alternator.

6. Disconnect the air injection pipes.

7. Remove the fan pulley, hub and bracket.

8. If so equipped, remove the air conditioning compressor drive belt.

9. If so equipped, remove the power steering pump drive belt.

10. Remove the crankshaft pulley and baffle plate.

11. Remove the upper, then the lower, belt covers.

12. Turn the crankshaft so the "A" mark on the camshaft pulley is at the top, aligned with the notch in the front housing.

13. Loosen the tensioner lock bolt and remove the tensioner spring.

14. Mark the forward rotation of the belt with paint to avoid confusion upon installation. Remove the belt.

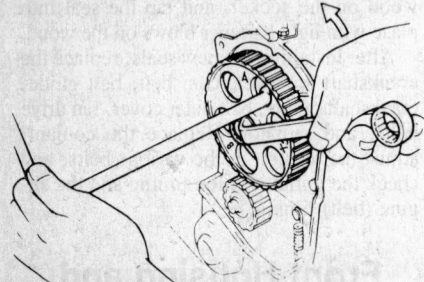

Camshaft pulley removal from the 1,998cc engine

15. Unbolt and remove the front housing.

16. Carefully, drive the camshaft seal from the housing.

17. Coat the outside of a new seal with clean engine oil and press it into place in the front housing.

18. Coat the seal lip with clean engine oil. Install the front housing, using a new gasket. Torque the bolts to 14–19 ft. lb.

19. Replace timing belt if it has been contaminated by oil or grease, or shows any sign of damage, wear, cracks or peeling.

20. To ease installation of belt, remove all the spark plugs.

21. Make sure the timing mark on the camshaft is aligned as described above, and that the timing mark (notch) on the crankshaft sprocket is aligned with the triangular shaped mark on the front housing.

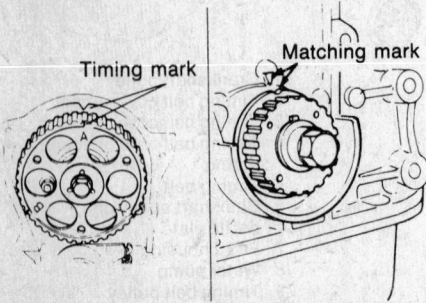

1,998cc engine timing mark alignment

22. Install the tensioner and spring, positioning the tensioner all the way to the intake manifold side and temporarily secure it there with the lock bolt.

23. Install the belt onto the sprockets from YOUR right side. If you are reusing the original belt, make sure you follow the directional mark previously made.

24. Loosen the lock bolt so that the tensioner applies tension to the belt.

25. Turn the crankshaft two full revolutions in the direction of normal rotation. This will apply equal tension to all points of the belt.

26. Make sure that the timing marks are still aligned. If not, repeat the belt installation procedure.

27. Tighten the tensioner lock bolt to 30–35 ft. lb.

28. Measure the timing belt tension by pressing on the belt at the midpoint of the longest straight run. Belt deflection should be 11–13mm. If not, repeat the belt adjustment procedure, above.

29. Installation of all other parts is the reverse of removal. Torque the belt cover bolts to 80 inch lb.; the fan bracket bolts to 40 ft. lb. When installing the drive belts on the various accessories, check the belt deflection.

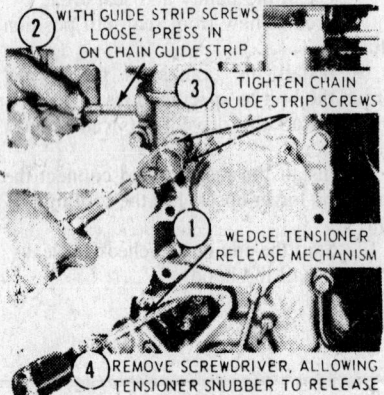

1,970cc engine timing chain tensioner adjustment

Timing Chain and Tensioner

REMOVAL & INSTALLATION

1979–84 1,970cc Engine

NOTE: Chain adjuster guide, 49 3953 260, is necessary for this procedure.

1. Remove the cylinder head and front cover. It is not necessary that the intake and exhaust manifolds be removed from the head.

2. Remove the oil pan, oil pump and pump drive chain.

3. Install the chain adjuster guide mentioned above.

4. Loosen the chain guide strip adjusting screws. Slightly rotate the timing chain in the direction of normal engine rotation. Press the top of the chain guide strip with a prybar and tighten the guide strip adjust-

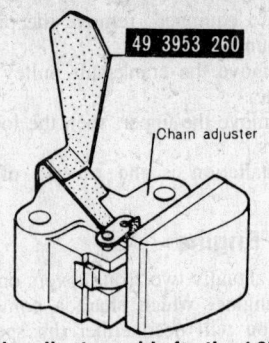

Chain adjuster guide for the 1,970cc engine

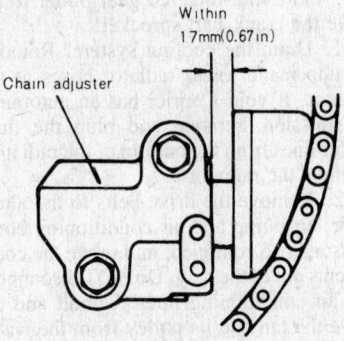

Checking chain adjuster head protrusion

ing screws. Check the protrusion of the chain adjuster head, as shown. If protrusion exceeds 17mm, replace the chain.

5. Remove the timing chain tensioner.

6. Remove the timing chain from the gears.

7. When installing the chain, make sure that the gears and chained are aligned as shown. The alignment marks on the gears must appear on the left, and fall between the nickel plated links.

8. Check the slack in the oil pump drive chain, after installation. Press on the chain, midway between the gears. If slack exceeds 4.0mm, install adjusting shims between the block and oil pump body. Shims are avail-

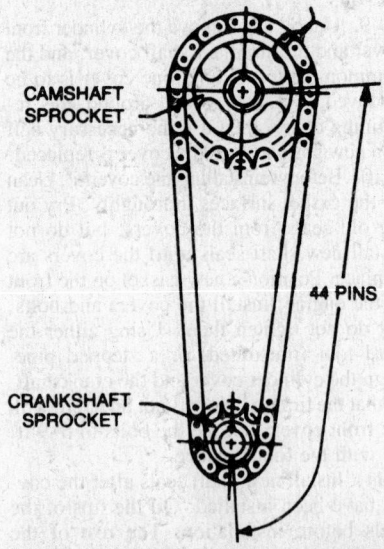

1,970cc engine timing chain alignment

able in thicknesses of 0.15mm. Tighten the oil pump sprocket bolt to 25 ft. lb.

9. Follow Step 4, above, and adjust the timing chain. Remove the guide tool.

10. Install the oil pan, using a new gasket and sealer.

11. Install all other parts in reverse order of removal.

Timing Belt

REMOVAL & INSTALLATION

1,998cc Engine

1. Disconnect the battery ground.
2. Drain the cooling system.
3. Remove the distributor.
4. Remove the fan shroud and fan.
5. Remove the alternator.
6. Disconnect the air injection pipes.
7. Remove the fan pulley, hub and bracket.
8. If so equipped, remove the air conditioning compressor drive belt.
9. If so equipped, remove the power steering pump drive belt.
10. Remove the crankshaft pulley and baffle plate.
11. Remove the upper, then the lower, belt covers.
12. Turn the crankshaft so that the "A" mark on the camshaft pulley is at the top, aligned with the notch in the front housing.
13. Loosen the tensioner lock bolt and remove the tensioner spring.
14. Mark the forward rotation of the belt with paint to avoid confusion upon installation. Remove the belt.
15. Replace the belt if it has been contaminated by oil or grease, or shows any sign of damage, wear, cracks or peeling.
16. To ease installation of the belt, remove all the spark plugs.
17. Make sure that the timing mark on the camshaft is aligned as described above, and that the timing mark (notch) on the crankshaft sprocket is aligned with the triangular shaped mark on the front housing.
18. Install the tensioner and spring, positioning the tensioner all the way to the intake manifold side and temporarily secure it there with the lock bolt.
19. Install the belt onto the sprockets from YOUR right side. If you are reusing the original belt, make sure you follow the directional mark previously made.
20. Loosen the lock bolt so that the tensioner applies tension to the belt.
21. Turn the crankshaft two full revolutions in the direction of normal rotation. This will apply equal tension to all points of the belt.
22. Make sure that the timing marks are still aligned. If not, repeat the belt installation procedure.
23. Tighten the tensioner lock bolt to 30–35 ft. lb.
24. Measure the timing belt tension by pressing on the belt at the midpoint of the

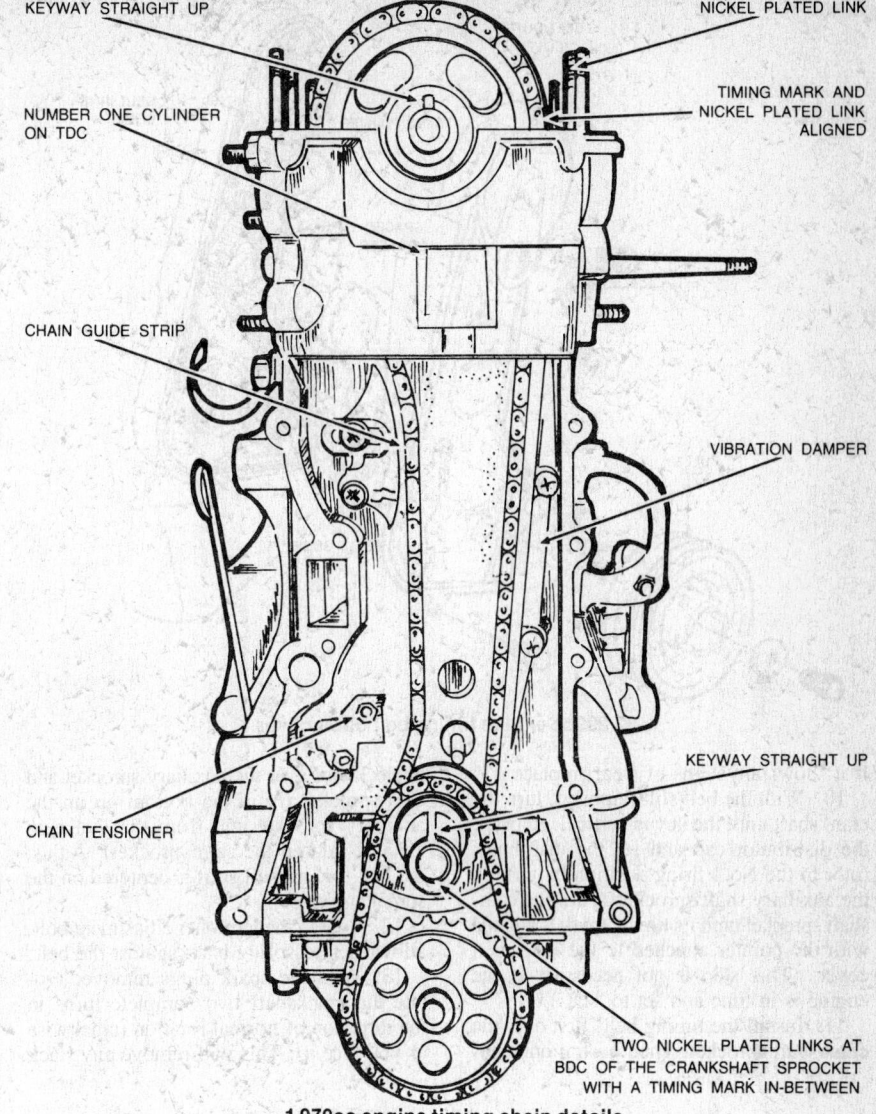

longest straight run. Belt deflection should be 11–13mm. If not, repeat the belt adjustment procedure, above.

25. Installation of all other parts is the reverse of removal. Torque the belt cover bolts to 80 inch lb.; the fan bracket bolts to 40 ft. lb. When installing the drive belts on the various accessories, check the belt deflection.

2,299cc Engines

1. Remove the access plug from the front of the belt cover.
2. Turn the crankshaft until the engine is at TDC, indicated when the timing pointer is aligned with the TDC notch on the crankshaft pulley. You can turn the crankshaft with a wrench on the pulley bolt; it is also easier with the spark plugs removed.

— **CAUTION** —

Always turn the engine in the direction of normal rotation (clockwise as you face it). Backward rotation will cause the belt to jump time or skip teeth.

3. Remove the distributor cap. The rotor should point to the No.1 cylinder plug wire tower.
4. Look through the access hole in the belt cover. The cam sprocket timing mark should be aligned with the timing pointer attached to the inner belt cover.
5. Loosen the adjustment bolts on the alternator, air pump, and a/c compressor (if equipped), and remove their drive belts. To provide clearance for removing the camshaft belt, remove the fan and its pulley from the water pump shaft.
6. Remove the belt outer cover. Note that there are tubular spacers underneath the two lower bolts.
7. Loosen the belt tensioner adjustment and pivot bolts. Lever the belt tensioner away from the belt and retighten the adjustment bolt to hold it away.
8. Remove the crankshaft pulley. It is held on by a single center bolt. Remove the belt guide behind it also.
9. Remove the camshaft drive belt. If it is not to be replaced, inspect it carefully for wear, cracks, or broken or missing teeth.

The illustration labels read: KEYWAY STRAIGHT UP, NUMBER ONE CYLINDER ON TDC, CHAIN GUIDE STRIP, CHAIN TENSIONER, NICKEL PLATED LINK, TIMING MARK AND NICKEL PLATED LINK ALIGNED, VIBRATION DAMPER, KEYWAY STRAIGHT UP, TWO NICKEL PLATED LINKS AT BDC OF THE CRANKSHAFT SPROCKET WITH A TIMING MARK IN-BETWEEN

1,970cc engine timing chain details

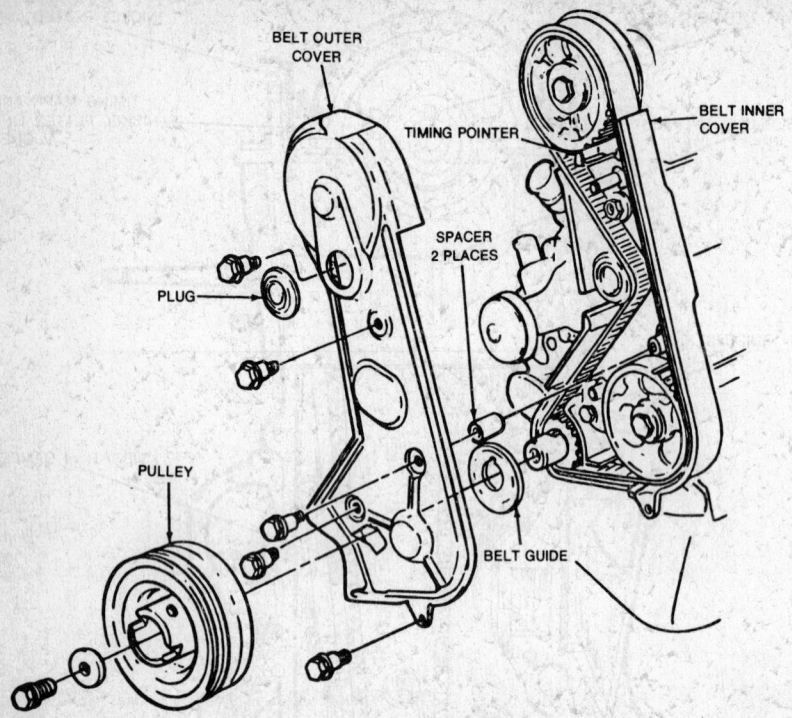

2,299cc engine timing belt outer cover

If it shows any signs of wear, replace it.

10. With the belt still removed, turn the crankshaft until the key is vertical. Remove the distributor cap and set the distributor rotor to the No.1 firing position by turning the auxiliary shaft sprocket. Turn the camshaft sprocket until its timing mark is aligned with the pointer attached to the inner belt cover. (This step is not necessary if the engine is in time and set to TDC).

11. Install the timing belt, first over the crankshaft sprocket. Then push it on counterclockwise over the auxiliary sprocket and cam sprocket (from the bottom up on the auxiliary sprocket and from the intake to exhaust side on the cam sprocket). Adjust the belt fore and aft so it is centered on the sprockets.

12. Loosen the tensioner adjustment bolt, allowing it to spring back against the belt.

13. With the spark plugs removed, rotate the crankshaft two complete turns in the direction of normal rotation (clockwise as you face it). This will remove any slack from the belt. Torque the tensioner adjustment bolt to 14–21 ft. lb., and pivot bolt to 28–40 ft. lb.

14. Replace the belt guide and crankshaft pulley.

15. Replace the spark plugs, distributor cap, outer cover, fan and pulley, and drive belts for the alternator and accessories. Adjust the drive belt tension. Start the engine and check the ignition timing; adjusting as necessary.

Timing Gears and Gear Case

REMOVAL & INSTALLATION

Diesel Engine

1. Rotate the engine so that #1 piston is on TDC of its firing stroke.
2. Disconnect the battery ground.
3. Drain the cooling system.
4. Remove the fan and fan shroud.
5. Remove all drive belts from the engine.

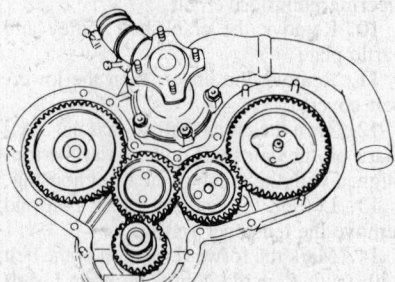

Diesel engine timing gear alignment

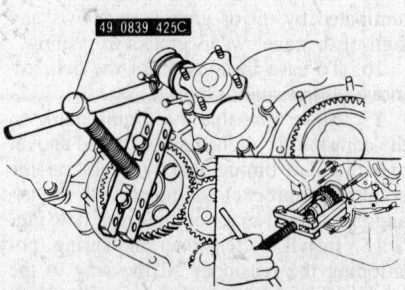

Use the puller illustrated or the equivalent, in the manner shown, to pull cam gear and crankshaft timing gear

6. Remove the power steering pump and its bracket, and position it out of the way, without disconnecting the hoses.
7. Remove the water pump.
8. Remove the crankshaft pulley bolt and, using a puller, remove the pulley.

Unbolt and remove the timing gear cover. Discard the gasket. The front cover seal may be replaced at this time. Coat the outer circumference of the new seal with sealer, and the sealing surface with clean engine oil.

9. Remove the oil pan.
10. Make sure that all timing marks are

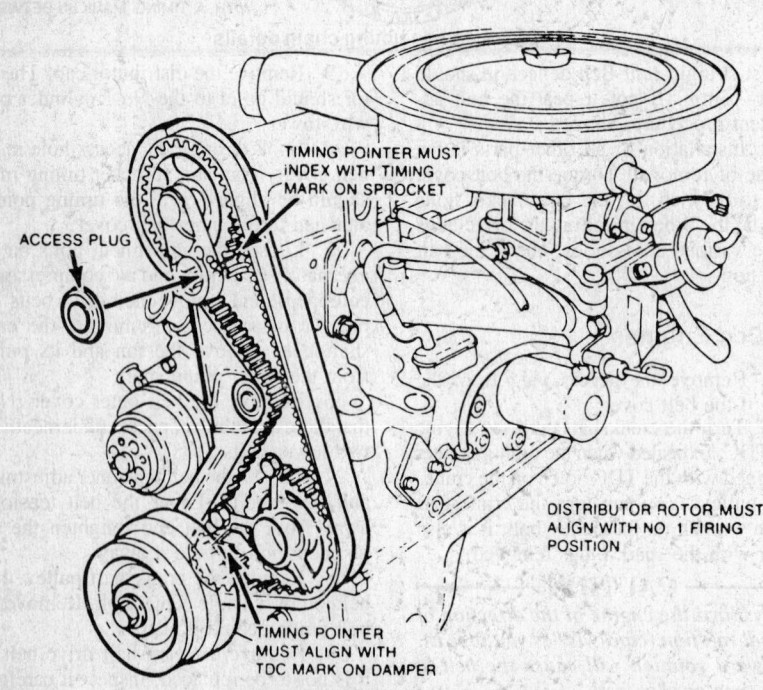

2,299cc engine timing belt alignment

aligned as illustrated. If not, rotate the engine to align them.

11. Remove the bolts from the camshaft gear, and remove the washer and friction gear.

12. Remove the bolts from the injection pump gear, and remove the washer and friction gear.

13. Using a puller, remove the camshaft, crankshaft and injection pump gears.

14. Matchmark the idler gears for installation reference and remove the nuts and gears.

15. Support the injection pump and remove the nuts attaching it to the timing gear case. Support the pump in this position for the rest of the procedure.

16. Remove the bolts attaching the gear case to the block and remove the case.

17. Discard all old gaskets and thoroughly clean the gasket mating surfaces.

18. Check all gears for wear and chipping. Replace any suspect parts.

19. Replace the front cover seal at this time.

20. Using a new gasket, coated with sealer, install the gear case. Torque the bolts to 12–17 ft. lb.

21. Aligning all timing marks, as shown in the accompanying illustration, install the gears in the following order:
   a. crankshaft and right idler
   b. camshaft
   c. left idler
   d. injection pump
   e. all friction gears and washers

22. Install all the nuts and bolts on the gears. Observe the following torques:
Camshaft gear, 45–50 ft. lb.
Idler gears, 17–23 ft. lb.
Injection pump gear, 40–50 ft. lb.

23. Using a new gasket coated with sealer, install the timing gear cover. Torque the bolts to 12–17 ft. lb.

24. Install all other parts in reverse order of removal.

Camshaft

REMOVAL & INSTALLATION

1,970cc Engine

Perform this operation on a cold engine only.

1. Scribe alignment marks on the hood hinges and remove the hood.

2. Remove the water pump.

3. Disconnect the coil wire and vacuum line from the distributor.

4. Rotate the crankshaft to place the No. 1 cylinder on TDC of the compression stroke. This can be determined by removing the spark plug and feeling compression with your thumb. When compression is felt, rotate the crankshaft until the pointer aligns with the TDC mark on the pulley.

5. Remove the plug wires and distributor cap. Remove the distributor.

6. Remove the valve cover.

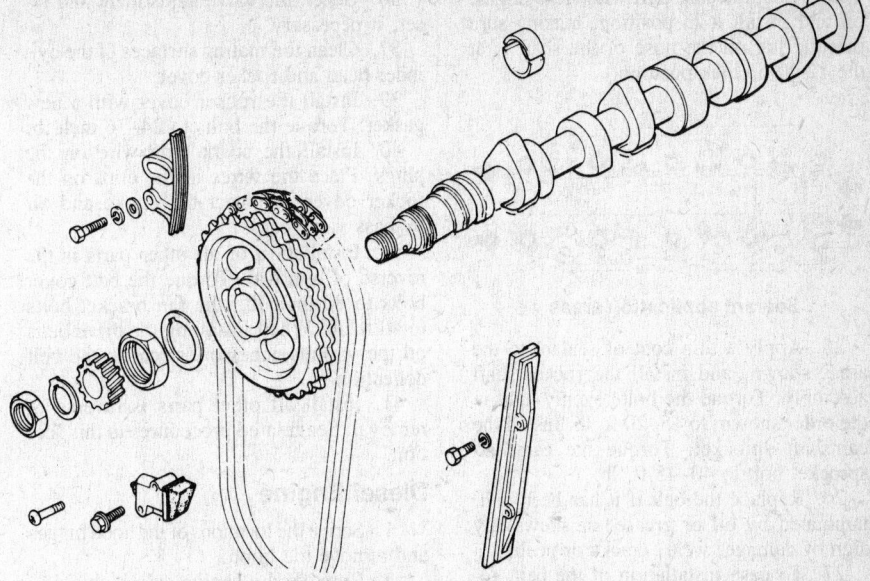

1,970cc camshaft components

7. Release the tension on the timing chain.

8. Remove the cylinder head bolts. Only do this on a cold engine.

9. Remove the rocker arm assembly.

10. Remove the nut, washer and distributor gear from the camshaft.

11. Remove the nut and washer holding the camshaft gear.

12. Remove the camshaft. Do not remove the camshaft gear from the timing chain. Be sure that the hear teeth and chain relationship is not disturbed. Wire the chain and cam gear to a place so that they will not fall into the front cover.

13. Clean all the gasket surfaces.

14. Clean the cylinder head bolt holes.

15. Install the camshaft on the head and install the camshaft gear.

16. Check the valve timing.

17. Install the rocker arm assembly.

18. Install and torque the head bolts.

19. Install the cam gear washer an nut.

20. Install the distributor gear, washer and nut.

21. Adjust the timing chain tension.

22. Check the camshaft end-play. It should be 0.001–0.007 in. If it exceeds 0.008 in., replace the thrust plate with a new one.

23. Install the distributor, distributor cap and plug wires.

24. Connect the vacuum line and coil wire.

25. Adjust the valve clearance cold. Install the valve cover and fill the cooling system.

26. Run the engine and check for leaks. When normal operating temperature is reached, adjust the hot valve clearance.

27. Adjust the carburetor and ignition.

28. Install the air cleaner and hood.

1,998cc Engine

1. Disconnect the battery ground.

2. Drain the cooling system.

3. Remove the distributor.

4. Remove the fan shroud and fan.

5. Remove the alternator.

6. Disconnect the air injection pipes.

7. Remove the fan pulley, hub and bracket.

8. If so equipped, remove the air conditioning compressor drive belt.

9. If so equipped, remove the power steering pump drive belt.

10. Remove the crankshaft pulley and baffle plate.

11. Remove the upper, then the lower, belt covers.

12. Turn the crankshaft so that the "A" mark on the camshaft pulley is at the top, aligned with the notch in the front housing.

13. Loosen the tensioner lock bolt and remove the tensioner spring.

14. Mark the forward rotation of the belt with paint to avoid confusion upon installation. Remove the belt.

15. Insert a bar through the hole in the camshaft sprocket to hold it in position and remove the sprocket bolt.

16. Disconnect the accelerator cable, if necessary.

17. If equipped, disconnect the air bypass valve cable.

18. Disconnect the spark plug wires. Remove the wires from the spark plug wire clips on the rocker covers and position them out of the way.

19. Remove the rocker cover and discard the gasket.

20. Remove the rocker arm shaft attaching bolts evenly in the order shown, and remove the rocker arm shafts.

21. Remove the camshaft rear seal cap.

22. Lift out the camshaft.

23. Inspect the camshaft for wear, heat scoring or obvious damage. Replace it if necessary. Check the lobes and journals for wear according to the specifications in the Camshaft Specification Chart.

24. Coat the camshaft with clean engine oil and install it in position, making sure that the lug on the nose of the shaft is at the 12:00 o'clock position.

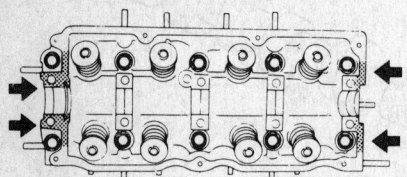

Sealant application areas

25. Apply a thin coat of sealant to the areas shown, and install the rocker shaft assembly. Torque the bolts evenly, and in the order shown, to 15–20 ft. lb. Install the camshaft sprocket. Torque the camshaft sprocket bolt to 40–45 ft. lb.

26. Replace the belt if it has been contaminated by oil or grease, or shows any sign of damage, wear, cracks or peeling.

27. To ease installation of the belt, remove all the spark plugs.

28. Make sure that the timing mark on the camshaft is aligned as described above, and that the timing mark (notch) on the crankshaft sprocket is aligned with the triangular shaped mark on the front housing.

29. Install the tensioner and spring, positioning the tensioner all the way to the intake manifold side and temporarily secure it there with the lock bolt.

30. Install the belt onto the sprockets from YOUR right side. If you are reusing the original belt, make sure you follow the directional mark previously made.

31. Loosen the lock bolt so that the tensioner applies tension to the belt.

32. Turn the crankshaft two full revolutions in the direction of normal rotation. This will apply equal tension to all points of the belt.

33. Make sure that the timing marks are still aligned. If not, repeat the belt installation procedure.

34. Tighten the tensioner lock bolt to 30–35 ft. lb.

35. Measure the timing belt tension by pressing on the belt at the midpoint of the longest straight run. Belt deflection should be 11–13mm. If not, repeat the belt adjustment procedure, above.

36. Check the valve adjustment and re-set, if necessary.

37. Clean the mating surfaces of the cylinder head and rocker cover.

39. Install the rocker cover with a new gasket. Torque the bolts to 24–36 inch lb.

40. Install the spark plug wire on the plugs. Place the wires in the clips on the rocker cover. Connect the choke and air by-pass valve cable.

41. Installation of all other parts is the reverse of removal. Torque the belt cover bolts to 80 inch lb.; the fan bracket bolts to 40 ft. lb. When installing the drive belts on the various accessories, check the belt deflection.

41. Install all other parts is reverse of removal. See related procedures in this section.

Diesel Engine

1. Scribe the locations of the hood hinges and remove the hood.
2. Remove the engine splash shield.
3. Drain the coolant.
4. Drain the engine oil.
5. Disconnect the battery cables and remove the battery.
6. Remove the air cleaner and the oil dipstick.
7. Remove the radiator shroud and the engine fan. Place the fan in an upright position to avoid fluid loss from the fan clutch.
8. Disconnect and tag all wires, hoses, cables, pipes and linkage from the engine.
9. Remove the clutch release cylinder.
10. Remove the oil cooler.
11. Remove the radiator.
12. Disconnect the exhaust pipe at the manifold and remove the exhaust manifold.
13. Dismount the air conditioning compressor and position it out of the way. Don't disconnect any refrigerant lines.
14. Dismount the power steering pump and position it out of the way without disconnecting any hoses.
15. Raise and support the truck on jackstands.
16. Attach a lifting sling and shop crane to the engine lifting eyes and take up the weight of the engine.
17. Support the transmission with a floor jack and remove the transmission-to-engine bolts.

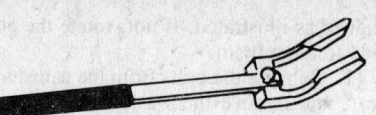

You will need a forked compressing tool like this to remove the 2.3 cam followers

2,299cc camshaft follower depressing tool

19. Remove the engine support plates and mounting nuts, push the engine forward to clear the transmission and lift it out of the truck.

20. Remove the rocker arm cover, rocker arm assemblies and pushrods, making sure you keep the pushrods in order of their removal. Remove the lifters, marking them also, for installation.

21. remove the timing gear case cover.
22. Remove the camshaft gear.
23. Remove the oil pan and oil pump.
24. Remove the camshaft thrust plate.
25. Carefully slide the camshaft from the block.
26. Inspect the camshaft for wear, damage or heat scoring. Check the dimensions of the shaft according to the specifications given in the Camshaft Specifications Chart.
27. Installation is the reverse of removal. Coat the camshaft journal and bearings with clean engine oil, and the lobes with polyethelene grease, prior to installation. See the appropriate related procedures in this chapter.

2,299cc Engine

You will need a special forked tool for this job. The forked tool is used to compress the hydraulic lash adjusters so that the cam followers may be removed and installed. You will also need a puller to remove the cam sprocket and a strip of teflon sealing tape (available at plumbing supply houses) for the cam sprocket bolt.

1. Turn the engine to TDC.
2. Remove the air cleaner. Number and remove the spark plug wires from the plugs. Remove the hose from the oil filler cap on the valve cover. Remove any other wires or hoses crossing the valve cover, and remove the valve cover.
3. Remove the camshaft belt outer cover.

Checking runout with V-blocks and a dial indicator. Rotate the cam after zeroing the indicator and look for a reading of less than .0003 in.

Checking end play

4. Loosen the belt tensioner adjuster and pivot bolts, and lever the tensioner away from the belt. Retighten the adjuster bolt to hold it away.

5. Slip the cam belt off the cam sprocket.

6. Remove the cam sprocket bolt and washer. Use a puller to remove the cam sprocket. Remove the belt guide.

7. Position the camshaft so that the base circle (low point) of the cam lobe is on the cam follower of the rearmost valve (#4 intake). Compress the lash adjuster with the forked tool, and withdraw the cam follower over the lash adjuster and out. It may also be necessary to compress the valve spring.

8. Repeat this procedure with all eight cam followers, working from the rear of the head to the front. Keep the followers in order; they must be returned to their original positions.

9. When all the followers have been removed, remove the two phillips head screws and washers and the retaining plate from the rear of the rear cam bearing tower.

10. Using a puller or a pointed piece of wood or plastic, pry the seal from the front of the camshaft. Slide the camshaft out through the front of the head.

11. Coat the camshaft and bearings thoroughly with engine oil. Slide the cam into the head.

12. Replace the retaining plate, screws and washers. Press a new front cam seal into place. Oil the lips of the seal before installation.

13. Coat the valve tips with Lubriplate® or its equivalent. Coat each cam follower with engine oil.

14. Working from the front to the rear, replace the cam followers. Rotate the cam so that the base circle of the cam lobe for the appropriate valve is facing the head. Use the forked tool to compress the lash adjuster, and install the follower over the lash adjuster and valve stem. It may also be necessary to compress the valve spring, using the same tool. Be sure the cam followers are returned to their original valves. Before rotating the camshaft to the proper position for the next valve, fully compress and release the lash adjuster of the cam follower just installed. It is imperative that this is done to prevent the adjusters from pumping up and providing incorrect clearance.

15. Slide the cam belt guide into place over the key. Install the cam sprocket over the key. Wrap the sprocket bolt threads with teflon tape, and install the bolt and washer.

16. Measure the camshaft end play. Push the cam to the rear of the head. Install a dial indicator so that the point is on the cam sprocket. By inserting a large prybar between the cam sprocket and head, lever the cam forward and release it. If the end play measures more than 0.009 in., the retaining plate must be replaced.

17. Turn the cam until its sprocket timing mark lines up with the indicator on the belt sprocket, pushing it on from the intake side to the exhaust side. Adjust the belt fore

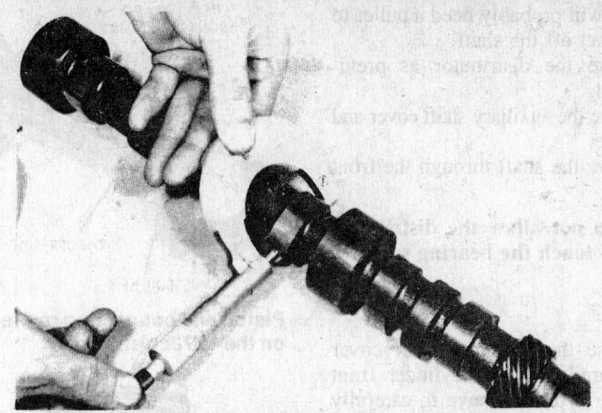

Checking cam height with a micrometer

and aft on the sprocket so that it is centered.

18. Release the cam belt tensioner so that it springs back against the belt. Remove the spark plugs. Using a wrench on the crankshaft pulley, rotate the engine two complete turns to remove slack from the belt. Rotate the engine in the normal direction of rotation (clockwise as you face it).

19. Tighten the belt tensioner adjuster bolt to 14–21 ft. lbs., the pivot bolt to 28–40 ft. lbs. Check the belt timing.

20. Replace the belt outer cover.

21. Install a new gasket on the valve cover with sealer, and install the valve cover.

Replace the spark plugs, spark plug wires, oil filler cap hose, air cleaner, and any other hoses or wires disconnected.

Auxiliary Shaft
REMOVAL & INSTALLATION

2,299cc Engine Only

1. Set the engine to TDC.

2. Remove the camshaft drive (timing) belt as previously outlined.

3. Remove the auxiliary shaft sprocket. It is held on by a single center bolt and

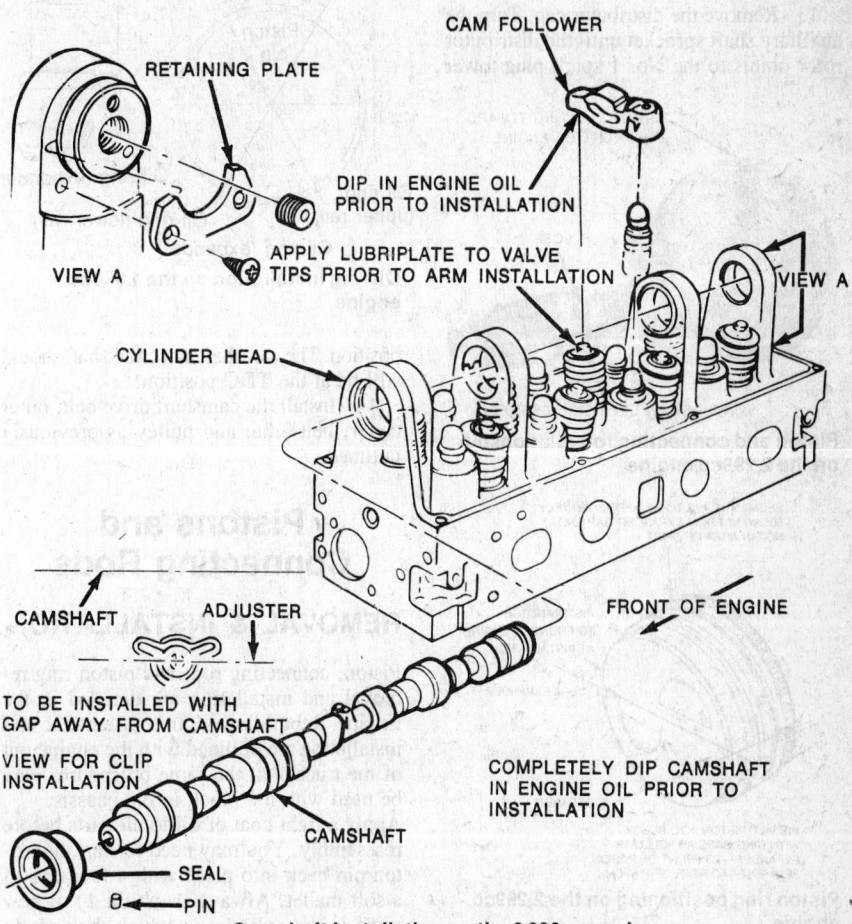

Camshaft installation on the 2,299cc engine

washer. You will probably need a puller to get the sprocket off the shaft.

4. Remove the distributor as previously outlined.

5. Remove the auxiliary shaft cover and retaining plate.

6. Remove the shaft through the front of the block.

NOTE: Do not allow the distributor drive gear to touch the bearing surfaces in the block.

7. Because the auxiliary shaft cover gasket is shared with the cylinder front (timing) cover, you will have to carefully cut off the old gasket around the marks left by the cover, if it did not tear and pull off when you removed the cover. Scrape off any traces of the old gasket, and cut a new gasket to fit. Do not install the new gasket yet.

8. Coat the auxiliary shaft, bearings, and gear with engine oil.

9. Slide the shaft into place, being careful not to bang the gear into the bearings.

10. Replace the retaining plate. Coat the new gasket with a thin layer of sealer and install onto the front of the block. Install the cover.

11. Install the distributor.

12. Install the auxiliary shaft sprocket, washer, and bolt.

13. Remove the distributor cap. Turn the auxiliary shaft sprocket until the distributor rotor points to the No. 1 spark plug tower

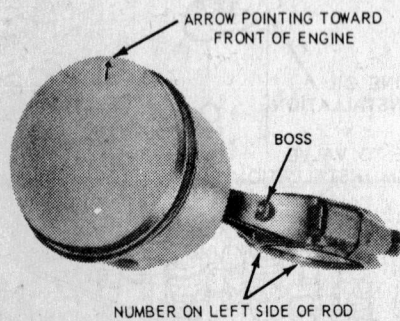

ARROW POINTING TOWARD FRONT OF ENGINE

BOSS

NUMBER ON LEFT SIDE OF ROD

Piston and connecting rod relationship on the 2,299cc engine

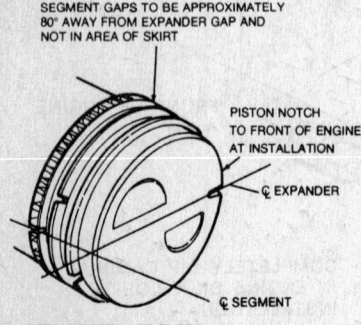

SEGMENT GAPS TO BE APPROXIMATELY 80° AWAY FROM EXPANDER GAP AND NOT IN AREA OF SKIRT

PISTON NOTCH TO FRONT OF ENGINE AT INSTALLATION

℄ EXPANDER

℄ SEGMENT

INSTALL PISTON INTO BLOCK WITH RING GAPS AS FOLLOWS EXPANDER—TO FRONT OF PISTON SEGMENT—TO REAR OF PISTON

Piston ring positioning on the 2,299cc engine

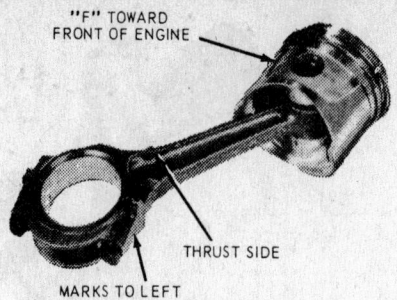

"F" TOWARD FRONT OF ENGINE

THRUST SIDE

MARKS TO LEFT

Piston and connecting rod relationship on the 1,970cc engine

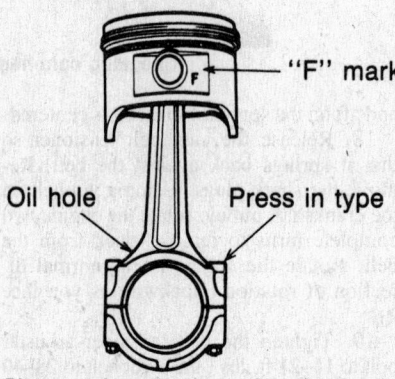

"F" mark

Oil hole   Press in type

Piston and connecting rod mating for the 1,998cc engine

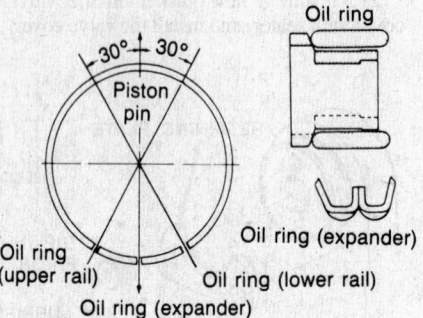

Oil ring

30° 30°

Piston pin

Oil ring (expander)

Oil ring (upper rail)   Oil ring (lower rail)

Oil ring (expander)

Oil ring installation on the 1,998cc engine

position. The camshaft and crankshaft should still be at the TDC position.

14. Install the camshaft drive belt, outer cover, belts, fan and pulley as previously outlined.

Pistons and Connecting Rods

REMOVAL & INSTALLATION

Piston, connecting rod, and piston ring removal and installation are detailed in the Engine Rebuilding section. Removal and installation are outlined with the engine out of the truck, but the same procedures may be used with the block in the chassis.

Apply a light coat of oil to all parts before reassembly. You may need to start the piston pin back into place with a few taps of a soft mallet. Always check the fit of new rings in the cylinder in which they are to

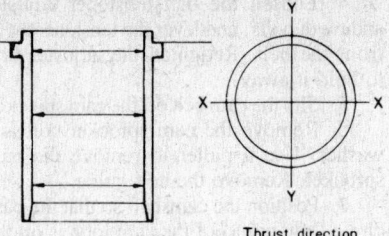

Y

X   X

Y

Thrust direction

Measure cylinder liners at locations and in directions shown

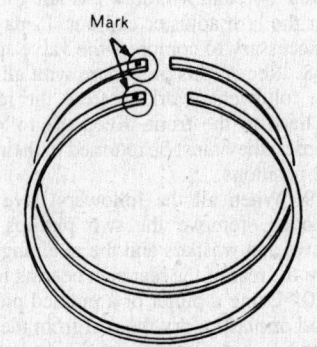

Mark

Install rings with stamped marks facing upward

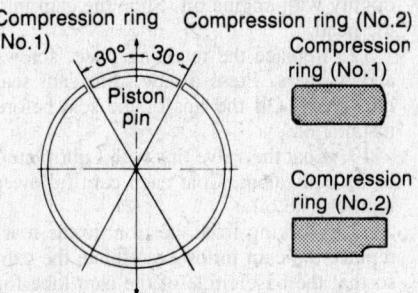

Compression ring (No.1)   Compression ring (No.2)

Compression ring (No.1)

30° 30°

Piston pin

Compression ring (No.2)

Compression ring installation on the 1,998cc engine

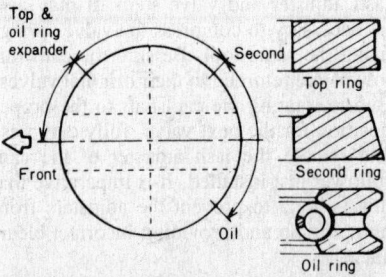

Top & oil ring expander   Second

Top ring

Front   Second ring

Oil   Oil ring

Identify and install top, second, and oil rings by cross-sections shown. Stagger ring gaps as shown

be used before installing. Press them square in the bore with a piston before measuring.

Piston installation position is shown in the illustrations. On the 1,970cc space the piston rings 120° apart, so that the gaps are not located on the thrust side or piston pin side.

On the 2,299cc, install the rings as shown. Be very careful when reinstalling pistons not to nick the crankshaft journals with the connecting rod bolts. Cover the bolts with a length of hose for protection.

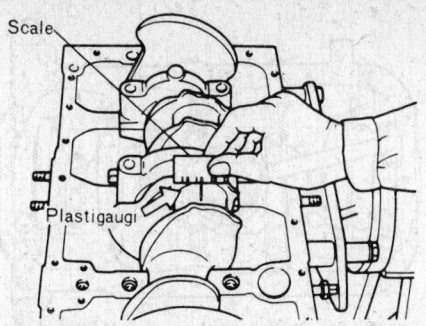

Checking rod bearing clearance with Plastigauge®

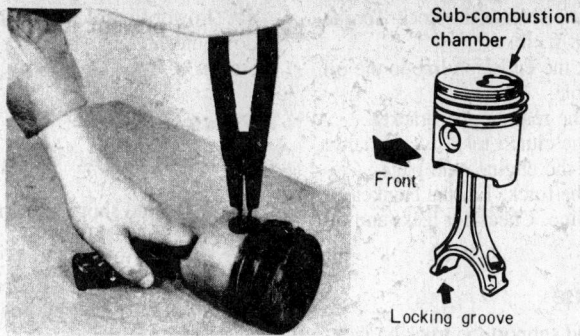

Diesel engine piston and connecting rod alignment

Crankshaft
REMOVAL & INSTALLATION

Main bearings may be replaced without removing the crankshaft. If the crankshaft must be removed, it is recommended that the engine be removed from the vehicle and mounted in a work stand. Refer to the Engine Rebuilding section.

Cylinder Liners
REMOVAL & INSTALLATION

Diesel Only

A hydraulic press and adapters are necessary for this procedure.

1. Remove the engine.
2. Remove the head.
3. Remove the pistons and connecting rods.
4. Remove the crankshaft and bearings.
5. Remove the camshaft.
6. Mount the block in the holding fixture under the press ram, bottom side up.
7. Drive the liners from the block.
8. Check the block bore for scratches. Remove them with an oil-soaked fine emery paper.
9. Invert the block and press the new liners in from the top.

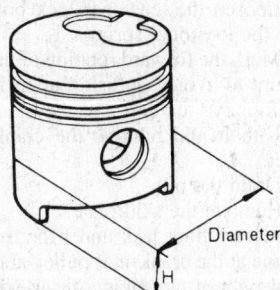

Measuring piston diameter. Measure distance H above the skirt

NOTE: Normal pressing pressure is 2,200–6,600 lb. Press pressure higher than 6,600 lb. will distort the liner; pressures lower than 2,200 lb. will result in a loose fit.

10. Once the liner is in place, check its protrusion above the head surface. Protrusion should be 0.026–0.031 in.

11. Check the liner bore. Bore for a new liner should be 3.5001 in. + 0.0019 in.

ENGINE LUBRICATION

Oil Pan
REMOVAL & INSTALLATION

2,299cc Engine

1. Raise and support the truck.
2. Remove the engine skid plate.
3. Drain the engine oil.
4. Remove the clutch release cylinder attaching nuts. Let the cylinder hang.

5. Remove the engine rear brace attaching bolts and loosen the bolts on the left side.
6. The oil pan is held on by 18, 16mm long bolts and washers around the perimeter, and 4, 20mm long bolts and washers. These must be returned to their original positions.
7. Remove the oil pump pickup tube from the pump.
8. Remove the oil pan.
9. Clean all the gasket surfaces.
10. Clean the oil pan, oil pump pickup tube and oil pump screen.
11. Install a new oil pan gasket with oil-resistant sealer. There are two side gaskets and a front and rear seal. The seals go on to the engine block, and the side pan gaskets go onto the pan. Refer to the illustration.
12. Install the oil pump pickup tube and screen.

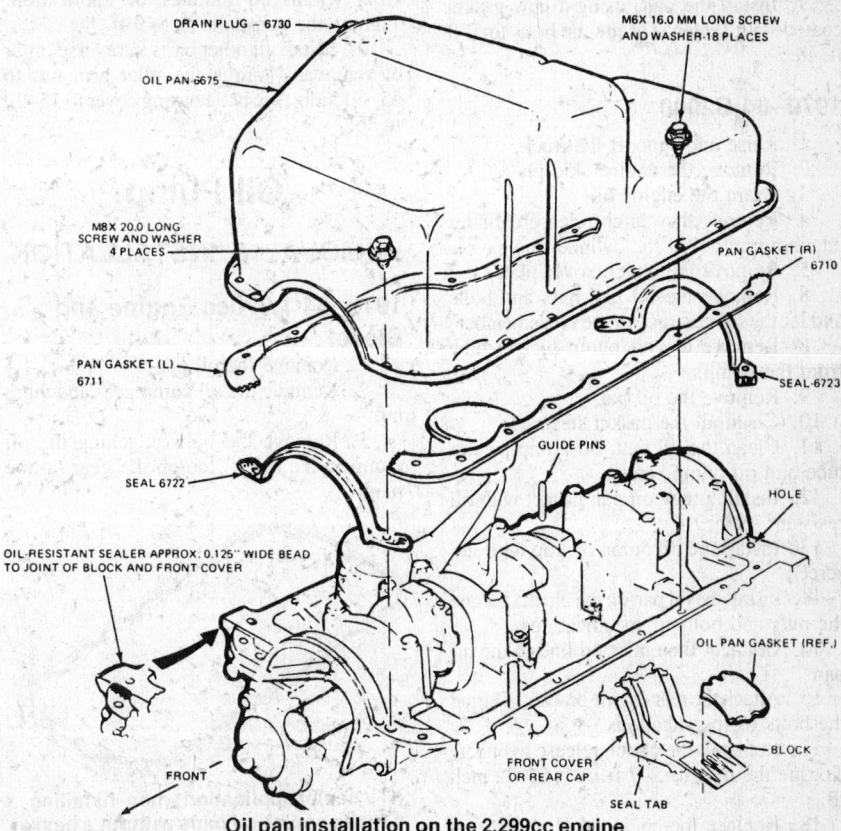

Oil pan installation on the 2,299cc engine

13. Install the oil pan on the block. Torque the bolts to 5–8 ft. lb.

14. Connect the emission line to the oil pan, if applicable.

15. Attach the rear engine bracket.

16. Install the clutch release cylinder.

17. Replace the engine skid plate.

18. Lower the truck. Fill the crankcase, and run the engine. Check for leaks and oil pressure.

Diesel Engine

1. Raise and support the truck on jackstands.

2. Remove the skid plate.

3. Drain the oil.

4. Unbolt and remove the oil pan.

5. Clean all the gasket surfaces. Straighten and portion of the pan rim that is bent.

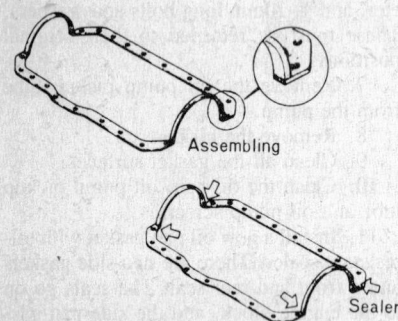

Assembling

Sealer

Apply sealer as shown

6. Clean the oil pan, oil pump pickup tube and oil pump screen.

7. Install the pan, using a new gasket coated with sealer. Torque the bolts to 5–9 ft. lb.

1979–84 B2000

1. Raise and support the truck.

2. Remove the engine skid plate.

3. Drain the engine oil.

4. Remove the clutch release cylinder attaching nuts. Let the cylinder hang.

5. Remove the clutch cover plate.

8. Remove the oil pan nuts and bolts and let the oil pan rest on the crossmember.

8 Remove the oil pump pickup tube from the pump.

9. Remove the oil pan.

10. Clean all the gasket surfaces.

11. Clean the oil pan, oil pump pickup tube and oil pump screen.

12. Install a new oil pan gasket with oil resistant sealer.

13. Install the oil pump pickup tube and screen.

14. Install the oil pan on the block. Torque the nuts and bolts to specifications.

15. Connect the emission line to the oil pan.

16. Attach the rear engine bracket. Torque the bolts to specifications.

17. Reinstall the clutch release cylinder. Torque the nuts to 5–7 ft. lb. (60–72 inch lb.

18. Replace the engine skid plate.

---

19. Lower the truck. Fill the crankcase, and run the engine. Check for leaks and oil pressure.

1,998cc Engine

1. Disconnect the battery ground cable.

2. Raise and support the truck on jackstands. Drain the oil.

3. Remove the skid plate.

4. Place a floor jack under the front of the engine at the crankshaft pulley and take up the weight of the engine. Or use a shop crane to support the engine.

5. Remove the crossmember.

6. Remove the cotter pin and nut and, with a puller, disconnect the idler arm from the center link.

7. Remove the engine mount gusset plates from the sides of the engine.

8. Remove the bell housing front cover.

9. Unbolt and remove the oil pan. A flat tipped screwdriver may be used to break the seal between the pan and block.

10. Clean all the gasket surfaces. Straighten and portion of the pan rim that is bent.

11. Clean the oil pan, oil pump pickup tube and oil pump screen.

12. If you are using a gasket, install a new oil pan gasket coated with oil resistant sealer. Place RTV silicone sealer at the points shown in the accompanying illustration. If you are using RTV silicone gasket material in place of a conventional gasket, run a $\frac{1}{8}$ inch bead around the rim of the pan, going inboard of each bolt hole. Tighten the pan bolts within 30 minutes of application. Tighten the pan bolts to 5–9 ft. lb.

13. Install all other parts in reverse order of removal. Torque the idler arm nut to 25–30 ft. lb.; the bell housing cover to 15–20 ft. lb.

Oil Pump

REMOVAL & INSTALLATION

1979–84 1,970cc Engine and Diesel

1. Remove the oil pan.

2. Remove the oil pump gear attaching nut.

3. Remove the bolts attaching the oil pump to the block. loosen the gear on the pump.

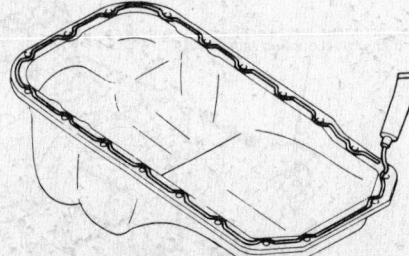

RTV sealer application when installing 1,998cc engine oil pans without a gasket

---

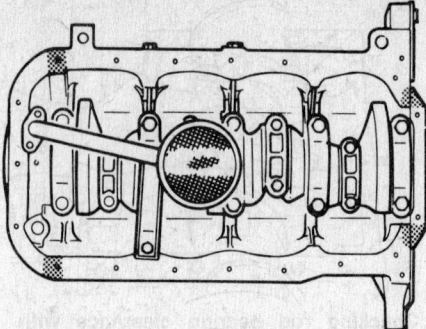

Sealer application points when using a gasket on 1,998cc engine oil pans

4. Remove the oil pump and gear

5. Install the oil pump gear in the chain.

6. Prime the oil pump and install it on the cylinder block. Install the bolts and tighten them securely.

7. Install the washer, gear and nut. Bend the locktab on the washer.

8. Install the oil pan. Fill the engine with oil. Start the engine and check for oil pressure. Check for leaks.

1986 B2000

1. Disconnect the battery ground.

2. Drain the cooling system.

3. Remove the distributor.

4. Remove the fan shroud and fan.

5. Remove the alternator.

6. Disconnect the air injection pipes.

7. Remove the fan pulley, hub and bracket.

8. If so equipped, remove the air conditioning compressor drive belt.

9. If so equipped, remove the power steering pump drive belt.

10. Remove the crankshaft pulley and baffle plate.

11. Remove the upper, then the lower, belt covers.

12. Turn the crankshaft so that the "A" mark on the camshaft pulley is at the top, aligned with the notch in the front housing.

13. Loosen the tensioner lock bolt and remove the tensioner spring.

14. Mark the forward rotation of the belt with paint to avoid confusion upon installation. Remove the belt.

15. Unbolt and remove the crankshaft sprocket.

16. Drain the oil.

17. Remove the skid plate.

18. Place a floor jack under the front of the engine at the crankshaft pulley and take up the weight of the engine. Or use a shop crane to support the engine.

19. Remove the crossmember.

20. Remove the cotter pin and nut and, with a puller, disconnect the idler arm from the center link.

21. Remove the engine mount gusset plates from the sides of the engine.

22. Remove the bell housing front cover.

23. Unbolt and remove the oil pan. A flat tipped screwdriver may be used to break the seal between the pan and block.

24. Remove the oil pick-up tube.

25. Unbolt and remove the oil pump.

26. Apply a thin coating of grease to the O-ring and install it in its recess in the pump body.

27. Apply a thin bead of RTV silicone sealer to the pump mounting surface.

28. Coat the oil seal lip with clean engine oil and install the pump. Torque the bolts to 14–19 ft. lb.

29. Clean all the gasket surfaces. Straighten and portion of the pan rim that is bent.

30. Clean the oil pan, oil pump pickup tube and oil pump screen.

31. If you are using a gasket, install a new oil pan gasket coated with oil resistant sealer. Place RTV silicone sealer at the points shown in the accompanying illustration. If you are using RTV silicone gasket material in place of a conventional gasket, run a 1/8 inch bead around the rim of the pan, going inboard of each bolt hole. Tighten the pan bolts within 30 minutes of application. Tighten the pan bolts to 5–9 ft. lb.

32. Install all other parts in reverse order of removal. Torque the idler arm nut to 25–30 ft. lb.; the bell housing cover to 15–20 ft. lb.

33. Replace timing the belt if it has been contaminated by oil or grease, or shows any sign of damage, wear, cracks or peeling.

34. To ease installation of the belt, remove all the spark plugs.

35. Make sure that the timing mark on the camshaft is aligned as described above, and that the timing mark (notch) on the crankshaft sprocket is aligned with the triangular shaped mark on the front housing.

36. Install the tensioner and spring, positioning the tensioner all the way to the intake manifold side and temporarily secure it there with the lock bolt.

37. Install the belt onto the sprockets from YOUR right side. If you are reusing the original belt, make sure you follow the directional mark previously made.

38. Loosen the lock bolt so that the tensioner applies tension to the belt.

39. Turn the crankshaft two full revolutions in the direction of normal rotation. This will apply equal tension to all points of the belt.

40. Make sure that the timing marks are still aligned. If not, repeat the belt installation procedure.

41. Tighten the tensioner lock bolt to 30–35 ft. lb.

42. Measure the timing belt tension by pressing on the belt at the midpoint of the longest straight run. Belt deflection should be 11–13mm. If not, repeat the belt adjustment procedure, above.

43. Installation of all other parts is the verse of removal. Torque the belt cover bolts to 80 inch lb.; the fan bracket bolts to 40 ft. lb. When installing the drive belts on the various accessories, check the belt deflection.

2,299cc Engine

The oil pump is mounted on the bottom of the engine block and is enclosed by the oil pan. To remove the pump, remove the oil pan, attaching bolts, and the pump. When installing, use a new gasket, and fill the pump with oil to prime it.

Oil Pump Chain

TENSION CHECK AND ADJUSTMENT

1,970cc Engines

Oil pump chain tension can be checked with a straightedge and a ruler. Lay the straightedge against the oil pump and crankshaft gears, alongside the chain. Depress the chain and measure the slack with a ruler. If slack exceeds 0.157 inches, the chain tension will have to be adjusted. Chain slack is reduced by the addition of shims between the oil pump and the cylinder block. The shims should be of equal thickness on each side of the pump.

Oil Cooler

REMOVAL & INSTALLATION

1,998cc Engine

1. Drain the cooling system.
2. Disconnect the coolant hoses at the oil cooler.
3. Remove the oil filter.
4. Remove the nut securing the cooler to the oil filter mounting stud.
5. Remove the cooler.
6. Installation is the reverse of removal. Coat the O-rings on the filter and cooler with clean engine oil prior to installation.

Rear Main Oil Seal

REMOVAL & INSTALLATION

1,970cc engine and the Diesel

If the rear main oil seal is being replaced independently of any other parts, it can be

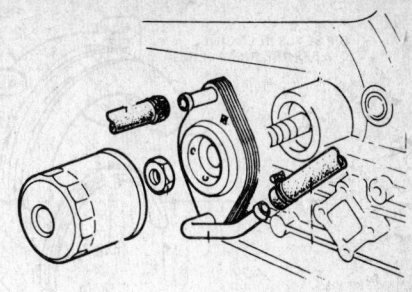

1.998cc engine oil cooler

done with the engine in place. If the rear main oil seal and the rear main bearing are being replaced, together, the engine must be removed from the truck.

1. Remove the transmission.
2. On trucks with a manual transmission, remove the clutch disc, pressure plate and flywheel. On trucks with an automatic transmission, remove the drive plate.
3. Using an awl, punch two holes in the crankshaft rear oil seal. They should be punched on opposite sides of the crankshaft, just above the bearing cap-to-cylinder block split line.
4. Install a sheet metal screw in each hole. Pry against both screws at the same time to remove the oil seal. Do not scratch the oil seal surface on the crankshaft.
5. Clean the oil recess in the cylinder block and bearing cap. Clean the oil seal surface on the crankshaft.
6. Coat the oil seal surfaces with oil. Coat the oil surface and the seal surface on the crankshaft with Lubriplate®. Install the oil seal and be sure that it is not cocked. Be sure that the seal surface was not damaged.
7. Install the flywheel. Coat the threads of the flywheel or drive plate attaching bolts with oil resistant sealer. Torque the bolts to specifications in sequence across from each other. Flywheel: Gasoline engine, 115–120 ft. lb.; Diesel engine, 100–140 ft. lb. Drive plate: 60–69 ft. lb.8.

Install the clutch, pressure plate and transmission.

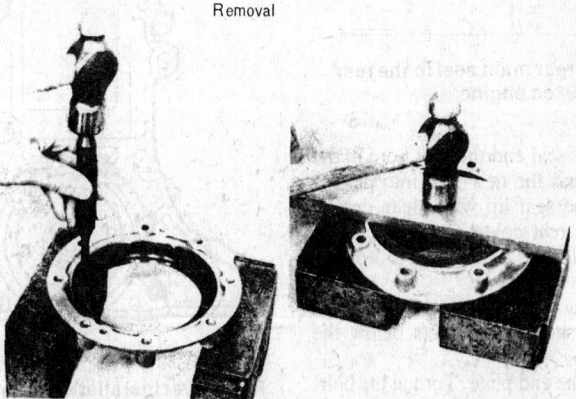

Removal

Diesel rear main seal removal and installation

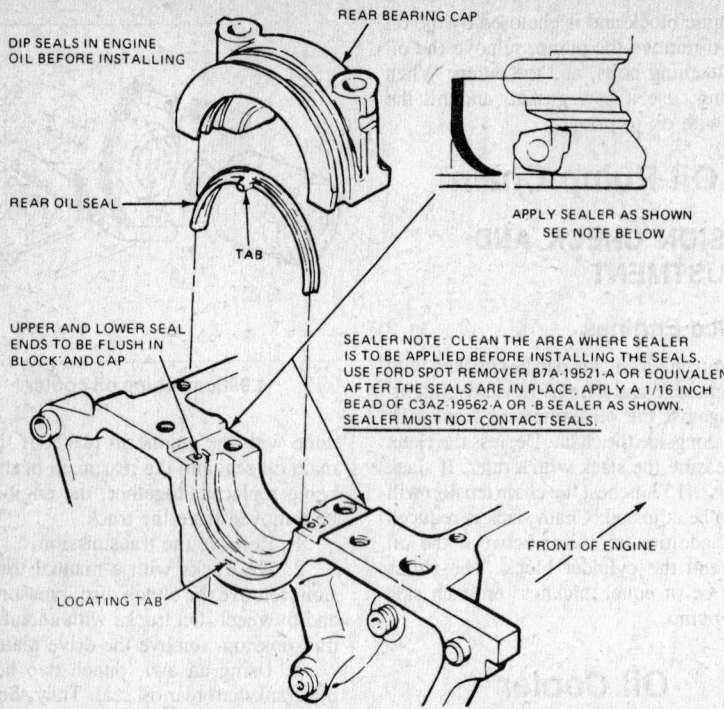

DIP SEALS IN ENGINE OIL BEFORE INSTALLING

REAR BEARING CAP

REAR OIL SEAL

TAB

APPLY SEALER AS SHOWN SEE NOTE BELOW

UPPER AND LOWER SEAL ENDS TO BE FLUSH IN BLOCK AND CAP

SEALER NOTE: CLEAN THE AREA WHERE SEALER IS TO BE APPLIED BEFORE INSTALLING THE SEALS. USE FORD SPOT REMOVER B7A-19521-A OR EQUIVALENT. AFTER THE SEALS ARE IN PLACE, APPLY A 1/16 INCH BEAD OF C3AZ-19562-A OR -B SEALER AS SHOWN. SEALER MUST NOT CONTACT SEALS.

LOCATING TAB

FRONT OF ENGINE

Rear main seal installation on the 2,299cc engine

1,998cc Engine

1. Remove the transmission.
2. Remove the clutch assembly.
3. Remove the flywheel.
4. Remove the end plate.
5. The seal is located in the rear cover. Remove the rear cover. Discard the gasket.
6. Drive the old seal from the rear cover.
7. Apply clean engine oil to the outer

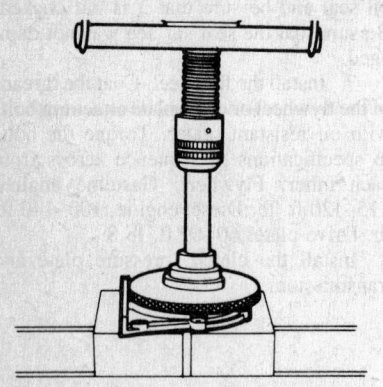

Installing the rear main seal in the rear cover on a 1,998cc engine

rim of the new seal and the seal bore in the rear cover. Press the new seal into place.

8. Coat the seal lip with clean engine oil. Install the rear cover and new gasket. Torque the bolts to 72–102 inch lb. (6–8.5 ft. lb.).

9. Using a sharp knife, cut away the part of the gasket that projects below the rear cover.

10. Install the end plate. Torque the bolts to 14–22 ft. lb.

11. Installation of other parts is the reverse of removal.

2,299cc Engines

1. Remove the oil pan. It may also be necessary to remove the oil pump to provide access to the main bearing cap bolts.

2. Loosen all the main bearing cap bolts, thereby lowering the crankshaft slightly, but not more than $\frac{1}{32}$ in.

3. Remove the rear main bearing cap, and remove the oil seal from the bearing cap and cylinder block. Install a small sheet metal screw in one end of the cylinder block half of the seal, and pull on the screw to remove the seal. Be careful not to scratch the seal surfaces.

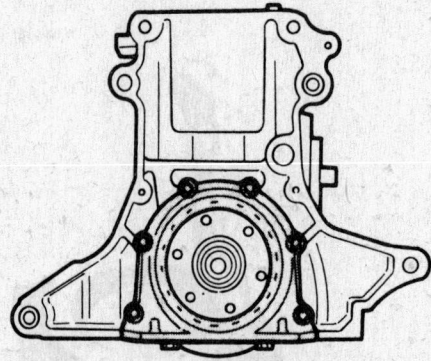

Rear cover installation on a 1,998cc engine

4. Clean the seal grooves in the cap and block with solvent (such as lacquer thinner). Use a brush to get behind the groove lip. Dry the area thoroughly. No solvent should remain to come in contact with the seal.

5. Dip the new seal halves in clean engine oil.

6. Carefully install the upper seal (block half) into its groove with the undercut side of the seal toward the front of the engine, by rotating it on the seal journal of the crankshaft until the ends are flush with the parting surface. Be sure that no rubber has been shaved off. Wipe the oil from the mating surface of the bearing cap and cylinder block.

7. Tighten the bearing cap bolts to 80–90 ft. lb.

8. Install the lower seal in the rear main bearing cap with the undercut side of the seal toward the front of the engine. Be sure that the seal ends are flush with the parting surface to mate with the upper seal when the cap is installed.

NOTE: Install the seals so that the locating tab faces the rear of the engine.

9. Apply a *small* amount of silicone sealer to the mating surface of the bearing cap. No sealer should come in contact with the rubber seals when the bearing cap is installed and tightened.

10. Install the rear main bearing cap and torque to 80–90 ft. lbs.

11. Install the oil pump (if removed) and oil pan. Fill the crankcase with oil, and operate the engine, checking for leaks.

ENGINE COOLING

Water Pump

REMOVAL & INSTALLATION

1,970cc Engine

1. Drain the cooling system.
2. Remove the lower hose from the water pump.
3. Disconnect the upper radiator hose from the engine and the lower radiator hose at the radiator.
4. Remove the radiator.
5. Remove the drive belts.
6. Remove the fan and pulley.
7. Remove the two small hoses from the water pump.
7. Unbolt and remove the water pump.
8. Clean the gasket surfaces of the water pump and cylinder block.
9. Install the water pump and new gasket on the block.

10. Install the lower hose on the water pump.

11. Install the fan and pulley. Install the crankshaft pulley.

12. Install the drive belts and adjust the tension.

13. Install the radiator.

14. Refill the cooling system with the specified amount and type of coolant. Install the radiator cap and start the engine. Check for leaks.

15. Install the hood.

1,998cc Engine

1. Disconnect the battery ground.

2. Drain the cooling system.

3. Remove the distributor.

4. Remove the fan shroud and fan.

5. Remove the alternator.

6. Disconnect the air injection pipes.

7. Remove the fan pulley, hub and bracket.

8. If so equipped, remove the air conditioning compressor drive belt.

9. If so equipped, remove the power steering pump drive belt.

10. Remove the crankshaft pulley and baffle plate.

11. Remove the upper, then the lower, belt covers.

12. Turn the crankshaft so that the "A" mark on the camshaft pulley is at the top, aligned with the notch in the front housing.

13. Loosen the tensioner lock bolt and remove the tensioner spring.

14. Mark the forward rotation of the belt with paint to avoid confusion upon installation. Remove the belt.

15. Remove the coolant inlet pipe and gasket from the pump.

16. Unbolt and remove the pump. Discard the gasket and O-ring.

17. Install the pump, using a new O-ring coated with clean coolant and a new gasket coated with sealer. Torque the bolts to 14–19 ft. lb.

18. Install the coolant inlet pipe, using a new gasket coated with sealer.

19. Replace the timing belt if it has been contaminated by oil or grease, or shows any sign of damage, wear, cracks or peeling.

20. To ease installation of the belt, remove all the spark plugs.

21. Make sure that the timing mark on the camshaft is aligned as described above, and that the timing mark (notch) on the crankshaft sprocket is aligned with the triangular shaped mark on the front housing.

22. Install the tensioner and spring, positioning the tensioner all the way to the intake manifold side and temporarily secure it there with the lock bolt.

23. Install the belt onto the sprockets from YOUR right side. If you are reusing the original belt, make sure you follow the directional mark previously made.

24. Loosen the lock bolt so that the tensioner applies tension to the belt.

25. Turn the crankshaft two full revolutions in the direction of normal rotation.

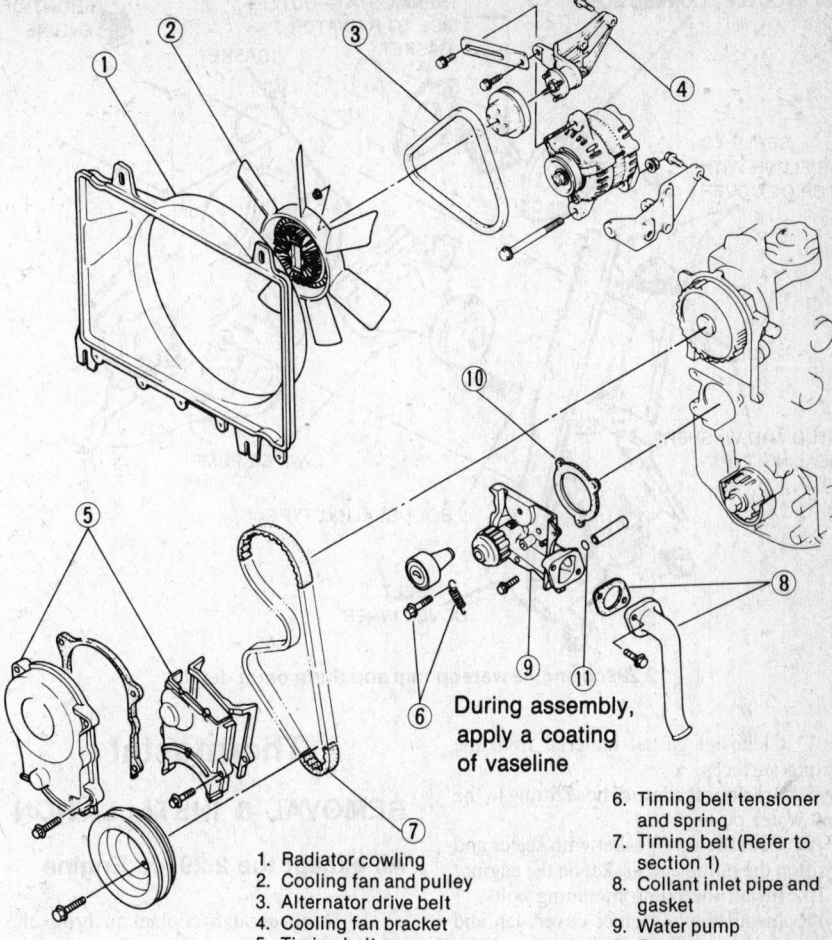

During assembly, apply a coating of vaseline

1. Radiator cowling
2. Cooling fan and pulley
3. Alternator drive belt
4. Cooling fan bracket
5. Timing belt cover, upper and lower
6. Timing belt tensioner and spring
7. Timing belt (Refer to section 1)
8. Collant inlet pipe and gasket
9. Water pump
10. Gasket
11. "O"-ring

Removing the 1,998cc engine water pump

This will apply equal tension to all points of the belt.

26. Make sure that the timing marks are still aligned. If not, repeat the belt installation procedure.

27. Tighten the tensioner lock bolt to 30–35 ft. lb.

28. Measure the timing belt tension by pressing on the belt at the midpoint of the longest straight run. Belt deflection should be 11–13mm. If not, repeat the belt adjustment procedure, above.

29. Installation of all other parts is the reverse of removal. Torque the belt cover bolts to 80 inch lb.; the fan bracket bolts to 40 ft. lb.

Diesel Engine

1. Drain the cooling system.

2. Remove the fan shroud.

3. Remove the fan, fan belt and pulley.

4. Disconnect the lower hose from the pump.

5. Remove the pump and discard the gasket.

6. Installation is the reverse of removal. Use a new water pump gasket coated with sealer.

2,299cc Engine

1. Drain the cooling system.

2. Disconnect the lower radiator hose and heater hose from the water pump.

3. Remove the alternator, air pump, and air conditioner drive belts. Remove the fan shroud, if your truck has one.

4. Remove the fan and pulley.

5. Remove the camshaft drive belt outer cover.

6. Remove the water pump attaching bolts and the water pump. It is not necessary to remove the belt inner cover.

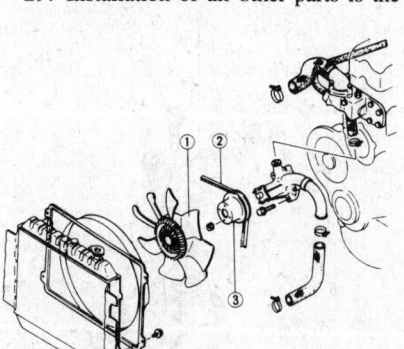

Diesel water pump removal. Remove the parts in the order shown

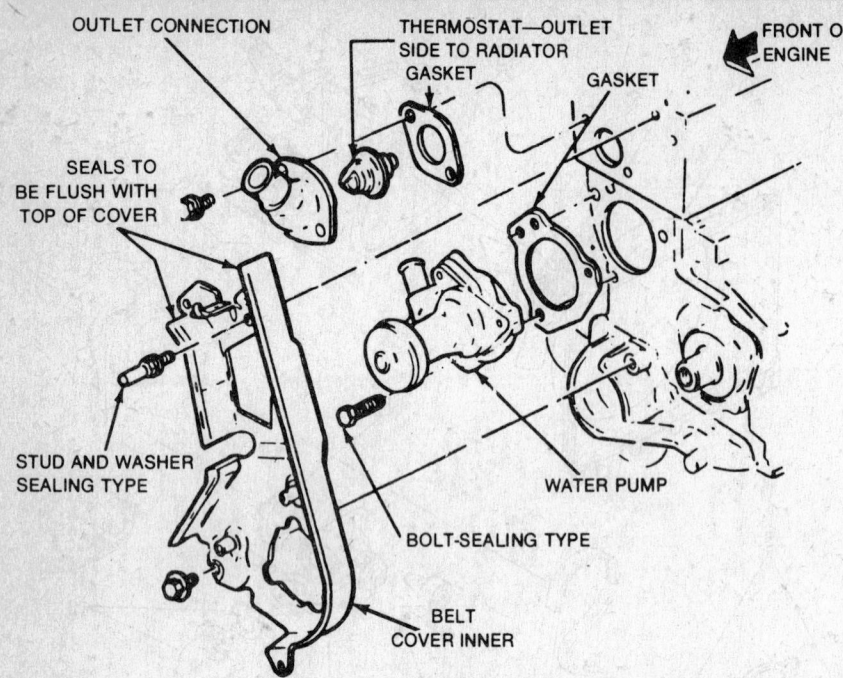

2,299cc engine water pump and thermostat details

7. Clean all gasket material from the mating surfaces.

8. Transfer the heater hose fitting to the new water pump.

9. Coat the new gasket with sealer and position the pump and gasket on the engine.

10. Install the pump mounting bolts.

11. Install the belt outer cover, fan and pulley, and accessory drive belts. Install the fan shroud, if equipped.

12. Connect the radiator hose.

13. Fill the cooling system. Turn the heater on, install the radiator and coolant recovery tank caps, and run the engine. Check for leaks. When the engine has cooled, check the coolant level, and add as necessary.

Radiator

REMOVAL & INSTALLATION

1. Drain the cooling system.

2. If equipped, remove the fan shroud.

3. Remove the fan. Don't lay the fan, if equipped with a fan clutch, on its side. Fluid will be lost and the fan clutch will have to be replaced.

4. Disconnect the upper and lower radiator hoses.

5. Unbolt and remove the radiator.

6. Install the radiator against the supports and tighten the mounting bolts.

7. Install the hoses on the radiator. Tighten the clamps.

8. Install the fan.

9. If equipped, install the fan shroud.

10. Refill the cooling system with the specified amount and type of coolant. Run the engine and check for leaks.

Thermostat

REMOVAL & INSTALLATION

All except the 2,299cc Engine

1. Drain enough coolant to bring the coolant level down below the thermostat housing. the thermostat housing is located on the left front side of the cylinder block. Disconnect the temperature sending unit wire.

2. Remove the coolant outlet elbow. If so equipped, position the vacuum control valve out of the way. The vacuum control valve is not used on California trucks.

3. Disconnect the coolant by-pass hose from the thermostat housing.

4. Remove the thermostat and housing from the engine.

5. Note the position of the jiggle pin and remove the thermostat from the housing.

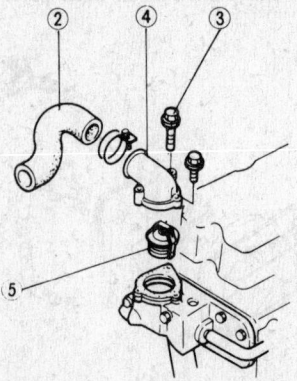

Diesel thermostat installation

6. Remove all gasket material from the parts.

7. Position the thermostat in the housing with the jiggle pin up. Coat a new gasket with sealer and install it on the thermostat housing.

8. Install the thermostat housing using a new gasket with water resistant sealer. Torque the bolts to 20 ft. lb.

9. Install the coolant outlet elbow and vacuum control valve (if equipped).

10. Connect the by-pass and radiator hoses.

11. Connect the temperature sending unit wire.

12. Fill the cooling system with the proper coolant. Operate the engine and check the coolant lever. Check for leaks.

2,299cc Engine

1. Drain the coolant so that the level is below the thermostat.

2. It is not necessary to remove the hose from the outlet connection, if you're careful. Remove the two bolts holding the outlet to the block and pull it away enough to provide access to the thermostat.

3. Remove the thermostat and gasket.

4. Clean the mounting surface and outlet housing of all old gasket material.

5. Coat a new gasket with silicone sealer. The gasket must go on before the thermostat.

6. Position the gasket against the engine, then place the thermostat on top of it with the outlet side towards the radiator.

7. Install the coolant outlet and the two retaining bolts. Torque the bolts to 14–21 ft. lb. Refill the cooling system, start the engine, and check for leaks and proper thermostat operation.

GASOLINE FUEL SYSTEM

Electric Fuel Pump

All models, through 1984, use an external electric fuel pump is mounted on the left frame rail adjacent to the fuel tank. Current is supplied to the pump through the ignition circuit and the pump will operate with the key in the RUN position.

TESTING THE FUEL PUMP

To determine that the fuel pump is in good operating condition, test for both volume and pressure should be performed. The tests are performed with the fuel pump installed, and the engine at normal operating temperature and idle speed. Be sure that the fuel filter has been changed within the specified interval. If in doubt, install a new filter.

Pressure Test

1. Remove the air cleaner.

2. Disconnect the fuel inlet line at the carburetor.

3. Connect a pressure gauge, a restrictor and a flexible hose between the fuel filter and the carburetor. Position the flexible hose and restrictor so that the fuel can be discharged into a suitable graduated container.

4. Before taking a pressure reading, operate the engine at idle speed and vent the system into the container by momentarily opening the hose restrictor.

5. Close the hose restrictor and allow the pressure to stabilize and note the reading. It should be 2.8–3.6 psi.

6. If the pump pressure is not within specifications, and the fuel filter and fuel lines are not blocked, the pump is malfunctioning and should be replaced.

7. If the pressure is within specifications, perform the volume test.

Volume Test

1. Open the hose restrictor and expel the fuel into the container, while observing the time required to discharge 1 pint. Close the restrictor. Fuel pump volume should be approximately 2 pints/minute.

2. If the pump volume is below specifications, repeat the test using an auxiliary fuel supply and a new filter. If the pump volume meets specifications while using an auxiliary fuel supply, check for a restriction in the fuel lines.

REMOVAL & INSTALLATION

1. Remove the fuel pump shield from the frame. Disconnect the electrical leads from the pump.

2. Disconnect the inlet and outlet lines from the pump. Plug the lines.

3. Unbolt and remove the pump from its mounting bracket.

4. Position the fuel pump on the mounting bracket and install the bolts. Be sure that both mounting surfaces are clean.

5. Connect the inlet and outlet hoses.

6. Connect the electrical leads to the pump.

7. Install the fuel pump shield.

Mechanical Fuel Pump

All 1986 models use a mechanically driven fuel pump, mounted on the left front of the cylinder head, driven by the camshaft.

PRESSURE TEST

1. Disconnect the pump-to-carburetor

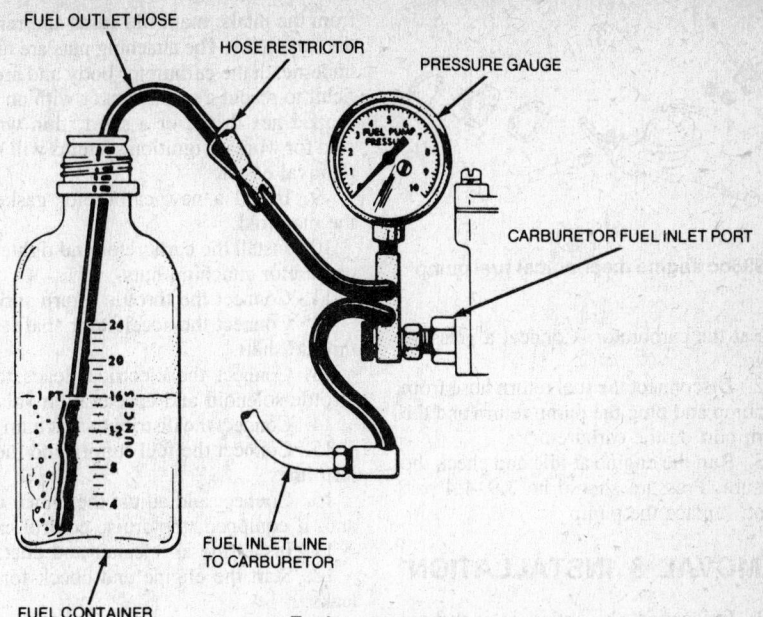

Fuel pump testing

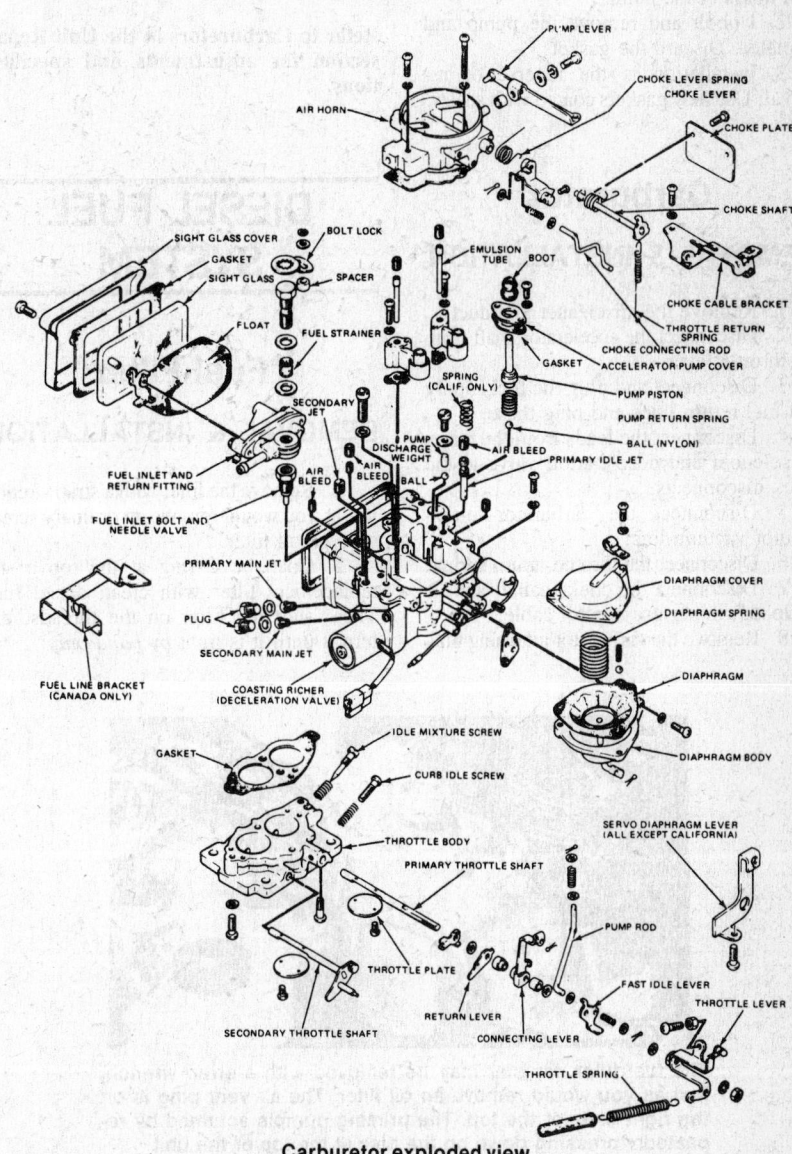

Carburetor exploded view

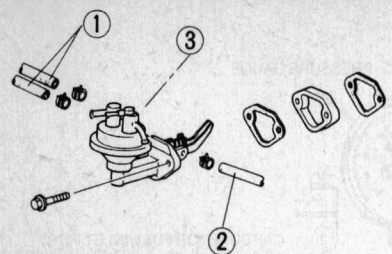

1,998cc engine mechanical fuel pump

hose at the carburetor. Connect a pressure gauge.

2. Disconnect the fuel return hose from the pump and plug the pump return and the return port on the carburetor.

3. Run the engine at idle and check the pressure. Pressure should be 3.9–4.4 psi. If not, replace the pump.

REMOVAL & INSTALLATION

1. Disconnect the outlet, inlet and return hoses at the pump.

2. Unbolt and remove the pump and insulator. Discard the gaskets.

3. Installation is the reverse of removal. Use new gaskets coated with sealer.

Carburetor

REMOVAL & INSTALLATION

1. Remove the air cleaner and duct.

2. Disconnect the accelerator shaft from the throttle lever.

3. Disconnect and plug the fuel supply and fuel return lines and plug these.

4. Disconnect the leads from the throttle solenoid and deceleration valve at the quick-disconnects.

5. Disconnect the carburetor-do-distributor vacuum line.

6. Disconnect the throttle return spring.

7. Disconnect the choke cable, and, if equipped, the cruise control cable.

8. Remove the carburetor attaching nuts

The fuel filter canister may be removed with a strap wrench, just as you would remove an oil filter. The air vent plug is on the right side, at the top. The priming pumpis actuated by repeatedly pressing down on the disc at the top of the unit

from the intake manifold studs and remove the carburetor. The attaching nuts are tucked underneath the carburetor body and are difficult to reach; a small socket with an "L" shaped hex drive, or a short, thin wrench sold for work on ignition systems will make removal easier.

9. Install a new carburetor gasket on the manifold.

10. Install the carburetor and tighten the carburetor attaching nuts.

11. Connect the throttle return spring.

12. Connect the accelerator shaft to the throttle shaft.

13. Connect the electrical leads to the throttle solenoid and deceleration valve.

14. Connect the distributor vacuum line.

15. Connect the fuel supply and fuel return lines.

16. Connect and adjust the choke cable and, if equipped, the cruise control cable.

17. Install the air cleaner and duct.

18. Start the engine and check for fuel leaks.

Refer to Carburetors in the Unit Repair section for adjustments and specifications.

DIESEL FUEL SYSTEM

Fuel Filter

REMOVAL & INSTALLATION

1. Remove the filter with a strap wrench, just as you would remove an ordinary screw-on type oil filter.

2. Coat the O-ring at the top of the replacement filter with clean diesel fuel. Then, start the filter on the threads, and turn it until it is tight *by hand only*.

3. You must now bleed the system by first loosening the vent plug, and then operating the priming pump until fuel without air bubbles is discharged from the vent. Tighten the vent plug back up and then run the engine to check for leaks.

Water Separator

REMOVAL & INSTALLATION

1. Disconnect the wiring connectors (1) and fuel lines (2).

2. Remove the bolts on either side, and remove the protective cover (3).

To remove the water separator, first disconnect the wiring electrical connectors (1), fuel lines (2) and shield (3). The separator itself is indicated by (4)

3. Remove the bolts on the bracket above the water separator, and remove the unit.

4. Bleed the system as described above at the end of the Fuel Filter Removal & Installation procedure.

Fuel Injection Lines

REMOVAL & INSTALLATION

1. Remove the bracket that fastens the four injection lines to the intake manifold.

2. Keeping track of which cylinder each line serves (lines are not interchangeable) and where each connects on the injection pump, unscrew injection line flare nuts at both ends and remove injection lines.

3. To install reverse the removal procedure torquing flare nuts to 18.1–21.7 ft. lb.

4. Start the engine and check for leaks. Note that until air is expelled from the system, the engine may run roughly.

Injection Timing

ADJUSTMENT

NOTE: A static timing gauge adapter and metric dial indicator are necessary for this procedure.

1. Disconnect the battery ground cables.

2. Remove the distributor head plug bolt from the injection pump.

3. Install the timing gauge adapter and metric dial indicator so that the indicator pointer is in contact with the injection pump plunger and the gauge reads 2.0mm.

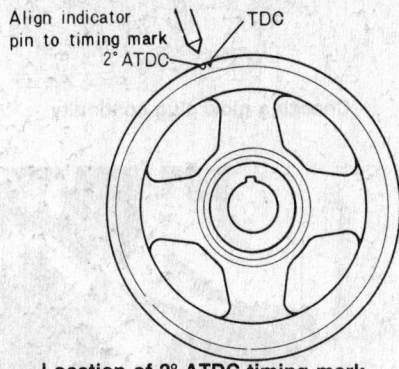

Location of 2° ATDC timing mark

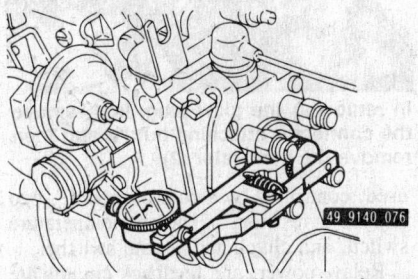

Install the special dial indicator as shown

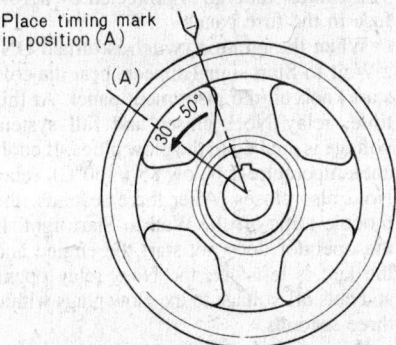

Turn the crankshaft pulley counterclockwise, as shown

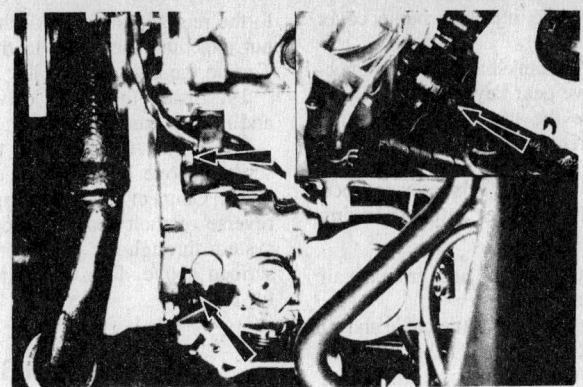

Arrows show locations of pump attaching nuts and bolt. Loosen them just slightly—so it's just possible to turn the pump

4. Align the 2°ATDC mark on the crankshaft pulley with the indicator on the timing gear case cover.

5. Slowly turn the engine counterclockwise until the dial indicator pointer stops moving (approximately 30°–50° pulley travel).

6. Adjust the dial indicator to 0. Confirm that the dial indicator does not move from 0, by rotating the crankshaft slightly from right to left.

7. Turn the crankshaft clockwise until the timing mark is once again aligned with the cover pointer. The dial indicator should read 1mm ± 0.02mm. If not, proceed to Step 8.

8. Loosen the injection pump mounting nuts and bolts.

9. Rotate the pump counterclockwise past the correct timing position, then clockwise until timing is correct.

10. Repeat the timing check to make sure the adjustment is correct.

Injection Pump

REMOVAL & INSTALLATION

1. Disconnect both battery ground cables.

2. Remove the radiator fan and shroud.

3. Remove the air conditioning compressor/power steering drive belt and idler pulley.

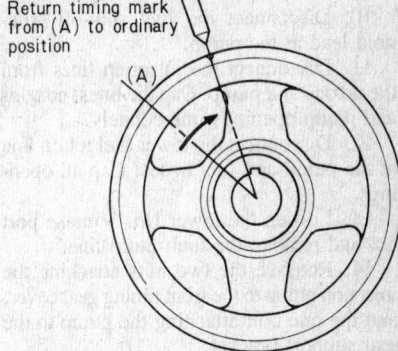

Then, return the pulley to its original position, turning it clockwise

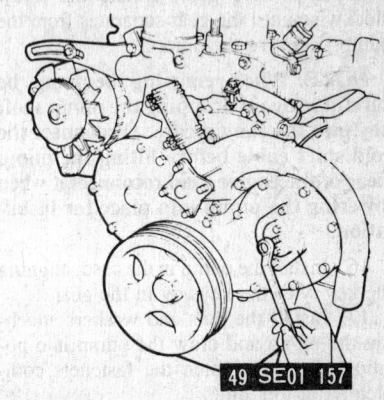

Using the special puller to slide the drive gear off the injection pump driveshaft

Items to be disconnected in injection pump removal (see text)

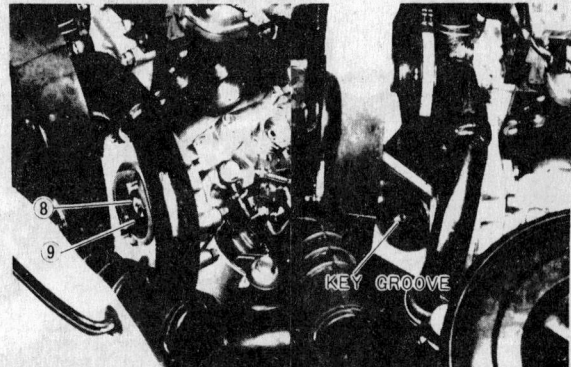

Injection pump drive gear lock nut and spring washer (8) and lock plate (9)

4. Remove the injection pump cover and gasket.

5. Turn the crankshaft until the injection pump drive gear keyway is at TDC.

6. Remove the large nut and washer attaching the drive gear to the injection pump.

NOTE: Be careful! It's easy to accidently drop the washer into the timing case.

7. Remove the intake hose from the air cleaner and manifold.

8. Disconnect the throttle cable and, if equipped, the cruise control cable, from the pump.

9. Disconnect the fuel inlet line from the pump. Cap the line and pump fitting immediately.

10. Disconnect the fuel shut-off solenoid lead at the pump.

11. Disconnect the injection lines from the nozzle and pump. Cap the lines, nozzles and pump openings immediately.

12. Disconnect the lower fuel return line at the pump and fuel hoses. Cap all openings.

13. Loosen the lower No. 3 intake port nut and remove the fuel return line.

14. Remove the two nuts attaching the injection pump to the front timing gear cover, and the one bolt attaching the pump to the rear support bracket.

15. Install a gear and hub remover in the drive gear cover and attach it to the injection pump drive gear. Rotate the screw clockwise until the gear separates from the pump. Remove the pump.

NOTE: When removing the pump, be careful to avoid dropping the pump shaft key into the timing case. Disconnect the cold start cable before lifting the pump clear of the engine, and reconnect it when lowering the pump into place for installation.

16. Install the pump in the case, aligning the key with the keyway in the gear.

17. Install the nuts and washers attaching the pump and draw the pump into position. Do not tighten the fasteners completely, at this time.

18. Install the bolt attaching the pump

to the rear support. Install the washer and nut attaching the pump to the drive gear. Torque the nut to 50 ft. lb.

19. Install the pump drive gear cover and new gasket.

20. Adjust the injection timing as described above.

21. Connect all other components in the reverse of their removal order. Bleed the system through the priming pump, as described above. Run the engine and check for leaks.

Injectors

REMOVAL & INSTALLATION

NOTE: A 27mm deep well socket is necessary for this procedure.

1. Disconnect both battery ground cables.

2. Disconnect the injection lines at the nozzle and pump. Cap all openings immediately.

3. Remove the fuel return line and gaskets.

4. Unbolt the fuel line heater from the head and position it out of the way.

5. Unscrew the nozzles.

6. Remove the copper washer and steel gasket from the nozzle. Discard them.

7. Clean the nozzles and seats with a cleaning kit made for diesel nozzles.

8. Using new gaskets and washers, install the nozzles in the head. Torque them to 50 ft. lb.

NOTE: The gaskets are installed with the blue side up.

9. Install all other parts in reverse of their removal order.

Glow Plugs

OPERATION

This engine uses a quick start glow plug system, enabling the operator to start the engine relatively quickly after the key-on sequence. One glow plug per cylinder is

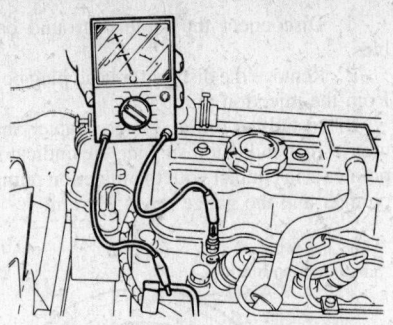

Checking glow plug continuity

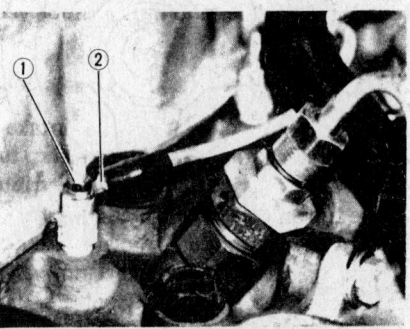

In removing the glow plug, first remove the connector attaching nut (1) and then remove the connector (2)

used, controlled by a control module, two relays, a resistor, a coolant temperature switch, and clutch and neutral switches.

Relay, power, and feedback circuits are protected by fusible links in the harness. The control module is protected by a 10A fuse in the fuse panel.

When the ignition switch is turned ON, a Wait-to-Start signal appears near the cold start knob on the instrument panel. At this time, relay No.1 closes and full system voltage is applied to the glow plugs. If coolant temperature is below 86°F (30°C), relay No.2 also closes. After three seconds, the module turns off the Wait-to-Start light. If the operator does not start the engine and the key is left ON, the No.1 relay opens and cuts off voltage to the glow plugs within three seconds.

However, if coolant temperature is below 86°F when the No. 1 relay opens, the No. 2 relay will remain closed, continuing reduced voltage to the glow plugs until the ignition switch is turned OFF.

When the engine is cranked, the control module cycles relay No. 1 intermittently, providing the glow plugs with between 4 and 12 volts, depending on which replay is closed.

Once the engine has started, the alternator output signals the control module to open the No. 1 relay and the afterglow function takes over, supplying between 4 and 5 volts to the glow plugs through the No. 2 relay as long as the coolant temperature remains below 86°F.

Once the truck is in motion, the clutch and neutral switches close, opening the No. 2 relay if the temperature switch hasn't already done so.

Removing injection nozzles (see text)

TESTING

1. Disconnect the leads from each glow plug. Connect one lead of an ohmmeter to the glow plug terminal and the other lead to a good ground. Set the ohmmeter on the X1 scale.

2. If the ohmmeter indicates the problem is not with the glow plug. If the ohmmeter indicates 1, replace the glow plug and retest.

REMOVAL & INSTALLATION

1. Disconnect the battery ground cables.

2. Disconnect the glow plug harness.

3. Using a 12mm deep well socket, unscrew the glow plugs.

4. Installation is the reverse of removal. Torque the glow plugs to 11–15 ft. lb.

Cold Start Device

ADJUSTMENT

1. Disconnect the cable from the advance lever on the pump.

2. Pull the control knob under the dash to the full out position.

3. Connect a tachometer to the engine.

4. Start the engine and push the advance lever all the way to the stopper. Connect the cable.

5. Turn the adjusting screw until engine speed is 1,150–1,250 rpm.

Fast Idle Control Device

TESTING

Trucks With Air Conditioning Only

When the air conditioning compressor cycles on, the vacuum pump signals the three-way solenoid valve, which in turn applies vacuum to a vacuum diaphragm unit connected to a control lever on the injection pump. Engine speed should be held at 700 rpm or increase to no more than 750 rpm. If engine speed drops below 700 rpm with the air conditioning compressor on, there is a leak in the vacuum circuit.

Fuel Tank

REMOVAL & INSTALLATION

NOTE: It is best to run the fuel tank as low as possible before removing the tank.

1. Raise and support the rear of the truck.

2. Remove the fuel tank drain plug and drain the gasoline into a metal container.

3. Install the drain plug.

NOTE: On models not equipped with a drain plug, disconnect the line which runs to the fuel pump at the tank and allow the tank to drain. Plug the line connection at the tank before proceeding.

4. Disconnect and plug the fuel pump line at the tank.

5. Disconnect the line from the condenser tank or vapor valve at the fuel tank.

6. If so equipped, disconnect the fuel return line.

7. Disconnect the fuel sending unit lead and the electrical connector.

8. Remove the fuel tank attaching bolts at the mounting bracket and lower the tank.

9. Raise the tank into position and install the attaching bolts securely.

10. Connect the fuel sending unit lead.

11. Connect the fuel return line if equipped.

12. Connect the condenser tank or vapor valve line.

13. Connect the fuel pump line.

MANUAL TRANSMISSION

REMOVAL & INSTALLATION

1979–84

1. Put the gearshift in Neutral.

2. Lift up the boot covering the shift lever and detach the gearshift tower from the extension housing. Remove the shift lever, tower and gasket as an assembly.

3. Cover the opening in the case with a heavy rag to keep dirt out.

4. Remove the negative battery cable. Raise and support the truck.

5. Disconnect the driveshaft at the rear axle.

6. Remove the driveshaft center bearing support and pull the driveshaft rearward to disconnect the driveshaft from the transmission. Install a plug in the extension housing to prevent lubricant from leaking out.

7. Remove the exhaust pipe brackets from the transmission case.

8. Disconnect the exhaust pipe hanger from the clutch housing.

9. Disconnect the exhaust pipe at the manifold and muffler and remove the exhaust pipe-resonator assembly or catalytic converter.

10. Unhook the clutch release lever return spring. Remove the clutch release cylinder and secure it out of the way.

11. Remove the speedometer cable from the extension housing.

12. Disconnect the starter motor and backup light wires.

13. Protect the oil pan with a block of wood and support the engine with a jack. Support the transmission with a separate jack.

14. Remove the starter.

15. Unbolt the transmission from the engine rear plate.

16. Unbolt the transmission mount from the crossmember.

17. Remove the crossmember.

18. Work the clutch housing off the locating dowels. Slide the transmission rearward until the input shaft spline clears the clutch disc.

19. Remove the transmission from the truck.

20. Be sure that all mating surfaces are free of dirt, burrs and paint.

21. Lift the transmission into place and start the input shaft into the clutch disc. Be sure that the splines align and move the transmission forward until the clutch housing seats on the locating dowels of the engine rear plate.

22. Bolt the clutch housing to the rear plate.

23. Install the starter motor.

24. Raise the engine and install the rear crossmember.

25. Install the rear transmission mount on the crossmember. Bolt the transmission to the rear mount.

26. Remove the jacks.

27. Install the driveshaft in the transmission extension housing. Install the center bearing.

28. Connect the driveshaft to the rear axle flange.

29. Install the exhaust pipe and resonator.

30. Connect the exhaust pipe to the flywheel housing and transmission brackets.

31. Connect the starter and back-up light wires.

32. Install the clutch release cylinder.

33. Adjust the clutch release lever free travel. Connect the return spring.

34. Connect the speedometer cable.

35. Fill the transmission with lubricant.

36. Lower the truck.

37. Install the shift tower and gasket. Install the boot.

38. Road test the truck and check for leaks.

1986

1. Disconnect the battery ground cable.

2. Raise and support the truck on jackstands.

3. Drain the transmission oil.

4. Remove the gearshift knob, remove the shift console attaching screws, and lift off the console.

5. Remove the shift lever-to-extension housing attaching bolts and remove the shift lever.

6. Remove the driveshaft.

7. Disconnect the speedometer cable from the transmission.

8. Disconnect the wiring at the starter and remove the starter.

9. Disconnect and tag all wiring at the transmission.

10. Disconnect the parking brake return spring, and disconnect the parking brake cables.

11. Remove the clutch slave cylinder.

12. Remove the transmission front support bracket.

13. Disconnect the exhaust pipe at the transmission and manifold.

14. Support the weight of the transmission with a floor jack or transmission jack.

15. Remove the transmission crossmember.

16. Lower the transmission to get access to the top bolts and remove the transmission-to-engine bolts.

17. Pull the transmission straight back and away from the engine. When clear, lower it and remove it from under the truck.

18. Installation is the reverse of removal. Torque the transmission-to-engine bolts to 60–65 ft. lb.; the gearshift lever bolts to 6–8 ft. lb.

LINKAGE ADJUSTMENT

The shifting mechanism of both Courier and Mazda manual transmissions is built into the transmission extension housing, therefore adjustments are not required.

For overhaul procedures, refer to the **Import Trucks portion of the Transmission Unit Repair Section.**

CLUTCH

The clutch is a dry single disc type, consisting of a clutch disc, clutch cover and pressure plate and a clutch release mechanism. It is hydraulically operated by a firewall mounted master cylinder and a clutch release slave cylinder mounted on the clutch housing.

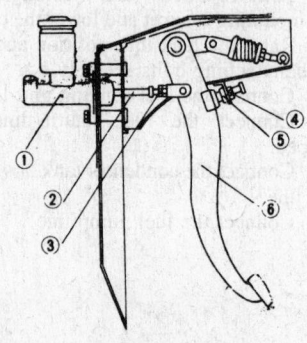

1. Master cylinder
2. Rod
3. Locknut
4. Adjusting bolt
5. Locknut
6. Clutch pedal

Clutch pedal height adjustment

Adjustments

CLUTCH PEDAL FREE-PLAY

The clutch pedal free-play is measured from the top of the pedal pad at rest to the point at which it stops when the pushrod hits the master cylinder piston. Free-play is adjusted by loosening the locknut on the pushrod and adjusting the pushrod length by rotating the rod. The clutch should have a free travel, measured at the pedal pad, of 0.025–0.121 in. for 1976–84 trucks, and 0.20–0.50 in. for 1986 trucks. Tighten the locknut when the adjustment is complete.

PEDAL HEIGHT

Pedal height is measured from the top of the pedal at rest, to the floor board, horizontally behind the pedal. Adjustment is made by loosening the locknut on the pedal stopper and turning the adjusting bolt. Pedal height should be:
   1979–81: 215mm
   1982–84 Gasoline engine: 205mm; Diesel engine: 215mm
   1986: 215mm

CLUTCH RELEASE LEVER ADJUSTMENT

No adjustment is possible. Instead, the stroke can be checked by raising the truck and moving the release rod. If the stroke measures less than 5mm (0.196 in.) the clutch pedal should be replaced.

Release Lever and Bearing

REMOVAL & INSTALLATION

1. The release lever (fork) is retained by a spring clip. Remove the spring clip and pull the fork from the pivot pin.

2. Remove the lever, dust cover boot and the release (throwout) bearing.

3. Inspect the parts carefully. Wipe off all the oil and dirt from the bearing, but do not soak it in solvent; it is prelubricated. Any burrs should be smoothed with crocus cloth. If burrs are present, inspect the transmission input shaft bearing retainer, and smooth any scoring with crocus cloth.

4. Coat the bearing retainer with a thin film of lithium base grease. Apply a thin film of this grease to both sides of the fork at contact points. Also lightly coat the release bearing surface where it contacts the pressure plate fingers.

5. Fill the grease groove inside the bearing hub with the lithium grease. Do not use polyethylene grease. Clean any excess grease from the bore of the hub, because excess grease will eventually work its way into the clutch disc.

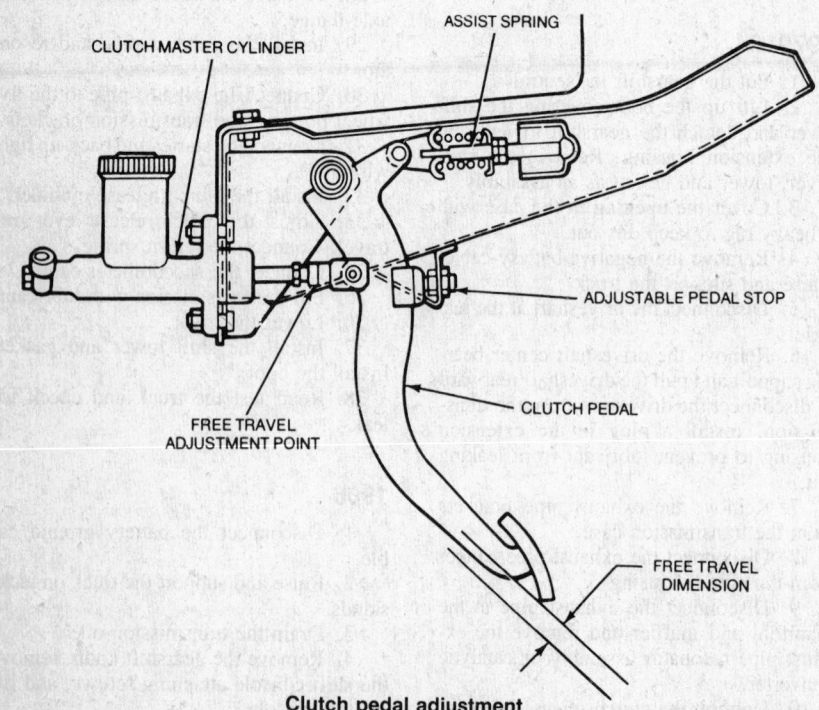

CLUTCH MASTER CYLINDER

ASSIST SPRING

ADJUSTABLE PEDAL STOP

CLUTCH PEDAL

FREE TRAVEL
ADJUSTMENT POINT

FREE TRAVEL
DIMENSION

Clutch pedal adjustment

6. Before installing the bearing, hold the inner race and rotate the outer race, applying pressure. If the rotation is noisy or rough, replace the bearing. Bearing failure is generally caused by improper free play settings at the release cylinder or pedal. Riding the pedal can reduce clearance, causing the bearing to constantly spin, increasing wear. The bearing can also fail due to release lever misalignment (bent out of plane or not centered on the housing bracket) or misalignment between the engine and transmission.

7. Apply a thin film of lithium grease to the input shaft bearing retainer portion of the clutch housing.

8. Dab the end of the pivot pin with grease, and drive the release lever onto it. Apply a thin film of grease to the contact points of the release lever, and install the release bearing. Hook the release collar spring back into place (if applicable).

9. Check the operation of the release bearing hub. It should slide freely on the input shaft bearing retainer.

10. Install the dust boot.

Clutch Unit

REMOVAL & INSTALLATION

All Models

1. Remove the transmission.
2. Remove the four attaching and two pilot bolts holding the clutch cover to the flywheel. Loosen the bolts evenly and a turn or two at a time. If the clutch cover is to be reinstalled, mark the flywheel and clutch cover to show the location of the two pilot holes.

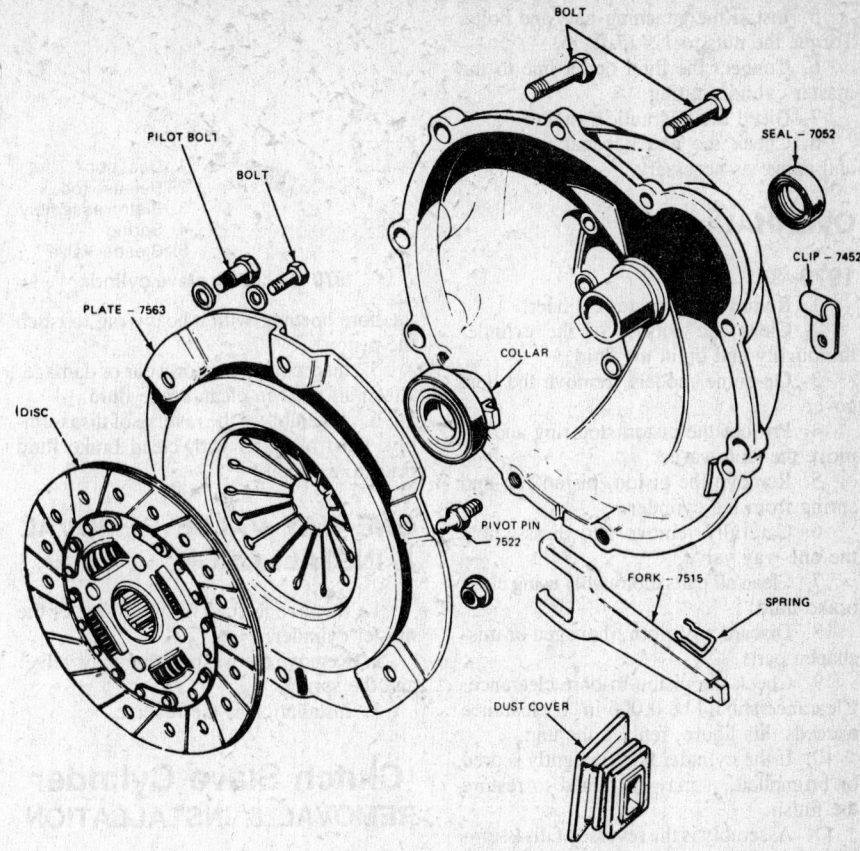

Typical clutch components

3. Remove the clutch disc.
4. Install the clutch disc on the flywheel. Do not touch the facing or allow the facing to come in contact with grease or oil. The clutch disc can be aligned using a tool made for that purpose, or with an old mainshaft.

5. Install the clutch cover on the flywheel and install the four standard bolts and the two pilot bolts.
6. To avoid distorting the pressure plate, tighten the bolts evenly a few turns at a time until they are all tight.
7. Torque the bolts to 13–20 ft. lbs. using a crossing pattern.
8. Remove the aligning tool.
9. Apply a light film of lubricant to the release bearing, release lever contact area on the release bearing hub and to the input shaft bearing retainer.
10. Install the transmission.
11. Check the operation of the clutch and if necessary, adjust the pedal free-play and the release lever.

Clutch Master Cylinder

REMOVAL & INSTALLATION

All Models

1. Disconnect and plug the fluid outlet line at the outlet fitting on the master cylinder one-way valve.
2. Remove the nuts and bolts attaching the master cylinder to the firewall.
3. Remove the cylinder straight out away from the firewall.
4. Start the pedal pushrod into the master cylinder and position the master cylinder on the firewall.

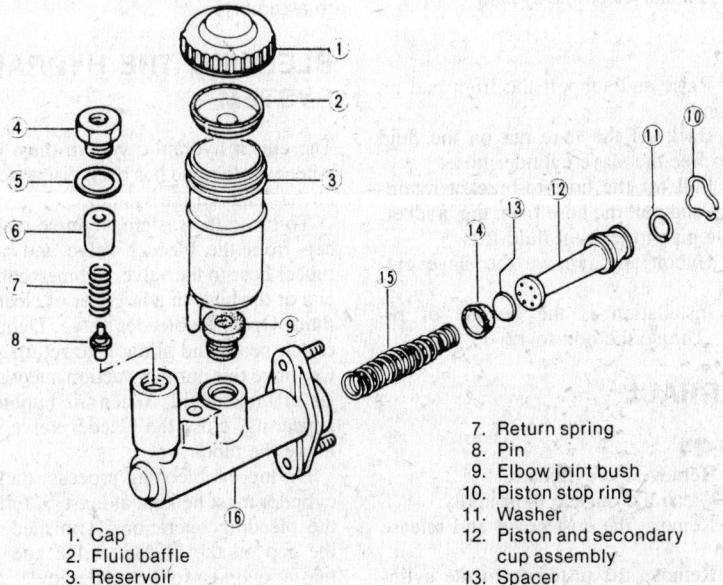

1. Cap
2. Fluid baffle
3. Reservoir
4. Joint bolt
5. Packing
6. Piston-oneway valve
7. Return spring
8. Pin
9. Elbow joint bush
10. Piston stop ring
11. Washer
12. Piston and secondary cup assembly
13. Spacer
14. Primary piston cup
15. Spring
16. Cylinder

1979–84 clutch master cylinder

5. Install the attaching nuts and bolts. Torque the nuts to 12–17 ft. lb.

6. Connect the fluid outlet line to the master cylinder fitting.

7. Bleed the hydraulic system.

8. Check the clutch pedal free-travel and adjust as necessary.

OVERHAUL

1979–84

1. Remove the master cylinder.

2. Clean the outside of the cylinder thoroughly and drain the fluid.

3. On some models, remove the dust cover.

4. Pry out the piston stop ring and remove the stop washer.

5. Remove the piston, piston cup and spring from the cylinder.

6. Carefully remove and disassemble the one-way valve.

7. Clean all parts thoroughly using clean brake fluid.

8. Discard any worn, damaged or mis-shapen parts.

9. Check the piston-to-bore clearance. Clearance should be 0.006 in. If clearance exceeds this figure, replace the unit.

10. If the cylinder bore is lightly scored or brinnelled, it may be honed to restore the finish.

11. Assembly is the reverse of disassembly. Coat all parts with clean brake fluid prior to assembly. Fill and bleed the system.

1986

1. Remove the master cylinder.

2. Using snapring pliers, press down on the piston and remove the snapring from the cylinder bore.

3. Remove the piston and secondary cup, primary cup protector, primary cup, return spring, reservoir and bushing.

4. The secondary piston and cup must be blown out with compressed air applied to the fluid pipe hole. Be careful to cover

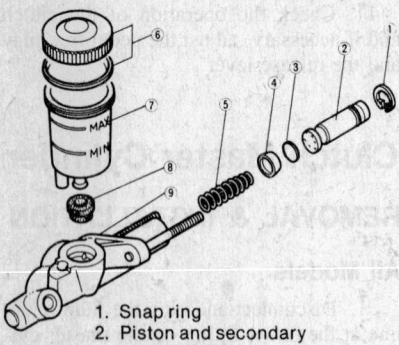

1. Snap ring
2. Piston and secondary cup assembly
3. Protector
4. Primary cup
5. Return spring
6. Tank cap and baffle
7. Reservoir tank
8. Bushing
9. Master cylinder body

1986 clutch master cylinder

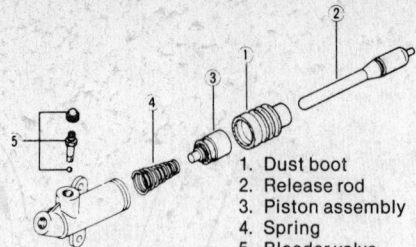

1. Dust boot
2. Release rod
3. Piston assembly
4. Spring
5. Bleeder valve

1979–84 clutch slave cylinder

the bore opening with a heavy rag to catch the piston.

5. Inspect all parts for wear or damage. Clean all parts in clean brake fluid.

6. Assembly is the reverse of disassembly. Coat all parts with clean brake fluid prior to assembly.

ONE-WAY VALVE REMOVAL & INSTALLATION

1. Remove the cap from the side of the master cylinder.

2. Remove the washer, one-way valve, and the spring.

3. Installation is the reverse.

Clutch Slave Cylinder
REMOVAL & INSTALLATION

1979–84

1. Disconnect and plug the line at the cylinder.

2. Unhook the lever from the pushrod.

3. Remove the nuts and washers attaching the slave cylinder to the clutch housing.

4. Installation is the reverse of removal. Torque the mounting nuts to 12–17 ft. lb. Fill and bleed the system.

1986

1. Raise and support the front end on jackstands.

2. Back off the flare nut on the fluid pipe to free the slave cylinder hose.

3. Pull off the hose-to-bracket retaining clip and pull the hose from the bracket. Cap the pipe to prevent fluid loss.

4. Unbolt and remove the slave cylinder.

5. Installation is the reverse of removal. Torque the bolt to 12–17 ft. lb.

OVERHAUL

1979–84

1. Remove the cylinder.

2. Clean the outside thoroughly.

3. Remove the dust cover and release rod.

4. Remove the piston from the cylinder.

5. Disassemble the bleeder valve.

6. Discard any worn, damaged or distorted parts.

7. Clean all parts in clean brake fluid.

8. The cylinder bore may be honed to remove slight surface damage.

9. Assembly is the reverse of disassembly. Coat all parts in clean brake fluid prior to assembly.

1986

1. Remove the cylinder.

2. Clean the outside thoroughly.

3. Remove the dust cover and release rod.

4. Remove the piston from the cylinder.

5. Remove the return spring.

6. Remove the bleeder screw and the small steel check ball underneath it.

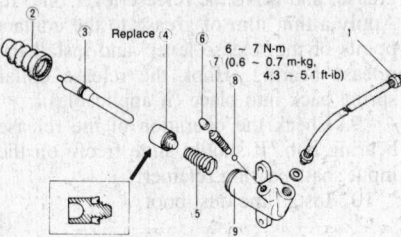

1. Flexible hose
2. Boot
3. Push rod
4. Piston and cup assembly
5. Return spring cylinder
6. Bleeder plug cap
7. Bleeder plug
8. Steel ball
9. Release cylinder

1986 clutch slave cylinder

7. Discard any worn, damaged or distorted parts.

8. Clean all parts in clean brake fluid.

9. The cylinder bore may be honed to remove slight surface damage.

10. Assembly is the reverse of disassembly. Coat all parts in clean brake fluid prior to assembly.

BLEEDING THE HYDRAULIC SYSTEM

The clutch hydraulic system must be bled whenever the line has been disconnected or air has entered the system.

To bleed the system, remove the rubber cap from the bleeder valve and attach a rubber hose to the valve. Submerge the other end of the hose in a large jar of clean brake fluid. Open the bleeder valve. Depress the clutch pedal and allow it to return slowly. Continue this pumping action and watch the jar of brake fluid. When air bubbles stop appearing, close the bleeder valve and remove the tube.

During the bleeding process, the master cylinder must be kept at least $3/4$ full. After the bleeding operation is finished, install the cap on the bleeder valve and fill the master cylinder to the proper level. Always use fresh brake fluid, and above all, do not use the fluid that was in the jar for bleeding, since it contains air. Install the master cylinder reservoir cap.

AUTOMATIC TRANSMISSION

Pan & Filter
REMOVAL & INSTALLATION

1. Raise and support the vehicle.
2. Place a drain pan under the transmission pan.
3. Remove the pan attaching bolts (except the two at the front). Loosen the two at the front slightly. Allow the fluid to drain.
4. Remove the pan.
5. Remove and discard the gasket.
6. Install a new pan gasket and install the pan on the transmission.
7. Lower the vehicle and fill the transmission with fluid. Check the transmission operation.

NOTE: The following adjustments should be performed in the order given. Be sure that the idle speed is set before performing any adjustments.

Adjustments
SHIFT LINKAGE ADJUSTMENT

1. Put the gearshift lever in Neutral.
2. Raise and support the truck.
3. Disconnect the clevis from the lower end of the selector lever operating arm.
4. Move the transmission manual lever to Neutral, the 3rd detent position from the rear of the transmission.
5. Loosen the two clevis retaining nuts and adjust the clevis so that it freely enters the hole of the lever. Tighten the retaining nuts to secure the adjustment.
6. Connect the clevis to the lever and attach it with the spring washer, flat washer and retaining clip.
7. Lower the truck and check the operation of the linkage. Be sure that all gears engage properly.

KICKDOWN SWITCH ADJUSTMENT

1. Turn the ignition switch to the ON position.
2. Loosen the kickdown switch attaching nut (the switch is located just above the accelerator pedal) and adjust the switch to engage when the accelerator pedal is depressed about $7/8$ of the way. The downshift solenoid will click when the switch engages.
3. Tighten the attaching nut and check the switch for proper operation.

SHIFT LEVER HANDLE INTERLOCK

The interlock should be adjusted when it does not perform as shown in the accompanying illustration, or, whenever the handle has been removed.

1. Back off the locknut below the handle.
2. Position the shifter in either N or D.
3. Screw in the handle until no play is felt at the interlock button.

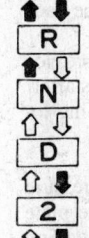

⇨ Button need not be depressed.

➡ Button must be pressed.

Automatic transmission shift interlock performance check

4. Turn the handle one additional turn, if necessary, to position the botton on the driver's side.
5. Depress the button and shift to P. If the lever cannot be moved to the P position, screw in the handle an additional turn, repeating the shift move and additional turn, until P can be engaged smoothly. From this point, shift through the various positions, confirming that the shifter works as shown in the illustration. If the lever can be shifted to R from either P or N, or into 2 from D, without depressing the button, back out on the handle.
6. When the adjustment is completed, check that the button protrudes 6.0mm from the handle in the N or P position. Recheck the shift pattern.
7. Tighten the locknut to 15 ft. lb.

NEUTRAL SAFETY SWITCH ADJUSTMENT

1. Adjust the manual linkage.
2. Place the transmission manual lever in Neutral (3rd detent from the rear of the transmission).
3. Remove the transmission manual lever retaining nut and lever.
4. Loosen the inhibitor switch attaching bolts. Remove the screw from the alignment pin hole at the bottom of the switch.
5. Rotate the switch and insert an alignment pin, 0.059 in. diameter into the alignment pin hole and internal rotor.

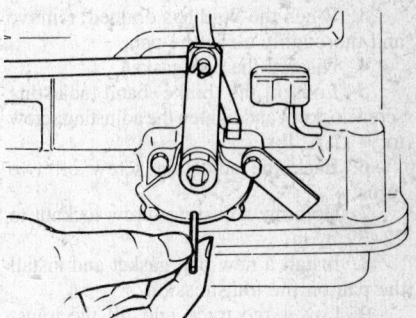

Neutral safety switch adjustment

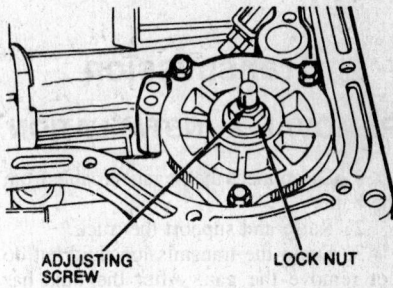

ADJUSTING SCREW   LOCK NUT

intermediate band adjustment

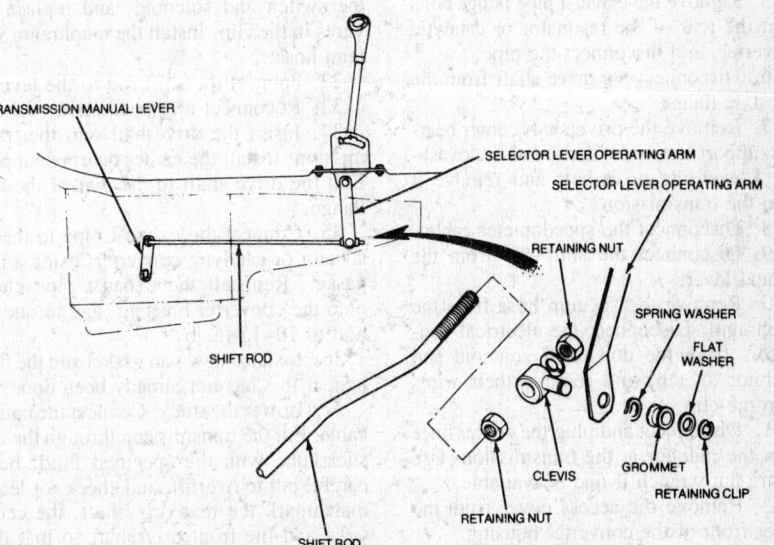

TRANSMISSION MANUAL LEVER

SELECTOR LEVER OPERATING ARM

SELECTOR LEVER OPERATING ARM

RETAINING NUT

SPRING WASHER

FLAT WASHER

SHIFT ROD

CLEVIS

GROMMET

RETAINING CLIP

RETAINING NUT

SHIFT ROD

Automatic transmission shift linkage adjustment

6. Tighten the two switch attaching bolts and remove the alignment pin.

7. Reinstall the alignment pin hole screw in the switch body.

8. Install the manual lever.

9. Check the operation of the switch. The engine should only start with the transmission selector lever in Neutral or Park.

INTERMEDIATE BAND ADJUSTMENT

1. Raise and support the truck.

2. Place a drain pan under the transmission and loosen the pan attaching bolts to drain the fluid. Finally remove all the bolts except the two at the front.

3. When the fluid has drained, remove and thoroughly clean the pan.

4. Discard the pan gasket.

5. Loosen the brake band adjusting screw locknut and tighten the adjusting screw to 9–11 ft. lb.

6. Back the adjusting screw off two turns.

7. Hold the adjusting screw locknut to 22–29 ft. lb.

8. Install a new pan gasket and install the pan on the transmission.

9. Lower the truck and fill the transmission with fluid.

Transmission

REMOVAL & INSTALLATION

1. Disconnect the negative cable from the battery.

2. Raise and support the truck.

3. Drain the transmission fluid but do not remove the pan. After the fluid has drained, install a few bolts to hold the pan in place, temporarily.

4. Remove the exhaust pipe bracket bolt from the right side of the converter housing.

5. Remove the exhaust pipe flange bolts from the rear of the resonator or catalytic converter, and disconnect the pipe.

6. Disconnect the drive shaft from the rear axle flange.

7. Remove the drive shaft center bearing support nuts, washers, and lockwashers. Lower the driveshaft and remove it from the transmission.

8. Disconnect the speedometer cable.

9. Disconnect the shift rod from the manual lever.

10. Remove the vacuum hose from the diaphragm. Disconnect the electrical connectors from the downshift solenoid and inhibitor switch, and remove their wires from the clip.

11. Disconnect and plug the cooler lines from the radiator at the transmission. Use a flare nut wrench if one is available.

12. Remove the access cover from the lower front of the converter housing.

13. Matchmark the drive plate (flywheel) and torque converter for reassembly. Remove the four bolts holding the torque converter to the drive plate.

14. Remove the bolts connecting the crossmember to the transmission.

15. Support the transmission with a jack. Remove the crossmember-to-frame bolts, and remove the crossmember.

16. Make sure that the transmission is securely supported. Secure it to the jack with a safety chain, if necessary.

17. Lower the transmission to provide working clearance, and remove the starter.

18. Remove the converter housing-to-engine bolts.

19. Remove the fluid filler tube.

20. With a pry bar, exert light pressure between the converter and the drive plate to prevent the converter from disengaging from the transmission as it is removed.

21. Lower the transmission and converter as an assembly. Be careful not to let the converter fall out.

22. Place the transmission on the jack. Be sure that the converter is properly installed.

23. Raise the transmission into place. Install the converter housing-to-engine bolts, and torque in two stages to 23–34 ft. lb.

24. Lower the transmission on the jack and install the starter.

25. Install the fluid filler tube with a new O-ring.

26. Raise the transmission slightly, and install the crossmember to the frame. Tighten the bolts to 23–34 ft. lb.

27. Lower the transmission and install the transmission-to-crossmember bolts. Tighten to 23–34 ft. lb.

28. Align the matchmarks made earlier on the torque converter and drive plate. Install the four attaching bolts and torque to 25–36 ft. lb. in three stages.

29. Install the access cover. Remove the jack.

30. Connect the cooler lines.

31. Install the electrical connectors to the switch and solenoid, and replace the wires in the clip. Install the diaphragm vacuum hose.

32. connect the shift rod to the lever.

33. Reconnect the speedometer cable.

34. Insert the driveshaft into the transmission. Install the center bearing support. Bolt the drive shaft to the rear of the axle flange.

35. Connect the exhaust pipe to the resonator or catalytic converter, using a new gasket. Reinstall the exhaust pipe clamp onto the converter housing, and torque the bolt to 10–15 ft. lb.

36. Install a new pan gasket and the fluid pan, if this has not already been done.

37. Lower the truck. Connect the battery cable. Fill the transmission through the dipstick tube with the specified fluid, being careful not to overfill, and check for leaks. matchmark the rear driveshaft, the center yoke and the front driveshaft so that they

DRIVE LINE

Driveshaft

REMOVAL & INSTALLATION

1. Matchmark the rear U-joint with the rear companion flange. Remove the bolts attaching the driveshaft to the rear companion flange.

2. On 2-piece units, remove the center support bearing bracket from the underbody.

3. Pull the driveshaft rearward and out of the transmission. Plug the rear seal opening.

4. Installation is the reverse of removal. Make sure that you align the matchmarks. Torque the rear companion flange bolts to 39–47 ft. lb.; the center bearing bracket nuts to 27–38 ft. lb.

U-Joint

For all information concerning U-joints, please refer to U-Joints/CV-Joints in the Unit Repair Section.

Center Bearing

REPLACEMENT

The center support bearing is a sealed unit which requires no periodic maintenance. The following procedure should be used if it becomes necessary to replace the bearing. You will need a pair of snap pliers for this job.

1. Remove the driveshaft assembly.

2. To maintain driveline balance, may be installed in their original positions.

3. Remove the center universal joint from the center yoke, leaving it attached to the rear driveshaft. See the following section for the correct procedure.

4. Remove the nut and washer securing the center yoke to the front driveshaft.

5. Slide the center yoke off the splines. The rear oil seal should slide off with it.

6. If the oil has remained on top of the snap ring, remove and discard the seal. Remove the snap ring from its groove. Remove the bearing.

7. Slide the center support and front oil seal from the front driveshaft. Discard the seal.

8. Install the new bearing into the center support. Secure it with the snap ring.

9. Apply a coat of grease to the lips of the new oil seals, and install them into the center support on either side of the bearing.

10. Coat the splines of the front driveshaft with grease. Install the center support

assembly and the center yoke onto the front driveshaft, being sure to match up the marks made during disassembly.

11. Install the washer and nut. Torque the nut to 116–130 ft. lb.

12. Check that the center support assembly rotates smoothly around the driveshaft.

13. Align the mating marks on the center yoke and the rear driveshaft, and assemble the center universal joint.

14. Install the driveshaft. Be sure that the rear yoke and the axle flange realigned properly.

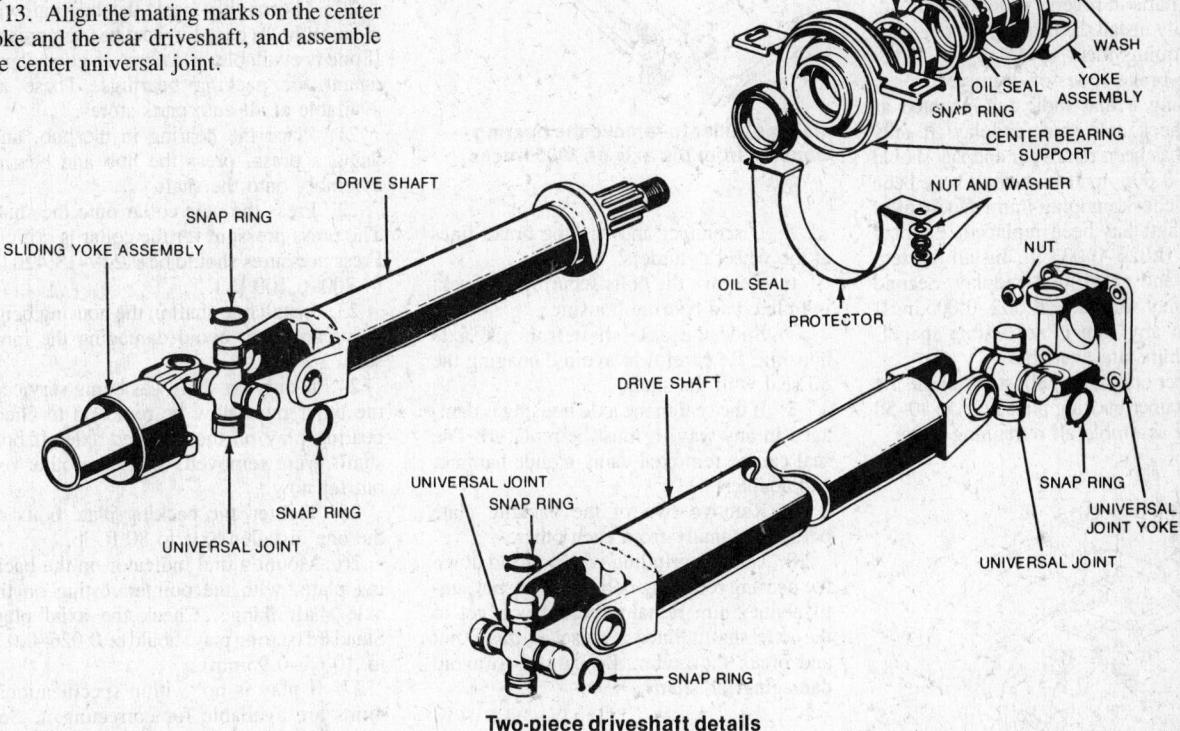

Two-piece driveshaft details

REAR AXLE

For rear axle overhaul, refer to the Import Truck portion of the Drive Axle Unit Repair Section.

Axle Shaft, Bearing and Seal

REMOVAL & INSTALLATION

1979–84

1. Raise and support the rear end on jackstands.

2. Remove the wheels and brake drums.

3. Remove the brake shoes.

4. Remove the parking brake cable retainer.

5. Disconnect and cap the brake lines at the wheel cylinders.

6. Remove the bolts securing the backing plate and bearing housing.

7. Slide the axle shaft from the axle housing.

8. Remove the oil seal from the axle housing and discard it. A puller may be necessary.

9. Straighten the tabs on the lockwasher and remove the nut and lockwasher from the axle shaft.

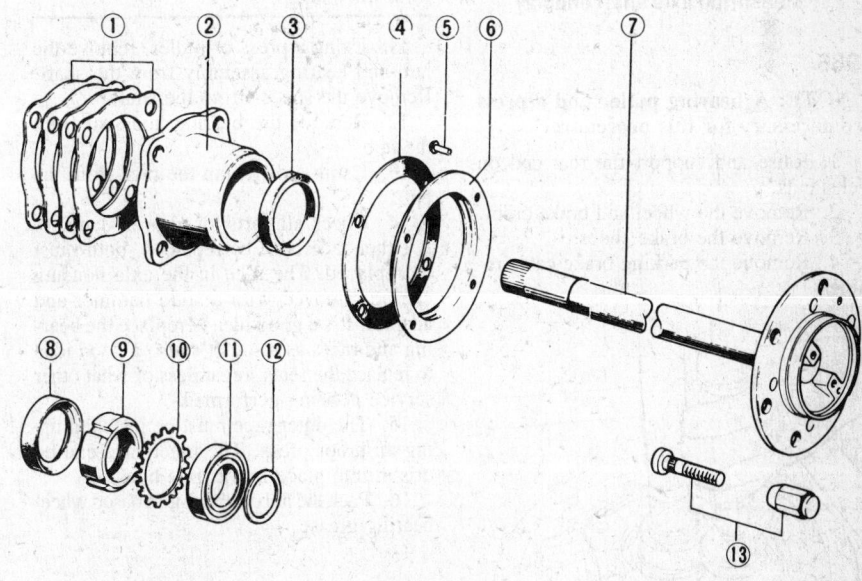

1. Shims	6. Baffle seal
2. Bearing housing	7. Axle shaft
3. Outer oil seal	8. Inner oil seal
4. Gasket	9. Lock nut
5. Rivet	10. Lock washer
	11. Bearing
	12. Spacer
	13. Hub bolt and lug nut

1979–84 rear axle shaft details

10. Remove the bearing and race from the shaft. A puller or press may be necessary. Discard the spacer.

11. Remove the outer seal from the bearing housing and discard it.

12. Discard the gasket from the baffle.

13. Using new seals and a new gasket, install all parts in reverse order of removal. Temporarily install the bearing/backing plate bolts, torquing them to 16 ft. lb. Don't install the brake shoes or drum yet.

14. Using a dial indicator mounted as shown, check axle shaft endplay. If only one shaft has been removed, endplay should be 0.002–0.006 in. If both shaft have been removed, check endplay immediately after the first shaft has been replaced. Endplay should be 0.026–0.033 in. Install the second shaft and check that endplay. Second shaft endplay should be 0.002–0.006 in. If endplay at any step is not within specifications, shims are available.

15. After endplay is adjusted, torque the bearing retainer/backing plate bolts to 40–50 ft. lb. and assemble all remaining parts.

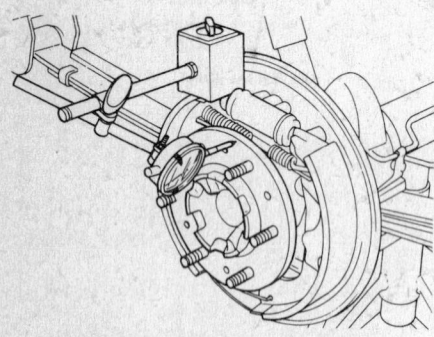

Measuring axle shaft endplay

1986

NOTE: A bearing puller and a press are necessary for this procedure.

1. Raise and support the rear end on jackstands.

2. Remove the wheel and brake drum.

3. Remove the brake shoes.

4. Remove the parking brake cable retainer.

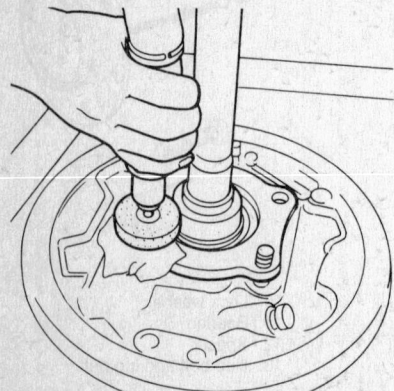

Grinding the bearing retainer collar on 1986 trucks

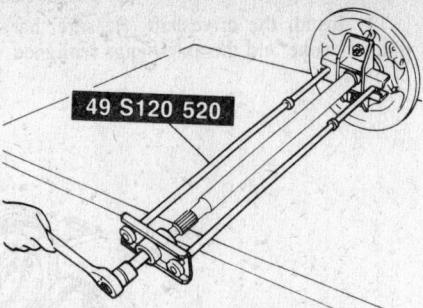

Using a puller to remove the bearing housing from the axle on 1985 trucks

5. Disconnect and cap the brake lines at the wheel cylinders.

6. Remove the bolts securing the backing plate and bearing housing.

7. Slide the axle shaft from the axle housing. Be careful to avoid damaging the oil seal with the shaft.

8. If the seal in the axle housing is damaged in any way, it must be replaced. The seal can be removed using a slide hammer and adapter.

9. Remove two of the backing plate bolts, diagonally from each other.

10. Using a grinding wheel, grind down the bearing retaining collar in one spot, until about 5mm remains before you get to the axle shaft. Place a chisel at this point and break the collar. Be careful to avoid damaging the shaft.

--- CAUTION ---

Wear some kind of protective goggles when grinding the collar and breaking the collar from the shaft!

11. Using a press or puller, remove the hub and bearing assembly from the shaft. Remove the spacer from the shaft.

12. Remove the bearing and seal from the hub.

13. Using a drift, tap the race from the hub.

14. Check all parts for wear or damage. If either race is to be replaced, both must be replaced. The race in the axle housing can be removed with a slide hammer and adapter. It's a good idea to replace the bearing and races as a set. It's also a good idea to replace the seals, regardless of what other service is being performed.

15. The outer race must be installed using an arbor press. The inner race can be driven into place in the axle housing.

16. Pack the hub with lithium based wheel bearing grease.

17. Tap a new oil seal into the axle housing until it is flush with the end of the housing. Coat the seal lip with wheel bearing grease.

18. Install a new spacer on the shaft with the larger flat surface up.

19. Install a new seal in the hub.

20. Thoroughly pack the bearing with clean, lithium based, wheel bearing grease. If one is available, use a grease gun adapter meant for packing bearings. These are available at all auto parts stores.

21. Place the bearing in the hub, and, using a press, press the hub and bearing assembly onto the shaft.

22. Press the new collar onto the shaft. The press pressure for the collar is critical. Press pressures should be 9,240–13,420 lb. (4,200–6,100 kg).

23. Install one shaft in the housing being very careful to avoid damaging the inner seal.

24. If only on shaft was being serviced, the other must now be removed to check bearing play on the serviced axle. If both shafts were removed, leave the other one out for now.

25. Tighten the backing plate bolts on the one installed axle to 80 ft. lb.

26. Mount a dial indicator on the backing plate, with the pointer resting on the axle shaft flange. Check the axial play. Standard bearing play should be 0.026–0.037 in. (0.65–0.95mm).

27. If play is not within specifications, shims are available for correcting it. See the table below.

28. Install the other shaft and torque the backing plate bolts. Check the play as on the first shaft. Play should be 0.002–0.010 in. If not, correct it with shims.

29. Install the brake drums and wheels. Bleed the brake system.

Differential
REMOVAL & INSTALLATION

1. Raise the vehicle and support it safely with jackstands.

2. Remove the differential drain plug and drain the lubricant from the differential. Install the plug after all of the fluid has drained.

3. Remove the axle shafts as previously outlined.

4. Remove the driveshaft(s) as previously outlined.

5. Remove the carrier-to-differential housing retaining fasteners and remove the carrier assembly from the housing.

SHIM SELECTION CHART

Part Number	Thickness mm (in.)
S083 26 165	0.10 (0.004)
S083 26 166	0.15 (0.006)
S083 26 167	0.50 (0.020)
S083 23 168	0.75 (0.030)

6. Clean the carrier and axle housing mating surfaces.

7. If the differential originally used a gasket between the carrier and the differential housing, replace the gasket. If the unit had no gasket, apply a thin film of oil-resistant silicone sealer to the mating surfaces of both the carrier and the housing and allow the sealer to set according to the manufacturer's instructions.

8. Place the carrier assembly onto the housing and install the carrier-to-housing fasteners. Torque the fasteners to 12–17 ft. lb.

9. Install the driveshaft(s) and axle shafts as previously outlined.

10. Install the brake drums and wheels.

11. Fill the differential with the proper amount of SAE 80W–90 fluid (see the Capacities Chart).

FRONT SUSPENSION

NOTE: On 1979–84 trucks, the front suspension consists of a wishbone-type, upper and lower control arm assembly with coil spring. Suspension travel is dampened by double acting shock absorbers.

1986 trucks use a torsion bar type front suspension, with upper and lower control arms. Conventional, double-acting shock absorbers are employed to dampen motion. A stabilizer bar is standard equipment.

Shock Absorber

TEST

The easiest way to check the performance of your shocks is to go to one corner of the truck and start it bouncing up and down. Get it going as much as you can and then release it. It should stop bouncing in less than two full bounces.

REMOVAL & INSTALLATION

1. Raise and support the front end on jackstands.

2. Remove the upper end nut, bushings and washers from the shock stem.

3. Remove the lower end attaching bolts.

4. Remove the shock from beneath the lower control arm.

5. Installation is the reverse of removal. Tighten the lower bolts to 25 ft. lb. on 1979–84 trucks, and 55–59 ft. lb. on 1986 trucks. Tighten the upper nut until $^1/_4$ inch of thread is visible above the locknut on 1979–84 trucks. On the 1986 B2000,

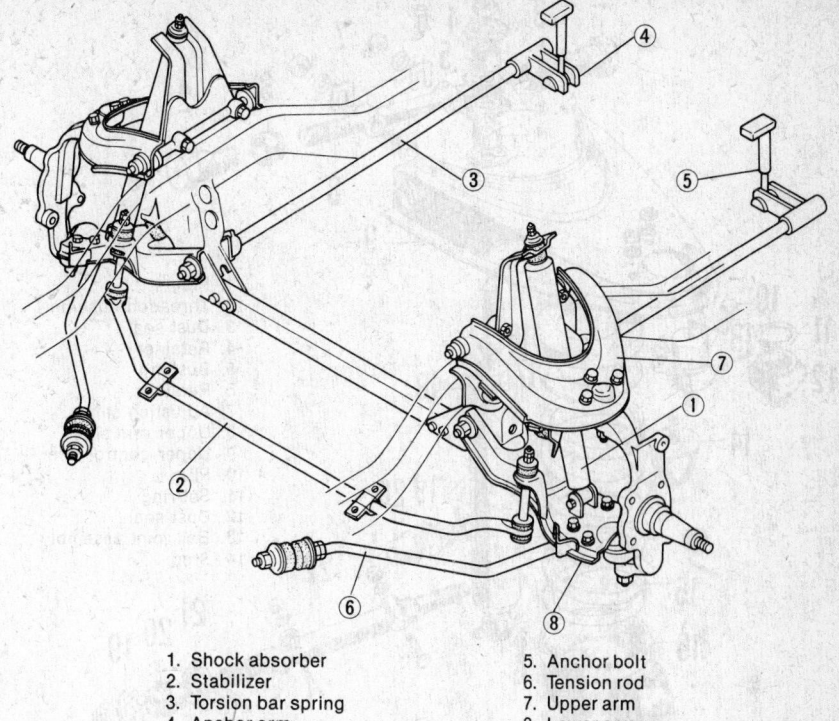

1. Shock absorber
2. Stabilizer
3. Torsion bar spring
4. Anchor arm
5. Anchor bolt
6. Tension rod
7. Upper arm
8. Lower arm

1986 front suspension

tighten the upper nut to 17–25 ft. lb. At this point, 7mm of thread should be visible above the nut.

Coil Spring

REMOVAL & INSTALLATION

1979–84

——— **CAUTION** ———

The spring is under a great deal of tension! It's best to use a coil spring compressor when removing the spring. Mishandling the spring could cause it to fly out of its mounting, causing a great deal of personal damage!

1. Raise and support the front end on jackstands under the frame.

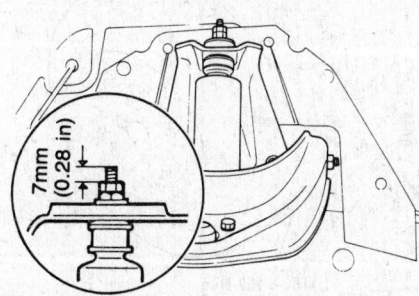

Tightening the upper nut on a 1986 front shock absorber

2. Remove the wheel.

3. Remove the shock absorber. Install the spring compressor.

4. Remove the stabilizer bar.

5. Support the lower arm with a floor jack.

6. Disconnect the upper and lower ball joints from the knuckle by removing the cotter pins and nuts and separating the ball joints with a ball joint separator tool.

7. Remove the upper control arm as described below.

8. Slowly lower the jack and remove the spring. Release the compressor to remove spring tension.

9. Installation is the reverse of re-

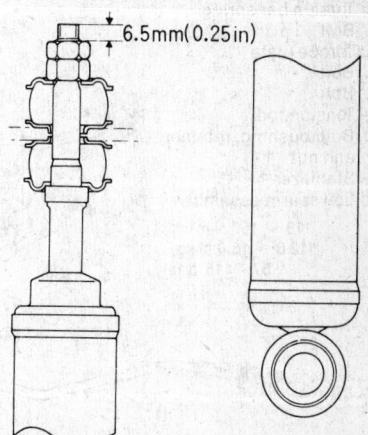

1979–84 front shock absorber upper nut installation

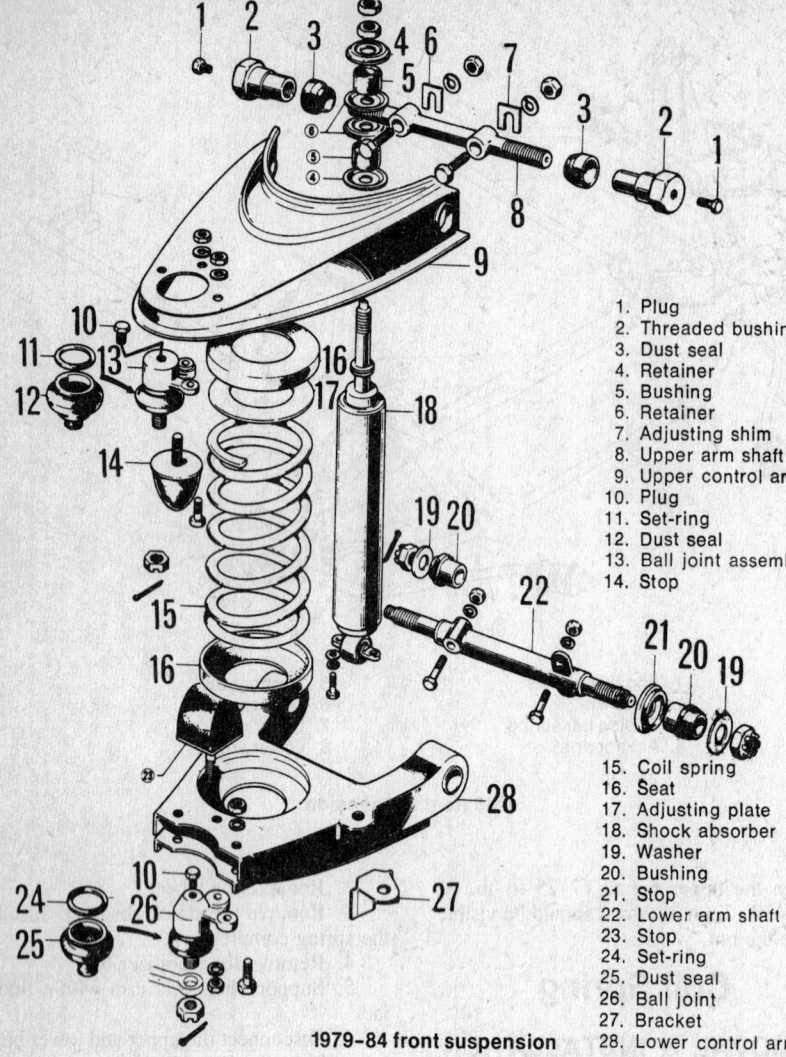

1. Plug
2. Threaded bushing
3. Dust seal
4. Retainer
5. Bushing
6. Retainer
7. Adjusting shim
8. Upper arm shaft
9. Upper control arm
10. Plug
11. Set-ring
12. Dust seal
13. Ball joint assembly
14. Stop
15. Coil spring
16. Seat
17. Adjusting plate
18. Shock absorber
19. Washer
20. Bushing
21. Stop
22. Lower arm shaft
23. Stop
24. Set-ring
25. Dust seal
26. Ball joint
27. Bracket
28. Lower control arm

1979–84 front suspension

1. Lug nut
2. Wheel and tire
3. Cotter pin
4. Nut
5. Bolt
6. Anchor bolt
7. Anchor swivel
8. Anchor arm
9. Torsion bar spring
10. Bolt
11. Torque plate
12. Bolt
13. Bolt
14. Tension rod
15. Bolt, bushing, retainer and nut
16. Stabilizer
17. Lower arm assembly

118 ~ 157 N-m
(12.0 ~ 16.0 m-kg,
87 ~ 115 ft-lb)

74 ~ 93 N-m
(7.6 ~ 9.5 m-kg,
54.9 ~ 68.6 ft-lb)

118 ~ 157 N-m
(12.0 ~ 16.0 m-kg, 87 ~ 115 ft-lb)

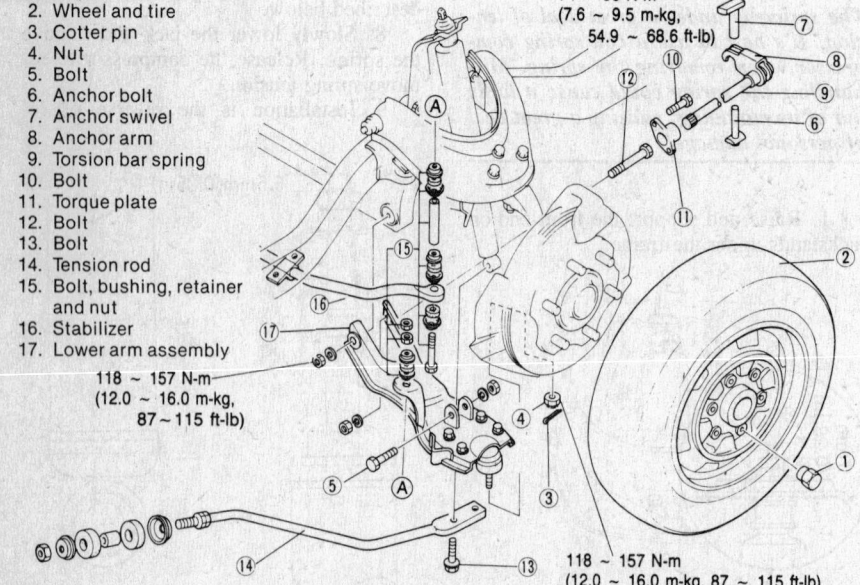

Torsion bar removal and installation for 1986 trucks

moval. It's best to replace springs in pairs, however, spacers are available to equalize road height.

Torsion Bar and Lower Control Arm

REMOVAL & INSTALLATION

1986 Only

NOTE: Special tools are necessary for this procedure.

1. Raise and support the front end on jackstands under the frame.

2. Remove the wheel.

3. Remove the cotter pin and nut from the lower ball joint.

4. Remove the lower shock absorber bolt.

5. Matchmark the anchor arm bolt and anchor swivel and remove the bolt and swivel.

6. Matchmark the torsion bar and anchor arm and the torsion bar and torque plate.

7. Remove the anchor arm and torsion bar from the torque plate. Separate the anchor arm from the torsion bar.

8. Unbolt and remove the torque plate.

9. Remove the lower arm-to-frame bolt. Separate the lower arm from the frame bracket with bushing puller/installer 49 0727 575.

10. Unbolt the tension rod from the lower arm and frame and remove it.

NOTE: Don't change the position of the double nut at the rear of the tension rod bushing, since it would affect caster.

11. Remove the stabilizer bar bolt, bushing, retainer and nut and remove the stabilizer bar.

12. Using a ball joint separator, separate the lower ball joint from the knuckle. Remove the lower control arm.

13. Inspect parts for wear or damage. Replace any suspect parts. Using a spring scale and adapter 49 0180 510B, check the ball joint preload. Pull scale reading should be 39.6 lb. or less. Measure the preload after first shaking the ball joint stud 3 or 4 times to make sure it is free.

14. Install the lower arm on the frame bracket and hand tighten the nut.

15. Install the lower ball joint on the knuckle and torque the nut to 115 ft. lb. Install the cotter pin.

16. Tighten the lower arm-to-frame nut to 115 ft. lb.

17. Position the torque plate and tighten the bolt to 68 ft. lb.

18. Coat the splines on the torsion bar with lithium based wheel bearing grease. Check the ends of the torsion bar. The bars are marked L for left and R for right. Don't confuse them. Align the matchmarks and install the torsion bar in the torque plate.

19. Coat the splines on the torsion bar with lithium based grease. Align the match-

marks and install the anchor arm on the torsion bar.

20. Install the anchor bolt and swivel and tighten the bolt until the matchmarks are mated.

21. Install the tension rod. Torque the bushing end nut to 90 ft. lb.; the lower arm end bolts to 85 ft. lb.

22. Install the stabilizer bar. Torque the bolt to 19 ft. lb.

23. Install the shock absorber bolt. Torque the bolt to 55–59 ft. lb.

24. Install the wheels and lower the truck to the ground.

25. Retorque the lower arm-to-frame bracket nut.

26. Check the front and rear tire pressures. Set the pressures to what are specified on the vehicle rating plate, except for P-metric radials. Set them at the maximum pressure shown on the side wall.

27. Measure the distance from the center of the wheel hub to the lip of the fender. This is the ride height. Proper ride height is obtained when the difference between the left and the right side is less than 10mm. Adjust the ride height by turning the anchor bolt.

NOTE: If, for some reason, you didn't matchmark the torsion bar anchor bolt, or the matchmarks were lost, or you're installing a new, unmarked torsion bar, here's a procedure to help you attain the correct ride height:

Install the anchor arm on the torsion bar so that there is 125mm between the lowest point on the arm and the crossmember directly above it.

Tighten the anchor bolt until the anchor arm contacts the swivel. Then, tighten the bolt an additional 45mm travel.

Stabilizer Bar

REMOVAL & INSTALLATION

1. Raise and support the front end on jackstands.

2. Unbolt the stabilizer bar-to-frame clamps.

3. Unbolt the stabilizer bar from the lower control arms. Keep all the bushings, washers and spacers in order.

4. Check all parts for wear or damage and replace anything which looks suspicious.

5. Installation is the reverse of removal. Tighten all fasteners lightly, then torque them to specifications with the wheels on the ground.

Stabilizer bar-to-control arm nut: 1979–84: 25 ft. lb.; 1986: 34 ft. lb.

Stabilizer-to-frame clamp bolts: 16 ft. lb.

Tension Rod
REMOVAL & INSTALLATION

1986 Only

1. Unbolt the tension rod from the lower arm and frame and remove it.

NOTE: Don't change the position of the double nut at the rear of the tension rod bushing, since it would affect caster.

2. Install the tension rod. Torque the bushing end nut to 90 ft. lb.; the lower arm end bolts to 85 ft. lb.

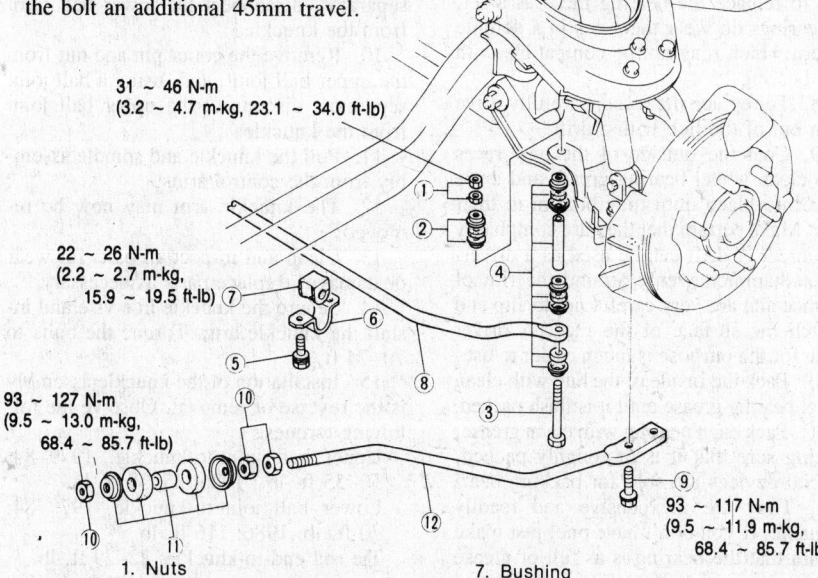

31 ~ 46 N-m
(3.2 ~ 4.7 m-kg, 23.1 ~ 34.0 ft-lb)

22 ~ 26 N-m
(2.2 ~ 2.7 m-kg,
15.9 ~ 19.5 ft-lb)

93 ~ 127 N-m
(9.5 ~ 13.0 m-kg,
68.4 ~ 85.7 ft-lb)

93 ~ 117 N-m
(9.5 ~ 11.9 m-kg,
68.4 ~ 85.7 ft-lb)

1. Nuts
2. Retainer and bushings
3. Bolt
4. Bushings, retainer and control link
5. Bolt
6. Stabilizer bracket
7. Bushing
8. Stabilizer
9. Bolt
10. Nut
11. Retainer, bushings and spacer
12. Tension rod

Stabilizer and tension rod removal, 1986 trucks

Upper Control Arm

REMOVAL & INSTALLATION

1979–84

1. Raise and support the front end on jackstands under the frame.

2. Using a floor jack, raise the lower control arm until the upper control arm is off the bumper stop.

3. Remove the wheel.

4. Place a chain through the coil spring as a safety measure, or install a spring compressor.

5. Remove the cotter pin and nut retaining the upper ball joint.

6. Using a ball joint separator, disconnect the ball joint from the spindle.

7. Working under the hood, remove the two upper arm retaining bolts and lift the arm from the truck. Note the number and position of any shims.

8. Installation is the reverse of removal. Place the shims in their original locations. Torque the two arm retaining bolts to 65–75 ft. lb.; the ball joint-to-arm bolts to 15–20 ft. lb.; the ball joint-to-spindle nut to 40–55 ft. lb.

1986

1. Raise and support the front end on jackstands placed under the frame.

2. Remove the wheels. Support the lower arm with a floor jack.

3. Remove the cotter pin and nut from the upper ball joint and separate the ball joint from the upper arm using a ball joint separator tool.

4. Remove the bushings and dust seals from the ends of the upper arm shaft.

5. Remove the nuts and bolts that retain the upper arm shaft to the support bracket. Note the number and location of the shims under the nuts. These must be installed in their exact locations for proper wheel alignment. Check all parts for wear or damage. Replace any suspect parts. Check the ball joint preload with a pull scale and adapter 49 0180 510B. Shake the ball joint stud a few times to make sure that it is free, then take the reading. The pull scale reading should be 40 lb. or less.6.

Installation is the reverse of removal. Torque the upper arm shaft mounting bolts to 60–68 ft. lb.; the ball joint nut to 30–37 ft. lb.

Lower Control Arm

REMOVAL & INSTALLATION

1979–84

1. Raise the front end and support it on jackstands under the frame.

2. Remove the wheels.

3. Remove the lower shock absorber bolts and push the shock up, out of the way.

T473

4. Disconnect the front stabilizer bar from the control arms.

5. Place a floor jack under the lower arm and raise the arm to compress the spring. Install a safety chain or spring compressor.

6. Unbolt the ball joint from the lower arm.

7. Pull the spindle and ball joint away from the arm.

8. Carefully lower the jack. The spring is under pressure, so be very careful that it is secured with the chain or spring compressor.

9. Remove the three lower control arm retaining bolts and lift the arm from the frame.

10. When installing the arm, safety chain the spring to the arm prior to installing the arm, or use a spring compressor. When the arm is in position, loosely install the ball joint bolts and remove the chain or compressor, then, tighten the ball joint nut to 70 ft. lb.; the three ball joint retaining nuts to 70 ft. lb. Install all other parts in reverse order of removal.

11. Have the front end alignment checked.

Ball Joints

INSPECTION

1. Inspect the dust seals. If cracked or brittle, replace them.

2. Check end play of both the upper and lower ball joints. If either exceeds 0.0039 in., it is defective.

REPLACEMENT

For replacement procedures, see the appropriate parts of either upper or lower control arm removal and installation.

Front Wheel Bearings

———— CAUTION ————
Brake shoes contain asbestos, which has been determined to be a cancer causing agent. Never clean the brake surfaces with compressed air! Avoid inhaling any dust from any brake surface! When cleaning brake surfaces, use a commercially available brake cleaning fluid.

ADJUSTMENT

1. Raise and support the front end on jackstands.

2. Remove the wheel. Remove the brake drum or disc brake caliper. Suspend the caliper out of the way. Don't disconnect the brake line.

3. Attach a spring scale to a wheel lug.

4. Pull the scale horizontally and check the force needed to start the wheel turning. The force should be 1.3–2.4 lbs. If the reading is not correct, proceed.

5. Remove the grease cap and cotter pin.

6. Tighten or loosen the hub nut until the correct pull rating is obtained.

7. Align the cotter pin holes and insert a new cotter pin. Replace the grease cap and wheel.

REMOVAL, REPACKING, INSTALLATION

1. Raise and support the front end on jackstands.

2. Remove the wheel.

3. Remove the grease cap, cotter pin, hub nut and flat washer.

4. On trucks with disc brakes, remove the caliper and suspend it out of the way without disconnecting the brake line. Slowly pull the hub from the spindle, positioning your hand to catch the outer bearing.

5. Remove the spacer, inner seal and inner bearing. Discard the seal.

6. Thoroughly clean the bearings and inside of the hub with a nonflammable solvent. Allow them to air dry.

7. Inspect the bearings for wear, damage, heat discoloration or other signs of fatigue. If they are at all suspect, replace them. When replacing bearings, it is a good idea to replace the bearing races as a set, as bearings do wear the races in a definite pattern which may not be compatible with new bearings.

8. To replace the races, carefully drive them out of the hub with a drift.

9. Coat the outside of the new races with clean wheel bearing grease and drive them into place until they bottom in their bore. Make certain that they are completely bottomed! A drift can be used as a driver, if you hammer evenly around the rim of the race and are very careful not to slip and scratch the surface of the race. A driver made for the purpose is much easier to use.

10. Pack the inside of the hub with clean wheel bearing grease until it is flush packed.

11. Pack each bearing with clean grease, making sure that it is thoroughly packed. Special devices are sold for packing bearings. They are inexpensive and readily available. If you don't have one, just make certain that the bearing is as full of grease

as possible by working it in with your fingers.

12. Install the inner bearing and seal. Drive the seal into place carefully until it is seated.

13. Install the spacer and the hub on the spindle.

14. Install the outer bearing, flat washer and hub nut.

15. Adjust the bearing as explained above.

16. Install the nut cap, cotter pin and grease cap. Install the wheel.

Knuckle and Spindle

REMOVAL & INSTALLATION

———— CAUTION ————
The coil spring on 1979–84 models is under great tension! Use a coil spring compressor for safety's sake.

1. Raise and support the front end on jackstands.

2. Remove the wheels.

3. Remove the brake drums or calipers. Suspend the calipers out of the way with a wire. Don't disconnect the brake line.

4. Remove the hub and bearings.

5. Remove the tie rod-to-knuckle nut, and, using a ball joint separator, remove the tie rod end from the knuckle.

6. On models through 1984, remove the shock absorber.

7. On models through 1984, install a spring compressor on the coil spring.

8. Support the lower arm with a floor jack.

9. Remove the cotter pin and nut from the lower ball joint, and, using a ball joint separator, disconnect the lower ball joint from the knuckle.

10. Remove the cotter pin and nut from the upper ball joint, and, using a ball joint separator, disconnect the upper ball joint from the knuckle.

11. Pull the knuckle and spindle assembly from the control arms.

12. The knuckle arm may now be removed.

13. Clean and inspect all parts for wear or damage. Replace parts as necessary.

14. Secure the knuckle in a vise and install the knuckle arm. Torque the bolts to 70–74 ft. lb.

15. Installation of the knuckle assembly is the reverse of removal. Observe the following torques:
Upper ball joint-to-knuckle, 1979–84: 50–55 ft. lb.; 1986: 35–38 ft. lb.
Lower ball joint-to-knuckle, 1979–84: 70 ft. lb.; 1986: 116 ft. lb.
Tie rod end-to-knuckle: 22–29 ft. lb.

REAR SUSPENSION

Leaf Springs

REMOVAL & INSTALLATION

1. Raise and support the truck, allowing the spring to hang freely.

───── **CAUTION** ─────
The rear leaf springs are under considerable tension. Be very careful when removing and installing them; they can exert enough force to cause serious injuries.

2. Place a floor jack under the rear axle to take up its weight.
3. Disconnect the lower end of the shock absorbers.
4. Remove the spring U-bolts and plate.
5. Remove the spring rear bolt.
6. Remove the front shackle nuts and the shackle.
7. Lift the spring from the truck.
8. Installation is the reverse of removal. Torque the spring rear shackle-to-frame nut to 58 ft. lb.; the rear shackle-to-spring nut to 72 ft. lb.; the U-bolt nuts to 58 ft. lb.; the front spring pin nut to 18 ft. lb.

Shock Absorbers

1979–84

1. Raise and support the truck on jackstands.
2. Unbolt the shock absorber at the top and bottom and remove it.
3. Installation is the reverse of removal. Tighten the nuts so that $1/4$ in. of thread is visible past the nut at each end of the shocks.

1986

1. Raise and support the rear end on jackstands.
2. Remove the wheels.
3. Unbolt the shock absorber at each end and remove it.
4. Installation is the reverse of removal. Torque each bolt to 57 ft. lb.

STEERING

Steering Wheel

REMOVAL & INSTALLATION

1. Disconnect the battery ground.
2. Pull the steering wheel pad straight up to remove it, then remove the horn button and contact.

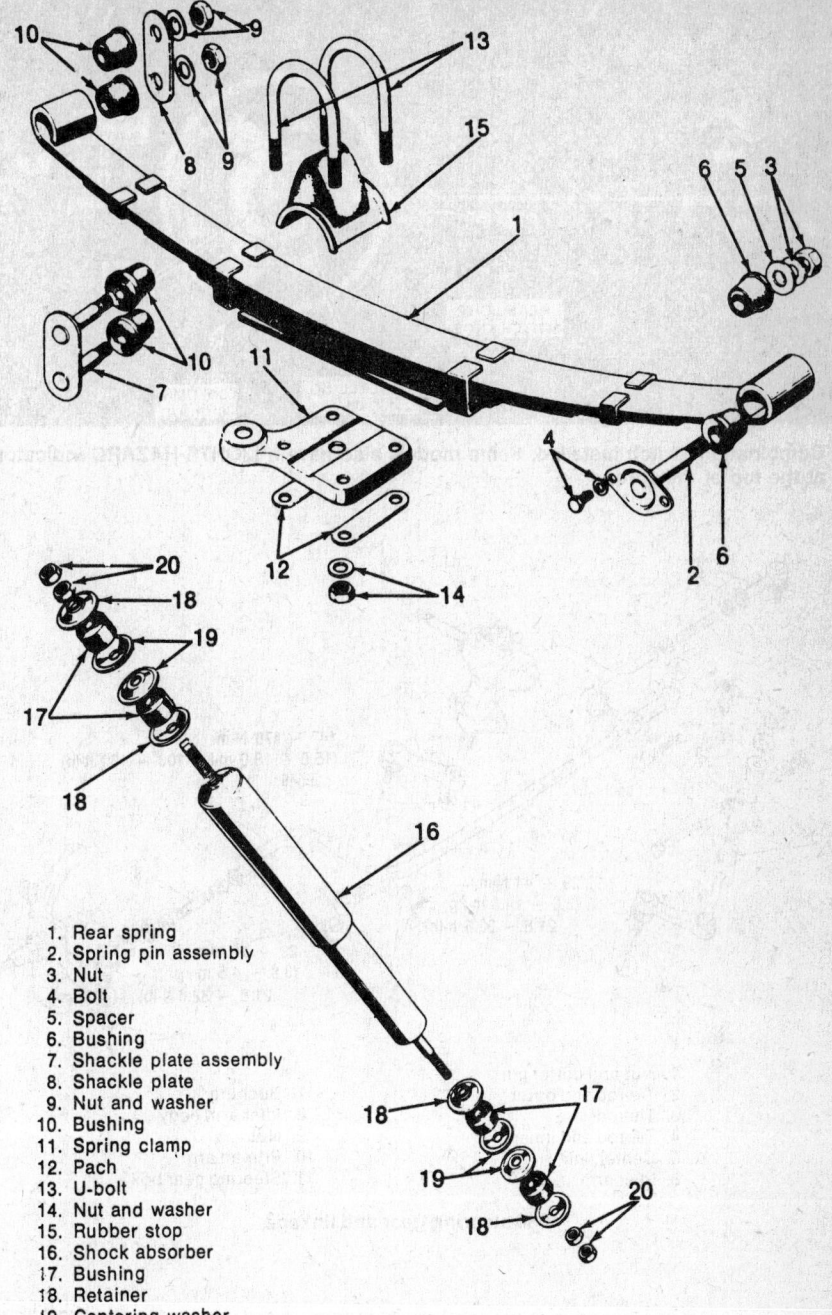

1. Rear spring
2. Spring pin assembly
3. Nut
4. Bolt
5. Spacer
6. Bushing
7. Shackle plate assembly
8. Shackle plate
9. Nut and washer
10. Bushing
11. Spring clamp
12. Pach
13. U-bolt
14. Nut and washer
15. Rubber stop
16. Shock absorber
17. Bushing
18. Retainer
19. Centering washer
20. Nut

1979–84 rear suspension

3. Remove the horn contact spring.
4. Matchmark the steering wheel and shaft.
5. Remove the wheel attaching nut and pull the wheel with a steering wheel puller.
6. Installation is the reverse of removal. Align the marks and tighten the nut to 25 ft. lb. on 1979–84 models; 35 ft. lb. on 1986 models.

Combination Switch

The combination turn signal, windshield wiper, and headlight switch is mounted on the steering column, and must be replaced as an assembly.

REMOVAL & INSTALLATION

1. Disconnect the negative battery cable.
2. Remove the steering wheel.
3. Remove the "Lights-Hazard" Indicator and the steering column shroud.
4. Unplug the electrical multiple connectors at the base of the steering column.
5. Pull the headlight knob from its shaft.

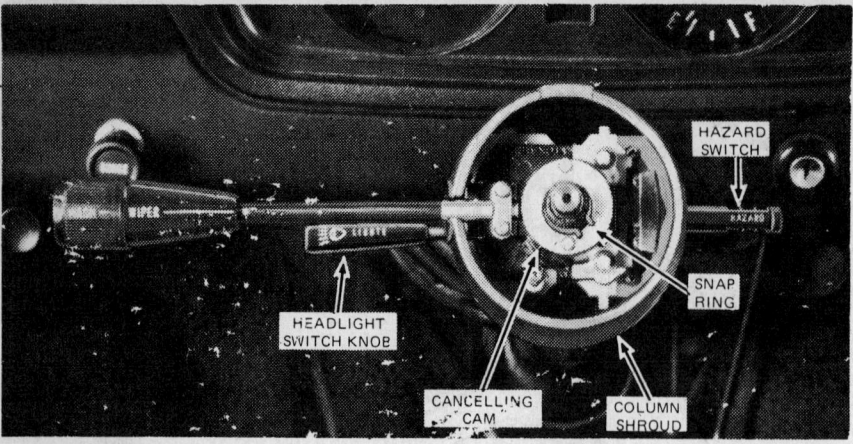

Combination switch installed. Some models also have a LIGHTS-HAZARD indicator at the top of the shroud.

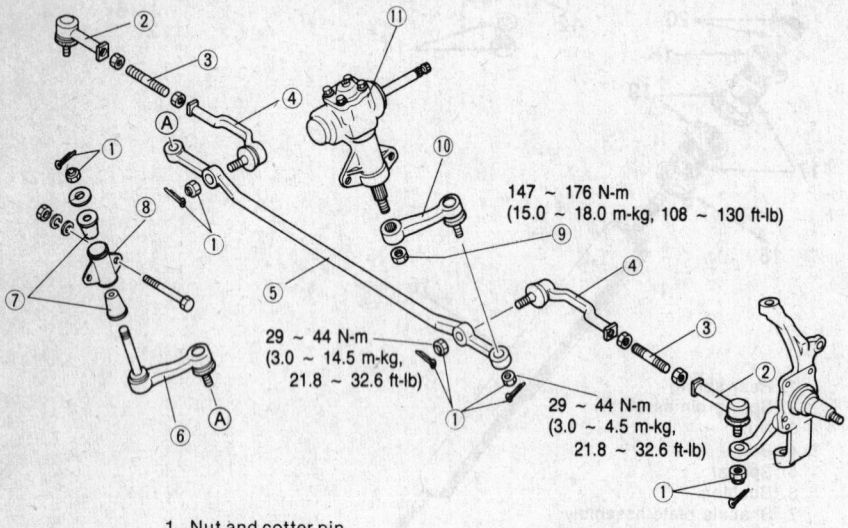

147 ~ 176 N·m
(15.0 ~ 18.0 m-kg, 108 ~ 130 ft-lb)

29 ~ 44 N·m
(3.0 ~ 14.5 m-kg,
21.8 ~ 32.6 ft-lb)

29 ~ 44 N·m
(3.0 ~ 4.5 m-kg,
21.8 ~ 32.6 ft-lb)

1. Nut and cotter pin
2. Tie-rod end outer
3. Tie-rod
4. Tie-rod end inner
5. Center link
6. Idler arm
7. Bushing
8. Idler arm body
9. Nut
10. Pitman arm
11. Steering gear box

1986 steering gear and linkage

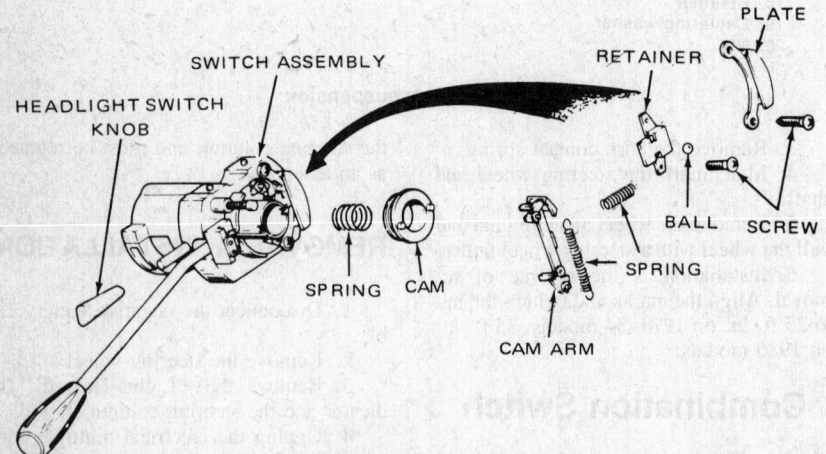

Turn signal components

6. Remove the snap ring, which retains the switch, from the steering shaft. Pull the turn indicator canceling cam from the shaft.

7. Remove the single retaining bolt near the bottom of the switch. Remove the complete switch from the column.

8. Installation is the reverse of removal. Check the operation of the switch before installing the steering wheel.

Removing Mazda steering column shroud

Idler Arm

REMOVAL & INSTALLATION

1. Raise and support the front end on jackstands.

2. Remove the idler arm-to-center link nut and cotter pin. Disconnect the center link from the idler arm using a ball joint separator.

3. Unbolt and remove the idler arm.

4. Installation is the reverse of removal. Torque the center link nut to 40 ft. lb.; the frame mounting nut to 58 ft. lb.

Pitman Arm

REMOVAL & INSTALLATION

1. Raise and support the front end on jackstands.

2. Remove the cotter pin and nut attaching the center link to the pitman arm.

3. Disconnect the center link from the pitman arm with a ball joint separator.

4. Matchmark the pitman arm and sector shaft.

5. Remove the pitman arm-to-sector shaft nut and remove the pitman arm. It may be necessary to use a puller.

6. Installation is the reverse of removal. Make sure you align the matchmarks. Tighten the pitman arm-to-sector shaft nut to 130 ft. lb.; the pitman arm-to-

center link nut to 32 ft. lb. If the cotter pin does not align, tighten to make it line up; never loosen it!

Center Link

REMOVAL & INSTALLATION

1. Raise and support the front end on jackstands.
2. Disconnect the center link at the tie rods, pitman arm and idler arm.
3. Installation is the reverse of removal. Tighten all of the nuts to 30 ft. lb.

Tie Rod Ends

REMOVAL & INSTALLATION

1. Loosen the tie rod jam nuts.
2. Remove and discard the cotter pin from the ball socket end, and remove the nut.

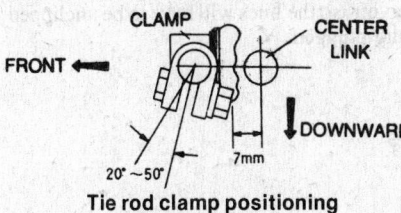

Tie rod clamp positioning

3. Use a ball joint puller to loosen the ball socket stud from the center link. Remove the stud from the kingpin steering arm in the same way.
4. Unscrew the tie rod end from the threaded sleeve, counting the number of threads until it's off. The threads may be left or right hand threads. Tighten the jam nuts to 58 ft. lb.
5. To install, lightly coat the threads with grease, and turn the new end in as many turns as were required to remove it. This will give the approximate correct toe-in.
6. Install the ball socket studs into center link and kingpin steering arm. Tighten the nuts to 30 ft. lb. Install a new cotter pin. You may tighten the nut to fit the cotter pin, but don't loosen it.
7. Check and adjust the toe-in, and tighten the tie rod clamps or jam nuts.

Manual Steering Gear

REMOVAL & INSTALLATION

1979–81

1. Remove the steering wheel as outlined above.
2. Remove the light switch knob.
3. Remove the steering column covers.
4. Remove the stop ring, cancelling cam and spring from the end of the column.
5. Disconnect the combination switch wiring.

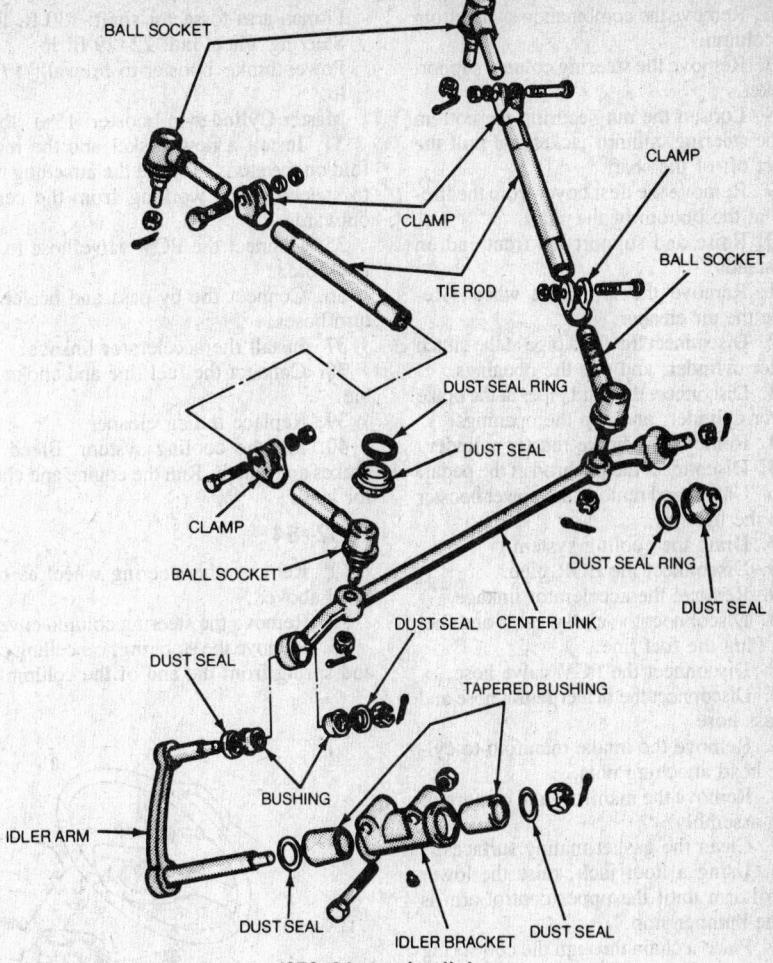

1979–84 steering linkage

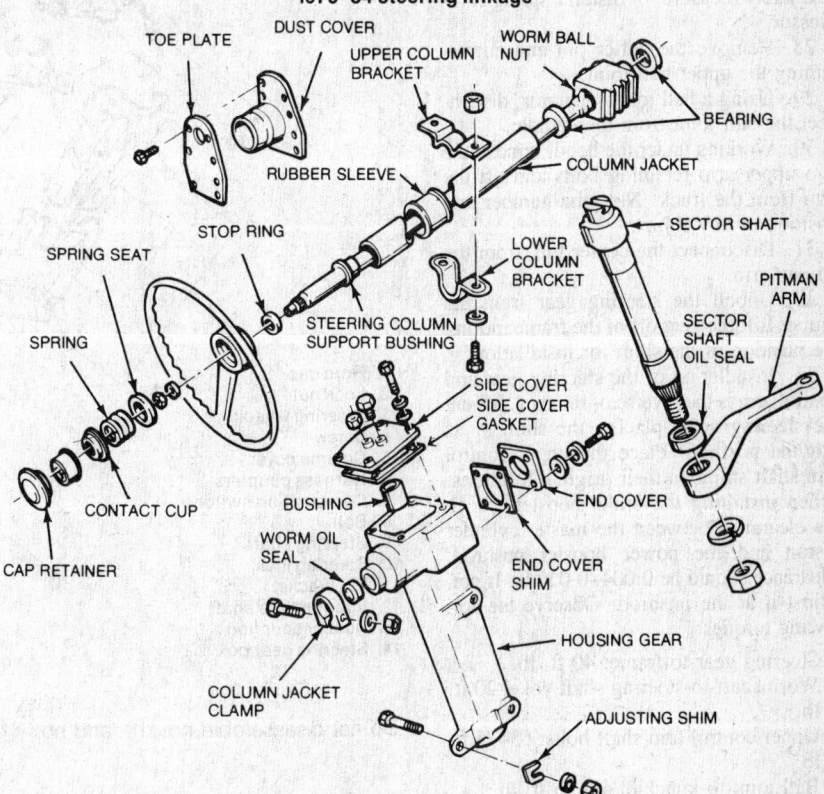

1979–81 steering column and gear

6. Remove the combination switch from the column.

7. Remove the steering column support bracket.

8. Loosen the nut securing the bottom of the steering column jacket and pull the jacket off of the shaft.

9. Remove the dust cover from the firewall at the bottom of the shaft.

10. Raise and support the front end on jackstands.

11. Remove the left front wheel. Remove the air cleaner.

12. Disconnect the fluid pipe at the clutch master cylinder, and cap the openings.

13. Disconnect the fluid pipes at the brake master cylinder, and cap the openings.

14. Remove the brake master cylinder.

15. Disconnect the pushrod at the pedal.

16. Unbolt and remove the power booster from the firewall.

17. Drain the cooling system.

18. Disconnect the EGR pipe.

19. Remove the accelerator linkage.

20. Disconnect the choke cable and fuel line. Plug the fuel line.

21. Disconnect the PCV valve hose.

22. Disconnect the heater return hose and by-pass hose.

23. Remove the intake manifold-to-cylinder head attaching nuts.

24. Remove the manifold and carburetor as an assembly.

25. Clean the gasket mating surfaces.

26. Using a floor jack, raise the lower control arm until the upper control arm is off the bumper stop.

27. Place a chain through the coil spring as a safety measure, or install a spring compressor.

28. Remove the cotter pin and nut retaining the upper ball joint.

29. Using a ball joint separator, disconnect the ball joint from the spindle.

30. Working under the hood, remove the two upper arm retaining bolts and lift the arm from the truck. Note the number and position of any shims.

31. Disconnect the center link from the pitman arm.

32. Unbolt the steering gear from the frame. Lift the gear off of the frame, noting the position of the shim for installation.

33. Installation of the steering gear and control arm is the reverse of removal. Mount the steering gear, placing the shim in its original position. Place the upper control arm shaft shims in their original locations. When installing the brake booster, check the clearance between the master cylinder piston and the power booster pushrod. Clearance should be 0.004–0.020 in. If not, adjust it at the pushrod. Observe the following torques:

Steering gear-to-frame: 40 ft. lb.
Wormshaft-to-steering shaft yoke: 20 ft. lb.
Upper control arm shaft bolts: 65–75 ft. lb.
Ball joint-to-knuckle: 40–55 ft. lb.
Center link-to-pitman arm: 30 ft. lb.
Pitman arm-to-sector shaft: 130 ft. lb.
Steering wheel nut: 22–29 ft. lb.
Power brake booster-to-firewall: 17 ft. lb.
Master Cylinder-to-booster: 15 ft. lb.

34. Install a new gasket and the manifold on the studs. Torque the attaching nuts to specification, working from the center outward.

35. Connect the PCV valve hose to the manifold.

36. Connect the by-pass and heater return hoses.

37. Install the accelerator linkage.

38. Connect the fuel line and choke cable.

39. Replace the air cleaner.

40. Fill the cooling system. Bleed the brakes and clutch. Run the engine and check for leaks.

1982–84

1. Remove the steering wheel as outlined above.

2. Remove the steering column covers.

3. Remove the stop ring, cancelling cam and spring from the end of the column.

4. Disconnect the combination switch wiring.

5. Remove the combination switch from the column.

6. Remove the steering column support bracket.

7. Loosen the nut securing the bottom of the steering column jacket and pull the jacket off of the shaft.

8. Remove the dust cover from the firewall at the bottom of the shaft.

9. Remove the bolt securing the yoke joint to the wormshaft and remove the steering shaft.

10. Remove the air cleaner.

11. On trucks with column shift, unbolt the lower bracket from the gear select rod and the shift rod.

12. Remove the lower bracket from the steering gear.

13. Remove the brakes lines from the master cylinder and cap the lines.

14. Unbolt and remove the master cylinder from the firewall or power booster. These trucks have a remotely mounted reservoir, so the lines will have to be unclipped and plugged.

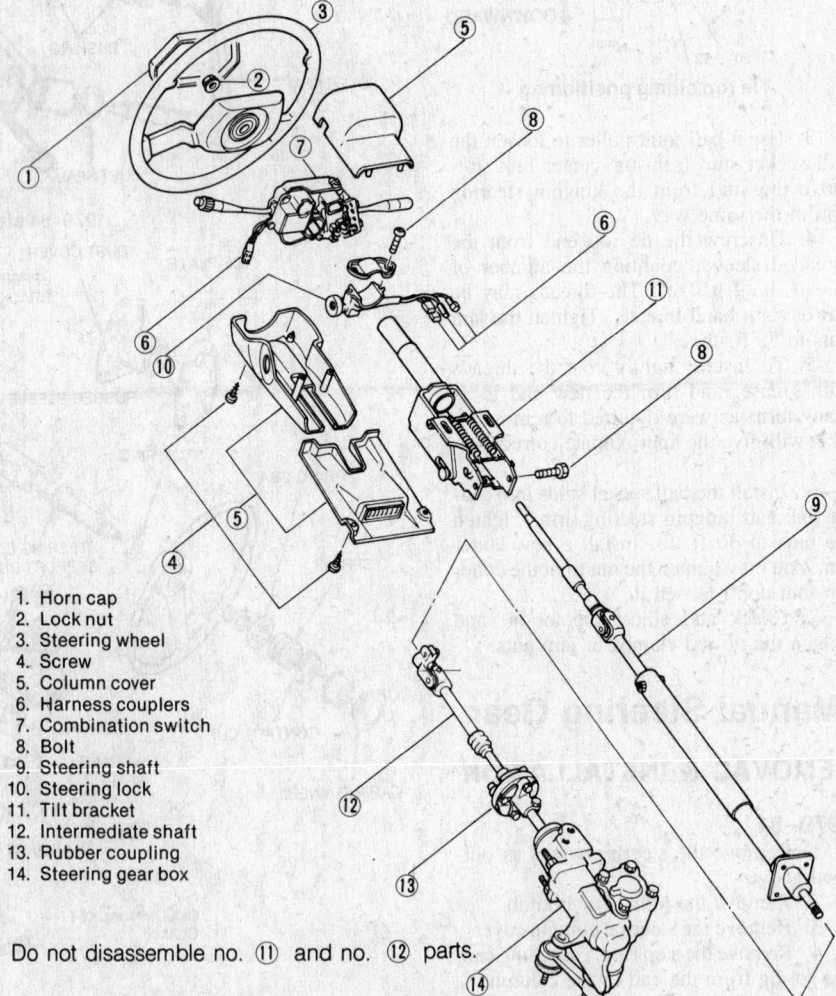

1. Horn cap
2. Lock nut
3. Steering wheel
4. Screw
5. Column cover
6. Harness couplers
7. Combination switch
8. Bolt
9. Steering shaft
10. Steering lock
11. Tilt bracket
12. Intermediate shaft
13. Rubber coupling
14. Steering gear box

Do not disassemble no. ⑪ and no. ⑫ parts.

1986 steering gear and column

15. Remove the cotter pin and nut and disconnect the center link from the pitman arm using a ball joint tool.

16. Remove the cotter pin and nut, matchmark the pitman arm and sector shaft and disconnect the pitman arm from the sector shaft using a ball joint tool.

17. Unbolt the steering gear from the frame, noting the position of any shim that might be installed.

18. Installation is the reverse of removal. Mount the steering gear, placing the shim in its original position. Observe the following torques:

Steering gear-to-frame: 40 ft. lb.
Center link-to-pitman arm: 30 ft. lb.
Pitman arm-to-sector shaft: 130 ft. lb.
Steering wheel nut: 22–29 ft. lb.
Master Cylinder-to-booster: 15 ft. lb.
Wormshaft-to-steering shaft yoke: 20 ft. lb.

1986

1. Raise and support the front end on jackstands.

2. Remove the pinch bolt securing the wormshaft to the steering shaft coupling.

3. Remove the cotter pin and nut securing the pitman arm to the center link and separate the pitman arm fro the link with a ball joint tool.

4. Unbolt the steering gear from the frame.

5. If the pitman arm is to be removed from the sector shaft, first matchmark their positions, relative to each other.

6. Installation is the reverse of removal. Observe the following torques:

Steering gear-to-frame: 40 ft. lb.
Wormshaft-to-steering shaft yoke: 28 ft. lb.
Pitman arm-to-sector shaft: 139 ft. lb.
Pitman arm-to-center link: 30 ft. lb.

Power Steering Gear

REMOVAL & INSTALLATION

This procedure is identical to the 1986 B2000 procedure for manual steering, with the exception that the power steering fluid hoses must be removed from the gear before the gear can be removed from the truck.

Power Steering Pump

REMOVAL & INSTALLATION

1. Raise and support the front end on jackstands.

2. Remove the power steering pump pulley nut.

3. Loosen the drive belt tensioner pulley and remove the belt.

4. Remove the pulley from the pump.

5. Position a drain pan under the pump and disconnect the hoses.

6. Remove the bracket-to-pump bolts and remove the pump from the truck.

7. Installation is the reverse of removal. Adjust the belt to give $1/2$ in. deflection along its longest straight run. Bleed the system.

BLEEDING THE SYSTEM

1. Raise and support the front end on jackstands.

2. Check the fluid level and fill it, if necessary.

3. Start the engine and let it idle. Turn the steering wheel lock-to-lock, several times. Recheck the fluid level.

4. Lower the truck to the ground.

5. With the engine idling, turn the wheel lock-to-lock several times again. If noise is heard in the fluid lines, air is present.

6. Put the wheels in the straight ahead position and shut off the engine.

7. Check the fluid level. If it is higher than when you last checked it, air is in the system. Repeat Step 5. Keep repeating Step 5 until no air is present.

BRAKES

For all brake system service and repair procedures not detailed below, please refer to "Brakes" in the Unit Repair section.

Adjustments

BRAKE PEDAL FREE-PLAY

Using the top of the pedal pad as a reference point, there should be 7.0–9.0mm on trucks through 1984 with power brakes, 4.0–7.0mm on 1986 trucks with power brakes, and, a little less than $1/8$ in. free-play before the pushrod contacts the master cylinder piston on models with non-power brakes.

1. Loosen the locknut on the master cylinder pushrod at the clevis.

2. Turn the pushrod to obtain the proper free-play, then tighten the nut.

BRAKE PEDAL HEIGHT

Pedal height is measured from the center of the pedal pad surface, horizontally to the firewall. On trucks through 1984, pedal height should be 8.1 inches. On 1986 trucks, pedal height should be 8.23–8.43 in. If not,

loosen the stop light switch locknut and turn the switch until the proper height is obtained. Tighten the locknut.

DRUM BRAKES

The rear drum brakes are self-adjusting. Manual adjustment is required only when

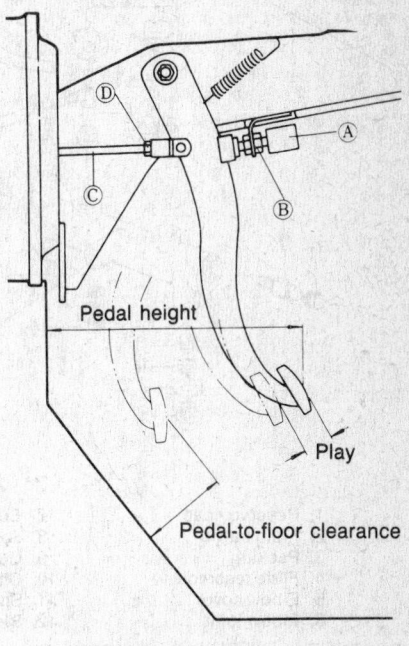

Pedal height adjustment

the brake shoes have been replaced, or when the length of the self-adjusting rod has been changes for some reason. The brakes should be cold (room temperature).

1. If the shoe retaining spring has been removed, first retract the pushrod fully (drum removed).

2. Raise and support the rear of the truck. The wheels must be free to turn.

3. Make sure the parking brake is fully released.

4. Remove the two adjusting hole plugs from the brake backing plate.

5. An arrow stamped on the backing plate indicates the direction to turn the adjuster starwheel to expand the shoes. Insert a brake spoon through the adjuster hole and turn the starwheel until the brakes are locked.

6. Insert a drift through the other adjuster hole. Use the drift to hold the pole lever of the self-adjuster firmly. Back off the starwheel three or four notches; the wheel should rotate freely (no drag).

7. Repeat the adjustment on the other wheel. Make sure the adjustment is exactly the same. Road test for equal brake action and readjust as necessary.

Disc Brakes

Disc brakes require no adjustments.

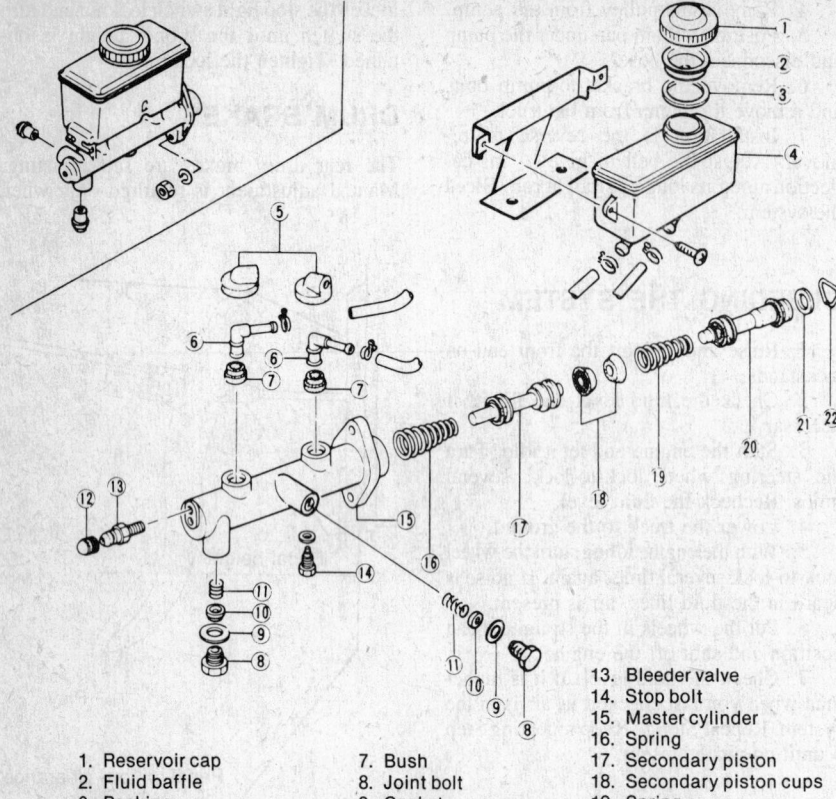

1. Reservoir cap
2. Fluid baffle
3. Packing
4. Fluid reservoir
5. Elbow cover
6. Elbow joint
7. Bush
8. Joint bolt
9. Gasket
10. Check valve
11. Spring
12. Bleeder cap
13. Bleeder valve
14. Stop bolt
15. Master cylinder
16. Spring
17. Secondary piston
18. Secondary piston cups
19. Spring
20. Primary piston
21. Stop washer
22. Stop ring

1979–84 brake master cylinder

1. Fluid-level sensor
2. Nut
3. Reserve tank cap
4. Reserve tank
5. Bushing
6. Stopper screw
7. O-ring
8. Primary piston assembly
9. Secondary piston assembly
10. Master cylinder body

1986 brake master cylinder

Master Cylinder

REMOVAL & INSTALLATION

1. Remove the brakes lines from the master cylinder and cap the lines.
2. On 1986 models, disconnect the fluid level sensor coupling.
3. Unbolt and remove the master cylinder from the firewall or power booster. 1979–84 models have a remotely mounted reservoir, so the lines will have to be unclipped and plugged.
4. Installation is the reverse of removal. Torque the mounting nuts to 15 ft. lb.
5. Bleed the system.

Power Booster

REMOVAL & INSTALLATION

1. Remove the master cylinder.
2. Disconnect the pushrod at the pedal.
3. Unbolt and remove the power booster from the firewall.
4. Installation is the reverse of removal. Check the clearance between the master cylinder piston and the power booster pushrod. Clearance should be 0.004–0.020 in. If not, adjust it at the pushrod. Torque the mounting nuts to 17 ft. lb.

Pressure Differential Valve

REMOVAL & INSTALLATION

Courier

1. Disconnect the warning light switch connector from the warning light switch.
2. Disconnect the brake inlet and outlet lines. Plug the lines.
3. Remove the valve assembly-to-cowl attaching bolt and remove the valve and switch assembly.
4. Position the valve and switch on the cowl. Install the retaining bolt.
5. Connect the brake lines to the valve.
6. Connect the warning light to the switch wiring connector.
7. Depress the brake pedal several times, then bleed the brake system.
8. Fill the master cylinder and check for proper operation.

Mazda

1. Disconnect the brake warning light switch connector, at the switch.
2. Disconnect the brake lines at the valve, and plug the lines.
3. Unbolt and remove the valve.
4. Installation is the reverse of removal. Bleed the system.

CENTRALIZING THE PRESSURE DIFFERENTIAL VALVE

After the brake system has been opened for repairs, or bled, the brake light will remain on. The pressure differential valve must be centered to make the light go off.

1. Turn the ignition switch ON, but don't start the engine.
2. Make sure that the master cylinder reservoirs are filled.
3. Slowly depress the brake pedal. The valve should center itself and the light go off. If not, bleed the brakes again and repeat the above procedure.

Parking Brake

ADJUSTMENT

1. Adjust the service brakes before attempting to adjust the parking brake.
2. Use the adjusting nut to adjust the length of the front cable so that the rear brakes are locked when the parking brake lever is pulled out 5–10 notches on trucks through 1984, and 11–13 notches on 1986 trucks.
3. After adjustment, apply the parking brake several times. Release the parking brake and make sure that the rear wheels rotate without dragging. If they drag, repeat the adjustment.

NOTE: If the parking brake cable is replaced, prestretch it by applying the parking brake hard three or four times before attempting adjustment.

Parking Brake Warning Light Switch

REMOVAL & INSTALLATION

1. Apply the parking brake to provide clearance between the switch assembly and the switch stop tab on the parking brake lever shaft.
2. Disconnect the switch wiring connector.
3. Remove the switch from its mounting bracket.
4. Install the switch on the mounting bracket.
5. Install the attaching screws.
6. Connect the switch wire connector.
7. Turn the ignition switch ON and check the operation of the switch. No adjustment to the switch is possible. If it is defective, replace the switch.

REMOVAL & INSTALLATION

Front Cable

1. Raise and support the front end on jackstands.
2. Remove the adjusting nut.

3. Separate the front cable from the equalizer and remove the jam nut.
4. Remove the return spring and boot from the housing.
5. Pull the lower housing forward and out of the slotted frame bracket. Slip the cable shaft sideways through the slot until the cable and housing are free of the bracket.
6. Disengage the upper cable connector from the brake lever by remove the clevis pin and retainer.
7. Remove the upper cable housing retaining clip and pull the upper cable and housing from the slotted bracket on the firewall.
8. Push the upper cable, cable housing and dust shield grommet through the firewall opening and into the engine compartment.
9. Remove the cable and housing.
10. Installation is the reverse of removal.

Rear Cable

1. Raise and support the rear end on jackstands.
2. Remove the pin and disconnect the equalizer from the clevis.
3. Disconnect the right hand cable from the left.
4. Remove the rear brake shoes.
5. Disengage the cables from the brake shoe levers.
6. Remove the cable housing retainer from the backing plate.
7. Pull the return spring to release the retainer plate from the end of the housing.
8. Loosen the cable housing-to-frame bracket locknut and remove the forward end of the cable housing from the frame bracket.
9. Remove the cable housing retaining clip bolts.
10. Disengage the cable housing-to-frame tension springs and pull the cable out of the backing plate.
11. Installation is the reverse of removal.

CHASSIS ELECTRICAL

Heater Case

REMOVAL & INSTALLATION

1. Disconnect the battery ground cable.
2. Drain the cooling system.
3. Remove the water valve shield at the left side of the heater.
4. Disconnect the two hoses from the left side of the heater.
5. At the heat-defroster door, at the water valve and at the outside recirculation door, disengage the control cable housing from the mounting clip on the heater. Disconnect

each of the three cable wires from the crank arms.
6. Disconnect the fan motor electrical lead.
7. Working inside the engine compartment, remove the two retaining nuts and the single bolt and washer which hold the heater to the firewall. A retaining bolt inside the passenger compartment must also be removed.
8. Disconnect the two defroster ducts from the heater and remove the heater.
9. Install the heater on the dash so that the heater duct indexes with the air intake duct and the two mounting studs enter their respective holes.
10. From the engine side of the firewall, install the nuts on the mounting studs. While an assistant holds the heater in position, install the mounting bolt.
11. Connect the defroster ducts.
12. Connect the heat-defrost door control cable to the door crank arm. Set the control lever (upper) in the HEAT position and turn the crank arm toward the mounting clip as far as it will go. Engage the cable housing in the clip and install the screw in the clip.
13. Connect the water valve control cable wire to the crank arm on the water valve lever. Locate the cable housing in the mounting clip. Set the control lever in the HOT position and pull the valve plunger and lever to the full outward position. This will move the lever crank arm toward the cable mounting clip as far as it will go. Tighten the clip and screw.
14. Insert the outside-recirculation door control cable into the hole in the door crank arm. Bend the wire over and tighten the screw. Set the center control lever in the REC position and turn the door crank arm toward the mounting clip as far as it will go. Engage the cable housing in the clip and install the screw in the clip.
15. Connect the fan motor electrical lead.
16. Connect the two hoses to the heater core tubes, at the left side of the heater, and tighten the clamp.
17. Install the water valve shield and tighten the three screws (left side of the heater).
18. Refill the cooling system and connect the battery ground cable.
19. Run the engine and check for leaks. Check the operation of the heater.

Heater Motor and Blower Fan

REMOVAL & INSTALLATION

1. Remove the heater assembly.
2. Remove the five screws and separate the halves of the heater assembly.
3. Loosen the fan retaining nut. Lightly tap on the nut to loosen the fan. Remove the fan and nut from the motor shaft.
4. Remove the three motor-to-case re-

taining screws and disconnect the bullet connector to the resistor and ground screw.

5. Rotate the motor and remove it from the case.

6. Install the motor in the case, rotating it slightly.

7. Install the retaining screws and connect the bullet connector and ground wire.

8. Install the fan on the shaft and install the nut.

9. Assemble the halves together and install the five retaining screws.

10. Install the heater in the truck. Check the operation of the heater.

Heater Core

REMOVAL & INSTALLATION

1. Remove the heater from the truck.

2. Remove the five screws and separate the halves of the case.

3. Loosen the hose clamps and slide the heater core from the case.

4. Slide the replacement core into the case. At the same time, connect the core tube to the water valve tube with the short hose and clamps.

5. Assemble the halves of the heater and install the five screws.

6. Install the heater in the truck. Check the operation of the heater.

Radio

REMOVAL & INSTALLATION

— CAUTION —

Never operate the radio with the speaker lead or antenna disconnected. Operation of the radio without a load will damage the amplifier's output transistors. If a replacement speaker is installed, be sure it is of the same impedance (resistance in ohms) as the original.

1. Disconnect the negative battery cable.

2. Pull off the heater control knobs, the instrument light brightness control knob, and the radio knobs.

3. Remove the ring nut and fiber washer for the brightness control. Remove the radio attaching nuts (shaft nuts). Remove the four screws for the meter hood (instrument trim panel) and remove the hood.

4. Slide the radio to the left until the rear support pin clears the support bracket. Pull the radio out from the instrument panel far enough to gain access to the wires at the rear of the radio chassis.

5. Disconnect the power lead, speaker leads and the antenna cable. Remove the radio.

6. To install, connect the wires to the radio.

7. Slide the radio into place. Move the radio to the left, engage the support pin with its bracket, then slide the radio to the right.

8. Install the meter hood, inserting the radio shafts and heater knobs through it as it is fitted into place. Install the four retaining screws but do not tighten them yet.

9. Install the washer and ring nut for the brightness control. Install the radio shaft nuts loosely. Tighten the meter hood screws, then tighten the radio shaft nuts.

10. Install the knobs. Connect the negative battery cable.

Windshield Wiper Switch

Because the steering wheel must be removed, the windshield wiper switch removal and installation procedure is in covered under steering column removal and installation.

Blade and Arm

REMOVAL & INSTALLATION

1. To remove the blade and arm, unscrew the retaining nut and pry the blade and arm from the pivot shaft. The shaft and arm are serrated to provide for adjustment of the wiper pattern on the glass.

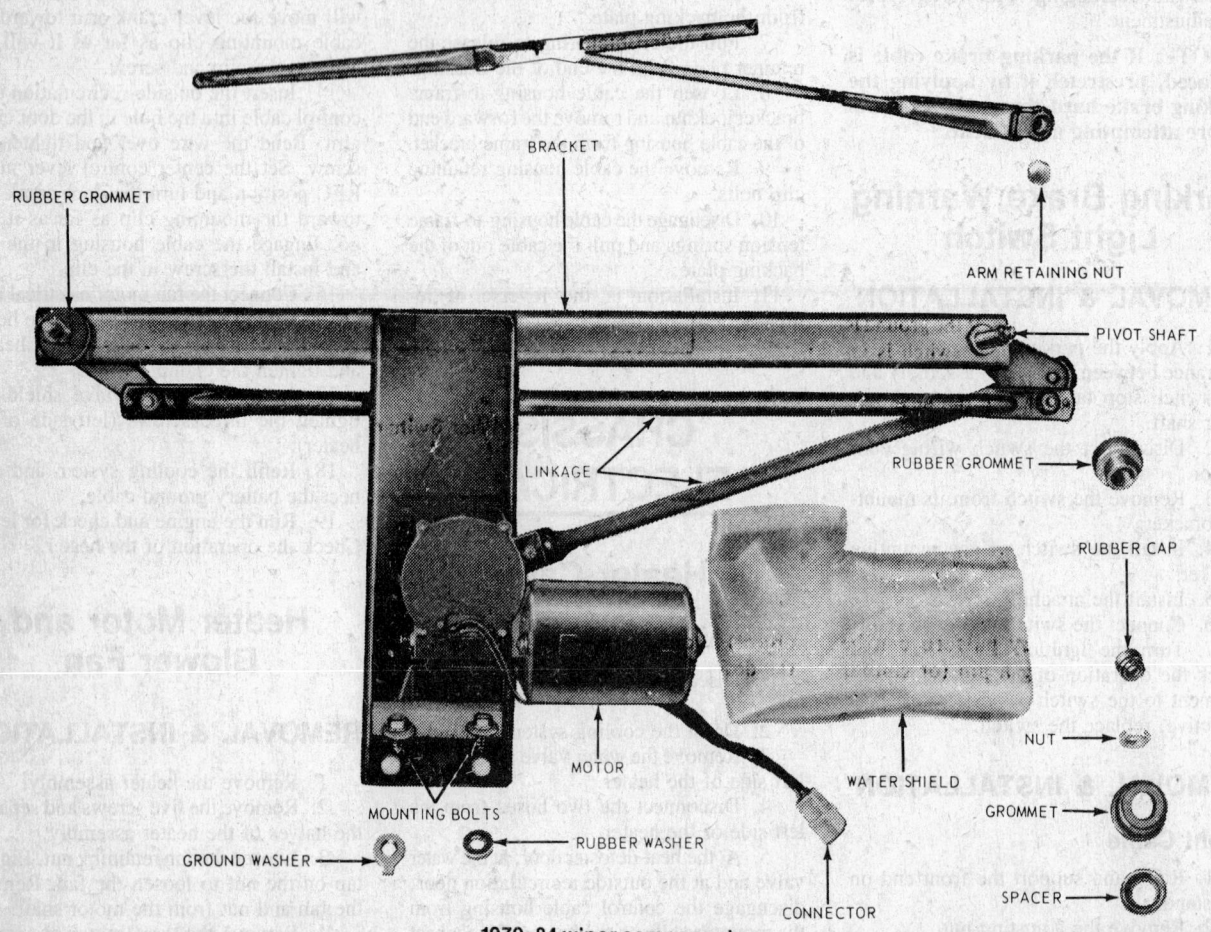

RUBBER GROMMET

BRACKET

ARM RETAINING NUT

PIVOT SHAFT

LINKAGE

RUBBER GROMMET

RUBBER CAP

NUT

MOTOR

WATER SHIELD

GROMMET

MOUNTING BOLTS

RUBBER WASHER

GROUND WASHER

CONNECTOR

SPACER

1979–84 wiper components

2. To set the arms back in the proper park position, turn the wiper switch on and allow the motor to cycle three or four times. Then turn off the wiper switch (do not turn off the wiper motor with the ignition key). This will place the wiper shafts in the proper park position.

3. Install the blade and arm on the shaft and install the retaining nut. The blades and arms should be positioned according to the illustration.

Wiper Motor, Linkage and Bracket

REMOVAL & INSTALLATION

1979–1984

1. Disconnect the battery ground cable.

2. Remove the wiper arms and blades by removing the retaining nuts.

3. Remove the rubber cap, nut, tapered spacer and rubber grommet from each pivot shaft.

4. Remove the two motor and bracket retaining bolts and washers.

5. Disconnect the wiper motor leads at the multiple connector.

6. Remove the motor and bracket assembly. Note the position of the ground washer and the rubber washer at the bracket mounting holes. Remove the plastic water shield.

7. To disconnect the motor from the bracket, remove the retaining clip that holds the linkage to the motor output arm. Note the position of the washers before removing the motor from the bracket.

8. Remove the four motor-to-bracket retaining bolts and remove the motor.

9. Install the wiper motor on the bracket and install the four retaining bolts.

10. Install the washers and position the linkage on the motor output arm. Install the retaining clip.

11. Install the plastic water shield.

12. Install the motor and bracket assembly in the truck.

13. Connect the multiple connector.

14. Install the washers, spacers and nuts on the pivot shafts.

15. Install the wiper arms and blades. Be sure the motor is in the Park position. This can be determined by cycling the motor several times. Adjust the position of the wipers. The clearance between the tips of the blades and the windshield moulding should be 20mm at park.

16. Connect the battery cable and check the operation of the wipers.

1986

1. Remove the wiper arm/blade assembly. Note that the arms are different. Don't confuse them.

2. Remove the rubber seal from the leading edge of the cowl.

3. Unbolt and remove the cowl.

4. Remove the access hole covers.

5. Remove the bolts holding the wiper shaft drives.

6. Matchmark the position of the wiper crank arm in relation to the face of the wiper motor. Disconnect the wiper linkage from the wiper motor crank arm.

7. Remove the wiper linkage.

8. Unbolt and remove the wiper motor. Disconnect the wiring harness.

9. Installation is the reverse of removal. Make sure that the parked height of the wiper arms, measured from the blade tips to the windshield moulding is 20mm. Torque the arm retaining nuts to 8–10 ft. lb.

Instrument Cluster

REMOVAL & INSTALLATION
1979–84

1. Disconnect the battery ground cable.

2. The meter hood (instrument cluster trim panel) must be removed for access to the cluster. See Steps 1–3 of the radio removal and installation procedure for details on meter hood removal.

3. Remove the screws holding the cluster to the instrument panel.

4. Pull the cluster rearward enough to gain access to the cluster assembly.

5. Reach behind the cluster and disconnect the speedometer cable.

6. Pull the multiple connector from the printed circuit.

7. Note the position of the two ammeter leads and disconnect them.

8. Remove the screw attaching the ground wire to the rear of the cluster. On trucks equipped with a coasting richer valve, remove the two connectors at the speedometer sensor switch.

9. Remove the instrument cluster.

10. Position the cluster assembly near the opening and connect the ground lead.

11. Connect the two ammeter leads to the ammeter.

12. Install the multiple connector at the rear of the cluster. On trucks equipped with a coasting richer valve, connect the two wires to the speedometer speed sensor.

13. Connect the speedometer cable to the speedometer head.

14. Install the four attaching screws.

15. On 1979 and later models, replace the meter hood.

16. connect the battery cable.

17. Run the engine and check the operation of all gauges.

1986

1. Disconnect the battery ground cable.

2. Reach behind the cluster and disconnect the speedometer cable.

3. Remove the screws attaching the cluster hood and carefully lift the hood off.

4. Remove the screw attaching the cluster pod to the dash panel and pull the pod out toward you, gradually. Reach behind the pod and disconnect the wiring connectors.

5. Remove the trip meter knob, and, on cluster w/tachometer, the clock adjust knob.

6. Remove the screws retaining the lens cover and lift off the cover.

7. Remove the screws retaining the cluster bezel and lift off the bezel.

8. Lift out the warning light plate.

9. On clusters wo/tachometer, remove, in order:
fuel gauge
speedometer
temperature gauge
printed circuit board

10. On cluster w/tachometer, remove, in order:
speedometer
digital clock
tachometer
fuel gauge
temperature gauge
printed circuit board

11. Installation is the reverse of removal.

Speedometer Cable

REMOVAL & INSTALLATION

1. Remove the instrument cluster.

2. Remove the old cable by pulling it out from the speedometer end of the cable housing. If the old cable is broken, the speedometer cable will have to be disconnected from the transmission and the broken piece removed from the transmission end.

3. Lubricate the lower $3/4$ of the new cable with speedometer cable lubricant, and feed the cable into the housing.

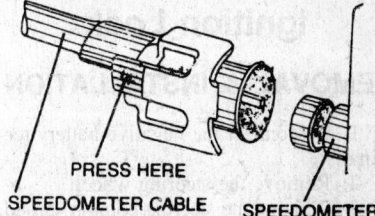

PRESS HERE
SPEEDOMETER CABLE   SPEEDOMETER
Speedometer cable connector

4. Connect the speedometer cable to the speedometer, and to the transmission if disconnected there.

5. Replace the instrument cluster.

Ignition Switch

REMOVAL & INSTALLATION

1979–81

1. Disconnect the battery ground cable.

2. Reach under the instrument panel and pull the wire connector from the rear of the switch.

3. Hold the switch body from behind the instrument panel and remove the black retaining nut by turning it counterclockwise.

4. Remove the switch from the rear of the instrument panel.

5. Position the switch in the instrument panel.

6. Hold the switch from behind the instrument panel. Install the retaining nut by turning it clockwise.

7. Plug the multiple connector into the back of the switch.

8. Connect the battery ground cable and check the operation of the switch.

1982–84

1. Disconnect the negative battery terminal.

2. Remove the steering wheel.

3. Remove the steering column shroud.

4. Disconnect the multiple connectors at the base of the combination switch.

5. Remove the switch retaining snap ring. Pull the turn signal indicator cancelling cam off the shaft.

6. Remove the switch retaining bolt and remove the complete switch from the column.8.

Installation is the reverse of removal.

1986

1. Disconnect the battery ground cable.

2. Remove the steering column covers.

3. Disconnect the wiring harness connector at the switch.

4. Remove the attaching screw and lift out the switch.

5. Installation is the reverse of removal.

Ignition Lock

REMOVAL & INSTALLATION

1. Disconnect the negative battery terminal.

2. Remove the steering wheel.

3. Remove the steering column shroud.

4. Disconnect the multiple connectors at the base of the combination switch.

5. Remove the switch retaining snap ring. Pull the turn signal indicator cancelling cam off the shaft.

6. Remove the switch retaining bolt and remove the complete switch from the column.

NOTE: Make a groove on the head of the bolts attaching the steering lock body to the column shaft using a saw. A screwdriver can be used to loosen the screws.

7. Remove the steering lock attaching bolts. Remove the steering lock.

8. Installation is the reverse of removal. During installation position a new steering lock on the column shaft. Tighten the bolts until the heads break off.

Headlights

REMOVAL & INSTALLATION

1. Remove the radiator grille attaching screws and remove the grille.

2. Remove the headlight bulb trim ring, by removing the three screws and rotating the ring clockwise. Support the headlight bulb and remove the trim ring.

NOTE: Do not disturb the headlight aim screws, which are installed in the housing next to the retaining screws.

3. Pull the plug connector from the rear of the bulb and remove the bulb.

4. Connect the plug connector to the rear of a new headlight.

5. Install the headlight in the housing, and locate the bulb tabs in the slots and the housing.

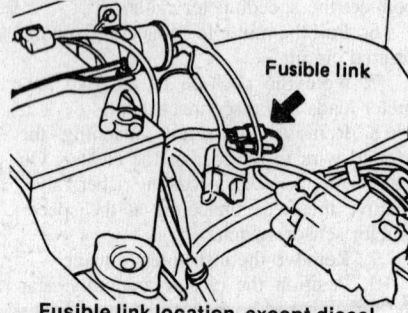

Fusible link

Fusible link location, except diesel

Fusible link

Fusible link location on the diesel

6. Position the trim ring over the bulb and loosely install the retaining screws. Rotate the ring counterclockwise to lock it in position. Tighten the three attaching screws. Check the headlight operation.

7. Install the grille.

8. Have the headlight aim checked.

Circuit Protection

The fuse box, on trucks through 1984, is located on the left side of the engine compartment near the windshield. On 1986 trucks, the fuse box is located under the instrument panel, on the left of the driver.

When a fuse blows out, inspect the electrical system for shorts or other faults. Fuses of specified capacity should be installed in their respective positions. Oversize fuses will allow excessive current to flow and should not be used. Spare fuses should be kept in a vinyl bag in the glove compartment.

1979–84 trucks have a fusible link which acts as a master fuse. The fusible link is a length of wire specially designed to melt under excessive electrical loads. It protects the entire electrical system. Replacements are made by splicing a new section into place. To replace the fusible link, first disconnect the battery negative cable. Then remove the old link and replace it with a link of similar capacity, available at your dealer.

1986 trucks have replaced the fusible link with an 80 amp fuse located under a protective cover on the right fender apron. This fuse is a push-in type block fuse.

Flashers and Relays

The hazard warning flasher is located to the left of the steering column, beneath the instrument panel, and is secured by a clamp and one screw. To remove it, simply unplug the electrical connector, loosen the screw, and slide the flasher out of the clamp. The turn signal flasher is located to the right of the steering column, beneath the instrument panel, and is secured in the same way as the hazard flasher.

The turn the signal relay is located to the immediate right of the hazard flasher, and is secured by two screws. To remove it, unplug the electrical connector and remove the screws.

Mitsubishi/D-50/Arrow

GENERAL ENGINE SPECIFICATIONS

Year	Engine Displacement cu in. (cc)	Fuel System Type	Horsepower @ rpm	Torque @ rpm (ft-lbs)	Bore × Stroke (in.)	Compression Ratio	Oil Pressure (psi)
'79–'86	121.7 (1995)	2-bbl	93 @ 5200 ① ⑤	108 @ 3000 ②	3.31×3.54	8.5:1	50–64
	155.92(2555)	2-bbl	105 @ 5000 ③ ⑥	139 @ 2500 ④	3.59×3.86	8.2:1	50–64
'83–'86	143.2 (2346)	Turbo Diesel	84 @ 4200	136 @ 2500	3.59×3.54	21.0:1	56 @ 2000 rpm

① Canada: 96 @ 5500 HP
② Canada: 109 @ 3500 torque
③ Canada: 108 @ 5000 HP (thru '80)
④ Canada: 140 @ 2500 torque (thru '80)
⑤ '82–'86 California 88 @ 5000
⑥ '82–'86 California 103 @ 5000

GASOLINE ENGINE TUNE-UP SPECIFICATIONS

(When analyzing compression test results, look for uniformity among cylinders, rather that specific pressures)

Year	Engine Displacement cc	Spark Plug Gap (in.)	Ignition Timing (deg) MT	Ignition Timing (deg) AT	Intake Valve Opens (deg) BTDC	Fuel Pump Pressure (psi)	Idle Speed (rpm)	Valve Clearance (in.) In	Valve Clearance (in.) Ex
'79–'86	1,995	0.039–0.043 ①	5B ③	5B ③	25	4.6–6.0	650 ± 50	0.006 ② Hot	0.010 Hot
	2,555	0.039–0.043 ①	7B ③	7B ③	25	4.6–6.0	750 ± 50	0.006 ② Hot	0.010 Hot

NOTE: The underhood specification sticker sometimes reflects tune-up specification changes made in production. Sticker figures must be used if they disagree with this chart.
① Canada: 0.028–0.031 in.
② Jet valve clearance: 0.006 (Hot) to 1982;
   1983 and later jet valve 0.010 in.
③ Canada: 1,995cc, 13B
          2,555cc, 7B
MT: manual transmission
AT: automatic transmission

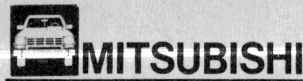

DIESEL ENGINE TUNE-UP SPECIFICATIONS

Year	Engine Displacement cc	Transmission	Curb Idle Speed (rpm)	Initial Injection Timing	Valve Clearance (hot) in
1983–86	2346	Manual	750 ± 100	2° ATDC @ 1mm (.0394 in.) plunger stroke	.010 intake .010 exhaust

CAPACITIES

Year	Model	Engine Displacement cc	Crankcase (qts) With Filter	Crankcase (qts) Without Filter	Transmission (qts) Manual 4-spd	Transmission (qts) Manual 5-spd	Transmission (qts) Automatic	Drive Axle (pts)	Gasoline Tank (gals)	Cooling System (qts) With AC	Cooling System (qts) Without AC
'79–'86	Pickups	1,995	4.5	4.0	2.2	—	6.8 ①	2.8 ④	15.8 ②	9.5	9.5
		2,555	4.5	4.0	—	2.4	6.8 ①	2.8 ①	15.8 ②	9.7	9.7
		2,346 ③	5.9	5.3	2.2	2.4	—	2.3 ④	15.1 ②	8.5	8.5
'83–'86	Montero	2,555	6.1	5.5	—	2.3	—	2.3	15.9	8.45	8.45

① '80–'82: 7.2 U.S. quarts
② '81 (2000) 15.1 gal
   (2555) 18.0 gal. also optional with 2000
   '82 and later: 18.0 optional on all
   pickup models
③ Turbo Diesel
④ Front axle; rear axle 3.8 pts.

FIRING ORDER

NOTE: To avoid confusion, always replace spark plug wires one at a time.

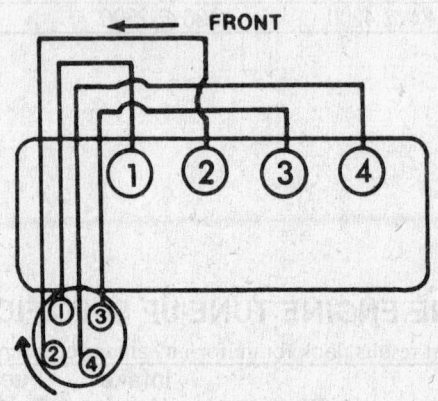

Gasoline engines

CRANKSHAFT AND CONNECTING ROD SPECIFICATIONS

(All measurements given in inches)

Year	Engine cc	Crankshaft Main Brg. Journal Dia.	Crankshaft Main Brg. Oil Clearance	Crankshaft Shaft End Play	Crankshaft Thrust on No.	Connecting Rod Journal Diameter	Connecting Rod Oil Clearance	Connecting Rod Side Clearance
'79–'86	1,995	2.2440	0.0008–0.0020	0.002–0.007	3	1.7720	0.0008–0.0020	0.004–0.010
'79–'86	2,555	2.3622	0.0008–0.0020	0.002–0.007	3	2.0866	0.0008–0.0024	0.004–0.010
'83–'86	2,346	2.598	0.008–0.0020	0.0008–0.0020	3	2.0866	0.0008–0.0024	0.004–0.010

VALVE SPECIFICATIONS

Year	Engine Displacement cc	Seat Angle (deg)	Face Angle (deg)	Spring Test Pressure (lbs @ in.)	Spring Installed Height (in.)	Stem-to-Guide Clearance (in.)		Stem Diameter (in.)	
						Intake	Exhaust	Intake	Exhaust
'79–'86	1,995	45	45	72 @ 1.59	1.591	0.0012–0.0024	0.0020–0.0035	0.3150	0.3150
	2,555	45	45	72 @ 1.59	1.591	0.0012–0.0024	0.0020–0.0035	0.3150	0.3150
	All (Jet valve)	45	45	5.5 @ .846	—	—	—	— 0.1693 —	
'83–'86	2,346	45	45	61 @ 1.59	1.591	0.0012–0.0024	0.0020–0.0035	0.3150	0.3150

NOTE: Jet valve is found in both 2.0L and 2.6L engines.

PISTON AND RING SPECIFICATIONS

Year	Engine cc	Piston to Bore Clearance	Ring Side Clearance			Ring Gap		
			Top Compression	Bottom Compression	Oil Control	Top Compression	Bottom Compression	Oil Control
'79–'86	1,995	0.0008–0.0016	0.0020–0.0035	0.0008–0.0024	snug	0.010–0.018	0.008–0.016	0.0078–0.028
'79–'86	2,555	0.0008–0.0016	0.0020–0.0035	0.0010–0.0020	snug	0.0120–0.018	0.010–0.015	0.012–0.024
'83–'86	2,346	.0016–.0024	0.001–0.002	0.001–0.003	0.001–0.003	0.001–0.016	0.001–0.016	0.001–0.016

TORQUE SPECIFICATIONS

Year	Engine Displacement cc	Cylinder Head Bolts	Rod Bearing Bolts	Main Bearing Bolts	Crankshaft Sprocket Bolt	Flywheel-to-Crankshaft Bolts	Manifolds	
							Intake	Exhaust
'79–'86	1,995	65–72 ①	33–34	37–39	80–94	94–101	11–14	11–14
'79–'86	2,555	65–72 ①	33–34	55–61	80–94	94–101	11–14	11–14
'83–'86	2,346	76–83 ②	33–34	55–61	123–137	94–101	11–14	11–14

① Cold engine; hot engine—73–79 ft. lbs.
② Cold engine; hot engine—84–90 ft. lbs.

ALTERNATOR AND REGULATOR SPECIFICATIONS

Model	Year	Alternator Identification Number	Rated Output @ 5000	Rated Output @ 2500	Brush Length (in.)	Brush Spring Tension (lbs.)	Regulated Voltage
All	'79	AQ2245G	41 amps	34 amps	0.669 ①	2.9–3.7 ②	14.1–14.7
	'80–'82	A2T16471	44 amps	37 amps	0.709	0.7–1	14.1–14.7
	'83–'86	A5T15470	—	45 amps	③	—	13.9–14.9
	'83–'86	A5T15370	—	45 amps	③	—	13.9–14.9

① Built-in type brush: 0.709 in.
② Built-in type spring: 0.71–1 lbs.
③ Replace brushes when worn down to wear limit line

BATTERY AND STARTER SPECIFICATIONS

Year	Engine Model cc	Battery Amp Hour Capacity	Starter Amps	Starter No Load Test Volts	Starter rpm	Brush Spring Tension (lbs)	Min. Brush Length (in.)	Type of Starter
'79–'82	1,995 & 2,555 w/MT	45, 60 ①	60	11.5	6,600	2.9–3.7	0.453	Direct Drive
	1,995 & 2,555 w/AT	45, 60 ①	90 ④	11.5 ③	②3,300	2.9–3.7	0.453	Gear Reduction
'83–'86	1,995 & 2,555	65	100	11.5	3,000	2.9–3.7	0.453	Gear Reduction
	1,995 & 2,555	65	100	11.5	3,000	2.9–3.7	0.453	Gear Reduction
	2,555 4 × 4	80	60	11.5	6,500	2.9–3.7	0.453	Direct Drive
	2,346	110	130	11.0	4,000	2.9–3.7	0.453	Gear Reduction

MT: Manual Transmission
AT: Automatic Transmission
TBD: Turbo Diesel
① 60 amp for Canada
② 1979: 4,500 rpm
③ 1979: 11 volts
④ 1979: 62 amps

BRAKE SPECIFICATIONS

(All measurements are given in inches unless noted)

Model	Year	Lug Nut Torque ft. lb.	Master Cylinder Bore	Brake Disc Thickness Std.	Min.	Runout	Brake Drum Diameter	Maximum Wear	Caliper Bore	Wheel Cylinder Bore
All	'79–'86	51–58	⅞	0.79	0.72	0.006	9.5 ②	9.579	2.125	0.750 ③

① Due to the variations in state inspection regulations, the minimum allowable lining thickness may be different from that recommended by the manufacturer.
② 4-wd: 9.99
③ 4-wd: 0.8125

TUNE-UP

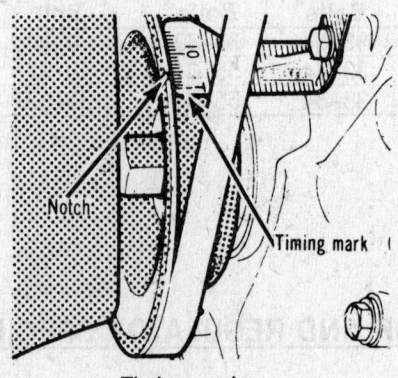

Timing marks

Electronic Ignition System (EIS)

For electronic ignition system component troubleshooting, refer to the Electrical section in Unit Repair.

Ignition Timing

ADJUSTMENT

1. Warm up the engine. Connect a tachometer and check the engine idle speed. Adjust it as outlined if not within specifications. If the timing mark on the front pulley is difficult to see, use chalk or a dab of paint to make it more visible.

2. Connect a timing light to the engine, as outlined in the instructions supplied by the manufacturer of the light.

3. Allow the engine to run at the specified idle speed with the gear shift in Neutral (Park, if automatic) and the air conditioning compressor and lights off.

—— CAUTION ——
Be sure the parking brake is firmly set and that the wheels are chocked.

4. Point the timing light at the timing marks indicated on the front timing chain cover. With engine at idle, timing should be at the specifications given in the tune-up chart at the beginning of this section. If it is not, loosen the attaching nut at the base of the distributor and rotate the distributor until the correct timing is achieved.

5. Stop the engine and retighten the attaching nut. Start the engine and recheck the timing.

6. Stop the engine and disconnect the timing light and tachometer.

Valve Lash

ADJUSTMENT

Gasoline Engines
Both the U engine (1,995cc) and the W engine (2,555cc) have a jet valve located beside the intake valve of each cylinder. The jet valve works off the intake valve rocker arm and injects a swirl of air into the combustion chamber to promote more complete burning of fuel.

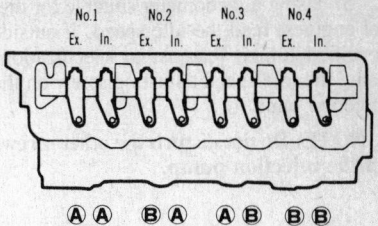

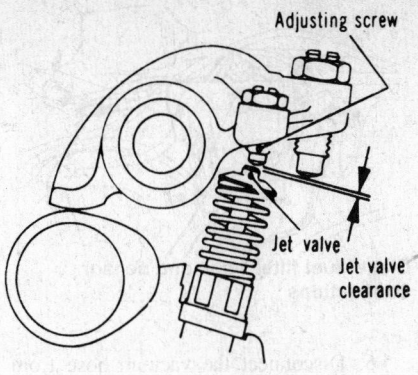

Jet valve adjustment

Valve adjustment

NOTE: When adjusting valve clearances, the jet valve must be adjusted before the intake valve.

1. Start the engine and allow it to reach normal operating temperature (170–190°F).

2. Stop the engine and remove the air cleaner and its hoses. Remove any other cables, hoses, wires, etc., which are attached to the valve cover, and remove the valve cover.

3. Disconnect the high tension coil-to-distributor wire at the coil.

4. Watch the rocker arms for No. 1 cylinder and rotate the crankshaft until the exhaust valve is closing and the intake valve has just started to open. At this point, No. 4 cylinder will be at top dead center (TDC) commencing its firing stroke.

5. Loosen the lock nut on cylinder No. 4 intake valve and back off the intake valve adjusting screw 2 or more turns.

6. Loosen the lock nut on the jet valve adjusting screw.

7. Turn the jet valve adjusting screw counterclockwise and insert a 0.006 in. feeler gauge between the jet valve stem and the adjusting screw.

8. Tighten the adjusting screw until it touches the feeler gauge. Take care not to press in the valve while adjusting because the jet valve spring is very weak.

NOTE: If the adjusting screw is tight, special care must be taken to avoid pressing down on the jet valve when adjusting the clearance or a false reading will result.

9. Tighten the lock nut securely while holding the rocker arm adjusting screw with a screwdriver to prevent it from turning.

10. Make sure that a 0.006 in. feeler gauge can be easily inserted between the jet valve and the rocker arm.

11. Adjust No. 4 cylinder's intake valve to 0.006 in. and its exhaust valve to 0.010 in. Tighten the adjusting screw locknuts and recheck each clearance.

12. Perform Step 4 in conjunction with the chart below to set up the remaining three cylinders for valve adjustments.

Exhaust Valve Closing	Adjust
No. 1 cylinder	No. 4 cylinder valves
No. 2 cylinder	No. 3 cylinder valves
No. 3 cylinder	No. 2 cylinder valves
No. 4 cylinder	No. 1 cylinder valves

13. Replace the valve cover and all other components. Run the engine and check for oil leaks at the valve cover.

Diesel Engines

1. Make sure the engine is at normal operating temperature.

2. Place the No. 1 piston at top dead center (TDC) on the compression stroke.

3. Loosen the locknuts on the rocker arms marked, A, in the illustration, and adjust the clearances to 0.010 in. with a feeler gauge. Retighten the locknuts.

4. After retightening the nuts, recheck clearances.

5. Place the No. 4 piston at TDC on the compression stroke.

6. Adjust valve clearances on the valves marked "B" in the illustration by following Step 3 above. Recheck clearances after tightening the locknuts.

7. Check the idle speed and readjust if necessary.

Idle Speed And Mixture

ADJUSTMENT

Gasoline Engine

1. Start and run the engine at idle until normal operating temperature is reached.

2. Check the tune-up specifications chart or the underhood decal for the correct curb idle speed.

3. Connect a tachometer (follow the instructions that came with the meter) and adjust the idle speed screw until the correct rpm is reached.

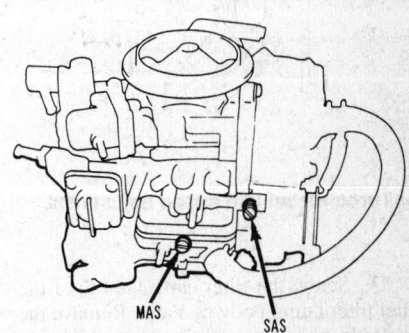

Diesel valve adjustment order

4. *Idle mixture adjustments should be made by an authorized garage using a CO meter.* However, a small amount of adjustment is possible (within the limits of the idle mixture screw limiter cap, which must not be removed).

NOTE: Some late model carburetors have a tamperproof, sealed idle mixture screw. These cannot be adjusted, except by an authorized service facility.

5. To adjust the idle mixture, first, adjust carb to correct curb idle speed. Next, watch the tachometer scale, listen to the engine and slowly turn the idle mixture screw clockwise. A drop in engine rpm or engine roughness will tell you when to stop. Then, slowly turn the mixture screw counterclockwise until once again you encounter rpm drop or engine roughness. A point, in between the clockwise or counterclockwise positions that gives you the highest rpm or smoothest running engine, is the best setting.

6. Check and readjust the curb idle speed, if necessary.

7. Have your adjustment checked with a CO meter as soon as possible.

Refer to the Carburetor section in Unit Repair for adjustments and specifications.

Carburetor idle adjustments

Diesel Engine

1. Make sure all accessories and lights are off, and the transmission is in Neutral.

2. Run the engine until the engine is at normal operating temperature.

3. Run the engine for more than 5 seconds at an engine speed of 2,000–3,000 rpm.

4. Run the engine at idle for 2 minutes.

5. Using a tachometer suitable for diesel engines, read the idle speed. If outside specified limits, readjust to specifications using the idle speed adjusting screw on the injection pump.

NOTE: Do not disturb the other screws on the injection pump.

Fuel Filter Element

REPLACEMENT

Diesel Engines Only

The fuel filter (and its maintenance) is crucial to the protection of the turbodiesel engine, keeping water and dirt out of the injection pump. The fuel filter also has a built in fuel heater, which prevents the waxing of diesel fuel in cold weather.

To remove the fuel filter element, the fuel filter canister must be removed. The fuel system must be bled after the new filter has been installed.

1. Disconnect the water level sensor connector, fuel heater connector, fuel temperature sensor connector (all electrical connections) and main fuel hoses at the fuel filter.

2. Remove the two bolts attaching the fuel filter to the vehicle, and remove the filter. Remove the protector and bracket.

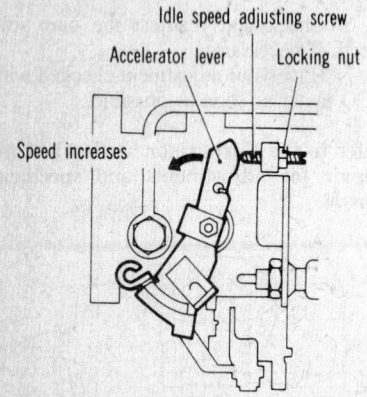

Idle speed adjusting screw

Accelerator lever — Locking nut

Speed increases

Turbo diesel idle speed adjustment

3. Screw the filter cartridge out of the fuel filter pump body by hand. Remove the water level sensor and the drain plug from the cartridge.

4. Check the filter for dirt deposits, and clean with kerosene if necessary.

5. Install the drain plug and water level sensor on the new cartridge. Torque the drain plug to 3–4 ft. lb.; torque the water level sensor to 9–10 ft. lb.

6. Whenever the fuel supply has run out, or the filter has been replaced, or the fuel lines have been disconnected, the fuel system must be bled of air. Loosen the air plug on the filter body. This is the plug

protruding out the top of the filter body on an angle.

7. Pull out the hand pump knob on the filter (on the side of the filter body) by turning it to the left. Pump the hand pump until the fuel coming out of the air plug hole has no air bubbles in it. Tighten the air plug.

DRAINING WATER FROM THE FUEL FILTER

When water accumulates in the fuel filter, the fuel water separator light will come on. Follow the procedure below as soon as possible.

1. Loosen the drain plug on the bottom of the fuel filter.

2. Drain the water by pumping the hand pump on the side of the filter body, then tighten the drain plug.

ENGINE MECHANICAL

Distributor

REMOVAL & INSTALLATION

1. Disconnect the battery ground cable.

2. Disconnect the wiring harness from the distributor control unit.

3. Mark the spark plug cables and pull them off the spark plugs.

NOTE: Always pull spark plug and coil cables at their boots to avoid breaking the wires inside the cables.

4. Remove the distributor cap by inserting a screwdriver into the two retaining screws, pushing in and turning clockwise.

5. Matchmark the distributor housing and the engine block. Mark the rotor position in the distributor as well. This will aid in correct positioning of the distributor during installation.

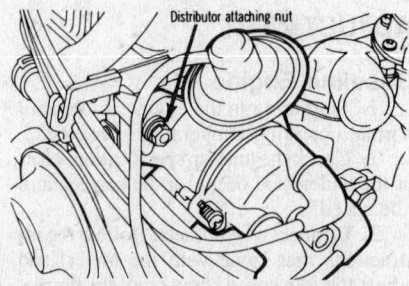

Distributor attaching nut

Distributor locknut location

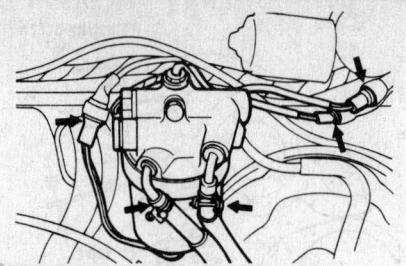

Diesel fuel filter hose and sensor connections

6. Disconnect the vacuum hose from the vacuum control unit.

7. Remove the distributor mounting nut and lift off the distributor assembly.

8. If the engine has been turned while the distributor is out: (Go to Step 12 if the engine has not been turned). Turn the engine crankshaft until the No. 1 cylinder is at top dead center on compression stroke. To find No. 1 cylinder, compression stroke, take off the distributor cap and turn the rotor and shaft until the rotor assembly is pointing toward the No. 1 cylinder lead in the distributor cap. Verify top dead center on the crankshaft pulley.

9. Align the mating mark (line) on the distributor housing with the mating mark (punch mark) on the distributor driven gear.

10. Install the distributor with the mating mark on the distributor attaching flange even with the center of the distributor retaining stud. Tighten the nut and replace the distributor cap, wires, and plug wires.

11. Set the ignition timing.

12. If the engine has not been turned while the distributor is out: Insert the distributor in the engine and align the marks made during removal.

13. Install the mounting nut, distributor cap, wires and plug wires, and vacuum line.

3. Start the engine and check the ignition timing.

Alternator

PRECAUTIONS

1. Always observe proper polarity of the battery connections. Be especially careful when jump starting the truck.

2. Never ground or short out any alternator or alternator regulator terminals.

3. Never operate the alternator with any of its or the battery's leads disconnected.

4. Always remove the battery or disconnect the cables while charging it.

5. Always disconnect the ground cable when replacing any electrical components.

6. Never subject the alternator to excessive heat or dampness if the engine is being steam cleaned.

7. Never use arc welding equipment with the alternator connected.

REMOVAL & INSTALLATION

1. Disconnect the battery ground cable.

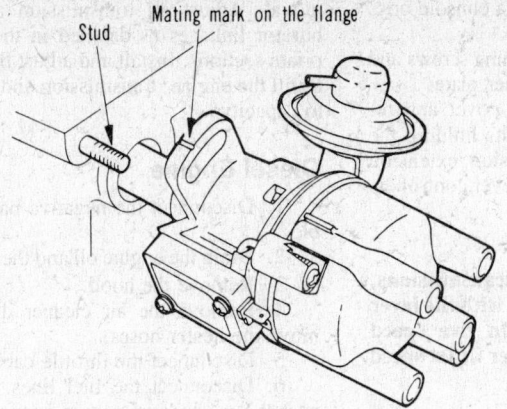

Align mark on flange with stud

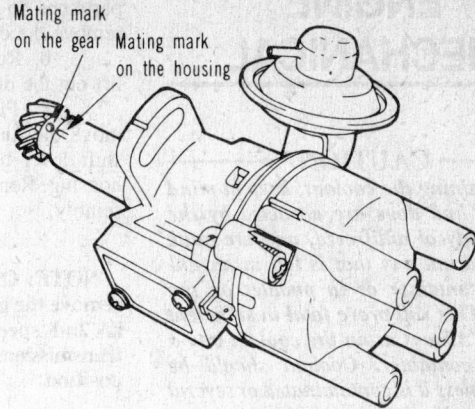

Mating mark alignment

2. Disconnect the cable from terminal "B" on the back of the alternator. Disconnect the other cables.

3. Remove the alternator brace bolt and the support bolt nut. Remove the drive belt.

4. Pull out the support bolt and remove the alternator assembly.

5. Align the hole in the alternator leg with the hole in the front case and insert the alternator support bolt from the front bracket side.

6. Install the brace bolt.

7. Install drive belt.

8. Push the alternator toward the front of the engine and check the clearance between the alternator leg and the front case. If the clearance is more than 0.008 in., insert spacers as required. 0.0078 in. thick spacers are available.

9. Adjust the belt tension as described below.

10. Tighten the alternator support bolt nut to 15–18 ft. lb. and the brace bolt to 9–11 ft. lb.

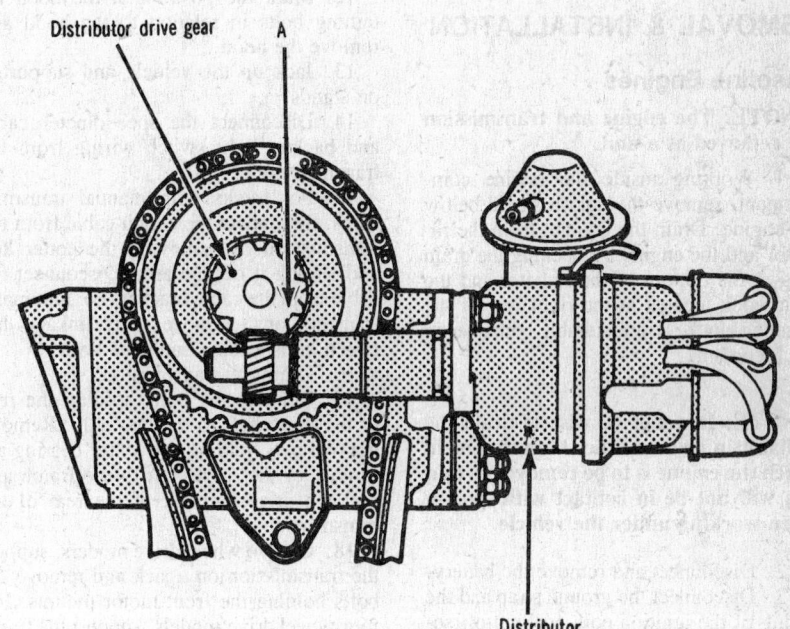

Distributor mounting

BELT ADJUSTMENT

Inspection and adjustment to the alternator drive belt should be performed every 15,000 miles. The belt should be replaced every 30,000 miles.

1. Inspect the drive belt to see that it is not cracked or worn. Be sure that its surfaces are free of grease or oil.

2. Pull the belt with a force of about 22 lbs. at a point halfway between the alternator pulley and the water pump pulley. The belt deflection should be $1/4$–$3/8$ in.

3. If the belt requires adjustment, loosen the alternator support bolt and alternator brace bolt and move the alternator to obtain specified deflection at 22 lbs. pressure.

4. After adjustment, tighten the alternator support bolt to 14–18 ft. lb., and the alternator brace bolt to 9–11 ft. lb.

── **CAUTION** ──
Do not overtighten the belt, or damage to the alternator bearings might result.

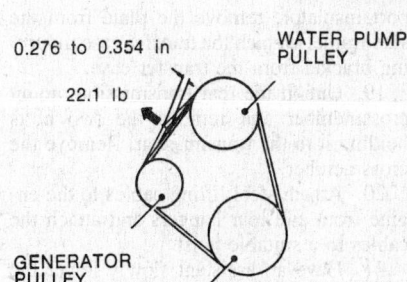

Drive belt tension adjustment

Regulator

REMOVAL & INSTALLATION

Both the 1,995cc and 2,555cc engines use an integrated circuit type regulator which is contained in the alternator. See Alternator Removal and Installation removal procedures. Adjustments of the regulator are confined to replacement.

Starter

REMOVAL & INSTALLATION

1. Disconnect the battery ground cable.

2. Disconnect the starter motor wiring.

3. Loosen and remove the two starter motor mounting bolts and remove the starter motor.

4. Installation is the reverse of removal.

For alternator and starter motor servicing, refer to the Electrical section of Unit Repair.

ENGINE MECHANICAL

CAUTION

When draining the coolant, keep in mind that cats and dogs are attracted by the ethelyne glycol antifreeze, and are quite likely to drink any that is left in an uncovered container or in puddles on the ground. This will prove fatal in sufficient quantity. Always drain the coolant into a sealable container. Coolant should be reused unless it is contaminated or several years old.

REMOVAL & INSTALLATION

Gasoline Engines

NOTE: The engine and transmission are removed as a unit.

1. Working inside the engine compartment, remove the splash shield below the engine. Drain the coolant from the radiator and the engine by opening the drain plug at the bottom of the radiator and the drain cock located at the right rear of the cylinder block. Use a suitable container to catch coolant.

NOTE: It would be wise to drain the radiator in an area other than the one in which the engine is to be removed so that you will not be in contact with coolant when working under the vehicle.

2. Disconnect and remove the battery.
3. Disconnect the ground strap and the wiring of the ignition coil, fuel cut-off solenoid valve, alternator, starter motor, water temperature gauge unit and oil pressure gauge unit.
4. Disconnect the air cleaner breather hose. Remove the air cleaner and disconnect the hot air duct and the vacuum hose.
5. Disconnect the accelerator control cable. For automatic transmissions, disconnect the transmission control rod.
6. Disconnect the radiator hoses by loosening their clips.
7. Disconnect the heater hose.
8. Disconnect the exhaust pipe from the exhaust manifold. The muffler pipe bracket should be detached at the transmission.
9. Disconnect the fuel hoses and vapor hose.
10. Remove the radiator and radiator cowl. Four bolts hold the radiator in place. On vehicles with automatic transmissions, remove and plug the two oil cooling pipes in the bottom of the radiator.
11. For trucks with four and five speed transmissions:
   a. Remove the lock screws and lift up the console box, inside the driver's compartment. In trucks without a console box, remove the carpet.
   b. Remove the attaching screws and lift out the dust cover retainer plate.
   c. Pull up the dust cover and remove the four attaching bolts holding the shift lever to the transmission extension housing. Remove the shift lever control assembly.

NOTE: On four speed transmissions, remove the gear shift lever with the lever in 2nd speed position. On five speed transmissions, place the lever in 1st speed position.

12. Mark the position of the hood retaining bolts in relation to the hood and remove the hood.
13. Jack up the vehicle and support it on stands.
14. Disconnect the speedometer cable and backup light switch wiring from the transmission.
15. For trucks with manual transmissions, disconnect the clutch cable from the transmission by removing the cotter key and sliding it off the arm. Disconnect the cable from the cable bracket. For automatic transmissions, remove shift linkage between transmission and shift lever.
16. Drain the transmission.
17. Remove the bolts holding the rear of the driveshaft to the rear axle. Remove the two nuts holding the center bearing assembly of the driveshaft to the frame and pull the driveshaft out of the rear of the transmission.
18. On two wheel drive models, support the transmission on a jack and remove the bolts holding the front motor mounts. For four wheel drive models, support the transfer case with a jack, and, after removing the transfer case mounting bracket and support insulator, remove the plate from the side frame. Detach the transfer case mounting bracket from the transfer case.
19. Unbolt the rear transmission mount crossmember and remove the two bolts holding it to the transmission. Remove the crossmember.
20. Attach steel lifting cables to the engine front and rear hangers and attach the cables to a suitable hoist.
21. Have an assistant slowly lower the jack under the transmission and pull the engine/transmission out of the vehicle by tilting it upwards and pulling forward.

NOTE: If the transmission will not clear the steering relay rod, raise it until the bell housing is above the rod, then remove the engine/transmission from the truck.

22. Installation is the reverse of removal. Adjust all transmission and carburetor linkages as detailed in the appropriate sections. Install and adjust the hood. Refill the engine, transmission and radiator to capacity.

Diesel Engine

1. Disconnect the negative battery cable.
2. Drain the engine oil and the coolant.
3. Remove the hood.
4. Remove the air cleaner duct. Remove the heater hoses.
5. Disconnect the throttle cable.
6. Disconnect the fuel lines. Disconnect the water level sensor connector and remove the fuel filter.
7. Remove the power steering pump, if equipped.
8. Disconnect the glow system cable, and the gauge unit harness connectors.
9. Disconnect the engine ground cable.
10. Disconnect the starter motor wiring harness.
11. Separate the clutch release cylinder from the transmission.
12. Disconnect the engine oil cooler hoses.
13. Disconnect the brake booster vacuum hose.
14. Disconnect the alternator wiring harness. Disconnect the oil pressure switch harness or the oil pressure gauge unit harness.
15. Remove the radiator assembly.
16. Remove the front exhaust pipe.
17. On 4 × 4 models, remove the engine and transfer case skid plates and under cover. On 2WD models, remove the under cover.
18. Disconnect the speedometer cable.
19. For 2WD models, disconnect the back-up light switch harness. On 4 × 4 models, disconnect the backup light switch harness and 4WD indicator light switch harness.
20. Remove the driveshafts.
21. Remove the gearshift lever assembly.
22. Support the transmission with a jack.
23. Detach the rear insulator from the transmission. Remove the No. 2 crossmember.
24. On 4 × 4 models, support the transfer case with a jack and, after removing the transfer case mounting bracket and support insulator, remove the plate from the side frame. Detach the transfer case mounting bracket from the transfer case.
25. Remove the engine mounting nuts from the front insulators. Using a chain block-and-tackle, raise and remove the engine and transmission assembly diagonally from the engine compartment.
26. Reverse the removal procedure for installation. Adjust the clutch and accelerator controls, align the hood and add coolant, engine oil and transmission lubricant. Check all controls for proper function before road testing.

Cylinder Head

REMOVAL & INSTALLATION

Gasoline Engine

———— CAUTION ————

Do not perform this operation on a warm engine. Remove the head bolts in the sequence shown. Loosen the head bolts evenly, not one at a time. Do not attempt to slide the cylinder head off the block, as it is located with dowel pins. Lift the head straight up and off the block.

1. Disconnect the battery and drain the cooling system. Disconnect the upper radiator hose.

———— CAUTION ————

When draining the coolant, keep in mind that cats and dogs are attracted by the ethelyne glycol antifreeze, and are quite likely to drink any that is left in an uncovered container or in puddles on the ground. This will prove fatal in sufficient quantity. Always drain the coolant into a sealable container. Coolant should be reused unless it is contaminated or several years old.

2. Remove the breather hoses and purge hose.
3. Remove the air cleaner and fuel line.
4. Remove the vacuum hose at the distributor and purge control valve.
5. Disconnect the spark plug wires after marking them for reinstallation.
6. Remove the distributor cap, and distributor by removing the retainer nut and pulling the unit out.
7. Disconnect the heater hose at the intake manifold.
8. Disconnect the water temperature gauge unit wire.
9. Place No. 1 piston in the Top Dead Center position to take pressure off the fuel pump rocker arm. Disconnect the fuel hoses and plug the line leading to the gas tank to prevent fuel leakage.
10. Remove the fuel pump mounting nuts or bolts and remove the fuel assembly. Remove the insulator and gaskets.
11. Disconnect the exhaust pipe at the exhaust manifold flange.
12. Remove the rocker cover.
13. Remove its breather and semicircular seal.
14. After slightly loosening the camshaft sprocket bolt, turn the crankshaft until No. 1 piston is at top dead center on compression stroke (both valves closed).

NOTE: Never turn the engine over using the camshaft bolt: it puts undue strain on the chain and other components.

15. Remove the camshaft sprocket bolt and distributor drive gear. Remove the camshaft sprocket and allow it to rest in the chain on the holder below.
16. Remove the cylinder head bolts in

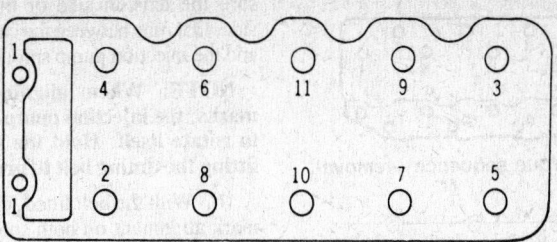

Gasoline engine head removal

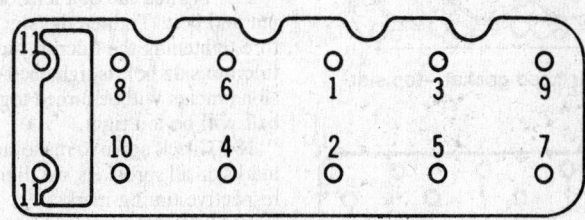

Gasoline engine head installation

the sequence shown in the illustration. Head bolts should be loosened in two or three stages to prevent head warpage.

NOTE: The cylinder head assembly is located with two dowel pins, front and rear, on the cylinder block. When removing, be careful not to slide it, or twist the camshaft sprocket and chain.

17. Remove the cylinder head assembly and cylinder head gasket.
18. Clean all gasket surfaces of cylinder block and cylinder head.
19. Install a new cylinder head gasket. Install the cylinder head assembly.

NOTE: Do not apply sealant to the head gasket and do not reuse an old head gasket.

20. Install the TEN cylinder head bolts. The two bolts at the front extend into the timing case, and are not included in the head bolt tightening sequence. In the tightening illustration, they are labeled No. 11. Starting at top center, tighten all ten cylinder head bolts to 35 ft. lb. in the sequence shown in the illustration. Repeat the tightening procedure, this time torque the bolts to 65–72 ft. lb. (cold engine), (72–80 ft. lb. hot engine).
21. Tighten the two front bolts (No. 11 in illustration) to 11–15 ft. lb.
22. Verify that No. 1 cylinder is at top dead center. Align the dowel pin in the end of the camshaft sprocket with the groove in the top of the front camshaft bearing cap and install the camshaft sprocket and chain while pulling up on the sprocket.
23. Install the distributor drive gear and the sprocket bolt.
24. Turn the crankshaft about 90° back, and tighten the camshaft sprocket bolt to 37–43 ft. lb. Very slowly turn the engine over two times to make sure the valve timing is correct. If the engine locks at a certain point in these two revolutions, the valve timing is not correct. Repeat Steps 22–24.

———— CAUTION ————

At this point, do not turn the engine over using the starter. If the valve timing is off, several of the valves could be bent.

25. Install the breather and semicircular seal to the cylinder head after applying sealant to surface contact points. Install the rocker cover with a new gasket.
26. Connect the exhaust pipe to the exhaust manifold flange. Tighten the bolts to 11–18 ft. lb.
27. Put No. 1 cylinder at top dead center and install the fuel pump with a new gasket and insulator. Connect all hoses.
28. Connect the water temperature gauge unit wire. Connect the heater hose to the intake manifold.
29. Install the distributor and spark plug cables. See Distributor Removal and Installation, above.
30. Connect the vacuum hose to the distributor and purge control valve. Connect the upper radiator hose and fill the cooling system with coolant.

NOTE: Many mechanics recommend that the engine oil be replaced after the head is removed to avoid water contamination from the coolant.

Diesel Engine

1. Observe the CAUTION at the beginning of the Gasoline Engine cylinder head removal procedure.
2. Disconnect the battery and drain the cooling system. Disconnect the upper radiator hose.
3. Label and disconnect any breather hoses which run over the cylinder head.
4. Disconnect the heater hose at the intake manifold. Disconnect the temperature sensor lead.
5. Unbolt the inlet piping and oil lines to the turbocharger. *Make sure you cover the turbo intake with a clean rag, as no*

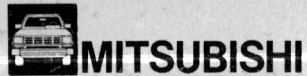
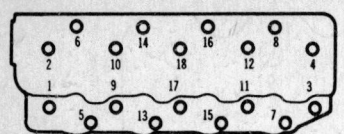
Diesel head torque sequence—removal

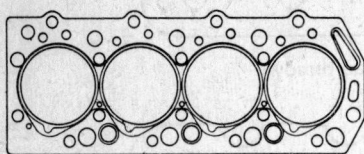
Diesel cylinder head gasket—top side

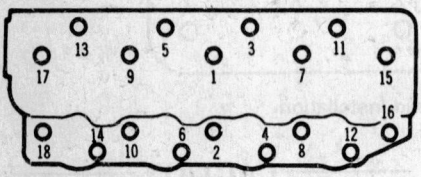
Diesel head torque sequence—installation

foreign material must enter any part of the turbo or turbo intake system.

6. Remove the fuel injection lines and glow plug connections. Remove the intake and exhaust manifolds.

7. Remove the timing belt front upper and lower covers. Set the No. 1 cylinder at TDC on the compression stroke. The timing marks should be lined up as shown in the illustration under Timing Belt. Remove the timing belt and remove injection pump.

NOTE: DO NOT disturb the crankshaft (if possible) while the timing belt is removed.

8. Remove the rocker cover. Loosen the cylinder head bolts in the sequence shown. Loosen the bolts gradually, making two or three passes.

9. Remove the cylinder head and gasket.

10. Clean all gasket surfaces of the cylinder block and head. *Always install a new gasket.*

11. Install the cylinder head gasket very carefully, as it has no identification marks. The accompanying illustration shows the top surface of the gasket.

12. Install the cylinder head to the block. Starting at No. 1 bolt, tighten all bolts to $1/2$ of their specified torque in the sequence shown. Finally, tighten all bolts to full specification in the same sequence. Proper bolt torque is 76–83 ft. lb.

13. Install the injection pump.

14. Move the timing belt tensioner fully toward the water pump and temporarily secure the tensioner.

14. Install the timing belt onto the injection pump sprocket and camshaft sprockets, making sure you don't turn the crankshaft sprocket while doing so. Make

sure the tension side of the belt (the top side, that runs between the camshaft sprocket and the injection pump sprocket) is *not slack*.

NOTE: When aligning the timing marks, the injection pump sprocket tends to rotate itself. Hold the sprocket while fitting the timing belt to prevent rotation.

16. With the belt fitted, check the timing mark alignment on both sprockets. Loosen the tensioner mounting bolts, which will apply tension to the belt.

17. Tighten the belt tensioner mounting nut and bolt. Tighten the slot side bolt before tightening the fulcrum side bolt. If the fulcrum side bolt is tightened first, the tension bracket will be turned together and the belt will be too tight.

18. Check again to make sure the timing marks on all sprockets are aligned with their respective timing marks.

19. Turn the crankshaft in the normal (clockwise) direction through two camshaft sprocket teeth and hold. Reverse the crankshaft to align the timing marks, and push down the belt with your finger at a point halfway, to check that the belt has about 0.16–0.20 in. deflection.

20. Install the rocker cover, along with a new gasket if necessary. Install the timing belt front upper and lower cover.

21. Install the intake and exhaust manifolds.

22. Reconnect the glow plugs and fuel injection lines.

23. Connect the turbocharger oil lines and inlet ducting.

NOTE: Be sure to prime the turbocharger with oil before installing the oil feed pipe.

24. Connect the heater hose and temperature sensor lead. Connect all breather hoses to their proper locations. Install the upper radiator hose and fill the cooling system. Connect the battery, start the engine and check timing. Check for water leaks.

Intake Manifold

─────── **CAUTION** ───────
When draining the coolant, keep in mind that cats and dogs are attracted by the ethelyne glycol antifreeze, and are quite likely to drink any that is left in an uncovered container or in puddles on the ground. This will prove fatal in sufficient quantity. Always drain the coolant into a sealable container. Coolant should be reused unless it is contaminated or several years old.

REMOVAL & INSTALLATION

Gasoline Engines

1. Drain the cooling system.

2. Remove the air cleaner assembly with its hoses from the engine.

3. Disconnect the fuel line and EGR lines.

4. Disconnect the accelerator linkage and, if so equipped, the automatic transmission shift cables at the carburetor.

5. Remove the water hose at the intake manifold. Remove the water hose at the carburetor.

6. Disconnect the water temperature sending unit.

7. Remove the manifold with the carburetor as a unit.

8. Installation is the reverse of removal. Tighten the manifold nuts to 11–14 ft. lb.

Exhaust Manifold

REMOVAL & INSTALLATION

Gasoline Engines

1. Remove the air cleaner.

2. Remove the heat shield from the exhaust manifold. Remove the EGR lines and reed valve, if equipped.

3. Unbolt the exhaust flange connection.

4. Remove the nuts holding the manifold to the cylinder head.

5. Remove the manifold.

6. Installation is the reverse of removal. Tighten the flange connection bolts to 11–18 ft. lb. Tighten the manifold bolts to 11–14 ft. lb.

Intake And Exhaust Manifolds

REMOVAL & INSTALLATION

Diesel Engines

NOTE: Although they are individual pieces, the intake and exhaust manifolds share the same gasket. Whenever one is removed, the other must be also, to replace the gasket.

1. Remove the turbocharger.

2. Remove the manifold connection from the intake manifold to the injection pump.

3. Loosen the attaching bolts and remove the intake manifold.

4. Loosen the attaching bolts and remove the exhaust manifold. Remove the gasket.

5. Installation is the reverse of removal. Clean all gasket mating surfaces before replacing the gasket. Torque the intake and exhaust manifold attaching bolts to 11–14 ft. lb.

NOTE: Be sure to prime the turbocharger with oil before installing the oil feed pipe.

Turbocharger

REMOVAL & INSTALLATION

Diesel Engines

1. Remove the exhaust pipe and gasket from the turbocharger. Remove the heat shield.

2. Remove the attaching bolts from the inlet fitting to the intake manifold. Loosen the hose clamps and remove the inlet fitting. Discard the gasket. Remove the rubber connecting hose.

3. Loosen the fitting and remove the oil feed pipe from the turbocharger assembly.

4. Loosen the hose clamps from the oil return line and remove the oil return line and flanged pipe from the turbocharger.

5. Remove the attaching bolts to the exhaust manifold flange and remove the turbocharger. Discard the mounting gasket; a new one should be fitted during assembly.

6. Installation is the reverse of removal. Clean all mounting surfaces before assembly, *and make sure no foreign material is allowed to enter the turbo manifolding or the turbo unit itself.*

NOTE: Before installing the oil feed pipe, pour clean engine oil into the turbocharger oil inlet. Make sure oil connections are tight and free from leaks. Replace all old gaskets with new ones.

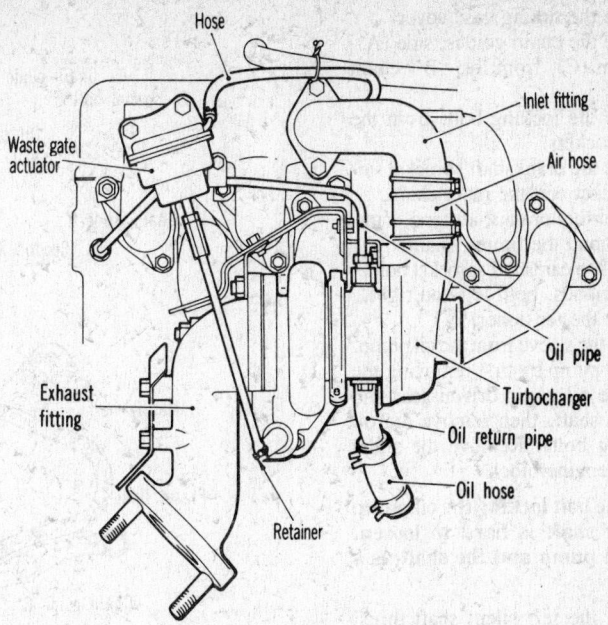

Turbocharger fittings and installation

Timing Chain, Cover, Silent Shafts and Tensioner

REMOVAL & INSTALLATION

Gasoline Engines

NOTE: All pickups are equipped with two Silent Shafts which cancel the vertical vibrating force of the engine and the secondary vibrating forces, which include the sideways rocking of the engine due to the turning direction of the crankshaft and other rolling parts. The secondary vibrating forces can be cancelled if forces equivalent in magnitude but opposite in direction are produced. In these engines, the opposite force is produced by silent shafts located in the upper left and lower right sides in the front of the cylinder block. The shafts are driven by a duplex chain and are turned by the crankshaft. The silent shaft chain assembly is mounted in front of the timing chain assembly and must be removed to service the timing chain.

1. Remove the battery cables.
2. Drain the radiator and remove it from the vehicle.
3. Remove the cylinder head (refer to cylinder head section for procedures).

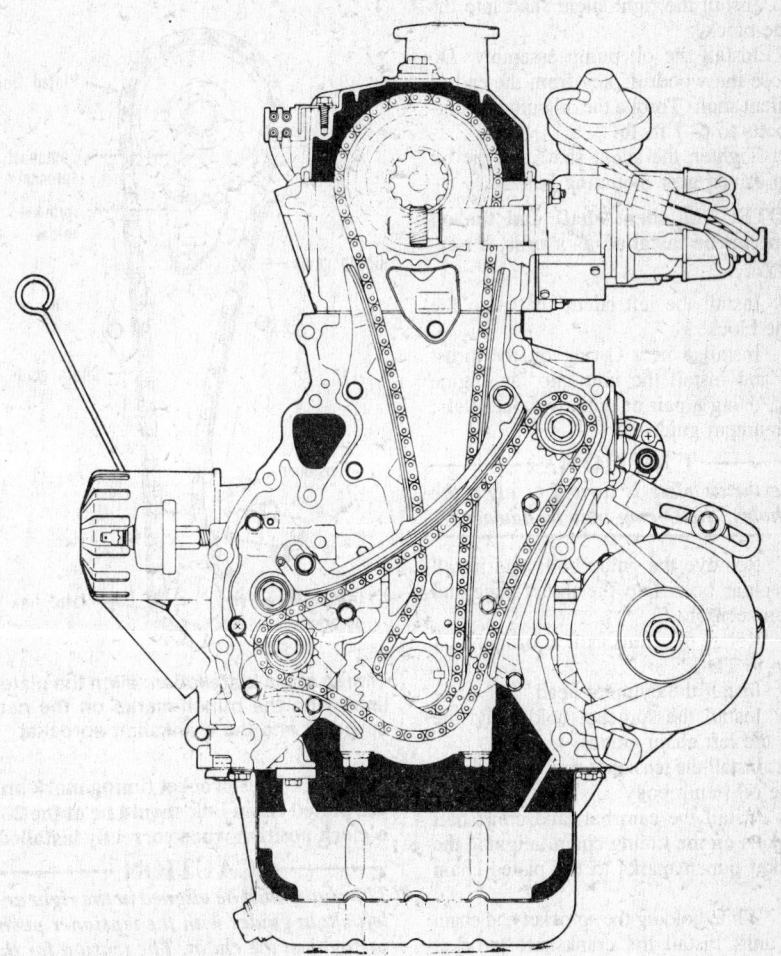

Front view of gasoline engine showing timing and silent shaft chains

4. Remove the cooling fan, spacer, water pump pulley and belt.
5. Remove the alternator. Remove the water pump.

6. Raise the front of the vehicle and support it on jackstands.
7. Remove the oil pan and screen. Remove the crankshaft pulley.

8. Remove the timing case cover.

9. Remove the chain guides, side (A), top (B), bottom (C), from the "B" chain (outer).

10. Remove the locking bolts from the "B" chain sprockets.

11. Remove the crankshaft sprocket, silent shaft sprocket and the outer chain.

12. Remove the crankshaft and camshaft sprockets and the timing chain.

13. Remove the camshaft sprocket holder and the chain guides, both left and right.

14. Remove the tensioner.

15. Remove the sleeve from the oil pump. Remove the oil pump by first removing the bolt locking the oil pump driven gear and the right silent shaft, then remove the oil pump mounting bolts. Remove the silent shaft from the engine block.

NOTE: If the bolt locking the oil pump and the silent shaft is hard to loosen, remove the oil pump and the shaft as a unit.

16. Remove the left silent shaft thrust washer and take the shaft from the engine block.

17. Install the right silent shaft into the engine block.

18. Install the oil pump assembly. Do not lose the woodruff key from the end of the silent shaft. Torque the oil pump mounting bolts to 6–7 ft. lb.

19. Tighten the silent shaft and the oil pump driven gear mounting bolt.

NOTE: The silent shaft and the oil pump can be installed as a unit, if necessary.

20. Install the left silent shaft into the engine block.

21. Install a new O-ring on the thrust plate and install the unit into the engine block, using a pair of bolts without heads, as alignment guides.

--- **CAUTION** ---

If the thrust plate is turned to align the bolt holes, the O-ring may be damaged.

22. Remove the guide bolts and install the regular bolts into the thrust plate and tighten securely.

23. Rotate the crankshaft to bring No. 1 piston to TDC.

24. Install the cylinder head.

25. Install the sprocket holder and the right and left chain guides.

26. Install the tensioner spring and sleeve on the oil pump body.

27. Install the camshaft and crankshaft sprockets on the timing chain, aligning the sprocket punch marks to the plated chain links.

28. While holding the sprocket and chain as a unit, install the crankshaft sprocket over the crankshaft and align it with the keyway.

29. Keeping the dowel pin hole on the camshaft in a vertical position, install the camshaft sprocket and chain on the camshaft.

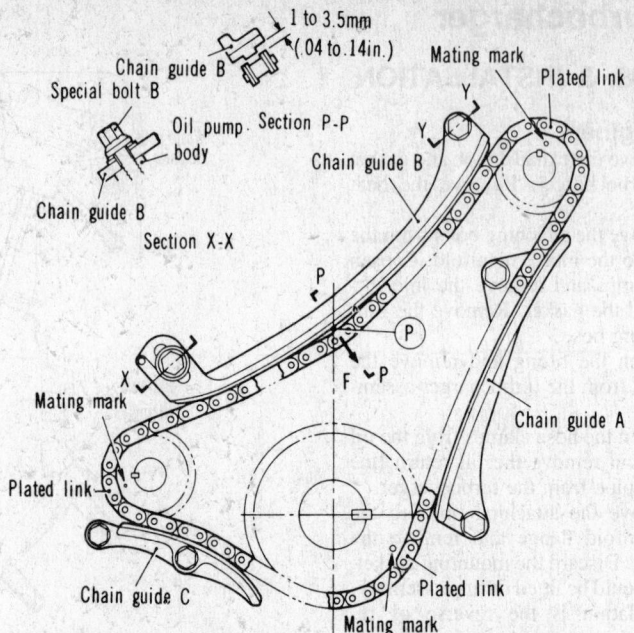

"Silent Shaft" balancing system—gasoline engine shown

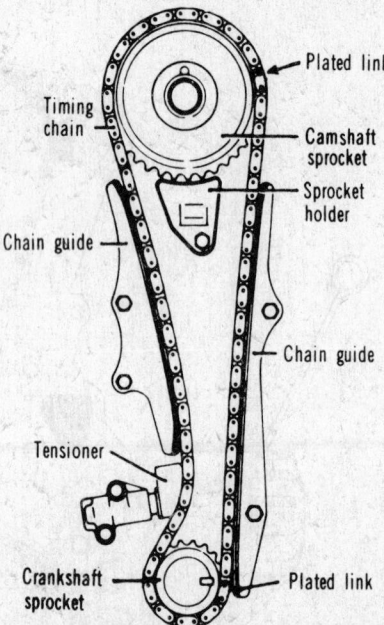

Timing chain installation: align the plated links with the punch-marks on the cam sprocket and the crankshaft sprocket

NOTE: The sprocket timing mark and the plated chain link should be at the 2–3 o'clock position when correctly installed.

--- **CAUTION** ---

The chain must be aligned in the right and left chain guides with the tensioner pushing against the chain. The tension for the inner chain is predetermined by spring tension.

30. Install the crankshaft sprocket for the outer or "B" chain.

31. Install the two silent shaft sprockets

and align the punched mating marks with the plated links of the chain.

32. Holding the two shaft sprockets and chain, install the outer chain in alignment with the mark on the crankshaft sprocket. Install the shaft sprockets on the silent shaft and the oil pump driver gear. Install the lock bolts and recheck the alignment of the punch marks and the plated links.

33. Temporarily install the chain guides, side (A), top (B), and bottom (C).

34. Tighten side (A) chain guide securely.

35. Tighten bottom (B) chain guide securely.

36. Adjust the position of the top (B) chain guide, after shaking the right and left sprockets to collect any chain slack, so that when the chain is moved toward the center, the clearance between the chain guide and the chain links will be approximately ⁹⁄₆₄ inch. Tighten the top (B) chain guide bolts.

37. Install the timing chain cover using a new gasket, being careful not to damage the front seal.

38. Install the oil screen and the oil pan, using a new gasket. Torque the bolts to 4.5–5.5 ft. lb.

39. Install the crankshaft pulley, alternator and accessory belts, and the distributor.

40. Install the oil pressure switch, if removed, and install the battery ground cable.

41. Install the fan blades, radiator, fill the system with coolant and start the engine.

Timing Belts

REMOVAL & INSTALLATION
Diesel Engines

1. Set the No. 1 cylinder at TDC on

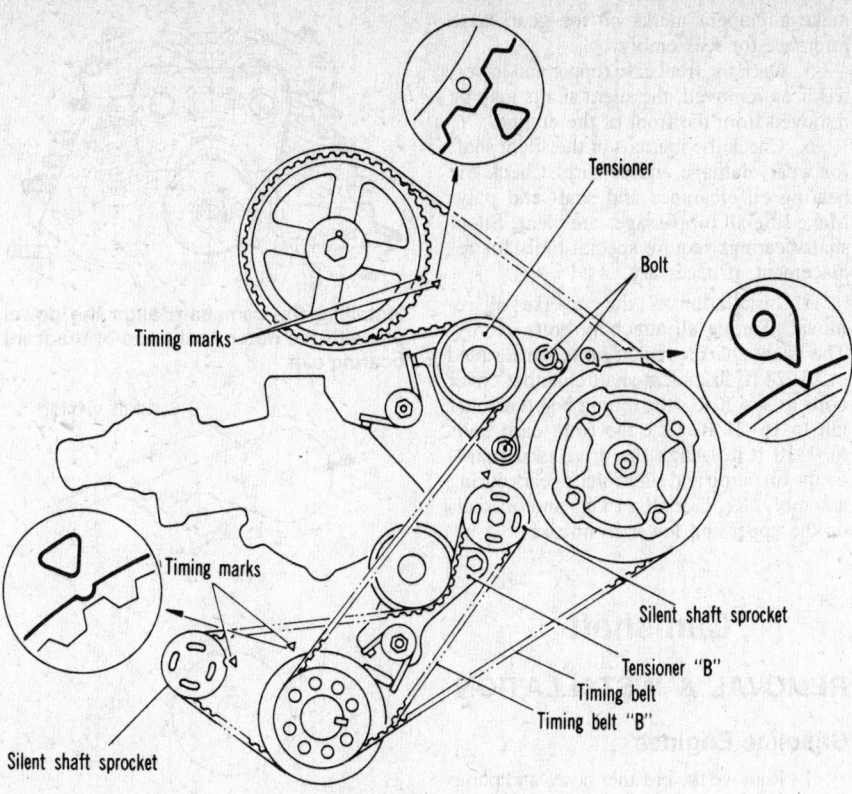

Diesel timing belt train

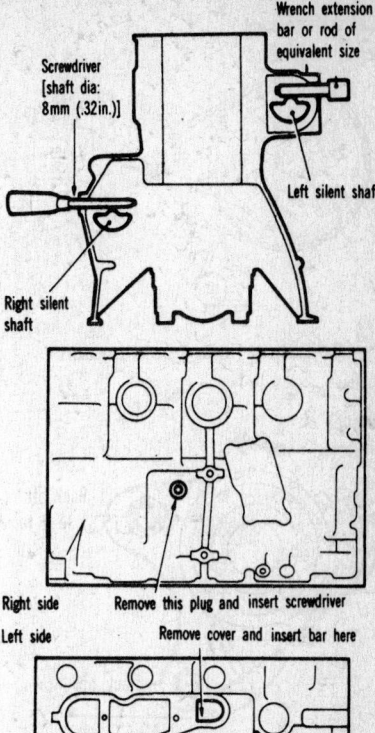

Preventing rotation of silent shafts

the compression stroke. Make sure all timing marks align.

2. If you will be reusing the belts, mark them both (the valve timing belt and the silent shaft belt) with an arrow indicating the direction of rotation before removal.

3. Remove the crankshaft pulley and both timing belt covers.

4. Slightly loosen both belt tensioner bolts, move the tensioners toward the water pump, then secure the tensioners by tightening the tensioner bolt temporarily.

5. Remove the camshaft sprocket and injection pump sprocket. *It is not necessary to remove the silent shaft sprockets or loosen the tensioner if the silent shaft timing belt is not being replaced.*

6. If the silent shaft belt is to be replaced, loosen the silent shaft sprockets after first locking the shafts in place with a small pry bar.

7. Remove the timing belt(s) and inspect for damage, wear or deterioration. *Do not use solvents or detergent to clean the timing belts, sprockets or tensioners.* Replace any component that is excessively contaminated with dirt, oil or grease.

8. Installation is the reverse of removal. Install the spacer to the left silent shaft with its chamfered edge toward the oil seal.

————— CAUTION —————
Installing this spacer incorrectly will result in damage to the oil seal.

9. Align the timing marks of the crank-

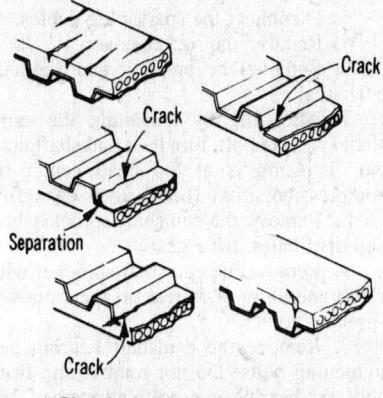

Check timing belt for wear

shaft and silent shaft drive sprockets when installing the silent shaft belt.

10. Loosen the silent belt tensioner and allow the spring tension to take up any belt slack. *Do not attempt to move the tensioner to obtain more tension.* Tighten the tensioner bolt to 25–28 ft. lb.

11. Make sure to install the belt so it rotates in the same direction as marked (if reusing the old belt). Standard belt deflection ("Play") when properly tensioned is 4–5mm (0.16–0.20 in.) as measured at midpoint between the upper silent shaft and crankshaft sprockets.

12. Align the timing marks for the camshaft, injection pump and crankshaft sprockets and install the timing belt. Install the timing belt first on the crankshaft

sprocket, then on the injection pump and camshaft sprockets.

NOTE: With the tension side kept taut, engage the timing belt teeth with each sprocket. The injection pump sprocket tends to to rotate itself, so it must be held while installing the belt.

13. Loosen the tensioner belt and allow the spring tension to take up the slack in the timing belt. *Do not attempt to apply more tension by manually moving the tensioner.* Once the belt is tensioned, tighten the tensioner bolt to 16–21 ft. lbs.

NOTE: Tighten the slot side bolt before tightening the fulcrum side bolt. If the fulcrum side is tightened earlier, the tension bracket will be turned together and the belt will be too tight.

14. Make sure the timing marks on all sprockets are still in correct alignment.

15. Turn the crankshaft in the normal direction of rotation through two camshaft sprocket teeth and hold it there. Reverse direction to align the timing marks, and push down on the belt at a point halfway between the camshaft and injection pump sprocket to check belt deflection. Normal deflection is 4–5mm (0.16–0.20 in.).

16. Reset the timing belt switch by depressing its knob until it is flush with the base and mount the timing belt covers.

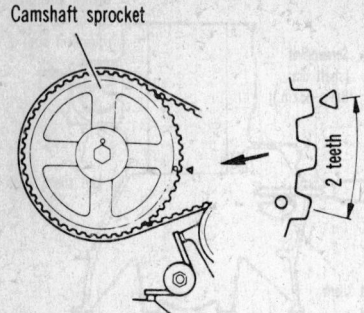

Adjusting timing belt tension

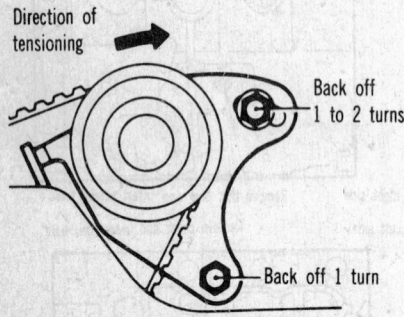

Loosening timing belt tensioner

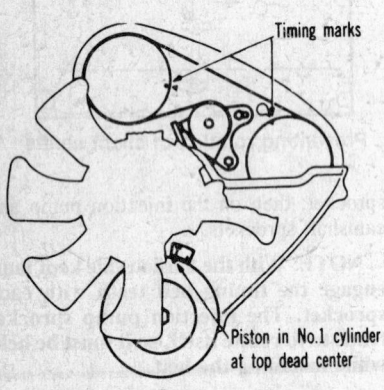

Checking timing belt alignment

Silent Shafts

REMOVAL & INSTALLATION

Diesel Engine

NOTE: This procedure is normally performed during a complete engine overhaul since it requires disassembly of most major engine components.

1. Remove the engine. Drain and remove the oil pan.
2. Remove the timing belts.
3. Remove the front case and oil pump assembly. When loosening the silent shaft driven gear attaching bolts, remove the plug on the right side of the cylinder block and insert a pry bar to prevent rotation of the shaft while loosening the bolt.
4. Before removing the oil pump gears,

make alignment marks on the gears as a reference for reassembly.

5. Once the front case (upper and lower) has been removed, the silent shafts may be removed from the front of the engine.
6. Check the journals of the silent shaft for wear, damage and seizure. Check the bearing oil clearance and shaft end play. Make sure all oil passages are clear. Silent shaft bearings require special tools for replacement, if necessary.
7. Installation is the reverse of removal. Torque all attaching bolts evenly. The silent shaft sprocket bolts are torqued to 25–28 ft. lb.; silent shaft chamber cover bolts to 3–4 ft.lb.; the timing belt tensioner nut to 16–21 ft. lbs.; the front case bolts to 9–10 ft.lb. Align the drive gear marks on the oil pump and silent shaft gears during assembly. Replace all gaskets and oil seals on the upper and lower front case.

Camshaft

REMOVAL & INSTALLATION

Gasoline Engines

1. Remove the breather hoses and purge hose.
2. Remove the air cleaner and fuel line.
3. Remove the fuel pump. Remove the distributor.
4. Diconnect the spark plug cables.
5. Remove the rocker cover.
6. Remove the breather and semicircular seal.
7. After slightly loosening the camshaft sprocket bolt, turn the crankshaft until No. 1 piston is at top dead center on compression stroke (both valves closed).
8. Remove the camshaft sprocket bolt and distributor drive gear.
9. Remove the camshaft sprocket with chain and allow it to rest on the camshaft sprocket holder.
10. Remove the camshaft bearing cap tightening bolts. Do not remove the front and rear bearing cap bolts altogether, but keep them inserted in the bearing caps so that the rocker assembly can be removed as a unit.
11. Remove the rocker arms, rocker shafts and bearing caps as an assembly.
12. Remove the camshaft.
13. Lubricate the camshaft lobes and bearings and fit camshaft into head.
14. Install the assembled rocker arm shaft assembly. The camshaft should be positioned so that the dowel pin on the front end of the cam is in the 12 o'clock position and in line with the notch in the top of the front bearing cap.
15. Install the bearing cap bolts. Starting at the center and working out, tighten the bolts to 7 ft. lb. Repeat the procedure, this time tightening them to 14–15 ft. lb.
16. Install the camshaft sprocket and distributor drive gear onto the camshaft while

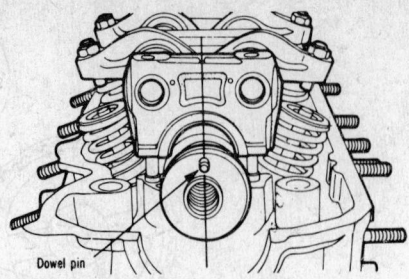

Installing the camshaft: align the dowel pin with the notch in the top of the front bearing cap

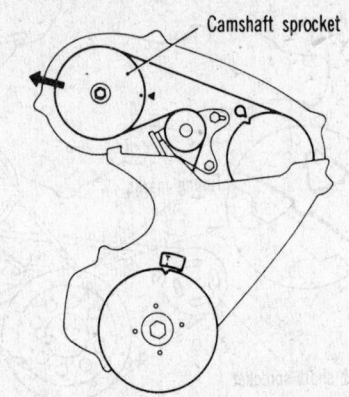

Camshaft timing sprocket installation showing timing marks

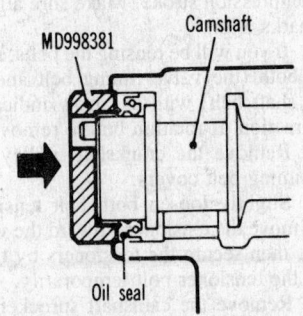

Camshaft oil seal installation—diesels

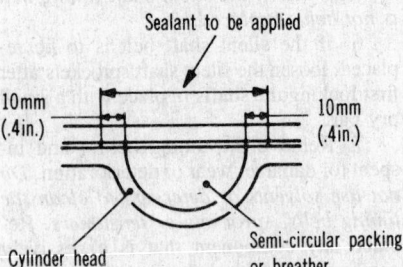

Apply sealant here—diesel camshaft installation

pulling it upward. Temporarily tighten the locking bolt.

17. Turn the crankshaft about 90° back and tighten the camshaft sprocket bolt to 37–43 ft. lb.
18. Temporarily set the valve clearance

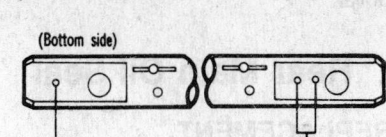

Front of engine

(Bottom side)

One oil hole   Two oil holes

Diesel rocker shaft orientation

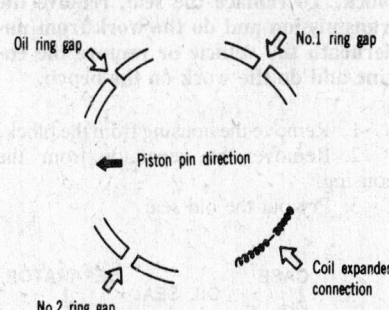

Oil ring gap   No.1 ring gap

Piston pin direction

No.2 ring gap   Coil expander connection

Piston ring end gap positioning—diesels

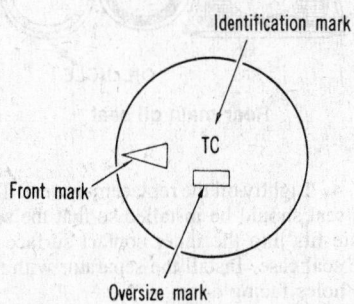

Identification mark

TC

Front mark

Oversize mark

Piston crown I.D. marking—diesels

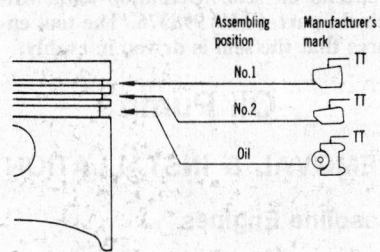

Assembling position   Manufacturer's mark

No.1   TT
No.2   TT
Oil   TT

Diesel piston ring installation

to cold engine specifications (see Valve Lash section).

19. Temporarily install the breather, semicircular seal and rocker cover and start the engine and run it at idle speed.

20. After the engine is at normal operational temperature, adjust the valves to hot engine specifications (see Valve Lash section).

21. Install breather and seal and apply sealant to the contact surfaces.

22. Install the rocker cover and tighten to 4–5 ft. lb.

23. Install distributor, fuel pump, air cleaner, fuel line, plug leads and other assemblies.

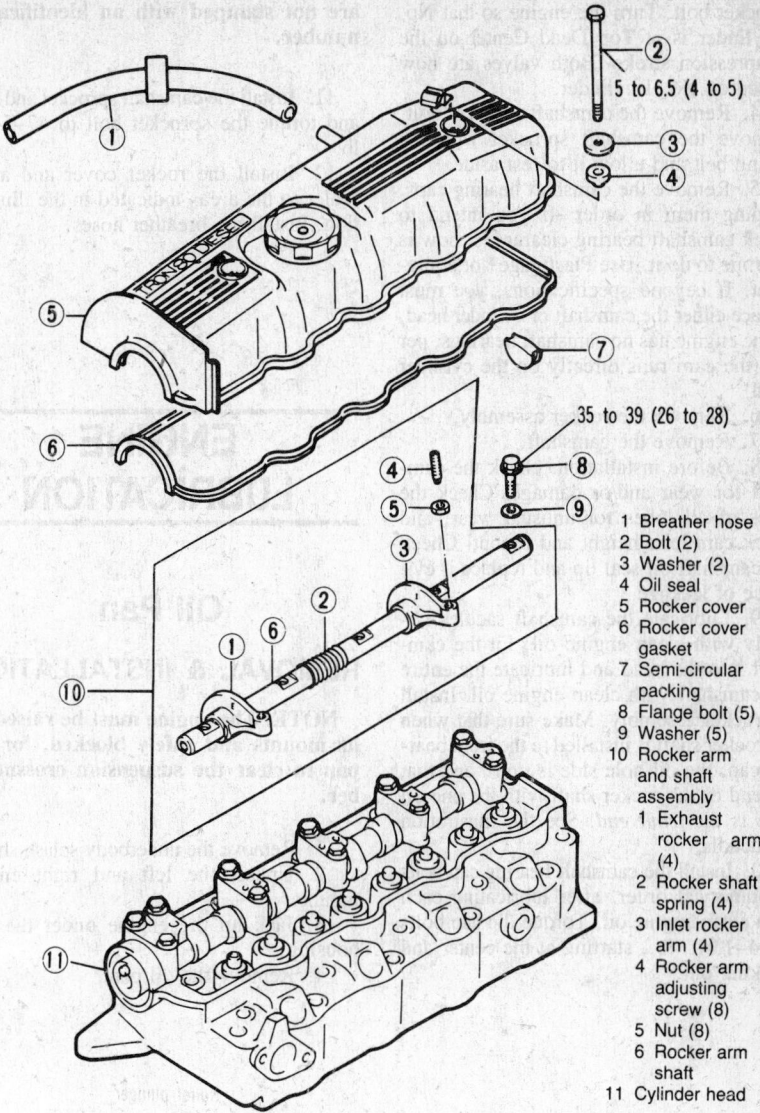

5 to 6.5 (4 to 5)

35 to 39 (26 to 28)

1 Breather hose
2 Bolt (2)
3 Washer (2)
4 Oil seal
5 Rocker cover
6 Rocker cover gasket
7 Semi-circular packing
8 Flange bolt (5)
9 Washer (5)
10 Rocker arm and shaft assembly
   1 Exhaust rocker arm (4)
   2 Rocker shaft spring (4)
   3 Inlet rocker arm (4)
   4 Rocker arm adjusting screw (8)
   5 Nut (8)
   6 Rocker arm shaft
11 Cylinder head

Diesel rocker assembly; torque specs in parentheses are ft. lbs.

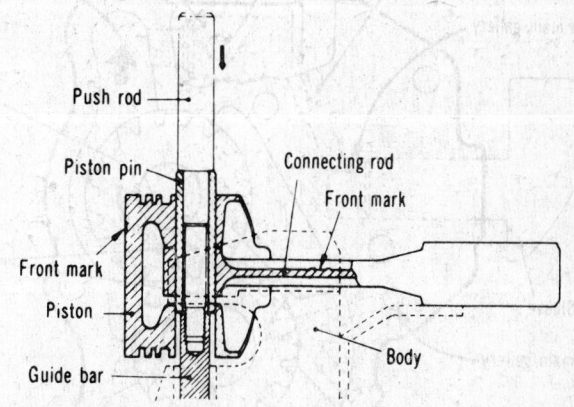

Push rod

Piston pin   Connecting rod

Front mark   Front mark

Piston

Guide bar   Body

Installing the piston pin. Front marks on piston and connecting rod also shown

Diesel Engines

1. Label and disconnect all breather hoses that are in the way of removing the valve rocker cover.

2. Remove the rocker cover.

3. Loosen, very slightly, the camshaft sprocket bolt. Turn the engine so that No. 1 cylinder is at Top Dead Center on the compression stroke. Both valves are now closed on No. 1 cylinder.

4. Remove the camshaft sprocket bolt. Remove the camshaft sprocket with the timing belt and allow it to rest aside.

5. Remove the camshaft bearing caps, keeping them in order. If you intend to check camshaft bearing clearances, now is the time to do it. Use Plastigage® or equivalent. If beyond specifications, you must replace either the camshaft or cylinder head, as the engine has no camshaft bearings, per se, (the cam runs directly on the cylinder head).

6. Remove the rocker assembly.

7. Remove the camshaft.

8. Before installation, check the camshaft for wear and/or damage. Check the faces of all lobes for unusual wear, and check cam lobe height and runout. Check the camshaft oil seal lip and replace if evidence of leakage.

9. Lubricate the camshaft saddles liberally with clean engine oil. Fit the camshaft into the head and lubricate the entire the camshaft with clean engine oil. Install the rocker assembly. Make sure that when the rocker shaft is installed to the front bearing cap, the oil hole side is *down* and that the end of the rocker shaft with the one oil hole is the *front end*. See the illustration for details.

10. Install the camshaft bearing cap bolts in numerical order, after lubricating each with clean engine oil. Torque the cap bolts to 14–15 ft. lb., starting at the center and working out.

NOTE: No. 1 and No.5 bearing caps are not stamped with an identification number.

11. Install the camshaft sprocket and belt, and torque the sprocket bolt to 47–54 ft. lb.

12. Install the rocker cover and apply sealer to the areas indicated in the illustration. Install the breather hoses.

ENGINE LUBRICATION

Oil Pan

REMOVAL & INSTALLATION

NOTE: The engine must be raised off its mounts and safely blocked, for the pan to clear the suspension crossmember.

1. Remove the underbody splash shield.

2. Unbolt the left and right engine mounts.

3. Jack up the engine under the bell housing.

4. Remove the oil pan.

5. Installation is the reverse of removal.

Rear Main Oil Seal

REPLACEMENT

NOTE: The rear main oil seal is located in a housing on the rear of the block. To replace the seal, remove the transmission and do the work from underneath the vehicle or remove the engine and do the work on the bench.

1. Remove the housing from the block.

2. Remove the separator from the housing.

3. Pry out the old seal.

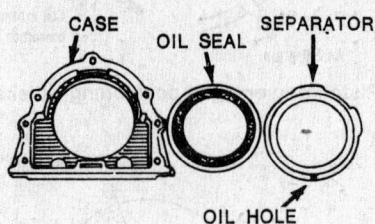

Rear main oil seal

4. Lightly oil the replacement seal. The oil seal should be installed so that the seal plate fits into the inner contact surface of the seal case. Install the separator with the oil holes facing down.

NOTE: To install the rear oil seal in the 2,346cc diesel engine, you must use a special oil seal installation tool, Mitsubishi part #MD 998376. The tool ensures that the seal is driven in evenly.

Oil Pump

REMOVAL & INSTALLATION

Gasoline Engines

See Timing Chain, Cover, Silent Shaft and Tensioner Removal and Installation procedure.

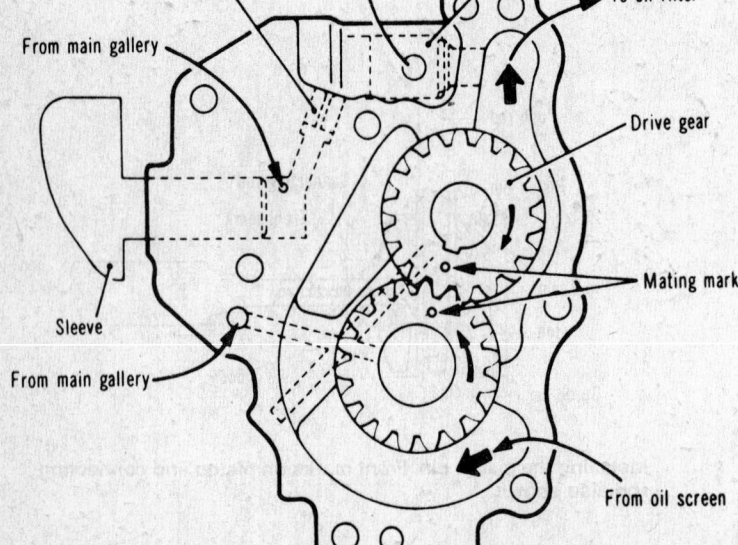

Gasoline engine oil pump layout. Notice mating marks on gears

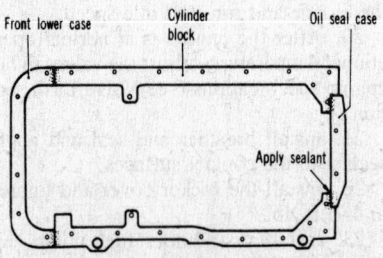

Apply oil pan sealant here—diesels

Diesel Engines

NOTE: The oil pump on the 2,346cc diesel is located behind the front lower case assembly and is driven directly by the crankshaft. Removal requires disassembly of the lower front case, and this is best done with the engine removed.

1. Drain the crankcase oil and remove the oil pan.
2. Remove the crankcase oil strainer.
3. Remove the timing belt assembly. Remove the belt drive gears.
4. Unbolt and remove the front upper case. The oil seal in the case can now be pried out carefully.

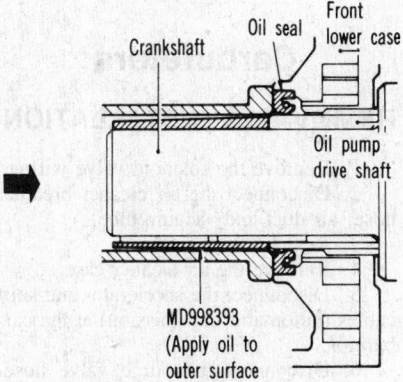

Installing diesel front cover oil seal

5. Remove the lower front case.
6. Before the oil pump inner and outer gears are removed, make alignment marks on both gears for reference during reassembly. Remove the oil pump.
7. Measure oil pump gear clearances. If side clearances are excessively large, or if the case or cover has evidence of step wear, replace the front case assembly.
8. Install the oil pump gears in the same direction as before disassembly, making sure the alignment marks made earlier line up. Lightly oil the gears with engine oil before installing the pump into the front case.
9. A special oil seal guide, Mitsubishi part #MD998385 or equivalent, is needed to install the oil seals into the front case. *New oil seals should always be installed.* A crankshaft seal installation tool and guide, Mitsubishi parts #MD998382 and MD998383 or equivalent are also needed.
10. Reverse the removal procedure for reassembly. You will need the above mentioned oil seal tools and guide to install the lower front case over the shafts. Also, use a new gasket behind the front case. Apply sealer to the four points illustrated on the oil pan flange.

1 Oil drain plug
2 Gasket
3 Bolt (24)
4 Oil pan gasket
5 Oil pan
6 Bolt (2)
7 Bolt (2)
8 Oil screen
9 Flange bolt (3)
10 Front upper case
11 Front upper case gasket
12 Oil seal
13 Left silent shaft
14 Plug cap
15 O-ring
16 Flange bolt
17 Flange bolt (7)
18 Front lower case assembly
  1 Flange bolt (3)
  2 Silent shaft gear cover
  3 Silent shaft driven gear
  4 Silent shaft drive gear
  5 Machine screw (5)
  6 Oil pump cover
  7 Oil pump outer gear
  8 Oil pump inner gear
  9 Plug
  10 Gasket
  11 Relief valve spring
  12 Relief valve
  13 Oil seal
  14 Front lower case
19 Front lower case gasket
20 Front oil seal
21 Oil pump gear drive shaft
22 Right silent shaft
23 Oil filter

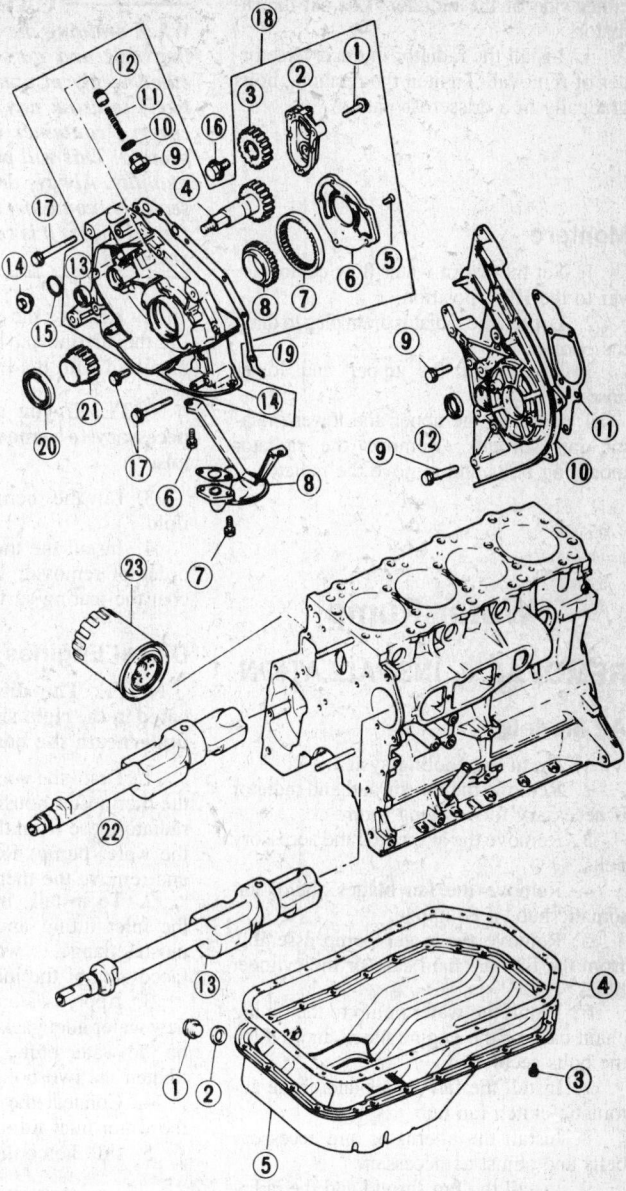

Diesel engine front cover assembly showing oil pump and silent shafts

ENGINE COOLING

CAUTION

When draining the coolant, keep in mind that cats and dogs are attracted by the ethelyne glycol antifreeze, and are quite likely to drink any that is left in an uncovered container or in puddles on the ground. This will prove fatal in sufficient quantity. Always drain the coolant into a sealable container. Coolant should be reused unless it is contaminated or several years old.

Radiator

REMOVAL & INSTALLATION

Pickup

1. Remove the splash panel from the bottom of the vehicle. Drain the radiator by opening the petcock. Remove the shroud on models so equipped.
2. Disconnect the radiator hoses at the engine. On automatic transmission vehicles, disconnect and plug the transmission lines to the bottom of the radiator.
3. Remove the two retaining bolts from

either side of the radiator. Lift out the radiator.

4. Install the radiator in the reverse order of removal. Tighten the retaining bolts gradually in a crisscross pattern.

Montero

1. Set the warm water flow control lever to the HOT position.
2. Loosen the radiator drain plug to drain the coolant.
3. Disconnect the upper and lower hoses.
4. Remove the upper and lower radiator fan shrouds. Remove the radiator mounting bolts and remove the radiator.

Water Pump

REMOVAL & INSTALLATION

All Models

1. Drain the cooling system.
2. Remove the fan shroud and radiator if necessary for working room.
3. Remove the alternator and accessory belts.
4. Remove the fan blades and/or automatic hub, if equipped.
5. Remove the water pump assembly from the timing chain case or the cylinder block.
6. Install the water pump to the timing chain case or the engine block and tighten the bolts securely.
7. Install the fan blades and/or the automatic clutch fan hub.
8. Install the alternator and accessory belts and adjust as necessary.
9. Install the fan shroud and the radiator, if removed.
10. Fill the cooling system, start the engine, and check for coolant leakage.

Thermostat

REMOVAL & INSTALLATION

Gasoline Engines

1. Drain the coolant below the level of the thermostat.

─────── CAUTION ───────

When draining the coolant, keep in mind that cats and dogs are attracted by the ethelyne glycol antifreeze, and are quite likely to drink any that is left in an uncovered container or in puddles on the ground. This will prove fatal in sufficient quantity. Always drain the coolant into a sealable container. Coolant should be reused unless it is contaminated or several years old.

---

2. Remove the two retaining bolts and lift the thermostat housing off the intake manifold with the hose still attached.

NOTE: If you are careful, it is not necessary to remove the upper radiator hose.

3. Lift the thermostat out of the manifold.
4. Install the thermostat in the reverse order of removal. Use a new gasket and coat the mating surfaces with sealer.

Diesel Engines

NOTE: The diesel thermostat is located in the right side of the water pump, underneath the hose inlet elbow.

1. Drain the coolant below the level of the thermostat housing. Remove the lower radiator hose from the water inlet fitting on the water pump. Remove the inlet fitting and remove the thermostat.
2. To install, insert the thermostat in the inlet fitting and make sure the thermostat flange is well seated in the spotfaced area of the inlet fitting.
3. Apply sealant to both sides of the new water inlet gasket and install the gasket on the water pump. Install the fitting and tighten the two bolts.
4. Connect the lower radiator hose to the water inlet fitting.
5. Fill the cooling system.

GASOLINE FUEL SYSTEM

Fuel Filter

REPLACEMENT

All models use an inline filter which should be replaced every 12,000 miles.

Fuel Pump

REMOVAL & INSTALLATION

1. Remove the fuel lines.
2. Unbolt the pump mounting bolts, and remove the pump, insulator, and gasket.
3. Coat both sides of a new insulator and gasket with sealer, and install the pump in the reverse order of removal.

TESTING

Disconnect the fuel line from the carburetor and attach a pressure tester to the end of the line. Crank the engine. The tester should show 4.6–6 psi.

Carburetors

REMOVAL & INSTALLATION

1. Remove the solenoid valve wiring.
2. Disconnect the air cleaner breather hose, air duct and vacuum tube.
3. Remove the air cleaner.
4. Remove the air cleaner case.
5. Disconnect the accelerator and shift cables (automatic transmission) at the carburetor.
6. Disconnect the purge valve hose. Remove the vacuum compensator, and fuel lines.
7. Drain the coolant.
8. Remove the water hose between the carburetor and the cylinder head.
9. Remove the carburetor.
10. Installation is the reverse of removal.

Refer to the Carburertor section in Unit Repair for adjustments and specifications.

DIESEL FUEL SYSTEM

Injection Pump

REMOVAL & INSTALLATION

1. Remove the timing belt upper cover.

DIESEL FUEL INJECTION SPECIFICATIONS

Injection Pump				Fuel Injectors	
Turning Direction	Injection Order	Injection Timing	Delivery Valve Opening Pressure	Injection Pressure	Injector Installation Torque
Clockwise, viewed from drive side	1-3-4-2	5°ATDC@ 1mm plunger stroke	306 psi	1,707 psi	44–50 ft. lbs.

2. Remove the nut and washer securing the injection pump sprocket.

NOTE: Be careful not to drop the nut and washer into the lower cover.

3. Turn the crankshaft to bring No. 1 piston to TDC on its compression stroke.

4. Use a suitable puller to loosen the sprocket from the taper section of the pump driveshaft. Do not remove the sprocket. Carefully set it in the timing belt lower cover with the belt engaged.

5. Remove the two water hoses from the wax element. Keep the end of the removed hose higher than the cylinder head to prevent coolant drainage.

6. Disconnect the boost compensator hose at the injection pump.

7. Remove the fuel injection pump. Make sure the delivery valves do not turn when loosening the pipe connections at the pump.

8. Remove the injection pump support bracket bolts.

9. Remove the injection pump mounting nuts and remove the injection pump from the engine.

───── **CAUTION** ─────

DO NOT rotate the crankshaft with the injection pump removed.

10. Installation is the reverse of removal. Make sure the timing marks on the camshaft sprocket and crankshaft pulley are aligned with their respective timing marks.

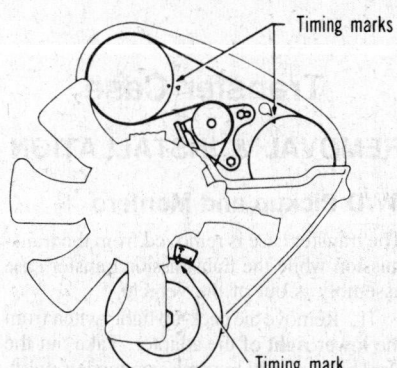

Timing marks set up for injection pump removal

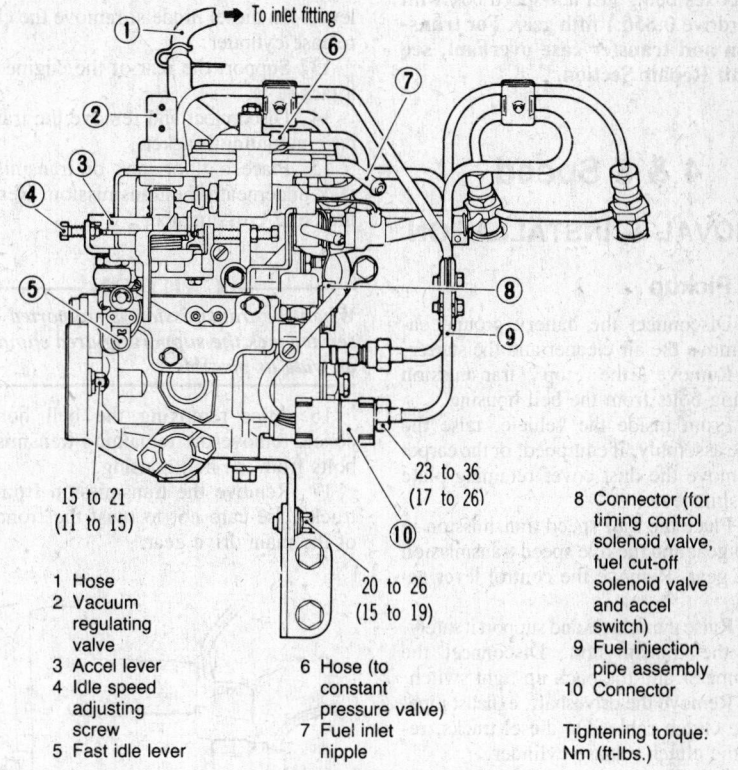

Diesel injection pump fittings

15 to 21
(11 to 15)

20 to 26
(15 to 19)

23 to 36
(17 to 26)

1 Hose
2 Vacuum regulating valve
3 Accel lever
4 Idle speed adjusting screw
5 Fast idle lever

6 Hose (to constant pressure valve)
7 Fuel inlet nipple

8 Connector (for timing control solenoid valve, fuel cut-off solenoid valve and accel switch)
9 Fuel injection pipe assembly
10 Connector

Tightening torque:
Nm (ft-lbs.)

→ To inlet fitting

11. After mounting the injection pump, carefully install the injection pump sprocket and belt. Make sure that the injection pump driveshaft key is not misplaced or dropped.

12. Adjust the injection timing and bleed the fuel system.

Fuel Injectors

REMOVAL & INSTALLATION

1. Remove the fuel delivery lines from the injectors and injection pump. Remove the lines as an assembly and cap all open fuel fittings on the injection pump immediately to prevent contamination with dirt or grease. *This is very important.*

2. Remove the fuel return pipe nuts, then remove the fuel return line.

3. Remove the injection nozzle assembly from the cylinder head with special tool

Removing injectors

(Mitsubishi part #MD998387 or a long, deepset socket).

4. Installation is the reverse of removal. Clean the area around each injector on the cylinder head before installing injectors, and fit a new nozzle tip gasket to each injector. Torque each injector to 44–50 ft. lb. Replace the heat shields.

NOTE: Exercise care when handling the fuel injector. It is a high precision part that is easily damaged by dirt or dropping.

MANUAL TRANSMISSION

The pickup uses both a 4-speed, the KM130, and a 5-speed, the KM132. The Montero uses only a 5-speed, the KM145. All three transmissions are virtually identical, the 5-speed boxes being just a 4-speed box with an overdrive 0.856:1 fifth gear. **For transmission and transfer case overhaul, see the Unit Repair Section.**

4 & 5 Speed

REMOVAL & INSTALLATION

2WD Pickup

1. Disconnect the battery ground cable, remove the air cleaner and the starter.
2. Remove the top transmission mounting bolts from the bell housing.
3. From inside the vehicle, raise the console assembly, if equipped, or the carpet and remove the dust cover retaining plate at the shift lever.
4. Place the four speed transmission in second gear and the five speed transmission in first gear. Remove the control lever assembly.
5. Raise the vehicle and support it safely. Drain the transmission. Disconnect the speedometer and the back up light switch.
6. Remove the driveshaft, exhaust pipe, and the clutch cable. On diesel trucks, remove the clutch release cylinder.
7. Support the transmission and remove the engine rear support bracket.
8. Remove the bell housing cover and bolts, move the transmission rearward, and lower it carefully to the floor. Remove the transmission from under the vehicle.
9. To install the transmission, reverse the removal procedure. Make sure the transmission is in the proper gear before installing the gear shift lever.

4WD Pickup

1. Disconnect the battery ground (negative) terminal.
2. Remove the air cleaner and starting motor.
3. Remove the transmission mounting bolts (two bolts on the upper side) from the bell housing.
4. In the cab, remove the lock screws and lift up the console box. In trucks without a console box, remove the carpet.
5. Remove the attaching screws and lift out the dust cover retaining plate.
6. Turn up the dust cover, and remove the control housing attaching bolts from the extension housing, and remove the control lever assembly.

7. Jack up the truck and support it with jackstands.
8. Drain the gearbox.
9. Remove the driveshafts. Disconnect the speedometer cable at the transmission.
10. Disconnect the back-up light switch harness and the 4WD indicator light switch harness, which is resting on the upper middle section of the transfer case.
11. Disconnect the front exhaust pipe.
12. On gasoline engined trucks, disconnect the clutch cable from the clutch control lever. On diesel models, remove the clutch release cylinder.
13. Support the rear of the engine with a jack.
14. Disconnect and remove the transfer case mounting bracket.
15. Place a floor jack or transmission jack underneath the transmission. Remove the No.2 crossmember.

CAUTION

When the transmission is supported on a service jack the supporting area should be as wide as possible.

---

16. After removing the bell housing cover, remove the remaining transmission bolts from the bell housing.
17. Remove the transmission from the truck. Use care not to twist the front end of the main drive gear.

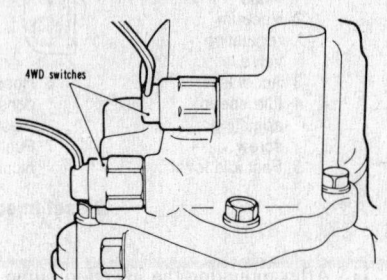

4WD switch location on manual transmissions

18. Reverse the above procedure for installation. Torque the transmission mounting bolts to 31–40 ft. lb.; the transmission-to-rear insulator bolts to 14–18 ft. lbs.; the transfer case support insulator-to-transfer case mounting bracket bolts to 14–18 ft. lb.; the transfer case mounting bracket bolts to 14–18 ft. lb. Set the transmission shift lever to the "neutral" position and the transfer case shift lever to the 4H position when installing the gear shift lever assembly. Adjust the clutch and fill the KM130 transmission with 2.2 qts. of hypoid gear oil; the KM132 with 2.4 qts.; the KM144 and KM145 transmission and transfer case with 2.3 qts., respectively.

Montero

1. Disconnect the negative cable from the battery.

2. Place the transmission lever in neutral, and the transfer case lever in 4H. Remove the gearshift lever assembly. Cover the opening with a shop rag to prevent any dirt from entering.
3. Jack up the truck and safely support it with jackstands. Remove the transfer case skid plate.
4. Drain the transfer case and transmission gearbox.
5. Remove the front and rear driveshafts.
6. Disconnect the speedometer cable, back-up light switch harness, and the 4WD indicator light harness, which rests on the top center of the transfer case.
7. Detach the clutch release cylinder from the transmission.
8. Remove the bell housing cover.
9. Detach the starter motor from the bell housing.

NOTE: On air conditioned Monteros, remove the front driveshaft, then lower the starting motor downward from underneath the vehicle to remove it.

10. Remove the front exhaust pipe mounting bracket.
11. Detach the engine support rear insulator from the No. 2 crossmember, and remove the No. 2 crossmember.
12. Support the transmission and transfer case assembly with a transmission jack or floor jack.
13. Unbolt the transfer case from the transfer case mounting bracket.
14. Remove the transmission mounting bolts from the engine.
15. Remove the transmission mounting bolts from the engine. With the jack still underneath the assembly and securely positioned, pull the transfer case and transmission assembly away from the engine.
16. Tilt the front of the transmission and transfer case assembly downward. Slowly lower the assembly with the jack, being careful that the rear of the transmission does not hit the No. 3 crossmember.

Transfer Case

REMOVAL & INSTALLATION

4WD Pickup and Montero

The transfer case is removed from the transmission while the transmission/transfer case assembly is out of the vehicle.

1. Remove the back-up light switch from the lower right of the adapter. Take out the steel ball (manual transmission models only).
2. Remove the plug from the right side of the transfer case and take out the select spring and the select plunger.
3. Remove the six bolts securing the control lever assembly and remove the control lever assembly and gasket.

4. On the four speed only, remove the plug from the top of the adaptor and take out the neutral return spring and plunger.

5. On five speed models, remove the plugs from the top of the adaptor and take out the resistance spring, steel ball, neutral springs and plungers.

6. On the four speed and automatic transmission transfer case, remove the bolts securing the transfer case adapter to the transmission.

7. On the four speed model only, with the change shifter tilted to the left, remove the control finger from the shift lug groove. Take out the transfer case assembly.

8. On the five speed model, drive out the lock pin from the change shifter using a ³/₁₆ in. punch. Remove the four bolts and two nuts securing the transfer case to the adapter, and remove the transfer case assembly from the adapter. Remove the change shifter from the control shaft.

9. Reverse the above procedure for installation. Make sure you mount the neutral return plungers and the springs in the hole on top of the adapter and tighten the plug until it is flush with the adapter surface.

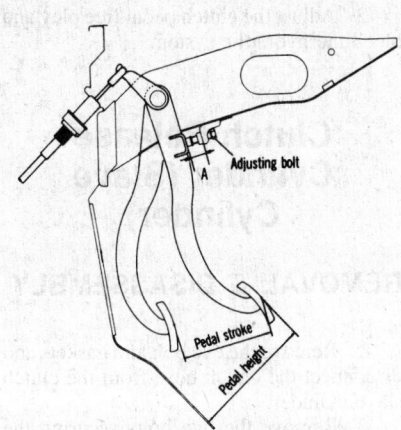

Adjusting clutch pedal height

CLUTCH

—— CAUTION ——

The clutch driven disc contains asbestos, which has been determined to be a cancer causing agent. Never clean clutch components with compressed air! Avoid inhaling any dust from any clutch surface! When cleaning clutch surfaces, use a commercially available brake cleaning fluid.

Clutch Cable

REMOVAL & INSTALLATION

All Gasoline Engine Pickups

1. Loosen the cable adjusting wheel inside the engine compartment while pulling the cable.

2. Loosen the clutch pedal adjusting bolt locknut and loosen the adjusting bolt.

3. Remove the cable end from the clutch throwout lever.

4. Remove the cable end from the clutch pedal.

5. Installation is the reverse of removal.

NOTE: Apply engine oil to the cable before replacing. Make sure the isolating pad is fitted on the cable after installation to keep the cable from rubbing the motor mount during operation.

Adjustment

PEDAL HEIGHT

1. Adjust the pedal height to the standard value with the adjusting bolt (see illustration), and check the pedal stroke and distance "A".

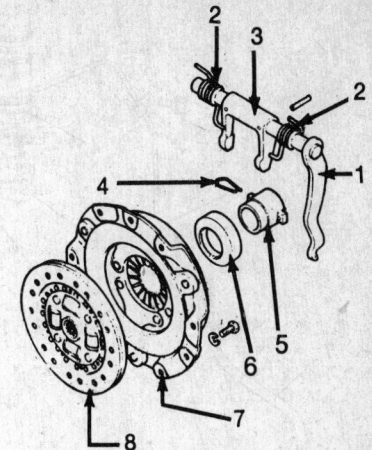

1. Clutch control shaft
2. Return spring
3. Clutch shift arm
4. Return clip
5. Release bearing carrier
6. Release bearing
7. Pressure plate assembly
8. Clutch disc

Exploded view of clutch assembly

NOTE: Insufficient pedal stroke results in only partial clutch release, causing hard gear shifting and gear grinding when shifting.

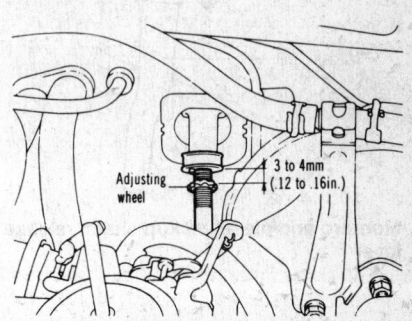

Adjusting the clutch cable

PEDAL HEIGHT ADJUSTMENT
Gasoline Engined Pickup Trucks

Description	Standard valve mm (in.)	
	2.0L	2.6L
Distance A	22 (.9)	20 (.8)
Pedal height	166 (6.5)	176 (6.9)
Pedal stroke	140 (5.5)	150 (5.9)

PEDAL HEIGHT ADJUSTMENTS
Diesel Pickups and Montero

Description	Standard value mm (in.)	
	Diesel pickups	Montero
Pedal height A	174 (6.8)	186–191 (7.3–7.5)
Pedal free play B	10 to 15 (.4 to .5)	5–10 (.2–.4)
Pedal to floorboard C	22 (.9) or more	—

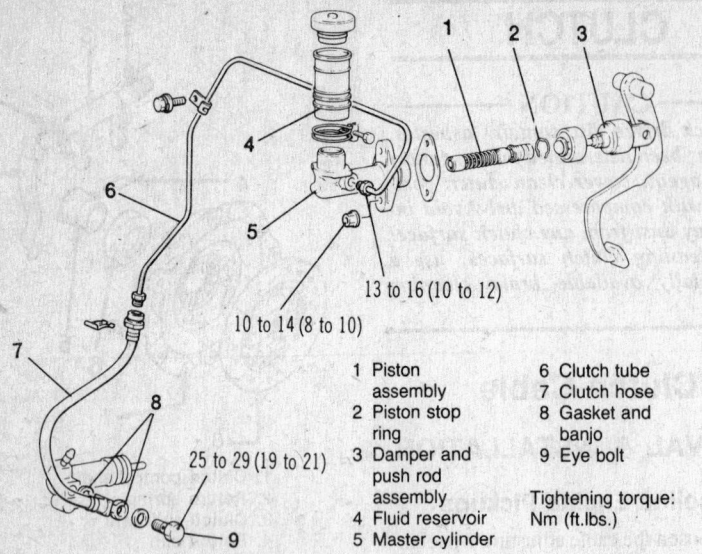

13 to 16 (10 to 12)

10 to 14 (8 to 10)

25 to 29 (19 to 21)

1 Piston assembly	6 Clutch tube
2 Piston stop ring	7 Clutch hose
3 Damper and push rod assembly	8 Gasket and banjo
4 Fluid reservoir	9 Eye bolt
5 Master cylinder	Tightening torque: Nm (ft.lbs.)

Hydraulic clutch assembly—diesel trucks and Montero

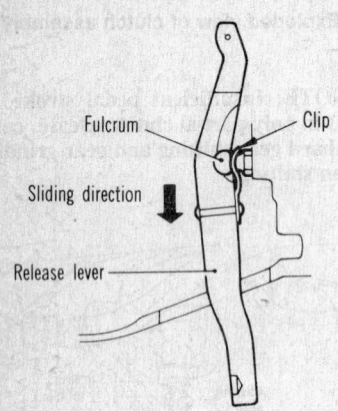

Montero and diesel pickup clutch release lever

2. In the engine compartment, at the fire wall, pull out the clutch cable a little and adjust the cable by turning the adjusting wheel until it is 0.12–0.16 in. from the insulator.

3. Clutch pedal free play should be within 0.8–1.4 in.

Hydraulic Clutch

All Turbo Diesel Models and Montero utilize a hydraulic clutch system. Clutch pedal pressure is converted, through the master cylinder, into fluid pressure, which operates a slave cylinder at the clutch. The slave cylinder operates the clutch control lever and shift arm, which moves the clutch release bearing and operates the clutch.

The master cylinder can be disassembled and rebuilt if worn and/or leaking. The slave cylinder should be replaced as a unit if defective.

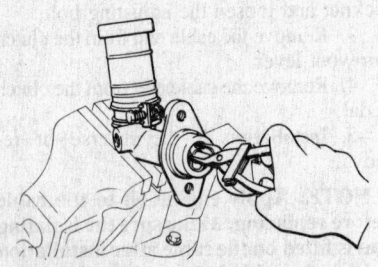

Removing master cylinder piston stop ring, hydraulic clutches

Master Cylinder

REMOVAL
1. Disconnect the clutch fluid tube from the master cylinder. Slowly depress the clutch pedal to drain the fluid.
2. Remove the cotter pin connecting the master cylinder pushrod to the clutch pedal.
3. Remove the two retaining nuts and remove the master cylinder.

DISASSEMBLY
1. Using snaping pliers, remove the piston stop ring. Pull out the piston assembly.
2. Wash the master cylinder, piston and cup in clean brake fluid. Never use solvents. Use care not to damage the cylinder, piston and piston cup.

NOTE: Do not disassemble the piston assembly.

INSPECTION
1. Check the inner surface of the master cylinder and the outer surface of the piston. Replace the piston if damaged or scored.
2. Check the master cylinder-to-piston

clearance. If clearance exceeds the service limit (0.0059 in.), replace either the piston or master cylinder unit. Standard piston diameter is 0.6231–0.6248 in.; standard master cylinder inside diameter is 0.6248 to 0.6265 in.

3. Check the piston cup for damage, deformation and wear. Replace the cup if at all defective. Also check the return springs for loss of tension and damage. Replace any part that is any related parts such as the return spring, piston cup and piston require replacement, it is necessary to replace the piston assembly.

ASSEMBLY

After checking clearances, assemble the master cylinder in the reverse order of disassembly. Always apply clean brake fluid to the master cylinder bore and piston cup prior to assembly.

INSTALLATION

1. Reverse the order of removal for installation. Tighten the master cylinder and clutch tube flare nut to 8–10 ft. lbs. and 10–12 ft. lb., respectively.
2. Adjust the clutch pedal free play and bleed the hydraulic system.

Clutch Release Cylinder (Slave Cylinder)

REMOVAL & DISASSEMBLY

1. Remove the eye bolt and gaskets and disconnect the clutch hose from the clutch slave cylinder.
2. Remove the two bolts securing the clutch slave cylinder and clutch housing, and remove the slave cylinder assembly.
3. Remove the rubber boot and push rod and take out the piston and spring.

INSPECTION

1. Check inside the slave cylinder body for scoring, rust or unusual wear.
2. Check the piston cup for wear or deformation.
3. Check the piston for rust or scoring.

REASSEMBLY & INSTALLATION

1. Insert the spring into the slave cylinder.
2. Liberally apply clean brake fluid to the outer surface of the piston, piston cup and cylinder bore.
3. Install the piston and piston cup into the release cylinder.
4. Install the push rod and boot.

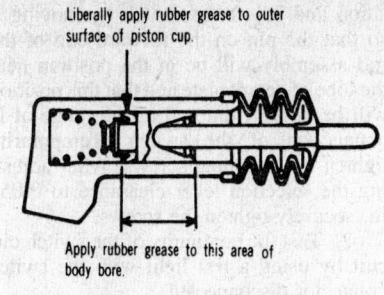

Liberally apply rubber grease to outer surface of piston cup.

Apply rubber grease to this area of body bore.

Slave cylinder assembly

5. Install the slave cylinder to the clutch housing and tighten the two bolts to 22–30 ft. lb.

6. Connect the clutch fluid line to the cylinder and tighten the eye bolt.

7. Bleed the clutch system.

HYDRAULIC CLUTCH BLEEDING

Whenever the clutch fluid line, clutch hose and/or the clutch master cylinder have been removed, or if the clutch pedal feels spongy (indicating air in the system), bleed the system. This is much the same as bleeding an hydraulic brake system.

1. Loosen the bleeder screw on the clutch slave cylinder. You'll have to jack up the truck and safely support it with jackstands for this.

2. Connect a length of rubber tubing to the bleeder nipple, with the other end of the tube submerged in a clear glass jar filled with clean brake fluid. Make sure the clutch master cylinder is topped up with fluid. Keep it topped up throughout the procedure.

3. Have an assistant push the clutch pedal down slowly until air bubbles stop coming out of the rubber hose in the jar. All air is now expelled from the system.

4. Have the assistant hold the clutch pedal down until while you retighten the bleeder screw.

5. Refill the clutch master cylinder with DOT 3 brake fluid.

—————— CAUTION ——————
Never mix brake fluids. Use only approved DOT 3 fluid.

CLUTCH PEDAL ADJUSTMENT

If clutch pedal height from floor or free play is not within the standard value, adjust as follows:

1. Turn the pedal adjusting bolt back to a position where it does not contact the pedal arm.

2. Loosen the push rod lock nut and adjust the pedal height to the standard value by turning the push rod.

3. Turn the pedal adjusting bolt until it comes into contact with the pedal arm, and then tighten the lock nut.

4. After making the adjustment depress the clutch pedal several times and check the clutch pedal-to-floorboard clearance. The pedal on diesel pickups should be within 22mm (0.9 in.) or more when the clutch is fully disengaged. The Montero pedal should be close to this same distance. If the pedal-to-floor clearance is less then this, air may be in the hydraulic system. If so, the system should be bled. Also, the clutch plates could be worn.

Clutch Disc

REPLACEMENT

1. Remove the transmission as outlined in Manual Transmission Removal and Installation.

NOTE: It is recommended that a clutch aligning tool be inserted in the clutch hub to prevent dropping of the clutch disc during disassembly.

2. Remove the pressure plate bolts, pressure plate and clutch disc.

3. From inside the transmission bell housing, remove the return spring clip and remove the release bearing assembly.

4. If necessary, remove the release control lever and spring pin with a $^3/_{16}$ inch punch. Remove the control lever shaft assembly and clutch shift arm, two felt packings and two return springs.

5. Installation is the reverse of removal.

AUTOMATIC TRANSMISSION

Pan and Filter

REMOVAL & INSTALLATION

1. Raise and support the vehicle.

2. Loosen the pan bolts from one end to the other allowing the fluid to drain out.

3. Unbolt the old filter from the pan.

4. Clean the pan and install a new filter. Tighten the filter bolts to 35 in.lb.

5. Install the pan and new gasket. Torque the pan bolts to 6–9 ft. lb.

6. Add four quarts of Dexron® II fluid, start the engine and move the lever through all positions, pausing momentarily in each. Add enough fluid to bring the level to the full mark on the dipstick.

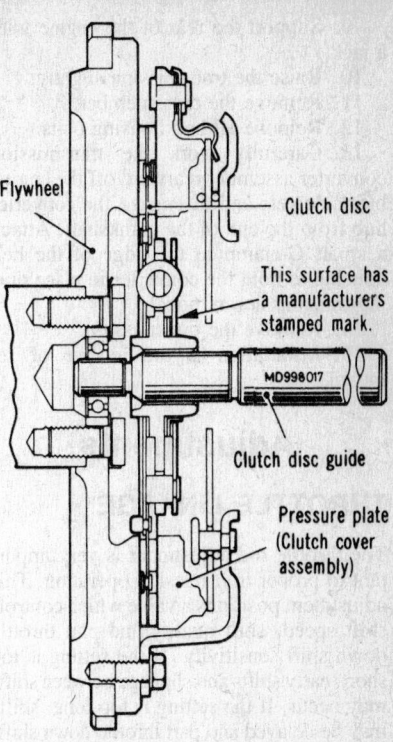

Flywheel

Clutch disc

This surface has a manufacturers stamped mark

MD998017

Clutch disc guide

Pressure plate (Clutch cover assembly)

Installing the clutch disc—use clutch disc guide as shown

Transmission

REMOVAL & INSTALLATION

NOTE: The transmission and converter must be removed as an assembly. Otherwise, the converter drive plate, pump bushing, or oil seal may be damaged. The drive plate will not support a load. Therefore, none of the weight of the transmission should be allowed to rest on the plate during removal.

1. Disconnect the battery ground cable, drain the transmission, and remove the cooler lines at the transmission.

2. Remove the starter and cooler line bracket.

3. Rotate the crankshaft clockwise and remove the bolts attaching the torque converter to drive plate.

4. Remove the driveshaft(s).

5. Disconnect the gearshift rod and torque shaft.

6. Disconnect the throttle rod from the lever at the left side of the transmission. Remove the linkage bellcrank from the transmission if so equipped.

7. Remove the oil filler tube and speedometer cable.

8. On 2WD models, disconnect the back-up light switch harness. On 4WD models, disconnect the back-up light switch harness and 4WD indicator light switch harness which are located near the upper middle of the transfer case.

9. Support the rear of the engine with a jack.

10. Raise the transmission slightly.

11. Remove the crossmember.

12. Remove all bell housing bolts.

13. Carefully work the transmission converter assembly rearward off the engine block dowels and disengage the converter hub from the end of the crankshaft. Attach a small C-clamp to the edge of the bell housing to hold the converter in place during transmission removal.

14. Remove the transmission.

15. Installation is the reverse of removal.

Adjustments

THROTTLE LINKAGE

The throttle rod adjustment is very important to proper transmission operation. This adjustment positions a valve which controls shift speed, shift quality and part throttle down shift sensitivity. If the setting is too short, early shifts and slippage between shifts may occur. If the setting is too long, shifts may be delayed and part throttle down shifts may be very sensitive.

1. Warm up the engine until it reaches normal operating temperature. With the carburetor automatic choke disengaged from the fast idle cam, adjust the engine idle speed by rotating speed adjusting screw (SAS). See Tune-Up Procedures at the beginning of this section.

2. Loosen the bolts on the linkage so that both rod "B" and "C" can slide properly.

3. Lightly push rod "A" or the transmission throttle lever and rod "C" toward the idle stopper, and set the rods to the idle position. Tighten the bolt securely that connects rods "B" and "C".

4. Make sure that when the carburetor throttle valve is wide open, the transmission throttle lever smoothly moves from the IDLE to the WIDE OPEN position (from 47.5° to 54°), and that there is some range in the lever stroke.

5. Also make sure that when the throttle linkage alone is slowly returned from the fully closed position, the transmission throttle lever completely returns to *idle* by return spring force.

KICKDOWN BAND

The kickdown band adjusting screw is located on the left side of the transmission case.

1. Loosen the locknut and back off approximately 5 turns. Test the adjusting screw for free turning in the transmission case.

2. Tighten the adjusting screw to 69 inch lb.

3. Back off the adjusting screw 3 $\frac{1}{2}$ turns from Step 2. Tighten the locknut to 35 ft. lb.

LOW & REVERSE BAND

1. Raise the vehicle, drain the transmission fluid and remove the pan.

2. This transmission has an allen socket adjustment screw at the servo end of lever. After removing the locknut this screw is tightened to 43 inch lb. torque then backed off 7 turns. Tighten the locknut to 30 ft. lb.

3. Reinstall the pan.

NEUTRAL SAFETY SWITCH

1. When testing the safety switch, check to see if the switch has been properly installed. Move the selector lever into N position and adjust the switch by moving it so that the pin on the forward end of the rod assembly will be in the position near the lobe of detent plate and that this position will be at the front end of the range of N connection of the switch. Temporarily tighten the attaching screws. After adjusting the selection lever clearance to 0.059 in. securely tighten the screws.

2. Test the continuity of the switch circuit by using a test light with the switch connector disconnected.

SHIFT LINKAGE ADJUSTMENT

NOTE: To adjust the shift linkage, the control cover must be removed.

1. Remove the shift handle assembly from the lever.

2. Take the position indicator assembly out upward. Remove the position indicator lamp.

3. Disconnect the control rod from the arm. Remove the lever bracket assembly.4.

Installation is the reverse of removal. If the proper turning effort (13–29 inch lb.) is not obtained, adjust it by using a selective wave washer of proper size. When the turning effort at the pivot A is checked, the pin at the forward end of the rod assembly must not slide with the detent plate. If the arm is loose, the bushing should be replaced.

DRIVE LINE

Driveshaft

REMOVAL & INSTALLATION

2WD Pickup Trucks

1. Make mating marks on the flange yoke and the differential companion flange.

2. Remove the bolts connecting the flange yoke to the differential companion flange, and remove the nuts attaching the center bearing assembly.

3. Remove the driveshaft by pulling it out. Installation is reverse of removal.

NOTE: When the sleeve yoke end of the driveshaft is pulled out from the transmission extension housing, transmission oil will flow out, if the front of the truck is raised higher than the rear.

—— CAUTION ——
When removing the driveshaft, be careful not to damage the oil seal lip and see that no foreign substance is present in the lip area.

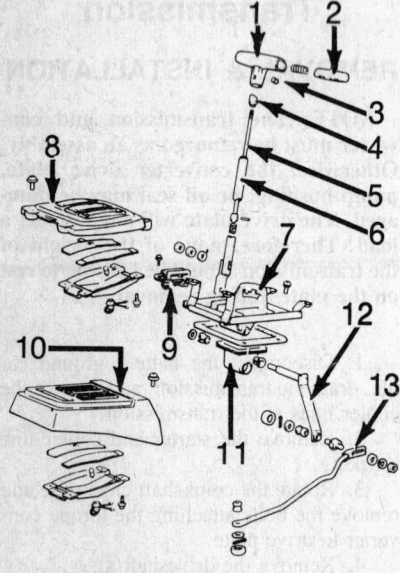

1. Selector handle
2. Push button
3. Set screw
4. Rod adjusting cam
5. Selector lever rod
6. Selector lever
7. Detent plate
8. Position indicator cover (for sports)
9. Inhibitor switch
10. Position indicator cover
11. Lever bracket cover
12. Control cover
13. Control rod

Exploded view of automatic transmission shift control

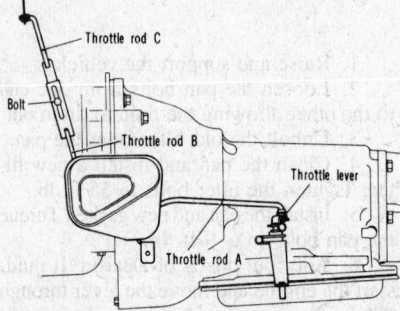

Throttle rod adjustment—automatic transmission

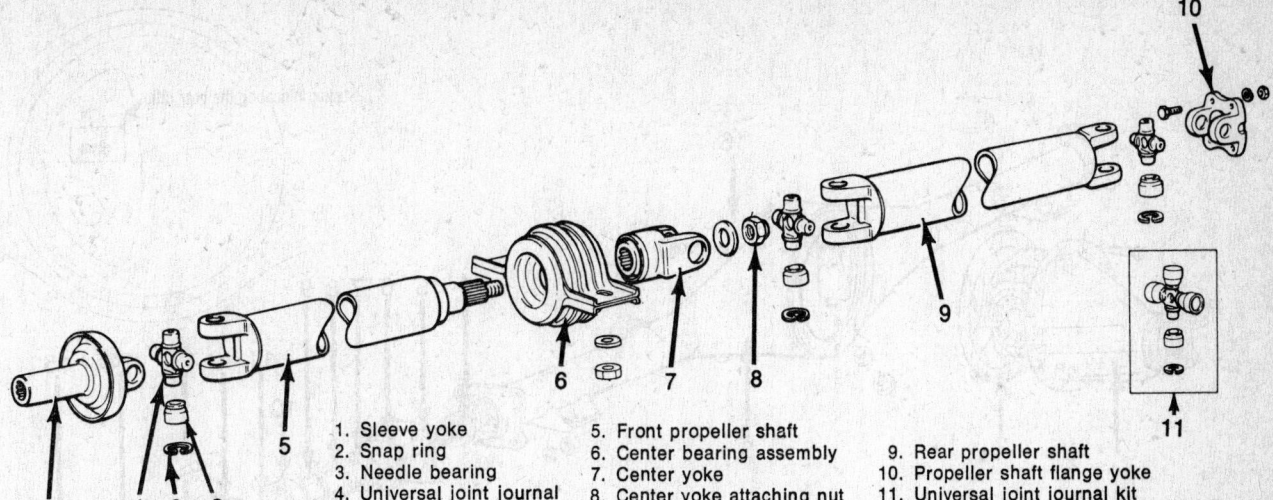

1. Sleeve yoke
2. Snap ring
3. Needle bearing
4. Universal joint journal

5. Front propeller shaft
6. Center bearing assembly
7. Center yoke
8. Center yoke attaching nut

9. Rear propeller shaft
10. Propeller shaft flange yoke
11. Universal joint journal kit

Exploded view of driveshaft

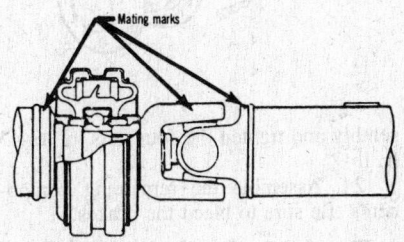

When installing center bearing align three mating marks

4WD Pickups and Montero

1. Set the free wheeling hubs to UN-LOCKED, and set the transfer case gear shift lever to the 2H position.

2. Make mating marks on the U-joint flange yoke and the companion flange on the differential and/or transfer case.

3. Remove the bolts connecting the flange yoke to the differential and/or transfer case companion flange.

4. Remove the driveshaft by pulling it out. Be careful not to damage the oil seal lip.

NOTE: When the driveshaft is pulled out of the transmission extension housing or the transfer case, transmission or transfer case oil may leak out depending on the angle of the vehicle. Be ready to catch any spills.

Center Bearing

REMOVAL & INSTALLATION

2WD Models

1. Remove driveshaft.
2. Disconnect the center universal joint.

3. Remove the nut holding the center yoke and remove the yoke. Remove the center bearing bracket from the bearing by prying on it.

4. Remove the center bearing using a gear puller.

NOTE: The center bracket and the mounting rubber are welded together and must be replaced as a unit.

5. Fill the bearing grease cavity with multipurpose grease.

6. Partially insert the center bearing into the shaft and install the bracket to the bearing.

7. Verify that the bracket mounting rubber is properly fitted in the bearing groove.

8. Refit the center yoke, making sure you align the notch on the yoke with the notch on the front driveshaft. Replace the attaching nut and tighten to 116–159 ft. lb.

9. Replace the center universal joint, making sure you align the notch on the rear driveshaft with the notch on the yoke.

10. Install the driveshaft.

NOTE: The manufacturer suggests that a new center yoke locking nut be used when the center bearing is removed.

REAR DRIVE AXLE

Axle Shaft, Bearing and Seal

REMOVAL & INSTALLATION

All Rear Axles

1. Jack up the vehicle and support it on stands.

2. Remove the rear wheels and brake drums.

3. Disconnect the brake line from the wheel cylinder and plug it to prevent fluid loss.

4. Remove the four nuts behind the brake backing plate holding the bearing case to the axle housing assembly.

5. Remove the backing plate, bearing case and the axle shaft as an assembly.

NOTE: It may be necessary to use a slide hammer to remove the assembly.

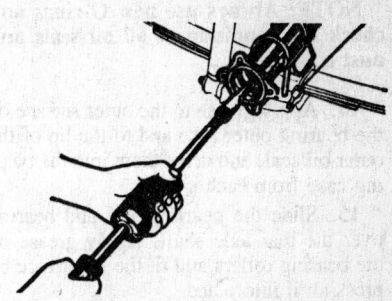

Removing the oil seal

6. Remove the O-ring and the bearing preload shims. Save the preload shims, as you will need them for reassembly.

7. Remove the oil seal with a hooked slide hammer.

8. To remove the axle shaft bearing, remove the notched locknut. This calls for a special tool, but you should be able to use a brass drift to knock it loose.

9. Remove the lock washer and plain washer.

10. Screw the lock nut back on to the axle shaft about three turns.

11. It will be necessary to fabricate a metal plate that fits over the axle shaft and butts the lock nut. Drill four holes in the plate that align with the four bearing case

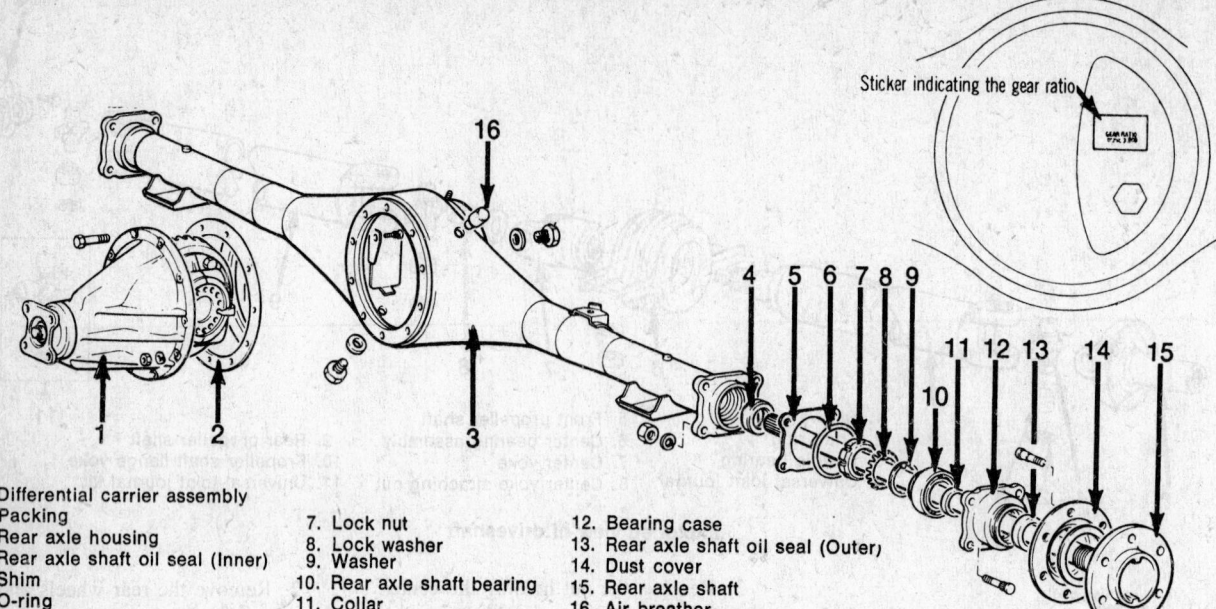

Sticker indicating the gear ratio

1. Differential carrier assembly
2. Packing
3. Rear axle housing
4. Rear axle shaft oil seal (Inner)
5. Shim
6. O-ring
7. Lock nut
8. Lock washer
9. Washer
10. Rear axle shaft bearing
11. Collar
12. Bearing case
13. Rear axle shaft oil seal (Outer)
14. Dust cover
15. Rear axle shaft
16. Air breather

Exploded view of drive axle assembly

studs and fit the plate. Refit two nuts and washers to the bearing case studs diagonally across from each other and tighten them evenly to free the bearing case and the bearing.

12. Use a hammer and drift to remove the bearing outer race from the bearing case.

13. Remove the outer oil seal from the bearing case.

NOTE: Always use new O-rings and check the condition of all oil seals and dust covers.

14. Apply grease to the outer surface on the bearing outer race and to the lip of the outer oil seal, and drive them into the bearing case from each side.

15. Slide the bearing case and bearing over the rear axle shaft. Apply grease on the bearing rollers and fit the inner race by pressing it into place.

─── **CAUTION** ───

Be careful not to damage or deform the dust cover.

16. Pack the bearing with grease.

17. Install the washer, the crowned lock washer and the lock nut in the order just given and tighten the lock nut to 130–159 ft. lb. if possible.

18. Bend the tab on the lock washer into the groove on the lock nut. If the tab and the groove do not line up, slightly tighten the lock nut until they do.

19. Drive the new inner oil seal into place after greasing it and refit the assembly. Be sure to fit the O-ring and shim and apply silicone rubber sealant to the bearing case face.

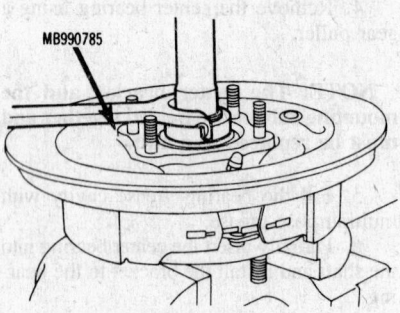

MB990785

Removing axle shaft locknut

NOTE: Be sure to bleed the brakes before road testing!

20. To adjust preload: Begin with the left side rear axle assembly and insert a 0.04 in. shim between the bearing case and the axle shaft housing. Tighten the four nuts to 36–43 ft. lb.

NOTE: Be sure to fit the O-ring and apply sealant.

21. Install the right side axle assembly into the right side housing without its shim and O-ring. Tighten the four nuts to 0.4 ft. lb.

22. Using a flat blade feeler gauge, measure the gap between the bearing case and the axle housing face. It should range between 0.002–0.008 in. Record the measurement.

23. Remove the axle shaft and select a shim that is the same thickness as the gap between the faces just measured, plus a shim with a thickness from 0.002–0.0079 in. and install them on the housing. Fit the O-ring and apply sealant. Fit the axle as-

sembly and tighten the four nuts to 36–43 ft. lb.

24. Assemble the remaining components. Be sure to bleed the brakes.

Rear Axle Assembly

REMOVAL & INSTALLATION

1. Loosen the rear wheel hub nuts and jack up the truck. Support the truck on jackstands placed forward of the rear spring front brackets.

2. Remove the rear wheels and remove the driveshaft.

NOTE: Support the differential housing with a jack to keep a slight amount of pressure on the springs.

3. Loosen the joint between the brake hose and the brake line and remove the stops to disconnect the brake hose. Plug the lines to prevent fluid loss.

4. Disconnect the rear cable of the parking brake at the balancer (refer to Brake section for procedure.)

5. Remove the shock absorbers, and the spring seats after removing the spring U-bolts.

6. Remove the spring shackle pin nuts and the shackle plate.

NOTE: The axle assembly will be supported solely by the jack under the differential case. Be careful not to allow it to drop.

7. With an assistant holding the axle assembly, slowly lower it to the ground.

8. Installation is the reverse of removal. Bleed the brakes after assembly.

AXLE SHAFT ASSEMBLY PRELOAD SHIMS

Part No.	Thickness of shim	
	mm	in.
MB092491	0.05 ± 0.005	.0020 ± .0002
MB092492	0.10 ± 0.010	.0040 ± .0004
MB092493	0.20 ± 0.015	.0079 ± .0006
MB092494	0.30 ± 0.020	.0118 ± .0008
MB092495	0.50 ± 0.025	.0197 ± .0010
MB092496	1.00 ± 0.040	.0394 ± .0016
MB092497	1.50 ± 0.050	.0591 ± .0020
MB092498	2.00 ± 0.055	.0787 ± .0022

FRONT DRIVE AXLE

Free Wheeling Hubs

REMOVAL & INSTALLATION

Manual Locking Hubs

1. Set the control handle to the FREE position. Unbolt the six cover bolts and remove the hub cover.
2. Using snapring pliers, remove the snapring from the driveshaft.
3. Remove the hub assembly from the front wheel.
4. To install, apply a semi-drying sealant to the front hub mounting surface of the hub body assembly, then torque the bolts to 36–43 ft. lb.
5. Measure driveshaft endplay using a dial indicator as shown in the illustration. If endplay exceeds 0.008–0.020 in., install a spacer on the shaft end so that the measurement will be within specifications. These are available from Dodge or Mitsubishi dealers.
6. Install the snapring. Install the hub cover.

Automatic Locking Hubs

1. Remove the hub cover. If the cover cannot be loosened by hand, protect the cover with a shop towel to avoid damaging it and use an oil filter strap wrench to loosen it.
2. Remove the O-ring from the hub cover.
3. Using snapring pliers, remove the snapring and spacer.
4. Remove the hub.
5. To install, apply a semi-drying sealer to the hub surface.

NOTE: Make sure there is no excess sealer on the outside of the hub.

6. Align the key of brake "B" in the illustration with the slot in the knuckle spindle. Loosely install the automatic hub assembly.

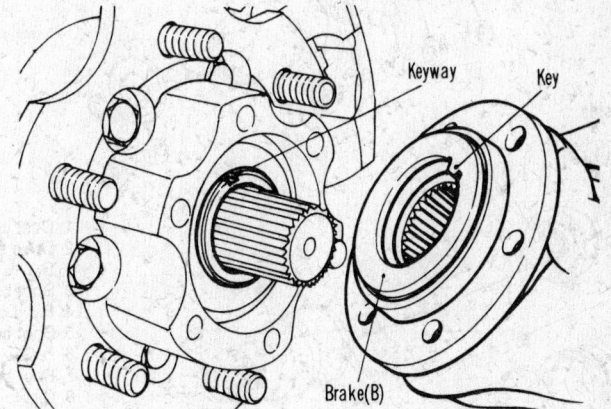

Free-wheeling hub installation. Notice keyway

7. Make sure that the hub and the free wheeling hub assembly are in close contact when the assembly is forced lightly against the hub. If not, turn the hub until close contact is obtained.
8. Tighten the free wheeling hub mounting bolts to 37–43 ft. lb.
9. Apply a suitable grease to the O-ring before mounting it into the cover.
10. Install the cover securely.

Outer Axle Shafts

REMOVAL

1. Remove the wheel and tire.
2. Remove the front brake caliper assembly. Do not disconnect the brake hose. Hang the caliper close by with a wire hook.
3. Remove the free wheeling hub cover assembly and remove the snapring from the driveshaft.
4. Remove the knuckle and the front hub together as a unit.
5. Remove the outer shaft as follows: for the left side on all 4WD pickups and Monteros, pull the shaft out of the differential carrier assembly. When pulling the left shaft from the differential carrier assembly, be careful that the shaft splines do not damage the oil seal.
6. On 4WD pickups jack up the right lower suspension arm. Remove the right shock absorber.

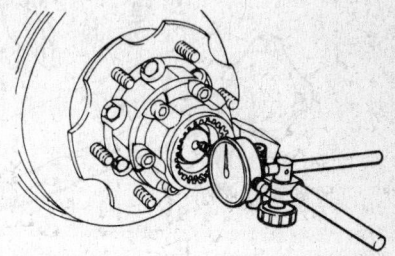

Use a dial indicator to measure axial drive shaft play—4WD models

NOTE: Do not lower the jack while disconnecting the shock absorber or after it is disconnected. The jack should not be removed until the upper part of the shock has been reconnected to the arm post of the side frame.

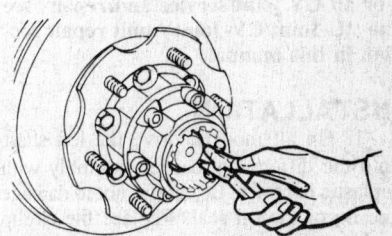

Removing snap-ring from front hub—4WD models

For left side

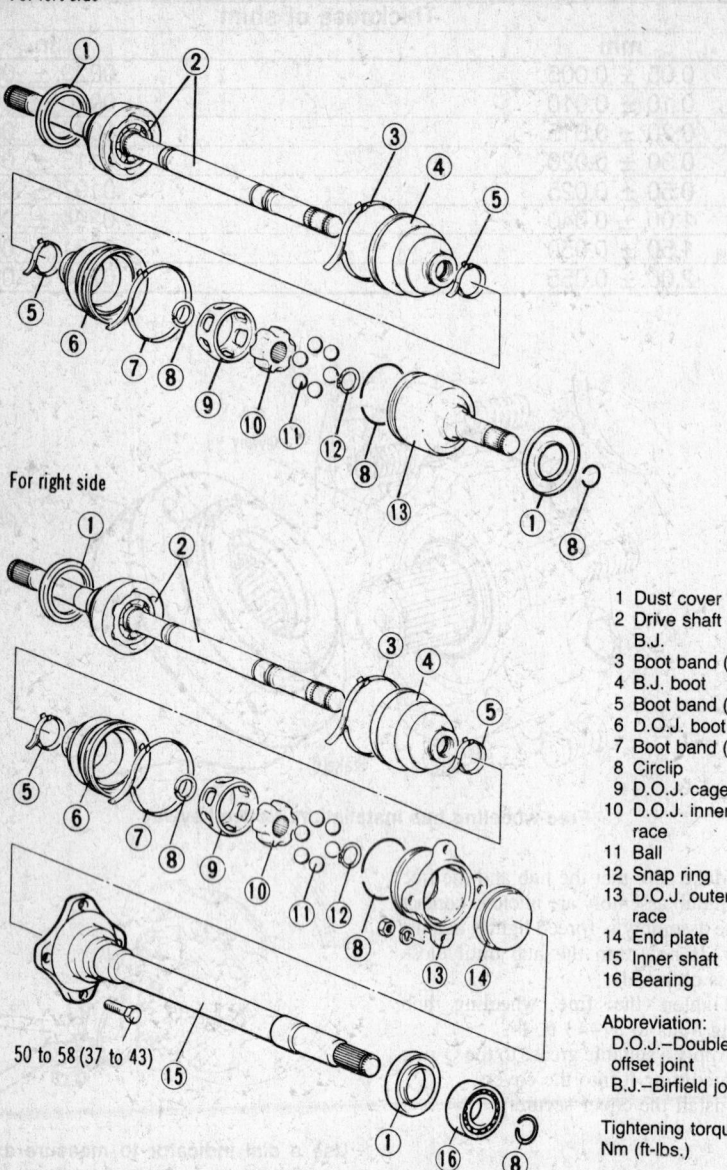

For right side

1 Dust cover
2 Drive shaft and B.J.
3 Boot band (A)
4 B.J. boot
5 Boot band (C)
6 D.O.J. boot
7 Boot band (B)
8 Circlip
9 D.O.J. cage
10 D.O.J. inner race
11 Ball
12 Snap ring
13 D.O.J. outer race
14 End plate
15 Inner shaft
16 Bearing

Abbreviation:
D.O.J.–Double offset joint
B.J.–Birfield joint

Tightening torque:
Nm (ft-lbs.)

50 to 58 (37 to 43)

Axle shaft and CV-joint assemblies—4WD models and Montero

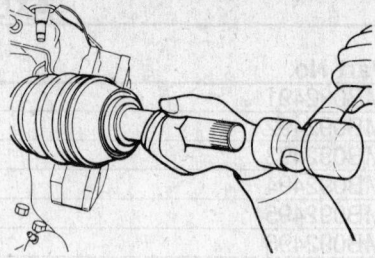

Use a plastic mallet to drive in left drive shaft

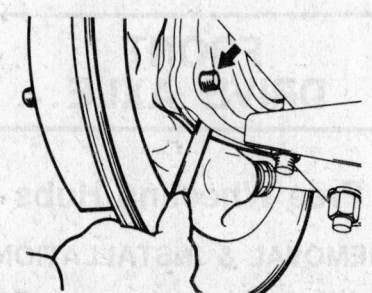

Measuring clearance between bearing case and axle housing face

7. On 4WD pickups, detach the right shaft from the inner shaft assembly and remove the shaft.

8. On Monteros, the right side shock absorber doesn't have to be removed. Detach the shaft from the differential carrier inner shaft, and remove it.

For all CV joint service and repair, see the "U-Joint/CV-Joint" unit repair section in this manual.

INSTALLATION

1. On all models, drive the left shaft into the differential carrier assembly with a plastic hammer. Be careful not to damage the lip of the oil seal. Replace the circlip on the spline with a new one.

2. Mount the knuckle together with the front hub assembly. Adjust the outer shaft

endplay using a dial indicator. Play should be within 0.008–0.020 in. Install a spacer (available in 0.012 in. increments from Dodge and Mitsubishi dealers) if endplay exceeds the above limits.

3. Install the right outer shaft to the inner shaft, and torque to 37–43 ft. lb. On 4WD pickup trucks, install the right shock absorber.

4. Install the knuckle and front hub assembly.

Inner Axle Shafts

REMOVAL & INSTALLATION

1. Slightly raise the lower suspension arm on a jack.

2. Remove the mounting nut from the top of the shock absorber and then detach the shock absorber from the crossmember. The shock must be removed when working on either side.

NOTE: When removing the shock absorber, do not lower the jack. Do not remove the jack until the top of the shock absorber is reattached to the crossmember.

3. Remove the outer shaft.

4. Attach a special driveshaft puller, Mitsubishi part #MB990906 or equivalent, to the inner shaft flange. Pull the inner shaft from the front differential carrier.

NOTE: When pulling the inner shaft from the carrier, be careful that the spline part of the inner shaft does not damage the oil seal.

5. If necessary, remove the housing tube.

6. Check for unusual wear or discoloration on the inner shaft.

7. Install the housing tube onto the front differential carrier and differential mounting bracket.

8. Drive the inner shaft into the front differential with the same special tool used to remove the shaft.

NOTE: Replace the circlip on the spline part of the inner shaft with a new one. Use care not to damage the lip of the oil seal.

9. Install the right outer shaft.

FRONT SUSPENSION— 2 WD

Coil Spring

REMOVAL & INSTALLATION

1. Raise the front of the vehicle and support it on jackstands.
2. Remove the wheel.
3. Remove the shock absorber (see below for procedures).
4. Remove stabilizer and strut bar (see below for procedures).
5. Compress the coil spring with a spring compressor.
6. Remove the relay rod from the steering arm.
7. Remove the upper knd lower ball joints using a ball joint remover.
8. Remove the coil spring.
9. Installation is the reverse of removal.

NOTE: The coil springs are color coded. The left side spring has a green band on it and the right side spring has a pink band on it. Do not mix the left and right springs.

10. Tighten the ball joint castle nuts to: upper, 43–65 ft. lb.; lower, 87–130 ft. lb.

Shock Absorbers

REMOVAL & INSTALLATION

1. Raise the vehicle and support it on jackstands.
2. Remove the wheel.
3. Remove the double lock nuts at the top of the shock absorber along with the rubber washer and its metal caps.
4. Remove the two bolts at the bottom of the shock absorber and withdraw the shock absorber through the bottom arm.
5. Installation is the reverse of removal. Be sure to refit all of the rubber cushion washers and their metal caps in the correct order. Tighten the upper shock absorber nut to 9–13 ft. lb. and install the lock nut. Tighten the two lower shock absorber bolts to 6–9 ft. lb.

Steering Knuckle

REMOVAL & INSTALLATION

1. Raise the vehicle and support it on jackstands.
2. Remove the wheel.
3. Remove the brake caliper assembly and the front hub assembly (see brake caliper and hub removal section, below).

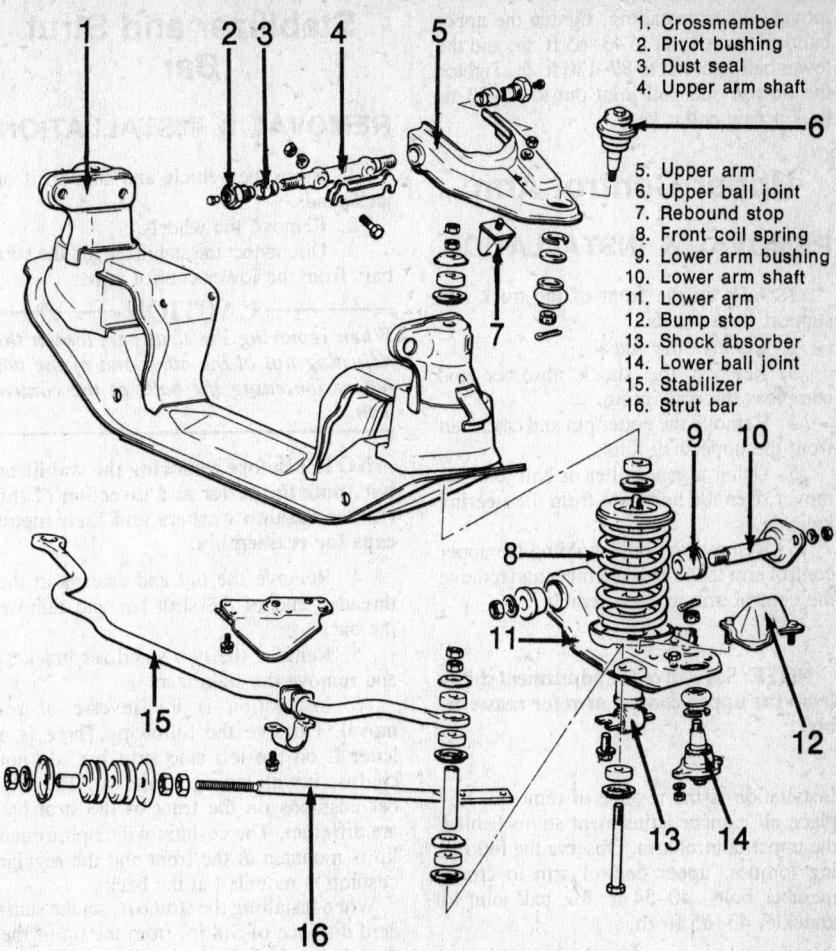

1. Crossmember
2. Pivot bushing
3. Dust seal
4. Upper arm shaft
5. Upper arm
6. Upper ball joint
7. Rebound stop
8. Front coil spring
9. Lower arm bushing
10. Lower arm shaft
11. Lower arm
12. Bump stop
13. Shock absorber
14. Lower ball joint
15. Stabilizer
16. Strut bar

Exploded view of front suspension

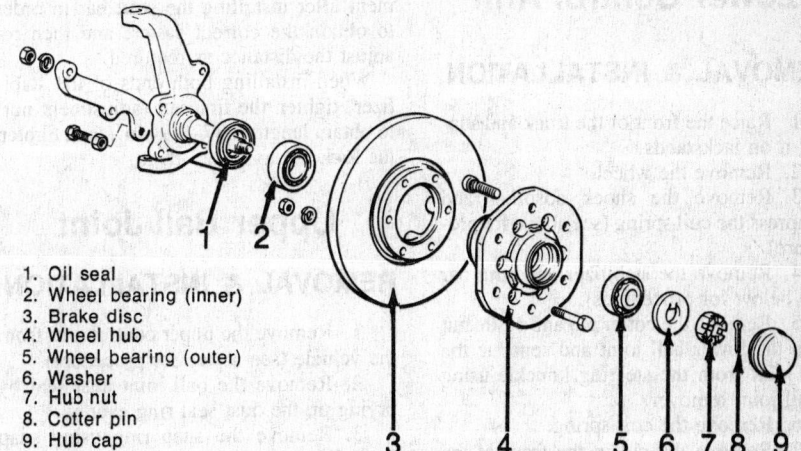

1. Oil seal
2. Wheel bearing (inner)
3. Brake disc
4. Wheel hub
5. Wheel bearing (outer)
6. Washer
7. Hub nut
8. Cotter pin
9. Hub cap

Steering knuckle and hub assembly

4. Disconnect the stabilizer and strut bar from the lower arm (see stabilizer and strut bar removal section, below).
5. Remove the shock absorber and compress the coil spring (see above for shock absorber removal).
6. Remove the relay rod from the steering arm using a ball joint remover.
7. Remove the cotter pins and castle nuts from the steering knuckle ball joints, and using either a gear puller or a ball joint remover, free the ball joints from the knuckle. Remove the knuckle.
8. Installation is the reverse of re-

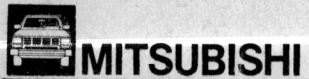

moval. When installing, tighten the upper ball joint castle nut to 43–65 ft. lb. and the lower ball joint nut to 87–130 ft. lb. Tighten the tie rod end ball joint nut to 25–33 ft. lb. Fit new cotter keys.

Upper Control Arm

REMOVAL & INSTALLATION

1. Jack up the front of the truck and support it on stands.
2. Remove the wheel.
3. Remove the shock absorber and compress the coil spring.
4. Remove the cotter pin and castle nut from the upper ball joint.
5. Using a gear puller or ball joint remover, free the ball joint from the steering knuckle.
6. Remove the bolts holding the upper control arm to the crossmember and remove the control arm as an assembly.

NOTE: Save all of the adjustment shims from the upper control arm for reassembly.

Installation is the reverse of removal. Replace all camber adjustment shims behind the upper control arm. Observe the following torques: upper control arm to crossmember bolts, 40–54 ft. lb., ball joint to knuckle, 43–65 ft. lb.

Lower Control Arm

REMOVAL & INSTALLATION

1. Raise the front of the truck and support it on jackstands.
2. Remove the wheel.
3. Remove the shock absorber and compress the coil spring (see above for procedure).
4. Remove the stabilizer and strut bar (see below for procedures).
5. Remove the cotter pin and castle nut from the lower ball joint and separate the ball joint from the steering knuckle using a ball joint remover.
6. Remove the coil spring.
7. Remove the nut in the front of the lower control arm mounting shaft. Remove the nuts at the rear of the shaft. Remove the shaft and remove the lower arm.
8. Installation is the reverse of removal. Tighten the front mounting shaft nut to 40–54 ft. lb. Tighten the rear nut to 6–9 ft. lb. Tighten the ball joint castle nut to 87–130 ft. lb. Tighten control arm shaft only after truck is on the ground.

Stabilizer and Strut Bar

REMOVAL & INSTALLATION

1. Raise the vehicle and support it on jackstands.
2. Remove the wheels.
3. Disconnect the stabilizer and the strut bars from the lower control arms.

———— CAUTION ————
When removing the strut bar, loosen the adjusting nut at the other end of the bar before loosening the bolts at the control arm.

NOTE: Before removing the stabilizer bar, note the order and direction of the rubber cushion washers and their metal caps for reassembly.

4. Remove the nut and spacers at the threaded end of the strut bar and remove the bar.
5. Remove the two stabilizer brackets and remove the stabilizer.
6. Installation is the reverse of removal. Observe the following. There is a letter L on the left side strut bar, do not confuse it with the right side bar. The rubber cushions on the front of the strut bar are different. The cushion with a protruded lip is mounted at the front and the regular cushion is mounted at the back.
   When installing the strut bar, set the standard distance of 3.8 in. from the tip of the threaded end of the bar to the rear face of the rear double nut. Lower the vehicle to the ground and tighten all nuts and bolts. Make sure you check the front wheel alignment after installing the strut bar in order to obtain the correct caster, and then readjust the distance as required.
   When installing both ends of the stabilizer, tighten the first nut (adjustment nut) to obtain length 0.87–0.94 in., then tighten the lock nut to 18–25 ft. lb.

Upper Ball Joint

REMOVAL & INSTALLATION

1. Remove the upper control arm from the vehicle (see above for procedure).
2. Remove the ball joint dust seal by prying up the dust seal ring evenly.
3. Remove the snap ring using snap ring pliers.
4. Using a ball joint remover and installer tool, press off the ball joint.

NOTE: A minimum of 2,200 lb. pressure will be required to remove the upper ball joint from the control arm.

5. To install the ball joint, press it into the burred hole, with the ball joint and upper arm mating marks aligned.

6. Make sure the ball joint snapring is a tight fit and install the dust cover.

Lower Ball Joint

REMOVAL & INSTALLATION

1. Jack up and safely support the vehicle. Remove the wheel.
2. Remove the coil spring (see above for procedures).
3. If you have not already done so, free the lower ball joint from the steering knuckle using a ball joint remover. Remove the dust cover from the ball joint.
4. Unbolt and remove the ball joint.
5. Installation is the reverse of removal. Install the ball joint with its tab side pointing to the rear of the vehicle. Tighten the ball joint to lower control arm bolts to 22–30 ft. lb.

FRONT SUSPENSION— 4 WD

Torsion Bars

REMOVAL & INSTALLATION

1. Jack up the front end of the vehicle and safely support it with jackstands.
2. Remove the front wheel, brake caliper assembly (but don't disconnect the brake hose) and the front hub assembly.
3. Support the lower arm from which the torsion bar is to be removed with a jack.
4. Detach the torsion bar dust covers from the torsion bar anchor arm assembly, and slide the covers a few inches down the torsion bar and out of the way.
5. Match mark the torsion bar to the anchor arm. This is important for later assembly. It is best to make scratch marks, as punching could damage the bar.
6. Loosen the adjusting nut and remove the torsion bar from the anchor arm.

NOTE: Remove the anchor arm assembly as necessary to make torsion bar removal easier.

7. To install, apply a multi-purpose grease to the torsion bar splines, the anchor arm splines and the inside of the dust boot and the anchor boot thread.
8. Make sure that the torsion bars are replaced on their proper sides respectively, if both bars were removed. Face the end of the bar with the identification mark forward, and align the mark on the anchor arm with the mating mark on the torsion bar when the bar is inserted in the anchor arm.

NOTE: When installing a new torsion bar, align the spline painted white with the mark on the anchor arm.

9. Select the relative position of the torsion bar splines and the anchor arm splines so that the dimension illustrated is as specified below, when the torsion bar and anchor arm are assembled, and with the rebound stop in contact with the side frame. Dimension A, left side, Montero: 5.43–5.73 in.; left side, pickup truck: 5.52–5.82 in; right side, Montero: 5.04–5.35 in.; right side, pickup truck: 5.32–5.62 in.

10. Tighten the adjusting nut so that the anchor bolt protrusion will become the dimensions shown in the illustration. Pickup truck dimensions are illustrated. Montero dimensions are: 2.17 in. left side; 2.68 in. right side.

11. Install the stabilizer bar to the lower arm. Tighten the nut so that the dimension from the bottom of the nut to the top of the bolt end is about 17mm.

12. Install the front hub assembly and brake caliper. Install the front wheel. Torque the lower arm-to-side frame bracket bolt to 108 ft. lb.

13. Measure the distance between the suspension bump stop bumper and bump stop bracket on the side frame with the vehicle unloaded. Distance should be 2.8 in. Tighten the adjusting nut on the anchor bolt if it is out of specification.

Shock Absorber

REMOVAL & INSTALLATION

The procedure for 4WD pickup trucks and Montero is similar to that for 2WD pickups.

Steering Knuckle

REMOVAL & INSTALLATION

1. Remove the tire and wheel assembly.
2. Remove the front hub assembly, including the brake rotor.
3. Using a tie rod tool, disconnect the tie rod from the steering knuckle.
4. Using ball joint tools, remove the upper and lower ball joints.
5. Remove the knuckle from the outer shaft.
6. If you intend to replace either the oil seal or needle bearing, refer to procedure below.
7. Reverse the removal procedure for installation.

Oil Seal and Needle Bearing

REMOVAL & INSTALLATION

1. Remove the steering knuckle.
2. Remove the oil seal then remove the

99 to 117 (73 to 86)

1 Upper arm shaft
2 Camber adjusting shim
3 Upper arm
4 Upper ball joint
5 Rebound stop
6 Snap ring
7 Ring
8 Dust cover
9 Joint cup (A)
10 Bushing
11 Shock absorber
12 Bushing (B)
13 Lower arm
14 Bushing (A)
15 Bump stop
16 Lower arm shaft
17 Anchor arm (B)
18 Lower ball joint
19 Torsion bar
20 Anchor bolt
21 Adjusting nut
22 Anchor arm assembly
23 Oil seal
24 Spacer
25 Needle bearing
26 Knuckle

8 to 11 (6 to 8)
59 to 88 (44 to 65)
12 to 17 (9 to 13)
69 to 98 (51 to 72)
8 to 11 (6 to 8)
118 to 176 (87 to 130)
94 to 117 (69 to 86)
138 to 156 (102 to 115)
138 to 156 (102 to 115)
20 to 29 (15 to 21)
53 to 73 (40 to 54)
40 to 49 (29 to 36)

Tightening torque : Nm (ft-lbs.)

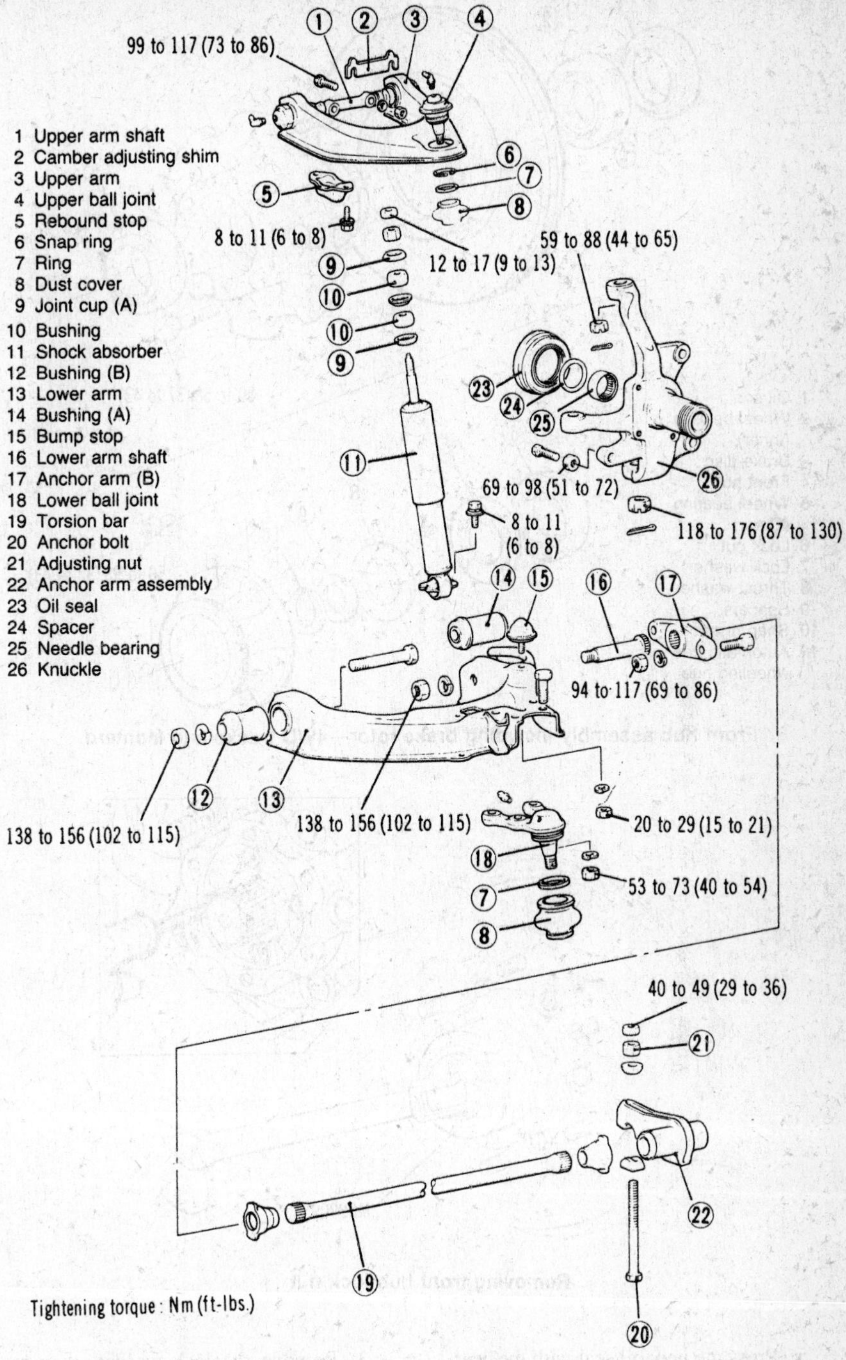

4WD pickup and Montero front suspension and knuckle assemblies

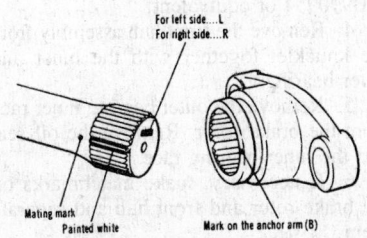

For left side....L
For right side....R

Mating mark
Painted white

Mark on the anchor arm (B)

Torsion bar installation—4WD pickups and Montero

spacer. Remove the needle bearing by tapping the needles uniformly.

NOTE: Once removed, the needle bearing cannot be reused and must be replaced.

3. To install, apply an SAE No. 2 EP grease to the roller surface of the new needle bearing. Press in the needle bearing using a bearing installation tool until it is flush with the knuckle end face. Be careful not to drive the bearing in too far.

4. Apply the SAE No. 2 EP grease to the knuckle contacting surface of the spacer. Install the spacer onto the knuckle with chamfered side toward the center of the vehicle.

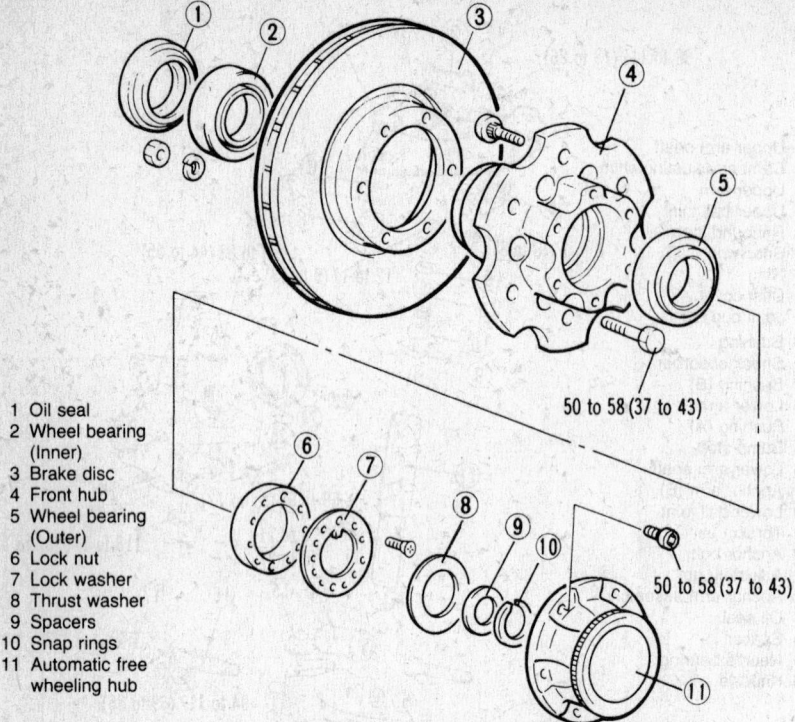

1 Oil seal
2 Wheel bearing (Inner)
3 Brake disc
4 Front hub
5 Wheel bearing (Outer)
6 Lock nut
7 Lock washer
8 Thrust washer
9 Spacers
10 Snap rings
11 Automatic free wheeling hub

50 to 58 (37 to 43)

Front hub assembly including brake rotor—4WD trucks and Montero

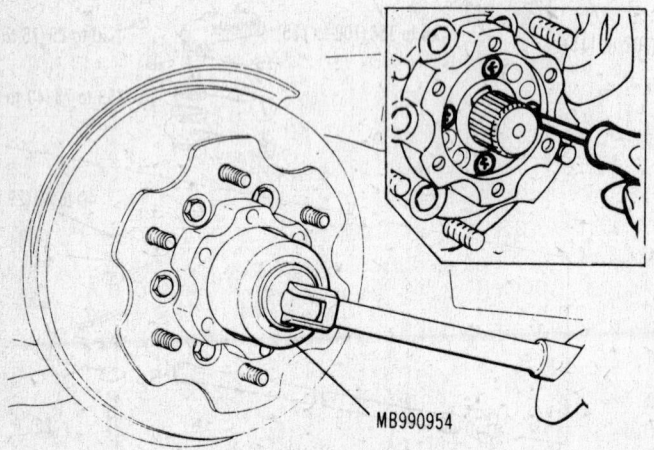

MB990954

Removing front hub lock nut

5. Press the new oil seal with the bearing tools until it is flush with the knuckle end face. Apply the grease to the inside and lip of the oil seal.
6. Install the steering knuckle.

Front Axle Hub and Bearings

REMOVAL & INSTALLATION

1. Remove the front tire assembly. Remove the front brake caliper, but do not disconnect the brake hose. Suspend the caliper with a wire out of the way.
2. Remove the free wheeling hub.

3. Remove the lock washer, then remove the lock nut with special tool MB990954 or equivalent.
4. Remove the front hub assembly from the knuckle, together with the inner and outer bearings.
5. Remove the outer bearing inner race from the brake rotor. Remove the oil seal and the inner bearing race.
6. If necessary, make matchmarks on the brake rotor and front hub and separate them.

BEARING REPLACEMENT

1. Carefully wipe all old grease from inside the front hub.

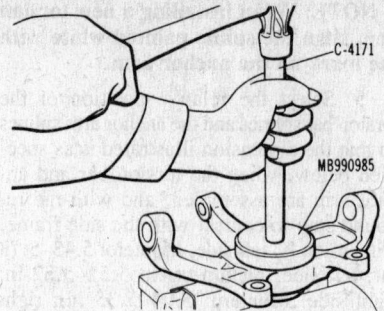

Press fitting oil seal into steering knuckle

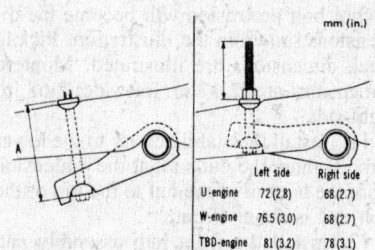

	Left side	Right side
U-engine	72 (2.8)	68 (2.7)
W-engine	76.5 (3.0)	68 (2.7)
TBD-engine	81 (3.2)	78 (3.1)

Torsion bar installation showing anchor arm and bolt adjustment

2. Using a brass drift, carefully tap out the inner and outer bearing races. Tap around each bearing uniformly.
3. Apply an SAE No. 2 EP grease to the outside surface of the new inner and outer bearing outer races.

NOTE: The bearing inner and outer race should be replaced as an assembly.

4. You'll need a special bearing installation tool to replace the inner and outer bearing outer races. The tools are, Mitsubishi part No. MB990938 and MB990933 or equivalent.

HUB INSTALLATION

1. Apply an SAE No. 2 EP multipurpose grease to the outer bearing outer race, oil seal lip and inside surface of the front hub.
2. Apply the grease to the inner bearing inner race and fit the inner race into the front hub.
3. Press the new oil seal into the front hub with the special tools until it is flush with the front hub end face.
4. Fit the knuckle into the front hub assembly. Using a special tool, part No. MB990954 or equivalent, which fits standard torque wrenches, torque the lock nut to 95–145 ft. lb. Loosen the lock nut to 0 ft. lb. (just loose), then torque to 18 ft. lb. Finally, loosen the nut 30°.
5. Install the lock washer. If the lock washer and lock nut holes do not align, align the holes by loosening the nut by not more than 20°.
6. Before installing the free wheeling hub assembly, measure the turning force of the front hub using a spring scale. If the measured value does not meet specifications (1–4 lbs.) retighten the lock nut to the specified torque given earlier.

7. Apply a semi-drying sealant to the free wheeling hub surface of the front hub and then tighten the front hub to the specified torque, 36–43 ft. lb.

8. Measure front driveshaft end play as described elsewhere in this chapter.

9. Assemble the remaining components in the reverse order of removal.

Upper Control Arm

REMOVAL & INSTALLATION

1. Jack up the front of the vehicle and safely support it with jackstands.

2. Loosen the anchor bolt of the torsion bar all the way.

3. Remove the shock absorber.

4. Discharge the brake fluid and disconnect the brake hose.

5. Loosen the nut holding the upper ball joint to the steering knuckle. Do not remove the nut, just loosen it.

6. Using a ball joint tool, disconnect the upper ball joint from the knuckle. Do not tear the rubber ball joint boot.

7. Remove the bolts connecting the upper arm shaft to the arm post of the side frame. Remove the upper control arm.

NOTE: The camber adjusting shims should be marked for later assembly.

8. Replace the upper ball joint if necessary.

9. Reassemble the upper control arm in the reverse order of disassembly, while noting the following:

a. When installing the upper arm assembly into the crossmember insert the upper arm shaft mounting bolts from the outside of the crossmember and put adjusting shims between the crossmember and the upper arm shaft in the order in which they were removed.

b. Tighten the torsion bar anchor bolts to the specs given under Torsion Bar Removal.

c. Tighten the upper arm shaft-to-crossmember bolts to 72–87 ft. lb.

d. Tighten the upper ball joint-to-knuckle nut to 43–65 ft. lb.

10. Check the front wheel alignment.

Lower Control Arm

REMOVAL & INSTALLATION

1. Remove the front skid plate and under cover if equipped.

2. Remove the torsion bar.

3. Remove the stabilizer bar.

4. Remove the shock absorber.

5. Remove the nut which retains the lower ball joint to the knuckle. Using a ball joint tool, carefully disconnect the lower ball joint from the knuckle.

6. Remove the front mounting bolts for the lower arm, and remove the lower arm assembly.

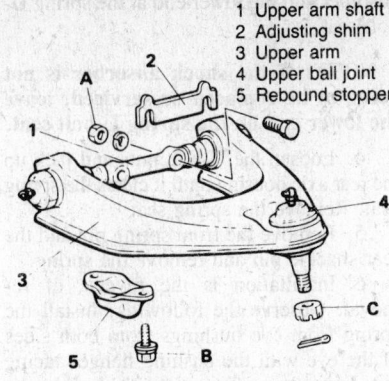

1 Upper arm shaft
2 Adjusting shim
3 Upper arm
4 Upper ball joint
5 Rebound stopper

	Nm	ft. lbs.
A	100-120	72-87
B	8-12	6-9
C	60-90	43-65

4WD and Montero upper control arm showing camber adjustment shim(s)

7. Replace the lower ball joint if necessary.

8. To install, temporarily mount the lower arm shaft to the crossmember.

NOTE: Assembly is easier if a solution of soapy water is applied to the lower arm shaft and to the rubber bushing.

9. Install the shock absorber and torsion bar.

10. Tighten the lower arm shaft to 101–116 ft. lb. with the vehicle lowered to the ground and unloaded.

11. Install the stabilizer bar.

12. Reinstall the remaining parts in the reverse order of removal.

Upper Ball Joint

REPLACEMENT

1. Jack up the front of the vehicle and safely support it with jackstands. Remove the tire.

2. Separate the upper ball joint from the steering knuckle.

3. Remove the rubber dust boot together with the ring.

4. Remove the snapring from the ball joint.

5. The upper ball joint must be pressed out of the arm. Depending on the type of press available, the control arm may or may not have to be removed.

6. The ball joint must be pressed into position when installing. Make sure the mating mark on the ball joint and the mark center of the control arm are aligned.

NOTE: Check to make sure there is no play between the ball joint groove and the snapring. If play exists, replace the snapring with a new one.

7. Apply an SAE No. 2 EP multipurpose grease to both the interior of the rubber

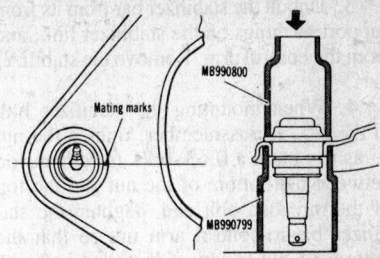

Press fitting upper ball joint—4WD and Montero

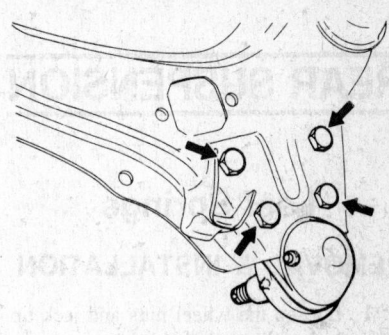

Removing lower ball joint

boot (fill the boot up) and the upper ball joint.

8. Apply a semi-drying sealant to the grooves in the upper ball joint. Secure the boot to the ball joint with the ring.

9. Install the remaining components in the reverse order of removal. Check front end alignment. Torque the ball joint-to-knuckle nut to 43–65 ft. lb.

Lower Ball Joint

REMOVAL & INSTALLATION

1. Remove the tire, after jacking up the front of the vehicle and supporting it with jackstands.

2. Unbolt the bottom of the shock absorber from the control arm.

3. Remove the nut which retains the lower ball joint to the steering knuckle.

4. Using a ball joint tool, disconnect the lower ball joint from the knuckle.

5. Unbolt the lower ball joint from the control arm.

6. Reverse the above procedure for installation. Fill the rubber ball joint boot with an SAE No. 2 multipurpose grease, and also apply the grease to the ball joint. Apply a semi-drying sealant to the grooves in the lower ball joint. Secure the boot to the joint with the ring.

Front Stabilizer Bar

REMOVAL & INSTALLATION

1. Jack up the front of the vehicle and safely support it with jackstands.

2. Remove the front skid plate.

3. Unbolt the stabilizer bar from its front support bushings on the stabilizer link, and from the control arm. Remove the stabilizer bar.

4. When mounting the stabilizer link to the No. 1 crossmember, tighten the nut so as to obtain a 0.63–0.71 in. dimension between the bottom of the nut and the top of the threaded bolt end. Tighten the stabilizer bar-to-control arm nut so that the bottom of nut-to-top of threaded bolt end dimension is 0.63–0.71 in.

REAR SUSPENSION

Leaf Springs

REMOVAL & INSTALLATION

1. Loosen the wheel nuts and jack up the vehicle. Support the frame on jackstands and lower the jack under the rear axle housing.

NOTE: Do not put jackstands under axle housing shafts.

2. Remove the parking brake cable clamp from the leaf spring.
3. Disconnect the upper end of the shock absorber and the lower end at the spring U-bolt seat.

NOTE: If the shock absorber is not going to be replaced or serviced, leave the lower end on the spring U-bolt seat.

4. Loosen the U-bolt nuts and jack up the rear axle housing until it clears the spring seat. Remove the spring seat.
5. Remove the front spring pin and the rear shackle pin and remove the spring.
6. Installation is the reverse of removal. Observe the following: Install the spring front eye bushings from both sides of the eye with the bushing flanges facing out. Insert the spring pin assembly from the wheel side and secure it to the hanger bracket with its bolt. Temporarily tighten the spring pin nut.
7. Repeat Step 6 on the rear spring mount.
8. Align the center of the U-bolt seat with the center bolt hole in the spring. Tighten the U-bolts to 47–54 ft. lb.

NOTE: Tighten the nuts on the U-bolts until all of the U-bolt threads protrude evenly.

9. Tighten the spring pins and shackle pins to 22–33 ft. lb.

Shock Absorbers

REMOVAL & INSTALLATION

1. Jack up the vehicle and remove the wheel.

2. Unbolt the top and bottom of the shock absorber and remove.
3. Installation is the reverse of removal. Tighten the shock absorber upper and lower mounting nuts to 13–18 ft. lb.

STEERING

Steering Wheel

REMOVAL & INSTALLATION

1. Pry off the steering wheel center foam pad.
2. Remove the steering wheel retaining nut.
3. Using a steering wheel puller, remove the wheel.
4. Be sure the front wheels are in a straight ahead position. Reverse the removal procedure for installation.

Steering Column

REMOVAL & INSTALLATION

1. Remove the air cleaner. Matchmark the column shaft and the steering gear shaft.
2. Remove the clamp bolt which holds the steering column shaft on the steering gear shaft.

NOTE: On vehicles with air conditioning, Step 2 must be done from under the truck.

3. Remove the horn pad and steering wheel retaining nut, then remove the steering wheel using a puller.
4. Loosen the tilt lock knob and lower the steering column fully.
5. Remove the steering column cover and disconnect the column wiring under the dashboard.
6. Remove the five bolts holding the base of the column at the fire wall.
7. Remove the four bolts holding the tilt column and remove the steering column from the vehicle.
8. Installation is the reverse of removal. Align the match marks on the steering column shaft and the steering gear shaft and couple the shafts before installing any bolts. Tighten the clamp bolt to 15–18 ft. lb.

Ignition Switch/Lock

REMOVAL & INSTALLATION

1. Remove the column cover.

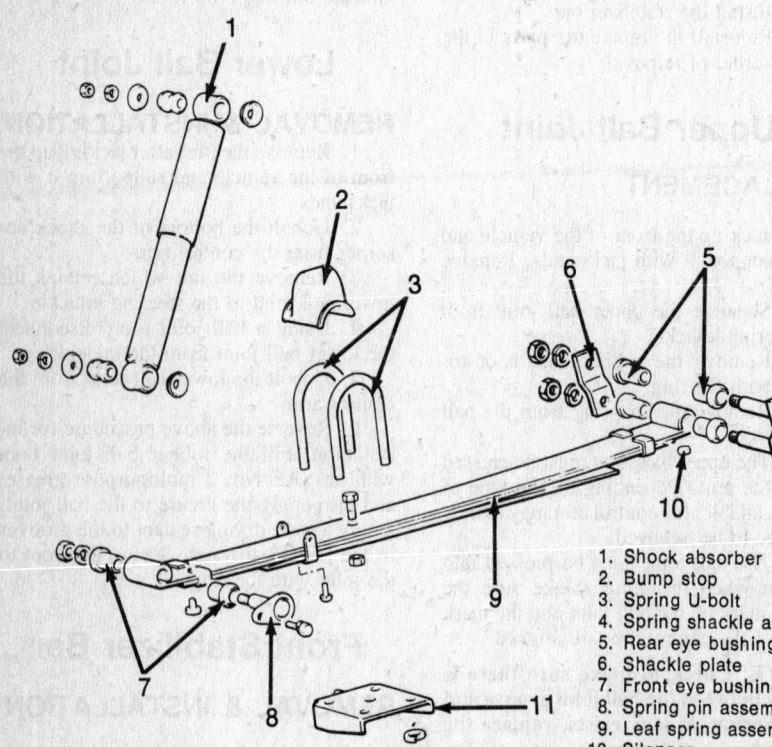

1. Shock absorber
2. Bump stop
3. Spring U-bolt
4. Spring shackle assembly
5. Rear eye bushing
6. Shackle plate
7. Front eye bushing
8. Spring pin assembly
9. Leaf spring assembly
10. Silencer
11. U-bolt seat

Rear suspension

2. Cut a notch in the lock bracket bolt head with a hacksaw.

3. Remove the lock bolts.

4. Disconnect the ignition harness and remove the switch/lock as a unit.

5. To remove the ignition switch, remove the screw holding it on the harness side and pull out the switch.

6. Installation is the reverse of removal.

NOTE: The steering wheel upper lock bracket and bolts should be replaced with new parts when the unit is installed. Before fully tightening the screw in the back of the ignition switch, insert the key and make sure the switch works smoothly.

Turn Signal Switch
REMOVAL & INSTALLATION

1. Remove the steering wheel (see above for procedure).

2. Put the tilt handle in its lowest position.

3. Remove the upper and lower column covers.

4. Remove the wiring harness band clip and disconnect the harness.

5. Remove the switch.

6. Installation is the reverse of removal with the following notes: Make sure the column switch aligns with the steering shaft center.

7. Place the wiring harness along the column tube as close as possible to the center line. Be sure to replace the adjustable wiring harness bands.

Steering Linkage
REMOVAL & INSTALLATION

1. Jack up the vehicle and support it on stands.

2. Remove the cotter pins and castle nuts holding the tie rod ends to the steering arms and the relay rod, and free the tie rods using either a suitable gear puller or a ball joint remover.

3. Unbolt and remove the relay rod in the same manner.

4. To remove the idler arm, remove the two bolts holding it to the frame and pull it out.

NOTE: The outer tie rod end has a left hand thread and the inner tie rod has a right handed thread on the driver's side.

5. Installation is the reverse of removal. Tighten all tie rod end nuts and relay rod nuts to 25–33 ft. lb.

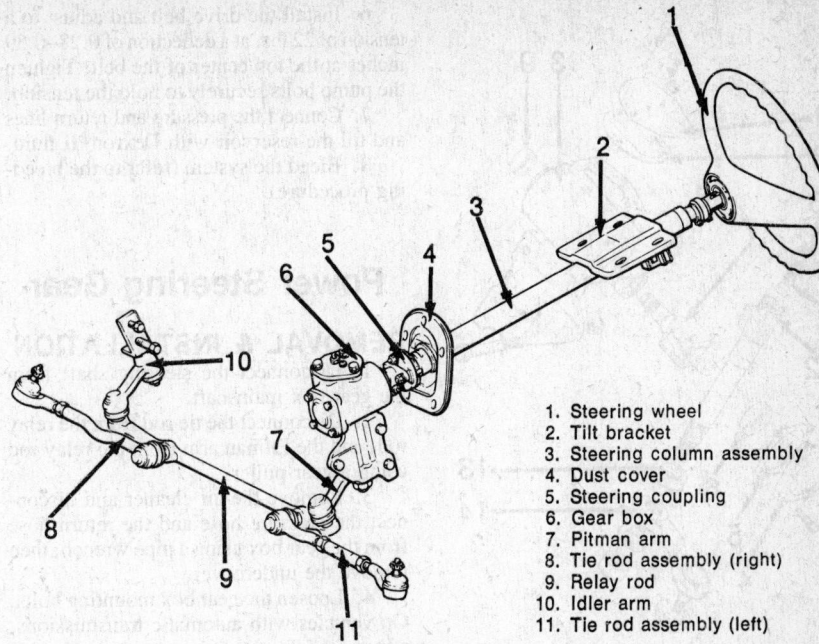

1. Steering wheel
2. Tilt bracket
3. Steering column assembly
4. Dust cover
5. Steering coupling
6. Gear box
7. Pitman arm
8. Tie rod assembly (right)
9. Relay rod
10. Idler arm
11. Tie rod assembly (left)

Manual steering column and gear assembly

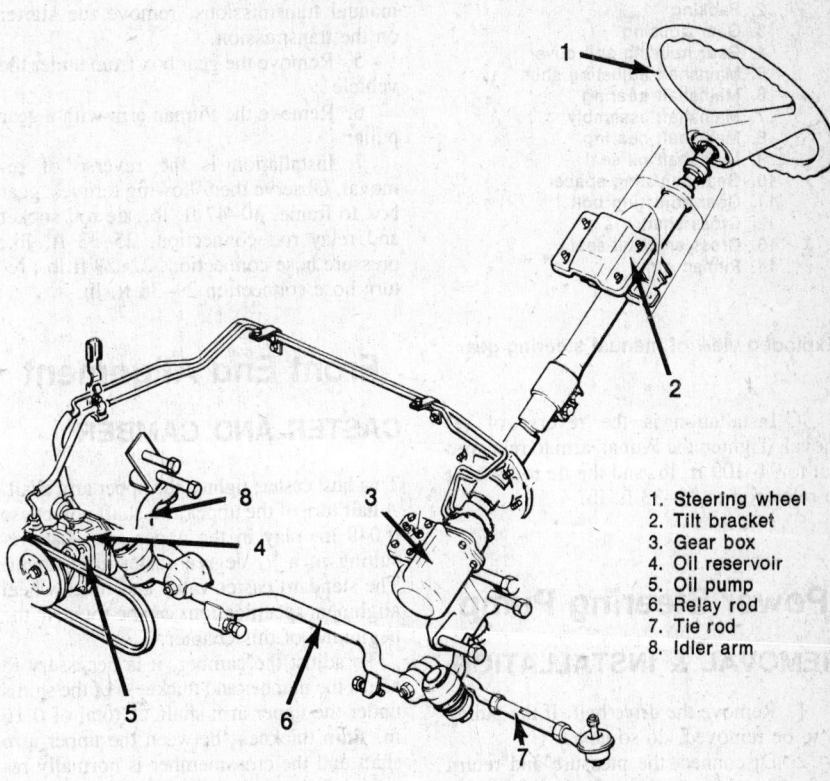

1. Steering wheel
2. Tilt bracket
3. Gear box
4. Oil reservoir
5. Oil pump
6. Relay rod
7. Tie rod
8. Idler arm

Power steering column and assembly

Manual Steering Gear
REMOVAL & INSTALLATION

1. Remove the clamp bolt connecting the steering shaft with the steering gear housing mainshaft.

2. Disconnect the tie rod and Pitman arm from the relay rod using a ball joint remover or gear puller.

3. Remove the three bolts holding the gear box to the frame and remove the gear box from under the vehicle.

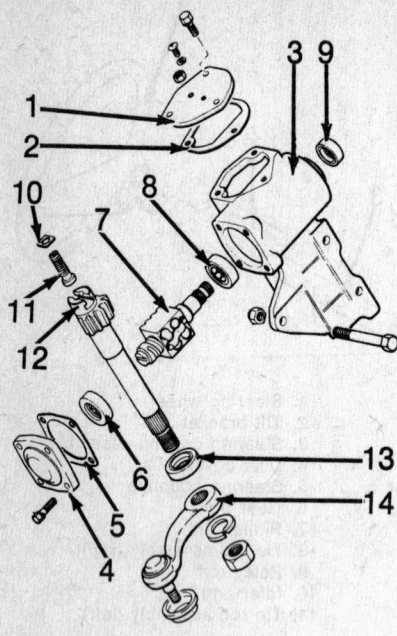

1. Gear housing upper cover
2. Packing
3. Gear housing
4. Gear housing end cover
5. Mainshaft adjusting shim
6. Mainshaft bearing
7. Mainshaft assembly
8. Mainshaft bearing
9. Mainshaft oil seal
10. Gear adjusting spacer
11. Gear adjusting bolt
12. Cross shaft
13. Cross shaft oil seal
14. Pitman arm

Exploded view of manual steering gear

4. Installation is the reverse of removal. Tighten the Pitman arm to relay rod nut to 94–109 ft. lb. and the tie rod socket to relay rod to 29–33 ft. lb.

Power Steering Pump

REMOVAL & INSTALLATION

1. Remove the drive belt. If the pulley is to be removed, do so now.
2. Disconnect the pressure and return lines. Catch any leaking fluid.
3. Remove the pump attaching bolts and lift the pump from the brackets.
4. Make sure the bracket bolts are tight and install the pump to the brackets.
5. If pulley had been removed, install it and tighten the nut securely. Bend the lock tab over the nut.

6. Install the drive belt and adjust to a tension of 22 lbs. at a deflection of 0.28–0.39 inches at the top center of the belt. Tighten the pump bolts securely to hold the tension.
7. Connect the pressure and return lines and fill the reservoir with Dexron®II fluid.
8. Bleed the system (refer to the bleeding procedure).

Power Steering Gear

REMOVAL & INSTALLATION

1. Disconnect the steering shaft from the gear box mainshaft.
2. Disconnect the tie rod from the relay rod, and the Pitman arm from the relay rod using a gear puller.
3. Remove the air cleaner and disconnect the pressure hose and the return hose from the gear box using a pipe wrench, then remove the undercover.
4. Loosen the gear box mounting bolts. On vehicles with automatic transmissions, remove the throttle linkage with the throttle linkage splash shield. On vehicles with manual transmissions, remove the starter on the transmission.
5. Remove the gear box from under the vehicle.
6. Remove the Pitman arm with a gear puller.
7. Installation is the reverse of removal. Observe the following torques: gear box to frame, 40–47 ft. lb.; tie rod socket and relay rod connection, 25–33 ft. lb.; pressure hose connection, 22–29 ft.lb.; return hose connection 29–36 ft. lb.

Front End Alignment

CASTER AND CAMBER

To adjust caster, tighten the upper arm shaft. A half turn of the upper arm shaft will cause 0.049 in. play in the upper arm shaft resulting in a $\frac{1}{4}$ degree caster adjustment. The standard caster value and other wheel alignment specifications can be found at the beginning of this chapter.

To adjust the camber, it is necessary to adjust the number and thickness of the shims under the upper arm shaft. A total of 0.16 in. shim thickness between the upper arm shaft and the crossmember is normally required for standard camber. A 0.024 in. adjustment in thickness of shims will provide about 8 minutes adjustment of camber.

TOE-IN

Toe-in can be adjusted by screwing the left tie rod turnbuckle in or out. One revolution

of the turnbuckle will vary in about 0.3 in. of toe-in adjustment. The toe-in may be increased or decreased by turning the tie rod turnbuckle toward the front or the rear of the vehicle respectively. After completion of the toe-in adjustment, check the difference in the length of the left and the right tie rods. If the difference exceeds 0.2 in., remove the right tie rod and adjust the length until the difference is reduced to 0.2 in. or less. An "L" stamped on the outer surface of the tie rod stands for left hand thread end.

BRAKES

---CAUTION---

Brake shoes contain asbestos, which has been determined to be a cancer causing agent. Never clean the brake surfaces with comporessed air! Avoid inhaling any dust from any brake surface! When cleaning brake surfaces, use a commercially available brake cleaning fluid.

Master Cylinder

REMOVAL & INSTALLATION

1. Remove all lines connected to the master cylinder. Slowly depress the brake pedal to remove the fluid.
2. Remove the master cylinder from the booster assembly.
3. Installation is the reverse of removal. Bleed the brakes.

Disc & Wheel Bearing

REMOVAL & INSTALLATION

1. Remove the caliper.
2. Pry off the dust cap. Tap out and discard the cotter pin. Remove the locknut.
3. Remove the brake disc and wheel hub.
4. Using a brass drift, carefully drive the outer bearing race out of the hub.
5. Remove the inner bearing seal and bearing.
6. Check the bearings for wear or damage and replace them if necessary. Drift the bearing race into place in the hub.
7. Pack the inner and outer wheel bearings with grease.
8. Install the inner bearing in the hub. Drive the seal on until its outer edge is even with the edge of the hub.
9. Install the hub/disc assembly on the spindle, being careful not to damage the oil seal.
10. Install the outer bearing, washer, and spindle nut. Adjust the bearing.

ADJUSTMENT

1. Tighten the spindle nut to 22 ft. lb. and then loosen it.
2. Tighten the nut to 6 ft. lb.
3. Install the cap on the nut. Do not back off the nut more than 30° for cotter pin hole-to-slot alignment.

Wheel Cylinder

REMOVAL & INSTALLATION

1. Remove brake shoes.
2. Disconnect the brake pipe from the rear of the wheel cylinder and plug it to prevent it from leaking fluid.
3. Remove the wheel cylinder from the brake backing plate.
4. Installation is the reverse of removal.

Power Brake Unit

REMOVAL & INSTALLATION

1. Remove the master cylinder.
2. Disconnect the vacuum hose from the power brake.
3. Remove the pin connecting the power brake operating rod to the pedal.
4. Loosen the nuts attaching the power brake to the fire wall and remove the power brake.
5. Installation is the reverse of removal. Apply sealer to all mounting surfaces before assembling.

Parking Brake

ADJUSTMENT

1. Release the parking brake.
2. Jack up the vehicle and support it on jackstands.
3. Make sure the balancer that the front of the cable rides in is parallel with the center line of the truck. The clearance between the balancer and the crossmember should be about 8 in.
4. Adjust the parking brake by turning the turnbuckle on the cable. The brake is properly adjusted when the parking brake handle can be pulled 16–17 notches (approx. 4.3 in.).
5. After adjusting the parking brake, make sure there is slack in the cable when the brake is in the off position. If the brake will not adjust correctly or fails on a hill, the rear brake shoes should be inspected for wear, oil or grease covered surfaces or malfunction.

CHASSIS ELECTRICAL

Blower Motor

REMOVAL & INSTALLATION

Pickups Without Air Conditioning

1. Remove the cluster panel.
2. Disconnect the cable between the motor and the heater unit.
3. Remove the three bolts holding the motor in the heater unit and pull out the fan.

NOTE: It may be necessary to unfasten the fan from the motor to remove them from under the dashboard.

4. Installation is the reverse of removal.

PickUps With Air Conditioning

— CAUTION —

The air conditioning system utilizes the blower motor assembly of the heater unit. However, it may be necessary to remove some of the air conditioning components to gain access to the motor. If this is the case, never attempt to loosen any of the air conditioning hoses during your work. They contain refrigerant under pressure, which could severely damage your eyes or skin on contact.

Montero

1. Disconnect the front wiring harness and blower motor coupling connectors.
2. Remove the lower mounting bolts of the blower assembly.
3. Remove the lap heater duct.
4. Remove the stopper of the glove box and push the glove box down.
5. Remove the "Recirc-Fresh" control wire and blower assembly mounting bolts.
6. Remove the blower assembly.
7. Reverse the above procedure for installation.

Heater Unit

— CAUTION —

When draining the coolant, keep in mind that cats and dogs are attracted by the ethelyne glycol antifreeze, and are quite likely to drink any that is left in an un-covered container or in puddles on the ground. This will prove fatal in sufficient quantity. Always drain the coolant into a sealable container. Coolant should be reused unless it is contaminated or several years old.

REMOVAL & INSTALLATION

Pickups Without Air Conditioning

1. Drain the cooling system.
2. Place the hot water flow control lever in the off position.
3. Remove the glove box, the center ventilation grille and duct, and the defroster duct.
4. Disconnect all control cables at the heater side.
5. Disconnect the water hoses.
6. Disconnect the harness from the heater fan motor.
7. Remove the top mounting bolts and the center mounting nuts, and remove the heater assembly.

Pickups With Air Conditioning

Removing the heater unit with air conditioning attached is similar to procedures used on units without air conditioning. It may be necessary to loosen or remove certain components of the air conditioning system to facilitate heater unit removal, however, *never* loosen the refrigerant pipes that lead into the air conditioning evaporator assembly. They are filled with a noxious fluid which, under certain conditions, could cause severe damage to your face or skin. Always leave all air conditioning work to skilled professionals.

Heater Core

REPLACEMENT

Montero

1. Remove the heater control lever arm and remove the water valve cover on the heater.
2. Drain the cooling system. Remove the heater pipe and water valve.
3. Disconnect the control arm linkage.
4. Remove the control arm.
5. Remove the heater core by moving it sideways.

NOTE: To prevent foreign material from getting in between the heater core and case, be careful not to remove the heater core felt when removing the heater core.

6. To install, after the center ventilator open/close damper has been placed in the fully closed position, turn the arm fully clockwise, then connect it to the link.

7. With the defroster/changeover damper in the fully closed defroster position, turn the arm fully counterclockwise, then connect it to the link.

8. With the water valve fully closed and the air intake damper fully closed, connect the arm to the link.

9. Connect each heater hose up to the proper inlet/outlet.

10. When installing the water hoses, apply a non-drying adhesive to the engine compartment side of the grommet. Tighten the clamps so that the bolt heads are accessible.

Radio
REMOVAL & INSTALLATION

1. Remove the instrument cluster bezel.

2. Remove the radio bracket attaching screws from the instrument panel, and remove the radio bracket.

3. Pull the radio out slightly, disconnect the antenna lead in, speaker connector and the power supply connector.

4. Take out the radio.

5. Installation is the reverse of removal.

Windshield Wiper Motor and Linkage
REMOVAL & INSTALLATION

1. Remove the wiper arms. Remove the arm shaft lock nuts and push in the shafts. Disconnect the electrical wiring.

2. Remove the bolts holding the motor bracket to the body and pull the wiper assembly outward and away from the body.

3. Hold the motor shaft and the linkage at right angles to each other and disconnect them. Remove the motor.

4. The linkages can be pulled from the opening in the front deck.

5. The installation is in the reverse of the removal, being sure to insert the linkage shaft bracket positioning boss positively in the hole provided in the body before tightening the wiper shaft nut.

6. Locate the wiper blades in the stopped position approximately $1/2$–$3/4$ in. above the bottom moulding or sealer of the windshield.

Instrument Panel
REMOVAL & INSTALLATION
Pickups

1. Disconnect the negative battery cable.

2. Remove the heater fan control knob, heater control knobs and the radio knobs.

3. Remove the ash tray and remove the two screws behind it holding the instrument panel bezel. Remove the two screws at the top of the bezel and remove the bezel.

4. Remove the four screws in the corners of the meter case.

5. Disconnect the speedometer cable and connectors from the back of the meter, and remove the meter assembly.

6. Installation is the reverse of removal.

Speedometer Cable
REMOVAL & INSTALLATION

1. Unfasten the speedometer cable from the rear of the speedometer. The instrument panel may have to be removed.

NOTE: The cable is fastened to the speedometer via a snap clip, which must be pressed down while the cable is being unfastened.

2. Unfasten the speedometer cable from the transmission.

3. Remove the bands holding the cable with the wiring harness and withdraw the cable through the engine compartment.

4. Installation is the reverse of removal. Always install the speedometer cable with the largest radius possible to prevent cable binding and noise.

Circuit Protection
FUSE BOX LOCATION

The fuse box is located below the hood release handle on the driver's side of the vehicle.

FUSIBLE LINK

The fusible link is located on the battery running from the positive (+) battery terminal. It is necessary to test the link for continuity with a circuit tester, since visual inspection is not enough to detect a melted fusible link. When the fusible link is melted, a dead short may be the cause.

Toyota

GENERAL ENGINE SPECIFICATIONS

Year	Engine Type	Engine Displacement Cu. In. (cc)	Carburetor Type	Horsepower (@ rpm)	Torque @ rpm (ft lbs)	Bore × Stroke (in.)	Compression Ratio	Oil Pressure @ rpm (psi)
'79–'80	20R	133.6 (2189)	2-bbl	90 @ 4800	122 @ 2400	3.48×3.50	8.4:1	64 @ 2500
'81–'86	22R	144.4 (2366)	2-bbl	96 @ 4800 ②④	93 @ 2800 ③⑤	3.62×3.50	9.0:1 ⑨	36–71 @ 3000
'84–'86	22R-E	144.4 (2366)	E.F.I.	106 @ 4800 ⑥	137 @ 2800 ⑦	3.62×3.50	9.0:1 ⑨	36–71 @ 3000
'79–'84	2F	258 (4230)	2-bbl	125 @ 3600	200 @ 1800	4.02×4.13	7.8:1	50–70 @ 2000
'81–'83	L	133.5 (2188)	D.F.I.	62 @ 4200 ①	93 @ 2400	3.54×3.39	21.5:1	60 @ 2000
'84–'86	2L	149.2 (2446)	D.F.I.	75 @ 4000	114 @ 2400	3.62×3.62	20.0:1	36–85 @ 3000
'85–'86	2L-T	149.2 (2446)	D.F.I. ⑧	84 @ 4000	137 @ 2400	3.62×3.62	9.0:1	36–85 @ 3000
'85–'86	22R-TE	144.4 (2366)	E.F.I. ⑧	135 @ 4800	173 @ 2800	3.62×3.50	9.3:1	36–71 @ 3000
'83–'86	3Y-EC	121.9 (1998)	E.F.I.	90 @ 4400	120 @ 3000	3.40×3.40	9.0:1	36–71 @ 3000

E.F.I. Electronic Fuel Injection
D.F.I. Diesel Fuel Injection
① '83: 59 @ 4200
② '83 Federal and '84 all: 100 @ 4800
③ '84: 130 @ 2800
④ '85: 103 @ 4800
⑤ '85: 133 @ 2800
⑥ '85: 116 @ 4800
⑦ '85: 140 @ 2800
⑧ Turbo
⑨ '86—9.3:1

TUNE-UP SPECIFICATIONS

Year	Engine Type	Spark Plugs Type	Spark Plugs Gap (in.)	Distributor Point Dwell (deg)	Distributor Point Gap (in.)	Ignition Timing (deg)▲ MT	Ignition Timing (deg)▲ AT	Fuel Pump Pressure (psi)	Manifold Vacuum at Idle* in. Hg	Compression Pressure (psi) @ 250 rpm**	Idle Speed (rpm) MT	Idle Speed (rpm) AT	Valve Clearance (in.) In	Valve Clearance (in.) Ex
'79–'84	2F	④	②	Electronic		7B	—	3.4–4.7	16.5	149	①	—	0.008	0.014
'79–'80	20R	W16 EX-U	0.031	52	0.018	8B	8B	2.1–4.3	15.75	156	800	850	0.008	0.012 ③
'81–'82	22R	W16 EXR-U	0.031	Electronic		8B	8B	2.1–4.3	15.8	156	700	750	0.008	0.012
'83–'86	22R	W16 EXR-U	0.031	Electronic		5B ⑥	5B ⑥	2.1–4.3	15.8	142–171	700 ⑤	750	0.008	0.012
'84–'86	22R-E	W16 EXR-U	0.031	Electronic		5B ⑦	5B ⑦	36–38	15.8	142–171	700 ⑤	750	0.008	0.012
'85–'86	22R-TE	W16 EXR-U	0.031	Electronic		5B ⑦	5B ⑦	36–38	15.8	120–149	—	—	0.008	0.012

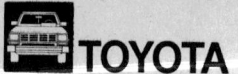
TUNE-UP SPECIFICATIONS

Year	Engine Type	Spark Plugs Type	Spark Plugs Gap (in.)	Distributor Point Dwell (deg)	Distributor Point Gap (in.)	Ignition Timing (deg) ▲ MT	Ignition Timing (deg) ▲ AT	Fuel Pump Pressure (psi)	Manifold Vacuum at Idle* in. Hg	Compression Pressure (psi) @ 250 rpm**	Idle Speed (rpm) MT	Idle Speed (rpm) AT	Valve Clearance (in.) In	Valve Clearance (in.) Ex
'83-'86	3Y-EC	P16R	0.043	Electronic		8B ⑧	8B ⑧	33-38	15.75	142-171	700	750	⑨	⑨

NOTE: If these figures do not correspond to information given on the engine compartment decal, use the figures on the decal. They are current for the engine in your truck.

▲ With automatic transmission in D (drive)
   and manual transmission in Neutral

* These are the minimum readings you must
   obtain

** Look for uniformity among cylinders rather
   than specific pressure

① '79-'80—800 '81-'82—650
② '79—0.039 '80-'82—0.031
③ 1979 20R exhaust valve 0.010 in.
④ '79—W14EX '80—W14EX-U '81-'82—W14EXR-U
⑤ EFI: 750 rpm
⑥ 1985-86: 0° @ 950 rpm, with vacuum advance OFF
⑦ T terminal shorted
⑧ Vacuum advance OFF
⑨ No adjustment is necessary

DIESEL ENGINE TUNE-UP SPECIFICATIONS

Year	Engine	Injector Opening Pressure (psi)	Idle Speed (rpm)	Valve Clearance (in.) Intake	Valve Clearance (in.) Exhaust	Cranking Compression Pressure @ 250 rpm	Maximum Compression Variance ③	Firing Order
'81-'83	L	1636-1778 ① 1493-1777 ②	700	0.010	0.014	284-455	71 psi	1-3-4-2
'84-'86	2L	2,276-2,389	700	0.010	0.014	284-455	71 psi	1-3-4-2
'85-'86	2L-T	2,276-2,389	700	0.010	0.014	284-455	71 psi	1-3-4-2

① New
② Used
③ Between highest and lowest cylinder

FIRING ORDERS

NOTE: To avoid confusion, always replace spark plug wires one at a time.

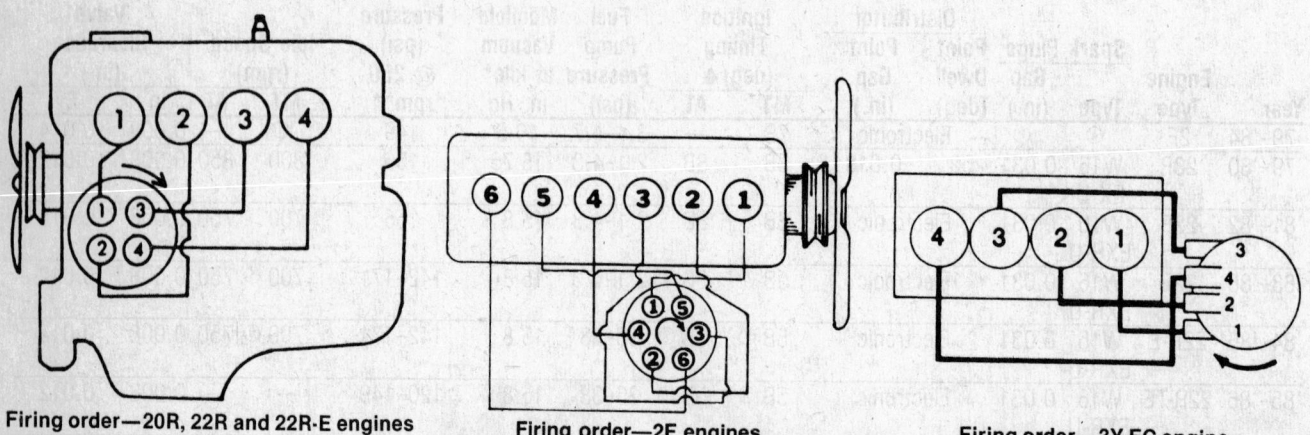

Firing order—20R, 22R and 22R-E engines

Firing order—2F engines

Firing order—3Y-EC engine

CAPACITIES

Year	Model	Engine Type	Crankcase▲ w/filter	Crankcase▲ wo/filter	Transmission▲ Manual	Transmission▲ Automatic	Trans. Case▲	Differential▲ Front	Differential▲ Rear	Fuel Tank■	Cooling System▲
'79	Pickup 2WD	20R	5.0	4.0	2.0 ①	7.0	—	—	1.6	12.1	8.5
'79½	Pickup 4WD	20R	5.0	4.0	2.0	6.7	1.7	2.0	2.1	—	7.4
'80	Pickup (all)	20R	5.0	4.0	2.0 ⑥	6.7 ④	1.7	2.4	2.3	—	8.9
'81–'82	Pickup (Gasoline)	22R	5.0	4.0	2.1	6.7	1.7	2.4	1.8 ⑦	13.7 ⑤	8.9
'81–'83	Pickup (Diesel)	L	6.1	5.1	1.9	—	—	—	1.8	16.0	11.1
'83	Pickup (Gasoline)	22R	5.0	4.0	⑪	6.9	1.7	2.3	1.8 ⑫	13.7 ⑤	8.9
'84	Pickup (Gasoline)	22R 22R-E	5.0	4.0	⑬	6.9	1.7	2.4	1.8 ⑫	13.7 ⑤	8.9
'84	Pickup (Diesel)	2L	6.0 ⑭	5.0 ⑭	⑮	—	1.7	2.3	1.8 ⑫	16.0	12.0
'79–'84	Land Cruiser	2F	8.4	7.4	⑧	—	2.6	2.2	2.2	22.2 ⑨	17.5 ⑩
'83–'86	Van	3Y-EC	3.7	3.2	2.3	6.9	—	—	1.3	15.9	7.5
'85–'86	4 Runner (Gasoline)	22R 22R-E	4.0	4.9	③	—	1.7	2.4	2.3	②	8.9
'85–'86	4 Runner (Diesel)	2L-T	6.1	5.1	3.2	—	1.7	2.4	2.3	②	10.4
'85–'86	Pickup (Gasoline)	22R 22R-E	4.9	4.0	④	6.9	—	—	1.9	⑯	8.9
'85–'86	Pickup (Diesel)	2L 2L-T	6.1	5.1	⑱	6.9	1.7	⑲	2.3	⑰	10.4

▲ Measurements in quarts
■ Measurements in gallons
① 5 speed trans.—2.8
② Std.—14.8; large—17.2
③ 22R engine—4.1; 22R-E—3.2
④ 4 speed—2.5; 5 speed (22R)—2.7; 5 speed (22R-E)—2.5
⑤ Long bed—16.1
⑥ 5 speed trans.—2.7
⑦ ¾ Ton—1.9; 4 × 4—2.3
⑧ 3 speed—1.8; 4 speed—3.3; 5 speed—4.7
⑨ Station Wagon—23.8
⑩ '79–'82 2-Door—19.9
   '79–'82 Wagon—18.3
⑪ 4 sp.: 2.1; 5 sp: 1.9
⑫ 4 × 4; 2.3
⑬ W52: 2.7
   W42: 2.9
   G52: 2.3
⑭ Calif.: 7 w/filter; 6 w/o filter
⑮ 2WD: 2.5
   4WD: 4.1
⑯ Short bed—13.7; long bed—17.2
⑰ Short bed—17.2; long bed (std.)—17.2; long bed (large)—19.3
⑱ 4 speed—2.5; 5 speed—2.3; 4WD—3.2
⑲ 4WD: 2.4

VALVE SPECIFICATIONS

Year	Engine Type	Seat Angle (deg.)	Face Angle (deg.)	Spring Test Pressure (lbs.)	Spring Installed Height (in.)	Stem-to-Guide Clearance (in.)▲ Intake	Stem-to-Guide Clearance (in.)▲ Exhaust	Stem Diameter (in.) Intake	Stem Diameter (in.) Exhaust
'79–'84	2F	45	44.5	71.6	1.693	0.0012–0.0024	0.0016–0.0028	0.3140	0.3137
'79–'80	20R	45	44.5	55.1	1.594	0.0008–0.0024	0.0012–0.0026	0.3138–0.3144	0.3136–0.3142
'81–'84	22R	45 ①	44.5	55.1	1.594	0.0008–0.0024	0.0012–0.0026	0.3145–0.3188	0.3136–0.3142
'81–'84	L	45 ①	44.5	56 ②	1.547	0.0008–0.0022	0.0016–0.0030	0.3336–0.3342	0.3328–0.3335

TOYOTA

VALVE SPECIFICATIONS

Year	Engine Type	Seat Angle (deg.)	Face Angle (deg.)	Spring Test Pressure (lbs.)	Spring Installed Height (in.)	Stem-to-Guide Clearance (in.)▲		Stem Diameter (in.)	
						Intake	Exhaust	Intake	Exhaust
'85–'86	22R 22R-E	45 ①	44.5	64	1.594	0.0008–0.0012	0.0012–0.0028	0.3145–0.3188	0.3136–0.3142
'85–'86	2L 2L-T	45 ①	44.5	64.4	1.547	0.0008–0.0022	0.0016–0.0030	0.3336–0.3342	0.3328–0.3335
'83–'86	3Y-EC	45 ①	44.5	64–77	1.598	0.0010–0.0024	0.0012–0.0026	0.3138–0.3144	0.3136–0.3142

▲ Valve guides are removable
① Blend the seat with 30° and 60° cutters to center the 45° portion on the valve face.
② 1984: 64 lbs.

CAMSHAFT SPECIFICATIONS

(All measurements in inches)

Year	Engine	Journal Diameter	Bearing Clearance	Maximum Journal Runout	Lobe Height▲		Maximum Thrust Clearance (End-Play)
					Intake	Exhaust	
'79–'80	20R	1.2982–1.2984	0.0004–0.0020	0.0080	1.6783–1.6819	1.6806–1.6842	0.0010 ①
'81–'82	22R	1.2982–1.2984	0.0004–0.0020	0.008	1.6783–1.6819	1.6806–1.6842	0.0010 ①
'79–'84	2F	②	0.0010–0.0030	0.0059	1.4960–1.5142	1.4920–1.5098	0.0080 ③
'81–'83	L (Diesel)	1.3767–1.3774	0.0009–0.0029	0.0020	1.6810–1.6948 ⑤	1.6900–1.7020 ⑥	0.0120 ④
'83–'86	3Y-EC	⑦	0.0010–0.0032	0.0024	1.5205–1.5244	1.5208–1.5248	0.012
'84–'86	2L (Diesel)	1.3767–1.3774	0.0009–0.0029	0.0020	1.8409	1.8602	0.012
'85–'86	2L-T (Diesel)	1.3767–1.3774	0.0009–0.0029	0.0020	1.8224	1.8602	0.012
'83–'86	22R	1.2984–1.2992	0.0004–0.0020	0.008	1.6783–1.6819	1.6807–1.6842	0.010 ①
'84–'86	22R-E	1.2984–1.2992	0.0004–0.0020	0.008	1.6783–1.6819	1.6807–1.6842	0.010 ①

▲ Measured from the bottom of the base circle to the top of the lobe (with the cam lobe pointing upward)
① Preferred—0.003–0.007
② Front—1.8880–1.8888
   Second—1.8289–1.8297
   Third—1.7699–1.7707
   Rear—1.7108–1.7116
③ Preferred—0.0035–0.0060
④ Preferred—0.0022–0.0061
⑤ 1984: 1.6809–1.6812
⑥ 1984: 1.6880–1.6900
⑦ Front—1.8291–1.8297
   Second—1.8192–1.8199
   Third—1.8094–1.8100
   Fourth—1.7996–1.8002
   Rear—1.7897–1.7904

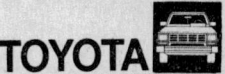

CRANKSHAFT AND CONNECTING ROD SPECIFICATIONS

(All measurements in inches)

Year	Engine Type	Crankshaft Main Brg Journal Dia	Crankshaft Main Brg • Oil Clearance	Crankshaft Shaft End-Play	Crankshaft Thrust on No.	Connecting Rod Journal Diameter	Connecting Rod Oil Clearance	Connecting Rod Side Clearance
'79–'84	2F	①	0.0008–0.0017	0.0024–0.0063	3	2.1252–2.1260	0.0008–0.0024	0.0043–0.0091
'79–'80	20R	2.3614–2.3622	0.0010–0.0022	0.0007–0.0079	3	2.0862–2.0866	0.0010–0.0022	0.0063–0.0102
'81–'84	22R	2.3614–2.3622	0.0006–0.0020	0.0008–0.0089	3	2.0862–2.0866	0.0008–0.0020	0.0008–0.0087
'81–'83	L (Diesel)	2.4402–2.4409	0.0012–0.0028	0.0016–0.0098	3	2.0858–2.0866	0.0012–0.0028	0.0031–0.0079
'85–'86	22R 22R-E	2.3616–2.3622	0.0010–0.0022	0.0008–0.0087	3	2.0861–2.0866	0.0010–0.0022	0.0008–0.0087
'85–'86	2L-T	2.4403–2.4409	0.0014–0.0025	0.0016–0.0098	3	2.1649–2.1654	0.0014–0.0025	0.0031–0.0079
'84–'86	2L (Diesel)	2.4403–2.4409	0.0014–0.0025	0.0016–0.0098	3	2.0861–2.0866	0.0014–0.0025	0.0031–0.0079
'83–'86	3Y-EC	2.2829–2.2835	0.0008–0.0020	0.0008–•0.0087	3	1.8892–1.8898	0.0008–0.0020	0.0063–0.0123

① No. 1—2.6367–2.6376; No. 2—2.6957–2.6967; No. 3—2.7548–2.7557; No. 4—2.8139–2.8148

TORQUE SPECIFICATIONS

(All readings in ft. lbs.)

Year	Engine Type	Cylinder Head Bolts	Rod Bearing Bolts	Main Bearing Bolts	Crankshaft Pulley Bolt	Flywheel-to-Crankshaft Bolts	Manifolds Intake	Manifolds Exhaust
'79–'84	2F	83–98	35–55	90–108 ①	116–145	59–62	28–37 ②	28–37 ②
'79	20R	52.1–63.7	39.1–47.7	68.7–83.2	79.6–94.0	61.5–68.7	10.6	28.9–36.2
'80	20R	53–63	40–47	69–83	120–130	73–86	13–19	29–36 ②
'81–'86	22R	53–63	40–47	69–83	120–130	73–86	13–19	29–36
'81–'83	L (Diesel)	84–90	37–43	71–81	69–75	84–90	8–11	11–15
'83–'86	3Y-EC	④	36	58	80	61 ③	36	36
'84–'86	2L (Diesel)	87	43	76	101	87	17	29
'85–'86	2L-T (Diesel)	87	43	76	101	87	17	38
'84–'86	22R-E	58	46	76	116	80	14	33

① Rear bearing—76–94 ft. lbs.
② California vehicles—37–51 ft. lbs.
③ 54 ft. lbs.: drive plate
④ 12 mm bolt: 14 ft. lbs.; 14 mm bolt: 65 ft. lbs.

PISTON AND RING SPECIFICATIONS

(All measurements in inches)

Year	Engine Type	Piston Clearance 68°F	Ring Gap Top Compression	Ring Gap Bottom Compression	Ring Gap Oil Control	Ring Side Clearance (Ring to Land) Top Compression	Ring Side Clearance (Ring to Land) Bottom Compression	Ring Side Clearance (Ring to Land) Oil Control
'79–'84	2F	0.0012–0.0020	0.0080–0.0160	0.0080–0.0160	snug	0.0012–0.0024	0.0008–0.0024	snug
'79–'80	20R	0.0012–0.0020	0.0040–0.0120	0.0040–0.0120	0.0040–0.0120	0.0012–0.0028	0.0012–0.0028	snug
'81–'83	22R	0.0020–0.0028	0.0094–0.0142	0.0071–0.0154	snug	0.0080	0.0080	snug

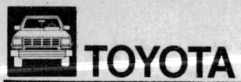

PISTON AND RING SPECIFICATIONS
(All measurements in inches)

Year	Engine Type	Piston Clearance 68°F	Ring Gap			Ring Side Clearance (Ring to Land)		
			Top Compression	Bottom Compression	Oil Control	Top Compression	Bottom Compression	Oil Control
'81–'83	L	0.0014–0.0022	0.0780–0.0157	0.0118–0.0197	0.0118–0.0197	0.0024–0.0039	0.0016–0.0031	0.0012–0.0028
'84	2L	0.0020–0.0028	0.0138–0.0232	0.0079–0.0213	0.0079–0.0193	0.0008–0.0026	0.0016–0.0039	0.0012–0.0028
'84–'86	22R	0.0012–0.0020	0.009–0.015	0.009–0.015	0.008–0.032	0.008	0.008	snug
'85–'86	22R-E	0.0012–0.0020	0.009–0.015	0.009–0.015	0.008–0.032	0.008	0.008	snug
'85–'86	2L-T	0.0020–0.0028	0.0138–0.0244	0.0079–0.0185	0.0079–0.0204	0.0008–0.0026	0.0016–0.0031	0.0012–0.0028
'85–'86	2L	0.0020–0.0028	0.0138–0.0244	0.0079–0.0185	0.0079–0.0204	0.0008–0.0026	0.0016–0.0031	0.0012–0.0028
'83–'86	3Y-EC	0.0030–0.0037	0.0087–0.0185	0.0059–0.0165	0.0079–0.0323	0.0012–0.0028	0.0012–0.0028	0.0012–0.0028

ALTERNATOR AND REGULATOR SPECIFICATIONS

Year	Engine	Alternator			Regulator					
					Field Relay				Regulator	
		Manufacturer	Output (amps)	Type	Contact Spring Deflection (in.)	Point Gap (in.)	Volts to Close (in.)	Air Gap (in.)	Regulator Point Gap (in.)	Volts
'79–'82	Exc. Diesel	Nippondenso	40	External	0.008–0.0018	0.016–0.047	4.5–5.8	0.008	0.010–0.018	13.8–14.8
'79–'84	Exc. Diesel	Nippondenso	①	Built-in	②	②	②	②	②	14.0–14.7
'81–'84	Diesel	Nippondenso	55	Built-in	②	②	②	②	②	13.8–14.8
'83–'86	3Y-EC	—	60	Built-in	②	②	②	②	②	13.5–15.1
'85–'86	Diesel	—	55	Built-in	②	②	②	②	②	13.8–14.4
'85–'86	Exc. Diesel	—	60	Built-in	②	②	②	②	②	13.5–15.1

① 40–45 Amp Standard, 55 Amp Optional
② Not Adjustable

STARTER SPECIFICATIONS

Year	Engine Type	Starter Type/ Rated Voltage	No Load Test			Brush Spring Tension	Minimum Brush Length
			Maximum Amps	Volts	Minimum rpm		
'79–'80	20R	Reduction Gear/12V	50	11.5	5000	21 ②	0.47
'84–'86	22R	Reduction Gear/12V, 1.4kw	90	11.5	3500	43–64 ②	0.39
'81–'83	L (Diesel)	Reduction Gear/12V	180	11.0	3500	7.1–8.8 ③	0.47
'79–'84	2F	Direct Drive/12V	50	11.5	5000	①	0.39
'81–'83	22R	Reduction Gear/12V, 1.0kw	90	11.5	3000	38–52 ②	0.39
'85–'86	22R-E	Reduction Gear/12V, 1.0kw	90	11.5	3000	43–64 ②	0.335
'85–'86	22R-E	Reduction Gear/12V, 1.4kw	90	11.5	3500	43–64 ②	0.394
'84–'86	2L (Diesel)	Reduction Gear/12V, 2.5kw	180	11.0	3500	7.1–8.8 ③	0.472
'85–'86	2L-T (Diesel)	Reduction Gear/12V, 2.5kw	180	11.0	3500	7.1–8.8 ③	0.472
'83–'86	3Y-EC	Reduction Gear/12V, 1.0kw	90	11.5	3000	43–64 ②	0.335
'83–'86	3Y-EC	Reduction Gear/12V, 1.4kw	90	11.5	3000	43–64 ②	0.394

① Not specified by manufacturer
② Ounces
③ Pounds

BRAKE SPECIFICATIONS
(All measurements in inches)

Year	Model	Lug Nut Torque (ft. lbs.)	Front Brake Disc		Maximum Drum Diameter		Minimum Lining Thickness	
			Minimum Thickness	Maximum Run-Out	Front	Rear	Front	Rear
'79–'80	Pick-up	65–86	②	0.006	—	10.08	—	0.04
'81–'83	Pick-up	65–86	②	0.006	—	10.08	—	0.04
'84	Pick-up	65–86	④	0.006	—	10.08	0.04	0.04
'79	Land Cruiser	65–86	0.740	0.005	11.70	11.70	③	0.06
'80–'84	Land Cruiser	66–86	0.750	0.005	—	11.70	—	0.06
'85–'86	4 Runner & Pickup	76	① ④	0.006	—	10.08	0.04	0.04
'83–'86	Van	76	0.748	0.006	—	10.08	0.04	0.04

① 2WD (PD 60 type): 0.945
② Except cab and chassis—0.453; cab and chassis—0.748
③ Drum brakes—0.06; disc brakes—0.04
④ 2WD (FS-17 type): 0.827; 4WD (S-12 + 8 type): 0.453

TUNE-UP

Breaker Points and Condenser

REPLACEMENT

Loosen the distributor cap-to-body clips and lift the cap straight up. Leave the leads connected to the cap. Remove the rotor and dust cover.

Clean the distributor cap and rotor with alcohol. Inspect them for cracks and other signs of wear or damage. Polish the points with a point file.

NOTE: DO NOT use emery cloth or sandpaper; these may leave particles on the points, causing them to arc.

1. Disconnect the point lead connector.
2. Remove the point retaining clip and remove the point hold-down screw.
3. Remove the point set.
4. Rotate the engine by hand or by using a remote starter switch, until the rubbing block is on the high point of the cam lobe.
5. Insert a 0.018 in. feeler gauge between the points; a slight drag should be felt.

NOTE: Some 20R engines have a plastic cap over the contact points, making it necessary to adjust the point gap at the rubbing block. Set the rubbing block between two cam lobes and adjust the clearance to 0.018 in. (See illustration). Always check the dwell angle after setting the points.

6. If no drag is felt or if the feeler gauge cannot be inserted, loosen but do not remove, the point hold-down screw.
7. Insert a screwdriver into the adjustment slot. Rotate the screwdriver until the proper point gap is attained.

NOTE: The point gap is increased by rotating the screwdriver counterclockwise and decreased by rotating it clockwise.

8. Tighten the point hold-down screw. Lubricate the cam lobes, the breaker arm, the rubbing block, the arm pivot and the distributor shaft with special high-temperature distributor grease.
9. Check the operation of the centrifugal advance mechanism by moving the rotor clockwise. Release the rotor; it should return to its original position. If it does not, check it for binding. Check the vacuum

advance unit by removing the cap and pressing in on the octane selector. Release the octane selector. It should snap back to its original position. Check for binding if it fails to do so.

10. Replace the condenser if it is suspect or as routine maintenance during the point replacement operation, in the following manner:

Remove the nut and the washer from the condenser lead terminal.

Remove the condenser mounting screw and withdraw the condenser.

To install, reverse the removal procedures.

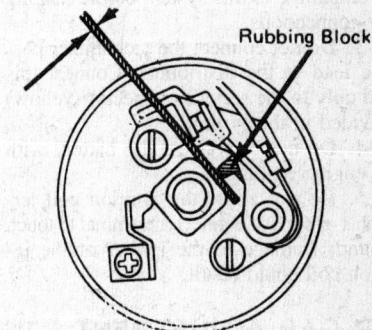

Some 20R engines are adjusted at the rubbing block rather than the contact points—see text

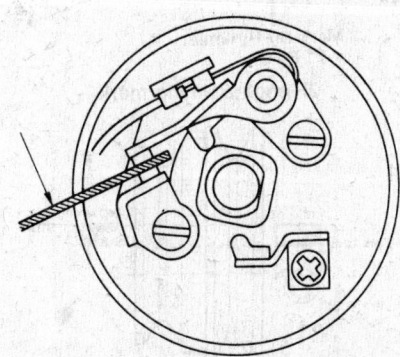

The arrow indicates the feeler gauge used to set the point gap. Make certain that the rubbing block rests on the high spot of the cam, as shown.

NOTE: The condenser is mounted on the outside of the distributor body on all models, except the Land Cruiser, which has it mounted inside the body.

Install the dust cover, the rotor and

the cap onto the distributor. Adjust the dwell and timing.

Dwell Angle

ADJUSTMENT

Connect a dwell/tachometer, in accordance with its manufacturer's instructions, between the distributor primary lead and a ground.

With the engine warmed up and running at the specified idle speed (see the tune-up chart), take a dwell reading. If the point dwell is not within specifications, shut the engine off and adjust the point gap, as outlined above.

NOTE: Increasing the point gap decreases the dwell angle and vice-versa.

Install the dust cover, the rotor and the cap. Check the dwell reading again.

Electronic Ignition

NOTE: The electronic ignition system is fully transistorized. A stationary magnetic pick-up coil (mounted in the distributor) and a toothed timing rotor (mounted on the distributor shaft) entirely replace the conventional breaker points and condenser. Because no mechanical contact exists between the pick-up coil and the timing rotor, the system is considered to be maintenance-free.

PRECAUTIONS—FULLY TRANSISTORIZED SYSTEM

1. If the engine will not start, DO NOT leave the ignition switch "ON" for more than 10 minutes.
2. Make sure that your test equipment is compatible to this system before making any connections.
3. Do not connect the tachometer positive lead to the distributor. Connect this lead only to the service connector (yellow) provided in the system.
4. Do not disconnect the battery with the engine running.
5. Do not allow the ignition coil terminals or service connector terminal to touch ground: Damage to the igniter or the ignition coil could result.

AIR GAP ADJUSTMENT

1. Remove the distributor cap, the rotor and the dust cover from the distributor.
2. Using the ignition switch, bump the engine until one of the timing rotor teeth aligns with the pick-up coil. Turn the ignition switch "OFF".
3. Using a flat, brass feeler gauge, check the gap between the timing rotor tooth and the pick-up coil. If adjustment is needed, loosen (but do not remove) the pick-up coil

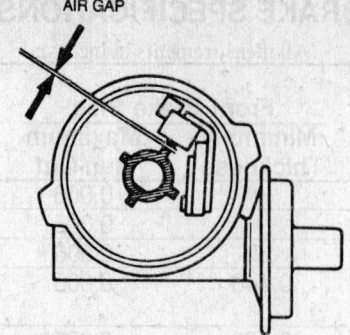

Check the air gap between the timing rotor and the pick-up coil

attaching screws and move the pick-up coil as necessary to attain the specified clearance. Tighten the pick-up coil attaching screws and recheck the clearance.

4. Reinstall the dust cover, the rotor and the distributor cap.

Refer to the Electrical section in Unit Repair for electronic ignition component troubleshooting procedures.

Ignition Timing

NOTE: The timing mark locations differ between the engines used in the Pickup and the 4-Runner (20R, 22R and 22R-E), the Van (3Y-EC) and the Land Cruiser (2F). The 20R, 22R, 22R-E and 3Y-EC timing marks are located on the crankshaft pulley (painted notch) and the timing cover (plate). The 2F timing marks are located on the flywheel (ball) and the bellhousing (pointer).

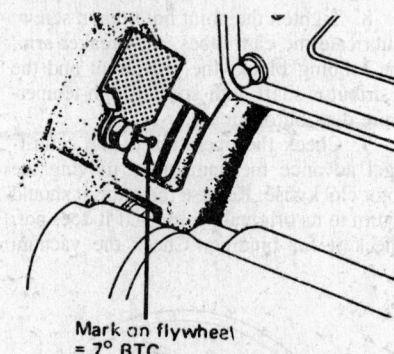

Mark on flywheel = 7° BTC

2F engine timing mark

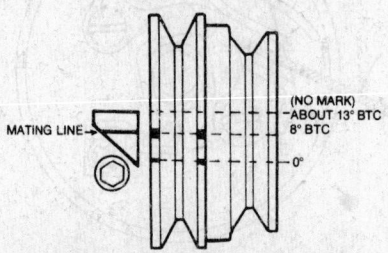

20R (1979) engine timing marks. The 8° notch is the larger and deeper of the two notches cut into the pulley. The 13° indication is only for 1979 trucks with High Altitude Compensation

1. Set the parking brake and block the wheels.
2. Clean off the timing marks and mark them with chalk or paint. The engine may have to bumped around (with the starter) to find the marks.
3. Warm the engine to operating temperatures. Connect a tachometer to the engine, then check and/or adjust the engine idle speed.

CAUTION

On the 20R, 22R and 22R-E engines, connect the positive (+) tachometer terminal either to the negative (−) ignition coil terminal (Type III) or to the yellow service connector (Type IV). On the 3Y-EC engine, connect the positive (+) tachometer to the service connector on the ignition coil/igniter (Type IIA) assembly. DO NOT connect it to the distributor side. Improper connections will damage the transistorized igniter.

---

4. Turn off the engine and connect a timing light according to the manufacturer's directions.

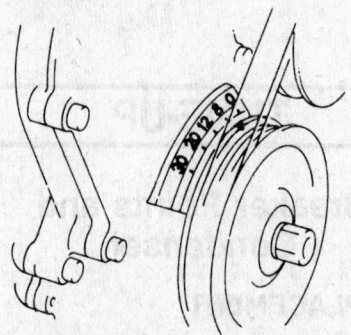

View of the timing marks used on the 3Y-EC engine—Van 1983 and later

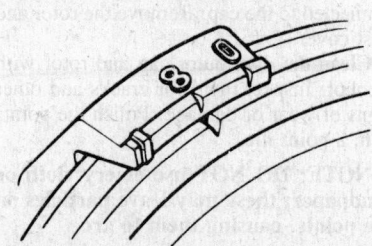

1980 and later, 20R, 22R and 22R-E engine timing marks—typical. The "8" mark is 8° before TDC. Some engines will also have a "5" mark which denotes 5° before top dead center

5. On the 22R-E engine, disconnect and short the "T" connector of the engine check harness (near the front of the vehicle). On all other models, disconnect and plug the vacuum hose(s) from the distributor vacuum unit.

NOTE: On equipped with a High Altitude Compensation (HAC) system there are two vacuum hoses which connect to the distributor. Both must be disconnected and plugged. These systems require an extra step in the timing procedure, found at the end of this section.

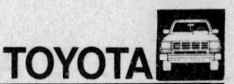
6. Be sure that the timing light wires are clear of the fan and pulleys, then start the engine.

CAUTION

Keep your fingers, clothes, hair, tools and wires clear of the fan and fan belts. Run the engine only in a well-ventilated area.

---

7. Allow the engine to run at the specified idle speed with the shift selector in Neutral (M/T) or Drive (A/T).

8. Point the timing light at the marks. With the engine at the specified idle, the marks should line up.

9. If the timing is incorrect, loosen the bolt at the base of the distributor just enough so that the distributor can be turned. Hold the distributor by its base and turn it slightly to advance or retard the timing as required. Once the marks are seen to align properly, tighten the bolt.

NOTE: On the 20R (1979) engine, minor timing corrections, may be adjusted with the octane selector, rather than by moving the distributor. See the Octane Selector section following for information.

10. After tightening the distributor bolt or adjusting the octane selector, recheck the timing. It is not unusual for it to change during the tightening process. It may take 2–3 tries to get it perfect. Turn off the engine, then disconnect the timing light and connect the vacuum line(s) at the distributor or the electrical "T" connector (22R-E), except on engines with HAC.

11. On engines with HAC (identified in the Note earlier) after setting the initial timing, reconnect the vacuum hoses at the distributor. Recheck the timing. It should now be about 20° BTDC (Van) or 12° BTDC (all other models).

12. If the advance is still low, pinch the hose between the HAC valve and the three way connector; it should now be to specifications. If not, the HAC valve should be checked for proper operation.

OCTANE SELECTOR ADJUSTMENT

20R Engine–1979

The octane selector is used as a fine adjustment to match the ignition timing to the grade of gasoline being used. It is located opposite the distributor vacuum advance unit, under a plastic cover. Normally the octane

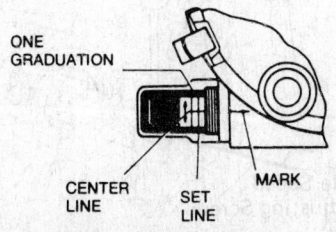

Octane selector setting

Valve clearance adjustment sequence—20R, 22R and 22R-E engines

selector should not require adjustment, however, adjustment is as follows:

1. Align the setting line with the threaded end of the housing and then align the center line with the setting mark on the housing.

2. Drive the truck at 16–22 miles per hour in High gear on a level road.

3. Depress the accelerator pedal all the way to the floor. A slight pinging sound should be heard; as the vehicle accelerates, the sound should go away.

4. If the pinging sound is loud or if it fails to disappear as the vehicle accelerates, retard the timing by turning the knob toward "R" (retard).

5. If there is no pinging sound at all, advance the timing by turning the knob toward "A" (advance).

6. When the adjustment is completed, replace the dust cover.

NOTE: One graduation of the octane selector is equal to about 10° of crankshaft angle.

Valve Adjustment

20R, 22R, 22R-E and 2F Engines

NOTE: If equipped with a Hot Air Intake (HAI) system or a Mixture Control (MC), disconnect and plug the hose(s) to prevent rough idling.

1. Start the engine and allow it to reach normal operating temperatures (above 175°F).

2. Stop the engine. Remove the air cleaner assembly, the hoses and the bracket, then any cables, hoses, wires and etc., which are attached to the valve cover. Remove the valve cover.

3. Set the No. 1 cylinder to TDC of the compression stroke. Place a wrench on the crankshaft pulley bolt and turn the engine until the notch on the crankshaft pulley is aligned with the 0° mark on the timing plate; the engine is at TDC.

NOTE: The rocker arms on cylinder No.1 should be loose and the rocker arms on cylinder No. 4 should be tight.

CAUTION

DO NOT start the engine. Valve clearances are checked with the engine stopped to prevent hot oil from being splashed out by the timing chain.

---

4. With the engine "Hot", the valve clearances are 0.008 in. (intake) and 0.012 in. (exhaust).

NOTE: The clearance is measured with a feeler gauge between the valve stem and the adjusting screw.

5. To adjust the valve clearance, loosen the locknut and turn the adjusting screw until the specified clearance is obtained. Tighten the locknut and check the clearance again. Adjust the intake valves of No.1 and 2 cylinders; the exhaust valves of No.1 and 3 cylinders.

6. Turn the crankshaft one revolution (360°). Adjust the intake valves of No.3 and 4 cylinders; the exhaust valves of No.2 and 4 cylinders.

7. To install the components, reverse the removal procedures.

NOTE: If you are assembling the engine after it has been dismantled, the valves must be adjusted with the engine cold. Obtain the valve clearance specification from the beginning of this section and add a minimum of .002" to these values. Remember: Excess clearance will not damage the engine whereas too little clearance could. After the initial engine start-up, allow the engine to reach normal operating temperature and adjust the valves as previously described to the "Hot" clearances listed in the specification chart.

3Y-EC Engines

Since this engine uses hydraulic lifters, no adjustment is necessary.

Diesel Engine

The valves are adjusted in basically the same manner as the gasoline engines, in that the engine must be off during the adjustment and that the clearance is checked with a feeler gauge between the rocker arm and the valve stem end.

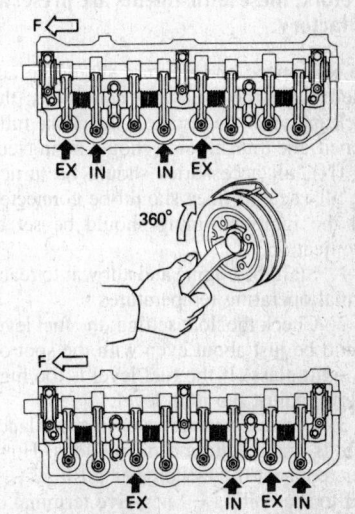

Valve clearance adjustment sequence—diesel engine

NOTE: The engine must be at normal operating temperature to obtain the proper valve clearances.

1. Remove the valve cover and rotate the crankshaft to align the TDC mark on the crankshaft pulley with the corresponding pointer. The valves of the number one cylinder should be closed (rocker arms should feel loose). If the rocker arms of the number one cylinder are tight, rotate the engine another 360° and again align the TDC marks.

2. With the engine "Hot", the valve clearances are 0.010 in. (intake) and 0.014 in. (exhaust).

3. Adjust the intake valves of No.1 and 2 cylinders; the exhaust valves of No.1 and 3 cylinders.

4. Rotate the crankshaft 360°, then place the crankshaft pulley notch on the 0° mark of the timing plate. Adjust the intake valves of No.3 and 4 cylinders; the exhaust valves of No.2 and 4 cylinders.

NOTE: Remember that the cylinders are numbered from the front of the engine to the rear and the valve arrangement from the front of the engine is E–I–E–I–E–I–E–I; "E" for exhaust valve and "I" for intake valve.

5. To install the components, reverse the removal procedures.

— CAUTION —
Never operate the engine with the valve cover removed, for "Hot" oil will be splashed everywhere.

Idle Speed and Mixture

CARBURETOR MODELS

20R (1979–80)2F (1979–84) Engines

NOTE: Idle mixture adjustments cannot be performed on 1981 and later carburetors; these adjustments are preset at the factory.

The idle speed and mixture should be adjusted under the following conditions: the air cleaner must be installed, the choke fully opened, the transmission should be in Neutral (N), all accessories should be turned off, all vacuum lines should be connected and the ignition timing should be set to specification.

1. Start the engine and allow it to reach normal operating temperatures.

2. Check the float setting; the fuel level should be just about even with the spot on the sight glass. If the fuel level is too high or low, adjust the float level.

3. Connect a tachometer in accordance with its manufacturer's instructions. However, connect the tachometer positive (+) lead to the coil's (−) negative terminal or to the igniter's service connector (if provided).

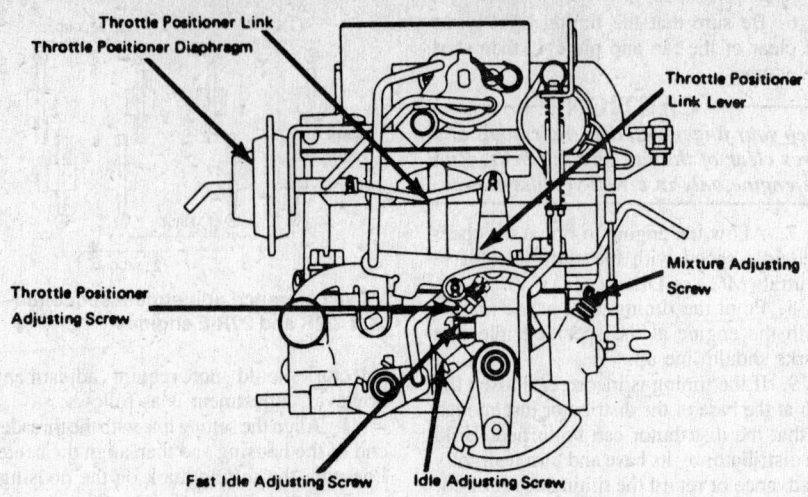

Carburetor adjusting screws—2F engine

— CAUTION —
DO NOT connect the tachometer to the distributor side; damage to the transistorized ignition could result. NEVER allow the tachometer terminal to touch ground for damage to the igniter or the ignition coil could result.

4. Using a pair of pliers, break the caps from the idle mixture screws. Turn the idle speed adjusting screw to obtain one of the following initial idle speeds: 20R–800 rpm (M/T) or 850 rpm (A/T): 2F–690 rpm (M/T).

5. Turn the idle mixture adjusting screw to increase the idle speed as much as is possible.

6. Next, turn the idle speed screw to again obtain the same idle speed figure given in Step 4.

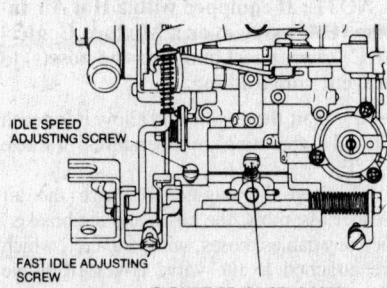

Carburetor adjusting screws—20R engines

7. If possible, turn the idle mixture screw to increase the idle speed again.

8. Repeat Steps 6 and 7 until the idle mixture adjusting screw will no longer increase the idle speed above the figure specified in Step 4.

9. Slowly turn the idle mixture screw clockwise, until the idle speed specified in the "Tune-Up Specifications" chart is reached (this makes the mixture leaner).

10. Disconnect the tachometer and install new idle mixture screw caps.

22R (1981 and Later) Engine

NOTE: The idle mixture screw is preset at the factory and no adjustment is needed or necessary.

The idle speed should be adjusted under the following conditions: the air cleaner must be installed, the choke fully opened, the transmission should be in Neutral (N), all accessories should be turned off, all vacuum lines should be connected and the ignition timing should be set to specification.

1. Start the engine and allow it to reach normal operating temperatures.

2. Check the float setting; the fuel level should be just about even with the spot on the sight glass. If the fuel level is too high or low, adjust the float level.

3. Connect a tachometer in accordance with its manufacturer's instructions. However, connect the tachometer positive (+) lead to the coil's (−) negative terminal or to the igniter's service connector (if provided).

— CAUTION —
DO NOT connect the tachometer to the distributor side; damage to the transistorized ignition could result. NEVER allow the tachometer terminal to touch ground for damage to the igniter or the ignition coil could result.

4. Using a pair of pliers, break the caps from the idle speed adjusting screw. Turn the idle speed adjusting screw to obtain the

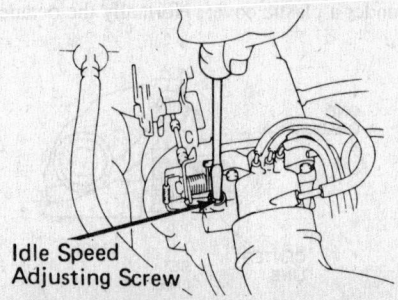

Adjusting the idle speed—22R-E engine

correct idle speed. On the 1981 models: 700 rpm (M/T and Federal 4-speed A/T) or 750 rpm (except, M/T or Federal 4-speed A/T); on the 1982 and later models: 700 rpm (M/T) or 750 rpm (A/T).

5. Disconnect the tachometer and install new idle speed adjusting screw cap.

FUEL INJECTION (GASOLINE)

22R-E (1984 and Later) Engine 3Y-EC (1983 and Later) Engine

The engines are equipped with a computer activated, electronic fuel injection system. Prior to adjusting the idle speed, make sure that: The air cleaner is installed. All vacuum hoses are connected. All pipes and hoses in the air intake system are connected and in good condition. All fuel injection system wiring is connected and in good condition. The engine is at normal operating temperature. All accessories are OFF. Transmission in Neutral.

——— CAUTION ———

Not all tachometers are compatible with the fuel injection system; consult the tachometer manufacturer's recommendations before installing the tachometer. NEVER allow the tachometer terminal to touch ground for damage to the igniter or the ignition coil could result.

1. Connect the tachometer positive (+) lead to the coil's (−) negative terminal or to the igniter's service connector (if provided).

2. Race the engine at 2,500 rpm for 2 minutes.

3. Run the engine at idle and turn the idle speed adjusting screw to obtain a the correct speed of: 700 rpm (M/T) or 750 rpm (A/T) for the 3Y-EC engine; 750 rpm for 22R-E engine.

4. Disconnect and remove the tachometer.

FUEL INJECTED (DIESEL)

L, 2L and 2L-T Engines

NOTE: The following adjustments are made with the transmission in Neutral, the parking brake applied fully, the accessories turned OFF and the air cleaner installed.

1. Warm the engine to normal operating temperatures and allow it to idle.

NOTE: If equipped with an L engine (1981–83), turn the idle adjustor knob counterclockwise; the knob should return to its unlocked position.

2. Remove the accelerator connector rod.

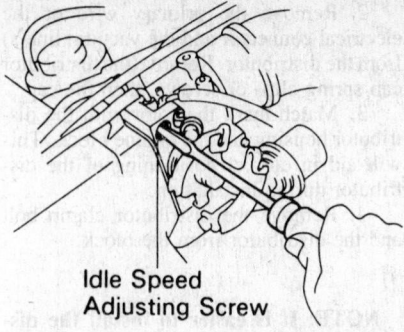

Idle Speed Adjusting Screw

Adjusting the idle speed—3Y-EC engine

3. Connect a tachometer to the engine according to the manufacturer's instructions and check the engine rpm at idle. The idle rpm should be 700 rpm.

4. If necessary, turn the idle adjusting screw on the fuel injection pump as required to obtain the 700 rpm idle speed.

5. Fully depress the injection pump lever, note the maximum engine speed and release the accelerator pedal immediately. The maximum rpm should be 4900.

6. If adjustment is necessary: Remove the wire seal of the maximum speed adjusting screw, if equipped. Using the Toyota tool No. 09275–54020, loosen the locknut of the maximum speed adjusting screw. Turn the maximum speed adjusting screw until the proper maximum rpm is obtained.

7. Install the accelerator connecting rod and adjust its length so that there is no slack in the accelerator cable.

8. Fully depress the accelerator pedal and make sure that the maximum speed is 4900 rpm; if not, adjust the stop bolt at the accelerator pedal.

9. Check that the idle speed increases as the idle adjustor knob is pulled outward and turned clockwise. Then turn the knob counterclockwise so that the rpm returns to the idle specification.

10. Turn the engine OFF and disconnect the tachometer from the engine.

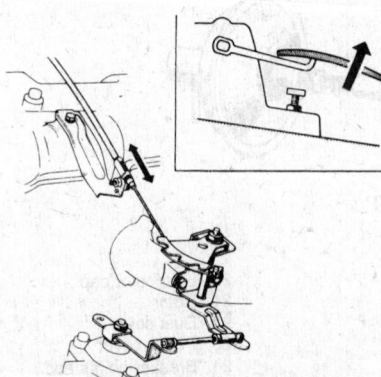

After setting the idle and maximum speeds on diesel engines, reinstall the accelerator cable and adjust the cable so that there is no slack—later models are similar

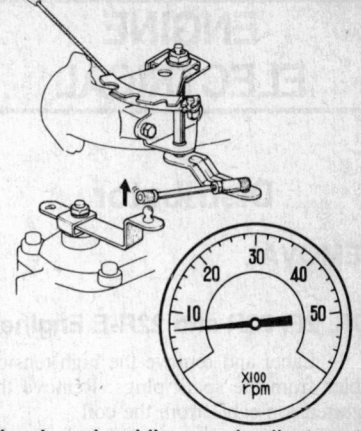

Diesel engine idle speed adjustment—disconnect the accelerator rod at the injection pump lever as shown, prior to adjustment

Diesel engine idle speed adjustment screw location

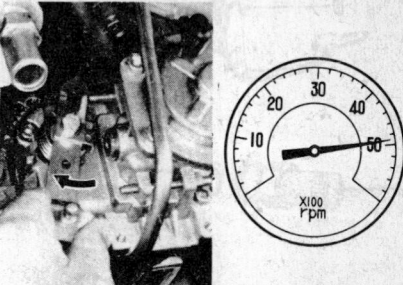

Diesel engine maximum speed adjustment—push the lever until it contacts the maximum speed adjusting screw

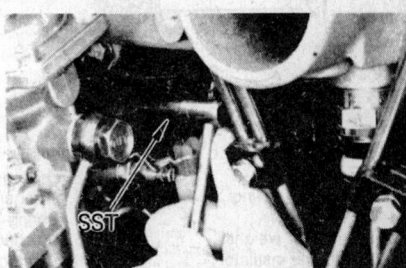

Diesel engine maximum speed adjustment—using the Toyota special service tool to loosen the maximum speed adjusting screw locknut

ENGINE ELECTRICAL

Distributor

REMOVAL

20R, 2F, 22R and 22R-E Engines

1. Label and remove the high tension cables from the spark plugs. Remove the high tension cable from the coil.

2. Remove the primary wire or the electrical connector and the vacuum line(s) from the distributor. Remove the distributor cap spring clips or screws, then the cap.

3. Match-mark the rotor with the distributor housing and the engine block. This will aid in correct positioning of the distributor during installation.

4. Remove the distributor clamp bolt and the distributor from the block.

NOTE: It is easier to install the distributor if the engine timing is not disturbed while it is removed. If the timing has been lost, see "Installation–Timing Disturbed".

3Y-EC Engine–Type IIA

1. Disconnect the negative battery cable.

2. Remove the front passenger seat from the vehicle.

3. Remove the engine service hole cover.

4. Disconnect the distributor vacuum advance hoses.

5. Disconnect the high tension cables from the spark plugs.

6. Remove the mounting bolt and the distributor from the engine.

INSTALLATION–TIMING NOT DISTURBED

20R, 2F, 22R and 22R-E Engines

1. Insert the distributor into the engine block and align the matchmarks made during removal.

2. Engage the distributor drive with the oil pump drive shaft.

3. Install the distributor clamp, the cap, the high tension wire, the primary wire or the electrical connector and the vacuum line(s).

4. Install the spark plugs cables.

5. Start the engine, check and/or adjust the timing, then adjust the octane selector (if equipped).

INSTALLATION–TIMING DISTURBED

20R, 2F, 22R and 22R-E Engines

If the engine has been cranked, dismantled or the timing otherwise lost, proceed as follows:

1. Determine the top dead center (TDC) of the No.1 cylinder's compression stroke by removing the spark plug from the No.1 cylinder and placing a finger or a compression gauge over the spark plug hole.

NOTE: Using a wrench, turn the crankshaft until the compression pressure starts to build up. Continue cranking the engine until the timing marks indicate TDC (0°).

2. Turn the crankshaft to align the timing marks on the 20R and 2F engine to 8° BTDC, on the 22R engine (1981–84), on the 22R-E engine (1984 and later) to 5° BTDC or on the 22R engine (1985 and later) to 0° TDC.

3. Temporarily install the rotor on the distributor shaft so that the rotor is pointing toward the No.1 terminal of the distributor cap.

4. Using a small screwdriver, align the slot on the distributor drive (oil pump drive-

1. Grease stopper	11. O-ring
2. Cam	12. Drive gear
3. Governor spring	13. Washer
4. E-ring	14. Spring
5. Governor weight	15. Washer
6. Terminal insulator	16. Bearing
7. Rubber plug	17. Washer
8. Hold-down clip for cap	18. Distributor shaft
9. Octane selector cap	19. Vacuum unit and octane selector assembly
10. Distributor housing	

20. Distributor cap
21. Rotor
22. Dust cover
23. Points cover
24. Breaker points and ground wire
25. Breaker plate
26. Damping spring

20R engine point type distributor

shaft) with the key on the bottom of the distributor shaft.

5. Install the distributor in the block by rotating it slightly (no more than one gear tooth in either direction) until the driven gear meshes with the drive.

NOTE: Oil the distributor drive gear and the oil pump driveshaft end before installation.

6. Temporarily tighten the lock bolt.

7. Remove the rotor, then install the dust cover, the rotor and the distributor cap.

8. Install the primary wire or the electrical connector and the vacuum line(s).

9. Install the No. 1 cylinder spark plug. Connect the cables to the spark plugs in the proper order by using the labels made during the removal procedures. Install the high tension wire on the coil.

10. Start the engine and adjust the ignition timing, then octane selector (if equipped).

3Y-EC Engine–Type IIA

1. Remove the No. 1 spark plug, place your finger over the opening and rotate the crankshaft (with a wrench) in the clockwise direction, until pressure is felt (this is TDC), then replace the spark plug.

NOTE: Make sure that the notch on the crankshaft pulley is aligned with the 0° mark on the timing plate.

2. Position the oil pump drive rotor slot 30° from the centerline.

3. On the distributor, align the groove on the housing with the pin of the driven gear (the drill mark side).

4. Insert the distributor by aligning the flange center with the bolt hole in the engine block.

5. Lightly tighten the hold-down bolt.

6. To complete the installation, reverse the removal procedures. Adjust the ignition timing.

Alternator

PRECAUTIONS

1. Always observe proper polarity of the battery connections; be especially careful when jump-starting the vehicle.

2. Never ground or short out any alternator or alternator regulator terminals.

3. Never operate the alternator with any of its or the battery's leads disconnected.

4. Always remove the battery or disconnect its output lead while charging it.

5. Always disconnect the ground cable when replacing any electrical components.

6. Never subject the alternator to excessive heat or dampness if the engine is being steam-cleaned.

7. Never use arc-welding equipment with the alternator connected.

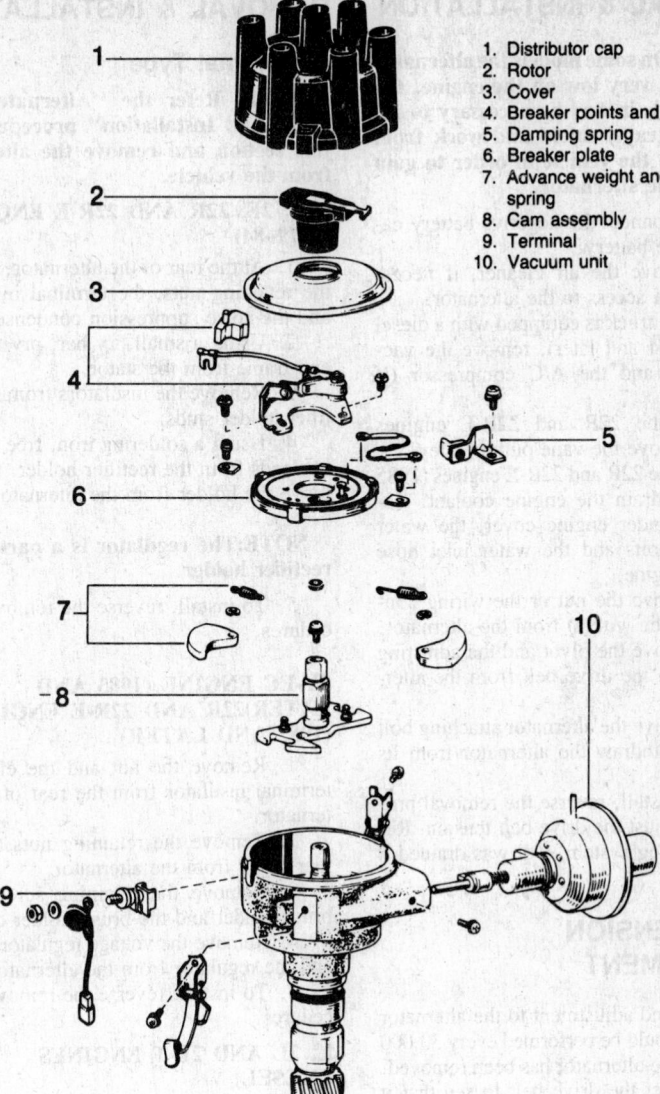

1. Distributor cap
2. Rotor
3. Cover
4. Breaker points and cover
5. Damping spring
6. Breaker plate
7. Advance weight and spring
8. Cam assembly
9. Terminal
10. Vacuum unit

2F engine point type distributor

1. Cam grease stopper
2. Signal rotor
3. Governor spring
4. Governor weight
5. Governor shaft
6. Plate washer
7. Compression coil spring
8. Thrust bearing
9. Washer
10. Dustproof packing
11. Steel plate washer
12. Rubber washer
13. Octane selector cap
14. Housing
15. O-ring
16. Spiral gear
17. Pin
18. Vacuum unit
19. Cord clamp
20. Breaker plate
21. Signal generator
22. Dustproof cover
23. Distributor rotor
24. Distributor cap
25. Rubber cap

Electronic ignition distributor used on 20R engines

REMOVAL & INSTALLATION

NOTE: On some models the alternator is mounted very low on the engine. On these models it may be necessary to remove the gravel shield and work from underneath the vehicle in order to gain access to the alternator.

1. Disconnect the negative battery cable from the battery.
2. Remove the air cleaner, if necessary, to gain access to the alternator.
3. If the truck is equipped with a diesel engine (1984 and later), remove the vacuum pump and the A/C compressor (if equipped).
4. On the 22R and 22R-E engines (1984), remove the vane pump pulley.
5. On the 22R and 22R-E engines (1985 and later), drain the engine coolant. Remove the under engine cover, the water inlet pipe bolts and the water inlet hose from the engine.
6. Remove the nut or the wiring connector and the wire(s) from the alternator.
7. Remove the pivot and the adjusting bolt(s), then the drive belt from the alternator.
8. Remove the alternator attaching bolt and then withdraw the alternator from its bracket.
9. To install, reverse the removal procedures. Adjust the drive belt tension. Refill the cooling system, if it was drained.

BELT TENSION ADJUSTMENT

Inspection and adjustment to the alternator drive belt should be performed every 30,000 miles or if the alternator has been removed.

1. Inspect the drive belt to see that it is not cracked or worn. Be sure that its surfaces are free of grease or oil.
2. Push down on the belt halfway between the fan and the alternator pulleys, (or crankshaft pulley) with thumb pressure. Belt deflection should be $\frac{3}{8}$–$\frac{1}{2}$ in.
3. If the belt tension requires adjustment, loosen the adjusting link bolt and move the alternator until the proper belt tension is obtained.

─────── CAUTION ───────
Do not overtighten the belt; damage to the alternator bearings could result.
─────────────────────────

4. Tighten the adjusting link bolt.

Voltage Regulator

Two types of voltage regulators are used: the Integrated Circuit (IC) and the Tirrill. The IC type is internally mounted in the alternator and is not adjustable. The Tirrill is externally mounted and is adjustable.

REMOVAL & INSTALLATION

IC—Internal Type

NOTE: Refer the "Alternator, Removal and Installation" procedures, in this section and remove the alternator from the vehicle.

20R, 2F, 22R AND 22R-E ENGINES (1979–84)

1. At the rear of the alternator, remove the retaining nuts, the terminal insulators and the noise suppression condenser.
2. Using a small pry bar, pry the rear end frame from the stator.
3. Remove the insulators from the rectifier holder studs.
4. Using a soldering iron, free the stator leads from the rectifier holder, then remove the holder from the alternator.

NOTE: The regulator is a part of the rectifier holder.

5. To install, reverse the removal procedures.

3Y-EC ENGINE (1983 AND LATER) 22R AND 22R-E ENGINES (1985 AND LATER)

1. Remove the nut and the electrical terminal insulator from the rear of the alternator.
2. Remove the retaining nuts and the rear cover from the alternator.
3. Remove the retaining screws, the brush holder and the brush holder cover.
4. Remove the voltage regulator screws and the regulator from the alternator.
5. To install, reverse the removal procedures.

L, 2L AND 2L-T ENGINES (DIESEL)

1. Remove the brush holder cover retaining nuts, the insulator, the rubber washer and the cover from the alternator.
2. Remove the screw and disconnect the blue wire from the brush holder/IC regulator assembly.
3. Lift the brush holder/IC regulator assembly from the rectifier holder, then remove the screw and the wire from the assembly.
4. Remove the screws and separate the IC regulator from the brush holder.
5. To install, reverse the removal procedures.

Tirrill—External Type (1979–84)

1. Disconnect the negative battery cable.
2. Disconnect the wiring harness.

NOTE: On Land Cruisers disconnect the leads from their screw terminals after noting their position for installation.

3. Remove the retaining hardware and remove the regulator.
4. To install, reverse the removal procedures.

VOLTAGE ADJUSTMENT

NOTE: Only external regulators used with gasoline engines are adjustable.

1. Disconnect the battery wire from the "B" terminal of the alternator. Remove the cover from the regulator assembly.
2. Using a voltmeter, connect the (+) positive lead to the "B" terminal of the alternator and the (−) negative lead to a ground.
3. Start the engine and gradually increase the engine speed to about 2000 rpm.
4. At this speed, the voltage reading should be 13.8–14.8 volts.
5. If the voltage does not fall within this range, a minor adjustment may be made to the adjusting arm. Disconnect the ground cable of the battery and remove the regulator cover. Bend the adjusting arm very slightly with a pair of needle nose pliers. Replace the cover and battery cable.
6. To install, reverse the removal procedures.

FIELD RELAY ADJUSTMENT

NOTE: Only external regulators used with gasoline engines are adjustable. This adjustment does not apply to Land Cruisers.

1. Remove the cover from the regulator assembly.
2. Clean off the points with emery cloth if they are dirty and wash them with solvent.
3. The relay actuating voltage is 4.5–5.8 volts. Adjust the point gap, as required, by bending the adjusting arm.
4. To install, reverse the removal procedures.

Starter

REMOVAL & INSTALLATION

1. Disconnect the negative battery cable at the battery end.

NOTE: On the 22R and 22R-E (1984 and later) engines equipped with an A/T, remove the transmission oil filler tube.

2. Disconnect all the wiring at the starter.

NOTE: If equipped with a 3Y-EC or an L-series (diesel) engine, disconnect the electrical connector from the starter.

3. Remove the mounting bolts and move the starter toward the front of the vehicle.
4. To install, reverse the removal procedures.

Refer to the Electrical section of Unit Repair for starter motor and alternator servicing information.

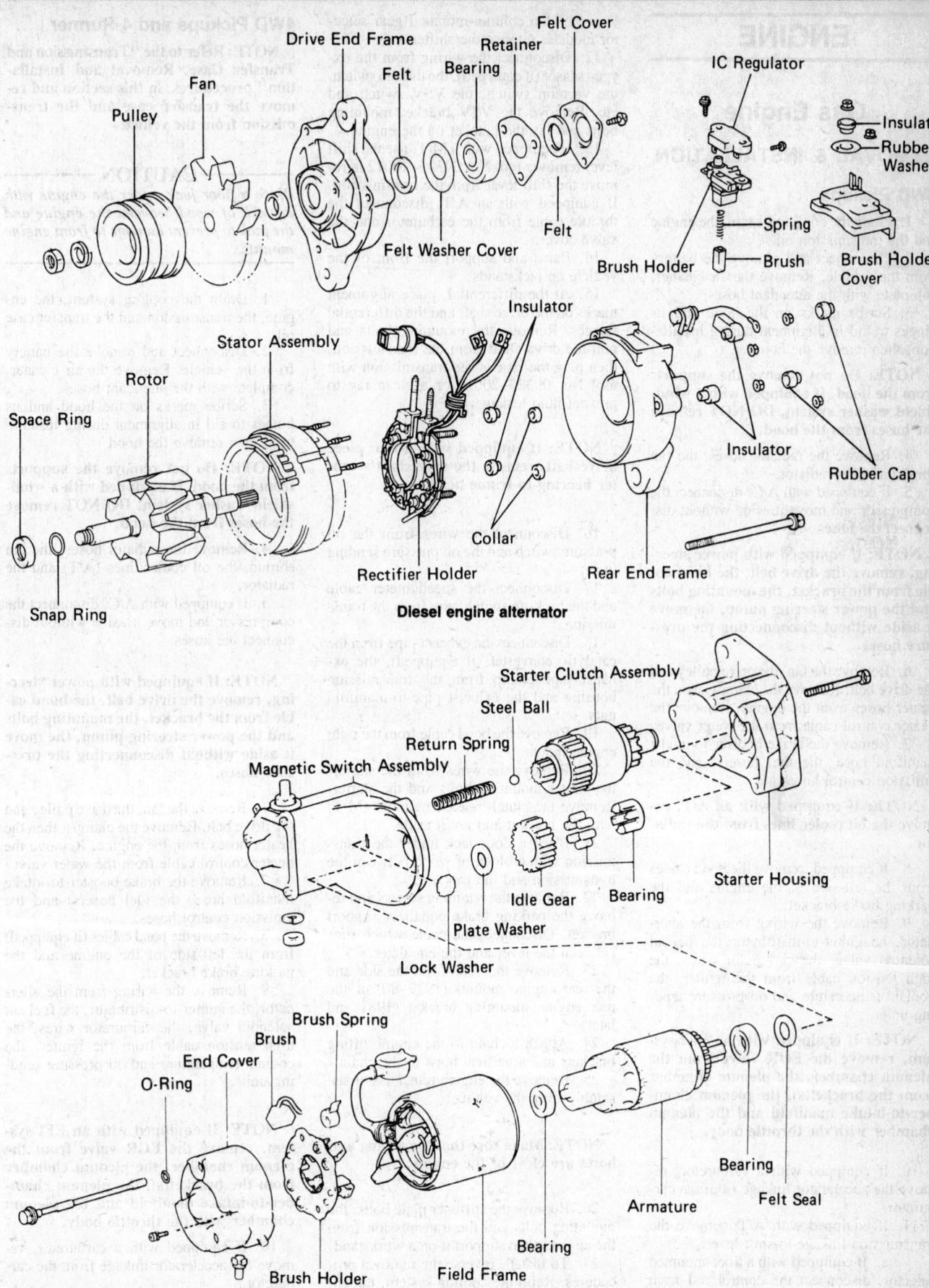

Pulley
Fan
Drive End Frame
Felt
Bearing
Retainer
Felt Cover
IC Regulator
Insulator
Rubber Washer
Spring
Brush Holder
Brush
Brush Holder Cover
Felt
Felt Washer Cover

Space Ring
Rotor
Stator Assembly
Insulator
Insulator
Rubber Cap
Snap Ring
Rectifier Holder
Collar
Rear End Frame

Diesel engine alternator

Starter Clutch Assembly
Steel Ball
Return Spring
Magnetic Switch Assembly
Idle Gear
Bearing
Plate Washer
Lock Washer
Starter Housing

Brush Spring
Brush
End Cover
O-Ring
Bearing
Armature
Bearing
Felt Seal
Brush Holder
Field Frame
Bearing

Diesel engine starter

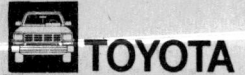
ENGINE

Gas Engine

REMOVAL & INSTALLATION

2WD Pickups

1. Drain the cooling system, the engine and the transmission oil.

2. Disconnect and remove the battery from the vehicle. Remove the air cleaner, complete with the attendant hoses.

3. Scribe marks on the hood and its hinges to aid in alignment during installation, then remove the hood.

NOTE: Do not remove the supports from the hood. If equipped with a windshield washer system, DO NOT remove the hoses from the hood.

4. Remove the radiator hoses, the fan shroud and the radiator.

5. If equipped with A/C, disconnect the compressor and move it aside without disconnect the hoses.

NOTE: If equipped with power steering, remove the drive belt, the bond cable from the bracket, the mounting bolts and the power steering pump, the move it aside without disconnecting the pressure hoses.

6. Remove the fan, the drive pulley and the drive belt. Remove the clamps, then the heater hoses from the engine. Remove the heater control cable from the water valve.

7. Remove the brake booster-to-intake manifold hose, the fuel hose(s) and the emission control hoses.

NOTE: If equipped with an A/T, remove the oil cooler lines from the radiator.

8. If equipped, remove the bond cables from the left-side of the engine and the parking brake bracket.

9. Remove the wiring from the alternator, the igniter-to-distributor, the fuel cut solenoid valve, the carburetor wires, the high tension cable from the igniter, the coolant temperature and oil pressure sending units.

NOTE: If equipped with an EFI system, remove the EGR valve from the plenum chamber, the plenum chamber from the bracket(s), the plenum chamber-to-intake manifold and the plenum chamber with the throttle body.

10. If equipped with a carburetor, remove the accelerator linkage from the carburetor.

11. If equipped with A/T, remove the transmission linkage-to-shift lever.

　a. If equipped with a floor-mounted selector, disconnect the control rod from the transmission.

　b. On column-mounted gear selector models, remove the shifter rod.

12. Disconnect the wiring from the oxygen sensor (if equipped), the thermo switch, the vacuum switch, the VSV switch and etc. Remove the VSV bracket mounting bolts and lay the bracket on the engine.

13. If equipped with a M/T, use the shift lever remover tool No. 09305–20012 to remove the shift lever from the transmission. If equipped with an A/T, disconnect the throttle cable from the carburetor and the valve cover.

14. Raise and support the front of the vehicle on jackstands.

15. At the differential, place alignment marks on the drive shaft and the differential flanges. Remove the mounting bolts and pull the drive shaft from the transmission, then plug the rear of the transmission with tool No. 09325–20010 or a clean rag to prevent fluid leakage.

NOTE: If equipped with a two piece driveshaft, remove the driveshaft's center bearing-to-frame bolts.

16. Disconnect the wires from the oil pressure switch and the oil pressure sending unit.

17. Disconnect the speedometer cable and the back-up switch wire from the transmission.

18. Disconnect the exhaust pipe from the catalytic converter (if equipped), the exhaust pipe clamp from the transmission housing and the exhaust pipe-to-manifold nuts.

19. Remove the bond cable from the right engine mount.

20. Remove the wires from the starter, the starter mounting bolts and the starter. Remove the clutch release cylinder (M/T) with the bracket and lay it aside.

21. Place a floor jack under the transmission with a block of wood between the transmission and the jack.

22. Remove the retaining screws and remove the parking brake equalizer support bracket. Disconnect the cable which runs between the lever and the equalizer.

23. Remove the bolts from the side and the rear engine mounts (1979–80) or the rear engine mounting bracket (1981 and later).

24. Attach a chain to the engine lifting brackets and a vertical hoist to the chain.

25. Remove the engine/transmission assembly from the vehicle.

NOTE: Make sure that the wiring and hoses are clear of the engine.

26. Remove the stiffener plate bolts, the mounting bolts and the transmission from the engine, then support it on a workstand.

27. To install, reverse the removal procedures. Refill the cooling system, the engine and the transmission.

4WD Pickups and 4-Runner

NOTE: Refer to the "Transmission and Transfer Case, Removal and Installation" procedures, in this section and remove the transfer case and the transmission from the vehicle.

——— CAUTION ———
Place a floor jack under the engine with a block of wood between the engine and the jack to prevent damage to front engine mounts.

1. Drain the cooling system, the engine, the transmission and the transfer case oil.

2. Disconnect and remove the battery from the vehicle. Remove the air cleaner, complete with the attendant hoses.

3. Scribe marks on the hood and its hinges to aid in alignment during installation, then remove the hood.

NOTE: Do not remove the supports from the hood. If equipped with a windshield washer system, DO NOT remove the hoses from the hood.

4. Remove the radiator hoses, the fan shroud, the oil cooler lines (A/T) and the radiator.

5. If equipped with A/C, disconnect the compressor and move it aside without disconnect the hoses.

NOTE: If equipped with power steering, remove the drive belt, the bond cable from the bracket, the mounting bolts and the power steering pump, the move it aside without disconnecting the pressure hoses.

6. Remove the fan, the drive pulley and the drive belt. Remove the clamps, then the heater hoses from the engine. Remove the heater control cable from the water valve.

7. Remove the brake booster-to-intake manifold hose, the fuel hose(s) and the emission control hoses.

8. Remove the bond cables (if equipped) from the left-side of the engine and the parking brake bracket.

9. Remove the wiring from the alternator, the igniter-to-distributor, the fuel cut solenoid valve, the carburetor wires, the high tension cable from the igniter, the coolant temperature and oil pressure sending units.

NOTE: If equipped with an EFI system, remove the EGR valve from the plenum chamber, the plenum chamber from the bracket(s), the plenum chamber-to-intake manifold and the plenum chamber with the throttle body.

10. If equipped with a carburetor, remove the accelerator linkage from the carburetor.

11. Disconnect the wiring from the ox-

ygen sensor (if equipped), the thermo switch, the vacuum switch, the VSV switch and etc. Remove the VSV bracket mounting bolts and lay the bracket on the engine.

12. If equipped with an A/T, disconnect the throttle cable from the carburetor and the valve cover.

13. Raise and support the front of the vehicle on jackstands.

14. Disconnect the wires from the oil pressure switch and the oil pressure sending unit.

15. Disconnect the exhaust pipe from the catalytic converter (if equipped) and the exhaust pipe-to-manifold nuts.

16. Remove the bond cable from the right engine mount.

17. Remove the wires from the starter, the starter mounting bolts and the starter. Remove the clutch release cylinder (M/T) with the bracket and lay it aside.

18. Remove the retaining screws and remove the parking brake equalizer support bracket. Disconnect the cable which runs between the lever and the equalizer.

19. Attach a chain to the engine lifting brackets and a vertical hoist to the chain.

20. Remove the engine from the vehicle, then support it on a workstand.

--- CAUTION ---

When removing the engine, make sure that the wiring and hoses are clear of the engine.

21. To install, reverse the removal procedures. Refill the cooling system, the engine, the transmission and the transfer case.

Land Cruiser & Wagon

1. Scribe alignment marks on the hood and hinges, then remove the hood. Drain the cooling system and engine oil.

2. Remove the radiator grille mounting bolts and the grille.

NOTE: On station wagon models, remove the parking light assembly and wiring first.

3. Remove the hood latch support rod. Detach the hood latch assembly from the radiator upper bracket, then remove the bracket.

4. Disconnect the heater and the radiator hoses from the radiator. Remove the radiator bolts and lift the radiator out of the vehicle.

5. Remove the heater hoses from the water valve and heater box. Disconnect the temperature control cable from the water valve.

6. Remove the the battery cables and the battery. Remove the wires from the starter solenoid terminal.

7. Remove the fuel lines from the pump and the fuel filter assembly.

8. Disconnect the primary wire from the ignition coil.

9. On the column shift models, detach both intermediate rods from the shifter shafts.

10. Remove the air cleaner assembly complete with hoses, from its bracket. Remove the emission control system cables and hoses. Remove the multi-connector from the alternator.

11. Disconnect the hand throttle, the accelerator and the choke linkages from the carburetor.

12. If equipped with vacuum-assisted 4WD engagement, remove the control unit vacuum hose from its manifold fitting.

13. Disconnect the oil pressure and water temperature gauge sender's wiring.

14. Remove the downpipe from the exhaust manifold. Detach the parking brake cable from the intermediate lever.

15. Unbolt and remove the front driveshaft from the flange on the transfer case output shaft.

16. Remove both the left and right engine stone shields, then the transmission skid-plate.

17. Remove the cotter pin and disconnect both the high and low-range shifter rods from their respective inner levers. Remove the high/low range shifter link lever and rod.

18. Disconnect the clutch release fork spring. Remove the clutch release cylinder from its mounting bracket at the rear of the engine.

19. If equipped with vacuum-assisted 4WD engagement, remove the clamp screws and withdraw the vacuum lines from the transfer case, control unit vacuum chamber.

20. Remove the 4WD indicator switch assembly.

21. Remove the speedometer cable from the transmission.

22. Disconnect the rear driveshaft from the transmission.

23. Detach the gearshift and the gear selector rods from the shift and the gear selector outer levers, respectively.

24. Remove the front and the rear engine mounts from the frame.

25. Install lifting hooks on the engine lift-points and connect to a vertical hoist.

26. Lift the engine slightly and toward the front, so the engine/transmission assembly clears the front of the vehicle.

27. To install, reverse the removal procedures. Refill the engine with coolant and lubricant.

Van

1. Disconnect the negative battery cable.

2. Remove the right seat and the engine service hole cover.

3. Drain the coolant from the radiator. Remove the reservoir tank, the heater hoses and the radiator.

4. Remove the air cleaner, the breather tube, the brake booster, the charcoal canister and the fuel hoses from the engine.

5. If equipped with power steering, remove the drive belt, the pulley, the wood-ruff key and the pump from the engine, then move the pump aside.

NOTE: When removing the power steering pump, DO NOT disconnect the pressure lines unless it is absolutely necessary.

6. Disconnect the accelerator cable with the bracket from the throttle body.

7. Disconnect the following wiring connectors from the: water temperature sender, oil pressure switch, IIA unit, A/C compressor, idle-up (A/C), VSV (A/C), A/T, water temperature switch (A/T), alternator connector and wire, air flow meter, solenoid resistor and etc.

8. Remove the fan shroud, the fan, the fluid coupling and the water pump pulley.

9. From inside the vehicle, remove the center pillar cover, the seat belt retractor and cover, then disconnect the electrical connectors from the ECU.

10. If equipped with A/C, remove the drive belt and the compressor mounting bolts, then move the compressor aside.

11. Raise and support the vehicle on jackstands, about 3 ft. off the floor.

12. Drain the engine oil. Remove the driveshaft and the front exhaust pipe.

13. Remove the transmission selector and shift cables, then the clutch release cylinder (M/T).

14. Disconnect the starter wires, the mounting bolts and the starter from the engine.

15. Remove the speedometer cable, the bond cable and the back-up light switch connector.

16. If equipped with a rear heater, disconnect the mode selector and the air mix damper cable from the damper. Disconnect the heater hoses to the rear heater unit.

17. Disconnect the bond cable(s) from the engine mount(s). Remove the engine under cover.

18. Disconnect the oil level sensor and the oil cooler hoses (A/T).

19. Place matchmarks on the front strut bar and the rear mounting nut. Remove the rear nut, the strut bar-to-lower control arm bolts and the strut.

20. Using an engine saddle, place it under the engine and support it. Place a floor jack under the transmission and support it.

NOTE: If equipped with a M/T, remove the engine rear mounting bracket from the body. If equipped with an A/T, remove the engine mounting member-to-transmission through bolt.

21. Remove the engine mounts-to-body nuts/bolts and lower the engine/transmission assembly, then remove the engine mounting member from the engine.

22. Remove the transmission from the engine.

23. To install, reverse the removal procedures. Refill the engine with oil and the cooling system with coolant.

Cylinder Head

——— CAUTION ———

DO NOT perform this operation on a warm engine. Remove the head bolts in the reverse of the tightening sequence. Loosen the head bolts evenly, not one at a time. Keep the pushrods in their original order. DO NOT attempt to slide the cylinder head off of the block, as it is located with dowel pins. Lift the head straight up and off the block.

REMOVAL & INSTALLATION

20R, 22R and 22R-E Engines

1. Disconnect the battery negative cable.

2. Drain the cooling system, both at the radiator and the block. The engine block drain is on the driver's side of the engine. The coolant, if good, may be reused.

——— CAUTION ———

Be sure to drain the engine oil, for it may become contaminated with coolant.

3. On the carburetor models, remove the air cleaner assembly complete with hoses. On the EFI models, remove the air cleaner hose.

NOTE: Cover the carburetor or throttle body opening with a clean cloth so that nothing can fall into it.

4. Remove the exhaust pipe-to-manifold nuts, then separate the pipe from the manifold. On the EFI models, disconnect the wire from the oxygen sensor.

5. Label and disconnect all of the various vacuum and emission hoses from the throttle body or the carburetor.

6. On the carburetor models (1984 and later), disconnect the following wire(s):

    a. The Vacuum Switching Valve (VSV) for A/C.

    b. The vacuum switch.

    c. The VSV for the Evaporative Emission Control (EVAP).

    d. The water temperature sender gauge.

    e. The cold mixture heater.

    f. The fuel cut solenoid valve.

    g. The Electronic Air Control Valve (EACV).

    h. The Vacuum Control Switch (VCS).

7. On the EFI models (1984 and later), disconnect the following wire(s):

    a. The cold start injector.

    b. The throttle position sensor.

    c. The water temperature sender gauge.

    d. The temperature sensor.

    e. The start injection time switch.

    f. The Overdrive (OD) thermo switch (for A/T).

    g. The injectors.

    h. The air valve.

8. At the cylinder head cover, remove the PCV hose, the spark plug wire holders, the distributor wiring connector and the throttle cable (if equipped with an A/T). Remove the mounting nuts and seals, then lift off the cylinder head cover.

——— CAUTION ———

Cover the oil return hole in the cylinder head to prevent objects from falling in.

9. Remove the upper radiator hose from the thermostat housing.

10. On the EFI models, remove the EGR valve from the chamber, the chamber-to-brace, the chamber-to-manifold nuts/bolts and the chamber with the throttle body.

11. At the distributor:

    a. Match-mark the rotor-to-distributor housing and the distributor housing-to-engine block.

    b. Disconnect the spark plug cables from the spark plugs (pull on the plug boot).

    c. Disconnect the primary ignition wire from the distributor cap and the electrical connector from the housing. d.

    Remove the distributor hold-down clamp, then the distributor from the cylinder head with the high tension cables attached.

12. Remove the fuel hoses from the fuel pump, the mounting bolts, the fuel pump and the gaskets from the cylinder head.

13. Disconnect the bond cables from the front and the rear of the cylinder head, then the wire(s) from the carburetor and the thermo-switch.

NOTE: If equipped with power steering, remove the drive belt, the pulley, the pump and the bracket, then move the pump aside without disconnecting the pressure hoses.

14. On the carburetor models, disconnect the following hoses:

    a. The water by-pass hose from the intake manifold.

    b. The heater inlet hose from the water valve.

    c. The brake booster hose from the intake manifold.

    d. The 2 fuel hoses from the pipes under the intake manifold.

    e. On Calif. models, the hose from the air injection tube.

    f. Label and disconnect the emission control hoses from the carburetor and the intake manifold that will interfere with the head removal.

15. On the EFI models, disconnect the following:

    a. The No.1 and No.2 hoses at the PCV.

    b. The air control valve hose.

    c. The actuator (cruise control) hose.

    d. The EGR vacuum modulator hose.

1. Rocker arm	9. Cam sprocket
2. Spring	10. Camshaft
3. Spacer	11. Camshaft bearing cap
4. Rocker shaft (intake)	12. Valve keeper
5. Head bolt	13. Spring retainer
6. Rocker stand	14. Valve spring
7. Rocker shaft (exhaust)	15. Valve seal
8. Distributor drive gear	16. Spring seat

17. Valve guide	
18. Half circle cam seal	
19. Cylinder head	
20. Intake valve	
21. Exhaust valve	
22. Rear cover (EGR cooler)	

Cylinder head components—20R, 22R and 22R-E engines

Supply part

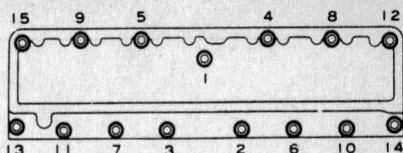

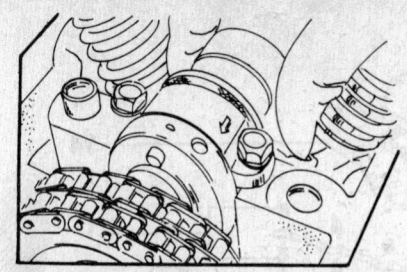

20R, 22R and 22R-E engines—rotate the camshaft so that the pin is at the top. Also note that the arrow on the camshaft bearing cap points to the front of the engine

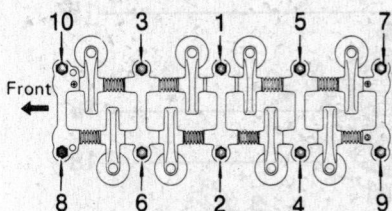

Cylinder head bolt tightening sequence—20R, 22R and 22R-E engines

e. The No. 1 air valve hose from the throttle body.

f. The No. 2 air valve hose from the chamber.

g. The No. 2 and No. 3 water by-pass hoses from the throttle body.

h. The actuator's air control valve hose.

i. The pressure regulator hose for the actuator.

j. The cold start injection tube.

k. The Bi-metal Vacuum Switch Valve (BVSV) hoses.

l. The brake booster hose.

16. On the carburetor models, disconnect the accelerator linkage from the carburetor. If equipped with an A/T, disconnect the throttle cable from the carburetor.

17. On the carburetor models (Federal and Canada), remove the air suction rear pipe.

18. On the EFI models, remove the pulsation damper, the fuel hose-to-delivery tube bolt and the fuel hose from the delivery tube. Disconnect the No. 4 by-pass hose from the air valve and the air valve from the intake manifold, then the by-pass hose from the intake manifold.

19. To remove the camshaft sprocket:

a. Place a wrench on the crankshaft bolt, then turn the crankshaft until the No. 1 piston is at the TDC of the compression stroke.

NOTE: The No. 1 piston is on TDC when the timing marks are on 0° and the valve of that cylinder are closed.

b. Place alignment marks on the timing chain and the camshaft sprocket.

c. Remove the half circle seal, the camshaft sprocket bolt, the distributor drive

gear, the fuel pump drive cam (20R and 22R), the camshaft thrust plate (22R-E) and the camshaft sprocket with the timing chain.

NOTE: When removing the camshaft sprocket and timing chain, allow the timing chain to remain on the crankshaft sprocket.

20. Remove the top timing chain cover bolt (in front of the cylinder head, before the cylinder head is removed.

21. Remove the cylinder head bolts (2–3 passes), starting from the outer ends and working toward the center.

— CAUTION —
If the cylinder head bolts are not removed in the correct order, warpage or cracking of the head may occur.

22. Remove the rocker arm assembly from the cylinder head. It may be necessary to use a small pry bar to loosen the rocker arm assembly.

23. Lift the cylinder head from the engine and place it on a workbench on 2 blocks of wood.

— CAUTION —
DO NOT pry between the cylinder head gasket and the engine block.

24. To install, use new gaskets and reverse the removal procedures. Torque the cylinder head-to-engine block to 58 ft. lbs., the timing chain cover-to-cylinder head bolt to 9 ft. lbs., the camshaft sprocket-to-camshaft bolt to 58 ft. lbs., the intake manifold bolts to 14 ft. lbs., the exhaust manifold bolts to 33 ft. lbs. and the rocker arm cover to 7–12 ft. lbs. Replace the cooling system

fluid and the engine oil. Adjust the valves, the drive belts, then check and/or adjust the timing.

Cylinder head bolt tightening sequence—2F engines

2F Engine

1. Disconnect the negative battery cable. Drain the cooling system.

2. Remove the air cleaner assembly, complete with its attendant hoses.

3. Detach the accelerator cable from the cylinder head cover support and the carburetor throttle arm.

4. Remove the choke cable and fuel lines from the carburetor. Remove the water hose bracket from the cylinder head cover.

5. Remove the clamps and the hoses from the water pump, then the water valve. Disconnect the heater temperature control cable from the water valve.

6. Disconnect the PCV line from the cylinder head cover. Disconnect the vacuum lines, from the vacuum switching valve to the various components of the emission control system.

7. Drain the engine oil. Remove the oil lines from the oil filter and the filter assembly from the manifold.

8. Detach the vacuum valve solenoid wire from the coil.

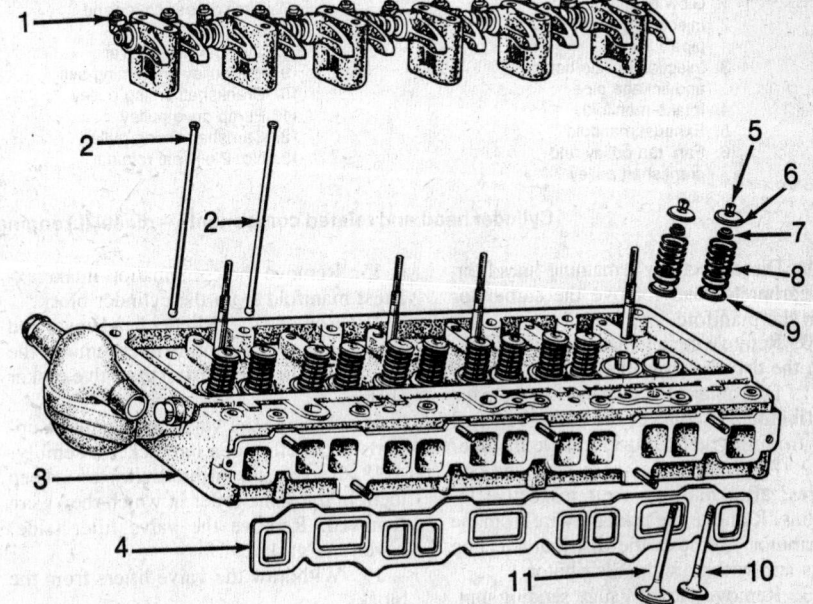

1. Rocker arm and shaft assembly
2. Pushrods
3. Cylinder head
4. Intake and exhaust manifold gasket
5. Valve keepers
6. Valve spring retainer
7. Valve seal
8. Valve spring
9. Valve spring seat
10. Exhaust valve
11. Intake valve

Cylinder head components—2F engine

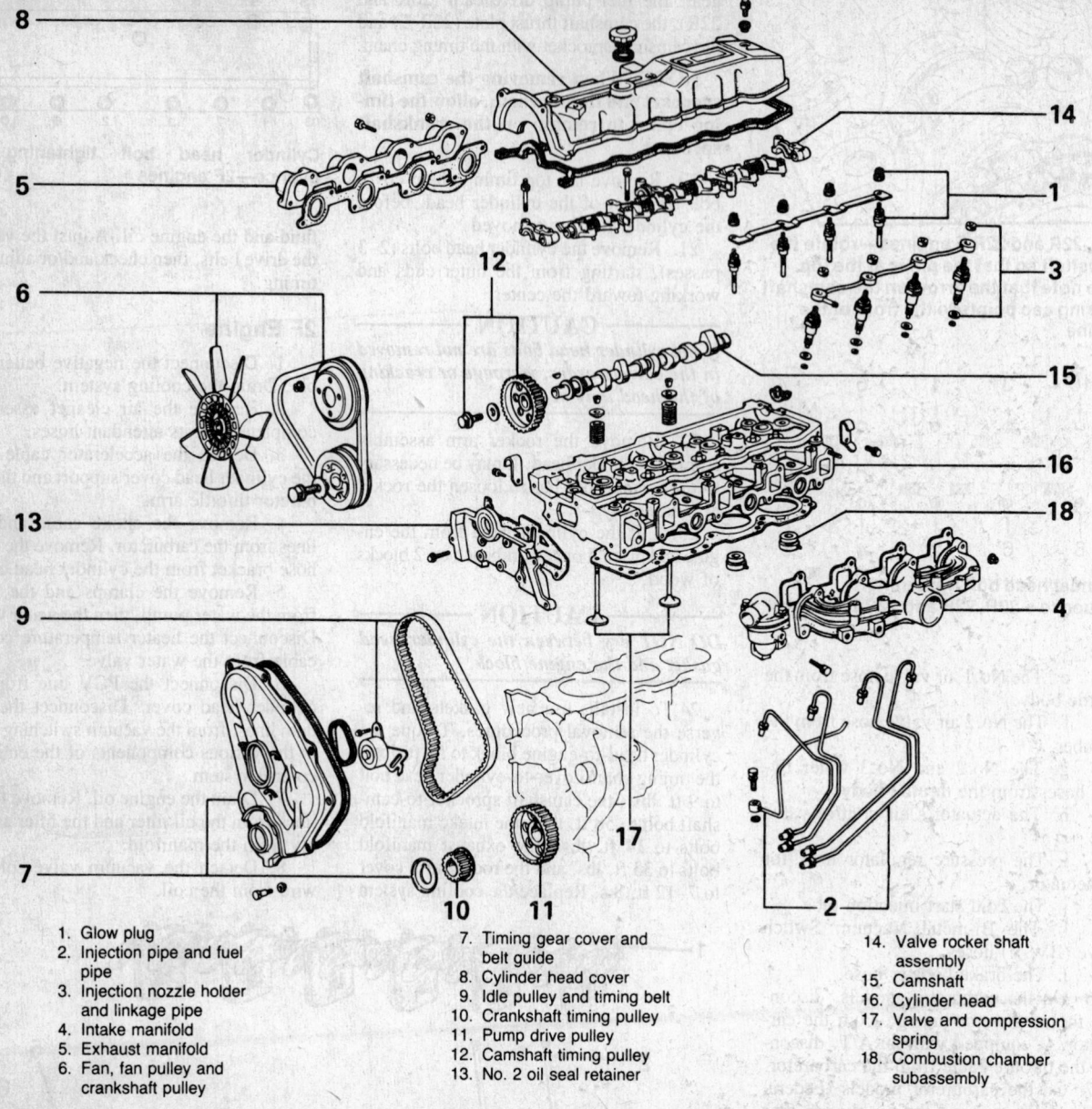

1. Glow plug
2. Injection pipe and fuel pipe
3. Injection nozzle holder and linkage pipe
4. Intake manifold
5. Exhaust manifold
6. Fan, fan pulley and crankshaft pulley
7. Timing gear cover and belt guide
8. Cylinder head cover
9. Idle pulley and timing belt
10. Crankshaft timing pulley
11. Pump drive pulley
12. Camshaft timing pulley
13. No. 2 oil seal retainer
14. Valve rocker shaft assembly
15. Camshaft
16. Cylinder head
17. Valve and compression spring
18. Combustion chamber subassembly

Cylinder head and related components—diesel (L) engine, 2L is similar

9. Disconnect any remaining lines from the carburetor and remove the carburetor from the manifold.

10. Remove the alternator adjusting link, then the drivebelt and the alternator.

11. Disconnect the vacuum line from the distributor. Disconnect the carburetor fuel line from the fuel pump. Remove the line.

12. Disconnect the spark plug and coil cables, after marking their respective locations. Remove the primary wire from the distributor. Remove the distributor clamp bolts and withdraw the distributor.

13. Remove the oil gauge sending unit. Remove the coil from its cylinder head bracket.

14. Remove the fuel pump and the oil filter tube clamping bolt from the valve lifter (side) cover. Drive the oil filler tube out of the cylinder block.

15. Remove the combination intake/exhaust manifold from the cylinder block.

16. Remove the cylinder head cover and gasket. Remove the oil delivery union, the spring and the sleeve from the valve rocker shafts.

17. Remove the valve rocker shaft supports nuts/bolts, then the rocker assembly.

18. Remove the pushrods, be sure to keep them in the same order in which they were removed. Remove the valve lifter (side) cover and gasket.

19. Withdraw the valve lifters from the block.

NOTE: The valve lifters should be kept, with their respective pushrods, in the sequence in which they were removed.

20. Disconnect the oil delivery union from the oil feed pipe.

21. Remove the cylinder head bolts in 2–3 stages and in the reverse order of installation.

22. Lift off the cylinder head and the gasket.

NOTE: Service procedures are covered in the Unit Repair Section.

23. To install, use new gaskets and reverse the removal procedures. Torque the cylinder head bolts to 83–98 ft. lbs. (in 3 steps), the rocker assembly nuts/bolts to 25–30 ft. lbs. (10mm) or 14–22 ft. lbs. (8mm). Adjust the valves. Refill the cooling system.

3Y-EC Engine

1. Disconnect the negative battery cable.

2. Remove the right-front seat and the engine service hole cover.

3. Drain the engine coolant (at the radiator and the left-side of the engine) and the engine oil.

4. If equipped with power steering, perform the following:

   a. Remove the air hoses from the air control valve.

   b. Drain the fluid from the reservoir tank.

   c. Disconnect the return hose from the pump.

   d. Using the tool No. 09631–22020, disconnect the pressure hose from the pump.

   e. Remove the drive belt, the pulley nut, the pulley, the woodruff key, the mounting bolts and the pump.

5. Remove the exhaust pipe and the bracket. Remove the air cleaner pipe and the hoses.

6. Disconnect the accelerator cable with the bracket from the throttle body.

7. Disconnect the water temperature sender gauge connector from the cylinder head.

8. Disconnect the following EFI connectors from the:

   a. The water thermo sensor.
   b. The start injector time switch.
   c. The cold start injector.
   d. The air valve.
   e. The throttle position sensor.
   f. The oxygen sensor.
   g. The water temperature switch (A/C equipped).

9. Disconnect the following hoses from the:

   a. The radiator inlet.
   b. The radiator breather.
   c. The reserve tank.
   d. The heater outlet.
   e. The PCV.
   f. The water by-pass.
   g. The brake booster vacuum.
   h. The charcoal canister.
   i. Label and disconnect the emission control.

10. Remove the throttle body from the air intake chamber.

11. Remove the EGR valve nuts from the intake chamber and the exhaust manifold-to-EGR valve union.

12. Disconnect the cold start injector pipe, the water by-pass hoses and the pressure regulator hose.

13. Using a 12mm offset box wrench, remove the air intake chamber brackets, then the chamber with the air valve.

14. Remove the wire clamp bolts and the injector connectors from the injectors.

15. Remove the exhaust manifold bracket, the heater pipe bracket, the fuel inlet pipe union bolt from the fuel filter and the fuel outlet hose.

16. Remove the spark plugs and the tubes.

17. Remove the cap nuts, the seal washers, the cylinder head cover and the gasket.

18. Remove the rocker arm shaft assembly nuts/bolts a little at a time, in 3–4 steps.

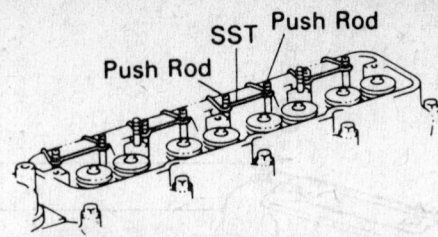

Using the special tools to hold the push rods in place, during the rocker arm installation

Remove the push rods, keeping them in order.

19. Remove the cylinder head bolts, a little at a time, in 3 passes. Lift the cylinder head off of the dowels and place it on 2 blocks of wood.

———— CAUTION ————

If the cylinder head bolts are not removed correctly, warpage or cracking of the cylinder head may occur. If necessary to pry the cylinder head from the block, pry between the cylinder head and the block projection.

20. Remove the valve lifters from the cylinder block.

21. Using a putty knife, clean the gasket mounting surfaces.

22. To install, use new gaskets and reverse the removal procedures. Torque the cylinder head bolts (in 3 passes) to 65 ft. lbs. (14mm) or 14 ft. lbs. (12mm), the rocker arm shaft-to-cylinder head bolts (in 3 passes) to 17 ft. lbs., the spark plugs to 13 ft. lbs., the air intake chamber bolts to 9 ft. lbs., the throttle body-to-intake chamber to 9 ft. lbs. and the exhaust pipe-to-exhaust manifold to 29 ft. lbs. Adjust the drive belts.

Refill the cooling system and the engine with oil. Check and/or adjust the timing.

NOTE: Use special tools No. 09270–71010 to hold the push rods in position, when installing the push rods.

Valve Rocker Shafts

REMOVAL & INSTALLATION

Gasoline Engines

The valve rocker shaft removal and installation is given as part of the cylinder head removal and installation procedure. Perform only the steps of the appropriate procedure necessary to remove or install the rocker shafts.

NOTE: On the 20R, 22R and 22R-E engines, the rocker arms are the same but the rocker shafts are different. Keep all parts in order so that they may be installed correctly. Lubricate all parts with engine oil prior to assembly.

VALVE ADJUSTMENT

20R, 22R and 22R-E Engines

1. Remove the cylinder head cover.
2. Turn the crankshaft (using a wrench) until the No.1 cylinder is on the TDC of the compression stroke. The valves of the No.1 cylinder must be loose.
3. To adjust the valve clearance, loosen the adjuster lock nut and turn the adjusting screw.
4. Using a feeler gauge, check and/or adjust the intake valve clearance of cylinders No.1 and 2, then the exhaust valve clearance of cylinders No.1 and 3.

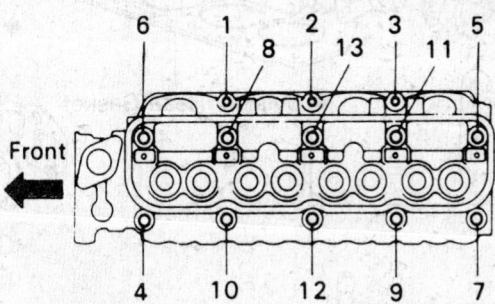

Bolt removal sequence—3Y-EC engine

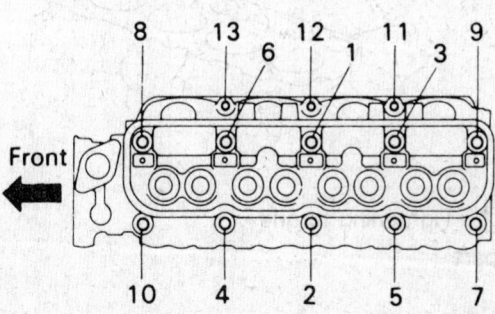

Cylinder head bolt torquing sequence— 3Y-EC

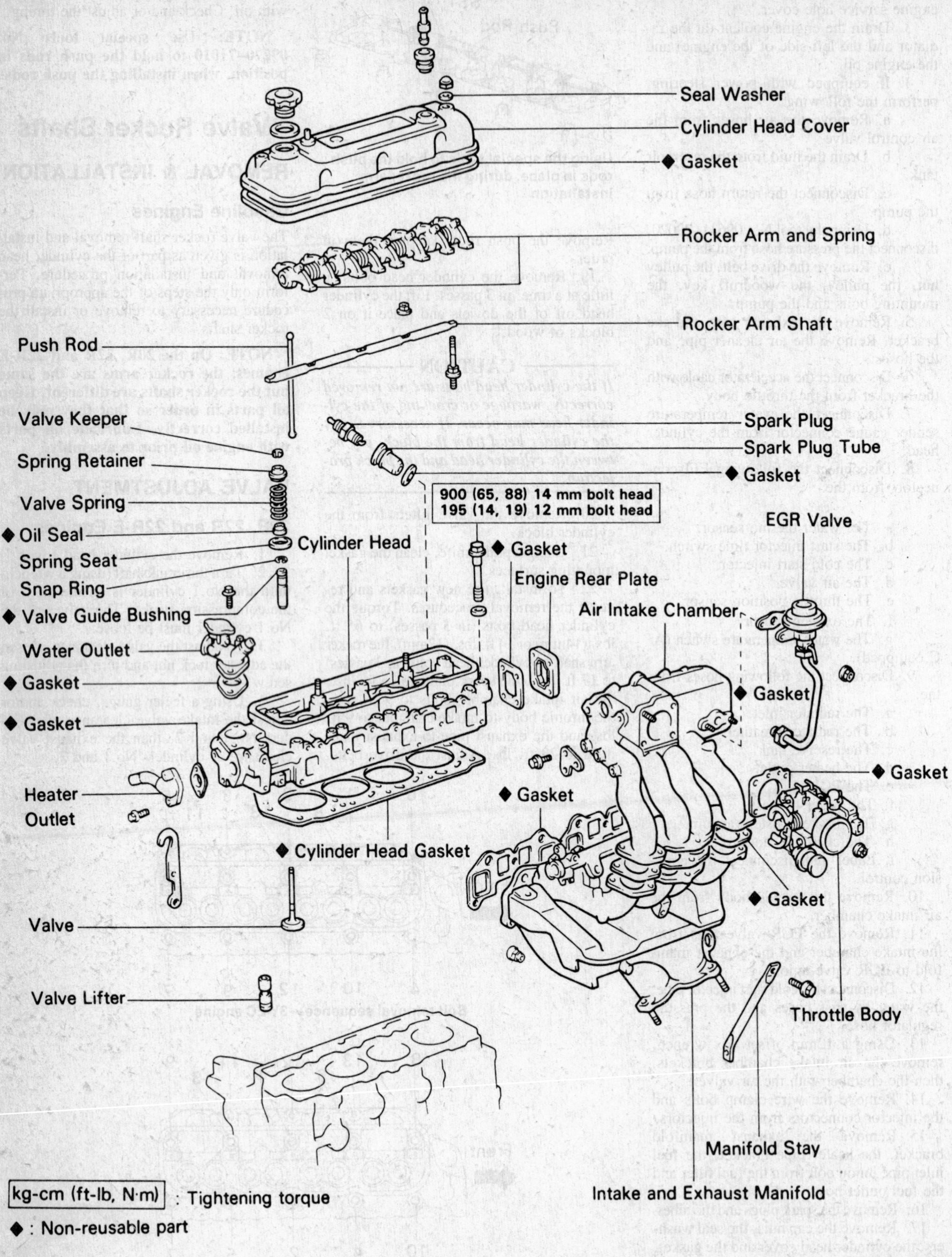

Seal Washer

Cylinder Head Cover

◆ Gasket

Rocker Arm and Spring

Rocker Arm Shaft

Push Rod

Spark Plug

Spark Plug Tube

◆ Gasket

Valve Keeper

Spring Retainer

Valve Spring

◆ Oil Seal

Cylinder Head

Spring Seat

Snap Ring

◆ Valve Guide Bushing

Water Outlet

◆ Gasket

◆ Gasket

Heater Outlet

◆ Cylinder Head Gasket

Valve

Valve Lifter

900 (65, 88) 14 mm bolt head
195 (14, 19) 12 mm bolt head

◆ Gasket

Engine Rear Plate

Air Intake Chamber

EGR Valve

◆ Gasket

◆ Gasket

◆ Gasket

◆ Gasket

Throttle Body

Manifold Stay

Intake and Exhaust Manifold

kg-cm (ft-lb, N·m) : Tightening torque

◆ : Non-reusable part

Exploded view of the 3Y-EC cylinder head

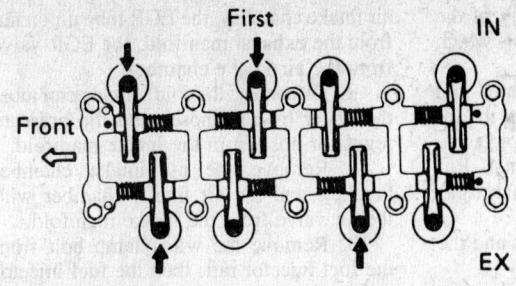

Adjusting the valve clearance (first step)—20R, 22R and 22R-E engines

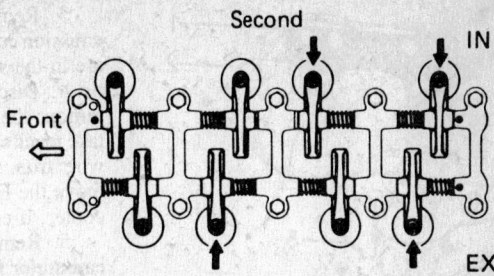

Adjusting the valve clearance (second step)—20R, 22R and 22R-E engines

NOTE: The intake valve clearance is 0.008 in. (cold); the exhaust valve clearance is 0.012 in. (cold).

5. Using a wrench on the crankshaft pulley bolt, turn the crankshaft one revolution, stopping on the 0° mark of the timing plate (TDC of No.4 cylinder).

6. Adjust the intake valves of cylinder No.3 and 4; the exhaust valves of cylinders No.2 and 4.

7. With the adjustment complete, install the removed components by reversing the removal procedures.

2F Engine

1. Operate the engine until it reaches normal operating temperatures.

2. Remove the cylinder head cover.

3. Turn the crankshaft (using a wrench) until the No.1 cylinder is on the TDC of the compression stroke. The valves of the No.1 cylinder must be loose.

4. To adjust the valve clearance, loosen the adjuster lock nut and turn the adjusting screw.

5. Using a feeler gauge, check and/or adjust the clearance of valve No.1, 2, 3, 5, 7 and 9 (as numbered from the front).

NOTE: The intake valve clearance is 0.008 in. (warm); the exhaust valve clearance is 0.014 in. (warm).

6. Using a wrench on the crankshaft pulley bolt, turn the crankshaft one revolution, aligning the 0° mark on the timing pulley with the timing pointer.

7. Adjust the clearance of valve No.4, 6, 8, 10, 11 and 12 (as numbered from the front).

8. With the adjustment complete, install the removed components by reversing the removal procedures. Check and/or adjust the timing.

3Y-EC Engine

The valves of this engine are hydraulic; no adjustment is necessary.

Intake Manifold

REMOVAL & INSTALLATION

20R and 22R Engines

1. Disconnect the battery negative ca-

ble. Drain the engine coolant to a level below the carburetor.

2. Remove the air cleaner assembly, complete with hoses.

3. Label and disconnect the vacuum lines from the manifold and the carburetor.

4. Remove the fuel line, the accelerator linkage, the electrical leads and the coolant hoses from the carburetor.

5. Remove the coolant by-pass hose from the manifold.

6. Remove the air valve from the intake manifold.

7. Unbolt and remove the intake manifold, complete with carburetor and EGR valve.

8. Cover the cylinder head intake ports with a clean cloth.

9. To install, use a new gasket and reverse the removal procedures. Torque the mounting nuts to 13–19 ft. lbs. Refill the cooling system.

NOTE: Tighten the bolts in several stages, working from the inside bolts outward.

22R-E Engine

1. Disconnect the battery negative cable.

2. Remove the air cleaner hose from the throttle body. Drain the cooling system to a level below the throttle body.

3. From the throttle body, remove the fuel lines from the distribution rail, the accelerator linkage, the throttle position sensor, the electrical leads, the coolant hoses, the PCV hose and the emission control hoses.

4. Remove the throttle body with the air intake chamber from the intake manifold.

5. Remove the coolant by-pass hose from the intake manifold.

6. Remove the intake manifold with the EGR valve.

7. Cover the cylinder head intake ports with a clean cloth.

8. Using a putty knife, clean the gasket mounting surfaces.

9. To install, use a new gaskets and reverse the removal procedures. Torque the mounting nuts to 13–19 ft. lbs. Refill the cooling system.

NOTE: Tighten the bolts in several stages, working from the inside bolts outward.

Exhaust Manifold

REMOVAL & INSTALLATION

20R, 22R and 22R-E Engines

1. Remove the exhaust pipe flange bolts and disconnect the exhaust pipe from the manifold.

NOTE: On the 22R-E engine, disconnect the oxygen sensor wiring connector.

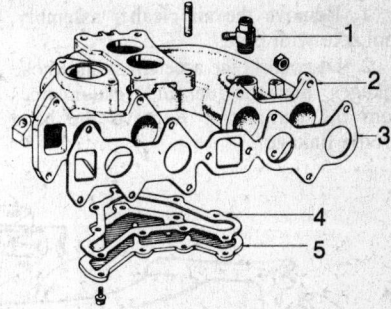

1. Vacuum fitting
2. Intake manifold
3. Gasket
4. Gasket
5. Cover

20R, 22R and 22R-E engine intake manifold components

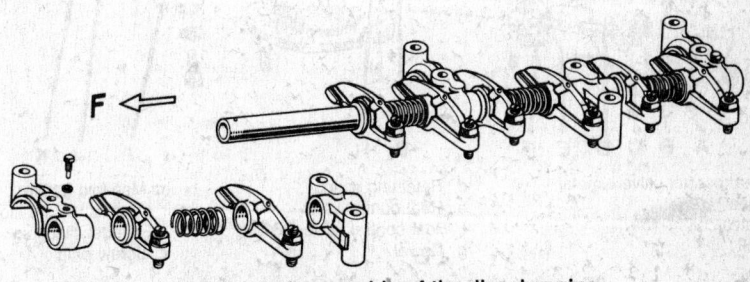

Rocker shaft assembly of the diesel engine

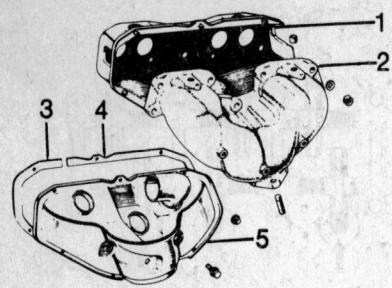

1. Inner heat stove
2. Exhaust manifold
3. Gasket
4. Gasket
5. Outer heat stove

20R, 22R and 22R-E engine exhaust manifold components

2. Remove the air cleaner tube from the heat stove. Remove the outer part of the heat stove.

3. Remove the mounting nuts, the exhaust insulator, the exhaust manifold with the air injection tube, then separate the inner portion of the heat stove from the manifold.

4. To install, use new gaskets and reverse the removal procedures. Torque the retaining nuts to 29–36 ft. lbs. (working from the inside out) and the exhaust pipe flange nuts to 25–32 ft. lbs.

Combination Manifold

REMOVAL & INSTALLATION

2F Engines

1. Remove the air cleaner assembly, complete with hoses.

2. Disconnect the accelerator and choke linkages, then the fuel and vacuum lines from the carburetor. Remove the hand throttle linkage.

3. Remove or move aside, any of the emission control system components which are in the way.

4. Disconnect the oil filter lines and remove the oil filter assembly from the intake manifold. Disconnect the solenoid valve wire from the ignition coil terminal. Remove the EGR tubes from the exhaust gas cooler, if equipped.

5. Remove the mounting bolts and the carburetor from the manifold.

6. Loosen the manifold retaining nuts, working from the inside out, in two or three stages.

7. Remove the intake/exhaust manifold assembly from the cylinder head as a complete unit.

8. Using a putty knife, clean the gasket mounting surfaces.

8. To install, use new gaskets and reverse the removal procedures. Tighten the bolts, working from the inside out.

NOTE: Tighten the bolts in two or three stages.

3Y-EC Engine

1. Disconnect the negative battery cable. Remove the right seat and the engine service hole cover.

2. Drain the engine coolant to a level below the throttle body.

3. Remove the air cleaner-to-throttle body hose.

4. Remove the accelerator cable with the bracket from the throttle body.

5. Disconnect the air valve connector, the throttle position sensor connector and the oxygen sensor connector.

6. Disconnect the PCV hose from the air intake chamber, the water by-pass hoses from the throttle body, the booster vacuum hose and the charcoal canister hose, then label and disconnect the emission control hoses.

7. Remove the throttle body from the air intake chamber, the EGR tube union nut from the exhaust manifold, the EGR valve from the air intake chamber.

8. Disconnect the cold start injector tube, the water by-pass hoses and the pressure regulator hose from the intake manifold.

9. Remove the air intake chamber brackets and the air intake chamber with the air valve from the intake manifold.

10. Remove the wire clamp bolt from the fuel injector rail, then the fuel injector rail from the fuel injectors.

11. Remove the exhaust manifold-to-intake manifold bracket, the exhaust manifold-to-engine bracket, the exhaust pipe from the exhaust manifold, the fuel inlet and outlet tubes union nut from the fuel rail.

12. Remove the spark plug wires, the spark plugs and the tubes.

13. Remove the retaining bolts, then intake and the exhaust manifolds as an assembly.

14. Using a putty knife, clean the gasket mounting surfaces.

15. To install, use new gaskets and reverse the removal procedures. Torque the manifold bolts to 36 ft. lbs., the air intake chamber-to-intake manifold bolts to 9 ft. lbs. and the throttle body bolts to 9 ft. lbs. Refill the cooling system.

Front Cover

REMOVAL & INSTALLATION

20R, 22R and 22R-E

1. Refer to the "Cylinder Head, Removal and Installation" procedures, in this section and remove the cylinder head.

2. Remove the radiator hoses, the fan shroud, the transmission oil cooler hoses from the radiator (A/T), the coolant reservoir tube, the mounting bolts and the radiator.

NOTE: On the 2WD models (1981–83), remove the idler arm bracket from the frame, the pitman arm from the selector shaft and the crossmember under the engine.

3. Remove the engine undercover and the engine mounting bolts, then place a floor jack under the transmission and raise it 0.98 in. Remove mounting nuts/bolts and the oil pan from the transmission.

4. Remove the fan from the water pump, the drive belts and the water pump pulley.

5. Remove the air pump, hoses and bracket, if equipped.

6. Remove the alternator adjuster bracket and move it towards the alternator.

7. Remove the center bolt on the crankshaft pulley. Using a gear puller tool No. 09213–31021, remove the crankshaft pulley.

8. Remove the mounting bolts, the water bypass tube and the heater tube.

9. On the 1985 and later models, remove the alternator adjusting bracket bolt and move the bracket toward the alternator.

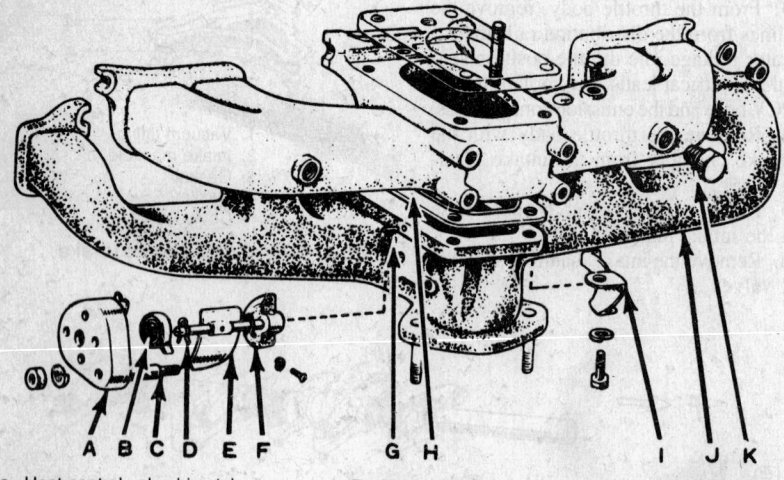

a. Heat control valve bimetal case
b. Valve coil
c. Bolt
d. Retaining spring
e. Heat control valve
f. Heat control valve shaft
g. Dowel
h. Manifold gasket
i. Counter weight stop
j. Exhaust manifold
k. Screw plug

2F engine combination manifold components

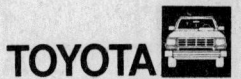
10. Remove the mounting bolts, the front cover and the gasket from the engine; it may be necessary to use a plastic hammer to loosen the front cover.

11. Using a putty knife, clean the gasket mounting surfaces.

12. To install, use new gaskets, sealant and reverse the removal procedures. Torque the front cover bolts to 9 ft. lbs. (8mm) or 29 ft. lbs. (10mm), the alternator adjusting bracket bolt to 9 ft. lbs., the oil pan bolts to 9 ft. lbs. and the crankshaft pulley bolt to 116 ft. lbs.

2F Engines

1. Drain the cooling system and the engine oil. Disconnect the negative battery cable.

2. Remove the air cleaner assembly, complete with hoses, from the bracket.

3. Remove the hood latch as well as its brace and support. Remove the headlight bezels and grille assembly.

4. Remove the upper and lower radiator hoses, the mounting bolts and the radiator.

NOTE: Remove the fan shroud, if equipped.

5. Loosen the drive belt adjusting link and remove the drive belt. Remove the alternator electrical connector, the retaining bolts and the alternator.

6. If equipped with an air injection pump, remove the hoses from the pump, the mounting bolts and the air pump.

7. Remove the fan and water pump as an assembly. Using a gear puller, remove the crankshaft pulley.

8. Remove the gravel shield from under the engine and the front driveshaft.

9. Remove the front oil pan bolts, to gain access to the bottom of the timing chain cover.

NOTE: It may be necessary to insert a thin knife between the pan and the gasket in order to break the pan loose. Use care not to damage the gasket.

10. To install, reverse the removal procedures. Adjust the drive belts and refill the cooling system.

3Y-EC

1. Disconnect the negative battery cable. Drain the cooling system.

2. If equipped with an A/T, remove and plug the oil cooler lines from the radiator.

3. Remove the fan shroud, the radiator hoses, the coolant reservoir hose and the upper radiator bolt. Raise the vehicle and support it on jackstands. Remove the engine under cover, the mounting bolts and the radiator.

4. Remove the drive belts, the fan, fluid coupling and the water pump pulley.

5. Using the tools No. 09213–70010 and 09330–00020, remove the crankshaft pulley center bolt. Using the wheel puller

tool No. 09213–31021, pull the crankshaft pulley from the crankshaft.

6. Remove the front cover mounting bolts. Using a small pry bar, lift the front cover from the engine.

7. Using a putty knife, clean the gasket mounting surfaces.

8. To install, use a new gasket, sealant and reverse the removal procedures. Using a soft faced hammer, drive the crankshaft pulley onto the crankshaft. Torque the crankshaft pulley bolt to 80 ft. lbs. (using the removal tools). Adjust the drive belts and refill the cooling system.

OIL SEAL REPLACEMENT

20R, 22R and 22R-E Engines

1. Refer to the "Front Cover, Removal and Installation" procedures, in this section and remove the crankshaft pulley.

2. Using a small pry bar, pry the oil seal from the oil pump housing.

3. Using the seal installation tool No. 09223–50010, drive the new seal into the oil pump housing. Apply multi-purpose grease to the lip of the new seal.

4. To complete the installation, reverse the removal procedures.

3Y-EC Engine

FRONT COVER REMOVED

1. Using a drive punch and a hammer, drive the oil seal from the front cover.

2. Using the seal installation tool No. 09223–22010 and a hammer, drive the new oil seal into the front cover.

3. Apply grease to the lip of the new seal.

FRONT COVER INSTALLED

1. Refer to the "Front Cover, Removal and Installation" procedures, in this section and remove the crankshaft pulley.

2. Using the seal removal tool No. 09308–10010, pull the oil seal from the front cover.

3. Apply multi-purpose grease to the lip of the new seal.

4. Using the seal installation tool No. 09223–22010 and a hammer, drive the new oil seal into the front cover.

5. To complete the installation, reverse the removal procedures.

Timing Chain and Tensioner

REMOVAL & INSTALLATION

20R, 22R and 22R-E Engines

1. Refer to the "Cylinder Head" and the "Front Cover, Removal and Installation" procedures, in this section, then remove the cylinder head and the front cover. Remove the oil pan.

2. Remove the chain from the damper and the cam sprocket together.

NOTE: If the chain and sprocket are worn, they will have to be replaced, along with the crankshaft sprocket.

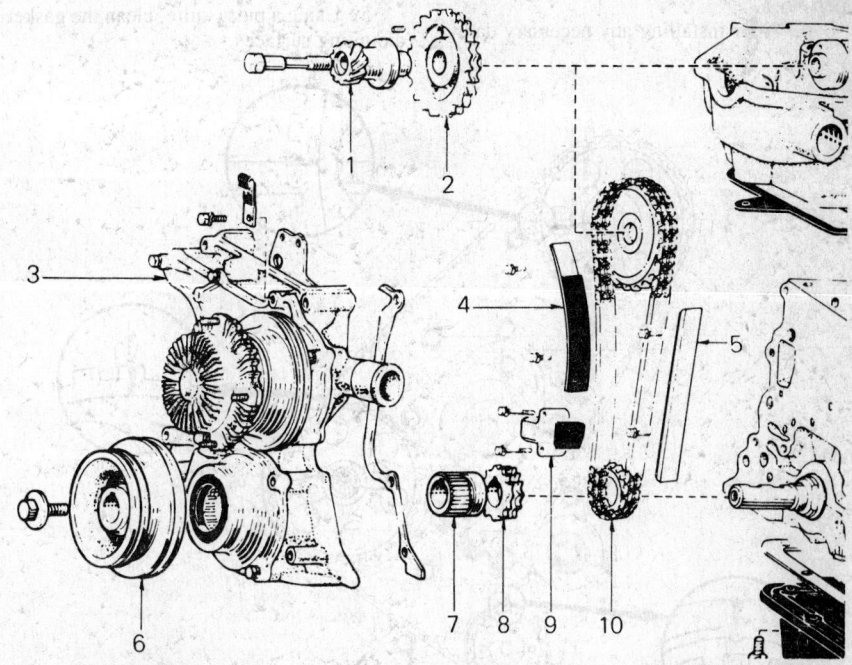

1. Distributor drive gear	5. Chain damper #1	8. Crankshaft sprocket
2. Cam sprocket	6. Crankshaft pulley	9. Chain tensioner
3. Timing chain cover	7. Pump drive spline	10. Timing chain
4. Chain damper #2		

Timing cover and related components—20R, 22R and 22R-E engines

T547

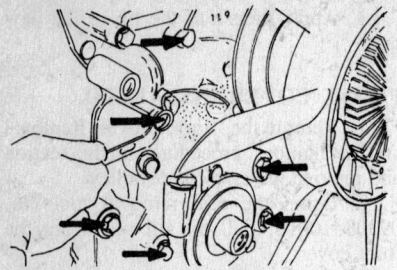

20R, 22R and 22R-E engines—only the six bolts indicated need to be removed to remove the timing cover

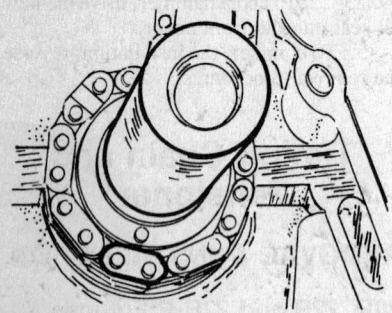

20R, 22R and 22R-E engines—align the crankshaft gear mark with the single bright link of the timing chain

3. Using the wheel removal tool No. 09213-36020, pull the crankshaft sprocket and pump drive spline as a unit.

4. Measure the chain tensioner for wear; if it is worn below 0.43 in., replace it as a unit.

5. Measure the chain dampers for wear; if measurements are below the limit: 0.20 in. for Damper No.1 and 0.18 in. for Damper No.2.

6. After installing any necessary dam-pers or a new tensioner, turn the crankshaft by hand until the key is at TDC. If removed, slide the crankshaft sprocket over the key. Place the chain on the sprocket so that the single bright link is over the mark on the sprocket.

7. Position the cam sprocket in the chain so that the timing mark on the sprocket is located between the two bright links (1979–84) or aligned with the bright link (1985 and later) of the chain.

8. Install the oil pump drive spline over the crankshaft key, if removed.

9. Turn the camshaft sprocket counterclockwise to take any slack out of the chain.

10. To complete the installation, use new gaskets, sealant and reverse the removal procedures.

3Y-EC

1. Refer to the "Front Cover, Removal and Installation" procedures, in this section and remove the front cover.

NOTE: Using a tension gauge, measure the slack of the timing chain, it should be 0.531 in. at 22 lbs. pressure.

2. Remove the mounting bolts and the timing chain tensioner.

3. Install the crankshaft pulley on the crankshaft. Using tools No. 09213-70010 and 09330-00020 to secure the crankshaft pulley, remove the camshaft mounting bolt with a socket wrench, then remove the crankshaft pulley.

4. Using the wheel puller tool No. 09950-20015, uniformly remove the camshaft sprocket with the crankshaft sprocket and chain.

5. Using a putty knife, clean the gasket mounting surfaces.

20R, 22R and 22R-E (1979–84) engines—align the camshaft sprocket mark between the two bright links of the timing chain

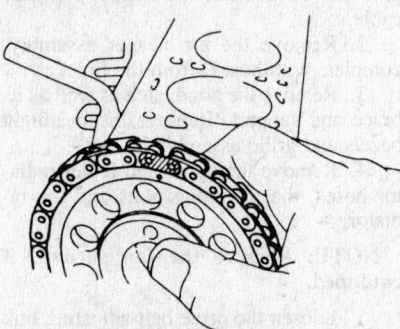

Aligning the timing chain to the camshaft sprocket—20R, 22R and 22R-E engines (1985 and later)

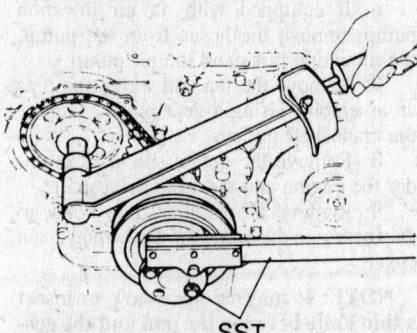

Removing and installing the camshaft sprocket bolt—3Y-EC engine

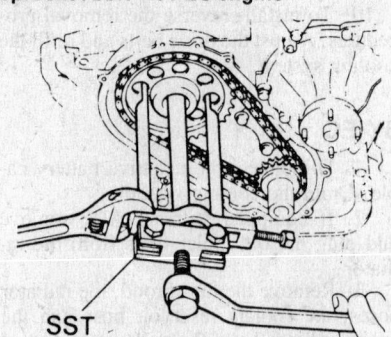

Removing the camshaft sprocket—3Y-EC engine

6. Upon installation, align the timing chain with the timing marks on the sprockets, then install the sprockets on their respective shafts.

7. To complete the installation, use new gaskets, sealant and reverse the removal

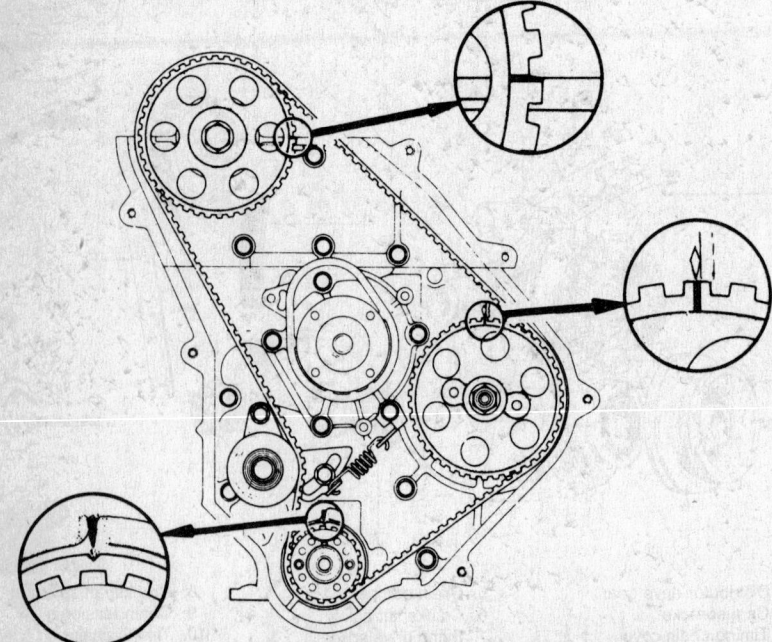

Final view of the timing marks after the crankshaft is turned twice—see text (diesel engine)

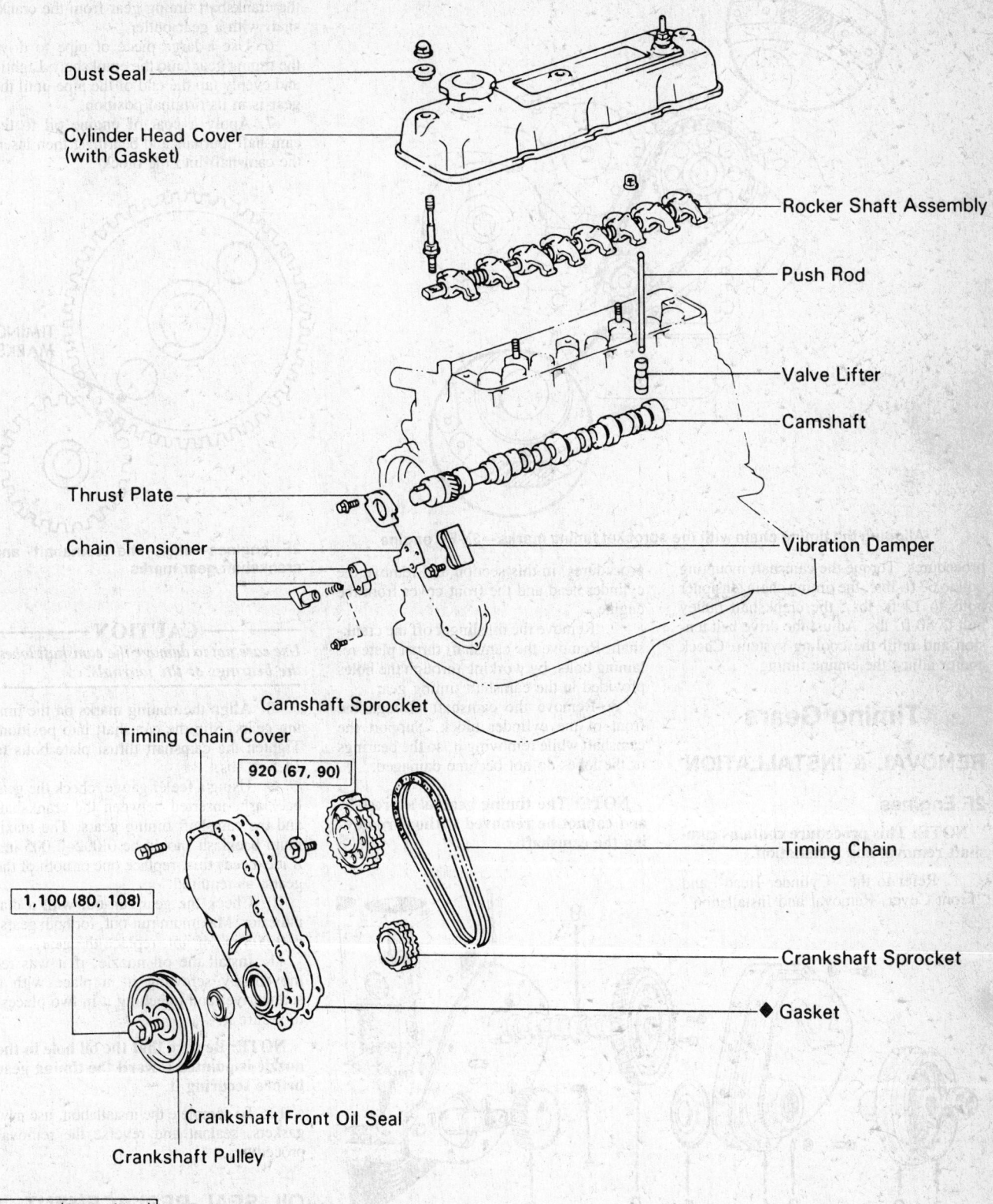

Dust Seal

Cylinder Head Cover
(with Gasket)

Rocker Shaft Assembly

Push Rod

Valve Lifter

Camshaft

Thrust Plate

Chain Tensioner

Vibration Damper

Camshaft Sprocket

Timing Chain Cover

920 (67, 90)

Timing Chain

1,100 (80, 108)

Crankshaft Sprocket

◆ Gasket

Crankshaft Front Oil Seal

Crankshaft Pulley

kg-cm (ft-lb, N·m) : Tightening torque

◆ : Non-reusable part

Exploded view of the timing chain and camshaft assembly—3Y-EC engine

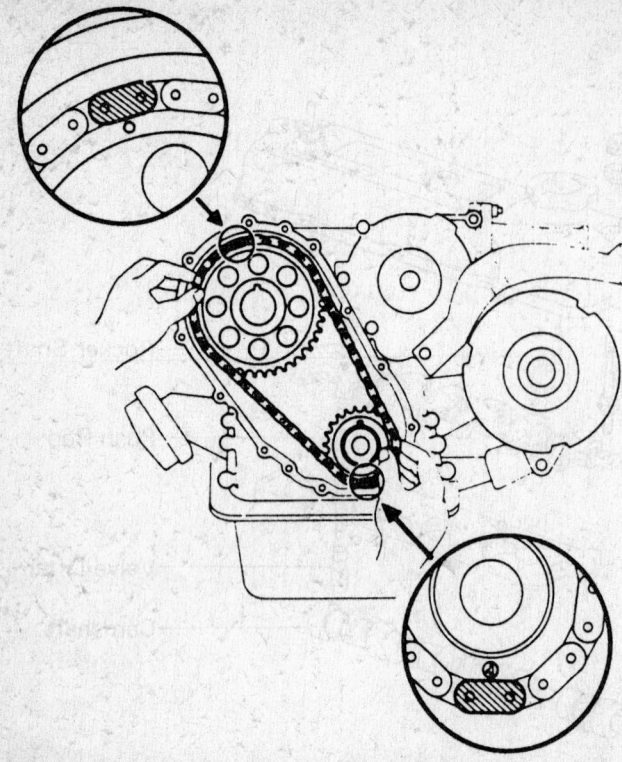

Aligning the timing chain with the sprocket timing marks—3Y-EC engine

procedures. Torque the camshaft mounting bolt to 67 ft. lbs., the timing chain tensioner bolts to 13 ft. lbs., the crankshaft pulley bolt to 80 ft. lbs. Adjust the drive belt tension and refill the cooling system. Check and/or adjust the engine timing.

Timing Gears

REMOVAL & INSTALLATION

2F Engines

NOTE: This procedure contains camshaft removal and installation.

1. Refer to the "Cylinder Head" and "Front Cover, Removal and Installation"

procedures, in this section and remove the cylinder head and the front cover from the engine.

2. Remove the oil slinger off the crankshaft. Remove the camshaft thrust plate retaining bolts, by working through the holes provided in the camshaft timing gear.

3. Remove the camshaft through the front of the cylinder block. Support the camshaft while removing it, so the bearings or the lobes do not become damaged.

NOTE: The timing gear is a press-fit and cannot be removed without removing the camshaft.

4. Inspect the crankshaft timing gear. Replace it if it has worn or damaged teeth.

5. Remove the sliding key, then pull the crankshaft timing gear from the crankshaft with a gear puller.

6. Use a large piece of pipe to drive the timing gear onto the crankshaft. Lightly and evenly tap the end of the pipe until the gear is in its original position.

7. Apply a coat of engine oil to the camshaft journals and bearings, then insert the camshaft into the block.

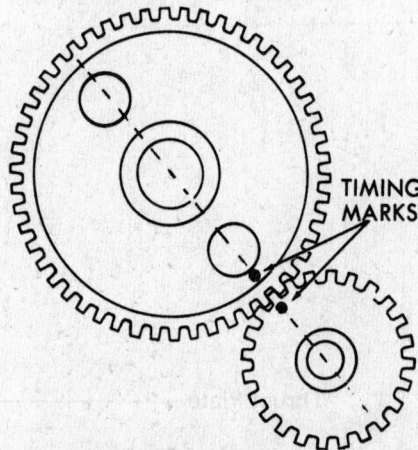

2F engines—align the camshaft and crankshaft gear marks

—— CAUTION ——
Use care not to damage the camshaft lobes, the bearings or the journals.

8. Align the mating marks on the timing gears. Slip the camshaft into position. Tighten the camshaft thrust plate bolts to 14.5 ft. lbs.

9. Using a feeler gauge, check the gear backlash, inserted between the crankshaft and the camshaft timing gears. The maximum backlash should be 0.002–0.005 in.; if it exceeds this, replace one or both of the gears, as required.

10. Check the gear run-out with a dial indicator. Maximum run-out, for both gears, is 0.008 in.; if not, replace the gear.

11. Install the oil nozzle, if it was removed, by screwing it in place with a screwdriver and punching it in two places, to secure it.

NOTE: Be sure that the oil hole in the nozzle is pointed toward the timing gear before securing it.

12. To complete the installation, use new gaskets, sealant and reverse the removal procedures.

OIL SEAL REPLACEMENT

2F Engine

1. Refer to the "Front Cover, Removal and Installation" procedures, in this section and remove the crankshaft pulley.

1. Crankshaft pulley
2. Balancer
3. Timing cover seal
4. Timing cover
5. Oil slinger
6. Crankshaft
7. Crankshaft key
8. Crankshaft gear
9. Camshaft

Timing gears, cover and related components—2F engines

2. Using a small pry bar, pry the oil seal from the front cover.

3. Using the seal installation tool No. 09515–35010, drive the new seal into the front cover. Apply multi-purpose grease to the lip of the new seal.

4. To complete the installation, reverse the removal procedures.

Camshaft

REMOVAL & INSTALLATION

Rotate the engine so that the notch on the crankshaft pulley aligns with 0° mark on the timing plate and the No.1 cylinder is at the TDC of the compression stroke. With the valve cover removed, the valves of the No.1 cylinder must be closed.

20R, 22R and 22R-E

1. Refer to the ''Cylinder Head, Removal and Installation'' procedures, in this section and remove the cylinder head from the engine.

2. Remove the rocker arm assembly from the cylinder head.

NOTE: It may be necessary to use a small pry bar to lift the rocker arm assembly from the cylinder head.

3. Using a feeler gauge, measure the thrust bearing clearance at the front of the camshaft; the standard clearance is 0.003–0.007 in, it should not exceed 0.0098 in.

4. Remove the camshaft bearing caps and lift out the camshaft. Keep the bearings in order so that they may be installed in their original position.

5. Check the camshaft journal caps for damage. Clean all of the bearing surfaces, including the caps, cam journal and the cylinder head.

6. With the camshaft in place on the cylinder head, lay small strips of plastigage® on each of the camshaft journals (at the tops of the journals, facing front-to-rear).

7. Reinstall the journal caps in their original locations (arrows facing forward) and torque the caps to 13–16 ft. lbs.

8. Remove the journal caps and gauge the width of the plastigage® against the chart on the plastigage® package. Maximum journal clearance is 0.004 in. If the journal clearance is greater than specified, measure the cam journal diameters with a micrometer. If the diameter of any cam journal is less than specified, obtain a new camshaft and recheck the journal clearance. If the clearance is still excessive, the cylinder head must be replaced.

9. To complete the installation, use new gaskets, sealant and reverse the removal procedures. Replenish the cooling system. Torque the camshaft bearing cap bolts to 14 ft. lbs., the cylinder head-to-engine block to 58 ft. lbs., the timing chain cover-to-cylinder head bolt to 9 ft. lbs., the camshaft sprocket-to-camshaft bolt to 58 ft. lbs., the intake manifold bolts to 14 ft. lbs., the exhaust manifold bolts to 33 ft. lbs. and the rocker arm cover to 7–12 ft. lbs. Replace the cooling system fluid and the engine oil. Adjust the valves, the drive belts, then check and/or adjust the timing.

NOTE: If a new cam is installed, use an assembly lube (available at most auto stores) on the cam lobes and engine oil on the journals. If the old cam is damaged excessively (lobes worn round, etc.) change the engine oil and filter.

2F Engine

NOTE: To service the 2F engine camshaft, refer to the previous Timing Gear Removal and Installation procedure.

3Y-EC Engine

1. Refer to the ''Timing Chain and Tensioner, Removal and Installation'' procedures, in this section and remove the timing chain from the engine.

2. Remove the right-front seat, the service hole cover and the distributor (type IIA).

3. Disconnect the cold start injector connector, place a shop towel under the injector tube, then remove the cold start injector union bolts, the injector and the gaskets.

Remove the valve cover, the mounting bolts and the rocker arm assembly.

4. Remove the push rods, keeping them in order. Using a wire hook or a magnetic finger, remove the valve lifters, keeping them in order.

5. Remove the thrust plate mounting bolts and the plate.

6. While turning the camshaft, slowly pull it out through the front of the engine, making sure not to damage the bearings, the camshaft lobes or the camshaft bearing surfaces.

7. Install the thrust plate, the camshaft sprocket and bolt onto the camshaft. Using a feeler gauge, measure the thrust bearing clearance, it should be 0.0028–0.0087 in.; if the clearance exceeds 0.012 in., replace the thrust plate.

8. Using a micrometer, check the bearing diameters of the camshaft. Using an internal micrometer, check the camshaft bearing diameters on the engine block.

9. Using a putty knife, clean the gasket bearing surfaces.

Using a putty knife, clean the gasket bearing surfaces.

NOTE: Before installing the valve lifters, coat them with oil.

10. To install, use new gaskets, sealant and reverse the removal procedures. Torque the camshaft thrust bearing plate bolts to 13 ft. lbs., the camshaft sprocket bolt to 67 ft. lbs., the timing chain tensioner bolts to 13 ft. lbs. and the crankshaft pulley bolt to 80 ft. lbs. Adjust the drive belts and refill the cooling system. Check and/or adjust the timing.

Pistons and Connecting Rods

REMOVAL & INSTALLATION

1. Refer to the ''Cylinder Head'' and the ''Oil Pan, Removal and Installation'' procedures, in this section, then remove the cylinder head and the oil pan from the engine.

2. Remove the oil pump (3Y-EC) and the strainer.

NOTE: On the 20R, 22R and 22R-E engines, it is not necessary to remove the oil pump.

3. Using a ridge reamer tool, remove the ridges from the top of the cylinder.

4. Measure the connecting rod side clearance.

5. Remove the connecting rod cap of the number one cylinder and check the oil clearance with plastigage®. Record the clearance and compare with the specification chart. Repeat this step for each connecting rod.

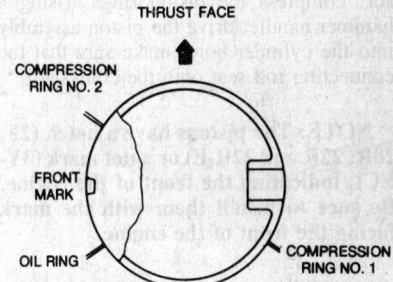

Ring gap staggering on the 20R, 22R and 22R-E engines

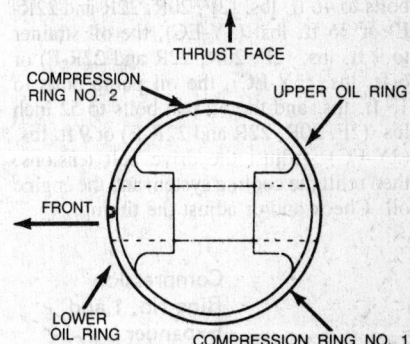

2F engines—after installing the piston rings, rotate each ring so that the ring gaps are positioned as shown. Failure to stagger the ring gaps will result in excessive oil consumption.

6. With the number one connecting rod cap removed, install a short piece of rubber hose onto each connecting rod bolt (the hose must completely cover the bolt).

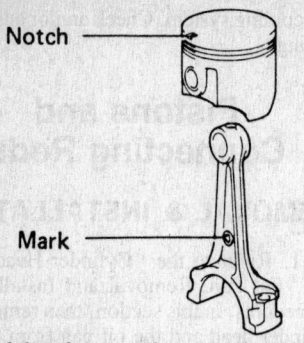

Notch

Mark

View of the piston and connecting rod—3Y-EC engine

7. Using a wooden or plastic handle (an old hammer handle works well), carefully tap the piston/connecting rod assembly out of the cylinder. Do not use excessive force as this could damage the connecting rods. Repeat this step for each cylinder.

NOTE: When removing the piston/connecting rod assemblies, be sure to keep the parts in order, for reassembly purposes.

8. To install the piston/connecting rod assemblies, place the assembly into the appropriate cylinders. Using a ring compressor, compress the piston rings. Using a hammer handle, drive the piston assembly into the cylinder bore; make sure that the connecting rod seat onto the crankshaft.

NOTE: The pistons have a notch (2F, 20R, 22R and 22R-E) or a dot mark (3Y-EC), indicating the front of the engine. Be sure to install them with the mark facing the front of the engine.

9. To complete the installation, use new gaskets, sealant and reverse the removal procedures. Torque the connecting rod cap bolts to 46 ft. lbs. (2F, 20R, 22R and 22R-E) or 36 ft. lbs. (3Y-EC), the oil strainer to 9 ft. lbs. (2F, 20R, 22R and 22R-E) or 6 ft. lbs. (3Y-EC), the oil pump bolts to 13 ft. lbs. and the oil pan bolts to 52 inch lbs. (2F, 20R, 22R and 22R-E) or 9 ft. lbs. (3Y-EC). Adjust the drive belt tensions, then refill the cooling system and the engine oil. Check and/or adjust the timing.

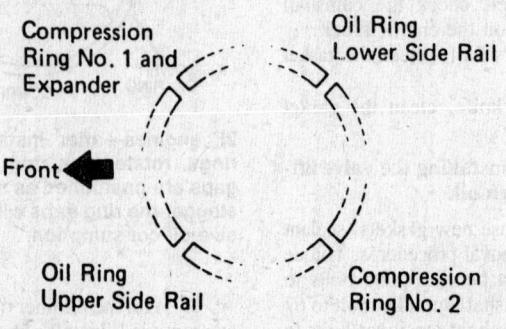

Compression Ring No. 1 and Expander

Oil Ring Lower Side Rail

Front

Oil Ring Upper Side Rail

Compression Ring No. 2

View of the compression ring positioning—3Y-EC engine

DIESEL ENGINE

Engine

REMOVAL & INSTALLATION

L Engine (1981–83)

1. Scribe match-marks on the hood and the hood supports, the remove the hood.
2. Remove the batteries from the vehicle.

NOTE: Some models may require the removal of both batteries from the vehicle.

3. Drain the cooling system and remove the radiator, shroud and radiator hoses.
4. If the vehicle is equipped with A/C, remove the compressor drive belt, unbolt the compressor and tie the compressor out of the way. DO NOT disconnect the refrigerant lines from the compressor.
5. Remove the drive belt, the water pump pulley and the cooling fan.
6. Disconnect the two heater hoses from the left side of the engine.
7. Disconnect the vacuum reservoir hose from the rear of the alternator.
8. Disconnect the vacuum hose from the "idle-up" unit, if the vehicle is equipped with air conditioning.
9. Disconnect the fuel hoses from the fuel injection pump.
10. Disconnect the wiring from the following components:
   a. Alternator.
   b. Thermo-switch.
   c. Oil pressure switch.
   d. No.1 glow plug relay (terminal +B).
   e. Starter.

NOTE: Mark these wires and tie them out of the way.

11. Disconnect the wiring from the left fender and the injection pump (accelerator wire). Also mark and tie these wires out of the way.
12. Using Toyota special service tool No. 09305–20012, remove the transmission shift lever from inside the vehicle.

13. Raise and support the vehicle on jackstands.
14. Drain the engine oil. Remove the engine under-cover and remove the backup light switch wire.
15. Remove the engine shock absorber and the drive shaft from the vehicle.

NOTE: Mark the drive shaft and the companion flange so that the shaft may be reinstalled in its original position.

16. Remove the speedometer cable and the exhaust pipe clamp from the transmission housing, then the exhaust pipe mounting nuts from the exhaust manifold.
17. Remove the clutch release cylinder and lay the cylinder along-side the frame.
18. Remove the engine mounting bolts from each side of the engine.
19. Place a jack under the transmission to support it. Remove the rear mount bracket at the crossmember and the crossmember.
20. Attach the engine lifting device to the engine.

NOTE: Check that all wiring and hoses are clear of the engine/transmission assembly.

21. Carefully, raise the engine/transmission assembly out of the engine compartment, being especially careful not to damage the air conditioning compressor, if equipped.
22. Remove the starter and the transmission from the engine, then mount the engine securely to a workstand.
23. To install, reverse the removal procedures. Refill the cooling system, the engine oil and/or the transmission.

2L and 2L-T Engines (1984 and Later)

NOTE: If equipped with 2WD, refer to the "Transmission, Removal and Installation" procedures, in this section and remove the transmission. If equipped with 4WD, refer to the "Transmission/Transfer Case, Removal and Installation" procedures, in this section and remove the transmission with the transfer case. DO NOT drain the fluid from the transmission or the transfer case.

1. Scribe match-marks on the hood and the hood supports, then remove the hood.
2. Remove the negative battery cable.
3. Drain the cooling system by opening the drain cocks at the radiator and the left-side of the engine block.
4. Remove the radiator hoses, the upper radiator shroud, the coolant reservoir hose, the mounting bolts and the radiator.

NOTE: If equipped with A/C, remove the lower radiator shroud.

5. If equipped with a 2L engine, remove the air cleaner.

6. Disconnect the accelerator cable from the fuel injection pump.

7. If equipped with A/C, remove the A/C vacuum hose from the VSV. Remove the drive belt, then the A/C compressor and move it aside. DO NOT disconnect the refrigerant lines from the compressor.

8. If equipped with a 2L-T engine, remove the turbocharger pressure hose from the pressure switch.

9. Disconnect the oil inlet hose from the vacuum pump.

10. Remove the inlet and the outlet fuel lines from the fuel injection pump.

11. Remove the wires or the electrical connectors from the following:
   a. The glow plug current sensor.
   b. The water temperature sensor.
   c. The glow plug resistor.
   d. The injection pump.
   e. The water temperature sender gauge.
   f. The starter.
   g. The engine ground cables.
   h. The oil pressure switch.
   i. The alternator.

NOTE: For California models, disconnect the throttle position sensor and the EVRV connectors. For 4WD models, disconnect the water temperature switch. Mark these wires and tie them out of the way.

12. Remove the engine splash shield.

13. If equipped with power steering, remove the drive belt, the mounting bolts and the pump, then move the pump aside. DO NOT disconnect the power steering pressure hoses.

14. If equipped with 2WD, remove the engine mounting shock absorber from the crossmember.

15. Remove the engine mounting insulator-to-crossmember nuts and bolts.

16. Remove the exhaust pipe mounting nuts from the exhaust manifold.

17. Attach the engine lifting device to the engine.

NOTE: Check that all wiring and hoses are clear of the engine.

18. Carefully, raise the engine out of the engine compartment, being especially careful not to damage the A/C compressor, if equipped.

19. Remove the starter and mount the engine securely to a workstand.

20. To install, reverse the removal procedures. Refill the cooling system and the engine oil. Adjust the drive belts.

Cylinder Head

—————— CAUTION ——————

Do not perform this operation on a warm engine. Remove the head bolts in the re-

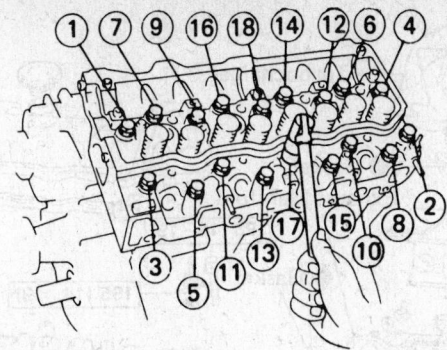

Cylinder head bolt removal sequence— diesel engine

verse of the tightening sequence. Loosen the head bolts evenly, not one at a time. Keep the pushrods in their original order. Do not attempt to slide the cylinder head off of the block, as it is located with dowel pins. Lift the head straight up and off the block.

REMOVAL & INSTALLATION

1. Using a sharp tool, scribe alignment marks of the hood brackets on the hood and remove the hood.

2. Remove the negative battery cables (L) or cable (2L and 2L-T).

3. Remove the air cleaner assembly (L), the air cleaner hose (2L) or the turbocharger intake pipe (2L-T).

4. Drain the cooling system. On the L engine, remove the radiator, the shroud and the radiator hoses. On the 2L and 2L-T engines, remove the inlet/outlet hoses, the shroud mounting bolts, the upper shroud, the lower shroud (A/C equipped), the radiator mounting bolts and the radiator.

5. On the 2L-T engine, disconnect the radiator reservoir tank hose from the water outlet and remove the turbocharger.

6. Disconnect the heater hoses at the engine and move the hoses aside.

7. On the L engine, disconnect the vacuum reservoir hose from the rear of the alternator, the alternator wire, the oil pressure switch wire, the B + terminal from the No.1 glow plug relay, the engine cable from the left fender and the starter wire. On the L and 2L engine, disconnect the accelerator

cable and the vacuum hose (A/C equipped) from the VSV.

8. If equipped with A/C, remove the drive belt and the compressor from the mounting brackets. Tie the compressor out of the way. DO NOT remove the refrigerant lines.

9. Disconnect the vacuum hoses, then the fuel hoses from the inlet and the outlet fuel pipes.

10. On the 2L and 2L-T engines, disconnect the following: the glow plug current sensor connector, the water temperature sensor connector, the glow plug resistor wires, the water temperature sender gauge connector and the water temperature switch (4WD).

11. Remove the fuel injection lines and the fuel injectors from the the engine. From the fuel injection pump, remove the nut, the gasket and the fuel inlet pipe. Arrange the injectors so that they may be reinstalled in their original locations.

12. Remove the PCV hose. Remove the engine cooling fan, pulley and drive belt. Remove the crankshaft pulley bolt, the pulley, the timing belt cover and the timing belt.

13. Using the wrench tool No. 09278–54011, remove the camshaft sprocket bolt. Using the wheel puller tool No. 09950–20015, pull the camshaft sprocket from the camshaft.

14. Remove the valve cover and upper front engine cover. Disconnect the accelerator connecting rod from the accelerator link.

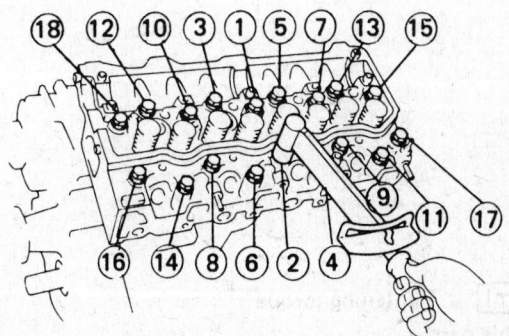

Cylinder head bolt installation sequence—diesel engine

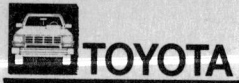
Turbocharger

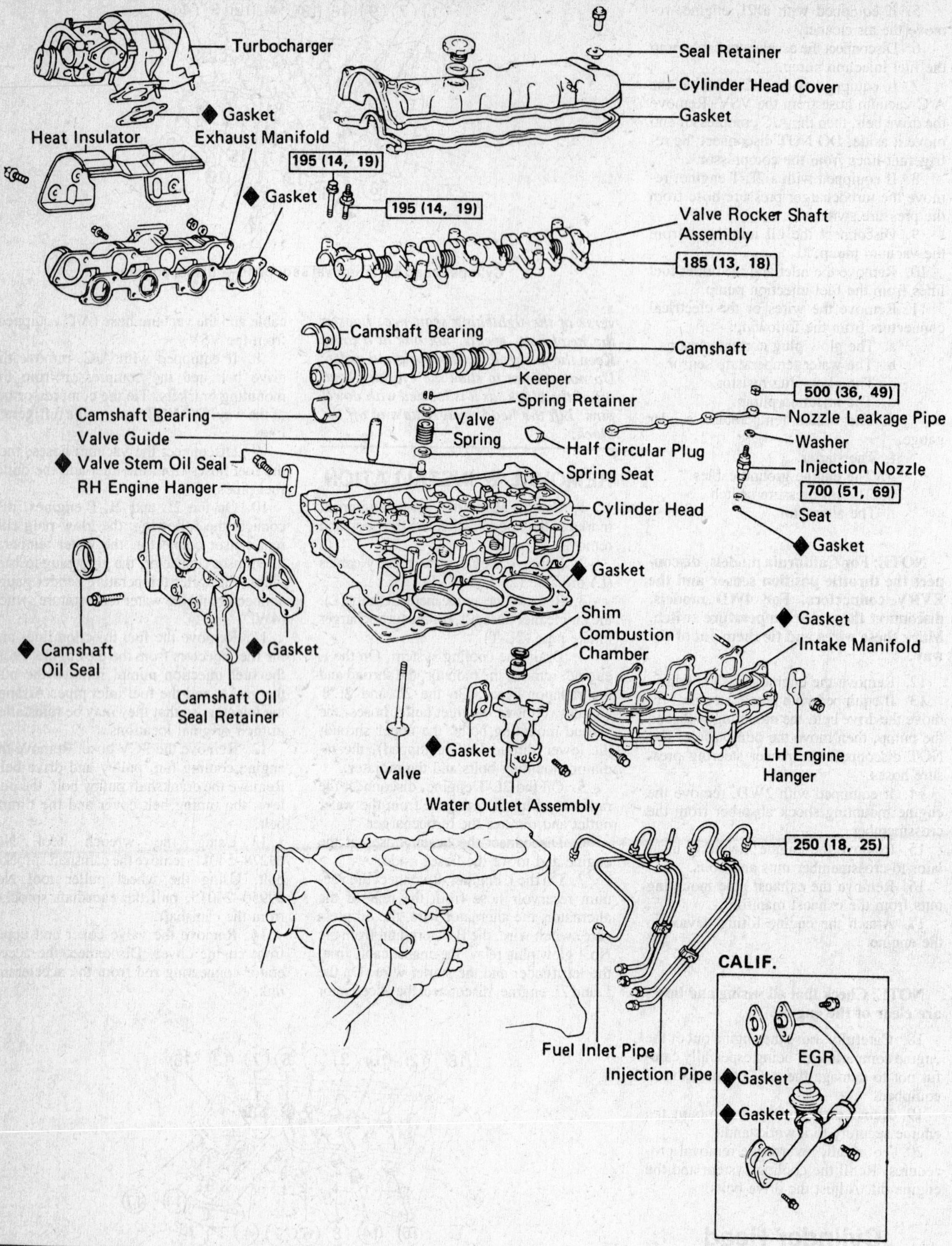

Seal Retainer

Cylinder Head Cover

Gasket

◆ Gasket
Exhaust Manifold

Heat Insulator

195 (14, 19)

◆ Gasket

195 (14, 19)

Valve Rocker Shaft
Assembly

185 (13, 18)

Camshaft Bearing

Camshaft

Keeper

Spring Retainer

500 (36, 49)

Camshaft Bearing

Valve Guide

Valve
Spring

Nozzle Leakage Pipe

Washer

◆ Valve Stem Oil Seal

Half Circular Plug

Injection Nozzle

RH Engine Hanger

Spring Seat

700 (51, 69)

Cylinder Head

Seat

◆ Gasket

◆ Camshaft
Oil Seal

◆ Gasket

◆ Gasket

Shim
Combustion
Chamber

◆ Gasket
Intake Manifold

Camshaft Oil
Seal Retainer

◆ Gasket

Valve

LH Engine
Hanger

Water Outlet Assembly

250 (18, 25)

Fuel Inlet Pipe

Injection Pipe

CALIF.

EGR

◆ Gasket

◆ Gasket

kg-cm (ft-lb, N·m) : Tightening torque

◆ : Non-reusable part

Exploded view of the 2L-T diesel engine—cylinder head

15. On Calif. models, remove the Electronic Vacuum Regulating Valve (EVRV), the EGR mounting bolts, the EGR valve and gasket.

16. Remove the intake manifold/left-hand hanger nuts/bolts, then the intake manifold and left-hand hanger.

17. Remove the mounting bolts and the water outlet assembly with the gasket. Remove the right-hand engine hanger and A/C bracket (if equipped).

18. On the turbocharger (2L-T), remove the clip, the union bolt and the 2 gaskets, then disconnect the oil pipe from the turbocharger.

19. Remove the manifold stay nuts/bolts and the stay. When removing the exhaust manifold, remove the mounting bolts and the heat insulator (2L-T) or the two heat insulators (2L), then the mounting nuts/bolts and the exhaust manifold.

20. Remove the mounting bolts, then camshaft oil seal retainer and gasket.

21. Using a feeler gauge, measure the camshaft thrust clearance. The standard clearance is 0.0022–0.0061 in.; the maximum is 0.012 in.

22. Gradually, remove the rocker arm shaft retaining bolts (starting at the outer ends and working toward the middle). Remove the rocker arm shaft assembly with the three upper bearing halves.

NOTE: When removing the camshaft bearings, be sure to arrange the bearings in order, for installation purposes.

23. Gradually, loosen and remove the head bolts in three passes. Lift the cylinder head (it may be necessary to use a pry bar to remove the cylinder head), remove it from the engine and place it on two blocks of wood.

24. Using a putty knife, clean the gasket mounting surfaces.

25. To install, use new gaskets and reverse the removal procedures. Torque the cylinder head bolts (in sequence) to 87 ft. lbs., the camshaft bearing bolts to 14 ft. lbs., the camshaft oil seal retainer bolts to 13 ft. lbs., the exhaust manifold-to-engine bolts to 29 ft. lbs. (2L) or 38 ft. lbs. (2L-T), the right-hand engine hanger bolts to 27 ft. lbs., the water outlet assembly bolts to 14 ft. lbs., the intake manifold/left-hand hanger bolts to 17 ft. lbs., the EGR valve bolts to 14 ft. lbs., the EGR valve nuts to 9 ft. lbs., the camshaft sprocket bolt to 72 ft. lbs., the injection nozzles to 51 ft. lbs., the injection pipe-to-nozzle leakage pipe to 36 ft. lbs., the injection pipe-to-nozzle to 18 ft. lbs. Adjust the drive belts and the valves. Refill the cooling system and the engine oil. Bleed the fuel injection system. Check and/or adjust the timing.

Valve Rocker Shafts

REMOVAL & INSTALLATION

1. Disconnect the cables which are po-
sitioned above the valve cover and move the cables aside.

2. Remove the valve cover.

3. Gradually loosen the rocker shaft support fasteners, working from the ends towards the center. Remove the rocker shaft and arms as an assembly.

NOTE: It is not necessary to remove the timing belt or related components.

4. To install the rocker shaft assembly, gradually tighten the support bolts, working from the center towards the ends. Finally tighten the bolts to 11–15 ft. lbs. Adjust the valves.

VALVE ADJUSTMENT

1. Remove the cylinder head cover.

2. Turn the crankshaft (using a wrench) until the No.1 cylinder is on the TDC of the compression stroke. The valves of the No.1 cylinder must be loose.

3. To adjust the valve clearance, loosen the adjuster lock nut and turn the adjusting screw.

4. Using a feeler gauge, check and/or adjust the intake valve clearance of cylinders No.1 and 2, then the exhaust valve clearance of cylinders No.1 and 3.

NOTE: The intake valve clearance is 0.011 in. (cold); the exhaust valve clearance is 0.015 in. (cold).

5. Using a wrench on the crankshaft pulley bolt, turn the crankshaft one revolution, stopping on the 0° mark of the timing plate (TDC of No.4 cylinder).

6. Adjust the intake valves of cylinder No.3 and 4; the exhaust valves of cylinders No.2 and 4.

7. With the adjustment complete, install the removed components by reversing the removal procedures.

Intake Manifold

REMOVAL & INSTALLATION

1. Disconnect the negative battery cables (L) or cable (2L and 2L-T).

2. Label and disconnect any wires or vacuum hoses which may be in the way.

3. On the 2L and 2L-T engines, remove the fuel injection tube clamp bolts from the intake manifold. If necessary, disconnect the fuel injection tubes from the fuel injectors and bend them out of the way.

4. On the L and 2L engines, disconnect the air inlet tube from the intake manifold. On the 2L-T engine, disconnect the turbocharger-to-intake manifold tube and the turbocharger-to-air cleaner tube.

5. Remove the mounting nuts/bolts (starting at the outer ends and working toward the center) and the intake manifold (with the left-hand engine hanger) from the engine.

6. Using a putty knife, clean the gasket mounting surfaces.

7. To install, use new gaskets and reverse the removal procedures. Torque the intake manifold nuts/bolts to 17 ft. lbs.

Exhaust Manifold

REMOVAL & INSTALLATION

1. Disconnect the negative battery cables (L) or cable (2L and 2L-T).

2. For the L and 2L engines, disconnect the exhaust pipe from the exhaust mainfold. For the 2L-T engine, refer to "Turbocharger, Removal and Installation" procedures, in this section and remove the turbocharger.

3. Remove the exhaust manifold-to-engine brace nuts/bolts and the brace.

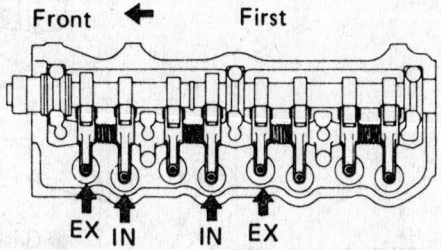

Adjusting the valve clearance (first step)—diesel engines

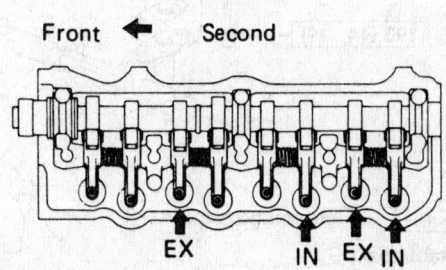

Adjusting the valve clearance (second step)—diesel engines

4. On the L and 2L engines, remove the exhaust manifold-to-engine bolts, the insulators and the manifold. On the 2L-T engine, remove the exhaust manifold-to-engine nuts/bolts, the insulators and the manifold.

NOTE: When removing the exhaust manifold bolts, loosen them (gradually) from the outer ends, working toward the center. When installing the manifold, work from the center toward the outer ends.

5. Using a putty knife, clean the gasket mounting surfaces.

6. To install, use new gaskets and reverse the removal procedures. Torque the exhaust manifold bolts to 29 ft. lbs. (L and 2L) or 38 ft. lbs. (2L-T), the insulator-to-exhaust manifold bolts to 9 ft. lbs.

Turbocharger

REMOVAL & INSTALLATION

2L-T Engine

1. Remove the negative battery cable. Drain the engine coolant.

2. Disconnect the vacuum hoses from the vacuum pump to the High Altitude Compensation valve (HAC) and from the Vacuum Switching Valve (VSV) to the idle-up actuator (A/C equipped), at the intake pipe.

3. Disconnect the accelerator cable from the intake manifold.

4. Disconnect the PCV hose. Remove the clamp and the No.1 air cleaner hose, then the clamp, the mounting bolts and the air cleaner pipe with the No.2 air hose.

5. Remove the mounting bolts and the heater hoses from the intake flange. Remove the clamps, the air intake pipe-to-valve cover bolts and the air intake pipe, then the No.1 air cleaner-to-compressor elbow hose.

6. Remove the bolts, the compressor elbow and the gasket from the turbocharger.

7. Remove the exhaust pipe-to-turbocharger flange nuts, the pipe and gasket from the turbocharger.

8. Remove the heat insulator from the turbocharger. Remove the clamps, the mounting nuts/bolts, then the No.2 and No.3 water by-pass tubes from the turbocharger.

9. Remove the mounting flange nuts and the oil pipe from the turbocharger.

10. Remove the turbocharger-to-exhaust manifold nuts, the turbocharger and the gaskets. If necessary, remove the turbine elbow from the turbocharger.

11. Using a putty knife, clean the gasket mounting surfaces.

NOTE: Before installing the turbocharger, pour about 1¼ cu. in. of oil into the turbocharger oil inlet, then spin the impeller wheel to splash oil on the bearing surface.

12. To install, use new gaskets and reverse the removal procedures. Torque the turbine elbow-to-turbocharger nuts to 19 ft. lbs., the turbocharger-to-exhaust manifold nuts to 38 ft. lbs., the oil pipe-to-turbocharger flange nuts to 14 ft. lbs., the turbine elbow-to-exhaust pipe nuts to 9 ft. lbs. and the compressor elbow-to-turbocharger bolts to 9 ft. lbs.

Front Cover

OIL SEAL REPLACEMENT

1. Refer to the "Timing Belt, Removal and Installation" procedures, in this section and remove the front cover.

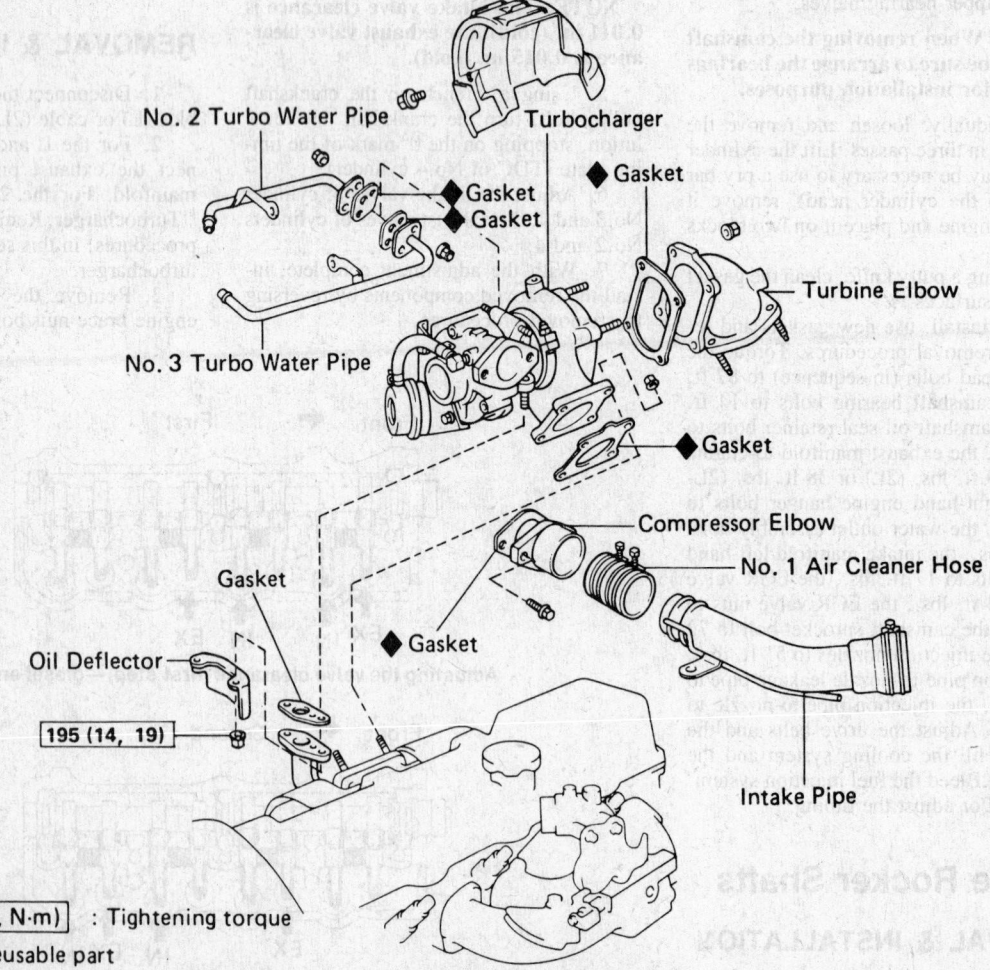

No. 2 Turbo Water Pipe
Turbocharger
◆ Gasket
◆ Gasket
◆ Gasket
◆ Gasket
No. 3 Turbo Water Pipe
Turbine Elbow
◆ Gasket
Compressor Elbow
No. 1 Air Cleaner Hose
Gasket
◆ Gasket
Oil Deflector
195 (14, 19)
Intake Pipe

kg-cm (ft-lb, N·m) : Tightening torque
◆ : Non-reusable part

Exploded view of the turbocharger— diesel engine

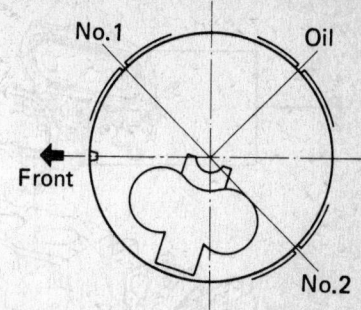

2. Remove the timing belt, the crankshaft sprocket and the oil pump from the engine.

NOTE: The oil seal is located in the oil pump which is attached to the front of the crankshaft.

3. Using a small pry bar, pry the oil seal from the front of the oil pump.

4. Apply multi-purpose grease to the seal lips.

5. Using the seal installation tool No. 09223–22010 and a hammer, drive the new oil seal into the oil pump housing.

6. To complete the installation, reverse the removal procedures.

Timing Belt and Tensioner

REMOVAL & INSTALLATION

1. Disconnect the negative battery cables (L) or cable (2L and 2L-T).

2. Drain the cooling system. Remove the radiator hoses, the upper and lower shrouds, the coolant reservoir hose and the radiator. Remove the drive belts, the fluid coupling and the pulley from the water pump fan.

3. If equipped with A/C, remove the mounting bolts and the A/C pulley from the crankshaft pulley; using tools 09213–54012 and 09330–00020, remove the set bolt and the plate washer from the crankshaft pulley. Using the wheel puller tool No. 09213–60016 (L) or 09213–60017 (2L and 2L-T), remove the crankshaft timing pulley.

4. If equipped with A/C, remove the bolts and the A/C idler pulley.

5. On the L and 2L engines, remove the mounting bolts and the front cover with the gasket. On the 2L-T engine, remove the clips and the turbo water hoses from the No. 1 turbo water tube, then the mounting bolts, the No. 1 turbo water tube, the front cover and the gasket.

6. Remove the timing belt guide.

7. On the 2L-T engine, remove the PCV hose from the air intake tube, the accelerator cable from the air intake tube, the clip and the injection pump, then the air cleaner tube and the No. 1 air cleaner hose.

8. Remove the grommet screws, the glow plug connector nuts, the current sensor, the glow plug connector and the glow plugs.

9. Using the crankshaft pulley set bolt and a wrench, turn the crankshaft until the TDC mark of the camshaft sprocket aligns with the top end of the cylinder head.

NOTE: If reusing the timing belt, mark an arrow on the belt in the direction of revolution and match-mark the belt with the sprockets, for installation purposes.

10. Using long pliers, remove the timing belt tension spring. Loosen the No. 1 idler pulley mount bolts and remove the timing belt.

11. Remove the mounting bolts, the plate washers and the tensioner pulley.

12. Using a putty knife, clean the gasket mounting surfaces.

NOTE: To install the timing belt, align the sprocket timing marks, install the timing belt, apply spring tension on the belt. Turn the crankshaft two clockwise revolutions, stopping at the TDC, then recheck the timing marks.

13. To complete the installation, use new gaskets and reverse the removal procedures. Torque the idler pulley bolts to 14 ft. lbs., the front cover bolts to 3–5 ft. lbs., the glow plugs and the current sensor to 9 ft. lbs., then the crankshaft pulley bolt to 69–75 ft. lbs. (L). Using tool No. 09214–60010 and a hammer, drive the crankshaft pulley on to the crankshaft; using tool Nos. 09213–54012 and 09330–00020, torque the crankshaft pulley set nut to 102 ft. lbs. (2L and 2L-T). Adjust the drive belts and refill the cooling system.

Camshaft

REMOVAL & INSTALLATION

Refer to the ''Cylinder Head, Removal and Installation'' procedures, in this section and remove the camshaft from the cylinder head. The journal clearances are checked in the same manner as the gasoline engines.

To install, use new gaskets, sealant and reverse the removal procedures. Torque the rocker shaft supports/camshaft bearing to 11–15 ft. lbs., the camshaft timing sprocket set bolt to 69–75 ft. lbs., the oil seal retainer to 8–12 ft. lbs. Adjust the drive belts and refill the cooling system.

NOTE: Refer to the ''Timing Belt, Removal and Installation'' procedures, in this section to properly install the timing belt.

Pistons and Connecting Rods

REMOVAL & INSTALLATION

1. Refer to the ''Cylinder Head'' and the ''Oil Pan, Removal and Installation'' procedures, in this section, then remove the cylinder head, the oil pan and the oil strainer.

2. Using a ridge reamer, remove the cylinder ridges.

3. Measure the connecting rod side clearance; the standard clearance is 0.0031–0.0079 in., the maximum clearance is 0.012 in.

4. Remove the connecting rod cap of the number one cylinder and check the oil clearance with plastigage®. Record the clearance and compare with the specifica-

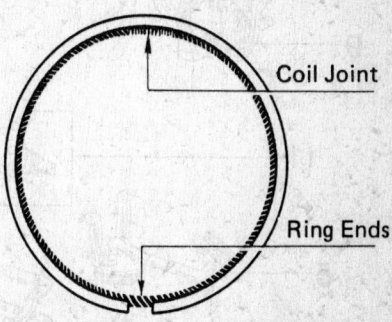

Piston ring gap staggering—diesel engine

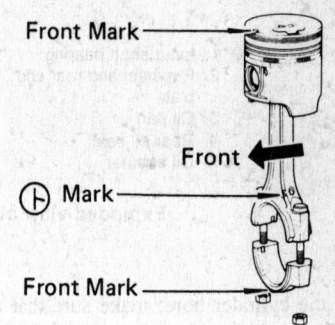

Oil ring installation on the diesel piston—position the ends of the expander opposite the ring gap

Piston and connecting rod installation—diesel engine

tion chart. Repeat this step for each connecting rod.

5. With the number one connecting rod cap removed, install a short piece of rubber hose onto each connecting rod bolt (the hose must completely cover the bolt).

6. Using a wooden or plastic handle (an old hammer handle works well), carefully tap the piston/connecting rod assembly out of the cylinder. Do not use excessive force as this could damage the connecting rods. Repeat this step for each cylinder.

NOTE: When removing the piston/connecting rod assemblies, be sure to keep the parts in order, for reassembly purposes.

7. To install the piston/connecting rod assemblies, place the assembly into the appropriate cylinders. Using a ring compressor, compress the piston rings. Using a hammer handle, drive the piston assembly

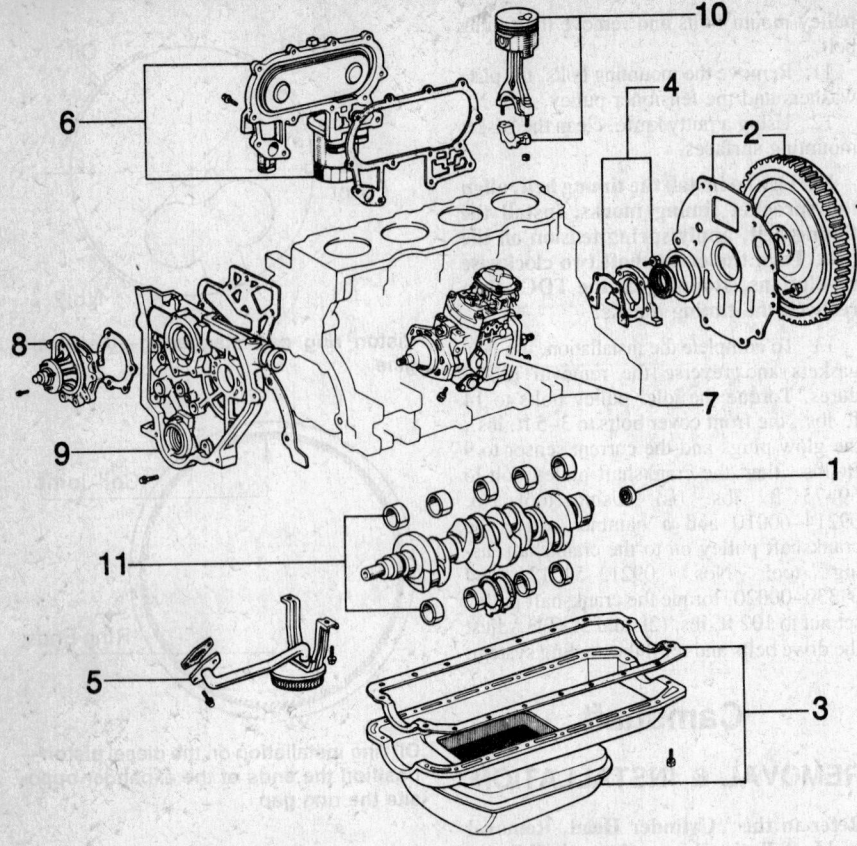

1. Input shaft bearing
2. Flywheel and rear end plate
3. Oil pan
4. Rear oil seal
5. Oil strainer
6. Oil filter bracket
7. Injection pump
8. Water pump
9. Timing belt case
10. Piston and connecting rod
11. Crankshaft

Exploded view of the diesel cylinder block

into the cylinder bore; make sure that the connecting rod seat onto the crankshaft.

NOTE: The pistons have a 0 mark (L and 2L) or an arrow (2L-T), indicating the front of the engine. Be sure to install them with the mark facing the front of the engine.

8. To complete the installation, use new gaskets, sealant and reverse the removal procedures. Torque the connecting rod cap bolts to 43 ft. lbs., the oil strainer to 9 ft. lbs., the oil pan bolts to 6 ft. lbs. or the nuts to 13 ft. lbs. and the engine crossmember bolts to 38 ft. lbs. Adjust the drive belt tensions, refill the cooling system and the engine oil. Check and/or adjust the fuel injection pump timing.

Crankshaft

For service information on the crankshaft and the main bearings, refer to "Engine Rebuilding" in the Unit Repair section.

GAS ENGINE LUBRICATION

Oil Pan

REMOVAL & INSTALLATION

Pickups (2WD and 4WD)

1. Remove the negative battery cable.
2. Raise and support the vehicle on jackstands.
3. Remove the under engine cover, then drain the engine oil.
4. On the 2WD (1979–83), remove the steering idler arm bracket, the pitman arm from the selector shaft and the crossmember.
5. On the 1984 and later models, support the front of the transmission with a floor jack, remove the engine-to-engine mount bolts and raise the engine 0.98 in.
6. Remove the oil pan nuts/bolts and the oil pan.

7. Using a putty knife, clean the gasket mounting surfaces.
8. To install, use a new gasket, sealant and reverse the removal procedures. Torque the oil pan bolts to 33–70 inch lbs. (1979–83) or 9 ft. lbs. (1984 and later). Refill the engine crankcase with oil.

Land Cruiser

1. Disconnect the negative battery cable.
2. Raise and support the vehicle on jackstands.
3. Remove the under-engine skid plates, then the flywheel side cover and undercover.
4. Disconnect the front driveshaft from the engine. Drain the engine oil.
5. Remove the oil pan bolts, the pan and gasket.
6. To install, use a new gasket, sealant and reverse the removal procedures. Refill the engine crankcase with oil.

Van

1. Disconnect the negative battery cable.
2. Raise and support the vehicle on jackstands.
3. Remove the stiffener plates from both sides of the engine. Drain the engine oil.
4. Disconnect the oil level sensor electrical connector.
5. Remove the oil pan bolts and the pan.

NOTE: If necessary, use tool No. 09032–00100, insert it between the oil pan and the cylinder block, then cut off the sealant.

6. Using a putty knife, clean the gasket mounting surfaces.
7. To install, use a new gasket, sealant and reverse the removal procedures. Torque the oil pan bolts to 9 ft. lbs.

Oil Pump

REMOVAL & INSTALLATION

Pickups (2WD and 4WD)

NOTE: When the oil pump has been removed from the engine, it is recommended that the new oil seal be installed.

1. Refer to the "Oil Pan, Removal and Installation" procedures, in this section and remove the oil pan and the oil strainer.
2. Remove the drive belts from the crankshaft pulley.
3. Remove the crankshaft pulley bolt. Using the wheel puller tool No. 09213–31021, remove the crankshaft pulley.
4. Remove the mounting bolts and the oil pump assembly. Refer to the oil pump

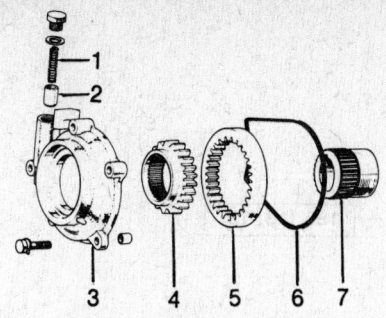

1. Relief valve spring
2. Relief valve
3. Pump body
4. Drive gear
5. Driven gear
6. O-ring
7. Drive spline

Exploded view of the 20R, 22R and 22R-E engine oil pump

specification chart to check the oil pump clearances.

NOTE: Check the timing chain cover for excessive wear or damage. If necessary, replace the gears or pump body or cover. Unbolt the relief valve (the vertical bolt on the pump body when attached to the engine), then check the piston, the oil passages and the sliding surfaces for burrs or scoring. Inspect the crankshaft front oil seal and replace if worn or damaged.

5. Using a putty knife, clean the gasket mounting surfaces.

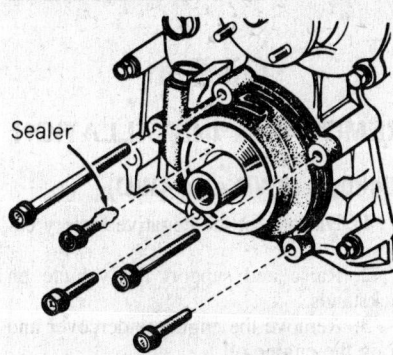

Sealer

Mounting of the 20R, 22R and 22R-E engine oil pump—apply sealer to the upper mounting bolt as shown

NOTE: Before installing the oil pump, be sure to pack the oil pump cavity with petroleum jelly.

6. To install, use a new oil pump-to-front cover O-ring, a new oil pan gasket, sealant and reverse the removal procedures. Adjust the drive belts and refill the crankcase with engine oil.

NOTE: When installing the oil pump bolts, apply sealant to the upper bolt. Be sure to apply sealer to the corners of the oil pan gasket before installing the pan.

Land Cruiser—2F Engine

NOTE: When the oil pump has been removed from the engine, it is recommended that the new oil seal be installed.

1. Refer to the "Oil Pan, Removal and Installation" procedures, in this section and remove the oil pan, the oil strainer and the union nuts on the oil pump pipe.

2. Remove the lock wire, the oil pump retaining bolt and the pipe from the engine.

3. Remove the oil pump cover and inspect the following parts for nicks, scoring, grooving and etc.: The pump cover, the drive/driven gears and the pump body.

4. If damage is excessive, replace the damaged parts or the complete pump. See the Oil Pump Specification chart to check the oil pump clearances.

5. To install, use new gaskets, sealant and reverse the removal procedures.

Van 3Y-EC Engine

1. Refer to the "Oil Pan, Removal and Installation" procedures, in this section and reverse the removal procedures.

2. Remove the oil pump mounting bolts, then pull out the pump assembly.

3. Using a putty knife, clean the gasket mounting surfaces.

4. To install, use new gaskets, sealant and reverse the removal procedures. Torque the oil pump bolts to 13 ft. lbs.

Rear Main Bearing Oil Seal

REMOVAL & INSTALLATION

20R, 22R and 22R-E Engines

1. Refer to the "Transmission" and/or "Clutch, Removal and Installation" procedures, in this section and remove the transmission (with the torque converter for A/T) and the clutch assembly (if equipped). Remove the transfer case, if equipped.

2. Remove the flywheel or the flex plate

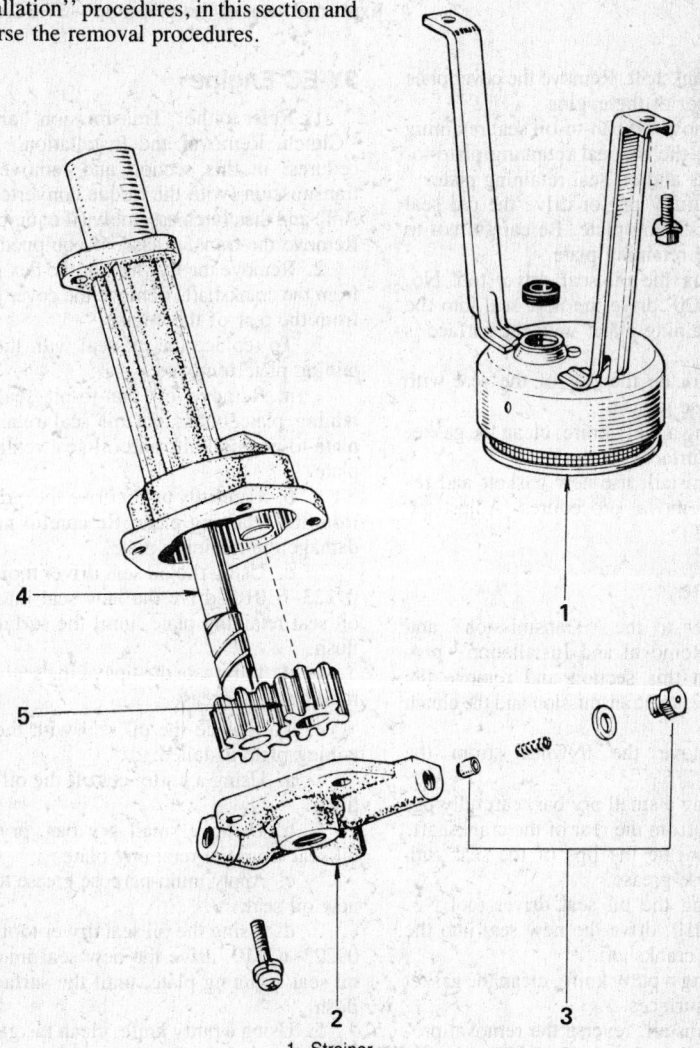

1. Strainer
2. Pump cover
3. Pressure relief valve
4. Drive gear
5. Driven gear

Exploded view of the 2F engine oil pump

kg-cm (ft-lb, N·m) : Tightening torque

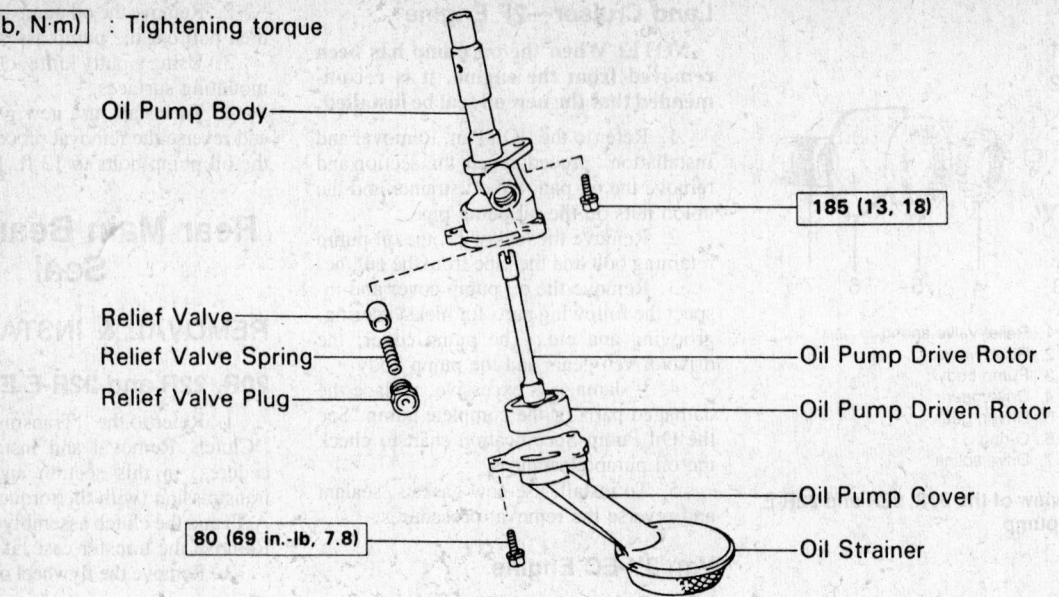

Oil Pump Body

Relief Valve
Relief Valve Spring
Relief Valve Plug

185 (13, 18)

Oil Pump Drive Rotor
Oil Pump Driven Rotor
Oil Pump Cover
Oil Strainer

80 (69 in.-lb, 7.8)

Exploded view of the oil pump—3Y-EC engine

from the crankshaft. Remove the cover plate from the rear of the engine.

3. Remove oil pan-to-oil seal retaining plate bolts, the oil seal retaining plate-to-engine bolts and oil seal retaining plate.

4. Carefully pry or drive the old seal from the retaining plate. Be careful not to damage the retaining plate.

5. Using the oil seal driver tool No. 09223–41020, drive the new seal into the oil seal retaining plate, until the surface is flush.

6. Lubricate the lips of the seal with multipurpose grease.

7. Using a putty knife, clean the gasket mounting surfaces.

8. To install, use new gaskets and reverse the removal procedures. Adjust the clutch (M/T).

2F Engine

1. Refer to the ''Transmission'' and ''Clutch, Removal and Installation'' procedures, in this section and remove the transfer case, the transmission and the clutch assembly.

2. Remove the flywheel from the crankshaft.

3. Using a small pry bar, carefully pry the oil seal from the rear of the crankshaft.

4. Lubricate the lips of the seal with multipurpose grease.

5. Using the oil seal driver tool No. 09223–60010, drive the new seal into the rear of the crankshaft.

6. Using a putty knife, clean the gasket mounting surfaces.

7. To install, reverse the removal procedures. Adjust the clutch (M/T).

3Y-EC Engine

1. Refer to the ''Transmission'' and/or ''Clutch, Removal and Installation'' procedures, in this section and remove the transmission (with the torque converter for A/T) and the clutch assembly (if equipped). Remove the transfer case, if equipped.

2. Remove the flywheel or the flex plate from the crankshaft. Remove the cover plate from the rear of the engine.

3. To replace the oil seal with the retaining plate removed:

  a. Remove oil pan-to-oil seal retaining plate bolts, the oil seal retaining plate-to-engine bolts and oil seal retaining plate.

  b. Carefully pry or drive the old seal from the retaining plate. Be careful not to damage the retaining plate.

  c. Using the oil seal driver tool No. 09223–63010, drive the new seal into the oil seal retaining plate, until the surface is flush.

  d. Lubricate the lips of the seal with multipurpose grease.

4. To replace the oil seal with the retaining plate installed:

  a. Using a knife, cut off the oil seal lip.

  b. Using a small pry bar, pry the oil seal from the retaining plate.

  c. Apply multi-purpose grease to the new oil seal.

  d. Using the oil seal driver tool No. 09223–63010, drive the new seal into the oil seal retaining plate, until the surface is flush.

5. Using a putty knife, clean the gasket mounting surfaces.

6. To complete the installation, reverse the removal procedures. Adjust the clutch (M/T).

DIESEL ENGINE LUBRICATION

Oil Pan

REMOVAL & INSTALLATION

Pickups (2WD and 4WD)

1. Disconnect the negative battery cable.

2. Raise and support the vehicle on jackstands.

3. Remove the engine undercover and drain the engine oil.

4. On the 2WD (1979–83), remove the steering idler arm bracket, the pitman arm from the selector shaft and the crossmember.

5. On the 2WD (1984 and later), remove the engine crossmember and the left-hand stiffener plate.

6. Remove the oil pan nuts/bolts and the oil pan.

7. Using a putty knife, clean the gasket mounting surfaces.

8. To install, use a new gasket, sealant and reverse the removal procedures. Torque the oil pan nuts to 13 ft. lbs., the oil pan

bolts to 72 inch lbs. and the crossmember bolts to 38 ft. lbs.

Oil Pump

REMOVAL & INSTALLATION

1. Refer to the "Timing Belt" and "Oil Pan, Removal and Installation" procedures, in this section and remove the timing belt, the oil pan and the oil strainer.

2. Remove the water pump and the timing pointer from the oil pump body.

3. Using tool No. 09278–54011, remove the fuel injection pump drive sprocket nut. Using the wheel puller tool No. 09213–60017, pull the injection pump sprocket from the injection pump.

4. Remove the No.1 and No.2 idler pulleys from the oil pump body.

5. Using the wheel puller tool No. 09213–60017, pull the crankshaft timing sprocket from the crankshaft.

6. Disconnect the water by-pass and the heater hoses from the union on the oil pump assembly.

NOTE: Before removing the injection pump-to-oil pump mounting bolts, check the injection period line

7. Remove the injection pump-to-oil pump mounting bolts, the oil pump-to-engine mounting bolts and the oil pump.

8. Remove the oil pump cover plate from the rear of the oil pump assembly to gain access to the oil pump gears.

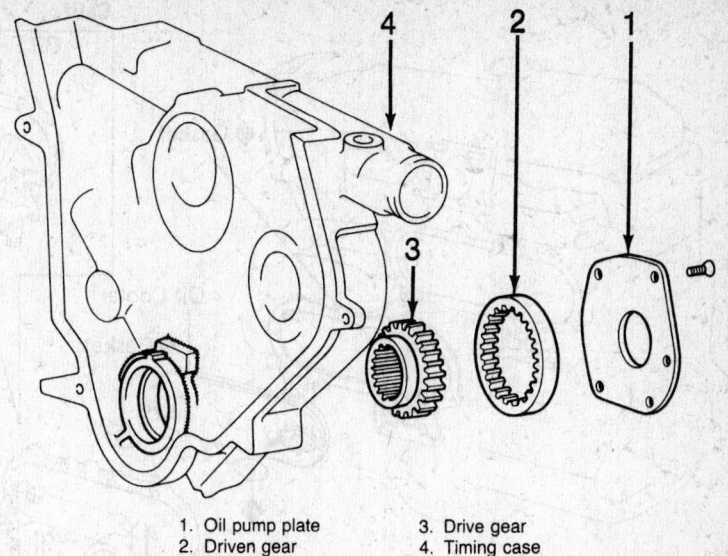

1. Oil pump plate
2. Driven gear
3. Drive gear
4. Timing case

Exploded view of the diesel oil pump

NOTE: Clearances are checked in the same manner as the pump used in the 20R, 22R and 22R-E engines.

9. Remove the pump gears and check the gears and timing case gear surfaces for damage or excessive wear.

10. Using a putty knife, clean the gasket mounting surfaces.

11. Install the gears with the triangular markings of each gear facing the pump plate side of the timing case. Install the pump cover plate.

NOTE: Before installing the oil pump to the engine, be sure to pack the pump cavity with petroleum jelly.

12. To install, use new gaskets, sealant and reverse the removal procedures. Torque the oil pump-to-engine bolts to 14 ft. lbs., the oil strainer bolts to 9 ft. lbs., the oil pan nuts to 13 ft. lbs., the oil pan bolts to 72 inch lbs., the left-hand stiffener plate-to-engine bolts to 29 ft. lbs. (2WD), the left-hand stiffener plate-to-upper transmission bolt to 53 ft. lbs. (2WD), the left-hand

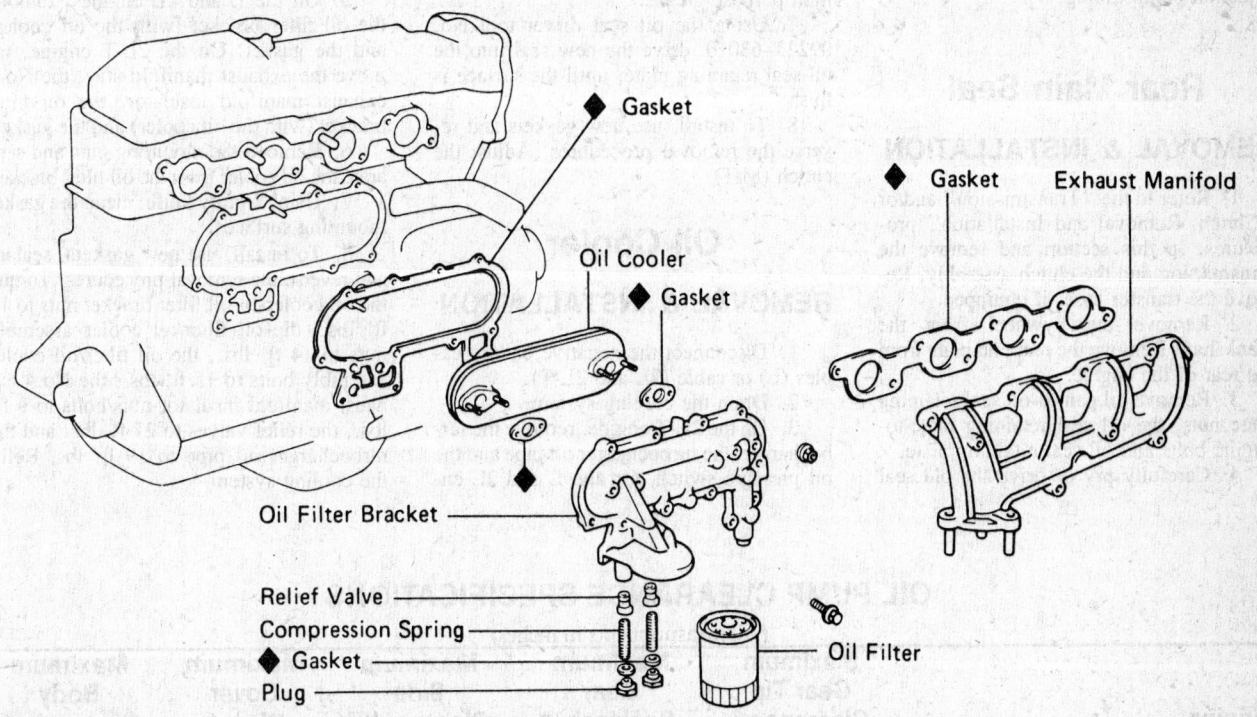

Gasket

Gasket Exhaust Manifold

Oil Cooler

Gasket

Oil Filter Bracket

Relief Valve

Compression Spring

Gasket

Plug

Oil Filter

Exploded view of the oil cooler assembly—L and 2L diesel engines

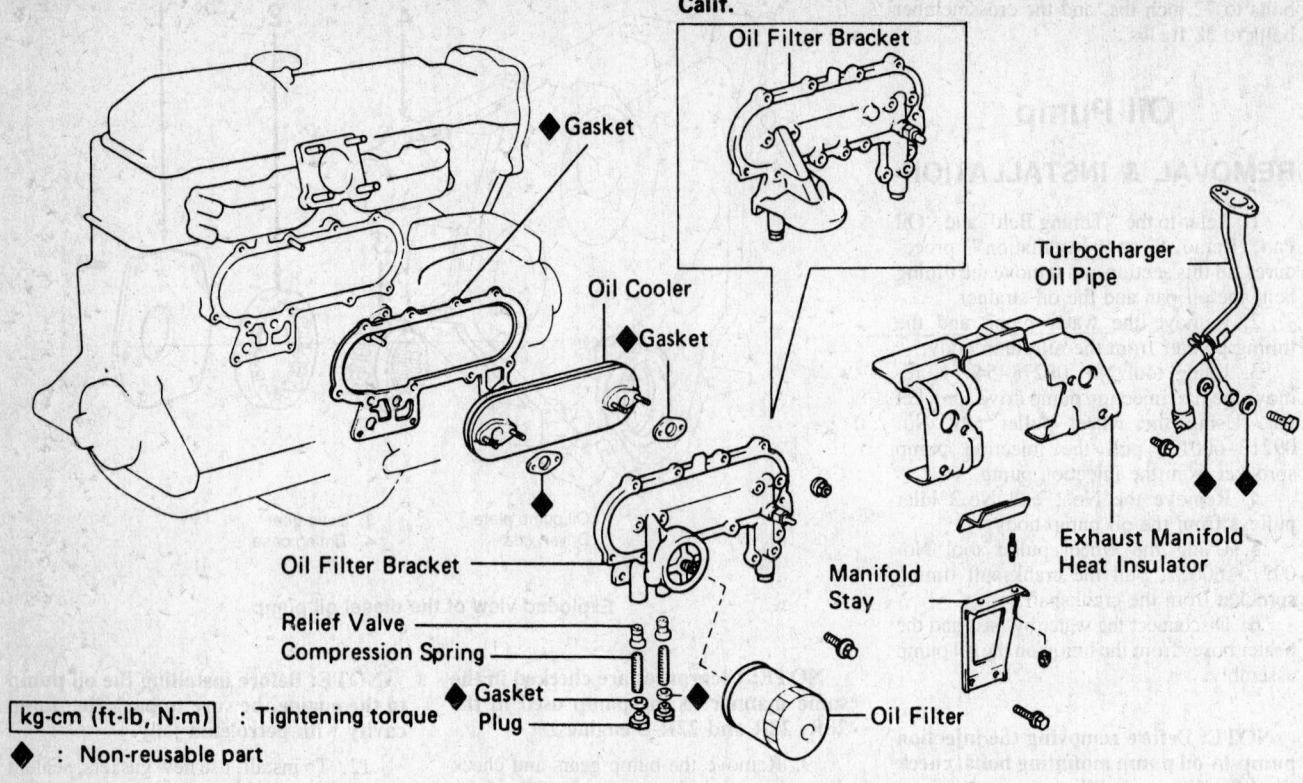

Gasket

Calif.
Oil Filter Bracket

Turbocharger
Oil Pipe

Oil Cooler

Gasket

Exhaust Manifold
Heat Insulator

Oil Filter Bracket

Manifold
Stay

Relief Valve

Compression Spring

Gasket
Plug

Oil Filter

kg-cm (ft-lb, N·m) : Tightening torque

◆ : Non-reusable part

Exploded view of the oil cooler assembly—2L-T diesel engine

stiffener plate-to-lower transmission bolt to 29 ft. lbs. (2WD) and the crossmember bolts to 38 ft. lbs. Adjust the drive belts. Refill the cooling system and the crankcase with engine oil. Check and/or adjust the fuel injection pump timing.

Rear Main Seal

REMOVAL & INSTALLATION

1. Refer to the ''Transmission'' and/or ''Clutch, Removal and Installation'' procedures, in this section and remove the transmission and the clutch assembly. Remove the transfer case, if equipped.
2. Remove the flywheel from the crankshaft. Remove the rear end plate from the rear of the engine.
3. Remove oil pan-to-oil seal retaining plate nuts, the oil seal retaining plate-to-engine bolts and oil seal retaining plate.
4. Carefully pry or drive the old seal

from the retaining plate. Be careful not to damage the retaining plate.
5. Using a putty knife, clean the gasket mounting surfaces.
6. Lubricate the lips of the seal with multipurpose grease.
7. Using the oil seal driver tool No. 09223–63010, drive the new seal into the oil seal retaining plate, until the surface is flush.
8. To install, use new gaskets and reverse the removal procedures. Adjust the clutch (M/T).

Oil Cooler

REMOVAL & INSTALLATION

1. Disconnect the negative battery cables (L) or cable (2L and 2L-T).
2. Drain the cooling system.
3. On the 2L-T engine, remove the turbocharger, the turbocharger oil pipe and the oil pressure switch. On the L and 2L en-

gines, disconnect the exhaust pipe from the exhaust manifold.
4. Remove the exhaust manifold insulator, the oil filter and the two relief valves (keep the disassembled parts in order).
5. On the L and 2L engines, remove the oil filter bracket (with the oil cooler) and the gasket. On the 2L-T engine, remove the exhaust manifold stay, the No.4 exhaust manifold insulator, the oil filter bracket (with the oil cooler) and the gasket.
6. Remove the mounting nuts and separate the oil cooler from the oil filter bracket.
7. Using a putty knife, clean the gasket mounting surfaces.
8. To install, use new gaskets, sealant and reverse the removal procedures. Torque the oil cooler-to-oil filter bracket nuts to 10 ft. lbs., the oil filter/oil cooler assembly nuts to 14 ft. lbs., the oil filter/oil cooler assembly bolts to 15 ft. lbs., the No.4 exhaust manifold insulator nuts/bolts to 9 ft. lbs., the relief valves to 27 ft. lbs. and the turbocharger oil pipe to 19 ft. lbs. Refill the cooling system.

OIL PUMP CLEARANCE SPECIFICATIONS

(All measurements in inches)

Engine	Maximum Gear Tip Clearance ①	Maximum Gear Backlash ②	Maximum Side Clearance ③	Maximum Cover Wear ④	Maximum Body Clearance ⑤
2F	0.008	0.037	0.006	0.006	—

OIL PUMP CLEARANCE SPECIFICATIONS
(All measurements in inches)

Engine	Maximum Gear Tip Clearance ①	Maximum Gear Backlash ②	Maximum Side Clearance ③	Maximum Cover Wear ④	Maximum Body Clearance ⑤
20R, 22R, 22R-E, L, 2L & 2L-T	0.012	—	0.006	—	0.008
3Y-EC	0.0079	—	0.0059	—	0.0079

① 2F Engines: Measured between the gear teeth of each gear and the pump body.
20R, 22R and L Engines: Measured between the gear teeth of each gear and the crescent.

② Measured between the gear teeth with the gears meshed together.
③ Measured between a straightedge positioned across the oil pump body and the gear faces.

④ Measured between a straightedge positioned across the cover and the cover wear (gear contact) surface.
⑤ Measured between the oil pump driven gear and the pump body.

GAS ENGINE COOLING

Radiator

REMOVAL & INSTALLATION

20R, 22R and 22R-E Engines

1. Disconnect the negative battery cable. Drain the cooling system.
2. Remove the clamps, then the upper and lower hoses from the radiator. If equipped with an automatic transmission, remove the oil cooler lines.
3. If necessary, remove the hood lock cable and the hood lock from the radiator upper support.

NOTE: It may be necessary to remove the grille in order to gain access to the hood lock/radiator support assembly.

4. If equipped, remove the fan shroud.
5. If equipped, disconnect the hose from the thermal expansion tank. If necessary, remove the tank from its bracket.
6. Remove the radiator upper support and the radiator.
7. To install, reverse the removal procedures. Check the transmission fluid level on vehicles with automatic transmissions. Refill the cooling system.

3Y-EC Engine

1. Drain the cooling system.
2. Remove the clamps, then the upper and lower hoses from the radiator.
3. If equipped with an automatic transmission, remove the oil cooler lines from the radiator, then the washer fluid tank and move it aside.
4. Remove the fan shroud and the upper radiator bolt.
5. Disconnect the hose from the thermal expansion tank.
6. Raise the front of the vehicle and support it on jackstands.

7. Remove the under engine cover and the radiator.
8. To install, reverse the removal procedures. Check the transmission fluid level on vehicles with automatic transmissions. Refill the cooling system.

Water Pump

REMOVAL & INSTALLATION

20R, 22R and 22R-E Engines

1. Drain the cooling system.
2. Unfasten the fan shroud securing bolts and remove the fan shroud, if equipped.
3. Loosen the alternator adjusting link bolt and remove the drive belt, then swing the alternator toward the engine.
4. If equipped with an air pump, air conditioning compressor or power steering pump drive belts, it may be necessary to loosen the adjusting bolt, remove the drive belt(s) and move the component out of the way.
5. Remove the fan from the fluid coupling, the fluid coupling and pulley from the water pump, then the water pump retaining bolts and the pump.

——— CAUTION ———
If the water pump is equipped with a fluid coupling, DO NOT tip the fluid coupling on its side, for the fluid will run out.

6. Using a putty knife, clean the gasket mounting surfaces.
7. To install, use a new gasket, sealant and reverse the removal procedures. Adjust the drive belt(s) tension. Refill the cooling system.

3Y-EC Engine

1. Drain the cooling system. Disconnect the drive belt from the water pump.
2. Remove the fan from the fluid coupling and the fluid coupling/pulley from the water pump.
3. Remove the drive belt adjusting bar (from the water pump), the mounting bolts and the water pump.

4. Using a putty knife, clean the gasket mounting surfaces.
5. To install, use a new gasket, sealant and reverse the removal procedures. Torque the water pump nuts/bolts to 13 ft. lbs., the drive belt adjusting bar to 29 ft. lbs., the pulley/fluid coupling-to-water pump nuts to 10 ft. lbs. and the fan-to-fluid coupling nuts to 10 ft. lbs. Adjust the drive belt tension. Refill the cooling system.

Thermostat

REMOVAL & INSTALLATION

20R, 22R and 22R-E Engines

1. Partially drain the cooling system to a level below the thermostat.

NOTE: Unless the upper radiator hose is positioned over one of the thermostat housing (water outlet) bolts, it is not necessary to detach the hose.

2. Remove the mounting bolts, the water outlet and the thermostat from the intake manifold.
3. Using a putty knife, clean the gasket mounting surfaces.
4. To install, use a new gasket, sealant and reverse the removal procedures. Refill the cooling system.

NOTE: When installing a new thermostat, be sure that the thermostat is positioned with the spring down.

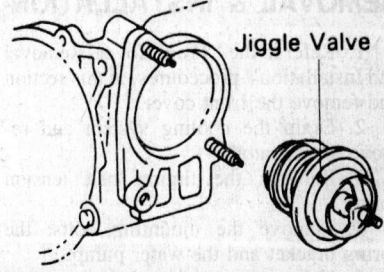

View of the thermostat installation position—3Y-EC engine

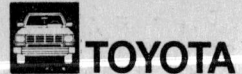

3Y-EC Engine

1. Drain the cooling system to a level below the thermostat.
2. Disconnect the radiator outlet hose from the thermostat housing.
3. Remove the mounting bolts, the thermostat housing and the thermostat.
4. Using a putty knife, clean the gasket mounting surfaces.
5. To install, use a new gasket, sealant and reverse the removal procedures, making sure that the jiggle valve is placed at the upper left position. Torque the thermostat housing to 9 ft. lbs. Refill the cooling system.

NOTE: When installing a new thermostat, be sure that the thermostat is positioned with the spring facing the engine block.

DIESEL ENGINE COOLING

Radiator

REMOVAL & INSTALLATION

1. Disconnect the negative battery cables (L) or cable (2L and 2L-T). Drain the cooling system.
2. Remove the clips or the clamps, then the inlet and outlet hoses from the radiator.
3. Remove the radiator fan shroud. If equipped with A/C, remove the lower radiator shroud.
4. Disconnect the coolant reservoir hose from the radiator.
5. Remove the mounting bolts and the radiator.
6. To install, reverse the removal procedures. Refill the cooling system.

Water Pump

REMOVAL & INSTALLATION

1. Refer to the "Front Cover, Removal and Installation" procedures, in this section and remove the front cover.
2. Drain the cooling system and remove the radiator.
3. Remove the timing belt tension spring.
4. Remove the mounting bolts, the spring bracket and the water pump.
5. Using a putty knife, clean the gasket mounting surfaces.

6. To install, use new gaskets, sealant and reverse the removal procedures. Torque the water pump bolts to 14 ft. lbs. Adjust the drive belts. Refill the cooling system.

Thermostat

REMOVAL & INSTALLATION

1. Drain the cooling system to a level below the thermostat.
2. Disconnect the radiator inlet hose from the thermostat housing.
3. Remove the mounting bolts, the thermostat housing and the thermostat.
4. Using a putty knife, clean the gasket mounting surfaces.
5. To install, use a new gasket, sealant and reverse the removal procedures. Refill the cooling system.

NOTE: When installing a new thermostat, be sure that the thermostat is positioned with the spring facing the engine block.

GAS FUEL SYSTEM

Fuel Filter

REMOVAL & INSTALLATION

——— CAUTION ———
When working on the fuel system, DO NOT smoke and be certain that there is no open flame in the area.

20R and 22R Engines

The fuel filter is located on the right-side of the fuel tank, in front of the right-rear wheel.
1. Raise and support the rear of the vehicle on jackstands.
2. Disconnect and plug the fuel lines at the fuel filter.
3. Remove the filter from the mounting bracket.
4. To install, use a new filter and reverse the removal procedures.

NOTE: When installing the fuel filter, be sure to install it in the correct direction of flow.

22R-E Engine

The fuel filter is located near the engine, at the rear left-side of air intake chamber.

1. Disconnect the negative battery cable.
2. Wrap a shop cloth around the fuel lines (to catch the excess fuel). Remove the fuel line-to-filter mounting bolts, the lines and the washers.
3. Remove the filter from the mounting bracket.
4. To install, use a new filter, gaskets and reverse the removal procedures.

NOTE: When installing the fuel filter, be sure to install it in the correct direction of flow.

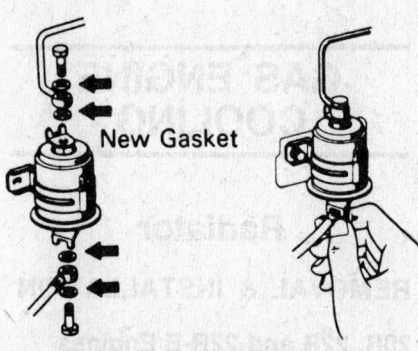
New Gasket

Exploded view of the fuel filter used on the 22R-E and 3Y-EC engines

3Y-EC Engine

The fuel filter is located at the lower right-side of the engine.
1. Disconnect the negative battery cable.
2. Wrap a shop cloth around the fuel lines (to catch the excess fuel). Remove the fuel line-to-filter mounting bolts, the lines and the washers.
3. Remove the filter from the mounting bracket.
4. To install, use a new filter, gaskets and reverse the removal procedures.

NOTE: When installing the fuel filter, be sure to install it in the correct direction of flow.

Mechanical Fuel Pump

A mechanical fuel pump is used on all engines except the 3Y-EC (1983 and later) and the 22R-E (1984 and later) engines. The mechanical fuel pump is actuated by an eccentric which is integral with the engine camshaft on 2F engines and bolted in front of the camshaft drive sprocket on 20R and 22R engines.

TESTING

Fuel pumps should always be tested on the

vehicle. The larger line between the pump and tank is the suction side of the system and the smaller line, between the pump and carburetor is the pressure side. A leak in the pressure side would be apparent because of dripping fuel. A leak in the suction side is usually only apparent because of a reduced volume of fuel delivered to the pressure side.

Leak Test

1. Tighten any loose line connections and look for any kinks or restrictions.

2. Disconnect the fuel line at the carburetor. Disconnect the distributor-to-coil primary wire. Place a container at the end of the fuel line and crank the engine a few revolutions. If little or no fuel flows from the line, either the fuel pump is inoperative or the line is plugged. Blow through the lines with compressed air and try the test again. Reconnect the line.

3. If fuel flows in good volume, check the fuel pump pressure to be sure.

Pressure Test

1. Attach a pressure gauge to the pressure side of the fuel line. On vehicles equipped with a vapor return system, squeeze off the return hose.

2. Run the engine at idle and note the reading on the gauge. Stop the engine and compare the reading with the specifications listed in the "Tune-Up Specifications" chart. If the pump is operating properly, the pressure will be as specified and will be constant at idle speed. If the pressure varies or is too high or low, the pump should be repaired or replaced, depending upon the pump type.

3. Remove the pressure gauge.

Flow Test

1. Disconnect the fuel line from the carburetor. Run the fuel line into a suitable measuring container.

2. Run the engine at idle until there is one pint of fuel in the container. One pint should be pumped in 30 seconds or less.

3. If the flow is below minimum, check for a restriction in the line.

REMOVAL & INSTALLATION

20R and 22R Engines

1. Disconnect the negative battery terminal.

———— CAUTION ————
When working on the fuel system, do not smoke or work near any fire hazard. Keep gasoline off rubber or leather parts.

2. Drain the cooling system to a level below the upper radiator hose, then remove the upper radiator hose.

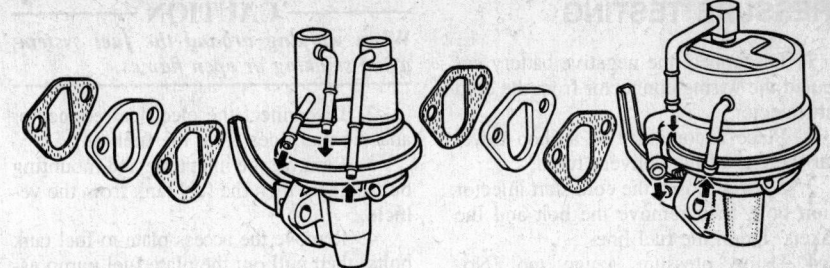

Mechanical fuel pumps used on 20R and 22R engines—arrows indicate the fuel flow in and out of the pump

3. Remove all three lines from the fuel pump, then the mounting bolts, the fuel pump and the gasket.

NOTE: The fuel pump is not repairable. It must be replaced as a complete unit.

4. To install, use a new gasket and reverse the removal procedures. Refill the cooling system.

2F Engines

1. Disconnect the negative battery cable.

2. Remove and plug the fuel lines at fuel pump. Remove the mounting bolts and the fuel pump.

NOTE: If any new rubber hose must be used to repair the fuel line, be sure that it is gasoline-resistant.

3. To install, use a new gasket and reverse the removal procedures.

NOTE: The 2F engine mechanical fuel pump is rebuildable, depending upon the parts availability. If the fuel pump is to be rebuilt rather than replaced, follow the instructions supplied with the rebuilding kit.

Electric Fuel Pump

An electric fuel pump is used on the 22R-E (1984 and later) and the 3Y-EC (1983 and later) engines. The fuel pump is wired into the ignition switch and oil pressure switch circuits. In the event of an oil pressure loss, the fuel pump is turned OFF so that the engine will stall, thus preventing engine damage due to the oil pressure loss. The fuel pump will operate only when the ignition switch is turned to the START position and when the oil pressure is normal.

OPERATION TESTING

1. Disconnect the electrical clip from the oil pressure switch.

2. Turn the ignition switch to the ON position (DO NOT start the engine).

3. Short the Fp and the +B terminals of the check connector. Check the cold start injector hose for pressure.

4. Check for a smooth flow of gasoline from the fuel filter outlet. If the pump is noisy, it is probably defective. If the pump

does not run, check the pump resistor and relay.

5. Disconnect the jumper wire and reconnect the check connector. Turn the ignition switch OFF.

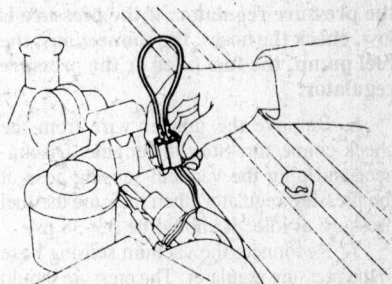

Shorting the Fp and +B terminals of the check connector—EFI system

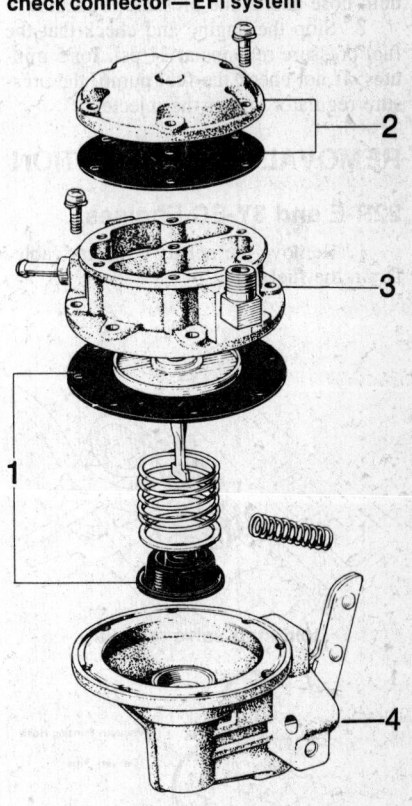

1. Diaphragm and spring
2. Cover and diaphragm
3. Upper body
4. Lower body

Mechanical fuel pump used on the 2F engine

PRESSURE TESTING

1. Disconnect the negative battery cable and the wiring connector from the cold start injector.

2. Place a container or a shop towel near the end of the delivery tube.

3. Slowly loosen the cold start injector union bolt, then remove the bolt and the gaskets. Drain the fuel line.

4. Using pressure gauge tool No. 09268–45011, connect it in line with the cold start injector. Reconnect the battery cable.

5. Short the Fp and +B terminals of the check connector wire. Turn the ignition switch to the ON position and measure to fuel pump pressure. It should be 33–38 psi. Turn the ignition switch OFF.

NOTE: If the pressure is high, replace the pressure regulator; if the pressure is low, check the hoses, the connections, the fuel pump, the fuel filter or the pressure regulator.

6. Remove the jumper wire from the check connector. Start the engine. Disconnect and plug the vacuum sensing hose at the pressure regulator, then measure the fuel pressure at idle. It should be 33–38 psi.

7. Reconnect the vacuum sensing hose to the pressure regulator. The pressure should now be 27–31 psi.; if not, check the vacuum hose and/or the pressure regulator.

8. Stop the engine and check that the fuel pressure remains at 21 psi. for 5 minutes. If not check the fuel pump, the pressure regulator and/or the injectors.

REMOVAL & INSTALLATION

22R-E and 3Y-EC Engines

1. Remove the negative battery cable. Drain the fuel tank.

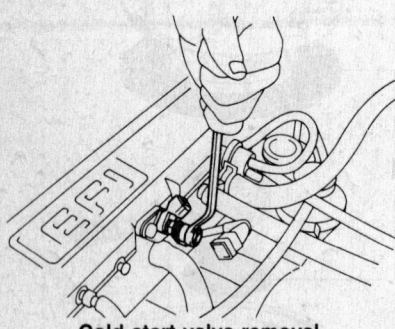

Cold start valve removal

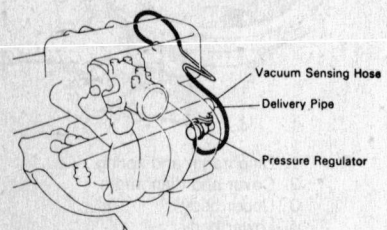

Removing the pressure regulator—3Y-EC engine

— CAUTION —
When working around the fuel system, avoid smoking or open flames.

2. Disconnect the electrical connector and the fuel lines from the fuel tank.

3. Remove the inlet tube and mounting bolts/straps, then the fuel tank from the vehicle.

4. Remove the access plate-to-fuel tank bolts, then pull out the plate/fuel pump assembly.

— CAUTION —
Do not operate the fuel pump unless it is immersed in gasoline and connected to its resistor.

5. Disconnect the electrical connectors from the fuel pump. Pull the bracket from the lower side of the fuel pump, then remove the fuel pump from the fuel hose.

6. Remove the rubber cushion, the clip and the fuel filter from the bottom of the fuel pump.

7. To install, use new gaskets and reverse the removal procedures. Torque the fuel pump bracket-to-fuel tank to 43 inch lbs. Refill the fuel tank.

Cold Start Injector

The EFI engines have a cold start injector located in the intake air chamber which aids in cold weather starting.

REMOVAL & INSTALLATION
22R-E and 3Y-EC Engines

1. Disconnect the negative battery cable and the cold start injector wire.

2. Place a shop towel or a container under the fuel delivery pipe and drain the fuel from the pipe.

3. Disconnect the fuel pipe from the cold start injector.

4. Remove the mounting bolts and the cold start injector from the intake air chamber.

5. To install, use new gaskets and reverse the removal procedures. Torque the injector bolts to 44–60 inch lbs.

Fuel Pressure Regulator

The fuel pressure regulator is located on the fuel delivery pipe of the EFI system, it maintains a constant fuel pressure in the injection system.

REMOVAL & INSTALLATION

22R-E and 3Y-EC Engines

NOTE: On the Van (3Y-EC) models, raise and support the vehicle on jack-stands.

1. Disconnect the vacuum sensing hose from the pressure regulator.

2. Place an shop towel or a container under the fuel hose connection and disconnect the fuel return hose from the regulator.

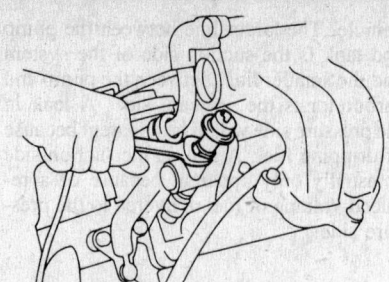

Removing the pressure regulator—22R-E engine

3. Remove the locknut (22R-E) or the mounting bolts (3Y-EC) and the pressure regulator from the fuel delivery pipe.

4. To install, reverse of removal. Torque the locknut to 22 ft. lbs. (22R-E) or bolts to 44–60 inch lbs. (3Y-EC).

Fuel Injector

For further information on the fuel injection system, refer to "Fuel Injection" in the "Unit Repair" section.

TESTING

Each injector may be tested for operation while on the engine, in two ways. One, is to obtain an inexpensive mechanic's stethoscope and listen for a clicking at the injector. The other way is by checking continuity at each injector terminal with a ohmmeter. Resistance should be 1.5–3.0 ohms.

REMOVAL & INSTALLATION

1. Refer the "Intake Air Chamber and Throttle Body, Removal and Installation" procedures, in this section and remove the air chamber.

2. Disconnect the wiring connector from the injector.

3. Disconnect the fuel hose from the delivery pipe.

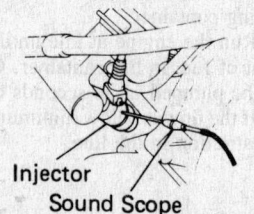

Injector
Sound Scope
Listening to injector with a mechanic's stethescope

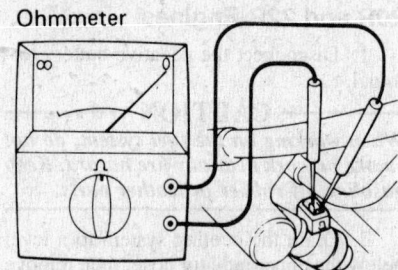

Ohmmeter

Testing the injector with an ohmmeter

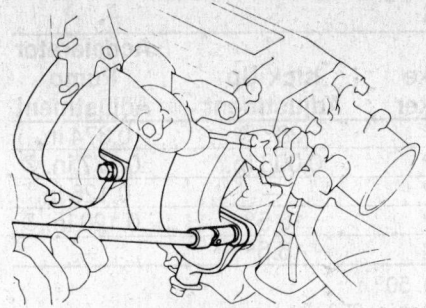

Injector removal

4. Remove the two mounting bolts (22R-E) or the nut/bolt (3Y-EC) and remove the injector along with the delivery pipe. Take care in handling the injectors. DO NOT DROP THEM! KEEP THEM AS CLEAN AS POSSIBLE!

NOTE: Injector performance tests are possible but expensive special tools are required. If these tools are unavailable, use the test procedures above. If the tools are available, they come with a set of instructions for their use.

5. To install, use new O-rings and reverse the removal procedures. Torque the hold down bolts to 12–16 ft. lbs. Check for fuel leakage.

NOTE: Each injector should have four insulators. Prior to installation, coat the O-rings with clean gasoline. Prior to tightening the hold down bolts, make sure that the injector rotates smoothly in its bore. If not, the O-rings are twisted.

Carburetors

REMOVAL & INSTALLATION

1. Disconnect the negative battery cable.

2. Label and disconnect the emission control hoses. Disconnect the air intake hose. Remove the mounting and butterfly nuts, then lift the air cleaner from the carburetor.

3. On the 20R engine, drain the engine coolant and disconnect the heater hose from the choke unit.

4. If equipped with an A/T, disconnect the throttle cable or rod. Disconnect the fuel hose, the emission control hose, the PCV hose and the wiring connector(s) from the carburetor.

5. Disconnect the accelerator linkage and the choke pipe (if equipped).

6. On the 2F engines, disconnect the magnetic valve wire from the coil terminal and the choke cable from the carburetor.

7. Remove the carburetor-to-manifold nuts/bolts and lift the carburetor off of the manifold.

8. Cover the open manifold with a clean cloth to prevent dirt and small objects from entering into the engine.

9. To install, reverse the removal procedures. After the engine has been started, check for fuel and vacuum leaks.

1. Pump jet
2. Spring
3. Outlet check ball
4. Secondary venturi
5. Primary venturi
6. Pump plunger
7. Spring
8. Ball retainer
9. Inlet check ball
10. Plug
11. Spring
12. AAP outlet check ball
13. Plug

14. AAP inlet check ball
15. Throttle positioner
16. Thermostatic valve cover
17. Thermostatic valve
18. Primary slow jet
19. Power valve
20. Power jet

21. Sight glass
22. Glass retainer
23. Diaphragm housing cap
24. Spring
25. Diaphragm
26. Housing
27. Fast idle cam

28. Solenoid valve
29. Carburetor body
30. Diaphragm
31. Spring
32. AAP housing
33. Secondary main jet
34. Primary main jet

Carburetor main body parts—20R engine

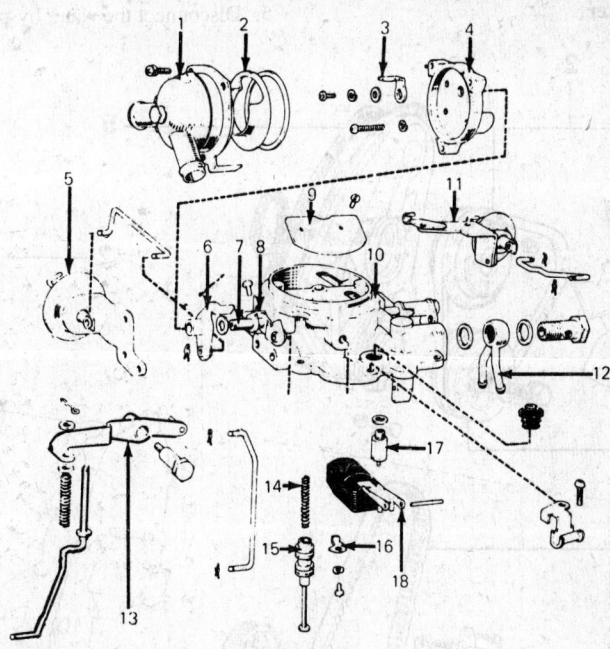

1. Choke coil water housing
2. Choke housing plate
3. Choke lever
4. Choke housing body
5. Choke breaker
6. Relief lever

7. Choke shaft
8. Connecting lever
9. Choke valve
10. Air horn
11. Choke opener
12. Union

13. Pump arm
14. Spring
15. Power piston
16. Piston retainer
17. Needle valve set
18. Float

Carburetor air horn parts—20R engine

CARBURETOR SPECIFICATIONS

Year	Engine	Float Level Adjustment	Fast Idle Adjustment	Choke Unloader	Choke Breaker	Kick Up Adjustment	Accelerator Pump Adjustment
'79–'84	2F	0.295 in. ④	—	50°	38°	28° ⑦	0.374 in.
'79–'80	20R	0.280 in.	0.047 in. ⑤	50°	38°	0.008 in.	0.177 in. ②
'81–'82	22R	0.413 in. ③	24°	45° ⑥	38°	—	0.126 in.
'83–'84	22R	0.386 ③	22°	50°	42°	16.5°	0.126 in. ⑧
'85–'86	22R	0.386 ③	23°	45°	42°	16.5°	—

② '80 20R 0.154 in.
③ Raised position; 1.89 lowered position
④ Raised position; 0.043 lowered position
⑤ '79 only. '80; 24°—see text

⑥ Except Canada. Canada: 50°
⑦ Except California. California: 25°
⑧ Calif.—0.150 in.

DIESEL FUEL SYSTEM

Injection Pump

REMOVAL & INSTALLATION

NOTE: Using a wrench on the center crankshaft pulley bolt, rotate the engine (clockwise only) until the TDC mark on the pulley is aligned with the pointer. Check that the valves of the number one cylinder are closed (rocker arms loose). If the valves are not closed, rotate the engine 360° and again align the TDC mark with the pointer.

1. Refer to the "Timing Belt, Removal and Installation" procedures, in this section and remove the timing belt.

NOTE: Using a piece of chalk or a crayon, mark the relationships between each of the timing gears and the timing belt, then the direction of the timing belt rotation.

2. Disconnect the boost compensator hose and the electrical connector from the fuel injection pump.
3. Disconnect the fuel hoses from the inlet and the outlet pipes.
4. Remove the accelerator connecting rod. On the 2L-T engine, remove the accelerator link.
5. Disconnect the water by-pass hoses.

Disconnect the fuel injection lines from the injection pump and the injectors, then remove the injection lines.

6. Disconnect and plug the fuel inlet and outlet lines at the injection pump.
7. Using the spanner wrench tool No. 09278–54011, remove the injection pump sprocket nut. Using the wheel puller tool No. 09213–60017, pull the injection pump sprocket from the injection pump.

NOTE: Check the factory-made alignment mark next to the outer pump fastener. This mark signifies the required relationship between the pump and the timing case assembly. Align this mark during installation.

8. Remove the mounting nuts/bolts and the injection pump.

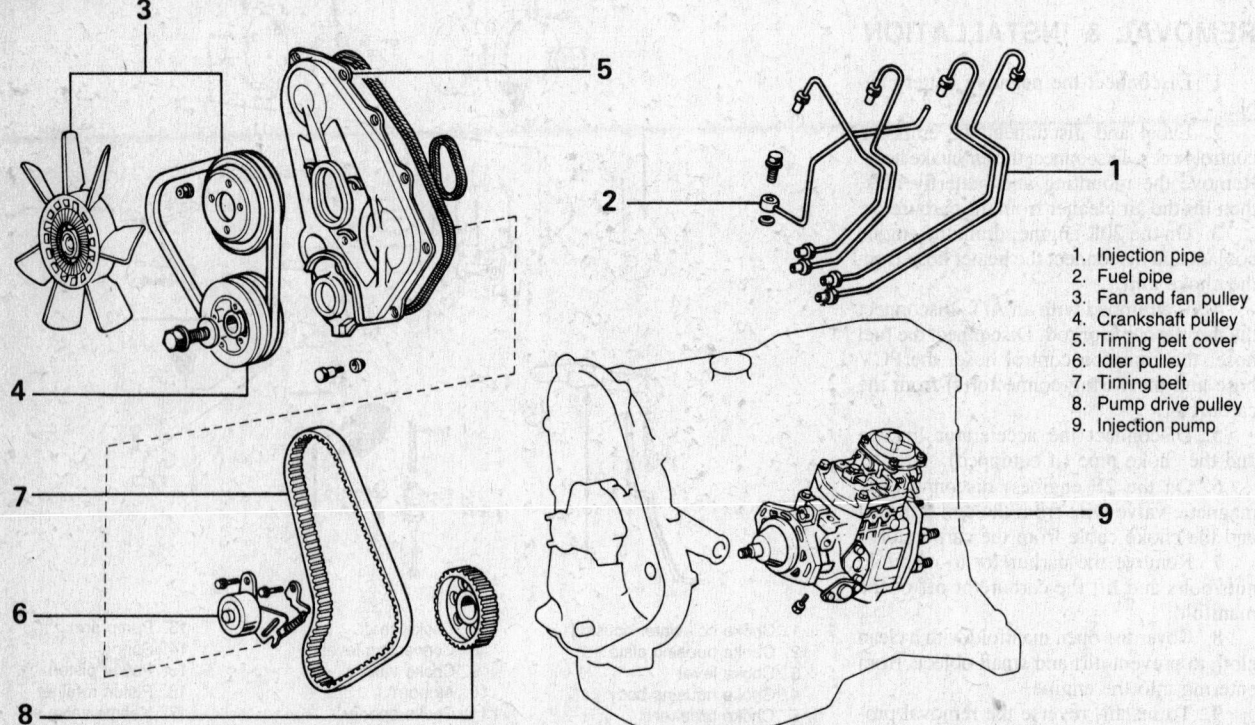

1. Injection pipe
2. Fuel pipe
3. Fan and fan pulley
4. Crankshaft pulley
5. Timing belt cover
6. Idler pulley
7. Timing belt
8. Pump drive pulley
9. Injection pump

Diesel fuel injection pump and related components

CAUTION

DO NOT disassemble the injection pump; only factory-authorized repair centers have the facilities to do so. No adjustments to the pump are possible.

9. To install, use new gaskets and reverse the removal procedures. Torque the injection pump-to-oil pump nuts to 15 ft. lbs. and the pump stay bolts to 13 ft. lbs. Adjust the timing belt, the drive belts and the ignition timing. Refill the cooling system.

NOTE: After any service is performed to the diesel fuel system, pump the priming handle on the fuel sedimenter assembly 30–40 times to purge air from the system.

Fuel Injector

REMOVAL

1. Disconnect the negative battery cables (L) or cable (2L and 2L-T).
2. On the 2L-T engines, remove the PCV hose, the No. 1 air cleaner hose, the air cleaner tube with the No. 2 hose, the heater hoses from the intake flange, the bracket bolts and the air intake tube.
3. Remove the injection line-to-manifold clamps (if equipped), the lines-to-injectors, the lines-to-injection pump and the injection lines from the vehicle.
4. Remove the leakage pipe from the injectors and note the location of each sealing washer.
5. Using the tool No. 09260–46012, remove the injector nozzle(s) from the cylinder head, noting the positions of the nozzle seats and seat gaskets.

CAUTION

DO NOT allow dirt to enter the engine through the nozzle holes.

NOTE: Remove accumulations of carbon from the nozzle holes.

6. Keep the injectors in order so that they may be installed in their original positions.

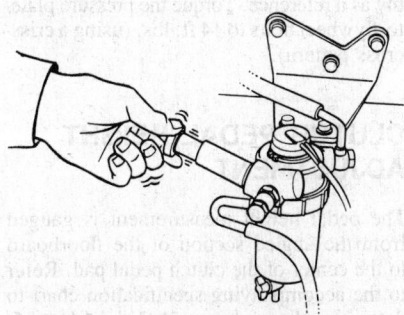

Location of the fuel system priming pump on the diesel engine—pump the handle 30–40 times after any fuel system service to purge air from the system— L engine

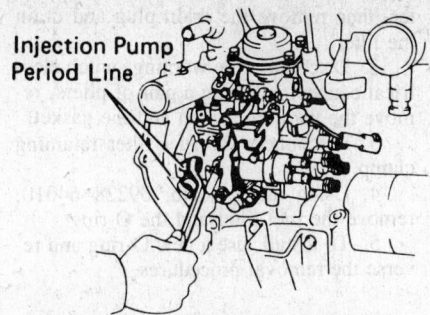

Inspecting the alignment of the fuel injection pump—diesel engines

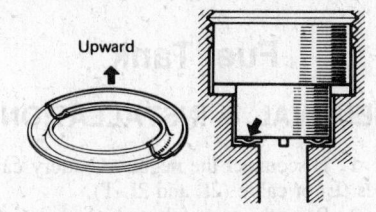

Correct installation of the injector nozzle seat—diesel engine

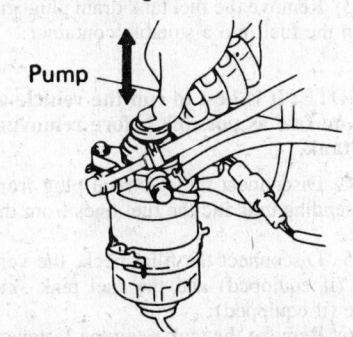

Using the sedimenter to prime the fuel system—2L and 2L-T diesel engines

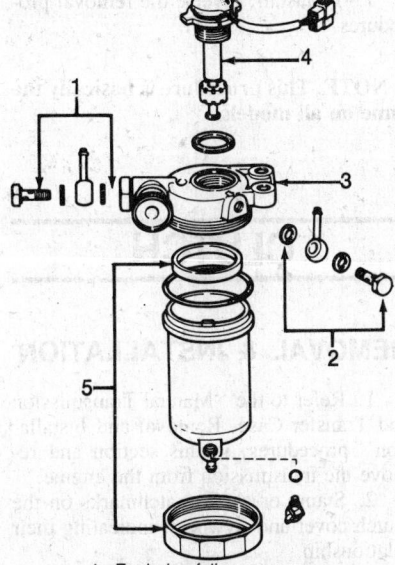

1. Fuel pipe follow screw
2. Fuel pipe follow screw
3. Fuel filter body
4. Level warning switch
5. Fuel sedimenter case and nut

Fuel sedimenter exploded view— L diesel engine

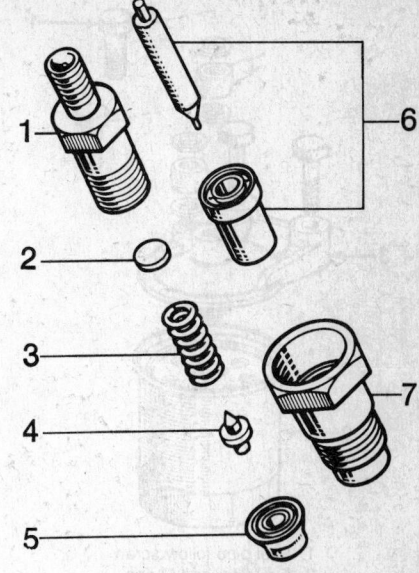

1. Nozzle holder retaining nut
2. Adjusting shim
3. Pressure spring
4. Pressure pin
5. Distance piece
6. Nozzle assembly
7. Nozzle holder body

Exploded view of the diesel fuel injection nozzle

7. If the engine exhibited any type of severe miss, excessive smoking or drastic decrease in power, it is best to have the nozzles professionally tested for opening pressure, leakage and spray pattern.
8. To install, use new gaskets and reverse the removal procedures. Torque the injector(s)-to-cylinder head to 51 ft. lbs., the leakage pipe-to-injector nuts to 36 ft. lbs., the injection pipes-to-intake manifold bolts to 18 ft. lbs.

NOTE: The nozzle seat is installed between the injector and the seat gasket; it must be positioned with the concave side of the seat towards the injector. After any service is performed to the diesel fuel system, pump the priming handle on the fuel sedimenter assembly several times until more resistance is felt.

CLEANING

1. Using the tool No. 09260–46012, remove the nozzle holder retaining nut from the nozzle holder body.
2. Remove the pressure spring, the shim, the pressure pin, the distance piece and the nozzle assembly.

CAUTION

CAUTION:

When disassembling the nozzle holder, be careful not to drop the inner parts.

3. Wash the nozzles in clean diesel fuel.
4. Remove carbon from the nozzle needle tip with a small, wooden stick. DO

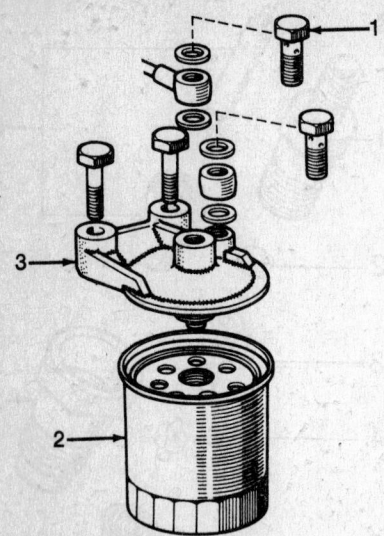

1. Fuel pipe follow screw
2. Fuel filter and O-ring
3. Fuel filter upper body

Fuel filter and filter body exploded view—L diesel engine

NOT use any metallic object to clean the nozzle tip.

NOTE: DO NOT touch the nozzle mating surfaces with your fingers.

5. Remove carbon from the exterior of the nozzle body with a brass bristled brush. DO NOT use a brush having regular, steel bristles.

6. Inspect all parts for damage and/or corrosion. If either of these conditions exist, the entire injector assembly must be replaced.

7. To assemble, reverse the disassembly procedures. Torque the nozzle holder retaining nut to 51 ft. lbs.

Fuel Filter/Sedimenter

REMOVAL & INSTALLATION

1. Place a container under the fuel fil-

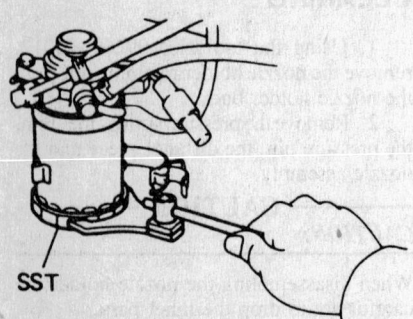

SST

Using tool 09228-64010, remove the fuel filter from the sedimenter—2L and 2L-T diesel engines

T570

ter, then remove the drain plug and drain the filter.

2. Disconnect the warning switch electrical connector. Using a pair of pliers, remove the warning switch and the gasket.

3. Disconnect the fuel filter retaining clamp.

4. Using the tool No. 09228-64010, remove the fuel filter and the O-ring.

5. To install, use a new O-ring and reverse the removal procedures.

NOTE: Lightly coat the O-ring with fuel before installation. Screw the filter on by hand, then tighten it the tool.

Fuel Tank

REMOVAL & INSTALLATION

1. Disconnect the negative battery cables (L) or cable (2L and 2L-T).

2. Raise the rear of the vehicle and support it on jackstands.

3. Remove the fuel tank drain plug and drain the fuel into a suitable container.

NOTE: It is best to run the vehicle as low on fuel as possible before removing the tank.

4. Disconnect the electrical plug from the sending unit and the fuel lines from the tank.

5. Disconnect the filler neck, the vent line (if equipped) and the fuel tank skid plate (if equipped).

6. Remove the tank retaining fasteners and carefully lower the tank from the vehicle.

7. To install, reverse the removal procedures.

NOTE: This procedure is basically the same on all models.

CLUTCH

REMOVAL & INSTALLATION

1. Refer to the "Manual Transmission and Transfer Case, Removal and Installation" procedures, in this section and remove the transmission from the engine.

2. Stamp or chalk matchmarks on the clutch cover and flywheel, indicating their relationship.

3. Loosen the clutch cover-to-flywheel retaining bolts one turn at a time. The pressure on the clutch disc must be released GRADUALLY.

4. Remove the clutch cover-to-flywheel bolts. Remove the clutch cover and the clutch disc.

5. If the clutch release bearing is to be replaced, perform the following:

a. Remove the bearing retaining clip(s), the bearing and hub.

b. Remove the release fork and the boot.

c. The bearing is press fitted to the hub.

NOTE: In some cases, the bearing is available with the hub from automotive suppliers. If this is not the case with your model, contact a machine shop and have the bearing replaced using a hydraulic press. Using other means to replace the bearing could result in personal injury.

d. Clean all parts and lightly grease the input shaft splines and all of the contact points.

e. Install the bearing/hub assembly, the fork, the boot and the retaining clip(s) in their original locations.

6. Inspect the flywheel surface for cracks, heat scoring (blue marks) and warpage. If oil is present on the flywheel surface, this indicates that either the engine rear oil seal or the transmission front oil seal is leaking. If necessary, refer to the appropriate section for seal replacement.

NOTE: Before installing any new parts, make sure that they are clean. During installation, do not get grease or oil on any of the components, as this will shorten clutch life considerably.

7. Using the alignment tool No. 09301-20020, position the clutch disc against the flywheel. (Pickups and Vans: The short side of the splined section faces the flywheel; Land Cruisers: The long side of the splined section faces the flywheel).

8. Install the clutch cover over the disc and install the bolts loosely. Align the pressure plate-to-flywheel matchmarks. If a new or rebuilt clutch cover assembly is installed, use the matchmark on the old cover assembly as a reference. Torque the pressure plate-to-flywheel bolts to 14 ft. lbs. (using a crisscross pattern).

CLUTCH PEDAL HEIGHT ADJUSTMENT

The pedal height measurement is gauged from the angled section of the floorboard to the center of the clutch pedal pad. Refer to the accompanying specification chart to determine the recommended pedal height.

If necessary, adjust the pedal height by loosening the locknut and turning the pedal stop bolt which is located above the pedal towards the drivers seat. Tighten the locknut after the adjustment.

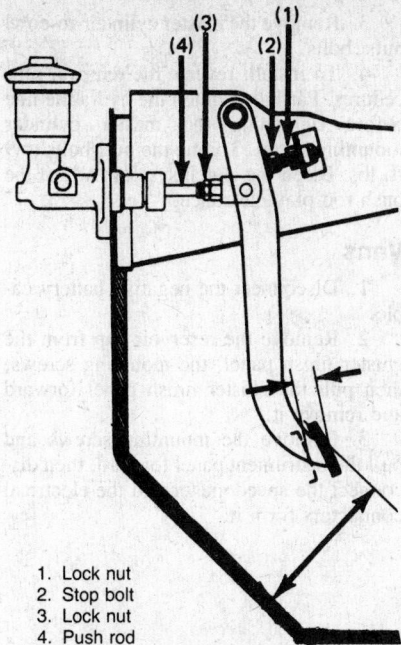

1. Lock nut
2. Stop bolt
3. Lock nut
4. Push rod

Clutch pedal adjustment points— typical. The distance between the ends of the long arrow is the pedal height. The distance between the two short arrows is the pedal free-play. Push rod play is only a small movement of the pedal— Pick-ups

CLUTCH PEDAL PUSH ROD PLAY ADJUSTMENT

The pedal push rod play is the distance between the clutch master cylinder piston and the pedal pushrod located above the pedal towards the firewall. Since it is nearly impossible to measure this distance at the source, it must be measured at the pedal pad, preferably with a dial indicator gauge. Refer to the accompanying specification chart to determine the recommended play.

If necessary, adjust the pedal play by loosening the pedal pushrod locknut and turning the pushrod. Tighten the locknut after the adjustment.

CLUTCH FORK TIP PLAY ADJUSTMENT

The fork tip play is the total amount of travel evident at the outer end of the clutch release fork where the fork comes in contact with the release cylinder pushrod. Refer to the accompanying specification chart to determine the recommended fork tip play.

The fork tip play is adjusted by loosening the release cylinder pushrod locknut and effectively increasing or decreasing the pushrod length as required.

NOTE: Some models do not have adjustable release cylinder pushrods. These models are identified by having no adjustment nuts on the pushrod.

CLUTCH PEDAL FREE-PLAY ADJUSTMENT

The free-play measurement is the total travel of the clutch pedal from the fully released position to where resistance is felt as the pedal is pushed downward. Refer to the accompanying specification chart to determine the recommended pedal free play.

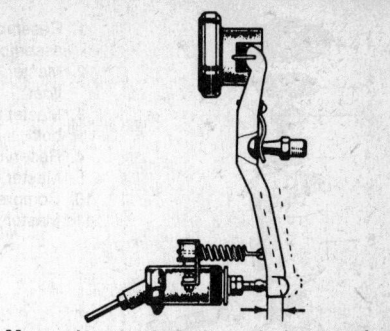

Measuring the fork tip end-play—typical

If the clutch pedal free play is incorrect, perform the previous clutch adjustments then bleed the system according to the procedure which follows. If a pedal free-play dimension is not listed for your model, perform the previous clutch adjustments and disregard the pedal free-play measurement.

Clutch Master Cylinder

REMOVAL & INSTALLATION

——— CAUTION ———
Brake fluid dissolves paint. DO NOT allow it to drip onto the body when removing the master cylinder.

Pickups and Land Cruiser

1. Disconnect the master cylinder pushrod pin from the top of the clutch pedal.
2. Using the tool No. 09751–36011, remove the hydraulic line from the master cylinder, being careful not to damage the compression fitting.

CLUTCH ADJUSTMENT SPECIFICATIONS
(All measurements in inches)

Year	Model	Pedal Height	Push Rod Play	Free-Play	Fork Tip Play
'79	Pickup	6.0–6.4	0.020–0.200	0.200–0.600	—
'80–'82	Pickup	6.0–6.4	0.040–0.200	0.200–0.600 ①	②
'79	Land Cruiser 2 dr.				
	w/P.B. ③	8.5	0.020–0.120	—	0.120–0.160
	wo/P.B. ④	7.9	0.020–0.120	—	0.120–0.160
'80–'84	Land Cruiser 2 dr.	8.5	0.040–0.200	—	0.157–0.197
'79	Land Cruiser Wagon				
	w/P.B. ③	7.3	0.020–0.200	—	0.120–0.160
	wo/P.B. ④	6.8	0.020–0.200	—	0.120–0.160
'80–'84	Land Cruiser Wagon	7.7	0.040–0.200	—	0.160–0.197
'83	Pickup	5.98–6.38	—	0.20–0.59	⑤
'84	Pickup	5.94	0.039–0.197	0.20–0.59	⑤
'85	Pickup	5.67 ⑥	0.039–0.197	0.20–0.59	⑤
'83–'86	Van	6.57–6.97 ⑥	0.039–0.197	0.20–0.59	⑦

① '80 4WD models—0.980–1.770
② '80 4WD models—0.079–0.118
③ With power brakes
④ Without power brakes
⑤ Non adjustable
⑥ From asphalt sheet
⑦ Self adjusting

1. Reservoir filler cap assembly
2. Master cylinder reservoir float
3. Master cylinder reservoir bolt
4. Reservoir bolt washer
5. Master cylinder reservoir
10. Compression spring
11. Master cylinder body
12. Master cylinder piston
13. Cylinder cup
14. Plate washer
15. Hole snap-ring
16. Master cylinder boot
17. Master cylinder pushrod
18. Nut
19. Master cylinder pushrod clevis

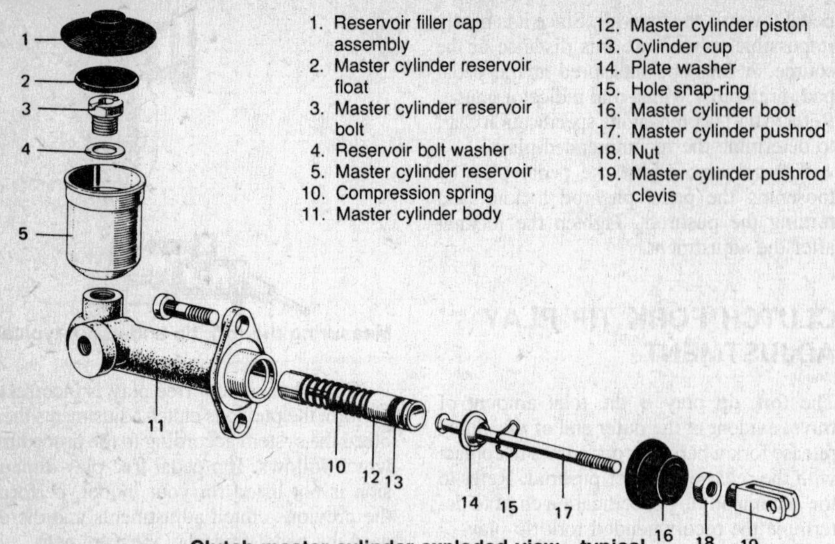

Clutch master cylinder exploded view—typical

3. Remove the master cylinder-to-cowl nuts/ bolts.

4. To install, reverse the removal procedures. Partially tighten the hydraulic line before tightening the master cylinder mounting nut(s). Torque the nuts/bolts to 9 ft. lbs. Bleed the clutch system. Adjust the push rod play clearance.

Vans

1. Disconnect the negative battery cable.

2. Remove the reservoir cap from the cluster finish panel, the mounting screws, then pull the cluster finish panel forward and remove it.

3. Remove the mounting screws and pull the instrument panel forward, then disconnect the speedometer and the electrical connectors form it.

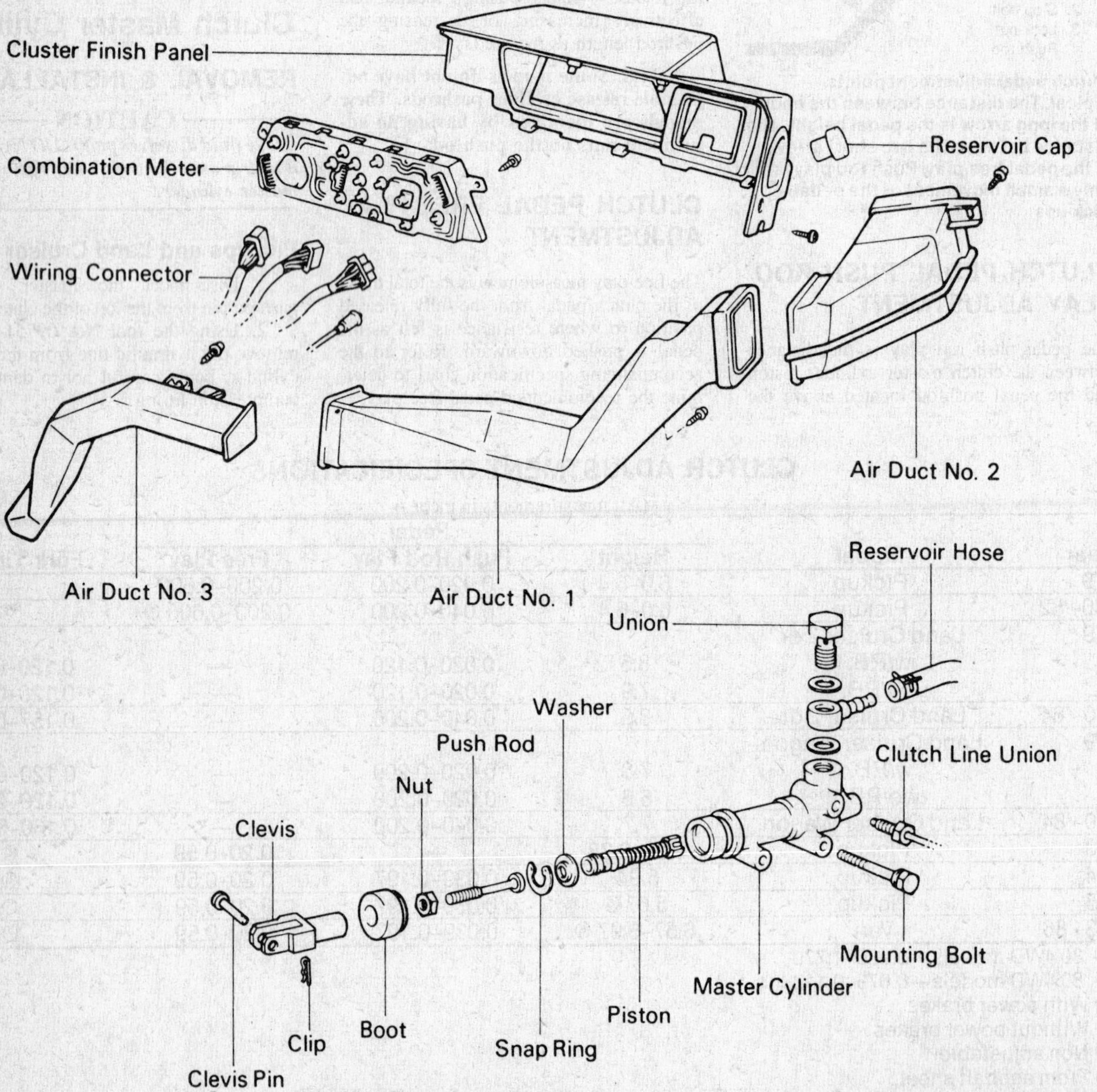

Cluster Finish Panel

Combination Meter

Wiring Connector

Reservoir Cap

Air Duct No. 2

Air Duct No. 3

Air Duct No. 1

Reservoir Hose

Union

Clutch Line Union

Washer

Push Rod

Nut

Clevis

Mounting Bolt

Master Cylinder

Clip

Boot

Piston

Snap Ring

Clevis Pin

Removing the clutch master cylinder— Vans

4. Remove the No.3, the No.1 and the No.2 air ducts.

5. Disconnect and plug the reservoir hose at the master cylinder. Using the tool No. 09751–36011, disconnect the clutch line union.

6. Remove the mounting bolts and the master cylinder.

7. To install, reverse the removal procedures. Bleed the clutch system. Adjust the clutch pedal.

Clutch Release Cylinder

REMOVAL & INSTALLATION

1. Raise and support the front of the vehicle on jackstands.

2. If equipped, remove the tension spring on the clutch fork.

3. Using the tool No. 09751–36011, remove the hydraulic line from the release cylinder. Be careful not to damage the fitting.

4. Turn the release cylinder pushrod in sufficiently to gain clearance from the fork.

5. Remove the mounting bolts and withdraw the cylinder.

6. To install, reverse the removal procedures. Bleed the clutch system. Adjust the fork tip clearance.

Hydraulic System Bleeding

NOTE: This procedure may be utilized when either the clutch master or release cylinder has been removed or if any of the hydraulic lines have been disturbed.

------ CAUTION ------
DO NOT spill brake fluid on the body of the vehicle as it will destroy the paint.

1. Fill the master cylinder reservoir with brake fluid.

2. Remove the cap and loosen the bleeder screw on the clutch release cylinder. Cover the hole with your finger.

3. Have an assistant pump the clutch pedal several times. Take your finger off the hole while the pedal is being depressed so that the air in the system can be released. Put your finger back on the hole and release the pedal.

4. When fluid pressure can be felt (with your finger) tighten the bleeder screw.

5. Place a short length of hose over the bleeder screw and the other end in a jar half full of clean brake fluid.

6. Depress the clutch pedal and loosen the bleeder screw. Allow the fluid to flow into the jar.

7. Tighten the plug, then release the clutch pedal.

8. Repeat this procedure until no air bubbles are visible in the bleeder tube.

9. When there are no more air bubbles in the system, tighten the plug fully with the pedal depressed. Replace the plastic cap.

10. Fill the master cylinder to the correct level with brake fluid. Check the system for leaks.

MANUAL TRANSMISSION

REMOVAL & INSTALLATION

2WD Pickup and Van

1. Disconnect the negative battery cables (L) or cable (2L and 2L-T).

2. If equipped with a floorshifter (Pickup), perform the following:

a. Remove the center floor console, if equipped.

b. Remove the shift lever handle, then the floor mat or carpet along with the shift lever boot in order to gain access to the shift lever.

c. Using the shift lever removal tool No. 09305–20011 (1979) or 09305–20012 (1980 and later), remove the shift lever.

NOTE: On the pickup (1984 and later) models, remove the boot and the shift lever from inside the vehicle.

3. Raise and support the vehicle on jackstands. Drain the transmission fluid.

4. Chalk matchmarks on the driveshaft flange and the differential pinion flange to indicate their relationships. These marks must be aligned during installation.

5. Remove the driveshaft flange bolts and the center support bearing-to-frame bolts (if equipped with a 2-piece driveshaft). Lower the driveshaft out of the vehicle. Using tool No. 09325–20010 (Pickup), insert it into the end of the transmission to prevent oil leakage.

6. On the van models, disconnect the shift and the select cables from the select outer levers, the clips and the cables.

7. Disconnect the back-up lamp switch electrical connector and the speedometer cable from the transmission, then tie the cable out of the way.

8. Disconnect the wiring at the starter. Remove the starter mounting bolts and lower the starter out of the vehicle.

9. Remove the exhaust pipe clamp and the exhaust pipe.

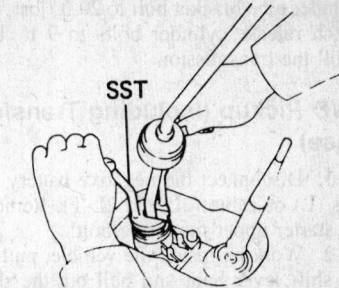

Using tool 09305-20012 to remove the shift lever

10. If the hydraulic line from the clutch release cylinder is clamped to the frame, remove the clamp retaining bolt. Remove the release cylinder mounting bolts and the fork spring (if equipped). Tie the release cylinder out of the way.

NOTE: It is not necessary to disconnect the hydraulic line from the release cylinder.

11. On column shift vehicles, disconnect the shift selector linkage at the transmission and remove the transmission cross shafts.

12. Support the rear of the transmission with a jack and remove the transmission-to-crossmember bolts, the crossmember-to-frame bolts and the crossmember from the vehicle.

NOTE: When removing the crossmember, raise the rear of the transmission SLIGHTLY, just enough to take the weight off of the crossmember.

13. Place a support under the engine with a wooden block (³⁄₄ in. thick) between the support and the engine oil pan.

------ CAUTION ------
The wooden block and support should be no more than about ¹⁄₄ away from the engine so that when the engine is lowered, damage will not occur to any underhood components. If possible, shim the support so that the wooden block touches the engine.

14. Remove the transmission-to-engine bolts, draw the transmission rearward and down, away from the engine.

NOTE: When removing the transmission, be careful not to damage the extension housing dust deflector.

15. To install, reverse the removal procedures. Torque transmission-to-engine bolts to 53 ft. lbs., the stiffener plate bolts to 27 ft. lbs., the transmission mount/bracket bolts to 19 ft. lbs., the rear engine mount bracket-to-crossmember bolts to 9 ft. lbs., the exhaust pipe-to-manifold bolts to 29 ft. lbs., the upper exhaust pipe bracket-to-clutch housing bolts to 27 ft. lbs., the lower exhaust pipe bracket-to-clutch housing bolts to 51 ft. lbs., the lower starter bolt/release

cylinder tube bracket bolt to 29 ft. lbs., the clutch release cylinder bolts to 9 ft. lbs. Refill the transmission.

4WD Pickup (Including Transfer Case)

1. Disconnect the negative battery cables (L) or cable (2L and 2L-T). Remove the starter upper mounting bolt.

2. Working inside the vehicle, pull up the shift lever boot and pull out the shift lever (using tool No. 09305–20012). If equipped with a 22R-E (1984 and later) engine, pull up the shift lever boot, then remove the mounting bolts and pull out the shift lever.

3. Using needle nose pliers, remove the transfer case shift lever snap ring and the shift lever.

4. Raise and support the vehicle on jackstands.

NOTE: Because of space limitations, it may be necessary to raise both the front and rear of the vehicle. If this is done, place jackstands under both axles as follows: On the outside of the U-bolts at the front axle; on the inside of the U-bolts at the rear axle.

5. Drain the lubricant from both the transmission and the transfer case.

6. Chalk matchmarks on the driveshaft flanges and the differential pinion flanges to indicate their relationships. These marks must be aligned during installation.

7. Remove the driveshaft mounting bolts and remove the front driveshaft assembly.

NOTE: DO NOT disassemble the front driveshaft to remove it.

8. Chalk matchmarks on the rear driveshaft and the slip yoke to indicate their relationships. These marks must be aligned during installation.

9. Remove the mounting bolts from the rearward flange of the rear driveshaft. Lower the driveshaft out of the vehicle. Remove the mounting bolts from the slip yoke flange then remove the flange and yoke assembly.

10. Unbolt the clutch release cylinder and tie it out of the way.

NOTE: It is not necessary to disconnect the hydraulic line from the clutch release cylinder.

11. Disconnect the starter motor electrical connectors. Remove the starter bolts and lower the starter from the vehicle.

12. At the transfer case, disconnect the speedometer cable (tie it out of the way), the back-up light switch connector and the 4WD indicator switch connector.

13. Disconnect the exhaust pipe clamp and the exhaust pipe from the transmission housing.

14. Remove the clutch release cylinder and the tube bracket, then move the cylinder aside.

NOTE: When removing the clutch release cylinder, DO NOT disassemble the hydraulic line from the cylinder.

15. Remove the crossmember-to-transfer case mounting bolts. Using a jack, raise the transmission and transfer case assembly SLIGHTLY off of the crossmember. Remove the crossmember-to-frame attaching bolts and remove the crossmember.

16. Place a support under the engine oil pan, with a wooden block (¾ in. thick) between the support and the engine oil pan.

— CAUTION —
The wooden block and support should be no more than about ¼ in. away from the engine so that when the engine is lowered, damage will not occur to any underhood components. If possible, shim the support so that the wooden block touches the engine.

17. Lower the jack until the engine rests on the support.

NOTE: For the next step, it is recommended that you have an assistant help you guide the transmission and transfer case assembly out of the vehicle.

18. Remove the exhaust pipe bracket and the stiffener plate bolts.

19. Remove the transmission-to-engine bolts, then draw the transmission/transfer case assembly rearward and down away from the engine.

20. Remove the transmission-to-transfer case adapter bolts and pull the transfer case from the transmission.

21. To install, use new gaskets and reverse the removal procedures. Torque transmission-to-engine bolts to 53 ft. lbs., the stiffener plate bolts to 27 ft. lbs., the transmission mount/bracket bolts to 19 ft. lbs., the rear engine mount bracket-to-crossmember bolts to 9 ft. lbs., the crossmember-to-frame bolts to 70 ft. lbs., the exhaust pipe-to-manifold bolts to 29 ft. lbs., the upper exhaust pipe bracket-to-clutch housing bolts to 27 ft. lbs., the lower exhaust pipe bracket-to-clutch housing bolts to 51 ft. lbs., the lower starter bolt/release cylinder tube bracket bolt to 29 ft. lbs., the clutch release cylinder bolts to 9 ft. lbs. Refill the transmission.

Land Cruiser
1979 MODELS

1. Disconnect the battery cables at the battery.

2. On 2-door models, perform the following:

    a. Remove the front seats, the seat tracks and the console box, if equipped.

    b. Remove the heater pipe clamp which is located on the transmission tunnel to the right of the transfer case shift lever.

    c. If the fuel tank is mounted beneath the passenger seat, drain the fuel, remove the fuel tank cover, disconnect the lines and etc., then remove the fuel tank.

3. Remove the shift lever knobs and the boots.

4. Using the tool No. 09305–60010, remove the transmission shift lever and the transmission tunnel cover.

5. Raise and support the vehicle on jackstands.

6. Drain the lubricant from the transmission and the transfer case.

7. Remove the engine undercover, located beneath the front driveshaft.

8. Chalk matchmarks on the driveshaft flanges and the differential pinion flanges to indicate their relationships. These marks must be aligned during installation.

9. Remove the driveshaft flanges, then the front and rear driveshafts.

10. Disconnect the speedometer cable from the transfer case and tie it out of the way.

11. Disconnect the parking brake cable at the parking brake lever. Leave the cable attached at the drum end; the cable will be removed with the transmission and transfer case assembly.

12. If equipped with a vacuum 4WD engagement system, mark and disconnect the following items at the transfer case:

    a. The indicator electrical connectors.

    b. The transfer switch wiring.

    c. The vacuum hoses.

13. Disconnect the back-up lamp switch connector. Unbolt the back-up lamp wiring harness clamp from the transfer case (if equipped).

14. On column shift models, disconnect the shift linkage from the transmission.

15. If equipped, remove the power take-off (PTO) lever.

16. Remove the crossmember-to-transfer case mounting bolts. Using a jack, raise the transmission and transfer case assembly SLIGHTLY off of the crossmember. Remove the crossmember-to-frame attaching bolts and remove the crossmember.

17. Place a support under the engine oil pan, with a wooden block (¾ in. thick) between the support and the engine oil pan.

— CAUTION —
The wooden block and support should be no more than about ¼ in. away from the engine so that when the engine is lowered, damage will not occur to any underhood components. If possible, shim the support so that the wooden block touches the engine.

18. Lower the jack until the engine rests on the support.

NOTE: For the next step, it is recommended that you have an assistant help you guide the transmission and transfer case assembly out of the vehicle.

19. Remove the exhaust pipe bracket and the stiffener plate bolts.

20. Remove the transmission-to-engine bolts, then draw the transmission/transfer case assembly rearward and down away from the engine.

21. To separate the transmission from the transfer case:

    a. Remove the 4WD engagement lever guide.

b. Remove the 4WD lever and rod as an assembly.

c. Remove the back-up lamp switch.

d. If equipped with a PTO, remove the PTO unit from the transmission. If not equipped with PTO, remove the left-side cover from the transfer case.

e. Remove the rear transfer case cover-to-transfer case bolts and the shaft nut located behind the cover.

NOTE: The nut is staked at the factory; to remove it, you must tap the staked portions outward to clear the shaft. Restake the nut after installation.

f. Remove the transfer case-to-transmission bolts.

NOTE: Two of the bolts are located at the left inner-side of the transfer case, where the PTO or cover was previously removed.

g. Using a puller (assembled to the transfer case and the transmission output shaft), separate the transfer case from the transmission.

22. To install, reverse the removal procedures. Torque transmission-to-engine bolts to 53 ft. lbs., the stiffener plate bolts to 27 ft. lbs., the transmission mount/bracket bolts to 19 ft. lbs., the rear engine mount bracket-to-crossmember bolts to 9 ft. lbs., the exhaust pipe-to-manifold bolts to 29 ft. lbs., the upper exhaust pipe bracket-to-clutch housing bolts to 27 ft. lbs., the lower exhaust pipe bracket-to-clutch housing bolts to 51 ft. lbs., the lower starter bolt/release cylinder tube bracket bolt to 29 ft. lbs., the clutch release cylinder bolts to 9 ft. lbs. Refill the transmission.

1980 AND LATER

1. Disconnect the negative battery cables.

2. Remove the entrance scuff plates from the floor of the interior.

3. Remove both side trim panels from beneath the instrument panel.

4. Remove the center heater duct and the front floor mat or carpet.

5. Remove the handles from both shift levers and the transmission tunnel cover along with the shift lever boots.

6. Disconnect the wiring from both the back-up lamp switch and the 4WD indicator (if equipped).

7. Using the tool No. 09305–55010, remove the transmission shift lever.

8. Raise and support the vehicle on jackstands. Remove the transfer case skid plate.

9. Disconnect the speedometer cable from the transfer case and tie it out of the way.

10. Chalk matchmarks on the driveshaft flanges and the differential pinion flanges

to indicate their relationships. These marks must be aligned during installation.

11. Remove the driveshaft flanges mounting bolts and the driveshaft assemblies.

12. Disconnect the starter electrical connectors, the mounting bolts and the starter from the vehicle.

13. Remove the clutch release cylinder and move it out of the way.

NOTE: It is not necessary to disconnect the hydraulic line from the release cylinder.

14. Drain the lubricant from both the transmission and the transfer case. Remove the tachometer sensor, if equipped.

15. Remove the crossmember-to-transfer case mounting bolts. Using a jack, raise the transmission and transfer case assembly SLIGHTLY off of the crossmember. Remove the crossmember-to-frame attaching bolts and remove the crossmember.

16. Place a support under the engine oil pan, with a wooden block (¾ in. thick) between the support and the engine oil pan.

--- CAUTION ---

The wooden block and support should be no more than about ¼ in. away from the engine so that when the engine is lowered, damage will not occur to any underhood components. If possible, shim the support so that the wooden block touches the engine.

17. Lower the jack until the engine rests on the support.

NOTE: For the next step, it is recommended that you have an assistant help you guide the transmission and transfer case assembly out of the vehicle.

18. Remove the exhaust pipe bracket and the stiffener plate bolts.

19. Remove the transmission-to-engine bolts, then draw the transmission/transfer case assembly rearward and down away from the engine.

20. To separate the transfer case from the transmission, remove the transfer case mounting bolts and slide the transfer case off of the transmission.

21. To install, reverse the removal procedures. Torque transmission-to-engine bolts to 53 ft. lbs., the stiffener plate bolts to 27 ft. lbs., the transmission mount/bracket bolts to 19 ft. lbs., the rear engine mount bracket-to-crossmember bolts to 9 ft. lbs., the exhaust pipe-to-manifold bolts to 29 ft. lbs., the upper exhaust pipe bracket-to-clutch housing bolts to 27 ft. lbs., the lower exhaust pipe bracket-to-clutch housing bolts to 51 ft. lbs., the lower starter bolt/release cylinder tube bracket bolt to 29 ft. lbs., the clutch release cylinder bolts to 9 ft. lbs. Refill the transmission.

SHIFT LINKAGE ADJUSTMENT

Pickup and Land Cruiser

COLUMN SHIFTER

The only adjustments which may be performed on the column shift linkages are for the length of the column-to-transmission rods. Adjust these so that the transmission operates smoothly.

FLOOR SHIFTER

All models equipped with a floor shifter have internally-mounted shift linkages. On older models, the linkage is contained in the side cover which is bolted on the transmission case.

Van

CABLE ADJUSTMENT

1. Remove the console box and loosen the adjusting lock nut.

2. Place the shift lever in the Neutral position.

3. Using a 0.20 in. dia. guide pin, insert it into the neutral adjust service hole; adjust the length of the cable by turning the adjusting nut.

4. After adjustment, remove the guide pin and reinstall the console box.

AUTOMATIC TRANSMISSION

DESCRIPTION

The A40 transmission, used on the 1979 models, is a fully automatic 3-speed but it does not use bands for gear changes, thus internal adjustments are not possible.

In 1980, the A40 was replaced by the A43 3-speed; internal adjustments are not required.

The A43D is a fully automatic 4-speed transmission first offered as an option on 1981 models and is available through the current model year. The 4th speed of this transmission is an overdrive ratio of 0.688 to 1, which offers improved gasoline mileage by lowering the engine rpm at highway speeds. The hydraulic circuit of the overdrive mode is electrically controlled. The main electrical components include the following:

1. A dash mounted overdrive control switch.

2. A dash mounted "OVERDRIVE-OFF" indicator lamp.

3. A transmission mounted solenoid.

4. An engine mounted thermo-switch which prevents overdrive engagement until

the engine coolant temperature reaches 131°F.

REMOVAL

A40 Transmission

1. Disconnect the negative battery cable(s) and the transmission throttle linkage from the carburetor.

2. Raise and support the vehicle on jackstands. Drain the transmission fluid.

3. Disconnect the wiring from the starter, then unbolt and lower the starter from the vehicle.

4. Disconnect the exhaust pipe from the exhaust manifold. Remove the exhaust clamp from the exhaust pipe.

5. Disconnect the shifting linkage from the driver's side of the transmission.

6. Disconnect the speedometer cable (tie it out of the way) and the parking brake cable (from the parking brake control lever).

7. Chalk matchmarks on the rear driveshaft flange and the differential pinion flange. These marks must be aligned during installation.

8. Unbolt the rear driveshaft flange.

NOTE: If the vehicle has a two-piece driveshaft, remove the center bearing bracket-to-frame bolts. Remove the driveshaft from the vehicle.

9. Using a floor jack and a wooden block (¾ in. thick), support the transmission at the pan. DO NOT raise the transmission; just raise the jack until the wooden block touches the transmission pan.

10. Place a support under the engine oil pan with a wooden block between the support and the engine.

── CAUTION ──

The wooden block and support should be no more than about ¼ in. away from the engine so that when the engine is lowered, damage will not occur to any underhood components.

11. Remove the transmission mount-to-crossmember bolts.

12. Raise the transmission SLIGHTLY, just enough to take the weight of the transmission off of the crossmember. Remove the crossmember-to-frame mounting bolts and the crossmember from the vehicle.

13. Slowly lower the transmission until the engine rests on the wooden support.

14. Disconnect and plug the two fluid cooler lines at the transmission to prevent the entry of dirt.

15. Remove the torque converter inspection plate and the torque converter-to-drive plate bolts, then slide the torque converter toward the transmission.

── CAUTION ──

Before performing step 19, place a drain pan under the torque convertor area of the transmission. Fluid leakage will occur as the transmission is uncoupled.

16. Remove the transmission-to-engine mounting bolts. Carefully pull the transmission to the rear and then lower it from the vehicle.

17. To install, reverse the removal procedures. Refill the transmission with Dexron II® fluid.

A43 and A43D Transmissions

1. Disconnect the negative battery cable. On the Pickup, remove the air cleaner assembly.

2. Disconnect the transmission throttle cable from the carburetor linkage (Pickup) or the throttle body (Van).

3. Raise and support the vehicle on jackstands. Drain the transmission fluid.

4. Disconnect the wiring connectors (near the starter) for the neutral start switch and the back-up light switch. If equipped, disconnect the solenoid (overdrive) switch wiring at the same location.

5. Disconnect the starter wiring at the starter. Remove the mounting bolts and the starter from the engine.

6. Chalk matchmarks on the rear driveshaft flange and the differential pinion flange. These marks must be aligned during installation.

7. Unbolt the rear driveshaft flange. If the vehicle has a two-piece driveshaft, remove the center bearing bracket-to-frame bolts. Remove the driveshaft from the vehicle.

8. Disconnect the speedometer cable (tie it out of the way) and the shift linkage from the transmission.

9. Disconnect the transmission oil cooler lines at the transmission.

10. Disconnect the exhaust pipe clamp and remove the oil filler tube.

11. Support the transmission, using a jack with a wooden block placed between the jack and the transmission pan. Raise the transmission, just enough to take the weight off of the rear mount.

12. On the Pickup models, remove the rear engine mount with the bracket and the engine under cover (pickups), to gain access to the engine crankshaft pulley. On the Van models, remove the fuel tank mounting bolts and support the fuel tank; remove the transmission mount through bolt.

13. Place a wooden block (or blocks) between the engine oil pan and the front frame crossmember.

── CAUTION ──

The wooden block and support should be no more than about ¼" away from the engine so that when the engine is lowered, damage will not occur to any underhood components.

14. Slowly, lower the transmission until the engine rests on the wooden block.

15. Remove the rubber plug(s) from the service holes located at the rear of the engine in order to gain access to the torque convertor bolts.

16. Rotate the crankshaft (to remove the torque convertor bolts) to access the bolts through the service holes.

17. Obtain a bolt of the same dimensions as the torque convertor bolts. Cut the head off of the bolt and hacksaw a screwdriver slot in the bolt opposite the threaded end.

NOTE: This modified bolt is used as a guidepin. Two guides pins are needed to properly install the transmission.

18. Thread the guide pin into one of the torque convertor bolt holes. The guide pin will help keep the convertor with the transmission.

19. Remove the stiffener plates from the transmission.

20. Remove the transmission-to-engine bolts, then carefully move the transmission rearward by prying on the guide pin through the service hole.

── CAUTION ──

As the transmission moves away from the engine about ⅛ in., feed wire through the front of the transmission and secure the wire in order to keep the convertor attached to the transmission. Also, try to keep the nose of the transmission pointed upward SLIGHTLY to help keep the convertor in place.

21. Pull the transmission rearward and lower it (front end down) out of the vehicle.

── CAUTION ──

Do not allow the attached cables to catch on any components during removal.

22. With the transmission out of the vehicle, remove the torque convertor as follows:

  a. Place a drain pan under the front of the transmission.

  b. Pull the convertor straight off of the transmission and allow the fluid to drain.

INSTALLATION

A40 Transmission

1. Apply a coat of multipurpose grease to the torque converter stub shaft and the pilot hole of the flywheel.

2. Assemble the torque converter into the transmission, so that it's output shaft and the transmission input shaft are aligned. Rotate the torque converter until the dowel pin is at the bottom.

3. Install a guide pin into the bottom bolt hole next to the dowel pin. Align the flywheel and the torque converter.

4. Tighten the torque converter bolts to 11–16 ft. lbs. Rotate the crankshaft ½ turn to reach all the bolts and tighten them evenly.

5. Bolt the transmission to the engine. Tighten the bolts to 37–51 ft. lbs.

6. To complete the installation, reverse the removal procedures. Fill the transmission with clean transmission fluid. Adjust the throttle and the shift linkages. Road test and check for leaks.

A43 and A43D Transmissions

1. Apply a coat of multi-purpose grease to the torque convertor stub shaft and the corresponding pilot hole in the flywheel.

2. Install the torque convertor into the front of the transmission. Push inward on the torque convertor while rotating it to completely couple the torque convertor to the transmission.

3. To make sure that the convertor is properly installed, measure the distance between the torque convertor mounting lugs and the front mounting face of the transmission. The proper distance is 0.080 in.

4. Install guide pins into two opposite mounting lugs of the torque convertor.

5. Raise the transmission to the engine, align the transmission with the engine alignment dowels and position the convertor guide pins into the mounting holes of the flywheel.

6. Install and tighten the transmission-to-engine mounting bolts. Torque the bolts to 47 ft. lbs.

7. Remove the convertor guide pins and install the convertor mounting bolts. Rotate the crankshaft as necessary to gain access to the guide pins and bolts through the service holes. Evenly, tighten the convertor mounting bolts to 13 ft. lbs. Install the rubber plugs into the access holes.

8. Install the engine undercover. Raise the transmission slightly and remove the wood block(s) from beneath the engine oil pan.

9. Install the transmission crossmember. Torque the crossmember-to-frame bolts to 26–36 ft. lbs.

10. Lower the transmission onto the crossmember and install the transmission mounting bolts. Torque the bolts to 19 ft. lbs.

11. Install the oil filler tube and connect the exhaust pipe clamp.

12. Connect the oil cooler lines to the transmission and torque the fittings to 25 ft. lbs.

13. To complete the installation, reverse the removal procedures. Adjust the transmission throttle cable. Refill the transmission with Dexron II® fluid. Road test the vehicle and check for leaks.

Transmission Pan and Filter

REMOVAL & INSTALLATION

1. Raise and support the front of the vehicle on jackstands.

2. Place a container under the transmission drain plug and drain the transmission fluid.

3. Remove the pan securing bolts, the pan and the gasket.

4. The pan may be washed in solvent for cleaning but must be absolutely dry when it is reinstalled. DO NOT wipe it out with a rag or you will risk leaving bits of lint inside the transmission.

5. Using a small pry bar, remove the oil tube (covering the oil strainer pan). Remove the oil strainer pan and the strainer. Clean the oil strainer.

6. Remove all traces of the old gasket from the pan and the transmission.

7. To install, use a new gasket(s) and reverse the removal procedures. Torque the oil strainer pan to 48 inch lbs. and the transmission pan to 39 inch lbs. Refill the transmission with Dexron II® fluid.

—— CAUTION ——
The pan bolts break easily if overtightened.

Adjustments

SHIFT LINKAGE

1979 Transmission

1. Check the shift linkage bushings for wear. Replace any that are excessively worn.

2. Set the manual valve lever on the transmission in the Neutral position.

3. Lock the connecting rod swivel with the locknut, so that the pointer, selector and manual valve lever are all in the Neutral position.

4. Check the operation by moving the selector through all the gears.

1980 and Later Transmissions

1. Loosen the adjustment nut on the transmission connecting rod (1980–83) or the shift cable (1984 and later).

2. Push the manual lever of the trans-

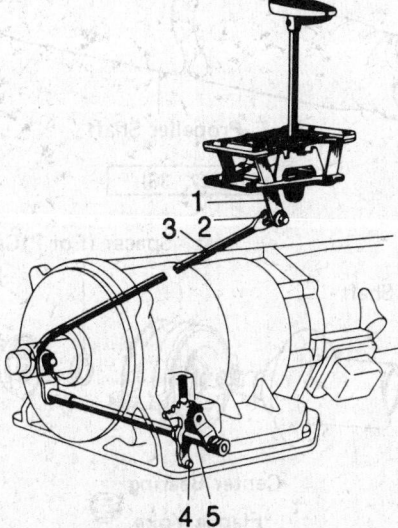

1979 floor shift automatic transmission linkage—adjust the connecting rod swivel at point 2

1. Shift lever
2. Connecting rod
3. Control rod
4. Manual valve lever
5. Manual valve lever shaft

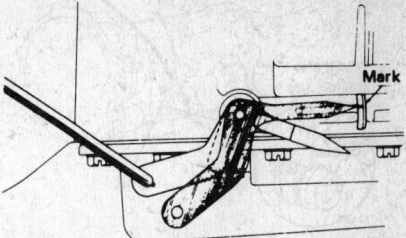

1979 floor shift automatic transmission—alignment marks on the transmission case and the throttle level

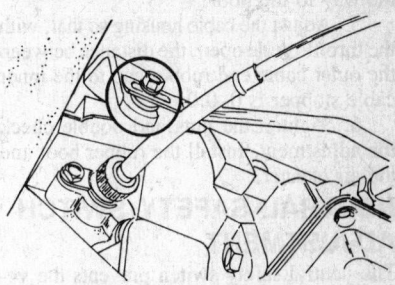

On 1980 and later automatics, loosen this nut to adjust the shift linkage

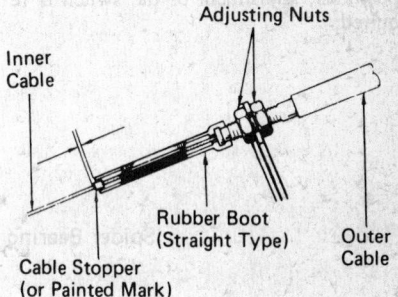

Throttle control cable adjustment—automatic transmission

mission fully forward (1980–83) or rearward (1984 and later).

3. Move the manual lever back three (1980–83) or two (1984 and later) notches, which is the NEUTRAL position.

4. Set the gearshift selector lever in it's NEUTRAL position.

5. Apply a slight amount of forward pressure on the selector lever (towards the Reverse position) and tighten the connecting rod (1980–83) or the shift cable (1984 and later) adjustment nut.

TRANSMISSION THROTTLE CONTROL ADJUSTMENT

1. Remove the air cleaner assembly.

2. Push the accelerator to the floor and check that the throttle valve opens fully; if not, adjust the accelerator link, so that it does.

3. Push back the rubber boot from the throttle cable which runs down to the transmission. Loosen the throttle cable adjustment nuts so that the cable housing can be adjusted.

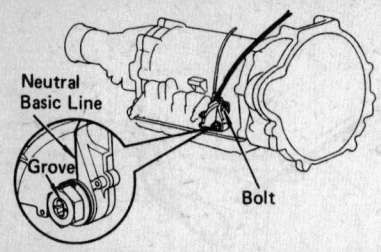

1980-83 neutral safety switch adjustment

4. Fully open the carburetor throttle by having an assistant press the accelerator all the way to the floor.

5. Adjust the cable housing so that, with the throttle wide open, the distance between the outer cable end rubber cap to the inner cable stopper is 0–0.04 in.

6. Tighten the nuts and double check the adjustment. Install the rubber boot and the air cleaner.

NEUTRAL SAFETY SWITCH ADJUSTMENT

The neutral safety switch prevents the vehicle from starting unless the gearshift selector is in either the PARK or NEUTRAL positions. If the vehicle will start in these positions, adjustment of the switch is required.

1979

1. Remove the center console screws, the electrical connector and the console from the vehicle.

2. Loosen the switch securing bolts.

3. Set the selector in the Drive position. Move the switch so that the arm just contacts the control shaft lever, then tighten the switch retaining bolts.

4. Check the operation of the switch; the truck should start only in Neutral or Park. The back-up lamps should only operate in the Reverse position.

5. If the switch cannot be adjusted (so that it functions properly), replace it with a new one.

6. Reinstall the console.

1980–83

1. Loosen the Neutral Start Switch bolt.

2. Place the gearshift selector lever in the Neutral position.

3. Align the shaft groove of the switch with the neutral Basic line. Hold the switch in this position and tighten the switch bolt to 35–60 inch lbs.

1984 and Later

1. Loosen the Neutral Start Switch bolt.

2. Place the selector lever in the Neutral position.

3. Disconnect the wires from the neutral start switch.

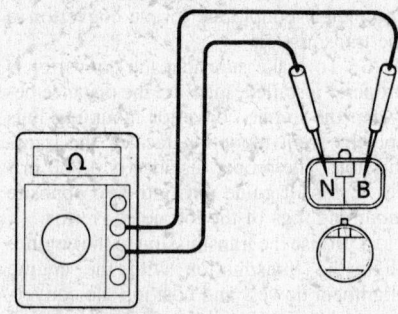

Checking the neutral start switch continuity—1984 and later

4. Connect an ohmmeter between the terminals of the switch.

5. Adjust the switch until there is continuity between the N and B terminals.

6. Reconnect the wires. Torque the bolt to 48 inch lbs.

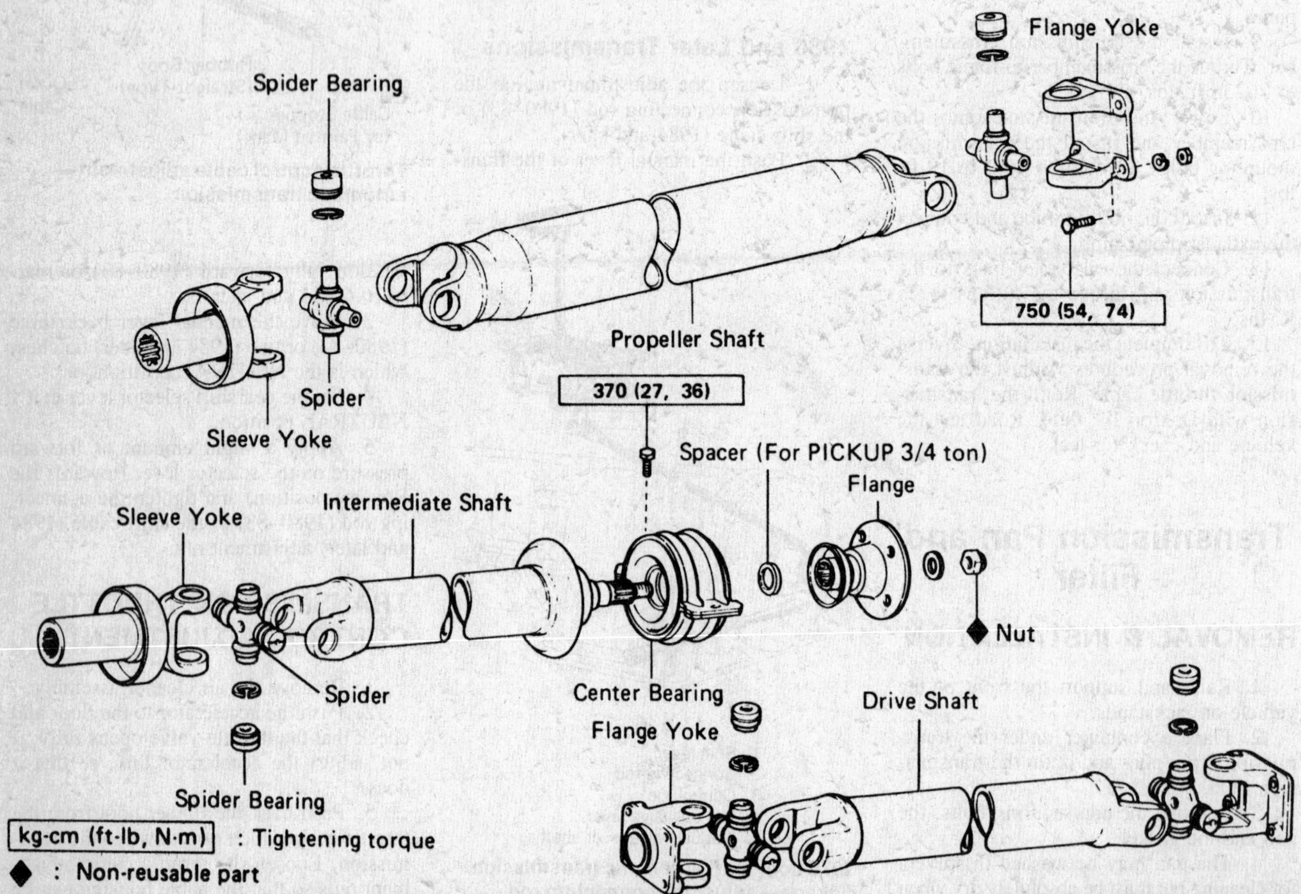

Spider Bearing

Spider

Sleeve Yoke

Propeller Shaft

Flange Yoke

750 (54, 74)

370 (27, 36)

Spacer (For PICKUP 3/4 ton)

Flange

Sleeve Yoke

Intermediate Shaft

Spider

Center Bearing

Flange Yoke

Drive Shaft

Nut

Spider Bearing

kg-cm (ft-lb, N·m) : Tightening torque

◆ : Non-reusable part

Exploded view of the driveshafts—2WD vehicles

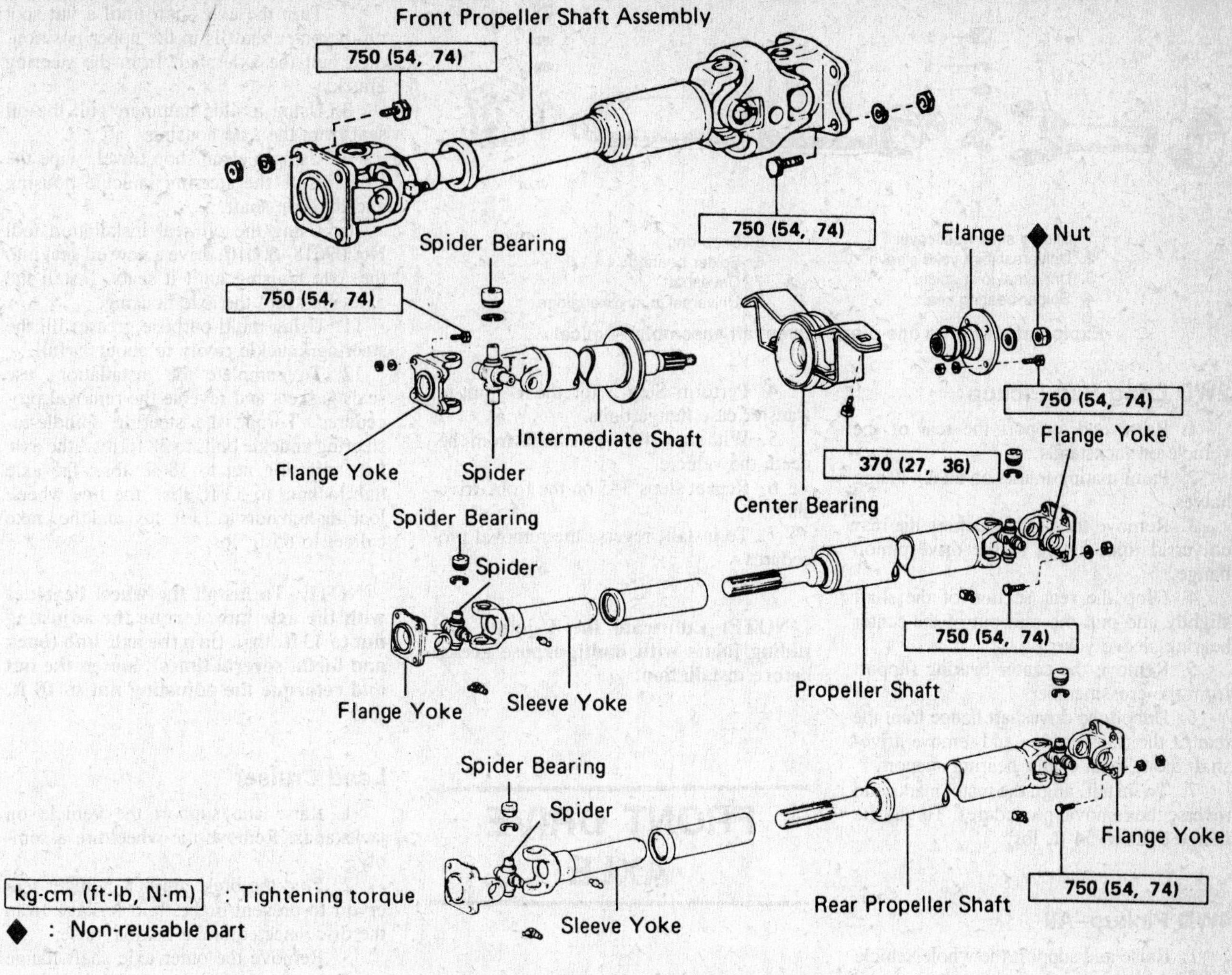

Front Propeller Shaft Assembly

750 (54, 74)

750 (54, 74)

Flange ◆ Nut

Spider Bearing

750 (54, 74)

750 (54, 74)

Flange Yoke

Flange Yoke   Spider   Intermediate Shaft

370 (27, 36)

Center Bearing

Spider Bearing

Spider

Flange Yoke   Sleeve Yoke

Propeller Shaft

750 (54, 74)

Spider Bearing

Spider

Flange Yoke

750 (54, 74)

Rear Propeller Shaft

750 (54, 74)

Sleeve Yoke

kg·cm (ft-lb, N·m) : Tightening torque

◆ : Non-reusable part

Exploded view of the driveshafts—4WD vehicles

DRIVE TRAIN

Driveshaft

REMOVAL & INSTALLATION

2WD Standard Bed Pickup and Van

1. Raise and support the rear of the vehicle on jackstands.

2. Paint a mating mark on the two halves of the rear universal joint flange.

3. Remove the bolts which hold the rear flange together.

4. Remove the splined end of the driveshaft from the transmission.

NOTE: Plug the end of the transmission with a rag or dummy flange to avoid losing transmission oil.

5. Remove the driveshaft from under the truck.

6. To install, reverse the removal procedures. Grease the splined end of the shaft before installing. Torque bolts to 31 ft. lbs. (Van) or 54 ft. lbs. (Pickup).

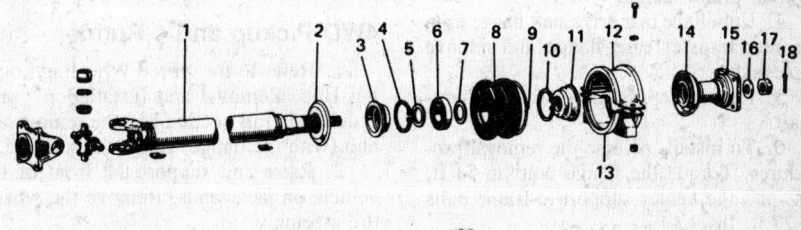

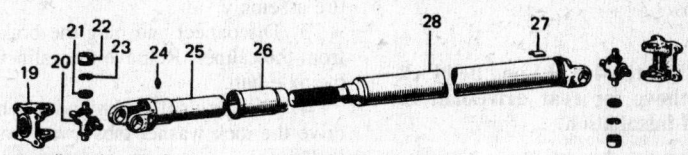

1. Intermediate driveshaft
2. Dust deflector No. I
3. Dust deflector No. 2
4. Hole snap-ring
5. Dust deflector No. 3
6. Radial ball bearing
7. Dust deflector No. 4
8. Center support bearing cushion
9. Set ring
10. Hole snap-ring
11. Dust deflector No. 2

12. Center support bearing housing No. 1
13. Center support bearing housing No. 2
14. Dust deflector No. 1
15. Universal joint flange
16. Plate washer
17. Castle nut
18. Cotter pin
19. Universal joint flange yoke
20. Universal joint spider

21. Universal joint spider bearing seal
22. Universal joint spider bearing
23. Hole snap-ring
24. Grease fitting
25. Universal joint sleeve yoke
26. Sliding shaft dust cover
27. Balance piece
28. Driveshaft

Exploded view of a two-piece driveshaft assembly—typical

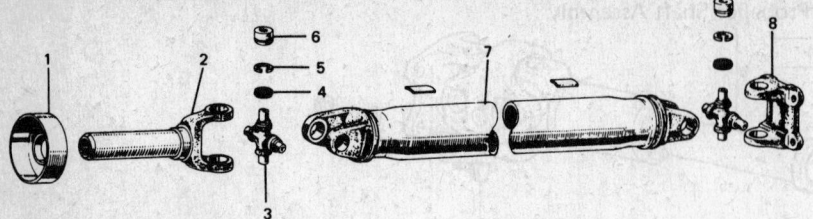

1. Sliding shaft dust cover
2. Universal joint yoke sleeve
3. Universal joint spider
4. Spider bearing seal
5. Snap-ring
6. Spider bearing
7. Driveshaft
8. Universal joint yoke falnge

Exploded view of a one-piece driveshaft assembly—typical

2WD Long Bed Pickup

1. Raise and support the rear of the vehicle on jackstands.
2. Paint mating marks on all six flange halves.
3. Remove the bolts attaching the rear universal joint flange to the drive pinion flange.
4. Drop the rear section of the shaft slightly and pull the unit out of the center bearing sleeve yoke.
5. Remove the center bearing support from the crossmember.
6. Unbolt the driveshaft flange from the rear of the transmission and remove driveshaft along with center bearing support.
7. To install, align the matchmarks and reverse the removal procedures. Torque the flange bolts to 54 ft. lbs.

4WD Pickup—All

1. Raise and support the whole vehicle off the ground and on jackstands.
2. Match-mark all driveshaft flanges BEFORE removing the bolts.
3. Unbolt the rear driveshaft flange from the rear pinion flange.
4. Unbolt the rear driveshaft flange from the rear transfer case flange and remove driveshaft.
5. Repeat steps 3 and 4 on front driveshaft.
6. To install, reverse the removal procedures. Torque the flange bolts to 54 ft. lbs. and the center support-to-frame bolts to 27 ft. lbs.

NOTE: For the 4 × 4 Long Bed Pickups, see above for rear driveshaft removal and installation.

Land Cruiser

1. Raise and support the vehicle on jackstands.
2. Match-mark all driveshaft flanges BEFORE removing the bolts.
3. Unfasten the bolts which secure the universal joint flange to the differential pinion flange.

4. Perform Step 2 for the U-joint-to-transfer case flange bolts.
5. Withdraw the driveshaft from beneath the vehicle.
6. Repeat steps 3–5 on the front driveshaft.
7. To install, reverse the removal procedures.

NOTE: Lubricate the U-joints and sliding joints with multipurpose grease before installation.

FRONT DRIVE AXLE

Axle Shaft

REMOVAL & INSTALLATION

4WD Pickup and 4-Runner

1. Refer to the "Free Wheeling/Locking Hub, Removal and Installation" procedures, in this section and the remove the hub (with the flange) from the axle hub.
2. Raise and support the front of the vehicle on jackstands. Remove the wheel/tire assembly.
3. Disconnect and plug the brake line from the caliper. Remove the caliper from the axle hub.
4. Using a drift punch and a hammer, drive the lock washer tabs away from the lock nut.
5. Using a 2 inch socket, remove the locknut from the axle shaft. Remove the lock washer, the adjusting nut, the thrust washer, the outer bearing and the axle hub/disc assembly from the vehicle.
6. Remove the knuckle spindle bolts, the dust seal and the dust cover. Using a brass bar and a hammer, tap the steering spindle from the steering knuckle.

7. Turn the axle shaft until a flat spot on the outer shaft is in the upper position, then pull the axle shaft from the steering knuckle.
8. Using a slide hammer, pull the oil seal from the axle housing.
9. Using a clean shop towel, wipe the from inside the steering knuckle housing and the axle shaft.
10. Using the oil seal installation tool No. 09618–60010, drive a new oil seal into the axle housing until it seats. Install the axle shaft into the axle housing.
11. Using multi-purpose grease, fill the steering knuckle cavity to about ¾ full.
12. To complete the installation, use seals/gaskets and reverse the removal procedures. Torque the steering spindle-to-steering knuckle bolts to 38 ft. lbs., the axle hub adjusting nut to 18 ft. lbs., the axle hub locknut to 33 ft. lbs., the free wheel/locking hub nuts to 23 ft. lbs. and the brake caliper to 65 ft. lbs.

NOTE: To install the wheel bearings with the axle hub, torque the adjusting nut to 43 ft. lbs., turn the axle hub (back and forth, several times), loosen the nut and retorque the adjusting nut to 18 ft. lbs.

Land Cruiser

1. Raise and support the vehicle on jackstands. Remove the wheel/tire assembly.
2. Plug the brake master cylinder reservoir to prevent brake fluid leakage from the disconnected brake flexible hose.
3. Remove the outer axle shaft flange cap (automatic locking hub) or the hub cover bolts and the cover (free wheel locking hub) and the shaft snap-ring from the axle hub.
4. Remove the outer axle shaft flange (automatic locking hub) or the hub ring (free wheel locking hub)-to-axle hub bolts, then alternately, screw two service bolts into the shaft flange or hub ring and remove the shaft flange or the hub ring with it's gasket.
5. Remove the brake drum set screws and the brake drum. If equipped with disc brakes, remove the caliper and disc.
6. Straighten the lockwasher and remove the front wheel bearing adjusting nuts with front wheel adjusting nut wrench or similar tool.
7. Remove the front axle hub together with its claw washer, bearings and oil seal.
8. Remove the clip and disconnect the brake flexible hose from the brake tube.
9. Cut and remove the lock wire, then remove the brake backing plate-to-steering knuckle bolts. Remove the brake backing plate together with the brake shoes, the tension springs and the wheel cylinder as an assembly.
10. Using a soft mallet, lightly, tap the steering knuckle spindle and remove the spindle with it's gasket.

NOTE: When removing the steering knuckle spindle on a vehicle equipped with the ball joint type axle shaft joint, be prepared for the disconnection of the outer axle shaft from the joint. The joint ball will fall from the joint. Try to cushion its fall or catch it if you can.

11. If equipped with the ball type axle shaft joint, slide the inner front axle shaft out of the axle housing. If equipped with the Birfield constant velocity joint type of axle shaft joint, remove the entire axle shaft assembly from the axle housing.

12. Using a bearing puller, remove the bushing from inside of knuckle spindle and the axle housing oil seal. Using a metal tube as a seating tool, drive oil seal into the axle housing and the new bushing into the knuckle spindle.

NOTE: If equipped with the ball joint type axle joint, install the inner axle with its proper spacer in position until the splines are fully meshed with the differential. If equipped with the Birfield constant velocity joint axle joint, install the axle into the housing and rotate the axle shaft until its splines mesh with the differential. Fill the steering knuckle about ¾ full with grease and place the joint ball on the inner shaft end.

13. To complete the installation, reverse the removal procedures. Adjust the wheel bearing preload.

Free-Wheeling and Automatic Locking Hubs

REMOVAL & INSTALLATION

1. If equipped with free-wheeling hubs, turn the hub control handle to the FREE position.
2. Remove the hub cover bolts and pull off the cover.
3. If equipped with automatic locking hubs, remove the axle bolt with the washer.
4. Using snap ring pliers, remove the snap-ring from the axle shaft.
5. Remove the hub body mounting nuts.
6. Remove the cone washers from the hub body mounting studs by tapping on the washer slits with a tapered punch.
7. Remove the hub body from the axle hub.
8. Apply multi-purpose grease to the inner hub splines.
9. To install, use new gaskets and reverse the removal procedures. Torque the hub body-to-axle hub nuts to 23 ft. lbs., the plate washer/bolt to 13 ft. lbs. (auto. locking hub) and the hub cover-to-hub body bolts to 7 ft. lbs.

NOTE: To install the snap ring onto the axle shaft, install a bolt into the axle shaft, pull it out and install the snap ring.

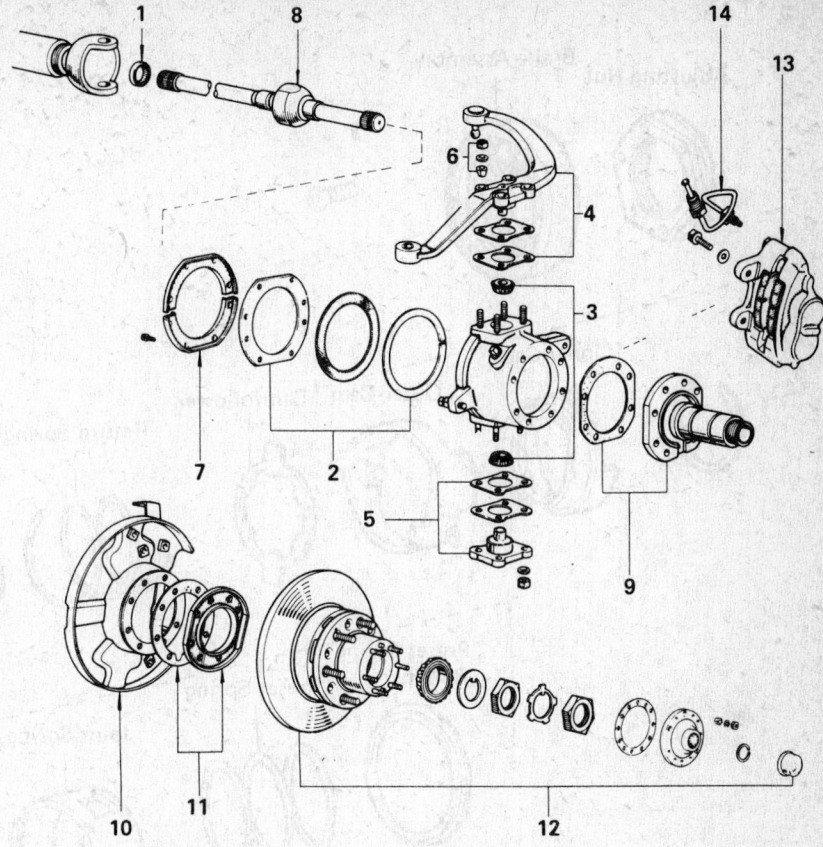

1. Oil seal
2. Oil seal set
3. Bearings
4. Steering knuckle
5. Bearing cup and shim
6. Nut, washer dowel
7. Oil seal retainer
8. Front axle shaft
9. Knuckle spindle and gasket
10. Dust cover
11. Dust seal and gasket
12. Front axle hub with disc
13. Brake caliper
14. Brake line

Front axle and steering knuckle (4WD)—pick-up illustrated, Land Cruiser similar

Front Wheel Bearings—4WD

REMOVAL & INSTALLATION

1. Refer to the "Free-Wheeling and Automatic Locking Hubs, Removal and Installation" procedures, in this section and remove the hubs.
2. Using a small pry bar, pry the grease seal from the rear of the disc/hub assembly, then remove the inner bearing from the assembly.
3. Using a shop cloth, wipe the grease from inside the disc/hub assembly.
4. Using a brass drift, drive the outer bearing races from each side of the disc/hub assembly.
5. Using solvent (NOT gasoline), clean all of the parts and blow dry with compressed air.
6. Using the bearing installation tool No. 09608-35013, drive the outer races into the disc/hub assembly until they seat against the shoulder.
7. Using multi-purpose grease, coat the area between the races and pack the bearings.
8. Place the inner bearing into the rear of the disc/hub assembly. Using the installation tool No. 09608-35013, drive a new grease seal into the rear of the disc/hub assembly until it is flush with the housing.
9. Install the disc/hub assembly onto the axle shaft, the outer bearing, the thrust washer and the adjusting nut.
10. To adjust the bearing preload, perform the following:
   a. Using tool No. 09607-60020, torque the adjusting nut to 43 ft. lbs.
   b. Turn the disc/hub assembly 2–3 times, from the left to the right.
   c. Loosen the adjusting nut until it can be turned by hand.
   d. Retorque the adjusting nut to 18 ft. lbs.
   e. Install the lock washer and the lock nut. Torque the lock nut to 33 ft. lbs.
   f. Check that the bearing has no play.
   g. Using a spring gauge, connect it to a wheel stud, the gauge should be held horizontal, then measure the rotating force, it should be 6–12 lbs.

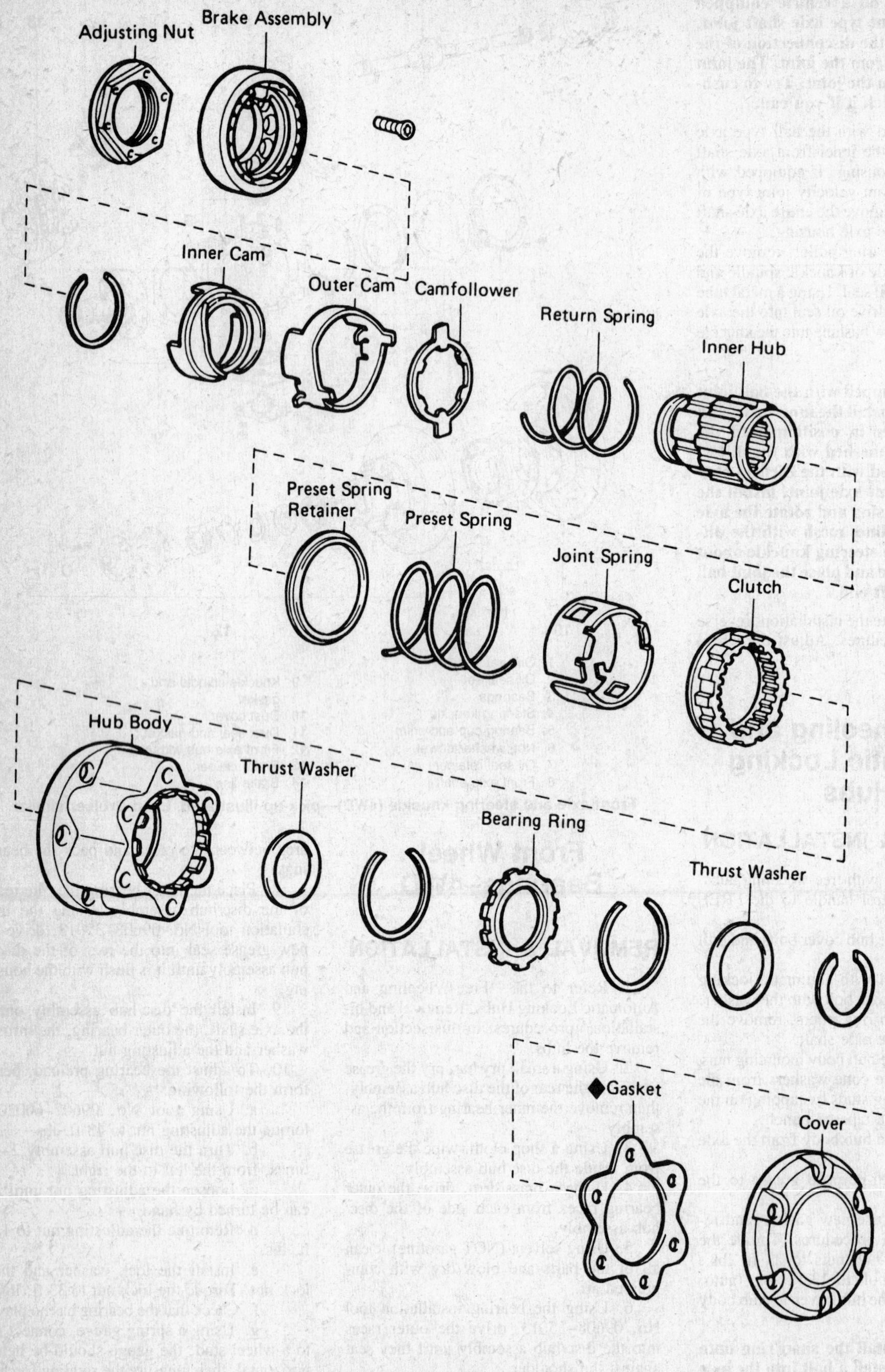

Adjusting Nut

Brake Assembly

Inner Cam

Outer Cam

Camfollower

Return Spring

Inner Hub

Preset Spring Retainer

Preset Spring

Joint Spring

Clutch

Hub Body

Thrust Washer

Bearing Ring

Thrust Washer

◆Gasket

Cover

Exploded view of the automatic locking hub—4WD vehicles

11. To complete the installation, reverse the removal procedures.

Differential

REMOVAL & INSTALLATION

1. Refer to the ''Front Axle Shaft, Removal and Installation'' procedures, in this section and remove the front axle shafts from the axle housing.

2. Drain the lubricant from the differential.

3. Match-mark the front driveshaft flange to the differential flange. Remove the mounting bolts and separate the driveshaft from the differential.

4. Remove the carrier retaining nuts and pull the carrier assembly out of the differential housing.

5. To install, use new gaskets and reverse the removal procedures. Torque the differential-to-axle nuts to 19 ft. lbs. and the front driveshaft flange-to-differential flange nuts/bolts to 54 ft. lbs. Refill the axle with 80W–90 gear oil to a level of ¼ inch below the fill hole.

NOTE: Before installing the carrier, apply a thin coat of liquid or silicone sealer to the carrier housing gasket and to the carrier side face of each carrier retaining nut.

For service information, refer to the Drive Axle section in Unit Repair.

Steering Knuckle

REMOVAL & INSTALLATION

2WD Pickup and Van

1. Raise and support the front of the vehicle on jackstands. Remove the wheel/tire assembly.

2. Remove the brake caliper (DO NOT disconnect the brake hose from the caliper) and suspend it on a wire.

3. Remove axle hub dust cap, the cotter pin, the nut lock, the adjusting nut, the thrust washer and the outer bearing, then pull the hub/disc assembly from the axle spindle.

4. Remove the backing plate cotter pins and the mounting nuts or bolts, then the backing plate.

5. Remove steering knuckle arm from the back of the steering knuckle.

6. Remove the nuts, the retainers and the bushings, then the shock absorber from the lower control arm.

7. Support the lower arm with a jack and raise to put pressure on spring.

──── **CAUTION** ────
Be careful not to unbalance vehicle support stands when jacking up lower arm.

8. Remove cotter pins, then the upper and lower ball joint nuts. Using the ball

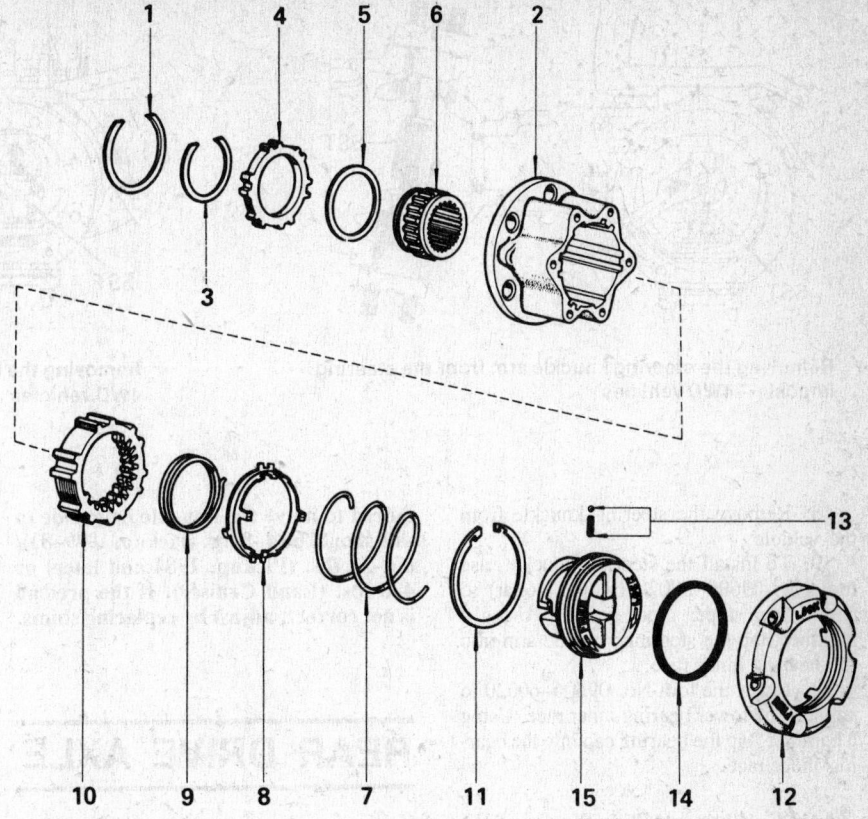

1. Snap-ring	6. Inner hub	11. Snap-ring
2. Free wheel hub body	7. Spring	12. Free wheel hub cover
3. Snap-ring	8. Pawl	13. Steel ball and spring
4. Free wheel hub ring	9. Spring	14. Seal
5. Spacer	10. Clutch	15. Control handle

Exploded view of a free-wheeling hub assembly

joint removal tool No. 09628–62010 (1979–83) or 09628–62011 (1984 and later), separate the ball joints from the steering knuckle.

9. Remove the steering knuckle from the vehicle.

NOTE: Whenever the hub/disc assembly is removed from the vehicle, it is good practice to replace the grease seal.

10. To install, reverse the removal procedures. Torque the upper ball joint nut to 80 ft. lbs. (Pickup) or 58 ft. lbs. (Van), the lower ball joint nut to 105 ft. lbs. (Pickup) or 76 ft. lbs., the steering knuckle arm-to-steering knuckle bolts to 80 ft. lbs. (Pickup) or 61 ft. lbs. (Van), the shock absorber-to-lower control arm nuts to 19 ft. lbs. and the backing plate-to-steering knuckle bolts to 80 ft. lbs. (Pickup) or 61 ft. lbs. (Van). Adjust the wheel bearing.

4WD Pickup and Land Cruiser

1. Refer to the ''Front Axle Shaft, Removal and Installation'' procedures, in this section and remove the front axle.

2. Remove the oil seal retainer and the oil seal set from the rear of the steering knuckle.

3. At the drag link end of the steering knuckle arm, remove the cotter pin. Using a screwdriver, remove the plug from the drag link, then disconnect the drag link from the steering knuckle arm.

4. Remove the tie-rod-to-steering knuckle, cotter pin and nut. Using the Ball Joint Removal tool No. 09611–22012, separate the tie-rod from the steering knuckle arm.

5. Remove the steering knuckle arm-to-steering knuckle (top) nuts and the steering knuckle-to-bearing cap (bottom) nuts. Using a tapered punch, tap the cone washers slits and remove the washers.

──── **CAUTION** ────
DO NOT tap on the bearings.

NOTE: DO NOT mix or lose the upper and lower bearing cap shims.

6. Using the bearing removal tool No. 09606–60020 (without a collar), press the steering knuckle arm with the shims from the steering knuckle.

7. Using the bearing removal tool No. 09606–60020 (without a collar), press the bearing cap with the shims from the steering knuckle.

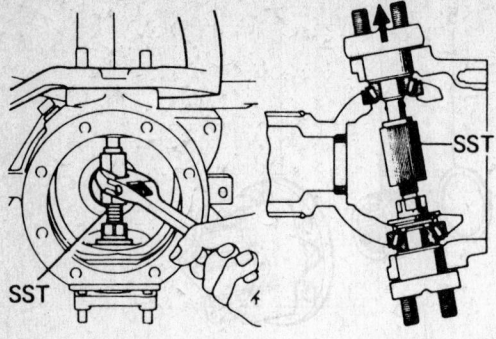

TOYOTA

Removing the steering knuckle arm from the steering knuckle—4WD vehicles

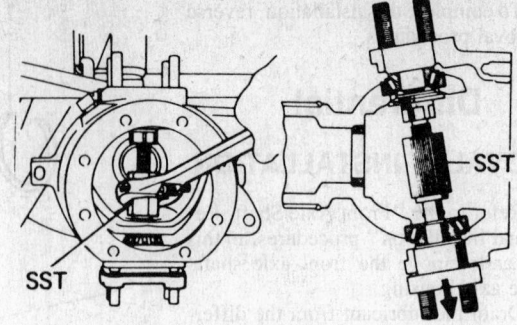

Removing the bearing cap from the steering knuckle—4WD vehicles

8. Remove the steering knuckle from the vehicle.

9. To install the steering knuckle, use tool No. 09606–60020 (with a collar) to support the upper inner bearing. Using a hammer, tap the steering knuckle arm into the bearing inner race.

10. Invert the tool No. 09606–60020 to support the lower bearing inner race. Using a hammer, tap the bearing cap into the bearing inner race.

NOTE: When installing the drag link-to-steering knuckle arm, torque the plug all the way, then loosen it 1⅓ turns and secure it with the cotter pin.

11. To install, use gaskets, seals, pack the steering knuckle with multi-purpose grease and reverse the removal procedures. Torque the steering knuckle arm-to-steering knuckle nuts to 71 ft. lbs., the bearing cap-to-steering knuckle nuts to 71 ft. lbs., the tie-rod-to-steering knuckle arm nut to 67 ft. lbs., the axle spindle-to-steering knuckle bolts to 38 ft. lbs. Adjust the wheel bearing preload.

NOTE: To test the knuckle bearing preload, attach a spring scale to the tie-rod end hole (at a right angle) in the steering knuckle arm. The force required to move the knuckle from side to side should be 4–8 lbs. (Pickup, 1979–83), 6.6–13 lbs. (Pickup, 1984 and later) or 4–5 lbs. (Land Cruiser). If the preload is not correct, adjust by replacing shims.

REAR DRIVE AXLE

Axle Shaft and Bearing

REMOVAL & INSTALLATION

PickUps

1. Loosen the rear wheel lug nuts, then raise and support the vehicle on jackstands. Remove the wheel/tire assembly.

2. Place a pan under the axle, remove the plug and drain the axle housing.

3. For 2WD models, remove the clip/clamp-to-frame bolts and disconnect the parking brake cable from the equalizer. For 4WD models, remove the pin and disconnect the rear parking brake cable from the bell crank.

4. Remove the brake drum securing screw and the drum.

5. Disconnect the brake line from the wheel cylinder and plug it, being careful not to damage the fitting.

6. Remove the brake backing plate-to-axle housing nuts and pull the backing plate with the axle from the axle housing.

— **CAUTION** —
When removing the axle shaft, be careful not to damage the oil seal.

7. Using a pair of snap ring pliers, remove the snap ring from the axle shaft.

8. Slip the tool No. 09521–25011 over the axle shaft and fasten it to the backing plate. Using two metal blocks and a press, press the axle from the backing plate assembly.

9. If necessary to remove the bearing from backing plate, perform the following:

   a. Remove the brake spring, the retracting spring clamp bolt, the lower springs, the shoe strut, the brake shoes and the parking brake lever.

   b. Using a slide hammer puller and the removal tool No. 09308–00010, pull the outer oil seal from the backing plate.

   c. Using the removal tools No. 09228–44010 and 09608–30011, press the bearing from the backing plate.

   d. Using the installation tools No. 09515–30010 and 09608–35013, press the new bearing into the backing plate.

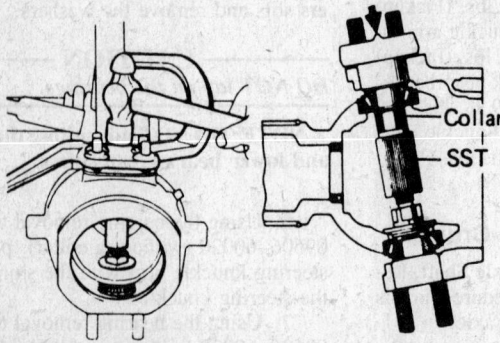

Installing the steering knuckle arm to the steering knuckle—4WD vehicles

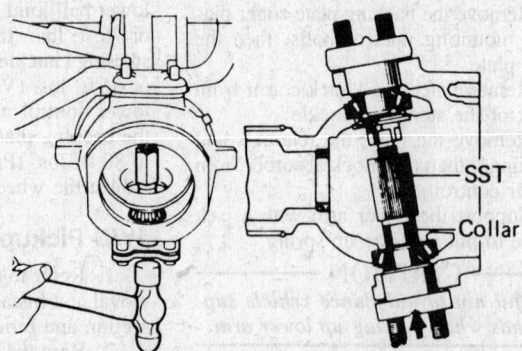

Installing the bearing cap to the steering knuckle—4WD vehicles

T584

e. Using the installation tool No. 09608–30011, press the new oil seal into the backing plate.

f. Reassemble the brake components to the backing plate.

10. Using a slide hammer and the removal tool No. 09308–00010, pull the oil seal from the axle housing.

11. Using the installation tool and a hammer, drive a new oil seal into the axle housing.

12. Using a press and the installation tool No. 09515–30010, press the axle shaft into the backing plate and the bearing retainer. Using snap ring pliers, install the snap ring onto the axle shaft.

13. Using a putty knife, clean the gasket mounting surfaces.

14. To complete the installation, reverse the removal procedures. Torque the backing plate-to-axle housing nuts to 51 ft. lbs. Adjust the brake shoe clearance and bleed the brake system. Refill the axle housing with SAE 90W GL5 gear oil.

Land Cruiser

SEMI-FLOATING TYPE DIFFERENTIAL

1. Loosen the rear wheel nuts. Raise and support the rear axle housing on jackstands. Remove the wheel/tire assembly.

2. Place a pan under the axle, remove the plug and drain the oil from the differential.

3. Remove the brake drum and related parts, as follows:

a. Remove the cover from the back of the differential housing.

b. Remove the pin from the differential pinion shaft.

c. Withdraw the pinion shaft and it's spacer from the case.

d. Use a mallet to tap the rear axle shaft toward the differential, then remove the C-lock from the axle shaft.

e. Withdraw the axle shaft from the housing.

4. Using a bearing puller, remove axle bearing and oil seal together from the axle housing. Using a metal tube and a hammer, drive the bearing and the seal into the housing until they seat.

CAUTION

DO NOT mix the parts of the left and right axle shaft assemblies.

---

5. To complete the installation, reverse the removal procedures. Refill the axle housing with SAE 90W GL5 gear oil.

NOTE: After installing the axle shaft, C-lock, spacer and pinion shaft, measure the clearance between the axle shaft and the pinion shaft spacer with a feeler gauge. The clearance should fall between 0.0024–0.0181 in. If the clearance is not within specifications, use one of the following spacers to adjust it:

a. 1.172–1.173 in.
b. 1.188–1.189 in.
c. 1.204–1.205 in.

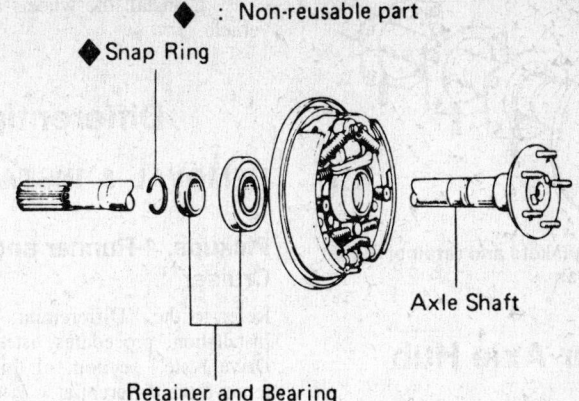

Exploded view of the rear axle assembly—Pick-ups and 4-Runner

FULL FLOATING TYPE DIFFERENTIAL

1. Loosen the rear wheel nuts. Raise and support the rear axle housing on jackstands. Remove the wheel/tire assembly.

2. Place a pan under the axle, remove the plug and drain the oil from the differential.

3. Remove the rear axle shaft plate nuts.

4. Remove the cone washers from the mounting studs by tapping the slits of the washers with a tapered punch.

5. Install bolts into the two unused holes of the axle shaft plate.

6. Tighten the bolts to draw the axle shaft assembly out of the housing.

7. To install, use a new gasket, sealant and reverse the removal procedures. Torque the axle shaft nuts to 21–25 ft. lbs.

Van

1. Loosen the rear wheel nuts. Raise and support the rear axle housing on jackstands. Remove the wheel/tire assembly.

2. Working through the hole in the axle flange, remove the backing plate-to-axle housing bolts.

3. Using a slide hammer puller and removal tool No. 09520–00031, pull the axle shaft from the housing.

4. Using a grinder, grind down the inner bearing retainer on the axle shaft. Using a chisel and a hammer, cut off the retainer and remove it from the shaft.

5. Using a arbor press and the removal tools 09527–21011, press the bearing from the axle shaft.

6. Using a slide hammer puller and the removal tool 09308–00010, pull the oil seal from the axle housing.

7. Lubricate the new oil seal with multipurpose grease. Using the installation tool No. 09517–30010 and a hammer, drive the new oil seal into the axle housing, to a depth of 0.236 in.

8. To install, use new gaskets and reverse the removal procedures. Torque the axle retainer-to-housing bolts to 48 ft. lbs.

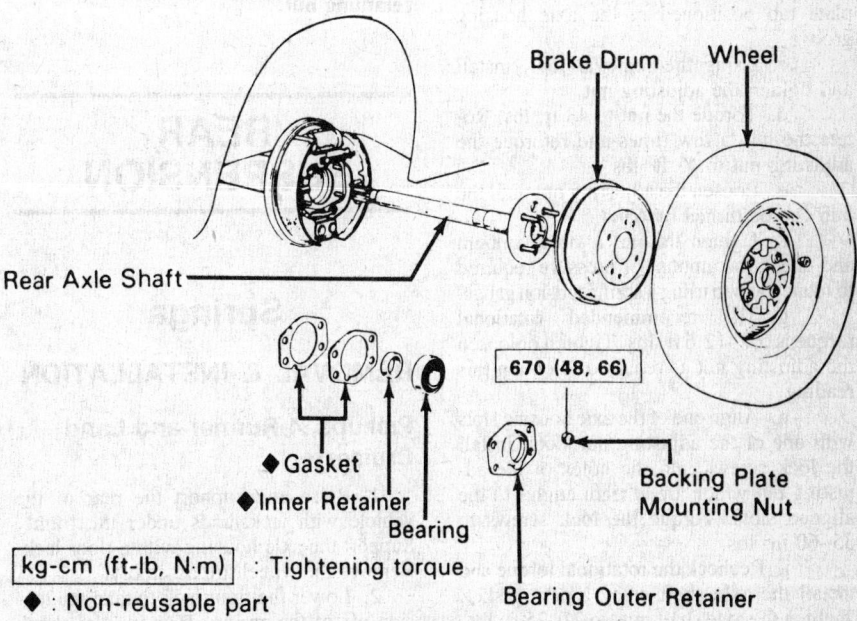

Exploded view of the rear axle assembly—Van

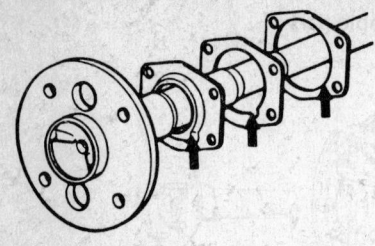

Aligning the gaskets and retainer of the axle shaft—Van

Rear Axle Hub

REMOVAL & INSTALLATION

Land Cruiser
FULL FLOATING TYPE DIFFERENTIAL

1. Refer to the "Axle Shaft, Removal and Installation" procedures, in this section and remove the axle shaft.

2. Loosen the lock screws. Using tool No. 09509–25011 and remove the adjusting nut from inside the hub.

3. Remove the hub from the axle housing. Inspect all parts for damage or excessive wear. Using a small pry bar, pry the oil seal from the hub.

4. Using a brass drift and a hammer, tap oil seal into the hub until it is firmly seated.

5. If the bearing race(s) needs replacement, drive the race(s) from the hub using a brass drift. Using the brass drift and a hammer, drive the new race(s) into the hub, until firmly seated.

6. To install the hub, perform the following procedures:

   a. Place the hub on the axle housing and install the outer bearing.

   b. Install the lock plate with the lock plate tab positioned in the axle housing groove.

   c. Using the removal tool, install and tighten the adjusting nut.

   d. Torque the nut to 43 ft. lbs. Rotate the hub a few times and retorque the adjusting nut to 43 ft. lbs.

   e. Loosen the adjusting nut until the hub can be turned by hand.

   f. Tighten the nut a small amount and check the amount of pressure required to rotate the hub using a spring tension gauge.

   g. The recommended rotational torque is 5.7–12.6 ft. lbs. Tighten or loosen the adjusting nut as required to obtain this reading.

   h. Align one of the axle housing slots with one of the adjusting nut slots. Install the lock screws into the holes of the adjusting nut which are at right angles to the aligned slots. Torque the lock screws to 35–60 in. lbs.

   i. Recheck the rotational torque and install the axle shaft using a new gasket. Tighten the axle shaft nuts to 21–25 ft. lbs.

j. Install the wheels and lower the vehicle.

Differential

REMOVAL & INSTALLATION

Pickups, 4-Runner and Land Cruiser

Refer to the "Differential, Removal and Installation" procedures, listed in the "Front Drive Axle" section, of this section and remove the differential.

Van

1. Refer to the "Front Axle Shaft, Removal and Installation" procedures, in this section and remove the front axle shafts from the axle housing.

2. Drain the lubricant from the differential.

3. Match-mark the front driveshaft flange to the differential flange. Remove the mounting bolts and separate the driveshaft from the differential.

4. Remove the carrier retaining nuts and pull the carrier assembly out of the differential housing.

5. To install, use new gaskets and reverse the removal procedures. Torque the differential-to-axle nuts to 23 ft. lbs. and the front driveshaft flange-to-differential flange nuts/bolts to 31 ft. lbs. Refill the axle with 80W–90 gear oil to a level of ¼ inch below the fill hole.

NOTE: Before installing the carrier, apply a thin coat of liquid or silicone sealer to the carrier housing gasket and to the carrier side face of each carrier retaining nut.

REAR SUSPENSION

Springs

REMOVAL & INSTALLATION

Pickups, 4-Runner and Land Cruiser

1. Raise and support the rear of the vehicle with jackstands under the frame. Support the axle housing with a floor jack. Remove the wheel/tire assembly.

2. Lower the floor jack to take the tension off of the spring. Remove the shock

absorber mounting nuts/bolts and the shock absorber.

3. On the Land Cruiser models, perform the following:

   a. Remove the cotter pins and the nuts from the lower end of the stabilizer link.

   b. Detach the link from the axle housing.

4. Remove the spring-to-axle housing U-bolt nuts, the spring seat (2WD) or spring bumper (4WD) and the U-bolt.

5. At the front of the spring, remove the hanger pin bolt. Disconnect the spring from the bracket.

6. Remove the spring shackle retaining nuts and the spring shackle inner plate, then carefully pry out the spring shackle with a pry bar.

7. Remove the spring from the vehicle.

--- **CAUTION** ---
Use care not to damage the hydraulic brake line or the parking brake cable.

8. To install, perform the following procedure:

   a. Install the rubber bushings in the eye of the spring.

   b. Align the eye of the spring with the spring hanger bracket and drive the pin through the bracket holes and rubber bushings.

NOTE: Use soapy water as lubricant (if necessary), to aid in pin installation. Never use oil or grease.

   c. Finger-tighten the spring hanger nuts/bolts.

   d. Install the rubber bushings in the spring eye at the opposite end of the spring.

   e. Raise the free end of the spring. Install the spring shackle through the bushings and the bracket.

   f. Install the shackle inner plate and finger-tighten the retaining nuts.

   g. Center the bolt head in the hole which is provided in the spring seat on the axle housing.

   h. Fit the U-bolts over the axle housing. Install the lower spring seat (2WD) or spring•bumper (4WD) and the nuts.

9. To complete the installation, reverse the removal procedures. Torque the U-bolt nuts to 72 ft. lbs. (2WD) or 90 ft. lbs. (4WD), the hanger pin-to-frame nut to 67 ft. lbs., the shackle pin nuts to 67 ft. lbs., the shock absorber bolts to 19 ft. lbs. (2WD) or 47 ft. lbs. (4WD).

NOTE: When installing the U-bolts, tighten the nuts so that the length of the bolts are equal.

Van

1. Raise and support the rear of the vehicle with jackstands under the frame. Support the axle housing with a floor jack. Remove the wheel/tire assembly.

2. Remove the shock absorber-to-axle housing bolt.

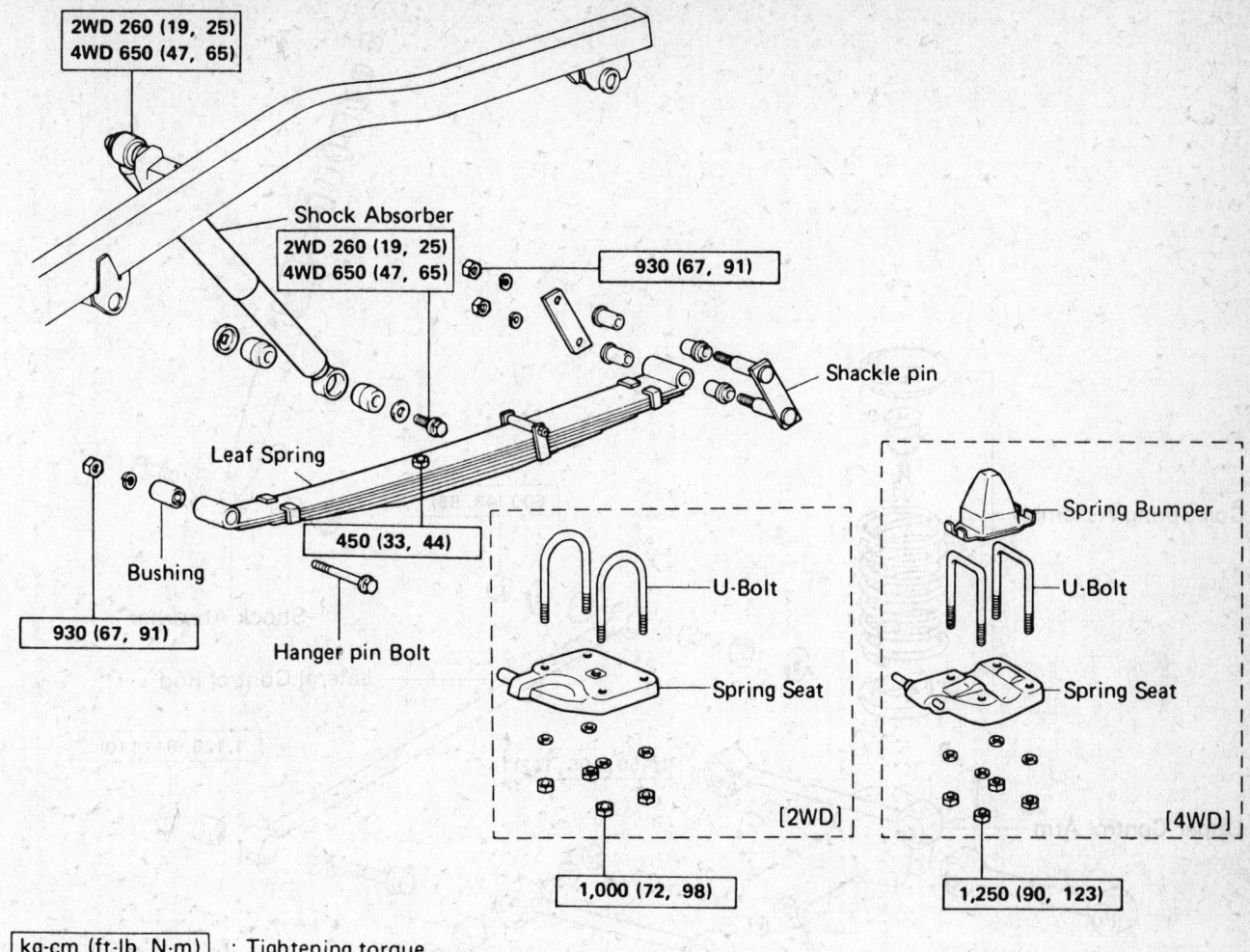

2WD 260 (19, 25)
4WD 650 (47, 65)

Shock Absorber
2WD 260 (19, 25)
4WD 650 (47, 65)

930 (67, 91)

Shackle pin

Leaf Spring

Spring Bumper

Bushing

450 (33, 44)

U-Bolt

U-Bolt

930 (67, 91)

Hanger pin Bolt

Spring Seat

Spring Seat

[2WD]

[4WD]

1,000 (72, 98)

1,250 (90, 123)

kg-cm (ft-lb, N·m) : Tightening torque

Exploded view of the rear suspension—Pick-ups, 4-Runner and Land Cruiser

3. Remove the stabilizer-to-axle housing bar bushing bracket bolts.

4. Remove the lateral control arm-to-axle housing nut and disconnect the lateral control arm.

5. Lower the floor jack, then remove the coil spring(s) and the insulators.

NOTE: While lowering the axle housing, be careful not to snag the brake line of the parking brake cable.

6. To install, reverse the removal procedures. Torque the shock absorber bolt to 27 ft. lbs., the lateral control arm-to-axle housing nut to 43 ft. lbs. and the stabilizer-to-axle housing bolts to 27 ft. lbs.

NOTE: Before tightening the lateral control arm and the stabilizer nuts/bolts, bounce the vehicle to stabilize the suspension.

Shock Absorbers

REMOVAL & INSTALLATION

Pickups, 4-Runner and Land Cruiser

1. Raise and support the rear of the vehicle on jackstands.

3. Remove the upper shock absorber retaining bolts from the upper frame member.

4. Remove the lower end bolt of the shock absorber from the spring seat.

5. Remove the shock absorber from the vehicle.

NOTE: Inspect the shock for wear, leaks or other signs of damage.

6. To install, reverse the removal procedures. Torque the upper bolt to 19 ft. lbs.

(2WD) or 47 ft. lbs. (4WD) and the lower bolt to 19 ft. lbs. (2WD) or 47 ft. lbs. (4WD).

Van

1. Raise and support the rear of the vehicle on jackstands.

2. Remove the shock absorber-to-axle housing bolt.

3. Working inside the vehicle, remove the lock nut, the retaining nut, the retainers and the rubber bushings from the top of the shock absorber.

NOTE: When removing the retaining nut, from the top of the shock absorber, it may be necessary to hold the top of the shock with a screwdriver, to keep it from turning.

4. Remove the shock absorber from the vehicle.

Coil Spring Assembly

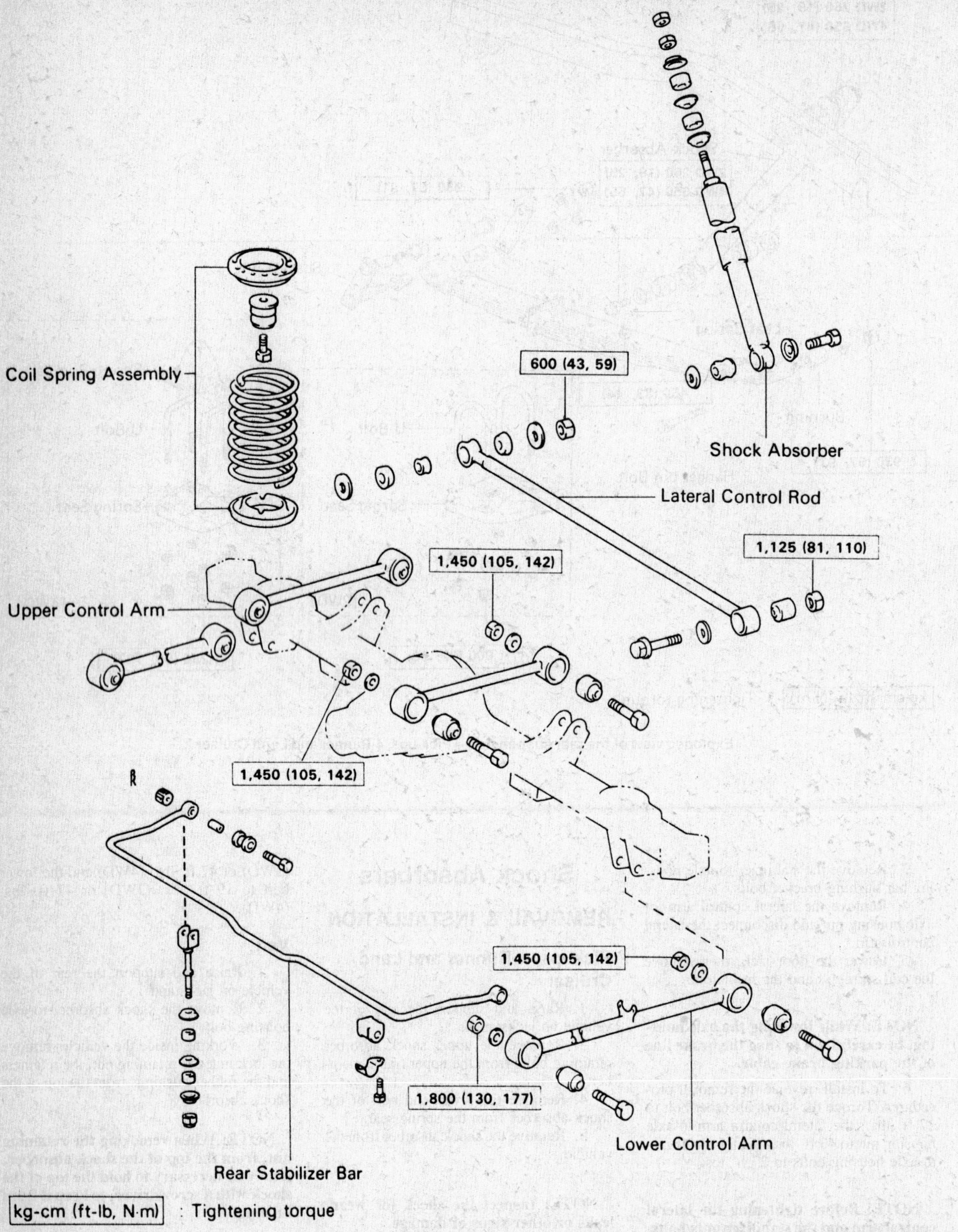

Upper Control Arm

600 (43, 59)

Shock Absorber

Lateral Control Rod

1,125 (81, 110)

1,450 (105, 142)

1,450 (105, 142)

1,450 (105, 142)

Lower Control Arm

1,800 (130, 177)

Rear Stabilizer Bar

kg-cm (ft-lb, N·m) : Tightening torque

Exploded view of the rear suspension—Van

5. To install, reverse the removal procedures. Torque the shock absorber-to-body nut to 16–24 ft. lbs. and the shock absorber-to-axle housing bolt to 27 ft. lbs.

Rear Control Arms

REMOVAL & INSTALLATION

Van

1. Raise and support the rear of the vehicle, with jackstands under the frame. Place a floor jack under the axle housing to support it.

2. Remove the upper control arm-to-body bolt, the upper control arm-to-axle housing bolt and the upper control arm from the vehicle.

3. Disconnect the brake line from the lower control arm.

4. Remove the lower control arm-to-body bolt, the lower control arm-to-axle housing bolt and the lower control arm from the vehicle.

5. Install the upper control arm to the body and to the axle housing with the nuts. DO NOT tighten the nuts.

6. Install the lower control arm to the body and to the axle housing with the nuts. DO NOT tighten the nuts.

7. Remove the jack and the supports from under the vehicle. Bounce the vehicle to stabilize the suspension.

8. Using the floor jack under the axle housing, raise the vehicle. Place the jackstands under the frame but DO NOT let them touch the frame.

9. To complete the installation, torque the upper control arm-to-body bolt to 105 ft. lbs., the upper control arm-to-axle housing bolt to 105 ft. lbs., the lower control arm-to-body bolt to 130 ft. lbs. and the lower control arm-to-axle housing bolt to 105 ft. lbs.

Lateral Control Rod

REMOVAL & INSTALLATION

Van

1. Raise and support the rear of the vehicle with jackstands under the frame. Place a floor jack under the axle housing and support it.

2. Remove the lateral control rod-to-axle housing nut.

3. Remove the lateral control rod-to-body nut and the control rod from the vehicle.

4. To install, raise the axle housing until the frame is just free of the jackstands.

5. Install the lateral control rod-to-body with the nut. DO NOT tighten the nut.

6. Install the lateral control rod-to-axle housing in the following order: washer, bushing, spacer, lateral control rod, bush-

ing, washer and nut. DO NOT tighten the nut.

7. Remove the jackstands, lower the vehicle to the floor and bounce it to stabilize the suspension.

8. Using the floor jack under the axle housing, raise the vehicle. Torque the lateral control rod-to-body nut to 81 ft. lbs. and the lateral control rod-to-axle housing nut to 43 ft. lbs.

FRONT SUSPENSION

Springs

REMOVAL & INSTALLATION

2WD Pickup and Van

These models are equipped with torsion bar front springs.

--- **CAUTION** ---

Great care must be taken to make sure springs are not mixed after removal. It is strongly suggested that before removal, each spring be marked with paint, showing front and rear of spring and from which

side of the truck it was taken. If the springs are installed backwards or on the wrong sides of the truck, they could fracture. If replacing the springs, it is not necessary to mark them.

1. Raise and support the front of the vehicle on jackstands.

2. Slide the boot from the rear of torsion bar spring, then paint an alignment mark from the torsion bar spring onto the anchor arm and the torque arm. There are right and left identification marks on the rear end of the torsion bar springs.

--- **CAUTION** ---

Be sure to mark the front of spring from back of spring.

3. On the rear torsion bar spring holder, there is a long bolt that passes through the arm of the holder and up through the frame crossmember. REMOVE THE LOCKING NUT ONLY FROM THIS BOLT.

4. Using a small ruler, measure the length from the bottom of the remaining nut to the threaded tip of the bolt and record this measurement.

5. Place a jack under the rear torsion bar spring holder arm and raise the arm to remove the spring pressure from the long bolt. Remove the adjusting nut from the long bolt.

6. SLOWLY lower jack.

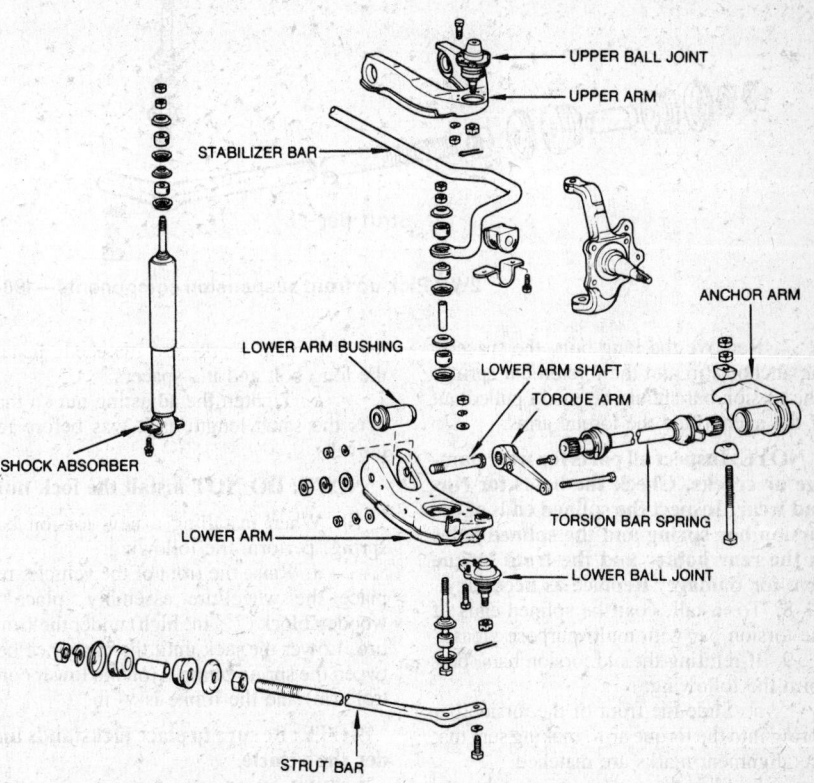

2WD Pick-up front suspension components—Pick-up (1979–83)

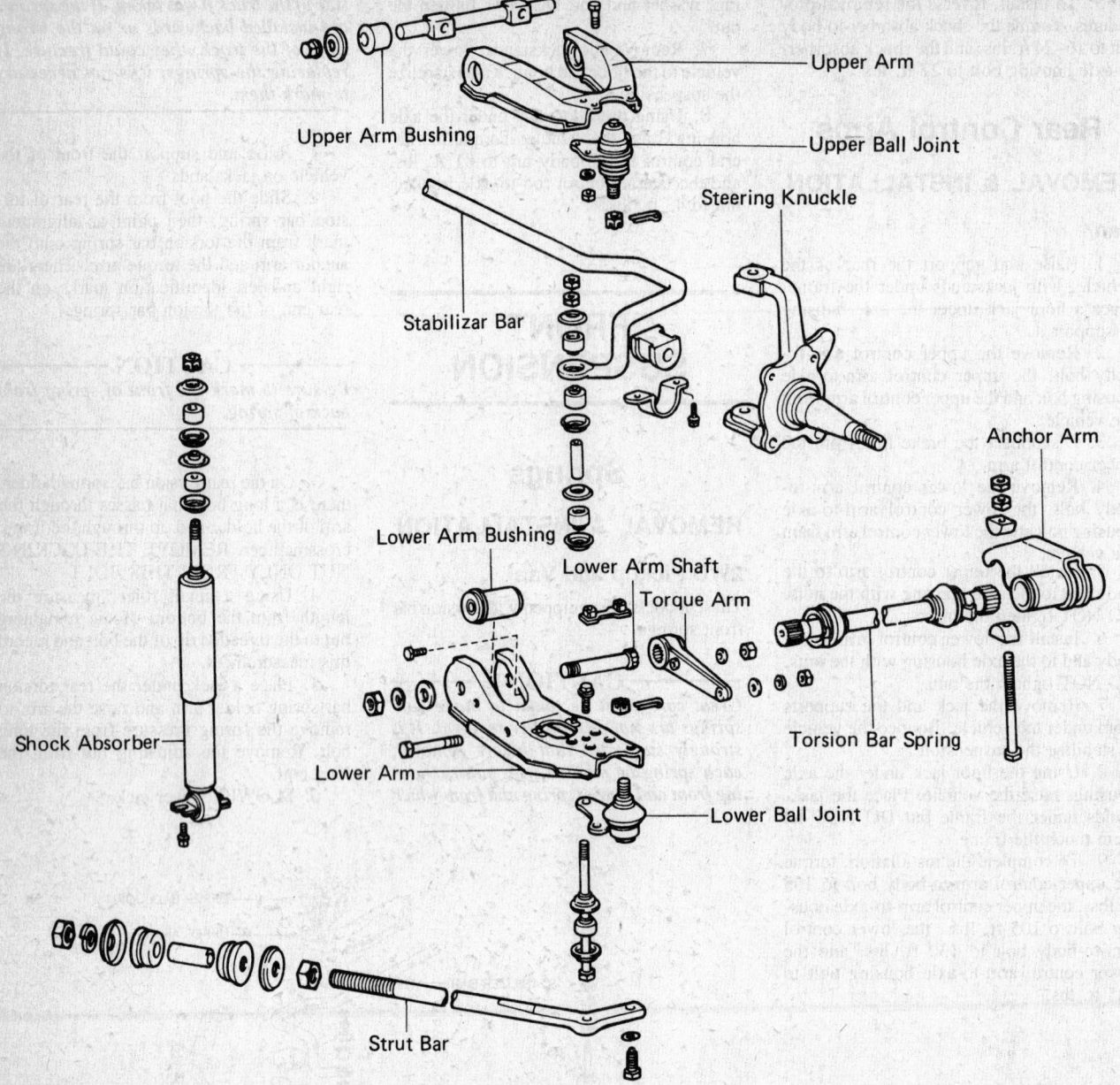

Upper Arm

Upper Arm Bushing

Upper Ball Joint

Steering Knuckle

Stabilizar Bar

Anchor Arm

Lower Arm Bushing

Lower Arm Shaft

Torque Arm

Shock Absorber

Torsion Bar Spring

Lower Arm

Lower Ball Joint

Strut Bar

2WD Pick-up front suspension components—1984 and later

7. Remove the long bolt, the spacers, the anchor arm and the torsion bar spring. The torsion bar should be easily pulled out of the anchor and the torque arms.

NOTE: Inspect all parts for wear damage or cracks. Check the boots for rips and wear. Inspect the splined ends of the torsion bar spring and the splined holes in the rear holder and the front torque arm for damage. Replace as necessary.

8. To install, coat the splined ends of the torsion bar with multi-purpose grease.

9. If refitting the old torsion bars, perform the following:

    a. Slide the front of the torsion bar spring into the torque arm, making sure that the alignment marks are matched.

    b. Slide the anchor arm onto the rear of the torsion bar spring, making sure that the alignment marks are matched. Install the long bolt and it's spacers.

    c. Tighten the adjusting nut so that it is the same length as it was before removal.

NOTE: DO NOT install the lock nut.

10. When installing a new torsion bar spring, perform the following:

    a. Raise the front of the vehicle, replace the wheel/tire assembly, place a wooden block (7½ in. high) under the front tire. Lower the jack until the clearance between the spring bumper (on the lower control arm) and the frame is ½ in.

NOTE: Be sure to place jackstands under the vehicle.

    b. **Slide the front of the torsion bar spring into the torque arm.**

Install the anchor arm into the rear of the torsion bar spring, then the long bolt and the spacers. the distance from the top of the upper spacer to the tip of the threaded end of bolt is 0.31–1.10 in. (½ ton vehicles) or 0.43–1.22 in. (¾ ton vehicles).

NOTE: Make sure the bolt and bottom spacer are snuggly in the holder arm while measuring.

    d. Remove the wooden block and lower the vehicle until it rests on the jackstands.

    e. Install and tighten the adjusting nut until the distance from the bottom of the nut to the tip of the threaded end of the bolt is 2.7–3.5 in.

NOTE: DO NOT install the lock nut.

11. Apply multi-purpose grease to the

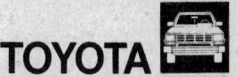
boot lips, then refit the boots to the torque and the anchor arms.

12. Lower the vehicle to the floor and bounce it several times to settle the suspension. With the wheels on the ground, measure the distance from the ground to the center of the lower control arm-to-frame shaft. Adjust the vehicle height using the adjusting nut on the anchor arm.

NOTE: If, after achieving the correct vehicle height, the distance from the bottom of the adjusting nut to the top of the threaded end of the long bolt is not within 2.7–3.5 in., change the position of the anchor arm-to-tension bar spring spline and reassemble.

13. Install and torque the lock nut on the long bolt to 61 ft. lbs.

CAUTION
Make sure the adjusting nut does not move when tightening lock nut.

4WD Pickup and 4-Runner

1. Raise and support the front of vehicle with jackstands under the frame.

VEHICLE HEIGHT

Year	Model	Pay Load	Tire Size	Front Height (in.) (Unloaded)
'79–'80	RN32L	½ Ton	185 SR 14-4PR	9.827
	RN42L		7.00-14-6PR	10.291
			E78-14 (B)	10.016
			ER78-14 (B)	9.866
	RN42L-KH	¾ Ton	7.50-14-6PR	10.961
	RN42L-3W (C & C)	¾ Ton	7.50-14-6PR	10.961
'81–'83	RN34	½ Ton	7.00-14-6PR	10.291
	RN44		E78-14 (B)	10.016
			ER78-14 (B)	9.866
			205/70 SR 14	9.512
	RN44L-KH	¾ Ton	7.50-14-6PR	10.961
	RN44L-3W C & C		7.50-14-6PR	10.961
'84	Short Bed (Std)		7.00-14-6PR	10.59
			ER78-14	10.04
	Long Bed (Std)		7.00-14-6PR	10.75
			ER78-14	10.20
	Long Bed (Soft Ride)		ER78-14	10.20
	Extra Cab (Soft Ride)		ER78-14	9.80
	Extra Cab (Std)		ER78-14	9.80
	¾ Ton		7.50-14-6PR	10.71
	C & C		7.50-14-6PR	10.83
	SR-5 (Short)		P195/75 R 14	9.76
			205/70 R 14	10.00
			ER78-14	10.00
	SR-5 (Long)		P195/75 R 14	9.96
			205/70 SR 14	10.20
			ER78-14	10.12
	Extra Cab SR5 (Long Bed)		P195/75 R 14	9.80
			205/70 SR 14	10.04
			ER78-14	9.96
'85–'86	Short Bed (Std)		7.00-14-6PR	10.63
	Long Bed (Std)		7.00-14-6PR	10.83
	Long Bed (Soft Ride)		P195/75 R 14	10.24
	Extra Cab (Soft Ride)		P195/75 R 14	9.84
	Extra Cab (Std)		P195/75 R 14	9.84
	1 Ton		185 R 14-LT8PR	10.31
	C & C		185 R 14-LT8PR	10.20
	Short Bed (SR-5)		P195/75 R 14	9.80
			205/70 SR 14	10.04
	Long Bed (SR-5)		P195/75 R 14	9.96
			205/70 SR 14	10.20
	Extra Cab SR-5 (Long Bed)		P195/75 R 14	9.84
			205/70 SR 14	10.08
'85–'86 (Diesel)	Short Bed (Std)		7.00-14-6PR	10.59
	Long Bed (Soft Ride)		P195/75 R 14	9.96
	Extra Cab (Soft Ride)		P195/75 R 14	9.80
	Extra Cab (Std)		P195/75 R 14	9.80
'83–'86	Van		P185/75 R 14	9.57

NOTE: DO NOT place the supports under the front axle housing.

2. Remove the wheel/tire assembly. Lower the axle housing until the tension is removed from the spring.

3. Remove the shock absorber-to-spring seat bolt and raise the shock up, out of the way.

4. If removing the driver's side front leaf spring, remove the cotter pin from the end of the steering drag link at the axle housing. Using a screwdriver, remove the plug from the end of the drag link.

5. Remove stabilizer bar-to-axle housing nut, bolt, spacer and washer assemblies. Remove the stabilizer bar-to-frame mounting clamps.

6. Disconnect brake line from the brake backing plate. Drive out shim holding brake line to holder and withdraw brake line. Plug end of brake line running to master cylinder to prevent fluid loss.

7. Place a jack under the front axle housing and raise to put pressure on the leaf spring. Remove the U-bolt nuts, the spring seat, the U-bolts and the spring bumper. Disconnect the drag link from the steering knuckle arm.

8. Lower the jack enough to take the pressure off the leaf spring but allow it to support the axle housing.

9. Remove the hanger pin nut/bolt (at the front of the spring) and the shackle pin nut/bolt (at the rear of the spring), then carefully pry the spring from retainers.

NOTE: It may be necessary to lower the jack under the axle housing to remove spring.

10. To install, reverse the removal procedures. Torque the U-bolt nuts to 90 ft.

lbs., the front hanger pin bolts to 8–11 ft. lbs., the front hanger pin nut to 67 ft. lbs., the rear shackle pin nuts to 67 ft. lbs., the shock absorber-to-body nuts to 19 ft. lbs., the shock absorber-to-spring seat bolt to 70 ft. lbs., the stabilizer bar clamps-to-frame bolts to 9 ft. lbs. and the stabilizer bar-to-axle housing nuts to 19 ft. lbs. Refill the brake master cylinder and bleed the system.

NOTE: Finger-tighten the hanger and the shackle pin nuts. Lower the vehicle to the floor and bounce it to stabilize the suspension.

Land Cruiser

Land Cruiser models are equipped with leaf springs in the front and rear. Thus, front spring removal is performed in almost the same manner as rear spring removal.

Refer to the "Rear Spring, Removal and Installation" procedures, in this section and remove the front spring in the same manner.

—————— CAUTION ——————
Be careful when raising or lowering the front suspension with a jack so as not to damage the steering system components.

Shock Absorber

REMOVAL & INSTALLATION

2WD Pickup and Van

1. Raise and support the front of the vehicle on jackstands. Remove the wheel/tire assembly.

2. Unfasten the double nuts at the top end of the shock absorber. Remove the cushions and the cushion retainers.

3. Remove the shock absorber-to-lower control arm bolts.

4. Compress the shock absorber and remove it from the vehicle.

5. To install, reverse the removal procedures. Torque the shock absorber-to-lower control arm bolts to 13 ft. lbs. and the shock absorber-to-body nuts to 19 ft. lbs.

4WD Pickup and 4-Runner

1. Raise and support the front of the vehicle on jackstands. Remove the wheel/tire assembly.

2. Unfasten the double nuts at the top end of the shock absorber. Remove the cushions and the cushion retainers.

3. Remove the shock absorber-to-axle housing bolt.

4. Compress the shock absorber and remove it from the vehicle.

5. To install, reverse the removal procedures. Torque the shock absorber-to-axle housing bolt to 33 ft. lbs. (1979–83) or 70 ft. lbs. (1984 and later) and the shock absorber-to-body nuts to 19 ft. lbs.

Land Cruiser

1. Raise and support the front of the vehicle on jackstands. Remove the wheel/tire assembly.

—————— CAUTION ——————
Be careful not to damage steering assembly when raising the front of the vehicle.

2. Remove mounting bolts from the top and the bottom of the shock and remove shock.

3. To install, reverse the removal procedures.

Stabilizer Bar

REMOVAL & INSTALLATION

2WD Pickup and Van

1. Refer to the "Spring, Removal and Installation" procedures, in this section and remove one of the torsion bar springs from the vehicle.

2. Remove the stabilizer bar-to-lower control arm nuts, the cushions and the bolts.

NOTE: Be sure to arrange the hardware as originally installed.

3. Remove the stabilizer bar-to-frame brackets and bushings and lower the stabilizer bar from the vehicle.

4. To install, reverse the removal procedures. Be sure to carefully inspect each bushing for damage and replace the bushing(s) (if necessary). Torque the stabilizer bar-to-lower control arm nuts to 9 ft. lbs. (Pickup) or 0.51–0.63 in. (Van) and the

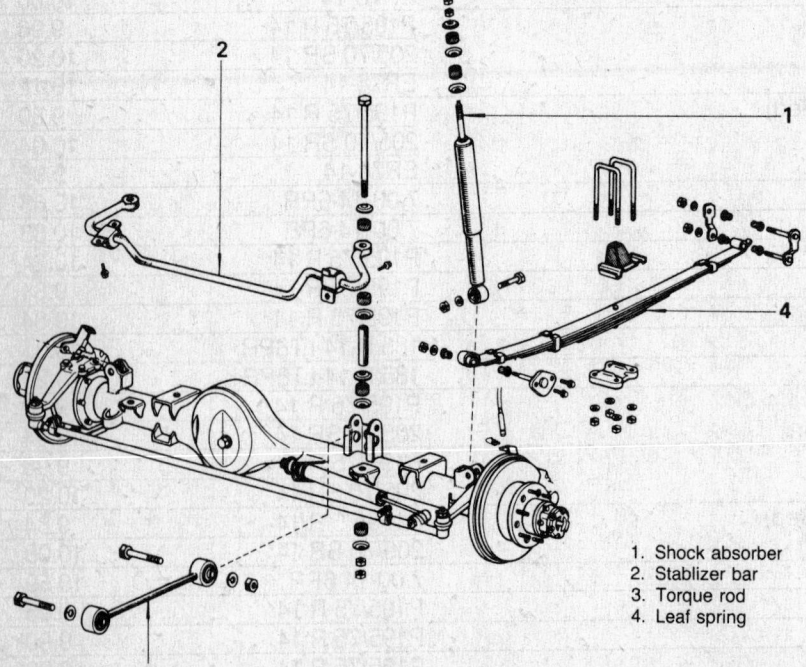

1. Shock absorber
2. Stablizer bar
3. Torque rod
4. Leaf spring

Front suspension components (4WD)—pick-up shown, Land Cruiser similar stabilizer bar-to-frame bracket bolts to 9 ft. lbs. (Pickup) or 14 ft. lbs. (Van). Adjust the tension bar spring and the vehicle heights.

NOTE: When torquing the stabilizer bar-to-lower control arm (on the Van), the distance from the end of the bolts to the top of the nut should be 0.51–0.63 in.

4WD Pickup, 4-Runner and Land Cruiser

1. Remove the stabilizer bar-to-axle housing nuts, the cushions and the bolts.

NOTE: Be sure to arrange the hardware as originally installed.

2. Remove the stabilizer bar-to-frame brackets and bushings and lower the stabilizer bar from the vehicle.

3. To install, reverse the removal procedures. Be sure to carefully inspect each bushing for damage and replace the bushing(s) (if necessary). Torque all stabilizer bar-to-frame bracket bolts to 9 ft. lbs. and the stabilizer bar-to-axle housing nuts to 19 ft. lbs.

Strut Bar

REMOVAL & INSTALLATION

2WD Pick-up and Van

1. Raise and support the vehicle on jackstands.

2. Place match-marks on the strut bar-to-inner mounting nut at the frame bracket.

3. Remove the front mounting nut from the strut bar.

4. Remove the strut bar-to-lower control arm.

5. To install, reverse the removal procedures. Torque the strut bar-to-lower control arm bolts to 70 ft. lbs. (Pickup) or 49 ft. lbs. (Van) and the strut bar-to-frame bracket nut to 90 ft. lbs. Check the front-end alignment.

Torque Rod

REMOVAL & INSTALLATION

4WD Pickup, 4-Runner and Land Cruiser

1. Raise and support the front of the vehicle on jackstands.

2. Remove the torque rod-to-axle housing nut/bolt.

3. Remove the torque rod-to-frame bracket nut/bolt.

4. Remove the torque rod from the vehicle.

5. To install, reverse the removal procedures. Torque the torque rod nuts and bolts to 105 ft. lbs.

Upper Control Arm

REMOVAL & INSTALLATION

2WD Pickup

1. Raise and support the vehicle on jackstands. Remove the wheel/tire assembly.

2. Remove the caliper and suspend it from the frame.

3. Raise the lower control arm with a jack.

4. Remove the nut from the upper ball joint stud.

5. Using the Ball Joint Removal tool No. 09628–62011, separate the ball joint from the steering knuckle. Be careful not to damage the ball joint boot.

6. Unbolt and remove the upper arm at the two bolts holding the inner shaft to the frame, taking note of the number and size of the aligning shims.

7. To install, reverse the removal procedures. Replace the shims in their original positions. Tighten fasteners but DO NOT torque them until the vehicle is on the ground.

8. Lower the vehicle to the ground and bounce it several times to align the suspension.

9. Torque the upper control arm-to-body bolts to 51–65 ft. lbs. (1979–83), 72 ft. lbs. (1984 and later) and the upper ball joint-to-steering knuckle nut to 80 ft. lbs.

Van

1. Refer to the "Spring, Removal and Installation" procedures, in this section and remove a torsion bar spring.

2. Remove the cool air intake duct.

3. Remove the upper ball joint-to-steering knuckle cotter pin and nut. Using the Ball Joint Removal tool No. 09628–62011, pull the ball joint from the steering knuckle.

NOTE: When separating the ball joint from the steering knuckle, be careful not to damage the ball joint boot.

4. Remove the upper control arm-to-frame bolts and the upper control arm from the vehicle.

5. To install, reverse the removal procedures. Torque the upper control arm-to-frame rear bolt to 112 ft. lbs., the upper control arm-to-frame front bolt to 65 ft. lbs. and the ball joint-to-steering knuckle nut to 58 ft. lbs. Check the front end alignment.

Lower Control Arm

REMOVAL & INSTALLATION

2WD Pickup

1. Refer to the "Spring, Removal and Installation" procedures, in this section and remove a torsion bar spring.

2. Remove the stablizer bar and the strut bar from the lower arm.

3. Remove the shock absorber from the lower arm.

4. Unbolt and remove lower ball joint.

NOTE: If the lower ball joint is not to be replaced, simply unbolt it from the lower control arm. It is not necessary to separate the ball joint from the steering knuckle.

5. Remove the lower control arm shaft nut. Remove the spring torque arm from the other side of the lower control arm, then remove the lower arm shaft bolt and the lower arm.

6. To install, reverse the removal procedures. Tighten the bolt(s) holding the lower control arm to the frame but do not torque them until the vehicle is on the ground. Torque the ball joint-to-lower control arm nuts/bolts to 18 ft. lbs. (8mm, 1979–83), 35 ft. lbs. (10mm, 1979–83) or 51 ft. lbs. (1984 and later), the strut bar-to-lower control arm bolts to 70 ft. lbs., the stabilizer bar-to-lower control arm bolts to 9 ft. lbs., the lower shock absorber bolt to 13 ft. lbs., upper shock absorber bolt to 18 ft. lbs. and the lower arm mounting nuts to 199 ft. lbs. Check the front end alignment.

CAUTION

DO NOT torque the control arm bolts fully until the vehicle is lowered and bounced several times; if the bolts are tightened with the control arm(s) hanging, excessive bushing wear will result.

Van

1. Raise and support the vehicle on jackstands.

2. Remove the stablizer bar and the strut bar from the lower arm.

3. Remove the shock absorber from the lower arm.

4. Unbolt and remove lower ball joint.

NOTE: If the lower ball joint is not to be replaced, simply unbolt it from the lower control arm. It is not necessary to separate the ball joint from the steering knuckle.

5. Match-mark the adjusting cam of the lower control arm.

6. Remove the adjusting cam, the nut and the lower control arm.

7. To install, reverse the removal procedures. Align the cam match-marks and finger tighten the nut. Torque the ball joint-to-lower control arm nuts/bolts to 49 ft. lbs., the strut bar-to-lower control arm bolts to 49 ft. lbs., the stabilizer bar-to-lower control arm bolts to 9 ft. lbs., the lower shock absorber bolt to 13 ft. lbs., upper shock absorber bolt to 19 ft. lbs. and the adjusting cam nut to 112 ft. lbs. Check the front end alignment.

CAUTION

DO NOT torque the control arm bolts fully until the vehicle is lowered and bounced several times.

Ball Joints

INSPECTION

To check the lower ball joint for wear, raise the lower control arm and check for excess play. If the loosness problem still exists, check the other suspension parts (wheel bearings, tie-rods and etc.). The bottom of the tire should not move more than 0.2 in. when the tire is pushed and pulled inward and outward. The tire should not move more than 0.09 in. up and down. If the play is greater than these figures, replace the ball joint. The upper ball joint should be replaced if a distinct looseness is felt when turning the ball joint stud with the steering knuckle removed.

REMOVAL & INSTALLATION

2WD Pickup and Van

1. Raise and support the vehicle on jackstands. Remove the wheel/tire assembly.
2. Support the lower control arm with a floor jack.
3. Remove the brake caliper and support it out of the way, with a wire.
4. Using the Ball Joint Removal tool No. 09611–22012, separate the tie-rod end from the knuckle arm.
5. Using the Ball Joint Removal tool No. 09628–62011, separate the upper or lower ball joint from the steering knuckle.

NOTE: Removal and installation will be easier if the bottom joint is removed first.

6. Remove the ball joint-to-control arm mounting bolts and separate the joint from the arm.
7. To install, reverse the removal procedures. Torque the ball joint-to-upper control arm bolts 20 ft. lbs. (Pickup) or 22 ft. lbs. (Van), the upper ball joint-to-steering knuckle nut to 80 ft. lbs. (Pickup) or 58 ft. lbs. (Van), the ball joint-to-lower control arm bolts to 15–21 ft. lbs. (8mm, 1979–83), 29–39 ft. lbs. (10mm, 1979–83), 51 ft. lbs. (Pickup, 1984 and later) or 49 ft. lbs. (Van), and the lower ball joint 105 ft. lbs. (Pickup) or 76 ft. lbs. (Van).

NOTE: Be sure to grease the ball joints before moving the vehicle.

Front Wheel Bearing

For the 4WD models, refer to the "Front Wheel Bearing" in this section to perform the wheel bearing the adjustments, the removal and the installation procedures.

ADJUSTMENT

2WD Pickup and Van

1. Raise and support the front of the vehicle on jackstands.
2. Remove the wheel bearing grease cap, the cotter pin, the lock nut.
3. Tighten the wheel bearing adjust nut to 25 ft. lbs.
4. Turn the disc/hub assembly 2–3 times from left to right.
5. Loosen the adjusting, so there is 0.020–0.039 in. axial play.
6. Install the lock nut, the cotter pin and the grease cap.
7. Lower the vehicle and road test.

REMOVAL & INSTALLATION

2WD Pickup and Van

1. Raise and support the front of the vehicle on jackstands. Remove the wheel/tire assembly.
2. Remove the brake caliper (DO NOT disconnect the brake hose from the caliper) and suspend it on a wire.
3. Remove axle hub dust cap, the cotter pin, the nut lock, the adjusting nut, the thrust washer and the outer bearing, then pull the hub/disc assembly from the axle spindle.
4. Using a small pry bar, pry the oil seal from the rear of the disc/hub assembly, then remove the inner bearing.
5. Using a brass drift and a hammer, drive the bearing races from both sides of the disc/hub assembly.
6. Clean the parts in solvent (NOT gasoline) and blow dry with compressed air (DO NOT use a rag).
7. Using the palm of the hand, force multi-purpose into the bearings.
8. Using the installation tool No. 09608–30011 (Pickup) or 09608–30021 (Van) and a hammer, drive the outer bearing race(s) into the disc/hub assembly until it seats.
9. Place some grease inside the disc/hub assembly (between the races) and install the bearing into the rear of the hub. Coat the new oil seal with grease.

10. Using the installation tool No. 09608–30011 (Pickup) or 09608–30021 (Van) and a hammer, drive the new seal into the rear of the hub until it is flush.
11. To complete the installation, reverse the removal procedures and adjust the wheel bearing. Torque the brake caliper to 65 ft. lbs. (Pickup) or 61 ft. lbs. (Van).

Front-End Alignment

Front-end alignment measurements require the use of special equipment. Before measuring alignment or attempting to adjust it, always check the following points:
1. Be sure that the tires are properly inflated.
2. See that the wheels are properly balanced.
3. Check the ball joints to determine if they are worn or loose.
4. Check front wheel bearing adjustment.
5. Be sure that the vehicle is on a level surface.
6. Check all suspension parts for tightness.

CASTER AND CAMBER ADJUSTMENTS

Measure the caster and camber angles. If they are not within specifications, adjust them by adding or subtracting the shims on the mounting bolts between the upper control arm and the suspension member:
1. To increase the camber, remove the shims equally from both of the control shaft mounting bolts. Do the reverse to decrease camber.
2. To increase the caster, add the camber adjusting shims to the rear mounting bolt or remove them from the front mounting bolt. Do the reverse to decrease caster.

NOTE: The caster and camber adjustments should always be performed in a single operation.

TOE-IN ADJUSTMENT

Measure the toe-in. Adjust it, if necessary, by loosening the tie-rod end clamping bolts and rotating the tie-rod adjusting tubes. Tighten the clamp bolts when finished.

NOTE: Both tie-rod ends should be the same length. If they are not, perform the adjustment until the toe-in is within specifications and the tie-rod ends are equal in length.

STEERING

Steering Wheel

REMOVAL & INSTALLATION

Three-Spoke

——— CAUTION ———

DO NOT attempt to remove or install the steering wheel by hammering on it. Damage to the energy-absorbing steering column could result.

1. Disconnect the negative battery cable.
2. Disconnect the horn and the turn signal multi-connector(s) at the base of the steering column shroud.
3. Loosen the trim pad retaining screws from the back side of the steering wheel.
4. Lift the trim pad and horn button assembly(ies) from the wheel.
5. Remove the steering wheel hub retaining nut.
6. Scratch match-marks on the hub and shaft to aid in the installation.
7. Using the steering wheel puller tool No. 09609–20010, pull the steering wheel from the steering column.
8. To install, reverse the removal procedures. Tighten the wheel retaining nut to 25 ft. lbs.

Two-Spoke

The two-spoke steering wheel is removed in the same manner as the three-spoke, except that the trim pad should be pried off with a screwdriver. Remove the pad by lifting it toward the top of the wheel.

Four-Spoke

——— CAUTION ———

DO NOT attempt to remove or install the steering wheel by hammering on it. Dam-

age to the energy absorbing steering column could result.

1. Disconnect the negative battery cable.
2. Disconnet the horn and the turn signal connectors at the base of the steering column shroud, underneath the instrument panel.
3. Gently pry the center emblem off the front of the steering wheel.
4. Insert a wrench through the hole and remove the steering wheel retaining nut.
5. Scratch match-marks on the hub and the shaft to aid installation.
6. Use the Steering Wheel Puller tool No. 09609–20010, pull the steering wheel from the steering column.
7. To install, reverse the removal procedures. Torque the steering wheel nut to 15–22 ft. lbs.

Turn Signal Switch

REMOVAL & INSTALLATION

1. Disconnect negative battery cable.
2. Remove the upper and the lower steering column shrouds.
3. Disconnect the combination switch electrical connector.
4. At the left-rear of the combination switch, remove the mounting screws and the turn signal switch.
5. If necessary to the remove the turn signal switch wires from the electrical connector, place a small screwdriver into the end of the connector, pry up on the retaining tab and pull the wire(s) from the connector.
6. To install, place the wire(s) into the electrical connector's slots, place a screwdriver behind the wire terminal and push the wire into the connector until the retaining tab locks it into place.
7. To complete the installation, reverse the removal procedures.

Combination Switch

The combination switch is composed of the turn signal, the light control, the hazard, the wiper and the washer switches.

REMOVAL & INSTALLATION

1. Refer to the "Steering Wheel, Removal and Installation" procedures, in this section and remove the steering wheel.
2. Remove the upper and lower steering column shroud screws.
3. Remove the combination switch screws and the switch from the column.
4. To install, reverse the removal procedures.

Ignition Lock/Switch

REMOVAL & INSTALLATION

The ignition lock/switch is located behind the combination switch on the steering column.

1. Refer to the "Combination Switch, Removal and Installation" procedures, in this section and remove the combination switch from the steering column.
2. Disconnect the ignition switch electrical connector from underneath the instrument panel.
3. Using the key in the ignition switch, turn it to the "ACC" position.
4. Using a thin rod, place it into the hole of the cylinder lock housing. Pushing down on the thin rod, pull out the cylinder lock.
5. To install, push the cylinder lock into the housing until the retaining tab locks it in place.
6. To complete the installation, reverse the removal procedures.

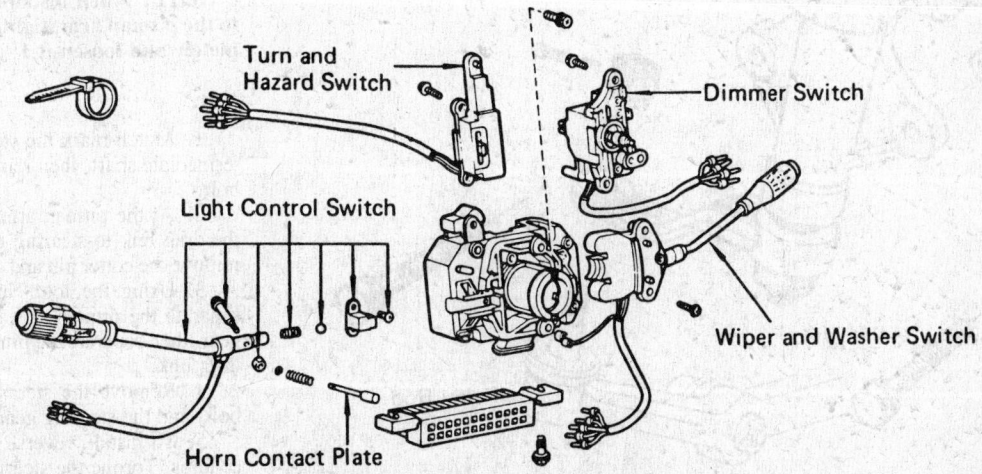

Exploded view of the combination switch

Power Steering Pump

REMOVAL & INSTALLATION

Pickup, Land Cruiser and 4-Runner

NOTE: On the 1984 and later models (except diesel), disconnect the air hoses from the air control valve and the high tension wires from the distributor. On the diesel models, remove the engine under cover.

1. Loosen the power steering pump pulley nut.

NOTE: Use the drive belt as a brake to keep the pulley from rotating.

2. Place a container under the pump. Disconnect the return line and the pressure tube, then drain the fluid into the container.

3. Loosen the idler pulley nut and the adjusting bolt, then remove the drive belt.

4. Remove the drive pulley and the woodruff key from the pump shaft.

5. Remove the mounting bolts and the power steering pump from the vehicle.

6. To install, reverse the removal procedures. Torque the pump pulley mounting bolt to 29 ft. lbs. (gasoline) or 45 ft. lbs. (diesel), the pump pulley nut to 32 ft. lbs. and the pressure hoses to 33 ft. lbs. Adjust the drive belt tension. Bleed the power steering system.

Van

1. Disconnect the air hoses from the air control valve of the power steering pump.

2. Drain the fluid from the power steering reservoir tank.

3. At the power steering pump, disconnect the return hose and the pressure tube.

4. Loosen the power steering pump adjusting bolt and remove the drive belt, the pulley and the woodruff key.

5. Remove the mounting bolts, the power steering pump and the bracket from the vehicle.

6. To install, reverse the removal procedures. Torque the power steering pump-to-engine bolts to 29 ft. lbs., the pulley set nut to 32 ft. lbs. and the pressure tube to 33 ft. lbs.

BLEEDING

1. Raise and support the front of the vehicle on jackstands.

2. Fill the pump reservoir with Dexron® automatic transmission fluid.

3. With the engine running, rotate the steering wheel from lock to lock several times. Add fluid as necessary.

NOTE: Perform the bleeding procedure until all of the air is bled from the system.

4. The fluid level should not have risen more than 0.20 in.; if it does, check the pump.

Manual Steering Gear

REMOVAL & INSTALLATION

2WD Pickup

1. Remove the pitman arm-to-relay rod cotter pin and nut. Using the tool No. 09611–22012, separate the relay rod from the pitman arm.

2. Match-mark the flexible steering coupling-to-steering gear, then remove the lock bolt and separate the steering coupling from the steering gear.

3. Remove the steering gear housing mounting bolts and the gear housing.

4. To install, reverse the removal procedures. Torque the housing-to-frame bolts to 26–36 ft. lbs. (1979–80), 37–43 ft. lbs. (1981–83) or 48 ft. lbs. (1984 and later), the pitman arm-to-relay rod nut 80–90 ft. lbs. (1979–83) or 67 ft. lbs. (1984 and later) and the steering gear-to-coupling yoke to 15–20 ft. lbs.

4WD Pickup and 4-Runner

1. Remove the stone shield from the gear housing, if equipped.

2. Match-mark the intermediate shaft-to-steering gear and disconnect them.

3. Remove the cotter pin and plug from the drag link.

4. Disconnect the drag link from the pitman arm.

5. Remove the pitman arm nut. Using the Puller tool No. 09610–55012, separate the pitman arm from the steering gear.

6. Remove the steering gear housing-to-frame bolts and the gear housing.

7. To install, reverse the removal procedures. Torque the steering gear-to-frame bolts to 42 ft. lbs., the steering gear-to-intermediate bolts to 29 ft. lbs., the pitman arm-to-steering gear nut to 127 ft. lbs.

NOTE: When installing the drag link to the pitman arm, tighten the plug completely and loosen it 1⅓ turns.

Van

1. Match-mark the steering gear-to-intermediate shaft, then remove the coupling bolt.

2. At the pitman arm-to-drag link and the drag link-to-steering gear connections, remove the cotter pin and the mounting nut.

3. Using the tool No. 09610–20012, separate the pitman arm from the steering gear, then separate the pitman arm from the drag link.

4. Remove the steering gear-to-frame bolts and the steering gear from the frame.

5. To install, reverse the removal procedures. Torque the steering gear-to-frame bolts to 70 ft. lbs., the steering gear-to-

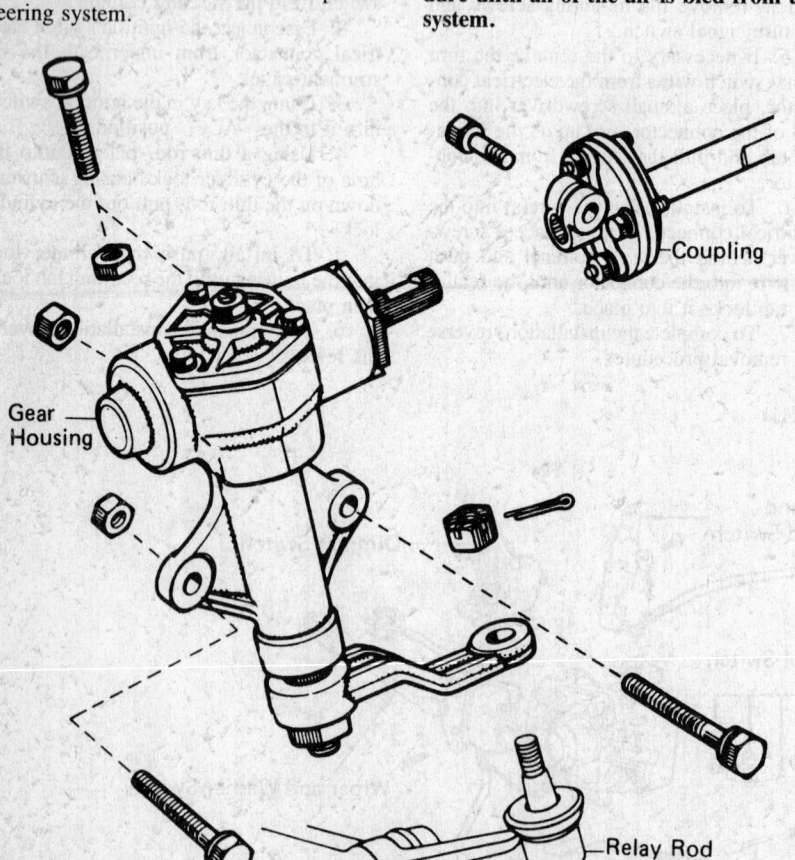

Gear Housing

Coupling

Relay Rod

2WD pick-up manual steering gear

coupling bolt to 18 ft. lbs., the pitman arm-to-steering gear nut to 90 ft. lbs. and the pitman arm-to-drag link nut to 67 ft. lbs.

Land Cruiser

55 SERIES

1. Remove the worm yokes from the worm and the main shaft.
2. Remove the intermediate shaft assembly.
3. Remove the Pitman arm from the sector shaft.
4. Remove the steering gear-to-frame bolts and the steering gear from the vehicle.
5. To install, reverse the removal procedures. Torque the Pitman arm to 119–141 ft. lbs.

NOTE: The intermediate shaft must be installed with the wheels in a straight ahead position and the steering wheel straight ahead.

40 SERIES

1. Remove the horn button assembly. Using a wheel puller, remove the steering wheel.
2. Remove the steering column jacket lower clamp and the turn signal switch assembly.
3. Remove the steering column access plate, then the carburetor and the oil filter.
4. Disconnect the No. 1 shift rod and select rod at the ends of the shift control and select levers.
5. Remove the lower shift control bracket clamp, the shift control lever, the select lever, the control shaft lower bracket, the control shaft low speed lever and the control shaft lower bracket.
6. Pull the control shaft out toward the driver's side.
7. Using a puller, remove the pitman arm from the steering gear.
8. Remove the steering gear box bracket cap and lift out the gear box.
9. To install, reverse the removal procedures. Torque the gear box bracket cap to 75–90 ft. lbs., the pitman arm to 120–140 ft. lbs. and the steering wheel nut to 30–50 ft. lbs.

ADJUSTMENTS

Adjustments to the manual steering gear are not necessary during normal service. Adjustments are performed only as part of overhaul, which is covered in the Unit Repair section.

Power Steering Gear

REMOVAL & INSTALLATION

Pickups (1979–80)

1. Disconnect and plug the hydraulic lines at the steering gear.
2. Mark-mark the intermediate shaft to steering gear shaft.

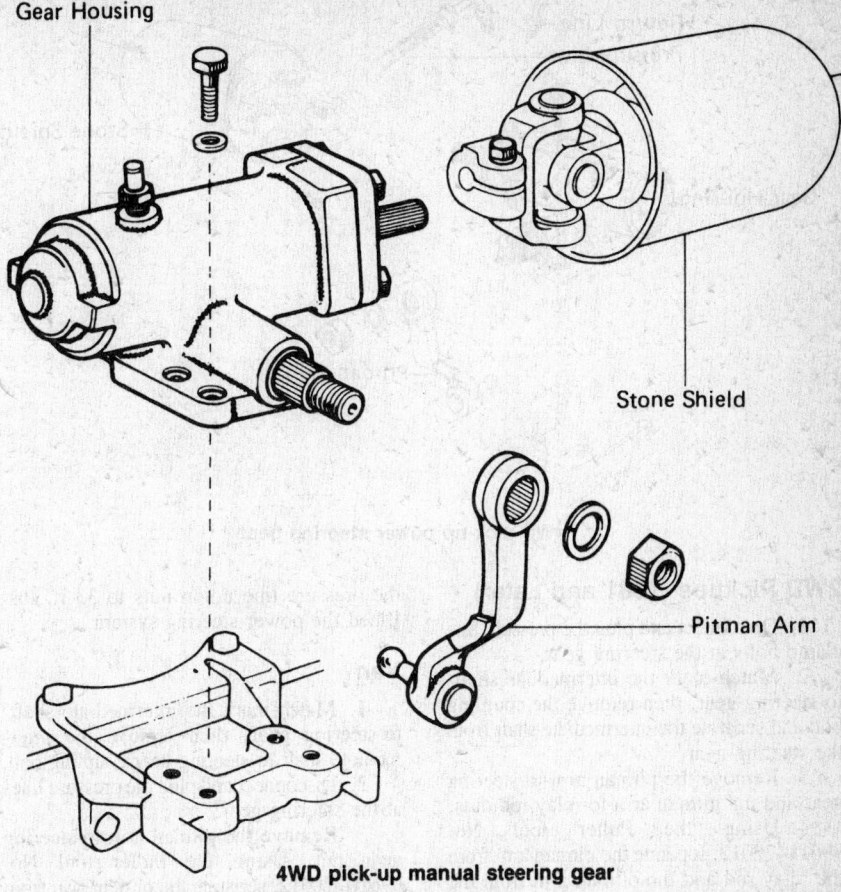

4WD pick-up manual steering gear

3. Remove the intermediate shaft-to-steering gear bolt and separate the intermediate shaft from the steering gear.
4. Using the Puller tool No. 09610–55012, separate the Pitman arm from the steering gear.
5. Remove the steering gear-to-frame bolts and the steering gear from the vehicle.
6. To install, reverse the removal procedures. Torque the steering gear-to-frame bolts to 37–47 ft. lbs., the pitman arm-to-steering gear nut to 116–137 ft. lbs., the intermediate shaft-to-steering gear bolt to 22–32 ft. lbs., the pressure hose fitting to 29–36 ft. lbs. and the return line fitting to 24–30 ft. lbs. Bleed the power steering system.

NOTE: During installation of the hydraulic lines, position each line clear of any surrounding components then tighten the fittings.

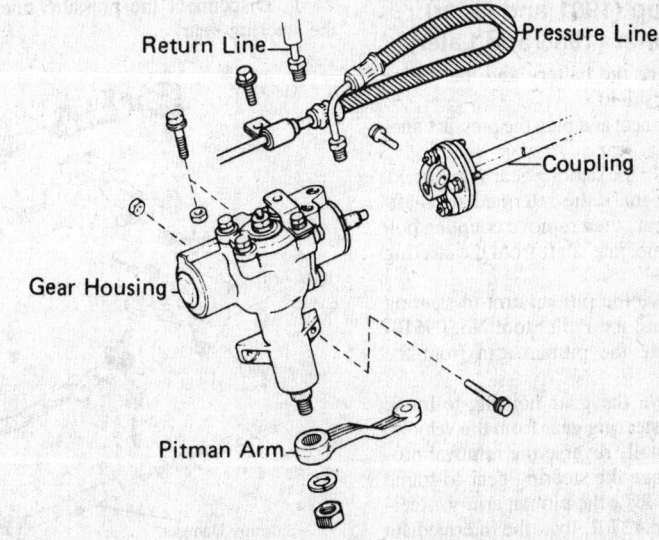

2WD pick-up power steering gear

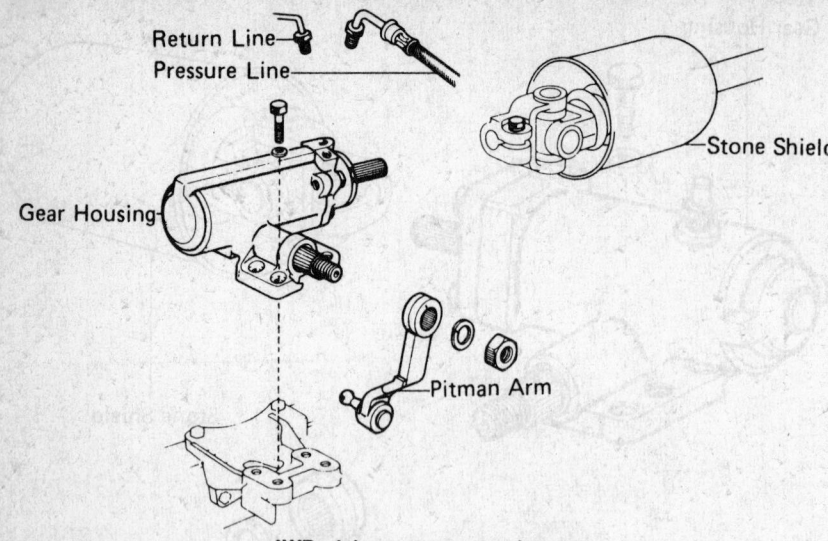

4WD pick-up power steering gear

2WD Pickups (1981 and Later)

1. Disconnect and plug the pressure line clamp bolts at the steering gear.
2. Match-mark the intermediate shaft-to-steering gear, then remove the coupling bolt and separate the intermediate shaft from the steering gear.
3. Remove the pitman arm-to-steering gear and the pitman arm-to-relay rod nuts.
4. Using the Puller tool No. 09611–22012, separate the pitman arm from the relay rod and the pitman arm from the steering gear.
5. Remove the steering gear-to-frame bolts and the steering gear from the vehicle.
6. To install, reverse the removal procedures. Torque the steering gear-to-frame bolts to 48 ft. lbs., the pitman arm-to-steering gear nut to 90 ft. lbs., the pitman arm-to-relay rod nut to 67 ft. lbs., the intermediate shaft-to-steering gear bolt to 19 ft. lbs. and the pressure line nuts to 33 ft. lbs. Bleed the power steering system.

4WD Pickup (1981 and Later) and 4-Runner (1985 and Later)

1. Remove the battery and the engine lower gravel shield.
2. Disconnect and plug the pressure lines at the steering gear.
3. Remove the steering gear stone shield.
4. Match-mark the intermediate shaft-to-steering gear, then remove coupling bolt and the intermediate shaft from the steering gear.
5. Remove the pitman arm-to-steering gear nut. Using the Puller tool No. 09610-5512, separate the pitman arm from the steering gear.
6. Remove the gear housing-to-frame bolts and the steering gear from the vehicle.
7. To install, reverse the removal procedures. Torque the steering gear-to-frame bolts to 42 ft. lbs., the pitman arm-to-steering gear nut to 127 ft. lbs., the intermediate shaft-to-steering gear bolt to 29 ft. lbs. and

the pressure line union nuts to 33 ft. lbs. Bleed the power steering system.

Van

1. Match-mark the intermediate shaft-to-steering gear, then remove the intermediate shaft-to-steering gear coupling bolt.
2. Disconnect and plug the pressure lines at the steering gear.
3. Remove the pitman arm-to-steering gear nut. Using the Puller tool No. 09610–20012, separate the pitman arm from the steering gear.
4. Remove the steering gear-to-frame bolts and the steering gear form the vehicle.
5. To install, reverse the removal procedures. Torque the steering gear-to-frame bolts to 78 ft. lbs., the pitman arm-to-steering gear nut to 90 ft. lbs., the intermediate shaft-to-steering gear bolt to 18 ft. lbs. and the pressure lines-to-steering gear to 33 ft. lbs. Bleed the power steering system.

Land Cruiser

1. Disconnect the pressure lines from the steering gear.

2. Remove the intermediate shaft-to-steering gear bolt and the steering column-to-firewall bolts.
3. Loosen the steering column-to-dash bolts. Remove the pitman arm-to-steering gear nut.
4. Using a puller, separate the relay rod from the Pitman shaft and the Pitman arm from the steering gear.
5. Pull the steering column towards the passenger compartment to uncouple the steering shaft from the steering gear.
6. Remove the steering gear-to-frame bolts and the steering gear from the vehicle.
7. To install, reverse the removal procedures. Torque the steering gear-to-frame bolts to 40–63 ft. lbs., the pitman arm-to-steering gear nut to 120–141 ft. lbs., the intermediate shaft-to-steering gear bolt to 22–32 ft. lbs., the pressure hose fitting to 29–36 ft. lbs. and the return hose fitting to 24–30 ft. lbs. Bleed the power steering system.

NOTE: During installation of the hydraulic lines, position each line clear of any surrounding components then tighten the fittings.

Steering Linkage

REMOVAL & INSTALLATION

2WD Pickup

1. Raise and support the front of the vehicle on jackstands.
2. Remove the front wheels.
3. Remove the pitman arm-to-relay rod nut. Using the Puller tool No. 09611–22012, separate the pitman arm from the relay rod.
4. Remove the idler arm-to-relay rod cotter pin and nut bolts and remove the idler arm-to-frame bolts.
5. Remove the tie-rod end-to-knuckle arm and the tie-rod end-to-relay rod cotter pin and nut.
6. Using the puller tool No. 09611–22012, separate the tie-rod from the knuckle arm and from the relay rod.

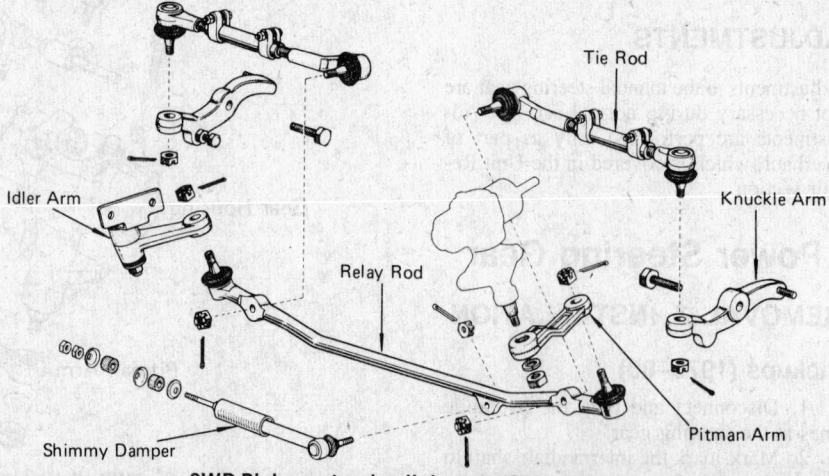

2WD Pick-up steering linkage—1979 and later

7. To install, reverse the removal procedures. Torque the tie-rod end-to-knuckle arm, the tie-rod end-to-relay rod nut to 67 ft. lbs., the relay rod-to-pitman arm nut to 67 ft. lbs. and the relay rod-to-idler arm to 43 ft. lbs.

4WD Pickup and 4-Runner

1. Raise and support the vehicle on jackstands. Remove the front wheels.

2. Remove tie-rod end-to-knuckle arm and the tie-rod-to-steering damper cotter pins and nuts. Remove the steering damper-to-axle housing nut, retainer and cushion; be sure to note the order of the cushions and retainers.

3. Using the Puller tool No. 09611–22012, separate the steering damper from the tie-rod, then the tie-rod ends from the knuckle arms. Remove the tie-rod and the steering damper from the vehicle.

4. At both ends of the drag link, remove the cotter pin. Using a screwdriver, remove the plug from both ends of the drag link.

NOTE: The cap may be tight, so you may have to use a wrench or pliers to turn the screw driver.

5. When the cap is removed, you should be able to dislodge the spring seat, spring and outer socket holder inside the drag link by working the steering knuckle back and forth. The steering knuckle socket in the drag link can now be removed.

NOTE: Be sure to note the order in which the spring seat, the spring and the outer socket are removed from the drag link.

6. To install, reverse the removal procedures. Torque the steering damper-to-axle housing nut to 9 ft. lbs., the tie-rod end-to-steering knuckle arm to 67 ft. lbs. and the steering damper-to-tie-rod end to 43 ft. lbs.

NOTE: Be sure to grease drag link ends at their grease nipples. When installing drag link end caps, tighten the plugs completely and loosen them 1⅓ turns.

Van

1. Raise and support the front of the vehicle on jackstands. Remove the wheel/tire assemblies.

NOTE: Before removing any component in the steering system, remove the cotter pin and the retaining nut first.

2. Using the Puller tool No. 09611–22012, separate the drag link from the pitman arm.

3. Using the Puller tool No. 09628–62011, separate the tie-rod ends from the relay rod, the relay rod from the center arm, the center arm from the center arm bracket and the tie-rod end from the knuckle arm.

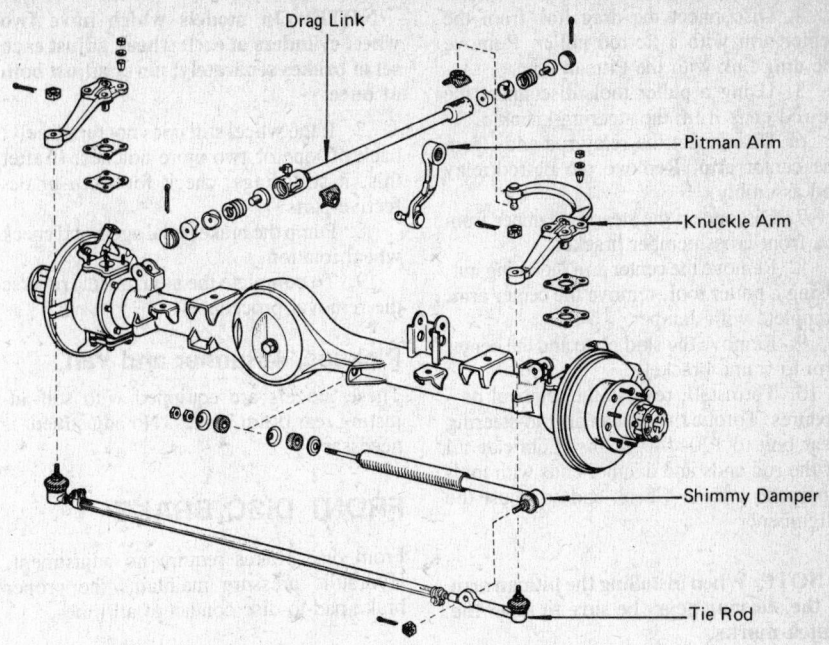

4WD Pick-up and 4-Runner steering linkage

4. Remove the idler arm-to-frame bolts and the idler arm from the vehicle.

5. To install, reverse the removal procedures. Torque the drag link-to-pitman arm nut to 67 ft. lbs., the drag link-to-center arm nut to 43 ft. lbs., the center arm-to-center arm bracket nut to 67 ft. lbs., the relay rod-to-idler arm nut to 43 ft. lbs., the relay rod-to-tie-rod end nut to 43 ft. lbs., the tie-rod end-to-knuckle arm nut to 43 ft. lbs., the idler arm-to-frame bolts to 58 ft. lbs. and the center arm bracket-to-frame bolts to 58 ft. lbs.

Land Cruiser

1. Raise and support the front of the vehicle on jackstands. Remove the wheels/tire assemblies.

2. Remove the pitman arm-to-steering gear nut.

NOTE: Punch match-marks on the Pitman arm-to-steering gear to aid reinstallation.

3. Using a puller tool, remove the pitman arm from the steering gear.

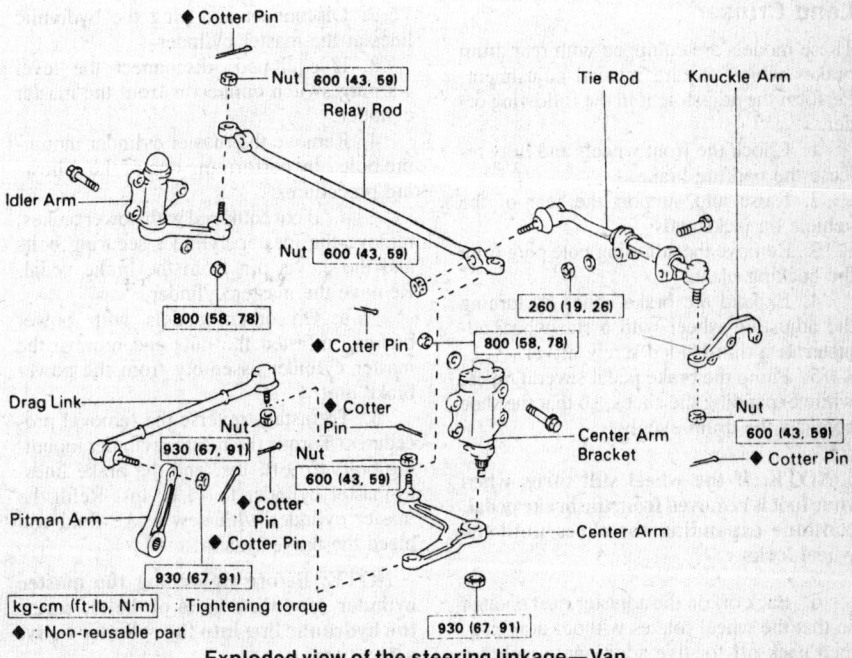

Exploded view of the steering linkage—Van

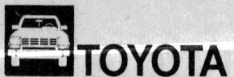

4. Disconnect the drag link from the center arm with a tie-rod puller. Remove the drag link with the Pitman arm.

5. Using a puller tool, disconnect the tie-rod ends from the steering knuckle.

6. Disconnect the relay rod ends from the center arm. Remove the tie-rod/relay rod assembly.

7. Disconnect the steering damper from the front crossmember bracket.

8. Remove the center arm mounting nut. Using a puller tool, remove the center arm, complete with damper.

9. Remove the skid plate and the center arm-to-frame bracket.

10. To install, reverse the removal procedures. Torque the pitman arm-to-steering gear bolt to 120–140 ft. lbs. Lubricate all of the rod ends and damper ends with multipurpose grease. Check and/or adjust the alignment.

NOTE: When installing the pitman arm to the steering gear, be sure to align the match-marks.

BRAKE SYSTEM

For servicing procedures not found in the following refer to the Brake section of Unit Repair.

Adjustments

REAR DRUM BRAKE

Land Cruiser

These models are equipped with rear drum brakes which require manual adjustment. Perform the adjustment in the following order:

1. Chock the front wheels and fully release the parking brake.

2. Raise and support the rear of the vehicle on jackstands.

3. Remove the adjusting hole plug from the backing plate.

4. Expand the brake shoes by turning the adjusting wheel with a star-wheel adjuster or a thin-bladed screw driver.

5. Pump the brake pedal several times, while expanding the shoes, so that the shoe contacts the drum evenly.

NOTE: If the wheel still turns when your foot is removed from the brake pedal, continue expanding the shoes until the wheel locks.

6. Back off on the adjuster, just enough so that the wheel rotates without dragging, then back off for five additional notches.

NOTE: On models which have two wheel cylinders at each wheel, adjust each set of brakes separately; never adjust both at once.

7. If the wheel still does not turn freely, back off one or two more notches. If after this, it still drags, check for worn or defective parts.

8. Pump the brake pedal again and check wheel rotation.

9. To complete the adjustment, reverse the removal procedures.

Pickups, 4-Runner and Van

These models are equipped with self-adjusting rear drum brakes. No adjustment is necessary.

FRONT DISC BRAKE

Front disc brakes require no adjustment. Hydraulic pressure maintains the proper brake pad-to-disc contact at all times.

NOTE: Because of this, the brake fluid level should be checked regularly.

Master Cylinder

REMOVAL & INSTALLATION

──────── CAUTION ────────
Be careful not to spill brake fluid on the painted surfaces of the vehicle; it will damage the paint.
────────────────────────

Pickups, 4-Runner and Land Cruiser

1. Using a syringe, remove the brake fluid from the master cylinder.

2. Disconnect and plug the hydraulic lines at the master cylinder.

3. If equipped, disconnect the level warning switch connector from the master cylinder.

4. Remove the master cylinder mounting bolts, by performing one of the following procedures:

   a. If not equipped with power brakes, remove the master cylinder securing bolts and the clevis pin from the brake pedal. Remove the master cylinder.

   b. On other models with power brakes, unfasten the nuts and remove the master cylinder assembly from the power brake unit.

5. To install, reverse the removal procedures. Torque the master cylinder mounting bolts to 9 ft. lbs. and the brake lines-to-master cylinder to 11 ft. lbs. Refill the master cylinder with new brake fluid and bleed the brake system.

NOTE: Before tightening the master cylinder mounting nuts or bolts, screw the hydraulic line into the cylinder body, a few turns.

Van

1. Disconnect the negative battery cable.

2. To expose the master cylinder, perform the following:

   a. Remove the master cylinder reservoir cap, located at the left side of the instrument panel.

   b. Remove the instrument cluster finish panel and the lower cluster finish panel. Disconnect the electrical connectors and the speedometer cable from the instrument panel, then remove the instrument panel.

   c. Remove the No.1, 2 and 3 air ducts.

3. Using a syringe, remove the brake fluid from the master cylinder reservoir.

4. Remove the reservoir hoses from the master cylinder. Disconnect and plug the brake lines at the master cylinder.

5. Remove the master cylinder mounting nuts, the vacuum check valve bracket and the master cylinder from the vehicle.

6. To install, reverse the removal procedures. Torque the master cylinder mounting nuts to 9 ft. lbs. and the brake lines-to-master cylinder to 11 ft. lbs. Refill the master cylinder with new brake fluid and bleed the brake system.

Load Sensing Proportioning Valve/By-Pass Valve

REMOVAL & INSTALLATION

1. Raise and support the vehicle on jackstands, so that it is level.

2. Disconnect the No.2 shackle from the bracket.

3. Disconnect and plug the brake lines from the load sensing valve.

4. Remove the load sensing valve bracket from the frame.

5. To install, reverse the removal procedures. Torque the load sensing valve-to-frame bolts to 14 ft. lbs. and the brake tubes to 11 ft. lbs. Bleed the brake system. Adjust the load sensing valve and the rear axle load. Check and/or adjust the length of the No.2 shackle (distance from the center of the No.2 shackle-to-shackle bracket bolt to the center of the No.1 shackle-to-spring bolt): 3.07 in. (2WD Pickup and Van) or 4.72 in. (4WD Pickup, 4-Runner and Land Cruiser).

ADJUSTMENT

1. Raise and support the vehicle on jackstands, so that it is level.

2. Check and/or adjust the rear axle load: 1,323 lbs. (¾ ton and C & C: 1979–83), 1,150 lbs. (FJ40, Land Cruiser), 1,200 lbs. (FJ60, Land Cruiser), 1,543 lbs. (2WD Pickup: 1984 and later, Van: 1983 and later), 1,433 lbs. (4WD Pick-up: 1979–83) or 1,653

lbs. (4WD Pickup: 1984 and later, 4-Runner: 1985 and later).

3. Using the Pressure Gauge tool No. 09705–29017 (1979–83) or 09709–29017 (1984 and later), install one (in the brake line) at the front wheel and one at the rear wheel.

4. Depress the brake pedal, raising the front pressure to 365 psi. (Land Cruiser), 711 psi. (Pickups, Van, 4-Runner), then check the rear brake pressure. The rear brake pressure should be 148–205 psi. (Land Cruiser), 398–540 psi. (Pickups: 1979–83, Van: 1983 and later), 455–597 psi. (2WD Pickup: 1984 and later) or 441–589 psi. (4WD Pickup: 1984 and later, 4-Runner: 1985 and later).

NOTE: When checking the fluid pressure, depress the pedal ONLY once and record the pressures within 2 seconds; NEVER depress it twice.

5. Depress the brake pedal, raising the front pressure to 835 psi. (Land Cruiser), 1,138 psi. (Pickups: 1979–83) or 1,422 psi. (Pickups: 1984 and later, Van: 1983 and later, 4-Runner: 1985 and later), then check the rear brake pressure. The rear brake pressure should be 312–411 psi. (Land Cruiser), 526–726 psi. (Pickups: 1979–83), 512–712 psi. (Van: 1983 and later), 696–896 psi. (2WD Pickup: 1984 and later) or 682–882 psi. (4WD Pickup: 1984 and later, 4-Runner: 1985 and later).

6. If the pressures do not fall within the specifications, perform the following:

a. Remove the No.2 shackle from the shackle bracket. Loosen the lock nut and adjust the length of the No.2 shackle (distance from the center of the No.2 shackle-to-shackle bracket bolt to the center of the No.1 shackle-to-spring bolt): 2.83–3.31 in. (2WD) or 4.49–4.96 in. (4WD).

NOTE: When adjusting the No.2 shackle length, lengthening the distance decreases the pressure and shortening the distance increases the pressure.

b. If the pressure cannot be adjusted with the No.2 shackle, raise or lower the valve body.

NOTE: When adjusting the valve body, lowering the valve lowers the pressure and raising the valve increases the pressure.

c. After adjusting the valve body, adjust the length of the No.2 shackle. If the adjustment cannot be accurately made, inspect the valve body.

NOTE: With the valve body installed in it's correct position, the distance between the valve body and the spring should be 0.04 in.

Power Brake Booster

REMOVAL & INSTALLATION

1. Refer to the "Master Cylinder, Removal and Installation" procedures, in this section and remove the master cylinder from the power brake booster.

2. Remove the vacuum hose form the power brake booster.

3. Working under the instrument panel, remove the brake pedal-to-brake booster rod clevis pin. Remove the power brake booster mounting bolts and the booster from the vehicle.

4. To install, reverse the removal procedures. Torque the power brake booster nuts to 9 ft. lbs. Check and/or adjust the brake pedal height.

NOTE: When installing a new booster, make sure there is a little clearance between the push rod end and the master cylinder piston.

Parking Brake

ADJUSTMENT

Pickups (1979–80)

NOTE: Adjust the rear brake shoes before attempting to adjust the parking brake.

1. Loosen the parking brake warning light switch bracket.

2. Push the parking brake lever in until it is stopped by the pawl.

3. Move the switch so that it will be "OFF" at this position but "ON" when the handle is pulled out.

4. Tighten the switch bracket and push the brake lever in again.

5. Working from under the vehicle, loosen the locknut on the parking brake cable equalizer.

6. Screw the adjusting nut IN, just enough so that the brake cables have no slack. Hold the adjusting nut in this position while tightening the locknut.

7. Rotate the rear wheels to make sure that the brakes are not dragging.

8. Pull out on the parking brake lever and count the number of notches needed to apply the parking brake; 7–10 notches is preferred.

Pickups (1981 and Later)
4-Runner (1985 and Later)

1. Make sure that the rear brakes are properly adjusted.

2. Pull the parking brake lever out as far as it will go, counting the number of notches heard in the travel: 10–16 notches (2WD) or 7–15 notches (4WD).

3. If these standards are not met, proceed as follows:

2WD Pickup

a. Working under the truck, tighten the adjusting nut at the equalizer until the travel is within limits and there is no drag at the rear shoes.

b. Apply the parking brake several times and again check that there is no drag with the brake released.

4WD Pickup and 4-Runner

a. Working under the truck, tighten the bellcrank stopper screw until the play at the rear brake links is gone, then loosen the nut one full turn. Tighten the locknut.

b. Tighten one of the adjusting nuts on the intermediate lever while loosening the other, until the travel is correct. Tighten the two locknuts.

c. Confirm that the bellcrank is in contact with the backing plate.

Van

NOTE: The rear brake shoe clearance should be adjusted before adjusting the parking brake.

1. Raise and support the rear of the vehicle on jackstands.

2. Remove the shift knob and the console box.

3. At the parking brake handle, loosen the cable locknut. Pull the hand brake UP 7–9 clicks.

4. Turn the adjust nut until the rear wheels can no longer be turned, then tighten the locknut.

5. Install the console and the shift knob.

Land Cruiser

Land Cruiser models use a separate drum brake assembly, operating on the driveshaft, to serve as a parking brake. Adjust it as follows:

1. Push the parking brake lever all the way in, so that the brake is released.

2. Raise and support the rear of the vehicle on jackstands.

3. Turn the parking brake adjustment shaft, located at the bottom of the parking brake backing plate, counterclockwise until the shoes seat against the drum.

4. Back the adjuster off one notch.

5. Apply the parking brake; the drum should be locked. Release the brake; the drum should rotate freely.

NOTE: If the drum does not rotate freely with the brake off, loosen the adjuster one more notch.

6. Adjust the turnbuckles on the parking brake intermediate levers and the adjusting nuts on the end of the parking brake cables, so that 7–12 notches are required to apply the parking brake.

CHASSIS ELECTRICAL

Heater

NOTE: On models equipped with A/C, the heater and the air conditioner are completely separate units. Be certain when working under the dashboard that only the heater hoses are disconnected. The air conditioning hoses are under pressure; if disconnected, the escaping refrigerant will freeze any surface with which it comes in contact, including your skin and eyes.

REMOVAL & INSTALLATION

Pickups, Van and 4-Runner

1. Disconnect the negative battery terminal.
2. Drain the cooling system.
3. Remove the glove box, the defroster hoses, the air damper, the air duct and the two side defroster ducts.
4. Remove the control unit from the instrument panel.
5. Disconnect the heater hoses from the core tubes.
6. Remove the retaining bolts and lift out the heater unit. At this point, the core may be pulled from the case.
7. To install, reverse the removal procedures.

Front Heater Core

NOTE: To service the heater core of Pickup, Van and 4-Runner models, refer to the "Heater, Removal and Installation" procedures, in this section.

REMOVAL & INSTALLATION

Land Cruiser–1979

1. Turn off the water valve.
2. Detach both hoses from the heater core.
3. Remove the air duct clamp.
4. Detach the defroster hoses from the heater box.
5. Remove the mounting bolts and withdraw the core.
6. To install, reverse the removal procedures.

1980 AND LATER

NOTE: The entire heater unit must be removed to gain access to the heater core. This procedure requires almost complete disassembly of the instrument panel and lowering of the steering column.

1. Note the following points before proceeding:
   a. Be sure to tag any wiring which must be disconnected so that it may be correctly installed.
   b. As the fasteners are removed, arrange them so that they may be installed in their original locations.
   c. Do not force any parts to remove them; if a part cannot easily be removed, remove any additional fasteners which may have been initially overlooked.
   d. When disconnecting the coolant hoses, be careful not to damage the heater core tubes. Place a drain pan under the coolant hose connections before disconnecting the hoses.
2. Disconnect the negative battery cable. Remove the glove box and the glove box door.
3. Remove the lower heater ducts (No.2 in the accompanying illustration). Remove the large heater duct from the passenger side of the heater unit (No.3 in the accompanying illustration).
4. Remove the ductwork from behind the instrument panel (No.4 in the accompanying illustration). If equipped, remove the radio.
5. Disconnect the wiring connector from the right side inner portion of the glove opening.
6. Remove the instrument panel pad. Remove the hood release lever. Disconnect the hand throttle control cable.
7. Remove the retaining screw from the left side of the fuse block.
8. Remove the steering column-to-instrument panel attaching nuts and carefully lower the steering column. Tag and disconnect the wiring as necessary in order to lower the column assembly.
9. Disconnect the electrical connector from the rheostat located to the left of the steering column opening.
10. Remove the center dual outlet duct which is attached to the upper portion of the heater unit (No.11 in the accompanying illustration).
11. Remove the lower instrument panel. The fasteners are located in the following places:
   a. Left side of the instrument panel: two at the left side end and two at the left lower end.
   b. Above the steering column: two.
   c. To the right of the steering column opening: two.
   d. Left upper corner of the glove box opening: two.
   e. Left lower corner of the glove box opening: one.
   f. Right side of the instrument panel: two at the right side end and two at the right lower end.
12. Tag and disconnect the hoses from the heater unit. Remove the heater unit-to-firewall fasteners and the heater unit.
13. Remove the heater core-to-heater unit pipe clamps and the heater core retaining clamp, then withdraw the heater core from the heater unit.

14. To install, reverse the removal procedures. Torque the steering column-to-instrument panel fasteners to 14–15 ft. lbs. Refill the cooling system.

Rear Heater Core

REMOVAL & INSTALLATION

Land Cruiser

1. Turn off the water valve and disconnect both hoses from the rear heater core.
2. Disconnect the wiring from the rear heater.
3. Remove the mounting bolts and lift out the core.
4. To install, reverse the removal procedures.

Heater Blower Motor

REMOVAL & INSTALLATION

Pickups (1979)

Refer to the "Heater, Removal and Installation" procedures, in this section and remove the blower motor.

Pickups (1980 and Later)Van (1983 and Later)4-Runner (1985 and Later)

1. Disconnect the electrical connector from motor.
2. Remove the three screws securing the motor and lift the motor from the case.
3. To install, reverse the removal procedures. Make sure that the seal around the motor flange is in good condition.

Land Cruiser

1979

1. Loosen the air duct clamping screws. Remove the ducts and the air duct screen.
2. Remove the mounting bolts and the blower motor with the fan.
4. To install, reverse the removal procedures.

1980 AND LATER

1. Disconnect the electrical connector from the blower motor.
2. Disconnect the flexible tube from the side of the blower motor.
3. Remove the blower motor fasteners and lower the blower motor out of the air inlet duct.
4. To install, reverse the removal procedures. During installation, be sure to position the motor so that the flexible tube can be attached to the motor.

Radio

———— CAUTION ————

Never operate the radio without a speaker; severe damage to the output transistors will result. If the speaker must be replaced, use a speaker of the correct impedance (ohms) or else the output transistors will be damaged and require replacement.

REMOVAL & INSTALLATION

Pickups, Van and 4-Runner

1. Disconnect the negative battery cable.
2. Remove the upper and lower steering column covers.
3. Remove the retaining screws holding the instrument cluster trim panel and trim panel.
4. Remove the radio knobs, the mounting nuts.
5. Remove the heater/air conditioner knobs from their control arms. DO NOT remove the blower fan control knob.
6. Remove the heater control dash light screws, the ash-tray and the center dash facade-to-dash screws. Pull the facade out, then disconnect the cigarette lighter and the blower fan control electrical connectors.
7. Remove any remaining radio screws and pull it out part way. Disconnect the power source, the speaker coupling and the antenna from the radio, then the radio through the dash.
9. To install, reverse the removal procedures.

Windshield Wiper Motor

REMOVAL & INSTALLATION

Pickups, Van and 4-Runner

1. Disconnect the wiring from the wiper motor. Remove the motor from the fire wall.
2. Remove the nut, then pry the wiper link from the crank arm.
3. Remove the motor.
4. To install, reverse the removal procedures.

Land Cruiser

EXCEPT 1980 AND LATER–STATION WAGON

1. Using a small pry bar, disconnect the wiper link from the motor.
2. Remove the rear motor bracket bolts.
3. Disconnect the wiper motor wiring.

4. Remove the wiper motor screws and withdraw the motor.
5. To install, reverse the removal procedures.

1980 AND LATER–STATION WAGON

NOTE: On these models, the wiper motor is removed with the linkage assembly.

1. Remove the wiper arm retaining nuts, then the wiper arm and blade assemblies.
2. Remove both wiper arm pivot covers and the pivot-to-cowl attaching screws.
3. Remove the two service hole covers from the cowl area of the engine compartment.
4. Disconnect the wiring from the wiper motor.
5. From the engine compartment, remove the wiper motor plate-to-cowl screws. Withdraw the wiper motor and the linkage from the cowl panel as an assembly.
6. Pry the linkage from of the wiper motor.
7. To install, reverse the removal procedures.

Windshield Wiper Linkage

REMOVAL & INSTALLATION

Pickups, Van and 4-Runner

1. Refer to the "Wiper Motor, Removal and Installation" procedures, in this section and remove the motor.
2. Remove the wiper arms by removing their retaining nuts and working them off their shafts.
3. Remove the wiper shafts nuts/spacers and push the shafts down into the body cavity. Pull the linkage out of the cavity through the wiper motor hole.
4. To install, reverse the removal procedures.

Land Cruiser–2 Door

1. Remove the wiper arm assemblies.
2. Remove the end plate from the pivot housing.
3. Remove the wiper motor with the linkage cable.
4. Separate the wiper motor and the transmission.
5. Remove the linkage cable.
6. To install, reverse the removal procedures.

Land Cruiser–Station Wagon

1979

1. Refer to the "Wiper Motor, Removal and Installation" procedures, in this section and remove the motor.
2. Remove the. wiper arm assemblies and the instrument cluster.
3. Loosen the throttle cable to improve access to the wiper linkage.
4. Remove the linkage attachment bolts and withdraw the linkage.
5. To install, reverse the removal procedures.

1980 AND LATER

Refer to the "Engine, Removal and Installation" procedures, in this section and remove the linkage.

Instrument Cluster

REMOVAL & INSTALLATION

Pickup, Van and 4-Runner

1. Disconnect the negative battery cable.
2. Remove the upper and lower steering column covers.
3. Remove the instrument trim panel screws and the panel.
4. Disconnect the speedometer cable from the speedometer.
5. Remove the instrument panel screws and pull the panel forward. Disconnect the electrical connectors from the back of the panel and remove the panel.
6. To install, reverse the removal procedures.

Land Cruiser

1. Disconnect the speedometer cable. Remove the instrument panel screws.
2. Loosen the steering column clamp by removing the attaching bolts.
3. Pull out the instrument panel and the speedometer, disconnect the electrical connectors and remove the panel.
4. To install, reverse the removal procedures.

Fuses and Fusible Links

The fuse box is located on the left-hand side (Pickups, 4-Runner and Land Cruiser) or behind the glove box (Van), underneath the dash. All models are equipped with fusible links on the battery cables running from the positive ($+$) battery terminal.

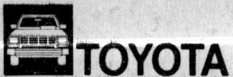
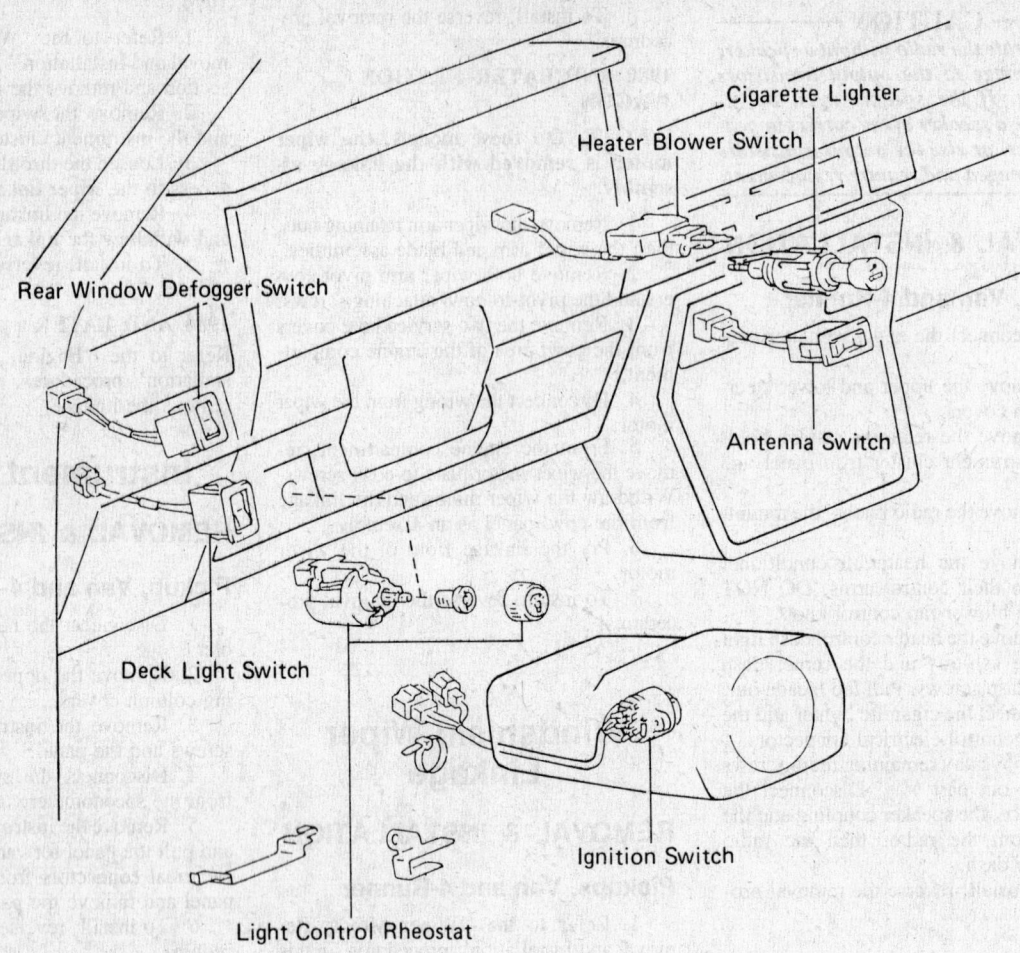

Cigarette Lighter

Heater Blower Switch

Rear Window Defogger Switch

Antenna Switch

Deck Light Switch

Ignition Switch

Light Control Rheostat

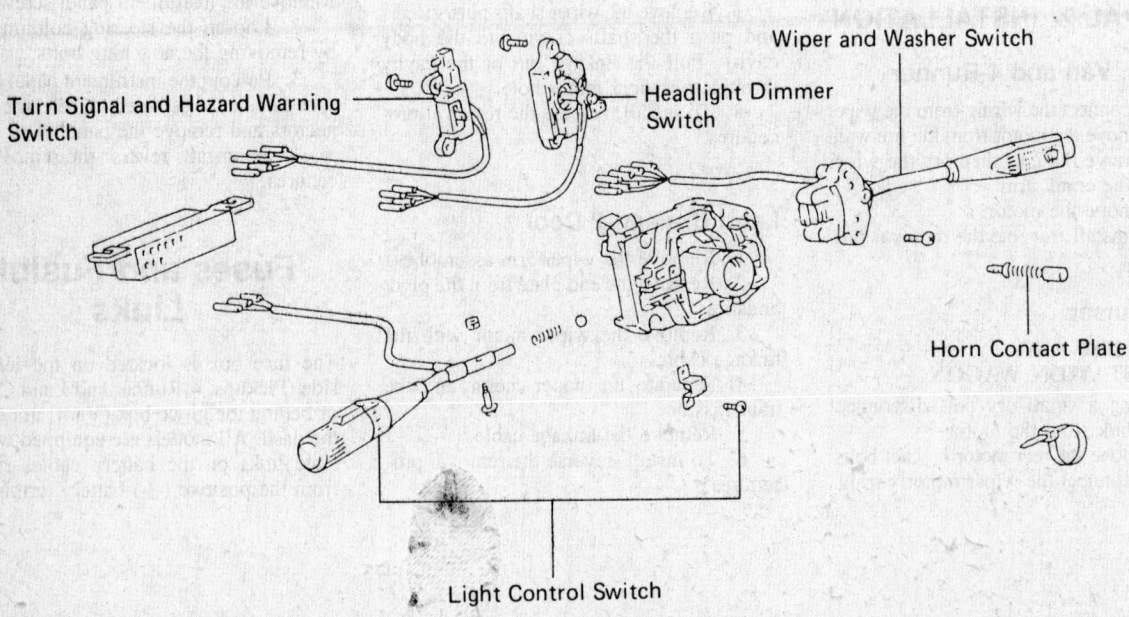

Wiper and Washer Switch

Turn Signal and Hazard Warning Switch

Headlight Dimmer Switch

Horn Contact Plate

Light Control Switch

Dash and steering column switches and relays

Volkswagen

GENERAL ENGINE SPECIFICATIONS

Year	Engine Displacement cu in. (cc)	Carburetor Type	Horsepower @ rpm (SAE)	Torque @ rpm (ft lbs) (SAE)	Bore × Stroke (in.)	Compression Ratio	Oil Pressure @ rpm (psi)
'80	88.9 (1457)	Fuel inj.	71 @ 5800	73 @ 3500	3.13 × 2.89	8:1	28 @ 2000
	89.7 (1471)	Diesel	48 @ 5000	56.5 @ 3000	3.01 × 3.15	23.5:1	27 @ 2000
'81–'82	105.0 (1715)	Fuel inj.	78 @ 5500①	88.2 @ 3100②	3.13 × 3.40	8.2:1	28 @ 2000
	97.1 (1588)	Diesel	52 @ 4800	71.5 @ 3000	3.01 × 3.40	23:1	28 @ 2000

① 74 @ 5000—California
② 89.6 @ 3000—California

GASOLINE TUNE-UP SPECIFICATIONS

Year	Engine Displacement cm³	Spark Plugs Type	Gap (in.)	Distributor Point Dwell (deg)	Point Gap (in.)	Ignition Timing (deg)	Intake Valve Opens (deg)	Compression Pressure (psi)	Idle Speed (rpm)	Valve Clearance (in.) In	Valve Clearance (in.) Ex
'80	1457	W7D N8Y	.024–.028	44–50	.016	3A @ idle	4B	142–184	850–1000	.008–.012	.016–.020
'81–'82	1715	W7D N8Y	.028	Electronic	—	3A @ idle	4B	142–184	850–1000①	.008–.012	.016–.020

NOTE: The underhood specifications sticker often reflects tune-up specification changes made in production. Sticker figures must be used if they disagree with those in this chart.
A—After Top Dead Center
B—Before Top Dead Center
① w/o idle stabilizer

DIESEL TUNE-UP SPECIFICATIONS

Year	Valve Clearance (cold) Intake (in.)	Exhaust (in.)	Intake Valve Opens (deg)	Injection Pump Setting (deg)	Injection Nozzle Pressure (psi) New	Used	Idle Speed (rpm)	Cranking Compression Pressure (psi)
'80	.008–.012	.016–.020	NA	Align marks	1849	1706	770–870	398 minimum
'81–'82	.008–.0012	.016–.020	NA	Align marks	1849	1706	800–850	398 minimum

NOTE: Valve clearance need not be adjusted unless it varies more than 0.002 in. from specification.
NA: Information not available

FIRING ORDERS

NOTE: Always remove spark plug wires one at a time.

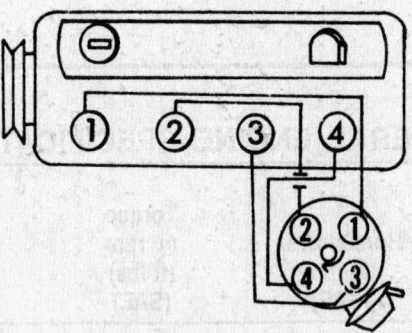

Firing order; 1-3-4-2

CAPACITIES

Year	Engine Displacement cu in. (cc)	Engine Crankcase (qts) With Filter	Without Filter	Transmission (pts) Manual	Automatic	Drive Axle (pts)	Gasoline Tank (gals)	Cooling System (pts)
'80	Gasoline	3.7	3.2	3.2①	6.4	1.6	15	9.8
	Diesel	3.7	3.2	2.6	—	1.6	15	12.6
'81–'82	Gasoline Diesel	4.7	4.2	3.2①	—	1.6	15	14.6

① 5-speed—4.2

CRANKSHAFT AND CONNECTING ROD SPECIFICATIONS

(All measurements are given in inches)

Year	Crankshaft Main Brg. Journal Dia.	Main Brg. Oil Clearance	Shaft End-Play	Thrust on No.	Connecting Rod Journal Diameter	Oil Clearance	Side Clearance (max.)
'80–'82	2.126	0.001–0.003	0.003–0.007	3	1.811	0.001–0.003	0.015

NOTE: Main and connecting rod bearings are available in three undersizes.

VALVE SPECIFICATIONS

Year	Seat Angle (deg)	Spring Test Pressure (lbs. @ in.)	Stem to Guide Clearance (in.)		Stem Diameter (in.)	
			Intake	Exhaust	Intake	Exhaust
'80–'82	45	96–106① @ 0.92 in.	0.001–0.002	0.001–0.002	0.314	0.314

NOTE: Exhaust valves must be grounded by hand.
① Outer spring, inner spring test pressure is 46–51 lbs. @ 0.72 in.

PISTON AND RING SPECIFICATIONS
(All measurements in inches)

Year, Model	Piston Clearance	Ring Gap			Ring Side Clearance		
		Top Compression	Bottom Compression	Oil Control	Top Compression	Bottom Compression	Oil Control
Gasoline Engine	0.001–0.003	0.012–0.018	0.012–0.018	0.010–0.016	0.001–0.002	0.001–0.002	0.001–0.002
Diesel Engine	0.001–0.003	0.012–0.020	0.012–0.020	0.010–0.016	0.002–0.400	0.002–0.003	0.001–0.002

NOTE: Three piston sizes are available to accommodate overbores up to 0.040 in.

TORQUE SPECIFICATIONS
(All readings in ft. lbs.)

Year	Cylinder Head Bolts	Rod Bearing Bolts	Main Bearing Bolts	Crankshaft Pulley Bolt	Flywheel To Crankshaft Bolts	Manifold	
						Intake	Exhaust
'80–'82	61①	33	47	56	54②	18	18

① 69 ft. lbs. warm
② Pressure plate to crankshaft bolts

BATTERY AND STARTER SPECIFICATIONS
(All models use 12 volt, negative ground system)

Year	Battery Amp Hour Capacity	Lock Test			No Load Test			Brush Spring Tension (oz.)	Minimum Brush Length (in.)
		Amps	Volts	Torque (ft. lbs.)	Amps	Volts	RPM		
'80–'82	45/54①	280–370	7.5	2.42	33–55	11.5	6000–8000	35.5	0.5

① w/AC

WHEEL ALIGNMENT

Year	CASTER		CAMBER		Toe-in (in.)	Steering Axis Inclination (deg)
	Range (deg)	Pref Setting (deg)	Range (deg)	Pref Setting (deg)		
'80–'82	+1°20'–2°20'	+1°50'	−10'–+50'	+20'	0.08	10°30'

TUNE-UP

VW recommends a tune-up, including new points and plugs, at 15,000 mile intervals. The only procedure required for diesel engines in this section is the valve lash adjustment and minimum/maximum engine speed checking and adjustment.

Breaker Points and Condenser

Snap off the two retaining clips on the distributor cap. Remove the cap and examine it for cracks, deterioration, or carbon tracking. Replace the cap, if necessary, by transferring one wire at a time from the old cap to the new one. Examine the rotor for corrosion or wear and replace it if questionable. Remove the dust shield. Check the points for pitting and burning. Slight imperfections on the contact surface may be filed off with a point file. It is best to replace the breaker point set. Always replace the condenser when you replace the point set.

1. Remove the rotor.
2. Unsnap the point connector from the terminal at the side of the distributor. Remove the retaining screw, and lift out the point set.
3. Install the new point set, making sure that the pin on the bottom engages the hole in the breaker plate.
4. Install the wire connector and the retaining screws (hand-tight).
5. Turn the engine with a wrench on the crankshaft pulley until the breaker arm rubbing block is in the high point of one of the cam lobes. Turn the engine only in the direction of normal rotation to avoid damage to the timing belt.
6. A 0.016 in. feeler gauge should just slip through the points. If the gap is incorrect, pivot a screwdriver in the point set notch and the two projections on the breaker plate to bring it within specifications.
7. When the gap is correct, tighten the retaining screw. Recheck the adjustment.
8. Lubricate the distributor cam with silicone grease.
9. Install the dust cover, rotor and distributor cap.
10. Check the dwell angle and the ignition timing.

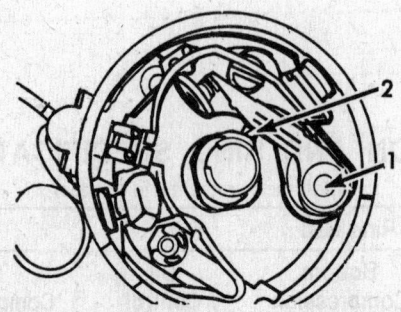

Breaker points and condenser. Lubricate at (1) with a drop of engine oil and at (2) with high melting point grease

11. The condenser is mounted on the outside of the distributor. Undo the mounting screw and the terminal block to replace.

Dwell Angle (Breaker-Point Gasoline Engines Only)

The dwell angle or cam angle is the number of degrees that the distributor cam rotates while the points are closed. There is an inverse relationship between dwell angle and point gap. Increasing the point gap will decrease the dwell angle and vice versa. Checking the dwell angle with a meter is a far more accurate method of measuring point opening than the feeler gauge method.

After setting the point gap to specification with a feeler gauge, check the dwell angle. Attach the dwell meter. The negative lead is grounded and the positive lead is connected to the primary wire. Terminal No.1 that runs from the coil to the distributor. Start the engine, let it idle and reach operating temperature, and observe the dwell on the meter. The reading should fall within the allowable range. If it does not, the gap will have to be reset. Dwell can also be checked with the engine cranking. In this case, dwell will vary between 0° and the dwell figure for that setting.

Hall Effect Electronic Ignition System

1981 and later trucks are equipped with the

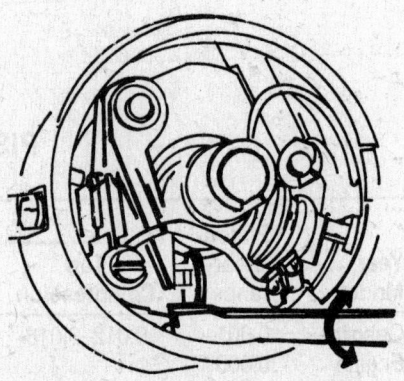

Adjusting the point gap

Hall effect electronic ignition system. The distributor contains the Hall Effect pickup assembly which replaces the breaker points assembly in conventional systems. The "Hall Effect" is a shift in magnetic field caused when one of the rotors on the distributor shaft passes the sensors mounted in the distributor. This shift performs the same function as breaker points, which is to allow the current (coil field) stored in the coil to collapse, causing a spark to run from the coil to the distributor and down to the spark plugs which make the current jump a gap between the two spark plug electrodes, causing a spark which ignites the air-fuel mixture in the combustion chamber. Since there are no breaker points and condenser to replace, the system should be maintenance free.

ELECTRONIC IGNITION PRECAUTIONS

When working on the Hall ignition, observe the following precautions to prevent damage to the ignition system.

1. Connect and disconnect test equipment only when the ignition switch is OFF.
2. Do not crank the engine with the starter for compression tests, etc., until the high tension coil wire is grounded.
3. Do not replace the original equipment coil with a conventional coil.
4. Do not install any kind of condenser to coil terminal.
5. Do not use a battery booster for longer than 1 (one) minute.
6. Do not tow cars with defective ignition systems without disconnect the plugs

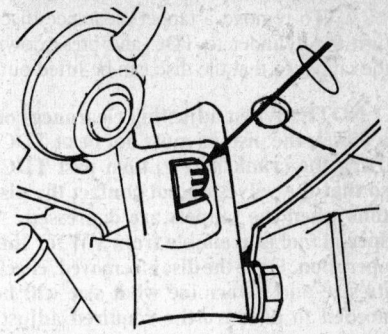
ELECTRONIC IGNITION TESTING SPECIFICATIONS

Component Tested	Specification
Rotor resistance	5000 ohms

Plug wire resistance
  With radio
  W/O radio
	
	800–1200 ohms
	0 ohms

Spark Plug Connector resistance
  Suppressed
  Not suppressed
	
	4000–6000 ohms
	800–1200 ohms

Air gap ①	0.25 mm (0.010 in.)
Inductive signal resistance ②	890–1285 ohms
Resistance from the coil tower to the negative coil terminal	5500–8000 ohms
Resistance from the coil positive terminal to the coil negative terminal	0.95–1.50 ohms
Vacuum retard	8°–10°
Vacuum advance	4°–8°
Centrifugal advance	6°–12°

① Air gap is adjustable by bending the teeth on the stator (reluctor)
② Measure resistance between the connectors to the control unit on the distributor (connects to the green and white wires)

Timing window

Timing mark aligned at 3° ATDC

on the idle stabilizer (if equipped) at the ignition control unit.

For electronic ignition component troubleshooting, refer to the Electrical section in Unit Repair.

Ignition Timing

BREAKER-POINT IGNITION SYSTEMS

1. Attach the timing light according to the manufacturer's instructions. Hook-up a dwell/tachometer since you'll need an rpm indication for correct timing.
2. Locate the timing mark opening in the clutch or torque converter housing at the rear of the engine directly behind the distributor. The OT mark stands for TDC or 0° advance. The other mark(s) identify advance and correct timing position. Mark them with chalk so that they will be more visible. Don't disconnect the vacuum line(s).
3. Start the engine and allow it to reach the normal operating temperature. The engine should be running at normal idle speed.
4. Shine the timing light at the marks.
5. The light should now be flashing when the timing mark and the V-shaped pointer are aligned.
6. If not, loosen the distributor hold-down bolt and rotate the distributor very slowly to align the marks.

7. Tighten the mounting nut when the ignition timing is correct.
8. Recheck the timing when the distributor is secured.

ELECTRONIC IGNITION SYSTEMS

1. Run the engine to operating temperature. Connect tachometer. See Electronic Ignition Precautions.
2. Stop the engine. Disconnect the plugs on the idle stabilizer (if equipped) at the control unit and plug them together.
3. Check the idle speed. It should be between 800–1000 rpm.
4. Attach the timing light according to its manufacturer's instruction. Shine the light at the timing marks (located on the flywheel; see Step 2 of Point Ignition Systems). The pointer in the hole must line up with the timing mark on the flywheel. To adjust the timing, loosen the distributor at its base and turn it until the marks are aligned.
5. Stop the engine and reconnect the plugs at the control unit.

Valve Lash

Check the valve clearance every 20,000 miles, with the engine at normal operating temperature.
1. Remove the camshaft cover and the distributor cap.

2. Set the engine at TDC on No. 1 cylinder by aligning the 0° T mark on the flywheel with the pointer and aligning the distributor rotor with the No. 1 cylinder mark on the rim of the distributor body.

NOTE: Always turn the crankshaft in the normal direction of rotation. Do not turn the engine by means of the timing belt (or camshaft bolt), because the belt will stretch or lose teeth.

3. The valve clearances of cylinder No. 1 should be checked when the valves of No. 4 cylinder overlap, i.e., when both No. 4 cylinder valves move in opposite directions simultaneously. It may be necessary to move the crankshaft slightly to find this position. When this happens, the exhaust valve is closing and the intake opening. Check and note the clearance of both the intake and exhaust valves for No. 1 cylinder.
4. Turn the crankshaft 180°. Check and note the valve clearances of cylinder No. 4 at the overlap position of cylinder No. 1.
5. Turn the crankshaft 180°. Check and note the valve clearances of cylinder No.2 at the overlap position of cylinder No. 3.
6. Compare the noted clearances with those listed in the Tune-Up Specifications. Adjustment is made by replacing the tappet clearance disc in the top of each tappet. These are available in 26 sizes ranging from 3.0mm (0.119) to 4.25mm (0.166 in.) in increments of 0.05mm (0.002 in.). The thickness of each disc is marked on the bottom.

NOTE: If a valve clearance deviates 0.002 in. or less from the specified clearance, it need not be adjusted.

7. To remove a tappet clearance disc, turn the cylinder to TDC and press down the tappet so that the disc can be lifted out.

NOTE: When adjusting clearances on a diesel, the pistons must not be at TDC. Turn the crankshaft 1/4 turn past TDC, so that the valves do not contact the pistons when the tappets are depressed. A special tool is available from VW for this operation. Once the disc is removed, check its size and determine what size will be needed to produce the required adjustment.

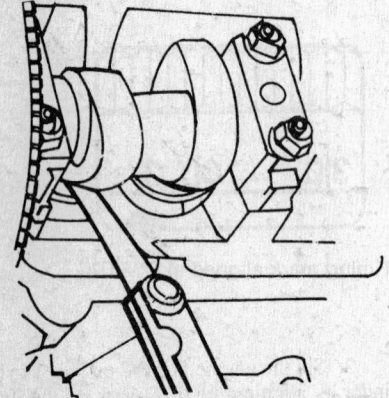

Checking valve clearance with a feeler guage

8. Install the required disc. When all the clearances have been corrected, recheck valve clearances.

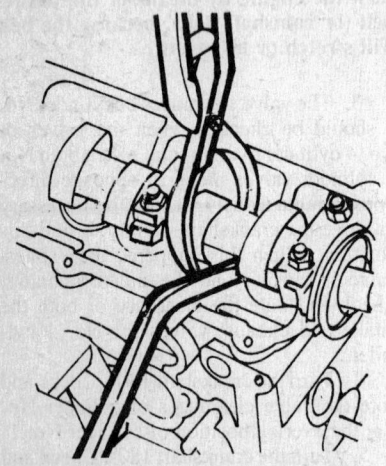

Using special tools to depress the tappet and remove the tappet clearance disc

CIS Fuel Injection

IDLE AND CO ADJUSTMENT

All Except 1980 California And 1981–82 Models

The following adjustments can be made only with a CO meter and the CO adjusting tool (VW–P377).

1. Run the engine until it reaches normal operating temperature.
2. Adjust the ignition timing to specification with the vacuum hoses connected and the engine at idle.
3. Adjust the idle speed to specification.
4. Remove the charcoal filter hose from the air cleaner except on Canadian models.
5. Turn on the headlight high beams.
6. Remove the plug from the CO adjusting hole and insert adjustment tool VW–P377. Turn the adjustment screw clockwise to raise the percentage of CO and counterclockwise to lower the percentage of CO.

CAUTION

Do not push the adjustment tool down or accelerate the engine with the tool n place.

---

7. Remove the tool after each adjustment and accelerate the engine briefly before reading the percentage of CO. The correct CO values are as follows:49 States: 1.5% M/T; 0.1% A/T 0.6 + 0.4%

1980 California, 1981–82 Models

1. The engine must be at operating temperature.
2. Disconnect crankcase breather hose at the cylinder head and plug the hose.
3. Disconnect the two plugs on the idle stabilizer at the control unit and plug them together.
4. Do not have any electrical accessories (air conditioner, lights, etc.) on.
5. Connect a tachometer and timing light. Check the timing. Adjust if necessary.
6. Check the idle speed against the specifications chart or your underhood sticker. Adjust the idle at the idle adjustment screw on the throttle chamber (880–1000 rpm).

NOTE: Only adjust the idle when the radiator fan is not on.

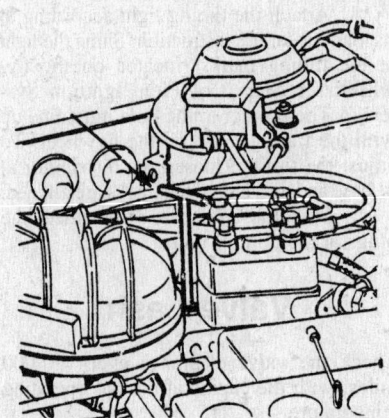

CO-adjusting tool installed—CIS fuel injection

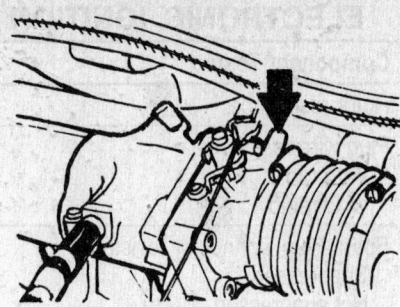

Fuel injection idle adjustment screw

Special adapter VW 1324 is necessary to use an external techometer on diesel engines

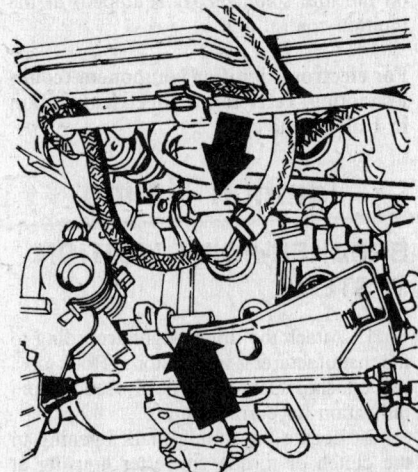

Diesel engine idle speed (upper) and maximum speed (lower) adjustment screws

Diesel Fuel Injection

IDLE SPEED/MAXIMUM SPEED ADJUSTMENTS

Volkswagen diesel engines have both an idle speed and a maximum speed adjustment. The maximum engine speed adjustment prevents the engine from over-revving

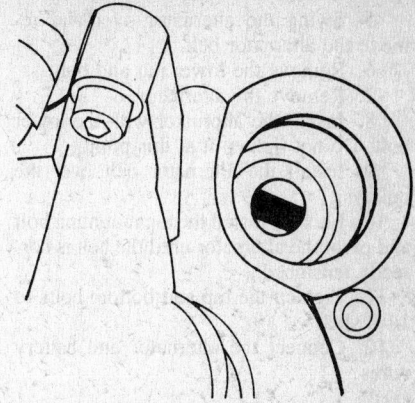
and swallowing itself whole. The adjusters are located side by side on top of the injection pump. The screw closest to the engine is the idle speed adjuster, while the outer screw is the maximum speed adjuster. The idle and maximum speed must be adjusted with the engine warm (normal operating temperature). Because the diesel engine has no conventional ignition, you will need a special adaptor (VW 1324) to connect your tachometer, or use the tachometer in the instrument panel, if equipped. You should check with the manufacturer of your tachometer to see if it will work with diesel engines. Adjust all engines to 770–870 rpm (through 1980) or 800–850 rpm (from 1981). When adjustment is correct, lock the locknut on the screw and apply a dab of paint of non-hardening thread sealer to prevent the screw from vibrating loose.

The maximum speed for all engines is between 5500–5600 rpm (through 1980) or 5300–5400 rpm (from 1981). If it is not in this range, loosen the screw and correct the speed (turning the screw clockwise decreases rpm). Lock the nut on the adjusting screw and apply a dab of paint in the same manner as you did on the idle screw.

CAUTION

Do not attempt to squeeze more power out of your engine by raising the maximum speed.

ENGINE ELECTRICAL

Distributor

REMOVAL & INSTALLATION

1. Disconnect the coil high tension wire.
2. Detach the primary wire.
3. Remove the distributor cap.
4. Turn the engine until the rotor aligns with the index mark on the outer edge of the distributor. This is the No. 1 position. Mark the bottom of the distributor housing and its mounting flange on the engine.
5. Remove the bolt and lift off the retaining flange. Lift the distributor straight out of the engine.
6. If the engine has not been disturbed while the distributor was out, i.e., the crankshaft was not turned, then reinstall the distributor in the reverse order of removal. Carefully align the marks.
7. If the engine has been rotated while the distributor was out, then proceed as follows:
8. Turn the crankshaft so that No.1 piston is on its compression stroke and the 0°T

timing mark is aligned with the V-shaped pointer.9.

Turn the distributor so that the rotor points approximately 15° before the No. 1 cylinder position on the distributor.

10. Insert the distributor into the engine block. If the oil pump drive doesn't engage, remove the distributor and, using a long screwdriver turn the pump shaft so that it is parallel to the centerline of the crankshaft.

11. Install the distributor, aligning the marks. Tighten the retaining nut.

12. Install the cap. Adjust the ignition timing.

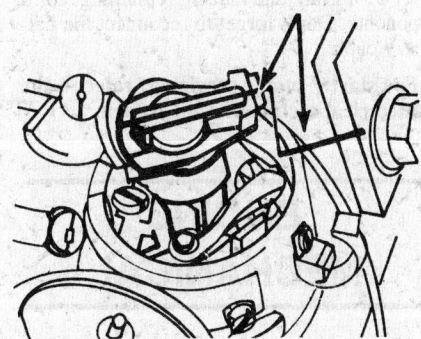

Rotor adjustment with notch for cylinder no. 1

7. Do not use test lamps of more than 12 volts (V) for checking diode continuity.

8. Do not short across or ground any of the terminals on the AC generator.

9. The polarity of the battery, generator, and regulator must be matched and

Alternator

ALTERNATOR PRECAUTIONS

An alternating current (AC) generator (alternator) is used. Unlike the direct current (DC) generators, there are several precautions which must be strictly observed in order to avoid damaging the unit.

1. Reversing the battery connections will result in damage to the diodes.
2. Booster batteries should be connected positive to positive and negative to a ground point of the vehicle being jumped.
3. Never use a fast charger as a booster to start cars with AC circuits.
4. When serving the battery with a fast charger, always disconnect the car battery cables.
5. Never attempt to polarize an AC generator.
6. Avoid long soldering times when replacing diodes or transistors. Prolonged heat is damaging to AC generators.

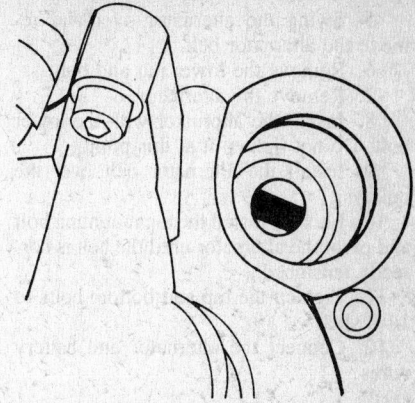

The oil pump drive should be parallel to the crankshaft

considered before making any electrical connections within the system.

10. Never operate the AC generator on an open circuit. Make sure that all connections within the circuit are clean and tight.

11. Disconnect the battery terminals when performing any service on the electrical system. This will eliminate the possibility of accidental reversal of polarity.

12. Disconnect the battery ground cable if arc welding is to be done on any part of the car.

Removing the lower alternator bolt through the timing cover

REMOVAL & INSTALLATION

The alternator and voltage regulator are combined in one housing. No voltage adjustment can be made with this unit. The regulator can be replaced without removing the alternator. Unbolt the regulator and remove from the rear.

1. Disconnect the battery cables.
2. Remove the multi-connector retaining bracket and unplug the connector from the rear of the alternator.
3. Loosen and remove the top mounting nut and bolt.
4. Using a socket inserted through the timing belt cover (it is not necessary to remove the cover), loosen the lower mounting bolt.

5. Swing the alternator over and remove the alternator belt.

6. Remove the lower nut and bolt.

7. Remove the alternator.

8. Install the alternator with the lower bolt. Do not tighten it at this point.

9. Install the alternator belt over the pulleys.

10. Loosely install the top mounting bolt and pivot the alternator until the belt is correctly tensioned.

11. Tighten the top and bottom bolts to 14 ft. lbs.

12. Connect the alternator and battery wires.

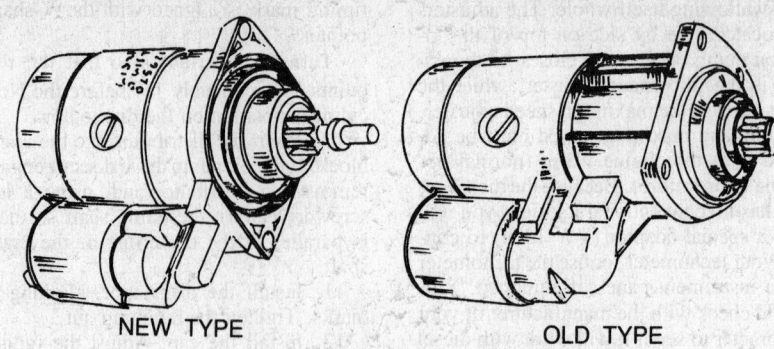

NEW TYPE   OLD TYPE

New and old type starters are not interchangeable

BELT REPLACEMENT AND TENSIONNG

1. Loosen the top alternator mounting bolt.

2. Using a socket inserted through the timing belt cover loosen the lower mounting bolt.

3. Use a wooden hammer handle or a broomstick to lever the alternator over and remove the belt.

4. Slip the new belt over the pulleys.

5. Pry the alternator over until the belt deflection midway between the crankshaft pulley and the alternator pulley is approx. ½ in.

6. Securely tighten the mounting bolts.

Starter

A new type of starter has been installed on some models that are equipped with a manual transmission. The new style starter is not interchangeable with the old design. The new starter does not need a rear support bracket.

REMOVAL & INSTALLATION

Old Style

1. Disconnect the battery ground cable.

2. Raise the front of the car and support on jackstands.

3. Mark with tape and then disconnect the wires from the starter solenoid.

4. Disconnect the large cable.

5. Remove the starter retaining nuts.

6. Unscrew the bolt. Remove the starter.

7. Installation of the starter is carried out in reverse order of removal.

New Style

1. Disconnect the battery ground cable.

2. Support the weight of the engine with either Volkswagen special tool 10–222 or use a jack with a block of wood under the oil pan. Don't jack the engine too high, just

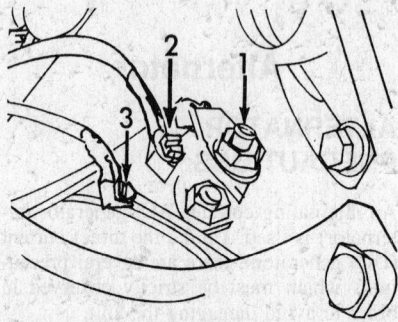

Starter electrical connections: (1) solenoid, (2) coil, (3) positive battery cable

take the weight off the motor mounts. Be careful not to bend the oil pan.

3. Remove the engine/transmission cover plate.

4. Unbolt and remove the starter side motor mount and carrier.

5. Disconnect and mark the starter wiring.

6. Remove the bolts holding the starter and remove the starter.

7. Install the starter and tighten the nuts and bolts to 14 ft. lbs.

8. Install the engine mount and carrier.

9. Install and attach remaining components. Don't forget to reconnect the battery cable.

For starter motor servicing, refer to the Electrical section of Unit Repair.

ENGINE MECHANICAL

Engine

REMOVAL & INSTALLATION

Gasoline Engines
MANUAL TRANSMISSION

NOTE: The engine and transmission are removed as an assembly.

1. Disconnect the battery ground cable.

2. Drain the coolant by unvolting the lower water pump flange or by removing the hoses.

——— CAUTION ———
Do not disconnect or loosen any refrigerant hose connections during engine removal on cars equipped with air conditioning.

3. On cars equipped with air conditioning: Loosen the compressor support bolts and remove the compressor. Remove the radiator cooling fan, air ducts and radiator. Remove the condenser. Place the air conditioning compressor and condenser out of the way without disconnecting any refrigerant lines.

4. Remove the radiator with the air ducts and fan.

5. Detach and label all the electrical wires connecting the engine to the body.

6. Disconnect and plug the fuel line at the fuel pump. Detach the coolant hoses at the left end of the engine. Disconnect the accelerator cable and remove the air cleaner.

7. Disconnect the speedometer cable from the transmission. Detach the clutch cable.

8. Remove the engine support to the right of the starter.

9. Remove the headlight caps inside the engine compartment.

10. Unbolt the driveshafts from the transmission and wire them up.

11. Unbolt the exhaust pipe from the manifold and unbolt the exhaust pipe brace.

12. Unbolt the transmission rear mount from the body (alongside the tunnel).

13. Detach the ground strap from the transmission and body.

14. Remove the shift linkage.

15. Attach a chain sling to the alternator bracket and the lifting eye at the left end of the engine. Lift the engine and transmission slightly.

16. Detach the engine carrier from the body and remove the left transmission carrier.

17. Lift the engine/transmission assembly carefully out of the car.

18. To separate the engine and transmission, turn the flywheel to align the lug on the flywheel (to the left of TDC) with the pointer in the opening. The engine and transmission can only be separated in this position. Remove the cover plate over the driveshaft flange and remove the engine to transmission bolts and the transmission housing cover plate.

19. To install the engine: Attach the transmission to the engine, the recess in the flywheel edge must be at 3:00 o'clock (facing the left end of the engine). Torque the engine to transmission bolts to 40 ft. lbs. Lift the engine/transmission assembly into place. Loosen the bolts for the engine and transmission mounts. Move the engine assembly from side to side until the rear transmission mount is straight. Center the left and right transmission mounts and tighten all transmission bolts. Push the front mount upward to center the rubber cone, then tighten the mount. Loosen the exhaust pipe clamps, release any strain, then tighten the clamps. Torque the 10 mm bolts to 29 ft. lbs. Torque the driveshaft flange bolts to 32 ft. lbs. Refill the cooling system.

AUTOMATIC TRANSMISSION

NOTE: The engine and transmission are removed as an assembly.

1. Shift the transmission into "Park." Disconnect both battery cables.
2. Drain the coolant by unbolting the lower water pump flange or by removing the hoses.

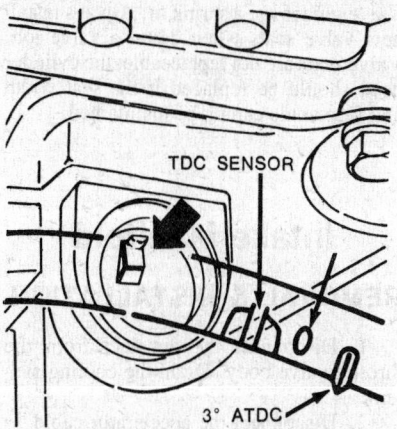

Aligning the flywheel for manual transmission and engine separation

—————— CAUTION ——————
Do not disconnect or loosen any refrigerant hose connections during engine removal on cars equipped with air conditioning.

3. On cars equipped with air conditioning, proceed as follows: Loosen the compressor support bolts and remove the compressor. Remove the radiator cooling fan, air ducts, and radiator. Remove the condenser. Place the air conditioning components out of the way without disconnecting any refrigerant lines.

4. Remove the radiator with the air ducts and fan.
5. Remove the air cleaner.
6. Detach the speedometer cable from the transmission.
7. Detach and label all electrical wires connecting the engine to the body. Detach the coolant hoses.
8. Remove the screws holding the accelerator cable bracket to the carburetor float bowl (do not disassemble linkage), detach the end of the gearshaft selector cable from the transmission, detach the accelerator cable and pedal cable at the transmission, and remove the two bracket bolts behind this linkage on the transmission.
9. Unbolt the exhaust pipe from the manifold.
10. Remove the rear transmission mount. Unbolt the driveshafts and wire them up out of the way.
11. Remove the converter cover plate and remove the three torque converter to drive plate bolts.
12. Attach a chain sling to the alternator bracket and the lifting eye at the left end of the engine. It may be necessary to remove the alternator. Lift the engine and transmission slightly.
13. Detach the engine front mounting support; remove the left transmission carrier and the right engine carrier.
14. Lift the engine/transmission assembly carefully out of the car.
15. The transmission can now be detached from the engine.
16. To install the engine: The engine to transmission bolts should be torqued to 40 ft. lbs. Lift the engine/transmission assembly into place and install the left transmission carrier, tightening first the body, then the transmission bolts. Lower the assembly to attach the engine carrier to the body, tightening the bolts to 40 ft. lbs. Install the engine mounting support. Check that all mounts and clamps are free of strain. Torque converter bolts should be torqued to 21 ft. lbs. and driveshaft bolts to 32 ft. lbs. Refill the cooling system. Check the adjustment of transmission and carburetor linkages.

Diesel Engines

NOTE: The diesel engine is removed with the transmission attached.

1. Disconnect the battery.
2. Disconnect the radiator hoses and drain the coolant. It can be saved for reuse, if it's not too old.
3. Remove the radiator complete with fan.
4. Remove the alternator.
5. Disconnect the fuel filter and set it aside near the windshield washer reservoir.
6. Detach the supply and return lines from the injection pump.

7. Disconnect the accelerator cable from the lever on the injection pump and remove the injection pump complete with bracket.
8. Disconnect the cold start cable from the pump.
9. Disconnect and label all electrical wires and leads.
10. Remove the front transmission mount.
11. Disconnect the clutch cable.
12. Remove the relay rod and connecting rod from the transmission and turn the relay lever shaft to the rear.
13. Disconnect the selector rod.
14. Unbolt the driveshafts and wire them up out of the way. Remove the rear support.
15. Disconnect the exhaust pipe at the manifold and remove the rear transmission mount.
16. Attach a lifting sling to the engine and take the weight from the engine mounts. Remove the left and right transmission mounts.
17. Carefully guide the engine out of the car while turning it slightly.
18. To separate the engine from the transmission, unscrew the plug from the TDC sensor opening and turn the flywheel to align the mark on the flywheel with the pointer. The engine/transmission can only be separated in this position.
19. Remove the cover plate over the driveshaft flange and remove the engine-to-transmission bolts.
20. Separate the engine and transmission.
21. Installation is the reverse of removal. Turn the flywheel so that the recess in the flywheel is level with the driveshaft flange. Lower the engine into the car and attach the left transmission mount to the transmission first. Align the rear transmission mount, center the engine/transmission and center the front transmission mount. Adjust the accelerator and cold start cables and bleed the injection system.

Cylinder Head

REMOVAL & INSTALLATION

The engine should be cold before the cylinder head can be removed. The head is retained by 10 socket head bolts. It can be removed without removing the intake and exhaust manifolds.

NOTE: 12 point socket head bolts are used. These should be used in complete sets only and need not be retorqued.

Gas Engines

1. Disconnect the battery ground cable.
2. Drain the cooling system.
3. Disconnect the air duct from the throttle valve assembly.
4. Disconnect the throttle valve assembly.

5. Remove the injectors and disconnect the line from the cold start valve.

6. Disconnect the radiator and heater hoses.

7. Disconnect the vacuum and PCV lines (label lines for installation).

8. Remove the auxiliary air regulator from the intake manifold.

9. Disconnect all electrical lines and remove the spark plugs (label all lines and wires for installation).

10. Separate the exhaust pipe from the exhaust manifold.

11. Remove the EGR line from the exhaust manifold.

12. Remove the intake manifold.

13. Remove the timing belt cover and belt.

14. Loosen the cylinder head bolts in the reverse of the tightening sequence.

15. Remove the bolts and lift the head straight off.

16. Check the flatness of the cylinder block.

17. Install the new cylinder head gasket with the word TOP or OBEN facing up.

18. Install bolts No. 10 and 8 first; these holes are smaller and will properly locate the gasket and cylinder head.

19. Install the remaining bolts. Tighten them in three stages using the sequence shown in the illustration. Cylinder head bolts

must be torqued cold to 55 ft. lbs., then tightened 1/4 turn more.

20. Install the remaining components in the reverse order of removal.

Diesel Engines

NOTE: The head is retained by Allen bolts. The engine should be cold when the head is removed. The word TOP or OBEN on the new gasket should face up.

1. Disconnect the battery ground cable.

2. Drain the cooling system.

3. Remove the air cleaner.

4. Disconnect the fuel lines. Disconnect and tag all electrical wires and leads.

5. Separate the exhaust pipe from the manifold. Disconnect the radiator and heater hoses.

6. Remove the timing cover and belt.

7. Loosen the cylinder head bolts in reverse order of the tightening sequence.

8. Remove the head. Do not lay the head on the gasket surface with the injectors installed. Support it at the ends on strips of wood.

9. Install the cylinder head with a new gasket. Be sure the new gasket has the same number of notches and the same identifying number as the old one, unless the pistons were also replaced. Install bolts No. 8 and 10 first and torque the bolts to the specification in the proper sequence.

OVERHAUL

Valve guides are a shrink fit. Always install new valve seals when doing a valve job. Valve seats are not replaceable; the cylinder head should be replaced if the seat width and face angle cannot be maintained.

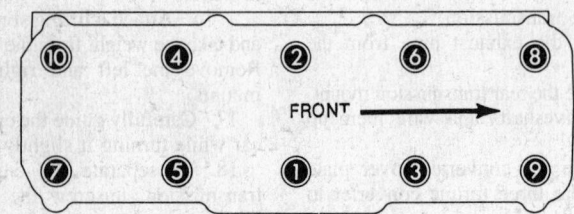

Cylinder head torque sequence

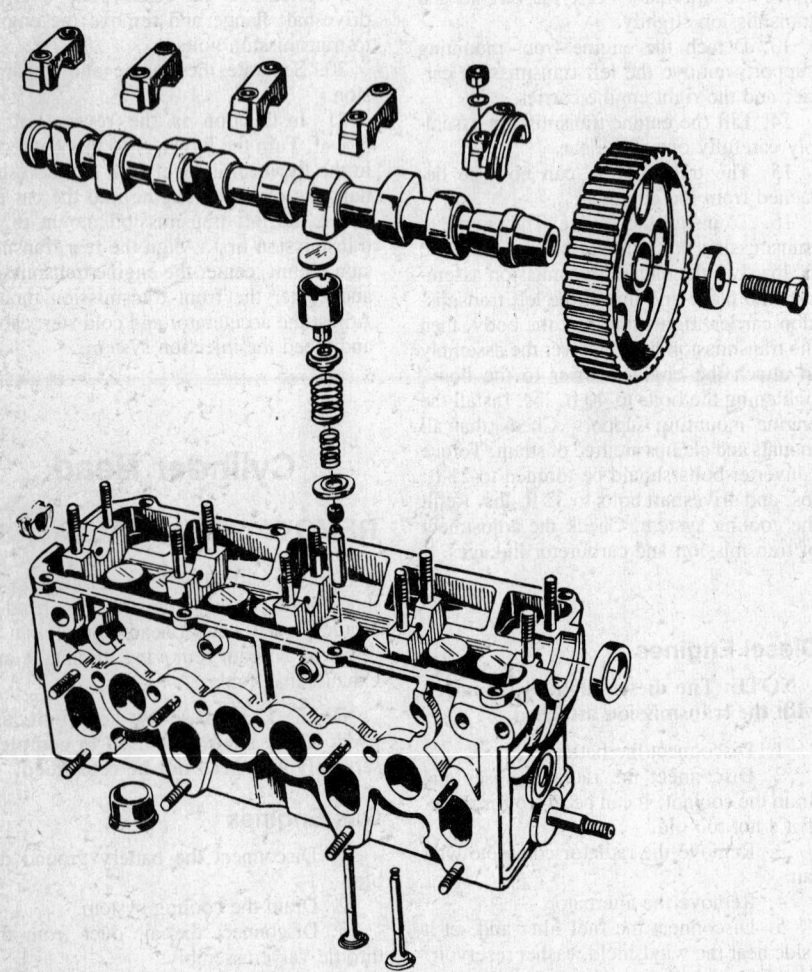

Exploded view of the cylinder showing valve train components

Intake Manifold

REMOVAL & INSTALLATION

1. Disconnect the air duct from the throttle valve body. Drain the cooling system.

2. Disconnect the accelerator cable.

3. Remove the injectors and disconnect the line from the cold start valve.

4. Disconnect all coolant hoses.

5. Disconnect all vacuum and emission control hoses (label all hoses for installation).

6. Remove the auxiliary air regulator.

7. Disconnect all electrical lines (label all wires for installation).

8. Disconnect the EGR line from the exhaust manifold.

9. Loosen and remove the retaining bolts and lift off the manifold.

10. Install a new gasket. Install the manifold and tighten the bolts to 18 ft. lbs.

11. Install the remaining components in the reverse order of removal.

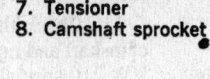

Exhaust Manifold

REMOVAL & INSTALLATION

1. Disconnect the EGR tube from the exhaust manifold.

2. Remove the interfering air pump components if so equipped.

3. Remove the air cleaner hose from the exhaust manifold.

4. Disconnect the intake manifold support.

5. Separate the exhaust pipe from the manifold.

6. Remove the retaining nuts and remove thee manifold.

7. Clean the cylinder head and manifold mating surfaces.

8. Install the exhaust manifold using a new gasket.

9. Tighten the nuts to 18 ft. lbs. Work from the inside out.

10. Install the remaining components in the reverse order of removal. Use a new manifold flange gasket.

Timing Belt Cover

REMOVAL & INSTALLATION

1. Loosen the alternator mounting bolts.

2. Pivot the alternator and slip the drive belt off the sprockets.

3. Unscrew the cover retaining nuts and remove the cover.

4. Reposition the spacers on the studs and then install the washers and nuts.

5. Install the alternator belt and adjust the tension.

Timing Belt

NOTE: The timing belt is designed to last for more than 60,000 miles and does not normally require tension adjustments. If the belt is removed or replaced, the basic valve timing must be checked and the belt retensioned.

REMOVAL, INSTALLATION AND TENSIONING

Gasoline Engines

Timing belt installation will be easier if the engine is set to TDC prior to belt removal. The 0°T mark will be aligned with the pointer on the bell housing, and the mark on the rear face of the camshaft pulley will align with the camshaft cover gasket on the left. Also, the V-notch in the crankshaft pulley should align with the dot mark on the intermediate shaft, and the distributor rotor should be pointing toward the mark on the rim of the housing. If the belt has broken and timing is off, remove the belt cover an belt, then set the engine to TDC before installing the belt, as outlined in Steps 5–7.

1. Alternator belt
2. Belt pulleys
3. Timing gear cover
4. Crankshaft sprocket
5. Intermediate sprocket
6. Drive belt
7. Tensioner
8. Camshaft sprocket

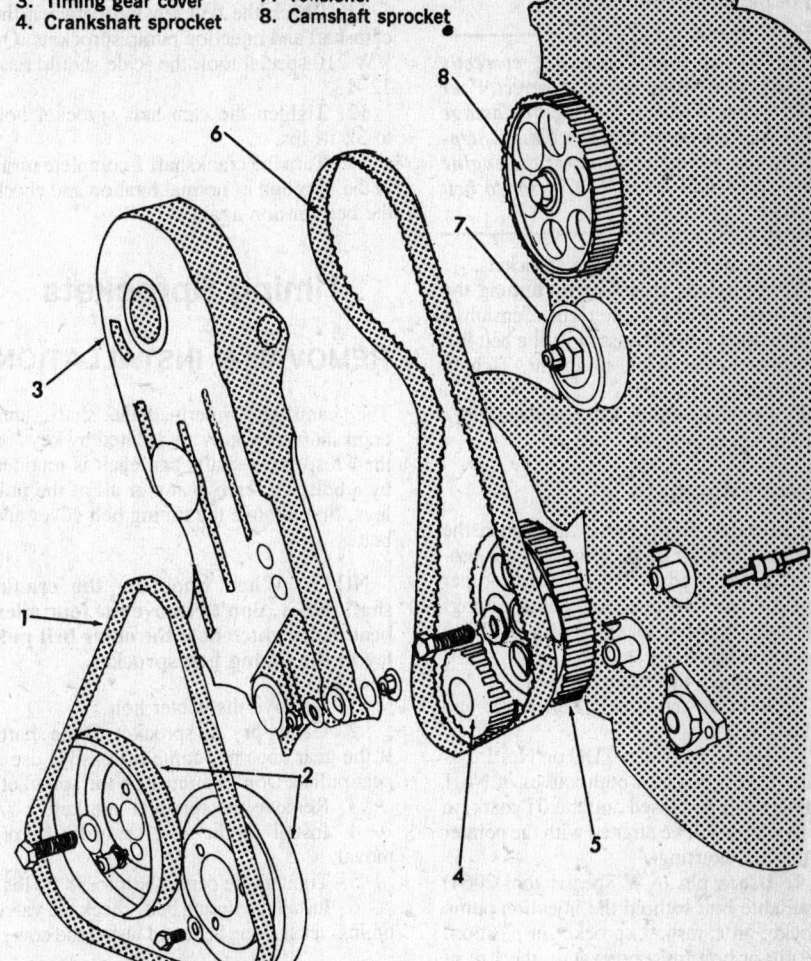

Exploded view of camshaft drive arrangement

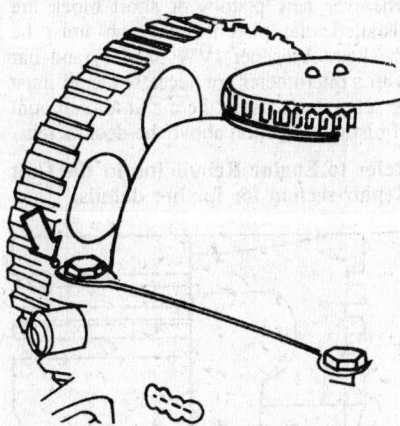

Camshaft sprocket alignment

1. Remove the timing belt cover.

2. While holding the large hex on the tension sprocket, loosen the pulley locknut.

3. Release the tensioner from the timing belt.

4. Slide the belt off the three toothed sprockets and remove it.

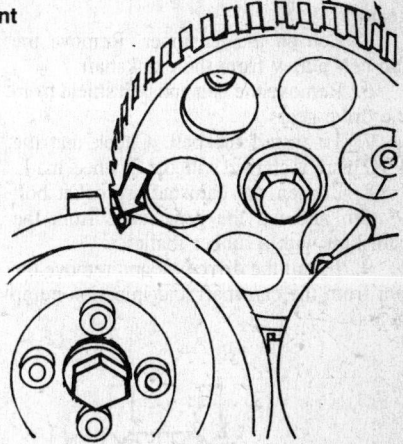

Crankshaft pulley and intermediate shaft alignment

5. Turn the crankshaft until No. 1 cylinder is at TDC. At this point, the 0°T mark will be aligned wit the pointer on the bell housing.

6. Align the timing mark on the rear face of the camshaft pulley with the camshaft cover gasket on the left.

7. Align the V-notch in the crankshaft pulley with the dot mark on the interme-

diate shaft. The distributor rotor should be pointing to the No. 1 cylinder mark on the rim of the distributor.

CAUTION

If the timing marks are not correctly aligned, valve timing will be incorrect. Poor performance and serious engine damage can result from improper valve timing. Steps 5–7 should not be necessary if the engine was in time and set to TDC prior to belt removal.

---

8. Install the belt on the sprockets.
9. Adjust the tensioner by turning the large tensioner hex to the right. Tension is correct when you can just twist the belt 90° with two fingers at the midpoint. Tighten the locknut to 32 ft. lbs.
10. Install the timing belt cover and check the ignition timing.

Diesel Engine

The drive belt on the diesel also drives the injection pump. It is necessary that this procedure is followed exactly to ensure proper valve timing and injection pump timing. You will also need special tool VW 210 to properly tension the belt.

1. Remove the alternator belt.
2. Remove the timing belt cover and rocker cover.
3. Set the engine at TDC on No. 1 cylinder. In this position both valves of No.1 cylinder will be closed and the 0T mark on the flywheel will be aligned with the pointer on the bell housing.
4. Use a pin (VW special tool 2064) or suitable bolt to hold the injection pump sprocket an camshaft sprocket in position. The pin or bolt must be exactly the size of the hole. There can be no "slop" in the gears.
5. Loosen the tensioner. Remove the fan belt pulley from the crankshaft.
6. Remove the belt and belt shield from the drive gears.
7. To install the belt. Check that the TDC mark is aligned with the flywheel mark.
8. Loosen the camshaft sprocket bolt ½ turn and tap the gear loose from the camshaft with a rubber mallet.
9. Install the drive belt and remove the pin from the camshaft and injection pump gears.

10. Tension the belt by turning the tensioner to the right.
11. Check the belt tension between the camshaft and injection pump sprockets. On VW 210 special tool, the scale should read 12–13.
12. Tighten the camshaft sprocket bolt to 32 ft. lbs.
13. Turn the crankshaft 2 complete turns in the direction of normal rotation and check the belt tension again.

Timing Sprockets

REMOVAL & INSTALLATION

The camshaft, intermediate shaft, and crankshaft sprockets are located by keys on their respective shafts and each is retained by a bolt. To remove any or all of the pulleys, first remove the timing belt cover and belt.

NOTE: When removing the crankshaft pulley, don't remove the four allen head bolts which hold the outer belt pulley to the timing belt sprocket.

1. Remove the center bolt.
2. Gently pry the sprocket off the shaft. If the gear does not come off easily, use a gear puller. Don't hammer on the sprocket.
3. Remove the sprocket and key.
4. Install in the reverse order of removal.
5. Tighten the center bolt to 58 ft. lbs.
6. Install the timing belt, check the valve timing, tension the belt, and install the cover.

DIESEL ENGINES

The same installation procedures apply to the diesel as to the gas engine. However, wherever new pistons or short block are installed, the piston projection must be checked. A spacer (VW 385/17) and bar with a micrometer are necessary, and must be set up to measure the maximum amount of piston projection above the deck height.

Refer to Engine Rebuilding in the Unit Repair section for further details.

Camshaft

REMOVAL & INSTALLATION

1. Remove the timing belt.
2. Remove the camshaft sprocket.
3. Remove the air cleaner
4. Remove the camshaft cover.
5. Unscrew and remove the Nos. 1, 3 and 5 bearing caps.
6. Unscrew the Nos. 2 and 4 bearing caps, diagonally and in increments.
7. Lift the camshaft out of the cylinder head.
8. Lubricate the camshaft out of the cylinder head.
9. Replace the camshaft oil seal with a new one whenever the cam is removed.
10. Install the Nos. 1, 3 and 5 bearing caps and tighten the nuts to 14 ft. lbs. Note that the bores are offset, and the numbers are not always on the same side.
11. Install the Nos. 2 and 4 bearing caps and diagonally tighten the nuts to 14 ft. lbs.

NOTE: If checking end-play, install a dial indicator so that the feeler touches the camshaft snout. End-play should be no more than 0.006 in.

12. Replace the seal in the No. 1 bearing cap. If necessary, replace the end plug in the cylinder head.
13. Install the camshaft cover.
14. Install the camshaft pulley and the timing belt.
15. Check the valve clearance.

Piston and Connecting Rods

GASOLINE ENGINES

The pistons must be installed in the block with the arrow at the edge of the crown facing toward the right front wheel. The connecting rod and cap alignment casting grooves must face the intermediate shaft. New connecting rod bolts must always be

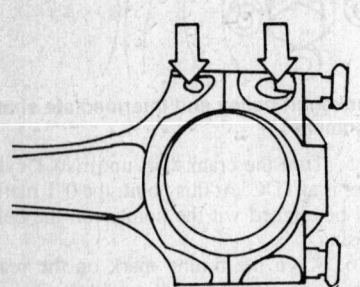

The connecting rod and cap alignment casting grooves must face the intermediate shaft

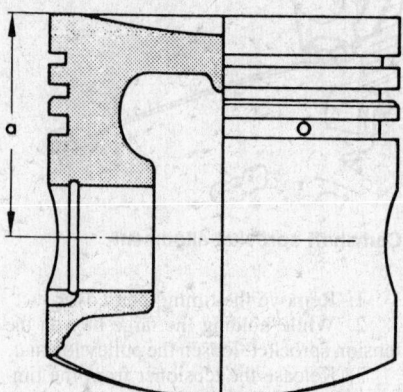

Piston height measurement, diesel engine

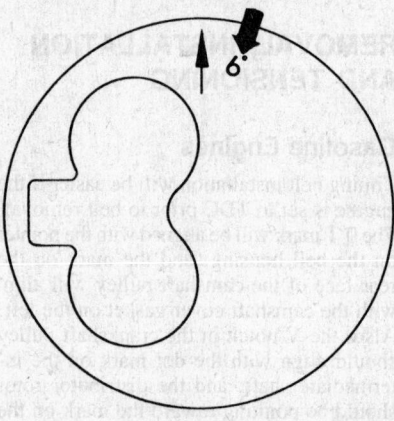

Diesel pistons are identified by the number "9" next to the arrow denoting installation direction

used. The pistons must be heated to 140°F in an oven before the piston pins can be pressed in. Three piston oversizes are available to accommodate overbores up to 0.040 in. There is a piston size code stamped on the cylinder block above the water pump.

ENGINE LUBRICATION

The lubrication system is conventional wet-sump design. The gear type oil pump is driven by the intermediate shaft. a pressure relief valve limits pressure and prevent extreme pressure from developing in the system. All oil is filtered by a full flow replaceable filter. A bypass valve assures lubrication in the event the filter becomes plugged. The oil pressure switch is located at the end of the cylinder head gallery (the end of the system) to assure accurate pressure readings.

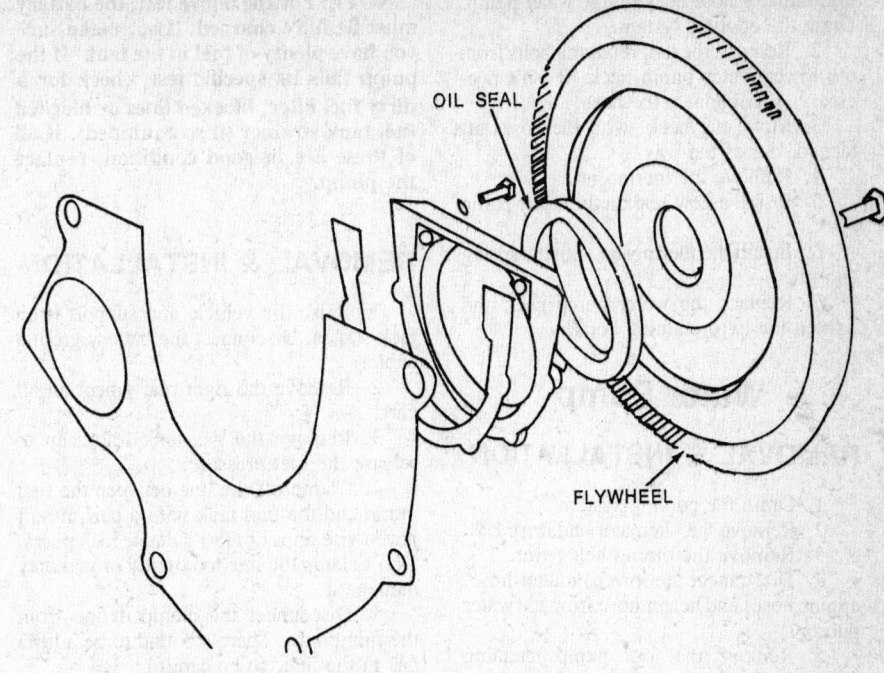

Rear main oil seal assembly

Oil Pan

REMOVAL & INSTALLATION

1. Drain the engine oil.
2. Loosen and remove the bolts retaining the oil pan.
3. Lower the pan from the car.
4. Install the pan using a new oil pan gasket.
5. Tighten the retaining bolt in a crisscross pattern. Tighten hex head bolts to 14 ft. lbs., or Allen head bolts to 7 ft. lbs.
6. Refill the engine with oil. Start the engine and examine the pan for leaks.

Rear Main Oil Seal

REPLACEMENT

The engine should be removed from the car. See Transmission Removal and Installation.

1. Remove the transmission and flywheel.
2. Using a small prybar, very carefully pry the old seal out of the support ring.
3. Remove the seal.
4. Lightly oil the replacement seal and then press it into place using a circular piece of flat metal. Be careful not to damage the seal or score the crankshaft.
5. Install the flywheel and transmission. Flywheel-to-engine bolts are tightened to 36 ft. lbs.

Oil Pump

REMOVAL & INSTALLATION

1. Remove the oil pan.
2. Remove the two mounting bolts.
3. Pull the oil pump down and out of the engine.
4. Unscrew the two bolts and separate the pump halves.
5. Remove the driveshaft and gear from the upper body.
6. Clean the bottom half in solvent. Pry up the metal edges to remove the filter screen for cleaning.
7. Examine the gears and driveshaft for wear or damage. Replace them if necessary.
8. Reassemble the pump halves.
9. Prime the pump with oil and install in the reverse order of removal.

ENGINE COOLING

The cooling system consists of a belt driven, external water pump, thermostat, radiator, and thermostatically controlled electric cooling fan. When the engine is cold the thermostat is closed and blocks the water from the radiator so the coolant is circulated only through the engine. When the engine warms up, the thermostat opens and the

radiator is included in thee coolant circuit. The thermostatic switch is in the bottom of the radiator and turns the electrical fan on at 199°F off at 186°F. This reduces power loss and engine noise.

—— CAUTION ——
The fan may run with the engine shut off. Keep fingers out of the way when the engine is warm.

Radiator and Fan

REMOVAL & INSTALLATION

1. Drain the cooling system.
2. Remove the inner shroud mounting bolts.
3. Disconnect the lower radiator hose.
4. Disconnect the thermostatic switch lead.
5. Remove the lower radiator shroud.
6. Remove the lower radiator mounting units.
7. Disconnect the upper radiator hose.
8. Detach the upper radiator shroud.
9. Disconnect the heater and intake manifold hoses.
10. Remove the side mounting bolts and lift the radiator and fan out as an assembly.
11. Installation is the reverse of removal.

Thermostat

REMOVAL & INSTALLATION

1. The thermostat is located in the bot-

tom radiator hose neck on the water pump. Drain the cooling system.

2. Remove the two retaining bolts from the lower water pump neck. It isn't necessary to disconnect the hose.

3. Move the neck, with the hoses attached, out of the way.

4. Remove the thermostat.

5. Install a new seal on the water pump neck.

6. Install the thermostat with the spring end up.

7. Replace the water pump neck and tighten the two retaining bolts.

Water Pump

REMOVAL & INSTALLATION

1. Drain the cooling system.
2. Remove the alternator and drive belt.
3. Remove the timing belt cover.
4. Disconnect the lower radiator hose, engine hose, and heater hose from the water pump.
5. Remove the four pump retaining bolts. Notice where the different length bolts are located.
6. Turn the pump slightly and lift it out of the engine block.
7. Installation is the reverse of removal. Use a new seal on the mating surface with the engine.

FUEL SYSTEM

Fuel Pump

TESTING

1. Have a helper operate the starter. Listen at the rear wheel to determine if the pump is running.

2. If the pump is not running, check the fuse on the front of the fuel pump relay.

3. If the fuse is good, replace the fuel pump relay.

4. If the fuel pump still does not operate, the fuel pump is faulty and must be replaced.

FUEL PUMP DELIVERY

1. Check the condition of the fuel filter, make sure it is clean.

2. Connect a jumper wire between the #1 terminal on the ignition coil and ground.

3. Disconnect the return fuel line and hold it in a measuring container with a capacity of 1 quart or 1000 cc.

4. Have a helper run the starter for 30 seconds while watching the quantity of fuel delivered. The minimum allowable flow is 900 cc (^{9}/$_{10}$ of a quart) in 30 seconds.

NOTE: For the above test, the battery must be fully charged. Also, make sure you have plenty of fuel in the tank. If the pump fails its specific test, check for a dirty fuel filter, blocked lines or blocked fuel tank strainer (if so equipped). If all of these are in good condition, replace the pump.

REMOVAL & INSTALLATION

1. Raise the vehicle and support it on jack stands. Disconnect the battery ground cable.

2. Remove the right rear wheel on all cars.

3. Remove the gas tank filler cap to release the fuel pressure.

4. Clamp off the line between the fuel pump and the fuel tank with a pair of soft jawed vise grips or other suitable lock pliers. Don't clamp the line too tightly or you may damage it.

5. Disconnect the clamped line from the fuel pump. There's bound to be a little gas in the line, so be careful.

6. If you vehicle has an accumulator mounted next to the fuel pump, disconnect the fuel lines from the accumulator. Disconnect the wiring from the fuel pump and remove all other lines after marking them for assembly.

7. Remove the nuts on the lower bracket, loosen the nut on the upper slotted bracket where it connects to the body and slide the pump out.

8. Install the new fuel pump in the reverse order of removal. Make sure that the new seal washers are installed on the fuel discharge line.

Diesel Fuel Injection

The diesel fuel system is an extremely complex and sensitive system. Very few repairs or adjustments are possible by the owner. Any service other than that listed here should be referred to an authorized VW dealer or diesel specialist. The injection pump itself is not repairable, it can only be replaced.

Any work done to the diesel fuel injection should be done with absolute cleanliness. Even the smallest specks of dirt will have a disastrous effect on the injection system. Do not attempt to remove the fuel injectors. They are very delicate and must be removed with a special tool to prevent damage. The fuel in the system is also under tremendous pressure (1700–1850 psi), so it's not wise to loosen any lines with the engine running. Exposing your skin to the spray from the injector at working pressure can cause fuel to penetrate the skin.

CHECKING INJECTION PUMP TIMING

Checking the injection pump timing also involves checking the valve timing. To alter the injection pump timing, the camshaft gear

must be removed and repositioned. This also changes the valve timing. Special tool (VW 210) is necessary to properly tension the injection pump drive belt on the diesel engine.

1. Set the engine at TDC on No. 1 cylinder. In this position, the TDC mark on the flywheel should be aligned with boss on the bell housing and both valves of No. 1 cylinder should be closed.

2. The marks on the pump and mounting plate should also be aligned.

3. If the valve timing is incorrect, set the valve timing as detailed in the engine section.

ACCELERATOR CABLE ADJUSTMENT

The ball pin on the pump lever should be pointing up and be aligned with the mark in the slot. The accelerator cable should be attached at the upper hole in the bracket. With the pedal in the full throttle position, adjust the cable so that the pump lever contacts the stop with no binding or strain.

COLD START CABLE ADJUSTMENT

When the cold start knob on the dash is pulled out, the fuel injection pump timing is advanced 2.5°. This improves cold starting and running until the engine warms up.

1. Insert the washer on the cable.
2. Insert the cable in the bracket with the rubber housing. Install the cable in the pin.
3. Install the lockwasher.
4. Move the lever to the zero position (direction of arrow). Pull the inner cable tight and tighten the clamp screw.

CHECKING GLOW PLUGS CURRENT SUPPLY

1. Connect a test light between No.4 cylinder glow plug and ground.
2. Turn the key to the heat position. The test light should light up.
3. If not, check the glow plug relay, ignition switch, or fuse box relay plate.

CHECKING GLOW PLUGS

1. Make this check after establishing that there is current to the glow plugs. Remove the wire and glow plug bus bar.

2. Connect the test light between the battery positive terminal and each glow plug in turn.

3. If the light lights, the glow plug is OK. If not, the glow plug is defective and must be replaced.

TRANSAXLE

Manual Transaxle

REMOVAL & INSTALLATION

The engine and transaxle may be re-

moved together as explained under Engine Removal and Installation or the transaxle may be removed alone, as explained here.

1. Disconnect the battery ground cable.

2. Support the left end of the engine at the lifting eye.

3. Remove the left transmission mount (between the transmission and the firewall).

4. Turn the engine until the lug on the flywheel (to the left of the TDC mark) aligns with the flywheel timing pointer.

5. Detach the speedometer drive cable, backup light wire, and clutch cable.

6. Remove the engine to transmission bolts.

7. Disconnect the shift linkage.

8. Detach the transmission ground strap.

9. Remove the starter.

10. Remove the engine mounting support near the starter.

11. Remove the rear transmission mount.

12. Unbolt and wire up the driveshafts.

13. From underneath, remove the bolts for the large cover plate, but don't remove it. Unbolt the small cover plate on the firewall side of the engine. Remove the engine to transmission nut immediately below the small plate.

14. Press the transmission off the dowels and remove it from below the car.

15. To install the transaxle, reverse the removal procedure: The recess in the flywheel edge must be at 3:00 o'clock. Tighten the engine to transmission bolts to 47 ft. lbs. Tighten the engine mounting support bolts to 47 ft. lbs. Tighten the driveshaft bolts to 32 ft. lbs.

16. Check the adjustment of the shift linkage.

SHIFT LINKAGE ADJUSTMENT

1. Align the holes of the lever housing plate with the holes of the lever bearing plate.

2. Loosen the shift rod clamp. Pull the boot off the lever housing and push it out of the way. It may be necessary to loosen the screws in the cover plate to free the boot.

3. Check that the shift finger is in the center of the stopping plate.

4. Adjust the shift rod end so that it is ¾ in. (⁹⁄₃₂ in. for five speed transmissions) from the right side of the lever housing. Tighten the shift rod clamp and check the shifter operation.

SELECTOR SHAFT LOCKBOLT ADJUSTMENT

Make this adjustment after linkage adjustment, if the linkage still feels spongy or jams.

1. Disconnect the shift linkage and put the transmission in Neutral.

2. Loosen the locknut and turn the adjusting sleeve in until the lockring lifts off the sleeve.

3. Turn the adjusting sleeve back until the lockring just contacts the sleeve. Tighten the locknut.

4. Turn the shaft slightly. The lockring should lift as soon as the shaft is turned.

5. Reconnect the linkage.

FIFTH GEAR LOCKBOLT ADJUSTMENT

This adjustment is made with the transmission in neutral. The fifth gear lockbolt is located on top of the transmission next to the selector shaft lockbolt. It has a large protective cap over it.

1. Remove the protective cap.

2. Loosen the locknut and tighten the adjusting sleeve until the detent plunger in the center of the sleeve just begins to move up.

3. Loosen the adjusting sleeve ⅓ of a turn and tighten the locknut. Make sure the transmission shifts in and out of fifth gear easily. Replace the protective cap.

CLUTCH

Pedal Freeplay

ADJUSTMENT

The clutch should have ⁹⁄₁₆ in. free play at the pedal. Pedal free play is the distance the pedal can be depressed before the linkage starts to act on the throwout bearing.

1. Adjust the clutch pedal free play by loosening the two nuts on the cable near the front of the transmission.

2 After obtaining the correct free play, tighten the adjusting nuts.

REMOVAL & INSTALLATION

1. Remove the transaxle.

2. Attach a toothed flywheel holder and gradually loosen the flywheel to pressure plate bolts one or two turns at a time in a crisscross pattern to prevent distortion.

3. Remove the flywheel and the clutch disc.

4. Remove the release plate retaining ring. Remove the release plate.

5. Lock the pressure plate in place and unbolt it from the crankshaft. Loosen the bolts one or two turns at a time in a crisscross pattern to prevent distortion.

6. On installation, use new bolts to attach the pressure plate to the crankshaft. Use a thread locking compound and torque the bolts in a diagonal pattern to 54 ft. lbs.

7. Lubricate the clutch disc splines with multi-purpose grease. Lubricate the release plate contact surface and pushrod socket with multi-purpose grease. Install the release plate, retaining ring, and clutch disc.

8. Install a dummy shaft to align the clutch disc.

9. Install the flywheel, tightening the bolts one or two turns at a time in a crisscross pattern to prevent distortion. Torque the bolts to 14 ft. lbs.

10. Install the transaxle.

Automatic Transaxle

REMOVAL & INSTALLATION

The engine and transaxle may be removed together as explained under Engine Removal and Installation or the transaxle may be removed alone, as explained here.

1. Disconnect both battery cables.

2. Disconnect the speedometer cable at the transmission.

3. Support the left end of the engine at the lifting eye. Attach a hoist to the transaxle.

4. Unbolt the rear transmission carrier from the body then from the transaxle. Unbolt the left side carrier from the body.

5. Unbolt the driveshafts and wire them up.

6. Remove the starter.

7. Remove the three converter to drive plate bolts.

8. Shift into P and disconnect the floorshift linkage at the transmission.

9. Remove the accelerator and carburetor cable bracket at the transmission.

10. Unbolt the left side transmission carrier from the transmission.

11. Unbolt the front transmission mount from the transmission.

12. Unbolt the bottom of the engine from the transmission. Lift the transaxle slightly, remove the rest of the bolts, pull the transmission off the mounting dowels, and lower the transaxle out of the car. Secure the converter so it doesn't fall out.

—— **CAUTION** ——
Don't tilt the torque converter.

13. To install: Be sure the torque converter is fully seated on the one-way clutch support. Push the transmission onto the mounting dowels and install two bolts. Lift the unit until the left driveshaft can be installed and install the rest of the bolts. Torque them to 39 ft. lbs.

14. Tighten the front transmission mount bolts to 39 ft. lbs. Install the left side transmission carrier to the transmission.

15. Connect the accelerator and carburetor cable bracket. Connect the floor shift linkage.

16. Tighten the torque converter to drive plate bolts to 22 ft. lbs. Torque the driveshaft bolts to 32 ft. lbs.

17. Install the rear transmission carrier and make sure that the left side carrier is aligned in the center of the body mount. Bolt the left side carrier to the body.

18. Connect the speedometer cable and the battery cables.

Pan and Filter

REMOVAL & INSTALLATION

1. Remove the drain plug and let the fluid drain into a pan. If the pan has no drain plug, loosen the pan bolts until a corner of the pan can be lowered to drain the fluid.

2. Remove the pan bolts and take off the pan.

3. Discard the old gasket and clean the pan out. Be very careful not to get any threads or lint from rags into the pan.

4. The filter needn't be replaced unless the fluid is dirty or smells burnt. The specified torque for the strainer screws is 2 ft.

5. Reinstall the pan with a new gasket and tighten the bolts, in a crisscross pattern, to 14 ft. lbs.

6. Using a long-necked funnel, pour in 2 ½ qts. of Dexron II automatic transmission fluid through the dipstick tube. Start the engine and shift through all the transmission ranges with the car stationary. Check the level on the dipstick with the lever in Neutral. It should be up to the lower end of the dipstick. Drive the car until it is warmed up and recheck the level.

Linkage

ADJUSTMENT

1. Check the cable adjustment as follows: Run the engine at 1000–1200 rpm with the parking brake on.

2. Select Reverse. A drop in engine speed should be noticed.

3. Select Park. Engine speed should increase. Pull the shift lever against Reverse, the engine speed shouldn't drop. Move the shift lever to engage Reverse. Engine speed should drop as the gear engages.

4. Move the shift lever to Neutral. An increase in engine speed should be noticed.

6. Shift into 1st. The lever must engage without having to overcome any resistance.

8. To adjust the cable, shift into Park. Loosen the cable, press the transmission lever all the way to the left and tighten the cable clamp.

TRANSMISSION CABLE ADJUSTMENT

1. Make sure the throttle is closed, and the choke and fast idle cam are off (carbureted models). Detach the cable end at the transmission.

2. Press the lever at the transmission into its closed throttle position.

3. You should be able to attach the cable end onto the transmission lever without moving the lever.

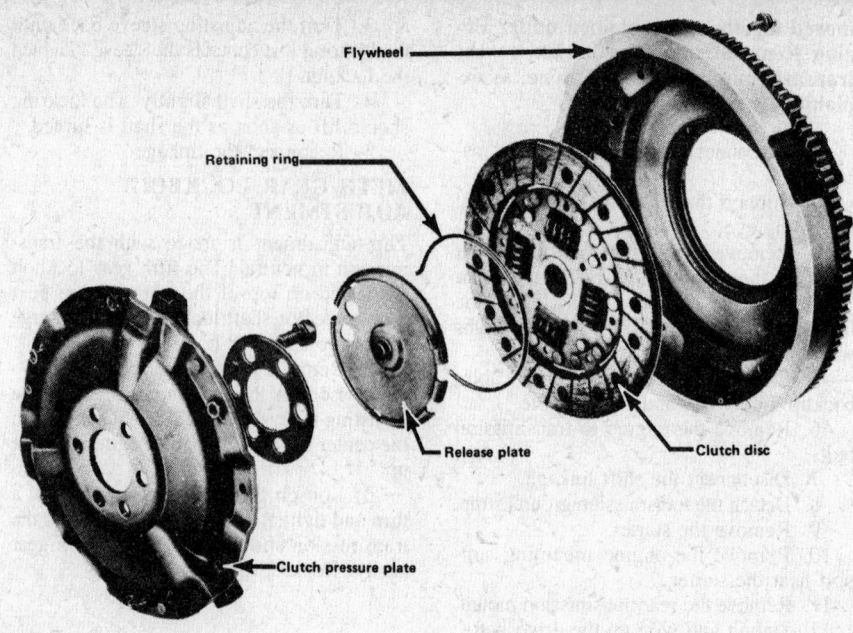

Clutch assembly: the pressure plate is bolted to the crankshaft and the clutch is actuated by a pushrod working on the release plate

4. Adjust the cable length to the correct setting.

SECOND GEAR (REAR) BAND ADJUSTMENT

NOTE: The transmission must be horizontal when band adjustments are performed.

1. Loosen the locknut on the adjusting screw, which is located on the front of the transmission.

2. Tighten the adjusting screw to 7 ft. lbs.

3. Loosen the screw and tighten it again to 4 ft. lbs.

4. Turn the screw out exactly 2 ½ turns and then tighten the locknut.

NEUTRAL START/BACKUP LIGHT SWITCH

The combination neutral start and backup light switch is mounted inside the shifter housing. The starter should operate in Park or Neutral only. Adjust the switch by moving it on its mounts. The backup lights should only come on when the shift selector is in the Reverse position.

DRIVE AXLES

Halfshafts

REMOVAL & INSTALLATION

1. With the car on the ground, remove the front axle nut.

2. Raise and support the front of the vehicle.

3. Remove the socket head bolts retaining the axle shaft to the transaxle.

4. Remove the bolt holding the ball joint to the steering knuckle and separate the knuckle from the ball joint.

5. Removing the ball joint from the knuckle should give enough clearance to remove the axle shaft. It pulls right out of the steering hub.

6. Installation is the reverse of removal. Tighten the axle shaft to transaxle bolts to 32 ft. lbs., the ball joint bolt to 21 ft. lbs. and the axle nut to 173 ft. lbs. Be sure to check the alignment after work is completed.

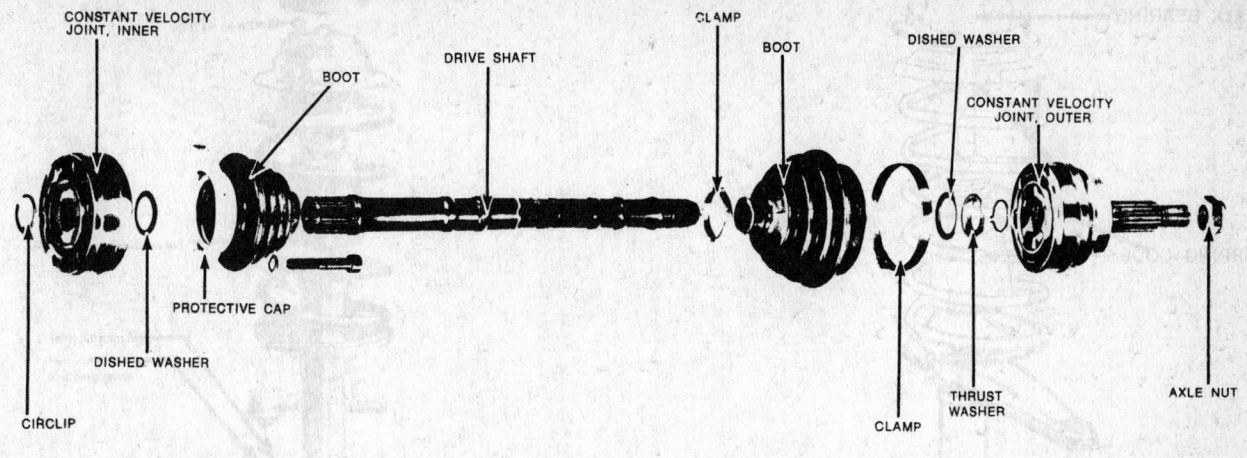

Axle shaft—exploded view

REAR SUSPENSION

Leaf Springs

REMOVAL & INSTALLATION

—————— *CAUTION* ——————
The springs are under a considerable amount of tension. Be very careful when removing or installing them; they can exert enough force to cause serious injuries.

1. Jack up the rear of the truck and support it with jackstands placed under the frame.
2. Disconnect the shock absorbers at their lower end.
3. Remove the nuts securing the U-bolts around the axle housing.
4. Place a jack under the rear axle housing and raise the housing to remove the weight off the springs.
5. Remove the nuts from the spring shackles, drive out the shackle pins and remove the spring from the vehicle.
6. Install the spring in the reverse order of removal. The weight of the truck must be on the rear wheels before tightening the front pin, shackle, and shock absorber attaching nuts. Tighten the front pin and shackle nuts, the U-bolt nuts and the shock absorber lower end nut.

Shock Absorbers

REMOVAL & INSTALLATION

The rear shock absorbers are removed simply by removing the upper and lower attaching nuts, and removing the component from the vehicle. They are installed in the

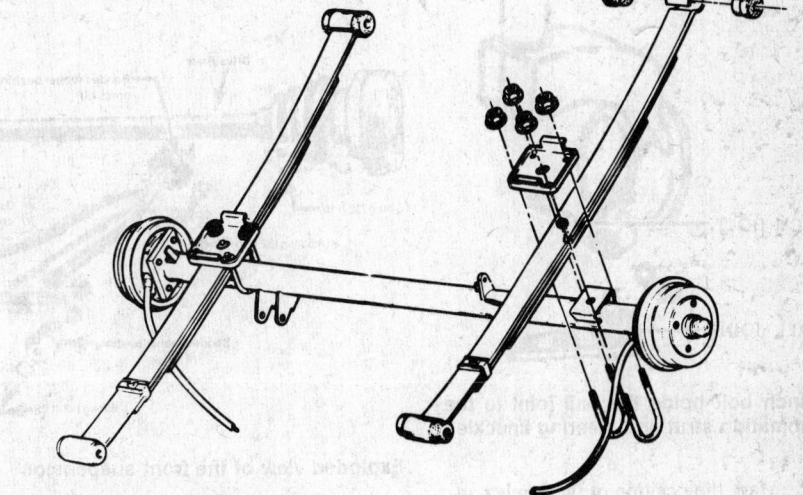

Rear suspension

reverse order. The weight of the vehicle must be on the rear wheels before tightening the shock absorber attaching nuts.

FRONT SUSPENSION

The front suspension is a simple strut design. It consists of a lower control arm, ball joint, and suspension strut. In a MacPherson strut design, such as this, the shock absorber strut serves as a locating member of the suspension as well as a damper. A shock absorber insert is located inside the strut. A coil spring is used.

Ball Joint

REMOVAL & INSTALLATION

1. Jack up the front of the truck and support it on stands.

2. Remove the retaining bolt and nut.
3. Pry the lower control arm and ball joint down and out of the strut.
4. Drill out the rivets; enlarge the holes to $^{21}/_{64}$ in.
5. Remove the ball joint assembly.
6. Bolt the new ball joint in place. Torque the bolts to 18 ft. lbs. Tighten the retaining bolt for the ball joint stud to 21 ft. lbs.

Shock Absorber

REMOVAL & INSTALLATION

Since the shock absorber cartridge is contained within the strut assembly, it is necessary to remove the strut and then compress the coil spring in order to remove the shock.

Strut

REMOVAL & INSTALLATION

1. Remove the brake hose from the strut clip.

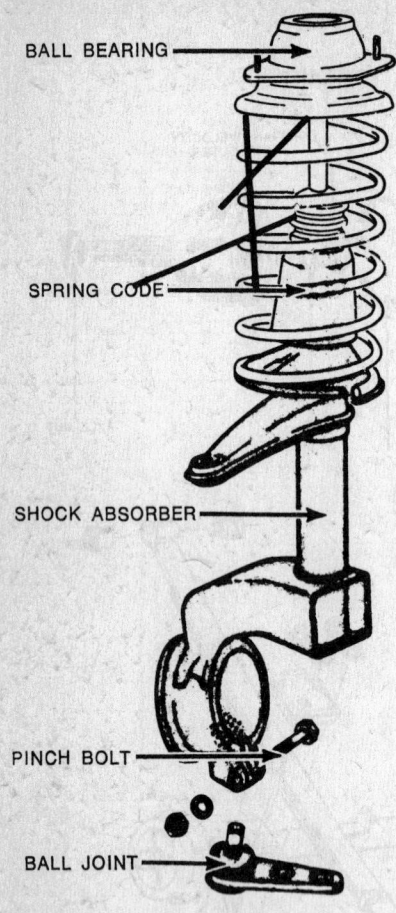

BALL BEARING

SPRING CODE

SHOCK ABSORBER

PINCH BOLT

BALL JOINT

A pinch bolt holds the ball joint to the combination strut and steering knuckle

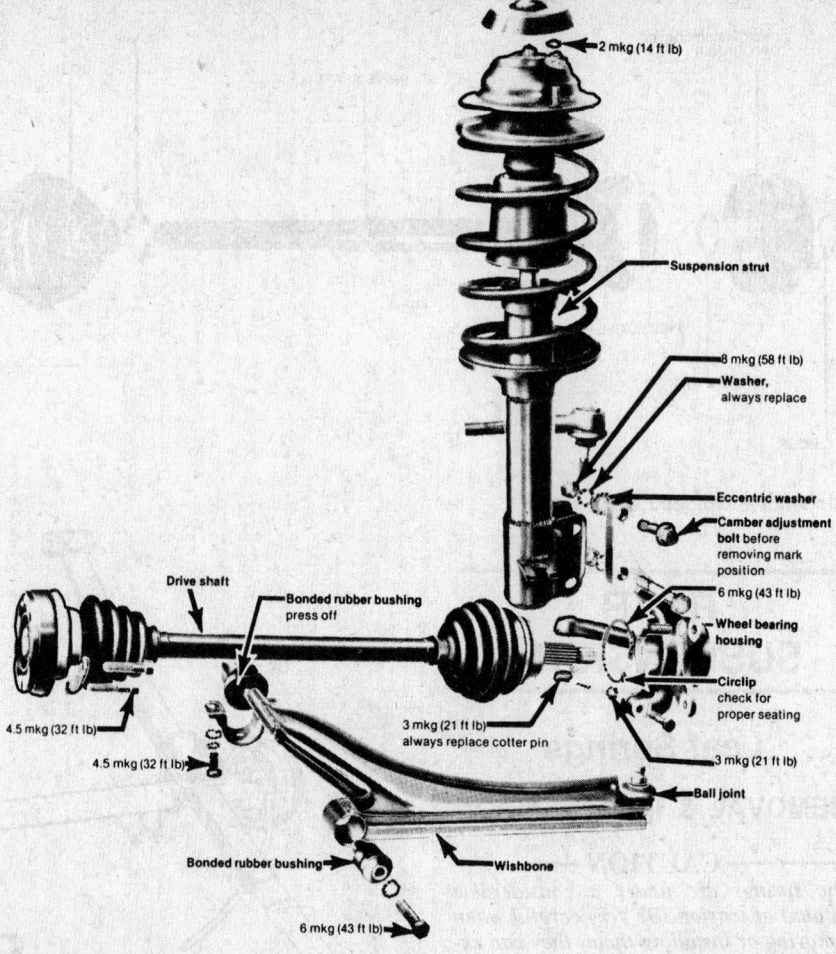

2 mkg (14 ft lb)

Suspension strut

8 mkg (58 ft lb)
Washer, always replace

Eccentric washer

Camber adjustment bolt before removing mark position

6 mkg (43 ft lb)

Wheel bearing housing

Circlip check for proper seating

Drive shaft

Bonded rubber bushing press off

4.5 mkg (32 ft lb)

4.5 mkg (32 ft lb)

3 mkg (21 ft lb) always replace cotter pin

3 mkg (21 ft lb)

Ball joint

Bonded rubber bushing

Wishbone

6 mkg (43 ft lb)

Exploded view of the front suspension

2. Mark the position of the camber adjustment bolts before removing them from the hub (wheel bearing housing).

3. Remove the upper mounting nuts and remove the strut from the car.

4. Installation is the reverse. The upper nuts are tightened to 14 ft. lbs., and the lower strut-to-hub bolts to 58 ft. lbs. Use new washers on the lower bolts. If the shock absorber was replaced, camber will have to be adjusted.

Coil Spring

REMOVAL & INSTALLATION

To remove the spring, the strut must be mounted in large vise, the spring compressed, the retaining nut and cover removed, and the spring slowly released. A special tool is needed to remove the shock absorber retainer, after which the shock absorber is easily removed. Assembly is the reverse of removal.

Front End Alignment

CAMBER ADJUSTMENT

Camber is adjusted by loosening the nuts

of the two bolts holding the top of the wheel bearing housing to the bottom of the strut, and turning the top eccentric bolt. The range of adjustment is 2°.

CASTER

Other than the replacement of damaged suspension components, caster is not adjustable.

TOE-IN ADJUSTMENT

Toe-in is checked with the wheels straight ahead. Only the right tie-rod is adjustable, but replacement left tie-rods are adjustable. Replacement left tie-rods should be set to the same length as the original. Toe-in should be adjusted only with the right tie-rod.

STEERING

The truck has rack and pinion steering with end-mounted tie-rods. No periodic maintenance is required on either rack and pinion steering system.

Steering Wheel

REMOVAL & INSTALLATION

1. Grasp the center cover pad and pull it from the wheel.

2. Loosen and remove the steering shaft nut.

3. Pull the wheel off the shaft. A puller isn't normally needed.

4. Disconnect the horn wire.

5. Replace the wheel in the reverse order of removal. Tighten the nut to 36 ft. lbs.

Turn Signal/Headlight Dimmer Switch

REPLACEMENT

1. Disconnect the battery ground cable.

2. Remove the steering wheel.

3. Remove the switch retaining screws.

4. Pry the switch housing off the column.

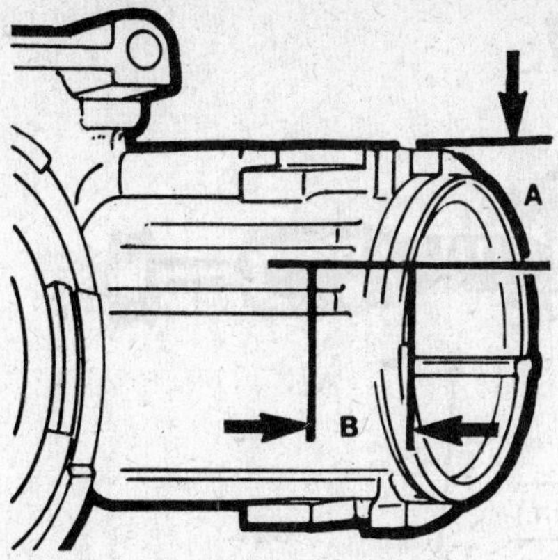
A
B

Dimensions for drilling the ignition lock cylinder hole (if not equipped)

5. Disconnect the electrical plugs at the back of the switch.
6. Remove the switch housing.
7. Replace in the reverse order of removal.

Ignition Switch/ Steering Lock

REMOVAL & INSTALLATION

NOTE: The access hole for removing the lock cylinder may be missing. Before the lock cylinder can be removed, drill a hole according to following dimensions: a= 12mm (0.472 in.) b = 10mm (0.393 in.) Drill the hole ⅛ in. deep.

1. Remove the steering wheel and turn signal switch. Remove the steering column shaft covers.
2. The lock is clamped to the steering column with special bolts whose heads shear off on installation. These must be drilled out in order to remove the switch.
3. On replacement, make sure that the lock tang is aligned with the slot in the steering column.

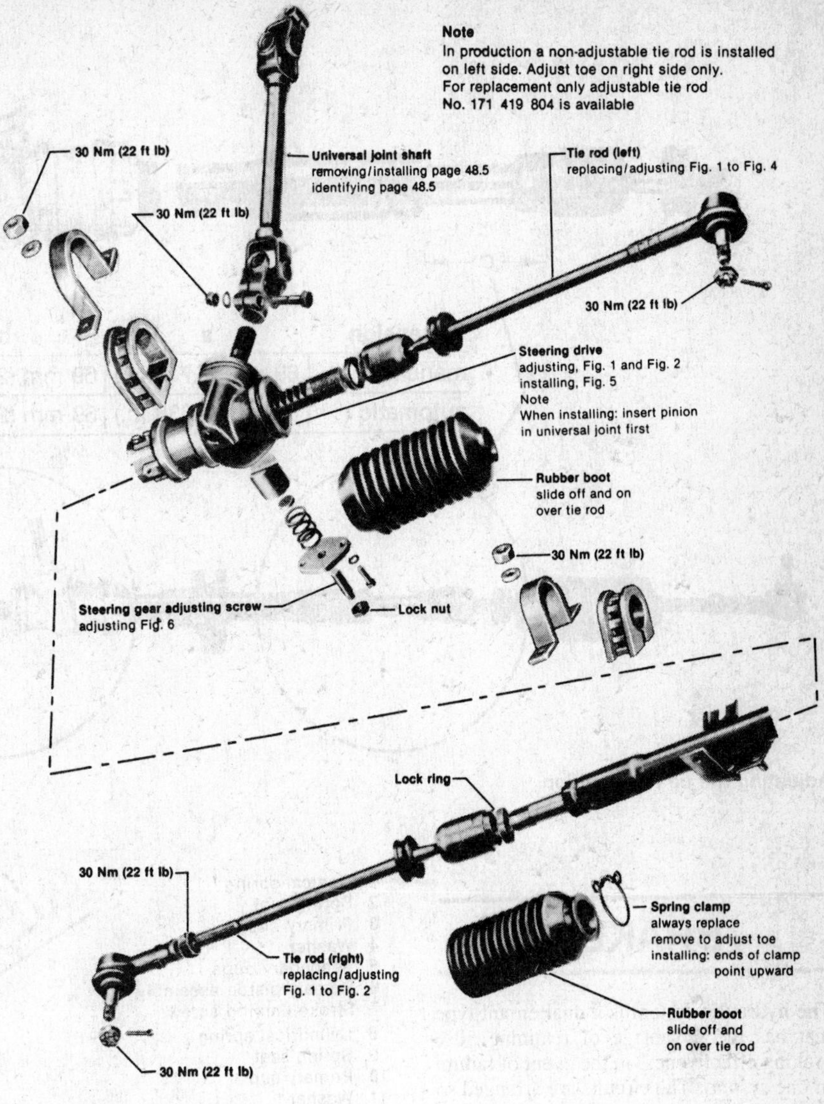
Note
In production a non-adjustable tie rod is installed on left side. Adjust toe on right side only.
For replacement only adjustable tie rod No. 171 419 804 is available
30 Nm (22 ft lb)
Universal joint shaft
removing/installing page 48.5
identifying page 48.5
Tie rod (left)
replacing/adjusting Fig. 1 to Fig. 4
30 Nm (22 ft lb)
30 Nm (22 ft lb)
Steering drive
adjusting, Fig. 1 and Fig. 2
installing, Fig. 5
Note
When installing: insert pinion in universal joint first
Rubber boot
slide off and on over tie rod
30 Nm (22 ft lb)
Steering gear adjusting screw
adjusting Fig. 6
Lock nut
Lock ring
30 Nm (22 ft lb)
Spring clamp
always replace
remove to adjust toe
installing: ends of clamp point upward
Tie rod (right)
replacing/adjusting
Fig. 1 to Fig. 2
Rubber boot
slide off and on over tie rod
30 Nm (22 ft lb)

Steering gear components

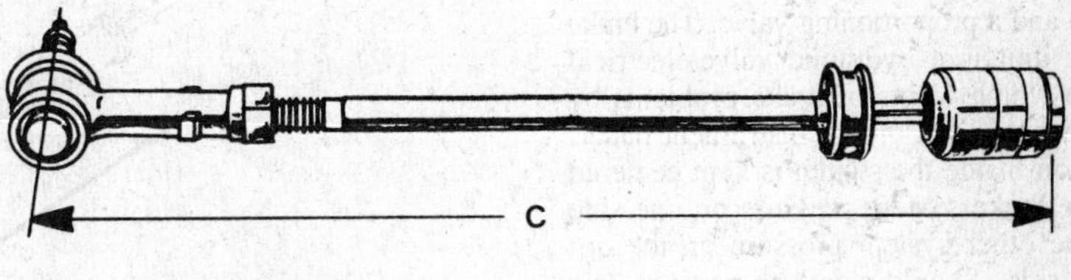
C

Right side tie rod adjustment

Steering Gear

REMOVAL & INSTALLATION

1. Disconnect the steering shaft universal joint and wire up out of the way.
2. Disconnect the tie rods at the steering rack and wire up and out of the way.
3. Remove the steering rack and drive.
4. Install the steering rack and drive and torque the attaching hardware to 14 ft. lbs.
5. Set the steering rack with equal distances between the housing on the right side and left side.
6. Install the tie rods and screw both sides in until an equal distance is reached on both rods.
7. Tighten the steering gear adjusting screw until it touches the thrust washer. Tighten the locknut.
8. Install the steering shaft.
9. Check the front end alignment.

Tie Rod End

REMOVAL & INSTALLATION

1. Center the steering rack.
2. Remove the cotter pin and nut from the tie rod end.
3. Disconnect the tie rod from the steering rack.
4. If the left side tie rod is being replaced, adjust it to 14.92 in. (379mm).
5. Adjust the steering rack and tie rods as outlined in Steps 5 and 6 of the "Steering Gear Removal and Installation."
6. Tighten the tie rod end retaining nut to 21 ft. lbs. and install a new cotter pin.

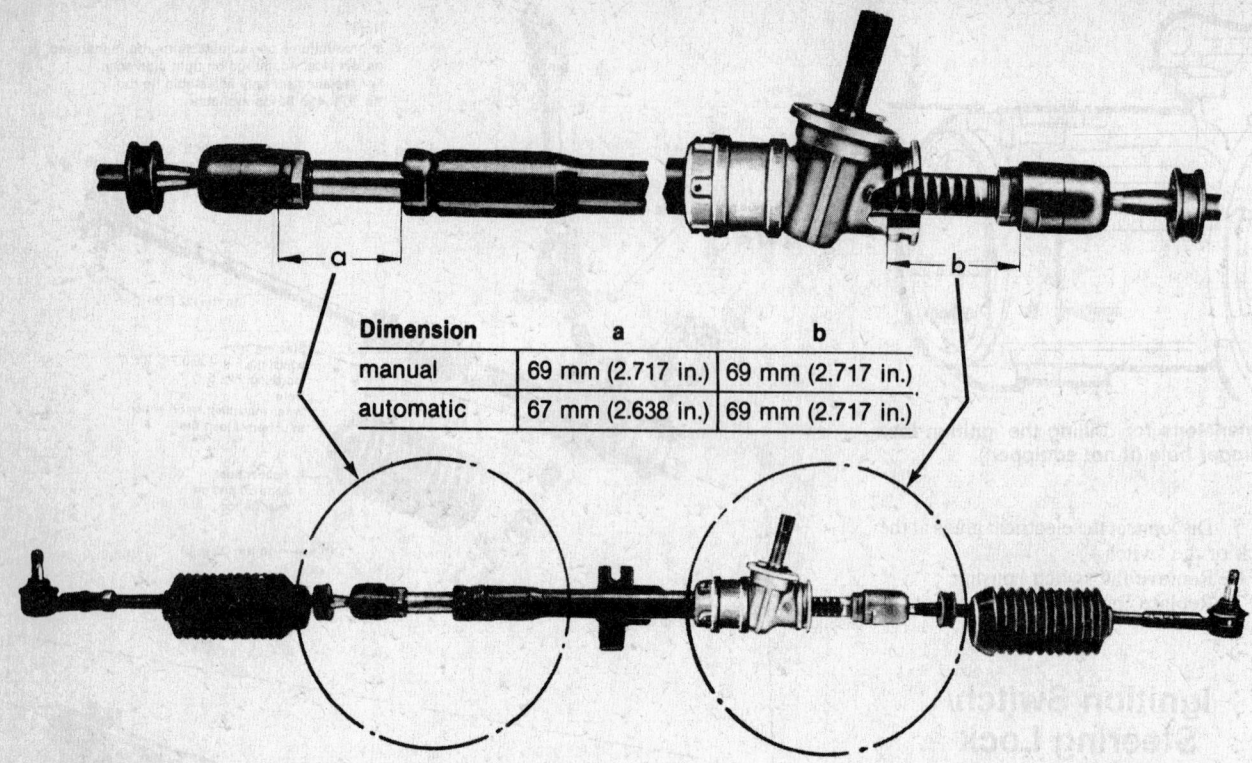

Dimension	a	b
manual	69 mm (2.717 in.)	69 mm (2.717 in.)
automatic	67 mm (2.638 in.)	69 mm (2.717 in.)

Adjusting the tie rod position

BRAKES

The hydraulic system is a dual circuit type that has the advantage of retaining 50% braking effectiveness in the event of failure in one system. The circuits are arranged so that you always have one front and one rear brake for a more controlled emergency stop. The right front and left rear are in one circuit; the left front and right rear are in the second circuit. There is also a brake failure switch and a proportioning valve. The brake failure unit is a hydraulic valve/electrical switch which warns of brake problems by the warning light on the instrument panel. A piston inside the switch is kept centered by one brake system pressure on one side and the other system pressure on the opposite side. Should a failure occur in one system, the piston would go to the "failed" side and complete an electrical circuit to the warning lamp. This switch also functions as a parking brake reminder light and will go out when the parking brake is released. The proportioning valve provides balanced front-to-rear braking during hard stops. Extreme brake line pressure will overcome the spring pressure on the piston within the valve causing it to proportionately restrict pressure to the rear brakes. In this manner, the rear brakes are kept from locking. The proportioner doesn't operate under normal braking conditions.

1. Conical spring
2. Spring seat
3. Primary cup
4. Washer
5. Secondary cups
6. Primary piston assembly
7. Stroke limiting screw
8. Cylindrical spring
9. Spring seat
10. Primary cup
11. Washer
12. Secondary cups
13. Circlip
14. Secondary piston assembly

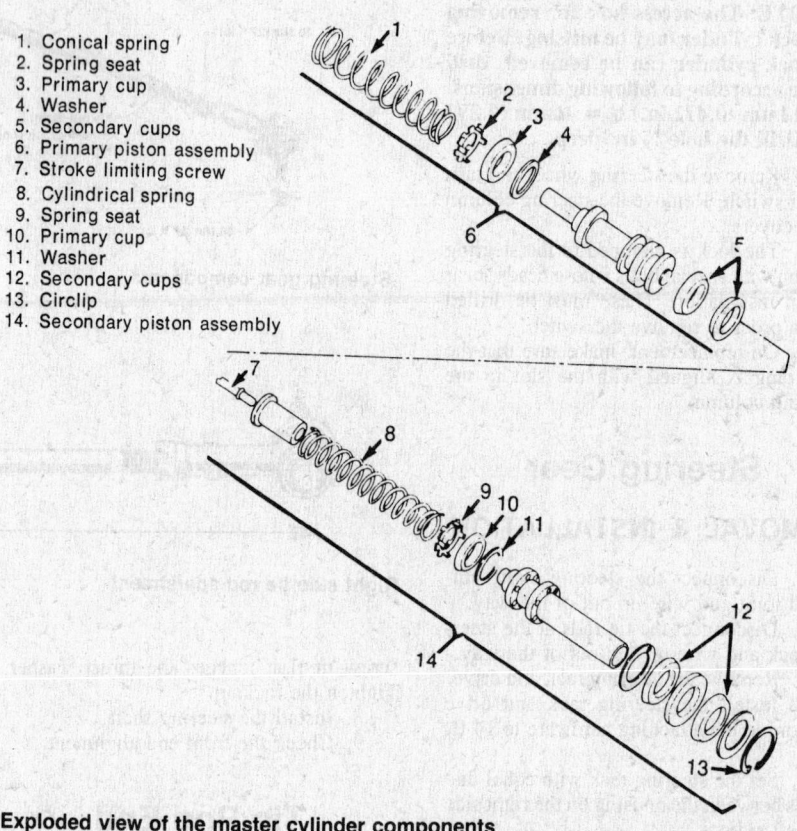

Exploded view of the master cylinder components

ADJUSTMENT

The front disc brakes require no adjust- ment, as disc brakes automatically adjust themselves to compensate for pad wear. The truck has self-adjusting rear drum brakes.

Master Cylinder

REMOVAL & INSTALLATION

1. Disconnect and plug the brake lines.
2. Disconnect the electrical plug from the sending unit for the brake failure switch.
3. Remove the two master cylinder mounting nuts.
4. Lift the master cylinder and reservoir out of the engine compartment being careful not to spill any fluid on the fender. Empty out and discard the brake fluid.

—————— CAUTION ——————
Do not depress the brake pedal while the master cylinder is removed.

5. Position the master cylinder and reservoir assembly onto the studs for the booster and install the washers and nuts. Tighten the nuts to no more than 9 ft. lbs.
6. Remove the plugs and connect the brake lines.
7. Bleed the entire brake system.

Parking Brake

ADJUSTMENT

1. On the Rabbit pick-up, adjustment is made at the cable end nuts on top of the handbrake lever. Block the front wheels. Raise the rear of the car and safely support.
2. Apply the parking brake so that the lever is on the second notch.
3. Tighten the compensator nut or adjusting nuts until both rear wheels cannot be turned by hand.
4. Release the parking brake lever and check that both wheels can be easily turned.

For brake system servicing other than described here, refer to Brakes in the Unit Repair section.

CHASSIS ELECTRICAL

Heater

The heater core and blower are contained in the heater assembly which is located in the passenger compartment under the center of the dash.

REMOVAL & INSTALLATION

1. Disconnect the battery ground cable.
2. Drain the cooling system.
3. Remove the windshield washer con-

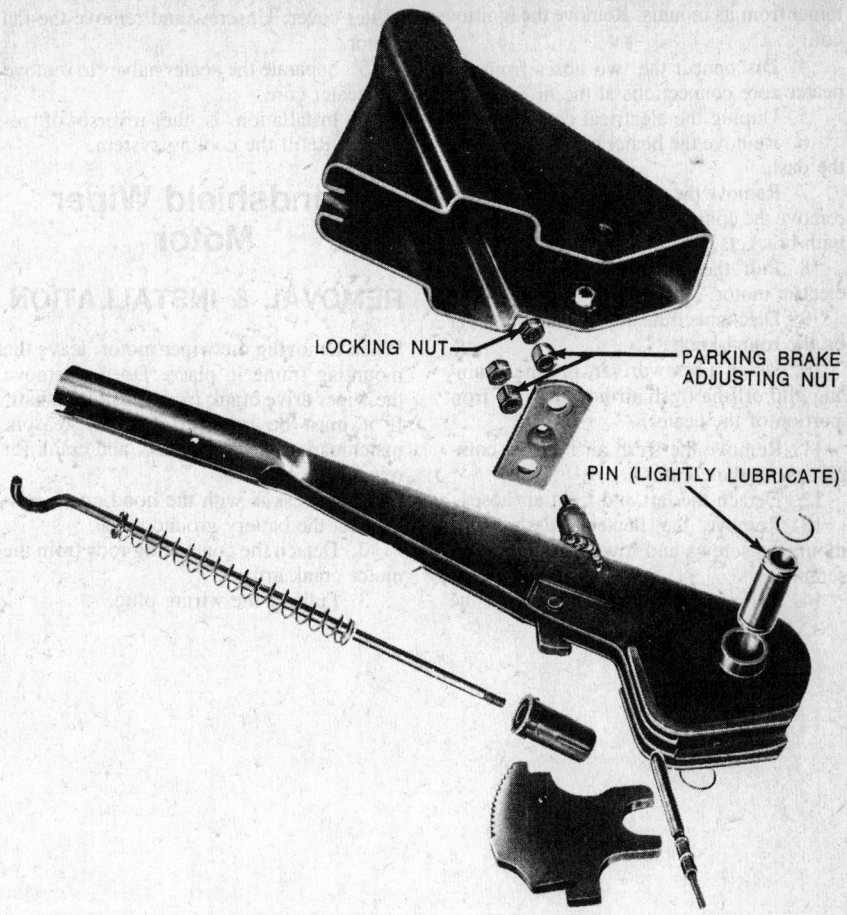

LOCKING NUT

PARKING BRAKE ADJUSTING NUT

PIN (LIGHTLY LUBRICATE)

Parking brakes: only one of the two cables is shown

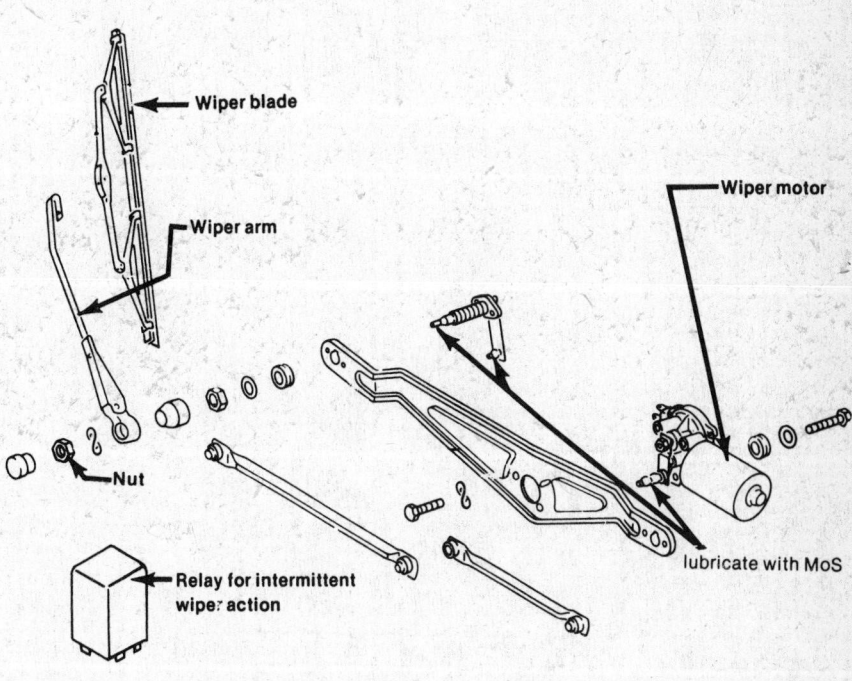

Wiper blade

Wiper arm

Wiper motor

Nut

Relay for intermittent wiper action

lubricate with MoS

Exploded view of wiper linkage

tainer from its mounts. Remove the ignition coil.

4. Disconnect the two hoses from the heater core connections at the firewall.

5. Unplug the electrical connector.

6. Remove the heater control knobs on the dash.

7. Remove the two retaining screws and remove the controls from the dash complete with brackets.

8. Pull the cable connection off the electric motor.

9. Disconnect the cable from the lever on the round knob.

10. Using a screwdriver, pry the retaining clip off the fresh air housing (the front portion of the heater).

11. Remove the fresh air housing complete with the controls.

12. Detach the left and right air hoses.

13. Remove the heater-to-dash panel mounting screws and lower the heater assembly.

14. Pull out the two pins and remove the heater cover. Unscrew and remove the fan motor.

15. Separate the heater halves to remove the heater core.

16. Installation is the reverse of removal. Refill the cooling system.

Windshield Wiper Motor

REMOVAL & INSTALLATION

When removing the wiper motor, leave the mounting frame in place. Do not remove the wiper drive crank from the motor shaft. If it must be removed for any reason, matchmark the shaft, motor, and crank for reinstallation.

1. Access is with the hood open. Disconnect the battery ground cable.

2. Detach the connecting rods from the motor crank arm.

3. Pull off the wiring plug.

4. Remove the 4 mounting bolts. You may have to energize the motor for access to the top bolt.

5. Remove the motor. Reverse the procedure for installation.

Instrument Cluster

REMOVAL & INSTALLATION

1. Disconnect the battery ground cable.

2. Remove the fresh air controls trim plate.

3. Remove the radio or glove box.

4. Unscrew the speedometer drive cable from the back of the speedometer. Detach the electrical plug.

5. Remove the attaching screw inside the radio/glove box opening.

6. Remove the instrument cluster. Reverse the procedure for installation.

UNIT REPAIR SECTION

Tools and Equipment

The service procedures in this book presuppose a familiarity with hand tools and their proper use. However, it is possible that you may have a limited amount of experience with the sort of equipment needed to work on an automobile. This section is designed to help you assemble a basic set of tools that will handle the majority of jobs you may undertake.

In addition to the normal assortment of screwdrivers and pliers, automotive service work requires an investment in wrenches, sockets and the handles needed to drive them, and various measuring tools such as torque wrenches and feeler gauges.

The best approach to gathering the required equipment is to proceed slowly, buying high-quality tools as they are needed. An initial investment should be made in a set of quality wrenches, ranging in size from ¼ inch to one inch, if your car has standard bolts, or from 5 mm to 19 mm if your car has metric fasteners. High quality forged wrenches are available in three styles: open end, box end, and combination open/box end. The combination tools are generally the most desirable as a starter set; the wrenches shown in the illustration are of the combination type.

The other set of tools inevitably required is a ratchet handle and socket set. This set should have the same size range as your wrench set. The ratchet, extension, and flex drives for the sockets are available in many sizes; it is advisable to choose a ⅜ inch drive set initially. One break in the inch/metric sizing war is that metric-sized sockets sold in the U.S. have inch-sized drive (¼, ⅜, ½, etc.). Sockets are available in six and twelve point versions; six point types are generally cheaper and are a good choice for a first set. The choice of a drive handle for the sockets should be made with some care. If this is your first set, take the plunge and invest in a flex-head ratchet; it will get into many places otherwise accessible only through a long chain of universal joints,

extensions and adapters. An alternative is a flex handle; such a tool is shown in the illustration, below the ratchet handle. In addition to the range of sockets mentioned, a rubber-lined spark plug socket should be purchased. Spark plugs have either a ¹³/₁₆ or a ⅝ inch hex; get the correct socket for the plugs in your car.

The most important thing to consider when purchasing hand tools is quality. Don't be misled by the low cost of "bargain" tools. Forged wrenches, tempered screwdriver blades, and fine tooth ratchets are a much better investment than their less expensive counterparts. The skinned knuckles and frustration inflicted by poor quality tools make any job an unhappy chore. Another consideration is that quality tools sold by reputable firms come with an on-the-spot replacement guarantee—if the tool breaks, you get a new one, no questions asked.

The tools needed for basic maintenance jobs, in addition to those just mentioned, include:

1. Jackstands, for support
2. Oil filter wrench
3. Oil filler spout or funnel
4. Grease gun
5. Battery hydrometer
6. Battery post and clamp cleaner
7. Container for draining oil
8. Lots of rags for the inevitable spills

In addition to these items there are several others which are not absolutely necessary, but handy to have around. These include a transmission funnel and filler tube, a drop (trouble) light on a long cord, an adjustable wrench (crescent wrench), and slip joint pliers.

A more extensive list of tools, suitable for tune-up work, can be drawn up easily. While the tools involved are slightly more sophisticated, they need not be outrageously expensive. For example, there are several inexpensive tach/dwell meters on the market that are every bit as good for the average mechanic as a $100.00 profes-

sional model. The key to these purchases is to make them with an eye towards adaptability and wide range. Using the tach/dwell meter example again, if the model you buy runs up to at least 1,500 rpm on the tachometer scale, the dwell meter works on 4, 6, or 8 cylinder engines, and the tachometer unit is adaptable to both conventional and electronic ignitions, it will serve for a long time on a variety of automobiles. A basic list of tune-up tools could include:

1. A tach/dwell meter
2. Spark plug gauge and gapping tool
3. Feeler blades
4. Timing light

In this list, the choice of a timing light should be made carefully. A light which works on the DC current supplied by the car battery is the best choice; it should have a xenon tube for brightness. If your car has electronic ignition, the light should have an inductive pick-up (the timing light illustrated has one of these), and since nearly all cars will have electronic ignition in the future, this feature is a reasonable one to look for.

In addition to these basic tools, there are several other tools and gauges you may find useful. These include:

1. A compression gauge. The screw-in type is slower to use, but eliminates the possibility of a faulty reading due to escaping pressure.
2. A manifold vacuum gauge
3. A test light
4. An induction meter. This is used to determine whether or not there is current flowing in a wire, and thus is extremely helpful in electrical troubleshooting.

Finally, you will probably find a torque wrench necessary for all but the most basic of work. The beam–type models are perfectly adequate, although the newer click (breakaway) type are more precise. Whichever type you choose, plan on having it recalibrated every once in a while.

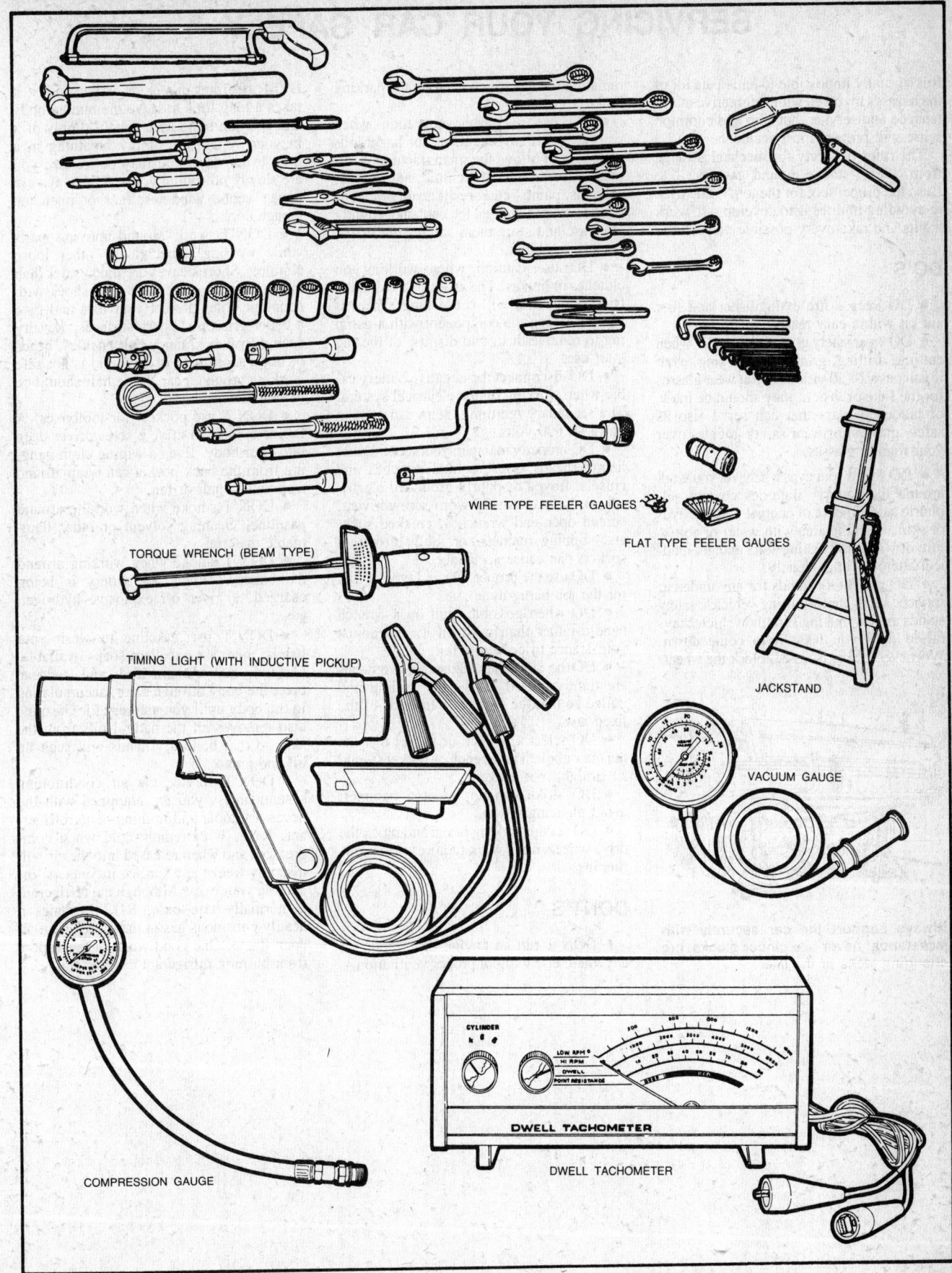

WIRE TYPE FEELER GAUGES

FLAT TYPE FEELER GAUGES

TORQUE WRENCH (BEAM TYPE)

JACKSTAND

TIMING LIGHT (WITH INDUCTIVE PICKUP)

VACUUM GAUGE

COMPRESSION GAUGE

DWELL TACHOMETER

A basic tool collection will handle almost any automotive repair work

SERVICING YOUR CAR SAFELY

It is virtually impossible to anticipate all of the hazards involved with automotive maintenance and service, but care and common sense will prevent most accidents.

The rules of safety for mechanics range from "don't smoke around gasoline," to "use the proper tool for the job." The trick to avoiding injuries is to develop safe work habits and take every possible precaution.

DO'S

• DO keep a fire extinguisher and first aid kit within easy reach.

• DO wear safety glasses or goggles when cutting, drilling, grinding, or prying, even if you have 20-20 vision. If you wear glasses for the sake of vision, they should be made of hardened glass that can serve also as safety glasses, or wear safety goggles over your regular glasses.

• DO shield your eyes whenever you work around the battery. Batteries contain sulphuric acid. In case of contact with the eyes or skin, flush the area with water or a mixture of water and baking soda and get medical attention immediately.

• DO use safety stands for any undercar service. Jacks are for raising vehicles; safety stands are for making sure the vehicle stays raised until you want it to come down. Whenever the car is raised, block the wheels

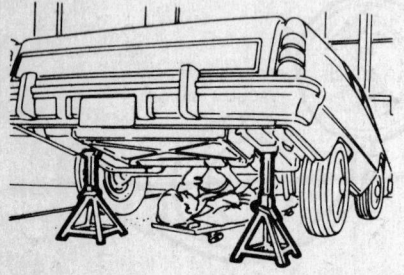

Always support the car securely with jackstands; never use cinder blocks, tire changing jacks or the like

remaining on the ground and set the parking brake.

• DO use adequate ventilation when working with any chemicals or hazardous materials. Follow the manufacturer's directions for usage. Brake fluid, anti-freeze, solvents, paints, etc. are all deadly poisons if taken internally. Seal the containers tightly after use and store them safely, out of the reach of children.

• DO use caution when working on clutches or brakes. The asbestos used in the friction material will cause lung cancer if inhaled. Wipe the component with a damp rag to remove dust, and dispose of the rag after use.

• DO disconnect the negative battery cable when working on the electrical system. The secondary ignition system can contain up to 40,000 volts.

• DO properly maintain your tools. Loose hammerheads, mushroomed punches and chisels, frayed or poorly grounded electrical cords, excessively worn screwdrivers, spread open-end wrenches, cracked sockets, slipping ratchets, or faulty droplight sockets can cause accidents.

• DO use the proper size and type of tool for the job being done.

• DO when possible, pull on a wrench handle rather than push on it, and adjust your stance to prevent a fall.

• DO be sure that adjustable wrenches are tightly closed on the nut or bolt and pulled so that the face is on the side of the fixed jaw.

• DO select a wrench or socket that fits the nut or bolt. The wrench or socket should sit straight, not cocked.

• DO strike squarely with a hammer; avoid glancing blows.

• DO set the parking brake and block the drive wheels if the work requires the engine running.

DONT'S

• DON'T run an engine in a garage or anywhere else without proper ventilation—

EVER! Carbon monoxide is poisonous; it takes a long time to leave the human body and you can build up a deadly supply of it in your system by simply breathing in a little every day. You may not realize you are slowly poisoning yourself. Always use power vents, windows, fans or open the garage doors.

• DON'T work around moving parts while wearing a necktie or other loose clothing. Short sleeves are much safer than long, loose sleeves; hard-toed shoes with neoprene soles protect your toes and give a better grip on slippery surfaces. Jewelry such as watches, fancy belt buckles, beads or body adornment of any kind is not safe working around a car. Long hair should be hidden under a hat or cap.

• DON'T use pockets for toolboxes. A fall or bump can drive a screwdriver deep into your body. Even a wiping cloth hanging from the back pocket can wrap around a spinning shaft or fan.

• DON'T smoke when working around gasoline, cleaning solvent or other flammable material.

• DON'T smoke when working around the battery. When the battery is being charged, it gives off explosive hydrogen gas.

• DON'T use gasoline to wash your hands; there are excellent soaps available. Gasoline may contain lead, and lead can enter the body through a cut, accumulating in the body until you are very ill. Gasoline also removes all the natural oils from the skin so that bone dry hands will suck up oil and grease.

• DON'T service the air conditioning system unless you are equipped with the necessary tools and training. The refrigerant, R-12, is extremely cold when compressed, and when released into the air will instantly freeze any surface it contacts, including your eyes. Although the refrigerant is normally non-toxic, R-12 becomes a deadly poisonous gas in the presence of an open flame. One good whiff of the vapors from burning refrigerant can be fatal.

Basic Maintenance

INTRODUCTION

Routine maintenance is probably the most important part of automobile care and the easiest to neglect. A regular program aimed at monitoring essential systems ensures that all components are in good and safe working order, and can prevent small problems from developing into major headaches. Routine maintenance also pays big dividends in keeping major repair costs at a minimum and extending the life of the car.

The owner's manual that came with your car includes a maintenance schedule, indicating service intervals in numbers of months or thousand of miles. This schedule should always be followed. We have provided, in each section, a guide to service intervals based on an averaging of manufacturer's recommendations. In most cases, the suggested interval offered here will be close to that given by the manufacturer of your car, but the manufacturer's schedule should always take precedence.

We have divided the maintenance work to be done into three categories: Under Hood, Under Car, and Exterior. The checks in each section require only a few minutes of attention every few weeks; the services to be performed can be easily accomplished in a morning. The most important part of any maintenance program is regularity. The few minutes or occasional morning spent on these seemingly trivial tasks will forestall or eliminate major problems later.

UNDER HOOD
Automatic Transmission, Automatic Transaxle

The fluid level in the automatic transmission or transaxle should be checked every three months or 6000 miles. All automatic transmissions have a dipstick for fluid level checks.

1. Drive the car until it is at normal operating temperature. The level should not be checked immediately after the car has been driven for a long time at high speed, or in city traffic in hot weather; in those cases, the transmission should be given a half hour to cool down.

2. Stop the car, apply the parking brake, then shift slowly through all gear positions, ending in Park. Leave the engine running.

3. Remove the dipstick, wipe it clean, then reinsert it, pushing it fully home.

4. Pull the dipstick again and, holding it horizontally, read the fluid level.

5. Cautiously feel the end of the dipstick to determine the temperature. Most dipsticks are marked with both cool and hot levels. If the fluid is not up to the correct level, more will have to be added.

6. Fluid is added through the dipstick tube. You will probably need the aid of a spout or a long-necked funnel. Be sure that whatever you pour through is perfectly clean and dry. Fluid recommendations can be found in the owner's manual.

Add fluid slowly, and in small amounts, checking the level frequently between additions. Do not overfill, which will cause foaming, fluid loss, slippage, and possible transmission damage.

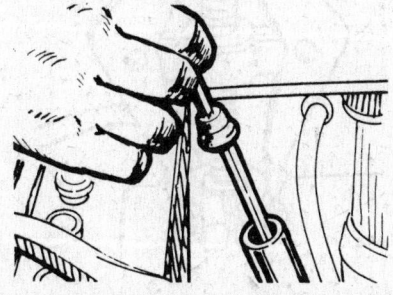

Check the automatic transmission fluid level with the dipstick provided

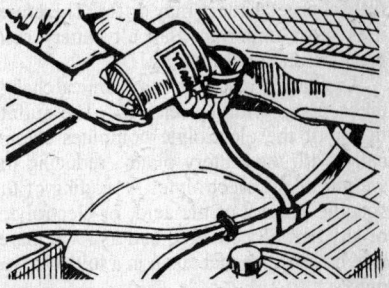

Fill the automatic transmission through the dipstick tube

Battery

FLUID LEVEL (EXCEPT "MAINTENANCE FREE" BATTERIES)

Check the battery electrolyte level at least once a month, or more often in hot weather or during periods of extended car operation. The level can be checked through the case on translucent polypropylene batteries; the cell caps must be removed on other models. The electrolyte level in each cell should be kept filled to the split ring inside, or the line marked on the outside of the case.

If the level is low, add only distilled water, or colorless, odorless drinking water, through

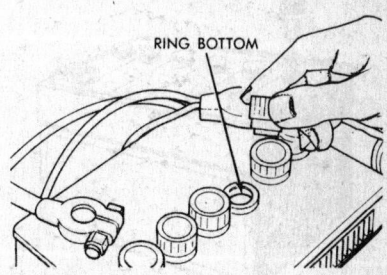

Fill the battery cell to the bottom of the split ring

U5

the opening until the level is correct. Each cell is completely separate from the others, so each must be checked and filled individually.

If water is added in freezing weather, the car should be driven several miles to allow the water to mix with the electrolyte. Otherwise, the battery could freeze.

SPECIFIC GRAVITY (EXCEPT "MAINTENANCE FREE" BATTERIES)

While not technically exact, a practical measurement of the chemical condition of the battery is indicated by measuring the specific gravity of the acid (electrolyte) contained in each cell. The electrolyte in a fully charged battery is usually between 1.260 and 1.280 times as heavy as pure water at the same temperature (80°F). Variations in the specific gravity readings for a fully charged battery may differ. Therefore, it is most important that all battery cells produce an equal reading.

As a battery discharges, a chemical change takes place within each cell. The sulfate factor of the electrolyte combines chemically with the battery plates, reducing the weight of the electrolyte. A reading of the specific gravity of the acid, or electrolyte, of any partially charged battery, will therefore be less than that taken in a fully charged one.

The hydrometer is the instrument used for determining the specific gravity of liquids. The battery hydrometer is readily available from many sources, including local auto replacement parts stores. The following chart gives an indication of specific gravity value, related to battery charge condition. If, after charging, the specific gravity between any two cells varies more than 50 points (.050), the battery is probably bad.

Specific Gravity Reading	Charged Condition
1.260–1.280	Fully charged
1.230–1.250	Three-quarter charged
1.200–1.220	One-half charged
1.170–1.190	One-quarter charged
1.140–1.160	Just about flat
1.110–1.130	All the way down

CABLES AND CLAMPS

Once a year, the battery terminals and the cable clamps should be cleaned. Loosen the clamps and remove the cables, negative cable first. On batteries with posts on top, the use of a puller specially made for the purpose is recommended. These are inexpensive, and available in auto parts stores. Side terminal battery cables are secured with a bolt.

Clean the cable clamps and the battery terminal with a wire brush until all corrosion, grease, etc. is removed and the metal is shiny. It is especially important to clean the inside of the clamp thoroughly, since a small deposit of foreign material or oxidation there will prevent a sound electrical connection and inhibit either starting or charging. Special tools are available for cleaning these parts, one type for conventional batteries and another type for side terminal batteries.

Before installing the cables, loosen the battery hold-down clamp or strap, remove the battery and check the battery tray. Clear it of any debris, and check it for soundness. Rust should be wire brushed away, and the metal given a coat of anti-rust paint. Replace the battery and tighten the hold-down clamp or strap securely, but be careful not to overtighten, which will crack the battery case.

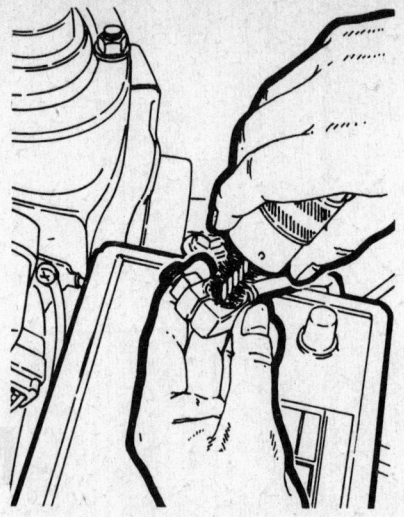

Clean the clamp with a wire brush

After the clamps and terminals are clean, reinstall the cables, negative cable last; do not hammer on the clamps to install. Tighten the clamps securely, but do not distort them. Give the clamps and terminals a thin external coat of grease after installation, to retard corrosion.

Check the cables at the same time that the terminals are cleaned. If the cable insulation is cracked or broken, or if the ends are frayed, the cable should be replaced with a new cable of the same length and gauge.

NOTE: Keep flame or sparks away from the battery; it gives off explosive hydrogen gas. Battery electrolyte contains sulphuric acid. If you should splash any on your skin or in your eyes, flush the affected area with plenty of clear water; if it lands in your eyes, get medical help immediately.

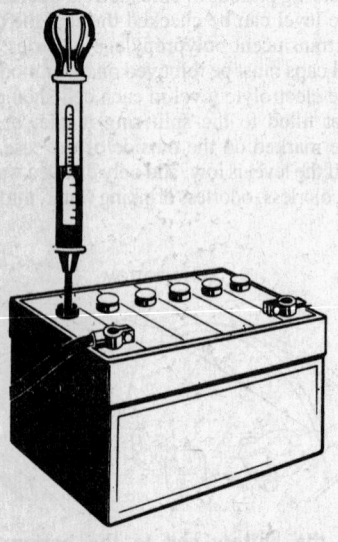

Testing battery specific gravity

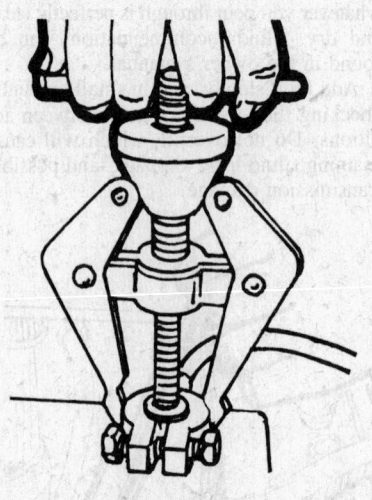

Use a puller to remove the clamp on post-type batteries

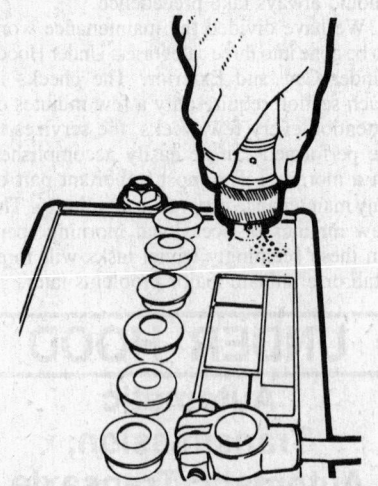

The posts are easily cleaned with a wire brush, or the battery post tool shown

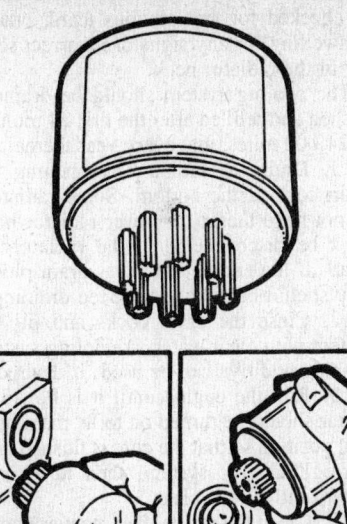

A special tool is required to clean the terminals and clamps on side terminal batteries

Brake Fluid

Once a month, the fluid level in the brake master cylinder should be checked.

1. Park the car on a level surface.

2. Clean off the master cylinder cover before removal. Some covers are retained by a bolt. Some of the newer master cylinders with plastic reservoirs have screw caps. Remove the cover, being careful not to drop or tear the rubber diaphragm which will probably be underneath. Be careful also not to drip any brake fluid on painted surfaces, as it eats paint.

NOTE: Brake fluid absorbs moisture from the air, which reduces effectiveness and will corrode brake parts once in the system. Never leave the master cylinder or the brake fluid container uncovered for any longer than necessary.

3. The fluid level should be about ¼ inch below the lip of the master cylinder well.

4. If fluid addition is necessary, use only extra heavy duty disc brake fluid meeting DOT 3 or DOT 4 specifications. The fluid should be reasonably fresh, because brake fluid deteriorates with age.

5. Replace the cover, making sure that the diaphragm is correctly seated.

If the brake fluid is constantly low, the system should be checked for leaks. However, it is normal for the fluid level to fall gradually as the disc brake pads wear; expect the fluid level to drop about ⅛ inch for every 10,000 miles of wear.

Belt Tension

Every six months or 12,000 miles, check

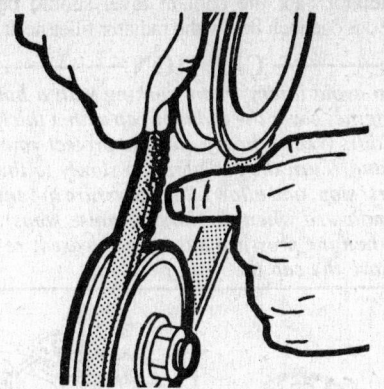

Check the belts for wear

the water pump, alternator, power steering pump, air pump, and air conditioning compressor drive belts for proper tension. Also look for signs of wear, fraying, separation, glazing and so on, and replace the belts as required.

Belt tension should be checked with a gauge made for the purpose. If a gauge is not available, tension can be checked with moderate thumb pressure applied to the belt at its longest span midway between pulleys. If the belt has a free span less than twelve inches, it should deflect approximately ⅛–¼ inch. If the span is longer than twelve inches, deflection can range between ⅛ and ⅜ inches.

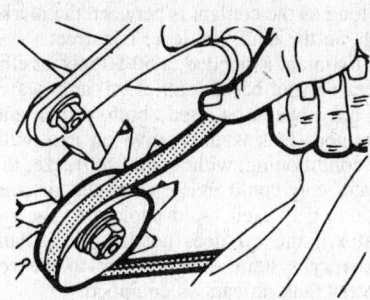

Check the belt tension at the middle of the longest span between pulleys

To adjust or replace belts:

1. Loosen the driven accessory's pivot and mounting bolts. Some air conditioning compressor belts are tensioned by an idler pulley; in this case, loosen the idler pulley and use a ½ in. drive ratchet in the square hole provided to lever the idler pulley up or down.

2. Move the accessory toward or away from the engine until the tension is correct. You can use a wooden hammer handle or broomstick as a lever, but do not use anything metallic.

3. Tighten the bolts and recheck the tension. If new bolts have been installed, run the engine for a few minutes, then recheck and readjust as necessary.

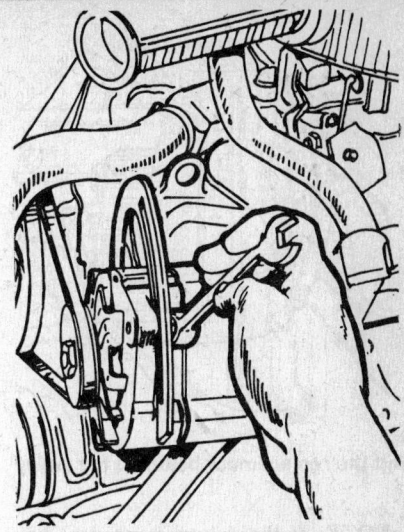

To either adjust or remove a belt, loosen the driven component's adjusting bolt

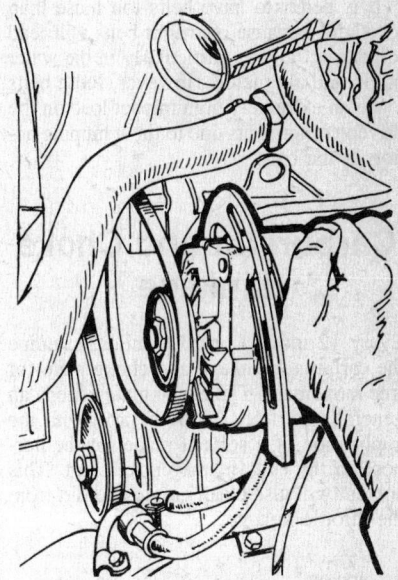

Push the component toward the engine to remove the belt

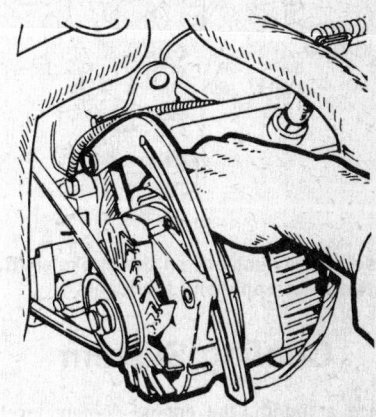

Pull outwards on the component to tension the belt, then tighten the bolts; recheck the belt tension after tightening

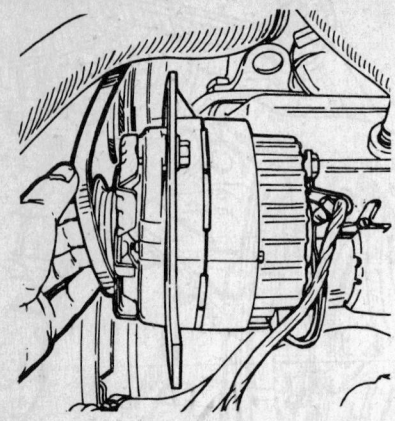

Slip the replacement belt over the pulley

NOTE: If the driven component has two drive belts, the belts should be replaced in pairs to maintain proper tension.

It is better to have belts too loose than too tight, because overtight belts will lead to bearing failure, particularly in the water pump and alternator. However, loose belts place an extremely high impact load on the driven components due to the whipping action of the belt.

Carburetor and Choke Linkage

Every 12 months or 6000 miles, examine the carburetor linkage and choke plate for free movement. The choke plate action can generally be freed, if necessary, with the application of a solvent made for the purpose to the ends of the choke shaft. This solvent will also clean grease and dirt from the throttle linkage.

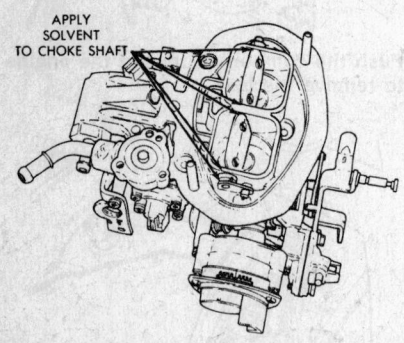

Use a spray solvent on the choke shaft, but do not apply any lubricants

Cooling System

Once a month, the engine coolant level should be checked. On cars without a coolant recovery system, this should only be done when the engine is cold. Remove the radiator cap; the coolant level should be about one inch below the radiator filler neck.

CAUTION

To avoid injury when working with a hot engine, cover the radiator cap with a thick cloth. Wear a heavy glove to protect your hand. Turn the radiator cap slowly to the first stop, and allow all the pressure to vent (indicated when the hissing noise stops). When the pressure has been released, remove the cap the rest of the way.

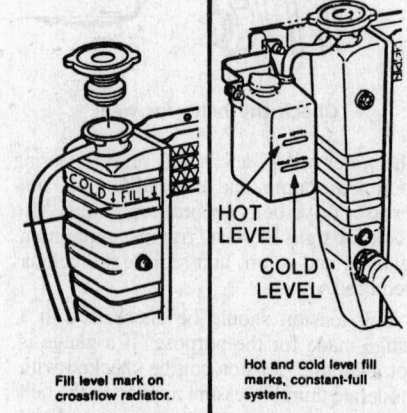

Proper coolant level is about one inch below the radiator neck, or between the lines on the recovery tank

On cars with a coolant recovery tank, coolant should be visible within the tank; as long as the coolant is between the markings on the tank, the level is correct.

If coolant is needed, a 50/50 mix of ethylene glycol-based antifreeze and water should always be used, both winter and summer. This is imperative on cars with air conditioning; without the antifreeze, the heater core could freeze when the air conditioning is used. Add coolant to the radiator if the car does not have a coolant recovery system. Add coolant to the recovery tank on cars so equipped.

The radiator hoses and clamps and the radiator cap should be checked at the same time as the coolant level. Hoses which are brittle, cracked, or swollen should be replaced. Clamps should be checked for tightness (screwdriver tight only—do not allow the clamp to cut into the hose or crush the fitting). The radiator cap gasket should

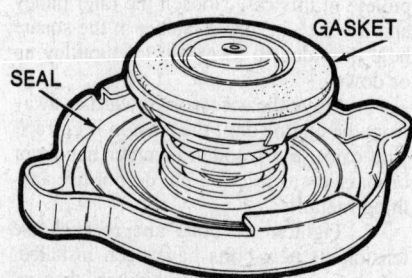

Check the radiator cap gasket and sealing surface

be checked for any obvious tears, cracks or swelling, or any signs of incorrect seating in the radiator neck.

The cooling system should be drained, flushed and refilled after the first 24 months or 24,000 miles, and every year thereafter.

1. Drain the radiator by opening the drain cock at the bottom. Some radiators do not have these; the lower radiator hose must be disconnected at the radiator instead. If the engine block has drain plugs, they should be opened to speed draining.

2. Close the drain cocks and fill the system with clear water. A cooling system flushing additive can be used, if desired.

3. Run the engine until it is hot. The heater should be turned on to its maximum heat position so that the core is flushed out.

4. Drain the system, then flush with water until it runs clear.

5. Clean out the coolant recovery tank, if equipped.

6. Fill the system with a 50/50 mix of ethylene glycol-based antifreeze and water. Fill the coolant recovery tank midway between the marks with this mixture also.

7. Run the engine until it is hot, then let it cool and top up the radiator or coolant recovery tank as necessary with the antifreeze/water mixture.

Heat Riser

The heat riser is a thermostatically or vacuum operated valve in the exhaust manifold (not all cars have one). It closes when the engine is warming up, in order to preheat the incoming fuel/air mixture. If it sticks open, the result will be frequent stalling during warmup, especially in cold and damp weather. If it sticks shut, the result will be a rough idle after the engine is warm.

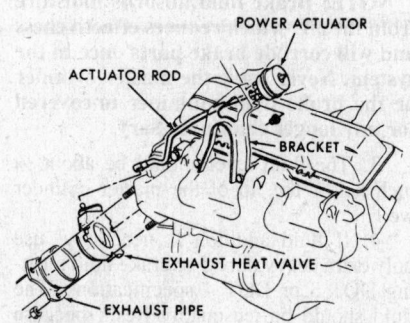

Exploded view of a vacuum-operated heat riser

The heat riser should move freely. It can be checked easily when the engine is cold by giving the counterweight on the valve shaft a twirl, or pulling the vacuum rod to open and shut the valve. If the valve is sticking or binding, a quick shot of solvent made for the purpose will free it up. This solvent should be applied every six months or 6000 miles to keep the valve free. If the valve is still stuck after application of the solvent, sometimes rapping the end of the

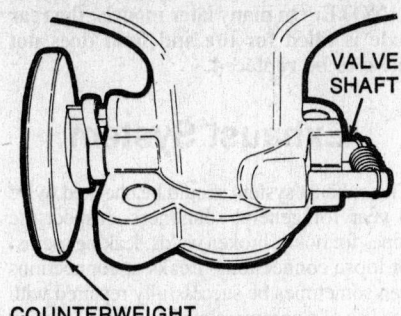

Thermostatically-operated heat control valve

shaft lightly with a hammer will break it loose. Otherwise, the components will have to be removed for further repairs.

Ignition Cables

The ignition system (points, condenser, rotor, spark plugs, etc.) receives regular attention in the form of a tune-up, and thus is not covered here. But one of the most commonly overlooked components is the ignition cable, or spark plug wire.

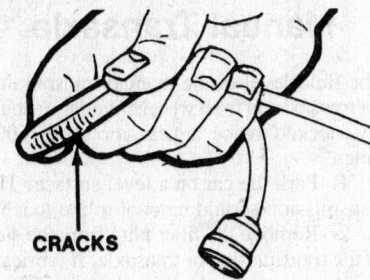

Inspect the ignition cables for cracks or breaks in the insulation

Although they rarely show any visible signs of deterioration, the ignition cables should be checked at every tune-up, and replaced at least every 50,000 miles. Cracking and embrittlement are of course obvious signs of wear, but most newer ca-

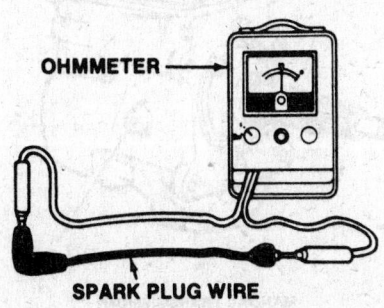

Test the ignition cables with an ohmmeter. Conventional ignition cables should be removed from the distributor cap, but electronic ignition wires should first be tested through the cap

bles have silicone insulation and thus are not prone to display these conditions.

The most reliable way to check the cables is with an ohmmeter. On conventional ignitions, the resistance should be less than 7,000 ohms per foot (wire removed). On cars with electronic ignitions, it is generally recommended to leave the wire attached to the distributor cap; test with one lead from the ohmmeter connected to the corresponding terminal in the distributor cap, the other lead touched to the disconnected end of the cable at the spark plug. Then, if resistance seems close to the limit, remove the wire from the cap and retest. In general, the spark plug wires on electronic ignitions should be replaced if the total resistance is over 36,000 ohms.

Always replace the cables with new ones of the same type. Replace the wires one at a time, working from the longest to the shortest.

Oil Level

The engine oil should be checked on a regular basis, ideally at each fuel stop, or once a week. It is best to check when the engine is at operating temperature, but checking the level immediately after shutting off the engine will give a false reading, because all of the oil will not yet have drained back into the crankcase. The car should be parked on a level surface to obtain an accurate reading.

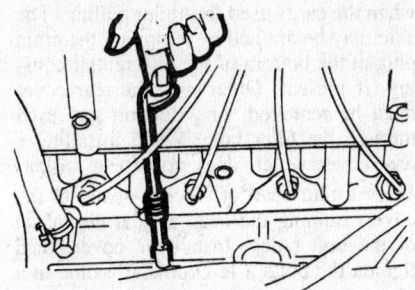

Check the engine oil level with the dipstick

1. Remove the oil dipstick. Wipe it clean, then replace it, seating it firmly.
2. Remove the dipstick again and hold it horizontally to prevent the oil from running. The level should be between the "Add" and "Full" marks on the dipstick. The dipstick may be marked "Add" and "Full," "Add" and "Safe," or may have lines scribed on it; in any case, the oil level should be above the lower marking.
3. If the oil is below the lower mark, enough oil should be added to the engine to raise the level to the upper mark. The markings are usually spaced so that one quart of oil will raise the level from the "Add" mark to the "Full" mark. Oil is added through the capped opening in the valve cover. Only oils labeled SF (gasoline engines) or CC (diesel engines) should be

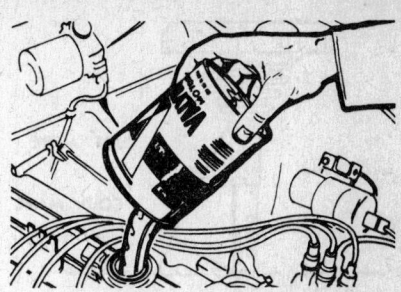

Add oil through the valve cover

used; select a viscosity that will be compatible with the temperatures expected until the next drain interval.

4. Replace the dipstick, then check the level again after any additions of oil. Be careful not to overfill, which will lead to leakage and seal damage.

Power Steering

The power steering fluid level is usually checked with a dipstick inserted into the pump reservoir. The dipstick may be attached to the reservoir cap, or inserted into a tube on the pump body. The level should be checked at every oil change. On some models, the power steering reservoir is translucent, allowing the level to be checked through the sides of the container without removing the cap. On others, the reservoir is a metal canister with a wingnut-attached

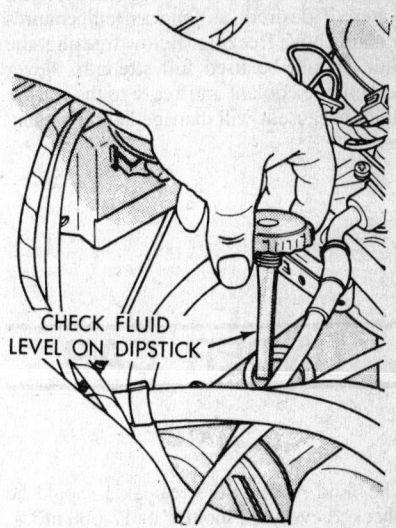

The power steering fluid level on many models is checked by means of a dipstick installed in the reservoir

cap. After the cap is removed, the level is checked with the scribed lines on the inside of the container.

On most models, the fluid level may be checked with the fluid either warm or cold. If checked with the fluid cold, the level will be slightly lower than with the fluid warm. If doubts arise about the specific procedures

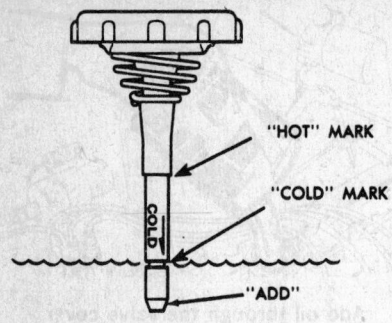

Typical power steering dipstick markings

"HOT" MARK
"COLD" MARK
"ADD"

for the car being checked, consult the owner's manual.

1. On all models, with the engine off, remove the dipstick, remove the cap or check the level through the side of the reservoir. If warm, the level should be between the "Hot" and "Cold" marks or even with the scribed line in the reservoir. If the fluid is cold, the level should be slightly lower.

2. If the level is low, add power steering fluid until the correct level is reached. Do not overfill the reservoir.

Windshield Washer Fluid

Check the fluid level in the windshield washer tank at every oil level check. The fluid can be mixed in a 50% solution with water, if desired, as long as temperatures remain above freezing. Below freezing, the fluid should be used full strength. Never add engine coolant antifreeze to the washer fluid, because it will damage the car's paint.

UNDER CAR

Axle

The fluid level in the rear axle should be checked every 12 months or 12,000 miles.

1. With the car parked on a level surface, remove the filler plug. The plug can be found either in the rear cover of the differential, or on the front of the pinion housing.

2. If lubricant trickles out when the plug is removed, the level is correct. If not, stick your finger in the hole (watch out for sharp threads); the fluid level should be even with edge of the filler hole.

3. If lubricant is needed, use SAE 80W-90 GL-5 gear oil (SAE 80W GL-5 in very cold climates) to fill standard axles. Limited

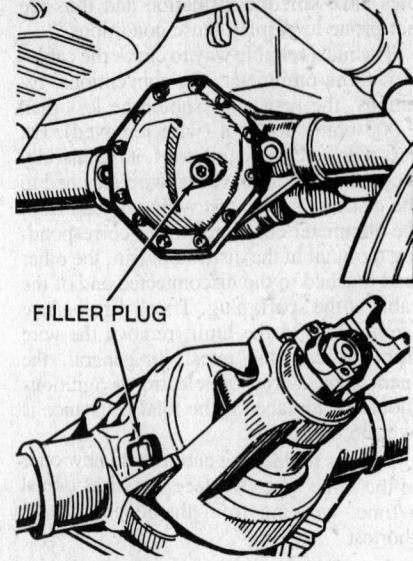

FILLER PLUG

Rear axle filler plug locations

slip axles require a special lubricant, available in auto parts stores.

4. When the level is correct, install the plug and tighten until snug. Do not overtighten.

Standard axles should be drained and refilled with fresh lubricant every 15,000 miles when the car is used to pull a trailer. Limited slip axles should be drained and refilled at the first 7500 miles; the limited slip lubricant should be changed every 7500 miles when the car is used for trailer pulling. The axle may be drained by removing the drain plug at the bottom of the differential housing, if present. Otherwise, the rear cover must be removed, or a suction gun used through the filler hole. When installing a rear cover which does not use a gasket, apply a thin bead of silicone sealer to the cover, running the bead around the inside of the bolt holes. Install the cover, then tighten the bolts a few turns at a time in a crisscross pattern.

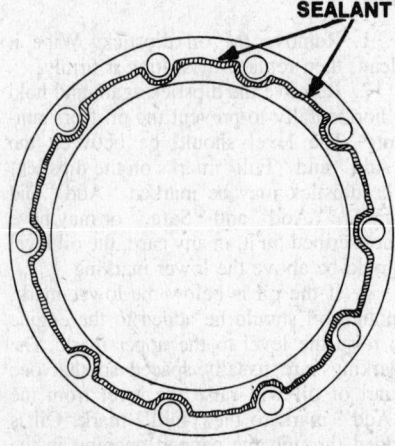

SEALANT

Apply a bead of silicone sealer to the rear cover if no gasket is used

NOTE: On many later models, the rear axle is filled for life and fluid does not have to be replaced.

Exhaust System

The exhaust system should be checked twice a year for general soundness. Inspect the pipes for holes, broken welds, leaking seams, or loose connections. Leaks at connections can sometimes be successfully repaired with the use of a commercial exhaust pipe sealer, but holes or breaks warrant replacement of the part. The exhaust pipe hangers and straps should be examined for any breaks or cracks; replace these as necessary. Some slight cracking of rubber hangers is normal, but deep cracks or cuts are cause for replacement.

— CAUTION —
Check the exhaust system only when it is cold. The temperature on an exhaust system using a catalytic converter can reach 1000°F after only a short period of engine operation.

Manual Transmission, Manual Transaxle

The fluid level in the manual transmission (or transaxle on front wheel drive cars) should be checked twice a year, or every 6000 miles.

1. Park the car on a level surface. The transmission should be cool to the touch.

2. Remove the filler plug from the side of the transmission or transaxle. If lubricant trickles out as the plug is removed, the fluid level is correct. If not, stick your finger into the hole (watch out for sharp threads); the lubricant should be right up to the edge of the filler hole.

3. If lubricant is needed, consult the owner's manual for the correct weight and type of fluid.

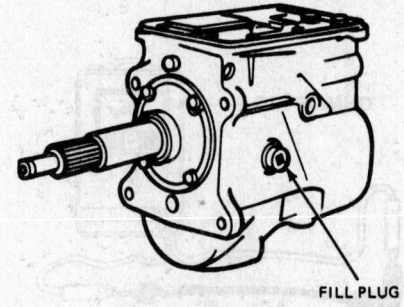

FILL PLUG

MANUAL TRANSMISSION
FILL TO BOTTOM OF
FILLER HOLE WITH
VEHICLE ON LEVEL
GROUND.

Typical manual transmission filler plug location

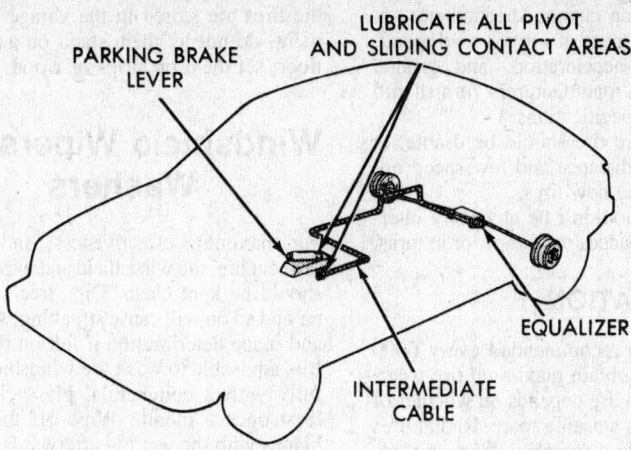

Lubricate the parking brake cable with white waterproof grease

NOTE: Some manual transmission/transaxle assemblies are filled with automatic transmission fluid rather than gear oil. Consult the owner's manual for lubricant information.

4. When the level is correct, install the filler plug and tighten until snug.

Parking Brake Linkage

The parking brake cable assembly should be inspected twice a year for fraying, kinks, and binding. A smooth white waterproof lubricant should be applied at the same time to all pivot points and areas in sliding contact.

Suspension Lubrication

Depending on the year of manufacture, there may be as many as twelve grease fittings on the suspension parts, or as few as two. Typical locations for grease nipples are on the ball joints, control arm pivot points, steering linkage, and the tie rod ends.

Lubricate these fittings with a small hand operated grease gun filled with EP chassis lubricant. Pump grease into the fitting slowly, until it begins to ooze out around the joint, or until the grease begins to expand the rubber boot around the fitting. Be extremely careful not to rupture any seals or boots, as this will lead to lubricant loss and contamination of the parts involved.

Occasionally, the grease nipples may become clogged with dirt or hardened grease. If so, unscrew them with a wrench of the proper size and clean them out with solvent. When reinstalled, they may be covered with plastic caps made for the purpose, or a piece of aluminum foil.

The chassis and suspension parts should be lubricated once a year, or every 7500 miles, whichever comes first.

Transfer Case

The transfer case on the four wheel drive Subaru shares a common lubricant supply with the transmission, therefore the transfer case lubricant supply does not have to be checked separately.

EXTERIOR

Drain Holes and Underbody

Most cars have drain holes spaced along the lower edge of the rocker panels and doors. These holes should be cleared of any debris or rust twice a year. A small screwdriver can be used to open plugged drain holes.

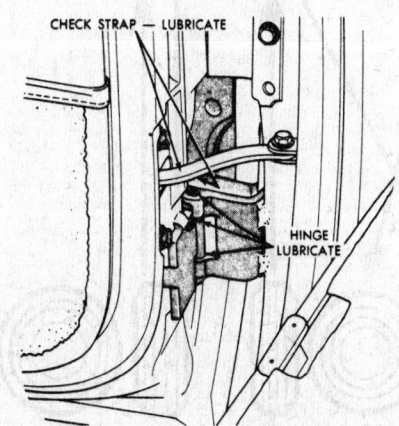

CHECK STRAP — LUBRICATE

HINGE LUBRICATE

Use engine oil to lubricate the door, hood, and trunk hinges

Every spring, the underbody should be flushed with clear water to remove deposits of mud, road salt, and debris. It is advisable to loosen any packed-in sediment before flushing to assure a more thorough cleaning.

Hinges and Locks

Once a year, the door, hood, and trunk hinges, and all locks should be lubricated to ensure smooth operation. The hinge points should be lightly oiled. Lock cylinders may be easily lubricated with a shot of silicone spray directed into the keyhole. Silicone lubricant also works well on the door latch mechanisms, and keeps the door, trunk, and window weatherseals pliable when applied in a light film.

Tires

Tires should be checked weekly for proper air pressure. A chart, located either in the glove compartment or on the driver's or passenger's door, gives the recommended inflation pressures. Maximum fuel economy and tire life will result if the pressure is maintained at the highest figure given on the chart. Pressures should be checked before driving since pressure can increase as

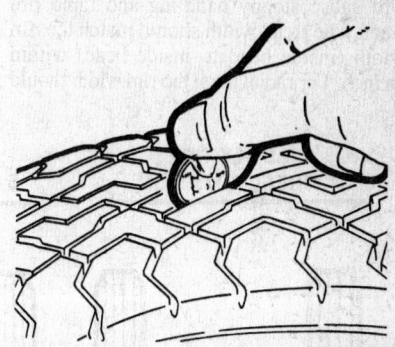

Tire tread depth can be checked with a penny. If the top of Lincoln's head is visible, the tires are due for replacement

much as six pounds per square inch (psi) due to heat buildup. It is a good idea to have your own accurate pressure gauge, because not all gauges on service station air pumps can be trusted. When checking pressures, do not neglect the spare tire. Note that some spare tires require pressures considerably higher than those used in the other tires.

While you are about the task of checking air pressure, inspect the tire treads for cuts, bruises and other damage. Check the air valves to be sure that they are tight. Replace any missing valve caps.

Check the tires for uneven wear that might indicate the need for front end alignment or tire rotation. Tires should be replaced when a tread wear indicator appears as a solid band across the tread.

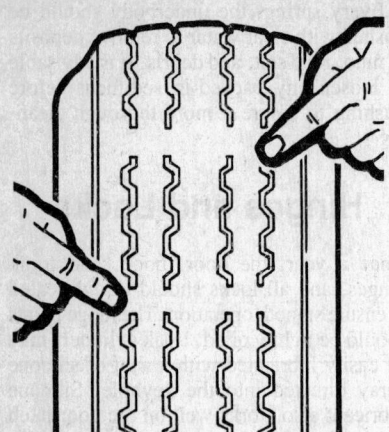

Tread wear indicators will appear as a band across the tire when the tread has worn out.

When buying new tires, give some thought to the following points, especially if you are considering a switch to larger tires or a different profile series:

1. All four tires must be of the same construction type. This rule cannot be violated. Radial, bias, and bias-belted tires must not be mixed.

2. The wheels should be the correct width for the tire. Tire dealers have charts of tire and rim compatibility. A mismatch will cause sloppy handling and rapid tire wear. The tread width should match the rim width (inside bead to inside bead) within an inch. For radial tires, the rim width should

be 80% or less of the tire (not tread) width.

3. The height (mounted diameter) of the new tires can change speedometer accuracy, engine speed at a given road speed, fuel mileage, acceleration, and ground clearance. Tire manufacturers furnish full measurement specifications.

4. The spare tire should be usable, at least for short distance and low speed operation, with the new tires.

5. There shouldn't be any body interference when loaded, on bumps, or in turns.

TIRE ROTATION

Tire rotation is recommended every 6000 miles or so, to obtain maximum tire wear. The pattern you use depends on whether or not your car has a usable spare. Radial tires should not be cross-switched (from one side of the car to the other); they last longer if their direction of rotation is not changed. Snow tires sometimes have directional arrows molded onto the side of their carcass; the arrow shows the direction of rotation. They will wear very rapidly if the rotation is reversed. Studded tires will lose their studs if their rotational direction is reversed.

NOTE: Mark the wheel position or direction of rotation on radial tires or studded snow tires before removing them.

STORAGE

Store the tires at proper inflation pressure

if they are mounted on wheels. Keep them in a cool dry place, laid on their sides. If the tires are stored in the garage or basement, do not let them stand on a concrete floor; set them on strips of wood.

Windshield Wipers and Washers

For maximum effectiveness, and longest element life, the windshield and wiper blades should be kept clean. Dirt, tree sap, road tar and so on will cause streaking, smearing and blade deterioration if left on the glass. It is advisable to wash the windshield carefully with a commercial glass cleaner at least once a month. Wipe off the rubber blades with the wet rag afterwards. For access to the blades on wiper systems which park below the hood line, turn the ignition key to "On" and run the wipers to the center of the windshield. Shut the wipers off with the ignition key, not the wiper switch. Do not attempt to move the wipers by hand; damage to the motor and drive mechanism will result.

If the blades are found to be cracked, broken or torn, they should be replaced immediately. Replacement intervals will vary with usage, although ozone deterioration usually limits blade life to about one year. If the wiper pattern is smeared or streaked, or if the blade chatters across the glass, the elements should be replaced. It is easiest and most sensible to replace the elements

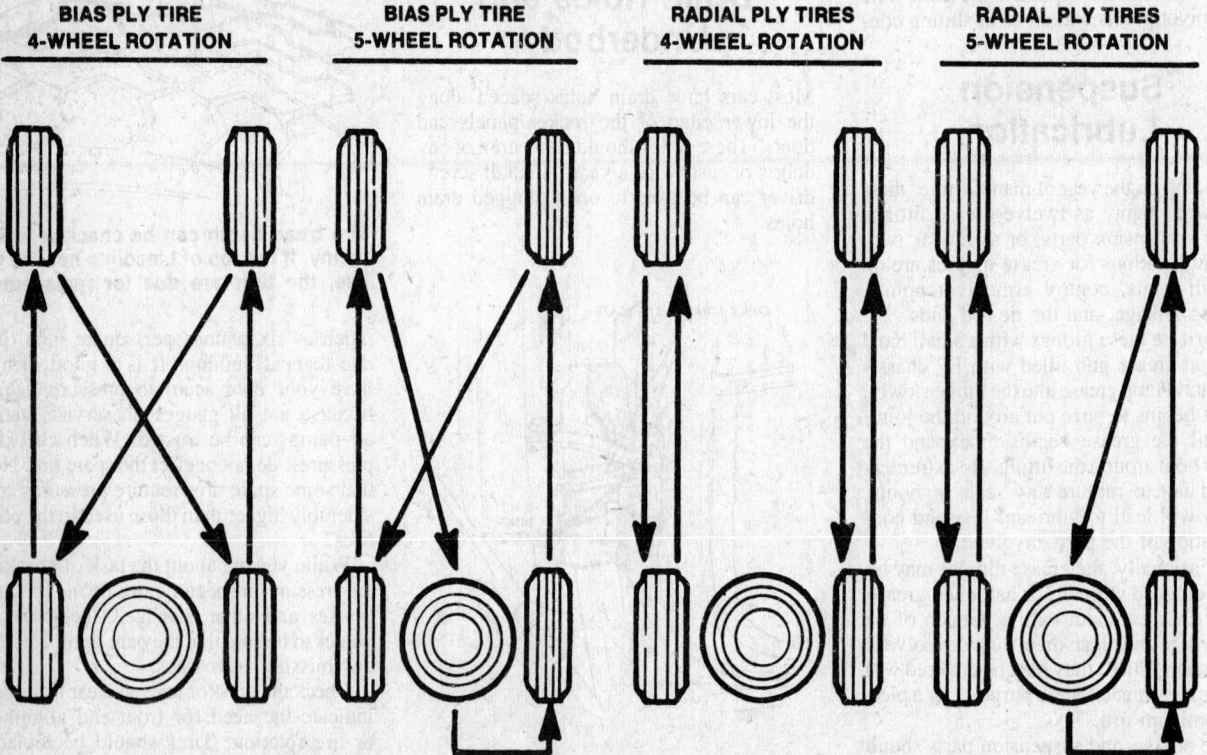

| BIAS PLY TIRE 4-WHEEL ROTATION | BIAS PLY TIRE 5-WHEEL ROTATION | RADIAL PLY TIRES 4-WHEEL ROTATION | RADIAL PLY TIRES 5-WHEEL ROTATION |

Tire rotation diagrams; note that radials should not be cross-switched

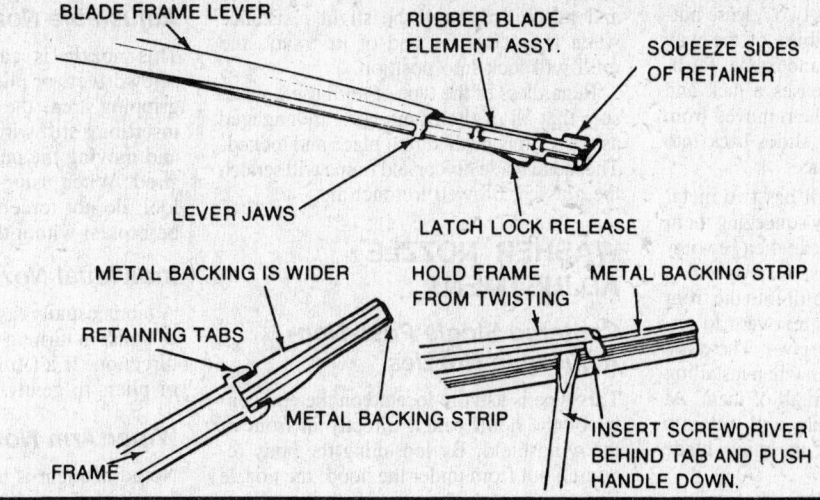

TRICO

BLADE FRAME LEVER

RUBBER BLADE ELEMENT ASSY.

SQUEEZE SIDES OF RETAINER

LEVER JAWS

LATCH LOCK RELEASE

METAL BACKING IS WIDER

HOLD FRAME FROM TWISTING

METAL BACKING STRIP

RETAINING TABS

METAL BACKING STRIP

FRAME

INSERT SCREWDRIVER BEHIND TAB AND PUSH HANDLE DOWN.

ANCO

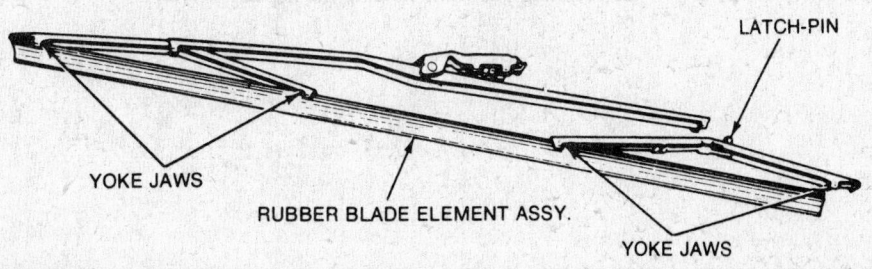

LATCH-PIN

YOKE JAWS

RUBBER BLADE ELEMENT ASSY.

YOKE JAWS

POLYCARBONATE

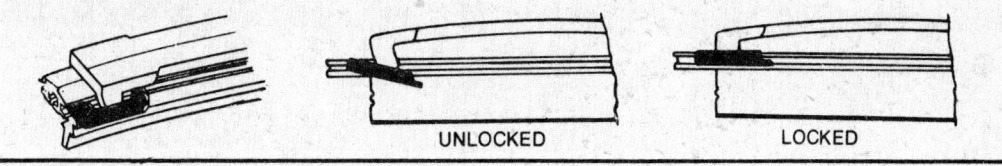

UNLOCKED

LOCKED

TRIDON

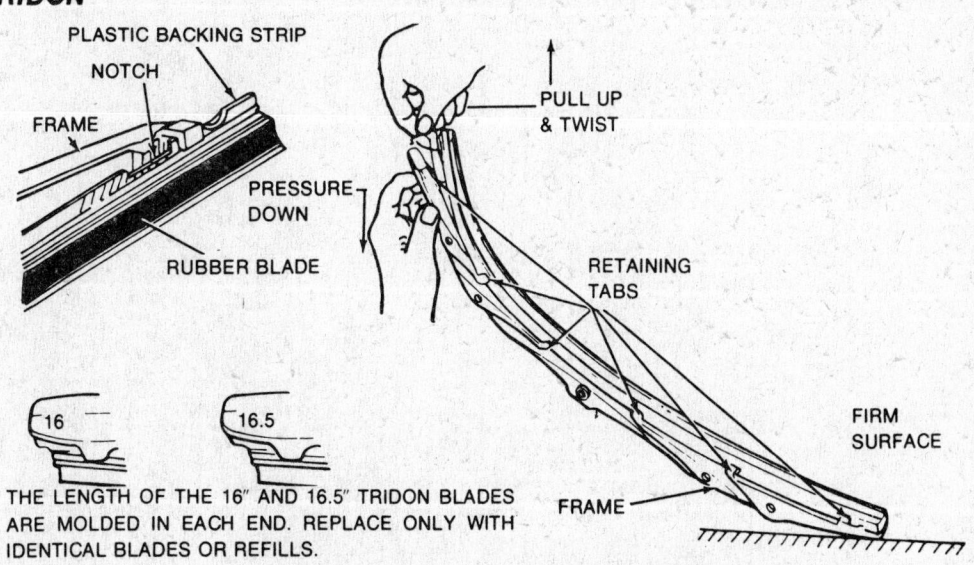

PLASTIC BACKING STRIP

NOTCH

FRAME

PULL UP & TWIST

PRESSURE DOWN

RUBBER BLADE

RETAINING TABS

FIRM SURFACE

THE LENGTH OF THE 16" AND 16.5" TRIDON BLADES ARE MOLDED IN EACH END. REPLACE ONLY WITH IDENTICAL BLADES OR REFILLS.

FRAME

Windshield wiper blade replacement methods

in pairs.

There are basically three different types of refills, which differ in their method of replacement. One type has two release buttons, approximately one-third of the way up from the ends of the blade frame. Pushing the buttons down releases a lock and allows the rubber filler to be removed from the frame. The new filler slides back into the frame and locks in place.

The second type of refill has two metal tabs which are unlocked by squeezing them together. The rubber filler can then be withdrawn from the frame jaws. A new refill is installed by inserting the refill into the front frame jaws and sliding it rearward to engage the remaining frame jaws. There are usually four jaws; be certain when installing that the refill is engaged in all of them. At the end of its travel, the tabs will lock into place on the front jaws of the wiper blade frame.

The third type is a refill made from polycarbonate. The refill has a simple locking device at one end which flexes downward out of the groove into which the jaws of the holder fit, allowing easy release. By sliding the new refill through all the jaws and pushing through the slight resistance when it reaches the end of its travel, the refill will lock into position.

Regardless of the type of refill used, make sure that all of the frame jaws are engaged as the refill is pushed into place and locked. The metal blade holder and frame will scratch the glass if allowed to touch it.

WASHER NOZZLE ADJUSTMENT

Centered Single Post—Non-Adjustable Nozzles

This type is usually located on the rear center of the hood panel, directly in front of the windshield. By loosening the body retaining nut from under the hood, the nozzle body can be turned to provide the best spray discharge to cover the windshield. Tighten the retaining nut while holding the nozzle in position.

Centered Single Post—Adjustable Nozzles

This nozzle is adjusted with a wrench, screwdriver, or pliers. If the nozzle has no gripping area, the adjustment is made by inserting a stiff wire into the nozzle opening and moving the nozzle in the direction desired. When using the wire as an adjuster tool, do not force the nozzle; the wire can be broken within the nozzle opening.

Individual Nozzles

A tab is usually fastened to the nozzle stem to assist in turning the nozzle in the desired direction. If a tab is not present, use a pair of pliers to gently move the nozzle.

Wiper Arm Nozzles

No adjustment is necessary on this type of nozzle, because the opening is centered on the wiper arm and moves along with the arm.

Air Conditioning

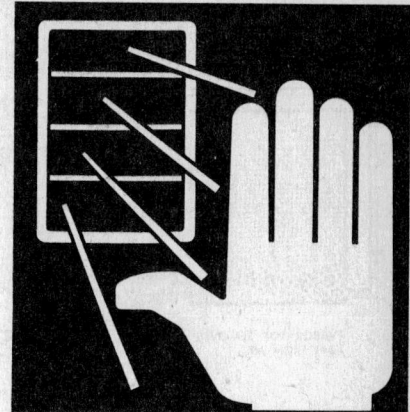

AIR CONDITIONING SYSTEMS

Automotive air conditioning systems are basic in design and operation, but many different components are used by the vehicle manufacturers to operate and control the systems to their specifications.

Basic System

The basic air conditioning system utilizes the compressor, condenser, evaporator, receiver-drier, expansion valve and a thermostatic or ambient type switch to control evaporator freeze-up. The controls are manually operated and the unit is basic in design. This system is usually installed as an add-on or after-market unit. A sight glass may be used in the system.

P.O.A. System

The P.O.A. (pilot operated absolute) suction throttling valve system contains the compressor condenser, evaporator, receiver-drier, expansion valve and a suction throttling valve. The suction throttling valve is used to keep the refrigerant gas in the evaporator at a pressure which will not allow the temperature of the evaporator core surface to go below 32 degrees F., thus preventing evaporator freeze-up. For the system to operate effectively, an equalizer line is connected between the suction side of the suction throttling valve and the ex-

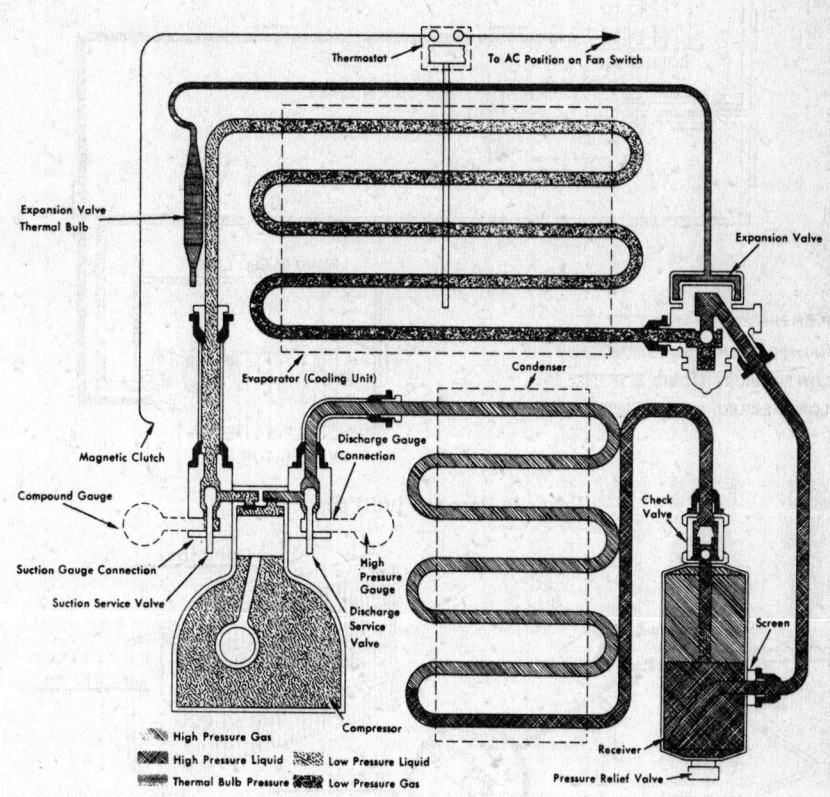

Basic air conditioning system

pansion valve diaphragm. This modifies the operation of the expansion valve which now is controlled by the evaporator outlet temperature and compression suction pressure.

When a crank type compressor is used with the P.O.A. system, an accumulator is placed between the evaporator and the com-

pressor. The accumulator operates as its name implies, accumulating any liquid refrigerant that may have passed from the evaporator and to prevent its moving to the compressor as a liquid, which may, in its form, cause internal compressor damage. A sight glass is normally used in this system.

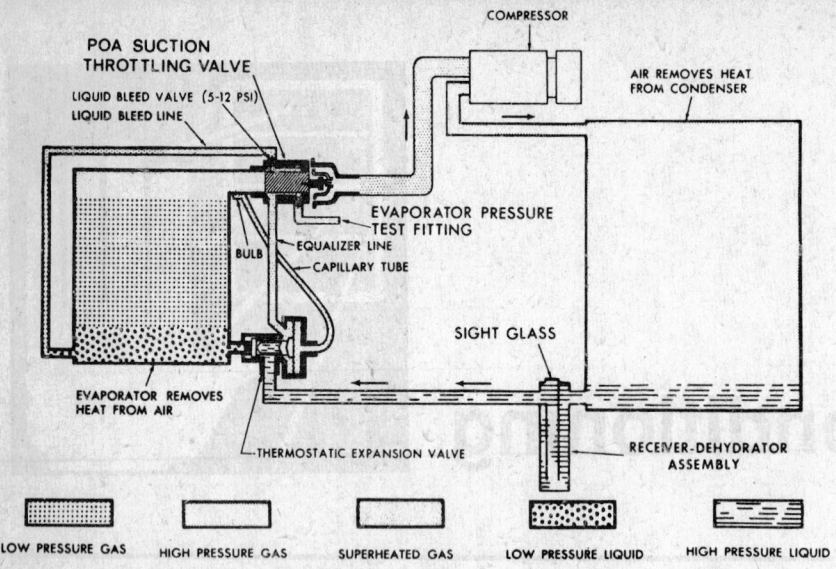

LOW PRESSURE GAS HIGH PRESSURE GAS SUPERHEATED GAS LOW PRESSURE LIQUID HIGH PRESSURE LIQUID

Pilot Operated Absolute (POA) system

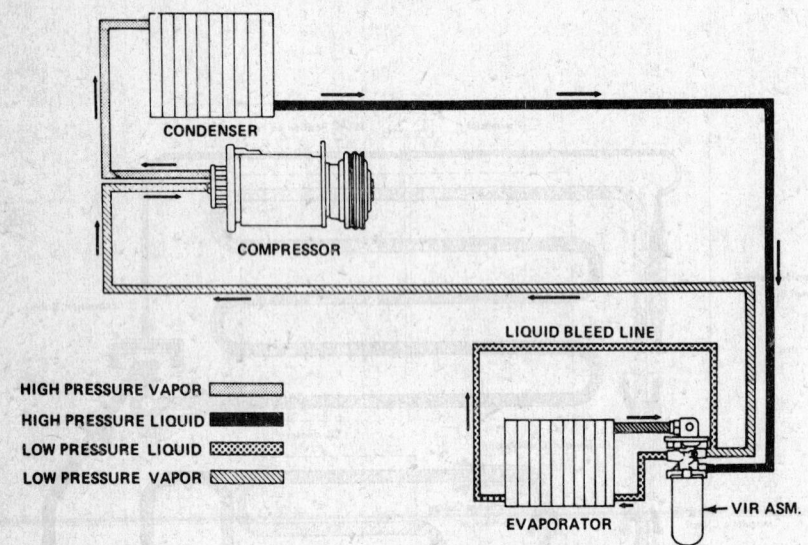

HIGH PRESSURE VAPOR
HIGH PRESSURE LIQUID
LOW PRESSURE LIQUID
LOW PRESSURE VAPOR

Valves In Receiver (VIR) system

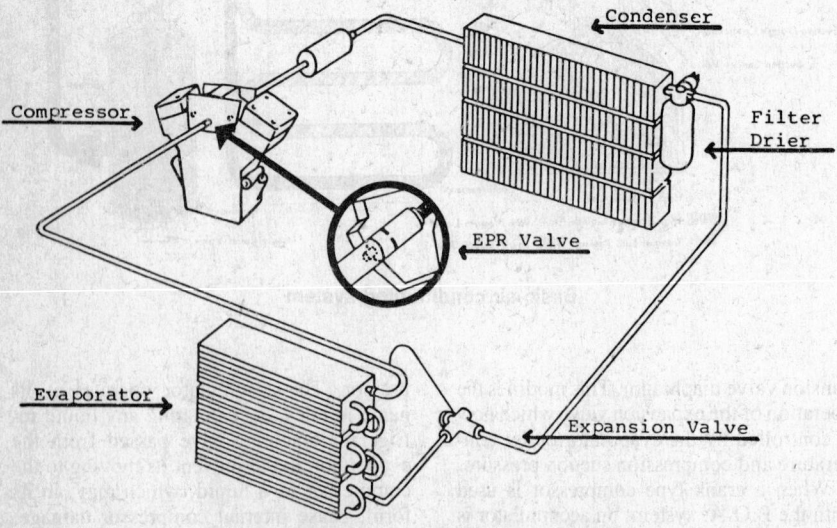

Evaporator Pressure Regulator (EPR) system

V.I.R. System

The V.I.R. system contains the compressor evaporator, condenser, muffler and a unit containing the P.O.A. valve, expansion valve and the receiver-drier. This unit is called the V.I.R. (valves in receiver) assembly. A muffler is normally used with this system and is located between the compressor and the condenser to absorb the compressor pulsations.

The V.I.R. assembly eliminates the outside equalizer line between the outlet of the P.O.A. valve and the expansion valve. The equalizer is now a drilled orfice in the wall between the P.O.A. valve and the expansion valve cavities of the V.I.R. housing. Should the valve prove defective during tests, the unit should be replaced, as it is not repairable or adjustable. A sight glass is normally used with this system.

E.P.R. System

The E.P.R. (evaporator pressure regulator) system includes the condenser, muffler, low pressure shut off valve receiver-drier, expansion valve, evaporator and a V-block, reciprocating crank type compressor. The E.P.R. valve is mounted on the suction side of the compressor and operates in conjunction with the expansion valve assembly, to regulate the flow of refrigerant from the evaporator to the compressor, under light air conditioning loads. By regulating the refrigerant flow, the evaporator temperature is controlled and freezing of the evaporator is prevented.

In contrast to other systems, the E.P.R. system uses the reheat procedure to control the temperature of the air, after it is cooled by passing through the evaporator fins. A manually controlled operating lever is connected to the heater water flow control valve and to a blend air door and the opening of the blend door proportions the amount of air around and through the heater core to control the mix of the cool and hot air for the desired inside temperature. A sight glass is used with this system.

Two types of expansion valves are used with this system. The first type has a capillary tube, mounted in a well on the suction line. The second type has no capillary tube, but senses the need to meter refrigerant into the evaporator by an internal sensing tube. This type of expansion valve is called the "H" type.

"H" Valve System

As was described in the E.P.R. system, the "H" expansion valve can be used with the E.P.R. valve, located in the V-block, reciprocating crank type compressor, to control the amount of refrigerant metered into the evaporator and to control the temperature of the evaporator coils to prevent freeze-up of the condensed moisture. However, when the "H" valve is used with the three piston, axial compressor, a cycling switch is used to control the temperature of the evaporator to prevent freeze-up, rather than the E.P.R. valve, as used with the reciprocating crank type compressor. This can be called the "H" valve system for explanation purposes only and should be recognized as such. The "H" system uses the same components as the other systems, basically the compressor (axial type), condenser, evaporator, expansion valve without a capillary tube ("H" type), receiver-drier, muffler and a low pressure shut off valve. The cycling clutch switch uses a capillary tube, attached to the surface of the suction line, to sense the need for refrigerant movement and compressor operation, therefore causing the electrical clutch pulley and coil to operate the compressor on demand from the cycling switch and to open the circuit to the coil when the demand is not needed. A sight glass is used with this system.

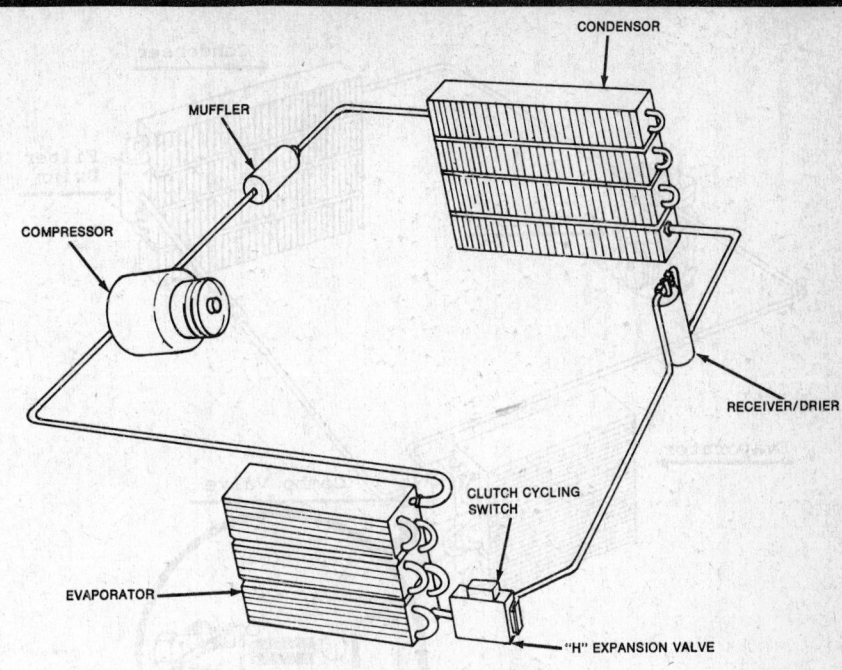

H type expansion valve system

CCOT System

The CCOT (cycling clutch orifice tube) system includes the compressor, condensor, evaporator, an accumulator-drier, a clutch cycling switch with a capillary tube, and a fixed orifce tube, mounted to the evaporator, replacing the expansion valve.

The clutch cycling switch with a temperature probing capillary tube, cycles the compressor clutch off and on as required to maintain a selected comfortable temperature within the vehicle, while preventing evaporator freeze-up. Full control of the system is maintained through the use of a selector control, mounted in the dash assembly. The selector control makes use of a vacuum supply and electrical switches to operate mode doors and the blower motor. A sight glass is not used in this system and one should not be installed. When charging the system, the correct quantity of refrigerant must be installed by measurement.

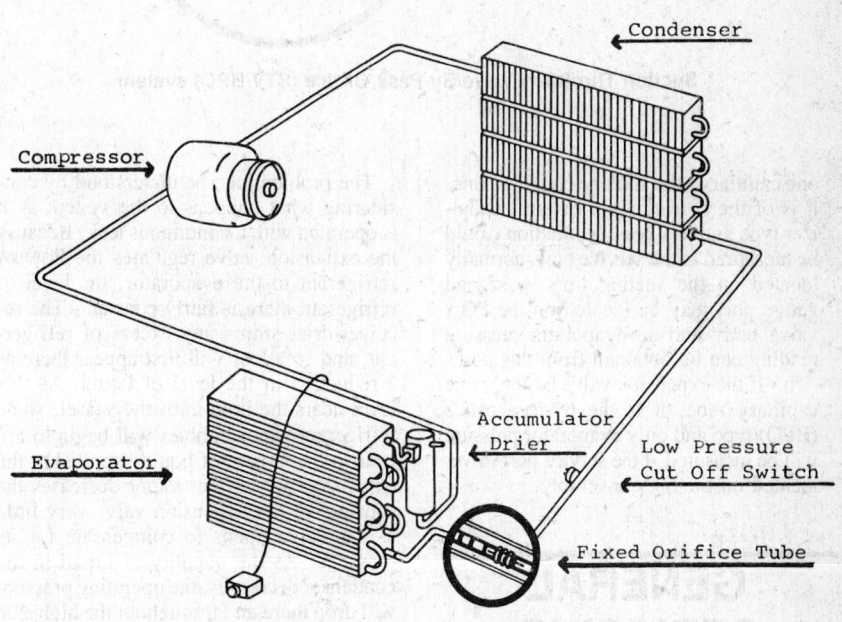

Cycling Clutch Orifice Tube (CCOT) system

STV/BPO System

The STV/BPO (suction throttling valve/bypass orifice) system uses either two types of external expansion valves or a mini-combination valve assembly contains an expansion valve, suction throttling valve and a service port. The expansion valve is of the "H" block design and is used to regulate the flow of refrigerant into the evaporator core. It is also the dividing point for the high and low pressure within the system. The suction throttling valve is used to control the evaporator pressure and to prevent coil freeze-up. The suction throttling starts when the compressor suction pressure decreases below the valve setting. The compressor suction pressure can continue to drop, but the evaporator pressure is held steady by the controlling or throttling action

of the STV. A pressure differential valve is used within the combination valve assembly, to allow oil–laden refrigerant to by-pass the restriction formed when the STV assembly is closed, to assure oil return to the compressor during times of reduced heat loads on the system. The by-pass valve remains closed under high heat loads since ample oil is moving through the system and compressor.

Evaporator pressure can only be measured on this system and a special type connector must be used to attach the high pressure gauge line to the service gauge port.

When either of the external type expansion valves are used, separate suction throttling valves are used. The operation of each is basically the same as the components of the combination valve assembly.

The type of external expansion valve used with the system will dictate either low suction or evaporator pressure measurements from the gauge service ports. To determine the pressure measurement that may be obtained from the system, examine the external expansion valve for one of the following conditions:

a. Should the expansion valve have

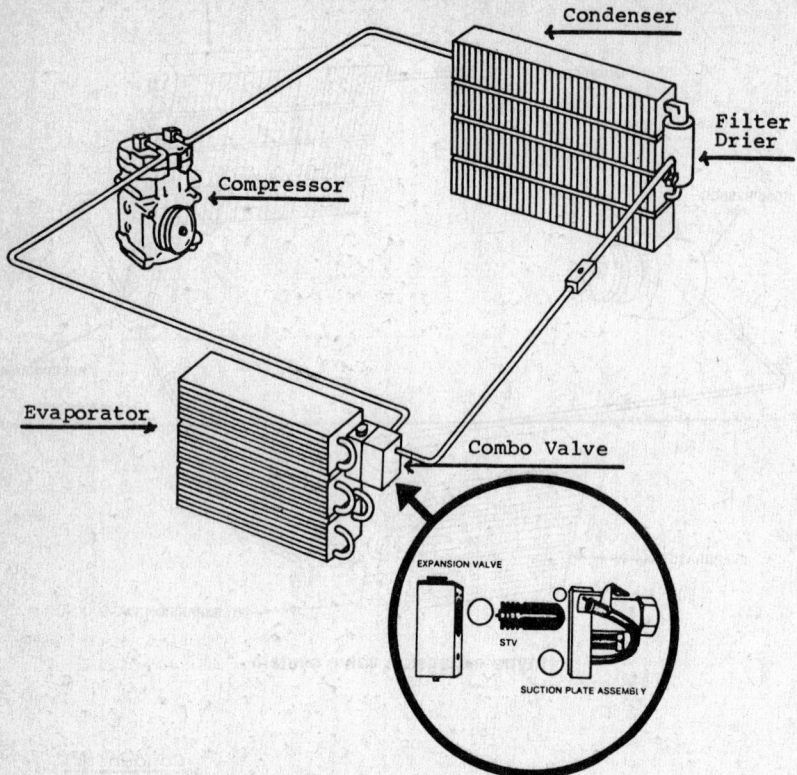

EXPANSION VALVE

STV

SUCTION PLATE ASSEMBLY

Suction Throttling Valve/By-Pass Orifice (STV/BPO) system

one capillary tube and one equalizer line, it is of the conventional external equalizer type and low pressure suction would be measured at the service port, normally located on the suction line. A second gauge port may be located on the POA valve body and an evaporator pressure reading can be obtained from this port.

b. If the expansion valve has only one capillary tube, it is the by-pass orfice (BPO) type and only evaporator pressure will be measured at the service port valve, located on the STV assembly.

GENERAL SERVICING PROCEDURES

The most important aspect of air conditioning service is the maintenance of a pure and adequate charge of refrigerant in the system. A refrigeration system cannot function properly if a significant percentage of the charge is lost. Leaks are common because the severe vibration encountered in an automobile can easily cause a sufficient cracking or loosening of the air conditioning fittings; as a result, the extreme operating pressures of the system force refrigerant out.

The problem can be understood by considering what happens to the system as it is operated with a continuous leak. Because the expansion valve regulates the flow of refrigerant to the evaporator, the level of refrigerant there is fairly constant. The receiver-drier stores any excess of refrigerant, and so a loss will first appear there as a reduction in the level of liquid. As this level nears the bottom of the vessel, some refrigerant vapor bubbles will begin to appear in the stream of liquid supplied to the expansion valve. This vapor decreases the capacity of the expansion valve very little as the valve opens to compensate for its presence. As the quantity of liquid in the condenser decreases, the operating pressure will drop there and throughout the high side of the system. As the R-12 continues to be expelled, the pressure available to force the liquid through the expansion valve will continue to decrease, and, eventually, the valve's orifice will prove to be too much of a restriction for adquate flow even with the needle fully withdrawn.

At this point, low side pressure will start to drop, and severe reduction in cooling capacity, marked by freeze-up of the evaporator coil, will result. Eventually, the operating pressure of the evaporator will be lower than the pressure of the atmosphere surrounding it, and air will be drawn into the system wherever there are leaks in the low side.

Because all atmospheric air contains at least some moisture, water will enter the system and mix with the R-12 and the oil. Trace amounts of moisture will cause sludging of the oil, and corrosion of the system. Saturation and clogging of the filter-drier, and freezing of the expansion valve orifice will eventually result. As air fills the system to a greater and greater extent, it will interfere more and more with the normal flows of refrigerant and heat.

From this description, it should be obvious that much of the repairman's time will be spent detecting leaks, repairing them, and then restoring the purity and quantity of the refrigerant charge. A list of general precautions that should be observed while doing this follows:

1. Keep all tools as clean and dry as possible.

2. Thoroughly purge the service gauges and hoses of air and moisture before connecting them to the system. Keep them capped when not in use.

3. Thoroughly clean any refrigerant fitting before disconnecting it in order to minimize the entrance of dirt into the system.

4. Plan any operation that requires opening the system beforehand, in order to minimize the length of time it will be exposed to open air. Cap or seal the open ends to minimize the entrance of foreign material.

5. When adding oil, pour it through an extremely clean and dry tube or funnel. Keep the oil capped whenever possible. Do not use oil that has not been kept tightly sealed.

6. Use only refrigerant 12. Purchase refrigerant intended for use in only automatic air conditioning systems. Avoid the use of refrigerant-12 that may be packaged for another use, such as cleaning, or powering a horn, as it is impure.

7. Completely evacuate any system that has been opened to replace a component, or that has leaked sufficiently to draw in moisture and air. This requires evacuating air and moisture with a good vacuum pump for at least one hour.

If a system has been open for a considerable length of time it may be advisable to evacuate the system for up to 12 hours (overnight).

8. Use a wrench on both halves of a fitting that is to be disconnected, so as to avoid placing torque on any of the refrigerant lines.

9. When overhauling a compressor, pour some of the oil into a clean glass and inspect it. If there is evidence of dirt or metal particles, or both, flush all refrigerant components with clean refrigerant before evacuating and recharging the system. In addition, if metal particles are present, the compressor should be replaced.

10. Schrader valves may leak only when under full operating pressure. Therefore, if leakage is suspected but cannot be located, operate the system with a full charge of refrigerant and look for leaks from all Schrader valves. Replace any faulty valves.

Additional Preventive Maintenance Checks

ANTIFREEZE

In order to prevent heater core freeze-up during A/C operation, it is necessary to maintain permanent type antifreeze protection of +15 degrees F. or lower. A reading of −15 degrees F. is ideal since this protection also supplies sufficient corrosion inhibitors for the protection of the engine cooling system.

NOTE: The same antifreeze should not be used longer than the manufacturer specifies.

RADIATOR CAP

For efficient operation of an air conditioned car's cooling system, the radiator cap should have a holding pressure which meets manufacturer's specifications. A cap which fails to hold these pressures should be replaced.

CONDENSER

Any obstruction of, or damage to, the condenser configuration will restrict the air flow which is essential to its efficient operation. It is therefore a good rule to keep this unit clean and in proper physical shape.

NOTE: Bug screens are regarded as obstructions.

CONDENSATION DRAIN TUBE

This single molded drain tube expels the condensation, which accumulates on the bottom of the evaporator housing, into the engine compartment.

If this tube is obstructed, the air conditioning performance can be restricted and condensation buildup can spill over onto the vehicle's floor.

Safety Precautions

Because of the importance of the necessary safety precautions that must be exercised when working with air conditioning systems and R-12 refrigerant, a recap of the safety precautions are outlined.

1. Avoid contact with a charged refrigeration system, even when working on another part of the air conditioning system or vehicle. If a heavy tool comes into contact with a section of copper tubing or a heat exchanger, it can easily cause the relatively soft material to rupture.

2. When it is necessary to apply force to a fitting which contains refrigerant, as when checking that all system couplings are securely tightened, use a wrench on both parts of the fitting involved, if possible. This will avoid putting torque on refrigerant tubing.

(It is advisable, when possible, to use tube or line wrenches when tightening these flare nut fittings.

3. Do not attempt to discharge the system by merely loosening a fitting, or removing the service valve caps and cracking these valves. Precise control is possible only when using the service gauges. Place a rag under the open end of the center charging hose while discharging the system to catch any drops of liquid that might escape. Wear protective gloves when connecting or disconnecting service gauge hoses.

4. Discharge the system only in a well ventilated area, as high concentrations of the gas can exclude oxygen and act as an anesthetic. When leak testing or soldering, this is particularly important, as toxic gas is formed when R-12 contacts any flame.

5. Never start a system without first verifying that both service valves are backseated, if equipped, and that all fittings throughout the system are snugly connected.

6. Avoid applying heat to any refrigerant line or storage vessel. Charging may be aided by using water heated to less than 125° to warm the refrigerant container. Never allow a refrigerant storage container to sit out in the sun, or near any other source of heat, such as a radiator.

7. Always wear goggles when working on a system to protect the eyes. If refrigerant contacts the eyes, it is advisable in all cases to see a physician as soon as possible.

8. Frostbite from liquid refrigerant should be treated by first gradually warming the area with cool water, and then gently applying petroleum jelly. *A physician should be consulted.*

9. Always keep refrigerant drum fittings capped when not in use. Avoid sudden shock to the drum, which might occur from dropping it, or from banging a heavy tool against it. *Never carry a drum in the passenger compartment of a car.*

10. Always completely discharge the system before painting the vehicle (if the paint is to be baked on), or before welding anywhere near refrigerant lines.

AIR CONDITIONING TOOLS AND GAUGES

Test Gauges

Most of the service work performed on any air conditioning system requires the use of a set of two gauges, one for the high (head) pressure side of the system, the other for the low (suction) side.

The low side gauge records both pressure and vacuum. Vacuum readings are calibrated from 0–30 inches and the pressure graduations read from 0 to no less than 60 psi.

The high side guage measures pressure from 0 to at least 600 psi.

Both gauges are threaded into a manifold that contains two hand shut-off valves. Proper manipulation of these valves and the use of the attached-test hoses allow the user to perform the following services:

1. Test high and low side pressures.
2. Remove air, moisture, and contaminated refrigerant.
3. Purge the system of (refrigerant).
4. Charge the system (with refrigerant).

The manifold valves are designed so they have no direct effect on gauge readings, but serve only to provide for, or cut off, flow of refrigerant through the manifold. During all testing and hook-up operations, the valves are kept in a closed position to

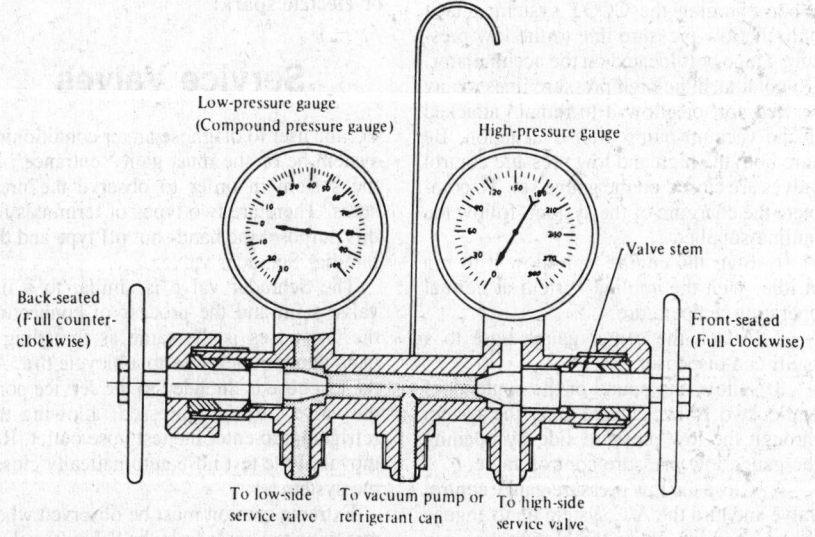

Low-pressure gauge
(Compound pressure gauge)

High-pressure gauge

Valve stem

Back-seated
(Full counter-clockwise)

Front-seated
(Full clockwise)

To low-side service valve   To vacuum pump or refrigerant can   To high-side service valve

Typical manifold gauge set

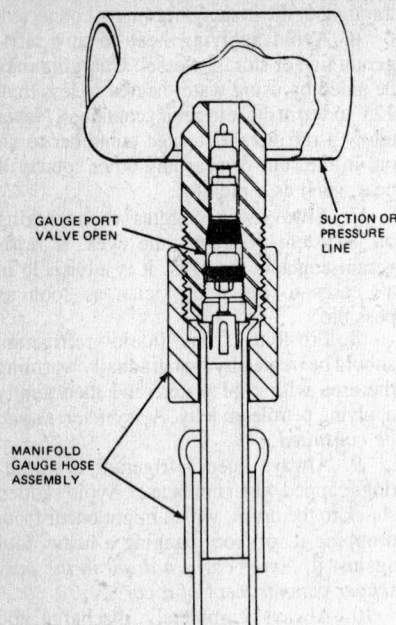

GAUGE PORT
VALVE OPEN

SUCTION OR
PRESSURE
LINE

MANIFOLD
GAUGE HOSE
ASSEMBLY

Manifold gauge hose connected to a Schraeder type service port

avoid disturbing the refrigeration system. The valves are opened only to purge the system of refrigerant or to charge it.

When purging the system, the center hose is uncapped at the lower end, and both valves are cracked open slightly. This allows refrigerant pressure to force the entire contents of the system out through the center hose. During charging, the valve on the high side of the manifold is closed, and the valve on the low side is cracked open. Under these conditions, the low pressure in the evaporator will draw refrigerant from the relatively warm refrigerant storage container into the system.

SYSTEMS WITH A SIGHT GLASS

Air conditioning systems that use a sight glass as a means to check the refrigerant level should be carefully checked to avoid under or over charging. The gauge set should be attached to the system for verification of pressures.

To check the system with the sight glass, clean the glass and start the vehicle engine. Operate the air conditioning controls on maximum for approximately five minutes to stabilize the system. The room temperature should be above 70 degrees. Check the sight glass for one of the following conditions:

1. If the sight glass is clear, the compressor clutch is engaged, the compressor discharge line is warm and the compressor inlet line is cool, the system has a full charge of refrigerant.

2. If the sight glass is clear, the compressor clutch is engaged and there is no significant temperature difference between

the compressor inlet and discharge lines, the system is empty or nearly empty. By having the gauge set attached to the system, a measurement can be taken. If the gauge reads less than 25 psi, the low pressure cut-off protection switch has failed.

3. If the sight glass is clear and the compressor clutch is disengaged, the clutch is defective, or the clutch circuit is open, or the system is out of refrigerant. Bypass the low pressure cut-off switch momentarily to determine the cause.

4. If the sight glass shows foam or bubbles, the system can be low on refrigerant. Occasional foam or bubbles is normal when the room temperature is above 110 degrees or below 70 degrees. To verify, increase the engine speed to approximately 1500 rpm and block the airflow through the condensor in order to increase the compressor discharge pressure to 225–250 psi. If the sight glass still shows bubbles or foam, the refrigerant level is low.

─── **CAUTION** ───

Do not operate the vehicle engine any longer than necessary with the condensor airflow blocked. This blocking action also blocks the cooling system radiator and will cause the system to overheat rapidly.

When the system is low on refrigerant, a leak is present or the system was not properly charged. Use a leak detector and locate the problem area and repair. If no leakage is found, charge the system to its capacity.

─── **CAUTION** ───

It is not advisable to add refrigerant to a system utilizing the suction throttling valve and a sight glass, because the amount of refrigerant required to remove the foam or bubbles will result in an overcharge and potentially damage system components.

CCOT SYSTEM

When charging the CCOT system, attach only the low pressure line to the low pressure gauge port located on the accumulator. Do not attach the high pressure lines to any service port or allow it to remain attached to the vacuum pump after evacuation. Be sure both the high and low pressure control valves are closed on the gauge set. To complete the charging of the system, follow the outline supplied.

1. Start the engine and allow it to run at idle, with the cooling system at normal operating temperature.

2. Attach the center gauge hose to a multi-can dispenser.

3. Allow one pound or the contents of one or two 14 oz. cans to enter the system through the low pressure side by opening the gauge low pressure control valve.

4. Close the low pressure gauge control valve and turn the A/C system on to engage the compressor. Place the blower motor in its high mode.

5. Open the low pressure gauge control valve and draw the remaining charge into the system.

6. Close the low pressure gauge control valve and the refrigerant source valve on the multi-can dispenser. Remove the low pressure hose from the accumulator quickly to avoid loss of refrigerant through the Schrader valve.

7. Install the protective cap on the gauge port and check the system for leakage.

8. Test the system for proper operation.

Leak Testing the System

There are several methods of detecting leaks in an air conditioning system; among them, the two most popular are (1) halide leak-detection or the "open flame method," and (2) electronic leak-detection.

The halide leak detection is a torch like device which produces a yellow-green color when refrigerant is introduced into the flame at the burner. A brilliant blue or violet color indicates the presence of large amounts of refrigerant at the burner. A small leak will cause the flame to turn a yellow-green color.

An electronic leak detector is a small portable electronic device with an extended probe. With the unit activated, the probe is passed along those components of the system which contain refrigerant. If a leak is detected, the unit will sound an alarm signal or activate a display signal depending on the manufacturer's design. It is advisable to follow the manufacturer's instructions as the design and function of the detection may vary significantly.

NOTE: Caution should be taken to operate either type of detector in well ventilated areas, so as to reduce the chance of personal injury, which may result from coming in contact with poisonous gases produced when R-12 is exposed to flame or electric spark.

Service Valves

For the user to diagnose an air conditioning system he or she must gain "entrance" to the system in order to observe the pressures. There are two types of terminals for this purpose, the hand shut off type and the familiar Schrader valve.

The Schrader valve is similar to a tire valve stem and the process of connecting the test hoses is the same as threading a hand pump outlet hose to a bicycle tire. As the test hose is threaded to the service port, the valve core is depressed, allowing the refrigerant to enter the test hose outlet. Removal of the test hose automatically closes the system.

Extreme caution must be observed when removing test hoses from the Schrader valves as some refrigerant will normally escape,

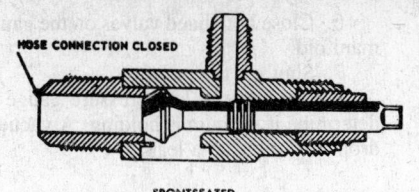

FRONTSEATED

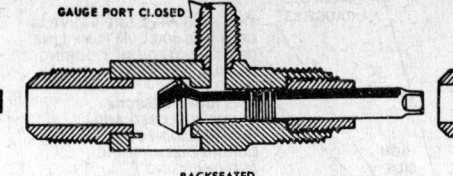

BACKSEATED

Manual service valve positions

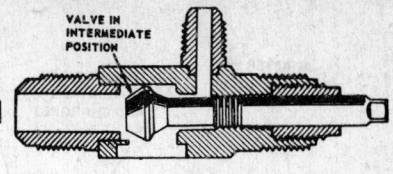

MID-POSITION (CRACKED)

usually under high pressure (observe safety precautions).

Some systems have hand shut-off valves (the stem can be rotated with a special racheting box wrench) that can be positioned in the following three ways:

1. FRONT SEATED—Rotated to full clockwise position.

a. Refrigerant will not flow to the compressor, but will reach the test gauge port. COMPRESSOR WILL BE DAMAGED IF SYSTEM IS TURNED ON IN THIS POSITION.

b. The compressor is now isolated and ready for service. However, care must be exercised when removing service valves from the compressor as a residue of refrigerant may still be present within the compressor. Therefore, remove service valves slowly, observing all safety precautions.

2. BACK SEATED—Rotated to full counterclockwise position. Normal position for system while in operation. Refrigerant flows to compressor but not to test gauge.

3. MID-POSITION (CRACKED)—Refrigerant flows to entire system. Gauge port (with hose connected) open for testing.

USING THE MANIFOLD GAUGES

The following are step-by-step procedures to guide the user to correct gauge usage.

1. WEAR GOGGLES OR FACE SHIELD DURING ALL TESTING OPERATIONS. BACKSEAT HAND SHUT-OFF TYPE SERVICE VALVES.

2. Remove caps from the high and low side of the service ports. Make sure both gauge valves are closed.

3. Connect the low side test hose to the service valve that leads to the evaporator (located between the evaporator outlet and the compressor).

4. Attach the high side test hose to the service valve that leads to the condenser.

5. Mid-position hand shutoff type service valves.

6. Start the engine and allow for warmup. All testing and charging of the system should be done after the engine and system have reached normal operation temperatures (except when using certain charging stations).

7. Adjust the air conditioner controls to maximum cold.

8. Observe the gauge readings. When

BAR GAUGE MANIFOLD AND COMPRESSOR SERVICE VALVE SETTINGS

Condition	Manifold Valves	Compressor Valves
Testing System	Both fully closed	Both cracked off backseat
Depressurizing System	Both cracked open	Both at mid position
Evacuating the system	Both wide open	Both at mid position
Charging in gas form with compressor running	High pressure valve closed	High pressure valve cracked off backseat
	Low pressure valve cracked	Low pressure valve at mid position
Charging in liquid form with compressor off	Low pressure valve closed	Both valves mid positioned
	High pressure valve wide open	

Note: A very small leak, causing system discharge about every two weeks, can be caused by a leaky Schrader type service valve. Check these valves with extra care when testing for a small leak.

the gauges are not being used it is a good idea to:

a. Keep both hand valves in the closed position.

b. Attach both ends of the high and low service hoses to the manifold if extra outlets are present on the manifold, or plug them if not. Also, keep the center charging hose attached to an empty refrigerant can. This extra precaution will reduce the possibility of moisture entering the gauges. If air and moisture have gotten into the gauges, purge the hoses by supplying refrigerant under pressure to the center hose with both gauge valves open and all openings unplugged.

DISCHARGING, EVACUATING AND CHARGING

Discharging the System

——— CAUTION ———

Perform this operation in a well-ventilated area.

When it is necessary to remove (purge)

the refrigerant pressurized in the system, follow this procedure:

1. Operate the air conditioner for at least 10 minutes.

2. Attach the gauges, shut off the engine and air conditioner.

3. Place a container or rag at the outlet of the center charging hose on the gauge. The refrigerant will be discharged there and this precaution will avoid its uncontrolled exposure.

4. Open the low side hand valve on the gauge slightly.

5. Open the high side hand valve slightly.

NOTE: Too rapid a purging process will be identified by the appearance of an oil foam. If this occurs, close the hand valves a little more until this condition stops.

6. Close both hand valves on the gauge set when the pressures read 0 and all the refrigerant has left the system.

Evacuating the System

Before charging any system it is necessary to purge the refrigerant and draw out the trapped moisture with a suitable vacuum pump. Failure to do so will result in ineffective charging and possible damage to the system.

Use this hook-up for the proper evacuation procedure:

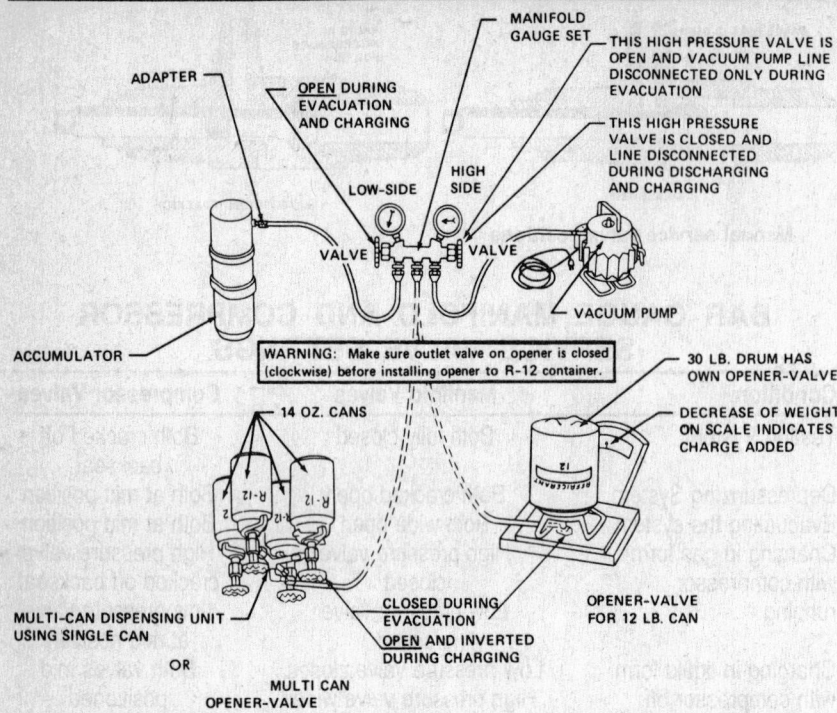

Typical gauge connections for discharge, evacuation and charging the system

1. Connect both service gauge hoses to the high and low service outlets.

2. Open both high and low side hand valves on the gauge manifold.

3. Open both service valves a slight amount (from back seated position), allow the refrigerant to discharge from the system.

4. Install the center charging hose of the gauge set to the vacuum pump.

5. Operate the vacuum pump for at least one hour (if the system has been subjected to open conditions for a prolonged period of time, it may be necessary to "pump the system down" overnight. Refer to the "System Sweep" procedure).

NOTE: If the low pressure gauge does not show at least 28" hg. within 5 minutes, check the system for a leak or loose gauge connectors.

6. Close both hand valves on the gauge manifold.

7. Shut off the pump.

8. Observe the low pressure gauge to determine if vacuum is holding. A vacuum drop may indicate a leak.

System Sweep

An efficient vacuum pump can remove all the air contained in a contaminated air conditioning system very quickly because of its vapor state. Moisture, however, is far more difficult to remove because the vacuum must force the liquid to evaporate before it will be able to remove it from the system. If a system has become severely contaminated, as, for example, it might become after all the charge was lost in conjunction with vehicle accident damage, moisture removal is extremely time consuming. A vacuum pump could remove all of the moisture only if it were operated for 12 hours or more.

Under these conditions, sweeping the system with refrigerant will speed the process of moisture removal considerably. To sweep, follow the following procedure:

1. Connect a vacuum pump to the gauges, operate it until vacuum ceases to increase, then continue operation for ten more minutes.

2. Charge the system with 50% of its rated refrigerant capacity.

3. Operate the system at fast idle for ten minutes.

4. Discharge the system.

5. Repeat twice the process of charging to 50% capacity, running the system for ten minutes, and discharging it, for a total of three sweeps.

6. Replace the drier.

7. Pump the system down as detailed in Step 1.

8. Charge the system.

Charging the System

─────── **CAUTION** ───────

Never attempt to charge the system by opening the high pressure gauge control while the compressor is operating. The compressor accumulating pressure can burst the refrigerant container, causing severe personal injuries.

BASIC SYSTEM

In this procedure the refrigerant enters the suction side of the system as a vapor while the compressor is running. Before proceeding, the system should be in a partial vacuum after adequate evacuation. Both hand valves on the gauge manifold should be closed.

1. Attach both test hoses to their respective service valve ports. Mid-position manually operated service valves, if present.

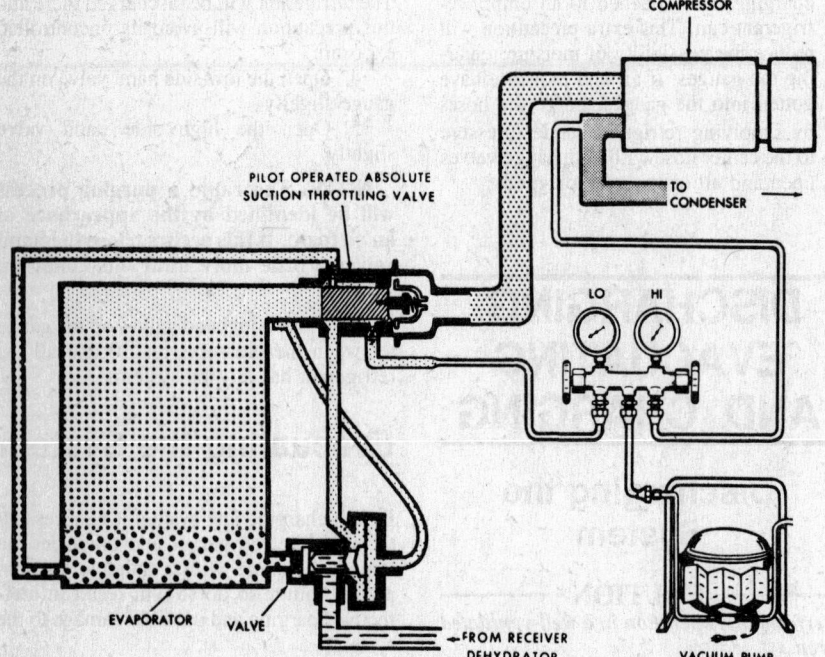

Schematic for evacuating the system

2. Install a dispensing valve (closed position) on the refrigerant container (single and multiple refrigerant manifolds are available to accommodate one to four 15 oz. cans).

3. Attach the center charging hose to the refrigerant container valve.

4. Open the dispensing valve on the refrigerant can.

5. Loosen the center charging hose coupler where it connects to the gauge manifold to allow the escaping refrigerant to purge the hose of contaminants.

6. Tighten the center charging hose connection.

7. Purge the low pressure test hose at the gauge manifold.

8. Start the engine, roll down the windows and adjust the air conditioner to maximum cooling. The engine should be at normal operating temperature before proceeding. The heated environment helps the liquid vaporize more efficiently.

9. Crack open the low side hand valve on the manifold. Manipulate the valve so that the refrigerant that enters the system does not cause the low side pressure to exceed 40 psi. Too sudden a surge may permit the entrance of unwanted liquid to the compressor. Since liquids cannot be compressed, the compressor will suffer damage if compelled to attempt it. If the suction side of the system remains in a vacuum, the system is blocked. Locate and correct the condition before proceeding any further.

NOTE: Placing the refrigerant can in a container of warm water (no hotter than 125° F) will speed the charging process. Slight agitation of the can is helpful too, but be careful not to turn the can upside down.

Some manufacturers allow for a partial charging of the A/C system in the form of a liquid (can inverted and compressor off) by opening the high side gauge valve only, and putting the high side compressor service valve in the middle position (if so equipped). The remainder of the refrigerant is then added in the form of a gas in the normal manner, through the suction side only.

SYSTEMS WITHOUT SIGHT GLASS, EXCEPT CCOT SYSTEM

The following procedure can be used to quickly determine whether or not an air conditioning system has the proper charge of refrigerant (providing ambient temperature is above 70° F. or 21° C.). This check can be made in a manner of minutes, thus facilitating system diagnosis by pinpointing the problem to the amount of charge in the system or by eliminating this possibility from the overall checkout.

1. Engine must be warm (thermostat open).

2. Hood and body doors open.

3. Selector lever set at NORM.

4. Temperature lever at COLD.

5. Blower on HI.

6. Normal engine idle.

7. Hand-feel the temperature of the evaporator inlet and outlet pipes with the compressor engaged.

a. Both same temperature or some degree cooler than ambient—proper condition: check for other problems.

b. Inlet pipe cooler than outlet pipe—low refrigerant charge.

• Add a slight amount of refrigerant until both pipes feel the same.

• Then add 15 oz. (1 can) additional refrigerant.

c. Inlet pipe has front accumulation—outlet pipe warmer: proceed as in Step b above.

If during the charging process the head pressure exceeds 200 psi, place an electric fan in front of the car and direct the turbulent air to the condenser. If no fan is available, repeatedly pour cool water over the top of the condenser. These cooling actions may be necessary on an extremely warm day to help dissipate the heat emitted by the engine during idle.

If this fails and pressure on the discharge side continues to rise, the system may be overcharged or the engine might be overheating. *Never* allow head pressure to go beyond 240 psi. during charging. If this condition occurs, stop the engine, find and correct the problem.

8. Continue dispensing refrigerant until the container is no longer cool to the touch. On a humid day, the outside of the container will frost. When the frost disappears the can is usually empty. To detach the dispensing can:

a. close the low pressure test gauge hand valve.

b. crack open the low pressure test hose at the manifold until the remaining pressure escapes.

c. tighten the hose coupler.

d. loosen the hose coupler connected to the refrigerant can.

e. discard the empty can and repeat Steps 2–8.

9. Continue to add refrigerant to the required capacity of the system. (Usually marked on the compressor).

CAUTION
DO NOT OVERCHARGE. This condition is usually indicated by an abnormally high side pressure reading and a noisy compressor resulting in ineffective cooling and damage to the system.

SYSTEMS WITH A SIGHT GLASS

Air conditioning systems that use a sight glass as a means to check the refrigerant level should be carefully checked to avoid under or over charging. The gauge set should be attached to the system for verification of pressures.

To check the system with the sight glass, clean the glass and start the vehicle engine. Operate the air conditioning controls on maximum for approximately five minutes

Amount of refrigerant Check item	Almost no refrigerant	Insufficient	Suitable	Too much refrigerant
Temperature of high pressure and low pressure lines.	Almost no difference between high pressure and low pressure side temperature.	High pressure side is warm and low pressure side is fairly cold.	High pressure side is hot and low pressure side is cold.	High pressure side is abnormally hot.
State in sight glass.	Bubbles flow continuously. **Bubbles will disappear and something like mist will flow when refrigerant is nearly gone.**	The bubbles are seen at intervals of 1 - 2 seconds.	Almost transparent. Bubbles may appear when engine speed is raised and lowered. **No clear difference exists between these two conditions.**	No bubbles can be seen.
Pressure of system.	High pressure side is abnormally low.	Both pressure on high and low pressure sides are slightly low.	Both pressures on high and low pressure sides are normal.	Both pressures on high and low pressure sides are abnormally high.
Repair.	**Stop compressor immediately** and conduct an overall check.	Check for gas leakage, repair as required, replenish and charge system.		Discharge refrigerant from service valve of low pressure side.

Using a sight glass to determine the relative refrigerant charge

to stabilize the system. The room temperature should be above 70 degrees. Check the sight glass for one of the following conditions:

1. If the sight glass is clear, the compressor clutch is engaged, the compressor discharge line is warm and the compressor inlet line is cool, the system has a full charge of refrigerant.

2. If the sight glass is clear, the compressor clutch is engaged and there is no significant temperature difference between the compressor inlet and discharge lines, the system is empty or nearly empty. By having the gauge set attached to the system, a measurement can be taken. If the gauge reads less than 25 psi, the low pressure cut-off protection switch has failed.

3. If the sight glass is clear and the compressor clutch is disengaged, the clutch is defective, or the clutch circuit is open, or the system is out of refrigerant. By-pass the low pressure cut-off switch momentarily to determine the cause.

4. If the sight glass shows foam or bubbles, the system can be low on refrigerant. Occasional foam or bubbles is normal when the room temperature is above 110 degrees or below 70 degrees. To verify, increase the engine speed to approximately 1500 rpm and block the airflow through the condenser in order to increase the compressor discharge pressure to 225–250 psi. If the sight glass still shows bubbles or foam, the refrigerant level is low.

— CAUTION —

Do not operate the vehicle engine any longer than necessary with the condenser airflow blocked. This blocking action also blocks the cooling system radiator and will cause the system to overheat rapidly.

When the system is low on refrigerant, a leak is present or the system was not properly charged. Use a leak detector and locate the problem area and repair. If no leakage is found, charge the system to its capacity.

— CAUTION —

It is not advisable to add refrigerant to a system utilizing the suction throttling valve and a sight glass, because the amount of refrigerant required to remove the foam or bubbles will result in an overcharge and potentially damaged system components.

CCOT SYSTEM

When charging the CCOT system, attach only the low pressure line to the low pressure gauge port located on the accumulator. Do not attach the high pressure line to any service port or allow it to remain attached to the vacuum pump after evacuation. Be sure both the high and the low pressure control valves are closed on the gauge set. To complete the charging of the system, follow the outline supplied.

1. Start the engine and allow it to run at idle, with the cooling system at normal operating temperature.

2. Attach the center gauge hose to a single or multi-can dispenser.

3. With the multi-can dispenser inverted, allow one pound or the contents of one or two 14 oz. cans to enter the system through the low pressure side by opening the gauge low pressure control valve.

4. Close the low pressure gauge control valve and turn the A/C system on to engage the compressor. Place the blower motor in its high mode.

5. Open the low pressure gauge control valve and draw the remaining charge into the system.

6. Close the low pressure gauge control valve and the refrigerant source valve, on the multi-can dispenser. Remove the low pressure hose from the accumulator quickly to avoid loss of refrigerant through the Schrader valve.

7. Install the protective cap on the gauge port and check the system for leakage.

8. Test the system for proper operation.

Leak Testing the System

There are several methods of detecting leaks in an air conditioning system; among them, the two most popular are (1) halide leak-detection or the "open flame method," and (2) electronic leak-detection.

The halide leak detection is a torch like device which produces a yellow-green color when refrigerant is introduced into the flame at the burner. A purple or violet color indicates the presence of large amounts of refrigerant at the burner.

An electronic leak detector is a small portable electronic device with an extended probe. With the unit activated, the probe is passed along those components of the system which contain refrigerant. If a leak is detected, the unit will sound an alarm signal or activate a display signal depending on the manufacturer's design. It is advisable to follow the manufacturer's instructions as the design and function of the detection may vary significantly.

— CAUTION —

Caution should be taken to operate either type of detector in well ventilated areas, so as to reduce the chance of personal injury, which may result from coming in contact with poisonous gases produced when R-12 is exposed to flame or electric spark.

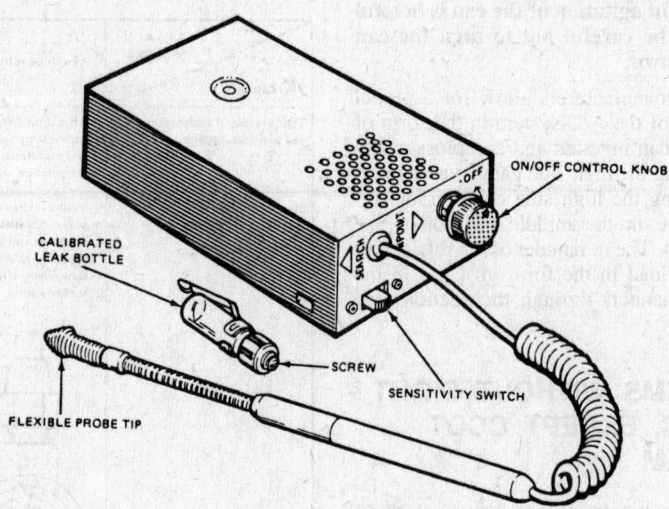

Electronic leak detector

Electrical

ELECTRICAL

TEST INSTRUMENTS

Ohmmeter

An ohmmeter is used to measure electrical resistance in a unit or circuit. The ohmmeter has a self-contained power supply. In use, it is connected across (or in parallel with) the terminals of the unit being tested.

Ammeter

An ammeter is used to measure current (amount of electricity) flowing through a unit, or circuit. Ammeters are always connected in the line (in series) with the unit or circuit being tested.

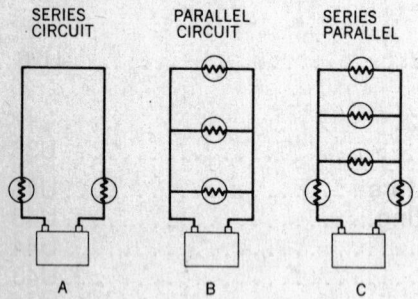

Basic electrical circuits

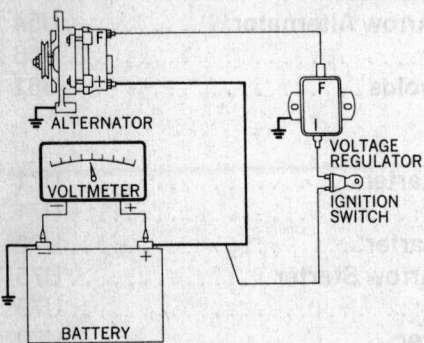

Voltmeter connected in parallel circuit

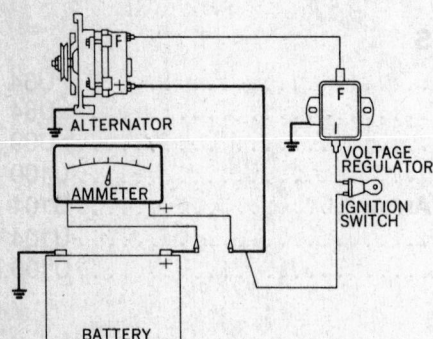

Ammeter connected to test wire

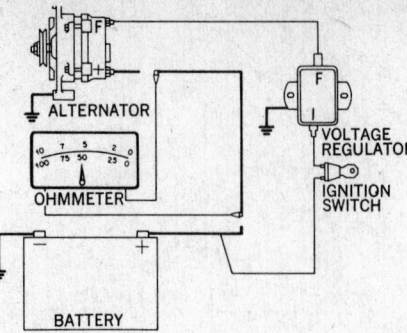

Ohmmeter connected to test wire resistance

Voltmeter

A voltmeter is used to measure voltage (electrical pressure) pushing the current through a unit, or circuit. The meter is connected across the terminals of the unit being tested.

Alternator Testing

IS IT THE ALTERNATOR OR THE VOLTAGE REGULATOR?

The first step in diagnosing troubles of the charging system, is to identify the source of failure. Does the fault lie in the alternator or the regulator? The next move depends upon preference or necessity; either repair or replace the offending unit.

Alternator output is controlled by the amount of current supplied to the field circuit of the system.

The alternator is capable of producing substantial current at idle speed. Higher maximum output is also a possibility. This presents a potential danger when testing. As a precaution, a field rheostat should be used in the field circuit when making the following isolation test. The field rheostat permits positive control of the amount of current allowed to pass through the field circuit during the isolation test. Unregulated alternator capacity could ruin the unit.

NOTE: Most manufacturers of precision gauges offer special test connectors, in sets, that will adapt to the leads and connections of any charging system.

There are certain precautionary measures that apply to alternator tests in general. These items are listed in detail to avoid repetition when testing each make of alternator, and to encourage a habit of good test procedure.

1. Check alternator drive belt for condition and tension.

2. Disconnect battery cables, check physical, chemical, and electrical condition of battery.

3. Be absolutely sure of polarity before connecting any battery in the circuit. Reversed polarity will ruin the diodes.

4. Never use a battery charger to start the engine.

5. Disconnect both battery cables when making a battery recharge hook-up.

6. Be sure of polarity hook-up when using a booster battery for starting.

7. Never ground the alternator output or battery terminal.

8. Never ground the field circuit between alternator and regulator.

9. Never run any alternator on an open circuit with the field energized.

10. Never try to polarize an alternator.

11. Do not attempt to motor an alternator.

12. The regulator cover must be in place when taking voltage limiter readings.

13. The ignition switch must be in off position when removing or installing the regulator cover.

14. Use insulated tools only to make adjustments to the regulator.

15. When making engine idle speed adjustments, always consider potential load factors that influence engine rpm. To compensate for electrical load, switch on the lights, radio, heater, air conditioner, etc.

Diagnosis

LOW OR NO CHARGING

1. Blown fuse.
2. Broken or loose fan belt.
3. Voltage regulator not working.
4. Brushes sticking.
5. Slip ring dirty.
6. Open circuit.
7. Bad wiring connections.
8. Bad diode rectifier.
9. High resistance in charging circuit.
10. Voltage regulator needs adjusting.
11. Grounded stator.
12. May be open rectifiers (check all three phases).
13. If rectifiers are found blown or open, check capacitor.

NOISY UNIT

1. Damaged rotor bearings.
2. Poor alignment of unit.
3. Broken or loose belt.
4. Open diode rectifiers.

REGULATOR POINTS BURNT OR STUCK

1. Regulator set too high.
2. Poor ground connections.
3. Shorted generator field.
4. Regulator air gap incorrect.

Chrysler Isolated Field Alternator (Electronic Regulator)

The Chrysler isolated field alternator derives its name from its construction. Both of the brushes are insulated from ground and there is no heat sink connection, thereby isolating the internal field.

TROUBLESHOOTING

Fusible Links

Chrysler Corporation trucks have a single fusible link which is connected between the starter relay and the junction block. Failure of this link will cause all electrical systems to stop functioning.

Charging System Operation

NOTE: If the current indicator is to give an accurate reading, the battery cables must be of the same gauge and length as the original equipment.

1. With the engine running and all electrical systems off, place a current indicator over the positive battery cable.
2. If a charge of about 5 amps is recorded, the charging system is working. If a draw of about 5 amps is recorded the system is not working. The needle moves toward the battery when a charge condition is indicated and away from the battery when a draw condition is indicated. If a draw is indicated, proceed to the next testing procedure. If an overcharge of 10–15 amps is indicated, check for a faulty regulator.

Ignition Switch-to-Regulator Circuit Check

1. Disconnect the regulator wires at the regulator.
2. Turn the key on but do not start the engine.
3. Using a voltmeter or test light, check for voltage across the I and F terminals. If there is current present, the circuit is good. If there is no current, check for bad connections, a bad ballast resistor, a bad ammeter, broken wires, or bad ground at the alternator or voltage regulator. Also, check for voltage from the I wire to ground; current should be present. Check for voltage from the F terminal to ground; current should not be present.

Isolation Test

This test determines whether the regulator or alternator is bad if everything else in the circuit was OK.

1. Disconnect, at the alternator, the wire that runs between one of the alternator field connections and the voltage regulator.
2. Run a jumper wire from the disconnected alternator terminal to ground.
3. Connect a voltmeter to the battery. The positive voltmeter lead connects to the positive battery terminal, and the negative lead goes to the negative terminal. Record the reading.
4. Make sure that all electrical systems are turned off. Start the engine. Do not race the engine.
5. Gradually raise engine speed to 1500–2000 rpm. There should be an increase of one to two volts on the voltmeter. If this is true, the alternator is good and the voltage regulator should be repaired. If there is no voltage increase, the alternator is faulty.

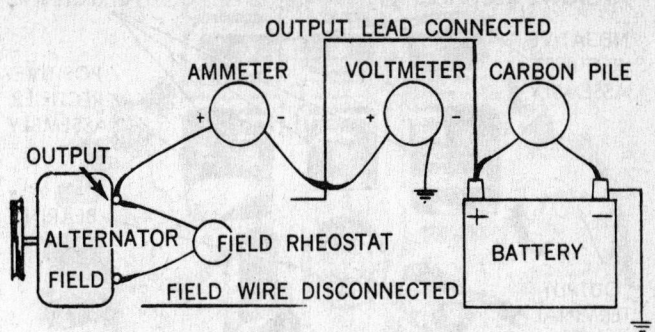

Checking current output of the charging system

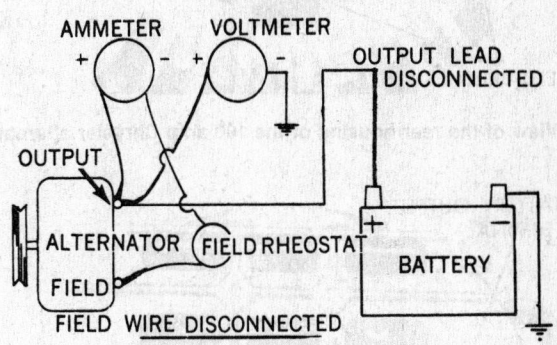

Checking field current draw

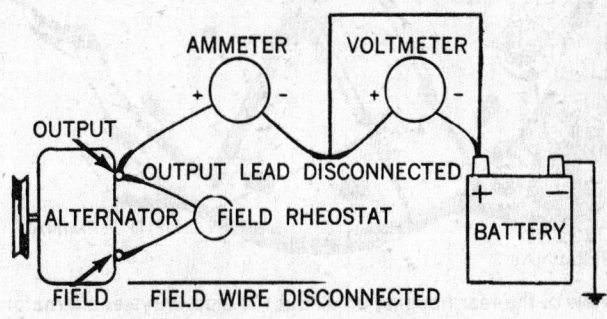

Checking charging system resistance

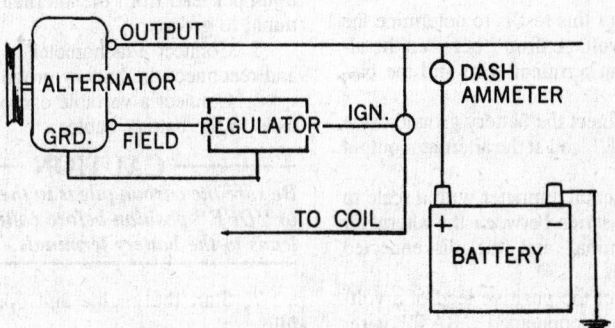

Alternator system with ammeter in the circuit

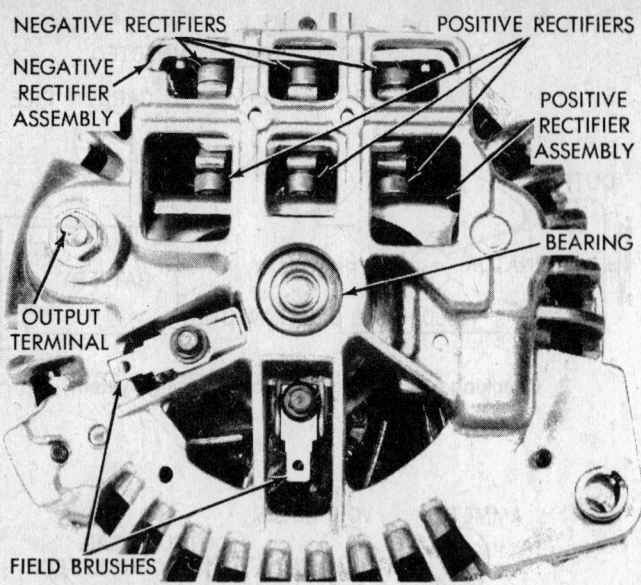

View of the rear housing of the 100 amp Chrysler alternator

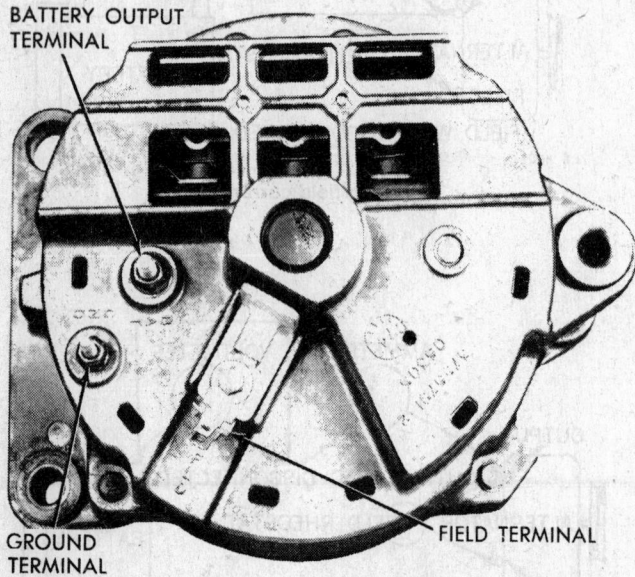

View of the rear housing on all but 100 amp Chrysler alternator

Charging Circuit Resistance Test

The purpose of this test is to determine the amount of "voltage drop" between the alternator output terminal wire and the battery.

1. Disconnect the battery ground cable and the "BAT" lead at the alternator output terminal.

2. Connect an ammeter with a scale to 100 amps in series between the alternator "BAT" terminal and the disconnected "BAT" wire.

3. Connect the positive lead of a voltmeter to the disconnected "BAT" wire. Connect the negative lead of the voltmeter to the negative post of the battery.

4. Disconnect the green colored regulator field wire from the alternator. Connect a jumper lead from the alternator field terminal to ground.

5. Connect a tachometer to the engine and reconnect the battery ground cable.

6. Connect a variable carbon pile rheostat to the battery cables.

——— CAUTION ———
Be sure the carbon pile is in the "OPEN" or "OFF" position before connecting the leads to the battery terminals.

7. Start the engine and operate at an idle.

8. Adjust the engine speed and carbon pile to maintain a flow of 20 amperes in the circuit. Observe the voltmeter reading which should not exceed .7 volts.

9. If a higher voltage reading is indicated, inspect, clean and tighten all connections in the charging system.

10. If necessary, a voltage drop test can be done at each connection until the excessive resistance is located.

11. If the charging system resistance is within specifications, reduce the engine speed, turn off the carbon pile rheostat and stop the engine. Remove battery ground cable.

12. Remove the test instruments from the electrical system and reconnect the charging system wiring. Reconnect the battery ground cable.

Current Output Test

This test determines if the alternator is capable of delivering its rated current output.

1. Disconnect the battery ground cable and the "BAT" lead wire at the alternator output terminal.

2. Connect an ammeter in series between the alternator output terminal and the disconnected "BAT" lead wire.

NOTE: The ammeter must have a scale of 100 amps.

3. Connect the positive lead of a voltmeter to the output terminal of the alternator and the negative lead to a good ground.

4. Disconnect the green colored wire at the voltage regulator and connect a jumper wire from the alternator field terminal to ground.

5. Connect a tachometer to the engine and reconnect the battery ground wire.

6. Connect a variable carbon pile rheostat between the positive and negative battery cables.

——— CAUTION ———
Be sure the rheostat control is in the "OPEN" or "OFF" position before connecting the leads to the battery cables.

7. Start the engine and operate at idle. Adjust the carbon pile rheostat control and the engine speed in increments until the voltmeter reading is 15 volts (13 volts for the 100 and 117 amp alternators) and the engine speed is 1250 rpm (900 rpm for the 100 and 117 amp alternators).

——— CAUTION ———
Do not allow the voltage to rise above 16 volts.

8. The ammeter readings must be within the following specifications.

Current Rating	Identification	Current Output
41 amp	Red or violet tag	40 amps min.
60 amp	Blue, natural or yellow	57 amps min.
100, 117 amp	Yellow	72 amps min.

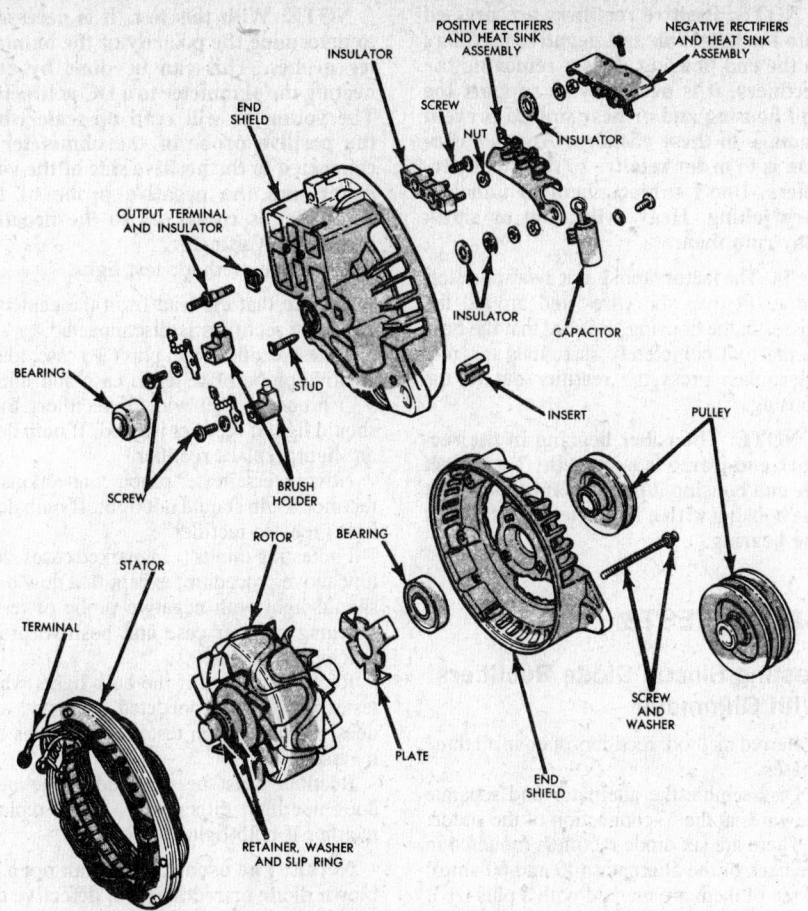

Ambient Temp. ¼ in. from Regulator	Voltage
20°F.	14.9 to 15.9
80°F.	13.9 to 14.6
140°F.	13.3 to 13.9
Above 140°F.	Less Than 13.6

Typical Chrysler alternator

5. If the voltage is below specifications, check the following: Voltage regulator ground: check voltage drop between regulator cover and ground. Harness wiring: disconnect regulator plug (ign. switch off), then turn on ign. switch and check for battery voltage at the terminals having the red and green leads. Wiring harness must be disconnected from the regulator when checking individual leads. If no voltage is present in either lead, the problem is in the truck wiring or alternator field.

6. If Step 5 tests showed no malfunctions, install a new regulator and repeat Step 4.

7. If voltage is above specifications, or fluctuates, check the following: Ground between regulator and body, and between body and engine. Ignition switch circuit between switch and regulator.

8. If voltage is still more than ½ volt above specifications, install a new regulator and repeat Step 4.

NOTE: If measured at the battery, current output will be approximately 5 amperes lower than specified.

9. If the readings are less than specified, the alternator should be removed and checked during a bench test.

10. After the current output test is completed, reduce the engine speed, turn the carbon pile rheostat off and then stop the engine.

11. Disconnect the battery ground cable, remove the ammeter, voltmeter and carbon pile. Remove the jumper wire from the field terminal and reconnect the green colored wire to the alternator field terminal.

12. Reconnect the battery cable, if no further testing is to be done to the charging circuit.

Rotor Field Coil Draw Test

The rotor field coil can be tested on or off the vehicle.

1. If on the vehicle, remove the drive belt and wiring connections from the alternator.

2. Connect a jumper wire from the negative terminal of the battery to one of the field terminals of the alternator.

3. Connect the test ammeter positive lead to the other field terminal of the alternator and the negative ammeter lead to the positive battery terminal.

4. Connect a jumper wire between the alternator end shield and the battery negative terminal.

5. Slowly rotate the alternator pulley by hand and observe the ammeter reading.

6. The field coil draw should be 4.5 to 6.5 amperes at 12 volts. (4.75 to 6.0 amperes at 12 volts on 100 and 117 amp alternators).

7. A low rotor coil draw is an indication of high resistance in the field coil circuit (brushes, slip rings or rotor coil). A higher rotor coil draw indicates possible shorted rotor coil or grounded rotor. No reading indicates an open rotor or defective brushes.

8. Remove the test equipment and jumper leads.

Electronic Voltage Regulator Test

1. Make sure battery terminals are clean and battery is charged.

2. Connect the positive lead of a test voltmeter to ignition Terminal No. 1 of the ballast resistor.

3. Connect the negative voltmeter lead to a good body ground.

4. Start engine and allow it to idle at 1250 rpm, all lights and accessories turned off. Voltage should be as follows:

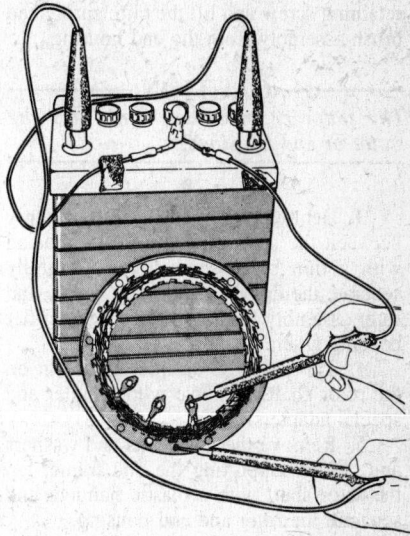

Stator test for ground on Chrysler systems

OVERHAUL & TESTING

Alternator disassembly, repair and assembly procedures are basically the same for all Chrysler alternators. Certain variations in design, or production modifications, could require slightly different procedures that should be obvious upon inspection of the unit being serviced.

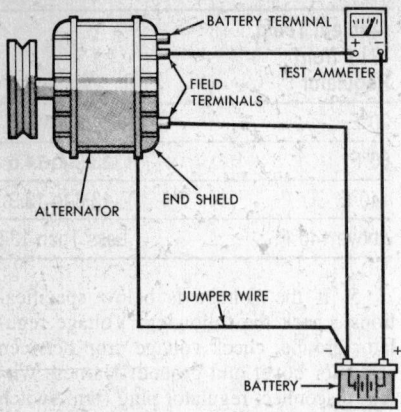

Rotor field coil current draw test

DISASSEMBLY

To prevent damage to the brush assemblies (100 and 117 amp), they should be removed before proceeding with the disassembly of the alternator. The brushes are mounted in a plastic holder that positions the brushes vertically against the slip rings.

1. Remove the retaining screw, flat washer, nylon washer and field terminal and carefully lift the plastic holder containing the spring and brush assembly from the end housing.

2. The ground brush (40 and 60 amp) is positioned horizontally against the slip ring and is retained in the holder that is integral with the end housing. Remove the retaining screw and lift the clip, spring and brush assembly from the end housing.

—————— CAUTION ——————
The stator is laminated, don't burr the stator or end housings.

3. Remove the through bolts and pry between the stator and drive end housing with a thin blade screwdriver. Carefully separate the drive end housing, pulley and rotor assembly from the stator and rectifier housing assembly.

4. The pulley is an interference fit on the rotor shaft. Remove with a puller and special adapters.

5. Remove the three nuts and washers and, while supporting the end frame, tap the rotor shaft with a plastic hammer and separate the rotor and end housing.

6. The drive end ball bearing is an interference fit with the rotor shaft. Remove the bearing with puller and adapters.

NOTE: Further dismantling of the rotor is not advisable, as the remainder of the rotor assembly is not serviced separately.

7. Remove the DC output terminal nuts and washers and remove terminal screw and inside capacitor (on units so equipped).

8. Remove the insulator. Cut rectifier wire at point of crimp. Support rectifier housing.

NOTE: Positive rectifiers are pressed into the heat sink and negative rectifiers in the end housing. When removing the rectifiers, it is necessary to support the end housing and/or heat sink to prevent damage to these castings. Another caution is in order relative to the diode rectifiers. Don't subject them to unnecessary jolting. Heavy vibration or shock may ruin them.

9. The factory tool is cut away and slotted to fit over the wires and around the bosses in the housing. Be sure that the bore of the tool completely surrounds the rectifier, then press the rectifier out of the housing.

NOTE: The roller bearing in the rectifier end frame is a press fit. To protect the end housing it is necessary to support the housing with a tool when pressing out the bearing.

BENCH TESTS

Testing Silicon Diode Rectifiers With Ohmmeter

Preferred method: rectifiers open in all three phases.

Disassemble the alternator and separate the wires at the Y-connection of the stator.

There are six diode rectifiers mounted in the back of the alternator (40 and 60 amp). Three of them are marked with a plus (+), and three are marked with a minus (−). These marks indicate diode case polarity.

NOTE: The 100 and 117 amp alternator has twelve silicone diodes. Six positive and six negative.

To test, set ohmmeter to its lowest range. If case is marked positive (+), place positive meter probe to case and negative probe to the diode lead. Meter should read between 4 and 10 ohms. Now, reverse leads of ohmmeter, connecting negative meter probe to positive case and positive meter probe to wire of rectifier. Set meter on a high range. Meter needle should move very little, if any (infinite reading). Do this to all positive diode rectifiers.

The diode rectifiers with minus (−) marks on their cases are checked the same way as above. Only now the negative ohmmeter probe is connected to the case for a reading of 4 to 10 ohms. Reverse leads as above for the other part to test.

If a reading of 4 to 10 ohms is obtained in one direction and no reading (infinity) is read on the ohmmeter in the other direction, diode rectifiers are good. If either infinity or a low resistance is obtained in both directions on a rectifier, it must be replaced.

If meter reads more than 10 ohms when ohmmeter positive probe is connected to positive on diode, and negative probe to negative, replace diode rectifier.

NOTE: With this test, it is necessary to determine the polarity of the ohmmeter probes. This can be done by connecting the ohmmeter to a DC voltmeter. The voltmeter will read up-scale when the positive probe of the ohmmeter is connected to the positive side of the voltmeter and the negative probe of the ohmmeter is connected to the negative side of the voltmeter.

Alternate method: test light.

Be sure that the lead from the center of the diode rectifiers is disconnected.

To test rectifiers with plus (+) case, touch positive probe of tester to case and minus (−) probe to lead wire of rectifier. Bulb should light if rectifier is good. If bulb does not light, replace rectifier.

Now reverse tester probe connections to rectifier. Bulb should not light. If bulb does light, replace rectifier.

For testing minus (−) marked cases, follow above procedure, except that now bulb should light with negative probe of tester touching rectifier case and positive probe touching lead wire.

Rectifier is good if the bulb lights when tester probes are connected one way, and does not light when tester connections are reversed.

Rectifier must be replaced if the bulb does not light either way. Also, replace rectifier if bulb lights both ways.

NOTE: The usual cause of an open or blown diode or rectifier is a defective capacitor or a battery that has been installed in reverse polarity. If the battery is installed properly and the diodes are open, test the capacitor.

Field Coil Draw

1. Connect a jumper between one FLD terminal and the positive terminal of a fully charged 12 volt battery.

2. Connect the positive lead of a test ammeter to the other field (FLD) terminal and the negative test lead to the negative battery terminal.

3. Slowly rotate the rotor by hand and observe the ammeter. The proper field coil draw is 2.3–2.7 amps at 12 volts.

NOTE: Field coil draw for the 100 and 117 ampere alternators should be 4.75 to 6.0 amperes at 12 volts.

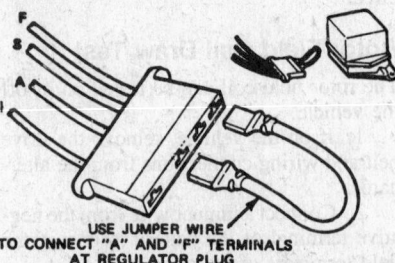

USE JUMPER WIRE
TO CONNECT "A" AND "F" TERMINALS
AT REGULATOR PLUG

USE OF JUMPER WIRE AT REGULATOR PLUG
TO TEST ALTERNATOR FOR NORMAL OUTPUT AMPS
AND FOR FIELD CIRCUIT WIRING CONTINUITY

Isolation test jumper wire

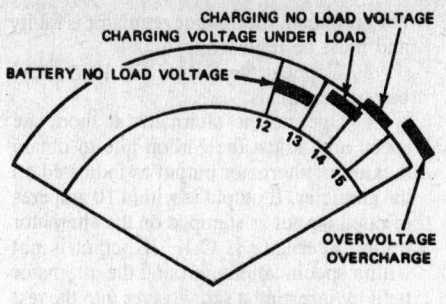

Voltmeter reading during isolation test

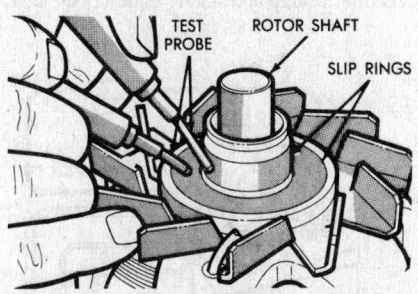

Rotor tests for short or open circuits on Chrysler systems

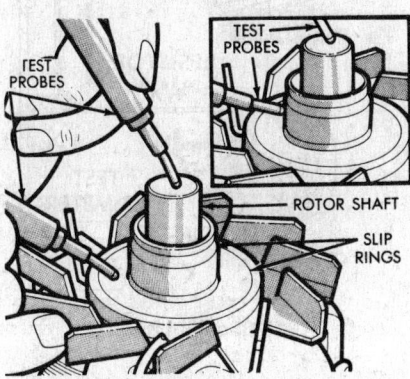

Rotor tests for ground on Chrysler systems

Field Circuit Ground Test

1. Touch one test lead of a 110 volt AC test bulb to one of the alternator brush (field) terminals and the other test lead to the end shield.
2. If the lamp lights, remove the field brush assemblies and separate the end housing by removing the three through-bolts.
3. Place one test lead on a slip ring and the other on the end shield.
4. If the lamp lights, the rotor assembly is grounded internally and must be replaced.
5. If the lamp does not light, the cause of the problem was a grounded brush.

Grounded Stator

1. Disconnect the diode rectifiers from the stator leads.

2. Test from stator leads to stator core, using a 110 volt test lamp. Test lamp should not light. If it does, stator is grounded and must be replaced.

Low Output

About 50% output accompanied with a growl-hum caused by a shorted phase or a shorted rectifier.

Perform Steps 1, 2 and 3 (rectifier open in all three phases). If the rectifiers are found to be within specifications, replace the stator assembly.

Current Output Too High

(No Control) Caused by Open Rectifier or Open Phase

Perform Steps 1, 2 and 3 (rectifier open in all three phases). If the rectifier tests satisfactorily, inspect the stator connections before replacing the stator.

ASSEMBLY

1. Support the heat sink or rectifier end housing on circular plate.
2. Check rectifier identification to be sure the correct rectifier is being used. The part numbers are stamped on the case of the rectifier. They are also marked, red for positive and black for negative.
3. Start the new rectifier into the casting and press it in squarely.

—————— CAUTION ——————
Do not start rectifier with a hammer or it will be ruined.

4. Crimp the new rectifier wire to the wires disconnected at removal, or solder (using a heat sink with rosin core solder).
5. Support the end housing on tool so that the notch in the support tool will clear the raised section of the heat sink, then press the bearing into position with tool SP3381, or equivalent.

NOTE: New bearings are pre-lubricated, additional lubrication is not required.

6. Insert the drive end bearing in the drive end housing and install the bearing plate, washers and nuts to hold the bearing in place.
7. Position the bearing and drive end housing on the rotor shaft and, while supporting the base of the rotor shaft, press the bearing and housing in position on the rotor shaft with an arbor press and arbor tool.

—————— CAUTION ——————
Be careful that there is no cocking of the bearing at installation; or damage will result. Press the bearing on the rotor shaft until the bearing contacts the shoulder on the rotor shaft.

8. Install pulley on rotor shaft. Shaft of rotor must be supported so that all pressing force is on the pulley hub and rotor shaft.

NOTE: Do not exceed 6,800 lbs. pressure. Pulley hub should just contact bearing inner race.

9. Some alternators will be found to have the capacitor mounted internally. Be sure the heat sink insulator is in place.
10. Install the output terminal screw with the capacitor attached through the heat sink and end housing.
11. Install insulating washers, lockwashers and locknuts.
12. Make sure the heat sink and insulator are in place and tighten the locknut.
13. Position the stator on the rectifier end housing. Be sure that all of the rectifier connectors and phase leads are free of interference with the rotor fan blades and that the capacitor (internally mounted) lead has clearance.
14. Position the rotor assembly in the rectifier end housing. Align the through bolt holes in the stator with both end housings.
15. Enter stator shaft in the rectifier end housing bearing, compress stator and both end housings manually and install through bolts, washers and nuts.
16. Install the insulated brush and terminal attaching screw.
17. Install the ground screw and attaching screw.
18. Rotate pulley slowly to be sure the rotor fan blades do not hit the rectifier and stator connectors.

Delcotron

NOTE: The internal alternator wiring is identical between the 10–SI, 15–SI and the 27–SI units, except the 10–SI uses a Wye stator winding while a Delta stator winding is used in the 15–SI and 27–SI alternators. The disassembly and assembly of the units remain basically the same.

Delcotron 10–SI and 15–SI

This system is an integrated AC generating system containing a built-in voltage regulator. Removal and replacement is essentially the same as for the standard AC generator.

The regulator is mounted inside the slip ring end frame. All regulator components are enclosed in an epoxy molding, and the regulator cannot be adjusted.

TROUBLESHOOTING

NOTE: See the "Alternator Test Plans" section before proceeding further. Make sure that the continuous running blower, if equipped, is disconnected. This blower will run with the key on even if the blower control is off, unless disconnected.

Low Charging Rate

1. After battery condition, drive belt tension, and wiring terminals and connections have been checked, charge the battery fully and perform the following test:

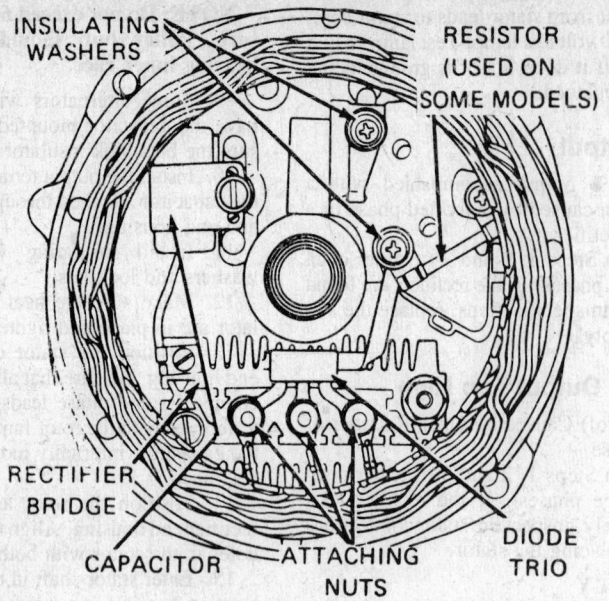

INSULATING WASHERS

RESISTOR (USED ON SOME MODELS)

RECTIFIER BRIDGE

CAPACITOR

ATTACHING NUTS

DIODE TRIO

Delcotron end-frame view

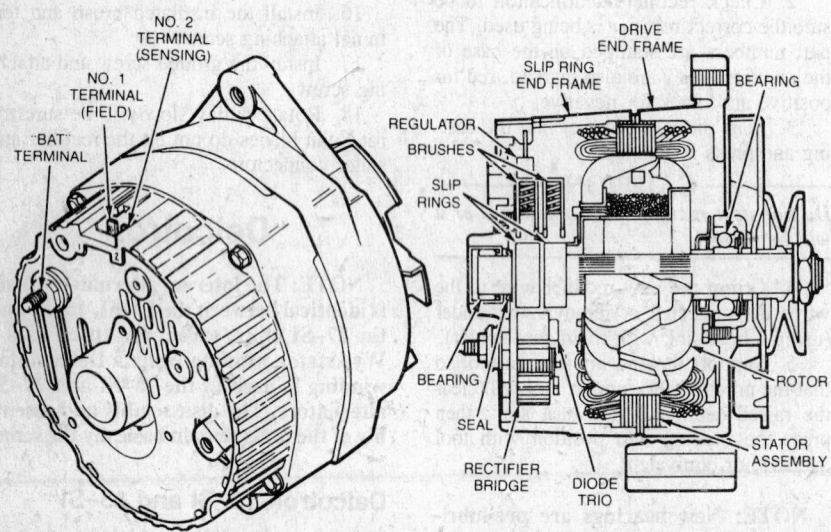

NO. 2 TERMINAL (SENSING)

NO. 1 TERMINAL (FIELD)

"BAT" TERMINAL

DRIVE END FRAME

SLIP RING END FRAME

BEARING

REGULATOR

BRUSHES

SLIP RINGS

BEARING

SEAL

RECTIFIER BRIDGE

DIODE TRIO

ROTOR

STATOR ASSEMBLY

10-SI Delcotron

2. Connect a test voltmeter between the alternator BAT. terminal and ground, ignition switch on. Connect the voltmeter in turn to alternator terminals No. 1 and No. 2, the other voltmeter lead being grounded as before. A zero reading indicates an open circuit between the battery and each connection at the alternator. If this test discloses no faults in the wiring, proceed to Step 3.

3. Connect the test voltmeter to the alternator BAT. terminal (the other test lead to ground), start the engine and run at 1,500–2,000 rpm with all lights and electrical accessories turned on. If the voltmeter reads 12.8 volts or greater, the alternator is good and no further checks need be made. If the voltmeter reads less than 12.8 volts, ground the field winding by inserting a screwdriver into the test hole in the end frame.

--- **CAUTION** ---

Do not force tab more than ¾ in. into end frame.

4. If the voltage increases to 13 volts or more, the regulator unit is defective. If voltage does not increase significantly, the alternator is defective.

Output Test

1. Connect a test voltmeter, ammeter and a 10 ohm, 6 watt resistor into the charging circuit. Do not connect the carbon pile to the battery posts at this time.

2. Increase alternator speed and observe voltmeter: if voltage is uncontrolled with speed and increases to 15.5 volts or more, check for a grounded brush lead clip as covered previously. If brush lead clip is

not grounded, the voltage regulator is faulty and must be replaced.

3. Connect the carbon pile load to the battery terminals.

4. Operate the alternator at moderate speed and adjust the carbon pile to obtain maximum alternator output as indicated on the ammeter. If output is within 10 amperes of rated output as stamped on the alternator frame, alternator is O.K. If output is not within specifications, ground the alternator field by inserting a screwdriver into the test hole in the end frame. If output now is within 10 amperes of rating, replace the voltage regulator; if still not within specifications, check field winding, diode trio, rectifier bridge and stator, as described later.

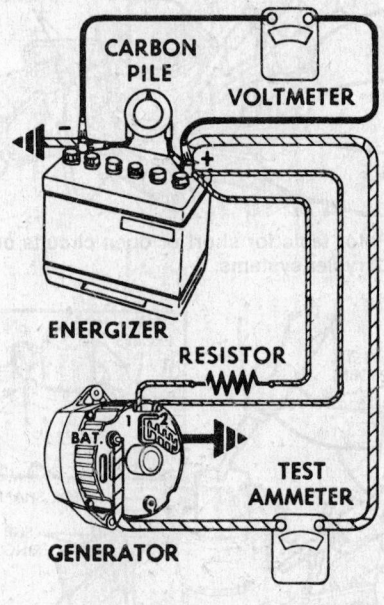

CARBON PILE

VOLTMETER

ENERGIZER

RESISTOR

BAT.

TEST AMMETER

GENERATOR

10-SI output test

SERVICING

1. Hold alternator in a vice, clamping the mounting flange lengthwise.

2. Make a scribe mark to help locate frame end parts in the same position during assembly.

3. Remove 4 through-bolts and separate the slip ring end frame from the drive end frame and rotor assembly.

NOTE: Prying at the stator slot will aid disassembly.

4. Remove 3 stator lead attaching nuts and separate stator from end frame.

5. Remove insulated screws and ground screw from brush holder. Remove diode trio, resistor, brush holder, and regulator from end frame.

6. Remove screws attaching capacitor to end frame and diode bridge; remove capacitor.

7. On 10–SI Series: Remove ground screw and battery terminal stud nut from rectifier bridge. Remove rectifier bridge,

terminal stud, and insulating washer from end frame. On 27–SI Series: Remove 2 ground screws, a connector strap screw, and the battery terminal stud nut. Remove rectifier bridge, connector, terminal stud, and insulating washers from end frame.

8. Press bearing from end frame using a tube slightly smaller OD than the bearing. Support the end frame from inside and press bearing from outside toward the inside.

NOTE: Some models may have a seal separate and in front of bearing. Discard the seal when replacing the bearing as the new bearing has an integral seal.

9. Separate drive end frame from rotor as follows: Place rotor in a vise and tighten only enough to permit pulley nut removal. Remove pulley nut, washer, pulley, fan, and collar from rotor shaft. Remove drive end frame from rotor shaft and remove rotor from vise.

10. Press bearing from drive end frame after removing bearing retainer plate: Remove screws attaching bearing seal and retainer assembly to housing. Support end frame from inside the housing on a metal tube with a slightly larger ID than the OD of the bearing. Press bearing and grease slinger (or flat washer used on some models) from end frame using a metal tube or collar against the grease slinger.

11. To assemble, reverse the order of the disassembly procedure. Torque the pulley nut to 50 ft. lbs.

— CAUTION —

During the assembly, do not interchange the ground screw (without insulator) for an insulated screw as this would cause uncontrolled or no alternator output.

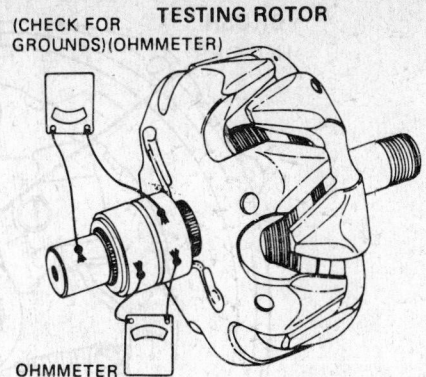

TESTING ROTOR
(CHECK FOR GROUNDS)(OHMMETER)

OHMMETER

Checking Delcotron rotor for grounds or open circuits

CHARGING SYSTEM TEST

High Charging Rate

1. With the battery fully charged, connect a voltmeter between alternator Terminal No. 2 and ground. If the reading is zero, No. 2 circuit from the battery is open.

2. If No. 2 circuit is OK, but an obvious overcharging condition still exists, proceed as follows: Remove the alternator and separate the end frames. Connect a low-range ohmmeter between the brush lead clip and the end frame, then reverse the lead connections. If both readings are zero, either the brush lead clip is grounded or the regulator is defective. A grounded brush lead clip can be due to a damaged insulating sleeve or omission of the insulating washer.

Diode Trio Testing

1. Before removing this unit, connect an ohmmeter between the brush lead clip and the end frame. The lowest reading scale should be used for this test.

2. After taking a reading, reverse the lead connections. If the meter reads zero, the brush lead clip is probably grounded, due to omission of the insulating sleeve or insulating washer.

3. Remove the three nuts which secure the stator.

4. Remove stator.

5. Remove the screw which secures the diode trio lead clip, then remove diode trio.

NOTE: The position of the insulating washer on the screw is critical; make sure it is returned to the same position on reassembly.

6. Connect an ohmmeter, on lowest range, between the single brush connector and one stator lead connector.

7. Observe the reading, then reverse the meter leads. Repeat this test with each of the other two stator lead connectors. The readings on each of these tests should NOT be identical, there should be one low and one high reading for each test. If this is not the case, replace the diode trio.

— CAUTION —

Do not use high voltage on the diode trio.

Rectifier Bridge Testing

1. Connect an ohmmeter between the heat sink (ground) and the base of one of the three terminals. Then, reverse the meter leads and take a reading. If both readings are identical, the bridge is defective and must be replaced.

2. Repeat this test with the remaining two terminals, then between the INSULATED heat sink (as opposed to the GROUNDED heat sink in previous test) and each of the three terminals. As before, if any two readings are identical, on reversing the meter leads, the rectifier bridge must be replaced.

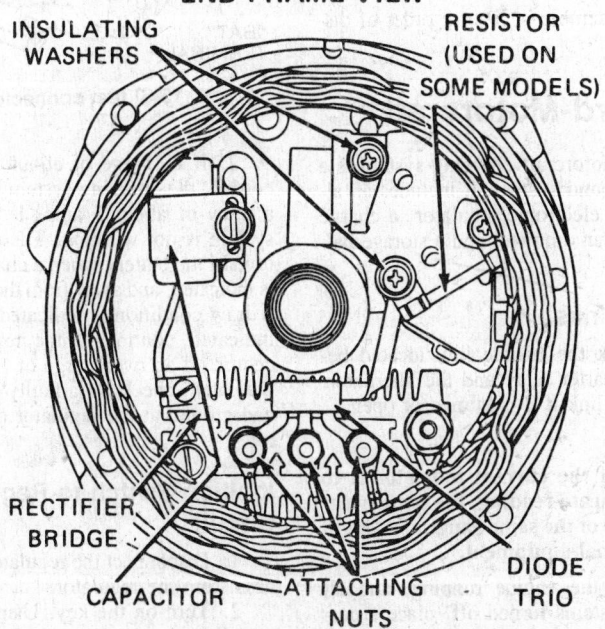

END FRAME VIEW

INSULATING WASHERS

RESISTOR (USED ON SOME MODELS)

RECTIFIER BRIDGE

CAPACITOR

ATTACHING NUTS

DIODE TRIO

10-SI brush lead clip ground test

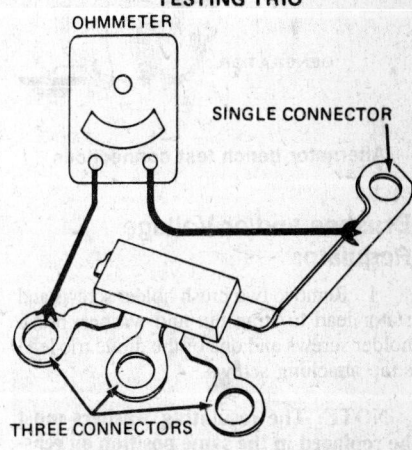

TESTING TRIO
OHMMETER

SINGLE CONNECTOR

THREE CONNECTORS

Testing the 10-SI diode trio

3. Remove the attaching screw and the BAT. terminal screw.

4. Disconnect the condenser lead.

5. Remove the rectifier bridge.

NOTE: The insulator between the insulated heat sink and the end frame is extremely important to the operation of the unit. It must be replaced in exactly the same position on reassembly.

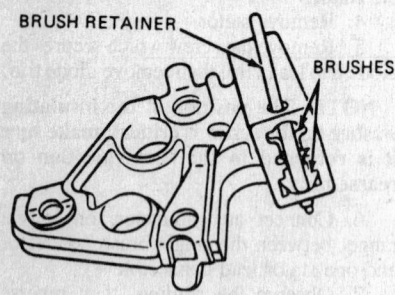

BRUSHES RETAINED IN HOLDER

10-SI brush holder

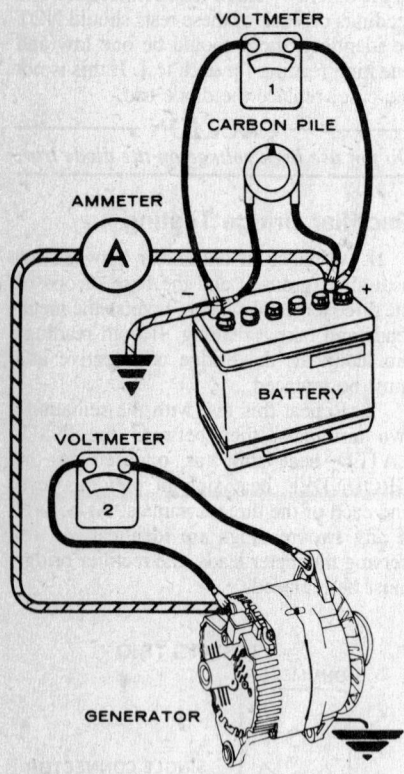

Alternator bench test connections

Brushes and/or Voltage Regulator

1. Remove two brush holder screws and stator lead to strap nut and washer, brush holder screws and one of the diode trio lead strap attaching screws.

NOTE: The insulating washers must be replaced in the same position on reassembly.

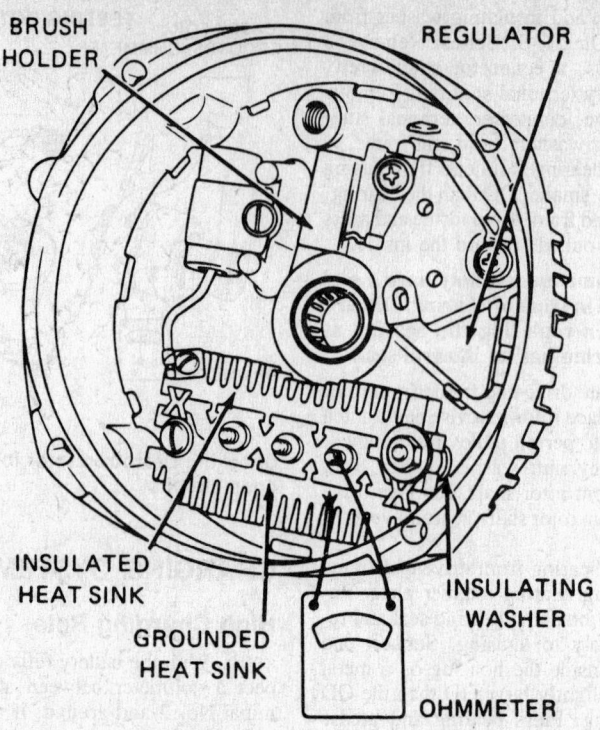

Testing the 10-SI rectifier bridge

2. Remove brush holder and brushes. The voltage regulator may also be removed at this time, if desired.

3. Brushes and brush spring must be free of corrosion and must be undamaged and completely free of oil or grease.

4. Insert spring and brushes into holder, noting whether they slide freely without binding. Insert wooden or plastic toothpick into bottom hole in holder to retain brushes.

NOTE: The brush holder is serviced as a unit; individual parts are not available.

5. Reassemble in reverse order of disassembly.

Ford-Motorcraft

The Ford-Motorcraft charging system is a negative ground system. It includes an alternator, an electronic regulator, a charge indicator or an ammeter and a storage battery.

Fusible Links

1. Check the fusible link located between the starter relay and the alternator. Replace the link if it is burned or open.

TESTING

NOTE: If the current indicator is to give an accurate reading, the battery cables must be of the same gauge and length as the original equipment.

1. With the engine running, and all electrical systems turned off, place a current indicator over the positive battery cable.

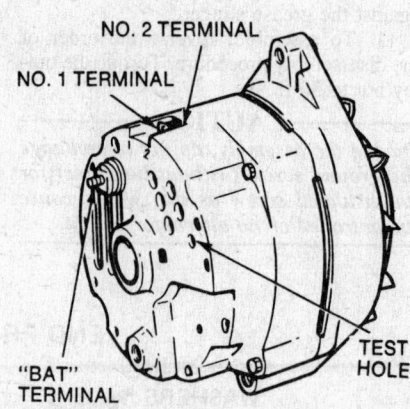

15-SI test connectors

2. If a charge of about 5 amps is recorded, the charging system is working. If a draw of about 5 amps is recorded, the system is not working. The needle moves toward the battery when a charge condition is indicated, and away from the battery when a draw condition is indicated. If a draw is indicated, continue to the next testing procedure. If an overcharge of 10–15 amps is indicated, check for a faulty regulator or a bad ground at the regulator or the alternator.

Ignition Switch to Regulator Circuit

1. Disconnect the regulator wiring harness from the regulator.

2. Turn on the key. Using a test light or voltmeter, check for voltage between the I wire and ground. Check for voltage be-

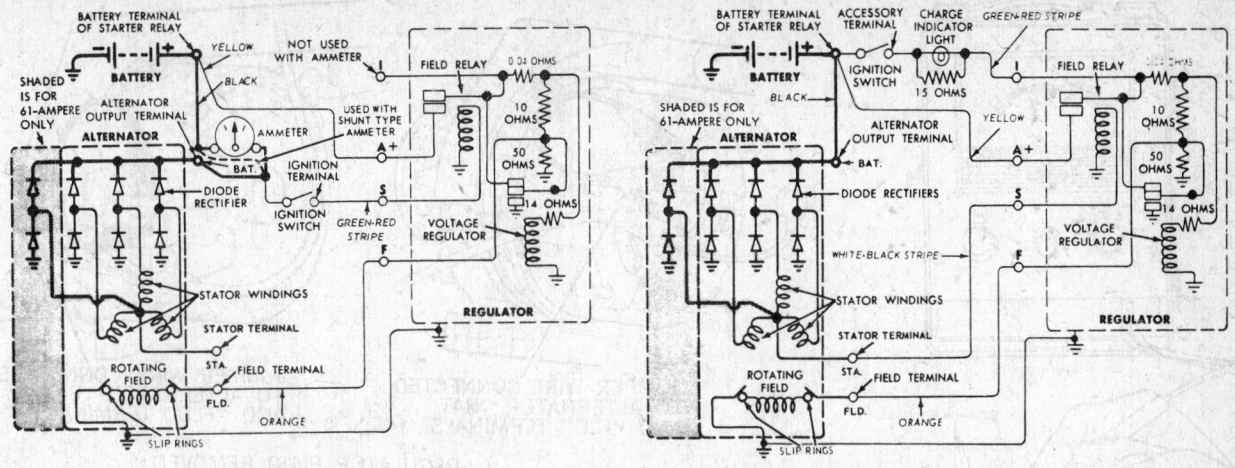

Charging system schematic with electro-mechanical regulator and charging light

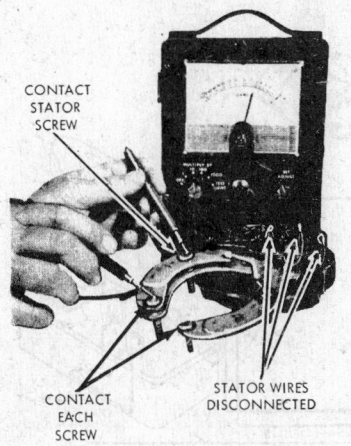

Testing the diodes on a 65 amp alternator

tween the A wire and ground. If voltage is present at this part of the system, the circuit is OK. If there is no voltage at the I wire, check for a burned-out charge indicator bulb, a burned-out resistor, or a break or short in the wiring. If there is no voltage present at the A wire, check for a bad connection at the starter relay or a break or short in the wire.

Isolation Test

This test determines whether the regulator or the alternator is faulty, after the rest of the circuit is found to be in good working order.

1. Disconnect the regulator wiring harness from the regulator.

2. Connect a jumper wire from the A wire to the F wire in the wiring harness plug.

3. Connect a voltmeter to the battery. The positive voltmeter lead goes to the positive terminal and the negative lead to the negative terminal. Record the reading on the voltmeter.

4. Turn off all of the electrical systems and start the engine. Do not race the engine.

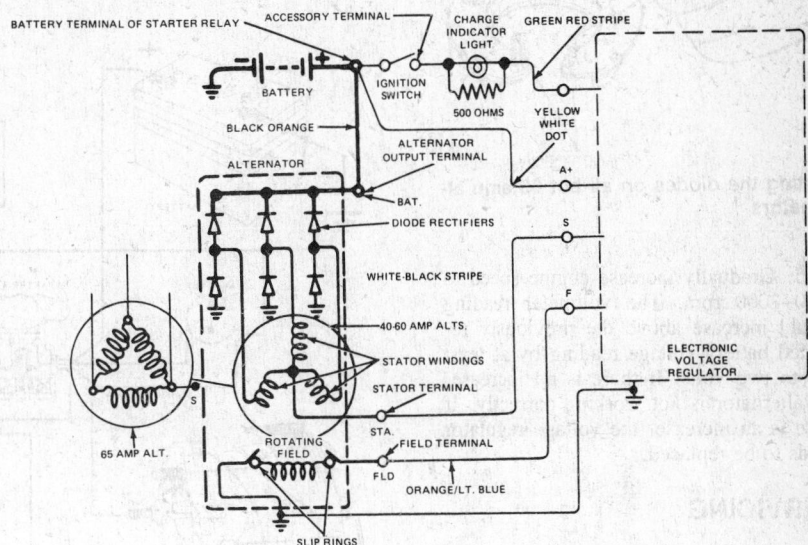

Alternator charging system with indicator lamp

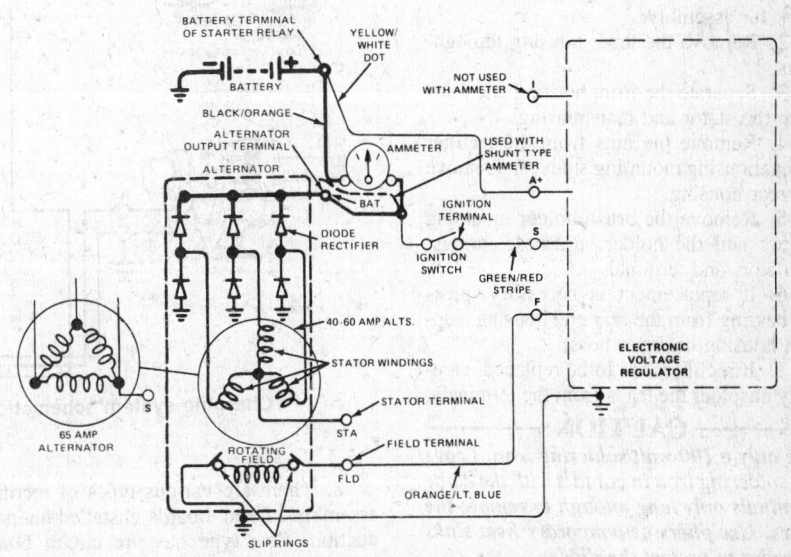

Alternator charging system with ammeter

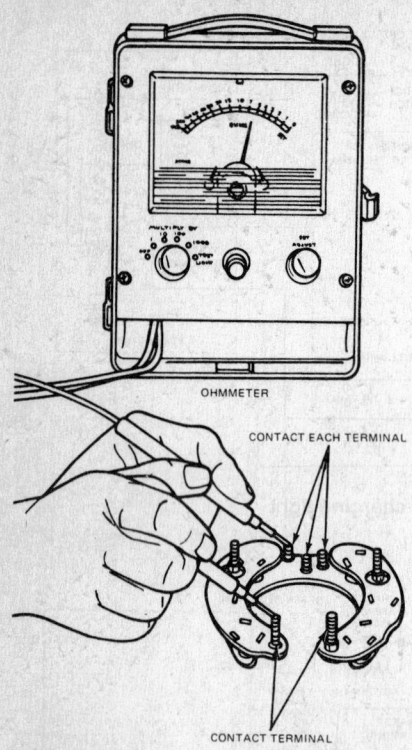

Testing the diodes on all but 65 amp alternators

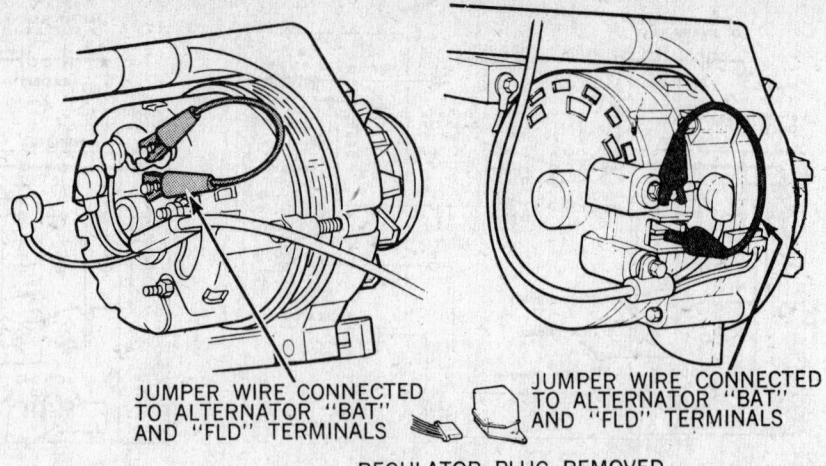

JUMPER WIRE CONNECTED TO ALTERNATOR "BAT" AND "FLD" TERMINALS

JUMPER WIRE CONNECTED TO ALTERNATOR "BAT" AND "FLD" TERMINALS

REGULATOR PLUG REMOVED FROM REGULATOR

Location of jumper wire for circuit tests on rear and side terminal alternators

5. Gradually increase engine speed to 1500–2000 rpm. The voltmeter reading should increase above the previously recorded battery voltage reading by at least one to two volts. If there is no increase, the alternator is not working correctly. If there is an increase, the voltage regulator needs to be replaced.

SERVICING

1979–80 Rear Terminal

DISASSEMBLY

1. Mark both end housings with a scribe mark for assembly.
2. Remove the three housing through-bolts.
3. Separate the front housing and rotor from the stator and rear housing.
4. Remove the nuts from the rectifier to rear housing mounting studs, and remove the rear housing.
5. Remove the brush holder mounting screws and the holder, brushes, springs, insulator, and terminal.
6. If replacement is necessary, press the bearing from the rear end housing, support housing on inner boss.
7. If rectifiers are to be replaced, carefully unsolder the leads from the terminals.

——— **CAUTION** ———

Use only a 100 watt soldering iron. Leave the soldering iron in contact with the diode terminals only long enough to remove the wires. Use pliers as temporary heat sinks in order to protect the diodes.

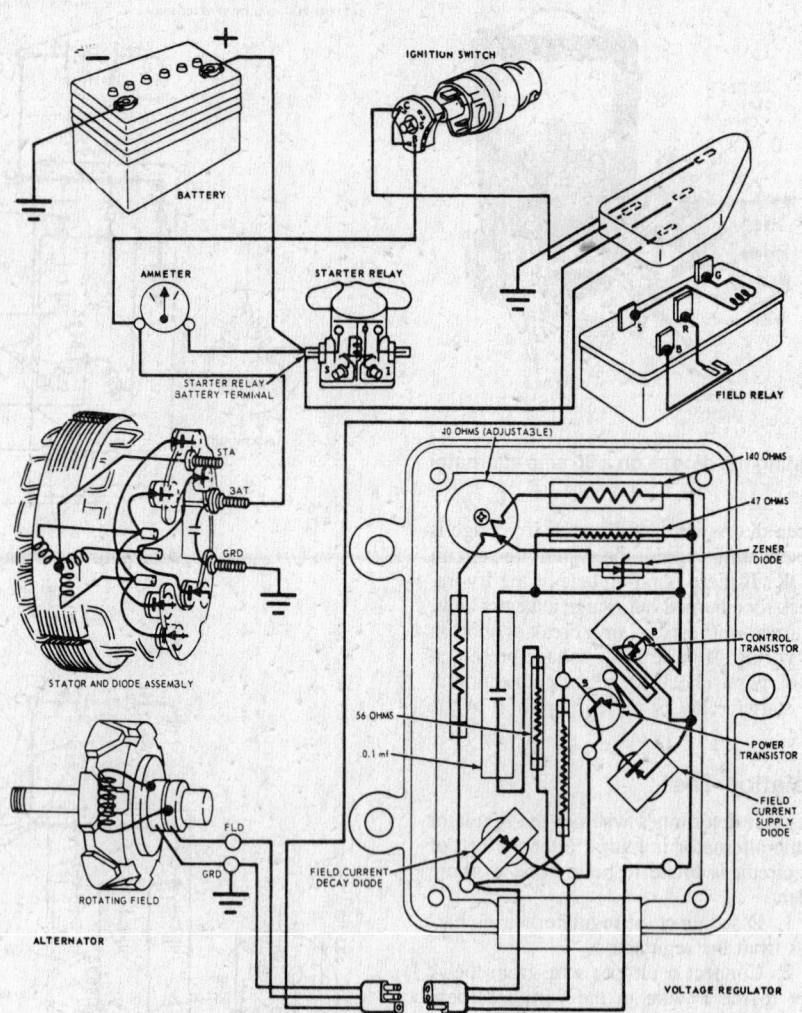

Charging system schematic with transistor regulator and ammeter

8. There are various types of rectifier assembly circuit boards installed in production. One type has the circuit board spaced away from the diode plates and the diodes are exposed. Another type consists of a single circuit board with integral diodes; and still another has integral diodes with an additional booster diode plate containing

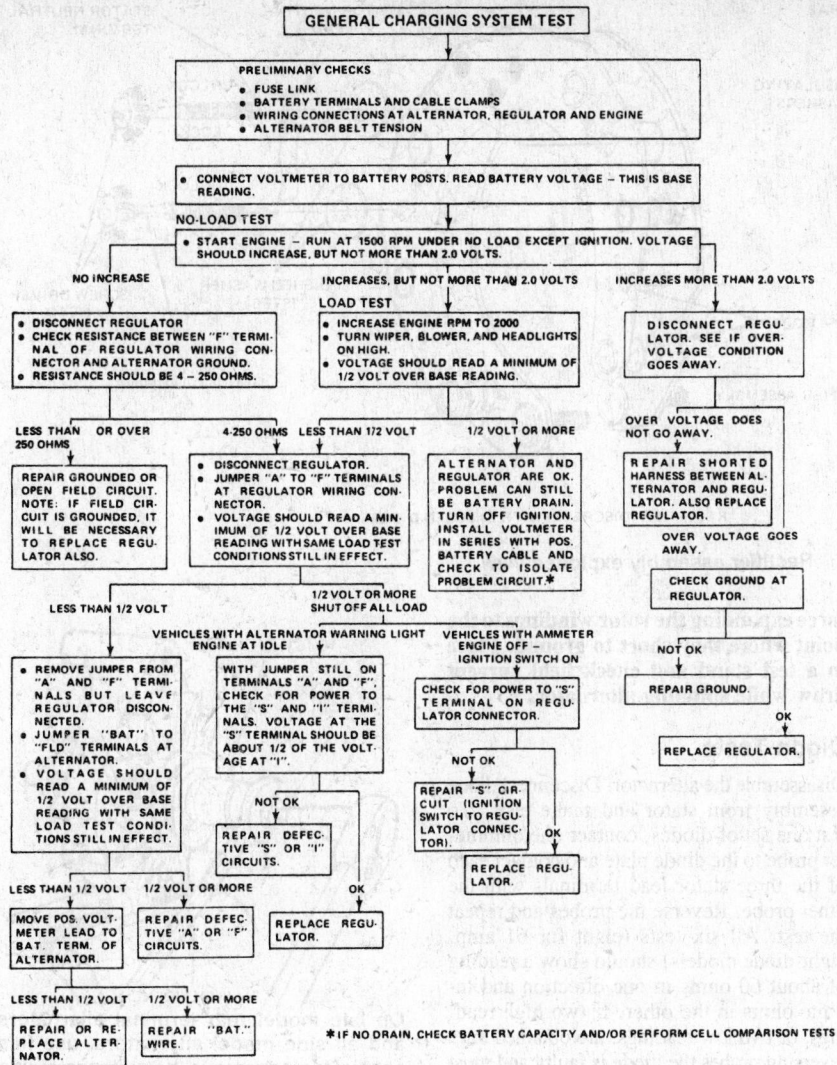

General charging system tests with ohmmeter and voltmeter

two diodes. This last type is used only on the eight-diode, 61 amp.

9. To disassemble, use the following procedures: Exposed Diodes: remove the screws from the rectifier by rotating bolt heads ¼ turn clockwise to unlock, then unscrewing. Integral Diodes: press out the stator terminal screw, making sure not to twist it while doing this. Do not remove grounded screw. Booster Diodes: press out the stator terminal screw about ¼ in., then remove the nut from the end of the screw and lift screw from circuit board, making sure not to twist it as it comes out.

10. Remove the drive pulley and fan. On alternator pulleys with threaded holes in the outer end of the pulley, use a standard puller for removal.

11. Remove the three screws that hold the front bearing retainer, and remove the front housing.

12. If the bearing is to be replaced, press from housing.

CLEANING & INSPECTION

1. The rotor, stator, diode rectifier assemblies, and bearings are not to be cleaned with solvent. These parts are to be wiped off with a clean cloth. Cleaning solvent may cause damage to the electrical parts or contaminate the bearing internal lubricant. Wash all other parts in solvent and dry them.

2. Rotate the front bearing on the driveshaft. Check for any scraping noise, looseness or roughness that indicates that the bearing is excessively worn. As the bearing is being rotated, look for excessive lubricant leakage. If any of these conditions exist, replace the bearing. Check rear bearing and rotor shaft.

3. Place the rear end housing on the slip ring end of the shaft and rotate the bearing on the shaft. Make a similar check for noise, looseness or roughness. Inspect the rollers and cage for damage. Replace the bearing if these conditions exist, or if the lubricant is missing or contaminated.

4. Check both the front and rear housings for cracks.

5. Check all wire leads on both the stator and rotor assemblies for loose soldered connections, and for burned insulation. Solder all poor connections. Replace parts that show burned insulation.

6. Check the slip rings for damaged insulation and runout. If the slip rings are more than 0.0005 in. out of round, take a light cut (minimum diameter limit 1.22 in.) from the face of the rings to true them. If the slip rings are badly damaged, the entire rotor will have to be replaced, as they are serviced as a complete assembly.

7. Replace any parts that are burned or cracked. Replace brushes that are worn to less than ⁵⁄₁₆ in. in length. Replace the brush spring if it had less than 7–12 oz. tension.

ASSEMBLY

1. Press the front bearing into the front housing boss, putting pressure on outer race only. Install bearing retainer.

2. If the stop ring on the driveshaft was damaged, install a new stop ring. Push the new ring onto the shaft and into the groove.

3. Position the front bearing spacer on the driveshaft against the stop ring.

4. Place the front housing over the shaft, with the bearing positioned in the front housing cavity.

5. Install fan spacer, fan, pulley, lockwasher and retaining nut and tighten nut to 60–100 ft. lbs. holding the drive shaft with an Allen key.

6. If rear bearing was removed, press a new one into rear housing.

7. Assemble brushes, springs, terminal and insulator in the brush holder, retract the brushes and insert a short length of ⅛ in. rod or stiff wire through the hole in the holder to hold the brushes in the retracted position.

8. Position the brush holder assembly in the rear housing and install mounting screws. Position brush leads to prevent shorting.

9. Wrap the three stator winding leads around the circuit board terminals and solder them using only rosin core solder and a 100-watt iron. Position the stator neutral lead eyelet on the stator terminal screw and install the screw in the rectifier assembly.

10. Exposed Diodes: insert the special screws through the wire lug, dished washers and circuit board. Turn ¼ turn counterclockwise to lock in place. Integral Diodes: insert the screws straight through the holes .Booster Diodes: position the stator wire terminal on the stator terminal screw, then position screw on rectifier. Position square insulator over the screw and into the square hole in the rectifier, rotate terminal screw until it locks, then press it in fingertight. Position the stator wire, then press the terminal screw into the rectifier and insulator with a vise.

NOTE: The dished washers are to be used on the molded circuit boards only. Using these washers on a fiber board will

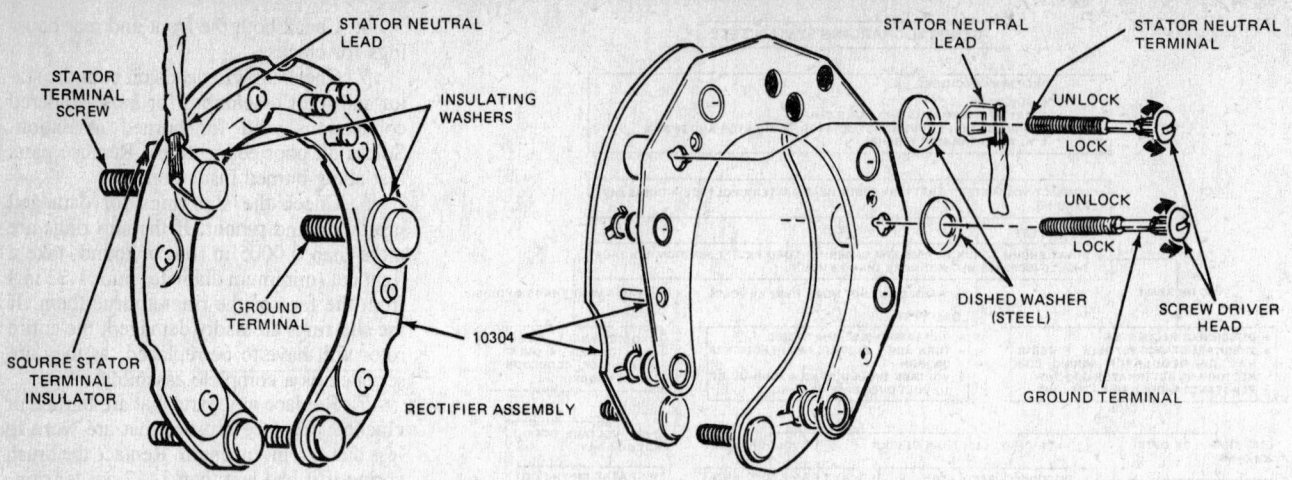

FLAT TYPE (INTEGRATED) RECTIFIER

RECTIFIER (DISCRETE) WITH EXPOSED DIODES

Rectifier assembly exploded view

result in a serious short circuit, as only a flat insulating washer between the stator terminal and the board is used on fiber circuit boards.

11. Place the radio noise suppression condenser on the rectifier terminals. With molded circuit board, install the STA and BAT terminal insulators. With fiber circuit board, place the square stator terminal insulator in the square hole in the rectifier assembly, then position BAT terminal insulator. Position the stator and rectifier assembly in the rear housing, making sure that all terminal insulators are seated properly in the recesses. Position STA, BAT and FLD insulators on terminal bolts; install nuts.

12. Clean the rear bearing surface of the rotor shaft with a rag, then position rear housing and stator assembly over rotor. Align matchmarks made during disassembly and install through-bolts. Remove brush retracting wire and place a dab of silicone sealer over the hole.

Field Current Draw Test

NOTE: Alternator must be removed from the truck.

1. Connect a test ammeter between the alternator frame and the positive post of a 12 volt test battery.

2. Connect a jumper wire between the negative test battery post and the alternator field terminal.

3. Observe the ammeter: Little or no current flow indicates high brush resistance, open field windings, or high winding resistance. Current in excess of specifications (approximately 2.9 amps. for most models) indicates shorted or grounded field windings, or brush leads touching.

NOTE: Sometimes the alternator produces current output at low engine speeds, but ceases to put out at higher speeds. This can be caused by centrifugal

force expanding the rotor windings to the point where they short to ground. Place in a test stand and check field current draw while spinning alternator.

Diode Tests

Disassemble the alternator. Disconnect diode assembly from stator and make tests. To test one set of diodes, contact one ohmmeter probe to the diode plate and contact each of the three stator lead terminals with the other probe. Reverse the probes and repeat the test. All six tests (eight for 61 amp. eight-diode models) should show a reading of about 60 ohms in one direction and infinite ohms in the other. If two high readings, or two low readings, are obtained after reversing probes the diode is faulty and must be replaced.

Stator Tests

Disassemble the stator from the alternator assembly and rectifiers. Connect test ohmmeter probes between each pair of stator leads. If the ohmmeter does not indicate equally between each pair of leads, the stator coil is open and must be replaced.

Connect test ohmmeter probes between one of the stator leads and the stator core. The ohmmeter should not show any reading. If it does show continuity, the stator winding is grounded and must be replaced.

1981 and Later Rear Terminal Alternators

DISASSEMBLY

1. Scribe the end housings and stator frame for alignment during reassembly, and then remove the three housing through bolts.

2. Separate the front housing and rotor assembly from the stator and rear housing. If there is resistance in pulling the front housing free, tap the front housing lightly with a plastic tipped hammer to break it loose.

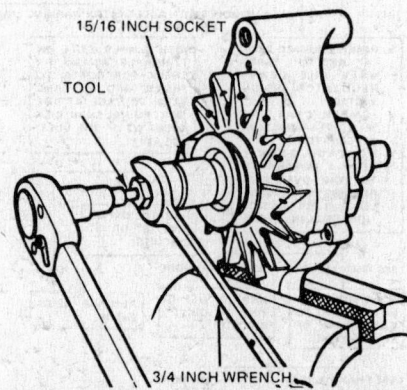

On late model rear terminal alternators and all side model alternators, use this special tool to remove the pulley retaining nut (© Ford Motor Co.)

3. Remove the brush springs from the brush holder, located in the rear housing.

4. Note the colors and locations of nuts, washers, and insulators on the back of the rear housing for reconnection. Then, remove all of them.

5. Remove the stator and rectifier assembly from the rear housing.

6. Remove the screws attaching the brush holder to the rear housing; then remove the brush holder, brushes, and brush terminal insulator.

7. Press the bearing out of the rear housing with an arbor press. Make sure the housing is supported as close as possible to the bearing boss to prevent damage to it.

8. Clamp the front housing in a vise equipped with protective jaws. Remove the drive pulley retaining nut. This requires a $5/16$ in. socket and $3/4$ in. wrench to drive it, and a special tool that passes through the center of the socket wrench, and locks onto the rotor shaft, and prevents it from turning as the nut is removed. Then, remove the lockwasher, drive pulley, fan, and fan spacer.

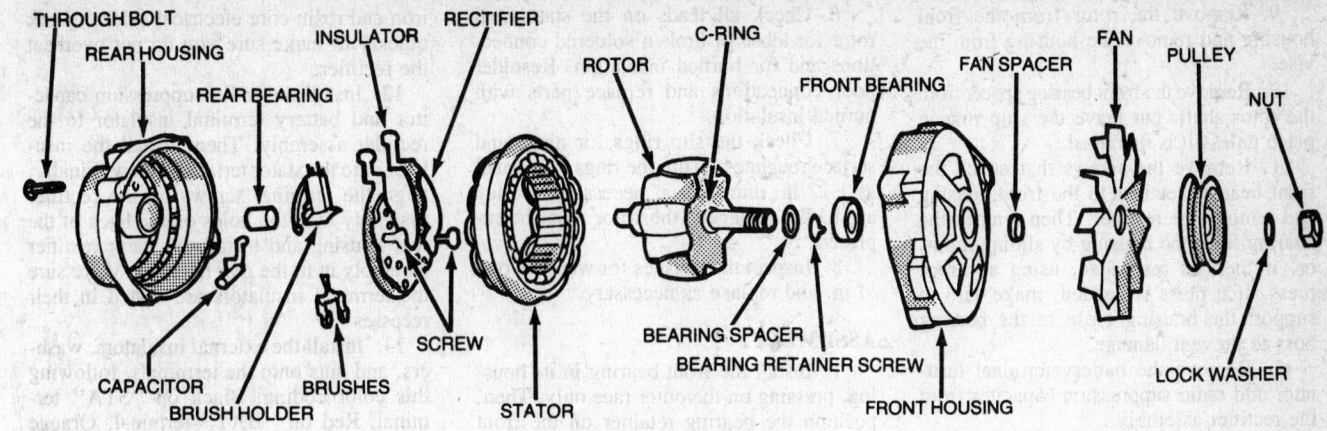

Side terminal alternator

THROUGH BOLT
REAR HOUSING
REAR BEARING
INSULATOR
RECTIFIER
ROTOR
C-RING
FRONT BEARING
FAN SPACER
FAN
PULLEY
NUT
CAPACITOR
BRUSH HOLDER
BRUSHES
SCREW
STATOR
BEARING SPACER
BEARING RETAINER SCREW
FRONT HOUSING
LOCK WASHER

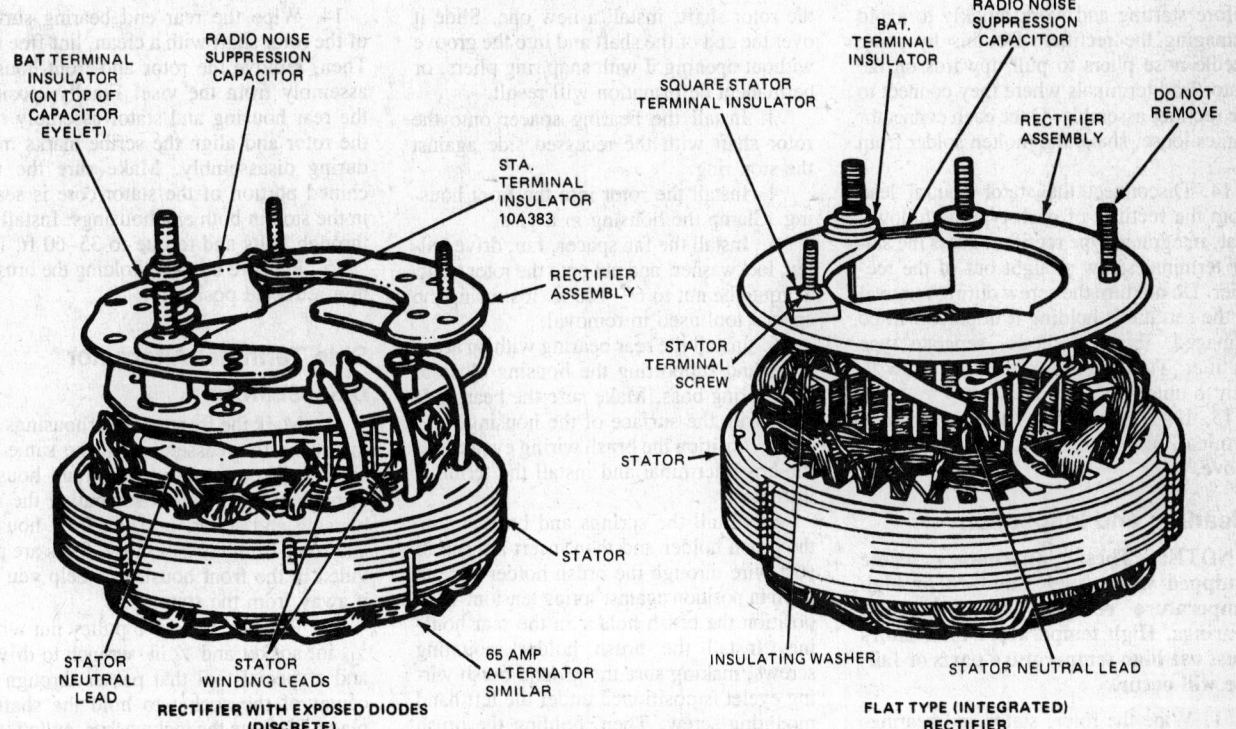

BAT TERMINAL INSULATOR (ON TOP OF CAPACITOR EYELET)
RADIO NOISE SUPPRESSION CAPACITOR
STA. TERMINAL INSULATOR 10A383
RECTIFIER ASSEMBLY
STATOR
STATOR NEUTRAL LEAD
STATOR WINDING LEADS
65 AMP ALTERNATOR SIMILAR

RECTIFIER WITH EXPOSED DIODES (DISCRETE)

BAT. TERMINAL INSULATOR
RADIO NOISE SUPPRESSION CAPACITOR
SQUARE STATOR TERMINAL INSULATOR
RECTIFIER ASSEMBLY
DO NOT REMOVE
STATOR TERMINAL SCREW
STATOR
INSULATING WASHER
STATOR NEUTRAL LEAD

FLAT TYPE (INTEGRATED) RECTIFIER

Stator and rectifier assemblies

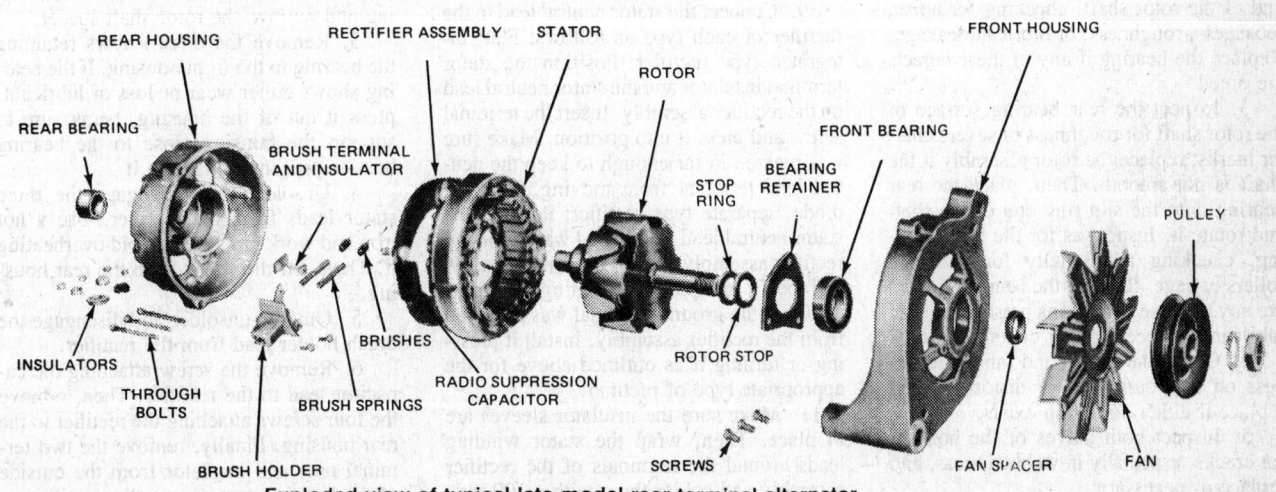

REAR HOUSING
REAR BEARING
BRUSH TERMINAL AND INSULATOR
INSULATORS
THROUGH BOLTS
BRUSH HOLDER
BRUSH SPRINGS
BRUSHES
RADIO SUPPRESSION CAPACITOR
RECTIFIER ASSEMBLY
STATOR
ROTOR
ROTOR STOP
SCREWS
STOP RING
BEARING RETAINER
FRONT BEARING
FRONT HOUSING
FAN SPACER
FAN
PULLEY

Exploded view of typical late model rear terminal alternator

9. Remove the rotor from the front housing and remove the housing from the vise.

10. Remove the front bearing spacer from the rotor shaft, but leave the stop ring in place unless it is damaged.

11. Remove the screws that attach the front bearing retainer to the front housing and remove the retainer. Then remove the bearing from the housing by sliding it out, or, if there is resistance, using an arbor press. If a press is needed, make sure to support the housing close to the bearing boss to prevent damage.

12. Remove the battery terminal insulator and radio suppression capacitor from the rectifier assembly.

13. Use a 100 watt soldering iron to unsolder the stator leads where they connect to the rectifier. Make sure the iron is hot before starting and work quickly to avoid damaging the rectifier. Do this by using needle nose pliers to pull upwards on the stator lead terminals where they connect to the rectifier assembly. Once each connector comes loose, shake the molten solder from it.

14. Disconnect the stator neutral lead from the rectifier of each type as follows: Flat, integrated type rectifier: Press the stator terminal screw straight out of the rectifier. Do not turn the screw during removal or the serrations holding it in place will be damaged. Exposed diode, separate type rectifier: Turn the stator terminal screw ¼ turn to unlock it and remove it.

15. If it is necessary to remove the ground terminal, follow the appropriate procedure above.

Cleaning and Inspection

NOTE: These alternators come equipped with either standard or high temperature rectifier assemblies and bearings. High temperature alternators must use high temperature parts or failure will occur.

1. Wipe the rotor, stator and bearings with a clean cloth. Do not use solvent.

2. Rotate the front bearing on the drive end of the rotor shaft, checking for noise, looseness, roughness, or lubricant leakage. Replace the bearing if any of these defects are noted.

3. Inspect the rear bearing surface of the rotor shaft for roughness or severe chatter marks; replace the rotor assembly if the shaft is not smooth. Then, place the rear bearing onto the slip ring end of the shaft and rotate it. Inspect as for the front bearing, checking additionally for damaged rollers or cage. Replace the bearing if there are any of these conditions present or if the lubricant has been lost or contaminated.

4. Check the pulley and fan for looseness on the rotor shaft or distortion, and replace if either condition exists.

5. Inspect both halves of the housing for cracks, especially in webbed areas, and replace as necessary.

6. Check all leads on the stator and rotor for loose or broken soldered connections and for burned insulation. Resolder poor connections and replace parts with burned insulation.

7. Check the slip rings for nicks and surface roughness. Turn the rings to as small as 1.22 in. diameter, if necessary. If they are badly damaged, the rotor must be replaced.

8. Inspect the brushes for wear beyond ¼ in. and replace as necessary.

ASSEMBLY

1. Install the front bearing in its housing, pressing on the outer race only. Then, position the bearing retainer on the front housing and install the attaching screws, torquing to 25–40 inch lbs.

2. If the stop ring was removed from the rotor shaft, install a new one. Slide it over the end of the shaft and into the groove without opening it with snap ring pliers, or permanent deformation will result.

3. Install the bearing spacer onto the rotor shaft with the recessed side against the stop ring.

4. Install the rotor into the front housing. Clamp the housing in a vise.

5. Install the fan spacer, fan, drive pulley, lockwasher, and nut onto the rotor shaft. Torque the nut to 60–100 ft. lbs. using the special tool used in removal.

6. Install the rear bearing with an arbor press and supporting the housing close to the bearing boss. Make sure the bearing is flush with the surface of the housing.

7. Position the brush wiring eyelet over the brush terminal and install the terminal insulator.

8. Install the springs and brushes into the brush holder and then insert a piece of stiff wire through the brush holder to hold them in position against spring tension. Then position the brush holder in the rear housing. Install the brush holder mounting screws, making sure the ground brush wiring eyelet is positioned under the left hand mounting screw. Then, holding the brush holder firmly against the housing, torque the screws to 17–25 inch lbs.

9. Connect the stator neutral lead to the rectifier of each type as follows: Flat, integrated type rectifier: Position the stator terminal insulator and the stator neutral lead on the rectifier assembly. Insert the terminal screw and press it into position. Make sure it is pressed in far enough to keep the neutral lead terminal from moving. Exposed diode, separate type rectifier: Position the stator neutral lead and dished washer on the rectifier assembly. Insert the terminal screw and lock it into place by rotating it ¼ turn.

10. If the ground terminal was removed from the rectifier assembly, install it pressing or turning it as outlined above for the appropriate type of rectifier.

11. Make sure the insulator sleeves are in place. Then, wrap the stator winding leads around the terminals of the rectifier assembly and solder them with a 100 watt iron and rosin core electrical solder. Work quickly to make sure you do not overheat the rectifier.

12. Install the radio suppression capacitor and battery terminal insulator to the rectifier assembly. Then, install the insulator onto the stator terminal screw. Finally, align the terminal screws on the rectifier assembly with the holes on the back of the rear housing and install the stator rectifier assembly in to the rear housing. Make sure the terminal insulators are seated in their recesses.

13. Install the external insulators, washers, and nuts onto the terminals, following this color coding: Black on "STA" terminal. Red on "BAT" terminal. Orange on "FLD" terminal. Torque the nut for the red lead to 30–55 inch lbs., and the other two to 25–35 inch lbs.

14. Wipe the rear end bearing surface of the rotor shaft with a clean, lint-free rag. Then, remove the rotor and front housing assembly from the vise. Finally, position the rear housing and stator assembly over the rotor and align the scribe marks made during disassembly. Make sure the machined portion of the stator core is seated in the stop in both end housings. Install the through bolts and torque to 35–60 ft. lbs.

15. Remove the wire holding the brushes in a retracted position.

Side Terminal Alternator
DISASSEMBLY

1. Mark the front and rear housings and the stator for reassembly in the same positions. Then, remove the four housing through bolts. Without separating the rear housing and stator, pull the front housing and rotor from the assembly. Slots are provided in the front housing to help you pry it away from the stator.

2. Remove the drive pulley nut with a ⁵⁄₁₆ in. socket and ¾ in. wrench to drive it and a special tool that passes through the center of the socket to hold the shaft in place. Remove the lockwasher, pulley, fan, and fan spacer from the rotor shaft. Finally, pull the rotor and shaft from the front housing and remove the rotor shaft spacer.

3. Remove the three screws retaining the bearing to the front housing. If the bearing shows either wear or loss of lubricant, press it out of the housing, being sure to support the housing close to the bearing boss to prevent damage to it.

4. Unsolder and disengage the three stator leads from the rectifier. Use a hot iron and work quickly to avoid overheating it. Then, lift the stator from the rear housing.

5. Quickly unsolder and disengage the brush holder lead from the rectifier.

6. Remove the screw attaching the capacitor lead to the rectifier. Then, remove the four screws attaching the rectifier to the rear housing. Finally, remove the two terminal nuts and insulator from the outside of the housing and remove the rectifier.

7. Remove the two screws attaching the brush holder to the housing and remove the brushes and holder. Remove the two rectifier insulators from the housing bosses.

8. Remove any sealing compound from the rear housing and brush holder.

9. Remove the screw attaching the capacitor to the rear housing and remove the capacitor.

10. Inspect the rear bearing for excessive wear, damage, or loss of lubricant and, if necessary press it out, supporting the housing close to the bearing boss to prevent damage to it.

ASSEMBLY

1. If the front bearing is being replaced, first press the new bearing into the housing, putting pressure on the outer race only. Then, install the bearing retaining screws and torque them to 25-40 inch lbs.

2. Place the inner spacer on to the rotor shaft and insert the rotor shaft into the center of the front bearing.

3. Install the fan spacer, fan, pulley, lockwasher, and nut onto the rotor shaft, in that order. Then, tighten the nut using the socket, open end wrench, and special tool used in removal.

4. If the rear bearing is being replaced, press a new one in by the outer race only

until the rear face is flush with the outer surface of the bearing boss.

5. Position the brush terminal on the brush holder. Then, install the springs and brushes in the holder, and insert a piece of stiff wire across in front of the brushes to retain them in a retracted position for assembly.

6. Position the brush holder in the rear housing and start the attaching screws. Make sure the wire retaining the brushes sticks

out far enough for you to pull it from the housing after the alternator is assembled. Poke any sealer that may be present out of the pin hole in the housing. Then, hold the brush holder firmly toward the brush enclosure opening while tightening the attaching screws. Finally, reseal the crack between the brush holder and the brush cavity in the rear housing with a body sealer. Don't use a silicone sealant.

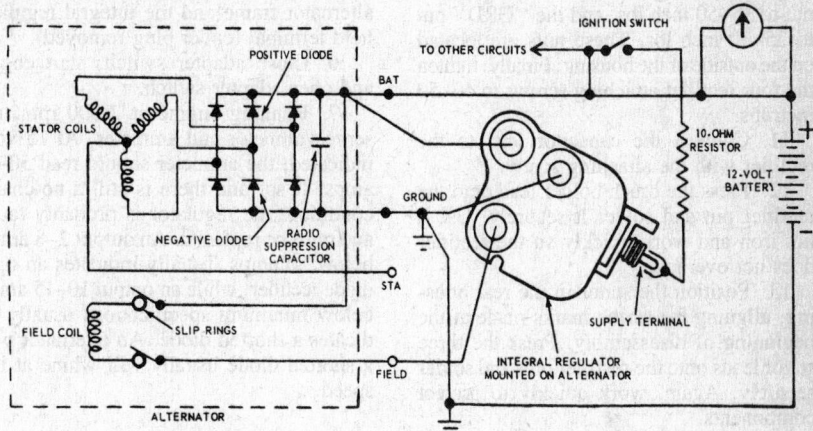

Charging system schematic with integral regulator

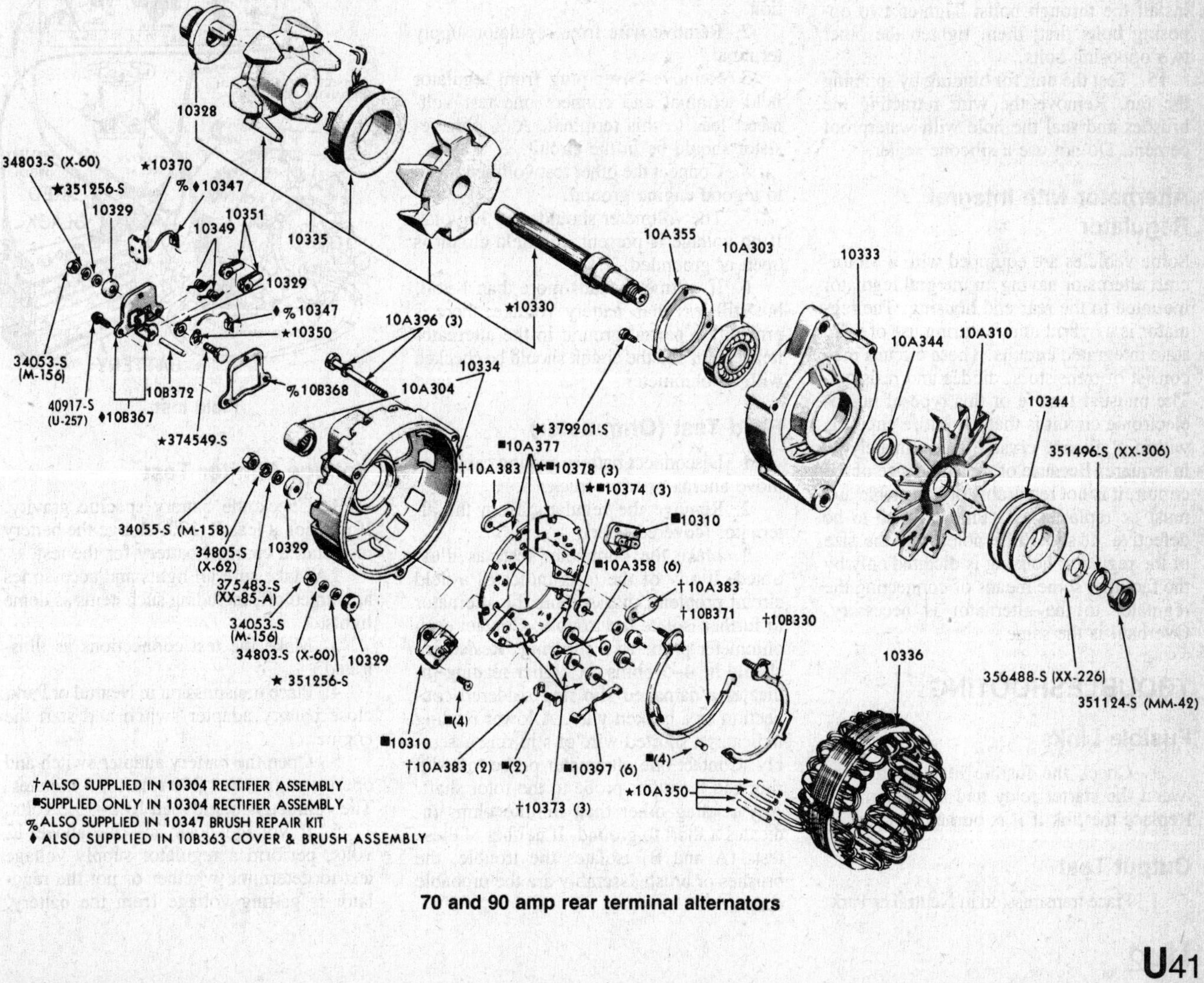

†ALSO SUPPLIED IN 10304 RECTIFIER ASSEMBLY
■SUPPLIED ONLY IN 10304 RECTIFIER ASSEMBLY
%ALSO SUPPLIED IN 10347 BRUSH REPAIR KIT
♦ALSO SUPPLIED IN 10B363 COVER & BRUSH ASSEMBLY

70 and 90 amp rear terminal alternators

7. Position the capacitor in the rear housing and start install the attaching screw.

8. Place the two rectifier bosses in the housing.

9. Place the insulator on the larger ("BAT") terminal of the rectifier and position the rectifier in the rear housing. Place the outside insulator on the "BAT" terminal install the nuts on both "BAT" and "GRD" terminals finger tight.

10. Start the four rectifier attaching screws. Then, tighten the "BAT" terminal nut to 35–50 inch lbs. and the "GRD" nut to 25–30 inch lbs. These nuts are located on the outside of the housing. Finally, tighten the four rectifier attaching screws to 40–50 inch lbs.

11. Connect the capacitor lead to the rectifier with the attaching screw.

12. Press the brush holder lead onto the rectifier pin and solder it securely. Use a hot iron and work quickly so the rectifier does not overheat.

13. Position the stator in the rear housing, aligning the scribe marks made at the beginning of disassembly. Press the three stator leads onto the rectifier pins and solder securely. Again, work quickly to protect components.

14. Position the rotor and front housing into/onto the stator and rear housing. Align the marks made in disassembly, and then install the through bolts. Tighten two opposing bolts first; then, tighten the other two opposing bolts.

15. Test the unit for binding by spinning the fan. Remove the wire retracting the brushes and seal the hole with waterproof cement. Do not use a silicone sealer.

Alternator with Integral Regulator

Some vehicles are equipped with a Motorcraft alternator having an integral regulator mounted to the rear end housing. The regulator is a hybrid unit featuring use of solid state integrated circuits. These circuits may consist of transistors, diodes and resistors. The unusual feature of this type of microelectronic circuit is that the entire circuit is within a silicone crystal approximately 1/8 in. square. Because of the small size of the circuit, it is not repairable or adjustable and must be replaced as a unit if found to be defective. It should be noted that the size of the regulator housing is dictated only by the fact that some means of connecting the regulator to the alternator is necessary. Overhaul is the same.

TROUBLESHOOTING

Fusible Links

1. Check the fusible link located between the starter relay and the alternator. Replace the link if it is burned or open.

Output Test

1. Place transmission in Neutral or Park.

2. Remove the positive battery cable and install a battery adapter switch in the line.

3. Attach one lead of a test voltmeter to the negative battery post and the other test lead to the circuit side of the adapter switch.

4. Connect a test ammeter to each side of the adapter switch, so that charging current will go through the ammeter when the switch is opened.

5. Connect a jumper wire between the alternator frame and the integral regulator field terminal (cover plug removed).

6. Close adapter switch, start engine and open adapter switch.

7. Running engine at 2,000 rpm, observe voltmeter and ammeter. At 15 volts indicated, the ammeter should read 50–57 amps. If so, and there is still a no-charge condition, the regulator is probably faulty and must be replaced. An output 2–8 amps. below 50 amps. usually indicates an open diode rectifier, while an output 10–15 amps. below minimum specifications usually indicates a shorted diode. An alternator with a shorted diode usually will whine at idle speed.

Field Test (Voltmeter)

1. Turn ignition switch to OFF position.

2. Remove wire from regulator supply terminal.

3. Remove cover plug from regulator field terminal and connect one test voltmeter lead to this terminal. A 1/4 ohm resistor should be in the circuit.

4. Connect the other test voltmeter lead to a good engine ground.

5. The voltmeter should read 12 volts. If no voltage is present, the field circuit is open or grounded.

6. If voltmeter reads more than 1 volt, but still less than battery voltage, there is probably a partial ground in the alternator field circuit and the circuit should be checked with an ohmmeter.

Field Test (Ohmmeter)

1. Disconnect battery ground cable; remove alternator from truck.

2. Remove the regulator from the alternator (covered later).

3. Make the ohmmeter tests as illustrated. If any of the tests indicates a field circuit problem, disassemble the alternator to further isolate the trouble. Contact each ohmmeter probe to a slip ring. Resistance should be 4–5 ohms. A higher reading indicates a damaged slip ring soldered connection or a broken wire. A lower reading indicates a shorted wire or slip ring assembly. Contact one ohmmeter probe to a slip ring and the other probe to the rotor shaft. Any reading other than infinite ohms indicates a short to ground. If neither of these tests (A and B) isolates the trouble, the brushes or brush assembly are the probable cause.

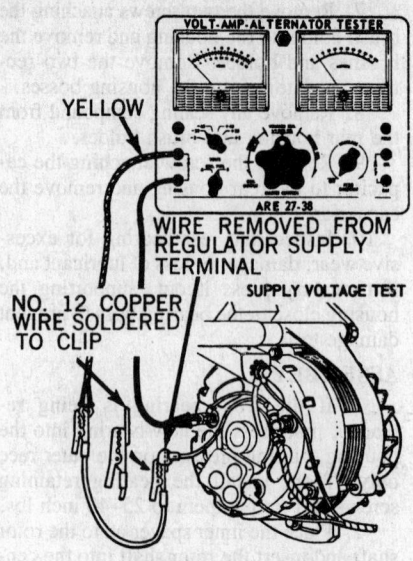

Supply voltage test

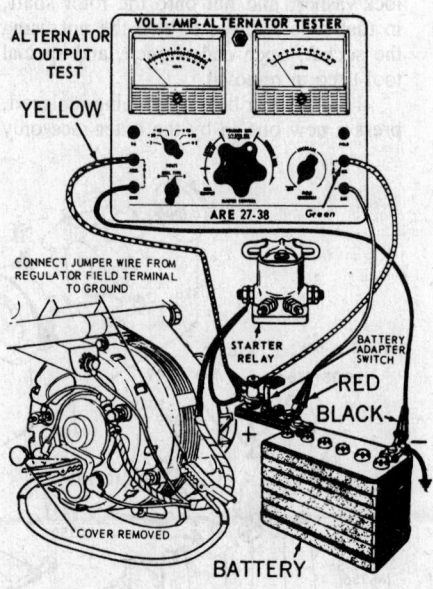

Field test

Voltage Limiter Test

1. Check the battery specific gravity. If it is not at least 1.230, charge the battery or install a charged battery for the test.

2. Make sure all lights and accessories are turned off, including such items as dome lights.

3. Make the test connections as illustrated.

4. Place transmission in Neutral or Park, close battery adapter switch and start the engine.

5. Open the battery adapter switch and operate engine at 2,000 rpm for 5 minutes. The voltmeter should read 13.3–15.3 volts.

6. If voltage does not rise above 12 volts, perform a regulator supply voltage test to determine whether or not the regulator is getting voltage from the battery.

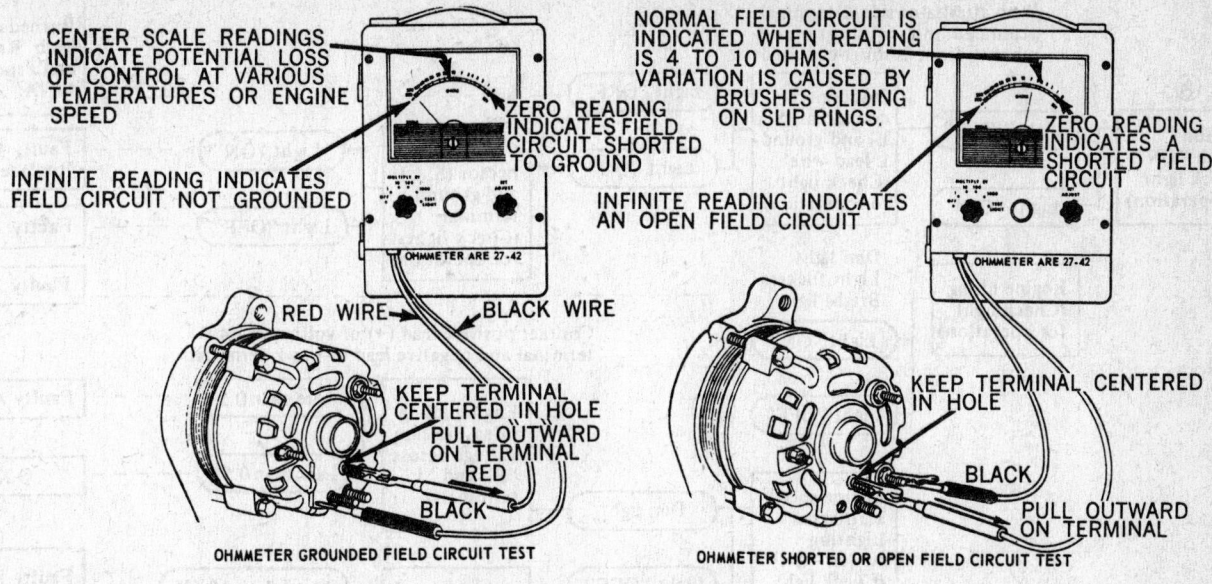

CENTER SCALE READINGS INDICATE POTENTIAL LOSS OF CONTROL AT VARIOUS TEMPERATURES OR ENGINE SPEED

INFINITE READING INDICATES FIELD CIRCUIT NOT GROUNDED

ZERO READING INDICATES FIELD CIRCUIT SHORTED TO GROUND

NORMAL FIELD CIRCUIT IS INDICATED WHEN READING IS 4 TO 10 OHMS. VARIATION IS CAUSED BY BRUSHES SLIDING ON SLIP RINGS.

ZERO READING INDICATES A SHORTED FIELD CIRCUIT

INFINITE READING INDICATES AN OPEN FIELD CIRCUIT

RED WIRE — BLACK WIRE

KEEP TERMINAL CENTERED IN HOLE PULL OUTWARD ON TERMINAL RED

BLACK

KEEP TERMINAL CENTERED IN HOLE

BLACK

PULL OUTWARD ON TERMINAL

OHMMETER GROUNDED FIELD CIRCUIT TEST

OHMMETER SHORTED OR OPEN FIELD CIRCUIT TEST

Field circuit test with ohmmeter

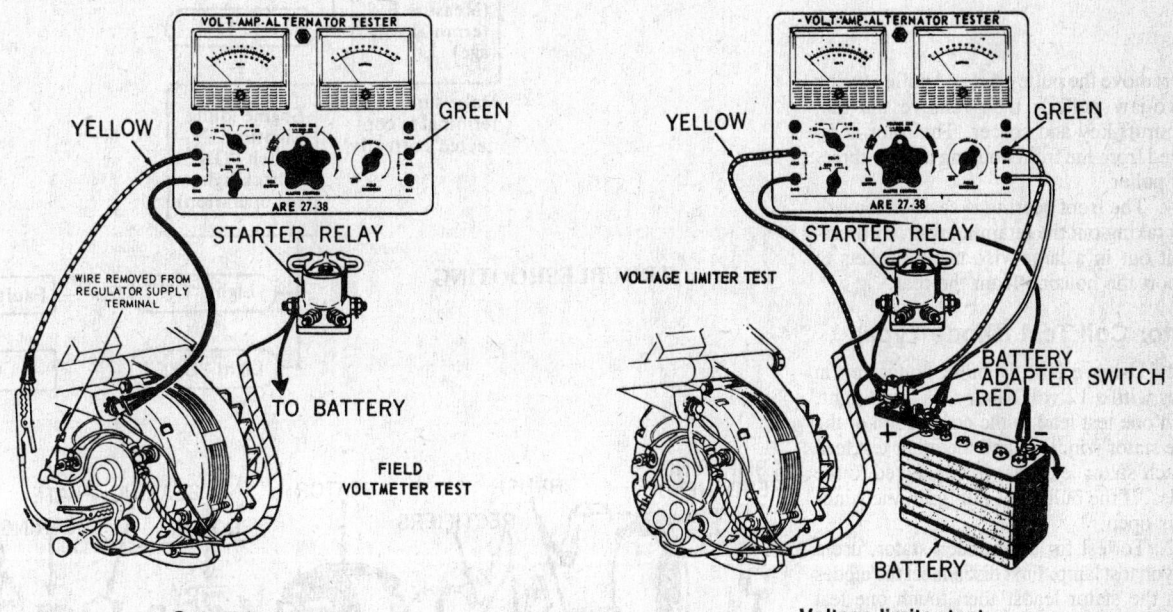

YELLOW

GREEN

WIRE REMOVED FROM REGULATOR SUPPLY TERMINAL

STARTER RELAY

TO BATTERY

FIELD VOLTMETER TEST

Output test

YELLOW

GREEN

VOLTAGE LIMITER TEST

STARTER RELAY

BATTERY ADAPTER SWITCH RED

BATTERY

Voltage limiter test

Before replacing a regulator, check the wiring of the entire charging system for shorts, opens, or high resistance connections.

Regulator Supply Voltage Test

The regulator is "turned on" by the application of battery voltage through a 10 ohm resistor wire. If the supply circuit is defective, the regulator will not function and the alternator will not put out current.

1. Connect a 12 volt test light or voltmeter between the regulator supply lead and ground.

2. Turn on the ignition switch. The test light should glow or the voltmeter indicate. If not, the supply circuit should be checked back to the battery, especially the resistance wire.

Prestolite System

Prestolite alternators incorporate an isolation diode, mounted as a component part of the internal positive heat sink assembly.

TROUBLESHOOTING

Ignition Switch-to-Regulator Circuit

1. Disconnect the regulator wires from the regulator.

2. Turn on the key. Using a test light or voltmeter, check for current between the I terminal and ground and the L terminal ground. If voltage is present, this part of

the system is OK. If no voltage is present, check for broken or shorted wires, a bad indicator bulb, a bad ammeter (if so equipped), or bad connections.

DISASSEMBLY

1. Remove the two brush mounting screws and cover, then tip the brush assembly away from the alternator and remove.

2. Matchmark the rear housing, stator and drive end housing, then remove the four retaining screws. The stator and rear housing are removed as a unit by tapping lightly with a fiber hammer to separate them from the front housing.

3. The rotor should not be removed unless it or the front bearing is defective. To remove the rotor under these conditions,

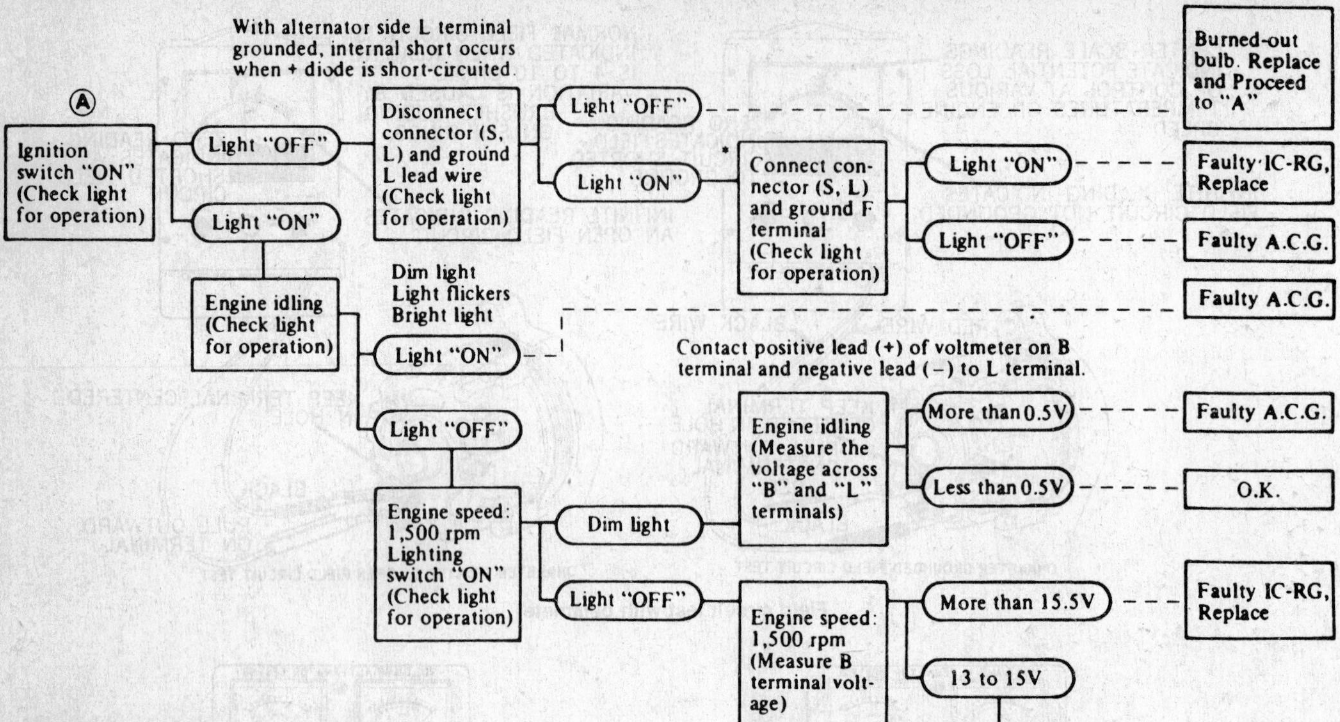

With alternator side L terminal grounded, internal short occurs when + diode is short-circuited.
A
Ignition switch "ON" (Check light for operation)
Light "OFF"
Light "ON"
Disconnect connector (S, L) and ground L lead wire (Check light for operation)
Light "OFF"
Light "ON"
Burned-out bulb. Replace and Proceed to "A"
Connect connector (S, L) and ground F terminal (Check light for operation)
Light "ON"
Light "OFF"
Faulty IC-RG, Replace
Faulty A.C.G.
Engine idling (Check light for operation)
Dim light
Light flickers
Bright light
Light "ON"
Faulty A.C.G.
Contact positive lead (+) of voltmeter on B terminal and negative lead (−) to L terminal.
Light "OFF"
Engine idling (Measure the voltage across "B" and "L" terminals)
More than 0.5V
Less than 0.5V
Faulty A.C.G.
O.K.
Engine speed: 1,500 rpm Lighting switch "ON" (Check light for operation)
Dim light
Light "OFF"
Engine speed: 1,500 rpm (Measure B terminal voltage)
More than 15.5V
Faulty IC-RG, Replace
13 to 15V

Make sure "S" terminal is connected correctly
Engine idling Lighting switch "ON" (Check light for operation)
Light "ON"
Faulty A.C.G.
Light "OFF"
O.K.

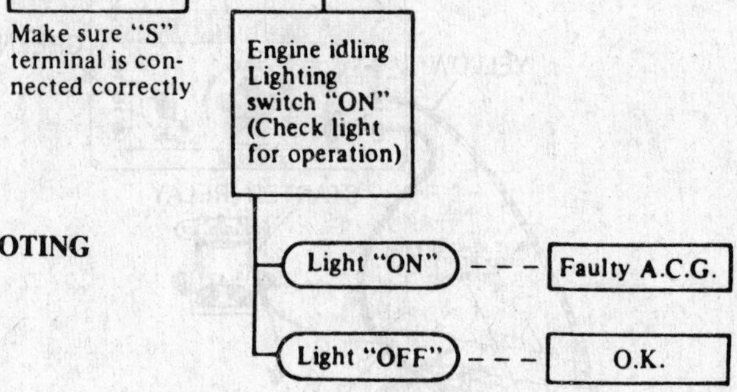

TROUBLESHOOTING

first remove the pulley nut and pulley (using a two-jaw puller), then remove the fan, Woodruff key and spacer. The rotor is removed from the front housing using a three-paw puller.

4. The front bearing is easily removed, after taking out the retaining ring, by pressing it out in a large vise using sockets to support the housing from the rear.

Stator Coil Test (Diode Type)

1. Using a No. 57 bulb, connected in series with a 12 volt battery, as a test light, touch one test lead to the connection of the three stator windings and the other test lead to each stator lead that is connected to the diodes. If the bulb does not light, the winding is open.

2. To test for a grounded stator, use a 110 volt test lamp. First disconnect the diodes from the stator leads, then touch one test lead to the stator core and the other test lead to each of the three stator leads. If the test lamp lights, the winding is grounded.

NOTE: If all other components are O.K. and alternator still does not work, it can be assumed that the stator windings are internally shorted. This type of short is impossible to detect by using the previous test. Diode tests are the same as for the Motorola alternator.

ASSEMBLY

1. Press the front bearing into the front housing, making sure the dust seal faces the rotor. Install the bearing retaining snapring, then press the shoulder of the shaft against the inner bearing race using a tool that fits over the shaft and against the race. Install the spacer, Woodruff key, fan and pulley, then install lockwasher and pulley nut.

SLIP RING END HEAD
BRUSH
STATOR
BEARING PLATE
RECTIFIERS
ROTOR
BEARING
BEARING
RECTIFIERS
CONNECTORS
DRIVE END HEAD

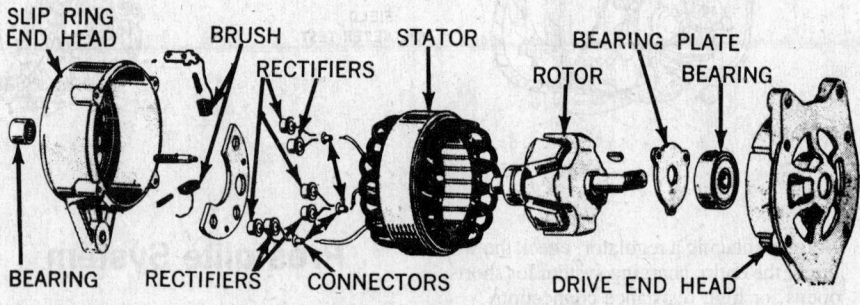

Prestolite alternator

2. Install the diode heat sink, negative diodes and stator. Solder any stator to diode connections that were unsoldered, using pliers as a heat sink to prevent overheating.

3. Install the rotor and front drive housing to stator and rear housing, aligning matchmarks made during disassembly. Install the four retaining screws, then the brush holder assembly and retaining screws.

4. Make sure the stator leads and brush holder assembly clear the rotor and that the rotor can be spun by hand without binding.

Datsun Alternator

DISASSEMBLY

1. On diesel engine models only, remove the vacuum pump.

2. Remove the through bolts and separate the front cover from the rear cover.

3. Place the rear cover side of the rotor in a vice and remove the pulley nut and pulley.

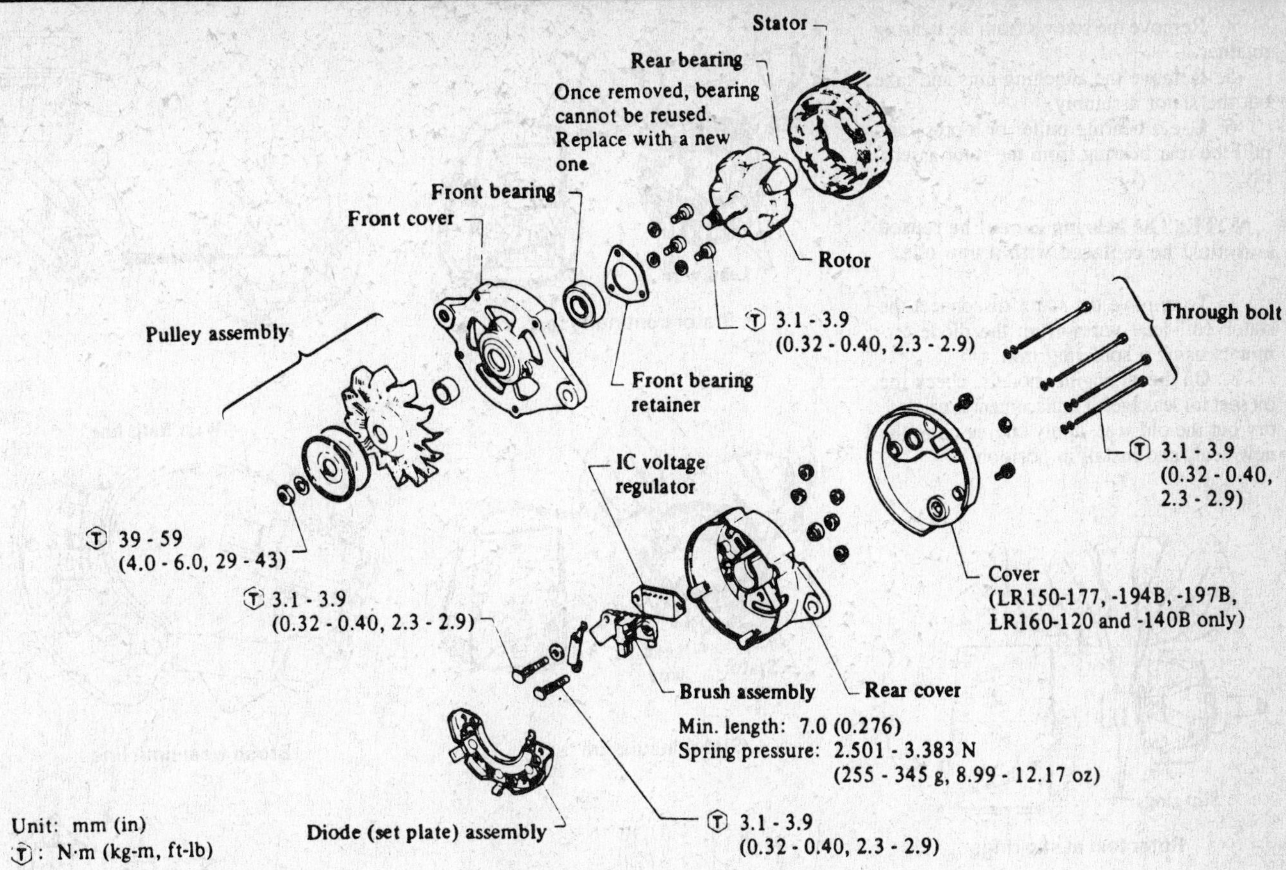

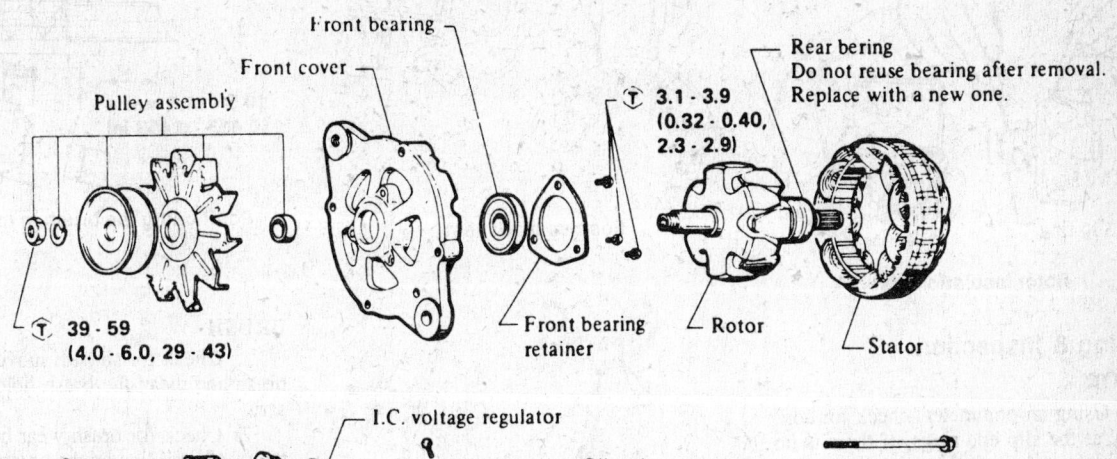

Stator

Rear bearing
Once removed, bearing cannot be reused. Replace with a new one

Front bearing

Front cover

Rotor

Pulley assembly

Ⓣ 3.1 - 3.9
(0.32 - 0.40, 2.3 - 2.9)

Front bearing retainer

Through bolt

Ⓣ 3.1 - 3.9
(0.32 - 0.40, 2.3 - 2.9)

IC voltage regulator

Ⓣ 39 - 59
(4.0 - 6.0, 29 - 43)

Ⓣ 3.1 - 3.9
(0.32 - 0.40, 2.3 - 2.9)

Cover
(LR150-177, -194B, -197B, LR160-120 and -140B only)

Brush assembly

Rear cover

Min. length: 7.0 (0.276)
Spring pressure: 2.501 - 3.383 N
(255 - 345 g, 8.99 - 12.17 oz)

Unit: mm (in)
Ⓣ : N·m (kg-m, ft-lb)

Diode (set plate) assembly

Ⓣ 3.1 - 3.9
(0.32 - 0.40, 2.3 - 2.9)

Datsun gasoline alternator—typical

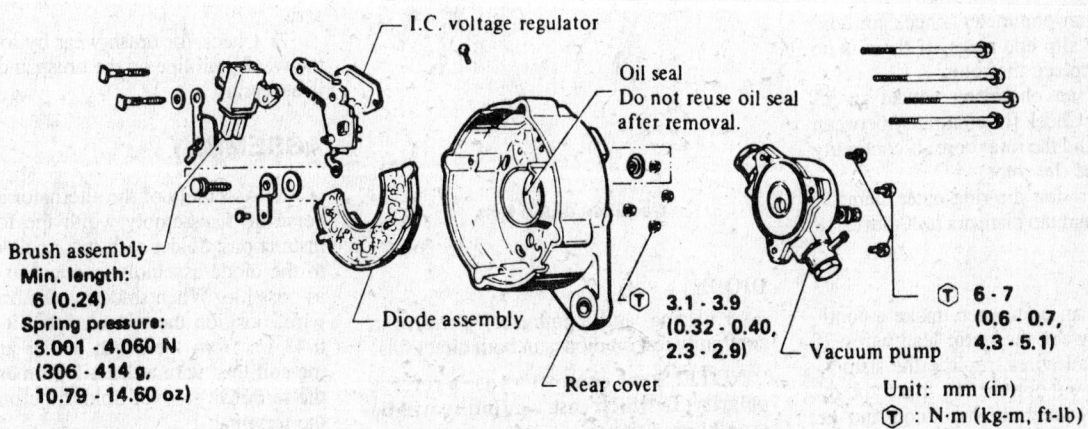

Front bearing

Front cover

Rear bering
Do not reuse bearing after removal. Replace with a new one.

Pulley assembly

Ⓣ 3.1 - 3.9
(0.32 - 0.40, 2.3 - 2.9)

Ⓣ 39 - 59
(4.0 - 6.0, 29 - 43)

Front bearing retainer

Rotor

Stator

I.C. voltage regulator

Oil seal
Do not reuse oil seal after removal.

Brush assembly
Min. length:
6 (0.24)
Spring pressure:
3.001 - 4.060 N
(306 - 414 g,
10.79 - 14.60 oz)

Diode assembly

Ⓣ 3.1 - 3.9
(0.32 - 0.40, 2.3 - 2.9)

Rear cover

Ⓣ 6 - 7
(0.6 - 0.7, 4.3 - 5.1)

Vacuum pump

Unit: mm (in)
Ⓣ : N·m (kg-m, ft-lb)

Datsun diesel engine alternator—typical

4. Remove the screws from the bearing retainer.

5. Remove the attaching nuts and take out the stator assembly.

6. Use a bearing puller or a press and pull the rear bearing from the rotor assembly.

NOTE: The bearing cannot be reused and must be replaced with a new one.

7. To remove the stator disconnect the stator coil lead wires from the diode terminals using a soldering iron.

8. On diesel engine models, check the oil seal for leakage. If replacement is needed, pry out the old seal, apply engine oil to the new seal, and install in position.

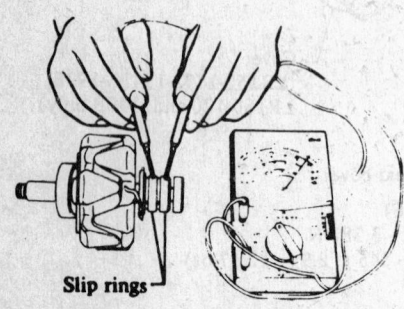

Rotor test at slip rings

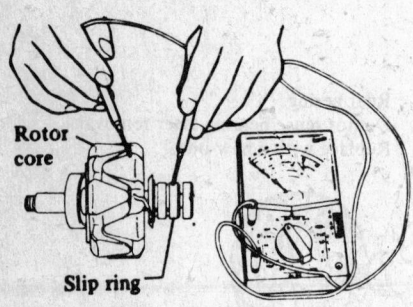

Rotor insulation test

Testing & Inspection
ROTOR

1. Using an ohmmeter, check for continuity at the slip end rings. If there is no continuity, replace the rotor.

2. Using an ohmmeter, make an insulation test. Check for continuity between the slip ring and the rotor core. If continuity exists, replace the rotor.

3. Measure the slip ring outer diameter for wear. Mimimum diameter is 30mm (1.18 in.).

STATOR

1. Using an ohmmeter, make a continuity test between the stator lead wires. If there is no lead wires, replace the stator.

2. Using an ohmmeter. make an insulation test between the stator core and the lead wire. If the continuity exists, replace the stator.

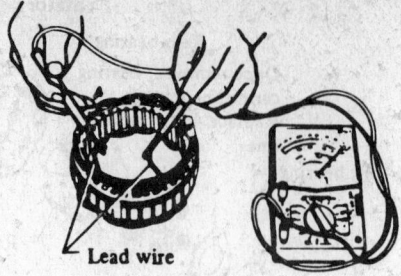

Stator continuity test

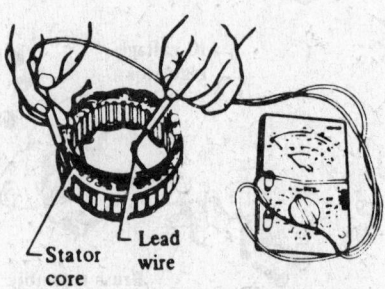

Stator insulation test

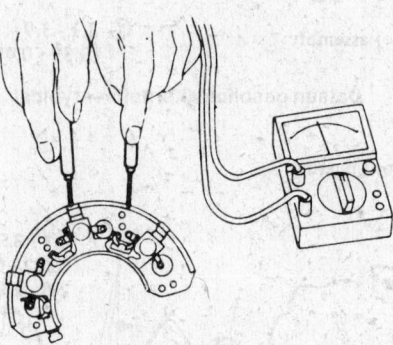

Positive diode test

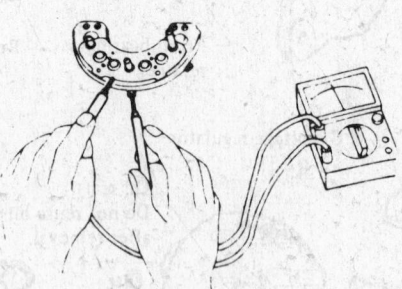

Negative diode test

DIODE

1. Using and ohmmeter, perform a continuity test on diodes in both directions.

NOTE: Some ohmmeters use a reverse polarity, in which case continuity will be exactly opposite.

2. Replace diodes as necessary.

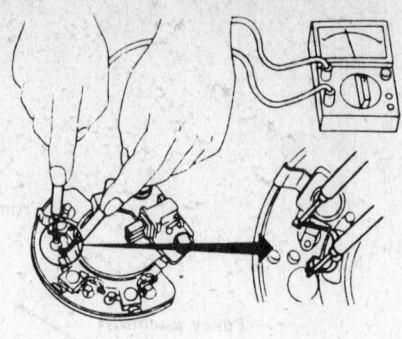

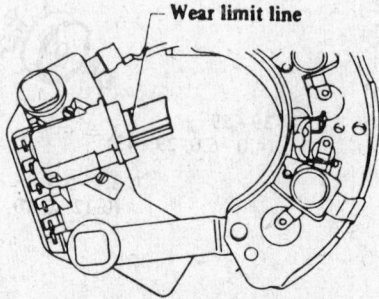

Brush wear limit line

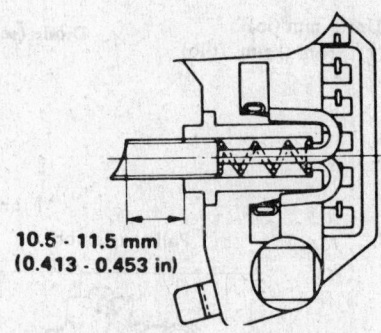

10.5 - 11.5 mm
(0.413 - 0.453 in)

Positioning the brush in the holder

BRUSH

1. Check for smooth movement of the brush and clean the brush holder if necessary.

2. Check for brush wear by looking at the wear limit line on the brush and replace if necessary.

ASSEMBLY

1. Assembly of the alternator is the reverse of disassembly wqith the following instructions: Solder each stator coil lead wire to the diode assembly terminal as quickly as possible. When soldering the brush lead wire, position the brush so that it extends 0.43 in. from the brush holder and wrap the coil lead wire at least 1.5 times around the terminal groove. Solder the outside of the terminal.

2. Tighten the pulley nut to 29–43 ft. lbs.

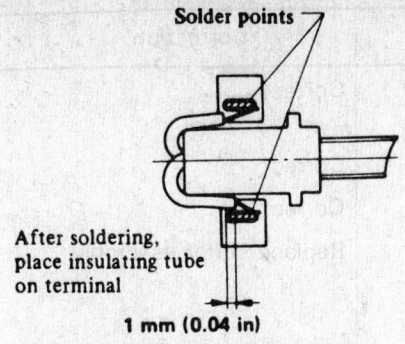

Solder points

After soldering,
place insulating tube
on terminal

1 mm (0.04 in)

Soldering the brush lead wires

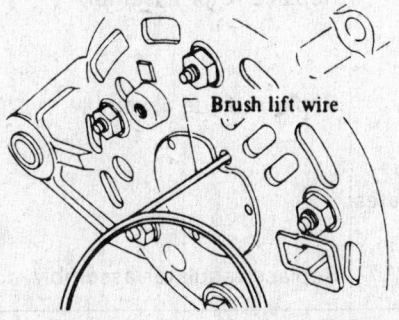

Brush lift wire

Insert a brush lift wire—gasoline engine

Use serration cap (Attach vinyl tape)
to prevent scratching oil seal

Brush lift wire

**Assembling the alternator—diesel
engine**

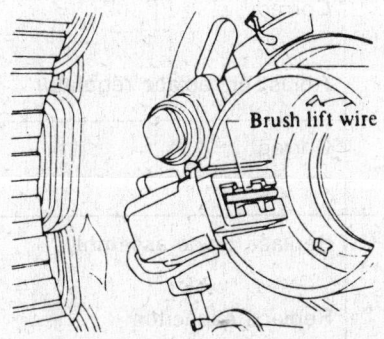

Brush lift wire

Internal view of the brush lift wire

3. Before installing the front and rear sides of the alternator, push the brush up with your fingers and retain the brush by inserting a wire from the outside into a lift hole. After installing the front and rear sides of the alternator, pull the brush lift by pushing towards the center.

NOTE: Do not pull brush lift by pushing towards the outside of the cover as it will damage the slip ring sliding

Isuzu/LUV Alternator

GAS ENGINE
Troubleshooting

1. Measure the resistance between F and E terminals (rotor coil resistance): The rotor coil circuit is normal if resistance measured across the terminals is 5 ohms. If resistance is higher than 5 ohms, the trouble is poor contact between the brushes and commutator. If no continuity exists between terminals F and E, the trouble is either an open coil rotor circuit, brush sticking or a broken lead wire. If resistance is lower than 5 ohms, it may be an indication of rotor coil layer short or the circuit being grounded.

2. Test the rectifying diodes in the following manner: Connect the positive (+) lead of a tester to the the alternator N terminal and the tester negative (—) lead to the alternator A terminal. If there exists a continuity between terminals, it indicates that one or more of the three diodes in the positive side are shorted. Connect the positive (+) lead of a tester to the alternator E terminal and the tester negative (-) lead to the alternator N terminal. If there exists a continuity, it indicates that one or more of the three diodes in the negative side are shorted.

DISASSEMBLY

1. Remove the through bolts and disconnect the lead wires at the connector.

2. Separate the alternator assembly into front and rear sections. The stator should be on the rear side.

3. Carefulley clamp the rotor in a vise and remove the pulley nut, then remove the pulley fan and rotor.

4. Remove the bearing retainer screws, then remove the ball bearing.

5. Remove the rear side nuts, then remove the stator from the rear cover together with the diodes, brush and capacitor.

6. Unsolder the diode to stator coil connections, then separate the diodes from the stator together with the brush and capacitor.

7. Remove the screws retaining the brush holder, then remove the diodes, brush and capacitor.

Testing & Inspection
ROTOR

1. Using an ohmmeter, check for continuity at the slip end rings. If there is no continuity, replace the rotor.

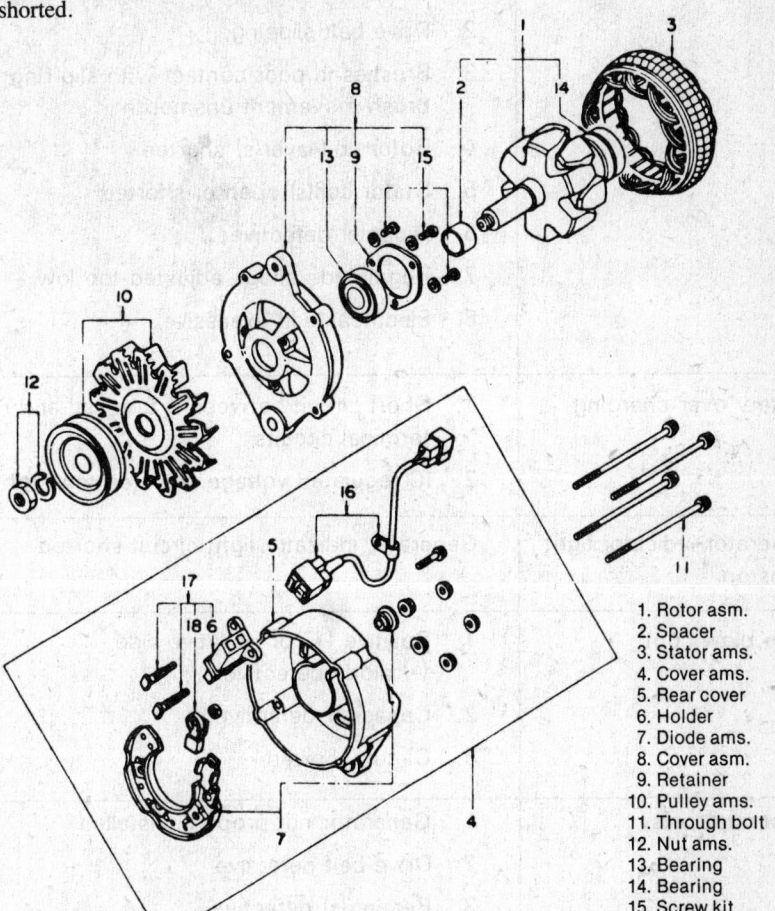

1. Rotor asm.
2. Spacer
3. Stator ams.
4. Cover ams.
5. Rear cover
6. Holder
7. Diode ams.
8. Cover asm.
9. Retainer
10. Pulley ams.
11. Through bolt
12. Nut ams.
13. Bearing
14. Bearing
15. Screw kit
16. Wire asm.
17. Brush and Condenser asm.
18. Condenser asm.

Isuzu/Luv alternator—gasoline engine

U47

Complaint	Cause	Correction
Battery not being charged	1. Cable between terminals broken or connector defective	Correct
	2. Charging system not properly grounded	Correct
	3. Brushes not in contact with slip ring	Correct
	4. Stator coil(s) open or burned Measure resistance across terminals at connectors between diode cover and stator coil leads.	Replace stator assembly
	5. Rotor coil(s) open or burned Measure resistance across F and E with connector at generator side disconnected	Replace rotor assembly
	6. Diode(s) defective Make a continuity test across B-N, B-E and N-E. Diodes are in good condition if tester indicates continuity only in one direction.	Replace diode assembly
	7. IC regulator defective	Replace regulator assembly
Battery under-charging	1. Wires between terminals poorly connected	Correct
	2. Drive belt slipping	Correct or replace belt
	3. Brushes in poor contact with slip ring or brush movement unsmooth	Correct or replace brush holder assembly
	4. Rotor coil layer(s) shorted	Replace rotor assembly
	5. Stator coil(s) open or shorted	Replace stator assembly
	6. Diode(s) defective	Replace diode assembly
	7. Regulated voltage adjusted too low	Adjust or replace regulator
	8. Electrical load excessive	Use higher capacity generator
Battery over-charging	1. Short circuit between B terminal and F terminal circuits	Correct
	2. IC regulated voltage adjusted too high	Adjust or replace regulator
Generator indicator light turns on	Generator indicator light circuit shorted	Correct
Fuse blows out	1. Positive (+) or negative side (−) diode defective	Replace diode assembly
	2. Capacitor defective	Replace capacitor
	3. Circuit shorted	Correct
Generator noisy	1. Generator not properly installed	Correct
	2. Drive belt defective	Replace belt
	3. Bearing(s) defective	Replace bearing(s)
	4. Diode(s) defective	Replace diode assembly
	5. Stator coil(s) shorted	Replace stator assembly

2. Using an ohmmeter, make an insulation test. Check for continuity between the slip ring and the rotor core. If continuity exists, replace the rotor.

3. Measure the slip ring outer diameter for wear. Mimimum diameter is 30mm (1.18 in.).

STATOR

1. Using an ohmmeter, make a continuity test between the stator lead wires. If there is no continuity, replace the stator.

2. Using an ohmmeter. make an insulation test between the stator core and the lead wire. If the continuity exists, replace the stator.

DIODE

1. Using and ohmmeter, perform a continuity test on diodes in both directions.

NOTE: Some ohmmeters use a reverse polarity, in which case continuity will be exactly opposite from the chart above.

2. Replace as necessary.

BRUSH

1. Check for smooth movement of the brush and clean the brush holder if necessary.

2. Check for brush wear by looking at the wear limit line on the brush and replace if necessary.

ASSEMBLY

1. Assembly of the alternator is the reverse of disassembly with the following instructions:

2. Before assembling the front and rear sections. insert a wire into the hole in the rear face of the rear cover from the outboard side to support the brush in the raised position , then insert the front section to which the rotor is assembled

DIESEL ENGINE

DISASSEMBLY

1. If so equipped, remove the vacuum pump attaching bolts, then hold the center plate and remove the vacuum pump in direction in line with the rotor shaft.

2. Remove the brush cover and the brush attaching bolts, then remove the brush from the holder.

3. Remove the through bolts and separate the body into front and rear sections.

NOTE: When separating, be careful so that the stator coils do not come off the rear cover. Also be careful you do not damage the oil seal when removing the rear cover. Taping the splines could provide some protection.

4. Carefully clamp the rotor assembly in a vise and remove the pulley nut.

5. Separate the pulley front cover and rotor, then remove the spacer and ball bearing.

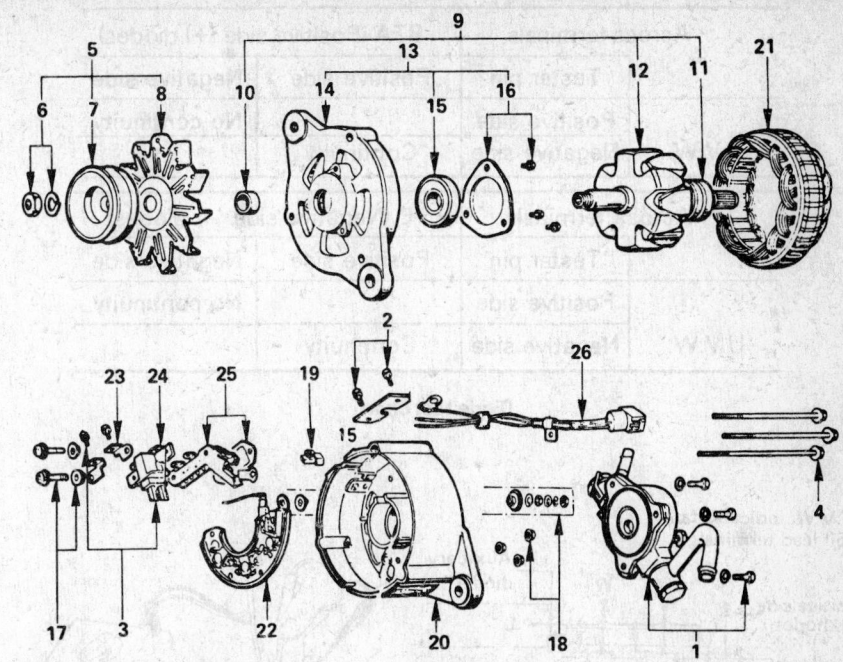

1. Vacuum pump
2. Cover
3. Brush
4. Through bolt
5. Pulley assembly
6. Pulley nut
7. Pulley
8. Fan
9. Rotor assembly
10. Spacer
11. Ball bearing
12. Rotor
13. Front cover assembly
14. Front cover
15. Ball bearing
16. Bearing retainer
17. Screw
18. Terminal bolt and nut
19. Lead wire
20. Rear cover
21. Stator
22. Diode
23. Holder plate
24. Brush holder
25. IC regulator assembly
26. Lead wire

Isuzu/Luv alternator—diesel engine

6. Remove the bearing retaining screws from the front cover , then remove the bearing.

7. Remove the terminal bolt and nut, then remove the lead wire.

8. Remove the nuts securing the B terminal and diode holder, then remove the screw inside the stator. Separate the stator and rear cover.

NOTE: Observe the position of the insulation washers for reassembly.

9. Remove the stator, then separate the diodes from the stator by melting the solder on the stator coil, diode and 'N' terminal leads. When melting the solder, hold the lead wire with long nose pliers to prevent heat from being transferred to the diodes.

10. Remove the holder plate and brush holder.

11. Melt away the solder on the IC holder plate terminal, then remove the IC regulator assembly.

12. If necessary, the vacuum pump may be disassembled by removing the center plate, exposing the rotor and vane.

Testing & Inspection
ROTOR

1. Using an ohmmeter, check for continuity at the slip end rings. If there is no continuity, replace the rotor.

2. Using an ohmmeter, make an insulation test. Check for continuity between the slip ring and the rotor core. If continuity exists, replace the rotor.

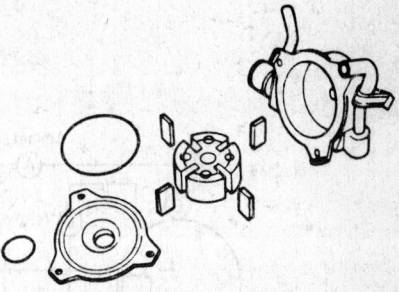

Disassembled view of vacuum pump

3. Measure the slip ring outer diameter for wear. Mimimum diameter is 30mm (1.18 in.).

STATOR

1. Using an ohmmeter, make a continuity test between the stator lead wires. If there is no continuity, replace the stator.

2. Using an ohmmeter. make an insulation test between the stator core and the lead wire. If the continuity exists, replace the stator.

DIODE

1. Using and ohmmeter, perform a continuity test on diodes in both directions.

NOTE: Some ohmmeters use a reverse polarity, in which case continuity will be exactly opposite from the chart above.

3. Replace as necessary.

Across terminals		BTA (Positive side (+) diodes)	
	Tester pin	Positive side	Negative side
U.V.W.	Positive side		No continuity
	Negative side	Continuity	

Across terminals		E (Negative side (–) diodes)	
	Tester pin	Positive side	Negative side
U.V.W.	Positive side		No continuity
	Negative side	Continuity	

Diode test chart

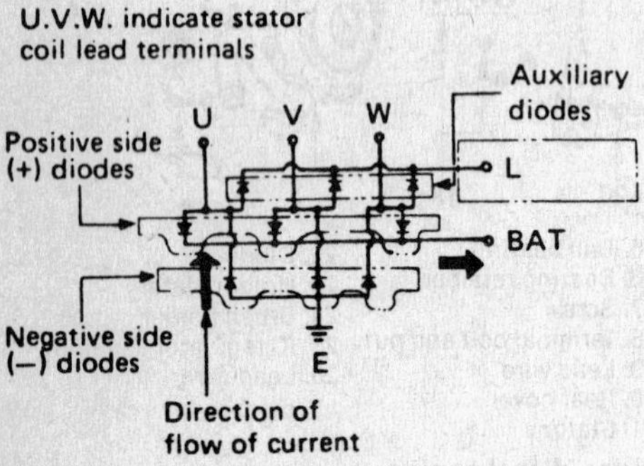

Testing diodes

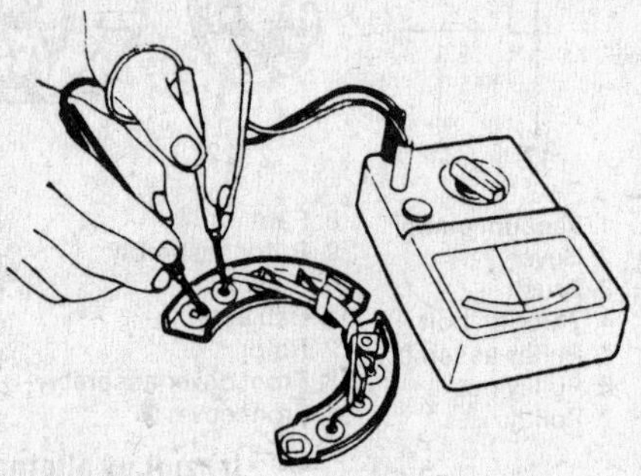

Testing diodes with an ohmmeter

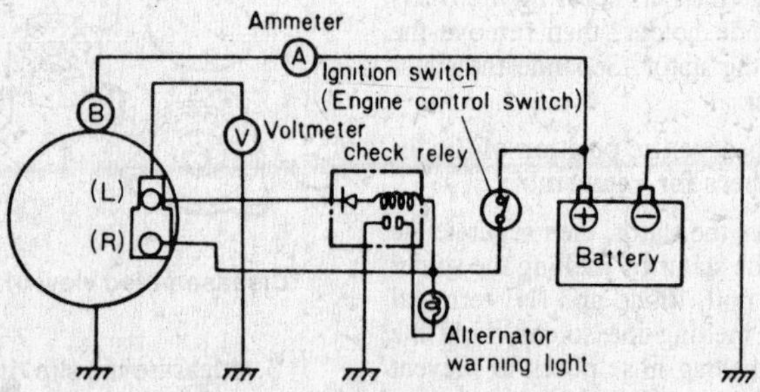

Checking charging system

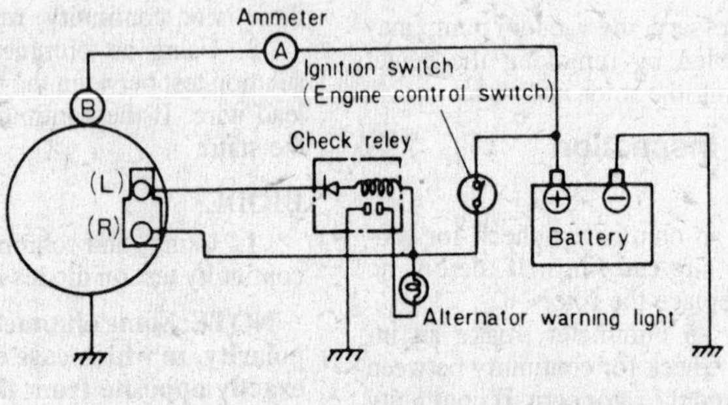

Checking regulated voltage with no-load

BRUSH

1. Check for smooth movement of the brush and clean the brush holder if necessary.
2. Check for brush wear by looking at the wear limit line on the brush and replace if necessary.

ASSEMBLY

1. Assembly of the alternator is the reverse of disassembly with the following instructions: Resolder the IC regulator lead wires. To prevent heat from being transfered to the diodes, use long nose pliers to hold the stator coil leads and diode leads and solder as quickly as posible.
2. Carefully clamp the rotor in a vise and torque the pulley nut to 33–43 ft. lbs.
3. Place some type of guide bar through the holes in the front cover and rear cover flange for alignment, then install the through bolts. Make sure the brush is installed in the brush holder correctly. If the vacuum pump was disassembled, position the rotor, with the serrated boss turned up, on the center plate and housing. Install the vanes into the slits in the rotor. The vanes should be installed with the camfered side turned outward. Install the housing, making sure the O-ring is not projected beyond the slot in the center plate. If the holes in the housing and center plate are not in alignment, adjust by turning the housing slightly, then tighten the three retaining bolts. Add engine oil (around 5 cc) through the filler port, then check that the pulley can be turned smoothly by hand.

Mazda/Courier Alternator

1979–84

Charging System Check

1. Connect an ammeter and voltmeter as shown in the illustration.
2. Turn the ignition switch "OFF".
3. Read the voltmeter connected between the "L" terminal and ground, if the alternator is normal the voltage should be zero (0V).
4. Turn the ignition switch "ON" and read the voltmeter. The voltage should be less than battery voltage, in the 1–3 V range. If the voltage is zero, either the alternator or wiring is bad. If the the voltage is near battery voltage, connect the "F" terminal to ground and read the voltmeter. if the voltage drops lower than the battery voltage, the IC regulator may be faulty.

Checking Regulated Voltage With No Load

1. Disconnect the wiring connected to alternator terminal "B".
2. Connect an ammeter (more than 40 A) between the alternator terminal "B" and the battery positive terminal.

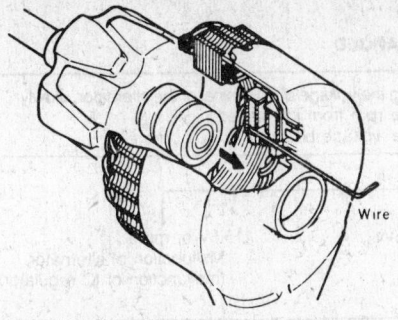

Insert a wire during assembly to avoid brush damage

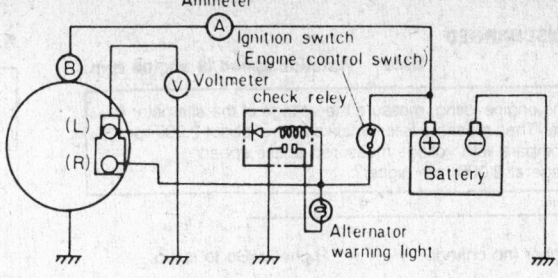

Checking output current

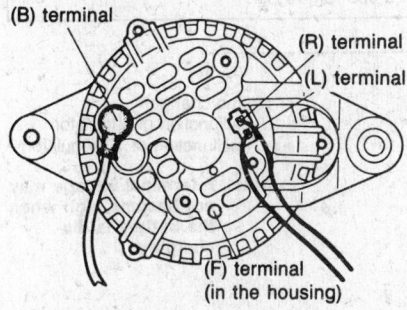

Mazda alternator terminal location

3. Connect a voltmeter between alternator terminal "L" and ground.

4. Start the engine and increase the engine speed to approximately 2,000 rpm. Turn off all unnecessary electric loads and read the value shown on the ammeter.

5. When the amperage in Step 4 is less than 5 amp., read the voltage (regulated voltage) of terminal "L". The regulated should be 14.1–14.7 V.

Checking Output Current

1. Disconnect the wiring connected to the alternator terminal "B".

2. Connect an ammeter (more than 40 amp) betwen the alternator terminal "B" and the battery positive terminal.

3. Start the engine and increase the engine speed to more than 2,500 rpm and read the maximum value shown on the ammeter. Apply all electrical loads, if the value shown on the ammeter is more than 90% of the rated output, the alternator is normal.

DISASSEMBLY

Diesel

1. Remove the vacuum pump.

2. Remove the retaining screw and remove the end cover from the top of the housing.

3. Remove the screws and remove the IC regulator and brush holder.

4. Remove the front and rear bracket assembly.

5. Carefully place the rotor in a vise and remove the pulley nut. Remove the pulley and fan assembly.

6. Remove the front bracket, rotor, rear bracket, and stator.

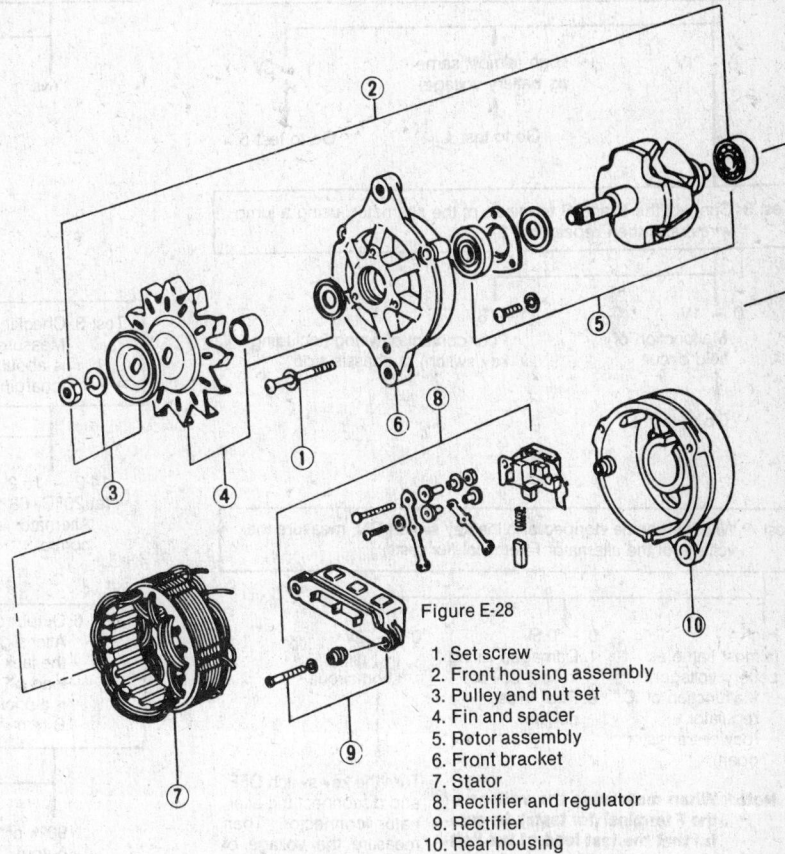

Figure E-28

1. Set screw
2. Front housing assembly
3. Pulley and nut set
4. Fin and spacer
5. Rotor assembly
6. Front bracket
7. Stator
8. Rectifier and regulator
9. Rectifier
10. Rear housing

Mazda Alternator—Gasoline engine—1979–84

Gas

1. Remove the set screw from the front bracket.

2. Separate the front housing assembly.

3. Carefully place the rotor in a vise and loosen the pulley nut, then remove the pulley and fan assembly.

4. Remove the rotor assembly, then the front bracket.

5. Disconnect the soldered portion of the stator lead wire with a soldering iron, then separate the stator and rectifier.

NOTE: This operation should be accomplished as quickly as possible as high temperature from the soldering iron could damage the rectifier.

6. Disconnect the brush holder and the IC regulator assembly from the rectifier with a soldering iron.

7. Assembly is the reverse order of disassembly. When mounting the rotor on the rear bracket, insert a wire through the hole in the rear bracket to avoid damage to the brushes.

1986

DISASSEMBLY

1. Place a soldering iron on the bearing box for approximately 3–4 minutes to heat it, then pull out the three bolts and insert a flat tip tool between the stator and front bracket and eleparate them.

ON-VEHICLE INSPECTION

A. BATTERY DISCHARGED

Note   Rotation speed is engine rpm.

Test 1: With the engine idling, measure the voltage of the alternator B terminal. Then measure B terminal voltage at about 2,000 rpm, and compare with voltage measured at idle speed.
Is voltage at 2,000 rpm higher?

Not higher (no change)   Higher →Go to test 5.

Test 2: With the engine stopped and the key switch ON, measure the voltage of the alternator L terminal. (To check for field current.)

0 ~ 1V   High (almost same   1 ~ 3V
                as battery voltage)
                        ↓   ↓
                 Go to test 4.   Go to test 5.

Test 3: Connect the B and R terminals of the alternator, using a jump wires, and then repeat test 2.

0 ~ 1V   1 ~ 3V
Malfunction of   Poor contact of wiring (including
field circuit   key switch) at chassis side
    ↓
Go to test 4

Test 4: With the engine stopped and the key switch ON, measure the voltage of the alternator F terminal (for tests).

High   0 ~ 0.5V   0.5 ~ 2V
(almost same   1. Damaged wiring   (+) diode
battery voltage)   in field coil or   short-circuit
  Malfunction of IC   2. Poor brush
  regulator   contact
  (power transistor
  open)

Note   When measuring the voltage of
the F terminal (for tests), be care-
ful that the test lead of the volt-
meter doesn't touch the
alternator housing.

Turn the key switch OFF, and disconnect the alternator connector. Then measure the voltage of the L terminal, and, if a voltage of about equal to battery voltage appears, the (+) diode is grounded.

Test 5: Output current test
Disconnect the wiring connected to the alternator B terminal, and connect an ammeter (60A or more) between the wiring and the terminal.
Start the engine, and then suddenly increase the rpm to 2,500 ~ 3,000 rpm and immediately read the maximum indication of the ammeter. (All electric loads should be left ON.)

70% or less of   70% ~ 90% of   90% or more of
nominal output   nominal output   nominal output
Malfunction of   Alternator is
alternator   normal
                        Go to test 6.

B. BATTERY OVERCHARGED

Test 1: While measuring the voltage of B terminal of the alternator, slowly increase engine rpm from idle speed.
Does B terminal voltage become 15.5V or more?

Less than 15.5V.   15.5V or more
                                Malfunction of alternator
                                (malfunction of IC regulator)

Test 2: Measure and compare the voltage at F terminal of the IC regulator during idling and at about 3,000 rpm.
Is the voltage at 3,000 rpm higher?

Yes   No (no change)
                                Malfunction of alternator
                                (malfunction of IC regulator)

                                Note   F terminal voltage may
                                not become high when
                                the battery is dis-
                                charging.

Test 3: Checking the adjustment voltage
Measure the alternator L terminal voltage when the engine speed is about 2,500 rpm. The battery must be fully charged (charging voltage 5A or less at 2,500 rpm).

14.2 ~ 15.2V   Abnormal
(at 20°C, 68°F)   Malfunction of
Alternator is   alternator.
normal.

Test 6: Output current test (re-check)
After slightly discharging the battery, repeat test 6. (In other words, the lack of output current flow may be because of a small flow load on load current.) In addition, also carefully check if there is a poor contact somewhere in the wiring between the alternator B terminal and the battery (+) terminal.

90% or less of nominal   90% or more of nominal
output   output
Remove the alterna-   Alternator is
tor from the engine   normal.
and inspect it
carefully.

Test 7: Adjustment voltage test (after test 6 or test 7, if needed)
Disconnect the cable from the battery (+) terminal and connect an ammeter (50A or more) between the cable and the battery.
Before starting the engine, connect the ammeter terminals so that starter current will flow to the ammeter.
Start the engine and read the ammeter indication (charging current) at about 2,500 rpm.
If it is less than 5A, measure the L terminal voltage at that time. This is the adjustment voltage.
If it is 5A or more, either charge for a while until it becomes less than 5A, or else replace with a fully charged battery and measure once again.

14.2 ~ 15.2V   Abnormal
(At 20°C, 68°F)   Malfunction of
IC regulator is   alternator
normal.

NOTE: The bearing box must be heated or the bearing cannot be pulled out.

2. Separate the front and rear sections being careful not to lose the stopper spring that fits around the circumference of the rear bearing.

3. Remove the pulley nut then disassemble the pulley, rotor and front bracket.

4. The rear bearing can be removed by using a bearing puller.

5. Remove the nut of the B terminal and the insulation bushing. Remove the rectifier retaining screws and the brush holder retaining screw, and then separate the rear bracket and stator.

6. Remove the IC regulator.

7. Remove the solder from the rectifier and stator leads.

NOTE: Do not use the soldering iron for more than 5 seconds as the rectifier may be damaged if overheated.

8. The brush may be removed by removing the solder from the pigtail.

Testing & Inspection

ROTOR

1. Using an ohmmeter, check for continuity at the slip end rings. If there is no continuity, replace the rotor.

2. Using an ohmmeter, make an insulation test. Check for continuity between the slip ring and the rotor core. If continuity exists, replace the rotor.

3. Measure the slip ring outer diameter for wear. Mimimum diameter is 30mm (1.18 in.).

STATOR

1. Using an ohmmeter, make a continuity test between the stator lead wires. If there is no continuity, replace the stator.

2. Using an ohmmeter. make an insulation test between the stator core and the lead wire. If the continuity exists, replace the stator.

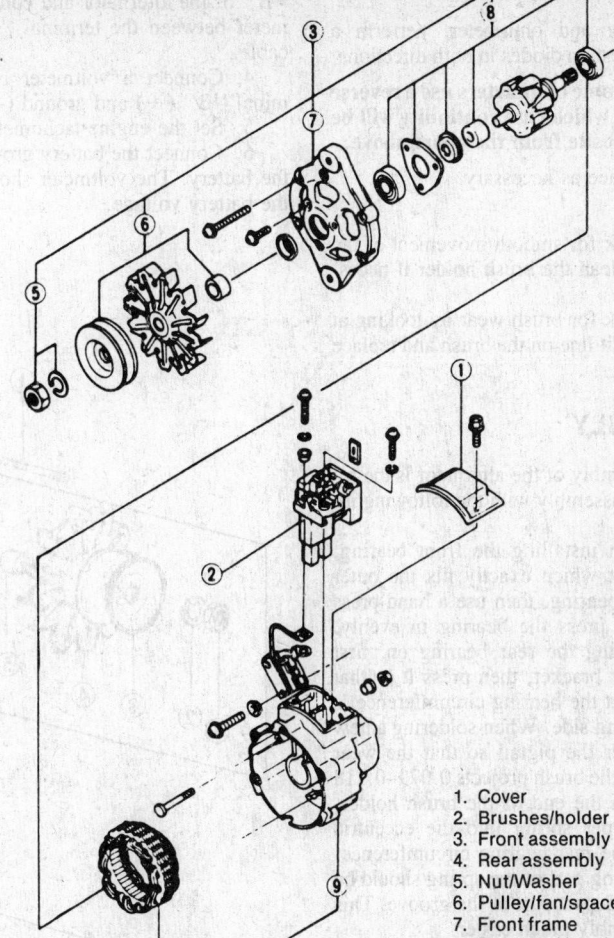

1. Cover
2. Brushes/holder
3. Front assembly
4. Rear assembly
5. Nut/Washer
6. Pulley/fan/spacer
7. Front frame
8. Rotor and bearings
9. Rear housing
10. Stator

Mazda Alternator—Diesel engine—1979–84

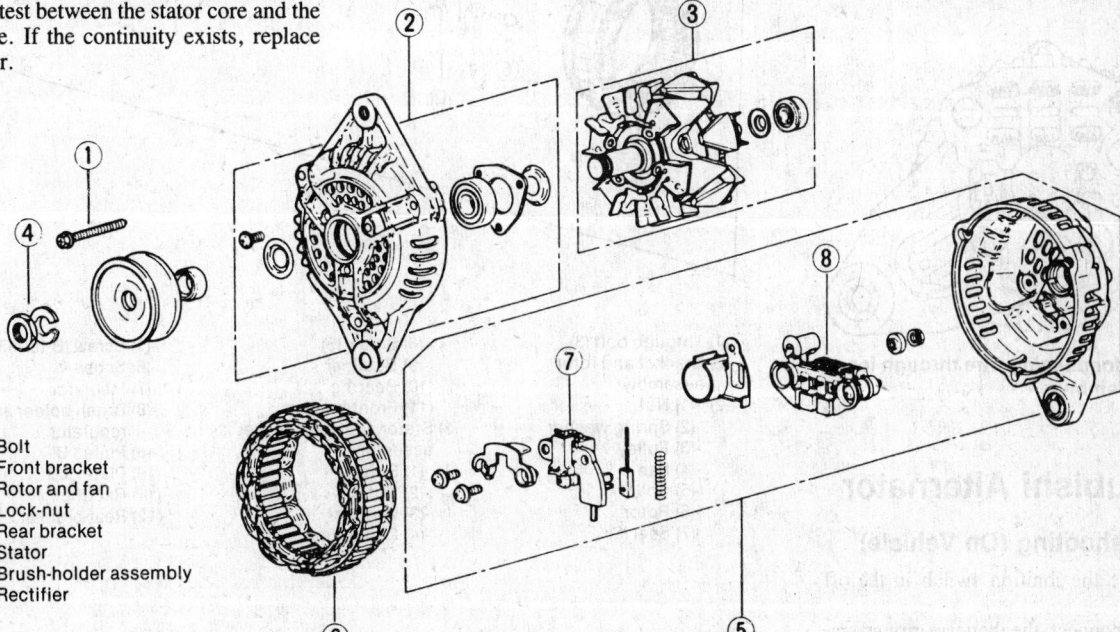

1. Bolt
2. Front bracket
3. Rotor and fan
4. Lock-nut
5. Rear bracket
6. Stator
7. Brush-holder assembly
8. Rectifier

Mazda alternator—1986

DIODE

1. Using and ohmmeter, perform a continuity test on diodes in both directions.

NOTE: Some ohmmeters use a reverse polarity, in which case continuity will be exactly opposite from the chart above.

2. Replace as necessary.

BRUSHES

1. Check for smooth movement of the brush and clean the brush holder if necessary.

2. Check for brush wear by looking at the wear limit line on the brush and replace if necessary.

ASSEMBLY

1. Assembly of the alternator is the reverse of disassembly with the following instructions:

2. When installing the front bearing, use a socket which exactly fits the outer race of the bearing, then use a hand press or vise and press the bearing in evenly. When pressing the rear bearing on, first heat the rear bracket, then press it so that the groove at the bearing circumference is at the slip ring side. When soldering a new brush, solder the pigtail so that the wear limit line of the brush projects 0.079–0.118 in. out from the end of the brush holder. Fit the stopper spring into the eccentric groove of the rear bearing circumference. The protruding part of the spring should be fit into the deepest part of the groove. This makes assembly much easier.

3. Before assembly, use a finger to push the brush into the brush holder, then pass a wire through the hole shown and secure the brush into position. After reassembly, manually turn the pulley to make sure that the rotor turns easily.

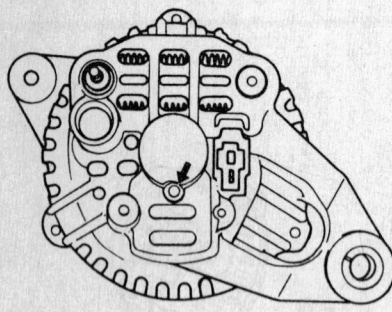

Hole for pushing wire through for reassembly

Mitsubishi Alternator

Troubleshooting (On Vehicle)

1. Place the ibnition switch in the off position.

2. Disconnect the battery ground cable.

3. Disconnect the cable from terminal "B" of the alternator and connect an ammeter between the terminal "B" and the cable.

4. Connect a voltmeter between terminal "B" (+) and ground (-).

5. Set the engine tachometer.

6. Connect the battery ground cable to the battery. The voltmeter should indicate the battery voltage.

7. Start the engine.

8. Turn on the lamps, accelerate the engine to the speed specified in the charts at the beginning of the Mitsubishi section and measure the output current. Check it against the chart.

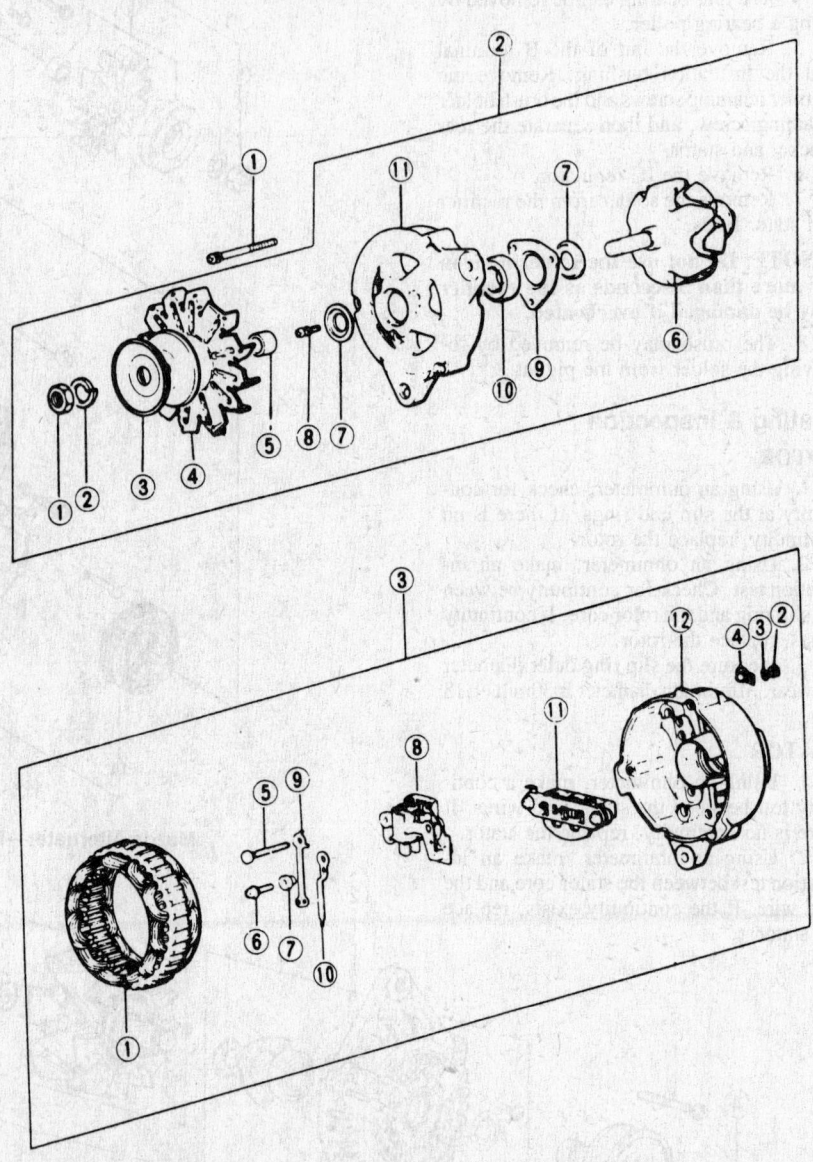

(1) Through bolt (3)	-(8) Screw (3)	-(5) Screw (B-terminal)
(2) Bracket and rotor	-(9) Retainer	-(6) Screw
assembly	-(10) Bearing	-(7) Insulator
(2) -(1) Nut	-(11) Front bracket	-(8) Brush holder and
-(2) Spring washer	(3) Stator and rear bracket	regulator
-(3) Pulley	assembly	-(9) Plate "B"
-(4) Fan	(3) -(1) Stator	-(10) Plate "L"
-(5) Collar	-(2) Nut	-(11) Rectifier
-(6) Rotor	-(3) Washer	-(12) Rear bracket
-(7) Seal (2)	-(4) Condenser	

Mitsubishi/D-50 Arrow alternator—gasoline engine

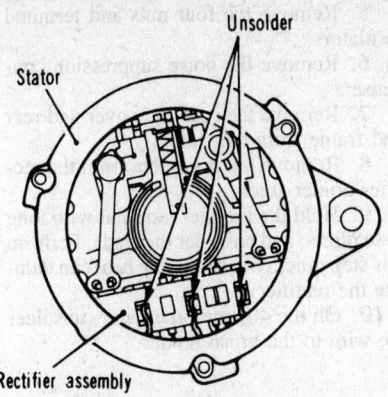

Unsolder the three wires to remove the stator assembly

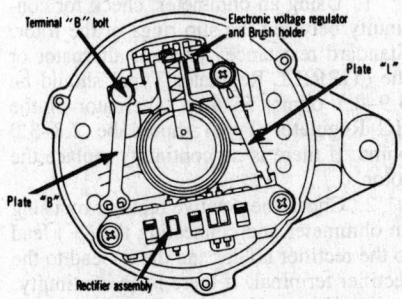

Location of rectifier and brush holder

DISASSEMBLY

Gas

1. Remove alternator from vehicle.
2. Remove the three through bolts from the alternator body.
3. Insert an appropriate pry tool between the front bracket and stator (see illustration). Pry the front bracket away from the stator. Remove the front bracket along with the rotor.

NOTE: If the Tool is inserted too deeply, the stator coil might be damaged.

4. Hold the rotor in a vise and remove the pulley nut. Then remove the pulley, fan, spacer and seal. Remove the rotor from the front bracket and remove the seal.
5. Unsolder the rectifier from the stator coil lead wires and remove the stator assembly.

NOTE: Make sure the solder is removed quickly (in less than five seconds). If a diode is heated to more than 150°C, it might be damaged.

6. Remove the condenser from terminal "B".
7. Unsolder the plates "B" and "L" from the rectifier assembly.
8. Remove the mounting screw and terminal "B" bolt and remove the electronic voltage regulator and brush holder. The regulator and brush holder cannot be separated.
9. Remove the rectifier assembly.

10. When only a brush or brush spring is to be replaced, it is not necessary to remove the stator, etc. Raise the brush holder assembly and unsolder the wire pigtail of the brush and remove the brush.

NOTE: Be very careful when bending the plates "B" and "L" so as not to disturb the rectifier moulding.

Testing & Inspection

1. Check the outside circumference of the slip ring for dirtiness and roughness. Clean or polish with armature paper, if required. A badly damaged slip ring or a slip ring worn down beyond the service limit should also be replaced. The service limit for the slip ring outside diameter is 1.268 in.
2. Check for continuity between the field coil and slip ring. If there is not continuity, the field coil is defective and the rotor must be replaced.
3. Check for continuity between the slip ring and the shaft (or core). If there is continuity, the rotor assembly must be replaced.
4. Check for continuity between the leads of the stator coil. If there is no continuity, the stator coil is defective.
5. Check for an open circuit between the stator coil leads and the stator core. If there is continuity between the stator core and the coil leads, the stator assembly must be replaced.

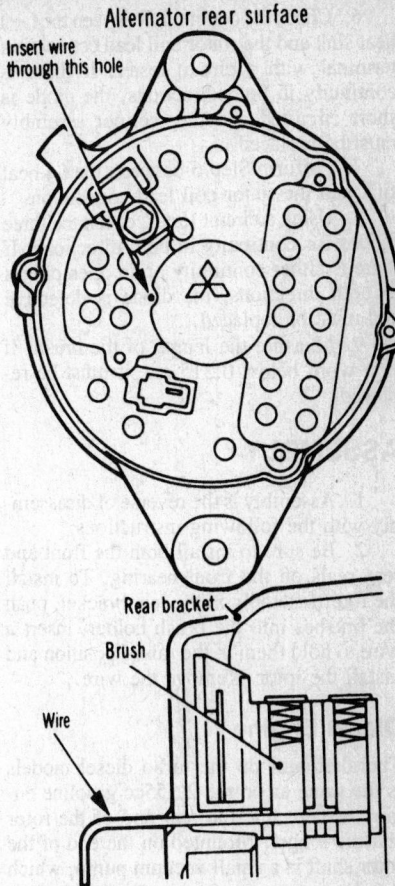

Insert a wire to hold the brushes

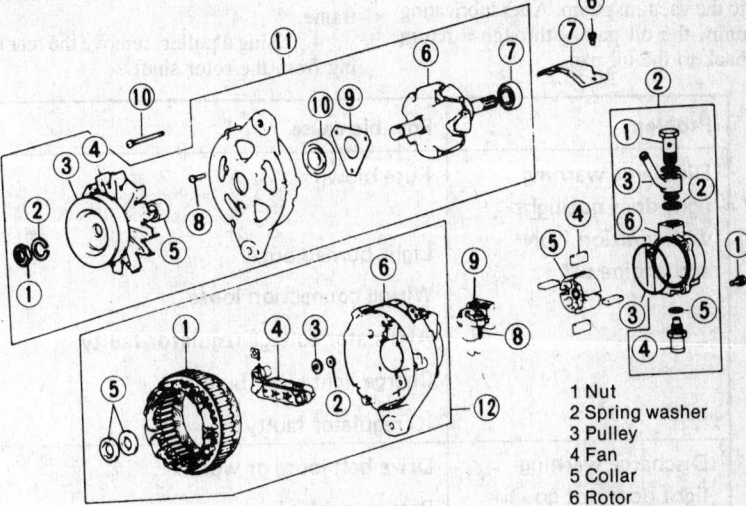

3. O-ring	1 Nut
4. Vane	2 Spring washer
1. Bolt (3)	5. Vacuum pump rotor	3 Pulley
2. Vacuum housing	6. Screw (2)	4 Fan
assembly	7. Cover	5 Collar
1 Check valve	8. Brush holder	6 Rotor
2 Gasket (2)	9. Electronic voltage	7 Rear ball bearing
3 Nipple	regulator	8 Screw (3)
4 Joint	10. Through bolt (3)	9 Bearing retainer
5 Gasket	11. Front bracket and rotor	10 Front ball bearing
6 Vacuum pump housing	assembly	11 Front bracket
		12. Stator and rear bracket
		1 Stator
		2 Nut (2)
		3 Insulator (2)
		4 Rectifier
		5 Oil seal set
		6 Rear bracket

Mitsubishi/D-50 Arrow alternator—diesel engine

6. Check for continuity between the (+) heat sink and the stator coil lead connection terminal with a circuit tester. If there is continuity in both directions, the diode is short circuited and the rectifier assembly must be replaced.

7. Perform Step 6 between the (-) heat sink and the stator coil lead connection.

8. Using a circuit tester, check the three diodes for continuity in both directions. If there is either continuity or an open circuit in both directions, the diode is defective and must be replaced.

9. Measure the length of the brush. If it is worn below 0.315 in., it must be replaced.

ASSEMBLY

1. Assembly is the reverse of disassembly with the following instructions:

2. Be sure to install both the front and rear seals on the front bearing. To install the rotor assembly in the rear bracket, push the brushes into the brush holder, insert a wire to hold them in the raised position and install the rotor. Remove the wire.

Diesel Engine

The alternator on the turbo diesel models is the same as on the 2,555cc gasoline engine, except that the rear end of the rotor shaft is longer. Mounted on the end of the rotor shaft is a small vacuum pump, which generates vacuum for the brake booster. The vacuum pump is a vane type. Oil for lubricating the pump passes through an oil hose from the oil filter bracket, and is supplied to the vacuum pump. After lubricating the pump, the oil passes through a return hose back to the oil pan.

Follow the Overhaul procedure above for gasoline engine alternators. The vacuum pump housing unbolts from the rear of the alternator housing. Always replace the O-ring. Grease the O-ring groove before installing the ring. Follow the following precautions below when assembling the vacuum pump.

1. Install the pump vanes with the rounded end outward. Any vane with worn or chipped ends should be replaced.

2. Make sure no foreign material is present when installing the pump. Any dirt can cause the pump to seize.

3. When the pump housing is installed, push it lightly in the direction of the arrow (illustration) to minimize clearance at point "A". Tighten the three bolts evenly. Pump performance is affected by the way the housing is installed.

4. Fill the pump with clean engine oil before bench testing or running the alternator. Do not operate the alternator without first lubing the pump.

Toyota Alternator

1979–84
DISASSEMBLY

1. Remove the three through bolts and pry the drive end from the stator. Do not pry the coil wires.

2. Place the drive end frame in a soft jaw vice, and remove the pulley nut spring washer, space collar, pulley, fan and space collar.

3. Remove the rotor from the drive end frame.

4. Using a puller, remove the rear bearing from the rotor shaft.

5. Remove the four nuts and terminal insulators.

6. Remove the noise suppression condenser.

7. Remove the rear end cover and rear end frame from the stator.

8. Remove the insulators from the rectifier holder studs.

9. Hold the rectifier terminal with long nose pliers, and unsolder the leads. Perform this step quickly as excessive heat can damage the rectifier.

10. On the 40 amp. alternator, unsolder the wire to the brush holder.

Testing & Inspection

1. Using an ohmmeter, check for continuity between the slip rings of the rotor. Standard resistance: 40 amp. alternator or the (TIRRILL Regulator Type) should be 3.9–4.1 ohms; 55 amp. alternator or the (IC Regulator Type) should be 2.8–3.0 ohms. If there is no continuity replace the rotor.

2. Check the positive rectifier by using an ohmmeter and connecting the (+) lead to the rectifier holder and the (-) lead to the rectifier terminal. If there is no continuity, replace the rectifier assembly with brush.

3. Check the negative rectifier by using an ohmmeter and connecting the (+) lead to the rectifier terminal and the (—) lead to the rectifier holder. If there is no continuity, replace the rectifier with brush.

4. Using an ohmmeter, connect the (+) lead to the (4) lead of the field diodes and the negative lead to the (1,2,3) leads of the field diodes (see illustration). If there is no continuity, replace the rectifier assembly with brush.

Problem	Possible cause	Remedy
Discharge warning light does not light with ignition "ON" and engine off	Fuse blown	Check "GAUGES" and "ENGINE"* fuses
	Light burned out	Replace light
	Wiring connection loose	Tighten loose connections
	Alternator voltage regulator faulty	Regulator
	Charge light relay faulty*	Check relay
	IC regulator faulty*	Replace IC regulator
Discharge warning light does not go out with engine running (battery requires frequent recharging)	Drive belt loose or worn	Adjust or replace drive belt
	Battery cables loose, corroded or worn	Repair or replace cables
	Fuse blown	Check "ENGINE" fuse
	Fusible link blown	Replace fusible link
	Alternator voltage regulator, charge light relay*, IC regulator* or alternator faulty	Check charging system
	Wiring faulty	Repair wiring

*IC Regulator Type only

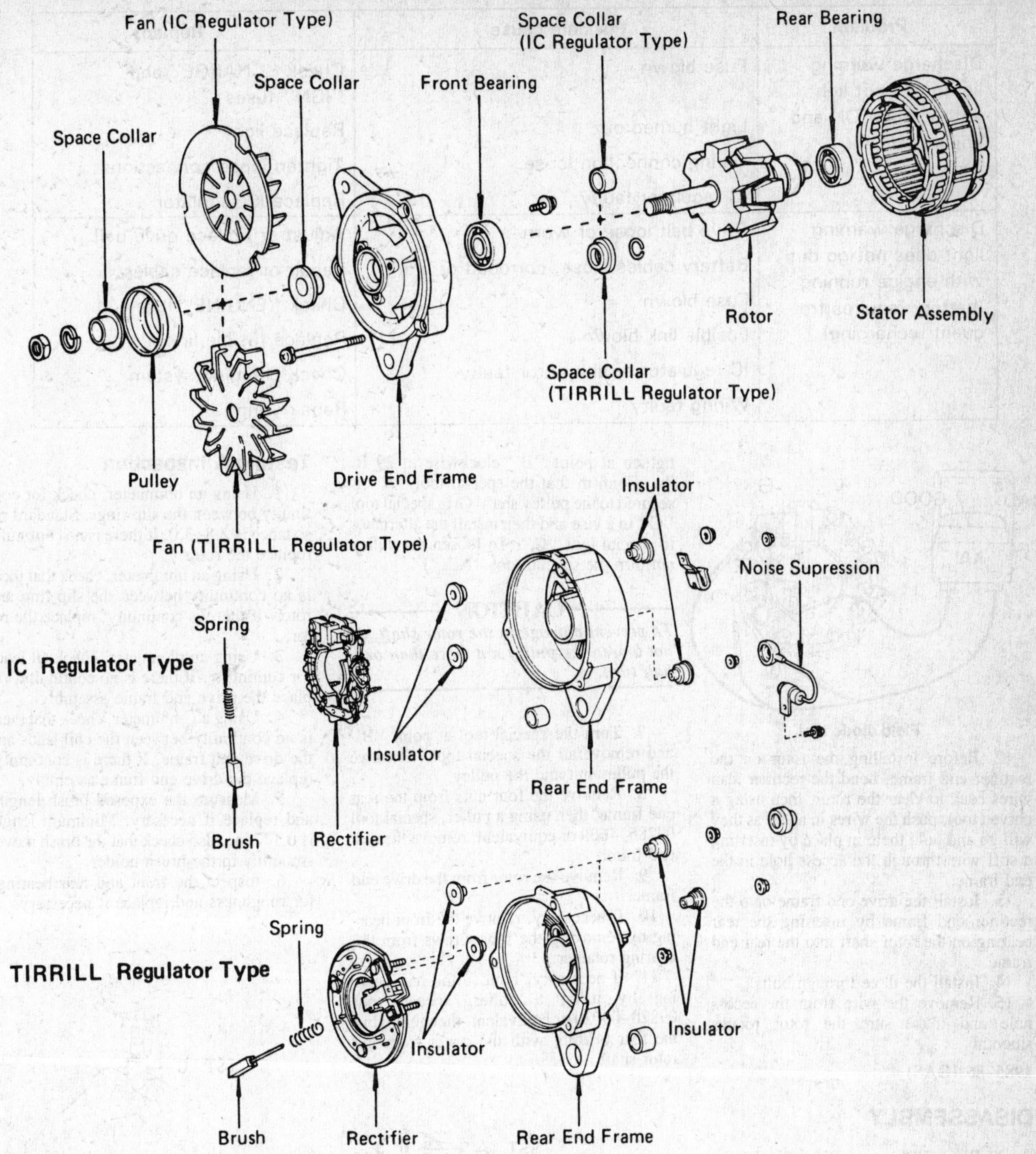

Fan (IC Regulator Type)

Space Collar

Space Collar
(IC Regulator Type)

Rear Bearing

Space Collar

Front Bearing

Space Collar

Rotor

Stator Assembly

Space Collar
(TIRRILL Regulator Type)

Pulley

Drive End Frame

Fan (TIRRILL Regulator Type)

Insulator

Noise Supression

IC Regulator Type

Spring

Insulator

Rear End Frame

Brush

Rectifier

Spring

TIRRILL Regulator Type

Insulator

Insulator

Brush

Rectifier

Rear End Frame

Toyota alternator—1979-84

ASSEMBLY

1. Hold the rectifier terminal with long nose pliers and solder each stator lead to the rectifier. Protect the rectifiers from heat.

2. Place the two insulators on the positive side of the rear end frame.

3. Install the rear end frame on the rectifier holder and check that the wires are not touching the case.

4. Place the rear end cover on the rear end frame.

5. Place the two insulators on the positive side studs.

6. Install the four nuts on the studs.

7. On the 55 amp. alternator or the (IC Regulator Type), mount the noise surpression condenser on the stud and connect the lead wire to the "B" terminal of the alternator.

8. Install the front bearing in the drive end frame, then install the three screws.

9. Press the rear bearing onto the rotor shaft.

10. Slide the spacer collar, then the drive end frame onto the rotor shaft.

11. Place the rotor in a soft jaw vice, then slide the spacer, fan, pulley and spacer collar on the rotor shaft. Tighten the nut to 37-47 ft. lbs.

Problem	Possible cause	Remedy
Discharge warning light does not light with ignition ON and engine off	Fuse blown	Check "CHARGE" and "IGN" fuses
	Light burned out	Replace light
	Wiring connection loose	Tighten loose connections
	IC regulator faulty	Replace IC regulator
Discharge warning light does not go out with engine running (battery requires frequent recharging)	Drive belt loose or worn	Adjust or replace drive belt
	Battery cables loose, corroded or worn	Repair or replace cables
	Fuse blown	Check "ENGINE" fuse
	Fusible link blown	Replace fusible link
	IC regulator or alternator faulty	Check charging system
	Wiring faulty	Repair wiring

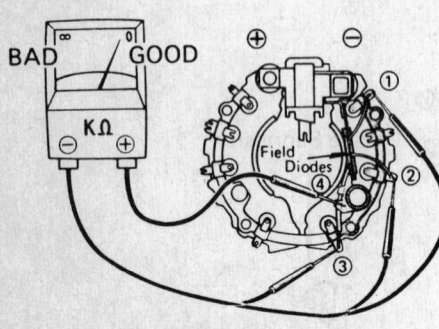

Field diode test

12. Before installing the rotor on the rectifier end frame, bend the rectifier lead wires back to clear the rotor, then using a curved tool, push the wires in as far as they will go and hold them in place by inserting a stiff wire through the access hole in the end frame.

13. Install the drive end frame onto the rectifier end frame by inserting the rear bearing on the rotor shaft into the rear end frame.

14. Install the three through bolts.

15. Remove the wire from the access hole and make sure the rotor rotates smoothly.

1985–86 (GAS)

DISASSEMBLY

1. Remove the nut and terminal insulator, then remove the three nuts and end cover.

2. Remove the two screws then remove the brush holder and brush holder cover.

3. Remove the three screws and remove the IC regulator.

4. Remove the four screws and remove the rectifier holder.

5. Remove the terminal insulator.

6. To remove the pulley and nut the manufacturer recommends the use of Special Tool 09820–63010 and to perform the following procedure: Hold the special tool at point "A" with a torque wrench, then

tighten at point "B" clockwise to 29 ft. lbs. Confirm that the special tool "A" is secured to the pulley shaft. Grip special tool "C" in a vise and then install the alternator to special tool "C". To loosen the pulley nut turn the special tool "A".

—— **CAUTION** ——
To prevent damage to the rotor shaft, do not loosen the pulley nut more than one-half turn.

7. Turn the special tool at point "B" and remove all the special tools. Remove the pulley nut and the pulley.

8. Remove the four nuts from the rear end frame, then using a puller, special tool 09286–46011 or equivalent, remove the rear end frame.

9. Remove the rotor from the drive end frame.

10. If necessary, remove the front bearing by removing the four screws from the bearing retainer.

11. If necessary, remove the rear bearing by using a puller, special tool 09820–00020 or equivalent, then removing the rear bearing with the cover from the rotor shaft.

Testing & Inspection

1. Using an ohmmeter, check for continuity between the slip rings. Standard resistance is 2.8–3.0. If there is no continuity replace the rotor.

2. Using an ohmmeter, check that there is no continuity between the slip ring and rotor. If there is continuity, replace the rotor.

3. Using an ohmmeter, check all leads for continuity. If there is no continuity, replace the drive end frame assembly.

4. Using an ohmmeter, check that there is no continuity between the coil leads and the drive end frame. If there is continuity, replace the drive end frame assembly.

5. Measure the exposed brush length, and replace if necesary. Minimum length is 0.177 in. Also check that the brush moves smoothly in the brush holder.

6. Inspect the front and rear bearings for roughness and replace if necessary.

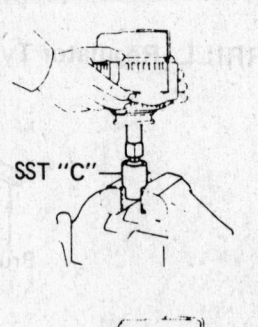

SST "C"

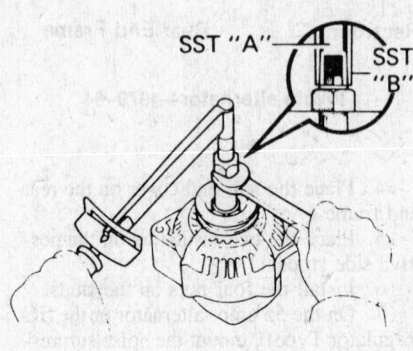

SST "A"   SST "B"

Pulley removal

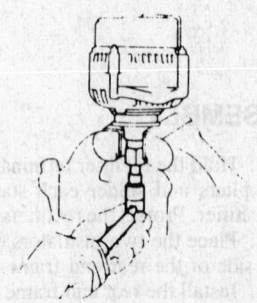

Pulley removal

Pulley Drive End Frame Front Bearing Rotor

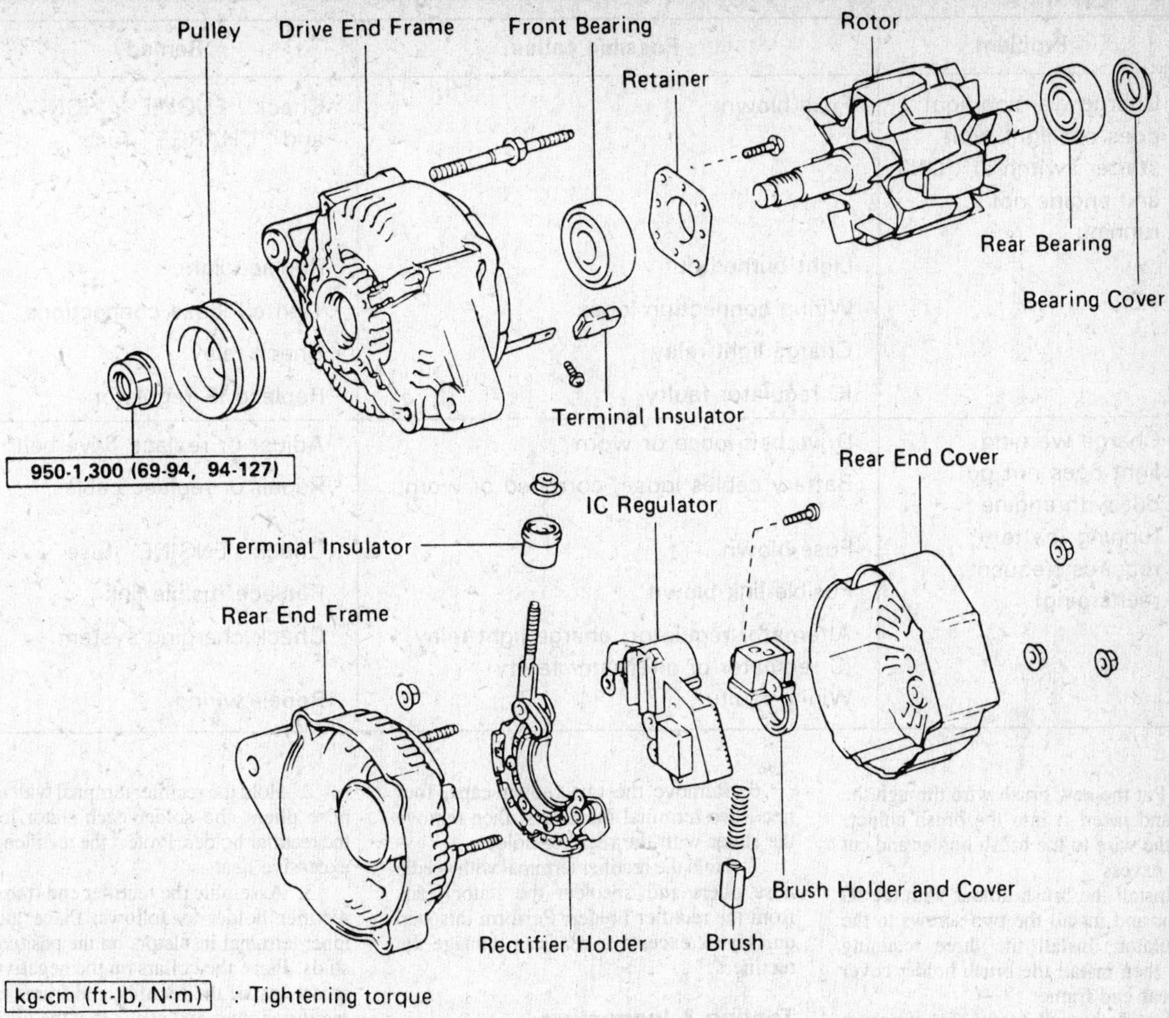

Retainer

Rear Bearing

Bearing Cover

950-1,300 (69-94, 94-127)

Terminal Insulator

Rear End Cover

Terminal Insulator

IC Regulator

Rear End Frame

Brush Holder and Cover

Rectifier Holder Brush

kg-cm (ft-lb, N·m) : Tightening torque

Toyota alternator—gas engine, 1985 and later

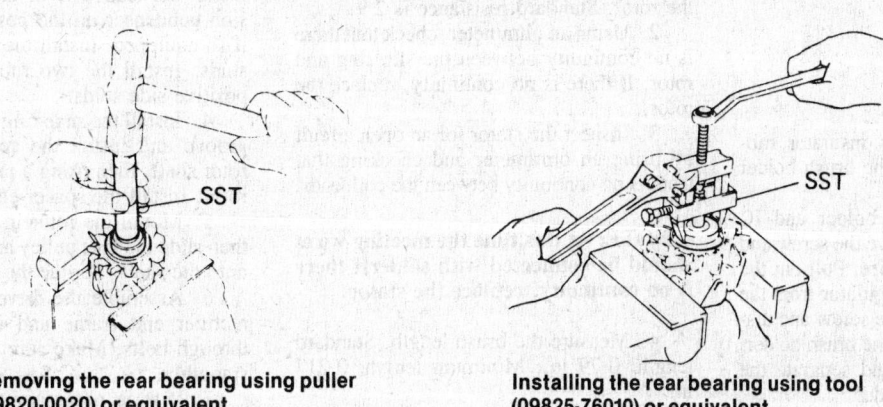

**Removing the rear bearing using puller
(09820-0020) or equivalent**

SST

**Installing the rear bearing using tool
(09825-76010) or equivalent**

SST

ASSEMBLY

1. If it is necessary to replace the rear bearing, use special tool 09285-76010 or equivalent and install the rear bearing and cover onto the rotor shaft.

2. Install the rotor to the drive end frame.

3. Using a plastic hammer, lightly tap the rear end frame on the drive end frame and install the four nuts.

4. The manufacturer recommends installing the pulley in the following manner using the special tools or their equivalents as indicated: Install the pulley to the rotor shaft and tighten the pulley nut by hand. Using special tool 09820-63010, hold at point "A" with a torque wrench and tighten tool "B" clockwise to 29 ft. lbs. Confirm tool "A" is secured to the pulley shaft. Grip special tool "C" in a vise and then

install the alternator to tool "C". To torque the pulley nut, turn tool "A" and tighten to 69–94 ft. lbs. Turn tool "B" and remove all the special tools.

5. Install the four terminal insulators on the lead wires.

6. Install the rectifier holder with the four screws.

7. If it is necessary to install a new brush, unsolder and remove the brush and

Problem	Possible cause	Remedy
Charge warning light does not light with starter switch at "ON" and engine not running	Fuse blown	Check "ENGINE", "IGN" and "CHARGE" fuse
	Light burned out	Replace light
	Wiring connection loose	Tighten loose connections
	Charge light relay	Check relay
	IC regulator faulty	Replace IC regulator
Charge warning light does not go out with engine running (battery requires frequent recharging)	Drive belt loose or worn	Adjust or replace drive belt
	Battery cables loose, corroded or worn	Repair or replace cables
	Fuse blown	Check "ENGINE" fuse
	Fusible link blown	Replace fusible link
	Alternator regulator, charge light relay, IC regulator or alternator faulty	Check charging system
	Wiring faulty	Repair wiring

spring. Put the new brush wire through the spring and insert it into the brush holder. Solder the wire to the brush holder and cut off any excess.

8. Install the brush holder with the IC regulator and install the two screws to the IC regulator. Install the three retaining screws, then install the brush holder cover to the rear end frame.

9. Install the end cover with the three retaining nuts and install the terminal insulator and nut. Make sure the rotor rotates smoothly.

1985–86 (DIESEL)

DISASSEMBLY

1. Remove the two nuts, insulator, rubber washer, then remove the brush holder cover.

2. Remove the brush holder and IC regulator as follows: Remove the screw and disconnect the blue lead wire. Pull out the brush holder with the IC regulator from the rectifier holder. Remove the screw and disconnect the lead wire from the brush holder. Remove the two screws and separate the IC regulator and brush holder.

3. Remove the three through bolts and remove the end frame with the rotor.

NOTE: It may be necessary to lightly tap the rotor shaft with a plastic hammer.

4. Mount the rotor in a soft jaw vice, and remove the pulley nut, spring washer, puley and fan.

5. Using a socket wrench and press, press out the rotor, spacer ring and collar. Remove the snap ring from the rotor shaft.

6. Remove the two rubber caps, four nuts, two terminal insulators, then remove the stator with the rectifier holder.

7. Hold the rectifier terminal with needle nose pliers and unsolder the stator leads from the rectifier holder. Perform this step quickly as excessive heat can damage the rectifiers.

Testing & Inspection

1. Using an ohmmeter, check that there is no continuity between the slip rings on the rotor. Standard resistance is 2.9.

2. Using an ohmmeter, check that there is no continuity between the slip ring and rotor. If there is no continuity, replace the rotor.

3. Inspect the stator for an open circuit by using an ohmmeter and checking that there is no continuity between the coil leads.

NOTE: At this time the meeting wires should be connected with solderIf there is no continuity, replace the stator.

4. Measure the brush length. Standard length: 0.79 in.; Minimum length: 0.217 in.

5. Inspect the bearing for roughness and replace if necessary.

ASSEMBLY

1. If it is necessary to replace the bearing remove the three retaining bolts, felt cover, felt, retainer, bearing, washer and felt. Insatll the bearing and related parts in the reverse order of removal and if necessary, lightly tap the bearing with a plastic hammer to install.

2. Hold the rectifier terminal with needle nose pliers and solder each stator lead to the rectifier holder. Protect the rectifiers from excessive heat.

3. Assemble the rectifier end frame and rectifier holder as follows: Place the two inner terminal insulators on the positive side studs. Place the collars on the negative side studs. Install the rectifier end frame on the rectifier holder and check that the wires are not touching the case. Place the two outer terminal insulators on the positive side studs. Install the lead wire of the noise suppression condenser on the positive side studs, if so equipped. Install the four nuts on the studs. Install the two rubber caps on the positive side studs.

4. Install the snap ring in the rotor shaft groove and install the space ring on the rotor shaft, then using a press, press in the rotor. Install the spacer collar.

5. Mount the rotor in a soft jaw vice, then slide the fan, pulley and spring washer onto the shaft. Torque the nut to 65 ft. lbs.

6. Assemble the drive end frame and rectifier end frame and install the three through bolts. Make sure the rotor rotates smoothly.

7. If it is necessary to install a new brush, unsolder and remove the brush and spring. Insert the new brush wire through the spring and install the brush in the brush holder. Solder the wire to the brush holder at the standard length of 0.79 in. and cut off the excess wire.

8. Install the IC regulator on the brush holder with the two retaining screws, and connect the white lead wire to the terminal of the IC regulator with the screw. Install the brush holder and blue lead wire with the screw.

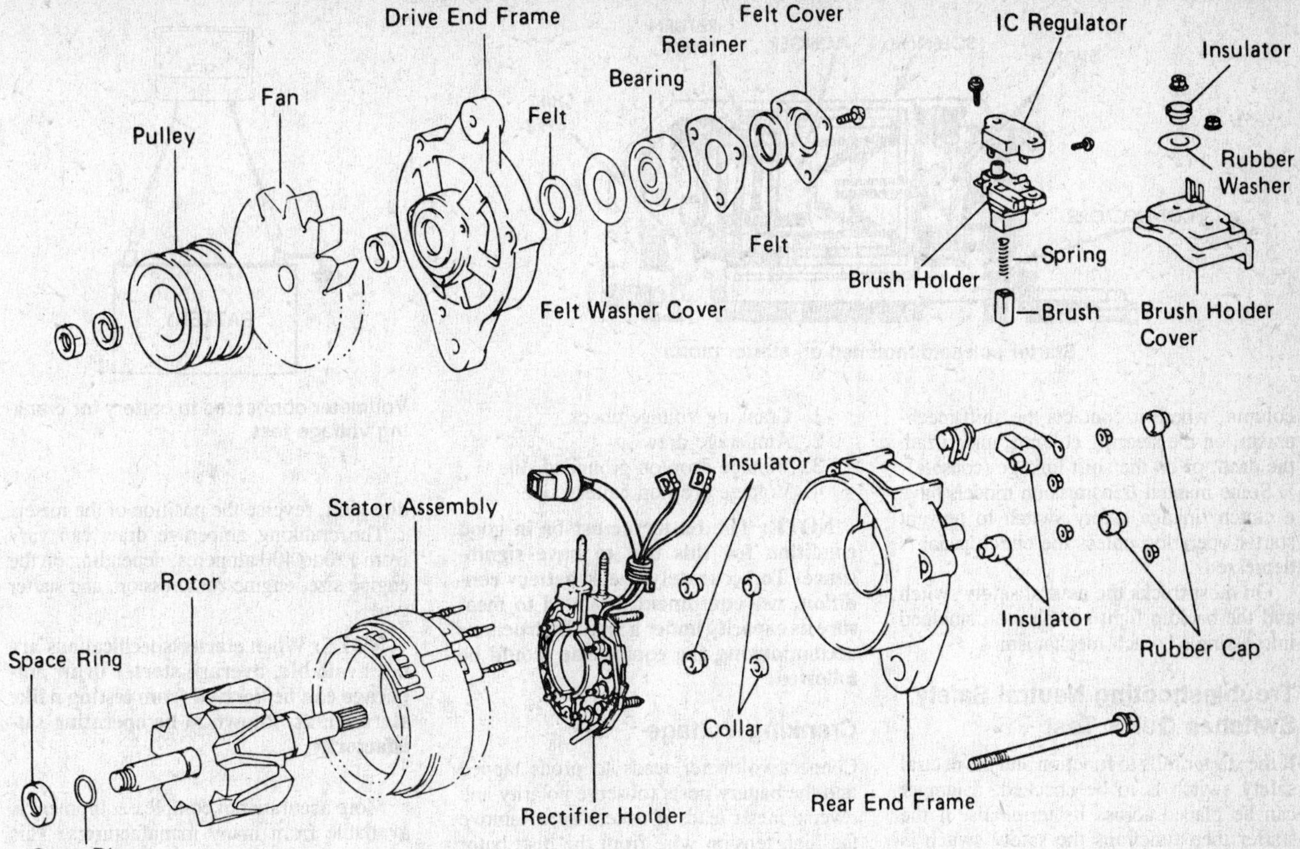

Pulley · Fan · Drive End Frame · Felt · Bearing · Retainer · Felt Cover · IC Regulator · Insulator · Rubber Washer · Spring · Brush Holder · Brush · Brush Holder Cover · Felt Washer Cover · Felt

Space Ring · Rotor · Stator Assembly · Insulator · Insulator · Rubber Cap · Snap Ring · Collar · Rectifier Holder · Rear End Frame

Toyota alternator—diesel engine, 1985 and later

9. Place the cover on the rectifier end frame and place the terminal insulator and rubber washer on terminal B. Install the two retaining nuts.

SWITCHES & SOLENOIDS

Magnetic Switches

Magnetic switches serve only to make contact for the starter motor. Usually, such switches are located on the inner fender panel, although they are found mounted on the starter in a few cases.

Magnetic Switches with Two Control Terminals

On this type of magnetic switch current is supplied from the ignition switch or transmission neutral button to one of the magnetic switch control terminals. The other control terminal is connected to the transmission neutral safety switch (on the transmission) where it is grounded.

Magnetic Switches with Ignition Resistor By-Pass Terminals

All normally use a magnetic switch with a single control terminal. The second terminal is an ignition resistor by-pass terminal.

SOLENOIDS WITHOUT RELAYS

This type of starter solenoid is always mounted on the starter. Makes electrical contact for the starter and pulls the starter and drive clutch into mesh with the flywheel. The Chrysler reduction gear starter has this solenoid embodied in the starter housing.

There is only one control terminal on the solenoid.

The ignition by-pass terminal is usually marked R or IGN, if it is used.

SOLENOIDS WITH SEPARATE RELAYS

The solenoid itself is always mounted on the starter. In addition to making contact for the starter, it also pulls the starter drive

clutch gear into mesh with the flywheel. A single control terminal is used on the solenoid itself. The relay is usually found mounted to the inner fender panel or on the firewall.

SOLENOIDS WITH BUILT-IN RELAYS

These units are always mounted on the starter and are connected, through linkage, to the starter drive clutch. The relay portion is a square box built into and integral with the front end of the solenoid assembly.

NEUTRAL SAFETY SWITCHES

The purpose of the neutral safety switch is to prevent the starter from cranking the engine except when the transmission is in neutral or park.

On some trucks the neutral safety switch is located on the transmission. It serves to ground the solenoid or magnetic switch, whichever is used.

On other trucks the neutral safety switch is located either at the bottom of the steering

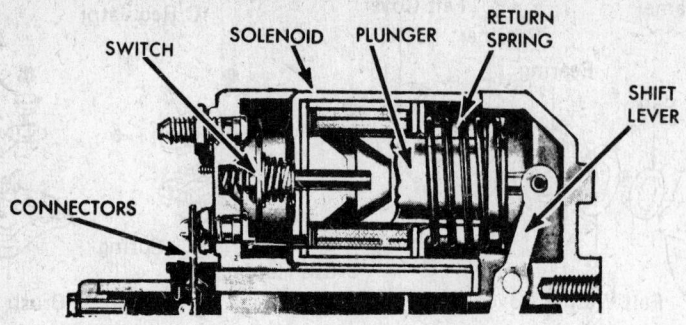

Starter solenoid mounted on starter motor

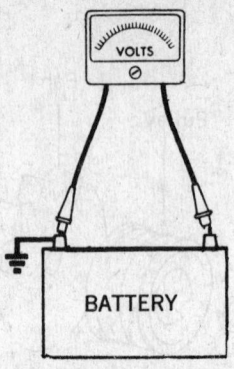

Voltmeter connected to battery for cranking voltage test

column, where it contacts the shift mechanism, on the steering column, underneath the dash, or on the shift linkage (console).

Some manual transmission models have a clutch linkage safety switch to prevent starter operation unless the clutch pedal is depressed.

On most trucks the neutral safety switch and the backup light switch are combined into a single switch mechanism.

Troubleshooting Neutral Safety Switches Quick Test

If the starter fails to function and the neutral safety switch is to be checked, a jumper can be placed across its terminals. If the starter then functions the safety switch is defective.

In the case of neutral safety switches with one wire, this wire must be grounded for testing purposes. If the starter works with the wire grounded, the switch is defective.

Neutral Safety Switch/Back-Up Light Switch

When the neutral safety switch is built in combination with the back-up light switch, the easiest way to tell which terminals are for the back-up lights is to take a jumper and cross every pair of wires. The pair of wires which light the back-up lamps should be ignored when testing the neutral safety switch. Once the back-up light wires have been located, jump the other pair of wires to test the neutral safety switch. If the starter functions only when the jumper is placed across these two wires, the neutral safety switch is defective or requires adjustment.

STARTING SYSTEMS

Starter Motor Testing

TESTING THE STARTER CIRCUIT

The starter circuit should be divided and tested in four separate phases:

1. Cranking voltage check
2. Amperage draw
3. Voltage drop on grounded side
4. Voltage drop on battery side

NOTE: The battery must be in good condition for this test to have significance. To accurately check battery condition, use equipment designed to measure its capacity under a load. Instructions accompanying the equipment should be followed.

Cranking Voltage

Connect voltmeter leads to prods tapped into the battery posts (observe polarity and reverse meter leads if necessary). Remove the high tension wire from the distributor cap and ground it to prevent starting. With electronic ignition, disconnect the control box harness from the distributor. Now, turn the key. Observe both voltmeter reading and cranking speed. The cranking speed should be even, and at a satisfactory rate of speed, with a voltmeter reading of at least 9.6 volts for 12 volt systems.

Amperage Draw

The amount of current the starter motor draws is usually (but not always) associated with the mechanical problems involved in cranking the engine. (Mechanical trouble in the engine, frozen or worn starter parts, misaligned starter or starter components, etc.) Because starter motor amperage draw is directly influenced by anything restricting the free turning of the engine, or starter, it is important that the engine and all components be at operating temperatures.

To measure starter current draw, remove the high tension wire from the center of the distributor cap and ground it. With electronic ignition, disconnect the control box harness from the distributor. A very simple and inexpensive starter current indicator is available at auto stores. This indicator is an induction type gauge and shows, without disconnecting any wires, starter current draw.

Place the yoke of the meter directly over the insulated starter supply cable (cable must be straight for a minimum of 2 in.). Close the starter switch for about 20 seconds, watch the meter dial and record the average reading. If the indicator swings in the wrong

direction, reverse the position of the meter.

The cranking amperage draw can vary from 150 to 400 amperes, depending on the engine size, engine compression, and starter type.

NOTE: When starter specifications are not available, average starter draw amperage can be derived from testing a like starter unit, known to be operating satisfactorily.

More accurate but complex equipment is available from many manufacturers. This equipment consists of a combination voltmeter, ammeter, and carbon pile rheostat. When using this equipment, follow the equipment manufacturer's procedures and recommendations.

High amperage and lazy performance would suggest an excessively tight engine, friction in the starter or starter drive, grounded starter field or armature.

Normal amperage and lazy performance suggest high resistance, or possibly poor connections somewhere in the starter circuit.

Low amperage and lazy or no performance suggest battery condition poor, bad cables or connections along the line.

Voltage Drop on Grounded Side

With a voltmeter on the 3 volt scale, without disconnecting any wires, connect negative test lead of the voltmeter to a prod secured in the grounded battery post. The positive test lead is connected to a cleaned, bare metal portion of the starter motor housing. Close the starter switch and note the voltmeter reading. If the reading is the same as battery reading, the ground circuit is open somewhere between the battery and the starter. In many cases the reading will be very small. The reading shown will indicate voltage drop (loss) between battery ground post and starter housing. The drop should not exceed 0.2 volt. If the voltage drop is above the specified amount, the next step is to isolate and correct the cause. It can be a bad cable or connection anywhere in the battery-to-starter ground circuit. A check

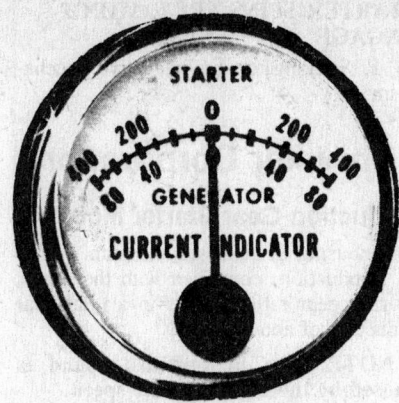

Starter current indicator

of this type should progress along the various points of possible trouble, between the battery ground post and the starter motor housing, until the trouble spot has been located.

Voltage Drop on Battery Side

Bad starter cranking may result from poor connections or faulty components of the battery or hot phase of the starter motor circuit. To check this phase of the circuit, without disconnecting any wires, connect one lead of a voltmeter to a prod secured in the hot post of the battery and the other voltmeter lead to the field terminal of the starting motor. The meter should be set to the 16–20 volt scale. Before closing the starter switch, the voltmeter reading will be that of the battery. After closing the starter switch, change the selector on the voltmeter to the 3 volt scale. With a jumper wire between the relay battery terminal and the relay starter switch terminal, crank the engine. If the starting motor cranks the engine, the relay (solenoid) is operating.

While the engine is being cranked, watch the voltmeter. It should not register more than 0.5 volt. If more than this, check each part of the circuit for voltage drop to isolate the trouble, (high resistance).

Without disturbing the voltmeter-to-battery hook-up, move the free voltmeter lead to the battery terminal of the relay (solenoid), and crank the engine. The voltmeter should show no more than 0.1 volt.

If this reading is correct, move the same voltmeter lead to the starting motor terminal of the relay (solenoid). While the engine is being cranked, the voltmeter should show no more than 0.3 volt. If it does, the trouble lies in the relay.

If the reading is correct, the trouble is in the cable or connections between the relay and the starting motor.

Diagnosis

STARTER WON'T CRANK ENGINE

1. Dead battery.
2. Open starter circuit, such as: Broken or loose battery cables. Inoperative starter

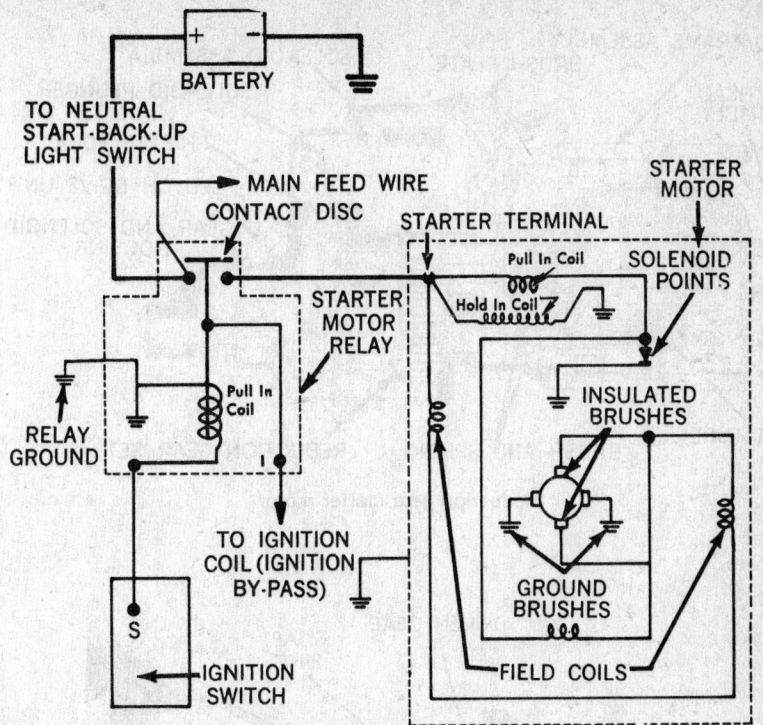

Positive engagement starter circuits

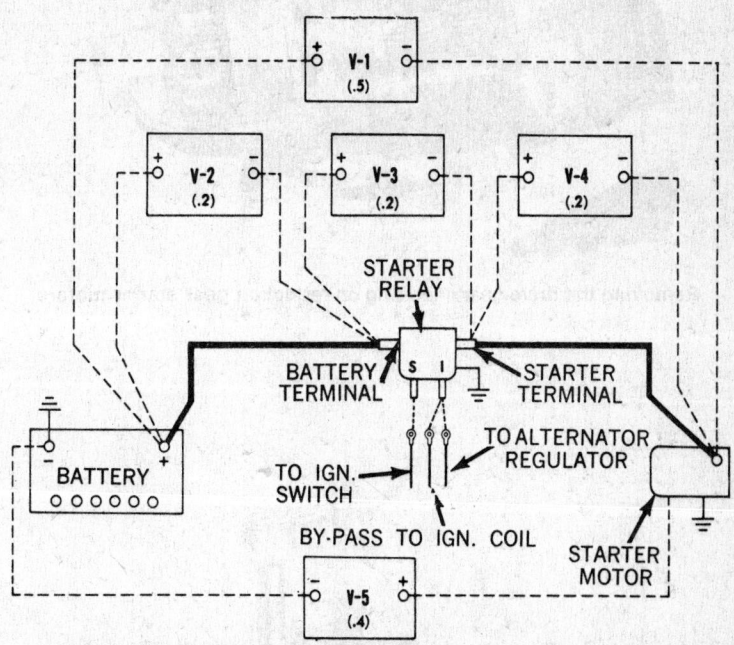

Starter cable resistance tests

motor solenoid. Broken or loose wire from starter switch to solenoid. Poor solenoid or starter ground. Bad starter switch.

3. Defective starter internal circuit, such as: Dirty or burnt commutator. Stuck, worn or broken brushes. Open or shorted armature. Open or grounded fields.

4. Starter motor mechanical faults, such as: Jammed armature end bearings. Bad bearing, allowing armature to rub fields.

Bent shaft. Broken starter housing. Bad starter worm or drive mechanism. Bad starter drive or flywheel driven gear.

5. Engine hard or impossible to crank such as: Hydrostatic lock, water in combustion chamber. Crankshaft seizing in bearings. Piston or ring seizing. Bent or broken connecting rod. Seizing of connecting rod bearing. Flywheel jammed or broken.

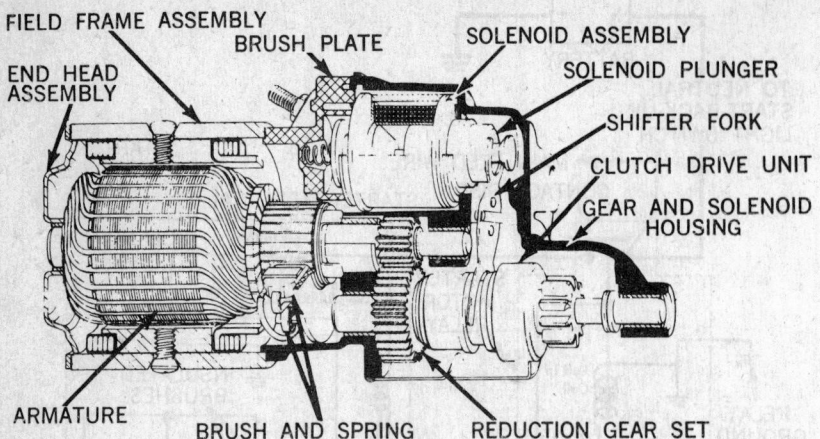

FIELD FRAME ASSEMBLY

BRUSH PLATE

SOLENOID ASSEMBLY

END HEAD ASSEMBLY

SOLENOID PLUNGER

SHIFTER FORK

CLUTCH DRIVE UNIT

GEAR AND SOLENOID HOUSING

ARMATURE

BRUSH AND SPRING

REDUCTION GEAR SET

Reduction gear starter motor

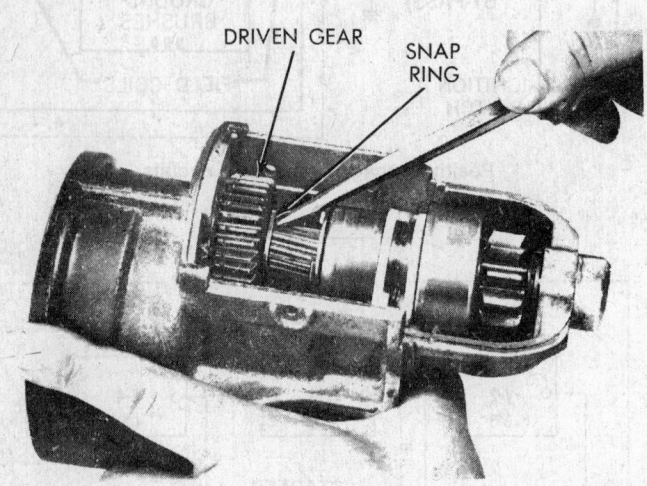

DRIVEN GEAR

SNAP RING

Removing the drive gear snap-ring on reduction gear starter motors

TERMINAL SCREW

Removing the terminal screw on reduction gear starter motors

STARTER SPINS FREE, WON'T ENGAGE

1. Sticking or broken drive mechanism.

Chrysler Corporation

Reduction Gear Starter Motor

The housing is die-cast aluminum. A 3.5 to 1 reduction, combined with the starter to ring gear ratio, results in a total gear reduction of about 45 to 1.

NOTE: The high-pitched sound is caused by the higher starter speed.

The positive shift solenoid is enclosed in the starter housing and is energized through the ignition switch. When ignition switch is turned to start, the solenoid plunger engages drive gear through a shifting fork. At the completion of travel, the plunger closes a switch to revolve the starter.

The tension of the spring-type shifting prevents a butt-tooth lock up and motor will not start before total shift.

An overrunning clutch prevents motor damage if key is held on after engine starts.

No lubrication is required due to Oilite bearings.

DISASSEMBLY

1. Support assembly in a vise equipped with soft jaws. Do not clamp. Care must be used not to distort or damage the die cast aluminum.

2. Remove the through-bolts and the end housing.

3. Carefully pull the armature up and out of the gear housing, and the starter frame and field assembly. Remove the steel and fiber thrust washer.

NOTE: On eight cylinder engines the starting motors have the wire of the shunt field coil soldered to the brush terminal. Six cylinder engines have the four coils in series and do not have a wire soldered to the brush terminal. One pair of brushes is connected to this terminal. The other pair of brushes is attached to the series field coils by means of a terminal screw. Carefully pull the frame and field assembly up just enough to expose the terminal screw and the solder connection of the shunt field at the brush terminal. Place two wood blocks between the starter frame and starter gear housing to facilitate removal of the terminal screw and unsoldering of the shunt field wire at the brush terminal.

4. Support the brush terminal with a finger behind terminal and remove screw.

5. On eight cylinder engine starters unsolder the shunt field coil lead from the brush terminal and housing.

6. The brush holder plate with terminal, contact and brushes is serviced as an assembly.

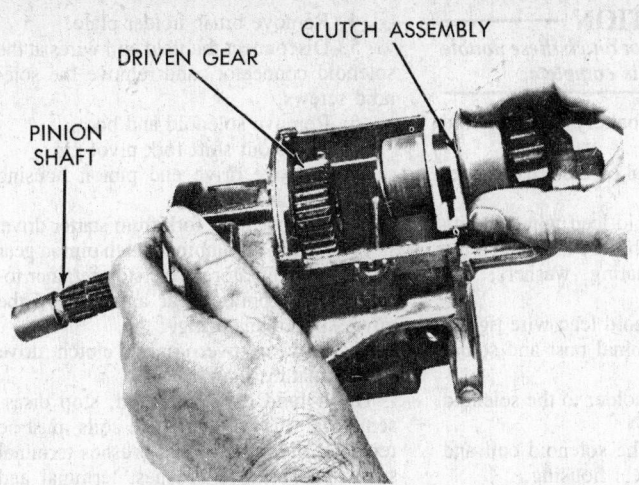

Removing the clutch assembly from a reduction gear starter

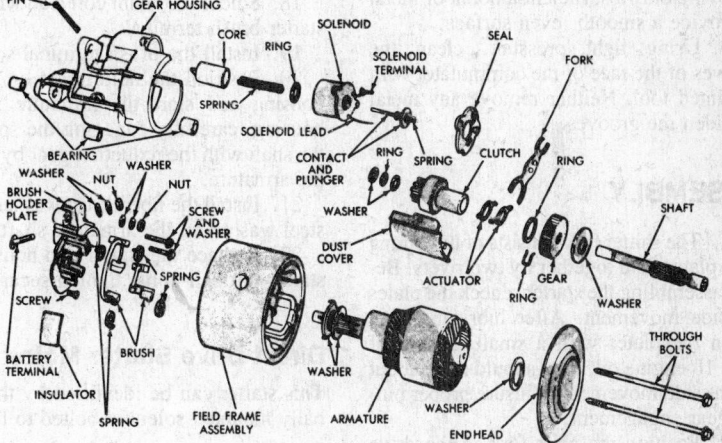

Shift fork and clutch arrangement on a reduction gear starter

7. Clean all old sealer from around plate and housing.

8. Remove the brush holder attaching screw.

9. On the shunt type, unsolder the solenoid winding from the brush terminal.

10. Remove $^{11}/_{32}$ in. nut, washer and insulator from solenoid terminal.

11. Remove brush holder plate with brushes as an assembly.

12. Remove gear housing ground screw.

13. The solenoid assembly can be removed from the well.

14. Remove nut, washer and seal from starter battery terminal and remove terminal from plate.

15. Remove solenoid contact and plunger from solenoid and remove the coil sleeve.

16. Remove the solenoid return spring, coil retaining washer, retainer and the dust cover from the gear housing.

17. Release the snap-ring that locates the driven gear on pinion shaft.

18. Release front retaining ring.

19. Push pinion shaft toward the rear and remove snap-ring, thrust washers, clutch and pinion, and two shift fork nylon actuators.

20. Remove driven gear and friction washer.

21. Pull shifting fork forward and remove moving core.

22. Remove fork retainer pin and shifting fork assembly. The gear housing with bushings is serviced as an assembly.

Replacement of Brushes

1. Brushes that are worn more than one-half the length of new brushes, or are oil-soaked, should be replaced.

2. When resoldering the shunt field and solenoid lead, make a strong, low-resistance connection using a high-temperature solder and resin flux. Do not use acid or acid-core solder. Do not break the shunt field wire units when removing and installing the brushes.

Reduction gear starter

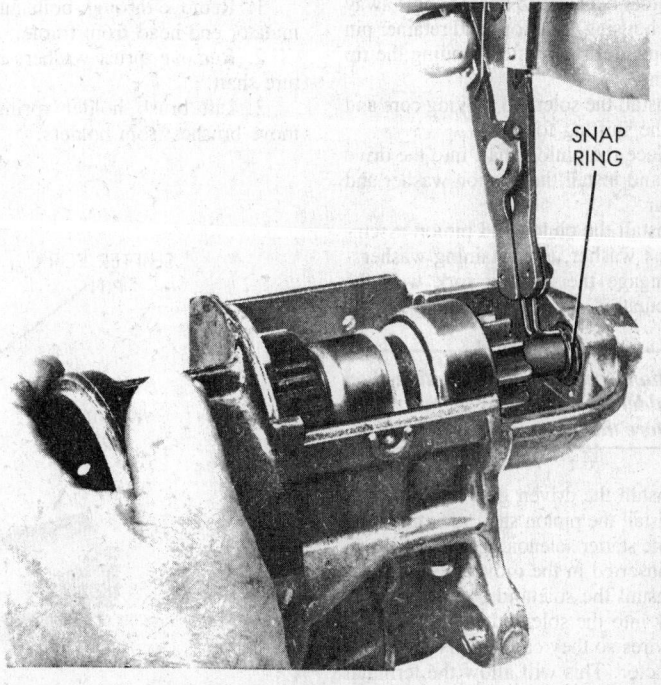

Removing the retaining ring from a reduction gear starter

Starter Clutch and Pinion Gear Inspection

1. Do not immerse the starter clutch unit in a cleaning solvent. The outside of the clutch and pinion must be cleaned with a cloth so as not to wash the lubricant from the inside of the clutch.

2. Rotate the pinion. The pinion gear should rotate smoothly and in one direction only. If the starter clutch unit does not function properly, or if the pinion is worn, chipped, or burred, replace the starter clutch unit.

Commutator Inspection

1. Inspect the commutator and the surface contacted by the brushes when the starter is assembled, for flat spots, out-of-roundness, or excessive wear.

2. Reface the commutator if necessary, removing only a sufficient amount of metal to provide a smooth, even surface.

3. Using light pressure, clean the grooves of the face of the commutator with a pointed tool. Neither remove any metal or widen the grooves.

ASSEMBLY

1. The shifter fork consists of two spring steel plates held together by two rivets. Before assembling the starter, check the plates for side movement. After lubricating between the plates with a small amount of SAE 10 engine oil, they should have about $\frac{1}{16}$ in. side movement to insure proper pinion gear engagement.

2. Position the shift fork in the drive housing and install the shifting fork retainer pin. One tip of the pin should be straight and the other bent at a 15 degree angle away from the housing. The fork and retainer pin should operate freely after bending the tip of the pin.

3. Install the solenoid moving core and engage the shifting fork.

4. Place the pinion shaft into the drive housing and install the friction washer and drive gear.

5. Install the clutch and pinion assembly, thrust washer, and retaining washer.

6. Engage the shifting fork with the clutch actuators.

——— CAUTION ———
The friction washer must be positioned on the shoulder of the splines of the pinion shaft before the driven gear is positioned.

7. Install the driven gear snap-ring.

8. Install the pinion shaft retaining ring.

9. The starter solenoid return spring can now be inserted in the moveable core.

10. Install the solenoid contact plunger assembly into the solenoid and reform the double wires so they can be curved around the contactor. This will allow the terminal stud to enter the brush holder properly.

——— CAUTION ———
The contactor must not touch these double wires after assembly is complete.

11. Assemble the battery terminal stud in the brush holder.

12. Position the seal on the brush holder plate.

13. Run the solenoid lead wire through the hole in the brush holder and attach the solenoid stud, insulating washers, flat washer, and nut.

14. Wrap the solenoid lead wire tightly around the brush terminal post and solder it.

15. Fix the brush holder to the solenoid attaching screws.

16. Gently lower the solenoid coil and brush plate into the gear housing.

17. Position the brush plate assembly into the starter gear housing, install the nuts, and tighten.

18. Solder the shunt coil lead wire to the starter brush terminal.

19. Install the brush terminal screw.

20. Position the field frame on the gear housing and start the armature into the housing, carefully engaging the splines on the shaft with the reduction gear by rotating the armature.

21. Install the fiber thrust washer and the steel washer on the armature shaft.

22. Replace the starter end housing and starter through-bolts; tighten securely.

Direct Drive Starter Motor

This starter can be identified by the externally mounted solenoid bolted to the case.

DISASSEMBLY

1. Remove through-bolts and tap commutator end head from frame.

2. Remove thrust washers from armature shaft.

3. Lift brush holder springs and remove brushes from holders.

4. Remove brush holder plate.

5. Disconnect the field coil wires at the solenoid connector, and remove the solenoid screws.

6. Remove solenoid and boot.

7. Drive out shift fork pivot pin.

8. Remove drive end pinion housing and spacer washer.

9. Remove shift fork from starter drive.

10. Slide overrunning clutch pinion gear toward commutator, drive stop retainer toward clutch pinion gear and remove the now-exposed snap-ring.

11. Remove overrunning clutch drive from armature shaft.

12. If field coils are good, stop disassembly at this point. If field coils must be replaced, remove ground brushes terminal screw and remove brushes, terminal and shunt wire. Remove pole shoe screws, using a ratchet-type impact driver and special wide screwdriver blade, then remove field coils.

13. Replacement of the brushes, inspection of the starter clutch and pinion, and inspection of the commutator procedures are the same as the reduction-gear starter procedures.

ASSEMBLY

1. Install field coils into frame, if removed.

2. Lubricate armature shaft and splines with engine oil.

3. Install starter drive, stop retainer, lock ring and spacer washer.

4. Install shift fork, with *narrow* leg of fork toward commutator.

5. Install pinion housing onto armature shaft, indexing shift fork with slot in housing.

6. Install shift fork pivot pin.

7. With clutch drive, shift fork, and pinion housing assembled onto the armature, slide armature into frame until pinion housing indexes with slot.

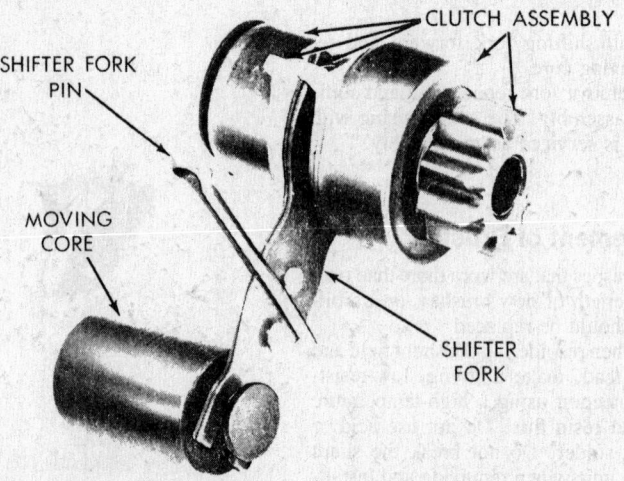

Removing the shift fork from a direct drive starter

8. Install solenoid and boot, tightening bolts to 60–70 inch lbs.

9. Connect field coil wires to solenoid connector, making sure they do not touch frame.

10. Install brush holder plate, indexing tang in frame hole.

11. Place brushes in holders, making sure field coil wires do not interfere.

12. Install thrust washers on commutator end of armature shaft to obtain a maximum of 0.010 in. end-play.

13. Install commutator end head and through-bolts. Tighten bolts to 40–50 inch lbs.

14. Measure drive gear pinion clearance; it should be ⅛ in. Adjust by moving solenoid fore and aft as required.

Ford Motor Co.

Motorcraft Positive Engagement Starter Motor

This starting motor is a series-parallel wound, four pole, four brush unit. It is equipped with an overrunning clutch drive pinion, which is engaged with the flywheel ring gear by an actuating lever, operated by a movable pole piece. This pole piece is hinged to the starter frame and can drop into position through an opening in the frame.

Three conventional field coils are located at three pole piece positions. The fourth field coil is designed to serve also as an engaging coil and a hold-in coil for the operation of the drive pinion.

When the ignition switch is turned to the start position, the starter relay is energized and current flows from the battery to the starter motor terminal. This prime surge of current first flows through the starter engaging coil, creating a very strong magnetic field. This magnetism draws the movable pole piece down toward the starter frame, which then causes the lever attached to it to move the starter pinion into engagement with the flywheel ring gear.

When the movable pole shoe is fully seated, it opens the field coil, grounding contacts, and the starter is then in normal operation. A holding coil is used to hold the movable pole shoe in the fully seated position during the engine cranking operation.

Trucks, equipped with automatic transmissions have a starter neutral switch circuit control. This is to prevent operation of the starter if the selector lever is not in Neutral or Park.

DISASSEMBLY

1. Remove brush cover band and starter drive gear actuating lever cover. Observe the brush lead locations for reassembly, then remove the brushes from their holders.

NOTE: Factory brush length is ½ in.; wear limit is ¼ in.

2. Remove the through-bolts, starter drive gear housing and the drive gear actuating lever return spring.

3. Remove the pivot pin retaining the starter gear actuating lever and remove the lever and the armature.

4. Remove the stop ring retainer. Remove and discard the stop ring holding the drive gear to the armature shaft; then remove the drive gear assembly.

5. Remove the brush end plate.

6. Remove the two screws holding the ground brushes to the frame.

7. On the field coil that operates the starter drive gear actuating lever, bend the tab up on the field retainer and remove the field coil retainer.

8. Remove the three coil retaining screws. Unsolder the field coil leads from the terminal screw, then remove the pole shoes and coils from the frame (use a 300 watt iron).

9. Remove the starter terminal nut, washer, insulator and terminal from the starter frame.

10. Check the commutator for runnout. If the commutator is rough, has flat spots, or is more than 0.005 in. out of round, reface the commutator. Clean the grooves in the commutator face.

11. Inspect the armature shaft and the two bearings for scoring and excessive wear. Replace if necessary.

12. Inspect the starter drive. If the gear teeth are pitted, broken, or excessively worn, replace the starter drive.

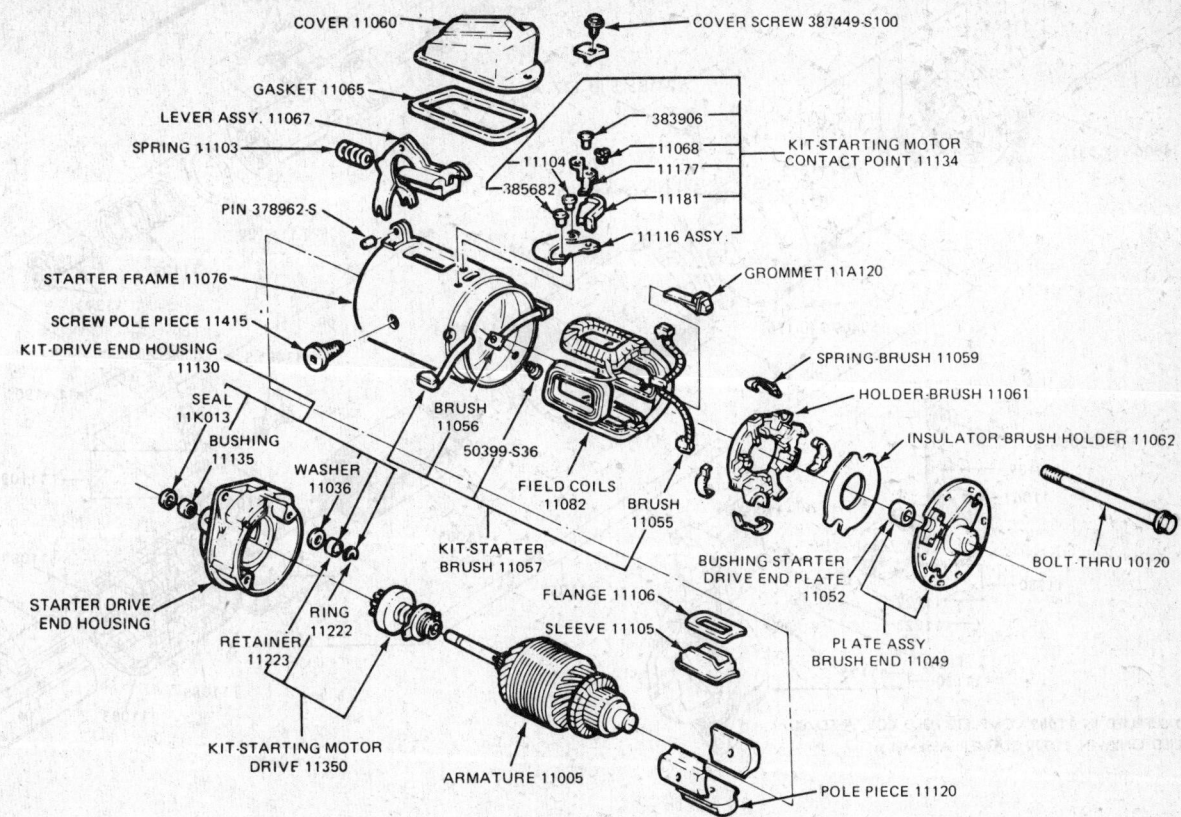

Ford starter

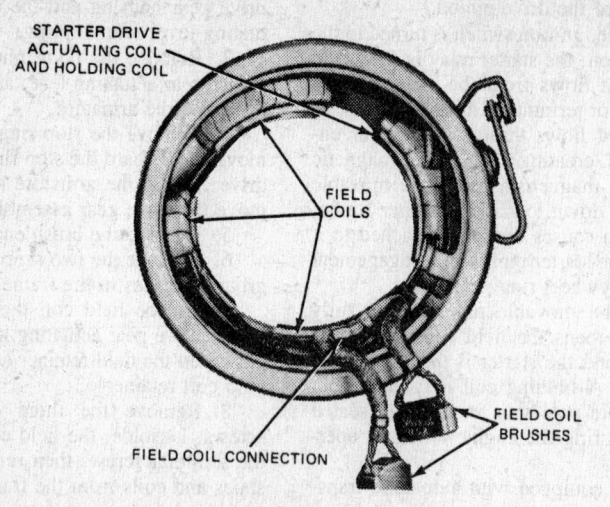

STARTER DRIVE
ACTUATING COIL
AND HOLDING COIL

FIELD
COILS

FIELD COIL
BRUSHES

FIELD COIL CONNECTION

Field coil assembly

ASSEMBLY

1. Install starter terminal, insulator, washers and retaining nut in the frame. (Be sure to position the slot in the screw perpendicular to the frame end surface.)

2. Position coils and pole pieces, with the coil leads in the terminal screw slot, then install the retaining screws. As the pole screws are tightened, strike the frame several sharp hammer blows to align the pole shoes. Tighten, then stake the screws.

3. Install solenoid coil and retainer and bend the tabs to hold the coils to the frame.

4. Solder the field coils and solenoid wire to the starter terminal, using resin-core solder and a 300 watt iron.

5. Check for continuity and ground connections in the assembled coils.

6. Position the solenoid coil ground terminal over the nearest ground screw hole.

7. Position the ground brushes to the starter frame and install retaining screws.

8. Position the brush end plate to the frame, with the end plate boss in the frame slot.

9. Lightly lubricate the armature shaft splines and install the starter drive gear assembly in the shaft. Install a new retaining stop ring and stop ring retainer.

11072
33924-S (M-28)
34052-S (M-155)
44710-S (X-10)
382136-S (MM-295-AG)
44727-S (X-24)
34079-S (M-163)
44713-S (X-13)
11390
11049
*50940-S
11068
11091
27185-S (U-177-A)
11059
34806-S (X-64)
11052
11126
34141-S (M-147)
11A122
11103
11393
50409-S (U-126)
11036
*43485-S
*11A120
11415
†11102
*305054-S
11005
†11083
11139
11067
†11083
11003
11350
11222
11223
11057
†11085
11036
11130
11135
†11083

†ALSO SUPPLIED IN 11082 COMPLETE FIELD COIL ASSEMBLY
♦SUPPLIED ONLY IN 11002 STARTER ASSEMBLY

Ford solenoid type starter

10. Position the fiber thrust washer on the commutator end of the armature shaft, then position the armature in the starter frame.

11. Position the starter drive gear actuating lever to the frame and starter drive assembly, and install the pivot pin.

NOTE: Fill drive gear housing bore ¼ full of grease.

12. Position the drive actuating lever return spring and the drive gear housing to the frame, then install and tighten the through-bolts. Do not pinch brush leads between brush plate and frame. Be sure that the stop ring retainer is properly seated in the drive housing.

13. Install the brushes in the brush holders and center the brush springs on the brushes.

14. Position the drive gear actuating lever cover on the starter and install the brush cover band with a new gasket.

15. Check starter no-load amperage draw.

Motorcraft Solenoid Actuated Starter Motor

This starter motor is a four-brush, four-field, four-pole wound unit. The frame encloses a wound armature, which is supported at the drive end by caged needle bearings and at the commutator end by a sintered copper bushing. The four pole shoes are retained to the frame by one pole screw apiece, and on each pole shoe is wound a ribbon-type field coil connected in series-parallel.

The solenoid is mounted to a flange on the starter drive housing, which encloses the entire shift mechanism and solenoid plunger. The solenoid utilizes two windings, a pull-in winding and a hold-in winding.

DISASSEMBLY

1. Disconnect the copper strap from the solenoid starter terminal, remove the remaining screws and remove the solenoid.

2. Loosen the retaining screw and slide the brush cover band back far enough to gain access to the brushes.

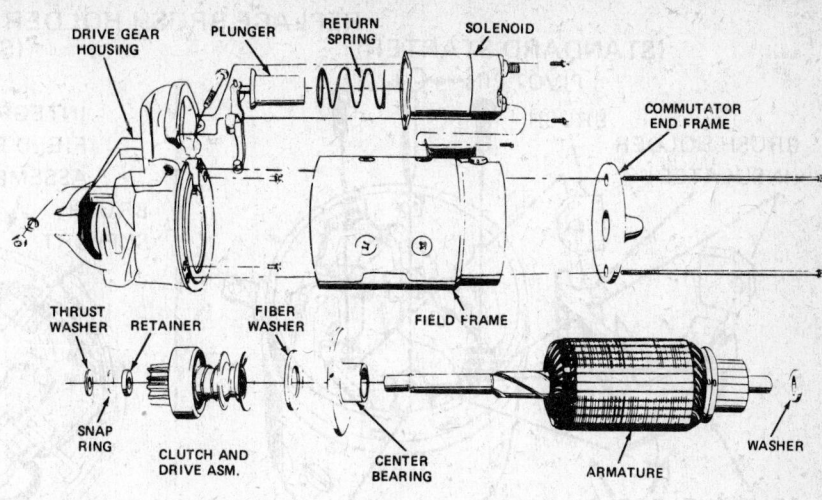

Delco starter

3. Remove the brushes from their holders, then remove the through-bolts and separate the drive end housing from the frame and brush end plate.

NOTE: Factory brush length is ½ in., wear limit ¼ in.

4. Remove the solenoid plunger and shift fork. These two items can be separated from each other by removing the roll pin.

5. Remove the armature and drive assembly from the frame. Remove the drive stop ring and slide the drive off the armature shaft.

6. Remove the drive stop ring retainer from the drive housing.

7. Inspection of the commutator, armature and bearings, and pinion gear procedures is the same as the positive engagement starter procedures.

ASSEMBLY

1. Lubricate the armature shaft splines with Lubriplate, then install drive assembly and a new stop ring.

2. Lubricate shift lever pivot pin with Lubriplate, then position solenoid plunger and shift lever assembly in the drive housing.

3. Place a new retainer in the drive housing. Apply a small amount of Lubriplate to the drive end of the armature shaft, then place armature and drive assembly into the drive housing, indexing the shift lever tangs with the drive assembly.

4. Apply a small amount of Lubriplate to the commutator end of the armature shaft, then position the frame and field assembly to the drive housing.

5. Position the brush plate assembly to the frame, making sure it properly indexes. Install through-bolts and tighten to 45–85 inch lbs.

6. Install brushes into their holders and make sure leads are not touching any interior starter components.

7. Place the rubber gasket between the solenoid mount and the frame surface.

8. Place the starter solenoid in position with metal gasket and spring, install heat shield (if so equipped) and install solenoid screws.

9. Connect copper strap and install cover band.

Delco-Remy Starter Motor

There are many different versions of the Delco-Remy starter, depending upon application. In general, six-cylinder engines use a unit having four field coils in series between the terminal and armature. Standard V8 engines use, depending on displacement, one of three types: one has two field coils in series with the armature and parallel to each other; another has two field coils in parallel between the field terminal and ground, and another has three field coils in series with the armature and one field connected between the motor terminal and ground. Heavy-duty starter motors, such as used on some of the largest G.M. high-output engines (over 400 cu. in.) have series compound windings. The relatively recent 20MT, 25MT, and 27MT starters are used with diesels only. They differ from the others mainly in the use of a center bearing. Most repair procedures are generally similar for them, too. Where additional procedures are required for the center bearing, they will be noted.

In spite of these differences, all Delco-Remy starters are disassembled and assembled in essentially the same manner.

DISASSEMBLY

1. Disconnect the field coil connectors from the motor solenoid terminal. Remove the solenoid mounting screws. Rotate the solenoid 90 degrees and remove it along with the plunger return spring.

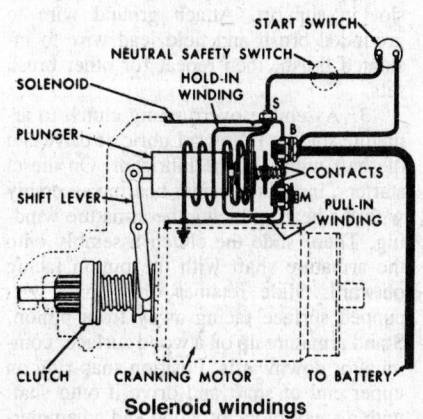

Solenoid windings

REPLACE BRUSH HOLDER

(STANDARD STARTER)   **(SMALL 5MT STARTER)**

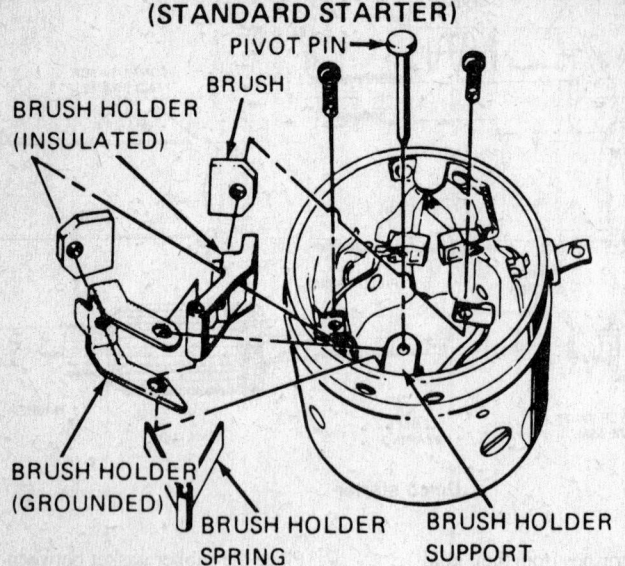

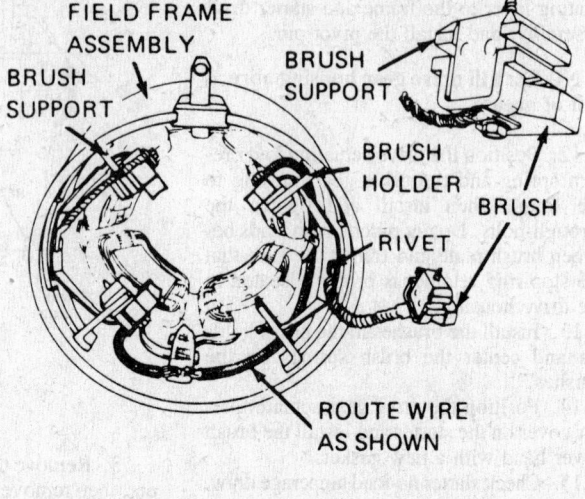

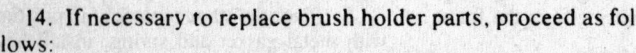

14. If necessary to replace brush holder parts, proceed as follows:

    a. Remove brush holder pivot pin which positions one insulated and one grounded brush.

    b. Remove brush spring.

    c. Replace brushes as necessary.

    a. Remove brush holder from brush support.

    b. Remove screw from brush holder and separate brush and holder.

    c. Inspect brush holder for wear or damage.

    d. Replace brushes and/or holders as necessary.

Brush holder removal—1981 and later Delco-Remy starters (© General Motors Corp.)

NOTE: On models so equipped, remove solenoid mounting screws.

2. Remove the through bolts.

3. Remove the commutator end frame. On diesel starters only, remove the insulator. On all starters, remove the washer, field frame assembly, and armature assembly from the drive housing. On diesel starters, the armature remains in the drive end frame.

4. On diesel starters only, remove the shift lever pivot bolt. On the 25MT only, remove the center bearing screws and remove the drive gear housing from the armature shaft. The shift lever/plunger assembly can now be separated from the starter clutch.

5. Remove the overrunning clutch from the armature shaft as follows: Slide the two-piece thrust collar off the end of the armature shaft. Slide a standard ½ in. pipe coupling or other spacer onto the shaft so that the end of the coupling butts against the edge of the retainer. Tap the end of the coupling with a hammer, driving retainer towards armature end of snap-ring. Remove snap-ring from its groove in the shaft using pliers. Slide retainer and clutch from armature shaft. On diesel starters, also remove the fiber washer and center bearing.

6. On 1980 and earlier starters, disconnect the brush assembly from the field frame by releasing the V-spring and removing the support pin. Pull the brush holders, brushes, and springs out as a unit and disconnect leads. On 1981 and later

starters (except 5MT), remove the pivot pin which holds the brush holder and one insulated and one grounded brush in place. Remove the brush spring. On the 5MT starter, remove the brush holder from the brush support. Remove the screw from the brush holder and separate the brush and holder.

7. On models so equipped, separate solenoid from lever housing.

Cleaning & Inspection

1. Clean parts with a rag, but do not immerse the parts in a solvent. Immersion in a solvent will dissolve the grease that is packed in the clutch mechanism and damage the armature and field coil insulation.

2. Test overrunning clutch action. The pinion should turn freely in the overrunning direction and must not slip in the cranking direction. Check pinion teeth to see that they have not been chipped, cracked, or excessively worn. Replace the unit if necessary.

3. Inspect the armature commutator. If the commutator is rough or out of round, it should be turned down. Some starter motor models use a molded armature commutator design and no attempt to undercut the insulation should be made or serious damage may result to the commutator.

ASSEMBLY

1. Install brushes into holders. Install solenoid, if so equipped.

2. Assemble insulated and grounded

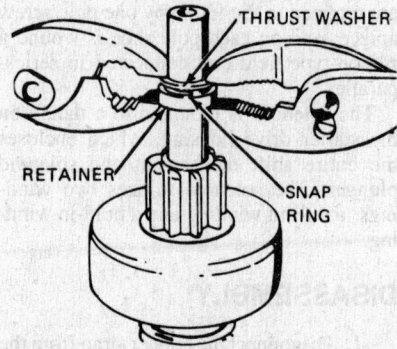

Forcing the snap-ring into the retainer

brush holder together using the V-spring and position the assembled unit on the support pin. Push holders and spring to bottom of support and rotate spring to engage the slot in support. Attach ground wire to grounded brush and field lead wire to insulated brush, then repeat for other brush sets.

3. Assemble overrunning clutch to armature shaft as follows: Lubricate drive end of shaft with silicone lubricant. On diesel starters, install the center bearing assembly with the bearing facing the armature winding. Then, slide the clutch assembly onto the armature shaft with the pinion facing outward. Slide retainer onto shaft with cupped surface facing away from pinion. Stand armature up on a wood surface, commutator downwards. Position snap-ring on upper end of shaft and drive it onto shaft with a small block of wood and a hammer.

Slide snap-ring into groove. Install thrust collar onto shaft with shoulder next to snap-ring. With retainer on one side of snap-ring and thrust collar on the other side, squeeze together with two sets of pliers until ring seats in retainer. On models without thrust collar, use a washer. Remember to remove washer before continuing.

4. Lubricate drive end bushing with silicone lubricant, then slide armature and clutch assembly into place, at the same time engaging shift lever with clutch. On 1981 and later non-diesel starters, the shift lever may be installed in the drive gear housing first. On the 25MT diesel starter only, install the center bearing screws and the shift lever pivot bolt. Tighten all securely.

5. Position field frame over armature and apply sealer (silicone) between frame and solenoid case. Position frame against drive housing, making sure brushes are not damaged in the process.

6. Lubricate the commutator end bushing with silicone lubricant. Place a leather brake washer on the armature shaft of gas engine starters. Place an insulator on the shaft of diesel engine starters. Slide the commutator end frame onto the shaft. Install the through bolts, making sure they pass through the bolt holes in the insulator on diesel starters. Install the through bolts and tighten to 65 inch lbs.

7. Reconnect field coil connector(s) to the solenoid motor terminal. Install solenoid mounting screws, if so equipped.

8. Check pinion clearance; it should be 0.010–0.140 in. on all models.

Prestolite Starter Motor

DISASSEMBLY

1. Remove the cover band and remove the brushes from their holders.
2. Remove the brush end plate mounting screws and the two through-bolts.
3. Remove the drive housing, end brush plate, and armature from the starter frame.
4. Compress the starter drive spring on the armature side of the shaft and remove the lock screw and remove the starter drive, center bearing plate and thrust washers.
5. Remove the four field pole shoes and remove the field coils from the frame.

NOTE: The positive brushes can be replaced on the field coils by soldering, and the negative brushes replaced on the brush end plate by riveting.

ASSEMBLY

1. Assemble the field coils and pole shoes into the frame and secure with screws.
2. Assemble the center bearing plate, thrust washers, and starter drive on the armature shaft and secure with the locking screw.
3. Place the armature assembly into the drive housing aligning the slot in the shaft center bearing support with the pin in the drive housing.
4. Install the end frame to the frame housing and install the six mounting screws.
5. Position the armature assembly into the frame housing and engage the frame dowel with the bolt of the drive frame. Install the two through-bolts and secure.
6. Install the brushes into the holders. Center the brush springs on the brushes and locate the insulated brush leads clear of the armature. Install the cover band.

Datsun

BRUSH REPLACEMENT

Non-Reduction Gear Type

1. With the starter out of the vehicle, remove the bolts holding the solenoid to the top of the starter and remove the solenoid.
2. To remove the brushes, remove the two thru-bolts, and the two rear cover attaching screws and remove the rear cover.
3. Disconnect the electrical leads and remove the brushes.
4. Install the brushes in the reverse order of removal.

Reduction Gear Type

1. Remove the starter. Remove the solenoid.
2. Remove the through bolts and the rear cover. The rear cover can be pried off with a screwdriver, but be careful not to damage the O-ring.
3. Remove the starter housing, armature, and brush holder from the center housing. They can be removed as an assembly.
4. Remove the positive side brush from its holder. The positive brush is insulated from the brush holder, and its lead wire is connected to the field coil.
5. Carefully lift the negative brush from the commutator and remove it from the holder.
6. Installation is the reverse.

STARTER DRIVE REPLACEMENT

Non-Reduction Gear Type

1. With the starter motor removed from the vehicle, remove the solenoid from the starter.
2. Remove the two thru-bolts and separate the gear from the yoke housing.
3. Remove the pinion stopper clip and the pinion stopper.
4. Slide the starter drive off the armature shaft.
5. Install the starter drive and reassemble the starter in the reverse order of removal.

Reduction Gear Type

1. Remove the starter.
2. Remove the solenoid and the shift lever.
3. Remove the bolts securing the center housing to the front cover and separate the parts.
4. Remove the gears and starter drive.
5. Installation is the reverse.

Isuzu

BRUSH REPLACEMENT

1. With the starter out of the vehicle, remove the bolts holding the solenoid to the top of the starter and remove the solenoid.
2. To remove the brushes, remove the two through-bolts and the two rear cover attaching screws and remove the rear cover.
3. Disconnect the brushes, electrical leads and remove the brushes.
4. Install the brushes in the reverse order of removal.

STARTER DRIVE REPLACEMENT

1. With the starter motor removed from the vehicle, remove the solenoid from the starter.
2. Remove the two through-bolts and separate the gear case from the yoke housing.
3. Remove the pinion stopper clip and the pinion stopper.

SPECIFICATIONS

PRESTOLITE STARTER

Vendor	Current Draw Under Normal Load (Amperes)	Minimum Stall Torque		Maximum Load (Amperes)	No-Load (Amperes)	Brushes				Through Bolt Torque (In-Lbs)	Mounting Bolt Torque (Ft-Lbs)
		(Ft-Lbs)	Volts			Mfg. Length (Inches)	Wear Limit (Inches)	Brush Spring Tension (Ounces)			
Prestolite	200	17.2	5	525	60	0.46–0.48	0.25	45–53		72–96	23–28

Maximum commutator runout in inches is 0.005. Maximum starting circuit voltage drop (battery + terminal to starter terminal) at normal engine temperature 0.5 volt.

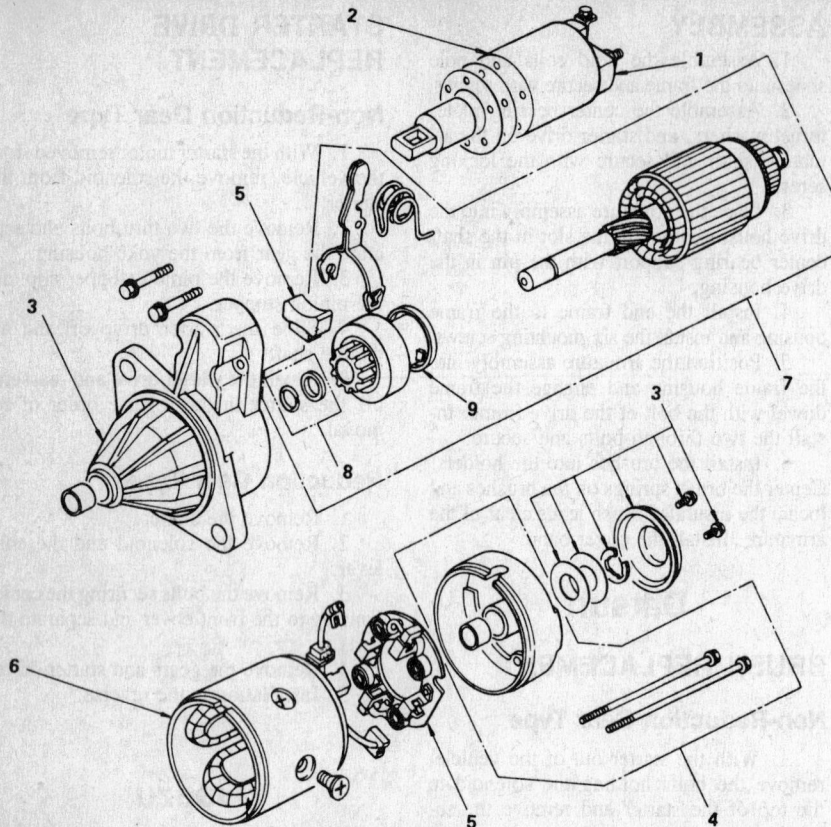

4. Slide the starter drive off the armature shaft.

5. Install the starter drive and reassemble the starter in the reverse order of removal.

Mazda/Courier

BRUSH REPLACEMENT

1. Remove the starter. Remove the two screws attaching the brush end bearing cover and remove the bearing cover.

2. Remove the through-bolts.

3. Remove the C-washer, washer and spring from the brush end of the armature shaft.

4. Pull the brush end cover from the starter frame.

5. Unsolder the two brushes from the field terminals an slide the brush holder from the armature shaft.

6. Cut the two brush wires at the brush holder and solder two new brushes to the brush holder.

7. Install the brush holder on the armature shaft and install the brushes in the brush holder.

8. Install the brush end cover on the starter frame and be sure that the ear tabs of the brush holder are aligned with the through-bolt holes.

9. Install the through-bolts.

10. Install the rubber gasket, spring, washer and C-washer on the armature shaft.

11. Install the brush end bearing cover on the brush end cover and install the two screws. If the brush holder tabs are not aligned with the through-bolts, the bearing cover screws cannot be installed.

1. Lead; M/terminal (magnetic switch)
2. Magnetic switch assembly
3. Dust cover and snap ring
4. Rear cover assembly and gear case assembly
5. Brush holder assembly
6. Field coil assembly and gasket
7. Armature assembly, with shift lever
8. Pinion stop clip
9. Pinion assembly

Isuzu/Luv starter—gasoline engine

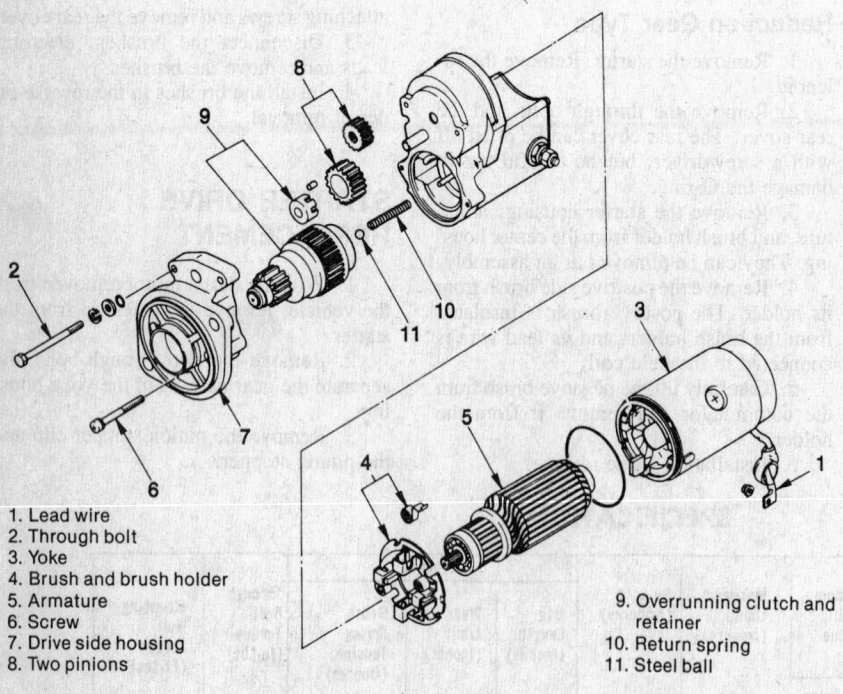

1. Lead wire
2. Through bolt
3. Yoke
4. Brush and brush holder
5. Armature
6. Screw
7. Drive side housing
8. Two pinions
9. Overrunning clutch and retainer
10. Return spring
11. Steel ball

Isuzu/Luv starter—diesel engine

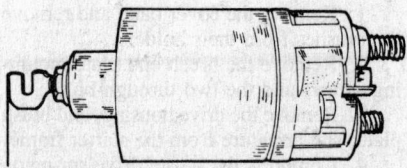

Starter solenoid plunger adjustment—adjust to 0.8 in.

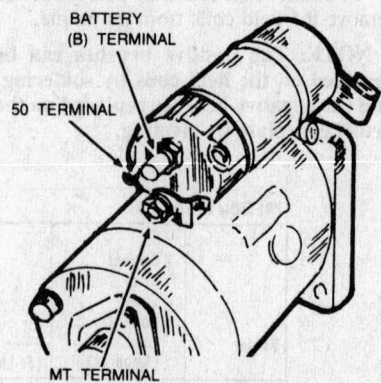

Starter solenoid terminals

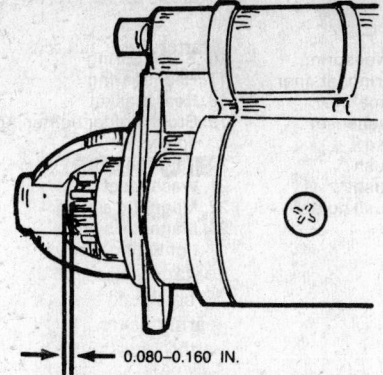

→ ← 0.080–0.160 IN.

Starter drive and clearance

SOLENOID REPLACEMENT

1. Remove the starter from the truck.
2. Disconnect the field strap from the solenoid terminal.
3. Remove the two solenoid attaching screws.
4. Disengage the solenoid plunger from the shift fork and remove the solenoid.
5. Install the solenoid on the drive end housing, making sure that the solenoid plunger hook is engaged with the shift fork.
6. Apply 12 volts to the solenoid S terminal and measure the clearance between the starter drive and the stop-ring retainer. It should be 0.080–0.020 in. If not, remove the solenoid and adjust the clearance by inserting an adjusting shim between the solenoid body and drive end housing.
7. Check the solenoid for proper operation and install the starter.
8. Check the operation of the starter.

STARTER OVERHAUL

1. Remove the starter from the truck.
2. Disconnect the field strap from the solenoid.
3. Remove the screws attaching the solenoid to the drive end housing. Disengage the solenoid plunger hook from the shift fork and remove the solenoid.
4. Remove the shift fork pivot bolt, nut and lockwasher.
5. Remove the through-bolts and separate the drive end housing from the starter frame. At the same time, disengage the shift fork from the drive assembly.
6. Remove the two screws attaching the brush end bearing cover to the brush end cover.
7. Remove the C-washer, washer and spring from the brush end of the armature shaft.
8. Pull the brush end cover from the starter frame.
9. Slide the armature from the starter frame and brushes.
10. Slide the drive stop-ring retainer toward the armature and remove the stop-ring. Slide the retainer and drive assembly off the armature shaft.

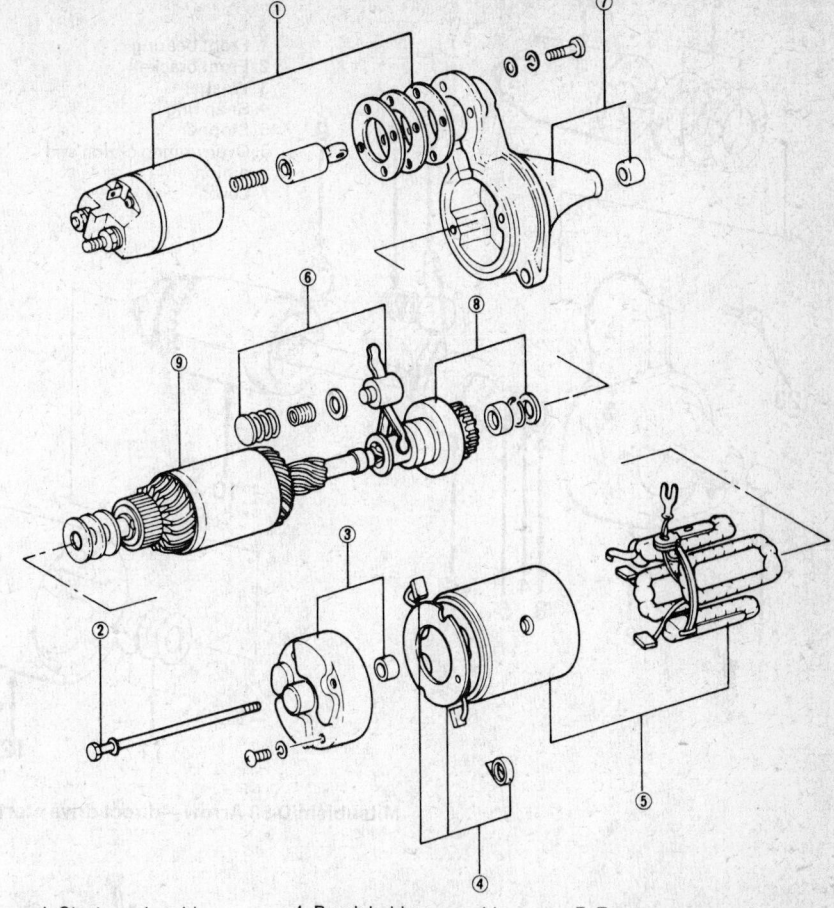

1. Starter solenoid
2. Bolts
3. Rear cover
4. Brush holder assembly
5. Yoke assembly
6. Lever assembly
7. Front cover assembly
8. Drive pinion
9. Armature set

Mazda starter—gasoline engine

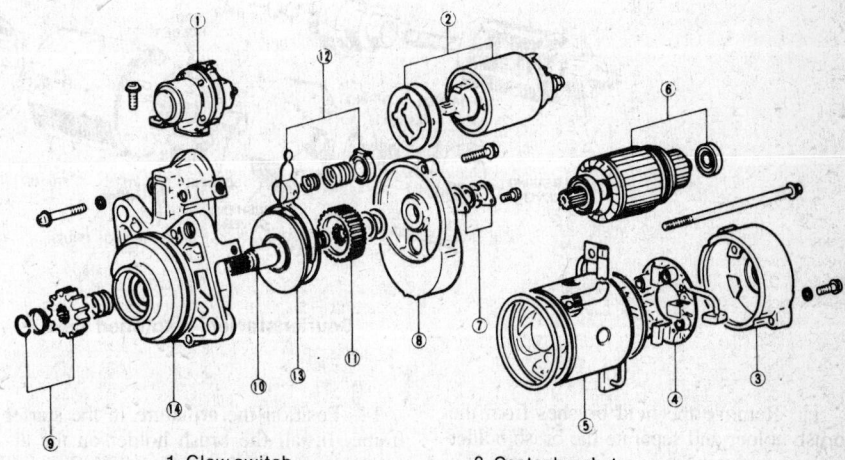

1. Glow switch
2. Starter solenoid
3. Rear housing
4. Brush holder assembly
5. Yoke assembly
6. Armature & bearing
7. Cover
8. Center bracket
9. Drive pinion
10. Pinion shaft
11. Gear
12. Lever & spring
13. Over running clutch
14. Front housing assembly

Mazda starter—diesel engine

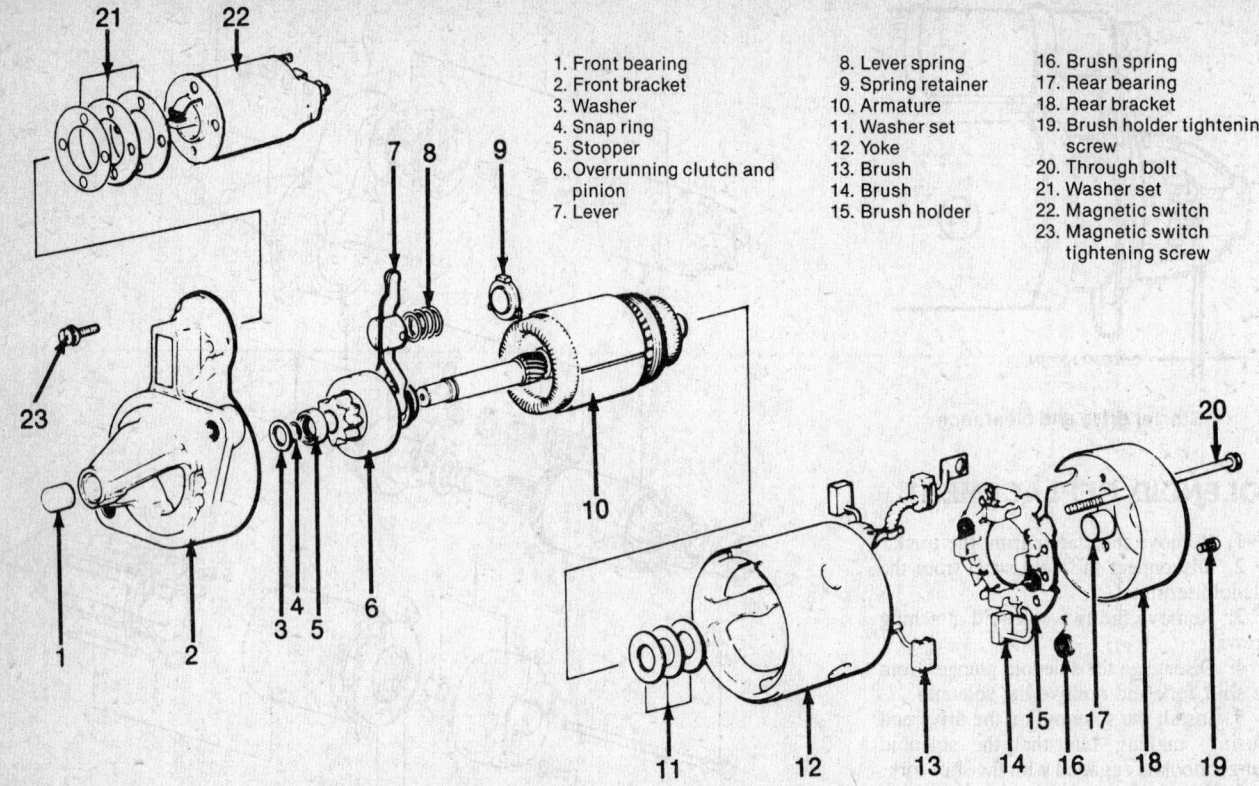

1. Front bearing
2. Front bracket
3. Washer
4. Snap ring
5. Stopper
6. Overrunning clutch and pinion
7. Lever
8. Lever spring
9. Spring retainer
10. Armature
11. Washer set
12. Yoke
13. Brush
14. Brush
15. Brush holder
16. Brush spring
17. Rear bearing
18. Rear bracket
19. Brush holder tightening screw
20. Through bolt
21. Washer set
22. Magnetic switch
23. Magnetic switch tightening screw

Mitsubishi/D-50 Arrow—direct drive starter

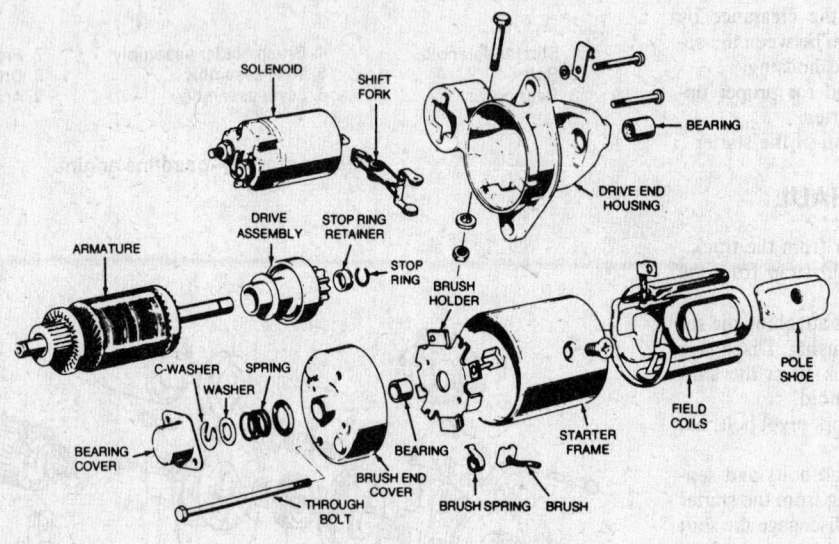

Courier starter—exploded view

11. Remove the field brushes from the brush holder and separate the brush holder from the starter frame.

12. Position the drive assembly on the armature shaft.

13. Position the drive stop-ring retainer on the armature shaft and install the drive stop-ring. Slide the stop-ring retainer over the stop-ring to secure the stop-ring on the shaft.

14. Position the armature in the starter frame. Install the brush holder on the armature and starter frame. Install the brushes in the brush holder.

15. Install the drive end housing on the armature shaft and starter housing. Engage the shift fork with the starter drive assembly as you move the drive end housing toward the starter frame.

16. Install the brush end cover on the

starter frame making sure that the rear tabs of the brush holder are aligned wit the through-bolt holes.

17. Install the through-bolts.

18. Install the rubber washer, spring, washer and C-washer on the armature shaft at the brush end. Install the brush end bearing cover on the brush end cover and install the attaching screws. If the brush end cover is not properly positioned, the bearing cover

screws cannot be installed.

19. Align the shift fork with the pivot bolt hole and install the pivot bolt, lockwasher and nut. Tighten the nut securely.

20. Position the solenoid on the drive end housing. Be sure that the solenoid plunger hook is engaged with the shift fork.

21. Install the two solenoid retaining screws and washers.

22. Apply 12 volts to the solenoid S terminal (ground the M terminal) and check the clearance between the starter drive and the stop-ring retainer. The clearance should be 0.080–0.020 in. If not, the solenoid plunger is not properly adjusted. The clearance can be adjusted by inserting an adjusting shim between the solenoid body and drive end housing.

23. Install the field strap and tighten the nut.

24. Install the starter. Check the operation of the starter.

Mitsubishi/D–50 Arrow

STARTER DRIVE, SOLENOID & BRUSH REPLACEMENT

Direct Drive Type

1. Remove the wire connecting the starter solenoid to the starter.

2. Remove the two screws holding the starter solenoid on the starter drive housing and remove the solenoid.

3. Remove the two long through bolts at the rear of the starter and separate the armature yoke from the armature.

4. Carefully remove the armature and the starter drive engagement lever from the front bracket, after making a mental note of the way they are positioned along with the attendant spring and spring retainer.

5. Loosen the two screws and remove the rear bracket.

6. Tap the stopper ring at the end of the drive gear engagement shaft in towards the drive gear to expose the snapring. Remove the snapring.

7. Pull the stopper, drive gear and overrunning clutch from the end of the shaft. For 1979 models with automatic transmissions, remove the center bracket, spring and spring retainer.

8. Inspect the pinion and spline teeth for wear or damage. If the engagement teeth are damaged, visually check the flywheel ring gear through the starter hole to insure that it is not damaged. It will be necessary to turn the engine over by hand to completely inspect the ring gear.

9. Check the brushes for wear. Their service limit length is 0.453 in. Replace if necessary.

10. For 1979 models with automatic transmissions, fit the spring retainer, spring and center bracket on the shaft.

11. Install the spring retainer and spring on the armature shaft.

12. Install the overrunning clutch assembly on the armature shaft.

13. Fit the stopper ring with its open side facing out on the shaft.

14. Install a new snapring and, using a gear puller, pull the stopper ring into place over the snapring.

15. Fit the small washer on the front end of the armature shaft.

16. Fit the engagement lever into the overrunning clutch and refit the armature into the front housing.

17. Fit the engagement lever spring and spring retainer into place and slide the armature yoke over the armature. Make sure you position the yoke with the spring retainer cut-out space in line with the spring retainer.

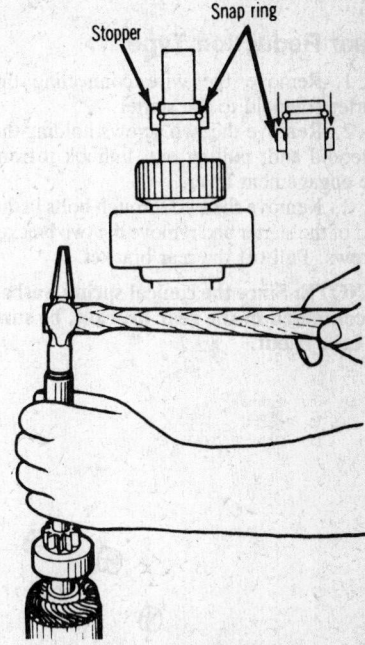

Removing snap ring

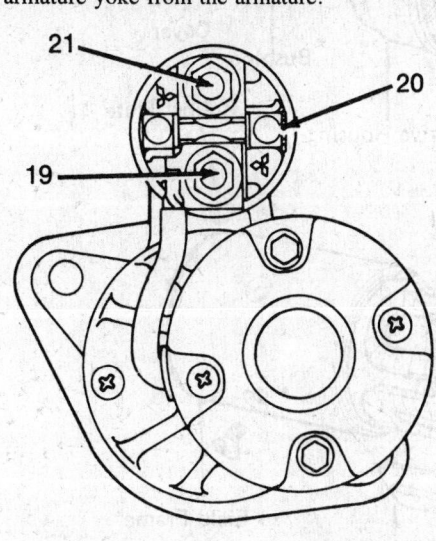

1. Lever spring
2. Packing
3. Lever
4. Front bracket
5. Pinion
6. Stopper
7. Ring
8. Pinion shaft assy.
9. Gear
10. Center bracket
11. Pole
12. Yoke
13. Field coil
14. Brush
15. Brush holder
16. Through bolt
17. Rear bracket
18. Magnetic switch
19. Terminal "M"
20. Terminal "S"
21. Terminal "B"

Mitsubishi/D-50 Arrow—gear reduction starter

NOTE: Make sure the brushes are seated on the commutator.

18. Replace the rear bracket and two retainer screws.

19. Install the two through bolts in the end of the yoke.

20. Refit the starter solenoid, making sure you fit the plunger over the engagement lever. Install the screws and connect the wire running from the starter yoke to the starter solenoid.

Gear Reduction Type

1. Remove the wire connecting the starter solenoid to the starter.

2. Remove the two screws holding the solenoid and, pulling out, unhook it from the engagement lever.

3. Remove the two through bolts in the end of the starter and remove the two bracket screws. Pull off the rear bracket.

NOTE: Since the conical spring washer is contained in the rear bracket, be sure to take it out.

4. Remove the yoke and brush holder assembly while pulling the brush upward.

5. Pull the armature assembly out of the mounting bracket.

6. In the side of the mounting bracket that the armature fits into, there is a small dust cap held by two screws. Remove it and remove the snapring and washer under it.

7. Remove the remaining bolts in the mounting bracket and split the reduction case.

NOTE: Several washers will come out when the case is split. These adjust the end play for the pinion shaft. Do not lose them.

8. Remove the reduction gear, lever and lever spring from the front bracket.

9. Using a brass drift or deep socket, knock the stopper ring on the end of the shaft in toward the pinion. Remove the snapring. Remove the stopper, pinion and pinion shaft assembly.

10. Remove the ball bearings at both ends of the armature.

NOTE: The ball bearings are pressed in the front bracket and are not replaceable. Replace them together with the bracket.

11. Inspect the pinion and spline teeth for wear or damage. If the engagement teeth are damaged, visually check the flywheel ring gear through the starter hole to insure that it is not damaged also. It will be necessary to turn the engine over by hand to completely inspect the ring gear.

12. Check the brushes for wear. Their service limit length is 0.0453 in. Replace if necessary.

13. Assembly is the reverse of disassembly. Be sure to replace all adjusting and thrust washers. When replacing the rear bracket, fit the conical spring pinion washer with its convex side facing out. Make sure that the brushes seat themselves on the commutator.

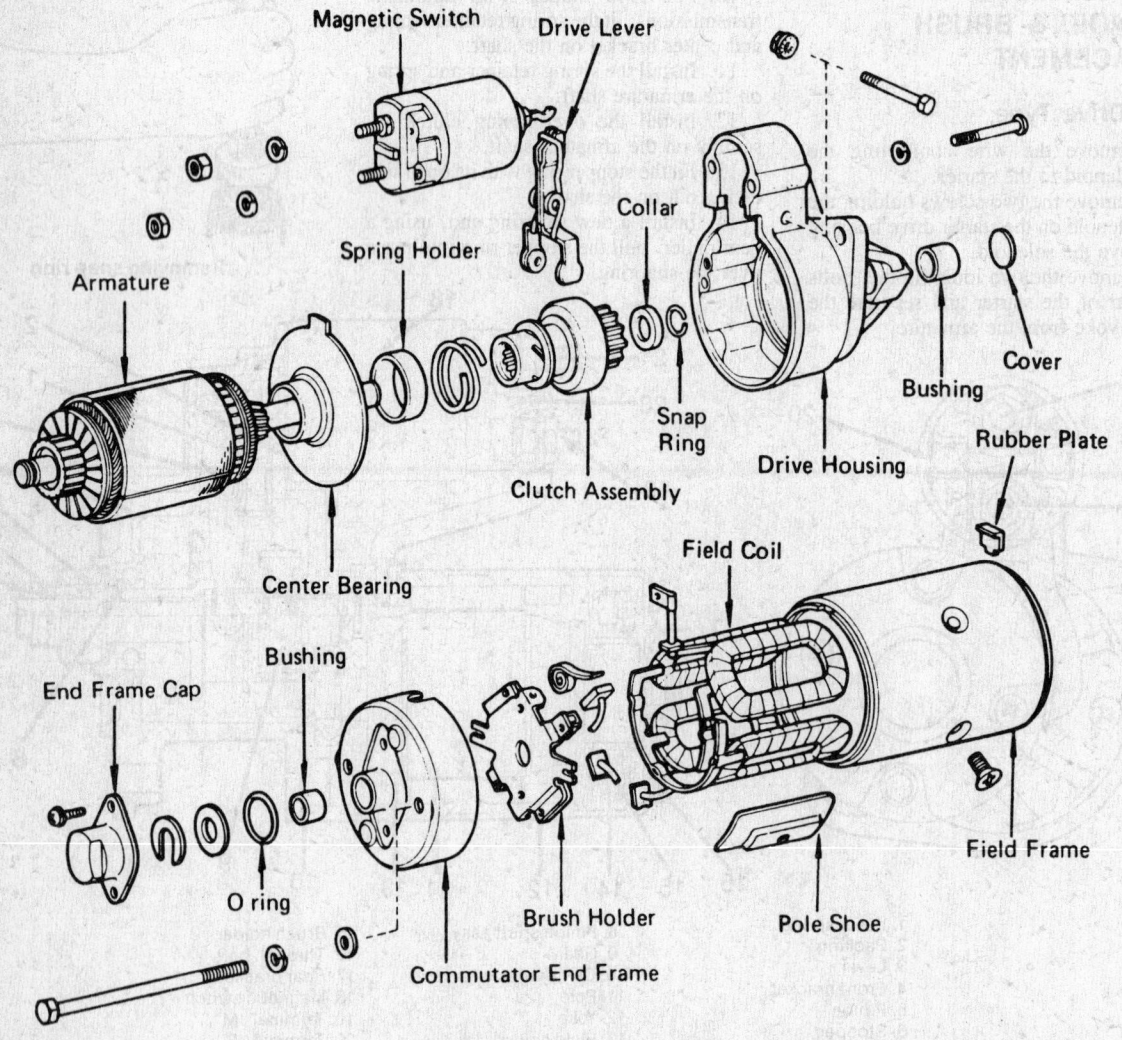

Toyota—direct drive starter—typical

1.0 kW Type

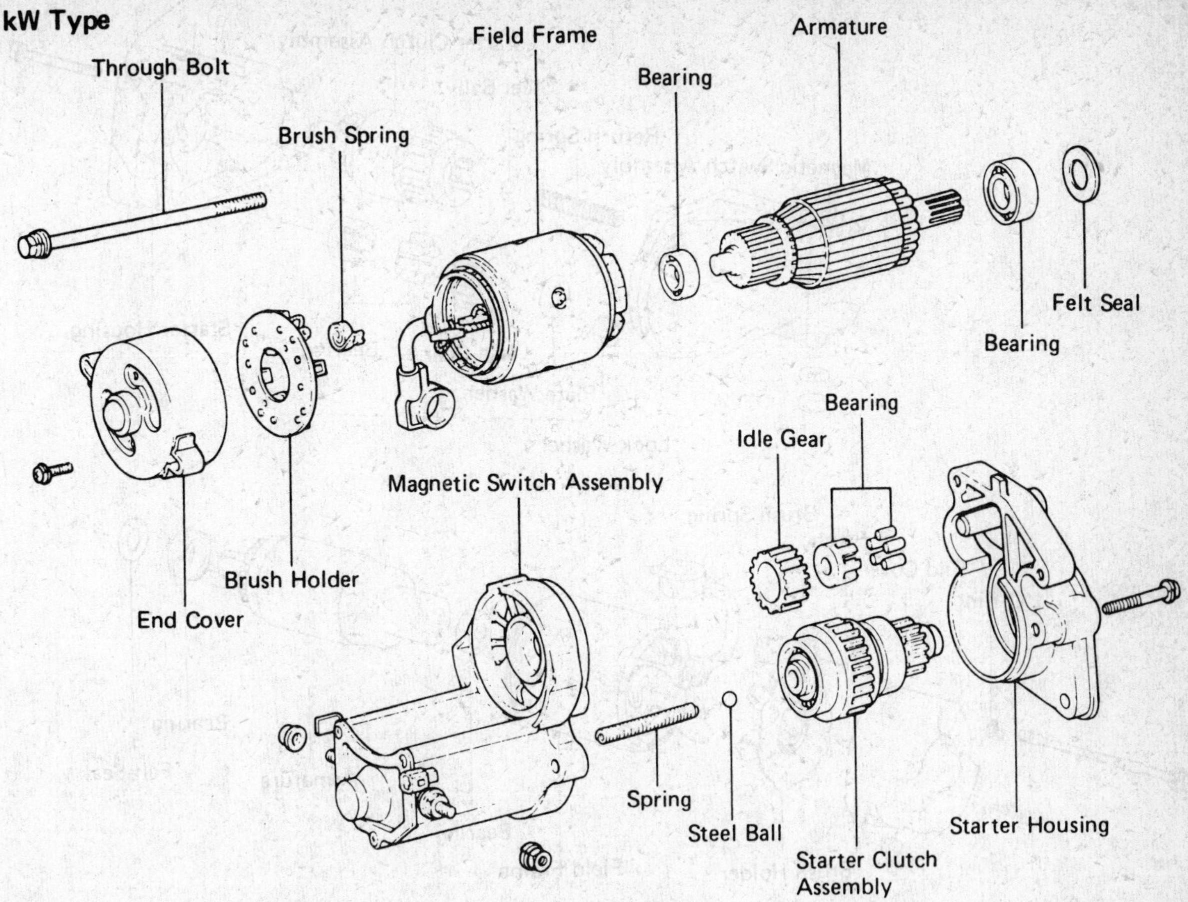

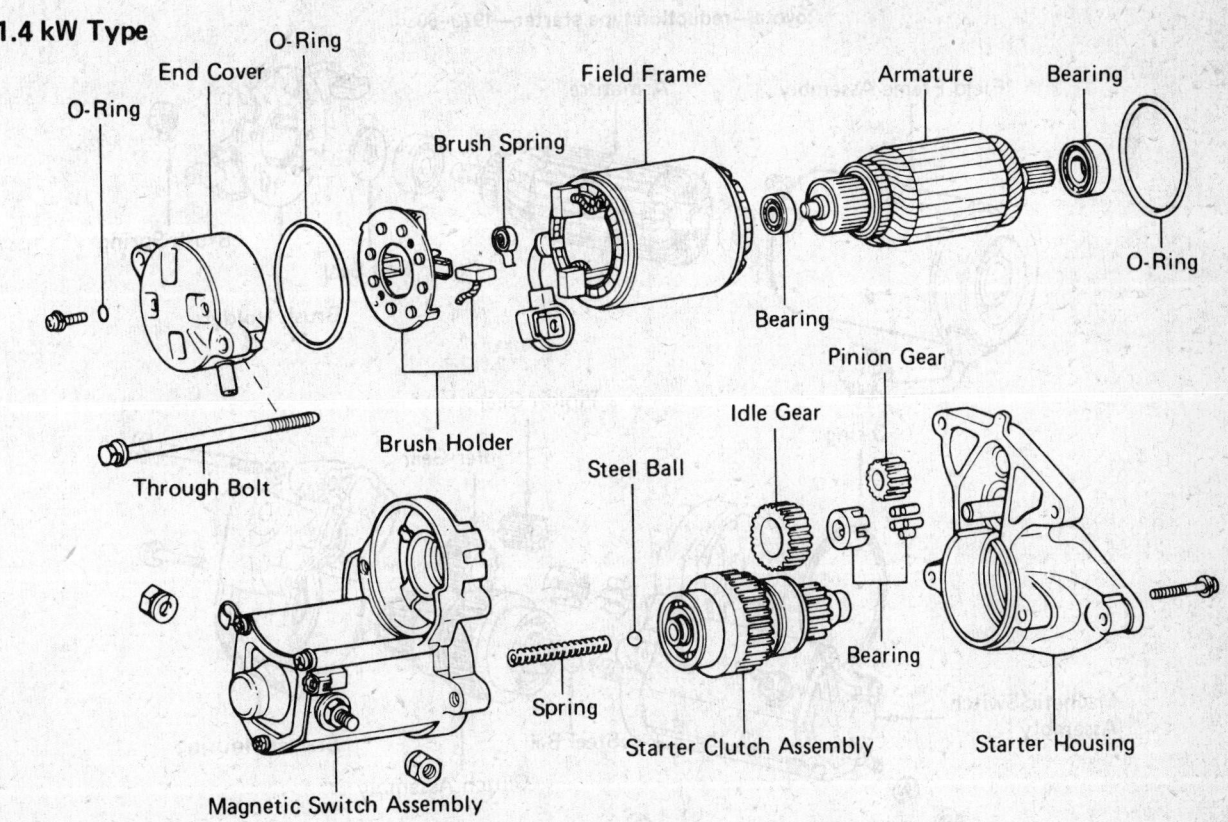

Through Bolt
Brush Spring
Field Frame
Bearing
Armature
Bearing
Felt Seal
Bearing
End Cover
Brush Holder
Magnetic Switch Assembly
Idle Gear
Bearing
Spring
Steel Ball
Starter Clutch Assembly
Starter Housing

1.4 kW Type

O-Ring
End Cover
O-Ring
Brush Spring
Field Frame
Armature
Bearing
O-Ring
Through Bolt
Brush Holder
Bearing
Pinion Gear
Idle Gear
Steel Ball
Bearing
Magnetic Switch Assembly
Spring
Starter Clutch Assembly
Starter Housing

Toyota—reduction type starter—1981 and later

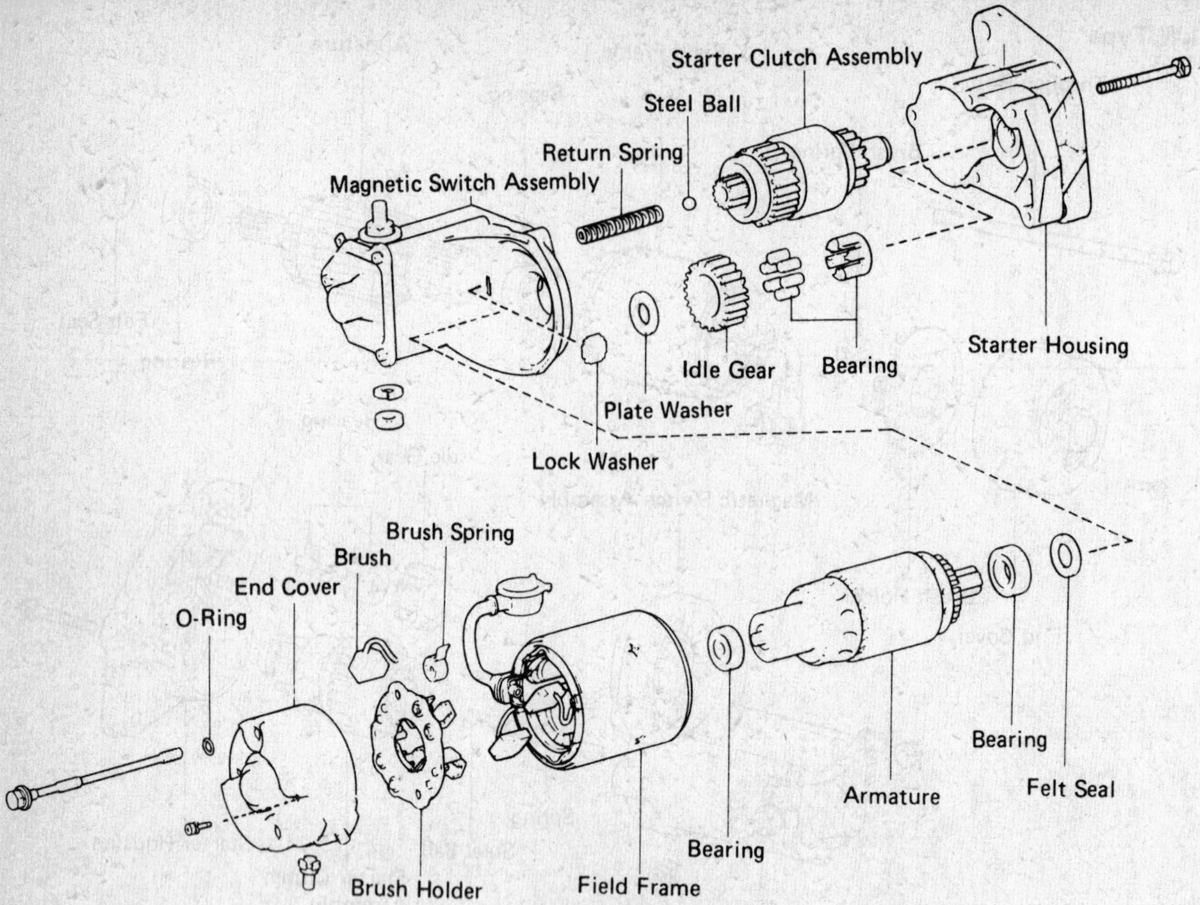

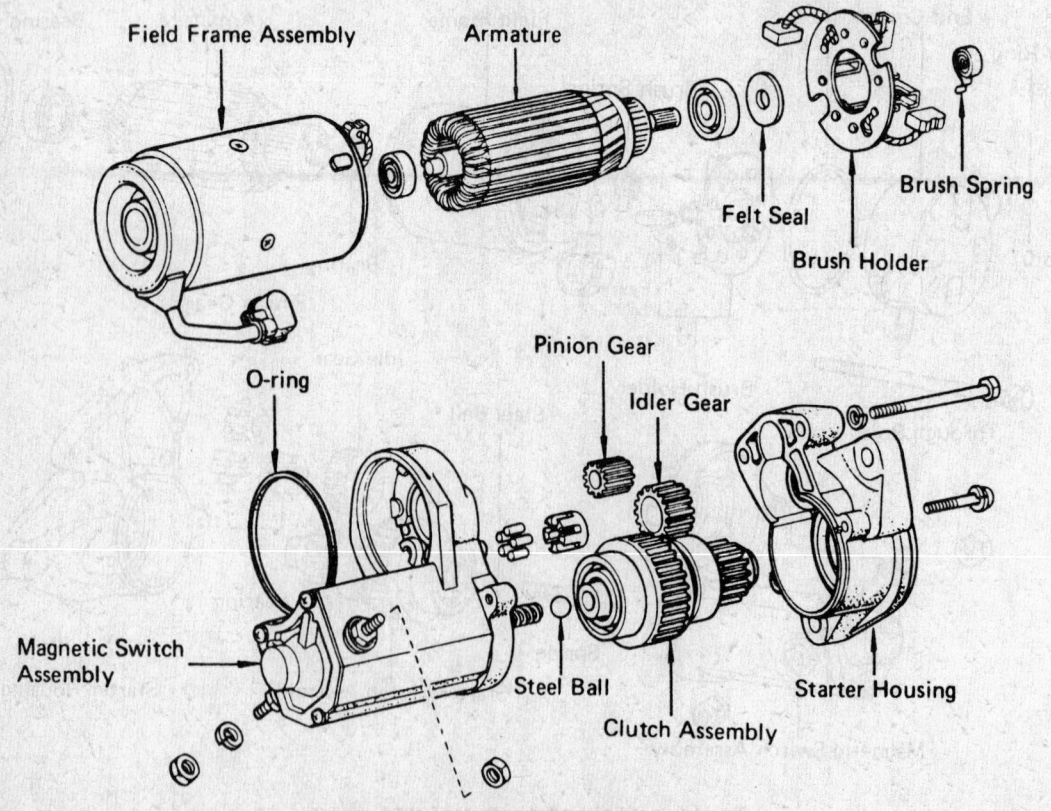

Starter Clutch Assembly

Steel Ball

Return Spring

Magnetic Switch Assembly

Idle Gear

Bearing

Starter Housing

Plate Washer

Lock Washer

Brush Spring

Brush

End Cover

O-Ring

Brush Holder

Field Frame

Bearing

Armature

Bearing

Felt Seal

Toyota—reduction type starter—1979–80

Field Frame Assembly

Armature

Brush Spring

Felt Seal

Brush Holder

Pinion Gear

Idler Gear

O-ring

Magnetic Switch Assembly

Steel Ball

Clutch Assembly

Starter Housing

Toyota—diesel engine starter

Toyota

STARTER DRIVE REPLACEMENT

Direct Drive Type

1. Remove the field coil lead from the solenoid terminal.

2. Remove the solenoid retaining screws. Remove the solenoid by tilting it upward and withdrawing it.

3. Remove the thru-bolts, then the drive housing from the field frame.

4. Remove the end-frame cap screws, the cap, the C-lock and the washer from the commutator end frame. Remove the commutator end frame from the field frame.

5. Withdraw the brushes from their holder if they are to be replaced.

NOTE: Check the brush length against the specification in the Battery and Starter Specifications chart. Replace the brushes with new ones if required.

6. Remove the armature assembly through the front of the field frame.

7. Place the armature vertically on a block of wood with the clutch assembly facing upwards.

8. Using a 14mm socket and a hammer, drive the stop collar (on the shaft) toward the armature to expose the snap ring.

9. Remove the snap ring, the stop collar and the drive assembly from the shaft.

10. To assemble, reverse the disassembly procedures. Pack the end bearing cover with multipurpose grease before installing it.

Reduction Type

NOTE: The starter must be removed from the vehicle, in order to perform this operation.

1. Remove the starter housing-to-field frame bolts and the starter housing-to-solenoid bolts, then separate the starter housing from the solenoid/field frame assembly.

2. On the 1.0 kw type, withdraw the clutch and idler gear. On the 1.4 kw, withdraw the pinion gear, the idler gear and the clutch.

3. To install, grease the moving parts and reverse the removal procedures.

VW

NOTE: A new type of starter has been installed on some models that are equipped with a manual transmission. The new style starter is not interchangeable with the old design.

OVERHAUL

Use the following procedure to replace brushes or starter drive.

1. Remove the solenoid.

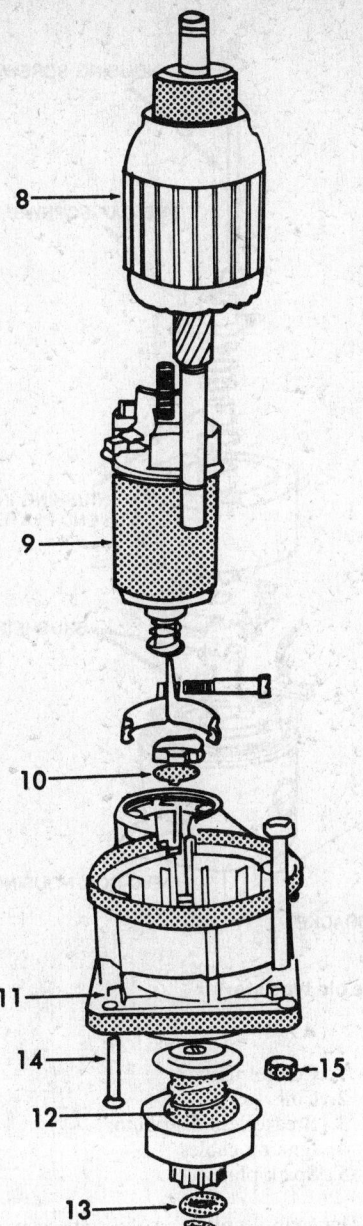

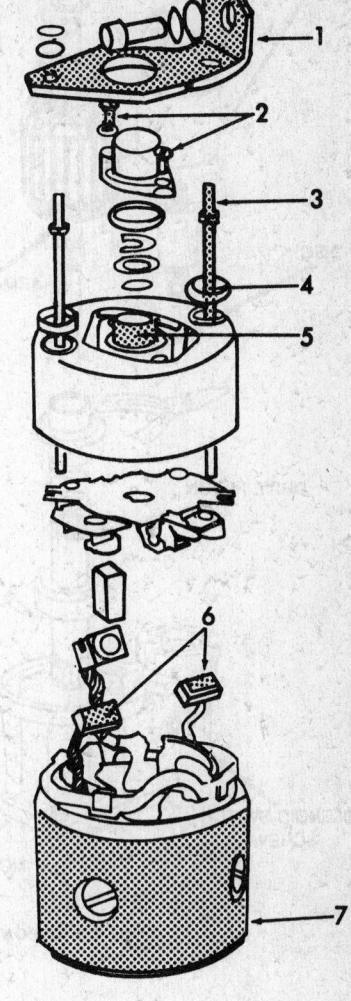

1. Mounting bracket	9. Solenoid
2. End cap screws	10. Disc
3. Housing screws	11. Mounting housing
4. Cupped washer	12. Drive pinion
5. End plate bushing	13. Stop ring
6. Brushes	14. Solenoid bolt
7. Field coil housing	15. Starter bolt and nut
8. Armature	16. Circlip

VW—exploded view of the new type starter

2. Remove the end bearing cap.

3. Loosen both of the long housing screws.

4. Remove the lockwasher and spacer washers.

5. Remove the long housing screws and remove the end cover.

6. Pull the two field coil brushes out of the brush housing.

7. Remove the brush housing assembly.

8. Loosen the nut on the solenoid housing, remove the sealing disc, and remove the solenoid operating lever.

9. Loosen the large screws on the side of the starter body and remove the field coil along with the brushes.

NOTE: If the brushes require replacement, the field coil and brushes and/or the brush housing and its brushes must be replaced as a unit.

10. If the starter drive is being replaced on the new type starter, push the stop-ring down and remove the circlip on the end of the shaft. Remove the stop-ring and remove the drive.

11. To remove the starter drive on old type starters, remove the armature and pull the drive unit off the end.

12. Assembly of the starter is carried out in the reverse order of disassembly. Use a gear puller to install the stop-ring in its groove (on models so equipped). Use a new circlip on the shaft.

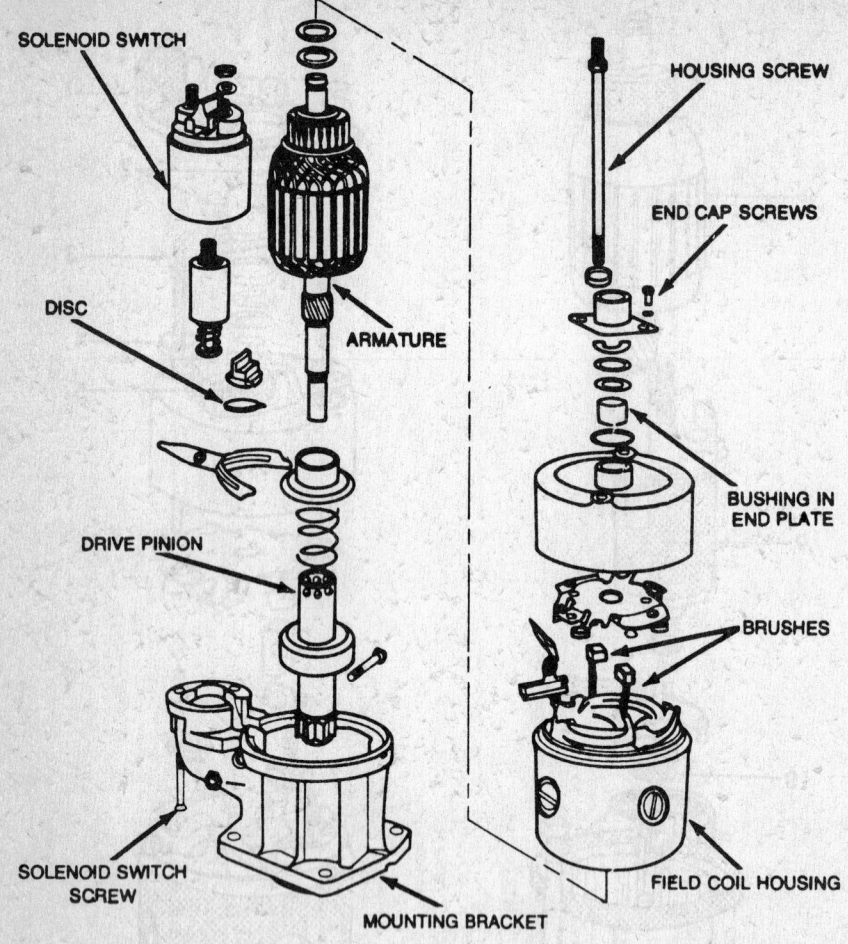

VW—exploded view of the old type starter

Labels: SOLENOID SWITCH, DISC, DRIVE PINION, SOLENOID SWITCH SCREW, MOUNTING BRACKET, ARMATURE, HOUSING SCREW, END CAP SCREWS, BUSHING IN END PLATE, BRUSHES, FIELD COIL HOUSING

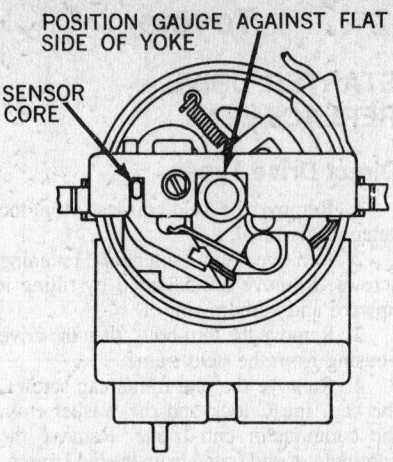

POSITION GAUGE AGAINST FLAT SIDE OF YOKE

SENSOR CORE

Using a special gauge to align the BID sensor coil

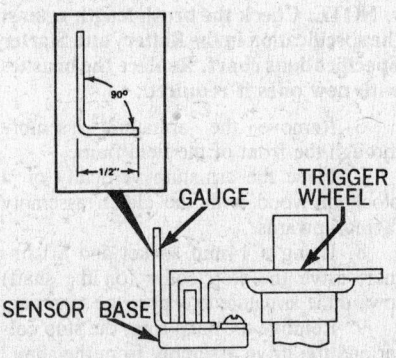

90°
1/2"
GAUGE
TRIGGER WHEEL
SENSOR BASE

Fabricating a BID wheel clearance gauge

SOLENOID REPLACEMENT

1. Remove the starter.
2. Remove the nut which secures the connector strip on the end of the solenoid.
3. Take out the two retaining screws on the mounting bracket and pull out the solenoid after it has been unhooked from the operating lever.
4. Installation is the reverse of removal. In order to facilitate engagement of the lever, the pinion should be pulled out as far as possible when inserting the solenoid.

IGNITION SYSTEM

AMC Breakerless Inductive Discharge (BID) Ignition

COMPONENTS

The AMC breakerless inductive discharge (BID) ignition system consists of five components:

1. Control unit
2. Coil
3. Breakerless distributor
4. Ignition cables
5. Spark plugs

The control unit is a solid-state, epoxy-sealed module with waterproof connectors. The control unit has a built-in current regulator, so no separate ballast resistor or resistance wire is needed in the primary circuit. Battery voltage is supplied to the ignition coil positive (+) terminal when the ignition key is turned to the "ON" or "START" position; low voltage is also supplied by the control unit.

The coil used with the BID system requires no special service. It works just like the coil in a conventional ignition system.

The distributor is conventional, except for the lack of points, condenser and cam. Advance is supplied by both a vacuum unit and a centrifugal advance mechanism. A standard cap, rotor, and dust shield are used.

In place of the points, cam, and condenser, the distributor has a sensor and trigger wheel. The sensor is a small coil which generates an electromagnetic field when excited by the oscillator in the control unit.

Standard spark plugs and ignition cables are used.

OPERATION

When the ignition switch is turned on, the control unit is activated. The control unit then sends an oscillating signal to the sensor which causes the sensor to generate a magnetic field. When one of the trigger wheel teeth enters this field, the strength of the oscillation in the sensor is reduced. Once the strength drops to a predetermined level, a demodulator circuit operates the control unit's switching transistor. The switching transistor is wired in series with the coil primary circuit; it switches the circuit off when it gets the demodulator signal.

From this point on, the BID ignition system works in the same manner as a conventional ignition system.

TROUBLESHOOTING

1. Check all of the BID ignition system electrical connections.
2. Disconnect the coil-to-distributor high tension lead.
3. Hold the end of the lead ½ in. away from a ground. Crank the engine. If there is a spark, the trouble is not in the ignition system.

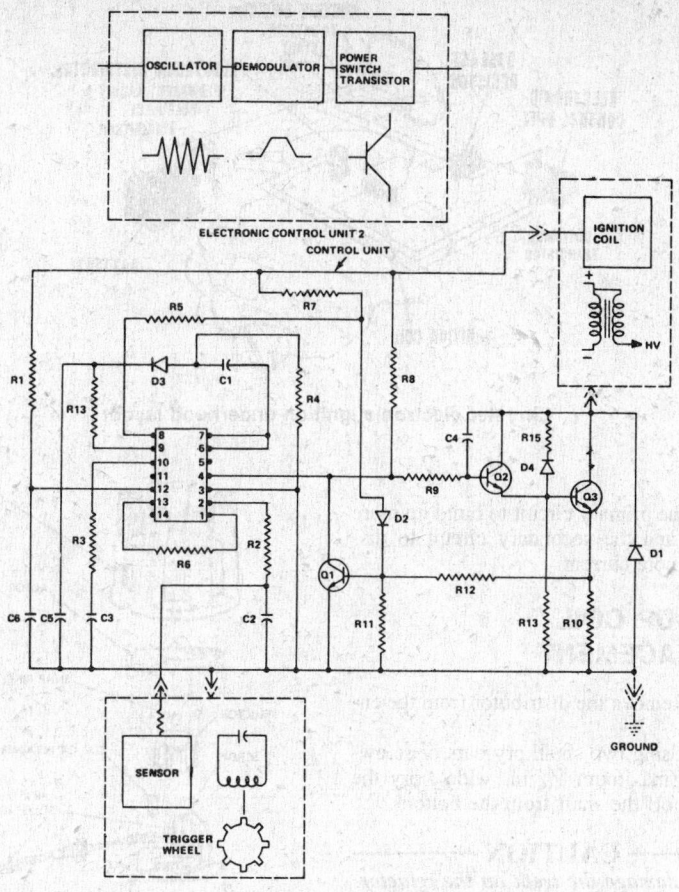

BID schematic

4. If there was no spark in Step 3, connect a test light with a No. 57 bulb between the positive coil terminal (+) and a good ground. Have an assistant turn the ignition switch to "ON" and "START" (Do not start the engine). The bulb should light in both positions; if it doesn't, the fault lies in the battery-to-coil circuit. Check the ignition switch and related wiring.

5. If the test light lit in Step 4, disconnect the coil-to-distributor leads at the connector and connect the test light between the positive (+) and negative (−) coil terminals.

6. Turn the ignition switch on. If the test light doesn't come on, check the control unit's ground lead. If the ground lead is in good condition, replace the control unit.

7. If the bulb lights in Step 6, leave the test light in place and short the terminals on the coil-to-distributor connector together with a jumper lead, (connector separated) at the coil side of the connector. If the light stays on, replace the control unit.

8. If the test light goes out, remove it. Check for a spark, as in Step 2, each time that the coil-to-distributor connector terminals are shorted together with the jumper lead. If there is a spark, replace the control unit; if there is no spark, replace the coil.

Coil Testing

Test the coil with a conventional coil checker or an ohmmeter. Primary resistance should be 1–2 ohms and secondary resistance should be 8–12 kilohms. The open output circuit should be more than 20 kilovolts. Replace the coil if it doesn't meet specifications.

Sensor Testing

Check the sensor resistance by connecting an ohmmeter to its leads. Resistance should be 1.8 ohms (±10%) at 77° F. Replace the sensor if it doesn't meet these specifications.

DISTRIBUTOR OVERHAUL

NOTE: If you must remove the sensor from the distributor for any reason, it will be necessary to have the special sensor positioning gauge in order to align it properly during installation.

1. Scribe matchmarks on the distributor housing, rotor, and engine block. Disconnect the leads and vacuum lines from the distributor. Remove the distributor. Unless the cap is to be replaced, leave it connected to the spark plug cables and position it out of the way.

2. Remove the rotor and dust cap.

3. Place a small gear puller over the trigger wheel, so that its jaws grip the inner shoulders of the wheel and not its arms. Place a thick washer between the gear puller and the distributor shaft to act as a spacer; do not press against the smaller inner shaft.

4. Loosen the sensor hold-down screw with a small pair of needle-nosed pliers; it has a tamperproof head. Pull the sensor lead grommet out of the distributor body and pull out the leads from around the spring pivot pin.

5. Release the sensor securing spring by lifting it. Make sure that it clears the leads. Slide the sensor off the bracket.

NOTE: A special gauge is required for sensor installation.

6. Remove the vacuum advance unit securing screw. Slide the vacuum unit out of the distributor. Remove it only if it is to be replaced.

7. Clean and dry the vacuum unit and sensor brackets. Lubrication of these parts is not necessary.

8. Install the vacuum unit, if it was removed.

9. Assemble the sensor, sensor guide, flat washer, and retaining screw. Tighten the screw only far enough to keep the assembly together; don't allow the screw to project below the bottom of the sensor.

NOTE: Replacement sensors come with a slotted-head screw to aid in assembly. If the original sensor is being used, replace the tamper-proof screw with a conventional one. Use the original washer.

10. Secure the sensor on the vacuum advance unit bracket, making sure that the tip of the sensor is placed in the notch on the summing bar.

11. Position the spring on the sensor and route the leads around the spring pivot pin. Fit the sensor lead grommet into the slot on the distributor body. Be sure that the lead can't get caught in the trigger wheel.

12. Place the special sensor positioning gauge over the distributor shaft, so that the flat on the shaft is against the large notch on the gauge. Move the sensor until the sensor core fits into the small notch on the gauge. Tighten the sensor securing screw with the gauge in place (through the round hole in the gauge).

13. It should be possible to remove and install the gauge without any side movement of the sensor. Check this and remove the gauge.

14. Position the trigger wheel on the shaft. Check to see that the sensor core is centered between the trigger wheel legs and that the legs don't touch the core.

15. Bend a piece of 0.050 in. gauge wire, so that it has a 90° angle and one leg ½ in. long. Use the gauge to measure the clearance between the trigger wheel legs and the sensor boss. Press the trigger wheel on the shaft until it just touches the gauge. Support the shaft during this operation.

16. Place 3 to 5 drops of SAE 20 oil on the felt lubricator wick.

17. Install the dust shield and rotor on the shaft.

18. Install the distributor on the engine using the matchmarks made during removal and adjust the timing. Use a new distributor mounting gasket.

Chrysler Electronic Ignition

COMPONENTS

This system consists of a special pulse-sending distributor, an electronic control unit, a two-element ballast resistor, and a special ignition coil.

The distributor does not contain breaker points or a condenser, these parts being replaced by a distributor reluctor and a pick-up unit.

OPERATION

The ignition primary circuit is connected from the battery, through the ignition switch, through the primary side of the ignition coil, to the control unit where it is grounded. The secondary circuit is the same as in conventional ignition systems: the secondary side of the coil, the coil wire to the distributor, the rotor, the spark plug wires, and the spark plugs.

The magnetic pulse distributor is also connected to the control unit. As the distributor shaft rotates, the distributor reluctor turns past the pick-up unit. As the reluctor turns past the pick-up unit, each of the eight teeth on the reluctor pass near the pick-up unit once during each distributor revolution (two crankshaft revolutions since the distributor runs at one-half crankshaft speed). As the reluctor teeth move close to the pick-up unit, the magnetic rotating reluctor induces voltage into the magnetic pick-up unit. This voltage pulse is sent to the ignition control unit from the magnetic pick-up unit. When the pulse enters the control unit, it signals the control unit to interrupt the ignition primary circuit. This causes the primary circuit to collapse and begins the induction of the magnetic lines of force from the primary side of the coil into the secondary side of the coil. This induction provides the required voltage to fire the spark plugs.

The advantages of this system are that the transistors in the control unit can make and break the primary ignition circuit much faster than conventional ignition points can, and higher primary voltage can be utilized, since this system can be made to handle higher voltage without adverse effects, whereas ignition breaker points cannot. The quicker switching time of this system allows longer coil primary circuit saturation time and longer induction time when the primary circuit collapses. This increased time

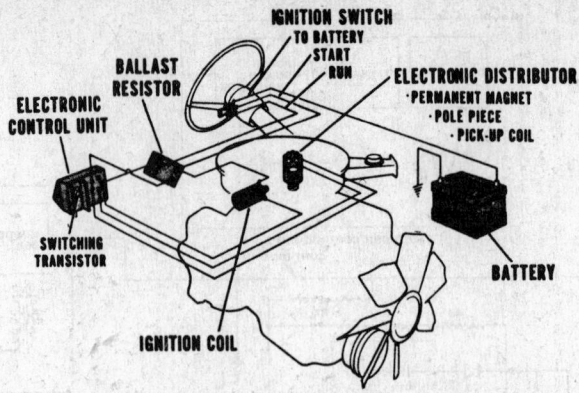

Chrysler electronic ignition underhood layout

allows the primary circuit to build up more current and the secondary circuit to discharge more current.

PICK-UP COIL REPLACEMENT

1. Remove the distributor from the engine.

2. Using two small pry-bars or screwdrivers (maximum $7/16$ in. wide), pry the reluctor off the shaft from the bottom.

─────── **CAUTION** ───────
Do not damage the teeth on the reluctor.

3. Unfasten the vacuum advance-to-distributor housing screws. Remove the vacuum unit, after disconnecting the arm from the upper plate.

4. Unfasten the pick-up coil wires from the distributor housing.

5. Unfasten the two screws which secure the lower plate to the distributor housing. Lift out the lower plate together with the upper plate and pick-up coil.

6. Separate the upper and lower plates by depressing the retaining clip on the underside of the plate and slide it away from the stud. The pick-up coil will come off with the upper plate; they cannot be separated; they must be serviced as an assembly.

7. Installation is the reverse of removal. Place a small amount of distributor grease on the support pins on the lower plate.

Air Gap Adjustment

1. Align one reluctor tooth with the pick-up coil tooth.

2. Loosen the pick-up coil hold-down screw.

3. Insert a 0.008 in. nonmagnetic feeler gauge between the reluctor tooth and the pick-up coil tooth.

4. Adjust the air gap so that contact is made between the reluctor tooth, the feeler gauge, and the pick-up coil tooth.

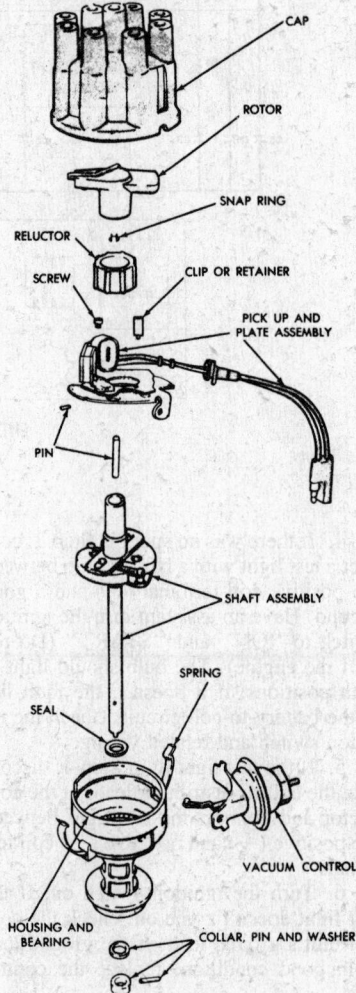

V8 distributor

5. Tighten the hold-down screw.

6. Remove the feeler gauge.

NOTE: No force should be required in removing the feeler gauge.

7. Check the air gap with a 0.008 in. feeler gauge. A 0.008 in. feeler gauge should not fit into the air gap.

CAUTION

A 0.008 in. feeler gauge can be forced into the air gap. Do not force the gauge into the air gap.

---

8. Apply vacuum to the vacuum unit and rotate the governor shaft. The pick-up pole should not hit the reluctor teeth. The gap was not properly adjusted if any hitting occurs. If hitting occurs on only one side of the reluctor, the distributor shaft is probably bent, and the governor and shaft assembly should be replaced.

TROUBLESHOOTING

Chrysler Corporation has an Electronic Ignition System Tester to be used when checking the system. However, many shops are not able to obtain this equipment, so an alternate method has been developed. The system may be tested using a voltmeter with a scale of 20 volts and an ohmmeter with a scale of 20,000 ohms.

When the ignition system is suspected of malfunctions, the following procedure should be used.

1. Inspect the secondary wires for cracks and tightness.

2. Check all primary wires and connections at the components for tightness.

3. Check and note the battery voltage.

4. Be sure the ignition switch is in the "OFF" position and remove the multi-wiring connector from the control unit.

5. Turn the ignition switch to the "ON" position and connect the negative lead of the voltmeter to a good ground.

6. Connect the positive lead of the voltmeter to the number 1 terminal of the wiring harness connector. The voltage should be within 1 volt of battery voltage. If more than 1 volt, the circuit must be checked for high resistance.

7. Connect the positive lead of the voltmeter to the wiring harness connector number 2 terminal. Available voltage should be within 1 volt of battery voltage. If more than 1 volt, the circuit should be checked for high resistance.

8. Connect the positive lead of the voltmeter to the wiring harness connector terminal number 3. Available voltage should be within 1 volt of battery voltage. If more than 1 volt, the circuit must be checked for high resistance.

9. Turn the ignition switch to the "OFF" position.

10. Connect an ohmmeter to the wiring harness connector terminal numbers 4 and 5. This checks the distributor pick-up coil. The ohmmeter resistance reading should be between 150 and 900 ohms. If the readings are higher or lower than specified, disconnect the dual lead connector coming from the distributor. Using the ohmmeter leads, check the resistance at the dual lead connector. If the reading is not between 150 and 900 ohms, replace the pick-up coil assembly in the distributor. If the reading is

within specifications at the dual lead connector, check the wiring harness between the control unit and the dual lead connector.

11. Connect one ohmmeter lead to a good ground and the other lead to either connector of the distributor. The ohmmeter should show an open circuit. If the ohmmeter shows a reading, the pickup coil in the distributor must be replaced.

12. When checking the electronic control unit, connect one ohmmeter lead to a good ground and the other lead to the control unit connector pin number 5. The ohmmeter should show continuity between

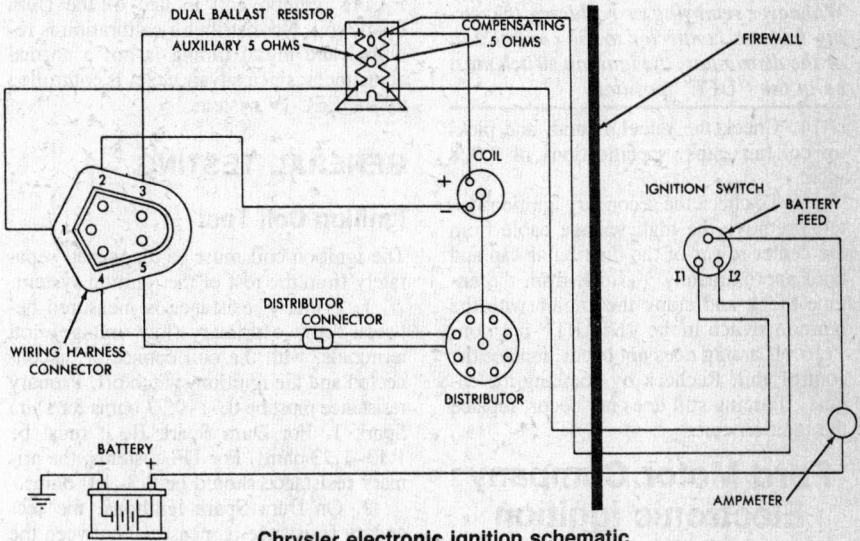

Chrysler electronic ignition schematic

TROUBLESHOOTING CHRYSLER ELECTRONIC IGNITION

Condition	Possible Cause	Correction
ENGINE WILL NOT START (Fuel and carburetion known to be OK)	a) Dual Ballast	Check resistance of each section: Compensating resistance: .50-.60 ohms @ 70°-80°F Auxiliary Ballast: 4.75-5.75 ohms Replace if faulty. Check wire positions.
	b) Faulty Ignition Coil	Check for carbonized tower. Check primary and secondary resistances: Primary: 1.41-1.79 ohms @ 70°-80°F Secondary: 9,200-11,700 ohms @ 70°-80°F Check in coil tester.
	c) Faulty Pickup or Improper Pickup Air Gap	Check pickup coil resistance: 400-600 ohms Check pickup gap: .010 in. feeler gauge should not slip between pickup coil core and an aligned reluctor blade. No evidence of pickup core striking reluctor blades should be visible. To reset gap, tighten pickup adjustment screw with a .008 in. feeler gauge held between pickup core and an aligned reluctor blade. After resetting gap, run distributor on test stand and apply vacuum advance, making sure that the pickup core does not strike the reluctor blades.
	d) Faulty Wiring	Visually inspect wiring for brittle insulation. Inspect connectors. Molded connectors should be inspected for rubber inside female terminals.
	e) Faulty Control Unit	Replace if all of the above checks are negative. Whenever the control unit or dual ballast is replaced, make sure the dual ballast wires are correctly inserted in the keyed molded connector.
ENGINE SURGES SEVERELY (Not Lean Carburetor)	a) Wiring	Inspect for loose connection and/or broken conductors in harness.
	b) Faulty Pickup Leads	Disconnect vacuum advance. If surging stops, replace pickup.
	c) Ignition Coil	Check for intermittent primary.
ENGINE MISSES (Carburetion OK)	a) Spark Plugs	Check plugs. Clean and regap if necessary.
	b) Secondary Cable	Check cables with an ohmmeter, or observe secondary circuit performance with an oscilloscope.
	c) Ignition Coil	Check for cabonized tower. Check in coil tester.
	d) Wiring	Check for loose or dirty connections.
	e) Faulty Pickup Lead	Disconnect vacuum advance. If miss stops, replace pickup.
	f) Control Unit	Replace if the above checks are negative.

the ground and the connector pin. If continuity does not exist, tighten the bolts holding the control unit to the vehicle panel and recheck. If continuity does not exist, the control unit must be replaced.

13. Reconnect the wiring harness at the control unit and distributor.

——————— CAUTION ———————

Whenever removing or replacing the wiring harness connector to the control unit or the distributor, the ignition switch must be in the "OFF" position.

14. Check the reluctor tooth and pickup coil air gaps, specifications of 0.008 inch.

15. To check the secondary ignition system, remove the high voltage cable from the center tower of the distributor cap and hold approximately $\frac{3}{16}$ inch from the engine block and crank the engine with the ignition switch in the "START" position.

16. If arcing does not occur, replace the control unit. Recheck by cranking the engine. If arcing still does not occur, replace the ignition coil.

Ford Motor Company Electronic Ignition Systems

Four different electronic ignition systems are used on Ford Motor Company vehicles, depending on year, engine and model. The four systems are:

1. Dura Spark I
2. Dura Spark II
3. Dura Spark III
4. TFI-IV (Thick Film Intgegrated)

Dura Spark I and Dura Spark II systems are nearly identical in operation, and virtually identical in appearance. The Dura Spark I uses a special control module which senses current flow through the ignition coil and adjust the dwell, or coil "on" time for maximum spark intensity. If the Dura Spark I module senses that the ignition is ON, but the distributor shaft is not turning, the current to the coil is turned OFF by the module. The Dura Spark II system does not have this feature, the coil is energized for the full amount of time that the ignition switch is ON. Keep this in mind when servicing the Dura Spark II system, as the ignition system could inadvertently "fire" while performing ignition system services (such as distributor cap removal) while the ignition is ON.

Dura Spark III is based on the previous systems, but the input signal is controlled by the EEC system, rather than as a function of engine timing and distributor armature position. The distributor, rotor cap, and control module are unique to this system; the spark plugs and plug wires are the same as those used with the Dura Spark II system. Although the Dura Spark II and III control modules are similar in appearance, they cannot be interchanged between systems.

The TFI-IV ignition system features a universal distributor using no centrifugal or vacuum advance. The distributor has a die cast base which incorporates an integrally mounted TFI (Thick Film Integrated) ignition module, a "Hall Effect" vane switch stator asasembly and provision for fixed octane adjustment. The TFI system uses an E-Core ignition coil in lieu of the Dura Spark coil. No distributor calibration is required and initial timing is not a normal adjustment, since advance etc. is controlled by the EEC-IV system.

GENERAL TESTING

Ignition Coil Test

The ignition coil must be diagnosed separately from the rest of the ignition system.

1. Primary resistance is measured between the two primary (low voltage) coil terminals, with the coil connector disconnected and the ignition switch off. Primary resistance must be 0.71–0.77 ohms for Dura Spark 1. For Dura Spark II, it must be 1.13–1.23 ohms. For TFI systems, the primary resistance should be 0.3–1.0 ohms.

2. On Dura Spark ignitions, the secondary resistance is measured between the BATT and high voltage (secondary) terminals of the ignition coil with the ignition off, and the wiring from the coil disconnected. Secondary resistance must be 7350–8250 ohms on Dura Spark I systems. Dura Spark II figure is 7700–9300 ohms. For TFI systems, the primary resistance should be 8000–11500 ohms.

3. If resistance tests are okay, but the coil is still suspected, test the coil on a coil tester by following the test equipment manufacturer's instructions for a standard coil. If the reading differs from the original test, check for a defective harness.

Resistance Wire Test

Replace the resistance wire if it doesn't show a resistance of 1.05–1.15 for Dura Spark II. The resistance wire isn't used on Dura Spark I or TFI systems.

Spark Plug Wire Resistance

Resistance on these wires must not exceed 5000 ohms per inch. To properly measure this, remove the wires from the plugs, and remove the distributor cap. Measure the resistance through the distributor cap at that end. Do not pierce any ignition wire for any reason. Measure only from the two ends.

NOTE: Silicone grease must be reapplied to the spark plug wires whenever they are removed. When removing the wires from the spark plugs, a special tool should be used. Do not pull on the wires. Grasp and twist the boot to remove the wire. Whenever the high tension wires are removed from the plugs, coil, or distributor, silicone grease must be applied to the boot before reconnection. Use a clean small screwdriver blade to coat the entire interior surface with Ford silicone grease D7AZ–19A33l–A, Dow Corning #111, or General Electric G–627.

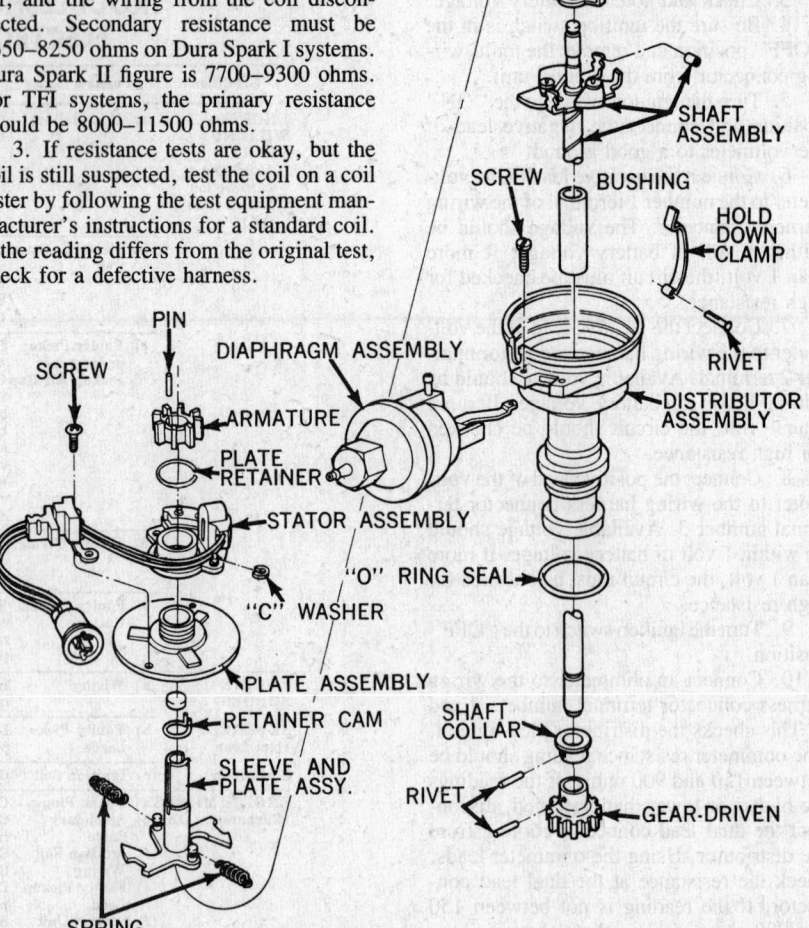

8-cylinder breakerless distributor

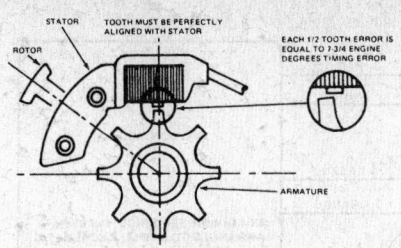

Armature alignment

Adjustments

The air gap between the armature and magnetic pick-up coil in the distributor is not adjustable, nor are there any adjustments for the amplifier module. Inoperative components are simply replaced. Any attempt to connect components outside the vehicle may result in component failure.

TROUBLESHOOTING DURA SPARK I

The following Dura Spark II troubleshooting procedures may be used on Dura Spark I systems with a few variations. The Dura Spark I module has internal connections which shut off the primary circuit in the run mode when the engine stalls. To perform the above troubleshooting procedures, it is necessary to by-pass these connections. However, with these connections by-passed, the current flow in the primary becomes so great that it will damage both the ignition coil and module unless a ballast resistor is installed in series with the primary circuit at the BAT terminal of the ignition coil. Such a resistor is available from Ford (Motorcraft part number DY–36). A 1.3 ohm, 100 watt wire-wound power resistor can also be used. To install the resistor, proceed as follows:

1. Release the BAT terminal lead from the coil by inserting a paper clip through the hole in the rear of the horseshoe coil connector and manipulating it against the locking tab in the connector until the lead comes free.

2. Insert a paper clip in the BAT terminal of the connector on the coil. Using jumper leads, connect the ballast resistor as shown.

3. Using a straight pin, pierce both the red and white leads of the module to short these two together. This will by-pass the internal connections of the module which turn off the ignition circuit when the engine is not running.

— **CAUTION** —

Pierce the wires only AFTER the ballast resistor is in place or the ignition coil and module will be damaged.

4. With the ballast resistor and by-pass in place, proceed with the Dura Spark II troubleshooting procedures. The resistor will become very hot during testing.

TROUBLESHOOTING DURA SPARK II

The following procedures can be used to determine whether the ignition system is working or not. If these procedures fail to correct the problem, a full troubleshooting procedure should be performed.

Preliminary Checks

1. Check the battery's state of charge and connections.

2. Inspect all wires and connections for breaks, cuts, abrasions, or burn spots. Repair as necessary.

3. Unplug all connectors one at a time and inspect for corroded or burned contacts. Repair and plug connectors back together.

DO NOT remove the dielectric compound in the connectors.

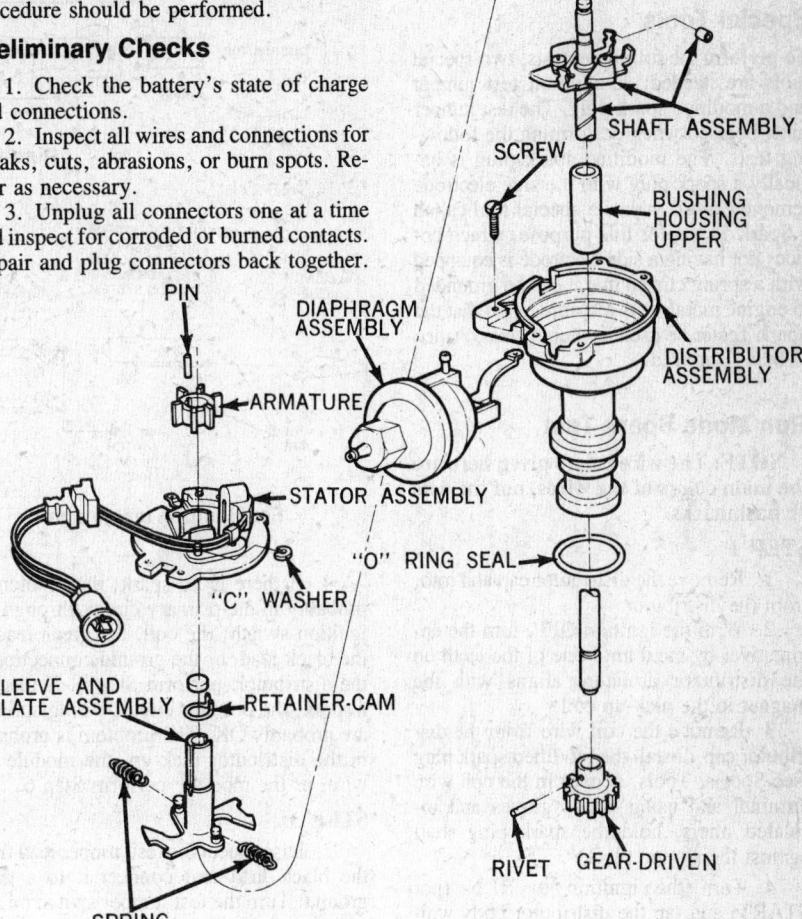

6-cylinder breakerless distributor

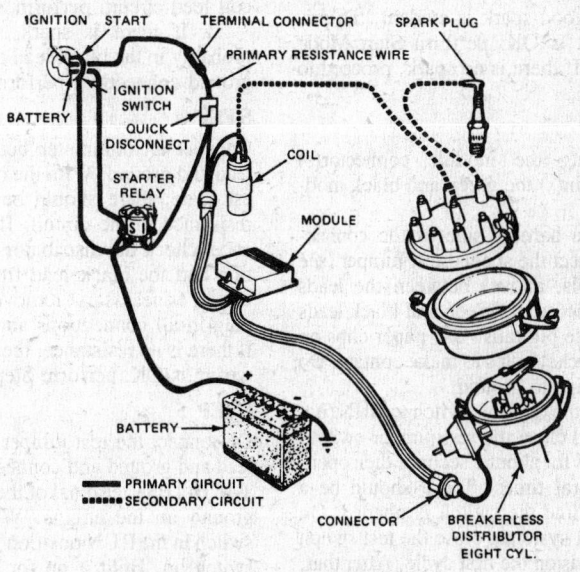

Breakerless ignition circuit routings

4. Check for loose or damaged spark plug or coil wires. A wire resistance check is given at the end of this section. If the boots or nipples are removed on 8mm ignition wires, reline the inside of each with new silicone di-electric compound (Motorcraft WA 10).

Special Tools

To perform the following tests, two special tools are needed; an ignition test jumper and a modified spark plug. The test jumper must be used when performing the following tests. The modified spark plug is basically a spark plug with the side electrode removed. Ford makes a special tool called a Spark Tester for this purpose, which besides not having a side electrode is equipped with a spring clip so that it can be grounded to engine metal. It is recommended that the Spark Tester be used as there is less chance of being shocked.

Run Mode Spark Test

NOTE: The wire colors given here are the main colors of the wires, not the dots or hashmarks.

STEP 1

1. Remove the distributor cap and rotor from the distributor.

2. With the ignition OFF, turn the engine over by hand until one of the teeth on the distributor armature aligns with the magnet in the pick-up coil.

3. Remove the coil wire from the distributor cap. Install the modified spark plug (see Special Tools, above) in the coil wire terminal and using heavy gloves and insulated pliers, hold the spark plug shell against the engine block.

4. Turn the ignition to RUN (not START) and tap the distributor body with a screwdriver handle. There should be a spark at the modified spark plug or at the coil wire terminal.

5. If a good spark is evident, the primary circuit is OK: perform Start Mode Spark Test. If there is no spark, proceed to Step 2.

STEP 2

1. Unplug the module connector(s) which contain(s) the green and black module leads.

2. In the harness side of the connector(s), connect the special test jumper (see Special Tools, above) between the leads which connect to the green and black leads of the module pig tails. Use paper clips on connector socket holes to make contact. Do not allow clips to ground.

3. Turn the ignition switch to RUN (not START) and close the test jumper switch. Leave closed for about 1 second, then open. Repeat several times. There should be a spark each time the switch is opened. On Dura Spark I systems, close the test switch for 10 seconds on the first cycle. After that, 1 second is adequate.

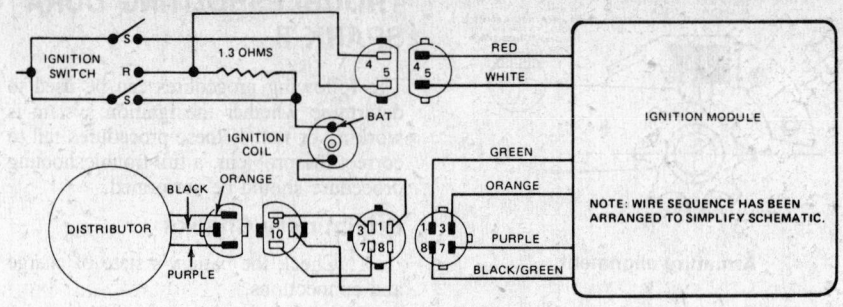

Dura-Spark schematic

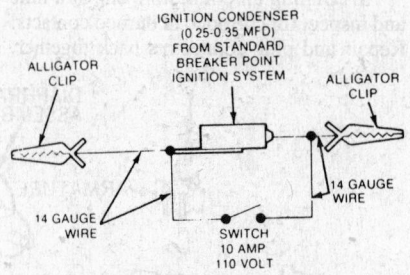

Ford ignition test jumper

4. If there is no spark, the problem is probably in the primary circuit through the ignition switch, the coil, the green lead or the black lead, or the ground connection in the distributor: perform Step 3. If there is a spark, the primary circuit wiring and coil are probably OK. The problem is probably in the distributor pick-up, the module red wire, or the module: perform Step 6.

STEP 3

1. Disconnect the test jumper lead from the black lead and connect it to a good ground. Turn the test jumper switch on and off several times as in Step 2.

2. If there is no spark, the problem is probably in the green lead, the coil, or the coil feed circuit: perform Step 5.

3. If there is spark, the problem is probably in the black lead or the distributor ground connection: perform Step 4.

STEP 4

Connect an ohmmeter between the black lead and ground. With the meter on its lowest scale, there should be no measurable resistance in the circuit. If there is resistance, check the distributor ground connection and the black lead from the module. Repair as necessary, remove the ohmmeter, plug in all connections and repeat Step 1. If there is no resistance, the primary ground wiring is OK: perform Step 6.

STEP 5

Disconnect the test jumper from the green lead and ground and connect it between the TACH-TEST terminal of the coil and a good ground on the engine. With the ignition switch in the RUN position, turn the jumper switch on. Hold it on for about 1 second then turn it off as in Step 2. Repeat several

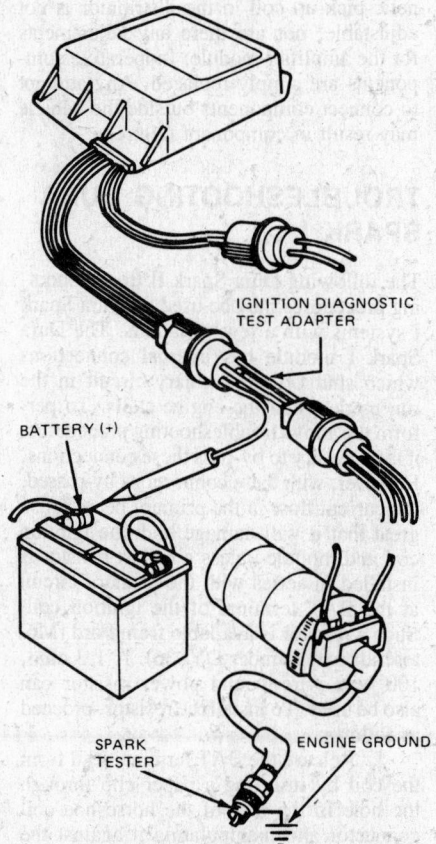

Connecting the Dura Spark III test adapter.

times. There should be a spark each time the switch is turned off. If there is no spark, the problem is probably in the primary circuit running through the ignition switch to the coil BAT terminal, or in the coil itself. Check coil resistance (test given later in this section), and check the coil for internal shorts or opens. Check the coil feed circuit for opens, shorts or high resistance. Repair as necessary, reconnect all connectors and repeat Step 1. If there is spark, the coil and its feed circuit are OK. The problem could be in the green lead between the coil and the module. Check for open or short, repair as necessary, reconnect all connectors and repeat Step 1.

STEP 6

To perform this step, a voltmeter which is

not combined with a dwell meter is needed. The slight needle oscillations ($\frac{1}{2}$ V) may not be detectable on the combined voltmeter/dwell meter unit.

1. Connect a voltmeter between the orange and purple leads on the harness side of the module connectors.

--- CAUTION ---

On the catalytic converter equipped cars, disconnect the air supply line between the Thermactor by-pass valve and the manifold before cranking the engine with the ignition off. This will prevent damage to the catalytic converter. After testing, run the engine for at least 3 minutes before reconnecting the by-pass valve, to clear excess fuel from the exhaust system.

2. Set the voltmeter on it lowest scale and crank the engine. The meter needle should oscillate slightly (about $\frac{1}{2}$ volt). If the meter does not oscillate, check the circuit through the magnetic pick-up in the distributor for open, shorts, shorts to ground and resistance. Resistance between the orange and purple leads should be 400–1000 ohms, and between each lead and ground should be more than 70,000 ohms. Repair as necessary, reconnect all connectors and repeat Step 1. If the meter oscillates, the problem is probably in the power feed to the module (red wire) or in the module itself: proceed to Step 7.

STEP 7

1. Remove all meters and jumpers and plug in all connectors.

2. Turn the ignition switch to the RUN position and measure voltage between the battery positive terminal and engine ground. It should be 12 volts.

3. Measure voltage between the red lead of the module and engine ground. To make this measurement, it will be necessary to pierce the red wire with a straight pin and connect the voltmeter to the straight pin and to ground. DO NOT ALLOW THE STRAIGHT PIN TO GROUND ITSELF.

4. The two readings should be within one volt of each other. If not within one volt, the problem is in the power feed to the red lead. Check for shorts, open, or high resistance and correct as necessary. After repairs, repeat Step 1. If the readings are within one volt, the problem is probably in the module. Replace with a good module and repeat Step 1. If this corrects the problem, reconnect the old module and repeat Step 1. If problem returns, permanently install the new module.

Start Mode Spark Test

NOTE: The wire colors given here are the main colors of the wires, not the dots or hashmarks.

1. Remove the coil wire from the distributor cap. Install the modified spark plug mentioned under "Special Tools" in the coil wire and ground it to engine metal either by its spring clip (Spark Tester) or by hold-ing the spark plug shell against the engine block with insulated pliers.

NOTE: See "CAUTION" under Step 6 of "Run Mode Spark Test", above.

2. Have an assistant crank the engine using the ignition switch and check for spark. If there is good spark, the problem is probably in the distributor cap, rotor, ignition cables or spark plugs. If there is no spark, proceed to Step 3.

3. Measure the battery voltage. Next, measure the voltage at the white wire of the module while cranking the engine. To make this measurement, it will be necessary to pierce the white wire with a straight pin and connect the voltmeter to the straight pin and to ground. DO NOT ALLOW THE STRAIGHT PIN TO GROUND ITSELF. The battery voltage and the voltage at the white wire should be within 1 volt of each other. If the readings are not within 1 volt of each other, check and repair the feed through the ignition switch to the white wire. Recheck for spark (Step 1). If the readings are within 1 volt of each other, or if there is still no spark after power feed to white wire is repaired, proceed to Step 4.

4. Measure the coil BAT terminal voltage while cranking the engine. The reading should be within 1 volt of battery voltage. If the readings are not within 1 volt of each other, check and repair the feed through the ignition switch to the coil. If the readings are within 1 volt of each other, the problem is probably in the ignition module. Substitute another module and repeat test for spark (Step 1).

TROUBLESHOOTING DURA SPARK III

PRELIMINARY CHECKOUT

NOTE: When making these tests, make sure to follow wiring back to the module for color coding purposes. All instrument checks for wiring integrity should be accompanied by visual inspection and by wiggling of connectors to check for either bad insulation or loose connections.

Begin troubleshooting by inspecting all vacuum hoses and both high tension and wiring harness wiring for proper routing and secure connections. Check also for damaged insulation and burned connectors.

Make sure the battery is fully charged and turn off all accessories before starting tests. The procedure requires the following equipment:

1. A precise volt/ohmmeter, preferably with digital readout.

2. A commercially available spark tester that can replace a spark plug in the circuit and provide visible evidence of a hot spark.

3. A diagnostic test adapter designed for the Dura Spark III system. This must plug into the three wire connector going to the ignition module and apply voltage to the right module circuit for testing it directly from the battery.

4. A 12 volt test light and a supply of straight pins for testing for voltage and resistance in the wiring.

TEST 1

1. Disconnect the three wire ignition module connector. Inspect the connector for damage, dirt, or corrosion. If none are present, proceed with Step 2. Otherwise, make repairs as necessary.

2. Install the test adapter between the two halves of the connector. Pull the coil high tension lead out of the distributor, install and ground the spark tester. Turn the ignition switch to the "RUN" position.

3. Touch the lead to the test adapter to the battery positive terminal while observing the spark tester. There should be a spark as the lead touches the terminal. Test repeatedly. If spark occurs consistently, go to Test 2. Otherwise, see Test 4.

TEST 2

1. Remove the diagnostic test adapter from the ignition module circuit, and reconnect the connector. Leave the spark tester in place, and observe it while cranking the engine. If there is no spark, go on to test three. If there is spark, follow the rest of the steps in this test.

2. Inspect the distributor cap, rotor, and adapter for cracks or carbon tracking and replace parts as necessary. Make sure there is silicone compound to protect the tip of the rotor.

3. If the inspections above do not reveal and cure the problem, check the distributor rotor alignment and correct as necessary.

TEST 3

1. If the starter relay has an "I" terminal, disconnect the cable from the starter relay to the starter motor. If it has no "I" terminal, disconnect the wire to the "S" terminal of the relay.

2. Insert a straight pin through the center of the white wire leading to the ignition module. Make sure it **does not** contact any ground.

3. Measure the voltage existing between the positive and negative battery terminals.

4. Connect the VOM negative lead to a good engine ground. Use a pin to get a connection to the white wire going to the ignition module (without grounding the pin). Connect the positive lead to the VOM to the pin. Finally, turn the ignition switch to the START position and read the voltage while wiggling the wiring.

5. Connect the positive lead of the VOM to the "Batt" terminal of the ignition coil and with the ignition switch in the START position, repeat the test.

6. If either reading was less than 90 percent of battery voltage, refer to a wiring diagram and repair wiring and connectors in the faulty circuit. Otherwise, replace the ignition switch.

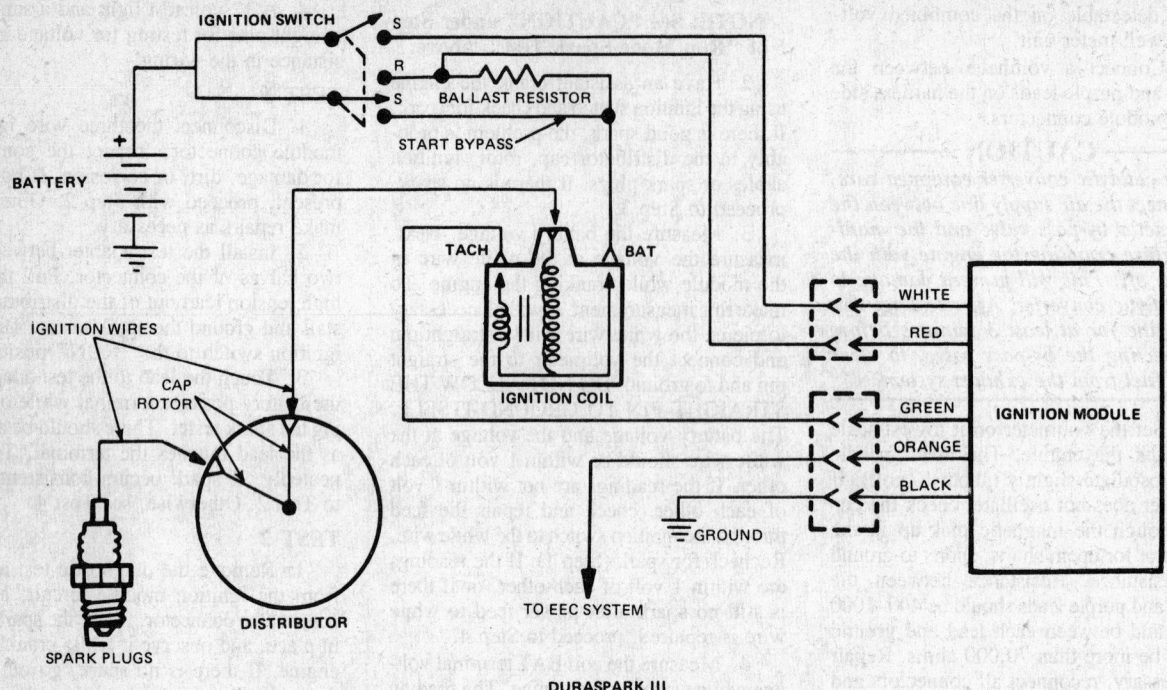

Wiring diagram of basic Dura Spark III ignition system

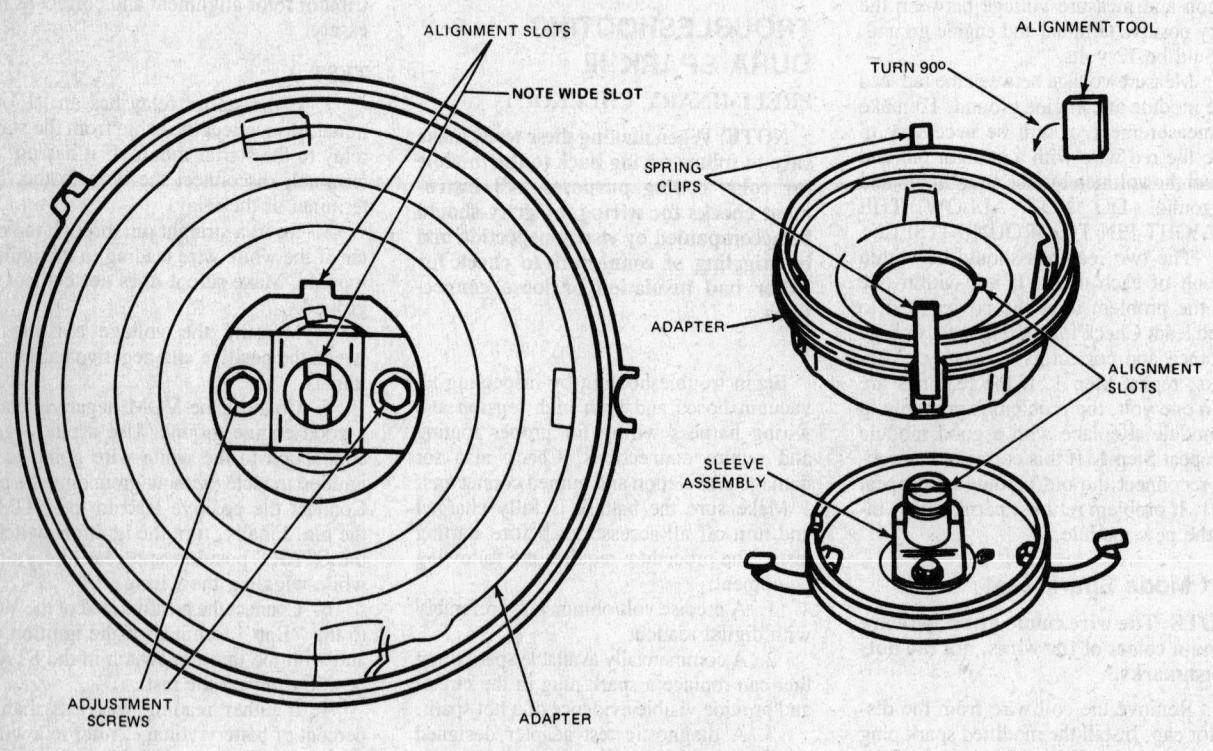

Rotor alignment—Dura Spark III

8. Remove the straight pin and reconnect the wiring to the starter relay. If the required repairs do not resolve the problem, or if both circuits passed the test, proceed to Test 4.

TEST 4

1. Connect a test lamp between the TACH terminal of the ignition coil and a ground on the engine. Connect the diagnostic test adapter into the ignition system as in the first test. Turn the ignition switch to the "RUN" position.

2. Touch the test lead for the adapter to the battery positive terminal repeatedly while observing the test light.

3. Remove the diagnostic connector and test light and reconnect the three prong connector. If the light flashes consistently, go to Test 5; if not, go to Test 6.

TEST 5

1. Disconnect the coil high tension wire and connector. Inspect the wire and connector for a burned appearance and replace parts as necessary. Inspect the coil for cracks or carbon tracking in the high tension connector, and replace it if necessary.

2. Test the high tension lead with an ohmmeter by removing it and connecting the meter to either end. Resistance should be 5,000 ohms per inch or less. If resistance is too high, replace the wire.

3. Measure the coil secondary circuit resistance by connecting the ohmmeter between the coil BATT terminal and the high voltage connector. Resistance must be between 7,700 ohms and 10,500 ohms. If resistance is either above or below this range, replace the coil.

4. Replace or reconnect parts as necessary.

TEST 6

1. If the starter relay has an "I" terminal, disconnect the cable from the relay to the starter motor. If there is no "I" terminal on the relay, disconnect the wire to the "S" terminal of the relay. Insert straight pins into the centers of the red and white module wires. DO NOT GROUND. Measure the exact voltage by connecting the voltmeter between the positive and negative battery terminals.

2. Connect the voltmeter negative lead between the base of the distributor and the pin passing through the red lead. Turn the ignition switch to the "RUN" position. Measure the voltage while wiggling the wiring.

3. Connect the voltmeter positive lead over to the pin running through the white wire. Turn the ignition switch to the START position and read the voltage while wiggling the wiring.

4. Repeat the step above, but with the positive lead attached to the BATT terminal on the ignition coil to test the ballast resistor bypass. Turn the ignition switch off, remove the straight pins, and reconnect the starter wiring.

5. If the voltage was more than 90 percent of battery voltage in every case, go on to the next test. If voltage was lower than 90 percent of battery voltage, inspect the wiring harness for faulty circuits and repair as necessary. Replace the ignition switch, or replace the radio interference capacitor on the coil as necessary until the system can pass the test.

TEST 7

1. Attach the negative lead of the voltmeter to an engine ground, and the positive lead to the BATT terminal on the coil.

2. With the ignition switch in the RUN position, measure the voltage.

3. Turn the ignition switch off and remove the voltmeter leads. If the voltage is 6 to 8 volts, proceed to Test 8; if it is less than 6 volts or greater than 8 volts, proceed to Test 9.

TEST 8

1. Disconnect the ignition module three wire connector and the connector at the coil for the wire from the ignition module. Repair or replace bad connectors as necessary.

2. Connect an ohmmeter between an engine ground ($-$) and the TACH terminal of the coil. Reconnect module and coil connectors. If resistance is greater than one ohm, replace the module. If it is one ohm or less, just inspect the wiring harness between the module and coil to make sure there are no wiring problems.

TEST 9

1. Disconnect the ignition coil primary connectors. Measure coil primary resistance by connecting an ohmmeter between the BATT and TACH terminals. Disconnect the ohmmeter and reconnect connectors.

2. If resistance is between .8 and 1.6 ohms, proceed to Test 10. If resistance is less than or greater than specification, replace the coil.

TEST 10

1. Insert a straight pin into the green wire at the module without grounding it.

2. Attach the negative lead of the voltmeter to an engine ground. Then, turn the ignition switch on and measure the voltage, first at the straight pin in the green wire and then at the TACH terminal of the ignition coil. If the difference in voltage is more than .5 volts, inspect and repair the green module-to-coil wire as necessary. If it is less than .5 volts, but more than 1.5, proceed to Test 12. If it is 1.5 volts or less, proceed to the test of the ballast resistor(Test 11).

3. Turn the ignition off and remove the leads and straight pin.

TEST 11

1. Disconnect the two wire connector to the ignition module. Repair any defects in the connectors. Disconnect the coil primary connector.

2. Use an ohmmeter to measure resistance of the circuit between the BATT

terminal of the ignition coil connector and the wiring harness connector mating with the red wire from the module. In both cases, measure on the wiring harness side not at the coil or at the wire leading to the module.

3. If the resistance is .8–1.6 ohms, replace the ignition module. If it is less than .8 or greater than 1.6 ohms, replace the ballast resistor.

4. Remove the ohmmeter and reconnect connectors as necessary.

TEST 12

1. Insert a straight pin into the black wire at the ignition module without grounding it.

2. Connect a voltmeter between the pin and an engine ground, with the positive lead at the pin. Turn the ignition switch to RUN position.

3. If the voltage read is less than .5 volt, replace the ignition module. If it is greater than .5 volts, proceed to the final test for further diagnosis of problems in the ground circuit.

4. Turn the ignition switch off, remove the voltmeter connectors and remove the straight pin.

TEST 13

1. Disconnect the three-wire connector to the module. Connect an ohmmeter between an engine ground and the *harness side* terminal mating with the black wire from the module. Read the resistance while wiggling the harness.

2. If resistance is less than one ohm, look for a problem in the wiring harness connectors and the module black wire. The problem is either intermittent or not in the ignition system. If resistance is greater than one ohm, inspect the harness and connectors between the module and ground connection. Repair wiring as necessary, disconnect the ohmmeter, and reconnect the connector.

Rotor Alignment Dura Spark III

1. Remove the distributor cap, noting the approximate location of the high tension wire going to No. 1 cylinder. Rotate the engine until No. 1 cylinder firing position is being approached by the rotor and the engine nears 0 degrees TDC. Remove the rotor and position a distributor alignment tool so that when the alignment slots reach an aligned position, the tool will slip into place.

2. Turn the engine very slowly until the tool just slips into the alignment slots, and then stop rotating the engine. Read the position of the timing mark on the vibration damper. If it is between 4 degrees Before Top Center and 4 degrees After Top Center, the rotor alignment meets specification. In this case, remove the tool, and reposition the rotor and cap. If the timing mark is outside these limits, proceed with the steps that follow.

3. Remove the alignment tool. Rotate the engine until the timing marks align at 0 degrees, TDC.

4. Loosen the two sleeve assembly adjustment screws. Then, rotate the sleeve assembly until the alignment tool can be inserted. With the tool in position, tighten the adjusting screws. Then, remove the tool, and install the rotor and cap.

TROUBLESHOOTING THE TFI–IV SYSTEM

NOTE: After performing any test which requires piercing a wire with a straight pin, remove the straight pin and seal the holes in the wire with silicone sealer.

Ignition Coil Secondary Voltage

1. Disconnect the secondary (high voltage) coil wire from the distributor cap and install a spark tester (see Special Tools, located with the Dura Spark Troubleshooting) between the coil wire and ground.
2. Crank the engine. A good, strong spark should be noted at the spark tester. If spark is noted, but the engine will not start, check the spark plugs, spark plug wiring, and fuel system. If there is no spark at the tester: Check the ignition coil secondary wire resistance; it should be no more than 5000 ohms per inch. Inspect the ignition coil for damage and/or carbon tracking. With the distributor cap removed, verify that the distributor shaft turns with the engine; if it does not, repair the engine as required. If the fault was not found proceed to the next test.

Ignition Coil Primary Circuit Switching.

1. Insert a small straight pin in the wire which runs from the coil negative (-) terminal to the TFI module, about one inch from the module.

─────── CAUTION ───────
The pin must not touch ground.
──────────────────────────

2. Connect a 12VDC test lamp between the straight pin and an engine ground.
3. Crank the engine, noting the operation of the test lamp. If the test lamp flashes, proceed to the next test. If the test lamp lights but does not flash, proceed to the Wiring Harness test. If the test lamp does not light at all, proceed to the Primary Circuit Continuity test.

Ignition Coil Resistance

Refer to the General Testing for an explanation of the resistance tests. Replace the ignition coil if the resistance is out of the specification range.

Wiring Harness

1. Disconnect the wiring harness connector from the TFI module; the connector tabs must be PUSHED to disengage the connector. Inspect the connector for damage, dirt, and corrosion.

2. Attach the negative lead of a voltmeter to the base of the distributor. Attach the other voltmeter lead to a small straight pin. With the ignition switch in the RUN position, insert the straight pin into the No. 1 terminal of the TFI module connector. Note the voltage reading. With the ignition switch in the RUN position, move the straight pin to the No. 2 connector terminal. Again, note the voltage reading. Move the straight pin to the No. 3 connector terminal, then turn the ignition switch to the START position. Note the voltage reading then turn the ignition OFF.
3. The voltage readings should all be at least 90% of the available battery voltage. If the readings are okay, proceed to the Stator Assembly and Module test. If any reading is less than 90% of the battery voltage, inspect the wiring, connectors, and/or ignition switch for defects. If the voltage is low only at the No. 1 terminal, proceed to the ignition coil primary voltage test.

Stator Assembly and Module

1. Remove the distributor from the engine.
2. Remove the TFI module from the distributor.
3. Inspect the distributor terminals, ground screw, and stator wiring for damage. Repair as necessary.
4. Measure the resistance of the stator assembly, using an ohmmeter. If the ohmmeter reading is 800–975 ohms, the stator is okay, but the TFI module must be replaced. If the ohmmeter reading is less than 800 ohms or more than 975 ohms; the TFI module is okay, but the stator module must be replaced.
5. Repair as necessary and reinstall the TFI module and the distributor.

TFI Module

1. Remove the distributor cap from the distributor, and set it aside (spark plug wires intact).
2. Disconnect the TFI harness connector.
3. Remove the distributor.
4. Remove the two TFI module retaining screws.
5. To disengage the modules terminals from the distributor base connector, pull the right side of the module down the distributor mounting flange and then back up. Carefully pull the module toward the flange and away from the distributor.

─────── CAUTION ───────
Step 5 must be followed EXACTLY; failure to do so will result in damage to the distributor module connector pins.
──────────────────────────

6. Coat the TFI module baseplate with a $\frac{1}{32}$ in. layer of silicone grease (Ford D7AZ–19A331–A or its equivalent).
7. Place the TFI module on the dis-

tributor base mounting flange. Position the module assembly toward the distributor bowl and carefully engage the distributor connector pins. Install and torque the two TFI module retaining screws to 9–16 inch lbs.
8. Install the distributor assembly.
9. Install the distributor cap and check the engine timing.

Primary Circuit Continuity

This test is performed in the same manner as the previous Wiring Harness test, but only the No. 1 terminal conductor is tested (ignition switch in RUN position). If the voltage is less than 90% of the available battery voltage, proceed to the coil primary voltage test.

Ignition Coil Primary Voltage

1. Attach the negative lead of a voltmeter to the distributor base.
2. Turn the ignition switch ON and connect the positive voltmeter lead to the negative (-) ignition coil terminal. Note the voltage reading and turn the ignition OFF. If the voltmeter reading is less than 90% of the available battery voltage, inspect the wiring between the ignition module and the negative (-) coil terminal, then proceed to the last test, which follows.

Ignition Coil Supply Voltage

1. Attach the negative lead of a voltmeter to the distributor base.
2. Turn the ignition switch ON and connect the positive voltmeter lead to the positive (+) ignition coil terminal Note the voltage reading then turn the ignition OFF. If the voltage reading is at least 90% of the battery voltage, yet the engine will still not run; first, check the ignition coil connector and terminals for corrosion, dirt, and/or damage; second, replace the ignition switch if the connectors and terminal are okay.
3. Connect any remaining wiring.

Delco-Remy High Energy Ignition (HEI) System

COMPONENTS

The Delco-Remy High Energy Ignition (HEI) System is a breakerless, pulse triggered, transistor controlled, inductive discharge ignition system.

There are only nine external electrical connections; the ignition switch feed wire, and the eight spark plug leads. On V8 engines, the ignition coil is located within the distributor cap, connecting directly to the rotor.

OPERATION

The magnetic pick-up assembly located inside the distributor contains a permanent magnet, a pole piece with internal teeth,

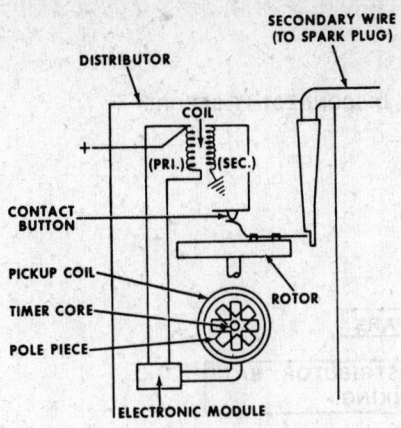

HEI schematic

and a pick-up coil. When the teeth of the rotating timer core and pole piece align, an induced voltage in the pick-up coil signals the electronic module to open the coil primary circuit. As the primary current decreases, a high voltage is induced in the secondary windings of the ignition coil, directing a spark through the rotor and high voltage leads to fire the spark plugs. The dwell period is automatically controlled by the electronic module and is increased with increasing engine rpm. The HEI System features a longer spark duration which is instrumental in firing lean and EGR diluted fuel/air mixtures. The condenser (capacitor) located within the HEI distributor is provided for noise (static) suppression purposes only and is not a regularly replaced ignition system component.

Some 1983 and later engines use an Electronic Spark Timing (EST) distributor. This unit replaces vacuum and centrifugal advance units with an electronic actuator. A vacuum sensor provides manifold vacuum information, and a reference pulse is generated by a pulse generator located near the engine vibration damper. These signals are fed to an Electronic Control Module which, in turn, sends a signal to the EST distributor for ignition timing determination.

Some 1983 and later 5.0 liter V8s also use an Electronic Spark Control (ESC) which responds to engine detonation with retardation of ignition timing. An engine mounted sensor detects the presence of vibration generated by detonation and signals the ESC controller to process the signal and adjust the timing via the actuator on the distributor. The system gradually retards the spark until detonation has disappeared. A failed sensor would produce occasional detonation.

These ESC equipped engines also have a "tip in" vacuum switch. When the throttle is suddenly opened and manifold vacuum is suddenly decreased, the switch contacts close to send a signal to the ESC controller to arbitrarily retard the spark to prevent knock. Thus, if the engine knocks for a very short time after a rapid opening of the throttle, this switch may be at fault.

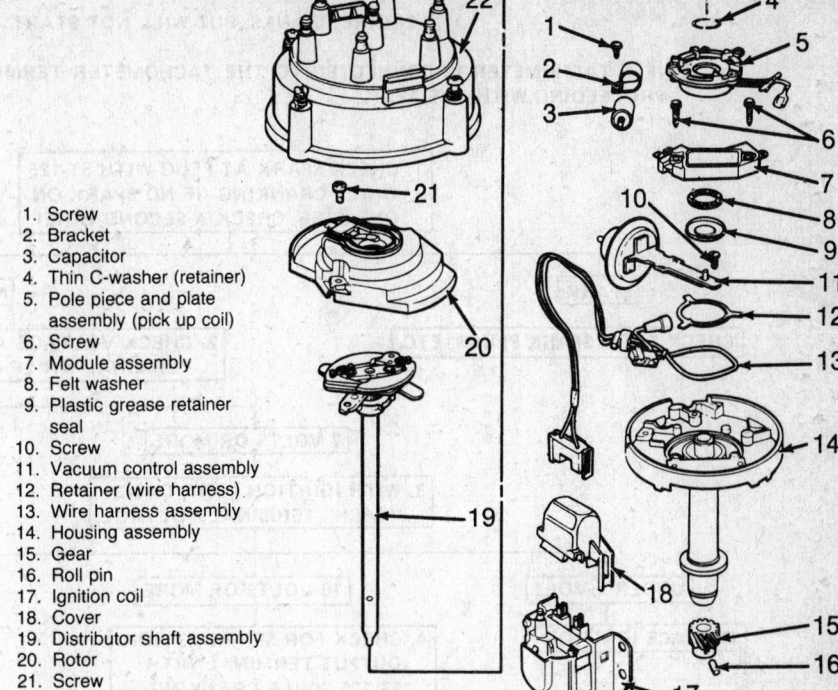

1. Screw
2. Bracket
3. Capacitor
4. Thin C-washer (retainer)
5. Pole piece and plate assembly (pick up coil)
6. Screw
7. Module assembly
8. Felt washer
9. Plastic grease retainer seal
10. Screw
11. Vacuum control assembly
12. Retainer (wire harness)
13. Wire harness assembly
14. Housing assembly
15. Gear
16. Roll pin
17. Ignition coil
18. Cover
19. Distributor shaft assembly
20. Rotor
21. Screw
22. Distributor cap

6-cylinder HEI distributor

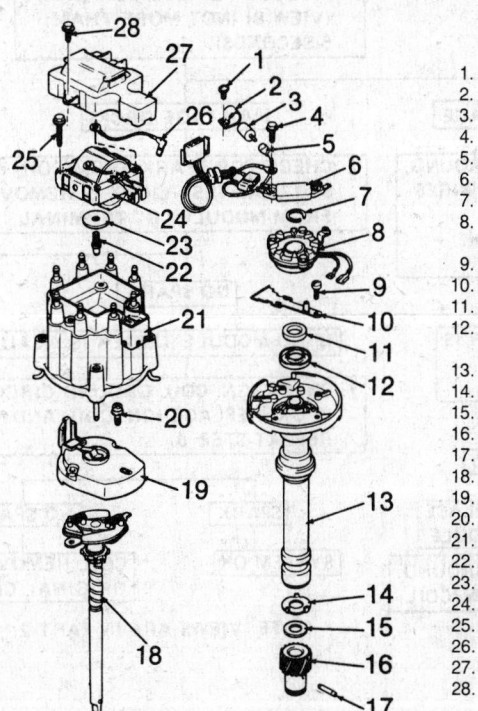

1. Screw
2. Bracket
3. Capacitor
4. Screw
5. Wiring harness assembly
6. Module assembly
7. Thin C-washer (retainer)
8. Pole piece and plate assembly (pick up coil)
9. Screw
10. Plastic retainer
11. Felt washer
12. Plastic grease retainer seal
13. Housing assembly
14. Thrust washer
15. Shim
16. Gear
17. Roll pin
18. Distributor shaft assembly
19. Rotor
20. Screw
21. Distributor cap
22. Resistor brush and spring
23. Seal
24. Ignition coil
25. Screw
26. Ground lead
27. Cover
28. Screw

V8 HEI distributor

ENGINE CRANKS, BUT WILL NOT START

NOTE: IF A TACHOMETER IS CONNECTED TO THE TACHOMETER TERMINAL, DISCONNECT IT BEFORE PROCEEDING WITH THE TEST.

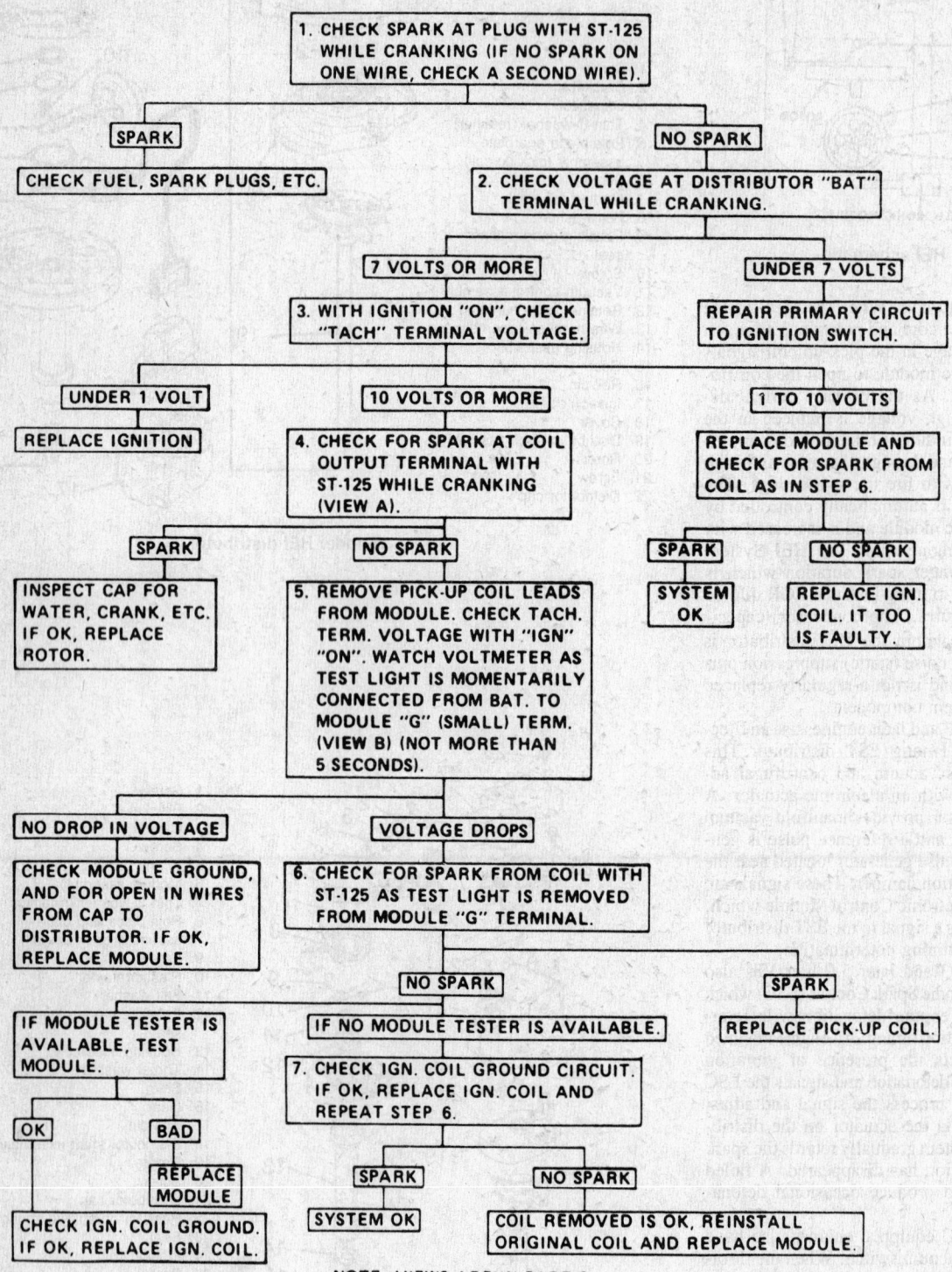

1. CHECK SPARK AT PLUG WITH ST-125 WHILE CRANKING (IF NO SPARK ON ONE WIRE, CHECK A SECOND WIRE).

SPARK → CHECK FUEL, SPARK PLUGS, ETC.

NO SPARK → 2. CHECK VOLTAGE AT DISTRIBUTOR "BAT" TERMINAL WHILE CRANKING.

7 VOLTS OR MORE → 3. WITH IGNITION "ON", CHECK "TACH" TERMINAL VOLTAGE.

UNDER 7 VOLTS → REPAIR PRIMARY CIRCUIT TO IGNITION SWITCH.

UNDER 1 VOLT → REPLACE IGNITION

10 VOLTS OR MORE → 4. CHECK FOR SPARK AT COIL OUTPUT TERMINAL WITH ST-125 WHILE CRANKING (VIEW A).

1 TO 10 VOLTS → REPLACE MODULE AND CHECK FOR SPARK FROM COIL AS IN STEP 6.

SPARK → INSPECT CAP FOR WATER, CRANK, ETC. IF OK, REPLACE ROTOR.

NO SPARK → 5. REMOVE PICK-UP COIL LEADS FROM MODULE. CHECK TACH. TERM. VOLTAGE WITH "IGN" "ON". WATCH VOLTMETER AS TEST LIGHT IS MOMENTARILY CONNECTED FROM BAT. TO MODULE "G" (SMALL) TERM. (VIEW B) (NOT MORE THAN 5 SECONDS).

SPARK → SYSTEM OK

NO SPARK → REPLACE IGN. COIL. IT TOO IS FAULTY.

NO DROP IN VOLTAGE → CHECK MODULE GROUND, AND FOR OPEN IN WIRES FROM CAP TO DISTRIBUTOR. IF OK, REPLACE MODULE.

VOLTAGE DROPS → 6. CHECK FOR SPARK FROM COIL WITH ST-125 AS TEST LIGHT IS REMOVED FROM MODULE "G" TERMINAL.

IF MODULE TESTER IS AVAILABLE, TEST MODULE.

NO SPARK → IF NO MODULE TESTER IS AVAILABLE.

SPARK → REPLACE PICK-UP COIL.

OK

BAD → REPLACE MODULE

CHECK IGN. COIL GROUND. IF OK, REPLACE IGN. COIL.

7. CHECK IGN. COIL GROUND CIRCUIT. IF OK, REPLACE IGN. COIL AND REPEAT STEP 6.

SPARK → SYSTEM OK

NO SPARK → COIL REMOVED IS OK, REINSTALL ORIGINAL COIL AND REPLACE MODULE.

NOTE: VIEWS ARE IN PART 2.

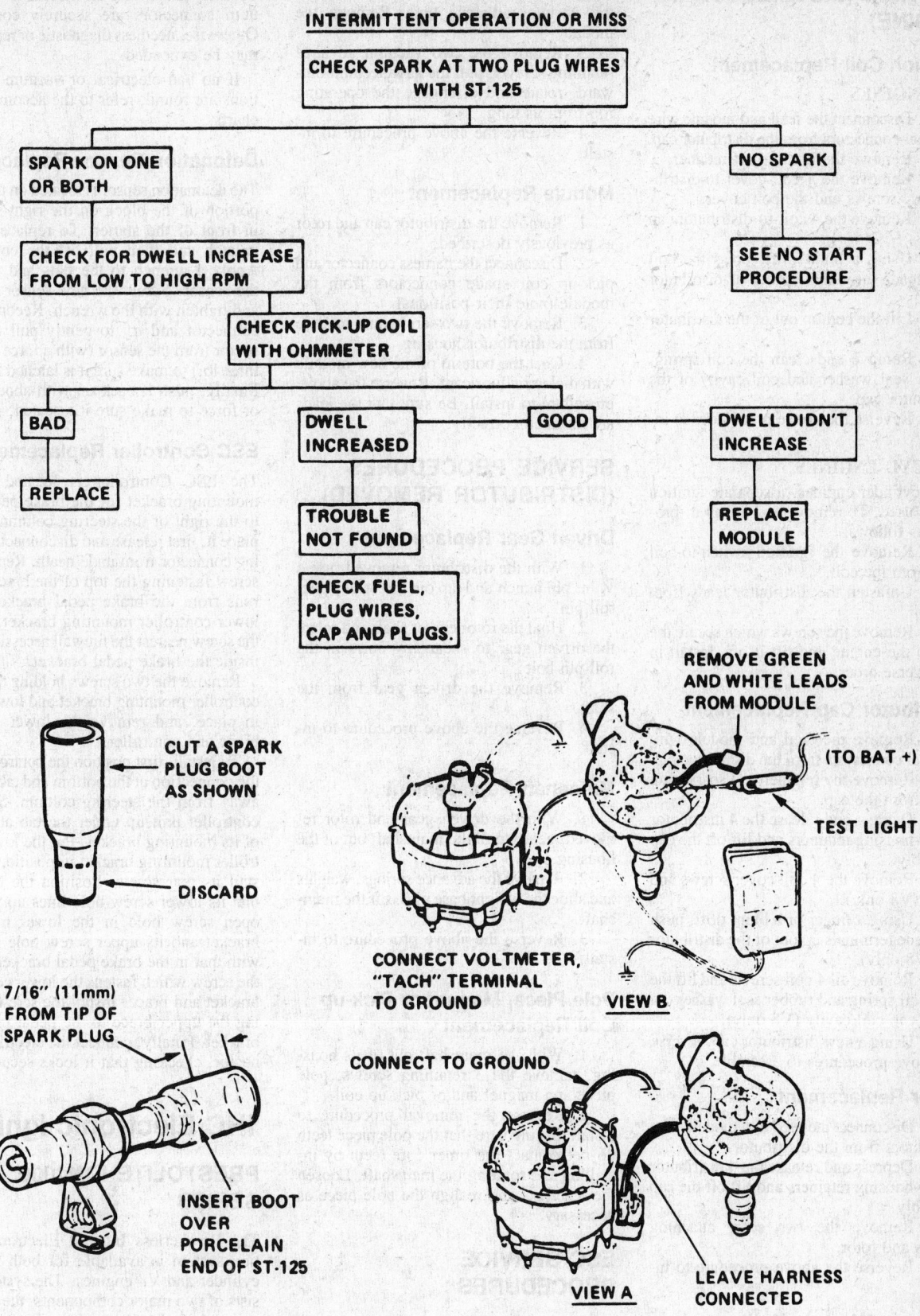

INTERMITTENT OPERATION OR MISS

CHECK SPARK AT TWO PLUG WIRES
WITH ST-125

SPARK ON ONE
OR BOTH

NO SPARK

SEE NO START
PROCEDURE

CHECK FOR DWELL INCREASE
FROM LOW TO HIGH RPM

CHECK PICK-UP COIL
WITH OHMMETER

BAD

REPLACE

DWELL
INCREASED

GOOD

DWELL DIDN'T
INCREASE

REPLACE
MODULE

TROUBLE
NOT FOUND

CHECK FUEL,
PLUG WIRES,
CAP AND PLUGS.

CUT A SPARK
PLUG BOOT
AS SHOWN

DISCARD

7/16" (11mm)
FROM TIP OF
SPARK PLUG

INSERT BOOT
OVER
PORCELAIN
END OF ST-125

REMOVE GREEN
AND WHITE LEADS
FROM MODULE

(TO BAT +)

TEST LIGHT

CONNECT VOLTMETER,
"TACH" TERMINAL
TO GROUND

VIEW B

CONNECT TO GROUND

VIEW A

LEAVE HARNESS
CONNECTED

REPAIRS (DISTRIBUTOR IN ENGINE)

Ignition Coil Replacement

V8 ENGINES

1. Disconnect the feed and module wire terminal connectors from the distributor cap.
2. Remove the ignition set retainer.
3. Remove the 4 coil cover-to-distributor cap screws and the coil cover.
4. Remove the 4 coil-to-distributor cap screws.
5. Using a blunt drift, press the coil wire spade terminals up out of distributor cap.
6. Lift the coil up out of the distributor cap.
7. Remove and clean the coil spring, rubber seal washer and coil cavity of the distributor cap.
8. Reverse the above procedures to install.

SIX CYL ENGINES

On 6 cylinder engines, a separate ignition coil is used. To remove and install it, proceed as follows:
1. Remove the ignition switch-to-coil lead from the coil.
2. Unfasten the distributor leads from the coil.
3. Remove the screws which secure the coil to the engine and lift it off. Install in the reverse order.

Distributor Cap Replacement

1. Remove the feed and module wire terminal connectors from the distributor cap.
2. Remove the retainer and spark plug wires from the cap.
3. Depress and release the 4 distributor cap-to-housing retainers and lift off the cap assembly.
4. Remove the 4 coil cover screws and cover (V8 only).
5. Using a finger or a blunt drift, push the spade terminals up out of the distributor cap (V8 only).
6. Remove all 4 coil screws and lift the coil, coil spring and rubber seal washer out of the cap coil cavity (V8 only).
7. Using a new distributor cap, reverse the above procedures to assemble.

Rotor Replacement

1. Disconnect the feed and module wire connectors from the distributor.
2. Depress and release the 4 distributor cap-to-housing retainers and lift off the cap assembly.
3. Remove the two rotor attaching screws and rotor.
4. Reverse the above procedure to install.

Vacuum Advance Unit Replacement

1. Remove the distributor cap and rotor as previously described.

2. Disconnect the vacuum hose from the vacuum advance unit. Remove the module.
3. Remove the two vacuum advance retaining screws, pull the advance unit outward, rotate and disengage the operating rod from its tang.
4. Reverse the above procedure to install.

Module Replacement

1. Remove the distributor cap and rotor as previously described.
2. Disconnect the harness connector and pick-up coil spade connectors from the module (note their positions).
3. Remove the two screws and module from the distributor housing.
4. Coat the bottom of the new module with dielectric lubricant. Reverse the above procedure to install. Be sure that the leads are installed correctly.

SERVICE PROCEDURES (DISTRIBUTOR REMOVED)

Driven Gear Replacement

1. With the distributor removed, use a $1/8$ in. pin punch and tap out the driven gear roll pin.
2. Hold the rotor end of shaft and rotate the driven gear to shear any burrs in the roll pin hole.
3. Remove the driven gear from the shaft.
4. Reverse the above procedure to install.

Mainshaft Replacement

1. With the driven gear and rotor removed, gently pull the mainshaft out of the housing.
2. Remove the advance springs, weights and slide the weight base plate off the mainshaft.
3. Reverse the above procedure to install.

Pole Piece, Magnet or Pick-up Coil Replacement

1. With the mainshaft out of its housing, remove the 3 retaining screws, pole piece and magnet and/or pick-up coil.
2. Reverse the removal procedure to install making sure that the pole piece teeth do not contact the timer core teeth by installing and rotating the mainshaft. Loosen the 3 screws and realign the pole piece as necessary.

ESC SERVICE PROCEDURES

Diagnosis

Before attempting to find an electrical or electronic problem with the Electronic Spark Control or Computer Command Control

System, check that all electrical and vacuum connectors are securely connected. Otherwise, needless diagnostic or repair time may be expended.

If no bad electrical or vacuum connections are found, refer to the accompanying charts.

Detonation Sensor Replacement

The detonation sensor is located on the lower portion of the block on the right side just in front of the starter. To replace it, first unlatch and then pull off the connector. Apply a wrench to the flats and unscrew the sensor. Screw the new sensor in place and tighten with the wrench. Reconnect the connector and try to gently pull the connector from the sensor (with a force of about three lb.) to make sure it is latched in place. Finally, push it back on with about 10 lb. of force to make sure it's seated.

ESC Controller Replacement

The ESC Controller is located on the mounting bracket for the brake pedal, just to the right of the steering column. To replace it, first release and disconnect the wiring connector from underneath. Remove the screw fastening the top of the brace which runs from the brake pedal bracket to the lower controller mounting bracket. This is the screw nearest the firewall accessible from inside the brake pedal bracket.

Remove the two screws holding the lower controller mounting bracket and lower brace in place, and remove the lower bracket, brace and controller.

To install, first position the controller with the connection at the bottom and tabs facing away from the steering column. Slide the controller unit up under the tab at the top of its mounting bracket. Put the lower controller mounting bracket in position and install its rear screw. Position the brace so that its lower screw hole lines up with the open screw hole in the lower mounting bracket and its upper screw hole lines up with that in the brake pedal bracket. Install the screw which fastens the lower controller bracket and brace. Install the screw fastening the upper brace to the pedal mounting bracket. Finally, connect the electrical connector, checking that it locks securely.

IHC Electronic Ignition

PRESTOLITE IDN4000 SERIES

The Breakerless Integral Electronic Ignition system is available for both the four cylinder and V8 engines. The system consists of two major components, the ignition coil and the distributor. The electronic control unit, consisting of a circuit board and sensor, is located within the distributor and is replaced as a complete unit, should service be required. Either a four or eight

HEI DIAGNOSIS CHART
Engine Cranks, But Will Not Start
1983 CCC
ENGINE CRANKS, BUT WILL NOT RUN
(WITH INTEGRAL IGNITION COIL)

PRELIMINARY

NOTE: Perform Diagnostic Circuit Check before using this procedure.
If a tachometer is connected to the tachometer terminal, disconnect it before proceeding with the test.
Intermittent no start may be caused by wrong pick-up or ignition coil.

1. Check spark at plug with ST-125 while cranking (if no spark on one wire, check a second wire). *

Spark

Check fuel, spark plugs, etc.

LEAVE HARNESS CONNECTED

CONNECT TO GROUND

VIEW A

No Spark

Disconnect 4 term. EST connector and see if engine will run.

7 16 (11mm) FROM TIP OF SPARK PLUG

INSERT BOOT OVER PORCELAIN END OF ST 125

Doesn't run

2. Check voltage at distributor "bat" terminal while cranking

Runs

See Code 42 chart.

---

7 volts or more

3. With ignition "on," check "tach" terminal voltage.

Under 7 volts

Repair primary circuit to ignition switch.

Under 1 Volt

It is faulty ign. coil connection or coil

Spark

Check color match of pick-up coil connector and ign. coil lead. ** Inspect cap for water, cracks, etc. If OK replace rotor.

10 Volts or More

4. Check for spark at coil output terminal with ST-125 while cranking. (View A)

No Spark

5. Remove pick-up coil leads from module Check tach. term. voltage with "ign" "on." Watch voltmeter as test light is momentarily connected from bat. + to module terminal "P". (View B) Not more than 5 seconds

1 to 10 Volts

Replace module and check for spark from coil as in Step 6.

Spark

System OK

No Spark

Replace ign. coil. It, too, is faulty.

REMOVE GREEN AND WHITE LEADS FROM MODULE

(TO BAT +)

TEST LIGHT

CONNECT VOLTMETER TACH TERMINAL TO GROUND

VIEW B

No Drop In Voltage

Check module grnd. and for open in wires from cap to distributor. If OK, replace mod.

Voltage Drops

6. Check for spark from coil with ST-125 as test light is removed from module terminal.

If module tester is available, test module

No Spark

If no module tester is available

7. Check ign. coil ground circuit. If OK, replace ign. coil and repeat Step 6.

Spark

It is pick-up coil or connections. Coil resistance should be 500-1500 ohms and not grounded.

OK | **Bad**

Replace module

Check ign. coil ground. If OK, replace ign. coil.

Spark

System OK

No Spark

Coil removed is OK, reinstall original coil and replace module.

RED WIRE — IGN. COIL — PICK-UP COIL — WHITE WIRE — CLEAR, BLACK.

P/N 1876209

YELLOW WIRE — IGN. COIL — PICK-UP COIL — RED WIRE — YELLOW

P/N 1875894

* A few sparks and then nothing, is considered no spark.

HEI DIAGNOSIS CHART
Continued
TROUBLE CODE 42
BYPASS OR EST PROBLEM

If vehicle will not start and run, check for grounded EST wire to ECM terminal "12." (Grounded and open EST circuit on 5.0L VIN "Y".)

A 1981 HEI module can cause a Code 42.

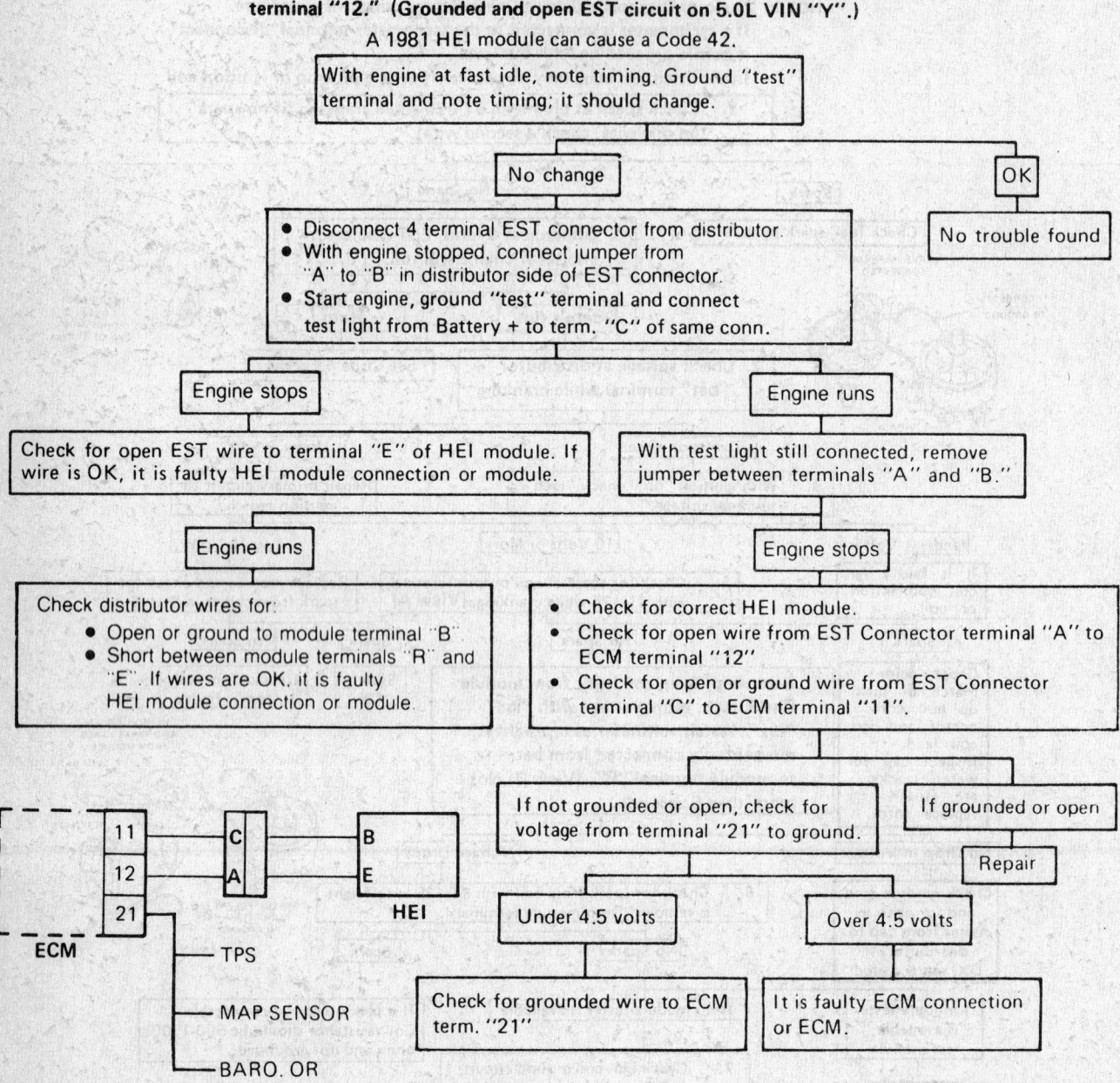

With engine at fast idle, note timing. Ground "test" terminal and note timing; it should change.

No change

OK

No trouble found

- Disconnect 4 terminal EST connector from distributor.
- With engine stopped, connect jumper from "A" to "B" in distributor side of EST connector.
- Start engine, ground "test" terminal and connect test light from Battery + to term. "C" of same conn.

Engine stops

Check for open EST wire to terminal "E" of HEI module. If wire is OK, it is faulty HEI module connection or module.

Engine runs

With test light still connected, remove jumper between terminals "A" and "B."

Engine runs

Check distributor wires for:
- Open or ground to module terminal "B".
- Short between module terminals "R" and "E". If wires are OK, it is faulty HEI module connection or module.

Engine stops

- Check for correct HEI module.
- Check for open wire from EST Connector terminal "A" to ECM terminal "12"
- Check for open or ground wire from EST Connector terminal "C" to ECM terminal "11".

If not grounded or open, check for voltage from terminal "21" to ground.

If grounded or open

Repair

Under 4.5 volts

Over 4.5 volts

Check for grounded wire to ECM term. "21"

It is faulty ECM connection or ECM.

ECM: 11, 12, 21 — C, A — B, E — HEI

— TPS

— MAP SENSOR

— BARO. OR VAC. SENSOR

---

toothed trigger wheel is located on the distributor shaft, depending upon the engine in which it is to be used.

The distributor is easily identified by the male type terminal on the distributor cap secondary system. The distributor incorporates a mechanical (centrifugal) spark advance system. Most distributors will have

a vacuum operated advance system to automatically provide the correct spark advance timing for the various engine speed and load conditions.

The sensor mounting plate configuration and vacuum advance unit location varies between distributors with clockwise and counterclockwise rotation.

Disassembly and Overhaul

The disassembly and overhaul of the distributor is similar to that of the conventional point type distributor. Certain specifications should be adhered to during the overhaul of the distributor. The distributor shaft side play should be between 0.002 and 0.004

ESC SYSTEMS DIAGNOSIS

ENGINE DETONATION ①

BEFORE ATTEMPTING ANY DIAGNOSIS, CHECK CONNECTION AT SENSOR AND INSURE THAT ALL CONNECTIONS ARE CLEAN AND TIGHT

WITH ENGINE RUNNING AT FAST IDLE SPEED & TRANS. IN N OR P (ENGINE SPEED ABOVE 1000 RPM), TAP EXHAUST MANIFOLD LIGHTLY & REPEATEDLY & CHECK FOR SPARK TIMING RETARD (WITH TIMING LIGHT).

RETARD → CHECK OTHER "ENGINE DETONATION" CAUSES

NO RETARD

DISCONNECT 10-PIN CONNECTOR FROM ESC CONTROLLER. MEASURE VOLTAGE FROM PIN B TO PIN K IN CONNECTOR WITH ENGINE OPERATING AT APPROX. 2000 RPM. VOLTAGE SHOULD BE GREATER THAN 80 MILLIVOLTS. (.08 VOLTS)

LOW → DISCONNECT SENSOR WIRE FROM SENSOR. MEASURE VOLTAGE FROM SENSOR TERMINAL TO GROUND. SHOULD BE GREATER THAN 80 MILLIVOLTS (.08 VOLTS) WITH ENGINE OPERATING AT APPROX 2000 RPM.

OK

HIGH OR LOW → REPLACE SENSOR.

OK

TRY TO START ENGINE (WITH 10-PIN CONNECTOR DISCONNECTED).

CHECK WIRES FROM PINS A, B, & K (IN 10-PIN CONNECTOR) FOR OPEN & SHORT CIRCUITS.

NO START / **START** → REPLACE DISTRIBUTOR MODULE.

OK → REPAIR SENSOR CONNECTOR.

NOT OK → REPAIR HARNESS.

RECONNECT 10-PIN CONNECTOR TO CONTROLLER. DISCONNECT SENSOR WIRE FROM SENSOR & INSERT A JUMPER WIRE INTO SENSOR WIRE CONNECTOR. WITH ENGINE RUNNING AT FAST IDLE SPEED, LAY WIRE ON TOP OF DISTRIBUTOR OVER IGNITION COIL & CHECK FOR SPARK TIMING RETARD.

NO RETARD / **RETARD** → REPLACE SENSOR.

MEASURE VOLTAGE FROM PIN H TO PIN K WITH IGNITION SWITCH ON. SHOULD BE MORE THAN 0.2 VOLT.

OK → REPLACE ESC CONTROLLER.

NOT OK → REPAIR OPEN CIRCUIT IN WIRE FROM PIN H IN ESC HARNESS.

① SOME OCCASIONAL TRACE-TO-LIGHT DETONATION IS ACCEPTABLE.

---

inch with a maximum of 0.006 inch. The distributor shaft end play should be 0.035 to 0.040 inch, except distributors with left hand (counterclockwise) rotation, which should be 0.004 to 0.018 inch. The distributor dwell should be 26 to 32 degrees, except distributors numbered IDN4001B, IDN4002R, IDN4001, IDN4010 and IDN4001A, which have a dwell of 28 to 34 degrees.

All distributors should have an air gap between the end of the trigger wheel tooth and the sensor of 0.008 inch.

Sensor to Trigger Wheel Air Gap Adjustment

1. Rotate the distributor shaft until one tooth of the trigger wheel is aligned with the center of the sensor.

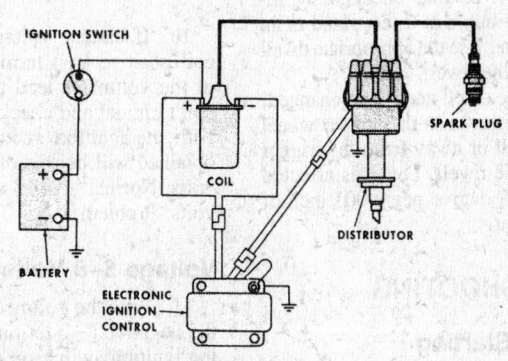

IHC electronic ignition system

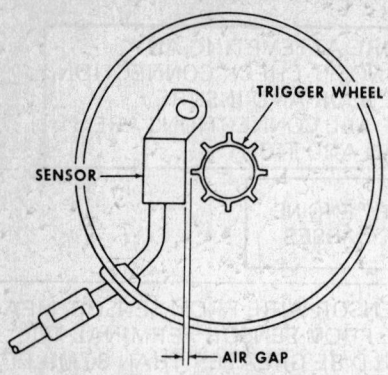

Sensor-to-trigger wheel gap

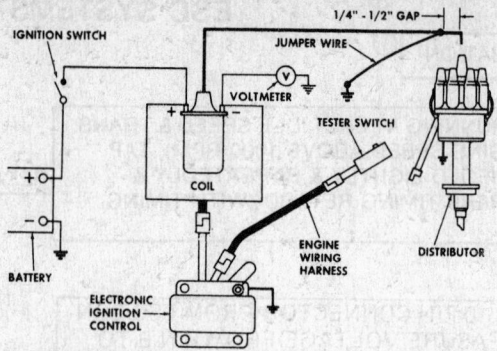

Voltmeter connected to coil negative terminal

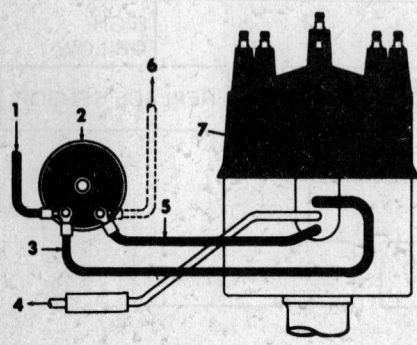

1. From ignition switch
2. Ignition coil
3. Red wire
4. To decelerate throttle modulator (where used)
5. Brown wire
6. To governor control unit (where used)
7. Distributor

Distributor primary wiring

NOTE: The trigger wheel tooth should be perpendicular to the flat surface of the sensor when properly aligned.

2. Using an appropriate feeler gauge, measure the air gap between the sensor and the end of the trigger wheel tooth.

3. Loosen and move the sensor as needed to obtain the specified air gap. Tighten the sensor mounting screw and recheck the air gap.

4. The dwell can be checked by installing the distributor in a test stand or in the vehicle engine. Use the appropriate dwell meter to check the dwell.

5. Should the dwell need to be changed, move the sensor towards the trigger wheel to decrease dwell or away from the trigger wheel to increase dwell. Dwell is affected approximately ½ degree per 0.001 inch of sensor movement.

TROUBLESHOOTING

Engine Not Starting

1. Battery voltage should be 12–13 volts before starting the tests.

2. Disconnect spark plug wire at the spark plug and hold the end terminal with an adapter, approximately ½ inch from a ground on the engine. Have an assistant crank the engine and observe for a spark from the wire and adapter to ground. Test at least two wires.

3. If a spark occurs, the electrical system is functioning and the no-start problem is elsewhere.

4. If no spark occurs, check for spark at the coil lead by disconnecting it from the distributor cap and holding it approximately ½ inch from a ground and again, have an assistant crank the engine.

5. If a spark occurs, the problem is in the distributor cap, rotor or spark plug cables.

6. If no spark occurs, check the distributor trigger wheel tooth to sensor air gap as previously outlined.

7. If the air gap is out of specifications, adjust and retest for spark. If still no spark occurs, "bump" the starter to position two trigger wheel teeth to straddle the sensor.

8. Connect a voltmeter between the coil positive (+) terminal and ground. Turn the ignition on and the voltage should read battery voltage (12–13 volts).

9. Should the voltage be noticeably lower than battery voltage, a high resistance exists between the battery and the coil. This resistance must be found and repaired before proceeding.

NOTE: Refer to the Primary Voltage Drop Test at the end of this troubleshooting outline.

10. If battery voltage is present at the coil positive (+) terminal, move the clip of the voltmeter lead to the negative (−) coil terminal and check the voltage present with the ignition switch on. The voltage obtained will be one of the following: 5–8 volts; Normal. 12–13 volts; Problem. 0–5 volts; Problem.

Voltage 5–8 Volts

1. With the voltmeter still connected to the negative (−) terminal of the coil, turn the ignition switch on and place the blade of a flat screwdriver against the face of the sensor while observing the voltmeter.

2. The voltage should increase to 12–13 volts.

3. Remove the screwdriver and the voltage should drop to 5–8 volts.

4. The voltage should switch up and down when the screwdriver blade is placed against and then removed from the sensor surface.

5. If the voltage does not switch up or down, the electronic control unit is defective and must be replaced.

6. To verify the secondary spark from the coil to ground as the voltage moves up or down, re-establish the ½ inch gap between the coil lead and engine ground.

7. With the ignition switch on, place the screwdriver blade against the face of the sensor and a spark should occur across the gap. If no spark occurs, the coil is defective and must be replaced.

8. After replacing the defective component(s), reassemble the shield, rotor, distributor cap and wiring. Recheck for spark at the spark plugs.

9. Check the dwell and the ignition timing. Adjust as required in this order. When the distributor is equipped with a vacuum advance, disconnect the vacuum hose before adjusting the timing.

Voltage 12–13 Volts

1. Connect a jumper wire between the distributor housing and the battery negative (−) terminal.

2. Observe the voltage reading. If the voltage remains at 12–13 volts, the electronic control unit is defective and must be replaced.

3. Should the voltage change to 5–8 volts with the jumper wire connected, a poor ground circuit exists between the distributor and the battery. All grounding straps should be examined, cleaned and tightened as required.

Voltage 0–5 Volts

1. Disconnect the voltmeter lead from the coil and disconnect the brown wire from the coil negative (−) terminal.

2. Reconnect the voltmeter lead to the coil negative (−) terminal, with the other voltmeter lead still connected to a ground.

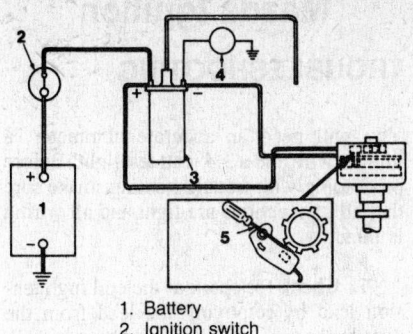

1. Battery
2. Ignition switch
3. Ignition coil
4. Voltmeter
5. Screwdriver

Testing the electronic control unit

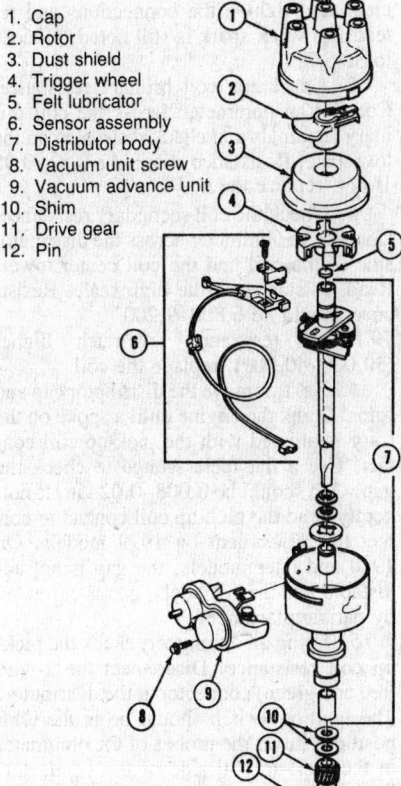

1. Cap
2. Rotor
3. Dust shield
4. Trigger wheel
5. Felt lubricator
6. Sensor assembly
7. Distributor body
8. Vacuum unit screw
9. Vacuum advance unit
10. Shim
11. Drive gear
12. Pin

BID distributor

3. With the ignition switch on, observe the voltage reading. If the voltage is still 0–5 volts, the coil is faulty and must be replaced.

4. If the voltage increases to 12–13 volts, the electronic control unit is defective and must be replaced.

5. After the necessary repairs are made, be sure to reconnect the brown wire to the coil negative (–) terminal.

PRIMARY VOLTAGE DROP TEST

1. Remove the distributor cap, rotor and shield. "Bump" the starter to position two teeth of the trigger wheel, straddling the sensor.

2. Connect the voltmeter positive (+) lead to the battery positive (+) post and connect the voltmeter negative (–) lead to the coil positive (+) terminal.

3. Turn the ignition on and observe the voltmeter reading. A reading of less than one volt should be obtained.

4. If a voltage reading higher than one volt exists, move the following components while observing the voltmeter scale. Battery cables. Starter solenoid battery terminal. Dash panel connector at the firewall (if used). Ammeter terminals. Ignition switch connectors.

5. If a fluctuation or upswing of the voltmeter is noted while flexing the connectors and cables, a poor connection or defective cable exists and must be corrected.

Datsun/Nissan Ignition

TROUBLESHOOTING

1. Make a check of the power supply circuit. Turn the ignition OFF. Disconnect the connector from the top of the IC unit. Turn the ignition ON. Measure the voltage at each terminal of the connector in turn by touching the probe of the positive lead of the voltmeter to one of the terminals, and touching the probe of the negative lead of the voltmeter to a ground, such as the engine. In each case, battery voltage should be indicated. If not, check all wiring, the ignition switch, and all connectors for breaks, corrosion, discontinuity, etc., and repair as necessary.

2. Check the primary windings off the ignition coil. Turn the ignition OFF. Disconnect the harness connector from the negative coil terminals. Use an ohmmeter to measure the resistance between the positive and negative coil terminals. If resistance is 0.84–1.02 ohms, the coil is OK. Replace if far from this range. If the power supply, circuits, wiring, and coil are in good shape, check the IC unit and pick-up coil, as follows:

3. Turn the ignition OFF. Remove the distributor cap and ignition rotor. Use an ohmmeter to measure the resistance between the two terminals of the pick-up coil, where they attach to the IC unit. Measure the resistance by reversing the polarity of the probes. If approximately 400 ohms are indicated, the pick-up coil is OK, but the IC unit is bad and must be replaced. If other than 400 ohms are measured, go to the next step.

4. Be certain the two pin connector to the IC unit is secure. Turn the ignition ON. Measure the voltage at the ignition coil negative terminal. Turn the ignition OFF.

—— **CAUTION** ——
Remove the tester probe from the coil negative terminal before switching the ignition OFF, to prevent burning out the tester.

5. If zero voltage is indicated, the IC unit is bad and must be replaced. If battery voltage is indicated, proceed.

6. Remove the IC unit from the distributor as follows: Disconnect the battery ground (negative) cable. Remove the distributor cap and ignition rotor. Disconnect

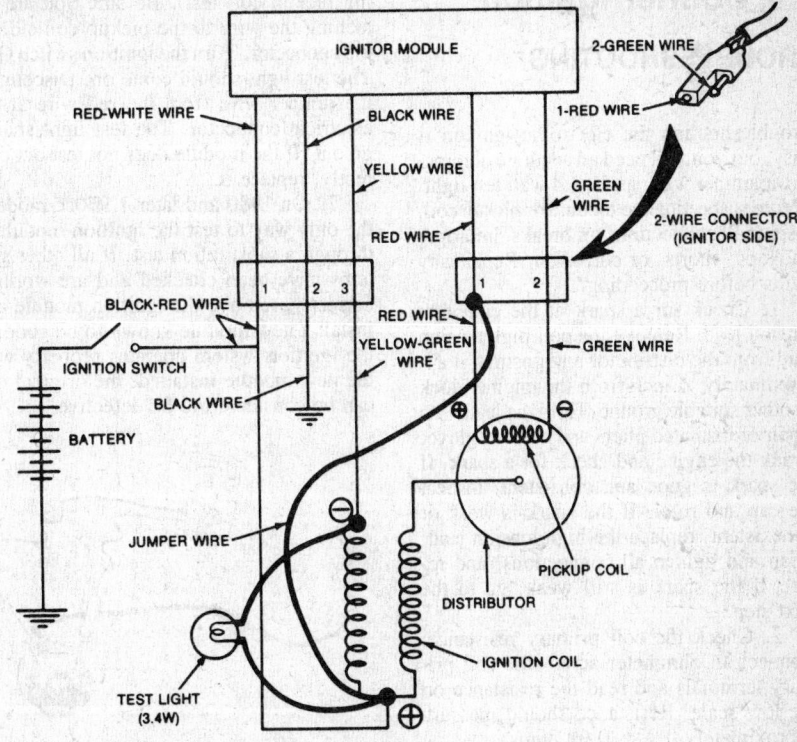

Electronic ignition troubleshooting

the harness connector at the top of the IC unit. Remove the two screws securing the IC unit to the distributor. Disconnect the two pick-up coil wires from the IC unit and remove the IC unit.

CAUTION

Pull the connectors free with a pair of needlenosed pliers. Do not pull on the wires to detach the connectors.

---

7. Measure the resistance between the terminals of the pick-up coil. It should be approximately 400 ohms. If so, the pick-up coil is OK, and the IC unit is bad. If not approximately 400 ohms, the pick-up coil is bad and must be replaced.

8. With a new pick-up coil installed, install the IC unit. Check for a spark at one of the spark plugs. If a good spark is obtained, the IC unit is OK. If not, replace the IC unit.

COLOR CODE

B   :BLACK
BW :BLACK WITH WHITE STRIPE
R   :RED
G   :GREEN
L   :BLUE

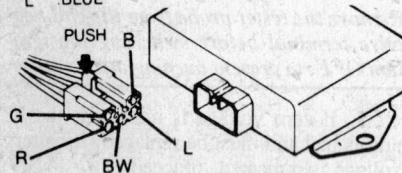

Electronic control unit connection

Courier Ignition

TROUBLESHOOTING

Troubleshooting the electronic ignition is easy, but you will need an accurate ohmmeter, a jumper wire, and a 3.4 watt test light. Before suspecting the module or pickup coil, inspect all connections for breaks, improper hookups, shorts, or corrosion. Repair any faults before proceeding.

1. Check for a spark at the coil high tension lead: Remove the coil high tension lead from the distributor and position it approximately ¼ inch from the engine block or other suitable ground. Hold the lead with a pair of insulated pliers and a heavy glove. Crank the engine and check for a spark. If the spark is good and consistent, inspect the cap and rotor. If the spark is weak or nonexistent, replace the high tension lead, clean and tighten all connections, and re-test. If the spark is still weak, go to the next step.

2. Check the coil primary resistance: Connect an ohmmeter across the coil primary terminals and read the resistance on the low scale. Resistance should measure approximately 0.9 ± 0.09 ohms 68°F. If the reading is far different, replace the coil.

3. Check the coil secondary resistance: Connect an ohmmeter across the distributor side of the coil and the coil center tower. Read the resistance on the high scale of the meter. Resistance should measure 6,800–9,200 ohms 70°F. If the resistance is much higher (30,000–40,000 ohms), replace the coil.

4. Next, remove the distributor cap and ignition rotor. Crank the engine around until a rotor spoke of the armature is aligned with the pickup coil. Use a flat feeler gauge to check the armature gap. It should measure 0.008–0.024 in. (0.20–0.60 mm). If it does not, gently bend the pickup coil to adjust.

NOTE: This does not apply to 1980 and later 1,970cc engines. The armature gap is not adjustable on these models.

5. Using the ohmmeter, measure the pickup coil resistance. Disconnect the 2 wire (red and green) connector at the distributor. The ignition switch should be OFF. Insert the probes of the ohmmeter into the pickup coil side of the connector. Resistance should be 760–840 ohms for the 2,299cc and 1979 1,970cc models. Resistance should be 1,050 ± 10% ohms 68°F for 1980 and later 1,970cc models. If resistance is not within specifications, replace the pickup coil.

6. Finally, test the ignition module. On 2,299cc and 1979 1,970cc models, connect the test light between the positive and negative terminals of the ignition coil. Connect a jumper wire between the positive coil terminal and the red wire of the pickup coil (at the connector unplugged in the preceding pickup coil test). Be sure you are attaching the wire to the pickup coil side of the connector. Turn the ignition switch ON. The test light should come on. Disconnect the jumper wire from the red wire at the electrical connector. The test light should go out. If the module does not test out correctly, replace it.

7. On 1980 and later 1,970cc models, the only way to test the ignition module is through a substitution test. If all other systems have been checked and are working correctly, remove the ignition module and install a new module known to be good. If the ignition system operates properly with the new module installed, the original one can be considered to be defective.

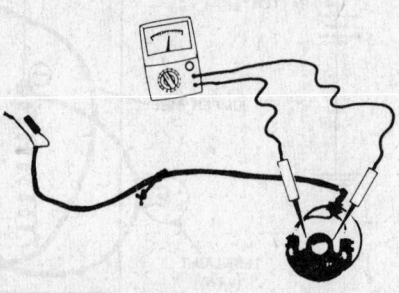

Checking pick-up coil resistance—Mazda

Mazda Ignition

TROUBLESHOOTING

You will need an accurate ohmmeter, a jumper wire and a 3.4 watt test light. Before proceeding with troubleshooting, make sure that all connections are tight and all wiring is intact.

1. Check for spark at the coil high tension lead by removing the lead from the distributor cap and holding it about ¼ inch from the engine block or other good ground. Use a heavy rubber glove or non-conductive clamp, such as a fuse puller or clothes pin, to hold the wire. Crank the engine and check for spark. If a good spark is noted, check the cap and rotor; if the spark is weak or nonexistent, replace the high tension lead, clean and tighten the connections and re-test. If a weak spark is still noted, proceed to Step 2.

2. Check the coil primary resistance. Connect an ohmmeter across the coil primary terminals and check resistance on the low scale. Resistance should be 0.81–0.98 If not, replace the coil.

3. Check the coil secondary resistance. Connect an ohmmeter across the distributor side of the coil and the coil center tower. Read resistance on the high scale. Resistance should be 6,800–9,200 70°F. If resistance is much higher (30,000–40,000), replace the coil.

4. Next, remove the distributor cap and rotor. Crank the engine until a spoke on the rotor is aligned with the pick-up coil contact. Use a flat feeler gauge to check the gap. Gap should be 0.008–0.024 in. If not, gently bend the pick-up coil contact to correct the adjustment on 1979 models. On 1980 and later models, the gap is not adjustable. On these models, gap is corrected by parts replacement.

5. Using an ohmmeter, check the pick-up coil resistance. Disconnect the 2-wire (red and green) connector at the distributor. The ignition switch should be in the OFF position. Insert the probes of the ohmmeter in the pick-up coil side of the connector. Resistance should be 760–840 for 1979 models, or 1,050 ± 10% for 1980 and later models. If not, replace the pick-up coil.

6. Finally, test the ignition module. On 1979 models, connect the test light between the positive and negative terminal of the ignition coil. Connect a jumper wire between the positive coil terminal and the red wire of the pick-up coil, at the connector that you unplugged in the previous test. Be sure that you are attaching the wire to the pick-up coil side of the connector. Turn the ignition switch ON. The test light should light. Disconnect the jumper wire from the connector. The light should go out. If not, replace the module.

7. On 1980 and later models, the only way to test the module is to substitute a known good module in its place.

Mitsubishi/D 50 Arrow Ignition

TROUBLESHOOTING

NOTE: If the engine will not start, go to Test 1 or Test 2, depending on the year of your truck.

If the engine will run:

1. Start the engine, allow to idle until the normal operating temperature is reached.

2. Check the ignition timing, adjust if necessary.

3. Visually check electrical connections for frayed insulation or bare wires. Make sure all plug-in connectors are clean and tight. Check the spark plug and coil wires for cracking, crossfiring, corroded terminals, continuity and resistance. Check the distributor cap for cracks or carbon tracking. Check any suspect parts. Check the spark plugs for foiling, nonfiring and correct gap.

4. If none of the checks have solved the problem, or the car fails to start, proceed to the following tests.

TEST 1: 1979–80

1. Remove the distributor cap by inserting a screwdriver in the ends of the two retaining screws, pushing in and turning the screws clockwise.

2. Remove the screws holding the rotor assembly and lift out the rotor.

3. Turn the ignition switch to the ON position.

4. Disconnect the high tension cable (coil wire) from the center terminal of the distributor cap and hold its end about a quarter of an inch away from a ground (cylinder block etc.).

NOTE: Use insulated pliers to hold the cable.

5. Insert a flatblade screwdriver between the reluctor and the stator. A spark should jump from the high tension wire to the ground. If a spark is not produced, a defective control unit, pick-up coil, ignition coil or faulty wiring may be the problem. Further service should be left to a qualified service technician. However, in the paragraphs that follow further tests are described.

TEST 2: FROM 1981

Remove the coil wire from the distributor cap tower. Hold the end of the wire with insulated pliers (prevents you from getting a shock). Locate the end of the wire about a quarter of an inch away from the cylinder head and have a friend crank the engine with the starter. Observe the spark or no spark condition. If a spark is produced, the IC igniter and ignition coil may be considered in good condition. Remove the distributor cap and check it for cracks, carbon tracking or dirt. Check the rotor for wear.

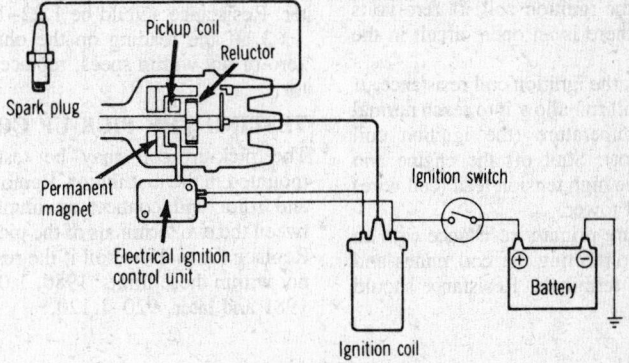

Electronic ignition—1979–80

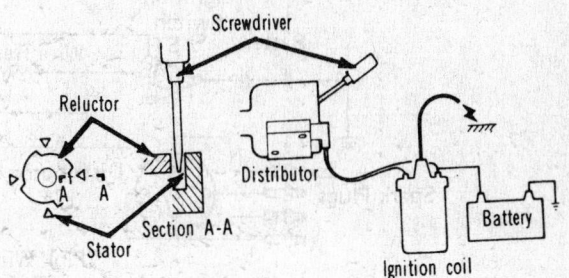

Testing ignition—1979-0

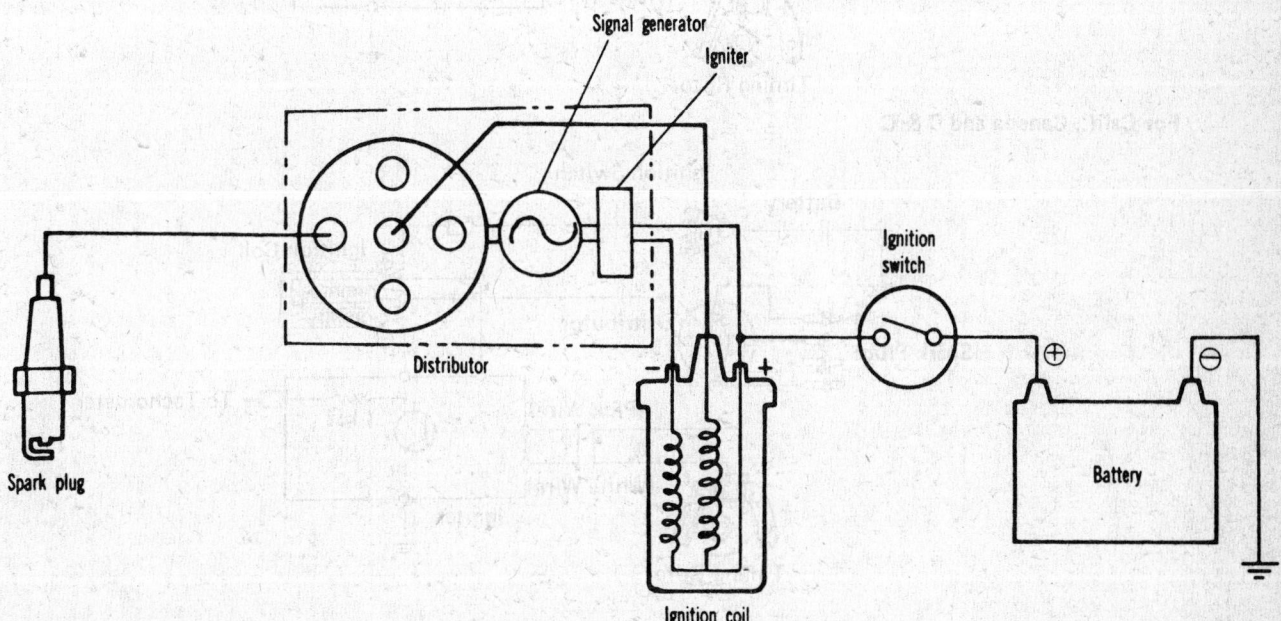

Electronic ignition system—1983 and later

Replace as necessary. If no sparks are produced, a defective control unit (internal), pick-up coil, ignition coil or faulty wiring may be the problem.

IGNITION COIL

If either Test 1 or Test 2 produces no spark, the ignition coil could be at fault. The fastest way to check is by substituting a known good coil. If a coil is not on hand, proceed with one or more of the following tests:

1. With the ignition switch in the ON position measure the voltage at the negative terminal of the ignition coil. If zero volts are shown, there is an open circuit in the coil.

2. Check the ignition coil resistance. If the engine will run allow it to reach normal operating temperature (the ignition coil should be hot). Shut off the engine and disconnect the high tension lead (coil wire) from the coil tower.

3. Measure primary resistance with an ohmmeter, connecting the coil minus and plus primary terminals. Resistance should be 0.7–0.85.

4. Measure the secondary resistance by connecting the ohmmeter between the contacts in the coil tower and the plus primary terminal. Resistance should be 9–11k.

5. Replace the coil if the voltage tests show zero volts or the resistances are not within specifications.

TESTING THE EXTERNAL RESISTOR

1. With the ignition switch off: connect an ohmmeter between the terminals of the external resistor.

2. Obtain a reading from the ohmmeter. Resistance should be 1.22–1.49.

3. If the reading on the ohmmeter is zero or not within specs, replace the resistor.

TESTING THE PICK-UP COIL

The pick-up coil may be tested while mounted in the distributor. Remove the cap and rotor and connect an ohmmeter between the two terminals of the pick-up coil. Replace the pick-up coil if the resistance is not within these limits: 1980, 1,050 ± 50; 1981 and later, 920–1.120,.

PICK-UP COIL REPLACEMENT

NOTE: The distributor must be removed from the engine.

1. Remove the distributor cap and rotor.

2. Remove the center mounting bolt (screw) and remove the governor assembly. Take care not to mix up the governor springs they must be installed in the same position. Remove the reluctor.

3. Remove the two mounting screws and take out the pick-up coil (79–80) or the pick-up coil and IC igniter ('81 and later). Carefully pull the igniter from the pick-up coil ('81 and later).

4. Installation of the new pick-up coil is the reverse of the removal.

RELUCTOR GAP CHECK

The reluctor gap can only be checked. Doing so is not a matter of routine maintenance, but usually is necessary only if the distributor is overhauled. However, too tight a fit between the signal rotor and the pickup stator could cause rotation of the distributor to damage the parts involved or produce an

For Federal except C & C

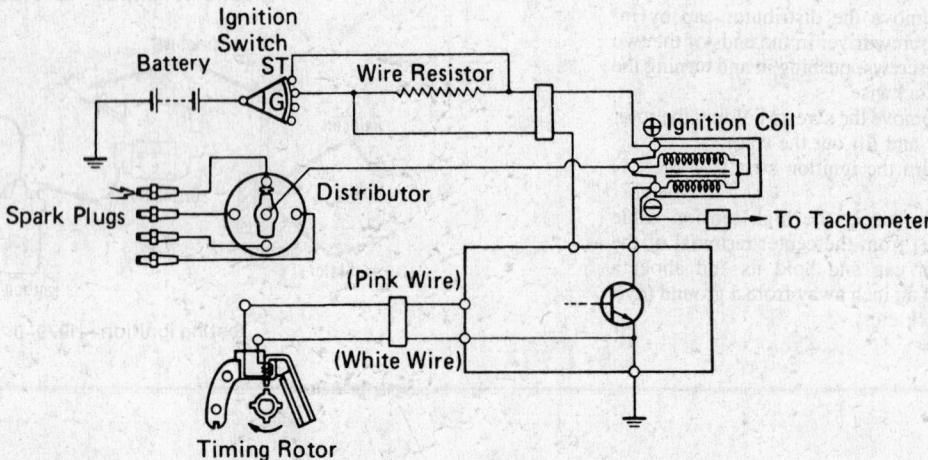

For Calif., Canada and C & C

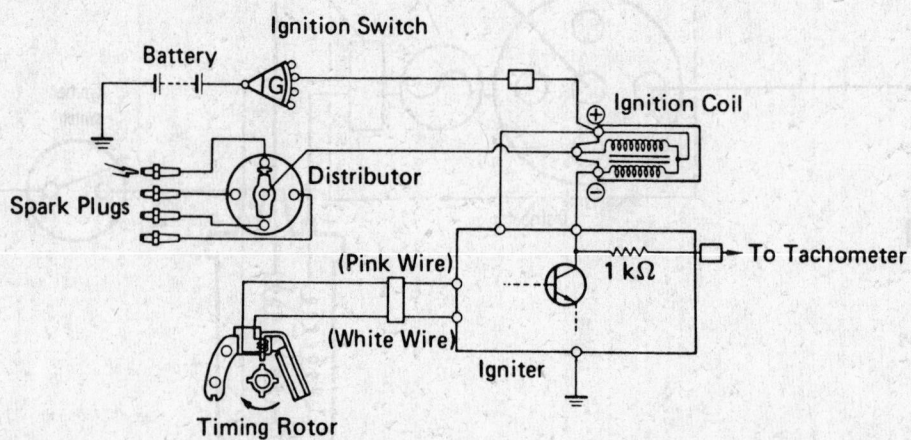

Ignition system circuit—1980

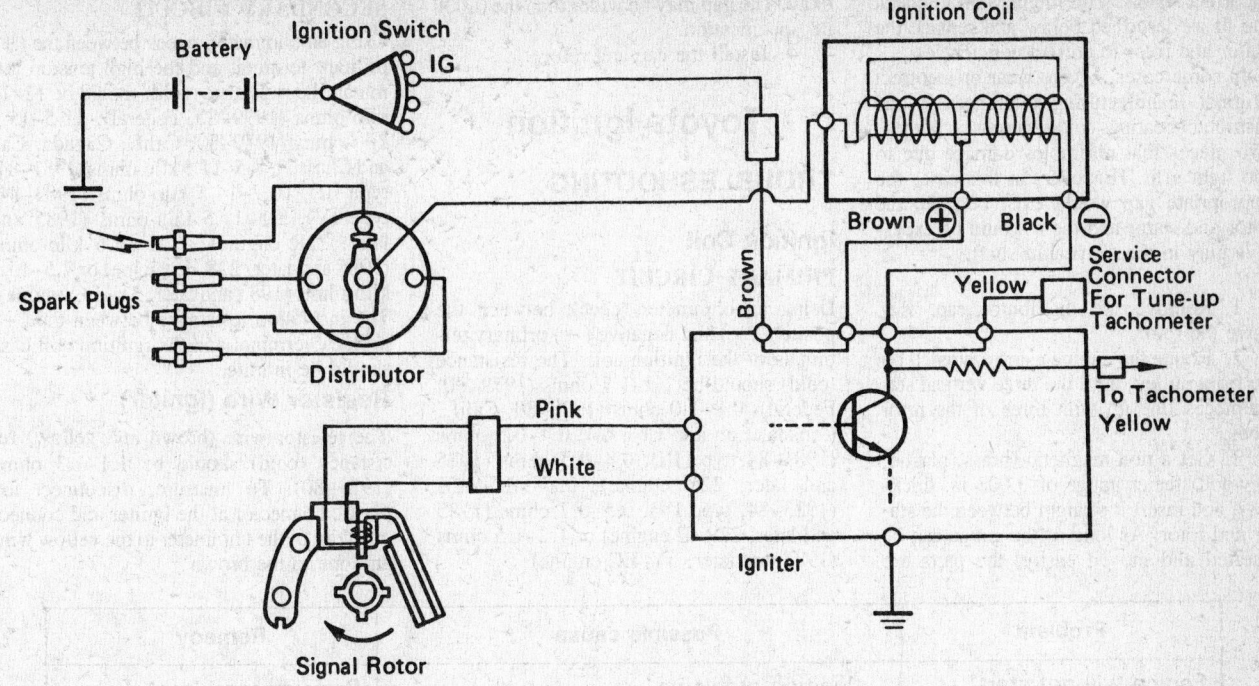

Ignition system circuit—1981 and later

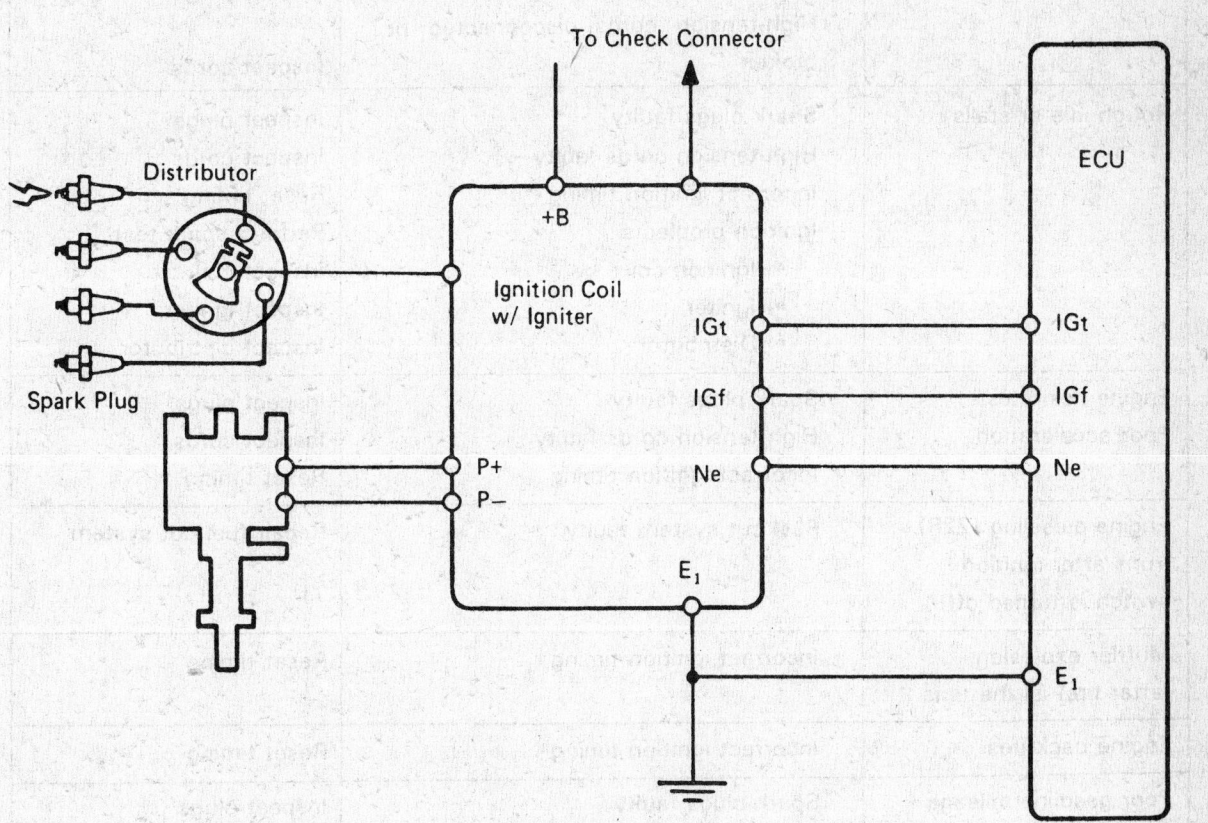

Electronic Spark Advance (ESA)—1985 and later 22RE engine

incorrect signal. You might want to check the fit as described below and replace the stator and rotor to correct deficiencies.

In some cases, severe wear or incorrect original manufacturing tolerances in the distributor bearings could cause wear of these two pieces that resembles damage due to too tight a fit. However, in this case, the appropriate gap would exist between the rotor and stator and there would be excessive play in the distributor shaft.

1. Remove the distributor cap. Remove the rotor.
2. Rotate the engine using a wrench on the front pulley, until the three vertical stator pieces line up with three of the rotor lobes.
3. Get a non-magnetic (brass, plastic, or wood) feeler gauge of 0.008 in. thickness, and insert it straight between the stator and rotor. As long as the gauge can be inserted and moved easily, the parts are

okay. The gap may be wider than the 0.008 in. specification.

4. Install the cap and rotor.

Toyota Ignition

TROUBLESHOOTING

Ignition Coil
PRIMARY CIRCUIT

Using an ohmmeter, check between the positive (+) and negative (−) primary terminals of the ignition coil. The resistance (cold) should be 1.3–1.7 ohms (1979–80, Federal), 0.8–1.0 ohms (1979–80, Calif., Canada, Cab and Chassis), 0.4–0.5 ohms (1981–84, type III), 0.4–0.5 ohms (1985 and later, 22R engine), 0.8–1.1 ohms (1983–84, type IV), 0.5–0.7 ohms (1985 and later, 22R–E engine) or 1.2–1.5 ohms (1983 and later, 3Y–EC engine).

SECONDARY CIRCUIT

Using an ohmmeter, check between the (+) primary terminal and the high tension terminal, the resistance (cold) should be 12–16 kilo-ohms (1979–80, Federal), 11.5–15.5 kilo-ohms (1979–80, Calif., Canada, Cab and Chassis), 8.5–11.5 kilo-ohms (1981–84, type III), 10.7–14.5 kilo-ohms (1983–84, type IV), 8.5–11.5 kilo-ohms (1985 and later, 22R engine), 11.4–15.6 kilo-ohms (1985 and later, 22R–E engine) or 7.5–10.5 kilo-ohms (1983 and later, 3Y–EC engine). The insulation resistance between the (+) primary terminal and the ignition coil case should be infinite.

Resistor Wire (Igniter)

The resistor wire (brown and yellow) resistance (cold) should be 1.1–1.3 ohms (1979–80). To measure, disconnect the plastic connector at the igniter and connect one wire of the ohmmeter to the yellow wire and one to the brown.

Problem	Possible cause	Remedy
Engine will not start/ Hard to start (cranks ok)	Ignition problems	Perform spark test
	• Ignition coil	Inspect coil
	• Igniter	Inspect igniter
	• Distributor	Inspect distributor
	Spark plugs faulty	Inspect plugs
	High-tension cords disconnected or broken	Inspect cords
Rough idle or stalls	Spark plugs faulty	Inspect plugs
	High-tension cords faulty	Inspect cords
	Incorrect ignition timing	Reset timing
	Ignition problems	Perform spark test
	• Ignition coil	Inspect coil
	• Igniter	Inspect igniter
	• Distributor	Inspect distributor
Engine hesitates/ Poor acceleration	Spark plugs faulty	Inspect plugs
	High-tension cords faulty	Inspect cords
	Incorrect ignition timing	Reset timing
Engine dieseling (22R) (runs after ignition switch is turned off)	Fuel cut system faulty	Repair fuel cut system
Muffler explosion (after fire) all the time	Incorrect ignition timing	Reset timing
Engine backfires	Incorrect ignition timing	Reset timing
Poor gasoline mileage	Spark plugs faulty	Inspect plugs
	Incorrect ignition timing	Reset timing
Engine overheats	Incorrect ignition timing	Reset timing

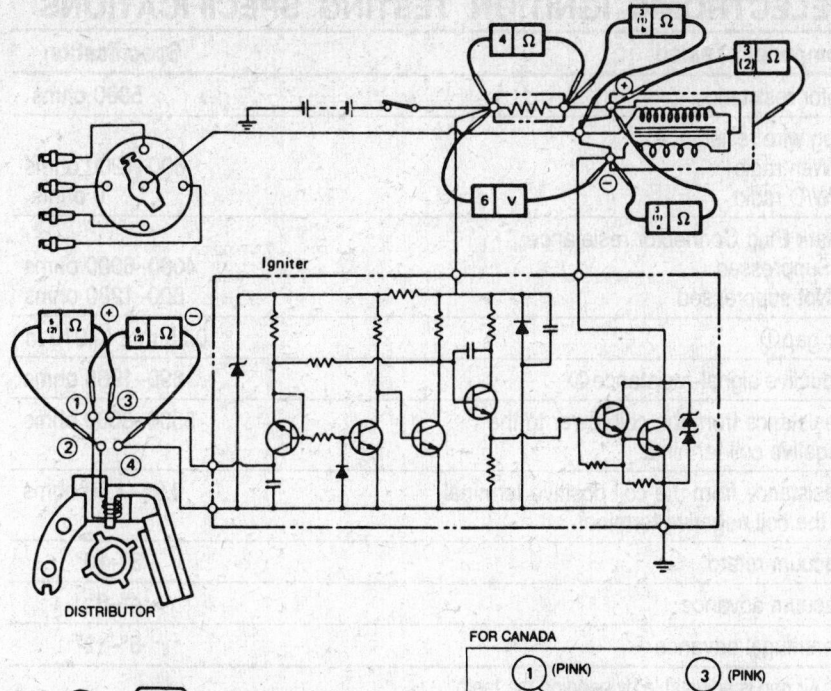

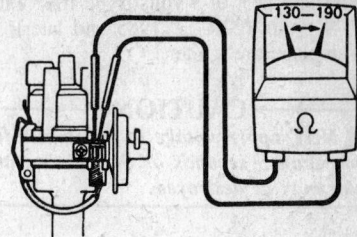

Measure the signal generator (pick-up coil) resistance at the pink and white wires

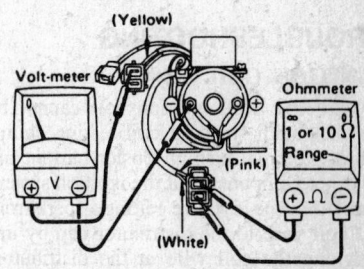

Use the ohmmeter as resistance at the igniter end of the distributor connector

DISTRIBUTOR
Igniter

(WHITE) **2**   **4** (WHITE)
(PINK) **1**   **3** (PINK)

DISTRIBUTOR   WIRING
CONNECTOR   CONNECTOR

FOR CANADA

1 (PINK)   **3** (PINK)
(BLACK)   (BLACK)
(WHITE) **2**   **4** (WHITE)

Fully transistorized ignition system troubleshooting

Air Gap (Distributor)

Remove the distributor cap and ignition rotor. Check the air gap between the timing rotor spoke and the pick-up coil. When aligned, the air gap should be 0.008–0.016 in. You will probably have to bump the engine around with the starter to line up the timing rotor.

Signal Generator

Unplug the electrical connector at the distributor, connect one ohmmeter lead to the white wire and the other lead to the pink wire. The resistance of the signal generator should be 130–190 ohms (1979–81), 140–180 ohms (1982 and later).

Power Transistor (Igniter)

FEDERAL (1979–80)

1. Unplug the electrical connector from the distributor and turn the ignition switch ON.

2. Using a 1.5v dry cell battery, connect the (+) terminal to the pink wire and the (−) terminal to the white wire. The voltage should be less than 5.0 volts; if not, replace the igniter.

3. Connect the (+) voltmeter lead to the (−) ignition coil primary terminal and the (−) voltmeter lead to the body ground. The voltage should measure 1–2 volts.

— CAUTION —

DO NOT apply voltage to the igniter for more than 5 seconds or the power transistor may be destroyed.

4. Reverse the dry cell terminals and check the voltage reading. The voltage should be 12 volts; if not, replace the igniter.

5. Turn the ignition switch OFF.

CALIF., CANADA, CAB AND CHASSIS (1979–80)

1. Turn the ignition ON.

2. Connect the (+) voltmeter lead to the (−) ignition coil primary terminal and the (−) voltmeter lead to the body ground. The voltage should measure 12 volts.

3. Unplug the wiring connector from the distributor.

4. Connect the (+) voltmeter lead to the (+) ignition coil terminal and the (−) lead to the body ground.

5. Using a 1.5v dry cell battery, connect the (+) terminal to the pink wire and the (−) terminal to the white wire. The voltage should be less than 1–5 volts; if not, replace the igniter.

— CAUTION —

DO NOT apply voltage to the igniter for more than 5 seconds or the power transistor may be destroyed.

6. Turn the ignition switch OFF.

ALL MODELS (1981–82) 22R ENGINE (1985 AND LATER)

1. Turn the ignition switch ON.

2. Connect the (+) voltmeter lead to the yellow connector of the igniter and the (−) voltmeter lead to the body ground. The voltage should measure 12 volts.

3. Unplug the electrical connector form the distributor.

4. Using a 1.5v dry cell battery, connect the (+) terminal to the pink wire and the (−) terminal to the white wire. The voltage should 1–5 volts (1981), 8–10 volts (1982) and (22R, 1985 and later); if not, replace the igniter.

— CAUTION —

DO NOT apply voltage to the igniter for more than 5 seconds or the power transistor may be destroyed.

5. Turn the ignition switch OFF.

TYPE III AND IV (1983–84) TYPE IIA, VAN (1983 AND LATER) 22R–E ENGINE (1985 AND LATER)

1. Turn the ignition switch ON.

2. Connect the (+) voltmeter lead to the ignition coil (−) terminal and the (−) voltmeter lead to the body ground. The voltage should measure 12 volts.

3. Unplug the electrical connector form the distributor.

NOTE: On the Van, Type IIA, DO NOT disconnect the electrical connector at the distributor.

4. Using a 1.5v dry cell battery, connect the (+) terminal to the pink wire and the (−) terminal to the white wire. The voltage should 8–10 volts (type III), 5–8

volts (type IV), 0–3 volts (type IIA, Van) or 5–8 volts (22R–E, 1985 and later); if not, replace the igniter.

CAUTION

DO NOT apply voltage to the igniter for more than 5 seconds or the power transistor may be destroyed.

---

5. Turn the ignition switch OFF.

VW

TROUBLESHOOTING

IGNITION COIL TEST

A defective Hall ignition coil cannot be checked with standard coil testing equipment. If there is no high tension current and all other components of the ignition system check out, see if you're getting a spark from the coil wire to the distributor cap by unplugging the coil wire at the distributor, holding the end of it with insulated pliers about ½ inch from ground (engine block, etc.) and turning over the engine. If a weak or no spark is obtained, try replacing the coil.

HALL PICKUP UNIT TEST

1. Check for voltage on positive terminal (15) of the ignition coil. There should be voltage with the ignition ON.
2. Ground a high tension coil wire.
3. Connect a test light (4 to 24 volts) between positive terminal (15) and negative terminal (1).
4. Crank the engine with the starter for approximately 5 seconds. The test light should flicker. If not, replace the ignition distributor.

IGNITION CONTROL UNIT TEST

1. Disconnect the plugs at the control unit and connect the plugs to each other.
2. Turn the ignition switch on and make sure there is current at positive terminal (15) of the ignition coil. Turn the ignition OFF.
3. Disconnect the high tension wire between the ignition coil and the distributor at the distributor.
4. Disconnect the wire (plug) between the control unit and the distributor at the distributor.
5. Connect the positive (+) terminal of the voltmeter to negative terminal (1) of the ignition coil and the negative (-) terminal to ground.

6. Turn the ignition ON. There should be a voltage reading of at least 12 volts. If voltage drops below 12 volts in one second, turn off the ignition. The control unit is defective and will have to be replaced.
7. Disconnect the green wire where it connects to the distributor and ground the wire. Turn the ignition switch ON. The voltmeter should read about 12 volts. Disconnect the ground wire. The voltage should drop to 6 volts. If not, replace the control unit. Turn off the ignition.
8. Connect the terminals of the voltmeter to the outer connector of the control unit. Connect the positive (+) lead to the red wire and the negative (-) lead to the brown wire. Switch on the ignition. The voltmeter should read about 10 volts. If not, replace the control unit.

IDLE STABILIZER

The idle stabilizer is located on top of the ignition control unit. The idle stabilizer controls idle speed by either advancing or retarding the distributor timing in accordance with engine load (air conditioner on, lights on, etc.). If idle speed is erratic or if the engine fails to start, try bypassing the idle stabilizer by disconnecting the two plugs at the idle stabilizer and plugging them together. If idle improves, the idle stabilizer should probably be replaced.

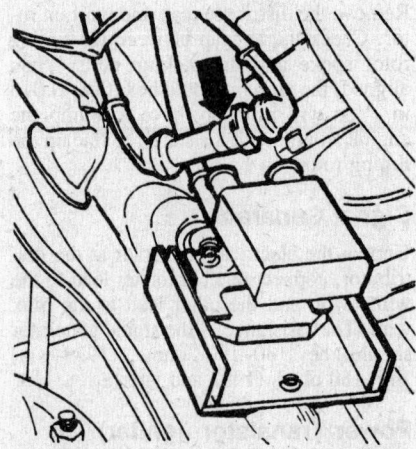

Disconnect the plugs on the idle stabilizer at the control unit and plug them together

ELECTRONIC IGNITION TESTING SPECIFICATIONS

Component Tested	Specification
Rotor resistance	5000 ohms
Plug wire resistance	
With radio	800–1200 ohms
W/O radio	0 ohms
Spark Plug Connector resistance	
Suppressed	4000–6000 ohms
Not suppressed	800–1200 ohms
Air gap ①	0.25 mm (0.010 in.)
Inductive signal resistance ②	890–1285 ohms
Resistance from the coil tower to the negative coil terminal	5500–8000 ohms
Resistance from the coil positive terminal to the coil negative terminal	0.95–1.50 ohms
Vacuum retard	8°–10°
Vacuum advance	4°–8°
Centrifugal advance	6°–12°

① Air gap is adjustable by bending the teeth on the stator (reluctor)

② Measure resistance between the connectors to the control unit on the distributor (connects to the green and white wires)

Emissions

GASOLINE ENGINE EMISSION CONTROL

Introduction

The emission control devices required by law on trucks are determined by weight classification and were considered either "light duty" or "heavy duty" applications, with the Gross Vehicle Weight (GVW) of 6000 lbs. as the dividing line. State and Federal Government regulations have now mandated a new weight standard from the 6000 lbs. GVW to a new GVW of 8500 lbs. or less as "light duty" and a GVW of 8500 lbs. or more as "heavy duty" applications.

The light duty emission devices are normally the same as used on the passenger cars.

During certain model years, passenger carrying vehicles, such as window vans with greater GVW of 6000 lbs. were also considered to be light duty models and must comply with the light duty emission control requirements.

Heavy duty truck models use fewer emission control devices than the light duty models, although more emission controls are being required in each succeeding year to comply with the changing emission control regulations and requirements.

The State of California remains stringent in their emission control standards and throughout this section, reference will be made to either the California, High Altitude or to the Federal engines. (Federal referring to the remaining 49 states, High Altitude referring to areas above 4000 ft. (1,219 meters).

Engine Modifications

Internal engine modifications have been made from year to year by redesigning the following:

1. Lowering the compression ratios to allow the use of low or nonleaded fuels.

2. Combustion chambers and piston modifications for a more efficient air/fuel flow rate and burning time.

3. Camshaft modification to improve valve timing and to increase valve overlap periods.

4. Higher engine operating temperatures and increased cooling areas.

5. Balanced fuel induction manifolds to properly balance the air/fuel flow to the cylinders.

6. Other modifications include changes in metals used in the construction of the engines and components to allow the operation of the engine with non-leaded, non-lubricating fuels.

External engine modifications have been made to the carburetors and distributors to provide the proper air/fuel mixture and to provide the proper timing of the ignition spark to insure the engine emission levels remain within the legislated limits, while providing the best engine performance and fuel economy at varying speeds and loads.

Jet Air System (Mitsubishi/Arrow D-50)

A jet air passage is provided in the car-

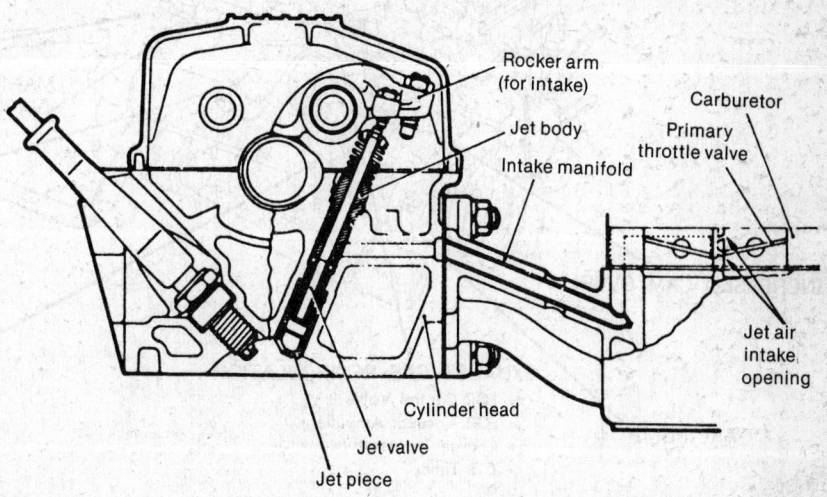

Mitsubishi/D50/Arrow jet valve air system

buretor, intake manifold, and cylinder head to direct air to a jet valve, operated simultaneously with the intake valve.

On the intake stroke, jet air is forced into the combustion chamber because of the pressure difference between the ends of the air jet passage.

This jet of air produces a strong swirl in the combustion chamber scavenging the residual gases around the spark plug.

The jet air volume lessens with increased throttle opening. It is at a maximum at idle.

MAINTENANCE

NOTE: Refer to Valve Lash Adjustment in the Mitsubishi/D–50/Arrow section for adjusting jet valve clearance.

No maintenance is required other than clearance adjustment during valve adjustment. The valve can be removed from the cylinder head for service or replacement.

Emission Control Systems

In order to control the engine crankcase, fuel and exhaust emissions, three major systems have been designated.

1. Crankcase controls are used to provide a more complete scavenging of the crankcase vapors and to route the vapors to the engine fuel induction system for burning with the air/fuel mixture.

2. Evaporation controls are used to prevent the emission of gasoline vapors from the fuel tank and carburetor, into the atmosphere. Charcoal canisters are used to store the gasoline vapors during periods of engine shutdown and during periods of engine operation, the gasoline vapors are drawn into the fuel induction system and burned with the air/fuel mixture.

3. Exhaust controls are used to limit the emission of Carbon Monoxide (CO), Hydrocarbons (HC) and Oxides of Nitrogen (NOx) from the engine exhaust. Numerous controls are used on the engines and the exhaust systems to perform this removal of pollutants.

MAINTENANCE

In order for the emission controls to function properly, maintenance must be performed at regular intervals, either by time or mileage increments. Owner manuals will normally contain a maintenance schedule for services to be done and should be followed for longer emission systems and vehicle life.

Emission Certification Label

An Emission Certification label is attached to either the engine or engine compartment sheet metal and should be consulted before any adjustments are made to the engine.

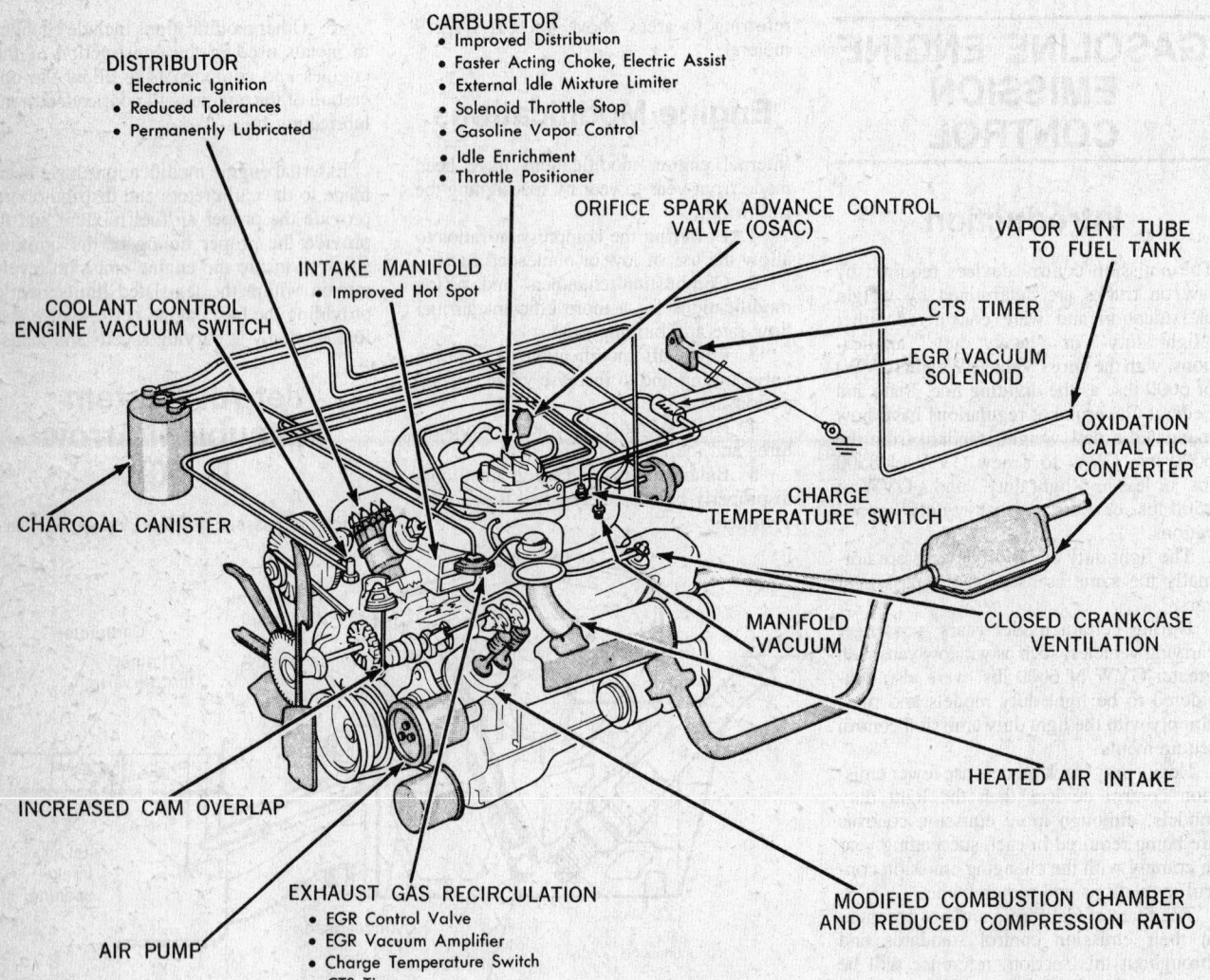

DISTRIBUTOR
- Electronic Ignition
- Reduced Tolerances
- Permanently Lubricated

CARBURETOR
- Improved Distribution
- Faster Acting Choke, Electric Assist
- External Idle Mixture Limiter
- Solenoid Throttle Stop
- Gasoline Vapor Control
- Idle Enrichment
- Throttle Positioner

INTAKE MANIFOLD
- Improved Hot Spot

COOLANT CONTROL ENGINE VACUUM SWITCH

CHARCOAL CANISTER

INCREASED CAM OVERLAP

AIR PUMP

EXHAUST GAS RECIRCULATION
- EGR Control Valve
- EGR Vacuum Amplifier
- Charge Temperature Switch
- CTS Timer

ORIFICE SPARK ADVANCE CONTROL VALVE (OSAC)

VAPOR VENT TUBE TO FUEL TANK

CTS TIMER

EGR VACUUM SOLENOID

OXIDATION CATALYTIC CONVERTER

CHARGE TEMPERATURE SWITCH

MANIFOLD VACUUM

CLOSED CRANKCASE VENTILATION

HEATED AIR INTAKE

MODIFIED COMBUSTION CHAMBER AND REDUCED COMPRESSION RATIO

Typical Emission Control System

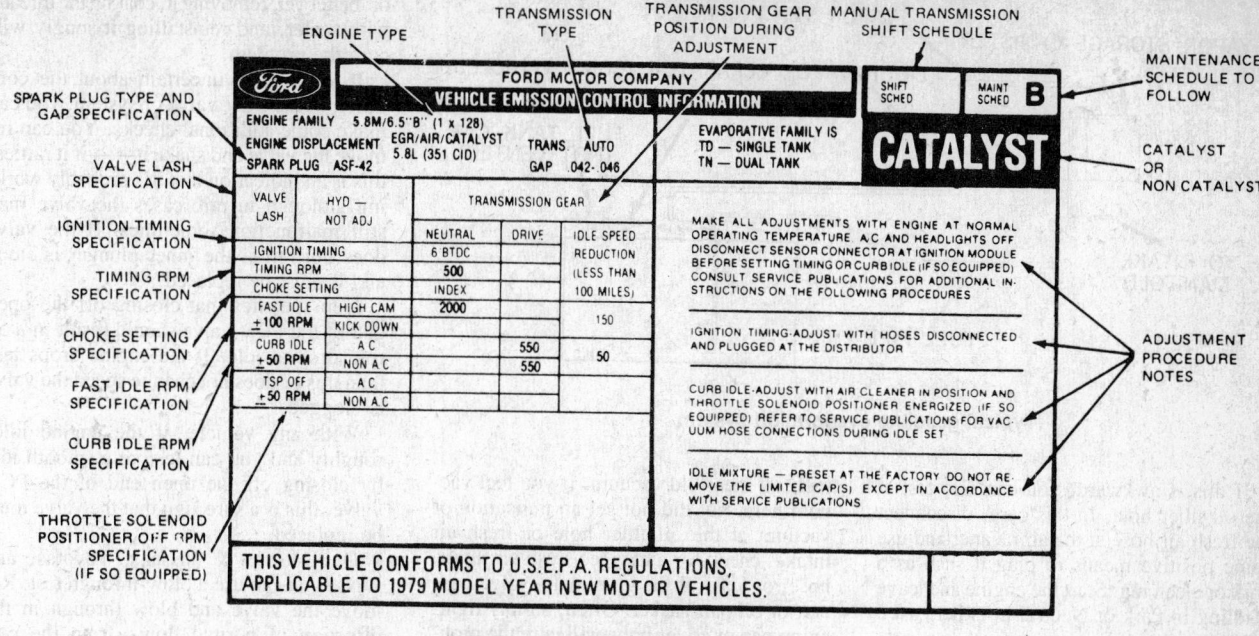

Labels around the diagram:

- ENGINE TYPE
- TRANSMISSION TYPE
- TRANSMISSION GEAR POSITION DURING ADJUSTMENT
- MANUAL TRANSMISSION SHIFT SCHEDULE
- SPARK PLUG TYPE AND GAP SPECIFICATION
- VALVE LASH SPECIFICATION
- IGNITION TIMING SPECIFICATION
- TIMING RPM SPECIFICATION
- CHOKE SETTING SPECIFICATION
- FAST IDLE RPM SPECIFICATION
- CURB IDLE RPM SPECIFICATION
- THROTTLE SOLENOID POSITIONER OFF RPM SPECIFICATION (IF SO EQUIPPED)
- MAINTENANCE SCHEDULE TO FOLLOW
- CATALYST OR NON CATALYST
- ADJUSTMENT PROCEDURE NOTES

FORD MOTOR COMPANY
VEHICLE EMISSION CONTROL INFORMATION

ENGINE FAMILY 5.8M/6.5"B" (1 x 128)
ENGINE DISPLACEMENT EGR/AIR/CATALYST 5.8L (351 CID) TRANS. AUTO
SPARK PLUG ASF-42 GAP .042-.046

EVAPORATIVE FAMILY IS
TD — SINGLE TANK
TN — DUAL TANK

SHIFT SCHED / MAINT SCHED B

CATALYST

VALVE LASH	HYD — NOT ADJ	TRANSMISSION GEAR		
		NEUTRAL	DRIVE	IDLE SPEED
IGNITION TIMING		6 BTDC		REDUCTION
TIMING RPM		500		(LESS THAN
CHOKE SETTING		INDEX		100 MILES)
FAST IDLE ±100 RPM	HIGH CAM	2000		
	KICK DOWN			150
CURB IDLE ±50 RPM	A C		550	
	NON A C		550	50
TSP OFF ±50 RPM	A C			
	NON A C			

MAKE ALL ADJUSTMENTS WITH ENGINE AT NORMAL OPERATING TEMPERATURE A.C AND HEADLIGHTS OFF DISCONNECT SENSOR CONNECTOR AT IGNITION MODULE BEFORE SETTING TIMING OR CURB IDLE (IF SO EQUIPPED) CONSULT SERVICE PUBLICATIONS FOR ADDITIONAL IN-STRUCTIONS ON THE FOLLOWING PROCEDURES

IGNITION TIMING-ADJUST WITH HOSES DISCONNECTED AND PLUGGED AT THE DISTRIBUTOR

CURB IDLE-ADJUST WITH AIR CLEANER IN POSITION AND THROTTLE SOLENOID POSITIONER ENERGIZED (IF SO EQUIPPED) REFER TO SERVICE PUBLICATIONS FOR VAC-UUM HOSE CONNECTIONS DURING IDLE SET

IDLE MIXTURE — PRESET AT THE FACTORY DO NOT RE-MOVE THE LIMITER CAP(S) EXCEPT IN ACCORDANCE WITH SERVICE PUBLICATIONS

THIS VEHICLE CONFORMS TO U.S.E.P.A. REGULATIONS, APPLICABLE TO 1979 MODEL YEAR NEW MOTOR VEHICLES.

Emission Certification Label (typical)

NOTE: It is a good practice to copy the information from the Emission Certification label and keep with the owner's manual, in case the label becomes mutilated or lost.

Crankcase Control System

POSITIVE CRANKCASE VENTILATION (PCV)

With the engine operating, crankcase ventilation air is drawn through an air cleaner mounted filter assembly, through a hose to the crankcase air inlet, down into the crankcase and up to the rocker arm chamber, out through a flow control valve and into a hose connected to the base of the carburetor or to the intake manifold. The crankcase vapors are then mixed with the air/fuel mixture and burned through the normal combustion process. The purpose of the flow control valve is to restrict the flow of crankcase vapors when the intake manifold vacuum is high (such as idle or coast modes), to avoid upsetting the air/fuel mixture at idle and causing roughness of the engine at low speeds or while idling.

With the flow control valve open at times of low engine vacuum and high air flow through the carburetor (such as having the throttle valves open as in the drive mode), the added crankcase vapor flow has no noticeable effect on the engine operation.

All late model Jeep 4–150 CID and 6–258 CID engines are equipped with the Computerized Emission Control (CEC) Fuel Feedback System. A PCV valve solenoid is installed in the PCV hose to close the crankcase ventilation system when the engine is operating at idle speed. The anti-dieseling relay system is used on the 4–150 CID engine to momentarily energize the PCV solenoid when the ignition key is turned off, to prevent air from entering the intake manifold below the throttle plates of the carburetor, thus preventing engine dieseling.

CRANKCASE CONTROL TESTING

Checking crankcase vacuum is the most effective way to test any PCV system. If there is vacuum in the crankcase, then the major part of the system has to be working.

On many models, you can effectively test the system with either a vacuum gauge designed for testing the PCV system or with a piece of paper. It is best to test the system at the fresh air intake——that is, where a large, open tube enters the air cleaner, often at a filter. Leave the PCV valve and its hose, and the oil filler cap, in place. Disconnect the fresh air hose at the air cleaner and test at the end of the hose.

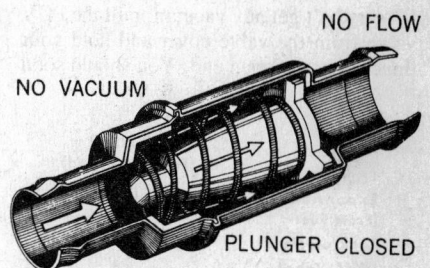

NO FLOW
NO VACUUM
PLUNGER CLOSED

PCV valve operation with engine off or during backfire

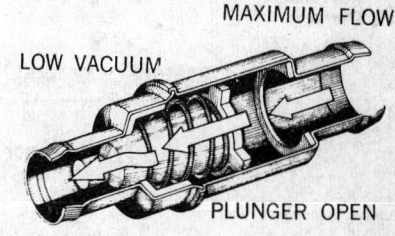

MAXIMUM FLOW
LOW VACUUM
PLUNGER OPEN

PCV valve operation at high engine speed

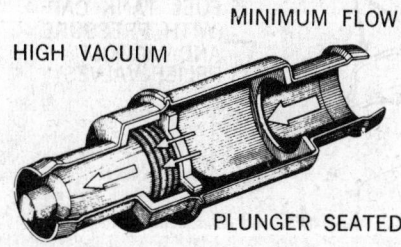

MINIMUM FLOW
HIGH VACUUM
PLUNGER SEATED

PCV valve at low engine speed or during idle

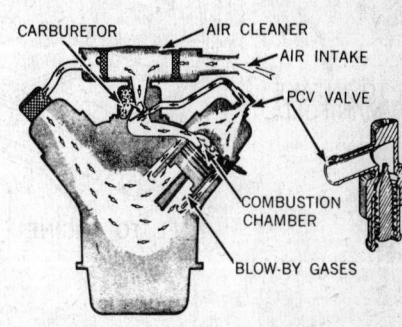

CARBURETOR
AIR CLEANER
AIR INTAKE
PCV VALVE
COMBUSTION CHAMBER
BLOW-BY GASES

Closed crankcase ventilation system

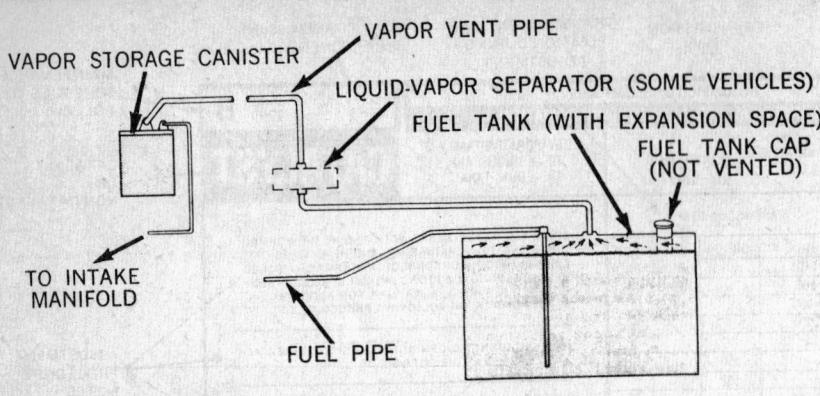

Typical gasoline evaporation system

If this is awkward, you can also test at the oil filler hole. In this case, disconnect the fresh air hose at the air cleaner and use some positive means to plug it such as a cork or clean rag. Start the engine and leave it idling in Park or Neutral in either case.

It may take a few seconds for the vacuum to build up enough to suck the paper against the opening. Make sure you allow time for the full force of the vacuum to take effect. If you don't get any vacuum, pull the PCV valve from the valve cover and hold your finger over the open end. You should soon

feel full manifold vacuum. If you feel vacuum here, but did not get an indication of vacuum at the oil filler hole or fresh air intake, check for a clogged fresh air intake hose, or leaks at the grommets, valve covers, or oil pan gasket. Often, simply tightening pan or cover bolts will cure the problem.

If there's no vacuum at the PCV valve, check for a clogged valved or hose, or a poor connection. In some cases, the hose is attached to a fitting that's screwed into the intake manifold. Tightening the fitting,

or, better yet, removing it, coating the threads with sealer, and reinstalling it snugly will cure the problem.

If you're still uncertain about the condition of the PCV valve or system, you can make some additional checks. You can remove the valve and shake it——if it rattles, this is an indication that it's probably working although in rare cases the valve may still malfunction sometimes. If the valve does not rattle, the inner plunger is stuck and it must be replaced.

Jeep specifies that closing off the open end of the PCV valve should result in a 50 rpm drop at idle. If the engine drops less than this or does not change speed the valve is clogged.

With any vehicle, if the engine idles roughly and you can restore a smooth idle by closing off the open end of the PCV valve, this is a sure sign that the valve must be replaced.

On Datsuns & Nissans, Toyotas, and Luvs, you can use a blow-through test. Remove the valve and blow through in the direction of normal flow—from the carburetor or intake manifold side. There should be a resistance to flow. Turn the valve around and blow through it from the valve cover side. The air should flow through it freely; if either test is failed, replace the valve.

On some foreign vehicles such as Isuzu and Luv, Dodge Arrow and D–50, Mitsubishi and VW, there is no PCV valve. Instead, an orifice serves to restrict the flow of air through the system under high vacuum conditions. Periodic maintenance includes checking the condition of hoses and connections, making sure hoses and especially the orifice are clean and, if the air cleaner contains a crankcase ventilation filter, cleaning or replacing it as necessary. If the filter consists of fiberglass within a plastic housing, replace it. If it consists of steel wool within a permanent housing (Mitsubishi, Arrow and D–50), clean it in a safe solvent and reinstall it. Hoses can be cleaned with solvent and air dried. Hoses must also be checked for cracks or leaks and replaced if defective.

Typical vapor canister storage

Evaporation Control System

To prevent the emission of gasoline vapors into the atmosphere from the gasoline tank and carburetor vents, vapors are routed by hoses to one or more charcoal filled canisters for storage while the engine is stopped and are routed from and/or through the canister(s) to the engine fuel induction system, when the engine is operating.

On single canister arrangements, the throttle valve is normally used as a purge valve, with the vapor hoses routed to the intake manifold or to the carburetor base. On some vehicle models, the purging of the canister is accomplished by air movement through the air cleaner snorkel and

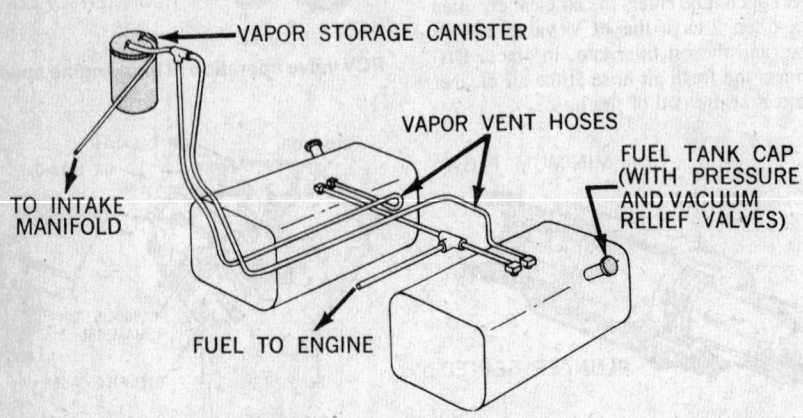

Vapor storage with twin fuel tanks (typical)

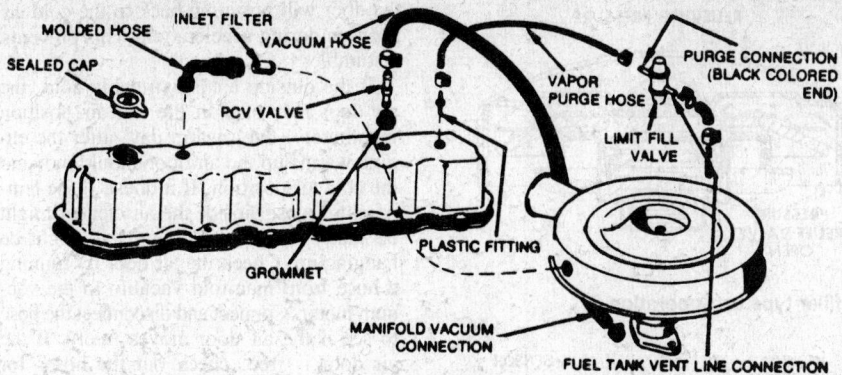

MOLDED HOSE   INLET FILTER   VACUUM HOSE
SEALED CAP   PURGE CONNECTION
                                        (BLACK COLORED END)
           PCV VALVE   VAPOR
                        PURGE HOSE   LIMIT FILL VALVE
   GROMMET   PLASTIC FITTING
        MANIFOLD VACUUM
        CONNECTION
                FUEL TANK VENT LINE CONNECTION

Jeep PCV system

into the engine, by having the purge hose connected from the canister to the snorkel.

On dual canister arrangements, the purging action of the secondary canister is triggered by a vacuum signal from the distributor vacuum hose to open the canister purge switch, which allows the vapors to purge through the PCV system and into the engine.

On some vehicles, a solenoid valve activated by the ignition switch allows intake vacuum to purge the canister. When the ignition is turned off, the solenoid valve closes to retain the vapors in the system. Valve failure occurs in the closed position and can result in lean running due to the existence of a vacuum in the float bowl.

Fuel filler caps that are used with the vapor Emission Control System normally have a pressure-vacuum valve assembly as part of the cap, to allow air to enter the tank as the fuel is consumed to avoid fuel tank collapse, when the vacuum is between 15 to 25 in. When the fuel tank internal pressure builds up from .75 to 2.0 psi (nominal) over atmospheric pressure, the pressure valve opens to relieve the excess internal pressures.

Larger trucks will normally have fuel caps with anti-surge mechanisms built into the caps to prevent fuel spillage during truck operation or will have non-vented caps with the fuel tanks vented through vapor storage canisters.

Vapor separators and anti-rollover valves are used with the vapor control systems, to avoid having raw fuel collect in the charcoal canister or to have fuel leakage in case of a vehicle rollover.

A carburetor external bowl fuel vapor vent, used on some vehicles (including Jeeps) provides an outlet for the fuel vapors when the engine is not in the operating mode, to prevent the vapors from entering the atmosphere.

To reduce the possibility of high temperature fuel vapor problems, certain 6–258 CID Jeep engine applications may be equipped with a fuel return system. This system consists of a hose connecting the fuel filter to the fuel tank, through a third nipple on the filter. The fuel filter must be positioned with the third nipple on the top

of the filter during its installation. During normal operation, a small amount of fuel returns to the tank, rather than entering the carburetor bowl. A one-way check valve is positioned in the return line, at the filter or the hose, to prevent fuel from returning to the carburetor through the fuel return line.

EVAPORATION CONTROL SYSTEM INSPECTION

The system inspection consists of examining the system rubber hoses, connections, metal lines, nylon lines, valves, separators and canisters. 10,000–12,000 miles is an adequate interval for inspection of the hoses——do it whenever spark plugs are replaced. The canister air filter must be replaced on some systems. The filter is located in the bottom of the canister and can be replaced by dismounting the canister and turning it over (always label the hoses going to the canister before disconnecting them.) On Mitsubishi trucks, the filter should be replaced every 24,000 miles.

Most systems require replacement of the entire canister at 50,000 mile intervals to replace the charcoal element, which becomes contaminated.

Checking the Datsun/Nissan Canister Purge Control Valve

To check the operation of the carbon can-

ister purge control valve, disconnect the rubber hose between the canister control valve and the T-fitting, at the T-fitting. Apply vacuum to the hose leading to the control valve. The vacuum condition should be maintained indefinitely. If the control valve leaks, remove the top cover of the valve and check for a dislocated or cracked diaphragm. If the diaphragm is damaged, a repair kit containing a new diaphragm, retainer, and spring is available and should be installed.

Thermostatic Air Cleaner

Fresh air supplied to the air cleaner comes either from the normal snorkel, or from a tube connected to an exhaust manifold stove. A door in the snorkel regulates the source of incoming air so that a warm engine always takes in warm air, approximately 100°F. The door may be controlled by a thermostatic spring or expansion bulb, or it may be vacuum operated. The vacuum operated designs use a thermostatic bimetal switch inside the air cleaner that bleeds off vacuum as the engine warms up, and regulates the position of the air door. On all late models, the snorkel is connected to a long tube so it takes in cooler air from out-

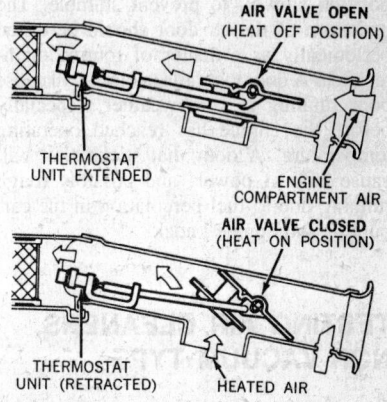

AIR VALVE OPEN (HEAT OFF POSITION)
THERMOSTAT UNIT EXTENDED
ENGINE COMPARTMENT AIR
AIR VALVE CLOSED (HEAT ON POSITION)
THERMOSTAT UNIT (RETRACTED)   HEATED AIR

Thermostatic controlled air cleaner operation

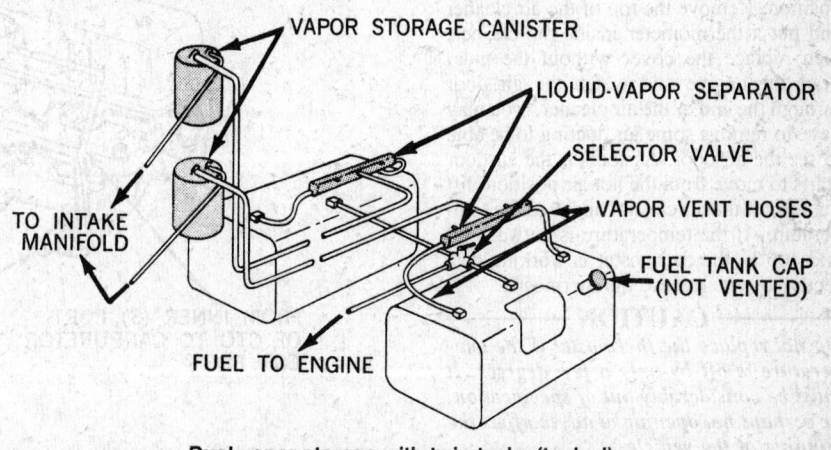

VAPOR STORAGE CANISTER
LIQUID-VAPOR SEPARATOR
SELECTOR VALVE
VAPOR VENT HOSES
FUEL TANK CAP (NOT VENTED)
TO INTAKE MANIFOLD
FUEL TO ENGINE

Dual vapor storage with twin tanks (typical)

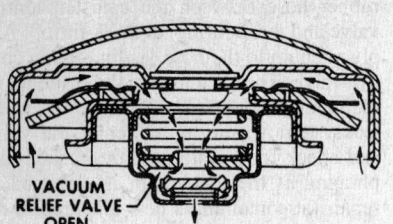

RELIEVING VACUUM

VACUUM
RELIEF VALVE
OPEN

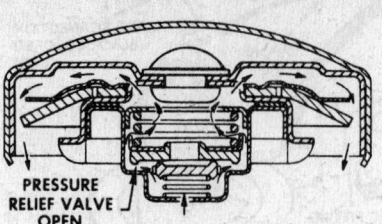

RELIEVING PRESSURE

PRESSURE
RELIEF VALVE
OPEN

Pressure-vacuum type fuel filler type cap operation

side the engine compartment. In hot climates the cool air tube is necessary because underhood air can easily reach 200°F.

Vacuum operated air doors are all designed so that the air cleaner takes in cold air when there is no vacuum. This means that an air door in the hot air position will switch to the cold position at wide open throttle because of the loss of manifold vacuum. The sudden switching of the door from hot to cold may cause a stumble or misfire in the engine, so some designs include a modulator valve mounted on the side of the air cleaner to block the vacuum and hold the door in the hot air position. A small thermostat inside the modulator opens it when the underhood temperatures reach normal. Other designs use a delay valve that allows the air door to move to the cold position slowly, to prevent stumble. The operation of the air door should be tested periodically as a matter of routine maintenance. A door that fails to close will cause poor running in cold weather, especially before the engine has reached operating temperature. A door that sticks shut will cause lack of power, and possibly rough running due to fuel percolation in the carburetor and engine knock.

TESTING AIR CLEANERS, NON-VACUUM TYPE

To test the non-vacuum type of heated air cleaner, start with an engine that is cold enough to have the air door in the hot air position. Remove the top of the air cleaner and put a thermometer inside the cleaner, then replace the cover without the nuts. Start the engine and watch the air door through the end of the air cleaner. You may have to remove some air ducting to be able to see the air door. As soon as the air door starts to move from the hot air position, lift the top off the air cleaner and read the temperature. If the temperature is between 80 and 100°F. the thermostat is working correctly. If not, replace the thermostat.

─────── **CAUTION** ───────
Do not replace the thermostat if the temperature is off by only a few degrees. It must be considerably out of specification, or perhaps not opening at all, to affect the running of the vehicle.

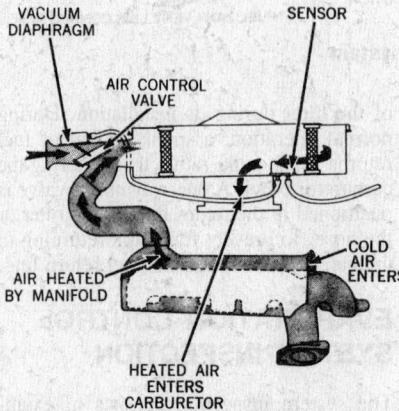

VACUUM
DIAPHRAGM

SENSOR

AIR CONTROL
VALVE

COLD
AIR
ENTERS

AIR HEATED
BY MANIFOLD

HEATED AIR
ENTERS
CARBURETOR

Vacuum operated air cleaner operation

TESTING AIR CLEANERS, VACUUM TYPE

To test the vacuum type of heated air cleaner, inspect the air door with the engine off. It should be in the cold air position. Start the engine. If the engine is cold, the air door should move to the hot air position. As the engine warms up, the air door should move to a mid position, depending on the outside air temperature.

the door will not jump back to the cold air position during acceleration. This prevents a stumble.

If the outside air is extremely cold, the air door may stay in the hot air position indefinitely. On a warm day, after the engine warms up the air door should move to the cold air position. If it doesn't, the temperature sensor inside the air cleaner might be faulty, or the air door itself might be hanging up. Check the air door by running a hose from manifold vacuum to the vacuum motor. Connect and disconnect the hose to see if the air door moves freely. If the air door is free, check out the hoses for leaks or blockage. If the hoses are okay, the trouble must be in the temperature sensor, and it should be replaced.

Modulators are used in the air cleaner vacuum line on some engines. The modulator mounts on the side of the air cleaner and has two hose connections, one to the air cleaner temperature sensor, and the other to the vacuum motor. Below 50–80°F. the modulator is a one-way check valve, which allows vacuum to move the air door to the hot air position, but traps the vacuum so

After the modulator warms up, the check valve unseats so that the vacuum can pass freely in either direction, and the air door then operates normally. The connections for the modulator are important. The connection in the center goes to the vacuum motor, and the connection on the edge goes to the vacuum source, which is the temperature sensor.

To test the modulator on a cold engine, apply enough vacuum to the edge port to move the air door to the hot position. Then remove the hose from the port, and the air door should stay in the hot position. Make the same test when the engine is warmed up, and the air door should move to the cold position when you pull off the hose.

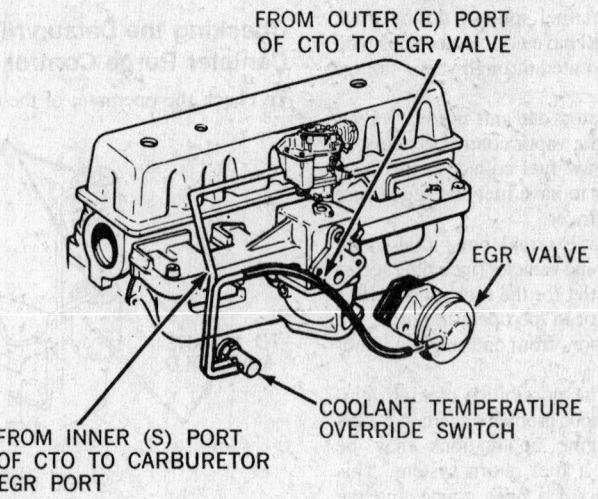

FROM OUTER (E) PORT
OF CTO TO EGR VALVE

EGR VALVE

COOLANT TEMPERATURE
OVERRIDE SWITCH

FROM INNER (S) PORT
OF CTO TO CARBURETOR
EGR PORT

Typical 6 cylinder engine EGR system

EVAPORATION CONTROL SYSTEM DIAGNOSIS CHART

Problem	Cause	Remedy
Persistent odor of fuel vapors	Canister saturated due to extend parking of vehicle.	Operate (idle) engine for several minutes to purge canister.
	Fuel tank cap not sealing.	Replace cap.
	Canister not purging:	
	a. Vacuum hose to intake manifold or tee obstructed or leaking.	Check for vacuum at canister end of hose. Blow through hose with compressed air. Replace hose if cracked, deteriorated or obstruction cannot be removed.
	b. Vacuum orifice in manifold fitting or tee obstructed.	Remove manifold fitting or tee and blow orally through fitting to check for obstruction. If orifice is plugged, soak fitting in solvent and blow out with compressed air.
	Loose vent hose connections or loose filler neck connections.	Pressure test vapor vent system for leakage. If leakage is indicated, visually inspect for damaged hoses or tubes, loose, damaged or missing clamps. Repair is needed.
Fuel leakage:		
a. From fuel tank cap	Fuel tank cap seal faulty.	Replace cap.
	Pressure relief valve in fuel tank cap faulty.	Test operation of pressure relief valve. If valve is faulty, replace cap.
	Valve vent hoses obstructed.	Remove fuel tank cap and blow hoses with compressed air. Replaced hoses if necessary.
b. From fuel tank liquid vapor separator or connecting tubes and hoses	Loose connections, cracked or broke tube or hose.	Pressure test vapor vent system for leakage. If leakage is indicated, repair as needed.
	Cracked or damaged fuel tank or liquid/vapor separator	Replace damaged components.
c. From vapor storage canister (through air flow filter)	Pressure relief valve in fuel tank cap faulty.	Test operation of valve. If faulty, replace cap.
Noisy fuel tank— wall fluctuation ("oilcanning")	Vacuum buildup in tank:	
	Vacuum relief valve in fuel tank cap faulty.	Test operation of valve. If faulty, replace cap.
	Pressure buildup in tank:	
	a. Pressure relief valve in fuel tank cap faulty.	Test operation of valve. If faulty, replace cap.
	b. Vapor vent hoses obstructed.	Remove fuel tank cap and blow out hoses with compressed air. Replace hoses if necessary.

Ignition Timing Control Systems

Ignition timing effects emissions significantly, especially when the throttle is nearly closed. Also, retarding timing under acceleration may reduce NO. For this reason, some emission systems may delay the application of vacuum spark advance under certain operating conditions.

Because of the effect such retardation has on engine heat, some of these systems may have controls which advance the spark via the distributor vacuum advance diaphragm only under conditions of high engine coolant temperature in order to prevent overheating.

MITSUBISHI/D–50/ARROW IGNITION TIMING CONTROL SYSTEM

When the engine is idling or operating at low speeds under light load or deceleration, the exhaust gas temperature is low, resulting in incomplete combustion of the air/fuel mixture. To prevent this, ignition timing is retarded under these conditions to maintain high exhaust gas temperature.

The units in the Ignition Timing Control system are as follow:

1. **Thermo Valve:** This valve is used to protect the engine from overheating. When coolant temperature reaches 203 degrees, the advance unit is allowed to operate, causing an increase in engine speed and a decrease in coolant temperature.

2. Single Diaphragm Distributor: This distributor has a single diaphragm vacuum advance unit, which advances the ignition timing as engine vacuum dictates. The single diaphragm distributor must not be interchanged with the dual diaphragm distributor. The distributor operating curves are different and would cause increased emissions. A thermo valve is not used with this type of distributor.

DATSUN/NISSAN SPARK TIMING CONTROL SYSTEM

A spark timing control system is added to manual transmission models sold in the U.S. in 1979. The system controls distributor vacuum advance, giving full vacuum advance when the transmission is in 4th or 5th, and partial advance in the first three gears. This provides better control of the combustion process, lowering emissions of HC and NO.

The system components include a top gear detecting switch, installed into the transmission, and a vacuum switching valve spliced into the distributor vacuum advance hose by means of a three way connector. When the transmission is shifted into either of the two top gears, the transmission switch goes on, thus activating the vacuum switching valve which closes its air bleed, giving full advance. Shifting into any gear but 4th or 5th turns the transmission switch off, deactivating the vacuum switching valve. The valve opens a vacuum leak, providing only partial vacuum advance to the distrib-

Checking the Transmission Switch

The switch can be checked easily with an ohmmeter. Connect the ohmmeter leads to the switch leads on the transmission. Shift back and forth between either 4 or 5 and one of the other gears. If the resistance does not change, replace the switch.

Vacuum Switching Valve

1. Disconnect the valve's electrical connectors. With the timing light installed, run the engine up to about 2,000 rpm and keep it there. Check the timing.

2. Connect the valve's electrical connectors directly to the battery with a pair of jumper wires. Be sure to observe correct polarity. If spark timing varies, the valve is ok. If not, replace it.

Deceleration Control Devices

Deceleration, especially sudden release of the throttle, produces more hydrocarbon and CO emissions than any other gas engine operating mode. Because of high manifold vacuum, both compression and turbulence in the cylinder are at a minimum. At the same time, fuel laying in the manifold un-

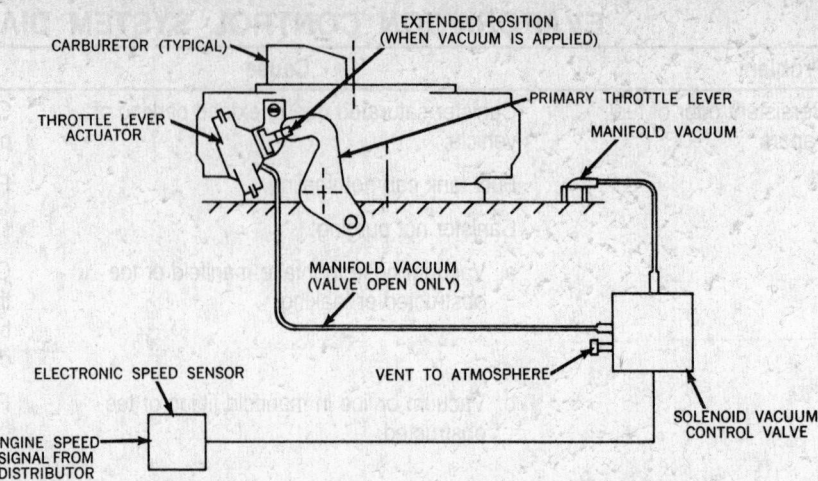

Typical throttle return control system used on G.M. vehicles

der cruise or acceleration conditions may suddenly be evaporated and pass through the engine practically unburned.

As a result of these combustion/emissions problems, many vehicles——especially those using high rpm engines and manual transmissions——are equipped with devices which slow or limit closing of the throttle when the vehicle is at speed, or admit extra mixture. In some cases, ignition timing may be advanced.

When improperly adjusted or malfunctioning, these systems may cause: engine racing and difficult shifting; inconsistent behavior of the engine when the throttle is released, also causing difficult shifting; or backfiring.

Since such a system can radically raise engine idle speed, you should make sure adjustments are to specification and that all elements of the system are snugly connected and working smoothly. Careless work could contribute to an accident!

CHEVROLET AND GMC THROTTLE RETURN CONTROL SYSTEM

1979–82

A throttle return control system (TRC) is used on some California truck engines. When the truck is coasting against the engine, the control valve is open to allow vacuum to operate the throttle lever actuator. The throttle lever actuator then pushes the throttle lever slightly open reducing the HC (hydrocarbon) emission level during coasting. When manifold vacuum drops below a predetermined level, the control valve closes, the throttle lever retracts, and the throttle lever closes to the idle position.

Control Valve Check and Adjustment

1. Disconnect the valve-to-carburetor

hose and connect it to an external vacuum source and a vacuum gauge.

2. Disconnect the valve-to-actuator hose at the connector and connect it to a vacuum gauge.

3. Place a finger firmly over the end of the bleed fitting.

4. Apply a minimum of 23 in. Hg vacuum to the control valve and seal off the vacuum source. The gauge on the actuator side should read the same as the gauge on the source side. If not, the valve needs adjustment. If vacuum drops off on either side (with the finger still on the bleed fitting), the valve is defective and should be replaced.

5. With a minimum of 23 in. Hg vacuum in the valve, remove the finger from the bleed fitting. The vacuum level in the actuator side will drop to zero and the reading on the source side will drop to a value that will be the valve set point of 21.5 in. Hg. If the valve is not within ½ in. Hg vacuum of the specified valve set point, adjust the valve.

6. Gently pry off the conical plastic cover.

7. Turn the adjusting screw in (clockwise) to raise the set point or out (counterclockwise) to lower the set point.

8. Recheck the valve set point.

9. If necessary, repeat the adjustment.

Throttle Valve Check and Adjustment

1. Disconnect the valve-to-actuator hose at the valve and connect it to an external vacuum source.

2. Apply 20 in. Hg vacuum to the actuator and seal the vacuum source. If the vacuum gauge reading drops, the valve is leaking and should be replaced.

3. Check the throttle lever, shaft, and linkage for freedom of operation.

4. Start the engine and warm it to operating temperature.

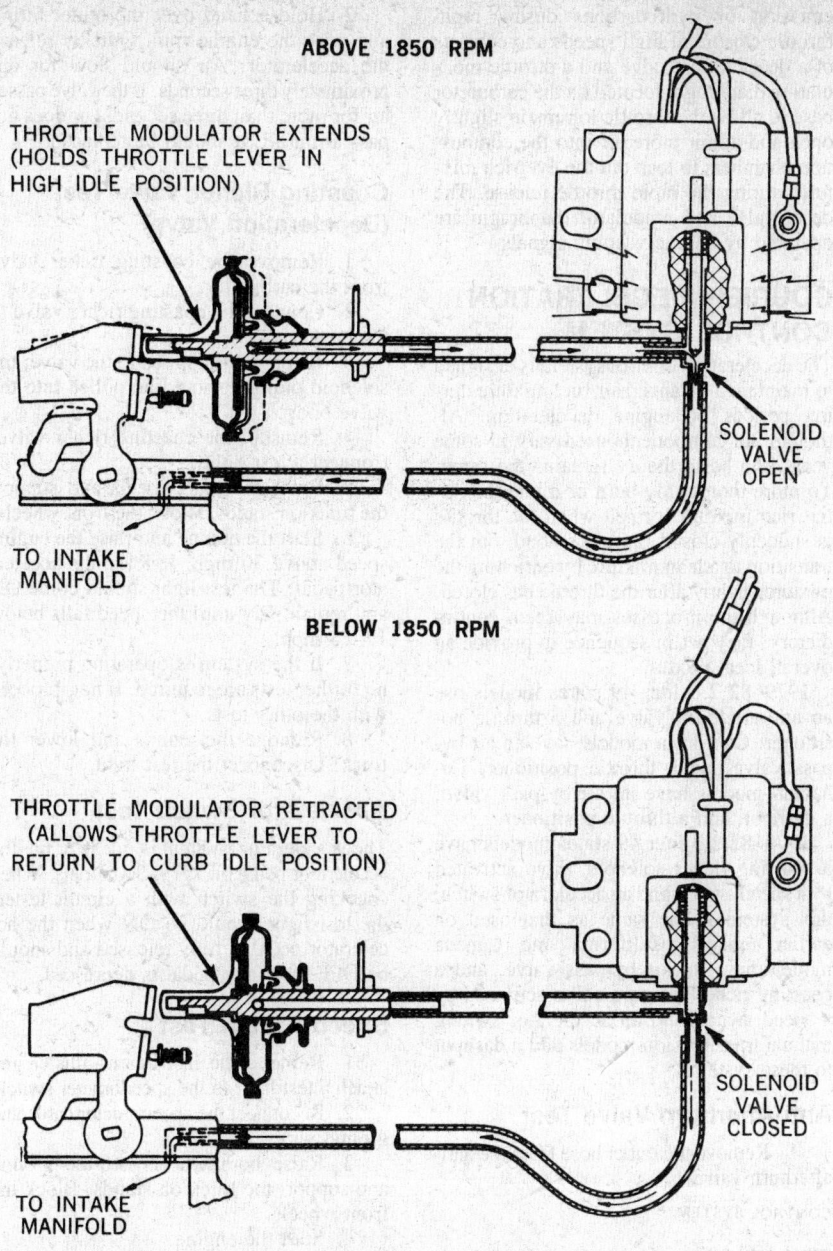

ABOVE 1850 RPM

THROTTLE MODULATOR EXTENDS
(HOLDS THROTTLE LEVER IN
HIGH IDLE POSITION)

SOLENOID
VALVE
OPEN

TO INTAKE
MANIFOLD

BELOW 1850 RPM

THROTTLE MODULATOR RETRACTED
(ALLOWS THROTTLE LEVER TO
RETURN TO CURB IDLE POSITION)

SOLENOID
VALVE
CLOSED

TO INTAKE
MANIFOLD

Typical throttle modulator operation used on IHC vehicles

5. Note the idle rpm.

6. Apply 20 in. Hg vacuum to the actuator and manually operate the throttle. Allow it to close against the extended actuator plunger. Note the engine rpm.

7. Release and reapply 20 in. Hg vacuum to the actuator and note the rpm at which the engine speed increases (do not assist the actuator).

8. If the engine speed obtained in Step 7 is not within 150 rpm of that obtained in Step 6, then the actuator may be binding. If the binding cannot be corrected, replace the actuator.

9. Release the vacuum from the actuator and the engine speed should return to within 50 rpm of the speed noted in Steps 4 and 5.

10. To adjust the actuator: Turn the screw on the actuator plunger until the specified TRC speed range (1475–1525 rpm) is obtained.

1983–84

This system is used on 1983–84 trucks with heavy duty emission controls. It consists of a vacuum operated throttle lever actuator, an electronic speed sensor, and a solenoid vacuum control valve, which converts the sensor's electronic signal to vacuum to actuate the system. The system opens the throttle slightly beyond its curb idle position to reduce hydrocarbons when the engine is overrunning.

Problems with the system would be in-

dicated by inconsistent idle speed or unaccustomed engine braking or backfiring. Troubleshoot the system as described below:

1. Inspect all vacuum hoses for cracks or bad connections and replace as necessary. Make sure wiring connectors are firmly attached at the distributor, speed switch, and vacuum solenoid.

2. Connect a tachometer sensitive enough to measure a change of 10 rpm to the distributor TACH terminal. Then, start the engine and run it at exactly 1890 rpm. Observe the throttle lever actuator. It should be extended.

3. Close the throttle until engine speed is exactly 1700 rpm. The throttle lever actuator should be retracted at this speed.

4. If either of these tests is failed, accelerate the engine within a wider range. If the actuator works this way, but not within the specs above, replace the electronic speed switch. Otherwise, proceed with the tests below.

5. Connect the negative probe of a voltmeter to an engine ground, and the positive probe to the hot connector of the wire to the vacuum solenoid. You do not have to disconnect the electrical connector to do this—the probe can be inserted into the connector body on the wire side to contact the metal terminal. Check for 12–14 volts here and, in a similar manner, at the speed switch. If voltage is present at only one of the two devices, repair the wiring harness as necessary. If there is no voltage at either device, repair connections at the distributor and bulkhead.

6. If proper voltage exists both places, ground the solenoid-to-switch connecting wire at the solenoid connector with a jumper. The throttle lever actuator should extend with the engine running.

7. If it does not extend, disconnect the hose at the side port of the solenoid that connects to the actuator. Check the orifice inside the port for plugging and clear it if necessary. If there is no plugging, replace the solenoid. Otherwise, retest it as in Step 6.

8. If the solenoid extended in Step 6, ground the solenoid-to-switch wire terminal at the speed switch. If it did not extend, repair the wire connecting the speed switch and solenoid. If it does extend, check the effectiveness of the speed switch ground connection. With the engine running, the voltmeter should read zero volts at the ground wire connection. If ground voltage is high, make repairs to the wiring.

9. Check the speed switch-to-distributor wire for proper connection. Make repairs as necessary. Then, repeat the test in which the engine is accelerated to 1890 rpm and the actuator is observed. If the actuator does not extend, replace the speed switch.

10. If the actuator remains extended at all speeds, remove the connector from the vacuum solenoid. If the actuator remains extended, check the orifice in the side port of the solenoid for plugging and, if nec-

essary, clear it out. Reconnect the vacuum line and run the test again. If the actuator now does not retract, disconnect the solenoid electrical connector. If the actuator still does not retract, replace the vacuum solenoid.

11. If the actuator retracts with the solenoid connector off, reconnect and then remove the speed switch connector. If the actuator now retracts, replace the speed switch. If the actuator does not retract, the solenoid-to switch wire is shorted to ground in the wiring harness. Make required repairs.

DODGE/PLYMOUTH VACUUM THROTTLE POSITIONER

Some 1979 and later models have a throttle positioner to hold the throttle slightly open on deceleration above 2000 rpm. This prevents excessive emission of unburned hydrocarbons. The system consists of an electronic speed switch, a vacuum solenoid valve, and a vacuum actuated throttle positioner. Adjust this device as follows:

1. Run the engine at about 2500 rpm.
2. Loosen the vacuum unit locknut and turn it until it just contacts the throttle lever.
3. Release the throttle. Adjust the unit to decrease engine speed until speed suddenly drops 1000 rpm or more. Back off the adjuster ¼ turn more and tighten the locknut.
4. Check the adjustment by accelerating the engine to about 2500 rpm and releasing the throttle. The engine should return to normal idle speed.

JEEP VACUUM THROTTLE MODULATING SYSTEM (VTM)

The VTM system is used to reduce the emission of hydrocarbons during rapid throttle closure at high speeds and consists of a deceleration valve and a throttle modulating diaphragm located on the carburetor base to allow the throttle to remain slightly open and admit more air into the combustion chambers to lean out the overrich mixture, during the rapid throttle release. The decel valve and modulator diaphragm are operated by engine vacuum signals.

COURIER DECELERATION CONTROL SYSTEM

The deceleration control system is designed to maintain a balanced air/fuel mixture during periods of engine deceleration. Although the components used vary in some years, the basic theory remains the same. To more thoroughly burn or dilute the initial rich mixture formed when the throttle is suddenly closed, and to smooth out the transition to a lean mixture by enriching the mixture slightly after the throttle has closed. Although the processes may seem contradictory, they act in sequence to provide an overall ideal mixture.

1979–82 2.0 liter 49 states models use an anti-afterburn valve and a throttle positioner. California models have an air bypass valve and a throttle positioner. Canadian models have an air by-pass valve, a dashpot, and a throttle positioner.

1979–82 2.3 liter 49 states models have a coasting richer solenoid valve activated by a speed switch and an accelerator switch; this system is the same as that used on earlier models. California and Canada models have an air by-pass valve, and a coasting richer solenoid valve activated by a speed switch and an accelerator switch; manual transmission models add a dashpot to these systems.

Anti-Afterburn Valve Test

1. Remove the outlet hose from the anti-afterburn valve.

2. Hold a hand over the outlet fitting and raise the engine rpm. Quickly release the accelerator. Air should flow for approximately three seconds. If the valve passes air for more than three seconds, or does not pass air at all, it should be replaced.

Coasting Richer Valve Test (Deceleration Valve)

1. Remove the coasting richer valve from the carburetor.
2. Connect the coasting richer valve to the battery.
3. As power is applied to the valve, the solenoid plunger should be pulled into the valve body.
4. Reinstall the coasting richer valve. Connect a test light.
5. Raise the rear wheels and support the truck on stands. Block the front wheels.
6. Start the engine and raise the engine speed above 30 mph. Release the accelerator pedal. The test light should come ON and remain ON until the speed falls below 17–23 mph.
7. If the system is operating properly, no further tests are required. If not, proceed with the other tests.
8. Remove the stands and lower the truck. Disconnect the test light.

Accelerator Switch Test

The accelerator switch is located on the accelerator pedal on 1979–80 models. When checking the switch with a circuit tester, the test light should be ON when the accelerator pedal is fully released and should be OFF when the pedal is depressed.

Speed Switch Test

1. Remove the instrument cluster and attach a test light to the speedometer switch.
2. Reconnect the speedometer cable and ground wire.
3. Raise both wheels off the ground and support the truck on stands. Block the front wheels.
4. Start the engine.
5. Depress the accelerator pedal to accelerate the engine and confirm that the speed switch is ON at speeds of 17–23 mph and OFF at speeds below 17–23 mph.
6. If not, replace the switch.
7. Lower the truck and remove the test light. Reinstall the instrument cluster.

Speed Switch Relay Test

Check the speed switch relay with a test light to be sure that it is operating at 17–23 mph.

Three Way Solenoid Valve Test

1979–82 2.0 ENGINES ONLY

1. Start the engine and allow it to reach normal operating temperature. Check the idle speed and adjust as necessary.
2. Disconnect the wire at the three way solenoid valve. This wire is coded either

THROTTLE RETURN CONTROL SYSTEM

CARBURETOR (TYPICAL)
EXTENDED POSITION (WHEN VACUUM IS APPLIED)
THROTTLE LEVER ACTUATOR
PRIMARY THROTTLE LEVER
MANIFOLD VACUUM
MANIFOLD VACUUM (VALVE OPEN ONLY)
ELECTRONIC SPEED SENSOR
VENT TO ATMOSPHERE
SOLENOID VACUUM CONTROL VALVE
ENGINE SPEED SIGNAL FROM DISTRIBUTOR

GMC/Chevrolet throttle return control system

black with a white stripe or brown with a red stripe on some models.

3. When the wire is disconnected, the engine speed should increase to 1000 rpm for 49 States trucks, or 1100 rpm for California models.

4. If the engine speed does not increase, the three way solenoid valve or the servo diaphragm is not operating correctly. Check the servo diaphragm using the following procedure; if the servo diaphragm is operating correctly, the three way valve is faulty and should be replaced.

Servo Diaphragm Test

1. Start the engine and set the idle speed to specification.

2. Stop the engine and disconnect the vacuum line between the vacuum control valve and the diaphragm at the diaphragm.

3. On 1979–82 models, disconnect the vacuum hose at the vacuum amplifier and the vacuum hose at the three way solenoid valve. Connect the vacuum hose from the servo diaphragm to the vacuum amplifier so that the intake manifold vacuum is applied directly to the servo diaphragm.

4. Disconnect and plug the vacuum line between the carburetor and the distributor.

5. Connect a tachometer and start the engine. The engine should idle at 900–1100 rpm (1000–1200 rpm for California models). If the engine speed is not correct, adjust by means of the servo diaphragm adjusting screw. If the correct speed is not obtainable, replace the diaphragm.

Vacuum Control Valve Test

1. Disconnect the vacuum hose between the vacuum control valve and the intake manifold at the manifold.

2. Attach a vacuum gauge in the line using a T-fitting.

3. Connect a tachometer to the engine. Start the engine and raise the speed to 300 rpm, then suddenly release the throttle. The vacuum reading should rise above 21.3 in., drop to that figure and hesitate there for one or two seconds, then drop to the normal idle vacuum of 16–18 in. Note that these readings are for sea level, and should be corrected accordingly.

4. If the vacuum reading is not within specification, adjust the vacuum control valve by turning the adjusting screw in the top of the valve. If the correct reading is unobtainable, replace the valve.

Vacuum Switch Test

1. Disconnect the vacuum hose between the vacuum switch and the vacuum control valve.

2. Using a T-fitting, connect a vacuum gauge between the vacuum switch and an external vacuum source.

3. Raise the vacuum reading above 8 in., then allow the vacuum to drop. The switch should click at approximately 6 in. If it does not, or if it clicks at a higher reading, replace the switch.

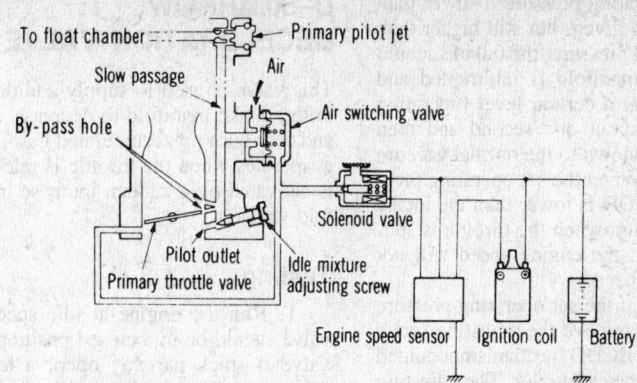

Schematic of D-50/Arrow deceleration fuel cutoff system

Air By-Pass Valve Test

1. Disconnect the air hose from the side of the air by-pass valve.

2. Connect a tachometer to the engine. Start the engine and raise the speed above 2000 rpm.

3. Release the throttle and check for air flow from the port on the side of the air by-pass valve. If there is no airflow, replace the valve.

Dashpot Adjustment

1. Check the engine idle speed and mixture, and adjust as necessary.

2. Remove the air cleaner.

3. With the tachometer still connected to the engine, loosen the dashpot locknut. Move the throttle lever and hold to maintain the engine speed at 2400–2600 rpm (2100–2300 rpm for California trucks).

4. Turn the dashpot until its rod contacts the throttle lever. Release the throttle lever and tighten the locknut.

5. Move the throttle lever until it contacts the dashpot rod and recheck the engine speed. Repeat the adjustment if necessary.

Accelerator Switch Adjustment

1979 AND LATER

The accelerator switch is mounted on the arm of the accelerator pedal.

1. Check the accelerator pedal to make sure that it moves freely.

2. Loosen the accelerator switch locknut.

3. Gradually turn the adjusting screw until the accelerator switch clicks.

4. Tighten the locknut.

Component Replacement

With the exception of the speed switch, replacement of the various components is simply a matter of disconnecting hoses, wires, etc., and dismounting the component. Mark all disconnected items so that they may be properly reconnected.

The speed switch is integrated with the speedometer assembly. To replace the speed switch, remove the speedometer as outlined later in the Courier section.

MAZDA BCDD SYSTEM

The BCDD consists of an independently operated auxiliary fuel system. This system functions when the engine is coasting to enrich the air-fuel mixture which minimizes the hydrocarbon content of the exhaust gases through more efficient combustion. This is accomplished without adversely affecting engine idle and the carbon monoxide content of the exhaust gases.

When intake manifold vacuum exceeds a predetermined value, a vacuum-actuated diaphragm opens an air passage allowing additional air to enter the intake manifold. When the additional air passage is opened, vacuum is brought to bear on another diaphragm which opens a fuel passage allowing additional fuel to enter the intake manifold.

Normally, the BCDD does not need adjustment. However, if the need should arise because of suspected malfunction of the system, proceed as follows:

1. Connect a tachometer to the engine.

2. Connect a quick-response vacuum gauge to the intake manifold.

3. Disconnect the BCDD solenoid valve electrical leads.

4. Start and warm up the engine until it reaches normal operating temperature.

5. Adjust the idle speed to the proper specification.

6. Raise the engine speed to 3,000–3,500 rpm under no-load (transmission in Neutral or Park), then allow the throttle to close quickly. Take notice as to whether or not the engine rpm returns to idle speed and if it does, how long the fall in rpm is interrupted before it reaches idle speed.

7. At the moment the throttle is snapped closed at high engine rpm, the vacuum in the intake manifold reaches −23.6 in. Hg and then gradually falls to about −16.5 in. Hg at idle speed. The process of the fall of intake manifold vacuum and engine rpm will take one of the following three forms:8.

When the operating pressure of the BCDD is too high, the system remains inoperative, and the vacuum in the intake manifold decreases without interruption just like that of an engine without a BCDD.

When the operating pressure is lower than that of the case given, but still higher than the properly set pressure, the fall of vacuum in the intake manifold is interrupted and kept constant at a certain level (operating pressure) for about one second and then gradually falls down to the normal vacuum at idle speed. When the set operating pressure of the BCDD is lower than the intake manifold vacuum when the throttle is suddenly released, the engine speed will not lower to idle speed.

9. To adjust the set operating pressure of the BCDD, remove the adjusting screw cover from the BCDD mechanism mounted on the side of the carburetor. The adjusting screw is a left-hand threaded screw. Turning the screw $1/8$ of a turn in either direction will change the operation pressure about 0.79 in. Hg. Turning the screw counterclockwise will increase the amount of vacuum needed to operate the mechanism and turning the screw clockwise will decrease the amount of vacuum needed to operate the mechanism.

10. The operating pressure for the BCDD is listed below. The decrease in intake manifold vacuum should be interrupted at these levels for about one second when the BCDD is operating correctly. 1979: -21.65 ± 0.75, all models. Don't forget to install the adjusting screw cover when the system is adjusted.

D–50/ARROW DECELERATION DEVICE

Closing of the throttle valve on deceleration is delayed in order to burn the air/fuel mixture more thoroughly. A vacuum controlled dashpot, attached to the carburetor linkage is used.

A servo valve detects intake manifold vacuum and closes if vacuum exceeds a preset value. Since the air in the dashpot diaphragm chamber cannot escape, the throttle linkage opening is temporarily retained. If the vacuum is below the preset value, the servo valve opens and the dashpot works normally.

Maintenance

Inspect the hoses for breaks and damage, and the valve body for cracks.

Adjustment

1. Have the engine running, brakes locked, and a tachometer attached.
2. Push the dashpot rod, connected to the carburetor arm, upward and into the dashpot until it stops.
3. Note the rpm of the engine under these conditions and, if necessary, adjust it to these specifications: U-engine: 2100–2300; W-engine: 1600–1800. Note the time that elapses between suddenly releasing the dashpot rod and return to normal idle speed. It should be 3–6 seconds; if not, replace the unit.

D–50/ARROW DECELERATION VALVE

This valve is used to supply additional air to the intake manifold to decrease vacuum and help burn up accumulated gasoline that evaporates when the throttle is released. It is activated by a sudden increase in manifold vacuum.

Testing

1. Run the engine at idle speed. The valve should be in a closed position. If the valve is stuck partway open, a lean idle mixture and poor idle will result. If the valve is sticking open, replace it.
2. Race the engine to 2,500 rpm and release the throttle suddenly. Watch the valve. It should open wide and then gradually close. If the valve does not operate smoothly and open quickly every time the engine is raced, replace it.

D–50/ARROW FUEL CUTOFF SYSTEM

Some vehicles have an air switching valve activated above 1700 rpm and at closed throttle. Under these conditions, carburetor ported vacuum opens the air switching valve and that valve, in turn, admits air to the idle and transition passages of the carburetor. The air pressure stops the flow of fuel, to improve fuel economy and eliminate emissions when the engine is overrunning.

Below 1700 rpm, a solenoid valve, activated through an engine speed sensor, deprives the system of vacuum so the engine will idle and operate at low speeds in the normal manner.

Testing

Start the engine and run it at idle. Then, disconnect the electrical connector to the solenoid valve. The engine should stall or nearly stall due to lean mixture. If not, check the vacuum passage for clogging. If the passage is clear, check the solenoid valve by disconnecting the vacuum line running to it. If disconnecting the vacuum line with the electrical connector disconnected makes the engine stall, replace the solenoid valve. If the vehicle has a stalling problem, make the following checks:

1. Test the voltage at the solenoid connector with the engine idling. If there is no voltage, check for wiring problems and replace or repair parts as necessary.
2. If there is voltage at the connector at idle, increase the engine speed to 1500 rpm. If voltage is lost before the engine reaches 1500 or at 1500, replace the engine speed sensor.
3. Increase the engine speed to 2500 rpm. If the voltage is not lost, replace the engine speed sensor.

DATSUN/NISSAN BOOST CONTROL DECELERATION DEVICE

The BCDD reduces hydrocarbon emissions during coasting conditions.

High manifold vacuum during coasting prevents the complete combustion of the air/fuel mixture because of the reduced amount of air. This condition will result in large HC emissions. Enriching the air/fuel mixture for a short time (during the high vacuum condition) will reduce the emission of HC in conjunction with the AIR system.

However, enriching the air/fuel mixture with only the mixture adjusting screw will cause poor engine idle, or invite an increase in the carbon monoxide (CO) content of the exhaust gases.

The BCDD consists of an independently operated auxiliary fuel system. This system functions when the engine is coasting to enrich the air-fuel mixture which minimizes the hydrocarbon content of the exhaust gases through more efficient combustion. This is accomplished without adversely affecting engine idle and the carbon monoxide content of the exhaust gases.

When intake manifold vacuum exceeds a predetermined value, a vacuum-actuated diaphragm opens an air passage allowing additional air to enter the intake manifold. When the additional air passage is opened, vacuum is brought to bear on another diaphragm which opens a fuel passage allowing additional fuel to enter the intake manifold.

When the engine changes from a coasting condition to that of idling, the transmission speed sensor closes an electrical circuit, energizing the vacuum control solenoid valve. When energized, the vacuum control solenoid valve vents the intake manifold vacuum to the atmosphere, thus causing the two diaphragms to return to their normal positions, closing off the additional air and fuel mixture. The transmission switch is not used on 1979 models.

LUV MIXTURE CONTROL VALVE

1979 and Later California and High Altitude Models

Disconnect the rubber hose connecting the mixture control valve with the intake manifold and plug the intake manifold side of the valve. If the mixture control valve is operating correctly, air will continue to blow out the mixture control valve for a few seconds after the accelerator pedal is fully depressed (engine running) and released quickly. If air continues to blow out for more than five seconds, replace the mixture control valve.

LUV COASTING ENRICHMENT SYSTEM

This system functions when the engine is coasting, to enrich the air/fuel mixture, which minimizes hydrocarbon content of the exhaust gases through efficient combustion.

A solenoid operated valve in the carburetor allows extra fuel to be drawn into the intake manifold during coasting. The solenoid is energized by the following switches.

1979 and later Federal with Manual Transmission:

1. Accelerator switch
2. Clutch switch
3. 3rd–4th gear switch

1979 and later California with Manual Transmission:

1. Accelerator switch
2. Clutch switch
3. Transmission neutral switch
4. Engine speed sensor

1979 and later California with Automatic Transmission:

1. Accelerator switch
2. Inhibitor switch
3. Engine speed sensor

When all of these switches turn on and the engine is coasting, the solenoid valve on the secondary side of the carburetor energizes and causes the valve to open. When the valve opens, the fuel is drawn out of the float chamber by engine vacuum and metered by the coasting jet below the secondary throttle valve. On acceleration the coasting richer circuit is opened causing the coasting richer valve to close, shutting off the supply of extra fuel.

Solenoid Valve Test

The valve should be open when the circuit is energized. A clicking noise should be heard when the valve operates.

Accelerator Switch Test

The accelerator switch should be closed when the pedal is not depressed and open when the pedal is depressed. Check with a test lamp.

Clutch Switch Test

1979 AND LATER

When the clutch pedal is depressed, the switch contacts are opened and the coasting richer system is de-energized. Test all years' clutch switches with a test lamp.

Transmission Switch Test

FEDERAL MANUAL TRANSMISSION MODELS

The transmission switch turns on when the

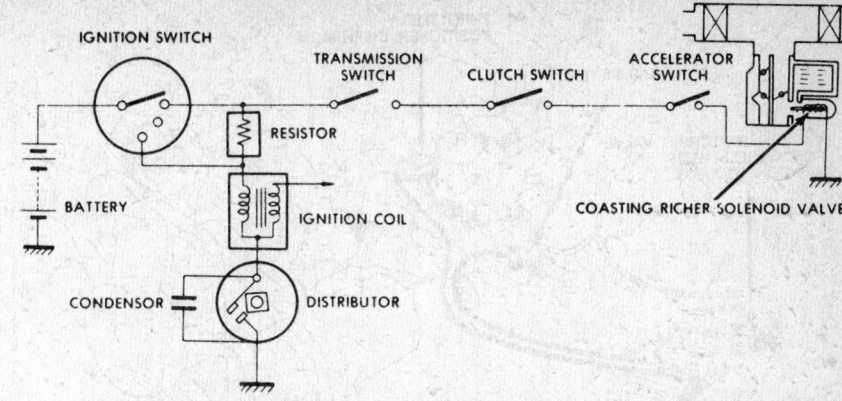

Schematic of LUV Coasting enrichment System—49 states

transmission is shifted into 3rd or 4th gear and energizes the coasting richer system. It de-energizes the system when the transmission is shifted into any other position.

CALIFORNIA MANUAL TRANSMISSION MODELS

The transmission switch turns on when the transmission is shifted into any gear and turns off when it is shifted into neutral position. The solenoid is energized when the switch is in the On position. Test with a test lamp.

Inhibitor Switch

CALIFORNIA AUTOMATIC TRANSMISSION MODELS

The inhibitor switch is installed on the shift linkage lever and energizes the coasting richer system in drive, low or second gear positions. Test the inhibitor switch with a test lamp.

MAZDA PISTON ENGINE THROTTLE OPENER

The throttle opener system consists of a servo-diaphragm connected to the throttle lever and a vacuum control valve which controls intake manifold vacuum through the servo-diaphragm.

Testing the System

SERVO-DIAPHRAGM

1. Start the engine and set the idle speed to 800 rpm. Stop the engine.
2. Disconnect the vacuum sensing tube between the servo-diaphragm and the vacuum control valve at the diaphragm.
3. Remove the intake manifold suction hole plug.
4. Connect the intake manifold and the servo-diaphragm with a tube so that the intake manifold vacuum goes directly to the servo-diaphragm.
5. Connect a tachometer and remove the vacuum sensing tube between the carburetor and distributor.
6. Start the engine and read the speed. If the engine is running 1300–1500 rpm,

the servo-diaphragm is operating normally. If the engine speed is 800–1300, adjust the speed with the throttle opening screw. If the engine speed remains normal, about 800 rpm, the servo-diaphragm is defective and should be replaced.

7. Remove the test equipment and reconnect all the lines.

Servo-Diaphragm Removal and Installation

1. Remove the air cleaner.
2. Disconnect the vacuum sensing tube from the diaphragm.
3. Remove the cotter pin and link.
4. Loosen the locknut and remove the servo-diaphragm.
5. Installation is the reverse of removal. Adjust the servo-diaphragm.

Vacuum Control Valve Removal and Installation

1. Remove the air cleaner.
2. Disconnect the vacuum sending tubes from the vacuum control valve.
3. Unbolt and remove the vacuum control valve.
4. Installation is the reverse of removal.

TOYOTA THROTTLE POSITIONER (TP) SYSTEM

During rapid deceleration large volumes of fuel/air mixture are drawn into the cylinders due to high manifold vacuum. If the throttle plates close completely there is not sufficient oxygen to permit complete combustion. Therefore a throttle positioner is installed to keep the throttle plates slightly open during deceleration. Vacuum is reduced under the throttle valve which, in turn, acts on the retard chamber of the distributor vacuum unit. This ignition retard compensates for the loss of engine braking caused by the partially-opened throttle.

Once the vehicle drops below a predetermined speed, the vacuum switching valve provides vacuum to the throttle positioner diaphragm. The throttle positioner then re-

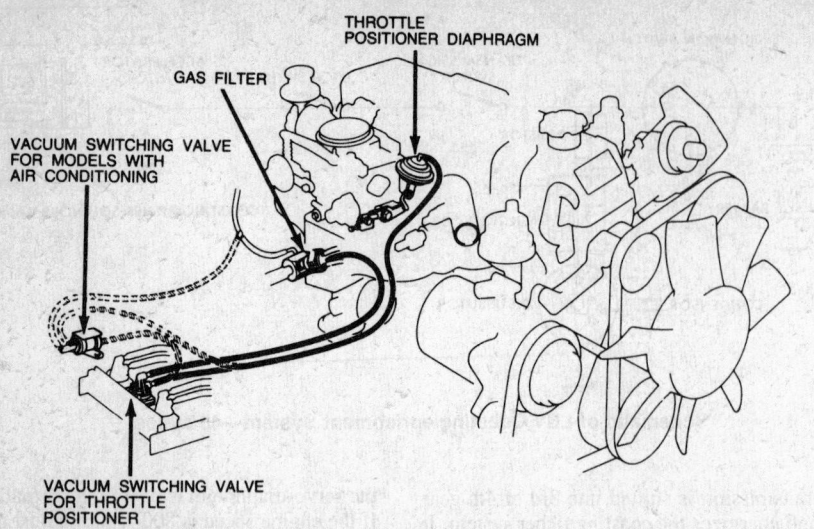

Schematic of a typical Toyota throttle positioner system

20R engine:	
manual transmission	1400
automatic transmission	1050
22R engine:	
all	1050
2F engine	
except '78–'80 Calif	1200
'78–'80 Calif	1400

tracts allowing the throttle valve to close completely. The distributor also returns to normal operation.

ADJUSTMENT

1. Start the engine and allow it to reach normal operating temperature.

2. Adjust the engine idle speed as previously outlined.

3. Detach the vacuum line from the throttle positioner diaphragm and plug the line.

4. Accelerate the engine slightly to set the throttle positioner.

5. Check the engine speed with a tachometer when the positioner is in place.

6. Check that the engine rpm reads as follows:

7. If necessary, adjust the engine rpm to the values listed in Step 6 by turning the throttle positioner screw.

8. Connect the vacuum hose to the positioner diaphragm. The engine idle should return to normal as soon as the vacuum line is connected.

9. If the throttle positioner fails to perform properly, check the linkage and diaphragm unit. If there are no defects in these components, the problem probably lies in the vacuum switching valve or in the speed sensor unit.

Exhaust Gas Recirculation

NOx (oxides of nitrogen) is a tailpipe emission caused by the oxidation of nitrogen in the combustion chamber. When the peak combustion temperatures go over 2500°F. NOx is formed in excessive amounts. To keep the combustion temperatures down, exhaust gas is recirculated.

Recirculation of the exhaust gases is accomplished by having a regulating valve between the exhaust and intake manifolds, and upon a predetermined demand, routing engine vacuum to the valve and opening the connecting port to allow the exhaust gases to flow into the intake manifold and mix with the air/fuel mixture.

In most systems, the exhaust gas passes through a filter on the way from the exhaust manifold to the EGR valve. There is no

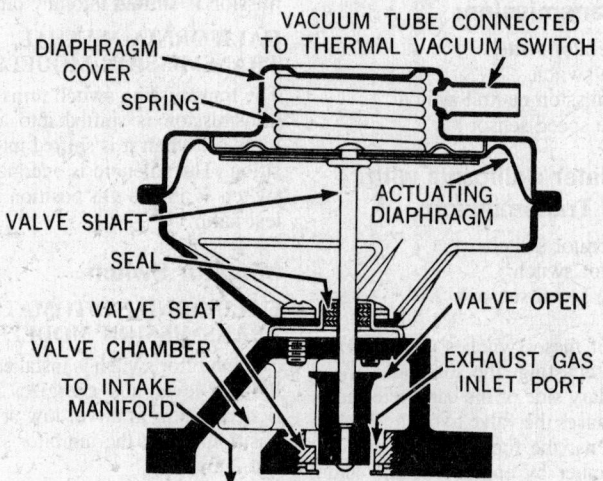

Cross section—ported vacuum signal EGR valve (typical)

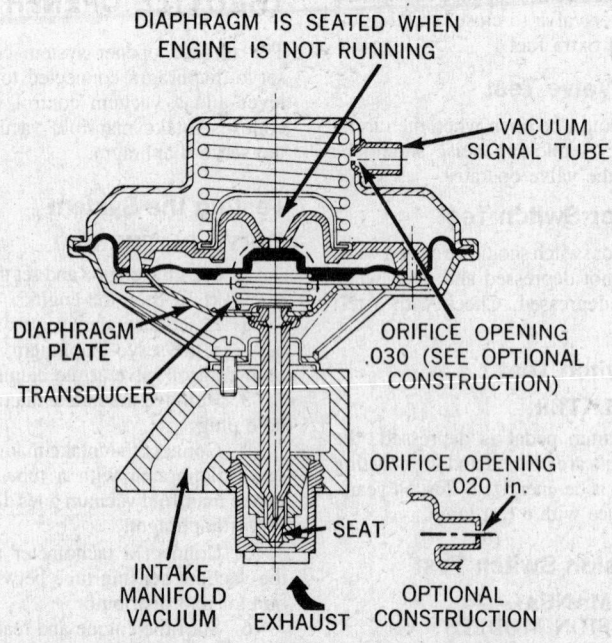

Cross section—typical negative back pressure EGR valve

EGR at idle or at full throttle, partial EGR at light throttle, and full EGR at medium throttle.

Three types of EGR valves are used, with the major differences in the method used to control the valve opening. The three types are as follows:

1. An EGR valve with no back pressure sensor and is controlled by ported vacuum.

2. An EGR valve with an integral back pressure sensor and is controlled by ported vacuum and exhaust gas back pressure. Both positive and negative type transducers are used to react to either high or low exhaust gas back pressures.

3. An EGR valve with an external, non-integral back pressure sensor and is controlled by ported vacuum and exhaust gas back pressure.

NOTE: Venturi vacuum is used as a triggering agent when a vacuum amplifier is used in the EGR system.

Several different types of controls are used to turn the vacuum to the EGR valve on and off. Most of them have to do with engine temperature, as described later.

When the EGR valve hose is connected to the base of the carburetor, without a separate amplifier, the system is operated by ported vacuum. The hose may not run directly from the EGR valve to the carburetor, but may go through a temperature control valve of some sort. In a ported vacuum system, the vacuum to operate the EGR valve is taken from a port that is above the throttle plate at idle, and thus not subject to vacuum. Because there is no vacuum, the spring in the EGR valve closes it, and the exhaust gas does not recirculate. As the throttle is opened, the port is exposed to vacuum, and the EGR valve opens.

Vacuum systems, with an amplifier, are the most complicated, because of the number of hoses. Manifold vacuum is connected to the amplifier by a hose, and then connects to the EGR valve. The amplifier also connects to venturi vacuum. At idle there is no venturi vacuum, but above idle the air moves through the carburetor venturi fast enough to create a vacuum. This slight amount of vacuum opens the amplifier, which then allows manifold vacuum to open the EGR valve.

Temperature controls for EGR systems come in many different designs. They are all made so that the EGR valve stays closed when the engine is cold. After the engine warms up, the temperature control allows the EGR valve to operate normally.

TESTING EGR SYSTEMS

Testing of the EGR systems should verify that the EGR valve is closed at idle, open above idle, and that the exhaust gas is actually recirculating. If the EGR valve sticks open at idle, the engine will run very rough, or may not even start. If this happens the valve should be removed and cleaned, or

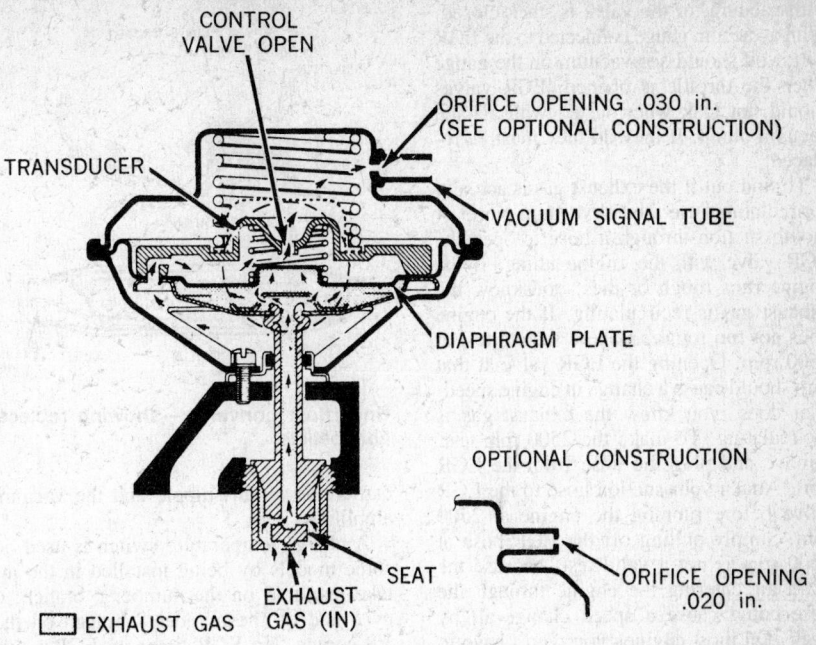

Cross section—typical positive back pressure EGR valve

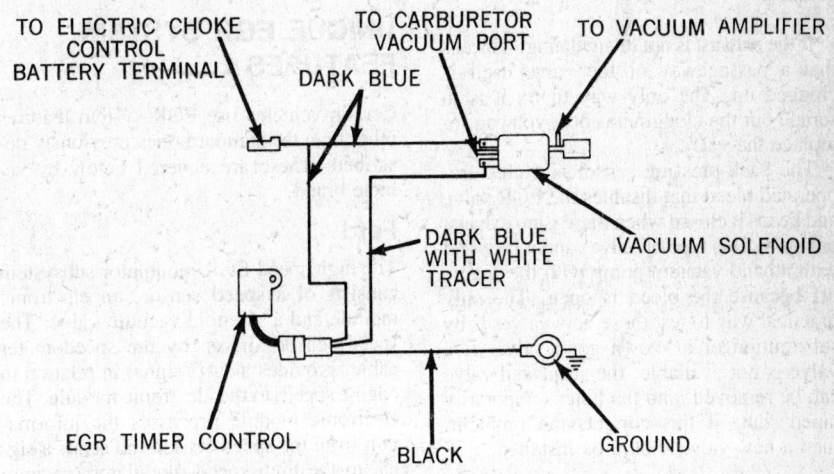

Typical time delay circuitry—EGR valve

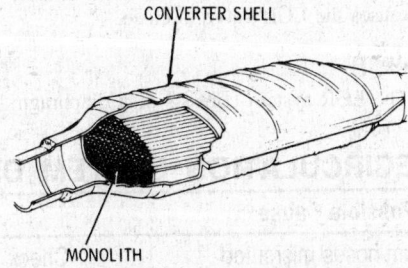

Cutaway of a monolith converter

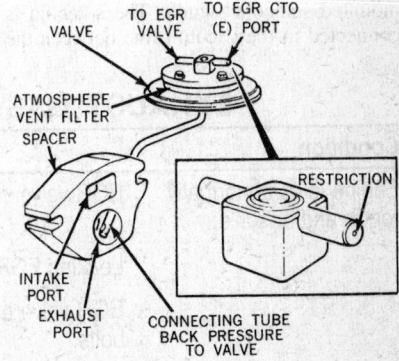

EGR valve with external, non-integral back pressure sensor

replaced. To check for valve opening above idle, check with a mirror or your fingers to see if the diaphragm or stem moves when the engine is at a fast idle in Park or Neutral. If the diaphragm does not move when the throttle is opened, there is either a problem with vacuum, or the valve is stuck closed. With a vacuum gauge connected to the EGR port, you should see vacuum on the gauge when the throttle is opened. EGR valves should not leak when tested with a hand vacuum pump. If they do they must be replaced.

To find out if the exhaust gas is actually recirculating, use a hand vacuum pump or mouth suction through a hose to open the EGR valve with the engine idling. If the engine runs rough or dies, you know the exhaust gas is recirculating. If the engine does not run rough, make a second test at 2500 rpm. Opening the EGR valve at that rpm should cause a change in engine speed. If it does, you know the exhaust gas is recirculating. To make the 2500 rpm test, remove and plug the hose from the EGR port. Attach your suction hose to the EGR valve before running the engine at 2500 rpm. Simply pulling off the EGR hose at 2500 rpm is not a valid test, because the extra air entering the engine through the hose could cause a speed change all by itself. On most engines you won't have to go this far, because opening the EGR valve at idle will prove that the exhaust is recirculating.

If the exhaust is not recirculating, it means that a passageway or the valve itself is clogged up. The only way to fix it is to scrape out the clogging as best you can, or replace the valve.

The back-pressure sensor is a pressure-operated bleed that disables the EGR valve and keeps it closed when there is no exhaust pressure. This type of valve cannot be tested with a hand vacuum pump with the engine off because the bleed is open. The only practical way to test these new valves is by substitution of a known good valve. If a valve is not available, the suspected valve can be removed, and the holes temporarily taped shut. If this corrects the problem, then a new valve should be installed.

EGR delay systems are used on some vehicles to prevent the recirculation of the exhaust gases for approximately 60 seconds after the ignition switch is turned on by an electrical timer, connected to an engine mounted solenoid switch. The solenoid is connected in the vacuum line between the

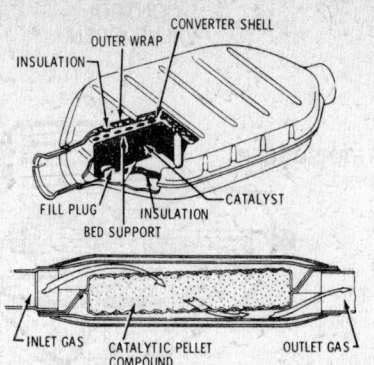

Underfloor converter—showing replaceable pellets

carburetor venturi nipple and the vacuum amplifier.

A charge temperature switch is used on some models by being installed in the intake manifold on the number 6 branch, 6 cyl., and on the number 8 branch on the V8 engine. No EGR timer on EGR valve operation is permitted when the air/fuel mixture temperature is below 60°F. (16°C).

UNIQUE EGR SYSTEM FEATURES

Certain vehicles use EGR system features other than the standard ones previously described. These are covered below by vehicle brand.

Ford

The high speed EGR modulator subsystem consists of a speed sensor, an electronic module and a solenoid vacuum valve. The speed sensor, driven by the speedometer cable, provides an AC signal in relation to engine speed, to the electronic module. The electronic module processes the information from the speed sensor and sends a signal to the high speed modulator (vacuum solenoid) valve. When the vehicle speed exceeds the module trigger speed, the solenoid vacuum valve closes which, in turn, causes the EGR valve to close.

Jeep

The EGR system consists of a diaphragm

actuated flow control valve (EGR valve), coolant temperature override switch (EGR CTO) and connecting hoses. In 1980, a Thermal Vacuum Switch, located in the air cleaner, was added to control the vacuum signal between the EGR and CTO.

EGR VALVE

The EGR valve is mounted on a machined surface at the rear of the intake manifold on V8 engines and on the side of the intake manifold on the Sixes. When a backpressure sensor is used, the EGR valve is mounted on a spacer which is an integral part of the backpressure sensor.

The valve is held in a normally closed position by a coil spring located above the diaphragm. A special fitting is provided at the carburetor to route ported (above the throttle plates) vacuum through the CTO and a TVS or BPS (when used) and hose connections to a fitting located above the diaphragm on the valve. A passage in the intake manifold directs exhaust gas from the exhaust crossover passage on V8s and from near the heat riser on the Sixes, to the EGR valve. When the diaphragm is actuated by vacuum, the valve opens and meters exhaust gas through special passages into the intake manifold below the carburetor.

COOLANT TEMPERATURE OVERRIDE SWITCH

This switch is located in the intake manifold at the coolant passage adjacent to the oil filler tube on V8s and on the left side of the cylinder block on the Sixes. The outer port of the switch is connected to either the EGR valve or BPS (when used). The inner port is connected by a hose to the EGR fitting at the carburetor. When the coolant temperature reaches 115°F the inner port of the switch opens and a vacuum signal is applied to the EGR valve. This vacuum signal is subject to regulation by the BPS when used.

THERMAL VACUUM SWITCH (TVS)

The TVS is located in the air cleaner on all 1980 and later engines, and functions as an on/off switch controlled by air cleaner air temperature. The TVS controls the vacuum passage between the EGR CTO valve and the EGR valve. At air temperature below

EXHAUST GAS RECIRCULATION SYSTEM DIAGNOSIS CHART

Condition	Possible Cause	Correction
Engine idles abnormally rough and/or stalls.	EGR valve vacuum hoses misrouted.	Check EGR valve vacuum hose routing. Correct as required.
	Leaking EGR valve.	Check EGR valve for correct operation.
	EGR valve gasket failed or loose EGR attaching bolts.	Check EGR attaching bolts for tightness. Tighten as required. If not loose, remove EGR valve and inspect gasket. Replace as required.
	EGR thermal control valve and/or EGR-TVS.	Check vacuum into valve from carburetor EGR port with engine at normal operating

EXHAUST GAS RECIRCULATION SYSTEM DIAGNOSIS CHART

Condition	Possible Cause	Correction
Engine idles abnormally rough and/or stalls.		temperature and at curb idle speed. Then check the vacuum out of the EGR thermal control valve to EGR valve. If the two vacuum readings are not equal within ± ½ in. Hg. (1.7 kPa), then proceed to EGR vacuum control diagnosis.
	Improper vacuum to EGR valve at idle.	Check vaccum from carburetor EGR port with engine at stabilized operating temperature and at curb idle speed. If vacuum is more than 1.0 in. Hg., refer to carburetor idle diagnosis.
Engine runs rough on light throttle acceleration, poor part load performance and poor fuel economy.	EGR valve vacuum hose misrouted.	Check EGR valve vacuum hose routing. Correct as required.
	Failed EGR vacuum control valve.	Same as listing in "Engine Idles Rough" condition.
	EGR flow unbalanced due to deposit accumulation in EGR passages or under carburetor.	Clean EGR passages of all deposits.
	Sticky or binding EGR valve.	Remove EGR valve and inspect. Clean or replace as required.
	Wrong or no EGR gaskets.	Check and correct as required.
Vehicle with back pressure EGR valve.	Control valve blocked or air flow restricted.	Check internal control valve function per service procedure.
Engine stalls on decelerations.	Restriction in EGR vacuum line.	Check EGR vacuum lines for kinks, bends, etc. Remove or replace hoses as required. Check EGR vacuum control valve function.
		Check EGR valve for excessive deposits causing sticky or binding operation. Clean or repair as required.
	Sticking or binding EGR valve.	Remove EGR valve and inspect, clean or repair as required.
Vehicle with a back pressure EGR valve.	Control valve blocked or air flow restricted.	Check internal control valve function per service procedure.
Part throttle engine detonation.	Insufficient exhaust gas recirculation flow during part throttle accelerations.	Check EGR valve hose routing. Check EGR valve operation. Repair or replace as required. Check EGR thermal control valve and/or EGR-TVS as listed in "Engine Idles Rough" section. Replace valve as required. Check EGR passage and valve for excessive deposit. Clean as required.
Vehicle with a back pressure EGR valve.	Control valve blocked or air flow restricted.	Check internal control valve function per service procedure.
Engine starts but immediately stalls when cold.	EGR valve hoses misrouted.	Check EGR valve hose routing.
	EGR system malfunctioning when engine is cold.	Perform check to determine if the EGR thermal control valve and/or EGR-TVS are operational. Replace as required.
Vehicle with a back pressure EGR valve.	Control valve blocked or air flow restricted.	Check internal control valve function per service procedure.

①Detonation can be caused by several other engine variables. Perform ignition and carburetor related diagnosis.

40–55 degrees F., the TVS prevents vacuum from opening the EGR valve, thus preventing EGR valve operation and improving cold engine driveability.

NOTE: A TVS valve is used for other engine related systems to control operations that require air cleaner intake air to be at the proper temperature before system operation is activated.

EXHAUST BACKPRESSURE SENSOR (BPS)

This device is used on some 1979–81 models in conjunction with the EGR system.

The BPS monitors exhaust backpressure and permits EGR operation only when engine operating conditions are favorable. The BPS units are variously calibrated, are not serviceable, and must be replaced with the identical part as a unit when necessary.

The BPS consists of a diaphragm valve and a spacer connected by a metal tube projecting into an exhaust port in the spacer body. The EGR valve mounts directly on the spacer. On some 1978–81 sixes, the sensor is integral with the EGR valve.

In operation, the metal tube connecting the diaphragm valve to the spacer routes exhaust backpressure from the particular exhaust port to the sensor. When the backpressure reaches a certain level the diaphragm valve spring pressure is overcome, permitting a vacuum signal to the EGR valve, providing that the CTO switch is open.

Thus, EGR operations is only permitted when the engine is warmed up sufficiently and exhaust backpressure relatively high, such as during acceleration and at some cruising speeds. When temperature or backpressure conditions are not met, the vacuum signal is vented to the atmosphere from a vent at the diaphragm valve.

Courier/Mazda

THREE-WAY SOLENOID VALVE

The Courier EGR system incorporates an EGR valve, vacuum amplifier, and water temperature thermal switch, all components typical of such systems. However, the signal produced by the thermal switch is communicated to the vacuum portion of the system by a three-way solenoid valve which is somewhat unique. To test the valve, proceed as follows:

1. Disconnect the electrical connectors from the thermo switch. Connect a jumper wire between the connectors to simulate a complete circuit.
2. Turn the ignition switch to ON.
3. Disconnect the vacuum hose from the EGR valve and blow into the hose. Air should be discharged from the three-way solenoid valve relief port. If it is not, replace the three-way valve.
4. Turn the ignition switch off, and remove the jumper wire from the thermo switch connectors. Disconnect the vacuum amplifier vacuum line from the three-way solenoid. Turn the ignition switch back to ON.

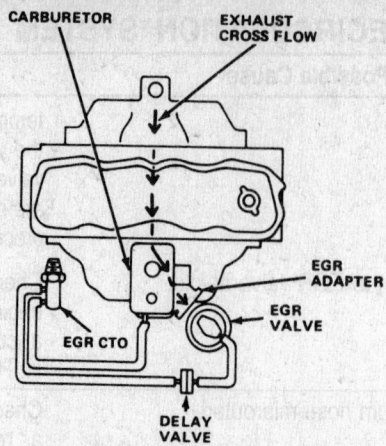

EGR system typical of 4-151

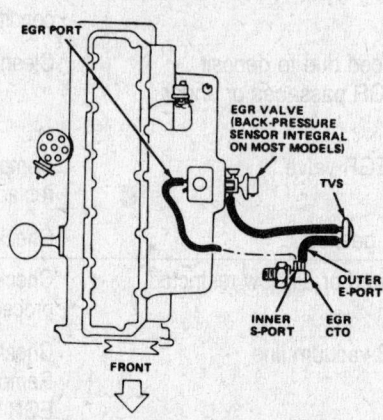

EGR system typical of 6-258

5. Blow into the vacuum line disconnected from the EGR valve, and check for air discharge from the vacuum amplifier port on the three-way solenoid valve. If there is no discharge, replace the solenoid valve.
6. After completion of all tests, reconnect the hoses to their original locations.

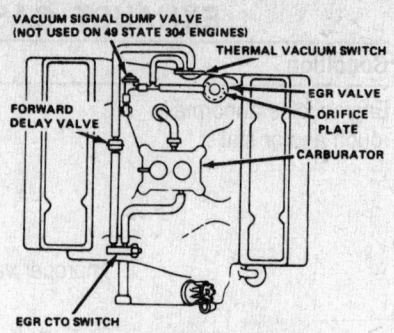

Typical V8 EGR system

Datsun/Nissan

EGR VALVE DIAPHRAGM LEAKAGE TEST

1. Remove the EGR valve and apply enough vacuum to the diaphragm to open the valve. Then, remove the source of vacuum and seal off the connection.
2. The valve should remain open for over 30 seconds after the vacuum is removed. Otherwise, the diaphragm leaks and the valve must be replaced.

THERMAL VACUUM VALVE

A thermal vacuum valve inserted in the engine thermostat housing controls the application of vacuum to the EGR valve. When the engine coolant reaches a predetermined temperature, the thermal vacuum valve opens and allows vacuum to be routed to the EGR valve. Below the predetermined temperature, the thermal vacuum valve closes and blocks vacuum to the EGR valve.

B.P.T. VALVE

1979 models have a B.P.T. valve installed between the EGR and the thermal vacuum valve. The B.P.T. valve has a diaphragm raised or lowered by exhaust back pressure. The diaphragm opens or closes an air bleed,

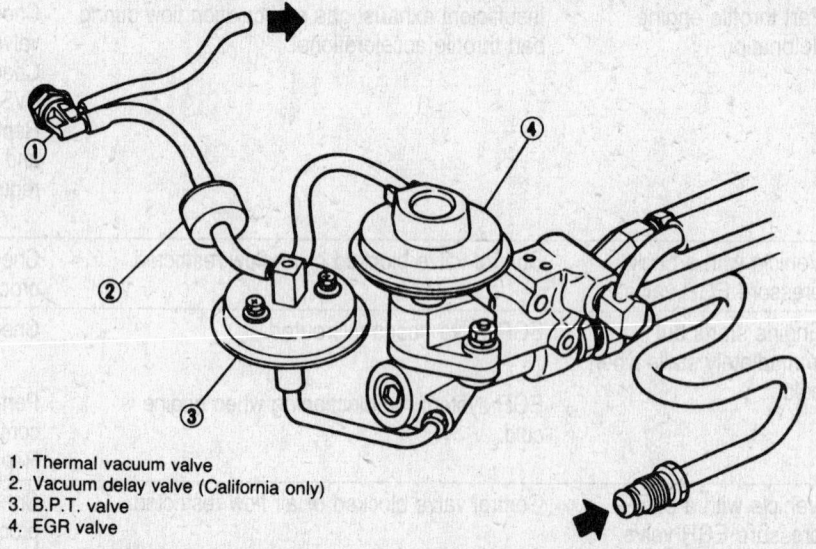

1. Thermal vacuum valve
2. Vacuum delay valve (California only)
3. B.P.T. valve
4. EGR valve

Datsun 1978 and later EGR system components. The hose at the top runs to the carburetor; the hose at the bottom runs to the exhaust manifold

which is connected into the EGR vacuum line. High pressure results in higher levels of EGR, because the diaphragm is raised, closing off the air bleed, which allows more vacuum to reach and open the EGR valve. Thus, the amount of recirculated exhaust gas varies with exhaust pressure.

VACUUM DELAY VALVE

1979 California models have a vacuum delay valve installed in the line between the thermal vacuum valve and the EGR valve. This valve delays rapid drops in vacuum in the EGR line, thus effecting a longer EGR time.

Mazda

EGR CONTROL VALVE TEST

1. Remove the air cleaner.
2. Run the engine at idle.
3. Disconnect the vacuum sensing tube from the EGR control valve.
4. Disconnect the vacuum sensing tube from the intake manifold vacuum control valve.
5. Connect this vacuum tube to the EGR control valve. The engine should stop. If not, clean or replace the EGR control valve.

REPLACING EGR CONTROL VALVE

1. Remove air cleaner.
2. Disconnect the vacuum sensing tube from the EGR control valve.
3. Disconnect the EGR control valve to exhaust manifold pipe.
4. Disconnect the pipe between the EGR control valve and the intake manifold.
5. Unbolt and remove the EGR control valve.
6. If old valve is to be reused, it should be cleaned with a wire brush before installation.
7. To install, reverse the above procedure.

TESTING THE THREE-WAY SOLENOID VALVE

The valve is located at the top center of the firewall in the engine compartment.

1. Disconnect the wiring to the water thermo switch and connect a jumper wire to the two connectors of the switch.
2. Disconnect the vacuum sensing tube from the EGR valve.
3. Disconnect the vacuum sensing tube which runs to the vacuum amplifier from the three way solenoid valve.
4. Turn the ignition switch on.
5. Blow through the solenoid valve from the tube disconnected from the EGR valve. Air should pass through the valve to the valve air filter.
6. Disconnect the jumper wire from the thermo switch.
7. Blow through the tube disconnected from the EGR valve and make sure the air passes through the opening to the vacuum amplifier tube.
8. If the solenoid valve does not operate properly, replace.

Mitsubishi/D–50/Plymouth

The following unique EGR system features are used on these trucks:

1. Thermo Valve: Used to stop EGR valve operation below approximately 131 degrees, in order to improve cold driveability and starting.
2. Dual EGR Control Valve: The EGR vacuum flow is suspended during idle and wide open throttle operation. A primary valve controls EGR flow when the throttle valve opening is relatively narrow, while a secondary control valve operates at wider openings.
3. Sub EGR Control Valve: This valve is linked to the throttle valve to closely modulate the EGR gas flow.
4. EGR Maintenance Warning Light: A light in the speedometer assembly to alert the driver to the need for EGR system maintenance. This device has a mileage sensor to light the visual signal at 15,000 mile intervals. Upon completion of the required EGR system maintenance, the warning light can be turned off by resetting the switch. It is in the speedometer cable, under the instrument panel.

MAINTENANCE

1. Check all vacuum hoses for cracks, breakage and correct installation.
2. Check EGR valve operation by applying vacuum to the EGR valve vacuum nipple with the engine idling. The idle should become rough.
3. Check the passages in the cylinder head and intake manifold for clogging. Clean as necessary.
4. Cold start the engine. The EGR port nipple should be open. When the coolant is warmed to over 131 degrees, the port should be closed.

Volkswagen with Carburetor-Type Fuel System

EGR DECELERATION VALVE TEST

1. Remove the hose from the deceleration valve. Plug the hose.
2. Run the engine for a few seconds at 3000 rpm.
3. Snap the throttle valve closed.
4. With your finger, check for suction at the hose connection.
5. Remove the hose from the connector.
6. Run the engine at about 3000 rpm. No suction should be felt.

RESETTING THE ELAPSED MILEAGE SWITCH

The EGR reminder light in the speedometer should light up every 15,000 miles as a reminder for maintenance. To reset the light switch, press the white button. The speedometer light should go out.

EGR VALVE REMOVAL AND INSTALLATION

1. Disconnect the EGR vacuum hose.
2. Unbolt the EGR line fitting on the

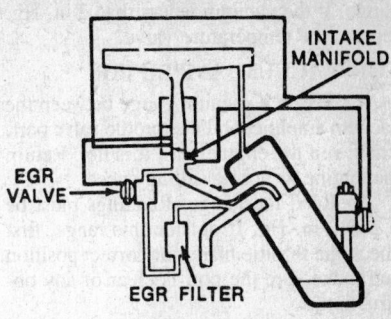

VW EGR system

side of the valve opposite to the hose connection.

3. Remove the two remaining EGR mounting bolts and then lift the EGR valve from the intake manifold.
4. Install the valve in reverse order, using a new gasket. Make sure all sealing surfaces are clean.

EGR DECELERATION VALVE

1. Remove the hose from the deceleration valve, and plug it. Install a tachometer.
2. Run the engine until it is warm. Then, speed it up until it is running at 3,000 rpm.
4. Continue running the engine at this speed for a few seconds and then feel for vacuum at the deceleration valve connection as you suddenly snap the throttle closed. There must be vacuum.
5. Run the engine at a steady 3,000 rpm and check again for vacuum. No vacuum should exist. If either test is failed, replace the valve.

Volkswagen with Fuel Injection

On fuel injected models, the EGR valve is controlled by a temperature control valve and a vacuum amplifier. The EGR valve itself is located on the front of the intake manifold.

TESTING THE EGR VALVE

Be sure the vacuum lines are not leaking. Replace any that are leaking or cracked.

1. Run the engine until it reaches operating temperature, and then allow it to idle. Install a tachometer and note the rpm.
2. Disconnect the vacuum hose at the EGR valve. Then, connect the line from the brake booster to the EGR valve (this can be done by installing a Tee in the vacuum line to the retard side of the distributor diaphragm and running a separate hose from there to the EGR valve.)
3. Reread the engine rpm. If it does not change, the EGR valve is clogged or damaged.

TESTING THE EGR TEMPERATURE VALVE

Run the engine until it reaches normal operating temperature.

1. With the engine idling, tee-in a vacuum gauge between the EGR temperature control valve and the EGR valve. Read the

gauge. If the vacuum is less than 2 in. Hg, replace the temperature valve.

EGR VACUUM AMPLIFIER

1. Tee-in a vacuum gauge between the vacuum amplifier and the throttle valve port. Then, run the engine until it is hot. Return the engine to normal idling speed.

2. Read the gauge. Readings must be 0.2–0.3 in. Hg. If outside this range, first check the throttle blade for correct position and make sure the port is clear of any obstruction.

3. Once there is vacuum to the amplifier from the throttle valve, move the vacuum gauge over to the line between the vacuum amplifier and temperature valve. Read the gauge in this position with the engine idling. If the gauge reads less than 2 in. Hg., replace the vacuum amplifier.

Catalytic Converter

A catalytic converter is a chamber in the exhaust system that contains a catalyst. When hydrocarbons or carbon monoxide pass over the catalyst they react with the oxygen in the exhaust and are converted into harmless water and carbon dioxide. The catalyst inside the converter is made in two forms. General Motors, and Jeep (up until the last two years or so) use the pellet form, in which loose pellets are packaged into the converter and can be emptied out and changed, if necessary. Ford, International and Chrysler use the honeycomb catalyst, which is built into the converter shell and is not replaceable. On Ford, International and Chrysler products the entire converter must be replaced if it goes bad.

On late model Jeep vehicles a dual bed monolithic-type or COC pellet type catalytic converter is used. The monolithic type is not serviceable, but the pellets can be replaced in the pellet type.

The stainless steel converter body is designed to last the life of the automobile, but excessive heat can cause premature converter failure. Although the excessive heat would be contained in the converter, the cause of the overheating would not be the fault of the converter, but from an outside source. A defective fuel system, air injection system or ignition system malfunction that permits unburned fuel to enter the converter will usually be the cause. If the converter is heat damaged, the cause must be located and repaired before a new converter is installed.

There is no positive way to test most converters in the field to see if they are actually working. Tailpipe readings may be used to set carburetor idle mixtures, when the car maker requires it, but taking a tailpipe reading to determine if the converter is working is not a positive check.

The one field check that is recommended in all cases is to inspect for mechanical damage. If a converter gets overheated, the catalyst can melt and block the exhaust. Pellets or pieces of the catalyst may even

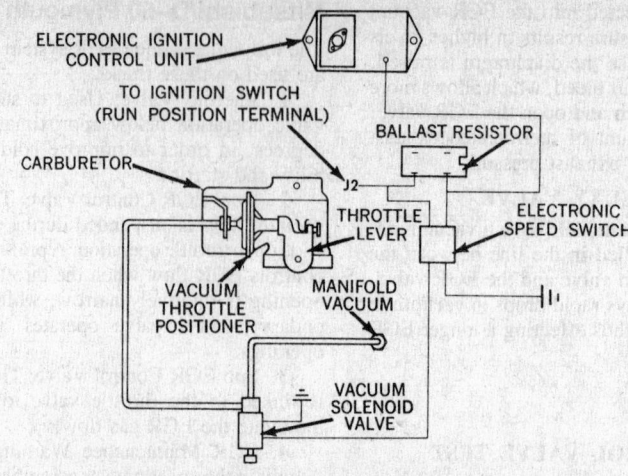

Typical throttle positioner system used on Chrysler Corp. vehicles having California Emission requirements

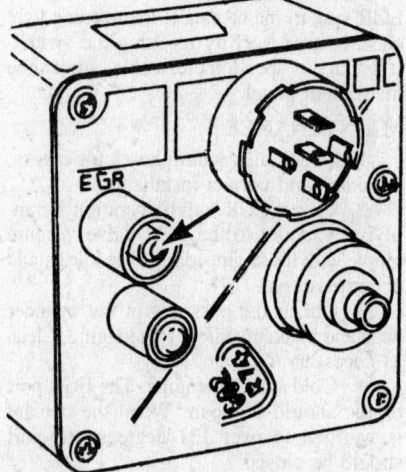

Resetting the VW EGR mileage counter

come flying out the tailpipe while the engine is running. If this happens, the pellets or the entire converter must be changed.

Checking for a melted converter that restricts the exhaust can be done with a vacuum gauge connected to the engine. Run the engine at about 2500 rpm in Park or Neutral. If the vacuum reading is steady, the exhaust is okay. If the vacuum reading slowly drops, it indicates a buildup of pressure in the exhaust.

The use of leaded fuel will slowly destroy the efficiency of the catalyst until finally, after several tanks full, it won't do its job any more. If used long enough, leaded fuel can even cause catalyst plugging to the point that the engine will not run. If you know that a vehicle has been run on several tanks of leaded fuel, then you can be sure that the catalyst has lost its ability to convert. But there is no way to test for this condition in the field.

Do not change the catalyst if the vehicle has been run on only one tank or less of leaded fuel. Switching back to lead free fuel

will allow the catalyst to recover and be almost as efficient as it was.

Checking the Converter on the VW Pickup

Most converters require very little or no direct maintenance. However, the VW pickup's converter must be inspected every 30,000 miles to check for damage to the ceramic insert due to heat cracking or possible road damage. To do this, proceed as described below.

— CAUTION —

Do not drop or strike the converter assembly or damage to the ceramic insert will result.

Damage and overheating of the catalytic converter will be indicated by flickering of the "Cat" warning light. This can be caused by:

1. Engine misfire caused by faulty spark plugs, faulty distributor cap or ignition wires.
2. Improper ignition timing.
3. CO valve set too high.
4. Faulty air pump diverter valve.
5. Faulty temperature sensor.
6. Engine under strain caused by trailer towing, high speed driving in hot weather, etc.

A faulty converter is indicated by one of the following symptoms:

1. Poor engine performance.
2. Engine stalling.
3. A rattle in the exhaust system.
4. A CO reading greater than .4% at the tail pipe at idle.

Check or replace the converter as follows:

1. Disconnect the temperature sensor.
2. Loosen and remove the bolts mounting the converter to the exhaust system and the chassis.
3. Remove the converter.
4. Hold the converter up to a strong light and look through both ends, checking for any blockage. If the converter is blocked, replace it.

5. Install in reverse order of removal.

6. Reset the elapsed mileage odometer by pushing the button marked ''Cat''.

Checking the Converter on Mitsubishi/D50/Arrow

1. Check the core for cracks and damage.

2. If the idle carbon monoxide and hydrocarbon content exceeds specifications and the ignition timing and idle mixture are correct, the converter must be replaced.

CONVERTER OVERHEAT PROTECTION

Engine controls are used to prevent the converter from being damaged by overheating due to overly rich fuel mixtures during periods of deceleration.

The controls are named differently by the manufacturers, but are all designed to accomplish the same purpose and to operate basically in the same manner. To prevent the engine from operating at a rich mode when the throttle plates are closed during deceleration, electrical and/or mechanical means are provided to hold the throttle plates open at predetermined engine speeds, in order to lean the air/fuel mixture as necessary to control the exhaust emissions. The engine control should be inoperative under engine speeds of 1800 to 2000 RPMs to avoid engine overrun or vehicle overspeed in slow traffic.

The various parts are as follows:

1. The throttle lever actuator is mounted as part of the carburetor assembly and operates when vacuum is applied to it from a separate solenoid vacuum control valve.

2. The solenoid vacuum control valve is controlled by a signal from the electronic speed sensor or a throttle modulator deceleration valve vacuum signal to allow vacuum to be routed to the throttle lever actuator.

3. Electronic speed sensor is mounted near or included with the distributor and senses the engine speed and sends a signal to the solenoid vacuum control valve as long as the preset speed is exceeded.

TESTING THE SYSTEM

To test the electrical speed sensor system, place the transmission in neutral or park and set the hand brake. With a tachometer attached to the engine, increase the speed to approximately 2000 RPM. The solenoid or modulator stem should extend to hold the carburetor throttle lever off curb idle setting. As the engine speed is reduced to below 1800 RPM, the solenoid or modulator stem should retract to the off position. A hand held vacuum pump and test lamp can be used to test the individual components of the system.

To test the vacuum operated system, without an electrical sensor, 21 to 22 in.

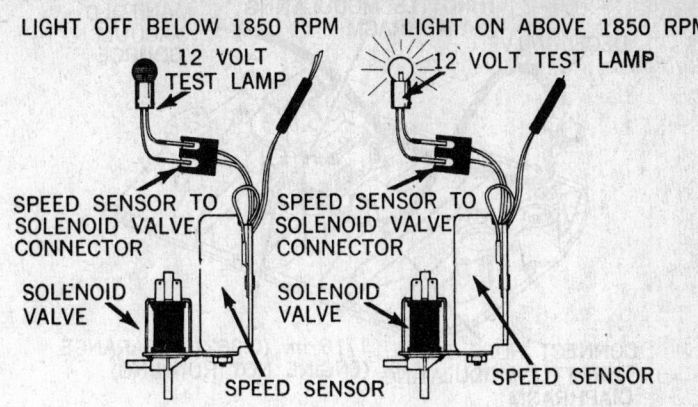

Testing an electric speed sensor switch

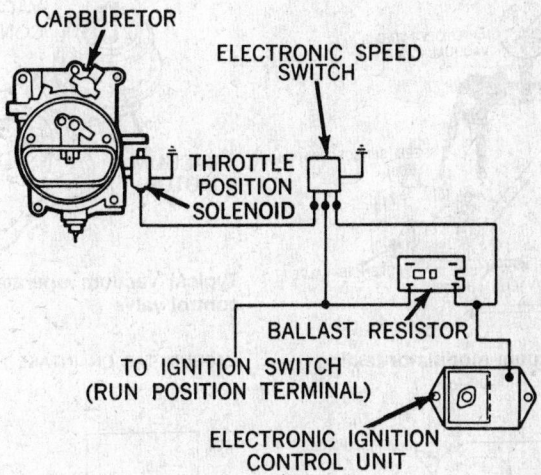

Typical throttle positioner electrical circuitry used on Chrysler Corp. vehicles

Hg. must be directed to the decel valve to open the port to direct vacuum to the throttle modulating diaphragm, located on the carburetor base. With the vacuum present, the stem of the modulating diaphragm will be extended. Release of the vacuum should allow the stem of the modulating diaphragm to retract.

Datsun/Nissan trucks with a catalytic converter also have a combined air control valve, which controls the amount of secondary air injected into the exhaust manifold. It is regulated by engine vacuum and air pump pressure, and works to keep the converter temperatures within proper limits. The combined air control valve replaces the air pump relief valve, found in the air pump system of earlier trucks not equipped with a catalytic converter.

Vacuum Operated Exhaust Heat Riser Valves

Exhaust heat riser valves have been used for many years to force part of the engine exhaust through a passageway under the intake manifold and preheat the fuel mixture. The heat valve was spring loaded into the closed position, but heat would make the spring relax so that during high speed operation or after warmup the exhaust would push it open.

Now, many engines use vacuum operated heat valves, controlled by a vacuum switch that is sensitive to engine temperature.

On these systems, manifold vacuum is used to close the valve, and force the exhaust gases through the crossover passage in the intake manifold. All the systems have a temperature valve that shuts the vacuum off when the engine warms up.

A simple coolant temperature-sensitive vacuum switch is mounted on the intake manifold coolant passage and has two hose connections. It actually does triple duty because it also controls the vacuum supply to the idle enrichment system and the air switching valve.

A second type vacuum switch has three hose connections, but one of them is a vent with a filter to keep the dirt out.

A third control uses either a coolant vacuum switch, or a vacuum solenoid con-

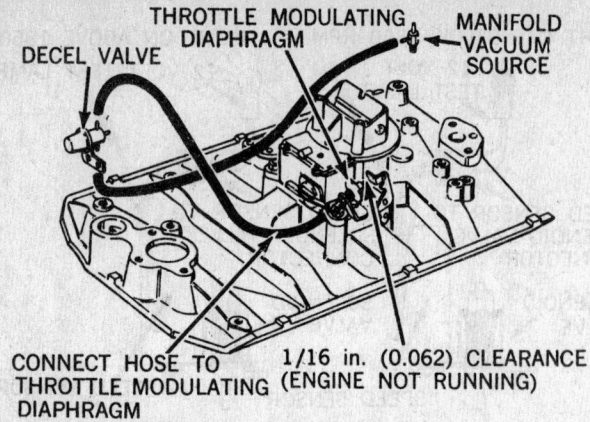

Typical vacuum throttle modulating system

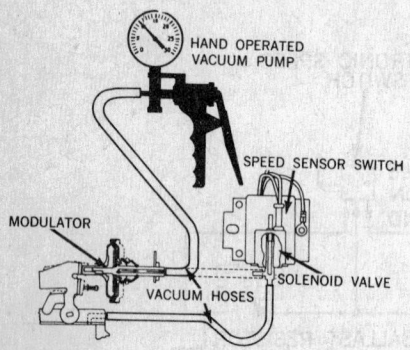

Typical vacuum modulator testing

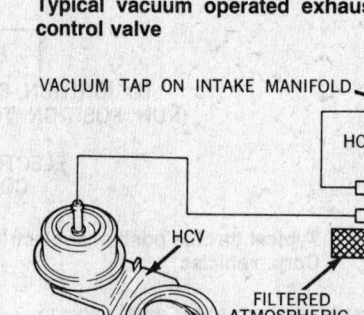

Typical vacuum operated exhaust heat control valve

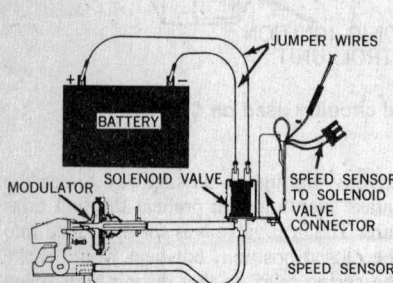

Testing an electric solenoid valve with the engine running

Typical exhaust heated control valve vacuum circuit using a ported vacuum coolant switch

nected to an oil temperature switch. The coolant vacuum switch has two hose connections and a vent when it controls the heat valve only. When it is tied into other emission control systems, it can have as many as five hose connections, and a vent. Some models also have a check valve in the hose so that vacuum will be trapped in the heat valve actuator when the engine is accelerated. This keeps the heat valve in the closed position and prevents a rattle.

TESTING VACUUM OPERATED EXHAUST HEAT RISER VALVES

Testing the vacuum operated heat riser valve is a matter of making sure it closes and

opens freely. You can move it to see if it works, on a warm engine. On a cold engine, the valve should be closed, and disconnecting the hose should allow it to open. On a cold engine, there should be vacuum at the vacuum actuator, and on a warm engine the vacuum should be shut off.

Chevrolet/GMC Early Fuel Evaporation System

This system is used on all light duty models. The six cylinder system consists of an EFE valve mounted at the flange of the exhaust manifold, an actuator, a thermal vacuum

switch (TVS), and a vacuum solenoid. The TVS is on the right side of the engine forward of the oil pressure switch. The TVS is normally closed and sensitive to oil temperature.

The V8 EFE system consists of an EFE valve at the flange of the exhaust manifold, an actuator, and a thermal vacuum switch. The TVS is located in the coolant outlet housing and directly controls vacuum.

In both systems, manifold vacuum is applied to the actuator, which in turn, closes the EFE valve. This routes hot exhaust gases to the base of the carburetor. When coolant (V8) or oil (six cylinder) temperatures reach a set limit, vacuum is denied to the actuator allowing an internal spring to return the actuator to its normal position, opening the EFE valve.

1983–84 4.8 L V8s use a manifold mounted thermostatic spring to actuate the EFE valve, much as was done with the traditional heat riser valve.

DIAGNOSIS

1. Allow the engine to cool until it is below 105°F. On some V8 engines, the EFE valve actuator arm is protected by a two-piece metal cover. Remove the cover if so equipped.

2. Start the engine while watching the EFE valve. The actuator link should be pulled into the diaphragm housing. If so, continue to run the engine to check that the valve opens before the engine is warmed up—— see the last two steps.

3. If the valve fails to close, disconnect the vacuum line running to it and apply a vacuum of at least 10 in. HG with a hand pump or by tapping directly to engine manifold vacuum with the engine idling. If the valve still does not close, it could be seized. Lubricate it with a manifold heat valve lubricant, and then retest it. If it does close, test it further by removing the source of vacuum and sealing off the vacuum line. The valve should remain closed for at least 20 seconds——if not, replace it. A valve which is not seized and still does not close has a bad vacuum diaphragm and must be replaced.

4. If the valve closes when external vacuum is applied, check the rest of the system for loose, kinked, pinched, plugged or cracked hoses and repair or replace parts as required. Also make sure vacuum is getting through the TVS or EFE vacuum solenoid (with engine still below 105°F.). Replace the thermostatic control device if necessary.

5. Warm up the engine to operating temperature. If the valve does not open, check the engine thermostat to ensure the engine reaches operating temperature; check the TVS or EFE solenoid to ensure vacuum is cut off before the engine reaches operating temperature; and check the valve to ensure it is not mechanically bound in the cold position. If these checks do not reveal the reason the valve won't open, there is

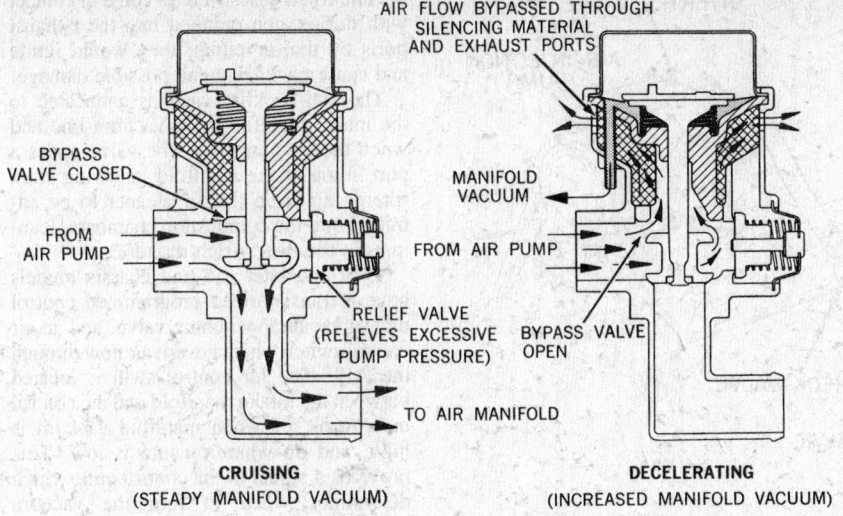

Diverter valve operation

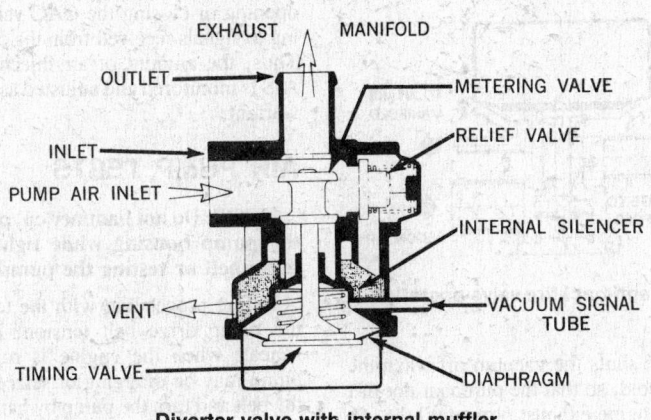

Diverter valve with internal muffler

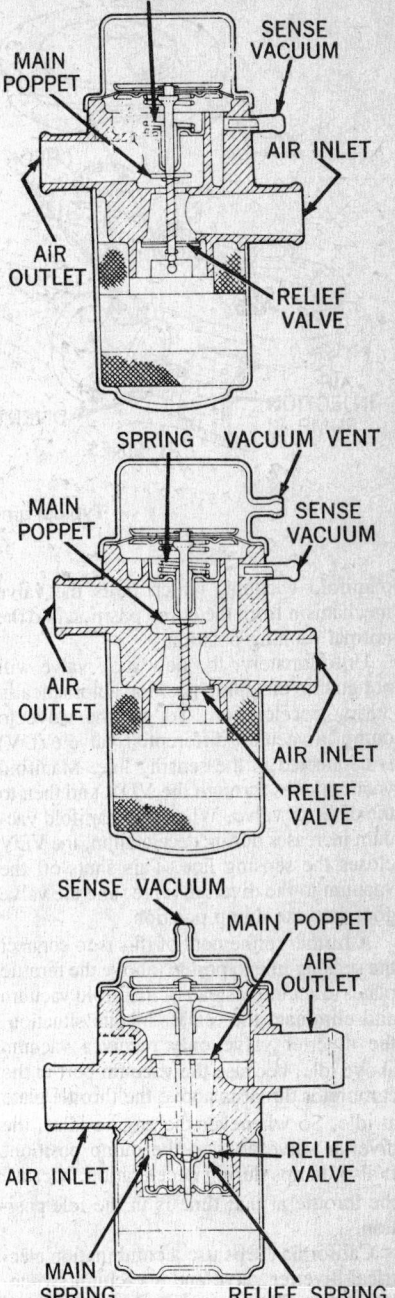

Three types of diverter valves: A. Air by-pass valve; B. Closed air by-pass valve; C. Timed air by-pass valve with vacuum vent

no air bleed effect in the diaphragm and the valve must be replaced.

Air Injection Systems

A belt-driven air pump supplies air to a small tube positioned in the exhaust port near each exhaust valve. The air mixes with unburned hydrocarbons in the exhaust and the hydrocarbons actually burn up in the exhaust system. Air injection systems are used on engines with catalytic converters, so that the converter gets enough air to keep the reaction going.

Plumbing on air injection systems varies considerably. A check valve is used between the pump and the exhaust port nozzle to keep hot exhaust gases from traveling up the plumbing and destroying the pump. V8s use two check valves.

An anti-backfire valve, also called by-pass valve or diverter valve, is used between the pump and the check valve. Usually, the diverter valve is mounted on the pump or near it. A small sensing hose connects the diverter valve to intake manifold vacuum. When the vacuum rises during de-

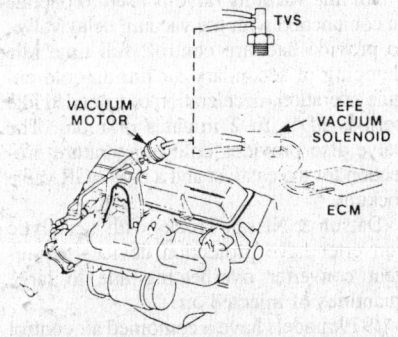

Chevrolet/GMC EFE systems use either a TVS or EFE vacuum solenoid to shut off the EFE valve when the engine warms up

celeration, the diverter valve opens, and sends the pump air into the atmosphere. This prevents the overrich deceleration mixture in the exhaust system from exploding or backfiring out the tailpipe. On some models of Jeeps, the diverter valve also functions as a pressure release valve for excessive air pump output. If the air pump belt squeals at high engine rpm or if you suspect the pump is taking too much power, you might want to inspect the valve.

Some models started using a diverter valve that has the small hose connection on the end instead of the side. The older diverter valve was normally in the running position, but the new one is normally in the dump position. In other words, the old valve allowed the air to pass through the engine exhaust ports regardless of whether the small sensing line was hooked up. The new valve, being normally in the dump position, must have the small sensing line hooked up to

U129

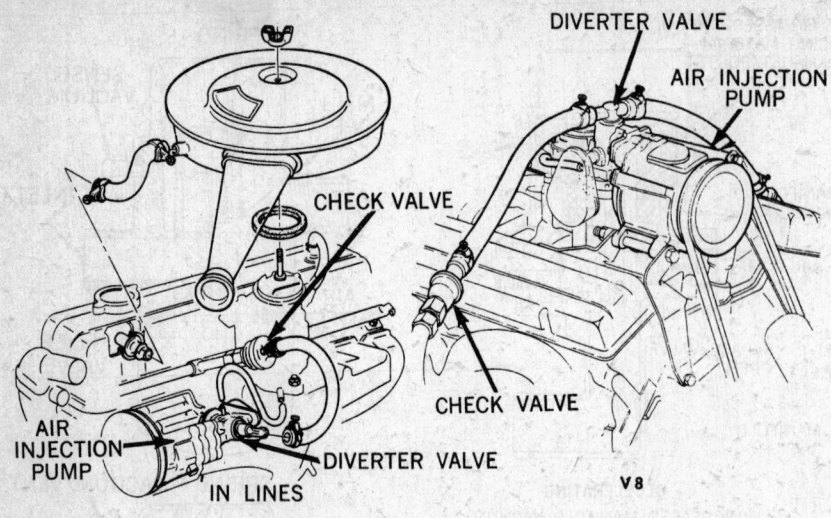

Typical air injection system

manifold vacuum, which pulls the valve mechanism from the dump position into the normal running position.

Unfortunately, the new style valve will not go into the dump position automatically during deceleration. To get the valve to dump, a vacuum differential valve (VDV) is connected in the sensing line. Manifold vacuum goes through the VDV and then to the diverter valve. When the manifold vacuum increases during deceleration, the VDV closes the sensing line. This shuts off the vacuum to the diverter valve, and the valve goes into the dump position.

A further refinement of this is to connect the sensing line to ported (above the throttle plates) vacuum instead of manifold vacuum and eliminate the VDV. In this situation, the diverter valve only receives vacuum above idle, because the vacuum port in the carburetor throat is above the throttle plate at idle. So whenever the engine idles, the diverter valve goes to the dump position. It also dumps during deceleration, because the throttle at that time is in the idle position.

California Jeeps use a combination electrical diverter valve and a vacuum switching valve. The diverter valve solenoid is controlled directly by the electronic control module (ECM) so that air is diverted to the air cleaner when the solenoid is de-energized at the following specified times: Engine not operating (electrical control). First five (5) seconds of any start up of the engine (electrical control). High electrical load on the engine control system. Closed throttle deceleration (vacuum control).

Some systems have a delay valve, similar to a spark delay valve, in the sensing hose. This delays for a few seconds the drop in vacuum when the throttle closes, so that the air is not dumped every time the driver takes his foot off the throttle in traffic.

Temperature controls are also used in the sensing hose hookup. Usually, the temper-

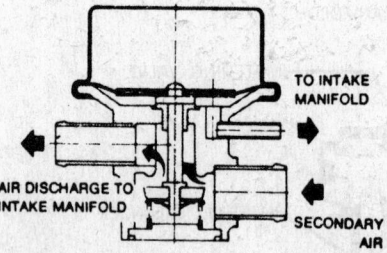

Datsun anti-backfire valve operation

ature valve shuts the vacuum off when the engine is cold, so that the pump air doesn't go to the engine exhaust ports until the engine warms up.

An idle vacuum valve is used to operate in conjunction with the vacuum delay valve, to provide backfire control, full time idle dumping of secondary air during cold engine operation, deceleration or extended idle periods of $\frac{1}{2}$ to 2 minutes or more. The valve also provides cold temperature protection for the catalyst and a cold EGR valve lockout.

Datsun & Nissan Trucks with a catalytic converter have protection devices to prevent converter overheating due to large quantities of injected air.

1979 models have a combined air control valve instead of the relief valve, emergency valve, and air control valve. The combined air control valve regulates the amount of injected air according to intake manifold and air pump discharge pressure, to prevent the converter from overheating.

An anti-backfire valve is installed in an air delivery hose. The purpose of the valve is to prevent backfiring in the exhaust manifold during deceleration. When the throttle closes suddenly, an overly rich air/fuel mixture exists in the intake manifold due to the lack of air getting past the throttle valves. This rich mixture will not completely burn in the combustion chamber. If

the unburned gases were to come in contact with the oxygen pumped into the exhaust ports by the air pump, they would ignite and cause backfiring and possible damage.

The anti-backfire valve is connected to the intake manifold by a vacuum line and when the vacuum rises, the valve opens a port in the intake manifold, allowing extra filtered air from the air cleaner to be admitted into the combustion chambers, leaning out the overly rich mixture.

1979 and later cab and chassis models have a transistorized programmed control unit, a vacuum switching valve, and an air control switch which govern air flow through the AIS. The air control switch, located between the intake manifold and the control unit, turns off when manifold vacuum is high, and on when vacuum is low. This provides a signal to the control unit, which determines when to turn the vacuum switching valve on or off accordingly. The vacuum switching valve controls the upper chamber of the CAC valve diaphragm, opening or closing the CAC valve according to signals received from the control unit. Thus, the amount of air injected into the AIS is monitored and adjusted as conditions warrant.

AIR PUMP TESTS

NOTE: Do not hammer on, pry or bend the pump housing while tightening the drive belt or testing the pump.

Before proceeding with the tests, check the pump drive belt tension. If the belt squeals when the engine is running, the pump may be dragging or seized. Remove the belt and turn the pump by hand to check for seizure. Disregard any chirping, squealing, or rolling sounds from inside the pump when turning it by hand, as these are normal.

Check the hoses and connections for leaks. Hissing or a blast of air is indicative of a leak. Soapy water, applied lightly around the area in question, is a good method for detecting leaks.

To test air output, disconnect the air hose from the pump wherever it is convenient. If you disconnect it from one check valve on a V8, the other hose should also be disconnected and plugged for the test. Run the engine at idle and feel the blast of air from the hose with your hand. Increase the engine speed to 1500 rpm and feel the blast of air again. If the blast increases, and is steady, the pump is okay.

PUMP NOISE DIAGNOSIS

The air pump is normally noisy; as engine speed increases, the noise of the pump will rise in pitch. The rolling sound the pump bearings make is normal. However, if this sound becomes objectionable at certain speeds, the pump may be defective and will have to be replaced.

A continual hissing sound from the air

AIR INJECTION REACTOR SYSTEM DIAGNOSIS CHART

Condition	Possible Cause	Correction
No air supply—accelerate engine to 1500 rpm and observe air flow from hoses. If the flow increases as the rpm's increase, the pump is functioning normally. If not, check possible cause.	Loose drive belt.	Tighten to specifications.
	Leaks in supply hose.	Locate leak and repair.
	Leak at fittings.	Tighten or replace clamps.
	Air expelled through by-pass valve:	
	a. Connect a vacuum line directly from engine manifold vacuum to by-pass valve	If this corrects the problem, go to step b. If not, replace air by-pass valve.
	b. Connect vacuum line from engine manifold vacuum source to by-pass valve through vacuum diffential valve directly, by passing the differential vacuum delay and separator valve.	If this corrects the problem, check differential vacuum, delay and separator valve and vacuum source line for plugging. Replace as required. If it doesn't, replace vacuum differential valve.
	Check valve inoperative.	Disconnect hose and blow through hose toward check valve. If air passes, function is normal. If air can be sucked from check valve, replace check valve.
	Pump failure.	Replace pump.
Excessive pump noise, chirping, rumbling, knocking, loss of engine performance.	Leak in hose.	Locate source of leak using soap solution and correct.
	Loose hose	Reassemble and replace or tighten hose clamp.
	Hose touching other engine parts.	Adjust hose position.
	Vacuum differential valve inoperative.	Replace vacuum differential valve.
	By-pass valve inoperative.	Replace by-pass valve.
	Pump mounting fasteners loose.	Tighten mounting screws as specified.
	Pump failure.	Replace pump.
	Check valve inoperative.	Replace check valve.
Excessive belt noise.	Loose belt.	Tighten to spec.
	Seized pump.	Replace pump.
Excessive pump noise. Chirping.	Insufficient break-in.	Run vehicle 10–15 miles at interstate speeds. Recheck.
Centrifugal filter fan damaged or broken.	Mechanical damage.	Replace centrifugal filter fan.
Exhaust tube bent or damaged.	Mechanical damage.	Replace exhaust tube.
Poor idle or driveability.	A defective A.I.R. system cannot cause poor idle or driveability.	Do not replace A.I.R. system.

pump pressure relief valve at idle indicates a defective valve. Replace the relief valve.

If the pump rear bearing fails, a continual knocking sound will be heard. Since the rear bearing is not separately replaceable, the pump will have to be replaced as an assembly.

ANTI-BACKFIRE VALVE TESTS

Detach the hose, which runs from the by-pass valve to the check valve.

Connect a tachometer to the engine. With the engine running at normal idle speed, check to see that air is flowing from the bypass valve hose connection.

Speed the engine up, so that it is running at 1,500–2,000 rpm. Allow the throttle to snap shut. The flow of air from the bypass valve at the check valve hose connection should stop momentarily and air should then flow from the exhaust port on the valve body or the silencer assembly.

Let the throttle snap shut several times. If the flow of air is not diverted into the atmosphere from the valve exhaust port or if it fails to stop flowing from the hose

connection, check the vacuum lines and connections. If these are tight, either the bypass valve or one of the accessory valves in the small sensing hose is defective and must be replaced.

A leaking diaphragm will cause the air to flow out both the hose connection and the exhaust port at the same time.

Late model systems should stop flowing at idle, as described earlier. If not, the bypass valve or accessory valve is defective.

To check the valve on Datsuns & Nissans, disconnect the hose from the air cleaner and place a finger on the end. Run the en-

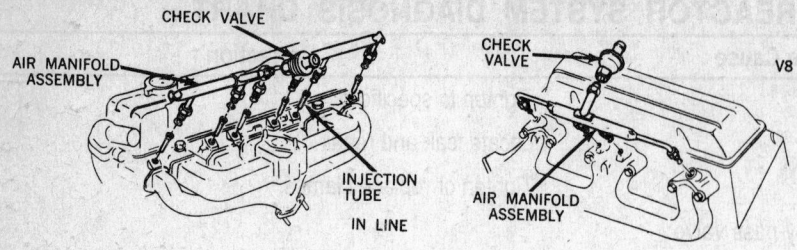

Typical air injection tubes

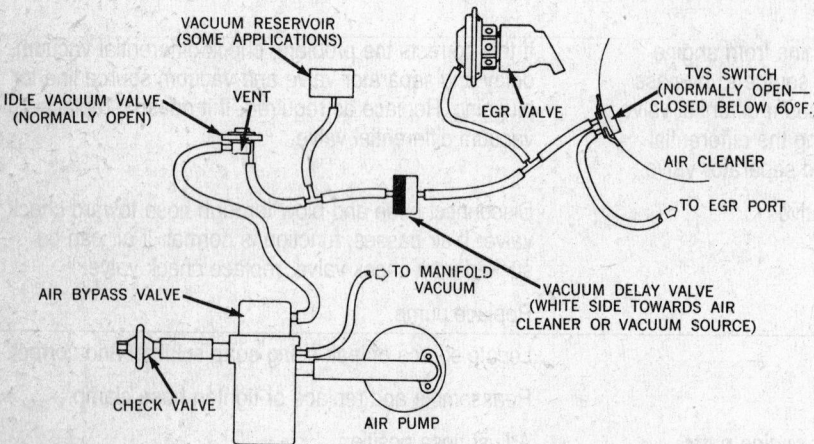

Typical air injection system with an idle vacuum valve used to control converter overheating

gine up to about 3,000 rpm, then quickly release the throttle. If the valve is performing correctly, suction should be felt at the end of the hose. If no suction is felt, replace the anti-backfire valve.

CHECK VALVE TEST

Remove the hose from the check valve. With the engine running at 1,500 rpm in Park or Neutral, hold the back of your hand near the check valve to test for exhaust gas leakage. If the valve leaks, it must be replaced.

NOTE: Vibration and flutter of the valve at idle is a normal condition caused by exhaust pulsations. It does not mean that the valve is defective.

DATSUN/NISSAN AIR PUMP RELIEF VALVE TEST

1. Disconnect the hoses leading to the check valve (on the air injection manifold) and the air control valve from the air hose connector. Plug the connector.
2. Start the engine and increase the engine speed to about 3,000 rpm. Place your finger on the outlet of the relief valve (inside the air cleaner housing) and check for air discharge. If you do not feel any air coming out, the relief valve is faulty, and must be replaced.

VACUUM DIFFERENTIAL VALVE TEST

Disconnect the small sensing hose at the bypass valve and connect a vacuum gauge to the hose. With the engine idling in Park or Neutral, the gauge should read full manifold vacuum.

Run the engine at a steady 2500 rpm in Park or Neutral, and release the throttle. As the engine decelerates, the vacuum gauge should drop close to zero, then return to full manifold vacuum as the engine speed drops to idle. If not, the VDV is defective and must be replaced.

NOTE: The small hose nozzle should be connected to manifold vacuum.

Datsun/Nissan Emergency Air Relief Valve

1. Warm up the engine.
2. Check all hoses for leaks, kinks, improper connections, etc.
3. Run the engine up to 2,000 rpm under no load. No air should be discharged from the valve.
4. Disconnect the vacuum hose from the valve. This is the hose which runs to the intake manifold. Run the engine up to 2,000 rpm. Air should be discharged from the valve. If not, replace it.

Datsun/Nissan Combined Air Control Valve

1. Check all hoses for leaks, kinks, and improper connections.
2. Thoroughly warm up the engine.
3. With the engine idling, check for air discharge from the relief opening in the air cleaner case.
4. Disconnect and plug the vacuum hose from the valve. Air should be discharged from the valve with the engine idling. If the disconnected vacuum hose is not plugged, the engine will stumble.
5. Connect a hand-operated vacuum pump to the vacuum fitting on the valve and apply 7.8–9.8 in. Hg of vacuum. Run the engine speed up to 3,000 rpm. No air should be discharged from the valve.
6. Disconnect and plug the air hose at the check valve, with the conditions as in the preceding step. This should cause the valve to discharge air. If not, or if any of the conditions in this procedure are not met, replace the valve.

AIR PUMP FILTER REPLACEMENT

Several American truck brands employ a standard type of air pump filter that requires periodic replacement. Proceed as follows:
1. Disconnect the air and vacuum hoses from the diverter valve.
2. Loosen the pump pivot and adjusting bolts and remove the drive belt.
3. Remove the pivot and adjusting bolts from the pump. Remove the pump and the diverter valve as an assembly.

--- **CAUTION** ---
Do not clamp the pump in a vise or use a hammer or pry bar on the pump housing.

4. To change the filter, break the plastic fan from the hub. It is seldom possible to remove the fan without breaking it.
5. Remove the remaining portion of the fan filter from the pump hub. Be careful that filter fragments do not enter the air intake hole.
6. Position the new centrifugal fan filter on the pump hub. Place the pump pulley against the fan filter and install the securing screws. Torque the screws alternately to 95 inch lbs. and the fan filter will be pressed onto the pump hub.
7. Install the pump on the engine and adjust its drive belt.

ISUZU AIR SWITCHING VALVE

This valve is designed to switch air flow from the air pump. It is operated by vacuum and air pump pressure which are switched by a vacuum switching valve.

While air pump pressure acts in the diaphragm chamber ''B'' (see illustration) through the vacuum switching valve, this valve flows the air from the air pump to the check valve. When manifold vacuum switched by the vacuum switching valve

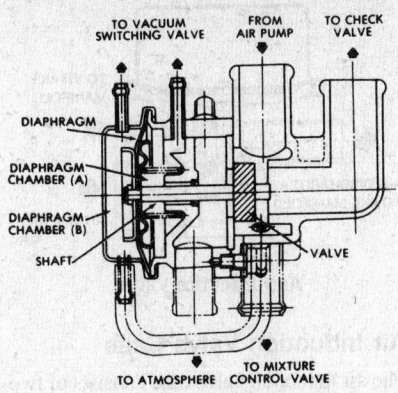

Isuzu air switching valve

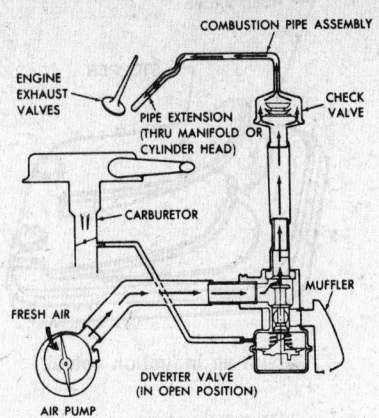

Typical air injection system operation

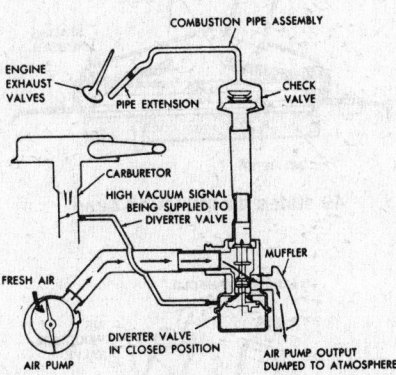

Typical air diverter valve operation

acts in the diaphragm chamber "B" the valve closes the air passage to the check valve, and at the same time opens the port to the atmosphere, so the air from the pump is diverted to the atmosphere.

AIR INJECTION MANIFOLDS

The air injection manifolds distribute the air via the diverter valve , to each of the exhaust manifold inlet ports. A check valve, incorporating a stainless steel spring plunger and an asbestos seat, is integral with each air injection manifold. The function of the check valve is to prevent the reverse flow of exhaust gases to the air pump during the pump or drive belt failure, or diverter valve bypass (vent) operation. The air injection tubes are mounted to the exhaust manifold and route the airflow into the inlet ports.

Pulse Air Injection Reactor Systems

GM PAIR

The Pulse Air Injection Reactor (PAIR) system is installed on the small inline 6 cylinder engine, used in General Motors light duty trucks, beginning in 1979. The PAIR system uses no air pump, but relies on the negative and positive exhaust gas impulses to draw fresh air into the exhaust manifold to assist in the further burning of the hydrocarbons (HC) before leaving the tailpipe.

Four individual check valves are used to prevent the exhaust gases from entering the fresh air intake chamber plenums. Two sets of pipes are used, one set in the front section of the exhaust manifold and the second set in the rear section of the exhaust manifold.

Two sets of plenum chambers are used and connected to the carburetor air cleaner by a common hose, for the fresh air intake.

During periods of high engine rpm, the check valves will remain closed to prevent the flow of exhaust gases to the engine air cleaner.

FAILURE DIAGNOSIS

1. Inspect the pulse air valve and pipes for leakage or defective operation, if a hissing noise is heard.
2. If one or more of the check valves are defective, exhaust gases will enter the carburetor area and cause poor drive-ability such as stalling, surge, or poor performance.

INSPECTION OF PULSE AIR INJECTION REACTOR SYSTEM

1. Burned off paint on the rocker arm plenum chambers indicates a defective pulse air valve. Rubber grommets and hoses will deteriorate and can cause a hissing noise.
2. Inspect the carburetor for pieces of rubber hoses or grommets, indicating an overheating of the components.
3. Inspect the operation of the pulse air valve by applying at least 17 in. Hg at the grommet end of the valve. A drop of 6 in. Hg in two seconds is allowed.

JEEP PULSE AIR SYSTEM

The Pulse Air Injection System utilizes the alternating positive and negative exhaust

pressure pulsations instead of an air pump to inject air into the exhaust system and produce exhaust gas oxidation. The air enters through the filtered side of the air cleaner to the air control valve. When opened by the air switch, the air control valve allows the air to continue to and through the air injection check valve. The air enters the exhaust system, either upstream or downstream from the check valve, air is injected either into the front exhaust pipe (upstream) or into the catalytic converter (downstream), depending upon the engine operating conditions. The CEC system micro computer unit (MCU) controls the switching operation.

Air Injection Check Valve

The air injection check valve is a one-way reed valve that is opened and closed by the negative and positive exhaust pressure pulsations. During the negative exhaust pulse (low Pressure), atmospheric pressure opens the check valve and forces air into the exhaust system. Being a one-way valve, the valve reed prevents exhaust from being forced back through the valve during the positive exhaust pressure pulsations (high pressure).

Air Control Valve

The air control valve controls the supply of filtered air routed to the air injection check valve. The valve is opened and closed by the air switch solenoid.

Air Switch Solenoid

The air switch solenoid controls the air control valve by switching the vacuum on and off. The solenoid is controlled by the micro computer unit (MCU).

Vacuum Storage Tank

Engine vacuum is stored in a reservoir tank until released by the air switch solenoid.

Micro Computer Unit (MCU)

The MCU switches air either upstream or downstream, depending upon the engine operating conditions, by energizing and de-energizing the air switch solenoids.

COURIER SECONDARY AIR INJECTION SYSTEM

Reed Valve and Air Pipe Assembly

This assembly is the main component of what is termed the secondary air injection system. Vehicles having this system are not equipped with air pumps.

The second system operates by sensing exhaust gas pulsations which in turn causes the reed valve to draw fresh air from the air cleaner and inject the fresh air into the exhaust system. The main advantage of this system is that it eliminates the parasitic

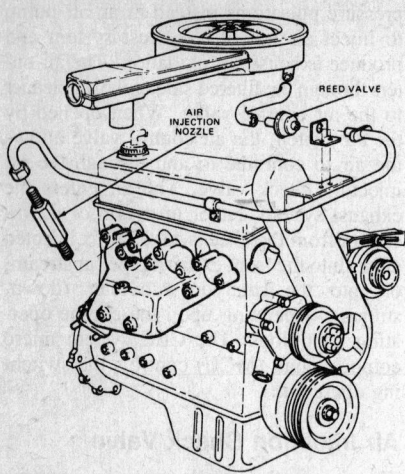

Courier secondary air injection system

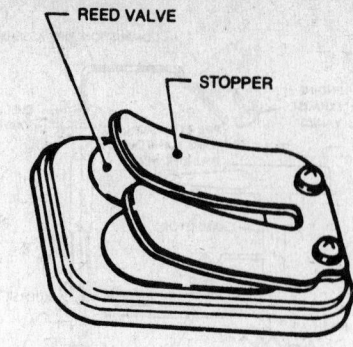

Datsun air induction valve

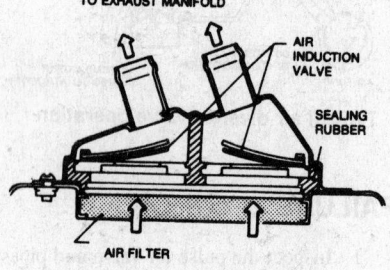

49 states air induction case

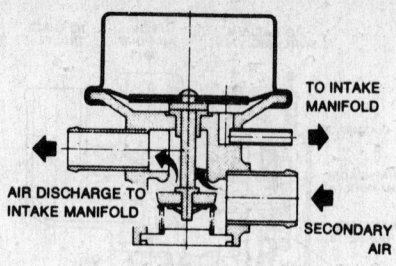

Anti-Backfire Valve

horsepower drain associated with systems using belt driven pumps.

Removal of the reed valve is simply a matter of disconnecting the hose at the valve and unbolting the valve from its mounting bracket.

REED VALVE TESTING

1. Warm the engine to normal operating temperature.
2. Disconnect the inlet hose from the reed valve.
3. Start the engine and allow it to idle.
4. Place your finger over the reed valve inlet. It should be felt that air is being drawn into the valve. Replace the valve if suction is not apparent.
5. Raise the engine rpm to 1500. Exhaust gas should not leak from the reed valve inlet; if it does, replace the valve.

Vacuum Delay Valve(s)

The description of this item is self-explanatory. The vacuum delay valve(s) "fine tunes" the vacuum operation of the Thermactor system for vehicles which are to be used in areas with stringent emission control regulations (e.g.——Calif.). Failure of a vacuum delay valve is extremely rare.

DATSUN/NISSAN AIR INDUCTION SYSTEM

Description

The air induction system is designed to send secondary air to the exhaust manifold, utilizing a vacuum caused by exhaust pulsation in the exhaust manifold.

The exhaust pressure in the exhaust manifold usually pulsates in response to the opening and closing of the exhaust valve and it decreases below atmospheric pressure periodically.

If a secondary air intake pipe is opened to the atmosphere under vacuum conditions, secondary air can be drawn into the

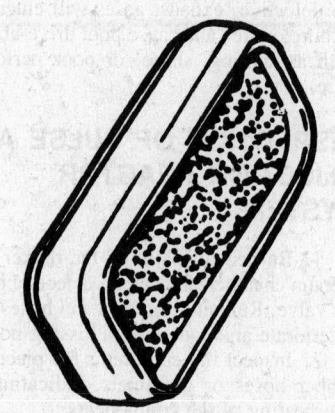

Air induction filter

exhaust manifold in proportion to the vacuum.

Therefore, the air induction system reduces CO and HC emissions in exhaust gases. The system consists of two air induction valves, a filter, hoses and E.A.I. (Exhaust Air Induction) tube(s).

Air Induction Valve Case

The air induction valve case consists of two reed valves, a rubber seal and a filter and is attached to the air cleaner. There are two types of air induction valve cases. Type-A is equipped with two hose connectors and is installed on California models, while Type-B is equipped with one connector and is installed on non-California models.

Air Induction Valve

Two reed valve type check valves are installed in the air cleaner. When the exhaust pressure is below atmospheric pressure (negative pressure), secondary air is sent to the exhaust manifold.

When the exhaust pressure is above atmospheric pressure, the reed valves prevent secondary air from being sent back to the air cleaner.

Air Induction Valve Filter

The air induction valve filter is installed at the dust side of the air cleaner. It purifies secondary air to be sent to the exhaust manifold.

Air Induction Pipe

The secondary air fed from the air induction valve goes through the E.A.I. pipe to the exhaust manifold.

Anti-Backfire (A.B.) Valve

This valve is actuated by intake manifold vacuum to prevent backfire in the exhaust system at the initial period of deceleration.

At this period, the mixture in the intake manifold becomes too rich to ignite and burn in the combustion chamber and burns easily in the exhaust system with injected air in the exhaust manifold.

The A.B. valve provides air to the intake manifold to make the air-fuel mixture leaner and prevents backfire.

The correct function of this valve reduces hydrocarbon emission during deceleration.

Component Removal and Installation

Air Induction Valve and Filter

Remove the valve and filter on the air cleaner. The air induction valve and valve filter can then be taken out easily. Installation is in the reverse sequence of removal.

Air Induction Pipe

Remove nut securing the pipe to the exhaust manifold. At the same time, remove the screws securing the bracket and rubber hose clamp.

The air induction pipe can then be taken out. Installation is in the reverse sequence of removal.

A.B. Valve

1. Remove air cleaner.
2. Remove air hoses and vacuum tube.

Then the A.B. valve can be taken out.

MITSUBISHI/D–50/ PLYMOUTH ARROW SECONDARY AIR SUPPLY SYSTEM

This system supplies air for the further combustion of unburned gases in the thermal reactor (California only) or exhaust manifold and consists of a reed valve, air hoses, and air passages built into the cylinder head.

The reed valve is operated by exhaust pulsations in the exhaust manifold. It draws fresh air through the air cleaner and supplies it to the exhaust ports.

Maintenance

Check for damage to the air hoses and air pipes. Make sure the air passages are open in the head.

Electronic Ignition System

A change has been made through the model years from the conventional distributors to the electronic ignition systems for more precise ignition control.

Different types are available from the manufacturers, but the operation of the systems are basically the same. Greater dependability, higher secondary voltages and less need for adjustments are the important factors considered in using this system for emission control.

Refer to the individual truck sections and to the Electrical section for expanded information.

DISTRIBUTOR CONTROLS

All distributor controls act in some way to change or eliminate vacuum advance during certain operating conditions. Usually, the control cuts down on the amount of vacuum advance, in effect retarding the spark, so that the exhaust will get hotter and burn up hydrocarbon and carbon monoxide emissions before they go out the tailpipe. This function also reduces NO emissions under acceleration.

The distributor vacuum advance unit might be connected, according to factory design,

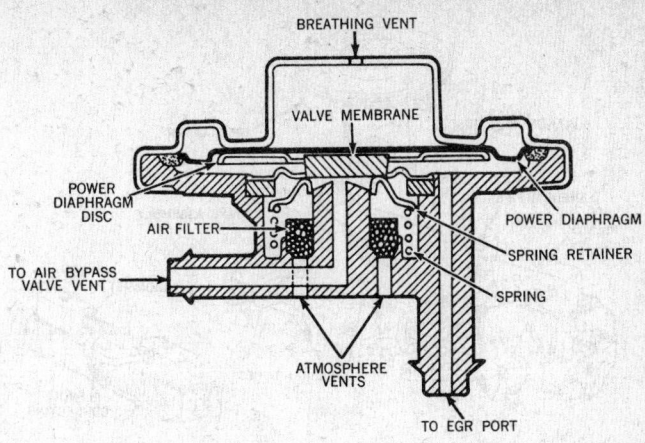

Cross section of an idle vacuum valve

to either manifold vacuum or ported (above the throttle plates) carburetor vacuum. Either way, the vacuum spark advance curve is approximately the same for all running conditions above idle. At idle, however, the manifold vacuum hookup results in full advance, while the ported hookup gives zero advance. If the hoses are hooked up the wrong way, the addition or lack of advance will affect idle speed, requiring a readjustment of the throttle position to bring the idle speed back to specifications. When this is done, emissions will usually be high, so it is important to keep the hoses hooked up correctly. Also, incorrectly connected hoses will have a critical effect on ignition timing. If the advance unit hose is connected to ported vacuum instead of manifold vacuum and timing is reset, actual ignition timing will be critically advanced and the engine will exhibit severe knock; if it is connected to manifold rather than ported vacuum, timing will be severely retarded and the engine will exhibit poor running, hesitation and overheating.

DUAL DIAPHRAGM DISTRIBUTORS

These distributors have two hose connections, one in the normal position, and the other closer to the distributor body. The hose fitting next to the body is for the retard diaphragm, and is connected to manifold vacuum. The retard diaphragm affects the spark only at idle, when there is no vacuum on the advance diaphragm. In effect, the retard diaphragm provides a movable resting place for the advance diaphragm. When ported vacuum is not acting on the advance diaphragm, it returns to the neutral or no-advance position against the retard diaphragm. At idle, manifold vacuum pulls the retard diaphragm to the retard position, and the advance diaphragm follows along to retard the spark. On the VW pickup, a temperature valve shuts off vacuum from the carburetor when the coolant temperature is below 130° F.

Testing Dual Diaphragm Distributors

To test a dual diaphragm distributor, connect a timing light to the engine. Remove the retard hose from the distributor and plug the hose. With the engine running, increase the speed to a fast idle and watch the timing marks. The timing should advance. If not, either the vacuum unit is faulty, the vacuum port is plugged, or there is a temperature control device that is shutting off the vacuum. Apply hand pump or mouth suction vacuum to the advance diaphragm and the timing should advance. If not, the distributor or advance unit must be repaired or replaced. Failure to advance could be caused by a faulty diaphragm or a sticky advance plate.

Remove the advance hose from the vacuum unit and read the timing at normal idle speed. Remove the plug that was inserted in the retard hose, and check for full manifold vacuum at the end of it. If there is no vacuum, temperature controls may be shutting it off.

Connect the hose to the retard diaphragm, or apply vacuum from another source. The timing should immediately retard several degrees. If not, the diaphragm is not working, and the unit must be replaced. Reconnect all hoses as they were originally.

DISTRIBUTOR VACUUM DECELERATION VALVE

Its purpose is to advance the spark during deceleration, by sending full manifold vacuum to the vacuum advance unit. At all other times the vacuum advance unit receives ported (above the throttle plates) carburetor vacuum.

Three checks should be made on the valve: the amount of vacuum at the distributor, any valve leaks, and the adjustment. To check the amount of vacuum at the distributor, use a T-fitting and a short length of vacuum hose to connect a vacuum gauge

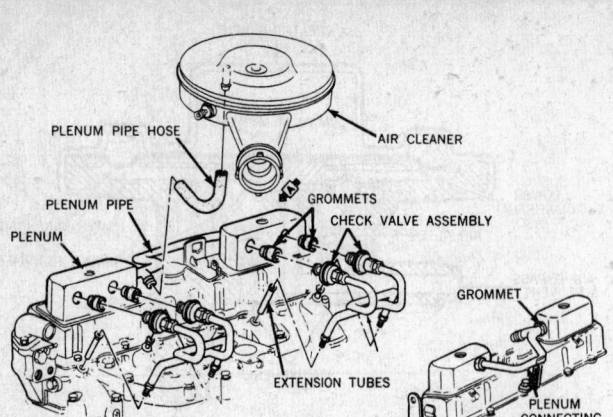

Pulse Air Injection System (PAIR)

into the distributor vacuum line near the distributor. At idle, with the engine fully warmed up, the vacuum on the gauge should be less than 1 Hg. If the gauge shows more than 1 Hg. the idle speed is too fast, or the valve is leaking. To check for a leak, remove the large manifold vacuum hose on the side of the valve. If the vacuum drops, the valve is leaking and must be replaced. If the vacuum stays high, reduce the engine idle speed so that the port in the carburetor is covered.

To check the valve adjustment, connect the manifold vacuum hose and run the engine at 2000 rpm for 5 seconds. Then release the throttle. The distributor vacuum should go over 16 in. Hg. and stay there for about one second. Within about three seconds after you release the throttle, the distributor vacuum should drop to below 6 in. Hg. If the carburetor is equipped with a dashpot to make the throttle close slowly, the time may be about one second longer. If the time is too long, remove the cover on the valve and turn the screw clockwise to reduce the time. To increase the time, turn the screw counterclockwise. If the valve will not adjust properly, it must be replaced, and the new valve adjusted to specifications.

SPARK DELAY VALVE

This small valve is connected between the carburetor and the distributor vacuum advance, so that the ported (above the throttle plates) vacuum to the distributor must pass through the valve. A restriction in the valve delays the vacuum applied to the vacuum advance unit so that the advance comes in slowly. When there is no vacuum at the carburetor port, as during idle or wide open throttle a check valve inside the spark delay valve opens and dumps the vacuum so that the vacuum advance unit returns to the no-advance position without any delay.

Spark delay valves can be tested for correct operation and leaks with a source of vacuum such as a hand vacuum pump or a running engine, and a vacuum gauge. Connect the vacuum gauge to the distributor

side of the valve, and the vacuum source to the other side. The gauge should rise slowly until it reads the amount of vacuum available. The time to rise to the maximum reading should be from one to 28 seconds. If the vacuum gauge does not read anything, the valve is plugged. If the vacuum reads instantly, without any delay, the valve is open. In either case, the spark delay valve must be replaced. To test the check valve part of the spark delay valve, remove the vacuum source and the vacuum gauge should drop instantly to zero without any delay. If there is any delay, the spark delay valve is defective and must be replaced.

TRANSMISSION CONTROLLED SPARK

The purpose of the transmission controlled spark is to eliminate vacuum spark advance in the lower gears. When the transmission is in high gear, vacuum spark advance is allowed for better gasoline milage and part throttle response. This system was originally used during the 1975–1976 model years on most light duty models and continued on some California and High Altitude vehicles in later years.

Some Jeep models use a Spark Coolant Temperature Override Switch. This system is used to override the TCS system to im-

prove driveability during the warmup period by providing full distributor vacuum advance operation until the temperature reaches 160°F within the cooling system. The system then reverts to the transmission controlled spark system.

A spark timing control system is added to manual transmission Datsun & Nissan models sold in the U.S. in 1979. The system controls distributor vacuum advance, giving full vacuum advance when the transmission is in 4th or 5th, and partial advance in the first three gears. This provides better control of the combustion process, lowering emissions of HC and NO.

The system components include a top gear detecting switch, installed into the transmission, and a vacuum switching valve spliced into the distributor vacuum advance hose by means of a three way connector. When the transmission is shifted into either of the two top gears, the transmission switch goes on, thus activating the vacuum switching valve which closes its air bleed, giving full advance. Shifting into any gear but 4th or 5th turns the transmission switch off, deactivating the vacuum switching valve. The valve opens a vacuum leak, providing only partial vacuum advance to the distributor.

Testing TCS Systems

Testing the system is done by connecting a vacuum gauge to the distributor vacuum line with a long hose so you can put it through the window into the front seat and see it while driving. There should be no vacuum in the lower gears on a warm engine, but after the transmission shifts into a gear that allows vacuum advance, you should see vacuum on the gauge. Engines that run their distributors on manifold vacuum will show vacuum at all times when in the proper gear. Engines that use ported (above the throttle plates) vacuum will show vacuum in the proper gear only when the throttle is open. If you don't get vacuum when you should, test the individual units in the system.

Vacuum solenoids can be tested by disconnecting all wiring and connecting hot

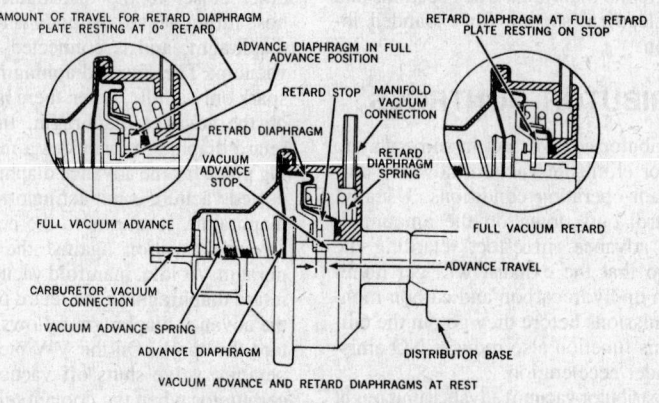

Dual diaphragm vacuum control

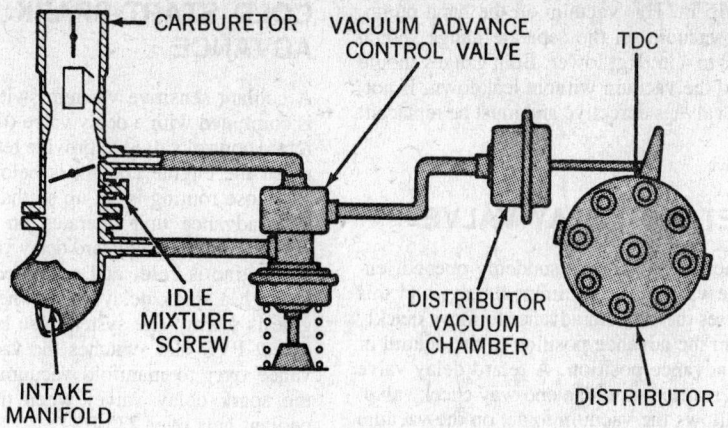

Carburetor-control valve-distributor relationship

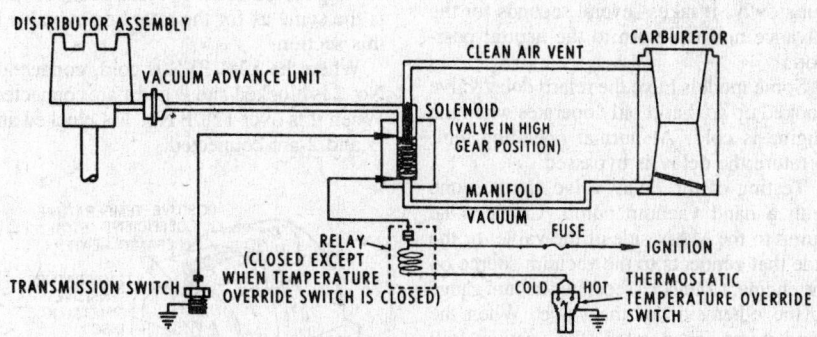

Typical transmission controlled spark system

and ground wires to the solenoid terminals, to make it open or close. You should be able to blow through the solenoid when it is open, but not when it's closed. Because solenoids exist in both normally open and normally closed designs, it is important to use the right solenoid. If the wrong solenoid is used, the system will work backwards, giving advance in the lower gears but not in high. The same goes for the transmission switch, which exists in both normally open and normally closed designs. The term "normally open" means that the solenoid or switch is open when it is not energized or activated. In the case of a vacuum solenoid, normally open means that if you were holding the solenoid in your hand without any wires connected to it, the vacuum passages would be open, allowing vacuum to pass. In the case of a transmission switch, the term "normally open" refers to the electrical path, which is "open" or "off" so that it will not conduct electricity. Normally closed, of course, means that the electric contacts are closed so that the current can pass. But normally closed on a vacuum solenoid means that the vacuum passage is blocked so the vacuum can't get through.

Datsun & Nissan Spark Timing Control System

1. Check all hoses and electrical wires

for proper connections, leaks or corrosion, and so on.

2. Check the distributor vacuum advance unit for proper operation. This can be checked by hooking up a timing light, starting the engine, then increasing engine speed and observing whether or not the timing marks advance. If not, the advance unit must be checked for binding or leaks.

3. With the timing light installed, increase the engine speed to 2,000 rpm. Have an assistant disengage the clutch, then shift between 3, 4, and 5, then back down and into neutral. Spark timing should vary when the transmission is in 4 or 5 (advance should

be greater). If this is not the case, check the vacuum switching valve.

TEMPERATURE ACTIVATED VACUUM (TAV) AND COLD TEMPERATURE ACTIVATED VACUUM (CTAV) SYSTEMS

This system switches the vacuum source back and forth between the carburetor spark port and EGR port, according to the air temperature. A 3-nozzle vacuum solenoid is used. Below approximately 55°F. outside air temperature, the temperature switch is open, and the solenoid is not energized. In this position, the solenoid connects the spark port to the vacuum advance unit. Above 55°F. the temperature switch closes, and energizes the solenoid. In this position, the solenoid connects the EGR port to the vacuum advance unit.

The temperature switch is located in the air cleaner, and a latching relay is on the firewall. Once the temperature switch has closed, the relay latches so that any sudden rush of cold air through the air cleaner will not cycle the solenoid on and off. The latching relay keeps the solenoid energized as long as the ignition switch is on. When the ignition switch is turned off, the relay unlatches and the system is ready for the next start, whether the air temperature is hot or cold. If the air at the temperature switch is over 55°F. the latching relay will come on when the ignition switch is turned on.

Testing

Test the system with a vacuum gauge connected to the vacuum advance hose at the distributor. With the temperature above 65°F. (to be sure the temperature switch has closed) you should be getting vacuum from the EGR port. If you disconnect the EGR port hose and the vacuum drops, you know the system is working. When making a cold test, the vacuum should come from the spark port hose, so disconnecting that hose should make the vacuum drop. Because both ports are above the throttle plate, the throttle must be opened slightly to get vacuum at the hose.

Identifying the spark port and EGR ports on the carburetor is easy if they are marked.

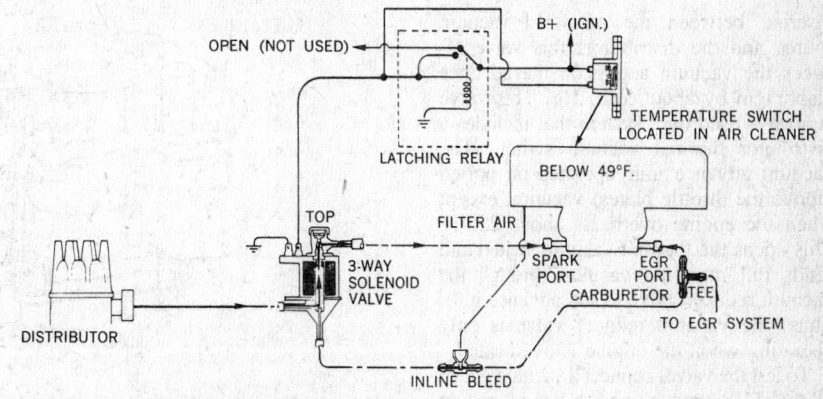

Cold temperature activated vacuum system

If there is no marking on the carburetor, connect two vacuum gauges, one to each port. At idle you should not have any vacuum. If you do have vacuum, it usually means the engine is idling too fast. Close the throttle slightly to slow down the idle and the vacuum should drop to almost zero.

When you open the throttle, you will see vacuum on one gauge before the other. The gauge that gets vacuum first is connected to the spark port.

ORIFICE SPARK ADVANCE CONTROL (OSAC)

It is a mechanism that delays the application of vacuum to the distributor vacuum advance unit. When the throttle is opened, the carburetor port is exposed to vacuum. This vacuum goes through a hose to the OSAC valve, and then to the distributor vacuum advance. The OSAC valve is sometimes mounted on the firewall, and sometimes on the air cleaner. Inside the OSAC valve is a calibrated orifice that delays the vacuum as much as 27 seconds, depending on the calibration of the valve.

Some OSAC valves have temperature control that senses the temperature inside the air cleaner or inside the plenum chamber behind the firewall, depending on where the valve is mounted. If the valve contains temperature control, it will be wide open below 60°F. bypassing the orifice and allowing vacuum advance without any delay. Above 60°F. the bypass closes and the delay takes over.

Testing

To test the valve, connect a vacuum gauge to the DIST connection on the valve. With the engine idling, you should have no reading on the gauge. If there is a reading, the engine is idling too fast. With the engine idling, open the throttle to a fast idle, and hold it steady. The vacuum on the gauge will rise slowly until it reaches a maximum reading. If not, there is something wrong with the system, and you should check out the hoses and the carburetor port, or replace the valve is necessary.

VACUUM REDUCER VALVE

Inserted between the manifold vacuum source and the distributor, this valve reduces the vacuum acting on the advance diaphragm by about 3 in. Hg. This valve is always used on a system that includes a distributor thermal vacuum switch. The vacuum advance unit operates on ported (above the throttle plates) vacuum, except when the engine overheats above 225°F. This opens the thermal vacuum switch and sends full manifold vacuum through the vacuum reducer valve to the advance unit. Thus, the vacuum reducer valve is only operating when the engine is overheated.

To test the valve, connect a vacuum gauge to the TVS nozzle, and a hand vacuum pump to the MAN nozzle. When you pump

up 15 in. Hg. vacuum on the hand pump, the vacuum on the separate gauge should be 3 to 4 in. Hg. lower. Both gauges should hold the vacuum without leakdown. If not, the valve is defective and must be replaced.

RETARD DELAY VALVE

When the throttle is suddenly opened, engine vacuum drops immediately, and this causes the vacuum advance to move quickly from the advance position to the neutral or no-advance position. A retard delay valve is a restriction with a one-way check valve. It allows the vacuum to act on the vacuum advance unit normally, but when the vacuum drops suddenly, the delay valve traps the vacuum in the advance unit and lets it out slowly. It takes several seconds for the advance unit to return to the neutral position.

Some models have the retard delay valve hooked up so that it only operates when the engine is cold. At normal operating temperature the delay is bypassed.

Testing of the delay valve can be done with a hand vacuum pump. Connect the pump to the MAN side of the valve, or the side that connects to the vacuum source on the engine. Connect a separate vacuum gauge to the other side of the valve. When the hand pump is operated, the vacuum will rise on both the pump gauge and the separate gauge equally. When the release is pulled, the pump gauge will drop to zero immediately, but the separate gauge will take several seconds to drop to zero. If it doesn't work that way, the delay valve is defective, and must be replaced.

COLD START SPARK ADVANCE

A coolant sensitive vacuum switch (PVS) is combined with a delay valve (distributor retard control valve) to provide retard delay when the engine coolant is below 128°F. The hose routing is set up so that the vacuum advance unit operates on manifold vacuum through the retard delay valve when the engine is cold, and on ported vacuum through a spark delay valve when the engine is warm. The system also has an overheat PVS that switches the vacuum advance over to manifold vacuum (through the spark delay valve) when the engine coolant gets over 235°F.

Testing the spark delay valve is covered in this section under Spark Delay Valve. Testing for the distributor retard control valve is the same as for the retard delay valve in this section.

When the 128° PVS is cold, connection No. 2 is blocked and D and 1 are connected. When it is over 128°F No. 1 is blocked and D and 2 are connected.

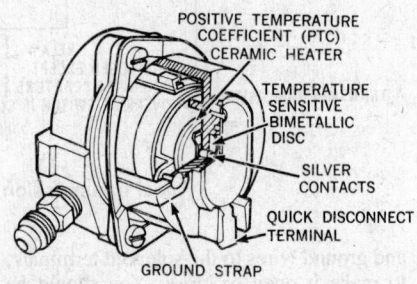

Electric choke assembly—Used with a choke stove

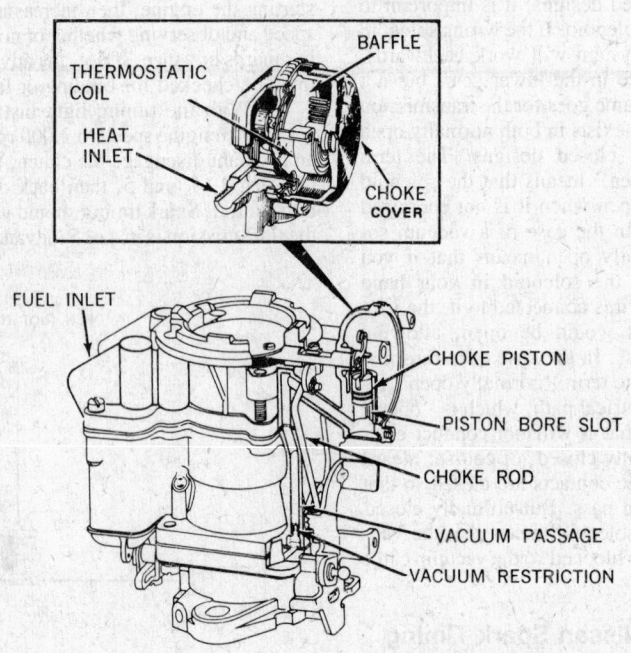

Non-electric choke assembly using a manifold heated choke stove

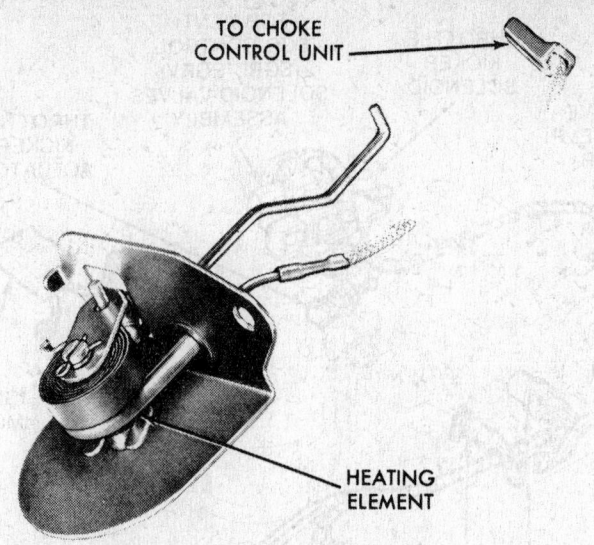

TO CHOKE
CONTROL UNIT

HEATING
ELEMENT

Manifold heat type choke assembly

Carburetor Choke Controls

NON-ELECTRIC CHOKE

A non-electric choke uses a "stove" on the exhaust manifold or a well on the intake manifold to provide heat. When the well is used, the choke coil is surrounded by the warm intake manifold, heated by the exhaust crossover passage. When the stove is used, the choke housing is connected to engine vacuum, and a long tube pulls the heated air from the stove into the choke housing to heat up the choke coil and cause the choke to open as the engine warms up. When an electric choke is used, it can be in addition to all the above, or it can be the only source of choke heat, depending on the design.

ELECTRIC CHOKE

The electric choke has a small heater next to the choke coil. This heater receives its current from different sources, depending on the car maker.

Ford Motor Company and Jeep chokes are powered from the alternator "center tap," which produces about 7 volts. As the alternator is only putting out voltage when the engine is running, the electric choke is automatically shut off when the engine is off. It is important that the choke is connected only to the special "center tap" provided on the alternator. The description "center tap" refers to the construction of the alternator wiring, and not to the location of the connection.

Inside is a thermostatic switch that turns on the heating element at approximately 80°F. Above that, the element stays on as long as the engine is running. The 80°F. figure was selected because the engine is warm enough at that temperature to keep running without the choke. When the heater comes on, the choke opens very quickly.

When the engine is shut off and cools down, the choke switch may stay on to as low as 65°F. at the choke housing. On a warm restart, where the choke switch was still on, the heating element would heat up the choke and open it shortly after the engine started.

Chrysler Corporation vehicles with an electric choke use a well type choke, which receives heat both from the intake manifold and the electric choke heater. A separate choke control unit is mounted on top of the intake manifold and connected to the heater with a wire. This wire disconnects at the choke control unit only, not at the heater.

Choke control units may be single and double stage. The double stage is recognized by the external resistor alongside the unit. The single stage unit turns on the choke heater at approximately 60°F. and off at 110°F. The double stage unit keeps the heater on below 60°F. but the current runs through the resistor. At approximately 60°F. the resistor is taken out of the circuit and the heater gets full current. At 110°F. the control unit turns the heater off.

Testing can be done with a non-powered test light on the choke terminal to find out if the heater is on or off. The ignition switch must be on. If the light glows, you know the control unit is on. On two-stage units, the light will glow dimly when the resistor is in the circuit, and brightly when the resistor is out. The current to the control unit comes from the ignition switch, and there is no fuse.

Chevrolet and GMC use an electric choke that is mounted on the carburetor. The choke has a dual element behind the coil spring. Whenever the engine is running, the choke heater is in operation. Below 50–70°F. a bimetal snap disc in the choke cover turns off the large section of the heating element so that only the small section gives off heat. Above 50–70°F. the disc switches on the

large heating element for faster choke opening.

Current to the choke is controlled by a three-terminal oil pressure switch. One of the terminals is a ground for the red oil pressure light on the instrument panel. The other two terminals are a switch in series between the ignition switch and the choke heater. Oil pressure operates the switch so that the choke gets current only when the engine is running. The circuit is fused through the backup light or transmission fuse in the fuse block.

NOTE: Failure of the choke heater circuit will cause the oil pressure light to go on.

IDLE ENRICHMENT SYSTEM

In order to reduce the cold engine stalling, a metering system is used to relate to the carburetor, rather than to the choke. The system enriches the carburetor mixtures in the curb idle and fast idle modes. The carburetor will have the complete idle system enriched during periods of cold to semi-cold operating conditions. The idle enrichment valve is manifold vacuum controlled and opens or closes a passageway that admits extra air to the idle circuit.

Some models have a coolant temperature control valve, mounted on a coolant passage and connected by hose between the manifold vacuum source and the idle enrichment valve. When the engine is cold, the valve is open and allows vacuum to operate the idle enrichment valve to richen the idle fuel mixture. When the engine warms up, the valve closes and cuts off the vacuum to the coolant temperature valve.

A vacuum solenoid may be included in the vacuum hose arrangement to provide EGR valve delay while the engine is cold.

Testing

Testing the system can be done on a cold engine by disconnecting the hose at the carburetor and connecting a vacuum gauge to the hose. Start the engine and note the length of time that vacuum appears on the gauge. At the end of the timed period, the gauge should drop to zero. Allow the engine to warm up to operating temperature and make the test again. This time you should not see any vacuum on the gauge, because the CCIE (coolant control idle enrichment), valve should be closed. If there is no timer, you will see vacuum for several minutes after a cold start, until the engine warms up.

To check the effect of the idle enrichment use a hand vacuum pump on the idle enrichment valve on the carburetor. With vacuum applied, the valve will be closed, richening the idle, and changing the idle speed. Release the vacuum and the speed should go back where it was. If there is no speed change, either the valve is not working, or a carburetor passage is blocked with dirt. The valve should also hold vacuum without leaking down.

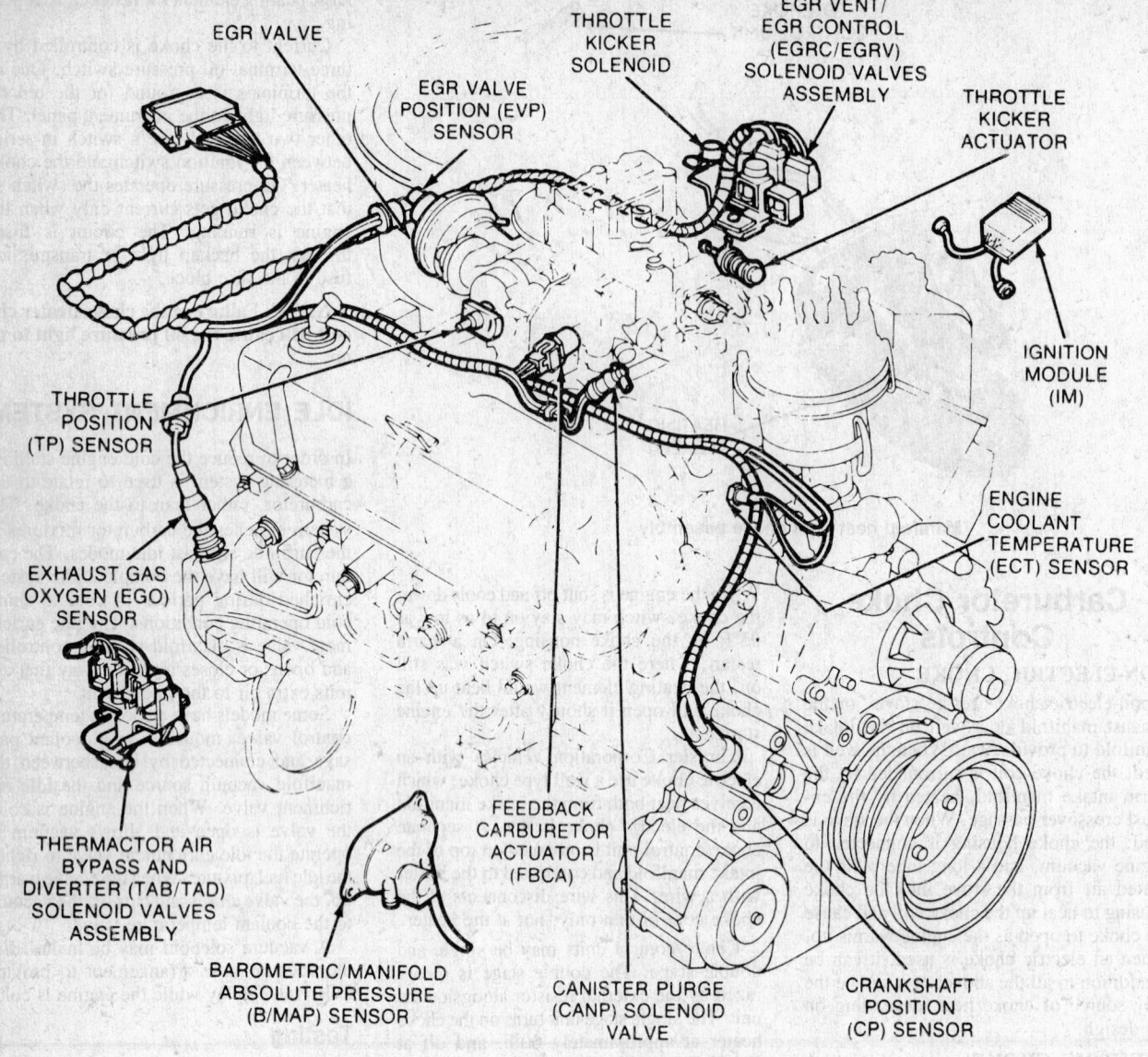

EGR VALVE

EGR VALVE POSITION (EVP) SENSOR

THROTTLE KICKER SOLENOID

EGR VENT/ EGR CONTROL (EGRC/EGRV) SOLENOID VALVES ASSEMBLY

THROTTLE KICKER ACTUATOR

IGNITION MODULE (IM)

THROTTLE POSITION (TP) SENSOR

EXHAUST GAS OXYGEN (EGO) SENSOR

ENGINE COOLANT TEMPERATURE (ECT) SENSOR

THERMACTOR AIR BYPASS AND DIVERTER (TAB/TAD) SOLENOID VALVES ASSEMBLY

FEEDBACK CARBURETOR ACTUATOR (FBCA)

BAROMETRIC/MANIFOLD ABSOLUTE PRESSURE (B/MAP) SENSOR

CANISTER PURGE (CANP) SOLENOID VALVE

CRANKSHAFT POSITION (CP) SENSOR

Typical feedback carburetor and electronic control system

Electronic Fuel Mixture Control Systems

The prime purpose of this system is to permit the use of a three-way catalyst, which improves driveability while maintaining emissions levels. The heart of the system is a catalyst composed of metals which not only oxidize unburnt hydrocarbons and CO, but reduce NO——nitrogen oxide emissions.

While the ordinary catalyst simply speeds the burning or combining with oxygen of CO and hydrocarbons, this system actually uses CO, a compound which has a strong affinity for oxygen to draw the oxygen from NO——the basic nitrogen oxide found in auto exhaust. Each molecule of CO becomes CO_2, by drawing one atom of oxygen from an NO molecule——a combi-

nation of one oxygen and one nitrogen. The system uses a catalyst to make this reaction occur much faster than it would under ordinary conditions.

The exact amount of oxygen necessary to oxidize all the CO produced in the combustion chambers must be present——and no more. Less will leave CO in the tailpipe, while more will leave NO. That's why the system must use an oxygen sensor. This is located in the tailpipe and, via a stepper motor, adds fresh air to the main circuit of the carburetor, reducing or increasing the added air as necessary to maintain perfect mixture. While most of these systems use an EGR valve, less exhaust gas must be recirculated because of what happens in the tailpipe. And, because the system uses both the oxygen present in nitrogen-oxides and oxygen from air added via an air pump to finish off the combustion process in the cat-

alyst, the mixture conditions are pretty rich in the combustion chambers. The result is smooth running and top performance if everything is working right.

While rough running may make you suspect that this system is performing erratically, you should not condemn it until a thorough check of all the car's basic running systems has been made. This is because the stepper motor control of mixture has only a very small leeway——it can only adjust mixture within a very small range. Dirt in the fuel filter or a carburetor passage, improperly set ignition timing or a bad plug or wire would be more than enough to create conditions for which the system could not possibly correct. In fact, if the symptoms are severe at all, it's almost a sure indication that the electronic mixture control system can't be at fault. Severe hesitation, misfire or hard starting don't come

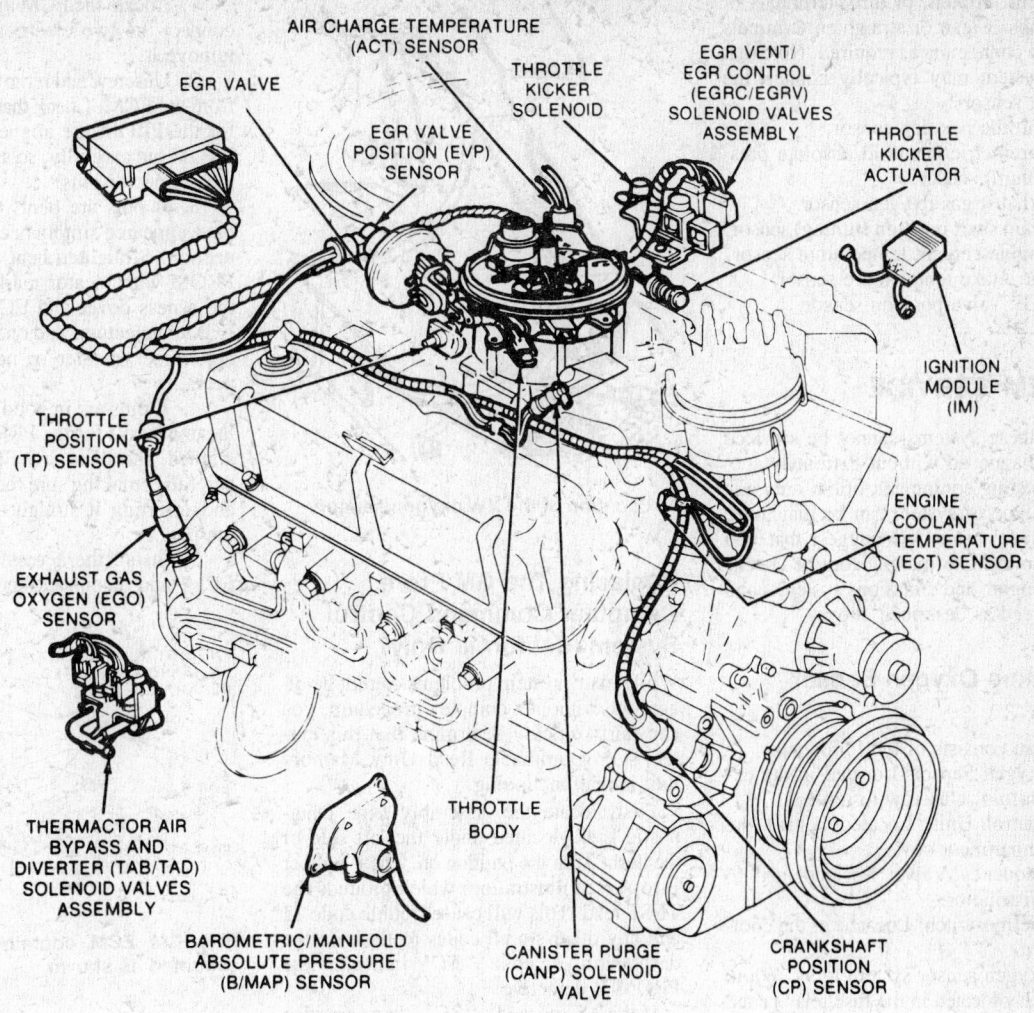

EGR VALVE

AIR CHARGE TEMPERATURE (ACT) SENSOR

EGR VALVE POSITION (EVP) SENSOR

THROTTLE KICKER SOLENOID

EGR VENT/ EGR CONTROL (EGRC/EGRV) SOLENOID VALVES ASSEMBLY

THROTTLE KICKER ACTUATOR

IGNITION MODULE (IM)

THROTTLE POSITION (TP) SENSOR

EXHAUST GAS OXYGEN (EGO) SENSOR

ENGINE COOLANT TEMPERATURE (ECT) SENSOR

THERMACTOR AIR BYPASS AND DIVERTER (TAB/TAD) SOLENOID VALVES ASSEMBLY

BAROMETRIC/MANIFOLD ABSOLUTE PRESSURE (B/MAP) SENSOR

THROTTLE BODY

CANISTER PURGE (CANP) SOLENOID VALVE

CRANKSHAFT POSITION (CP) SENSOR

EFI wiring and vacuum diagram

from mixture that's no more than 10% off optimum settings. And, under cold engine conditions or heavy throttle, the system often does not control mixture anyway.

If you're satisfied that all the basic systems are working right, what symptoms might indicate that this system is at fault?

1. Excessive CO reading at the tailpipe.

2. Excessive NO.

3. Catalyst odor (slightly rich mixture).

4. Moderately poor fuel economy (slightly rich mixture).

5. Moderate surge or hesitation (slightly lean mixture).

GM suggests that the first thing to do with their California Computer Command Control System is to very carefully go over every electrical and vacuum connection in the system. A loose connection of either kind can readily duplicate the symptoms of a bad sensor or other component. This is excellent procedure for all such emission controls.

Electrical connectors should be disconnected and inspected for improper instal-

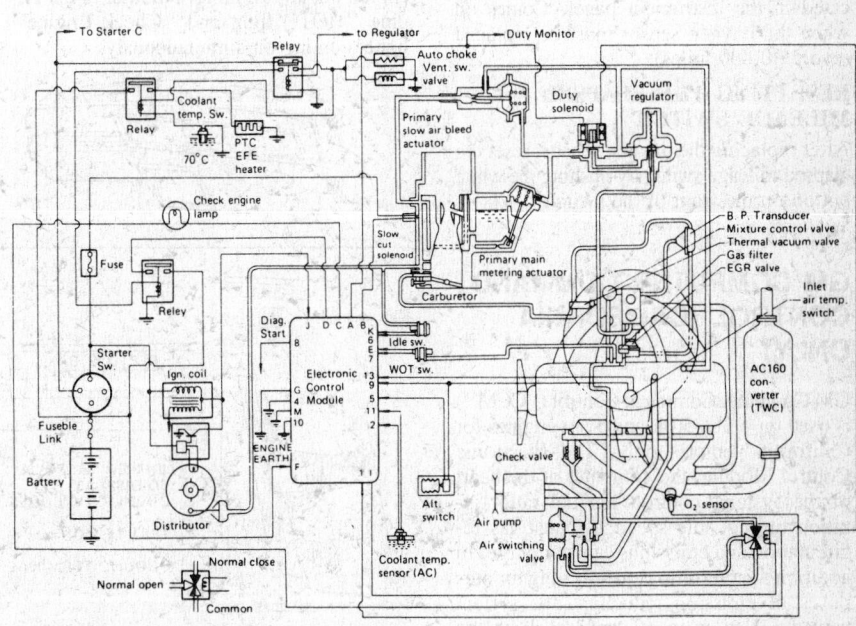

The 1983 Isuzu Closed Loop Emission Control System for California, incorporating a typical electronic mixture control system

lation, bent, broken, or dirty terminals or mating tabs. Clean or straighten terminals or replace connectors as required. Note that such a system may typically contain the following sensors:

1. Throttle position sensor.
2. Barometric/manifold absolute pressure (vacuum) sensor.
3. Exhaust gas oxygen sensor.
4. Crankshaft position (timing) sensor.
5. Engine coolant temperature sensor.
6. Air intake temperature sensor.
7. EGR valve position sensor.

SYSTEM SERVICE

Most of these systems cannot be serviced or even diagnosed without extremely specialized testing equipment; often, only one brand or type of system can be diagnosed with a given tester. We suggest that you confine yourself to making routine checks of basic engine and emissions systems, and then proceed as described above.

VW Pickup Oxygen Sensor System

The system consists of the following:

1. Oxygen Sensor: Located in the exhaust manifold. Unscrew to replace.
2. Control Unit: Located behind the glove compartment cover.
3. Frequency Valve: Located next to the fuel distributor.
4. Thermoswitch: Located in the cooling system.
5. Oxygen sensor system relay: White colored relay located in the fuse/relay panel.
6. Elapsed mileage switch: Located on the firewall.
7. Warning light: Marked OXS and located in the instrument panel. Comes on when the oxygen sensor must be replaced (every 30,000 miles).

RESETTING THE ELAPSED MILEAGE SWITCH

After replacing the oxygen sensor, reset the elapsed mileage switch by pushing the white button on the front of the switch.

GM COMPUTER COMMAND CONTROL (CALIFORNIA ONLY)

GM Computer Command Control ("CCM") is used on 4.3L, 5.0L and 5.7L engines for California vehicles only. The Electronic Control Module, the computer at the heart of the system, controls air/fuel ratio, ignition timing, idle speed, exhaust gas recirculation and emissions canister purge. In addition to controlling spark timing for purposes of emission reduction, the system controls it so as to prevent knock on the 4.3L engine. It controls transmission converter lockup as well.

Location of the VW oxygen sensor

Replacing The GM Prom, Computer Command Control System (California Only)

While many system problems cannot be diagnosed without a complex procedure, you can fairly quickly determine that this system's Programmable Read Only Memory unit is malfunctioning.

First, locate the Assembly Line Diagnostic Link, located under the left side of the dash. Turn the ignition on. Then, jumper B to A (see illustration) which grounds the TEST lead. This will cause trouble code 12 and any other stored codes to flash on the dashboard. A code "51" indicates the PROM is defective.

If there is no code "51", there are also two other indications of a defective PROM. One is a 0 degree dwell reading with the engine hot and running. Another is the engine "HOT" light and "Check Engine" lights coming on simultaneously.

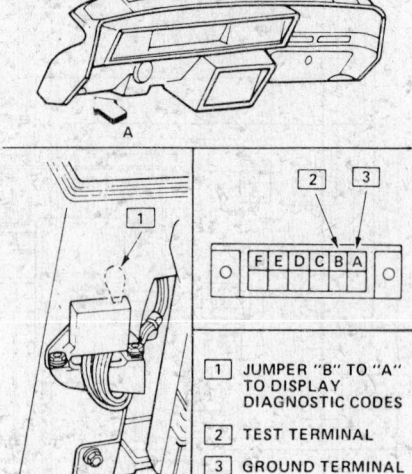

1 JUMPER "B" TO "A" TO DISPLAY DIAGNOSTIC CODES
2 TEST TERMINAL
3 GROUND TERMINAL

Grounding the "TEST" lead of the Assembly Line Diagnostic link

1. Locate the ECM under the dash, disconnect the two electrical connectors and remove it.
2. Unscrew and remove the access cover from the ECM. Check that the locator marks for the PROM are aligned. Then, pull the PROM out carefully, so as to avoid bending the connector pins.
3. If pins are bent, the Code 51 may have appeared simply because of a bad connection. Straighten bent pins, reinstall the PROM with locator marks aligned, install the access cover and ECM, reconnect the ECM connectors, and operate the engine to determine whether or not Code 51 reappears.
4. If pins are in good condition and the locator marks for the PROM were properly aligned, install a new PROM. Connect it carefully, making sure locating marks align and inserting it straight to avoid bending the pins.
5. Install the access cover, install the ECM, and reconnect the ECM connectors.

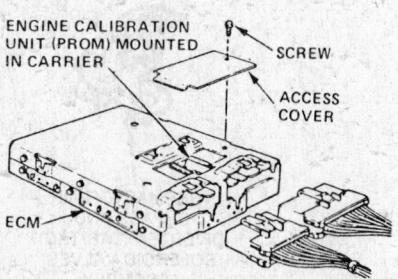

The GM ECM contains a PROM unit mounted as shown

JEEP COMPUTERIZED EMISSION CONTROL (CEC) SYSTEM

Some vehicles are equipped with the Computerized Emission Control (CEC) System which is an electronically controlled fuel feedback system that controls undesirable emissions to the atmosphere and maintains the ideal air/fuel ratio to provide an optimum balance between the emissions and engine performance. The system uses a micro computer unit (MCU), numerous signal input sensors and several output components. Based on the engine operating conditions, relayed to the MCU by input sensor signals, the MCU generates output signals to provide the proper air/fuel mixture, proper ignition timing and engine idle speed.

The system operates in either of two modes, the closed loop mode or the open loop mode. Closed loop is when the air/fuel ratio is varied, according to the oxygen content tin the exhaust gases. In the open loop mode, the air/fuel ratio is predetermined by the MCU for a number of engine operating conditions, such as engine start-up, cold engine operation, or wide open throttle (WOT) position. When the engine is started, the MCU then determines in which mode of operation , (closed or open loop), the engine should be operating. The MCU determines this by monitoring the input signals from the various input components, such as the air and coolant temperature information, engine rpm information and vacuum levels.

The MCU operates the system in the open loop mode based on a priority rating for the various predetermined engine operating conditions. It continues to operate the system and the MCU output components in the open loop mode until such time as a closed loop mode of operation is indicated. At this time, the MCU shifts the operation to the closed loop mode. Based on the oxygen content in the exhaust gases, and other inputs, it continues to operate the system in the closed loop mode, constantly varying the air/fuel ratio to maintain the optimum 14.7:1 ratio.

The engine operating conditions are constantly being monitored by the MCU and any changes that occur during the engine operation are quickly detected by the MCU, which places the system back in the appropriate open mode of operation.

Jeep CEC System Components

1. Micro computer unit (MCU)
2. Oxygen sensor
3. Thermal electric switch (TES)
4. Coolant temperature switch (CTS)
5. Four in. vacuum switch
6. Ten in. vacuum switch
7. Wide open throttle switch
8. Engine rpm (TACH) voltage
9. Stepper motor (in carburetor)
10. Dual bed catalytic converter
11. Altitude jumper wire
12. Knock sensor
13. Electronic control unit to advance or retard ignition timing
14. Idle relay
15. SOLE-VAC Throttle positioner
16. Idle solenoid
17. Upstream and downstream air switch solenoids
18. PCV valve shut-off solenoid
19. Bowl vent solenoid
20. Intake manifold heater switch.

The CEC system controls the air/fuel ratio with movable air metering pins, visible from the top of the carburetor air horn, that are driven by an MCU controlled stepper motor. The stepper motor moves the metering pins in increments or small steps via electrical impulses generated by the MCU. The MCU causes the stepper motor to drive the metering pins to a richer or leaner position in reaction to voltage input from the oxygen sensor, located in the exhaust manifold. The oxygen sensor voltage varies in reaction to changes in the exhaust gas content of oxygen. Because the content of oxygen in the exhaust gas indicates the completeness of the combustion process, it is a reliable indicator of the air/fuel mixture that is entering the combustion chamber.

Because the oxygen sensor only reacts to oxygen, any air leak or malfunction between the carburetor and the sensor may cause the sensor top provide erroneous voltage output. This could be caused by a manifold air leak or malfunctioning secondary air checks valve.

The engine operation characteristics never quite permit the MCU to compute a single metering pin position that constantly provides the optimum air/fuel mixture. Therefore, closed loop operation is characterized by constant movement of the metering pins because the MCU is forced constantly to make small corrections in the air/fuel mixture, in an attempt to create the optimum air/fuel mixture ratio of 14.7:1.

Computer Command Control (CCC OR C3) System

Late model Jeep Vehicles equipped with a V6 cylinder engine and a California emissions package, have a self-diagnostic system with a CHECK ENGINE light mounted in the instrument panel cluster.

The self-diagnostic system detects the troubles most likely to occur. The diagnostic system illuminates the CHECK ENGINE light when a trouble is detected. When a jumper wire is connected between trouble code TEST terminals 6 and 7 of the 15 terminal diagnostic connector, the CHECK ENGINE light will flash a trouble code or codes that indicate the trouble area.

For a bulb and system check, the CHECK ENGINE light will illuminate when the ignition switch is turned ON and the engine not started. If the test terminals are then grounded, the light will flash a code 12 that indicates the self- diagnostic system is operational. A code 12 consists of one (1) flash, followed by a short pause, then two (2) flashes in quick succession. After a longer pause, the code will repeat its self two (2) more times.

When the engine is started, the CHECK ENGINE light will remain on momentarily and then will go off. If the CHECK ENGINE light remains on, the self-diagnostic system has detected a trouble in the operational components. If the test terminals are then grounded, the trouble code will be flashed three (3) times. If more than one trouble has been detected, each trouble code will be flashed three (3) times. The trouble codes will flash in numeric order (lowest number code first). The trouble code series will repeat as long as the test terminals are grounded.

A trouble code indicates a trouble in a particular circuit or component. Trouble code 14, for example, indicates a trouble in the coolant sensor circuit, which includes the coolant sensor, connector, harness and the ECM. The procedure for locating trouble is in accompanying chart 14. Similar diagnostic charts are provided for each code.

The absence of a code does not mean a system is trouble free, because the self-diagnostic system does not detect all possible troubles. To determine if their is a

system problem, a System Performance Test is necessary. This test is made when the CHECK ENGINE light and the self-diagnostic system do not indicate a problem, but the system is suspected because no other reason can be found for the complaint.

Trouble Code Memory

When a problem develops in the feedback system, the CHECK ENGINE light will illuminate and a trouble code will be stored in the ECM memory. If the fault is intermittent, the CHECK ENGINE light will be turned off after ten (10) seconds when the problem ceases. However, the trouble code will be retained in the ECM memory until the battery voltage to the ECM is removed. To accomplish this, remove the negative battery cable for at least ten (10) seconds to erase all stored trouble codes.

System Performance

The system should be considered as a possible source of trouble for engine performance, fuel economy and exhaust emission complaints, ONLY AFTER normal engine diagnosis has been completed. In many cases, the feedback system has been blamed for engine performance problems, when only simple, basic engine problems have been the cause.

The system performance test verifies the system is functioning correctly and each step should be followed completely. Do not skip steps in order to short cut the tests.

Trouble Code Identification

NOTE: The trouble code will be flashed on the CHECK ENGINE light only if a trouble exist that pertains to the following codes. Any codes stored in the ECM from a problem that has ceased to exist, will be erased from the ECM memory after fifty (50) engine starts.

1. Trouble Code 12 — No distributor reference pulses to the ECM. This code is not stored in memory and will only flash while the trouble exists. This is a normal code with the ignition ON and the engine not operating.

2. Trouble Code 13 — Refers to oxygen sensor circuit. The engine must operate up to five (5) minutes at part throttle , under road load, before this code will appear.

3. Trouble Code 14 — Coolant sensor circuit has short circuit. The engine must operate up to five (5) minutes before this code will appear.

4. Trouble Code 15 — Coolant sensor circuit has open circuit. The engine must operate up to five (5) minutes before this code will appear.

5. Trouble Code 21 — Throttle positioner sensor circuit problem. The engine must operate up to twenty five (25) seconds at specified curb idle speed before this code will appear.

6. Trouble Code 23 — The Mixture

control solenoid has a short circuit to ground or an open circuit.

7. Trouble Code 34 — Vacuum sensor circuit problem. The engine must operate up to five (5) minutes at the specified curb idle speed before this code will appear.

8. Trouble Code 41 — No distributor reference pulses to the ECM at the specified engine manifold vacuum. This code will be stored in the ECM memory.

9. Trouble Code 42 — Electronic Spark Timing (EST) bypass circuit or EST circuit has short circuit to ground or an open circuit.

10. Trouble Code 44 — Lean exhaust indication. The engine must operate up to five (5) minutes, be in closed loop operation mode and at part throttle before this code will appear.

11. Trouble Code 44 and 45 at the same time — Indicates a faulty oxygen sensor.

12. Trouble Code 45 — Rich exhaust indication. The engine must operate up to five (5) minutes, be in closed loop and at part throttle before this code will appear.

13. Trouble Code 51 — Faulty calibration unit (PROM) or installed improperly. It requires up to thirty (30) seconds before this code swill appear.

14. Trouble Code 54 — Mixture Control solenoid circuit has a short circuit and/or a faulty ECM.

15. Trouble Code 55 — Voltage reference has short circuit to ground (terminal 21), a faulty oxygen sensor or ECM.

DIESEL ENGINE EMISSION CONTROLS

Crankcase Depression Regulator

Diesels use a crankcase ventilation system that provides for positive removal and re-burning of blowby gases, much as a PCV system does. The system is different, however, because of the very low vacuum that is present in the unthrottled diesel intake system. Because of the minimal vacuum, no fresh air is drawn in to scavenge the crankcase gases. The gases are drawn out of the crankcase via a line to the oil filler or a valve cover and drawn, on some diesels, through a crankcase depression regulator and into the intake manifold. The regulator maintains a constant vacuum——on GM 6.2 liter engines it's 3–4 in. of water. Under conditions of high manifold vacuum, such as high rpm operation, the regulator closes to keep crankcase vacuum from getting too high. Under ideal conditions, it opens to maintain it. The unit will also attempt to compensate for varying in-

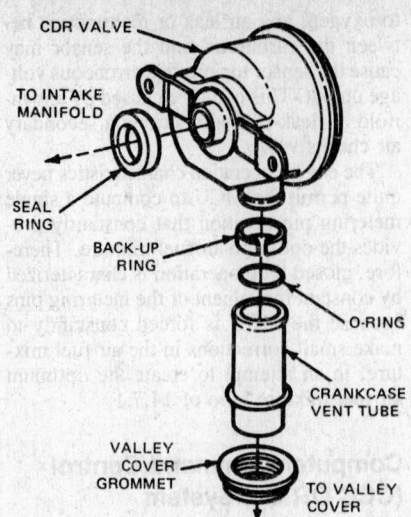

Exploded view of Ford and IH 6.9 liter CDR valve

take restriction, for example from a clogged air cleaner.

A clogged CDR valve, or a valve stuck in the closed position would produce high crankcase pressure and oil leaks or even a blown front seal. A valve stuck in the open position could draw air and dust into the crankcase, especially if there are leaks in valve cover or crankcase gaskets.

To test the system, adapt a suitable gauge to the oil filler tube in place of the cap, making certain there are no air leaks. Make sure the air cleaner is in reasonable condition and that there are no restrictions in the intake system. Idle and accelerate the engine while watching the gauge. A constant low vacuum that varies very little with engine rpm indicates the system is working.

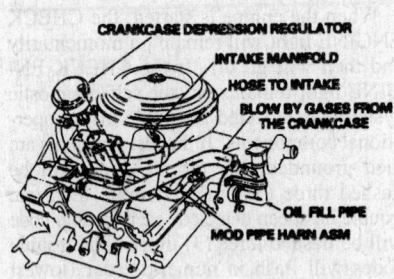

Schematic of GM 6.2 liter CDR system

Replacement is simply a matter of removing the air cleaner and unbolting the valve, and then bolting on a new one. Make sure, however, that you replace all seals and gaskets and that they are properly seated.

Mazda Intake Shutter Valve System

On Mazda diesel pickups, an intake shutter valve system is used in place of the crankcase depression regulator. The intake shut-

ter valve system may require checking of the vacuum it generates and adjustment of the diaphragm that operates the air shutter. To check, remove the plug in the intake manifold. Install an adapter with metric threads in place of the plug, and connect a vacuum gauge. Disconnect the electrical connector at the 3-way valve. Then, start the engine and run it at idle. Now, read the vacuum gauge. It should read approx. 16.92 in. of mercury.

If the reading is not correct, adjust the adjusting screw on the shutter valve until it reaches the specified value. Then, stop the engine, reconnect the electrical connector, and replace the plug in the intake manifold.

Mazda 3-Way Solenoid Valve

1. Disconnect the vacuum line leading to the intake shutter valve diaphragm. Plug a vacuum gauge into the open end of the vacuum tube. Disconnect the electrical connector at the 3-way valve.

2. Run the engine at idle and check for vacuum. If vacuum is present, the 3-way valve is ok. If not, replace it.

3. Remove the vacuum gauge and reconnect the vacuum line and electrical connector.

EGR Systems

GM 6.2L DIESEL

This diesel engine uses an EGR valve located under the air cleaner cover to recirculate exhaust gas and reduce peak combustion temperatures. This reduces NOx much as in a gas engine.

There are several differences in the way the EGR valve is energized, however. And, because there is little intake system vacuum in a diesel, a special EPR valve increases exhaust backpressure to aid EGR flow at idle. This valve, too requires a unique actuation system.

One unique feature of the system is that a throttle position switch on the injection pump actuates the EGR valve and EPR valve at different throttle positions.

IDLE OPERATION

Both valves are energized by vacuum produced via the engine's vacuum pump. Vacuum to each valve is controlled via a solenoid——so there's an EGR solenoid valve and an EPR solenoid valve.

At idle, both are open passing vacuum through and opening both the EGR and EPR valves. Just off idle, the EPR solenoid is de-energized, and the EPR valve closes. EGR is maintained under these conditions because of an increase in intake vacuum and exhaust backpressure.

At high throttle openings, diesels do not require EGR as the fuel mixture becomes richer, leaving less excess oxygen for NOx formation. Under these conditions, at a fairly advanced throttle opening, the Throttle Position Switch sends current to the EGR solenoid closing the valve, vacuum to the EGR valve is cut off, and the EGR valve closes. This minimizes diesel smoke at full or near-full throttle conditions.

TROUBLESHOOTING THE SYSTEM

If the engine gives heavy exhaust smoke under full throttle conditions only, suspect the EGR system, and proceed as follows:

1. Operate the engine until the thermostat opens. Then, remove the air cleaner cover and watch the operation of the EGR valve with the engine idling. The valve should be open. This is indicated if the valve head is in the up position and there is noticeable exhaust noise in the intake. If the valve is not open, check for loose electrical or vacuum connections. Confirm that the EGR valve is actually open by pulling off the vacuum hose to the top of the valve. The valve stem should drop and there should be a noticeable reduction in noise. Reconnect the hose. If the valve is working, skip to Step 4.

2. Check the vacuum to the EGR valve with a vacuum gauge. It should be 20 in. of water. If not, check for this vacuum at the outlet of the vacuum pump, and replace the pump or replace repair vacuum lines as necessary.

3. If the EGR valve is actually receiving vacuum but does not open and close as the hose is connected and disconnected, replace the EGR valve.

4. Operate the throttle lever through 20 degrees of travel. The EGR valve should close when the throttle position switch reaches the calibration point. If not, proceed with the following steps.

5. Check the pink wire leading to the Throttle Position Switch. If there is no voltage at the connector, check for loose connections, damaged wiring, or a blown 20 amp gauge circuit fuse.

6. Once voltage is supplied to the TPS, check that the blue wire from the TPS (leading to the EPR solenoid) also has voltage. If the contact at the TPS has no voltage at idle with the pink wire hot, replace the TPS.

7. With the engine off, key on operate the throttle through 20 degrees of travel. Measure voltage at the blue wire. It should cut out at approximately 15 degrees of throttle opening. Then measure voltage at the yellow wire. It should cut in at 20 degrees of throttle opening.

8. Check that all electrical connections are good and correctly wired, and that all vacuum connections are good and properly routed. If wiring and vacuum lines are good, and operation of the throttle through 20 degrees does not first close the EPR valve and then close the EGR valve, replace the appropriate solenoid.

GM DIESEL ELECTRONIC CONTROL SYSTEM (CALIFORNIA ONLY)

1985 and later California trucks equipped with the 6.2L diesel use an electronically controlled EGR system and transmission converter clutch. The Diesel Electronci Control (DEC) system uses the following components:

1. Diesel Electronic Control Module (ECM)
2. Absolute Pressure (MAP) sensor
3. EGR Solenoid
4. EGR vent solenoid
5. EGR valve
6. EPR solenoid
7. EPR valve
8. Throttle position sensor
9. Engine speed sensor
10. ALCL connector
11. TCC solenoid
12. Vehicle speed sensor

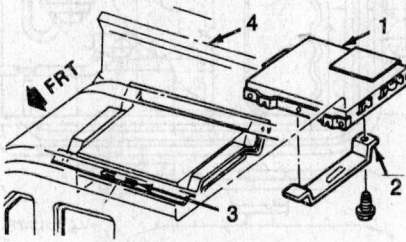

1. Electronic control module
2. Retainer
3. ECM mounting housing
4. Plenum panel

Location of the electronic control module on conventional cab Chev./GMC trucks

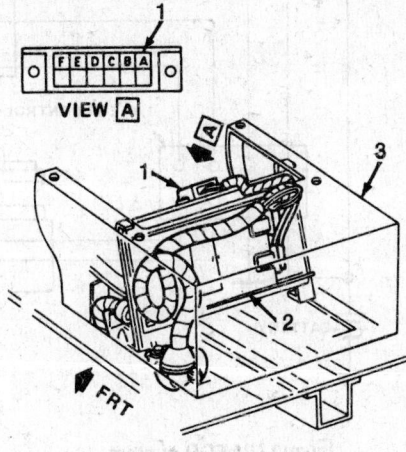

1. ALCL connector used with diesel diagnostic tool J34750 or equivalent
2. ECM
3. Seat riser

Location of the electronic control connector on Chev./GMC vans

The Diesel Electronic Control module is a computer that controls:

1. EGR system
2. Exhaust Pressure Regulation (which is related to EGR control)
3. EGR system diagnosis
4. Engagement of the Transmission Converter Clutch

The ECM monitors: engine rpm through the engine speed sensor; EGR vacuum through the MAP sensor; throttle position through the throttle sensor; and vehicle speed through the vehicle speed sensor to control the EGR and TCC.

A special Diesel Diagnostic Check tool is required to check out this system—— there is no automatic flashing of trouble codes when a problem occurs. However, the first step in any situation where trouble begins is to make a careful check for bad connections at the many connecting points for vacuum hoses and electrical wires under the hood. Proceed as follows:

1. Inspect all vacuum hoses for incorrect routing, pinches, disconnected connections, and make repairs as necessary. Be sure to track hoses under the air cleaner and all accessories to ensure proper vacuum flow.

2. Inspect all wires for incorrect or loose connections, burning or chafing, pinching or contact with sharp edges or hot parts such as exhaust manifolds. Make repairs here, as well.

ISUZU/LUV DIESEL EGR (CALIFORNIA ONLY)

This system incorporates an ordinary EGR valve mounted on the intake manifold. The valve is vacuum actuated, the vacuum being produced by the engine's vacuum pump and sent to the valve through a vacuum switching valve. The solenoid operated vacuum switching valve interrupts the vacuum to the EGR valve under conditions determined by an EGR controller. This device, in turn, receives signals indicating engine rpm and throttle lever position. At light throttle, the EGR valve receives vacuum between 1600 and 2200 rpm. Under heavy throttle, it receives vacuum only above 2200 rpm.

A thermal vacuum switch measures coolant temperature and is connected in series in the vacuum line between the vacuum pump and the EGR valve. It prevents EGR from occurring at coolant temperatures below 122 degrees F.

TROUBLESHOOTING THE SYSTEM

1. Apply about 14 in. HG vacuum to the EGR valve with a vacuum pump. The valve should open all the way and should not leak down. Otherwise, replace it.

2. Remove the Thermal Vacuum Valve from the cooling system (drain coolant first, as necessary). Try to blow through the valve (make sure it has cooled if removed from a hot engine). No air should pass. Then, immerse the wet portion of the valve in water and heat it to about 130 degrees F. Now, air should pass through the valve. If either test is failed, replace the valve.

3. Disconnect the connectors to the Vacuum Switching Valve, and run jumper wires so battery voltage can be applied directly. If you can hear the plunger operating when current is applied, the valve is good. Otherwise, replace it.

4. Connect a voltmeter positive lead to the light green/black terminal of the EGR controller, and the negative lead to black/yellow terminal. Run the engine at over 1600 rpm. There should be just under 12 volts across the leads. Otherwise, replace the unit.

5. Connect a voltmeter positive lead to the blue/yellow terminal of the control lever position sensor, and connect the negative lead to the blue/red terminal. Do not disconnect the connector. Move the control lever until there is a clearance of .275 in: between the lever and the idle adjusting screw. The voltage reading should be 3.3–4.4 volts. Otherwise, replace the sensor.

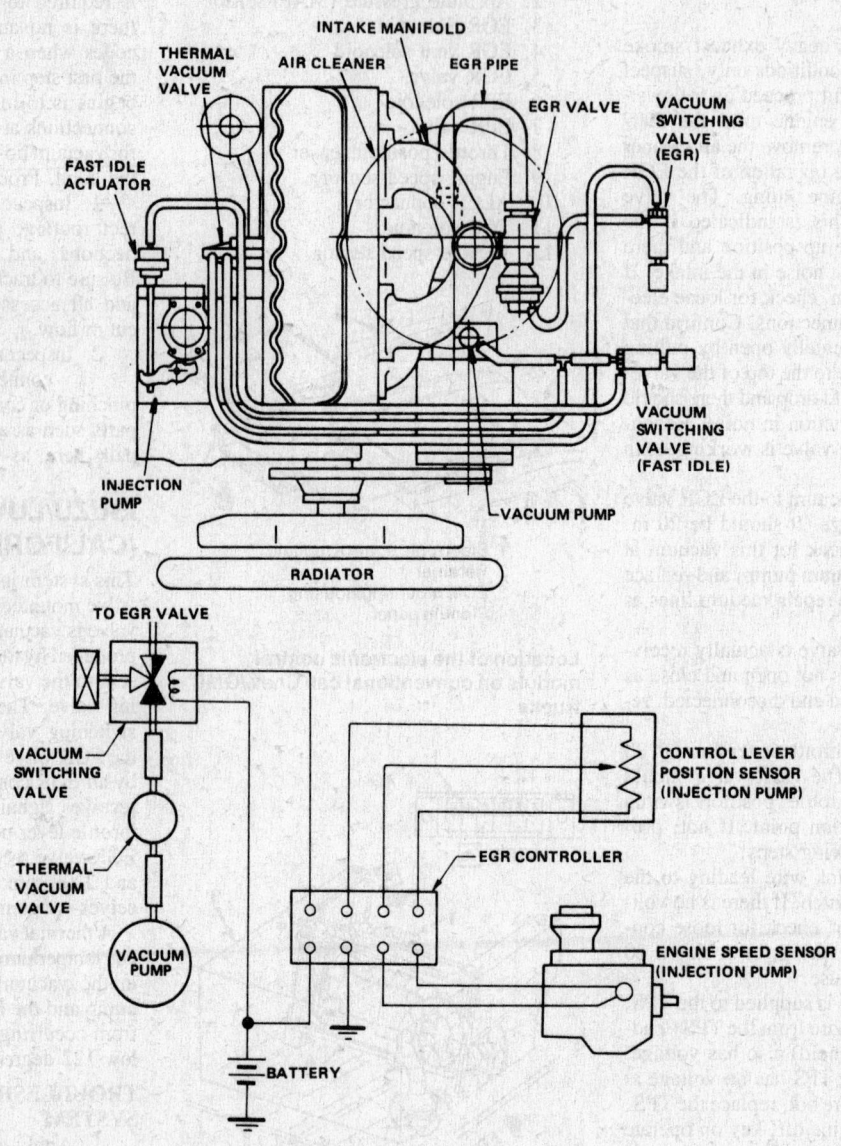

Isuzu/LUV EGR system

Carburetors

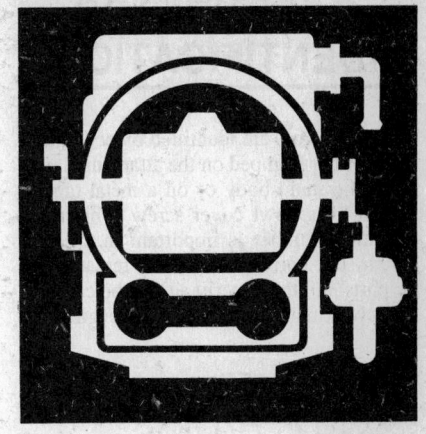

INDEX

CARBURETOR IDENTIFICATION

All carburetors are identified by code numbers, either stamped on the attaching flange side, the main body or on a metal tag retained by a bowl cover screw. This identification number is important in order to obtain the correct carburetor replacement or parts and to properly adjust the carburetor when matched to a specific engine.

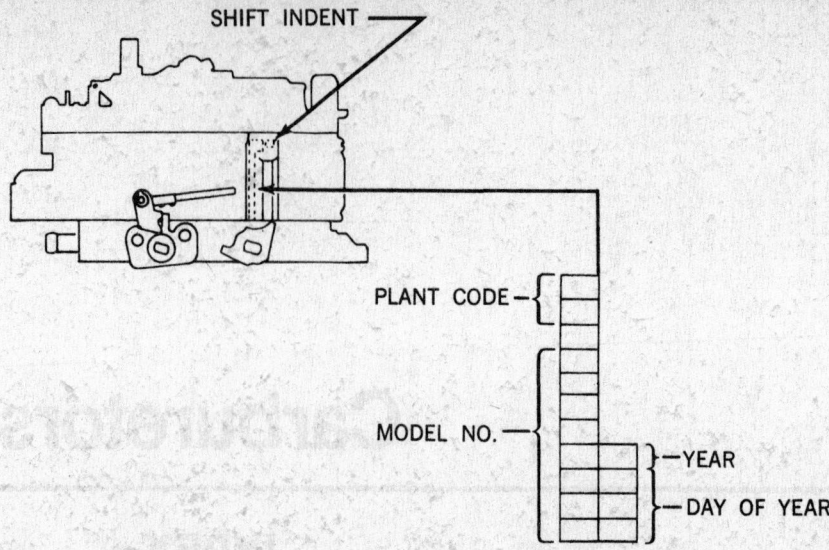

Rochester four barrel models—typical (© General Motors Corp.)

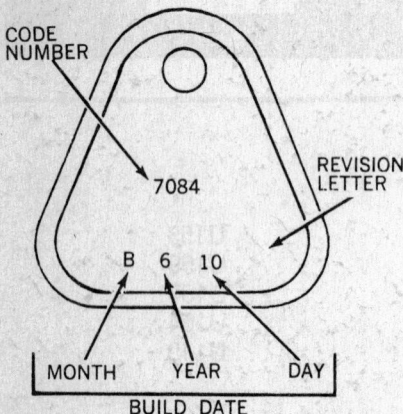

Carter carburetors for Jeep usage—typical (© Jeep Corp.)

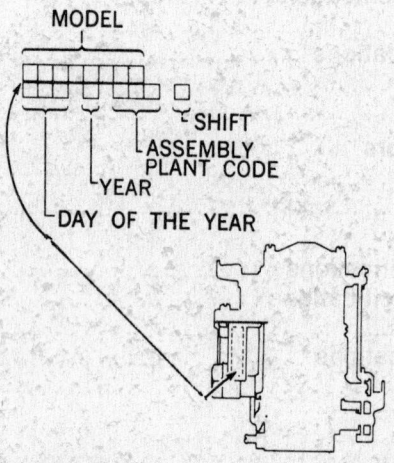

Rochester one barrel models—typical (© General Motors Corp.)

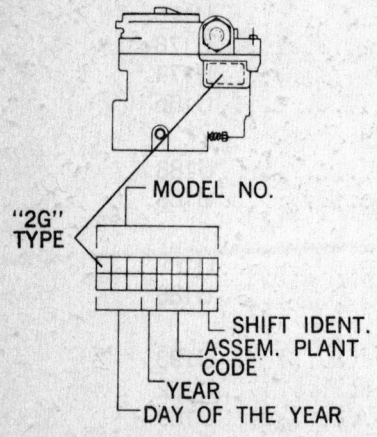

Rochester two barrel models—typical (© General Motors Corp.)

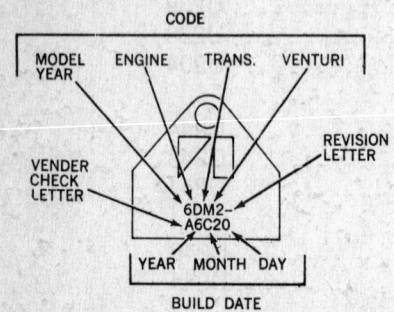

Motorcraft carburetors for Jeep usage—typical (© Jeep Corp.)

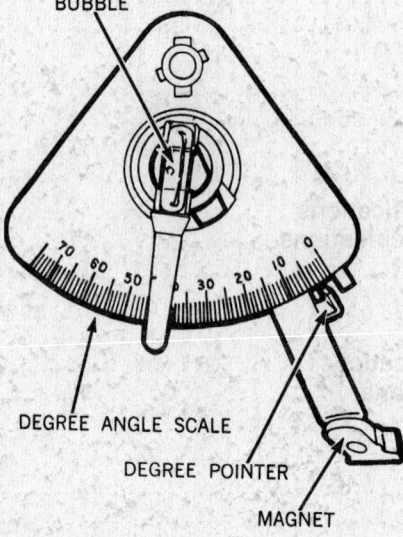

Degree angle tool—typical (© Kent-Moore Tools)

Special Tools

An angle degree tool is recommended by Rochester Products Division for use to confirm adjustments to the choke valve and related linkages on late model two and four barrel carburetors in place of the plug type gauges. Decimal and degree conversion charts are provided for use with the angle degree tool. To use the angle gauge, rotate the degree scale until zero (0) is opposite the pointer. With the choke valve completely closed, place the gauge magnet squarely on top of the choke valve and rotate the bubble until it is centered. Make the necessary adjustments to have the choke valve at the specified degree angle opening as read from the degree angle tool. The carburetor may be off the engine for adjustments, but make sure the carburetor is held firmly during the use of the angle gauge.

A variety of other special adjustment tools may be necessary during the overhaul of different carburetors covered in this section. When required, the tools are illustrated and tool numbers given for reference. Most carburetor overhaul kits contain the float level gauges and specifications nec-

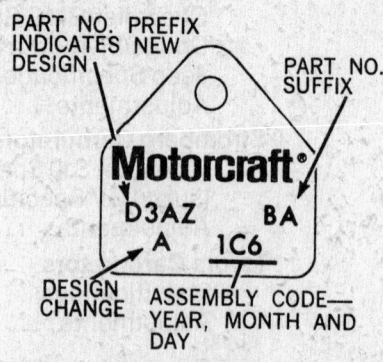

Motorcraft carburetors for Ford usage—typical (© Ford Motor Co.)

essary for complete rebuilding, and if specifications differ from those given in the following charts, use the values listed in the overhaul instructions with a specific kit. Before beginning any overhaul procedures, read through each section to make sure all required special tools are on hand in order to complete the repair.

OVERHAUL TIPS

When the carburetor is disassembled, wash all parts (except diaphragms, electric choke units, pump plunger, and any other plastic, leather, fiber, or rubber parts) in clean carburetor solvent. Do not leave parts in the solvent any longer than is necessary to sufficiently loosen the deposits. Excessive cleaning may remove the special finish from the float bowl and choke valve bodies, leaving these parts unfit for service. Rinse all parts in clean solvent and blow them dry with compressed air or allow them to air dry. Wipe clean all cork, plastic, leather, and fiber parts with a clean, lint-free cloth.

Blow out all passages and jets with compressed air and be sure that there are no restrictions or blockages. Never use wire or similar tools to clean jets, fuel passages, or air bleeds. Clean all jets and valves separately to avoid accidental interchange. Check all parts for wear or damage. If wear or damage is found, replace the defective parts. Especially check the following:

1. Check the float needle and seat for wear. If wear is found, replace the complete assembly.

2. Check the float hinge pin for wear and the float(s) for dents or distortion. Replace the float if fuel has leaked into it.

3. Check the throttle and choke shaft bores for wear or an out-of-round condition. Damage or wear to the throttle arm, shaft, or shaft bore will often require replacement of the throttle body. These parts require a close tolerance of fit. Wear may allow air leakage, which could affect starting and idling.

NOTE: Throttle shafts and bushings are not included in overhaul kits. They can be purchased separately.

4. Inspect the idle mixture adjusting needles for burrs or grooves. Any such condition requires replacement of the needle, since you will not be able to obtain a satisfactory idle.

5. Test the accelerator pump check valves. They should pass air one way but not the other. Test for proper seating by blowing and sucking on the valve. Replace the valve if necessary. If the valve is satisfactory, wash the valve again to remove breath moisture.

6. Check the bowl cover for warped surfaces with a straight edge.

7. Closely inspect the valves and seats for wear and damage, replacing as necessary.

8. After the carburetor is assembled, check the choke valve for freedom of operation.

Carburetor overhaul kits are recommended for each overhaul. These kits contain all gaskets and new parts to replace those that deteriorate most rapidly. Failure to replace all parts supplied with the kit (especially gaskets) can result in poor performance later.

After cleaning and checking all components, reassemble the carburetor, using new parts and referring to the exploded view. When reassembling, make sure that all screws and jets are tight in their seats, but do not overtighten as the tips will be distorted. Tighten all screws gradually, in rotation. Do not tighten needle valves into their seats. Uneven jetting will result. Always use new gaskets. Be sure to adjust the float level, following the instructions contained in the rebuilding kit, when reassembling.

CARTER CARBURETORS

IDLE SPEED ADJUSTMENT

Model BBD Carburetor

NOTE: The carburetor choke and intake manifold heater must be off. This occurs when the engine coolant heats to approximately 160°F.

1. Have the engine at normal operating temperature. Connect a tachometer to the ignition coil negative (TACH) terminal.

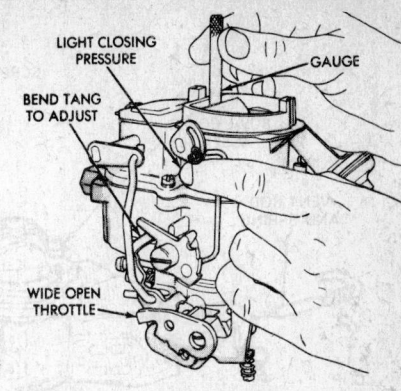

Adjusting choke unloader — BBD carburetor

2. Remove the vacuum hose to the SOL-VAC vacuum actuator unit. Plug the vacuum hose. Disconnect the holding solenoid wire connector.

3. Adjust the curb (slow) idle speed screw to obtain the correct curb idle speed. Refer to the specifications under Idle Speed or refer to the Emission Information label, under the hood, for the correct curb idle engine rpm.

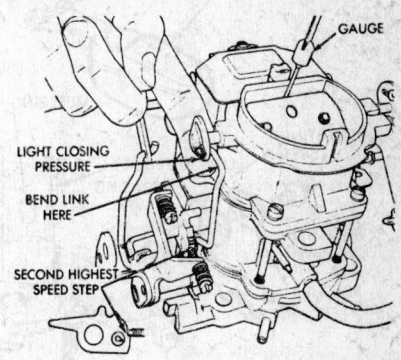

Fast idle cam position adjustment—BBD carburetor

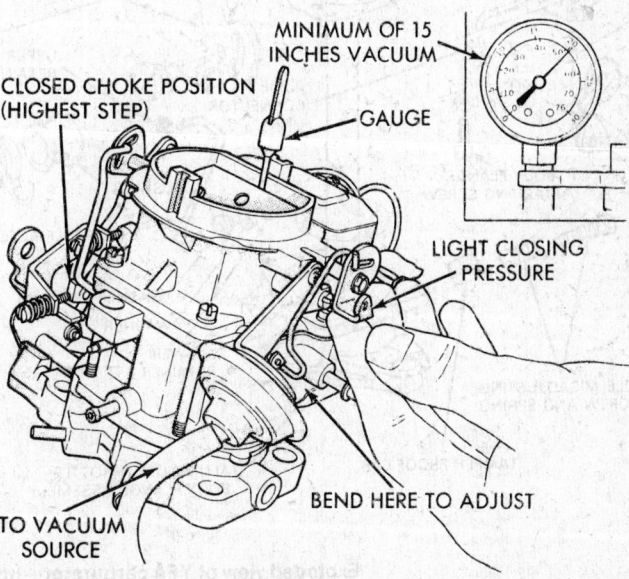

Adjustment of initial opening (vacuum kick)—BBD carburetor

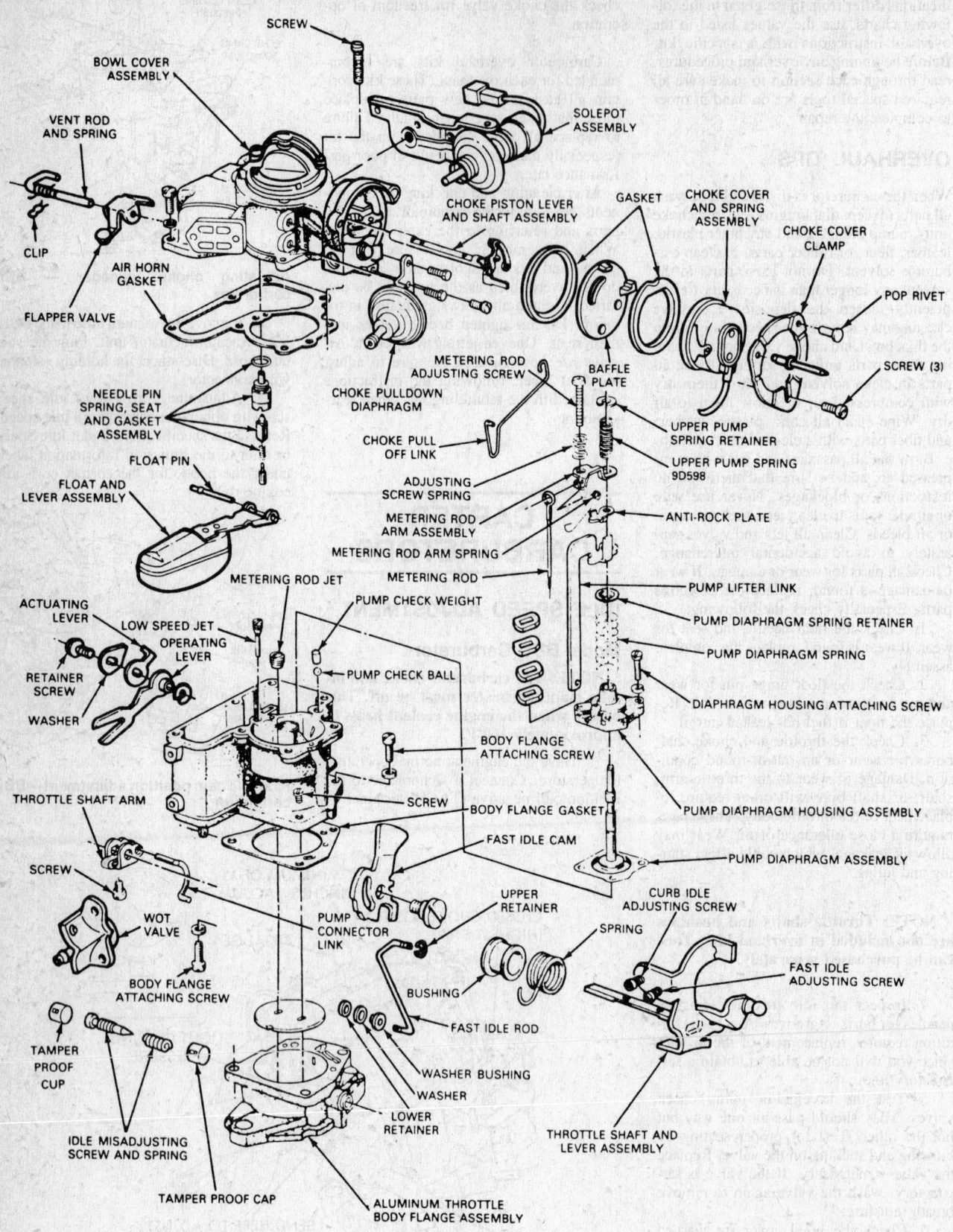

Exploded view of YFA carburetor—non-feedback type

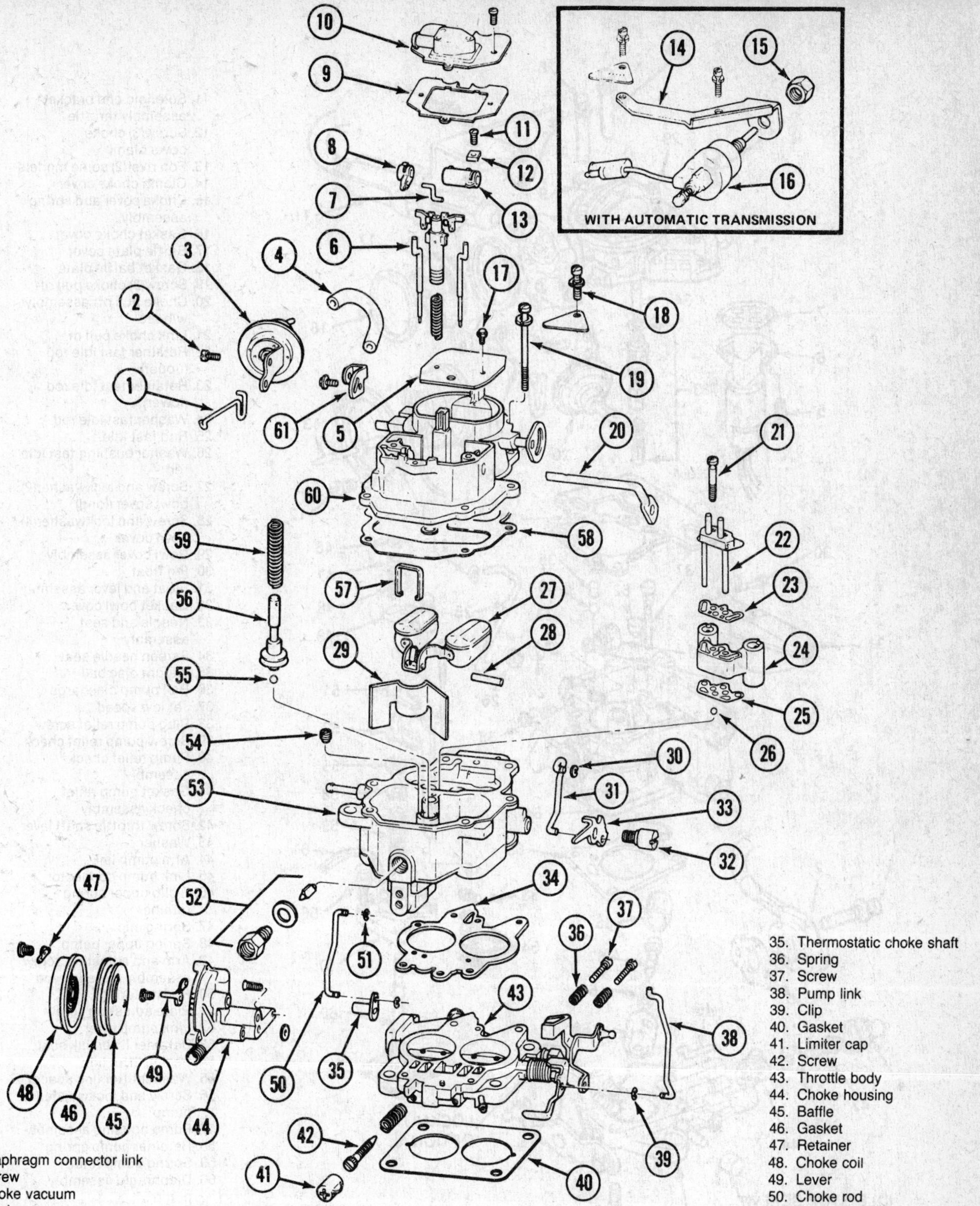

WITH AUTOMATIC TRANSMISSION

35. Thermostatic choke shaft
36. Spring
37. Screw
38. Pump link
39. Clip
40. Gasket
41. Limiter cap
42. Screw
43. Throttle body
44. Choke housing
45. Baffle
46. Gasket
47. Retainer
48. Choke coil
49. Lever
50. Choke rod
51. Clip
52. Needle and seat assembly
53. Main body
54. Main metering jet
55. Check ball (large)
56. Accelerator pump plunger
57. Fulcrum pin retainer
58. Gasket
59. Spring
60. Air horn
61. Lever

1. Diaphragm connector link
2. Screw
3. Choke vacuum diaphragm
4. Hose
5. Valve
6. Metering rod
7. S-Link
8. Pump arm
9. Gasket
10. Rollover check valve
11. Screw
12. Lock
13. Rod lifter
14. Bracket

15. Nut
16. Solenoid
17. Screw
18. Air horn retaining screw (short)
19. Air horn retaining screw (long)
20. Pump lever
21. Venturi cluster screw
22. Idle fuel pick-up tube
23. Gasket

24. Venturi cluster
25. Gasket
26. Check ball (small)
27. Float
28. Fulcrum pin
29. Baffle
30. Clip
31. Choke link
32. Screw
33. Fast idle cam
34. Gasket

Carter BBD two barrel carburetor—typical

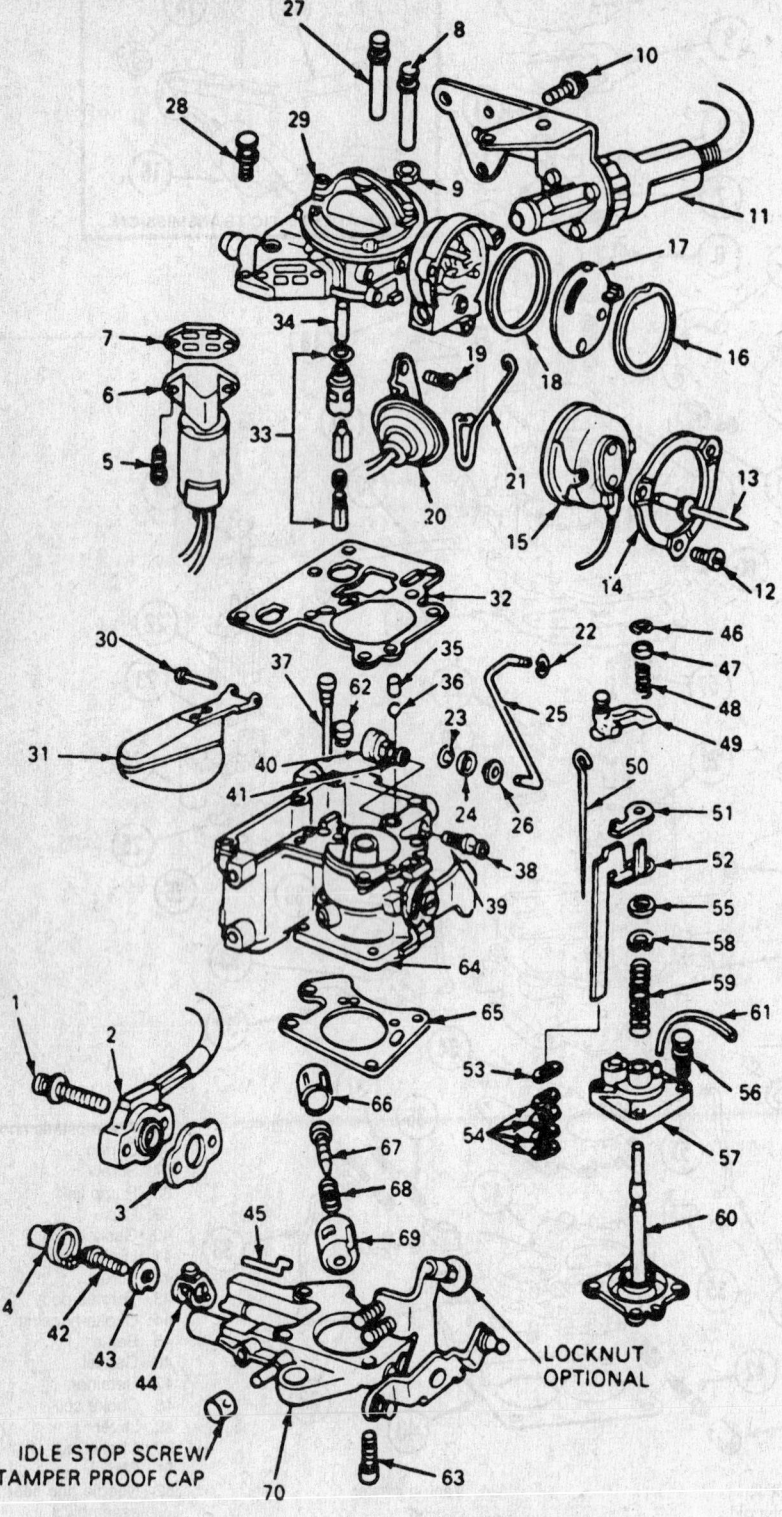

11. Solenoid and bracket assembly throttle
12. Screw(3) choke cover clamp
13. Pop rivet(2) some models
14. Clamp choke cover
15. Choke cover and spring assembly
16. Gasket choke cover
17. Baffle plate cover
18. Gasket baffle plate
19. Screw(2) choke pull off
20. Choke pull off assembly w/hose
21. Link choke pull off
22. Retainer fast idle rod (upper)
23. Retainer fast idle rod (lower)
24. Washer fast idle rod
25. Rod fast idle
26. Washer bushing fast idle rod
27. Screw and lockwasher(2) bowl cover (long)
28. Screw and lockwasher(4) bowl cover
29. Bowl cover assembly
30. Pin float
31. Float and lever assembly
32. Gasket bowl cover
33. Needle and seat assembly
34. Screen needle seat
35. Weight disc ball
36. Ball pump discharge
37. Jet low speed
38. Plug pump relief screw
39. Screw pump relief check
40. Pump relief check assembly
41. Gasket pump relief check assembly
42. Screw throttle shaft lever
43. Washer
44. Arm pump link
45. Link pump connector
46. E-clip upper spring retainer
47. Spring clip
48. Spring upper pump
49. Arm and adjusting screw assembly metering rod
50. Rod metering
51. Plate adjusting screw
52. Link pump lifter
53. Retainer lifter link seal
54. Seal(4) lifter link
55. Washer lifter link spacer
56. Screw and lockwasher(4) pump
57. Pump housing assembly
58. Retainer pump spring
59. Spring pump return
60. Diaphragm assembly pump
61. Tube pump passage
62. Jet main
63. Screw(4) throttle body
64. Bowl assembly
65. Gasket throttle body
66. Cap idle needle
67. Needle idle adjusting
68. Spring idle adjusting needle
69. Clip idle needle
70. Throttle body assembly

IDLE STOP SCREW/ TAMPER PROOF CAP

LOCKNUT OPTIONAL

1. Screw and lockwasher(2) throttle sensor
2. Throttle sensor assembly
3. Plate sensor
4. Drive coupler sensor
5. Screw(2) feedback solenoid
6. Feedback solenoid assembly
7. Gasket feedback solenoid
8. Screw and lockwasher solenoid bracket
9. Locknut bracket screw
10. Screw(3) bracket

Exploded view of YFA carburetor—feedback type

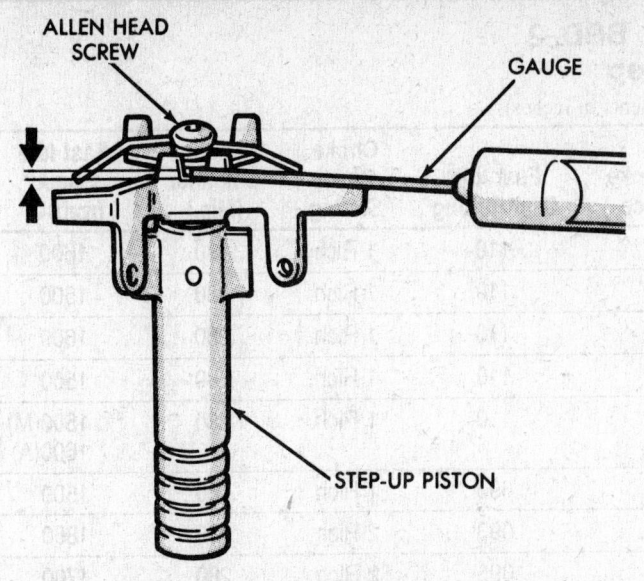

Step-up piston clearance adjustment—BBD carburetor

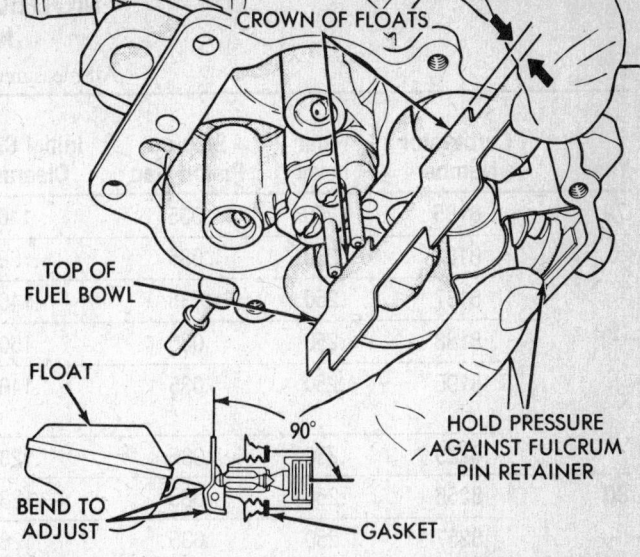

Adjusting float level with bowl inverted—BBD carburetor

MODEL BBD
Dodge/Plymouth
(All measurements in inches)

Year	Carburetor Number	Float Level	Choke Unloader	Fast Idle Cam Setting	Choke Valve Initial Opening w/Vacuum Kick	Fast Idle Speed (rpm)	Accelerator Pump Setting	Step-up Piston Gap
'79	8214S	.250	.280	.070	.110	1400	.500	—
	8215S	.250	.280	.070	.110	1600	.500	—
	8249S	.250	.280	.070	.110	1400	.500	—
	8232S	.250	.280	.070	.110	1500	.500	—
	8210S	.250	.280	.070	.110	1400	.500	—
	8211S	.250	.280	.070	.110	1500	.500	—
'80–'82	8146S	.250	.310	.070	.070	1500	.500	.035
	8147S	.250	.310	.110	.150	1500	.500	.035
'82	8348S	.250	.310	.070	.130	1600	.500	.035
	8352S	.250	.310	.070	.130	1600	.500	.035
'83	8146S	.250	.310	.070	.070	1500	.470	.035
	8147S	.250	.310	.110	.150	1500	.470	.035
	8371S	.250	.280	.070	.130	1600	.470	.035
	8374S	.250	.280	.070	.130	1400	.470	.035
	8359S	.250	.280	.070	.130	1400	.470	.035
	8358S	.250	.280	.070	.130	1400	.470	.035
'84	8387S	.250	.310	.110	.150	1700	.470	.035
	8386S	.250	.310	.070	.070	1500	.470	.035
	8374S	.250	.280	.070	.130	1400	.470	.035
	8359S	.250	.280	.070	.130	1400	.470	.035
	8358S	.250	.280	.070	.130	1400	.470	.035

Note: Choke is fixed on all models

MODEL BBD–2
Jeep
(All measurements in inches)

Year	Carburetor Number	Float Level	Step-up Piston Gap	Initial Choke Clearance	Fast Idle Cam Setting	Choke Cover Setting	Choke Unloader (Min.)	Fast Idle Speed (rpm)①
'79	8185	.250	.035	.140	.110	1 Rich	.280	1600
	8186	.250	.035	.150	.110	1 Rich	.280	1500
	8187	.250	.035	.140	.110	1 Rich	.280	1600
	8188	.250	.035	.150	.110	1 Rich	.280	1500
	8195	.250	.035	.140	.110	1 Rich	.280	1500(M) 1600(A)
	8229	.250	.035	.128	.095	1 Rich	.280	1500
'80	8256	.250	.035	.128	.093	2 Rich	.280	1850
	8257	.250	.035.	.128	.095	2 Rich	.280	1700
	8253	.250	.035	.128	.095	2 Rich	.280	1850
	8254	.250	.035	.120	.086	2 Rich	.280	1700
	8255	.250	.035	.140	.093	2 Rich	.280	②
	8277	.250	.035	.116	.081	1 Rich	.280	1700
'81	8302	.250	.035	.140	.095	1 Rich	.280	1850
	8303	.250	.035	.140	.095	1 Rich	.280	1700
	8311	.250	.035	.120	.085	1 Rich	.280	1700
	8306	.250	.035	.140	.095	1 Rich	.280	1700
	8312	.250	.035	.140	.095	1 Rich	.280	②
	8307	.250	.035	.140	.095	1 Rich	.280	1700
'82–'83	8338	.250	.035	.140	.095	1 Rich	.280	1850
	8339	.250	.035	.140	.095	1 Rich	.280	1700
	8340	.250	.035	.150	.110	1 Rich	.280	1700
	8341	.250	.035	.150	.150	1 Rich	.280	1700
	8349	.250	.035	.128	.095	2 Rich	.280	②
	8351	.250	.035	.130	.095	Index	.280	1700
'84–'85	8383	.250	.035	.140	.095	1 Rich	.280	1850
	8384	.250	.035	.140	.095	1 Rich	.280	1700

①On second step of fast idle cam with TCS solenoid and EGR disconnected.
②Manual transmission 1700 rpm and automatic transmission 1850 rpm

4. Apply a direct source of vacuum to the vacuum actuator, using a hand vacuum pump or its equivalent. When the SOL-VAC throttle positioner is fully extended, turn the vacuum actuator adjustment screw on the throttler lever until the specified engine rpm is obtained. Disconnect the vacuum source from the vacuum actuator.

5. With a jumper wire, apply battery voltage (12 volts) to energize the holding solenoid.

NOTE: The holding wire connector can be installed and either the rear window defroster or the air conditioner (with the compressor clutch wire disconnected) can be turned on to energize the holding solenoid.

6. Hold the throttle open manually to allow the throttle positioner to fully extend.

NOTE: Without the vacuum actuator, the throttle must be opened manually to allow the SOL-VAC throttle positioner to fully extend.

7. If the holding solenoid idle speed is not within specifications, adjust the idle using the ¼ in. hex-headed adjustment screw on the end of the SOL-VAC unit. Adjust to specifications.

8. Disconnect the jumper wire from the SOL-VAC holding solenoid wire connector, if used. Connect the wire connector to the SOL-VAC unit, if not connected. Install the original vacuum hose to the vacuum actuator.

9. Remove the tachometer and if disconnected, connect the compressor clutch wire. Install any other component that was previously removed.

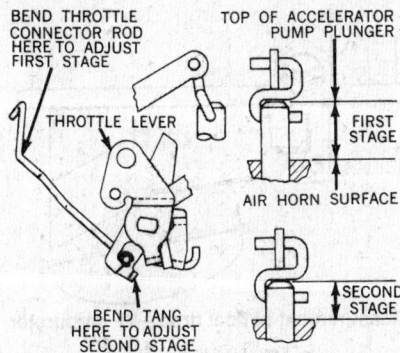

Adjustment of the primary and secondary accelerator pump— Thermo-Quad® carburetor

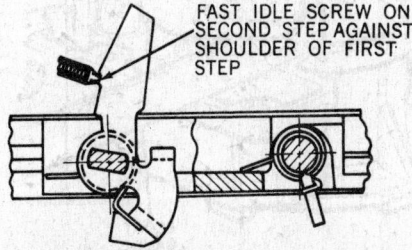

Carter Thermo-Quad® fast idle speed adjustment cam position

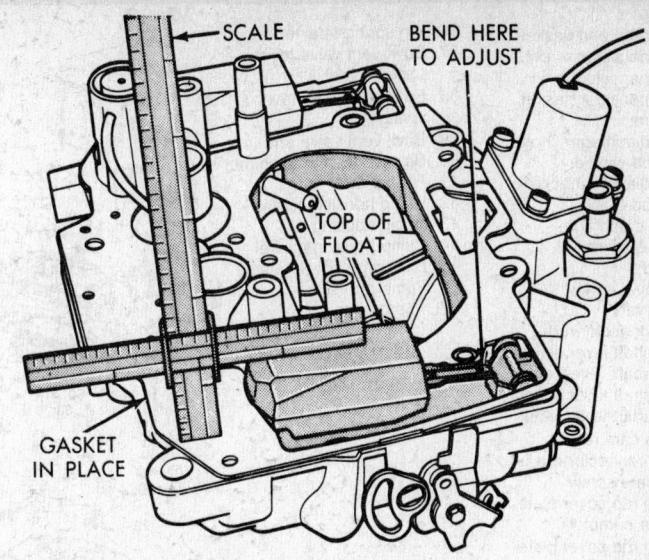

Carter TQ float height measurement

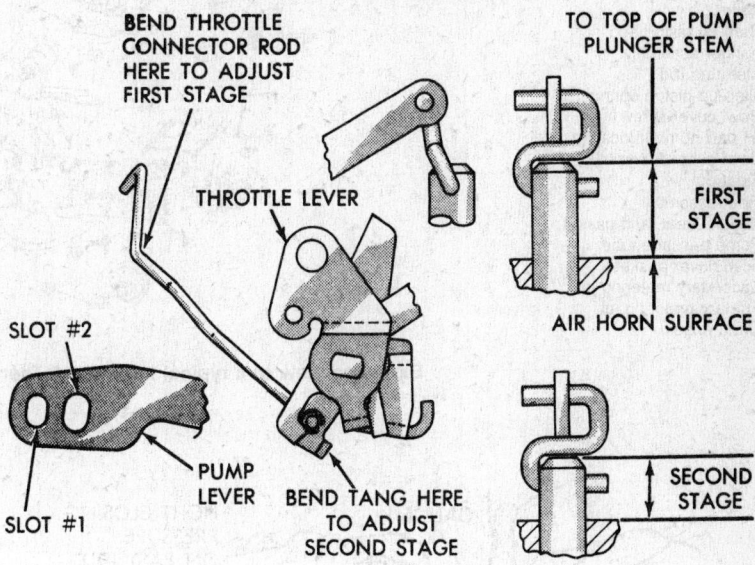

Accelerator pump stroke adjustment—Thermo-Quad® carburetor

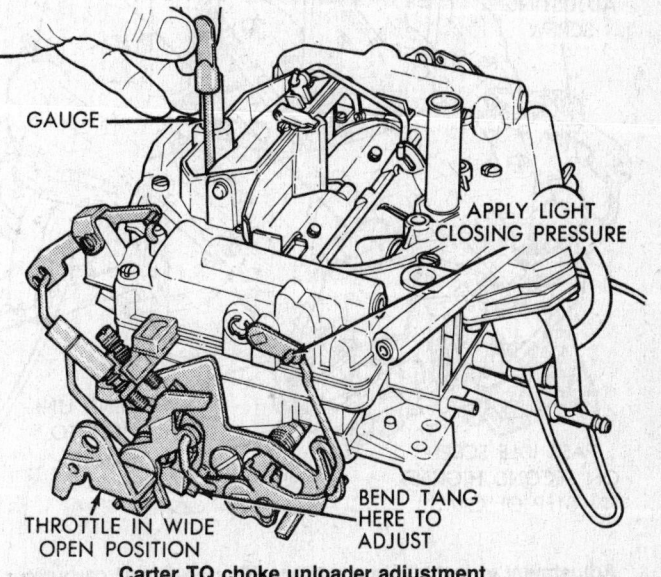

Carter TQ choke unloader adjustment

1. Fuel inlet nut and gasket
2. Idle compensator screw
3. Idle compensator
4. Idle compensator gasket
5. "E" retainer
6. Primary diaphragm choke pull-off rod washer
7. Primary diaphragm choke pull-off rod
8. Auxiliary diaphragm choke pull-off rod (if equipped)
9. Choke lever screw
10. Choke lever
11. Choke connector rod
12. Countershaft lever screw
13. Countershaft, lever, outer
14. Countershaft lever spring
15. Countershaft lever, inner
16. Fast idle cam rod
17. Throttle connector rod
18. Cover plate screw
19. Metering rod cover plate (opposite pump)
20. Metering rod cover plate (pump side)
21. Step-up piston cover plate
22. Step-up piston and hanger assembly
23. Metering rod
24. Step-up piston spring
25. Bowl cover screw
26. IH part number location
27. Bowl cover assembly
28. Float pin
29. Float assembly
30. Needle, seat, and gasket
31. Pump passage tube
32. Bowl cover gasket
33. Secondary metering jet
34. Primary metering jet
35. Quad rings

36. Pin spring retainer
37. Bowl vent valve lever, upper
38. Bowl vent valve lever spring
39. Bowl vent valve arm
40. Bowl vent valve grommet
41. Rivet plug
42. Pump housing screw
43. Pump housing
44. Pump housing gasket
45. Discharge check needle
46. Pump arm screw
47. Pump arm

48. Pump "S" link
49. Air valve lock plug
50. Air valve adjustment plug
51. Air valve spring
52. Pump intake check assembly
53. Plunger assembly
54. Plunger spring
55. Main body
56. Main body gasket
57. Step-up piston lifter
58. Step-up piston lifter lever pin
59. Solenoid and diaphragm choke pull-off bracket screw
60. Solenoid
61. Solenoid operating lever screw
62. Curb idle speed screw and lever
63. Bowl vent lever, lower
64. Throttle shaft washer
65. Hose
66. Primary diaphragm choke pull-off bracket
67. Auxiliary choke pull-off and dashpot
68. Auxiliary choke pull-off and bracket (if equipped)
69. Dashpot and bracket
70. Limiter cap
71. Idle mixture screw
72. Idle mixture screw spring
73. Throttle body assembly
74. Carter part number location
75. Low idle speed screw

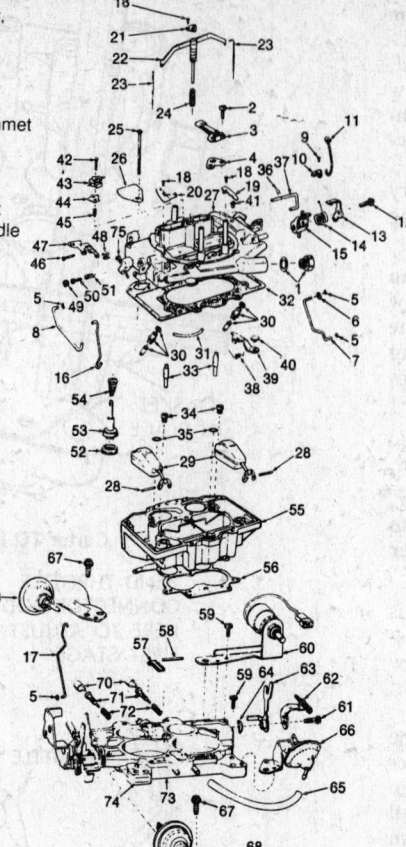

Exploded view of a typical late-model Thermo-Quad®

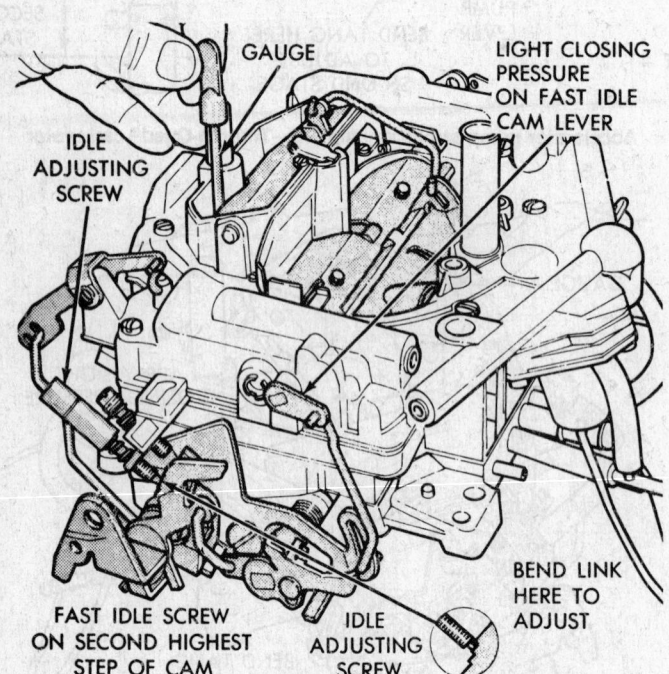

GAUGE

LIGHT CLOSING PRESSURE ON FAST IDLE CAM LEVER

IDLE ADJUSTING SCREW

BEND LINK HERE TO ADJUST

FAST IDLE SCREW ON SECOND HIGHEST STEP OF CAM

IDLE ADJUSTING SCREW

Adjustment of fast idle cam setting—Thermo-Quad® carburetor

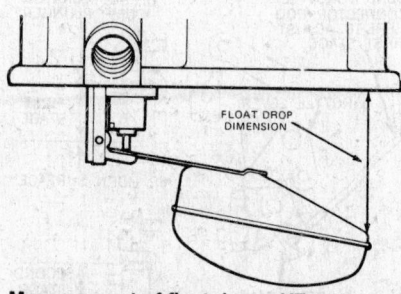

FLOAT DROP DIMENSION

Measurement of float drop—YF carburetor

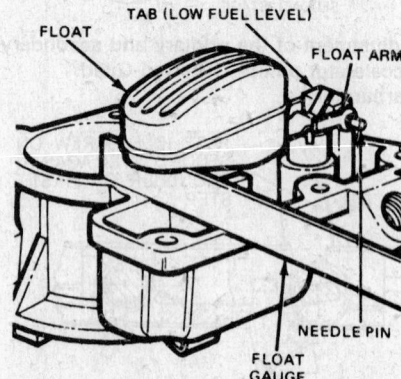

TAB (LOW FUEL LEVEL)

FLOAT

FLOAT ARM

NEEDLE PIN

FLOAT GAUGE

Measurement of float drop—YF/YFA carburetor

MODEL TQ
International
(All measurements in inches)

Year	Carburetor Number	Float Level	Fast Idle Speed (rpm)	Auto Choke Setting	Fuel Bowl Vent Clearance	Accelerator Pump Stroke Adjustment	Secondary Throttle Lock-Out Adjustment	Metering Rod Adjustment	Vacuum Kick Adjustment	Vacuum Pull-off Choke Adjustment	Fast Idle Cam and Linkage Adjustment	Choke Unloader Adjustment
'79–'81	TQ91285	.91 ± .030 (Old Needles) .88 ± .030 (New Needles)	1600	¼ Rich	.800–.830	Primary① .328–.358 Secondary① .120–.260	.060–.090	.468 ± .031	Vac. High .440–.460 Vac. Low .235–.255	.840–.880	.089–.109	.280–.320
	TQ6591S, 6550S	1.06	1550–1600	1 Rich	.800–.830	Primary .328–.358 Secondary .120–.260	.060–.090	15/32	Vac. High .335–.355 Vac. Low .250–.270	.840–.880 (6550S only)	.089–.109	.280–.320
	TQ6590S, 6552S, 6551S	1.06	1550–1600	1 Rich	.800–.830	Primary .328–.358	.060–.090	15/32	Vac. High .335–.355 Vac. Low .250–.270	.840–.880 (6551S only)	.089–.109	.280–.320

① Rod in inner hole

MODEL TQ
Dodge/Plymouth
(All measurements in inches)

Year	Carburetor Number	Float Level	Fast Idle Speed (rpm)	Accelerator Pump Stroke Adjustment①	Secondary Throttle Lock-Out Adjustment	Vacuum Kick Adjustment	Choke Diaphragm Rod Adjustment	Fast Idle Cam and Linkage Adjustment③	Choke Unloader Adjustment②
'79	9228S	29/32	1600	11/32 (9/64)	.060–.090	.100	.040	.100	.500
	9229S	29/32	1600	11/32 (9/64)	.060–.090	.100	.040	.100	.500
	9223S	29/32	1600	11/32 (9/64)	.060–.090	.100	.040	.100	.500
	9227S	29/32	1600	11/32 (9/64)	.060–.090	.100	.040	.100	.500
	9224S	29/32	1600	5/16 (3/16)	.060–.090	.100	.040	.100	.500
	9225S	29/32	1600	5/16 (3/16)	.060–.090	.100	.040	.100	.500
	9207S	29/32	1600	31/64 (23/64)	.060–.090	.150	.040	.100	.310
	9208S	29/32	1600	31/64 (23/64)	.060–.090	.150	.040	.100	.310
	9209S	29/32	1600	31/64 (23/64)	.060–.090	.150	.040	.100	.310
	9210S	29/32	1600	31/64 (23/64)	.060–.090	.150	.040	.100	.310
	9211S	29/32	1400	31/64 (23/64)	.060–.090	.100	.040	.100	.500
	9212S	29/32	1400	31/64 (23/64)	.060–.090	.100	.040	.100	.500
	9247S	29/32	1400	31/64 (23/64)	.060–.090	.100	.040	.100	.500
	9248S	29/32	1400	31/64 (23/64)	.060–.090	.100	.040	.100	.500
'80	9279S	29/32	1600	.340 (.190)	.060–.090	.130	.040	.130	.310
	9288S	29/32	1600	.340 (.190)	.060–.090	.120	.040	.120	.310
	9296S	29/32	1500	.340 (.140)	.060–.090	.100	.040	.100	.310
	9254S	29/32	1500	.340 (.190)	.060–.090	.130	.040	.100	.310

MODEL TQ
Dodge/Plymouth
(All measurements in inches)

Year	Carburetor Number	Float Level	Fast Idle Speed (rpm)	Accelerator Pump Stroke Adjustment①	Secondary Throttle Lock-Out Adjustment	Vacuum Kick Adjustment	Choke Diaphragm Rod Adjustment	Fast Idle Cam and Linkage Adjustment③	Choke Unloader Adjustment②
'80	9265S	29/32	1600	.340(.140)	.060–.090	.150	.040	.100	.310
	9255S	29/32	1600	.340(.190)	.060–.090	.120	.040	.100	.310
	9252S	29/32	1600	.340(.190)	.060–.090	.120	.040	.120	.310
	9251S	29/32	1600	.340(.190)	.060–.090	.120	.040	.120	.310
	9292S	29/32	1600	.340(.190)	.060–.090	.150	.040	.130	.310
	9298S	29/32	1600	.340(.140)	.060–.090	.150	.040	.100	.310
	9299S	29/32	1600	.340(.140)	.060–.090	.150	.040	.100	.310
	9281S	29/32	1600	.340(.190)	.060–.090	.180	.040	.130	.310
	9261S	29/32	1600	.340(.190)	.060–.090	.130	.040	.130	.310
'81	9311S	29/32	1500	.340	.060–.090	.150	.040	.100	.310
	9314S	29/32	1500	.340	.060–.090	.150	.040	.100	.310
	9325S	29/32	1500	.340	.060–.090	.120	.040	.100	.310
	9329S	29/32	1600	.340	.060–.090	.130	.040	.100	.310
	9330S	29/32	1500	.340	.060–.090	.130	.040	.100	.310
	9331S	29/32	1700	.340	.060–.090	.110	.040	.100	.310
	9332S	29/32	1700	.340	.060–.090	.110	.040	.100	.310
	9357S	29/32	1800	.340	.060–.090	.130	.040	.100	.310
	9358S	29/32	1700	.340	.060–.090	.180	.040	.100	.310
	9359S	29/32	1500	.340	.060–.090	.130	.040	.100	.310
'82	9342S	29/32	1600	.340	.060–.090	.130	.040	.100②	.310
	9375S	29/32	1800	.340	.060–.090	.130	.040	.130②	.310
	9376S	29/32	1700	.340	.060–.090	.130	.040	.130②	.310
	9379S	29/32	1500	.340	.060–.090	.130	.040	.130②	.310
'83	9342S	29/32	1600	.340(.390)	.060	.130	.040	.100②	.310
	9375S	29/32	1800	.340(.390)	.060	.130	.040	.130②	.310
	9379S	29/32	1500	.340(.390)	.060	.130	.040	.130②	.310
	9376S	29/32	1700	.340(.390)	.060	.180	.040	.100②	.310
'84	9386S	29/32	1600	.340(.190)	.060–.090	.170	.040	.100	.310
	9387S	29/32	1500	.340(.190)	.060–.090	.150	.040	.100	.310
	9379S	29/32	1500	.340(.190)	.060–.090	.130	.040	.130	.310
	9376S	29/32	1700	.340(.190)	.060–.090	.180	.040	.100	.310

Note: Choke is fixed on all models
① Stage I (Stage II). No Stage II adjustment on 1981–82 models
② Measure at the lowest edge of the choke valve on the throttle lever side
③ Set the linkage with idle on the second highest step of the cam

MODELS YF/YFA
Ford
(All measurements in inches)

Year	Carburetor Number	Float Level	Float Drop	Choke Unloader Setting	Choke Setting	Dash Pot Plunger	Initial Choke Opening
'79	D8TE						
	BVA	25/32	1 19/32	.280	Index	—	.230
	CKB	25/32	1 19/32	.280	Index	.070	.230
	BWA	25/32	1 19/32	.280	Index	.070	.230
	BUA	25/32	1 19/32	.280	Index	.070	.230
	BUB	25/32	—	—	—	—	—
	CNA	25/32	1 19/32	.280	Index	.070	.230
	AAA	25/32	1 19/32	.280	Index	.070	.230
	UA	23/32	1½	—	Manual	—	—
	CDA	23/32	1½	—	Manual	—	—
	D8UE						
	AAA	—	—	—	—	—	—
	ZA	25/32	1 19/32	.280	Index	.070	.230
	D6TE						
	ZA	23/32	—	.280	Index		.290
	D6UE						
	MA	23/32	—	.280	Index		.290
	D2UE						
	EA	25/32	1 19/32	—	—	—	—
'80	E0TE–9510						
	ABA,FA, LA,KA	.69	1.53	.28	①	—	.290
	ACA,ARA	.69	1.53	.28	①	—	.320
	AEA,AFA, ALA,AKA, ATA,CA, GA	.69	1.53	.28	①	—	.230

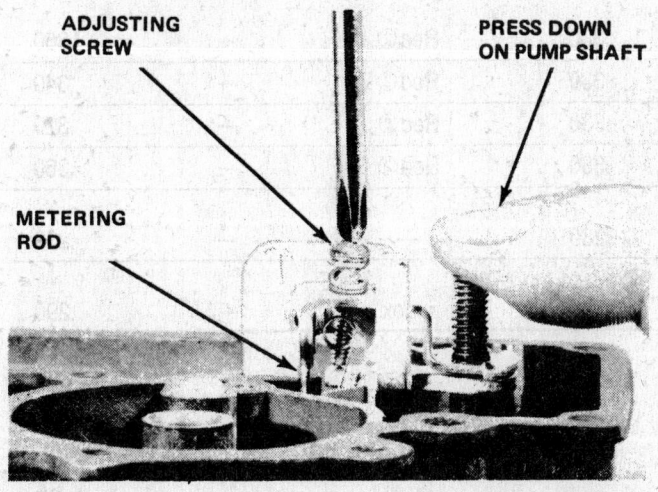

Metering rod adjustment—YF carburetor with electric choke

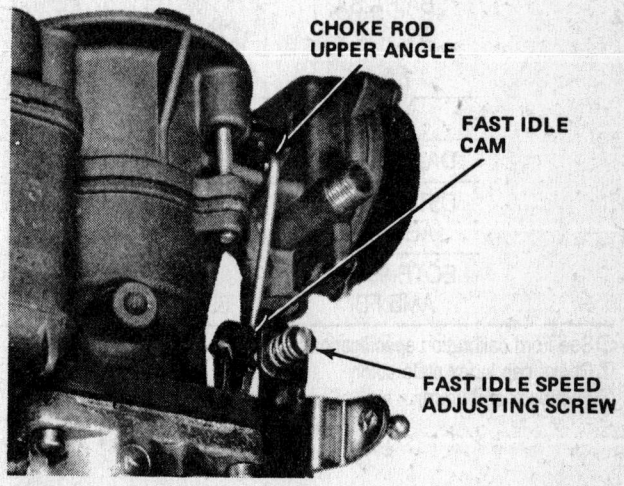

Fast idle cam and linkage adjustment—YF carburetor

MODELS YF/YFA
Ford
(All measurements in inches)

Year	Carburetor Number	Float Level	Float Drop	Choke Unloader Setting	Choke Setting	Dash Pot Plunger	Initial Choke Opening
'81	E0TE–9510 AMA,FA	.69	—	.28	Index	—	.290
	D5TE–9510 CA,VA	.69	—	.28	Index	—	.290
	AGB	⅜	—	.28	1 Rich	—	.230
	E1TE–9510 UA,ARA, ARB	.78	—	.28	Index	—	.230
	AUA,VA	.78	—	.28	2 Rich	—	.300
	AZA,GA	.78	—	.330	2 Rich	—	.320
'82	E2TE–AMA E2UE–EA	.78	—	.28	Index	—	.230
	E2TE BZA,BVA	.78	—	.28	Index	—	.270
	CEA,JA	.78	—	.320	2 Rich	—	.330
	YA,AAA	.78	—	.280	Index	—	.300
	MA,ANA	.78	—	.280	2 Rich	—	.300
	KA	.78	—	.330	2 Rich	—	.320
	AAA	.78	—	.280	Index	—	.300
	EZUE–DA	.78	—	.330	2 Rich	—	.320
'83–'84	E37E–9510 LB,NB, RB,TB	.65	—	.270	Gray②	—	.320
	E37E–9510 BB	.65	—	.270	Yellow②	—	.320
'85	E5TE-9510 DA	.65	—	.270	Gray②	—	.320
	VA,UA,TA, BA,RA,SA, JA	.78	—	.330	Red②	—	.360
	FA	.78	—	.330	Red②	—	.340
	HA	.78	—	.330	Red②	—	.320
	DA,MA,CA③	.78	—	.330	Red②	—	.360
	D5TE-9510 AGB	⅜	—	.280	—	—	.230
	E0TE-9510 AMB,FB	.69	—	.280	index	—	.290

① See Ford calibration specifications
② Choke cap index plate color
③ Feedback carburetor

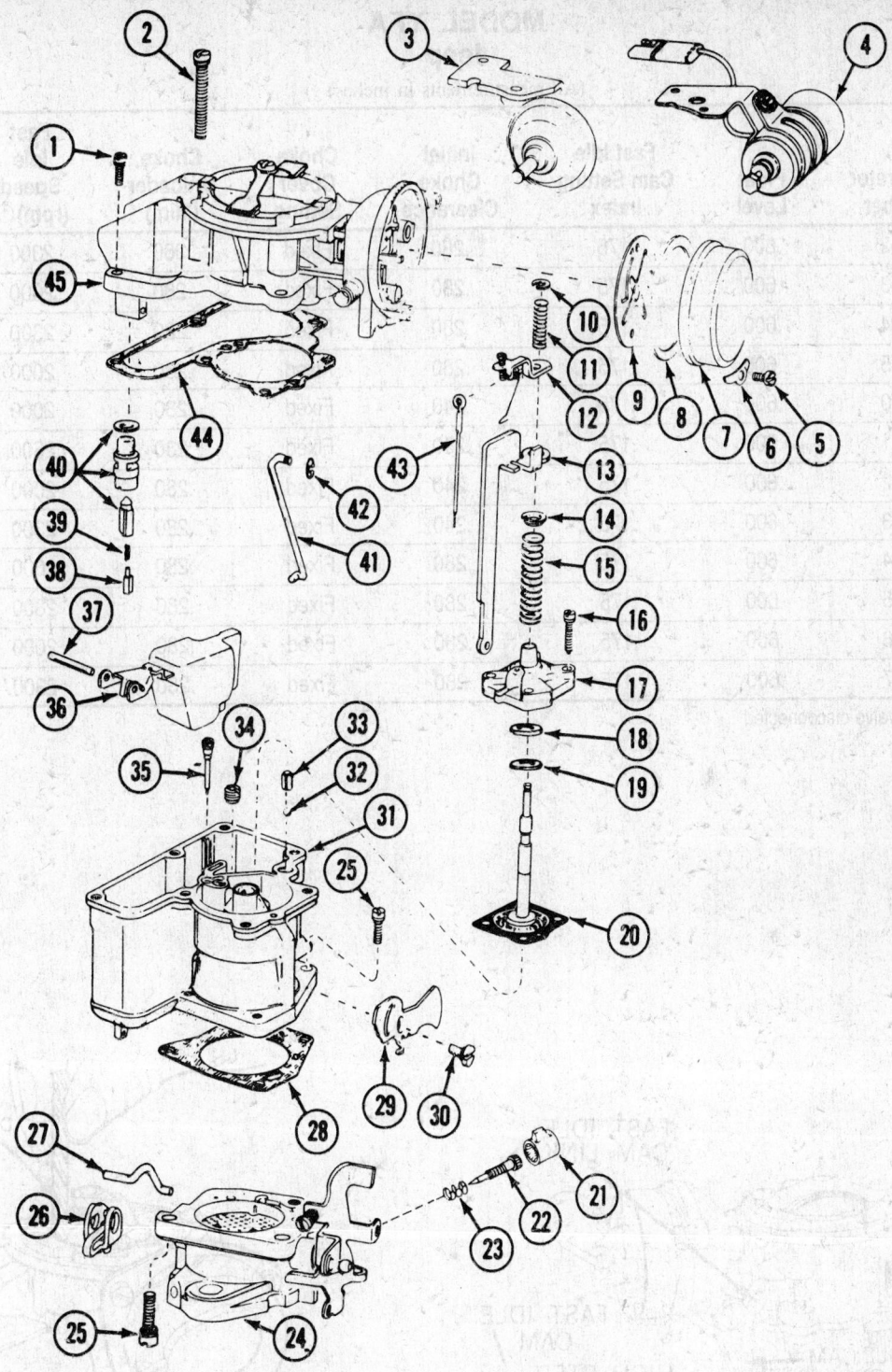

1. Air horn screw (short)
2. Air horn screw (long)
3. Dashpot and bracket
4. Solenoid and bracket
5. Coil housing screw
6. Coil housing retainer
7. Choke cover
8. Coil housing gasket
9. Coil housing baffle plate
10. Upper pump spring retainer
11. Upper pump spring

12. Metering rod arm
13. Diaphragm lifter link
14. Washer
15. Diaphragm spring
16. Diaphragm housing screw (4)
17. Diaphragm housing
18. Washer
19. Spacer
20. Diaphragm
21. Idle screw limiter cap
22. Idle mixture screw

23. Spring
24. Throttle body
25. Body flange screw (3)
26. Throttle shaft arm
27. Pump connector link
28. Body gasket
29. Fast idle cam
30. Fast idle cam screw
31. Main body
32. Discharge ball
33. Discharge ball weight
34. Metering jet

35. Low speed jet
36. Float
37. Float pin
38. Needle pin
39. Needle spring
40. Needle, needle seat, gasket
41. Choke connector rod
42. Choke connector rod retainer
43. Metering rod
44. Air horn gasket
45. Air horn

Exploded view of YF carburetor—typical

MODEL YFA
Jeep
(All measurements in inches)

Year	Carburetor Number	Float Level	Fast Idle Cam Setting Index	Initial Choke Clearance	Choke Cover Setting	Choke Unloader (Min.)	Fast Idle Speed (rpm)①	Bowl Vent Opens
'83	7452	.600	.175	.280	Fixed	.280	2300	2 Step
	7453	.600	.175	.280	Fixed	.280	2000	2 Step
	7454	.600	.175	.280	Fixed	.280	2300	2 Step
	7455	.600	.175	.280	Fixed	.280	2000	2 Step
'84	7700	.600	.175	.240	Fixed	.280	2000	—
	7701	.600	.175	.240	Fixed	.280	2300	—
	7702	.600	.175	.240	Fixed	.280	2000	—
	7703	.600	.175	.240	Fixed	.280	2300	—
'85	7704	.600	.175	.280	Fixed	.280	2000	—
	7705	.600	.175	.280	Fixed	.280	2300	—
	7706	.600	.175	.280	Fixed	.280	2000	—
	7707	.600	.175	.280	Fixed	.280	2300	—

① Engine hot, EGR valve disconnected

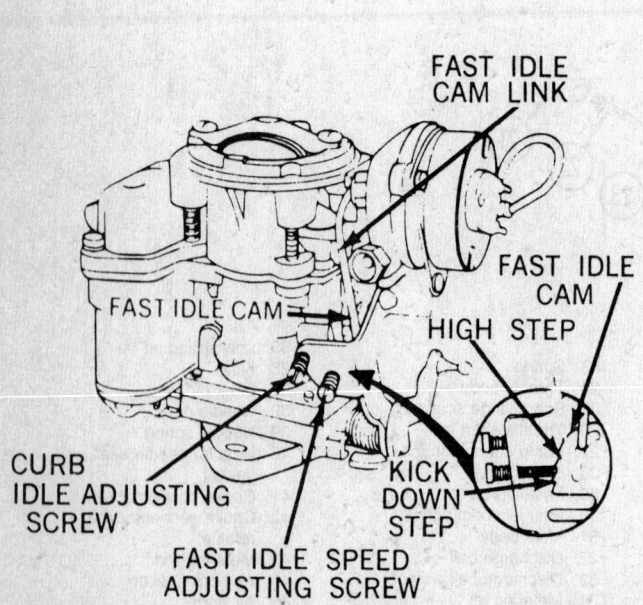

Typical adjustment points—YF carburetor with electric choke

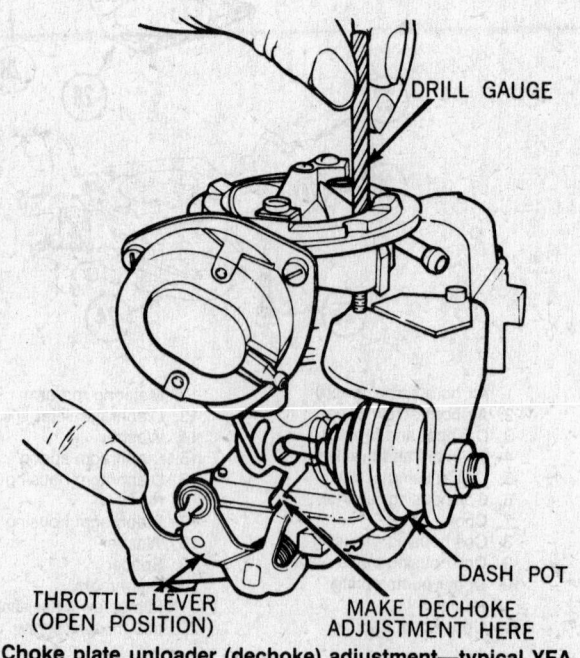

Choke plate unloader (dechoke) adjustment—typical YFA carburetor

IDLE MIXTURE ADJUSTMENT

Model YFA Carburetor

The idle mixture is preset at the time of manufacture and should normally not require re-adjustment. To prevent easy access to the idle mixture screw, a tamper resistant plug is set into the carburetor assembly to cover the screw. Should adjustment be required due to system diagnosis, contamination, replacement of components or tampering, the following procedure may be used to bring the adjustment into compliance with specifications.

1. Connect a tachometer to the TACH terminal of the ignition coil wire connector and a dwell meter to the mixture solenoid test terminals in the diagnosis connector (D2–14 and D2–7) and adjust the dwell meter to the 6 cylinder scale.

2. If the idle mixture screw tamper resistant plug has not been removed, the carburetor must be removed from the engine for access to the plug. With the carburetor off the engine, invert the carburetor and place it in a suitable holding device and remove the plug by drilling a ⅛ in. hole in the center, installing a self-tapping screw and pulling the plug from the carburetor.

3. Reinstall the carburetor on the engine, connect all lines and wires.

4. Place the transmission in the NEUTRAL position and apply the parking brake.

5. Disconnect and plug the canister purge vacuum hose at the charcoal canister.

6. Start the engine and operate at fast idle speed to bring the engine and coolant to normal operating temperature, thus allowing the CEC (feedback) system to operate in the CLOSED LOOP mode of operation.

7. Return the engine to idle speed and adjust the carburetor for an idle speed 700 rpm for A/T equipped vehicles in DRIVE and 750 rpm in NEUTRAL for M/T equipped vehicles.

8. Adjust the idle mixture screw to obtain an average dwell reading of between 25 and 35 degrees, with 30 degrees preferred.

9. If the dwell is too low, turn the idle mixture screw counterclockwise (out). If the dwell is too high, turn the idle mixture screw clockwise (in).

NOTE: Allow time for the system to react and stabilize after each movement of the adjusting screw. The feedback system is very sensitive to adjustments.

10. Observe the final dwell indication with the adjusting tool removed. If the specified dwell cannot be obtained by adjustment, inspect the carburetor idle circuits for air leaks, restrictions and etc. Do any necessary repairs.

11. When the adjustment is complete, connect the canister purge hose and adjust the idle speed to specifications.

12. Stop the engine and remove the tachometer and dwell meter.

13. Plug the idle mixture adjusting screw opening with RTV sealant.

14. Install the gasket and the air cleaner assembly on the carburetor.

DATSUN/NISSAN CARBURETORS

IDLE MIXTURE ADJUSTMENT

1. Remove the carburetor.

2. Drill a small hole carefully through the idle mixture plug and remove the plug with a sheet metal screw and pliers. Be careful not to drill into the adjusting screw and remove all metal shavings with compressed air.

3. Install the carburetor on the engine.

4. Start the engine and allow it to reach normal operating temperature. Race the engine two or three times to make sure the choke is completely off.

5. Connect a suitable emission test device and adjust the idle % CO by turning the mixture adjusting screw. Check the underhood sticker for specifications.

6. Once the mixture is set to specifications, install a new mixture screw plug into the carburetor.

AUTOMATIC CHOKE ADJUSTMENT

1. With the engine cold, make sure the choke is fully closed. Press the gas pedal all the way to the floor and release, or pull the choke knob out on early models with that system.

2. Check the choke linkage for binding. The choke plate should be easily opened and closed with your finger. If the choke sticks or binds, it can usually be freed with a liberal application of a carburetor cleaner made for the purpose. A couple of quick shots from a spray can of this stuff normally does the trick. If not, the carburetor will have to be disassembled for repairs.

3. The choke is correctly adjusted when the index mark on the choke housing (notch) aligns with the center mark on the carburetor body. If the setting is incorrect, loosen the three screws clamping the choke body in place and rotate the choke cover left or right until the marks align. Tighten the screws carefully to avoid cracking the housing.

THROTTLE LINKAGE ADJUSTMENT

When the primary throttle valve is opened to an angle of 50° from its closed position,

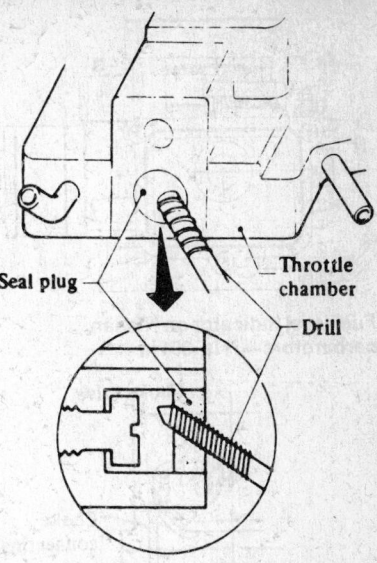

Removing idle mixture screw plug—Nissan carburetors

the adjust plate which is integral with the primary throttle valve, is brought into contact with portion A (see illustration) of the return plate. When the primary throttle valve is opened farther, the return plate is pulled apart from the stopper (B in the illustration), allowing the secondary throttle valve to open. To adjust the linkage:

1. Measure the clearance between the primary throttle valve and the wall of the throttle chamber at the center of the throttle valve when the adjust plate is brought into contact with portion A of the return plate. Standard clearance is 0.26–0.32 in.

2. If necessary, make the adjustment by bending the portion A of the return plate.

FLOAT LEVEL ADJUSTMENT

The fuel level is normal if it is within the lines on the window glass of the float chamber when the vehicle is resting on level

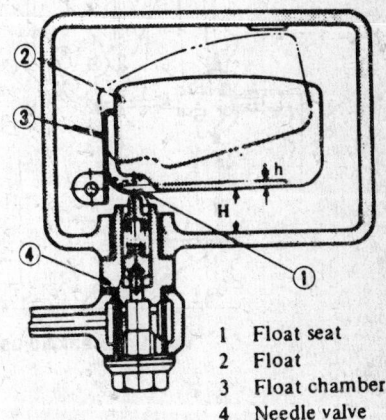

1	Float seat
2	Float
3	Float chamber
4	Needle valve

Float level adjustment—Nissan carburetors

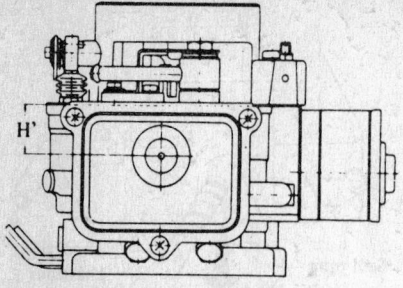

Fuel level indicator on Nissan carburetors—H is .091 in.

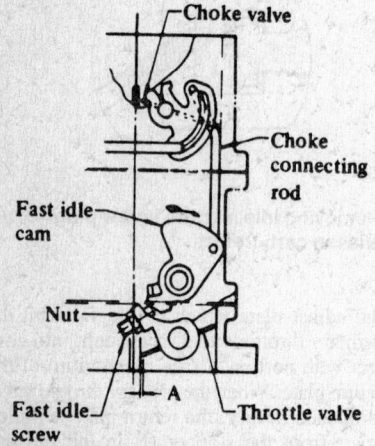

Adjusting throttle valve gap—Nissan carburetors

ground and the engine is off. If the fuel level is outside the lines, remove the float housing cover. Have an absorbent cloth under the cover to catch the fuel from the fuel bowl. Adjust the float level by bending the needle seat on the float. The needle valve should have an effective stroke of about .051–.067 in. When necessary, the needle

valve stroke can be adjusted by bending the float stopper.

NOTE: Be careful not to bend the needle valve rod when installing the float and baffle plate, if removed.

FAST IDLE ADJUSTMENT

1. With the carburetor removed from the vehicle, place the upper side of the fast idle screw on the second step of the fast idle cam and measure the clearance between the throttle valve and the wall of the throttle valve chamber at the center of the throttle valve (A in the illustration). Refer to the specification chart for proper clearance. Adjust by turning the fast idle screw.

2. Install the carburetor on the engine. Place the fast idle screw on the second step of the cam.

3. Start the engine and measure the fast idle rpm with the engine at operating temperature. Refer to the underhood emission sticker for correct fast idle speed specifications.

4. To adjust the fast idle speed, loosen the locknut and turn the fast idle adjusting screw.

NOTE: The first step of the fast idle adjustment procedure is not absolutely necessary; it should be used as a guide to correct adjustment at overhaul.

CHOKE UNLOADER ADJUSTMENT

1. Close the choke valve completely.
2. Hold the choke valve closed by stretching a rubber band between the vacuum break lever and a stationary part of the carburetor.

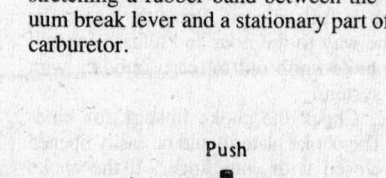

3. Open the throttle lever fully.
4. With the throttle lever fully open, adjust the clearance between the choke valve and the carburetor body to specifications by bending the unloader tongue. Make sure that the throttle valve opens completely when the carburetor is mounted on the engine.

VACUUM BREAK ADJUSTMENT

1. Close the choke valve completely.
2. Hold the choke vavle by stretching a rubber band between the vacuum break lever and stationary part of the carburetor.
3. Push the vacuum break stem with pliers, then gently pull it fully straight.
4. Adjust the gap between the choke valve and carburetor body by bending the unloader tongue.

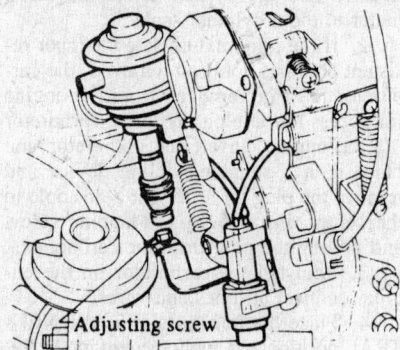

Dash pot adjustment—Nissan carburetors

DASHPOT ADJUSTMENT

NOTE: Air conditioner should be off for this adjustment.

1. Adjust the idle speed and mixture before making adjustments to the dashpot. Warm the engine to operating temperature, and connect a tachometer to the engine.

2. Move the throttle lever by hand, and note the engine speed when the dashpot plunger just touches the throttle lever.

3. The engine speed should be 1650–1850 rpm with automatic transmission, or 1900–2100 rpm with manual transmission through 1980. On 1981 and later models the speed is 1600 rpm.

4. If not, loosen the locknut and turn the adjusting screw until the engine speed is in the proper range. Tighten the locknut. On 1979 models with air conditioning, a different dashpot is used. Adjustment is made by turning the screw on the throttle lever which contacts the plunger.

5. Open the throttle and allow it to close by itself. The dashpot should smoothly reduce the idling speed from 2000 to 1000 rpm in about three seconds.

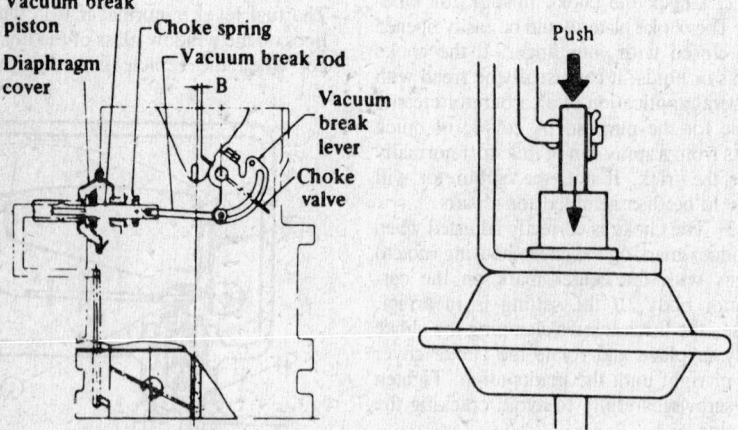

Vacuum break adjustment—Nissan carburetors

DATSUN (NISSAN) CARBURETORS

(All measurements in inches)

Engine	Carburetor Number	Fuel Level	Throttle Valve Gap ①		Vacuum Break ②	Choke Unloader	Dashpot Adjustment (rpm)
			MT	AT			
Z24	DFP384-3③	.910	.0280–.0335	.0343–.0398	.1220–.1457	.0807–.1122	1700–1900
	DFP384-4③	.910	.0280–.0335	.0343–.0398	.1220–.1457	.0807–.1122	1400–1600
	DFP384-12③	.910	.0280–.0335	.0343–.0398	.1220–.1457	.0807–.1122	1700–1900
	DCR384-3	.910	.0280–.0335	.0343–.0398	.0965–④.1201	.0807–.1122	1400–1600
	DCR384-4	.910	.0280–.0335	.0343–.0398	.0965–.1201	.0807–.1122	1400–1600
	CDR384-5⑤	.910	.0280–.0335	.0343–.0398	.0913–.1150	.0807–.1122	1400–1600
	DCR384-6⑤	.910	.0280–.0335	.0343–.0398	.0913–.1150	.0807–.1122	1400–1600
	DCR384-11A	.910	.0280–.0335	.0343–.0398	.0965–.1201	.0807–.1122	1400–1600
	DCR384-15⑤	.910	.0280–.0335	.0343–.0398	.0965–.1201	.0807–.1122	1400–1600
	DCR384-21A	.910	.0280–.0335	.0343–.0398	.0965–.1201	.0807–.1122	1400–1600
	DCR384-25⑤	.910	.0280–.0335	.0343–.0398	.0965–.1201	.0807–.1122	1400–1600
Z20	CDR342-8	.910	.0299–.0354	.0343–.0398	.0965–④.1201	.0807–.1122	1400–1600

MT Manual Transmission
AT Automatic Transmission
① Second step of fast idle cam
② Above 68°F
③ California Models
④ MPG Models: .1031–.1268
⑤ Canadian Models

HOLLEY CARBURETORS

MIXTURE ADJUSTMENT
225 Engine w/Holley 6145
Electronic Feedback Carburetor

Note: To perform this procedure you will need propane enrichment equipment and a precisely regulated vacuum supply.

1. Remove the concealment plug which gives access to the mixture screw. Connect a tachometer according to manufacturer's instructions. With the parking brake on and transmission in Neutral, start the engine and operate it on the second step of the fast idle cam until it is hot. Then, open the throttle to bring the engine to normal idle speed.

2. Disconnect and plug the EGR valve vacuum line at the EGR valve. Jumper the carburetor switch to a good ground. Leave the air cleaner in place.

3. Trace the vacuum hose leading to the choke diaphragm back to the Tee and

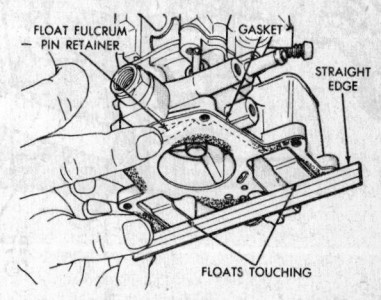

Float adjustment—6145 carburetor

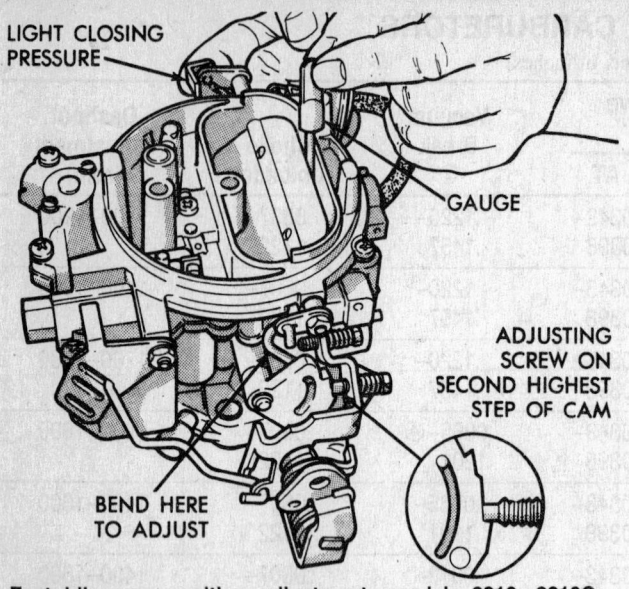

LIGHT CLOSING PRESSURE

GAUGE

ADJUSTING SCREW ON SECOND HIGHEST STEP OF CAM

BEND HERE TO ADJUST

Fast idle cam position adjustment—models 2210, 2210C, and 2245 carburetors

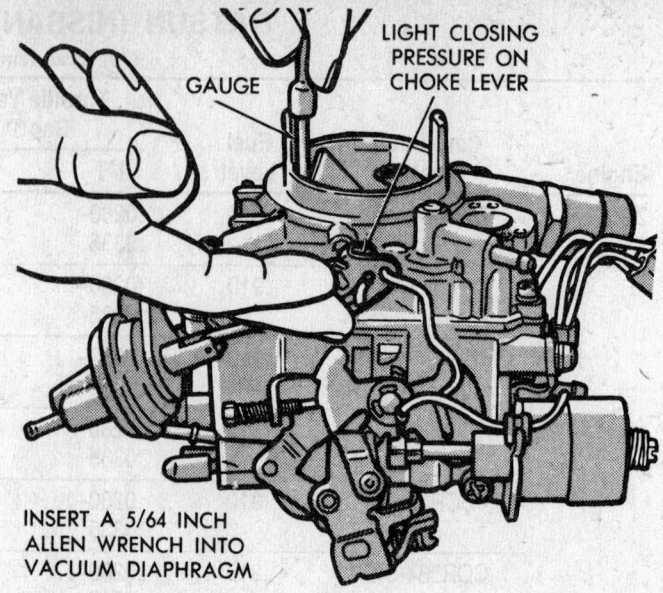

GAUGE

LIGHT CLOSING PRESSURE ON CHOKE LEVER

INSERT A 5/64 INCH ALLEN WRENCH INTO VACUUM DIAPHRAGM

Choke valve initial setting (vacuum kick)—model 1945 carburetor

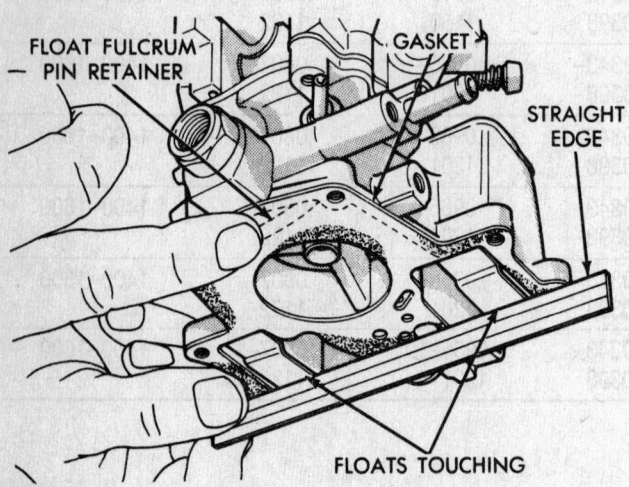

FLOAT FULCRUM — PIN RETAINER

GASKET

STRAIGHT EDGE

FLOATS TOUCHING

Adjusting the fuel level with the fuel bowl inverted—model 1945 carburetor

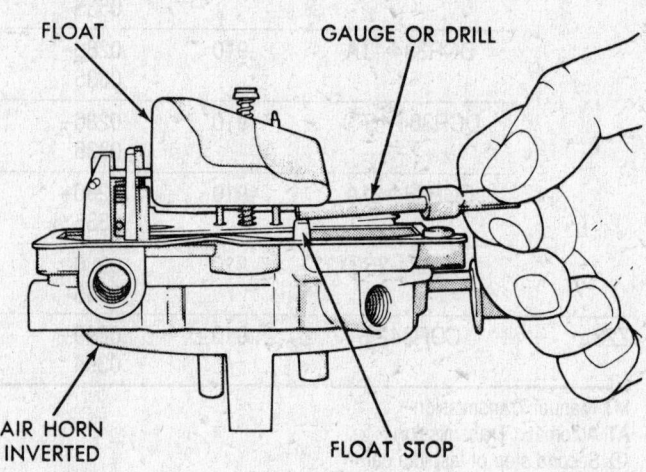

FLOAT

GAUGE OR DRILL

AIR HORN INVERTED

FLOAT STOP

Adjusting the float level—models 2210C and 2245 carburetors

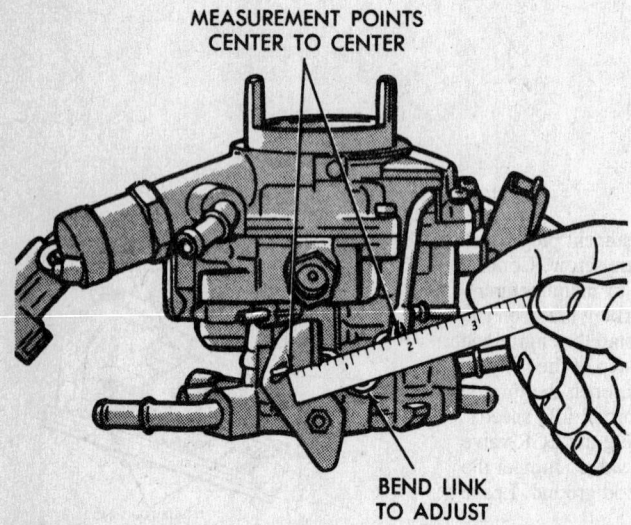

MEASUREMENT POINTS CENTER TO CENTER

BEND LINK TO ADJUST

Accelerator pump piston stroke adjustment—model 1945 carburetor

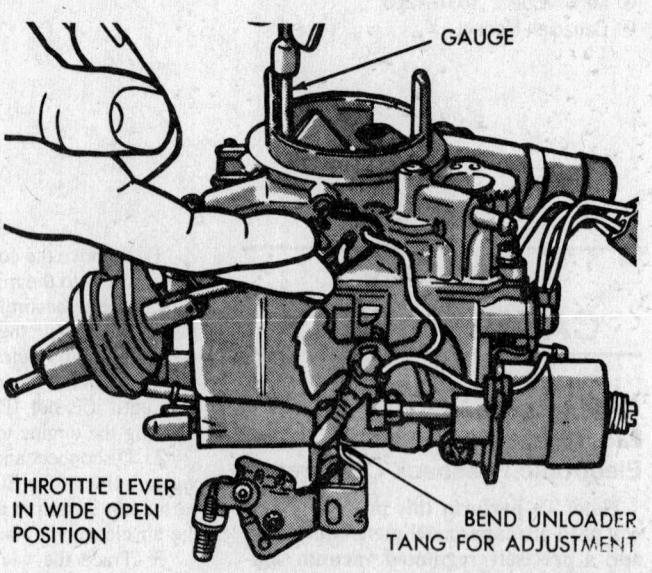

GAUGE

THROTTLE LEVER IN WIDE OPEN POSITION

BEND UNLOADER TANG FOR ADJUSTMENT

Choke valve unloader adjustment—model 1945 carburetor

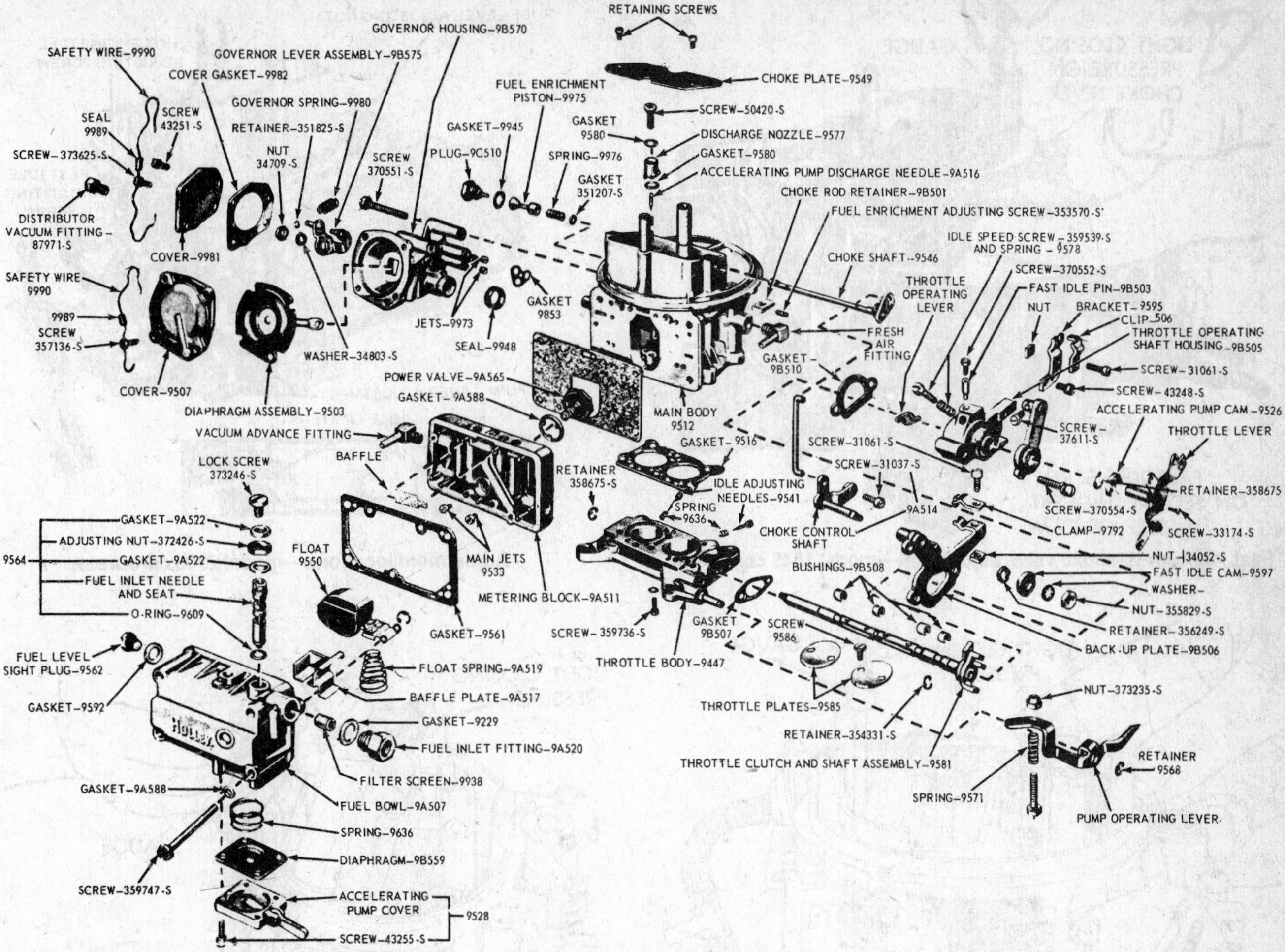

RETAINING SCREWS

SAFETY WIRE—9990
GOVERNOR LEVER ASSEMBLY—9B575
COVER GASKET—9982
GOVERNOR HOUSING—9B570
GOVERNOR SPRING—9980
CHOKE PLATE—9549

SEAL 9989
SCREW 43251-S
RETAINER—351825-S
GASKET 9945
GASKET 9580
SCREW—50420-S
SCREW—373625-S
NUT 34709-S
SCREW 370551-S
PLUG—9C510
SPRING—9976
DISCHARGE NOZZLE—9577
GASKET—9580
DISTRIBUTOR VACUUM FITTING—87971-S
FUEL ENRICHMENT PISTON—9975
GASKET 351207-S
ACCELERATING PUMP DISCHARGE NEEDLE—9A516
CHOKE ROD RETAINER—9B501
FUEL ENRICHMENT ADJUSTING SCREW—353570-S

SAFETY WIRE 9990
CHOKE SHAFT—9546
IDLE SPEED SCREW—359539-S AND SPRING—9578
9989
SCREW 357136-S
JETS—9973
GASKET 9853
THROTTLE OPERATING LEVER
FRESH AIR FITTING
SCREW—370552-S
FAST IDLE PIN—9B503
NUT
BRACKET—9595
CLIP—506
THROTTLE OPERATING SHAFT HOUSING—9B505

COVER—9507
WASHER—34803-S
SEAL—9948
GASKET 9B510
SCREW—31061-S
SCREW—43248-S
ACCELERATING PUMP CAM—9526

DIAPHRAGM ASSEMBLY—9503
POWER VALVE—9A565
MAIN BODY 9512
GASKET—9516
SCREW—37611-S
THROTTLE LEVER

VACUUM ADVANCE FITTING
GASKET—9A588
SCREW—31061-S
RETAINER—358675

LOCK SCREW 373246-S
BAFFLE
RETAINER 358675-S
SCREW—31037-S
IDLE ADJUSTING NEEDLES—9541
9A514
SCREW—370554-S
SCREW—33174-S

GASKET—9A522
FLOAT 9550
SPRING 9636
CLAMP—9792

ADJUSTING NUT—372426-S
CHOKE CONTROL SHAFT
NUT—34052-S

9564
GASKET—9A522
MAIN JETS 9533
BUSHINGS—9B508
FAST IDLE CAM—9597
WASHER

FUEL INLET NEEDLE AND SEAT
METERING BLOCK—9A511
NUT—355829-S

O-RING 9609
GASKET 9B507
RETAINER—356249-S
BACK-UP PLATE—9B506

FUEL LEVEL SIGHT PLUG—9562
GASKET—9561
SCREW—359736-S
SCREW 9586

GASKET 9592
THROTTLE BODY—9447
NUT—373235-S

FLOAT SPRING—9A519
THROTTLE PLATES—9585
RETAINER 9568

GASKET—9229
BAFFLE PLATE—9A517
RETAINER—354331-S

GASKET—9A588
FUEL INLET FITTING—9A520
THROTTLE CLUTCH AND SHAFT ASSEMBLY—9581

FILTER SCREEN—9938
SPRING—9571
PUMP OPERATING LEVER

SCREW—359747-S
FUEL BOWL—9A507
SPRING—9636
DIAPHRAGM—9B559

ACCELERATING PUMP COVER
9528
SCREW—43255-S

Holley two barrel—typical

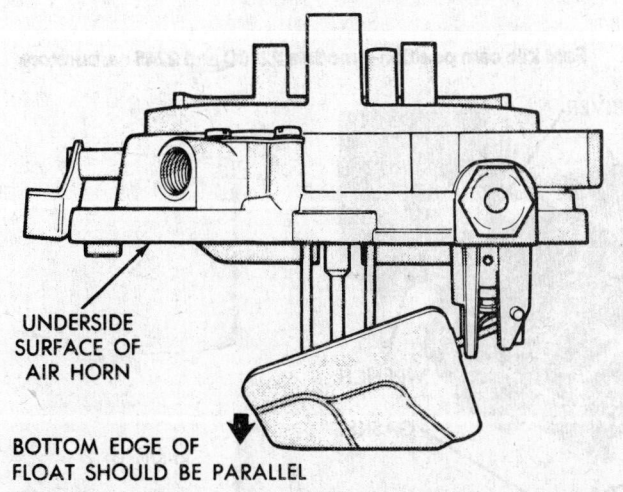

UNDERSIDE SURFACE OF AIR HORN

BOTTOM EDGE OF FLOAT SHOULD BE PARALLEL

Adjusting the float drop—models 2210, 2210C and 2245 carburetors

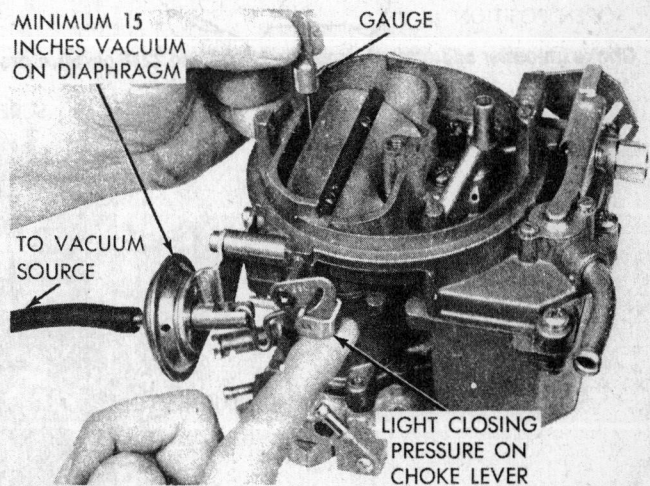

MINIMUM 15 INCHES VACUUM ON DIAPHRAGM

GAUGE

TO VACUUM SOURCE

LIGHT CLOSING PRESSURE ON CHOKE LEVER

Adjusting the initial choke valve setting—models 2210, 2210C and 2245 carburetors

Fast idle cam-to-choke valve adjustment—model 1945 carburetor

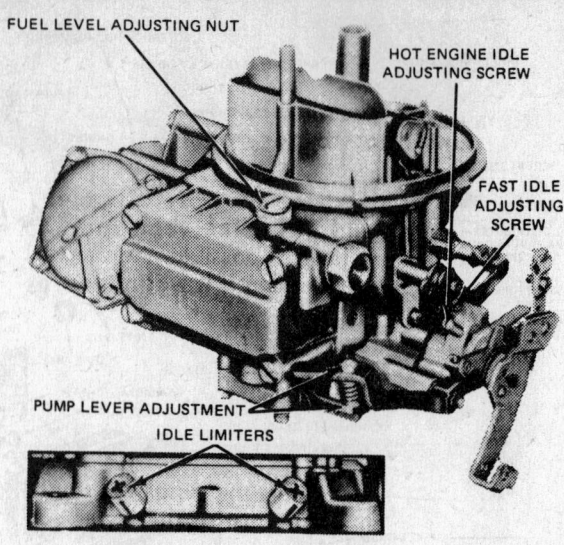

Adjustment locations—model 2300 carburetor

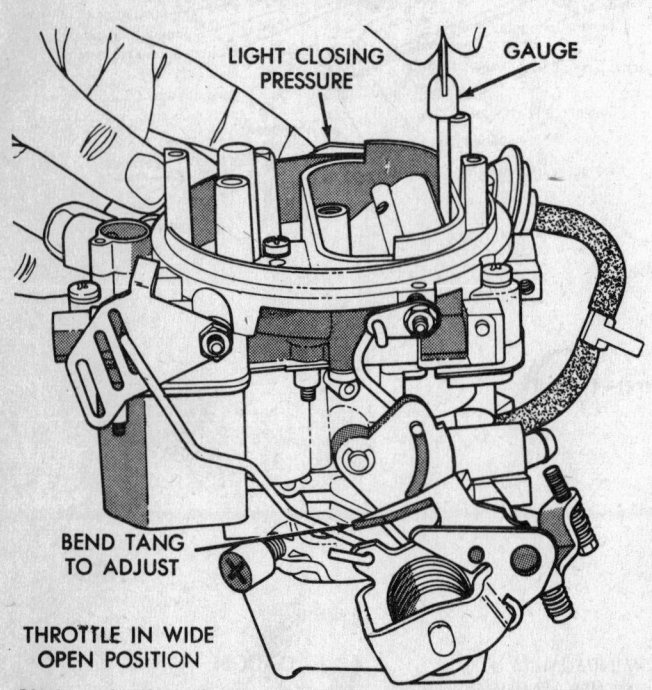

Choke unloader adjustment—models 2210C and 2245 carburetors

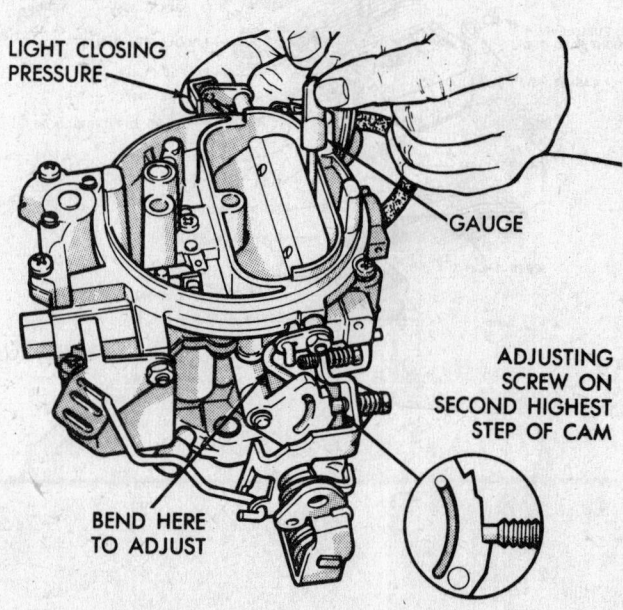

Fast idle cam position—models 2210C and 2245 carburetors

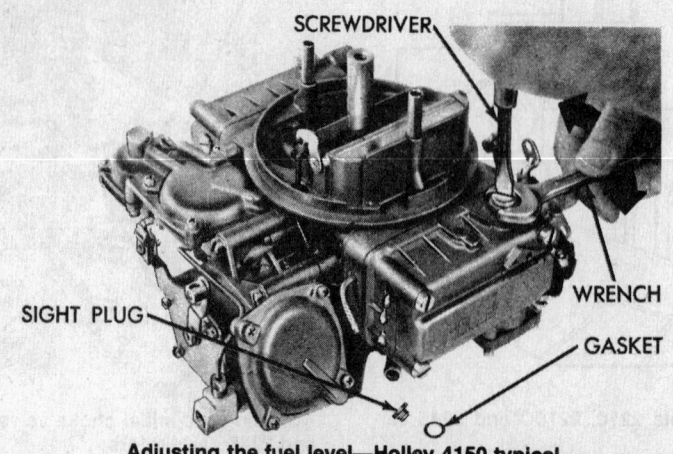

Adjusting the fuel level—Holley 4150 typical

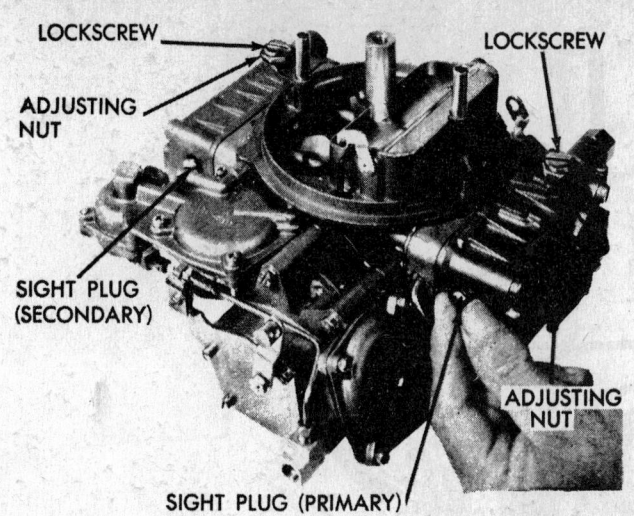

LOCKSCREW

LOCKSCREW

ADJUSTING
NUT

SIGHT PLUG
(SECONDARY)

ADJUSTING
NUT

SIGHT PLUG (PRIMARY)

Fuel level sight plug location—Holley 4150 typical

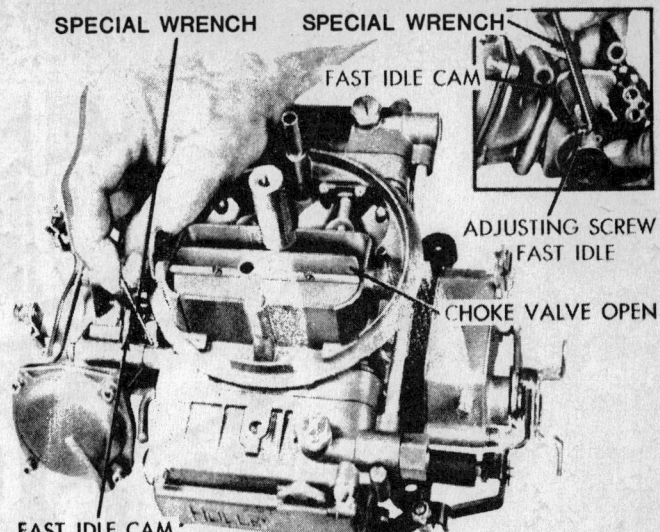

SPECIAL WRENCH

SPECIAL WRENCH

FAST IDLE CAM

ADJUSTING SCREW
FAST IDLE

CHOKE VALVE OPEN

FAST IDLE CAM

Fast idle speed adjustment—Holley 4150 typical

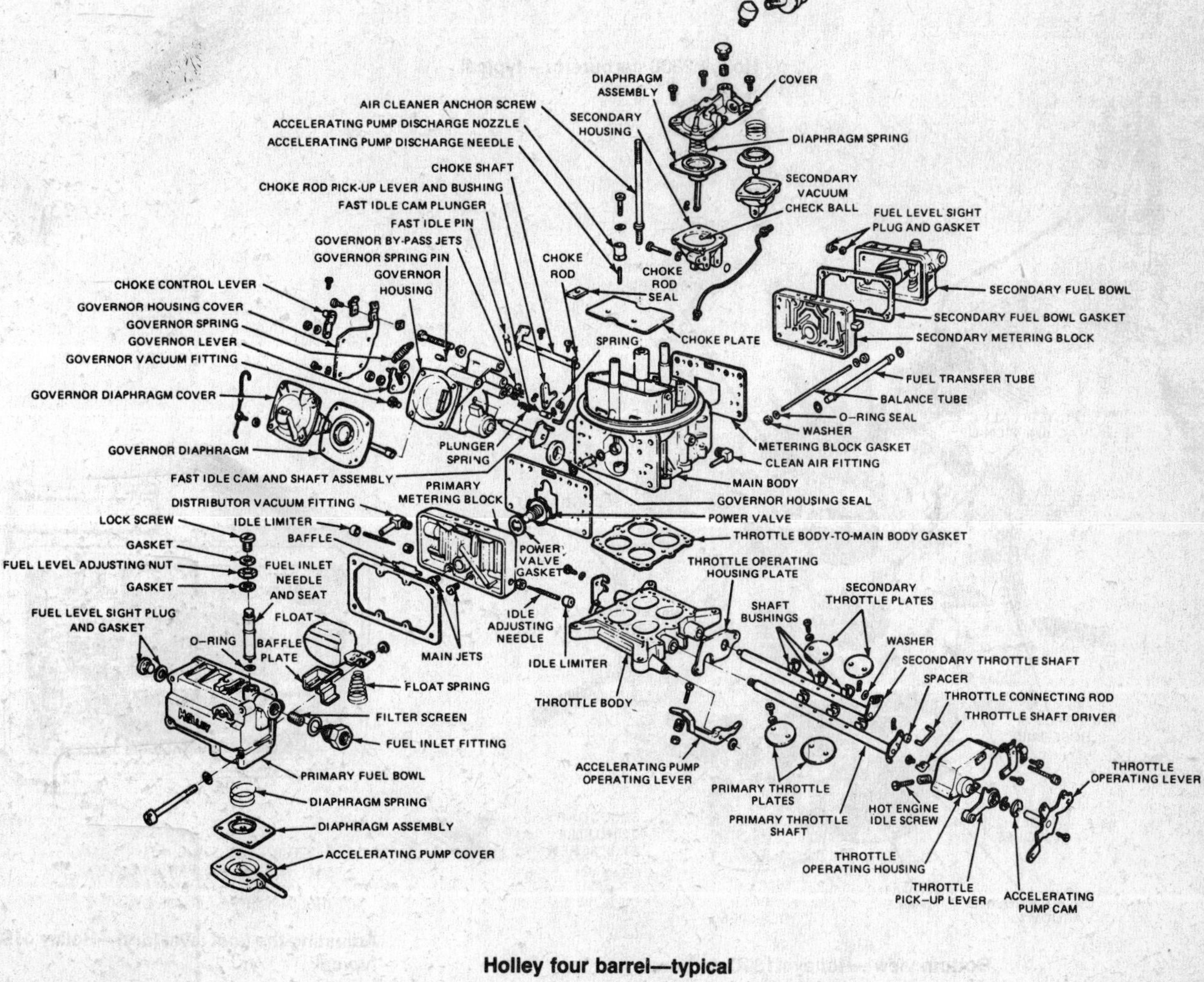

DIAPHRAGM
ASSEMBLY

COVER

AIR CLEANER ANCHOR SCREW

SECONDARY
HOUSING

ACCELERATING PUMP DISCHARGE NOZZLE

ACCELERATING PUMP DISCHARGE NEEDLE

DIAPHRAGM SPRING

CHOKE SHAFT

CHOKE ROD PICK-UP LEVER AND BUSHING

SECONDARY
VACUUM
CHECK BALL

FAST IDLE CAM PLUNGER

FUEL LEVEL SIGHT
PLUG AND GASKET

FAST IDLE PIN

GOVERNOR BY-PASS JETS

CHOKE
ROD

GOVERNOR SPRING PIN

SECONDARY FUEL BOWL

GOVERNOR
HOUSING

CHOKE CONTROL LEVER

SECONDARY FUEL BOWL GASKET

GOVERNOR HOUSING COVER

CHOKE
ROD
SEAL

SECONDARY METERING BLOCK

GOVERNOR SPRING

GOVERNOR LEVER

FUEL TRANSFER TUBE

GOVERNOR VACUUM FITTING

SPRING

CHOKE PLATE

BALANCE TUBE

GOVERNOR DIAPHRAGM COVER

O-RING SEAL

WASHER

METERING BLOCK GASKET

GOVERNOR DIAPHRAGM

CLEAN AIR FITTING

PLUNGER
SPRING

MAIN BODY

FAST IDLE CAM AND SHAFT ASSEMBLY

GOVERNOR HOUSING SEAL

PRIMARY
METERING BLOCK

DISTRIBUTOR VACUUM FITTING

POWER VALVE

LOCK SCREW

IDLE LIMITER

THROTTLE BODY-TO-MAIN BODY GASKET

GASKET

BAFFLE

POWER
VALVE
GASKET

THROTTLE OPERATING
HOUSING PLATE

FUEL LEVEL ADJUSTING NUT

FUEL INLET
NEEDLE
AND SEAT

SECONDARY
THROTTLE PLATES

GASKET

IDLE
ADJUSTING
NEEDLE

SHAFT
BUSHINGS

FUEL LEVEL SIGHT PLUG
AND GASKET

FLOAT

O-RING

BAFFLE
PLATE

WASHER

IDLE LIMITER

SECONDARY THROTTLE SHAFT

MAIN JETS

THROTTLE BODY

SPACER

FLOAT SPRING

THROTTLE CONNECTING ROD

FILTER SCREEN

THROTTLE SHAFT DRIVER

FUEL INLET FITTING

ACCELERATING PUMP
OPERATING LEVER

THROTTLE
OPERATING LEVER

PRIMARY FUEL BOWL

PRIMARY THROTTLE
PLATES

DIAPHRAGM SPRING

PRIMARY THROTTLE
SHAFT

HOT ENGINE
IDLE SCREW

DIAPHRAGM ASSEMBLY

THROTTLE
OPERATING HOUSING

ACCELERATING PUMP COVER

THROTTLE
PICK-UP LEVER

ACCELERATING
PUMP CAM

Holley four barrel—typical

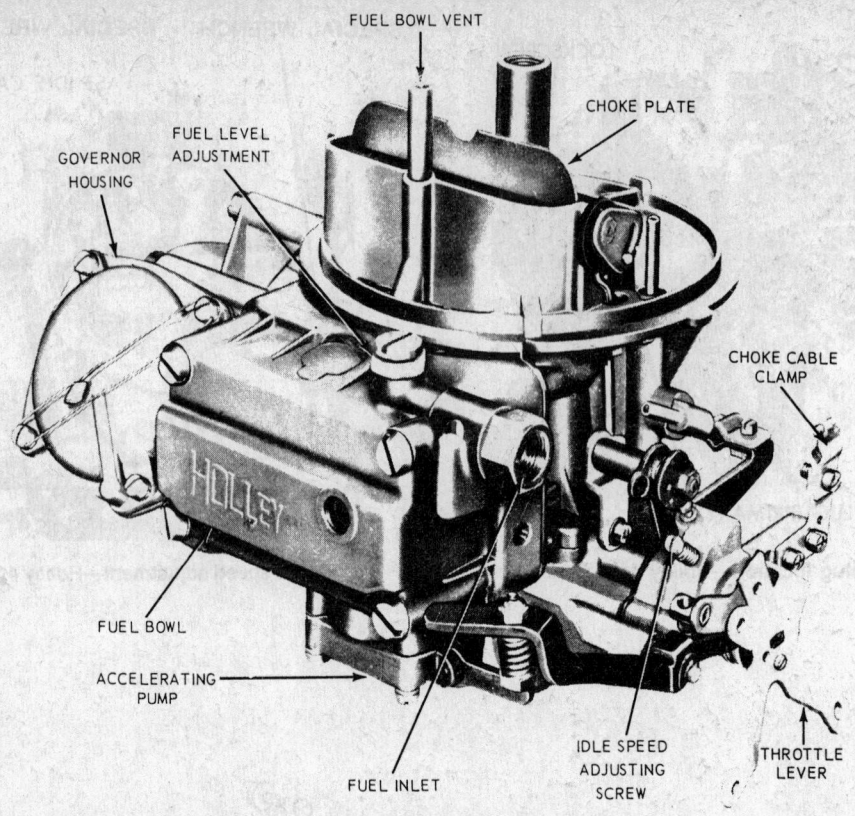

FUEL BOWL VENT

CHOKE PLATE

GOVERNOR HOUSING

FUEL LEVEL ADJUSTMENT

CHOKE CABLE CLAMP

HOLLEY

FUEL BOWL

ACCELERATING PUMP

FUEL INLET

IDLE SPEED ADJUSTING SCREW

THROTTLE LEVER

Holley 2300 carburetor—typical

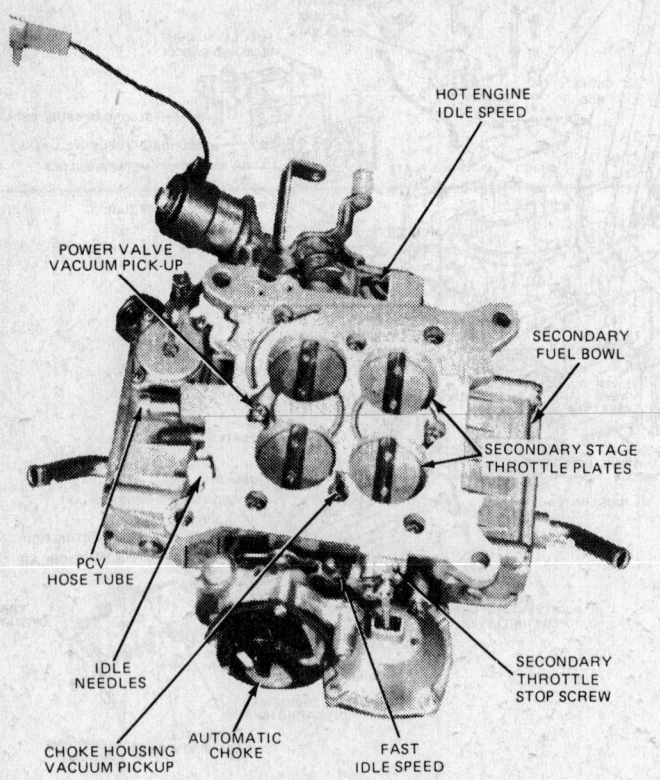

HOT ENGINE IDLE SPEED

POWER VALVE VACUUM PICK-UP

SECONDARY FUEL BOWL

SECONDARY STAGE THROTTLE PLATES

PCV HOSE TUBE

IDLE NEEDLES

SECONDARY THROTTLE STOP SCREW

CHOKE HOUSING VACUUM PICKUP

AUTOMATIC CHOKE

FAST IDLE SPEED

Bottom view—Holley 4180C carburetor

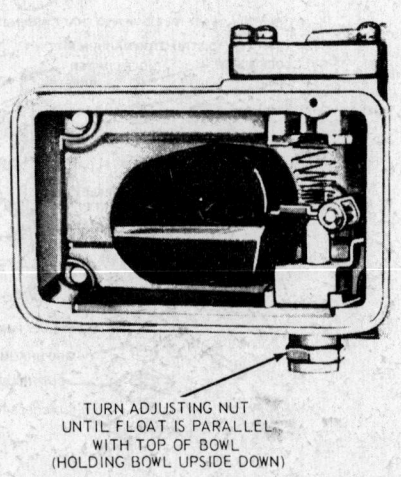

TURN ADJUSTING NUT UNTIL FLOAT IS PARALLEL WITH TOP OF BOWL (HOLDING BOWL UPSIDE DOWN)

Adjusting the float level (dry)—Holley 4150 typical

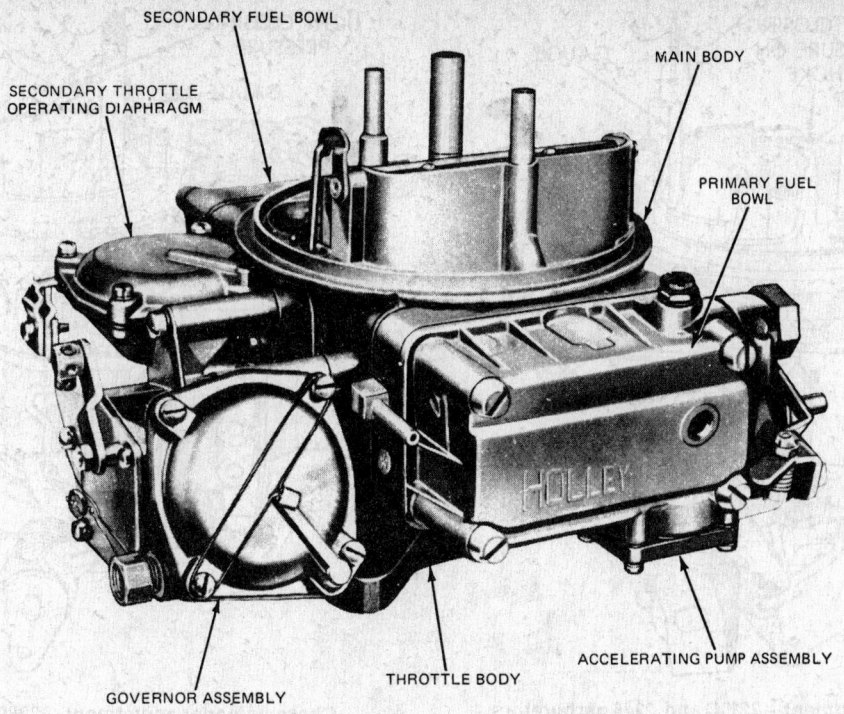

SECONDARY FUEL BOWL

SECONDARY THROTTLE
OPERATING DIAPHRAGM

MAIN BODY

PRIMARY FUEL
BOWL

GOVERNOR ASSEMBLY

THROTTLE BODY

ACCELERATING PUMP ASSEMBLY

Holley 4150 carburetor—typical

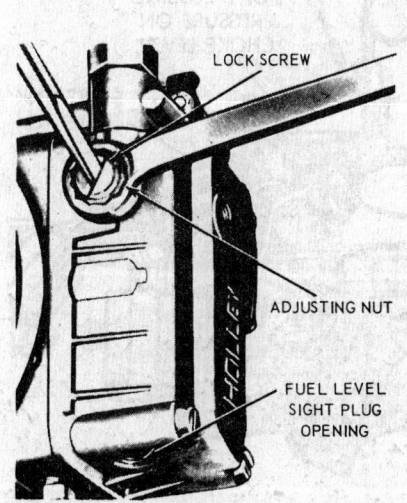

LOCK SCREW

ADJUSTING NUT

FUEL LEVEL
SIGHT PLUG
OPENING

**Adjusting the fuel level (wet)—Holley 4150
typical**

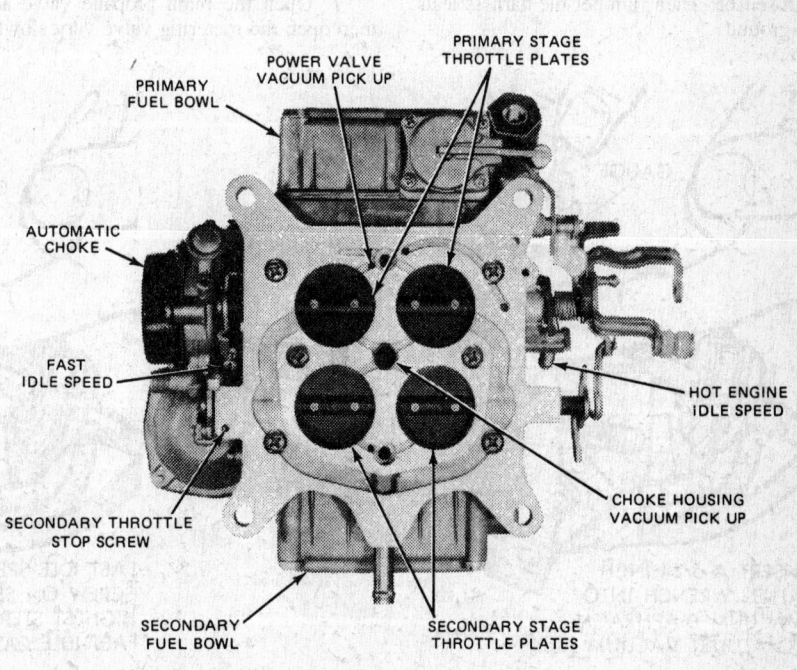

PRIMARY
FUEL BOWL

POWER VALVE
VACUUM PICK UP

PRIMARY STAGE
THROTTLE PLATES

AUTOMATIC
CHOKE

FAST
IDLE SPEED

HOT ENGINE
IDLE SPEED

SECONDARY THROTTLE
STOP SCREW

CHOKE HOUSING
VACUUM PICK UP

SECONDARY
FUEL BOWL

SECONDARY STAGE
THROTTLE PLATES

Bottom view—Holley 4160C carburetor

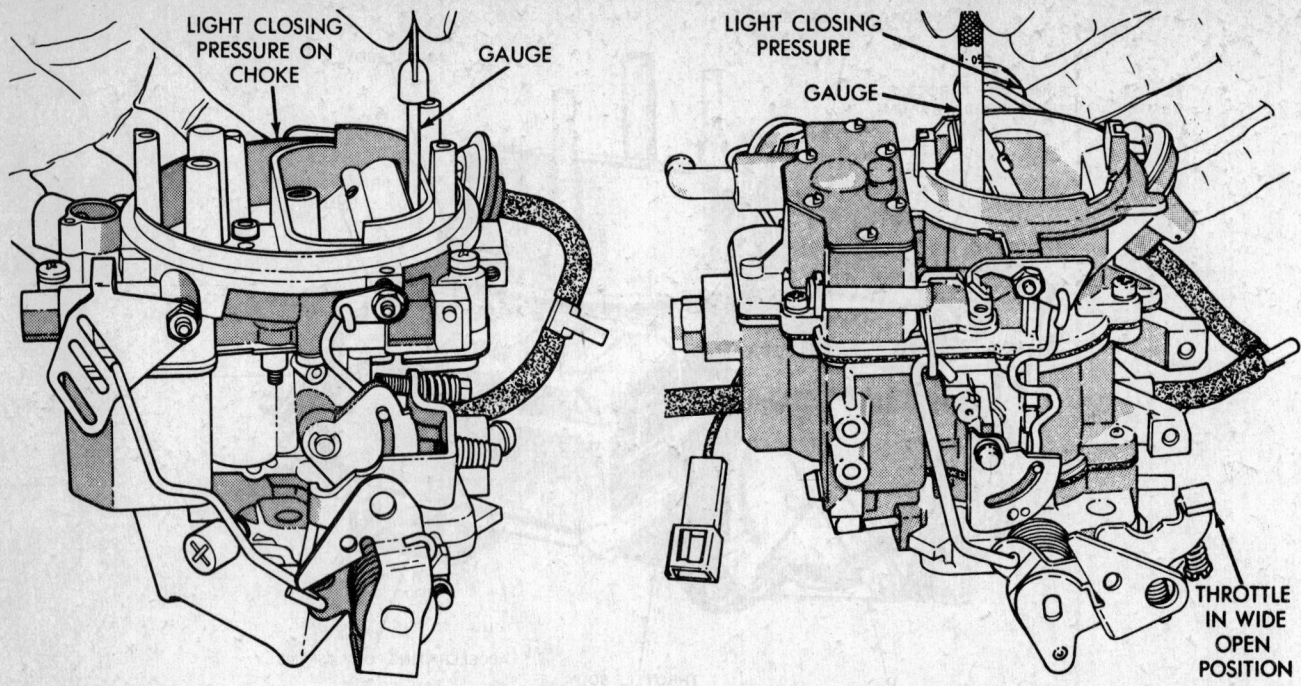

Vacuum kick adjustment—2210C and 2245 carburetors

Choke unloader adjustment—2280 carburetor

disconnect this hose only; then connect the propane supply hose in its place.

4. Make sure the propane bottle is securely upright and in a safe location. Then, pull the PCV valve from the valve cover and allow it to draw underhood air. Disconnect the control hose (which is ³⁄₁₆ in. in diameter) from the charcoal canister and plug it.

5. Being careful not to touch the hot exhaust manifold, and pulling directly on the bullet connector only, disconnect the oxygen sensor harness lead from the oxygen sensor. Then, jumper the harness lead to ground.

CAUTION

Make sure you don't put any stress on the wire to the oxygen sensor as you do this.

---

6. Wait two minutes to allow the effect of disconnecting the oxygen sensor to take full effect. As you wait, disconnect the vacuum line at the vacuum transducer on the Spark Control Computer. Then, connect an auxiliary vacuum supply to the vacuum transducer and set it at 61 in. of vacuum. When the two minutes have elapsed, proceed to the next step.

7. Open the main propane valve and then open the metering valve very slowly,

as adding too much propane will cause the engine rpm to suddenly decrease by running too rich. You'll find a point where turning the metering valve open farther will just begin slowing the rpm back down. Optimize the valve opening very carefully, allowing time for the engine to adjust, so the engine runs at the highest rpm possible.

8. Referring to the engine compartment sticker, adjust the idle speed screw (located on top of the carburetor solenoid) to get the rpm specified for propane enrichment. Repeat the last part of Step 7 in order to optimize the rpm. If this raises rpm, repeat the idle speed adjustment. Go

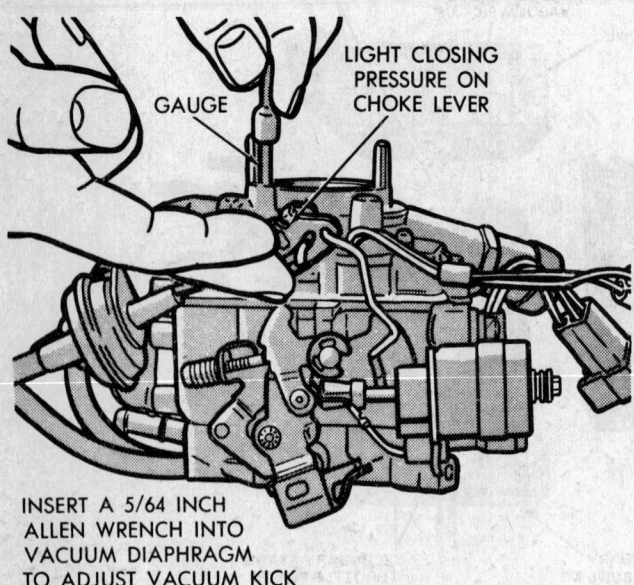

Vacuum kick adjustment—6145 carburetor

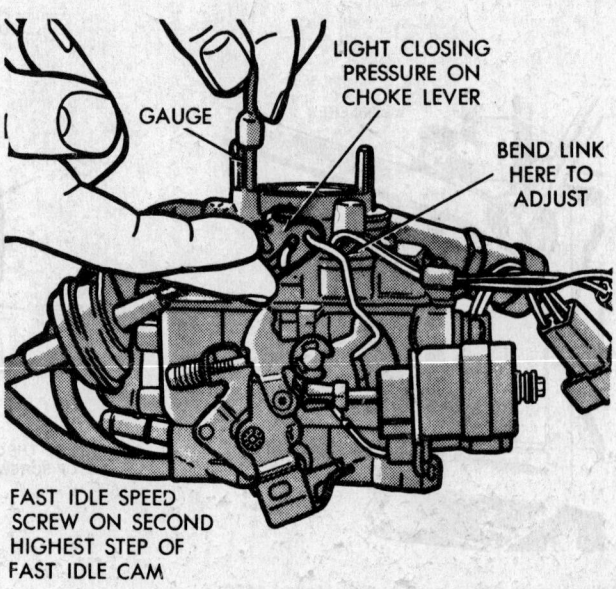

Fast idle cam adjustment—6145 carburetor

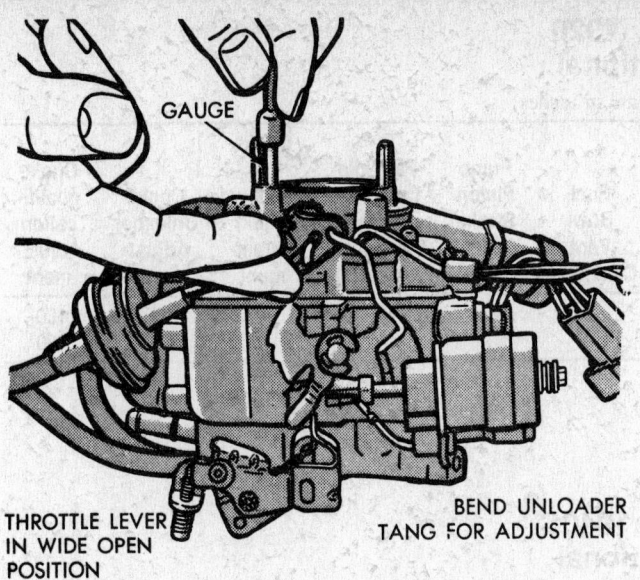

Choke unloader adjustment—6145 carburetor

THROTTLE LEVER IN WIDE OPEN POSITION

GAUGE

BEND UNLOADER TANG FOR ADJUSTMENT

LIGHT CLOSING PRESSURE

GAUGE

STEM FULLY DEPRESSED

TO VACUUM SOURCE

APPLY 15 INCHES OF VACUUM ON DIAPHRAGM

Vacuum kick adjustment—2280 carburetor

back and forth in this way until the engine runs at the specified rpm with the propane adjustment at the optimum level.

9. Turn off the propane metering valve and main valve. Allow the engine to run for one minute in order to allow the rpm to stabilize. Then, adjust the mixture screw very slowly, pausing after each change to allow the engine to stabilize, until you achieve the specified idle rpm and, at the same time, the smoothest possible idle.

10. Again, open the main propane valve and then adjust the metering valve carefully to optimize engine rpm without changing the throttle setting. Measure the rpm, it should be no more than 25 rpm more than in Step 9. If it is, repeat Steps 7–9, as you have failed to get the optimum propane mixture level at the specified propane rpm. Retest as necessary.

11. Turn off both propane valves. Disconnect the propane line and restore all vacuum and electrical connections changed at the beginning of the procedure. Install a new concealment plug over the mixture adjusting screw.

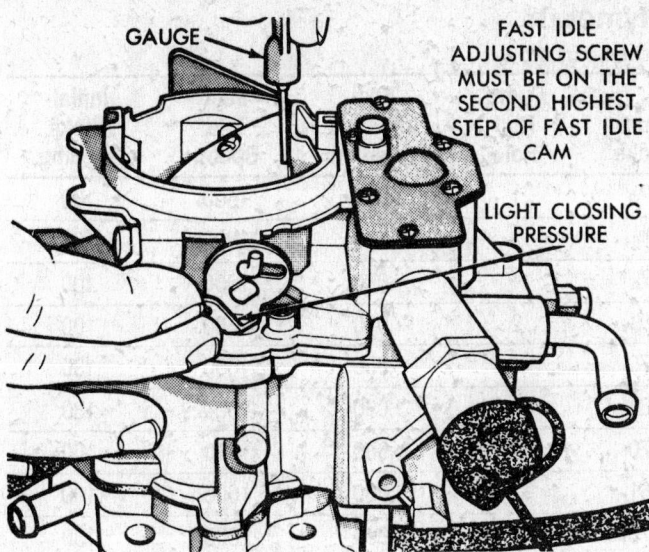

GAUGE

FAST IDLE ADJUSTING SCREW MUST BE ON THE SECOND HIGHEST STEP OF FAST IDLE CAM

LIGHT CLOSING PRESSURE

Fast idle adjustment—2280 carburetor

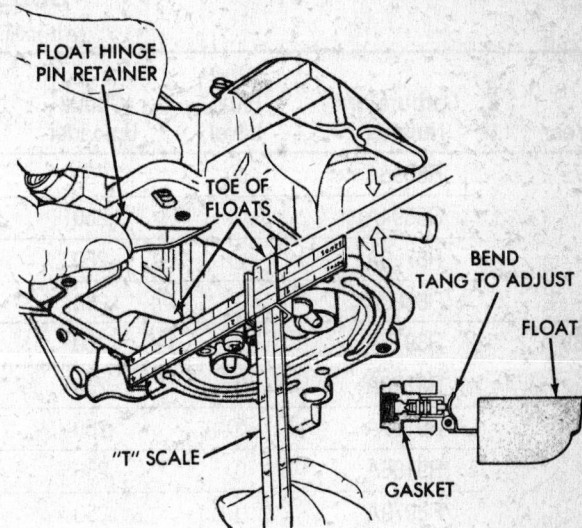

FLOAT HINGE PIN RETAINER

TOE OF FLOATS

"T" SCALE

BEND TANG TO ADJUST

FLOAT

GASKET

Float level adjustment—2280 carburetor

MODEL 1920
International
(All measurements in inches)

Year	Carburetor Number	Float Level	Fuel Level	Fast Idle Speed (rpm)	Auto Choke Setting	Dash-Post Setting	Fuel Bowl Vent Clearance	Pump Piston Stroke Adjustment	Fast Idle Cam Pos. Adjustment (Top Step—Hot)	Choke Vac. Pulldown (Kick) Adjustment	Choke Unloader Adjustment	Choke qualification Adjustment
'79–'81	7771	①	11/16 ± 1/32	2200	1 Lean	—	—	25/32	—	—	.235–.295	.150–.180

① Flush with top edge of bowl and with fuel inlet valve held closed

MODEL 1940C
International
(All measurements in inches)

Year	Carburetor Number	Float Level	Fuel Level	Automatic Choke	Dash Pot Setting	Choke Unloader	Fast Idle (rpm)	Curb Idle (rpm)	Pump Stroke	Idle %CO
'79–'81	7771	①	11/16 ± 1/32	1 Rich	.100–.130	.235–.295	2200	675–725	25/32	0.3–1.5

① Flush with top edge of bowl, with fuel inlet valve held closed

MODEL 1945
Dodge/Plymouth
(All measurements in inches)

Year	Carburetor Number	Dry Float Level	Choke Unloader	Pump Stroke	Pump Rod Hole	Fast Idle Cam Position	Fast Idle Speed	Initial Choke Opening
'79	R8593A	①	.250	2 7/32	1	.080	1600	.100
	R8594A	①	.250	2 21/64	2	.080	1600	.100
	R8799A	①	.250	2 7/32	1	.080	1600	.100
	R8800A	①	.250	2 21/64	2	.080	1600	.100
'80	R8978A	①	.250	1.70	1	.080	1600	.130
	R8720A	①	.250	1.61	2	.090	1600	.130
	R9107A	①	.250	1.70	1	.080	1600	.100
	R9106A	①	.250	1.61	2	.080	1600	.100
	R8979A	①	.250	1.70	1	.080	1600	.130
	R8721A	①	.250	1.61	2	.080	1600	.130
'81	R9131A	①	.250	1.61	2	.090	1600	.130
	R9132A	①	.250	1.61	2	.090	1600	.130
	R9134A	①	.250	1.61	2	.090	1600	.130
	R9152A	①	.250	1.61	2	.080	1800	.130
	R9153A	①	.250	1.61	2	.080	1800	.130
	R9399A	①	.250	1.61	2	.080	1800	.130

MODEL 1945
Dodge/Plymouth
(All measurements in inches)

Year	Carburetor Number	Dry Float Level	Choke Unloader	Pump Stroke	Pump Rod Hole	Fast Idle Cam Position	Fast Idle Speed	Initial Choke Opening
'82	R9132A	①	.250	—	2	.090	1600	.130
	R9134A	①	.250	—	2	.090	1600	.130
	R9153A	①	.250	—	2	.080	1800	.130
	R9399A	①	.250	—	2	.080	1800	.130
	R9762A	①	.250	—	2	.090	1600	.130
	R9765A	①	.250	—	2	.080	1800	.130
'83	R40055A	①	.250	—	2	.080	1600	.130
	R40056A	①	.250	—	2	.080	1600	.130
	R9399-1A	①	.250	—	2	.080	1800	.130
	R9134-1A	①	.250	—	2	.090	1800	.130
'84	R40088A	①	.250	1.61	2	.080	1600	.130
	R40089A	①	.250	1.61	2	.080	1600	.130
	R40102A	①	.250	1.70	1	.080	1800	.130
	R40103A	①	.250	1.61	2	.090	1600	.130
'85	R40102A	①	.250	1.70	1	.080	1800	.130
	R40244A	①	.250	1.61	2	.090	1600	.130
	R40159	①	.250	1.61	2	.080	1600	.130
	R40160	①	.250	1.61	2	.090	1600	.130

Note: Choke setting is fixed and non-adjustable.
① Flush with top of bowl cover gasket, carb inverted

MODEL 2210C
International
(All measurements in inches)

Year	Carburetor Number	Float Level	Fuel Level	Fast Idle Speed	Choke Setting	Choke Unloader	Choke Qualification Adjustment
'79–'81	6620-1 7309	.180	½①	2000	②	.198–.258	.040–.070
	7214, 7214-1	.180	½①	1800	Preset	.228 ± .030	.135 ± .015
	7309, 6620-2, 7133, 7940 8241	.180	½①	2200	②	.198–.258	.040–.070
	7657, 7217, 8244	.180	½①	1800	Preset	.198–.258	.120–.150

① @ 5.5 psi
② Choke with index marks—1 notch lean (restrained)
   Choke without index marks—preset (unrestrained)

MODEL 2245C
International
(All measurements in inches)

Year	Carburetor Number	Float Level	Fuel① Level	Automatic Choke	Dash Pot Setting	Choke Unloader	Fast Idle (rpm)	Curb Idle (rpm)	Idle % CO
'79	7773	.180	½	Preset	.105–.135	.198–.258	2200	675–725②	0.1–0.8

① 5.5 psi of fuel pressure
② Transmission in Neutral and air conditioning off

MODEL 2245
Dodge/Plymouth
(All measurements in inches)

Year	Carburetor Number	Float Level (Dry)	Choke Unloader	Choke Setting	Pump Rod Location (Hole)	Fast Idle Cam Position	Fast Idle Speed (rpm)	Initial Choke Opening
'79	R8597A	.200②	.170	①	1	.110	1600	.110
	R8598A	.200②	.170	①	1	.110	1600	.110
	R8925A	.200②	.170	①	1	.110	1600	.110
'80–'81	R7871A	3⁄16	.170	Fixed	1	.110	1700	.150
'82	R9816A	3⁄16	.170	Fixed	3	.110	1700	.150
'83	R9816A	5⁄32–7⁄32	.170	Fixed	2	.110	1700	.150

① Fixed setting
② Float drop—bottom of float to be parallel with air horn bottom

MODEL 2280, 2280G
Dodge/Plymouth
(All measurements in inches)

Year	Carburetor Number	Float Level①	Choke Vacuum Kick	Fast Idle Cam	Fast Idle (rpm)	Choke Unloader	Bowl Vent Valve
'80	R8999A	5⁄16	.130	.070	1600	.310	.030
	R9000A	5⁄16	.130	.070	1600	.310	.030
	R9001A	5⁄16	.150	.070	1600	.310	.030
	R9209A	5⁄16	.150	.070	1600	.310	.030
	R9224A	5⁄16	.150	.070	1600	.310	.030
'81	R9135A	9⁄32	.110	.070	1500	.310	—
	R9136A	9⁄32	.130	.070	1500	.310	—
	R9151A	9⁄32	.110	.070	1500	.310	—
	R9437A	9⁄32	.130	.070	1500	.310	—

MODEL 2280, 2280G
Dodge/Plymouth
(All measurements in inches)

Year	Carburetor Number	Float Level ①	Choke Vacuum Kick	Fast Idle Cam	Fast Idle (rpm)	Choke Unloader	Bowl Vent Valve
'82	R9491A	7/32–10/32	.140	.052	1500	.200	—
	R9493A	7/32–10/32	.140	.052	1500	.200	—
	R9572A	7/32–10/32	.140	.052	1500	.200	—
'83	R9499A	9/32	.140	.052	1500	.200	—
	R9951A	9/32	.140	.052	1500	.200	—
'84	R9951A	9/32	.140	.070	1500	.250	—
	R40093A	9/32	.140	.052	1500	.200	—
'85	R40164	9/32	.140	.070	1600	.150	.035
	R40167	9/32	.140	.070	1600	.150	.035
	R40172A	9/32	.140	.070	1450	.200	.035

① Measured from surface of fuel bowl to the toe of each float

MODEL 2300, 2300C, 2300G, 2300EG
International
(All measurements in inches)

Year	Carburetor Number	Fuel Level	Fast Idle (rpm)	Governor No load (rpm)	Governor Full load (rpm)	Fast Idle (rpm)	Curb Idle (rpm)	Idle % CO
'79–'81	8736	3/8	2200	4000	3800	1450 ± 50	625–675	1.0–3.0
	7922	3/8	2400	3800	3600	1350 ± 50	525–575	0.5–2.5
	8741	3/8	2000	3400	3200	1350 ± 50	500–550	1.0–2.0
	9072	3/8	2000	3800	3600	1350 ± 50	625–675	1.0–3.0
	8242	3/8	2000	4000	3800	1450 ± 50	625–675	1.0–3.0
	9076	3/8	2400	3800	3600	1350 ± 50	525–575	0.5–2.0
	8245	3/8	2400	3800	3600	1350 ± 50	525–575	0.5–2.0
	8248	3/8	2000	3400	3200	1350 ± 50	500–550	1.0–2.0
	8232	3/8	2200	4000	3800	1350 ± 50	650–700	2.0
	8180	3/8	2200	4000	3800	1300 ± 50	650–700	1.5
	8236	3/8	2400	3800	3600	1350 ± 50	525–575	0.5–2.5
	8235	3/8	2400	3800	3600	1350 ± 50	525–575	0.5–2.5
	8238	3/8	2000	3400	3200	1350 ± 50	500–550	1.5–3.0

MODEL 2300 EG
Ford
(All measurements in inches)

Year	Carburetor Number	Float Level	Float Drop	Choke Unloader Setting	Choke Setting	Dash Pot Plunger	Initial Choke Setting
'80	E0TE–9510 PA, PB, BCA, BCB, EVA, EYA	①	—	—	—	—	.375
'81	D9TE–9510 ABC, APA	①	—	—	Manual	—	.350–.400

① Check float level to the bottom of the sight plug in the carburetor

MODEL 2300EG
Ford
(All measurements in inches)

Year	Carburetor Number	Fuel Level (Wet)	Pump Lever Location	Power Valve Timing	Pulldown Setting
'82	D9TE APA AGA	①	# 2	9.5-7.0/3.0-.5	.350–.400
	E0TE-PA E2TE-NA	①	# 1	9.5-7.0/3.0-.5	.350–.400

MODEL 2300EG
Ford
(All measurements in inches)

Year	Carburetor Number	Fuel Level (Wet)	Pump Lever Location	Enrichment Valve Ident.	Main Jet Ident.
'83	D9TE-APA E2TE-AGA E0TE-PA E2TE-NA	①	—	12	58
'84–'85	D9TE-9510 APA, PA	①	#2	12	58
	E2TE-9510 AGA, NA	①	#2	12	58

① At bottom of sight plug

MODEL 2300C
Ford
(All measurements in inches)

Year	Carburetor Number	Fuel Level (Wet)	Pump Cam Location	Pulldown Setting	Choke Setting	Choke Unloader
'83	E2TE-9510-DPA	①	1	.210–.230	3 Rich	.300

MODEL 4150EG
Chevrolet/GMC
(All measurements in inches)

Year	Carburetor Number	Float Level (Dry)	Accelerator Pump (Min.)	Fast Idle (rpm)	Air Vent Clearance	Fast Idle Mechanical Clearance
'79	R8278A	②	.015	2200	.045–.075	.031
	R8280A	②	.015	2200	.045–.075	.031
	R8282A	②	.015	2200	.045–.075	.031
	R8283A	②	.015	2200	.045–.075	.031
	R8444A	②	.015	2200	.045–.075	.031
	R8279A	②	.015	2200	.045–.075	.031
	R8281A	②	.015	2200	.045–.075	.031
'80–'84	R8848A	②	.015	2200	.045–.075	.031
	R8849A	②	.015	2200	.045–.075	.031
	R8852A	②	.015	2200	.045–.075	.031
	R8853A	②	.015	2200	.045–.075	.031
	R8856A	②	.015	2200	.045–.075	.031
	R8850A	②	.015	2200	—	.031
	R8851A	②	.015	2200	—	.031
	R8854A	②	.015	2200	—	.031
	R8855A	②	.015	2200	—	.031
	R8857A	②	.015	2200	—	.031
'85	All Numbers (Federal and Canadian)	②	#1	③	.045–.075	.031
	All Numbers (California)	②	#1	③	—	.031

Note: Secondary set screw should be ½ turn open
③ See underhood specifications sticker

① Primary bowl—.197
   Secondary bowl—.166
② Primary bowl—.194
   Secondary bowl—.213

MODELS 4150G, 4150EG
International
(All measurements in inches)

Year	Carburetor Number	Fuel Level	Fast Idle Setting (rpm)	Governor Speed No-Load (rpm)	Governor Speed Full-Load (rpm)	Curb Idle Speed (rpm)	Idle Mixture Setting % CO
'79–'80 (Federal)	6803-3	①	2000	3800	3600	650–700	2.0 Max
	7215	①	2400	3800	3600	525–575	0.5–2.5
	7251	①	2400	3800	3600	525–575	0.5–2.5
	6911	①	2000	3400	3200	500–550	1.5–3.0
'79–'80 (Calif.)	7028, 7529	①	②	3800	3600	625–675	0.5–1.5
	7218, 7218-1	①	2400	3800	3600	525–575	0.5–2.5
	7581	①	2400	3800	3600	525–575	0.5–2.5
	7921	①	2400	3800 ± 50	—	525–575	0.5–2.5
'79–'80 (50 States)	7029, 7029-1	①	2400	3800	3600	525–575	0.5–2.5
	6974	①	2000	3400	3200	500–550	1.0–2.0

① Primary ⅜ in., Secondary ⅝ in.
② Mechanical setting-.015–.020 in.

MODEL 4152EG
Chevrolet/GMC
(All measurements in inches)

Year	Carburetor Number	Dry Float Level Primary	Dry Float Level Secondary	Secondary Set Screw	Fast Idle	Pump Cam Position
'85	All	.194	.213	½ turn open	.031 ①	#1 hole

① Mechanical setting

MODEL 4180EG
Ford
(All measurements in inches)

Year	Carburetor Number	Float Level (Dry)	Float Drop	Choke Unloader Setting	Choke Setting	Dash Pot Plunger	Initial Choke Setting
'80	E0TE-9510 ETA, JB, EUA, RB, JA, RA, ERA, MB, SB, SA, ESA, MA	①	—	—	—	—	.210
	D9TE-9510 AHE, EBA, ETA, EUA	①	—	—	—	—	.210

MODEL 4180EG
Ford
(All measurements in inches)

Year	Carburetor Number	Float Level (Dry)	Float Drop	Choke Unloader Setting	Choke Setting	Dash Pot Plunger	Initial Choke Setting
'81	D9TE-9510 EBA, AHE, ETA, EUA	①	—	—	Manual	—	.185–.235
	E0TE-9510 RA, JA, MA, SA	①	—	—	Manual	—	.185–.235

① Check float level to the bottom of the sight plug in the carburetor

MODEL 4180EG
Ford
(All measurements in inches)

Year	Carburetor Number	Fuel Level (Wet)	Choke Pulldown Setting	Pump Lever Location	Power Valve Timing
'82	E2TE-CSA, AJA CUA, CTA, AKA, CVA E0TE-SA495 D9TE-EVA495	①	.180–.440	#1	9.5-7.0/3.0-.5

① At bottom of sight plug

MODEL 4180C
Ford
(All measurements in inches)

Year	Carburetor Number	Fuel Level (Wet)	Choke Pulldown Setting	Choke Unloader Setting	Choke Setting	Pump Lever Location	Enrichment Valve Indent.
'80	D9TE-9510 DKA	①	—	.300–.330	—	—	②
'81	D9TE-9510 BKA	①	—		5 Rich	—	③
'82	E1UE-RA	①	.200–.220	.295–.335	2 Rich	#1	④
'83	E3TE-9510-PC 9510-PC	①	.210–.230	.300–.330	3 Rich	#1	11
	E3TE-9510-SB 9510-TB	①	.210–.230	.300–.330	3 Rich	#1	8

MODEL 4180C
Ford
(All measurements in inches)

Year	Carburetor Number	Fuel Level (Wet)	Choke Pulldown Setting	Choke Unloader Setting	Choke Setting	Pump Lever Location	Enrichment Valve Indent.
'84–'85	E4TE-9510-ARA	①	.185	.300	Index	#1	13
	E3TE-9510 PD	①	.220	.295–.335	3 Rich	#1	8
	TC	①	.210–.230	.300–.330	3 Rich	#1	8
	RD	①	.200	.315	3 Rich	#1	8A
	SC	①	.210–.230	.315	3 Rich	#1	8A

① At bottom of sight plug
② Initial choke setting .210 in.
③ Initial choke setting .195–.225
④ Power valve timing 9.5–7.0/3.0–.5

MODEL 4190EG

Year	Carburetor Number	Fuel Level (Wet)	Enrichment Valve Ident.	Pump Cam Location	Primary Main Jet	Secondary Main Jet
'83	E2TE-CMA CNA	①	12	#1	62	68
'84–'85	E2TE-9510 AGA, NA	①	12	#2	58	—
	CMA, CNA	①	12	#1	62	683
	CPA, CRA	①	12	#1	62	683
	E3HE-9510 AA, BA, CA, DA	①	12	#1	62	683

① At bottom of sight plug

MODEL 6145
Dodge/Plymouth
(All measurements in inches)

Year	Carburetor Number	Float Setting	Choke Vacuum Kick	Choke Unloader Adjustment	Fast Idle Cam Position	Fast Idle (rpm)	Pump Piston Stroke
'83	R40029A	①	.150	.250	.090	1600	1.70
	R40030A	①	.150	.250	.090	1600	1.61

MODEL 6145
Dodge/Plymouth
(All measurements in inches)

Year	Carburetor Number	Float Setting	Choke Vacuum Kick	Choke Unloader Adjustment	Fast Idle Cam Position	Fast Idle (rpm)	Pump Piston Stroke
'84	R40098A	①	.150	.250	.070	1600	1.61
	R40099A	①	.150	.250	.070	1600	1.61
'85	R40161	①	.150	.250	.060	1600	1.75
	R40162	①	.150	.250	.070	1600	1.75

① With bowl inverted, float lungs just touch a straightedge run along gasket surface.
② 1450 rpm with manual transmission

MODEL 6280
Dodge/Plymouth
(All measurements in inches)

Year	Carburetor Number	Float Level	Choke Vacuum Kick	Fast Idle Cam	Choke Unloader	Accelerator Pump Stroke	Fast Idle Speed	Propane Idle Speed
'85	R40132	①	.150	.070	.250	②	1400	710
	R40133	①	.130	.070	.150	②	1625③	775④

① Measured from surface of fuel bowl to top of each float
② Flush with top of bowl vent
③ 1450 rpm with manual transmission
④ 740 rpm with manual transmission

MAZDA CARBURETORS

FAST IDLE ADJUSTMENT

2299cc Engine

1. Close the choke fully.
2. Place the fast idle adjusting screw on the first step of the fast idle cam.
3. Check the clearance between the throttle plate and the inside of the throttle wall with a wire feeler gauge. It should measure 0.058–0.066 in. (0.061–0.071 in. for 1981 and later California models).
4. If the clearance is incorrect, adjust by means of the fast idle screw, clockwise to increase or counterclockwise to decrease.

1970cc Engine

1. Close the choke fully.
2. Place the fast idle adjusting screw on the first step of the fast idle cam.
3. Check the clearance between the throttle plate and the inside of the throttle wall with a wire feeler gauge. It should measure 0.051–0.059 in.
4. If the clearance is incorrect, adjust by means of the fast idle screw, clockwise to increase or counterclockwise to decrease.

1998cc Engine

1. Set the fast idle cam so that the fast idle lever rests on the second step of the cam.
2. Adjust the clearance between the air horn wall and the lower edge of the throttle plates by turning the fast idle adjusting screw. Clearance should be 0.029–0.044 in.

FLOAT AND FUEL LEVEL ADJUSTMENT

2299cc Engine

1. With the engine running, check the fuel level in the sight glass (in the fuel bowl).
2. If the fuel level is not at the specified mark on the sight glass, remove the carburetor from the truck.
3. Remove the fuel bowl cover.
4. Invert the carburetor and lower the float until the tang on the float just contacts the needle valve.
5. Measure the clearance between the float and the edge of the bowl. It should be 0.236 in.
6. If the clearance is not as specified, bend the float tang until the proper clearance is obtained.

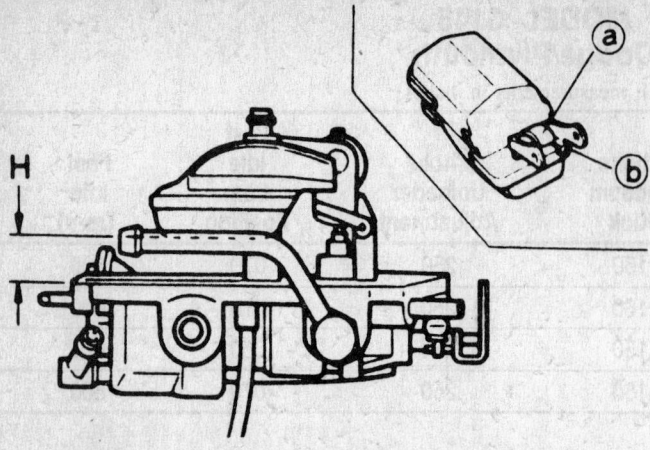

Float level adjustment—Mazda carburetor

7. Install the fuel bowl cover.

8. Reinstall the carburetor on the truck.

9. Recheck the fuel level at the sight glass.

1970cc Engine

1. With the engine running, check the fuel level in the sight glass (in the fuel bowl).

2. If the fuel level is not at the specified mark on the sight glass, remove the carburetor from the truck.

3. Remove the fuel bowl cover.

4. Invert the carburetor and lower the float until the tang on the float just contacts the needle valve.

5. Measure the clearance between the float and the edge of the bowl. Clearance should be 0.0335 in.

6. If the clearance is not as specified, bend the float seal lip until the proper clearance is obtained.

7. Turn the carburetor right side up and measure the clearance between the bottom of the float and the bowl. Clearance should be 0.039 in.

8. If not, bend the float stopper until the correct clearance is obtained.

9. Install the fuel bowl cover.

10. Reinstall the carburetor on the truck.

11. Recheck the fuel level at the sight glass.

1986 B2000

1. Remove the air horn from the carburetor.

2. Turn the air horn upside down on a level surface. Allow the float to hang under its own weight.

3. Measure the clearance between the float and the air horn gasket surface. The gap should be 11.5–12.5mm. If not, bend the float seat lip to obtain the correct gap.

4. Turn the air horn right side up and allow the float to hang under its own weight.

5. Measure the gap between the BOTTOM of the float and the air horn gasket surface. The gap should be 46.0–47.0mm. If not, bend the float stopper until it is.

CHOKE ADJUSTMENT

1970cc and 2299cc Engines

There are four adjustments to be made to the choke in these years; choke/throttle valve opening adjustment; choke diaphragm adjustment; choke unloader adjustment; and choke thermostat (bi-metal) adjustment.

CHOKE/THROTTLE VALVE OPENING ANGLE

1. Adjust the fast idle cam as previously outlined before making this adjustment.

2. Place the fast idle screw on the second step of the fast idle cam.

3. Measure the clearance between the edge of the choke plate and the throttle bore with a wire gauge. Clearance should be 0.016–0.028 in. for the 1970cc, Federal models (0.024–0.036 in. for California models); 0.051–0.071 in. for 1979–80 2299cc Engines; 0.039–0.051 in. for 1981 and later 2299cc Federal models (0.041–0.067 in. for California models).

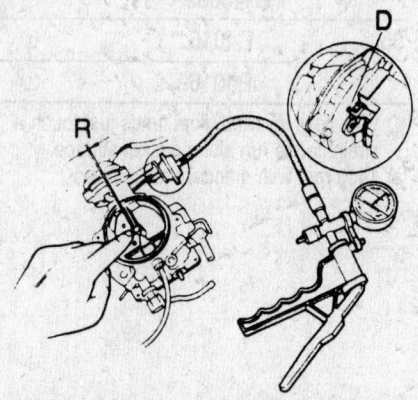

Choke diaphragm adjustment—Mazda carburetor

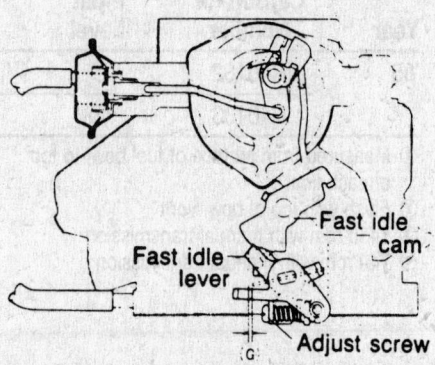

Fast idle cam adjustment—Mazda carburetor

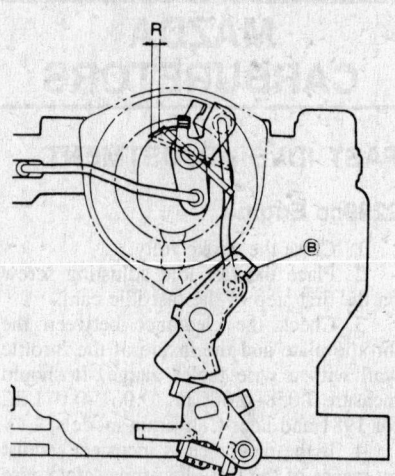

Choke valve clearance adjustment—Mazda carburetor

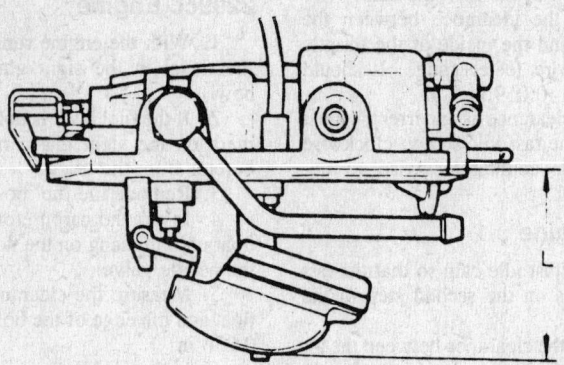

Float drop measurement—Mazda carburetor

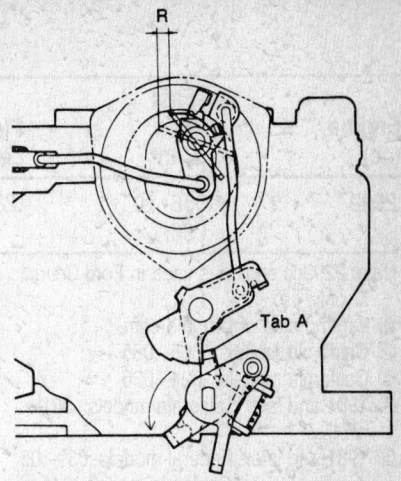

4. If the clearance is incorrect, adjust by bending the starting arm. If a large adjustment is required, the choke rod can be bent slightly.

CHOKE DIAPHRAGM

1. Remove the vacuum hose from the choke diaphragm. Attach a vacuum pump to the diaphragm fitting and apply approximately 15.6 in. Hg vacuum to the diaphragm.

2. Check to see that the fast idle screw is on the first step of the fast idle cam.

3. Press on the choke plate slightly to settle it. Measure the clearance between the edge of the choke plate an the throttle bore. Clearance should be 0.047–0.067 in. for 1970cc Federal models; 0.065–0.085 in. for California models; 0.051–0.071 for Federal 2299cc engines and 0.063–0.079 in. for California models.

4. If the clearance is incorrect, adjust by bending the choke lever.

CHOKE UNLOADER

1. Close the choke plate fully. Open the throttle plate fully.

2. Measure the clearance between the edge of the choke plate and the throttle bore with a wire gauge. Clearance should be 0.079–0.099 in. for 1970cc engines; 0.090–0.110 in. for the 2299cc engine.

3. If the clearance is incorrect, bend the choke unloader adjusting nail (tang).

THERMOSTAT

1. The index mark on the thermostat cover should be aligned with the center mark on the choke housing.

2. Adjust by loosening the thermostat cover retaining screws slightly and shifting the position of the cover. Tighten the screws after adjustment.

1998cc Engine

Three choke related adjustments are performed on these units.

CHOKE DIAPHRAGM

1. Disconnect the vacuum line from the choke diaphragm unit.

2. Using a vacuum pump, apply 15.7 in. Hg to the diaphragm.

3. Using light finger pressure, close the choke valve. Check the clearance at the upper edge of the choke valve. Clearance should be 0.066–0.084 in.

4. If not, bend the tab on the choke lever to adjust it.

CHOKE VALVE CLEARANCE

1. Position the fast idle lever on the second step of the fast idle cam.

2. The leading edge of the choke valve should be 0.60–1.0mm from fully closed. If not, bend either the tab on the cam or the choke rod to adjust. The tab will give smaller adjustment increments.

CHOKE UNLOADER

1. Open the primary throttle valve all the way and hold it in this position.

2. The leading edge of the choke valve should be 2.74–3.60mm from fully closed. If not, bend the tab on the throttle lever.

THROTTLE LINKAGE ADJUSTMENT

1979–84

1. Loosen the locknuts on the longer linkage rod and rotate both ends in the sockets until the proper accelerator travel from idle to wide-open throttle is obtained.

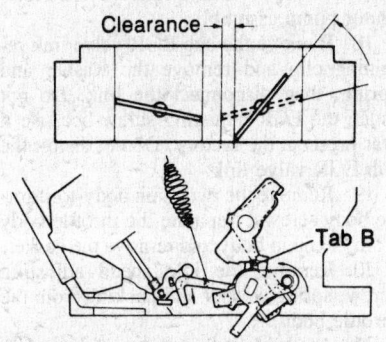

Secondary throttle valve adjustment—Mazda carburetor

2. Tighten the locknuts to set the adjustment.

1986

1. Check the cable deflection at the carburetor. Deflection should be 1.0–3.0mm. If not, adjust it by turning the adjusting nut at the bracket near the carburetor.

2. Depress the accelerator pedal to the floor. The throttle plates should be vertical. If not, adjust their position by turning the adjusting nut on the accelerator pedal bracket.

SECONDARY THROTTLE VALVE ADJUSTMENT

All Models

The clearance between the primary throttle valve and the air horn wall, when the secondary throttle valve just starts to open should be: 1979–84 Courier 1970cc and Mazda B2000: 0.256 ± 0.015 in. 1986 B2000: 0.289–0.325 in. If the clearance is not as specified, bend the fast idle link (1979–84) or the tab B (1986).

Choke unloader adjustment—Mazda carburetor

MAZDA CARBURETORS

(All measurements in inches)

Engine (cc)	Fast Idle Cam	Float Level	Choke Valve Opening	Choke Diaphragm	Choke Unloader	Secondary Throttle Valve
1970	.051–.059	.0335	.016 ③ .028	.047–② .067	.079–.099	.256
1998	.029–.044	.453–.492	.023–.039	.066–.084	.107–.141	.289–.325

MAZDA CARBURETORS

(All measurements in inches)

Engine (cc)	Fast Idle Cam	Float Level	Choke Valve Opening	Choke Diaphragm	Choke Unloader	Secondary Throttle Valve
2299	.058–④ .066	.236	.051–⑤ .071	.051–⑥ .071	.090– .110	.256

Note: 2299cc engine is used in Ford Courier only

① On 2nd step of fast idle cam
② California models: .065–.085
③ California models: .024–.036
④ 1981 and later California models: .061–.071
⑤ 1981 and later: Federal models-.039–.051
   California models-.041–.067
⑥ California models: .063–.079

MITSUBISHI CARBURETORS

OVERHAUL

1979–80

1. Disconnect the water hose.
2. Remove the throttle return spring and damper spring.
3. Remove the throttle adjuster lever spring and the secondary return spring.
4. Remove the choke unloader link retaining clip and disconnect the choke unloader link.
5. Disconnect the vacuum hose.
6. Disconnect the lower end of the diaphragm chamber link and remove the diaphragm chamber. Do not immerse the diaphragm chamber assembly in cleaner.
7. Remove the two screws and remove the air switching valve (ASV).
8. Remove the six float chamber cover screws. Separate the float chamber cover from the carburetor main body by tapping with a plastic hammer. Do not pry it off.
9. Remove the float chamber cover gasket.
10. Remove the float lever pin and the float.
11. Remove the needle valve assembly, gasket and filter.
12. Do not remove the automatic choke system because the factory setting will be disturbed.
13. Turn the main body upside down and remove the pump discharge check ball and weight.
14. Remove the fuel cut-off solenoid.

15. Remove the main jets and pilot jets. Do not tamper with the screws with white paint on heads.
16. Remove the enrichment assembly.
17. Disconnect the pump rod from the throttle shaft lever and remove the accelerator pump assembly.
18. Remove the sub EGR valve link retaining clip and remove the washer and spring, then disconnect the link. Do not touch the EGR adjusting screw because it was preset at the factory. Do not distort the sub EGR valve link.
19. Remove the two main body-to-throttle body screws. Separate the throttle body from the main body and remove the gasket.
20. Remove the idle speed adjusting screw, spring washer and packing from the throttle body.
21. Assembly is the reverse of disassembly.

OVERHAUL

1981 and Later

NOTE: A thorough test and check of minor carburetor adjustments should precede any major carburetor overhauls. Sometimes poor performance can be the result of loose, misadjusted or malfunctioning engine or electrical components.

1. Pull the water hose off the nipple of throttle body and off the nipple of wax element portion.
2. Grind down the head of choke cover lock screws (in 2 positions) by using a hand grinder or other instruments.
3. Disconnect the ground line of the fuel cut-off solenoid at the float chamber cover.
4. Remove the throttle return spring and the damper spring.
5. Pull off the vacuum hose connecting the depression chamber and the throttle body.
6. Remove the accelerator pump rod from the throttle lever.
7. Remove the dashpot rod (manual transmission) or throttle opener rod (automatic transmission) from the free lever.
8. Remove the depression chamber rod from the secondary throttle lever.
9. Remove the float chamber cover screws (6). Four screws connect the float chamber cover to the main body. Two screws connect the cover to the throttle body.
10. Remove only the main body by lifting the float chamber cover (the cover can not be removed because the choke unloader rod is connected to the throttle shaft.) Don't turn the carburetor up side down, during the procedure. Turning the carburetor causes accelerator pump check weight, ball, and steel ball of anti-overfill device to drop.

NOTE: When lifting the chamber cover, the venturi retaining spring may drop down.

11. Remove the E-ring at the lower end of the choke unloader rod, and disconnect the rod from the lever.
12. Do not remove any device from the float chamber unless necessary, especially the autochoke system.
13. The float can be removed by pulling the pin off.
14. The needle valve can be removed by removing the screw and retainer.
15. Remove the accelerator pump and fuel cut off solenoid.

16. Don't remove parts without necessity. Don't remove the throttle valves and don't touch SAS and dashpot adjusting screws.

17. Assembly is the reverse of disassembly.

FLOAT LEVEL ADJUSTMENT

1. Invert the float chamber cover assembly without the gasket.

2. Position a universal float level gauge or similar depth gauge in the float. The distance from the bottom of the float to the surface of the float chamber cover should be 0.787 ± 0.0394 in.

3. If the reading is not within this range, the shim under the needle seat must be changed. Shim kits are available from the dealer. Adding or removing a shim will change the float level by three times the shim thickness.

CHOKE VALVE SETTING

1. Fit the strangler spring to the float lever.

2. Assemble the choke pinion, aligning the inscribed line or black painted line on the tooth of the choke pinion with the inscribed line of the cam lever.

3. Temporarily tighten the new lockscrews.

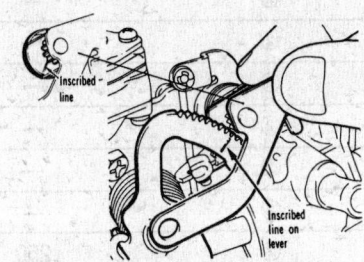

Choke valve setting—Mitsubishi carburetor

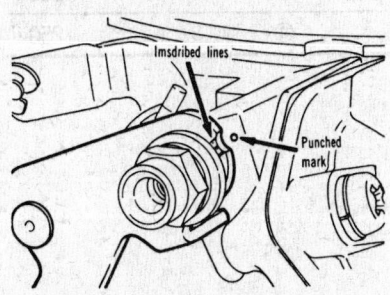

Setting choke pinion—Mitsubishi carburetor

4. Set the choke valve by moving the pinion arm up or down. Align a punched mark on the float chamber cover at the center of the three inscribed lines and secure the pinion arm with the lock screws.

FAST IDLE OPENING ADJUSTMENT

Before adjustment, the carburetor must be at room temperature (73°F) for about one hour. Adjust the fast idle opening to specifications by turning the fast idle adjusting screw. On the G63B with manual transmission, the opening adjustment should be

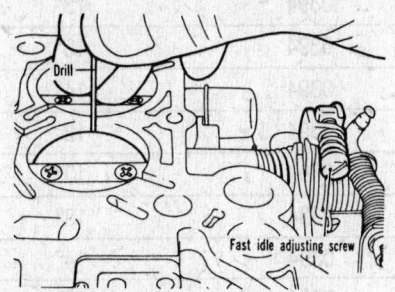

Fast idle opening adjustment— Mitsubishi carburetor

0.025 in. On the G63B with automatic and G54B with manual transmission, the opening should be 0.028 in. On the G54B with automatic transmission, the opening should be 0.031 in.

THROTTLE POSITION SWITCH (TPS) ADJUSTMENT

1. Start the engine and allow it to reach normal operating temperature. Make sure the fast idle cam is released.

2. Make sure the throttle valve is fully shut.

3. Connect a digital voltmeter between terminals 2 and 3 of the TPS connector. Do not separate the TPS connector and body side harness.

4. Turn the ignition ON, then adjust the TPS adjustment screw so the output voltage reads 250 mV.

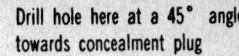

Drill hole here at a 45° angle towards concealment plug

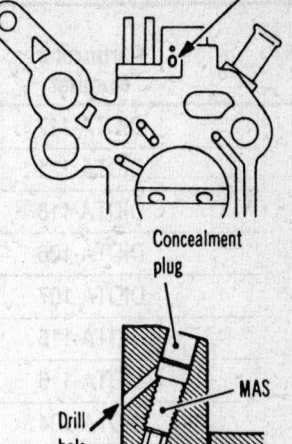

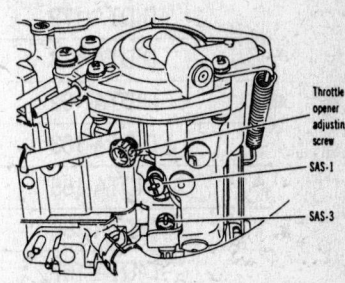

Removing mixture screw plug— Mitsubishi carburetor

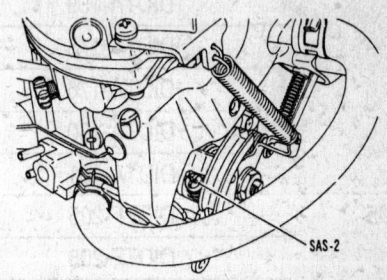

Location of adjustment screws— Mitsubishi carburetor

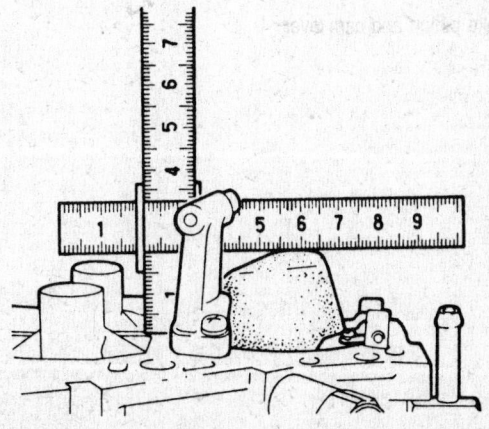

Idle speed adjusting screw—Mitsubishi carburetor

Float adjustment on non-feedback Mitsubishi carburetor

MITSUBISHI CARBURETORS
(All measurements in inches or degrees)

Year	Carburetor Number	Float Level	Fast Idle Opening ①	Choke Valve	TPS Voltage (mV)
'83	DIDTA-117	.0394	12°	②	—
	DIDTA-121	.0394	12°	②	—
	DIDTA-118	.0394	13°	②	—
	DIDTA-106	.0394	13°	②	—
	DIDTA-107	.0394	14°	②	—
	DIDTA-115	.0394	12°	②	—
	DIDTA-116	.0394	13°	②	—
	DIDTA-104	.0394	13°	②	—
	DIDTA-105	.0394	14°	②	—
'84	DIDTA-177	.0394	12°	②	—
	DIDTA-179	.0394	12°	②	—
	DIDTA-178	.0394	13°	②	—
	DIDTA-182	.0394	13°	②	—
	DIDTA-183	.0394	14°	②	—
	DIDTA-165	.0394	12°	②	—
	DIDTA-166	.0394	13°	②	—
	DIDTA-167	.0394	12°	②	—
	DIDTA-168	.0394	13°	②	—
	DIDTA-169	.0394	14°	②	—
	DIDTA-175	.0394	12°	②	—
	DIDTA-176	.0394	13°	②	—
	DIDTA-180	.0394	13°	②	—
	DIDTA-181	.0394	14°	②	—
'85	DIDTF-205	.0394	.025	②	250
	DIDTF-206	.0394	.028	②	250
	DIDTF-207	.0394	.028	②	250
	DIDTF-208	.0394	.031	②	250

① @ 73°F
② Align marks on choke pinion and cam lever

MOTORCRAFT CARBURETORS

IDLE MIXTURE ADJUSTMENT

Model 2150 Carburetor

NOTE: The idle mixture adjustment screws are concealed by tamper resistant caps. The idle mixture should be adjusted only if the mixture adjustment screws were removed or altered during major carburetor overhaul or tampering.

1. Connect a tachometer to the ignition coil negative terminal.
2. Start the engine and allow it to reach normal operating temperature.

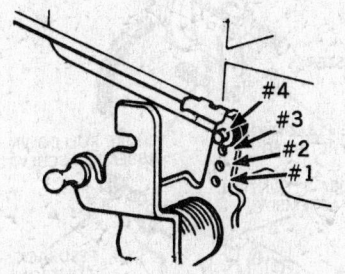

Accelerator pump stroke hole location—typical models 2100 and 2150 carburetors

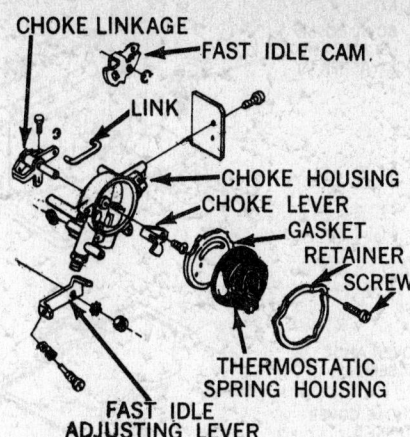

Automatic choke assembly (typical)—model 2100 carburetor

3. Set the parking brake and chock the wheels. Position the automatic transmission in the DRIVE detent.
4. Be sure choke is completely off and the idle speed is set to specifications.
5. Turn the idle mixture adjusting screws clockwise (leaner) until a perceptible loss of engine speed (rpm) is noted on the tachometer.
6. Turn the idle mixture adjusting screws counterclockwise (richer) until the highest engine speed (rpm) is obtained.
7. This position of the idle mixture adjusting screws is referred to as the LEAN BEST IDLE.
8. For the final adjustment, turn both idle mixture adjusting screws clockwise in small, equal amounts until the specified idle speed drop is noted on the tachometer.

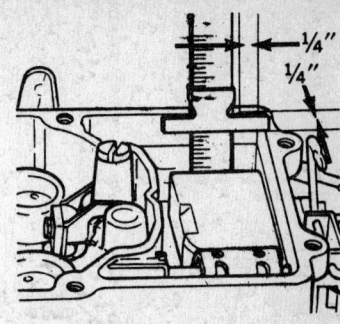

Fuel level adjustment (wet)—models 2100 and 2150 carburetors

Indexing marks for automatic choke thermostatic spring housing and choke housing—typical models 2100 and 2150 carburetors

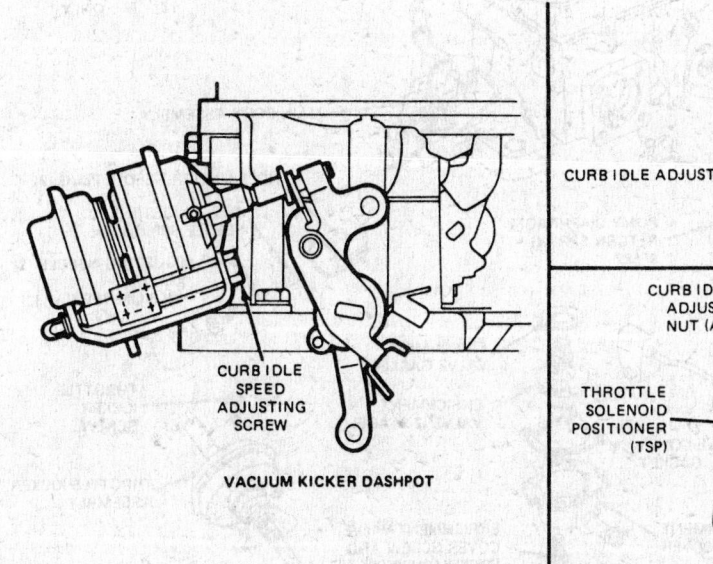

VACUUM KICKER DASHPOT

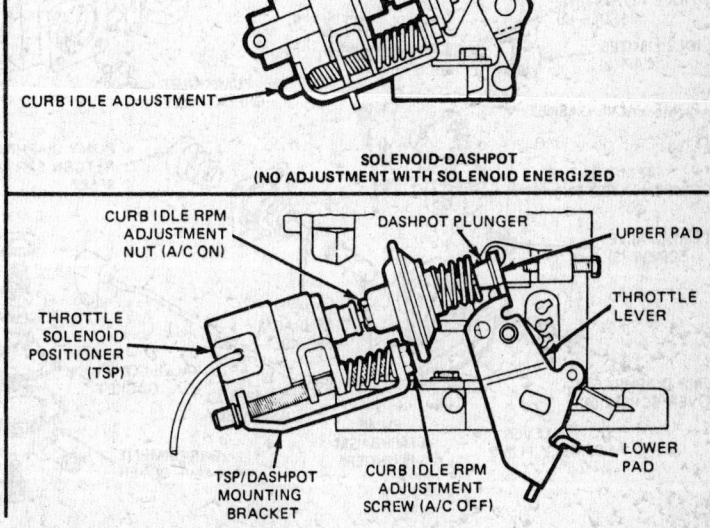

Curb idle adjustment with throttle positioner—Motorcraft 2150 carburetor

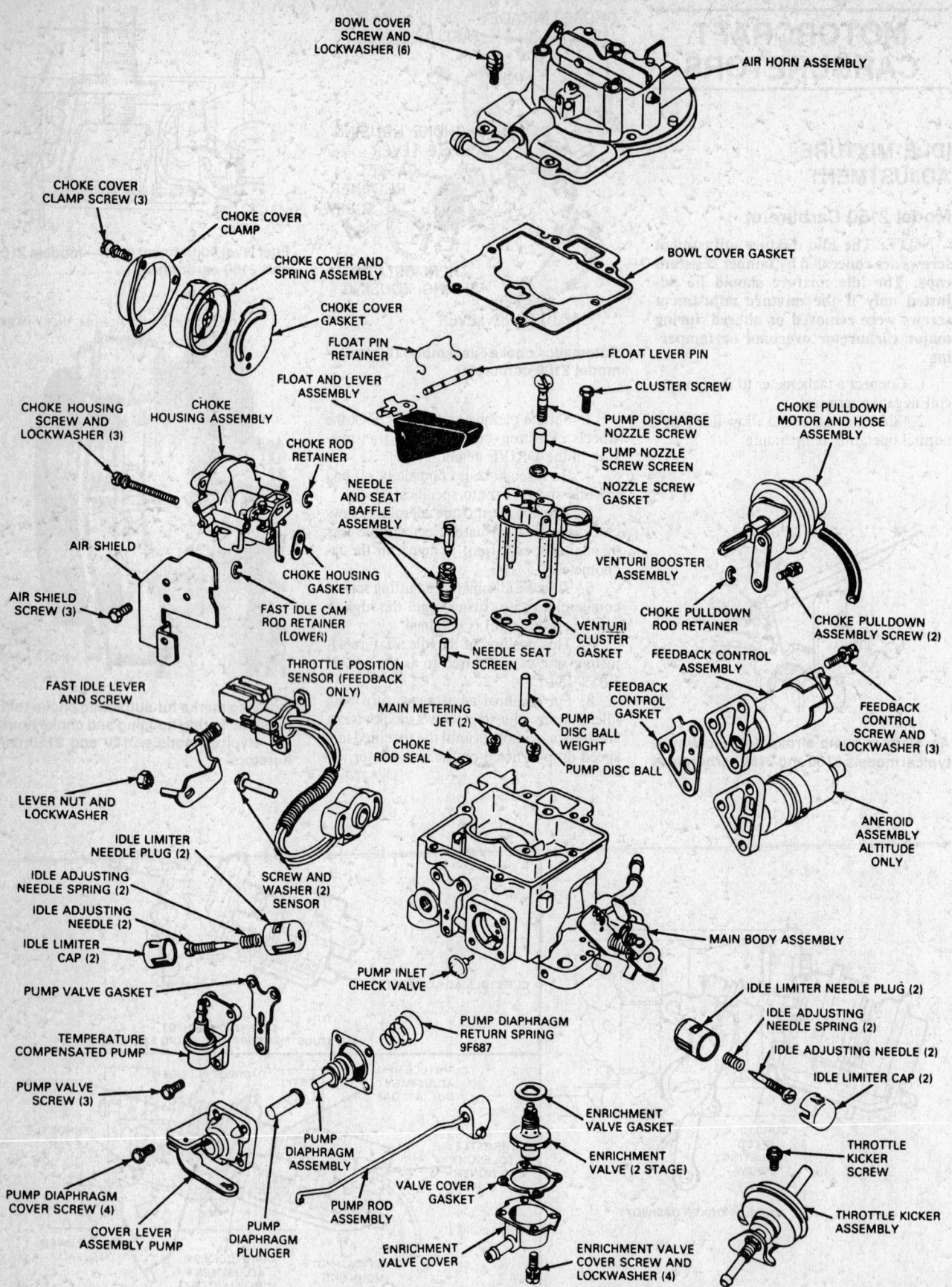

BOWL COVER
SCREW AND
LOCKWASHER (6)

AIR HORN ASSEMBLY

CHOKE COVER
CLAMP SCREW (3)

CHOKE COVER
CLAMP

CHOKE COVER AND
SPRING ASSEMBLY

BOWL COVER GASKET

CHOKE COVER
GASKET

FLOAT PIN
RETAINER

FLOAT LEVER PIN

FLOAT AND LEVER
ASSEMBLY

CLUSTER SCREW

CHOKE PULLDOWN
MOTOR AND HOSE
ASSEMBLY

CHOKE HOUSING
SCREW AND
LOCKWASHER (3)

CHOKE
HOUSING ASSEMBLY

PUMP DISCHARGE
NOZZLE SCREW

CHOKE ROD
RETAINER

PUMP NOZZLE
SCREW SCREEN

NOZZLE SCREW
GASKET

NEEDLE AND SEAT
BAFFLE
ASSEMBLY

AIR SHIELD

VENTURI BOOSTER
ASSEMBLY

CHOKE PULLDOWN
ROD RETAINER

CHOKE PULLDOWN
ASSEMBLY SCREW (2)

AIR SHIELD
SCREW (3)

CHOKE HOUSING
GASKET

FAST IDLE CAM
ROD RETAINER
(LOWER)

VENTURI
CLUSTER
GASKET

FEEDBACK CONTROL
ASSEMBLY

FEEDBACK CONTROL
SCREW AND
LOCKWASHER (3)

FAST IDLE LEVER
AND SCREW

NEEDLE SEAT
SCREEN

FEEDBACK
CONTROL
GASKET

THROTTLE POSITION
SENSOR (FEEDBACK
ONLY)

MAIN METERING
JET (2)

PUMP
DISC BALL
WEIGHT

FEEDBACK
CONTROL
SCREW AND
LOCKWASHER (3)

LEVER NUT AND
LOCKWASHER

CHOKE
ROD SEAL

PUMP DISC BALL

ANEROID
ASSEMBLY
ALTITUDE
ONLY

IDLE LIMITER
NEEDLE PLUG (2)

IDLE ADJUSTING
NEEDLE SPRING (2)

SCREW AND
WASHER (2)
SENSOR

IDLE ADJUSTING
NEEDLE (2)

IDLE LIMITER
CAP (2)

MAIN BODY ASSEMBLY

IDLE LIMITER NEEDLE PLUG (2)

IDLE ADJUSTING
NEEDLE SPRING (2)

PUMP VALVE GASKET

PUMP INLET
CHECK VALVE

IDLE ADJUSTING
NEEDLE (2)

TEMPERATURE
COMPENSATED PUMP

PUMP DIAPHRAGM
RETURN SPRING
9F687

IDLE LIMITER CAP (2)

PUMP VALVE
SCREW (3)

ENRICHMENT
VALVE GASKET

THROTTLE
KICKER
SCREW

PUMP DIAPHRAGM
COVER SCREW (4)

PUMP
DIAPHRAGM
ASSEMBLY

ENRICHMENT
VALVE (2 STAGE)

COVER LEVER
ASSEMBLY PUMP

PUMP ROD
ASSEMBLY

VALVE COVER
GASKET

THROTTLE KICKER
ASSEMBLY

PUMP
DIAPHRAGM
PLUNGER

ENRICHMENT
VALVE COVER

ENRICHMENT VALVE
COVER SCREW AND
LOCKWASHER (4)

Exploded view of 2150A feedback carburetor

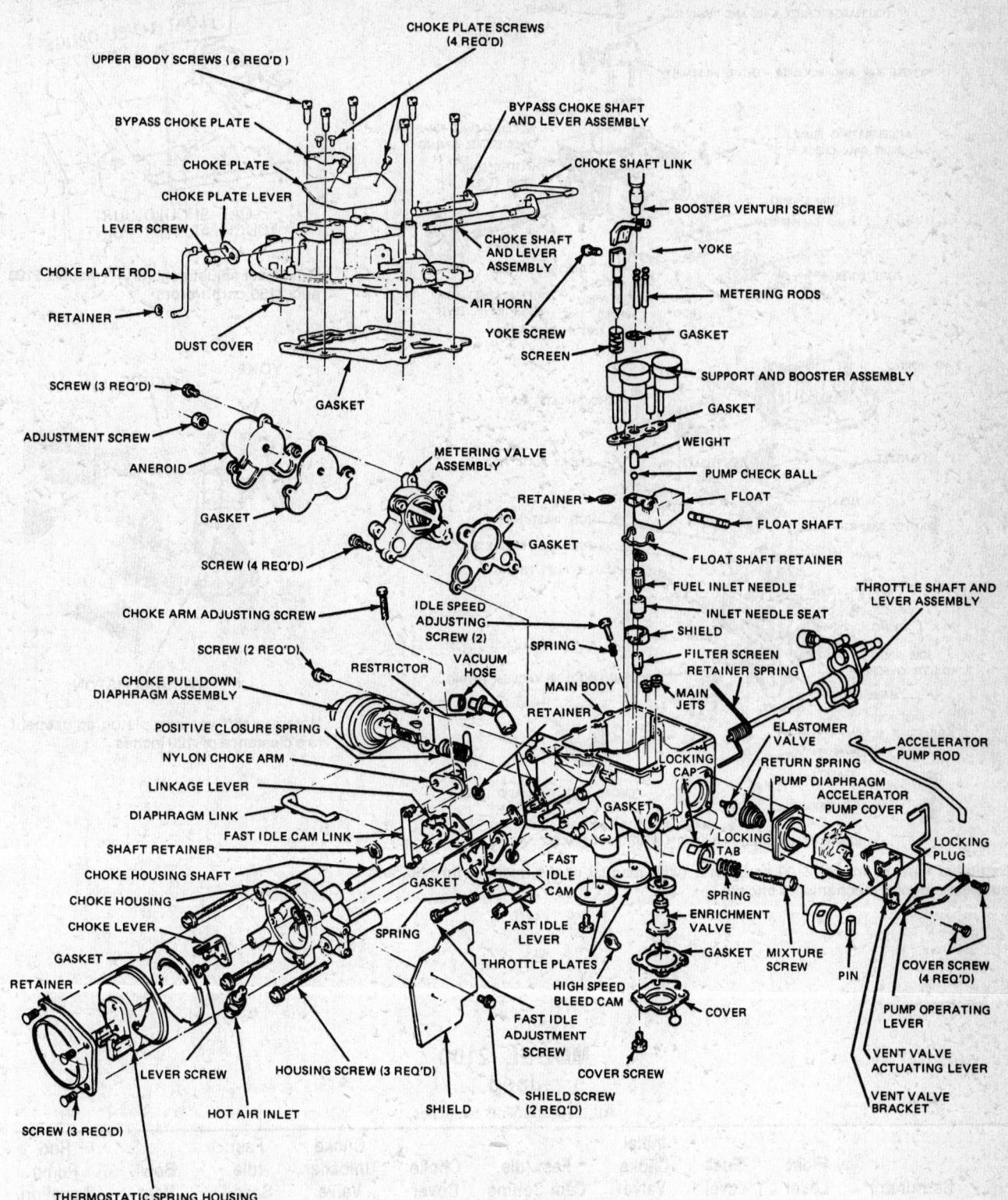

Typical late model 2150 carburetor

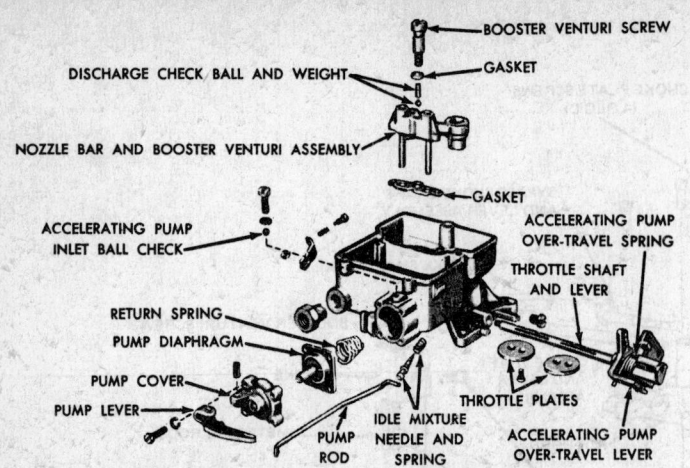

BOOSTER VENTURI SCREW
GASKET
DISCHARGE CHECK BALL AND WEIGHT
NOZZLE BAR AND BOOSTER VENTURI ASSEMBLY
GASKET
ACCELERATING PUMP INLET BALL CHECK
ACCELERATING PUMP OVER-TRAVEL SPRING
THROTTLE SHAFT AND LEVER
RETURN SPRING
PUMP DIAPHRAGM
PUMP COVER
PUMP LEVER
PUMP ROD
IDLE MIXTURE NEEDLE AND SPRING
THROTTLE PLATES
ACCELERATING PUMP OVER-TRAVEL LEVER

LEFT FRONT VIEW

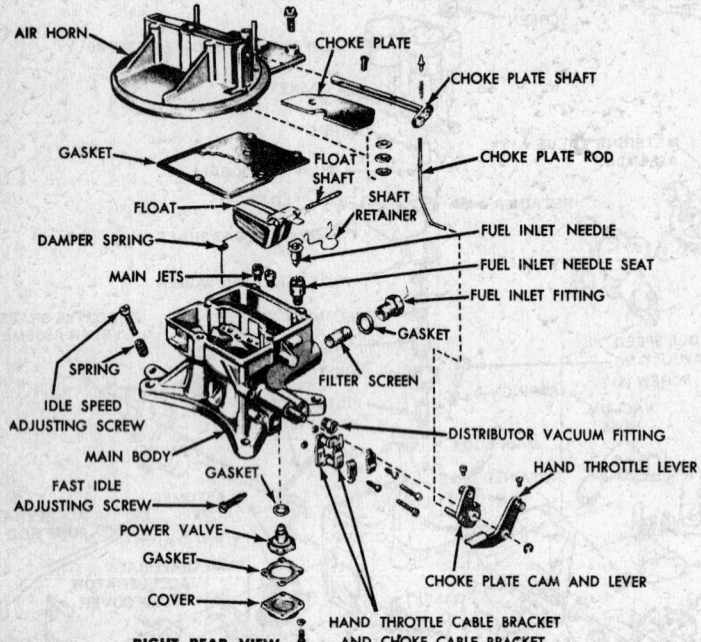

AIR HORN
CHOKE PLATE
CHOKE PLATE SHAFT
GASKET
FLOAT SHAFT
CHOKE PLATE ROD
SHAFT RETAINER
FLOAT
DAMPER SPRING
FUEL INLET NEEDLE
FUEL INLET NEEDLE SEAT
MAIN JETS
FUEL INLET FITTING
GASKET
SPRING
FILTER SCREEN
IDLE SPEED ADJUSTING SCREW
MAIN BODY
DISTRIBUTOR VACUUM FITTING
FAST IDLE ADJUSTING SCREW
GASKET
HAND THROTTLE LEVER
POWER VALVE
GASKET
COVER
CHOKE PLATE CAM AND LEVER
HAND THROTTLE CABLE BRACKET AND CHOKE CABLE BRACKET

RIGHT REAR VIEW

Exploded view of model 2100 carburetor (with manual choke, manual throttle, and automatic choke mechanisms shown)

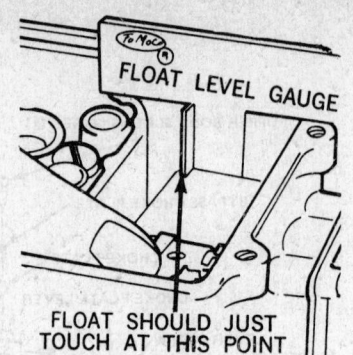

FLOAT LEVEL GAUGE

FLOAT SHOULD JUST TOUCH AT THIS POINT

Float level adjustment (dry)—models 2100 and 2150 carburetors

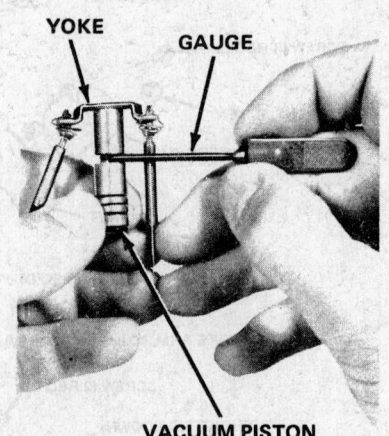

YOKE
GAUGE
VACUUM PISTON

Metering rod vacuum piston adjustment to a clearance of .120 inches

MODEL 2100
Jeep
(All measurements in inches)

Year	Carburetor Number	Float Level (Dry)	Fuel Level (Wet)	Initial Choke Valve Clearance	Fast Idle Cam Setting ②	Choke Cover Setting	Choke Unloader Valve Clearance	Fast Idle Speed ①	Bowl Vent Clearance	Rod Pump Location Hole
'79	9DM2	.555	.930	.125	.120	2 Rich	.250	1500	.120	—
	9DM2C	.555	.930	.132	.120	1 Rich	.250	1500	.120	—
	9DA2J	.555	.930	.128	.113	1 Rich	.250	1600	.120	—
	9DM2H	.555	.930	.140	.125	Index	.250	1500	.120	—

MODEL 2100
Jeep
(All measurements in inches)

Year	Carburetor Number	Float Level (Dry)	Fuel Level (Wet)	Initial Choke Valve Clearance	Fast Idle Cam Setting ②	Choke Cover Setting	Choke Unloader Valve Clearance	Fast Idle Speed ①	Bowl Vent Clearance	Rod Pump Location Hole
'80	ODMJ12	.375	.093	.125	.113	2 Rich	.300	1500	.120	3
	ODM2JC	.375	.093	.120	.106	2 Rich	.300	1500	.120	3
	ODA2J2	.375	.093	.120	.106	2 Rich	.300	1600	.120	3
	ODA2J	.375	.093	.128	.113	2 Rich	.300	1600	.120	3
	ODM2A	.375	.093	.128	.113	2 Rich	.360	1500	.120	3

① TCS solenoid and EGR disconnected, fast idle screw on 2nd cam step.

② Measured between choke valve and air horn, fast idle screw on 2nd cam step.

MODEL 2150
Ford
(All measurements in inches)

Year	Carburetor Number	Float Level (Dry)	Choke Unloader Setting	Choke Setting	Accelerator Pump Rod Location	Fuel Level (Wet)	Choke Pulldown Setting (Min)
'79	D8TE						
	LA	31/64	—	Index	3	7/8	.130
	ARA	31/64	—	Index	3	7/8	.145
	DA	31/64	—	Index	4	7/8	.130
	CTA	31/64	—	Index	4	7/8	.130
	DBA	31/64	—	2 Rich	4	7/8	.130
	BLA	31/64	—	2 Rich	4	7/8	.130
	CRA	31/64	—	2 Rich	4	7/8	.130
	BJA	31/64	—	3 Rich	3	7/8	.175
	ATA	31/64	—	2 Rich	3	7/8	.145
	BEA	31/64	—	3 Rich	2	7/8	.200
	BA	31/64	—	Index	3	7/8	.140
	DP7E						
	AGA	31/64	—	2 Rich	3	7/8	.145
	D7UE						
	APA	31/64	—	2 Rich	3	7/8	.170
	D8UE						
	VA	31/64	—	Index	4	7/8	.145
	DA	31/64	—	3 Rich	3	7/8	.185
	KA	7/16	—	Index	3	13/16	.185
	GA	31/64	—	1 Rich	2	7/8	.205

MODEL 2150
Ford
(All measurements in inches)

Year	Carburetor Number	Float Level (Dry)	Choke Unloader Setting	Choke Setting	Accelerator Pump Rod Location	Fuel Level (Wet)	Choke Pulldown Setting (Min)
'79	HA	$^{31}/_{64}$	—	Index	2	$^7/_8$	.215
	MA	$^{31}/_{64}$	—	Index	2	$^7/_8$	.215
	MB	$^{31}/_{64}$	—	Index	2	$^7/_8$	.215
	SA	$^{31}/_{64}$	—	3 Rich	3	$^7/_8$	.180
'80	E0TE-9510 BGA, CYA, GZA, ABA, BEA	.810	—	.20	—	—	.140
	BHA	.810	—	.20	—	—	.135
	BRA, DDA	.810	—	.20	—	—	.128
	CFA, EAA	.810	—	.25	—	—	.128
	CVA, NA	.875	—	.20	—	—	.105
	BYA	.875	—	.115	—	—	.140
	BSA Calibration Number: 0-59J-R0 0-59G-R10	.875	—	.115	—	—	.140
	0-59J-R10 0-59H-R10	.875	—	.20	—	—	.140
	DCA	.875	—	.25	—	—	.140
	AAA, PA, SA, TA, VA	.810	—	.25	—	—	.185
	AA	.810	—	.25	—	—	.105
	CLA	.875	—	.20	—	—	.140
	BLA, BFA, BZA	.875	—	.25	—	—	.148
	CCA, CBA	.875	—	.25	—	—	.159
	EDA, DGA	.875	—	.25	—	—	.155
	EEA, EFA	.875	—	.20	—	—	.160
	DEA, ECA	.875	—	.25	—	—	.175
	DFA	.875	—	.25	—	—	.185
'81	E1TE-9510 BJA, CHA, BCA	$^{31}/_{64}$	—	.250	V notch	—	.148
	BTA	$^{31}/_{64}$	—	.250	V notch	—	.130
	BVA	$^7/_{16}$	—	.200	V notch	—	.130
	CEA, CFA	$^{31}/_{64}$	—	.200	V notch	—	.160
	CAA, BYA	$^{31}/_{64}$	—	.250	V notch	—	.175
	BSA	$^{31}/_{64}$	—	.250	V notch	—	.130

MODEL 2150
Ford

(All measurements in inches)

Year	Carburetor Number	Float Level (Dry)	Choke Unloader Setting	Choke Setting	Accelerator Pump Rod Location	Fuel Level (Wet)	Choke Pulldown Setting (Min)
'81	CCA	31/64	—	.250	V notch	—	.155
	BZA, CBA	31/64	—	.250	V notch	—	.180
	CNA, CMA, CPA, CRA, CSA, CKA, CLA	7/16	—	.200	V notch	—	.125
	E1UE-9510 KA	31/64	—	.250	V notch	—	.180
	GA	7/16	—	.200	V notch	—	.130
	HA	31/64	—	.200	V notch	—	.125
'82	E2TE- BNA, CGA	7/16	.200	V notch	2	.810	.125
	BMA, CFA	7/16	.250	V notch	2	.810	.125
	DAA, CYA	7/16	.250	V notch	2	.810	.115–.135
	BEA	7/16	.200	V notch	2	.810	.130
	BLA	7/16	.200	V notch	2	.810	.125
	AYA, BEA, CJA, BFA	7/16	.200	V notch	2	.810	.130
	E2UE-JA	31/64	.200	V notch	2	.875	.130
	E2TE-BAA, BBA	7/16	.200	V notch	2	.810	.130
	CKA	7/16	.200	V notch	2	.810	.120
	E2UE-FA E1UE-JA	31/64	.200	V notch	3	.875	.120
	E2UE-KA	31/64	.200	V notch	2	.875	.120
	E2TE-BPA,BRA	31/64	.200	V notch	2	.875	.120
	E2UE-ANA AAA AKA E2UE-RA	31/64	.250	V notch	3	.875	.170–.190
	E2UE-SA	31/64	.250	V notch	3	.875	.182
	E2UE-HA ABA E1UE-KA	31/64	.250	V notch	3	.875	.170–.190
	E2TE-BHA BGA	31/64	.250	V notch	4	.875	.180
	E2TE-DCA E2TE-DBA	31/64	.250	V notch	3	.875	.170–.190
	E2TE-BKA BJA	31/64	.250	V notch	4	.875	.175
	E2TE-DDA DEA	31/64	.250	V notch	3	.875	.170–.190

MODEL 2150
Ford

(All measurements in inches)

Year	Carburetor Number	Float Level (Dry)	Choke Unloader Setting	Choke Setting	Accelerator Pump Rod Location	Fuel Level (Wet)	Choke Pulldown Setting (Min)
'83	E3TE-9510						
	BCA						
	BFA						
	BBA						
	BGA	—	.25	V notch	3	.810	.115–135
	AUA	7/16	.20	V notch	3	.810	.142
	BHA	7/16	.20	V notch	3	.810	.152
	AYA	7/16	.25	V notch	4	.810	.137
	AVA	7/16	.25	V notch	3	.810	.149
	BJA	7/16	.20	V notch	4	.810	.157
	BLA	7/16	.20	V notch	3	.810	.157
	BEA	7/16	.20	V notch	3	.810	.149
	BMA	7/16	.20	V notch	4	.810	.150
	E3UE-9510						
	CA,FA	31/64	.20	V notch	3	.875	.120
	BA,KA	31/64	.20	V notch	2	.875	.120
	E3TE-9510						
	BAA,BPA	31/64	.25	V notch	3	.875	.130
	E2UE-9510						
	DA						
	EA						
	ANA						
	AKA	31/64	.25	V notch	3	.875	.180
	E37E-9510						
	LB	.650	.27	—	—	—	.320
	E37E-9510						
	ABA						
	AAA						
	ADA	7/16	.250	V notch	4	.810	.126–.146
'84	E37E-9510						
	AEA	7/16	.200	V-notch	4	.810	.136
	E4TE-9510						
	AUA	31/64	.200	V-notch	4	.875	.150
	AFA	31/64	.200	V-notch	3	.875	.144
	ADA	7/16	.200	V-notch	4	.810	.152
	ACA	7/16	.200	V-notch	4	.810	.155
	ATA	31/64	.200	V-notch	3	.875	.130
'85	E57E-9510				4	.810④	.136
	BA,CA	1/16	.250	3-Rich			
	E5TE-9510			V-notch	4	.875	.150
	YA	9/32	.220				
	ACA	9/32	.200	3 Rich	4	.875	.150
	AAA	1/4	.200	V-notch	4	.810	.155
	PA	1/4	.200	3 Rich	4	.810	.152

MODEL 2150
Ford
(All measurements in inches)

Year	Carburetor Number	Float Level (Dry)	Choke Unloader Setting	Choke Setting	Accelerator Pump Rod Location	Fuel Level (Wet)	Choke Pulldown Setting (Min)
'84	E3UE-9510 EA,DA	31/64	.250	V-notch	3	.875	.180

① Wet float setting 1980 only
② For choke settings see the Ford Calibration Specifications 1980 only
③ For 1979 carburetor specifications see Ford Calibration Specifications

MODEL 2150
Jeep
(All measurements in inches)

Year	Carburetor Number	Float Level (Dry)	Fuel Level (Wet)	Initial Choke Valve Clearance	Fast Idle Cam Setting ②	Choke Cover Setting	Choke Unloader Valve Clearance	Fast Idle Speed ①	Bowl Vent Clearance	Rod Pump Location Hole
'79	9RHM2	.555	.930	.104	.086	2 Rich	.348	1500	—	3
	9RHA2	.555	.930	.113	.093	2 Rich	.350	1600	—	3
'80	ORHM2	.575	.930	.104	.081	2 Rich	.348	1500	.120	3
	ORHA2	.575	.930	.104	.081	2 Rich	.350	1600	.120	3
'81	DMJ2	.375	.930	.125	.113	2 Rich	.300	1500	.120	3
	DA2J	.375	.930	.128	.113	1 Rich	.300	1600	.120	3
	DM2A	.375	.930	.128	.113	1 Rich	.360	1500	.120	3
	RHM2	.575	.930	.104	.081	2 Rich	.348	1500	.120	3
	RHA2	.575	.930	.113	.086	2 Rich	.350	1600	.120	3
'82	2RHM2	19/64–23/64 ③	.930	.116	.076	1 Rich	.350	1500	.120	—
	2RHA2	19/64–23/64 ③	.930	.116	.076	1 Rich	.350	1600	.120	—
'84–'85	4RHA2	.575	.930	.136	.086	2 Rich	.350	1600	.120	—
	5RHA2	.328	.930	.118	.076	Y	.420	1600	—	—

① TCS solenoid and EGR disconnected, fast idle screw on 2nd cam step
② Measured between choke valve and air horn fast idle screw on 2nd cam step
③ Measured from machined bowl surface to a point ⅛ inch from float tip with the needle seated

MODEL 7200
1981 Ford
(All measurements in inches)

Year	Carburetor Number	Float Level (Dry)	Float Drop	Choke Unloader Setting	Choke Setting	Dash Pot Plunger	Initial Choke Setting
'81	E1TE-9510						
	YA,AHA	1.455①	—	—	Index	—	—
	ZA	1.040①	—	—	Index	—	—

① ± .025 inches

MODEL 7200
1982–'83 Ford
(All measurements in inches)

Year	Carburetor Number	Float Setting	Float Drop	Control Vacuum Regulator	Pulldown Timing	Pump lever Lash
'82	E1TE-2A, AHA E2TE-CDA, CCA	1.070–1.010	1.490–1.430	.245–.255	2–5 sec.	.010①
'83	E3TE-9510– BVA, BYA	1.070–1.010	1.490–1.430	.245–.255	2–5 sec.	.010①

① Plus one turn counter-clockwise.

FORD CALIBRATION NUMBERS LIGHT TRUCK ONLY

Year	Emission Calibration Number	Choke Setting	Fast Idle (RPM) High Cam	Kick Down①	Choke Valve Pull Down
'77	7-53G-RO	3 Rich	2000	—	—
	7-53H-RO	3 Rich	2000	—	—
	7-53S-RO	3 Rich	2000	—	—
	7-54K-RO	3 Rich	2100	—	—
	7-54S-RO	3 Rich	2100	—	—
	7-54T-RO	3 Rich	2100	—	—
	7-59G-RO	Index	1900	—	—
	7-60G-RO	Index	1900	—	—
	7-62G-RO	Index	1900	—	—
	7-64U-RO	3 Rich	—	1500	—

FORD CALIBRATION NUMBERS LIGHT TRUCK ONLY

Year	Emission Calibration Number	Choke Setting	Fast Idle (RPM) High Cam	Kick Down①	Choke Valve Pull Down
'77	7-65G-RO	3 Rich	—	1500	—
	7-65H-RO	3 Rich	—	1500	—
	7-65U-RO	Index	—	1500	—
	7-71-RO	3 Rich	—	1250	—
	7-71J-RO	3 Rich	—	1250	—
	7-72-RO	3 Rich	—	1500	—
	7-72J-RO	3 Rich	—	1500	—
	7-73-RO	3 Rich	—	1200	—
	7-74-RO	3 Rich	—	1500	—

① Kickdown—2nd step of fast idle cam

FORD CALIBRATION NUMBERS LIGHT AND HEAVY TRUCKS

Year	Emission Calibration Number	Choke Setting	Fast Idle (RPM) High Cam	Fast Idle (RPM) Kick Down ①	Choke Valve Pull Down
'78	5-81A-RO	None	—	—	—
	5-81B-RO	None	2200	—	—
	5-82-RO	None	2200	—	—
	5-82A-RO	None	2200	—	—
	5-85A-R1	None	2200	—	—
	5-85J-R6	None	—	—	—
	5-86A-RO	None	2200	—	—
	5-86J-R3	None	2500	—	—
	5-89-R2	None	2000	—	—
	5-89J-R3	None	2200	—	—
	5-90A-R1	None	2200	—	—
	5-90B-R1	None	2200	—	—
	5-90J-R3	None	2200	—	—
	6-51A-RO	Index	—	1600	—
	6-51E-RO	Index	—	1600	—
	6-93-R7	None	2500	—	—
	6-94-R4	None	2500	—	—
	6-95-R6	None	2500	—	—
	6-95-95	None	2500	—	—
	7-60E-R11	Index	1900	—	—
	7-71-R10	2 Rich	—	1250	—
	7-71J-R10	2 Rich	—	1450	—
	7-72-R11	3 Rich	—	1500	—
	7-72J-R11	3 Rich	—	1500	—
	7-73-R10	2 Rich	—	1250	—
	7-74-R10	3 Rich	—	1500	—
	7-74J-R11	3 Rich	—	1500	—
	7-75A-R16	1 Rich	—	1250	—
	7-76A-R10	2 Rich	—	1500	—
	7-76J-R11	3 Rich	—	1700	—
	7-77-R10	1 Rich	—	1500	—
	7-77A-R10	None	—	1500	—
	7-77J-R10	1 Rich	—	1500	—
	7-77M-R10	None	—	1500	—
	7-78-R10	1 Rich	—	1500	—
	7-78J-R10	1 Rich	—	1500	—
	7-79-R1	3 Rich	—	1250	—
	7-80-RO	3 Rich	—	1500	—
	7-81K-RO	None	1400	—	—
	7-93J-RO	None	2500	—	—

FORD CALIBRATION NUMBERS LIGHT AND HEAVY TRUCKS

Year	Emission Calibration Number	Choke Setting	Fast Idle (RPM) High Cam	Fast Idle (RPM) Kick Down ①	Choke Valve Pull Down
'78	7-95J-RO	None	2500	—	—
	7-96J-90	None	2500	—	—
	8-51J-RO	Index	—	1600	—
	8-51K-RO	Index	—	1600	—
	8-51L-RO	Index	—	1600	—
	8-51M-RO	Index	—	1600	—
	8-51S-RO	Index	—	1600	—
	8-51T-RO	Index	—	1600	—
	8-52K-RO	Index	—	1600	—
	8-52L-R10	Index	—	1600	—
	8-52U-RO	Index	—	1600	—
	8-53A-RO	1 Rich	2000	—	—
	8-53G-RO	3 Rich	2000	—	—
	8-53S-RO	3 Rich	2000	—	—
	8-54A-RO	3 Rich	2000	—	—
	8-54G-RO	3 Rich	2000	—	—
	8-54S-RO	3 Rich	2100	—	—
	8-54T-R10	3 Rich	2100	—	—
	8-59G-RO	Index	2100	—	—
	8-59T-R2	Index	2100	—	—
	8-60A-RO	2 Rich	2200	—	—
	8-60J-RO	Index	2100	—	—
	8-60S-R11	Index	1900	—	—
	8-60S-R12	Index	1900	—	—
	8-62T-RO	—	—	—	—
	8-62J-RO	Index	2100	—	—
	8-64G-RO	3 Rich	—	1750	—
	8-64S-RO	Index	—	1500	—
	8-65A-RO	Index	2100	—	—
	8-65G-RO	1 Rich	2000	—	—
	8-65S-RO	Index	—	1500	—
	8-65U-RO	Index	—	1500	—
	R-66U-RO	Index	—	1500	—
	8-97-RO	Index	—	1200	—
	7-97J-RO	Index	—	1200	—
'79	9-51G-RO	Index	—	1600	.230
	9-51J-RO	Index	—	1600	.230
	9-51K-RO	Index	—	1600	.230
	9-51L-RO	Index	—	1600	.230
	9-51M-RO	Index	—	1600	.230

FORD CALIBRATION NUMBERS LIGHT AND HEAVY TRUCKS

Year	Emission Calibration Number	Choke Setting	Fast Idle (RPM) High Cam	Kick Down①	Choke Valve Pull Down
'79	9-51S-RO	Index	—	1600	.230
	9-51T-RO	Index	—	1600	.230
	9-52G-RO	Index	—	1600	.230
	9-52J-RO	Index	—	1600	.230
	9-52L-RO	Index	—	1600	.230
	9-52M-RO	Index	—	1600	.230
	9-53G-RO	3 Rich	2000	—	.140
	9-53H-RO	3 Rich	2000	—	.140
	9-54A-RO	—	—	—	.145
	9-54G-RO	3 Rich	2000	—	.145
	9-54H-RO	3 Rich	2000	—	.145
	9-54J-RO	2 Rich	2000	—	.145
	9-54R-RO	3 Rich	2000	—	.145
	9-54S-RO	1 Rich	2400	—	.136
	9-54T-RO	3 Rich	2000	—	.145
	9-54U-RO	1 Rich	2400	—	.136
	9-59H-RO	Index	2000	—	.135
	9-59J-RO	Index	2000	—	.145
	9-59K-RO	Index	2000	—	.145
	9-59S-RO	Index	2000	—	.135
	9-60T-RO	—	—	—	.150
	9-60G-RO	Index	2000	—	.145
	9-60H-RO	Index	2000	—	.150
	9-60J-RO	Index	2000	—	.140
	9-60L-RO	Index	2000	—	.150
	9-60M-RO	Index	2000	—	.150
	9-60S-RO	3 Rich	2100	—	.150
	9-61G-RO	Index	2000	—	.145
	9-61H-RO	Index	2000	—	.135
	9-62A-RO	—	—	—	.145
	9-62B-RO	—	—	—	.145
	9-62J-RO	Index	1900	—	.145
	9-62M-RO	Index	1900	—	.145
	9-63H-RO	Index	—	1500	.190
	9-64G-RO	Index	2200	—	.200
	9-64H-RO	Index	2200	—	.200
	9-64S-RO	Index	2200	—	.200
	9-66G-RO	5 Rich	—	1600	.210
	9-72J-RO	3 Rich	2000	—	.150
	9-77J-RO	Index	—	1600	.290

FORD CALIBRATION NUMBERS LIGHT AND HEAVY TRUCKS

Year	Emission Calibration Number	Choke Setting	Fast Idle (RPM) High Cam	Kick Down①	Choke Valve Pull Down
'79	9-77M-RO	Manual	2550	—	—
	9-78J-RO	Index	—	1600	.290
	9-83G-RO	Manual	2200	—	—
	9-83H-RO2	Manual	2500	—	—
	9-87G-RO	Manual	2700	—	—
	9-97J-RO	5 Rich	—	1600	.210
	7-76J-R11	3 Rich	—	1700	.180
	7-93J-RO	Manual	2500	—	—
	7-95J-RO	Manual	2500	—	—
	9-71J-RO	3 Rich	1750	—	—
	9-73J-RO	2 Rich	1750	—	—
	9-74J-RO	3 Rich	2000	—	—
	9-97J-R11	5 Rich	—	1600	—
'80	0-51F-RO	Index	—	1600	—
	0-51G-RO	2 Rich	—	1400	—
	0-51H-RO	2 Rich	—	1400	—
	0-51L-RO	Index	—	1400	—
	0-51M-RO	Index	—	1400	—
	0-51S-RO	Index	—	1600	—
	0-51T-RO	Index	—	1600	—
	0-52H-RO	2 Rich	—	1400	—
	0-52J-RO	2 Rich	—	1400	—
	0-52L-RO	2 Rich	—	1400	—
	0-52M-RO	2 Rich	—	1400	—
	0-52S-RO	Index	—	1600	—
	0-53D-RO	3 Rich	2000	—	—
	0-53G-RO	3 Rich	2000	—	—
	0-53H-RO	3 Rich	2000	—	—
	0-53K-RO	3 Rich	2000	—	—
	0-53L-RO	3 Rich	2000	—	—
	0-53N-RO	3 Rich	2500	—	—
	0-53Q-RO	3 Rich	2500	—	—
	0-53S-RO	3 Rich	2500	—	—
	0-54D-RO	3 Rich	2000	—	—
	0-54F-RO	3 Rich	2000	—	—
	0-54G-RO	Index	2000	—	—
	0-54H-RO	3 Rich	2000	—	—
	0-54K-RO	Index	2000	—	—
	0-54L-RO	3 Rich	2000	—	—
	0-54M-RO	3 Rich	2100	—	—

FORD CALIBRATION NUMBERS LIGHT AND HEAVY TRUCKS

Year	Emission Calibration Number	Choke Setting	Fast Idle (RPM) High Cam	Kick Down①	Choke Valve Pull Down
'80	0-54N-RO	1 Rich	2400	—	—
	0-54P-RO	3 Rich	2000	—	—
	0-54Q-RO	1 Rich	2400	—	—
	0-54R-RO	3 Rich	2100	—	—
	0-54T-RO	3 Rich	2000	—	—
	0-54V-RO	3 Rich	2100	—	—
	0-60A-RO	3 Rich	2000	—	—
	0-59C-RO	3 Rich	2000	—	—
	0-59G-RO	3 Rich	2000	—	—
	0-59G-R10	3 Rich	2000	—	—
	0-59H-RO	3 Rich	2000	—	—
	0-59H-R10	3 Rich	2000	—	—
	0-59J-RO	3 Rich	2000	—	—
	0-59J-R10	3 Rich	2000	—	—
	0-59S-RO	3 Rich	2000	—	—
	0-60B-RO	3 Rich	2000	—	—
'81	1-57G-R1	—	2200	—	—
	1-58G-RO	—	2000	—	—
	1-51D-RO	—	—	1400	—
	1-51D-R10	—	—	1400	—
	1-51E-RO	—	—	1400	—
	1-51F-RO	—	—	1400	—
	1-51G-RO	—	—	1400	—
	1-51H-RO	—	—	1400	—
	1-51K-RO	—	—	1400	—
	1-51L-RO	—	—	1400	—
	1-51E-R10	—	—	1400	—
	1-51F-R10	—	—	1400	—
	1-51G-R10	—	—	1400	—
	1-51H-R10	—	—	1400	—
	1-51K-R10	—	—	1400	—
	1-51L-R10	—	—	1400	—
	1-51S-RO	—	—	1400	—
	1-51S-R10	—	—	1400	—
	1-51T-RO	—	—	1400	—

FORD CALIBRATION NUMBERS LIGHT AND HEAVY TRUCKS

Year	Emission Calibration Number	Choke Setting	Fast Idle (RPM) High Cam	Kick Down①	Choke Valve Pull Down
'81	1-52G-RO	—	—	1400	—
	1-52H-RO	—	—	1400	—
	1-52K-RO	—	—	1400	—
	1-52L-RO	—	—	1400	—
	1-52G-R10	—	—	1400	—
	1-52H-R10	—	—	1400	—
	1-52K-R10	—	—	1400	—
	1-52L-R10	—	—	1400	—
	1-52S-RO	—	—	1400	—
	1-52T-RO	—	—	1400	—
	1-53D-RO	—	2200	—	—
	1-53F-RO	—	2200	—	—
	1-53G-RO	—	2200	—	—
	1-53H-RO	—	2200	—	—
	1-53K-RO	—	2200	—	—
	1-53D-R10	—	2200	—	—
	1-53G-R10	—	2200	—	—
	1-53K-R10	—	2200	—	—
	1-59A-RO	—	2000	—	—
	1-59B-RO	—	2000	—	—
	1-59H-RO	—	2000	—	—
	1-59K-RO	—	2000	—	—
	1-60A-RO	—	2200	—	—
	1-60B-RO	—	2200	—	—
	1-60H-R1	—	2000	—	—
	1-60J-RO	—	2000	—	—
	1-60K-RO	—	2000	—	—
	1-63T-RO	—	1700	—	—
	1-64A-RO	—	2000	—	—
	1-64G-R1	—	2000	—	—
	1-64H-R2	—	2000	—	—
	1-64R-R1	—	1650	—	—
	1-64S-RO	—	1650	—	—
	1-64T-RO	—	1650	—	—

①Kickdown—2nd step of fast idle cam

FORD CALIBRATION NUMBERS LIGHT TRUCK ONLY

Year	Emission Calibration Number	Choke Pulldown Setting	Choke Unloader Setting	Choke Cap Setting	Float Setting (Dry)	Float Level (Wet)	Pump Lever Location
'82	2-55D-R0	.125	.200	V-Notch	7/16	.810	#2
	2-56D-R0	.125	.250	V-Notch	7/16	.810	#2
	2-56D-R10	.115–.135	.250	V-Notch	7/16	.810	#2
	2-57G-R0	.130	.200	V-Notch	7/16	.810	#2
	2-58H-R0	.125	.200	V-Notch	7/16	.810	#2
	2-51D-R0	.270	.28	Index	.78	—	—
	2-51E-R0	.270	.28	Index	.78	—	—
	2-51F-R0	.270	.28	Index	.78	—	—
	2-51G-R0	.270	.28	Index	.78	—	—
	2-51K-R0	.230	.28	Index	.78	—	—
	2-51L-R0	.230	.28	Index	.78	—	—
	2-51P-R0	.320	.330	2 Rich	.78	—	—
	2-51P-R10	.320	.330	2 Rich	.78	—	—
	2-51S-R0	.320	.330	2 Rich	.78	—	—
	2-51T-R0	.320	.330	2 Rich	.78	—	—
	2-51X-R0	.300	.28	Index	.78	—	—
	2-51Y-R0	.300	.28	Index	.78	—	—
	2-52G-R0	.300	.28	2 Rich	.78	—	—
	2-52H-R0	.300	.28	2 Rich	.78	—	—
	2-52K-R0	.300	.28	2 Rich	.78	—	—
	2-52L-R0	.300	.28	2 Rich	.78	—	—
	2-52S-R0	.320	.330	2 Rich	.78	—	—
	2-52T-R0	.320	.330	2 Rich	.78	—	—
	2-52Y-R0	.300	.28	Index	.78	—	—
	2-53D-R0	.130	.200	V-Notch	7/16	.810	#2
	2-53F-R0	.130	.200	V-Notch	7/16	.810	#2
	2-53G-R0	.130	.200	V-Notch	7/16	.810	#2
	2-53H-R0	.130	.200	V-Notch	7/16	.810	#2
	2-53K-R0	.130	.200	V-Notch	7/16	.810	#2
	2-53X-R1	.130	.200	V-Notch	7/16	.810	#2
	2-54D-R0	.125	.200	V-Notch	7/16	.810	#2
	2-54F-R0	.130	.200	V-Notch	31/64	.875	#2
	2-54G-R0	.125	.200	V-Notch	7/16	.810	#2
	2-54H-R0	.130	.200	V-Notch	31/64	.875	#2
	2-54K-R0	.125	.200	V-Notch	7/16	.810	#2
	2-54L-R0	.125	.200	V-Notch	7/16	.810	#2
	2-54P-R0	—	—	Index	1.070–1.010	—	—
	2-54R-R0	—	—	Index	1.070–1.010	—	—
	2-54X-R1	.120	.200	V-Notch	7/16	.810	#2
	1-63T-R0	—	—	Index	1.070–1.010	—	—
	1-63T-R10B	—	—	Index	1.070–1.010	—	—
	1-64H-R2	.120	.200	V-Notch	31/64	.875	#3
	1-64R-R1	—	—	—	1.070–1.010	—	—
	1-64S-R0	—	—	—	1.070–1.010	—	—
	1-64T-R0	—	—	—	1.070–1.010	—	—
	1-64T-R10	—	—	—	1.070–1.010	—	—

FORD CALIBRATION NUMBERS LIGHT TRUCK ONLY

Year	Emission Calibration Number	Choke Pulldown Setting	Choke Unloader Setting	Choke Cap Setting	Float Setting (Dry)	Float Level (Wet)	Pump Lever Location
'82	2-63Y-R10B	—	—	Index	1.070–1.010	—	—
	2-64X-R0	.120	.200	V-Notch	31/64	.875	#2
	2-64Y-R10B	—	—	Index	1.070–1.010	—	—

FORD CALIBRATION NUMBERS MEDIUM AND HEAVY TRUCK

Year	Emission Calibration Number	Choke Pulldown Setting	Choke Unloader Setting	Choke Cap Setting	Float Setting (Dry)	Float Level (Wet)	Pump Lever Location
'82	5-77-R1	.230	.28	1 Rich	3/8	—	—
	5-78-R1	.230	.28	1 Rich	3/8	—	—
	5-77G-R12	.290	.28	Index	.69	—	—
	9-77S-R10	.290	.28	Index	.69	—	—
	9-78J-R0	.290	.28	Index	.69	—	—
	9-78J-R11	.290	.28	Index	.69	—	—
	7-79-R1	.130	.250	V-Notch	31/64	.875	#3
	7-80-R0	.130	.250	V-Notch	31/64	.875	#3
	2-75J-R17	.170–.190	.250	V-Notch	31/64	.875	#3
	2-76J-R17	.170–.190	.250	V-Notch	31/64	.875	#3
	7-75A-R11	.173	.250	V-Notch	31/64	.875	#3
	7-75A-R12	.173	.250	V-Notch	31/64	.875	#3
	7-75J-R14	.170–.190	.250	V-Notch	31/64	.875	#3
	7-76A-R11	.182	.250	V-Notch	31/64	.875	#3
	7-76A-R12	.182	.250	V-Notch	31/64	.875	#3
	7-76J-R11	.170–.190	.250	V-Notch	31/64	.875	#3
	7-76J-R13	.170–.190	.250	V-Notch	31/64	.875	#3
	7-76J-R14	.170–.190	.250	V-Notch	31/64	.875	#3
	7-76J-R15	.170–.190	.250	V-Notch	31/64	.875	#3
	9-83G-R12	.350–.400	—	—	—	①	②
	9-83H-R11	.180–.440	—	—	—	①	#1
	9-83H-R14	.180–.440	—	—	—	①	#1
	9-73J-R11	.180	.250	V-Notch	31/64	—	#4
	9-73J-R12	.180	.250	V-Notch	31/64	—	#4
	9-73J-R13	.170–.190	.250	V-Notch	31/64	—	#3
	9-73J-R14	.170–.190	.250	V-Notch	31/64	—	#3
	9-74J-R11	.175	.250	V-Notch	31/64	—	#4
	9-74J-R12	.175	.250	V-Notch	31/64	.875	#4
	9-74J-R13	.165–.185	.250	V-Notch	31/64	.875	#3
	9-74J-R14	.165–.185	.250	V-Notch	31/64	.875	#3
	9-87G-R11	.180–.440	—	—	—	①	#1
	9-97J-R12	.200–.220	.295–.335	2 Rich	—	①	#1

① At bottom of sight plug
② #1-E0TE-PA, E2TE-NA; #2-E2TE-AGA, D9TE-APA

FORD CALIBRATION NUMBERS LIGHT TRUCK ONLY

Year	Emission Calibration Number	Choke Pulldown Setting	Choke Bi-Metal I.D.	Cam Index Setting	Choke Unloader	Float Level (Dry)
'83	3-41D-R01	.320	REL	.140	.22	.650
	3-41D-R10	.320	REL	.140	.22	.650
	3-41P-R02	.320	REL	.140	.22	.650
	3-41P-R11	.320	REL	.140	.22	.650
	3-41P-R12	.320	REL	.140	.22	.650
	3-49S-R01	.320	REL	.140	.22	.650
	3-49S-R10	.320	REL	.140	.22	.650
	3-49S-R11	.320	REL	.140	.22	.650
	3-49X-R01	.320	REL	.140	.22	.650
	3-49X-R11	.320	REL	.140	.22	.650
	3-50S-R01	.320	REL	.140	.22	.650
	3-50S-R11	.320	REL	.140	.22	.650
	3-50X-R10	.320	REL	.140	.22	.650
	3-50X-R11	.320	REL	.140	.22	.650
	3-55D-R00	.115–.135	2080ME350	V-Notch	.25	.810①
	3-56D-R00	.115–.135	2080ME350	V-Notch	.25	.810①
	3-51D-R00	.270	EC	.140	.28	.780
	3-51E-R01	.270	EB	.140	.28	.780
	3-51F-R00	.270	EC	.140	.28	.780
	3-51G-R00	.270	EC	.140	.28	.780
	3-51H-R00	.270	EB	.140	.28	.780
	3-51K-R00	.270	EB	.140	.28	.780
	3-51L-R00	.270	EB	.140	.28	.780
	3-51P-R00	.320	EC	.140	.330	.780
	3-51R-R00	.320	EC	.140	.330	.780
	3-51R-R10	.320	EC	.140	.330	.780
	3-51S-R00	.320	EC	.140	.330	.780
	3-51S-R10	.320	EC	.140	.330	.780
	3-51T-R00	.320	EC	.140	.330	.780
	3-51T-R10	.320	EC	.140	.330	.780
	3-51V-R00	.300	EB	.140	.28	.780
	3-51X-R00	.300	EB	.140	.28	.780
	3-51Z-R00	.300	EB	.140	.28	.780
	3-52E-R00	.270	EC	.140	.28	.780
	3-52F-R00	.300	EC	.140	.28	.780
	3-52G-R00	.300	EC	.140	.28	.780
	3-52K-R00	.300	EC	.140	.28	.780

FORD CALIBRATION NUMBERS LIGHT TRUCK ONLY

Year	Emission Calibration Number	Choke Pulldown Setting	Choke Unloader Setting	Choke Cap Setting	Float Setting (Dry)	Float Level (Wet)	Pump Lever Location
'83	3-52R-R00	.320	EC	.140	.330		.780
	3-52R-R10	.320	EC	.140	.330		.780
	3-52S-R00	.320	EC	.140	.330		.780
	3-52S-R10	.320	EC	.140	.330		.780
	3-52T-R00	.320	EC	.140	.330		.780
	3-52T-R10	.320	EC	.140	.330		.780
	3-52V-R00	.300	EB	.140	.28		.780
	3-52Y-R00	.300	EB	.140	.28		.780
	3-52Z-R00	.300	EB	.140	.28		.780
	3-53F-R00	.142	2100MF	V-Notch	.20		7/16
	3-53G-R00	.142	2100MF	V-Notch	.20		7/16
	3-53K-R00	.142	2100MF	V-Notch	.20		7/16
	3-53L-R00	.142	2100MF	V-Notch	.20		7/16
	3-53W-R00	.152	2100MF400	V-Notch	.20		7/16
	3-53Y-R00	.152	2100MF400	V-Notch	.20		7/16
	3-53Z-R00	.152	2100MF400	V-Notch	.20		7/16
	3-54E-R00	.137	2100MF	V-Notch	.25		7/16
	3-54F-R00	.149	2100KH	V-Notch	.25		7/16
	3-54J-R00	.137	2100MF	V-Notch	.25		7/16
	3-54L-R00	.149	2100KH	V-Notch	.25		7/16
	3-54P-R00	.157	2080ME400	V-Notch	.20		7/16
	3-54R-R00	.157	2080ME400	V-Notch	.20		7/16
	3-54T-R00	.157	2100ME400	V-Notch	.20		7/16
	3-54W-R00	.149	2120KH400	V-Notch	.20		7/16
	3-54Y-R00	.150	2100MF400	V-Notch	.20		7/16
	3-54Z-R00	.150	2100MF400	V-Notch	.20		7/16
	1-63T-R12	②	③	.355–.365④	.245–.255⑤		1.010–1.070
	1-63T-R13	②	③	.355–.365④	.245–.255⑤		1.010–1.070
	1-64H-R02	.120	2150ME	V-Notch	.20		31/64
	1-64T-R12	②	③	.355–.365	.245–.255⑤		1.010–1.070
	1-64T-R13	②	③	.355–.365	.245–.255⑤		—
	2-63Y-R11	②	③	.355–.365④	.245–.255⑤		1.010–1.070
	2-63Y-R12	②	③	.355–.365④	.245–.255⑤		1.010–1.070
	3-64X-R00	.120	2100ME	V-Notch	.20		31/64
	2-64Y-R11	②	③	.355–.365	.245–.255⑤		1.010–1.070
	2-64Y-R12	②	③	.355–.365	.245–.255⑤		1.010–1.070

ROCHESTER CARBURETORS

IDLE MIXTURE ADJUSTMENT

Model 1ME Carburetor
LEAN IDLE DROP PROCEDURE

1. Set the parking brake and block the drive wheels of the vehicle.

2. Remove the air cleaner for access to the carburetor, but leave all vacuum hoses connected.

3. Disconnect and plug any vacuum hoses as directed on the underhood emission control sticker.

4. Place the transmission in Park (A/T) or Neutral (M/T).

5. Start the engine and allow it to reach normal operating temperature. The choke should be fully open and all accessories switched off.

6. Connect an accurate tachometer to the engine according to the manufacturer's instructions.

7. Disconnect the vacuum advance and plug the hose. Check the ignition timing and, if necessary, adjust to the specifications listed on the underhood sticker. Reconnect the vacuum advance.

8. Carefully remove the cap from the idle mixture screw. Exercise caution so as not to bend the screw. Lightly seat the screw, then back it out just enough so the engine will run.

9. Back the screw out ⅛ turn at a time until maximum idle speed is obtained, then set the idle speed to the higher value shown on the underhood specifications sticker. Repeat the step to make sure you have maximum idle speed.

10. Turn the mixture screw in ⅛ turn at a time until the idle speed reaches the lower value shown on the underhood specifications sticker.

11. Reset the idle speed to specification shown on the underhood sticker, then check and adjust the fast idle as described on the sticker. Reconnect all vacuum hoses.

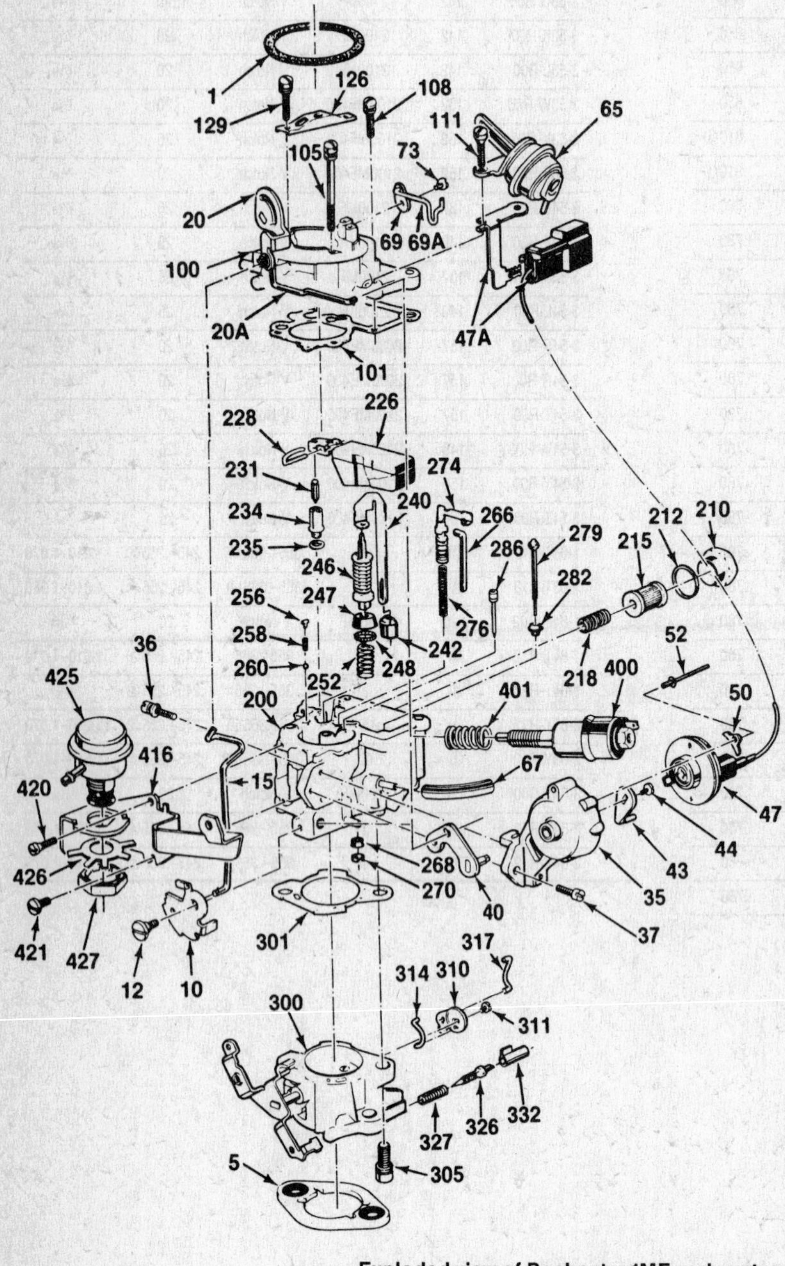

1. Gasket—air cleaner
5. Gasket—flange
10. Cam—fast idle
12. Screw—fast idle cam attaching
15. Link—fast idle cam
20. Choke shaft, lever & link assembly
20A. Link—choke
35. Choke housing & bearing assembly
36. Screw assembly—choke housing attaching
37. Screw—choke housing attaching
40. Choke shaft & lever assembly
43. Lever—choke stat
44. Screw—stat lever attaching
47. Electric choke cover & stat assembly
47A. Connector & bracket assembly
50. Retainer—choke cover
52. Rivet—choke cover attaching
65. Vacuum break assembly—bowl side
67. Hose—vacuum break
69. Vacuum break lever & link assembly
69A. Link—vacuum break
73. Screw—lever attaching
100. Air horn assembly
101. Gasket—air horn to float bowl
105. Screw assembly—air horn to float bowl (long)
108. Screw assembly—air horn to float bowl
111. Screw—air horn to float bowl (countersunk)
126. Bracket—air cleaner
129. Screw assembly—air cleaner bracket attaching
200. Float bowl assembly
210. Nut—fuel inlet
212. Gasket—fuel inlet nut
215. Filter—fuel inlet
218. Spring—fuel filter
226. Float
228. Hinge pin—float
231. Needle—float
234. Seat—float needle
235. Gasket—float needle seat
240. Rod—pump
242. Seal—pump rod
246. Pump assembly
247. Cup—pump plunger
248. Spring—pump plunger
252. Spring—pump return
256. Guide—pump discharge spring
258. Spring—pump discharge ball
260. Ball—pump discharge
266. Rod—power piston
268. Seal—power piston rod
270. Retainer—power piston rod seal
274. Power valve piston assembly
276. Spring—power piston
279. Metering rod & spring assembly
282. Jet—main metering
286. Idle tube assembly
300. Throttle body assembly
301. Gasket—float bowl to throttle body
305. Screw assembly—float bowl to throttle body
310. Lever—pump & power rod
311. Screw—pump lever attaching
314. Link—power rod
317. Link—pump
326. Needle—idle mixture
327. Spring—idle mixture needle
332. Limiter—idle mixture needle
400. Solenoid—idle stop
401. Spring—idle stop solenoid
416. Bracket—throttle kicker
420. Screw—bracket attaching (countersunk)
421. Screw—bracket attaching
425. Throttle kicker assembly
426. Washer-tap locking
427. Nut-throttle kicker assembly attaching

Exploded view of Rochester 1ME carburetor—1981-84 models

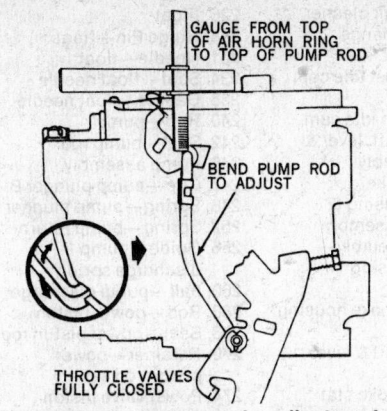

GAUGE FROM TOP OF AIR HORN RING TO TOP OF PUMP ROD

BEND PUMP ROD TO ADJUST

THROTTLE VALVES FULLY CLOSED

Accelerator pump rod adjustment— Rochester model 2G

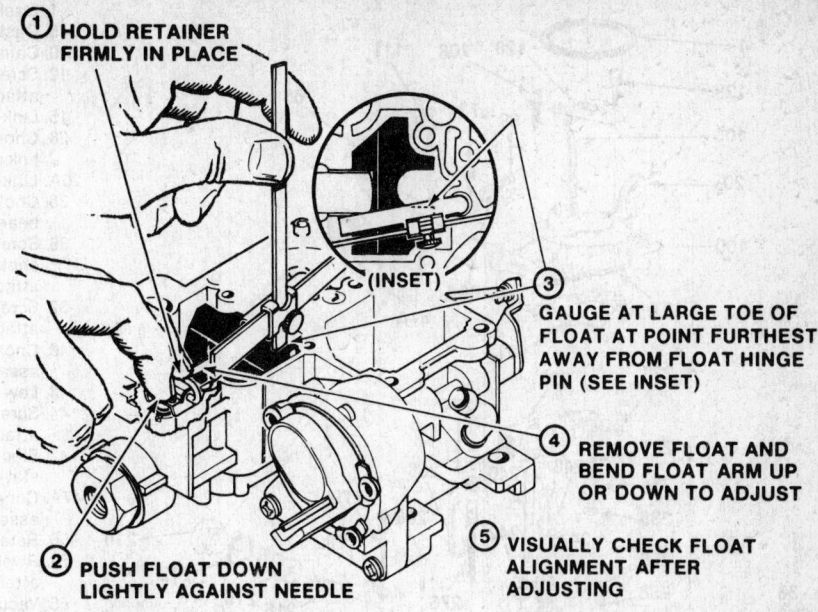

① **HOLD RETAINER FIRMLY IN PLACE**

(INSET)

③ **GAUGE AT LARGE TOE OF FLOAT AT POINT FURTHEST AWAY FROM FLOAT HINGE PIN (SEE INSET)**

④ **REMOVE FLOAT AND BEND FLOAT ARM UP OR DOWN TO ADJUST**

② **PUSH FLOAT DOWN LIGHTLY AGAINST NEEDLE**

⑤ **VISUALLY CHECK FLOAT ALIGNMENT AFTER ADJUSTING**

Setting float level—model 2SE carburetor

IDLE MIXTURE ADJUSTMENT

Model 2SE Carburetor

PROPANE ENRICHMENT PROCEDURE

NOTE: The idle mixture screws have been adjusted at the time of manufacture and are sealed. Only after major carburetor overhaul, throttle body replacement or if high emissions are occurring, should any attempt be made to adjust the mixture screws. Mixture adjustment requires artificial enrichment by adding propane through the air cleaner assembly using suitable propane equipment and adapters. Refer to the underhood emission control sticker for instructions and propane enrichment rpm specifications before attempting adjustment.

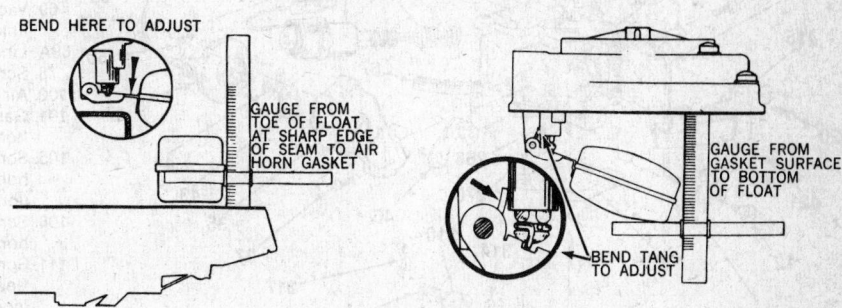

BEND HERE TO ADJUST

GAUGE FROM TOE OF FLOAT AT SHARP EDGE OF SEAM TO AIR HORN GASKET

Float level adjustment—Rochester model 2G

GAUGE FROM GASKET SURFACE TO BOTTOM OF FLOAT

BEND TANG TO ADJUST

Float drop adjustment—Rochester model 2G

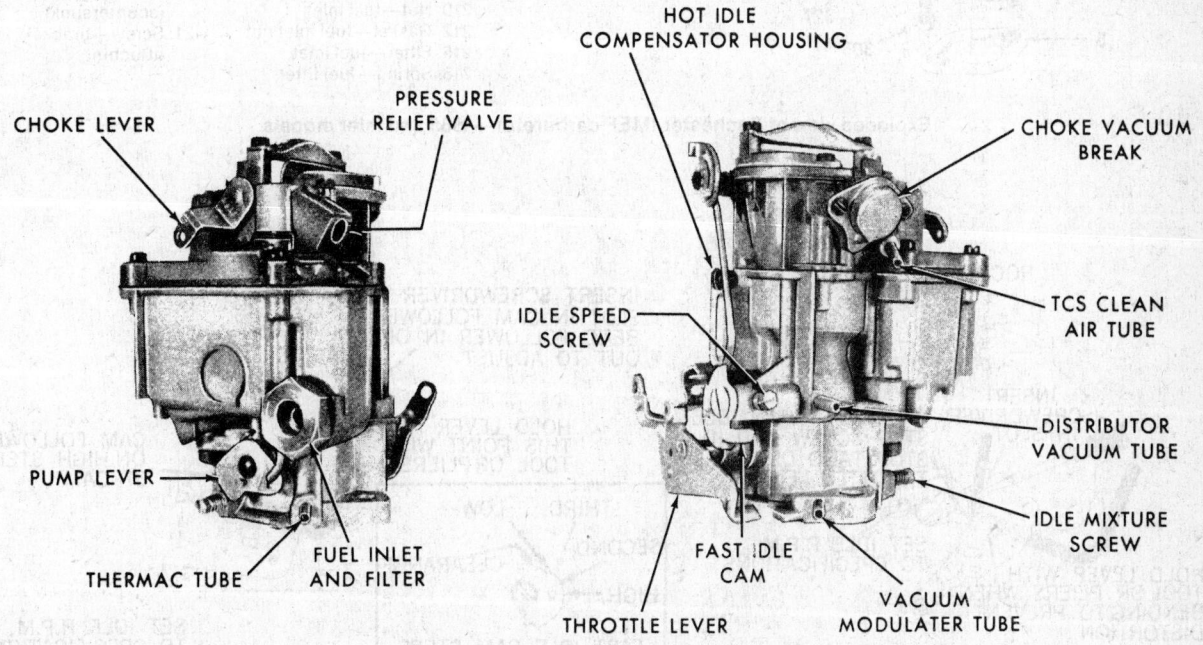

HOT IDLE COMPENSATOR HOUSING

CHOKE LEVER

PRESSURE RELIEF VALVE

CHOKE VACUUM BREAK

IDLE SPEED SCREW

TCS CLEAN AIR TUBE

DISTRIBUTOR VACUUM TUBE

IDLE MIXTURE SCREW

PUMP LEVER

THERMAC TUBE

FUEL INLET AND FILTER

FAST IDLE CAM

THROTTLE LEVER

VACUUM MODULATER TUBE

Rochester Monojet® carburetor—typical

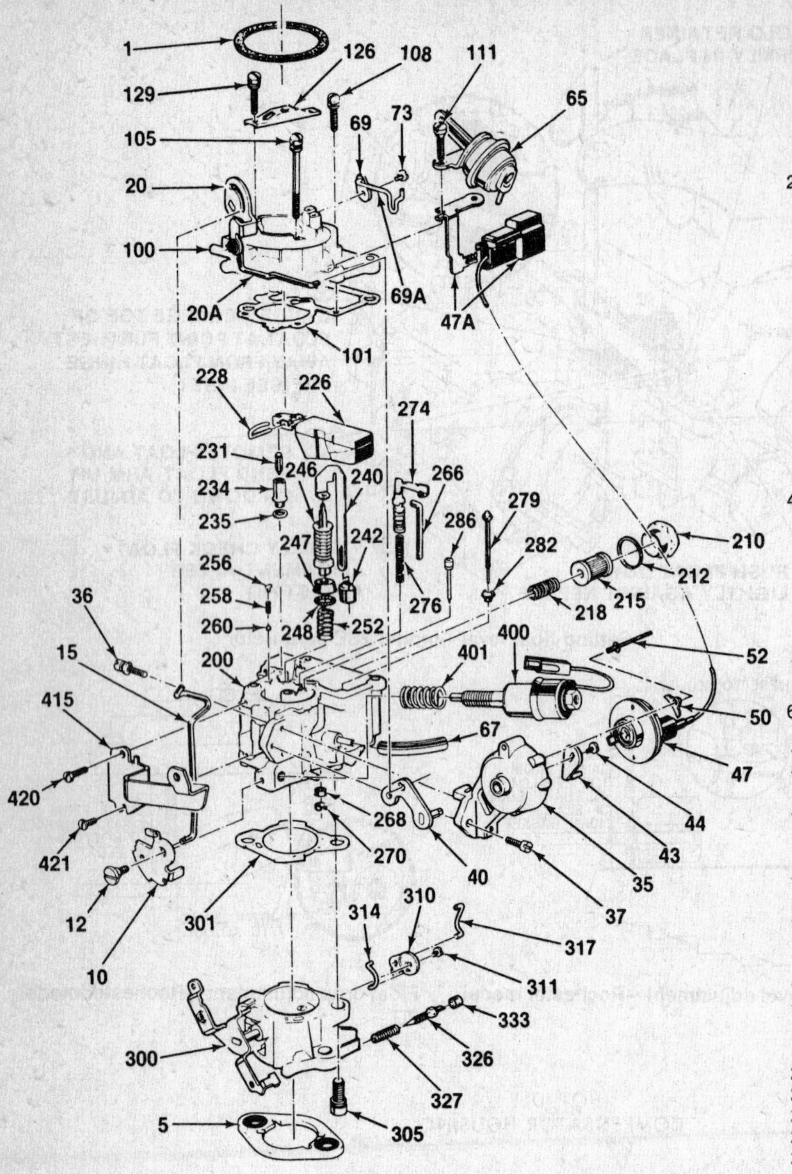

1. Gasket—air cleaner
5. Gasket—flange
10. Cam—fast idle
12. Screw—fast idle cam attaching
15. Link—fash idle cam
20. Choke shaft, lever & link assembly
20A. Link—choke
35. Choke housing & bearing assembly
36. Screw assembly—choke housing attaching
37. Screw—choke housing attaching
40. Choke shaft & lever assembly
43. Lever—choke stat
44. Screw—stat lever attaching
47. Electric choke cover & stat assembly
47A. Connector & bracket assembly
50. Retainer—choke cover
52. Rivet—choke cover attaching
65. Vacuum break assembly—bowl side
67. Hose—vacuum break
69. Vacuum break lever & link assembly
69A. Link—vacuum break
73. Screw—lever attaching
100. Air horn assembly
101. Gasket—air horn to float bowl
105. Screw assembly—air horn to float bowl (long)
108. Screw assembly—air horn to float bowl
111. Screw—air horn to float bowl (countersunk)
126. Bracket—air cleaner
129. Screw assembly—air cleaner bracket attaching
200. Float bowl assembly
210. Nut—fuel inlet
212. Gasket—fuel inlet nut
215. Filter—fuel inlet
218. Spring—fuel filter

226. Float
228. Hinge Pin—float
231. Needle—float
234. Seat—float needle
235. Gasket—float needle
240. Rod—pump
242. Seal—pump rod
246. Pump assembly
247. Cup—pump plunger B
248. Spring—pump plunger
252. Spring—pump return
256. Guide—pump discharge spring
260. ball—pump discharge
266. Rod—power piston
268. Seal—power piston rod
270. Retainer—power piston rod seal
274. Power valve piston assembly
276. Spring—power piston
279. Metering rod & spring assembly
282. Jet—main metering
286. Idle tube assembly
300. Throttle body assembly
301. Gasket—float bowl to throttle body
305. Screw assembly—float bowl to throttle body
310. Lever—pump & power rod
311. Screw—pump lever attaching
314. Link—power rod
317. Link—pump
326. Needle—idle mixture
327. Spring—idle mixture needle
332. Limiter—idle mixture needle
333. Plug—idle mixture needle
400. Solenoid—idle stop
401. Spring—idle stop solenoid
415. Bracket—throttle return spring anchor
420. Screw—bracket attaching (countersunk)
421. Screw—bracket attaching

Exploded view of Rochester 1MEF carburetor—1985 and later models

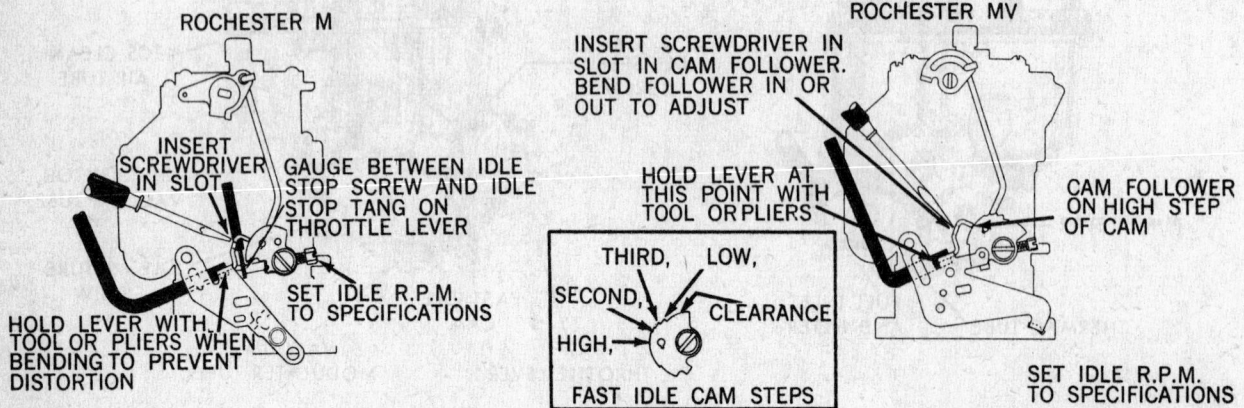

Fast idle adjustment—Monojet® carburetor

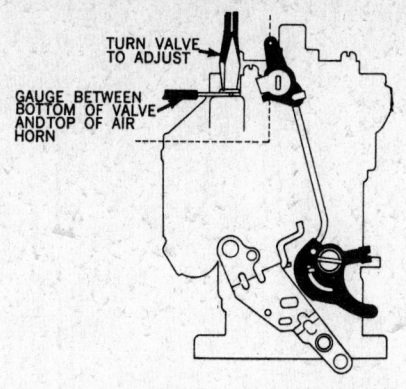

Idle vent adjustment—Monojet® carburetor

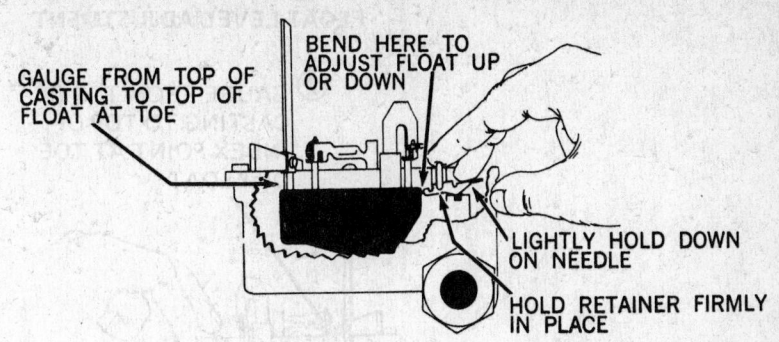

Float level adjustment—Monojet® carburetor

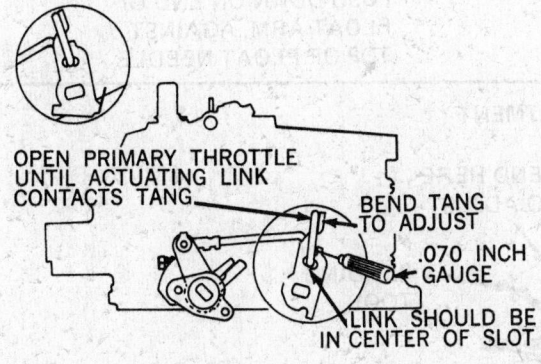

Secondary opening adjustment—typical Quadrajet® carburetor

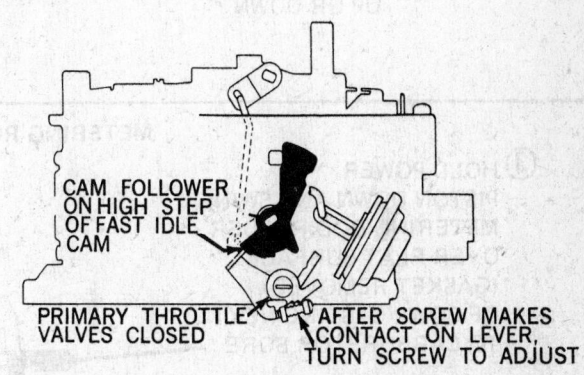

Fast idle adjustment—typical Quadrajet® carburetor

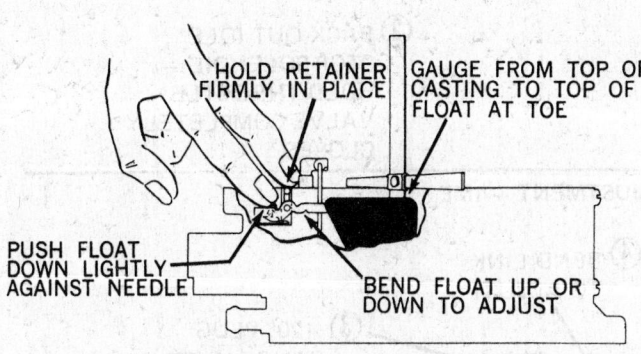

Float level adjustment—typical Quadrajet® carburetor

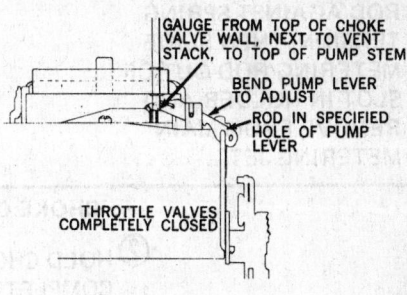

Pump rod adjustment—typical Quadrajet® carburetor

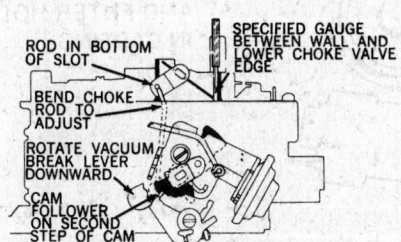

Choke rod adjustment—typical Quadrajet® carburetor

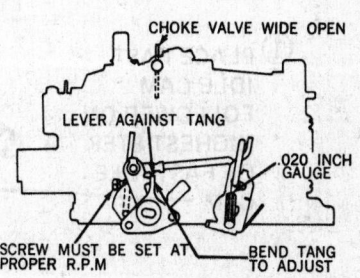

Secondary closing adjustment—typical Quadrajet® carburetor

FLOAT LEVEL ADJUSTMENT

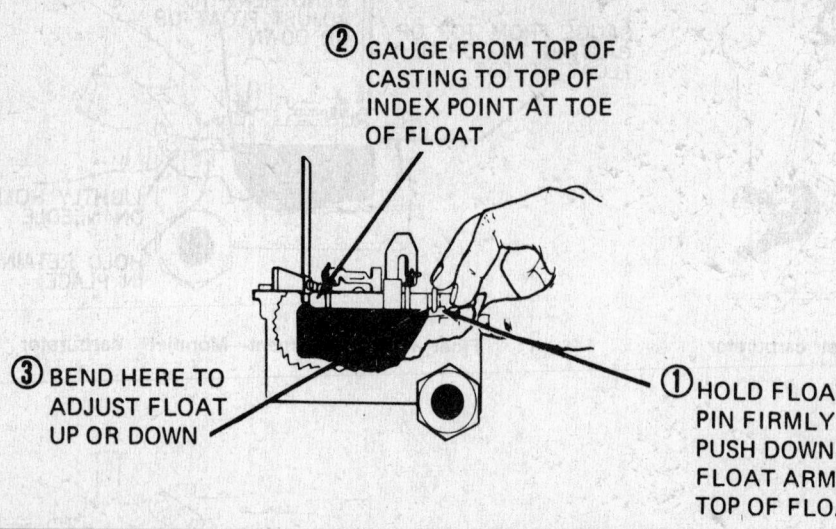

② GAUGE FROM TOP OF CASTING TO TOP OF INDEX POINT AT TOE OF FLOAT

③ BEND HERE TO ADJUST FLOAT UP OR DOWN

① HOLD FLOAT RETAINING PIN FIRMLY IN PLACE — PUSH DOWN ON END OF FLOAT ARM, AGAINST TOP OF FLOAT NEEDLE

METERING ROD ADJUSTMENT

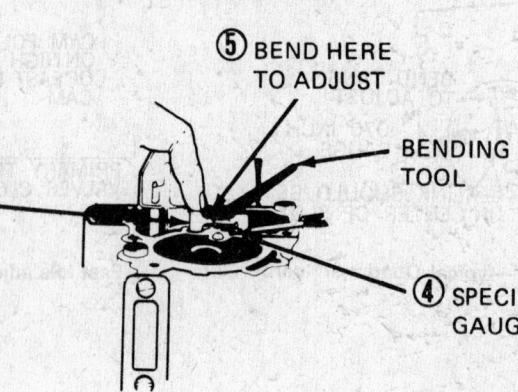

③ HOLD POWER PISTON DOWN AND SWING METERING ROD HOLDER OVER FLAT SURFACE (GASKET REMOVED) OF BOWL CASTING NEXT TO CARBURETOR BORE

① REMOVE METERING ROD BY HOLDING THROTTLE VALVE WIDE OPEN. PUSH DOWNWARD ON METERING ROD AGAINST SPRING TENSION, THEN SLIDE METERING ROD OUT OF SLOT IN HOLDER AND REMOVE FROM MAIN METERING JET.

⑤ BEND HERE TO ADJUST

BENDING TOOL

④ SPECIFIED PLUG GAUGE — SLIDE FIT

② BACK OUT IDLE STOP SOLENOID — HOLD THROTTLE VALVE COMPLETELY CLOSED

CHOKE COIL LEVER ADJUSTMENT — 1ME

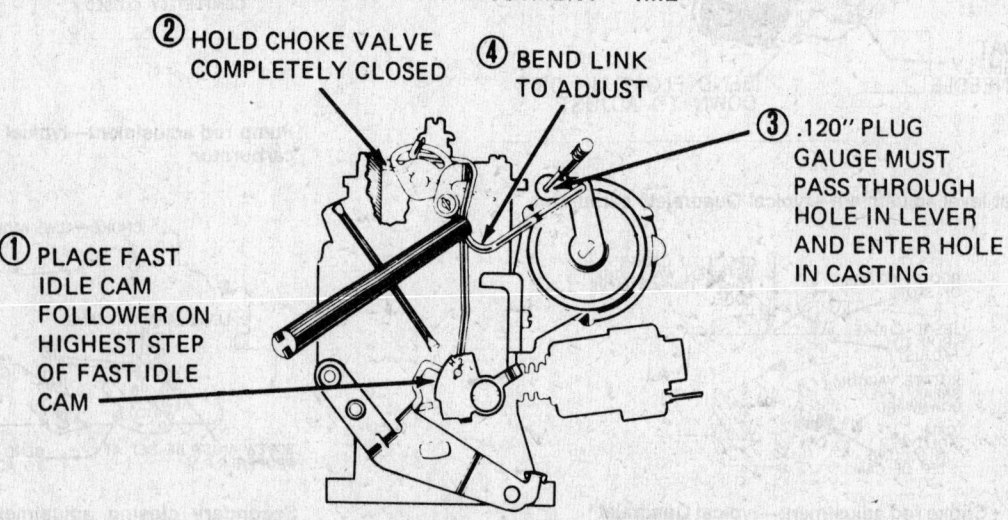

② HOLD CHOKE VALVE COMPLETELY CLOSED

④ BEND LINK TO ADJUST

③ .120" PLUG GAUGE MUST PASS THROUGH HOLE IN LEVER AND ENTER HOLE IN CASTING

① PLACE FAST IDLE CAM FOLLOWER ON HIGHEST STEP OF FAST IDLE CAM

Carburetor adjustments—1ME, 1MEF models

CHOKE ROD (FAST IDLE CAM) ADJUSTMENT (2ND STEP)

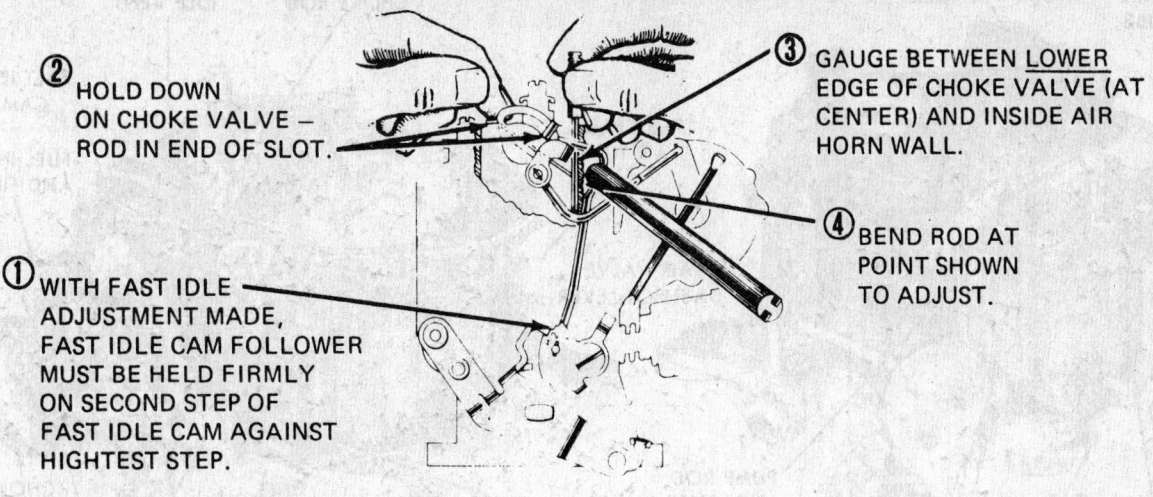

② HOLD DOWN ON CHOKE VALVE – ROD IN END OF SLOT.

③ GAUGE BETWEEN LOWER EDGE OF CHOKE VALVE (AT CENTER) AND INSIDE AIR HORN WALL.

④ BEND ROD AT POINT SHOWN TO ADJUST.

① WITH FAST IDLE ADJUSTMENT MADE, FAST IDLE CAM FOLLOWER MUST BE HELD FIRMLY ON SECOND STEP OF FAST IDLE CAM AGAINST HIGHTEST STEP.

VACUUM BREAK ADJUSTMENT – 1ME (BOWL SIDE)

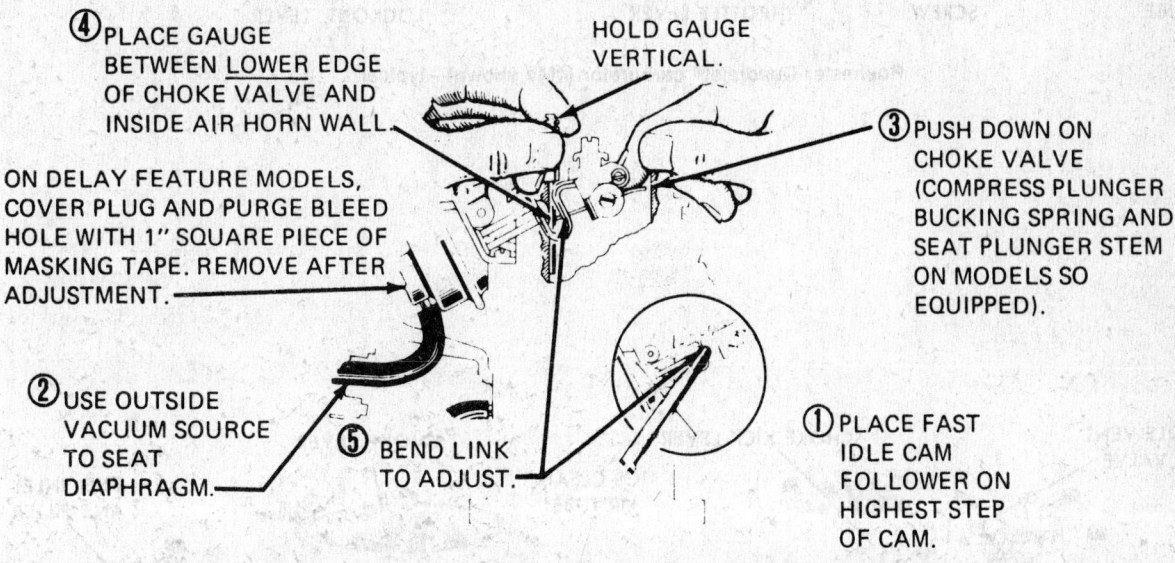

④ PLACE GAUGE BETWEEN LOWER EDGE OF CHOKE VALVE AND INSIDE AIR HORN WALL.

HOLD GAUGE VERTICAL.

③ PUSH DOWN ON CHOKE VALVE (COMPRESS PLUNGER BUCKING SPRING AND SEAT PLUNGER STEM ON MODELS SO EQUIPPED).

ON DELAY FEATURE MODELS, COVER PLUG AND PURGE BLEED HOLE WITH 1″ SQUARE PIECE OF MASKING TAPE. REMOVE AFTER ADJUSTMENT.

② USE OUTSIDE VACUUM SOURCE TO SEAT DIAPHRAGM.

⑤ BEND LINK TO ADJUST.

① PLACE FAST IDLE CAM FOLLOWER ON HIGHEST STEP OF CAM.

UNLOADER ADJUSTMENT – 1ME (WIDE OPEN KICK)

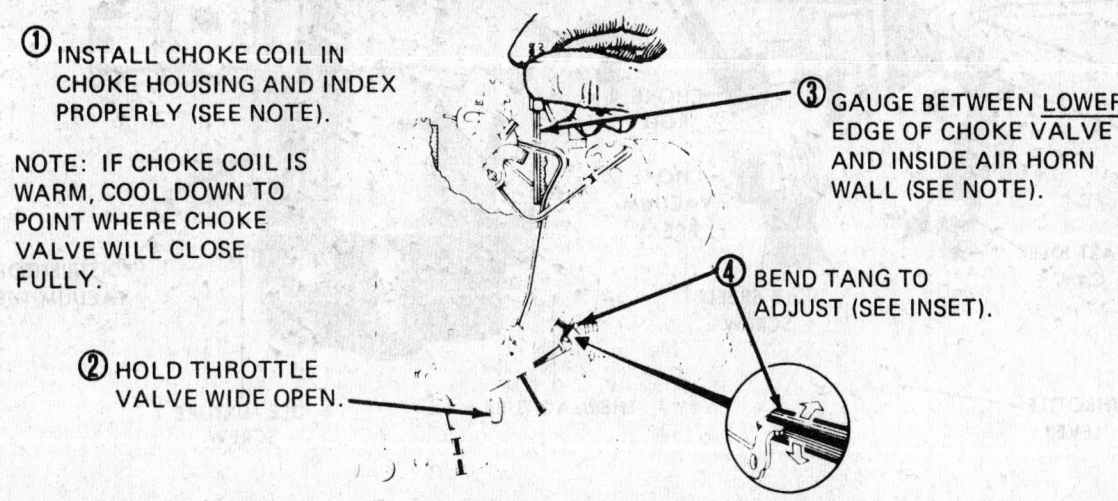

① INSTALL CHOKE COIL IN CHOKE HOUSING AND INDEX PROPERLY (SEE NOTE).

NOTE: IF CHOKE COIL IS WARM, COOL DOWN TO POINT WHERE CHOKE VALVE WILL CLOSE FULLY.

③ GAUGE BETWEEN LOWER EDGE OF CHOKE VALVE AND INSIDE AIR HORN WALL (SEE NOTE).

④ BEND TANG TO ADJUST (SEE INSET).

② HOLD THROTTLE VALVE WIDE OPEN.

Carburetor adjustments—1ME, 1MEF models

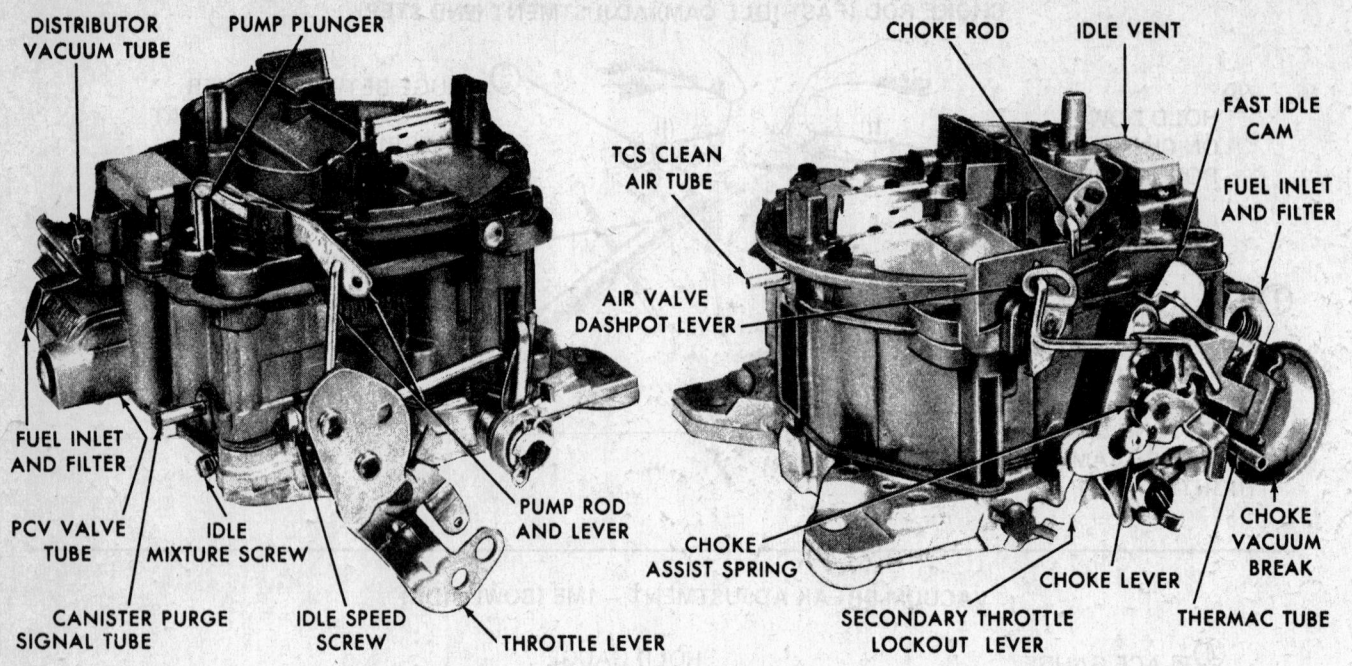

DISTRIBUTOR VACUUM TUBE

PUMP PLUNGER

CHOKE ROD

IDLE VENT

TCS CLEAN AIR TUBE

FAST IDLE CAM

FUEL INLET AND FILTER

AIR VALVE DASHPOT LEVER

FUEL INLET AND FILTER

PCV VALVE TUBE

IDLE MIXTURE SCREW

PUMP ROD AND LEVER

CHOKE VACUUM BREAK

CANISTER PURGE SIGNAL TUBE

IDLE SPEED SCREW

THROTTLE LEVER

CHOKE ASSIST SPRING

CHOKE LEVER

THERMAC TUBE

SECONDARY THROTTLE LOCKOUT LEVER

Rochester Quadrajet® carburetor (4MV shown)—typical

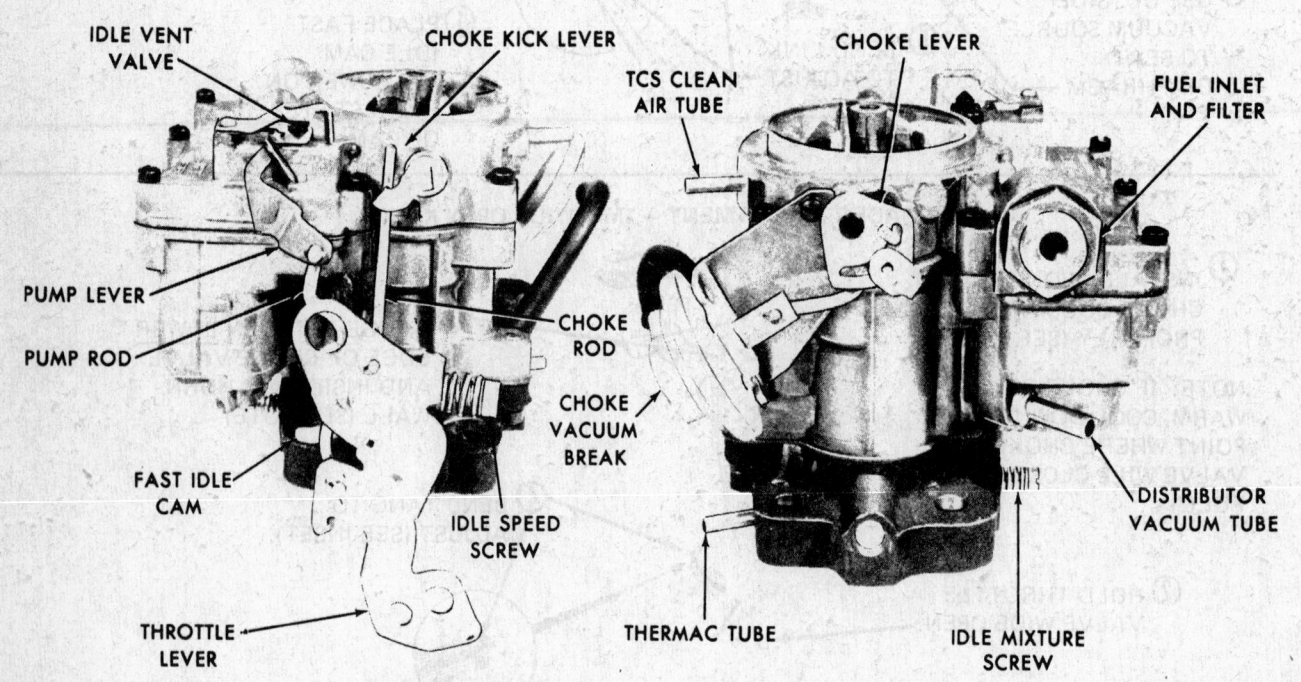

IDLE VENT VALVE

CHOKE KICK LEVER

CHOKE LEVER

TCS CLEAN AIR TUBE

FUEL INLET AND FILTER

PUMP LEVER

PUMP ROD

CHOKE ROD

CHOKE VACUUM BREAK

FAST IDLE CAM

IDLE SPEED SCREW

DISTRIBUTOR VACUUM TUBE

THROTTLE LEVER

THERMAC TUBE

IDLE MIXTURE SCREW

Rochester model 2GV—typical

1. Gasket—air cleaner
5. Gasket—flange
10. Cam—fast idle
12. Screw—fast idle cam attaching
15. Link—choke
18. Choke lever & swivel assembly
19. Screw—choke rod swivel
30. Choke control bracket assembly
100. Air horn assembly
101. Gasket—air horn to float bowl
105. Screw assembly—air horn to float bowl
107. Screw assembly—air horn to float bowl (long)
120. Nut—fuel inlet
122. Gasket—fuel inlet nut
125. Filter—fuel inlet
128. Spring—fuel filter
130. Link—pump
131. Retainer—pump link
134. Pump shaft & outside lever assembly
136. Washer—pump shaft
138. Seal—pump shaft
140. Lever—pump, inside
141. Screw—pump lever attaching
149. Pump assembly
150. Cup—pump plunger
151. Spring—pump plunger
152. Retainer—pump assembly
160. Float
162. Baffle—fuel inlet
165. Pin—float lever hinge
170. Needle—float
175. Seat—float needle
176. Gasket—float needle seat
180. Power valve piston assembly
200. Float bowl assembly
210. Spring—pump return
211. Ball—pump inlet check
230. Venturi cluster assembly
231. Gasket—venturi cluster to float bowl
235. Screw—venturi cluster—bowl attaching
236. Lockwasher—venturi cluster—bowl screw
238. Screw—venturi cluster—bowl attaching (center)
239. Gasket—venturi cluster to bowl pump discharge screw
245. Guide—pump discharge spring
247. Spring—pump discharge ball
248. Ball—pump discharge check
253. Jet—main metering
258. Power valve assembly—vacuum
259. Gasket—power valve assembly
262. power valve assembly (pump plunger-actuated)

263. Gasket—power valve assembly
300. Throttle body assembly
301. Gasket—float bowl to throttle body
305. Screw—float bowl to throttle body
306. Lockwasher—float bowl to throttle body screw
310. Needle—idle mixture
311. Spring—idle mixture needle
315. Limiter—idle mixture needle
318. Screw—(throttle stop)

319. Spring—(throttle stop screw)
400. Bracket—idle speed device
401. Screw—solenoid bracket assembly attaching
405. Bracket—idle speed device
406. Rivet—bracket attaching
420. Throttle kicker assembly
421. Washer—tab locking
422. Nut—throttle kicker assembly attaching
425. Solenoid—idle stop
427. Nut—solenoid attaching

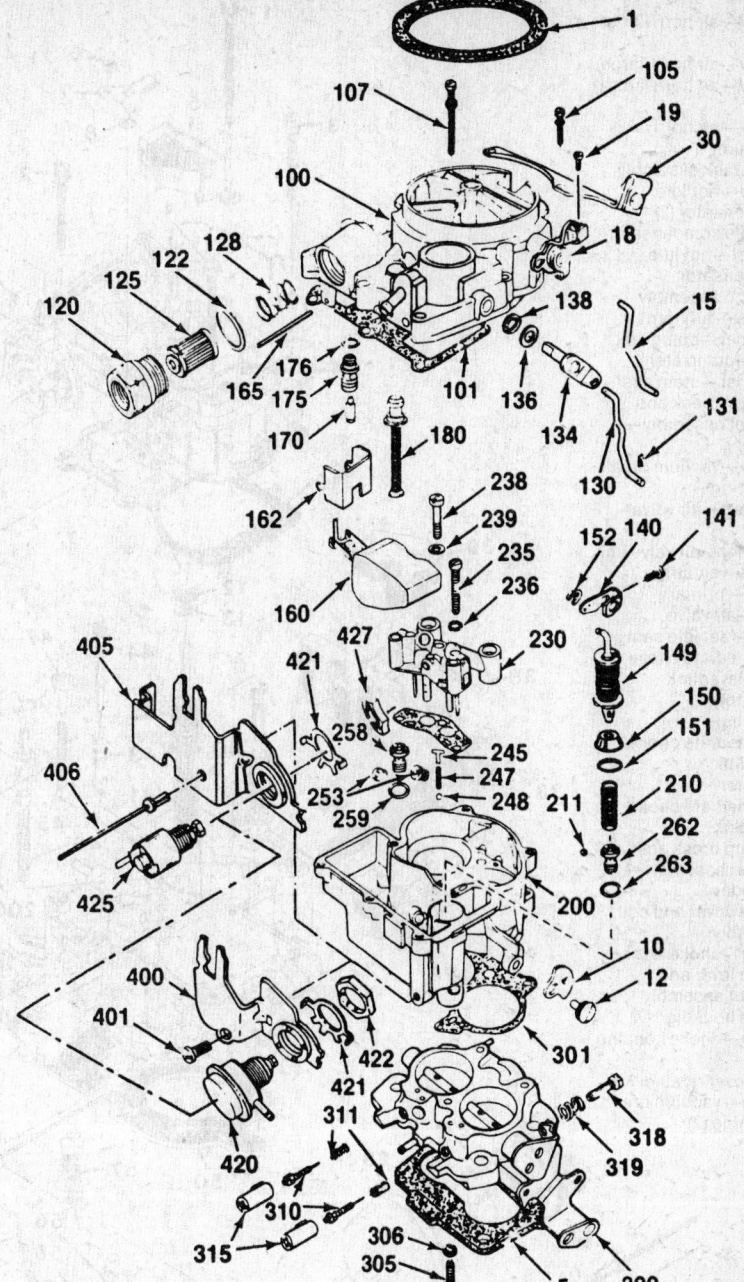

Exploded view of Rochester 2G, 2GF carburetor

1. Screw—air horn (long) (2)
2. Screw—air horn (large)
3. Screw—air horn (short) (3)
4. Screw—air horn (medium)
5. Vent stack assembly
6. Screw—hot idle compensator (2)
7. Hot idle compensator
8. Gasket—hot idle compensator
9. Air horn assembly
10. Gasket—air horn
11. Retainer—pump link
12. Seal—pump stem
13. Retainer—stem seal
14. Vacuum break and bracket assembly—primary
15. Screw—vacuum break attaching
16. Bushing—air valve link
17. Retainer—air valve link
18. Hose—vacuum break—primary
19. Link—air valve
20. Link—fast idle cam
21. Intermediate choke shaft/lever/link assembly
22. Bushing—intermediate choke shaft link
23. Retainer—intermediate choke shaft link
24. Vacuum break and bracket assembly—secondary
25. Choke cover and coil assembly
26. Screw—choke lever
27. Choke lever and contact assembly
28. Choke housing
29. Screw—choke housing (2)
30. Stat cover retainer kit
31. Screw—vacuum break attaching (2)

32. Float bowl assembly
33. Nut—fuel inlet
34. Gasket—fuel inlet nut
35. Filter—fuel inlet
36. Spring—fuel filter
37. Float assembly
38. Hinge pin—float
39. Insert—float bowl
40. Needle and seat assembly
41. Spring—pump return
42. Pump—assembly
43. Jet—main metering
44. Rod—main metering assembly
45. Ball—pump discharge
46. Spring—pump discharge

47. Retainer—pump discharge spring
48. Power piston assembly
49. Spring—power piston
50. Gasket—throttle body
51. Throttle body assembly
52. Pump rod
53. Clip—cam screw
54. Screw—cam
55. Spring—throttle stop screw
56. Screw—throttle stop
57. Idle needle and spring
58. Screw—throttle body attaching (4)
59. Nut—idle solenoid
60. Retainer—idle solenoid
61. Idle solenoid

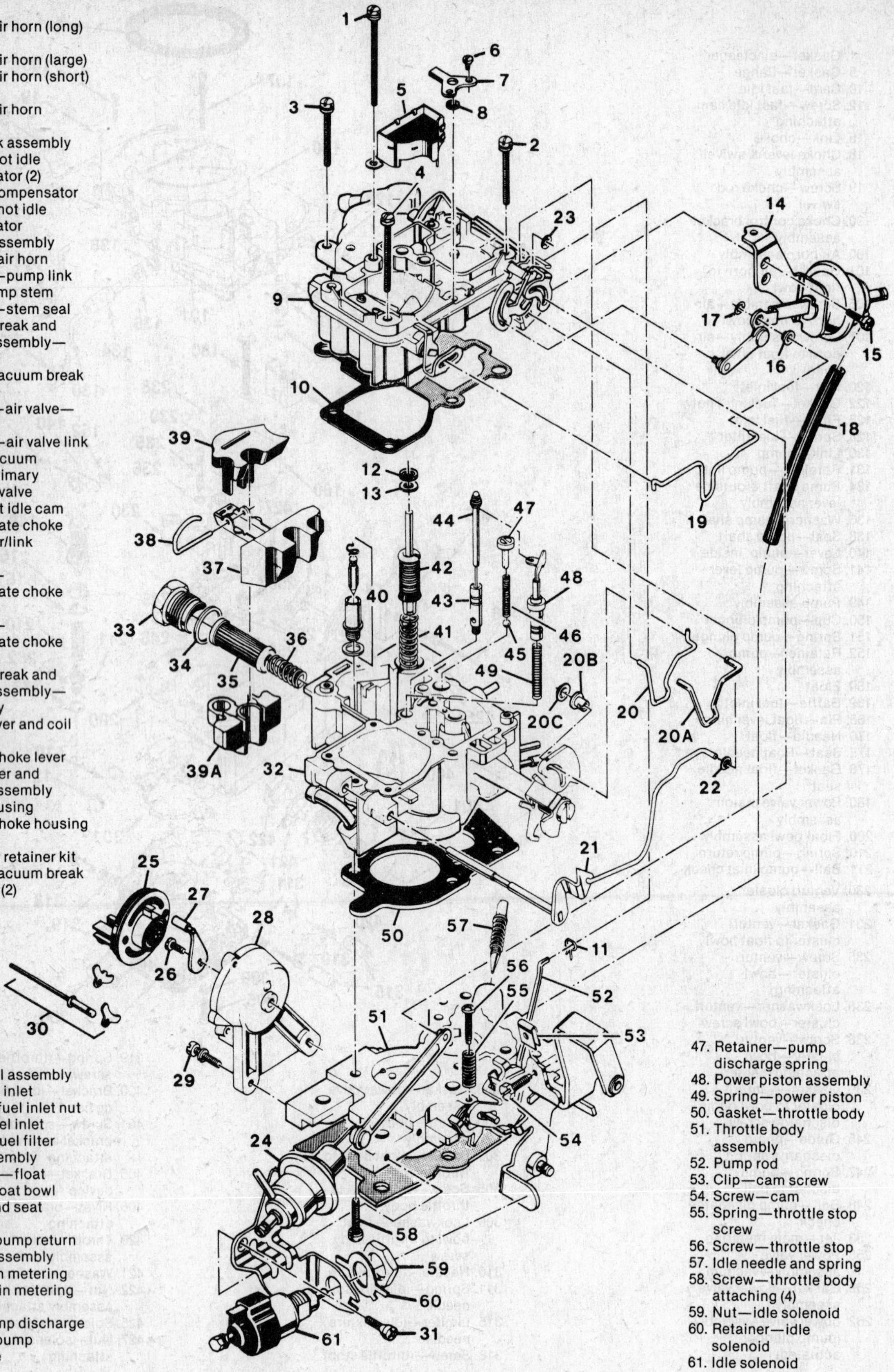

Exploded view of Rochester 2SE carburetor

1. Set the parking brake and block the wheels. Attach a calibrated tachometer to the engine.

2. Disconnect and plug the canister purge line to the carburetor.

3. Disconnect the vacuum advance hose and plug it. Adjust the ignition, timing, if necessary.

4. Adjust the carburetor idle speed to specifications. Disconnect the crankcase ventilation hose from the air cleaner.

5. Insert the hose with the rubber stopper from the propane valve and special adapter tool, into the air cleaner crankcase ventilation hose hole.

NOTE: The propane cylinder must be vertical during the adjustment procedure.

6. With the engine at normal operating temperature and running, the A/T equipped vehicles in the DRIVE position and the M/T equipped vehicles in the NEUTRAL po-

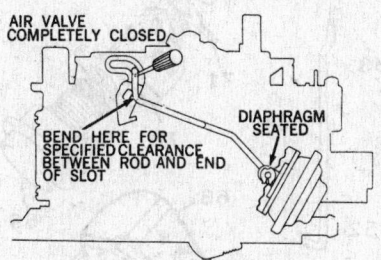

Air valve dashpot adjustment—typical Quadrajet® carburetor

sition, slowly open the propane cylinder control valve to allow propane to enter the carburetor.

7. Continue to add propane until the idle speed increases to the maximum enriched idle rpm and then, because of over richness, will drop. Note the maximum enriched idle rpm.

NOTE: If a rich rpm cannot be obtained, check for an empty propane cylinder or propane system leaks.

8. The propane enrichment is the difference between curb idle speed and the maximum enrichment idle rpm.

9. The maximum enrichment idle rpm is the curb idle rpm plus the propane enrichment rpm.

10. If the maximum enrichment idle rpm is within the specifications, the idle mixture is correct. If so, remove the propane tube and install the crankcase ventilation tube in the air cleaner assembly.

11. If the maximum enriched idle rpm is not within the specifications, remove the carburetor from the engine to gain access to the tamper resistant plugs covering the idle mixture screw.

NOTE: A portion of the throttle base must be cut and the plugs crushed in order to expose the idle mixture adjusting screws.

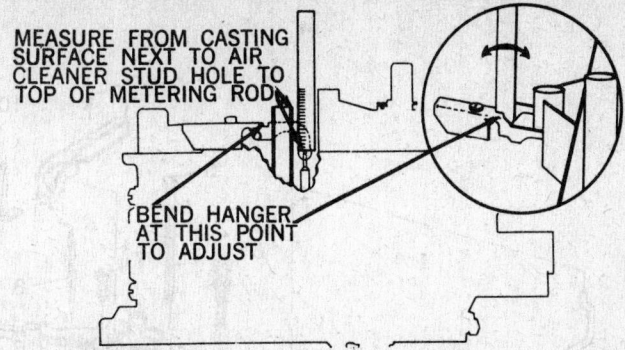

Secondary metering adjustment—typical Quadrajet® carburetor

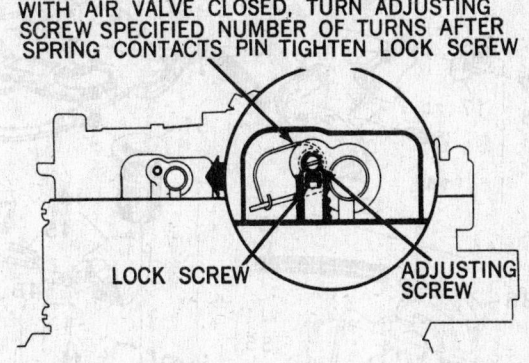

Air valve spring adjustment—typical Quadrajet® carburetor

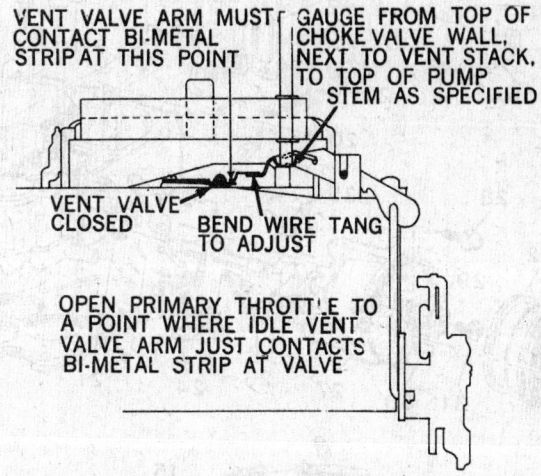

Idle vent adjustment—typical Quadrajet® carburetor

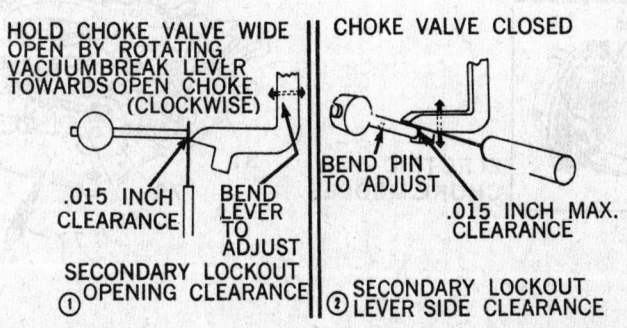

Secondary lockout adjustment—typical Quadrajet® carburetor

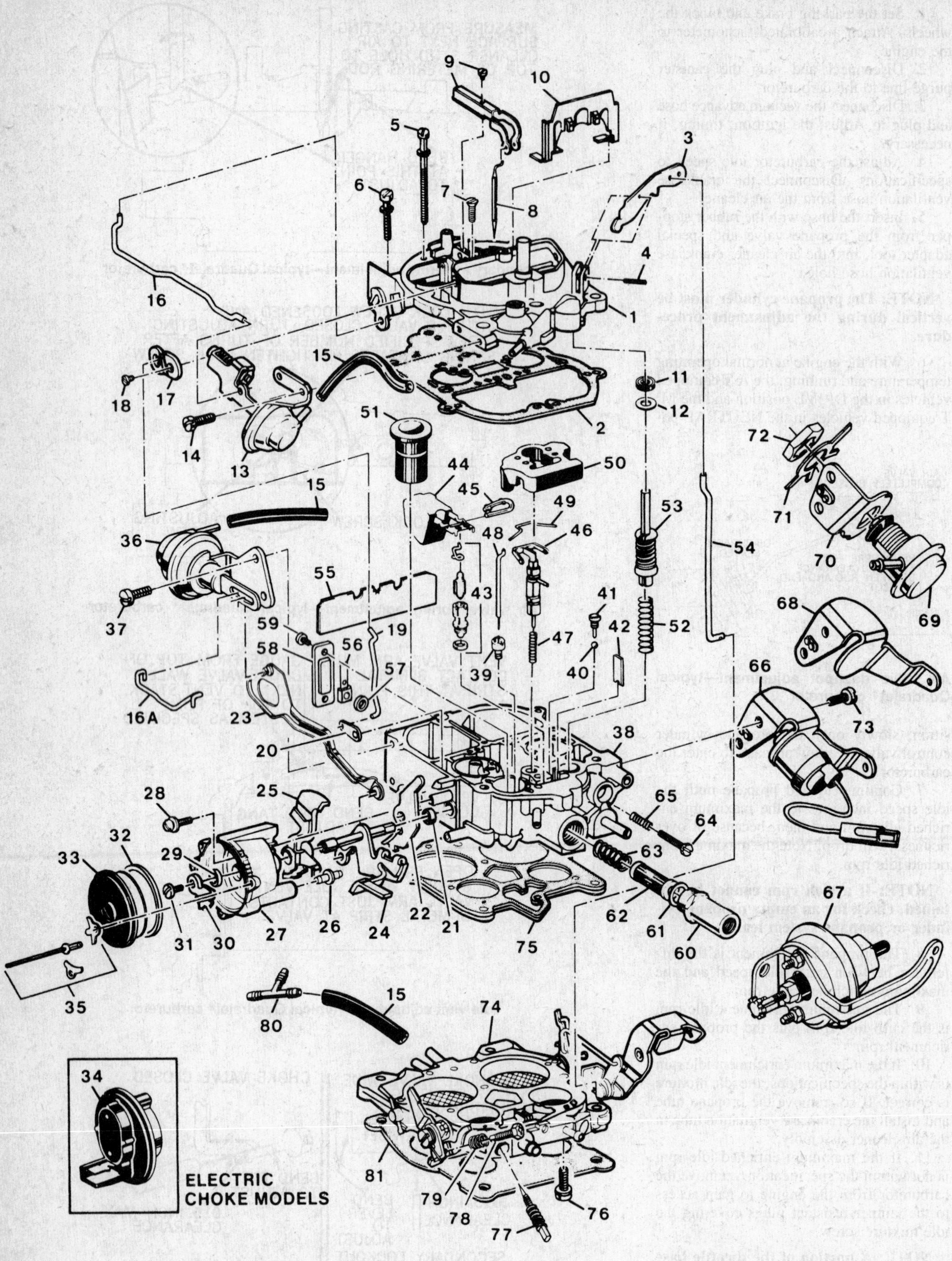

ELECTRIC
CHOKE MODELS

M4MC/M4ME carburetor exploded view →

1. Air Horn Assy.
2. Gasket—Air Horn
3. Lever—Pump Actuating
4. Roll Pin—Pump Lever Hinge
5. Screw—Air Horn Long (2)
6. Screw—Air Horn Short ( )
7. Screw — Air Horn Countersunk (2)
8. Metering Rod—Secondary (2)
9. Holder and Screw—Secondary Metering Rod
10. Baffle—Secondary Air
11. Seal—Pump Plunger
12. Retainer—Pump Seal
13. Vac. Break Control & Bracket—Front
14. Screw—Control Attaching (2)
15. Hose—Vacuum
16. Rod—Air Valve
16A. Rod—Air Valve (Truck)
17. Lever—Choke Rod (Upper)
18. Screw—Choke Lever
19. Rod—Choke
20. Lever—Choke Rod (Lower)

21. Seal—Intermediate Choke Shaft
22. Lever—Secondary Lockout
23. Link—Rear Vacuum Break
24. Int. Choke Shaft & Lever
25. Cam—Fast Idle
26. Seal—Choke Housing to Bowl (Hot Air Choke)
27. Kit—Choke Housing
28. Screw—Choke Housing to Bowl
29. Seal—Intermediate Choke Shaft (Hot Air Choke)
30. Lever—Choke Coil
31. Screw—Choke Coil Lever
32. Gasket—Stat Cover (Hot Air Choke)
33. Stat Cover & Coil Assy. (Hot Air Choke)
34. Stat Cover & Coil Assy. (Electric Choke)
35. Kit — Stat Cover Attaching
36. Rear Vacuum Break Assembly
37. Screw—Vacuum Break Attaching (2)
40. Ball—Pump Discharge

41. Retainer—Pump Discharge Ball
42. Baffle—Pump Well
43. Needle & Seat Assembly
44. Float Assembly
45. Hinge Pin — Float Assembly
46. Power Piston Assembly
47. Spring—Power Piston
48. Rod—Primary Metering (2)
49. Spring—Metering Rod Retainer
50. Insert—Float Bowl
51. Insert—Bowl Cavity
52. Spring—Pump Return
53. Pump Assembly
54. Rod—Pump
55. Baffle—Secondary Bores
56. Idle Compensator Assembly
57. Seal—Idle Compensator
58. Cover—Idle Compensator
59. Screw—Idle Compensator Cover (2)
60. Filter Nut—Fuel Inlet
61. Gasket—Filter Nut
62. Filter—Fuel Inlet
63. Spring—Fuel Filter

64. Screw—Idle Stop
65. Spring — Idle Stop Screw
66. Idle Speed Solenoid & Bracket Assembly
67. Idle Load Compensator & Bracket Assembly
68. Bracket—Throttle Return Spring
69. Actuator—Throttle Lever (Truck Only)
70. Bracket—Throttle Lever Actuator (Truck Only)
71. Washer—Actuator Nut (Truck Only)
72. Nut—Actuator Attaching (Truck Only)
73. Screw—Bracket Attaching (2)
74. Throttle Body Assembly
75. Gasket—Throttle Body
76. Screw—Throttle Body (3)
77. Idle Mixture Needle & Spring Assy. (2)
78. Screw — Fast Idle Adjusting
79. Spring — Fast Idle Screw
80. Tee—Vacuum Hose
81. Gasket—Flange

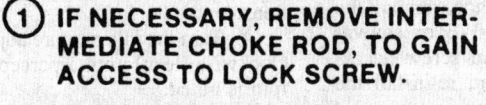

① **IF NECESSARY, REMOVE INTERMEDIATE CHOKE ROD, TO GAIN ACCESS TO LOCK SCREW.**

② **LOOSEN LOCK SCREW USING 3/32" (2.381mm) HEX WRENCH.**

③ **TURN TENSION-ADJUSTING SCREW CLOCKWISE UNTIL AIR VALVE OPENS SLIGHTLY.**

TURN ADJUSTING SCREW COUNTERCLOCKWISE UNTIL AIR VALVE JUST CLOSES. CONTINUE COUNTERCLOCKWISE SPECIFIED NUMBER OF TURNS.

④ **TIGHTEN LOCK SCREW.**

⑤ **APPLY LITHIUM BASE GREASE TO LUBRICATE PIN AND SPRING CONTACT AREA.**

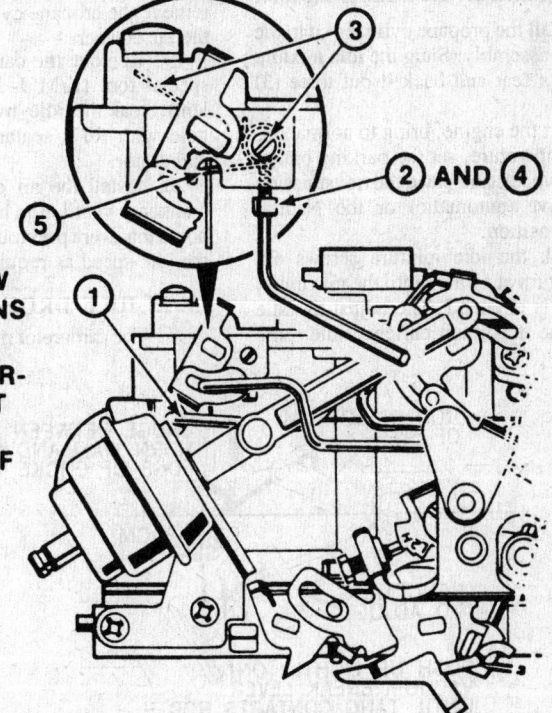

Air valve spring adjustment—2SE carburetor

① IF RIVETED, DRILL OUT AND REMOVE RIVETS. REMOVE CHOKE COVER AND COIL ASSEMBLY.

② PLACE FAST IDLE SCREW ON HIGH STEP OF FAST IDLE CAM.

③ PUSH ON INTERMEDIATE CHOKE LEVER UNTIL CHOKE VALVE IS CLOSED.

④ INSERT .085" (2.18mm) PLUG GAGE IN HOLE.

⑤ EDGE OF LEVER SHOULD JUST CONTACT SIDE OF GAGE.

⑥ SUPPORT AT "S" AND BEND INTERMEDIATE CHOKE ROD TO ADJUST.

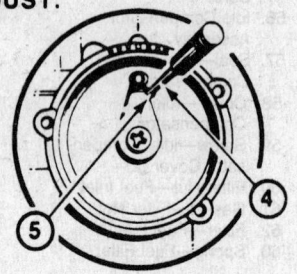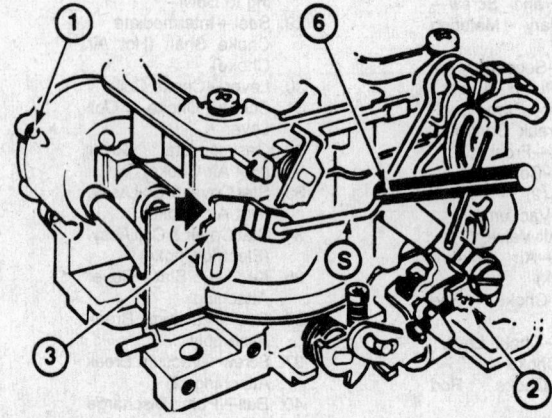

Choke coil lever adjustment—2SE carburetor

12. Remove the screws; a special tool (Kent Moore J–29030–B or equivalent) is used to attach to the heads of the screws to gain access for movement of the screws. Re-install the carburetor on the engine and connect all components.

NOTE: Modify special Kent Moore tool J–29030–B by grinding ⅛ inch off the rear and ¼ inch off the front of the tool.

13. Install the propane cylinder kit to the air cleaner assembly. Snug the idle mixture screw to its seat and back it out three (3) turns.

14. Start the engine, bring to normal operating temperature, set the parking brake, chock the wheels and place the transmission in the Drive (automatic) or the Neutral (manual) position.

15. Back the idle mixture screws out (richer, ⅛ turn at a time) until the maximum idle speed is obtained. The adjust the idle speed to the maximum enriched idle rpm.

16. Turn the mixture screws in clockwise (⅛ turn at a time) until the idle speed attains the specified curb idle speed.

17. Check the maximum enriched rpm with the propane too. If not within specifications, refer to Step 5 and repeat the procedure.

18. When the mixture and idle speed have been properly adjusted, stop the engine, remove the propane cylinder and hose from the air cleaner.

19. Remove the carburetor, remove the special tool (K/M J–29030–B or equivalent), seal the idle mixture screw access hole with RTV sealant and re-install the carburetor.

20. Install the air cleaner, connect the crankcase ventilation hose and any vacuum hoses that were previously removed. Adjust the idle speed as required.

LEAN IDLE DROP PROCEDURE

1. The carburetor must be removed from the vehicle and the tamper resistant plugs removed from the throttle body as noted under the idle mixture adjustment with propane outline.

2. Connect a tachometer to the ignition system, have the engine at normal operating temperature, apply the parking brakes securely, chock the wheels and position the A/T gear selector in DRIVE and the M/T in NEUTRAL.

3. Adjust the idle speed to specifications.

4. Turn the idle mixture adjusting screw clockwise (lean) until a perceptible loss of rpm is noted.

5. Turn the idle mixture adjusting screw counterclockwise (rich) until the highest engine rpm is attained, Do not turn the screw any further than the point at which the highest engine rpm is first attained. This is referred to as LEAN BEST IDLE.

NOTE: The engine speed will increase above curb idle speed an amount that corresponds approximately to the lean drop specifications of 20 rpm.

6. As a final adjustment, turn the idle mixture screw clockwise in increments until the specified drop (20 rpm) is attained.

NOTE: If the final rpm differs more than ± 30 rpm from the original set curb idle speed, adjust the curb idle speed to the specified engine rpm and repeat the above steps.

7. Remove the air cleaner, remove the carburetor and the modified special tool or its equivalent. Install RTV sealer in the idle mixture screw access hole.

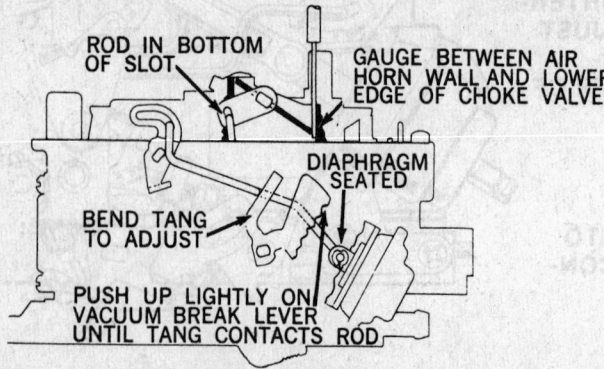

ROD IN BOTTOM OF SLOT

GAUGE BETWEEN AIR HORN WALL AND LOWER EDGE OF CHOKE VALVE

DIAPHRAGM SEATED

BEND TANG TO ADJUST

PUSH UP LIGHTLY ON VACUUM BREAK LEVER UNTIL TANG CONTACTS ROD

Vacuum break adjustment—typical Quadrajet® carburetor

8. Install the carburetor and air cleaner assembly. Adjust the engine idle speed to specifications as required.

IDLE MIXTURE ADJUSTMENT

Model E2SE Carburetor (Feedback Type)

NOTE: Each carburetor has been calibrated at the factory and should not normally need adjustment in the field. However, should a diagnosis indicate the need for adjustment due to emission failure or replacement of critical components, the idle mixture can be adjusted using the following procedure.

1. Remove the carburetor from the engine and remove the tamper resistant plug in order to gain access to the idle mixture adjusting screw.

2. Modify special Kent Moore tool J–29030–B or its equivalent, by grinding ⅛ inch off the rear and ¼ inch off the front of the tool. Place the modified tool onto the idle mixture adjusting screw.

3. Turn the idle mixture screw in until it is lightly seated and back out four (4) turns.

NOTE: If the seal in the air horn concealing the idle air bleed has been removed, replace the air horn. If the seal is still in place, do not remove the seal.

4. Remove the vent stack screen assembly to gain access to the lean mixture screw.

5. Turn the lean mixture screw in until lightly bottomed and then back out 2½ turns.

NOTE: Some resistance should be felt. If not, remove the screw and inspect for the presence of the spring.

6. Install the carburetor on the engine with the modified tool installed on the mixture adjusting screw. Do not install the air cleaner and gasket.

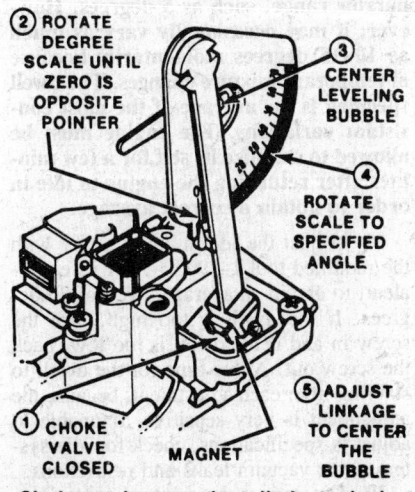

② ROTATE DEGREE SCALE UNTIL ZERO IS OPPOSITE POINTER

③ CENTER LEVELING BUBBLE

④ ROTATE SCALE TO SPECIFIED ANGLE

① CHOKE VALVE CLOSED MAGNET

⑤ ADJUST LINKAGE TO CENTER THE BUBBLE

Choke angle gauge installed—typical

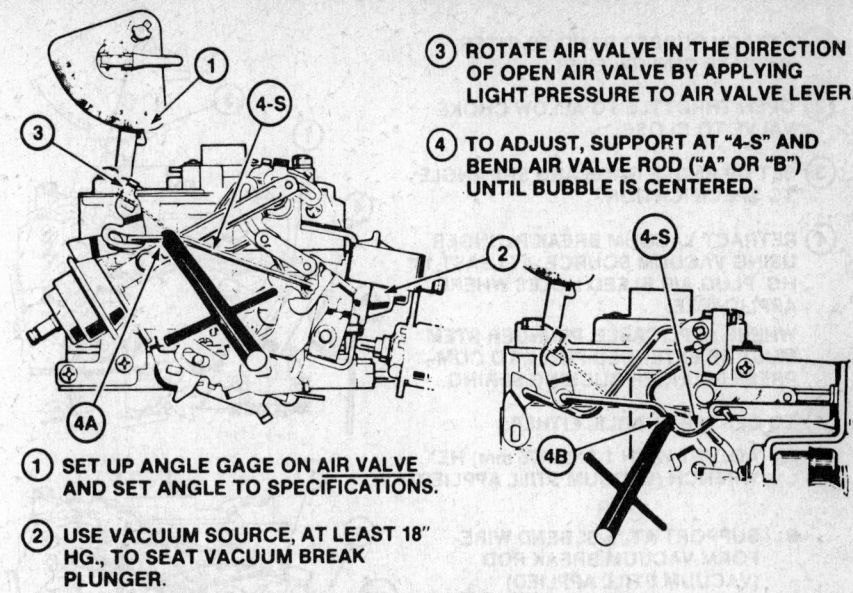

③ ROTATE AIR VALVE IN THE DIRECTION OF OPEN AIR VALVE BY APPLYING LIGHT PRESSURE TO AIR VALVE LEVER.

④ TO ADJUST, SUPPORT AT "4-S" AND BEND AIR VALVE ROD ("A" OR "B") UNTIL BUBBLE IS CENTERED.

① SET UP ANGLE GAGE ON AIR VALVE AND SET ANGLE TO SPECIFICATIONS.

② USE VACUUM SOURCE, AT LEAST 18″ HG., TO SEAT VACUUM BREAK PLUNGER.

Air valve rod adjustment—2SE, E2SE carburetors

① ATTACH RUBBER BAND TO INTERMEDIATE CHOKE LEVER.

② OPEN THROTTLE TO ALLOW CHOKE VALVE TO CLOSE.

③ SET UP ANGLE GAGE AND SET ANGLE TO SPECIFICATIONS.

④ PLACE FAST IDLE SCREW ON SECOND STEP OF CAM AGAINST RISE OF HIGH STEP.

⑤ PUSH ON CHOKE SHAFT LEVER TO OPEN CHOKE VALVE AND TO MAKE CONTACT WITH BLACK CLOSING TANG.

⑥ SUPPORT AT "S" AND ADJUST BY BENDING FAST IDLE CAM ROD UNTIL BUBBLE IS CENTERED.

FAST IDLE CAM

Choke rod and fast idle cam adjustment—2SE, E2SE carburetors

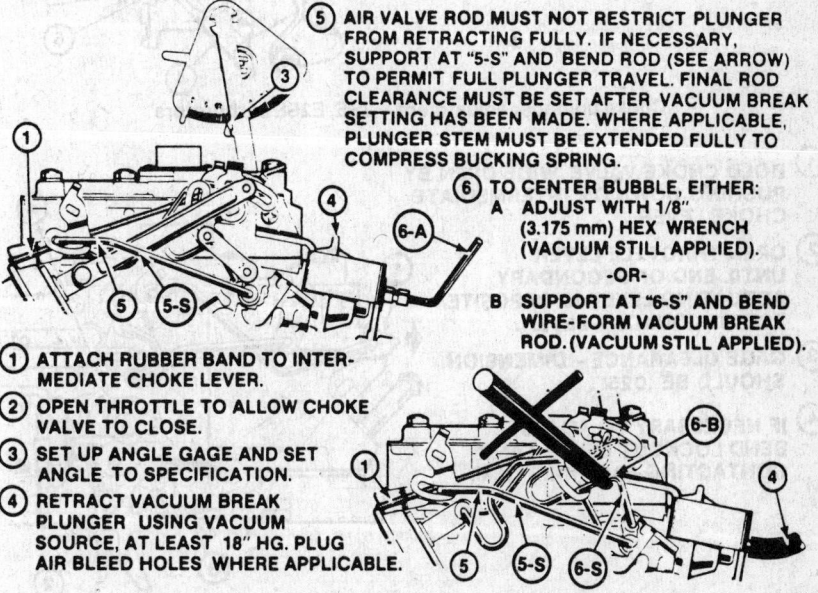

⑤ AIR VALVE ROD MUST NOT RESTRICT PLUNGER FROM RETRACTING FULLY. IF NECESSARY, SUPPORT AT "5-S" AND BEND ROD (SEE ARROW) TO PERMIT FULL PLUNGER TRAVEL. FINAL ROD CLEARANCE MUST BE SET AFTER VACUUM BREAK SETTING HAS BEEN MADE. WHERE APPLICABLE, PLUNGER STEM MUST BE EXTENDED FULLY TO COMPRESS BUCKING SPRING.

⑥ TO CENTER BUBBLE, EITHER:
A ADJUST WITH 1/8″ (3.175 mm) HEX WRENCH (VACUUM STILL APPLIED).
-OR-
B SUPPORT AT "6-S" AND BEND WIRE-FORM VACUUM BREAK ROD. (VACUUM STILL APPLIED).

① ATTACH RUBBER BAND TO INTERMEDIATE CHOKE LEVER.

② OPEN THROTTLE TO ALLOW CHOKE VALVE TO CLOSE.

③ SET UP ANGLE GAGE AND SET ANGLE TO SPECIFICATION.

④ RETRACT VACUUM BREAK PLUNGER USING VACUUM SOURCE, AT LEAST 18″ HG. PLUG AIR BLEED HOLES WHERE APPLICABLE.

Primary vacuum break adjustment—2SE, E2SE carburetors

① ATTACH RUBBER BAND TO INTER-MEDIATE CHOKE LEVER.

② OPEN THROTTLE TO ALLOW CHOKE VALVE TO CLOSE.

③ SET UP ANGLE GAGE AND SET ANGLE TO SPECIFICATION.

④ RETRACT VACUUM BREAK PLUNGER USING VACUUM SOURCE, AT LEAST 18" HG. PLUG AIR BLEED HOLES WHERE APPLICABLE.

WHERE APPLICABLE, PLUNGER STEM MUST BE EXTENDED FULLY TO COM-PRESS PLUNGER BUCKING SPRING.

⑤ TO CENTER BUBBLE, EITHER:

A. ADJUST WITH 1/8" (3.175 mm) HEX WRENCH (VACUUM STILL APPLIED)

-OR

B. SUPPORT AT "5-S", BEND WIRE-FORM VACUUM BREAK ROD (VACUUM STILL APPLIED)

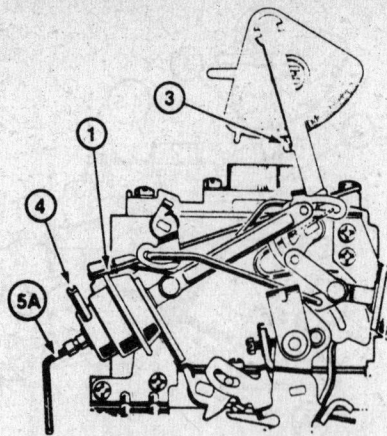

Secondary vacuum break adjustment—2SE, E2SE carburetors

① ATTACH RUBBER BAND TO INTER-MEDIATE CHOKE LEVER.

② OPEN THROTTLE TO ALLOW CHOKE VALVE TO CLOSE.

③ SET UP ANGLE GAGE AND SET ANGLE TO SPECIFICATIONS.

④ HOLD THROTTLE LEVER IN WIDE OPEN POSITION.

⑤ PUSH ON CHOKE SHAFT LEVER TO OPEN CHOKE VALVE AND TO MAKE CONTACT WITH BLACK CLOSING TANG.

⑥ ADJUST BY BENDING TANG UNTIL BUBBLE IS CENTERED.

Choke unloader adjustment—2SE, E2SE carburetors

① HOLD CHOKE VALVE WIDE OPEN BY PUSHING DOWN ON INTERMEDIATE CHOKE LEVER.

② OPEN THROTTLE LEVER UNTIL END OF SECONDARY ACTUATING LEVER IS OPPOSITE TOE OF LOCKOUT LEVER.

③ GAGE CLEARANCE - DIMENSION SHOULD BE .025".

④ IF NECESSARY TO ADJUST, BEND LOCKOUT LEVER TANG CONTACTING FAST IDLE CAM.

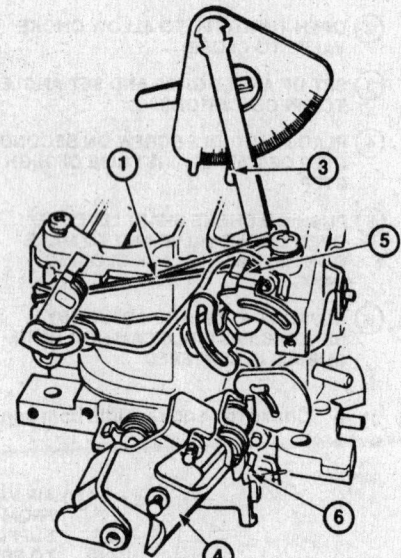

Secondary lockout adjustment—2SE, E2SE carburetors

7. Disconnect the bowl vent line at the carburetor, disconnect the EGR valve hose and the canister purge hose at the carburetor. Cap the carburetor ports.

8. Refer to the Vehicle Emission Control Information label diagram, located under the vehicle hood, and locate the hose from port 'D' on the carburetor to the temperature sensor and the secondary vacuum break thermal vacuum switch.

9. Disconnect the hose at the temperature sensor on the air cleanser and plug the hose.

10. Connect a dwell meter positive probe to the mixture control solenoid dwell test wire with a green connection.

11. Connect the negative probe to ground and set the meter at the 6 cylinder scale position.

12. Connect a tachometer to the ignition system, set the parking brake and chock the wheels.

13. Place the transmission in Park (automatic) or Neutral (manual).

14. Start and operate the engine until normal operating temperature is reached and the Electronic Engine Control System is in the closed loop mode of operation.

15. Operate the engine at 3000 rpm and adjust the lean mixture screw slowly in small increments, allowing time for the dwell to stabilize after turning the screw to obtain an average dwell of 35 degrees.

16. If the dwell is too low, back the screw out and if too high, turn the screw in. If unable to adjust to specifications, inspect the main metering system for leaks, restriction, etc.

17. Return the engine to idle speed. Allow the engine to stabilize before the dwell is recorded.

NOTE: The mixture control (MC) solenoid dwell is an indication of the ratio of ON to OFF time. The dwell of the MC solenoid is used to determine the calibration and is sensitive to changes in the fuel mixture caused by heat, air leaks, etc. While the engine is idling, it is normal for the dwell to increase and decrease fairly constant over a relativity narrow range, such as 5 degrees. However, it may occasionally vary as much as 10–15 degrees momentarily because of temporary mixture changes. The dwell specified is the average of the most consistant variations. The engine must be allowed to stabilize its self for a few minutes after returning the engine to idle in order to obtain a correct average.

18. Adjust the idle mixture screw with the modified tool J–29030–A or its equivalent, to obtain an average dwell of 25 degrees. If the dwell is too high, turn the screw in and if the dwell is too low, back the screw out. Allow time for the dwell to stabilize after each adjustment, because the adjustment is very sensitive. If unable to adjust to specifications, check for idle system air or vacuum leaks and restrictions.

19. Disconnect the mixture control so-

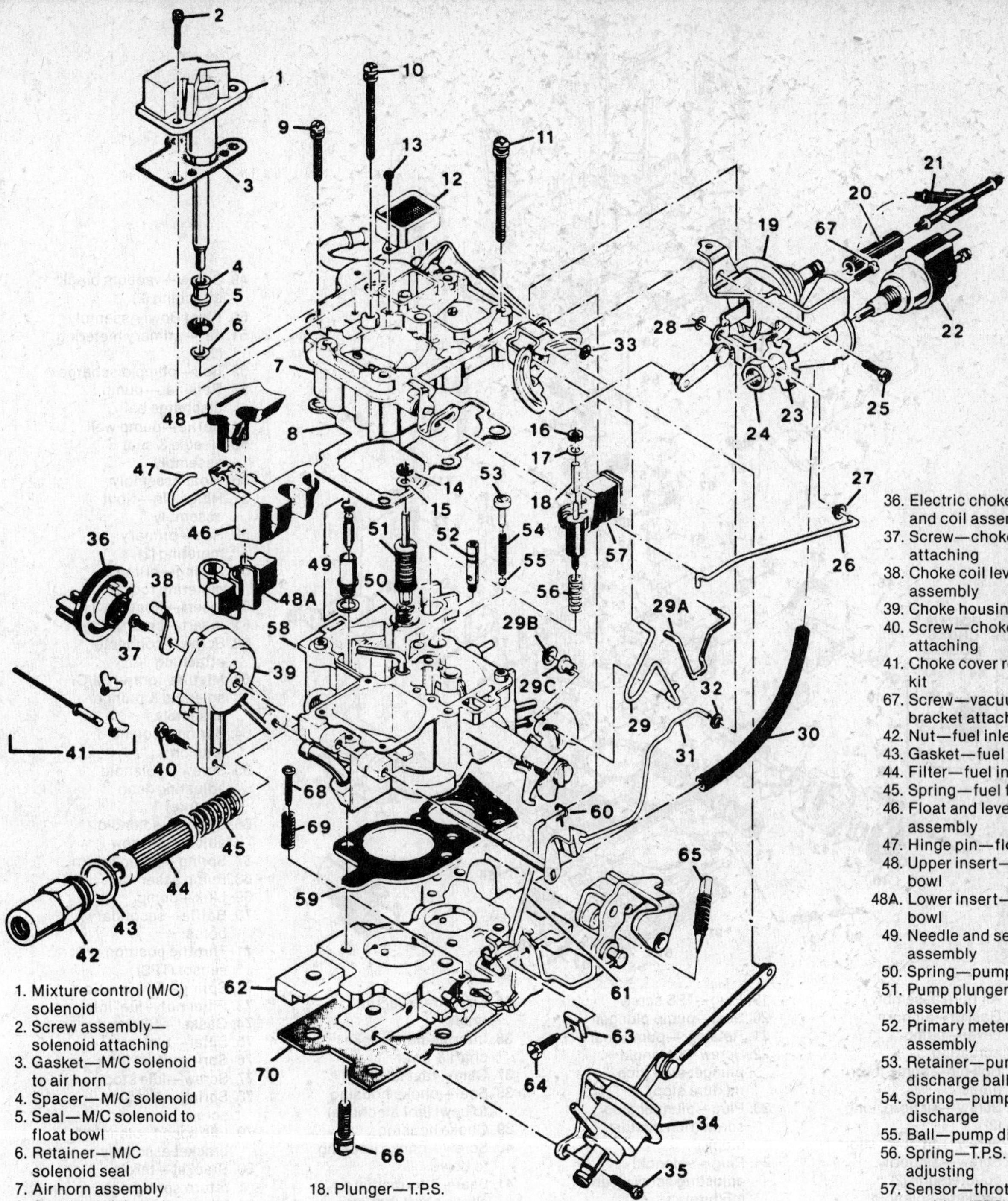

1. Mixture control (M/C) solenoid
2. Screw assembly—solenoid attaching
3. Gasket—M/C solenoid to air horn
4. Spacer—M/C solenoid
5. Seal—M/C solenoid to float bowl
6. Retainer—M/C solenoid seal
7. Air horn assembly
8. Gasket—air horn to float bowl
9. Screw—air horn to float bowl (short)
10. Screw—air horn to float bowl (long)
11. Screw—air horn to float bowl (large)
12. Vent stack and screen assembly
13. Screw—vent stack attaching
14. Seal—pump stem
15. Retainer—pump stem seal
16. Seal—T.P.S. plunger
17. Retainer—T.P.S. plunger seal
18. Plunger—T.P.S. actuator
19. Vacuum break and bracket assembly—primary
20. Hose—vacuum break primary
21. Tee—vacuum break
22. Solenoid—idle speed
23. Retainer—idle speed solenoid
24. Nut—idle speed solenoid attaching
25. Screw—vacuum break bracket attaching
26. Link—air valve
27. Bushing—air valve link
28. Retainer—air valve link
29. Link—fast idle cam
29A. Link—fast idle cam
29B. Retainer—link
29C. Bushing—link
30. Hose—vacuum break
31. Intermediate choke shaft/lever/link assembly
32. Bushing—intermediate choke link
33. Retainer—intermediate choke link
34. Vacuum break and link assembly—secondary
35. Screw—vacuum break attaching

36. Electric choke—cover and coil assembly
37. Screw—choke lever attaching
38. Choke coil lever assembly
39. Choke housing
40. Screw—choke housing attaching
41. Choke cover retainer kit
67. Screw—vacuum break bracket attaching
42. Nut—fuel inlet
43. Gasket—fuel inlet nut
44. Filter—fuel inlet
45. Spring—fuel filter
46. Float and lever assembly
47. Hinge pin—float
48. Upper insert—float bowl
48A. Lower insert—float bowl
49. Needle and seat assembly
50. Spring—pump return
51. Pump plunger assembly
52. Primary metering jet assembly
53. Retainer—pump discharge ball
54. Spring—pump discharge
55. Ball—pump discharge
56. Spring—T.P.S. adjusting
57. Sensor—throttle position (TPS)
58. Float bowl assembly
59. Gasket—float bowl
60. Retainer—pump link
61. Link—pump
62. Throttle body assembly
63. Clip—cam screw
64. Screw—fast idle cam
65. Idle needle and spring assembly
66. Screw—throttle body to float bowl
68. Screw—idle stop
69. Spring—idle stop screw
70. Gasket—insulator flange

Exploded view of Rochester E2SE carburetor

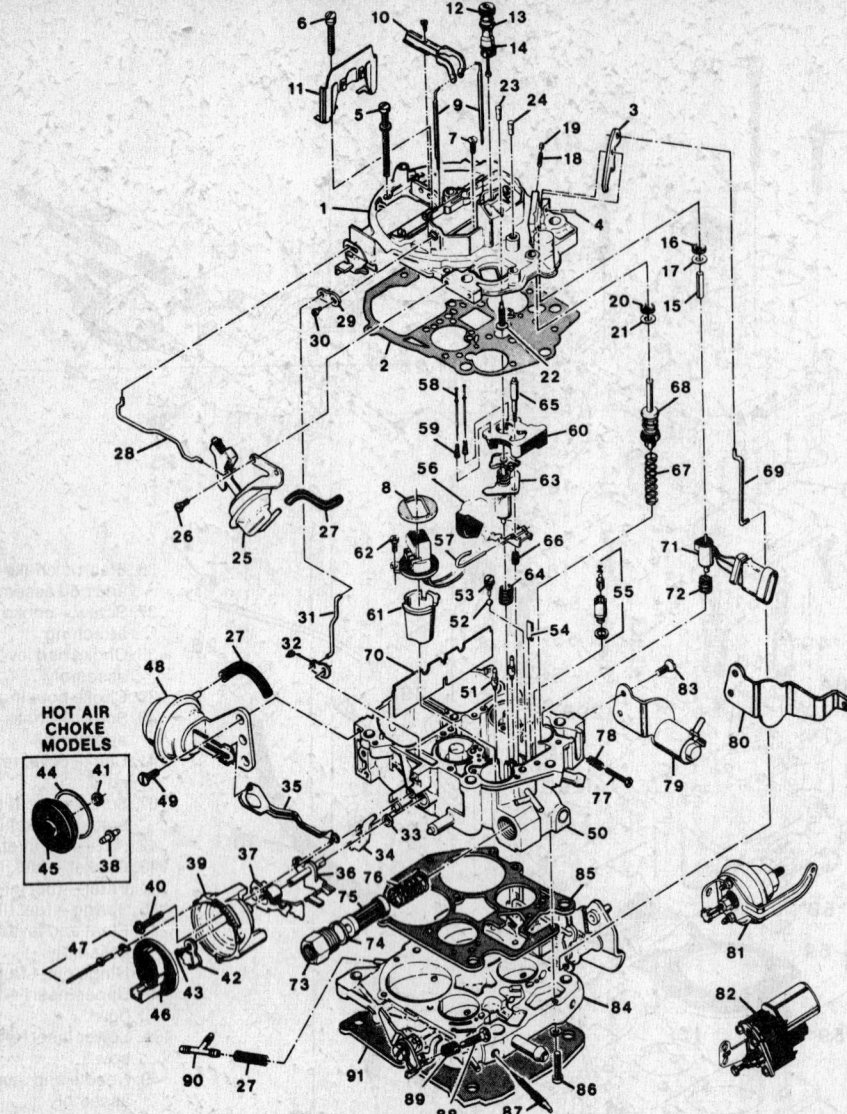

HOT AIR CHOKE MODELS

49. Screw—vacuum break attaching (2)
50. Float Bowl Assembly
51. Jet—primary metering (2)
52. Ball—pump discharge
53. Retainer—pump discharge ball
54. Baffle—pump well
55. Needle & seat assembly
56. Float assembly
57. Hinge pin—float assembly
58. Rod—primary metering (2)
59. Spring—primary metering rod (2)
60. Insert—float bowl
61. Insert—bowl cavity
62. Screw—connector attaching
63. Mixture control (M/C) solenoid & plunger assembly
64. Spring—solenoid tension
65. Screw—solenoid adjusting (lean mixture)
66. Spring—solenoid adjusting screw
67. Spring—pump return
68. Pump assembly
69. Link—pump
70. Baffle—secondary bores
71. Throttle position sensor (TPS)
72. Spring—TPS Tension
73. Filter nut—fuel inlet
74. Gasket—filter nut
75. Filter—fuel inlet
76. Spring—fuel filter
77. Screw—idle stop
78. Spring—idle stop screw
79. Idle speed solenoid & bracket assembly
80. Bracket—throttle return spring
81. Idle load compensator & bracket assembly
82. Idle speed control & bracket assembly
83. Screw—bracket attaching
84. Throttle body assembly
85. Gasket—throttle body
86. Screw—throttle body
87. Idle needle & spring assembly (2)
88. Screw—fast idle adjusting
89. Spring fast idle screw
90. Tee—vacuum hose
91. Gasket—flange

1. Air horn assembly
2. Gasket—air horn
3. Lever—pump actuating
4. Roll pin—pump lever hinge
5. Screw—air horn, long (2)
6. Screw—air horn, short
7. Screw—air horn, countersunk (2)
8. Gasket—solenoid connector to air horn
9. Metering rod—secondary (2)
10. Holder & screw—secondary metering rod
11. Baffle—secondary air
12. Valve—idle air bleed
13. "O" ring (thick)—idle air bleed valve
14. "O" ring (thin)—idle air bleed valve
15. Plunger—TPS actuator
16. Seal—TPS plunger
17. Retainer—TPS seal
18. Screw—TPS adjusting

19. Plug—TPS screw
20. Seal—pump plunger
21. Retainer—pump seal
22. Screw—solenoid plunger stop (rich mixture stop)
23. Plug—plunger stop screw (rich mixture stop)
24. Plug—solenoid adjusting screw (lean mixture)
25. Vacuum break & bracket—front
26. Screw—vacuum break attaching (2)
27. Hose—vacuum
28. Rod—air valve
29. Lever—choke rod (upper)
30. Screw—choke lever
31. Rod—choke
32. Lever—choke rod (lower)
33. Seal—intermediate choke shaft
34. Lever—secondary lockout

35. Link—rear vacuum break
36. Intermediate choke shaft & lever
37. Cam—fast idle
38. Seal—choke housing to bowl (hot air choke)
39. Choke housing
40. Screw—choke housing to bowl
41. Seal—intermediate choke shaft (hot air choke)
42. Lever—choke coil
43. Screw—choke coil lever
44. Gasket—Stat cover (hot air choke)
45. Stat cover & coil assembly (hot air choke)
46. Stat cover & coil assembly (electric choke)
47. Kit—stat cover attaching
48. Vacuum break assembly—rear

Exploded view of Rochester E4ME carburetor

lenoid and check for and engine speed change of at least 50 rpm. If the rpm does not change enough, inspect the idle air bleed circuit for restrictions, leaks, etc.

20. Increase the engine speed to 3000 rpm and operate for a few minutes. Note the dwell which should be varying with an average indications of 35 degrees.

21. If the average dwell is not at 25 degrees, adjust the lean mixture screw.

22. After adjusting the lean mixture screw, adjust the idle mixture screw to obtain 25 degrees dwell.

23. If at an average dwell of 25 degrees, remove the carburetor from the engine, remove the modified tool J–29030–A or equivalent from the idle mixture screw and seal the access hole with RTV sealant.

24. Install the carburetor, connect all disconnected components and install the vent screen. Verify the idle speed is within specifications.

THROTTLE POSITION SWITCH (TPS) ADJUSTMENT

1. Drill a $5/64$ in. hole in the TPS adjustment screw plug and remove the plug with a small slide hammer. Be careful when drilling so as not to damage the adjustment screw head.

2. Disconnect the TPS connector and jumper all three terminals.

3. Connect a digital voltmeter from TPS center terminal (B) to bottom terminal (C).

4. With the ignition ON (engine off), turn the TPS adjustment screw to obtain the specified voltage at curb idle position with the A/C off and the ISC fully retracted. See the TPS Adjustment Chart for voltage specifications.

5. After adjustment, install a new adjustment screw plug into the air horn.

IDLE SPEED AND MIXTURE ADJUSTMENT

Model E4ME, M4ME, M4MC Carburetors

1. Set the parking brake and block the drive wheels. Disconnect and plug the hoses as directed on the underhood emission control sticker. Connect a dwell meter to the mixture control solenoid dwell terminal and a tachometer to the engine.

2. Start the engine and allow it to reach

normal operating temperature; a varying dwell should be noted on the dwell meter.

3. Check the engine idle speed and adjust to specifications listed on the underhood sticker if necessary.

4. With the engine idling in Drive (Neutral for manual transmissions), check the dwell readings on the six cylinder scale. If varying within the 10–50° range, mixture is correct. If not, continue with adjustment procedure.

5. Remove the idle air bleed valve cover by drilling out the rivets with a No. 35 drill bit. Use care when drilling to prevent damage to the air horn casting. Cover the carburetor intake ports with masking tape to prevent metal filings from falling into the engine. With the cover removed, look for the presence or absence of an identification letter on top of the idle air bleed valve. If the valve doesn't have an identification letter, use Procedure A below. If the valve does have an identification letter, use Procedure B to continue adjustment.

PROCEDURE A (WITHOUT ID LETTER)

1. Install idle air bleed valve gauging tool J–33815–2, BT–8253–B, or equivalent in throttle side D-shaped vent hole in the air horn casting. The upper end of the tool should be positioned over the open cavity next to the idle air bleed valve.

2. While holding the gauging tool down lightly so that the solenoid plunger is against the solenoid stop, adjust the idle air bleed valve so tha the gauging tool will pivot over and just contact the top of the valve. This presets the valve for on-vehicle adjustment. Remove the gauging tool.

3. Start the engine and allow it to reach normal operating temperature.

4. While idling in Drive (Neutral for manual transmission), use a screwdriver to slowly turn the idle air bleed valve until the dwell varies within the 25–35° range, attempting to get as close to 30° as possible. Perform this step carefully, as the bleed valve is very sensitive and should be turned in $1/8$ turn increments only. If the dwell cannot be set within the range, it will be necessary to remove the plugs and adjust the idle mixture needles.

5. Remove the carburetor and place on a suitable holding fixture with the manifold side up. Be careful not to damage any linkage, tubes or parts protruding from the air horn. Make two parallel cuts in the throttle body with a hacksaw, one on each side of the locator points beneath the idle mixture needle plug. The cuts should reach down

to the steel plugs, but should not extend more than $1/8$ in. beyond the locator points.

6. Place a flat punch near the ends of the saw marks, hold the punch at a 45° angle, then drive it into the throttle body until the casting breaks away, exposing the steel plug. Remove the plug.

7. Using tool J–29030–B, BT–7610–B, or equivalent, turn both mixture screws in until lightly seated, then out the number of turns listed in the Specifications Chart.

8. Install the carburetor on the engine using a new flange mounting gasket.

9. Readjust the idle air bleed valve as described above. If the dwell is still below 25°, turn both mixture screws counterclockwise an additional turn and try again. If above 35°, turn both mixture screws clockwise an additional turn.

10. After adjustments are complete, seal the mixture screw openings with silicone sealer to discourage further adjustment and prevent a possible fuel vapor leak. On vehicles without an ISC, adjust curb idle speed if necessary. Check and adjust the fast idle speed as described on the underhood emission sticker.

PROCEDURE B (WITH ID LETTER)

1. Install air bleed valve gauging tool J–33815–2, BT–8253–B, or equivalent in throttle side D-shaped vent hole in the upper air horn casting. The upper end of the tool should be positioned over the open cavity next to the idle air bleed valve.

2. While holding the gauging tool down lightly so that the solenoid plunger is against the solenoid stop, adjust the idle air bleed valve so that the gauging tool will pivot over and just contact the top of the valve. The valve is now set properly and no further adjustment is necessary.

3. If the dwell readings are still incorrect, remove the idle mixture plugs as described in Steps 5–8, above.

4. While idling in Drive (Neutral on manual transmission), adjust both mixture screws equally in $1/8$ turn increments until the dwell reading varies within the 25–35° range, attempting to get as close to 30° as possible. If the reading is too low, turn the mixture screws counterclockwise and vice-versa. Allow time for dwell readings to stabilize after each adjustment.

5. Once all adjustments are complete, seal the idle mixture screws with silicone sealant to discourage further adjustment and prevent a fuel vapor leak. On vehicles without and ISC, adjust the curb idle speed if necessary and adjust the fast idle speed as described on the underhood sticker.

MODEL 1ME/1M/1MEF
Chevrolet/GMC
(All measurements in inches)

Year	Carburetor Number	Float Level	Choke Unloader Setting	Choke Setting	Fast Idle Speed (rpm)	Metering Rod Setting	Fast Idle Cam 2nd Step	Choke Vacuum Break
'79	17058009	¼	—	Index	2400	.065	—	—
	17058011	¼	—	Index	2400	.065	—	—
	17059009	⁵⁄₁₆	.520	2 Rich	2400	.065 ①	.275	.400
	17059309	⁵⁄₁₆	.521	2 Rich	2400	.065	.275	.400
	17059359	⁵⁄₁₆	.521	2 Rich	2400	.065	.275	.400
'80	17080009	¹¹⁄₃₂	.520	②	2400	.090	.275	.400
	17080309	¹¹⁄₃₂	.520	②	2400	.090	.275	.400
	17080359	¹¹⁄₃₂	.520	②	2400	.090	.275	.400
'81–'83	17081009	¹¹⁄₃₂	.520	②	③	.090	.275	.400
	17081309	¹¹⁄₃₂	.520	②	③	.090	.275	.400
	17081329	¹¹⁄₃₂	.520	②	③	.090	.275	.400
'84–'86	17081009	¹¹⁄₃₂	.520	②	③	.090	.275	.400
	17084329	¹¹⁄₃₂	.520	②	③	.090	.275	.400
	17085009	¹¹⁄₃₂	.520	②	③	.090	.275	.400
	17085036	¹¹⁄₃₂	.520	②	③	.090	.275	.400
	17085044	¹¹⁄₃₂	.520	②	③	.090	.275	.400
	17085045	¹¹⁄₃₂	.520	②	③	.090	.275	.400
	17086096	¹¹⁄₃₂	.520	②	③	.090	.275	.400
	17086101	¹¹⁄₃₂	.520	②	③	.090	.275	.400
	17086102	¹¹⁄₃₂	.520	②	③	.090	.275	.400

① .090 inches on medium duty truck applications
② Not adjustable
③ See emission label under hood
④ Lower edge of choke valve

MODEL 2G/2GV/2GC/2GE/2GF
Chevrolet/GMC
(All measurements in inches)

Year	Carburetor Number	Float Level	Float Drop	Choke Setting	Pump Rod Location	Fast Idle Speed (rpm)	Choke Vacuum Break
'79	7044133	¹⁹⁄₃₂	1⁹⁄₃₂	Index	1⁹⁄₁₆	①	—
	7044134	¹⁹⁄₃₂	1⁹⁄₃₂	Index	1⁷⁄₁₆	①	—
	17059126	⅝	1⁹⁄₃₂	Index	1¹⁵⁄₃₂	①	—
	17059127	¹⁷⁄₃₂	1⁹⁄₃₂	Index	1¹⁵⁄₃₂	①	—

MODEL 2G/2GV/2GC/2GE/2GF
Chevrolet/GMC
(All measurements in inches)

Year	Carburetor Number	Float Level	Float Drop	Choke Setting	Pump Rod Location	Fast Idle Speed (rpm)	Choke Vacuum Break
'79	17059423	⅝	1⁹/₃₂	Index	1²¹/₃₂	①	—
	17059424	¹⁷/₃₂	1⁹/₃₂	Index	1¹⁵/₃₂	①	—
	17059420	¹⁷/₃₂	1⁹/₃₂	Index	1¹⁵/₃₂	①	—
'80–'86	7044133	¹¹/₁₆	1⁹/₃₂	Manual	1⁹/₁₆	①	—
	7044134	¹¹/₁₆	1⁹/₃₂	Manual	1⁷/₁₆	①	—
	17058120	¹¹/₁₆	1⁹/₃₂	Manual	1²¹/₃₂	①	—
	17080120	⅝	1⁹/₃₂	Manual	1²¹/₃₂	①	—
	17080126	⅝	1⁹/₃₂	Manual	1²¹/₃₂	①	—
	17080127	⅝	1⁹/₃₂	Manual	1²¹/₃₂	①	—
	17080129	⅝	1⁹/₃₂	Index	1²¹/₃₂	①	.130
	17080420	⅝	1⁹/₃₂	Manual	1²¹/₃₂	①	—
	17080423	⅝	1⁹/₃₂	Manual	1²¹/₃₂	①	—
	17080424	⅝	1⁹/₃₂	Manual	1²¹/₃₂	①	—
	17082129	⅝	1⁹/₃₂	Manual	1²¹/₃₂	①	—
	17082420	⅝	1⁹/₃₂	Manual	1¹⁵/₃₂	①	—
	17084432	⅝	1⁹/₃₂	Manual	1¹⁵/₃₂	①	—
	17084433	⅝	1⁹/₃₂	Manual	1²¹/₃₂	①	—
	17085120	⅝	1⁹/₃₂	Manual	1¹⁵/₃₂	①	—
	17085126	⅝	1⁹/₃₂	Manual	1¹⁵/₃₂	①	—
	17085464	⅝	1⁹/₃₂	Manual	1¹⁵/₃₂	①	—
	17085465	⅝	1⁹/₃₂	Manual	1²¹/₃₂	①	—

① See Tune-Up Specifications or underhood sticker

MODEL 2SE
Chevrolet/GMC
(All measurements in inches)

Year	Carburetor Number	Float Level	Choke Unloader①	Choke Setting	Pump Rod Adj.③	Fast Idle (rpm)	Fast Idle Cam 2nd step①	Choke Vacuum Break①
'79	17059640	⅛	49°	②	⁹/₁₆	2000	—	20°
	17059641	⅛	49°	②	⁹/₁₆	1800	—	23.5°
	17059643	⅛	49°	②	⁹/₁₆	1800	—	23.5°
	17059740	⅛	49°	②	⁹/₁₆	2000	—	20°
	17059741	⅛	49°	②	⁹/₁₆	2100	—	20°
	17059764	⅛	49°	②	⁹/₁₆	2100	—	20°
	17059765	⅛	49°	②	⁹/₁₆	2100	—	23.5°
	17059767	⅛	49°	②	⁹/₁₆	2100	—	23.5°

MODEL 2SE
Chevrolet/GMC
(All measurements in inches)

Year	Carburetor Number	Float Level	Choke Unloader①	Choke Setting	Pump Rod Adj.③	Fast Idle (rpm)	Fast Idle Cam 2nd step①	Choke Vacuum Break①
'80	17080621	⅛	41°	⑤	9/16	④	17°	22°
	17080622	⅛	41°	⑤	9/16	④	17°	22°
	17080623	⅛	41°	⑤	9/16	④	17°	22°
	17080626	⅛	41°	⑤	9/16	④	17°	22°
	17080720	⅛	41°	⑤	9/16	④	17°	22°
	17080721	⅛	41°	⑤	9/16	④	17°	23.5°
	17080722	⅛	41°	⑤	9/16	④	17°	20°
	17080723	⅛	41°	⑤	9/16	④	17°	23.5°
'81	17081621	3/16	38°	⑤	5/8	④	15°	38°
	17081622	3/16	38°	⑤	5/8	④	15°	38°
	17081623	3/16	38°	⑤	5/8	④	15°	38°
	17081624	3/16	38°	⑤	5/8	④	15°	38°
	17081625	3/16	38°	⑤	5/8	④	15°	38°
	17081626	3/16	38°	⑤	5/8	④	15°	38°
	17081627	3/16	38°	⑤	5/8	④	15°	38°
	17081629	3/16	41°	⑤	5/8	④	15°	38°
	17081630	3/16	38°	⑤	5/8	④	15°	38°
	17081633	3/16	38°	⑤	5/8	④	15°	38°
	17081720	3/16	41°	⑤	5/8	④	15°	38°
	17081721	3/16	41°	⑤	5/8	④	15°	38°
	17081725	3/16	41°	⑤	5/8	④	15°	38°
	17081726	3/16	41°	⑤	5/8	④	15°	38°
	17081727	3/16	41°	⑤	5/8	④	15°	38°

① Use angle degree tool or change over to decimal equivalent on the conversion chart at the end of this section
② 1 notch counterclockwise
③ Measure distance from air horn casting
④ See emissions label underhood for exact rpm specification
⑤ Riveted choke cap is not adjustable under normal circumstances

MODEL 2SE
Chevrolet/GMC
(All measurements in inches or degrees)

Year	Carburetor Number	Float Level	Choke Coil Lever	Choke Rod①	Primary Vacuum Break	Secondary Vacuum Break	Air Valve Rod	Choke Unloader
'82	17082334	3/16	.085	15°	26°	38°	1°	42°
	17082335	3/16	.085	15°	26°	38°	1°	42°
	17082336	3/16	.085	15°	26°	38°	1°	42°
	17082337	3/16	.085	15°	26°	38°	1°	42°
	17082338	3/16	.085	15°	26°	38°	1°	42°
	17082339	3/16	.085	15°	26°	38°	1°	42°
	17082341	3/16	.085	15°	30°	37°	1°	42°
	17082342	3/16	.085	15°	30°	37°	1°	42°
	17082344	3/16	.085	15°	30°	37°	1°	42°
	17082345	3/16	.085	15°	30°	37°	1°	42°
	17082431	3/16	.085	15°	24°	38°	1°	42°
	17082433	3/16	.085	15°	24°	38°	1°	42°
	17082480	3/16	.085	15°	26°	38°	1°	42°
	17082481	3/16	.085	15°	26°	38°	1°	42°
	17082482	3/16	.085	15°	23°	38°	1°	42°
	17082483	3/16	.085	15°	26°	38°	1°	42°
	17082484	3/16	.085	15°	26°	38°	1°	42°
	17082485	3/16	.085	15°	26°	38°	1°	42°
	17082486	3/16	.085	15°	28°	38°	1°	42°
	17082487	3/16	.085	15°	28°	38°	1°	42°
	17082488	3/16	.085	15°	28°	38°	1°	42°
	17082489	3/16	.085	15°	28°	38°	1°	42°
	17082348	7/16	.085	22°	26°	32°	1°	40°
	17082349	7/16	.085	22°	28°	32°	1°	40°
	17082350	7/16	.085	22°	26°	32°	1°	40°
	17082351	7/16	.085	22°	28°	32°	1°	40°
	17082353	7/16	.085	22°①	28°	35°	1°	30°
	17082355	7/16	.085	22°	28°	35°	1°	30°
'83	17083410	3/16	.085	15°	23°	38°	1°	42°
	17083411	3/16	.085	15°	26°	38°	1°	42°
	17083412	3/16	.085	15°	23°	38°	1°	42°
	17083413	3/16	.085	15°	26°	38°	1°	42°
	17083414	3/16	.085	15°	23°	38°	1°	42°
	17083415	3/16	.085	15°	26°	38°	1°	42°
	17083416	3/16	.085	15°	23°	38°	1°	42°
	17083417	3/16	.085	15°	26°	38°	1°	42°
	17083419	3/16	.085	15°	28°	38°	1°	42°

MODEL 2SE
Chevrolet/GMC
(All measurements in inches or degrees)

Year	Carburetor Number	Float Level	Choke Coil Lever	Choke Rod①	Primary Vacuum Break	Secondary Vacuum Break	Air Valve Rod	Choke Unloader
'83	17083421	3/16	.085	15°	26°	38°	1°	42°
	17083423	3/16	.085	15°	28°	38°	1°	42°
	17083425	3/16	.085	15°	26°	38°	1°	42°
	17083427	3/16	.085	15°	26°	38°	1°	42°
	17083429	3/16	.085	15°	28°	38°	1°	42°
	17083560	3/16	.085	15°	28°	38°	1°	42°
	17083562	3/16	.085	15°	28°	38°	1°	42°
	17083565	3/16	.085	15°	28°	38°	1°	42°
	17083569	3/16	.085	15°	28°	38°	1°	42°
	17083348	7/16	.085	22°	30°	32°	1°	40°
	17083349	7/16	.085	22°	30°	32°	1°	40°
	17083350	7/16	.085	22°	30°	32°	1°	40°
	17083351	7/16	.085	22°	30°	32°	1°	40°
	17083352	7/16	.085	22°	30°	35°	1°	40°
	17083353	7/16	.085	22°	30°	35°	1°	40°
	17083354	7/16	.085	22°	30°	35°	1°	40°
	17083355	7/16	.085	22°	30°	35°	1°	40°
	17083360	7/16	.085	22°	30°	32°	1°	40°
	17083361	7/16	.085	22°	28°	32°	1°	40°
	17083362	7/16	.085	22°	30°	32°	1°	40°
	17083363	7/16	.085	22°	28°	32°	1°	40°
	17083364	7/16	.085	22°	30°	35°	1°	40°
	17083365	7/16	.085	22°	30°	35°	1°	40°
	17083366	7/16	.085	22°	30°	35°	1°	40°
	17083367	7/16	.085	22°	30°	35°	1°	40°
	17083390	13/32	.085	28°	30°	35°	1°	38°
	17083391	13/32	.085	28°	30°	35°	1°	38°
	17083392	13/32	.085	28°	30°	35°	1°	38°
	17083393	13/32	.085	28°	30°	35°	1°	38°
	17083394	13/32	.085	28°	30°	35°	1°	38°
	17083395	13/32	.085	28°	30°	35°	1°	38°
	17083396	13/32	.085	28°	30°	35°	1°	38°
	17083397	13/32	.085	28°	30°	35°	1°	38°
'84	17084348	11/32	.085	22°	30°	32°	1°	40°
	17084349	11/32	.085	22°	30°	32°	1°	40°
	17084350	11/32	.085	22°	30°	32°	1°	40°
	17084351	11/32	.085	22°	30°	32°	1°	40°

MODEL 2SE
Chevrolet/GMC
(All measurements in inches or degrees)

Year	Carburetor Number	Float Level	Choke Coil Lever	Choke Rod ①	Primary Vacuum Break	Secondary Vacuum Break	Air Valve Rod	Choke Unloader
'84	17084352	11/32	.085	22°	30°	35°	1°	40°
	17084353	11/32	.085	22°	30°	35°	1°	40°
	17084354	11/32	.085	22°	30°	35°	1°	40°
	17084355	11/32	.085	22°	30°	35°	1°	40°
	17084360	5/32	.085	22°	30°	32°	1°	40°
	17084362	5/32	.085	22°	30°	32°	1°	40°
	17084364	5/32	.085	22°	30°	35°	1°	40°
	17084366	5/32	.085	22°	30°	35°	1°	40°
	17084390	7/16	.085	28°	30°	38°	1°	38°
	17084391	7/16	.085	28°	30°	38°	1°	38°
	17084392	7/16	.085	28°	30°	38°	1°	38°
	17084393	7/16	.085	28°	30°	38°	1°	38°
	17084394	7/16	.085	28°	30°	40°	1°	38°
	17084395	7/16	.085	28°	30°	40°	1°	38°
	17084396	7/16	.085	28°	30°	40°	1°	38°
	17084397	7/16	.085	28°	30°	40°	1°	38°
	17084410	11/32	.085	15°	23°	38°	1°	42°
	17084412	11/32	.085	15°	23°	38°	1°	42°
	17084425	11/32	.085	15°	26°	36°	1°	40°
	17084427	11/32	.085	15°	26°	36°	1°	40°
	17084560	11/32	.085	15°	24°	34°	1°	38°
	17084562	11/32	.085	15°	24°	34°	1°	38°
	17084569	11/32	.085	15°	24°	34°	1°	38°
'85	17085348	5/32	.085	22°	32°	36°	1°	40°
	17085350	5/32	.085	22°	32°	36°	1°	40°
	17085351	11/32	.085	22°	32°	36°	1°	40°
	17085352	5/32	.085	22°	30°	34°	1°	40°
	17085354	5/32	.085	22°	30°	34°	1°	40°
	17085355	11/32	.085	22°	30°	34°	1°	40°
	17085360	5/32	.085	22°	32°	36°	1°	40°
	17085362	5/32	.085	22°	32°	36°	1°	40°
	17085363	11/32	.085	22°	32°	36°	1°	40°
	17085364	5/32	.085	22°	30°	34°	1°	40°
	17085366	5/32	.085	22°	30°	34°	1°	40°
	17085367	11/32	.085	22°	30°	34°	1°	40°
	17085372	5/32	.085	22°	32°	36°	1°	40°
	17085374	5/32	.085	22°	32°	36°	1°	40°

Note: Specified angle for use with angle degree tool.
① Adjust with fast idle cam on 2nd step.

MODEL 2SE/E2SE
Jeep
(All measurements in inches or degrees)

Year	Carburetor Number	Float Level	Pump Stem Height	Fast② Idle Cam	Fast Idle (rpm)	Air① Valve Link	Primary Vacuum Break	Choke Unloader	Choke Setting
'81	17081790	.208	.128	25°	2400	2°	19°	32°	③
	17081791	.256	.128	25°	2600	2°	19°	32°	③
	17081796	.208	.128	25°	2400	2°	—	19°	③
	17081797	.208	.128	25°	2600	2°	—	19°	③
'82	17082380	.169	1.28	18°	2400	2°	21°	34°	③
	17082381	.169	1.28	18°	2400	2°	21°	34°	③
	17082389	.169	.128	18°	2400	2°	19°	34°	③

Note: Specified angle for use with angle
degree tool
① Maximum degree setting
② 2nd step on cam
③ Tamper resistant—riveted cover

MODEL 2SE/E2SE
Jeep
(All measurements in inches or degrees)

Year	Carburetor Number	Float Level	Air Valve Windup	Choke Coil Lever	Fast Idle Cam 2nd Step	Primary Vacuum Break	Secondary Vacuum Break	Air Valve Rod	Choke Unloader
'83–'84	17084581	5/32	1	.085	22°	26°	32°	1°	40°
	17084580	5/32	1	.085	22°	26°	32°	1°	40°
	17084582	5/32	1	.085	22°	26°	32°	1°	40°
	17084583	5/32	1	.085	22°	26°	32°	1°	40°
	17084384	1/8	1	.085	22°	25°	30°	1°	40°
'85–'86	17085380	5/32	1	.085	22°	26°	32°	1°	40°
	17085381	5/32	1	.085	22°	26°	32°	1°	40°
	17085382	5/32	1	.085	22°	26°	32°	1°	40°
	17085383	5/32	1	.085	22°	26°	32°	1°	40°
	17085384	1/8	1	.085	22°	25°	30°	1°	40°

Note: Specified angle for use with angle
degree tool

MODEL E2SE
Chevrolet/GMC
(All measurements in inches or degrees)

Year	Carburetor Number	Float Level	Choke Coil Lever	Choke Rod ①	Primary Vacuum Break	Secondary Vacuum Break	Air Valve Rod	Choke Unloader
'83	17083356	13⁄32	.085	22°	25°	35°	1°	30°
	17083357	13⁄32	.085	22°	25°	35°	1°	30°
	17083358	13⁄32	.085	22°	25°	35°	1°	30°
	17083359	13⁄32	.085	22°	25°	35°	1°	30°
	17083368	1⁄8	.085	22°	25°	35°	1°	30°
	17083370	1⁄8	.085	22°	25°	35°	1°	30°
	17083450	1⁄8	.085	28°	27°	35°	1°	45°
	17083451	1⁄4	.085	28°	27°	35°	1°	45°
	17083452	1⁄8	.085	28°	27°	35°	1°	45°
	17083453	1⁄4	.085	28°	27°	35°	1°	45°
	17083454	1⁄8	.085	28°	27°	35°	1°	45°
	17083455	1⁄4	.085	28°	27°	35°	1°	45°
	17083456	1⁄8	.085	28°	27°	35°	1°	45°
	17083630	1⁄4	.085	28°	27°	35°	1°	45°
	17083631	1⁄4	.085	28°	27°	35°	1°	45°
	17083632	1⁄4	.085	28°	27°	35°	1°	45°
	17083633	1⁄4	.085	28°	27°	35°	1°	45°
	17083634	1⁄4	.085	28°	27°	35°	1°	45°
	17083635	1⁄4	.085	28°	27°	35°	1°	45°
	17083636	1⁄4	.085	28°	27°	35°	1°	45°
	17083650	1⁄8	.085	28°	27°	35°	1°	45°
	17083430	11⁄32	.085	15°	26°	38°	1°	42°
	17083431	11⁄32	.085	15°	26°	38°	1°	42°
	17083434	11⁄32	.085	15°	26°	38°	1°	42°
	17083435	11⁄32	.085	15°	26°	38°	1°	42°
'84	17072683	9⁄32	.085	28°	25°	35°	1°	45°
	17074812	9⁄32	.085	28° ①	25°	35°	1°	45°
	17084356	9⁄32	.085	22°	25°	30°	1°	30°
	17084357	9⁄32	.085	22°	25°	30°	1°	30°
	17084358	9⁄32	.085	22°	25°	30°	1°	30°
	17084359	9⁄32	.085	22°	25°	30°	1°	30°
	17084368	1⁄8	.085	22°	25°	30°	1°	30°

CARBURETORS ROCHESTER

MODEL E2SE
Chevrolet/GMC
(All measurements in inches or degrees)

Year	Carburetor Number	Float Level	Choke Coil Lever	Choke Rod①	Primary Vacuum Break	Secondary Vacuum Break	Air Valve Rod	Choke Unloader
'84	17084370	1/8	.085	22°	25°	30°	1°	30°
	17084430	11/32	.085	15°	26°	38°	1°	42°
	17084431	11/32	.085	15°	26°	38°	1°	42°
	17084434	11/32	.085	15°	26°	38°	1°	42°
	17084435	11/32	.085	15°	26°	38°	1°	42°
	17084452	5/32	.085	28°	25°	35°	1°	45°
	17084453	5/32	.085	28°	25°	35°	1°	45°
	17084455	5/32	.085	28°	25°	35°	1°	45°
	17084456	5/32	.085	28°	25°	35°	1°	45°
	17084458	5/32	.085	28°	25°	35°	1°	45°
	17084532	5/32	.085	28°	25°	35°	1°	45°
	17084534	5/32	.085	28°	25°	35°	1°	45°
	17084535	5/32	.085	28°	25°	35°	1°	45°
	17084537	5/32	.085	28°	25°	35°	1°	45°
	17084538	5/32	.085	28°	25°	35°	1°	45°
	17084540	5/32	.085	28°	25°	35°	1°	45°
	17084542	1/8	.085	28°	25°	35°	1°	45°
	17084632	9/32	.085	28°	25°	35°	1°	45°
	17084633	9/32	.085	28°	25°	35°	1°	45°
	17084635	9/32	.085	28°	25°	35°	1°	45°
	17084636	9/32	.085	28°	25°	35°	1°	45°
'85	17085356	4/32	.085	22°	25°	30°	1°	30°
	17085357	9/32	.085	22°	25°	30°	1°	30°
	17085358	4/32	.085	22°	25°	30°	1°	30°
	17085359	9/32	.085	22°	25°	30°	1°	30°
	17085368	4/32	.085	22°	25°	30°	1°	30°
	17085369	9/32	.085	22°	25°	30°	1°	30°
	17085370	4/32	.085	22°	25°	30°	1°	30°
	17085371	9/32	.085	22°	25°	30°	1°	30°
	17085452	5/32	.085	28°	25°	35°	1°	45°
	17085453	5/32	.085	28°	25°	35°	1°	45°
	17085458	5/32	.085	28°	25°	35°	1°	45°

Note: Specified angle for use with angle
degree tool
① All models: Lean mixture screw–2½ turns
Idle mixture screw–4 turns

MODEL M2MC/M2ME
Chevrolet/GMC

(All measurements in inches ro degrees)

Year	Carburetor Number	Float Level	Choke Unloader	Choke Setting	Pump① Rod Adj.	Fast② Idle (rpm)	Fast Idle Cam Setting	Choke Vacuum Break
'79	17059100	15/32	—	1 Lean	13/32	1600	38°	29°
	17059101	15/32	—	1 Lean	13/32	1600	38°	29°
	17059102	15/32	—	1 Lean	13/32	1600	38°	29°
	17059103	15/32	—	1 Lean	13/32	1600	38°	29°
	17059142	15/32	—	1 Lean	13/32	1600	38°	29°
	17059143	15/32	—	1 Lean	13/32	1600	38°	29°
	17059144	15/32	—	1 Lean	13/32	1600	38°	29°
	17059145	15/32	—	1 Lean	13/32	1600	38°	29°
'80	17080100	7/16	38°	—	9/32	③	38°	29°
	17080102	7/16	38°	—	9/32	③	38°	29°
	17080142	7/16	38°	—	9/32	③	38°	29°
	17080143	7/16	38°	—	9/32	③	38°	29°
	17080145	7/16	38°	—	9/32	③	38°	29°
'81	17081101	13/32	38°	—	5/16	③	38°	29°
	17081103	13/32	38°	—	5/16	③	38°	29°
	17081142	13/32	38°	—	5/16	③	38°	29°
	17081143	13/32	38°	—	5/16	③	38°	29°
	17081144	13/32	38°	—	5/16	③	38°	29°
	17081145	13/32	38°	—	5/16	③	38°	29°

Note: Specified angle for use with angle
degree tool
① Rod installed in the inner hole of the pump
lever (nearest the carburetor)
② Manual transmission—1300 rpm in neutral
③ See underhood emissions label for idle
speed specifications

MODEL E4ME
Chevrolet/GMC
(All measurements in inches or degrees)

Year	Carburetor Number	Float Level	Rich Mixture Screw	Idle Mixture Needle Turns	Air Valve Spring Turns	Choke Rod	Front Vacuum Break	Rear Vacuum Break	Air Valve Rod	Choke Unloader	Idle Air Bleed Valve
'83	17083202	11/32	—	3⅜	⅞	20°	—	27°	—	38°	①
	17083203	11/32	—	3⅜	⅞	38°	—	27°	—	38°	①
	17083204	11/32	—	3⅜	⅞	20°	—	27°	—	38°	①
	17083207	11/32	—	3⅜	⅞	38°	—	27°	—	38°	①
	17083216	11/32	—	3⅜	⅞	20°	—	27°	—	38°	①
	17083218	11/32	—	3⅜	⅞	20°	—	27°	—	38°	①
	17083236	11/32	—	②	⅞	20°	—	27°	—	38°	1.756
	17083506	7/16	—	②	⅞	20°	27	36°	—	36°	1.756
	17083508	7/16	—	②	⅞	20°	27	36°	—	36°	1.756
	17083524	7/16	—	②	⅞	20°	25	36°	—	36°	1.756
	17083526	7/16	—	②	⅞	20°	25	36°	—	36°	1.756
'84	17084201	11/32	4/32	3⅜	⅞	20°	27°	—	.025	38°	①
	17084205	11/32	4/32	3⅜	⅞	38°	27°	—	.025	38°	①
	17084208	11/32	4/32	3⅜	⅞	20°	27°	—	.025	38°	①
	17084209	11/32	4/32	3⅜	⅞	38°	27°	—	.025	38°	①
	17084210	11/32	4/32	3⅜	⅞	20°	27°	—	.025	38°	①
	17084507	7/16	4/32	②	1	20°	27°	36°	.025	36°	①
	17084509	7/16	4/32	②	1	20°	27°	36°	.025	36°	①
	17084525	7/16	4/32	②	1	20°	25°	36°	.025	36°	①
	17084527	7/16	4/32	②	1	20°	25°	36°	.025	36°	①
'85	17085202	11/32	4/32	3⅜	⅞	20°	27°	—	.025	38°	①
	17085203	11/32	4/32	3⅜	⅞	20°	27°	—	.025	38°	①
	17085204	11/32	4/32	3⅜	⅞	20°	27°	—	.025	38°	①
	17085207	11/32	4/32	3⅜	⅞	38°	27°	—	.025	38°	①
	17085218	11/32	4/32	3⅜	⅞	20°	27°	—	.025	38°	①
	17085502	7/16	—	②	⅞	20°	26°	36°	.025	39°	①
	17085503	7/16	—	②	⅞	20°	26°	36°	.025	39°	①
	17085506	7/16	—	②	1	20°	27°	36°	.025	36°	①
	17085508	7/16	—	②	1	20°	27°	36°	.025	36°	①
	17085524	7/16	—	②	1	20°	25°	36°	.025	36°	①
	17085526	7/16	—	②	1	20°	25°	36°	.025	36°	①

Note: Specified angle for use with angle
degree tool
Lean mixture screw-1.304 gauge
Choke stat lever-.120 gauge
① Preset with 1.756 gauge, final adjustment
on vehicle
② Preset 3 turns, final adjustment on vehicle

MODEL M4MC/4MV QUADRAJET
Chevrolet/GMC
(All measurements in inches)

Year	Carburetor Number	Float Level	Air Valve Dashpot	Pump Rod Adj.	Pump Rod Hole	Initial Choke Valve Opening	Vacuum Break	Choke Unloader	Air Valve Spring Turns
'79	17059212	7/16	.015	9/32	Inner	.314	.136	.260	3/4
	17059512	13/32	.015	9/32	Inner	.314	.136	.260	3/4
	17059061	15/32	.015	13/32	Inner	.314	.129	.277	7/8
	17059201	15/32	.015	13/32	Inner	.314	.129	.277	7/8
	17059065	15/32	.015	13/32	Inner	.314	.129	.277	7/8
	17059205	15/32	.015	13/32	Inner	.314	.129	.277	7/8
	17059066	15/32	.015	13/32	Inner	.314	.129	.277	7/8
	17059206	15/32	.015	13/32	Inner	.314	.129	.277	7/8
	17059068	15/32	.015	13/32	Inner	.314	.129	.277	7/8
	17059208	15/32	.015	13/32	Inner	.314	.129	.277	7/8
	17059069	15/32	.015	13/32	Inner	.314	.129	.277	7/8
	17059209	15/32	.015	13/32	Inner	.314	.129	.277	7/8
	17059076	15/32	.015	13/32	Inner	.314	.129	.277	7/8
	17059226	15/32	.015	13/32	Inner	.314	.129	.277	7/8
	17059077	15/32	.015	13/32	Inner	.314	.129	.277	7/8
	17059227	15/32	.015	13/32	Inner	.314	.129	.277	7/8
	17059213	15/32	.015	9/32	Inner	.234	.129	.260	1
	17059215	15/32	.015	9/32	Inner	.234	.129	.260	1
	17059363	15/32	.015	13/32	Inner	.314	.149	.277	7/8
	17059503	15/32	.015	13/32	Inner	.314	.149	.277	7/8
	17059506	15/32	.015	13/32	Inner	.314	.149	.277	7/8
	17059368	15/32	.015	13/32	Inner	.314	.149	.277	7/8
	17059508	15/32	.015	13/32	Inner	.314	.149	.277	7/8
	17059377	15/32	.015	9/32	Outer	.314	.149	.277	7/8
	17059527	15/32	.015	9/32	Outer	.314	.149	.277	7/8
	17059378	15/32	.015	9/32	Outer	.314	.149	.277	7/8
	17059528	15/32	.015	9/32	Outer	.314	.149	.277	7/8
	17059509	15/32	.015	13/32	Inner	.314	.179	.277	7/8
	17059515	15/32	.015	9/32	Inner	.234	.129	.260	1
	17059510	15/32	.015	9/32	Inner	.314	.179	.277	7/8
	17059529	15/32	.015	9/32	Inner	.234	.129	.260	1
	17059513	15/32	.015	9/32	Inner	.234	.129	.260	1
	17059586	15/32	.015	13/32	Inner	.314	.179	.277	7/8
	17059588	15/32	.015	13/32	Inner	.314	.179	.277	7/8
	17059229	15/32	.015	9/32	Inner	.234	.129	.260	1
	17059520	3/8	.015	9/32	Inner	.324	.164	.277	7/8
	17059521	3/8	.015	9/32	Inner	.314	.164	.277	7/8

MODEL M4MC/4MV QUADRAJET
Chevrolet/GMC
(All measurements in inches)

Year	Carburetor Number	Float Level	Choke Unloader	Choke Setting	Pump① Rod Adj.	Fast Idle Cam 2nd Step	Choke Vacuum Break
'80	17080201	15/32	42°	②	9/32	46°	23°
	17080205	15/32	42°	②	9/32	46°	23°
	17080206	15/32	42°	②	9/32	46°	23°
	17080224	15/32	42°	②	9/32	46°	23°
	17080290	15/32	42°	②	9/32	46°	26°
	17080291	15/32	42°	②	9/32	46°	26°
	17080292	15/32	42°	②	9/32	46°	26°
	17080295	15/32	42°	②	9/32	46°	23°
	17080297	15/32	42°	②	9/32	46°	23°
	17080503	15/32	42°	②	9/32	46°	26°
	17080506	15/32	42°	②	9/32	46°	26°
	17080508	15/32	42°	②	9/32	46°	26°
	17080523	15/32	42°	②	9/32	26°	23°
	17080524	15/32	42°	②	9/32	46°	23°
	17080525	15/32	42°	②	9/32	46°	23°
	17080526	15/32	42°	②	9/32	46°	23°
	17080226	15/32	42°	②	9/32	46°	23°
	17080227	15/32	42°	②	9/32	46°	23°
	17080527	15/32	42°	②	9/32	46°	23°
	17080528	15/32	42°	②	9/32	46°	23°
	17080213	3/8	40°	②	9/32	37°	30°
	17080215	3/8	40°	②	9/32	37°	30°
	17080513	3/8	40°	②	9/32	37°	30°
	17080515	3/8	40°	②	9/32	37°	30°
	17080229	3/8	40°	②	9/32	37°	30°
	17080529	3/8	40°	②	9/32	37°	30°
	17080225	15/32	42°	②	9/32	46°	23°
	17080212	3/8	40°	②	9/32	30°	24°
	17080512	3/8	40°	②	9/32	30°	24°
'81	17080212	3/8	40°	②	9/32	30°	24°
	17080213	3/8	40°	②	9/32	30°	23°
	17080215	3/8	40°	②	9/32	30°	23°
	17080298	3/8	40°	②	9/32	30°	23°
	17080507	3/8	40°	②	9/32	30°	23°
	17080512	3/8	40°	②	9/32	30°	24°
	17080513	3/8	40°	②	9/32	30°	23°
	17081200	15/32	42°	②	9/32	23°	24°

MODEL M4MC/4MV QUADRAJET
Chevrolet/GMC
(All measurements in inches)

Year	Carburetor Number	Float Level	Choke Unloader	Choke Setting	Pump① Rod Adj.	Fast Idle Cam 2nd Step	Choke Vacuum Break
'81	17081201	15/32	42°	②	9/32	23°	23°
	17081205	15/32	42°	②	9/32	23°	23°
	17081206	15/32	42°	②	9/32	23°	23°
	17081220	15/32	42°	②	9/32	23°	23°
	17081226	15/32	42°	②	9/32	23°	24°
	17081227	15/32	42°	②	9/32	—	24°
	17081290	13/32	42°	②	9/32	24°	23°
	17081291	13/32	42°	②	9/32	24°	23°
	17081292	13/32	42°	②	9/32	24°	23°
	17081506	13/32	36°	②	9/32	36°	23°
	17081508	13/32	36°	②	9/32	36°	23°
	17081524	13/32	36°	②	5/16③	36°	25°
	17081526	13/32	36°	②	5/16③	36°	25°

Note: Specified angle for use with angle degree tool

① Place the pump arm linkage in the inner hole of the arm, except on carburetors with a 5/16 pump rod height (see ③)

② 1980 and 1981 choke cover are riveted in position and are not adjustable under normal conditions

③ On carburetors with 5/16 pump rod height, place the pump arm linkage in the outer hole of the arm

MODEL M4MC/M4ME QUADRAJET
Chevrolet/GMC
(All measurements in inches or degrees)

Year	Carburetor Number	Float Level	Pump Rod Hole	Pump Rod Setting	Choke Rod① Setting	Air Valve Rod	Vacuum Break Front	Vacuum Break Rear	Air Valve Turns	Choke Unloader	Propane Enrichment (rpm)
'82	17080212	3/8	inner	9/32	46°	.025	24°	30°	3/4	40°	②
	17080213	3/8	inner	9/32	37°	.025	23°	30°	1	40°	②
	17080215	3/8	inner	9/32	37°	.025	23°	30°	1	40°	②
	17080298	3/8	inner	9/32	37°	.025	23°	30°	1	40°	②
	17080507	3/8	inner	9/32	37°	.025	23°	30°	1	40°	②
	17080512	3/8	inner	9/32	46°	.025	24°	30°	3/4	40°	②
	17080513	3/8	inner	9/32	37°	.025	23°	30°	3/4	40°	②
	17082213	3/8	inner	9/32	37°	.025	23°	30°	1	40°	②

MODEL M4MC/M4ME QUADRAJET
Chevrolet/GMC
(All measurements in inches or degrees)

Year	Carburetor Number	Float Level	Pump Rod Hole	Pump Rod Setting	Choke Rod① Setting	Air Valve Rod	Vacuum Break Front	Vacuum Break Rear	Air Valve Turns	Choke Unloader	Propane Enrichment (rpm)
'82	17082220	$^{13}/_{32}$	inner	$^9/_{32}$	46°	.025	24°	34°	$^7/_8$	39°	②
	17082221	$^{13}/_{32}$	inner	$^9/_{32}$	46°	.025	24°	34°	$^7/_8$	39°	150
	17082222	$^{13}/_{32}$	inner	$^9/_{32}$	46°	.025	24°	34°	$^7/_8$	39°	50
	17082223	$^{13}/_{32}$	inner	$^9/_{32}$	46°	.025	24°	34°	$^7/_8$	39°	100
	17082224	$^{13}/_{32}$	inner	$^9/_{32}$	46°	.025	24°	34°	$^7/_8$	39°	50
	17082225	$^{13}/_{32}$	inner	$^9/_{32}$	46°	.025	24°	34°	$^7/_8$	39°	150
	17082226	$^{13}/_{32}$	inner	$^9/_{32}$	46°	.025	24°	34°	$^7/_8$	39°	50
	17082227	$^{13}/_{32}$	inner	$^9/_{32}$	46°	.025	24°	34°	$^7/_8$	39°	50
	17082230	$^{13}/_{32}$	inner	$^9/_{32}$	46°	.025	26°	36°	$^7/_8$	39°	②
	17082231	$^{13}/_{32}$	inner	$^9/_{32}$	46°	.025	26°	36°	$^7/_8$	39°	②
	17082234	$^{13}/_{32}$	inner	$^9/_{32}$	46°	.025	26°	36°	$^7/_8$	39°	②
	17082235	$^{13}/_{32}$	inner	$^9/_{32}$	46°	.025	26°	36°	$^7/_8$	39°	②
	17082290	$^{13}/_{32}$	inner	$^9/_{32}$	46°	.025	24°	34°	$^7/_8$	39°	②
	17082291	$^{13}/_{32}$	inner	$^9/_{32}$	46°	.025	24°	34°	$^7/_8$	39°	②
	17082292	$^{13}/_{32}$	inner	$^9/_{32}$	46°	.025	24°	34°	$^7/_8$	39°	②
	17082293	$^{13}/_{32}$	inner	$^9/_{32}$	46°	.025	24°	34°	$^7/_8$	39°	100
	17082506	$^{13}/_{32}$	inner	$^9/_{32}$	46°	.025	23°	36°	$^7/_8$	39°	50
	17082508	$^3/_8$	inner	$^9/_{32}$	46°	.025	23°	36°	$^7/_8$	39°	50
	17082513	$^{13}/_{32}$	inner	$^9/_{32}$	46°	.025	23°	30°	$^3/_4$	40°	②
	17082524	$^{13}/_{32}$	outer	$^5/_{16}$	46°	.025	25°	36°	$^7/_8$	39°	20
	17082526	$^{13}/_{32}$	outer	$^5/_{16}$	46°	.025	25°	36°	$^7/_8$	39°	20
'83	17080201	$^{15}/_{32}$	inner	$^9/_{32}$	46°	.025	—	23°	$^7/_8$	42°	②
	17080205	$^{15}/_{32}$	inner	$^9/_{32}$	46°	.025	—	23°	$^7/_8$	42°	②
	17080206	$^{15}/_{32}$	inner	$^9/_{32}$	46°	.025	—	23°	$^7/_8$	42°	②
	17080213	$^3/_8$	inner	$^9/_{32}$	37°	.025	23°	30°	1	40°	②
	17080290	$^{15}/_{32}$	inner	$^9/_{32}$	46°	.025	—	26°	$^7/_8$	42°	②
	17080291	$^{15}/_{32}$	inner	$^9/_{32}$	46°	.025	—	26°	$^7/_8$	42°	②
	17080292	$^{15}/_{32}$	inner	$^9/_{32}$	46°	.025	—	26°	$^7/_8$	42°	②
	17080298	$^3/_8$	inner	$^9/_{32}$	37°	.025	23°	30°	1	40°	②
	17080507	$^3/_8$	inner	$^9/_{32}$	37°	.025	23°	30°	1	40°	②
	17080513	$^3/_8$	inner	$^9/_{32}$	37°	.025	23°	30°	1	40°	②
	17082213	$^9/_{32}$	inner	$^9/_{32}$	37°	.025	23°	30°	1	40°	②
	17083234	$^{13}/_{32}$	inner	$^9/_{32}$	46°	.025	—	26°	$^7/_8$	39°	20
	17083235	$^{13}/_{32}$	inner	$^9/_{32}$	46°	.025	—	26°	$^7/_8$	39°	100
	17083290	$^{13}/_{32}$	inner	$^9/_{32}$	46°	.025	—	24°	$^7/_8$	39°	40
	17083291	$^{13}/_{32}$	inner	$^9/_{32}$	46°	.025	—	24°	$^7/_8$	39°	100
	17083292	$^{13}/_{32}$	inner	$^9/_{32}$	46°	.025	—	24°	$^7/_8$	39°	40

MODEL M4MC/M4ME QUADRAJET
Chevrolet GMC
(All measurements in inches or degrees)

Year	Carburetor Number	Float Level	Pump Rod Hole	Pump Rod Setting	Choke Rod① Setting	Air Valve Rod	Vacuum Break Front	Vacuum Break Rear	Air Valve Turns	Choke Unloader	Propane Enrichment (rpm)
'83	17083293	¹³⁄₃₂	inner	⁹⁄₃₂	46°	.025	—	24°	⅞	39°	100
	17083298	⅜	inner	⁹⁄₃₂	37°	.025	23°	30°	1	40°	②
	17083507	⅜	inner	⁹⁄₃₂	37°	.025	23°	30°	1	40°	②
	17080212	⅜	inner	⁹⁄₃₂	46°	.025	24°	30°	¾	40°	②
	17080512	⅜	inner	⁹⁄₃₂	46°	.025	24°	30°	¾	40°	②
	17083220	¹³⁄₃₂	inner	⁹⁄₃₂	46°	.025	—	24°	⅞	39°	150
	17083221	¹³⁄₃₂	inner	⁹⁄₃₂	46°	.025	—	24°	⅞	39°	150
	17083222	¹³⁄₃₂	inner	⁹⁄₃₂	46°	.025	—	24°	⅞	39°	50
	17083223	¹³⁄₃₂	inner	⁹⁄₃₂	46°	.025	—	24°	⅞	39°	150
	17083224	¹³⁄₃₂	inner	⁹⁄₃₂	46°	.025	—	24°	⅞	39°	50
	17083225	¹³⁄₃₂	inner	⁹⁄₃₂	46°	.025	—	24°	⅞	39°	150
	17083226	¹³⁄₃₂	inner	⁹⁄₃₂	46°	.025	—	24°	⅞	39°	50
	17083227	¹³⁄₃₂	inner	⁹⁄₃₂	46°	.025	—	24°	⅞	39°	50
	17083230	¹³⁄₃₂	inner	⁹⁄₃₂	46°	.025	—	26°	⅞	39°	20
	17083231	¹³⁄₃₂	inner	⁹⁄₃₂	46°	.025	—	26°	⅞	39°	100
'84	17084200	¹³⁄₃₂	inner	⁹⁄₃₂	46°	.025	—	26°	⅞	39°	②
	17084206	¹³⁄₃₂	inner	⁹⁄₃₂	46°	.025	—	26°	⅞	39°	20
	17084211	¹³⁄₃₂	inner	⁹⁄₃₂	46°	.025	—	26°	⅞	39°	②
	17084220	¹³⁄₃₂	inner	⁹⁄₃₂	46°	.025	—	26°	⅞	39°	80
	17084221	¹³⁄₃₂	inner	⁹⁄₃₂	46°	.025	—	26°	⅞	39°	80
	17084226	¹³⁄₃₂	inner	⁹⁄₃₂	46°	.025	—	24°	⅞	39°	30
	17084227	¹³⁄₃₂	inner	⁹⁄₃₂	46°	.025	—	24°	⅞	39°	30
	17084228	¹³⁄₃₂	inner	⁹⁄₃₂	46°	.025	—	26°	⅞	39°	80
	17084229	¹³⁄₃₂	inner	⁹⁄₃₂	46°	.025	—	26°	⅞	39°	80
	17084230	¹³⁄₃₂	inner	⁹⁄₃₂	46°	.025	—	26°	⅞	39°	20
	17084231	¹³⁄₃₂	inner	⁹⁄₃₂	46°	.025	—	26°	⅞	39°	40
	17084234	¹³⁄₃₂	inner	⁹⁄₃₂	46°	.025	—	26°	⅞	39°	20
	17084235	¹³⁄₃₂	inner	⁹⁄₃₂	46°	.025	—	26°	⅞	39°	80
	17084290	¹³⁄₃₂	inner	⁹⁄₃₂	46°	.025	—	24°	⅞	39°	30
	17084291	¹³⁄₃₂	inner	⁹⁄₃₂	46°	.025	—	26°	⅞	39°	100
	17084292	¹³⁄₃₂	inner	⁹⁄₃₂	46°	.025	—	24°	⅞	39°	30
	17084293	¹³⁄₃₂	inner	⁹⁄₃₂	46°	.025	—	26°	⅞	39°	100
	17084294	¹³⁄₃₂	inner	⁹⁄₃₂	46°	.025	—	26°	⅞	39°	30
	17084298	¹³⁄₃₂	inner	⁹⁄₃₂	46°	.025	—	26°	⅞	39°	30

MODEL M4MC/M4ME QUADRAJET
Chevrolet/GMC
(All measurements in inches or degrees)

Year	Carburetor Number	Float Level	Pump Rod Hole	Pump Rod Setting	Choke Rod① Setting	Air Valve Rod	Vacuum Break Front	Vacuum Break Rear	Air Valve Turns	Choke Unloader	Propane Enrichment (rpm)
'85	17084500	12/32	inner	9/32	37°	.025	23°	30°	1	40°	②
	17084501	12/32	inner	9/32	37°	.025	23°	30°	1	40°	②
	17084502	12/32	inner	9/32	46°	.025	24°	30°	7/8	40°	②
	17085000	12/32	inner	9/32	46°	.025	24°	30°	7/8	40°	②
	17085001	12/32	inner	9/32	46°	.025	23°	30°	1	40°	②
	17085003	12/32	inner	9/32	46°	.025	23°	—	7/8	35°	②
	17085004	13/32	inner	9/32	46°	.025	23°	—	7/8	35°	②
	17085205	13/32	inner	9/32	20°	.025	26°	38°	7/8	39°	②
	17085206	13/32	inner	9/32	46°	.025	—	26°	7/8	39°	20
	17085208	13/32	inner	9/32	20°	.025	26°	38°	7/8	39°	10
	17085209	13/32	outer	3/8	20°	.025	26°	36°	7/8	39°	50
	17085210	13/32	inner	9/32	20°	.025	26°	38°	7/8	39°	10
	17085211	13/32	outer	3/8	20°	0.25	26°	36°	7/8	39°	50
	17085212	13/32	inner	9/32	46°	.025	23°	—	7/8	35°	②
	17085213	13/32	inner	9/32	46°	.025	23°	—	7/8	35°	②
	17085215	13/32	inner	9/32	46°	.025	—	26°	7/8	32°	②
	17085216	13/32	inner	9/32	20°	.025	26°	38°	7/8	39°	②
	17085217	13/32	inner	9/32	20°	.025	26°	36°	1/2	39°	②
	17085219	13/32	inner	9/32	20°	.025	26°	36°	1/2	39°	②
	17085220	13/32	outer	3/8	20°	.025	—	26°	7/8	32°	75
	17085221	13/32	outer	3/8	20°	.025	—	26°	7/8	32°	75
	17085222	13/32	inner	9/32	20°	.025	26°	36°	1/2	39°	20
	17085223	13/32	outer	3/8	20°	.025	26°	36°	1/2	39°	50
	17085224	13/32	inner	9/32	20°	.025	26°	36°	1/2	39°	20
	17085225	13/32	outer	3/8	20°	.025	26°	36°	1/2	39°	50
	17085226	13/32	inner	9/32	20°	.025	—	24°	7/8	32°	20
	17085227	13/32	inner	9/32	20°	.025	—	24°	7/8	32°	20
	17085228	13/32	inner	9/32	46°	.025	—	24°	7/8	39°	30
	17085229	13/32	inner	9/32	46°	.025	—	24°	7/8	39°	30
	17085230	13/32	inner	9/32	20°	.025	—	26°	7/8	32°	20
	17085231	13/32	inner	9/32	20°	.025	—	26°	7/8	32°	40
	17085235	13/32	inner	9/32	46°	.025	—	26°	7/8	39°	80
	17085238	13/32	outer	3/8	20°	.025	—	26°	7/8	32°	75
	17085239	13/32	outer	3/8	20°	.025	—	26°	7/8	32°	75
	17085290	13/32	inner	9/32	46°	.025	—	24°	7/8	39°	30
	17085291	13/32	outer	3/8	46°	.025	—	26°	7/8	39°	100
	17085292	13/32	inner	9/32	46°	.025	—	24°	7/8	39°	30

MODEL M4MC/M4ME QUADRAJET
Chevrolet/GMC
(All measurements in inches or degrees)

Year	Carburetor Number	Float Level	Pump Rod Hole	Pump Rod Setting	Choke Rod① Setting	Air Valve Rod	Vacuum Break Front	Vacuum Break Rear	Air Valve Turns	Choke Unloader	Propane Enrichment (rpm)
'85	17085293	13/32	outer	3/8	46°	.025	—	26°	7/8	39°	100
	17085294	13/32	inner	9/32	46°	.025	—	26°	7/8	39°	②
	17085298	13/32	inner	9/32	46°	.025	—	26°	7/8	39°	②

Note: Specified angle for use with angle degree tool. Choke coil lever setting is .120 in. for all carburetors.
① Second step of fast idle cam
② See Underhood Specifications sticker

QUADRAJET MODELS
Dodge/Plymouth

Year	Carburetor Number	Float Level	Air Valve Spring Turns	Fast Idle cam	Choke Rod	Vacuum Kick	Air Valve Rod	Choke Unloader	Propane rpm
'85	1785408	13/32	1/2	20°	.143	27°①	.025	38°②	800
	1785409	13/32	5/8	20°	.143	27°①	.025	38°②	750
	1785415	13/32	1/2	20°	.143	27°①	.025	38°②	800
	1785416	13/32	3/4	20°	.143	27°①	.025	38°②	800

Note: Specified angle for use with angle gauge tool
① Plug gauge-.214 in.
② Plug gauge-.345 in.

ANGLE DEGREE TO DECIMAL CONVERSION
Model M2MC, M2ME and M4MC Carburetor

Angle Degrees	Decimal Equiv. Top of Valve	Angle Degrees	Decimal Equiv. Top of Valve
5	.023	33	.203
6	.028	34	.211
7	.033	35	.220
8	.038	36	.227
9	.043	37	.234
10	.049	38	.243
11	.054	39	.251
12	.060	40	.260
13	.066	41	.269
14	.071	42	.277
15	.077	43	.287
16	.083	44	.295
17	.090	45	.304
18	.096	46	.314
19	.103	47	.322

ANGLE DEGREE TO DECIMAL CONVERSION
Model M2MC, M2ME and M4MC Carburetor

Angle Degrees	Decimal Equiv. Top of Valve	Angle Degrees	Decimal Equiv. Top of Valve
20	.110	48	.332
21	.117	49	.341
22	.123	50	.350
23	.129	51	.360
24	.136	52	.370
25	.142	53	.379
26	.149	54	.388
27	.157	55	.400
28	.164	56	.408
29	.171	57	.418
30	.179	58	.428
31	.187	59	.439
32	.195	60	.449

ANGLE DEGREE TO DECIMAL CONVERSION
Model 4MV Carburetor

Angle Degrees	Decimal Equiv. Top of Valve	Angle Degrees	Decimal Equiv. Top of Valve
5	.019	33	.158
6	.022	34	.164
7	.026	35	.171
8	.030	36	.178
9	.034	37	.184
10	.038	38	.190
11	.042	39	.197
12	.047	40	.204
13	.051	41	.211
14	.056	42	.217
15	.060	43	.225
16	.065	44	.231
17	.070	45	.239
18	.075	46	.246
19	.080	47	.253
20	.085	48	.260
21	.090	49	.268
22	.095	50	.275
23	.101	51	.283
24	.106	52	.291
25	.112	53	.299
26	.117	54	.306
27	.123	55	.314
28	.128	56	.322
29	.134	57	.329
30	.140	58	.337
31	.146	59	.345
32	.152	60	.353

TPS ADJUSTMENT SPECIFICATIONS
Chevrolet/GMC Models

Year	Engine Code	TPS Voltage
'83	H	.51
	G	.51
	F	.40
	L	.40
'84	B	.255
	X	.31
	Z	.255
	G	.48
	H	.48
	G	.48
	F	.41
	L	.41
'85	G	.48
	H	.48
	G	.48
	F	.41
	L	.41
	N	.25

Note: Measure voltage with throttle at curb idle position, ignition ON, engine and A/C OFF. All values ± 0.1 volt.

STROMBERG CARBURETORS

IDLE SPEED AND MIXTURE ADJUSTMENT

1979–80 Models

1. Start the engine and run it until it reaches operating temperature.

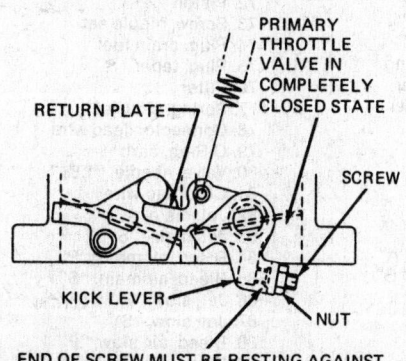

RETURN PLATE — PRIMARY THROTTLE VALVE IN COMPLETELY CLOSED STATE — SCREW — KICK LEVER — NUT

END OF SCREW MUST BE RESTING AGAINST RETURN PLATE

Kick lever adjustment—DCH340

2. If it hasn't already been done, check and adjust the ignition timing. After you have set the timing, turn off the engine.

3. Attach tachometer to the engine.

4. Remove the air cleaner.

5. Start the engine and, with transmission in Neutral, check the idle speed on the tachometer. If the reading is correct, turn off the engine and remove the tachometer. If it is not correct, proceed to the following steps.

6. Turn the idle adjusting screw with a screwdriver clockwise to increase idle speed and counterclockwise to decrease it.

7. If the vehicle is equipped with air conditioning:

  a. Turn on the A/C to maximum cold and high blower. Disconnect the vacuum line to the air cleaner housing air compensator and plug the intake manifold.

  b. Open the throttle approximately 1/3 and allow the throttle to close. This will allow the speed-up solenoid to reach full travel.

  c. Adjust the fast idle screw to set the idle speed to 900 rpm.

  d. Open the throttle about 1/3 and allow it to close. Read the idle rpm. Repeat Step C until the correct reading is obtained. Shut off the engine.

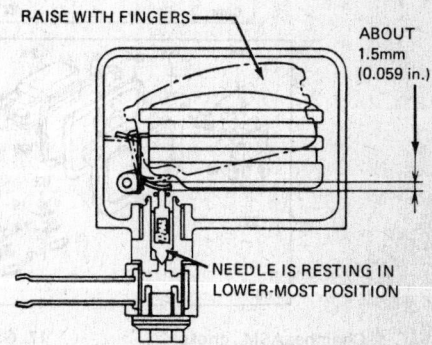

RAISE WITH FINGERS — ABOUT 1.5mm (0.059 in.) — NEEDLE IS RESTING IN LOWER-MOST POSITION

DCH340 needle valve stroke adjustment

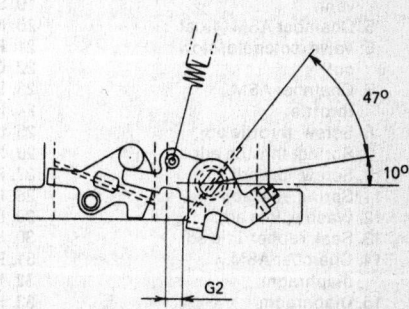

47° — 10° — G2

Secondary throttle opening point clearance—DCH340

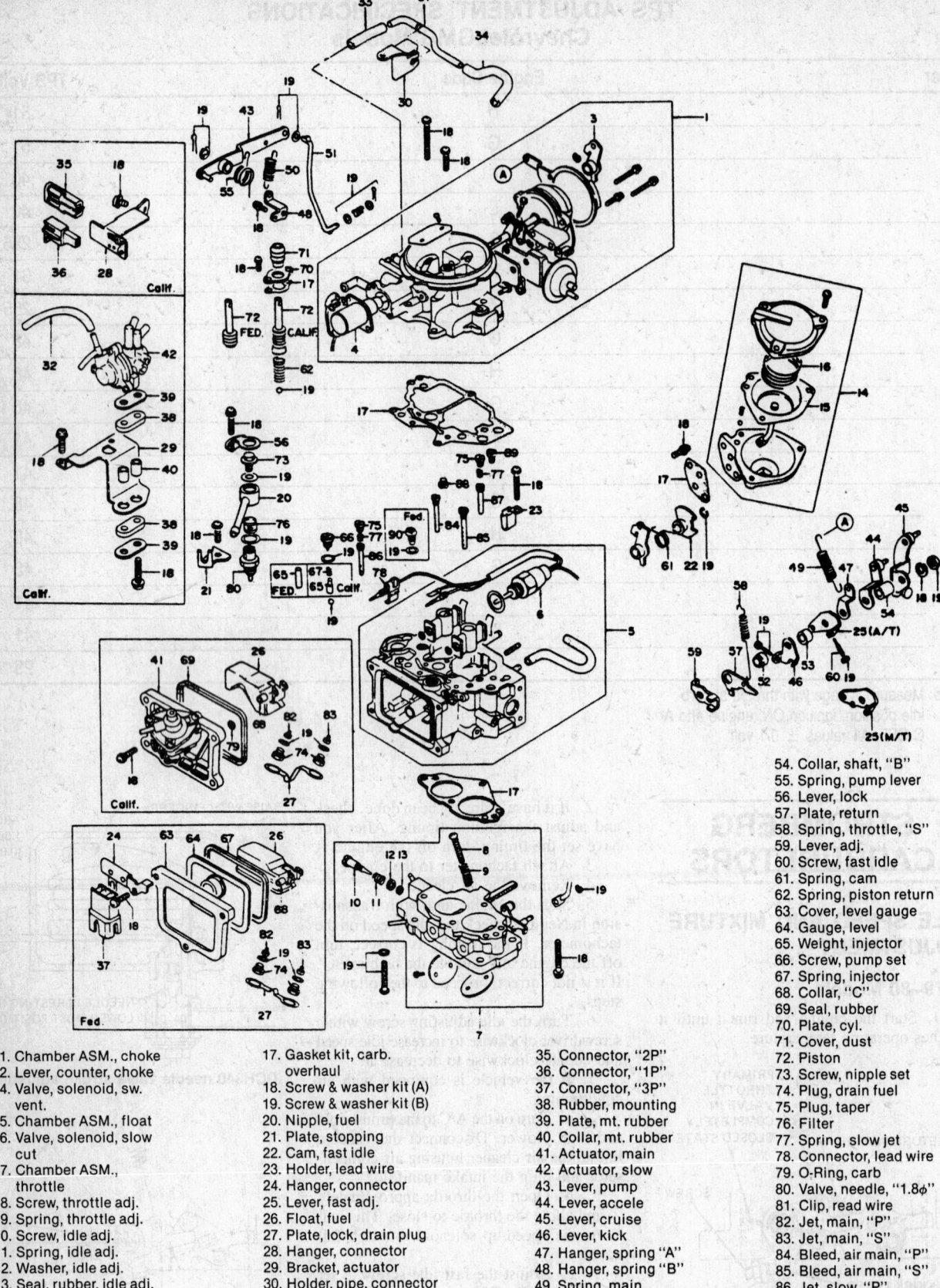

1. Chamber ASM., choke
2. Lever, counter, choke
3. Valve, solenoid, sw. vent.
4. Chamber ASM., float
5. Valve, solenoid, slow cut
6. Chamber ASM., throttle
7. Screw, throttle adj.
8. Spring, throttle adj.
9. Screw, idle adj.
10. Spring, idle adj.
11. Washer, idle adj.
12. Seal, rubber, idle adj.
13. Chamber ASM., diaphragm
14. Diaphragm
15. Spring, diaphragm

17. Gasket kit, carb. overhaul
18. Screw & washer kit (A)
19. Screw & washer kit (B)
20. Nipple, fuel
21. Plate, stopping
22. Cam, fast idle
23. Holder, lead wire
24. Hanger, connector
25. Lever, fast adj.
26. Float, fuel
27. Plate, lock, drain plug
28. Hanger, connector
29. Bracket, actuator
30. Holder, pipe, connector
31. Pipe, connector
32. Hose, rubber, "C"
33. Hose, rubber, "D"
34. Hose, rubber, "E"

35. Connector, "2P"
36. Connector, "1P"
37. Connector, "3P"
38. Rubber, mounting
39. Plate, mt. rubber
40. Collar, mt. rubber
41. Actuator, main
42. Actuator, slow
43. Lever, pump
44. Lever, accele
45. Lever, cruise
46. Lever, kick
47. Hanger, spring "A"
48. Hanger, spring "B"
49. Spring, main
50. Spring, assist
51. Rod, pump
52. Sleeve
53. Collar, shaft, "A"

54. Collar, shaft, "B"
55. Spring, pump lever
56. Lever, lock
57. Plate, return
58. Spring, throttle, "S"
59. Lever, adj.
60. Screw, fast idle
61. Spring, cam
62. Spring, piston return
63. Cover, level gauge
64. Gauge, level
65. Weight, injector
66. Screw, pump set
67. Spring, injector
68. Collar, "C"
69. Seal, rubber
70. Plate, cyl.
71. Cover, dust
72. Piston
73. Screw, nipple set
74. Plug, drain fuel
75. Plug, taper
76. Filter
77. Spring, slow jet
78. Connector, lead wire
79. O-Ring, carb
80. Valve, needle, "1.8φ"
81. Clip, read wire
82. Jet, main, "P"
83. Jet, main, "S"
84. Bleed, air main, "P"
85. Bleed, air main, "S"
86. Jet, slow, "P"
87. Jet, slow, "S"
88. Bleed, air, slow, "P"
89. Bleed, air, slow, "S"
90. Valve, power

Exploded view of Stromberg DCH340 carburetor

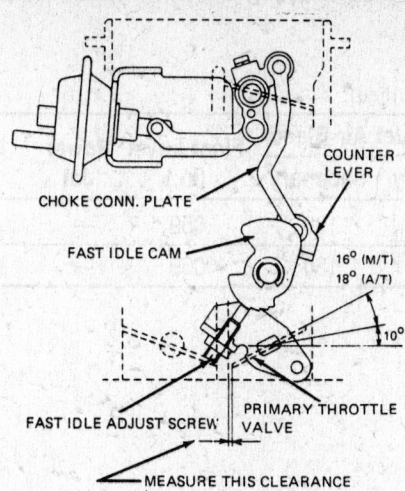

Primary throttle valve adjustment—
DCH340

8. Turn the mixture adjusting screw all the way in. Seat the needle tip lightly to avoid damaging the tip. Back the screw out 3½ turns.

9. Start the engine. Turn the mixture screw until the maximum engine rpm is achieved.

10. Reset the engine idle speed.

11. Turn the idle mixture screw clockwise (lean) until the engine speed drops 50 rpm.

12. Reset the idle mixture screw ½ turn counterclockwise (rich) from Step 11 position.

13. Reset the throttle adjusting screw to required idle speed.

14. Unplug and reconnect any vacuum lines that may have been disconnected.

1981 and Later Models

The idle mixture adjustment is the same as previously described with the following exceptions:

1. Make the idle speed adjustment with the engine at normal operating temperature. Be sure the choke is fully opened, air conditioning off and the air cleaner installed.

2. Disconnect and plug the distributor vacuum, the canister purge and EGR vacuum lines. Shut off the vacuum to the idle compensator by bending the rubber hose.

3. Adjust to required idle speed with throttle adjusting screw.

4. If equipped with air conditioning, turn AC to max cold and high blower.

5. Open throttle to approx. ⅓ and allow throttle to close, the speed up solenoid should activate. Adjust speed up solenoid screw until 900 rpm is reached.

6. In order to adjust the idle mixture you must first remove the plug that covers the mixture screw. To remove the plug, first remove the carburetor and turn it upside down.

7. Knock out the plug carefully with a hammer and screwdriver (see illustration).

8. Reinstall the carburetor.

9. Turn the mixture screw all the way in (seated lightly) and then back it out 2 turns for Federal models and 1 turn for California models. Readjust the idle if necessary.

THROTTLE LINKAGE ADJUSTMENT

When the primary throttle valve is opened to an angle of 50° from its closed position, the adjust plate which is interlocked with the primary throttle valve is brought into contact with the return plate. When the primary throttle valve is opened farther, the return plate is pulled apart from the stopper allowing the secondary throtttle valve to open. To adjust the linkage:

1. Measure the clearance between primary throttle valve and the wall of the throttle chamber at the center of the throttle valve when the adjust plate is brought into contact with the return plate. Standard clearance is 0.26–0.32 in.

2. If necessary, make the adjustment by bending the tab of the return plate.

FLOAT LEVEL ADJUSTMENT

The fuel level is normal if it is within the lines on the window glass of the float chamber when the vehicle is resting on level ground and the engine is off. If the fuel level is outside the lines, remove the float housing cover. Have an absorbent cloth under the cover to catch the fuel from the fuel bowl. Adjust the float level by bending the needle seat on the float. The needle valve

should have an effective stroke of about 0.059 in. When necessary, the needle valve stroke can be adjusted by bending the float stopper.

NOTE: Be careful not to bend the needle valve rod when installing the float and baffle plate, if removed.

KICK LEVER ADJUSTMENT

1. Bring the primary side throttle valve into the complete closed position, by turning the throttle adjustment screw.

2. On manual transmission models, with the throttle valve completely closed, loosen the lock nut on the kick lever screw and turn the screw until it is in contact with the return plate and tighten the lock nut.

3. On automatic transmission models with the throttle valve completely closed, bend the end of the kick lever until it is in contact with the return plate.

ELECTRIC AUTOMATIC CHOKE

1. Install the thermostat cover by fitting the end of the choke lever into the bimetal hook.

2. Align the thermostat housing line (thickest one) with the line on the thermostat cover.

3. Measure the clearance between the choke valve edge and the choke chamber wall when the choke is fully closed. The standard clearance should be 0.11–0.29 in. This clearance is equal to a bimetal lever angle of 30°.

4. If the measured value is not correct, adjust it by bending the bimetal lever as necessary.

ELECTRIC CHOKE ADJUSTMENT

Align the thickest line on the thermostat housing with the line on the thermostat cover. Measure clearance between the cover side stopper and the bimetal level side stopper when the diaphragm is fully stroked with negative pressure or finger pressure. If the measured value deviates from the standard clearance of 0.28–0.29 in. or the equivalent bimetal lever angle of 20°, adjust with the adjusting screw.

MODEL DCH340, DFP340
Isuzu/LUV

(All numbers are OEM jet numbers unless otherwise specified)

Carburetor Number	Main Jet		Main Air Bleed		Slow Jet		Slow Jet Air Bleed		Float Level (in.)	Power Jet
	Primary	Secondary	Primary	Secondary	Primary	Secondary	Primary	Seconary		
DCH340-227	114	170	120	70	50	100	150	100	.059	50
DCH340-228	114	170	120	70	50	100	150	100	.059	50

MODEL DCH340, DFP340
Isuzu/LUV

(All numbers are OEM jet numbers unless otherwise specified)

Carburetor Number	Main Jet		Main Air Bleed		Slow Jet		Slow Jet Air Bleed		Float Level (in.)	Power Jet
	Primary	Secondary	Primary	Secondary	Primary	Secondary	Primary	Seconary		
DFP340-3	93	180	100	90	53	125	120	130	.059	—
DFP340-4	93	180	100	90	53	125	120	130	.059	—

Note: Throttle valve opening angle is 16 degrees on all models

MODEL DCH340
Chevrolet S-10

(All measurements in inches)

Float Needle Valve Stroke	Primary Throttle Valve Adjustment	Secondary Throttle Opening Point	Kick Lever Adjustment
.059	.050–.059①	.24–.30	②

① Applies to manual trans. automatic trans- .059–.069
② Zero clearance between kick lever screw and return plate—throttle fully closed.

TOYOTA CARBURETORS

FLOAT LEVEL ADJUSTMENT

20R and 22R Engines

With the engine idling, check the fuel level in the carburetor sight glass. If the fuel level is even with the line, no adjustment is necessary. If it is not, adjust the float as follows:

1. Remove and invert the air horn so that the float hangs by its own weight towards

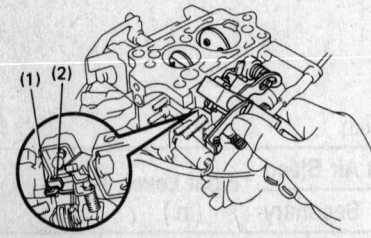

Adjust the throttle valve opener by bending primary (1) or secondary (2) levers—Toyota carburetors

the air horn. With the air horn gasket removed, measure the distance between the tip of the float and the air horn. The distance should be 0.276 for 1979–80 engines, 0.413 for 1981–82 (22R) engines and 0.386 for 1983 and later engines. If the distance is incorrect, adjust the float by bending the center tab (marked "A" in the illustration) on all models.

2. 20R engines: Raise the float away from the air horn and measure the distance between the needle valve push pin and the float lip. The distance should be 0.040 in. If adjustment is necessary floats, bend the single center tab which resembles the tab marked "A" in the illustration.

3. 22R Engines: Raise the float away from the air horn, then measure the distance between the air horn and the tips of the float, furthest away from the air horn. The distance should be 1.89 in. If adjustment is necessary, remove the float from the air horn and bend the tab which is furthest away from the floats (centered between the hinge points). Recheck the setting after the adjustment has been made.

2F Engines

1. Remove the carburetor air horn. Invert the air horn and allow the float to hang towards the air horn.

2. With the air horn gasket removed,

measure the distance between the float and the air horn, at the end of the float opposite the needle valve. The distance should be 0.295 in. If adjustment is necessary, remove the float and bend the tab which is centered between the hinge pivot points. After the adjustment is completed, reinstall the float and recheck the setting.

3. Lift upward on the float and measure the distance between the needle valve push pin and the lip of the float. The distance should be 0.043 in. If adjustment is necessary, remove the float and bend the tabs located just inside of the hinge points. After the adjustment is completed, reinstall the float and recheck the setting.

FAST IDLE ADJUSTMENT–OFF VEHICLE

20R Engines (1979) and 2F Engines

1. Refer to the "Carburetor, Removal and Installation" procedures, in this section and remove the carburetor.

2. Close the choke valve completely and invert the carburetor.

3. Using a wire-type feeler gauge, check the clearance between the upper half of the

primary throttle blade and the throttle bore. The clearance should be 0.047 for 20R engines or 0.051 for 2F engines. If necessary, adjust the clearance by turning the fast idle screw.

4. To install, reverse the removal procedures.

20R Engines (1980) and 22R Engines (1981 and Later)

NOTE: A special blade angle tool must be obtained to properly make this adjustment.

1. Refer to the "Carburetor, Removal and Installation" procedures, in this section and remove the carburetor.

2. Close the choke valve completely and set the throttle shaft lever to the first step of the fast idle cam.

3. Attach the blade angle tool to the primary throttle blade. Adjust the primary throttle blade angle to the correct angle from horizontal by turning the fast idle screw.

4. Remove the angle tool from the carburetor.

5. To install, reverse the removal procedures.

FAST IDLE ADJUSTMENT—ON VEHICLE

20R Engines (1979)

1. Start the engine and allow it to reach normal operating temperature.

2. Stop the engine and disconnect the vacuum hose from the EGR valve. Connect a tachometer to the engine.

3. Open the throttle valve slightly and close the choke plate, which will set the fast idle cam.

4. Disconnect the vacuum hose(s) from the distributor vacuum unit. Plug the vacuum hose end(s).

5. Without touching the accelerator pedal, start the engine and read the tachometer. If necessary, adjust the fast idle speed to 2400 rpm by turning the fast idle screw.

6. Reconnect the vacuum hoses to both the EGR valve and the distributor vacuum unit. Disconnect the tachometer from the engine.

20R Engines (1980) and 22R Engines (1981 and Later)

1. Start the engine and allow it to reach normal operating temperatures.

2. Stop the engine and connect a tachometer to the engine.

3. Remove the air cleaner assembly.

4. Disconnect and plug the vacuum hose at the fast idle cam breaker (if equipped).

5. Disconnect the vacuum hose(s) from the distributor vacuum unit.

6. Disconnect the vacuum hose from the EGR valve.

7. Open the throttle valve slightly and

fully pull up on the fast idle linkage. Release the throttle.

8. Without touching the accelerator pedal, start the engine and read the tachometer. If necessary, adjust the fast idle speed to 2400 rpm (1980–82) or 2600 rpm (1983 and later), by turning the fast idle screw.

9. Reconnect the vacuum hoses, disconnect the tachometer and reinstall the air cleaner.

2F Engines

1. Start the engine and allow it to reach normal operating temperatures.

2. Stop the engine and connect a tachometer to the engine.

3. Remove the air cleaner assembly.

4. Disconnect the vacuum hoses from both the EGR valve and the distributor vacuum unit.

5. Pull the dash mounted choke control knob fully outward.

6. Open the choke plate and prevent it from closing using a screwdriver.

--- CAUTION ---
DO NOT jam the screwdriver into place.

7. Start the engine and read the tachometer. If necessary, adjust the fast idle speed to 1800 rpm by turning the fast idle screw.

8. Remove the screwdriver from the choke, disconnect the tachometer and reconnect the vacuum hoses.

9. Install the air cleaner assembly.

IDLE MIXTURE PRESET ADJUSTMENT

NOTE: Perform this adjustment on any rebuilt carburetor prior to installation of the carburetor.

1. Carefully turn the idle mixture adjusting screw clockwise (in) until the screw seats lightly.

--- CAUTION ---
Do not force the screw. The tip of the screw is easily damaged.

2. Turn the screw counterclockwise (out) 1½ turns for 20R and 2F engines; 2½ turns for 1981 22R and 4½ turns, 1982 and later engines.

3. Reset the idle mixture adjustment after the engine is running.

AUTOMATIC CHOKE INSPECTION AND ADJUSTMENT

NOTE: Steps 1–4 must be performed with the engine cold and turned OFF.

1. Remove the air cleaner lid.

2. Depress the accelerator pedal. The choke plate should close. If the choke plate closes, proceed to Step 5.

3. If the choke plate does not close, loosen the three screws around the thermostat case.

--- CAUTION ---
Do not loosen the center housing screw; coolant leakage will occur.

4. Rotate the case just until the choke plate closes and tighten the case screws.

5. Start the engine and allow it to reach normal operating temperature. If the choke plate opens fully, the choke adjustment is correct. If it does not, loosen the three case screws and rotate the case just until the choke plate is fully open. Tighten the case screws.

CHOKE UNLOADER ADJUSTMENT

NOTE: Perform the following adjustment on a cold engine. A special blade angle gauge is needed for this adjustment.

1. Remove the air cleaner lid.

2. Depress the accelerator pedal to set the choke plate in the fully closed position.

3. Attach the blade angle gauge to the choke plate.

4. Open the primary throttle valve completely. The unloader should open the choke plate to the figure shown in the carburetor specification chart. On 2F and 20R engines, bend the fast idle lever to adjust the angle. On 22R engines, bend the first throttle arm to adjust the angle.

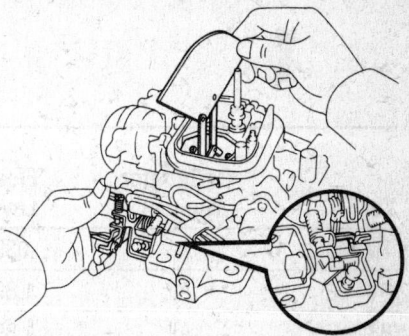

Choke unloader adjustment—Toyota carburetor

5. Remove the blade angle gauge and reinstall the air cleaner lid.

CHOKE BREAKER ADJUSTMENT

NOTE: A special blade angle gauge is needed for this adjustment.

1. Remove the air cleaner assembly.

2. Attach the blade angle gauge to the choke plate.

3. With the choke blade closed, push the choke breaker link towards the choke

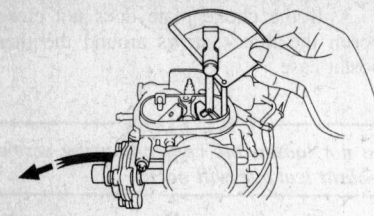

Choke breaker adjustment—Toyota carburetor

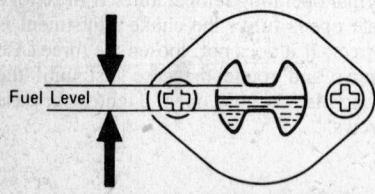

Fuel level check—Toyota carburetor

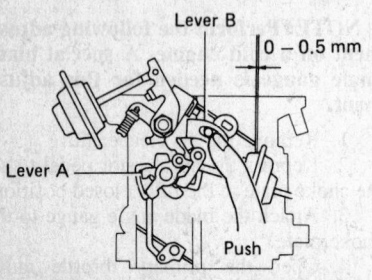

Secondary throttle valve adjustment—Toyota carburetor

breaker until resistance is felt. The choke plate should open to the angle shown in the carburetor specification chart. Bend the choke breaker link to adjust the opening.

4. Detach the blade angle gauge and install the air cleaner assembly.

SECONDARY KICK-UP ADJUSTMENT

This adjustment is not necessary on 22R engines.

20R Engines

1. Remove the carburetor as previously outlined.

2. With the carburetor inverted, open the primary throttle blade fully. Measure the distance between the uppermost portion of the secondary throttle blade and the secondary throttle bore, using a wire-type feeler gauge. The clearance should be 0.008 in. If adjustment is necessary, bend the secondary throttle lever as required to attain the proper clearance.

3. Install the carburetor as previously outlined.

2F Engines

NOTE: A special blade angle gauge is needed for this adjustment.

1. Remove the carburetor as previously outlined.

2. Attach the blade angle gauge to the secondary throttle blade.

3. Open the primary throttle blade fully and read the blade angle gauge. The secondary throttle blade should open slightly to an angle of 28° (except California) or 25° (California). If adjustment is necessary, bend the secondary throttle lever as required to attain the proper angle.

4. Detach the blade angle gauge and install the carburetor as previously outlined.

ACCELERATOR PUMP ADJUSTMENT

NOTE: This adjustment is not required on 22R engines.

1. Remove the air cleaner assembly.

2. With the choke plate fully open, measure the accelerator pump stroke at the top of the accelerator pump rod by fully opening the throttle from the closed position.

3. The total pump stroke should be 0.177 in. for 20R (1979) engines, 0.154 in. for 20R (1980) engines or 0.374 in. for 2F engines. Adjust the stroke by bending the linkage ("A") which is attached to the accelerator pump arm opposite the accelerator pump.

4. Install the air cleaner assembly.

TOYOTA CARBURETORS

(All measurements in inches or degress)

Year	Carburetor Number	Float Level	Throttle Valve Angle	Secondary Touch Angle	Fast Idle Angle	Choke Unloader Angle	Idle Mixture Screw①
'79–'80	All	.039	90	59	24	50	1½
'81–'82	All	1.89	90	50	24	45	2½
'83	All	1.89	90	59	22	50	4
'84	All	1.89	90	59	22	50	4½
'85	All	1.89	90	59	23	45	3½

Note: Angle specifications for use with angle degree tool
① Turns out

Fuel Injection

INDEX

FUEL INJECTION

General Information

There are two basic types of fuel injection systems currently in production, Port and Throttle Body. Constant Injection Systems (CIS) are a mechanically controlled, Port type of fuel injection that uses linkage between an air flow sensor and fuel distributor to regulate the air/fuel mixture by moving a piston up and down. Air Flow Controlled (AFC or EFI), Port and Throttle Body (TBI) injection systems use an intake air flow (or air mass) sensor and an electronic control unit that monitors engine conditions and then regulates the fuel mixture by sending electrical impulses to the injector(s). By varying the length of time the solenoid-type fuel injector is open, more or less fuel is delivered. A fuel injection systems is called PORT type when the injectors are mounted in the cylinder head and spray the fuel charge directly behind the intake valve. Throttle Body Injection (TBI) uses one or two injectors mounted atop the intake manifold (much like a conventional carburetor) that spray the fuel charge down through a throttle body butterfly valve. The fuel charge is drawn into the intake manifold and distributed to the cylinders in the conventional manner.

Fuel injection combined with electronics and various engine sensors provides a precise fuel management system that meets all the demands for improved fuel economy, increased performance and lower emissions more precisely and reliably than is possible with a conventional carburetor. A fuel injected engine generally averages ten percent more power and fuel economy with lower emissions than a carbureted engine. Even "feedback" carburetors with computer controls cannot achieve the accurate fuel metering necessary to meet the lower emission levels required by upcoming Federal standards. Because of its precise control, fuel injection allows the engine to operate with a stoichiometric or optimum fuel ratio of 14.7 parts air to one part fuel throughout the entire engine rpm range, under all operating conditions. This 14.7:1 fuel mixture assures that all the carbon and hydrogen is burned in the combustion chamber during the power stroke. This produces the lowest combination of emissions from unburned hydrocarbons, carbon monoxide and oxides of nitrogen in the exhaust gases. By using an oxygen sensor to measure the oxygen content of the exhaust gases, the on-board computer (control unit) can constantly adjust and "fine tune" the fuel mixture in response to changing temperature, load and altitude conditions. In addition, this degree of fuel control allows the catalytic converter to function at peak efficiency for emission control.

It's important to identify all system components, how they work and their relationship to one another in operation before attempting any maintenance or repair procedures on any particular type of fuel injection. All fuel injection systems are delicate and vulnerable to damage from rust, dirt, water and careless handling. The shock of hitting a cement floor when dropped from about waist height can ruin a control unit, for example. Because of the close tolerances involved in fuel injector construction (25 millionths of an inch on some), any rust or dirt particles in the fuel lines can ruin machined surfaces or block the nozzle completely. Water in the fuel can do more damage than a well-placed grenade and some fuel additives or chemicals may damage fuel lines or components like the oxygen sensor. Many fuel injection components, while similar in both appearance and function, can vary from one manufacturer to another so it is very important to identify exactly which type of injection system is being used on a particular engine, and to become familiar with all related sensors and equipment. In addition, there are manufacturer's modifications within the major groups of CIS, AFC and TBI type systems that utilize different sensors and electronic control units that are similar in appearance but are not interchangeable with any other system. Each individual system has some sensors, capabilities and characteristics unique to its own design. Although they may look pretty much the same to the casual observer, each PROM or memory unit is programmed differently for each type of engine it is supposed to regulate. For this reason, the careful recording of part numbers or engine serial numbers and the correct identification of vehicle make, model and year is vital to insure the correct replacement parts are obtained. Always double check all numbers before installing any new component.

COMPARING CARBURETOR CIRCUITS TO FUEL INJECTION COMPONENTS

It makes fuel injection a little easier to understand by comparing the functions of major systems and components in the carburetor to those in a typical port fuel injection system. Fuel injection is basically just a more precise way of doing the same thing—delivering the correct air/fuel mixture to the combustion chambers at the correct time under all engine operating conditions.

One major difference is that the accelerator pedal actually feeds fuel to the intake manifold when depressed (via the accelerator pump linkage) on the carburetor, even if the engine is stopped and the ignition switch is turned OFF. On a fuel injection system, the accelerator pedal merely opens or closes the butterfly-type throttle plate, allowing more air to flow into the intake system. There is no direct, mechanical linkage connection that provides fuel for the engine. The amount of fuel delivered by the port fuel injection system is regulated either by movement of the fuel distributor plunger (CIS systems), or by the electronic control unit that sets the fuel mixture according to the output signals from various sensors (AFC system). It's impossible to flood a fuel injected engine by pumping the accelerator pedal with the ignition switch off since the pedal only opens and closes an air valve. No fuel can be delivered unless the control unit is energized and sends impulses to the injectors.

To assist cold starting, modern carburetors incorporate an automatic choke assembly which consists of mechanical linkage operating a butterfly valve that reduces the amount of incoming air, thereby increasing the amount of vacuum that pulls on the main fuel discharge nozzle in the carburetor. This increased vacuum draws a larger quantity of fuel from the float bowl, providing the richer fuel mixture necessary for cold starting. The butterfly valve (choke plate) movement is controlled either manually, or by a bimetal coil spring that responds to temperature and is usually heated either by hot water from the cooling system, hot air from the exhaust manifold, or by an electric element in the choke spring housing. Because it is a mechanical system, it is vulnerable to malfunctions due to loose or sticking linkage, broken springs, and incorrect adjustment, causing hard starting or flooding problems on cold engines. All fuel injection systems incorporate a cold start system that does essentially the same thing as the carburetor choke system, but in a much more precise way with different components. Some fuel injection systems use an extra fuel injector called a "cold start valve" that is mounted in the intake manifold downstream from the throttle plate. This solenoid-type injector provides an extra fuel charge when the ignition is switched ON with the engine cold and is regulated through a temperature sensitive switch designed to cut off the signal to the cold start injector after about 12 seconds. This temperature sensitive switch, called a "thermo-time switch", is also designed to lock out the cold start system operation when the engine is warm and the enriched fuel mixture is not necessary for starting. Some computer-controlled fuel injection systems are designed to enrich the fuel mixture by providing longer signal to the injectors (increasing the amount of time the nozzle is open) without the need for a separate cold start injector. The longer the solenoid-type injectors are open, the more fuel is delivered to the combustion chambers.

NOTE: Because some cold start injectors are designed to operate when the key is switched ON, it is possible to flood a cold engine by turning the key on and off a few times. Although the effect is the same as pumping the accelerator on a carbureted engine, the cause is totally different. Some manufacturers design

the cold start valve to spray only when the starter is cranking to avoid this problem

SAFETY PRECAUTIONS

——————— CAUTION ———————

Whenever working on or around any fuel injection system, always observe these general precautions to prevent the possibility of personal injury or damage to fuel injection components.

1. Never install or remove battery cables with the key ON or the engine running. Jumper cables should be connected with the key OFF to avoid power surges that can damage electronic control units. Engines equipped with computer controlled fuel injection systems should avoid both giving and getting jump starts due to the possibility of serious damage to components due to arcing in the engine compartment.

2. Always remove the battery cables before charging the battery. Never use a high-output charger on an installed battery or attempt to use any type of "hot shot" (24 volt) starting aid.

3. Never remove or attach wiring harness connectors with the ignition switch ON, especially to the electronic control unit.

4. When checking compression on engines with AFC injection systems, unplug the cable from the battery to the relays.

5. Always depressurize the fuel system before attempting to disconnect any fuel lines.

6. Always use clean rags and tools when working on an open fuel injection system and take care to prevent any dirt from entering the system. Wipe all components clean before installation and prepare a clean work area for disassembly and inspection of components. Use lint-free cloths to wipe components and avoid using any caustic cleaning solvents.

7. Do not drop any components during service procedures and never apply 12 volts directly to a fuel injector unless instructed specifically to do so. Some injectors windings are designed to safely handle only 4 or 5 volts and can be destroyed in seconds if 12 volts are applied directly to the connector.

8. Remove the electronic control unit if the vehicle is to be placed in an environment where temperatures exceed approximately 176 degrees F (80 degrees C), such as a paint spray booth or when arc or gas welding near the control unit location in the car.

PROBLEM DEFINITIONS

1. STALLS–engine stops running at idle or when driving. Determine if the stalling condition is only present when the engine is either hot or cold, or if it happens consistently regardless of operating temperature.

2. LOADS UP–engine misses due to excessively rich mixture. This usually occurs during cold engine operation and is characterized by black smoke from the tailpipe.

3. ROUGH IDLE–engine runs unevenly at idle. This condition can range from a slight stumble or miss up to a severe shake.

4. TIP IN STUMBLE–a delay or hesitation in engine response when accelerating from idle with the car at a standstill. Some slight hesitation conditions are considered normal when they only occur during cold operation and gradually vanish as the engine warms up.

5. MISFIRE–rough engine operation due to a lack of combustion in one or more cylinders. Fouled spark plugs or loose ignition wires are the most common cause.

6. HESITATION–a delay in engine response when accelerating from cruise or steady throttle operation at road speed. Not to be confused with the tip in stumble described above.

7. SAG–engine responds initially, then flattens out or slows down before recovering. Severe sags can cause the engine to stall.

8. SURGE–engine power variation under steady throttle or cruise. Engine will speed up or slow down with no change in the throttle position. Can happen at a variety of speeds.

9. SLUGGISH–engine delivers limited power under load or at high speeds. Engine loses speed going up hills, doesn't accelerate as fast as normal, or has less top speed than was noted previously.

10. CUTS OUT–temporary complete loss of power at sharp, irregular intervals. May occur repeatedly or intermittently, but is usually worse under heavy acceleration.

11. POOR FUEL ECONOMY–significantly lower gas mileage than is considered normal for the model and drivetrain in question. Always perform a careful mileage test under a variety of road conditions to determine the severity of the problem before attempting corrective measures. Fuel economy is influenced more by external conditions, such as driving habits and terrain, than by a minor malfunction in the fuel injection system that doesn't cause another problem (like rough operation).

Ford Multi-Point (Port) Fuel Injection

The Electronic Fuel Injector System (EFI) is classified as a multi-point, pulse time, mass air flow fuel injection system. Fuel is metered into the intake air steam in accordance with engine demand through the injectors mounted on a tuned intake manifold. An on-board vehicle electronic engine control (EEC) computer accepts inputs from various engine sensors to compute the required fuel flow rate necessary to maintain a programmed air/fuel ratio throughout the entire engine operational range. The

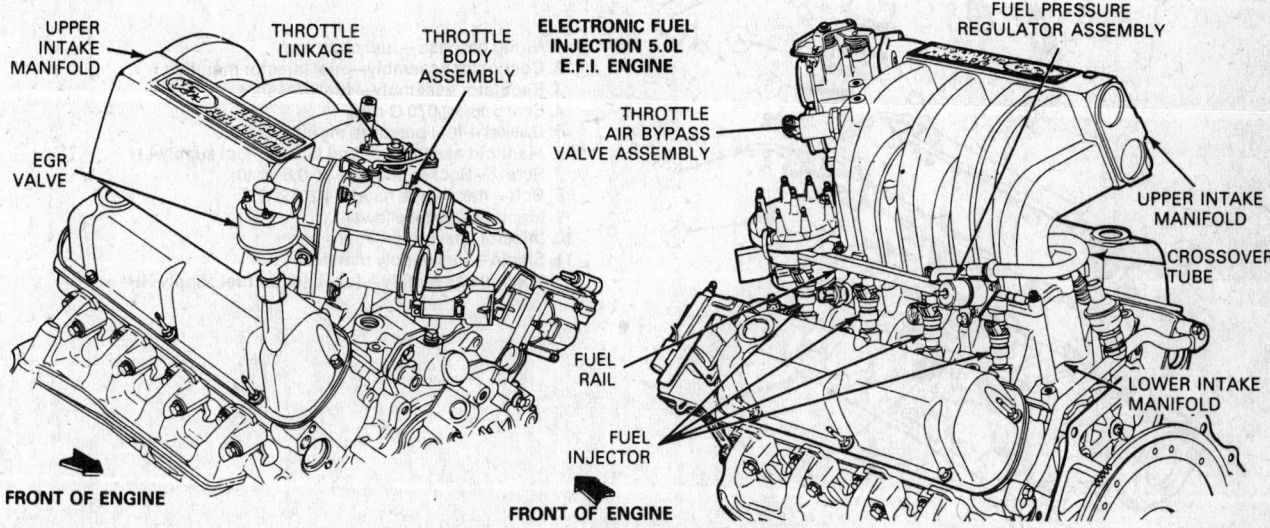

Ford EFI-V8 engine component location

computer then outputs a command to the fuel injectors to meter the approximate quantity of fuel.

The fuel delivery subsystem consists of a high pressure, chassis mounted, electric fuel pump and a low pressure in-tank pump. The pumps deliver fuel from the fuel tank through a 20 micron fuel filter to a fuel charging manifold assembly. The fuel charging manifold assembly incorporates electrically actuated fuel injectors directly above each of the engine's intake ports. The injectors, when energized, spray a metered quantity of fuel into the intake air stream. A constant fuel pressure drop is maintained across the injector nozzles by a pressure regulator. The regulator is connected in series with the fuel injectors and positioned down stream from them. Excess fuel supplied by the pump, but not required by the engine, passes through the regulator and returns to the fuel tank through a fuel return line.

All injectors (4 cylinder) or one bank (8 cylinder) are energized simultaneously, once every crankshaft revolution. The period of time that the injectors are energized (injector "on time" or the pulse width) is controlled by the vehicles' Engine Electronic Control (EEC) computer. Air entering the engine is measured by a speed density meter. This air flow information and input from various other engine sensors is used to compute the required fuel flow rate. The computer determines the needed injector pulse width and outputs a command to the injector to meter the exact quantity of fuel.

The fuel injector nozzles are electromechanical devices which both meter and atomize fuel delivered to the engine. The injectors are mounted in the lower intake manifold and are positioned so that their tips are directing fuel just ahead of the engine intake valves. The injector bodies consist of a solenoid actuated pintle and needle valve assembly. An electrical control signal from the Electronic Engine Control unit activates the injector solenoid causing the pintle to move inward off the seat, allowing fuel to flow. Since the injector flow orifice is fixed and the fuel pressure drop across the injector tip is constant, fuel flow to the engine is regulated by how long the solenoid is energized.

The fuel pressure regulator is attached to the fuel supply manifold assembly downstream of the fuel injectors. It regulates the fuel pressure supplied to the injectors. The regulator is a diaphragm operated relief valve in which one side of the diaphragm senses fuel pressure and the other side is subjected to intake manifold pressure. The nominal fuel pressure is established by a spring pre-load applied to the diaphragm. Balancing one side of the diaphragm with manifold pressure maintains a constant fuel pressure drop across the injectors. Fuel, in excess of that used by the engine, is bypassed through the regulator and returns to the fuel tank.

The throttle body assembly controls air flow to the engine through a butterfly-type valve. The body is a single piece die casting made of aluminum. It has a bore with an air bypass channel around the throttle plate. This bypass channel controls both cold and warm engine idle airflow control as regulated by an air bypass valve assembly. The valve assembly is an electromechanical device controlled by the EEC computer. It incorporates a linear actuator which positions a variable area metering valve. Other features, on the V8 throttle body assembly include: A preset stop tp locate the WOT position. A throttle body mounted throttle position sensor and canister purge ports for evaporative emission control.

The fuel supply manifold assembly is the component that delivers high pressure fuel from the vehicle fuel supply line to the fuel injectors. The assembly consists of a single preformed tube or stamping with four injector connectors on four cylinder engines; or two banks of stamped fuel rails connected by a crossover. A mounting flange for the fuel pressure regulator, a pressure relief valve for diagnostic testing or field service fuel system pressure bleed down (on some models) and mounting attachments which locate the fuel manifold assembly and provide fuel injector retention. The crossover on V8 models uses a push connect fitting.

The air intake manifold is a two piece (upper and lower intake manifold) aluminum casting. Runner lengths are tuned to optimize engine torque and power output.

Fuel Charging Assembly

NOTE: If any of the sub-assemblies are to be serviced and/or removed, with the fuel charging assembly mounted to the engine, the following steps must be taken.

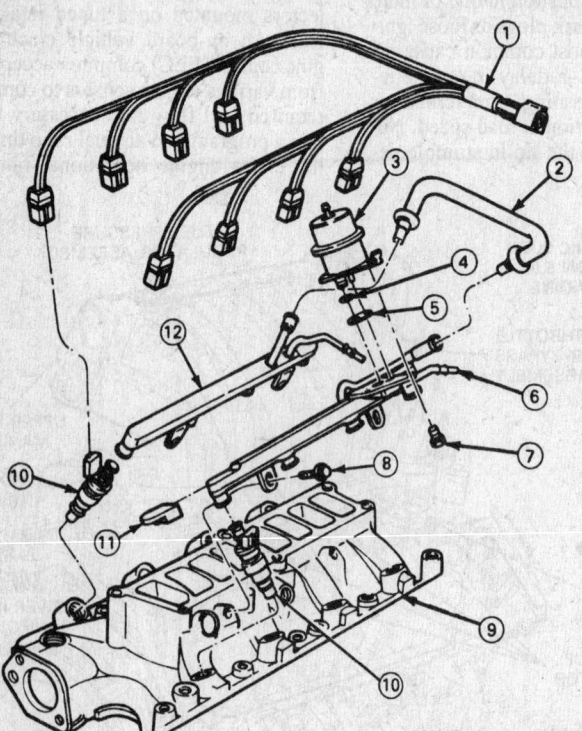

1. Wiring harness—fuel charging
2. Connector assembly—fuel injector manifold
3. Regulator assembly—fuel pressure
4. Seal 5 16 × .070 O-ring
5. Gasket—fuel pressure regulator
6. Manifold assembly—fuel injector fuel supply-LH
7. Screw—Socket head 5.0 × 0.8 × 10
8. Bolt—hex flange head 1 4-20 × .75
9. Manifold—intake lower
10. Injector assembly—fuel
11. Shield—fuel supply manifold
12. Manifold assembly—fuel injector fuel supply-RH

Ford EFI-V8 fuel charging manifold assembly

1. Make sure the ignition key is in the OFF position.

2. Drain the coolant from the radiator (on four cylinder models).

3. Disconnect the negative battery cable (on 4 cylinder engines).

4. Remove the fuel cap to relieve fuel tank pressure.

5. On four cylinder engines: Relieve the pressure from the fuel system at the pressure relief valve (If a pressure relief valve is not present, follow the procedure for V8). Special tool T80L-9974-A or its equal is needed for this procedure. Disconnect the fuel supply line. Identify and disconnect the fuel return lines and vacuum connections. Disconnect the injector wiring harness by disconnecting the ECT sensor in the heater supply tube, under the lower intake manifold. Disconnect the air by-pass connector from EEC harness.

6. On V8 engines: Locate and disconnect the electrical connection to either the fuel pump relay, the inertia switch or the in-line high pressure fuel pump. Crank the engine for about ten seconds. The engine may start and run briefly. After it stops, crank again.

7. Disconnect the negative battery cable.

NOTE: Not all assemblies may be serviceable while on the engine. In some cases, removal of the fuel charging assembly may facilitate service of the various sub-assemblies. To remove the entire fuel charging assembly, the following should be observed.

Upper Intake Manifold—4 Cylinder

REMOVAL & INSTALLATION

1. Disconnect the air cleaner outlet tube from the air intake throttle body.

2. Unplug the throttle position sensor from the wiring harness.

3. Unplug the air by-pass valve connector.

4. Remove three upper manifold retaining bolts.

5. Remove upper manifold assembly.

6. Remove and discard the gasket from the lower manifold assembly. If scraping is necessary be careful not to damage gasket surfaces, or allow any material to drop into the lower manifold.

7. Installation is the reverse of removal. Tighten the upper intake manifold bolts 15–22 ft. lbs. Use a new gasket between the manifolds.

Fuel Charging Assembly—4 Cylinder

NOTE: If any of the sub-assemblies are to be serviced and/or removed, with the fuel charging assembly mounted to the engine, the following steps must be taken.

1. Make sure the ignition key is the the OFF position.

2. Drain the coolant from the radiator.

3. Disconnect the negative battery cable.

4. Remove the fuel cap to relieve fuel tank pressure.

5. Relieve the pressure from the fuel system at the pressure relief valve. Special tool T80L-9974-A or its equal is needed for this procedure.

6. Disconnect the fuel supply line.

7. Identify and disconnect the fuel return lines and vacuum connections.

8. Disconnect the injector wiring harness by disconnecting the ECT sensor in the heater supply tube, under the lower intake manifold.

9. Disconnect the air by-pass connector from EEC harness.

NOTE: Not all assemblies may be serviceable while on the engine. In some cases, removal of the fuel charging assembly may facilitate service of the various sub-assemblies. To remove the entire fuel charging assembly, the following should be observed.

REMOVAL & INSTALLATION

1. Remove the engine air cleaner outlet tube between the vane air meter and air throttle body by loosening two clamps.

2. Disconnect and remove the accelerator and speed control cables (if so equipped) from the accelerator mounting bracket and throttle lever.

3. Disconnect the rear vacuum line to the dash panel vacuum tree and the front vacuum line to the air cleaner and fuel pressure regulator.

4. Disconnect the PCV system by removing the following:

   a. Two large forward facing connectors on the throttle body and intake manifold.

   b. Throttle body port hose at the straight plastic connector.

   c. Canister purge line at the straight plastic connector.

   d. PCV hose at the valve cover.

   e. Unbolt the PCV separator support bracket from cylinder head and remove PCV system.

5. Disconnect the EGR vacuum line at the EGR valve.

6. Disconnect the EGR tube from the upper intake manifold by removing the two flange nuts.

7. Remove the dipstick and its tube.

8. Remove the fuel return line.

9. Remove six manifold mounting nuts.

10. Remove the manifold with wiring harness and gasket.

11. Installation is the reverse of removal. Tighten the manifold bolts 12–15 ft. lbs.

Fuel Pressure Regulator—4 Cylinder

REMOVAL & INSTALLATION

——————— CAUTION ———————

Before attempting this procedure depressurize the fuel system.

1. Remove the vacuum line at the pressure regulator.

2. Remove the three Allen retaining screws from the regulator housing.

3. Remove the pressure regulator, gasket and O-ring. Discard gasket and inspect O-ring for deterioration.

NOTE: If scraping is necessary be careful not to damage the gasket surface.

4. Installation is the reverse of removal. Lubricate the O-ring with light oil prior to installation. Tighten the three screws 27–40 inch lbs.

Injector Manifold Assembly—4 Cylinder

REMOVAL & INSTALLATION

1. Remove the fuel tank cap. Release the pressure from the fuel system.

2. Disconnect the fuel supply and return lines.

3. Disconnect the wiring harness from the injectors.

4. Disconnect the vacuum line from the fuel pressure regulator valve.

5. Remove the two fuel injector manifold retaining bolts.

6. Carefully disengage the manifold from the fuel injectors. Remove the manifold.

7. Installation is the reverse of removal. Torque the fuel manifold bolts 15–22 ft. lbs.

Pressure Relief Valve—4 Cylinder

REMOVAL & INSTALLATION

1. If the fuel charging assembly is mounted to the engine, the fuel system must be depressurized.

2. Using an open end wrench or suitable deep well socket, remove the pressure relief valve from the injection manifold.

3. Installation is the reverse of removal. Torque the valve to 4–7 ft. lbs.

Throttle Position Sensor—4 Cylinder

REMOVAL & INSTALLATION

1. Disconnect the throttle position sensor from the wiring harness.

2. Remove the two retaining screws.

3. Remove the throttle position sensor.

4. Installation is the reverse of removal. Torque the sensor screws to 11–16 inch lbs.

Air Bypass Valve Assembly—4 Cylinder

REMOVAL & INSTALLATION

1. Disconnect the air bypass valve assembly connector from the wiring harness.

2. Remove the two air bypass valve retaining screws.

3. Remove the air bypass valve and gas-

ket. If necessary to remove the gasket by scraping, be careful not to damage the gasket surface.

4. Installation is the reverse of removal. Torque the air bypass valve assembly to 6–8 ft. lbs.

Air Intake Throttle Body–4 Cylinder

REMOVAL & INSTALLATION

1. Remove four throttle body nuts. Make sure that the throttle position sensor connector and air by-pass valve connector have been disconnected from the harness. Disconnect air cleaner outlet tube.

2. Identify and disconnect vacuum hoses.

3. Remove throttle bracket.

4. Carefully separate the throttle body from the upper intake manifold.

5. Remove and discard the gasket between the throttle body and the upper intake

manifold. If scraping is necessary be careful not to damage gasket surfaces, or allow any material to drop into the manifold.

6. Installation is the reverse of removal. Tighten the throttle body to upper intake manifold nuts 12–15 ft. lbs.

Fuel Injector–4 Cylinder

REMOVAL & INSTALLATION

NOTE: The fuel system must be depressurized prior to starting this procedure.

1. Disconnect the fuel supply and return lines.

2. Remove the vacuum line from the fuel pressure regulator.

3. Disconnect the wiring harness.

4. Remove the fuel injector manifold assembly.

5. Carefully remove the connectors from the individual injectors.

6. Grasping the injectors body, pull up

while gently rocking the injector from side to side.

7. Inspect the injector O-rings (two per injector) for signs of deterioration. Replace as needed.

8. Inspect the injector "plastic hat" (covering the injector pintle) and washer for signs of deterioration. Replace as needed. If a hat is missing, look for it in the intake manifold.

9. Installation is the reverse of removal. Lubricate all O-rings with a light oil. Carefully seat the fuel injector manifold assembly on the four injectors and secure the manifold with the attaching bolts. Torque the bolts 15–22 ft. lbs.

Upper & Lower Intake Manifold–V8

REMOVAL & INSTALLATION

────── CAUTION ──────
Discharge fuel system pressure before starting any work that involves disconnecting fuel system lines.

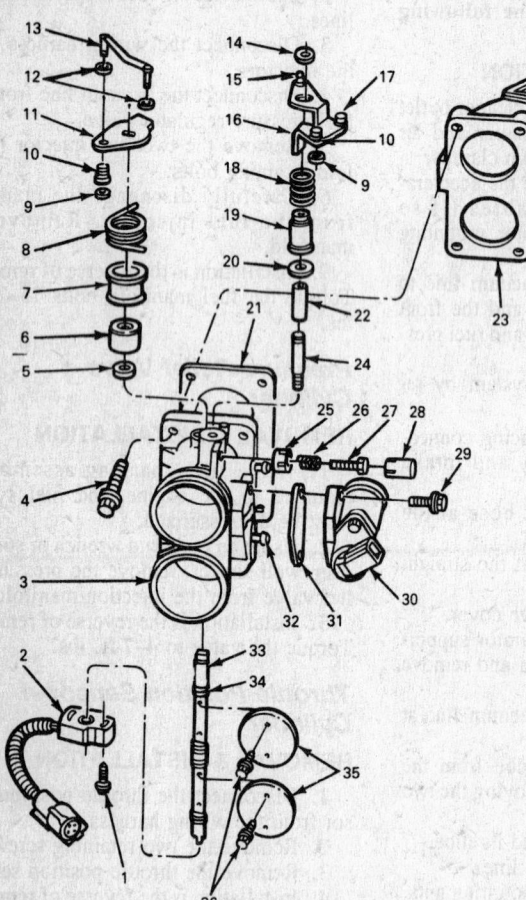

AIR INTAKE CHARGE THROTTLE BODY AND UPPER MANIFOLD ASSEMBLY

1. Screw and washer assembly—M4 × 22
2. Throttle position sensor
3. Body—air intake charge throttle
4. Bolt—5 16–18 × 1.25
5. Seal—throttle control shaft
6. Dust cover—engine throttle shaft
7. Bushing—engine throttle shaft
8. Spring—engine throttle return
9. Retaining ring
10. Bushing—throttle control lever
11. Lever—engine throttle
12. Washer—flat
13. Throttle control rod
14. Ring—external retaining
15. Ball—engine throttle lever
16. Lever—throttle control
17. Pin—engine transmission linkage
18. Spring—secondary throttle return
19. Bushing—throttle control shaft
20. Washer—8.65 × 18.25 × .023 flat steel
21. Gasket—air intake charge throttle
22. Bearing—throttle control linkage
23. Manifold—intake upper
24. Hub—throttle control
25. Plug—engine idle adjusting screw locking
26. Spring—engine idle adjusting screw
27. Screw—10.32 × 1/50 hex head slotted
28. Cap—engine idle adjusting screw
29. Bolt—M6 × 20
30. Air bypass valve assembly
31. Gasket—air bypass
32. Tube
33. Shaft—air intake charge throttle
34. Seal—Air intake charge throttle shaft
35. Plate—throttle
36. Screw—M4 × 07 × 9.0 × 4.9 hex head wash

Ford EFI-V8 air intake throttle body and upper manifold

1. To remove the upper manifold: Remove the air cleaner. Disconnect the electrical connectors at the air bypass valve, throttle position sensor and EGR position sensor.

2. Disconnect the throttle and transmission linkage at the throttle ball and from the throttle body. Remove the bolts that secure the bracket to the intake and position the bracket and cables out of the way.

3. Disconnect the upper manifold vacuum fitting connections by removing all the vacuum lines at the vacuum tree (label lines for position identification). Remove the vacuum lines to the EGR valve and fuel pressure regulator.

4. Disconnect the PCV system by disconnecting the hose from the fitting at the rear of the upper manifold.

5. Remove the two canister purge lines from the fittings at the throttle body.

6. Disconnect the EGR tube from the EGR valve by loosening the flange nut.

7. Remove the bolt from the upper intake support bracket to upper manifold. Remove the upper manifold retaining bolts and remove the upper intake manifold and throttle body as an assembly.

8. Clean and inspect all mounting surfaces of the upper and lower intake manifolds.

9. Position a new mounting gasket on the lower intake manifold and install the upper manifold in the reverse order of removal. Mounting bolts are torqued to 12–18 ft. lbs.

10. To remove the lower intake manifold: Upper manifold and throttle body must be removed first.

11. Drain the cooling system. Remove the distributor assembly, cap and wires.

12. Disconnect the electrical connectors at the engine coolant temperature sensor and sending unit, at the air charge temperature sensor and at the knock sensor.

13. Disconnect the injector wiring harness from the main harness assembly. Remove the ground wire from the intake manifold stud. The ground wire must be installed at the same position it was removed from.

14. Disconnect the fuel supply and return lines from the fuel rails.

15. Remove the upper radiator hose from the thermostat housing. Remove the bypass hose. Remove the heater outlet hose at the intake manifold.

16. Remove the air cleaner mounting bracket. Remove the intake manifold mounting bolts and studs. Pay attention to the location of the bolts and studs for reinstallation. Remove the lower intake manifold assembly.

17. Clean and inspect the mounting surfaces of the heads and manifold.

18. Apply a ¹⁄₁₆ inch bead of RTV sealer to the ends of the manifold seals (at the junction point of the seals and gaskets). Install the end seals and intake gaskets on the cylinder heads. The gaskets must interlock with the seal tabs.

19. Install locator bolts at opposite ends of each head and carefully lower the intake manifold into position. Install and tighten the mounting bolts and studs, in sequence, to 23–25 ft. lbs. Wait ten minutes and once again torque in sequence. Install the remaining components in the reverse order of removal.

Air Intake Throttle Body–V8
REMOVAL & INSTALLATION

1. Disconnect the air intake hose.
2. Disconnect the throttle position sensor and air by-pass valve connectors.

3. Remove the four throttle body mounting nuts and carefully separate the air throttle body from the upper intake manifold.

4. Remove and discard the mounting gasket. Clean all mounting surfaces using care not to damage the gasket surfaces of the throttle body and manifold. Do not allow any material to drop into the intake manifold.

5. Install the throttle body in the reverse order of removal. The mounting nuts are tightened to 12–15 ft. lbs.

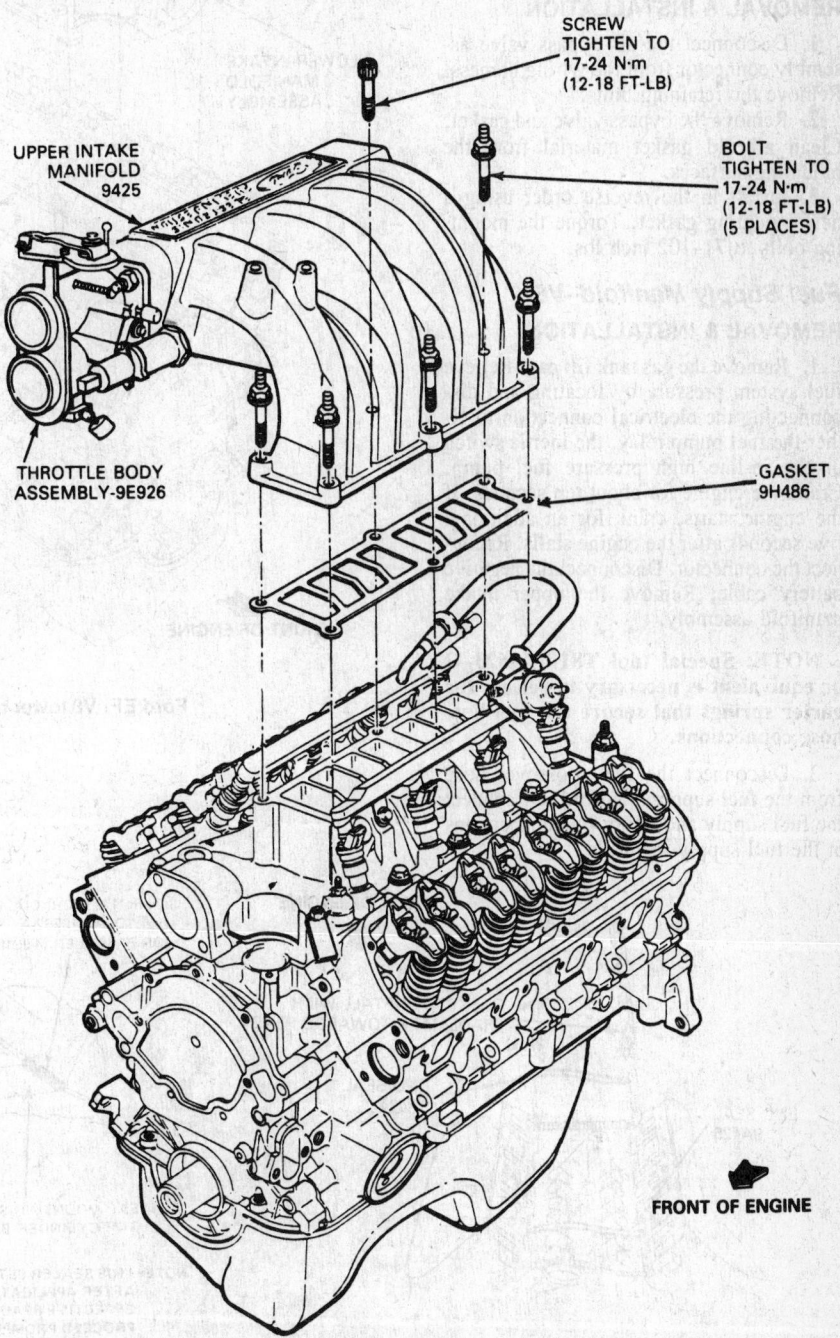

SCREW
TIGHTEN TO
17-24 N·m
(12-18 FT-LB)

BOLT
TIGHTEN TO
17-24 N·m
(12-18 FT-LB)
(5 PLACES)

UPPER INTAKE
MANIFOLD
9425

THROTTLE BODY
ASSEMBLY-9E926

GASKET
9H486

FRONT OF ENGINE

Ford EFI-V8 upper manifold and throttle body assembly

Throttle Position Sensor–V8
REMOVAL & INSTALLATION

1. Disconnect the wiring harness. Scribe a reference mark across the edge of the sensor and throttle body for correct reinstallation.

2. Remove the retaining screws and the position sensor.

3. Install in reverse order. When thje throttle position sensor is installed on the throttle body, the wiring harness should be pointing directly to the air bypass valve.

Air Bypass Valve Assembly–V8
REMOVAL & INSTALLATION

1. Disconnect the air bypass valve assembly connector from the wiring harness. Remove the retaining bolts.

2. Remove the bypass valve and gasket. Clean all old gasket material from the mounting surfaces.

3. Install in the reverse order using a new mounting gasket. Torque the mounting bolts to 71–102 inch lbs.

Fuel Supply Manifold–V8
REMOVAL & INSTALLATION

1. Remove the gas tank fill cap. Relieve fuel system pressure by locating and disconnecting the electrical connection to either the fuel pump relay, the inertia switch or the in-line high pressure fuel pump. Crank the engine for about ten seconds. If the engine starts, crank for an additional five seconds after the engine stalls. Reconnect the connector. Disconnect the negative battery cable. Remove the upper intake manifold assembly.

NOTE: Special tool T81P–19623–G or equivalent is necessary to release the garter springs that secure the fuel line/hose connections.

2. Disconnect the fuel crossover hose from the fuel supply manifold. Disconnect the fuel supply and return line connections at the fuel supply manifold.

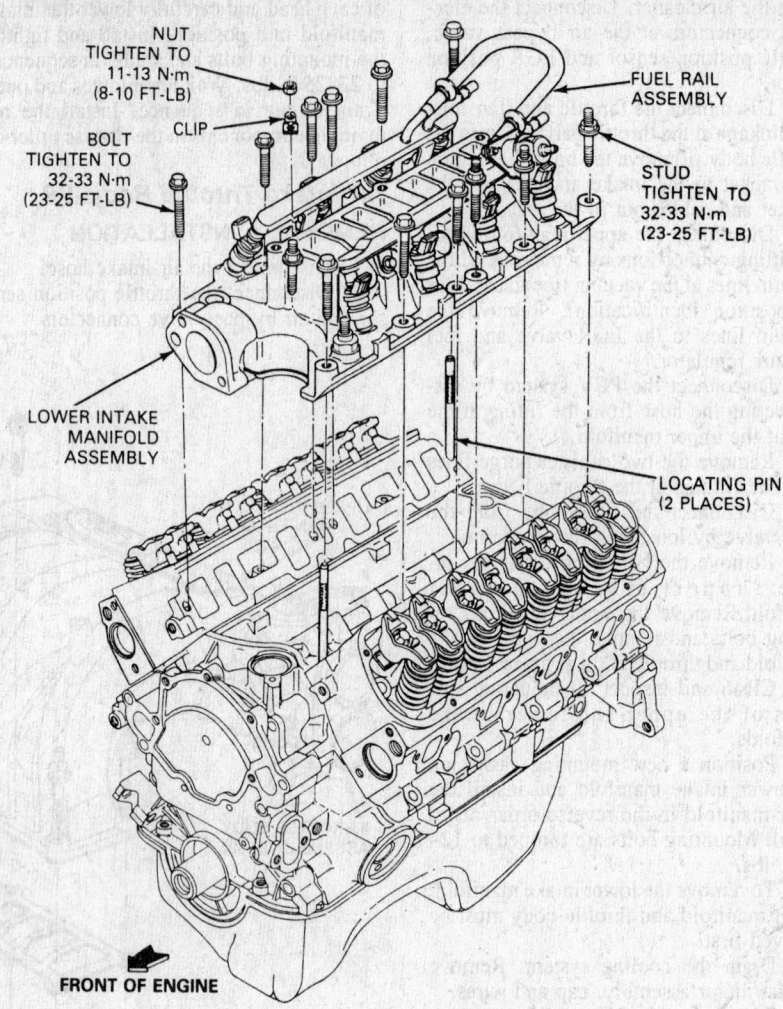

Ford EFI-V8 lower manifold installation

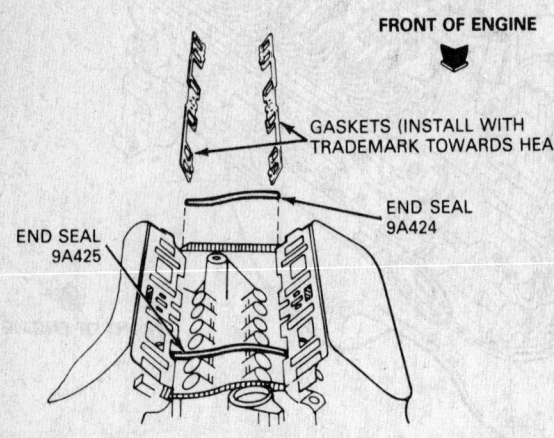

Ford EFI-V8 lower manifold gasket installation

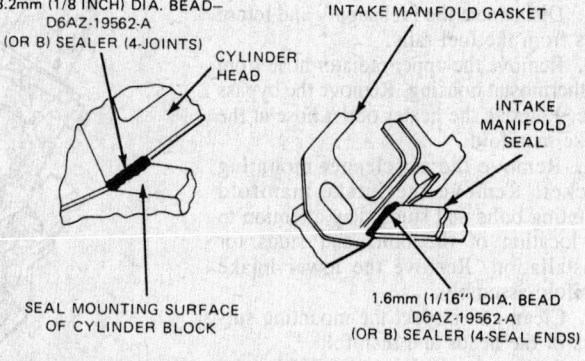

NOTE: THIS SEALER SETS UP WITHIN 15 MINUTES AFTER APPLICATION. TO ASSURE EFFECTIVE SEALING, ASSEMBLY SHOULD PROCEED PROMPTLY.

Ford EFI-V8 lower manifold seal installation

3. Remove the two fuel supply manifold retaining bolts. Carefully disengage the manifold from the fuel injectors and remove the manifold.

4. When installing: Make sure the injector caps are clean and free of contamination. Place the fuel supply manifold over each injector and seat the injectors into the manifold. Make sure the caps are seated firmly.

5. Torque the fuel supply manifold retaining bolts to 15–22 ft. lbs. Install the remaining components in the reverse order of removal.

NOTE: Fuel injectors may be serviced after the fuel supply manifold is removed. Grasp the injector and pull up on it while gently rocking injector from side to side. Inspect the mounting O-rings and replace any that show deterioration.

Fuel Pressure Regulator–V8
REMOVAL & INSTALLATION

1. Discharge the fuel pressure.
2. Remove the vacuum line. Remove the retaining screws from the regulator housing.
3. Remove the regulator, gasket and O-ring. Clean all mounting surfaces.
4. Install in reverse order.

Jeep (4 Cylinder) Throttle Body Injection

Throttle Body
REMOVAL & INSTALLATION

1. Remove the upper cover assembly from the TBI.
2. Remove the lower cover assembly retaining bolts and lower cover assembly.
3. Remove the throttle cable and return spring.
4. Disconnect the wire harness connector from the injector.
5. Disconnect the wire harness connector from the wide open throttle (WOT) switch.
6. Disconnect the wire harness connector from the ISC motor.
7. Disconnect the fuel supply pipe from the throttle body.
8. Disconnect the fuel return pipe from the throttle body.
9. Disconnect the vacuum hoses from the throttle body assembly. Identify the tag and hoses for installation reference.
10. Disconnect the potentiometer wire connector.
11. Remove the throttle body-to-manifold retaining nuts from the studs.

12. Remove the throttle body assembly from the intake manifold.
13. If the throttle body assembly is being replaced, transfer the following components to the replacement throttle body: idle speed control (ISC) motor/WOT switch and bracket.
14. Install the replacement throttle body assembly on the intake manifold. Use a replacement gasket between the components.
15. Install the throttle body-to-manifold retaining nuts on the studs.
16. Connect the vacuum hoses.
17. Connect the fuel return pipe to the throttle body.
18. Connect the fuel supply pipe to the throttle body.
19. Connect the wire harness connector to the injector.
20. Connect the wire harness connector to the WOT switch.
21. Connect the wire harness connector to the ISC motor.
22. Connect the potentiometer wire connector.
23. Install the throttle cable and return spring.
24. Install the lower cover assembly and retaining nuts.
25. Install the upper cover assembly.

FORD ELECTRONIC FUEL INJECTION TROUBLESHOOTING

Symptom	Possible Problem Areas
Surging, backfire, misfire, runs rough	1. EEC distributor rotor registry 2. EGR solenoid(s) defective 3. Distributor, cap, body, rotor, ignition wires, plugs, coil defective 4. Pulse ring behind vibration damper misaligned or damaged 5. Spark plug fouling
Stalls on deceleration	1. EGR solenoid(s) or valve defective 2. EEC distributor rotor registry
Stalls at idle	1. Idle speed wrong 2. Throttle kicker not working
Hesitates on acceleration	1. Acceleration enrichment system defective 2. Fuel pump ballast bypass relay not working
Fuel pump noisy	1. Fuel pump ballast bypass relay not working
Engine won't start	1. Fuel pump power relay defective, no spark, EGR system defective, no or low fuel pressure 2. Crankshaft position sensor not seated, clearance wrong, defective 3. Pulse ring behind vibration damper misaligned, sensor tabs damaged 4. Power and ground wires open or shorted, poor electrical connections 5. Inertia switch tripped
Engine starts and stalls or runs rough	1. Fuel pump ballast wire defective 2. Manifold absolute pressure (MAP) sensor circuit not working 3. Low fuel pressure 4. EGR system problem 5. Microprocessor and calibration assembly faulty
Starts hard when cold	1. Cranking signal circuit faulty

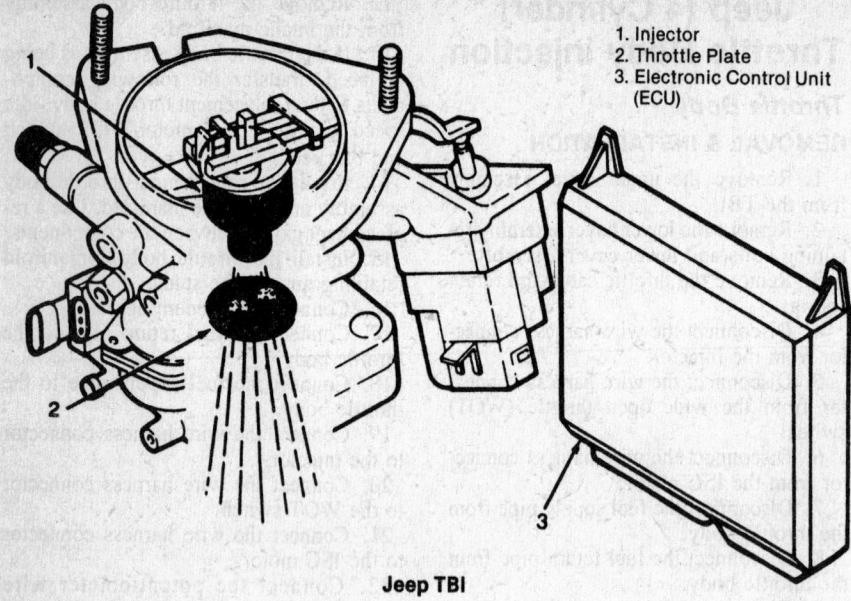

1. Injector
2. Throttle Plate
3. Electronic Control Unit (ECU)

Jeep TBI

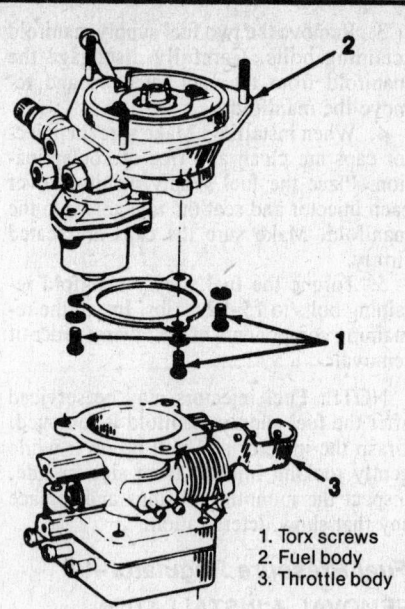

1. Torx screws
2. Fuel body
3. Throttle body

Jeep fuel body assembly

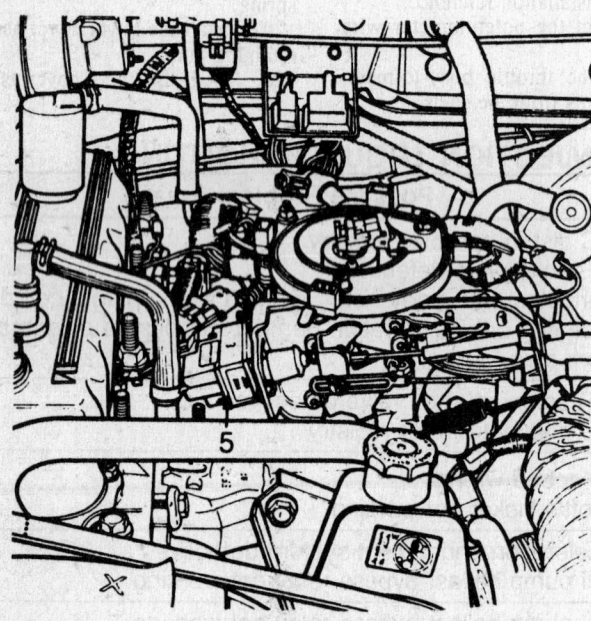

Jeep ISA motor location

Fuel Body Assembly

REMOVAL & INSTALLATION

1. Remove the throttle body assembly from the intake manifold.
2. Remove the three Torx head screws that retain the fuel body to the throttle body.
3. Remove the original gasket and discard.
4. Install the replacement fuel body on the throttle body using a replacement gasket.

5. Install the three fuel body-to-throttle body retaining Torx head screws and tighten securely.
6. Install the throttle body assembly on the intake manifold.

Fuel Pressure Regulator

REMOVAL & INSTALLATION

1. Remove the throttle body assembly.
2. Remove the three retaining screws that secure the pressure regulator to the fuel body.

3. Remove the pressure regulator assembly. Note the location of the components for assembly reference.
4. Discard the gasket.
5. Position the pressure regulator assembly with a replacement gasket.
6. Install the three retaining screws to secure the pressure regulator to the fuel body.
7. Operate the engine and check for leaks.
8. Install the throttle body assembly.

Fuel Injector

REMOVAL & INSTALLATION

1. Remove the air inlet cover and hose.
2. Remove the injector wire connector.
3. Remove the injector retainer clip screws.
4. Remove the injector retainer clip.
5. Using a pair of small pliers, gently grasp the center collar of the injector (between electrical terminals) and carefully remove the injector with a lifting-twisting motion.
6. Discard the upper and lower O-rings. Note that the backup ring fits over the upper O-ring.
7. Lubricate with light oil and install a replacement lower O-ring in the housing bore.
8. Lubricate with light oil and install replacement upper O-ring in the housing bore.
9. Install the backup ring over the upper O-ring.
10. Position the replacement injector in the fuel body and center the nozzle in the lower housing bore. Seat the injector using a pushing-twisting motion.
11. Align the wire terminals in the proper orientation.
12. Install the retainer clip and screws.

13. Install the injector wire connector.
14. Install the air inlet cover and hose.

Throttle Position Sensor (TPS)
REMOVAL & INSTALLATION

1. Remove the upper and lower air inlet covers.
2. Remove the throttle body assembly.
3. Remove the two Torx retaining screws.
4. Remove the throttle position sensor from the throttle shaft lever.
5. Position the throttle position sensor over the throttle shaft lever.
6. Install the two Torx head screws to retain the sensor.
7. Install the throttle body assembly.
8. Install the upper and lower cover assemblies.

Idle Speed Actuator (ISA) Motor
REMOVAL & INSTALLATION

NOTE: The closed throttle (die) switch is integral with the motor.

1. Disconnect the throttle return spring.
2. Disconnect the wire harness connector from the motor.
3. Remove the motor-to-bracket retaining nuts. Use a backup wrench as not to remove the ISA studs which hold the ISA motor together.
4. Remove the motor from the bracket.
5. Install the ISA motor on the bracket.
6. Install the motor-to-bracket retaining nuts.
7. Connect the wire harness connector to the ISA motor.
8. Connect the throttle return spring.

EGR Valve
REMOVAL & INSTALLATION

NOTE: The EGR valve is located on the intake manifold below the throttle body.

1. Disconnect the vacuum hose from the EGR valve.
2. Remove the two EGR valve-to-intake manifold retaining bolts.
3. Remove the EGR valve from the intake manifold.
4. Discard the original gasket.
5. Clean the EGR valve and intake manifold mating surfaces.
7. Position the replacement EGR valve on the intake manifold with a replacement gasket.
8. Install the EGR valve-to-intake manifold bolts.
9. Connect the vacuum hose to the EGR valve.

Manifold Absolute Pressure (MAP) Sensor
REMOVAL & INSTALLATION

1. Disconnect the wire harness connector.
2. Disconnect the vacuum hose.

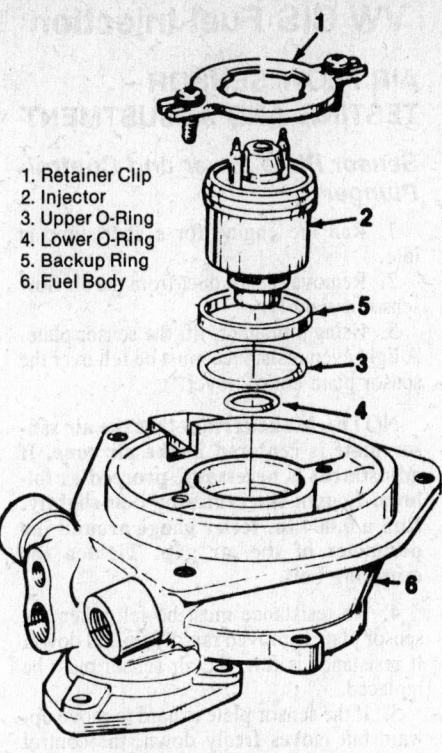

1. Retainer Clip
2. Injector
3. Upper O-Ring
4. Lower O-Ring
5. Backup Ring
6. Fuel Body

Jeep injector assembly

3. Remove the retaining nuts.
4. Remove the MAP sensor from the dash panel.
5. Install in the reverse order.

Manifold Air/Fuel Temperature (MAT) Sensor
REMOVAL & INSTALLATION

1. Disconnect the wire harness connector from the MAT sensor harness.
2. Remove the MAT sensor from the intake manifold.
3. Clean the threads in the manifold.
4. Install the replacement MAT sensor in the intake manifold.
5. Connect the wire harness connector to the MAT sensor harness.

Coolant Temperature Sensor (CTS)
REMOVAL & INSTALLATION

1. Remove the wire harness connector from the coolant temperature sensor (CTS). (Located at the rear of the intake manifold).

------- CAUTION -------
DO NOT remove the CTS with the cooling system hot and under pressure because serious burns from coolant can occur.

2. Remove the CTS from the intake manifold and rapidly plug the hole to prevent loss of coolant.

3. Install the replacement CTS in the intake manifold.
4. Connect the wire harness connector to the CTS harness connector.

Power (B+) Relay/Fuel Pump and Compressor Clutch Relay
REMOVAL & INSTALLATION

1. Remove the power (B+) relay, the fuel pump relay, or the clutch relay from the group of relays on the right inner fender panel.
2. Install in reverse order and connect the replacement relay to the wire harness connector.

Canister Purge/EGR Solenoid
REMOVAL & INSTALLATION

1. Disconnect the vacuum hoses from the solenoid.
2. Disconnect the electrical connection from the solenoid.
3. Remove the solenoid assembly from the relay bracket.
4. Install the solenoid to the relay bracket.
5. Connect the electrical connection to the solenoid.
6. Connect the vacuum hoses to the solenoid.

Wide Open Throttle (WOT) Switch
REMOVAL & INSTALLATION

1. Remove the air inlet cover.
2. Disconnect the throttle return spring.
3. Disconnect the throttle cable.
4. Disconnect the wire harness connector from the WOT switch.
5. Remove the two WOT switch-to-bracket screws.
6. Remove the WOT switch.
7. Position the replacement of WOT switch on the bracket and secure with the two switch-to-bracket screws.
8. Connect the wire harness connector.
9. Connect the throttle cable.
10. Connect the throttle return spring.
11. Install the air inlet bonnet.

Fuel Filter
REMOVAL & INSTALLATION

1. Raise and safely support the vehicle.
2. Remove the filter retaining strap bolt.
3. Install a pinch clamp on the inlet hose side of the filter to prevent gravity leakage of gas from the line when the filter is disconnected.
4. Remove the fuel line mounting clamps and remove the filter.
5. Install the filter into position and install the inlet and outlet fuel hoses and clamps. Tighten the clamps securely.
6. Install the filter retaining strap and bolt.
7. Remove the pinch clamp.
8. Lower vehicle.

Toyota (EFI) Fuel Injection

System Description

There are three basic systems used by the Toyota EFI: Fuel, Air Induction and Electronic Controls.

FUEL SYSTEM

An electric fuel pump supplies sufficient fuel, under constant pressure, to the injectors. The injectors "inject" a metered amount of fuel into the intake manifold as controlled and measured by signals from the ECU (electrical control unit).

AIR INDUCTION SYSTEM

The air induction system consists of the air intake, an air flow meter, an air valve, an air intake chamber and other components. All of which contribute to the supply of the proper amount of air to the intake manifold as controlled electronically.

ELECTRONIC CONTROLS

The ECU receives signals from various sensors indicating various changes in engine operation and conditions which include: Intake air volume, Intake air temperature, Coolant temperature, Engine rpm, Acceleration/deceleration, Exhaust content, etc.

The signals are utilized by the ECU to determine the injection duration necessary for an optimum air/fuel ratio.

Other functions of the ECU are: Electronic spark advance, Turbocharger overpressure, Transmission converter lock-up, Engine trouble diagnostics and a Fail-safe (back-up) system to provide minimal drivability in case of sensor malfunction.

Troubleshooting

Engine troubles are not usually caused by the EFI system. When troubleshooting, always first check the condition of all related systems.

Many times the most frequent cause of the problem is a bad contact in a wiring connector, so always make sure that the connections are secure. When inspecting the connector, pay particular attention to the following points:
1. Check to see that the terminals are not bent.
2. Check to see that the connector is pushed all the way in and locked.
3. Check that there is no change in the signal when the connector is tapped or wiggled.
Actual troubleshooting of the EFI system and the EFI computer is a complex process which requires the use of expensive and hard to find tools. Other than checking the function of the main components individually, specific testing is possible only if the required equipment is available.

VW CIS Fuel Injection

AIR FLOW SENSOR— TESTING AND ADJUSTMENT

Sensor Plate Lever and Control Plunger

1. Run the engine for a short time at idle.
2. Remove the air duct from the air flow sensor assembly.
3. Using a magnet, lift the sensor plate. A light even resistance must be felt over the sensor plate entire travel.

NOTE: Make certain that the air sensor plate is centered in the air cone. If adjustment is necessary, proceed as follows: Loosen the centering bolt slightly. Run a 0.004 in. feeler gauge around the perimeter of the air gap. Tighten the centering bolt.

4. No resistance must be felt when the sensor plate is moved rapidly up and down. If resistance is felt, the air sensor must be replaced.
5. If the sensor plate is hard to move upward but moves freely down, the control plunger is sticking. Remove the fuel distributor and clean the control plunger in solvent. If after installation the plunger is still sticking, the fuel distributor must be replaced.

Sensor Plate Height Adjustment

The height adjustment of the sensor plate must be checked under fuel pressure. You will also need a bridging adaptor (US 4480/3).

1. Install a pressure gauge in the line between the fuel distributor and control pressure regulator.
2. Remove the rubber elbow from the air flow sensor housing.
3. Remove the fuel pump relay from the fuse panel and install a bridging adapter (US 4480/3).
4. Switch the adapter ON and wait until pressure reads 49–54 psi.
5. Switch the bridging adapter OFF; the pressure should fall to 28–37 psi.
6. The upper edge of the sensor plate must be even with the bottom of the air cone taper or no more than 0.020 in. below the bottom of the taper.

7. Bend the clip to adjust the height.
8. Recheck the pressure readings after adjusting.
9. Remove the pressure gauge, reconnect the fuel lines and install the fuel pump relay.

Thermo-Time Switch

NOTE: To properly perform the following tests, the engine must be cold with the water temperature below 95°F (35°C).

1. Disconnect the electrical connector from the cold start valve on the end of the intake manifold.
2. Connect a test light across the cold start valve terminals.
3. Connect a jumper wire from the #1 terminal on the ignition coil to a good ground.
4. Have an assistant operate the starter. If the test light fails to light after 8 seconds, the thermo-time switch is defective and should be replaced.

Cold Start Valve

1. Remove the electrical connector from the cold start valve.

NOTE: Do not remove the fuel line from the cold start valve.

2. Remove the cold start valve from the manifold and point the nozzle into a measuring container.
3. Connect a jumper wire from one terminal of the cold start valve to terminal #15 on the ignition coil.
4. Connect a second jumper wire from the other cold start valve terminal to ground.
5. Remove the fuel pump relay and bridge the relay plate terminals #L13 and #L14 with a fused (8 amp) jumper wire.
6. Have an assistant turn the ignition switch on while observing the fuel spray pattern from the cold start valve. The spray pattern from the nozzle must be cone-shaped and steady, if not replace the valve.
7. Turn the ignition switch off.
8. Wipe the nozzle dry with a clean rag and check for leakage. If drops form within one minutes, the valve is defective and must be replaced.

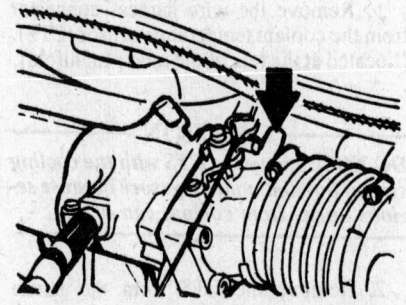

CIS-idle adjustment screw location

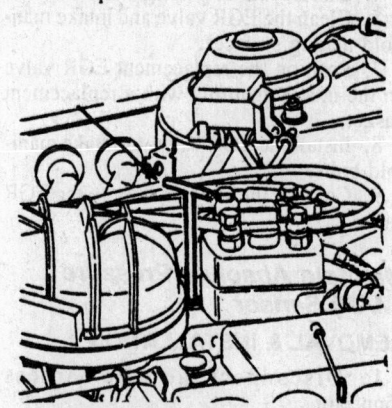

CIS-CO adjusting tool installed

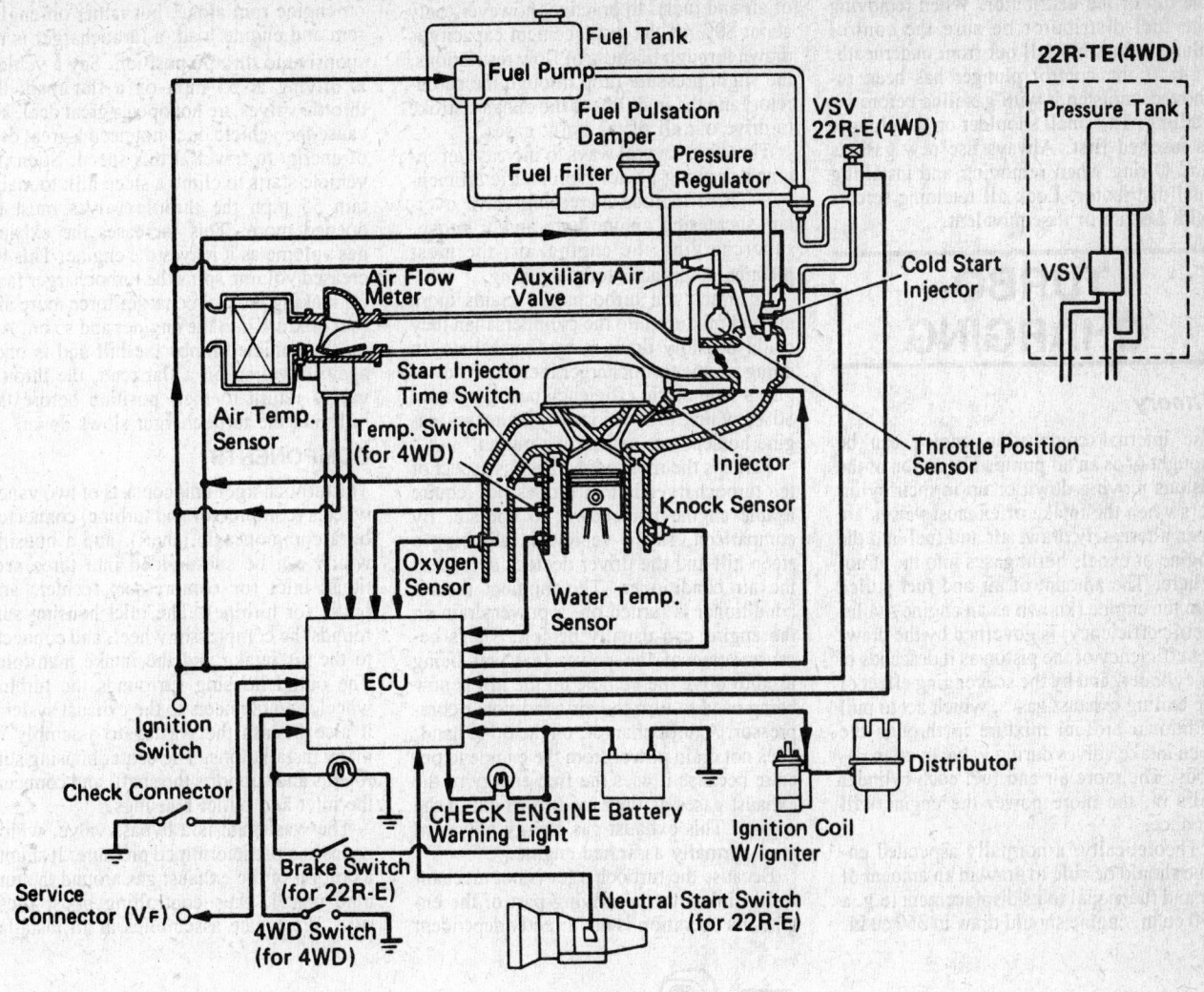

22R-TE(4WD)

Fuel Tank
Fuel Pump
Fuel Pulsation Damper
Fuel Filter
Pressure Regulator
VSV 22R-E(4WD)
Pressure Tank
VSV
Air Flow Meter
Auxiliary Air Valve
Cold Start Injector
Air Temp. Sensor
Start Injector Time Switch
Temp. Switch (for 4WD)
Injector
Throttle Position Sensor
Knock Sensor
Oxygen Sensor
Water Temp. Sensor
ECU
Ignition Switch
Check Connector
CHECK ENGINE Warning Light
Battery
Distributor
Ignition Coil W/igniter
Service Connector (VF)
Brake Switch (for 22R-E)
4WD Switch (for 4WD)
Neutral Start Switch (for 22R-E)

Toyota EFI

Auxiliary Air Regulator

NOTE: The engine must be cold to perform this test.

1. Unplug the electrical connector from the auxiliary air regulator.
2. Start the engine and check the idle.
3. Pinch the hose between the auxiliary air regulator and the intake manifold. The engine idle should drop. Reconnect the electrical connector.
4. Repeat the test on the engine at operating temperature. The idle speed should remain constant when the hose is pinched. If not, replace the auxiliary air regulator.

Control Pressure Regulator

The system is under considerable constant pressure. The only practical test that should be attempted by the owner is one using an ohmmeter. Be sure the engine is at normal operating temperature. There should be not loose fuel fittings or other fire hazards when the electrical connections are disengaged.

1. Remove the electrical connector from the control pressure regulator and auxiliary air regulator.

2. Start the engine and run it at idle.
3. Check the terminals of the control pressure regulator wiring harness for voltage. It should be at least 11.5 volts.
4. Connect an ohmmeter across the terminals of the control pressure regulator socket. Resistance should be between 16 and 22 ohms. If there is no resistance, replace the control pressure regulator.

Fuel Injectors

1. Remove the injector but leave it connected to the fuel line.
2. Point the injector into a measuring container.
3. Remove the fuel pump relay and bridge the relay plate terminals #L13 and L14 with a fused (8 amp) jumper wire.
4. Remove the air duct from the air flow sensor.
5. Have an assistant turn the ignition switch on.
6. Lift the air flow sensor plate with a magnet and observe the injector nozzle spray pattern. The spray pattern must be cone-shaped and even, if not replace the injector.

7. Turn the ignition off and hold the injector horizontally. It should not drip.

NOTE: One or more injectors may be checked at the same time.

8. Moisten the rubber seals on the injectors with fuel before installing.
9. Press the injectors firmly into place.

Fuel Distributor

REMOVAL & INSTALLATION

1. Release the pressure in the system by loosening the fuel line on the control pressure regulator (large connector). Have a rag ready to catch the fuel that escapes.
2. Mark the fuel lines in the top of the distributor so that you will be able to put them back in their correct positions.

NOTE: Using different colored paints is usually a good marking device. When you mark each line, be sure to mark the spot where it connects to the distributor.

3. Clean the fuel lines, then remove them from the distributor. Remove the little looped wire plug (the CO adjusting screw plug). Remove the two retaining screws in

the top of the distributor. When removing the fuel distributor be sure the control plunger does not fall out from underneath.

4. If the control plunger has been removed, moisten it with gasoline before installing. The small shoulder on the plunger is inserted first. Always use new gaskets and O-ring when removing and installing fuel distributor. Lock all retaining screws with Loctite or its equivalent.

TURBO-CHARGING

Theory

The internal combustion engine can be thought of as an air pump. The action of the pistons moving down or up in their cylinders when the intake or exhaust valves are open alternately draws air and fuel into the engine or expels burnt gases into the atmosphere. The amount of air and fuel pulled into the engine (known as an engine's volumetric efficiency) is governed by the drawing efficiency of the piston as it descends in its cylinder, and by the scavenging effect of the exiting exhaust gases, which act to pull additional air/fuel mixture in through the open intake valves during valve overlap periods. The more air and fuel each cylinder pulls in, the more power the engine will produce.

Theoretically, a normally asperated engine should be able to draw in an amount of air and fuel equal to its displacement (e.g. a 350 cu in. engine should draw in 350 cu in.

of air and fuel). In practice, however, only about 80% of the displacement capacity is drawn through because of flow restrictions, the slight pressure drop through the carburetor, and the inability of the exhaust stroke to drive out all of the burnt gases.

There are several ways to increase an engine's drawing power (volumetric efficiency). These include increasing valve overlap, increasing engine bore and/or stroke, supercharging the engine, or, the most practical approach, turbocharging.

In effect, the turbocharger crams more air/fuel mixture into the cylinders than they could possibly draw in by themselves. In doing so, the turbocharger increases the engine's volumetric efficiency past its normal 80%, which proportionately increases engine horsepower and torque output.

Perhaps the most advantageous aspect of the turbocharger is that it does not require usable engine horsepower to operate. By comparison, say a vehicle is climbing a steep hill and the driver decides to turn on the air conditioner. The moment the air conditioner is turned on, a power drain on the engine can usually be felt. That's because some of the power that was being used to drive the vehicle up the hill is now being used to turn the air conditioner compressor. A turbocharger, on the other hand, does not drain power from the engine to operate because it uses the free energy of the exhaust gases as they are blown out of the engine. This exhaust gas energy is wasted on a normally aspirated engine.

Because the turbocharger is not mechanically linked to the driving part of the engine, its operation is not directly dependent

on engine rpm alone, but rather on engine rpm and engine load: a turbocharger is responsive to throttle position. Say a vehicle is driving at 55 mph on a flat road: the throttle valves are not open a great deal, because the vehicle does not need a great deal of energy to travel at this speed. Soon the vehicle starts to climb a steep hill: to maintain 55 mph the throttle valves must be opened more. This increases the exhaust gas volume as it leaves the engine. This increased volume spins the turbocharger faster, making the turbocharger force more air/fuel mixture into the engine, and so on. After the vehicle climbs the hill and is once again traveling on a flat road, the throttle valves return to their position before the hill, and the turbocharger slows down.

COMPONENTS

The turbocharger unit consists of two vaned wheels (compressor and turbine) connected by a common axle (shaft), and a housing which can be sub-divided into three sections: inlet (or compressor), center, and outlet (or turbine). The inlet housing surrounds the compressor wheel, and connects to the air intake and the intake manifold. The outlet housing surrounds the turbine wheel, and connects to the exhaust system; it also houses the wastegate assembly in many installations. The center housing surrounds and supports the shaft, and connects the inlet and outlet housings.

The wastegate is a bypass valve, which opens at a predetermined pressure. It shunts a portion of the exhaust gas around the turbine wheel, thus controlling boost pressure. Wastegate assemblies in all installa-

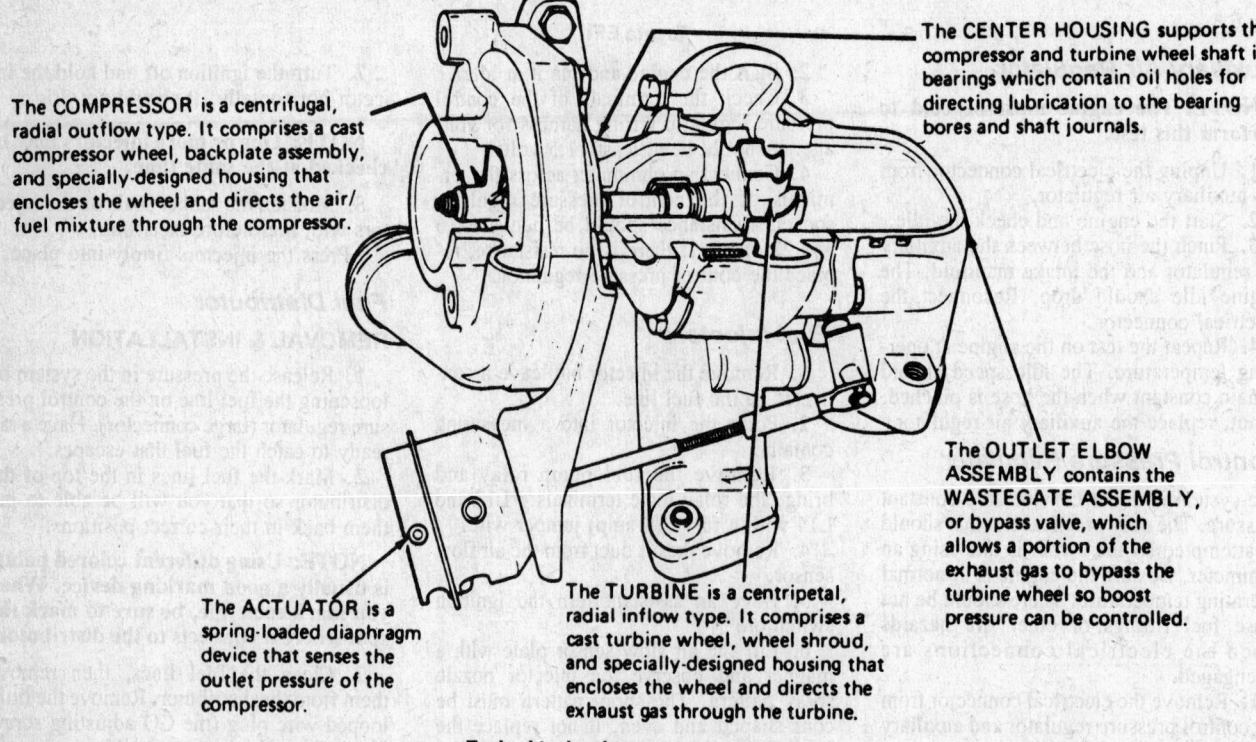

The COMPRESSOR is a centrifugal, radial outflow type. It comprises a cast compressor wheel, backplate assembly, and specially-designed housing that encloses the wheel and directs the air/fuel mixture through the compressor.

The CENTER HOUSING supports the compressor and turbine wheel shaft in bearings which contain oil holes for directing lubrication to the bearing bores and shaft journals.

The ACTUATOR is a spring-loaded diaphragm device that senses the outlet pressure of the compressor.

The TURBINE is a centripetal, radial inflow type. It comprises a cast turbine wheel, wheel shroud, and specially-designed housing that encloses the wheel and directs the exhaust gas through the turbine.

The OUTLET ELBOW ASSEMBLY contains the WASTEGATE ASSEMBLY, or bypass valve, which allows a portion of the exhaust gas to bypass the turbine wheel so boost pressure can be controlled.

Typical turbocharger components

tions covered in this book are installed in the outlet housing.

OPERATION

Turbocharger operation is remarkably simple. The turbine wheel is installed in the path of the engine's exhaust gas, and the compressor wheel is installed in the intake path. Exhaust gas is directed through the turbine housing, causing the turbine wheel to spin. This spinning motion is transferred by the connecting shaft to the compressor wheel. As the compressor wheel spins, it packs the intake charge (which is being drawn through the carburetor in all installations covered in this book) into a dense mass, which is fed into the engine. Combustion converts the charge into exhaust. The exhaust charge is directed through the turbine housing, where it spins the turbine wheel, and then out through the turbine housing discharge into the exhaust system.

Thus, turbocharger operation is self-perpetuating. However, unchecked turbocharger operation will increase compressor pressure (called boost pressure) beyond the design limits of the engine, and will seriously damage internal engine components. Boost pressure is controlled by the wastegate. When boost pressure rises to a predetermined value, the wastegate opens, bypassing exhaust flow around the turbine.

Greater volumetric efficiency is a benefit of the turbocharging process, but increased cylinder pressure is a drawback, because it raises the engine's octane requirement. The two are inseparable, so a method must be devised to compensate for the increased octane requirement to avoid detonation (spark knock). Water injection, alcohol injection, low boost pressures, charge intercoolers, ignition spark retardation, and alcohol fuels have all been used to control detonation, with varying degrees of success.

One method of controlling detonation is by limiting boost and by spark retardation. Wastegate operation begins at at a fixed psi, and enough exhaust gas is routed around the turbine to limit boost to a fixed maximum. The electronic ignition system has been modified in the turbocharged en-

TURBOCHARGER TROUBLESHOOTING

Problem	Cause	How To Check	Solution
No boost	Gasket leak, hole in exhaust system	Temporarily block tailpipe with engine running. Any exhaust leaks in the system will be heard.	Repair leaks (usually at gasket surfaces)
	Dirty air filter	Remove air filter and check	Replace or clean filter
	Blocked air intake	Visually inspect for blockage	Clear intake
	Worn valves or rings	Compression test engine	Repair
	Throttle valves not opening completely	Manually operate throttle linkage, check valve movement	Adjust linkage, repair carburetor
	Exhaust blockage	Check catalytic converter for melted and blocked catalyst, check muffler and exhaust pipes for debris	Replace catalytic converter, repair exhaust system
	Wastegate stuck open	Test wastegate operation	Repair or replace wastegate assembly
Fuel odor under boost	Leak at compressor or intake manifold	Look for fuel stains at fittings	Tighten fittings or replace gaskets
Ignition miss at high speed, under load	Spark plug gap too large	Remove spark plugs, measure gap	Reduce gap
	Faulty coil	Test Coil	Replace
Ignition miss (often)	Excessive resistance in ignition cables	Check cable resistance (see Tune-Up Unit Repair section)	Replace cables as necessary
Oil leaks into turbine	Blocked oil return hose	Remove hose and check for blockage or crimps	Repair or replace hose
Detonation	Fuel octane rating too low	Check octane rating of fuel used against that recommended by manufacturer (consult owner's manual)	Switch to higher octane unleaded fuel
	Faulty sensor	Check G.M. as instructed here; have Ford system checked by qualified technician	Replace as necessary
	Faulty ignition retard unit	Refer to qualified technician	Repair or replace as necessary
	Engine overheating	Check coolant level, debris clogged radiator, no coolant circulation, blocked thermostat	Repair or replace as necessary
Poor idle	Air leak between compressor and carburetor	Listen at joints for hissing sound while the engine idles	Repair

gine to include two spark retardation points. When boost pressure reaches approximately one-half to one psi, a switch in the intake manifold sends a signal to the ignition module, which retards ignition timing six degrees. A second manifold switch sends its signal when boost reaches a certain psi, resulting in additional degrees of retard.

A slightly different way of controlling detonation is to limit boost maximum psi. A detonation sensor is installed in the engine block or intake manifold. Vibrations caused by detonation are transmitted to the sensor, which sends a signal to the Electronic Spark Control (ESC) module. The module processes this signal, and sends a command signal to the distributor to retard timing.

Testing Wastegate Operation

As noted before, the wastegate is a safety valve for the engine. If the wastegate sticks shut, boost pressure will build until the air/fuel mixture charge becomes too powerful for the mechanical components (pistons, bearings, etc.) and causes engine damage.

If the wastegate sticks open, little or not boost will be received from the turbocharger, which translates into mediocre engine performance. The simplest wastegate test is to to remove the pressure hose at the wastegate diaphragm unit, connect a pressure pump (such as the type used for cooling system testing) and apply pressure. At the specified opening pressure, the link between the wastegate and its diaphragm unit will just move (about .015 in). The movement is not great, but it should be easy to see.

If the wastegate does not move, try to operate the linkage by hand. It should move under moderate hand pressure. If it move, the problem is probably in the diaphragm unit (broken diaphragm). To test the diaphragm, remove the vacuum hose from the diaphragm, hook up a manual vacuum pump and apply 25 in. Hg of vacuum to the diaphragm unit. If the vacuum drops below 18 in. Hg within one minute, replace the diaphragm unit.

NOTE: Some turbos use a diaphragm which opens the wastegate during idle and part throttle, when there's no boost, to reduce engine backpressure and improve fuel economy. To test this type of unit, apply about 20 in. Hg of vacuum to the diaphragm unit: the wastegate link should move slightly. This unit operates solely with plenum vacuum and can be identified by the absence of a boost pressure signal line on the diaphragm unit.

Testing Operation of Detonation Sensor

Connect a tachometer and timing light to the engine, run the engine at 1800–2500 rpm and tap on the intake manifold next to the detonation sensor.

NOTE: Be careful to keep all wires, clothing and tools away from moving engine parts.

Rap continuously, quickly and moderately hard. This should trigger the detonation sensor. When it triggers, engine speed should drop at least 200 rpm and timing should retard at least 2–4°, probably more.

LUBRICATION

The turbocharger shaft spins in bearings lubricated by engine oil. Turbine speeds routinely reach 120,000–140,000 rpm, making an adequate and well-filtered oil supply critical for proper operation. Any interruption or contamination of the oil supply will result in engine damage as well. Accelerating the engine to top rpm immediately after starting can result in engine and turbocharger damage (due to the lack of oil pressure). Immediately shutting down the engine after it has been operated at high rpm for an extended period can also result in turbocharger damage, since oil pressure will be shut off, but the turbine will continue to spin for a few moments. (Shutting the throttle abruptly when the engine is at high speed can also cause extensive damage, but for a different reason: sudden closed throttle operation causes the mixture to become very lean, resulting in detonation, high engine temperature, and consequent damage.)

Before starting the engine when changing the oil and filter, or performing any operation which results in oil drainage or loss:

1. Disconnect the ignition to distributor harness connector from the distributor module.

2. Crank the engine several times until the oil light goes out. Do not crank the engine for more than twenty seconds at a time.

3. Reconnect the harness and start the engine.

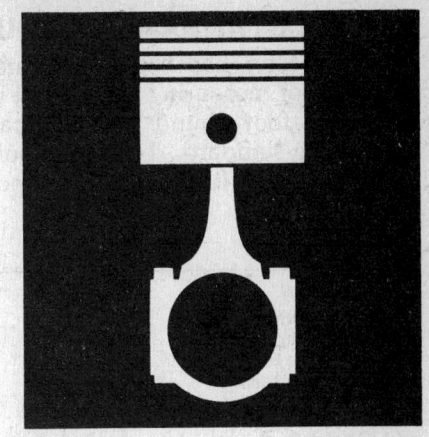

Engine Rebuilding

This section describes, in detail, the procedures involved in rebuilding a typical engine. The procedures are basically identical to those used in rebuilding engines of nearly all design and configurations.

The section is divided into two parts. The first, Cylinder Head Reconditioning, assumes that the cylinder head is removed from the engine, all manifolds are removed, and the cylinder head is on a workbench. The camshaft should be removed from overhead cam cylinder heads. The second section, Cylinder Block Reconditioning, covers the block, pistons, connecting rods and crankshaft. It is assumed that the engine is mounted on a work stand, and the cylinder head and all accessories are removed.

Procedures are identified as follows:

Unmarked—Basic procedures that must be performed in order to successfully complete the rebuilding process.

Starred (*)—Procedures that should be performed to ensure maximum performance and engine life.

Double starred (**)—Procedures that may be performed to increase engine performance and reliability.

In many cases, a choice of methods is also provided. Methods are identified in the same manner as procedures. The choice of method for a procedure is at the discretion of the user.

The tools required for the basic rebuilding procedure should, with minor exceptions, be those included in a mechanic's tool kit. An accurate torque wrench, and a dial indicator (reading in thousandths) mounted on a universal base should be available. Special tools, where required, all are readily available from the major tool suppliers. The services of a competent automotive machine shop must also be readily available.

When assembling the engine, any parts that will be in frictional contact must be prelubricated, to provide protection on initial start-up. Any product specifically formulated for this purpose may be used. NOTE: *Do not use engine oil.* Where semi-permanent (locked but removable) installation of bolts or nuts is desired, threads should be cleaned and coated with Loctite® or a similar product (non-hardening).

Aluminum has become increasingly popular for use in engines, due to its low weight and excellent heat transfer characteristics. The following precautions must be observed when handling aluminum engine parts:
—Never hot-tank aluminum parts.
—Remove all aluminum parts (identification tags, etc.) from engine parts before hot-tanking (otherwise they will be removed during the process).
—Always coat threads lightly with engine oil or anti-seize compounds before installation, to prevent seizure.
—Never over-torque bolts or spark plugs in aluminum threads. Should stripping occur, threads can be restored using any of a number of thread repair kits available (see next section).

Magnaflux and Zyglo are inspection techniques used to locate material flaws, such as stress cracks. Magnafluxing coats the part with fine magnetic particles, and subjects the part to a magnetic field. Cracks cause breaks in the magnetic field, which are outlined by the particles. Since Magnaflux is a magnetic process, it is applicable only to ferrous materials. The Zyglo process coats the material with a fluorescent dye penetrant, and then subjects it to blacklight inspection, under which cracks glow brightly. Parts made of any material may be tested using Zyglo. While Magnaflux and Zyglo are excellent for general inspection, and locating hidden defects, specific checks of suspected cracks may be made at lower cost and more readily using spot check dye. The dye is sprayed onto the suspected area, wiped off, and the area is then sprayed with a developer. Cracks then will show up brightly. Spot check dyes will only indicate surface cracks; therefore, structural cracks below the surface may escape detection. When questionable, the part should be tested using Magnaflux or Zyglo.

REPAIRING DAMAGED THREADS

Several methods of repairing damaged threads are available. Heli-Coil® (shown here), Keenserts® and Microdot® are among the most widely used. All involve basically the same principle—drilling out stripped threads, tapping the hole and installing a prewound insert— making welding, plugging and oversize fasteners unnecessary.

Two types of thread repair inserts are usually supplied—a standard type for most Inch Coarse, Inch Fine, Metric Coarse and Metric Fine thread sizes and a spark plug type to fit most spark plug port sizes. Consult the individual manufacturer's catalog to determine exact applications. Typical thread repair kits will contain a selection of prewound threaded inserts, a tap (corresponding to the outside diameter threads of the insert) and an installation tool. Most manufacturers also supply blister-packed thread repair inserts separately and a master kit with a variety of taps and inserts plus installation tools.

Before effecting a repair to a threaded hole, remove any snapped, broken or damaged bolts or studs. Penetrating oil can be used to free frozen threads; the offending item can be removed with locking pliers or with a screw or stud extractor. After the hole is clear, the thread can be repaired as follows.

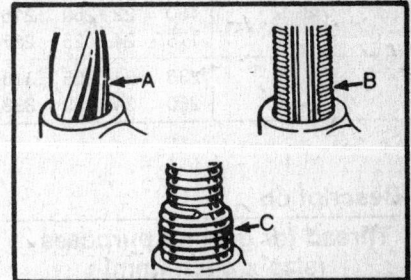

A. Drill out the damaged threads with the specified drill. Drill completely through the hole or to the bottom of a blind hole.

B. With the tap supplied tap the hole to receive the threaded insert. Keep the tap well oiled and back it out frequently to avoid clogging the threads.

C. Screw the threaded insert onto the installation tool until the tang engages the slot. Screw the insert into the tapped hole until it is ¼–½ turn below the top surface. After installation, break the tang off with a hammer and punch.

STANDARD TORQUE SPECIFICATIONS AND CAPSCREW MARKINGS

Newton-Meter has been designated as the world standard for measuring torque and will gradually replace the foot-pound and kilogram-meter torque measuring standard. Torquing tools are still being manufactured with foot-pounds and kilogram-meter scales, along with the new Newton-Meter standard. To assist the repairman, foot-pounds, kilogram-meter and Newton-Meter are listed in the following charts, and should be followed as applicable.

U.S. BOLTS

SAE Grade Number	1 or 2			5			6 or 7			8		
Capscrew Head Markings: Manufacturer's marks may vary. Three-line markings on heads below indicate SAE Grade 5. 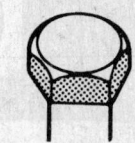				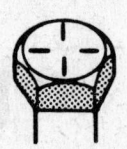								
Usage	Used Frequently			Used Frequently			Used at Times			Used at Times		
Quality of Material	Indeterminate			Minimum Commercial			Medium Commercial			Best Commercial		
Capacity Body Size	Torque			Torque			Torque			Torque		
(inches)–(thread)	Ft-Lb	kgm	Nm	Ft-Lb	kgm	Nm	Ft-Lb	kgm	Nm	Ft-Lb	kgm	Nm
1/4–20	5	0.6915	6.7791	8	1.1064	10.8465	10	1.3630	13.5582	12	1.6596	16.2698
–28	6	0.8298	8.1349	10	1.3830	13.5582				14	1.9362	18.9815
5/16–18	11	1.5213	14.9140	17	2.3511	23.0489	19	2.6277	25.7605	24	3.3192	32.5396
–24	13	1.7979	17.6256	19	2.6277	25.7605				27	3.7341	36.6071
3/8–16	18	2.4894	24.4047	31	4.2873	42.0304	34	4.7022	46.0978	44	6.0852	59.6560
–24	20	2.7660	27.1164	35	4.8405	47.4536				49	6.7767	66.4351
7/16–14	28	3.8132	37.9629	49	6.7767	66.4351	55	7.6065	74.5700	70	9.6810	94.9073
–20	30	4.1490	40.6745	55	7.6065	74.5700				78	10.7874	105.7538
1/2–13	39	5.3937	52.8769	75	10.3725	101.6863	85	11.7555	115.2445	105	14.5215	142.3609
–20	41	5.6703	55.5885	85	11.7555	115.2445				120	16.5860	162.6960
9/16–12	51	7.0533	69.1467	110	15.2130	149.1380	120	16.5960	162.6960	155	21.4365	210.1490
–18	55	7.6065	74.5700	120	16.5960	162.6960				170	23.5110	230.4860
5/8–11	83	11.4789	112.5329	150	20.7450	203.3700	167	23.0961	226.4186	210	29.0430	284.7180
–18	95	13.1385	128.8027	170	23.5110	230.4860				240	33.1920	325.3920
3/4–10	105	14.5215	142.3609	270	37.3410	366.0660	280	38.7240	379.6240	375	51.8625	508.4250
–16	115	15.9045	155.9170	295	40.7985	399.9610				420	58.0860	568.4360
7/8–9	160	22.1280	216.9280	395	54.6285	535.5410	440	60.8520	596.5520	605	83.6715	820.2590
–14	175	24.2025	237.2650	435	60.1605	589.7730				675	93.3525	915.1650
1–8	236	32.5005	318.6130	590	81.5970	799.9220	660	91.2780	894.8280	910	125.8530	1233.7780
–14	250	34.5750	338.9500	660	91.2780	849.8280				990	136.9170	1342.2420

METRIC BOLTS

Description	Torque ft-lbs. (Nm)			
Thread for general purposes (size x pitch (mm))	Head Mark 4		Head Mark 7	
6 x 1.0	2.2 to 2.9	(3.0 to 3.9)	3.6 to 5.8	(4.9 to 7.8)
8 x 1.25	5.8 to 8.7	(7.9 to 12)	9.4 to 14	(13 to 19)
10 x 1.25	12 to 17	(16 to 23)	20 to 29	(27 to 39)
12 x 1.25	21 to 32	(29 to 43)	35 to 53	(47 to 72)
14 x 1.5	35 to 52	(48 to 70)	57 to 85	(77 to 110)
16 x 1.5	51 to 77	(67 to 100)	90 to 120	(130 to 160)
18 x 1.5	74 to 110	(100 to 150)	130 to 170	(180 to 230)
20 x 1.5	110 to 140	(150 to 190)	190 to 240	(160 to 320)
22 x 1.5	150 to 190	(200 to 260)	250 to 320	(340 to 430)
24 x 1.5	190 to 240	(260 to 320)	310 to 410	(420 to 550)

CAUTION: Bolts threaded into aluminum require much less torque

NOTE: This engine rebuilding section is a guide to accepted rebuilding procedures. Typical examples of standard rebuilding procedures are illustrated.

CYLINDER HEAD RECONDITIONING

Procedure	Method
Identify the valves:	Invert the cylinder head, and number the valve faces front to rear, using a permanent felt-tip marker.
Remove the rocker arms (OHV engines only):	Remove the rocker arms with shaft(s) or balls and nuts. Wire the sets of rockers, balls and nuts together, and identify according to the corresponding valve.
Remove the camshaft (OHC engines only):	See the engine service procedures earlier in this book for details concerning specific engines.
Remove the valves and springs:	Using an appropriate valve spring compressor (depending on the configuration of the cylinder head), compress the valve springs. Lift out the keepers with needlenose pliers, release the compressor, and remove the valve, spring, and spring retainer.
Remove glow plugs and fuel injectors (Diesel engines only):	Label and remove all fuel injectors and glow plugs from the head. Glow plugs unscrew. See the appropriate car section for injector removal. Inspect glow plugs for bulges, cracks or signs of melting. Clean injector tips with a steel brush, then inspect for evidence of melting.
**Remove pre-combustion chamber inserts (Diesel engines only):	**Remove the pre-combustion chambers using a hammer and a thin, blunt brass drift, inserted through the injector hole (or glow plug hole, whichever is more convenient). If chamber is to be reused, carefully remove all carbon from it. NOTE: *Remove chamber only if being replaced, if a glow plug tip has broken off and must be removed, or if chamber is obviously damaged or loose.*

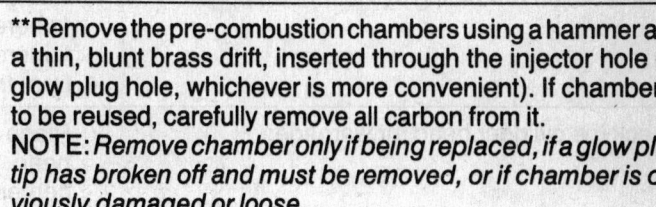

Removing pre-combustion chamber with a drift (© G.M. Corp.)

Check the valve stem-to-guide clearance:	Clean the valve stem with lacquer thinner or a similar solvent to remove all gum and varnish. Clean the valve guides using solvent and an expanding wire-type valve guide cleaner. Mount a dial indicator so that the stem is at 90° to the valve stem, as close to the valve guide as possible. Move the valve off its seat, and measure the valve guide-to-stem clearance by rocking the stem back and forth to actuate the dial indicator. Measure the valve stems using a micrometer, and compare to specifications, to determine whether stem or guide wear is responsible for excessive clearance.

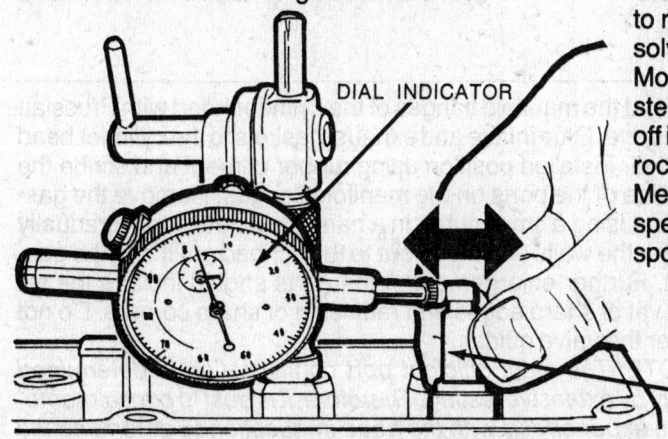

DIAL INDICATOR

VALVE STEM

Checking the valve stem-to-guide clearance

CYLINDER HEAD RECONDITIONING

Procedure	Method

De-carbon the cylinder head and valves:

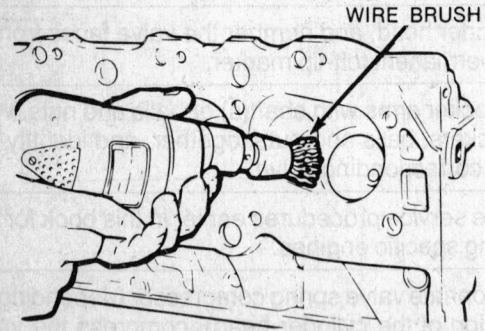

WIRE BRUSH

Removing carbon from the cylinder head

Chip carbon away from the valve heads, combustion chambers, and ports, using a chisel made of hardwood. Remove the remaining deposits with a stiff wire brush.
NOTE: *Ensure that the deposits are actually removed, rather than burnished.*

Hot-tank the cylinder head (cast iron heads only):
CAUTION: *Do not hot-tank aluminum parts.*

Have the cylinder head hot-tanked to remove grease, corrosion, and scale from the water passages.
NOTE: *In the case of overhead cam cylinder heads, consult the operator to determine whether the camshaft bearings will be damaged by the caustic solution.*

Degrease the remaining cylinder head parts:

Using solvent (i.e., Gunk), clean the rockers, rocker shaft(s) (where applicable), rocker balls and nuts, springs, spring retainers, and keepers. Do not remove the protective coating from the springs.

Check the cylinder head for warpage:

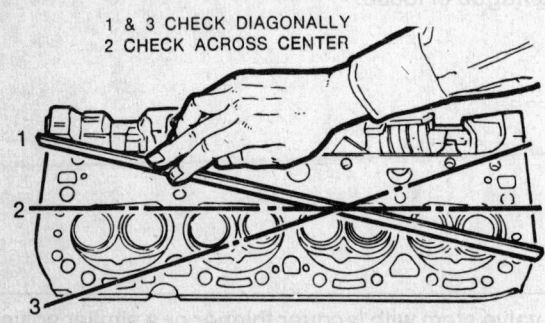

1 & 3 CHECK DIAGONALLY
2 CHECK ACROSS CENTER

Checking cylinder head for warpage

Place a straight-edge across the gasket surface of the cylinder head. Using feeler gauges, determine the clearance at the center of the straight-edge. Measure across both diagonals, along the longitudinal centerline, and across the cylinder head at several points. If warpage exceeds .003′ in a 6′ span, or .006′ over the total length, the cylinder head must be resurfaced.
NOTE: *If warpage exceeds the manufacturer's maximum tolerance for material removal, the cylinder head must be replaced.*
When milling the cylinder heads of V-type engines, the intake manifold mounting position is altered, and must be corrected by milling the manifold flange a proportionate amount.

**Porting and gasket matching:

**Coat the manifold flanges of the cylinder head with Prussian blue dye. Glue intake and exhaust gaskets to the cylinder head in their installed position using rubber cement and scribe the outline of the ports on the manifold flanges. Remove the gaskets. Using a small cutter in a hand-held power tool gradually taper the walls of the port out to the scribed outline of the gasket. Further enlargement of the ports should include the removal of sharp edges and radiusing of sharp corners. Do not alter the valve guides.
NOTE: *The most efficient port configuration is determined only by extensive testing. Therefore, it is best to consult someone experienced with the head in question to determine the optimum alterations.*

CYLINDER HEAD RECONDITIONING

Procedure	Method

*Knurling the valve guides:

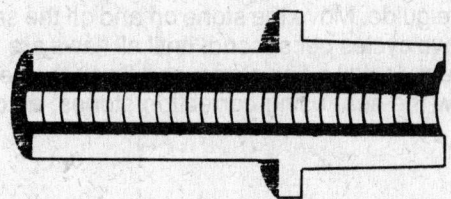

Cut-away view of a knurled valve guide

*Valve guides which are not excessively worn or distorted may, in some cases, be knurled rather than replaced. Knurling is a process in which metal is displaced and raised, thereby reducing clearance. Knurling also provides excellent oil control. The possibility of knurling rather than replacing valve guides should be discussed with a machinist.

Replacing the valve guides:
NOTE: *Valve guides should only be replaced if damaged or if an oversize valve stem is not available.*

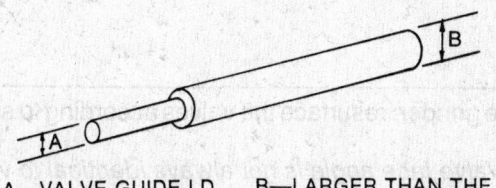

A—VALVE GUIDE I.D.   B—LARGER THAN THE VALVE GUIDE O.D.
Valve guide removal tool

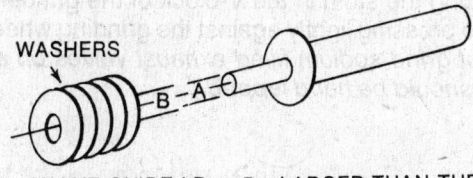

WASHERS

A—VALVE GUIDE I.D.   B—LARGER THAN THE VALVE GUIDE O.D.
Valve guide installation tool (with washers used for installation)

Depending on the type of cylinder head, valve guides may be pressed, hammered, or shrunk in. In cases where the guides are shrunk into the head, replacement should be left to an equipped machine shop. In other cases, the guides are replaced as follows: Press or tap the valve guides out of the head using a stepped drift (see illustration). Determine the height above the boss that the guide must extend, and obtain a stack of washers, their I.D. similar to the guide's O.D., of that height. Place the stack of washers on the guide, and insert the guide into the boss.
NOTE: *Valve guides are often tapered or beveled for installation.*
Using the stepped installation tool (see illustration), press or tap the guides into position. Ream the guides according to the size of the valve stem.

Replacing valve seat inserts:

Replacement of valve seat inserts which are worn beyond resurfacing or broken, if feasible, must be done by a machine shop.

Resurfacing the valve seats using reamers:

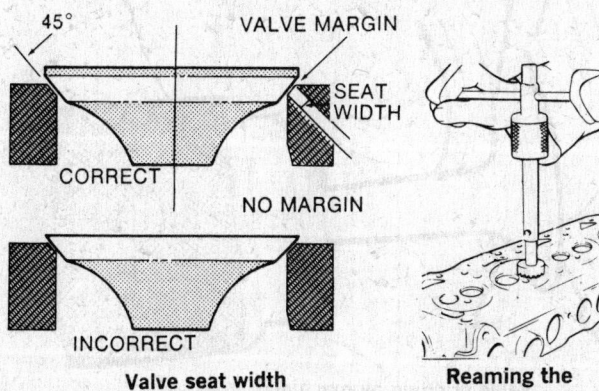

Valve seat width and centering

Reaming the valve seat

Select a reamer of the correct seat angle, slightly larger than the diameter of the valve seat, and assemble it with a pilot of the correct size. Install the pilot into the valve guide, and using steady pressure, turn the reamer clockwise.
CAUTION: *Do not turn the reamer counterclockwise.*
Remove only as much material as necessary to clean the seat. Check the concentricity of the seat (see below). If the dye method is not used, coat the valve face with Prussian blue dye, install and rotate it on the valve seat. Using the dye marked area as a centering guide, center and narrow the valve seat to specifications with correction cutters.
NOTE: *When no specifications are available, minimum seat width for exhaust valves should be $5/64''$, intake valves $1/16''$.*
After making correction cuts, check the position of the valve seat on the valve face using Prussian blue dye.
NOTE: *Do not cut induction hardened seats; they must be ground.*

CYLINDER HEAD RECONDITIONING

Procedure	Method

*Resurfacing the valve seats using a grinder:

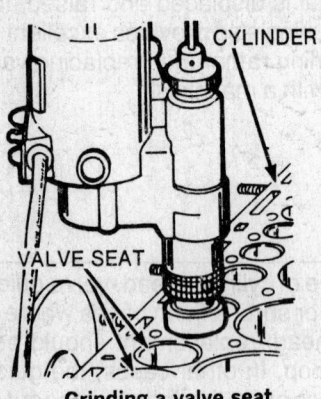

Grinding a valve seat

*Select a pilot of the correct size, and a coarse stone of the correct seat angle. Lubricate the pilot if necessary, and install the tool in the valve guide. Move the stone on and off the seat at approximately two cycles per second, until all flaws are removed from the seat. Install a fine stone, and finish the seat. Center and narrow the seat using correction stones, as described above.

Resurfacing (grinding) the valve face:

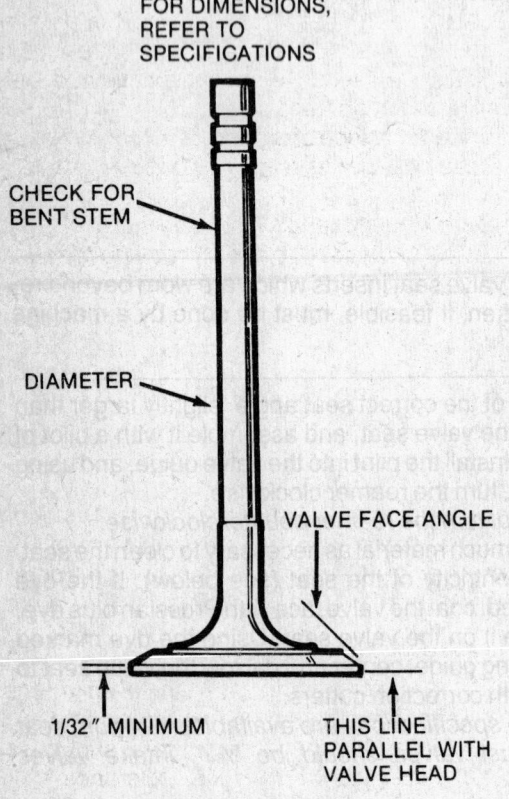

FOR DIMENSIONS,
REFER TO
SPECIFICATIONS

CHECK FOR
BENT STEM

DIAMETER

VALVE FACE ANGLE

1/32" MINIMUM

THIS LINE
PARALLEL WITH
VALVE HEAD

Critical valve dimensions

Using a valve grinder, resurface the valves according to specifications.
CAUTION: *Valve face angle is not always identical to valve seat angle.*
A minimum margin of 1/32" should remain after grinding the valve. The valve stem top should also be squared and resurfaced, by placing the stem in the V-block of the grinder, and turning it while pressing lightly against the grinding wheel.
NOTE: *Do not grind sodium filled exhaust valves on a machine. These should be hand lapped.*

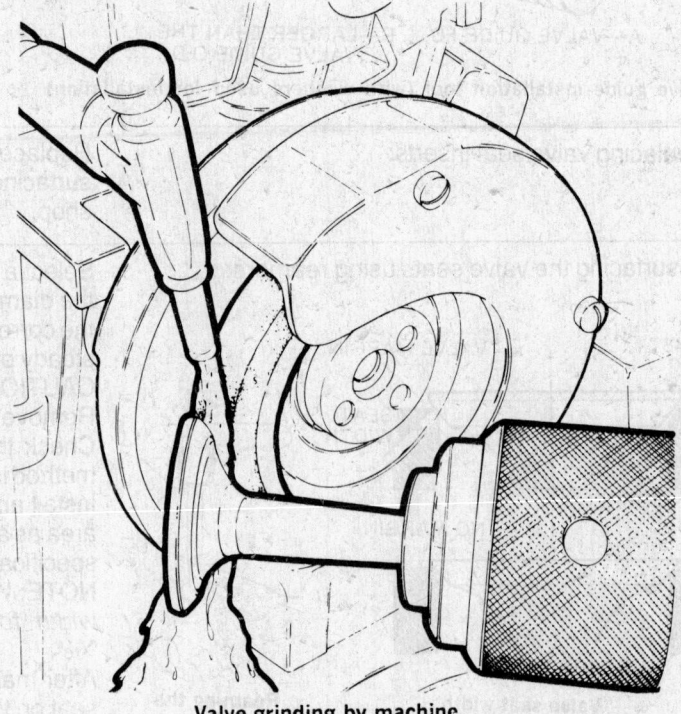

Valve grinding by machine

CYLINDER HEAD RECONDITIONING

Procedure	Method

Checking the valve seat concentricity:

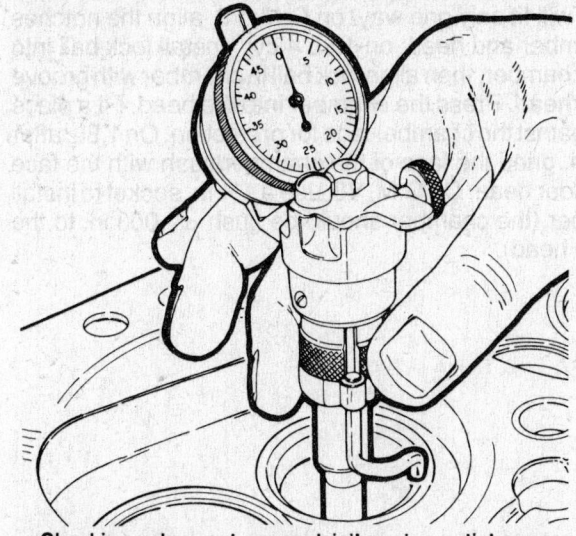

Checking valve seat concentricity using a dial gauge

Coat the valve face with Prussian blue dye, install the valve, and rotate it on the valve seat. If the entire seat becomes coated, and the valve is known to be concentric, the seat is concentric.
*Install the dial gauge pilot into the guide, and rest the arm on the valve seat. Zero the gauge, and rotate the arm around the seat. Run-out should not exceed .002″.

*Lapping the valves:
NOTE: *Valve lapping is done to ensure efficient sealing of resurfaced valves and seats.*

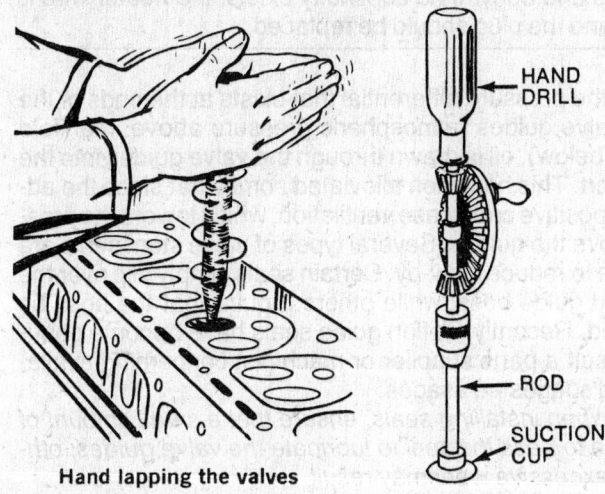

Hand lapping the valves

HAND DRILL
ROD
SUCTION CUP

Home made mechanical valve lapping tool

*Invert the cylinder head, lightly lubricate the valve stems, and install the valves in the head as numbered. Coat valve seats with fine grinding compound, and attach the lapping tool suction cup to a valve head.
NOTE: *Moisten the suction cup.*
Rotate the tool between the palms, changing position and lifting the tool often to prevent grooving. Lap the valve until a smooth, polished seat is evident. Remove the valve and tool, and rinse away all traces of grinding compound.
**Fasten a suction cup to a piece of drill rod, and mount the rod in a hand drill. Proceed as above, using the hand drill as a lapping tool.
CAUTION: *Due to the higher speeds involved when using the hand drill, care must be exercised to avoid grooving the seat.* Lift the tool and change direction of rotation often.

Check the valve springs:

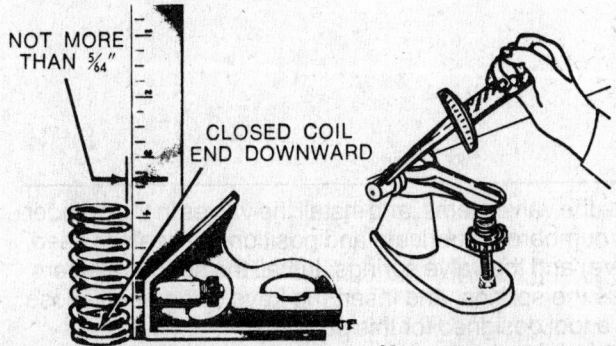

NOT MORE THAN 5/64″
CLOSED COIL END DOWNWARD

Checking valve spring free length and squareness

Measuring valve spring test pressure

Place the spring on a flat surface next to a square. Measure the height of the spring, and rotate it against the edge of the square to measure distortion. If spring height varies (by comparison) by more than 1/16″ or if distortion exceeds 1/16″, replace the spring.
**In addition to evaluating the spring as above, test the spring pressure at the installed and compressed (installed height minus valve lift) height using a valve spring tester. Springs used on small displacement engines (up to 3 liters) should be ∓ 1 lb. of all other springs in either position. A tolerance of ∓ 5 lbs. is permissible on larger engines.

CYLINDER HEAD RECONDITIONING

Procedure	Method

Install pre-combustion chambers (Diesel engines only)

Pre-combustion chambers are press-fit into the head. The chambers will fit only one way: on G.M. V8, align the notches in the chamber and head; on 1.8L 4 cyl., install lock ball into groove in chamber, then align lock ball in chamber with groove in cylinder head. Press the chamber into the head. Fit a piece of metal against the chamber face for protection. On 1.8L, after installation, grind the face of the chamber flush with the face of the cylinder head. On G.M. V8, use a 1¼ in. socket to install the chamber (the chamber should be flush ± .003 in. to the face of the head).

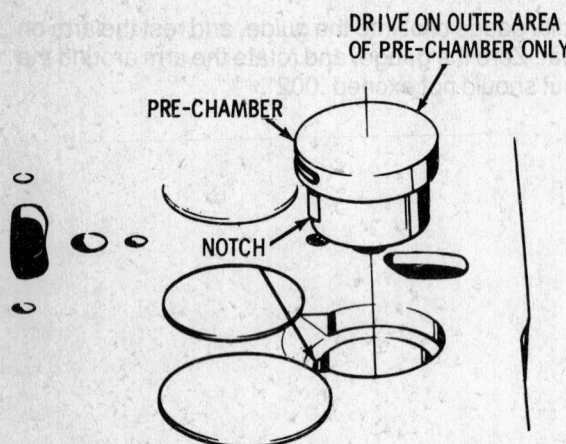

DRIVE ON OUTER AREA OF PRE-CHAMBER ONLY

PRE-CHAMBER

NOTCH

Align the notches to install the pre-combustion chamber (© G.M. Corp.)

Install fuel injectors and glow plugs (Diesel engines)

Before installing glow plugs, check for continuity across plug terminals and body. If no continuity exists, the heater wire is broken and the plug should be replaced.

---

*Install valve stem seals:

*Due to the pressure differential that exists at the ends of the intake valve guides (atmospheric pressure above, manifold vacuum below), oil is drawn through the valve guides into the intake port. This has been alleviated somewhat since the addition of positive crankcase ventilation, which lowers the pressure above the guides. Several types of valve stem seals are available to reduce blow-by. Certain seals simply slip over the stem and guide boss, while others require that the boss be machined. Recently, Teflon guide seals have become popular. Consult a parts supplier or machinist concerning availability and suggested usages.

NOTE: *When installing seals, ensure that a small amount of oil is able to pass the seal to lubricate the valve guides; otherwise, excessive wear may result.*

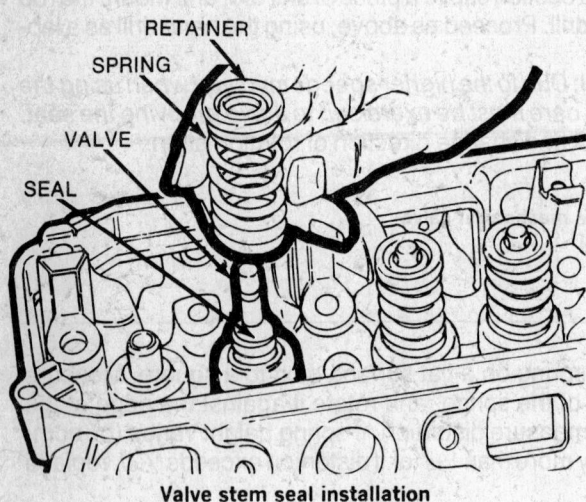

RETAINER

SPRING

VALVE

SEAL

Valve stem seal installation

Install the valves:

Lubricate the valve stems, and install the valves in the cylinder head as numbered. Lubricate and position the seals (if used, see above) and the valve springs. Install the spring retainers, compress the springs, and insert the keys using needlenose pliers or a tool designed for this purpose.

NOTE: *Retain the keys with wheel bearing grease during installation.*

CYLINDER HEAD RECONDITIONING

Procedure	Method

Check valve spring installed height:

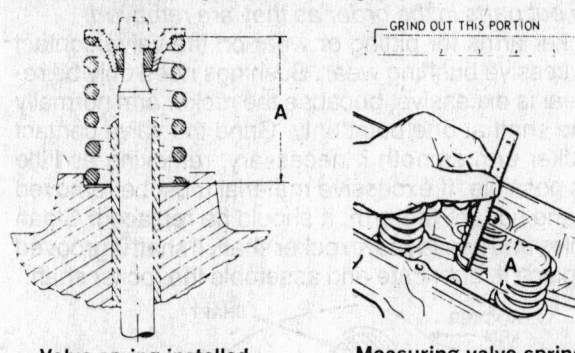

Valve spring installed
height dimension

Measuring valve spring
installed height

Measure the distance between the spring pad and the lower edge of the spring retainer, and compare to specifications. If the installed height is incorrect, add shim washers between the spring pad and the spring.
CAUTION: *Use only washers designed for this purpose.*

Install the camshaft (OHC engines only) and check end play:

See the engine service procedures earlier in this book for details concerning specific engines.

Inspect the rocker arms, balls, studs, and nuts (OHV engines only):

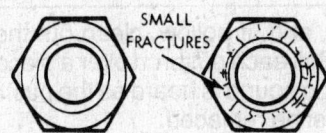

Stress cracks in the rocker nuts

Visually inspect the rocker arms, balls, studs, and nuts for cracks, galling, burning, scoring or wear. If all parts are intact, liberally lubricate the rocker arms and balls, and install them on the cylinder head. If wear is noted on a rocker arm at the point of valve contact, grind it smooth and square, removing as little material as possible. Replace the rocker arm if excessively worn. If a rocker stud shows signs of wear, it must be replaced (see below). If a rocker nut shows stress cracks, replace it. If an exhaust ball is galled or burned, substitute the intake ball from the same cylinder (if it is intact), and install a new intake ball.
NOTE: *Avoid using new rocker balls on exhaust valves.*

Replacing rocker studs (OHV engines only):

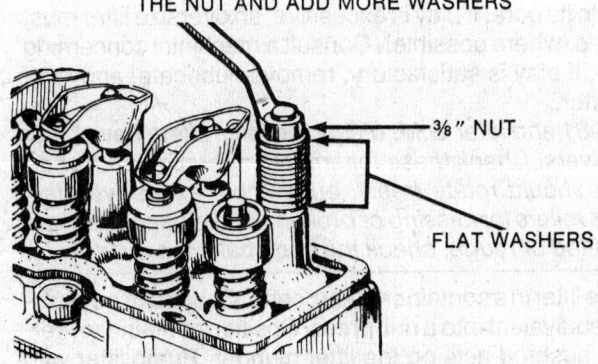

AS STUB BEGINS TO PULL UP,
IT WILL BE NECESSARY TO REMOVE
THE NUT AND ADD MORE WASHERS

⅜″ NUT

FLAT WASHERS

Extracting a pressed-in rocker stud

In order to remove a threaded stud, lock two nuts on the stud, and unscrew the stud using the lower nut. Coat the lower threads of the new stud with Loctite®, and install.
Two alternative methods are available for replacing pressed in studs. Remove the damaged stud using a stack of washers and a nut (see illustration). In the first, the boss is reamed .005–.006″ oversize, and an oversize stud pressed in. Control the stud extension over the boss using washers, in the same manner as valve guides. Before installing the stud, coat it with white lead and grease. To retain the stud more positively drill a hole through the stud and boss, and install a roll pin. In the second method, the boss is tapped, and a threaded stud installed. Retain the stud using Loctite® Stud and Bearing Mount.

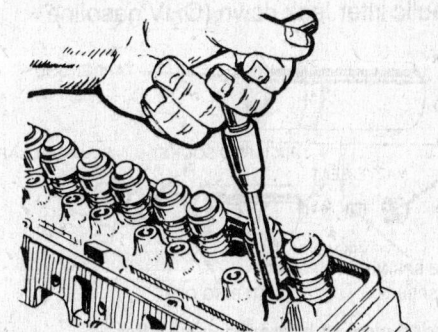

Reaming the stud bore for oversize rocker studs

CYLINDER HEAD RECONDITIONING

Procedure	Method

Inspect the rocker shaft(s) and rocker arms (OHV engines only):

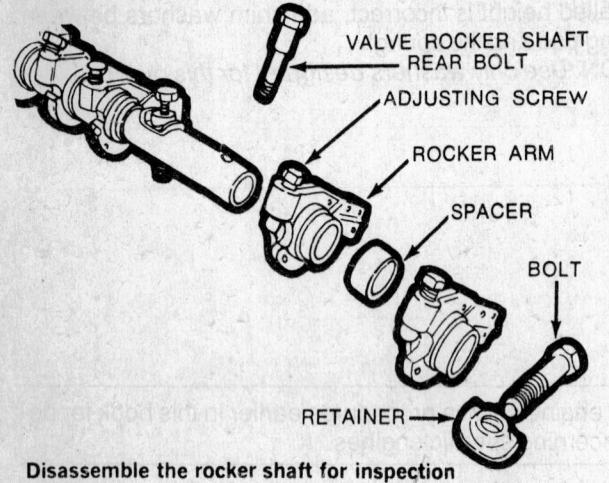

Disassemble the rocker shaft for inspection

(labels: VALVE ROCKER SHAFT REAR BOLT, ADJUSTING SCREW, ROCKER ARM, SPACER, BOLT, RETAINER)

Remove rocker arms, springs and washers from rocker shaft. NOTE: *Lay out parts in the order as they are removed.* Inspect rocker arms for pitting or wear on the valve contact point, or excessive bushing wear. Bushings need only be replaced if wear is excessive, because the rocker arm normally contacts the shaft at one point only. Grind the valve contact point of rocker arm smooth if necessary, removing as little material as possible. If excessive material must be removed to smooth and square the arm, it should be replaced. Clean out all oil holes and passages in rocker shaft. If shaft is grooved or worn, replace it. Lubricate and assemble the rocker shaft.

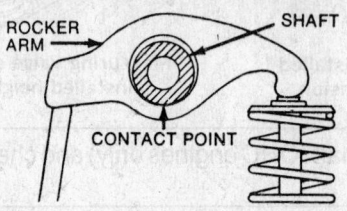

Rocker arm-to-rocker shaft contact area

(labels: ROCKER ARM, SHAFT, CONTACT POINT)

Inspect the camshaft bushings and the camshaft (OHC engines):

See next section.

Inspect the pushrods (OHV engines only):

Remove the pushrods, and, if hollow, clean out the oil passages using fine wire. Roll each pushrod over a piece of clean glass. If a distinct clicking sound is heard as the pushrod rolls, the rod is bent, and must be replaced.

*The length of all pushrods must be equal. Measure the length of the pushrods, compare to specifications, and replace as necessary.

Inspect the valve lifters (OHV engines only):

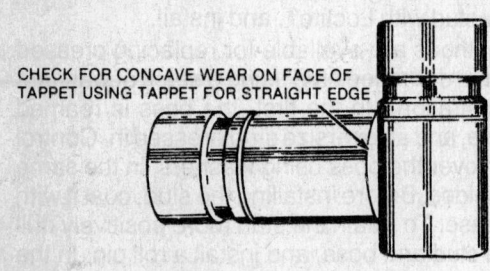

CHECK FOR CONCAVE WEAR ON FACE OF TAPPET USING TAPPET FOR STRAIGHT EDGE

Checking the lifter face

Remove lifters from their bores, and remove gum and varnish, using solvent. Clean walls of lifter bores. Check lifters for concave wear as illustrated. If face is worn concave, replace lifter, and carefully inspect the camshaft. Lightly lubricate lifter and insert it into its bore. If play is excessive, an oversize lifter must be installed (where possible). Consult a machinist concerning feasibility. If play is satisfactory, remove, lubricate, and reinstall the lifter.
NOTE: *1981 and later G.M. diesel V8 valve lifters have roller cam followers. Check these for smooth operation and wear. The roller should rotate freely, but without excessive play. Check the rollers for missing or broken needle bearings. If the roller is pitted or rough, check the camshaft lobe for wear.*

---

*Testing hydraulic lifter leak down (OHV gasoline engines only):

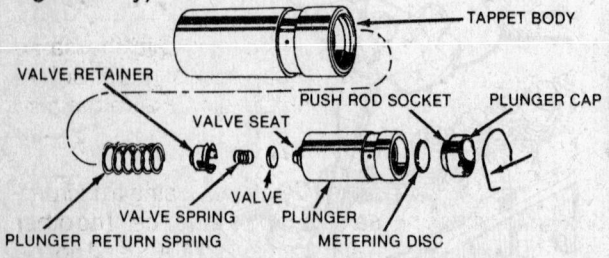

Typical exploded view of hydraulic valve lifter

(labels: TAPPET BODY, PUSH ROD SOCKET, PLUNGER CAP, VALVE RETAINER, VALVE SEAT, VALVE, PLUNGER, METERING DISC, VALVE SPRING, PLUNGER RETURN SPRING)

Submerge lifter in a container of kerosene. Chuck a used pushrod or its equivalent into a drill press. Position container of kerosene so pushrod acts on the lifter plunger. Pump lifter with the drill press, until resistance increases. Pump several more times to bleed any air out of lifter. Apply very firm, constant pressure to the lifter, and observe rate at which fluid bleeds out of lifter. If the fluid bleeds very quickly (less than 15 seconds), lifter is defective. If the time exceeds 60 seconds, lifter is sticking. In either case, recondition or replace lifter. If lifter is operating properly (leak down time 15–60 seconds), lubricate and install it.

CYLINDER HEAD RECONDITIONING

Procedure	Method
Bleed the hydraulic lifters (diesel engines only):	After the cylinder heads are installed on G.M. V8 diesels, the valve lifters must be bled down before the crankshaft is turned. Failure to bleed down the lifters will cause damage to the valve train. See diesel engine rocker arm replacement procedure in Oldsmobile 88, 98, etc. car section for procedures. NOTE: *When installing new lifters, prime by working the lifter plunger while submerged in clean kerosene or diesel fuel.*

CYLINDER BLOCK RECONDITIONING

Procedure	Method
Checking the main bearing clearance: 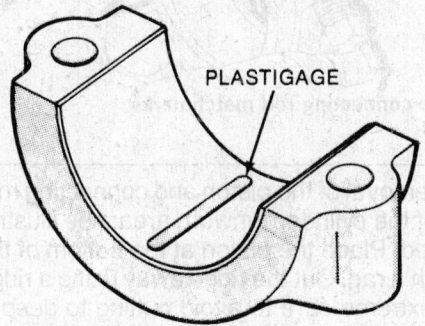 Plastigage® installed on the lower bearing shell 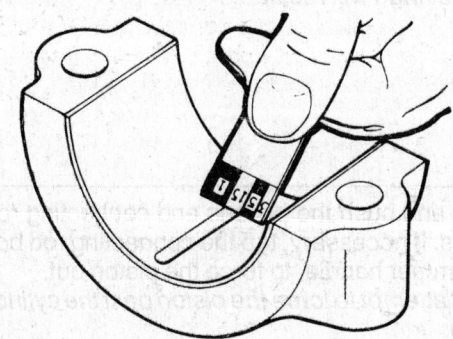 Measuring Plastigage® to determine bearing clearance	Invert engine, and remove cap from the bearing to be checked. Using a clean, dry rag, thoroughly clean all oil from crankshaft journal and bearing insert. NOTE: *Plastigage is soluble in oil; therefore, oil on the journal or bearing could result in erroneous readings.* Place a piece of Plastigage along the full length of journal, reinstall cap, and torque to specifications. Remove bearing cap, and determine bearing clearance by comparing width of Plastigage to the scale on Plastigage envelope. Journal taper is determined by comparing width of the Plastigage strip near its ends. Rotate crankshaft 90° and retest, to determine journal eccentricity. NOTE: *Do not rotate crankshaft with Plastigage installed.* If bearing insert and journal appear intact, and are within tolerances, no further main bearing service is required. If bearing or journal appear defective, cause of failure should be determined before replacement. *Remove crankshaft from block (see below). Measure the main bearing journals at each end twice (90° apart) using a micrometer, to determine diameter, journal taper and eccentricity. If journals are within tolerances, reinstall bearing caps at their specified torque. Using a telescope gauge and micrometer, measure bearing I.D. parallel to piston axis and at 30° on each side of piston axis. Subtract journal O.D. from bearing I.D. to determine oil clearance. If crankshaft journals appear defective, or do no meet tolerances, there is no need to measure bearings; for the crankshaft will require grinding and/or undersize bearings will be required. If bearing appears defective, cause for failure should be determined prior to replacement.
Checking the connecting rod bearing clearance:	Connecting rod bearing clearance is checked in the same manner as main bearing clearance, using Plastigage. Before removing the crankshaft, connecting rod side clearance also should be measured and recorded. *Checking connecting rod bearing clearance, using a micrometer, is identical to checking main bearing clearance. If no other service is required, the piston and rod assemblies need not be removed.

CYLINDER BLOCK RECONDITIONING

Procedure	Method
Removing the crankshaft:	Using a punch, mark the corresponding main bearing caps and saddles according to position (i.e., one punch on the front main cap and saddle, two on the second, three on the third, etc.). Using number stamps, identify the corresponding connecting rods and caps, according to cylinder (if no numbers are present). Remove the main and connecting rod caps, and place sleeves of plastic tubing over the connecting rod bolts, to protect the journals as the crankshaft is removed. Lift the crankshaft out of the block.

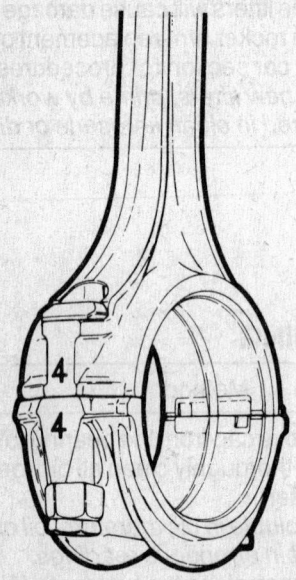

Connecting rod matched to cylinder with a number stamp

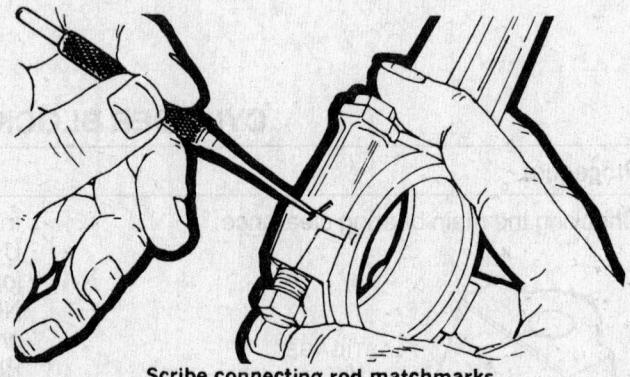

Scribe connecting rod matchmarks

Procedure	Method
Remove the ridge from the top of the cylinder:	In order to facilitate removal of the piston and connecting rod, the ridge at the top of the cylinder (unworn area; see illustration) must be removed. Place the piston at the bottom of the bore, and cover it with a rag. Cut the ridge away using a ridge reamer, exercising extreme care to avoid cutting to deeply. Remove the rag, and remove cuttings that remain on the piston. CAUTION: *If the ridge is not removed, and new rings are installed, damage to rings will result.*

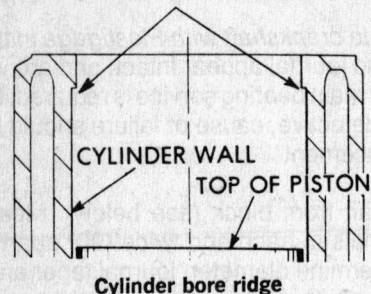

RIDGE CAUSED BY CYLINDER WEAR

CYLINDER WALL

TOP OF PISTON

Cylinder bore ridge

Procedure	Method
Removing the piston and connecting rod:	Invert the engine, and push the pistons and connecting rods out of the cylinders. If necessary, tap the connecting rod boss with a wooden hammer handle, to force the piston out. CAUTION: *Do not attempt to force the piston past the cylinder ridge (see above).*

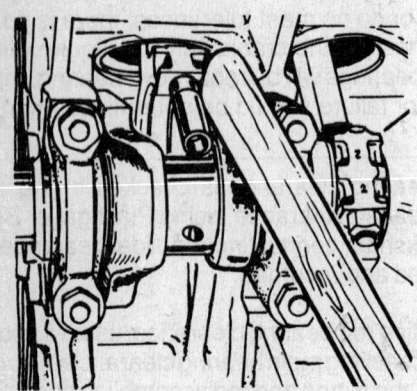

Removing the piston

CYLINDER BLOCK RECONDITIONING

Procedure	Method
Service the crankshaft:	Ensure that all oil holes and passages in the crankshaft are open and free of sludge. If necessary, have the crankshaft ground to the largest possible undersize. **Have the crankshaft Magnafluxed, to locate stress cracks. Consult a machinist concerning additional service procedures, such as surface hardening (e.g., nitriding, Tuftriding) to improve wear characteristics, cross drilling and chamfering the oil holes to improve lubrication, and balancing.
Removing freeze plugs:	Drill a small hole in the middle of the freeze plugs. Thread a large sheet metal screw into the hole and remove the plug with a slide hammer.
Remove the oil gallery plugs:	Threaded plugs should be removed using an appropriate (usually square) wrench. To remove soft, pressed in plugs, drill a hole in the plug, and thread in a sheet metal screw. Pull the plug out by the screw using pliers.
Hot-tank the block: NOTE: *Do not hot-tank aluminum parts.*	Have the block hot-tanked to remove grease, corrosion, and scale from the water jackets. NOTE: *Consult the operator to determine whether the camshaft bearings will be damaged during the hot-tank process.*

Check the block for cracks:	Visually inspect the block for cracks or chips. The most common locations are as follows:
	    Adjacent to freeze plugs.
	    Between the cylinders and water jackets.
	    Adjacent to the main bearing saddles.
	    At the extreme bottom of the cylinders.
	Check only suspected cracks using spot check dye (see introduction). If a crack is located, consult a machinist concerning possible repairs.
	
	**Magnaflux the block to locate hidden cracks. If cracks are located, consult a machinist about feasibility of repair.

Install the oil gallery plugs and freeze plugs:	Coat freeze plugs with sealer and tap into position using a piece of pipe, slightly smaller than the plug, as a driver. To ensure retention, stake the edges of the plugs. Coat threaded oil gallery plugs with sealer and install. Drive replacement soft plugs into block using a large drift as a driver. *Rather than reinstalling lead plugs, drill and tap the holes, and install threaded plugs.
*Check the deck height:	*The deck height is the distance from the crankshaft centerline to the block deck. To measure, invert the engine, and install the crankshaft, retaining it with the center main cap. Measure the distance from the crankshaft journal to the block deck, parallel to the cylinder centerline. Measure the diameter of the end (front and rear) main journals, parallel to the centerline of the cylinders, divide the diameter in half, and subtract it from the previous measurement. The results of the front and rear measurements should be identical. If the difference exceeds .005″, the deck height should be corrected. NOTE: *Block deck height and warpage should be corrected at the same time.*

CYLINDER BLOCK RECONDITIONING

Procedure	Method

Check the block deck for warpage:

Using a straightedge and feeler gauges, check the block deck for warpage in the same manner that the cylinder head is checked (see Cylinder Head Reconditioning). If warpage exceeds specifications, have the deck resurfaced.
NOTE: *In certain cases a specification for total material removal (Cylinder head and block deck) is provided. This specification must not be exceeded.*

Check the bore diameter and surface:

Measuring the cylinder bore with a dial gauge

Visually inspect the cylinder bores for roughness, scoring, or scuffing. If evident, the cylinder bore must be bored or honed oversize to eliminate imperfections, and the smallest possible oversize piston used. The new pistons should be given to the machinist with the block, so that the cylinders can be bored or honed exactly to the piston size (plus clearance). If no flaws are evident, measure the bore diameter using a telescope gauge and micrometer, or dial guage, parallel and perpendicular to the engine centerline, at the top (below the ridge) and bottom of the bore. Subtract the bottom measurements from the top to determine taper, and the parallel to the centerline measurements from the perpendicular measurements to determine eccentricity. If the measurements are not within specifications, the cylinder must be bored or honed, and an oversize piston installed. If the measurements are within specifications the cylinder may be used as is, with only finish honing (see below).
NOTE: *Prior to boring, check the block deck warpage, height and bearing alignment.*
CAUTION: *The 4 cyl. 140 G.M. engine cylinder walls are impregnated with silicone. Boring or honing can be done only by a shop with the proper equipment.*

Measuring cylinder bore with a telescope gauge

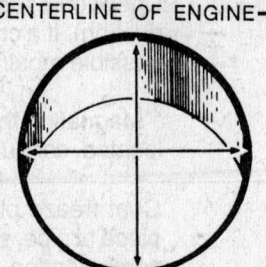

A—AT RIGHT ANGLE TO CENTERLINE OF ENGINE
B—PARALLEL TO CENTERLINE OF ENGINE
Cylinder bore measuring points

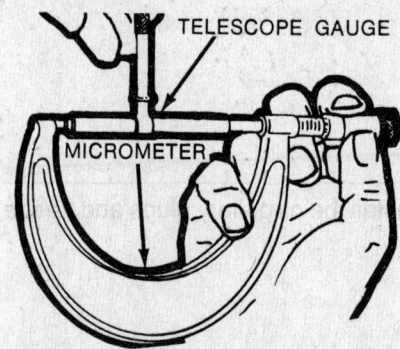

Determining cylinder bore by measuring telescope gauge with a micrometer

Check the cylinder block bearing alignment:

Checking main bearing saddle alignment

Remove the upper bearing inserts. Place a straightedge in the bearing saddles along the centerline of the crankshaft. If clearance exists between the straightedge and the center saddle, the block must be alignbored.

CYLINDER BLOCK RECONDITIONING

Procedure	Method

Clean and inspect the pistons and connecting rods:

Using a ring expander, remove the rings from the piston. Remove the retaining rings (if so equipped) and remove piston pin.

NOTE: *If the piston pin must be pressed out, determine the proper method and use the proper tools; otherwise the piston will distort.*

Clean the ring grooves using an appropriate tool, exercising care to avoid cutting too deeply. Thoroughly clean all carbon and varnish from the piston with solvent.

CAUTION: *Do not use a wire brush or caustic solvent on pistons.*

Inspect the pistons for scuffing, scoring, cracks, pitting, or excessive ring groove wear. If wear is evident, the piston must be replaced. Check the connecting rod length by measuring the rod from the inside of the large end to the inside of the small end using calipers (see illustration). All connecting rods should be equal length. Replace any rod that differs from the others in the engine.

*Have the connecting rod alignment checked in an alignment fixture by a machinist. Replace any twisted or bent rods.

*Magnaflux the connecting rods to locate stress cracks. If cracks are found, replace the connecting rod.

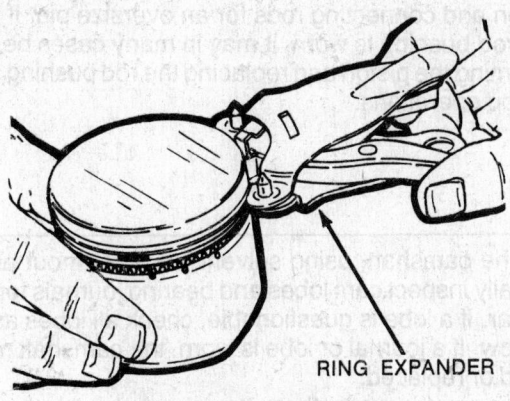

RING EXPANDER

Removing the piston rings

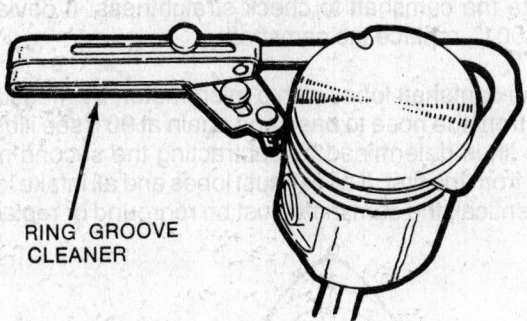

RING GROOVE CLEANER

Cleaning the piston ring grooves

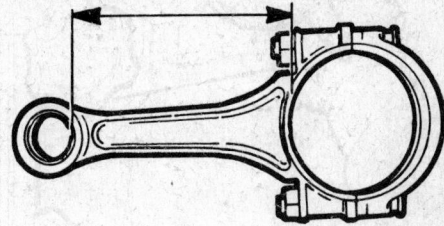

Check the connecting rod length (arrow)

Fit the pistons to the cylinders:

Using a telescope gauge and micrometer, or a dial gauge, measure the cylinder bore diameter perpendicular to the piston pin, 2½° below the deck. Measure the piston perpendicular to its pin on the skirt. The difference between the two measurements is the piston clearance. If the clearance is within specifications or slightly below (after boring or honing), finish honing is all that is required. If the clearance is excessive, try to obtain a slightly larger piston to bring clearance within specifications. Where this is not possible, obtain the first oversize piston, and hone (or if necessary, bore) the cylinder to size.

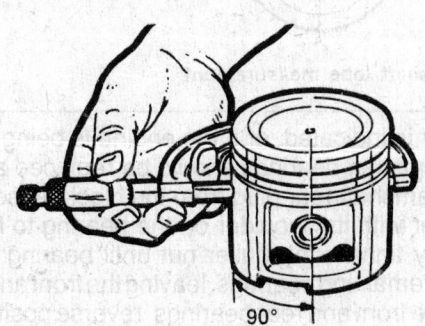

90°

Measuring the piston prior to fitting

Assemble the pistons and connecting rods:

Inspect piston pin, connecting rod small end bushing, and piston bore for galling, scoring, or excessive wear. If evident, replace defective part(s). Measure the I.D. of the piston boss and connecting rod small end, and the O.D. of the piston pin. If within specifications, assemble piston pin and rod.

CAUTION: *If piston pin must be pressed in, determine the proper method and use the proper tools; otherwise the piston will distort.*

CYLINDER BLOCK RECONDITIONING

Procedure	Method

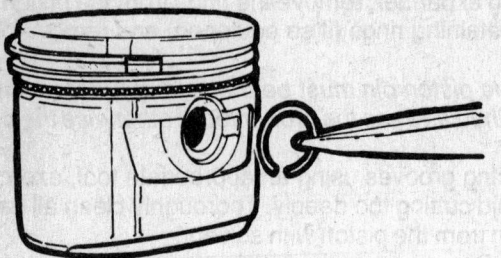

Installing piston pin lock rings

Install the lock rings; ensure that they seat properly. If the parts are not within specifications, determine the service method for the type of engine. In some cases, piston and pin are serviced as an assembly when either is defective. Others specify reaming the piston and connecting rods for an oversize pin. If the connecting rod bushing is worn, it may in many cases be replaced. Reaming the piston and replacing the rod bushing are machine shop operations.

Clean and inspect the camshaft:

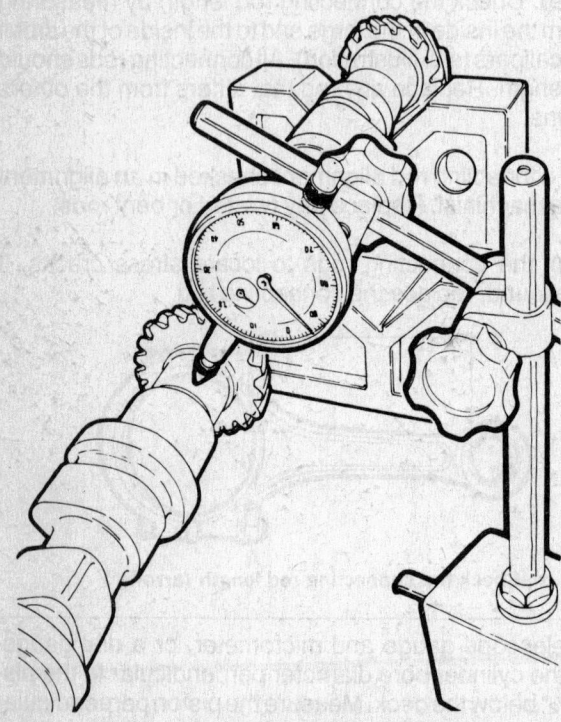

Checking the camshaft for straightness

Degrease the camshaft, using solvent, and clean out all oil holes. Visually inspect cam lobes and bearing journals for excessive wear. If a lobe is questionable, check all lobes as indicated below. If a journal or lobe is worn, the camshaft must be reground or replaced.

NOTE: *If a journal is worn, there is a good chance that the bushings are worn.*

If lobes and journals appear intact, place the front and rear journals in V-blocks, and rest a dial indicator on the center journal. Rotate the camshaft to check straightness. If deviation exceeds .001°, replace the camshaft.

*Check the camshaft lobes with a micrometer, by measuring the lobes from the nose to base and again at 90° (see illustration). The lift is determined by subtracting the second measurement from the first. If all exhaust lobes and all intake lobes are not identical, the camshaft must be reground or replaced.

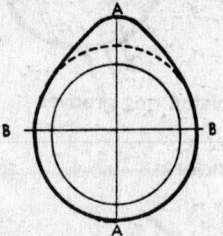

Camshaft lobe measurement

Replace the camshaft bearings (OHV engines only):

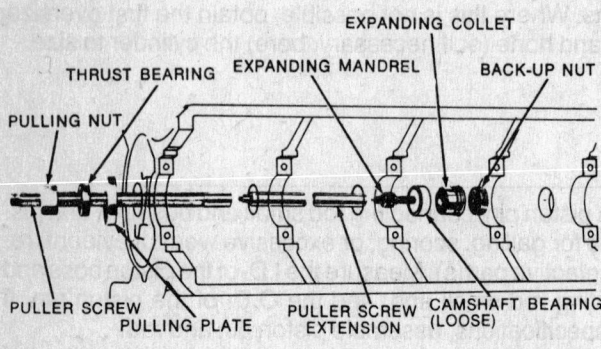

Camshaft removal and installation tool (typical)

If excessive wear is indicated, or if the engine is being completely rebuilt, camshaft bearings should be replaced as follows: Drive the camshaft rear plug from the block. Assemble the removal puller with its shoulder on the bearing to be removed. Gradually tighten the puller nut until bearing is removed. Remove remaining bearings, leaving the front and rear for last. To remove front and rear bearings, reverse position of the tool, so as to pull the bearings in toward the center of the block. Leave the tool in this position, pilot the new front and rear bearings on the installer, and pull them into position: Return the tool to its original position and pull remaining bearings into postion.

NOTE: *Ensure that oil holes align when installing bearings.*

Replace camshaft rear plug, and stake it into position to aid retention.

CYLINDER BLOCK RECONDITIONING

Procedure	Method

Finish hone the cylinders:

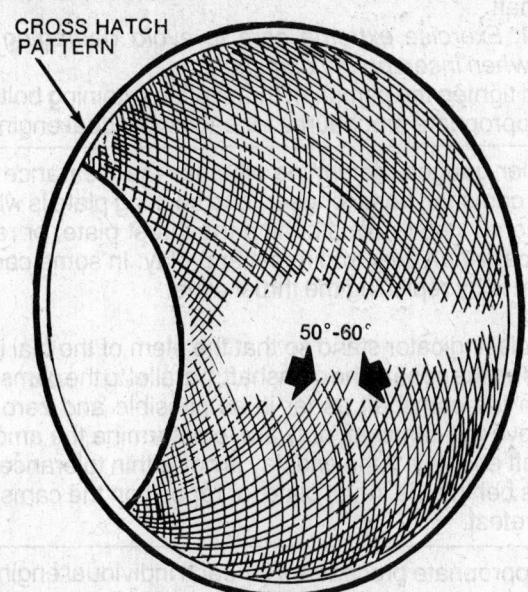

CROSS HATCH PATTERN

50°-60°

Chuck a flexible drive hone into a power drill, and insert it into the cylinder. Start the hone, and move it up and down the cylinder at a rate which will produce approximately a 60° cross-hatch pattern (see illustration).
NOTE: *Do not extend the hone below the cylinder bore.*
After developing the pattern, remove the hone and recheck piston fit. Wash the cylinders with a detergent and water solution to remove abrasive dust, dry, and wipe several times with a rag soaked in engine oil.

Check piston ring end-gap:

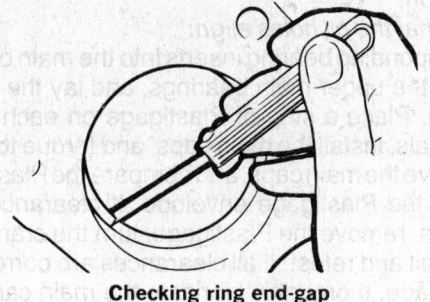

Checking ring end-gap

Compress the piston rings to be used in a cylinder, one at a time, into that cylinder, and press them approximately 1" below the deck with an inverted piston. Using feeler gauges, measure the ring end-gap, and compare to specifications. Pull the ring out of the cylinder and file the ends with a fine file to obtain proper clearance.
CAUTION: *If inadequate ring end-gap is utilized, ring breakage will result.*

Install the piston rings:

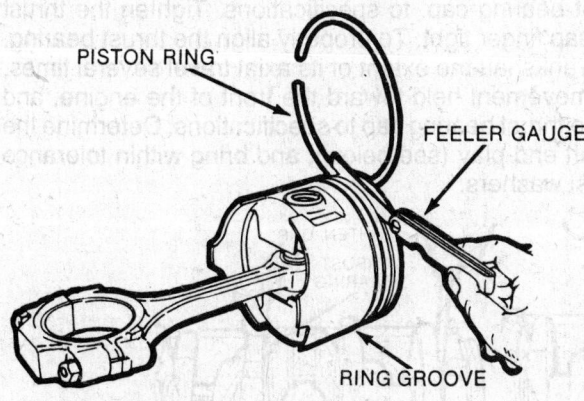

PISTON RING

FEELER GAUGE

RING GROOVE

Checking ring side clearance

Inspect the ring grooves in the piston for excessive wear or taper. If necessary, recut the groove(s) for use with an over-width ring or a standard ring and spacer. If the groove is worn uniformly, overwidth rings, or standard rings and spacers may be installed without recutting. Roll the outside of the ring around the groove to check for burrs or deposits. If any are found, remove with a fine file. Hold the ring in the groove, and measure side clearance. If necessary, correct as indicated above.
NOTE: *Always install any additional spacers above the piston ring.*
The ring groove must be deep enough to allow the ring to seat below the lands (see illustration). In many cases, a "go-no-go" depth gauge will be provided with the piston rings. Shallow grooves may be corrected by recutting, while deep grooves require some type of filler or expander behind the piston. Consult the piston ring supplier concerning the suggested method. Install the rings on the piston, lowest ring first, using a ring expander.
NOTE: *Position the ring markings as specified by the manufacturer (see car section).*

CYLINDER BLOCK RECONDITIONING

Procedure	Method
Install the camshaft (OHV engines only):	Liberally lubricate the camshaft lobes and journals, and install the camshaft. CAUTION: *Exercise extreme care to avoid damaging the bearings when inserting the camshaft.* Install and tighten the camshaft thrust plate retaining bolts. See the appropriate procedures for each individual engine.

Check camshaft end-play (OHV engines only):

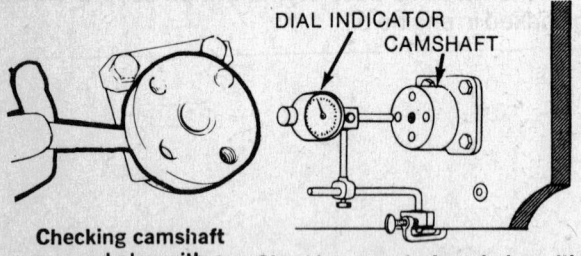

DIAL INDICATOR
CAMSHAFT

Checking camshaft end-play with a feeler gauge

Checking camshaft end-play with a dial indicator

Using feeler gauges, determine whether the clearance between the camshaft boss (or gear) and backing plate is within specifications. Install shims behind the thrust plate, or reposition the camshaft gear and retest end-play. In some cases, adjustment is by replacing the thrust plate.

*Mount a dial indicator stand so that the stem of the dial indicator rests on the nose of the camshaft, parallel to the camshaft axis. Push the camshaft as far in as possible and zero the gauge. Move the camshaft outward to determine the amount of camshaft endplay. If the endplay is not within tolerance, install shims behind the thrust plate, or reposition the camshaft gear and retest.

Install the rear main seal (where applicable):	See the appropriate procedures for each individual engine.

Install the crankshaft:

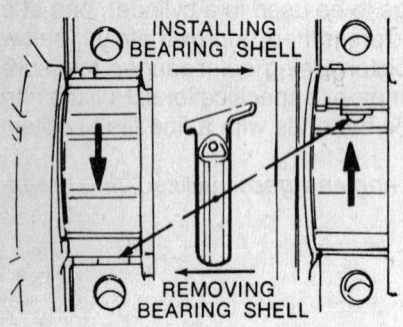

INSTALLING
BEARING SHELL

REMOVING
BEARING SHELL

Removal and installation of upper bearing insert using a roll-out pin

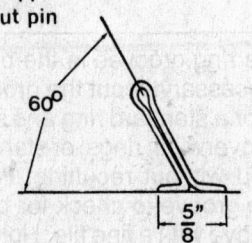

60°

5"/8

Home-made bearing roll-out pin

Thoroughly clean the main bearing saddles and caps. Place the upper halves of the bearing inserts on the saddles and press into position.
NOTE: *Ensure that the oil holes align.*
Press the corresponding bearing inserts into the main bearing caps. Lubricate the upper main bearings, and lay the crankshaft in position. Place a strip of Plastigage on each of the crankshaft journals, install the main caps, and torque to specifications. Remove the main caps, and compare the Plastigage to the scale on the Plastigage envelope. If clearances are within tolerances, remove the Plastigage, turn the crankshaft 90°, wipe off all oil and retest. If all clearances are correct, remove all Plastigage, thoroughly lubricate the main caps and bearing journals, and install the main caps. If clearances are not within tolerance, the upper bearing inserts may be removed, without removing the crankshaft, using a bearing roll out pin (see illustration). Roll in a bearing that will provide proper clearance, and retest. Torque all main caps, excluding the thrust bearing cap, to specifications. Tighten the thrust bearing cap finger tight. To properly align the thrust bearing, pry the crankshaft the extent of its axial travel several times, the last movement held toward the front of the engine, and torque the thrust bearing cap to specifications. Determine the crankshaft end-play (see below), and bring within tolerance with thrust washers.

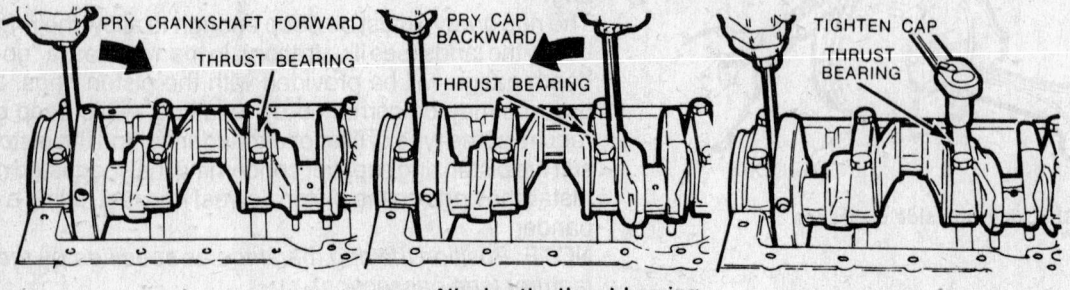

PRY CRANKSHAFT FORWARD
THRUST BEARING

PRY CAP BACKWARD
THRUST BEARING

TIGHTEN CAP
THRUST BEARING

Aligning the thrust bearing

CYLINDER BLOCK RECONDITIONING

Procedure	Method

Measure crankshaft end-play:

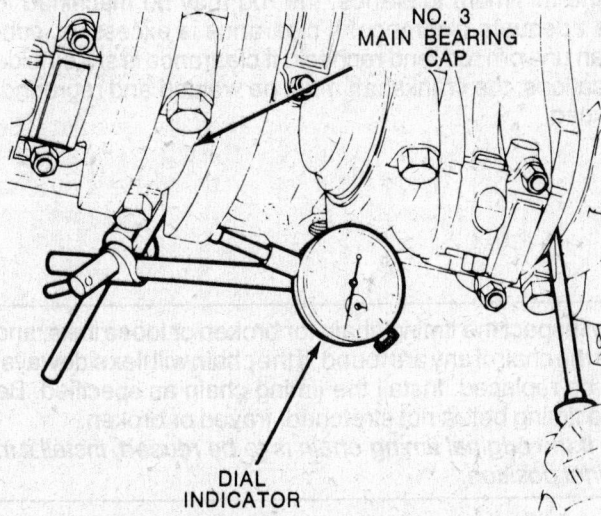

Checking crankshaft end-play with a dial indicator

Mount a dial indicator stand on the front of the block, with the dial indicator stem resting on the nose of the crankshaft, parallel to the crankshaft axis. Pry the crankshaft the extent of its travel rearward, and zero the indicator. Pry the crankshaft forward and record crankshaft end-play.

NOTE: *Crankshaft end-play also may be measured at the thrust bearing, using feeler gauges* (see illustration).

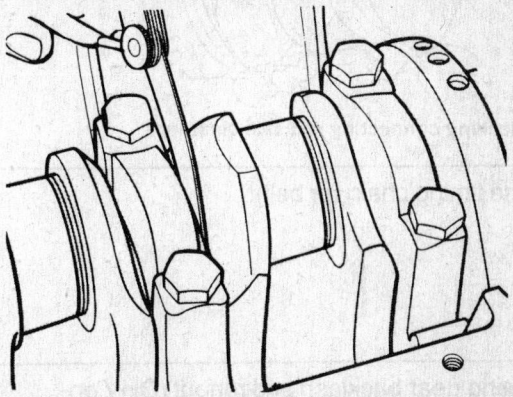

Checking crankshaft end-play with a feeler gauge

Install the pistons:

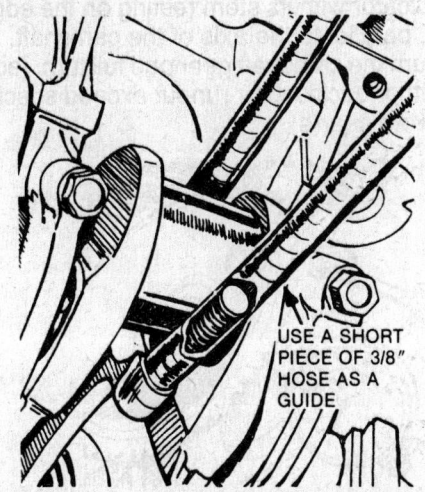

Tubing used to protect crankshaft journals and cylinder walls during piston installation

Press the upper connecting rod bearing halves into the connecting rods, and the lower halves into the connecting rod caps. Position the piston ring gaps according to specifications (see car section), and lubricate the pistons. Install a ring compressor on a piston, and press two long (8″) pieces of plastic tubing over the rod bolts. Using the tubes as a guide, press the pistons into the bores and onto the crankshaft with a wooden hammer handle. After seating the rod on the crankshaft journal, remove the tubes and install the cap finger tight. Install the remaining pistons in the same manner. Invert the engine and check the bearing clearance at two points (90° apart) on each journal with Plastigage.

NOTE: *Do not turn the crankshaft with Plastigage installed.*

If clearance is within tolerances, remove *all* Plastigage, thoroughly lubricate the journals, and torque the rod caps to specifications. If clearance is not within specifications, install different thickness bearing inserts and recheck.

CAUTION: *Never shim or file the connecting rods or caps.*

Always install plastic tube sleeves over the rod bolts when the caps are not installed, to protect the crankshaft journals.

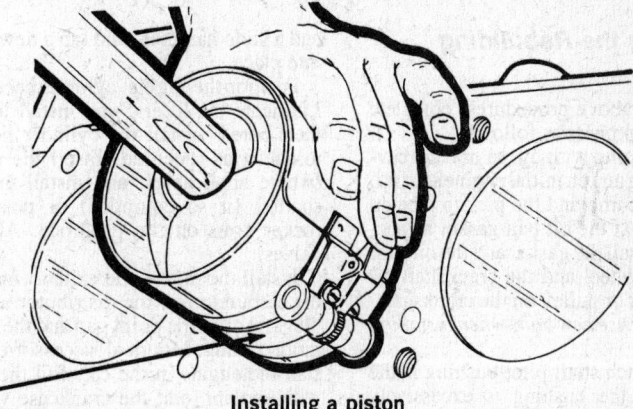

Installing a piston

CYLINDER BLOCK RECONDITIONING

Procedure	Method
Check connecting rod side clearance: 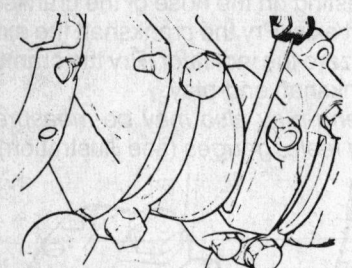 Checking connecting rod side clearance	Determine the clearance between the sides of the connecting rods and the crankshaft, using feeler gauges. If clearance is below the minimum tolerance, the rod may be machined to provide adequate clearance. If clearance is excessive, substitute an unworn rod, and recheck. If clearance is still outside specifications, the crankshaft must be welded and reground, or replaced.
Inspect the timing chain (or belt):	Visually inspect the timing chain for broken or loose links, and replace the chain if any are found. If the chain will flex sideways, it must be replaced. Install the timing chain as specified. Be sure the timing belt is not stretched, frayed or broken. NOTE: *If the original timing chain is to be reused, install it in its original position.*
Check timing gear backlash and runout (OHV engines): 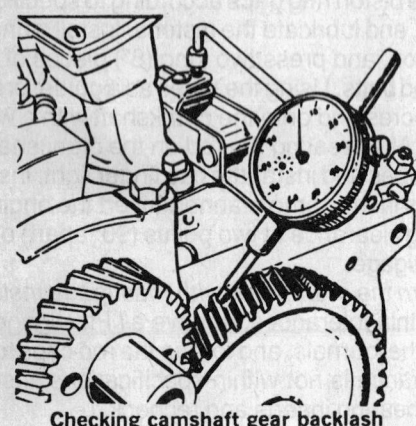 Checking camshaft gear backlash	Mount a dial indicator with its stem resting on a tooth of the camshaft gear (as illustrated). Rotate the gear until all slack is removed, and zero the indicator. Rotate the gear in the opposite direction until slack is removed, and record gear backlash. Mount the indicator with its stem resting on the edge of the camshaft gear, parallel to the axis of the camshaft. Zero the indicator, and turn the camshaft gear one full turn, recording the runout. If either backlash or runout exceed specifications, replace the worn gear(s). 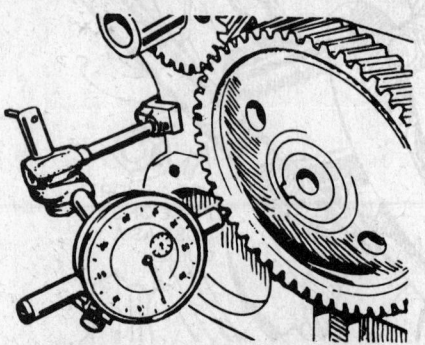 Checking camshaft gear runout

Completing the Rebuilding Process

Following the above procedures, complete the rebuilding process as follows:

Fill the oil pump with oil, to prevent cavitating (sucking air) on initial engine start up. Install the oil pump and the pickup tube on the engine. Coat the oil pan gasket as necessary, and install the gasket and the oil pan. Mount the flywheel and the crankshaft vibration damper or pulley on the crankshaft. NOTE: *Always use new bolts when installing the flywheel.*

Inspect the clutch shaft pilot bushing in the crankshaft. If the bushing is excessively worn, remove it with an expanding puller and a slide hammer, and tap a new bushing into place.

Position the engine, cylinder head side up. Lubricate the lifters, and install them into their bores. Install the cylinder head, and torque it as specified. Insert the pushrods (where applicable), and install the rocker shaft(s) (if so equipped) or position the rocker arms on the pushrods. Adjust the valves.

Install the intake and exhaust manifolds, the carburetor(s), the distributor and spark plugs. Adjust the point gap and the static ignition timing. Mount all accessories and install the engine in the car. Fill the radiator with coolant, and the crankcase with high quality engine oil.

Break-In Procedure

Start the engine, and allow it to run at low speed for a few minutes, while checking for leaks. Stop the engine, check the oil level, and fill as necessary. Restart the engine, and fill the cooling system to capacity. Check the point dwell angle and adjust the ignition timing and the valves. Run the engine at low to medium speed (800–2500 rpm) for approximately ½ hour, and retorque the cylinder head bolts. Road test the car, and check again for leaks.

Follow the manufacturer's recommended engine break-in procedure and maintenance schedule for new engines.

Steering Gear

DOMESTIC TRUCKS

Manual Steering Gear
STEERING GEAR ALIGNMENT

Before any steering gear adjustments are made, it is recommended that the front end of the truck be raised and a thorough inspection be made for stiffness or lost motion in the steering gear, steering linkage and front suspension. Worn or damaged parts should be replaced, since a satisfactory adjustment of the steering gear cannot be obtained if bent or badly worn parts exist.

It is also very important that the steering gear be properly aligned in the truck. Misalignment of the gear places a stress on the steering worm shaft, therefore a proper adjustment is impossible. To align the steering gear, loosen the steering gear-to-frame mounting bolts to permit the gear to align itself. Check the steering gear to frame mounting seat. If there is a gap at any of the mounting bolts, proper alignment may be obtained by placing shims where excessive gap appears. Tighten the steering gear-to-frame bolts. Alignment of the gear in the truck is very important and should be done carefully so that a satisfactory, trouble-free gear adjustment may be obtained.

Ford Steering Gear – Recirculating Ball Type

1979–83
STEERING WORM AND SECTOR GEAR ADJUSTMENTS

The ball nut assembly and the sector gear must be adjusted properly to maintain a minimum amount of steering shaft end play and a minimum amount of backlash between the sector gear and the ball nut. There are only two adjustments that may be done on this steering gear and they should be done as given below:

1. Remove the steering gear from the vehicle.

2. Loosen the locknut on the sector shaft adjustment screw and turn the adjusting screw counterclockwise about three turns.

3. Measure the worm bearing preload by attaching an inch lbs. torque wrench to the input shaft. Note the reading required to rotate input shaft about 1½ turns either side of center. If the torque reading is not about 4–5 inch lbs., adjust the gear as given in the next step.

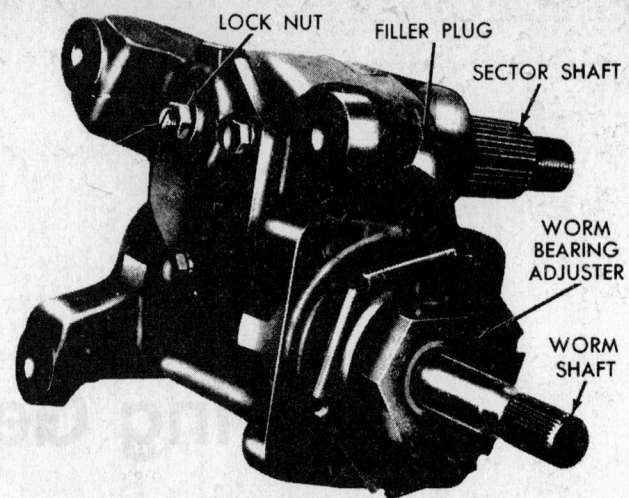

Steering gear adjustments

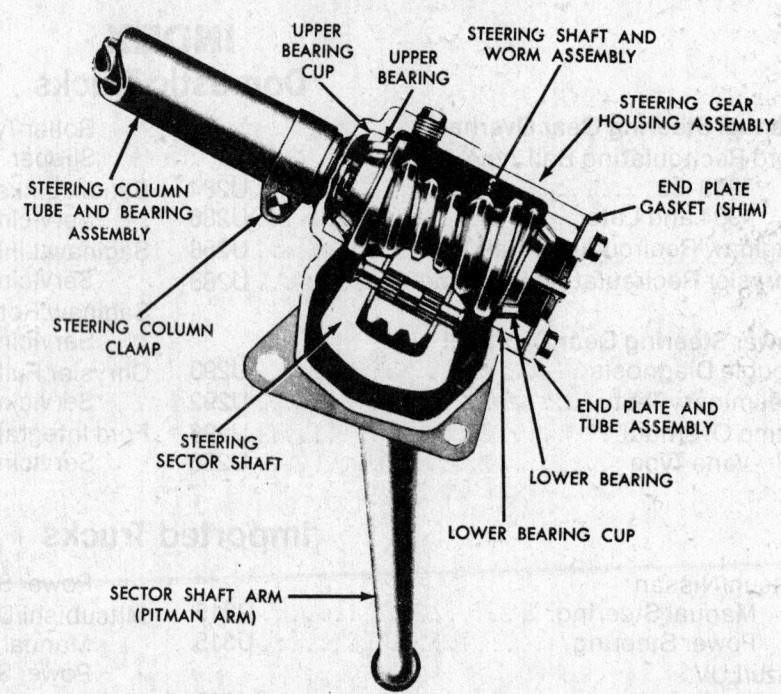

Worm and roller type steering gear

4. Loosen the steering shaft bearing adjuster lock nut and tighten or back off the bearing adjusting screw until the preload is within the specified limits.

5. Tighten the steering shaft bearing adjuster lock nut, and recheck the preload torque.

6. Turn the input shaft slowly to either stop. Turn gently against the stop to avoid possible damage to the ball return guides. Then rotate the shaft three turns to center the ball nut.

7. Turn the sector adjusting screw clockwise until the proper torque (9–10 inch lbs.) is obtained that is necessary to rotate the worm gear past its center (high spot).

8. With the input shaft centered, hold the sector shaft and check the lash between the ball nuts, balls, and worm shaft by applying 15 lbs. torque to the steering input shaft in both right and left turn directions. The total travel of the wrench should not exceed 1¼ in.

9. Tighten the sector adjusting screw locknut, and recheck the backlash. Install the steering gear.

Disassembly

1. Rotate the steering shaft three turns from either stop.

2. Remove the sector shaft adjusting screw locknut and loosen the screw one

STEERING TROUBLE DIAGNOSIS

Condition	Possible Cause	Correction
Excessive Play or Looseness in the Steering	1. Steering gear shaft adjusted too loose or shaft and/or bushing badly worn.	1. Replace worn parts and adjust according to instructions.
	2. Excessive steering gear worm end play due to bearing adjustment.	2. Adjust according to instructions.
	3. Steering linkage loose or worn.	3. Replace worn parts.
	4. Front wheel bearings improperly adjusted.	4. Adjust wheel bearings.
	5. Steering arm loose on steering gear shaft.	5. Inspect for damage to the gear shaft and steering arm, replace parts as necessary.
	6. Steering gear housing attaching bolts loose.	6. Tighten the attaching bolts to specifications.
	7. Steering arms loose at steering knuckles.	7. Tighten according to specifications.
	8. Working pins or bushings.	8. Replace king pins and bushings.
	9. Loose spring shackles.	9. Adjust or replace parts as necessary.
Hard Steering	1. Low or uneven tire pressure.	1. Inflate the tires to recommended pressures.
	2. Insufficient lubricant in the steering gear housing or in steering linkage.	2. Lubricate as necessary.
	3. Steering gear shaft adjusted too tight.	3. Adjust according to instructions.
	4. Improper caster or toe-in.	4. Align the wheels.
	5. Steering column misaligned.	5. See "Steering Gear Alignment."
Wheel Tramp (Excessive Vertical Motion of Wheels)	1. Incorrect tire pressure.	1. Inflate the tires to recommended pressures.
	2. Improper balance of wheels, tires and brake drums.	2. Balance as necessary.
	3. Loose tie rod ends or steering connections.	3. Inspect and repair as necessary.
	4. Worn or inoperative shock absorbers.	4. Replace the shock absorbers.
	5. Excessive run-out of brake drums, wheels or tires.	5. Repair or replace as required.
Shimmy	1. Badly worn and/or unevenly worn tires.	1. Rotate tires or replace if necessary.
	2. Wheels and tires out of balance.	2. Balance wheel and tire assemblies.
	3. Worn or loose steering linkage parts.	3. Replace parts are required.
	4. Worn king pins and bushings.	4. Replace king pins and bushings.
	5. Loose steering gear adjustments.	5. Adjust steering gear as necessary.
	6. Loose wheel bearings.	6. Adjust wheel bearings.
	7. Improper caster setting.	7. Adjust caster to specifications.
	8. Weak or broken springs.	8. Replace as required.
	9. Incorrect tire pressure or tire sizes not uniform.	9. Check tire sizes and inflate tires to recommended pressure
	10. Faulty shock absorbers.	10. Replace as necessary.
Pull to One Side (Tendency of the Vehicle to Veer in one Direction Only)	1. Incorrect tire pressure or tires not uniform.	1. Check tire sizes and inflate the tires to recommended pressures.
	2. Wheel bearings improperly adjusted.	2. Adjust wheel bearings.
	3. Dragging brakes.	3. Inspect for weak, or broken brake shoe spring, binding pedal.
	4. Improper caster, camber or toe-in.	4. Adjust to specifications.
	5. Grease, dirt, oil or brake fluid on brake linings.	5. Inspect, replace and adjust as necessary.
	6. Broken or sagging rear springs.	6. Replace the rear springs.
	7. Bent front axle, linkage or steering knuckle.	7. Replace the parts as necessary.
	8. Worn or tight king pin bushings.	8. Lubricate or replace as necessary.
Wander or Weave	1. Improper caster, camber or toe-in.	1. Adjust to specifications.
	2. Worn king pin and bushings.	2. Replace parts as required.
	3. Worn or improperly adjusted front wheel bearings.	3. Adjust or replace parts as necessary.
	4. Loose spring shackles.	4. Adjust or replace parts as necessary.
	5. Incorrect tire pressure or tire sizes not uniform.	5. Check tire sizes and inflate tires to recommended pressure.
	6. Loose steering gear mounting bolts.	6. Tight to specifications.
	7. Tight king pin bushings.	7. Lubricate or ream to proper fit.
	8 Tight king pin thrust bearings.	8. Adjust to .001 to .005 inch clearance.

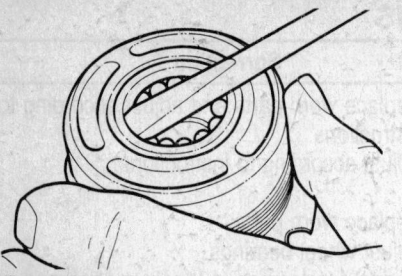

Removing bearing retainer from the adjuster plug

turn. Remove the steering shaft bearing adjuster, and the housing cover bolts and remove the sector shaft. Remove the shaft by turning the screw clockwise. Keep the shim with the screw.

3. Remove the sector shaft from the housing.

4. Carefully pull the steering shaft and ball nut from the housing, and remove the steering shaft lower bearing. Do not run the ball nut to either end of the worm gear to prevent damaging the ball return guides. Disassemble the ball nut only if there are signs of binding or tightness.

5. To disassemble the ball nut, remove the ball return guide clamp and the ball return guides from the ball nut. Keep ball nut clamp side up until ready to remove the ball bearings.

6. Turn the ball nut over and rotate the worm shaft from side to side until all 50 balls have dropped out into a clean pan. With all balls removed, the nut will slide off the wormshaft.

7. Remove the upper bearing cup from the bearing adjuster and the lower cup from the housing. It may be necessary to tap the housing or the adjuster on a wooden block to jar the bearing cups loose.

Inspection

1. Carefully clean and inspect all parts. If the inspection shows bearing damage, the sector shaft bearing and the oil seal should be pressed out.

2. If the sector shaft bearing and oil seals were removed, press new bearings and oil seals into the housing. Do not clean, wash, or soak seals in cleaning solvent.

3. Apply the recommended steering gear lubricant to the housing and seals, filling the pocket between sector shaft bearings.

Assembly

1. Install the bearing cup in the lower end of the housing and a bearing cup in the adjuster nut. Install a new seal in the bearing adjuster if the old seal was removed.

2. Apply gear lube to the outside of the worm shaft and the inside of the ball nut. Lay the steering shaft down, and position the ball nut on the shaft with the guide holes upward and the shallow end of the teeth to the left of the steering wheel position. Align the grooves in worm and ball

nut by sighting through the guide holes.

3. Insert the ball guides into the holes in the ball nut, lightly tapping them, if necessary, to seat them.

4. Insert 25 balls into the hole in the top of each ball guide. If necessary, rotate the shaft slightly to distribute the balls evenly in the circuit.

5. Install the ball guide clamp, tightening the screws to the proper torque. Check that the worm shaft rotates freely.

6. Coat the threads of the steering shaft bearing adjuster, the housing cover bolts, and the sector adjusting screw with a suitable oil-resistant sealing compound. Do not apply sealer to female threads and do not get sealer on the steering shaft bearings.

7. Coat the worm bearings, sector shaft bearings, and gear teeth with steering gear lubricant.

8. Clamp the housing in a vise, with the sector shaft axis horizontal, and place the steering shaft lower bearing in its cup. Place the steering shaft and ball nut assemblies in the housing.

9. Position the steering shaft upper bearing on top of the worm gear and install the steering shaft bearing adjuster, adjuster nut, and the bearing cup. Leave the nut loose.

10. Adjust the worm bearing preload according to the instructions given earlier.

11. Position the sector adjusting screw and adjuster shim, and check for a clearance of not more than 0.002 in. between the screw head and the end of the sector shaft. If the clearance exceeds 0.002 in., add enough shims to reduce the clearance to under 0.002 in. clearance.

12. Start the sector shaft adjusting screw into the housing cover. Install a new gasket on the cover.

13. Rotate the steering shaft until the ball nut teeth mesh with the sector gear teeth, tilting the housing so the ball will tip toward the housing cover opening.

14. Lubricate the sector shaft journal and install the sector shaft and cover. With the cover moved to one side, fill the gear with lubricant (about 0.97 lb.). Push the cover and the sector shaft into place, and install the two top housing bolts. Do not tighten the bolts until checking to see that there is some lash between the ball nut and the sector gear teeth. Hold or push the cover away from the ball nut and tighten the bolts to the proper torque (30–40 ft. lbs.).

15. Loosely install the sector shaft adjusting screw lock nut and adjust the sector shaft mesh load as given earlier. Tighten the adjusting screw lock nut.

1984 and Later (Koyo)

PRELOAD AND MESHLOAD ADJUSTMENT (GEAR REMOVED FROM VEHICLE)

1. Tighten the sector cover bolts to 40 ft. lbs.

2. Loosen the preload adjuster locknut and tighten the worm bearing adjuster nut

until all end play has been removed. Lubricate the wormshaft seal with a drop of Type F automatic transmission fluid.

3. Using a $^{11}/_{16}$ in., 12 point socket and an inch lbs. torque wrench, turn the wormshaft all the way to the right. Measure the left turn torque required to rotate the wormshaft at a constant speed for approximately $1\frac{1}{2}$ turns. This torque reading is preload.

4. Tighten or loosen the adjuster nut as required until the correct preload of 7–9 inch lbs. is obtained. Tighten the adjuster locknut to 187 ft. lbs.

5. Rotate the wormshaft from stop to stop, counting the total number of turns, then turn back halfway, placing the gear at the center. (Approximately 7 turns stop to stop.)

6. Again, using the tools in Step 3, observe the highest reading while the wormshaft is turned approximately 90° either way across center. If the highest reading (meshload) is within 12–14 in lbs. and at least 4 inch lbs. over the preload, turn the sector shaft adjusting screw as required.

7. Hold the sector shaft adjusting screw and tighten the locknut to 25 ft. lbs.

Disassembly

1. Rotate the steering shaft from stop to stop, counting the total number of turns. Then turn exactly half-way back, placing the gear on center.

2. Remove the sector adjusting cover bolts, then remove the sector shaft with the cover. Remove the cover from the shaft by turning the screw clockwise. Keep the shim with the screw.

3. Using a special locknut wrench, loosen the worm bearing adjuster locknut and remove the adjuster plug and wormshaft thrust bearing.

4. Carefully pull the wormshaft and ball nut assembly from the housing and remove the upper thrust bearing.

NOTE: To avoid damage to the return guides, keep the ball nut from running down to either end of the worm.

5. Pry out the sector shaft and wormshaft seals and discard them.

NOTE: Individual parts for this manual steering gear are not availabe for service. Do not disassemble. If the worm cannot rotate freely in the ball nut, replace the entire assembly.

6. The adjuster/plug bearing cup can be removed using a puller tool and slide hammer.

7. The housing bearing cup can be removed from the housing using a hammer and a suitable size bearing driver or socket.

8. The sector cover bushing is not serviceable. If found to be defective the entire sector cover including the bushing must be replaced.

9. The sector shaft needle bearing is serviced only as part of the housing unit and is

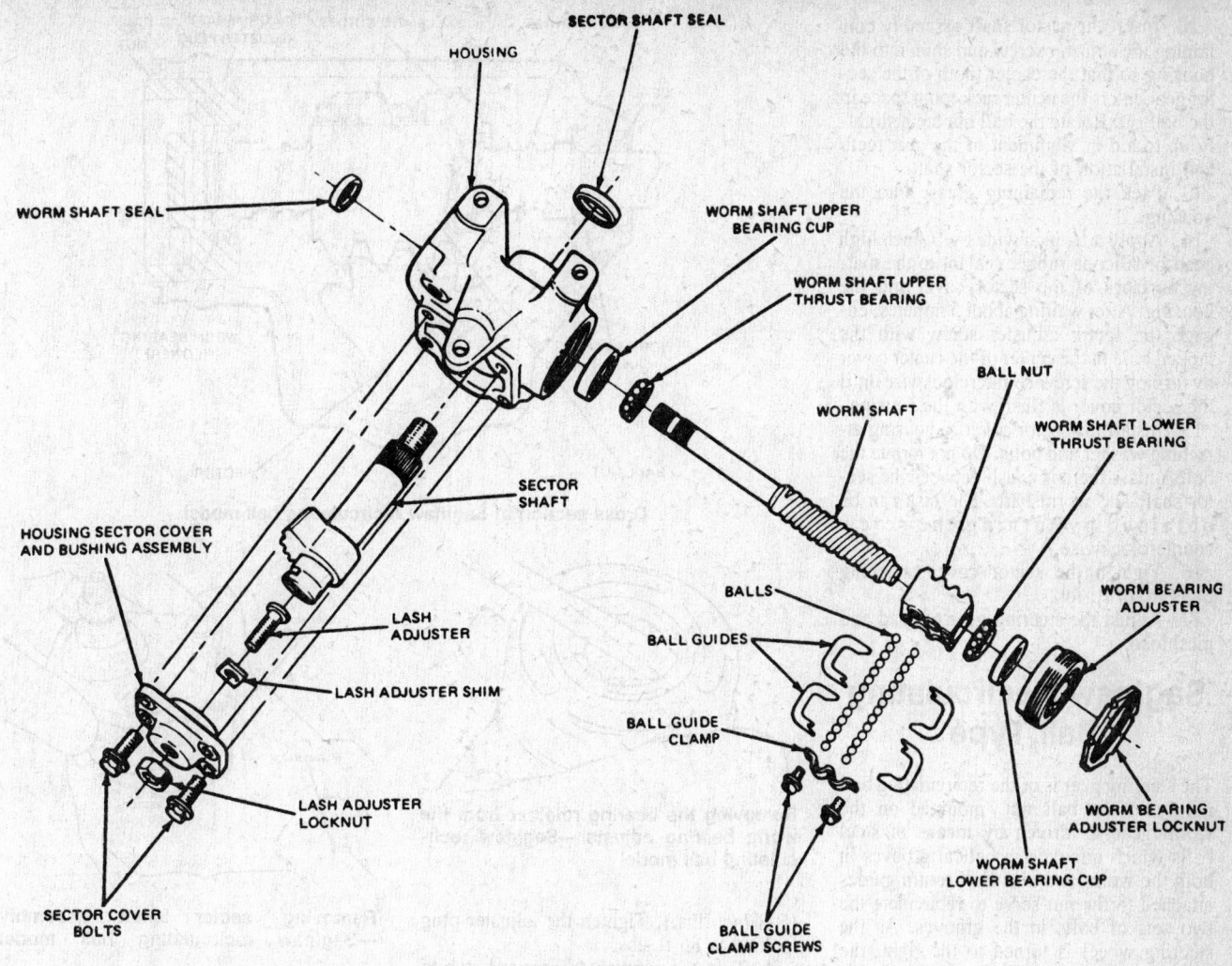

Ford recirculating ball steering gear (Koyo) 1984 and later

not serviced separately. If one or more needles fall out, they may be cleaned and put back using steering gear lube to hold them in place.

Inspection

1. Wash all parts in a cleaning solvent and dry thoroughly with air.
2. Inspect all bearings and bushings for wear.
3. Inspect the ball nut gear for chipping.
4. Inspect the ball nut and wormshaft for tightness and binding.
5. Inspect the housing for cracks.
6. Inspect the sector gear teeth for chipping.
7. Check the clearance between the sector adjusting screw head and the bottom of the sector shaft T-slot. If the clearance is more than 0.004 in. install a new shim as required to reduce the clearance to 0.004 in. or less. A steering gear lash adjuster kit is available containing five different size shims. While holding the sector adjusting screw, turn the sector shaft back and forth. The sector shaft must turn freely. If the sector shaft does not turn freely, increase the

T-slot clearance using an appropriate shim from the lash adjuster kit. Make sure the resulting clearance is not more than 0.004 in.

Assembly

1. If the wormshaft bearing cup was removed from the housing, install a new cup Using Tool T82T-3504-AH, or equivalent.
2. If the adjuster plug bearing cup was removed, install a new cup using Tool T82T-3504-AH or equivalent.
3. Install the sector shaft seal in the housing using Tool T82T-3504-AH, or equivalent. Press the seal until it bottoms out.
4. Tap the womshaft seal in the housing, using a suitable size socket and a hammer. Assemble the seal flush with the housing surface.
5. Clamp the steering gear housing in a vise with the wormshaft bore horizontal and the sector cover opening up.
6. Apply steering gear grease to the wormshaft bearings, sector shaft needle bearing in the housing and the sector cover bushing.

7. Slip one of the wormshaft bearings over the wormshaft splined end. Insert the wormshaft and ball nut assembly into the housing. Feed the splined end of the wormshaft through the bearing cup and seal. Place the remaining wormshaft thrust bearing in the adjuster plug bearing cup.
8. Install the adjuster plug and locknut into the housing opening being careful to guide the wormshaft end into the bearing until nearly all end play has been removed from the wormshaft.
9. Position the sector adjusting screw and shim into the sector shaft slot. Check the clearance between the screw head and the sector shaft T-slot. Refer to Step 7 under Inspection above.
10. Lubricate the steering gear with 14.8 ounces by weight of steering gear grease. Rotate the wormshaft until the ball nut is near the end of its travel. Pack as much grease into the housing as possible without loosing it out at the sector shaft opening. Rotate the wormshaft to move the ball nut near the other end of its travel and pack more grease into the housing.
11. Rotate the wormshaft until the ball nut is in the center of its travel.

12. Insert the sector shaft assembly containing the adjuster screw and shim into the housing so that the center tooth of the sector gear enters the center rack tooth space in the ball nut. Rotate the ball nut teeth slightly up to aid in alignment of the gear teeth and installation of the sector shaft.

13. Pack the remaining grease into the housing.

14. Apply a $\frac{1}{8}$ inch wide by $\frac{1}{8}$ inch high bead of silicone rubber sealant to the mating surfaces of the sector cover and the housing. After waiting about 5 minutes, engage the sector adjuster screw with the tapped hole in the center of the center cover by turning the screw counterclockwise until the sector cover is flush with the housing.

15. Install the sector cover to housing attaching washer and bolts. Do not torque the bolts unless there is a lash between the sector shaft and wormshaft. The lash can be obtained by turning the screw counterclockwise.

16. Tighten the sector cover attaching bolts to 40 ft. lbs.

17. Adjust the steering gear preload and meshload.

Saginaw Recirculating Ball Type

The steering gear is of the recirculating ball nut type. the ball nut, mounted on the worm gear, is driven by means of steel balls which circulate in helical grooves in both the worm and nut. Ball return guides attached to the nut serve to recirculate the two sets of balls in the grooves. As the steering wheel is turned to the right, the ball nut moves upward. When the wheel is turned to the left, the ball nut moves downward.

The sector teeth on the pinion shaft and the ball nut are designed so that they fit the tightest when the steering wheel is straight ahead. This mesh action is adjusted by an adjusting screw which moves the pinion shaft endwise until the teeth mesh properly. The worm bearing adjuster provides proper preloading of the upper and lower bearings. Before doing the adjustment procedures given below, ensure that the steering problem is not caused by faulty suspension components, bad front end alignment, etc. Then, proceed with the following adjustments.

STEERING WORM AND SECTOR ADJUSTMENT

1. Tighten the worm bearing adjuster plug until all end play has been removed, then loosen $\frac{1}{4}$ turn.

2. Use an $\frac{11}{16}$ in. 12 point socket to carefully turn the wormshaft all the way into the right corner then turn back about $\frac{1}{2}$ turn.

3. Tighten the adjuster plug until the proper thrust bearing preload is obtained

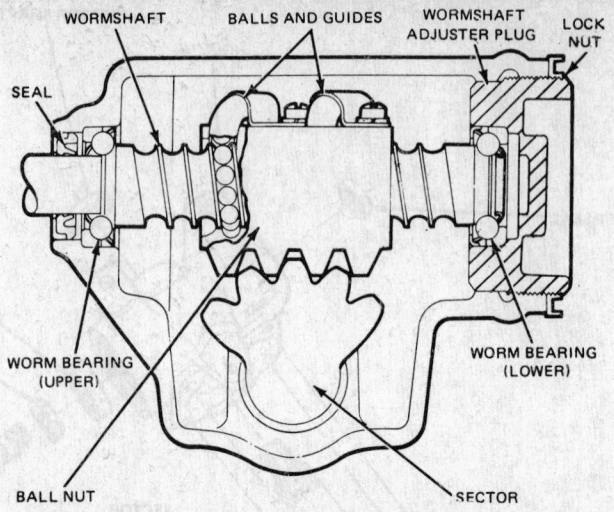

Cross section of Saginaw recirculating ball model

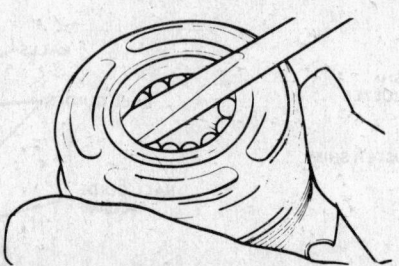

Removing the bearing retainer from the worm bearing adjuster—Saginaw recirculating ball model

(5–8 inch lbs.). Tighten the adjuster plug locknut to 85 ft. lbs.

4. Turn the wormshaft from one stop to the other counting the number of turns. Then turn the shaft back exactly half the number of turns to the center position.

5. Turn the lash (sector shaft) adjuster screw clockwise to remove all lash between the ball nut and sector teeth. Tighten the locknut to 25 ft. lbs.

6. Using an $\frac{11}{16}$ in. 12 point socket and an inch lb. torque wrench, observe the highest reading while the gear is turned through the center position. It should be 16 inch lbs. or less.

7. If necessary repeat Steps 5 and 6.

Disassembly

1. Place the steering gear in a vise, clamping onto one of the mounting tabs. The wormshaft should be in a horizontal position.

2. Rotate the wormshaft from stop to stop and count the total number of turns. Turn back exactly halfway, placing the gear on center.

3. Remove the three self locking bolts which attach the sector cover to the housing.

4. Using a plastic hammer, tap lightly on the end of the sector shaft and lift the sector cover and sector shaft assembly from the gear housing.

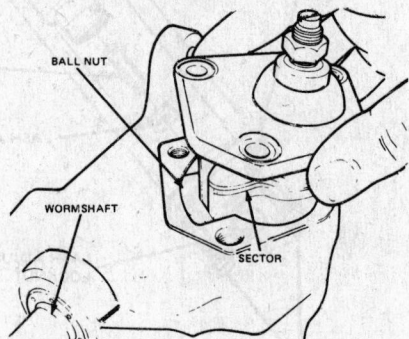

Removing sector shaft assembly —Saginaw recirculating ball model

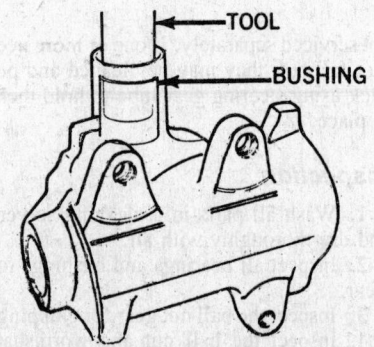

Removing sector shaft bushing—Saginaw recirculating ball model

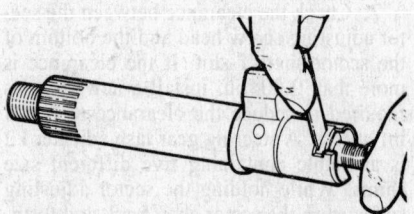

Checking lash adjuster end clearance —Saginaw recirculating ball model

NOTE: It may be necessary to turn the wormshaft by hand until the sector will pass through the opening in the housing.

5. Remove the locknut from the adjuster plug and remove the adjuster plug assembly.

6. Pull the wormshaft and ball nut assembly from the housing.

NOTE: Damage may be done to the ends of the ball guides if the ball nut is allowed to rotate to the end of the worm.

7. Remove the worm shaft upper bearing from inside the gear housing.

8. Pry the wormshaft lower bearing retainer from the adjuster plug housing and remove the bearing.

9. Remove the locknut from the lash adjuster screw in the sector cover. Turn the lash adjuster screw clockwise and remove it from the sector cover. Slide the adjuster screw and shim out of the slot in the end of the sector shaft.

10. Pry out and discard both the sector shaft and wormshaft seals.

Inspection

1. Wash all parts in cleaning solvent and blow dry with an air hose.

2. Use a magnifying glass and inspect the bearings and bearing caps for signs of indentation, or chipping. Replace any parts that show signs of damage.

3. Check the fit of the sector shaft in the bushings in the sector cover and housing. If these bushings are worn, a new sector cover and bushing assembly or housing bushing should be installed.

4. Check steering gear wormshaft assembly for being bent or damaged.

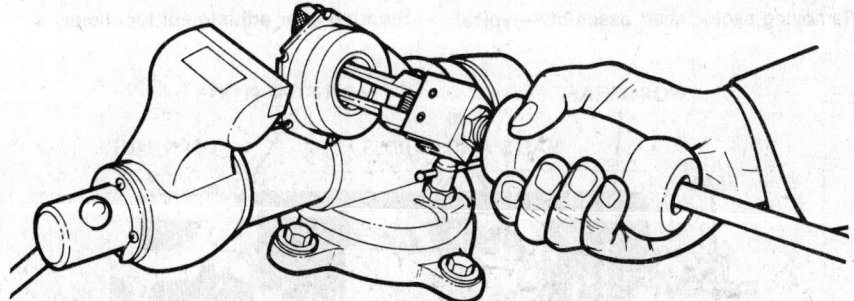

Removing worm shaft lower bearing cup from the adjuster plug—Saginaw recirculating ball model

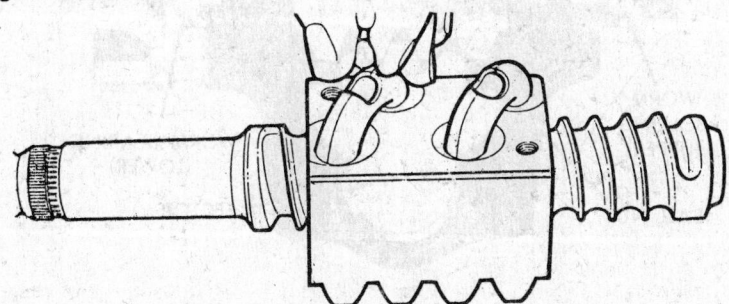

Filling the ball circuits—Saginaw recirculating ball model

SHAFT SEAL REPLACEMENT

1. Remove the old seal from the pump body.

2. Install the new seal by pressing the outer diameter of the seal with a suitable size socket.

NOTE: Make sure the socket is large enough to avoid damaging the external lip of the seal.

SECTOR SHAFT BUSHING REPLACEMENT

1. Place the steering gear housing in an arbor press.

2. Press the sector shaft bushing from the housing.

NOTE: Service bushings are bored to size and require no further reaming.

SECTOR COVER BUSHING REPLACEMENT

1. The sector cover bushing is not serviced separately. The entire sector cover assembly including the bushing must be replaced as a unit.

BALL NUT SERVICE

If there is any indication of binding or tightness when the ball nut is rotated on the worm the unit should be disassembled, cleaned and inspected as follows:

Ball Nut Disassembly

1. Remove the screws and clamp retaining the ball guides in the ball nut. Pull the guides out of the ball nut.

2. Turn the ball nut upside down and rotate the wormshaft back and forth until all the balls have dropped out of the ball nut. The ball nut can now be pulled endwise off the worm.

3. Wash all parts in solvent and dry them with air. Use a magnifying glass and inspect the worm and nut grooves and the surface of all balls for signs of indentation. Check all ball guides for damage at the ends. Replace any damaged parts.

Ball Nut Assembly

1. Slip the ball nut over the worm with the ball guide holes up and the shallow end of the ball nut teeth to the left from the steering wheel position. Sight through the ball guide to align the grooves in the worm.

2. Place two ball guide halves together and insert them in the upper circuit in the ball nut. Place the two remaining guides together and insert them in the lower circuit.

3. Count out 25 balls and place them in a suitable container. This is the proper number of balls for one circuit.

4. Load the 25 balls into one of the guide holes while turning the wormshaft gradually away from that hole.

5. Fill the remaining ball circuit in the same manner.

6. Assemble the ball guide clamp to the ball nut and tighten the screws to 18–24 inch lbs.

7. Check the assembly by rotating the ball nut on the worm to see that it moves freely.

NOTE: Do not rotate the ball nut to the end of the worm threads as this may damage the ball guides.

Assembly

1. Coat the threads of the adjuster plug, sector cover bolts and lash adjuster with a non-drying oil resistant sealing compound.

NOTE: Do not apply compound to the female threads. Use extreme care when applying compound to the bearing adjuster so that it does not come in contact with the wormshaft bearing.

2. Place the steering gear housing in a vise with the wormshaft bore horizontal and the sector cover opening up.

3. Make sure that all seals, bushings and bearing cups are installed in the gear housing and that the ball nut is installed on the wormshaft.

4. Slip the wormshaft upper bearing assembly over the wormshaft and insert the wormshaft and ball nut assembly into the housing, feeding the end of the shaft through the upper ball bearing cup and seal.

5. Place the wormshaft lower bearing assembly in the adjuster plug bearing cup and press the stamped retainer into place with a suitable size socket.

6. Install the adjuster plug and locknut into the lower end of the housing while carefully guiding the end of the wormshaft into the bearing until nearly all end play has been removed from the wormshaft.

7. Position the lash adjuster including the shim in the slotted end of the sector shaft.

NOTE: End clearance should not be greater than .002. If the end clearance is greater than .002 a shim package is available with thicknesses of .063, .065, .067, .069.

8. Lubricate the steering gear with 11 oz. of steering gear grease. Rotate the wormshaft until the ball nut is at the other end of its travel and then pack as much new lubricant into the housing as possible without losing out the sector shaft opening. Rotate the wormshaft until the ball nut is at the other end of its travel and pack as much lubricant into the opposite end as possible.

9. Rotate the wormshaft until the ball nut is in the center of travel. This is to make sure that the sector shaft and ball nut will engage properly with the center tooth of the sector entering the center tooth space in the ball nut.

10. Insert the sector shaft assembly including lash adjuster screw and shim into the housing so that the center tooth of the sector enters the center tooth space in the ball nut.

11. Pack the remaining portion of the lubricant into the housing and also place some in the sector cover bushing hole.

12. Place the sector cover gasket on the housing.

13. Install the sector cover onto the sector shaft by reaching through the sector cover with a screwdriver and turning the lash adjuster screw counterclockwise until the screw bottoms, then back the screw off one-half turn. Loosely install a new lock nut onto the adjuster screw.

14. Install and tighten the sector cover bolt to 30 ft. lbs.

Chrysler Recirculating Ball Type

The steering gear is of the recirculating ball nut type. The ball nut, mounted on the worm gear, is driven by means of steel balls which circulate in helical grooves in both the worm and nut. Ball return guides attached to the nut serve to recirculate the two sets of balls in the grooves. As the steering wheel is turned to the right, the ball nut moves upward. When the wheel is turned to the left, the ball nut moves downward.
The sector teeth on the pinion shaft and the ball nut are designed so that they fit the tightest when the steering wheel is straight ahead. This mesh action is adjusted by an adjusting screw which moves the pinion shaft endwise until the teeth mesh properly. The worm bearing adjuster provides proper preloading of the upper and lower bearings.

WORM BEARING PRE-LOAD ADJUSTMENT

1. Remove the steering gear arm and lockwasher from the sector shaft, using a suitable gear puller.

2. Remove the horn button or horn ring.

3. Loosen the sector-shaft adjusting screw locknut, and back out the adjusting screw about two turns.

4. Turn the steering wheel two complete turns from the straight ahead position, and place an inch lb. torque wrench on the steering shaft nut.

5. Rotate the steering shaft at least one turn toward the straight ahead position while measuring the torque on the torque wrench. The torque should be between 1⅛ and 4½ inch lbs. to move the steering wheel. If torque is not within these limits, loosen the worm shaft bearing adjuster locknut and turn the adjuster clockwise to increase the preload or counterclockwise to decrease the preload. When the preload is correct, hold the adjuster screw steady and tighten the locknut. Recheck preload.

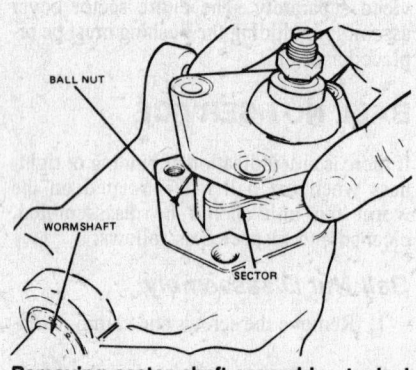

Removing sector shaft assembly—typical

BALL NUT RACK AND SECTOR MESH ADJUSTMENT

NOTE: This adjustment can be accurately made only after proper preloading of worm bearing.

1. Turn steering wheel gently from one stop to the other, counting the number of turns. Turn the steering wheel back exactly half way, to the center position.

2. Turn the sector-shaft adjusting screw clockwise to remove all lash between ball nut rack and the sector gear teeth, then tighten adjusting screw locknut to 35 ft. lbs.

3. Turn the steering wheel about ¼ turn away from the center or high spot position. With the torque wrench on the steering wheel nut measure the torque required to turn the steering wheel through the high spot at the center position. The reading should be between 8 and 11 inch lbs. This

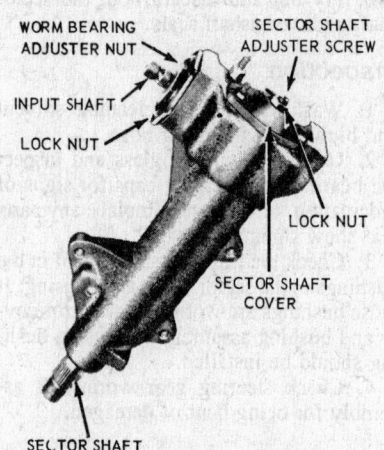

Steering gear adjustment locations

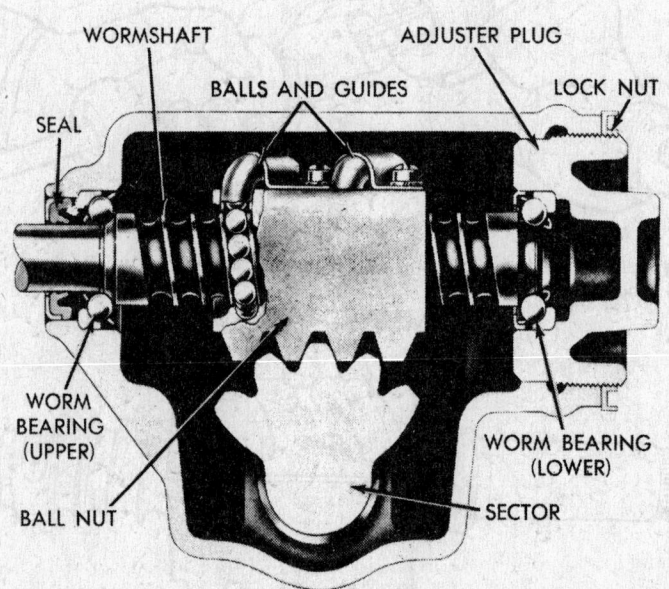

is the total of the worm shaft bearing pre-load and the ball nut rack and sector gear mesh load. Readjust the sector-shaft adjustment screw if necessary to obtain a correct torque reading.

4. After completing the adjustments, place the front wheels in a straight ahead position, and with the steering wheel and steering gear centered, install the steering arm on sector-shaft. Tighten the steering arm retaining nut to 180 ft. lbs.

Disassembly & Assembly

1. Attach the steering gear assembly to a holding fixture and put the holding fixture in a bench vise. Thoroughly clean the outside surface before disassembly.

2. Loosen the sector-shaft adjusting screw locknut, and back out the adjusting screw about two turns to relieve the mesh load between the ball nut rack and the sector gear teeth.

3. Position the steering gear worm shaft in a straight ahead position.

4. Remove the attaching bolts from the sector-shaft cover and slowly remove the sector-shaft while sliding an arbor tool into the housing. Remove the locknut from the adjusting screw and remove the screw from the cover by turning it clockwise. Slide the adjustment screw and its shim out of the slot in the end of the sector-shaft.

5. Loosen the worm shaft bearing adjuster locknut with a brass drift (punch) and remove the locknut. Hold the worm shaft steady while unscrewing the adjuster. Slide the worm adjuster off the shaft.

--- CAUTION ---

Handle the adjuster carefully to avoid damaging the aluminum threads. Also, do not run the ball nut down to either end of the worm shaft to avoid damaging the ball guides.

6. Carefully remove the worm and ball nut assembly. This assembly is serviced as a complete assembly only and is not to be disassembled or the ball return guides removed or disturbed.

7. Remove the sector-shaft needle bearing by placing the gear housing in an arbor press; insert a tool in the lower end of the housing and press both bearings through the housing. The sector-shaft cover assembly, including a needle bearing or bushing, is serviced as an assembly.

8. Remove the worm shaft oil seal from the worm shaft bearing adjuster by inserting a blunt punch behind the seal and tapping alternately on each side of the seal until it is driven out of the adjuster.

9. Remove the worm shaft in the same manner as that given in Step 8. Be careful not to cock the bearing cup and distort the adjuster counter bore.

10. Remove the lower cup if necessary. Pull the bearing cup out.

11. Wash all parts in clean solvent and dry thoroughly. Inspect all parts for wear, scoring, pitting, etc. Test operation of the worm shaft and ball nut assembly. If ball nut does not travel smoothly and freely on the worm shaft or if there is binding, replace the assembly.

NOTE: Extreme care must be taken when handling the aluminum worm bearing adjuster to avoid thread damage. Also, be careful not to damage the threads in the gear housing. Always lubricate the worm bearing adjuster before screwing it into the housing.

12. Inspect the sector-shaft for wear and check the fit of the shaft in the housing bearings. Inspect the fit of the shaft pilot bearing in the housing. Be sure the worm shaft is not bent or damaged.

13. Install the sector-shaft lower needle bearing. Press the bearing into the housing about $\frac{7}{16}$ in. below the end of the bore to leave space for the new oil seal.

14. Install the upper needle bearing in the same manner and press it into the inside end of the housing bore flush with the inside end of the bore surface.

15. Install the worm shaft bearing cups (upper and lower) by placing them and their spacers in the adjuster nut and press them into place.

16. Install the worm shaft oil seal by placing the seal in the worm shaft adjuster with the metal seal retainer up. Drive the seal into place with a suitable sleeve until it is just below the end of the bore in the adjuster.

NOTE: Apply a coating of steering gear lubricant to all moving parts during assembly. Also, put lubricant on and around oil seal lips.

17. Clamp the holding fixture and housing in a bench vise with the bearing adjuster opening upward. Place a thrust bearing in the lower cup in the housing.

18. Hold the ball nut from turning and insert the worm shaft and ball nut assembly into the housing with the end of the worm shaft resting in the thrust bearing. Place the upper thrust bearing on the worm shaft. Thoroughly lubricate the threads on the adjuster and the threads in the housing.

19. Place a protective sleeve of tape over the splines on the worm shaft to avoid damaging the seal. Slide the adjuster assembly over the shaft.

20. Thread the adjuster into the housing and tighten the adjuster to 50 ft. lbs. while rotating the worm shaft to seat the bearings.

21. Loosen the adjuster so no bearing preload exists. Tighten the adjuster for a worm shaft bearing preload of $1\frac{1}{8}$ to $4\frac{1}{2}$ inch lbs. Tighten the bearing adjuster locknut and recheck the preload.

22. Before installing the sector-shaft, pack the worm shaft cavities in the housing above and below the ball nut with steering gear lubricant. A good grade of multi-purpose lubricant may be used if steering gear lubricant is not available. Do not use gear oil. Pack enough lubricant into the worm cavities to cover the worm.

23. Slide the sector-shaft adjusting screw and shim into the slot in the end of the shaft. Check the end clearance for no more than 0.004 in. clearance. If the clearance is not within the limit, remove old shim and install a new shim, available in three different thicknesses, to get the proper clearance.

24. Start the sector-shaft and adjuster screw into the bearing in the housing cover. Using a screwdriver through the hole in the cover, turn the screw counterclockwise to pull the shaft into the cover. Install the adjusting screw locknut, but do not tighten at this time.

25. Rotate the worm shaft to center the ball nut.

26. Place a new gasket on the housing cover and install the sector-shaft and cover aasembly into the steering gear housing. Be sure to coat the sector-shaft and sector teeth with steering gear lubricant before installing the sector-shaft in the housing. Allow some lash between the sector-shaft sector teeth and the ball nut rack. Install and tighten the cover bolts to 25 ft. lbs.

27. Place the sector-shaft seal on the cross-shaft with the lip of the seal facing the housing. Press the seal in place.

28. Turn the worm shaft about $\frac{1}{4}$ turn away from the center of the high spot position. Using a torque wrench and a $\frac{3}{4}$ inch socket on the worm shaft spline, check the torque needed to rotate the shaft through the high spot. The reading should be between 8 and 11 inch lbs. Readjust the sector-shaft adjusting screw until the proper reading is obtained. Tighten the locknut to 35 ft. lbs. and recheck sector-shaft torque.

SECTOR-SHAFT OIL SEAL

Replacement

1. Remove the steering gear arm retaining nut and lockwasher.

2. Remove seal with a seal puller or other appropriate tool.

3. Place a new oil seal onto the splines of the sector-shaft with the lip of the seal facing the housing.

4. Remove the tool, and install the steering gear arm, lockwasher, and retaining nut. Tighten the nut to 180 ft. lbs. torque.

Models	Sector Shaft Diameter
254, 301	1¼ inch
376, 378	1⅜ inch
408	1½ inch
504	1¾ inch

POWER STEERING

GENERAL INFORMATION

The procedures for maintaining, adjusting, and repairing the power steering systems and components discussed in this chapter

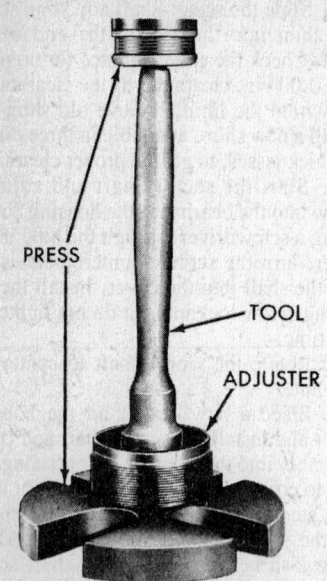

Installing the wormshaft upper bearing cup

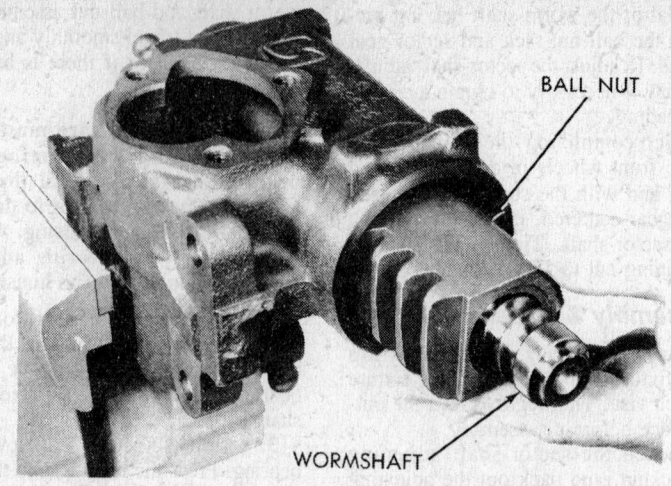

Removing the wormshaft and ballnut assembly

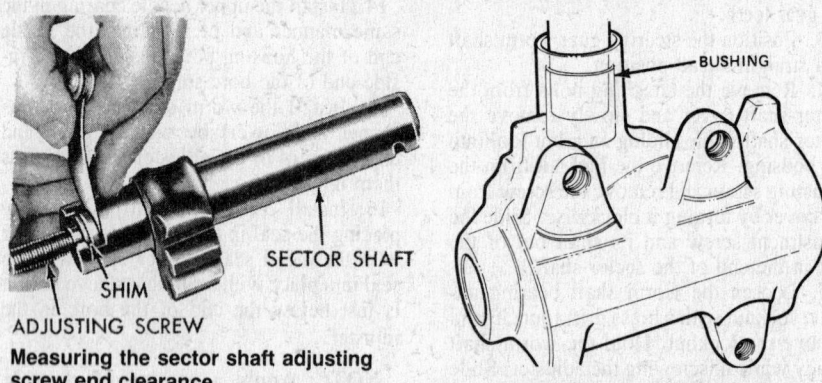

Measuring the sector shaft adjusting screw end clearance

Removing sector shaft bushing

are to be done only after determining that the steering linkages and front suspension systems are correctly aligned and in good condition. All worn or damaged parts should be replaced before attempting to service the power steering system. After correcting any condition that could affect the power steering, do the preliminary tests of the steering system components.

PRELIMINARY TESTS

Lubrication

Proper lubrication of the steering linkage and the front suspension components is very important for the proper operation of the steering systems of trucks equipped with power steering. Most all power steering systems use the same lubricant in the steering gear box as in the power steering pump reservoir, and the fluid level is maintained at the pump reservoir.

With power cylinder-assist power steering, the steering gear is of the standard mechanical type and the lubricating oil is self contained within the gear box and the level is maintained by the removal of a filler plug on the gear box housing. The control valve assembly is mounted on the gear box and is lubricated by power steering oil from the power steering pump reservoir, where the level is maintained.

Air Bleeding

Air bubbles in the power steering system must be removed from the fluid. Be sure the reservoir is filled to the proper level and the fluid is warmed up to operating temperature. Then, turn the steering wheel through its full travel three or four times until all the air bubbles are removed. Do not hold the steering wheel against its stops. Recheck the fluid level.

Removing wormshaft lower bearing cup using special tool

Fluid Level Check

1. Run the engine until the fluid is at the normal operating temperature. Then, turn the steering wheel through its full travel three or four times, and shut off the engine.
2. Check the fluid level in the steering reservoir. If the fluid level is low, add enough fluid to raise the level to the Full mark on the dipstick or filler tube.

Pump Belt Check

1. Inspect the pump belt for cracks, glazing, or worn places. Using a belt tension gauge, check the belt tension for the proper range of adjustment. The amount of tension varies with the make of truck and the condition of the belt. New belts (those belts used less than 15 minutes) require a higher figure. The belt deflection method of adjustment may be used only if a belt tension gauge is not available. The belt should be adjusted for a deflection of $\frac{1}{4}$ to $\frac{3}{8}$ in..

Fluid Leaks

Check all possible leakage points (hoses, power steering pump, or steering gear) for loss of fluid. Turn engine on and rotate the steering wheel from stop to stop several times. Tighten all loose fittings and replace any defective lines or valve seats.

STEERING TROUBLE DIAGNOSIS
Power Steering

Condition	Possible Cause	Correction
Intermittent or No Power Assist	1. Belt slipping and/or low fluid level.	1. Adjust or replace belt. Add fluid as necessary.
	2. Piston or rod binding in power cylinder. (Linkage type).	2. Repair or replace piston and rod.
	3. Sliding sleeve stuck in control valve. (Linkage type).	3. Free-up or replace sleeve.
	4. Improper pump operation.	4. Refer to "Power Steering Pump."
Poor or No Recovery from Turns	1. Improper caster setting.	1. Adjust to specifications.
	2. Steering gear adjustments too tight.	2. Adjust according to instructions.
	3. Improper spool nut adjustment. (Linkage type).	3. Adjust according to instructions.
	4. Valve spool installed backwards. (Linkage type).	4. Install valve spool correctly.
	5. Low tire pressure.	5. Inflate tires to recommended pressure.
	6. Tight steering linkage.	6. Lubricate as necessary.
	7. King pins frozen.	7. Lubricate as necessary.
Lack of Effort (Both Turns)	1. Improper sector shaft adjustment.	1. Adjust Sector Shaft.
	2. Pressure plates on wrong side of reactions rings.	2. Gear Recondition.
Lack of Effort (Left Turn Only)	1. Left turn reaction seal "O" ring worn, damaged or missing.	1. Gear Recondition.
	2. Left turn reaction oil passageway not drilled in housing or cylinder head.	2. Replace parts as required.
	3. Left turn reaction ring sticking in cylinder head.	3. Replace parts as required.
Lack of Effort (Right Turn Only)	1. Right turn U-shaped reaction seal worn, damaged, or missing.	1. Gear Recondition.
	2. Right turn reaction oil passageway not drilled in housing head, or ferrule pin.	2. Replace parts as required.
	3. Right turn reaction ring sticking in housing head.	3. Replace parts as required.
Lack of Assist (Left Turn Only)	1. Left turn reaction seal "O" ring worn, damaged, or missing.	1. Gear Recondition.
Lack of Assist (Right Turn Only)	1. Right turn U-shaped reaction seal worn, damaged, or missing.	1. Gear Recondition.
	2. Worm sealing ring (teflon) worm sleeve seal, ferrule pin "O" ring damaged or worn.	2. Gear Recondition.
	3. Excessive internal leakage thru piston end plug and/or side plugs.	3. Replace worm-piston assembly.
Lack of Assist (Both Turns)	1. Low oil level in pump reservoir (usually accompanied by pump noise).	1. Fill to proper level.
	2. Loose pump belt.	2. Adjust belts.
	3. Pump output low.	3. Pressure test pump.
	4. Engine idle too low.	4. Adjust engine idle.
	5. Excessive internal leakage thru piston end plug and/or side plugs.	5. Replace worm-piston assembly.

STEERING TROUBLE DIAGNOSIS
Power Steering Noise

Condition	Possible Cause	Correction
Objectionable "Hiss"	1. Noisy valve	1. Do not replace valve unless "hiss" is extremely objectionable. A replacement valve will also exhibit sight noise and is not always a cure for the objection.
Rattle or Chuckle Noise in Steering Gear	1. Gear loose on frame.	1. Check gear mounting bolts. Torque bolts to specifications.
	2. Steering linkages looseness.	2. Check linkage pivot points for wear. Replace if necessary.
	3. Pressure hose touching other parts of truck.	3. Adjust hose position. Do not bend tubing by hand.
	4. Loose Pitman shaft over center adjustment. **NOTE:** A slight rattle may occur on turns because of increased clearance off the "high point". This is normal and clearance must not be reduced below specified limits to eliminate this slight rattle.	4. Adjust
	5. Loose Pitman arm.	5. Torque Pitman arm pinch bolt.
Squawk Noise in Steering Gear When Turning or Recovering From a Turn	1. Dampener O-ring on valve spool cut.	1. Replace dampener O-Ring.
	2. Loose or worn valve.	2. Replace valve.
Chirp Noise in Steering Gear	1. Gear relief valve.	1. Replace relief valve.
Chirp Noise in Steering Pump	1. Loose belt.	1. Adjust belt tension.
Belt Squeal (Particularly Noticeable at Full Wheel Travel and Standstill Parking)	1. Loose belt.	1. Adjust belt tension.
Growl Noise in Steering Pump	1. Excessive back pressure in hoses or steering gear caused by restriction.	1. Locate restriction and correct. Replace part if necessary.
Growl Noise in Steering Pump (Particularly Noticeable at Standstill Parking)	1. Scored pressure plates, thrust plate or rotor.	1. Replace parts and flush system.
	2. Extreme wear of cam ring.	2. Replace parts.
Groan Noise in Steering Pump	1. Low oil level.	1. Fill reservoir to proper level.
	2. Air in the oil. Poor pressure hose connection.	2. Torque connector. Bleed system.
Rattle or Knock Noise in Steering Pump	1. Loose pump pulley nut.	1. Torque nut.
Rattle Noise in Steering Pump	1. Vanes not installed properly.	1. Install properly.
	2. Vanes sticking in rotor slots.	2. Repair or replace.
Swish Noise in Steering Pump	1. Defective flow control valve.	1. Replace part.
Whine Noise in Steering Pump	1. Pump Shaft bearing scored.	1. Replace housing and shaft. Flush and bleed system.

STEERING TROUBLE DIAGNOSIS
Power Steering Pump

Condition	Possible Cause	Correction
Intermittent Assist	1. Flow control valve sticking. 2. Slipping belt. 3. Low fluid level. 4. Low pump efficiency.	1. Pressure test pump and service as necessary. 2. Adjust belt. 3. Inspect and correct fluid level. 4. Pressure test pump and service as necessary.
No Assist	1. Pump seizure. 2. Broken slipper spring(s). 3. Flow control bore plug ring not in place. 4. Flow control valve sticking.	1. Replace pump. 2. Recondition pump or replace as necessary. 3. Replace snap ring. Inspect groove for depth. 4. Pressure test pump and service as necessary.
No Assist When Parking Only	1. Wrong pressure relief valve. 2. Broken "O" ring on flow control bore plug. 3. Loose pressure relief valve. 4. Low pump efficiency	1. Install proper relief valve. 2. Replace "O" ring. 3. Tighten valve. DO NOT ADJUST. 4. Pressure test pump and service as necessary.
Noisy Pump	1. Low fluid level. 2. Belt noise. 3. Foreign material blocking pump housing oil inlet hole.	1. Inspect and correct fluid level. 2. Inspect for pulley alignment, paint or grease on pulley and correct. Adjust belt. 3. Remove reservoir, visually check inlet oil hole and service as necessary.
Pump Vibration	1. Pump hose interference with sheet metal or brake lines. 2. Belt loose. 3. Pulley loose or out of round. 4. Crankshaft pulley loose or damaged. 5. Bracket pivot bolts loose.	1. Reroute hoses. 2. Adjust belt. 3. Replace pulley. 4. Replace crankshaft pulley. 5. If unable to tighten, replace bracket.
Pump Leaks	1. Cap or filler neck leaks. 2. Reservoir solder joints leak. 3. Reservoir "O" ring leaking. 4. Shaft seal leaking. 5. Loose rear bracket bolts. 6. Loose or faulty high pressure ferrule. 7. Rear bolt holes stripped or casting cracked.	1. Correct fluid level. 2. Resolder or replace reservoir as necessary. 3. Inspect sealing area of reservoir. Replace "O" ring or reservoir as necessary. 4. Replace seal. 5. Tighten bolts. 6. Tighten fitting to 24 foot-pounds or replace as necessary. 7. Repair, if possible, or replace pump.

Turning Effort

Check the turning effort required to turn the steering wheel after aligning the front wheels and inflating the tires to the proper pressure.

1. With the vehicle on dry pavement and the front wheel straight ahead, set the parking brake and turn the engine on.

2. After a short warm-up period for the engine, turn the steering wheel back and forth several times to warm the steering fluid.

3. Attach a spring scale to the steering wheel rim and measure the pull required to turn the steering wheel one complete revolution in each direction. The effort needed to turn the steering wheel should not exceed the limits specified.

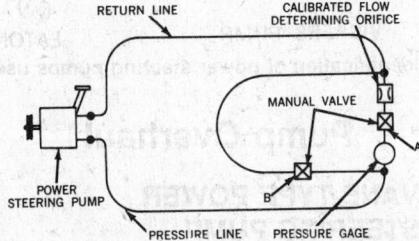

Power steering pump test circuit diagram

NOTE: This test may be done with the steering wheel removed and a torque wrench applied on the steering wheel nut.

Power Steering Hose Inspection

1. Inspect both the input and output hoses of the power steering pump for worn spots, cracks, or signs of leakage. Replace hose if defective, being sure to reconnect the replacement hose properly. Many power steering hoses are identified as to where they are to be connected by special means, such as fittings that will only fit on the correct pump fitting, or hoses of special lengths.

Test Driving Truck to Check the Power Steering

When test driving to check power steering, drive at a speed between 15 and 20 mph. Make several turns in each direction. When

STEERING TROUBLE DIAGNOSIS
Power Steering

Condition	Possible Cause	Correction
Hard Steering	1. Low or uneven tire pressure.	1. Inflate the tires to recommended pressures.
	2. Insufficient lubricant in the steering gear housing or in steering linkage.	2. Lubricate as necessary.
	3. Steering gear shaft adjusted too tight.	3. Adjust according to instructions.
	4. Improper caster or toe-in.	4. Align the wheels.
	5. Steering column misaligned.	5. See "Steering Gear Alignment."
	6. Loose, worn or broken pump belt.	6. Adjust or replace belt.
	7. Air in system.	7. Bleed air from system.
	8. Low fluid level in the pump reservoir.	8. Fill to correct level.
	9. Pump output pressure low.	9. See "Pressure Test."
	10. Leakage at power cylinder piston rings. (Linkage type).	10. Replace piston rings and repair as required.
	11. Binding or bent cylinder linkage. (Linkage type).	11. Replace or repair as required.
	12. Valve spool and/or sleeve sticking. (Linkage type).	12. Free-up or replace as required.

a turn is completed, the front wheels should return to the straight ahead position with very little help from the driver.

If the front wheels fail to return as they should and yet the steering linkage is free, well oiled and properly adjusted, the trouble is probably due to misalignment of the power cylinder or improper adjustment of the spool valve.

The power steering pump supplies all the power assist used in power steering systems of all designs. There are various designs of pumps used by the truck manufacturers but all pumps supply power to operate the steering systems with the least effort. All power steering pumps have a reservoir tank built onto the oil pump. These pumps are driven by belt turned by pulleys on the engine, normally on the front of the crankshaft.

During operation of the engine at idle speed, there is provision for the power steering pump to supply more fluid pressure. During driving speeds or when the truck is moving straight ahead, less pressure is needed and the excess is relieved through a pressure relief and flow control valve. The pressure relief part of the valve is inside the flow control and is basically the same for all pumps. The flow control valve regulates, or controls, the constant flow of fluid from the pump as it varies with the demands of the steering gear. The pressure relief valve limits the hydraulic pressure built up when the steering gear is turned against its stops.

During pump disassembly, make sure all work is done on a clean surface. Clean the outside of the pump thoroughly and do not allow dirt of any kind to get inside. Do not immerse the shaft oil seal in solvent.

If replacing the rotor shaft seal, be extremely careful not to scratch sealing surfaces with tools.

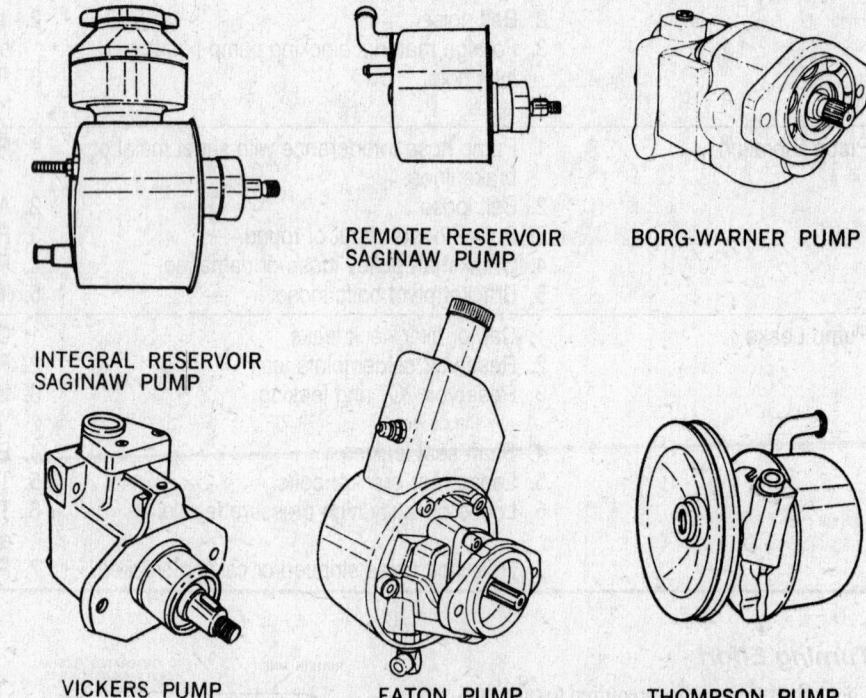

REMOTE RESERVOIR SAGINAW PUMP

BORG-WARNER PUMP

INTEGRAL RESERVOIR SAGINAW PUMP

VICKERS PUMP

EATON PUMP

THOMPSON PUMP

Identification of power steering pumps used on General Motors Trucks—typical

Pump Overhaul

VANE TYPE POWER STEERING PUMP

The vane type power steering pump is used in Saginaw steering systems. The operation is basically the same as that of the roller type pumps. Centrifugal force moves a number of vanes outward against the pump ring, causing a pumping action of the fluid to the control valve.

Disassembly

1. Clean the outside of the pump in a non-toxic solvent before disassembling.
2. Mount the pump in a vise, being careful not to squeeze the front hub too tight.
3. Remove the union and seal.
4. Remove the reservoir retaining studs and separate the reservoir from the housing.
5. Remove the mounting bolt and union O-rings.
6. Remove the filter and filter cage; discard the element.

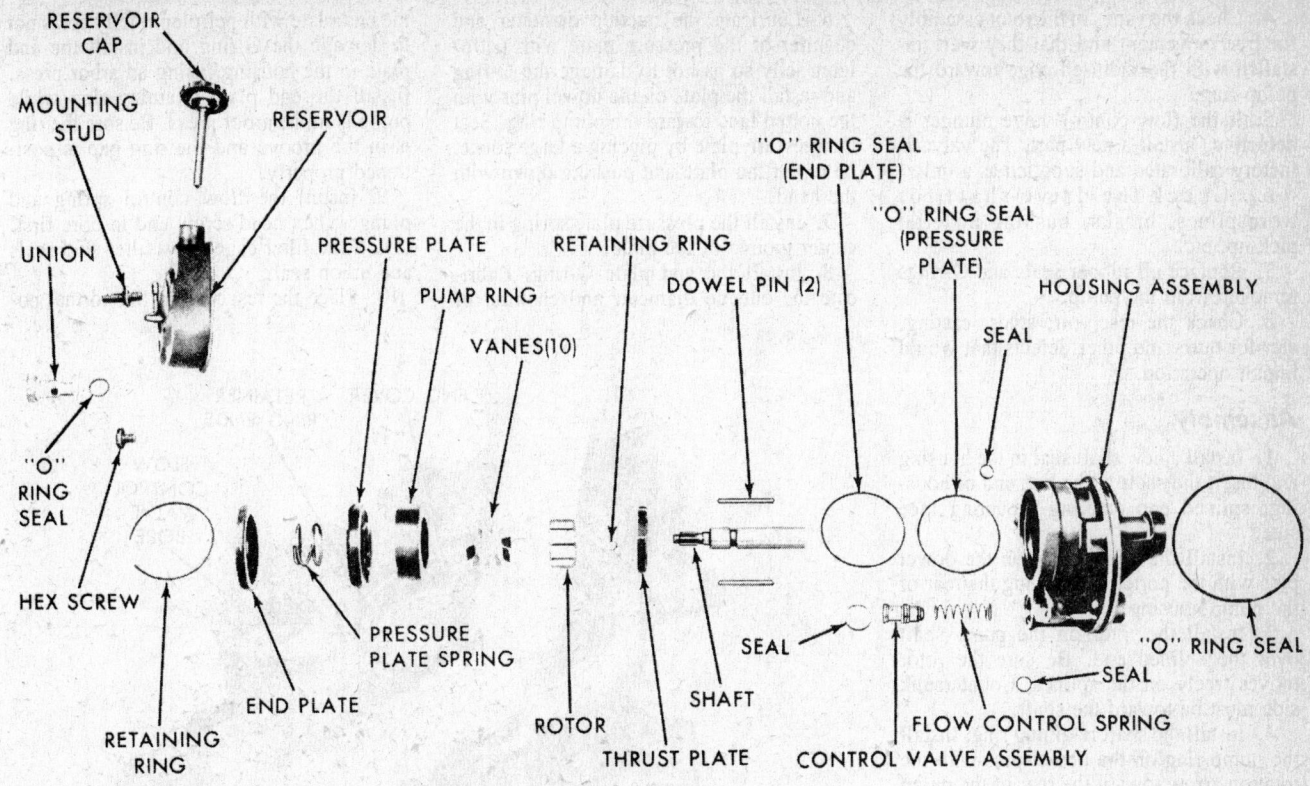

Exploded view of vane type pump

7. Remove the end plate retaining ring by compressing the retaining ring and then prying it out with a removal tool. The retaining ring may be compressed by inserting a small punch in the ⅛ in. diameter hole in the housing and pushing in until the ring clears the groove.

8. Remove the end plate. The end plate is spring-loaded and should rise above the housing level. If it is stuck inside the housing, a slight rocking or gentle tapping should free the plate.

9. Remove the shaft woodruff key and tap the end of the shaft gently to free the pressure plate, pump ring, rotor assembly, and thrust plate. Remove these parts as one unit.

10. Remove the end plate O-ring. Separate the pressure plate, pump ring, rotor assembly, and thrust plate.

Inspection

Clean all metal parts in a non-toxic solvent and inspect them as given below:

1. Check the flow control valve for free movement in the housing bore. If the valve is sticking, see if there is dirt or a rough spot in the bore.

2. Check the cap screw in the end of the flow control valve for looseness. Tighten if necessary being careful not to damage the machined surfaces.

3. Inspect the pressure plate and the pump plate surfaces for flatness and check that there are no cracks or scores in the parts. Do not mistake the normal wear marks for scoring.

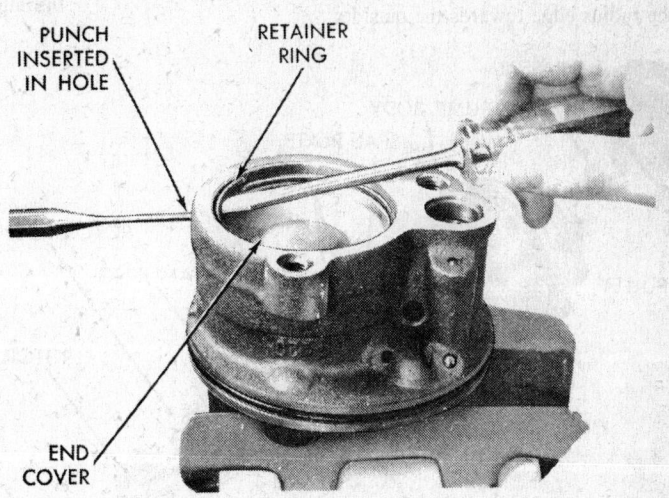

Removing end cover retaining ring

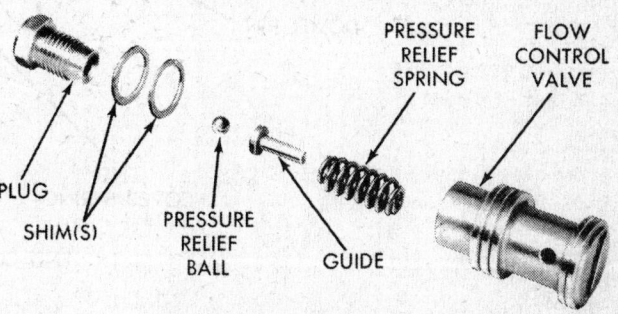

Flow control valve, vane type pump

4. Check the vanes in the rotor assembly for free movement and that they were installed with the radiused edge toward the pump ring.

5. If the flow control valve plunger is defective, install a new part. The valve is factory calibrated and supplied as a unit.

6. Check the drive shaft for worn splines, breaks, bushing material pick-up, etc.

7. Replace all rubber seals and O-rings removed from the pump.

8. Check the reservoir, studs, casting, etc. for burrs and other defects that would impair operation.

Assembly

1. Install a new shaft seal in the housing and insert the shaft at the hub end of housing, splined end entering mounting face side.

2. Install the thrust plate on the dowel pins with the ported side facing the rear of the pump housing.

3. Install the rotor on the pump shaft over the splined end. Be sure the rotor moves freely on the splines. Countersunk side must be toward the shaft.

4. Install the shaft retaining ring. Install the pump ring on the dowel pins with the rotation arrow toward the rear of the pump housing. Rotation is clockwise as seen from the pulley.

5. Install the vanes in the rotor slots with the radius edge towards the outside.

6. Lubricate the outside diameter and chamfer of the pressure plate with petroleum jelly so as not to damage the O-ring and install the plate on the dowel pins with the ported face toward the pump ring. Seat the pressure plate by placing a large socket on top of the plate and pushing down with the hand.

7. Install the pressure plate spring in the center groove of the plate.

8. Install the end plate O-ring. Lubricate the outside diameter and chamfer of the end plate with petroleum jelly so as not to damage the O-ring and install the end plate in the housing, using an arbor press. Install the end plate retaining ring while pump is in the arbor press. Be sure the ring is in the groove and the ring gap is positioned properly.

9. Install the flow control spring and plunger, hex head screw end in bore first. Install the filter cage, new filter stud seals and union seal.

10. Place the reservoir in the normal po-

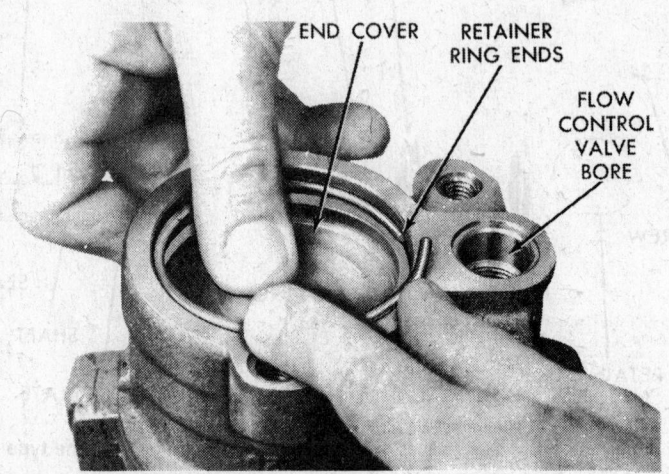

Installing end cover plate and retainer ring

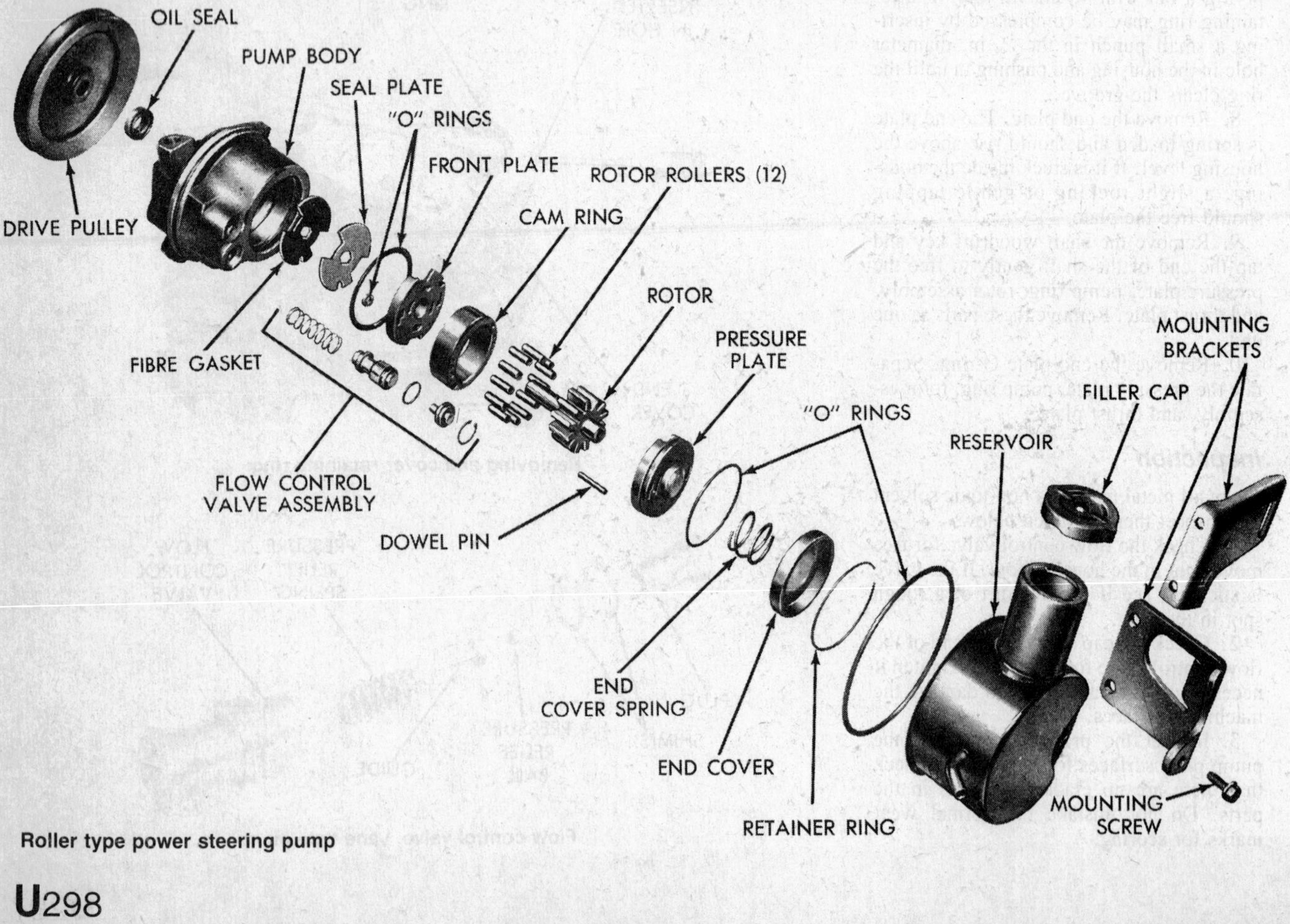

END COVER — RETAINER RING ENDS — FLOW CONTROL VALVE BORE

OIL SEAL
PUMP BODY
SEAL PLATE
"O" RINGS
FRONT PLATE
ROTOR ROLLERS (12)
CAM RING
ROTOR
PRESSURE PLATE
"O" RINGS
MOUNTING BRACKETS
FILLER CAP
RESERVOIR
DOWEL PIN
END COVER SPRING
END COVER
RETAINER RING
MOUNTING SCREW
DRIVE PULLEY
FIBRE GASKET
FLOW CONTROL VALVE ASSEMBLY

Roller type power steering pump

sition and press down until the reservoir seats on the housing. Check the position of the stud seals and the union seal.

11. Install the studs, union, and drive shaft woodruff key. Support the shaft on the opposite side of the key when tapping the key into place.

ROLLER TYPE POWER STEERING PUMP

The roller type power steering pump is designed similar to other constant flow centrifugal force pumps. A star-shaped rotor forces 12 steel rollers against the inside surface of a cam ring. As the rollers follow the eccentric pattern of the cam ring, oil is drawn into the inlet ports and exhausted through the discharge ports while the rollers are moved into vee shaped cavities of the rotor, forcing oil into the high pressure circuit. A flow control valve permits a regulated amount of fluid to return to the intake side of the pump when excess output is produced during high speed operation. This reduces the power needs to drive the pump and minimizes temperature build-up.

The flow control valve used in one make of pump is a two-stage valve. Fluid under high pressure passes through two holes into a metering circuit located in a sealed passage. At low speed, about 2.7 gpm. passes to the gear. As speed increases and the valve moves, excess fluid is bypassed to the inlet and the valve blocks flow through one hole. This drops the flow to about 1.6 gpm. at high speeds.

When steering conditions produce excessive pressure needs (such as turning the wheels against the stops), the pressure built up in the steering gear exerts force on the spring end of the flow control valve.

This end of the valve contains the pressure relief valve. High pressure lifts the relief valve ball from its seat, allowing fluid to flow through a trigger orifice located in the front land of the flow control valve. This reduces pressure on the spring end of the valve which then opens and allows the fluid to return to the intake side of the pump. This action limits the maximum pressure output of the pump to a safe level. Normally, the pressure needs of the pump are below the maximum limits, causing the pressure relief ball and the flow control valve to remain closed.

Disassembly

1. Remove pump from engine, drain reservoir, and clean outside of pump. Clamp the pump in a vise at the mounting bracket.

2. Remove the drive pulley.

3. Remove the shaft seal by installing the seal remover adapter over the end of the drive shaft with the large end toward the pump. Place the seal remover tool over the shaft and through the adapter. Then, screw the tapered thread well into the metal por-

Installing pressure plate

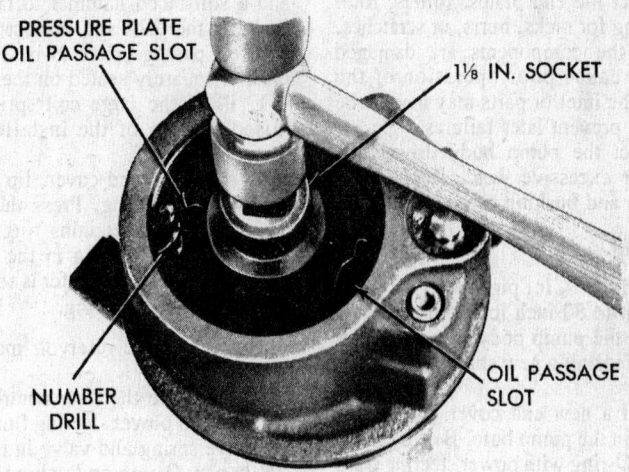

Seating pressure plate

tion of the seal. Tighten the large drive nut and remove the seal.

4. Remove the pump from the vise and remove the bracket mounting bolts. Remove the bracket.

5. Remove the reservoir and place the pump in a soft-faced vise with the shaft down. Discard the mounting bolt and the reservoir O-rings.

6. Move the end cover retaining ring around until one end of the ring lines up with the hole in the pump body. Insert a small punch in the hole and push it in far enough to bend the ring so a screwdriver can be inserted between the ring and the housing. Remove the ring.

7. Remove the end cover and spring from the housing. It may be necessary to tap the cover gently to loosen it in the housing.

8. Remove the pump from the vise and turn the pump over so the rotating pump

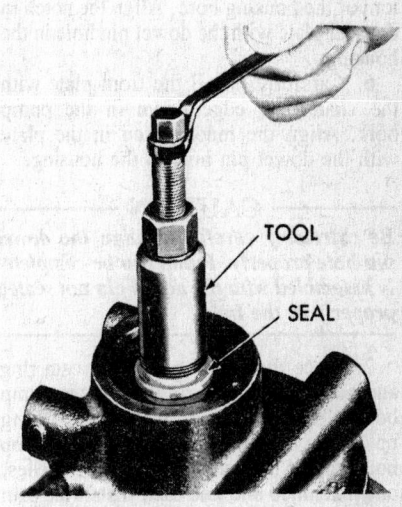

Removing shaft seal

may come out of the housing. Tap the end of the drive shaft to loosen these parts. Lift the pump body off the rotating group. Check that the seal plate is removed from the bottom of the housing bore.

9. Discard the O-rings from the pressure plate and end cover.

10. Remove the snap-ring, bore plug, flow control valve and spring from the housing. Discard the O-ring. If necessary to dismantle the flow control valve for cleaning, see the procedure for disassembly.

Inspection

1. Remove the clean out plug with an Allen wrench.

2. Wash all metal parts in clean, nontoxic solvent. Blow out all passages with compressed air and air dry all cleaned parts.

3. Inspect the drive shaft for excessive wear and the seal area for nicks or scoring. Replace if necessary.

4. Inspect the end plates, rollers, rotor and cam ring for nicks, burrs, or scratches. If any of the components are damaged enough to cause poor operation of the pump, all the interior parts may have to be replaced to prevent later failures.

5. Inspect the pump body drive shaft bushing for excessive wear. Replace the pump body and bushing as one assembly.

Assembly

1. Install the ⅛ in. pipe clean out plug, tightening it to 80 inch lbs. torque.

2. Place the pump body on a clean flat surface and install a new shaft seal into the bore.

3. Install a new end cover O-ring into the groove in the pump bore. Be sure to lubricate the O-ring with power steering fluid before installing it.

4. Lubricate and install a new O-ring in the groove on the pump body where the reservoir fits snugly.

5. Install the brass seal plate to the bottom of the housing bore. Align the notch in the seal plate with the dowel pin hole in the housing.

6. Carefully install the front plate with the chamfered edge down in the pump bore. Align the index notch in the plate with the dowel pin hole in the housing.

CAUTION

Be extremely careful to align the dowel pin hole properly. Pump can be completely assembled with the dowel pin not seated properly in the hole.

---

7. Place the dowel pin in the cam ring and position the cam ring inside the pump bore. Notch in the cam ring must be facing up (away from the pulley end of pump housing). If the cam ring has two notches, one machined and one cast, install the cam ring with the machined notch up. Check the amount of dowel pin extending above the

cam ring surface. If more than ³/₁₆ in. is showing, the dowel pin is not seated in the index hole in the housing.

8. Install the rotor and shaft in the cam ring and carefully install the 12 steel rollers in the cavities of the rotor. Lubricate the rotor, rollers, and the inside surface of cam ring with power steering fluid. Rotate the shaft by hand to be sure all the rollers are seated parallel with the shaft and are not sticking or binding.

9. Position the pressure plate by carefully aligning the index notch on the plate with the dowel pin and inserting a clean drill (number 13 to 16) in the cam ring oil hole next to the dowel pin notch until it bottoms on the housing floor.

10. Lubricate and install a new O-ring on the pressure plate. Position the pressure plate in the pump bore so that the dowel pin is in the index notch on the plate and the drill extends through the oil passage in the pressure plate. Seat the pressure plate on the cam ring using a clean 1⅛ in. socket and a soft-faced hammer to tap it gently. Remove the drill and inspect the plate at both oil passage slots to be sure that the plate is squarely seated on the cam ring.

11. Place the large coil spring over the raised portion of the installed pressure plate.

12. Place the end cover, lip edge facing up, over the spring. Press the end cover down below the retaining ring groove. Install the retaining ring in the groove. Be sure the end cover chamfer is squarely seated against the snap-ring.

13. Replace the reservoir mounting bolt seal.

14. Lubricate the flow control valve assembly with power steering fluid and insert the valve spring and valve in the bore. Install a new O-ring on the bore plug, lubricate with fluid, and carefully install in the bore. Install the snap-ring with the sharp edge up. Do not depress the bore plug more than ¹/₁₆ in. below the snap-ring groove.

15. Place the reservoir on the pump body and visually align the mounting bolt hole. Tap the reservoir down on the pump with a plastic-faced hammer.

16. Remove the pump from the vise and install the mounting brackets with the mounting bolts on the pump. Tighten the bolts to 18 ft. lbs. torque.

17. Install the drive pulley by using the installer tool as follows: place the pulley on the end of the shaft and thread the installer tool into the ⅜ in. threaded hole in the end of the shaft. Put the installer shaft in a vise and tighten the drive nut against the thrust bearing, pressing the pulley on the shaft until it is flush. Do not try to press the pulley on the shaft without the special installer tool since the pump interior will be damaged by any other installation procedure. A small amount of drive shaft end play will be seen when the pulley is installed. This end play is necessary and will be minimized by a thin coat of oil between the ro-

tor and the end plates when the pump is operating.

18. Install the pump assembly on the engine, install the drive belt and hoses (use new O-ring on pressure hose), and check for leaks.

FLOW CONTROL VALVE

Disassembly

1. After removing the pump from the engine and the reservoir from the pump, remove the snap-ring and plug from the flow bore. Discard the O-ring.

2. Depress the control valve against the spring pressure and allow the valve to spring out of the bore. If the valve is stuck in the bore or it did not come out of the bore far enough, it may be necessary to tap the housing lightly to remove it.

3. If the valve has dirt or foreign particles on it or in its bore, the rest of the pump needs cleaning. The hoses should be flushed and the steering gear valve body reconditioned. If the valve bore is badly scored, replace the pump body and the flow control valve.

4. Remove any nicks or burrs by gently rubbing the valve with crocus cloth. Clamp the valve land in a vise with soft-jaws and remove the hex head ball seat and shims. Note the number and gauge (thickness) of the shims on the ball seat. They must be reinstalled for the same shim thickness to keep the same value of relief pressure.

5. Remove the valve from the vise and remove the pressure relief ball, guide, and spring.

Assembly

1. Insert the spring, guide and pressure relief ball in the end of the flow control valve.

2. Install the hex head plug using the exact number and thickness shims that were removed. Tighten the plug to 80 inch lbs. torque.

3. Lubricate the valve with power steering fluid and insert the flow control valve spring and valve in the housing bore. Install a new O-ring on the bore plug, lubricate with fluid and carefully install into the bore. Install the snap-ring. Do not depress the bore plug more than ¹/₁₆ in. beyond the snap-ring groove.

SLIPPER TYPE POWER STEERING PUMP

The slipper type power steering pump is a belt-driven constant displacement assembly that uses a number of spring-loaded slippers in the pump rotor to force fluid from the inlet side to the flow control valve. Openings in the metering pin allow a flow of about two gpm. of fluid to the steering gear before the flow control valve directs the excess fluid to the inlet side of the pump again. Maximum pressure in the pump is limited by the pressure relief valve which opens when the pressure exceeds the maximum limits.

The slipper type power steering pump discussed in this section is used on Ford trucks and is called the Ford-Thompson power steering pump.

Disassembly

1. Drain as much fluid from the pump as possible after removing the pump from the truck.

2. Install a $\frac{3}{8}$x16 capscrew in the end of the pump shaft to avoid damaging the shaft end with the pulley remover tool. Install the pulley remover tool on the pulley hub and place the pump and remover tool in a vise. Hold the pump steady and turn the tool nut counterclockwise to draw the pulley off the shaft. The pulley must be removed without in and out pressure on the pump shaft to avoid damaging the internal thrust washers.

3. Remove the pump reservoir by installing the pump in a holding fixture with an adapter plate in a vise with the reservoir facing up.

4. Remove the outlet fitting hex nut and any other attaching parts from the reservoir case.

5. Invert the pump so the reservoir is now facing down. Using a wooden block, remove the reservoir by tapping around the flange until the reservoir is loose. Remove the reservoir O-ring seal and the outlet fitting gasket from the pump.

6. Again invert the pump assembly in the vise, remove the pump housing holding bolts and the pump housing.

7. Remove the housing cover, the O-ring seal and the pressure springs from inside the pump housing. Remove the pump cover gasket and discard it.

8. Remove the retainer end plate and upper pressure plate. In some pumps, the end plate and the upper pressure plate are made as one unit.

9. Remove the loose fitting dowel pin. Be careful not to bend the fixed dowel pin which remains in the housing plate assembly.

10. Remove the rotor assembly being careful not to let the slippers and springs fall out of the rotor. It may not be necessary to disassemble the rotor assembly unless the lower pressure plate, housing plate, rotor shaft and/or seal is to be replaced. However, the rotor assembly may be disassembled by removing the slippers and springs from the cam ring.

11. Remove any rust, dirt, burrs, or scoring from the pulley end of the rotor shaft before removing the shaft from the housing plate. The shaft must come out without restrictions to avoid scoring or damaging the bushing. Remove the pump rotor shaft.

12. Remove the lower pressure plate.

13. Remove the rotor shaft seal after first wrapping a piece of 0.005 in. shim stock around the shaft and pushing it into the inside of the seal until it touches the bushing.

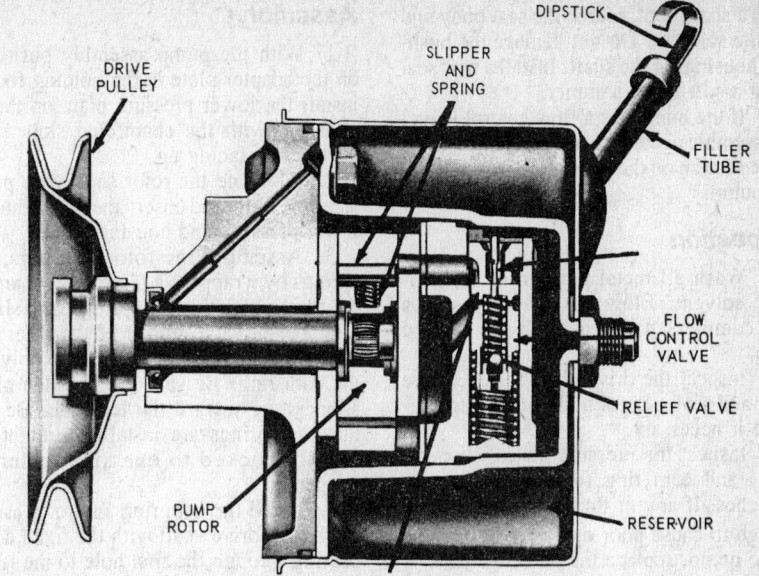

Ford Thompson power steering pump—sectional view

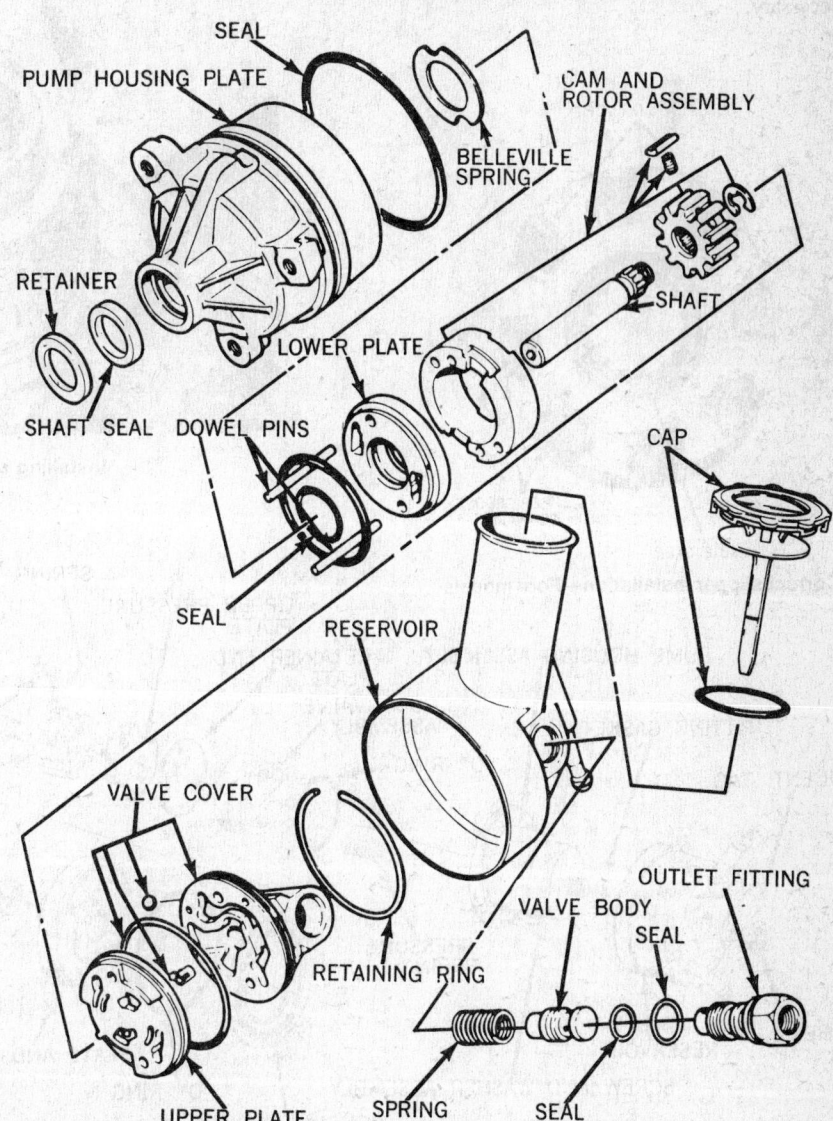

Exploded view of model C-11 slipper type pump

With a sharp tool, pierce the seal body and pry the seal out. Do not damage the bushing, housing, or the shaft. Install a new seal using a soft-faced hammer.

14. If the pump has a flow control valve, disassemble according to instructions given in the section on the roller type power steering pump.

Inspection

1. Wash all metal parts in clean, non-toxic solvent. Blow out all oil passages with compressed air and air dry all cleaned parts.

2. Inspect the drive shaft for excessive wear and seal area for nicks or scoring. Replace if necessary.

3. Inspect the pressure plates, slippers, rotor, and cam ring for nicks, burrs, or scratches. If any of the parts are damaged enough to cause poor operation or binding of the pump, replace the defective part.

4. Inspect the pump body drive shaft bushing for excessive wear. Replace if necessary.

Assembly

1. With the pump assembly positioned on the adapter plate in the holding fixture, install the lower pressure plate on the anchor pin with the chamfered slots at the center hole facing up.

2. Lubricate the rotor shaft with power steering fluid and insert the shaft into the lower pressure and housing plates.

3. Assemble the rotor, slippers, and springs by wrapping a piece of wire around the rotor, installing the springs, and sliding a slipper in each groove of the rotor over the springs. Then, insert the assembly into the cam ring. Be sure the flat side of the slippers are toward the left side. Be sure that the springs are installed straight and are not cocked to one side under the slippers.

4. Install the cam ring and rotor assembly on the drive shaft with the fixed dowel passing through the first hole to the left of the cam notch when the arrow on the cam

outside diameter is pointing toward the lower pressure plate. If the cam and rotor assembly does not seat properly, turn the rotor shaft slightly until the spline teeth mesh, allowing the cam and rotor to drop into position.

5. Insert the loose fitting dowel through the cam insert and lower pressure plate into the hole in the housing plate assembly. When both dowels are installed properly, they will be the same height.

6. Install the upper pressure plate so the tapered notch is facing down against the cam insert. The fixed dowel should pass through the round dowel hole and the loose dowel through the long hole. The slot between the ears on the outside of the pressure plate should match the notch on the cam insert.

7. Install the retainer end plate so the slot on the end plate matches the notches on the upper pressure plate and the cam insert.

8. Install the pump valve assembly O-

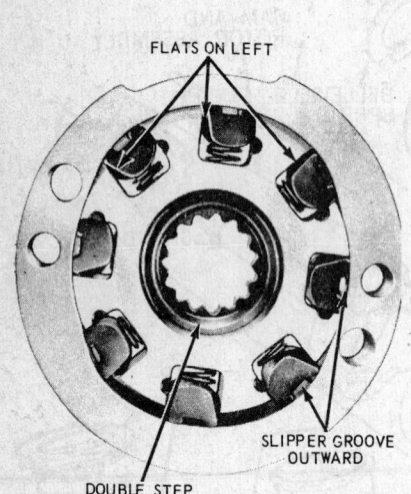

FLATS ON LEFT

SLIPPER GROOVE OUTWARD

DOUBLE STEP

Correct slipper installation—Ford models

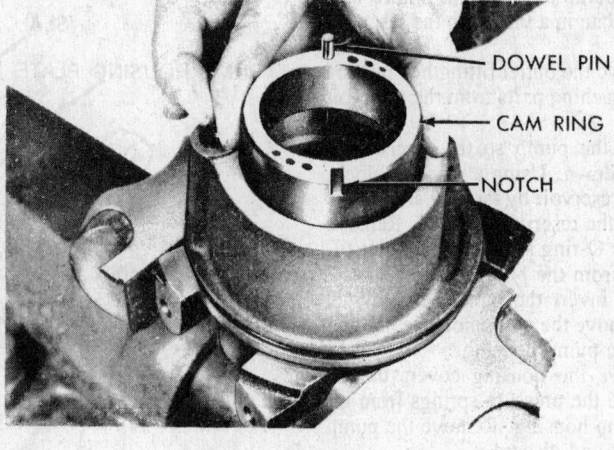

DOWEL PIN

CAM RING

NOTCH

Installing cam ring

PUMP HOUSING ASSEMBLY

FITTING GASKET

IDENT. TAG

"O" RING

PUMP VALVE ASSEMBLY

RETAINER END PLATE

UPPER PRESSURE PLATE

SPRING

ROTOR SHAFT

LOWER PRESSURE PLATE

DOWEL PIN

CAM AND ROTOR ASSEMBLY

PULLEY

PUMP SHAFT SEAL

PLATE AND BUSHING ASSEMBLY

"O" RING

HOUSING GASKET

PRESSURE SPRING

SCREW AND WASHER ASSEMBLY

RESERVOIR

HEX NUT

Ford Thompson power steering pump—sectional view

ring seal on the pump valve assembly. Do not twist the seal.

9. Place the pump valve assembly on top of the retainer end plate with the large exhaust slot on the pump valve in line with the outside notches of the cam, upper pressure plate, and retainer end plate. All parts must be fully seated. If correctly installed, the relief valve stem will be in line with the lube return hole in the pump housing plate.

10. Put small amounts of vaseline on the pump housing plate to hold the cover gasket in place. Install the cover gasket in place.

11. Insert the pressure plate springs into the pockets in the pump valve assembly.

12. Plug the intake hole in the housing.

13. Lubricate the inside of the housing and the housing cover seal with power steering fluid. Install two studs for use as positioning guides, one in the bolt hole nearest the drain hole and the other in the bolt hole on the opposite side of the housing plate.

14. Align the small lube hole in the housing rim and the lube hole in the housing plate. Install the housing, using a steady, even, downward pressure. Do not jar the pressure spring out of position. Remove the guide studs and loosely install the housing retaining bolts finger tight.

15. Tighten the retaining bolts evenly to 28–32 ft. lbs. until the housing flange contacts the gasket.

16. Install a $\frac{3}{8}$x16 hex head screw into the end of the rotor shaft and put a torque wrench on it. Check the amount of torque needed to rotate the rotor shaft. If the torque is more than 15 inch lbs., loosen the retaining bolts slightly and rotate the rotor shaft. Then, retighten the retaining bolts evenly. Do not use the pump if the shaft torque exceeds 15 inch lbs.

17. Release the pin in the bench holding fixture and shake the pump assembly back and forth. If there is a rattle, the pressure springs have fallen out of their seats and must be reinstalled.

18. Install the reservoir O-ring seal on the housing plate without twisting it. Lubricate the seal and install the reservoir, aligning the notch in the reservoir flange with the notch in the outside edge of the pump housing plate and bushing assembly. Using only a soft-faced hammer, tap at the rear outer corners of the reservoir. Inspect the assembly to be sure the reservoir is fully seated on the housing plate.

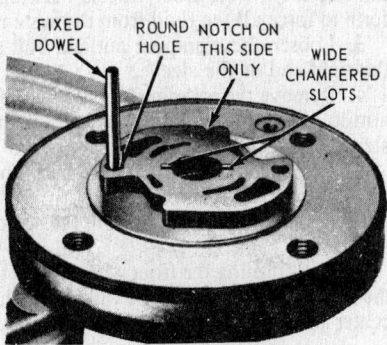

Lower pressure plate installed

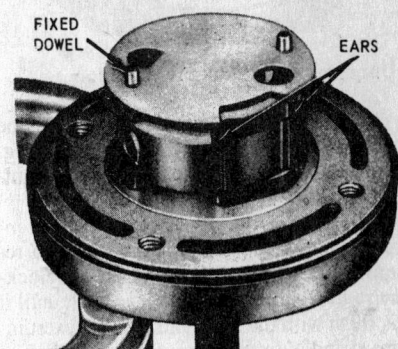

Retainer end plate installation

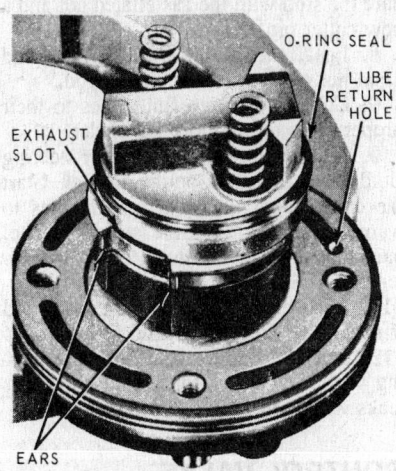

Valve and pressure spring installation

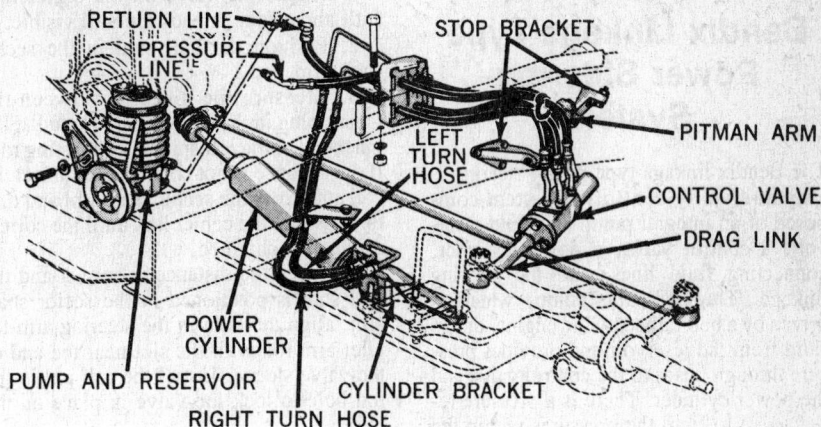

Linkage type power steering installation—typical

Cam and rotor installation

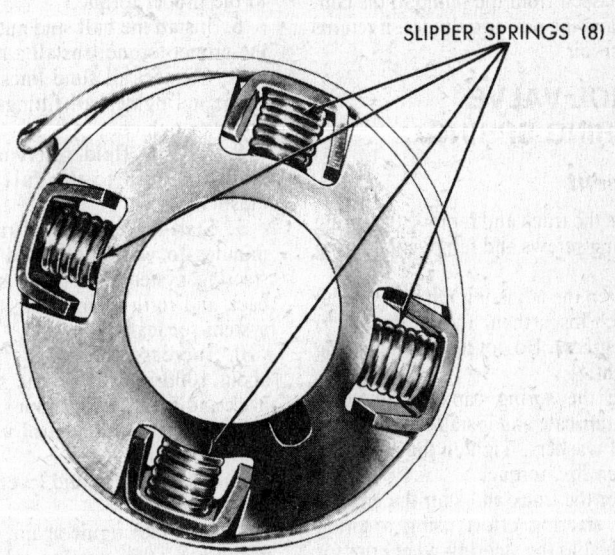

Correct slipper installation—Chrysler models

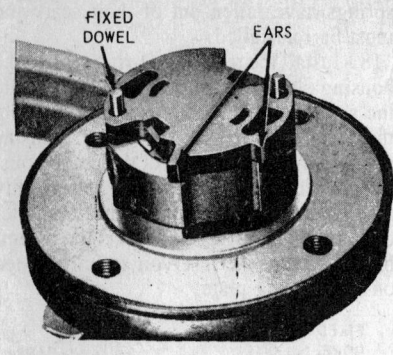

Upper pressure plate installation

19. Install the identification tag (if one was removed) on the outlet valve fitting. Install the outlet valve fitting nut and tighten to 48–45 ft. lbs. torque.

20. Turn the pump assembly over and install the pulley using the tool used to remove the pulley. Turn the tool nut clockwise to draw the pulley on the shaft until it is flush with the shaft end. Do not exert inward and outward pressures on the shaft to avoid damaging the internal thrust areas. Remove the tool.

Bendix Linkage Type Power Steering System

The Bendix linkage-type power steering is a hydraulically controlled system composed of an integral pump and fluid reservoir, a control valve, a power cylinder, connecting fluid lines, and the steering linkage. The hydraulic pump, which is driven by a belt turned by the engine, draws fluid from the reservoir and provides pressure through hoses to the control valve and the power cylinder. There is a pressure relief valve to limit the pressures within the steering system to a safe level. After the fluid has passed from the pump to the control valve and the power cylinder, it returns to the reservoir.

CONTROL VALVE CENTERING SPRING

Adjustment

1. Raise the truck and remove the spring cap attaching screws and remove the spring cap.

2. Tighten the adjusting nut snug (about 90–100 inch lbs.); then, loosen the nut ¼ turn (90 degrees). Do not turn the adjusting nut too tight.

3. Place the spring cap on the valve housing. Lubricate and install the attaching screws and washers. Tighten the screws to 72–100 inch lbs. torque.

4. Lower the truck and start the engine. Check the steering effort using a spring scale attached to the steering wheel rim for a pull of no more than 12 lbs.

POWER STEERING CONTROL VALVE

Removal

1. Raise the truck. If a two post hoist is used, be sure to place the hoist adapters under the front suspension steering arms. Do not allow the hoist adapters to contact the steering linkage.

2. Disconnect the four fluid line fittings at the control valve and drain the fluid from the lines. Turn the front wheels back and forth to force all the fluid from the system.

3. Loosen the clamping nut and bolt at the right end of the sleeve.

4. Remove the roll pin from the steering arm-to-idler arm rod through the slot in the sleeve.

5. Remove the control valve ball stud nut.

6. Remove the ball stud from the sector shaft arm.

7. After turning the front wheels fully to the left, unthread the control valve from the center link steering arm-to-idler arm rod.

Installation

1. Thread the valve on the center link until about four threads are still visible.

2. Position the ball stud in the sector shaft arm.

3. Measure the distance between the grease plug in the sleeve and the stud at the inner end of the left spindle connecting rod. If the distance is not correct, disconnect the ball stud from the sector shaft arm and turn the valve on the center link until the correct distance is obtained.

4. When the distance is correct and the ball stud is positioned in the sector shaft arm, align the hole in the steering arm-to-idler arm rod with the slot near the end of the valve sleeve. Install the roll pin in the rod hole to lock the valve in place on the rod.

5. Tighten the valve sleeve clamp bolt to the proper torque.

6. Install the ball stud nut and tighten to the proper torque. Install a new cotter pin.

7. Connect all fluid lines to the control valve and tighten all fittings securely. Do not over-tighten.

8. Fill the fluid reservoir with power steering fluid to the full mark on the dipstick.

9. Start the engine and run it for a few minutes to warm the fluid in the power steering system. Turn the steering wheel back and forth to the stops and check the system for leaks.

10. Increase the engine idle speed to about 1000 rpm. Turn the steering wheel back and forth several times, then stop the engine. Check the control valve and hose connections for leaks.

11. Recheck the fluid level and add fluid if necessary.

12. Start the engine again, and check the position of the steering wheel when the front wheels are straight ahead. Do not make any adjustments until toe-in is checked.

13. With engine running, check front wheel toe-in.

14. Check steering wheel turning effort which should be equal in both directions.

POWER STEERING POWER CYLINDER

Removal and Installation

1. Disconnect the two fluid lines from the power cylinder and drain the fluid.

2. Remove the pal nut, attaching nut, washer and the insulator from the end of the power cylinder rod. Remove the cotter pin and castellated nut holding the power cylinder stud to the center link.

3. Disconnect the power cylinder stud from the center link.

4. Remove the insulator sleeve and washer from the end of the power cylinder.

5. Inspect the tube fittings and seats in the power cylinder for nicks, burrs, or other damage. Replace the seats or tubes if damaged.

6. Install the washer, sleeve and the insulator on the end of the power cylinder rod.

7. While extending the rod as far as possible, insert the rod in the bracket on the frame and then, compress the rod so the stud may be inserted in the center link. Secure the stud with the castellated nut and a new cotter pin.

8. Install the insulator, washer, nut, and a pal nut on the power cylinder rod.

9. Connect the two fluid lines to their proper ports on the power cylinder.

10. Fill the reservoir with power steering fluid to the full mark on the dipstick. Start the engine and run for a few minutes to warm the fluid. Turn the steering wheel back and forth to the stops to fill the system. Stop the engine.

11. Recheck the fluid level and add fluid if necessary. Check for fluid leaks.

12. Start the engine again, turn the steering wheel back and forth, and check for leaks while the engine is running.

CONTROL VALVE

Disassembly

1. Clean the outside of the control valve of dirt and fluid.

2. Remove the centering spring cap from the valve housing. The control valve should be put in a soft-faced bench vise during disassembly. Clamp the control valve around the sleeve flange only to avoid damaging the valve housing, spool, or sleeve.

3. Remove the nut from the end of the valve spool bolt. Remove the washers, spacer, centering spring, adapter, and the bushing from the bolt and valve housing.

4. Remove the two bolts holding the valve housing and the sleeve together. Sep-

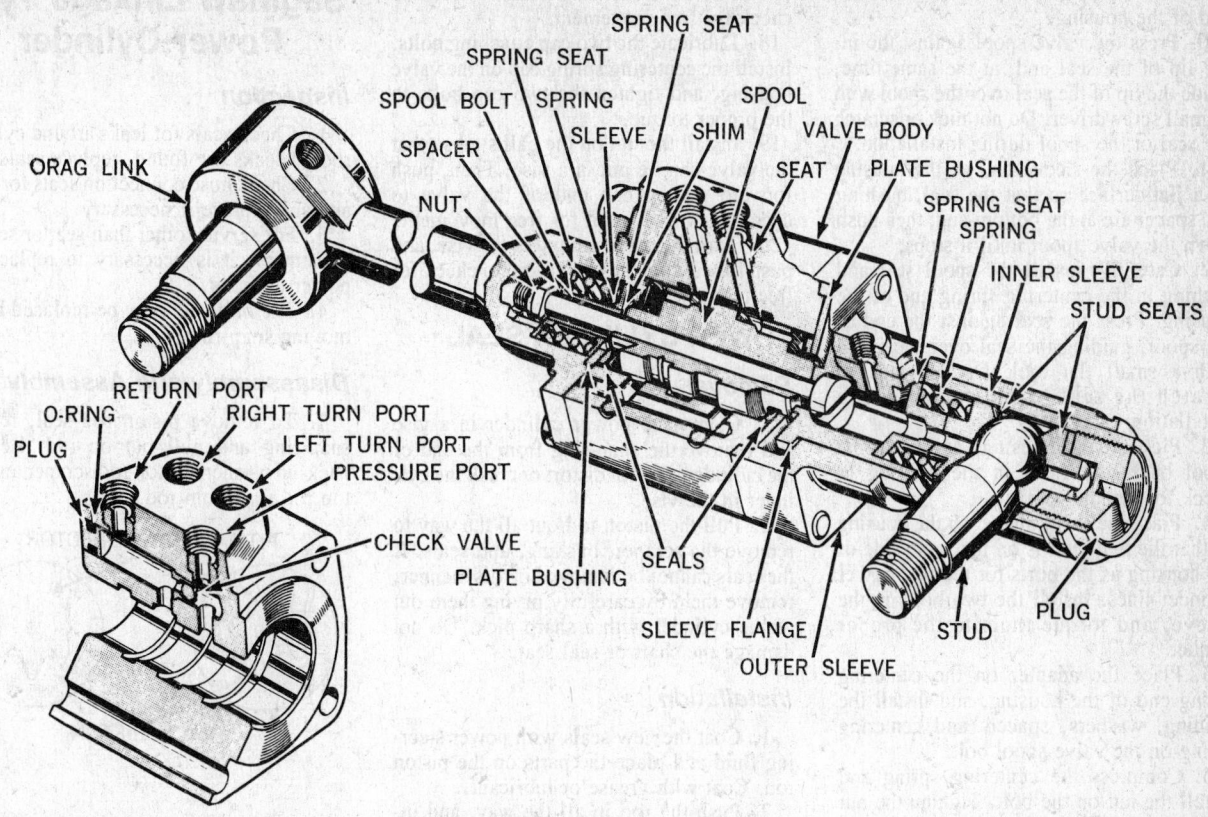

Control valve cross section—typical

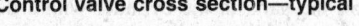

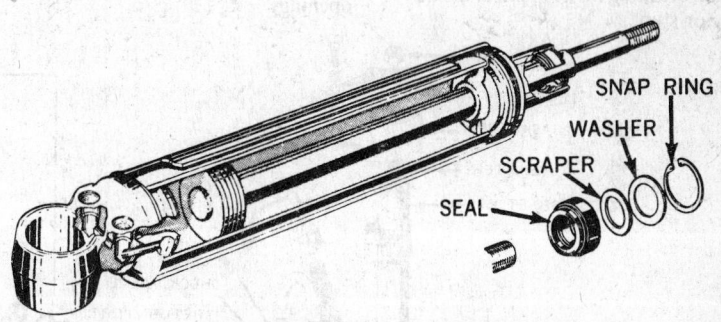

Phantom view of power cylinder—typical

arate the valve housing and the sleeve.

5. Remove the plug from the sleeve. Push the valve spool out of the centering spring end of the valve housing, and remove the seal from the spool.

6. Remove the spacer, bushing and valve housing.

7. Drive the pin out of the travel regulator stop with a punch and hammer. Pull the head of the valve spool bolt tightly against the travel regulator stop before driving the pin out of the stop.

8. Turn the travel regulator stop counterclockwise in the valve sleeve to remove the stop from the sleeve.

9. Remove the valve spool bolt, spacer, and rubber washer from the travel regulator stop.

10. Remove the rubber boot and clamp from the valve sleeve. Slide the bumper, spring, and ball stud seat out of the valve sleeve and remove the ball stud socket from the sleeve.

11. Remove the return port hose seat and the return port relief valve.

12. Remove the spring plug and O-ring. Then remove the reaction limiting valve.

13. Replace all worn or damaged hose seats by using an Easy-Out screw extractor or a bolt of proper size as a puller. Tap the existing hole in the hose seat, using a starting tap of the correct size. Remove all metal chips from the hose seat after tapping. Place a nut and washer on a bolt of the same size as the tapped hole. The washer must be large enough to cover the hose seat

port. Insert the bolt in the tapped hole and remove the hose seat by turning the nut clockwise and drawing the bolt out. Install a new hose seal in the port, and thread a bolt of the correct size in the port. Tighten the bolt enough to bottom the seal in the port.

Assembly

1. Coat all parts of the control valve assembly with power steering fluid. Seals should be coated with lubricant before installation.

2. Install the reaction limiting valve, spring and plug. Install the return port relief valve and the hose seat.

3. Insert one of the ball stud seats (flat end first) into the ball stud socket, and insert the threaded end of the ball stud into the socket.

4. Place the socket in the control valve sleeve so that the threaded end of the ball stud can be pulled out through the slot.

5. Place the other ball stud seat, spring, and bumper in the socket. Install and securely tighten the travel regulator stop.

6. Loosen the stop just enough to align the nearest hole in the stop with the slot in the ball stud socket and install the stop pin in the ball stud socket, travel regulator stop, and valve spool bolt.

7. Install the rubber boot, clamp, and the plug on the control valve sleeve. Be sure the lubrication fitting is turned on tightly and does not bind on the ball stud socket.

8. Insert the valve spool in the valve housing, rotating it while installing it.

9. Move the spool toward the centering spring end of the housing, and place the

10. Press the valve spool against the inner lip of the seal and, at the same time, guide the lip of the seal over the spool with a small screwdriver. Do not nick or scratch the seal or the spool during installation.

11. Place the sleeve end of the housing on a flat surface so that the seal, bushing and spacer are at the bottom end; then push down the valve spool until it stops.

12. Carefully install the spool seal and bushing in the centering spring end of the housing. Press the seal against the end of the spool, guiding the seal over the spool with a small flat tool. Do not nick or scratch the seal or the spool during installation.

13. Pick up the housing, and slide the spool back and forth in the housing to check for free movement.

14. Place the valve sleeve on the housing so that the ball stud is on the same side of the housing as the ports for the two power cylinder lines. Install the two bolts in the sleeve, and torque them to the proper torque.

15. Place the adapter on the centering spring end of the housing, and install the bushing, washers, spacers and centering spring on the valve spool bolt.

16. Compress the centering spring and install the nut on the bolt. Tighten the nut snug (about 90–100 inch lbs.); then, loosen it not more than 1/4 turn. Do not over-tighten to avoid breaking the stop pin at the travel regulator stop.

17. Move the ball stud back and forth to check for free movement.

18. Lubricate the two cap attaching bolts. Install the centering spring cap on the valve housing, and tighten the two cap bolts to the proper torque.

19. Install the nut on the ball stud so that the valve can be put in a vise. Then, push forward on the cap end of the valve to check the valve spool for free movement.

20. Turn the valve around in the vise, and push forward on the sleeve end to check for free movement.

POWER CYLINDER SEAL

Removal

1. Clamp the power cylinder in a vise and remove the snap-ring from the end of the cylinder. Do not distort or crack the cylinder in the vise.

2. Pull the piston rod out all the way to remove the scraper, bushing, and seals. If the seals cannot be removed in this manner, remove them by carefully prying them out of the cylinder with a sharp pick. Do not damage the shaft or seal seat.

Installation

1. Coat the new seals with power steering fluid and place the parts on the piston rod. Coat with grease or lubricant.

2. Push the rod in all the way, and install the parts in the cylinder with a deep socket slightly smaller than the cylinder opening.

Saginaw Linkage Type Power Cylinder

Inspection

1. Check seals for leaks around cylinder rod. If leaks are found, replace seals.

2. Check hose connection seats for damage and replace if necessary.

3. For service other than seat or seal replacement, it is necessary to replace the power cylinder.

4. The ball stud may be replaced by removing snap-ring.

Disassembly and Assembly

1. To remove piston rod seal, remove snap-ring and pull out on rod. Remove back-up washer, piston rod scraper and piston rod seal from rod.

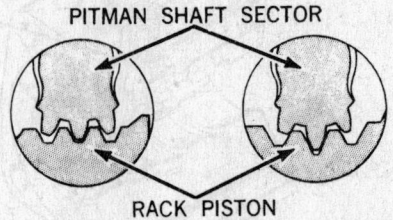

Comparison of constant ratio and variable ratio steering shaft teeth

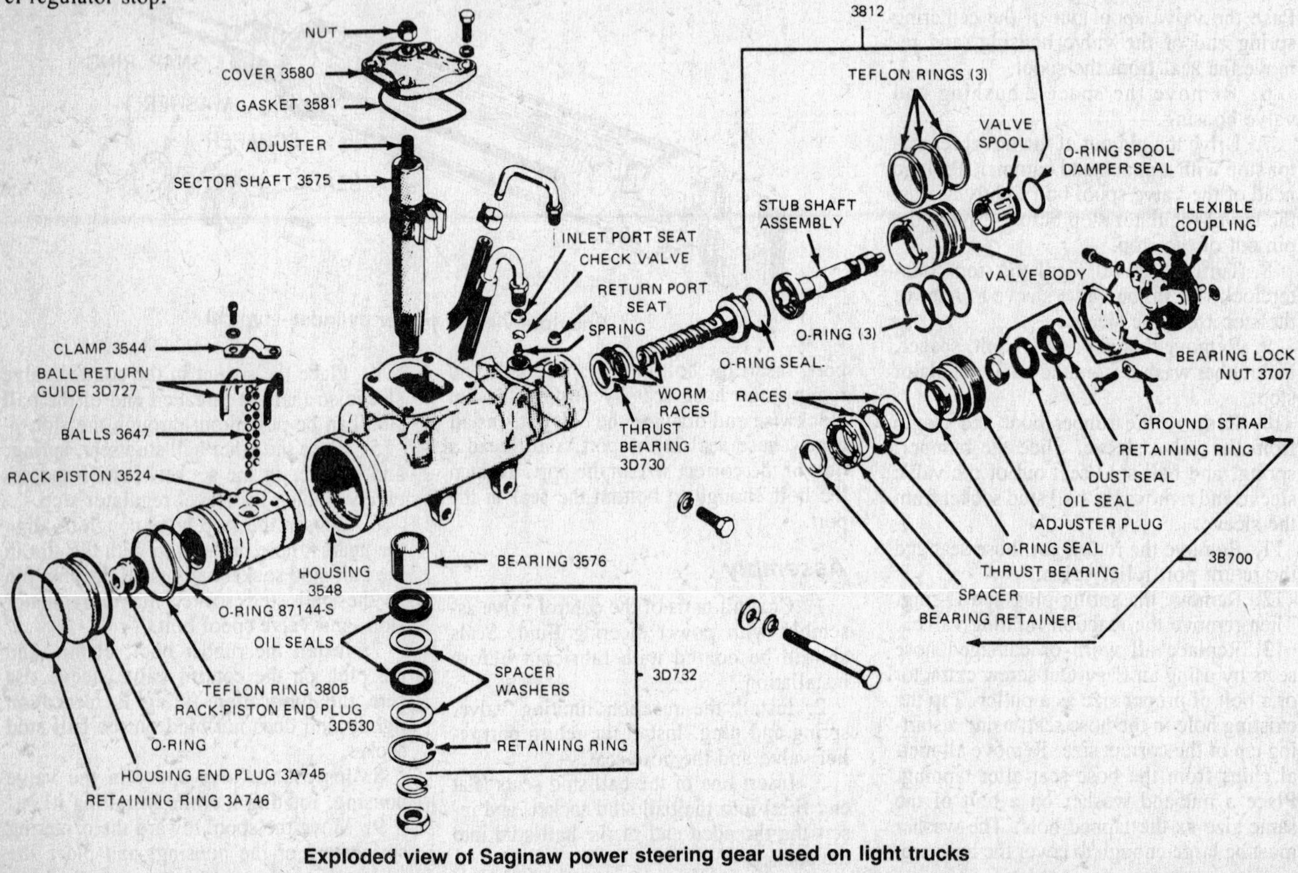

Exploded view of Saginaw power steering gear used on light trucks

2. To remove the ball stud, depress the end plug and remove the snap-ring. Push on the end of the ball stud and the end plug, spring, spring seat, ball stud and seal may be removed. If the ball seat is to be replaced, it must be pressed out.

3. Reverse disassembly procedure. Be sure snap-ring is properly seated.

Saginaw Rotary Type Power Steering

The rotary-type power steering gear is designed with all components in one housing. The power cylinder is an integral part of the gear housing. A double-acting type piston allows oil pressure to be applied to either side of the piston. The one-piece piston and power rack is meshed to the sector shaft.

The hydraulic control valve is composed of a sleeve and valve spool. The spool is held in the neutral position by the torsion bar and spool actuator. Twisting of the torsion bar moves the valve spool, allowing oil pressure to be directed to either side of the power piston, depending upon the directional rotation of the steering wheel, to give power assist.

On many trucks of the General Motors Corporation, a modified version of the rotary valve power steering system provides variable ratio steering to assist the driver to steer the truck easier and safer. The steering gear ratio will vary from a high ratio of about 16:1 while steering straight ahead to a lower gear ratio of about 12.1:1 while making a full turn to either side.

POWER STEERING UNIT

Fluid Used

This unit uses Dexron II automatic transmission fluid. The fluid capacity is $4\frac{1}{2}$ pints.

Bleeding the System

Fill the pump reservoir to within $\frac{1}{2}$ in. of the top. Start and run the engine to attain normal operating temperatures. Now, turn the steering wheel through its entire travel three or four times to expel air from the system, then recheck the fluid level.

Checking Steering Effort

Run the engine to attain normal operating temperatures. With the wheels on a dry floor, hook a pull scale to the spoke of the steering wheel at the outer edge. The effort required to turn the steering wheel should be $3\frac{1}{2}$–5 lbs. If the pull is not within these limits, check the hydraulic pressure.

Worm Bearing Preload and Sector Mesh Adjustments

Disconnect the pitman arm from the sector shaft, then back off on the sector shaft adjusting screw on the sector shaft cover.

Center the steering on the high point, then attach a pull scale to the spoke of the steering wheel at the outer edge. The pull required to keep the wheel moving for one complete turn should be $\frac{1}{2}$–$\frac{2}{3}$ lbs.

If the pull is not within these limits, loosen the thrust bearing locknut and tighten or back off on the valve sleeve adjuster locknut to bring the preload within limits.

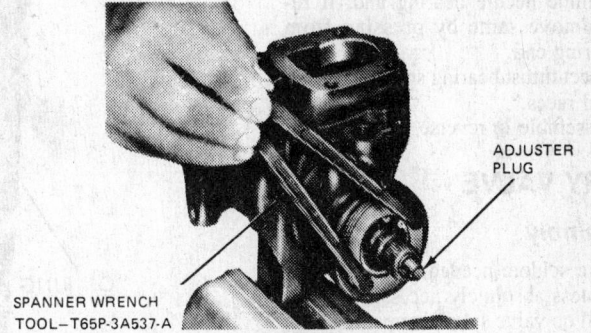

SPANNER WRENCH
TOOL—T65P-3A537-A

Removing adjuster plug

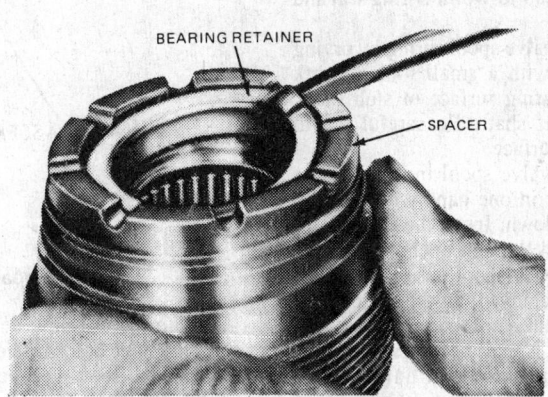

BEARING RETAINER

SPACER

Removing thrust bearing retainer

Tighten the thrust bearing locknut and recheck the preload.

Slowly rotate the steering wheel several times, then center the steering on the high point. Now, turn the sector shaft adjusting screw until a steering wheel pull of 1–$1\frac{1}{2}$ lbs. is required to move the worm through the center point. Tighten the sector shaft adjusting screw locknut and recheck the sector mesh adjustment.

Install the pitman arm and draw the arm in position with the nut.

Service Operations

ADJUSTER PLUG AND ROTARY VALVE

Removal

1. Thoroughly clean exterior of gear assembly. Drain by holding valve ports down

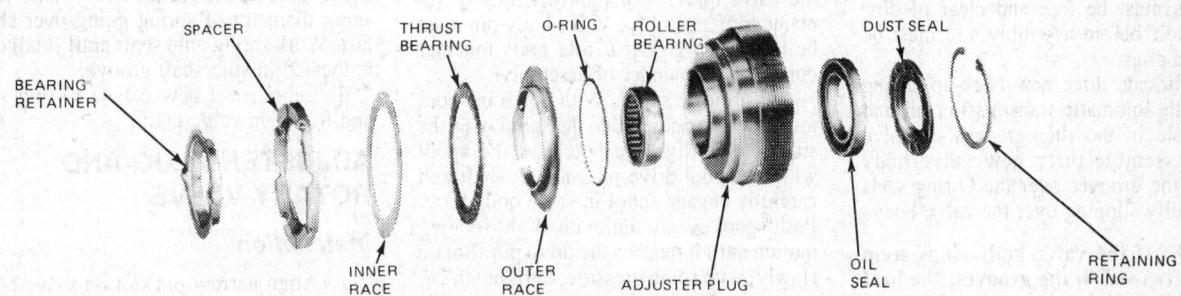

BEARING RETAINER · SPACER · THRUST BEARING · O-RING · ROLLER BEARING · DUST SEAL

INNER RACE · OUTER RACE · ADJUSTER PLUG · OIL SEAL · RETAINING RING

Adjuster plug disassembled

and rotating worm back and forth through entire travel.

2. Place gear in vise.

3. Loosen adjuster plug locknut with punch. Remove adjuster plug with spanner.

4. Remove rotary valve assembly by grasping stub shaft and pulling it out.

ADJUSTER PLUG

Disassembly & Assembly

1. Remove upper thrust bearing retainer with screwdriver. Be careful not to damage bearing bore. Discard retainer. Remove spacer, upper bearing and races.

2. Remove and discard adjuster plug O-ring.

3. Remove stub shaft seal retaining ring (Truarc pliers will help) and remove and discard dust seal.

4. Remove stub shaft seal by prying out and discard.

5. Examine needle bearing and, if required, remove same by pressing from thrust bearing end.

6. Inspect thrust bearing spacer, bearing rollers and races.

7. Reassemble in reverse of above.

ROTARY VALVE

Disassembly

Repairs are seldom needed. Do not disassemble unless absolutely necessary. If the O-ring seal on valve spool dampener needs replacement, perform this portion of operation only.

1. Remove cap-to-worm O-ring seal and discard.

2. Remove valve spool spring by prying on small coil with a small tool to work spring onto bearing surface of stub shaft. Slide spring off shaft. Be careful not to damage shaft surface.

3. Remove valve spool by holding the valve assembly in one hand with the stub shaft pointing down. Insert the end of pencil or wood rod through opening in valve body cap and push spool until it is out far enough to be removed. In this procedure, rotate to prevent jamming. If spool becomes jammed it may be necessary to remove stub shaft, torsion bar and cap assembly.

Assembly

All parts must be free and clear of dirt, chips, etc., before assembly and must be protected after.

1. Lubricate three new back-up O-ring seals with automatic transmission oil and reassemble in the ring grooves of valve body. Assemble three new valve body rings in the grooves over the O-ring seals by carefully slipping over the valve body.

NOTE: If the valve body rings seem loose or twisted in the grooves, the heat of the oil during operation will cause them to straighten.

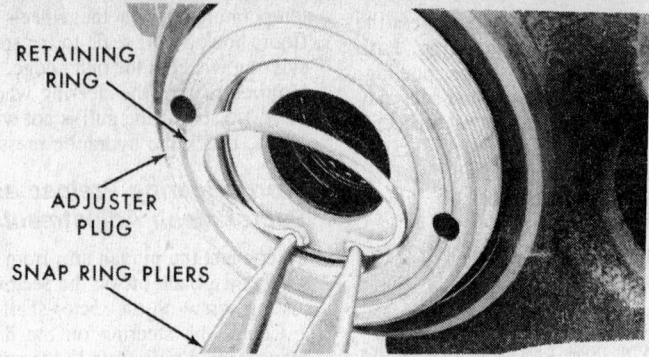

Removing adjuster plug retaining seal ring

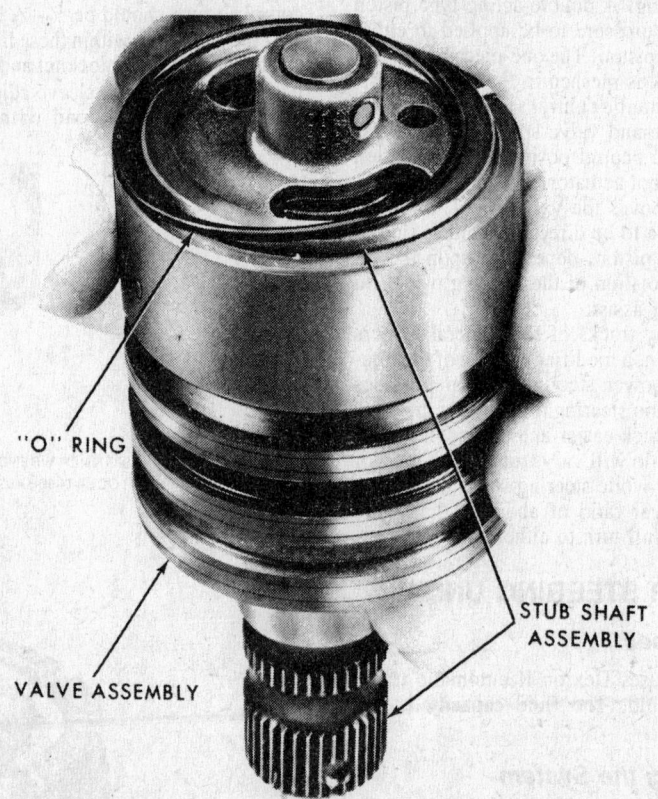

Separating valve spool (center) from valve body

2. Lubricate a new dampener O-ring with automatic transmission fluid and install in valve spool groove.

3. Assemble stub shaft torsion bar and cap assembly in the valve body, aligning the groove in the valve cap with the pin in the valve body. Tap lightly with soft remainder of assembly. Valve body pin must be in the cap groove. Hold parts together during the remainder of assembly.

4. Lubricate spool. With notch in spool toward valve body, slide the spool over the stub shaft. Align the notch on the spool with the spool drive pin on stub shaft and carefully engage spool in valve body bore. Push spool evenly and with slight rotating motion until it reaches the drive pin. Rotate slowly, with some pressure, until notch engages pin. Be sure dampener O-ring seal is evenly distributed in the spool groove.

— CAUTION —

Use extreme care because spool to valve body clearance is very small. Damage is easily caused.

5. With seal protector over stub shaft, slide valve spool spring over shaft, with small diameter of spring going over shaft last. Work spring onto shaft until small coil is located in stub shaft groove.

6. Lubricate a new cap to O-ring seal and install in valve body.

ADJUSTER PLUG AND ROTARY VALVE

Installation

1. Align narrow pin slot on valve body with valve body drive pin on the worm. Insert the valve assembly into gear housing

by pressing against valve body with finger tips. Do not press on stub shaft or torsion bar. The return hole in the gear housing should be fully visible when properly assembled.

CAUTION

Do not press on stub shaft as this may cause shaft and cap to pull out of valve body, allowing the spool dampener O-ring seal to slip into valve body oil grooves.

2. With seal protector over end of stub shaft, install adjuster plug assembly into gear housing snugly with spanner, then back plug off approximately one-eighth turn. Install plug locknut but do not tighten. Adjust preload as described in the adjustment section.

3. After adjustment, tighten locknut.

PITMAN SHAFT

Removal & Installation

1. Completely drain the gear assembly and thoroughly clean the outside.

2. Place gear in vise.

3. Rotate stub shaft until pitman shaft gear is in center position. Remove side cover retaining bolts.

4. Tap end of pitman shaft with soft hammer and slide shaft out of housing.

5. Remove and discard side cover O-ring seal.

6. The seals, washers, retainers and bearings may now be removed and examined.

7. Examine all parts for wear or damage and replace as required.

8. Install in reverse of above. Make proper adjustment as described in adjustment section.

RACK-PISTON NUT AND WORM ASSEMBLY

Removal

1. Completely drain the gear assembly and thoroughly clean the outside.

2. Remove pitman shaft assembly as previously described.

3. Rotate housing end plug retaining ring so that one end of ring is over hole in gear housing. Spring one end of ring so pin punch can be inserted to lift it out.

4. Rotate stub shaft to full left turn position to force end plug out of housing.

5. Remove and discard housing end plug O-ring seal.

6. Remove rack-piston nut end plug with ½ in. square drive.

7. Insert special tool in end of worm. Turn stub shaft so that rack-piston nut will go into tool and then remove rack-piston nut from gear housing.

8. Remove adjuster plug and rotary valve assemblies as previously described.

9. Remove worm and lower thrust bearing and races.

10. Remove cap O-ring seal and discard.

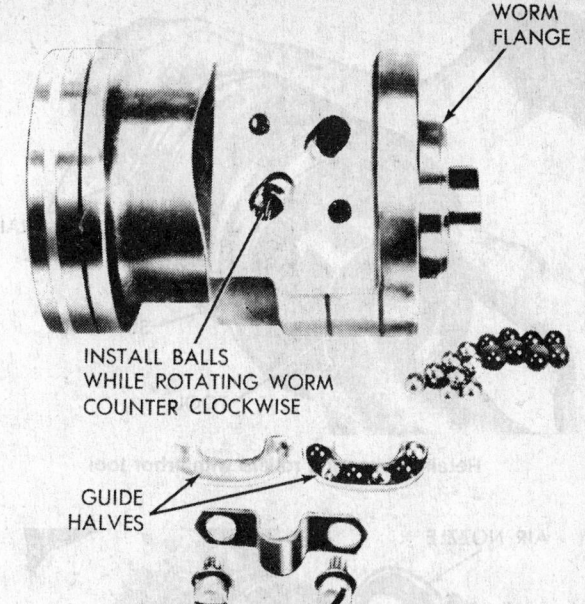

Installing balls in rack piston

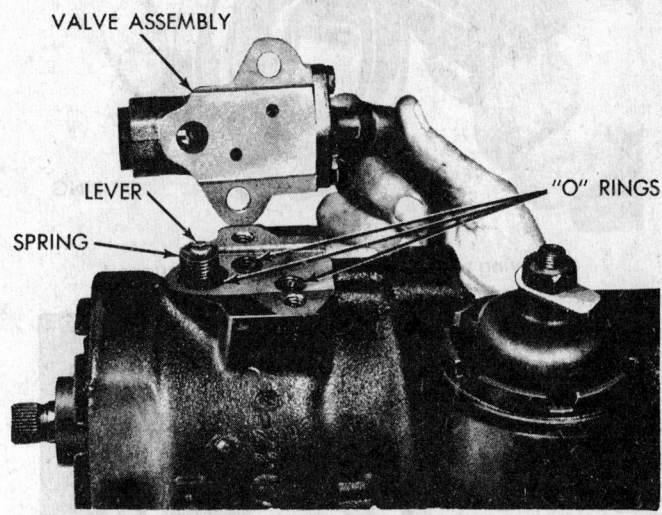

Removing valve body assembly

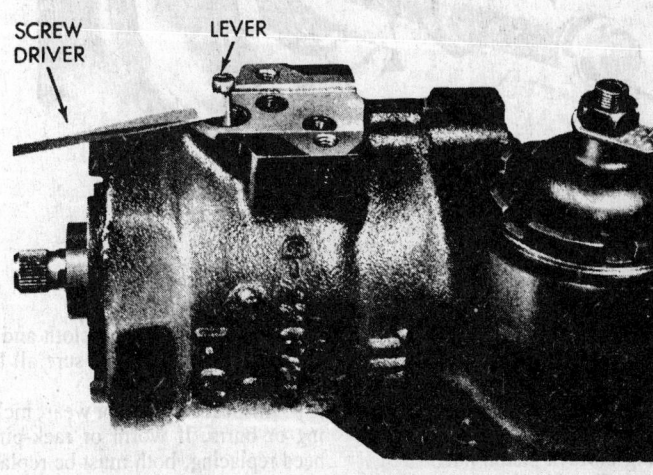

Removing pilot lever

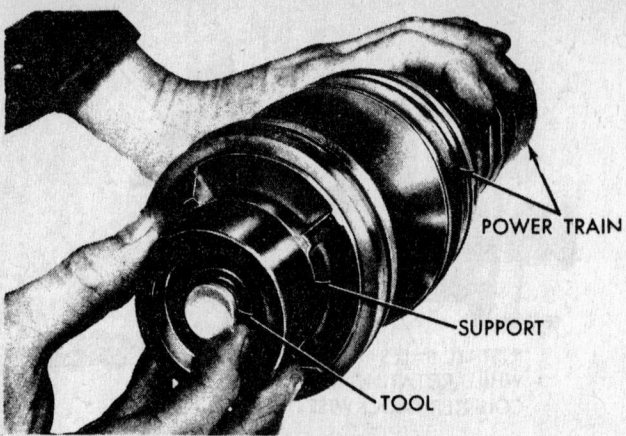

POWER TRAIN

SUPPORT

TOOL

Retaining bearing rollers with arbor tool

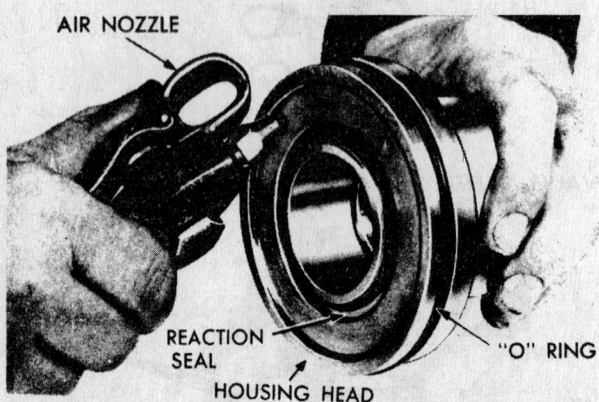

AIR NOZZLE

REACTION SEAL

HOUSING HEAD

"O" RING

Removing reaction seal from wormshaft support

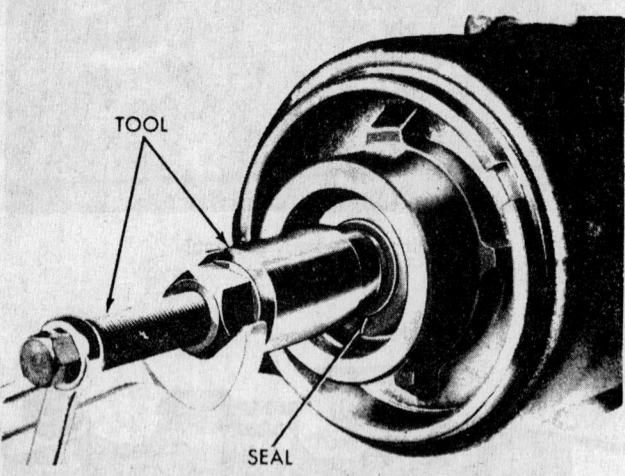

TOOL

SEAL

Removing worm shaft oil seal

RACK-PISTON NUT AND WORM

Disassembly & Assembly

1. Remove and discard piston ring and back-up O-ring on rack piston nut.

2. Remove ball guide clamp and return guide.

3. Place nut on clean cloth and remove ball retaining tool. Make sure all balls are removed.

4. Inspect all parts for wear, nicks, scoring or burrs. If worm or rack-pinion nut need replacing, both must be replaced as a matched pair.

5. In reassembling reverse the above.

NOTE: When assembling, alternate black and white balls, and install guide and clamp. Packing with grease helps in holding during assembly. When new balls are used, various sizes are available and a selection must be made to secure proper torque when making the high point adjustment.

RACK-PISTON NUT AND WORM ASSEMBLY

Installation

1. Install in reverse of removal procedure.

2. In all cases use new O-ring seals.

3. Make adjustments as previously described.

Chrysler Full-Time Power Steering (Constant Control Type)

The Chrysler Corporation Constant Control Type Power Steering Gear System consists of a hydraulic pressure pump, a power steering gear and connecting hoses.

The power steering gear housing contains a gear shaft and sector gear, a power piston with gear teeth milled into the side of the piston which is in constant mesh with the gear shaft sector teeth, a worm shaft which connects the steering wheel to the power piston through a coupling. The worm shaft is geared to the piston through recirculating ball contact.

A pivot lever is fitted into the spool valve at the upper end and into a drilled hole in the center thrust bearing race at the lower end. The center thrust bearing race is held firmly against the shoulder of the worm shaft by two thrust bearings, bearing races and an adjusting nut. The pivot lever pivots in the spacer which is held in place by the pressure plate.

When the steering wheel is turned to the left the worm shaft moves out of the power piston a few thousandths of an inch, the center thrust bearing race moves the same distance since it is clamped to the worm shaft. The race thus tips the pivot lever and moves the spool valve down, allowing oil under pressure to flow into the left-turn power chamber and force the power piston down. As the power piston moves, it rotates the cross-shaft sector gear and, through the steering linkage, turns the front wheels.

On a right turn the worm shaft moves into the power piston, the center thrust bearing race thus tips the pivot lever and moves the spool valve up, allowing oil under pressure to flow into the right power chamber and force the power piston up.

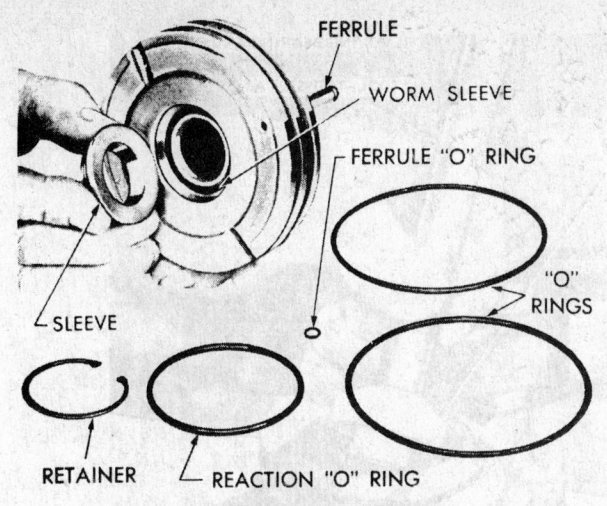

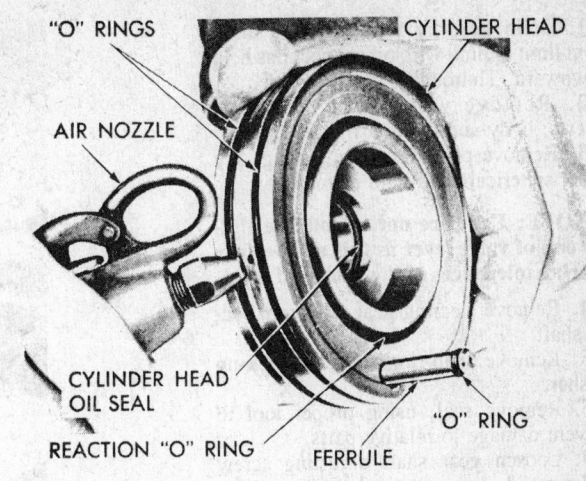

Removing cylinder head oil seal

Removing reaction seal from cylinder head

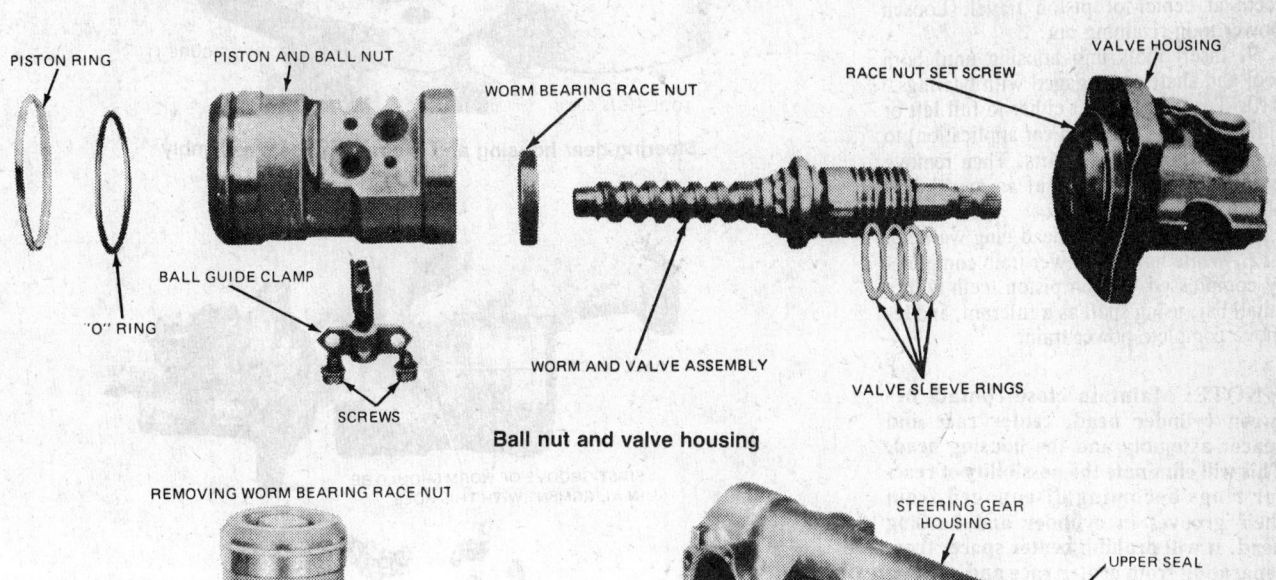

Ball nut and valve housing

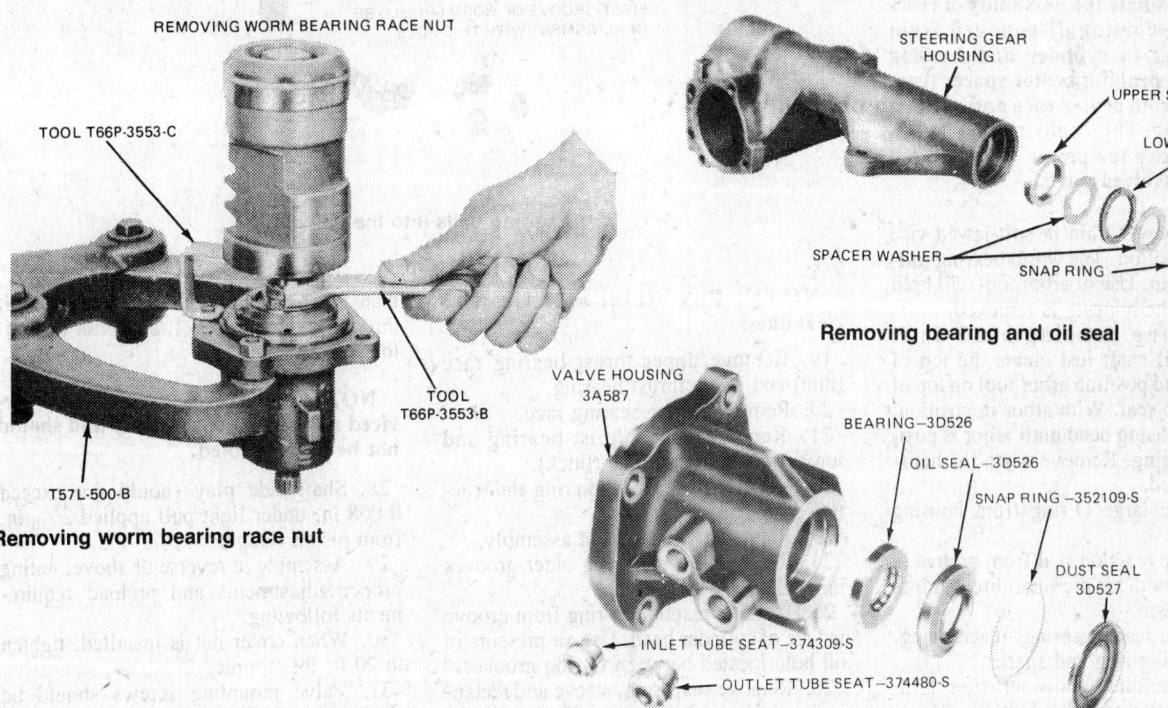

Removing worm bearing race nut

Removing bearing and oil seal

Valve housing disassembled

Reconditioning

1. Drain gear by turning worm shaft from limit to limit with oil connections held downward. Thoroughly clean outside.

2. Remove valve body attaching screws, body and three O-rings.

3. Remove pivot lever and spring. Pry under spherical head with a small bar.

NOTE: Use care not to collapse slotted end of valve lever as this will destroy bearing tolerances of the spherical head.

4. Remove steering gear arm from sector shaft.

5. Remove snap-ring and seal back-up washer.

6. Remove seal, using proper tool to prevent damage to relative parts.

7. Loosen gear shaft adjusting screw locknut and remove gear shaft cover nut.

8. Rotate wormshaft to position sector teeth at center of piston travel. Loosen power train retaining nut.

9. Insert tools into housing until both tool and shaft are engaged with bearings.

10. Turn worm shaft either to full left or full right (depending on car application) to compress power train parts. Then remove power train retaining nut as mentioned above.

11. Remove housing head tang washer.

12. While holding power train completely compressed, pry on piston teeth with a small bar, using shaft as a fulcrum, and remove complete power train.

NOTE: Maintain close contact between cylinder head, center race and spacer assembly and the housing head. This will eliminate the possibility of reactor rings becoming disengaged from their grooves in cylinder and housing head. It will prohibit center spacer from separating from center race and cocking in the housing. This could make it impossible to remove the power train without damaging involved parts.

13. Place power train in soft-jawed vise in vertical position. The worm bearing rollers will fall out. Use of arbor tool will hold roller when the housing is removed.

14. Raising housing head until wormshaft oil shaft just clears the top of wormshaft and position arbor tool on top of shaft and into seal. With arbor in position, pull up on housing head until arbor is positioned in bearing. Remove when the housing is removed.

15. Remove large O-ring from housing head groove.

16. Remove reaction seal from groove in face of head with air pressure directed into ferrule chamber.

17. Remove reactor spring, reactor ring, worm balancing ring and spacer.

18. While holding wormshaft from turning, turn nut with enough force to release staked portions from knurled section and remove nut.

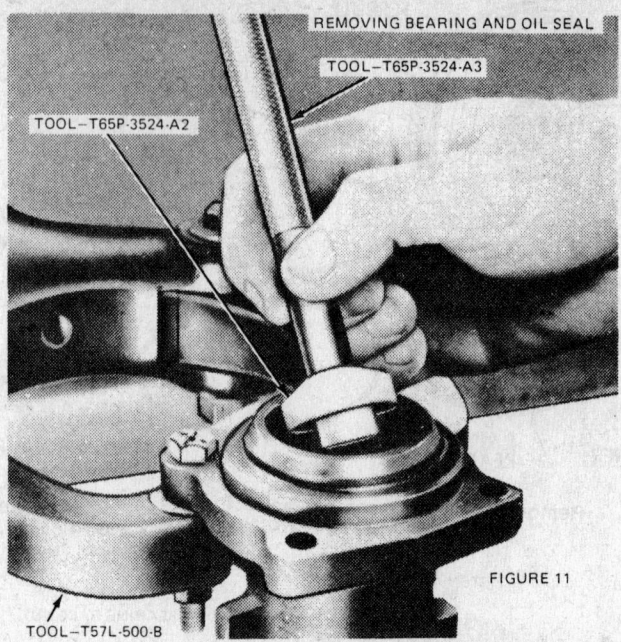

FIGURE 11

Steering gear housing and sector shaft seal assembly

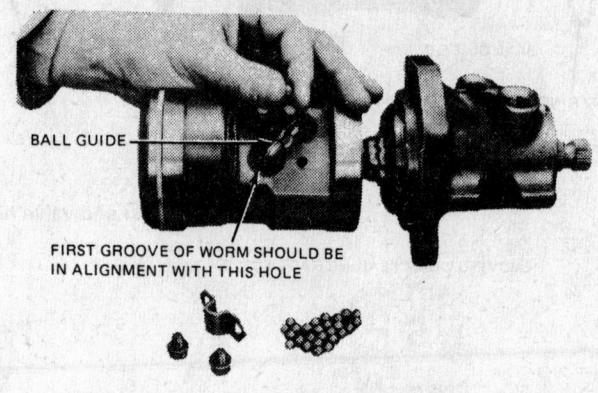

Loading balls into the ball guide

NOTE: Pay strict attention to cleanliness.

19. Remove upper thrust bearing race (thin) and upper thrust bearing.

20. Remove center bearing race.

21. Remove lower thrust bearing and lower thrust bearing race (thick).

22. Remove lower reaction ring and reaction spring.

23. Remove cylinder head assembly.

24. Remove O-rings from outer grooves in head.

25. Remove reaction O-ring from groove in face of cylinder head. Use air pressure in oil hole located between O-ring grooves.

26. Remove snap-ring, sleeve and rectangular oil seal from cylinder head counterbore.

27. Test wormshaft operation. Not more than 2 inch lbs. should be required to turn it through its entire travel, and with a 15 ft. lb. side load.

NOTE: The worm and piston is serviced as a complete assembly and should not be disassembled.

28. Shaft side play should not exceed 0.008 in. under light pull applied $2\frac{5}{16}$ in. from piston flange.

29. Assemble in reverse of above, noting proper adjustments and preload requirements following.

30. When cover nut is installed, tighten to 20 ft. lbs. torque.

31. Valve mounting screws should be tightened to 200 inch lbs. torque.

32. With hoses connected, system bled, and engine idling roughly, center valve unit

until not self-steering. Tap on head of valve body attaching screws to move valve body up, and tap on end plug to move valve body down.

33. With steering gear on center, tighten gear shaft adjusting screw until lash just disappears.

34. Continue to tighten ⅜ to ½ turn and tighten locknut to 50 ft. lbs.

Ford Integral Power Steering Gear

The Ford integral power steering unit is a torsion-bar type.

The torsion bar power steering unit includes a worm and one-piece rack piston, which is meshed to the gear teeth on the steering sector shaft. The unit also includes a hydraulic valve, valve actuator, input shaft and torsion bar assembly which are mounted on the end of the worm shaft and operated by the twisting action of the torsion bar.

The torsion-bar type of power steering gear is designed with the one piece rack-piston, worm and sector shaft in one housing and the valve spool in an attaching housing. This makes possible internal fluid passages between the valve and cylinder, thus eliminating all external lines and hoses, except the pressure and return hoses between the pump and gear assembly.

The power cylinder is an integral part of the gear housing. The piston is double acting, in that fluid pressure may be applied to either side of the piston.

A selective metal shim, located in the valve housing of the gear is for the purpose of tailoring steering gear efforts. If efforts are not within specifications they can be changed by increasing or decreasing shim thickness as follows:

1. Efforts heavy to the left; increase shim thickness.

2. Efforts light to the left; decrease shim thickness.

Adjustments

The only adjustment which can be performed is the total over center position load, to eliminate excessive lash between the sector and rack teeth.

1. Disconnect the Pitman arm from the sector shaft.

2. Disconnect the fluid return line at the reservoir, at the same time cap the reservoir return line pipe.

3. Place the end of the return line in a clean container and cycle the steering wheel in both directions as required, to discharge the fluid from the gear.

4. Turn the steering wheel to 45 degrees from the left stop.

5. Using an in. lb. torque wrench on the steering wheel nut, determine the torque required to rotate the shaft slowly through an approximately ⅛ turn from the 45 degree position.

6. Turn the steering gear back to center, then determine the torque required to rotate the shaft back and forth across the center position. Loosen the adjuster nut, and turn the adjuster screw until the reading is 11–12 inch lbs. greater than the torque 45 degrees from the stop. Tighten the lock nut while holding the screw in place.

7. Recheck the readings and replace the Pitman arm and the steering wheel hub cover.

8. Correct the fluid return line to the reservoir and fill the reservoir with specified lubricant to the proper level.

VALVE CENTERING SHIM

Removal & Installation

1. Hold the steering gear over a drain pan in an inverted position and cycle the input shaft several times to drain the remaining fluid from the gear.

2. Mount the gear in a soft-jawed vise.

3. Turn the input shaft to either stop then, turn it back approximately 1¾ turns to center the gear.

4. Remove the two sector shaft cover attaching screws, the brake line bracket and the identification tag.

5. Tap the lower end of the sector shaft with a soft-faced hammer to loosen it, then lift the cover and shaft from the housing as an assembly. Discard the O-ring.

6. Remove the four valve housing attaching bolts. Lift the valve housing from the steering gear housing while holding the piston to prevent it from rotating off the worm shaft.

7. Remove the valve housing and the lube passage O-rings and discard them.

8. Place the valve housing, worm and piston assembly in the bench mounted holding fixture with the piston on the top.

9. Rotate the piston upward (back off) 3½ turns.

10. Insert tool T66P–3553–C or equivalent (with the arm facing away from the piston) into a bolt hole in the valve housing. Rotate the arm into position under the piston.

11. Loosen the Allen head race nut set screw from the valve housing.

12. Using tool T66P–3553–B or equivalent, loosen the worm bearing race nut.

13. Lift the piston-worm assembly from the valve housing. During removal hold the piston to prevent it from spinning off at the shaft.

14. Change the power steering valve centering shim.

15. Install the piston-worm assembly into the valve housing. Hold the piston worm to prevent it from spinning off of the shaft.

16. Install the worm bearing race nut and torque to 2–8 inch lbs. using tool T66P–3553–B or equivalent.

17. Install the race nut set screw (Allen head) through the valve housing.

18. Rotate the piston upward (back off)

½ turn and remove tool T66P–3553–C or equivalent.

19. Remove the valve housing, worm, and piston assembly from the holding fixture.

20. Position a new lube passage O-ring in the counterbore of the gear housing.

21. Apply vaseline to the teflon seal on the piston.

22. Place a new O-ring on the valve housing.

23. Slide the piston and valve into the gear housing being careful not to damage the teflon seal.

24. Align the lube passage in the valve housing with the one in the gear housing, and install but do not tighten the attaching bolts.

25. Rotate the ball nut so that the teeth are in the same place as the sector teeth. Tighten the four valve housing attaching bolts to 35–45 ft. lbs.

26. Position the sector shaft cover O-ring in the steering gear housing. Turn the input shaft as required to center the piston.

27. Apply vaseline to the sector shaft journal; then, position the sector shaft and cover assembly in the gear housing. Install the brake line bracket, steering gear identification tag and the two sector shaft cover attaching studs.

28. Position an inch lb. torque wrench on the gear input shaft and adjust the meshload to approximately 4 inch lbs. Then, torque the sector shaft cover attaching studs to 55–70 ft. lbs.

29. After the cover attaching bolts have been tightened to specification, adjust the mesh load to 17 inch lbs. with an inch lb. torque wrench.

STEERING GEAR DISASSEMBLY

1. Hold the steering gear over a drain pan in an inverted position and cycle the input shaft several times to drain the remaining fluid from the gear.

2. Mount the gear in a soft-jawed vise.

3. Remove the lock nut from the adjusting screw.

4. Turn the input shaft to either stop then, turn it back approximately 1¾ turns to center the gear.

5. Remove the two sector shaft cover attaching studs, the brake line bracket and the identification tag.

6. Tap the lower end of the sector shaft with a soft-hammer to loosen it, then lift the cover and shaft from the housing as an assembly. Discard the O-ring.

7. Turn the sector shaft cover counterclockwise off the adjuster screw.

8. Remove the four valve housing attaching bolts. Lift the valve housing from the steering gear housing while holding the piston to prevent it from rotating off the worm shaft. Remove the valve housing and the lube passage O-rings and discard them.

9. Stand the valve body and piston on end with the piston end down. Rotate the input shaft counterclockwise out of the piston allowing the ball bearings to drop into the piston.

10. Place a cloth over the open end of the piston and turn it upside down to remove the balls.

11. Remove the two screws that attach the ball guide clamp to the ball nut and remove the clamp and the guides.

12. Install the valve body assembly in the holding fixture (do not clamp in a vise) and loosen the race nut screw (Allen head) from the valve housing and remove the worm bearing race nut.

13. Carefully slide the input shaft, worm and valve assembly out of the valve housing. Due to the close diametrical clearance between the spool and housing, the slightest cocking of the spool may cause it to jam in the housing.

14. Remove the shim from the valve housing bore.

Valve Housing

1. Remove the dust seal from the rear of the valve housing and discard the seal.

2. Remove the snap-ring from the valve housing.

3. Turn the fixture to place the valve housing in an inverted position.

4. Insert special tool in the valve body assembly opposite the seal end and gently tap the bearing and seal out of the housing.

Discard the seal. Caution must be exercised when inserting and removing the tool to prevent damage to the valve bore in the housing.

5. Remove the fluid inlet and outlet tube seats with an EZ-out if they are damaged.

6. Coat the fluid inlet and outlet tube seats with vaseline and position them in the housing. Install and tighten the tube nuts to press the seats to the proper location.

7. Coat the bearing and seal surface of the housing with a film of vaseline.

8. Seat the bearing in the valve housing. Make sure that the bearing is free to rotate.

9. Dip the new oil seal in gear lubricant; then, place it in the housing with the metal side of the seal facing outward. Drive the seal into the housing until the outer edge of seal does not quite clear the snap-ring groove.

10. Place the snap-ring in the housing; then, drive on the ring until the snap-ring seats in its groove to properly locate the seal.

11. Place the dust seal in the housing with the dished side (rubber side) facing out. Drive the dust seal into place. The seal must be located behind the undercut in the input shaft when it is installed.

Worm and Valve

1. Remove the snap-ring from the end of the actuator.

2. Slide the control valve spool off the actuator.

3. Install the valve spool evenly and slowly with a slight oscillating motion into the flanged end of valve housing with the valve identification groove between the valve spool lands outward, checking for freedom of valve movement within the housing working area. The valve spool should enter the housing bore freely and fall by its own weight.

4. If the valve spool is not free, check for burrs at the outward edges of the working lands in the housing and remove with a hard stone.

5. Check the valve for burrs and if burrs are found, stone the valve in a radial direction only. Check for freedom of the valve again.

6. Remove the valve spool from the housing.

7. Slide the spool onto the actuator making sure that the groove in the spool annulus is toward the worm.

8. Install the snap-ring to retain the spool. The beveled ID of the snap-ring must be assembled toward the spool.

9. Check the clearance between the spool and the snap-ring. The clearance should be between .0005–.035 in.. If the clearance is not within these limits, select a snap-ring that will allow a clearance of .002 in..

Piston and Ball Nut

1. Remove the teflon ring and the O-ring from the piston and ball nut.

2. Dip a new O-ring in gear lubricant and install it on the piston and ball nut.

3. Install a new teflon ring on the piston and ball nut being careful not to stretch it any more than necessary.

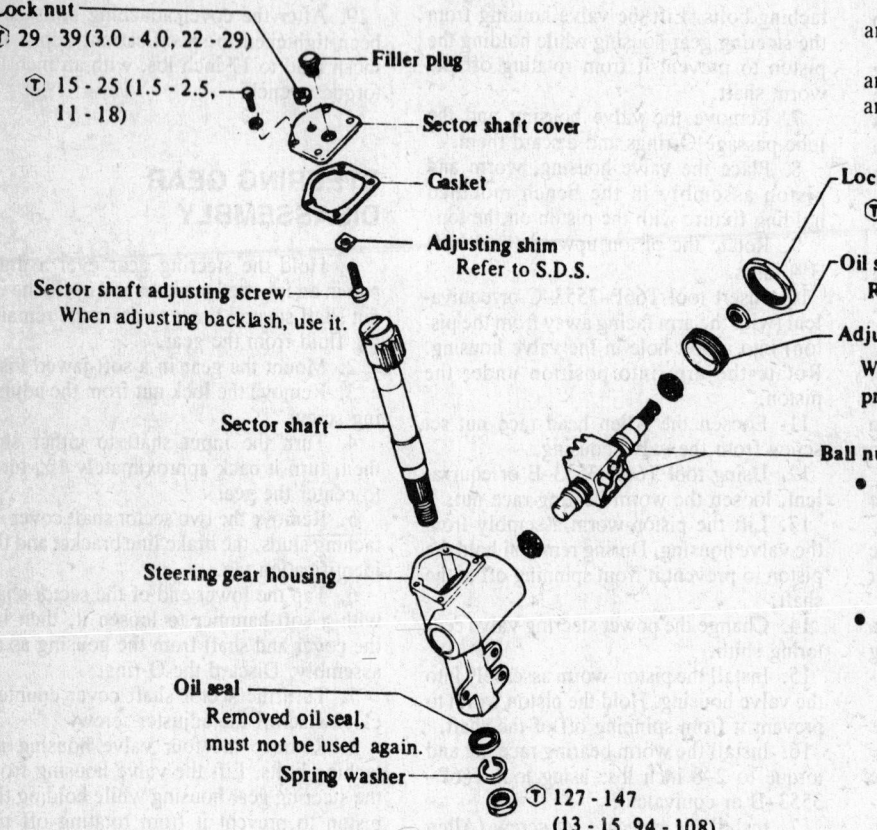

Lock nut
🔧 29 - 39 (3.0 - 4.0, 22 - 29)
🔧 15 - 25 (1.5 - 2.5, 11 - 18)
Filler plug
Sector shaft cover
Gasket
Adjusting shim
Refer to S.D.S.
Sector shaft adjusting screw
When adjusting backlash, use it.
Sector shaft
Steering gear housing
Oil seal
Removed oil seal, must not be used again.
Spring washer
🔧 127 - 147 (13 - 15, 94 - 108)

Lock nut
🔧 245 - 314 (25 - 32, 181 - 231)
Oil seal
Removed oil seal, must not be used again.
Adjusting plug
When adjust worm bearing preload, use it.
Ball nut and worm assembly
• Be careful not to allow ball nut to run down to either end of worm.
The ends of ball guides will be damaged if nut is rotated until it stops at the end of worm.
• Do not detach ball nut from worm shaft assembly.
If necessary, replace ball nut and worm assembly with sector shaft.

🔧 : N·m (kg-m, ft-lb)

Datsun manual steering gear—typical

Steering Gear Housing

1. Remove the snap-ring and the spacer washer from the lower end of the steering gear housing.

2. Remove the lower seal from the housing. Lift the spacer washer from the housing.

3. Remove the upper seal in the same manner as the lower seal. Some housings require only one seal and one spacer.

4. Dip both sector shaft seals in gear lubricant.

5. Apply lubricant to the sector shaft seal bore of the housing and position the sector shaft inner seal into the housing with the lip facing inward. Press the seal into place. Place a spacer washer (0.090 in.) on top of the seal and apply more lubricant to the housing bore.

6. Place the outer seal in the housing with the lip facing inward and press it into place. Then, place a 0.090 in. spacer washer on top of the seal.

7. Position the snap-ring in the housing. Press the snap-ring into the housing to properly locate the seals and engage the snap-ring in the groove.

STEERING GEAR ASSEMBLY

Do not clean, wash, or soak seals in cleaning solvent.

1. Mount the valve housing in the holding fixture with the flanged end up.

2. Place the required thickness valve spool centering shim in the housing.

3. Carefully install the worm and valve in the housing.

4. Install the race nut in the housing and torque it to 42 ft. lbs.

5. Install the race nut set screw (Allen head) through the valve housing and torque to 20–25 inch lbs.

6. Place the piston on the bench with the ball guide holes facing up. Insert the worm shaft into the piston so that the first groove is in alignment with the hole nearest to the center of the piston.

7. Place the ball guide in the piston. Place the 27 to 29 balls, depending on the piston design, in the ball guide turning the worm in a clockwise direction as viewed from the input end of the shaft. If all of the balls have not been fed into the guide upon reaching the right stop, rotate the input shaft in one direction and then in the other while installing the balls. After the balls have been installed, do not rotate the input shaft or the piston more than 3½ turns off the right stop to prevent the balls from falling out of the circuit.

8. Secure the guides in the ball nut with the clamp.

9. Position a new lub passage O-ring in the counterbore of the gear housing.

10. Apply petroleum jelly to the teflon seal on the piston.

11. Place a new O-ring on the valve housing.

12. Slide the piston and valve into the gear housing being careful not to damage the teflon seal.

13. Align the lube passage in the valve housing with the one in the gear housing and install but do not tighten the attaching bolts.

14. Rotate the ball nut so that the teeth are in the same plane as the sector teeth. Tighten the four valve housing attaching bolts to 35–45 ft. lbs.

15. Position the sector shaft cover O-ring in the steering gear housing. Turn the input shaft as required to center the piston.

16. Apply vaseline to the sector shaft journal then position the sector shaft and cover assembly in the gear housing. Install the brake line bracket, the steering identification tag and two sector shaft cover attaching bolts. Torque the bolts to 55–70 ft. lbs.

17. Attach an in. lb. torque wrench to the input shaft. Adjust the mesh load to 17 inch lbs.

IMPORTED TRUCKS

Datsun/Nissan

Manual Steering Gear
DISASSEMBLY

1. Place the steering gear in a vise using tool KV48100301 or its equivalent.

2. Set the worm gear in a straight ahead position.

3. Remove the sector shaft cover retaining bolts.

4. Pull the sector shaft out, being careful not to damage the oil seal. Make sure the worm gear is set in a straight ahead position and do not remove the shaft needle bearings from the steering gear housing.

5. Remove the sector shaft cover, then if necessary the sector shaft oil seal.

6. Loosen the adjusting plug lock nut using tool K48101500 or its equivalent.

7. Draw out the worm gear with the worm bearing.

--- CAUTION ---
Be careful not to allow the ball nut to run down to either end of the worm or damage to the ball guides may result.
---

8. Remove the seal from the adjusting plug using the appropriate tool.

ASSEMBLY & ADJUSTMENT

NOTE: **Before reassembly, clean and lubricate all parts with gear fluid, and fill the space between the sealing lips of the new sector shaft and adjusting plug oil seals with the recommended multi-purpose grease.**

1. Position the worm gear assembly with the worm bearing assembly in the gear housing.

2. Install the adjusting plug using tool KV4801400.

3. Adjust the worm bearing preload.

   a. Rotate the worm shaft a few turns in both directions to set worm bearing.

   b. Adjust the worm bearing preload by turning the adjusting plug in the tighten direction to 1.7–5.2 inch lbs.

4. Apply suitable sealant around the lock nut inner surface.

5. Tighten the locknut using the appropriate tools. Recheck the worm bearing preload.

6. Select a suitable adjusting shim and adjust the end play between the sector shaft and the adjusting screw (0.0004–0.0012 in.).

7. Coat the seal face with gear fluid, then press the oil seal to the steering gear housing using the appropriate tool.

8. Install the sector cover on the adjusting screw with the sector shaft.

9. Set the worm gear in a straight ahead position.

10. Insert the sector shaft and sector cover assembly with gasket into the gear housing, using care not to scratch the oil seal.

11. Tighten the sector cover to gear housing bolts to 11–18 ft. lbs.

12. Pour the recommended gear oil into through the filler hole and install the filler plug.

13. Tighten the adjusting screw so that the gear preload is within specification.

Power Steering Gear
TURNING TORQUE MEASUREMENT

NOTE: **Before disassembly, measure the turning torque.**

1. Measure the turning torque at a 360° position as follows:

   a. Install the steering gear in a vice using tool KV48100301 or its equivalent.

   b. Turn the stub shaft all the way to the right and left several times.

   c. Measure the turning torque at 360° position from the straight ahead position. (6.1–10.4 inch lbs.). If it is beyond specification, the gear must be replaced as an assembly.

2. Measure the turning torque at a straight ahead position as follows:

   a. Set the worm gear in a straight ahead position. This position is where the stub shaft is turned 1.9 turns from the lock position.

   b. Measure the turning torque at this position it should be 0.9–3.5 inch lbs., (higher than at 360°).

3. After the adjustment is completed, tighten the lock nut. If the turning torques are not within specifications, replace the gear assembly.

DISASSEMBLY

1. Remove the adjusting screw locknut and replace the O-ring.

2. Install the gear on tool KV48100301.

3. Set the stub shaft in a straight ahead position. The straight ahead position is where the stub shaft is turned two turns by 45° from the lock position.

4. Disconnect the sector shaft cover bolt.

NOTE: Do not loosen the adjusting screw locknut. Turning the lock nut at this time could cause damage to the O-ring resulting in an oil leak.

5. Knock out the end of the sector shaft approximately 0.79 in..

6. Connect a roll of plastic film to the sector shaft. (1 mm thick, length and width approximately 8 in.).

7. Attach the plastic film to the two bearings located inside the gear housing while simultaneously pulling out the sector shaft so that the bearings will not drop into the housing.

8. Carefully pry out the dust seal so that it will not damage the inner side of the gear housing.

9. Remove the snap ring.

10. Carefully remove the special washer and oil seal.

11. Remove the O-ring.

12. Remove the torx screws, then remove the rear housing together with the worm gear assembly.

NOTE: When the worm assembly is removed, the piston may turn and come off under its own weight. Hold the piston to prevent it from turning. If piston to rear housing clearance exceeds 1.38 in. by loosening, recirculating ball will be out of groove of worm; do not reinstall the piston but replace the entire assembly.

13. Remove the O-rings.

14. Remove the snap ring, then the rear cover.

15. Remove the O-ring then the oil seal.

ASSEMBLY

1. Apply a thin coat of vaseline to a new adjusting screw lock nut O-ring and insert it into the groove.

2. Apply a thin coat of vaseline to the new oil seal and dust seal then press in the new seal and the special washer.

3. Install a new snap ring into the gear housing. Turn the snap ring to make sure it fits into the groove and always install the snap ring with its rounded edges facing the oil seal.

4. Press in a new dust seal.

5. Apply a thin coat of vaseline to a new O-ring then fit the O-ring into the sector shaft cover.

6. Set the piston rack at a straight ahead position. Turn the piston rack about 10° to

15° toward yourself with your finger. This is for smooth insertion of the sector gear.

7. Wrap vinyl tape around serration area of sector shaft to prevent damage to the oil seal lip during insertion.

8. With the plastic film wraped around the sector shaft as in Step 7 of disassembly, insert the sector shaft into the gear housing. Gradually remove the plastic film being careful not to drop the bearings into the gear housing.

9. Tighten the sector shaft cover bolts to 20–24 ft. lbs. in a criss-cross pattern.

10. At this time check the turning torque and steering gear preload as described earlier. If there is a great difference between the values before and after disassembly, it will be necessary to replace the assembly.

11. Apply a thin coat of vaseline to new rear housing O-ring then install.

12. Gradually insert the worm gear and rear housing assembly into the gear housing, being careful not to damage the oil seal and O-rings.

13. Install and tighten the torx screws in a criss-cross pattern to 20–24 ft. lbs.

14. Install a new O-ring and oil seal.

15. Install the rear cover then the snap ring. Turn the snap ring to make sure it fits into the groove and always install the snap ring with its rounded edge facing the rear cover.

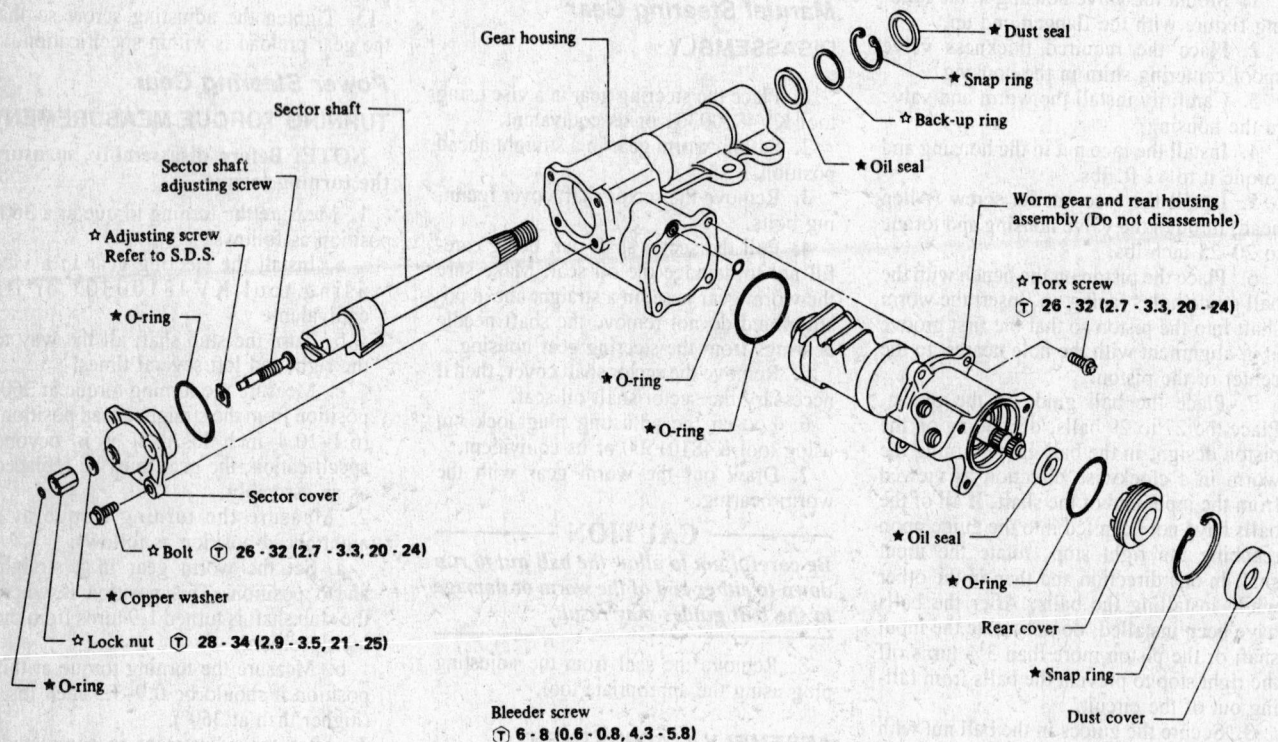

Datsun power steering gear—1984 and later

16. Select suitable adjusting shims and adjust the end play between the sector shaft and adjusting screw. End play should be 0.0004–0.0012 in.

Power Steering Pump

1982–83

1. Remove the pulley and use new washer upon reassembly.

2. Remove the cap assembly.

3. Remove the tank and O-ring.

4. Remove the connector bolt, washers and joint.

5. Remove the connector then remove the O-ring.

6. Remove the rear O-ring as follows: With the tank and tank O-ring removed, remove the bracket, snap ring, rear cover and spring then remove the O-ring.

NOTE: Do not face the rear cover side of the housing downwards, otherwise the side plate may fall. If dropped do not attempt to reassemble them, rather replace the oil pump assembly.

7. Replace the pulley shaft oil seal as follows: With the pulley removed remove the snap ring, then remove the pulley shaft assembly. Remove the oil seal.

ASSEMBLY

1. Clean all disassembled parts in a suitable cleaning solvent.

2. Always use new washers, seals and O-rings, and coat all O-rings with a thin coat of vaseline before installing.

3. Install a new O-ring then install the tank. Temporarily tighten the bolt and then after the pump is installed on the vehicle then tighten the bolt securely.

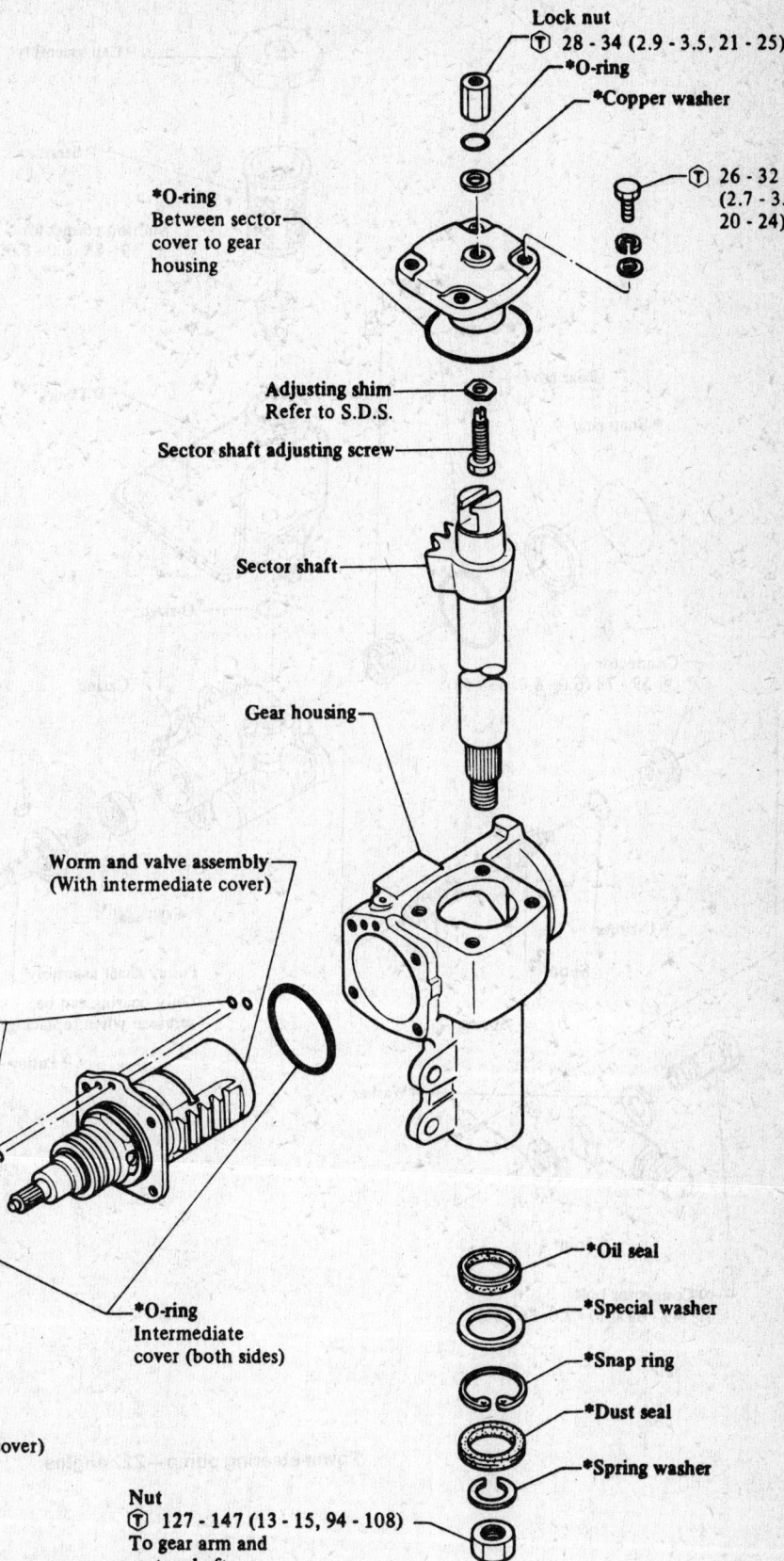

Lock nut
Ⓣ 28 - 34 (2.9 - 3.5, 21 - 25)
*O-ring
*Copper washer
Ⓣ 26 - 32 (2.7 - 3.3, 20 - 24)

*O-ring
Between sector cover to gear housing

Adjusting shim
Refer to S.D.S.

Sector shaft adjusting screw

Sector shaft

Gear housing

Worm and valve assembly (With intermediate cover)

*O-ring

*O-ring
Intermediate cover (both sides)

Rear housing

Dust cover

*Oil seal
(In rear cover)

Ⓣ 26 - 32
(2.7 - 3.3, 20 - 24)

* : Do not reuse once removed.
Ⓣ : N·m (kg-m, ft-lb)

*Oil seal
*Special washer
*Snap ring
*Dust seal
*Spring washer

Nut
Ⓣ 127 - 147 (13 - 15, 94 - 108)
To gear arm and sector shaft

Datsun power steering gear—1982-83

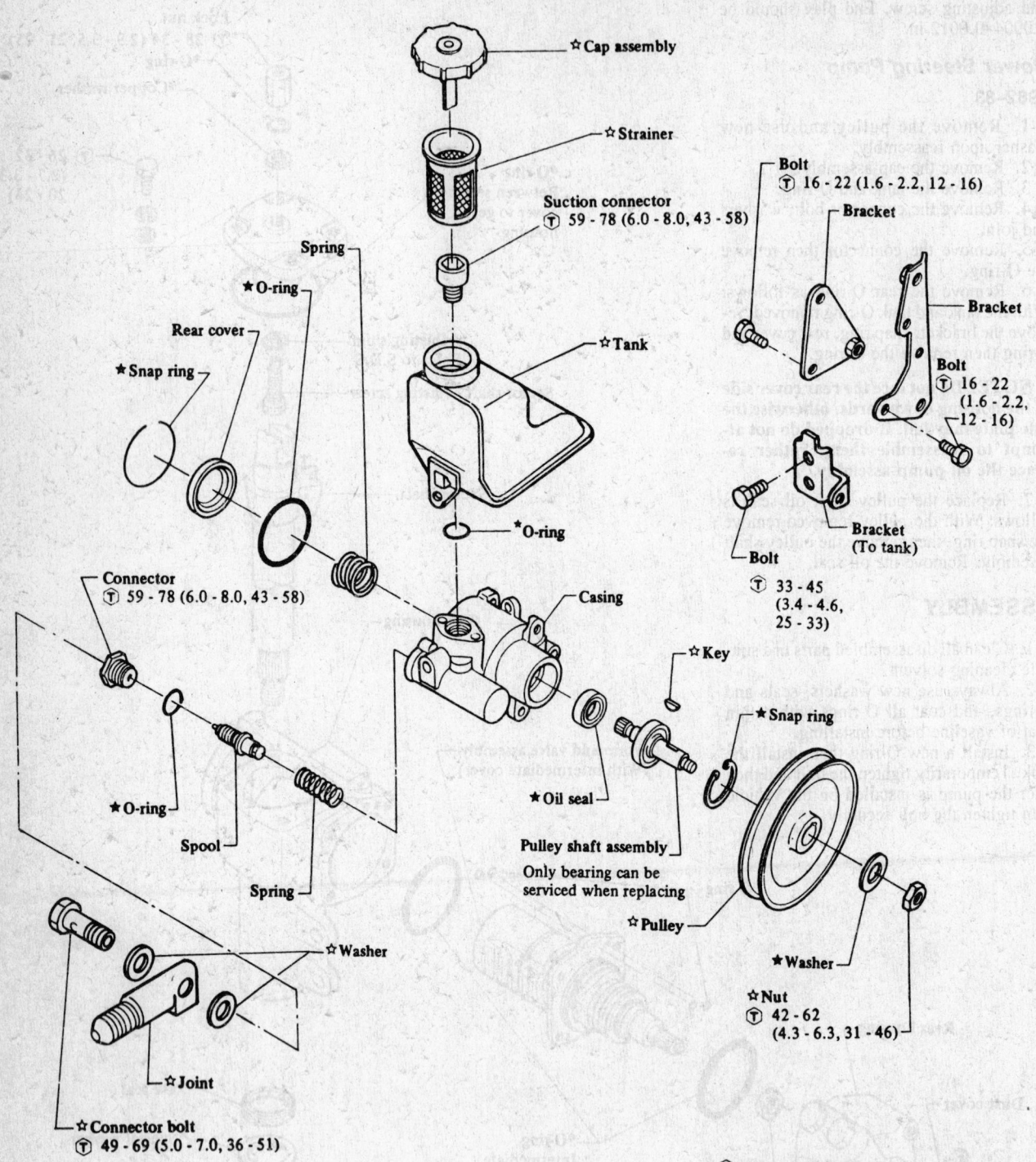

☆Cap assembly

☆Strainer

Suction connector
Ⓣ 59 - 78 (6.0 - 8.0, 43 - 58)

Spring

★O-ring

Rear cover

★Snap ring

☆Tank

★O-ring

Connector
Ⓣ 59 - 78 (6.0 - 8.0, 43 - 58)

Casing

★O-ring

Spool

Spring

☆Washer

☆Joint

☆Connector bolt
Ⓣ 49 - 69 (5.0 - 7.0, 36 - 51)

Bolt
Ⓣ 16 - 22 (1.6 - 2.2, 12 - 16)

Bracket

Bracket

Bolt
Ⓣ 16 - 22
(1.6 - 2.2,
12 - 16)

Bracket
(To tank)

Bolt

Ⓣ 33 - 45
(3.4 - 4.6,
25 - 33)

☆Key

★Snap ring

★Oil seal

Pulley shaft assembly
Only bearing can be
serviced when replacing

☆Pulley

★Washer

☆Nut
Ⓣ 42 - 62
(4.3 - 6.3, 31 - 46)

Ⓣ : N·m (kg-m, ft-lb)
★ or ☆ : are available for service replacement.
★ : always replace when disassembled.

Power steering pump—Z22 engine

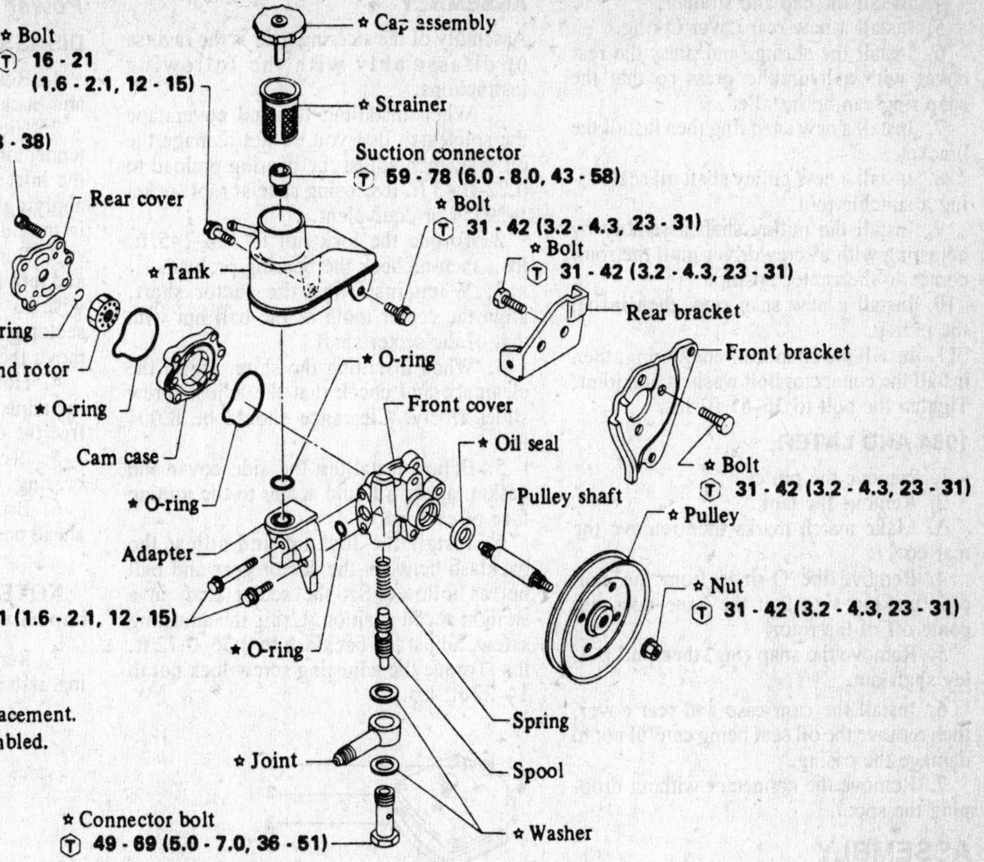

☆ Bolt
Ⓣ 16 - 21
(1.6 - 2.1, 12 - 15)

★ Cap assembly

★ Bolt
Ⓣ 38 - 52 (3.9 - 5.3, 28 - 38)

★ Strainer

Suction connector
Ⓣ 59 - 78 (6.0 - 8.0, 43 - 58)

Rear cover

☆ Bolt
Ⓣ 31 - 42 (3.2 - 4.3, 23 - 31)

★ Tank

☆ Bolt
Ⓣ 31 - 42 (3.2 - 4.3, 23 - 31)

Rear bracket

★ Snap ring

Vane and rotor

★ O-ring

★ O-ring

Front cover

Front bracket

Cam case

★ Oil seal

☆ Bolt
Ⓣ 31 - 42 (3.2 - 4.3, 23 - 31)

★ O-ring

Pulley shaft

☆ Pulley

Adapter

☆ Bolt
Ⓣ 16 - 21 (1.6 - 2.1, 12 - 15)

☆ Nut
Ⓣ 31 - 42 (3.2 - 4.3, 23 - 31)

★ O-ring

Spring

★ or ☆: available for service replacement.
★: always replace when disassembled.
Ⓣ : N·m (kg-m, ft-lb)

★ Joint

Spool

★ Connector bolt
Ⓣ 49 - 69 (5.0 - 7.0, 36 - 51)

★ Washer

Power steering pump—Z24 engine

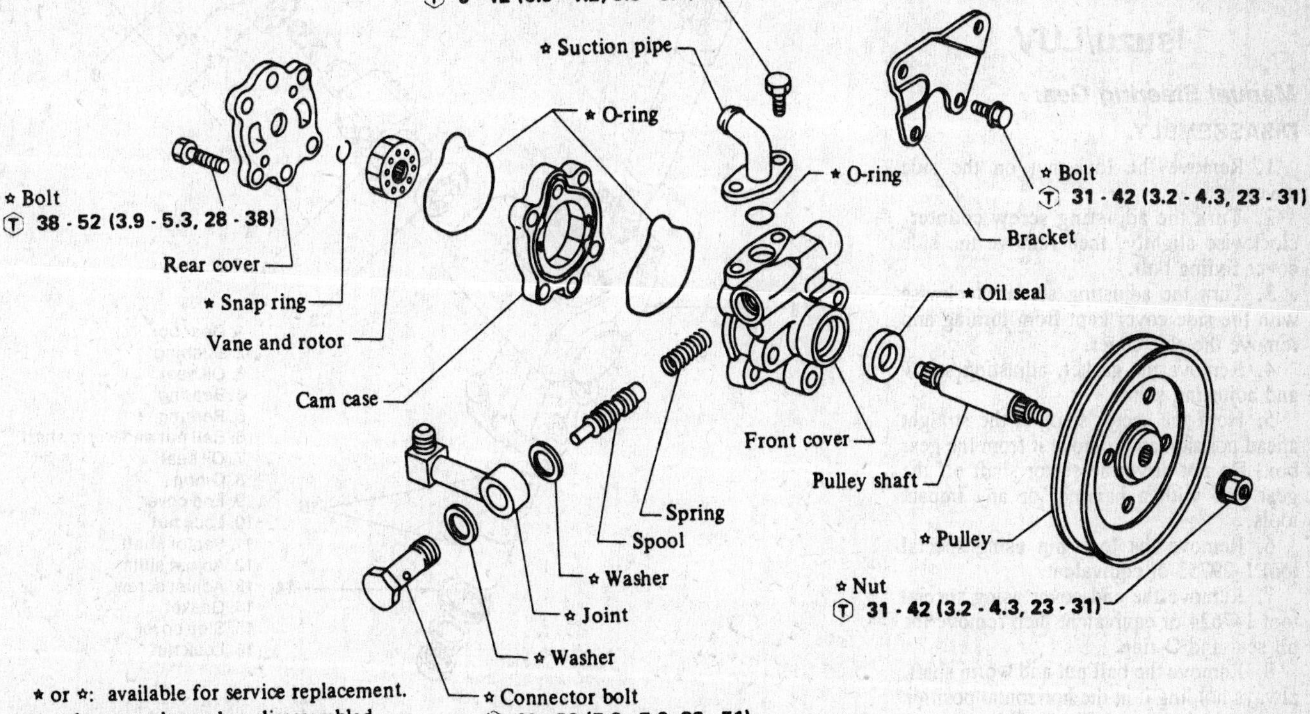

☆ Bolt
Ⓣ 9 - 12 (0.9 - 1.2, 6.5 - 8.7)

★ Suction pipe

★ O-ring

★ Bolt
Ⓣ 38 - 52 (3.9 - 5.3, 28 - 38)

★ O-ring

☆ Bolt
Ⓣ 31 - 42 (3.2 - 4.3, 23 - 31)

Rear cover

Bracket

★ Snap ring

Vane and rotor

★ Oil seal

Cam case

Front cover

Pulley shaft

Spring

☆ Pulley

Spool

★ Washer

☆ Nut
Ⓣ 31 - 42 (3.2 - 4.3, 23 - 31)

★ Joint

★ Washer

★ or ☆: available for service replacement.
★: always replace when disassembled.
Ⓣ : N·m (kg-m, ft-lb)

★ Connector bolt
Ⓣ 49 - 69 (5.0 - 7.0, 36 - 51)

Power steering pump—SD25 engine

4. Install the cap and strainer.

5. Install a new rear cover O-ring.

6. Install the spring, and press the rear cover with a hydraulic press so that the snap ring can be installed.

7. Install a new snap ring then install the bracket.

8. Install a new pulley shaft oil seal, using a suitable tool.

9. Install the pulley shaft assembly by adjusting with a screwdriver until the rotor comes to the center position.

10. Install a new snap ring, then install the pulley.

11. Install the connector and O-ring, then install the connector bolt washers and joint. Tighten the bolt to 36–51 ft. lbs.

1984 AND LATER

1. Remove the pulley.

2. Remove the tank.

3. Make match marks then remove the rear cover.

4. Remove the O-rings from the cam case making sure that the vane does not come off of the rotor.

5. Remove the snap ring, then pull pulley shaft out.

6. Install the cam case and rear cover, then remove the oil seal being careful not to damage the casing.

7. Remove the connector without dropping the spool.

ASSEMBLY

1. Assembly of the oil pump is the reverse of disassembly. Before installing O-rings and oil seals, apply a thin coat of power steering fluid to them.

Isuzu/LUV

Manual Steering Gear
DISASSEMBLY

1. Remove the lock nut on the side cover.

2. Turn the adjusting screw counterclockwise slightly, then remove the side cover fixiing bolt.

3. Turn the adjusting screw clockwise with the side cover kept from turning and remove the side cover.

4. Remove the gasket, adjusting screw and adjusting shim.

5. Hold the sector shaft in the straight ahead position and remove it from the gear box. Do not drive the sector shaft off the gear box with a hammer or any impact tools.

6. Remove the lock nut using special tool J–29753 or equivalent.

7. Remove the end cover using special tool J–7624 or equivalent then remove the oil seal and O-ring.

8. Remove the ball nut and worm shaft, always holding it in the horizontal position so that the ball-nut will not slide out.

9. Remove the bearings oil seal and bushings.

ASSEMBLY

Assembly of the steering gear is the reverse of disassembly with the following instructions:

1. When installing the end cover tape the splines so that you do not damage the oil seal, then adjust the bearing preload to 0.22–0.43 ft. lbs.,using special tool socket J–29754 or equivalent.

2. Torque the lock nut to 116–145 ft. lbs., then recheck the bearing preload.

3. When installing the sector shaft, align the center tooth of the ball nut with that of the sector shaft.

4. When installing the shim, adjust the clearance and check that the adjust screw slides freely. Clearance should be 0.004 in..

5. Before installing the side cover and gasket, apply a liquid sealer to the joining face of each part.

6. Install the locknut and adjust the backlash between the sector gear and ball nut as follows: Set the sector shaft in a straight ahead position. Using the adjusting screw, adjust the backlash to 0.36–0.72 ft. lbs. Torque the adjusting screw lock nut to 14–22 ft. lbs.

Power Steering Gear
DISASSEMBLY

1. Remove the dust cover, retaining ring and buck up ring.

2. Clean the faces of the stub shaft extended outward and plug the hose fitting on the inlet side, then remove the oil seal by applying compressed air through the hole in the outlet side.

3. Remove the adjusting screw Lock nut and turn the adjusting screw counter-clockwise to remove the preload between the sector gear and the rack piston, then remove the top cover bolts.

4. Hold the top cover stationary, turn the adjusting screw clockwise to raise and free the cover, then remove the cover.

5. Remove the O-ring and needle bearing.

6. Bring the stub shaft into a straight ahead position and remove the sector shaft.

NOTE: Do not use a hammer or any impact tools to remove the sector shaft.

7. Remove the ball nut and valve housing assembly.

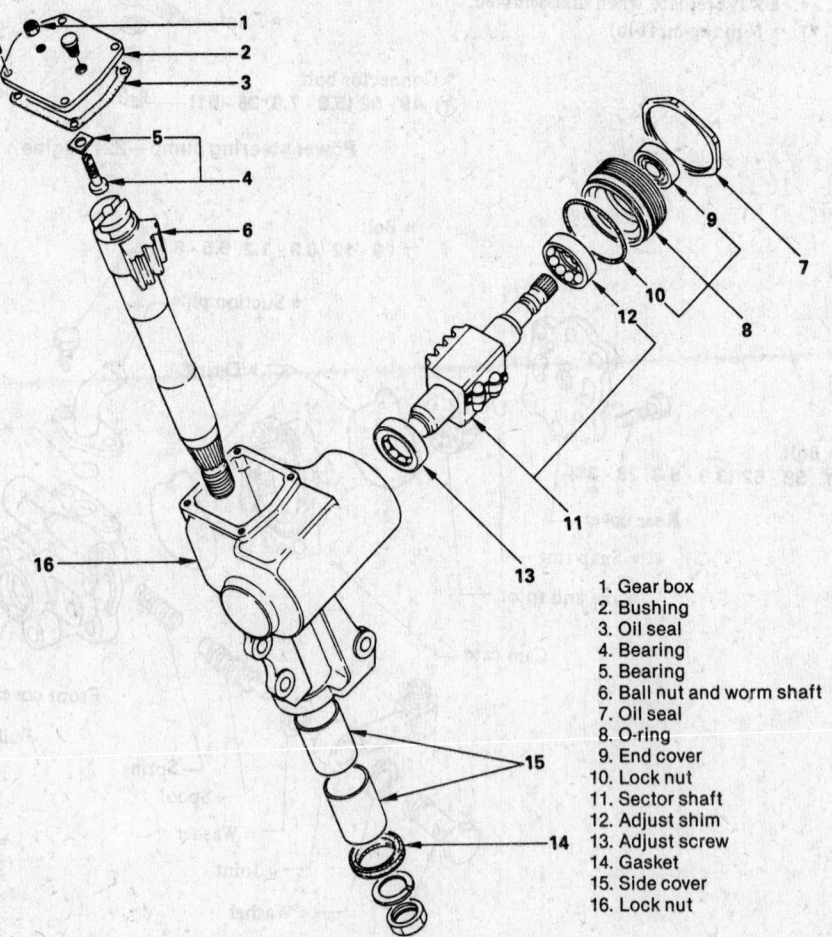

1. Gear box
2. Bushing
3. Oil seal
4. Bearing
5. Bearing
6. Ball nut and worm shaft
7. Oil seal
8. O-ring
9. End cover
10. Lock nut
11. Sector shaft
12. Adjust shim
13. Adjust screw
14. Gasket
15. Side cover
16. Lock nut

Isuzu/Luv manual steering gear

NOTE: Always keep the ball nut and valve housing assembly in the horizontal position, or the rack piston will fall off onto the end of the worm, causing the rack piston to slip out of the worm shaft and the balls to fall out.

4. Remove the remainder of the seals and bearings.

ASSEMBLY

NOTE: Before reassembly apply a thin coat of grease to the lips of all seals and O-rings.

1. Install the seal ring and dust seal the the gear box.

2. Install the needle bearing into the housing.

3. Install the two O-rings and the seal ring onto the ball nut and valve assembly.

4. Install the ball nut and valve assembly and tighten the valve housing retaining bolts to 27–30 ft. lbs.

NOTE: Be careful not to drop the O-ring fitted to the oil passage in the valve housing. Refer to the note in Step 7 of disassembly.

5. Install the oil seal using special tool J–26508, then install the back up ring and retaining ring. Turn the face with the rounded edge (outer circumference) to the oil seal.

6. Install the dust cover.

7. Tape the sector shaft serration to protect the seal ring from damage, then align the center tooth of the ball nut with that of the sector shaft and install the sector shaft.

8. Install the needle bearing, O-ring and top cover. Tighten the top cover bolts to 27–30 ft. lbs.

9. Install the lock nut and adjust the backlash between the sector gear and ball nut as follows: Set the sector shaft in a straight ahead position. Using the adjusting screw, adjust the backlash to 3.4–5.6 ft. lbs. Tighten the lock nut to 25–35 ft. lbs.

Power Steering Pump

DISASSEMBLY

1. Remove the pulley.
2. Remove the end plate retaining ring then remove the end plate.
3. Remove the pressure plate spring.
4. Remove the pump cartridge and shaft.
5. Remove the pressure plate.
6. Remove the cam.
7. Remove the retaining ring, then remove the rotor vane thrust plate and dowel pins.
8. Remove the O-rings from the pump housing.
9. Remove the control valve assembly.
10. Remove the oil seal from the housing.

ASSEMBLY

1. Assembly is the reverse of removal with the following instructions:

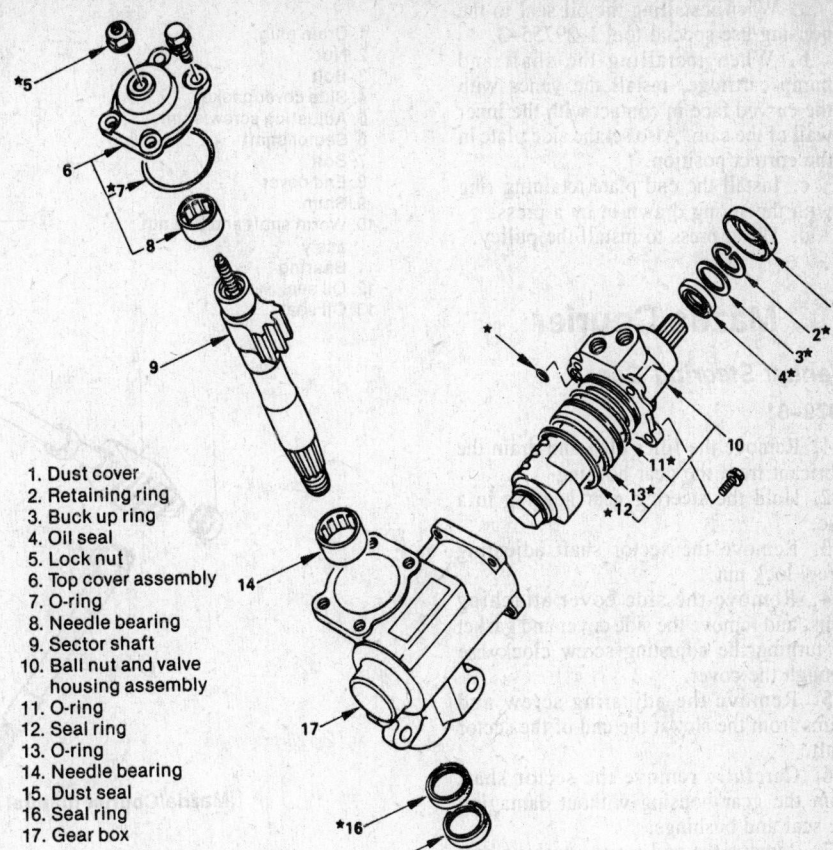

1. Dust cover
2. Retaining ring
3. Buck up ring
4. Oil seal
5. Lock nut
6. Top cover assembly
7. O-ring
8. Needle bearing
9. Sector shaft
10. Ball nut and valve housing assembly
11. O-ring
12. Seal ring
13. O-ring
14. Needle bearing
15. Dust seal
16. Seal ring
17. Gear box

★ Repair kit

Isuzu/Luv power steering gear

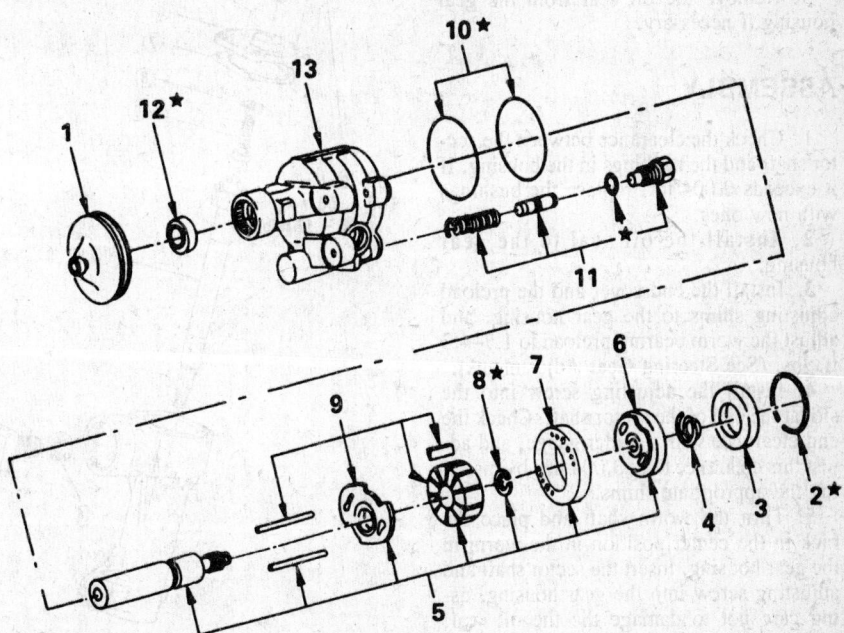

1. Pulley
2. End plate retaining ring
3. End plate
4. Pressure plate spring
5. Pump cartridge and shaft
6. Pressure plate
7. Cam
8. Retaining ring
9. Rotor, vane, thrust plate and dowel pins
10. O-rings
11. Control valve assembly
12. Oil seal
13. Pump housing

Isuzu/Luv power steering pump

a. When installing the oil seal to the housing use special tool J–29755–3.

b. When installing the shaft and pump cartridge, install the vanes with the curved face in contact with the inner wall of the cam. Also set the side plate in the correct position.

c. Install the end plate retaining ring with the spring drawn in by a press.

d. Use a press to install the pulley.

Mazda/Courier

Manual Steering Gear

1979–81

1. Remove the filler plug and drain the lubricant from the gear housing.

2. Hold the steering gear housing in a vise.

3. Remove the sector shaft adjusting screw lock nut.

4. Remove the side cover attaching bolts, and remove the side cover and gasket by turning the adjusting screw clockwise through the cover.

5. Remove the adjusting screw and shims from the slot at the end of the sector shaft.

6. Carefully remove the sector shaft from the gear housing without damaging the seal and bushings.

7. Remove the end cover attaching bolt and remove the end cover and shims.

8. Remove the worm shaft and ball nut assembly from the gear housing.

9. Remove the oil seal from the gear housing if necessary.

ASSEMBLY

1. Check the clearance between the sector shaft and the bushings in the housing. If it exceeds 0.004 in., replace the bushings with new ones.

2. Install the oil seal to the gear housing.

3. Install the end cover and the preload adjusting shims to the gear housing, and adjust the worm bearing preload to 1.7–4.3 ft. lbs. (See Steering Gear Adjustments).

4. Install the adjusting screw into the slot at the end of the sector shaft. Check the end clearance with a feeler gauge, and adjust this clearance to 0–0.004 in., by inserting the appropriate shims.

5. Turn the worm shaft and place the rack in the center position in the worm in the gear housing. Insert the sector shaft and adjusting screw into the gear housing, using care not to damage the the oil seal. Make sure the center of the sector gear is in alignment with the center of the rack.

6. Place the side cover and gasket onto the adjusting screw and turn the adjusting screw counter-clockwise until it is screwed into the proper position.

7. Install the side cover attaching bolts and tighten the bolts.

1. Drain plug
2. Nut
3. Bolt
4. Side cover/gasket
5. Adjusting screw/shim
6. Sector shaft
7. Bolt
8. End cover
9. Shim
10. Worm shaft and ball nut ass'y
11. Bearing
12. Oil seal
13. Oil seal

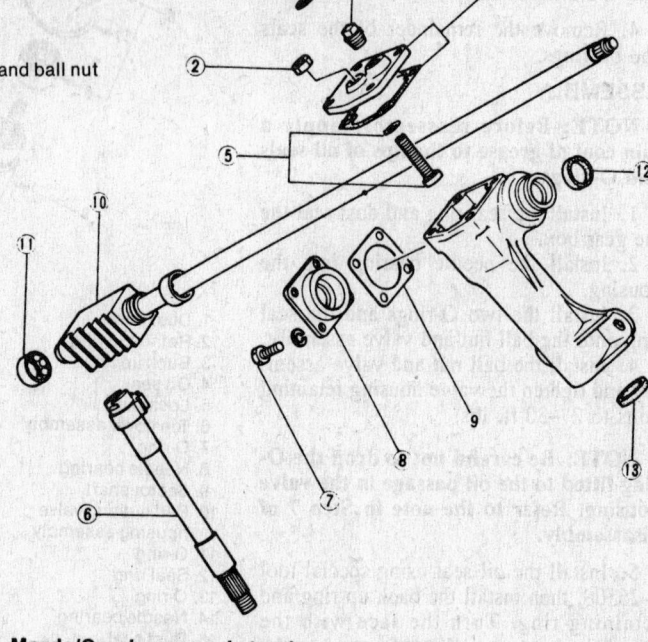

Mazda/Courier manual steering gear—1979–81

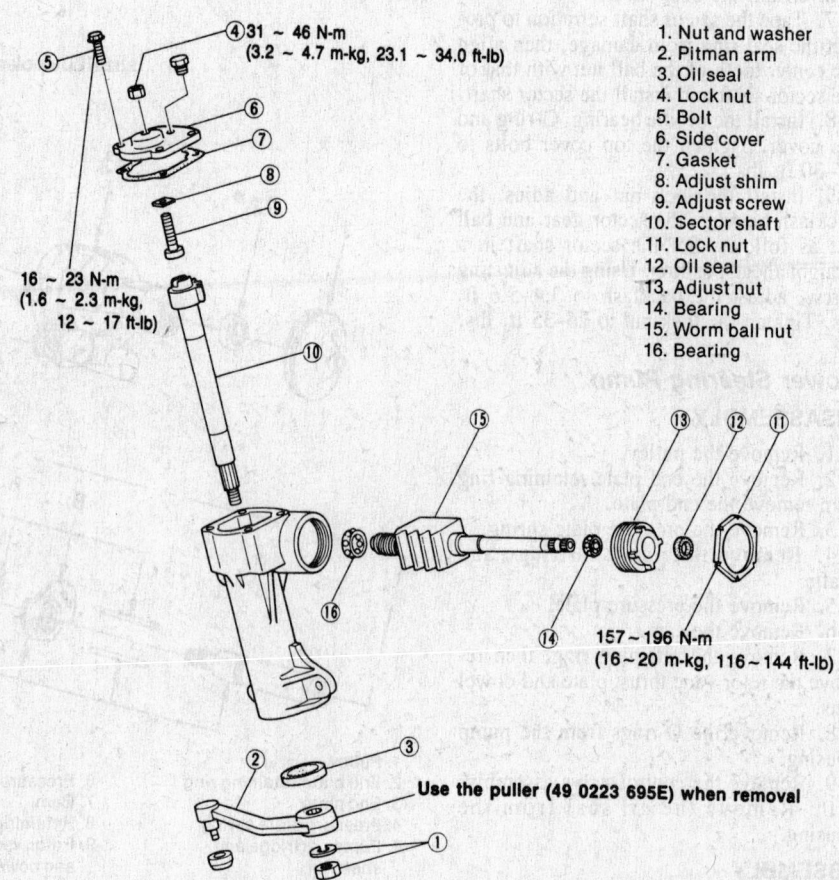

4 31 ~ 46 N·m
(3.2 ~ 4.7 m-kg, 23.1 ~ 34.0 ft-lb)

16 ~ 23 N·m
(1.6 ~ 2.3 m-kg,
12 ~ 17 ft-lb)

157 ~ 196 N·m
(16 ~ 20 m-kg, 116 ~ 144 ft-lb)

1. Nut and washer
2. Pitman arm
3. Oil seal
4. Lock nut
5. Bolt
6. Side cover
7. Gasket
8. Adjust shim
9. Adjust screw
10. Sector shaft
11. Lock nut
12. Oil seal
13. Adjust nut
14. Bearing
15. Worm ball nut
16. Bearing

Use the puller (49 0223 695E) when removal

Mazda/Courier manual steering gear—1982 and later

8. Adjust the backlash between the sector gear and rack. (See Steering Gear Adjustments).

Worm Bearing Preload

1. Disconnect the pitman arm from the center link and remove the horn cap if the gear is still installed in the truck.

2. With a torque wrench, rotate the worm shaft and check the rotating torque. The rotating torque (preload) should be between 5.2–7.8 inch lbs. If the reading is not within specifications, adjust the preload as follows: Remove the end cover attaching bolts and the end cover together with the shims. If the preload is less than 5.2 in. lb., reduce the size of the shim. If the preload is more than 7.8 in. lb. add to the shim size. Install the end cover and recheck the worm bearing preload.

Sector Gear And Ball Nut Backlash

NOTE: Adjust the worm bearing preload before adjusting the backlash.

1. Turn the worm shaft gently and stop it at the center position.

2. Loosen the lock nut of the adjusting screw and turn the screw in or out until the correct adjustment is made. The standard backlash is 0.

3. Tighten the adjusting screw lock nut.

1982 AND LATER

1. Remove the pitman arm nut and washer and remove the pitman arm using a puller, then remove the seal.

2. Remove the lock nut and bolt, then remove the side cover and gasket.

3. Remove the adjusting shim and adjusting screw, then pull out the sector shaft while keeping it in the middle position.

4. Remove the lock nut with special tool 49 1391 580 or its equivalent.

5. Remove the oil seal and the adjusting nut with special tool 49 UB39 585 or its equivalent.

6. Remove the bearing then the worm ball nut and the bearing at the end of the ball nut.

ASSEMBLY

1. Assemble in the reverse order of disassembly with the following instructions:

a. Apply grease or oil to the lips of all oil seals.

b. Preload of the worm ball nut is 0.5–1.0 lb., using a pull scale (without sector). If an adjustment is to be made loosen the locknut and adjust with the adjusting screw. Tighten the locknut to 116–144 ft. lbs.

c. Set the adjusting screw and shim, in the "T" groove on the top of the sector shaft and measure the clearance in the axial direction. If the clearance is more than 0.004 in. adjust by selecting the proper size shim.

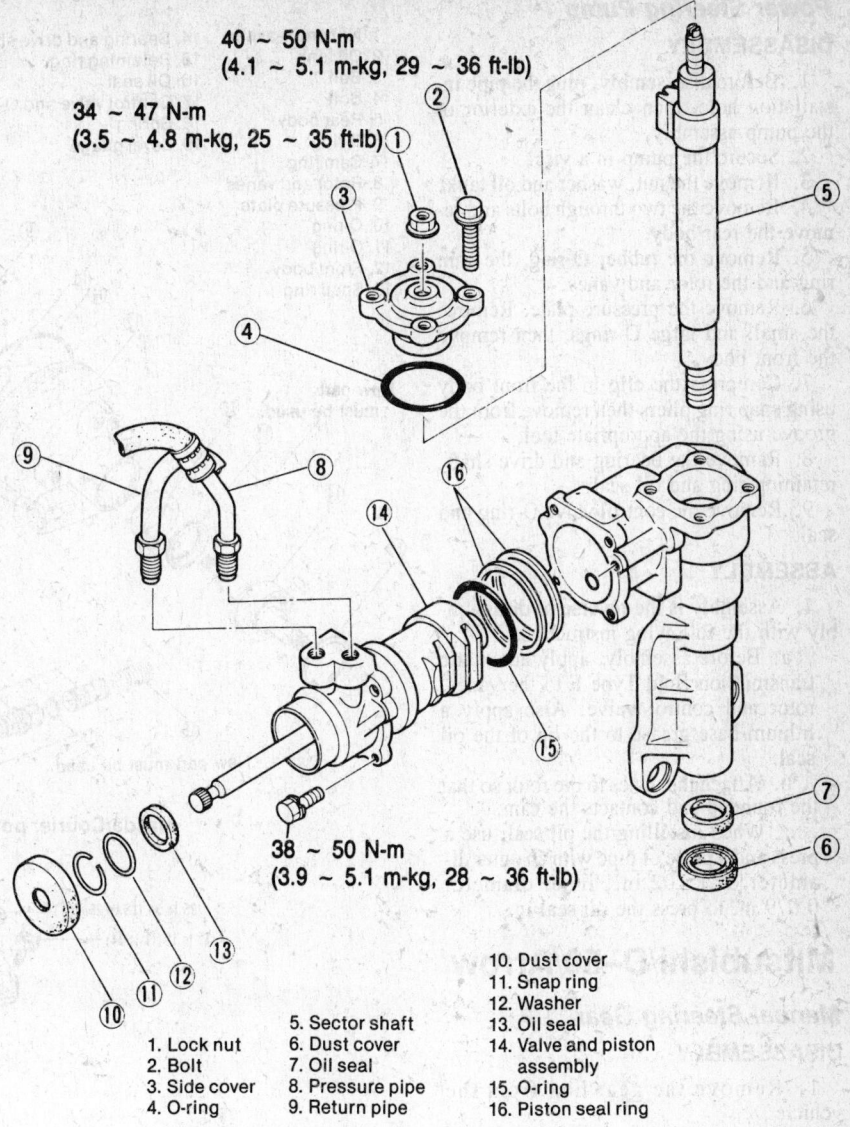

40 ~ 50 N-m
(4.1 ~ 5.1 m-kg, 29 ~ 36 ft-lb)

34 ~ 47 N-m
(3.5 ~ 4.8 m-kg, 25 ~ 35 ft-lb) ①

38 ~ 50 N-m
(3.9 ~ 5.1 m-kg, 28 ~ 36 ft-lb)

1. Lock nut
2. Bolt
3. Side cover
4. O-ring
5. Sector shaft
6. Dust cover
7. Oil seal
8. Pressure pipe
9. Return pipe
10. Dust cover
11. Snap ring
12. Washer
13. Oil seal
14. Valve and piston assembly
15. O-ring
16. Piston seal ring

Mazda/Courier power steering gear

d. When installing the sector shaft, make sure the teeth of the sector shaft mesh with the center part of the teeth of the worm ball nut. The side cover tightening torque is 12–17 ft. lbs.

e. The pitman arm attaching nut torque is 108–130 ft. lbs.

f. Adjust the backlash while keeping the steering gear in the straight forward position. Backlash: 0 mm

Power Steering Gear
DISASSEMBLY

1. Plug the openings of all pipe installation fittings, and clean the exterior of the gear assembly.

2. Remove the lock nut and bolts from the side cover, then remove the O-ring.

3. Remove the sector shaft dust cover and oil seal.

4. Remove the dust cover, snap ring, washer and oil seal then remove the valve and piston assembly.

5. Remove the O-ring and piston seal ring.

ASSEMBLY

1. Assemble in the reverse order of disassembly with the following instructions.

a. Apply grease or oil to the lips of all oil seals.

b. Its a good idea to apply a sealing agent to the adjusting screws after adjustments are made.

c. Position the worm shaft with the vehicle in the straight ahead position, and then set the sector shaft adjustment screw so that the preload at that position is 2.2 lb. or less. Tighten the adjustment screw lock nut to 25–35 ft. lbs. The preload at the straight ahead position must be 0.4–0.9 lb. higher than when the steering wheel is turned 360° to the left or right.

d. Tighten the side cover bolts to 29–36 ft. lbs.

Power Steering Pump

DISASSEMBLY

1. Before disassembly, plug the pipe installation hole, then clean the exterior of the pump assembly.

2. Secure the pump in a vise.

3. Remove the nut, washer and oil tank.

4. Remove the two through bolts and remove the rear body.

5. Remove the rubber O-ring, the cam ring and the rotor and vanes.

6. Remove the pressure plate. Remove the small and large O-rings, then remove the front body.

7. Compress the clip in the front body using snap ring pliers then remove from the groove using the appropriate tool.

8. Remove the bearing and drive shaft, retaining ring and oil seal.

9. Remove the control valve O-ring and seal.

ASSEMBLY

1. Assembly is the reverse of disassembly with the following instructions:

   a. Before assembly, apply automatic transmission fluid Type F to the vanes, rotor and control valve. Also apply a lithium base grease to the lip of the oil seal.

   b. Attach the vanes to the rotor so that the rounded end contacts the cam.

   c. When installing the oil seal, use a press and a piece of pipe with an outer diameter of 1.102 in., inner diameter 0.079 in. to press the oil seal in.

Mitsubishi/D–50/Arrow

Manual Steering Gear

DISASSEMBLY

1. Remove the gear box from the vehicle.

2. Remove the nut holding the Pitman arm on the cross shaft and using a gear puller, pull the arm from the shaft.

3. Before disassembling any further, record the starting preload of the mainshaft as a guide for reassembly.

4. Loosen the lock nut on the cross shaft adjusting bolt and turn the bolt slightly counterclockwise. Remove the cover bolts.

5. Lift the cover up slightly and turn the adjusting bolt in until it unfastens from the cover and remove the cover.

6. Turn the cross shaft until its teeth will fit through the cover hole and pull it out of the gear housing.

NOTE: Use care not to damage the cross shaft splines and the oil seal when removing the cross shaft.

7. Measure the main shaft starting preload with the cross shaft removed.

8. Loosen the end cover attaching bolts and remove the end cover and shim.

NOTE: Keep the shim for reassembly.

1. Nut and washer
2. Oil tank
3. Bolt
4. Bolt
5. Rear body
6. O-ring
7. Cam ring
8. Rotor and vanes
9. Pressure plate
10. O-ring
11. O-ring
12. Front body
13. Snap ring
14. Bearing and drive shaft
15. Retaining ring
16. Oil seal
17. Control valve and O-ring
18. Spring
19. Level gauge

20 ~ 29 N-m (2.0 ~ 3.0 m-kg, 14 ~ 22 ft-lb)

17 ~ 20 N-m (1.7 ~ 2.0 m-kg, 12 ~ 14 ft-lb)

New part must be used.

New part must be used.

New part must be used.

Mazda/Courier power steering pump

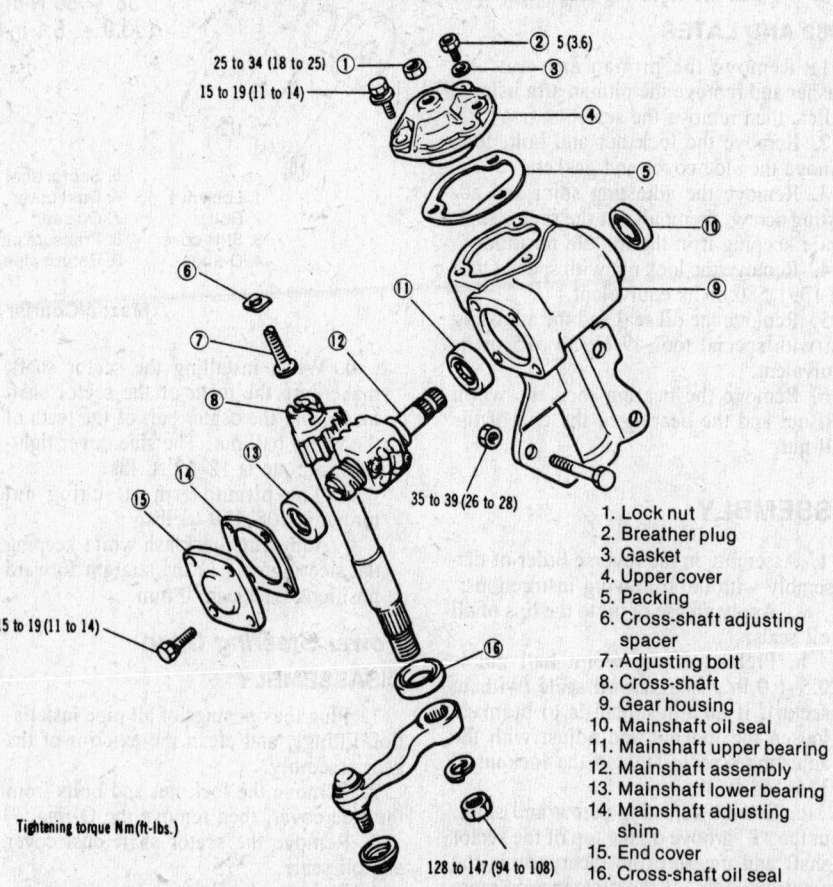

25 to 34 (18 to 25)
5 (3.6)
15 to 19 (11 to 14)
35 to 39 (26 to 28)
15 to 19 (11 to 14)
128 to 147 (94 to 108)

Tightening torque Nm(ft-lbs.)

1. Lock nut
2. Breather plug
3. Gasket
4. Upper cover
5. Packing
6. Cross-shaft adjusting spacer
7. Adjusting bolt
8. Cross-shaft
9. Gear housing
10. Mainshaft oil seal
11. Mainshaft upper bearing
12. Mainshaft assembly
13. Mainshaft lower bearing
14. Mainshaft adjusting shim
15. End cover
16. Cross-shaft oil seal

Mitsubishi/D-50 Arrow manual steering gear

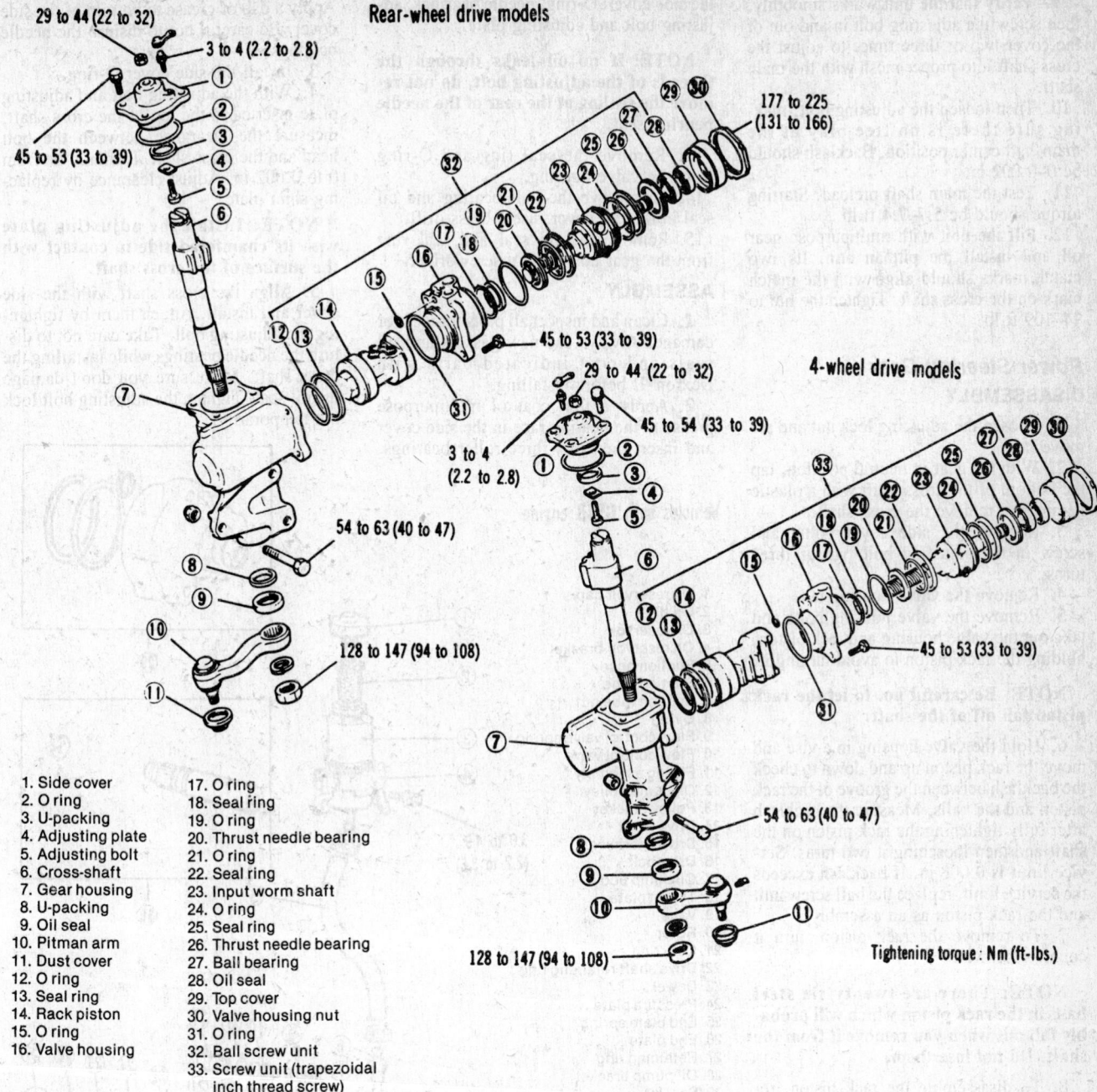

Rear-wheel drive models

29 to 44 (22 to 32)
3 to 4 (2.2 to 2.8)
45 to 53 (33 to 39)
177 to 225 (131 to 166)
45 to 53 (33 to 39)
54 to 63 (40 to 47)
128 to 147 (94 to 108)

4-wheel drive models

29 to 44 (22 to 32)
3 to 4 (2.2 to 2.8)
45 to 54 (33 to 39)
45 to 53 (33 to 39)
54 to 63 (40 to 47)
128 to 147 (94 to 108)

Tightening torque : Nm (ft-lbs.)

1. Side cover
2. O ring
3. U-packing
4. Adjusting plate
5. Adjusting bolt
6. Cross-shaft
7. Gear housing
8. U-packing
9. Oil seal
10. Pitman arm
11. Dust cover
12. O ring
13. Seal ring
14. Rack piston
15. O ring
16. Valve housing
17. O ring
18. Seal ring
19. O ring
20. Thrust needle bearing
21. O ring
22. Seal ring
23. Input worm shaft
24. O ring
25. Seal ring
26. Thrust needle bearing
27. Ball bearing
28. Oil seal
29. Top cover
30. Valve housing nut
31. O ring
32. Ball screw unit
33. Screw unit (trapezoidal inch thread screw)

Mitsubishi/D-50 Arrow power steering gear

9. Gently pull out the main shaft, ball nut assembly and the bearings.

NOTE: Never attempt to disassemble the main shaft and ball nut assembly.

ASSEMBLY

1. Check the component parts for wear or damage. Make sure the ball nut slides easily on the mainshaft. There should not be excessive free play. Never allow the ball nut to run entirely to the end of its travel, or it could be damaged.

2. Insert the main shaft assembly into

the gear housing. Hold the main shaft horizontally.

3. Install the oil seal after applying a small amount of grease to its lip.

4. Install the gasket, shim and gasket end cover to the housing. Tighten the four end cover bolts to 11–45 ft.lb. Use sealant on both the cover gasket and the bolt threads.

5. Measure the main shaft preload. It should be between 3–4.8 ft.lb. If not, adjust it replacing the shim with a thicker or thinner shim. Shims come in thicknesses

from 0.0020 to 0.0200 in.

6. Fit the adjusting bolt and shim in the top of the cross shaft and, using a feeler gauge, check the clearance between the adjusting bolt head and the cross shaft. Clearance should be 0–0.002. If not, replace the shim.

7. Insert the cross shaft into the gear housing. Be sure to align the teeth on both shafts in the center of their travel.

8. Install the cover and torque the bolts to 11–14 ft.lb. Apply sealant to the cover gasket and the threads of the bolts.

9. Verify that the unit works smoothly, then screw the adjusting bolt in and out of the cover two or three times to adjust the cross shaft into proper mesh with the main shaft.

10. Then loosen the adjusting bolt, making sure there is no free play at the mainshaft center position. Backlash should be 0–0.002 in.

11. Test the main shaft preload. Starting torque should be 5.7–7.4 ft.lb.

12. Fill the unit with multipurpose gear oil and install the pitman arm. Its two match marks should align with the match mark on the cross shaft. Tighten the nut to 94–109 ft.lb.

Power Steering Gear
DISASSEMBLY

1. Loosen the adjusting lock nut and remove it.

2. With the gear in neutral position, tap the bottom of the cross shaft with a plastic hammer to remove the cross shaft.

3. Remove the side cover bolts, and screw in the adjusting bolt two or three turns.

4. Remove the valve housing nut.

5. Remove the valve housing bolts and take out the valve housing and rack piston, holding the rack piston to avoid turning it.

NOTE: Be careful not to let the rack piston fall off of the shaft.

6. Hold the valve housing in a vise and move the rack piston up and down to check the backlash between the groove of the rack piston and the balls. Measure the backlash after fully tightening the rack piston on the shaft and then loosening it two turns. Service limit is 0.008 in. If backlash exceeds the service limit, replace the ball screw unit and the rack piston as an assembly.

7. To remove the rack piston, turn it counterclockwise.

NOTE: There are twenty-six steel balls in the rack piston which will probably fall out when you remove it from the shaft. Do not lose them.

8. To disassemble the rack piston, remove the circular holder, the circulator, the steel balls, the seal ring and the O-ring. Do not disassemble the rack piston end cap.

9. Loosen the top cover and remove it and the input worm shaft from the valve housing.

10. Remove worm shaft thrust plate, thrust needle roller bearing, two seal rings and two O-rings.

11. Screw in the adjusting bolt at the tip of the cross shaft and remove the side cover.

NOTE: There are thirty-three needle bearing rollers which may fall out when you remove the cross shaft. Do not lose them.

12. Remove the following parts from the side cover: O-ring, needle bearings, adjusting bolt and adjusting plate.

NOTE: If no oil leaks through the threads of the adjusting bolt, do not remove the sealing at the rear of the needle bearing seat.

13. Remove the seal ring and O-ring from the valve housing.

14. To remove the ball bearing and oil seal in the top cover, use a brass drift.

15. Remove the oil seal and seal ring from the gear box using a screwdriver.

ASSEMBLY

1. Clean and inspect all parts for wear or damage. Always use new gaskets and oil seals and coat indicated parts with Dexron®II before installing.

2. Apply a thin coat of multipurpose grease to the bearing race in the side cover and insert the thirty-three roller bearings.

Apply a dab of grease to bottom of the side cover. Be careful not to disturb the needle bearings.

3. Install the side cover O-ring.

4. With the adjusting bolt and adjusting plate inserted in the top of the cross shaft, measure the clearance between the bolt head and the cross shaft. It should be from 0 to 0.002 in. Adjust clearance by replacing shim plate.

NOTE: Install the adjusting plate with its chamfered side in contact with the surface of the cross shaft.

5. Align the cross shaft with the side cover and install. Attach them by tightening the adjusting bolt. Take care not to disturb the needle bearings while installing the cross shaft. Make sure you don't damage the oil seal. Tighten the adjusting bolt lock nut temporarily.

Vehicles with G54B engine

1. Oil reservoir cap
2. Oil filter
3. Oil reservoir
4. Oil reservoir bracket
5. Suction hose
6. Suction tube
7. Suction tube nut
8. O ring
9. Flow control valve spring
10. Flow control valve
11. Fitting assembly
12. Oil pump pulley
13. Pulley bracket
14. Drive shaft
15. Drive shaft seal
16. Drive belt
17. Oil pump body
18. Thrust plate
19. Vane
20. Rotor
21. Cam ring
22. Drive shaft retaining ring
23. Dowel
24. Pressure plate
25. End plate spring
26. End plate
27. Retaining ring
28. Oil pump bracket
29. Seal kit
30. Drive shaft kit

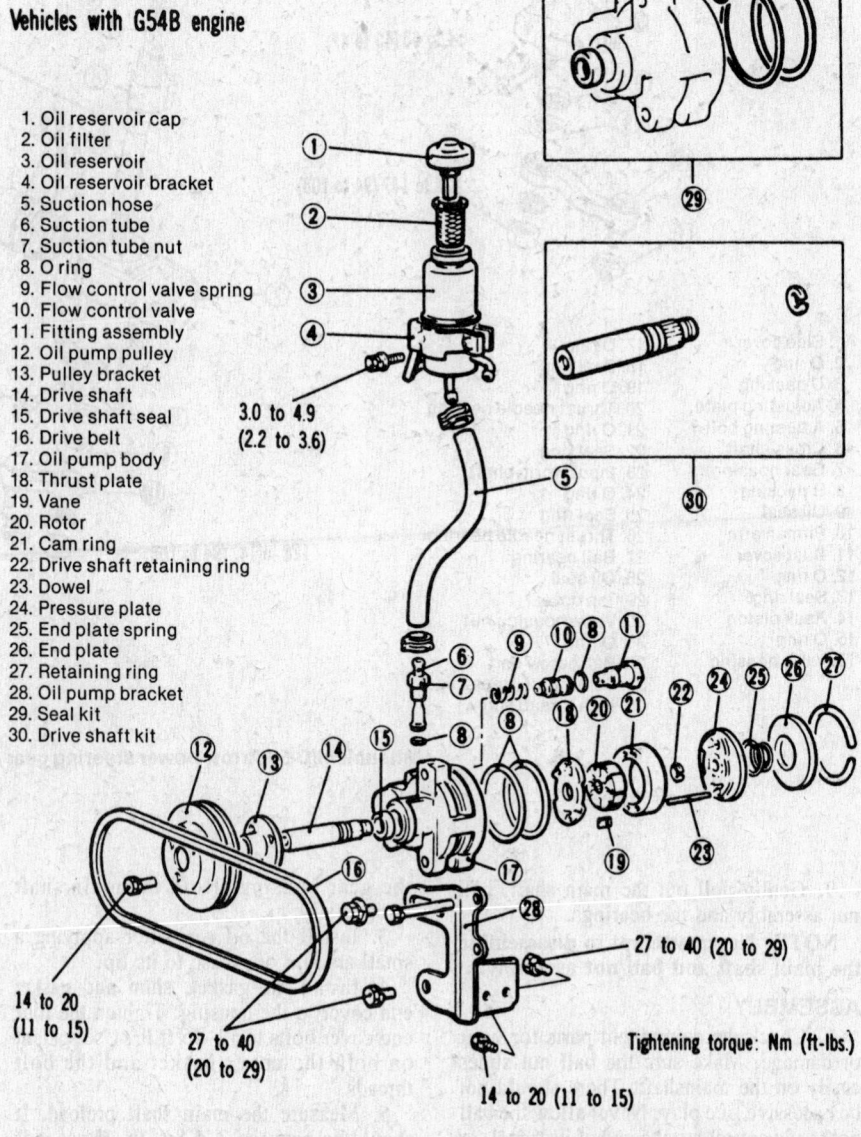

Tightening torque: Nm (ft-lbs.)

Power steering pump—G548 (155.9) engine

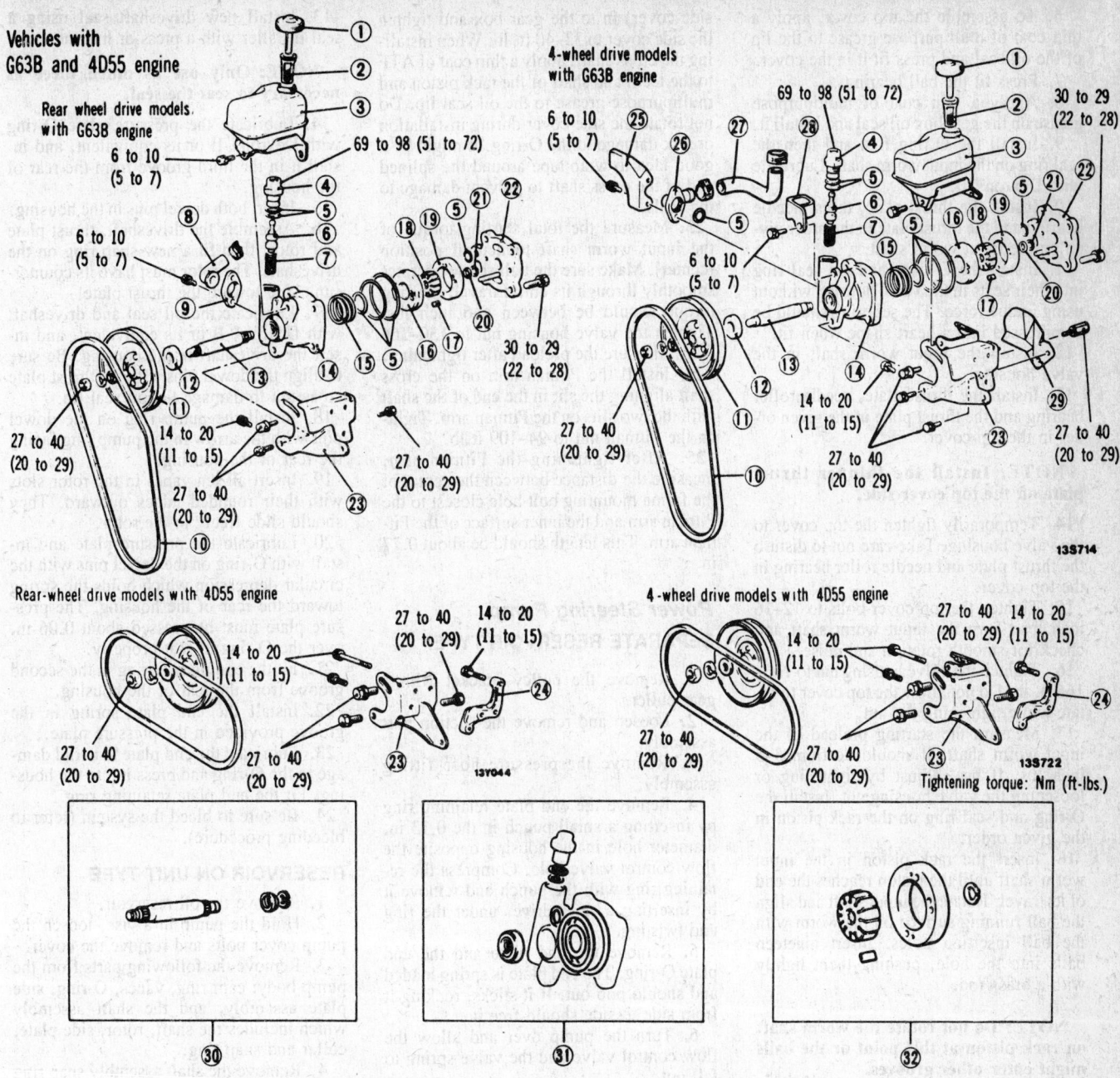

Vehicles with G63B and 4D55 engine

Rear wheel drive models with G63B engine

69 to 98 (51 to 72)
6 to 10 (5 to 7)
6 to 10 (5 to 7)
30 to 29 (22 to 28)
27 to 40 (20 to 29)
14 to 20 (11 to 15)
27 to 40 (20 to 29)

4-wheel drive models with G63B engine

6 to 10 (5 to 7)
69 to 98 (51 to 72)
30 to 29 (22 to 28)
6 to 10 (5 to 7)
27 to 40 (20 to 29)
14 to 20 (11 to 15)
27 to 40 (20 to 29)
27 to 40 (20 to 29)

13S714

Rear-wheel drive models with 4D55 engine

27 to 40 (20 to 29)
14 to 20 (11 to 15)
14 to 20 (11 to 15)
27 to 40 (20 to 29)
27 to 40 (20 to 29)

13Y044

4-wheel drive models with 4D55 engine

27 to 40 (20 to 29)
14 to 20 (11 to 15)
14 to 20 (11 to 15)
27 to 40 (20 to 29)
27 to 40 (20 to 29)

13S722
Tightening torque: Nm (ft-lbs.)

1. Oil reservoir cap
2. Oil filter
3. Oil reservoir
4. Connector
5. O ring
6. Flow control valve
7. Flow control valve spring
8. Suction tube
9. Suction plate
10. Drive belt
11. Oil pump pulley
12. Drive shaft seal
13. Drive shaft
14. Oil pump body
15. Side plate spring
16. Side plate
17. Vane
18. Rotor
19. Cam ring
20. Collar
21. Drive shaft retaining ring
22. Pump cover
23. Oil pump bracket
24. Oil pump bracket stay
25. Guide bracket
26. Suction connector
27. Suction hose
28. Wrench holder
29. Heat insulator
30. Drive shaft kit
31. Seal kit
32. Cartridge kit

Power steering pump—G635 (121.9) and 4D55 (143.2) engines

6. To assemble the top cover, apply a thin coat of multipurpose grease to the lip of the oil seal and press fit it in the cover.

7. Press fit the ball bearing.

8. Apply a thin coat of multipurpose grease on the gear box oil seal and install it.

9. Install the O-ring first and then the seal ring on the input worm shaft. Lubricate with Dexron® II.

10. Install the thrust plate, thrust needle bearing and the thrust plate in the order given on the input worm shaft.

11. Install the O-rings and the seal ring into their seats in the valve housing without using undue force. The seal ring should be compressed into a heart shape when fit.

12. Install the input worm shaft in the valve housing.

13. Install the thrust plate, needle roller bearing and the thrust plate in the given order in the top cover.

NOTE: Install the thinner thrust plate on the top cover side.

14. Temporarily tighten the top cover to the valve housing. Take care not to disturb the thrust plate and needle roller bearing in the top cover.

15. Tighten the top cover bolts to 12–16 inch lbs. Turn the input worm shaft and check for smooth rotation and noise.

16. Tighten the valve housing nut to 130–166 ft. lb. Do not allow the top cover to rotate while tightening the nut.

17. Measure the starting preload of the input worm shaft. It should be from 3–5 inch lbs. If not, adjust by tightening or loosening the valve housing nut. Install the O-ring and seal ring on the rack piston in the given order.

18. Insert the rack piston in the input worm shaft until the piston reaches the end of its travel. Rotate the input shaft and align the ball running surface on the worm with the ball insertion holes. Insert nineteen balls into the hole, pushing them lightly with a brass rod.

NOTE: Do not rotate the worm shaft on rack piston at this point or the balls might enter other grooves.

19. After installing all nineteen of the balls, make sure the last ball is about half an in. below the end of the rack piston. If there is more than a half in. clearance, it probably means one or more of the balls has fallen into a different worm groove. Remove the assembly and begin again.

20. Insert the remaining seven balls in the circulator, holding them in place with grease. Fit the circulator in place and tighten the screws.

21. Hold the gear box in a vise and install the ball screw unit. Tighten the valve housing to 33–40 ft. lb. After installation, rotate the input worm shaft to move the rack piston to the neutral (center) position. Be careful not to damage the seal ring when installing the rack piston.

22. Install the cross shaft assembly (with side cover) in to the gear box and tighten the side cover to 33–40 ft. lb. When installing the cross shaft, apply a thin coat of ATF to the teeth and shaft of the rack piston and multipurpose grease to the oil seal lip. Do not rotate the side cover during installation or risk damage to the O-ring. It might be a good idea to wrap tape around the splined end of the cross shaft to prevent damage to the seals.

23. Measure the total starting torque of the input worm shaft to neutral position (center). Make sure the ball screw operates smoothly through its entire travel. Starting torque should be between 4–6 inch lbs. Tighten the valve housing nut to 130–166 ft. lb. Measure the preload after tightening.

24. Install the Pitman arm on the cross shaft aligning the slit in the end of the shaft with the two slits on the Pitman arm. Tighten the Pitman nut to 94–109 ft. lb.

25. After tightening the Pitman arm, measure the distance between the center of the frame mounting bolt hole closest to the Pitman arm and the inner surface of the Pitman arm. This length should be about 0.77 in.

Power Steering Pump
SEPARATE RESERVOIR TYPE

1. Remove the pulley bracket with a gear puller.

2. Loosen and remove the suction port assembly.

3. Remove the pressure hose fitting assembly.

4. Remove the end plate retaining ring by inserting a small punch in the 0.13 in. diameter hole in the housing opposite the flow control valve hole. Compress the retaining ring with the punch and remove it by inserting a screwdriver under the ring and twisting.

5. Remove the end plate and the end plate O-ring. The end plate is spring loaded and should pop out. If it sticks, rocking it from side to side should free it.

6. Turn the pump over and allow the flow control valve and the valve spring to fall out.

7. With the end cover O-ring removed, tap lightly on the end of the driveshaft to free the pressure plate.

8. Remove the pressure plate, driveshaft, pump ring, vanes and rotor.

NOTE: Do not remove the welch plug. If it is cracked or otherwise damaged, replace the whole housing body.

9. Remove the driveshaft retaining ring.

10. Remove the rotor and thrust plate from the driveshaft and both dowel pins from the housing.

11. Pry the driveshaft seal out of the housing, being careful not to damage the housing, discard the shaft seal.

12. Clean all parts and inspect them for wear or damage.

13. Install new driveshaft seal using a seal installer with a press or hammer.

NOTE: Only use as much force as necessary to seat the seal.

14. Lubricate the pressure plate O-ring with Dexron® II or its equivalent, and install it in the third groove from the rear of the housing.

15. Insert both dowel pins in the housing.

16. Assemble the driveshaft, thrust plate and rotor, then fit a new snap ring on the driveshaft. The rotor must have its countersunk side toward the thrust plate.

17. Lubricate the oil seal and driveshaft with Dexron® II or its equivalent, and insert the driveshaft in the housing. Be sure to align the dowel pins with the thrust plate so as not to damage the oil seal lip.

18. Install the pump ring on the dowel pins with the arrow in the pump ring facing the rear of the housing.

19. Insert all ten vanes in the rotor slots with their rounded edges outward. They should slide freely in the rotor.

20. Lubricate the pressure plate and install with O-ring on the dowel pins with the circular depression which holds the spring toward the rear of the housing. The pressure plate must be pressed about 0.06 in. over the O-ring to seat properly.

21. Fit the end plate O-ring in the second groove from the rear of the housing.

22. Install the end plate spring in the groove provided in the pressure plate.

23. Lubricate the end plate to avoid damage to the O-ring and press it into the housing. Fit the end plate retaining ring.

24. Be sure to bleed the system (refer to bleeding procedure).

RESERVOIR ON UNIT TYPE

1. Remove the oil reservoir.

2. Hold the pump in a vise, loosen the pump cover bolts and remove the cover.

3. Remove the following parts from the pump body: cam ring, vanes, O-ring, side plate assembly, and the shaft assembly which includes the shaft, rotor, side plate, collar and snap ring.

4. Remove the shaft assembly snap ring and remove the collar, rotor side plate.

5. Remove the oil seal with a screwdriver. Remove the suction connector.

6. Remove the connector at the top of the pump body and remove the flow control valve assembly and flow control spring.

7. Assembly is the reverse of disassembly with the following instructions:

   a. Clean and check all parts for wear. Always use new gaskets and lubricate all parts with Dexron® II before assembling.

   b. Pay close attention to the illustrations for the installing direction of the side plate, rotor and collar.

   c. When installing the cam ring, the countersunk holes at the end of the vanes face toward the cover.

   d. Fit the vanes with their rounded sides pointed out.

   e. Bleed the system.

Toyota

Manual Steering Gear (2WD)

DISASSEMBLY

1. Remove the oil filler plug and drain the gear oil.
2. Using a puller, remove the pitman arm from the sector shaft.
3. Remove the adjusting screw lock nut and three retaining bolts from the end cover.
4. Remove the end cover by tightening the adjusting screw.
5. Pull the sector shaft from the housing.

NOTE: On some later models remove the needle bearings from the housing at this time.

6. Using special tool 09617-30040 or equivalent, remove the locknut.
7. Using special tool 09616-22010 or equivalent, remove the bearing adjusting screw.
8. Carefully pull the worm shaft out of the gear housing.

NOTE: Never attempt to disassemble the ball nut from the steering worm shaft.

ASSEMBLY

1. Inspect the worm bearings, bearing races and oil seal and if a problem is suspected replace them.
2. If it is necessary to replace the oil seal, remove it with an appropriate prying tool, then using special tool 09620-30010 or equivalent, install a new seal.
3. If it is necessary to replace the outer race in the gear housing use special tool 09612-65013 or equivalent to remove it, then use special tool 09620-30010 or equivalent to install a new one.
4. If it is necessary to replace the outer race from the adjusting nut, perform the following: Remove the oil seal with an appropriate prying tool. Using special tool 09612-30012 or equivalent remove the outer race from the nut. Using special tool 09620-30010 or equivalent, install a new race and then a new seal into the nut.
5. If necessary remove the inner races from the shaft with a press, then using special tool 09620-30010 or equivalent, press new ones into the shaft.
6. Measure the shaft thrust clearance with a feeler gauge. The maximum clearance should be less than 0.0020 in. If necessary, install a new thrust washer between the sector shaft and the adjusting screw to provide the minimum clearance.
7. Apply MP grease to the bushing needle, roller bearings and oil seals.
8. Place the worm bearings on the shaft and insert the shaft into the housing.
9. Install the bearing adjusting screw and using special tool 09616-22010 or equivalent, gradually tighten the adjusting screw until it is snug.

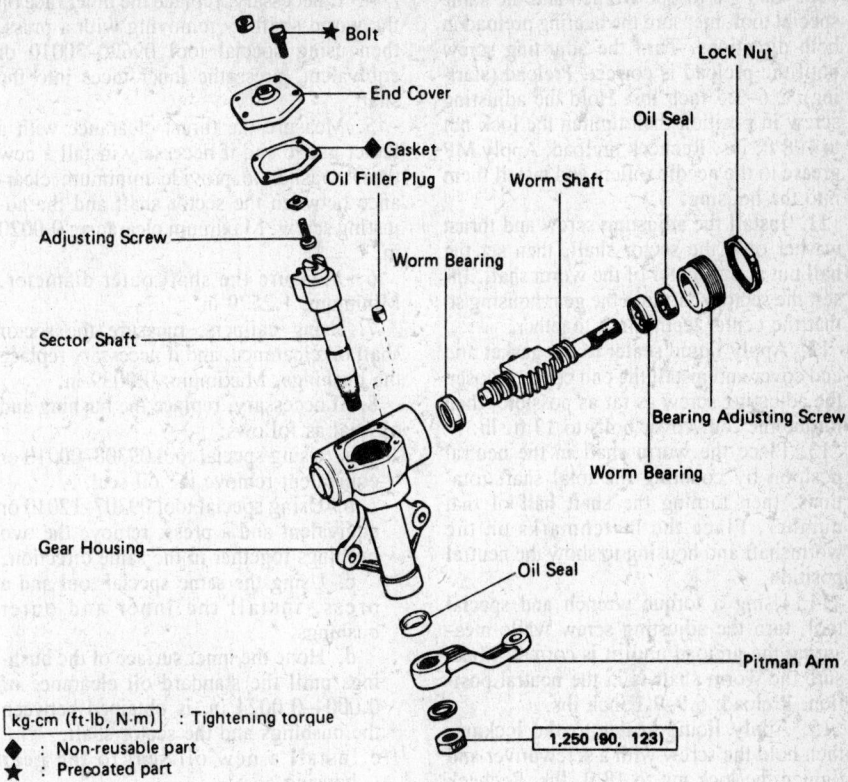

kg-cm (ft-lb, N·m) : Tightening torque
◆ : Non-reusable part
★ : Precoated part

Toyota manual steering gear—2WD

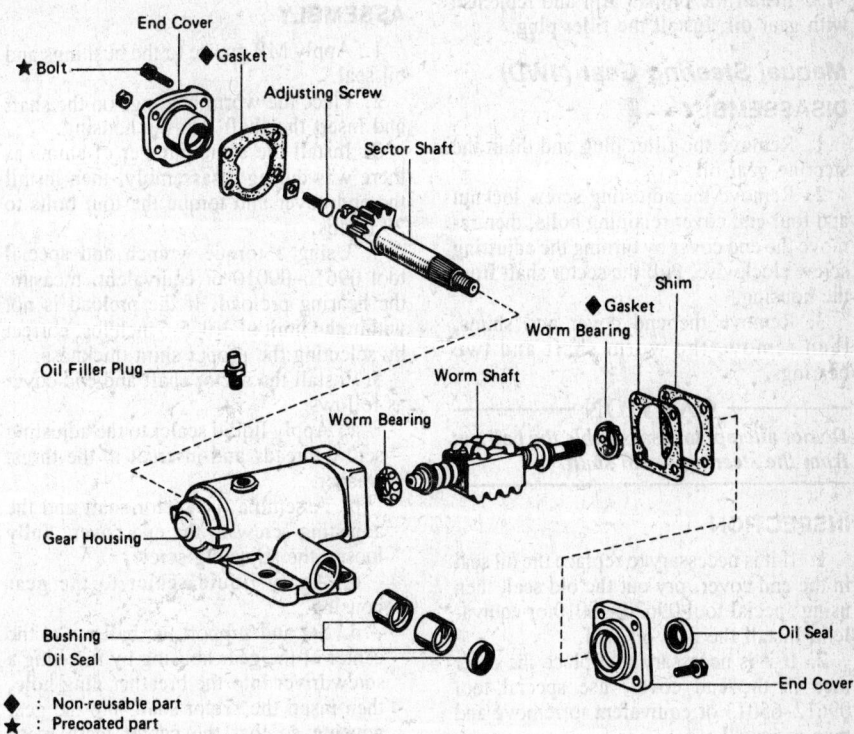

◆ : Non-reusable part
★ : Precoated part

Toyota manual steering gear—4WD

10. Using a torque wrench and the same special tool, measure the bearing preload in both directions. Turn the adjusting screw until the preload is correct. Preload (starting): 2.6–3.5 inch lbs. Hold the adjusting screw in position then tighten the lock nut to 108 ft. lbs. Recheck preload. Apply MP grease to the needle rollers and install them into the housing.

11. Install the adjusting screw and thrust washer onto the sector shaft, then set the ball nut at the center of the worm shaft. Insert the sector shaft into the gear housing so that the center teeth mesh together.

12. Apply liquid sealer to the gasket and end cover and install the end cover. Loosen the adjusting screw as far as possible, then torque the four cover bolts to 13 ft. lb.

13. Place the worm shaft in the neutral position by counting the total shaft rotations, then turning the shaft half of that number. Place the matchmarks on the wormshaft and housing to show the neutral position.

14. Using a torque wrench and special tool, turn the adjusting screw while measuring the preload until it is correct. Make sure the worm shaft is in the neutral position. Preload: 6.9–9.1 inch lbs.

15. Apply liquid sealer to the locknut, then hold the screw with a screwdriver and tighten the lock nut to 18 ft. lbs. Recheck the preload.

16. Measure the sector shaft backlash. There should be no backlash within 100 degrees of the right and left sides from the neutral position.

17. Install the pitman arm and replenish with gear oil. Install the filler plug.

Manual Steering Gear (4WD)
DISASSEMBLY

1. Remove the filler plug and drain the steering gear oil.

2. Remove the adjusting screw locknut and four end cover retaining bolts, then remove the end cover by turning the adjusting screw clockwise. Pull the sector shaft from the housing.

3. Remove the end cover and shims, then remove the worm shaft and two bearings.

——— CAUTION ———
Do not attempt to disassemble the ball nut from the steering worm shaft.

INSPECTION

1. If it is necessary to replace the oil seal in the end cover, pry out the old seal, then using special tool 09620–30010 or equivalent, install the new oil seal.

2. If it is necessary to replace the outer race in the end cover, use special tool 09612–65013 or equivalent to remove and then to install a new one.

3. If it is necessary to replace the outer race in the gear housing, use special tool 09612–65013 to remove and then install a new one.

4. If necessary, replace the inner race on the worm shaft by removing with a press, then using special tool 09620–30010 or equivalent, press the inner races into the shaft.

5. Measure the thrust clearance with a feeler gauge and if necessary install a new thrust washer to provide minimum clearance between the sector shaft and the adjusting screw. Maximum clearance: 0.0020 in.

6. Measure the shaft outer diameter. Minimum: 1.2579 in.

7. Using calipers, measure the sector shaft oil clearance, and if necessary replace the bushings. Maximum: 0.0039 in.

8. If necessary, replace the bushing and oil seal as follows:

   a. Using special tool 09308–00010 or equivalent remove the oil seal.

   b. Using special tool 09307–12010 or equivalent and a press, remove the two bushings together in the same direction.

   c. Using the same special tool and a press, install the inner and outer bushings.

   d. Hone the inner surface of the bushings until the standard oil clearance of 0.0004–0.0024 in. is obtained between the bushings and the sector shaft.

   e. Install a new oil seal to the gear housing.

9. Check the end cover bushing for wear or damage. Using calipers measure the sector shaft oil clearance. Maximum clearance is : 0.0039 in., if found to be excessive replace the end cover.

ASSEMBLY

1. Apply MP grease to the bushings and oil seal.

2. Place the worm bearings on the shaft and insert the shaft into the housing.

3. Install the same number of shims as there was during disassembly, then install the end cover and torque the four bolts to 29 ft. lbs.

4. Using a torque wrench and special tool 09616–00010 or equivalent, measure the bearing preload. If the preload is not within the limit of 3.0–5.6 inch lbs. correct by selecting the proper shim thickness.

5. Install the sector shaft and end cover as follows:

   a. Apply liquid sealer to the adjusting screw threads and insert it in the thrust washer.

   b. Assemble the sector shaft and the adjusting screw to the end cover. Fully loosen the adjusting screw.

   c. Apply liquid sealer to the gear housing.

   d. Set and support the ball nut at the center of the gear housing by inserting a screwdriver into the breather plug hole, then insert the sector shaft into the gear housing so that the center teeth mesh together.

   c. Torque the four bolts to 29 ft. lbs.

6. Place the worm shaft in the neutral position by counting the total shaft rotations then turning the shaft back half of that number. Place matchmarks on the worm shaft and housing to show neutral position.

7. Using the special tool and a torque wrench, turn the adjusting screw while measuring the preload until it is correct. Preload: 6.9–9.5 inch lb.

NOTE: Make sure the wormshaft is in the neutral position.

8. Hold the adjusting screw with a screwdriver and tighten the lock nut to 31 ft. lbs. Recheck preload.

9. Measure the sector shaft backlash (0–0.0106 in.) within 100 degrees of the left and right sides from the neutral position. Refill with gear oil.

Power Steering Gear (2WD)
DISASSEMBLY

1. Remove the adjusting screw lock nut from the end cover and the four bolts, then screw in the adjusting screw until the cover comes off.

2. Tap on the cross shaft end with a plastic hammer and pull out the cross shaft.

3. Remove the four cap screws from the housing. Hold the power piston nut with your thumb so it cannot move, and turn the wormshaft clockwise, then withdraw the valve body and power piston assembly.

——— CAUTION ———
Make sure the power piston nut does not come off the worm shaft.

INSPECTION

1. Mount the valve body in a vise, then using a dial indicator, check the ball clearance. Move the worm gear up and down. The maximum ball clearance is 0.0059 in. If the clearance is excessive, the power control valve assembly must be replaced.

2. Clamp the cross shaft in a vise, then using a dial indicator check the end play. The end play should be 0.0012–0.0020 in. If necessary, adjust end play as follows: Use a chisel and a hammer and remove the lock nut stake. Using special tool 09630–00011 or its equivalent, loosen the lock nut. Adjust the adjusting screw for the correct end play and tighten the lock nut. Stake the lock nut.

3. Replace the Teflon ring and needle roller bearings as follows:

   a. Pry out the old seal from the pitman arm end of the housing.

   b. Remove the snap ring using snap ring pliers.

   c. Remove the metal spacer, Teflon ring and O-ring.

   d. Using special tool 09630–00011 or equivalent, drive out the bearings.

   e. Using the same special tool, install the top bearing with the long flange out. Drive the bearing in flush with the inside casting surface.

   f. Install the lower bearing with the long flange out. The special tool will

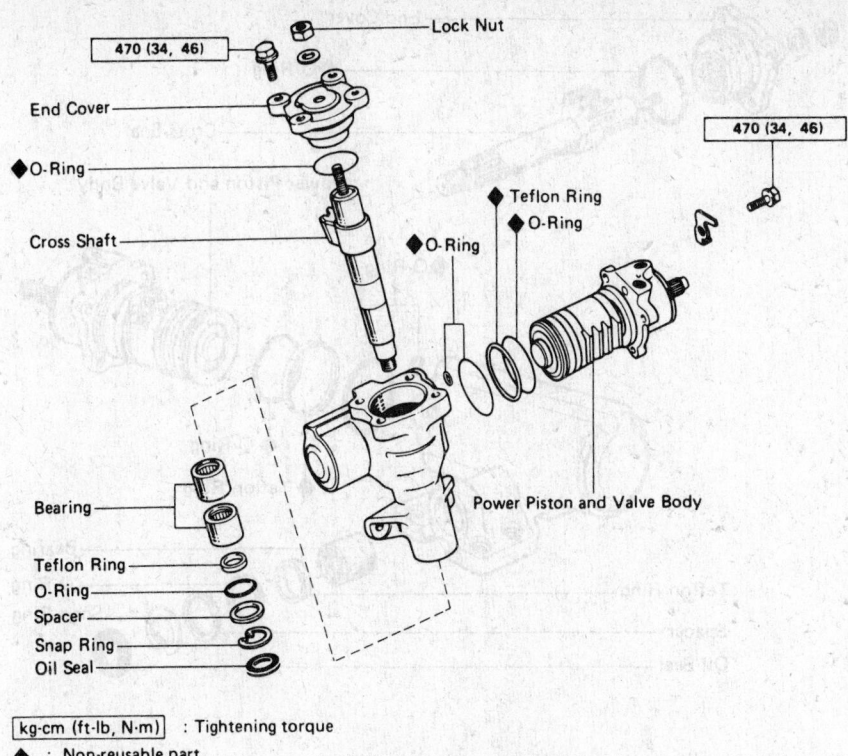

470 (34, 46) — Lock Nut

End Cover

◆O-Ring

Cross Shaft

◆O-Ring

470 (34, 46)

◆Teflon Ring
◆O-Ring

Bearing

Teflon Ring
O-Ring
Spacer
Snap Ring
Oil Seal

Power Piston and Valve Body

| kg-cm (ft-lb, N·m) | : Tightening torque
◆ : Non-reusable part

Toyota power steering gear—2WD

bottom and correctly position the bearing.

g. Install the O-ring and metal spacer.

h. Using snap ring pliers, install the snap ring.

i. Form the seal into a heart shape and install it by hand, then using the same special tool as above form the seal, then drive the oil seal into the gear housing.

ASSEMBLY

1. Install two new O-rings and insert the valve body into the housing. Torque the bolts in a diagonal pattern to 34 ft. lbs.

2. Inspect the worm shaft bearing as follows: Using special tool 09630–00011 or its equivalent, remove the lock nut and the bearing cap. Remove the worm bearing and O-ring, then install a new O-ring and bearing cap.

3. Adjust the worm bearing preload as follows: Using special tool 09630–00011, tighten the bearing cap until the preload is correct. Using special tool 09616–00010 or its equivalent, and a torque wrench, check the preload of the bearing. Preload should be 3.5–5.6 in. lb. Hold the power piston nut to prevent it from turning. Tighten the lock nut while holding the bearing cap to 36 ft. lbs. Recheck the preload.

4. Install a new O-ring on the end cover.

5. Assemble the cross shaft to the end cover, and fully loosen the adjusting screw.

6. Set the worm gear at the center of the gear housing, then insert and push the cross

shaft into the gear housing so that the center teeth mesh together.

7. Install the four cap bolts and torque the four cap bolts in a diagonal pattern to 34 ft. lbs.

8. Turn the worm shaft to full lock in both directions and determine the exact center.

9. Adjust the total preload as follows: Install special tool 09616–00010 or its equivalent, and a torque wrench, on the center worm shaft. Turn the adjusting screw while measuring the preload until it is correct. Total preload: (Add worm preload), 1.7–2.6 inch lbs.

10. Install a new washer, then install and tighten the lock nut to 34 ft. lbs.

11. Recheck the total preload.

Power Steering Gear (4WD)
DISASSEMBLY

1. Remove the adjusting screw lock nut from the end cover and the four bolts, then screw in the adjusting screw until the cover comes off.

2. Tap on the cross shaft end with a plastic hammer and pull out the cross shaft.

3. Remove the four cap screws from the housing. Hold the power piston nut with your thumb so it cannot move, and turn the wormshaft clockwise, then withdraw the valve body and power piston assembly. Make sure the power piston nut does not come off the worm shaft.

INSPECTION

1. Mount the valve body in a vise, then

using a dial indicator, check the ball clearance. Move the worm gear up and down. The maximum ball clearance is 0.0059 in. If the clearance is excessive, the power control valve assembly must be replaced.

2. Clamp the cross shaft in a vise, then using a dial indicator check the end play. The end play should be 0.0012–0.0020 in. If necessary, adjust end play as follows: Use a chisel and a hammer and remove the lock nut stake. Using special tool 09630–00011 or its equivalent, loosen the lock nut. Adjust the adjusting screw for the correct end play and tighten the lock nut. Stake the lock nut.

3. Replace the Teflon ring and needle roller bearings as follows: Pry out the old seal from the pitman arm end of the housing. Remove the snap ring using snap ring pliers. Remove the metal spacer, Teflon ring and O-ring. Using special tool 09630–00011 or equivalent, drive out the bearing.

Using special tool 09631–60010 or its equivalent, install a new bearing so that it is positioned 0.929 in. away from the housing inner end surface. Install a new Teflon ring together with a new O-ring to the special tool above, then install to the gear housing using the tool. Install the metal spacer. Using snap ring pliers, install the snap ring. Using tool 09630–00011 form the seal, then drive the seal into the gear housing, using tool 09631–600010.

ASSEMBLY

1. Install two new O-rings and insert the valve body into the housing. Torque the bolts in a diagonal pattern to 34 ft. lbs.

2. Inspect the worm shaft bearing as follows: Using special tool 09630–00011 or its equivalent, remove the lock nut and the bearing cap. Remove the worm bearing and O-ring, then install a new O-ring and bearing cap.

3. Adjust the worm bearing preload as follows: Using special tool 09630–00011, tighten the bearing cap until the preload is correct. Using special tool 09616–00010 or its equivalent, and a torque wrench, check the preload of the bearing. Preload should be 3.5–5.6 inch lb. Hold the power piston nut to prevent it from turning. Tighten the lock nut while holding the bearing cap to 36 ft. lbs. Recheck the preload.

4. Install a new O-ring on the end cover.

5. Assemble the cross shaft to the end cover, and fully loosen the adjusting screw.

6. Set the worm gear at the center of the gear housing, then insert and push the cross shaft into the gear housing so that the center teeth mesh together.

7. Install the four cap bolts and torque the four cap bolts in a diagonal pattern to 34 ft. lbs.

8. Turn the worm shaft to full lock in both directions and determine the exact center.

9. Adjust the total preload as follows: Install special tool 09616–00010 or its equivalent, and a torque wrench, on the center worm shaft. Turn the adjusting

screw while measuring the preload until it is correct. Total preload: (Add worm preload), 1.7–2.6 inch lbs.

10. Install a new washer, then install and tighten the lock nut to 34 ft. lbs.

11. Recheck the total preload

Power Steering Pump

1979–80

1. Clamp the pump in a vise.

2. Remove the union from the rear housing.

3. Mark the front and rear housings and fixed ring for correct reassembly purposes.

4. Remove the six front housing bolts, then using a plastic hammer, tap off the front housing.

CAUTION

Do not allow the shaft to come out, otherwise the slippers and springs can fly out.

5. Hold your hand over the fixed ring and carefully remove the shaft.

6. Separate the rear housing and fixed ring, and if necessary tap with a plastic hammer to do so.

7. Remove the O-ring and side plate from the front and rear housings housing.

8. Remove the flow control valve as follows: Remove the screw and lock plate. Using snap ring pliers, remove the snap ring. Temporarily install the screw. Using needle nose pliers, grasp the screw and pull out the plug. Remove the spring and flow control valve by hand.

ASSEMBLY

1. Install the flow control valve as follows: Make sure the letter inscribed on the flow control valve matches the letter stamped on the rear of the pump body. Lubricate the flow control valve and spring with automatic transmission fluid. Install the flow control valve, spring, plug, snap ring, snap ring keeper, and screw to the flow control valve housing.

2. Install the side plate with the chamfer facing down on the rear housing, then lubricate and install the O-ring.

3. Install the side plate and O-ring on the front housing.

4. Select and use the fixed ring, rotor shaft and slippers inscribed with the same mark.

5. Align the match marks and place the fixed ring on the rear housing. Install the two bolts as guides.

6. Using a plastic hammer, tap the fixed ring into place, then remove the guide bolts.

7. Lubricate and install the rotor into the fixed ring.

8. Assemble the spring seat, four springs and slipper. Check that there is no more than 0.020 in. difference in the spring length for each set.

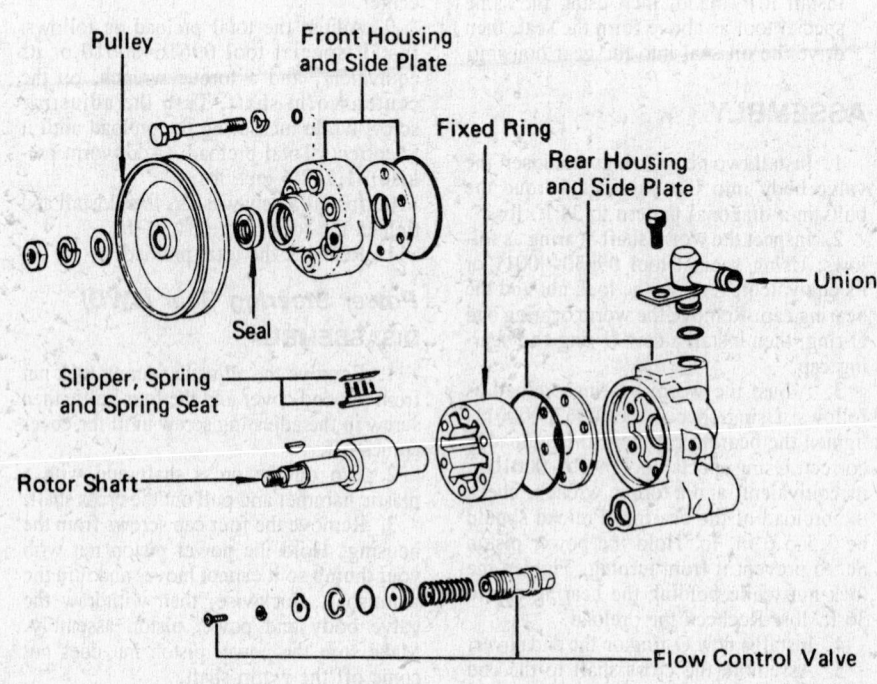

Toyota power steering gear—4WD

◆ : Non-reusable part

Toyota power steering pump—1979–80

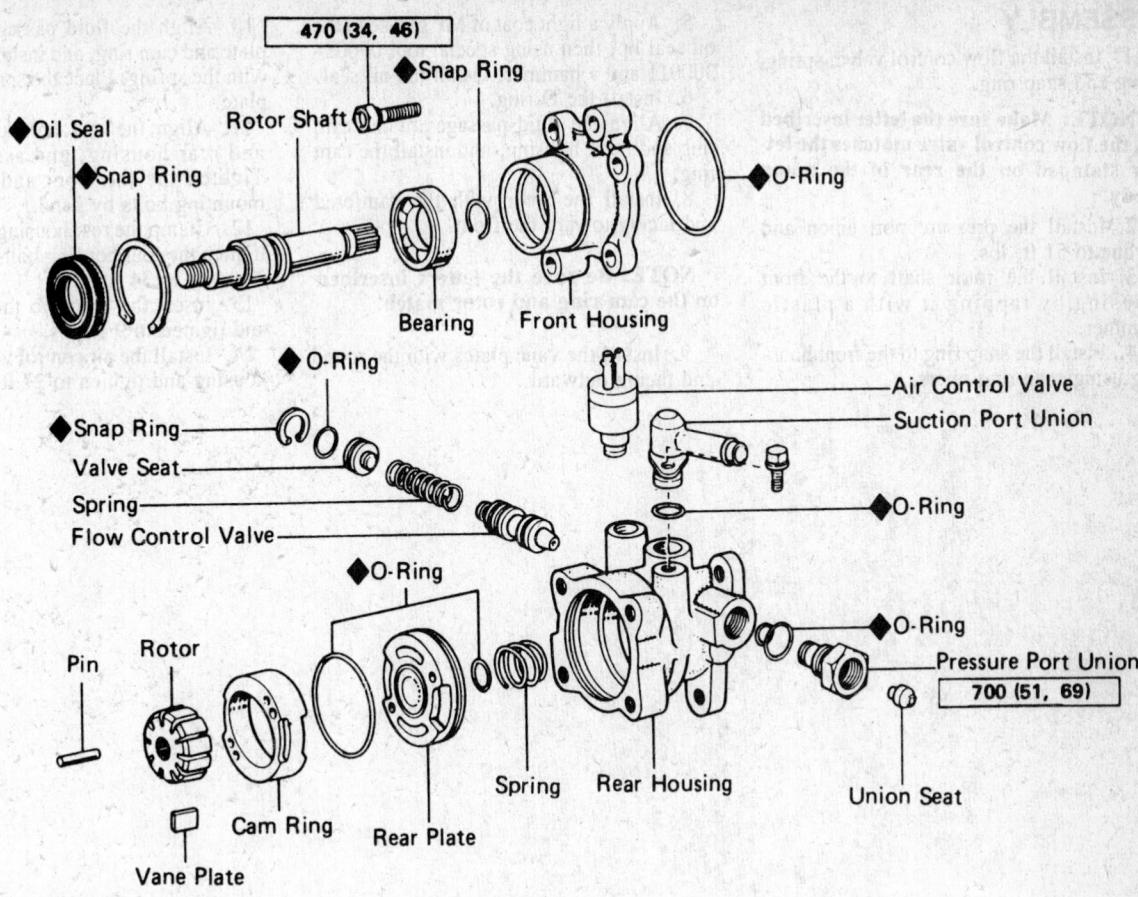

470 (34, 46)

◆Oil Seal
◆Snap Ring

◆Snap Ring
Rotor Shaft
Bearing
Front Housing
◆O-Ring

◆O-Ring
◆Snap Ring
Valve Seat
Spring
Flow Control Valve

Air Control Valve
Suction Port Union
◆O-Ring

◆O-Ring
◆O-Ring
Pressure Port Union
700 (51, 69)

Pin
Rotor
Vane Plate
Cam Ring
Rear Plate
Spring
Rear Housing
Union Seat

kg-cm (ft-lb, N·m)

◆ : Non-reusable part

Toyota power steering pump—1981 and later

9. Install the slipper assembly with the notched or open side facing counterclockwise. Install all the slippers toward the rotor

10. Align the match marks on the front housing with the fixed ring and rear housing and install the two guide bolts, then using a plastic hammer, tap the housing into position.

11. Install the six bolts with lubricated O-rings into the housing, then tighten the bolts evenly 3 or 4 times to 24–30 ft. lbs.

1981 AND LATER

1. Clamp the pump in a vise.

2. Remove the air control valve from the rear housing on the 22R–E engine only.

3. Remove the suction port union from the rear housing.

4. Place matchmarks on the front and rear housing.

5. Remove the four front housing bolts, then using a plastic hammer, tap off the front housing. Be careful that the vane plates, rotor and cam ring do not fall out.

6. Remove the cam ring, rotor, and vane plates without scratching them.

7. Remove the rotor shaft as follows: Clamp the front housing in a vise. Pry off the oil seal using a chisel and a hammer. Remove the snap ring using snap ring pliers. Using a plastic hammer, lightly tap the rotor shaft out of the front housing.

8. Using a plastic hammer, tap the bottom end of the rear housing, and remove the rear plate and spring.

9. Remove the flow control valve as follows: Temporarily a bolt to the plug. Push the bolt and remove the snap ring with snap ring pliers. Pull out the bolt and remove the plug. Remove the spring and flow control valve by hand. Remove the pressure port union.

INSPECTION

1. Check the front housing bushing for wear or damage. If wear or damage is found the entire housing must be replaced.

2. Check the oil clearance between the

bushing and rotor shaft. Maximum oil clearance is 0.0028 in.

3. If necessary, replace the rotor shaft bearing as follows: Using snap ring pliers, remove the snap ring. Using a press, press out the old bearing and press in the new one. Using the snap ring pliers, install the snap ring.

4. Measure the cam ring thickness. Check that the difference between the rotor and cam ring is less than the maximum 0.0024 in. If the difference is excessive, replace the cam ring with one having the same letter as the rotor.

5. Check the vane plates for wear or scratches.

6. Measure the clearance between the vane plate and the rotor groove. Maximum clearance is 0.0024 in.

7. Check the flow control valve for wear or damage. If necessary, replace the valve with one having the same letter on the rear housing.

8. Check that the flow control valve spring is within 1.85–1.97 in. If not replace it.

ASSEMBLY

1. Install the flow control valve, spring, plug and snap ring.

NOTE: Make sure the letter inscribed on the flow control valve matches the letter stamped on the rear of the pump body.

2. Install the pressure port union and torque to 51 ft. lbs.

3. Install the rotor shaft to the front housing by tapping it with a plastic hammer.

4. Install the snap ring to the front housing using snap ring pliers.

5. Apply a light coat of MP grease to the oil seal lip, then using special tool 09608–300011 and a hammer, install the oil seal.

6. Install the O-ring.

7. Align the fluid passages of the cam ring and front housing, and install the cam ring.

8. Install the rotor with the camfered end facing toward the front.

NOTE: Be sure the letters inscribed on the cam ring and rotor match.

9. Install the vane plates with the round end facing outward.

10. Align the fluid passages of the rear plate and cam ring, and install the rear plate with the spring. Place the spring on the rear plate.

11. Align the matchmarks on the front and rear housing, and assemble them. Tighten the the front and rear housing mounting bolts by hand.

12. Clamp the rear housing in a vise, then tighten the four housing bolts evenly in 3 or 4 passes to 34 ft. lbs.

13. Insert the union to the rear housing and tighten to 9 ft. lbs.

14. Install the air control valve to the rear housing and tighten to 27 ft. lbs.

Wheel Alignment

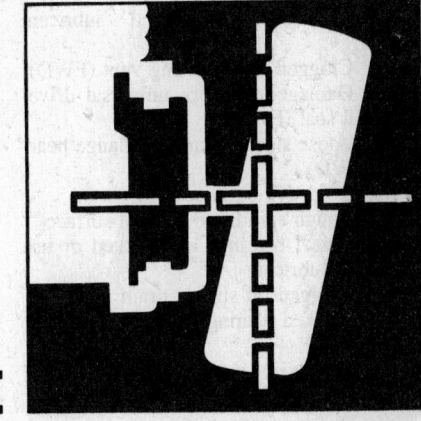

SUSPENSION & ALIGNMENT TROUBLE DIAGNOSIS

UNEVEN TIRE WEAR

1. Tire pressures low
2. Excessive camber
3. Tires out of balance
4. Tires overloaded
5. Out of round tires and rims
6. Caster incorrect
7. Toe-in incorrect
8. High speed driving into turns
9. Unequal tire size
10. Improper tracking
11. Bent or worn steering and suspension components

STEERING WHEEL SPOKE POSITION NOT PROPERLY CENTERED

1. Steering gear set off ''high-spot''
2. Improper toe-in
3. Relationship between lengths of tie-rods not equal
4. Bent steering components
5. Steering wheel improperly placed on steering shaft

HARD STEERING

1. Tire pressure low
2. Wheel spindle bent
3. Steering assembly binding or maladjusted
4. Tie rod ends tight

5. Caster excessive
6. Kingpins or ball joints too tight
7. Lack of lubrication to steering and suspension units

SHIMMY

1. Tire pressure incorrect
2. Tires of unequal size
3. Loose wheel bearings
4. Loose steering arms or steering gear adjustment
5. Steering gear loose on frame
6. Loose or broken steering linkage rods or internal adjustment parts
7. Spring shackles loose
8. Ball joints or kingpins and bushings worn
9. Front end alignment out of specifications
10. Wheels and tires out of balance
11. Wheels and tires out of round or loose on hub
12. Shock absorbers worn out
13. U-bolts loose on axle to spring
14. Worn or out-of-round brake drum or rotor (shimmy felt upon brake application)

WANDER OR WEAVE

1. Tire pressure incorrect
2. Tires of unequal size
3. Bent spindle
4. Wheel bearings loose or worn
5. Kingpins worn or bent
6. Kingpins tight in steering knuckle or bushings
7. Steering gear assembly too tight or too loose
8. Too little caster
9. Too much or too little camber
10. Too much or too little toe-in
11. Front axle bent or shifted
12. Springs broken

13. Frame diamond shaped
14. Rear axle housing shifted or bent
15. Steering linkage tight or binding
16. Lack of lubrication to front suspension or steering linkage
17. Defective power steering assembly

FRONT END RIDES HARD

1. Improper tire pressure
2. Springs broken or too stiff
3. Shock absorbers too stiff or malfunctioning
4. Front end alignment incorrect
5. Loose suspension components

VEHICLE STEERS TO ONE SIDE AT ALL TIMES

1. Incorrect caster setting
2. Incorrect camber setting
3. Incorrect kingpin inclination or wheel support angle
4. Unequal tire pressure or tire size
5. One side brake drag
6. Unequal shock absorber control
7. Bent or damaged steering and suspension components
8. Uneven or weak spring condition, front or rear
9. Broken center or shackle bolts
10. Frame bent causing improper tracking

NOISY FRONT END

1. Lack of, or improper lubrication
2. Loose steering linkage
3. Loose suspension parts
4. Loose brake parts
5. Worn universal (FWD)
6. Worn differential (FWD)
7. Loose sheet metal

LUBRICATION LEAKING INTO DRUM OR ON ROTOR

1. Excessive differential lubricant (FWD)
2. Clogged axle housing vent (FWD)
3. Damaged or worn universal drive-shaft oil seal (FWD)
4. Loose steering knuckle flange bearings (FWD)
5. Defective outer seal
6. Rough spindle to oil seal surface
7. Wheel bearings overpacked or use of wrong lubricate
8. Clogged oil slinger drain
9. Cracked steering knuckle outer flange

EXCESSIVE TIRE WEAR

1. Incorrect wheel alignment
2. Failure to rotate tires
3. Improper tire inflation
4. Overload or improperly loaded vehicle
5. High tire temperature operation
6. Excessive speed, quick starts and quick stops
7. Bent suspension, frame or wheel parts
8. Tires out of balance
9. Uneven brake application

Wheel Alignment

For a truck to have safe steering control with a minimum of tire wear, certain established rules must be followed. These rules fix the values of planes, angles and radii relative to each other and to truck and tire dimensions. Some factors are built in, with no provision for adjustment; others are adjustable within limits. The entire system depends upon all value factors, separately and combined. It is therefore difficult to change some of the established settings without influencing others.

This system is called steering geometry or wheel alignment and requires a complete check of all the factors involved. Definitions of these factors and the effect each one has on the truck are given in the following paragraphs.

STEERING WHEEL POSITION

Always check steering wheel alignment in conjunction with and at the same time as toe-in. In fact, the steering wheel spoke position, with the truck on a straight section of highway, may be the first indication of front end misalignment.

If the truck has been wrecked, or indicates any evidence of steering gear or linkage disturbance, the Pitman arm should be disconnected from the sector shaft. The steering wheel (or gear) should be turned from extreme right to extreme left to determine the halfway point in its turning scope.

This will be the spot on the gear that is in action during straight ahead driving and in which position the steering gear should be adjusted. With the steering wheel in the straight-ahead position and the steering gear adjusted to zero lash status, reconnect the Pitman arm.

Steering Geometry

CAMBER ANGLE

Camber is the amount that the front wheels are inclined outward or inward at the top. Camber is spoken of, and measured, in degrees from the perpendicular.

The purpose of the camber angle is to take some of the load off the spindle outboard bearing.

CASTER ANGLE

Caster is the amount that the kingpin (or in the case of trucks without king-pins, the knuckle support pivots) is tilted towards the back or front of the truck. Caster is usually spoken of, and measured, in degrees. Positive caster means that the top of the kingpin is tilted toward the back of the truck. Positive caster is indicated by the sign +.

Negative caster is exactly the opposite; the top of the kingpin is tilted toward the front of the truck. This is generally indicated by the sign −. Negative caster is sometimes referred to as reverse caster.

The effect of positive caster is to cause the truck to steer in the direction in which it tends to go. Positive caster in the front wheels may cause the truck to steer down off a crowned road or steer in the direction of a cross wind. For this reason, a number of our modern trucks are arranged with negative caster so that the opposite is true; the truck tends to steer up a crowned road and into a cross wind.

CASTER ANGLE CORRECTION

Caster angle specifications are based on the vehicle load limits, which will usually result in a level frame.

Since load requirements may vary, the frame does not always remain level and must be considered when determining the correct caster angle.

To measure the frame angle, the vehicle should be on a smooth and level surface. Place a bubble protractor on the frame rail and measure the degree of frame tilt and in what direction, either front or rear.

Two methods of determining caster angles are used. The first method is to determine the caster angle from the wheel with alignment equipment, and the second method is to obtain the desired caster angle from the specification charts. The frame angle is then added to or subtracted from the caster angles as necessary. The two methods are outlined.

First Method

1. Determine the frame angle
   a. Frame high at rear—frame angle is negative.
   b. Frame low at rear—frame angle is positive.
2. Determine the caster angle at the wheel with the alignment checking equipment.
3. Add or subtract frame angle from or to the determined caster angle.
   a. Negative frame angle is added to positive caster angle.
   b. Positive frame angle is subtracted from positive caster angle.
   c. Negative frame angle is subtracted from negative caster angle.
   d. Positive frame angle is added to negative caster angle.
4. Determine the corrected caster angle and the specified caster angle and correct on the vehicle. Use the following examples as guides.

EXAMPLE NO. 1 (FRAME LOWER AT REAR—POSITIVE)

Measured wheel caster angle	+2°
Frame angle	3°
Actual caster angle (Frame at zero degrees)	−1°
Specifications (desired)	+2°
Necessary degrees to change	+3°

EXAMPLE NO. 2 (FRAME HIGHER AT REAR—NEGATIVE)

Measured wheel caster angle	+2°
Frame angle	2°
Actual caster angle (Frame at zero degrees)	+4°
Specifications (desired)	+3°
Necessary degrees to change	−1°

Second Method

1. Measure the frame angle.
   a. Front of frame down—frame angle positive.
   b. Front of frame up—frame angle negative.
2. From the specifications, determine the specified or desired caster setting.
3. Add or subtract the frame angle from the specified caster setting.
   a. Positive frame angle is subtracted from the specified setting.
   b. Negative frame angle is added to the specified caster setting.
4. Using wheel alignment equipment, obtain the measured caster angle from the wheel and determine the corrected specified setting, using the following examples as guides.

EXAMPLE NO. 3 (FRONT OF FRAME LOWER—POSITIVE)

Specified setting	+1°
Frame angle (positive—subtract)	−1°
Corrected specified setting	0°
Reading obtained from wheel	−1°
Adjust wheel to corrected specified setting of—	0°

EXAMPLE NO. 4 (FRONT OF FRAME HIGHER—NEGATIVE)

Specified setting	+2°
Frame angle (negative—add)	1°
Corrected specified setting	+3°
Reading obtained from wheel	+2°
Adjust to corrected specified setting of—	+3°

Angle of Kingpin Inclination

In addition to the caster angle, the kingpins (or knuckle support pivots) are also inclined toward each other at the top. This angle is known as kingpin inclination and is usually spoken of, and measured, in degrees.

The effect of kingpin inclination is to cause the wheels to steer in a straight line regardless of outside forces such as crowned roads, cross winds, etc., which may tend to make it steer at a tangent. As the spindle is moved from extreme right to extreme left it apparently rises and falls. Notice that it reaches its highest position when the wheels are in the straight-ahead position. In actual operation, the spindle cannot rise and fall because the wheel is in constant contact with the ground.

Therefore, the truck itself will rise at the extreme right turn and come to its lowest point at the straight-ahead position, and again rise for an extreme left turn. The weight of the truck will tend to cause the wheels to come to the straight-ahead position, which is the lowest position of the truck itself.

INCLUDED ANGLE

Included angle is the name given to that angle which includes kingpin inclination and camber. It is the relationship between the centerline of the wheel and the centerline of the kingpin (or the knuckle support pivots). This angle is built into the knuckle (spindle) forging and will remain constant throughout the life of the truck, unless the spindle itself is damaged.

When checking a truck on the front end stand, always measure kingpin inclination as well as camber unless some provision is made on the stand for checking condition of the spindle. Where no such provision is made, add the kingpin inclination to the camber for each side of the truck. These totals should be exactly the same, regardless of how far from the norm the readings may be.

For example the left side of the truck checks 5½° kingpin inclination and 1° positive camber—total 6½°. Since both sides check exactly the same for the included angle, it is unlikely that both spindles, in this instance, are bent. Adjusting to correct for camber will automatically set correct kingpin inclination.

A bent spindle would show up like this: left side of the truck has ¾° positive camber. 5¼° kingpin inclination—total 6° included angle. Right side of truck has 1¼° positive camber, 6° kingpin inclination—total 7¼° included angle. One of these spindles is bent and if adjustments are made to correct camber, the kingpin inclination will be incorrect due to the bent spindle.

Since the most common cause of a bent spindle is striking the curb when parking, which causes the spindle to bend upward, the side having the greater included angle usually has the bent spindle. It will be found impossible to achieve good alignment and minimum tire wear unless the bent spindle is replaced.

Toe-in

Toe-in is the amount that the front wheels are closer together at the front than they are at the back. This dimension is usually spoken of, and measured, in inches or fractions of inches.

Generally speaking, the wheels are toed-in because they are cambered. When a truck operates with 0° camber it will be found to operate with zero toe-in. As the required camber increases, so does the toe-in. The reason for this is that the cambered wheel tends to steer in the direction in which it is cambered. Therefore it is necessary to overcome this tendency of the wheel by compensating very slightly in the direction opposite to that in which it tends to roll. Caster and camber both have an effect on toe-in. Therefore toe-in is the last thing on the front end which should be corrected.

TOE-OUT STEERING RADIUS

When a truck is steered into a turn, the outside wheel of the vehicle scribes a much larger circle than the inside wheel. Therefore, the outside wheel must be steered to a somewhat less angle than the inside wheel. This difference in angle is often called toe-out.

The change in angle from toe-in in the straight-ahead position to toe-out in the turn is caused by the relative position of the steering arms to the kingpin and to each other.

If a line were drawn from the center of the kingpin through the center of the steering arm-tie rod attaching hole at each wheel, these lines would be found to cross almost exactly in the center of the rear axle.

If the front end angles, including toe-in, are set correctly, and the toe-out is found to be incorrect, one or both of the steering arms are bent.

Tracking

While tracking is more a function of the rear axle and frame, it is difficult to align the front suspension when the truck does not track straight. Tracking means that the centerline of the rear axle follows exactly the path of the centerline of the front axle when the truck is moving in a straight line.

On trucks that have equal tread, front and rear, the rear tires will follow in exactly the tread of the front tires, when moving in a straight line. However, there are many trucks whose rear tread is wider than the front tread. On such trucks, the rear axle tread will straddle the front axle tread an equal amount on both sides, when moving in a straight line.

Perhaps the easiest way to check a truck for tracking is to stand directly in back of it and watch it move in a straight line down the street. If the observer will stand as near to the center of the truck as possible, he can readily observe, even with the difference in perspective between the front and rear wheels, whether or not they are tracking properly. If the truck is found to track incorrectly, the difficulty will be found in either the frame or in the rear axle alignment.

Another more accurate method to check tracking is to park the truck on a level floor and drop a plumb-line from the extreme outer edge of the front suspension lower A-frame. Use the same drop point on each side of the truck. Make a chalk line where the plumb-line strikes the floor. Do the same with the rear axle, selecting a point on the rear axle housing for the plumb-line.

Measure diagonally from the left rear mark to the right front mark and from the right rear mark to the left front mark. These two diagonal measurements should be exactly the same. A ¼″ variation is acceptable.

If the diagonal measurements taken are different, measure from the right rear mark to the right front mark and from the left rear to the left front. These two measurements should also be the same within ¼″.

If the diagonal measurements are different, but the longitudinal measurements are the same, the frame is swayed (diamond shaped).

However, in the event that the diagonal measurements are unequal and the longitudinal measurements are also unequal, and the truck is tracking incorrectly, the rear axle is misaligned.

If the diagonal and longitudinal measurements are both unequal, but the truck appears to track correctly on the street, a kneeback is indicated.

NOTE: A kneeback means that one complete side of the front suspension is bent back. This is often caused by crimping the front wheels against the curb when parking the vehicle, then starting up without straightening the wheels out.

DOMESTIC TRUCKS
Chevrolet and GMC

Vehicle Identification: Year / Chevrolet Model (A)	GMC Model (A)	Adj. Ill. No.	Caster @ Height Measurement (Degrees) — Suspension Height Measurement (M): 1½	2	2½	3	3½	3¾	4	Camber (Degrees): Min	Pref.	Max.	Toe-in (Inches)	Toe-in (Millimeters)
1985-86														
Astrovan		(1)	②	②	②	②	②	②	②	1/8	15/16	1¾	5/32	4.0
1982-86														
S-10(4×2)	S-15(4×2)	(1)	①	①	①	①	①	①	①	0	13/16	1⅝	1/8	3.2
S10(4×4)	S-15(4×4)	(2)	①	①	①	①	①	①	①	0	13/16	1⅝	1/8	3.2
1982-86														
C-10	C-1500	(3)	—	—	3⅝	3⅛	2⅝	2⅜	2	0	11/16	1⅜	3/16 ± 1/8	4.8 ± 3.2
C-20,30	C-2500,3500	(3)	—	—	1½	15/16	5/16	1/8	0	−½	3/16	7/8	3/16 ± 1/8	4.8 ± 3.2
1982-86														
K-10,20	K-1500,2500	N/A	—	—	8	8	8	8	8	5/16 ③	1 ③	1 11/16 ③	3/16 ± 1/8 ④	4.8 ± 3.2 ④
K-30	K-3500	N/A	—	—	8	8	8	8	8	−3/16	½	1 3/16	3/16 ± 1/8	4.8 ± 3.2
1982-84														
G-10,20	G-1500,2500	(3)	3½	3⅛	2 11/16	2⅜	2⅛	1 15/16	1⅞	−3/16	½	1 3/16	3/16 ± 1/8	4.8 ± 3.2
G-30	G-3500	(3)	2⅞	2 3/16	1⅝	1	½	3/16	0	−½	3/16	7/8	3/16 ± 1/8	4.8 ± 3.2
1982-84														
P-10	P-1500	(3)	—	—	2 5/16	1 11/16	1 3/16	15/16	5/8	−½	3/16	7/8	3/16 ± 1/8	4.8 ± 3.2
P-20,30	P-2500,3500	(3)	—	—	2 5/16	1 11/16	1 3/16	15/16	5/8	−½	3/16	7/8	3/16 ± 1/8	4.8 ± 3.2
1982-84														
P30 Motor Home	P-3500 Motor Home	(3)	—	—	5½	5	4⅜	4⅛	3⅞	−½	3/16	7/8	5/16 ± 1/8	7.9 ± 3.2
1980-81														
G-10,20	G-1500,2500	(3)	3½	3⅛	2 11/16	2 13/32	2⅛	1 13/16	—	−3/16	½	1 3/16	3/16 ± 1/8	4.8 ± 3.2
G-30	G-3500	(3)	2 3/16	2 3/16	1⅝	1	½	0	—	−½	3/16	7/8	3/16 ± 1/8	4.8 ± 3.2
1979-81														
C-10	C-1500	(3)	—	—	2 3/32	1 13/16	1 3/16	11/16	3/16	−½	3/16	1⅞	3/16 ± 1/8	4.8 ± 3.2
C-20,30	C-2500,3500	(3)	—	—	1½	29/32	1 5/16	0	−11/16	−½	3/16	7/8	3/16 ± 1/8	4.8 ± 3.2
K-10,20	K-1500,2500	N/A	—	—	8	8	8	8	8	5/16	1	1 11/16	0 ± 1/8	0 ± 3.2
K-30	K-3500	N/A	—	—	8	8	8	8	8	−3/16	½	1 3/16	0 ± 1/8	0 ± 3.2
P-10	P-1500	(3)	—	—	2 5/16	1 11/16	1 3/16	5/8	1/8	−½	3/16	7/8	3/16 ± 1/8	4.8 ± 3.2
P-20,20(B)	P-2500,3500(B)	(3)	—	2 29/32	2 11/16	1 11/16	1 3/16	5/8	3/16	−½	3/16	7/8	3/16 ± 1/8	4.8 ± 3.2
P-30 Motor Home(B)	P-3500 Motor Home(B)		—	—	5½	5	4 13/32	3 13/16	3 5/16	−½	3/16	7/8	5/16 ± 1/8	7.9 ± 3.2
1979														
G-10,20	G-1500,2500	(3)	2 29/32	2 5/16	2	1⅝	1 5/16	29/32	—		N/A		3/16 ± 1/8	4.8 ± 3.2
G-30	G-3500	(3)	3 13/32	2 11/16	2⅛	1½	1	13/32	—		N/A		3/16 ± 1/8	4.8 ± 3.2

With vehicle level, measure frame angle with a bubble protractor. Record the suspension height measurement.
a. Subtract an up-in-rear frame angle from a positive caster specification.
b. Subtract a down-in-rear frame angle from a negative caster specification.
c. Add an up-in-rear frame angle to a negative caster specification.
d. Add a down-in-rear frame angle to a positive caster specification.
(A) Vehicle height must be checked and corrected before alignment is performed.
① 1° Min. 2° Pref. 3° Max.
② 1 11/16 Min. 2 11/16 Pref. 3 11/16 Max.
③ ¾ Min. 1½ Pref. 2 Max. for '85–'86 models
④ 0 ± 1/8 (0 ± 3.2mm) for '85–'86 models

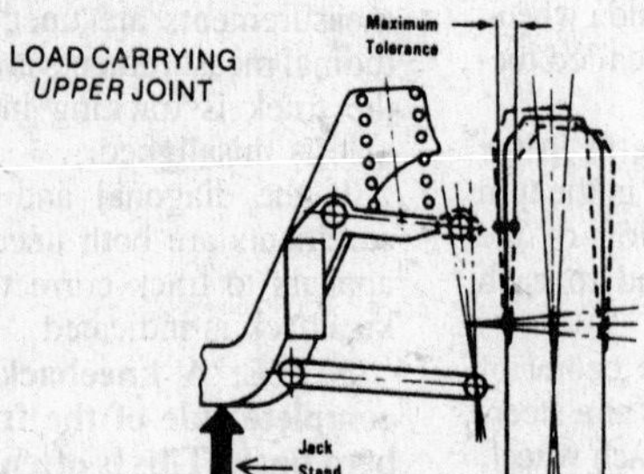
LOAD CARRYING
UPPER JOINT
Maximum Tolerance
Jack Stand

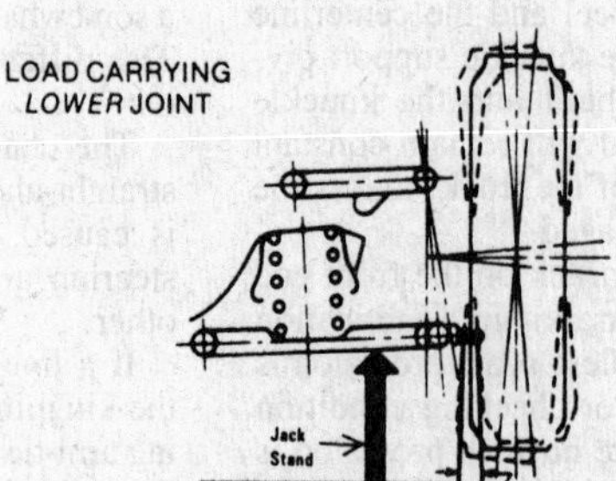
LOAD CARRYING
LOWER JOINT
Jack Stand
Maximum Tolerance

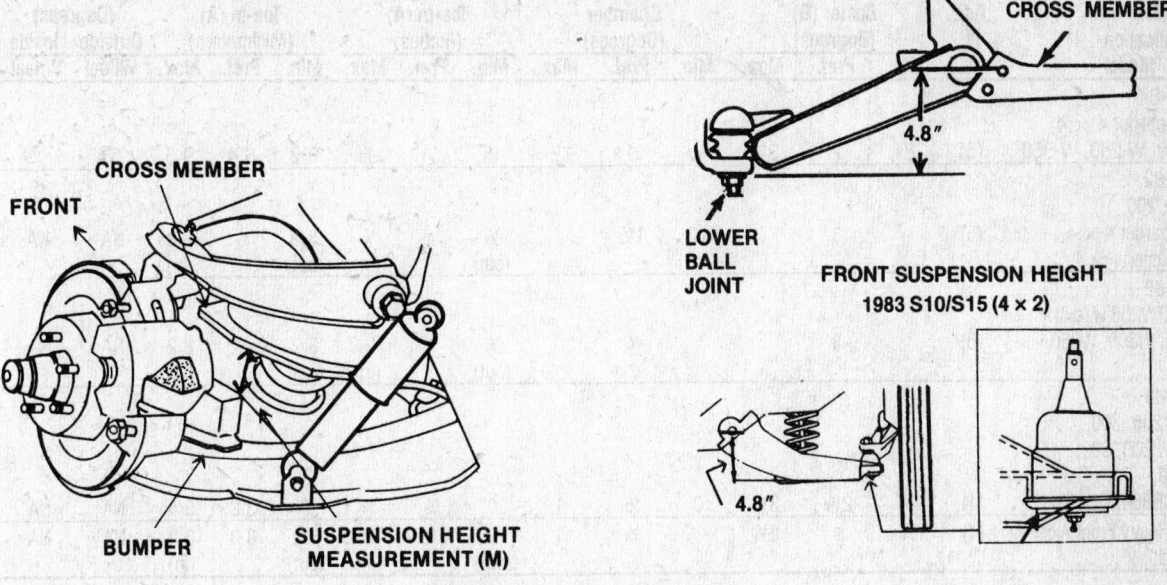

FRONT SUSPENSION HEIGHT
S10/S15 (4 × 4)

CROSS MEMBER

4.8″

LOWER
BALL
JOINT

FRONT SUSPENSION HEIGHT
1983 S10/S15 (4 × 2)

4.8″

CROSS MEMBER

FRONT

BUMPER

SUSPENSION HEIGHT
MEASUREMENT (M)

Chevrolet/GMC measuring points

CHRYSLER CORP. LIGHT TRUCKS

Vehicle Identification Year Model	Adj. Ill. No.	Caster (B) (Degrees) Min.	Pref.	Max.	Chamber (Degrees) Min.	Pref.	Max.	Toe-In (A) (Inches) Min.	Pref.	Max.	Toe-In (A) (Millimeters) Min.	Pref.	Max.	Toe-Out On Turns (Degrees) Outside Wheel	Inside Wheel	Strg. Axis Incl. (Deg.)
'85–'86 Caravan, Mini Ram Van, Voyager																
Front	—	——— Fixed ———			-¼	5/16	¾	⅛ (out)	1/16 (in)	7/32 (in)	3.2 (out)	1.6 (in)	5.6 (in)	NA	NA	NA
Rear	—	——— Fixed ———			−1⅛	−½	−⅛	¼ (out)	0	¼ (in)	6.3 (out)	0	6.3 (in)			
'85–'86 Caravan, Mini Ram, Van, Voyager																
Front	—	——— Fixed ———			1/16	5/16	1 1/16	⅛ (out)	1/16 (out)	0	3.2 (out)	1.6 (out)	0.0	NA	NA	NA
Rear	—	——— Fixed ———			⅛	⅝	1⅛	5/32 (out)	3/32 (in)	11/32 (in)	4.0 (out)	2.5 (in)	8.8 (in)			
1981–86 B100, 200 300; PB100, 200, 200; MB300, 400 CB300, 400	(4)	1¼	2¼	3¼	0	½	1	0	⅛	¼	0.0	3.2	6.0	NA	NA	NA
'85–'86 Ram-Charger 4 × 2	(2)	−½	½	1½	0	½	1	0	3/16	½	0	4.8	12.7	33	33	NA
1983–84 Ramcharger 4 × 2, D-150, D-250, D350	(2)	−½	½	1½	0	½	1	1/16	⅛	3/16	0.0	3.2	4.8	33	33	NA
1979–82 Ramcharger 4 × 2 Trail Duster 4 × 2, D100, 150, 200, 250, 300, 350	(2)	−½	½	1½	0	¼	1	0	⅛	¼	0	3.2	6.4	33	33	8½

CHRYSLER CORP. LIGHT TRUCKS

Vehicle Identification Year Model	Adj. Ill. No.	Caster (B) (Degrees) Min.	Pref.	Max.	Chamber (Degrees) Min.	Pref.	Max.	Toe-In (A) (Inches) Min.	Pref.	Max.	Toe-In (A) (Millimeters) Min.	Pref.	Max.	Toe-Out On Turns (Degrees) Outside Wheel	Inside Wheel	Strg. Axis Incl. (Deg.)
1983–84 Ramcharger 4 × 4, W-150, W-250, W-350	(5)	½	2	3½	½	1	1½	⅛	¼	⅜	3.2	6.4	9.5	①	①	8.5
1979–82 W150, 200, Ramcharger 4 × 4, Trail Duster 4 × 4	(5)		3			1½		⅛ (out)		⅛ (in)	3.2 (out)		3.2 (in)	NA	NA	8½
1979–82 W200, W250 w/extra equip., W300, W350, W400	(5)		3			½		⅛ (out)		⅛ (in)	3.2 (out)		3.2 (in)	NA	NA	8½
1979–80 B100, 200, 300, PB100, 200, 300; MB300, 400; CB300, CB400	(4)	1¼	2¼	3¼	0	½	1	0	⅛	¼	0	3.2	6.4	NA	NA	NA
W/Heavy Front Axle Load	(4)	¼		2¾	–¼	⅜	1	⅛ (out)	0	⅛ (in)	3.2 (out)	0.0	3.2 (in)	NA	NA	NA

INTERNATIONAL

Vehicle Identification Year Model	Adj. Ill. No.	Caster (Degrees) Min.	Pref.	Max.	Camber (Degrees) Min.	Pref.	Max.	Toe-In (Inches) Min.	Pref.	Max.	Toe-In (Millimeters) Min.	Pref.	Max.	Toe-Out On Turns (Degrees) Outside Wheel	Inside Wheel
80 Scout, All Models	(5)		2½			½		3/32		5/16	1.6		7.9	NA	NA
79 Scout, All Models	(5)		0			1		0		3/16	0		5.1	NA	NA

AMC—JEEP

Vehicle	Ill. No.	Caster (Degrees) Min.	Pref.	Max.	Camber (Degrees) Min.	Pref.	Max.	Toe-In (Inches) Min.	Max.	Toe-In (Millimeters) Min.	Max.	Toe-Out On Turns (Degrees) Outside Wheel	Inside Wheel	Strg. Axis Incl. (Deg.)
'86–'84 Sportwagon, Cherokee, Wagoneer	(5)	7	7½	8	–½	0	½	1/32 0 1/32 (out) (in)		0.8 0 0.8 (out) (in)		——NA——		NA
83–82 CJ-5	(5)	6	6	7	0	0	½	3/64	3/32	1.2	2.4	——29——		8½
85–82 CJ-7 and Scrambler	(5)	6	6	7	0	0	½	3/64	3/32	1.2	2.4	——32——		8½
85–82 Cherokee, Wagoneer and Pickup Truck	(5)	4	4	5	0	0	½	3/64	3/32	1.2	2.4	——36-37——		8½
81 "CJ" Models	(5)	6	6	7	1½	1½	2	3/64	3/32	1.2	2.4	——(A) 31——		8½
81–80 Cherokee, Wagoneer and Pickup Trucks	(5)		4			0		3/64	3/32	1.2	2.4	——(A) 37——		8½
80–79 "CJ" Models	(5)		3			1½		3/64	3/32	1.2	2.4	——(A) 31½——		8½
79 Cherokee, Wagoneer and Pickup Trucks	(5)		4			1½		3/64	3/32	1.2	2.4	——37½——		8½

1981 CJ models 31°–32°; 1980–81 except CJ 37°–38°.

FORD

Year/Model	Axle Part Number Ride Height Min.	Max.	Adj. Ill. No.	Caster (Degrees) Min.	Max.	Camber (Degrees) Min.	Max.	Toe-in (Inches)	Toe-in (Millimeters)
1984–86 Ranger, 4×2(1)(3)(6)	Forged Axle		(6)	(4)	(4)				
	3¼	3½		5¼	8¼	-2	-½	1/32	0.8
	3½	3¾		4½	7½	-1⅝	⅛	1/32	0.8
	3¾	4		3½	6½	-½	1	1/32	0.8
	4	4¼		3	6	½	1¼	1/32	0.8
	4½	4¾		1⅞	4⅞	1¼	2¾	1/32	0.8
	Stamped Axle		(6)						
	3	3¼		5¼	8¼	-2	-½	1/32	0.8
	3¼	3½		4½	7½	-1⅝	⅛	1/32	0.8
	3½	3¾		3½	6½	-½	1	1/32	0.8
	3¾	4		3	6	½	1¼	1/32	0.8
	4¼	4½		1⅞	4⅞	1¼	2¾	1/32	0.8
1983 Ranger, 4×2(1)(3)(6)			(6)	(4)	(4)				
	2¾	3¼		4½	7	-1	1	1/32	0.8
	3¼	3½		4	6½	-½	1¾	1/32	0.8
	3½	4		3⅜	5⅞	0	2⅜	1/32	0.8
	4	4¼		2⅝	5⅛	¾	3	1/32	0.8
	4¼	4¾		2	4½	1½	3¾	1/32	0.8
1984–86 Ranger, & Bronco II 4×4(1)(3)(6)			(6)	(4)	(4)				
	2¾	3		5½	8½	-2	-½	1/32	0.8
	3¼	3½		4	7	-1	½	1/32	0.8
	3½	3¾		3	6	0	1½	1/32	0.8
	4	4¼		2	5	1	2½	1/32	0.8
	4¼	4½		1	4	2	3½	1/32	0.8
1983 Ranger, 4×4(1)(3)(6)			N/A	(4)	(4)	(4)	(4)		
	2¾	3¼		5	8	-1	½	1/32	0.8
	3¼	3½		4	7	0	1½	1/32	0.8
	3½	4		3	6	½	2	1/32	0.8
	4	4¼		2½	5½	1¼	2¾	1/32	0.8
	4¼	4¾		1¾	5	2	3¾	1/32	0.8
1985–86 F250, 350, 4×2(1)(2)			N/A	(4)	(4)				
	2½	2¾		5¼	7¼	-1	½	3/32	2.4
	2¾	3		5	7	-½	¾	3/32	2.4
	3¼	3½		4½	6½	-⅛	1¼	3/32	2.4
	3¾	4		4	6	½	1¾	3/32	2.4
	4	4¼		3½	5½	¾	2¼	3/32	2.4
1983–84 F250, 350, 4×2(1)(3)(A)			N/A	(4)	(4)				
	2½	2¾		5½	7	-½	½	1/32	0.8
	2¾	3¼		5	6	-½	1½	1/32	0.8
	3¼	3½		4	5	¼	2¼	1/32	0.8
	3½	4		3	4	1	3	1/32	0.8
	4	4¼		(7)	(7)	2	3½	1/32	0.8
1981–82 F-250, 350, 4×2(1)(3)			N/A	(4)	(4)	(4)	(4)		
	2	2¼		5¾	9	-2½	0	1/32	0.8
	2¼	2¾		4¾	8	-1½	1	1/32	0.8
	2¼	3¼		3¾	7	-¼	1¾	1/32	0.8
	3¼	3¾		2¾	6	¼	2¾	1/32	0.8
	3½	4		1¾	5	1	3½	1/32	0.8
	4	4¼		¾	4	2	4½	1/32	0.8

FORD

Year/Model	Axle Part Number Ride Height Min.	Max.	Adj. Ill. No.	Caster (Degrees) Min.	Max.	Camber (Degrees) Min.	Max.	Toe-in (Inches)	Toe-in (Millimeters)
1984–86 F250, 350, 4×4(1)(3)(6)			N/A						
	5	5¼		3	5	-¼	-1¾	1/32	0.8
	5½	5¾		3⅛	5⅛	¾	-¾	1/32	0.8
	6	6¼		3¼	5¼	1½	½	1/32	0.8
	6¼	6½		3⅜	5⅜	3	1½	1/32	0.8
	6¾	7		3½	5½	4	2½	1/32	0.8
1983 F250, 350, 4×4(1)(3)(A)			N/A	(4)	(4)	(4)	(4)		
	5	5½		3 1/16	5⅛	-1¾	¾	1/32	0.8
	5½	6		3⅛	5¼	-¾	1¾	1/32	0.8
	6	6¼		3¼	5⅜	¼	3	1/32	0.8
	6¼	6¾		3⅜	5½	1½	4	1/32	0.8
	6¾	7		3½	5½	2½	4¼	1/32	0.8
1981–82 F250, 350 4×4(1)(3)			N/A	(4)	(4)				
	4¾	5		3	5	4¾	5	1/32	0.8
	5	5½		3⅛	5⅛	-1¾	¾	1/32	0.8
	5½	6		3⅛	5⅛	-¾	1¾	1/32	0.8
	6	6¼		3¼	5¼	¼	2¾	1/32	0.8
	6¼	6¾		3⅜	5⅜	1¼	4	1/32	0.8
	6¾	7		3½	5½	2½	5	1/32	0.8
1985–86 E150(1)(3)			N/A	(4)	(4)	(4)	(4)		
	3½	3¾		—	—	—	—	1/32	0.8
	4	4½		7½	9½	-1¼	¼	1/32	0.8
	4½	5		6¼	8¼	-⅛	1¼	1/32	0.8
	5	5½		5	7	⅞	2¼	1/32	0.8
	5½	5¾		3¼	5¼	1¾	3¼	1/32	0.8
1984 E150(1)(3)			N/A	(4)	(4)	(4)	(4)		
	3½	3¾		—	—	—	—	1/32	0.8
	4	4¼		4⅝	6¼	½	-¾	1/32	0.8
	4½	4¾		3¼	5¼	1¾	½	1/32	0.8
	5	5¼		2	4	2¾	1½	1/32	0.8
	5½	5¾		¼	2¼	3⅝	2⅜	1/32	0.8
1983 E100, 150(1)(3)			(6)						
	3¾	4		—	—	-¾	¾	1/32	0.8
	4	4½		4½	5¼	-⅝	1⅝	1/32	0.8
	4½	5		3¼	4	⅜	2⅝	1/32	0.8
	5	5½		2	2¾	1¼	3⅝	1/32	0.8
	5½	5¾		¾	2¼	2¼	4⅛	1/32	0.8
1981–82 E-100, 150(1)(3)			N/A						
	3¼	3½		6½	8	-1¾	-¼	1/32	0.8
	3½	3¾		5¾	7¼	-1½	¼	1/32	0.8
	3¾	4		5	6¾	-1	¾	1/32	0.8
	4	4¼		4½	5¾	-½	1¼	1/32	0.8
	4¼	4½		4	5¼	0	1¾	1/32	0.8
	4½	4¾		3¼	4½	½	2¼	1/32	0.8
	4¾	5		2½	4	1	2¾	1/32	0.8
	5	5¼		2	3¼	1½	3¼	1/32	0.8
	5¼	5½		1½	2¾	2	3¾	1/32	0.8
1984–86 E150 4×2(1)(3)(A)			(6)	(4)	(4)				
	3¼	3½		5	7	-¾	¾	1/32	0.8
	3½	4		4	6	¼	1¼	1/32	0.8
	4	4¼		3¼	5¼	½	2	1/32	0.8
	4¼	4¾		2½	4½	2	3½	1/32	0.8
	4¾	5		1½	3½	3	4½	1/32	0.8

FORD

Year/ Model	Axle Part Number Ride Height Min.	Max.	Adj. Ill. No.	Caster (Degrees) Min.	Max.	Camber (Degrees) Min.	Max.	Toe-in (Inches)	Toe-in (Millimeters)
1983 F-100, 150, 4×2(1)(3)(A)			(6)	(4)	(4)				
	2¾	3¼		5½	7½	−1½	¾	1/32	0.8
	3¼	3½		5	6	−¾	1½	1/32	0.8
	3½	4		4¼	5¼	¼	2½	1/32	0.8
	4	4¼		3¼	4¼	1	3½	1/32	0.8
	4¼	4¾		2½	3½	2	4½	1/32	0.8
1981–82 F-100, 150 4×2(1)(3)			N/A	(4)	(4)	(4)	(4)		
	2¼	2¾		6	10	−3	−½	1/32	0.8
	2¾	3¼		5	9	−2	½	1/32	0.8
	3¼	3½		4	8	−1¼	1¼	1/32	0.8
	3½	4		3	7	−¼	2¼	1/32	0.8
	4	4¼		2	6	½	3	1/32	0.8
	4¼	4¾		1	5	1½	4	1/32	0.8
1980 F-100, 150 4×2(1)(2)			N/A	(4)	(4)	(4)	(4)		
	2¼	2¾		6	10	−3	−½	3/32	2.4
	2¾	3¼		5	9	−2	½	3/32	2.4
	3¼	3½		4	8	−1¼	1¼	3/32	2.4
	3½	4		3	7	−¼	2¼	3/32	2.4
	4	4¼		2	6	½	3	3/32	2.4
	4¼	4¾		1	5	1½	4	3/32	2.4
1981–82 F-150, Bronco 4×4(1)(3)			N/A	(4)	(4)				
	2¾	3¼		6	9	−2½	−¼	1/32	0.8
	3¼	3½		5	8	−1¾	½	1/32	0.8
	3½	4		4	7	−¾	1½	1/32	0.8
	4	4¼		3	6	0	2¼	1/32	0.8
	4¼	4¾		2	5	1	3¼	1/32	0.8
	4¾	5		1	4	1¾	4	1/32	0.8
1985–86 F150, Bronco 4×4(1)(3)			(6)	(4)	(4)				
	3¼	3½		6	8	−1¾	−¼	1/32	0.8
	3½	3¾		5	7	−¾	¾	1/32	0.8
	4	4¼		4	6	¼	1¾	1/32	0.8
	4¼	4½		3	5	1¼	2¾	1/32	0.8
1983 F-150, Bronco 4×4 4×4(1)(3)			(6)	(4)	(4)				
	3¼	3½		6	7	−1½	¾	1/32	0.8
	3½	4		5	6	−¾	1¾	1/32	0.8
	4	4¼		4	5	¼	2¾	1/32	0.8
	4¼	4¾		3	4	1¼	3½	1/32	0.8
1980 F-150, Bronco 4×4(1)(2)			N/A	(4)	(4)				
	2¾	3¼		6	9	−2½	−¼	3/32	2.4
	3¼	3½		5	8	−1¾	½	3/32	2.4
	3½	4		4	7	−¾	1½	3/32	2.4
	4	4¼		3	6	0	2¼	3/32	2.4
	4¼	4¾		2	5	1	3¼	3/32	2.4
	4¾	5		1	4	1¾	4	3/32	2.4
1984–86 E250, 350 (1)(3)			N/A	(4)	(4)	(4)	(4)		
	3¼	3½		—	—	—	—	1/32	0.8
	3¾	4		7⅝	9⅝	−¾	½	1/32	0.8
	4¼	4½		6¼	8¼	¼	1½	1/32	0.8
	4¾	5		5	7	1¼	2½	1/32	0.8
	5¼	5½		3¾	5¾	2¼	3½	1/32	0.8

FORD

Year/ Model	Axle Part Number Ride Height Min.	Max.	Adj. Ill. No.	Caster (Degrees) Min.	Max.	Camber (Degrees) Min.	Max.	Toe-in (Inches)	Toe-in (Millimeters)
1983 E-250, 350(1)(3)			N/A						
	3¾	4		—	—	−⅝	⅞	1/32	0.8
	4	4½		7¼	8	−½	1⅞	1/32	0.8
	4½	5		6	6¾	½	2⅞	1/32	0.8
	5	5½		4¾	5½	1½	3⅞	1/32	0.8
	5½	5¾		3¼	4	2½	4⅜	1/32	0.8
1980–82 E-250, 350(1)(3)			N/A						
	3¼	3½		9	10½	−1¾	−¼	1/32	0.8
	3½	3¾		8½	9¾	−1½	¼	1/32	0.8
	3¾	4		7⅞	9	−1	¾	1/32	0.8
	4	4¼		7⅛	8½	−½	1¼	1/32	0.8
	4¼	4½		6½	7¾	0	1¾	1/32	0.8
	4½	4¾		5¾	7	½	2¼	1/32	0.8
	4¾	5		5¼	6½	1	2¾	1/32	0.8
	5	5¼		4⅝	6	1½	3¼	1/32	0.8
	5¼	5½		4	5½	2	3¾	1/32	0.8
1980 F-250, 350 4×2(1)(3)			N/A	(4)	(4)	(4)	(4)		
	2	2¼		5¼	9	−2½	0	1/32	0.8
	2¼	2¾		4¾	8	−1½	1	1/32	0.8
	2¾	2¼		3¾	7	−¾	1¾	1/32	0.8
	3¼	3½		2¾	6	¼	2¾	1/32	0.8
	3½	4		1¾	5	1	3½	1/32	0.8
	4	4¼		¾	4	2	4½	1/32	0.8
1980 F-250, 350 4×4(1)(2)			(6)	(4)	(4)				
	4¼	4¾		3½	5¾	−4	−1¼	3/32	2.4
	4¾	5		3¼	5½	−2¾	0	3/32	2.4
	5	5½		3	5¼	−1½	1¼	3/32	2.4
	5½	6		2¾	5	−¼	2½	3/32	2.4
	6	6¼		2½	4¾	1	3¾	3/32	2.4
	6¼	6¾		2¼	4½	2¼	5	3/32	2.4
1980 E-100, 150(1)(5)			N/A						
	3¼	3½		6¼	8	−1¾	−¼	¼	6.4
	3½	3¾		5¾	7¼	−1½	¼	¼	6.4
	3¾	4		5	6¾	−1	¾	¼	6.4
	4	4¼		4½	5¾	−½	1¼	¼	6.4
	4¼	4½		4	5¼	0	1¾	¼	6.4
	4½	4¾		3¼	4½	½	2¼	¼	6.4
	4¾	5		2½	4	1	2¾	¼	6.4
	5	5¼		2	3¼	1½	3¼	¼	6.4
	5¼	5½		1½	2¾	2	3¾	¼	6.4
1979 F-100, F-150, F-250 (D) (6200-6800 GVW) (4×2)			N/A						
	2¾	3		8⅜	9⅝	−2	−¼	3/32	2.4
	3	3¼		7¾	9	−1½	⅛	3/32	2.4
	3¼	3½		7	8⅜	−1⅛	½	3/32	2.4
	3½	3¾		6¼	7⅝	−¾	1	3/32	2.4
	3¾	4		5⅞	7⅛	−⅜	1¼	3/32	2.4
	4	4¼		5⅛	6½	0	1⅝	3/32	2.4
	4¼	4½		4½	5⅞	⅜	2	3/32	2.4
	4½	4¾		3¾	5¼	¾	2⅜	3/32	2.4
	4¾	5		3¼	4⅝	1¼	1¾	3/32	2.4
	5	5¼		2½	4	1⅞	3⅛	3/32	2.4

FORD

Year/ Model	Axle Part Number Ride Height Min.	Max.	Adj. Ill. No.	Caster (Degrees) Min.	Max.	Camber (Degrees) Min.	Max.	Toe-in (Inches)	Toe-in (Millimeters)
1979			N/A						
F-250 R/C (7700-7900) (4×2) (D)	2¾	3		7¾	9	−1⅞	−¼	3/32	2.4
F-250 S/C (6300-7800) (D8TA-BA) (D)									
	3	3¼		7	7⅜	−1½	⅛	3/32	2.4
	3¼	3½		6⅜	7¾	−1⅛	½	3/32	2.4
	3½	3¾		5⅞	7⅛	−¾	⅞	3/32	2.4
	3¾	4		5⅛	6½	−⅜	1¼	3/32	2.4
	4	4¼		4½	5⅞	0	1⅝	3/32	2.4
	4¼	4½		3⅞	5¼	⅜	2	3/32	2.4
	4½	4¾		3¼	4⅝	⅞	2½	3/32	2.4
	4¾	5		2⅝	4	1¼	2¾	3/32	2.4
	5	5¼		2	3⅜	1⅝	3⅛	3/32	2.4
1979			N/A						
F-350, F-250 S/C (3100) (D)	2¾	3		9⅝	11	−2⅛	−⅝	3/32	2.4
(4×2) F-250 S/C RPO Suspension (D8TA-DA)(D)									
	3	3¼		8⅞	10⅜	−1¾	−¼	3/32	2.4
	3¼	3½		8⅜	9¾	−1⅜	⅛	3/32	2.4
	3½	3¾		7¾	9	−1⅛	½	3/32	2.4
	3¾	4		7	8⅜	−¾	¾	3/32	2.4
	4	4¼		6⅜	7¾	−1¼	1¼	3/32	2.4
	4¼	4½		5¾	7⅛	0	1⅝	3/32	2.4
	4½	4¾		5⅛	6½	⅜	2	3/32	2.4
	4¾	5		5⅛	6½	¾	2⅜	3/32	2.4
	5	5¼		3⅞	5¼	1¼	2¼	3/32	2.4
1979			N/A						
E-100/150 (D7UA-BB) (C) (4×2)	3¼	3½		6¼	8	−1¾	−¼	1/32	0.8
	3½	3¾		5¾	7¼	−1½	¼	1/32	0.8
	3¾	4		5	6¾	−1	¾	1/32	0.8
	4	4¼		4½	5¾	−½	1¼	1/32	0.8
	4¼	4½		4	5¼	0	1¾	1/32	0.8
	4½	4¾		3¼	4½	½	2¼	1/32	0.8
	4¾	5		2½	4	1	2¾	1/32	0.8
	5	5¼		2	3¼	1½	3¼	1/32	0.8
	5¼	5½		1½	2¾	2	3¾	1/32	0.8
1979			N/A						
E-250/350 (D5UA-AB) (C) (4×2)	3¼	3½		9	10½	−1¾	−¼	1/32	0.8
	3½	3¾		8½	9¾	−1½	¼	1/32	0.8
	3¾	4		7⅞	9	−1	¾	1/32	0.8
	4	4¼		7⅛	8½	−½	1¼	1/32	0.8
	4¼	4½		6½	7¾	0	1¾	1/32	0.8
	4½	4¾		5¾	7	½	2¼	1/32	0.8
	4¾	5		5¼	6½	1	2¾	1/32	0.8
	5	5¼		4⅝	6	1½	3¼	1/32	0.8
	5¼	5½		4	5½	2	3¾	1/32	0.8

(A) Right side height measurement shown; left side height should be 0″ to 7/16″ higher on any one vehicle.
(1) All vehicles with normal operating attitude.
(2) Nominal toe setting is 2.5 mm (3/32 inch). Range is 0.8 mm (3/32 inch) out to 5.6 mm (7/32 inch) in.
(3) Nominal toe setting is 8 mm (1/32 inch). Range is 2.5 mm (3/32 inch) out to 4 mm (5/32 inch) in.
(4) Not Adjustable.
(5) Toe range is 0″ to ¼″, ⅛″ nominal.
(6) Side-to-side variation; Caster 1½°, Camber 23/32°.
(7) Vehicle height is too high.

If caster or camber are not with specifications, find the true ride height by installing wooden blocks as pictured below. This will aid in determining which wheel is out of specification.

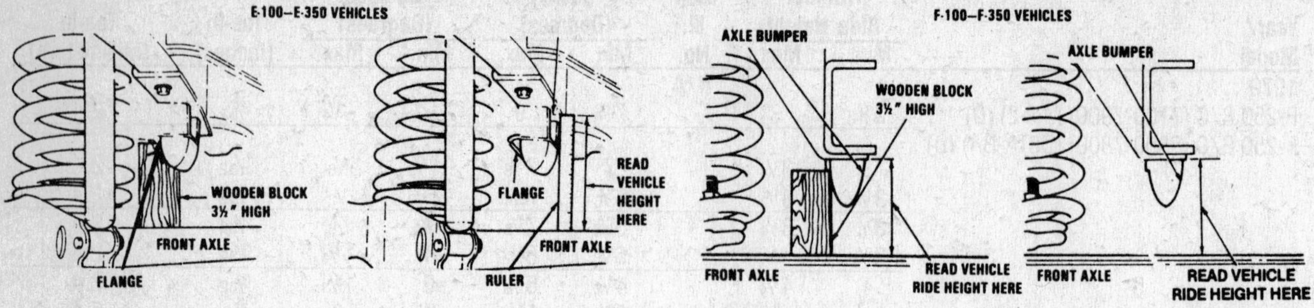
E-100—E-350 VEHICLES
WOODEN BLOCK 3½" HIGH
FRONT AXLE
FLANGE
FLANGE
READ VEHICLE HEIGHT HERE
FRONT AXLE
RULER
F-100—F-350 VEHICLES
AXLE BUMPER
WOODEN BLOCK 3½" HIGH
FRONT AXLE
READ VEHICLE RIDE HEIGHT HERE
AXLE BUMPER
FRONT AXLE
READ VEHICLE RIDE HEIGHT HERE

To find ride height on vehicles equipped with front leaf springs:
Measure from bottom of frame to top of front axle.

Ford measuring points

IMPORTED TRUCKS

Vehicle Identification Year Model	Adj. Ill. No.	Caster (Degrees)			Camber (Degrees)			Toe-In (Inches)			Toe-In (Millimeters)			Toe-Out on Turns (Degrees)	
		Min.	Pref.	Max.	Min.	Pref.	Max.	Min.	Pref.	Max.	Min.	Pref.	Max.	Outside Wheel	Inside Wheel
CHEVROLET LUV															
82–81 Series 12, 11(4×2)	(1)	0	1/2	1	0	1/2	1	0	1/16	1/8	0.0	2.0	4.0	33	37
82–81 Series 12, 11(4×4)	(7)	−3/16	5/16	13/16	1/16	9/16	1 1/16	1/16 (out)	0	1/16 (in)	2.0 (out)	0	2.0 (in)	33	35
80–79 Series 10, 9(4×4)	(7)	−1 1/16	5/16	15/16	−3/16	9/16	15/16	1/8 (out)	0	1/8 (in)	3.2 (out)	0	3.2 (in)	30	39
80–79 Series 10,9,8,6(4×2)	(1)	−13/16	−3/16	13/16	−1/4	1/2	1 1/4	1/8 (out)	0	1/8 (in)	3.2 (out)	0	3.2 (in)	30	39
DATSUN/NISSAN															
'86–'85 Pick-up (4×2)	(7)	13/16	—	1 13/16	0	—	1							18	20
w/Radial Tires								1/16	—	5/32	2.0	—	4.0		
w/Bias Tires								3/16	—	9/32	5.0	—	7.0		
84 720 Pick-up Truck (4×2)															
w/Radial Tires	(7)	13/16	—	23/16	0	—	1	1/16	—	5/32	2.0	—	4.0	18	20
w/Bias Tires	(7)	13/16	—	23/16	0	—	1	3/16	—	9/32	5.0	—	7.0	18	20
83–82 Pick-up Truck (4×2)	(7)	13/16		1 13/16	0		1	7/32		9/32	5.1		7.1	18	20
86–84 720 Pick-up Truck (4×4)	(7)	15/16		1 15/16	3/16		1 3/16	1/16		5/32	2.0		4.0	18	20
83–81 Pick-up Truck (4×4)	(7)	13/16		23/16	0		1	13/64		9/32	5.6		7.1	18	18 1/2
81–80 Pick-up Truck	(7)	13/16		1 13/16	0		1	13/64		9/32	5.6		7.1	18	20
79 Pick-up Truck	(7)	1/2		2	−1/4		1 1/4	7/32		9/32	5.6		7.1	30 1/2	35+
+ Plus or minus 1° is considered acceptable.															
DODGE D50, PLYMOUTH ARROW, MITSUBISHI (A)(B)															
86–83 D-50 Pick-up Truck (4×4)	(1)	1	2	3	1/2	1	1 1/2	3/32		11/32	2.4		8.7	28	30 3/4
86-80 D-50 Pick-up Truck (4×2)	(1)	1 1/2	2 1/2	3 1/2	1/2	1	1 1/2	3/32		11/32	2.4		8.7	30 1/2	37
79 D-50 Pick-up Truck	(1)	2	3	4	1/2	1	1 1/2	3/32		11/32	2.4		8.7	30 1/2	37
82–80 Arrow Pick-up Truck	(1)	1 1/2	2 1/2	3 1/2	1/2	1	1 1/2	3/32		11/32	2.4		8.7	30 1/2	37
79 Arrow Pick-up Truck	(1)	2	3	4	1/2	1	1 1/2	3/32		11/32	2.4		8.7	30 1/2	37
84–83 Mitsubishi (4×2)	(1)	1 1/2	2 1/2	3 1/2	1/2	1	1 1/2	3/32		11/32	2.0		9.0	30 1/2	37
84–83 Mitsubishi (4×4)	(1)	1	2	3	1/2	1	1 1/2	3/32		11/32	2.0		9.0	28	30 3/4

IMPORTED TRUCKS

Vehicle Identification Year Model	Adj. Ill. No.	Caster (Degrees) Min.	Pref.	Max.	Camber (Degrees) Min.	Pref.	Max.	Toe-In (Inches) Min.	Pref.	Max.	Toe-In (Millimeters) Min.	Pref.	Max.	Toe-Out on Turns (Degrees) Outside Wheel	Inside Wheel
ISUZU															
81–86 Pick-up Truck (4×2)	(8)(9)	0	½	1	0	½	1	0	5/64	5/32	0.0	2.0	4.0	33	37
81–86 Pick-up Truck (4×4)	(7)	−3/16	5/16	13/16	1/16	9/16	1 1/16	5/64 (out)	0	5/64 (in)	2.0 (out)	0.0	2.0 (in)	33	35
MAZDA															
79–86 B2000, B2200	(3)	11/16	1	1 5/16	7/16	¾	1¼	0	1/8	¼	0	3.0	6.0	30 11/16	32½
Maximum variation wheel, caster 1 1/16°, camber ½°															
TOYOTA															
86–85 4-Runner	N/A	2	3	4	¼	1	1¾	0	1/32	1/16	0	1	2	N/A	N/A
85 Pick-up (4×2)															
½ Ton Long Bed	(4)	11/16	1 3/16	1 11/16	0	½	1							30	34
w/Bias Tires								3/16	¼	9/32	5.0	6.0	7.0		
w/Radial Tires								1/16	1/8	5/32	2.0	3.0	4.0		
1 Ton	(1)	1/16	9/16	1 1/16	0	½	1	1/8	5/32	3/16	3.0	4.0	5.0	30	34
Cab & Chassis	(1)	−7/16	1/16	9/16	0	½	1	1/8	5/32	3/16	3.0	4.0	5.0	30	34
86–84 Pick-up (4×4)	N/A	1¼	2¼	3¼	¼	1	1¾							29	30½
w/Radial Tires								0	1/32	1/16	0	1.0	2.0		
w/Bias Tires								1/8	5/32	3/16	3.0	4.0	5.0		
86–84 Pick-up (4×2)															
½ Ton Short Bed		1/16	11/16	1 3/16	0	½	1							30	40
w/Bias Tires	(1)							1/8	5/32	3/16	3.0	4.0	5.0		
w/Radial Tires	(1)							0	1/32	1/16	0	1.0	2.0		
84 Pick-up (4×2)															
½ Ton Long Bed	(1)	11/16	1 3/16	1 11/16	0	½	1							30	34
¾ Ton	(1)	1/16	11/16	1 3/16	0	½	1							30	34
Cab & Chassis	(1)	11/16	1 3/16	1 11/16	0	½	1							30	34
All w/Bias Tires								3/16	¼	9/32	5.0	6.0	7.0		
All w/Radial Tires								1/16	1/8	5/32	2.0	3.0	4.0		
83–84 Hilux (4×4)	N/A	2¾	3½	4¼	¼	1	1¾							29	30½
w/Radial Tires							0	3/64	5/64	0	1.0	2.0			
w/Bias Tires								1/8	5/32	13/64	3.0	4.0	5.0		
83–82 Hilux (4×2) Pick-up															
RN½ Ton	(1)	½	1	1½	9/16	1 1/16	1 9/16							29	36
w/Bias Tires							5/32	13/64	15/64	4.0	5.0	6.0			
w/Radial Tires								3/64	5/64	1/8	1.0	2.0	3.0		
RN ¾ Ton, Cab & Chassis	(1)	0	½	1	9/16	1 1/16	1 9/16							29	36
w/Bias Tires								5/32	13/64	15/64	4.0	5.2	6.0		
w/Radial Tires								3/64	5/64	1/8	1.2	2.0	3.1		
81–80 Hilux (4×2)	(1)	0	½	1	9/16	1 1/16	1 9/16							29	36
w/Radial Tires								3/64	5/64	1/8	1.2	2.0	3.2		
w/Bias								5/32	13/64	¼	4.0	5.2	6.4		
79 Hilux (4×2)	(1)	0	½	1	9/16	1 1/16	1 9/16	5/32	13/64	15/64	4.0	5.2	6.0	29	36
79 Hilux (4×4)	N/A		3½			1									
w/HR78×15B								0	3/64	5/64	0	1.2	2.0	29	36
w/H78×15B								7/64	5/32	15/64	2.8	4.0	6.0	29	30½
VOLKSWAGEN															
83–81 Rabbit Pick-up Front	(10)	13/16	1 5/16	1 13/16	−3/16	5/16	13/16	7/32 (out)	1/8 (out)	1/32 (out)	5.6 (out)	3.2 (out)	0.8 (out)	18½	20
Rear	N/A				−1	0	1	½ (out)	0	½ (in)	12.7 (out)	0.0	12.7 (in)		
80 Rabbit Pick-up Front	(10)	1 5/16	1 13/16	2 5/16	−3/16	5/16	13/16	7/32 (out)	1/8 (out)	1/32 (out)	5.6 (out)	3.2 (out)	0.8 (out)	18½	20
Rear	N/A				−9/16	0	9/16	0	5/32	5/16	0	4.0	7.9		

(A) Variation between wheel ½°.
(B) Side to side variation ½° or less.

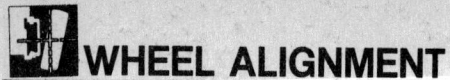

LUV SUSPENSION HEIGHT TABLES

Model & Year	Measurement at	Front Suspension		Rear Suspension	
		Inches	MM	Inches	MM
82-81 (4 × 4)		5	127	7 11/16	195
82-81 (4 × 2)	Base Model, Soft Ride	4	102	6 1/8	155
	Base Model, Cab & Chassis Comp.	4	102	7 1/2	190
	Cab & Chassis	4	102	8 5/16	210
	Long Wheel Base Model	4	102	7 1/2	190
80-79 4 × 4		4 13/16	122	7 11/16	195
80-78 4 × 2	Base Model, Soft Ride	4 5/8	116.8	6 1/8	155
	Base Model, Cab & Chassis Comp.	4 5/8	116.8	7 1/2	190
	Cab & Chassis	4 5/8	116.8	8 5/16	210
	Long Wheel Base Model	4 5/8	116.8	7 1/2	190
77-76 4 × 2	at Curb	4 5/8	116.8	6	152.4
	at G.V.W.	3 5/8	92.0	5 1/32	127.8
75-74 4 × 2	at Curb	4 5/8	116.8	7 13/32	188.0
	at G.V.W.	3 5/8	92.0	5 1/32	127.8

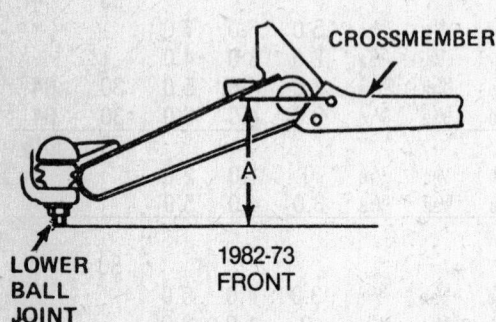

CROSSMEMBER

LOWER BALL JOINT

1982-73 FRONT

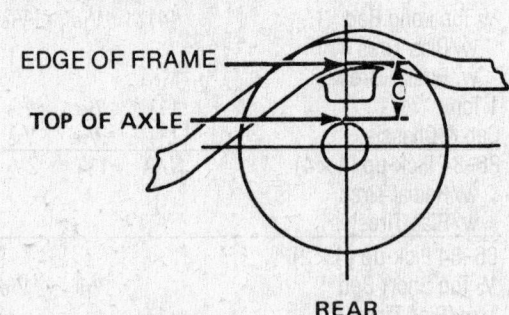

EDGE OF FRAME

TOP OF AXLE

REAR

TRIM HEIGHTS

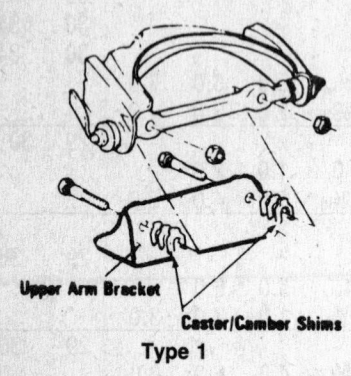

Upper Arm Bracket

Caster/Camber Shims

Type 1

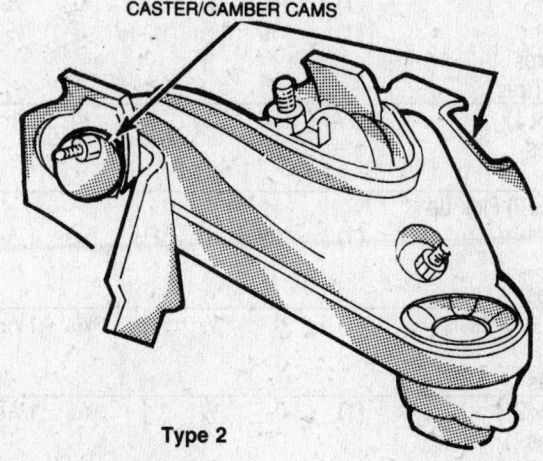

CASTER/CAMBER CAMS

Type 2

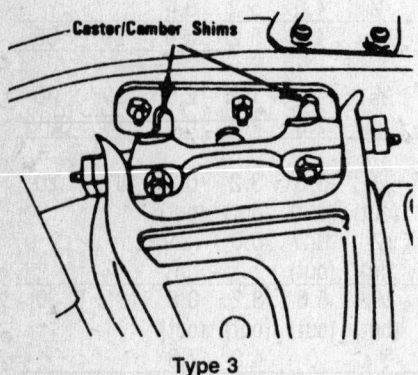

Caster/Camber Shims

Type 3

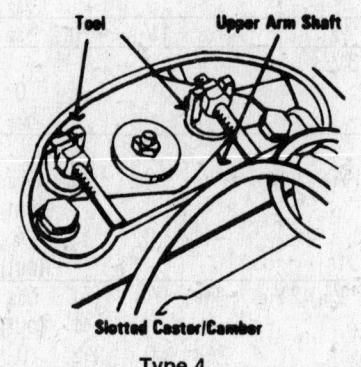

Tool

Upper Arm Shaft

Slotted Caster/Camber

Type 4

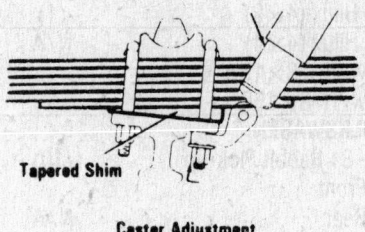

Tapered Shim

Caster Adjustment

Type 5

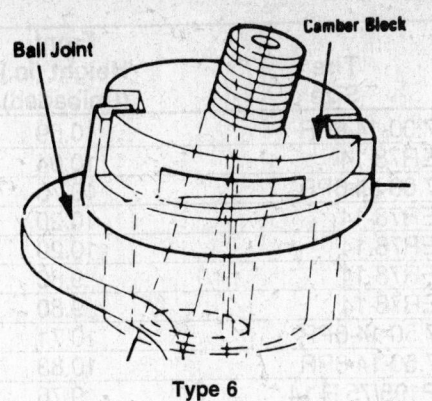

Type 6

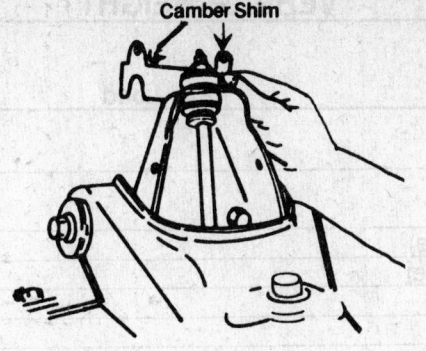

Type 8

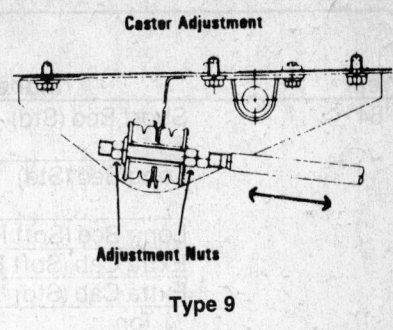

Type 9

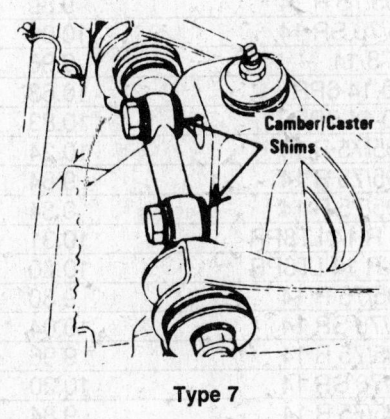

Type 7

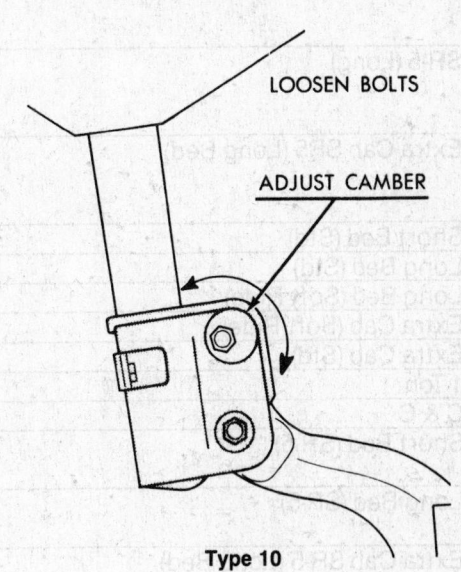

Type 10

TOYOTA PICK-UPS

VEHICLE HEIGHT

Year	Model	Pay Load	Tire Size	Front Height (in.) (Unloaded)
'79–'80	RN32L	½ Ton	185 SR 14-4PR	9.827
	RN42L		7.00-14-6PR	10.291
			E78-14 (B)	10.016
			ER78-14 (B)	9.866
	RN42L-KH	¾ Ton	7.50-14-6PR	10.961
	RN42L-3W (C & C)	¾ Ton	7.50–14–6PR	10.961
'81–'83	RN34	½ Ton	7.00-14-6PR	10.291
	RN44		E78-14 (B)	10.016
			ER78-14 (B)	9.866
			205/70 SR 14	9.512
	RN44L-KH	¾ Ton	7.50-14-6PR	10.961
	RN44L-3W C & C		7.50-14-6PR	10.961

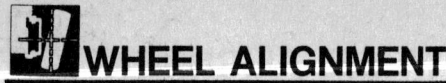

VEHICLE HEIGHT

Year	Model	Pay Load	Tire Size	Front Height (in.) (Unloaded)
'84	Short Bed (Std)		7.00-14-6PR	10.59
			ER78-14	10.04
	Long Bed (Std)		7.00-14-6PR	10.75
			ER78-14	10.20
	Long Bed (Soft Ride)		ER78-14	10.20
	Extra Cab (Soft Ride)		ER78-14	9.80
	Extra Cab (Std)		ER78-14	9.80
	¾ Ton		7.50-14-6PR	10.71
	C & C		7.50-14-6PR	10.83
	SR-5 (Short)		P195/75 R 14	9.76
			205/70 R 14	10.00
			ER78-14	10.00
	SR-5 (Long)		P195/75 R 14	9.96
			205/70 SR 14	10.20
			ER78-14	10.12
	Extra Cab SR5 (Long Bed)		P195/75 R 14	9.80
			205/70 SR 14	10.04
			ER78-14	9.96
'85–'86	Short Bed (Std)		7.00-14-6PR	10.63
	Long Bed (Std)		7.00-14-6PR	10.83
	Long Bed (Soft Ride)		P195/75 R 14	10.24
	Extra Cab (Soft Ride)		P195/75 R 14	9.84
	Extra Cab (Std)		P195/75 R 14	9.84
	1 Ton		185 R 14-LT8PR	10.31
	C & C		185 R 14-LT8PR	10.20
	Short Bed (SR-5)		P195/75 R 14	9.80
			205/70 SR 14	10.04
	Long Bed (SR-5)		P195/75 R 14	9.96
			205/70 SR 14	10.20
	Extra Cab SR-5 (Long Bed)		P195/75 R 14	9.84
			205/70 SR 14	10.08
'85-'86 (Diesel)	Short Bed (Std)		7.00-14-6PR	10.59
	Long Bed (Soft Ride)		P195/75 R 14	9.96
	Extra Cab (Soft Ride)		P195/75 R 14	9.80
	Extra Cab (Std)		P195/75 R 14	9.80
'83–'86	Van		P185/75 R 14	9.57

Manual Transmission Overhaul

INDEX

MANUAL TRANSMISSION SERVICE

Sequence of Diagnosis

In order to determine the problems that may exist in a transmission, a systematic diagnosis procedure should be followed to locate and repair the malfunction.

1. Consult with the owner or operator to identify the problem.

2. Road test, whenever possible with the owner or operator, to verify the problem is within the transmission and not caused by a related component.

3. Verify that all controls are operating properly and in good condition.

4. With the unit removed from the vehicle, inspect it prior to the disassembly.

5. During the disassembly, inspect the varied parts to locate the source of the problem.

6. Replace companion gears to defective or worn gears. Do not re-install a part that does not have a long service life remaining.

7. make any modifications or changes as recommended by the manufacturer.

Diagnosis & Troubleshooting

Noises with Transmission in Neutral

1. Misalignment of transmission.
2. Worn flywheel pilot bearing.
3. Worn or scored countershaft bearings.
4. Worn or rough reverse idler gear.
5. Sprung or worn countershaft.
6. Excessive backlash in gears.
7. Worn mainshaft pilot bearing.
8. Scuffed gear tooth contact surface.
9. Insufficient lubrication.
10. Use of incorrect grade of lubricant.

Noises with Transmission in Gear

1. Worn or rough mainshaft rear bearing.
2. Rough, chipped or tapered sliding gear teeth.
3. Noisy speedometer gears.
4. Excessive end play of mainshaft gears.
5. Refer to conditions listed above under noises with transmission in neutral.

Growling, Humming and Grinding

1. Pitted, chipped or cracked gears.

2. Damaged gears or chips in lubricant from failed power-take-off.
3. Excessive gear wear from high mileage or overloading.

Hissing, Thumping and Bumping

1. Bad bearings on way to failure.
2. Broken bearings and retainers.

Metallic Rattles

1. Engine torsional vibration.
2. Clutch disc assembly worn or without torsional vibration dampers.
3. Engine idle speed too low.
4. Rough engine idle.
5. Excessive backlash in power-take-off mounting.

Squealing, Gear Whine and Gear Seizure

1. One of the free-running gears seizing on thrust-face or fluted diameter momentarily, then letting go.
2. Whine of excessive backlash in mating gears or improper shimming of power-take-off unit.

Walking or Jumping out of Gear

CAUSES OUTSIDE TRANSMISSION

1. Improperly positioned forward remote control which limits full travel forward and backward from the remote neutral position.
2. Improper adjustment or length shift rods or linkage that limits travel of forward remote from neutral position.
3. Loose bell cranks, sloppy ball and socket joints.
4. Shift rods, cables, etc., too spongy, flexible, or not secured properly at both ends.
5. Worn or loose engine mounts if forward unit is mounted to frame.
6. Forward remote mount too flimsy, loose on frame, etc.
7. Set screws loose at remote control joints.
8. Air shift system partially inoperative.
9. Transmission and engine out of alignment either vertically or horizontally.

CAUSES INSIDE TRANSMISSION

1. Shift tower or cover loose or interlock balls or pins worn or springs broken.
2. Shift fork pads not square with shift rod bore.
3. Shift rod poppet springs broken.
4. Shift rod poppet notches worn.
5. Shift rod bent or sprung out of line.
6. Shift fork pads or groove in sliding gear or collar worn excessively.
7. Shift fork pads not square with rod bore.
8. Worn taper on gear teeth, spacers or bearings.
9. Backing rings or retaining rings not installed properly on rear unit curvic rings on gears.

Hard Shifting

PRELIMINARY INVESTIGATION

1. Not enough clutch pedal free-play.
2. Worn or inoperative clutch hydraulic cylinder.
3. Worn or loose clutch shaft, levers.
4. Worn or loose throwout bearing or carrier.
5. Low air pressure to main auxiliary unit shift cylinder.
6. Air leaks in cylinders, control lines or cab control valve.
7. Improper remote control function.
8. No lubricant in remote control units.
9. No lubricant in (or grease fittings on) U-joints or swivels of remote controls.

UNSYNCHRONIZED (CONSTANT MESH) TRANSMISSIONS

1. Lack of lubricant or wrong lubricant used causing buildup of sticky varnish and sludge deposits on splines of shaft and gears.
2. Sliding clutch gears tight on splines of shaft.
3. Clutch teeth burred over, chipped or badly mutilated due to improper shifting.
4. Driver not familiar with proper shifting procedure for this transmission. Also includes proper shifting if used with 2-speed axle, auxiliary, etc.
5. Clutch or drive gear pilot bearing seized, rough, or dragging.
6. Clutch brake engaging too soon when clutch pedal is depressed.

SYNCHRONIZED TRANSMISSION

1. Badly worn or bent shift rods.
2. Loose or flimsy remote controls, spongy or flexible rods and/or cables preventing full application of force to hold and synchronize gears.
3. Further, driver may not be able to feel the synchronizer action which usually results in a snap-type shift.
4. Synchronizer bronze or aluminum rings worn or steel chips imbedded in rings prevent proper synchronization.
5. Damaged synchronizer such as broken poppet springs, poppets jammed, loose or broken blocker pins.
6. Free running gears, seized or galled on either the thrust face or diameters.

Sticking in Gear

1. Clutch not releasing.
2. Inoperative slave power units.
3. Sliding clutch gears tight on splines.
4. Chips wedged between or under splines of shaft and gear.
5. Improper adjustment, excessive wear or lost motion in shifter linkage.
6. Clutch brake set too high on clutch pedal locking gears behind hopping guard.

Crash Shifting or Raking Gears

SYNCHRONIZED TRANSMISSIONS

1. Raking of gears during manual shift may be caused by a defective synchronizer

or improper shifting technique for synchronized transmission.

2. Occurs with cold, heavy oil, but synchronizer begins to work properly when transmission oil reaches normal operating temperature.

3. Heavy oil prevents the synchronizer cone from breaking through oil film and doing job properly.

4. Glazing of synchronizer cones due to use of E.P. addition in multi-purpose axle lubricant.

5. Synchronizer cones worn smooth causing loss of clutching action: Failure to control engine speed drop-off during up-shift. Failure to bring engine speed nearly up to governor speed when driver shifting. Attempted shifting without using clutch.

6. Blocker pin detents worn resulting in loss of blocker action.

7. Blocker pins loose, broken or turned over.

Oil Leaks

1. Oil level too high.
2. Wrong lubricant in unit.
3. Non-shielded bearing used at front or rear bearing cap (where applicable.)
4. Seals (if used) defective or omitted from bearing cap, wrong type seal used, etc.
5. Screwback threads in bearing caps off location, worn out, or filled with varnish, sludge, dirt, etc.
6. Transmission breather omitted, plugged internally, etc.
7. Capscrews loose, omitted or missing from remote control, shifter housing, bearing caps, P.T.O. or covers, etc.
8. Welch "seal" plugs loose or missing entirely from machine openings in case.
9. Oil drain-back openings in bearing caps or case plugged with varnish, dirt, covered with gasket material, etc.
10. Broken gaskets, gaskets shifted or squeezed out of position, pieces still under bearing caps, clutch housing, P.T.O. and covers, etc.
11. Cracks or holes in castings.
12. Drain plug loose.
13. Also possibility that oil leakage could be from engine.
14. Internal O-ring worn in air cylinders, leaking air into transmission, pressurizing transmission.

Vibration

ORIGINATING IN TRANSMISSION

1. Sprung mainshafts and countershaft.
2. Gears that have seized to shaft and broken loose.
3. Bearings that are extremely worn allowing rotating shafts to oscillate from intended centers.

ORIGINATING ELSEWHERE BUT APPARENTLY IN TRANSMISSION

1. Drive lines out of static or dynamic balance.

2. Out of phase, wrong drive line working angles.

3. Worn crosses and bearings in U-joints.

4. Loose mounting or worn center bearings.

5. Worn and pitted teeth on ring gear and pinion of driving axle(s).

6. Wheels out of balance.

7. Warped parking brake drum or disc.

Bearing Failure

1. Dirt, always abrasive enters through seals, breathers, dirty containers.

2. Lapping action of fine steel particles from balls and raceways.

3. Entry of chips from hammers, chisels, punches during disassembly and assembly.

4. Bearing jammed with chip(s) may turn on shaft or in housing.

5. Brinnelling, ball depressions, spalling.

6. Excessive looseness under load scrubs shaft and bearing bore.

7. Failure due to heat: Failure of lubricant circulation. Lubricant deterioration or low level. Radically tight bearing caused by expansion of inner race when mounted on shaft or compression of outer race when pressed into housing. Off-square mounting producing heat at retainers.

Transfer Case

DIAGNOSIS & TROUBLESHOOTING

Slips Out of Gear (High-Low)

1. Shifting poppet spring weak.
2. Bearing broken or worn.
3. Shifting fork bent.
4. Improper control rod adjustment.

Slips Out of Front Wheel Drive

1. Shifting poppet spring weak or broken.
2. Bearing worn or broken.
3. Excessive shaft end-play.
4. Shifting fork bent.

Hard Shifting

1. Lack of lubricant.
2. Shift lever binding on shaft.
3. Shifting poppet ball scored.
4. Shifting fork bent.
5. Low tire pressure.

Backlash

1. Companion yoke loose.
2. Transfer case loose on mounts.
3. Internal parts excessively worn.

Noisy

1. Low lubricant level.
2. Bearings improperly adjusted or excessively worn.

3. Gears worn or damaged.
4. Improper alignment of driveshafts or U-joints.

Oil Leakage

1. Excessive amount of lubricant in case.
2. Vent clogged.
3. Gaskets or seals leaking.
4. Bearings loose or damaged.
5. Driveshaft yoke mating surfaces scored.

Overheating

1. Excessive or insufficient amount of lubricant.
2. Bearing adjustment too tight.

CLEANING COMPONENTS

Cleanliness of parts, tools, and work area is of the utmost importance. All transmission components (except bearing assemblies) should be cleaned in cleaning solvent and dried with compressed air before any inspection or work is begun. Great care should be taken when cleaning bearings. Bearings should always be cleaned separately from other parts in clean cleaning solvent and not gasoline. They must never be cleaned in a hot solution tank. It is advisable that they be soaked in cleaning fluid and then tapped against a block of wood in order to free any solidified lubricant that may be trapped inside. Rinse bearings thoroughly in clean solvent and then dry them with moisture-free compressed air being careful not to spin the bearings with the air stream. Rotate each bearing slowly and inspect rollers or balls for any signs of excessive wear, roughness, or damage. Those bearings not in excellent condition must be replaced. If they pass this inspection, they should be dipped in clean oil and wrapped in clean lintless cloth to protect them until installation.

INSPECTION OF COMPONENTS

All parts must be completely and carefully inspected and replaced for any signs of wear, stress, discoloration or warpage due to excessive heat. Whenever available, the magna flux process should be used on all parts except roller and ball bearings, to detect small cracks unseen by the eye. Inspect the breather assembly to see that it is not clogged or damaged and check all threaded parts for stripped or cross threads. Oil passages must be cleared of obstructions by the use of air pressure or brass rods and all gaskets, oil seals, lock wires, cotter pins, and snap rings are to be replaced. Small nicks or burrs in gears or splines can be removed with a fine abrasive stone. It is important that any housings or covers having cracks or other damage should be replaced and not welded. Synchronizers, not in excellent condition, must be replaced. The bronze synchronizer cone should be checked for wear or for any steel chips that may have become imbedded in it. Springs must be inspected for free length, compressed length, distortion, or collapsed coils.

NOTE: The splines on many clutch gears, mainshafts, etc., are equipped with a machined relief called a "hopping guard". With the clutch gear engaged, the mating gear is free to slip into this notch, preventing the two gears from separating or "walking out of gear" under various load conditions. This is not a worn or chipped gear. Do not grind or discard the gear.

Check all shafts for spline wear or damage. If the mainshaft 1st and reverse sliding gear or clutch hub have worn into the sides of the splines, the shaft should be replaced. Shift forks, shift rods, interlock balls and pins must be replaced if scored, worn, distorted or damaged.

DOMESTIC TRUCKS

AMC/Jeep

Model AX 4 is a 4-speed manual transmission while AX 5 is a 5-speed manual transmission. Both transmissions have synchromesh engagement in all forward gears controlled by a floor shift mechanism integrated into the transmission top cover.

NOTE: The following components and materials must be replaced whenever the transmission is overhauled. Lip-type oil seals. Lock nuts. All roll pins. All snap-rings. Loctite Thread Lock or Loctite 242 Sealer.

DISASSEMBLY

1. Remove the clutch housing.
2. Remove the straight screw plug, spring and ball using a Torx bit to remove the screw plug, and a magnet to remove spring and ball.
3. Remove five adapter housing bolts and one nut.
4. Remove the shift lever housing set bolt and lock plate.
5. Remove the plug at the rear of the shift fork shaft.
6. Remove the large magnet to pull the shaft out.
7. Remove the select lever from the top while rotating.
8. Remove the five adapter housing bolts two studs and one nut.
9. Using a plastic hammer, tap and remove the extension housing. Leave the gas-

ket attached to the intermediate plate.
10. Remove the front bearing retainer and outer snap-rings from the two front bearings.
11. Separate the intermediate plate from the transmission case using a small plastic hammer and remove the case.
12. Mount the intermediate plate in a vise. Be careful not to damage the plate.

NOTE: Before placing the intermediate plate in a vise, insert bolts, washers, and nuts in the open holes at the bottom of plate. Tighten vise against these bolts to prevent damage to the plate.

13. Remove the straight screw plug, locking balls and springs using a Torx bit and magnet.
14. Remove the five slotted spring pins using a hammer and punch and then remove the two E-rings from the shift rails.

—————— CAUTION ——————
The locking ball from the reverse shift head and locking ball and pin from the intermediate housing will fall from the holes so be sure to catch them. If they do not come out, remove them with a magnet.

15. Pull out the shift fork shaft No. 4 from the intermediate plate and catch the locking ball.
16. Remove shift fork shaft No. 4 and the 5th gear fork.
17. Pull out shift fork shaft No. 5 from the intermediate plate, and remove it with the reverse shift head.

—————— CAUTION ——————
The interlock pins will fall from their hole. If they do not come out, remove them with a magnet.

18. Remove the shift fork shaft No. 3 from the intermediate plate and catch the interlock pins.

—————— CAUTION ——————
The interlock pin will fall from the hole so be sure to catch it. If it does not come out, remove it with a magnet.

19. Remove shift fork shaft No. 1 from the intermediate plate being careful not to drop the interlock pin.
20. Remove shift fork shaft No. 2, shift fork No. 2 and shift fork No. 1.
21. Remove the reverse idle gear shaft stopper, reverse idler gear and shaft.
22. Remove the reverse shift arm from the reverse shift arm bracket.
23. Using a feeler gauge, measure the counter 5th gear thrust clearance. Standard Clearance: 0.004–0.012 in.
24. Engage two gears to lock the output shaft. Using a hammer and chisel, loosen the staked part of the nut on the countershaft.
25. Remove the lock nut. Disengage the gears.
26. Remove the gear spline piece No. 5, synchronizer ring, needle roller bearing

and counter fifth gear using tool J–22888 or equivalent.
27. Remove the spacer and use a magnet to remove the ball.
28. Remove the reverse shift arm bracket.
29. Remove the rear bearing retainer bolts with a Torx bit and the snap-ring using snap-ring pliers.
30. Remove the output shaft, counter gear and input shaft as a unit from the intermediate plate by pulling on the counter gear and tapping on the intermediate plate with a plastic hammer.
31. Remove the input shaft with fourteen needle roller bearings from the output shaft.
32. Remove the counter rear bearing from the intermediate plate.
33. Measure the thrust clearance of each gear. Standard Clearance: 0.004–0.10 in.
34. Using two awls and a hammer, tap out the snap-ring.
35. Using a press, remove the fifth gear, rear bearing, first gear and the inner race.
36. Remove the needle roller bearing.
37. Remove the synchronizer ring and locking ball.
38. Using a press, remove hub sleeve No. 1 assembly, synchronizer ring, second gear.
39. Remove the needle roller bearing.
40. Remove the snap-ring from hub sleeve No. 2.
41. Using a press, remove the hub sleeve, synchronizer ring, and third gear.
42. Remove the needle roller bearing.

Component Inspection

OUTPUT SHAFT & INNER RACE

1. Check the output shaft and inner race for wear or damage.
2. Using calipers, measure the output shaft flange thickness. Minimum Thickness: 0.189 in.
3. Using calipers, measure the inner face flange thickness. Minimum Thickness: 0.157 in.
4. Using a micrometer, measure the outer diameter of the output shaft journal surface. 2nd Gear Minimum: 1.495 in. 3rd Gear Minimum: 1.377 in.
5. Using a micrometer, measure the outer diameter of the inner race. Minimum Diameter: 1.535 in.
6. Using a dial indicator, measure the shaft runout. Maximum Runout: 0.002 in.

FIRST GEAR OIL CLEARANCE

1. Using a dial indicator, measure the oil clearance between the gear and inner race with the needle roller bearing installed. Standard Clearance: 0.0004–0.0013 in.
2. Using a dial indicator, measure the oil clearance between the gear and shaft with the needle roller bearing installed. Standard Clearance: 2nd and 3rd Gears: 0.0004–0.0013 in. Counter 5th Gear: 0.0004–0.0013 in.

Synchronizer Ring Inspection

1. Check for wear or damage. Turn the ring and push it in to check the braking action.
2. Measure the clearance between the synchronizer ring back and the gear spline end. Standard Clearance: 0.040–0.078 in. Minimum Clearance: 0.031 in.

Shift Fork and Hub Sleeve Clearance

1. Using a feeler gauge, measure the clearance between the hub sleeve and shift fork. Maximum Clearance: 0.039 in.

Input Shaft and Bearing Inspection and Removal

1. Check for wear or damage. If necessary, remove the bearing snap-ring using snap-ring pliers and remove the bearing.
2. Using a press, remove the bearing.
3. Using a press and tool J–34603 or equivalent, install the new bearing.
4. Select a snap-ring that will allow minimum axial play and install it on the shaft.

Counter Gear and Bearing Inspection

1. Check the gear teeth for wear or damage.
2. Check the bearing for wear or damage.

Counter Gear Front Bearing Replacement

1. Using snap-ring pliers, remove the snap-ring.
2. Press out the bearing using tool J–22912–01 or equivalent.
3. Replace the side race.
4. Using tool J–28406 or equivalent, press in the bearing and inner race.
5. Select a snap-ring that will allow minimum axial play and install it on the shaft.

Front Bearing Retainer Inspection

1. Check retainer for damage.
2. Check the oil seal lip for wear or damage.

Oil Seal Replacement

1. Using a awl, pry the old seal out of the housing.
2. Press in the new oil seal using tool J–34602 or equivalent.
3. The oil seal depth is 0.441–0.480 in. from the housing-to-transmission surface to the top edge of the seal.

Reverse Restrict Pin Replacement

1. Check for wear or damage.
2. Using a Torx bit, remove the screw plug.

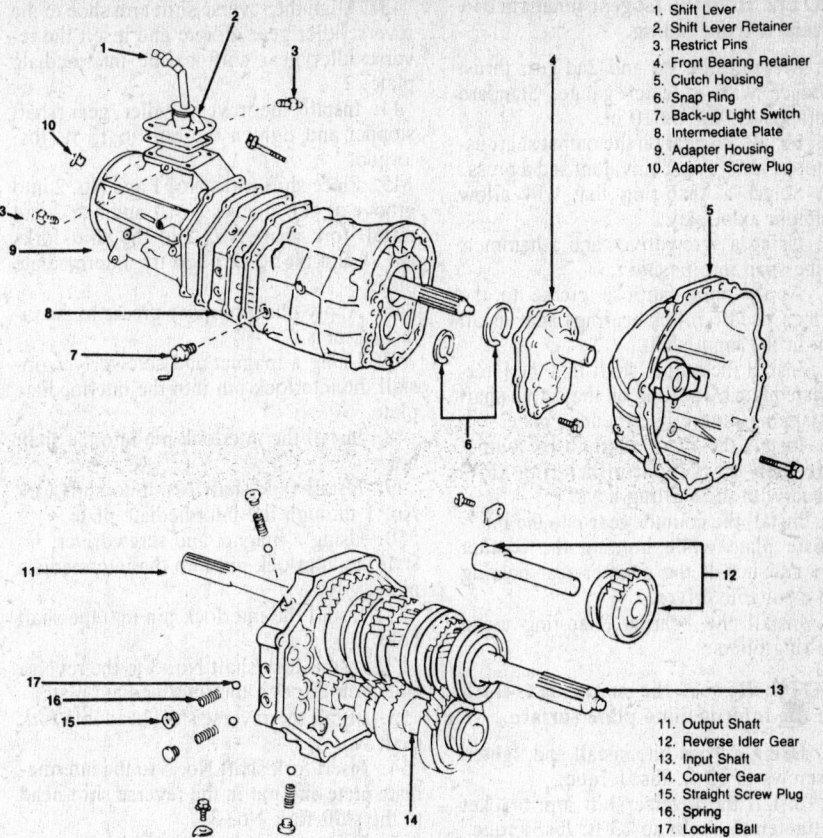

1. Shift Lever
2. Shift Lever Retainer
3. Restrict Pins
4. Front Bearing Retainer
5. Clutch Housing
6. Snap Ring
7. Back-up Light Switch
8. Intermediate Plate
9. Adapter Housing
10. Adapter Screw Plug

11. Output Shaft
12. Reverse Idler Gear
13. Input Shaft
14. Counter Gear
15. Straight Screw Plug
16. Spring
17. Locking Ball

Jeep 4-5 speed

3. Using a hammer and pin punch, drive out the slotted spring pin.
4. Pull off the lever housing and slide out the shaft.
5. Install the lever housing.
6. Using a hammer and pin punch, drive out the slotted spring pin.
7. Using a Torx bit, install and torque the screw plug to 27 ft. lbs. torque.

Adapter Housing & Oil Seal Inspection & Replacement

1. Check the adapter housing for wear or damage.
2. Replace the oil seal with tool J–29184 or equivalent.

ASSEMBLY

1. Install the clutch hub No. 1 and No. 2 into hub sleeves along with the shifting keys.

----- CAUTION -----
Install the key springs so their gaps are not in line.

2. Install the shifting springs under the shifting keys.
3. Apply gear oil on the output shaft and 3rd gear needle roller bearing.
4. Place the 3rd gear synchronizer ring on the gear and align the ring slots with the shifting keys.

5. Install the needle roller bearing in the 3rd gear and hub sleeve No. 2.
6. Select a new snap-ring (2) that will allow minimum axial play and install it on the shaft.
7. Using a feeler gauge, measure the 3rd gear thrust clearance. Standard Clearance: 0.004–0.010 in.
8. Apply gear oil on the output shaft and 2nd gear needle bearing.
9. Place the 2nd gear synchronizer ring on the 2nd gear and align the ring slots with the shifting keys.
10. Install the needle roller bearing in the 2nd gear.
11. Using a press install the 2nd gear and hub sleeve No. 1.
12. Install the first gear locking ball in the output shaft.
13. Apply gear oil to the needle roller bearing.
14. Assemble the first gear, synchronizer ring, needle roller bearing and bearing inner race.
15. Install the assembly on the output shaft, with the synchronizer ring slots aligned with the shifting keys.
16. Turn the inner race to align it with the locking ball.
17. Install the output shaft rear bearing using tool J–34603 or equivalent and a press.
18. Install the bearing on the output shaft with the outer race snap-ring groove toward the rear.

NOTE: Hold the 1st gear inner race to prevent it from falling.

19. Measure the 1st and 2nd gear thrust clearance with a feeler gauge. Standard Clearance: 0.004–0.010 in.

20. Install 5th gear on the output shaft using tool J–34603 or equivalent and a press.

21. Select a snap-ring that will allow minimum axial play.

22. Using a screwdriver and a hammer, tap the snap into position.

23. Apply multi-purpose grease to the fourteen needle roller bearings and install them in the input shaft.

24. Install the output shaft into the intermediate plate by pulling on the output shaft and tapping on the intermediate plate.

25. Install the input shaft to the output shaft with the synchronizer ring slots aligned with the shifting keys.

26. Install the counter gear into the intermediate plate while holding the counter gear, and install the counter rear bearing with a suitable driver.

27. Install the bearing snap-ring using snap-ring pliers.

NOTE: Be sure the snap-ring is flush with the intermediate plate surface.

28. Using a Torx bit, install and tighten the screws to 13 ft. lbs. torque.

29. Install the reverse shift arm bracket and tighten the bolts to 13 ft. lbs. torque.

30. Install the ball and spacer.

31. Install the shifting keys and hub sleeve No. 3 onto the counter 5th gear.

— CAUTION —

Install the key springs positioned so the end gaps are not in line.

32. Install shifting key springs under the shifting keys.

33. Apply gear oil to the needle roller bearing and install the counter 5th gear with hub sleeve No. 3 and needle roller bearings.

34. Install the synchronizer ring on gear spline piece.

35. Using tool J–28406 or equivalent drive in gear spline piece No. 5 with the synchronizer ring slots aligned with the shifting keys.

NOTE: When installing gear spline piece No. 5, support the counter gear in front with a 3–5 lb. hammer or equivalent.

36. Engage two gears to lock the output shaft.

37. Install and tighten the lock nut to 90 ft. lbs. torque on the counter shaft.

38. Stake the lock nut.

39. Disengage the gears.

40. Measure the counter fifth gear thrust clearance using a feeler gauge. Standard Clearance: 0.004–0.012 in.

41. Install the reverse shift arm to the pivot of the reverse shift arm bracket.

42. Install the reverse idler gear on the shaft.

43. Align the reverse shift arm shoe to the reverse idler gear groove and insert the reverse idler gear shift to the intermediate plate.

44. Install the reverse idler gear shaft stopper and tighten the bolt to 13 ft. lbs. torque.

45. Place shift forks No. 1 and No. 2 into groove of hub sleeves No. 1 and No. 2 and install fork shaft No. 2 to the shift forks No. 1 and No. 2 through the intermediate plate.

46. Apply multi-purpose grease to the interlock pins.

47. Using a magnet and screwdriver, install the interlock pin into the intermediate plate.

48. Install the interlock pin into the shaft hole.

49. Install fork shaft No. 1 to shift fork No. 1 through the intermediate plate.

50. Using a magnet and screwdriver, install the interlock pin into the intermediate plate.

51. Install the interlock pin into the shaft hole.

52. Install fork shaft No. 3 to the reverse shift arm through the intermediate plate.

53. Install the reverse shift head into fork shaft No. 5.

54. Insert fork shaft No. 5 to the intermediate plate and put in the reverse shift head to the shift fork No. 3.

55. Using a magnetic finger and screwdriver, install the locking ball into the reverse shift head hole.

56. Shift hub sleeve No. 3 to the 5th speed position.

57. Place shift fork No. 3 into the groove of hub sleeve No. 3 and install fork shaft No. 4 to shift fork No. 3 and reverse shift arm.

58. Using a magnet and screwdriver, install the locking ball into the intermediate plate and insert fork shaft No. 4 to the intermediate plate.

59. Check the interlock by positioning the shift fork shaft No. 1 to the 1st speed position.

60. Fork shafts No. 2, No. 3, No. 4 and No. 5 should not move.

61. Using a pin punch and a hammer, drive in new slotted spring pins in each shift fork, reverse shift arm and reverse shift head.

62. Install two fork shaft E-rings.

63. Apply liquid sealer to the screw plugs.

64. Install the locking balls, springs and screw plugs with a Torx bit and tighten to 14 ft. lbs. torque.

NOTE: Install the short spring into the tower of the intermediate plate.

65. Remove the intermediate plate from the vise.

66. Remove the bolts, nuts, washers and gasket.

Case Installation

1. Align each bearing outer race, each fork shaft end and reverse idler gear with the holes in the case and install the case on the intermediate plate. If necessary, tap on the case with a plastic hammer.

2. Install two new bearing snap-rings.

3. Install front bearing retainer with a new gasket.

4. Apply liquid sealer to the bolts.

5. Install and tighten the bolts to 12 ft. lbs. torque.

6. Install the new gasket to the intermediate plate.

7. Install the adapter housing.

8. Install and tighten the adapter bolts to 27 ft. lbs. torque.

9. Install the shift lever housing.

10. Insert the shift lever into the adapter and shift lever housing.

11. Install and tighten shift lever housing bolt with a lock plate to 28 ft. lbs. torque. Lock the lock plate.

12. Install and tighten the adapter screw plug to 13 ft. lbs. torque.

13. Apply liquid sealer to the plug.

14. Install the locking ball, spring and screw plug and tighten the plug to 14 ft. lbs. torque.

15. Check to see that the input shaft and output shafts rotate smoothly.

16. Check to see that shifting can be done smoothly to all positions.

17. Install the black restrict pin on the reverse gear/5th gear side.

18. Install the remaining pin and tighten the pins to 20 ft. lbs. torque.

19. Install the shift lever retainer with a new gasket and tighten the bolts to 13 ft. lbs. torque.

20. Install the back-up light switch and tighten to 27 ft. lbs. torque.

21. Install the clutch housing and tighten the bolts to 27 ft. lbs. torque.

GMC S–Series

DISASSEMBLY

77mm 4 Speed

1. Remove drain plug and drain lubricant from transmission.

2. Thoroughly clean the exterior of the transmission assembly.

3. Using a hammer and punch, remove the roll pin that attaches the offset lever to shift rail.

4. Remove extension housing attaching bolts. Separate the extension housing from the transmission case and remove housing and offset lever as an assembly.

5. Remove detent ball and spring from offset lever and remove roll pin from extension housing or offset lever.

6. Remove transmission shift cover attaching bolts. Using a screwdriver, pry shift cover loose and remove cover from transmission case.

7. Remove clip that retains reverse lever to reverse lever pivot bolt.

8. Remove reverse lever pivot bolt and

remove reverse lever and fork as an assembly.

9. Using a hammer and punch, mark position of front bearing cap to transmission case. Remove front bearing cap bolts and remove bearing cap.

10. Remove small retaining and large locating snap rings from front drive gear bearing.

11. Install bearing puller J–22912–01 on front bearing and puller J–8433–1 with two bolts on end of drive gear and remove and discard bearing. A new bearing must be used when assembling the transmission.

12. Remove retaining and locating snap rings from rear bearing and mainshaft. Install puller J–22912–01 on bearing and puller J–8433–1 with two bolts (J–33171) on end of mainshaft and remove and discard used bearing. A new bearing must be used when assembling transmission.

13. Remove drive gear from mainshaft and transmission case.

14. Remove mainshaft from transmission case by tipping mainshaft down at the rear and lifting shaft out through shift cover opening.

15. Using a hammer and punch, remove roll pin retaining reverse idler gear shaft in transmission case. Remove idler gear and shaft from case.

16. Remove countershaft from rear of case using loading tool J–26624. Remove countershaft gear and loading tool as an assembly from case along with thrust washers.

Mainshaft Disassembly

1. Scribe alignment mark on third-fourth synchronizer hub and sleeve for reassembly. Remove retaining snap ring and remove third-fourth synchronizer assembly from mainshaft.

2. Slide third gear off mainshaft.

3. Remove second gear retaining snap ring. Remove tabbed thrust washer, second gear and blocker ring from mainshaft.

4. Remove first gear thrust washer and roll pin from mainshaft. Use pliers to remove roll pin.

5. Remove first gear and blocker ring from mainshaft.

6. Scribe alignment mark on first-second synchronizer hub and sleeve for reassembly.

7. Remove synchronizer springs and keys from first-second sleeve and remove sleeve from shaft.

NOTE: Do not attempt to remove the first-second hub from the mainshaft. The hub and mainshaft are assembled and machined as a unit.

8. Remove loading Tool J–26624, roller bearings, spacers and thrust washers from the countershaft gear.

Drive Gear Disassembly

1. Remove roller bearings from cavity of drive gear.

2. Wash parts in a cleaning solvent.
3. Inspect gear teeth for wear.
4. Inspect drive shaft pilot for wear.

Cover Disassembly

1. Place selector arm plates and shift rail in neutral position (centered).

2. Rotate shift rail until selector arm disengages from selector arm plates and roll pin is accessible.

3. Remove selector arm roll pin using a pin punch and hammer.

4. Remove shift rail, shift forks, selector arm plates, selector arm, interlock plate and roll pin.

5. Remove shift cover to extension housing "O" ring seal using a screwdriver.

6. Remove nylon inserts and selector arm plates from shift forks. Note position of inserts and plates for assembly reference.

ASSEMBLY

1. Install nylon inserts and selector arm plates in shift forks.

2. If removed, install shift rail plug. Coat edges of plug with sealer before installing.

3. Coat shift rail and rail bores with light weight grease and insert shift rail in cover. Install rail until flush with inside edge of cover.

4. Place first-second shift fork in cover with fork offset facing rear of cover and push shift rail through fork. The first-second shift fork is the larger of the two forks.

5. Position selector arm and C-shaped interlock plate in cover and insert shift rail through arm. Widest part of interlock plate must face away from cover, and selector arm roll pin hole must face downward and toward rear of cover.

6. Position third-fourth shift fork in cover with fork offset facing rear of cover. Third-fourth shift fork selector arm plate must be under first-second shift fork selector arm plate.

7. Push shift rail through third-fourth shift fork and into front bore in cover.

8. Rotate shift rail until selector arm plate at forward end of rail faces away from, but is parallel to cover.

9. Align roll pin holes in selector arm and shift rail and install roll pin. Roll pin must be flush with surface of selector arm to prevent pin from contacting selector arm plates during shifts.

10. Install a new shift cover to extension housing "O" ring seal. Coat "O" ring seal with transmission lubricant.

Drive Gear Assembly

1. Coat roller bearings and drive gear bearing bore with light weight grease. Install roller bearings into bore of drive gear.

Mainshaft Assembly

1. Coat mainshaft and gear bores with transmission lubricant.

2. Install first-second synchronizer sleeve on mainshaft, aligning marks previously made.

3. Install synchronizer keys and springs into the first-second synchronizer sleeve. Engage tang end of springs into the same synchronizer key but position open ends of springs so they face away from one another.

4. Place blocking ring on first gear and install gear and ring on mainshaft. Be sure synchronizer keys engage notches in first gear blocking ring.

5. Install first gear roll pin in mainshaft.

6. Place blocking ring on second gear and install gear and ring on mainshaft. Be sure synchronizer keys engage notches in second gear blocking ring. Install second gear thrust washer and snap ring on mainshaft. Be sure thrust washer tab is engaged in mainshaft notch.

7. Measure second gear end play using feeler gauge. Insert gauge between gear and thrust washer. End play should be 0.004–0.014 in. If end play is over 0.014, replace thrust washer and snap ring and inspect synchronizer hub for excessive wear.

8. Place blocking ring on third gear and install gear and ring on mainshaft.

9. Install third-fourth synchronizer sleeve on hub, aligning marks previously made.

10. Install synchronizer keys and springs in third-fourth synchronizer sleeve. Engage tang end of each spring in same key but position open ends of springs so they face away from one another.

11. Install third-fourth synchronizer assembly on the mainshaft with machined groove in hub facing forward. Install snap ring on mainshaft. Be sure synchronizer keys are engaged in notches in third gear blocker ring.

12. Install Tool J–26624 into countershaft gear. Using a light weight grease, lubricate roller bearings and install into bores at front and rear of countershaft gear. Install roller bearing retainers on Tool J–26624.

Transmission Assembly

1. Coat countershaft gear thrust washers with grease and position washer in case.

2. Position countershaft gear in case and install countershaft from rear of case. Be sure that thrust washers stay in place during installation of countershaft and gear.

3. Position reverse idler gear in case with shift lever groove facing rear of case and install reverse idler shaft from rear of case. Install roll pin in shaft and center pin in shaft.

4. Install mainshaft assembly into the case. Do not disturb position of synchronizer assemblies during installation.

5. Install fourth gear blocking ring in third-fourth synchronizer sleeve. Be sure synchronizer keys engage in notches in blocker ring.

6. Install drive gear into case and engage with mainshaft.

7. Position mainshaft first gear against rear of case. Using a new bearing, start front bearing onto drive gear. Align bearing with bearing bore in case and drive bearing onto drive gear and into case using Tool J-25234.

8. Install front bearing retaining and locating snap rings.

9. Apply a ⅛ in. bead of RTV sealant, #732 or equivalent, on case mating surface of front bearing cap. Install bearing cap aligning marks previously made. Apply nonhardening sealer on attaching bolts and install bolts. Torque bolts to specifications.

10. Install first gear thrust washer with oil groove facing first gear on mainshaft, aligning slot in washer with first gear roll pin.

11. Using a new bearing, position rear bearing on mainshaft. Align bearing with bearing bore in case and drive bearing into case using Tool J–25234.

12. Install locating and retaining snap rings on rear bearing.

13. Install speedometer gear and retaining clip on mainshaft.

14. Apply nonhardening sealer to threads of reverse lever pivot bolt and start bolt into case. Engage reverse lever fork in the reverse idler gear and reverse lever on pivot bolt. Tighten bolt to specifications and install retaining clip.

15. Rotate drivegear and mainshaft gears. If blocker rings tend to stick on gears, release the rings by gently prying them off the cones.

16. Apply a ⅛ in. bead of RTV Sealant, #732 or equivalent, on the cover mating surface of transmission. Place reverse lever in neutral, and position cover on case.

17. Install 2 dowel type bolts first to align cover on case. Install remaining cover bolts and torque to specifications. The offset lever to shift rail roll pin hole must be in the vertical position after cover installation.

18. Apply a ⅛ in. bead of RTV Sealant, #732 or equivalent, on the extension housing to transmission case mating surface.

19. Place extension housing over mainshaft to a position where shift rail is in shift cover opening.

20. Install detent spring in offset lever. Place ball in neutral guide plate detent position. Apply pressure on the offset lever, slide offset lever onto shift rail and seat extension housing to transmission case.

21. Install extension housing retaining bolts. Torque bolts to specifications.

22. Align hole in offset lever and shift rail and install roll pin.

23. Fill transmission to its proper level with recommended lubricant.

5 Speed

DISASSEMBLY

1. Remove drain bolt on transmission case and drain lubricant.

2. Thoroughly clean the exterior of the transmission assembly.

3. Using pin punch and hammer, remove roll pin attaching offset lever to shift rail.

4. Remove extension housing-to-transmission case bolts and remove housing and offset lever as an assembly.

NOTE: Do not attempt to remove the offset lever while the extension housing is still bolted in place. The lever has a positioning lug engaged in the housing detent plate which prevents moving the lever far enough for removal.

5. Remove detent ball and spring from offset lever and remove roll pin from extension housing or offset lever.

6. Remove plastic funnel, thrust bearing race and thrust bearing from rear of countershaft.

NOTE: The countershaft rear thrust bearing, bearing washer and plastic funnel may be found inside the extension housing.

7. Remove bolts attaching transmission cover and shift fork assembly and remove cover.

NOTE: Two of the transmission cover attaching bolts are alignment-type dowel bolts. Note the location of these bolts for assembly reference.

8. Using a punch and hammer, drive the roll pin from the fifth gearshift fork while supporting the end of the shaft with a block of wood.

9. Remove fifth synchronizer gear snap ring, shift fork, fifth gear synchronizer sleeve, blocking ring and fifth speed drive gear from rear of countershaft.

10. Remove snap ring from fifth speed driven gear.

11. Using a hammer and punch, mark both bearing cap and case for assembly reference.

12. Remove front bearing cap bolts and remove front bearing cap. Remove front bearing race and end play shims from front bearing cap.

13. Rotate drive gear until flat surface faces countershaft and remove drive gear from transmission case.

14. Remove reverse lever C-clip and pivot bolt.

15. Remove mainshaft rear bearing race and then tilt mainshaft assembly upward and remove assembly from transmission case.

16. Unhook overcenter link spring from front of transmission case.

17. Rotate fifth gear-reverse shift rail to disengage rail from reverse lever assembly. Remove shift rail from rear of transmission case.

18. Remove reverse lever and fork assembly from transmission case.

19. Using hammer and punch, drive roll pin from forward end of reverse idler shaft and remove reverse idler shaft, rubber "O" ring and gear from the transmission case.

20. Remove rear countershaft snap ring and spacer.

21. Insert a brass drift through drive gear opening in front of transmission case and, using an arbor press, carefully press countershaft rearward to remove rear countershaft bearing.

22. Move countershaft assembly rearward, tilt countershaft upward and remove from case. Remove countershaft front thrust washer and rear bearing spacer.

23. Remove countershaft front bearing from transmission case using an arbor press.

Mainshaft Disassembly

1. Remove thrust bearing washer from front end of mainshaft.

2. Scribe reference mark on third-fourth synchronizer hub and sleeve for reassembly.

3. Remove third-fourth synchronizer blocking ring, sleeve, hub and third gear as an assembly from mainshaft.

4. Remove snap ring, tabbed thrust washer, and second gear from mainshaft.

5. Remove fifth gear with Tool J-22912–01 or its equal and arbor press. Slide rear bearing off mainshaft.

6. Remove first gear thrust washer, roll pin, first gear and synchronizer ring from mainshaft.

7. Scribe reference mark on first-second synchronizer hub and sleeve for reassembly.

8. Remove synchronizer spring and keys from first-reverse sliding gear and remove gear from mainshaft hub. Do not attempt to remove the first-second-reverse hub from mainshaft. The hub and shaft are assembled and machined as a matched set.

Drive Gear Disassembly

1. Remove bearing race, thrust bearing, and roller bearings from cavity of drive gear.

2. Using Tool J-22912–01 or its equal and arbor press, remove bearing from drive gear.

3. Wash parts in a cleaning solvent.

4. Inspect gear teeth and drive shaft pilot for wear.

Drive Gear Assembly

1. Using Tool J-22912–01 or its equal with an arbor press, install bearing on drive gear.

2. Coat roller bearings and drive gear bearing bore with grease. Install roller bearings into bore of drive gear.

3. Install thrust bearing and race in drive gear.

Mainshaft Assembly

1. Coat mainshaft and gear bores with transmission lubricant.

2. Install first-second synchronizer sleeve on mainshaft hub aligning marks made at disassembly.

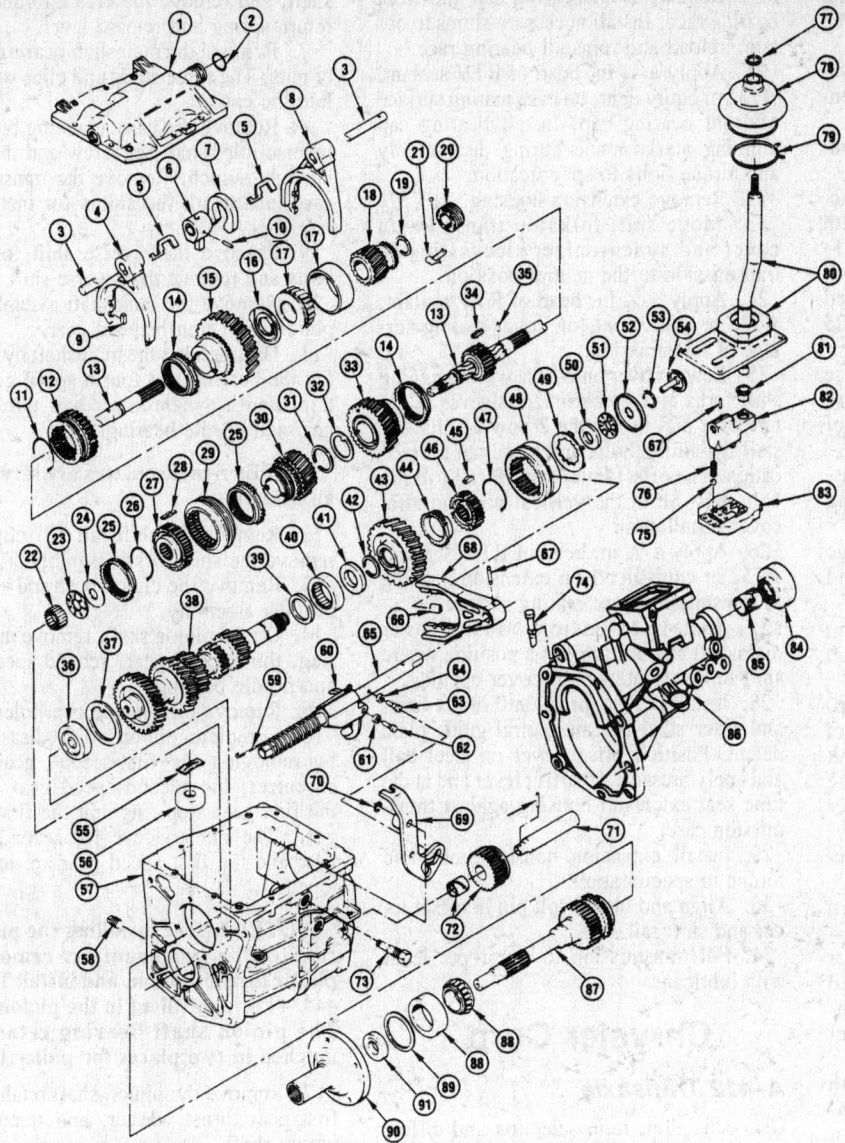

33. GEAR, 2nd Speed
34. KEY, 1 & 2 Syn
35. PIN, 1st Spd Gr Thr Wa Ret
36. BEARING, Cntr Gr Frt
37. WASHER, Cntr Gr Frt Thr
38. GEAR, Counter
39. SPACER, Counter Gr Brg Frt
40. BEARING, Cntr Gr Rr
41. SPACER, Counter Gr Brg Rr
42. RING, Snap
43. GEAR, 5th Spd Drive
44. RING, 5th Syn
45. KEY, 5th Syn
46. HUB, 5th Syn
47. SPRING, 5th Syn
48. SLEEVE, 5th Syn
49. RETAINER, 5th Syn Key
50. RACE, 5th Syn Thr Brg Frt
51. BEARING, 5th Syn Ndl Thr
52. RACE, 5th Syn Thr Brg Rr
53. RING, Snap
54. FUNNEL, Trans Oiling
55. NUT, Magnet
56. MAGNET
57. CASE, Trans
58. PLUG, Fill & Drain
59. SPRING, Rev Lock
60. FORK, Rev Shift
61. ROLLER, Fork
62. PIN, Rev Fork
63. PIN, Shift Rail
64. ROLLER, Rail pin
65. RAIL, 5th & Rev Shft
66. INSERT, Shift Fork
67. PIN, Roll
68. FORK, 5th Shift
69. LEVER, 5th & Rev Relay
70. RING, Rev. Relay Lever Ret
71. SHAFT, Rev Idler Gr
72. GEAR, Rev Idler (Incl Bshg)
73. PIN, 5th Spd Shft Lvr Piv
74. VENTILATOR, Ext
75. BALL, Steel
76. SPRING, Detent
77. RETAINER, Cont Lvr Boot
78. BOOT, Cont Lvr
79. RETAINER, Cont Lvr Boot Lwr
80. CONTROL, Trans Lvr & Hsg
81. SLEEVE, Shft Lvr Dmpr
82. LEVER, Offset Shift
83. PLATE, Detent & Guide
84. SEAL, Ext Rr Oil
85. BUSHING, Extension Housing
86. HOUSING, Extension
87. GEAR, Main Drive
88. BEARING, Front
89. SHIM, Brg Adj
90. RETAINER, Drive Gr Brg
91. SEAL, Drive Gr Brg Oil

1. COVER, Trans
2. SEAL, "O" Ring, Cvr to Ext.
3. SHAFT, Shift
4. FORK, 3rd & 4th Shift
5. PLATE, Shift Fork
6. ARM, Control Selector
7. PLATE, Gear Sel Intlk
8. FORK, 1st & 2nd Shift
9. INSERT, Shift Fork
10. PIN, Roll
11. SPRING, Syn
12. GEAR, Rev Sldg
13. SHAFT, Output, W/1 & 2 Syn
14. RING, 1 & 2 Syn Blkg
15. GEAR, 1st Speed
16. WASHER, 1st Spd Gr Thrust

17. BEARING, Rear
18. GEAR, 5th Spd Drvn
19. RING, Snap
20. GEAR, Speedo Dr
21. CLIP, Speedo Dr Gr
22. BEARING, Main Shf Rlr
23. BEARING, Main Dr Gr Thr Ndl
24. RACE, Main Dr Gr Ghr Brg
25. RING, 3 & 4 Syn
26. SPRING, 3 & 4 Syn
27. HUB, 3 & 4 Syn
28. KEY, 3 & 4 Syn
29. SLEEVE, 3 & 4 Syn
30. GEAR, 3rd Speed
31. RING, Snap
32. WASHER, 2nd Spd Gr Thr

77mm 5-speed exploded view

3. Install first-second synchronizer keys and springs. Engage tang end of each spring in same synchronizer key but position open end of springs opposite of each other.

4. Install blocker ring and second gear on mainshaft. Install tabbed thrust washer and second gear retaining snap ring on mainshaft. Be sure washer tab is properly seated in mainshaft notch.

5. Install blocker ring and first gear on mainshaft. Install first gear roll pin and then first gear thrust washer.

6. Slide rear bearing on mainshaft.

7. Install fifth speed gear on mainshaft using Tool J–22912–01 and arbor press. Install snap ring on mainshaft.

8. Install third gear, third-fourth synchronizer assembly and thrust bearing on mainshaft. Synchronizer hub offset must face forward.

ASSEMBLY

1. Coat countershaft front bearing bore

with Loctite 601, or equivalent, and install front countershaft bearing flush with facing of case using an arbor press.

2. Coat countershaft tabbed thrust washer with grease and install washer so tab engages depression in case.

3. Tip transmission case on end and install countershaft in front bearing bore.

4. Install countershaft rear bearing spacer. Coat countershaft rear bearing with grease and install bearing using Tool J–29895 and sleeve J–33032, or its equivalent. The bearing when correctly installed will extend beyond the case surface 0.125 in.

5. Position reverse idler gear in case with shift lever groove facing rear of case and install reverse idler shaft from rear of case. Install roll pin in idler shaft.

6. Install assembled mainshaft in transmission case. Install rear mainshaft bearing race in case.

7. Install drive gear in case, and engage in third-fourth synchronizer sleeve and blocker ring.

8. Install front bearing race in front bearing cap. Do not install shims in front bearing cap at this time.

9. Temporarily install front bearing cap.

10. Install fifth speed-reverse lever, pivot bolt and retaining clip. Coat pivot bolt threads with nonhardening sealer. Be sure to engage reverse lever fork in reverse idler gear.

11. Install countershaft rear bearing spacer and retaining snap ring.

12. Install fifth speed gear on countershaft.

13. Insert fifth speed-reverse rail in rear of case and install in to reverse fifth speed lever. Rotate rail during installation to simplify engagement with lever. Connect spring to front of case.

14. Position fifth gear shift fork on fifth gear synchronizer assembly and install synchronizer on countershaft and shift fork on shift rail. Make sure roll pin hole in shift fork and shift rail are aligned.

15. Support fifth gear shift rail and fork on a block of wood and install roll pin.

16. Install thrust race against fifth speed synchronizer hub and install snap ring. Install thrust bearing against race on countershaft. Coat both bearing and race with petroleum jelly.

17. Install lipped thrust race over needle-type thrust bearing and install plastic funnel into hole in end of countershaft gear.

18. Temporarily install extension housing and attaching bolts. Turn transmission case on end, and mount a dial indicator on extension housing with indicator on the end of mainshaft.

19. Rotate mainshaft and zero dial indicator. Pull upward on mainshaft until end play is removed and record reading. Mainshaft bearings require a preload of 0.001–0.005 in. To set preload, select a shim pack measuring 0.001–0.005 in. greater than the dial indicator reading recorded.

20. Remove front bearing cap and front bearing race. Install necessary shims to obtain preload and reinstall bearing race.

21. Apply a $\frac{1}{8}$ in. bead of RTV sealant, #732 or equivalent, on case mating surface of front bearing cap. Install bearing cap aligning marks made during disassembly and torque bolts to specification.

22. Remove extension housing.

23. Move shift forks on transmission cover and synchronizer sleeves inside transmission to the neutral position.

24. Apply a $\frac{1}{8}$ in. bead of RTV sealant, #732 or equivalent, on cover mating surface of transmission.

25. Lower cover onto case while aligning shift forks and synchronizer sleeves. Center cover and install the 2 dowel bolts. Install remaining bolts and torque to specification. The offset lever to shift rail roll pin hole must be in the vertical position after cover installation.

26. Apply a $\frac{1}{8}$ in. bead of RTV Sealant, #732 or equivalent, on extension housing to transmission case mating surface.

27. Install extension housing over mainshaft and shift rail to a position where shift rail just enters shift cover opening.

28. Install detent spring into offset lever and place steel ball in neutral guide plate detent. Position offset lever on steel ball and apply pressure on offset lever and at the time seat extension housing against transmission case.

29. Install extension housing bolts and torque to specification.

30. Align and install roll pin in offset lever and shift rail.

31. Fill transmission to its proper level with lubricant.

Chrysler Corp.

A–412 Transaxle

Gear reduction, ratio selection and differential functions are combined in a single unit. The transaxle assembly is housed in a two piece magnesium case. One piece is the transmission housing and the other piece is the clutch and differential assembly housing.

DISASSEMBLY

1. Remove the clutch push rod.

2. Remove the drive flange dust plug, snap-ring, cone washer drive flange and drive flange oil seal.

3. Remove the selector shaft cover, push out the selector shaft and remove the selector shaft oil seal.

4. Remove the mainshaft bearing retaining nut rubber plugs, and remove the clutch release bearing end cover. While removing, hold the clutch release lever in upward position to avoid loading end cover and damaging case threads.

5. Remove the release bearing and the sleeve.

6. Remove the circlips from the torque shaft, and remove the clutch torque shaft, return spring and release lever.

7. Remove the mainshaft bearing retainer nuts. The three studs and clips will drop into the case.

8. Remove the case attaching bolts, the reverse idler shaft set screw and the back-up light switch. Remove the transmission case, and mark the shims for installation reference.

9. Remove the reverse shift fork supports and remove the reverse shift fork.

10. Remove the mainshaft assembly and pinion shaft fourth speed gear.

11. Disassemble the mainshaft by removing the bearing and fourth speed gear, the third-fourth synchronizer and third speed gear and needle bearing.

NOTE: Synchronizers are serviced as an assembly.

12. Remove the shift rail "E" clips, and remove the shift forks assembly.

13. Remove the clutch push rod seal and bushing assembly.

14. On the pinion shaft, remove the snapring, third speed gear, second speed gear and needle bearing.

15. Remove the reverse gear idler shaft.

16. Complete pinion shaft disassembly by removing the first-second gears synchronizer, the second speed gear sleeve, the first gear stop ring and the first speed gear. The inner sleeve for second speed gear and the first speed gear are removed together.

NOTE: Before installing the puller to remove the synchronizer, remove the plastic thrust bottom, and install Tool L–4443–4 or equivalent in the pinion shaft. The pinion shaft bearing retainer is notched in two places for puller jaws.

17. Remove the pinion shaft retainer and first gear thrust washer, and remove the pinion shaft.

Differential Repair

1. Remove the axle shaft circlips.

2. Remove differential bearing cone and cup.

NOTE: Bearing cones and cups are matched sets and must be replaced as assemblies.

3. Remove the side gears.

4. Remove the pinion shaft snap-ring, and drive out the pinion shaft. Pinion shaft gears and plastic thrust washer can now be removed from differential case. When installing pinion shaft be sure to align plastic thrust washer with case to avoid damage to thrust washer holes.

5. Drill out the ring gear rivets. The new ring gear is installed with bolts and nuts.

Differential Bearing Preload Adjusting

NOTE: Differential Bearing Preload adjustment is necessary after replace-

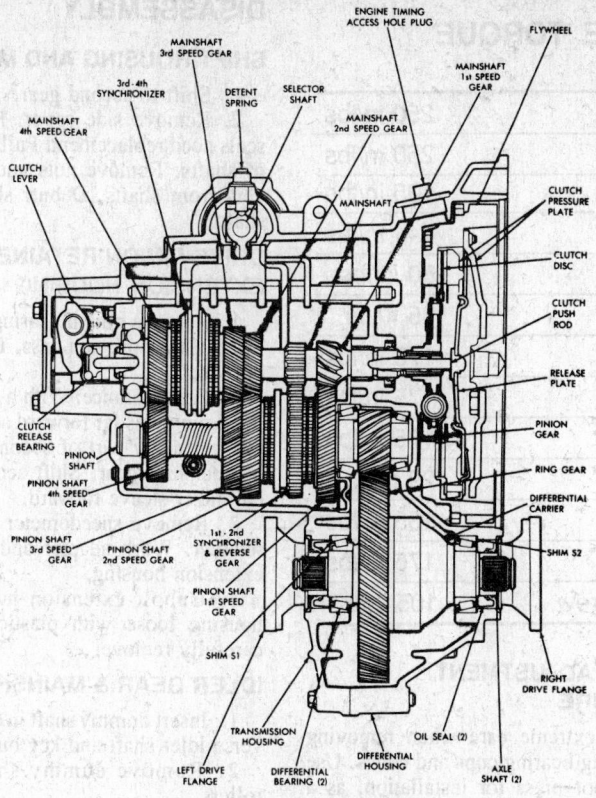

Chrysler A-412 transaxle

ment of the transmission case, clutch housing, differential case or differential bearings.

1. Install cup of bearing (opposite ring gear) with shim S2 in the clutch/differential housing. Shim S2 is always 0.039 in. thick.

2. Install outer race on ring gear side without shim S1 in transmission housing.

3. Install differential in its housing.

4. Place transmission housing in position with gasket and tighten five bolts to 14 ft. lbs.

5. Install a dial indicator, and move the differential up and down for measurement reading.

NOTE: Do not turn differential when measuring because bearings will settle and give incorrect reading.

6. Correct bearing preload is obtained by adding a constant figure (0.40 mm)(.015 in.) to measured reading. For example:

	mm.	inch
Measured reading	0.90	.035
Plus preload (constant figure)	0.40	.015
Shim Thickness (S1) =	1.30	.050

7. Remove the transmission case, and drive out the outer bearing cup.

8. Insert selected shim S1; the thickest shim first. Shims are available in sizes ranging from 0.006–0.031 in.

9. Drive in bearing cup and install transmission housing with gasket and tighten. Before installing transmission housing, remove one axle shaft to check turning torques.

Checking Turning Torque

1. Lubricate bearings with transmission oil, and check for the following turning torque: New bearings; 10.4–21.7 inch lbs. Used bearings; minimum 2.7 inch lbs.

Pinion Shaft Bearing Preload Adjustment

1. If clutch housing, ring and pinion gears or differential bearing are changed, it is necessary to adjust preload on pinion shaft bearing.

2. Place a 0.65 mm shim in bearing housing, and press in the small bearing cup.

3. Install pinion shaft and tighten cover nuts to 14 ft. lbs.

4. Mount a dial gauge and move pinion shaft up and down for measurement reading. Do not turn pinion shaft when measuring because bearings will settle and give incorrect measurement.

5. Specified bearing preload is obtained by adding a constant figure (0.20 mm) to measured reading and shim thickness (0.65 mm). Shims are available in sizes from 0.65 mm (.025 in.) to 1.40 mm (.055 in.).

6. Remove ball bearing retainer, pinion

For example:

Shim installed	0.65 mm (.025 inch)
Measured reading	0.30 mm (.012 inch)
Preload (constant figure)	0.20 mm (.008 inch)
Shim	= 1.15 mm

shaft and small bearing cup and 0.65 mm shim. Install correct shim.

ASSEMBLY

1. Install the pinion shaft.

2. Install first gear thrust washer flat side up, and install pinion shaft retainer.

3. Install first gear and stop ring and first/second gears synchronizer. The lowest thrust collar on the hub must go toward second gear. Install the sleeve with reverse teeth nearest fifth gear. Slots in synchronizer ring must be aligned with struts in first/second gears synchronizer assembly to avoid damage to stop ring on assembly.

4. Install second gear bearing race.

5. Install and correctly align reverse idler gear shaft.

6. Install needle bearing and second speed gear.

7. Install third speed gear. Select and install a retaining snap-ring which will provide 0.000 to less than 0.004 in. end-play.

8. Install the mainshaft and third/fourth gears synchronizer assembly.

9. Install fourth speed gear mainshaft needle bearing and mainshaft fourth speed gear.

10. Install fourth speed gear and snap-ring on pinion shaft.

11. Use Tool L–4442 to correctly adjust mainshaft position to specifications.

12. Adjust mainshaft end-play only if transmission case, clutch housing or mainshaft has been changed.

13. Install shift forks and "E" clips.

14. Install reverse shift fork and support brackets.

15. Use guide pins to install transmission case on clutch housing. Be sure pinion shaft is aligned with pinion shaft needle bearing in transmission case.

16. Install mainshaft bearing snap-ring.

17. Install reverse idler shaft bolt and install selector shaft assembly.

18. Install mainshaft bearing retainer and washer.

19. Install clutch torque shaft, return spring, release lever and circlips.

20. Install release bearing and sleeve.

21. Install the clutch release bearing end cover and mainshaft bearing retainer nut rubber plug.

22. Install selector shaft cover.

23. Install back-up light switch.

24. Install and adjust detent plunger: Loosen locknut. Tighten adjusting sleeve until gap can be seen between lock ring and adjusting sleeve. Loosen adjusting sleeve

A-412 MANUAL TRANSAXLE TORQUE SPECIFICATIONS

Clutch Housing Case Bolt	250 in/lbs
Clutch Housing Case Stud	250 in/lbs
Release Bearing End Cover Screw	105 in/lbs
Back-up Light Switch	144 in/lbs
Electronic Timing Probe Retainer	80 in/lbs
Gearshift Selector Shaft Cover	35 ft/lbs
Gearshift Detent Body Lock Nut	175 in/lbs
Drain Plug	175 in/lbs
Fill Plug	175 in/lbs
Pinion Shaft Bearing Retainer Bolt	29 ft/lbs
Mainshaft Ball Bearing Retaining Nut	155 in/lbs
Reverse Idler Shaft Set Screw	175 in/lbs
Reverse Idler Fork Bracket—Clutch Housing Screw	105 in/lbs

¼ turn. Hold adjusting sleeve in this position and tighten locknut.

25. Install the clutch push rod and the selector shaft boot seal.

A-460 Transaxle

INTERMEDIATE SHAFT ASSEMBLY

The 1-2 and 3-4 shift forks and synchronizer stop rings are interchangeable. However, if parts are to be reused, reassemble them in their original position. When assembling the intermediate shaft, make sure all gears turn freely and have a minimum of 0.003 in. end-play.

INPUT SHAFT

Shim thickness calculation need only be done if any of the following parts are replaced:
1. Transaxle case.
2. Input shaft seal retainer.
3. Bearing retainer plate.
4. Rear end cover.
5. Input shaft.
6. Input shaft bearings.

NOTE: Refer to Bearing Adjustment Procedure to determine the proper shim thickness for correct bearing preload and proper bearing turning torque.

DIFFERENTIAL

Shim thickness calculation need only be done if any of the following parts are replaced:
1. Transaxle case.
2. Differential bearing retainer.
3. Extension housing.
4. Differential case.
5. Differential bearings.

NOTE: Refer to Bearing Adjustment Procedure to determine the proper shim thickness for correct bearing preload and proper bearing turning torque.

BEARING ADJUSTMENT PROCEDURE

1. Take extreme care when removing and installing bearing cups and cones. Use only an arbor press for installation, as a hammer may not properly align the bearing cup or cone. Burrs or nicks on the bearing seat will give a false end play reading while gauging for proper shims. Improperly seated bearing cups and cones are subject to low mileage failure.

2. Bearing cups and cones should be replaced if they show signs of pitting or heat distress. If distress is seen on either the cup or bearing rollers, both cup and cone must be replaced.

3. Bearing end play and drag torque specifications must be maintained to avoid premature bearing failures. Used (original) bearing may lose up to 50% of the original drag torque after break-in.

NOTE: All bearing adjustments must be made with no other component interference or gear inter-mesh.

4. Replace bearings as a pair. For example, if one differential bearing is defective, replace both differential bearings. If one input shaft bearing is defective, replace both input shaft bearings.

5. Bearing cones must be reused if removed.

6. Turning torque readings should be obtained while smoothly rotating in either direction (breakaway reading is not indicative of the true turning torque).

7. Replace the oil baffle, if damaged.

A-230 3 Speed

The A-230 is a three-speed transmission equipped with two synchronizer units to assist in the engagement of all forward gears. Lubricant capacity is 5 pints.

DISASSEMBLY

SHIFT HOUSING AND MECHANISM

1. Shift to second gear.
2. Remove side cover. If shaft O-ring seals need replacement: Pull shift-forks out of shafts. Remove nuts and operating levers from shafts. Deburr shafts. Remove shafts.

DRIVE PINION RETAINER & EXTENSION HOUSING

1. Remove pinion bearing retainer from front of transmission case. Pry off retainer oil seal.
2. For clearance: With a brass drift, tap drive pinion as far forward as possible. Rotate cut away part of second gear next to countershaft gear. Shift second/third synchronizer sleeve forward.
3. Remove speedometer pinion adapter retainer. Work adapter and pinion out of extension housing.
4. Unbolt extension housing. Break housing loose with plastic hammer and carefully remove.

IDLER GEAR & MAINSHAFT

1. Insert dummy shaft in case to push reverse idler shaft and key out of case.
2. Remove dummy shaft and idler rollers.
3. Remove both tanged idler gear thrust washers.
4. Remove mainshaft assembly through rear of case.

COUNTERSHAFT GEAR & DRIVE PINION

1. Using a mallet and dummy shaft, tap the countershaft rearward enough to remove key. Drive countershaft out of case, being careful not to drop the washers.
2. Lower countershaft gear to bottom of case.
3. Remove snap-ring from pinion bearing outer race (outside front of case).
4. Drive pinion shaft into case with plastic hammer. Remove assembly through rear of case.
5. If bearing is to be replaced, remove snap-ring and press off bearing.
6. Lift counter shaft gear and dummy shaft out through rear of case.

MAINSHAFT

1. Remove snap-ring from front end of mainshaft along with second gear stop ring and second gear.
2. Spread snap-ring in mainshaft bearing retainer. Slide retainer back off the bearing race.
3. Remove snap-ring at rear of mainshaft. Support front side of reverse gear. Press bearing off mainshaft.
4. Remove from press. Remove mainshaft bearing and reverse gear from shaft.
5. Remove snap-ring and first-reverse synchronizer assembly from shaft. Remove stop ring and first gear rearward.

1. 1st speed gear
2. Ring
3. Spring
4. Sleeve
5. Struts
6. Clutch gear
7. Spring
8. Snap-ring
9. Reverse gear bushing
10. Reverse gear
11. Output shaft bearing
12. Snap-ring
13. Snap-ring
14. Bearing retainer
15. Extension housing gasket
16. Extension seal
17. Seal
18. Extension bushing
19. Snap-ring
20. Synchronizer ring
21. Spring
22. Sleeve
23. Clutch gear
24. Struts
25. Spring
26. Ring
27. 2nd speed gear
28. Output shaft
29. Snap-ring
30. Roller
31. Main drive gear (pinion)
32. Drive pinion bearing
33. Snap-ring
34. Snap-ring
35. Drive pinion seal

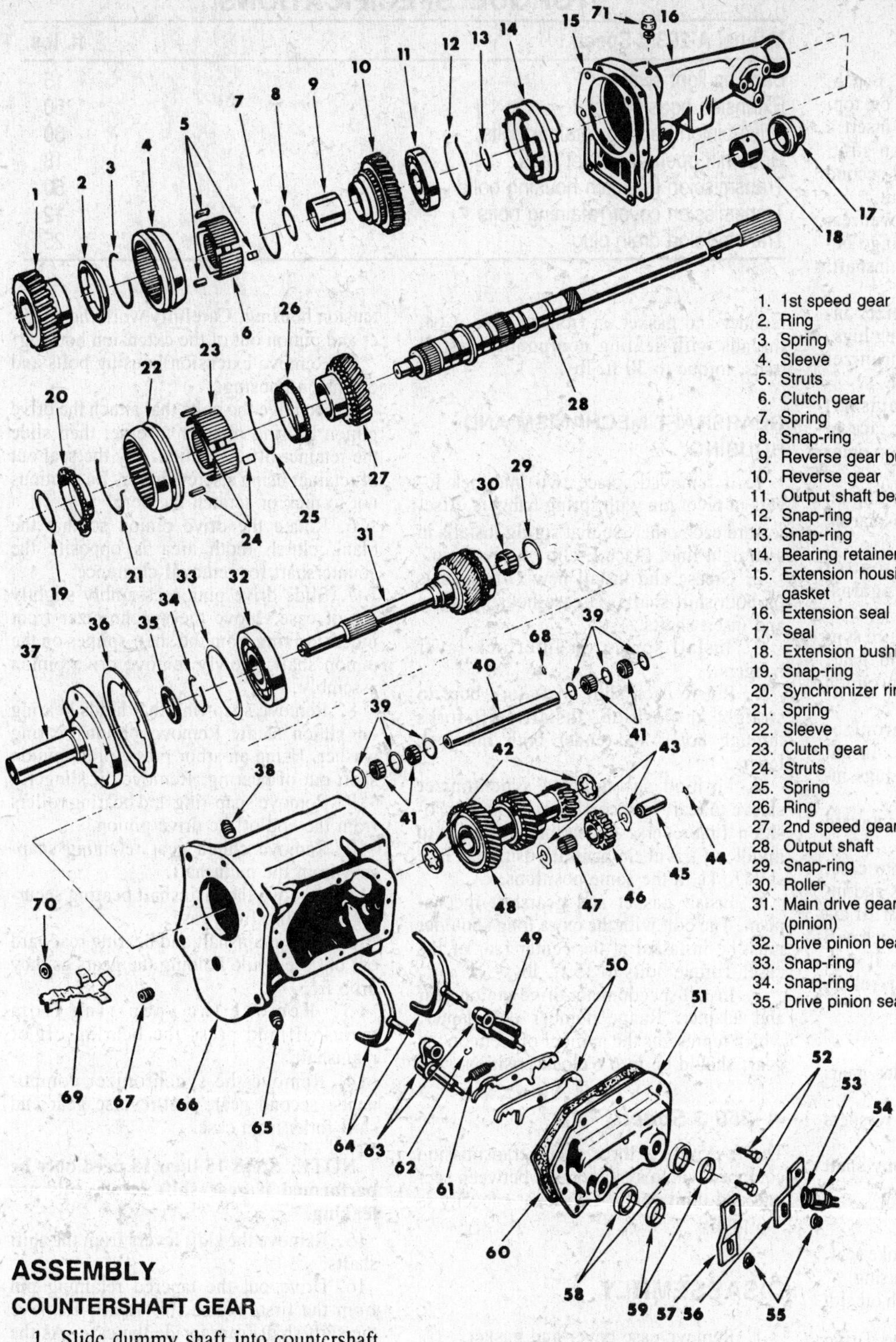

36. Retainer gasket
37. Bearing retainer
38. Plug
39. Countershaft bearing washer
40. Countershaft
41. Roller
42. Thrust washer
43. Rev. idler gear shaft
44. Key
45. Rev. idler gear shaft
46. Reverse idler gear
47. Rev. roller
48. Countershaft gear
49. Low and rev. lever
50. Interlock lever
51. Housing
52. Bolt
53. Gasket
54. Back-up light switch
55. Locking nut
56. Lever
57. Lever
58. Seal
59. Retainer
60. Housing gasket
61. Interlock spring
62. Snap-ring
63. Lever w/shaft
64. Gearshift fork
65. Plug
66. Case
67. Expansion plug
68. Countershaft key
69. Clip
70. Magnet
71. Extension vent

A-230 3-speed

ASSEMBLY

COUNTERSHAFT GEAR

1. Slide dummy shaft into countershaft gear.

2. Slide one roller thrust washer over dummy shaft and into gear, followed by 22 greased rollers.

3. Repeat Step 2, adding one roller thrust washer on end.

4. Repeat Steps 2 and 3 at other end of countershaft gear. There is a total of 88 rollers and 6 thrust washers.

5. Place greased front thrust washer on dummy shaft against gear with tangs forward.

6. Grease rear thrust washer and stick it in place in the case, with tangs rearward. Place countershaft gear assembly in bottom of transmission case until drive pinion is installed.

PINION GEAR

1. Press new bearing on pinion shaft with snap-ring groove forward. Install new snap-ring.

2. Install 15 rollers and retaining ring in drive pinion gear.

3. Install drive pinion and bearing assembly into case.

4. Position countershaft gear assembly by positioning it and thrust washers so countershaft can be tapped into position. Be careful to keep the countershaft against the dummy shaft to keep parts from falling between them. Install key in countershaft.

MAINSHAFT

1. Place a stop ring flat on the bench. Place a clutch gear and a sleeve on top. Drop the struts in their slots and insert a strut spring with the tang inside on strut. Turn the assembly over and install second strut spring, tang in a different strut.

2. Slide first gear and stop ring over rear of mainshaft and against thrust flange between assembly over rear of mainshaft, first and second gears on shaft.

3. Slide first/reverse synchronizer indexing hub slots to first gear stop ring lugs.

4. Install first/reverse synchronizer clutch gear snap-ring on mainshaft.

5. Slide reverse gear and mainshaft bearing on shaft, supporting inner race of bearing. Be sure snap-ring groove on outer race is forward.

6. Install bearing retaining snap-ring on mainshaft. Slide snap-ring over the bearing and seat it in groove.

7. Place second gear over front of mainshaft with thrust surface against flange.

8. Install stop ring and second/third synchronizer assembly against second gear. Install second/third synchronizer clutch gear snap-ring on shaft.

9. Move second/third synchronizer sleeve forward as far as possible. Install front stop ring inside sleeve with lugs indexed to struts.

10. Rotate cut out on second gear toward countershaft gear for clearance.

11. Insert mainshaft assembly into case. Tilt assembly to clear cluster gears and insert pilot rollers in drive pinion gear. If assembly is correct, the bearing retainer will bottom to the case without force. If not, check for a misplaced strut, pinion roller, or stop ring.

REVERSE IDLER GEAR

1. Place dummy shaft into idler gear. Insert 22 greased rollers.

2. Position reverse idler thrust washers in case with grease.

3. Position idler gear and dummy shaft in case. Install idler shaft and key.

EXTENSION HOUSING

1. Remove extension housing yoke seal. Drive bushing out from inside housing.

2. Align oil hole in bushing with oil slot in housing. Drive bushing into place. Drive new seal into housing.

3. Install extension housing and gasket to hold mainshaft and bearing retainer in place.

DRIVE PINION BEARING RETAINER

1. Install outer snap-ring on drive pinion bearing. Tap assembly back until snap-ring contacts case.

2. Using seal installer tool or equivalent, install a new seal in retainer bore.

3. Position main drive pinion bearing

retainer and gasket on front of case. Coat threads with sealing compound, install bolts, torque to 30 ft. lbs.

GEARSHAFT MECHANISM AND HOUSING

1. If removed, place two interlock levers in pivot pin with spring hangers offset toward each other, so that spring installs in a straight line. Place E-clip on pivot pin.

2. Grease and install new O-ring seals on both shift shafts. Grease housing bores and insert shafts.

3. Install spring on interlock lever hangers.

4. Rotate each shift shaft fork bore to straight up position. Install shift forks through bores and under both interlock levers.

5. Position second/third synchronizer sleeve to rear, in second gear position. Position first/reverse synchronizer sleeve to middle of travel, in neutral position. Place shift forks in the same positions.

6. Install gasket and gearshift mechanism. The bolt with the extra long shoulder must be installed at the center rear of the case. Torque bolts to 15 ft. lbs.

7. Install speedometer drive pinion gear and adapter. Range number on adapter, which represents the number of teeth on the gear, should be in 6 o'clock position.

A-250 3 Speed

The A-250 is a three speed transmission equipped with a synchronizer between second and third gears. Lubricant capacity is 4½ pints.

DISASSEMBLY

1. Remove case cover and gasket.

2. Measure the synchronizer "float" with a pair of feeler gauges. Measurement is made between the synchronizer outer ring pin and the opposite synchronizer outer ring. This measurement must be made on two pins 180 degrees apart with equal gap on both ends for "float" determination. The measurement should be between 0.060–0.117 in. A snug fit should be maintained between feeler gauge and pins.

3. Remove the bolt and retainer holding the speedometer pinion adapter in the ex-

TORQUE SPECIFICATIONS

Manual A-203 3-Speed	ft. lbs.
Back up light switch	15
Extension housing bolts	50
Drive pinion bearing retainer bolts	30
Gearshift operating lever nuts	18
Transmission to clutch housing bolts	50
Transmission cover retaining bolts	12
Transmission drain plug	25

tension housing. Carefully work the adapter and pinion out of the extension housing.

4. Remove extension housing bolts and extension housing.

5. Remove the bolts that attach the drive pinion bearing retainer to case, then slide the retainer off the pinion. Pry the seal out of retainer using a screwdriver. Be cautious not to nick or scratch the bore.

6. Rotate the drive pinion so that the blank clutch tooth area is opposite the countershaft for removal clearance.

7. Slide drive pinion assembly slightly out of case. Move the synchronizer front inner stop ring from the short splines on the pinion shaft. Slowly remove drive pinion assembly.

8. Remove snap-ring that holds bearing on pinion shaft. Remove pinion bearing washer. Using an arbor press, press pinion shaft out of bearing. Remove oil slinger.

9. Remove snap-ring and bearing rollers from the end of the drive pinion.

10. Remove clutch gear retaining snap-ring from the mainshaft.

11. Remove the mainshaft bearing securing snap-ring from case.

12. Slide mainshaft and bearing rearward out of case while holding the gears as they drop free.

13. Remove the snap-ring from mainshaft and press the bearing off of mainshaft.

14. Remove the synchronizer components, second gear, first/reverse gear and shift forks from case.

NOTE: Steps 15 thru 18 need only be performed if gear shift lever seals are leaking.

15. Remove the shift levers from the shift shafts.

16. Drive out the tapered retaining pin from the first/reverse shift shaft. Remove the shift shaft from inside the case. As the detent balls are spring loaded, when the shafts are removed the balls will drop to the bottom of the transmission case.

17. Remove the interlock sleeve, spring and both detent balls from case. Drive tapered retaining pin out of second/third shaft and remove shaft from case.

18. Drive shift shaft seals out of case with a suitable drift.

19. Check end play of countershaft gear with a feeler gauge. The end play should be between 0.005–0.022 in. This measure-

ment is used to determine if a new thrust washer is necessary during reassembly.

20. Using a countershaft bearing arbor, drive the countershaft towards the rear of the case until the small key can be removed from the countershaft.

21. Drive the countershaft the rest of the way out of the case, keeping the arbor tight against the end of the countershaft. This will prevent loss of roller bearings.

22. Remove the countershaft gear, front thrust washer and rear thrust washer from the case.

23. Remove the bearing rollers, spacer ring and center spacer from the countershaft gear.

24. Drive the reverse idler gear shaft out of the transmission case using a suitable drift. Remove the Woodruff key from the end of the reverse idler shaft.

25. Remove the reverse idler gear and thrust washers out of the case. Remove the bearing rollers from the gear.

ASSEMBLY

1. Slide the countershaft gear bearing roller spacer over arbor tool. Coat the bore of gear with lubricant and slide tool and spacer into gear bore.

2. Lubricate the bearing rollers with heavy grease and install two rows of 22 rollers each in both ends of gear in area around arbor. Cover with heavy grease and install bearing spacer rings in each end of gear and between roller rows.

3. If countershaft gear end play was found to be excessive during disassembly, install new thrust washers. Cover with heavy grease and install thrust washer and thrust needle bearing and cap at each end of countershaft gear and over arbor. Install gear and arbor in the case, and make sure that tabs on rear thrust washer slide into grooves in the case.

4. Drive the arbor forward out of the countershaft gear and through the bore in the front of the case using the countershaft and a soft faced hammer. When the countershaft is almost in place, make certain the keyway in the countershaft is aligned with the key slot in the rear of the case. Insert the shaft key and continue to drive the countershaft into the case until the key is bottomed in the slot.

5. Position special arbor tool in the reverse idler gear and install the 22 roller bearings using a heavy grease.

6. Place the front and rear thrust washers at each end of the reverse idler gear. Position the assembly in the transmission case with the chamfered end of the gear teeth towards the front. Make sure that the thrust washer tabs engage the slots in case.

7. Insert reverse idler shaft into the bore at rear of case with keyway to the rear, pushing the arbor towards the front of the case.

8. When the keyway is aligned with the slot in the case, insert the key in the keyway. Drive the shaft forward until the key is seated in the recess.

NOTE: Steps 9 through 14 need only be performed if the shift levers have been disassembled.

9. Place new shift shaft seals in the case and drive it into position with suitable drift.

10. Carefully slide the first/reverse shift shaft into the case and lock into place with a tapered retaining pin. Position the lever so that the center detent is aligned with the interlock bore.

11. Install the interlock sleeve into the bore followed by a detent ball, spring and pin.

12. Install remaining detent ball and hold in place with detent ball holding tool.

13. Depress the detent ball and carefully install the second/third shift shaft. Align center detent with detent ball and secure lever with tapered retaining pin.

14. Install shift levers and tighten retaining nuts to 18 ft. lbs.

15. Press the bearing on the mainshaft and select and install snap-ring that gives minimum end play.

16. Move shift lever to reverse position, and then place the first/reverse gear and shift fork in the case.

NOTE: Both shift forks are offset toward the rear of the transmission case.

17. Assemble the synchronizer parts with shift fork and second gear.

18. Place the second gear assembly in the transmission case and insert the shift fork into its lever.

19. Install the mainshaft carefully through the gear assembly until it bottoms in rear of case.

20. Install synchronizer clutch gear snap-ring on mainshaft.

21. Select and install mainshaft bearing snap-ring in case.

22. If "float" measurement was found to be outside specifications, install or remove shims to place "float" within range.

23. Install oil slinger on drive pinion shaft and slide against the gear.

24. Slide the bearing over the pinion shaft with snap-ring groove away from gear, then seat bearing on shaft using an arbor press.

25. Install keyed washer between bearing and retaining snap-ring groove.

26. Secure bearing and washer with selected thickness snap ring. If large snap-ring around bearing was removed, install it at this time.

27. Place drive pinion shaft in a vise with soft faced jaws and install the 14 roller bearings in the shaft cavity. Coat the roller bearings with a heavy grease and install retaining ring in groove.

28. Rotate the drive pinion so that the blank clutch tooth area is next to the countershaft. Guide the drive pinion through the front of case and engage the inner stop ring with the clutch teeth. Then seat pinion bearing. The pinion shaft is fully seated when the snap-ring is in full contact with the case.

29. Install a new seal in the pinion bearing retainer.

30. Position retainer assembly and new gasket on the case. Use sealing compound on bolts and tighten to 30 ft. lbs.

31. Slide the extension housing and a new gasket over mainshaft. Guide shaft through bushing and oil seal. Use sealing compound on the bolt used in the hole tapped through the transmission case. Install remaining bolts and tighten all to 50 ft. lbs.

32. Install the transmission cover and gasket and tighten cover bolts to 12 ft. lbs.

33. Rotate the speedometer pinion gear and adapter assembly so that the number on the adapter corresponding to the number of teeth on the gear is in the 6 o'clock position as the assembly is installed.

34. Fill the transmission with the proper lubricant and install the drain plug and tighten to 25 ft. lbs. Install the back-up light switch and tighten to 15 ft. lbs.

35. Rotate the drive pinion shaft and check operation of transmission by running the transmission through all gear ranges.

A–390 3 Speed

The A–390 is a three speed synchromesh transmission. Lubricant capacity is $4\frac{1}{2}$ pints.

DISASSEMBLY

1. Remove the bolts that attach the cover to the case. Remove the cover and gasket.

2. Remove the long spring that retains the detent plug in the case. Remove the detent plug with a small magnet.

3. Remove the bolt and retainer securing the speedometer pinion adapter to the transmission case. Carefully work the adapter and pinion out of the extension housing.

4. Remove the bolts that attach the extension housing to the transmission case. Slide the extension housing off the output shaft.

5. Remove the bolts that attach the input shaft bearing retainer to the case. Slide the retainer off the shaft. Using a suitable tool, pry the seal out of the retainer. Be careful not to nick or scratch the bore in which the seal is pressed or the surface on which the seal is bottomed.

6. Remove the lubricant fill plug from the right side of the case. Working through the fill plug opening, drive the roll pin out of the countershaft with a $\frac{1}{4}$ in. punch.

7. Working with the countershaft bearing arbor and a soft faced hammer, tap the countershaft toward the front of the case with the arbor tool to remove the expansion plug from the countershaft bore at the front of the case. The countershaft is a loose fit in the case and will slide easily.

8. Insert the arbor tool through the front of the case and push the countershaft out of

the rear of the case so the roll pin hole in the countershaft does not travel through the roller bearings. The countershaft gear will drop to the bottom of the case. Remove the countershaft from the rear of the case.

9. Place both shift levers in neutral (center) position.

10. Remove the input shaft assembly and stop ring from the front of the case.

11. Remove the set screw that secures the first/reverse shift fork to the shift rail. Slide the first/reverse shift rail out through the rear of the case.

12. Move the second/third shift fork rearward for access to the set screw. Remove the set screw from the fork. Using a suitable tool, rotate the shift rail one quarter ($\frac{1}{4}$) turn.

13. Lift the interlock plug from the case with a magnet.

14. Tap on the inner end of the second/third shift rail to remove the expansion plug from the front of the case. Remove the shift rail through the front of the case.

15. Remove the second/third shift rail detent plug and spring from the detent bore with a magnet.

16. Tap the output shaft assembly rearward until the output shaft bearing clears the case. Remove both shift forks. Remove the snap-ring that retains the output shaft bearing to the output shaft.

17. Assemble the output shaft bearing removal tool over the output shaft and bearing. Remove the output shaft bearing.

18. Remove the output shaft assembly through top of the case.

19. Using a suitable drift, drive the reverse idler gear shaft toward the rear, and out of the transmission case.

20. Lift the reverse idler gear and thrust washer out of the case.

21. Remove the countershaft gear, arbor assembly, and thrust washers from the bottom of the case.

22. Remove the countershaft roll pin from the bottom of the case.

23. Remove the snap-ring that retains the second/third synchronizer clutch gear and sleeve assembly on the output shaft. Slide the second/third synchronizer assembly off the end of the output shaft.

NOTE: Do not separate the second-third synchronizer clutch gear, sleeve, struts, or spring unless inspection reveals that a replacement is necessary.

24. Slide the second gear and stop ring off the output shaft.

25. Remove the snap-ring and thrust washer retaining the first gear. Slide the first gear and stop ring off the output shaft.

26. Remove the snap-ring that retains the first/reverse synchronizer hub on the output shaft. The first/reverse synchronizer hub is a press fit on the output shaft. To avoid damage to the synchronizer, remove the synchronizer hub using an arbor press. Do not attempt to remove or install the hub by hammering or prying.

SHIFT LEVERS AND SEALS

1. Remove the operating levers from their respective shafts. Remove any burrs from the shafts to avoid damage to the case.

2. Push the shift levers out of the transmission case. Remove and discard the O-ring seal from each shaft.

3. Lubricate the new seals with transmission oil and install them on the shafts.

4. Install the shift levers in the case.

5. Install the operating levers and tighten the retaining nuts to 18 ft. lbs.

INPUT SHAFT BEARING AND ROLLERS

1. Remove the snap-ring securing the bearing on the input shaft. Carefully press the input shaft out of the bearing with an arbor press.

2. Remove the fifteen bearing rollers from the cavity in the end of the input shaft.

3. Install the 15 bearing rollers in the cavity of the input shaft. Coat the rollers with a thin film of grease to retain them during installation.

4. Slide the input shaft bearing over the input shaft, snap-ring groove away from the gear end. Seat the bearing assembly on the input shaft with an arbor press.

5. Secure the bearing with the snap-ring. Be sure the snap-ring is properly seated. If a large snap-ring around the bearing was removed, be sure to install it at this time.

SYNCHRONIZERS

NOTE: If either synchronizer is to be disassembled, mark all parts so that they will be reassembled in the same position. Do not mix parts from the two synchronizers.

1. Push the synchronizer hub off each synchronizer sleeve.

2. Separate the struts and springs from the hubs.

3. Install the spring on the front side of the first/reverse synchronizer hub, making sure that all three strut slots are fully covered. Hang the three struts on the spring and in the slots with the wide end of the strut inside the hub.

4. With the alignment marks on the hub and sleeve aligned, push the sleeve down on the hub until the struts are in the neutral detent. Place the stop ring on top of the synchronizer assembly.

5. With the alignment marks on the second/third synchronizer sleeve and hub aligned, slide the sleeve on the hub. Drop in the three struts in the strut slots. Install the spring with the hump in the center, into the hollow of the strut. Turn the assembly over and install the other spring so that the hump in the center of the spring is inserted in the same strut. Place the stop ring on each end of the synchronizer assembly.

COUNTERSHAFT GEAR AND BEARING

1. Remove the countershaft bearing arbor, the roller bearings and the two bearing retainers from the countershaft gear.

2. Coat the bore in each end of the countershaft gear with grease.

3. Insert the countershaft arbor and install twenty five roller bearings and the retainer washer in each end of the countershaft gear.

4. Position the countershaft gear and arbor assembly in the transmission case. Align the gear bore and the thrust washers with the bores in the case and install the countershaft.

5. Using a feeler gauge, check the countershaft gear end play. The end play should be within 0.004–0.018 in. If the clearance is not within limits, replace the thrust washers.

6. After establishing the correct end play, install the arbor tool in the countershaft gear and lower the gear and tool out of the bottom of the transmission case.

ASSEMBLY

1. Coat the countershaft gear thrust surfaces in the case with a thin film of grease and position the two thrust washers in place. Place the countershaft gear and arbor assembly in the proper position in the bottom of the transmission case. The countershaft gear will remain in the bottom of the case until the output and input shafts are installed.

2. Coat the reverse idler gear thrust surfaces in the case with a thin film of grease and position the two thrust washers in place. Install the reverse idler gear in the case and align the gear bore with the thrust washers in the case bore. Install the reverse idler shaft.

3. Measure the reverse idler gear end play with a feeler gauge. End play should be 0.004–0.018 in. If the clearance is not within limits, replace the thrust washers. If the end-play is correct, leave the reverse idler gear in place.

4. Lubricate the output shaft splines and the machined surfaces with transmission oil.

5. Slide the first reverse synchronizer onto the output shaft with the fork groove toward the front. The first/reverse synchronizer hub is a press fit on the output shaft. To eliminate the possibility of damage to the hub, install the hub using an arbor press.

NOTE: Do not attempt to install the hub by hammering or driving.

6. Secure the hub on the output shaft with the snap-ring. Slide the first gear and stop ring onto the output shaft, aligning the slots in the stop ring with the struts. Install the thrust washer and snap-ring.

7. Slide the second gear and stop ring on the output shaft.

8. Install the second/third synchronizer assembly on the output shaft. Rotate the second gear to index the struts with the

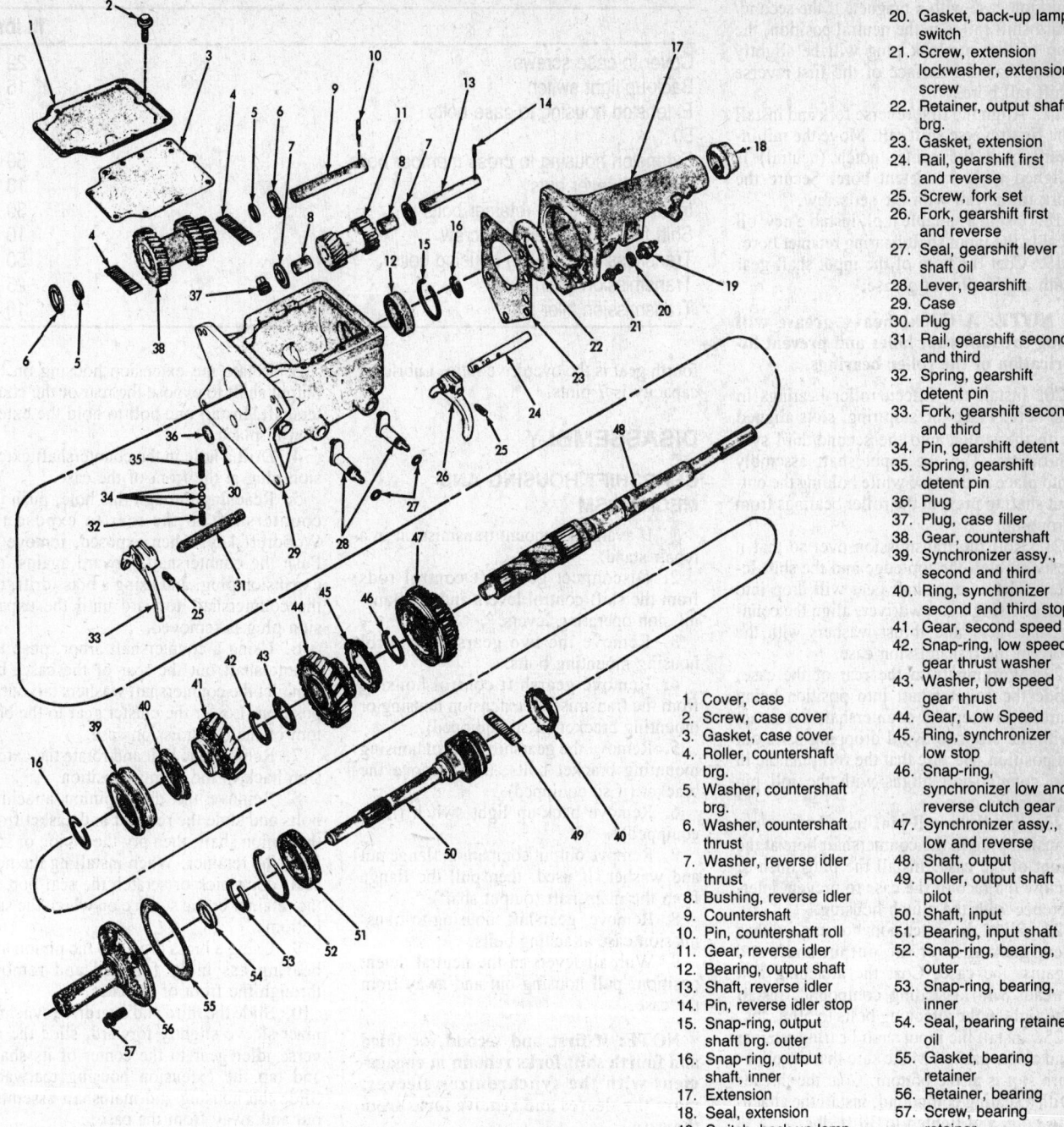

1. Cover, case
2. Screw, case cover
3. Gasket, case cover
4. Roller, countershaft brg.
5. Washer, countershaft brg.
6. Washer, countershaft thrust
7. Washer, reverse idler thrust
8. Bushing, reverse idler
9. Countershaft
10. Pin, countershaft roll
11. Gear, reverse idler
12. Bearing, output shaft
13. Shaft, reverse idler
14. Pin, reverse idler stop
15. Snap-ring, output shaft brg. outer
16. Snap-ring, output shaft, inner
17. Extension
18. Seal, extension
19. Switch, back-up lamp
20. Gasket, back-up lamp switch
21. Screw, extension lockwasher, extension screw
22. Retainer, output shaft brg.
23. Gasket, extension
24. Rail, gearshift first and reverse
25. Screw, fork set
26. Fork, gearshift first and reverse
27. Seal, gearshift lever shaft oil
28. Lever, gearshift
29. Case
30. Plug
31. Rail, gearshift second and third
32. Spring, gearshift detent pin
33. Fork, gearshift second and third
34. Pin, gearshift detent
35. Spring, gearshift detent pin
36. Plug
37. Plug, case filler
38. Gear, countershaft
39. Synchronizer assy., second and third
40. Ring, synchronizer second and third stop
41. Gear, second speed
42. Snap-ring, low speed gear thrust washer
43. Washer, low speed gear thrust
44. Gear, Low Speed
45. Ring, synchronizer low stop
46. Snap-ring, synchronizer low and reverse clutch gear
47. Synchronizer assy., low and reverse
48. Shaft, output
49. Roller, output shaft pilot
50. Shaft, input
51. Bearing, input shaft
52. Snap-ring, bearing, outer
53. Snap-ring, bearing, inner
54. Seal, bearing retainer oil
55. Gasket, bearing retainer
56. Retainer, bearing
57. Screw, bearing retainer

A-390 3-speed

slots in the stop ring. Secure the synchronizer with a snap-ring.

9. Position the output shaft assembly in the transmission case. Place the transmission in a vertical position with the front of the case flat on the work bench. Place a 1¼ in. block of wood under the end of the output shaft. The block of wood will hold the output shaft assembly up during installation of the output shaft bearing.

10. Install the large snap-ring on the output shaft bearing. Place the bearing on the output shaft with the large snap-ring up.

Drive the bearing on the shaft until it is seated on the shaft. Secure the bearing on the output shaft with the snap-ring. Return the transmission to a horizontal position.

11. Insert both shift forks in the case and in their proper sleeves. Push the output shaft assembly into position and tap it forward until the output shaft bearing is seated in the transmission case.

12. Install the shortest detent spring followed by a detent plug into the case. Place the second/third synchronizer assembly in the second gear position.

13. Align the second/third shift fork and install the second/third shift rail. The second/third shift rail is the shortest of the two shift rails. It will be necessary to depress the detent plug to enter the shift rail in the bore. Move the rail inward until the detent plug engages the forward notch (second gear position).

14. Secure the fork to the rail with the set screw. Move the synchronizer to the neutral position.

15. Install a new expansion plug in the transmission case.

16. Install the interlock plug in the transmission case with a magnet. If the second/third shift rail is in the neutral position, the top of the interlock plug will be slightly lower than the surface of the first/reverse shift rail bore.

17. Align the first/reverse fork and install the first/reverse shift rail. Move the rail inward until the center notch (neutral) is aligned with the detent bore. Secure the fork to the rail with the set screw.

18. Using a suitable tool, install a new oil seal in the input shaft bearing retainer bore.

19. Coat the bore of the input shaft gear with a thin film of grease.

NOTE: A thick, heavy grease will plug the lubricant holes and prevent lubrication of the roller bearings.

20. Install the fifteen roller bearings in the bore. Place the stop ring, slots aligned with the struts, into the second/third synchronizer. Tap the input shaft assembly into place in the case while holding the output shaft to prevent the roller bearings from dropping.

21. Roll the transmission over so that it rests on both the top edge and the shift levers. The countershaft gear will drop into place. Using a screwdriver, align the countershaft gear and thrust washers with the bore in the transmission case.

22. Working from the rear of the case, slide the countershaft into position being careful to keep the countershaft in contact with the arbor to avoid dropping parts out of position. Be sure that the roll pin hole in the countershaft aligns with the roll pin hole in the case.

23. Install the roll pin. Install a new expansion plug in the countershaft bore at the front of the case. Install the plug flush or below the face of the case to prevent interference with the clutch housing.

24. Slide the extension housing, with a new gasket, over the output shaft and against the case. Coat the attaching bolt threads with a sealing compound. Install and tighten the attaching bolts to 50 ft. lbs.

25. Install the input shaft bearing retainer and a new gasket. Make sure that the oil return slot is at the bottom. Coat the threads with a sealing compound, install the attaching bolts and tighten to 30 ft. lbs.

26. Install the remaining detent plug into the case followed by the detent spring.

27. Install the filler plug and the back-up light switch. Pour lubricant over the entire gear train while rotating the input shaft and the output shaft.

28. Place the cover and a new gasket on the transmission. Coat the attaching screw threads with a sealing compound. Install and tighten the attaching screws to 22 ft. lbs.

Overdrive—4 Speed

The Overdrive–4 Speed transmission is a four speed unit with all forward gears synchronized. Third gear is direct, while the

TORQUE SPECIFICATIONS

	ft. lbs.
Cover to case screws	22
Back-up light switch	15
Extension housing to case bolts	50
Extension housing to cross member bolts	50
Gearshift lever nuts	18
Input shaft bearing retainer bolts	30
Shift fork to shift rail set screw	10
Transmission to clutch housing bolts	50
Transmission drain plug	25
Transmission filler plug	15

fourth gear is the overdrive ratio. Lubricant capacity is 7 pints.

DISASSEMBLY

GEARSHIFT HOUSING AND MECHANISM

1. If available, mount transmission in a repair stand.
2. Disconnect gearshift control rods from the shift control levers and the transmission operating levers.
3. Remove the two gearshift control housing mounting bolts.
4. Remove gearshift control housing from the transmission extension housing or mounting bracket (if so equipped).
5. Remove the gearshift control housing mounting bracket bolts, then remove the bracket (if so equipped).
6. Remove back-up light switch (if so equipped).
7. Remove output companion flange nut and washer, if used, then pull the flange from the mainshaft (output shaft).
8. Remove gearshift housing-to-transmission case attaching bolts.
9. With all levers in the neutral detent position, pull housing out and away from the case.

NOTE: If first and second, or third and fourth shift forks remain in engagement with the synchronizer sleeves, move the sleeves and remove forks from the case.

10. Remove nuts, lock washers and flat washers that hold first/second, and third/fourth speed shift operating levers to the shafts.
11. Disengage shift levers from the flats on the shafts and remove levers. Remove the E-ring on the overdrive four speed.

EXTENSION HOUSING, MAINSHAFT & MAIN DRIVE PINION

1. Remove the bolt and retainer holding the speedometer pinion adapter in the extension housing, then remove the pinion adapter.
2. Remove the bolts attaching the extension housing to the transmission case.

3. Rotate the extension housing on the output shaft to expose the rear of the countershaft. Install one bolt to hold the extension in place.
4. Drill a hole in the countershaft extension plug at the front of the case.
5. Reaching through this hole, push the countershaft to the rear to expose the Woodruff key; when exposed, remove it. Push the countershaft forward against the expansion plug, and using a brass drift, tap the countershaft forward until the expansion plug is removed.
6. Using a countershaft arbor, push the countershaft out the rear of the case, but don't let the countershaft washers fall out of position. Lower the cluster gear to the bottom of the transmission case.
7. Remove the bolt and rotate the extension back to the normal position.
8. Remove the drive pinion attaching bolts and slide the retainer and gasket from the pinion shaft, then pry the pinion or seal from the retainer. When installing the new seal, don't nick or scratch the seal bore in the retainer or the surface on which the seal bottoms.
9. Using a brass drift, tap the pinion and bearing assembly forward and remove through the front of the case.
10. Slide the third and overdrive synchronizer sleeve slightly forward, slide the reverse idler gear to the center of its shaft, and tap the extension housing rearward. Slide the housing and mainshaft assembly out and away from the case.
11. Remove the snap-ring holding the third and overdrive synchronizer clutch gear and sleeve assembly to the mainshaft, then remove the synchronizer assembly.
12. Slide the overdrive gear and stop ring off the mainshaft. Using pair of long nose pliers, compress the snap ring holding the mainshaft bearing in the extension housing. With it compressed, pull the mainshaft assembly and bearing out of the extension housing.
13. Remove the snap-ring holding the mainshaft on the shaft. The bearing is removed by inserting steel plates on the front side of the first speed gear, then pressing the mainshaft through the bearing being careful not to damage the gear teeth.

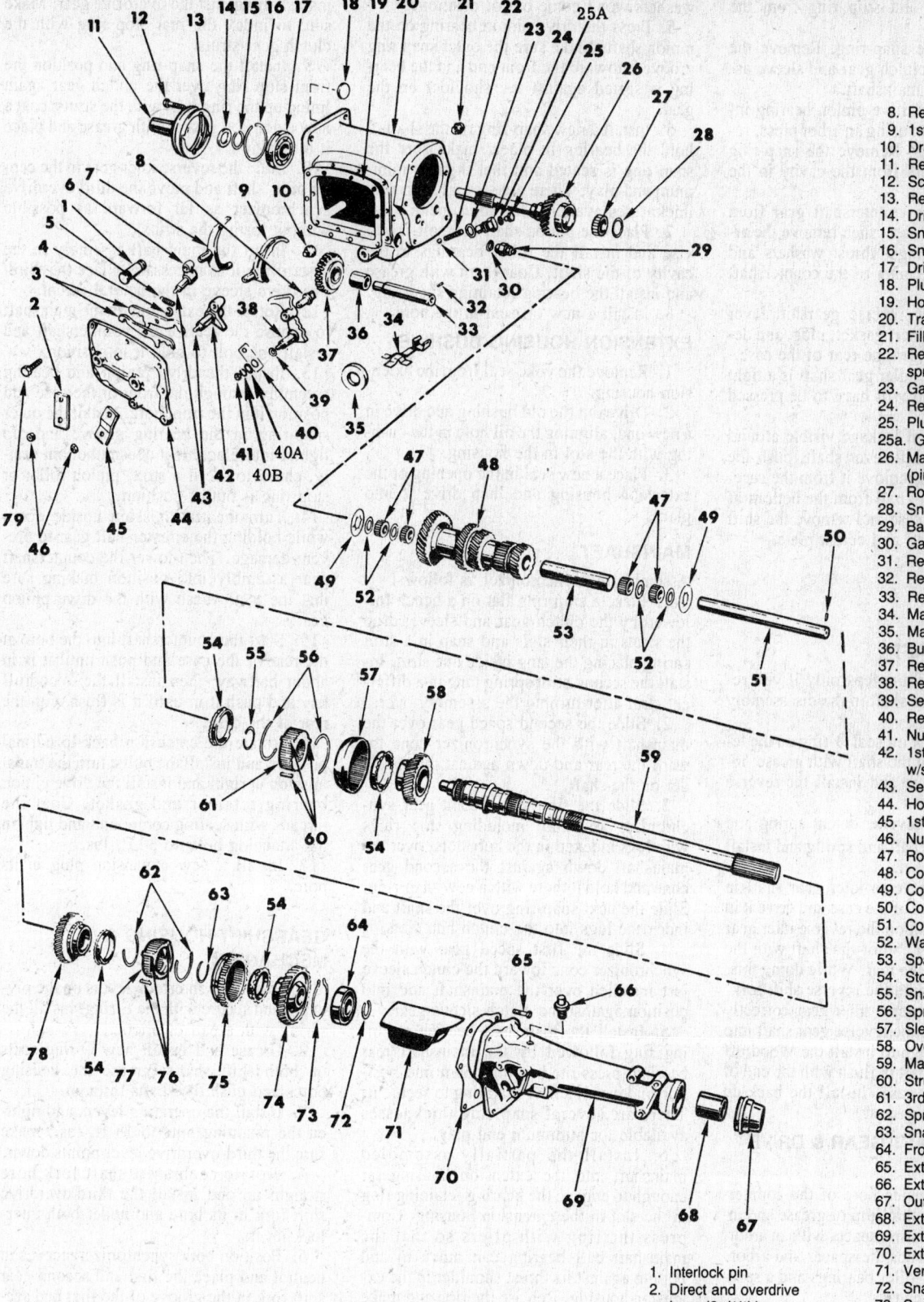

8. Reverse fork
9. 1st and 2nd fork
10. Drain plug
11. Retainer
12. Screw
13. Retainer gasket
14. Drive pinion seal
15. Snap-ring
16. Snap-ring
17. Drive pinion bearing
18. Plug
19. Housing gasket
20. Transmission case
21. Filler plug
22. Rev. detent ball spring
23. Gasket
24. Retainer
25. Plug
25a. Gasket
26. Main drive gear (pinion)
27. Roller
28. Snap-ring
29. Back-up light switch
30. Gasket
31. Rev. Lever detent ball
32. Rev. idler shaft key
33. Rev. idler gear shaft
34. Magnet clip
35. Magnet
36. Bushing
37. Reverse idler gear
38. Reverse lever w/shaft
39. Seal
40. Reverse lever
41. Nut
42. 1st and 2nd lever w/shaft
43. Seal
44. Housing
45. 1st and 2nd lever
46. Nut
47. Rollers
48. Countershaft gear
49. Countershaft washer
50. Countershaft key
51. Countershaft
52. Washer
53. Spacer
54. Stop ring
55. Snap-ring
56. Spring
57. Sleeve
58. Overdrive gear
59. Main or output shaft
60. Struts
61. 3rd and direct gear
62. Spring
63. Snap-ring
64. Front bearing
65. Extension bolt
66. Extension vent
67. Output shaft seal
68. Extension bushing
69. Extension
70. Extension gasket
71. Vent baffle
72. Snap-ring
73. Snap-ring
74. 1st speed gear
75. 1st and 2nd clutch sleeve gear
76. Synchronizer struts
77. Clutch gear
78. 2nd speed gear
79. Case housing bolt

1. Interlock pin
2. Direct and overdrive lever (3rd/4th)
3. Interlock lever
4. Snap-ring
5. Interlock spring
6. 3rd and overdrive (4th) lever
7. 3rd and overdrive (4th) fork

A-833 4-speed

14. Remove the bearing, retainer ring, first speed gear and stop ring from the shaft.

15. Remove the snap-ring. Remove the first and second clutch gear and sleeve assembly from the mainshaft.

16. Remove the drive pinion bearing inner snap-ring, then using an arbor press, remove the bearing. Remove the snap-ring and bearing rollers from the cavity in the drive pinion.

17. Remove the countershaft gear from the bottom of the case, then remove the arbor, needle bearings, thrust washers and spacers from the center of the countershaft gear.

18. Remove the reverse gearshift lever detent spring retainer, gasket, plug, and detent ball spring from the rear of the case.

19. The reverse idler gear shaft is a tight fit in the case and will have to be pressed out.

20. If there is oil leakage visible around the reverse gearshift lever shaft, push the lever shaft in and remove it from the case. Remove the detent ball from the bottom of the transmission case and remove the shift fork from the shaft and detent plate.

ASSEMBLY

REVERSE SHAFT

Follow the first four steps only if you removed the reverse shaft in the disassembly procedure.

1. Install a new oil seal O-ring on the lever shaft and coat the shaft with grease; insert it into its bore and install the reverse fork in the lever.

2. Install the reverse detent spring and gasket; insert the ball and spring and install the plug and gasket.

3. Place the reverse idler gear shaft in position in the end of the case and drive it in far enough to position the reverse idler gear on the protruding end of the shaft with the fork slot toward the rear. While doing this, engage the slot with the reverse shift fork.

4. With the reverse idler gear correctly positioned, drive the reverse gear shaft into the case far enough to install the Woodruff key. Drive the shaft in flush with the end of the transmission case. Install the back-up light switch and gasket.

COUNTERSHAFT GEAR & DRIVE PINION

1. Coat the inside bore of the countershaft gear with a thin film of grease and install the roller bearing spacer with an arbor, into the gear; center the spacer and arbor.

2. Install the roller bearings and a spacer ring on each end.

3. Replace worn thrust washers; coat the new ones with grease and install them over the arbor with the tang side toward the case boss.

4. Install the countershaft assembly into the case and allow the gear assembly to sit on the bottom of the case so that the thrust washers won't come out of position.

5. Press the drive pinion bearing on the pinion shaft. Make sure the outer snap ring groove is toward the front end and the bearing is seated against the shoulder on the gear.

6. Install a new snap-ring on the shaft to hold the bearing in place; make sure the snap-ring is seated and that there is minimum end play. There are several snap-ring thicknesses available for adjustment.

7. Place the pinion shaft in a soft-jawed vise and install the roller bearings in the cavity of the shaft. Coat them with grease and install the bearing retaining snap-ring.

8. Install a new oil seal in the bore.

EXTENSION HOUSING BUSHING

1. Remove the yoke seal from the extension housing.

2. Drive out the old bushing and drive in a new one, aligning the oil hole in the bushing with the slot in the housing.

3. Place a new seal in the opening of the extension housing and then drive it into place.

MAINSHAFT

Assemble the synchronizer as follows:

1. Place a stop ring flat on a bench followed by the clutch gear and sleeve; drop the struts in their slots and snap in a strut spring placing the tang inside one strut. Install the second strut spring tang in a different strut after turning the assembly over.

2. Slide the second speed gear over the mainshaft with the synchronizer cone toward the rear and down against the shoulder on the shaft.

3. Slide the first and second gear synchronizer assembly including stop rings with lugs indexed in the hub slots, over the mainshaft down against the second gear cone and hold it there with a new snap-ring. Slide the next snap-ring over the shaft and index the lugs into the clutch hub slots.

4. Slide the first speed gear with the synchronizer cone toward the clutch sleeve just installed over the mainshaft and into position against the clutch sleeve gear.

5. Install the mainshaft bearing retaining ring followed by the mainshaft rear bearing; press the bearing down into position and install a new snap ring to secure it. There are several snap-ring thicknesses available for minimum end play.

6. Install the partially assembled mainshaft into the extension housing far enough to engage the bearing retaining ring in the slot in the extension housing. Compress the ring with pliers so that the mainshaft ball bearing can move in and bottom against its thrust shoulder in the extension housing. Release the ring and make sure that it is seated.

7. Slide the overdrive gear over the mainshaft with the synchronizer cone toward the front followed by the gear's snap-ring.

8. Install the third-overdrive gear synchronizer clutch gear assembly on the mainshaft against the overdrive gear. Make sure to index the rear stop ring with the clutch gear struts.

9. Install the snap-ring and position the front stop ring over the clutch gear again lining up the ring lugs with the struts; coat a new extension gasket with grease and place it in position.

10. Slide the reverse idler gear to the center of its shaft and move the third/overdrive synchronizer as far forward as possible without losing the struts.

11. Insert the mainshaft assembly in the case tilting it as necessary. Place the third/overdrive sleeve in the neutral detent.

12. Rotate the extension on the mainshaft to expose the rear of the countershaft and install one bolt to hold it in position.

13. Install the drive pinion and bearing assembly through the front of the case and position it in the front bore. Install the outer snap-ring in the bearing groove and tap lightly into place. If it doesn't bottom easily, check to see if a strut, pinion roller or stop ring is out of position.

14. Turn the transmission upside down while holding the countershaft gear to prevent damage. Then lower the countershaft gear assembly into position making sure that the teeth mesh with the drive pinion gear.

15. Start the countershaft into the bore at the rear of the case and push until it is in about halfway; then install the Woodruff key and push it in until it is flush with the rear of the case.

16. Rotate the extension back to normal position and install the bolts; turn the transmission upright and install the drive pinion bearing retainer and gasket. Coat the threads with sealing compound and tighten the attaching bolts to 30 ft. lbs.

17. Install a new expansion plug in its bore.

GEARSHIFT HOUSING & MECHANISM

1. Install the interlock levers on the pivot pin and secure with the E-ring. Install the spring with a pair of pliers.

2. Grease and install new O-ring seals on both shift shafts; grease the housing bores and push the shafts through.

3. Install the operating levers and tighten the retaining nuts to 18 ft. lbs.; make sure the third-overdrive lever points down.

4. Rotate each shift shaft fork bore straight up and install the third/overdrive shift fork in its bore and under both interlock levers.

5. Position both synchronizer sleeves in neutral and place the first and second gear shift fork in the groove of the first and second gear synchronizer sleeve. Slide the reverse idler gear to neutral. Turn the transmission on its right side and place the gearshift housing gasket in place holding it there with grease. Install the reverse detent ball and spring into the case bore.

6. As the shift housing is lowered in place, guide the third–overdrive shift fork into its synchronizer groove then lead the shaft of the first and second shift lever.

7. Raise the interlock lever with a screwdriver to allow the first and second shift fork to slip under the levers. The shift housing will now seat against the case.

8. Install the bolts lightly and shift through all the gears to check for proper operation.

9. The reverse shift lever and the first and second gear shift lever have cam surfaces which mate in reverse position to lock the first and second lever, the fork and synchronizer in the neutral position.

10. To check for proper operation, put the transmission in reverse, and, while turning the input shaft, move the first and second lever in each direction. If it locks up or becomes harder to turn, select a new shift lever size with more or less clearance. If there is too little cam clearance, it will be difficult or impossible to shift into reverse.

11. Grease the reverse shaft, install the operating lever and nut, and install the speedometer drive pinion gear and adapter, making sure the range number is in the straight down position.

Ford

3.03 3 Speed

The Ford 3.03 is a fully synchronized three speed transmission. All gears except reverse are in constant mesh. Forward speed gear changes are accomplished with synchronizer sleeves.

DISASSEMBLY

1. Drain the lubricant by removing the lower extension housing bolt.

2. Remove the case cover and gasket.

3. Remove the long spring that holds the detent plug in the case and remove the detent plug with a small magnet.

4. Remove the extension housing and gasket.

5. Remove the front bearing retainer and gasket.

6. Remove the filler plug on the right side of the transmission case. Working through the plug opening, drive the roll pin out of the case and countershaft with a ¼ in. punch.

7. Hold the countershaft gear with a hook. Install dummy shaft and push the countershaft out of the rear of the case. As the countershaft comes out, lower the gear cluster to the bottom of the case. Remove the countershaft.

8. Remove the snap-ring that holds the speedometer drive gear on the output shaft. Slip the gear off the shaft and remove the gear lock ball.

9. Remove the snap-ring that holds the output shaft bearing. Using a special bearing puller, remove the output shaft bearing.

TORQUE SPECIFICATIONS

	ft. lbs.
Back up light switch	15
Drive pinion bearing, retainer bolts	30
Extension housing to case bolts	50
Gearshift to mounting plate	24
Gearshift mounting plate to extension	12
Shift lever nuts	18
Transmission to clutch housing bolts	50
Transmission drain plug	25

10. Place both shift levers in the neutral (center) position.

11. Remove the set screw that holds the first/reverse shift fork to the shift rail. Slip the first/reverse shift rail out through the rear of the case.

12. Move the first/reverse synchronizer forward as far as possible. Rotate the first/reverse shift fork upwards and lift it out of the case.

13. Place the second/third shift fork in the second position. Remove the set screw. Rotate the shift rail 90 degrees.

14. Lift the interlock plug out of the case with a magnet.

15. Remove the expansion plug from the second/third shift rail by lightly tapping the end of the rail. Remove the second/third shift rail.

16. Remove the second/third shift rail detent plug and spring from detent bore.

17. Remove the input gear and shaft from the case.

18. Rotate the second/third shift fork upwards and remove from case.

19. Using caution, lift the output shaft assembly out through top of case.

20. Lift the reverse idler gear and thrust washers out of case. Remove the countershaft gear, thrust washer and dummy shaft from case.

21. Remove the snap-ring from the front of the output shaft. Slip the synchronizer and second gear off shaft.

22. Remove the second snap-ring from output shaft and remove the thrust washer, first gear and blocking ring.

23. Remove the third snap-ring from the output shaft. The first/reverse synchronizer hub is a press fit on the output shaft. Remove the synchronizer hub with an arbor press.

— CAUTION —
Do not attempt to remove or install the synchronizer hub by prying or hammering.

SHIFT LEVERS & SEALS

1. Remove shift levers from the shafts. Slip the levers out of case. Discard shaft sealing O-rings.

2. Lubricate and install new O-rings on shift shafts.

3. Install the shift shafts in the case and secure shift levers.

INPUT SHAFT BEARINGS

1. Remove the snap-ring securing the input shaft bearing. Using an arbor press, remove the bearing.

2. Press the input shaft bearing onto shaft using correct tool.

SYNCHRONIZERS

1. Scribe alignment marks on synchronizer hubs before disassembly. Remove each synchronizer hub from the synchronizer sleeves.

2. Separate the inserts and insert springs from the hubs.

— CAUTION —
Do not mix parts from the separate synchronizer assemblies.

3. Install the insert spring in the hub of the first/reverse synchronizer. Be sure that the spring covers all the insert grooves. Start the hub on the sleeve making certain that the scribed marks are properly aligned. Place the three inserts in the hub, small ends on the inside. Slide the sleeve and reverse gear onto hub.

4. Install one insert spring into a groove on the second/third synchronizer hub. Be sure that all three insert slots are covered. Align the scribed marks on the hub and sleeve and start the hub into the sleeve. Position the three inserts on the top of the retaining spring and push the assembly together. Install the remaining retainer spring so that the spring ends cover the same slots as the first spring. Do not stagger the springs. Place a synchronizer blocking ring on the ends of the synchronizer sleeve.

COUNTERSHAFT GEAR BEARINGS

1. Remove the dummy shaft, needle bearings and bearing retainers from the countershaft gear.

2. Coat the bore in each end of the countershaft gear with grease.

3. Hold the dummy shaft in the gear and install the needle bearings in the case.

4. Place the countershaft gear, dummy shaft, and needle bearings in the case.

5. Place the case in a vertical position. Align the gear bore and the thrust washers with the bores in the case and install the countershaft.

6. Place the case in a horizontal position. Check the countershaft gear end play with a feeler gauge. Clearance should be

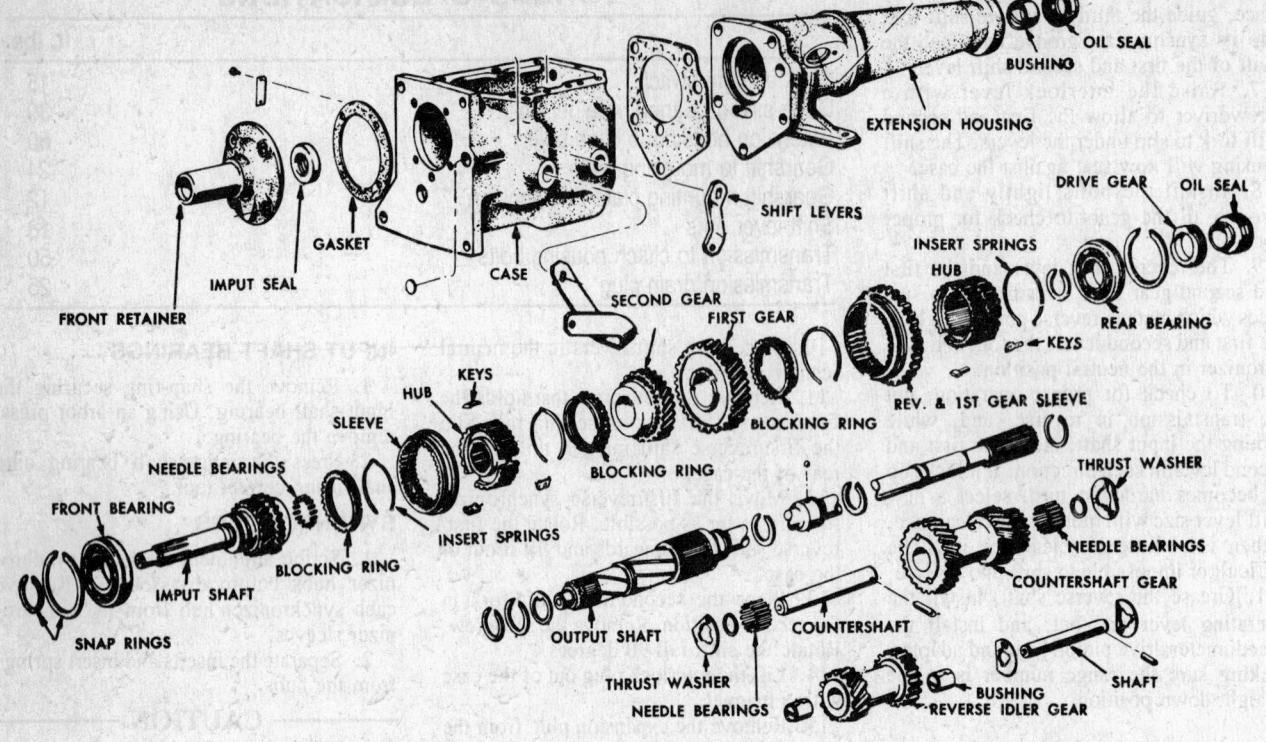

Ford 3.03 3-speed

between 0.004–0.018 in. If clearance does not come within specifications, replace the thrust washers.

7. Install the dummy shaft in the countershaft gear and leave the gear at the bottom of the transmission case.

ASSEMBLY

1. Cover the reverse idler gear thrust surfaces in the case with a thin film of lubricant, and install the two thrust washers in the case.

2. Install the reverse idler gear and shaft in the case. Align the case bore and thrust washers with gear bore and install the reverse idler shaft.

3. Measure the reverse idler gear end play with a feeler gauge; clearance should be between 0.004–0.018 in. If end play is not within specifications, replace the thrust washers. If clearance is correct, leave the reverse idler gear in case.

4. Lubricate the output shaft splines and machined surfaces with transmission oil.

5. The first/reverse synchronizer hub is a press fit on the output shaft. Hub must be installed in an arbor press. Install the synchronizer hub with the teeth-end of the gear facing towards the rear of the shaft.

— CAUTION —

Do not attempt to install the first/reverse synchronizer with a hammer.

6. Place the blocking ring on the tapered surface of the first gear.

7. Slide the first gear on the output shaft with the blocking ring toward the rear of the shaft. Rotate the gear as necessary to engage the three notches in the blocking ring with the synchronizer inserts. Install the thrust washer and snap-ring.

8. Slide the blocking ring onto the tapered surface of the second gear. Slide the second gear with blocking ring and the second/third synchronizer on the mainshaft. Be sure that the tapered surface of second gear is facing the front of the shaft and that the notches in the blocking ring engage the synchronizer inserts. Install the snap-ring and secure assembly.

9. Cover the core of the input shaft with a thin coat of grease.

— CAUTION —

A thick film of grease will plug lubricant holes and cause damage to bearings.

10. Install bearings. Install the input shaft through the front of the case and insert snap-ring in the bearing groove.

11. Install the output shaft assembly in the case. Position the second/third shift fork on the second/third synchronizer.

12. Place a detent plug spring and a plug in the case. Place the second/third synchronizer in the second gear position (toward the rear of the case). Align the fork and install the second/third shift rail. It will be necessary to depress the detent plug to install the shift rail in the bore. Move the rail forward until the detent plug enters the forward notch (second gear).

13. Secure the fork to the shift rail with a

set screw and place the synchronizer in neutral.

14. Install the interlock plug in the case.

15. Place the first/reverse synchronizer in the first gear position (towards the front of the case). Place the shift fork in the groove of the synchronizer. Rotate the fork into position and install the shift rail. Move the shift rail inward until the center notch (neutral) is aligned with the detent bore. Secure shift fork with set screw.

16. Install a new shift rail expansion plug in the front of the case.

17. Hold the input shaft and blocking ring in position and move the output shaft forward to seat the pilot in the roller bearings on the input gear.

18. Tap the input gear bearing into place while holding the output shaft. Install the front bearing retainer and gasket. Torque attaching bolts to specifications.

19. Install the large snap-ring on the rear bearing. Place the bearing on the output shaft with the snap-ring end toward the rear of the shaft. Press the bearing into place using a special tool. Secure the bearing to the shaft with the snap-ring.

20. Hold the speedometer drive gear lock ball in the detent and slide the speedometer drive gear into position. Secure with snap-ring.

21. Place the transmission in the vertical position. Working with a screwdriver through the drain hole in the bottom of the case, align the bore of the countershaft gear and the thrust washer with the bore in the case.

22. Working from the rear of the case,

push the dummy shaft out of the countershaft gear with the countershaft. Align the roll pin hole in the countershaft with the matching hole in the case. Drive the shaft into place and install the roll pin.

23. Position the new extension housing gasket on the case with sealer. Install the extension housing and torque to specification.

24. Place the transmission in gear and pour gear oil over entire gear train while rotating the input shaft.

25. Install the remaining detent plug and long spring in case.

26. Position cover gasket on case with sealer and install cover. Torque cover bolts to specifications.

27. Check operation of transmission in all gear positions.

4 Speed Overdrive

The Ford 4 speed overdrive transmission is fully synchronized in all forward gears. The 4 speed shift control is serviced as a unit and should not be disassembled. The lubricant capacity is 4.5 pints.

DISASSEMBLY

1. Remove retaining clips and flat washers from the shift rods at the levers.

2. Remove shift linkage control bracket attaching screws and remove shift linkage and control brackets.

3. Remove cover attaching screws. Then lift cover and gasket from the case. Remove the long spring that holds the detent plug in the case. Remove the plug with a magnet.

4. Remove extension housing attaching screws. Then, remove extension housing and gasket.

5. Remove input shaft bearing retainer attaching screws. Then, slide retainer from the input shaft.

6. Working a dummy shaft in from the front of the case, drive the countershaft out the rear of the case. Let the countergear assembly lie in the bottom of the case. Remove the set screw from the first/second shift fork. Slide the first/second shift rail out of the rear of the case. Use a magnet to remove the interlock detent from between the first/second and third/fourth shift rails.

7. Locate first/second speed gear shift lever in neutral. Locate third/fourth speed gear shift lever in third speed position.

NOTE: On overdrive transmissions, locate third/fourth speed gear shift-lever in the fourth speed position.

8. Remove the lockbolt that holds the third/fourth speed shift rail detent spring and plug in the left side of the case. Remove spring and plug with a magnet.

9. Remove the detent mechanism set screw from top of case. Then, remove detent spring and plug with a small magnet.

10. Remove attaching screw from the third/fourth shift fork. Tap lightly on the inner end of the shift rail to remove the expansion plug from front of case. Then, withdraw the third/fourth speed shift rail from the front. Do not lose the interlock pin from rail.

11. Remove attaching screw from the first and second speed shift fork. Slide the first/second shift rail from the rear of case.

12. Remove the interlock and detent plugs from the top of the case with a magnet.

13. Remove the snap-ring or disengage retainer that holds the speedometer drive gear to the output shaft, then remove speedometer gear drive ball.

14. Remove the snap-ring used to hold the output shaft bearing to the shaft. Pull out the output shaft bearing.

15. Remove the input shaft bearing snap-rings. Use a press to remove the input shaft bearing. Remove the input shaft and blocking ring from the front of the case.

16. Move output shaft to the right side of the case. Then, maneuver the forks to permit lifting them from the case.

17. Support the thrust washer and first-speed gear to prevent sliding from the shaft, then lift output shaft from the case.

18. Remove reverse gear shift fork attaching screw. Rotate the reverse shift rail 90°, then, slide the shift rail out the rear of the case. Lift out the reverse shift fork.

19. Remove the reverse detent plug and spring from the case with a magnet.

20. Using a dummy shaft, remove the reverse idler shaft from the case.

21. Lift reverse idler gear and thrust washers from the case. Be careful not to drop the bearing rollers or the dummy shaft from the gear.

22. Lift the countergear, thrust washers, rollers and dummy shaft assembly from the case.

23. Remove the next snap-ring from the front of the output shaft. Then, slide the third/fourth synchronizer blocking ring and the third speed gear from the shaft.

24. Remove the next snap-ring and the second speed gear thrust washer from the shaft. Slide the second speed gear and the blocking ring from the shaft.

25. Remove the snap-ring, then slide the first/second synchronizer, blocking ring and the first speed gear from the shaft.

26. Remove the thrust washer from rear of the shaft.

CAM & SHAFT SEALS

1. Remove attaching nut and washers from each shift lever, then remove the three levers.

2. Remove the three cams and shafts from inside the case.

3. Replace the old O-rings with new ones that have been well-lubricated.

4. Slide each cam and shaft into its respective bore in the transmission.

5. Install the levers and secure them with their respective washers and nuts.

SYNCHRONIZERS

1. Push the synchronizer hub from each synchronizer sleeve.

2. Separate the inserts and springs from the hubs. Do not mix parts of the first/second with parts of third/fourth synchronizers.

3. To assemble, position the hub in the sleeve. Be sure the alignment marks are properly indexed.

4. Place the three inserts into place on the hub. Install the insert springs so that the irregular surface (hump) is seated in one of the inserts. Do not stagger the springs.

COUNTERSHAFT GEAR

1. Dismantle the countershaft gear assembly.

2. Assemble the gear by coating each end of the countershaft gear bore with grease.

3. Install dummy shaft in the gear. Then install 21 bearing rollers and a retainer washer in each end of the gear.

REVERSE IDLER GEAR

1. Dismantle reverse idler gear.

2. Assemble reverse idler gear by coating the bore in each end of reverse idler gear with grease.

3. Hold the dummy shaft in the gear and install the 22 bearing rollers and the retainer washer into each end of the gear.

4. Install the reverse idler sliding gear on the splines of the reverse idler gear. Be

TORQUE SPECIFICATIONS

	ft. lbs.
Input shaft gear bearing retainer to transmission case	30–36
Transmission to flywheel housing	37–42
Transmission cover to transmission case	14–19
Speedometer cable retainer to transmission extension	3–4.5
Transmission extension to transmission case	42–50
Flywheel housing to engine	40–50
Gear shift lever to cam and shaft assembly lock nuts	18–23
U-Joint flange to output shaft	60–80
Filler plug	10–20
Shifter fork set screws	10–18
T.R.S. switch to case	15–20

sure the shift fork groove is toward the front.

INPUT SHAFT SEAL

1. Remove the seal from the input shaft bearing retainer.

2. Coat the sealing surface of a new seal with lubricant, then press the new seal into the input shaft bearing retainer.

ASSEMBLY

1. Grease the countershaft gear thrust surfaces in the case. Then, position a thrust washer at each end of the case.

2. Position the countershaft gear, dummy shaft, and roller bearings in the case.

3. Align the gear bore and thrust washers with the bores in the case. Install the countershaft.

4. With the case in a horizontal position, countershaft gear end-play should be from 0.004–0.018 in. Use thrust washers to obtain play within these limits.

5. After establishing correct endplay, place the dummy shaft in the countershaft gear and allow the gear assembly to remain on the bottom of the case.

6. Grease the reverse idler gear thrust surfaces in the case, and position the two thrust washers.

7. Position the reverse idler gear, sliding gear, dummy, etc., in place. Make sure that the shift fork groove in the sliding gear is toward the front.

8. Align the gear bore and thrust washers with the case bores and install the reverse idler shaft.

9. Reverse idler gear end-play should be 0.004–0.018 in. Use selective thrust washers to obtain play within these limits.

10. Position reverse gear shift rail detent spring and detent plug in the case. Hold the reverse shift fork in place on the reverse idler sliding gear and install the shift rail from the rear of the case. Lock the fork to the rail with the Allen head set screws.

11. Install the first/second synchronizer onto the output shaft. The first and reverse synchronizer hub are a press fit and should be installed with gear teeth facing the rear of the shaft.

NOTE: On overdrive transmissions, first and reverse synchronizer hub is a slip fit.

12. Place the blocking ring on second gear. Slide second speed gear onto the front of the shaft with the synchronizer coned surface toward the rear.

13. Install the second speed gear thrust washer and snap-ring.

14. Slide the fourth gear onto the shaft with the synchronizer coned surface front.

15. Place a blocking ring on the fourth gear.

16. Slide the third/fourth speed gear synchronizer onto the shaft. Be sure that the inserts in the synchronizer engage the notches in the blocking ring. Install the snap-ring onto the front of the output shaft.

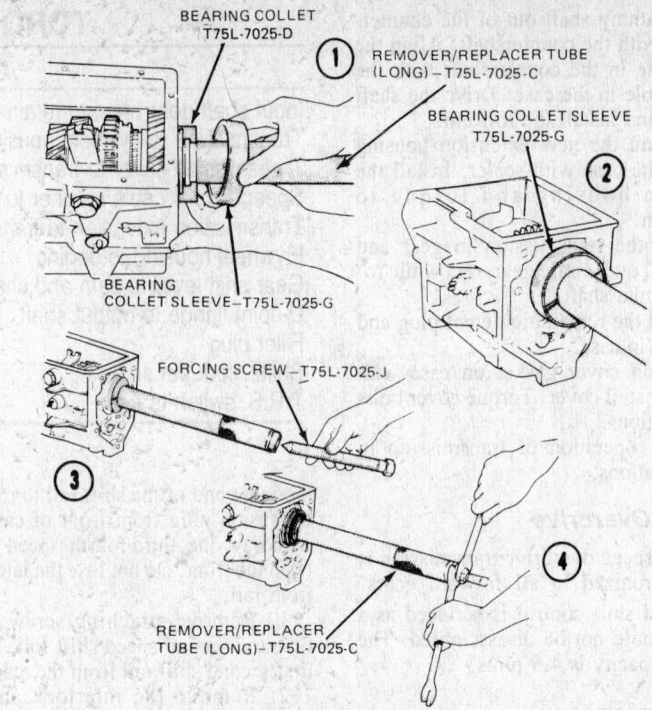

Removing output shaft bearing

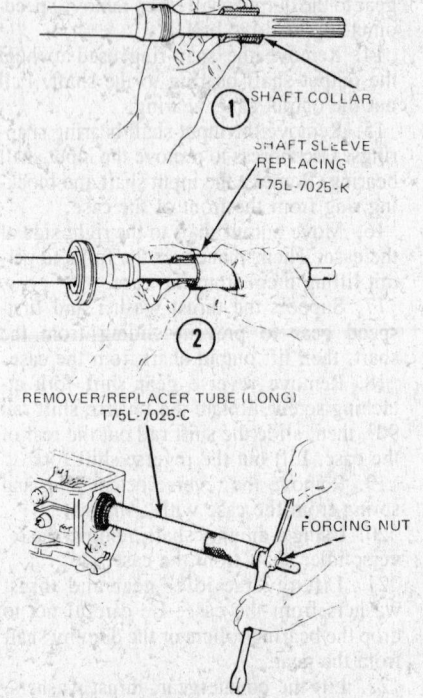

Installing output shaft bearing

17. Put the blocking ring on the first gear.

18. Slide the first gear onto the rear of the output shaft. Be sure that the inserts engage the notches in the blocking ring and that the shift fork groove is toward the rear.

19. Install heavy thrust washer onto the rear of the output shaft.

20. Lower the output shaft assembly into the case.

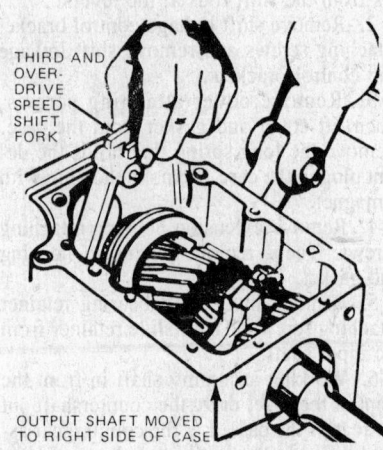

Removing shift fork from case

21. Position the first/second speed shift fork and the third/fourth speed shift fork in place on their respective gears. Rotate them into place.

22. Place a spring and detent plug in the detent bore. Place the reverse shift rail into neutral position.

23. Coat the third/fourth speed shift rail interlock pin (tapered ends) with grease, then position it in the shift rail.

24. Align the third/fourth speed shift fork with the shift rail bores and slide the shift rail into place. Be sure that the three detents are facing the outside of the case. Place the front synchronizer into fourth-speed position and install the set screw into the third/fourth speed shift fork. Move the synchronizer to neutral position. Install the third/fourth speed shift rail detent plug, spring

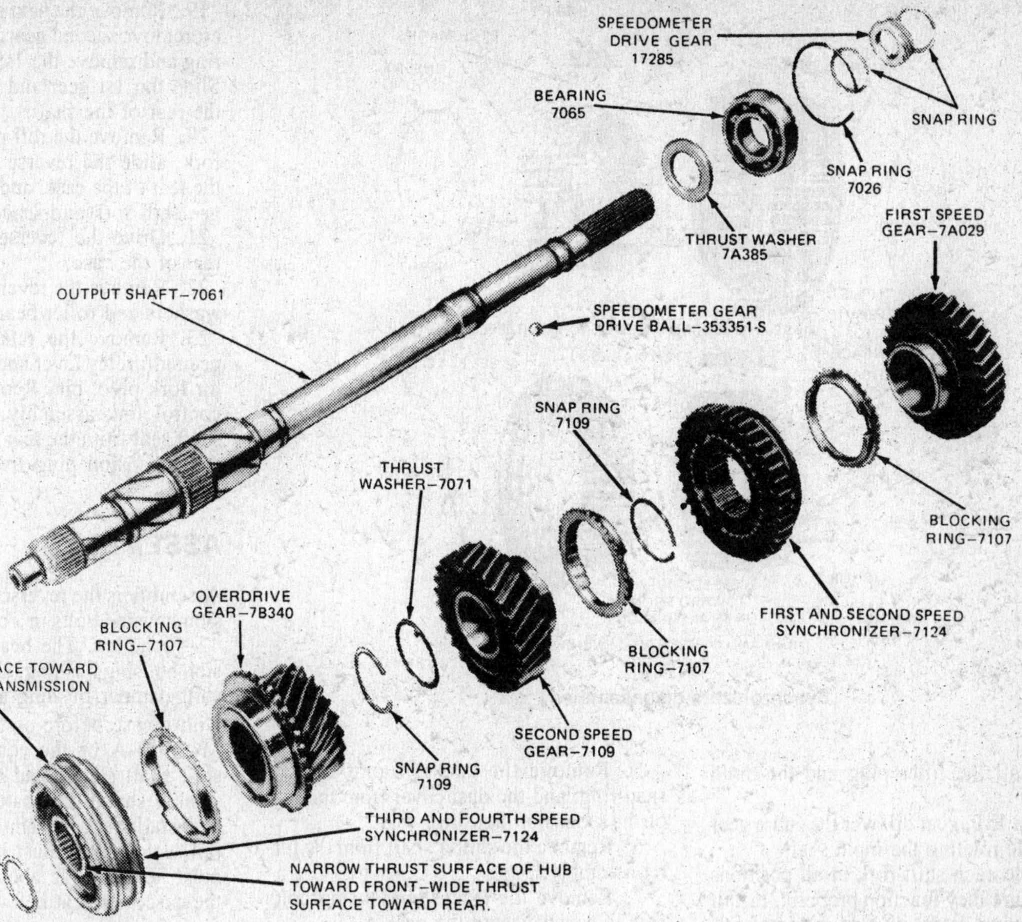

SPEEDOMETER
DRIVE GEAR
17285

BEARING
7065

SNAP RING

SNAP RING
7026

FIRST SPEED
GEAR—7A029

THRUST WASHER
7A385

SPEEDOMETER GEAR
DRIVE BALL—353351-S

OUTPUT SHAFT—7061

SNAP RING
7109

THRUST
WASHER—7071

BLOCKING
RING—7107

FIRST AND SECOND SPEED
SYNCHRONIZER—7124

OVERDRIVE
GEAR—7B340

BLOCKING
RING—7107

STEPPED SURFACE TOWARD
FRONT OF TRANSMISSION

BLOCKING
RING—7107

SNAP RING
7109

SECOND SPEED
GEAR—7109

SNAP RING
7109

THIRD AND FOURTH SPEED
SYNCHRONIZER—7124

NARROW THRUST SURFACE OF HUB
TOWARD FRONT—WIDE THRUST
SURFACE TOWARD REAR.

Output shaft

and bolt into the left side of the transmission case. Place the detent plug (tapered ends) in the detent bore.

25. Align first/second speed shift fork with the case bores and slide the shift rail into place. Lock the fork with the set screw.

26. Coat the input gear bore with a small amount of grease. Then install the 15 bearing rollers.

27. Put the blocking ring in the third/fourth synchronizer. Place the input shaft gear in the case. Be sure that the output shaft pilot enters the roller bearing of the input shaft gear.

28. With a new gasket on the input bearing retainer, dip attaching bolts in sealer, install bolts and torque to 30–36 ft. lbs.

29. Press on the output shaft bearing, then install the snap-ring to hold the bearing.

30. Position the speedometer gear drive ball in the output shaft and slide the speedometer drive gear into place. Secure gear with snap-ring.

31. Align the countershaft gear bore and thrust washers with the bore in the case. Install the countershaft.

32. With a new gasket in place, install and secure the extension housing. Dip the extension housing screws in sealer, then torque screws to 42–50 ft. lbs.

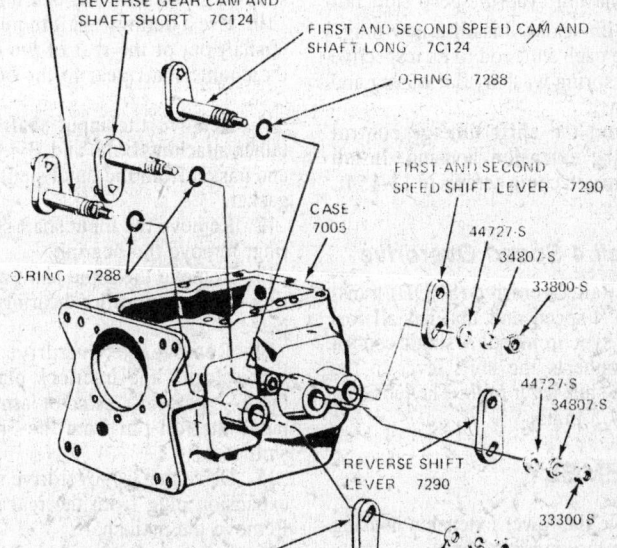

THIRD AND OVERDRIVE
CAM AND SHAFT
(SHORT) 7C124

REVERSE GEAR CAM AND
SHAFT SHORT 7C124

FIRST AND SECOND SPEED CAM AND
SHAFT LONG 7C124

O-RING 7288

FIRST AND SECOND
SPEED SHIFT LEVER 7290

CASE
7005

44727-S

34807-S

33800-S

O-RING 7288

44727-S

34807-S

REVERSE SHIFT
LEVER 7290

33300 S

THIRD AND OVERDRIVE
SHIFT LEVER 7285

44727-S

34807 S

33800 S

Cams, shafts and shift levers

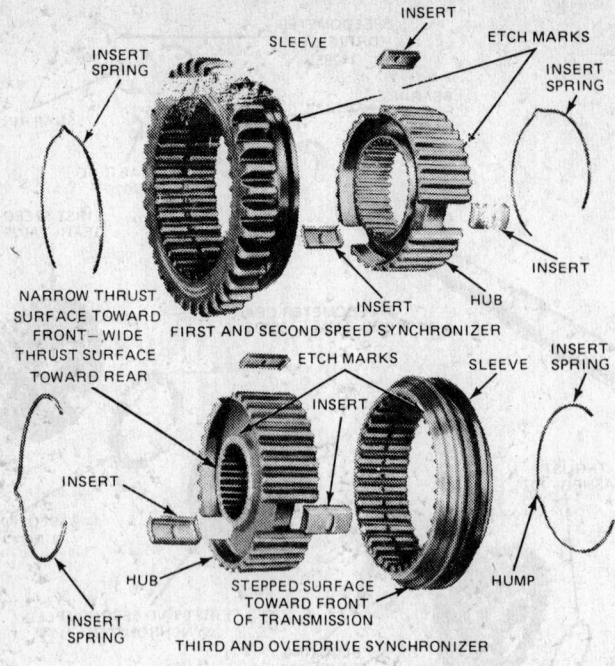

INSERT

SLEEVE

INSERT

ETCH MARKS

INSERT
SPRING

INSERT
SPRING

INSERT

NARROW THRUST
SURFACE TOWARD
FRONT—WIDE
THRUST SURFACE
TOWARD REAR

INSERT

HUB

FIRST AND SECOND SPEED SYNCHRONIZER

ETCH MARKS

INSERT

SLEEVE

INSERT
SPRING

INSERT

INSERT

HUB

STEPPED SURFACE
TOWARD FRONT
OF TRANSMISSION

HUMP

INSERT
SPRING

THIRD AND OVERDRIVE SYNCHRONIZER

Synchronizers disassembled

33. Install the filler plug and the drain plug.

34. Pour E.P. gear oil over the entire gear train while rotating the input shaft.

35. Place each shift fork in all positions to make sure they function properly. Install the remaining detent plug in the case, followed by the spring.

36. With a new cover gasket in place, install the cover. Dip attaching screws in sealer, then torque screws to 14–19 ft. lbs.

37. Coat the third/fourth speed shift rail plug bore with sealer. Install a new plug.

38. Secure each shift rod to its respective lever with a spring washer, flat washer and retaining pin.

39. Position the shift linkage control bracket to the extension housing. Install and torque the attaching screws to 12–15 ft. lbs.

Single Rail 4 Speed Overdrive

The Single Rail Overdrive (SROD) transmission is a 4-speed unit that has all forward speeds synchronized. A single control rod (rail) connects the shift lever to the transmission shift lever rails. The lubricant capacity is 4.5 pints.

DISASSEMBLY

1. Remove the lower extension housing bolt to drain the transmission.

2. Remove the cover screws; remove the cover and discard the gasket.

3. Remove the screw, detent spring and plug from the case; a magnetized rod will aid in removal.

4. Drive the roll pin from the shifter shaft.

5. Remove the backup lamp switch, snap-ring, and the dust cover from the rear of the extension housing.

6. Remove the shifter shaft from the turret assembly.

7. Remove the extension housing bolts and housing; discard the gasket.

8. Remove the speedometer gear snap-ring; slide the gear from the shaft and remove the drive ball.

9. Remove the output shaft bearing snap-ring. Remove the bearing.

10. Use a dummy shaft to push the countershaft out of the rear of the case. Lower the countershaft gear to the bottom of the case.

11. Remove the input shaft bearing retainer attaching bolts and slide the retainer and gasket from the input shaft; discard the gasket.

12. Remove the input shaft bearing snap-ring; remove the bearing.

13. Remove the input shaft and blocking ring (including roller bearings) from the case.

14. Remove the overdrive shift pawl, gear selector and interlock plate. Remove the 1–2 gearshift selector arm plate. Remove the roll pin from the 3rd–overdrive shift fork.

15. Drive the 3rd–overdrive shift rail and expansion plug from the rear of the case. Remove the mainshaft.

16. Remove the 1st and 2nd gear shift fork; remove the 3rd–overdrive shift fork.

17. Remove the countershaft gear and thrust washers from the case.

18. Remove the snap-ring from the front of the output shaft. Slide the 3rd gear and O.D. synchronizer, blocking ring, and gear from the shaft.

19. Remove the next snap-ring and washer; remove second gear. Remove next snap-ring and remove the 1st–2nd synchronizer. Slide the 1st gear and blocking ring from the rear of the shaft.

20. Remove the roll pin from the reverse fork, slide the reverse shifter rail through the rear of the case, and remove the reverse gearshift fork and spacer.

21. Drive the reverse gear shaft out the rear of the case.

22. Remove the reverse idler gear, thrust washers and roller bearings.

23. Remove the retaining clip, reverse gearshift relay lever and reverse gear selector fork pivot pin. Remove the O.D. shift control link assembly. Remove the shift shaft seal from the rear of the case; remove the expansion plug from the front of the case.

ASSEMBLY

Assembly is the reverse. Tighten the extension housing bolts in a criss-cross pattern to 42–50 ft. lbs. The bearing rollers, extension housing bushing, shifter shaft and gear shift damper bushing are to be lubricated with grease before assembly (Ford ESW–M1C109–A or the equivalent). The gear shift shaft sleeve and the turret cover assembly should be coated with sealer prior to installation. The intermediate and high rail welch plug must be seated firmly; it must not protrude above the front face of the case, nor seat below 0.6 in. below the front face.

With the 1st gear thrust washer clamped tightly against the output shaft shoulder, 1st gear endplay must be 0.005–0.024 in. 2nd gear endplay must be 0.003–0.021 in. O.D. endplay must be 0.009–0.023 in. Countershaft gear endplay, checked after installation between the thrust washers, must be 0.004–0.018 in.

When the gearshift selector arm plate is seated in the 1st–2nd shift fork plate slot, the shifter shaft must pass freely through the bore without binding.

Ranger & Bronco II

4 Speed (Gasoline Engines)
DISASSEMBLY

1. Remove the nuts attaching the bell housing to the transmission case. Remove the bell housing and gasket.

2. Remove the drain plug and drain lubricant from the transmission. Clean the metal filings from the magnet of the drain plug (if necessary). Install the drain plug.

3. Place transmission in neutral.

4. Remove the four 12mm bolts attaching the gearshift lever retainer to the extension housing. Remove the gearshift lever retainer and gasket.

5. Remove the six 14mm bolts attaching the extension housing to the transmission case.

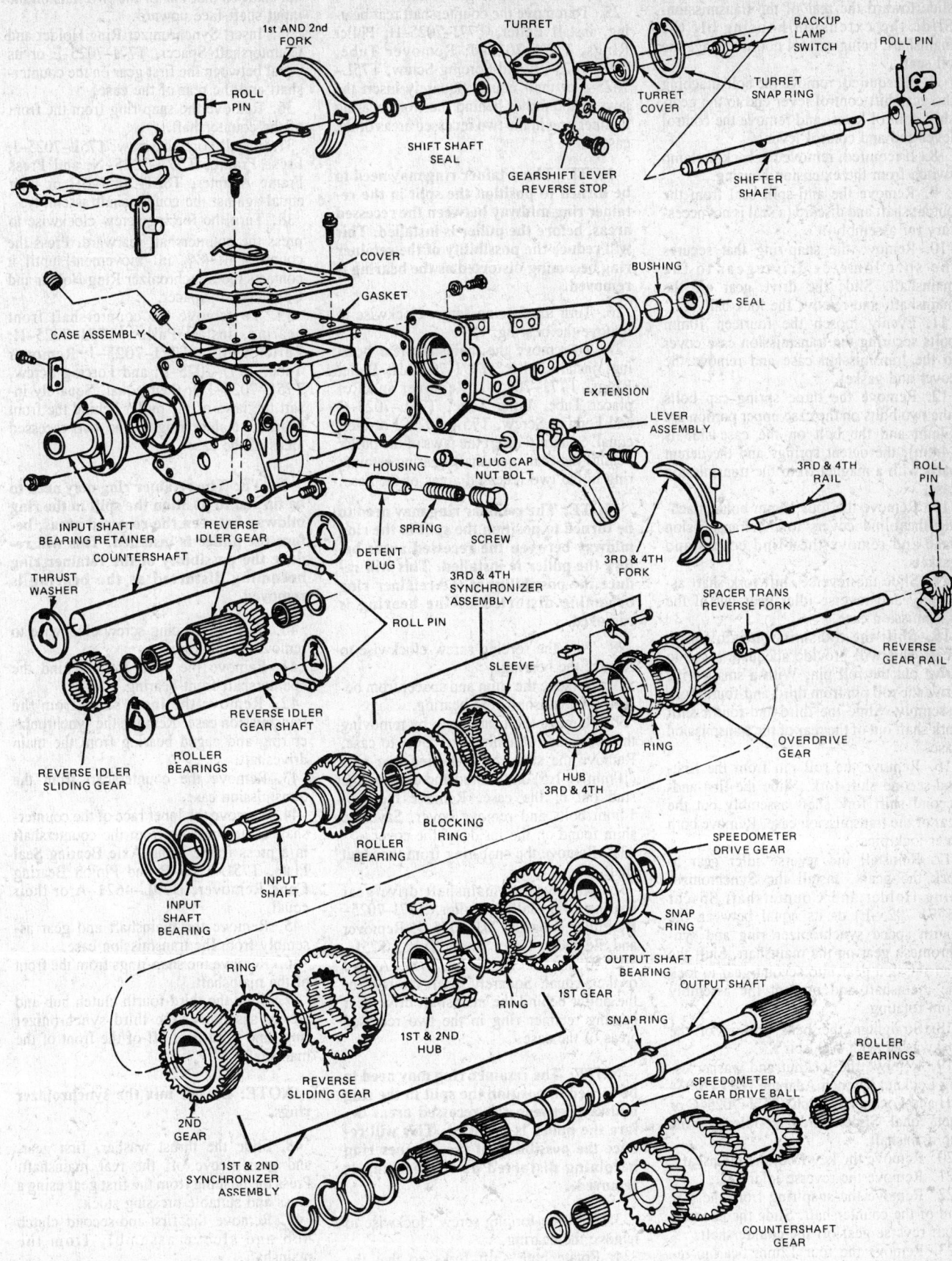

1st AND 2nd FORK

ROLL PIN

SHIFT SHAFT SEAL

TURRET

TURRET COVER

GEARSHIFT LEVER REVERSE STOP

BACKUP LAMP SWITCH

TURRET SNAP RING

ROLL PIN

SHIFTER SHAFT

COVER

GASKET

BUSHING

SEAL

CASE ASSEMBLY

EXTENSION

HOUSING

LEVER ASSEMBLY

PLUG CAP NUT BOLT

3RD & 4TH RAIL

ROLL PIN

INPUT SHAFT BEARING RETAINER

COUNTERSHAFT

REVERSE IDLER GEAR

SPRING

SCREW

DETENT PLUG

3RD & 4TH FORK

3RD & 4TH SYNCHRONIZER ASSEMBLY

SPACER TRANS REVERSE FORK

REVERSE GEAR RAIL

THRUST WASHER

ROLL PIN

SLEEVE

RING

OVERDRIVE GEAR

REVERSE IDLER GEAR SHAFT

REVERSE IDLER SLIDING GEAR

ROLLER BEARINGS

HUB 3RD & 4TH

BLOCKING RING

ROLLER BEARING

INPUT SHAFT

SPEEDOMETER DRIVE GEAR

INPUT SHAFT BEARING

SNAP RING

RING

OUTPUT SHAFT BEARING

RING

1ST GEAR

OUTPUT SHAFT

ROLLER BEARINGS

SNAP RING

1ST & 2ND HUB

REVERSE SLIDING GEAR

SPEEDOMETER GEAR DRIVE BALL

2ND GEAR

1ST & 2ND SYNCHRONIZER ASSEMBLY

COUNTERSHAFT GEAR

Single rail overdrive transmission

6. Raise the control lever to the left and slide toward the rear of the transmission. Slide the extension housing off the mainshaft, being careful not to damage the oil seal.

7. If required, remove the bolt attaching the gearshift control lever end to the gearshift control lever, and remove the control lever end and control lever.

8. If required, remove the back-up lamp switch from the extension housing.

9. Remove the anti-spill seal from the output shaft and discard (a seal is not necessary for assembly).

10. Remove the snap-ring that secures the speedometer drive gear to the mainshaft. Slide the drive gear off the mainshaft, and remove the lock ball.

11. Evenly loosen the fourteen 10mm bolts securing the transmission case cover to the transmission case and remove the cover and gasket.

12. Remove the three spring cap bolts (the two bolts on the case upper portion are 17mm and the bolt on the case side is 14mm), the detent springs and the detent balls with a magnet from the transmission case.

13. Remove the four 10mm bolts attaching the blind covers to the transmission case and remove the blind covers and gaskets.

14. Slide the reverse shift fork shaft assembly and reverse idler gear out of the transmission case.

15. Shift the transmission into fourth gear. This will provide adequate space to drive out the roll pin. With a small drift, drive the roll pin from third-and-fourth fork assembly. Slide the third-and-fourth shift fork shaft out of the rear of the transmission case.

16. Remove the roll pin from the first-and-second shift fork. Slide the first-and-second shift fork shaft assembly out the rear of the transmission case. Remove both inter-lock pins.

17. Reinstall the reverse idler gear to lock the gears. Install the Synchronizer Ring Holder and Countershaft Spacer (T77J-7025-E) or its equal between the fourth speed synchronizer ring and synchromesh gear on the mainshaft. Shift the transmission gear into second gear to lock the mainshaft and prevent the assembly from rotating.

18. Straighten the bent portion of the lockwasher with a chisel.

19. Remove the locknut and washer using Locknut Wrench Adapter T82T-7003-CH and Locknut Wrench, T77J-7025-C or their equal. Slide the reverse-idler gear off the mainshaft.

20. Remove the key from the mainshaft.

21. Remove the reverse idler gear.

22. Remove the snap-ring from the rear end of the countershaft. Slide the countershaft reverse gear off the countershaft.

23. Remove the four 12mm bearing retainer attaching bolts.

24. Remove the bearing retainer together with the reverse idler gear shaft.

25. To remove the countershaft rear bearing, install Puller, T77J-7025-H; Puller Rings, T77J-7025-J; Remover Tube, T77J-7025-B; and Forcing Screw, T75L-7025-J or their equal. Squarely insert the jaws of the puller behind the front bearing retainer ring in the two recessed areas of the case.

NOTE: The retainer ring may need to be turned to position the split in the retainer ring midway between the recessed areas, before the puller is installed. This will reduce the possibility of the retainer ring becoming distorted as the bearing is removed.

26. Turn the forcing screw clockwise to remove the bearing.

27. To remove the mainshaft rear bearing, install Puller, T77J-7025-H; Puller Rings, T77J-7025-J; Remover and Replacer Tube, Long Tube, T75L-7025-C and Forcing Screw, T75L-7025-J or their equal. Squarely insert the jaws of the puller behind the rear mainshaft bearing retainer ring in the two recessed areas of the case.

NOTE: The retainer ring may need to be turned to position the split in the ring midway between the recessed areas before the puller is installed. This will reduce the possibility of the retainer ring becoming distorted as the bearing is removed.

28. Turn the forcing screw clockwise to remove the bearing.

29. Remove the shim and spacer from behind the mainshaft rear bearing.

30. Remove the front cover by removing the four studs attaching the cover to case. Remove the studs by installing two nuts (10mm × 1.5) on the stud and drawing the stud out of the case. Remove the four 14mm bolts and remove cover. Save the shim found on the inside of the cover.

31. Remove the snap-ring from the input shaft.

32. Remove the mainshaft drive gear bearing by installing Puller, T77J-7025-H; Puller Rings, T77J-7025-J; Remover and Replacer Tube, Short Tube, T75L-7025-B; and Forcing Screw, T75L-7025-J or their equal. Squarely insert the jaws of the puller behind the mainshaft drive gear bearing retainer ring in the two recessed areas of the case.

NOTE: The retainer ring may need to be turned to position the split in the ring midway between the recessed areas before the puller is installed. This will reduce the possibility of the retainer ring becoming distorted as the bearing is removed.

33. Turn the forcing screw clockwise to remove the bearing.

34. Rotate both shift forks so that the main geartrain will fall to the bottom of the case. Remove the shift forks. Rotate the input shaft so that one of the two flats on the input shaft face upward.

35. Insert Synchronizer Ring Holder and Countershaft Spacer, T77J-7025-E or its equal between the first gear on the countershaft and the rear of the case.

36. Remove the snap-ring from the front of the countershaft.

37. Install Forcing Screw, T75L-7025-J; Press Frame, T77J-7025-N; and Press Frame Adapter, T82T-7003-BH or their equal against the countershaft assembly.

38. Turn the forcing screw clockwise to press the countershaft rearward. Press the countershaft ($\frac{3}{16}$ in. movement) until it contacts the Synchronizer Ring Holder and Countershaft Spacer.

39. To remove the countershaft front bearing, install Puller, T77J-7025-H; Puller Rings, T77J-7025-J; Remover Tube, T77J-7025-B; and Forcing Screw, T75L-7025-J or their equal. Squarely insert the jaws of the puller behind the front bearing retainer ring in the two recessed areas of the case.

NOTE: The retainer ring may need to be turned to position the split in the ring midway between the recessed areas, before the puller is installed. This will reduce the possibility of the retainer ring becoming distorted as the bearing is removed.

40. Turn the forcing screw clockwise to remove the bearing.

41. Remove the shim from behind the countershaft front bearing.

42. Remove the input shaft from the transmission case. Remove the synchronizer ring and caged bearing from the main driveshaft.

43. Remove the countershaft from the transmission case.

44. Remove the inner race of the countershaft center bearing from the countershaft in a press frame using Axle Bearing Seal Plate, T75L-1165-B and Pinion Bearing Cone Remover, D79L-4621-A or their equal.

45. Remove the mainshaft and gear assembly from the transmission case.

46. Remove the snap-rings from the front of the mainshaft.

47. Slide the third-fourth clutch hub and sleeve assembly, the third synchronizer ring, and third gear off of the front of the mainshaft.

NOTE: Do not mix the synchronizer rings.

48. Slide the thrust washer, first gear, and gear sleeve off the rear mainshaft. Press the bushing from the first gear using a press and suitable pressing stock.

49. Remove the first-and-second clutch hub and sleeve assembly from the mainshaft.

50. Clean and inspect transmission case, gears, bearings, and shafts.

ASSEMBLY

Before beginning the assembly procedure, three measurements must be performed: Mainshaft Thrust Play, Countershaft Thrust Play and Mainshaft Bearing Clearance.

Mainshaft Thrust Play

Check the mainshaft thrust play by measuring the depth of the mainshaft bearing bore in the transmission rear cage by using a depth micrometer (D80P-4201-A) or its equal. Then measure the mainshaft rear bearing height. The difference between the two measurements indicates the required thickness of the adjusting shim. The standard thrust play is 0 to 0.0039 in. Adjusting shims are available in 0.0039 in. and 0.0118 in. sizes.

Countershaft Thrust Play

Check the countershaft thrust play by measuring the depth of the countershaft front bearing bore in the transmission case by using a depth micrometer (D80P-4201-A) or its equal. Then measure the countershaft front bearing height. The difference between the two measurements indicates the required thickness of the adjusting shims. The standard thrust play is 0 to 0.0039 in. Adjusting shims are available in 0.0039 in. and 0.0118 in. sizes.

Mainshaft Bearing Clearance

Check the main driveshaft bearing clearance by measuring the depth of the bearing bore in the clutch adapter plate with a depth micrometer, D80P-4201-A or its equal. Make sure the micrometer is on the second step of the plate. Measure the bearing height. The difference between the two measurements indicates the required adjusting shim thickness. The standard clearance is 0 to 0.0039 in. If an adjusting shim is required, select one to bring the clearance to within specifications.

1. Assemble the first-and-second synchromesh mechanism by installing the clutch hub to the sleeve, placing the three synchronizer keys into the clutch hub key slots and installing the key springs to the clutch hub.

NOTE: When installing the key springs, the open end tab of the springs should be inserted into the hub holes with the springs turned in the same direction. This will keep the spring tension on each key uniform.

2. Assemble the third-and-fourth synchromesh mechanisms in the same manner as first-and-second synchromesh mechanism.
3. Place the synchronizer ring on the third gear and slide the third gear to the front of the mainshaft with the synchronizer ring toward the front.
4. Slide the third-and-fourth clutch hub and sleeve assembly to the front of the mainshaft, making sure that the three synchronizer keys in the synchromesh mechanism engage the notches in the synchronizer ring.

NOTE: The direction of the third-and-fourth clutch hub and sleeve assembly should be as shown.

5. Install the snap-ring to the front of the mainshaft.
6. Place the synchronizer ring on the second gear and slide the second gear to the mainshaft with the synchronizer ring toward the rear of the shaft.
7. Slide the first-and-second clutch hub and sleeve assembly to the mainshaft with the oil grooves of the clutch hub toward the front of the mainshaft. Make sure that the three synchronizer keys in the synchromesh mechanism engage the notches in the second synchronizer ring.
8. Insert the first gear sleeve in the mainshaft.
9. Press the bushing in the first gear using a press and suitable press stock.
10. Place the synchronizer ring on the first gear and slide the first gear onto the mainshaft with the synchronizer ring facing the front of the shaft. Rotate the first gear as necessary to engage the three notches in the synchronizer ring with the synchronizer keys.
11. Install the original thrust washer on the mainshaft.
12. Position the mainshaft and gears assembly in the case.
13. Position the caged bearing in the front end of the mainshaft.
14. Place the synchronizer ring on the input shaft (fourth gear), and install the input shaft to the front end of the mainshaft, making sure that the three synchronizer keys in the third-and-fourth synchromesh mechanism engage the notches in the synchronizer ring.
15. Position the first-and-second shift fork and third-and-fourth shift fork in the groove of the clutch hub and sleeve assembly.
16. Press the inner race of the countershaft rear bearing onto the countershaft using Center Bearing Replacer, T77J-7025-K or its equal.
17. Position the countershaft gear in the case, making sure that the countershaft gear engages each gear of the mainshaft assembly.
18. Install the correct shim in the mainshaft rear bearing bore as determined in the Mainshaft Thrust Play Measurement.
19. Position the main drive gear bearing and the mainshaft rear bearing into the proper bearing bores. Be sure the synchronizer and shifter forks have not been moved out of position.
20. Install the Dummy Bearing Replacer, T75L-7025-Q; Mainshaft Front Bearing Replacer, T82T-7003-DH; Replacer Tube, T77J-7025-M; Press Frame Adapter, T82T-7003-BH; and Press Frame, T77J-7025-N or their equal on the case.

Position the Synchronizer Ring Holder and Countershaft Spacer, T77J-7025-E or their equal between the mainshaft drive gear and synchronizer ring. Turn the forcing screw on the press frame until both bearings are properly seated.

21. Install the main drive gear bearing snap-ring.
22. Place the correct shim in the countershaft front bearing bore as determined by the Countershaft Thrust Play Measurement.
23. Position the countershaft front and rear bearings in the bores and install the tools. Turn the forcing screw until the bearing is properly seated. Use the rear bearing as a pilot.
24. Install the snap-ring to secure the countershaft front bearing.
25. Install the bearing retainer together with the reverse idler gear shaft to the transmission case and tighten the four 12mm attaching bolts.
26. Slide the counter reverse gear onto the countershaft with the chamfer to the rear. Install the snap-ring to secure the counter reverse gear.
27. Install the key on the mainshaft.
28. Slide the reverse gear and lockwasher (tab facing outward) onto the mainshaft (chamfer on teeth should be to rear). Install a new locknut and hand tighten.
29. Shift into second gear and reverse gear to lock rotation of the mainshaft. Tighten the locknut to 145-203 ft. lbs. torque using the Locknut Wrench (T77J-7025-C) and Locknut Adapter, T82T-7003-CH or their equal.
30. Place the fourth-and-third clutch sleeve in third gear using Synchronizer Ring Holder and Countershaft Spacer, T77J-7025-E or its equal.
31. Check the clearance between the synchronizer key and the exposed edge of the synchronizer ring with a feeler gauge. If the measurement is greater than 0.079 in., the synchronizer key can pop out of position. To correct this, change the thrust washer (selective fit) between the mainshaft rear bearing and the first gear. Available thrust washer sizes are 0.098, 0.118, and 0.138.
32. Check the clearance again with a feeler gauge. If the clearance is within specifications, bend the tab of the lockwasher.
33. Slide the first-and-second shift fork shaft assembly into the case (from rear of case). Install the roll pin. Secure the first-and-second shift fork to the fork shaft by staking the roll pin.

NOTE: Be sure to use a new roll pin.

34. Insert the inter-lock pin into the transmission using the lockout pin replacer tool.
35. Slide the third-and-fourth shift fork shaft assembly into the case (from rear of case). Secure the third-and-fourth shift fork to the fork shaft by staking the roll pin. Place transmission in neutral.

NOTE: Be sure to use a new roll pin.

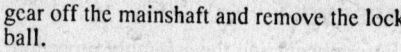

36. Insert the inter-lock pin into the transmission.

37. Slide the reverse fork shaft assembly and reverse idler gear into the transmission case from the rear of the case with the gear chamfer forward. Secure the reverse shift fork to the fork shaft by staking the roll pin.

NOTE: Be sure to use a new roll pin.

38. Position the three detent balls and three springs into the case. Place copper washers on the top two bolts and install the three spring cap bolts.

39. Install the two blind covers and gaskets. Tighten the 10mm attaching bolts.

40. Install the lock ball, speedometer drive gear, and snap-ring onto the mainshaft.

41. Apply a thin coat of sealing agent, Gasket Maker, E2AZ-19562-A (ESE-M4G234-A2) or equivalent to the contacting surfaces of the transmission case and extension housing.

42. Position the extension housing with the gearshift control lever end laid down to the left as far as it will go. Tighten the four 14mm attaching bolts.

NOTE: The lower two bolts must be coated with Loctite or equivalent.

43. If removed, insert the speedometer driven gear assembly to the extension housing and secure it with the bolt.

44. Check to ensure the gearshift control lever operates properly.

45. Install the transmission case cover gasket and cover with drain plug to rear. Install and tighten the fourteen 10mm attaching bolts.

46. Position the gasket and gearshift lever retainer to the extension housing, and tighten the four attaching bolts.

47. Install the correct size shim on the second step of the clutch adapter plate as determined by the Mainshaft Bearing Clearance Measurement.

48. Coat the clutch adapter plate with sealer, Gasket Maker, E2AZ-19562-A (ESE-M4G234-A2) or equivalent. Install the clutch adapter plate to the transmission case and tighten the four bolts and four studs.

49. Remove the filler plug and install 3.0 pints of Ford Manual Transmission Lube, D8DZ-19C547-A (ESP-M2C83-C) or equivalent. Reinstall filler plug and tighten to 18-29 ft. lbs.

4 Speed (Diesel Engines)
DISASSEMBLY

1. If not already drained, remove the drain plug and drain the transmission fluid into a suitable container. Remove the fork and release bearing from the clutch housing.

2. Install transmission in Holding Fixture T57L-500-B. Remove six bolts (12mm) attaching the front cover to the transmission case and remove the front cover shim and gasket.

3. Remove front cover oil seal.

4. Remove the input shaft snap-ring.

5. Remove outer snap-ring on input shaft bearing. Install Bearing Collet Tool T75L-7025-E or its equal on main input shaft front bearing, Remover Tube T75L-7025-B and Forcing Screw T75L-7025-J or their equal. Slide Bearing Collet Sleeve T75L-7025-G or its equal over remover tube and bearing collet, and turn forcing screw to remove input shaft bearing.

6. Remove the eight bolts (12mm) attaching the extension housing to the transmission case. Slide the extension housing off the mainshaft, with the control lever end laid down and to the left as far as it will go.

7. Remove the bolt (10mm) attaching the control lever end to the control rod and remove the control lever end and rod from the extension housing.

8. Remove the speedometer driven gear assembly from the extension housing.

9. Remove the back-up lamp switch and neutral sensing switch.

10. Remove the snap-ring that secures the speedometer drive gear on the mainshaft. Slide the speedometer drive gear off the mainshaft and remove the lock ball.

11. Install Bearing Pusher Tool T83T-7111-A or its equal over countershaft front bearing. Turn forcing screw to force countershaft, together with the countershaft front bearing, from the transmission housing.

12. Slide Bearing Holder and Gear Shaft Assembly from the transmission housing.

13. Remove three spring cap bolts, three springs and shift locking balls.

NOTE: Reverse spring is shortest. Take care, the spring-loaded lower ball will pop out.

14. Remove the reverse shift rod and shift fork assembly and reverse gear from the bearing housing.

15. Remove roll pins fixing shift forks to the rods. Push each of the shift rods rearward through the fork and bearing housing and remove the shift rods and forks.

NOTE: Mark 3rd-4th and 1st-2nd shift forks before removal to simplify installation.

16. Remove the lower reverse shift rod locking ball and spring, and the interlock pins from the bearing housing.

17. Straighten the tab of the lockwasher. Lock transmission synchronizers into any two gears and remove the mainshaft lock nut using Adapter Tool T83T-7025-A and Tool Shaft T77J-7025-C or its equal.

18. Remove the snap-ring from the rear end of the countershaft and slide off the counter reverse gear.

19. Remove the five bearing cover bolts (12mm) and cover, and the reverse idler gear shaft from the bearing housing.

20. With a soft hammer, tap the rear end of the mainshaft and countershaft in turn, being careful not to damage the shafts, and remove these shafts from the bearing housing.

21. Carefully separate the input shaft and caged needle roller bearing from the mainshaft.

22. Remove rear countershaft bearing from the bearing housing using Remover Tube Tool T77J-7025-B or its equal.

23. Remove rear mainshaft bearing from the bearing housing using Bearing Remover Tool T77F-4222-A and Remover Tube Tool T77J-7025-B or its equal.

24. Remove the thrust washer, first gear, sleeve and synchronizer ring from the rear of the mainshaft.

25. Using snap-ring pliers, remove the snap-ring from the front of the mainshaft.

26. Using a press and Remover Tool T71P-4621-B or its equal, remove the third-and-fourth clutch hub, sleeve, synchronizer ring and third gear from the front of the mainshaft.

27. Using a press and Remover Tool T71P-4621-B or its equal, remove the first-and-second clutch hub and sleeve assembly synchronizer ring, and second gear

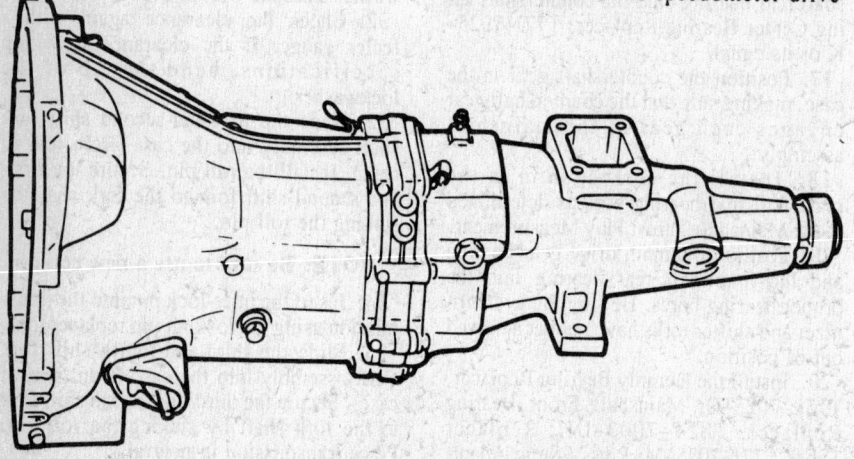

Four speed manual (diesel engines)

from the rear of the mainshaft in the same manner as described in the previous step.

28. Press front bearing from countershaft using Remover Tool D79L–4621–A or T71P–4621–B or its equal and a suitable stock piece.

29. Perform cleaning and inspection procedures described in a separate section of this manual.

ASSEMBLY

1. Assemble the third-and-fourth clutch by installing the clutch hub and synchronizer into the sleeve, placing the three keys into the clutch hub slots and installing the springs onto the hub.

NOTE: When installing the key springs, the open end tab of the springs should be inserted into the hub holes. This will keep the spring tension on each key uniform.

2. Assemble the first-and-second clutch hub and sleeve in the same manner as described in Step 1 above.

3. Install the third gear and synchronizer ring onto the front section of the mainshaft.

4. Install the third-and-fourth clutch hub assembly onto the mainshaft by using a press. Hold assembly together and slowly press into place.

NOTE: Make sure the clutch hub assembly is facing in the correct direction.

5. Fit the snap-ring on the mainshaft.

6. Install the second gear, synchronizer ring onto the rear section of the mainshaft.

7. Install the first-and-second clutch hub assembly onto the mainshaft by using a press.

8. Install the synchronizer ring, first gear with sleeve, and thrust washer onto the mainshaft.

9. Install the input shaft and the needle roller bearing to the mainshaft.

10. Check the countershaft rear bearing clearance. Measure the depth of the countershaft bearing bore in the bearing housing using a depth micrometer (D80P–4201–A or equivalent). Then, measure the countershaft bearing height. The difference between the two measurements indicates the required thickness of the adjusting shim. The clearance should be less than 0.0039 in. The adjusting shims are available in the following thickness: 0.0039 in. and 0.0118 in.

11. Check the mainshaft bearing clearance in the same manner as for the countershaft rear bearing clearance. The clearance should be less than 0.0039 in. The adjusting shims are available in the following thickness: 0.0039 in. and 0.0118 in.

12. Position proper shim on countershaft rear bearing and press into bearing housing using Installer Tool T77J–7025–B.

13. Position proper shim on mainshaft bearing and press into bearing housing using Installer Tool T77J–7025–K or its equal.

14. Position front bearing on countershaft and press into place using Bearing Replacer Tool T71P–7025–A or its equal.

15. Mesh countershaft and mainshaft assembly and position the two on the bearing housing. Make certain that thrust washer is installed on mainshaft assembly at the rear of the first gear.

16. While holding mainshaft assembly in place, press countershaft assembly into bearing housing using Replacer Tool T71P–7025–A or its equal to hold rear countershaft bearing in housing.

17. Install the bearing cover and reverse idle gear shaft to the bearing housing. The cover must be seated in the groove on the idle gear shaft.

18. Install the reverse gear with the key onto the mainshaft. Install the lock nut on the mainshaft and hand tighten.

NOTE: When installing the mainshaft reverse gear and the countershaft reverse gear, both gears should be fitted so that the chamfer on the teeth faces rearward.

19. Install the countershaft reverse gear and secure it with the snap-ring.

NOTE: After installing reverse gears, lock transmission in any two gears.

20. Insert the short spring and locking ball into the reverse bore of the bearing housing.

21. While holding down the ball with a punch or other suitable tool, install the reverse shift rod and shift lever assembly with the reverse idle gear at the same time.

22. Using the dummy shift rails (Tool Number T72J–7280 or its equal), install each shift fork rod and interlock pins.

23. Install the first-and-second shift fork and third-and-fourth shift fork to their respective clutch sleeves.

24. Align the roll pin holes of each shift fork and rod. Install the new roll pins.

NOTE: When assembling the shift fork and control end, a new roll pin should be installed with a pin slit positioned in the direction of the shift rod axis.

25. Install the shift locking balls and springs into their respective positions and install the spring cap bolt.

NOTE: The short spring and ball are installed in the reverse bore.

26. Apply a thin coat of Silicone Sealer D6AZ–19562–B, or equivalent, on both contacting surfaces of the bearing housing.

27. Install the bearing housing assembly to the transmission case.

28. Temporarily attach the bearing housing to the transmission with two top and two bottom bolts and tightened extension housing mounting bolts to position the countershaft front bearing in the bore.

NOTE: If necessary, remove plugs from bell housing shift rod bores to align shift rods. After installation of bearing housing assembly is complete, reinstall plugs using a silicone sealer (D6AZ–19562–B or equivalent).

29. Tighten the mainshaft locknut to 116–174 ft. lbs. using Adapter T83T–7025–A and Tool Shaft T77J–7025–C or equivalent.

30. Bend a tab on the lockwasher using Staking Tool T77J–7025–F or equivalent.

31. Install the speedometer drive gear with the lock ball onto the mainshaft and secure it with a snap-ring.

32. With the outer snap-ring in place on the main driveshaft front bearing, place bearing, shim 389117–2S, and Adapter Tool T75L–7025–N or equivalent over the input shaft.

33. Thread the Replacer Shaft T75L–7025–K or equal onto the Adapter Tool. Install the Replacer Tube T75L–7025–B or equal over the Replacer Shaft and install the nut and washer on the forcing screw.

34. Slowly tighten the nut until the adapter is secure on the input shaft. Make certain that all tools are aligned.

35. Tighten the nut on the forcing screw until the bearing outer snap-ring is seating against the housing. Remove the installation tools.

36. Install the input shaft snap-ring.

37. Install the speedometer driven gear assembly to the extension housing and attach with the bolt and lock plate.

38. Insert the shift control lever through the holes from the front side of the extension housing.

39. Install the control lever end to the control lever and tighten the attaching bolt (10mm) to 20–25 ft. lbs.

40. Install the back-up lamp switch and neutral sensing switch to the extension housing and tighten the switches to 20–25 ft. lbs.

41. Remove the bolts installed previously to temporarily hold the bearing housing.

42. Apply a thin coat of Silicone Sealer D6AZ–19562–B or equivalent on the contacting surface of the bearing housing and extension housing.

43. Install the extension housing to the bearing housing with the control lever and laid down to the left as far as it will go. Tighten the eight attaching bolts (12mm). Check to ensure that the control rod operates properly.

5 Speed Overdrive (Gasoline Engines)

DISASSEMBLY

1. Remove the nuts attaching the bell housing to the transmission case. Remove the bell housing gasket.

2. Remove the drain plug and drain lubricant from the transmission into a suitable container. Clean the metal filings from

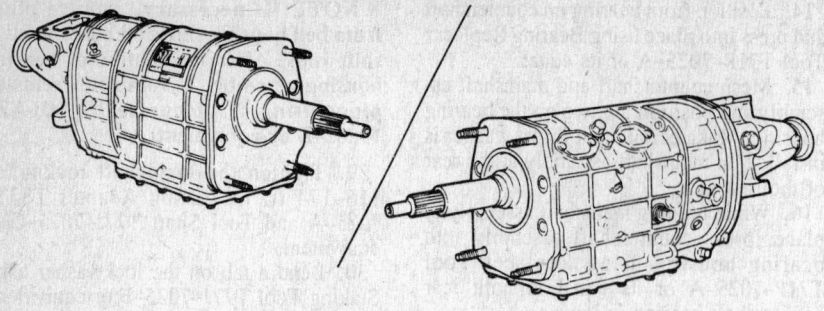

Ranger 5-speed overdrive

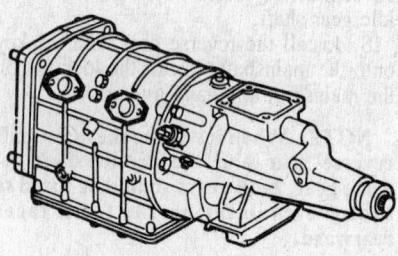

Four speed manual (gasoline engines)

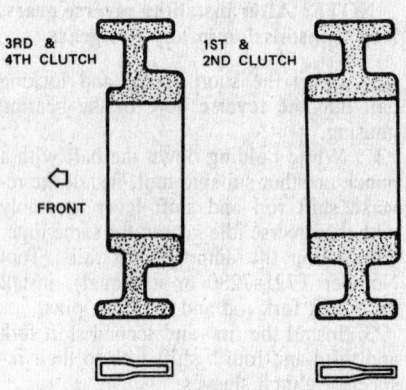

Clutch hub assembly direction

the magnet of the drain plug, if necessary. Install the drain plug.

3. (Optional) Position the Bench Mount Holding Fixture (T57L–500–B) or its equal to the studs on the right side of the transmission housing. Secure in place with the Bench Holding Fixture Adapter (T77J–7025–D) or its equal to prevent damage to the metric stud threads.

4. Place the transmission in neutral.

5. Remove the speedometer sleeve and driven gear assembly from the extension housing.

6. Remove the three bolts (14mm) and four nuts (14mm) attaching the extension housing to the transmission case. There are two longer outer bolts and one short center (bottom) bolt used.

7. Raise the control lever to the left and slide toward the rear of the transmission. Slide the extension housing off the mainshaft, being careful not to damage the oil seal.

8. Pull the control lever and rod out the front end of the extension housing.

9. If required, remove the back-up lamp switch from the extension housing.

10. Remove the anti-spill seal from the mainshaft and discard. (A seal is not necessary for assembly.)

11. Remove the snap-ring that secures the speedometer drive gear to the mainshaft. Slide the drive gear off the mainshaft, and remove the lock ball.

12. Evenly loosen the fourteen 10mm bolts securing the transmission case cover to the transmission case. Remove the cover and gasket.

13. Mark the shift rails and forks to aid during transmission assembly. Remove the roll pins attaching the shift rod ends to the shift rod and remove the shift rod ends.

14. Gently pry the bearing housing away from the transmission case with a screwdriver, being careful not to damage the housing or case. Slide the bearing housing off the mainshaft.

15. Remove the snap-ring and washer retaining the mainshaft rear bearing to the mainshaft.

16. Assemble the Bearing Puller Ring Tool (T77J–7025–J), Bearing Puller Tool (T77J–7025–H), and Forcing Screw (T75L–7025–J) on the Remover and Replacer Tube Tool (T75L–7025–B) or their equal. Slide the tool assembly over the mainshaft and engage the puller jaws behind the rear bearing. Tighten the jaws evenly onto the bearing with a wrench, then turn the forcing screw to remove the mainshaft rear bearing.

17. Remove the snap-ring from the rear end of the countershaft. Assemble the Bearing Puller Tool (T77J–7025–H), Bearing Puller Ring (T77J–7025–J) and Forcing Screw (T75L–7025–J) onto the Remover Tube (T77J–7025–B) or their equal. Slide the tool assembly over the countershaft and engage the puller jaws behind the countershaft rear bearing. Tighten the jaws evenly onto the bearing with a wrench, then turn the forcing screw to remove the bearing.

18. Remove the counter fifth gear and spacer from the rear of the countershaft.

19. Tap the housing with a plastic hammer, if necessary, and remove center housing. Remove the reverse idler gear and two spacers with housing.

20. Remove the cap screw (12mm) from center housing and remove idler gear shaft.

21. Remove the three spring cap bolts. The two bolts on the case upper portion are 17mm and the bolt on the case side is 14mm. Remove the detent springs and the detent balls with a magnet from the transmission case.

22. Remove the four 10mm bolts attaching the blind covers to the transmission case and remove the blind covers and gaskets.

23. Remove the roll pin from the fifth-and-reverse shift fork. Slide the fifth-and-reverse shift fork shaft out of the transmission case.

24. Shift the transmission into fourth gear. This will provide adequate space to drive out the roll pin. With a small drift, drive the roll pin from third-and-fourth shift fork. Slide the third-and-fourth shift fork shaft out of the rear of the transmission case.

25. Remove the roll pin from the first-and-second shift fork. Slide the first-and-second shift fork shaft assembly out the rear of the transmission case. Remove both inter-lock pins.

26. Remove the snap-ring that secures the fifth gear to the mainshaft.

27. Remove the thrust washer and lock ball, fifth gear and synchronizer ring from the rear of the mainshaft.

28. Install the Synchronizer Ring Holder and Countershaft Spacer (T77J–7025–E) or its equal between the fourth-speed synchronizer ring and synchromesh gear on the mainshaft. Shift the transmission into second gear to lock the mainshaft and prevent the assembly from rotating.

29. Straighten the staked portion of the mainshaft bearing locknut with the Staking Tool (T77J–7025–F) or its equal. Using the Locknut Wrench (T77J–7025–C) or its equal remove the mainshaft bearing locknut.

30. Slide the reverse gear and clutch hub assembly off the mainshaft.

31. Remove the counter reverse gear from the countershaft.

32. If installed, remove the transmission from the holding fixture and set on a workbench.

33. Remove the bolts (12mm) attaching the mainshaft center bearing cover to the transmission and remove the bearing cover.

34. To remove the countershaft center bearing, install Puller T77J–7025–H. Puller Rings T77J–7025–J, Remover Tube T77J–7025–B, and Forcing Screw T75L–7025–J or their equal. Squarely insert the jaws of the puller behind the center bearing retainer ring in the two recessed areas of the case.

NOTE: The retainer ring may need to be turned to position the split in the retainer ring midway between the recessed areas before the puller is installed. This will reduce the possibility of the retainer ring becoming distorted as the bearing is removed.

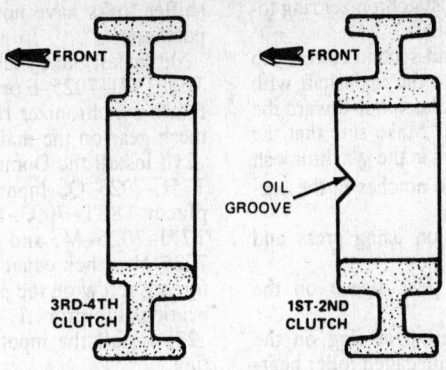
35. Turn the forcing screw to remove the bearing.

36. To remove the mainshaft center bearing, install Puller T77J–7025–H, Puller Rings T77J–7025–J, Long Remover Tube T75L–7025–C and Forcing Screw T75L–7025–J or their equal. Squarely insert the jaws of the puller behind the rear mainshaft bearing retainer ring in the two recessed areas of the case.

37. Turn the forcing screw clockwise to remove the bearing.

38. Remove the shim and spacer from behind the mainshaft rear bearing along with the bearing.

39. Remove the front cover by first removing the four studs attaching the cover to case. Remove the studs by installing two nuts (10mm × 1.5) on the stud and drawing the stud out of the case. Remove the four 14mm bolts and remove the cover. Save the shim found on the inside of the cover.

40. Remove the snap-ring from the input shaft.

41. Remove the input shaft bearing by installing Puller T77J–7025–H, Puller Rings T77J–7025–J, Remover Tube T75L–7025–B, and Forcing Screw T75L–7025–J or their equal. Squarely insert the jaws of the puller behind the input shaft bearing retainer ring in the two recessed areas of the case.

NOTE: The retainer ring may need to be turned to position the split in the ring midway between the recessed areas before the puller is installed.

42. Turn the forcing screw clockwise to remove the bearing.

43. Rotate both shift forks so that the main gear train will fall to the bottom of the case. Remove the shift forks. Rotate the input shaft so that one of the two flats on the input shaft faces upward.

44. Remove the snap-ring from the front of the countershaft.

45. Remove Synchronizer Ring Holder T77J–7025–E or its equal from the front of the case and insert between the first gear on the countershaft and the rear of the case.

46. Install Forcing Screw T75L–7025–J, Press Frame T77J–7025–N, and Press Frame Adapter T82T–7003–BH or their equal against the countershaft assembly.

47. Turn the forcing screw clockwise to press the countershaft rearward. Press the countershaft ($\frac{3}{16}$ in. movement) until it contacts the Synchronizer Ring Holder and Countershaft Spacer.

48. To remove the countershaft front bearing, first remove the press frame. Then, install Puller T77J–7025–H, Puller Rings T77J–7025–J, Remover Tube T77J–7025–B, and Forcing Screw T75L–7025–J or their equal. Squarely insert the jaws of the puller behind the front bearing retainer ring in the two recessed areas of the case.

NOTE: The retainer ring may need to be turned to position the split in the ring

midway between the recessed areas before the puller is installed.

49. Turn the forcing screw clockwise to remove the bearing.

50. Remove the shim from behind the countershaft front bearing.

51. Remove the countershaft from the transmission case.

52. Remove the input shaft from the transmission case. Remove the synchronizer ring and caged bearing from the mainshaft.

53. Remove the mainshaft and gear assembly from the transmission case.

54. Remove the inner race of the countershaft center bearing from the countershaft in a press frame using Axle Bearing Seal Plate T75L–1165–B and Pinion Bearing Cone Remover D79L–4621–A or their equal.

55. Remove first gear and first-and-second synchronizer ring. Remove snap-ring retainer from mainshaft.

NOTE: Do not mix synchronizer rings.

56. Install Bearing Remover Tool T71P–4621–B or its equal between second and third gear.

57. Press the mainshaft out of third gear and third-and-fourth clutch hub sleeve.

58. Press the first-and-second clutch hub and sleeve assembly, and first gear sleeve from the mainshaft.

59. Clean and inspect the case, gears, bearings and shafts.

ASSEMBLY

NOTE: As each part is assembled, coat the part with manual transmission oil D8DZ–19C547–A (ESP–M2C83–C) or equivalent.

Before beginning the assembly procedure, three measurements must be performed: Mainshaft Thrust Play, Countershaft Thrust Play and Mainshaft Bearing Clearance.

Mainshaft Thrust Play

Check the mainshaft thrust play by measuring the depth of the mainshaft bearing bore

in the transmission rear cage by using a depth micrometer (D80P–4201–A). Then measure the mainshaft rear bearing height. The difference between the two measurements indicates the required thickness of the adjusting shim. The standard thrust play is 0 to 0.0039 in. Adjusting shims are available in 0.0039 in. and 0.0118 in. sizes.

Countershaft Thrust Play

Check the countershaft thrust play by measuring the depth of the countershaft front bearing bore in the transmission case by using a depth micrometer (D80P–4201–A). Then measure the countershaft front bearing height. The difference between the two measurements indicates the required thickness of the adjusting shims. The standard thrust play is 0 to 0.0039 in. Adjusting shims are available in 0.0039 in. and 0.0118 in. sizes.

Mainshaft Bearing Clearance

Check the mainshaft bearing clearance by measuring the depth of the bearing bore in the clutch adapter plate with a depth micrometer, D80P–4201–A. Make sure the micrometer is on the second step of the plate. Measure the bearing height. The difference between the two measurements indicates the required adjusting shim thickness. The standard clearance is 0 to 0.0039 in. If an adjusting shim is required, select one to bring the clearance to within specifications.

1. Assemble the first-and-second synchromesh mechanism and the third-and-fourth synchromesh mechanism by installing the clutch hub to the sleeve. Place the three synchronizer keys into the clutch hub key slots and install the key springs to the clutch hub.

NOTE: When installing the key springs, the open end tab of the springs should be inserted into hub holes with springs turned in the same direction. This will keep the spring tension on each key uniform.

2. Place the synchronizer ring on the second gear and position the second gear to

Clutch hub assembly direction

the mainshaft with the synchronizer ring toward the rear of the shaft.

3. Slide the first-and-second clutch hub and sleeve assembly to the mainshaft with the oil grooves of the clutch hub toward the front of the mainshaft. Make sure that the three synchronizer keys in the synchromesh mechanism engage the notches in the second synchronizer ring.

4. Press into position using press and suitable replacer tool.

5. Insert the first gear sleeve on the mainshaft.

6. Place the synchronizer ring on the third gear along with the caged roller bearing and slide the third gear to the front of the mainshaft with the synchronizer ring toward the front.

7. Press the third-and-fourth clutch hub and sleeve assembly to the front of the mainshaft. Make sure that the three synchronizer keys in the synchromesh mechanism engage the notches in the synchronizer ring.

NOTE: Make sure the installed direction of the third-and-fourth clutch hub and sleeve assembly are correct.

8. Install the snap-ring to the front of the mainshaft.

9. Slide the needle bearing for the first gear to the mainshaft.

10. Place the synchronizer ring on the first gear. Slide the first gear onto the mainshaft with the synchronizer ring facing the front of the shaft. Rotate the first gear, as necessary, to engage the three notches in the synchronizer ring with the synchronizer keys.

11. Install the original thrust washer to the mainshaft.

12. Position the mainshaft and gear assembly in the case.

13. Position the first-and-second shift fork and third-and-fourth shift fork in the groove of the clutch hub and sleeve assembly.

14. Position the caged bearing in the front end of the mainshaft.

15. Place the synchronizer ring on the input shaft (fourth gear) and install the input shaft to the front end of the mainshaft. Make sure that the three synchronizer keys in the third-and-fourth synchromesh mechanism engage the notches in the synchronizer ring.

16. Press the inner race of the countershaft rear bearing onto the countershaft using Center Bearing Replacer T77J-7025-K or its equal.

17. Position the countershaft gear in the case, making sure that the countershaft gear engages each gear of the mainshaft assembly.

18. Install the correct shim on the mainshaft center bearing as determined in the Mainshaft Thrust Play Measurement in this Section.

19. Position the input shaft bearing and the mainshaft center bearing to the proper bearing bores. Be sure the synchronizer and

shifter forks have not been moved out of position.

20. Install the Synchronizer Ring Holder Tool T77J-7025-E or its equal between the fourth synchronizer ring and the synchromesh gear on the mainshaft.

21. Install the Dummy Bearing Replacer T75L-7025-Q, Input Shaft Bearing Replacer T82T-7003-DH, Replacer Tube T77J-7025-M, and Press Frame T77J-7025-N or their equal on the case. Turn the forcing screw on the press frame until both bearings are properly seated.

22. Install the input shaft bearing snap-ring.

NOTE: Be sure that the synchronizer and shift forks are properly positioned during seating of bearings. After bearings are seated, make certain that both synchronizers operate freely.

23. Place the correct shim in the countershaft front bearing bore.

24. Position the countershaft front and center bearings in the bores and install the tools. Turn the forcing screw until the bearing is properly seated. Use the center bearing as a pilot.

25. Install the snap-ring to secure the countershaft front bearing.

26. Remove the synchronizer ring holder.

27. Install the bearing cover to the transmission case and tighten the four attaching bolts. Tighten to 41–59 ft. lbs.

28. Install the reverse idler gear and shaft with a spacer on each side of shaft.

29. Slide the counter reverse gear (chamfer side forward) and spacer onto the countershaft.

30. Slide the thrust washer, reverse gear, caged roller bearings and clutch hub assembly onto the mainshaft. Install a new locknut (hand tight).

31. Shift into second gear and reverse gear to lock the rotation of the mainshaft. Tighten the locknut to 115 to 172 ft. lbs. using the Locknut Wrench T77J-7025-C or its equal.

32. Stake the locknut into the mainshaft keyway using the staking tool.

33. Place the fourth-and-third clutch sleeve in third gear using Synchronizer Ring Holder and Countershaft Spacer T77J-7025-E.

34. If new synchronizers have been installed, check the clearance between the synchronizer key and the exposed edge of the synchronizer ring with a feeler gauge. If the measurement is greater than 0.079 in., the synchronizer key can pop out of position. To correct this, change the thrust washer (selective fit) between the mainshaft center bearing and the first gear. Available thrust washer sizes are 0.098, 0.118 and 0.138 in.

35. If new synchronizers were installed, check the clearance again with a feeler gauge. If the clearance is within specifications, bend the tab of the lockwasher.

36. Position the fifth synchronizer ring on the fifth gear. Slide the fifth gear onto the mainshaft with the synchronizer ring toward the front of the shaft. Rotate the fifth gear, as necessary, to engage the three notches in the synchronizer ring with the synchronizer keys in the reverse and clutch hub assembly.

37. Install the lock ball and thrust washer on the rear of the fifth gear.

38. Install the snap-ring on the rear of the thrust washer. Check the clearance between the thrust washer and the snap-ring. If the clearance is not within 0.0039 to 0.0118 in., select the proper size thrust washer to bring the clearance within specifications.

39. Slide the first-and-second shift fork shaft assembly into the case (front rear of case). Secure the first-and-second shift fork to the fork shaft with the roll pin.

NOTE: Be sure to use a new roll pin.

40. Insert the inter-lock pin into the transmission using the lockout pin replacer tool.

41. Shift transmission into fourth gear. Slide the third-and-fourth shift fork shaft into the case, from rear of case. Secure the third-and-fourth shift fork to the fork shaft with the roll pin. Insert inter-lock pin.

NOTE: Be sure to use a new roll pin.

42. Shift synchronizer hub into fifth gear. Position reverse and fifth fork on the clutch hub and slide the reverse-and-fifth fork shaft into the case (from rear of case). Secure the reverse-and-fifth shift fork to the fork shaft with the roll pin.

NOTE: Be sure to use a new roll pin.

43. Install the two blind covers and gaskets. Tighten the attaching bolts (10mm) to 23–34 ft. lbs.

44. Position the three detent balls and three springs into the case and install the spring cap bolts (12mm and 17mm).

45. Apply a thin coat of Gasket Maker E2AZ-19562-A (ESE-M4G234-A2) or equivalent to the contacting surfaces of the center housing and transmission case.

46. Position the center housing on the case. Align the reverse idler gear shaft boss with the center housing attaching bolt boss. Install and tighten the idler shaft capscrew (12mm) and tighten to 41–59 ft. lbs.

47. Slide the counter fifth gear to the countershaft.

48. Position the countershaft rear bearing on the countershaft. Press into position using the Adjustable Press Frame T77J-7025-N and Forcing Screw T75L-7025-J or their equal.

49. Install the thrust washer and snap-ring to the rear of the countershaft rear bearing. Check the clearance between the thrust washer and the snap-ring using a feeler gauge.

50. If the clearance is not within 0.0000 to 0.0059 in., select the proper size thrust washer to bring the clearance within speci-

fications, 0.0748, 0.0787, 0.0827, or 0.0866 in.

51. If installed, remove filler plugs. Position the mainshaft rear bearing on the mainshaft. Press into place using the Adjustable Press Frame T77J-7025-N, Dummy Bearing T75L-7025-Q1 and Forcing Screw T75L-7025-J or their equal.

52. Install the thrust washer and snapring to the rear of the mainshaft rear bearing. Check the clearance between the thrust washer and the snap-ring. The clearance should be 0.0000 to 0.0039 in. If the clearance is not within specifications, replace the thrust washer to bring the clearance within specifications, 0.0787, 0.0846, or 0.0906 in.

53. Apply a thin coat of Gasket Maker E2AZ-19562-A (ESE-M4G234-A2) or equivalent to the contacting surfaces of the bearing housing and center housing.

54. Position the bearing housing on the center housing.

55. Install each shift fork shaft end onto the proper shift fork shaft. (Note the scribe marks made during disassembly) and secure with roll pins.

56. Install the lock ball, speedometer drive gear, and snap-ring onto the mainshaft.

57. If removed, install control lever and rod in extension housing.

58. Apply a thin coat of Gasket Maker E2AZ-19562-A (ESE-M4G234-A2) or equivalent to the contacting surfaces of the bearing housing and extension housing.

59. Position the extension housing in the bearing housing with the gearshift control lever end laid down to the left as far as it will go. Tighten the attaching bolts and nuts (14mm) to 60–80 ft. lbs. There are two longer outer bolts and one shorter center (bottom) bolt used.

60. If removed, insert the speedometer driven gear assembly to the extension housing and secure it with the bolt.

61. Check to ensure the gearshift control lever operates properly.

62. Install the transmission case cover gasket and cover with drain plug to the rear. Install and tighten the fourteen 10mm attaching bolts to 23–34 ft. lbs.

63. Install the correct size shim on the second step of the front cover as determined by the mainshaft bearing clearance measurement.

64. Coat the front cover with Gasket Maker E2AZ-19562-A (ESE-M4G234-A2) or equivalent. Install the front cover to the transmission case and tighten the four bolts and four studs.

65. Install 3.0 pints of Ford Manual Transmission Lube D8DZ-19C547-A (ESP-M2C83-C) or equivalent. Re-install the filler plugs and tighten to 18–29 ft. lbs.

5 Speed Overdrive (Diesel Engines)

DISASSEMBLY

1. If not already drained, remove the

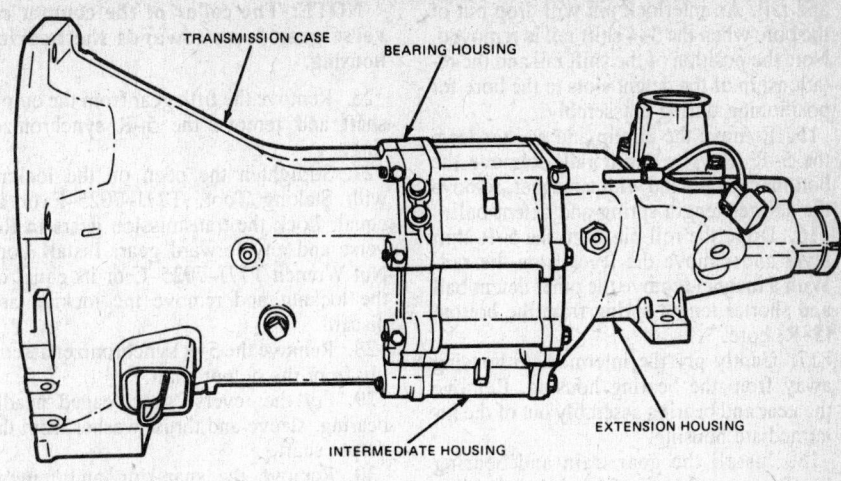

5-speed manual overdrive (diesel engines)

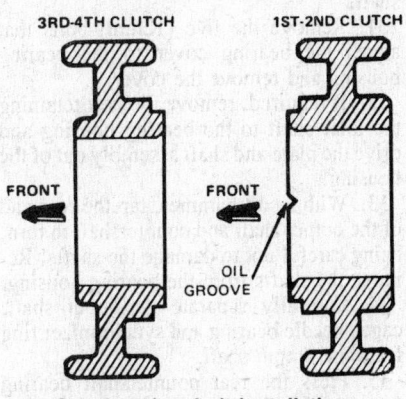

Synchronizer hub installation

drain plug and drain the transmission fluid into a suitable container. Remove the fork and release bearing from the transmission case.

2. Install the transmission in Bench Mounted Holding Fixture, T57L-500-B or its equal. Remove the six bolts (12mm) attaching the front cover to the transmission case and remove the front cover, shim (located in cover) and gasket.

3. Remove the front cover oil seal using Inner Seal Removal Tool, T75P-3504-G and Impact Slide Hammer, T50T-100-A or their equal.

4. Remove the input shaft snap-ring.

5. If installed, remove the gearshift lever. Remove the four bolts (12mm) and remove the retainer and gasket from the extension housing.

6. Remove the outer retaining ring on the input shaft bearing. Install Bearing Collet Tool, T75L-7025-E on the input shaft bearing, and Remover Tube, T75L-7025-B and Forcing Screw T75L-7025-J or their equal. Slide Bearing Collet Sleeve, T75L-7025-G or its equal over the Remover Tube and Bearing Collet, and turn the forcing screw to remove the input shaft bearing.

7. Remove the bolt (12mm) that attaches the control lever end to the control rod

and remove the control lever end and rod from the extension housing.

8. Remove the eight (12mm) bolts attaching the extension housing to the intermediate housing and transmission housing. Slide the extension housing off the output shaft with the control lever end laid down and to the left as far as it will go.

9. Remove the speedometer driven gear assembly from the extension housing.

10. Remove the back-up lamp switch and the neutral safety switch.

11. Remove the grommet from the end of the output shaft. Remove the snap-ring that secures the speedometer drive gear on the output shaft. Slide the speedometer drive gear off the output shaft and remove the lock ball.

12. Install Bearing Pusher Tool, T83T-7111-A or its equal over the countershaft front bearing. Turn the forcing screw to force the countershaft (together with the countershaft front bearing) from the transmission case. Remove the pusher tool assembly.

NOTE: The countershaft front bearing may remain in the transmission case. Remove the bearing with a suitable driver.

13. Remove and discard the roll pin from the 1-2 shift fork. Remove the cir-clip from the rail. Remove the upper cap bolt and with a magnet, remove the spring and detent ball from the bore. Remove the 1-2 shift rail and 1-2 shift fork. Note the position of the 1-2 shift fork in relation to the 3-4 shift fork for positioning during reassembly. The shift forks and rails are not interchangeable. Also note the position of the shift rail and the relationship of the detent slots to the bore for positioning during reassembly. The three detent slots in the shift rails face towards the cap bolts.

14. Remove the roll pin from the 3-4 shift fork. Remove the circlip from the rail. Remove the middle cap bolt and with a magnet, remove the spring and detent ball from the bore. Remove the 3-4 shift fork

and rail. An interlock pin will drop out of the bore when the 3–4 shift rail is removed. Note the position of the shift rail and the relationship of the detent slots to the bore for positioning during reassembly.

15. Remove the circlip and washer from the 5–R (Reverse) shift rail. Remove the bottom cap bolt and with a magnet, remove the shorter length spring and detent ball.

16. Drive the roll pin from the 5–R shift lever and remove the lever from the rail. With a magnet remove the other detent ball and shorter length spring from the bottom (5–R) bore.

17. Gently pry the intermediate housing away from the bearing housing. Remove the gear and bearing assembly out of the intermediate housing.

18. Install the gear train and bearing housing assembly in a fabricated holding tool positioned in a vise. A soft-jawed vise may be used in place of the holding tool.

19. Remove the bottom cap bolt and with a magnet, remove the shorter length spring and detent ball from the bore. Drive the roll pin out of the 5–R shift fork and discard. Remove the 5–R shift rail. An inter-lock pin will drop out of the bore when the 5–R shift rail is removed. Note the position of the 5–R shift fork in relation to the bearing housing for positioning during reassembly. Also note the position of the shift rail and the relationship of the detent slots to the bore for positioning during reassembly. The three detent slots in the shift rail face towards the cap bolt.

20. Remove the retaining ring from the output shaft ball bearing. Remove the thrust washer.

21. To remove the output shaft rear bearing, place Shaft Protectors, D80L–625–2 and D80L–625–3 or their equal on the end of the output shaft. It may be necessary to hold the shaft protectors in place with putty. Install Puller, T77J–7025–H, Collet (2), T77J–7025–J or their equal against the bearing so the jaws of the puller are against the rear of the bearing. Place Tube (Long), T75L–7025–C or its equal over the output shaft. Install Forcing Screw, T75L–7025–J or its equal into the tube and turn the forcing screw clockwise to remove the bearing. Discard the bearing and install a new one during reassembly.

22. Remove the snap-ring from the countershaft rear bearing. Install Puller, T77J–7025–H, Collet (2), T77J–7025–J, Tube (Short, T77J–7025–B) and Forcing Screw, T75L–7025–J or their equal. Turn the forcing screw clockwise and remove the bearing. Discard the bearing and install a new one during reassembly.

23. Remove the retaining ring, thrust washer and lock ball from the output shaft.

24. Remove the fifth gear and sleeve from the countershaft.

NOTE: The collar of the fifth gear faces towards the bearing housing.

25. Remove the reverse gear from the countershaft.

NOTE: The collar of the counter-reverse gear faces towards the bearing housing.

26. Remove the fifth gear from the output shaft and remove the 5–R synchronizer ring.

27. Straighten the peen on the locknut with Staking Tool, T77J–7025–F or its equal. Lock the transmission gears in Reverse and any forward gear. Install Lock Nut Wrench T77J–7025–C or its equal on the locknut and remove the locknut and discard.

28. Remove the 5–R synchronizer assembly from the output shaft.

29. Pry the reverse gear, caged needle bearing, sleeve and thrust washer from the output shaft.

30. Remove the snap-ring and remove the reverse idler gear from the idler shaft. Remove the keyed thrust washer from the shaft.

31. Remove the five (12mm) bolts that attach the bearing cover to the bearing housing and remove the cover.

32. If required, remove the bolt retaining the idler shaft to the bearing housing and drive the plate and shaft assembly out of the housing.

33. With a soft hammer, tap the rear end of the output shaft and countershaft in turn, being careful not to damage the shafts. Remove the shafts from the bearing housing.

34. Carefully separate the input shaft, caged needle bearing and synchronizer ring from the output shaft.

35. Press the rear countershaft bearing from the bearing housing using Remover Tube Tool, T77J–7025–B or its equal.

36. Press the rear output shaft bearing from the bearing housing using Bearing Remover Tool, T77F–4222–A and Remover Tube Tool, T77J–7025–B or their equal.

37. Remove the thrust washer, first gear, sleeve and synchronizer ring from the rear of the output shaft.

38. Using snap-ring pliers, remove the snap-ring from the front of the output shaft.

39. Using a press and Remover Tool, T71P–4621–B or its equal, remove the third-and-fourth hub, sleeve, synchronizer ring and third gear from the front of the output shaft.

40. Using a press and Remover Tool, T71P–4621–B or its equal, remove the first-and-second hub and sleeve assembly synchronizer ring, and second gear from the rear of the output shaft in the same manner as described in the previous step.

41. Press the front bearing from the countershaft using Remover Tool D79L–4621–A or T71P–4621–B and suitable press stock.

42. Inspect all parts.

ASSEMBLY

1. Assemble the 3–4 synchronizer assembly by installing the keys in the hub and sliding the sleeve over the hub and keys. Install the springs onto the hub.

NOTE: When installing the springs, the open end tab of the springs should be inserted into the hub holes. This will keep the spring tension on each key uniform.

2. Assemble the 1–2 synchronizer assembly in the same manner as described in Step 1. Assemble the 5–R synchronizer assembly as also described in Step 1 and install the retaining ring in the 5–R assembly.

3. Install the third gear and synchronizer ring onto the front section of the output shaft.

4. Install the 3–4 synchronizer assembly onto the output shaft by using a press. Hold the assembly together and slowly press in place. Make sure the three recesses in the synchronizer ring are aligned with the three keys in the synchronizer hub.

NOTE: The direction of the hub is as shown. The recesses in each synchronizer sleeve must face each other.

5. Fit the snap-ring on the output shaft.

6. Install the second gear, synchronizer ring onto the rear section of the output shaft.

7. Install the 1–2 synchronizer assembly onto the output shaft by using a press. Hold the assembly together and slowly press in place. Make sure the three keys recesses in the synchronizer ring are aligned with the three keys in the synchronizer hub.

NOTE: The direction of the hub is as shown. The recesses in the synchronizer sleeve must face each other.

8. Install the synchronizer ring, first gear with sleeve, and thrust washer onto the output shaft.

9. Install the input shaft and the needle roller bearing to the output shaft.

10. Check the countershaft rear bearing clearance. Measure the depth of the countershaft bearing bore in the bearing housing with a depth micrometer (D80P–4201–A or equivalent). Install the retaining ring on the bearing and with a depth micrometer measure the distance between the inside edge of the ring and the end of the bearing. The difference between the two measurements indicates the required thickness of the adjusting shim. The clearance should be less than 0.0039 in. The adjusting shims are available in 0.0039 in. and 0.0118 in. sizes.

11. Check the output shaft bearing clearance. Measure the depth of the bearing bore with a depth micrometer. Measure the width of the bearing with a micrometer. The difference between the two measurements indicates the required thickness of the adjusting shim. The clearance should be less than 0.0039 in. Adjusting shims are available in 0.0039 in. and 0.0118 in. sizes.

12. Position the proper shim on the countershaft rear bearing and press into the bearing housing using Installer Tool, T77J–7025–B or its equal.

13. Position the proper shim on the out-

put shaft bearing and press the bearing into the bearing housing using Installer Tool, T77J–7025–K or its equal.

14. Position the front bearing on the countershaft and press the bearing into place using Bearing Replacer Tool, T71P–7025–A or its equal.

15. Mesh the countershaft and the output shaft assembly and position the two in the bearing housing. Make sure that the thrust washer is installed on the mainshaft assembly at the rear of the first gear. Make sure the three recesses in the synchronizer ring are aligned with the three keys in the synchronizer hub.

16. While holding the mainshaft assembly in place, press the countershaft assembly into the bearing housing using Replacer Tool T71P–7025–A or its equal to hold the rear countershaft bearing in the housing.

17. Position the bearing cover on the bearing housing. Install the five (12mm) bolts and tighten.

18. If removed, drive the reverse idler shaft into the bearing housing. Install the bolt and tighten.

19. Install the thrust washer, sleeve, caged needle bearing and reverse gear on the output shaft.

20. Install the reverse gear on the countershaft. The offset on the gear must face the bearing housing.

21. Place the keyed thrust washer so the tab is in the groove in the bearing housing. Install the reverse idler gear so the squared portion of the gear faces the bearing housing. Make sure the reverse idler gear and reverse gear are in mesh. Install the spacer and snap-ring on the idler shaft.

22. Install the 5–R synchronizer assembly on the output shaft.

23. Lock the transmission in Reverse and any forward gear. Install a new locknut on the output shaft and tighten to 94–152 ft. lbs. using Locknut Wrench, T77J–7025–C or its equal.

24. Bend the tab on the locknut with Staking Tool, T77J–7025–F or its equal.

25. Install the 5–R synchronizer ring and gear on the output shaft. Make sure the three recesses in the synchronizer ring are aligned with the three keys in the synchronizer hub.

26. Install the sleeve and the counter-fifth gear on the countershaft.

27. Install the lock ball in the output shaft and position the thrust washer so the slot in the washer is over the lock ball. Install the retaining ring.

28. Position the output shaft assembly in a press and press the output shaft bearing on the shaft using Dummy Bearing Replacer, T75L–7025–Q or its equal and an appropriate length of press stock. Install the thrust washer and retaining ring.

29. Position the countershaft in a press and press the countershaft rear bearing on the shaft using Dummy Bearing Replacer, T75L–7025–Q or its equal and an appropriate length of press stock. Install the thrust washer and retaining ring.

30. Position all synchronizers in the neutral position. Install the shorter length spring and detent ball in the bottom (5–R) bore. Compress the ball and spring with Dummy Shift Rail Tools, T72J–7280 or its equal and install the dummy shift rail in the bore. Install the 5–R shift rail in the bottom bore and make sure the three detent slots in the rail face the cap bolt and the interlock slot in the 5–R rail faces towards the 1–2 bore. Install the interlock pin through the top bore so it is positioned in the channel between the 5–R rail bore and 3–4 rail bore. Install the 3–4 rail in the housing and make sure the three detent slots in the rail face the middle bore. Insert the interlock pin in the channel the 3–4 rail and the 1–2 rail bore. Install the 1–2 shift rail in the housing so the three detent slots in the rail face the top bore.

NOTE: The interlock pins are identical and all four detent balls are identical. The springs for the 5–R or bottom bore are of a shorter length than the other two springs.

31. Install the first-and-second shift fork and the third-and-fourth shift forks to their respective sleeves.

32. Align the roll pin holes of each shift fork and rod. Install new roll pins.

NOTE: When installing the shift fork and control end, a new roll pin should be installed with a pin slit positioned in the direction of the shift rod axis. If not removed, remove the shift levers from the shift rails. Remember from which rail each lever was removed for correct installation upon assembly.

33. Install the detent balls and springs into their respective bores and install the three cap bolts.

NOTE: The shorter length spring is installed in the bottom (reverse) bore.

34. Install the circlips on the 1–2 and 3–4 shift rails. Install the circlip and washer on the 5–R shift rail.

35. Apply a thin coating of Silicone Sealer, D6AZ–19562–B or equivalent to the mating surfaces of the transmission case and the bearing housing. Install the transmission case on the bearing housing.

36. Apply a thin coating of Silicone Sealer, D6AZ–19562–B or equivalent to the mating surfaces of the bearing housing and intermediate housing. Install the intermediate housing to the bearing housing.

37. Position the shift lever gates on the appropriate shift rails. Install new roll pins.

38. Place the lock ball in the output shaft and position the speedometer drive gear over the ball. Install the snap-ring. Install the grommet on the end of the output shaft.

39. Apply a thin coating of Silicone sealer, D6AZ–19562–B or equivalent to the extension housing and the intermediate housing. Slide the extension housing over the output shaft (the control lever must be moved to the far left) and onto the extension housing. Install the bolts and tighten.

NOTE: If necessary, remove the plugs from the transmission case shift rod bores to align the shift rods. After the installation of the bearing housing assembly, reinstall the plugs using Silicone Sealer, D6AZ–19562–B or equivalent.

40. With the outer snap-ring in place on the input shaft front bearing, place the bearing, shim and Adapter Tool, T75L–7025–N or its equal over the input shaft.

41. Thread the Replacer Shaft, T75L–7025–K or its equal onto the Adapter Tool. Install the Replacer Tube, T75L–7025–B or its equal over the Replacer Shaft and install the nut and washer on the forcing screw.

42. Slowly tighten the nut until the adapter is securely on the input shaft. Make sure the tools are aligned.

43. Tighten the nut on the forcing screw until the bearing outer snap-ring is seated. Remove the installation tools.

NOTE: The input shaft bearing retaining ring must be flush with the transmission case. If not flush, it will be necessary to tap on the end of the input shaft with a soft hammer until the bearing is seated.

44. Install the input shaft snap-ring.

45. Measure the distance between the end of the installed input bearing in the transmission case with a depth micrometer. Measure the distance between the bearing cover gasket and the bottom of the bearing bore in the cover. The difference between the two measurements is the clearance between the outer bearing race and the front cover. The clearance should be less than 0.0039 in. Clearance can be adjusted by installing an adjusting shim. Shims are available in sizes of 0.006 in. and 0.012 in.

46. Install a new oil seal in the front cover using Installer Tool, T71P–7050–A. Install the shim in the recess in the front cover.

47. Apply gear lubricant to the lip of the oil seal inside the front cover and install the front cover to the transmission case. Install the six (12mm) bolts and tighten.

48. Install the control lever end to the control lever and tighten the attaching bolt (10mm) to 20–25 ft. lbs.

49. Install the back-up lamp switch and the neutral safety switch to the extension housing and tighten the switches to 20–25 ft. lbs.

50. Install the gearshift lever retainer and gasket to the extension housing. Install the four (12mm) bolts and tighten to 20–27 ft. lbs. If required, install the gearshift lever.

51. Install the release bearing and fork.

Muncie

Model SM330 3 Speed (83MM)

The G.M. Corporation Model SM 330

(83MM) (Muncie) is a three-speed transmission using helical constant mesh gears. The engagement of all gears except reverse is assisted by synchronizers.

DISASSEMBLY

1. Remove side cover and shift forks.
2. Unbolt extension and rotate to line up groove in extension flange with reverse idler shaft. Drive reverse idler shaft and key out of case with a brass drift.
3. Move second-third synchronizer sleeve forward. Remove extension housing and mainshaft assembly.
4. Remove reverse idler gear from case.
5. Remove third speed blocker ring from clutch gear.
6. Expand snap-ring which holds mainshaft rear bearing. Tap gently on end of mainshaft to remove extension.
7. Remove clutch gear bearing retainer and gasket.
8. Remove snap-ring. Remove clutch gear from inside case by gently tapping on end of clutch gear.
9. Remove oil slinger and 16 mainshaft pilot bearings from clutch gear cavity.
10. Slip clutch gear bearing out front of case. Aid removal with a screwdriver between case and bearing outer snap-ring.
11. Drive countershaft and key out to rear.
12. Remove countergear and two tanged thrust washers.

MAINSHAFT

1. Remove speedometer drive gear. Some speedometer drive gears, made of metal, must be pulled off.
2. Remove rear bearing snap-ring.
3. Support reverse gear. Press on rear of mainshaft to remove reverse gear, thrust washer, and rear bearing. Be careful not to cock the bearing on the shaft.
4. Remove first and reverse sliding clutch hub snap-ring.
5. Support first gear. Press on rear of mainshaft to remove clutch assembly, blocker ring, and first gear.
6. Remove second and third speed sliding clutch hub snap-ring.
7. Support second gear. Press on front of mainshaft to remove clutch assembly, second speed blocker ring, and second gear from shaft.

CLUTCH KEYS & SPRINGS

Keys and springs may be replaced if worn or broken, but the hubs and sleeves must be kept together as originally assembled.
1. Mark hub and sleeve for reassembly.
2. Push hub from sleeve. Remove keys and springs.
3. Place three keys and two springs, one on each side of hub, so all three keys are engaged by both springs. The tanged end of the springs should not be installed into the same key.
4. Slide the sleeve onto the hub, aligning the marks.

EXTENSION OIL SEAL & BUSHING

1. Remove seal.
2. Using bushing remover and installer, or other suitable tool, drive bushing into extension housing.
3. Drive new bushing in from rear. Lubricate inside of bushing and seal. Install new oil seal with extension seal installer or suitable tool.

CLUTCH BEARING RETAINER OIL SEAL

1. Pry old seal out.
2. Install new seal using seal installer or suitable tool. Seat seal in bore.

ASSEMBLY

MAINSHAFT

1. Lift front of mainshaft.
2. Install second gear with clutching teeth up; the rear face of the gear butts against the mainshaft flange.
3. Install a blocking ring with clutching teeth downward. All three blocking rings are the same.
4. Install second and third synchronizer assembly with fork slot down. Press it onto mainshaft splines. Both synchronizer assemblies are identical but are assembled differently. The second-third speed hub and sleeve is assembled with the sleeve fork slot toward the thrust face of the hub; the first-reverse hub and sleeve, with the fork slot opposite the thrust face. Be sure that the blocker ring notches align with the synchronizer assembly keys.
5. Install synchronizer snap-ring. Both synchronizer snap-rings are the same.
6. Turn rear of shaft up.
7. Install first gear with clutching teeth upward; the front face of the gear butts against the flange on the mainshaft.
8. Install a blocker ring with clutching teeth down.
9. Install first and reverse synchronizer assembly with fork slot down. Press it onto mainshaft splines. Be sure blocker ring notches align with synchronizer assembly keys and synchronizer sleeves face front of mainshaft.
10. Install snap-ring.
11. Install reverse gear with clutching teeth down.
12. Install steel reverse gear thrust washer with flats aligned.

13. Press rear ball bearing onto shaft with snap-ring slot down.
14. Install snap-ring.
15. Install speedometer drive gear and retaining clip.

ASSEMBLY

1. Start the transmission unit assembly: Place a row of 29 roller bearings, a bearing washer, a second row of 29 bearings, and a second bearing washer at each end of the countergear. Hold in place with grease.
2. Place countergear assembly through rear case opening with a tanged thrust washer, tang away from gear, at each end. Install countershaft and key from rear of case. Be sure that thrust washer tangs are aligned with notches in case.
3. Place reverse idler gear in case. Do not install reverse idler shaft yet.

NOTE: The reverse idler gear bushing may not be replaced separately—only as a unit.

4. Expand snap-ring in extension. Assemble extension over mainshaft and onto rear bearing. Seat snap-ring.
5. Load 16 mainshaft pilot bearings into clutch gear cavity. Assemble third speed blocker ring onto clutch gear clutching surface with teeth toward gear.
6. Place clutch gear assembly, without front bearing, over front of mainshaft. Make sure that blocker ring notches align with keys in second-third synchronizer assembly.
7. Stick gasket onto extension housing with grease. Assemble clutch gear, mainshaft, and extension to case together. Make sure that clutch gear teeth engage teeth of countergear anti-lash plate.
8. Rotate extension housing. Install reverse idler shaft and key.
9. Torque extension bolts to 45 ft. lbs.
10. Install oil slinger with inner lip facing forward. Install front bearing outer snap-ring and slide bearing into case bore.
11. Install snap-ring to clutch gear stem. Install bearing retainer and gasket and torque to 20 ft. lbs. Retainer oil return hole must be at 6 o'clock.
12. Shift both synchronizer sleeves to neutral positions. Install side cover, insert-

TORQUE SPECIFICATIONS
Muncie-83MM

	ft. lbs.
Extension to case attaching	45
Drain plug	30
Filler plug	15
Side cover attaching bolts	22
Main drive gear retainer bolts	22
Transmission case to clutch housing bolts	45

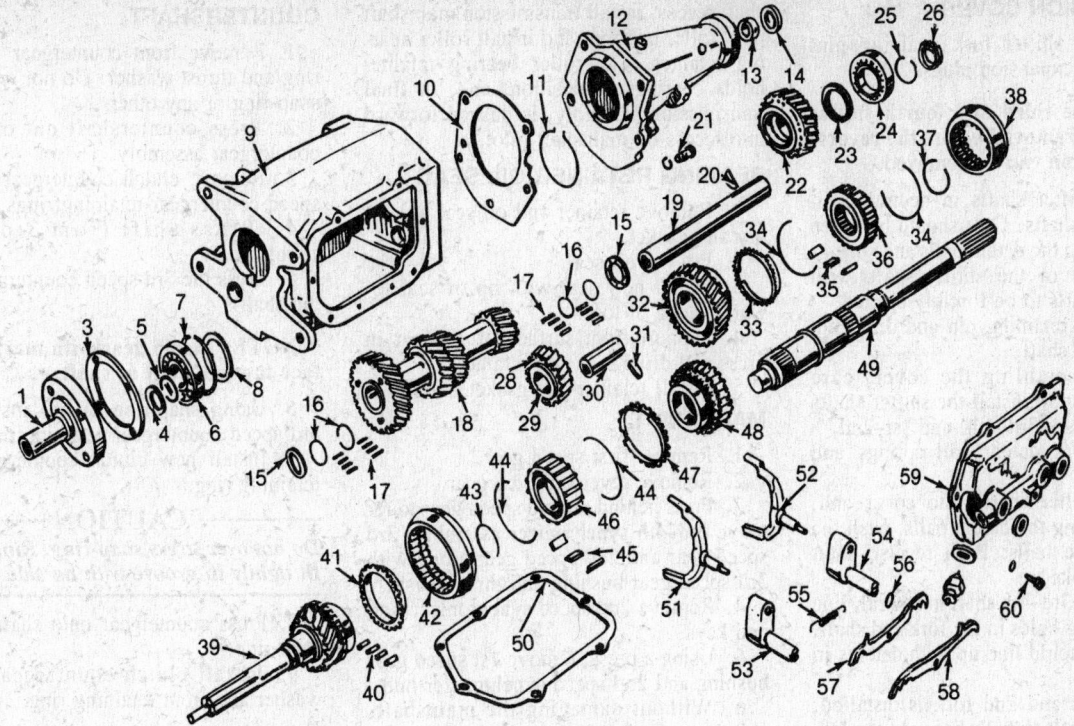

1. Bearing retainer
2. Bolt and lock washer
3. Gasket
4. Oil seal
5. Snap-ring (bearing-to-main drive gear)
6. Main drive gear bearing
7. Snap-ring bearing
8. Oil slinger
9. Case
10. Gasket
11. Snap-ring (rear bearing-to-extension)
12. Extension
13. Extension bushing

14. Oil seal
15. Thrust washer
16. Bearing washer
17. Needle bearings
18. Countergear
19. Countershaft
20. Woodruff key
21. Bolt (extension-to-case)
22. Reverse gear
23. Thrust washer
24. Rear bearing
25. Snap-ring
26. Speedometer drive gear
27. Retainer clip

28. Reverse idler gear
29. Reverse idler bushing
30. Reverse idler shaft
31. Woodruff key
32. 1st speed gear
33. 1st speed blocker ring
34. Synchronizer key spring
35. Synchronizer keys
36. 1st and reverse synchronizer hub assembly
37. Snap-ring
38. 1st and reverse synchronizer collar
39. Main drive gear

40. Pilot bearings
41. 3rd speed blocker ring
42. 2nd and 3rd synchronizer collar
43. Snap-ring
44. Synchronizer key spring
45. Synchronizer keys
46. 2nd and 3rd synchronizer hub
47. 2nd speed blocker ring
48. 2nd speed gear
49. Mainshaft
50. Gasket
51. 2nd and 3rd shifter

fork
52. 1st and reverse shifter fork
53. 2-3 shifter shaft assembly
54. 1st and reverse shifter shaft assembly
55. Spring
56. O-ring seal
57. 1st and reverse detent cam
58. 2nd and 3rd detent cam
59. Side cover
60. Bolt and lock washer

Muncie 83 mm 3-speed

ing shifter forks in synchronizer sleeve grooves.

13. Torque side cover bolts to 20 ft. lbs.

Model SM465 4 Speed (117mm)

Muncie model SM465 transmission is a four speed transmission using helical gears. The action of all gears except reverse is aided by synchronizers.

DISASSEMBLY

1. Remove transmission cover assembly.

NOTE: Move reverse shifter fork so that reverse idler gear is partially engaged before attempting to remove cover. Forks must be positioned so rear edge of the slot in the reverse fork is in line with the front edge of the slot in the forward forks as viewed through tower opening.

2. Lock transmission into two gears. Remove the universal joint flange nut, universal joint front flange and brake drum assembly. On 4WD models, use a special tool to remove mainshaft rear lock nut.

3. Remove parking brake and brake flange plate assembly on those vehicles having a drive-shaft parking brake.

4. Remove rear bearing retainer and gasket.

5. Slide speedometer drive gear off mainshaft.

6. Remove clutch gear bearing retainers and gasket.

7. Remove countergear front bearing cap and gasket.

8. Using a prybar, pry off countershaft front bearing.

9. Remove countergear rear bearing snap-rings from shaft and bearing. Using special tool, remove countergear rear bearings.

10. Remove clutch gear bearing outer race to case retaining ring.

11. Remove clutch gear and bearing by tapping gently on bottom side of clutch gear shaft and prying directly opposite against the case and bearing snap-ring groove at the same time. Remove fourth gear synchronizer ring.

CAUTION

Index cut out section of clutch gear in down position with countergear to obtain clearance for removing clutch gear.

12. Remove rear mainshaft bearing snap-ring and, using special tools, remove bearing from case. Slide 1st speed gear thrust washer off mainshaft.

13. Lift mainshaft assembly from case. Remove synchronizer cone from shaft.

14. Slide reverse idler gear rearward and move countergear rearward, then lift to remove from case.

15. To remove reverse idler gear, drive reverse idler gear shaft out of case from front to rear using a drift. Remove reverse idler gear from case.

TRANSMISSION COVER

1. Remove shifter fork retaining pins and drive out expansion plugs.

NOTE: The third and fourth shifter fork must be removed before the reverse shifter head pin can be removed.

2. With shifter shafts in neutral position, remove shafts. Care should be taken when removing the detent balls and springs since removal of the shifter shafts will cause these parts to be forcibly ejected.

3. Remove retaining pin and drive out reverse shifter shaft.

4. In reassembling the cover, care should be taken to install the shifter shafts in order; reverse, 3rd–4th, and 1st–2nd.

5. Place fork detent ball springs and balls in cover.

6. Start shifter shafts into cover and, while depressing the detent balls, push the shafts over the balls. Push reverse shaft through the yoke.

7. With the 3rd–4th shaft in neutral, line up the retaining holes in the fork and shaft. Detent balls should line up with detents in shaft.

8. After 1st and 2nd fork is installed, place two inner-lock balls between the low speed shifter shaft and the high speed shifter shaft in the crossbore of the front support boss. Grease the interlock pin and insert it in the 3rd–4th shifter shaft hole. Continue pushing this shaft through cover bore and fork until retainer hole in fork lines up with hole in shaft.

9. Place two interlock balls in crossbore in front support boss between reverse, and 3rd and 4th shifter shaft. Then push remaining shaft through fork and cover bore, keeping both balls in position between shafts until retaining holes line up in fork and shaft. Install retaining pin.

10. Install 1st/2nd fork and reverse fork retaining pins. Install new shifter shaft hole expansion plugs.

CLUTCH GEAR & SHAFT

1. Remove mainshaft pilot bearing rollers from clutch gear if not already removed, and remove roller retainer. Do not remove snap-ring on inside of clutch gear.

2. Remove snap-ring securing bearing on steam of clutch gear.

3. To remove bearing, position a special tool to the bearing and, with an arbor press, press gear and shaft out of bearing.

4. Press bearing and new oil slinger onto clutch gear shaft using a special tool. Slinger should be located flush with bearing shoulder on clutch gear. Be careful not to distort oil slinger.

5. Install bearing snap-ring on clutch gear shaft.

6. Install bearing retainer ring in groove on O.D. of bearing. The bearing must turn freely on the shaft.

7. Install snap-ring on I.D. of mainshaft pilot bearing bore in clutch gear.

8. Lightly grease bearing surface in shaft recess, install transmission mainshaft pilot roller bearings and install roller bearing retainer. The roller bearing retainer holds bearings in position and, in final transmission assembly, is pushed forward into recess by mainshaft pilot.

BEARING RETAINER OIL SEAL

1. Remove retainer and oil seal assembly and gasket.

2. Pry out oil seal.

3. Install new seal with lip of seal toward flange of tool.

4. Support front surface of retainer in press and drive seal into retainer.

5. Install retainer and gasket on case.

MAINSHAFT

1. Remove first speed gear.

2. Remove reverse driven gear.

3. Press behind second speed gear to remove 3rd–4th synchronizer assembly, 3rd speed gear and 2nd speed gear along with 3rd speed gear bushing and thrust washer.

4. Remove 2nd speed synchronizer ring and keys.

5. Using a press, remove 1st speed gear bushing and 2nd speed synchronizer hub.

6. Without damaging the mainshaft, chisel out the 2nd speed gear bushing.

ASSEMBLY

1. Lubricate with E.P. oil and press onto mainshaft.

--------- **CAUTION** ---------
1st, 2nd and 3rd speed gear bushings are sintered iron, exercise care when installing.

2. Press 1st and 2nd speed synchronizer hub onto mainshaft with annulus toward rear of shaft.

3. Install 1st and 2nd synchronizer keys and springs.

4. Press 1st speed gear bushing onto mainshaft until it bottoms against hub.

NOTE: Lubricate all bushings with E.P. oil before installation of gears.

5. Install synchronizer blocker ring and 2nd speed gear onto mainshaft and against synchronize hub. Align synchronizer key slots with keys in synchronizer hub.

6. Install 3rd speed gear thrust washer onto mainshaft inserting washer tang in slotted shaft. Then press 3rd speed gear bushing onto mainshaft against thrust washer.

7. Install 3rd speed gear and synchronizer blocker ring against 3rd speed gear thrust washer.

8. Align synchronizer key ring slots with synchronizer assembly keys and drive 3rd and 4th synchronizer assembly onto mainshaft. Secure assembly with snap-ring.

9. Install reverse driven gear with fork groove toward rear.

10. Install 1st speed gear against 1st and 2nd synchronizer hub. Install 1st speed gear thrust washer.

COUNTERSHAFT

1. Remove front countergear retaining ring and thrust washer. Do not re-use this snap-ring or any others.

2. Press countershaft out of clutch countergear assembly.

3. Remove clutch countergear and 3rd speed countergear retaining rings.

4. Press shift from 3rd speed countergear.

5. Press the 3rd speed countergear onto the shaft.

NOTE: Install gear with marked surface toward front of shaft.

6. Using snap-ring pliers, install new 3rd speed countergear retaining ring.

7. Install new clutch countergear rear retaining ring.

--------- **CAUTION** ---------
Do not over stress snap-ring. Ring should fit tightly in groove with no side play.

8. Press countergear onto shaft against snap-ring.

9. Install clutch countergear thrust washer and front retaining ring.

ASSEMBLY

1. Lower the countergear into the case.

2. Place reverse idler gear in transmission case with gear teeth toward the front. Install idler gear shaft from rear to front, being careful to have slot in end of shaft facing down and flush with case.

3. Install mainshaft assembly into case with rear of shaft protruding out rear bearing hole in case. Rotate case onto front end.

NOTE: Install 1st speed gear thrust washer on shaft, if not previously installed.

4. Install snap-ring on bearing O.D. and place rear mainshaft bearing on shaft. Drive bearing onto shaft and into case.

5. Install synchronizer cone on mainshaft and slide rearward to clutch hub.

--------- **CAUTION** ---------
Make sure three cut-out sections of 4th speed synchronizer cone align with three clutch keys in clutch assembly.

6. Install snap-ring on clutch gear bearing O.D. Index cut out portion of clutch gear teeth to obtain clearance over countershaft drive gear teeth, and install into case.

7. Install clutch gear bearing retainer and gasket and torque to 15–18 ft. lbs.

8. Rotate case onto front end.

9. Install snap-ring on countergear rear bearing O.D., and drive bearing into place. Install snap-ring on countershaft at rear bearing.

10. Tap countergear front bearing assembly into case.

11. Install countergear front bearing cap and new gasket and torque to 20–30 inch lbs.

12. Slide speedometer drive gear over mainshaft to bearing.

13. Install rear bearing retainer with new gasket. Be sure snap-ring ends are in lube slot and cut out in bearing retainer. Install bolts and tighten to 15–18 ft. lbs. Install brake backing plate assembly on those models having driveshaft brake.

NOTE: On models equipped with 4-wheel drive, install rear lock nut and washer and torque to 120 ft. lbs. and bend washer tangs to fit slots in nut.

14. Install parking brake drum and/or universal joint flange.

15. Lock transmission in two gears at once. Install universal joint flange locknut and tighten to 90–120 ft. lbs.

16. Move all transmission gears to neutral except the reverse idler gear which should be engaged approximately ³/₈ of an inch (leading edge of reverse idler gear taper lines up with the front edge of the 1st speed gear). Install cover assembly and gasket. Shifting forks must slide into their proper positions on clutch sleeves and reverse idler gear. Forks must be positioned as in removal.

17. Install cover attaching bolts and gearshift lever and check operation of transmission.

New Process

NP435 4 Speed

1. Mount the transmission in a holding fixture. Remove the parking brake assembly, if one is installed.

2. Shift the gears into neutral by replacing the gear shift lever temporarily, or by using a bar or screw driver.

3. Remove the cover screws, the second screw from the front on each side is shouldered with a split washer for installation alignment.

4. While lifting the cover, rotate slightly counterclockwise to provide clearance for the shift levers. Remove the cover.

5. Lock the transmission in two gears and remove the output flange nut, the yoke, and the parking brake drum as a unit assembly.

NOTE: The drum and yoke are balanced and unless replacement of parts are required, it is recommended that the drum and yoke be removed as a assembly.

6. Remove the speedometer drive gear pinion and the mainshaft rear bearing retainer.

7. Before removal and disassembly of the drive pinion and mainshaft, measure the end play between the synchronizer stop ring and the third gear. Clearance should be within 0.050–0.070 in. If necessary, add corrective shims during assembly.

NOTE: Record this reading for reference during assembly.

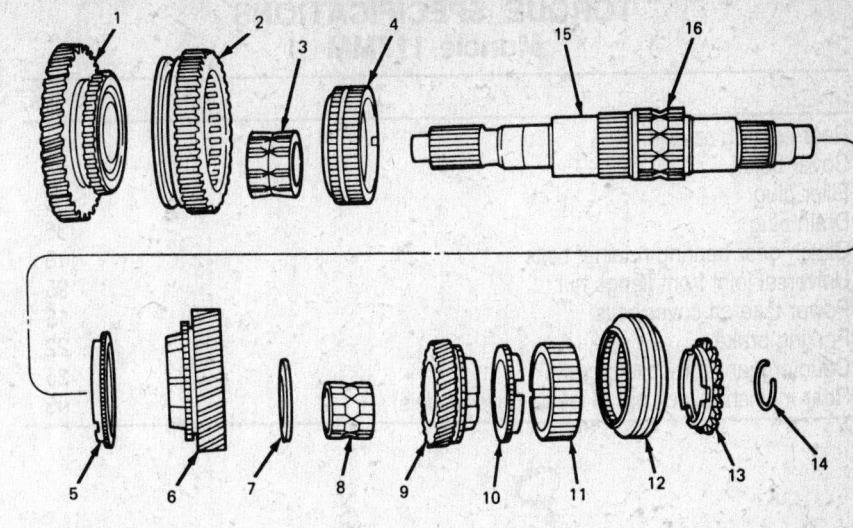

1. 1st speed gear
2. Reverse driven gear
3. 1st gear bushing
4. 1st-2nd gear synchronizer hub assembly
5. 2nd speed blocker ring
6. 2nd speed gear
7. Thrust washer
8. 3rd speed bushing
9. 3rd speed gear
10. 3rd speed blocker ring
11. 3rd-4th speed synchronizer hub assembly
12. 3rd-4th speed synchronizer sleeve
13. 4th speed blocker ring
14. Snap-ring
15. Mainshaft
16. 2nd speed gear bushing

Muncie 117 mm 4-speed mainshaft

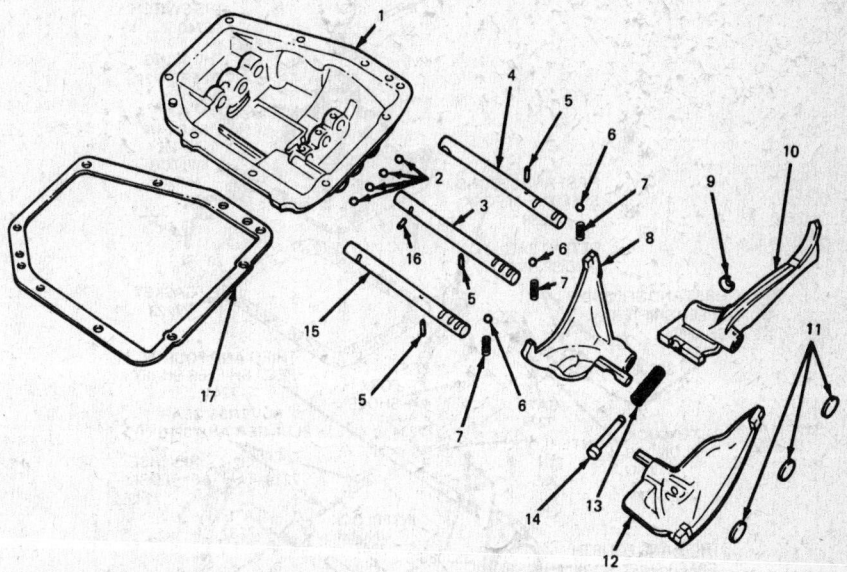

1. Transmission cover
2. Interlock balls
3. 3rd-4th shifter shaft
4. Reverse shifter shaft
5. Fork retaining pin
6. Detent ball
7. Detent spring
8. 3rd-4th shifter fork
9. C-ring lock clip
10. Reverse shifter fork
11. Shifter shaft hole plugs
12. 1st-2nd shifter fork
13. Interlock plunger spring
14. Reverse interlock plunger
15. 1st-2nd shifter shaft
16. Interlock pin
17. Cover gasket

Muncie 117 mm 4-speed cover assembly

8. Remove the drive pinion bearing retainer.

9. Rotate the drive pinion gear to align the space in the pinion gear clutch teeth with the countershaft drive gear teeth. Remove the drive pinion gear and the tapered roller bearing from the transmission by pulling on the pinion shaft, and rapping the face of the case lightly with a brass hammer.

10. Remove the snap-ring, washer, and the pilot roller bearings from the recess in the drive pinion gear.

11. Place a brass drift in the front center of the mainshaft and drive the shaft rearward.

TORQUE SPECIFICATIONS
Muncie-117MM

	ft. lbs.
Rear bearing retainer	18
Cover bolts	25
Filler plug	35
Drain plug	35
Clutch gear bearing retainer bolts	18
Universal joint front flange nut	95
Power take off cover bolts	18
Parking brake	22
Countergear front cover screws	25
Rear mainshaft lock nut (4 wheel drive models)	95

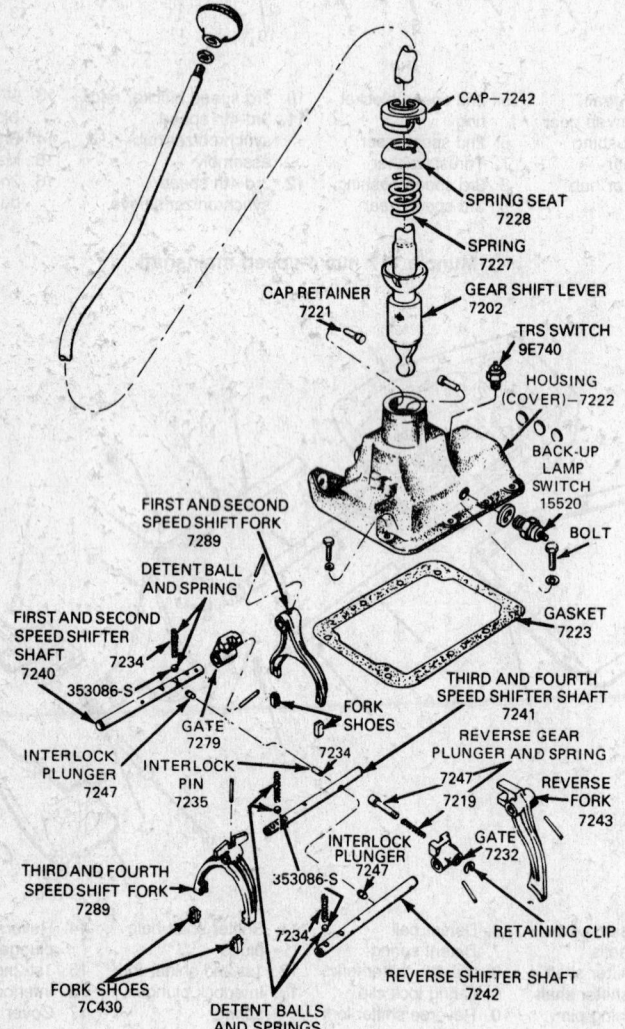

NP 435 gearshift housing

12. When the mainshaft rear bearing has cleared the case, remove the rear bearing and the speedometer drive gear with a suitable gear puller.

13. Move the mainshaft assembly to the rear of the case and tilt the front of the mainshaft upward.

14. Remove the roller type thrust washer.

15. Remove the synchronizer and stop rings separately.

16. Remove the mainshaft assembly.

17. Remove the reverse idler lock screw and lock plate.

18. Using a brass drift held at an angle, drive the idler shaft to the rear while pulling.

19. Lift the reverse idler gear out of the case.

NOTE: If the countershaft gear does not show signs of excessive side play or end play and the teeth are not badly worn or chipped, it may not be necessary to replace the countershaft gear.

20. Remove the bearing retainer at the rear end of the countershaft. The bearing assembly will remain with the retainer.

21. Tilt the cluster gear assembly and work it out of the transmission case.

22. Remove the front bearings from the case with a suitable driver.

MAINSHAFT

1. Remove the clutch gear snap-ring.
2. Remove the clutch gear, the synchronizer outer stop ring to third gear shim, and the third gear.
3. Remove the special split lock ring with two screw drivers. Remove the second gear and synchronizer.
4. Remove the first-reverse sliding gear.
5. Drive the old seal out of the bearing retainer.
6. Place the mainshaft in a soft-jawed vise with the rear end up.
7. Install the first-reverse gear. Be sure the two spline springs, if used, are in place inside the gear as the gear is installed on the shaft.
8. Place the mainshaft in a soft-jawed vise with the front end up.
9. Assemble the second speed synchronizer spring and synchronizer brake on the second gear. Secure the brake with a snapping ring making sure that the snap-ring tangs are away from the gear.
10. Slide the second gear on the front of the mainshaft. Make sure that the synchronizer brake is toward the rear. Secure the gear to the shaft with the two piece lock ring. Install the third gear.
11. Install the shim between the third gear and the third-fourth synchronizer stop ring. Refer to the measurements of end play made during disassembly to determine if additional shims are needed.

NOTE: The exact determination of end-play must be made after the complete assembly of the mainshaft and the main drive pinion is installed in the transmission case.

REVERSE IDLER GEAR

Do not disassemble the reverse idler gear. If it is no longer serviceable, replace the assembly complete with the integral bearings.

COVER & SHIFT FORK UNIT

NOTE: The cover and shift fork assembly should be disassembled only if inspection shows worn or damaged parts, or if the assembly is not working properly.

1. Remove the roll pin from the first-

second shift fork and the shift gate with an "easy out".

NOTE: A square type or a closely wound spiral "easy out" mounted in a tap is preferable for this operation.

2. Move the first-second shift rail forward and force the expansion plug out of the cover. Cover the detent ball access hole in the cover with a cloth to prevent it from flying out. Remove the rail, fork, and gate from the cover.

3. Remove the third-fourth shift rail, then the reverse rail in the manner outlined in Steps 1 and 2 above.

4. Compress the reverse gear plunger and remove the retaining clip. Remove the plunger and spring from the gate.

5. Install the spring on the reverse gear plunger and hold it in the reverse shift gate. Compress the spring in the shift gate and install the retaining clip.

6. Insert the reverse shift rail in the cover and place the detent ball and spring in position. Depress the ball and slide the shift rail over it.

7. Install the shift gate and fork on the reverse shift rail. Install a new roll pin in the gate and the fork.

8. Place the reverse fork in the neutral position.

9. Install the two interlock plungers in their bores.

10. Insert the interlock pin in the third-fourth shift rail. Install the shift rail in the same manner as the reverse shift rail.

11. Install the first-second shift rail in the same manner as outlined above. Make sure the interlock plunger is in place.

12. Check the interlocks by shifting the reverse shift rail into the Reverse position. It should be impossible to shift the other rails with the reverse rail in this position.

13. If the shift lever is to be installed at this point, lubricate the spherical ball seat and place the cap in place.

14. Install the back-up light switch.

15. Install new expansion plugs in the bores of the shift rail holes in the cover. Install the rail interlock hole plug.

DRIVE PINION & BEARING RETAINER

1. Remove the tapered roller bearing from the pinion shaft with a suitable tool.

2. Remove the snap-ring, washer, and the pilot rollers from the gear bore, if they have not been previously removed.

3. Pull the bearing race from the front bearing retainer with a suitable puller.

4. Remove the pinion shaft seal with a suitable tool.

6. Position the drive pinion in an arbor press.

7. Place a wood block on the pinion gear and press it into the bearing until it contacts the bearing inner race.

8. Coat the roller bearings with a light film of grease to hold the bearings in place,

and insert them in the pocket of the drive pinion gear.

9. Install the washer and snap-ring.

10. Press a new seal into the bearing retainer. Make sure that the lip of the seal is toward the mounting surface.

11. Press the bearing race into the retainer.

ASSEMBLY

1. Press the front countershaft roller bearings into the case until the cage is flush with the front of the transmission case. Coat the bearings with a light film of grease.

2. Place the transmission with the front of the case facing down. If uncaged bearings are used, hold the loose rollers in place in the cap with a light film of grease.

3. Lower the countershaft assembly into the case placing the thrust washer tangs in the slots in the case, and inserting the front end of the shaft into the bearing.

4. Place the roller thrust bearing and race on the rear end of the countershaft. Hold the bearing in place with a light film of grease.

5. While holding the gear assembly in alignment, install the rear bearing retainer gasket, retainer, and bearing assembly. Install and tighten the cap screws.

6. Position the reverse idler gear and bearing assembly in the case.

7. Align the idler shaft so that the lock plate groove in the shaft is in position to install the lock plate.

8. Install the lock plate, washer, and cap screw.

9. Make sure the reverse idler gear turns freely.

10. Lower the rear end of the mainshaft assembly into the case, holding the first gear on the shaft. Maneuver the shaft through the rear bearing opening.

NOTE: With the mainshaft assembly moved to the rear of the case, be sure the third-fourth synchronizer and shims remain in position.

11. Install the roller type thrust bearing.

12. Place a wood block between the front of the case and the front of the mainshaft.

13. Install the rear bearing on the mainshaft by carefully driving the bearing onto the shaft and into the case, snap-ring flush against the case.

14. Install the drive pinion shaft and bearing assembly. Make sure that the pilot rollers remain in place.

15. Install the spacer and speedometer drive gear.

16. Install the rear bearing retainer and gasket.

17. Place the drive pinion bearing retainer over the pinion shaft, without the gasket.

18. Hold the retainer tight aginst the bearing and measure the clearance between the retainer and the case with a feeler gauge.

NOTE: End play in Steps 19 and 20 below allows for normal expansion of parts during operation, preventing seizure and damage to bearings, gears, synchronizers, and shafts.

19. Install a gasket shim pack 0.010–0.015 in. thicker than measured clearance between the retainer and case to obtain the required 0.007–0.017 in. pinion shaft end play. Tighten the front retainer bolts and recheck the end play.

20. Check the synchronizer end play clearance (0.050–0.070 in.) after all mainshaft components are in position and properly tightened. Two sets of feeler gauges are used to measure the clearance. Care should be used to keep both gauges as close as possible to both sides of the mainshaft for best results.

NOTE: In some cases, it may be necessary to disassemble the mainshaft and change the thickness of the shims to keep the end play clearance within the specified limits, 0.050–0.070 in. Shims are available in two thicknesses.

21. Install the speedometer drive pinion.

22. Install the yoke flange, drum, and drum assembly.

23. Place the transmission in two gears at once, and tighten the yoke flange nut.

24. Shift the gears and/or synchronizers into all gear positions and check for free rotation.

25. Cover all transmissions components with a film of transmission oil to prevent damage during start up after initial lubricant fill-up.

26. Move the gears to the neutral position.

27. Place a new cover gasket on the transmission case, and lower the cover over the transmission.

28. Carefully engage the shift forks into their proper gears. Align the cover.

29. Install a shouldered alignment screw with split washer in the screw hole second from the front of the cover. Try out gear operation by shifting through all ranges. Make sure everything moves freely.

30. Install the remaining cover screws.

NP445 4 Speed

DISASSEMBLY

1. Place the transmission in a holding fixture and drain the lubricant.

2. Shift the transmission gears into neutral. Remove the gearshift cover attaching bolts. Note that the two bolts opposite the tower are shouldered to properly position the cover. Lift the cover straight up and remove.

3. Lock the transmission in two gears at once and remove the mainshaft nut and yoke.

4. Loosen and remove the extension housing bolts. Remove the mainshaft extension housing and the speedometer drive pinion.

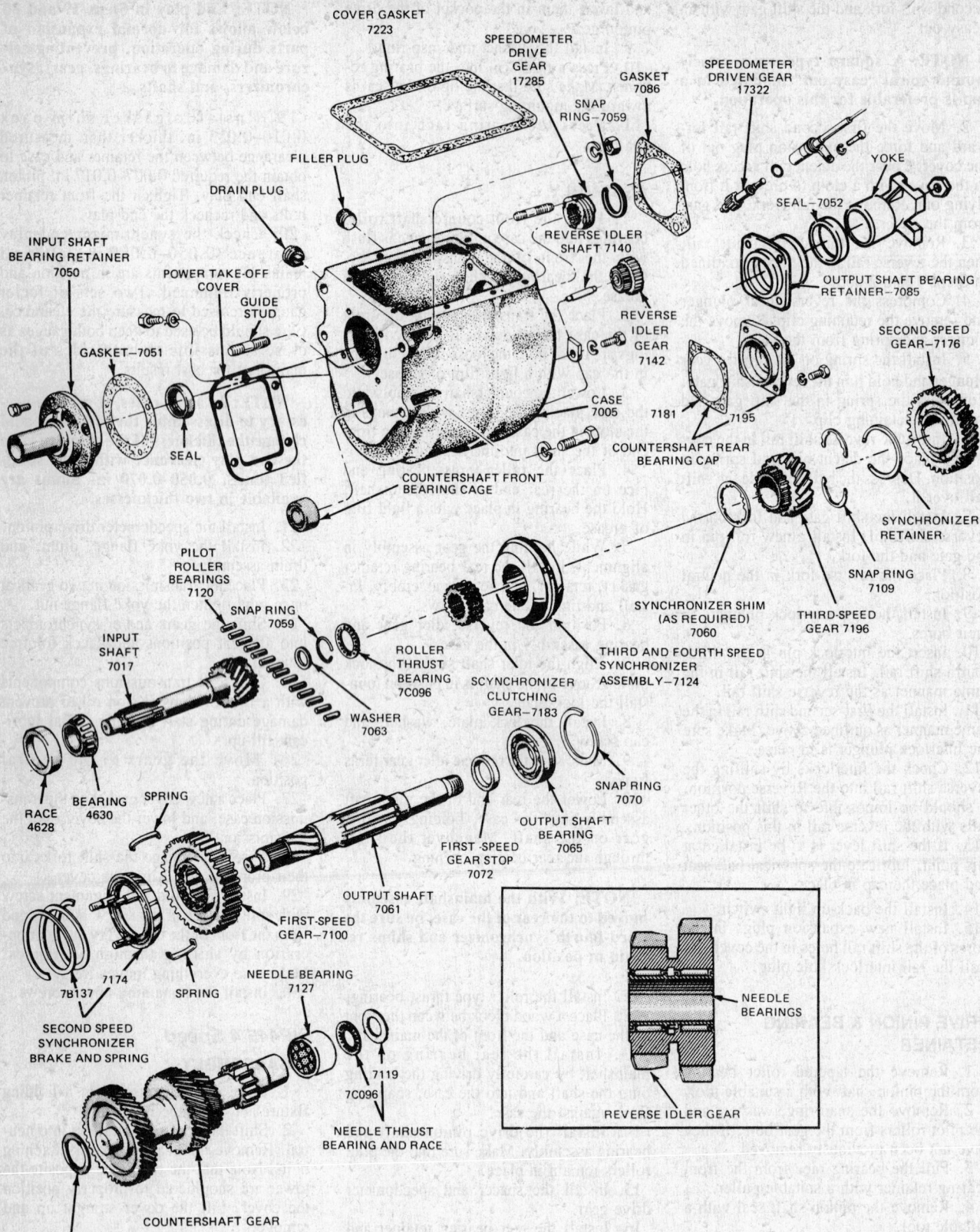

COVER GASKET
7223

SPEEDOMETER
DRIVE
GEAR
17285

GASKET
7086

SPEEDOMETER
DRIVEN GEAR
17322

SNAP
RING—7059

YOKE

SEAL—7052

FILLER PLUG

DRAIN PLUG

REVERSE IDLER
SHAFT 7140

OUTPUT SHAFT BEARING
RETAINER—7085

INPUT SHAFT
BEARING RETAINER
7050

POWER TAKE-OFF
COVER

GUIDE
STUD

REVERSE
IDLER
GEAR
7142

SECOND-SPEED
GEAR-7176

GASKET—7051

SEAL

CASE
7005

7181

7195

COUNTERSHAFT REAR
BEARING CAP

SYNCHRONIZER
RETAINER

COUNTERSHAFT FRONT
BEARING CAGE

SNAP RING
7109

PILOT
ROLLER
BEARINGS
7120

SNAP RING
7059

SYNCHRONIZER SHIM
(AS REQUIRED)
7060

THIRD-SPEED
GEAR 7196

INPUT
SHAFT
7017

ROLLER
THRUST
BEARING
7C096

SCYNCHRONIZER
CLUTCHING
GEAR—7183

THIRD AND FOURTH SPEED
SYNCHRONIZER
ASSEMBLY—7124

WASHER
7063

SNAP RING
7070

RACE
4628

BEARING
4630

SPRING

FIRST-SPEED
GEAR STOP
7072

OUTPUT SHAFT
BEARING
7065

OUTPUT SHAFT
7061

FIRST-SPEED
GEAR—7100

7174

7B137

SPRING

NEEDLE BEARING
7127

SECOND SPEED
SYNCHRONIZER
BRAKE AND SPRING

7119

7C096

NEEDLE THRUST
BEARING AND RACE

NEEDLE
BEARINGS

REVERSE IDLER GEAR

THRUST WASHER

COUNTERSHAFT GEAR

New Process 435 4-speed

TORQUE SPECIFICATIONS
New Process 435

	ft. lbs.
Cover screws	20–40
Drive gear retaining screw	15–25
Front countershaft retainer screw	15–25
Front countershaft bearing washer screw	12–22
Flange nut	125
Mainshaft rear retainer screw	15–25
Rear countershaft retainer screw	15–25
PTO cover screws	8–12
Filler and drain plugs	25–45
Reverse idler shaft lock screw	20–40
Brake link shoulder screw	20–40

5. Remove the bolts from the drive pinion front bearing retainer and pull the bearing retainer and gasket off.

6. Rotate the drive pinion gear to align the pinion gear flat with the countershaft drive gear teeth. Remove the drive pinion gear and the tapered roller bearing from the transmission.

7. Remove the mainshaft thrust bearing.

8. Push the mainshaft assembly to the rear of the transmission and tilt the front of the mainshaft up.

9. Remove the mainshaft assembly from the transmission case.

10. Remove the reverse idler lock screw and lock plate.

11. Using a suitable size brass drift, carefully drive the reverse idler shaft out the REAR of the case.

─────── CAUTION ───────

Do not attempt to drive the reverse idler shaft forward! This will damage the transmission case and the reverse idler shaft.

12. Remove the countershaft rear bearing retainer.

13. Slide the countershaft to the rear, then up and out of the case.

14. Drive the countershaft forward, out of the bearing and the case.

MAINSHAFT

1. Place the mainshaft in a soft-jawed vise with the front end up.

2. Lift the third-fourth synchronizer and high speed clutch off the mainshaft.

3. Remove the third gear.

4. Remove the second gear snap-ring. Lift off the thrust washer.

5. Remove the second gear.

6. Remove the first-reverse synchronizer and clutch gear.

7. Install the mainshaft in the vise rear end up.

8. Remove the tapered bearing from the shaft with a suitable gear puller.

9. Remove the first gear snap-ring and thrust washer.

10. Remove the first gear.

11. Lubricate all parts with transmission lubricant prior to assembly.

12. Place the mainshaft in a soft-jawed vise with the rear end up.

13. Slide the first gear over the mainshaft, with the clutch gear facing down. Install the thrust washer and snap-ring.

14. Install the reverse gear over the end of the mainshaft with the fork groove facing down.

15. Install the mainshaft rear bearing on the mainshaft with a sleeve of suitable size. Press the bearing on its inner race.

16. Install the mainshaft in the vise with the front end facing up.

17. Install the first-reverse synchronizer.

18. Install the second gear on the mainshaft.

19. Install the keyed thrust washer, ground side toward the second gear and secure with the snap-ring.

20. Install the third gear and one shim on the mainshaft.

21. Install the third fourth synchronizer over the mainshaft. Make sure that the slotted end of the clutch gear is positioned toward the third gear.

COVER & SHIFT FORK

NOTE: **The cover and shift fork assembly should be disassembled only if inspection shows worn or damaged parts, or if the assembly is not working properly.**

1. Remove the roll pin from the first-second shift fork and the shift gate. Use a square-type or spiral wound "easy-out" mounted in a tap handle for these operations.

2. Move the first-second shift rail rearward and force the expansion plug out of the cover. Cover the detent ball access hole in the cover with a cloth to prevent it from flying out. Remove the rail fork, and gate from the cover.

3. Remove the third-fourth shift rail, then the reverse rail in the manner outlined in Steps 1 and 2 above.

4. Compress the reverse gear plunger and remove the retaining clip. Remove the plunger and spring from the gate.

5. Apply a thin film of grease on the interlock slugs and slide them into the openings in the shift rail supports.

6. Install the reverse shift rail through the reverse shift fork plate and the reverse shift fork.

7. Secure the reverse shift plate and the shift fork with the roll pins. Install the interlock pin in the third-fourth shift rail. Hold in place with a thin film of grease.

8. Slide the third-fourth shift rail into the rail support from the rear of the cover. Slide the rail through the third-fourth shift fork and poppet ball and spring. Secure the third-fourth shift fork with the roll pin.

9. Install the interlock pin in the first-second shift rail and secure with a light coat of grease. Slide the first-second shift rail into the case, through the shift fork and shift gate. Hold the poppet ball and spring down until the shift rail passes.

10. Secure the first-second shift rail and gate with the roll pins.

ASSEMBLY

1. Install the countershaft front bearing in the case using a $1\frac{3}{8}$ in. socket as a driver. Grease the needle bearings prior to installation. Hold the bearings in place with a socket of suitable size while seating the bearing retainer. Drive the retainer in until it is flush with the case.

2. Install the tanged thrust washer on the countershaft with the tangs facing out. Install the countershaft in the transmission case.

3. Install the countershaft rear bearing retainer over the rear bearing. Use a new washer and position the retainer with the curved segment toward the bottom of the case.

4. Install the reverse idler gear into the case with the chamfered section facing the rear. Hold the thrust washer and needle bearings in position.

5. Slide the reverse idler shaft into the case, from the rear, and through the reverse idler gear. Make sure that the lock notch is down and at the rear of the case.

6. Install the reverse idler shaft lock and bolt.

7. Place the mainshaft in a soft-jawed vise with the front end facing up.

8. Install the drive gear on top of the mainshaft.

9. Measure the clearance between the high-speed synchronizer and the drive gear with two feeler gauges. If the clearance is greater than 0.043–0.053 in., install synchronizer shims between the third gear and the synchronizer brake drum. After the required shims have been installed, remove the drive gear from the mainshaft.

10. Install the mainshaft into the transmission case. Place the thrust washer over the pilot end of the mainshaft.

11. Position the drive gear so that the cutaway portion of the gear is facing down. Slide the drive gear into the front of the case and engage the mainshaft pilot in the pocket of the drive gear.

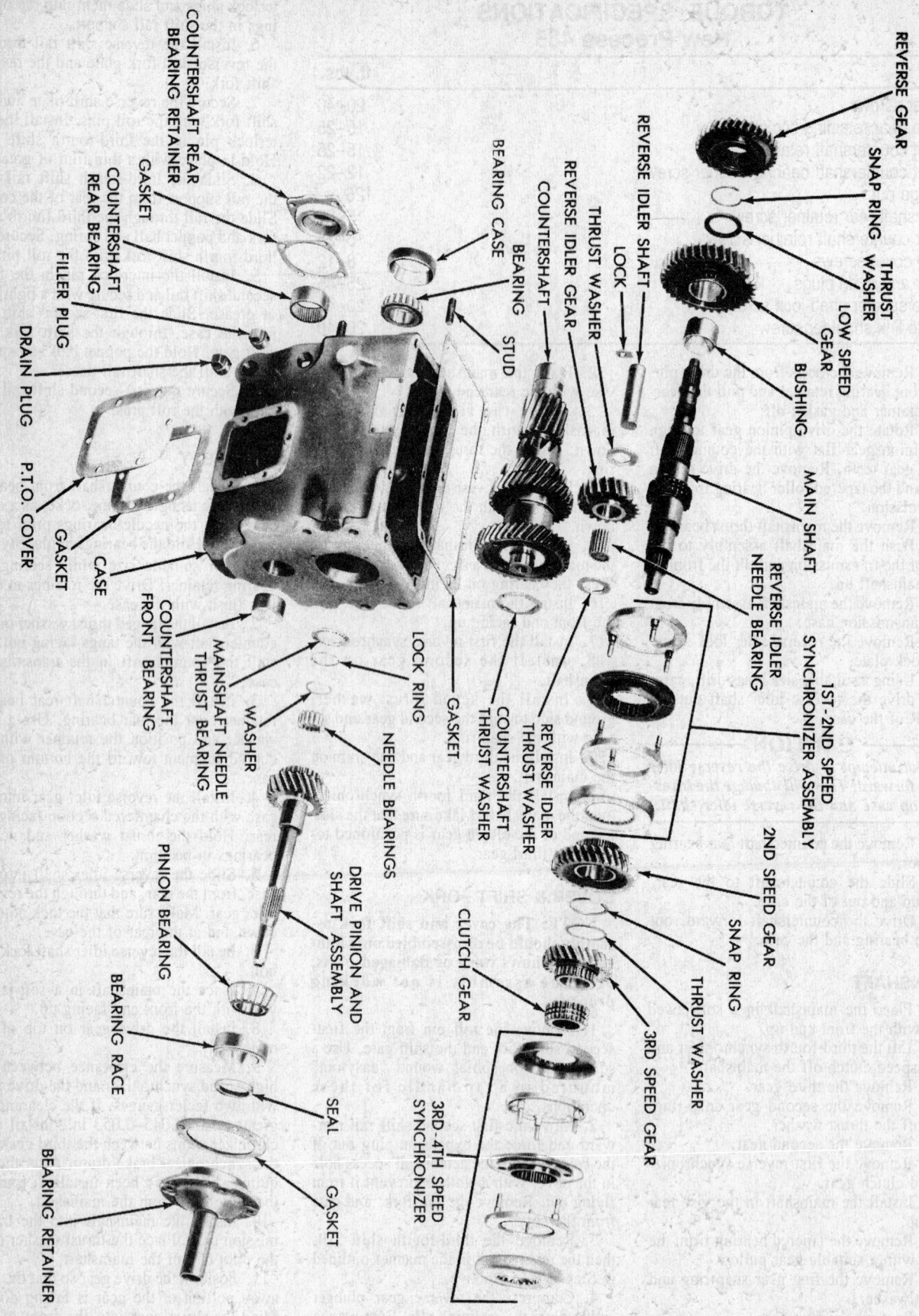

COUNTERSHAFT REAR
BEARING RETAINER

GASKET

COUNTERSHAFT
REAR BEARING

FILLER PLUG

DRAIN PLUG

P.T.O. COVER

GASKET

CASE

BEARING CASE

BEARING

COUNTERSHAFT

REVERSE IDLER GEAR

REVERSE IDLER SHAFT

THRUST WASHER

LOCK

STUD

REVERSE GEAR

SNAP RING

THRUST
WASHER

LOW SPEED
GEAR

BUSHING

MAIN SHAFT

REVERSE IDLER
NEEDLE BEARING

1ST-2ND SPEED
SYNCHRONIZER ASSEMBLY

2ND SPEED GEAR

SNAP RING

THRUST WASHER

3RD SPEED GEAR

CLUTCH GEAR

3RD-4TH SPEED
SYNCHRONIZER

COUNTERSHAFT
FRONT BEARING

MAINSHAFT NEEDLE
THRUST BEARING

WASHER

NEEDLE BEARINGS

LOCK RING

GASKET

COUNTERSHAFT
THRUST WASHER

REVERSE IDLER
THRUST WASHER

PINION BEARING

DRIVE PINION AND
SHAFT ASSEMBLY

BEARING RACE

SEAL

GASKET

BEARING RETAINER

New Process 445

TORQUE SPECIFICATIONS
New Process 445

	ft. lbs.
Cover screws	20–40
Drive gear retaining screw	15–25
Front countershaft retaining screw	15–25
Front countershaft bearing washer screw	12–22
Flange nut	125
Mainshaft rear retainer screw	15–25
Rear countershaft retainer screw	15–25
PTO cover screws	8–12
Filler and drain plugs	25–45
Reverse idler shaft lock screw	20–40
Brake link shoulder screw	20–40

12. Slip the drive gear front bearing retainer over the shaft on gasket, and do not secure with bolts.

13. Install the mainshaft rear bearing retainer. Tighten the screws to specifications.

14. Hold the retainer against the front of the transmission case and measure the clearance between the front bearing retainer and the front of the case with a feeler gauge. Record the measurement and remove the bearing retainer.

15. Install a gasket pack on the front bearing retainer which is 0.010–0.015 in. thicker than the clearance measured in Step 14. Install the front bearing retainer and torque attaching screws to specification.

16. The end play float of the front synchronizer must be checked before installation of the transmission cover assembly. Measure the end play "float" by inserting two feeler gauges opposite one another between the third gear and the synchronizer stop ring. Accurate measurement can be made only after all mainshaft parts are in place and torqued to specification.

17. If the front synchronizer end play "float" does not fall between 0.050–0.070 in., shims should be added or removed as required, from between the third gear and the synchronizer stop ring.

18. Install the yoke retaining nut on the rear of the mainshaft. Shift the transmission into two gears at the same time and torque the yoke nut to 125 ft. lbs.

19. Shift the transmission into neutral.

20. Install the cover gasket.

21. Shift the transmission into second gear. Shift the cover into second.

22. Carefully lower the cover into position. It may be necessary to position the reverse gear to permit the fork to engage its groove.

23. Install the cover aligning screws (shouldered) and tighten with fingers only.

24. Install the remaining cover screws and tighten to specifications.

Saginaw

SM326 76mm 3 Speed

The Saginaw Model SM326 is a synchro-mesh three-speed transmission using helical constant mesh gears. The engagement of all gears except reverse is assisted by synchronizers.

DISASSEMBLY

1. Remove side cover assembly and shift forks.

2. Remove clutch gear bearing retainer.

3. Remove clutch gear bearing to gear stem snap-ring. Pull clutch gear outward until a screwdriver can be inserted between bearing and case. Remove clutch gear bearing.

4. Remove speedometer driven gear and extension bolts.

5. Remove reverse idler shaft snap-ring. Slide reverse idler gear forward on shaft.

6. Remove mainshaft and extension assembly.

7. Remove clutch gear and third-speed blocker ring from inside case. Remove 14 roller bearings from clutch gear.

8. Expand the snap-ring which retains the mainshaft rear bearing. Remove the extension.

9. Using a dummy shaft, drive the countershaft and key out the rear of the case. Remove the gear, two tanged thrust washers, and dummy shaft. Remove bearing washer and 27 roller bearings from each end of countergear.

10. Use a long drift to drive the reverse idler shaft and key through the rear of the case.

11. Remove reverse idler gear and tanged steel thrust washer.

MAINSHAFT

1. Remove second and third speed sliding clutch hug snap-ring from mainshaft. Remove clutch assembly, second speed blocker ring, and second speed gear from front of mainshaft.

2. Depress speedometer drive gear retaining clip. Remove gear. Some units have a metal speedometer drive gear which must be pulled off.

3. Remove rear bearing snap-ring.

4. Support reverse gear. Press on rear of mainshaft. Remove reverse gear, thrust washer, spring washer, rear bearing, and snap-ring. When pressing off the rear bearing, be careful not to cock the bearing on the shaft.

5. Remove first and reverse sliding clutch hub snap-ring. Remove clutch assembly, first speed blocker ring, and first gear.

CLUTCH KEYS & SPRINGS

Keys and springs may be replaced if worn or broken, but the hubs and sleeves are matched pairs and must be kept together.

1. Mark hub and sleeve for reassembly.

2. Push hub from sleeve. Remove keys and springs.

3. Place three keys and two springs, one on each side of hub, in position, so all three keys are engaged by both springs. The tanged end of the springs should not be installed into the same key.

4. Slide the sleeve onto the hub, aligning the marks.

NOTE: A groove around the outside of the synchronizer hub marks the end that must be opposite the fork slot in the sleeve when assembled.

EXTENSION OIL SEAL & BUSHING

1. Remove seal.

2. Using bushing remover and installer tool, or other suitable tool, drive bushing into extension housing.

3. Drive new bushing in from the rear. Lubricate inside of bushing and seal. Install new oil seal with extension seal installer tool or other suitable tool.

CLUTCH BEARING RETAINER OIL SEAL

1. Pry old seal out.

2. Install new seal using seal installer or suitable tool. Seat seal in bore.

ASSEMBLY

1. Turn front of mainshaft up.

2. Install second gear with clutching teeth up; the rear face of the gear butts against the flange on the mainshaft.

3. Install a blocker ring with clutching teeth down. All three blocker rings are the same.

4. Install second and third speed synchronizer assembly with fork slot down. Press it onto mainshaft splines. Both synchronizer assemblies are the same. Be sure that blocker ring notches align with synchronizer assembly keys.

5. Install synchronizer snap-ring. Both synchronizer snap-rings are the same.

6. Turn rear of shaft up.

7. Install first gear with clutching teeth up; the front face of the gear butts against the flange on the mainshaft.

8. Install a blocker ring with clutching teeth down.

9. Install first and reverse synchronizer assembly with fork slot down. Press it onto mainshaft splines. Be sure blocker ring

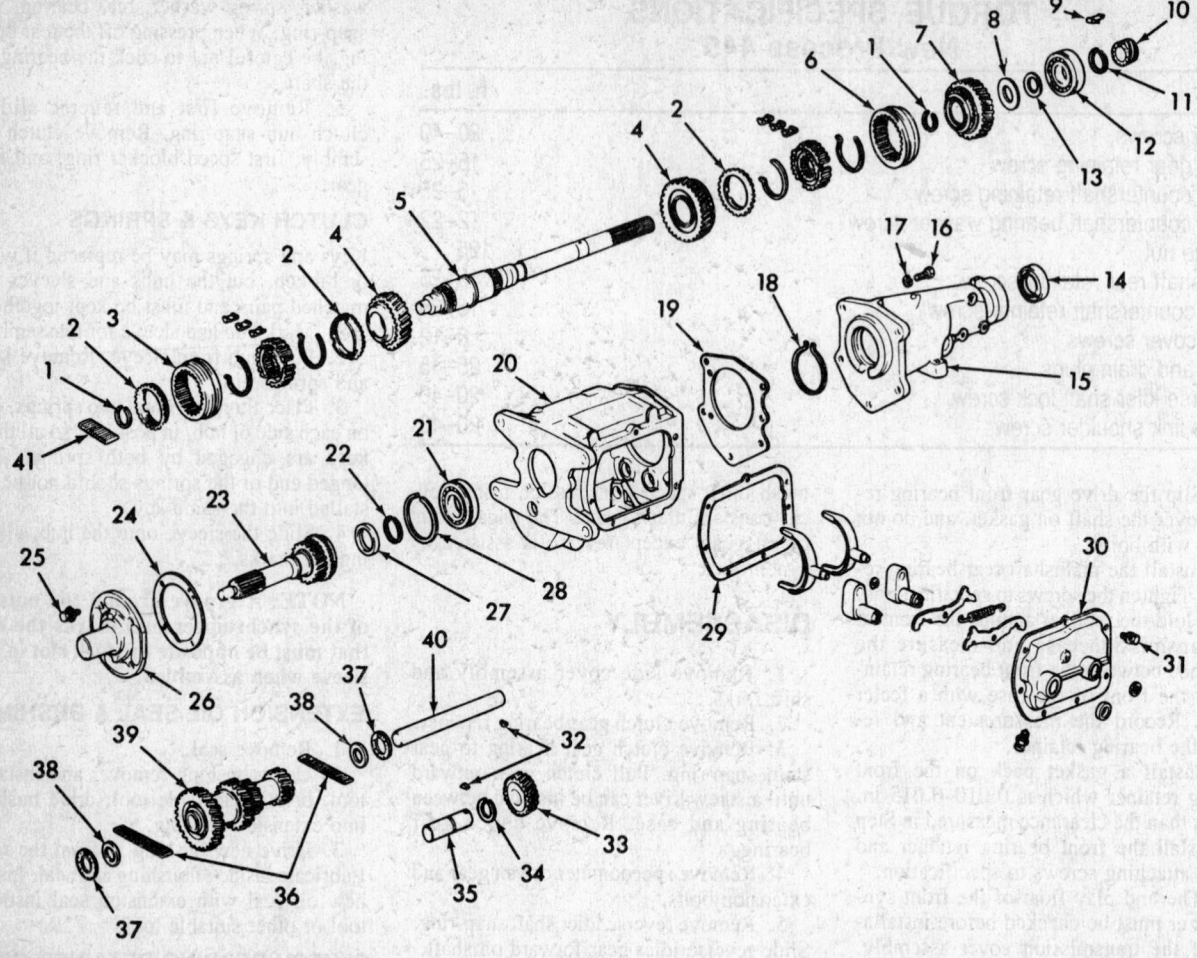

1. Synchronizer retainer ring
2. Synchronizer blocking ring
3. Synchronizer assembly
4. Second speed gear
5. Main shaft
6. Synchronizer assembly
7. Gear assembly
8. Thrust washer
9. Retainer clip
10. Speedometer drive gear
11. Ring
12. Mainshaft bearing
13. Washer
14. Seal
15. Extension housing
16. Bolt
17. Washer
18. Rear bearing location ring
19. Gasket
20. Case
21. Bearing assembly
22. Ring
23. Clutch gear
24. Gasket
25. Bolt and lockwasher
26. Retainer assembly
27. Ring
28. Clutch gear bearing locating ring
29. Cover gasket
30. Cover assembly
31. Bolt and lockwasher
32. Woodruff keys
33. Gear assembly
34. Retaining ring
35. Shaft
36. Roller
37. Washer
38. Washer
39. Gear assembly
40. Counter gear shaft
41. Mainshaft bearing roller

Saginaw transmission

notches align with synchronizer assembly keys.

10. Install snap-ring.

11. Install reverse gear with clutching teeth down.

12. Install steel reverse gear thrust washer and spring washer.

13. Press rear ball bearing onto shaft with snap-ring slot down.

14. Install snap-ring.

15. Install speedometer drive gear and retaining clip. Press on metal speedometer drive gear.

CASE ASSEMBLY

1. Using dummy shaft load a row of 27 roller bearings and a thrust washer at each end of countergear. Hold in place with grease.

2. Place countergear assembly into case through rear. Place a tanged thrust washer, tang away from gear at each end. Install countershaft and key, making sure that tangs align with notches in case.

3. Install reverse idler gear thrust washer, gear, and shaft with key from rear of case. Be sure thrust washer is between gear and rear of case with tang toward notch in case.

NOTE: The reverse idler gear bushing may not be replaced separately, only as a unit with the gear.

4. Expand snap-ring in extension. Assemble extension over rear of mainshaft and onto rear bearing. Seat snap-ring in rear bearing groove.

5. Install 14 mainshaft pilot bearings into clutch gear cavity. Assemble third speed blocker ring onto clutch gear clutching surface with teeth toward gear.

6. Place clutch gear, pilot bearings, and third speed blocker ring assembly over front of mainshaft assembly. Be sure blocker rings align with keys in second-third synchronizer assembly.

7. Stick extension gasket to case with grease. Install clutch gear, mainshaft, and extension together. Be sure clutch gear engages teeth of countergear anti-lash plate. Torque extension bolts to 45 ft. lbs.

8. Place bearing over stem of clutch gear and into front case bore. Install front bearing to clutch gear snap-ring.

9. Install clutch gear bearing retainer and gasket. The retainer oil return hole must be at the bottom. Torque to 10 ft. lbs.

10. Install reverse idler gear shaft E-ring.

11. Shift synchronizer sleeves to neutral

TORQUE SPECIFICATIONS
Saginaw-76mm

	ft. lbs.
Extension to case attaching bolts	35–55
Drain and filler plugs	10–15
Side cover attaching bolts	18–24
Clutch gear retainer bolts	18–24

TORQUE SPECIFICATIONS
Tremec T-150

	ft. lbs.
Back-up light switch	15–20
Fill and drain plugs	10–20
Front bearing cap bolt	30–36
Shift control housing bolts	20–25
Transfer case drive gear locknut	150
Transfer case to transmission bolts	30
TCS switch	18

positions. Install cover, gasket, and forks, aligning forks with synchronizer sleeve grooves. Torque side cover bolts to 10 ft. lbs.

12. Install speedometer driven gear.

Tremec

T–150 3 Speed Transmission (77 mm)

The Tremec T–150 (77 mm) transmission is used in varied vehicle applications, with or without transfer cases. The gear selection is controlled by either a top shift housing or by a remote control shift lever assembly. Although some of the gears and case applications are not interchangeable, the gear arrangement is basically the same.

DISASSEMBLY

1. Remove the bolts securing the transfer case to the transmission. Remove the transfer case.

2. Remove the transfer case drive gear locknut, flat washer, and drive gear. Remove the large fiber washer from the rear bearing adapter. Move the second-third clutch sleeve forward and the first-reverse sleeve to the rear before removing the locknut.

3. Remove the transmission oil plug and drive the countershaft out of the case with a suitable size drift. Do not lose the countershaft access plug when removing the countershaft. With the countershaft removed the countershaft gear will lie at the bottom of the case, leave it there until the mainshaft is removed.

4. Punch alignment marks in the front bearing cap and the transmission case for assembly reference.

5. Remove the front bearing cap and gasket.

6. Remove the large lock ring from the front bearing.

7. Remove the clutch shaft, front bearing, and the second-third synchronizer assembly. A special tool is required for this operation.

8. Remove the rear bearing and adapter assembly with a brass drift and hammer. Drive the adapter out the rear of the case with light blows from the hammer.

9. Remove the mainshaft assembly. Tilt the spline end of the shaft downward and lift the front end up and out of the case.

10. Remove the countershaft tool and arbor as an assembly. Remove the countershaft thrust washers, countershaft roll pin, and any pilot roller bearings that may have fallen into the case.

11. Remove the reverse idler shaft. Insert a brass drift through the clutch shaft bore in the front of the case and tap the shaft until the end with the roll pin clears the counter bore in the rear of the case. Remove the shaft.

12. Remove the reverse idler gear and thrust washers from the case.

13. Remove the retaining snap-ring from the front of the mainshaft. Remove the second-third synchronizer assembly and second gear. Mark the hub and sleeve for reference during assembly.

NOTE: Observe the position of the insert springs and the inserts during removal for correct assembly.

14. Remove the insert springs from the second-third synchronizer, remove the three inserts, and separate the sleeve from the synchronizer hub retaining snap-ring.

15. Remove the snap-ring and the tabbed thrust washer from the mainshaft and remove the first gear blocking ring.

16. Remove the first-reverse synchronizer hub snap-ring.

NOTE: Observe the position of the insert springs and the inserts during removal for correct assembly.

17. Remove the first-reverse sleeve, insert spring and the three insert from the hub. Remove the spacer from the rear of the mainshaft.

─── CAUTION ───

Do not attempt to remove the press fit hub by hammering. Hammer blows will damage the hub and mainshaft.

18. Remove the front bearing retaining snap-ring and any remaining roller bearings from the clutch shaft.

19. Press the front bearing off the clutch shaft with an arbor press.

─── CAUTION ───

Do not attempt to remove the bearing by hammering. Hammer blows will damage the bearing and the clutch shaft.

20. Clamp the rear bearing adapter in a soft-jawed vise. Do not over-tighten.

21. Remove the rear bearing retaining snap-ring. Remove the bearing adapter from the vise.

22. Press the rear bearing out of the adapter with an arbor press.

23. Thoroughly wash all parts in clean solvent and dry with compressed air. Do not dry the bearings with compressed air, use a clean shop cloth. Clean the needle and clutch shaft bearings by placing them in a shallow parts cleaning tray and covering them with solvent. Allow the bearings to air dry on a clean shop cloth. Check the case for the following: Cracks in the bores, bosses, or bolt holes. Stripped threads in bolt holes. Nicks, burrs, rough surfaces in the shaft bores or on the gasket surfaces.

24. Check the gear and synchronizer assemblies for the following: Broken, chipped, or worn gear teeth. Damaged splines on the synchronizer hubs or sleeves. Bent or damaged inserts. Damaged needle bearings or bearing bores in the countershaft gear. Broken or worn teeth or excessive wear of the blocking rings. Wear of galling of the countershaft, clutch shaft, or reverse idler shaft. Worn thrust washers. Nicked, broken, or worn mainshaft or clutch shaft splines. Bent, distorted, or weak snap-rings. Worn bushings in the reverse idler gear. Replace the gear if the bushings are worn. Rough, galled, or broken front or rear bearings.

ASSEMBLY

1. Lubricate the reverse idler shaft bore and bushings with transmission oil.

2. Coat the transmission case reverse idler gear thrust washer surfaces with petroleum jelly and install the thrust washers in the case.

NOTE: Make sure the locating tangs on the thrust washers are aligned in the slots in the case.

3. Install the reverse idler gear. Align the gear bore, thrust washers, and case bore. Install the reverse idler shaft from the rear of the transmission case. Be sure to align and seat the roll pin in the shaft into the counter bore in the rear of the case.

4. Measure the reverse idler gear end-play by inserting a feeler gauge between the thrust washer and the gear. End-play should be 0.004–0.018 in. If end play ex-

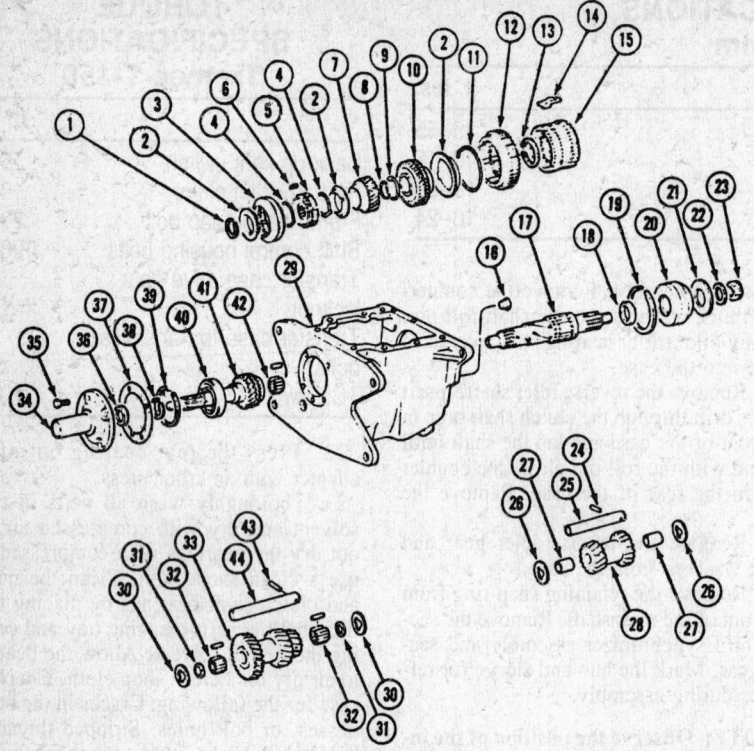

1. Mainshaft retaining snap-ring
2. Synchronizer blocking rings (3)
3. Second-third synchronizer sleeve
4. Second-third synchronizer insert spring (2)
5. Second-third hub
6. Second-third synchronizer insert (3)
7. Second gear
8. First gear retaining snap-ring
9. First gear tabbed thrust washer
10. First gear
11. First-reverse synchronizer insert spring
12. First-reverse sleeve
13. First-reverse hub retaining snap-ring
14. First-reverse synchronizer insert (3)
15. First-reverse hub
16. Countershaft access plug
17. Mainshaft
18. Mainshaft spacer
19. Rear bearing adapter lock ring
20. Rear bearing and adapter assembly
21. Fiber washer
22. Flat washer
23. Locknut
24. Roll pin
25. Reverse idler gear shaft
26. Thrust washer
27. Bushing (part of idler and gear
28. Reverse idler gear
29. Transmission case
30. Thrust washer (2)
31. Bearing retainer (2)
32. Countershaft needle bearings (50)
33. Countershaft gear
34. Front bearing cap
35. Bolt (4)
36. Front bearing cap oil seal
37. Gasket
38. Front bearing retainer snap-ring
39. Front bearing lockring
40. Front bearing
41. Clutch shaft
42. Mainshaft pilot roller bearings
43. Roll pin
44. Countershaft

T-150

ceeds 0.018 in., remove the reverse idler gear and replace the thrust washers.

5. Coat the needle bearing bores in the countershaft gear with petroleum jelly. Insert the arbor tool in the bore of the gear and install the (25) needle bearings and the retainer washers at each end of the countershaft gear.

6. Coat the countershaft gear thrust washer surface with petroleum jelly and position the thrust washers in the case.

NOTE: Make sure the locating tangs on the thrust washers are aligned in the slots in the case.

7. Insert the countershaft into the bore at the rear of the case just far enough to hold the thrust washer in place.

8. Install the countershaft gear in the case. Do not install the roll pin at this time. Align the gear bore, thrust washers, the bores in the case, and install the countershaft.

NOTE: Do not remove the arbor tool completely.

9. Measure the countershaft gear endplay by inserting a feeler gauge between the washer and the countershaft gear. Endplay should be 0.004–0.018 in. If the endplay exceeds 0.018 in., remove the gear and replace the thrust washer.

10. When the correct countershaft gear end-play has been obtained, install the countershaft arbor and remove the countershaft. Allow the countershaft gear to re-

main at the bottom of the case, leave the countershaft in the case enough to hold the thrust washer in place.

11. Coat the splines and machined surfaces on the mainshaft with transmission oil. Install the first-reverse synchronizer on the output shaft splines by hand. The end of the hub with the slots should face the front of the shaft. Use an arbor press to complete the hub installation. Install the retaining snap-ring in the groove farthest to the rear.

CAUTION

Do not attempt to drive the hub on the shaft with a hammer.

12. Coat the splines of the first-reverse hub with transmission oil and install the first reverse sleeve and gear halfway onto the hub, with the gear end of the sleeve facing the rear of the shaft. Align the marks made during disassembly.

13. Install the insert spring in the first-reverse hub. Make sure the spring bottoms in the hub and covers all three insert slots. Position the three "T" shaped inserts in the hub with the small ends in the hub slots and the large ends inside the hub. Push the inserts fully into the hub so they seat on the insert spring, slide the first-reverse sleeve and gear over the inserts until the inserts engage in the sleeve.

14. Coat the bore and the blocking ring surface of first gear with transmission oil and place blocking ring on the tapered surface of the gear.

15. Install the first gear on the output shaft. Rotate the gear until the notches in the blocking ring engage the inserts in the first-reverse synchronizer assembly. Install the tanged thrust washer, sharp end facing out, and retaining snap-ring on the mainshaft.

16. Coat the bore and blocking ring surface of the second gear with transmission oil. Place the second gear blocking ring on the tapered surface of second gear.

17. Install the second gear on the output shaft with the tapered surface of the gear facing the front of the mainshaft.

18. Install one insert spring into the second-third synchronizer hub. Be sure that the spring covers all three insert slots in the hub. Align the second-third sleeve with the hub using the marks made during disassembly. Start the sleeve onto the hub.

19. Place the three inserts into the hub slots and on top of the insert spring. Push the sleeve fully onto the hub to engage the inserts in the sleeve. Install the remaining insert spring in the exact position as the first spring. The ends of both springs must cover the same slot in the hub and not be staggered.

NOTE: The inserts have a small lip on each end. When they are correctly installed, this lip will fit over the insert spring.

20. Install the second-third synchronizer assembly on the mainshaft. Rotate the sec-

ond gear until the notches in the blocking ring engage the inserts in the second-third synchronizer assembly.

21. Install the retaining snap-ring on the mainshaft and measure the end-play between the snap-ring and the second-third synchronizer hub. The end-play should be 0.040–0.014 in. If the end-play exceeds the limit, replace the thrust washer and all the snap-rings on the mainshaft assembly. Install the spacer on the rear of the mainshaft.

22. Install the mainshaft assembly in the case. Be sure that the first-reverse sleeve and gear is in the neutral (centered) position.

23. Press the rear bearing into the rear bearing adapter with an arbor press. Install the rear bearing retaining ring and the bearing adapter lockring.

24. Support the mainshaft assembly and install the rear bearing and adapter assembly in the case. Use a soft faced hammer to seat the adapter in the case.

25. Install the large fiber washer in the rear bearing adapter. Install the transfer drive gear, flat washer, and locknut. Tighten the locknut to 150 ft. lbs. torque.

26. Press the front bearing onto the clutch shaft. Install the bearing retaining snap-ring on the clutch shaft and the lockring into its groove.

27. Coat the bore of the clutch shaft assembly with petroleum jelly and install the (15) roller bearings in the clutch shaft bore.

—— CAUTION ——

Do not use chassis grease or a similar heavy grease in the clutch shaft bore. Heavy grease will plug the lubricant holes in the shaft and prevent proper lubrication of the roller bearings.

28. Coat the blocking ring surface of the clutch shaft with transmission oil. Position the blocking ring on the clutch shaft.

29. Support the mainshaft assembly and insert the clutch shaft through the front bearing bore in the case. Seat the mainshaft pilot in the clutch shaft roller bearings. Tap the bearings into place with a soft faced hammer.

30. Apply a thin film of sealer to the front bearing cap gasket and position the gasket on the case. Be sure the cutout in the gasket is aligned with the oil return hole in the case.

31. Remove the front bearing cap oil seal with a suitable tool. Install a new seal with a suitable driver.

32. Install the front bearing cap and tighten the bolts to 33 ft. lbs. Be sure that the marks on the cap and the transmission case are aligned and the oil return slot in the cap lines up with the oil return hole in the case.

33. Make a wire loop about 18–20 inches long and pass the wire under the countershaft gear assembly. The wire loop should raise and support the countershaft gear assembly when it is pulled upward.

34. Raise the countershaft gear with the wire. Align the bore in the countershaft gear with the front thrust washer and the countershaft. Start the countershaft into the gear with a soft faced hammer.

35. Align the roll pin hole in the countershaft with the roll pin holes in the case and complete the installation of the countershaft. Install the countershaft access plug in the rear of the case and seat with a soft faced hammer.

36. Install the countershaft roll pin in the case. Use a magnet or needle nose pliers to insert and start the pin in the case. Use a ½ in. punch to seat the pin. Install the transmission filler plug.

37. Shift the synchronizer sleeves through all gear ranges and check their operation. If the clutch shaft and mainshaft appear to bind in the neutral position, check for blocking rings sticking on the first or second gear tapers.

38. Install the transfer case on the transmission. Tighten the attaching bolts to 30 ft. lbs.

SHIFT CONTROL HOUSING

1. Remove the back-up light switch and the transmission controlled spark switch (TCS) if so equipped.

2. Remove the shift control housing cap, gasket, spring retainer, and the shift lever spring as an assembly.

3. Invert the housing and mount in a soft-jawed vise.

4. Move the second-third shift rail to the rear of the housing, rotate the shift fork toward the first-reverse rail until the roll pin is accessible. Drive the roll pin out of the fork and rail with a pin punch. Remove the shift fork and the roll pin.

NOTE: The roll pin hole in the shift fork is offset. Mark the position of the shift fork for assembly reference.

5. Remove the second-third shift rail using a brass drift or hammer. Catch the shift rail plug as the rail drives it out of the housing. Cover the shift and poppet ball holes in the cover to prevent the poppet ball from flying out. Mark the location of the shift rail for assembly reference.

6. Rotate the first-reverse shift fork away from the notch in the housing until the roll pin is accessible. Drive the roll pin out of the fork and rail using a pin punch. Remove the shift fork and roll pin.

NOTE: The roll pin hole in the shift fork is offset. Mark the position of the shift fork for assembly reference.

7. Remove the first-reverse shift rail using a brass drift or hammer. Catch the shift rail plug as the rail drives it out of the housing. Cover the shift and poppet ball holes in the cover to prevent the poppet ball from flying out. Mark the location of the shift rail for assembly reference.

8. Remove the poppet balls, springs, and the interlock plunger from the housing.

9. Install the poppet springs and the detent plug in the housing.

10. Insert the first-reverse shift rail into the housing, and install the shift fork on the shift rail.

11. Install the poppet ball on the top of the spring in the first-reverse rail.

12. Using a punch or wooden dowel, push the poppet ball and spring downward into the housing bore and install the first-reverse shift rail.

13. Align the roll pin holes in the first-reverse shift fork and install the roll pin. Move the shift rail to the neutral (center) detent.

14. Insert the second-third shift rail into the housing and install the poppet ball on top of the spring in the shift rail bore.

15. Using a punch or wooden dowel, push the poppet ball and spring downward into the housing bore and install the second-third shift rail.

16. Align the roll pin holes in the second-third shift rail and the shift fork and install the roll pin. Move the shift rail to the neutral (center) position.

17. Install the shift rail plugs in the housing, and remove the shift control cover from the vise.

18. Install the shift lever, shift lever spring, spring retainer, gasket, and the shift control housing cap as an assembly. Tighten the cap securely.

11. Install the back-up light switch and the TCS switch if so equipped.

Warner

T–4 & T–5 4 & 5 Speed

For T–5 procedures, refer to the GM S–Series

DISASSEMBLY

1. Drain the transmission lubricant. 2WD models are not equipped with a drain plug; the fluid must be siphoned from the transmission.

2. Use a pin punch and hammer to remove the offset lever-to-shift rail roll pin.

3. Remove the extension housing (2WD) or the adapter (4WD). Remove the housing and the offset lever as an assembly.

4. Remove the detent ball and spring from the offset lever. Remove the roll pin from the extension housing or adapter.

5. Remove the countershaft rear thrust bearing and race.

6. Remove the transmission cover and shift fork assembly. Two of the transmission cover bolts are alignment type dowel pins. Mark their location so that they may be reinstalled in their original locations.

7. Remove the reverse lever to reverse lever pivot bolt C-clip.

8. Remove the reverse lever pivot bolt. Remove the reverse lever and fork as an assembly.

9. Mark the position of the front bearing

cap to case, then remove the bearing cap bolts and cap.

10. Remove the front bearing race and the shims from the bearing cap. Use a small pry bar and remove the front seal from the bearing cap.

11. Rotate the main drive gear shaft until the flat portion of the gear faces the countershaft, then remove the main drive gear shaft assembly.

12. Remove the thrust bearing and 15 roller bearings from the clutch shaft. Remove the output shaft bearing race. Tap the output shaft with a plastic hammer to loosen it if necessary.

13. Tilt the output shaft assembly upward and remove the assembly from the case.

14. Carefully pull off the countershaft rear bearing with the proper puller after marking the position for reinstallation.

15. Move the countershaft rearward and tilt it upward to remove it from the transmission case. Remove the countershaft bearing spacer.

16. Remove the reverse idler shaft roll pin, then remove the reverse idler shaft and gear.

17. Press off the countershaft front bearing. Use the appropriate pullers and remove the bearing from the main drive gear shaft.

18. Remove the extension housing or adapter oil seal and remove the back-up light switch from the case.

OUTPUT SHAFT DISASSEMBLY

1. Remove the thrust bearing washer from the front of the output shaft.

2. Scribe matchmarks on the hub and sleeve of the 3rd-4th synchronizer so that these parts may be reassembled properly.

3. Remove the 3rd-4th synchronizer blocking ring, sleeve and hub as an assembly.

4. Remove the insert springs and the inserts from the 3rd-4th synchronizer and separate the sleeve from the hub.

5. Remove the 3rd speed gear from the shaft.

6. Remove the 2nd speed gear to output shaft snapring, the tabbed thrust washer and the 2nd speed gear from the shaft.

7. Use an appropriate puller and remove the the output shaft bearing.

8. Remove the 1st gear thrust washer, the roll pin, the 1st speed gear and the blocking ring.

9. Scribe matchmarks on the 1st-2nd synchronizer sleeve and the output shaft.

10. Remove the insert spring and the inserts from the 1st-reverse sliding gear, then remove the gear from the output hub.

OUTPUT SHAFT ASSEMBLY

1. Coat the output shaft and the gear bores with transmission lubricant.

2. Align the matchmarks and install the 1st-2nd synchronizer sleeve on the output shaft hub.

3. Install the three inserts and two springs into the 1st-reverse synchronizer sleeve.

NOTE: The tanged end of each spring should be positioned on the same insert but the open face of each spring should be opposite each other.

4. Install the blocking ring and the 2nd speed gear onto the output shaft.

5. Install the tabbed thrust washer and 2nd gear snapring in the output shaft; be sure that the washer is properly seated in the notch.

6. Install the blocking ring and the 1st speed gear onto the output shaft, then install the 1st gear roll pin.

7. Press the rear bearing onto the shaft.

8. Install the remaining components onto the output shaft: The 1st gear thrust washer. The 3rd speed gear. The 3rd-4th synchronizer hub inserts and the sleeve (the hub offset must face forward). The thrust bearing washer on the rear of the countershaft.

COVER & FORKS DISASSEMBLY

1. Place the selector arm plates and the shift rail centered in the Neutral position.

2. Rotate the shift rail counterclockwise until the selector arm disengages from the selector arm plates; the selector arm roll pin should now be accessible.

3. Pull the shift rail rearward until the selector contacts the 1st-2nd shift fork.

4. Use a $\frac{3}{16}$ in. pin punch and remove the selector arm roll pin and the shift rail.

5. Remove the shift forks, the selector arm, the roll pin and the interlock plate.

6. Remove the shift rail oil seal and O-ring.

7. Remove the nylon inserts and the selector arm plates from the shift forks.

NOTE: Mark the position of the parts so that they may be properly installed.

COVER & FORK ASSEMBLY

1. Attach the nylon inserts to the selector arm plates and through the shift forks.

2. If removed, coat the edges of the shift rail plug with sealer and install the plug.

3. Coat the shift rail and the rail bores with petroleum jelly, then slide the shift rail into the cover until the end of the rail is flush with the inside edge of the cover.

4. Position the 1st-2nd shift fork into the cover; with the offset of the shift fork facing the rear of the cover. Push the shift rail through the fork. The 1st-2nd fork is the larger of the two forks.

5. Position the selector arm and the C-shaped interlock plate into the cover, then push the shift rail through the arm. The widest part of the interlock plate must face away from the cover and the selector arm roll pin must face downward, toward the rear of the cover.

6. Position the 3rd-4th shift fork into the cover with the fork offset facing the rear of the cover. The 3rd-4th shift selector arm

plate must be positioned under the 1st-2nd shift fork selector arm plate.

7. Push the shift rail through the 3rd-4th shift fork and into the front cover rail bore.

8. Rotate the shift rail until the forward selector arm plate faces away from parallel to the cover.

9. Align the roll pin holes of the selector arm and the shift rail and install the roll pin. The roll pin must be installed flush with the surface of the selector arm to prevent selector arm plate to pin interference.

10. Install the O-ring into the groove of the shift rail oil seal, then install the oil seal carefully after lubricating it.

CASE ASSEMBLY

1. Apply a coat of Loctite® 601, or equivalent, to the outer cage of the front countershaft bearing, then press the bearing into the bore until it is flush with the case.

2. Apply petroleum jelly to the tabbed countershaft thrust washer and install the washer with the tab engaged in the corresponding case depression.

3. Tip the transmission case on end and install the countershaft into the front bearing bore.

4. Install the rear countershaft bearing spacer and coat the rear bearing with petroleum jelly. Install the rear countershaft bearing using the appropriate tools. The rear bearing is properly installed when 0.125 in. is extended beyond the case surface.

5. Position the reverse idler into the case (the shift lever groove must face rearward) and install the reverse idler shaft into the case. Install the shaft retaining pin.

6. Install the output shaft assembly into the transmission case.

7. Install the main drive gear bearing onto the main drive shaft using the appropriate tools. Coat the roller bearings with petroleum jelly and install them in the main drive gear recess. Install the thrust bearing and race.

8. Install the 4th gear blocking ring onto the output shaft. Install the rear output shaft bearing race.

9. Install the main drive gear assembly into the case, engaging the 3rd-4th synchronizer blocking ring.

10. Install a new seal in the front bearing cap and in the rear extension or adapter.

11. Install the front bearing into the front bearing cap but do not (at this time) install the shims. Temporarily install the cap to the transmission without applying sealer.

12. Install the reverse lever, the pivot pin (coat the threads with non-hardening sealer) and the retaining C-clip. Be sure the reverse lever fork is engaged with the reverse idler gear.

13. Coat the countershaft rear bearing race and the thrust bearing with petroleum jelly, then install the parts into the extension housing or adapter.

14. Temporarily install the extension housing or adapter without sealer, tighten

the retaining bolts slightly, but do not final torque them.

15. Turn the transmission case on end and mount a dial indicator in position to measure output shaft end play. To eliminate end play the bearings must be preloaded from 0.001–0.005 in. Check the endplay. Select a shim pack that measures 0.001–0.005 in. thicker than the measured endplay.

16. Install the shims under the front bearing cap. Apply an ⅛ inch bead of RTV sealer to the cap. Align the reference marks and install the cap on the front of the transmission. Torque the mounting bolts to 15 ft. lbs. Recheck the output shaft end play, none should exist. Adjust if necessary.

17. Remove the extension housing or adapter. Move the shift forks and synchronizer sleeves to their neutral position. Apply an ⅛ in. bead of RTV sealer to the cover to case mounting surface. Align the forks with their sleeves and carefully lower the cover into position. Center the cover and install the alignment dowels. Install the mounting bolts and tighten to 9 ft. lbs.

NOTE: The offset lever to shift rail roll pin must be position vertically; if not, repeat Step 17.

18. Apply a ⅛ in. bead of RTV sealer to the extension housing or adapter and install over the output shaft.

NOTE: The shift rail must be positioned so that it just enters the shift cover opening.

19. Install the detent spring into the offset lever and place the steel ball into the Neutral guide plate detent. Apply pressure to the detent spring and offset lever, then slide the offset lever on the shift rail and seat the extension housing or adapter plate against the transmission case. Install and tighten the mounting bolts to 25 ft. lbs.

20. Install the roll pin into the offset lever and shift rail. Install the damper sleeve in the offset lever. Coat the back up lamp switch threads with sealer and install the switch, tighten to 15 ft. lbs.

T-14A, T-15A 3 Speed

The Warner T-14A, T-15A are fully synchronized three-speed transmissions having helical drive gears throughout. Lubricant capacity is 2½ pints.

DISASSEMBLY

1. Separate transfer case from transmission by removing five capscrews.

2. Remove gearshift housing and disassembly by removing shift rails, poppet balls, springs, and shift forks.

3. Remove nut, flat washer, transfer case drive gear, adapter, and spacer.

4. Remove main drive gear bearing retainer gasket.

5. Remove main drive gear and mainshaft bearing snap-rings and bearings.

6. Remove main drive gear and mainshaft assembly.

NOTE: The T-15A transmission must be shifted into second gear to allow removal of the mainshaft and gear assembly.

7. On remote shift models, remove roll pins from lever shafts and housing. From inside case, slide levers and interlock assembly out. Remove forks and lever assemblies.

8. Remove lock plate from reverse idler shaft and countershaft.

9. Drive countershaft out to rear with dummy shaft. Remove countergear and two thrust washers. Remove spacer washers, rollers, and spacer from gear.

10. Drive reverse idler shaft out to rear. Remove gear, washers, and roller bearings.

11. Remove clutch hub snap-ring and second-third synchronizer assembly.

12. Remove second and reverse gears.

13. Remove clutch hub snap-ring and low synchronizer assembly.

14. Remove low gear.

SYNCHRONIZER

1. Remove springs. Low synchronizer has only one spring; second-third, two.

2. Mark sleeve and hub before separating.

3. Remove hub.

4. Remove three shifter plates from hub.

5. Inspect all parts for wear.

6. Assembly in reverse order of disassembly. On second-third unit, make sure that spring openings are 120 degrees from each other, with spring tension opposed.

NOTE: If a synchronized assembly is replaced on a floor shift unit, the shift fork operating the synchronizer being replaced must have the letter A just under the shaft hole on the side opposite the pin.

7. Wash all parts in solvent.

8. Air dry but do not spin bearings with air pressure.

9. Check case bearing and shaft bores for cracks or burrs.

10. Check all gears and bronze blocking rings for cracks, and chipped, worn, or cracked teeth. If any gears are replaced, also replace the meshing gears.

11. Check all bearings and bushings for wear or damage.

12. Check that synchronizer sleeves slide freely on clutch hubs.

ASSEMBLY

1. Place reverse idler gear with dummy shaft, roller bearing, and thrust washers in case. Install reverse idler shaft.

2. Assemble countershaft center spacer, four bearing spacers, and bearing rollers in countershaft gear.

3. Install large countergear thrust washer in front of case. Position small thrust washer on countergear hub with lip facing groove in case. Holding countergear in position, push in countershaft from rear.

4. Install lock plate in slots of reverse idler shaft and countershaft.

5. Install to mainshaft: Low gear. Bronze blocking ring. Low synchronizer assembly. Largest snap-ring that fits in groove. Second gear. Bronze blocking ring. Second-third synchronizer assembly. Largest snap-ring that fits in groove. Reverse gear.

6. Install mainshaft assembly through top of case.

7. Install bronze blocking ring to second-third synchronizer assembly.

8. On remote shift units, install shifter shafts, with new O-rings, into case.

NOTE: T-15 interlock levers are marked as to location. T-14 levers have no marks and are interchangeable.

9. Depress interlock lever while installing shift fork into shift lever and synchronizer clutch sleeve. Install poppet spring. Install tapered pins securing shafts in case.

10. Install main drive gear roller bearings.

11. Install main drive gear and oil slinger into case with cutaway portion of gear toward countergear. Install main drive gear to mainshaft.

12. Using bearing installer and thrust yoke tool, install main drive gear and mainshaft bearings and drive into position. The thrust yoke is needed to prevent damage to the synchronizer clutch.

13. Install main drive gear and mainshaft bearing snap-rings. The mainshaft bearing snap-ring is 0.010 thicker than main drive gear bearing snap-ring.

14. Install mainshaft rear bearing adapter, spacer, transfer case drive gear, flat washer, and nut. Torque nut to 130–170 ft. lbs.

15. Install main drive gear bearing retainer (with new oil seal) and gasket. Align oil drain holes in retainer and gasket.

16. Install case cover gasket. On remote shift units, install cover gasket with vent holes to left side.

17. Position gear train and floor-shift assembly in neutral. Insert shifter forks into clutch sleeves and torque to 8–15 ft. lbs.

T-18, T-18A & T-19 4 Speed

The Warner T-18, T-18A and T-19 transmissions have four forward speeds and one reverse. A power take-off opening is provided on certain transmissions, depending upon the models and applications and can be located on either the right or left sides of the case. The T-18 and T-18A transmissions are synchronized in second, third and fourth speeds only, while the T-19 transmission is synchronized in all forward gears. The disassembly and assembly remains basically the same for the transmission models.

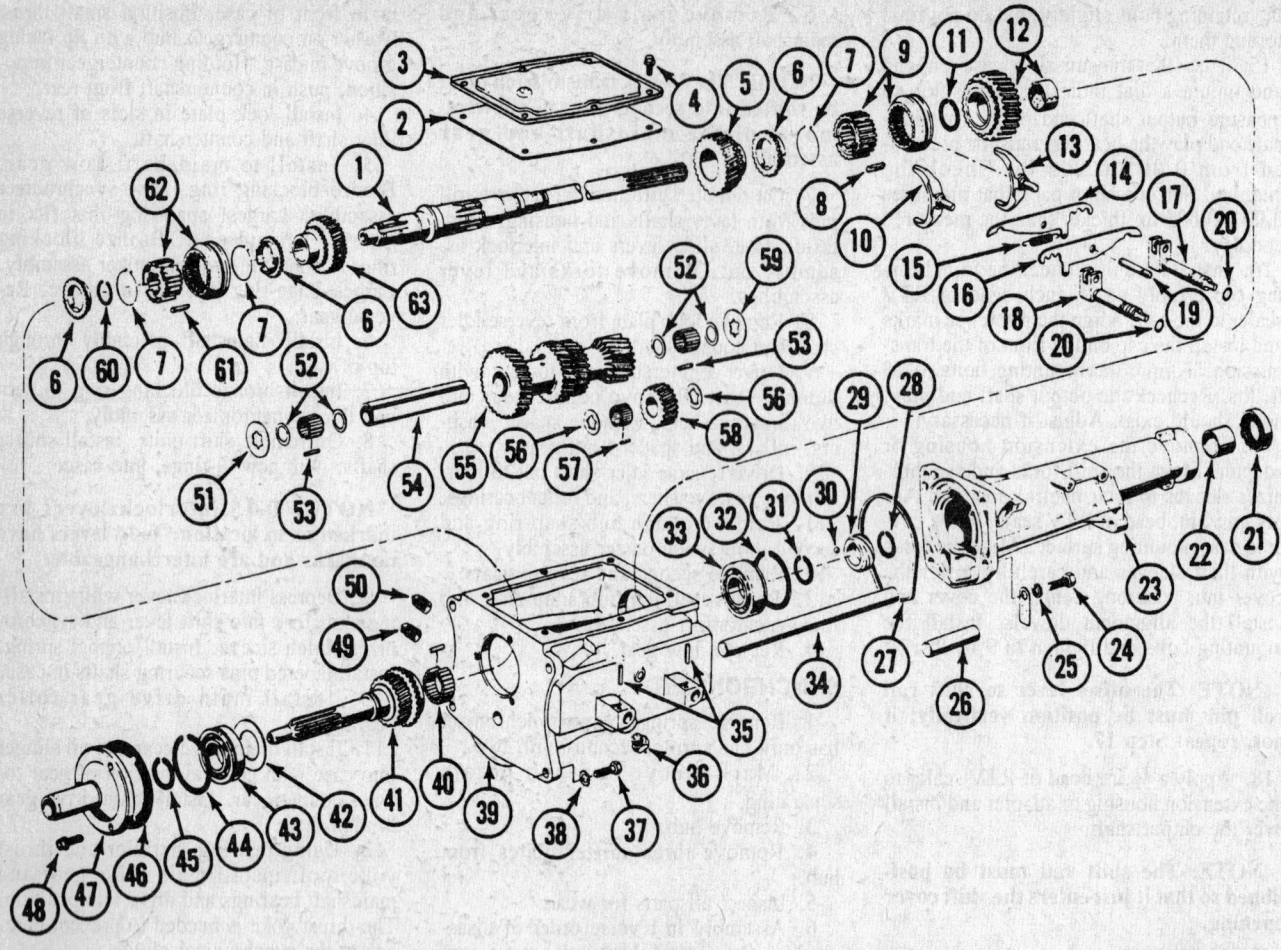

Warner T-14 or T-15

1. Spline shaft
2. Gasket
3. Case cover
4. Bolt
5. First gear
6. Clutch friction ring set
7. Shaft plate retaining spring
8. Clutch shaft first and reverse plate
9. First and reverse clutch assembly
10. Shifter second and high fork
11. Clutch first and reverse gear snapring
12. Reverse gear
13. Shifter first and reverse R fork
14. Shifter interlock first and reverse lever
15. Speed finder interlock poppet spring
16. Shifter interlock second and third lever
17. Shifter fork first and reverse shaft
18. Shifter fork second and third shaft
19. Shifter fork interlock lever pivot pin
20. Shifter fork shaft seal
21. Rear bearing cap oil seal
22. Rear bearing cap bushing
23. Rear bearing cap
24. Bolt
25. Lock washer
26. Idler gear shaft
27. Rear bearing cap gasket
28. Speedometer drive gear ring
29. Speedometer drive gear
30. Speedometer drive gear ball
31. Rear ball bearing lockring
32. Rear ball bearing lockring
33. Rear ball bearing
34. Countershaft
35. Shifter fork retaining pin
36. Solenoid control switch
37. Bolt
38. Lock washer
39. Case
40. Spline shaft pilot bearing roller
41. Clutch shaft
42. Front ball bearing washer
43. Front ball bearing
44. Front ball bearing lockring
45. Front ball bearing snap-ring
46. Gasket
47. Front bearing cap
48. Bolt
49. Drain plug
50. Filler pipe plug
51. Front countershaft gear thrust washer
52. Countershaft gear
53. Countershaft gear bearing roller
54. Countershaft gear roller bearing spacer
55. Countershaft gear
56. Reverse idler gear bearing roller washer
57. Reverse idler gear bearing roller
58. Reverse idler gear
59. Rear countershaft thrust washer (less lip)
60. Clutch second and third snap-ring
61. Clutch shaft second and third plate
62. Second and third clutch assembly
63. Second gear

TORQUE SPECIFICATIONS

Location	N.m.	ft. lbs.
Front bearing retainer to case	14–20	10–15
Cover to case	14–24	10–18
Control levers to lever shafts	20–34	15–25
Rear bearing retainer to case	31–37	23–27
Companion flange to mainshaft	122–163	90–120
Control lever housing bolt	14–20	10–15

DISASSEMBLY

1. After draining the transmission and removing the parking brake drum (or shoe assembly), lock the transmission in two gears and remove the U-joint flange, oil seal, speedometer driven gear and bearing assembly. Lubricant capacity is 6½ pints.

2. Remove the output shaft bearing retainer and the speedometer drive gear and spacer.

3. Remove the output shaft bearing snap-ring, and remove the bearing.

4. Remove the countershaft and idler shaft retainer and the power take-off cover.

5. After removing the input shaft bearing retainer, remove the snap-rings from the bearing and the shaft.

6. Remove the input shaft bearing and oil baffle.

7. Drive out the countershaft (from front). Keep the dummy shaft in contact with the countershaft to avoid dropping any rollers.

8. After removing the input shaft and the synchronizer blocking ring, pull the idler shaft.

9. Remove the reverse gear shifter arm, the output shaft assembly, the idler gear, and the cluster gear. When removing the cluster, do not lose any of the rollers.

OUTPUT SHAFT

1. Remove the third- and high-speed synchronizer hub snap-ring from the output shaft, and slide the third- and high-speed synchronizer assembly and the third-speed gear off the shaft. Remove the synchronizer sleeve and the inserts from the hub. Before removing the two snap-rings from the ends of the hub, check the end play of the second-speed gear (0.005–0.024 in.).

2. Remove the second-speed synchronizer snap-ring. Slide the second-speed synchronizer hub gear off the hub. Do not lose any of the balls, springs, or plates. Pull the hub off the shaft, and remove the second-speed synchronizer from the second-speed gear. Remove the snap-ring from the rear of the second-speed gear, and remove the gear, spacer, roller bearings, and thrust washer from the output shaft. Remove the remaining snap-ring from the shaft.

CLUSTER GEAR

Remove the dummy shaft, pilot bearing rollers, bearing spacers, and center spacer from the cluster gear.

REVERSE IDLER GEAR

Rotate the reverse idler gear on the shaft, and if it turns freely and smoothly, disassembly of the unit is not necessary. If any roughness is noticed, disassemble the unit.

GEAR SHIFT HOUSING

1. Remove the housing cap and lever. Be sure all shafts are in neutral before disassembly.

2. Tap the shifter shafts out of the housing while holding one hand over the holes in the housing to prevent loss of the springs and balls. Remove the two shaft lock plungers from the housing.

CLUSTER GEAR ASSEMBLY

Slide the long bearing spacer into the cluster gear bore, and insert the dummy shaft in the spacer. Hold the cluster gear in a vertical position, and install one of the bearing spacers. Position the 22 pilot bearing rollers in the cluster gear bore. Place a spacer on the rollers, and install 22 more rollers and another spacer. Hold a large thrust washer against the end of cluster gear and turn the assembly over. Install the rollers and spacers in the other end of the gear.

REVERSE IDLER GEAR ASSEMBLY

1. Install a snap-ring in one end of the idler gear, and set the gear on end, with the snap-ring at the bottom.

2. Position a thrust washer in the gear on top of the snap-ring. Install the bushing on top of the washer, insert the 37 bearing rollers, and then a spacer followed by 37 more rollers. Place the remaining thrust washer on the rollers, and install the other snap-ring.

OUTPUT SHAFT ASSEMBLY

1. Install the second speed gear thrust washer and snap-ring on the output shaft. Hold the shaft vertically, and slide on the second speed gear. Insert the bearing rollers in the second-speed gear, and slide the spacer into the gear. (The T–18 model does not contain second speed gear rollers or spacer). Install the snap-ring on the output shaft at the rear of the second-speed gear. Position the blocking ring on the second-speed gear. Do not invert the shaft because the bearing rollers will slide out of the gear.

2. Press the second-speed synchronizer hub onto the shaft, and install the snap-ring. Position the shaft vertically in a soft-jawed vise. Position the springs and plates in the second-speed synchronizer hub, and place the hub gear on the hub.

3. With the T–19 model, press the first and second speed synchronizer onto the shaft and install the snap-ring. Install the first speed gear and snap-ring on the shaft and press on the reverse gear. For the T–19, ignore Steps 2 and 4.

4. Hold the gear above the hub spring and ball holes, and position one ball at a time in the hub, and slide the hub gear downward to hold the ball in place. Push the plate upward, and insert a small block to hold the plate in position, thereby holding the ball in the hub. Follow these procedures for the remaining balls.

5. Install the third speed gear and synchronizer blocking ring on the shaft.

6. Install the snap-rings at both ends of the third and high-speed synchronizer hub. Stagger the openings of the snap-rings so that they are not aligned. Place the inserts in the synchronizer sleeve, and position the sleeve on the hub.

7. Slide the synchronizer assembly onto the output shaft. The slots in the blocking ring must be in line with the synchronizer inserts. Install the snap-ring at the front of the synchronizer assembly.

GEAR SHIFT HOUSING

1. Place the spring on the reverse gear shifter shaft gate plunger, and install the spring and plunger in the reverse gate. Press the plunger through the gate, and fasten it with the clip. Place the spring and ball in the reverse gate poppet hole. Compress the spring and install the cotter pin.

2. Place the spring and ball in the reverse shifter shaft hole in the gear shift housing. Press down on the ball, and position the reverse shifter shaft so that the reverse shifter arm notch does not slide over the ball. Insert the shaft part way into the housing.

3. Slide the reverse gate onto the shaft, and drive the shaft into the housing until the ball snaps into the groove of the shaft. Install the lock screw lock wire to the gate.

4. Insert the two interlocking plungers in the pockets between the shifter shaft holes. Place the spring and ball in the low and second shifter shaft hole. Press down on the ball, and insert the shifter shaft part way into the housing.

5. Slide the low and second shifter shaft gate onto the shaft, and install the corresponding shifter fork on the shaft so that the offset of the fork is toward the rear of the housing. Push the shaft all the way into the housing until the ball engages the shaft groove. Install the lock screw and wire that fastens the fork to the shaft. Install the third and high shifter shaft in the same manner. Check the interlocking system. Install new expansion plugs in the shaft bores.

CASE ASSEMBLY

1. Coat all parts, especially the bearings, with transmission lubricant to prevent scoring during initial operation.

2. Position the cluster gear assembly in the case. Do not lose any rollers.

3. Place the idler gear assembly in the case, and install the idler shaft. Position the slot in the rear of the shaft so that it can engage the retainer. Install the reverse shifter arm.

4. Drive out the cluster gear dummy shaft by installing the countershaft from the rear. Position the slot in the rear of the shaft so that it can engage the retainer. Use thrust washers as required to get 0.006 to 0.020 in. cluster gear end play. Install the countershaft and idler shaft retainer.

5. Position the input shaft pilot rollers and the oil baffle, so that the baffle will not rub the bearing race. Install the input shaft and the blocking ring in the case.

6. Install the output shaft assembly in the case, and use a special tool to prevent jamming the blocking ring when the input shaft bearing is installed.

7. Drive the input shaft bearing onto the shaft. Install the thickest select-fit snap-ring that will fit on the bearing. Install the input shaft snap-ring.

8. Install the output shaft bearing.

9. Install the input shaft bearing without a gasket, and tighten the bolts only enough to bottom the retainer on the bearing snap-ring. Measure the clearance between the retainer and the case, and select a gasket (or gaskets) that will seal in the oil and prevent

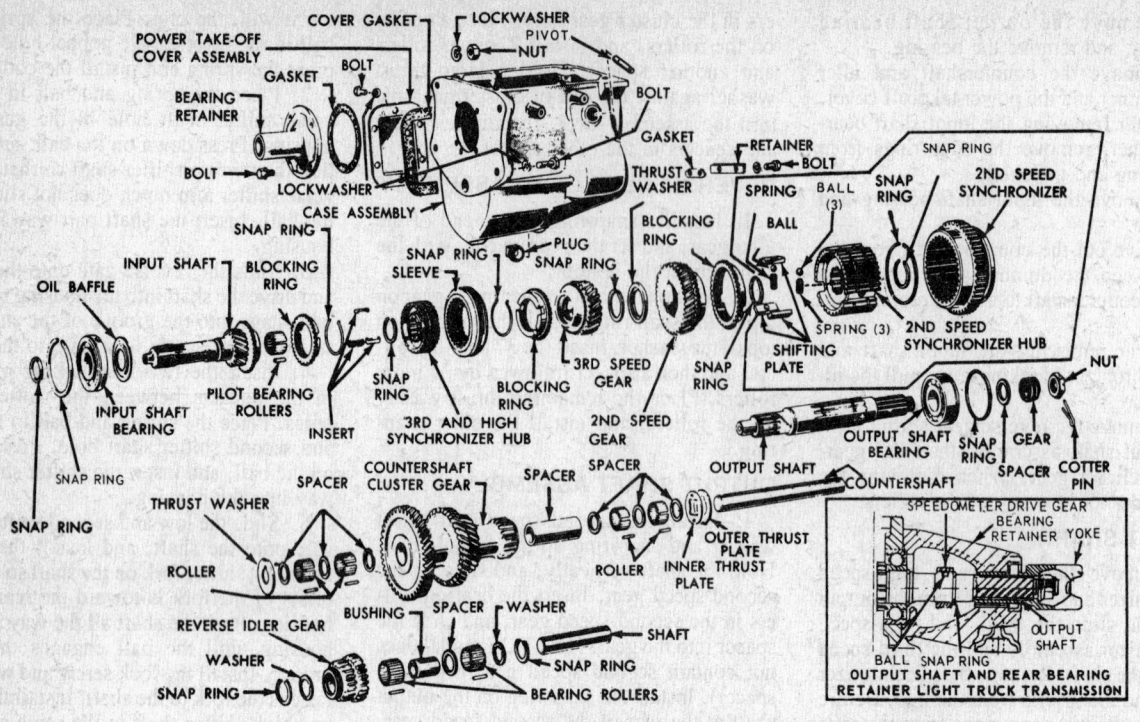

Warner T-18

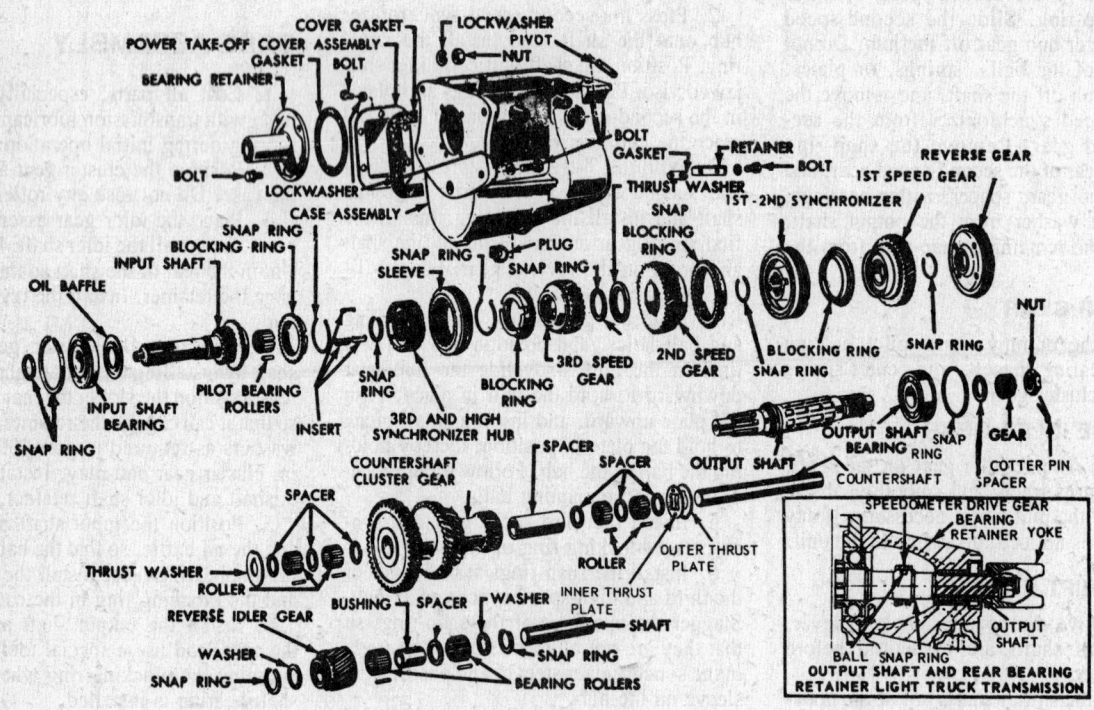

Warner T-19

end play between the retainer and the snapring. Torque the bolts to specification.

10. Position the speedometer drive gear and spacer, and install a new output shaft bearing retainer seal.

11. Install the output shaft bearing retainer. Torque the bolts to specification, and install safety wire.

12. Install the brake shoe (or drum), and torque the bolts to specification. Install the U-joint flange. Lock the transmission in two gears and torque the nut to specification.

13. Install the power take-off cover plates with new gaskets. Fill the transmission according to specifications.

T-15D 3 Speed

The Warner T-15D transmission has three synchronized forward speeds and one reverse. The transmission has either a remote controlled shift lever on the steering column or a top cover shift lever assembly. This transmission can be used with or without a transfer case in the drive line with the use of different extension housing or bearing retainer designs.

DISASSEMBLY

1. Drain the transmission of its lubricant and remove either the top cover or the shift lever assembly from the top of the transmission. Remove the front bearing retainer.

2. Remove the front main drive gear bearing snap-rings and remove the bearing from the shaft and transmission case with the aid of a bearing puller or its equivalent.

3. Remove the main drive gear from the transmission case by having the cutaway portion of the gear teeth positioned downward towards the cluster gear. As the gear is removed from the mainshaft, do not lose the needle roller bearing from the bearing pocket.

4. Remove the rear extension housing or the bearing retainer from the rear of the transmission case.

5. Remove the mainshaft rear bearing snap-rings and remove the bearing from the mainshaft and transmission case with a bearing puller or its equivalent.

6. Column shift: Position the gears in second speed, move the mainshaft to the left and remove the shift forks.

7. Remove the mainshaft by tilting the front of the assembly upward and lifting it through the top of the case.

IDLER GEAR, CLUSTER GEAR & SHAFTS

1. Tap the reverse idler gear shaft and the countershaft rearward to allow the removal of the lockplate from the slots in both shafts.

2. Using a brass drift, drive the reverse idler gear shaft towards the rear and out of the transmission case. Avoid losing the needle roller bearings from the gear bore.

3. Using a dummy countershaft or its equivalent, drive the countershaft from the rear of the transmission case. Lift the cluster gear assembly from the transmission case. Mark the thrust washer locations.

MAINSHAFT

1. Remove the second/third speed synchronizer snap-ring from the front of the mainshaft. Remove the synchronizer from the mainshaft, after matchmarking the sleeve and hub.

2. Remove the second speed gear from the mainshaft.

3. Remove the reverse gear from the rear of the mainshaft. Remove the rear synchronizer (first/reverse) hub snap-ring.

4. Remove the first/reverse synchronizer unit from the rear of the mainshaft.

NOTE: Only one blocker ring is used with the first/reverse synchronizer assembly as the reverse speed gear is not a synchromesh unit.

5. Remove the first speed gear from the mainshaft.

6. Clean the transmission case with solvent and inspect for cracks, worn bearing bores or other damages.

7. Clean and inspect all gears and bronze blocking rings for cracks, chipped or cracked teeth or excessive wear on the teeth. Should a gear require replacement, the meshing gear should be replaced also.

8. Inspect all bearings and bushings for wear or damage. The thrust washers should be renewed upon transmission assembly, if grooved or distorted.

9. Inspect the synchronizer clutch sleeves for abnormal wear and ease of operation.

10. Lubricate all internal transmission components before installation.

ASSEMBLY

1. The assembly of the mainshaft gears is in the reverse of the removal procedure. During the assembly, the snap-rings are of the selective thickness type and should be selected to obtain the following end play measurements: Second/third speed synchronizer; 0.004 to 0.020 in. measured between the snap-ring and the second speed synchronizer hub. First/reverse synchronizer; 0.005 to 0.020 in. measured between the first speed gear and the collar on the mainshaft.

IDLER GEAR & COUNTERSHAFT

1. Using the dummy countershaft or its equivalent, install the needle roller bearings, spacers and thrust washers in the cluster gear bore.

NOTE: Use vaseline type lubricant to hold the needle, roller bearings in place.

2. Place the cluster gear assembly into the transmission case and install the countershaft from the rear to the front of the case, through the cluster gear, forcing the dummy shaft out the front shaft bore of the case.

3. During the installation of the countershaft, be sure to maintain alignment of the spacers and thrust washers.

4. Install the needle roller bearings into the bore of the reverse idler gear and hold in place with a vaseline type lubricant.

5. Position the thrust washers on the gear and place the assembly between the transmission case web and the rear inner surface of the case.

6. Carefully drive the reverse idler gear shaft through the case bore and into the reverse idler gear assembly. Be sure to keep the thrust washers aligned to avoid damage to them.

7. With both the countershaft and the reverse idler gear shaft in Place, install the lock plate with the tabs on the top side, into the slots of each shaft. Drive the shafts forward until the lock plate is flush against the case surface.

MAINSHAFT

1. Tilt the mainshaft assembly and install the rear of the shaft assembly into the case. Lower the mainshaft assembly into the case. If the transmission is controlled by a steering column shift lever, move the mainshaft and install the shifting forks into place on the clutch sleeves and shift mechanism.

2. Using a mainshaft support or equivalent, block and support the front of the mainshaft. Install the rear mainshaft bearing with a bearing installer tool or equivalent.

3. Install the large and small snap-rings on the rear bearing and mainshaft. Remove the front shaft support.

4. Install the main drive gear with the oil baffle, into the case. Position the cutaway portion of the gear downward towards the countershaft/cluster gear assembly, to aid in the installation of the drive gear. Use caution to avoid dropping the needle roller bearings as the mainshaft front stub enters the main drive gear bearing pocket.

5. Install the main drive gear bearing and the retaining snap-rings. Be sure the oil baffle is in place.

6. Install the front bearing retainer with a new gasket.

7. Install the rear bearing retainer or extension housing, using a new gasket. Install the oil seal as required.

8. Floor Shift: Place the gears in a neutral position and the shift lever housing components in neutral. Place the shifting levers in their respective sliding sleeve grooves and bolt the cover to the transmission case. Column Shift: Install the top cover with a new gasket and bolt into place on the transmission.

9. Fill the transmission with lubricant to its proper level (3 pints) and move the gear shifting mechanism by hand to be assured of proper gear selection before installation of the transmission into the vehicle.

SR–4 4 Speed

The Warner SR–4 transmission is a four speed, constant mesh unit, providing synchromesh engagement in all forward gears.

DISASSEMBLY

1. Separate the transmission from the transfer case, if attached.

2. Drain the lubricant from the transmission by removing the lower adapter housing bolt.

3. If the shift lever housing has not been removed, place the shift lever in the neutral position, remove the retaining bolts and lift the shift lever housing from the transmission.

4. Remove the flanged nut holding the offset lever to the shift rail. Remove the offset lever.

5. Remove the adapter housing retain-

ing bolts and the housing from the transmission case.

6. Remove the shift control housing retaining bolts and remove the cover and gasket. Mark the location of the two dowel bolts to reinstall in their original position.

7. Remove the spring clip holding the reverse lever to the reverse lever pivot bolt. Remove the reverse lever pivot bolt, allowing the removal of the reverse lever and reverse lever fork as an assembly.

8. Match mark the front bearing retainer to the transmission case and remove the bearing retainer and gasket.

9. Remove the large and small snap-rings from the front and rear ball bearings on the input and output shafts.

10. With the aid of a bearing puller tool or equivalent, remove the input shaft ball bearing and remove the input shaft from the case.

11. Remove the rear (output shaft) bearing from the shaft with the aid of a bearing puller tool or equivalent.

12. Remove the output shaft assembly as a unit from the transmission case. Do not allow the synchronizer sleeves to separate from the hubs during the removal.

13. Push the reverse idler gear shaft rearward and remove the shaft and gear from the case.

14. Using a dummy countershaft, push the countershaft to the rear of the case. Remove the cluster gear assembly and dummy countershaft as a unit, from the transmission case.

15. Separate the dummy countershaft and remove the 50 needle roller bearings, spacers and thrust washers from the cluster gear.

NOTE: The cluster gear front thrust washer is of a plastic material, while the rear thrust washer is metal.

COUNTERSHAFT GEAR BEARING

1. Remove the dummy shaft, bearing retainer washers and needle bearings from the countershaft gear. Clean and inspect the parts.

2. Coat the bore at each end of the countershaft gear with grease to retain the needle bearings.

3. While holding the dummy shaft in the gear, install the needle bearings and retainer washers in each end of the gear.

4. Slide first gear off the output shaft, and remove the first speed blocker ring. Take care not to lose the sliding gear from the first and second speed synchronizer assembly.

5. Clean and inspect all parts.

6. Place a blocker ring on the cone of first gear, and slide the gear and ring assembly onto the output shaft. Make sure that the inserts in the synchronizer engage in the blocker ring notches.

7. Install the spring pin retaining first gear to the output shaft.

8. Install a blocker ring on the cone of second gear, and slide the gear and ring assembly onto the output shaft. Make sure that the inserts in the synchronizer engage in the blocker ring notches.

9. Install the second gear thrust washer and new snap-ring on the shaft.

10. Install a blocker ring on the cone of third gear, and slide the gear and ring assembly onto the output shaft. Install the third and fourth speed synchronizer. Make sure that the inserts in the synchronizer engage in the blocker ring notches.

11. Install a new third and fourth gear synchronizer snap-ring.

12. Place the first gear thrust washer (oil slinger) on the shaft and on the spring pin retaining first gear.

13. Assembly end play measurements are as follows: Second gear; 0.004 to 0.014 in., measured between the second speed gear and the thrust washer. Third/fourth synchronizer hub; 0.004 to 0.014 in., measured between the output shaft snap-ring and the third speed synchronizer hub.

COVER ASSEMBLY

1. Remove the detent screw, spring and plunger.

2. Pull the shifter shaft rod rearward, rotating it counterclockwise.

3. Remove the spring pin retaining the manual selector and interlock to the shifter shaft.

4. Remove the shifter shaft from the cover taking care not to damage the seal.

5. Remove the manual selector and interlock plate.

6. Remove the first and second speed shifter fork. Remove the third and fourth speed shifter fork.

7. Clean and inspect all parts. Replace the shifter shaft seal and welch plug, if damaged.

8. Assemble the two plastic inserts to each shift fork; the two projections on the inside of the inserts fit into the blind holes in the ends of the shift forks. Insert the selector arm plates into the shift forks.

9. Install the third and fourth speed shifter fork into the cover.

10. Install the first and second speed shifter fork into the cover. Lubricate the shifter shaft bore with grease.

11. Install the manual selector arm through the interlock plate, and position the two pieces into the cover, with the wide leg of the interlock plate towards the inside of the transmission case.

12. Align the shifter shaft in the cover, and insert the shaft through the shifter forks and manual selector. Coat the shifter shaft with a light coating of grease. Make sure the detent grooves face the plunger side of the cover.

13. Align the pin holes in the manual selector arm and shifter shaft. Install the spring pin flush with the surface of the selector arm.

14. Install the detent plunger, spring, and plug. Tighten the plug to 8–12 ft. lbs.

8. Check the operation of the shift forks in each gear position.

OUTPUT SHAFT

1. Scribe alignment marks on the synchronizer and blocker rings. Remove the snap-ring from the front of the output shaft. Slide the third and fourth speed synchronizer assembly, blocker rings and third gear off the shaft.

2. Remove the next snap-ring and the second gear thrust washer from the shaft. Slide second gear and the blocker ring off the shaft, taking care not to lose the sliding gear from the first and second speed synchronizer assembly. The first and second speed synchronizer hub cannot be removed from the output shaft.

3. Remove the first gear thrust washer (oil slinger) from the rear of the output shaft. Remove the spring pin retaining first gear onto the shaft.

SYNCHRONIZER

1. Scribe reference marks on the hub and sleeve of the synchronizer.

2. Push the sleeve from the hub of each synchronizer.

3. Separate the inserts and insert springs from the hubs. Do not mix the parts between the first/second speed synchronizer and the third/fourth speed synchronizer. Clean and inspect all parts.

NOTE: The first/second speed synchronizer hub is not to be removed from the shaft. They have been assembled and machined as a matched unit during manufacturing to assure concentricity.

4. To assemble, position the sleeve on the hub, aligning the previously marked reference points.

5. Position the three inserts per hub and install the insert springs, being sure that the bent end of the springs are seated in one of the inserts. The springs on each side of the hubs must face in opposite directions and the openings be 180 degrees apart.

ASSEMBLY

1. Coat the countershaft thrust washers with a vaseline type lubricant and position the plastic type washer at the front of the case and the metal washer at the rear of the case.

2. With the 50 needle roller bearings in place in the cluster gear and the dummy countershaft in place, install the countershaft/cluster gear assembly into the case.

CAUTION

Be sure the thrust washers are not displaced during the gear installation.

---

3. Align the cluster gear bore with the case bores and install the countershaft from the rear to the front of the case, pushing the dummy countershaft from the gear and case.

4. Position the reverse idler gear with the shift lever groove facing to the front and install the shaft from the rear of the case.

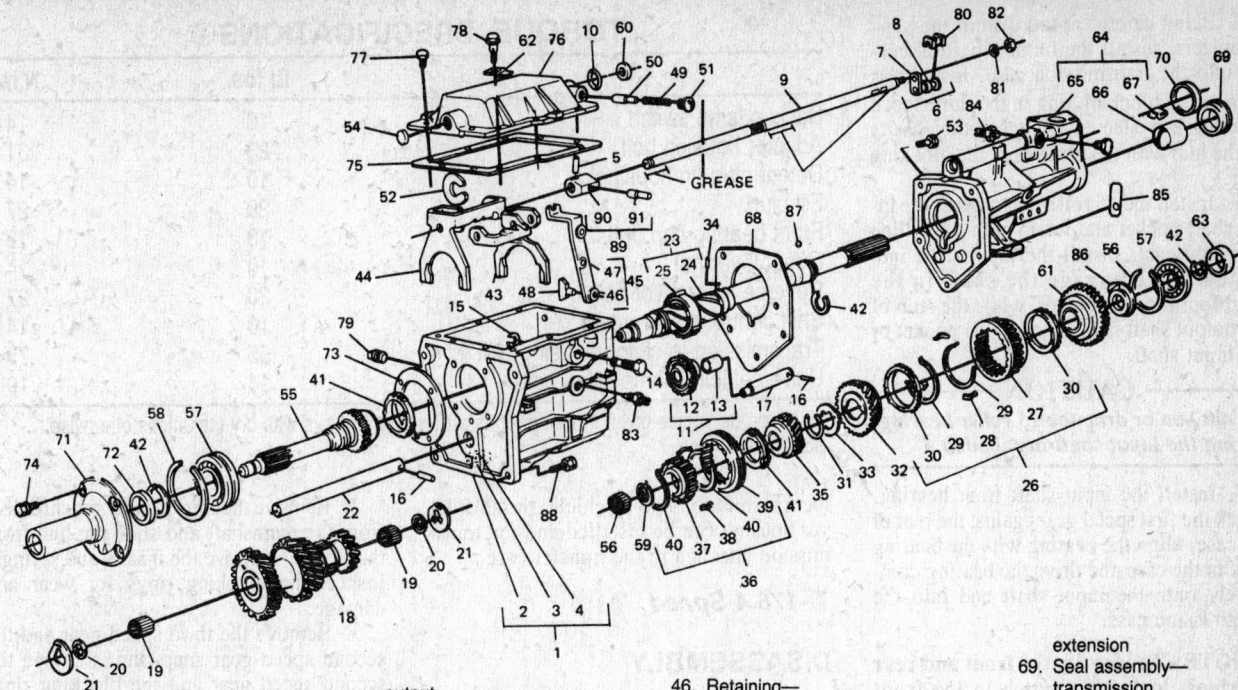

1. Case assembly—transmission
2. Case—transmission
3. Magnet—transmission case chip
4. Nut spring $9/64$
5. Pin—$3/16$ diameter × $13/16$ rolled spring
6. Lever assembly—transmission gearshift shaft offset
7. Lever transmission gearshift shaft offset
8. Pin—transmission gearshift shaft offset lever
9. Shaft—transmission shifter
10. Seal—O-ring
11. Gear and bush assembly—transmission reverse idler sliding
12. Gear—transmission reverse idler sliding
13. Bushing—transmission reverse idler gear
14. Pin—transmission reverse gear selector fork pivot
15. Ring—$7/16$ retaining
16. Pin—$1/4$ x 1 spring
17. Shaft—transmission reverse idler gear
18. Gear—transmission countershaft
19. Roller—transmission countershaft bearing
20. Washer—208/.918 flat
21. Wasner—transmission countershaft gear thrust
22. Countershaft—transmission
23. Shaft assembly—transmission output
24. Shaft—transmission output
25. Hub—transmission synchronizer 1st and 2nd gear cluster
26. Shaft and gear assembly—transmission output
27. Gear—transmission reverse sliding
28. Insert—transmission synchronizer hub
29. Spring—transmission synchronizer retaining
30. Ring—transmission synchronizer blocking
31. Ring—transmission 2nd speed gear retaining snap
32. Gear—transmission 2nd speed
33. Washer—transmission 2nd speed gear thrust
34. Pin—$1/8$ × $1/4$ rolled spring
35. Gear—transmission 3rd speed
36. Synchronizer assembly—3rd and 4th speed
37. Hub—transmission synchronizer
38. Insert—transmission synchronizer hub
39. Sleeve—transmission 3rd and 4th gear clutch hub
40. Spring—transmission synchronizer retaining
41. Ring—transmission synchronizer blocking
42. Ring—transmission m/d gear bearing shaft snap
43. Fork—transmission 1st & 2nd gear shift
44. Fork—transmission 3rd & 4th gear shift
45. Lever assembly—transmission reverse gear shaft relay
46. Retaining—transmission reverse gear shaft relay lever
47. Lever—transmission reverse gear shaft relay
48. Fork—transmission reverse gear shift
49. Spring—transmission shifter interlock
50. Plunger—transmission meshlock
51. Screw—m12 x 10 round head flat
52. Plate—transmission gear selector interlock
53. Screw & washer assembly—m10 x 30 hex head
54. Plug—$3/4$ diameter welch type
55. Shaft—transmission input
56. Roller—transmission mainshaft bearing
57. Bearing assembly—transmission m/d gear ball
58. Ring—m/d gear bearing retaining snap
59. Ring—1.00 retaining
60. Seal—transmission shift shaft
61. Gear—transmission 1st speed
62. Clip—spark control switch wire retaining
63. Gear—speedometer drive
64. Extension assembly—transmission
65. Extension—transmission
66. Bushing—transmission extension
67. Stop—transmission gear shift lever reverse
68. Gasket—transmission
69. Seal assembly—transmission extension oil
70. Plug—transmission extension
71. Retainer—transmission input shaft gear bearing
72. Seal assembly—transmission input shaft oil
73. Gasket—transmission input shaft bearing retainer
74. Bolt—M8 x 20 hex head-lock
75. Gasket—transmission case cover
76. Cover—transmission case
77. Screw—m6 x 20 hex head
78. Bolt—m6 x 32 hex washer HD shoulder
79. Plug—$1/2$-14 pipe (filler)
80. Bushing—transmission gear shift damper
81. Washer—spring lock
82. Nut—hexagon
83. Switch assembly—back-up lamp
84. Switch assembly—transmission seat belt warning sensor
85. Tag—transmission service identification
86. Washer—transmission 1st gear thrust
87. Ball—.25 diameter
88. Screw and lockwasher assembly—m12 x 40
89. Arm assembly—transmission control selector
90. Arm—transmission control selector
91. Pin—transmission gear shift

Warner SR-4

5. Being careful not to disturb the synchronizers, install the output shaft assembly into the transmission case. Install the fourth gear blocking ring in the third speed synchronizer sleeve, engaging the inserts on the hub with the grooves of the blocking ring.

6. Install the 15 roller bearings in the input shaft pocket and retain with a vaseline type lubricant. Install the input shaft into the case and engage the shaft in the third/fourth synchronizer, while the stub of the output shaft is installed in the pocket of the input shaft.

CAUTION

Do not jam or drop the 15 roller bearings during the input shaft installation.

7. Install the input shaft front bearing. Block the first speed gear against the rear of the case, align the bearing with the bearing bore in the case and drive the bearing completely onto the input shaft and into the transmission case.

NOTE: To identify the front and rear bearings, look for a notch in the front bearing race. The rear bearing has no notch.

8. Install the front bearing retaining and locating snap-rings.
9. Install the front bearing cap oil seal and install the cap (bearing retainer) with a new gasket to the transmission case. Install the retaining bolts.
10. Install the first speed thrust washer on the output shaft with the oil grooves facing the first speed gear. Install the rear bearing onto the output shaft and into the case bearing bore.

NOTE: Be sure the first gear thrust washer is engaged on the first gear roll pin before installing the rear bearing.

11. Install the retaining and locating snap-rings on the rear bearing and output shaft.
12. Position the reverse lever in the case, on the pivot bolt and install the retaining clip. Tighten the pivot bolt. Be sure the reverse lever fork is engaged in the reverse idler gear.
13. Rotate the input shaft and output shaft gears and blocking rings to insure freeness of movement. Blocking ring to gear clutch tooth face should have a clearance of 0.030 in.
14. Place the reverse lever in the neutral position and install the cover assembly on the transmission case. Place the two dowel bolts in their original positions and install the remaining retaining bolts.
15. Install a new oil seal in the adapter housing and, using a new gasket, install the adapter housing to the transmission case.
16. Install 3 pints of lubricant into the transmission.
17. Install the offset lever and retain with the flanged nut.
18. Depending upon the installation of

TORQUE SPECIFICATIONS ①

	ft. lbs.	N.m.
Backup lamp switch	10	14
Adapter housing bolt	23	31
Detent plug (in housing)	10	14
Fill plug	20	27
Front bearing cap bolt	13	18
Offset lever nut	10	14
Reverse lever pivot bolt	20	27
Shift control housing bolt	10	14
Transmission-to-clutch housing bolt	55	75
Universal joint clamp strap bolt	14	19

① All torque values given in foot-pounds and newton-meters with dry fits unless otherwise specified.

the transmission into a vehicle, the shift lever housing can be installed and the transmission attached to the transfer case.

T-176 4 Speed

DISASSEMBLY

1. Remove the transfer case from the rear of the transmission.
2. Remove the shift control housing. Mark the location of the two dowel bolts in the housing.
3. Drain the lubricant from the transmission, if not previously done. Remove the rear adapter housing.
4. With a dummy countershaft tool, remove the countershaft from the transmission, front to rear. Allow the cluster gear to lay on the bottom of the case.
5. Remove the rear bearing locating and retaining snap-rings. Remove the rear bearing with a bearing remover tool or equivalent.
6. Match mark the front bearing retainer to the case for easier installation, remove the retaining bolts and the retainer.
7. Remove the locating and retaining snap-rings from the front bearing. Remove the front bearing and the input shaft using a puller tool or equivalent.
8. Remove the mainshaft pilot bearing rollers from the input shaft pocket. Engage the third speed synchronizer.
9. Remove the mainshaft assembly by lifting the front of the shaft upward and out.
10. Remove the cluster gear assembly from the case. Locate and remove any thrust washers and needle roller bearings from the case.
11. Tap the reverse idler gear shaft from the case and remove the reverse idler gear and thrust washers.
12. Separate the reverse idler gear from the sliding gear. Do not lose the needle roller bearings.

MAINSHAFT

1. Remove the third/fourth speed synchronizer snap-ring from the front of the mainshaft.

2. Remove the third/fourth synchronizer from the mainshaft and slide the hub from the sleeve. Remove the inserts and springs. Inspect the blocking rings for wear and damage.
3. Remove the third speed gear and the second speed gear snap-ring. Remove the second speed gear and the blocking ring. Remove the tabbed thrust washer.
4. Remove the snap-ring from the first/second synchronizer hub. Remove the hub and the reverse gear with sleeve as an assembly. Match mark the hub and sleeve for assembly references. Remove the inserts and springs as the sleeve is removed.
5. Remove the first speed gear thrust washer from the rear of the shaft and remove the first speed gear and the blocking ring.
6. Assemble the first/second synchronizer hub, inserts and springs. Install the clutch sleeve. Be sure to position the spring ends 180 degrees apart.
7. Install the assembled first/second speed synchronizer hub and the reverse gear with sleeve, on the mainshaft. Secure with a new snap-ring.
8. Install the first speed gear and blocking ring on the rear of the mainshaft and install the first gear thrust washer.
9. Install a new tabbed thrust washer on the mainshaft with the tab seated in the mainshaft tab bore.
10. Install the second speed gear and the blocking ring on the mainshaft and secure with a new snap-ring.
11. Install the third speed gear and blocking ring on the mainshaft.
12. Assemble the third/fourth speed synchronizer hub, inserts, and springs. Be sure the spring ends are 180 degrees apart.
13. Install the assembled third/fourth speed synchronizer on the mainshaft and secure with a new snap-ring.
14. The measured end play between the snap-ring and the third/fourth speed synchronizer should be 0.004 to 0.014 in.

ASSEMBLY

1. Load the reverse idler gear with the 44 needle roller bearings and a bearing re-

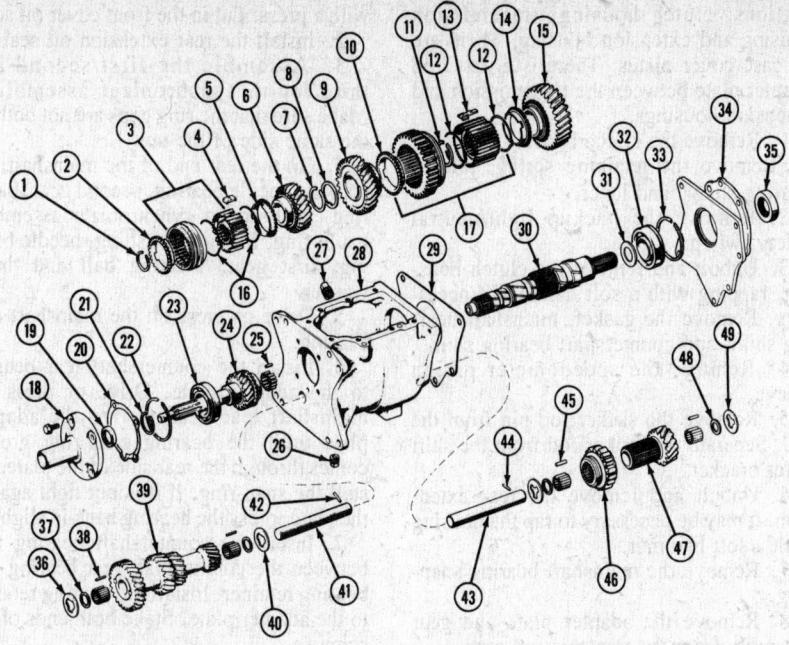

1. Third-fourth gear snap-ring
2. Fourth gear synchronizer ring
3. Third-fourth gear clutch assembly
4. Third-fourth gear plate
5. Third gear synchronizer ring
6. Third speed gear
7. Second gear snap-ring
8. Second gear thrust washer
9. Second speed gear
10. Second gear synchronizer ring
11. Main shaft snap-ring
12. First-second synchronizer spring
13. Low-second plate
14. First gear synchronizer ring
15. First gear
16. Third-fourth synchronizer spring

17. First-second gear clutch assembly
18. Front bearing cap
19. Oil seal
20. Gasket
21. Snap-ring
22. Lock ring
23. Front ball bearing
24. Clutch shaft
25. Roller bearing
26. Drain plug
27. Fill plug
28. Case
29. Gasket
30. Spline shaft
31. First gear thrust washer
32. Rear ball bearing
33. Snap-ring
34. Adapter plate
35. Adapter seal
36. Front countershaft gear thrust washer
37. Roller washer
38. Rear roller bearing
39. Countershaft gear
40. Rear countershaft thrust washer
41. Countershaft
42. Pin
43. Idler gear shaft
44. Pin
45. Idler gear roller bearing
46. Reverse idler sliding gear
47. Reverse idler gear
48. Idler gear washer
49. Idler gear thrust washer

Warner T-176

tainer on each end of the gear. Install the sliding gear on the reverse idler gear. Install lubricated thrust washers into the case.

2. Install the reverse idler assembly into the case and install the reverse idler gear shaft.

3. Be sure to engage the thrust washer locating tabs in the case locating slots.

4. Seat the reverse idler gear shaft roll pin into the counterbore in the case. The reverse idler gear end play should be 0.004 to 0.018 in.

5. Install the 42 needle roller bearings in the cluster gear, using the dummy countershaft as a bearing holder. Use of a vaseline type lubricant is suggested to hold the bearings in place.

6. Position the lubricated thrust washers in place on the inside of the transmission case. Position the thrust washer tabs in the tab slots of the case.

7. Insert the countershaft into the rear case bore, just far enough to hold the rear thrust washer. Lower the cluster gear assembly into the case and align the gear bore with the case bore. Push the countershaft into the cluster gear, displacing the dummy countershaft out the front case bore hole.

Do not completely remove the dummy countershaft.

8. Measure the cluster gear end play which should be 0.004 to 0.018 in. Correct

as required and reinstall the dummy countershaft into the cluster gear, pushing the countershaft from the gear.

9. Allow the cluster gear to remain at

TORQUE SPECIFICATIONS①

	ft. lbs.	N.m.
Backup lamp switch	15	20
Drain and fill plugs	15	20
Front bearing cap bolts	13	18
Shift housing-to-transmission case bolts	13	18
Support plate bolts	18	24

①All torque values given in foot-pounds and newton-meters with dry fits unless otherwise specified.

T-176 SPECIFICATIONS
Lubricant Capacity and End-Play Tolerances

End-Play Tolerances:
Countershaft Gear to Case .... 0.004 to 0.018 inch (0.10 to 0.45 mm)
Reverse Idler Gear to Case .... 0.004 to 0.018 inch (0.10 to 0.45 mm)
Mainshaft Gear Train .... 0.004 to 0.018 inch (0.10 to 0.45 mm)
Lubricant Capacity .... 3.5 pints (1.7 liters)
Lubricant Type .... SAE 85W-90, APJ GL5

the bottom of the case until the input and mainshaft has been installed to provide the necessary assembly clearance.

10. With the synchronizers in the neutral position, install the mainshaft assembly into the case.

11. Install the front bearing part way on the input shaft and install the 15 roller bearing in the shaft pocket.

NOTE: Do not use a heavy grease to hold the bearings in the pocket as the grease can plug the lubrication holes. Use only a vaseline type lubricant.

12. Position the blocking ring on the third/fourth synchronizer. Support the mainshaft assembly and insert the input shaft through the front bearing bore of the case. Seat the mainshaft pilot hub into the bearing pocket of the input shaft and tap the front bearing and input shaft into the case, using a soft faced hammer.

13. When the bearing is fully seated, install the bearing retainer housing, but not the snap-rings at this time.

14. Install the rear bearing on the mainshaft and the bearing bore of the case. It will be necessary to seat the rear bearing further than the locating snap-ring would allow, so do not install the locating snap-ring until after the retaining snap-ring is installed.

15. Remove the front bearing retainer housing and fully seat the front bearing on the input shaft. Install the retaining and locating snap-rings. Install a new oil seal in the retainer housing and install on the transmission case.

16. Install the locating snap-ring on the rear bearing, if not previously done.

17. To install the cluster gear and countershaft, turn the transmission case on end with the input shaft down. Align the cluster gear bore and thrust washers with the case bores. Tap the countershaft into place and displace the dummy countershaft out the front of the case. Do not allow the dummy shaft to drop to the floor.

18. Level the transmission case and install the extension adapter housing with a new gasket.

19. Shift the synchronizer sleeves by hand to insure correct operation. Install 3.5 pints of lubricant into the case and install a new gasket on the shift housing flange. With the gears in the neutral position and the shift lever forks in their neutral position, install the shift lever housing in place on the transmission case.

IMPORTED TRUCKS

Datsun/Nissan

4 Speed
DISASSEMBLY

This transmission is constructed in three sections: clutch housing, transmission housing and extension housing. There are no case cover plates. There is a cast iron adapter plate between the transmission and extension housings.

1. Remove the clutch housing dust cover. Remove the retaining spring, release bearing sleeve and lever.

2. Remove the backup light/neutral safety switch.

3. Unbolt and remove the clutch housing, rapping with a soft hammer if necessary. Remove the gasket, mainshaft bearing shim, and countershaft bearing shim.

4. Remove the speedometer pinion sleeve.

5. Remove the striker rod pin from the rod. Separate the striker rod from the shift lever bracket.

6. Unbolt and remove the rear extension. It may be necessary to rap the housing with a soft hammer.

7. Remove the mainshaft bearing snap-ring.

8. Remove the adapter plate and gear assembly from the transmission case.

9. Punch out the shift fork retaining pins. Remove the shift rod snap-rings. Remove the detent plugs, springs and balls from the adapter plate. Remove the shift rods, being careful not to lose the interlock balls.

10. Remove the snap-ring, speedometer drive gear and locating ball.

11. Remove the nut, lockwasher, thrust washer, reverse hub and reverse gear.

12. Remove the snap-ring and countershaft reverse gear. Remove the snap-ring, reverse idler gear, thrust washer and needle bearing.

13. Support the gear assembly while rapping on the rear of the mainshaft with a soft hammer.

14. Remove the setscrew from the adapter plate. Remove the shaft nut, spring washer, plain washer and reverse idler shaft.

15. Remove the bearing retainer and the mainshaft rear bushing.

16. To disassemble the mainshaft (rear section), remove the front snap-ring, third/fourth synchronizer assembly, third gear and needle bearing. From the rear, remove the thrust washer, locating ball, first gear, needle bearing, first gear bushing, first/second synchronizer assembly, second gear, and needle bearing.

17. To disassemble the clutch shaft, remove the snap-ring and bearing spacer and press off the bearing.

18. To disassemble the countershaft, press off the front bearing. Press off the rear bearing, press off the gears and remove the keys.

19. Remove the retaining pin, control arm pin and shift control arm from the rear of the extension housing.

ASSEMBLY

1. Place the O-ring in the front cover. Install the front cover to the clutch housing with a press. Put in the front cover oil seal.

2. Install the rear extension oil seal.

3. Assemble the first/second and third/fourth synchronizer assemblies. Make sure that the ring gaps are not both on the same side of the unit.

4. On the rear end of the mainshaft, install the needle bearing, second gear, baulk ring, first/second synchronizer assembly, baulk ring, first gear bushing, needle bearing, first gear, locating ball and thrust washer.

5. Drive or press on the mainshaft rear bearing.

6. Install the countershaft rear bearing to the adapter plate. Drive or press the mainshaft rear bearing into the adapter plate until the bearing snap-ring groove comes through the rear side of the plate. Install the snap-ring. If it is not tight against the plate, press the bearing back in slightly.

7. Insert the countershaft bearing ring between the countershaft rear bearing and bearing retainer. Install the bearing retainer to the adapter plate. Stake both ends of the screws.

8. Insert the reverse idler shaft from the rear of the adapter plate. Install the spring washer and plain washer to the idler shaft.

9. Place the two keys on the countershaft and oil the shaft lightly. Press on third gear and install a snap-ring.

10. Install the countershaft into its rear bearing.

11. From the front of the mainshaft, install the needle bearing, third gear, baulk ring, third/fourth synchronizer assembly and snap-ring. Snap-rings are available in thicknesses from 0.0561 to 0.0640 in. to adjust gear end-play.

12. Press the main drive bearing onto the clutch shaft. Install the main drive gear spacer and a snap-ring. Snap-rings are available in thicknesses from 0.0710 to 0.0820 in. to adjust gear end-play.

13. Insert a key into the countershaft drive gear with fourth gear and drive on the countershaft fourth gear with a drift. The rear end of the countershaft should be held steady while driving on the gear, to prevent rear bearing damage.

14. Install the reverse hub, reverse gear, thrust washer, and lock tab on the rear of the mainshaft. Install the shaft nut temporarily.

15. Install the needle bearing, reverse idler gear, thrust washer, and snap-ring.

16. Place the countershaft reverse gear and snap-ring on the rear of the countershaft. Snap-rings are available in thicknesses from 0.0433 to 0.0590 in. to adjust gear end-play.

17. Engage both first and second gears to lock the shaft.

18. On the rear of the mainshaft, install the snap-ring, locating ball, speedometer drive gear, and snap-ring. Snap-rings are available in thicknesses from 0.0433 to 0.0590 in.

19. Recheck end-play and backlash of all gears.

20. Place the reverse shift fork on the reverse gear and install the reverse shift rod. Install the detent ball, spring and plug. Install the fork retaining pin. Place two interlock balls between the reverse shift rod and the third/fourth shift rod location. Install the third/fourth shift fork and rod. Install the detent ball, spring and plug. This plug is shorter than the other two. Install the fork retaining pin. Place two interlock balls between the first/second shift rod location and the third/fourth shift rod. Install the first/second shift fork and rod. Install the detent ball, spring and plug.

21. Install the shift rod snap-rings.

22. Apply sealant sparingly to the adapter plate and transmission housing. Install the transmission housing to the adapter plate and bolt it down temporarily.

23. Drive in the countershaft front bearing with a drift. Place the snap-ring in the mainshaft front bearing.

24. Apply sealant sparingly to the adapter plate and extension housing. Align the shift rods in the neutral positions. Position the striker rod to the shift rods and bolt down the extension housing.

25. Insert the striker rod pin, connect the rod to the shift lever bracket and install the striker rod pin retaining ring. Replace the shift control arm.

26. To select the proper mainshaft bearing shim, first measure the amount the bearing protrudes from the front of the transmission case. Then measure the depth of the bearing recess in the rear of the clutch housing. Required shim thickness is found by subtracting, the difference is required shim size. Shims are available in thicknesses of 0.0551 and 0.0630 in.

27. To select the proper countershaft front bearing shim, measure the amount that the bearing is recessed into the transmission case. Shim thickness should equal this measurement. Shims are available in thicknesses from 0.0157 to 0.0394 in.

28. Apply sealant sparingly to the clutch and transmission housing mating surfaces.

29. Replace the clutch operating mechanism.

30. Install the shift lever temporarily and check shifting action.

5 Speed

This transmission is similar to the 4 speed transmission (Model F4W71B). The overhaul can be accomplished by following the outline for the disassembly and assembly of the 4 speed.

Servo type synchromesh is used, instead of the Borg Warner type in the four speed. Shift linkage and interlock arrangements are the same, except the reverse shift rod also operates fifth gear. Most service procedures are identical to those for the four speed unit.

Those unique to the five-speed follow:

DISASSEMBLY

To disassemble the synchronizers, remove the circlip, synchronizer ring, thrust block, brake band, and anchor block. Be careful not to mix parts of the different synchronizer assemblies.

ASSEMBLY

1. The synchronizer assemblies for second, third, and fourth are identical. When assembling the first gear synchronizer, be sure to install the 0.0866 in. thick brake band at the bottom.

2. When assembling the mainshaft, select a third gear synchronizer hub snap-ring

Apply sealant to mating surface of transmission case and rear extension.

Oil seal
Apply gear oil to oil seal.

T 4 - 5 (0.4 - 0.5, 2.9 - 3.6)

Upper cover

Gasket

Rear extension (4-speed)

Dowel pin

Dust cover

Bearing

Adapter plate

Oil gutter (4-speed)

T 16 - 21 (1.6 - 2.1, 12 - 15)

T 16 - 23 (1.6 - 2.3, 12 - 17)

Top gear switch
T 20 - 29 (2.0 - 3.0, 14 - 22)
Reverse lamp switch
T 20 - 29 (2.0 - 3.0, 14 - 22)

Transmission case

Bearing retainer

Gasket

Front cover

Filler plug
T 25 - 34 (2.5 - 3.5, 18 - 25)

Apply sealant to thread of bolts.

Oil seal
Apply gear oil to oil seal.

Drain plug
T 25 - 34 (2.5 - 3.5, 18 - 25)

T 20 - 34 (2.0 - 3.5, 14 - 25)

T 16 - 21 (1.6 - 2.1, 12 - 15)
Apply sealant to thread of bolts.

T : N·m (kg-m, ft-lb)

Datsun/Nissan 4 speed transmission case assembly

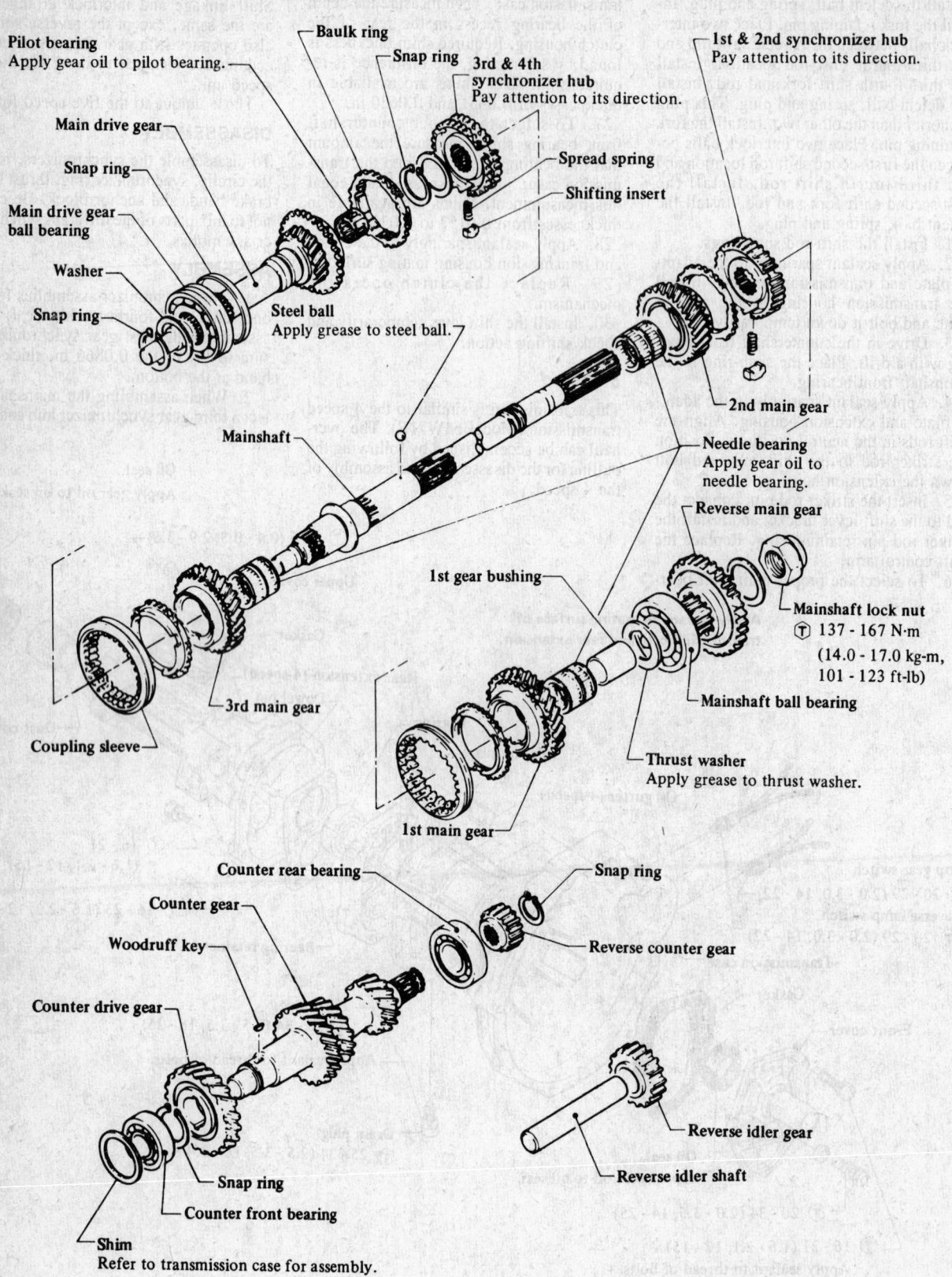

Pilot bearing
Apply gear oil to pilot bearing.

Main drive gear

Snap ring

Main drive gear ball bearing

Washer

Snap ring

Steel ball
Apply grease to steel ball.

Mainshaft

Baulk ring

Snap ring

3rd & 4th synchronizer hub
Pay attention to its direction.

Spread spring

Shifting insert

1st & 2nd synchronizer hub
Pay attention to its direction.

2nd main gear

Needle bearing
Apply gear oil to needle bearing.

Reverse main gear

Mainshaft lock nut
Ⓣ 137 - 167 N·m
(14.0 - 17.0 kg-m,
101 - 123 ft-lb)

Mainshaft ball bearing

Thrust washer
Apply grease to thrust washer.

3rd main gear

Coupling sleeve

1st gear bushing

1st main gear

Counter rear bearing

Counter gear

Woodruff key

Counter drive gear

Snap ring

Counter front bearing

Shim
Refer to transmission case for assembly.

Snap ring

Reverse counter gear

Reverse idler gear

Reverse idler shaft

Datsun/Nissan 4 speed transmission gear assemblies

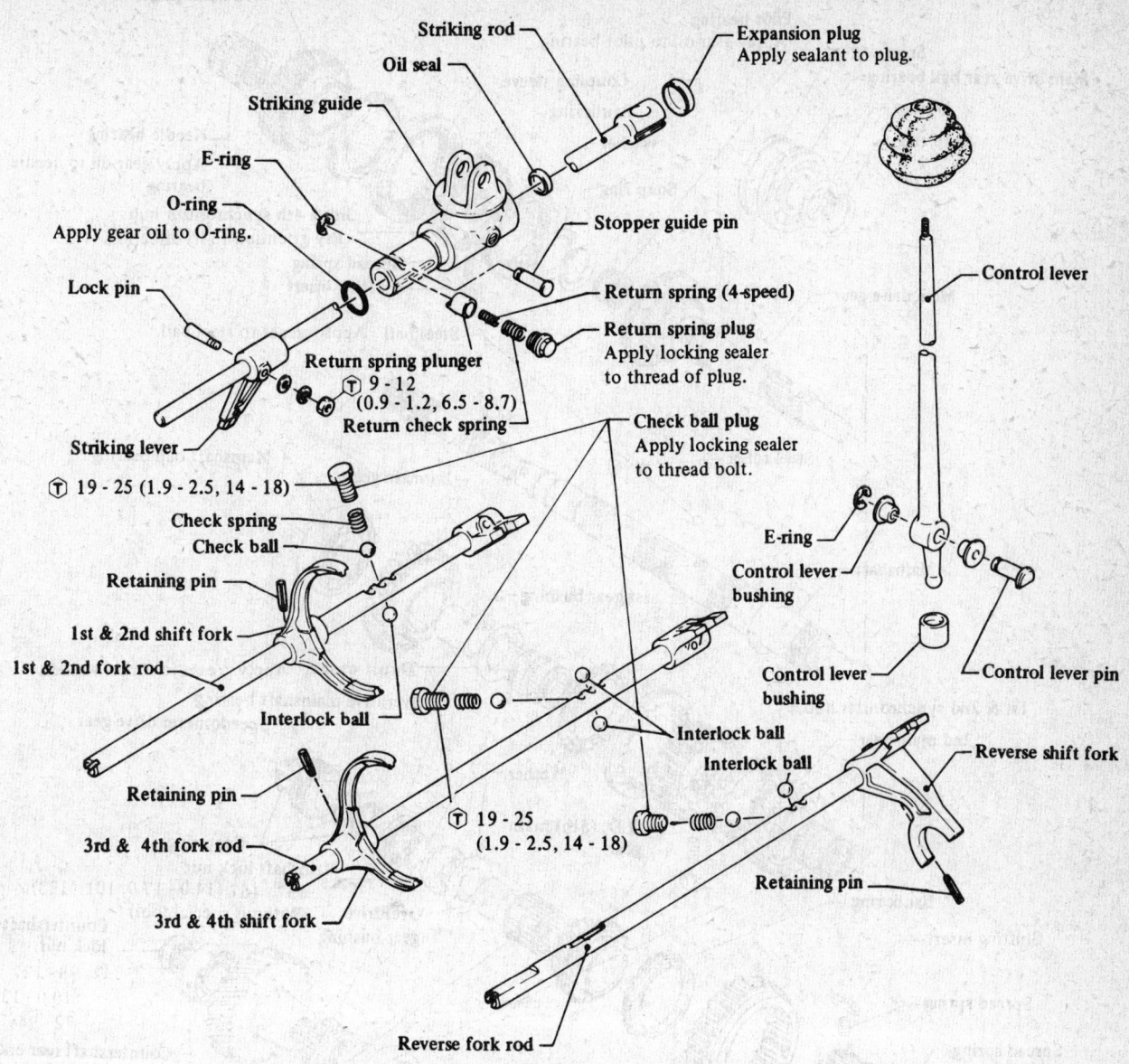

Striking rod

Oil seal

Expansion plug
Apply sealant to plug.

Striking guide

E-ring

O-ring
Apply gear oil to O-ring.

Lock pin

Stopper guide pin

Return spring (4-speed)

Return spring plug
Apply locking sealer
to thread of plug.

Return spring plunger
Ⓣ 9 - 12
(0.9 - 1.2, 6.5 - 8.7)
Return check spring

Striking lever

Check ball plug
Apply locking sealer
to thread bolt.

Ⓣ 19 - 25 (1.9 - 2.5, 14 - 18)

Check spring

Check ball

E-ring

Retaining pin

Control lever
bushing

1st & 2nd shift fork

1st & 2nd fork rod

Control lever
bushing

Control lever pin

Interlock ball

Interlock ball

Interlock ball

Ⓣ 19 - 25
(1.9 - 2.5, 14 - 18)

Reverse shift fork

Retaining pin

3rd & 4th fork rod

Retaining pin

3rd & 4th shift fork

Control lever

Reverse fork rod

Datsun/Nissan 4 speed transmission forks and shifter assemblies

to minimize hub end-play. Snap-rings are available in thicknesses of 0.061–0.630 in., 0.0591–0.0610 in. and 0.0571–0.0591 in. The synchronizer hub must be installed with the longer boss to the rear.

3. When reassembling the gear train, install the mainshaft, countershaft, and gears to the adapter plate. Hold the rear nut and force the front nut against it to a torque of 217 ft. lbs. for '79 models and 123 ft. lbs. for '80 and later models. Select a snap-ring to minimize end-play of the fifth gear bearing at the rear of the mainshaft. Snap-rings are available in thicknesses from 0.0433 to 0.0551 in.

Isuzu/LUV

4 Speed
DISASSEMBLY

1. Remove the boot, clutch fork and throwout bearing.
2. Remove bearing retainer, gasket and spring washer.
3. Remove the speedometer gear and bushing.
4. Remove the shifter cover and gasket.
5. Remove the back-up switch on Cali-

fornia vehicles and both back-up and CRS switches on all others.
6. Remove the rear extension and gasket.
7. Remove the thrust washers and reverse idler gear.
8. Remove the snap-rings, speedometer drive gear and key from the mainshaft.
9. Remove the spring pin from the reverse shifter fork and reverse gear.
10. Remove the snap-ring from the outer circumference of the clutch gear shaft ball bearing.
11. Remove the center support assembly from the transmission case.

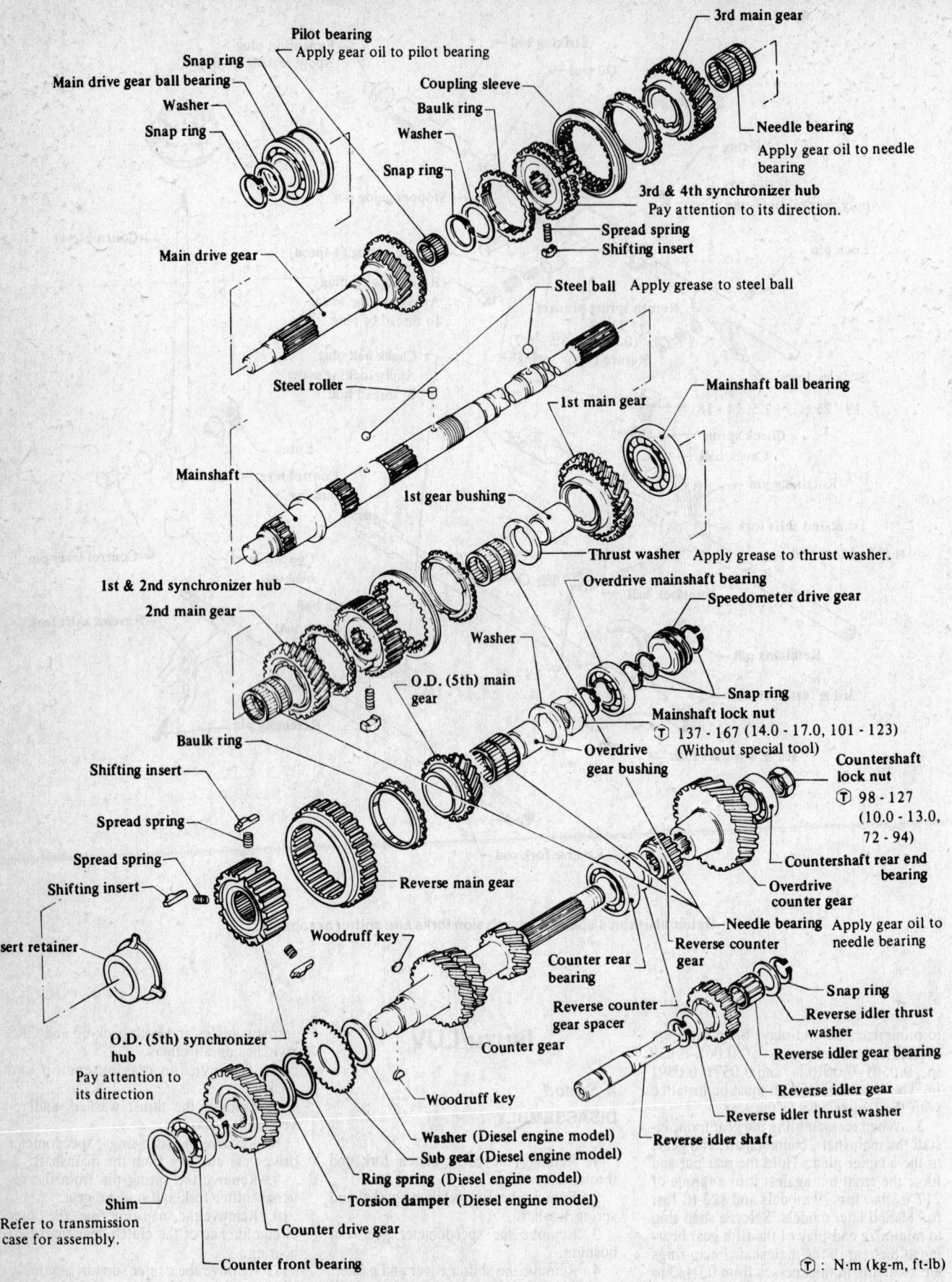

Datsun/Nissan 5 speed transmission gear assemblies

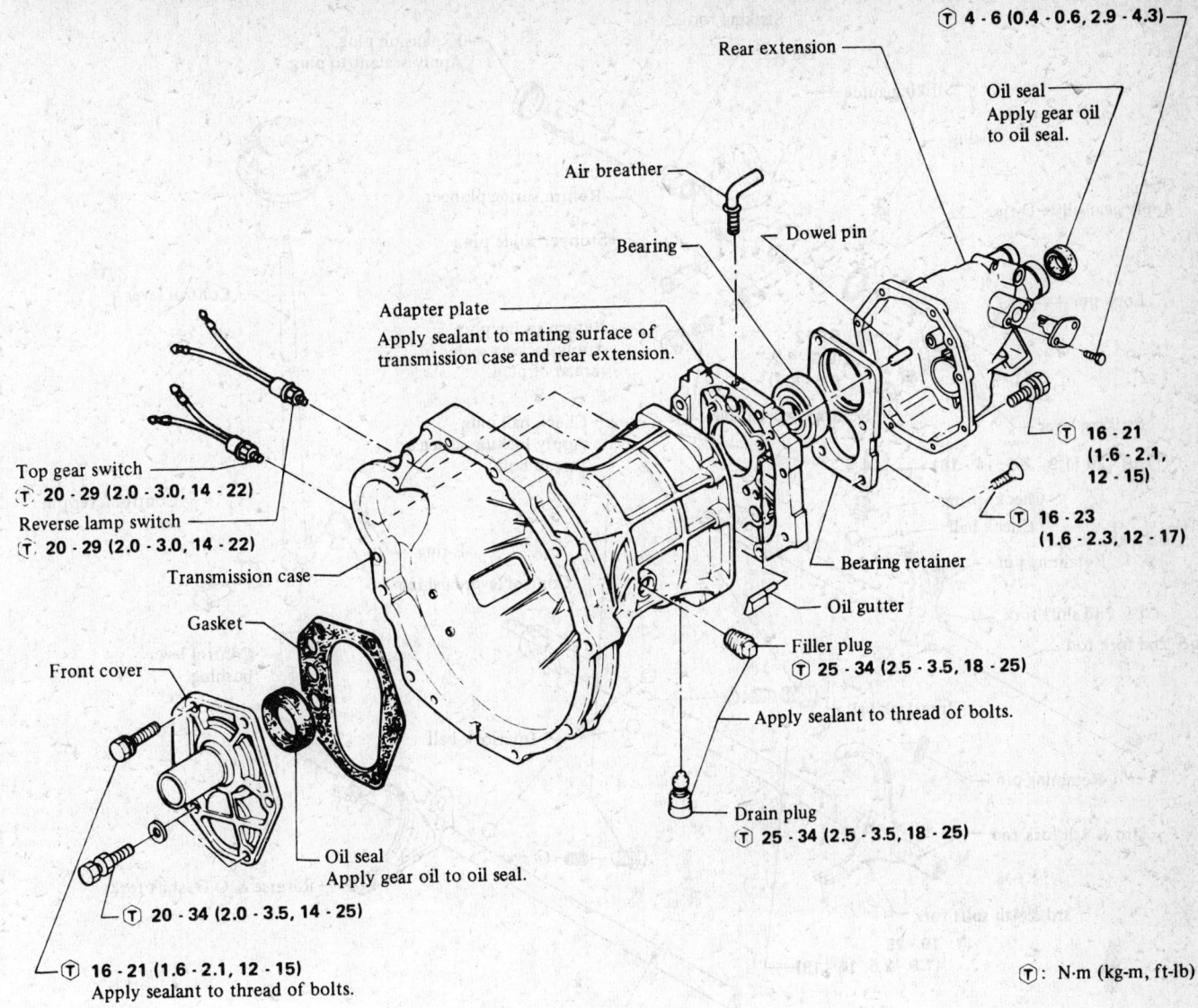

Ⓣ 4 - 6 (0.4 - 0.6, 2.9 - 4.3)

Rear extension

Oil seal
Apply gear oil
to oil seal.

Air breather

Bearing

Dowel pin

Adapter plate
Apply sealant to mating surface of
transmission case and rear extension.

Ⓣ 16 - 21
(1.6 - 2.1,
12 - 15)

Ⓣ 16 - 23
(1.6 - 2.3, 12 - 17)

Top gear switch
Ⓣ 20 - 29 (2.0 - 3.0, 14 - 22)

Reverse lamp switch
Ⓣ 20 - 29 (2.0 - 3.0, 14 - 22)

Transmission case

Bearing retainer

Oil gutter

Filler plug
Ⓣ 25 - 34 (2.5 - 3.5, 18 - 25)

Gasket

Front cover

Apply sealant to thread of bolts.

Drain plug
Ⓣ 25 - 34 (2.5 - 3.5, 18 - 25)

Oil seal
Apply gear oil to oil seal.

Ⓣ 20 - 34 (2.0 - 3.5, 14 - 25)

Ⓣ 16 - 21 (1.6 - 2.1, 12 - 15)
Apply sealant to thread of bolts.

Ⓣ : N·m (kg-m, ft-lb)

Datsun/Nissan 5 speed transmission case assembly

12. Drive out the spring pins from the third and fourth and first and second shift forks.

NOTE: When removing the spring pin, hold a round bar against the end of the shifter rods to prevent damage.

13. Remove the detent spring plate from the center support, then remove the detent springs and balls.

14. Remove the first and second and the third and fourth shifter rods from the center support, then remove the shifter forks.

15. Remove the reverse shifter rod forward as it is fitted with a stopper pin.

NOTE: Be careful not to lose the detent interlock plugs located between the shifter rods in the center support.

16. Move both synchronizers rearward to prevent turning of the mainshaft.

NOTE: It may be necessary to tap the synchronizers with the hammer handle to get them engaged.

17. Remove the locknut and washer from the mainshaft.

18. Remove the nut, washer countershaft reverse gear and collar from the rear of the countergear.

19. Remove the center support countergear bearing snap-ring.

20. Remove the center support.

21. Separate the clutch gear, needle bearings and blocker ring from the mainshaft assembly.

22. Press the rear bearing from the mainshaft.

23. Remove the thrust washer, 1st speed gear, needle roller bearing, a collar and blocker ring.

24. Remove the 1st and 2nd gear synchronizer assembly.

25. Remove the 2nd gear, blocker ring and needle roller bearing from the mainshaft.

26. Remove the snap-ring, 3rd and 4th synchronizer assembly and blocker ring from the mainshaft.

27. Remove the 3rd gear and needle bearings.

28. Remove the snap-ring and press off the clutch bearing and countergear bearing from the shaft.

ASSEMBLY

1. Stand the front of the mainshaft upward and install the 3rd speed gear and needle roller bearing with the tapered side of the gear facing the front of the mainshaft.

2. Install a blocker ring with the clutching teeth upward over the synchronizing surface of the 3rd speed gear.

3. If it is necessary to reassemble the synchronizer assembly, turn the face of the

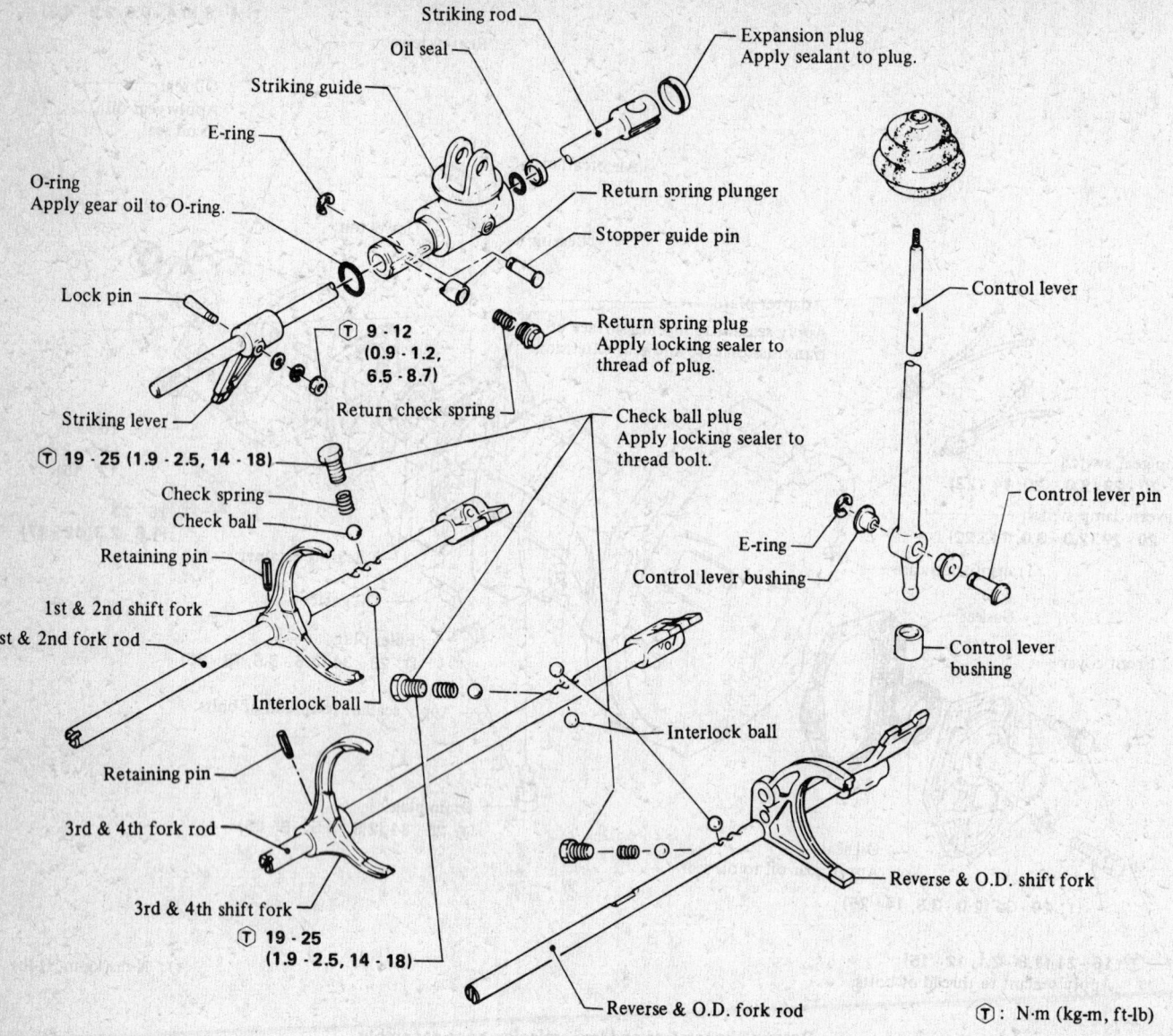

Striking rod

Oil seal

Striking guide

Expansion plug
Apply sealant to plug.

E-ring

O-ring
Apply gear oil to O-ring.

Return spring plunger

Stopper guide pin

Control lever

Lock pin

Ⓣ 9 - 12
(0.9 - 1.2,
6.5 - 8.7)

Return spring plug
Apply locking sealer to
thread of plug.

Striking lever

Return check spring

Check ball plug
Apply locking sealer to
thread bolt.

Control lever pin

Ⓣ 19 - 25 (1.9 - 2.5, 14 - 18)

E-ring

Control lever bushing

Check spring

Check ball

Retaining pin

Control lever
bushing

1st & 2nd shift fork

1st & 2nd fork rod

Interlock ball

Interlock ball

Retaining pin

3rd & 4th fork rod

Reverse & O.D. shift fork

3rd & 4th shift fork

Ⓣ 19 - 25
(1.9 - 2.5, 14 - 18)

Reverse & O.D. fork rod

Ⓣ : N·m (kg-m, ft-lb)

Datsun/Nissan 5 speed transmission forks and shifter assemblies

synchronizer hub with the heavy boss to the face of the sleeve with the light chamfering on the outer rim.

4. Fit the keys into the key groove and position the synchronizer springs into the hole in the side face of the hub.

5. Install the 3rd and 4th synchronizer assembly on the mainshaft with the face of the sleeve with the light chamfer rearward.

6. Turn the rear of the mainshaft upward and install the 2nd speed gear and needle roller bearing on the mainshaft with the tapered surface of the gear facing the rear of the mainshaft.

7. Install a blocker ring with the clutching teeth downward over the synchronizing surface of the 2nd speed gear.

8. Install the 1st and 2nd synchronizer assembly with the chamfer on the sleeve facing the front of the mainshaft.

9. Install a blocker ring with the clutching teeth rearward.

10. Install the collar, needle roller bearing and 1st speed gear on the mainshaft.

NOTE: The tapered side of the gear should be facing the front of the mainshaft.

11. Install the 1st speed gear thrust washer on the mainshaft with the grooved side facing 1st gear.

12. Press the rear bearing on the mainshaft with the snap-ring groove facing the front of the mainshaft.

13. If removed, press the ball bearing on the clutch gearshaft with the snap-ring groove on the bearing facing the front of the transmission. Install the snap-ring on the clutch gear shaft.

14. Assemble the needle roller bearing,

blocker ring and clutch gear to the front of the mainshaft.

15. If removed, press on the countergear ball bearing with the snap-ring groove facing the rear of the transmission.

16. If removed, install the snap-rings in the inner circumference of the mainshaft and countergear holes of the center support.

17. If removed, insert the idler gear shaft with the lock plate groove side into the center support from the rear, then install the lock plate.

18. Mesh the countergear with the mainshaft assembly and install a holding tool on the mainshaft and countergear.

19. Install the center support.

20. Press the center support onto the shaft until the countergear bearing is brought into contact with its snap-ring.

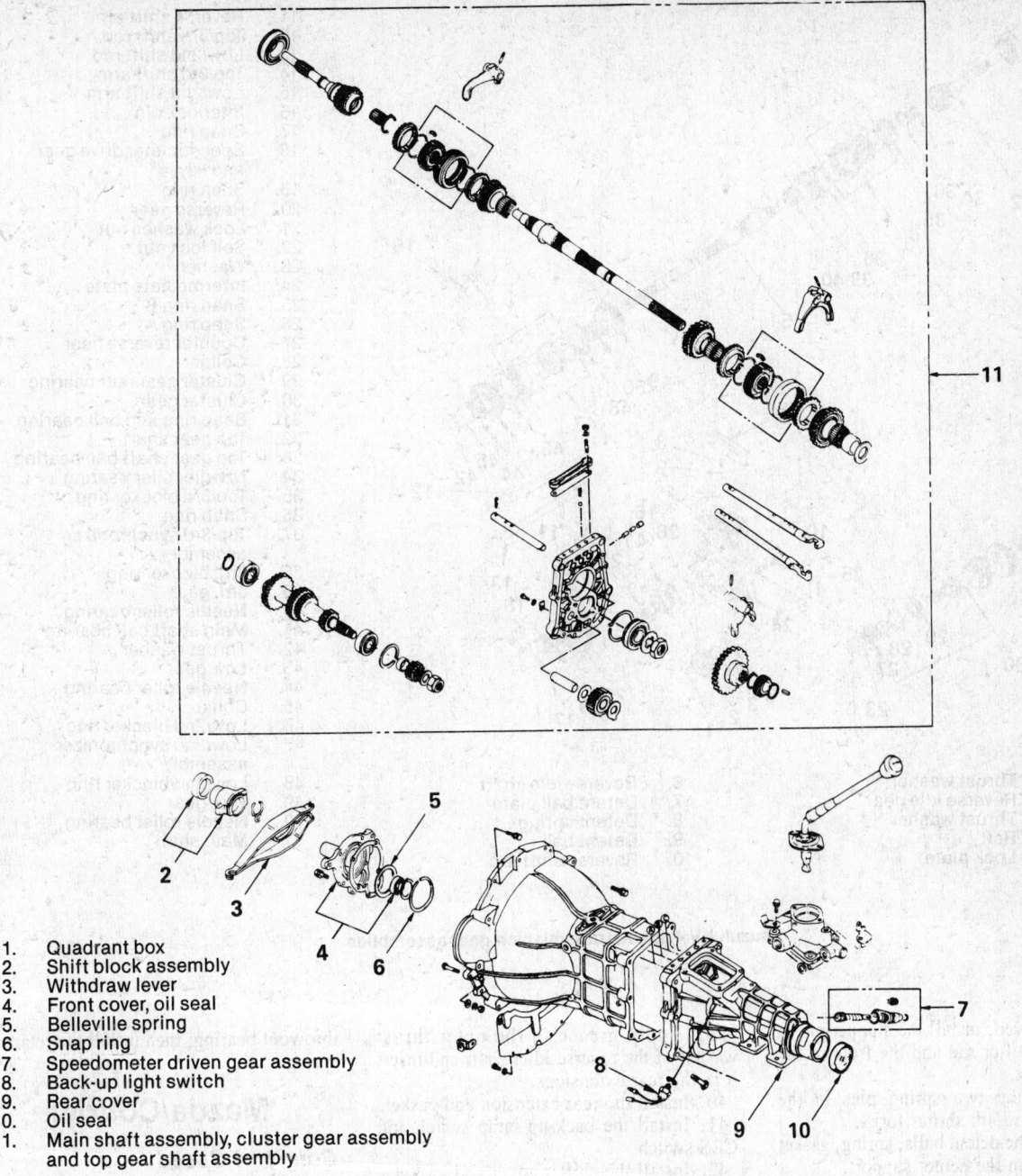

1. Quadrant box
2. Shift block assembly
3. Withdraw lever
4. Front cover, oil seal
5. Belleville spring
6. Snap ring
7. Speedometer driven gear assembly
8. Back-up light switch
9. Rear cover
10. Oil seal
11. Main shaft assembly, cluster gear assembly and top gear shaft assembly

Isuzu/LUV 4 speed transmission major components

21. Expand the countergear bearing snapring and press the center support further until the mainshaft and countergear snaprings are fitted into their grooves.

22. Remove the holding tool from the mainshaft and countergear.

23. Move both synchronizers rearward to prevent turning of the mainshaft.

24. Install the collar, countershaft reverse gear, washer and nut on the rear of the countergear.

NOTE: Install the locknut so that the chamfered side is facing the lock washer.

25. Install the locknut and lock washer on the mainshaft.

NOTE: Install the locknut so that the chamfered side is facing the lockwasher.

26. Apply grease to the two detent plugs and insert them into their detent holes from the middle hole of the center support.

27. Install the 1st and 2nd shifter forks and the 3rd and 4th into their grooves in the synchronizer assembly.

28. Install 3rd and 4th shifter rod from the rear of the center support through the

middle hole and into the 1st and 2nd, 3rd and 4th shifter forks. Align the spring pin hole in the shifter fork with the hole in the shifter rod.

NOTE: Identify the 3rd and 4th shifter rod by the two detent grooves on the side of the rod.

29. Install the 1st and 2nd shifter rod from the rear of the center support through the 1st and 2nd shifter fork and align the hole in the rod to the hole in the shifter fork.

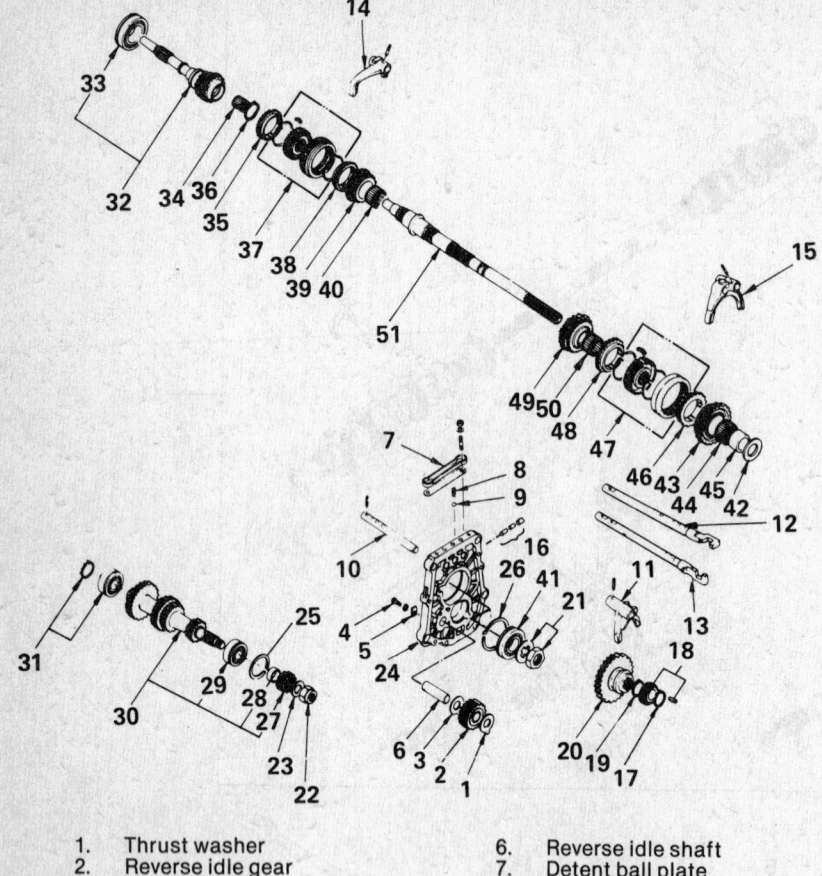

11.	Reverse shift arm
12.	Top/3rd shift rod
13.	Low/2nd shift rod
14.	Top/3rd shift arm
15.	Low/2nd shift arm
16.	Interlock pin
17.	Snap ring
18.	Speedometer drive gear and key
19.	Snap ring
20.	Reverse gear
21.	Lock washer, nut
22.	Self lock nut
23.	Washer
24.	Intermediate plate
25.	Snap ring B
26.	Snap ring A
27.	Counter reverse gear
28.	Collar
29.	Cluster gear rear bearing
30.	Cluster gear
31.	Snap ring and ball bearing
32.	Top gear shaft
33.	Top gear shaft ball bearing
34.	Needle roller bearing
35.	Top/3rd blocker ring
36.	Snap ring
37.	Top/3rd synchronizer assembly
38.	3rd. blocker ring
39.	3rd. gear
40.	Needle roller bearing
41.	Main shaft ball bearing
42.	Thrust washer
43.	Low gear
44.	Needle roller bearing
45.	Collar
46.	Low/2nd blocker ring
47.	Low/2nd synchronizer assembly
48.	Low/2nd blocker ring
49.	2nd. gear
50.	Needle roller bearing
51.	Main shaft

1.	Thrust washer	6.	Reverse idle shaft
2.	Reverse idle gear	7.	Detent ball plate
3.	Thrust washer	8.	Detent spring
4.	Bolt	9.	Detent ball
5.	Lock plate	10.	Reverse shift rod

Isuzu/LUV 4 speed transmission gear assemblies

30. If removed, install the stopper pin in the reverse shifter rod and the front of the center support.

31. Install the two spring pins in the 1st/2nd and 3rd/4th shifter forks.

32. Install the detent balls, spring, gasket and retainer on the center support.

33. Install the center support assembly and gasket.

34. Assemble the reverse shifter fork to the reverse gear and install these parts into position from the rear side of the mainshaft, then connect them to the reverse shifter rod.

35. Install the spring pin in the reverse shifter fork.

36. Install the thrust washer and reverse idler gear on the idler shaft.

NOTE: The reverse idler gear should be installed with undercut teeth forward.

37. Install the speedometer drive gear snap-ring and key on the mainshaft.

38. Install a new oil seal in the rear extension.

39. Apply grease to the outer thrust washer of the reverse idler shaft and insert it in the rear extensions.

40. Install the rear extension and gasket.

41. Install the back-up lamp switch and CRS switch.

42. Install the shifter cover and gasket.

43. Install the oil O-ring to the speedometer drive gear and install the gear.

44. Install the front bearing retainer seal.

45. Install a snap-ring in the outer circumference of the clutch gear bearing.

46. Apply grease to the bearing retainer spring washer and place it in the bearing retainer with the dished face turned to the bearing outer race.

47. Install the bearing retainer to the front of the transmission case.

NOTE: The shorter bolts are used on countergear front bearing side of the bearing retainer.

48. Install the ball stud to the bearing retainer.

49. Install the boot clutch fork and

throwout bearing, then install the retaining spring.

Mazda/Courier

Courier 4 Speed
DISASSEMBLY

The clutch housing, split transmission case, and extension housing are all made of aluminum.

1. Drain the oil.

2. Remove the clutch housing, bearing retainer, release bearing, and release fork.

3. Remove the speedometer shaft sleeve and driven gear.

4. Remove the extension housing.

5. Remove the back-up light switch.

6. Separate the case halves. Do not pry apart.

7. Measure gear backlash. The backlash for all gears should be 0.004–0.008 in.

8. Remove the countergear set from the right-hand half of the case.

9. Use a magnet to remove the ball from the second countergear bearing.

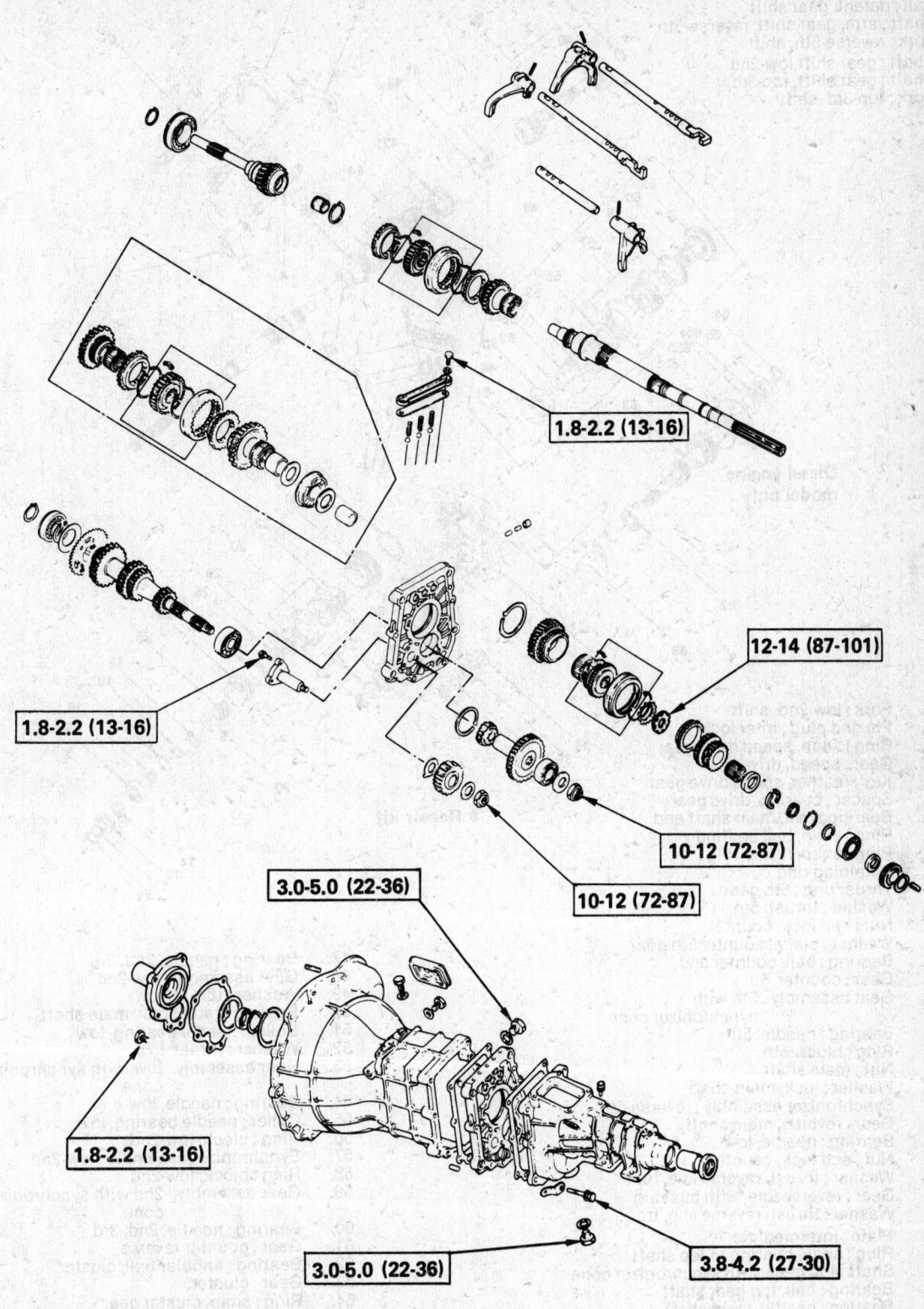

1.8-2.2 (13-16)

1.8-2.2 (13-16)

12-14 (87-101)

10-12 (72-87)

10-12 (72-87)

3.0-5.0 (22-36)

1.8-2.2 (13-16)

3.0-5.0 (22-36)

3.8-4.2 (27-30)

Isuzu/LUV 5 speed transmission gear assemblies torque values

1. Plate ; detent spring
2. Spring ; detent ball
3. Ball ; detent, gear shift
4. Shaft ; arm, gear shift, reverse-5th
5. Fork ; reverse-5th, shift
6. Shaft ; gear shift low-2nd
7. Shaft ; gear shift, top-3rd
8. Fork ; top-3rd, shift

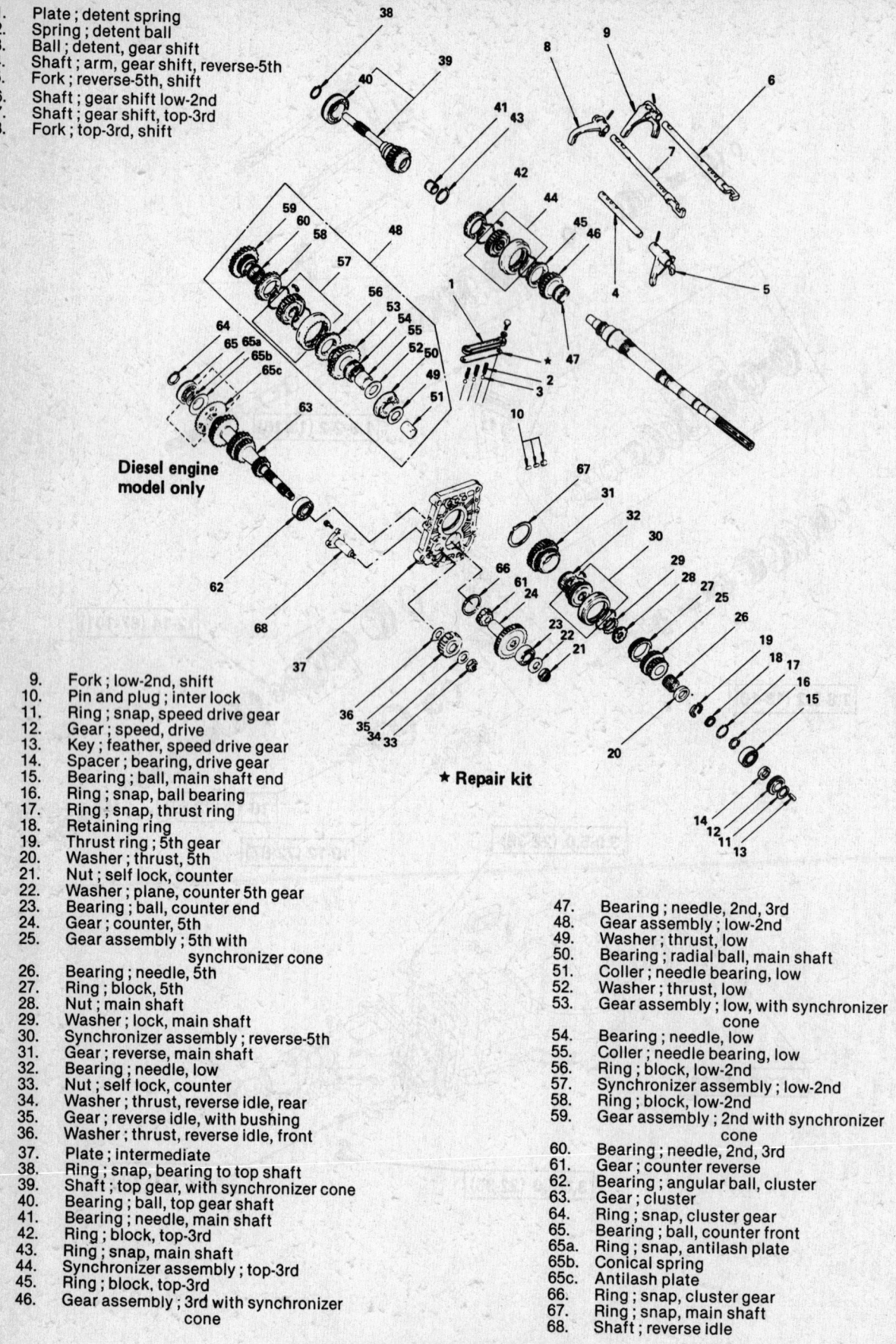

Diesel engine model only

★ Repair kit

9. Fork ; low-2nd, shift
10. Pin and plug ; inter lock
11. Ring ; snap, speed drive gear
12. Gear ; speed, drive
13. Key ; feather, speed drive gear
14. Spacer ; bearing, drive gear
15. Bearing ; ball, main shaft end
16. Ring ; snap, ball bearing
17. Ring ; snap, thrust ring
18. Retaining ring
19. Thrust ring ; 5th gear
20. Washer ; thrust, 5th
21. Nut ; self lock, counter
22. Washer ; plane, counter 5th gear
23. Bearing ; ball, counter end
24. Gear ; counter, 5th
25. Gear assembly ; 5th with
    synchronizer cone
26. Bearing ; needle, 5th
27. Ring ; block, 5th
28. Nut ; main shaft
29. Washer ; lock, main shaft
30. Synchronizer assembly ; reverse-5th
31. Gear ; reverse, main shaft
32. Bearing ; needle, low
33. Nut ; self lock, counter
34. Washer ; thrust, reverse idle, rear
35. Gear ; reverse idle, with bushing
36. Washer ; thrust, reverse idle, front
37. Plate ; intermediate
38. Ring ; snap, bearing to top shaft
39. Shaft ; top gear, with synchronizer cone
40. Bearing ; ball, top gear shaft
41. Bearing ; needle, main shaft
42. Ring ; block, top-3rd
43. Ring ; snap, main shaft
44. Synchronizer assembly ; top-3rd
45. Ring ; block, top-3rd
46. Gear assembly ; 3rd with synchronizer
    cone

47. Bearing ; needle, 2nd, 3rd
48. Gear assembly ; low-2nd
49. Washer ; thrust, low
50. Bearing ; radial ball, main shaft
51. Coller ; needle bearing, low
52. Washer ; thrust, low
53. Gear assembly ; low, with synchronizer
    cone
54. Bearing ; needle, low
55. Coller ; needle bearing, low
56. Ring ; block, low-2nd
57. Synchronizer assembly ; low-2nd
58. Ring ; block, low-2nd
59. Gear assembly ; 2nd with synchronizer
    cone
60. Bearing ; needle, 2nd, 3rd
61. Gear ; counter reverse
62. Bearing ; angular ball, cluster
63. Gear ; cluster
64. Ring ; snap, cluster gear
65. Bearing ; ball, counter front
65a. Ring ; snap, antilash plate
65b. Conical spring
65c. Antilash plate
66. Ring ; snap, cluster gear
67. Ring ; snap, main shaft
68. Shaft ; reverse idle

Isuzu/LUV 5 speed transmission gear assemblies

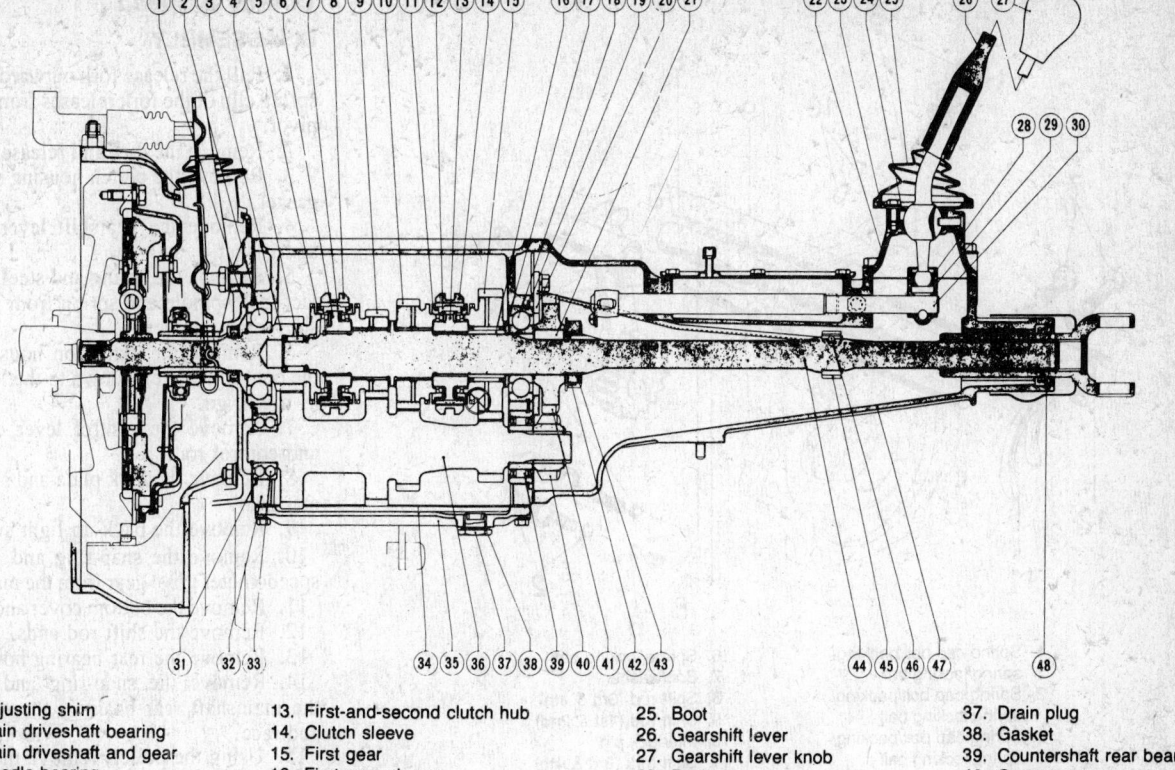

1. Adjusting shim
2. Main driveshaft bearing
3. Main driveshaft and gear
4. Needle bearing
5. Synchronizer ring
6. Third-and-fourth clutch hub
7. Synchronizer key
8. Clutch sleeve
9. Third gear
10. Second gear
11. Synchronizer ring
12. Synchronizer key
13. First-and-second clutch hub
14. Clutch sleeve
15. First gear
16. First gear sleeve
17. Thrust washer
18. Mainshaft bearing
19. Adjusting shim
20. Bearing cover plate
21. Key
22. Gearshift lever retainer
23. Cover
24. Shim
25. Boot
26. Gearshift lever
27. Gearshift lever knob
28. Bush
29. Control lever end
30. Gearshift control lever
31. Adjusting shim
32. Transmission case
33. Countershaft front bearing
34. Gasket
35. Transmission under cover
36. Countershaft
37. Drain plug
38. Gasket
39. Countershaft rear bearing
40. Counter reverse gear
41. Reverse gear
42. Lock washer
43. Locknut
44. Mainshaft
45. Speedometer drive gear
46. Lock ball
47. Extension housing
48. Mainshaft oil seal

Cross section of Mazda/Courier 4-speed without intermediate housing

10. Withdraw the input and the output shafts as a unit.

11. Use a punch to drive the three slotted spring pins out of the shift forks and shift fork shafts.

NOTE: The slotted pin cannot always be fully removed from the first/second shift fork; however, the shift fork can still be withdrawn. Do not try to force the pin out, as damage to the transmission case could result.

12. Remove the case cover and the three detent balls and springs.

13. Remove the shift fork shafts in the following order: First/second shaft. Pin. Reverse shift fork shaft. Third/fourth shaft. Pin.

14. Measure the thrust clearance of the reverse idler gear. The specified clearance is 0.002–0.020 in.

15. Remove the idler shaft. Remove the gear and washer.

16. Measure the thrust clearance of the gears on the output shaft: 1st, 2nd, 5th; 0.006–0.010 in. 3rd; 0.006–0.012 in. Reverse; 0.008–0.012 in. 5th gear is optional.

17. Disassemble the components of the output shaft. Five-speed transmissions have an extra gearset and related parts. Replace the front bearing, if it is rough or noisy. Use a drift and a press. Remove the snap-ring first. For bearing installation, replacement snap-rings are available in a range of sizes 0.0925–0.1024 in. to obtain minimum axial play between the input shaft and the bearing.

ASSEMBLY

1. Assemble the components of the synchronizer hubs, and the output shaft.

2. Install the rear bushing on the output shaft, being careful to install it in the proper direction.

3. Install the ball into the groove of the bushing and slide the bushing over the shaft.

4. Install the needle roller bearing, reverse gear, the ball and the reverse gear synchronizer hub.

5. Install the following items on the output shaft of the four-speed transmission, in the order indicated: Large-diameter reverse gear spacer. Long spacer. Shims.

6. Install the following items on the output shaft of the five-speed transmission in the order indicated: Ball. Fifth gear synchronizer ring. Fifth gear. Needle roller bearing. Bushing. Rear support ball bearing.

7. Install the shims and the nut on the end of the output shaft.

NOTE: If the original nut is being used, change the number of shims to alter the locking portion of the nut.

8. Check the thrust clearance of each gear.

9. Working from the rear of the output shaft, install 3rd gear, synchronizer, spacer, and 3rd/4th synchronizer hub (should face forward).

10. Select a snap-ring to obtain a thrust clearance of less than 0.002 in. for the 3rd/4th synchronizer hub.

11. Assemble the following from the rear of the output shaft: Snap-ring. Key. Speedometer drive gear. Snap-ring.

12. Check thrust clearance of gears.

13. Install the fork and shaft assembly in the transmission case.

14. Insert the straight pins in the grooves on either side of the third/fourth shift shaft.

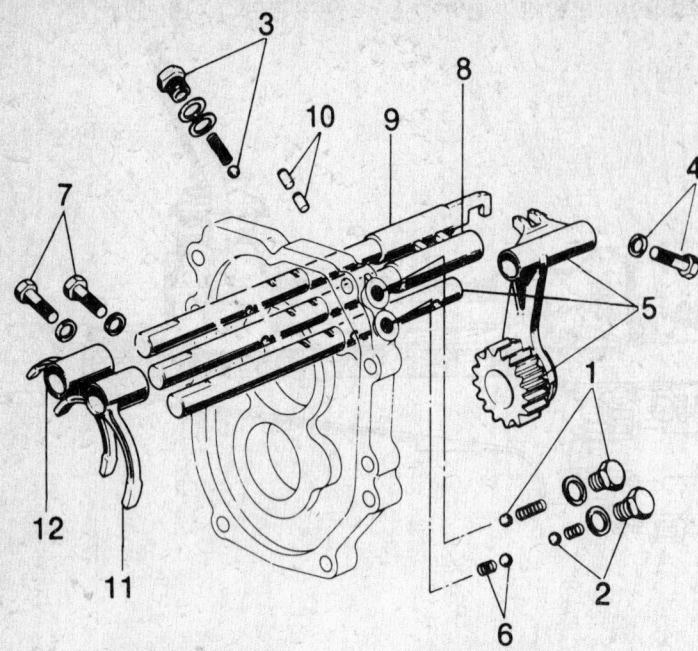

1. Spring cap bolt/packing/
   spring/locking ball
2. Spring cap bolt/packing/
   spring/locking ball
3. Spring cap bolt/packing/
   spring/locking ball
4. Bolt/washer
5. Shift fork (Reverse)/rod/
   reverse idler gear

6. Spring/locking ball
7. Bolt/washer
8. Shift rod (3rd & 4th)
9. Shift rod (1st & 2nd)
10. Interlock pin
11. Shift fork (3rd & 4th)
12. Shift fork (1st & 2nd)

Exploded view of shift selector rods and forks, 4-speed with intermediate housing

15. Assemble the first/second gear shaft and fork.
16. Perform Step 11 for the reverse shift fork shaft.
17. Insert the three detent balls, followed by their springs.
18. Place the cover gasket on the case and install the cover.
19. Use a punch to drive a slotted spring pin into each shift fork to secure it.
20. Assemble the input and output shafts.
21. Install the shift forks into their respective grooves on the input/output shaft assembly.
22. Install the shaft assembly in the right-hand half of the transmission case, so that the snap-ring is positioned firmly against the front surface of the transmission case.
23. Apply grease to the countergear rear bearing lockball. Insert the ball into the hole in the rear bearing outer race.
24. Place the countergear assembly into the right-hand half of the transmission case. Mate the lockball with the hole in the transmission case. Place the bearing snapring firmly against the front surface of the transmission case.
25. Install the reverse idler gear.
26. Install the washers, so that their protrusions align with the grooves in the transmission case.
27. Install the shaft into the case and through the gears and washers.

28. Align the grooves in the idler shaft with the hole in the shaft boss. Install the retaining bolt and washer into the boss.
29. Apply a light coating of liquid sealer over the joint surfaces of the transmission case halves.

CAUTION
Do not apply sealer to the ½ in. hole for the back-up light switch.

30. Align the transmission case locating pins with their holes and assemble the halves of the case.

NOTE: There are four different bolt lengths, do not install the wrong bolt in the wrong hole.

31. Insert the ball, spring, and washer in the back-up light switch hole. Screw in the switch assembly.
32. Install the gasket and bolt the extension housing to the rear of the transmission.
33. Install the speedometer shaft sleeve and drive gear.
34. Apply grease to the conical springs. Install one spring over the input shaft bearing and the other over the countershaft bearing. Install the spacer over the countershaft bearing spring, after coating the spacer with grease.
35. Install the gasket and the clutch housing.

Courier 5 Speed
DISASSEMBLY

1. Pull the release fork outward until the spring clip of the fork releases from the ball pivot.
2. Remove the fork and release bearing.
3. Remove the clutch housing shim and gasket.
4. Remove the gearshift lever retainer and gasket.
5. Remove the spring and steel ball, select lock spindle and spring from the gearshift lever retainer.
6. Remove the extension housing with the control lever end down to the left as far as it will go.
7. Remove the control lever end, key and control rod.
8. Remove the lock plate and speedometer gear.
9. Remove the back-up light switch.
10. Remove the snap-ring and slide the speedometer drive gear from the mainshaft.
11. Remove the bottom cover and gasket.
12. Remove the shift rod ends.
13. Remove the rear bearing housing.
14. Remove the snap-ring and remove the mainshaft rear bearing, thrust washer and race.
15. Using the puller, remove the washer and countershaft rear bearing.
16. Remove the counter fifth gear.
17. Remove the intermediate housing.
18. Remove the springs and shift locking balls.
19. Remove the two blind covers and gaskets from the case.
20. Remove the reverse/fifth shift rod, fork and interlock pin.
21. Remove the first/second and third/fourth shift forks, rods and interlock pins.
22. Remove the snap-ring and slide the washer, fifth gear and synchronizer ring from the mainshaft. Also, remove the steel ball and needle bearing.
23. Lock the rotation of the mainshaft with second and reverse.
24. Remove the locknut and slide the reverse/fifth clutch hub and sleeve assembly, synchronizer ring, reverse gear and needle bearing from the mainshaft.
25. Remove the spacer and counter reverse gear from the countershaft.
26. Remove the reverse idler gear, thrust washers and shaft from the transmission case.
27. Remove the bearing rear cover plate.
28. Remove the snap-ring from the front end of the countershaft and install Mazda tool number 49 0839 445 synchronizer ring holder or its equivalent between the fourth synchronizer ring and the synchromesh gear on the main driveshaft.
29. Remove the countershaft front bearing.
30. Remove the adjusting shim from the countershaft front bearing bore.
31. Remove the countershaft center bearing outer race.

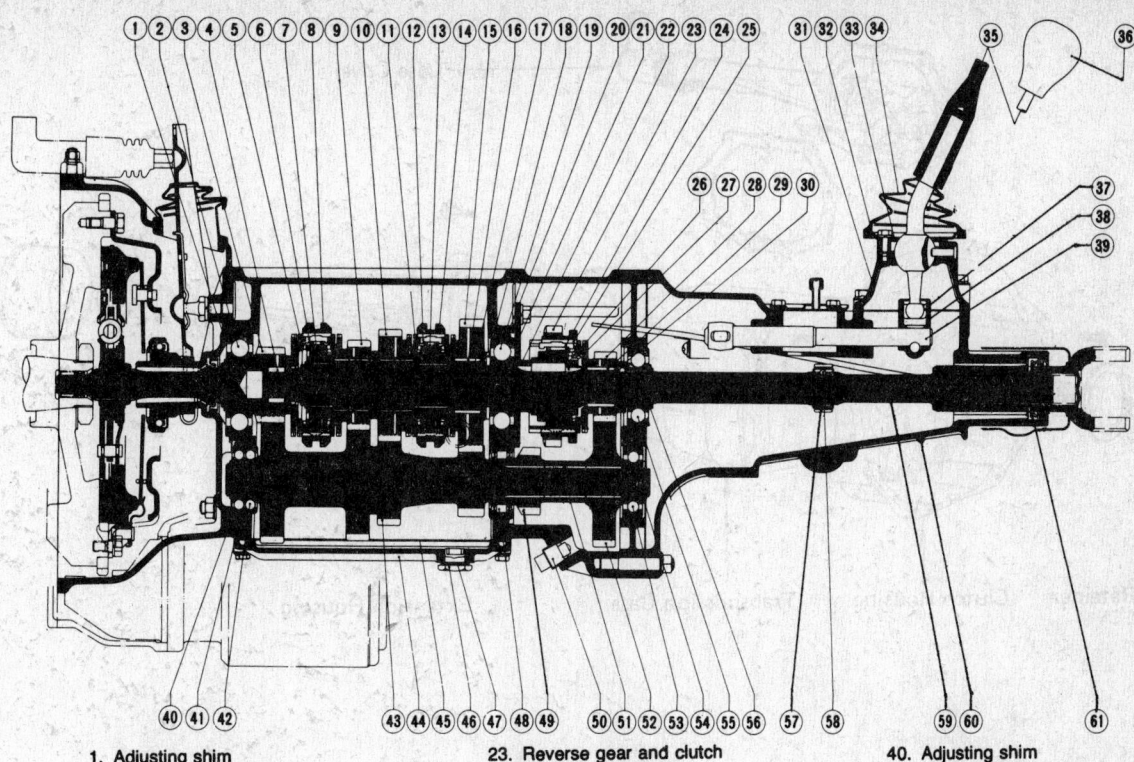

1. Adjusting shim
2. Main driveshaft bearing
3. Main driveshaft gear
4. Needle bearing
5. Synchronizer ring
6. Synchronizer key
7. 3rd-and-4th clutch hub
8. Clutch sleeve
9. 3rd gear
10. 2nd gear
11. Synchronizer ring
12. Synchronizer key
13. 1st-and-2nd clutch hub
14. Clutch sleeve
15. 1st gear
16. Needle bearing
17. Needle bearing inner race
18. Thrust washer
19. Mainshaft front bearing
20. Adjusting shim
21. Bearing cover plate
22. Spacer

23. Reverse gear and clutch sleeve assembly
24. Synchronizer key
25. Synchronizer ring
26. Lock washer
27. Locknut
28. 5th gear
29. Needle bearing
30. Thrust washer
31. Gearshift lever retainer
32. Cover
33. Gasket
34. Boot
35. Gearshift lever
36. Gearshift lever knob
37. Bush
38. Gearshift control lever end
39. Gearshift control lever

40. Adjusting shim
41. Transmission case
42. Countershaft front bearing
43. Countershaft
44. Transmission under cover
45. Gasket
46. Drain plug
47. Gasket
48. Countershaft center bearing
49. Counter reverse gear
50. Drain plug
51. Spacer
52. Counter 5th gear
53. Countershaft rear bearing
54. Thrust washer
55. Mainshaft rear bearing
56. Thrust washer
57. Speedometer drive gear
58. Lock ball
59. Mainshaft
60. Extension housng
61. Mainshaft oil seal

Cross section Mazda/Courier 5-speed

32. With a special puller and attachment, remove the mainshaft front bearing, thrust washer and inner race along with the adjusting shim from the mainshaft front bearing bore.

33. Remove the snap-ring, and remove the main driveshaft bearing.

34. Remove the countershaft center bearing inner race with the puller.

35. Separate the input shaft from the mainshaft and remove the input shaft.

36. Remove the synchronizer ring and needle bearing from the input shaft.

37. Remove the mainshaft assembly.

38. Remove the first/second and third/fourth shift forks from the case.

39. Remove the snap-ring and slide the

third/fourth clutch hub and sleeve assembly, synchronizer ring and third gear from the mainshaft.

40. Remove the thrust washer, first gear and needle bearing from the rear of the mainshaft.

41. Press out the needle bearing inner race, synchronizer ring, first and second clutch hub, sleeve assembly, synchronizer ring and second gear from the mainshaft.

ASSEMBLY

1. Install the third/fourth clutch hub into the sleeve, place the three keys into the clutch hub slots and install the springs onto the hub.

2. Assemble the first/second and

reverse/fifth clutch hub and sleeve.

3. Install the needle bearing, second gear, synchronizer ring, and first/second clutch assembly on the rear section of the mainshaft.

4. Press on the first gear needle bearing inner race.

5. Install the third gear and synchronizer ring onto the front section of the mainshaft.

6. Install the third/fourth clutch assembly onto the mainshaft.

7. Install the snap-ring on the mainshaft.

8. Install the needle bearing, synchronizer ring, first gear and thrust washer on the mainshaft.

9. Install the mainshaft assembly.

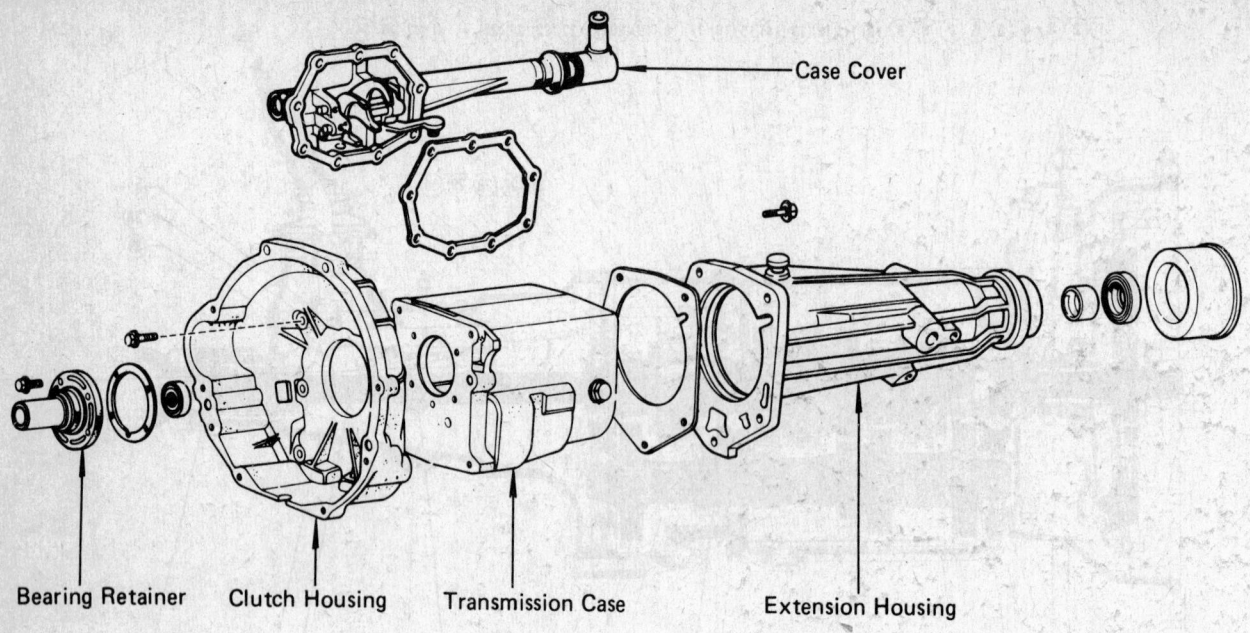

Case Cover

Bearing Retainer   Clutch Housing   Transmission Case   Extension Housing

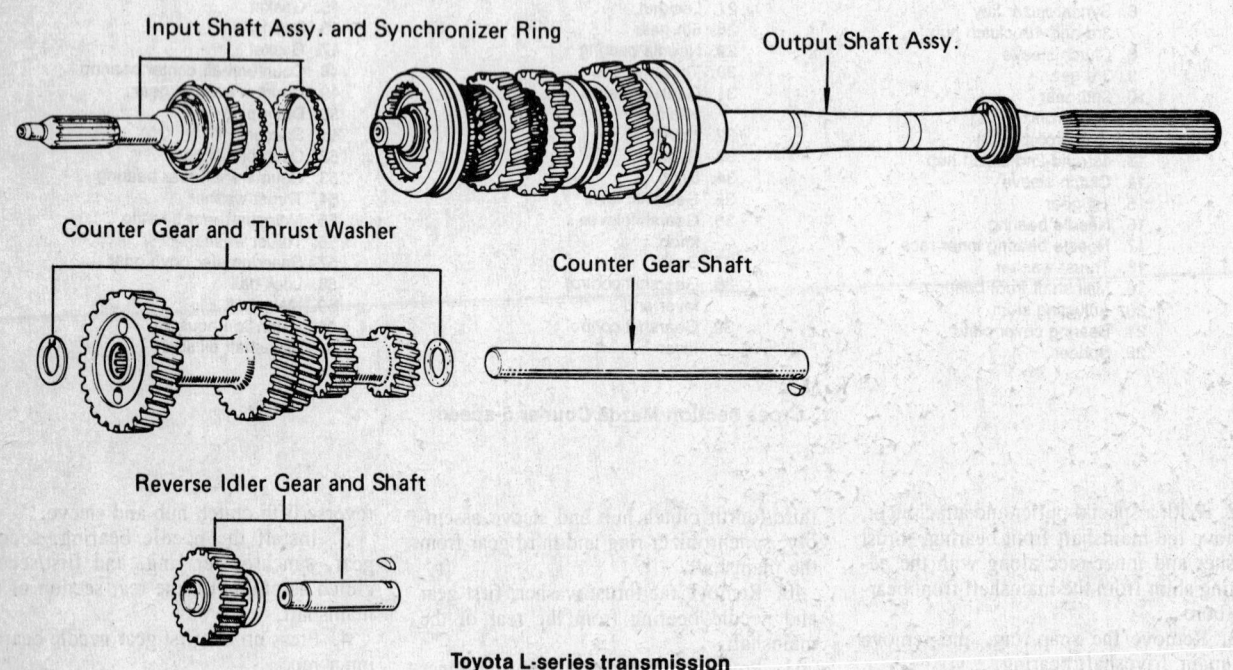

Input Shaft Assy. and Synchronizer Ring

Output Shaft Assy.

Counter Gear and Thrust Washer

Counter Gear Shaft

Reverse Idler Gear and Shaft

Toyota L-series transmission

10. Install the needle bearing on the front end of the mainshaft.

11. Install the first/second and third/fourth shift forks in their respective clutch sleeves.

12. Check the mainshaft bearing end-play. Check the depth of the mainshaft bearing bore in the case. Measure the mainshaft bearing height. The difference

indicates the required adjusting shim to give a total end-play of less than 0.0039 in.

13. Install the synchronizer ring holder tool between the fourth synchronizer ring and the synchromesh gear on the input shaft.

14. Position the shims and mainshaft bearing in the bore and install with a press.

15. Install the input shaft bearing in the

same way.

16. Check the countershaft front bearing end-play in the same way as the mainshaft bearing end-play.

17. Install the front bearing snap-ring.

18. Press the countershaft center bearing into position.

19. Install the bearing cover plate.

20. Install the reverse idler gearshaft,

thrust washers and reverse idler gear.

21. Install the counter reverse gear and spacer on the rear end on the countershaft.

22. Install the thrust washer and press the needle bearing inner race of the reverse gear on the mainshaft.

23. Install the needle bearing, reverse gear, synchronizer ring, reverse/fifth clutch assembly and new mainshaft locknut on the mainshaft.

24. Lock the mainshaft with the second and reverse gears. Tighten the locknut.

25. Install the needle bearing, synchronizer ring and fifth gear on the mainshaft.

26. Install the thrust washer, steel ball and snap-ring on the mainshaft.

27. Check the thrust washer-to-snap-ring clearance. It should be 0.0039–0.0118 in.

28. Install the first/second shift rod through the holes in the case and fork.

29. Install the interlock pin with a special installer and guide.

30. Install the third/fourth shift rod through the holes in the case and fork.

31. Align the holes and install the lockbolts of each shift fork and rod.

32. Install the interlock pin as above.

33. Position the reverse/fifth shift fork on the clutch sleeve and install the shift rod.

34. Tighten the lockbolt.

35. Install the three shift locking balls, springs and cap bolts.

36. Place the third/fourth clutch sleeve in third gear.

37. Check the clearance between the synchronizer key and the exposed edge of the synchronizer ring with a feeler gauge. The gap should be 0.026–0.079 in. Adjust by varying thrust washers.

38. Install the two blind covers and gaskets.

39. Install the undercover and gasket.

40. Apply a thin coat of sealer to the mating edges and install the intermediate housing on the transmission case. Align the lockbolt holes of the housing and reverse idler gearshaft, install and tighten the lockbolt.

41. Position the counter fifth gear and bearing to the rear end of the countershaft and install with a press.

42. Install the thrust washer and snap-ring.

43. Check the clearance between the washer and snap-ring. Clearance should be less than 0.0039 in.

44. Install the mainshaft rear bearing.

45. Install the thrust washer and snap-ring.

46. Check the thrust washer-to-snap-ring clearance. Clearance should be less than 0.0059 in.

47. Apply a thin coat of sealing agent to the mating surfaces and install the bearing housing on the intermediate housing.

48. Install the shift rod ends on their respective rods.

49. Install the speedometer drive gear and steel ball on the mainshaft. Secure it with a snap-ring.

50. Install a speedometer driven gear assembly on the extension housing and secure it with the bolt and lock plate.

51. Insert the control rod through the holes from the front side of the extension housing.

52. Align the key and insert the control lever end in the control rod.

53. Install the bolt and tighten it to 20–30 ft. lbs.

54. Install the back-up light switch.

55. Place the gasket on the case and install the extension housing with the control lever end down and as far to the left as it will go.

56. Insert the select lock spindle and spring from the underside of the shift lever retainer.

57. Install the steel ball and spring in alignment with the spindle groove and install the spring cap bolt.

58. Install the gearshift lever retainer and gasket on the extension housing.

59. Check the bearing end-play. Measure the depth of the bearing bore in the housing. Measure the height of the bearing protrusion. The difference indicates the thickness of the shim needed. The end-play should be less than 0.0039 in.

60. Place the gasket on the front side of the case. Apply lubricant to the lip of the oil seal and install the clutch housing on the case.

61. Install the release bearing and fork on the clutch housing.

Mazda 4 & 5 Speed (Thru '84)

On models equipped with a 5 speed transmission, the disassembly and assembly of the rear extension housing, selector levers and forks are completed in the same manner as the 4 speed transmission. After this has been done, the added housing can be removed by taking out the retaining bolts. The housing will have to be lightly tapped with a soft-faced hammer. The removal of the housing exposes the 5th/reverse synchronizer assembly, the reverse countergear, the countershaft and mainshaft bearings. The bearings are pulled from the shafts and then the gears can be removed. The assembly is in the reverse of the removal procedure.

DISASSEMBLY

1. Remove the throwout bearing and fork from the clutch housing. Remove the front transmission cover, shim and gasket. Remove the snap-ring from the maindrive shaft.

2. Remove the gearshift lever retainer and gasket from the extension housing. Unbolt and remove the extension housing by sliding the housing off the mainshaft with the control lever end laid down to the left as far as possible.

3. Unbolt and remove the control lever end and rod from the extension housing.

4. Remove the speedometer driven gear and the back-up light switch.

5. Remove the snap-ring retaining the speedometer drive gear on the main shaft. Slide the gear off of the mainshaft. Remove the lock ball.

6. Use an appropriate pusher tool and remove the transmission case from the intermediate bearing housing plate.

7. Remove the bearing from the case. remove the front bearing from the countershaft using a suitable puller.

8. Remove the three spring cap bolts and remove the springs and shift locking balls. Remove the reverse shift rod, fork assembly and reverse gear from the bearing housing. Remove each shift fork mounting spring pin and push the shift rods rearward out of the housing. Remove the rods and forks. Remove the reverse shift rod locking ball and retaining pins from the bearing housing.

9. Remove the reverse gear and key from the mainshaft. Remove the snap ring from the rear end of the countershaft and slide the reverse counter gear off.

10. Remove the bearing cover and the reverse idler gear from the bearing housing.

11. Use a "soft" brass or copper hammer and tap the rear ends of the mainshaft and countershaft lightly and in turn until they can be removed from the bearing housing.

12. Remove the bearings from the housing. Remove the thrust washer, first gear, sleeve and synchronizer ring from the mainshaft.

13. Remove the snap-ring from the front of the mainshaft. Press the third and fourth clutch hub and sleeve assembly, synchronizer ring and third gear from the front of the mainshaft. Press the first and second clutch hub and sleeve assembly, synchronizer ring and second gear from the rear of the mainshaft.

14. Clean and inspect all parts. Replace as necessary.

ASSEMBLY

1. Assemble the third and fourth, first and second clutches by installing the clutch hub into the sleeve, positioning the three keys in the slots and installing the retainer springs.

2. Install the third gear and synchronizer ring onto the front of the mainshaft. Install the third and fourth clutch hub assembly onto the mainshaft. Be sure the hub is facing in the proper direction. Fit the retaining snap-ring in position.

3. Install the second gear and synchronizer ring onto the rear section of the mainshaft. Press the first and second clutch assembly onto the mainshaft. Install the synchronizer ring, first gear with sleeve and the thrust washer onto the mainshaft.

4. Install the main driveshaft and needle bearing to the mainshaft.

5. Check the countershaft rear bearing and mainshaft bearing clearances. Clearance should be less than 0.0039 in. Shims are available in 0.0039 and 0.118 in. sizes. Press each bearing and shim into position in the bearing housing.

6. Press the countershaft and mainshaft

1. Bolt
2. Bearing cover
3. Reverse idle gear shaft
   (4 speed transmission)
4. Ball bearing
5. Adjustment shim
6. Ball bearing
7. Adjustment shim
8. Bearing housing

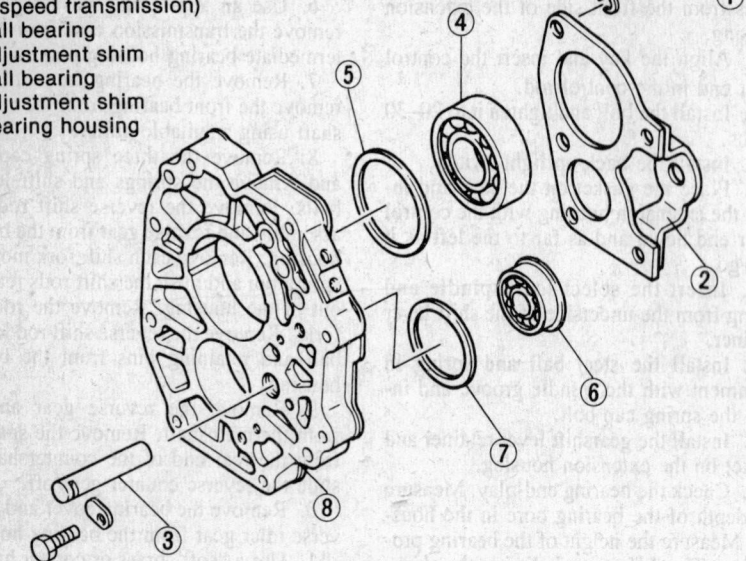

Mazda 4 & 5 speed bearing housing parts

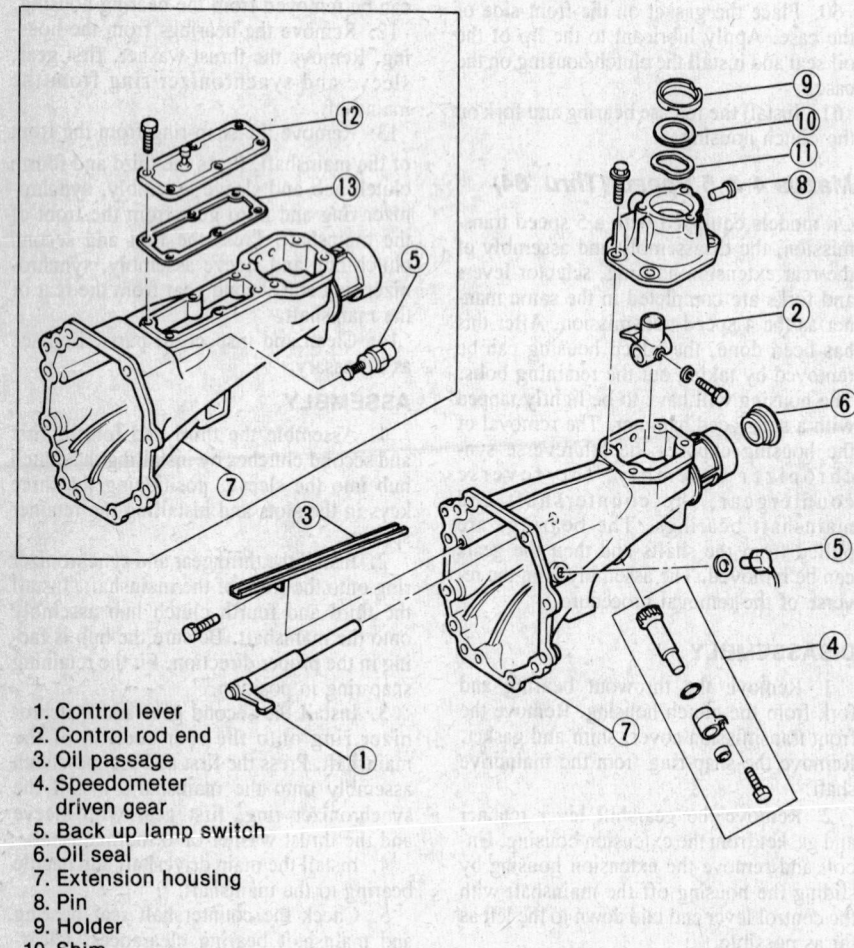

1. Control lever
2. Control rod end
3. Oil passage
4. Speedometer
   driven gear
5. Back up lamp switch
6. Oil seal
7. Extension housing
8. Pin
9. Holder
10. Shim
11. Wave washer
12. Cover } 4 speed
13. Gasket } transmission

Mazda 4 & 5 speed extension housing parts

assemblies in position on the bearing housing.

7. Install the reverse idler gear shaft and bearing cover to the bearing housing. Install the reverse gear and key onto the mainshaft. When installing the reverse gears, both gears should be fitted so that the chamfer on the teeth face rearward. Tighten the locknut to 125–150 ft. lbs. Bend the lock tab on the washer. Install the countershaft reverse gear and secure with snapring.

8. Insert the spring and ball into the bearing housing. Push down on the ball with a suitable tool and install the reverse shift rod, shift lever and reverse idler gear at the same time.

9. Install the first and second shift fork and the third and fourth shift fork over the clutch sleeves. Install the shift fork rods, align the mounting holes and secure with new spring pins. Install the pins with the split parallel to the shift fork mounting rod.

10. Install the shift locking balls and springs in their respective positions and secure with the retaining bolt.

11. Apply a thin coat of sealer to both sides of the bearing housing and mount the bearing housing in position on the transmission case.

12. Install the speedometer drive gear and steel ball on the mainshaft, secure them with the snap-ring.

13. Install the mainshaft and countershaft ball bearings.

14. Install the speedometer driven gear assembly to the extension housing. Insert the shift control lever through the holes from the front side of the extension housing. Install the control lever end to the control lever and tighten the mounting bolt to 20–25 ft. lbs.

15. Mount the back-up light switch on the extension housing. Apply a thin coat of sealer to the extension housing mounting flange and install the housing onto the bearing housing. Be sure the control lever is laid over to the left as far as possible when installing the extension housing. Install the gearshift lever retainer and gasket.

16. Lubricate the front oil seal and install the front cover on the transmission. The front bearing should have less than 0.004 play when the front cover is installed. Shims of 0.006 and 0.012 in. are available for adjustment purposes.

17. Install the throwout bearing fork and bearing and install the transmission.

Mazda 4 & 5 Speed (1986)

The 4 and 5 speed transmissions are basically the same, with an added housing located between the adapter plate and the rear extension housing, to carry the 5th and reverse gears. Added roller bearings are used in the housing to prevent shaft misalignment.

DISASSEMBLY

1. Remove the throwout bearing return

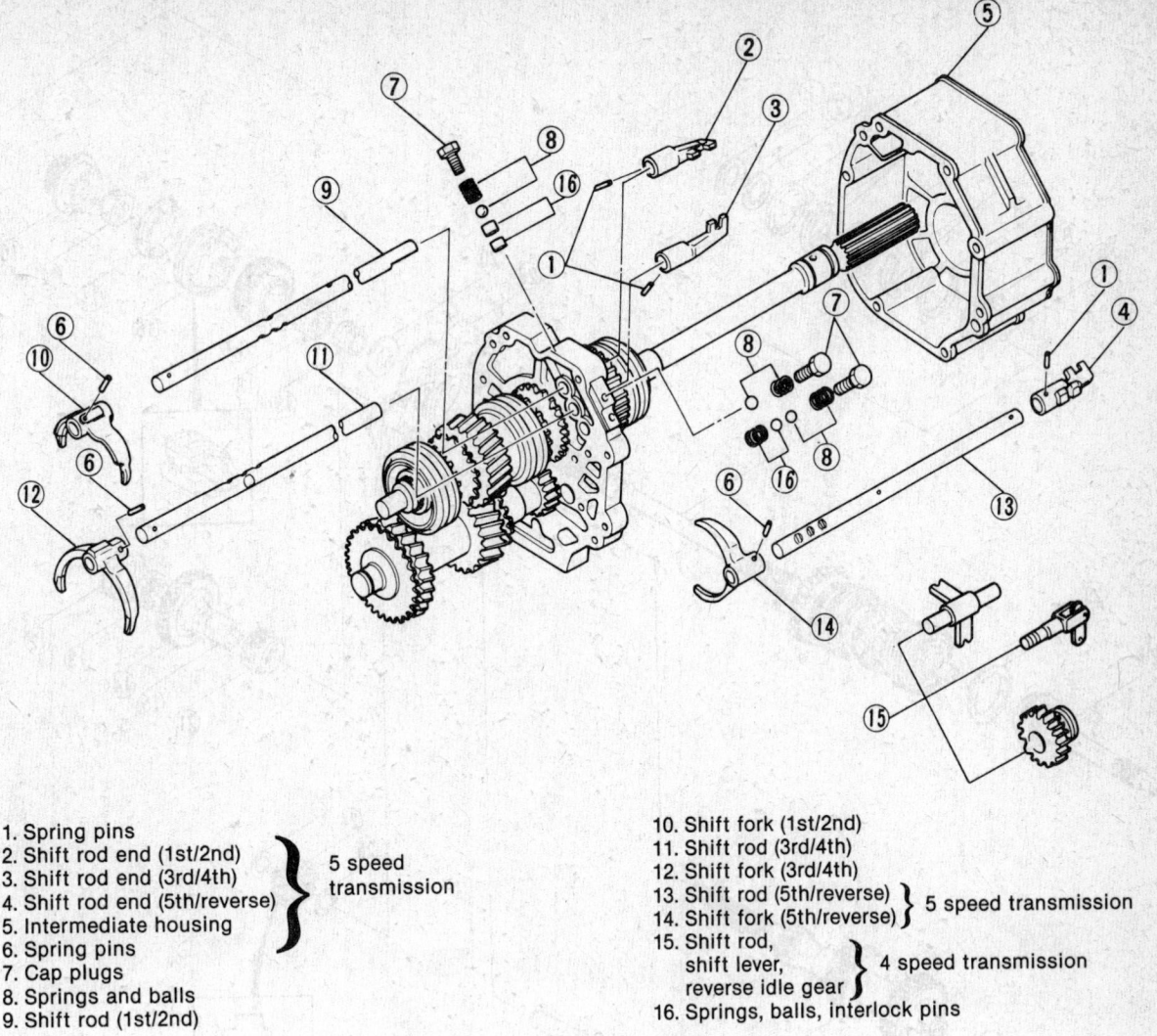

1. Spring pins
2. Shift rod end (1st/2nd)
3. Shift rod end (3rd/4th)
4. Shift rod end (5th/reverse)
5. Intermediate housing
6. Spring pins
7. Cap plugs
8. Springs and balls
9. Shift rod (1st/2nd)

} 5 speed transmission

10. Shift fork (1st/2nd)
11. Shift rod (3rd/4th)
12. Shift fork (3rd/4th)
13. Shift rod (5th/reverse)
14. Shift fork (5th/reverse)

} 5 speed transmission

15. Shift rod, shift lever, reverse idle gear

} 4 speed transmission

16. Springs, balls, interlock pins

Mazda 4 & 5 speed shift forks and rods

spring, throwout bearing, and the release fork.

2. Remove the bearing housing.

3. Remove the input shaft and countershaft snap-rings.

4. Remove the floorshift lever retainer, complete with gasket.

5. Unfasten the cap bolt and withdraw the spring, steel ball, select lock pin and spring from the retainer.

6. Remove the extension housing. Turn the control lever as far left as it will go and slide the extension housing off the output shaft.

7. Remove the spring seat and spring from the end of the shift control lever.

8. Loosen the spring cap and withdraw the spring and plunger from their bore.

9. Remove the control rod and boss from the extension housing.

10. Remove the speedometer driven gear. Remove the back-up light switch.

11. Remove the speedometer drive gear.

12. Tap the front ends of the input shaft and countershaft with a plastic hammer;

then remove the intermediate housing assembly from the transmission case.

13. Remove the three cap bolts; then withdraw the springs and lockballs.

14. Remove the reverse shift rod, reverse idler gear, and shift lever.

15. Remove the setscrews from all the shift forks and push the shift rods rearward to remove them. Remove the shift forks.

16. Withdraw the reverse shift rod lockball, spring, and interlock pins from the intermediate housing.

17. Remove reverse gear and key from the output shaft.

18. Remove the reverse countergear.

19. Remove the countershaft and output shaft from the intermediate housing.

20. Remove the bearings from the intermediate housing and transmission case.

21. Remove the snap-ring from the output shaft.

22. Slide the third/fourth clutch hub, sleeve, synchronizer ring, and third gear off the output shaft.

23. Remove the thrust washer, first gear,

sleeve, synchronizer ring, and second gear from the rear of the output shaft.

ASSEMBLY

1. Install the third/fourth synchronizer clutch hub on the sleeve. Place the three synchronizer keys in the clutch hub key slots. Install the key springs with their open ends 120° apart.

2. Install third gear and the synchronizer ring on the front of the output shaft. Install the third/fourth clutch hub assembly on the output shaft. Be sure that the larger boss faces the front of the shaft.

3. Secure the gear and synchronizer with the snap-ring.

4. Perform Step 1 to the first/second synchronizer assembly.

5. Position the synchronizer ring on second gear. Slide second gear on the output shaft so that the synchronizer ring faces the rear of the shaft.

6. Install the first/second clutch hub assembly on the output shaft so that its oil grooves face the front of the shaft. Engage

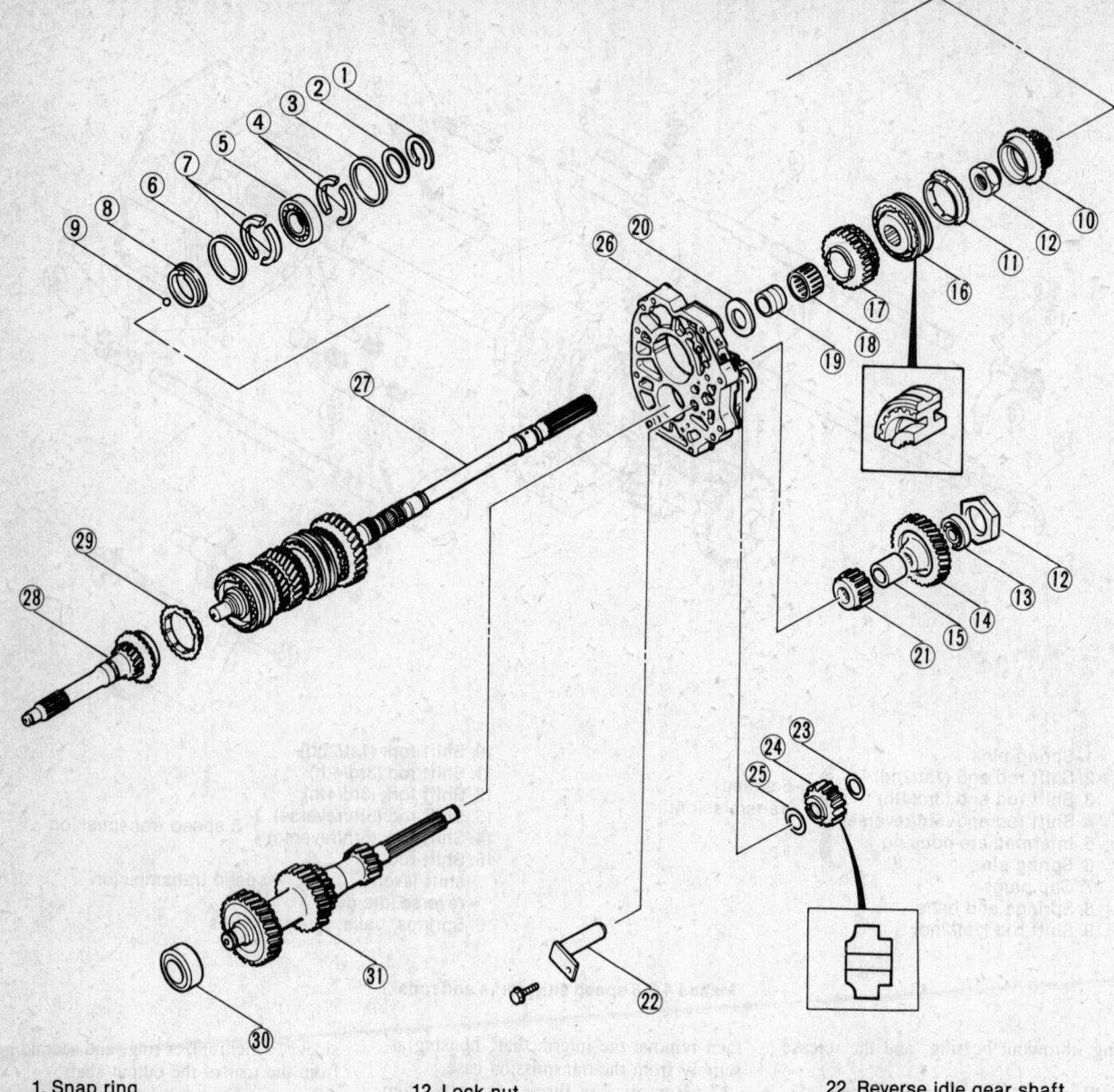

Mazda 5 speed main and countershaft assemblies

1. Snap ring
2. Washer
3. Retaining ring
4. C washers
5. Ball bearing
6. Retaining ring
7. C washers
8. Thrust lock washer
9. Ball
10. 5th gear
11. Synchronizer ring

12. Lock nut
13. Ball bearing
14. Counter gear
15. Spacer
16. Clutch hub assembly
    (5th/reverse)
17. Reverse gear
18. Needle bearing
19. Inner race
20. Washer
21. Counter reverse gear

22. Reverse idle gear shaft
23. Washer
24. Reverse idle gear
25. Washer
26. Bearing housing assembly
27. Main shaft and gear
    assembly
28. Main drive gear
29. Synchronizer ring
30. Ball bearing
31. Counter shaft gear

the keys in the notches on the second gear synchronizer ring.

7. Slide the first gear sleeve onto the output shaft. Position the synchronizer ring on first gear. Install the first gear on the output shaft so that the synchronizer ring faces frontward. Rotate the first gear as required to engage the notches in the syn-

chronizer ring with the keys in the clutch hub.

8. Slip the thrust washer on the rear of the output shaft. Install the needle bearing on the front of the output shaft.

9. Install the synchronizer ring on fourth gear and install the input shaft on the front of the output shaft.

10. Press the countershaft rear bearing and shim into the intermediate housing, then press the countershaft into the rear bearing.

11. Keep the thrust washer and first gear from falling off the output shaft by supporting the shaft. Install the output shaft on the intermediate housing. Be sure that each

1. Lock nut
2. Plain washer
3. Reverse gear
4. Woodruff key
5. Snap ring
6. Counter reverse gear
7. Bearing housing assembly
8. Main drive gear
9. Synchronizer ring
10. Counter shaft gear
11. Ball bearing
12. Main shaft and gear assembly

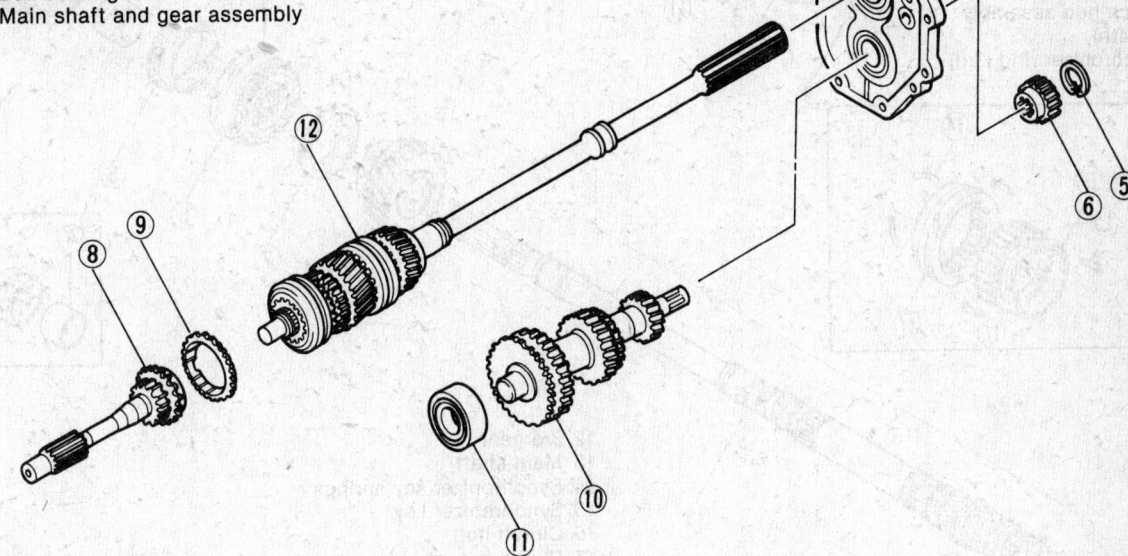

Mazda 4 speed main and countershaft assemblies

output shaft gear engages with its opposite number on the countershaft.

12. Tap the output shaft bearing and shim into the intermediate housing with a plastic hammer. Install the cover.

13. Install reverse gear on the output shaft and secure it with its key.

NOTE: The chamfer on the teeth of both the reverse gear and the reverse countergear should face rearward.

14. Install the reverse countergear.

15. Install the lockball and spring into the bore in the intermediate housing. Depress the ball with a screwdriver.

16. Install the reverse shift rod, lever, and idler gear at the same time. Place the reverse shift rod in the neutral position.

17. Align the bores and insert the shift interlock pin.

18. Install the third/fourth shift rod into the intermediate housing and shift bores. Place the shift rod in Neutral.

19. Install the next interlock pin in the bore.

20. Install the first/second shift rod.

21. Install the lockballs and springs in their bores. Install the cap bolt.

22. Install the speedometer drive gear and lockball on the output shaft, and install its snap-ring.

23. Apply scaler to the mating surfaces of the intermediate housing. Install the intermediate housing in the transmission case.

24. Install the input shaft and countershaft front bearings in the transmission case.

25. Secure the speedometer driven gear.

26. Install the control rod through the holes in the front of the extension housing.

27. Align the key with the keyway and install the yoke on the end of the control rod. Install the yoke lockbolt.

28. Fit the plunger and spring into the extension housing bore and secure with the spring cap.

29. Turn the control rod all the way to the left and install the extension housing on the intermediate housing.

30. Insert the spring and select lockpin inside the gearshift retainer. Align the steel ball and spring with the lockpin slot, and secure it with the spring cap.

31. Install the spring and spring seat in the control rod yoke.

32. Install the gearshift lever retainer over its gasket on the extension housing.

33. Lubricate the lip of the front bearing cover oil seal and secure the cover on the transmission case.

34. Check the clearance between the front bearing cover and bearing. It should

be less than 0.006 in. If it is not within specifications insert additional adjusting shims. The shims are available in 0.006 in. or 0.012 in. sizes.

35. Install the throwout bearing, return spring and release fork.

NOTE: On 5 Speed transmissions, the disassembly and assembly of the rear extension housing, selector levers and forks are completed in the same manner as the 4 speed transmission. After this has been done, the added housing can be removed by taking out the retaining bolts. The housing will have to be lightly tapped with a soft-faced hammer. The removal of the housing exposes the 5th/reverse synchronizer assembly, the reverse countergear, the countershaft and mainshaft bearings. The bearings are pulled from the shafts and then the gears can be removed. The assembly is in the reverse of the removal procedure.

Mitsubishi/D 50 Arrow

4 Speed (KM130)
DISASSEMBLY

1. Remove the undercover.

1. Washer
2. 1st gear
3. Gear sleeve
4. Needle bearing and inner
5. Synchronizer ring (1st)
6. Clutch hub assembly (1st/2nd)
7. Synchronizer ring (2nd)
8. 2nd gear
9. Snap ring
10. Clutch hub assembly (3rd/4th)
11. Synchronizer ring (3rd)

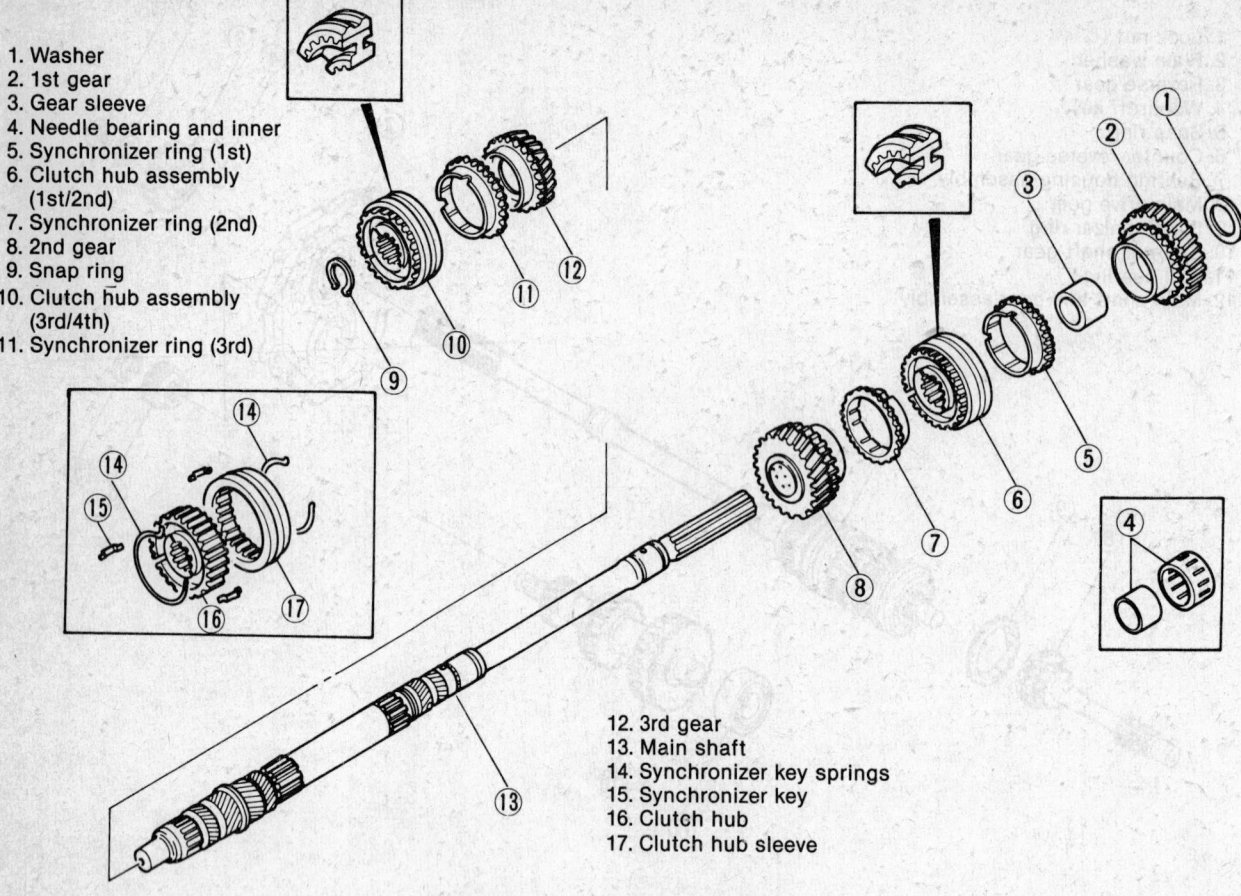

12. 3rd gear
13. Main shaft
14. Synchronizer key springs
15. Synchronizer key
16. Clutch hub
17. Clutch hub sleeve

Mazda 4 & 5 speed mainshaft parts

2. Remove the back-up light switch. Be careful not to lose the steel ball.

3. Remove the speedometer gear sleeve clamp and remove the speedometer driven gear and sleeve assembly from the extension housing assembly.

4. Remove the extension housing bolts. Turn the shift lever to the left and pull off the extension housing.

5. Loosen the three poppet plugs, then remove the three poppet springs and the three steel balls.

6. Place the 1st/2nd speed shift rod in Neutral position.

7. Remove the reverse shift rail and fork assembly together with the reverse idler gear.

8. Using a $^3/_{16}$ in. punch, drive off the 3rd/4th and 1st/2nd speed shift fork spring pins. Push each shift rod toward the rear of the transmission case and remove the shift forks. Remember to remove the interlock plunger.

9. Remove the snap-ring from the rear end of the counter gear and then remove the reverse counter gear.

10. Unlock the mainshaft locknut and remove the locknut. The locknut can be loosened by double-engaging the 3rd speed gear and the 1st speed gear.

11. Remove the reverse gear from the mainshaft.

12. Remove the five attaching screws and then remove the rear bearing retainer.

13. Remove the front bearing retainer.

14. With the counter gear pressed to the rear, remove the rear bearing snap-ring. Then using a bearing puller remove the rear counter bearing.

15. Remove the snap-ring from the front counter bearing. Pull off the bearing with a bearing puller.

16. Pull the counter gear out of the case.

17. Remove the main drive pinion from the front of the case. To remove the bearing from the main drive pinion, remove the two snap-rings and then remove the bearing with a bearing puller.

18. Remove the mainshaft bearing snap-ring and remove the bearing using a dual post bearing puller (special tool MD998056-10 and MD998056 or equivalents).

19. Remove the mainshaft assembly by lifting it up through the case.

20. Disassemble the mainshaft assembly in the following order: Pull off the 1st speed gear, the 1st/2nd speed synchronizer and the 2nd speed gear toward the rear of the mainshaft. Remove the snap-ring from

the forward end of the mainshaft, then remove the 3rd/4th speed synchronizer and the 3rd speed gear.

21. If removing the shift control shaft assembly, remove the pin locking the gear shifter using a $^3/_{16}$ in. punch. To remove the lock pin, press the gear shifter forward and drive the lock pin off, being careful not to bend the control shaft. Inspect the parts after cleaning. Replace any worn, damaged or defective.

ASSEMBLY

1. If the main drive pinion bearing has been removed, replace it using a pipe fit over the end of the pinion shaft. Make sure the pipe does not apply pressure on the ball bearings but only on the bearing race, or bearing damage could result.

2. Fit a snap-ring which gives a clearance of no more than 0–0.002 in. and install it on the drive pinion.

3. Assemble the mainshaft in the following order: Assemble the 3rd/4th speed and the 1st/2nd speed synchronizers. The front and rear ends of the synchronizer sleeve and hub can be identified as shown in the illustration. The synchronizer spring is installed as shown. Install the needle bearing, 3rd speed gear, synchronizer ring

and the 3rd/4th speed synchronizer assembly onto the mainshaft from the front end. Be careful not to confuse the front and the rear of the synchronizer assembly.

4. Select and install a snap-ring that will give the 3rd/4th speed synchronizer hub an end-play from 0.0 to 0.003 in.

5. Third speed gear end-play should be from 0.002 to 0.008 in.

6. Install the needle bearing, the 2nd speed gear, the synchronizer assembly, the bearing sleeve, the needle bearing, the 1st speed gear, and the bearing spacer onto the mainshaft from the rear end.

7. Push the bearing spacer forward and check the 1st and 2nd speed gear end-play. Clearance should be within 0.002–0.008 in.

8. Insert the mainshaft assembly into the transmission case and fit the mainshaft center bearing using a bearing driver. Hold the forward end of the mainshaft by hand at the front of the case.

9. Install the needle bearing and the synchronizer ring, then insert the main drive pinion assembly into the case from the front.

10. Insert the countershaft gear into the case. With a snap-ring fitted to the countershaft front needle bearing, drive the bearing into the case by hammering on the outer race of the bearing.

11. Fit a snap-ring to the countershaft rear ball bearing and then install the bearing with a bearing installer.

12. Install the front bearing retainer. When installing the retainer, install a spacer that will give a clearance of 0.0–0.004 in. Apply sealant to both sides of the front bearing retainer packing and apply gear oil to the oil seal lip. Install packing and oil seal.

13. Install the rear bearing retainer and its five screws. It is suggested that each screw head be staked with a pointed punch to prevent them from coming loose.

14. Install the reverse gear on the mainshaft and tighten the locknut to 73–94 ft. lbs. Lock the nut at the notch of the mainshaft.

15. Install the spacer and counter reverse gear to the counter gear rear end.

16. Install a snap-ring of the proper size so that the reverse counter gear and play will be from 0.0 to 0.003 in.

17. Install the 3rd/4th and 1st/2nd speed shift forks into their respective synchronizer sleeves. Insert each shift rod from the rear of the case. Lock the shift forks and rod with spring pins, install the interlock plunger between the shift rods.

NOTE: The spring pins should be installed with the slits parallel to the shift rod.

18. Install the reverse shift rod and fork assembly together with the reverse idler gear.

19. Insert the ball and poppet spring with the small end on the ball side into each shift rod. Tighten the plugs to the specified positions. After installation, seal each plug head with sealant.

20. Apply sealant to both sides of the extension housing packing and fit the packing into the housing.

21. Turn the gear shift control down to the left and install the extension to the transmission case.

22. Make sure the forward end of the control finger is snug in the slot of the shift lug and fit the extension housing bolts after coating their threads with sealant.

23. Apply gear oil to the speedometer driven gear and install the gear and sleeve assembly in the extension housing. Make sure the sleeve flange and its mating areas on the extension housing are free of dirt, or it will cause the gears to be misaligned and could damage them.

24. Rotate the speedometer driven gear and sleeve assembly so that the number on the sleeve, which is the same as the number of teeth on the gear, is in the "U" mark position as the assembly is installed.

25. Install the speedometer gear clamp with its tongs in the sleeve positioning slots.

26. Install the backup light switch with its steel ball.

27. Refit the under cover and torque the bolts to 6–7 ft. lbs.

28. Install the transmission control lever assembly and fill the gear shifter area with grease. Fill the transmission with lubricant.

5 Speed (KM132)

DISASSEMBLY

1. Remove the clutch release bearing and carrier.

2. Remove the spring pin and the clutch control shaft. Remove the felt, return spring and clutch shift arm.

3. Remove the case cover.

4. Remove the back-up light switch.

5. Remove the extension housing.

6. Remove the speedometer drive gear.

7. Remove the ball bearing from the mainshaft rear end.

8. Loosen three poppet spring plugs, then remove three poppet springs and three balls.

9. Remove the 3–4 and 1–2 speed shift fork spring pins. Pull off each shift rail toward the rear of the transmission case, then remove the shift fork. Remove the interlock plunger.

10. Remove the overtop and reverse shift forks spring pins, shift rails and forks.

11. Loosen the locknuts (mainshaft and countershaft rear ends).

12. Pull off the counter overtop gear and the ball bearing at the same time using a puller. Remove the spacer and the counter reverse gear.

13. Remove the overtop gear and sleeve from the mainshaft. Remove the overtop synchronizer assembly and spacer.

14. Remove the reverse idler gear.

15. Remove the rear bearing retainer.

16. Drive the reverse idler gear shaft from inside the case.

17. Remove the front bearing retainer.

18. With the counter gear pressed to the rear, remove the rear bearing snap-ring. Remove the counter rear bearing.

19. Remove the counter front bearing.

20. Remove the counter gear from the inside of the case.

21. Remove the main drive pinion from the front of the case. Remove the main drive pinion bearing.

22. Remove the mainshaft bearing snapring. Remove the ball bearing.

23. Pull the mainshaft assembly from the case.

24. Disassemble the mainshaft in the following order: Remove the 1st gear, the 1–2 speed synchronizer and the 2nd speed gear toward the rear of the mainshaft. Remove the snap-ring from the forward end of the mainshaft. Remove the 3–4 speed synchronizer and the 3rd gear.

25. Disassemble the extension housing: Remove the lock plate and the speedometer driven gear. Remove the plug, spring and neutral return plunger.

When removing the control shaft assembly, pull off the lock pin locking the gear shifter. To remove the lock pin, press the gear shifter forward and pull it off.

ASSEMBLY

1. Install the ball bearing on the main drive pinion. Install a selective snap-ring so that there will be 0–0.002 in. clearance between the snap-ring and the bearing. Thickness of snap-ring– .0906 in.; .0925 in.; .0945 in.; .0965 in.; .0984 in. Identification color– White; None; Red; Blue; Yellow.

2. Install the mainshaft in the following order: Assemble the 3–4 speed and 1–2 speed synchronizers. Be sure the synchronizer assemblies are installed facing in the proper direction. Install the needle bearing, the 3rd speed gear, the synchronizer ring, and the 3–4 speed synchronizer assembly on to the mainshaft from the front end. Select and install a snap-ring of proper size so that the 3–4 speed synchronizer hub end-play will be 0–0.003 in. Check the 3rd gear end-play (.0016–.0079 in.). Install the needle bearing, the 2nd speed gear, the synchronizer assembly, the bearing sleeve, the needle bearing, the 1st speed gear, and the bearing spacer on the mainshaft from the rear. With the bearing spacer pressed forward, check the 2nd and 1st gear end-play (0.0016–0.0079 in.).

3. Install the mainshaft into the transmission case and drive in the mainshaft center bearing.

4. Install the needle bearing and the synchronizer ring. Install the main drive pinion assembly into the case from the front.

5. Install the countershaft gear into the case. Drive the front bearing into the case.

6. Install the snap-ring on the countershaft rear bearing.

7. Install the front bearing retainer. Select and install a spacer of proper size so that the clearance will be 0–0.0039 in. Re-

1. Transmission case
2. Main drive pinion
3. Synchronizer assy (3-4 speed)
4. 3rd speed gear
5. 2nd speed gear
6. Synchronizer assy (1-2 speed)
7. 1st speed gear
8. Rear bearing retainer
9. Synchronizer assy (overtop)
10. Overtop gear
11. Control finger
12. Neutral return finger
13. Control shaft
14. Control lever cover
15. Control lever assy
16. Stopper plate
17. Control housing
18. Change shifter
19. Mainshaft
20. Speedometer drive gear
21. Extension housing
22. Counter overtop gear
23. Counter reverse gear
24. Reverse idler gear
25. Reverse idler gear shaft
26. Case cover
27. Counter gear
28. Front bearing retainer
29. Clutch shift arm
30. Release bearing carrier
31. Clutch control shaft
32. Return spring

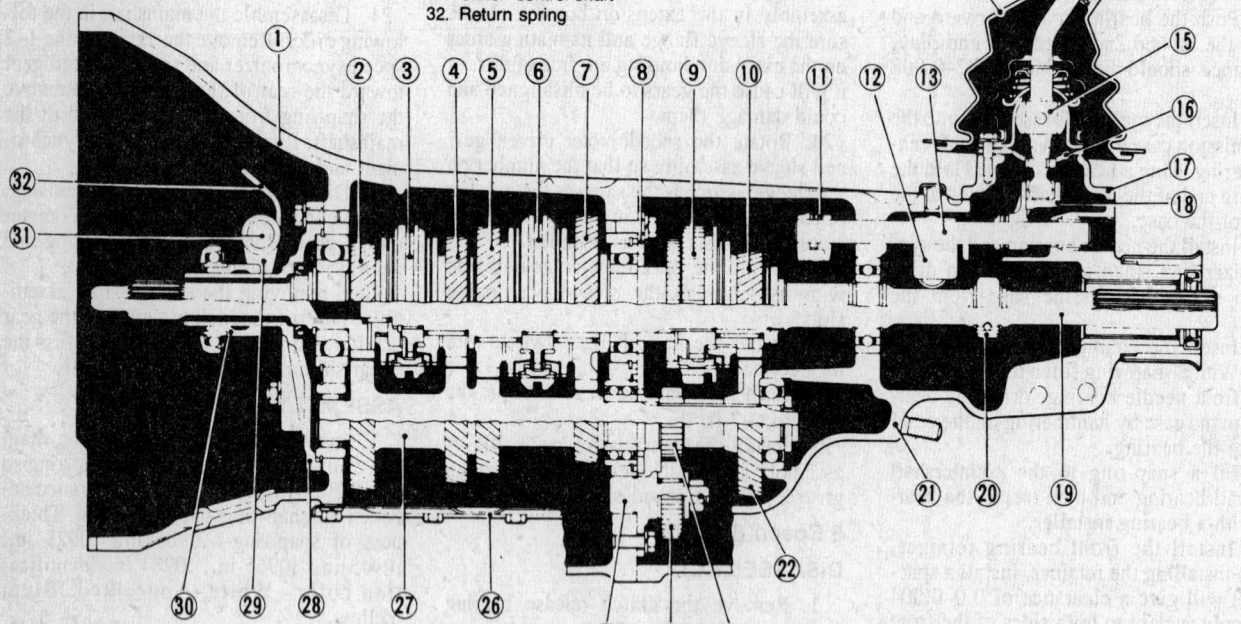

Cross section of 5-speed transmission, model KM 132

place the front bearing retainer oil seal.

8. Install the rear bearing retainer.

9. Install the reverse idler gear shaft.

10. Install the needle bearing, the reverse idler gear and the thrust washer. Check the reverse idler gear end-play (.0047–.0110 in.). Install the thrust washer with the ground side toward the gear side.

11. Assemble the overtop synchronizer.

12. Install the spacer, the stop plate, the overtop synchronizer assembly, the overtop gear bearing sleeve, the needle bearing, the synchronizer ring and the overtop gear in the written order on to the mainshaft from the rear end. Check the overtop gear end-play.

13. Install the spacer, the counter reverse gear, the spacer, the counter overtop gear and the ball bearing on to the countershaft gear from the rear end.

14. Insert the 3–4 and 1–2 speed shift forks into respective synchronizer sleeves. Insert each shift rail from the rear of the case. Lock the shift forks and rails with spring pins. Install an interlock plunger between shift rails. The pin should be in-

stalled with the slit in the axial direction of the shift rail.

15. Insert the ball and poppet spring into each shift rail. Install the poppet spring with the small end on the ball side.

16. Install the ball bearing on to the rear end of the mainshaft.

17. Install the speedometer drive gear.

18. Install the extension housing. Turn the change shifter fully down to the left. Make sure the forward end of the control finger is snugly fitted in the slot of the shift lug.

19. Install the neutral return plungers, the spring, and resistance spring and ball. Tighten each plug till its top is flush with the boss top surface.

20. Install the speedometer driven gear sleeve into the extension housing and into mesh with the drive gear.

21. Install the back-up light switch. Remember the steel ball.

22. Install the under cover.

23. Insert the clutch control shaft. Install the packing (felt), the return spring and the clutch shift arm. The spring pin should be

installed in such a manner that the slip will be at right angles with the axis of the control shaft.

24. Install the transmission control lever assembly. Fill the gear shifter area with grease.

25. After reassembly, rotate the drive pinion to see if it rotates smoothly.

Toyota

4 & 5 Speed (L Series)

The following procedures are for a L Series 4 speed, other L Series transmissions use similar procedures.

DISASSEMBLY

NOTE: The shifting controls are mounted in a side cover on the transmission case. The remote controls are either column or floor mounted.

1. Remove speedometer drive unit in the extension housing (Except 4 × 4 Pickup).

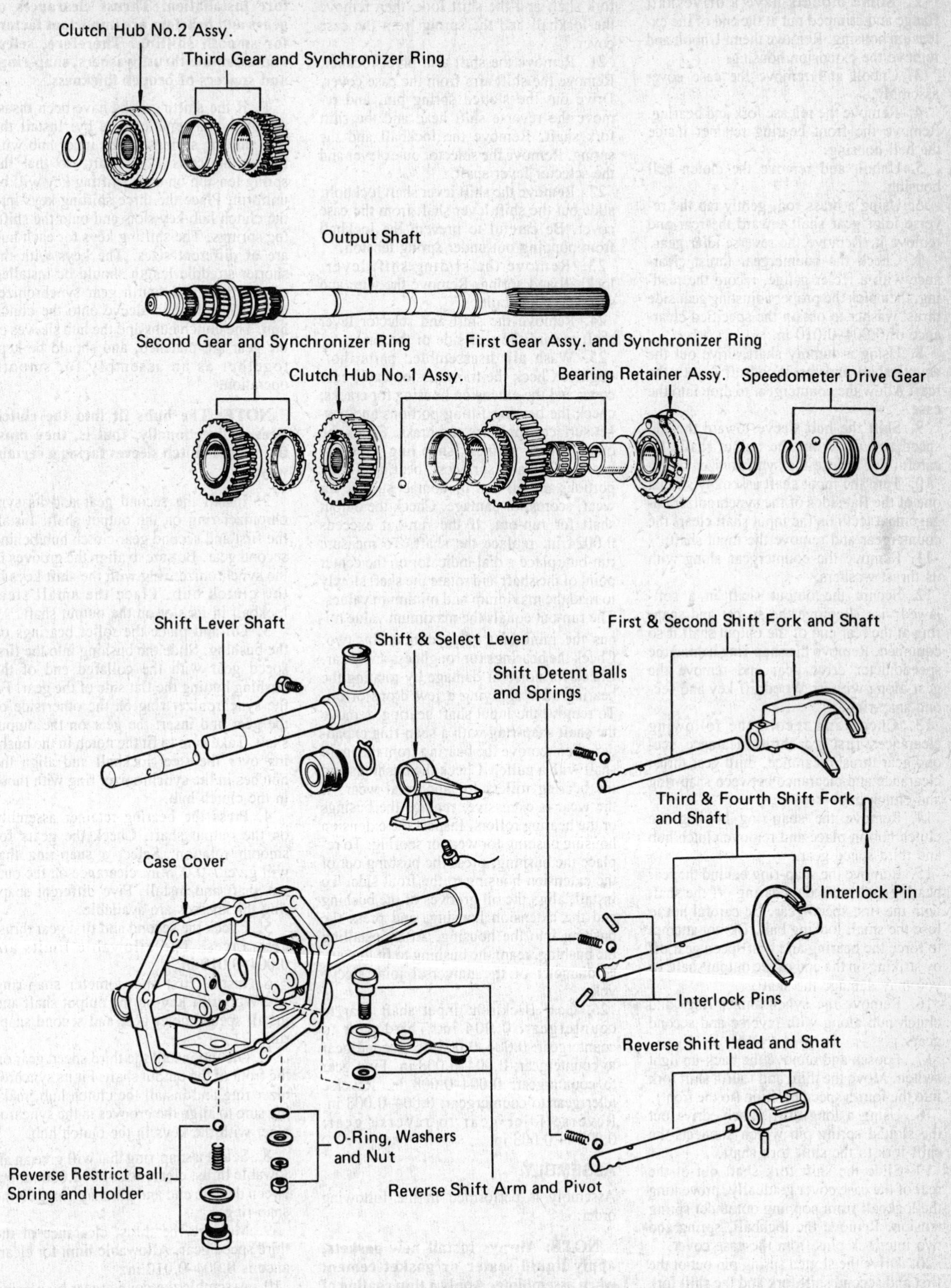

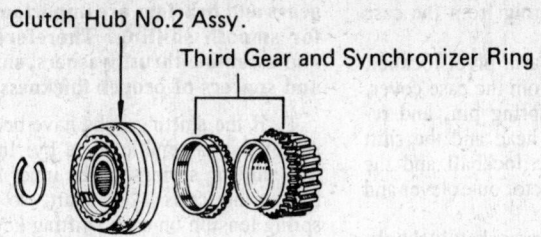

Clutch Hub No.2 Assy.

Third Gear and Synchronizer Ring

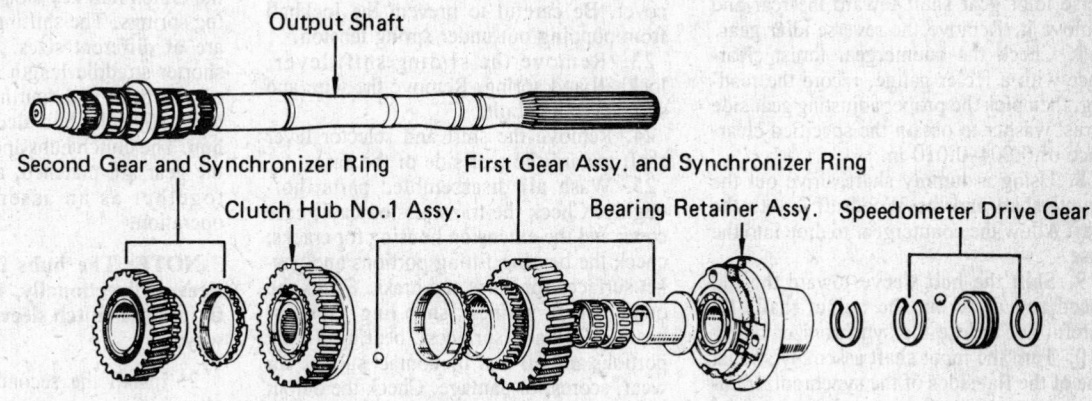

Output Shaft

Second Gear and Synchronizer Ring

First Gear Assy. and Synchronizer Ring

Clutch Hub No.1 Assy.

Bearing Retainer Assy.

Speedometer Drive Gear

Shift Lever Shaft

Shift & Select Lever

First & Second Shift Fork and Shaft

Shift Detent Balls and Springs

Third & Fourth Shift Fork and Shaft

Interlock Pin

Case Cover

Interlock Pins

Reverse Shift Head and Shaft

O-Ring, Washers and Nut

Reverse Shift Arm and Pivot

Reverse Restrict Ball, Spring and Holder

Toyota L-series transmission

2. Some models have a driveshaft flange and crimped nut at the end of the extension housing. Remove them. Unbolt and remove the extension housing.

3. Unbolt and remove the case cover assembly.

4. Remove the release fork and bearing. Remove the front bearing retainer inside the bell housing.

5. Unbolt and remove the clutch bell housing.

6. Using a brass rod, gently tap the reverse idler gear shaft toward the rear and remove it. Remove the reverse idler gear.

7. Check the countergear thrust clearance with a feeler gauge, record the reading, then pick the proper adjusting gear side thrust washer to obtain the specified clearance of 0.004–0.010 in.

8. Using a dummy shaft, drive out the countershaft and the Woodruff key to the rear. Allow the countergear to drop into the case.

9. Shift the hub sleeve toward the top speed and draw out the output shaft. Be careful not to lose the synchronizer ring.

10. Turn the input shaft assembly so that one of the flat sides of the synchronizer engagement teeth on the input shaft clears the countergear and remove the input shaft.

11. Remove the countergear along with its thrust washers.

12. Secure the output shaft in a soft-jawed vise. Remove the spacer and snap-ring at the rear end of the output shaft if so equipped. Remove the snap-ring behind the speedometer drive gear and remove the gear along with its Woodruff key and second snap-ring.

13. Check and record the following clearances: first gear thrust clearance, second gear thrust clearance, third gear thrust clearance and clearance between snap-ring and clutch hub.

14. Remove the snap-ring holding the clutch hub in place and remove clutch hub and third speed gear.

15. Remove the snap-ring behind the rear bearing and press the bearing off the shaft with the first speed gear. Be careful not to lose the small locking ball. Do not attempt to force the bearing and first speed gear off by striking on the end of the output shaft, or you may damage the shaft.

16. Remove the synchronizer rings and clutch hub along with reverse and second gears.

17. Loosen and remove the back-up light switch. Move the third and fourth shift fork into the fourth speed position (to the front).

18. Using a long drift punch, drive out the slotted spring pin which connects the shift fork to the shift fork shaft.

19. Slide the shift fork shaft out of the rear of the case cover gradually, preventing the lockball from popping out under spring tension. Remove the lockball, spring and two interlock pins from the case cover.

20. Drive the slotted spring pin out of the first and second shift fork and the shift fork shaft in the same manner. Remove the shift

fork shaft and the shift fork, then remove the lockball and the spring from the case cover.

21. Remove the shift arm pivot locknut. Remove the shift arm from the case cover. Drive out the slotted spring pin, and remove the reverse shift head and the shift fork shaft. Remove the lockball and the spring. Remove the selector outer lever and the selector lever shaft.

22. Remove the shift lever shaft lockbolt, slide out the shift lever shaft from the case cover. Be careful to prevent the lockball from popping out under spring tension.

23. Remove the sliding shift lever, lockball and spring. Remove the wire and shift lever lockbolt.

24. Remove the shift and selector lever shaft toward the rear side of the case.

25. Wash all disassembled parts thoroughly. Check the transmission case, case cover and the extension housing for cracks; check the bearing fitting portions and gasket surfaces for burrs and nicks. Check the output shaft splines, snap-ring grooves, bearing contact surfaces, bearing fitting portions and oil seal lip contact surface for wear, scores, or damage. Check the output shaft for run-out. If the run-out exceeds 0.0024 in. replace the shaft. To measure run-out, place a dial indicator on the center point of the shaft and rotate the shaft slowly to read the maximum and minimum values. The run-out equals the maximum value minus the minimum value divided by two. Check the bearings for roughness and wear. Check for noise or damage by rotating the bearing after applying a few drops of oil. To remove the input shaft bearing, remove the shaft snap-ring with a snap-ring expander, then remove the bearing from the input shaft with a puller. Check the bushings and the bearing rollers for abnormal wear. If the wear is excessive, replace the busings or the bearing rollers. Inspect the extension housing bushing for wear or scoring. To replace the bushing, press the bushing out of the extension housing to the front side. To install, align the oil grooves of the bushing and the extension housing, and press the bushing into the housing. After installing the bushing, ream the bushing to fit the outer diameter of the universal joint sleeve yoke.

26. Gear Backlash: Input shaft gear to countergear: 0.004 in. Third gear to countergear: 0.004–0.008 in. Second gear to countergear: 0.004–0.008 in. First gear to countergear: 0.004–0.008 in. Reverse idler gear to countergear: 0.004–0.008 in. Reverse idler gear to reverse gear: 0.004– +0.008 in.

ASSEMBLY

Assembly is performed in the following order:

NOTE: Always install new gaskets, apply liquid sealer or gasket cement when assembling. Apply a thin coating of transmission lubricant on all parts be-

fore installation. Thrust clearances of gears and bearings are important factors for smooth shifting. Therefore, select and assemble thrust washers, snap-rings and spacers of proper thickness.

1. If the shifting hubs have been disassembled, reassemble them by: Install the two shifting springs in the inner hub with the spring ends 120° apart, so that the spring tension on each shifting key will be uniform. Place the three shifting keys into the clutch hub key slots and onto the shifting springs. The shifting keys for each hub are of different sizes. The keys with the shorter straddle length should be installed on the third and fourth gear synchronizer unit. Slide the hub sleeve onto the clutch hub. The clutch hubs and the hub sleeves of the gear are matched, and should be kept together as an assembly for smooth operation.

NOTE: The hubs fit into the clutch sleeves directionally, that is, they must fit into the clutch sleeves facing a certain way.

2. Install the second gear and its synchronizer ring on the output shaft. Install the first and second gear clutch hub behind second gear. Be sure to align the grooves in the synchronizer ring with the shift keys in the clutch hub. Place the small steel lockball in its slot on the output shaft.

3. Oil and place the roller bearings on the bushing. Slide the bushing into the first speed gear with the collared end of the bushing butting the flat side of the gear. Fit the synchronizer ring on the other side of the gear and insert the gear on the output shaft. Take care to fit the notch in the bushing over the steel lockball and align the notches in the synchronizer ring with those in the clutch hub.

4. Press the bearing retainer assembly on the output shaft. Check the gears for smooth rotation. Select a snap-ring that will give 0–0.006 in. clearance on the output shaft and install. Five different snap-ring thicknesses are available.

5. Check the second and first gear thrust clearances. The allowable limits are 0.004–0.010 in.

6. Install first speedometer snap-ring and Woodruff key on the output shaft and install speedometer gear and second snap-ring.

7. Oil and install the third speed gear on the front of the output shaft. Fit its synchronizer ring and install the clutch hub, making sure to align the grooves in the synchronizer with the keys in the clutch hub.

8. Select a snap-ring that will give an allowable thrust clearance of 0–0.006 in. between the hub end and the snap-ring. Fit the snap-ring.

9. Measure the thrust clearance of the third speed gear. Allowable limit for clearance is 0.004–0.010 in.

10. Assemble the countergear by placing the spacers and greased roller bearings in

both ends and fitting a tube or rod that is the exact length of the countergear with the thrust washers on its ends and fit it into the countergear.

11. Referring to the thrust measurement on the countergear when disassembling, select thrust washers that give an allowable clearance of 0.004–+0.010 in. Adjust clearance by replacing the rear thrust washer.

12. Install the thrust washers on the ends of the countergear. Rear thrust washers are identified by Number 1, 2, 3 or 4 stamped on their outside face. Be sure to install the washers with their protruding groove facing out.

13. Place the countergear assembly into the case with the notches in the thrust washers facing up and align the notches with the case grooves.

14. Allow the countergear to drop down into the case. Do not replace the countergear shaft at this point.

15. Install the input shaft in the same manner of removal, aligning one of the flat sides of the clutch hub with the large countershaft gear and, using a brass drift, drive the input shaft bearing into the case, tapping it around the outer ball race.

16. Fit the synchronizer ring on the input shaft. Shift the clutch hub on the end of the output shaft forward.

17. Fit the output shaft into the case, making sure the slotted spring pin in the bearing retainer aligns with its groove in the case, and align the grooves in the synchronizer ring with the clutch hub keys.

18. Oil the countershaft. Raise the countergear assembly so that it aligns with the case holes and, making sure the thrust washers are in place, insert the countergear shaft from the rear of the case with the Woodruff key slot at the rear. The countergear shaft will push the tube out of the case. Install the Woodruff key and fit it into its slot in the case.

19. Install the reverse idler gear with its toothed end facing forward. Slide the reverse idler gear shaft in with its Woodruff key slot end facing the rear. Install the Woodruff key and tap the shaft in so that the key is in its seat in the case.

20. Install the front bearing retainer. Be sure to align the oil seal slot in the retainer with the oil hole in the case. Coat the bolt threads with sealer and tighten to 5–6 ft. lbs.

21. Install the clutch bell housing. Coat the bolt threads with sealer and tighten to 37–50 ft. lbs.

22. Install clutch release fork assembly.

23. Apply grease to the extension housing rear oil seal and install the extension housing. Be careful not to damage the oil seal on the output shaft when assembling.

24. Coat the extension housing bolts with sealer and tighten to 22–32 ft. lbs.

25. Oil the speedometer gear assembly and fasten to the extension housing.

26. To assemble the transmission case cover, install the shift arm pivot onto the reverse shift arm, and insert into the case.

27. Assemble the shift and selector lever shaft together with the shift and selector lever, and secure the bolt with a wire.

28. Insert the reverse shift fork shaft compression spring and lockball into the case, and insert the fork shaft from the rear side, then secure the shift head with a new slotted spring pin.

29. Align the fork shaft positioning groove with the shift interlock pin groove.

30. Align the reverse shift fork shaft knob with the reverse shift fork shaft, and install the O-ring, washer and nut onto the shift arm pivot. Insert the shift interlock pin into the rear side of the case cover and the compression spring and lockball into the front side, and assemble the shift fork together with the first and second shift fork shaft. Secure the shift fork with a new slotted spring pin.

31. Align the shift fork shaft positioning groove with the shift interlock pin groove. Insert the two shift interlock pins into the front side of the case cover.

32. Insert the compression spring and the lockball, and assemble the shift fork together with the third and fourth shift fork shaft, then secure the shift fork with a new slotted spring pin.

33. Install the lockball, compression spring and reverse restricting ball holder.

34. Check all shift forks for smooth movement. Tighten to 27–32 ft. lbs.

35. Install the back-up light switch on the case cover.

36. Align each shift fork and the reverse shift arm with the respective gears, and install the transmission case cover, within the gasket, onto the transmission. Torque the case cover retaining bolts to 11–16 ft. lbs.

37. To adjust the shift arm pivot, loosen the locknut on the shift arm pivot, turn the shift arm pivot clockwise until friction is felt, when the reverse idler gear contacts with the first gear and/or the countergear.

38. Next from this position, turn the shift arm pivot counterclockwise approximately 90 degrees. Tighten the pivot locknut securely.

39. With the input shaft rotating, make sure that there is no noise and that the reverse idler gear does not contact other gears in the transmission.

40. If no friction is felt when the shift arm pivot is turned clockwise, set the pivot line mark at 60 degrees rearward from its horizontal position to the case cover surface.

41. If necessary, replace the oil seal in the extension housing after assembling the transmission using the oil seal puller, and pull out the oil seal together with the dust seal.

4 & 5 Speed (W Series)

The following procedures are for a W Series 5 speed transmission, other W Series transmissions use similar procedures.

DISASSEMBLY

1. Drain the oil.
2. Remove the clutch housing, with the release fork, bearing and hub still attached.
3. Remove the back-up light switch.
4. Remove the gearshift lever retainer.
5. Rotate the shift rod housing counterclockwise (viewed from behind) and then disconnect the rod from the shift fork shafts.
6. Unbolt and remove the extension housing.
7. Drive out the slotted pin and separate the shift rod, housing and spring.
8. Remove the front bearing retainer.
9. Take off both of the front countershaft covers, and the spacer.
10. Remove the snap-rings from the input and countershaft bearings.
11. Remove the intermediate plate.
12. When removing the intermediate plate, leave all the gears and other parts attached.
13. Remove the speedometer driven gear. There are two reverse restrictor pins. The pins are located underneath plugs on the extension housing.
14. Remove the straight screw plugs from the shift forks and withdraw the springs.
15. Drive the slotted spring pins out of each shift fork.
16. Slide the gear shift fork shafts back and remove the forks.
17. Remove the speedometer drive gear snap-ring and remove the drive gear.
18. Remove the output shaft bearing.
19. Remove the countershaft bearing.
20. Remove the fifth and reverse gears from the countershaft.
21. Remove the snap-ring, fifth gear, its synchronizer ring, needle roller bearing, and fifth gear bearing inner race from the output shaft.
22. Remove the reverse gear and clutch hub from the output shaft.
23. Loosen the bolt and remove the reverse idler gear stop from the rear cover. Withdraw the reverse idler shaft from the rear; remove the reverse idler gear and spacer.
24. Remove the output shaft rear bearing retainer. Remove the rear bearing snapring.
25. Push the countergear bearing outer race rearward, and remove the bearing. Separate the countergear from the intermediate plate.
26. Separate the input shaft and synchronizer ring from the output shaft.
27. Remove the output shaft from the intermediate plate.
28. Remove the hub and synchronizer ring, followed by third gear.
29. Press off the rear bearing.
30. Remove the following items from the output shaft, in the order listed: First gear. Roller bearing with inner race. Synchronizer ring. Reverse gear. Clutch hub. Second gear. Synchronizer ring.

ASSEMBLY

1. Install the sleeve over the third gear synchronizer hub. Insert the three shift keys

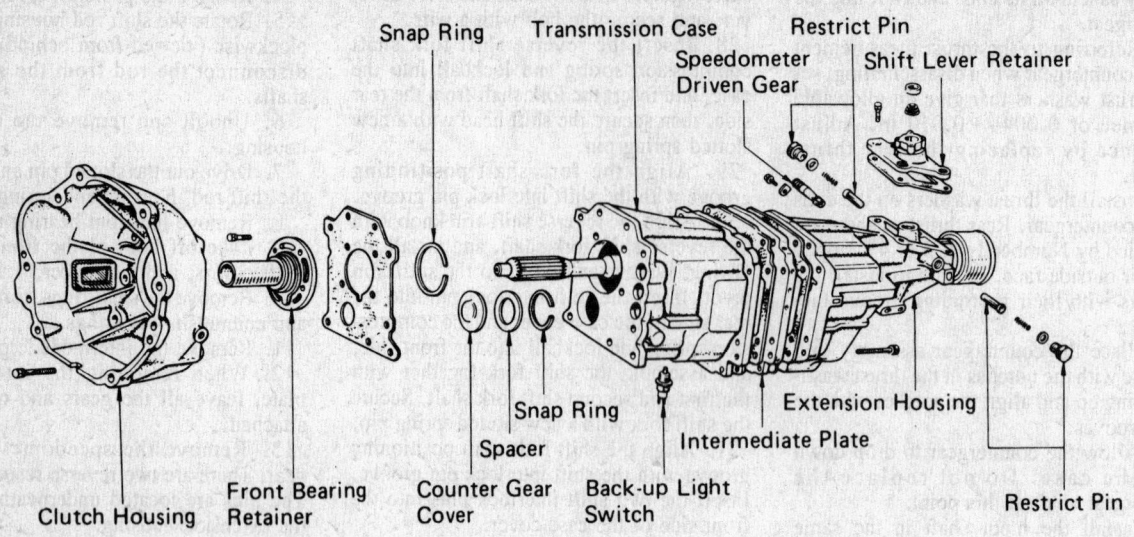

Snap Ring

Transmission Case

Restrict Pin

Speedometer Driven Gear

Shift Lever Retainer

Snap Ring

Spacer

Counter Gear Cover

Back-up Light Switch

Extension Housing

Intermediate Plate

Restrict Pin

Clutch Housing

Front Bearing Retainer

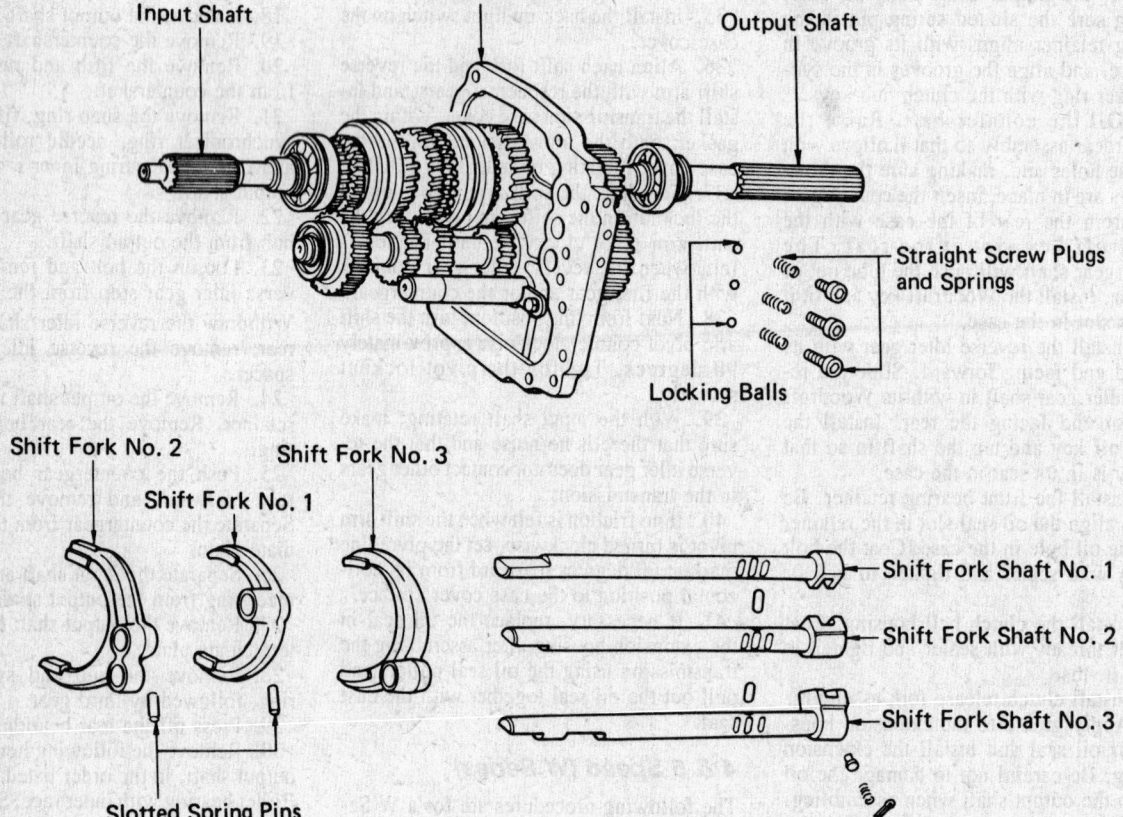

Input Shaft

Intermediate Plate

Output Shaft

Straight Screw Plugs and Springs

Locking Balls

Shift Fork No. 2

Shift Fork No. 1

Shift Fork No. 3

Shift Fork Shaft No. 1

Shift Fork Shaft No. 2

Shift Fork Shaft No. 3

Slotted Spring Pins

Toyota W-series transmission

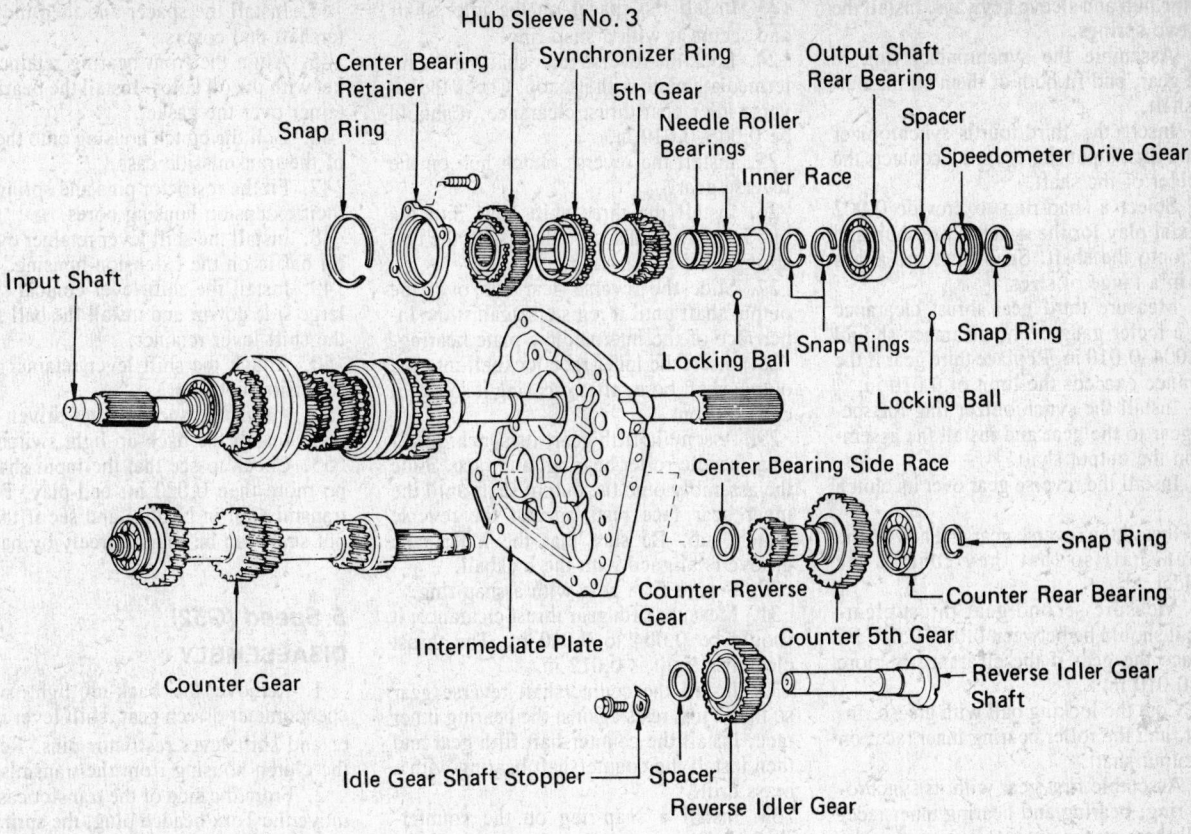

Input Shaft

Snap Ring

Center Bearing Retainer

Hub Sleeve No. 3

Synchronizer Ring

5th Gear

Needle Roller Bearings

Inner Race

Output Shaft Rear Bearing

Spacer

Speedometer Drive Gear

Locking Ball

Snap Rings

Snap Ring

Locking Ball

Center Bearing Side Race

Counter Reverse Gear

Counter 5th Gear

Counter Rear Bearing

Snap Ring

Reverse Idler Gear Shaft

Intermediate Plate

Counter Gear

Idle Gear Shaft Stopper

Spacer

Reverse Idler Gear

Snap Ring

Synchronizer Ring

Hub Sleeve No. 2

3rd Gear

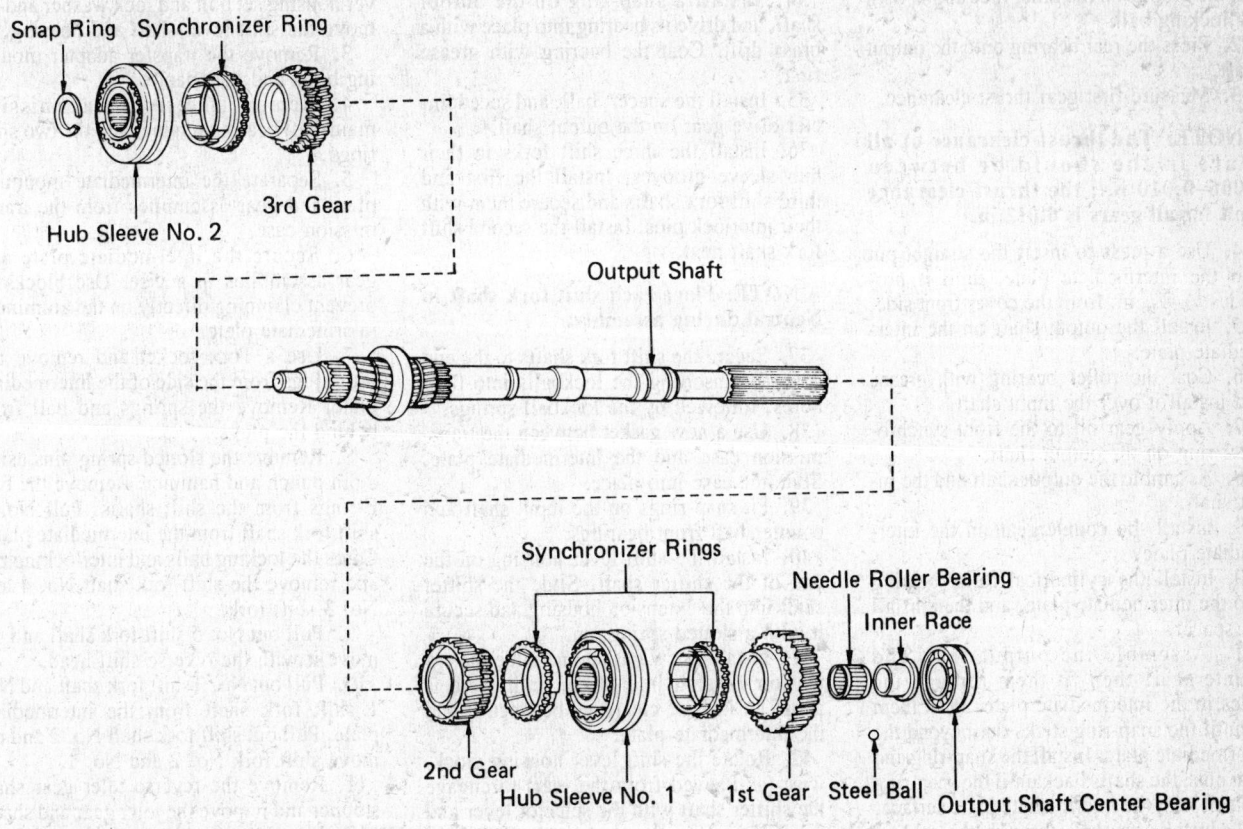

Output Shaft

Synchronizer Rings

Needle Roller Bearing

Inner Race

2nd Gear

Hub Sleeve No. 1

1st Gear

Steel Ball

Output Shaft Center Bearing

Toyota W-series transmission

into the hub and sleeve keyways. Install the hub two springs.

2. Assemble the synchronizer ring to third gear, and fit both of them on the output shaft.

3. Insert the third/fourth synchronizer hub on the output shaft, until it contacts the shoulder of the shaft.

4. Select a snap-ring to provide 0.002 in. axial play for the synchronizer hub and fit it onto the shaft. Snap-rings are available in a range of sizes.

5. Measure third gear thrust clearance with a feeler gauge. The clearance should be 0.004–0.010 in. Replace third gear if the clearance exceeds the limit of 0.010 in.

6. Install the synchronizer ring for second gear to the gear and install the assembly on the output shaft.

7. Install the reverse gear over its clutch hub.

8. Install the reverse gear and hub on the output shaft so that they contact the shoulder.

9. Measure second gear thrust clearance; it should be between 0.004–0.010 in. Replace the gear if the clearance is more than 0.010 in.

10. Coat the locking ball with grease. Insert it, and the roller bearing inner race, on the output shaft.

11. Assemble first gear with its synchronizer ring, bearing and bearing inner race. Install them on the output shaft, so that the end of the inner race contacts the clutch hub and the groove on the inner race aligns with the locking ball.

12. Press the rear bearing onto the output shaft.

13. Measure first gear thrust clearance.

NOTE: The thrust clearance of all gears in the should be between 0.006–0.010 in.; the thrust clearance limit for all gears is 0.012 in.

14. Use a press to insert the straight pin into the intermediate plate, until it protrudes $1/4$–$5/16$ in. from the cover front side.

15. Install the output shaft on the intermediate plate.

16. Coat the roller bearing with grease and install it over the input shaft.

17. Apply gear oil to the front synchronizer ring on the output shaft.

18. Assemble the output shaft and the input shaft.

19. Install the countergear on the intermediate plate.

20. Install the cylindrical roller bearing into the intermediate plate, and then install the spacer.

21. Assemble the output shaft and countergear, then fit them through the holes in the intermediate plate. Push them in until the snap-ring sticks out beyond the intermediate plate. Install snap-ring and then push the shafts back until the snap-ring is flush with the intermediate plate surface.

22. Install the shaft through the reverse idler gear. Insert the end of the shaft into the end of the intermediate plate.

23. Install the spacer on the idler shaft and secure it with a snap-ring.

24. Lock the reverse idler shaft on the intermediate plate with its stop. Check the reverse idler gear thrust clearance, it should be 0.006–0.010 in.

25. Install the reverse clutch hub on the reverse gear.

26. Install the three shift keys into the hub keyways and secure them with the two springs and a snap-ring.

27. Slide the reverse gear hub over the output shaft until it registers against the inner race of the intermediate plate bearing.

28. Insert the inner race lockball into the output shaft bore, after greasing it so that it can't fall out.

29. Assemble fifth gear, its synchronizer ring, needle roller bearing, and race. Slide the assembly onto the output shaft until the inner bear face rests against the reverse clutch hub. Be sure that the inner race groove is aligned with the lockball.

30. Secure fifth gear with a snap-ring.

31. Measure fifth gear thrust clearance; it should be 0.004 to 0.010 in. The thrust clearance limit is 0.012 in.

32. Install the countershaft reverse gear so that it just rests against the bearing inner race. Install the countershaft fifth gear and then install the countershaft bearing with a brass drift.

33. Install a snap-ring on the countershaft; select a snap-ring from one of the four available sizes.

34. Install a snap-ring on the output shaft, and drive its bearing into place with a brass drift. Coat the bearing with grease first.

35. Install the spacer, ball, and speedometer drive gear on the output shaft.

36. Install the three shift forks in their hub sleeve grooves. Install the first and third shift shafts and secure them with their interlock pins. Install the second shift fork shaft next.

NOTE: Place each shift fork shaft in Neutral during assembly.

37. Secure the shift fork shafts to the end cover by inserting the lockballs into their bores, followed by the lockball springs.

38. Use a new gasket between the transmission case and the intermediate plate. Slide the case into place.

39. Fit snap-rings on the input shaft and countershaft front bearings.

40. Install the shift lever housing on the end of the shifter shaft. Slide the shifter shaft into the extension housing and secure it with a slotted spring pin.

41. Install a new gasket and slide the extension housing into place, until there is about an inch of clearance between it and the intermediate plate.

42. Rotate the shift lever housing clockwise (as viewed from the rear) to engage the shifter shaft with the selector lever and the shift fork shaft.

43. Slide the extension housing the rest of the way.

44. Install the spacer and then the countershaft end covers.

45. Align the front bearing retainer gasket with the oil holes. Install the bearing retainer over the gasket.

46. Bolt the clutch housing onto the front of the transmission case.

47. Fit the restrictor pins and springs into their extension housing bores.

48. Install the shift lever retainer over the oil baffle on the extension housing.

49. Install the shift lever conical spring, large side down, and install the ball seat in the shift lever retainer.

50. Attach the shift lever retainer to the extension housing.

51. Install the speedometer driven gear.

52. Install the back-up light switch.

53. Check to see that the input shaft has no more than 0.020 in. end-play. Put the transmission in Neutral and see if the output shaft can be rotated freely by hand.

5 Speed (G52)
DISASSEMBLY

1. Remove the back-up light switch, speedometer driven gear, shift lever retainer and shift lever restrictor pins. Remove the clutch housing from the transmission.

2. From the side of the transfer case, remove the Torx headed plug, the spring and ball. Remove the Torx plug from the back of the transfer adapter. Remove the shift lever housing set bolt and lock washer and remove the shift lever shaft and housing.

3. Remove the transfer adapter mounting bolts and the adapter.

4. Remove the front transmission mainshaft bearing retainer and the two snap rings.

5. Separate the intermediate mounting plate and gear assemblies from the transmission case.

6. Secure the intermediate plate and gear assemblies in a vise. Use blocks to prevent clamping directly on the aluminum intermediate plate.

7. Use a Torx socket and remove the four plugs from the side of the intermediate plate. Remove the springs and ball from behind the plugs.

8. Remove the slotted spring pins using a pin punch and hammer. Remove the two E-rings from the shift shafts. Pull No. 4 shift fork shaft from the intermediate plate. Catch the locking balls and interlocking pin and remove the shift fork shaft No. 4 and No. 3 shift fork.

9. Pull out No. 5 shift fork shaft and remove it with the reverse shift head.

10. Pull out No. 3 shift fork shaft and No. 1 shift fork shaft from the intermediate plate. Pull out shift fork shaft No. 2 and remove shift fork No. 2 and No. 1.

11. Remove the reverse idler gear shaft stopper and remove the idler gear and shaft.

12. Remove the reverse shift arm from the reverse shift arm bracket.

13. Measure the 5th gear clearance on the

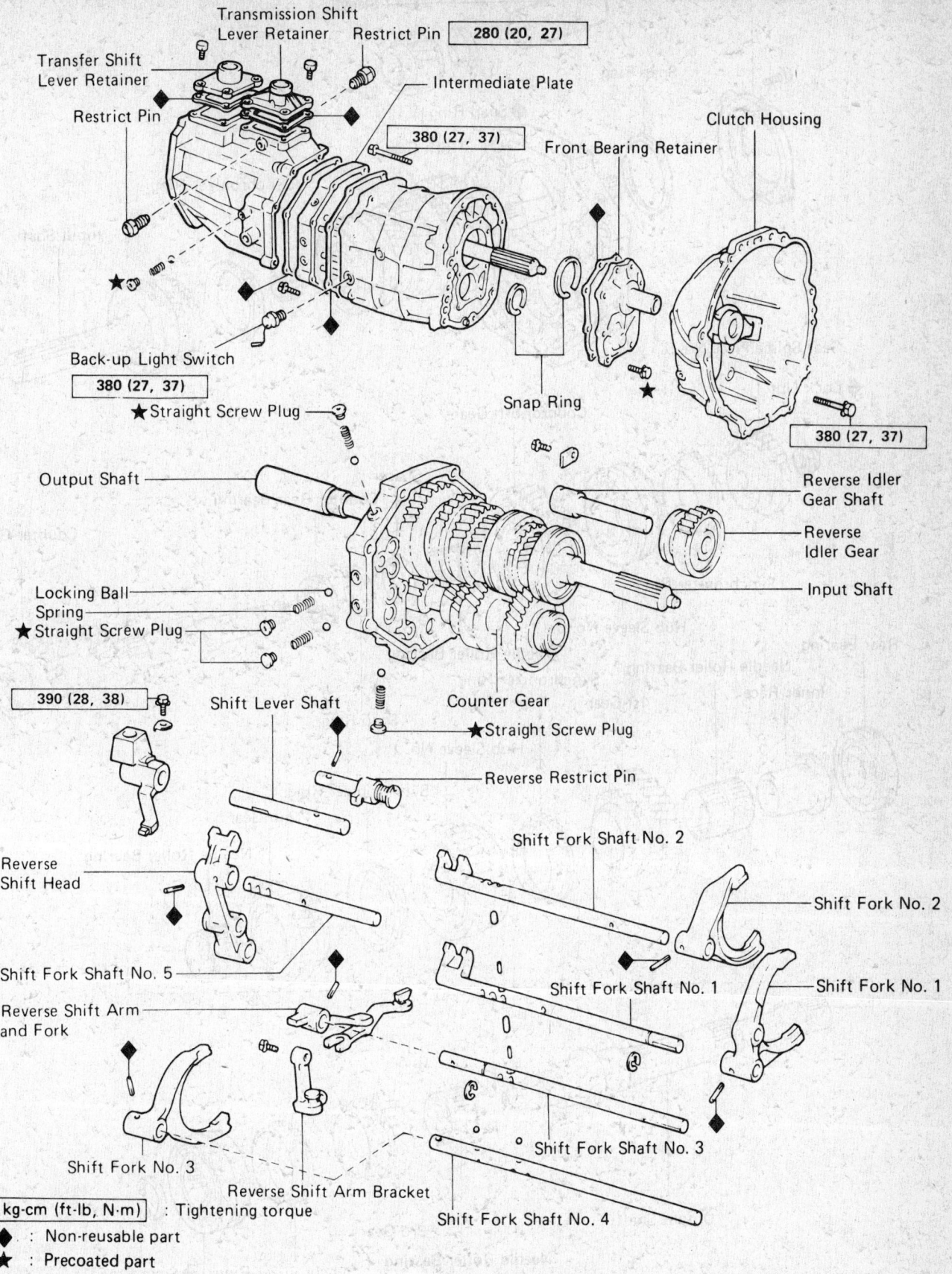

Transmission Shift
Lever Retainer   Restrict Pin   **280 (20, 27)**

Transfer Shift
Lever Retainer

Intermediate Plate

Restrict Pin

380 (27, 37)

Front Bearing Retainer

Clutch Housing

Restrict Pin

Back-up Light Switch

380 (27, 37)

★ Straight Screw Plug

Snap Ring

380 (27, 37)

Output Shaft

Reverse Idler
Gear Shaft

Reverse
Idler Gear

Locking Ball
Spring

★ Straight Screw Plug

Input Shaft

390 (28, 38)

Shift Lever Shaft

Counter Gear

★ Straight Screw Plug

Reverse Restrict Pin

Reverse
Shift Head

Shift Fork Shaft No. 2

Shift Fork No. 2

Shift Fork Shaft No. 5

Reverse Shift Arm
and Fork

Shift Fork Shaft No. 1

Shift Fork No. 1

Shift Fork No. 3

Reverse Shift Arm Bracket

Shift Fork Shaft No. 3

Shift Fork Shaft No. 4

kg-cm (ft-lb, N·m) : Tightening torque

◆ : Non-reusable part

★ : Precoated part

Toyota G-series transmission

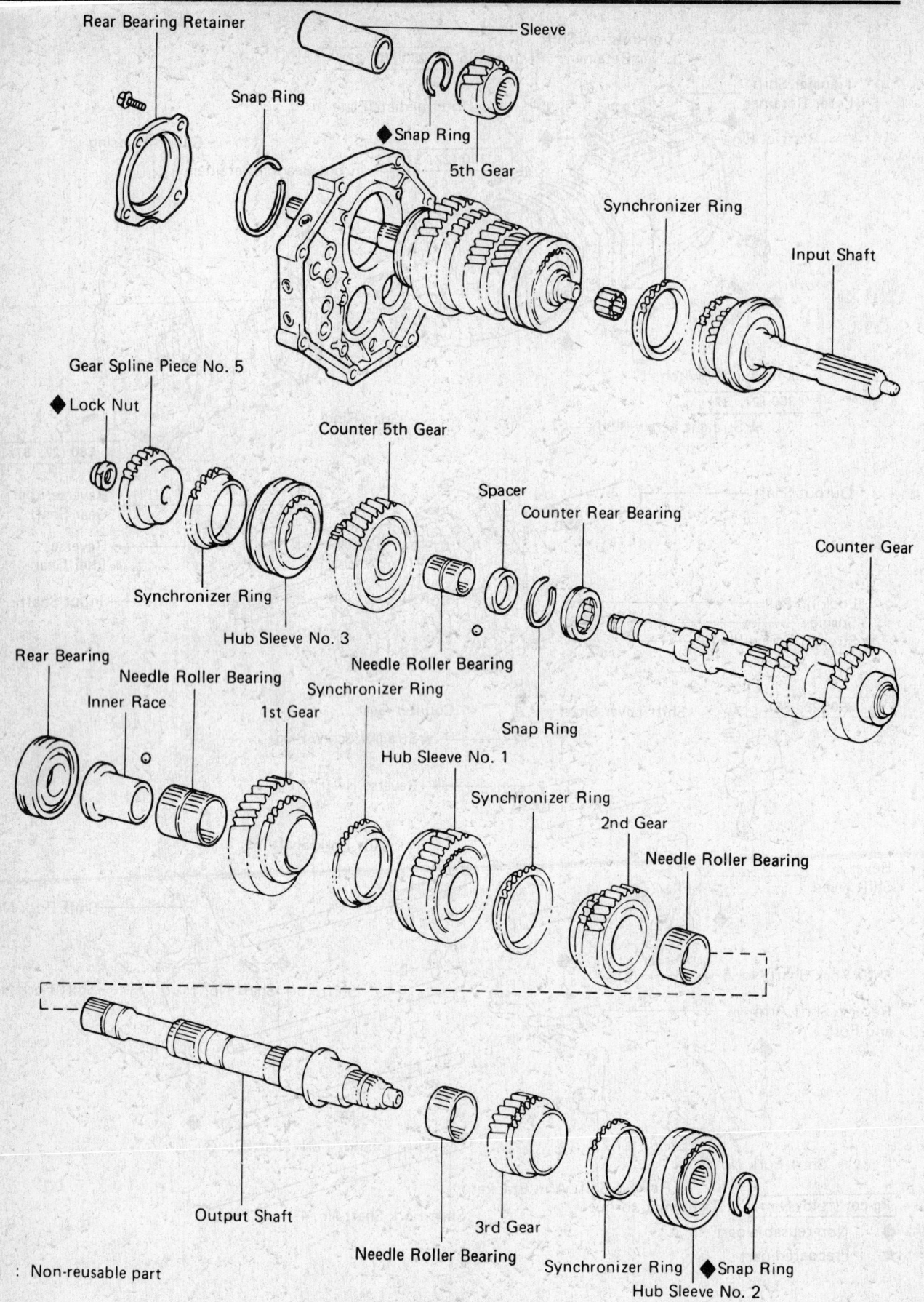

Rear Bearing Retainer

Sleeve

Snap Ring

◆ Snap Ring

5th Gear

Synchronizer Ring

Input Shaft

Gear Spline Piece No. 5

◆ Lock Nut

Counter 5th Gear

Spacer

Counter Rear Bearing

Counter Gear

Synchronizer Ring

Hub Sleeve No. 3

Needle Roller Bearing

Snap Ring

Rear Bearing

Needle Roller Bearing

Synchronizer Ring

Inner Race

1st Gear

Hub Sleeve No. 1

Synchronizer Ring

2nd Gear

Needle Roller Bearing

Output Shaft

3rd Gear

Needle Roller Bearing

Synchronizer Ring

◆ Snap Ring

Hub Sleeve No. 2

◆ : Non-reusable part

Toyota G-series transmission

countershaft. Clearance should be 0.0039–0.118 in.

14. Remove the 5th gear synchronizer ring, needle bearings and fifth gear from the countershaft by engaging the gear and using a hammer and chisel to loosen the staked nut. Remove the nut, disengage the gear and, using a puller, remove the gear splined No. 5 synchronizer assembly, needle bearing and 5th gear from the countershaft.

15. Remove the spacer and ball from the countershaft. Remove the two bolts and the reverse shift arm bracket. Use a Torx socket and remove the rear bearing retainer.

16. Remove the snap ring from the main output shaft and remove the output shaft, countershaft and input shaft as an assembly. Separate the output shaft and needle bearings (14) from the input shaft. Remove the countershaft rear bearing from the intermediate plate. Remove the sleeve from the output shaft using an appropriate puller. Measure each gear thrust clearance, maximum clearance is 0.0098 in.

17. Press off the fifth gear, rear bearing and first gear assemblies after removing the retaining snap ring. Remove the locking ball.

18. Press off the No. 1 synchronizer assembly and second gear. Remove the needle bearing. Remove the snap ring and press off No. 2 synchronizer assembly and 3rd gear. Remove the needle bearing.

19. Inspect all parts and replace as necessary. Required clearances follow: Output flange thickness; 0.1890 in. Inner race flange thickness; 0.1571 in. Output shaft journal thickness–2nd gear; 1.4954 in. 3rd gear; 1.3773 in.: Shaft out-of-round; 0.0020 in.: Synchronizer ring clearance; 0.0039–0.079 in.: Shift forks to hub sleeves; 0.039 in. max.

ASSEMBLY

1. Replace the grease retainers as necessary. Check and replace the input shaft and the countershaft front bearings if necessary. Select a snap ring that will allow the minimum axial play.

2. Assemble the gears and synchronizers in the reverse order of removal. Make sure the synchronizers are installed facing in the proper direction. Use snap rings that will allow a minimum of axial play. Check clearances of all gears, synchronizers and bearings.

3. After the input, output and countershafts are ready for installation (reverse order of removal): Install the output shaft into the intermediate plate by pulling on the output shaft and tapping on the intermediate plate with a soft hammer.

4. Apply MP grease to the needle bearings and install them in position on the input shaft. Install the input shaft to the output shaft with the synchronizer ring slots aligned with the shifting keys.

5. Install the countershaft and bearing to the intermediate plate. Install the output shaft snap ring flush with the intermediate

plate. Install the rear bearing retainer on the intermediate plate. Tighten the retainers to 13 ft. lbs. Install the reverse shift arm bracket. Install the ball and spacer on the countershaft and install the fifth gear with hub and needle bearings. Install the synchronizer assembly with the ring slots aligned with the shifting keys. Engage the gear and install the locknut. Torque the locknut to 87 ft. lbs. and stake the locknut. Disengage the gears. Check fifth gear thrust clearance.

6. Install the reverse shift arm to the pivot of the reverse shift arm bracket.

7. Install the reverse idler gear and shaft. Align the reverse shift arm shoe with the reverse idler gear groove and insert the reverse idler gear shaft into the intermediate plate. Install the gear shaft stopper and tighten the retaining bolt to 13 ft. lbs.

8. Position No. 1 and No. 2 shift forks into the grooves of their respective hub sleeves and install the No. 2 fork shaft through the forks and intermediate plate. Apply MP grease to the interlock pins and install the pins into the forks and intermediate plate.

9. Install the interlock pin in No. 1 shift shaft hole. Install the shift shaft through fork No. 1 and the intermediate plate. Install the interlock pin into the intermediate plate.

10. Install the interlock pin to the head of shift shaft No. 3. Install the shaft through the reverse shift arm and the intermediate plate.

11. Install the reverse shift head into fork shaft No. 5. Install fork shaft No. 5 to the intermediate plate and put the reverse shift head onto shift fork shaft No. 3.

12. Install the locking ball into the reverse shift head hole. Shift the No. 3 synchronizer hub sleeve to the 5th speed position. Place shift fork No. 3 into the groove of hub sleeve No. 3 and install fork shaft No. 4 to shift fork No. 4 and the reverse shift arm. Install the locking ball into the intermediate plate and insert fork shaft No. 4 to the intermediate plate.

13. Shift fork shaft No. 1 to the 1st speed position. Forks No. 2, No. 3, No. 4 and No. 5 should not move. Install the slotted retainer spring pins into each shift fork, reverse shift arm and reverse shift head. Install the two fork shaft E-rings. Apply sealer to the locking ball cover screw plugs. Install the locking balls, springs and screw plugs with a Torx socket and tighten to 14 ft. lbs. The short spring goes in the bottom of the intermediate plate.

14. Mount the intermediate plate and gear assemblies to the transmission. Align each bearing outer race, each fork end and reverse idler shaft with the case installation holes, tap with a plastic hammer if necessary to help with alignment. Install the two bearing snap rings.

15. Install the front bearing retainer with a new gasket after applying sealer to the mounting bolts. Tighten the bolts to 12 ft. lbs.

16. Install a new gasket on the intermediate plate. Install the transfer adapter and eight mounting bolts, torque the bolts to 27 ft. lbs. Insert the shift lever housing to the transfer adapter and connect the shift fork shafts. Insert the shift lever shaft to the transfer adapter and shift lever housing. Install and torque the shift lever housing bolt, torque to 28 ft. lbs. Install the front Torx plug cover, tighten to 27 ft. lbs.

17. Install the side locking ball, spring and screw plug. Apply sealer to the plug before installation, torque to 14 ft. lbs.

18. After installing the extension housing or transfer adapter, check to see that the input and output shafts rotate smoothly. Check to see that shifting can be made smoothly in all positions.

19. Install the black restrict pin on the reverse gear/5th gear side. Install the other restrict pin and torque both to 20 ft. lbs.

20. Install the clutch housing and torque the bolts to 27 ft, lbs. Install the shift lever retainer and new gasket. Tighten to 13 ft. lbs.

21. Install the back-up light switch (27 ft. lbs.). Install the release fork and bearing and install the transmission.

Land Cruiser 3 Speed
DISASSEMBLY

1. Remove the transfer case shift lever guide, cotter pin, and lockbolt. Remove the shift lever and linkage; do not lose the link lever shoe.

2. Remove the back-up light switch and gasket from the transmission cover.

3. Remove the transfer case cover, complete with gasket. Remove the power take-off cover and gasket.

4. Straighten out the tabs on the input shaft nut lockwasher and remove the nut. Slide the spacer off.

NOTE: Lock the power take-off drive gear with a wooden block or brass drift to keep the shaft from turning while the nut is being removed.

5. Loosen the five bolts which secure the transfer case to the transmission case and separate the cases with a puller. Hold the power take-off drive gear, spacer, and input gear, so that they don't drop out.

6. Unfasten the bolt and remove the gear selector outer lever.

7. Unfasten the bolts and remove the transmission case cover, complete with gasket.

8. Loosen the bolts and remove the front bearing retainer, with gasket, from the transmission case.

9. Drive the shift fork shaft out toward the front of the case with a hammer and a brass drift. Use care not to lose the fork balls, springs and pin.

10. Withdraw the first/reverse shift fork and the second/third shift fork from the transmission case. Remove the lockballs and springs.

11. Drive the countershaft rearward with a brass drift. Remove the countershaft Woodruff key.

NOTE: The countergear should remain in the case.

12. Remove the input shaft and bearing with a puller. Install the puller on the front of the input shaft.

13. Use a hammer and a brass drift to drive the output shaft rearward until the output shaft bearing clears the case. Do not pound on the output shaft; tap it lightly.

14. Separate the bearing from the output shaft with a puller.

15. Remove the output shaft and the related components from the transmission case.

16. Use a snap-ring expander to remove the snap-ring from the front of the output shaft. Slide the synchronizer clutch hub, sleeve, synchronizer ring, second gear, and the first/reverse gearset off of the shaft.

17. Remove the countershaft drive gear, spacer, roller bearing, and washers. Note the placement of the gear thrust washers. Use care not to lose the rollers.

18. Drive the reverse idler gear shaft rearward and remove its Woodruff key.

19. Remove the reverse idler gear, rollers and thrust washers from the case.

Gear Backlash: Input shaft gear-to-countershaft drive gear: 0.004 in. Second-to-countergear: 0.004 in. First/reverse gear-to-countergear: 0.008 in. Countergear-to-reverse idler gear: 0.008 in. Reverse gear-to-reverse idler gear: 0.008 in. Synchronizer ring-to-gear: 0.039 in.

ASSEMBLY

NOTE: Use new gaskets, oil seals, and dust seals. Coat the gaskets with sealer.

1. Apply a light coating of gear oil to all components, prior to assembly.

2. Grease the bore of the reverse idler gear. Insert the bearing rollers and washer in the bore.

3. Install the reverse idler gear and the two thrust washers into the case. Drive the reverse idler gear shaft through the case, by gently tapping it into place from behind. Lock the shaft into place with the Woodruff key.

4. Grease the bore of the countershaft drive gear and fit its spacer. Install the rollers in the bore and hold them in place with a heavy coating of grease. Install the washers in the bore.

5. Place the countershaft drive gear, thrust washer, and side thrust washers in the case.

6. If the bearing was removed from the input shaft, press it into place on the shaft.

7. Select a snap-ring that will give the input shaft minimum end-play, and install it on the shaft.

8. Grease bore of the input shaft, install the rollers, and then fit the snap-ring.

9. Carefully drive the input shaft assembly and bearing into the transmission case.

10. Lift the countershaft drive gear up and install the countershaft from the rear of the case. Secure the countershaft with its Woodruff key.

11. Use a feeler gauge to measure the countershaft thrust clearance; it should be 0.002–0.008 in. Select a countergear side thrust washer of the proper size to obtain the specified thrust clearance.

12. Install the two synchronizer shifting key springs into the clutch hub, so that the open ends are 120° apart. Place the three shifting keys into the clutch hub key slots.

13. Slide the clutch hub sleeve into the clutch hub.

14. Fit second gear, its synchronizer ring, and the synchronizer assembly on the output shaft. Check second gear thrust clearance with a feeler gauge. It should be 0.003–0.009 in.

15. Working from the rear, slide the first/reverse gearset on the output shaft.

16. Install the output shaft assembly in the transmission case. Using a suitable brass drift, install the rear bearing over the output shaft and into the case.

17. Install both shift forks and retain them with their balls and lockpins.

18. Depress the shift fork lockballs with a screwdriver, then drive the shift fork shaft through the case, and into shift forks.

19. Install a new O-ring on the shift fork shaft and lock it in place with its pin.

20. Coat the front bearing retainer with liquid sealer and install the bearing retainer over it.

21. Install a suitable size pipe over the transmission output shaft. Place the transfer case input gear, power take-off drive gear, and the spacers over the pipe, which should be projecting through the transfer case.

22. Coat a new gasket with liquid sealer and install.

23. Install the transfer case on the transmission case. Be sure to install the two short bolts from the inside of the transfer case.

24. Install the bearing over the end of the transmission output shaft and into the transfer case with a drift.

25. Install the transfer case input shaft spacer.

26. Install the transfer case cover over its gasket.

27. Coat the gasket with liquid sealer and install the power take-off cover.

28. Install the back-up light switch and gasket.

29. Install the transfer front drive fork and its gasket on the transfer case extension housing.

4 Speed
DISASSEMBLY

1. Remove the transfer case shift lever guide, cotter pin, and lockbolt. Remove the shift lever and linkage, do not lose the link lever shoe.

2. Remove the back-up light switch and gasket from the transmission cover.

3. Remove the transfer case cover, complete with gasket. Remove the power take-off cover and gasket.

4. Straighten out the tabs on the input shaft nut lockwasher and remove the nut. Slide the spacer off.

NOTE: Lock the power take-off drive gear with a wooden block or brass drift to keep the shaft from turning while the nut is being removed.

5. Loosen the five bolts which secure the transfer case to the transmission case and separate the cases with a puller. Hold the power take-off drive gear, spacer, and input gear, so that they don't drop out.

6. Remove the transfer case input shaft gear stop from the transmission rear bearing retainer.

7. Remove the rear bearing retainer and gasket.

8. Remove the front bearing retainer and gasket.

9. Remove the input shaft and front bearing from the case with a puller.

NOTE: Prior to removing the input shaft, align the slot in the input shaft with the countershaft drive gear.

10. Remove the synchronizer ring.

11. Use an expander to remove the snap-ring on the output shaft rear bearing. Remove the bearing with a puller.

12. Withdraw the output shaft and gearset from the transmission case.

13. Straighten out the tabs on the countershaft front bearing retainer and remove the bolts; then remove the retainer and lockwasher.

14. Remove the front countershaft bearing. Remove the bearing spacer. Remove the rear countershaft bearing.

15. Remove the countershaft.

16. Remove the reverse shift arm pivot and shift arm.

17. Install a puller on the reverse idler shaft, and pull the gear shaft and key from the transmission case. Remove the reverse idler gear from the case.

18. Slide the first gear and its thrust washer off the back end of the output shaft.

19. Remove the snap-ring from the front of the output shaft; then remove the clutch hub, sleeve, synchronizer ring, and third gear from the shaft.

20. Remove the snap-ring and slide second gear and thrust washer off the shaft.

21. Slide the reverse gear synchronizer ring off the output shaft.

22. Press the countershaft drive gear off of the countershaft. Remove the Woodruff key and spacer from the shaft.

23. Repeat Step 22 for the third gear and Woodruff key.

24. Move the shift forks into neutral. Drive the slotted spring pin out of the third/fourth shift fork with a long drift.

25. Without using excessive force, drive

the third/fourth shift fork shaft and plug forward with a brass drift. Remove the shift fork, lockball, and spring.

26. Drive the spring pins out of the reverse shift fork and boss with a long pin punch.

27. Remove the plug from the rear of the transmission case. Drive the first/second shift fork shaft forward with a brass drift. Remove the shift fork, boss, lockball, and spring.

28. Drive the spring pins out of the reverse shift fork and boss. Loosen and remove the tapered screw plugs from the transmission cover; then push the interlock rollers out with a long drift.

29. Remove the cotter pin, spring, and lockball from the shift fork boss.

30. Remove the C-washer, then withdraw the reverse return spring and plunger from the boss.

Gear Backlash: Input gear-to-countershaft drivegear: 0.004 in. Third gear-to-countershaft gear: 0.004 in. Second gear-to-countershaft gear: 0.004 in. First gear-to-countershaft gear: 0.004 in. Reverse gear-to-reverse idler gear: 0.005 in.

ASSEMBLY

1. Apply a light coating of gear oil to all components, prior to assembly.

2. Place the reverse idler gear in the case with its fork groove facing forward.

3. Align the Woodruff key, groove, and slot. Carefully drive the reverse idler shaft through the holes in the case, and into the gear.

NOTE: If you install new bushings in the idler gear, be sure that their openings are at least 90° apart.

4. Adjust the reverse idler gear position by trunging the shaft arm pivot to obtain a distance of 4.49 in. between the outer rear of the transmission case and the reverse idler gear. Move the gear to neutral, the distance between the front end of the gear and outer rear of the transmission case should now be 2.71 in. Adjust by rotating the shift arm pivot.

5. Install the Woodruff key into the groove of the countershaft and align its keyway with the countershaft third gear. Place the gear on the shaft with the long hub facing forward.

6. Slide the spacer on the countershaft and press its drivegear on, in a similar manner to third gear (Step 5), with its long hub facing rearward.

7. Place the countershaft assembly into the transmission case and install the countershaft rear bearing with a press. Install the snap-ring over the end of the countershaft.

8. Install the front bearing spacer, protruded end forward, on the countershaft. Press the front bearing on, until its snap-ring registers firmly against the transmission case.

─────── CAUTION ───────

Apply pressure to the outer bearing race only, to prevent damage to the bearing.

9. Place the bearing retainer and lockwasher on the front of the countershaft.

10. Slide second gear on the output shaft so that its synchronizer outer ring faces rearward. Select a snap-ring so that second gear has a thrust clearance of 0.004–0.012 in.

11. Slide third gear on the output shaft, so that its synchronizer cone faces forward.

12. Install the two synchronizer shifting key springs into the clutch hub, so that the open ends are 120° apart. Place the three shifting keys into the clutch hub key slots.

13. Slide the clutch hub sleeve into the clutch hub.

14. Install the synchronizer ring and slide the synchronizer assembly over the output shaft. The grooves should face the rear.

15. Select a snap-ring to provide 0.008 in. thrust clearance for the clutch hub.

16. Slide the reverse gear synchronizer ring over the output shaft. Check the ring for smooth movement.

17. Slide first gear on the output shaft.

18. Place the output shaft assembly in the transmission case.

19. Fit the first gear thrust washer on the output shaft. Align the slot in the thrust washer with the output shaft pin.

20. Install the output shaft rear bearing with a press. Be sure to apply pressure on the bearing outer race only. Install the rear bearing and its gasket.

21. Press-fit the bearing on the input shaft. Grease the bearing rollers and install all 18 in the input shaft.

22. Place the input shaft in the transmission case, so that the synchronizer ring keyways align with the shift keys. Install the output shaft bearing spacer in the output shaft.

─────── CAUTION ───────

Be sure the bearing rollers or the spacer do not fall into the transmission case during installation.

23. Install the input shaft bearing retainer and gasket.

NOTE: Make sure that all shift fork shafts are in neutral during the shift fork shaft assembly steps, below.

24. Install the spring and return plunger in the reverse shift boss. Secure them with the C-washer. Install the ball and spring then secure them with a cotter pin.

25. Place the reverse shift fork and boss in the transmission cover, its bore, shift fork and into its boss, while depressing the lockball with a screwdriver.

26. Slide the reverse shift fork shaft through the front of the cover. Install the fork lock-spring and ball in the cover.

27. Align the holes and drive the slotted spring pin through the reverse shift fork and boss to secure them to the shaft.

28. Coat the roller with grease and install it into the interlock hole in the cover.

29. Place the first/second shift boss and fork into the cover. Install the first/second shift fork shaft and spacer from the front of the cover, after fitting its pin, and while depressing the lockball with a screwdriver.

30. Secure the first/second shift fork and boss with their slotted spring pins. Install another roller into the hole in the cover.

31. Install the third/fourth shift fork lockspring and ball in the shift fork. Install the shift fork and shaft in the cover and secure them with a slotted spring pin.

32. Check the shift fork assembly for smooth operation. Install the plugs.

33. Install a suitable size pipe over the transmission output shaft. Place the transfer case input gear, power take-off drive gear, and the spacers over the pipe, which should be projecting through the transfer case.

34. Coat a new gasket with liquid sealer and install.

35. Install the transfer case on the transmission case. Be sure to install the two short bolts from the inside of the transfer case.

36. Remove the pipe from the output shaft and transfer case.

37. Install the bearing over the end of the transmission output shaft and into the transfer case with a drift.

38. Install the transfer case input shaft spacer.

39. Install the transfer case cover over its gasket.

40. Coat the gasket with liquid sealer and install the power take-off cover.

41. Install the back-up light switch and gasket.

42. Install the transfer front drive fork and its gasket on the transfer case extension housing.

Manual Transmission Troubleshooting

Condition	Probable Cause
Jumping out of high gear	1. Misalignment of transmission case or clutch housing. 2. Worn pilot bearing in crankshaft. 3. Bent transmission shaft. 4. Worn high speed sliding gear. 5. Worn teeth in clutch shaft. 6. Insufficient spring tension on shifter rail plunger. 7. Bent or loose shifter fork. 8. End-play in clutch shaft. 9. Gears not engaging completely. 10. Loose or worn bearings on clutch shaft or mainshaft.
Sticking in high gear	1. Clutch not releasing fully. 2. Burred or battered teeth on clutch shaft. 3. Burred or battered transmission mainshaft. 4. Frozen synchronizing clutch. 5. Stuck shifter rail plunger. 6. Gearshift lever twisting and binding shifter rail. 7. Battered teeth on high speed sliding gear or on sleeve. 8. Lack of lubrication. 9. Improper lubrication. 10. Corroded transmission parts. 11. Defective mainshaft pilot bearing.
Jumping out of second gear	1. Insufficient spring tension on shifter rail plunger. 2. Bent or loose shifter fork. 3. Gears not engaging completely. 4. End-play in transmission mainshaft. 5. Loose transmission gear bearing. 6. Defective mainshaft pilot bearing. 7. Bent transmission shaft. 8. Worn teeth on second speed sliding gear or sleeve. 9. Loose or worn bearings on transmission mainshaft. 10. End-play in countershaft.
Sticking in second gear	1. Clutch not releasing fully. 2. Burred or battered teeth on sliding sleeve. 3. Burred or battered transmission mainshaft. 4. Frozen synchronizing clutch. 5. Stuck shifter rail plunger. 6. Gearshift lever twisting and binding shifter rail. 7. Lack of lubrication. 8. Second speed transmission gear bearings locked will give same effect as gears stuck in second. 9. Improper lubrication. 10. Corroded transmission parts.
Jumping out of low gear	1. Gears not engaging completely. 2. Bent or loose shifter fork. 3. End-play in transmission mainshaft. 4. End-play in countershaft. 5. Loose or worn bearings on transmission mainshaft. 6. Loose or worn bearings in countershaft. 7. Defective mainshaft pilot bearing.
Sticking in low gear	1. Clutch not releasing fully. 2. Burred or battered transmission mainshaft. 3. Stuck shifter rail plunger. 4. Gearshift lever twisting and binding shifter rail. 5. Lack of lubrication. 6. Improper lubrication. 7. Corroded transmission parts.

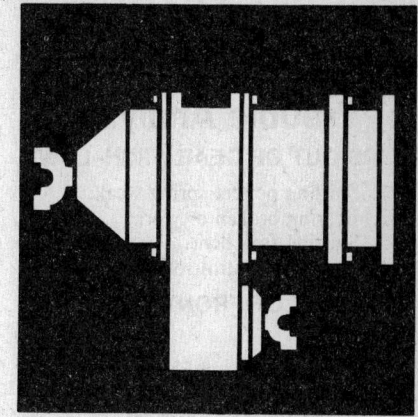

Transfer Cases

INDEX

TRANSFER CASE SERVICE

Trouble Analysis

SLIPS OUT OF GEAR (HIGH–LOW)

1. Shifting poppet spring weak.
2. Bearing broken or worn.
3. Shifting fork bent.
4. Improper control rod adjustment.

SLIPS OUT OF FRONT WHEEL DRIVE

1. Shifting poppet spring weak or broken.
2. Bearing worn or broken.
3. Excessive shaft end-play.
4. Shifting fork bent.

HARD SHIFTING

1. Lack of lubricant.
2. Shift lever binding on shaft.
3. Shifting poppet ball scored.
4. Shifting fork bent.
5. Low tire pressure.

BACKLASH

1. Companion yoke loose.
2. Transfer case loose on mounts.
3. Internal parts excessively worn.

NOISY

1. Low lubricant level.
2. Bearings improperly adjusted or excessively worn.
3. Gears worn or damaged.
4. Improper alignment of driveshafts or U-joints.

OIL LEAKAGE

1. Excessive amount of lubricant in case.
2. Vent clogged.
3. Gaskets or seals leaking.
4. Bearings loose or damaged.
5. Driveshaft yoke mating surfaces scored.

OVERHEATING

1. Excessive or insufficient amount of lubricant.
2. Bearing adjustment too tight.

CLEANING & INSPECTION

During overhaul, all components of the transfer case (except bearing assemblies) should be thoroughly cleaned with solvent and dried with air pressure prior to inspection and reassembly. Be sure all gasket sealing material is cleaned off of the case, cover plates and mounting flanges.

Bearing Cleaning

Proper cleaning of bearings is of utmost importance. Bearings should always be cleaned separately from other parts.

Soak all bearing assemblies in clean solvent or fuel oil. Bearings should never be cleaned in a hot solution tank. Wash the bearings in solvent until all old lubricant is loosened. Hold races so that bearings will not rotate; then clean bearings with a soft bristled brush until all dirt has been removed. Remove loose particles of dirt by tapping bearing flat against a block of wood. Rinse bearings in clean solvent; then blow bearings dry with air pressure.

CAUTION
Do not spin bearings while drying.

After drying, rotate each bearing slowly while examining balls or rollers for roughness, damage, or excessive wear. Replace all bearings that are not in first class condition. After cleaning and inspecting bearings lubricate generously with recommended lubricant, then wrap each bearing in clean paper until ready for reassembly.

INSPECTION

1. Inspect all parts for discoloration or warpage.
2. Examine all gears and splines for chipped, worn, broken or nicked teeth. Small nicks or burrs may be removed with a fine abrasive stone.
3. Inspect the breather assembly to make sure that it is open and not damaged.
4. Check all threaded parts for damaged, stripped, or crossed threads.
5. Replace all gaskets, oil seals and snap-rings.
6. Inspect housings, retainers and covers for cracks or other damage. Replace the damaged parts.
7. Inspect keys and keyways for condition and fit.
8. Inspect shift forks for wear, distortion or any other damage.

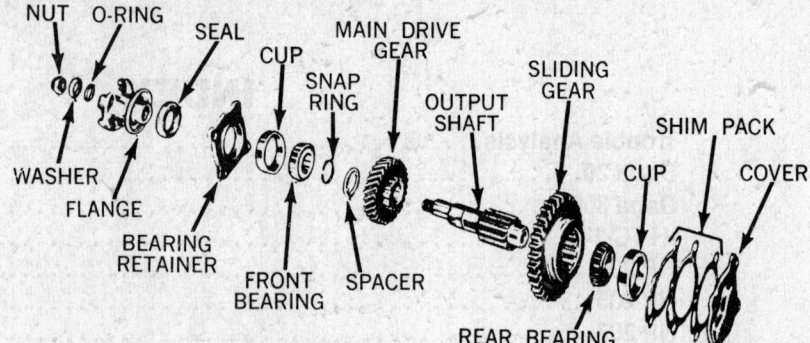

Exploded view of the front output shaft of a Dana 20 transfer case (© Ford Motor Co.)

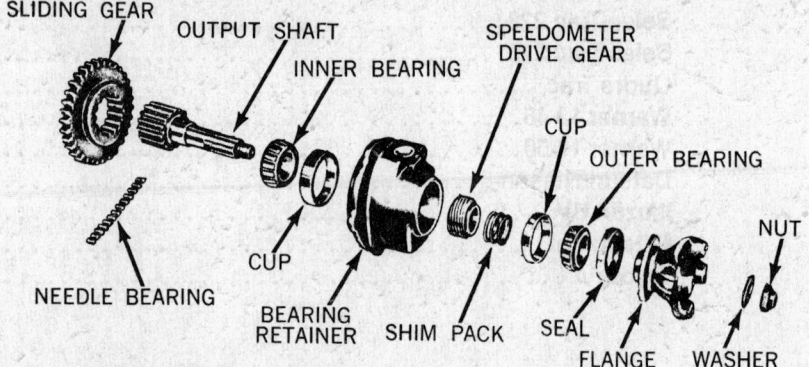

Exploded view of the rear output shaft of a Dana 20 transfer case (© Ford Motor Co.)

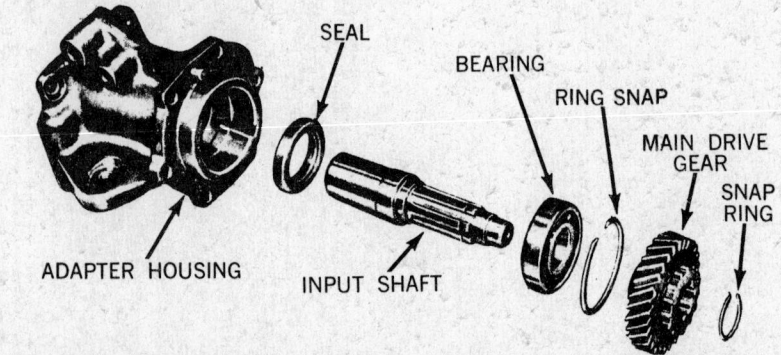

Exploded view of the input shaft of a Dana 20 transfer case (© Ford Motor Co.)

9. Check detent ball springs for free length, compressed length, distortion or collapsed coils.

10. Check bearing fit on their respective shafts and in their bores or cups. Inspect bearings, shafts and cups for wear.

NOTE: If either the bearings or cups are worn or damaged, it is advisable to replace both parts.

11. Inspect all bearing rollers or balls for pitting or galling.

12. Examine detent balls for corrosion or brinneling. If shift bar detents show wear, replace them.

13. Replace all worn or damaged parts. When assembling the transfer case, coat all moving parts with recommended lubricant.

Dana Model 20

The Dana Model 20 is a two-speed gearbox that controls the power from the transmission to the front and rear driving axles. Positions of the transfer case are: four-wheel-drive low (4L), neutral (N), two-wheel-drive high (2H) and four-wheel-drive high (4H).

DISASSEMBLY

Transfer Case

1. Clean any dirt from the transfer case and remove the bottom cover plate.

2. Remove the retaining plug, flat washer, detent spring and ball which engages the front drive shift rail detent rod. Then, remove plug from front drive detent rod access hole.

3. Remove the retaining plug, detent spring and ball which engages the rear drive shift rail detent rod.

4. Remove the idler shaft lockplate.

5. Using a hammer and soft drift, drive the idler shaft rearward and out of the case; then lift out the thrust washers and idler gear.

NOTE: When removing the idler gear, do not lose any of the rollers.

6. Remove the flange retaining nuts from the front and rear output shafts.

7. Remove the flange from the front and rear output shafts. Discard the O-ring.

8. Remove the bolts securing the adapter housing to the case; then remove the adapter as an assembly.

9. Remove the bolts which attach the rear output shaft bearing retainer to the case; then remove the retainer and output shaft as an assembly.

NOTE: Be sure not to lose any of the rollers.

10. Disconnect the shift rail link from the two shift rails.

11. Lift out the rear output shaft sliding gear.

1 Input shaft
2 Transfer case
3 Input gear
4 Snap ring
5 Sliding clutch gear
6 Rear input shaft needle bearing
7 Rear output shaft front bearing

8 Rear Output shaft front bearing cup
9 Rear output shaft housing gasket
10 Rear output shaft housing breather
11 Speedometer driven gear
12 Rear output shaft housing
13 Rear output shaft rear bearing cup
14 Rear output shaft rear bearing
15 Rear output shaft yoke
16 Rear output shaft locknut
17 Washer
18 Rear output shaft
19 Rear output shaft seal
20 Shims
21 Speedometer drive gear
22 Intermediate shaft lock plate bolt
23 Intermediate shaft lock plate
24 Intermediate shaft bearing spacer
25 Intermediate shaft
26 Intermediate shaft needle bearings
27 Intermediate shaft tanged thrust washer

28 Intermediate gear
29 Front output shaft rear cover
30 Front output shaft rear bearing
31 Front output shaft rear cover shim pack
32 Front output shaft rear bearing cup
33 Front output shaft sliding clutch gear
34 Drain Plug
35 Front output shaft drive gear
36 Spacer
37 Front output shaft front bearing
38 Front output shaft front bearing cup
39 Spacer
40 Front output shaft seal
41 Front output shaft bearing
42 Front output shaft yoke
43 Rubber "O" ring
44 Washer
45 Front Output shaft locknut

FRONT

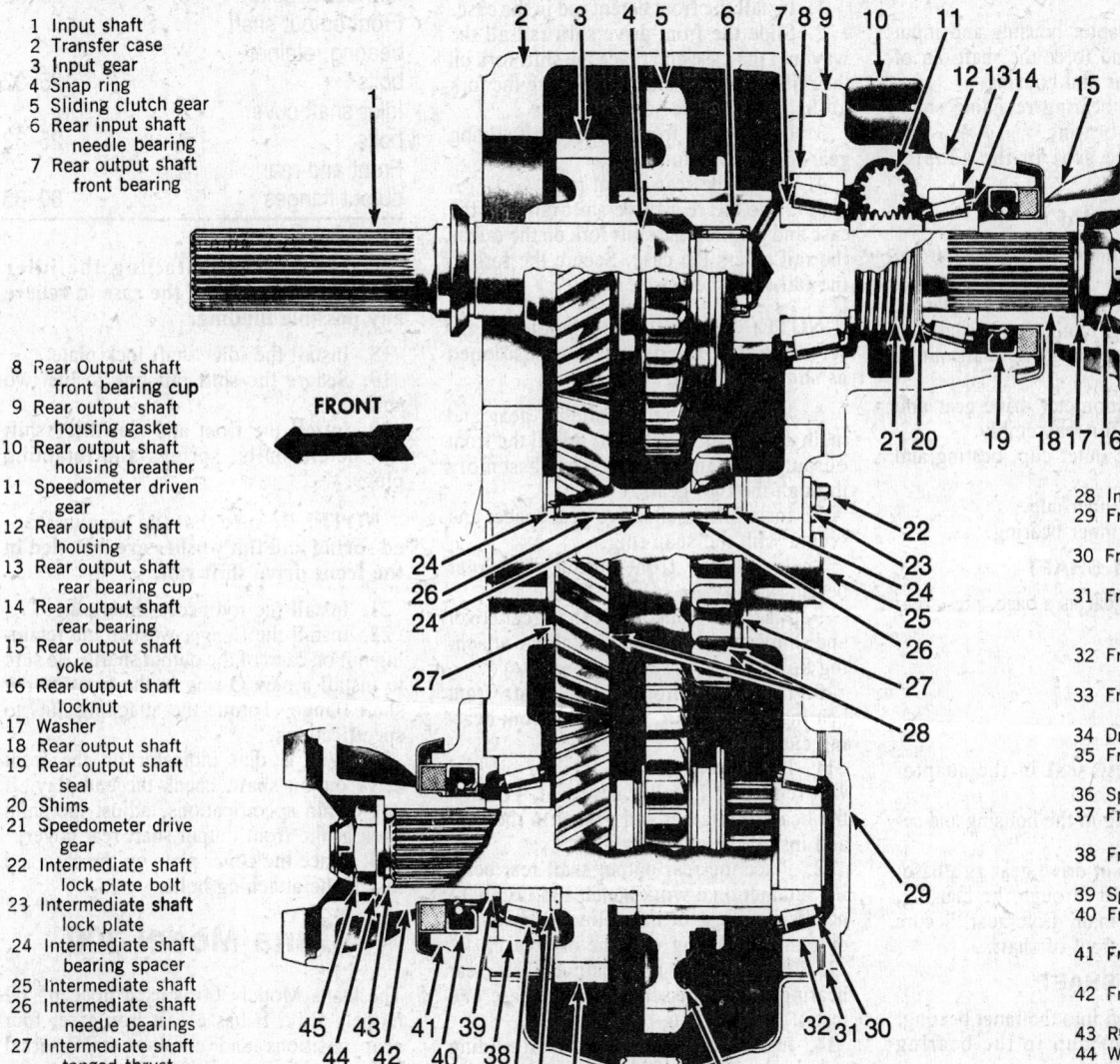

Cross section of Dana model 20 transfer case (© General Motors Corp.)

12. Remove the setscrew securing the rear fork to the shift rail; then remove the rear drive shift rail and fork.

13. Remove the front output shaft rear cover and shims. Fasten the shims together.

14. Remove the front output shaft bearing retainer and gasket.

15. Tap the threaded end of the front output shaft; then remove the rear cup.

16. Angle the front output shaft front bearing away from the main drive gear to allow removal of the snap-ring; then tap the shaft and rear bearing out of the case.

17. Lift out the sliding gear, main drive gear, front bearing, spacer and snap-ring.

18. Remove the front cup.

19. Remove the setscrew securing the front shift fork to the shift rail; then remove the rail and fork.

20. Remove the detent rods.

21. Remove shift rail oil seal.

INPUT SHAFT

1. Remove the snap-ring from the front of the shaft.

2. Place the adapter housing and input shaft on a press and force the shaft out of the main drive gear and housing.

3. Remove the bearing retaining snap-ring; then remove bearing.

4. Remove the seal in the adapter housing.

REAR OUTPUT SHAFT

1. Remove needle bearings from bore of shaft.

2. Remove speedometer driven gear.

3. Place bearing retainer and shaft assembly in a press; then force shaft out of retainer.

4. Lift off speedometer drive gear and shims. Tag shims for reassembly.

5. Press out the outer cup, bearing and seal.

6. Remove the inner cup.

7. Remove the inner bearing.

FRONT OUTPUT SHAFT

Using the sliding gear as a base, press rear bearing off shaft.

ASSEMBLY

INPUT SHAFT

1. Install a new seal in the adapter housing.

2. Install bearing in the housing and secure with snap-ring.

3. Using the main drive gear as a base, force the input shaft through the housing, seal, bearings and main drive gear. Secure with snap-ring on front of shaft.

REAR OUTPUT SHAFT

1. Press the shaft into the inner bearing.

2. Install outer cup in the bearing retainer.

3. Install the inner cup.

4. Position the outer bearing in the retainer; then place the shims and speedometer drive gear on the shaft. Install shaft in

the bearing retainer housing.

5. Place the bearing retainer and shaft in a vise. Install the output shaft flange and torque the retaining nut to specifications.

6. With a dial indicator on the flange end of the shaft, measure end-play. Adjust shim pack between the speedometer drive gear and outer bearing to achieve correct clearance.

7. After setting correct end-play, remove flange and press bearing retainer seal into housing.

8. Install the flange, washer and nut. Tighten the nut to specifications.

FRONT OUTPUT SHAFT

Using a press, force front output rear bearing on shaft.

SHIFT RAIL OIL SEALS

Install the two shift rail oil seals with appropriate tools.

Transfer Case

1. Install the front detent rod in the case.

2. Slide the front drive shift rail all the way into the case and place the shift fork on the rail as it enters the case. Secure the fork to the rail with the setscrew.

3. Position the front output shaft sliding gear in the shift fork.

4. Install the rear detent rod.

5. Slide the rear drive shift rail into the case and position the shift fork on the rail as the rail enters the case. Secure the fork to the rail with the setscrew.

NOTE: The shift rails should be inserted so that the detents are positioned as shown in illustration.

6. While holding the sliding gear and main drive gear in position, install the front output shaft and rear bearing assembly through the two gears.

7. Install the main drive gear spacer and secure with the snap-ring.

8. Install the front output shaft rear bearing cup.

9. Place the front output shaft rear cover and shims on the case and install the attaching bolts.

10. Install the front output shaft front bearing on the shaft. Install the front bearing cup.

11. If the front bearing retainer oil seal was removed, install a new seal. Position the bearing retainer and gasket to the case and install the attaching bolts.

12. Place the rear output shaft rear bearing retainer on a work bench and install 13 needle bearings in the splined hub of the output shaft, using vaseline or grease.

13. Position the rear output shaft rear bearing retainer assembly to the case and install the attaching bolts.

14. Install the rear output shaft sliding gear in the shifting fork and on the splines of the output shaft.

15. Position the adapter housing assembly on the rear output shaft and case. Install the attaching bolts.

16. Install the roller bearings in the bore of the idler shaft gear with vaseline or grease.

17. Position the idler gear and thrust washers in the case; then drive the idler shaft into the rear of the case through the idler gear and thrust washers.

SPECIFICATIONS

END PLAY (IN.)

Front output shaft	0.001–0.005
Rear output shaft	0.001–0.005

TORQUE LIMITS (FT. LBS.)

Transfer case to transmission extension bolts	20–30
Transfer case to transmission output shaft nut	60–80
Front output shaft rear cover bolts	25–32
Front output shaft bearing retainer bolts	25–32
Idler shaft cover bolts	25–32
Front and rear output flanges	80–85

NOTE: After installing the idler shaft, tap the sides of the case to relieve any possible binding.

18. Install the idler shaft lock plate.

19. Secure the shift rail link to the two shift rails.

20. Install the front and rear drive shift rail detent balls, springs and retaining plugs.

NOTE: Be sure that the heavier loaded spring and flat washer are installed in the front drive shift rail.

21. Install the rod access hole plug.

22. Install the flange, washer and retaining nut on each of the output shafts. Be sure to install a new O-ring in the front output shaft flange. Torque the attaching nuts to specifications.

23. With a dial indicator on the front drive output shaft, check the end-play. If not within specifications, adjust the shim pack at the front output shaft rear cover.

24. Place the cover plate on the case and install the attaching bolts.

Dana Model 300

The Dana Model 300 is used in Jeep® CJ models only. It has a cast iron case, four gear positions and employs an external floor mounted gearshift linkage for range control. It is a part time, 2 speed unit with undifferentiated high and low ranges. It is used with both manual and automatic transmission. Low range reduction is 2.6:1.

DISASSEMBLY

1. Drain the unit and remove the shift lever assembly.

2. Remove the bottom cover.

NOTE: The bottom cover has been coated with a sealant. Use a putty knife to break the seal and work the knife around the bottom of the cover to break it loose. Don't try to wedge the cover off.

3. With a puller, remove the front and rear yokes.

4. Unbolt and remove the input shaft support from the case. The rear output shaft gear and input shaft will come with it as an assembly.

NOTE: The support has been coated with sealant, remove it as you did the bottom cover.

5. Remove the rear output shaft clutch sleeve from the case.

6. Remove and discard the snap ring retaining the rear output shaft gear on the input shaft and remove the gear.

7. Remove and discard the input bearing snapring.

8. Remove the input shaft bearing from the support. Tap the end of the shaft with a soft mallet to aid removal.

9. Remove the input shaft bearing and end-play shims from the shaft with an arbor press.

10. Remove the input shaft oil seal from the support.

11. Unbolt and remove the intermediate shaft lockplate.

12. Remove the intermediate shaft. Tap the shaft out of the case using a brass punch and plastic mallet.

13. Remove and discard the intermediate shaft O-ring seal.

14. Remove the intermediate gear assembly and thrust washers.

NOTE: The thrust washers have locating tabs which must fit into notches in the case at assembly.

15. Remove the needle bearings and spacers from the intermediate gear. There are 48 needle bearings and three spacers.

16. Remove the rear bearing cap attaching bolts and remove the cap. A plastic mallet will aid in removal.

NOTE: The rear bearing cap has been coated with sealant.

17. Remove the end play shims and speedometer drive gear from the rear output shaft.

18. Remove and discard the rear output shaft oil seal. Remove the bearings and races from the rear cap.

19. Unbolt and remove the front and rear output shaft shift forks from the shift rods.

20. Remove the shift rods. Insert a punch through the clevis pin holes in the rods and rotate the rods while pulling them out of the case.

NOTE: When the shift rods are free of the case, take care to avoid losing the shift rod poppet balls and springs.

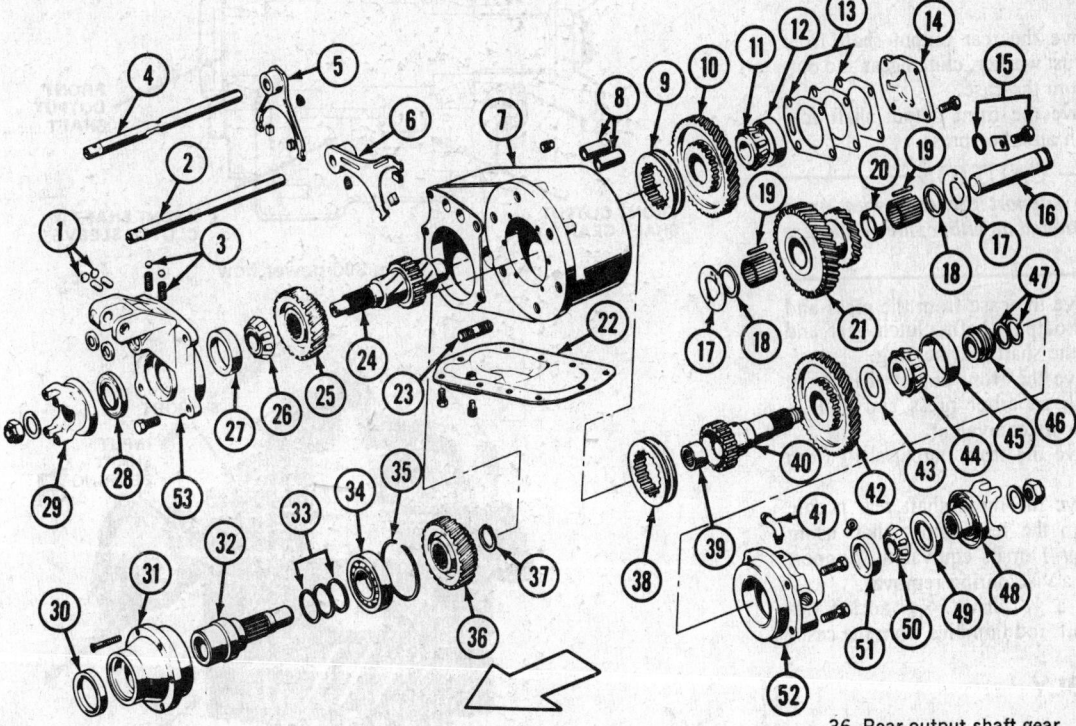

1 Interlock plugs and interlocks	19 Intermediate gear shaft needle bearings	36 Rear output shaft gear
2 Shift rod—rear output shaft fork		37 Snap ring
3 Poppet balls and springs	20 Bearing spacer (thick)	38 Clutch sleeve—rear output shaft
4 Shift rod—front output shaft fork	21 Intermediate gear	39 Input shaft rear bearing (needle) (or pilot bearing)
5 Front output shaft shift fork	22 Bottom cover	
6 Rear output shaft shift fork	23 Stud (case-to-trans.)	40 Rear output shaft
7 Transfer case	24 Front output shaft	41 Vent
8 Thimble covers	25 Front ouptut shaft gear	42 Clutch gear—rear output shaft
9 Clutch sleeve—front output shaft	26 Front output shaft bearing (front)	43 Thrustwasher
10 Clutch gear—front output shaft	27 Front output shaft bearing race	44 Bearing—rear output shaft front
11 Bearing—front output shaft rear	28 Oil seal	45 Race—rear output shaft bearing
12 Race—front output shaft bearing	29 Front yoke	46 Speedometer drive gear
13 End play shims—front output shaft	30 Seal	47 End play shims
14 Cover plate	31 Support—input shaft	48 Rear yoke
15 Lock plate, bolt and washer	32 Input shaft	49 Rear output shaft oil seal
16 Intermediate gear shaft	33 Shims	50 Bearing—rear output shaft rear
17 Thrust washer	34 Input shaft bearing	51 Bearing race
18 Bearing spacer (thin)	35 Input shaft bearing snap ring	52 Rear bearing cap
		53 Front bearing cap

Exploded view of Dana 300

21. Remove the shift forks from the case.
22. Remove the bolts attaching the front cap to the case and remove the cap.

NOTE: The front cap has been coated with sealant.

23. Remove the front output shaft and shift rod oil seals from the front cap.
24. Remove the bearing race from the front cap.
25. Remove the cover plate bolts and remove the plate and end play shims from the case. Keep the shims together for assembly.
26. Move the front output shaft toward the front of the case.
27. Remove the front output shaft rear bearing race.
28. Remove the rear output shaft front bearing. Position the case on wood blocks. Seat the clutch gear on the case interior surface and tap the shaft out of the bearing with a soft mallet.

NOTE: It the bearing is difficult to remove, an arbor press may have to be used.

29. Remove the rear output shaft front bearing, thrust washer, clutch gear and output shaft from the case.
30. Remove the front output shaft rear bearing with an arbor press.

--- **CAUTION** ---

Be sure to support the case with wood blocks positioned on either side of the case bore.

31. Remove the case from the press and remove the output shaft, clutch gear and sleeve and the shaft rear bearing.
32. Remove the front output shaft front bearing with an arbor press and tool J–22912–01 or its equivalent.
33. Remove the front output shaft from the gear.
34. Remove the input shaft rear needle bearing from the rear output shaft using tool J–29369–1 or its equivalent. Support the shaft in a vise during removal.
35. Using a $\frac{3}{8}$″ drive, $\frac{7}{16}$″ socket, remove the shift rod thimbles from the case.

ASSEMBLY

Coat all parts with SAE 85W–90 oil before assembly.

1. Apply Loctite® 220 or its equivalent to the thimbles and install them in the case.
2. Install the front output shaft gear on the front output shaft. Be sure that the clutch teeth on the gear face the shaft gear teeth.
3. Install the front bearing on the front output shaft using an arbor press. Be sure that the bearing is seated against the gear.
4. Install the front output shaft in the case and install the clutch sleeve and gear on the shaft.
5. Install the front output shaft rear bearing using an arbor press.

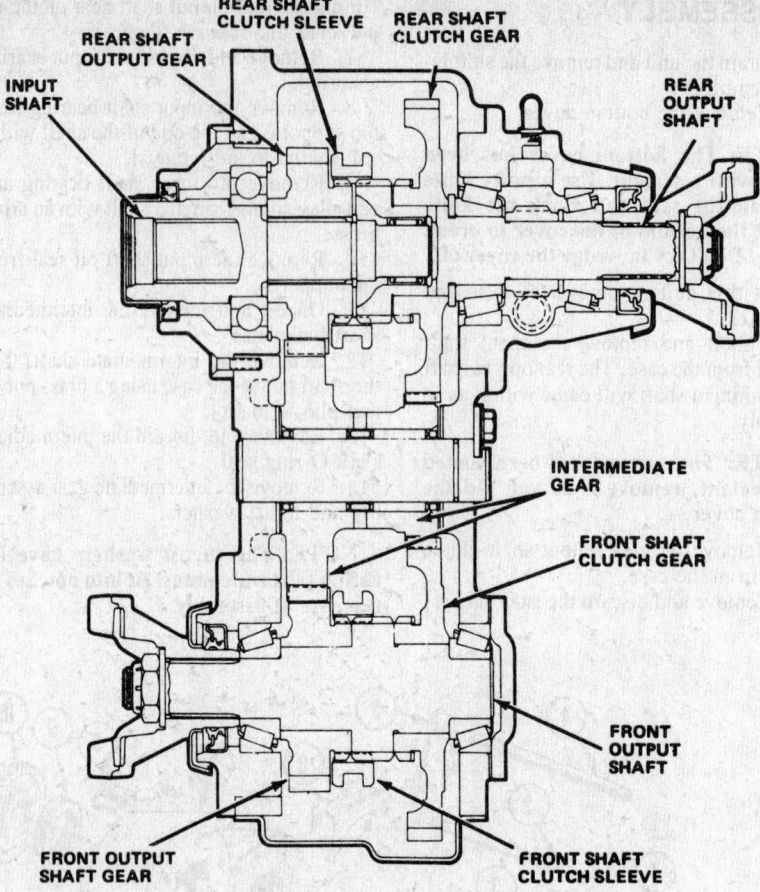

Dana 300 power flow

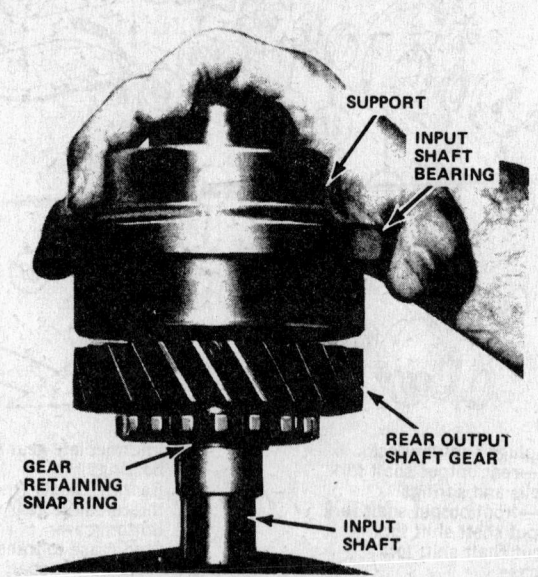

Front support, input shaft and rear output shaft gear removal

NOTE: Install an old yoke nut on the shaft to avoid damage to the threads.

6. Install the input shaft needle bearings in the rear output shaft with tool J–29179 or its equivalent.
7. Position the rear output shaft clutch gear in the case and insert the rear output shaft into the gear.
8. Install the thrust washer and front bearing on the rear output shaft using an arbor press.
9. Install the shims and bearing on the input shaft using an arbor press.

10. Install a new input shaft seal.

11. Using a new snap-ring, install the input shaft and bearing in the support.

12. Install the rear output shaft gear on the input gear and install a new gear retaining ring.

13. Measure the clearance between the input gear and the gear retaining snap-ring using a feeler gauge. Clearance should not exceed .003 in. If clearance is beyond tolerance, add shims between the input shaft and bearing.

14. Install the clutch sleeve on the rear output shaft.

15. Apply Loctite® 515 or equivalent to the mating surfaces of the input shaft support and install the support assembly, shaft and gear in the case. Use two support bolts to align the support on the case and tap the support into position with a soft mallet. Torque the support bolts to 10 ft. lbs.

16. Install the rear bearing cap front bearing race.

17. Install the rear bearing cap rear bear-

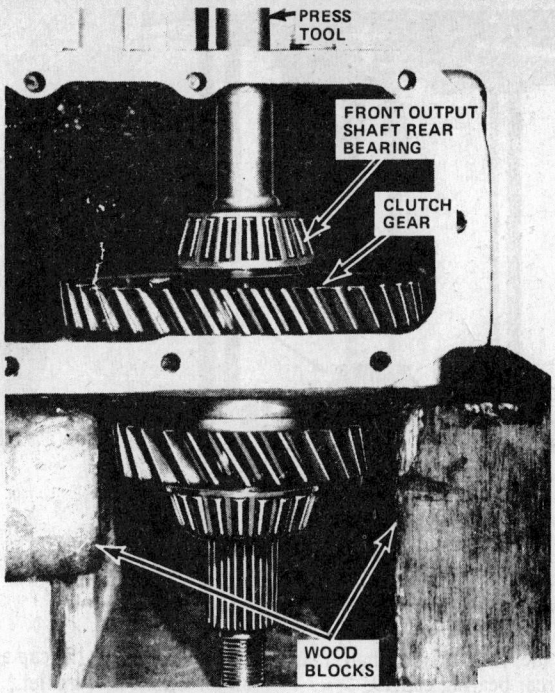

Front output shaft rear bearing removal

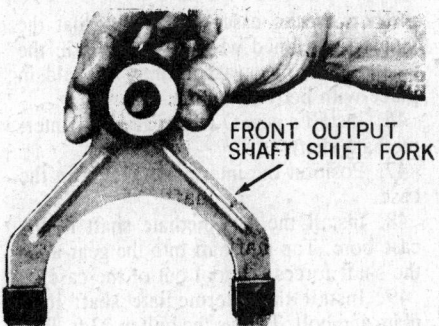

Shift fork installation

Shift fork set screw removal

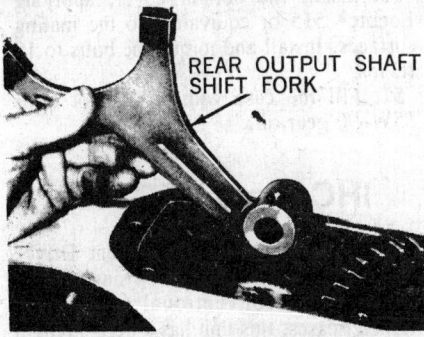

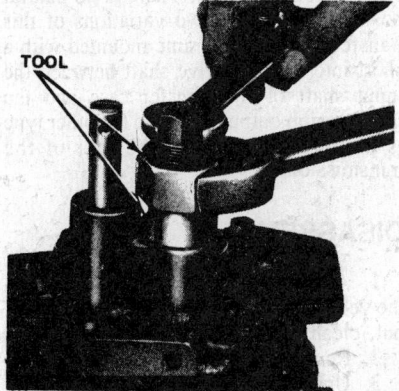

Shift rod oil seal removal

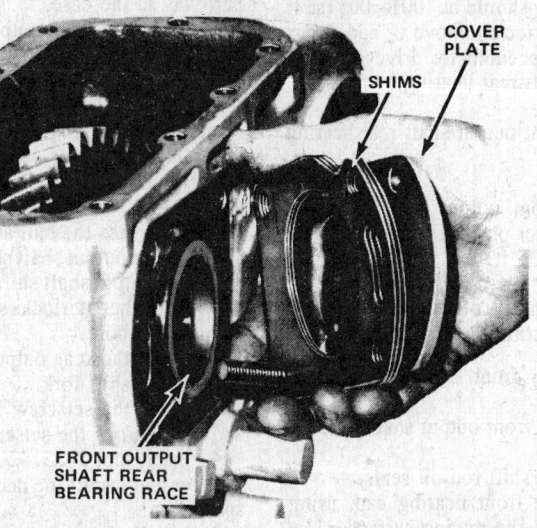

Cover plate, shims and front output shaft rear bearing race

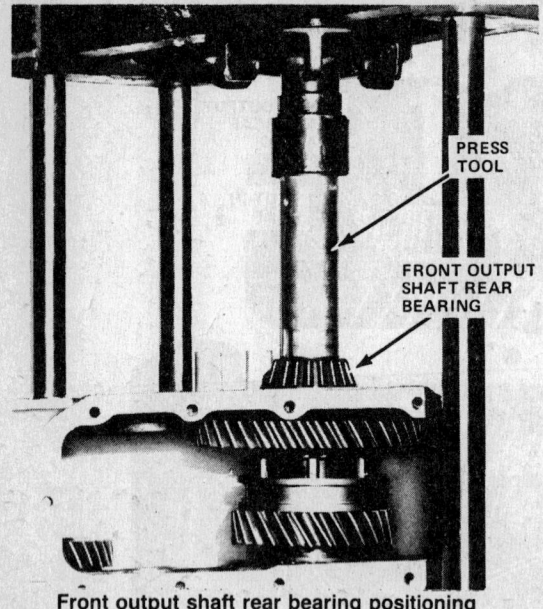

Front output shaft rear bearing positioning

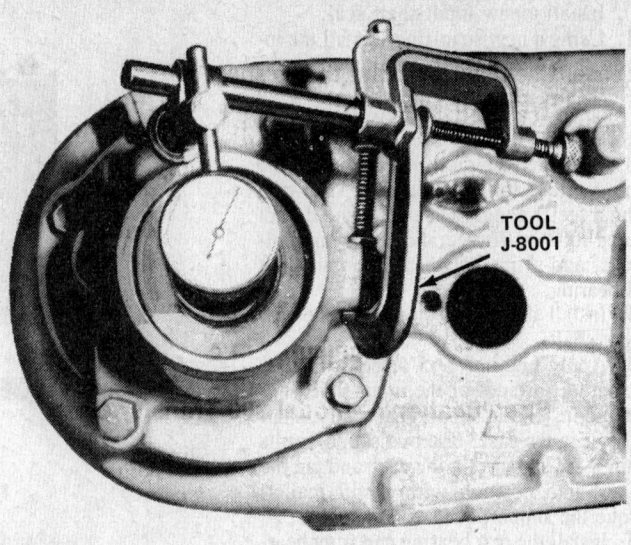

Checking rear output shaft end-play

18. Position the rear output shaft rear bearing in the rear bearing cap.

19. Install the rear output shaft yoke oil seal.

20. Install the speedometer gear and end-play shims on the rear output shaft.

21. Apply Loctite® 515 or equivalent to the mating surfaces of the cap and install the rear bearing cap. Use two cap bolts to align the cap and tap it into place with a soft mallet.

22. Tighten the cap bolts to 35 ft. lbs.

23. Install the rear output shaft yoke. Torque a new locknut to 120 ft. lbs.

24. Clamp a dial indicator on the rear output shaft bearing cap. Position the indicator stylus so that it contacts the end of the shaft.

25. Pry the shaft back and forth to check end-play. End-play should be .001–.005 in. If play is not correct, remove or add shims between the speedometer drive gear and the output shaft rear bearing.

26. Install the front output shaft rear bearing race.

27. Install the front output shaft end play shims and cover plate. Tighten the cover plate bolts to 35 ft. lbs.

NOTE: Apply Loctite® 220 to the bolts before installation.

28. Install the front output shaft front bearing race.

29. Install the front output shaft yoke oil seal.

30. Install the shift rod oil seals.

31. Install the front bearing cap, using Loctite® 515 on the mating surfaces. Use two bolts to align the cap and tap it into position with a soft mallet.

32. Install and tighten the bearings cap bolts to 35 ft. lbs.

33. Seat the rear bearing cup against the cover plate by tapping the end of the front output shaft with a plastic mallet. Mount a dial indicator on the front bearing cap and position the stylus against the end of the output shaft. Pry the shaft back and forth to check end-play. End-play should be .001–.005 in. If the play is not correct, add or remove shims between the cover plate and case. If shims are added seat the rear bearing cup again before checking.

34. Install the front output shaft yoke. Tighten the new locknut to 120 ft. lbs.

35. Install the front and rear output shaft shift forks.

36. Install the front output shaft shift rod poppet ball and spring in the front bearing cap.

37. Compress the poppet ball and spring and install the front output shaft shift rod part way in the case.

38. Insert the front output shaft shift rod through the shift fork.

39. Align the setscrew hole in the shift fork and rod. Install and tighten the set-screw to 14 ft. lbs.

40. Install the rear output shaft shift rod poppet ball and spring in the front bearing cap.

41. Compress the ball and spring and install the rear output shaft shift rail part way. The front output shaft shift rod should be in neutral and the interlocks seated in the front bearing cap bore.

42. Insert the rear output shaft shift rod through the shift fork.

43. Align the setscrew holes in the fork and rod. Torque the setscrew to 14 ft. lbs.

44. Insert tool J–25142 in the intermediate gear and install the needle bearings and spacer.

45. Install the intermediate gear thrust washers in the case. Make sure that the tangs are aligned with the grooves in the case. The thrust washers may be held in place with petroleum jelly.

46. Install a new O-ring seal on the intermediate shaft.

47. Position the intermediate gear in the case.

48. Install the intermediate shaft in the case bore. Tap the shaft into the gear until the shaft forces the tool out of the case.

49. Install the intermediate shaft lock plate and bolt. Torque the bolt to 23 ft. lbs.

50. Install the bottom cover, applying Loctite® 515 or equivalent to the mating surfaces. Install and torque the bolts to 15 ft. lbs.

51. Fill the case with 4 pints of SAE 85W–90 gear oil.

IHC Model TC–143

The IHC Model TC–143 "Silent Drive" transfer case is a chain driven single speed unit. Unlike conventional gear driven transfer cases, this unit has a high-strength link-belt type loop of chain driving two broad-faced sprockets. There is no neutral position. There are two variations of this transfer case; one is frame mounted with a short intermediate drive shaft between the input shaft of the transfer case and the transmission output shaft and the other type is mounted directly to the rear of the transmission.

DISASSEMBLY

1. After removing the transfer case from the vehicle and draining all of the lubricant out, clean the outside of the case.

2. Remove the shift cover.

3. Unscrew and remove the indicator light switch from the case.

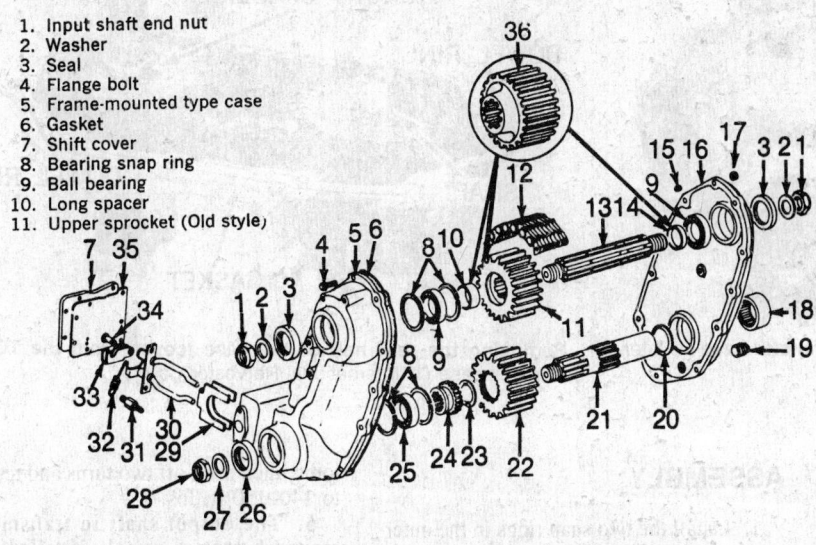

1. Transmission shaft coupling
2. Washer
3. Transmission shaft end nut
4. Input shaft coupling
5. Speedometer gear
6. Gasket
7. Transmission mounted type case
8. Flange bolt
9. Gasket
10. Snap input shaft ring
11. Bearing snap ring
12. Ball bearing
13. Long spacer
14. Upper sprocket (Old style)
15. Chain
16. Input shaft
17. Short spacer
18. Dowel ring
19. Cover
20. Flange nut
21. Seal
22. Washer
23. End nut
24. Roller bearing
25. Drain plug
26. Thrust washer
27. Output shaft
28. Lower sprocket
29. Thrust washer
30. Sliding clutch
31. Ball bearing
32. Seal
33. Washer
34. Output shaft end nut
35. Shift shoe
36. Assembly shifter
37. Spring stud
38. Shift spring
39. Shift clevis
40. Clevis pin
41. Shift cover gasket
42. Shift cover
43. Upper sprocket (new style)

Exploded view of a transmission mounted type International Harvester TC-143 transfer case (© International Harvester Co.)

1. Input shaft end nut
2. Washer
3. Seal
4. Flange bolt
5. Frame-mounted type case
6. Gasket
7. Shift cover
8. Bearing snap ring
9. Ball bearing
10. Long spacer
11. Upper sprocket (Old style)
12. Chain
13. Input shaft
14. Short spacer
15. Dowel ring
16. Cover
17. Flange nut
18. Roller bearing
19. Drain plug
20. Thrust washer
21. Outout shaft
22. Lower sprocket
23. Thrust washer
24. Sliding clutch
25. Ball bearing
26. Seal
27. Washer
28. Output shaft end nut
29. Shift shoe
30. Assembly shifter
31. Spring stud
32. Shift spring
33. Shift clevis
34. Clevis pin
35. Shift cover gasket
36. Upper sprocket (new style)

Exploded view of a frame mounted type International Harvester TC-143 transfer case (© International Harvester Co.)

4. With the rear output shaft flange clamped in a soft jawed vise, remove the flange retaining nut. Remove the flange from the vise and remove the flange from the rear output shaft. Use a puller, if necessary.

5. Turn the case so it rests on the flanges and remove the bolts securing the two halves of the case. Lift the top (rear) half of the case from the assembly and discard the gasket.

6. If present, remove the short spacer from the rear side of the input shaft. Also, if the thrust washer did not stay with the cover when removed, remove it now from the front output shaft.

7. With the case again secured in a soft jawed vise, place your thumbs on the ends of the shafts with your fingers under the sprockets. Pull the sprockets together with the chain off the shafts and out of the case as an assembly.

8. Unhook and remove the shift spring.

9. Remove the shift assembly mounting bolt and spring stud from the case.

10. Pull the shift cranks from the bosses inside the case.

11. If present, remove the long spacer from the input shaft and the washer from the front output shaft.

12. Lift the sliding clutch and its shift shoe from the front output shaft.

13. The two shafts are removed from the case with a press or by tapping them out with a soft hammer.

NOTE: If the transfer case is a transmission mounted unit, a snap ring on the input shaft must be removed before the shaft can be removed from the case.

14. After the shafts are removed, remove the oil seals and the bearing snap rings, and press or tap out the two ball bearings.

15. Pry and remove the thrust washer from the boss for the front output shaft roller bearing in the other half of the case.

16. Press the roller bearing cage from the inside and out of the cover.

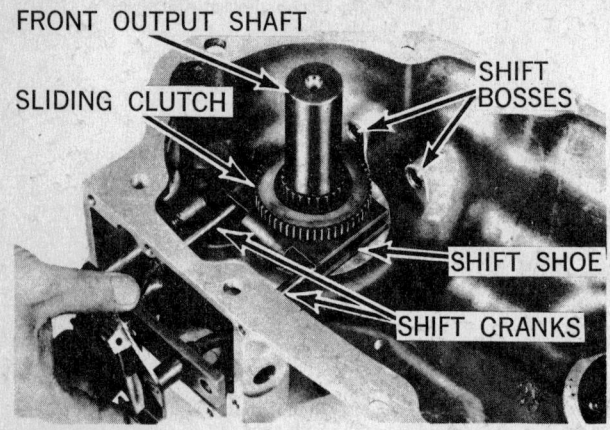

Removing the shifter mechanism from the TC-143 transfer case (© International Harvester Co.)

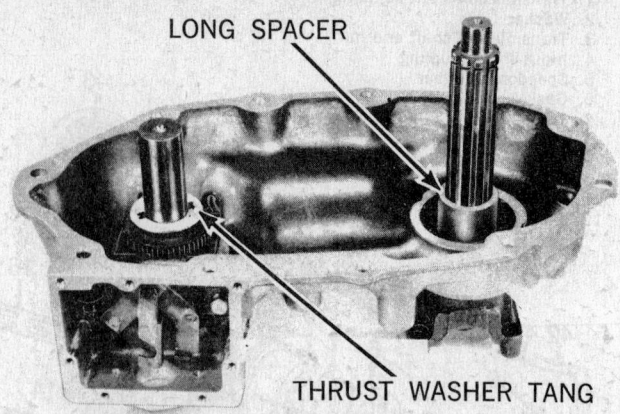

Installation of the thrust washer and long spacer, if so equipped, on the TC-143 transfer case (© International Harvester Co.)

Removing the shifter cover on a TC-143 transfer case (© International Harvester Co.)

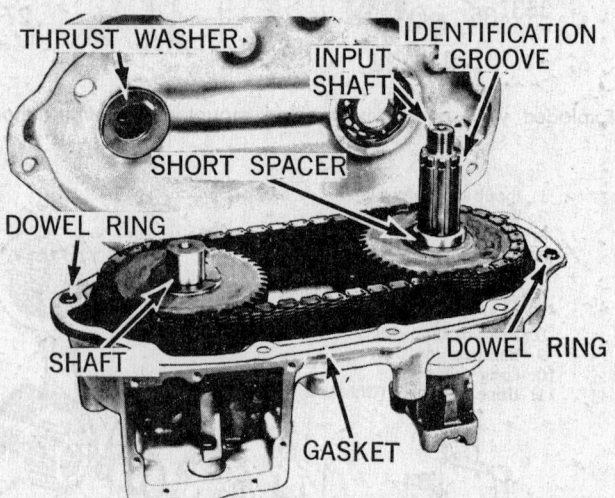

Removing the rear half of the case (cover) from the TC-143 transfer case (© International Harvester Co.)

17. Press the ball bearing on the outer race from the outside and out of the cover.

Cleaning and Inspection

1. Clean all parts in solvent, removing all traces of old gaskets, sealants and lubricants, and dry the parts with compressed air.
2. Examine all the ball and roller bearings for wear or damage and replace as necessary.
3. Inspect the sprocket teeth and bores for damage and wear. Check the internal splines and clutch teeth for chipped surfaces. Small nicks or burrs can be removed with a file.
4. Check the smooth and splined surfaces of the shafts for wear or damage. The sliding clutch must move freely on the output shaft, but excessive clearance is remedied by replacement of parts.
5. Examine the chain for bent or broken lines. If either condition exists, replace the chain.

ASSEMBLY

1. Install the two snap rings in the outer grooves of both bearing bores in the front half of the case.
2. Coat both bearing bores with lubricant and press or tap with a soft hammer both ball bearing assemblies into the case.
3. Install the two snap rings in the inner grooves of the two bearing bores in the front case half.
4. Install the shaft oil seals from the outside of the front case half. They are best installed with a press.
5. On a frame mounted transfer case, position the lightly lubricated shafts in the bearings, then, pull the shafts into the bearings by tightening the flange attaching nuts with the flanges installed. The input shaft is installed with the identification groove toward the rear of the transfer case. Tighten the flange attaching nut on the front output shaft to 200–250 ft. lbs., and the flange attaching nut on the input shaft until it bottoms, then back off two turns and retighten to 140–150 ft. lbs.
6. The output shaft in transmission mounted transfer case is installed in the same manner as outlined for frame mounted units in Step 5.
7. The input shaft in transmission mounted transfer cases has a snap-ring to be installed on the front end of the shaft prior to installation in the bearing. The shaft is then pressed or tapped into the bearing with a soft hammer until the snap-ring is bottomed against the bearing.
8. Assemble the shift shoe to the sliding clutch and install the sliding clutch to the front output shaft.
9. Insert the shifter assembly into the transfer case so the shift cranks of the shifter pass through the shift shoe before being guided into the shift bosses. Make sure the shifter operates the sliding clutch and then secure the assembly with the bolt and spring stud.

NOTE: The flange bolt must be in-

stalled before the spring stud because the position of the spring stud when installed prevents the installation of the flange bolts.

10. Install the thrust washer to the front output shaft and the long spacer to the input shaft, if so equipped with a long spacer.

NOTE: Be sure the thrust washer tangs fit down into the splines on the shaft.

11. Lightly lubricate the bores of both sprockets and secure the case in a soft jawed vise by the end of the input shaft.
12. Install the upper sprocket to the outer end of the input shaft. Do not slide the sprocket completely into the case.
13. Position the chain over the sprocket and place the lower sprocket inside the chain.
14. Pull down on the lower sprocket enough to slide the lower sprocket onto the output shaft.
15. Slide both sprockets and chain onto the sprockets as far as they will go.
16. Install the short spacer on the input shaft (not required on later models).
17. Install the shift spring between the spring post and the shifter assembly.
18. Press the ball bearing into the input shaft bore of the rear cover from the inside surface.
19. Press the roller bearing into the output shaft bore of the rear cover from the outside surface until it is flush with the bearing bore.
20. Press the seal for the rear of the input shaft (rear output) into the cover until it is flush.
21. Place the cover on a bench, inside facing up. Coat the back side of the thrust washer with a thin coat of sealant and position it on the roller bearing boss with the tang on the washer mated with the oil passage in the boss.

NOTE: Use the sealant sparingly and make sure none of it enters the bearing or blocks the two oil passages.

22. Position a new gasket and two dowel rings on the mating surface of the case.
23. Position the cover onto the case, guiding the two shafts into their respective bearings. Secure the cover to the case with the attaching nuts and bolts, tightening them to 29–38 ft. lbs.
24. Install the indicator light switch to the case and check its operation with a test light.
25. Install the shift cover gasket and cover on the case and secure it with bolts and lockwashers tightened to 4–6 ft. lbs. Do not overtighten.
26. Install the rear output flange on the rear of the input shaft so both the input flange and the output flange are on the same plane. The flanges must be assembled in this manner to prevent vibration.
27. Install the washer and a nylon insert locknut on the shaft tightened to 140–150 ft. lbs. to secure the flange.

New Process Model 203

The New Process Model 203 transfer case is a full-time 4WD unit that operates in 4WD at all times. The unit incorporates a differential similar to axle differentials; compensating for different speeds of the front and rear axles resulting from varying speeds while turning and operating over different surfaces.

There are five shift positions with this transfer case; Neutral, High and High Lock, and Low and Low Lock. The Lock positions are used under low traction conditions. In the Lock position, the differential action of the transfer case is eliminated, by locking the front and rear output shafts together. In this mode, neither the front or rear axle can rotate independently of the other.

DISASSEMBLY

1. Loosen rear output shaft flange retaining nut and remove front output shaft flange and washer.
2. Tap the front output shaft dust seal away from case assembly. Remove front output shaft bearing retainer and gasket.
3. Position transfer case assembly on blocks with input shaft facing downward.
4. Remove rear output shaft assembly from transfer case. Slide the differential carrier off the shaft.
5. Place a 1½ in. to 2 in. band type hose clamp on input shaft to retain bearings.
6. Lift shift rail and driveout pin retaining shift fork.
7. Remove shift rail poppet ball plug, gasket and ball from case. Use a magnet to remove poppet ball.
8. Push shift rail down, lift up on lockout clutch and remove shift fork from clutch assembly.
9. Remove front output shaft rear bearing retainer. It may be necessary to gently tap front of shaft or cautiously pry retainer from case. Make certain that no roller bearings are lost from rear cover.
10. When necessary, remove rear bearing by pressing.
11. Pry front output shaft front bearing from lower side of case.
12. Remove front output shaft assembly from case.
13. Lift intermediate housing from range box, after removing bolts.
14. Remove chain from intermediate housing.
15. Remove lockout clutch, drive gear and input shaft from range box.
16. Install a 1½ to 2″ band type hose clamp on end of input shaft to retain roller bearings.
17. Pull up on shift rail and remove rail from link.
18. Lift input shaft assembly from range box.

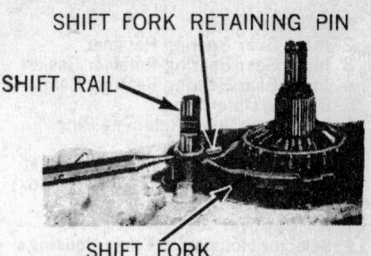

Removing the pin retaining the shift fork to the shift rail in a New Process 203 transfer case (© Ford Motor Co.)

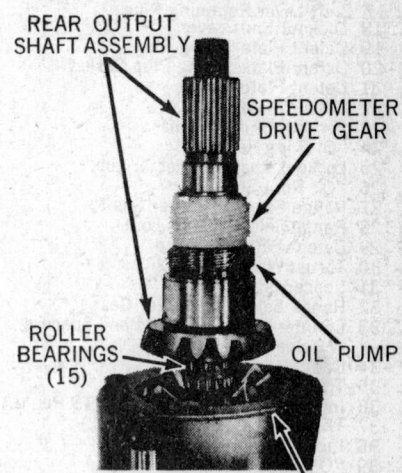

Removing the rear output shaft. The differential carrier assembly is removed next—New Process 203 transfer case (© Ford Motor Co.)

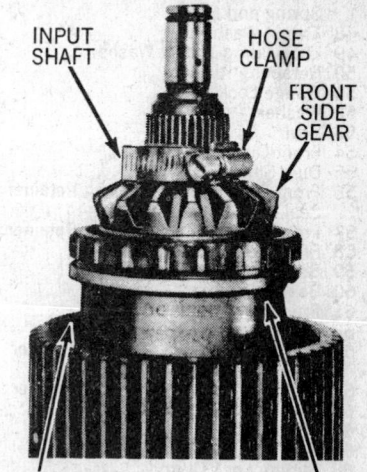

Place a hose clamp around the end of the input shaft to prevent losing the 123 roller bearings out of the clutch assembly—New Process 203 transfer case (© Ford Motor Co.)

ASSEMBLY

1. Position range box with input gear side down, on wood blocks.

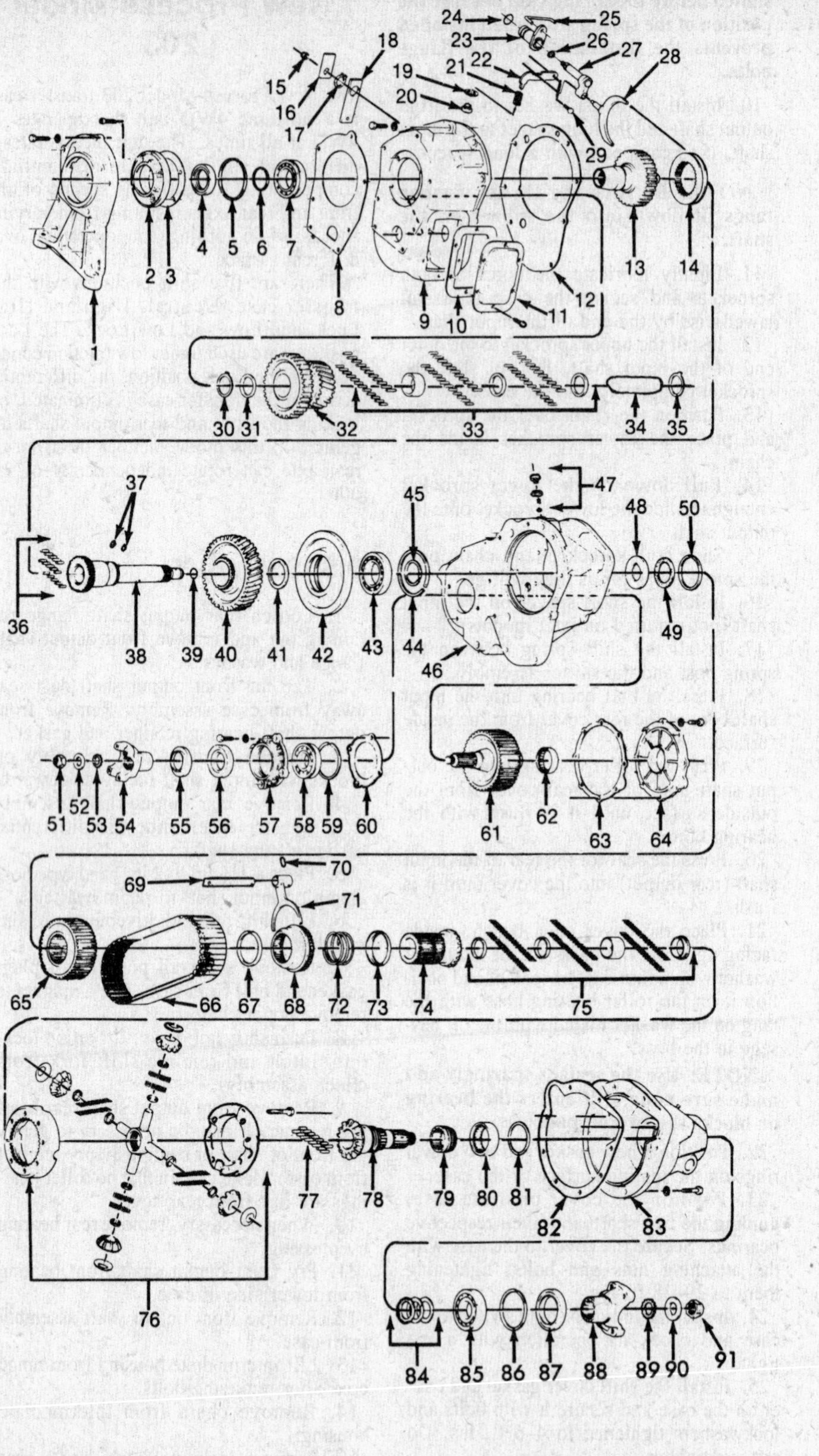

1 Adapter
2 Input Gear Bearing Retainer
3 Input Gear Bearing Retainer Gasket
4 Input Gear Bearing Retainer Seals
5 Bearing Outer Ring
6 Bearing to Shaft Retaining Ring
7 Input Gear Bearing
8 Adapter to Selector Housing Gasket
9 Range Selector Housing (Range Box)
10 P.T.O. Cover Gasket
11 P.T.O. Cover
12 Selector Housing to Chain Housing Gasket
13 Main Drive Input Gear
14 Range Selector Sliding Clutch
15 Shift Lever Lock Nut
16 Range Selector Shift Lever
17 Shift Lever Retaining Ring
18 Lockout Shift Lever
19 Detent Plate Spring Plug
20 Detent Plate Spring Plug Gasket
21 Detent Plate Spring
22 Detent Plate
23 Lockout Shifter Shaft
24 "O" Ring Seal
25 Lockout Shaft Connector Link
26 "O" Ring Seal
27 Range Selector Shifter Shaft
28 Range Selector Shift Fork
29 Detent Plate Pivot Pin
30 Thrust Washer
31 Spacer (short)
32 Range Selector Counter Gear
33 Counter Roller Bearings and Spacers (72 Bearings Req'd.)
34 Countergear Shaft
35 Thrust Washer
36 Input Shaft Roller Bearings (15 Req'd.)
37 Thrust Washer Pins (2 Req'd.)
38 Input Shaft
39 "O" Ring Seal
40 Low Speed and Bushing
41 Thrust Washer
42 Input Shaft Bearing Retainer
43 Input Shaft Bearing
44 Input Shaft Bearing Retaining Ring (Large)
45 Input Shaft Bearing Retaining Ring
46 Chain Drive Housing
47 Lockout Shift Rail Poppet Plug, Gasket, Spring and Ball
48 Thrust Washer
49 Lubricating Thrust Washer
50 Retaining Ring
51 Flange Lock Nut
52 Washer
53 Seal
54 Front Output Yoke
55 Dust Shield
56 Front Output Shaft Bearing Retainer Seal
57 Front Output Shaft Bearing Retainer
58 Front Output Shaft Bearing
59 Bearing Outer Ring
60 Bearing Retainer Gasket
61 Front Output Shaft
62 Front Output Shaft Rear Bearing
63 Front Output Rear Bearing Retainer Cover Gasket
64 Front Output Rear Bearing Retainer
65 Drive Shaft Sprocket
66 Drive Chain
67 Retaining Ring
68 Sliding Lock Clutch
69 Lockout Shift Rail
70 Shift Fork Retaining Pin
71 Lockout Shift Fork
72 Lockout Clutch Spring
73 Spring Washer Cup
74 Front Side Gear
75 Front Side Gear Bearing and Spaces (123 Bearings Req'd.)
76 Differential Carrier Assembly (132 Bearings Req'd.)
77 Rear Output Shaft Roller Bearings (15 Req'd.)

78 Rear Output Shaft
79 Speedometer Drive Gear
80 Rear Output Shaft Front Roller Bearing
81 Oil Pump "O" Ring Seal
82 Rear Output Housing Gasket
83 Rear Output Housing
84 Shim Pack

85 Rear Output Rear Bearing
86 Bearing Retainer
87 Rear Output Shaft Seal
88 Rear Output Flange
89 Rear Output Shaft Rubber Seal
90 Washer
91 Flange Nut

Exploded view of a New Process 203 full time transfer case (© General Motors Corp.)

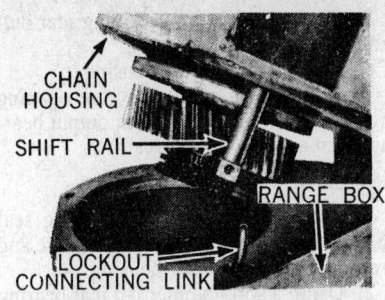

Disengaging the shift rail from the locknut connecting link in a New Process 203 transfer case (© Ford Motor Co.)

2. Place gasket on input housing.

3. Install lockout clutch and drive sprocket on input shaft assembly. Install a 2″ band type hose clamp on end of input shaft to prevent loss of bearings during installation.

4. Place input shaft, lockout clutch and drive sprocket in range box. Align tab on bearing retainer with notch in gasket.

5. Engage lockout clutch shift rail to the connector link. Position rail in housing bore and turn shifter shaft lowering rail into the housing. This will prevent the link and rail from becoming disconnected.

6. Place the drive chain in housing with the chain around the outer wall.

7. Secure the chain housing to the range box. Be sure that the shift rail engages the channel of the housing. Place the chain on the input drive sprocket.

8. Place the front output sprocket in transfer case. Turn the clutch drive gear to assist in positioning chain on sprocket.

9. Position the shift fork and rail on the clutch assembly. Install the clutch assembly completely into the drive sprocket. Insert retaining pin in shift fork and rail.

10. Install front output bearing, gasket, retainer, bolts, flange, gasket, seal, washer and retaining nut.

11. If rear bearing was removed from front output shaft, press a new bearing into the outside face of cover until bearing is flush with opening.

12. Install front output shaft, rear bearing, retainer, gasket and bolts.

13. Slip differential carrier assembly on the input shaft. Bolts on carrier must face rear of shaft.

14. Load bearings in pinion shaft, install rear output housing assembly, gasket and bolts.

15. Install a dial indicator on the rear housing. The indicator must contact the end of the output shaft. While holding the rear flange, rotate the front output shaft and find the highest point of gear hop. Reset indicator and with rear output shaft at high point, pull up on the end of the shaft to determine end-play. Remove indicator and install shim pack to control end-play to between 0 and .005″. The shim pack is positioned on the shaft in front of the rear bearing. Check for binding of rear output shaft.

SPECIFICATIONS
TORQUE (FT. LBS.)

Adapter to transfer case bolts	38
Adapter to transmission bolts	40
Transfer case to frame nuts (upper)	50
Transfer case to frame nuts (lower)	65
Shift lever attaching nuts	25
Shift lever rod swivel locknuts	50
Shift lever locking arm nut	150 in. lbs.
Skid plate bolt retaining nuts	45
Crossmember bolt retaining nuts	45
Adapter mount bolts	25
Intermediate case to range box bolts	30
Front output bearing retainer bolts	30
Output shaft yoke nuts	150
Front output rear bearing retainer bolts	30
Differential assembly screws	45
Rear output shaft housing	30
Poppet ball retainer nut	15
PTO cover bolts	15
Front input bearing retainer bolts	20
Filler plug	25

16. Insert lockout clutch shift rail poppet ball, spring and screw plug in transfer case.

17. Install poppet plate spring, gasket and plug, if they were not previously installed.

18. Install shift levers on the range box, if these were not left on vehicle.

19. Torque all bolts, locknuts and plugs to specifications.

20. Fill transfer case with specified lubricant until the proper level is reached. Secure filler plug.

SUBASSEMBLIES OVERHAUL

Lockout Clutch
DISASSEMBLY

1. Remove front side gear from input shaft assembly.

2. Remove thrust washer, roller bearings and spacers from front side gear bore. The position of the spacers must be noted.

3. Remove the snap-ring which holds the drive sprocket to clutch assembly. Slip the drive sprocket from the front side gear.

4. Remove the lower snap-ring.

5. Remove sliding gear, spring and spring cup washer from the front side gear.

6. Thoroughly clean and inspect all component parts. Replace any component that is worn or defective.

ASSEMBLY

1. Place spring cup washer, spring and sliding clutch gear on front side gear.

2. Secure sliding clutch to front side gear with a snap-ring.

3. Spread petroleum jelly on front side gear and install roller bearings and spacers.

4. Place thrust washer in gear end of front side gear.

5. Slide drive sprocket on clutch splines and secure with snap-ring.

Differential Carrier
DISASSEMBLY

1. Separate differential carrier sections and lift out pinion gear and spider assembly.

2. Note that undercut side of pinion gear spider faces toward front of side gear.

3. Remove pinion thrust washers, pinion roller washer gears and roller bearings from spider unit.

4. Thoroughly clean and inspect all component parts. Replace any component that is worn or damaged.

ASSEMBLY

1. Spread petroleum jelly on pinion gears and install roller bearings.

2. Position on the leg of each spider, pinion roller washer, pinion gear and thrust washer.

3. Position the spider assembly in front half of the carrier. The undercut surface of the spider thrust surface face downward or toward teeth.

4. Secure carrier halfs together. Make certain the marks are aligned. Torque all bolts to specifications.

Input Shaft
DISASSEMBLY

1. Remove thrust washer and spacer from shaft.

2. Remove bearing retainer assembly from input shaft.

3. Hold low speed gear and lightly tap shaft from gear. Note the position of the thrust washer pins in input shaft.

4. Remove snap-ring holding input bearing in retainer using a screw driver. Lightly tap rear bearing out of retainer.

5. Remove pilot roller bearing and O-ring from end of input shaft.

6. Thoroughly clean and inspect all component parts. Replace any component that is worn or damaged.

ASSEMBLY

1. Tap or press input bearing into retainer. Be sure that ball loading slots are toward concave side of retainer. Install securing snap ring. Make certain that selective

snap-ring of proper thickness is used to provide tightest fit.

2. Position low speed gear on shaft, clutch end facing gear end of input shaft.

3. Place thrust washers on input shaft, align slot in washer with pin in shaft. Slide or tap washers into position.

4. Place input bearing retainer on shaft and secure with snap-ring. Snap-rings are selective. Use snap-ring that provides tightest fit.

5. Slip spacer and thrust washer on shaft and align with locating pin.

6. Spread heavy grease on end of shaft and install roller bearings.

7. Install rubber O-ring at end of shaft.

Range Box

DISASSEMBLY

1. Remove poppet plate spring, plug and gasket.

2. Remove clutch fork and sliding gear by disengaging sliding clutch gear from input gear.

3. Remove upper shift lever from shifter shaft.

4. Remove snap-ring and lower shift lever.

5. Push shifter shaft assembly down and remove lockout clutch connector link. The long end of connector link engages poppet plate.

6. Remove shifter shaft assembly and separate shafts. Remove O-rings.

7. When necessary to remove poppet plate, drive pivot shaft out and remove plate and spring from bottom of case.

8. Remove input gear bearing retainer and seal assembly. Release snap-ring from retainer and tap bearing out of assembly.

9. Release snap-ring holding input shaft bearing to shaft and remove bearing.

10. Remove countershaft from cluster gear and case assembly from intermediate case side. Remove cluster gear assembly from range box.

11. Remove cluster gear thrust washers from case.

12. Thoroughly clean and inspect all component parts. Replace any component that is worn or damaged.

ASSEMBLY

1. Spread heavy grease in cluster bore and using proper tool install roller bearings and spacers.

2. Spread heavy grease on case and install thrust washers. Engage tab on thrust washers with slot in case.

3. Place cluster gear assembly in case and install countershaft through front of range box and into gear assembly. Flat face of countershaft must be aligned with case gasket.

4. Place bearing on input gear shaft with snap-ring groove facing out, install a new retaining ring. Insert input gear and bearing in housing. The retaining ring used in this operation is a select fit. Use ring that provides the tightest fit.

5. Secure input gear and bearing with a snap-ring.

6. Match up oil slot in retainer with drain hole in case and insert input gear bearing retainer and gaskets. Install bolts and torque to specifications.

7. Spread sealant on pin and install poppet pin and pivot pin in housing.

8. Lubricate and install new O-rings on inner and outer shifter shafts.

9. Insert shifter shafts in housing and engage long end of lockout clutch connector link with outer shifter shaft. Complete this operation before assembly bottoms out.

10. Install lower shift lever and retaining ring.

11. Install upper shift lever and shaft retaining nut.

12. Install shift fork and sliding clutch gear. Push fork up into shifter shaft and engage poppet plate. Move sliding clutch gear onto input shaft gear.

13. Insert poppet plate spring, gasket and plug in housing top. Make certain that spring engages poppet plate.

Input Gear Bearing

REPLACEMENT

1. Remove bearing retainer and gasket from housing.

2. Remove and discard snap-ring holding bearing in retainer.

3. Pry bearing from case and remove from shaft.

4. Inspect input gear and bearing retainer for damage or wear. Replace if necessary.

5. Place new bearing and snap-ring on input gear. Using a soft hammer, tap bearing into position. Secure snap-ring.

6. Insert bearing retainer into housing and secure with attaching bolts. Tighten bolts to specifications.

Input Gear Seal

REPLACEMENT

1. Remove bearing retainer from housing.

2. Remove seal from retainer by prying.

3. Place new seal on retainer and install with proper seal driver.

4. Install bearing retainer in housing and secure with attaching bolts. Tighten bolts to specifications.

Rear Output Shaft Housing

DISASSEMBLY

1. Remove speedometer driven gear from housing.

2. Remove rear output flange and washer, if they have not been removed previously.

3. Using a soft hammer tap on flange end of pinion and remove the pinion. If speedometer drive gear does not come off with pinion reach into case and remove.

4. Remove old seal from bore with suitable prying tool.

5. Remove snap ring retaining rear output rear bearing.

6. Tap bearing out of housing.

7. Install a long drift into rear opening of housing and drive out front output bearing. Remove seal and discard.

ASSEMBLY

1. Spread grease on front bearing seal and place in bore. Place bearing in bore and press until it bottoms in housing.

2. Using a soft hammer tap rear bearing into place. Secure with proper snap-ring. Snap-rings are selective, use the one that provides the tightest fit.

3. Place rear seal in bore and drive into position with suitable tool. When seal is in position it should be approximately $1/8$" to $3/16$" below housing face.

4. Place speedometer drive gear on output shaft with shims of approximately .050" thickness. Insert output shaft into carrier through housing front opening.

5. Install flange and washer on output shaft. Leave retaining nut loose until shim requirements are known.

6. Install speedometer driven gear.

Front Output Shaft Bearing Seal

REPLACEMENT

1. Remove old seal from retainer bore.

2. Inspect and clean retainer.

3. Spread sealer on outer edge of new seal.

4. Place new seal in retainer bore and drive into position with proper tool.

Front Output Bearing

REPLACEMENT

1. Remove rear cover from case assembly and discard gasket.

2. Press old bearing from cover.

3. Place new bearing on outside face of cover. Cover bearing with a wood block and press into cover until bearing is flush with opening.

4. Place gasket on transfer case and tap cover into position using a soft hammer. Secure cover with attaching bolts and tighten to specifications.

New Process Model 205

The New Process Model 205 transfer case is a two-speed gearbox mounted between the main transmission and the rear axle. The gearbox transmits power from the transmission and engine to the front and rear driving axles.

DISASSEMBLY

Transfer Case

1. Clean the exterior of the case.

2. Remove the nuts from the universal joint flanges.

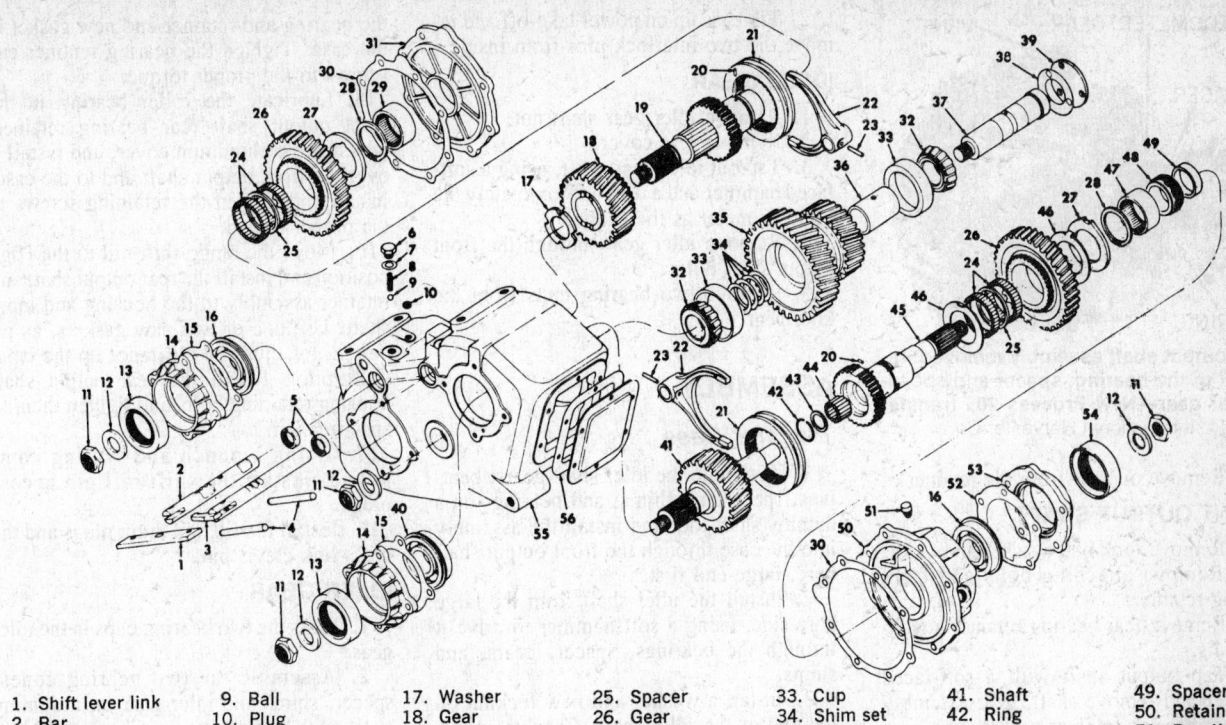

1. Shift lever link	9. Ball	17. Washer	25. Spacer	33. Cup	41. Shaft	49. Spacer
2. Bar	10. Plug	18. Gear	26. Gear	34. Shim set	42. Ring	50. Retainer
3. Bar	11. Nut	19. Shaft	27. Washer	35. Gear	43. Washer	51. Breather
4. Plunger	12. Washer	20. Pin	28. Ring	36. Spacer	44. Bearing	52. Gasket
5. Seal	13. Seal	21. Clutch	29. Bearing	37. Shaft	45. Gear	53. Retainer
6. Screw	14. Retainer	22. Fork	30. Gasket	38. Gasket	46. Washer	54. Seal
7. Gasket	15. Gasket	23. Pin	31. Retainer	39. Cover	47. Bearing	55. Case
8. Spring	16. Bearing	24. Bearing	32. Cone	40. Bearing	48. Gear	56. Gasket

Exploded view of a New Process 205 transfer case (© International Harvester Co.)

3. Remove the front output shaft rear bearing retainer, front bearing retainer and drive flange.

4. Tap the front output shaft assembly from the case with a soft hammer. Remove the sliding clutch, front output high gear, washer and bearing from the case.

5. Remove the rear output shaft housing attaching bolts and remove the housing, output shaft, bearing retainer and speedometer gear.

6. Slide the rear output shaft from the housing.

NOTE: Be careful not to lose the 15 needle bearings that will be loose when the rear output shaft is removed.

7. Drive the two 1/4 in. shift rail pin access hole plugs into the transfer case with a punch and hammer.

8. Remove the two shift rail detent nuts and springs from the case. Use a magnet to remove the two detent balls.

9. Position both shift rails in neutral and remove the shift fork retaining roll pins with a long punch.

10. Remove the clevis pin from one shift rail and rail link.

11. Remove the range shift rail first, then the 4WD shift rail.

12. Remove the shift forks and and sliding clutch from the case. Remove the input shaft bearing retainer, bearing and shaft.

13. Remove the cup plugs and rail pins, if they were driven out, from the case.

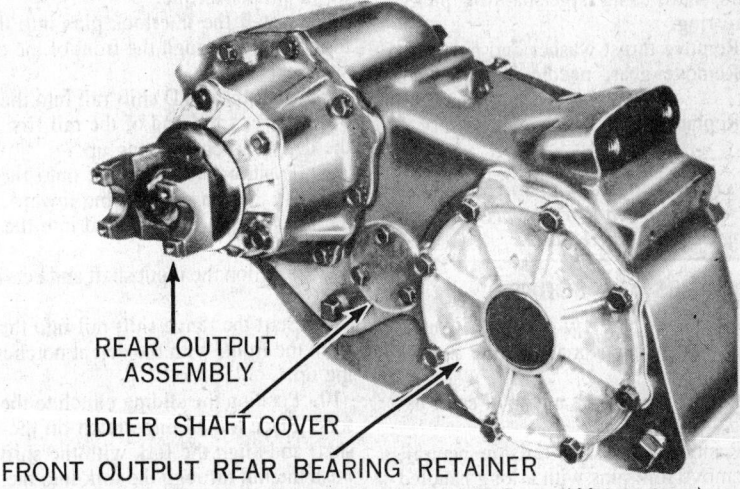

Rear view of a New Process 205 transfer case (© General Motors Corp.)

14. Remove the locknut from the idler gear shaft.

15. Remove the idler gear shaft rear cover.

16. Remove the idler gear shaft, using a soft hammer and a drift.

17. Roll the idler gear assembly to the front output shaft hole and remove the assembly from the case.

REAR OUTPUT SHAFT AND YOKE

1. Loosen rear output shaft yoke nut.

2. Remove shaft housing bolts, then remove the housing and retainer assembly.

3. Remove retaining nut and yoke from the shaft, then remove the shaft assembly.

4. Remove and discard snap-ring.

5. Remove thrust washer and pin.

6. Remove tanged bronze washer. Remove gear needle bearings, spacer and second row of needle bearings.

7. Remove tanged bronze thrust washer.

8. Remove pilot rollers, retainer ring and washer.

9. Remove oil seal retainer, ball bearing, speedometer gear and spacer. Discard gaskets.

10. Press out bearing.

SPEEDOMETER GEAR — VENT
SPACER
BEARING

Rear output shaft assembly removed: removal of the bearing, spacer and speedometer gear—New Process 205 transfer case (© International Harvester Co.)

11. Remove oil seal from the retainer.

FRONT OUTPUT SHAFT

1. Remove lock nut, washer and yoke.
2. Remove attaching bolts and front bearing retainer.
3. Remove rear bearing retainer attaching bolts.
4. Tap output shaft with a soft-faced hammer and remove shaft, gear assembly and rear bearing retainer.
5. Remove sliding clutch, gear, washer and bearing from output high gear.
6. Remove sliding clutch from the high output gear; then remove gear, washer and bearing.
7. Remove gear retaining snap-ring from the shaft, using large snap-ring picks. Discard ring.
8. Remove thrust washer and pin.
9. Remove gear, needle bearings and spacer.
10. Replace rear bearing, if necessary.

———— CAUTION ————
Always replace the bearing and retainer as an assembly. Do not try to press a new bearing into an old retainer.

SHIFT RAILS AND FORKS

1. Remove the two poppet nuts, springs, and using a magnet, the poppet balls.
2. Remove cup plugs on top of case, using a ¼" punch.
3. Position both shift rails in neutral, then remove fork pins with a long handled screw extractor.
4. Remove clevis pins and shift rail link.
5. Lower shift rails; upper rail first and then lower.
6. Remove shift forks and sliding clutch.
7. Remove the front output high gear, washer and bearing. Remove the shift rail cup plugs.

INPUT SHAFT

1. Remove snap-ring in front of bearing. Tap shaft out rear of case and bearing out front of case, using a soft-faced hammer or mallet.

2. Tilt case up on power take-off and remove the two interlock pins from inside.

IDLER GEAR

1. Remove idler gear shaft nut.
2. Remove rear cover.
3. Tap out idler gear shaft, using a soft-faced hammer and a drift approximately the same diameter as the shaft.
4. Remove idler gear through the front output shaft hole.
5. Remove two bearing cups from the idler gear.

ASSEMBLY

Transfer Case

1. Assemble the idler shaft gears, bearings, spacer and shims, and bearings on a dummy shaft tool and install the assembly into the case through the front output shaft bore, large end first.
2. Install the idler shaft from the large bore side, using a soft hammer to drive it through the bearings, spacer, gears, and shims.
3. Install a washer and new locknut on the end of the idler shaft. Check to make sure the idler gear rotates freely. Tighten the locknut to specification.
4. Install the idler shaft cover with a new gasket so the flat side faces the rear bearing retainer of the front output shaft. Install and tighten the two retaining screws to the proper torque.
5. Install the interlock pins into the interlock bore through the front of the output shaft opening.
6. Start the 4WD shift rail into the front of the case, solid end of the rail first, with the detent notches facing up.
7. Position the shift fork onto the shift rail with the long end facing inward. Push the rail through the fork and into the Neutral position.
8. Position the input shaft and bearing in the case.
9. Start the range shift rail into the case from the front, with the detent notches facing up.
10. Position the sliding clutch to the shift fork. Place the sliding clutch on the input shaft and align the fork with the shift rail. Push the rail through the fork into the Neutral position.
11. Install the roll pins that lock the shift forks to the shift rails with a long punch.
12. Position the front wheel drive high gear and its thrust washer in the case. Position the sliding clutch in the shift fork. Shift the rail and fork into the front wheel drive (4WD-Hi) position, while at the same time, meshing the clutch with the mating teeth on the front wheel drive high gear.
13. Align the thrust washer, high gear and sliding clutch with the bearing bore in the case and insert the front output shaft and low gear into the high gear assembly.
14. Install a new seal in the front bearing retainer of the front output shaft, and install

the bearing and retainer and new gasket in the case. Tighten the bearing retainer cap screws to the proper torque.
15. Lubricate the roller bearing in the front output shaft rear bearing retainer, which is the aluminum cover, and install it over the front output shaft and to the case. Install and tighten the retaining screws to the proper torque.
16. Move the range shift rail to the High position and install the rear output shaft and retainer assembly to the housing and input shaft. Use one or two new gaskets, as required, to adjust the clearance on the input shaft pilot. Install the rear output shaft housing retaining bolts and tighten them to specification.
17. Using a punch and sealing compound, install the shift rail pin access plugs.
18. Install the fill and drain plugs and the cross-link clevis pin.

IDLER GEAR

1. Press the two bearing cups in the idler gear.
2. Assemble the two bearing cones, spacer, shims and idler gear on a dummy shaft, with bore facing up. Check end-play.
3. Install idler gear assembly (with dummy shaft) into the case, large end first, through the front output shaft bore.
4. Install idler shaft from large bore side, driving it through with a soft-faced hammer or mallet.

SPECIFICATIONS

END PLAY (IN.)

Idler gear	0.000–0.002
Rear output shaft	0.002–0.027

TORQUE LIMITS (FT. LBS.)

Idler shaft locknut	150
Idler shaft cover	20
Front output shaft front bearing retainer	30–35
Front output shaft yoke locknut	130–150
Rear output shaft bearing retainer and housing	30–35
Rear output shaft yoke locknut	130–150
P.T.O. cover	15
Front output shaft rear bearing retainer	30–35
Filler and drain plugs	30
Case to frame	130
Case to adapter	25
Adapter mount	75
Case bracket to frame	
Upper	30
Lower	65
Adapter to transmission	
Manual transmission	30–35
Automatic transmission	30–35

5. Install washer and new locknut. Check for free rotation and measure end-play. Torque locknut to specifications.

6. Install idler shaft cover and new gasket. Torque cover bolts to specifications.

NOTE: Flat side of cover must be positioned towards front output shaft rear cover.

SHIFT RAILS AND FORKS

1. Press the two rail seals into the case.

NOTE: Install seals with metal lip outward.

2. Install interlock pins from inside case.

3. Insert slotted end of front output specifications.

CASE

1. Install power take-off cover and gasket. Torque attaching bolts to specifications.

2. Install cup plugs at rail pin holes.

NOTE: After installing, seal the cup plugs.

3. Install drain and filler plugs. Torque to specifications.

4. Install shift rail cross link, clevis pins and lock pins.

New Process Model 207

The 207 transfer case is an aluminum case, chain drive, four position unit providing four-wheel drive high and low ranges, a two-wheel high range, and a neutral position. It is a part-time four-wheel drive unit. Torque input in four-wheel high and low ranges is undifferentiated. The range positions on the 207 transfer case are selected by a floor mounted gearshift lever.

The 207 case is a two-piece aluminum case containing front and rear output shafts, two drive sprockets, a shift mechanism and a planetary gear assembly. The drive sprockets are connected and operated by the drive chain. The planetary assembly which consists of a three pinion carrier and an annulus gear provide the four-wheel drive low range when engaged.

DISASSEMBLY

1. Remove fill and drain plugs.

2. Remove front yoke. Discard yoke seal washer and yoke nut.

3. Turn transfer case on end and position front case on wood blocks.

4. Shift transfer case to 4 Lo.

5. Remove extension housing attaching bolts. Using a hammer, tap the shoulder on the extension housing to break sealer loose.

6. Remove the snap-ring for the rear bearing from the mainshaft and discard.

7. Remove the rear retainer attaching

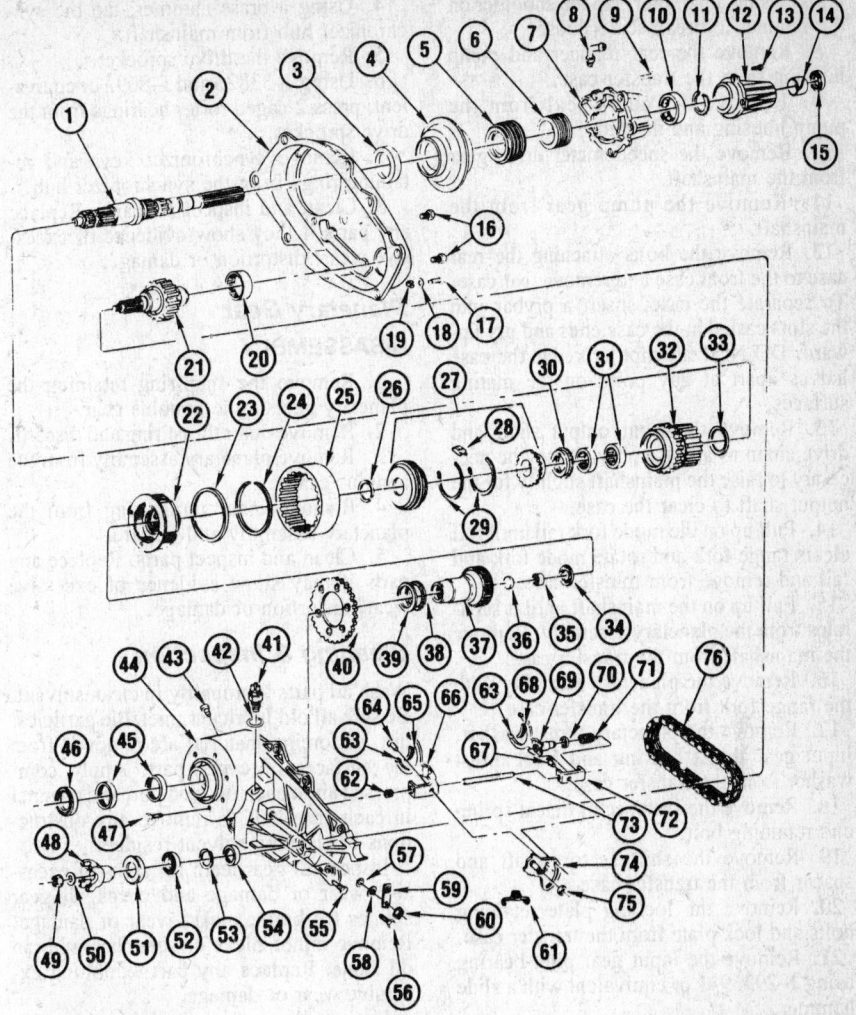

1. SHAFT, Main Drive
2. HOUSING, Case
3. SEAL, Oil Pump Hsg.
4. HOUSING, Oil Pump
5. PUMP, Oil
6. GEAR, Speedo Drive
7. RETAINER, Main Shf. Rr. Brg.
8. CONNECTOR, Case Vent
9. BOLT
10. BEARING, Main Shf. Rr.
11. RING, Main Shf. Rr. Brg. Ret.
12. EXTENSION, Main Shf.
13. BOLT, Hex
14. BUSHING, Case Main Shf. Ext.
15. SEAL, Main Shf. Ext.
16. PLUG, Case Oil
17. BOLT, Hex (M10 × 1.5 × 35)(2 req'd)
18. WASHER, Hsg. Alignment Dowel
19. DOWEL, Hsg. Alignment
20. BEARING, Frt. Otpt. Shf. Pilot
21. SHAFT, Frt. Otpt.
22. CARRIER ASM, Planet Gear
23. WASHER, Planet Gr. Carr. Ret. Rg. Thrust
24. RING, Planet Gr. Carr. Ret.
25. GEAR, Planet Gr. Carr. Annulus
26. RING, Main Dr. Shf. Syn.

27. SYNCHRONIZER ASM, Main Dr. Shf.
28. STRUT, Syn.
29. SPRING, Syn. Strut
30. RING, Syn. Stop
31. BEARING, Dr. Chain Sprocket
32. SPROCKET, Dr. Chain
33. WASHER, Dr. Chain Sprocket Thrust
34. WASHER, Input Main Dr. Gr. Thrust
35. BEARING, Input Dr. Gr. Pilot
36. PLUG, Cup
37. GEAR ASM, Input Main Dr.
38. BEARING, Input Dr. Gr. Thrust
39. WASHER, Input Dr. Gr. Thrust Brg.
40. PLATE, Low Range Lock
41. SWITCH, Four Whl. Dr. Ind. Light
42. SEAL, Four Whl. Dr. Ind. Light Switch
43. PLUG, Oil Access Hole
44. HOUSING, Case (Frt. Half)
45. BEARING, Input Dr.
46. SEAL, Input Dr. Gr.
47. BOLT, Hex
48. YOKE, Frt. Otpt. Prop. Shf.
49. NUT, Frt. Otpt. Prop. Shf. Yoke
50. WASHER, Frt. Otpt. Prop.

Shf. Yoke (Rubber)
51. DEFLECTOR, Frt. Otpt. Prop. Shf. Yoke
52. SEAL, Frt. Otpt. Shf.
53. RING, Frt. Otpt. Shf. Brg. Ret.
54. BEARING, Frt. Otpt. Shf.
55. SCREW, Shift Sector Spr.
56. SCREW
57. SEAL, Shift Sector & Shf. Oil
58. RETAINER, Shift Sector & Shf.
59. LEVER, Shifter Shf.
60. NUT, Shift Shf. Lvr.
61. SPRING ASM, Shift Sector
62. BUSHING, Range Fork
63. PAD, Fork End
64. PIN, Range Shift Fork
65. PAD, Range Shift Fork Center
66. FORK ASM, Range Shift
67. PIN, Mode Shift Fork Brkt.
68. PAD, Mode Shift Fork Center
69. FORK ASM, Mode Shift
70. CUP, Mode Shift Fork Spr.
71. SPRING, Mode Shift Fork
72. BRACKET ASM, Mode Shift Fork
73. SHAFT, Shift Fork
74. SECTOR, W/Shf., Shift
75. SPACER, Shift Sector Shf.
76. CHAIN, Drive

Exploded view new process 207

bolts. Using a hammer, tap the shoulder on the retainer to break sealer loose.

8. Remove the rear retainer and pump housing from the transfer case.

9. Remove the pump seal from the pump housing and discard.

10. Remove the speedometer drive gear from the mainshaft.

11. Remove the pump gear from the mainshaft.

12. Remove the bolts attaching the rear case to the front case and remove rear case. To separate the case, insert a prybar into the slots casted in the case ends and pry upward. DO NOT attempt to wedge the case halves apart at any point on the mating surfaces.

13. Remove the front output shaft and drive chain as an assembly. It may be necessary to raise the mainshaft slightly for the output shaft to clear the case.

14. Pull up on the mode fork rail until rail clears range fork and rotate mode fork and rail and remove from transfer case.

15. Pull up on the mainshaft until it separates from the planetary assembly. Remove the mainshaft from the transfer case.

16. Remove the planetary assembly with the range fork from the transfer case.

17. Remove the planetary thrust washer, input gear thrust bearing and front thrust washer from the transfer case.

18. Remove the shift sector detent spring and retaining bolt.

19. Remove the shift sector, shaft and spacer from the transfer case.

20. Remove the locking plate retaining bolts and lock plate from the transfer case.

21. Remove the input gear pilot bearing using J–29369–1 or equivalent with a slide hammer.

22. Remove the front output shaft seal, input shaft seal and the rear extension seal using a brass drift.

23. Using J–33841 with J–8092 or equivalent, press the 2 caged roller bearings for the front input shaft gear from the transfer case.

24. Using J–29369–2 with J–33367 or a slide hammer, remove the rear bearing for the front output shaft.

25. Using a hammer and drift, remove the rear mainshaft bearing from the rear retainer.

26. Using an awl, remove the snap-ring retaining the front output shaft bearing. Using a hammer and drift, remove the bearing from the case.

27. Remove the bushing from the extension housing using J–33839 with J–8092 or equivalent. Press bushing from the extension housing.

Mainshaft
DISASSEMBLY

1. Remove the speedometer gear.

2. Using an awl, pry off the pump gear from the mainshaft.

3. Remove the snap-ring retaining the synchronizer hub from the mainshaft.

4. Using a brass hammer, tap the synchronizer hub from mainshaft.

5. Remove the drive sprocket.

6. Using J–33826 and J–8092 or equivalent, press 2 caged roller bearings from the drive sprocket.

7. Remove synchronizer keys and retaining rings from the synchronizer hub.

8. Clean and inspect all parts. Replace any parts if they show evidence of excessive wear, distortion or damage.

Planetary Gear
DISASSEMBLY

1. Remove the snap-ring retaining the planetary gear in the annulus gear.

2. Remove outer thrust ring and discard.

3. Remove planetary assembly from the annulus gear.

4. Remove inner thrust ring from the planetary assembly and discard.

5. Clean and inspect parts. Replace any parts if they show evidence of excessive wear, distortion or damage.

Cleaning & Inspection

Wash all parts thoroughly in clean solvent. Be sure all old lubricant, metallic particles, dirt, or foreign material are removed from the surfaces of every part. Apply compressed air to each oil feed port and channel in each case half to remove any obstructions or cleaning solvent residue.

Inspect all gear teeth for signs of excessive wear or damage and check all gear splines for burrs, nicks, wear or damage. Remove minor nicks or scratches with an oil stone. Replace any part exhibiting excessive wear or damage.

Inspect all snap-rings and thrust washers for evidence of excessive wear, distortion or damage. Replace any of these parts if they exhibit these conditions.

Inspect the two case halves for cracks, porosity damaged mating surfaces, stripped bolt threads, or distortion. Replace any part that exhibits these conditions. Inspect the low range lock plate in the front case. If the lock plate teeth or the plate hub is cracked, broken, chipped, or excessively worn, replace the lock plate and the lock plate attaching bolts.

Inspect the condition of all needle, roller and thrust bearings in the front and rear case halves and the input gear. Also, check the condition of the bearing bores in both cases and in the input gear, rear output shaft and rear retainer. Replace any part that exhibits signs of excessive wear or damage.

Planetary Gear
ASSEMBLY

1. Install the inner thrust ring on planetary assembly.

2. Install the planetary assembly into the annulus gear.

3. Install the outer thrust ring and then the snap-ring.

Mainshaft
ASSEMBLY

1. Using J–33828 and J–8092 or equivalent, install the front drive sprocket bearing. Press bearing until tool bottoms out. Bearing should be flush with front surface. Reverse tool on J–8092 or equivalent and press rear bearing into sprocket until tool bottoms out. The rear bearing should be recessed after installation.

2. Install thrust washer on the mainshaft.

3. Install drive sprocket on the mainshaft.

4. Install blocker ring and synchronizer hub on the mainshaft. Seat hub on mainshaft and install a new snap-ring to retain.

5. Install pump gear on the mainshaft. Tap the gear with a hammer to seat on mainshaft.

6. Install speedometer gear on the mainshaft.

ASSEMBLY

All of the bearings used in the transfer case must be correctly positioned to avoid covering the bearing oil feed holes. After installation of bearings, check the bearing position to be sure the feed hole is not obstructed or blocked by a bearing.

1. Install the lock plate in the transfer case. Coat case and lock plate surfaces around bolt holes with Loctite 515 or equivalent.

2. Position the lock plate to the case and align bolt holes in lock plat with case. Install attaching bolts and torque to specification.

3. Install the roller bearings for the input shaft into the transfer case using J–33830 and J–8092 or equivalent. Press bearings until tool bottoms in bore.

4. Install the front output shaft rear bearing, using J–33832 and J–8092 or equivalent. Press bearing until tool bottoms in case.

5. Install the front output shaft front bearing using J–33833 and J–8092 or equivalent. Press bearing until tool bottoms in bore.

6. Install the snap-ring that retains the front output shaft bearing in case.

7. Install the front output shaft seal using J–33834 or equivalent.

8. Install the input shaft seal using J–33831 or equivalent.

9. Install spacer on shift sector shaft and install sector in transfer case. Install shift lever and retaining nut. Torque to specification.

10. Install shift sector detent spring and retaining bolt.

11. Install the pilot bearing into the input gear using J–33829 and J–8092 or equivalent. Press bearing until tool bottoms out.

12. Install the input gear front thrust bearing and input gear in transfer case.

13. Install the planetary gear thrust wash-

TORQUE SPECIFICATIONS
Model 207

Description	N•m	Ft. lbs.
Bolt Locking Plate to Transfer Case	27–40	20–30
Nut-Front Output Yoke	122–176	90–130
Switch Vacuum	20–34	15–25
Nut-Shift Lever	20–27	15–20
Bolt-Transfer Case	27–34	20–25
Bolt-Rear Retainer	20–27	15–20
Bolt-Extension Housing	27–34	20–25
Bolt-Drain-Fill	40–54	30–40
Bolt-Adapter to Transfer Case	26–40	19–29
Bolt-Shift Bracket	65–85	47–62
Bolt-Shift Lever Pivot	120–140	88–103
Bolt-Shift Lever Adjusting	34–48	25–35

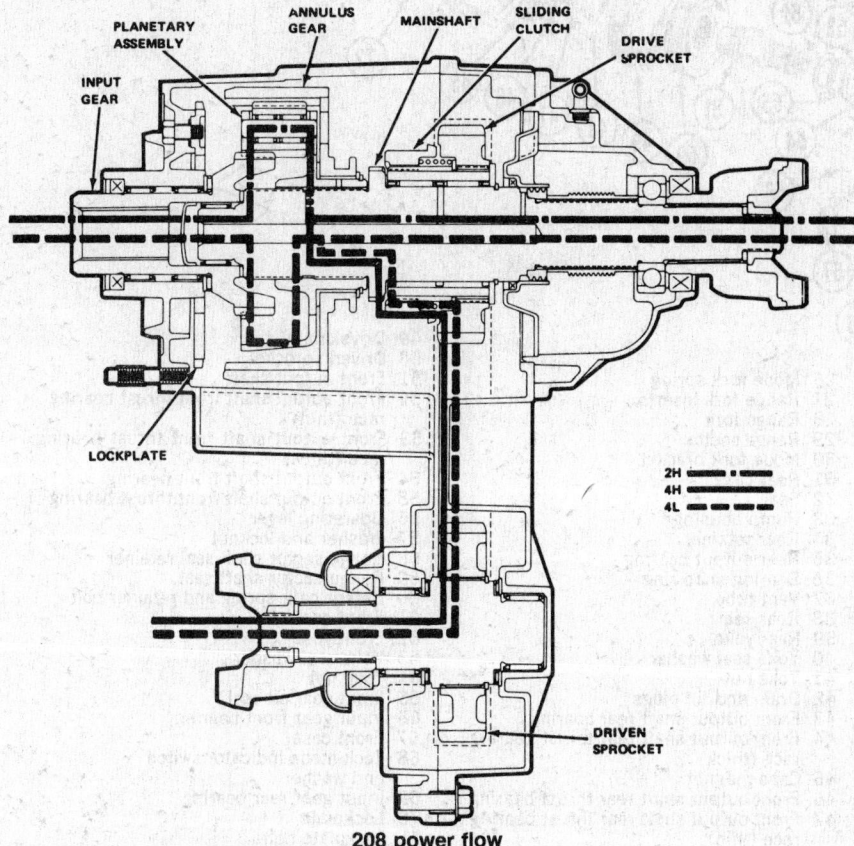

208 power flow

er on the input gear. Position range fork on planetary assembly and install planetary assembly into the transfer case.

14. Install the mainshaft into the transfer case. Make sure the thrust washer is aligned with the input gear and planetary assembly before installing mainshaft.

15. Install mode fork on synchronizer sleeve and rotate until mode fork is aligned with range fork. Slide mode fork rail down through range fork until rail is seated in bore of transfer case.

16. Position drive chain on front output shaft and install chain on drive sprocket. Install front output shaft in the transfer case. It may be necessary to slightly raise the mainshaft to seat the output shaft in the case.

17. Install the magnet into pocket of transfer case.

18. Apply ⅛" bead of Loctite 515® or equivalent to the mating surface of the front case. Install rear case on the front case aligning dowel pins. Install bolts and torque to 20–25 ft. lbs. Install the two bolts with washers into the dowel pin holes.

19. Install the output bearing into the rear retainer using J–33833 and J–8092 or equivalent. Press bearing until seated in bore.

20. Install pump seal in pump housing using J–33835 or equivalent. Apply petroleum jelly to pump housing tabs and install housing in rear retainer.

21. Apply ⅛" bead of Loctite 515® or equivalent to mating surface of rear case and install retaining bolts. Torque bolts to specification 15–20 ft. lbs.

22. Using a new snap-ring, install snapring on mainshaft. Pull up on mainshaft and seat snap-ring in its groove.

23. Install bushing in extension housing using J–33826 and J–8092 or equivalent. Press bushing until tool bottoms in bore.

24. Install a new seal in the extension housing using J–33843 or equivalent.

25. Apply ⅛" bead of Loctite® 515 or equivalent to mating surface of extension housing. Align extension housing to the rear retainer and install attaching bolts. Torque bolts to specification 20–25 ft. lbs.

26. Install front yoke on output shaft. Install a new yoke seal washer with a new nut and torque to specification.

27. Install drain plug and torque to specification. Install fill plug.

New Process Model 208

The New Process Model 208 is a part-time unit with a two piece aluminum housing. On the front case half, the front output shaft, front input shaft, four wheel drive indicator switch and shift lever assembly are located. On the rear case half, the rear output shaft, bearing retainer and drain and fill plugs are located.

DISASSEMBLY

1. Drain the fluid from the case.

2. Remove the attaching nuts from the front and rear output yokes. Remove the yokes and sealing washers.

3. Remove the four bolts and separate the rear bearing retainer from the rear case half.

4. Remove the retaining ring, speedometer drive gear nylon oil pump housing, and oil pump gear from the rear output shaft.

5. Remove the eleven bolts and separate the case halves by inserting a screw driver in the pry slots on the case.

6. Remove the magnetic chip collector from the bottom of the rear case half.

7. Remove the thick thrust washer, thrust bearing and thin thrust washer from the front output shaft assembly.

8. Remove the drive chain by pushing

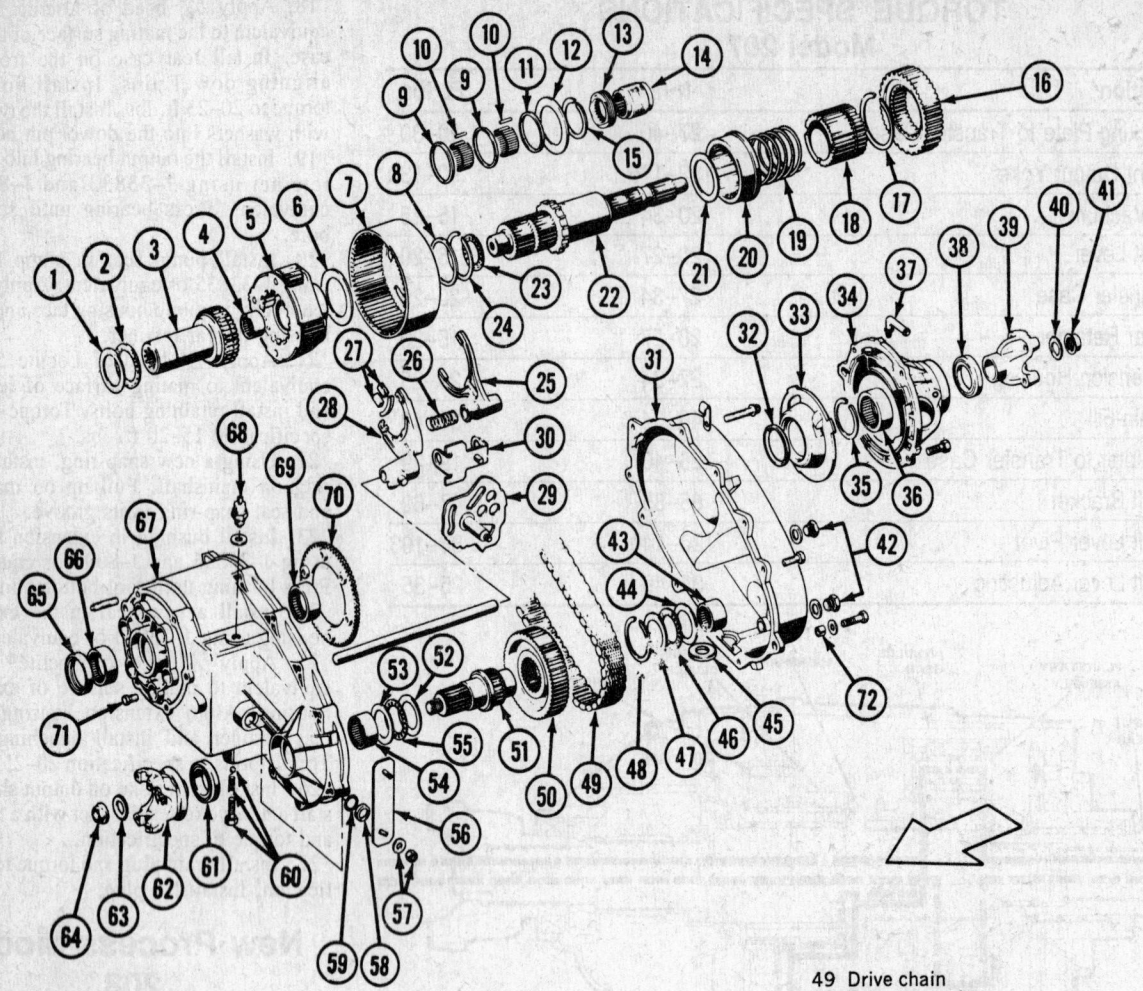

1 Input gear thrust washer
2 Input gear thrust bearing
3 Input gear
4 Mainshaft pilot bearing
5 Planetary assembly
6 Planetary thrust washer
7 Annulus gear
8 Annulus gear thrust washer
9 Needle bearing spacers
10 Mainshaft needle bearings (120)
11 Needle bearing spacer
12 Thrust washer
13 Oil pump
14 Speedometer gear
15 Drive sprocket retaining ring
16 Drive sprocket
17 Sprocket carrier stop ring
18 Sprocket carrier
19 Clutch spring
20 Sliding clutch
21 Thrust washer
22 Mainshaft
23 Mainshaft thrust bearing
24 Annulus gear retaining ring
25 Mode fork

26 Mode fork spring
27 Range fork inserts
28 Range fork
29 Range sector
30 Mode fork bracket
31 Rear case
32 Seal
33 Pump housing
34 Rear retainer
35 Rear output bearing
36 Bearing snap ring
37 Vent tube
38 Rear seal
39 Rear yoke
40 Yoke seal washer
41 Yoke nut
42 Drain and fill plugs
43 Front output shaft rear bearing
44 Front output shaft rear thrust bearing
    race (thick)
45 Case magnet
46 Front output shaft rear thrust bearing
47 Front output shaft rear thrust bearing
    race (thin)
48 Driven sprocket retaining ring

49 Drive chain
50 Driven sprocket
51 Front output shaft
52 Front output shaft front thrust bearing
    race (thin)
53 Front output shaft front thrust bearing
    race (thick)
54 Front output shaft front bearing
55 Front output shaft front thrust bearing
56 Operating lever
57 Washer and locknut
58 Range sector shaft seal retainer
59 Range sector shaft seal
60 Detent ball, spring and retainer bolt
61 Front seal
62 Front yoke
63 Yoke seal washer
64 Yoke nut
65 Input gear oil seal
66 Input gear front bearing
67 Front case
68 Lock mode indicator switch
    and washer
69 Input gear rear bearing
70 Lockplate
71 Lockplate bolts
72 Case alignment dowels

Exploded view of 208

the front input shaft inward and by angling the gear slightly to obtain adequate clearance to remove the chain.

9. Remove the output shaft from the front case half and slide the thick thrust washer, thrust bearing and thin thrust washer off the output side of the front output shaft.

10. Remove the screw, poppet spring and check ball from the front case half.

11. Remove the four wheel drive indicator switch and washer from the front case half.

12. Position the front case half on its face and lift out the rear output shaft, sliding clutch and clutch shift fork and spring.

13. Place a shop towel on the shift rail. Clamp the rail with a vise grip pliers so that they lay between the rail and the case edge. Position a pry bar under the pliers and pry out the shift rail.

14. Remove the snap ring and thrust washer from the planetary gear set assembly in the front case half.

Oil pump removal

Front output shaft rear thrust bearing removal

Mainshaft thrust bearing and input gear

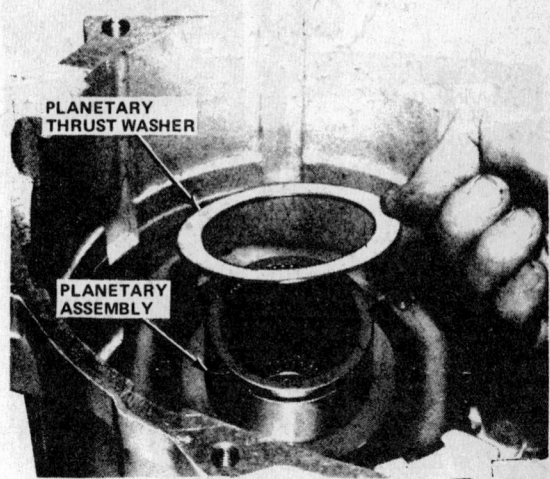

Planetary thrust washer and planetary assembly

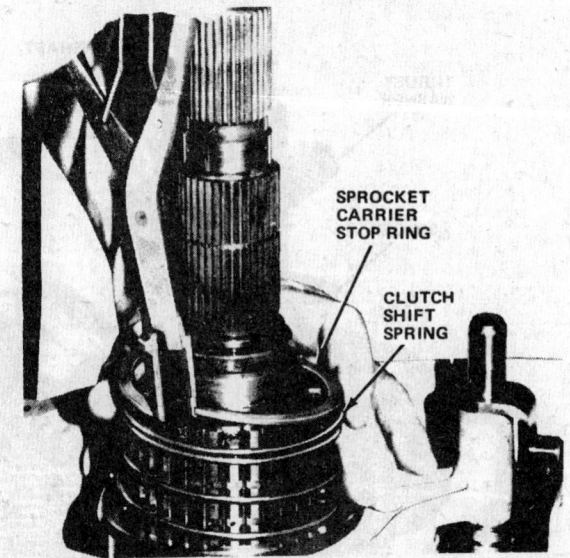

Sprocket carrier clutch ring and spring removal

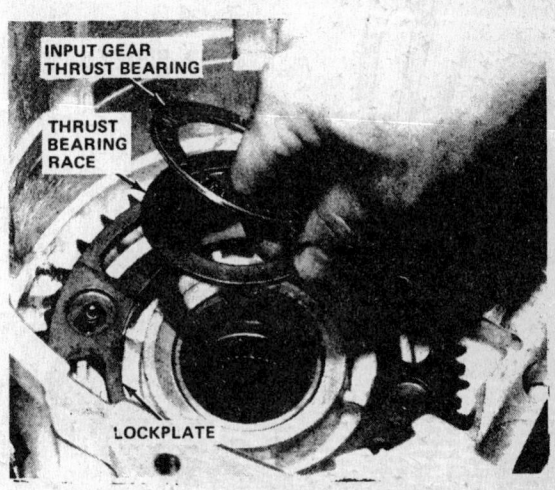

Input gear thrust bearing and race removal

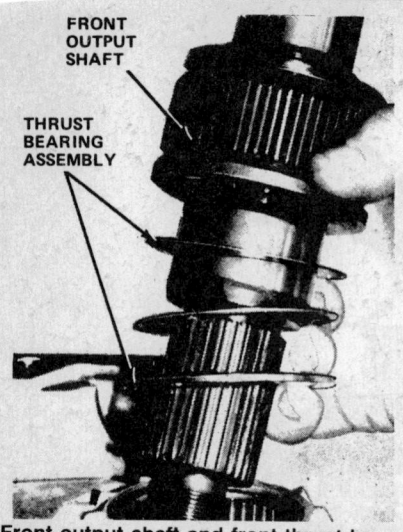

Front output shaft and front thrust bearing assembly removal

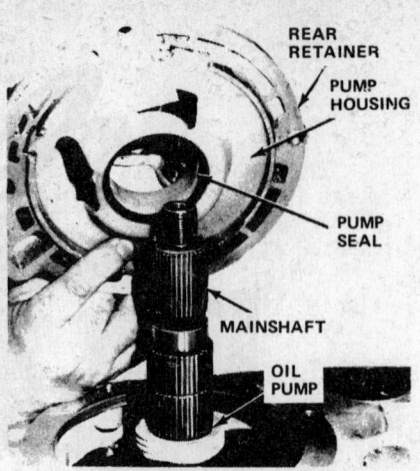

Rear retainer removal

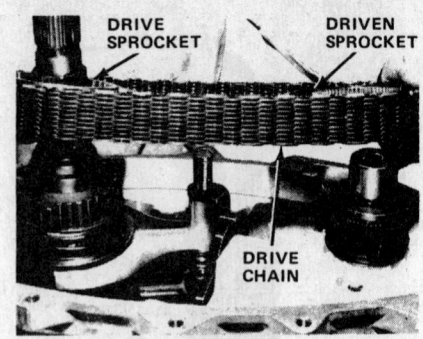

Sprocket and chain removal

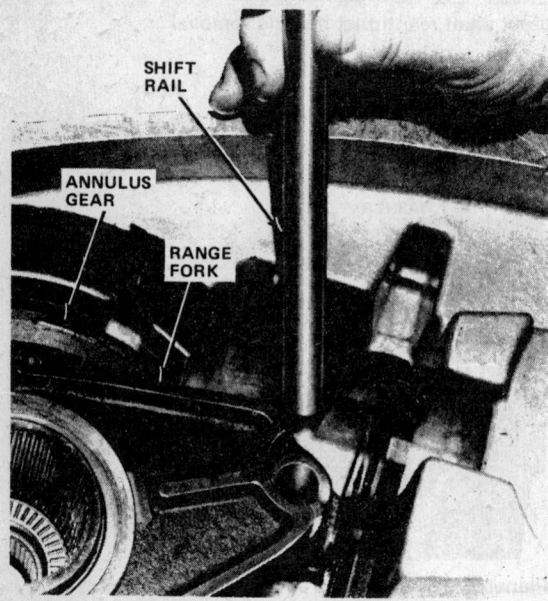

Annulus gear and shift rail installation

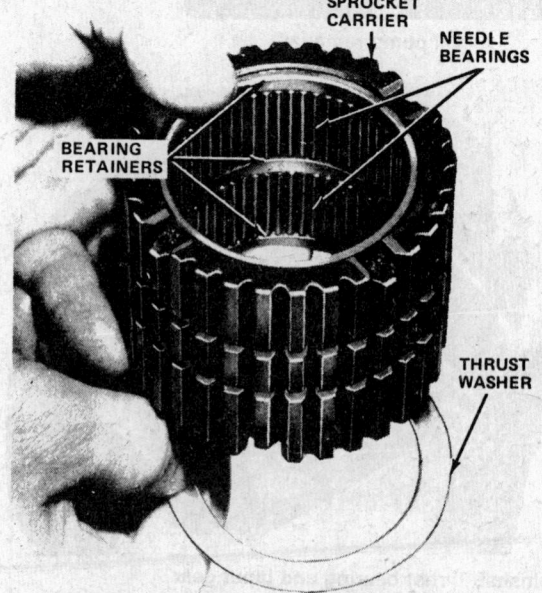

Assembling sprocket carrier components

Driven sprocket retaining snap-ring removal

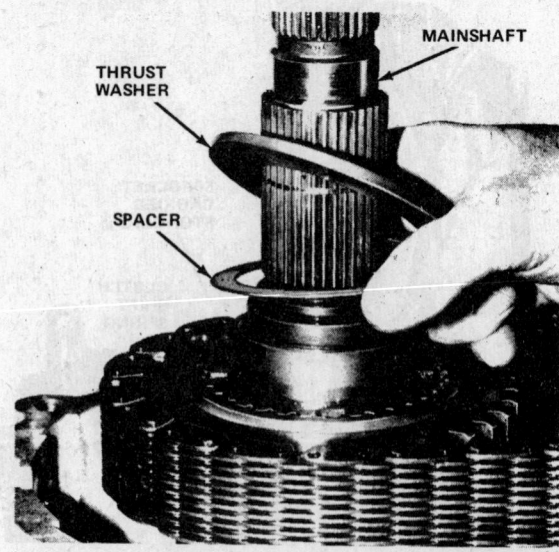

Drive sprocket thrust washer and spacer removal

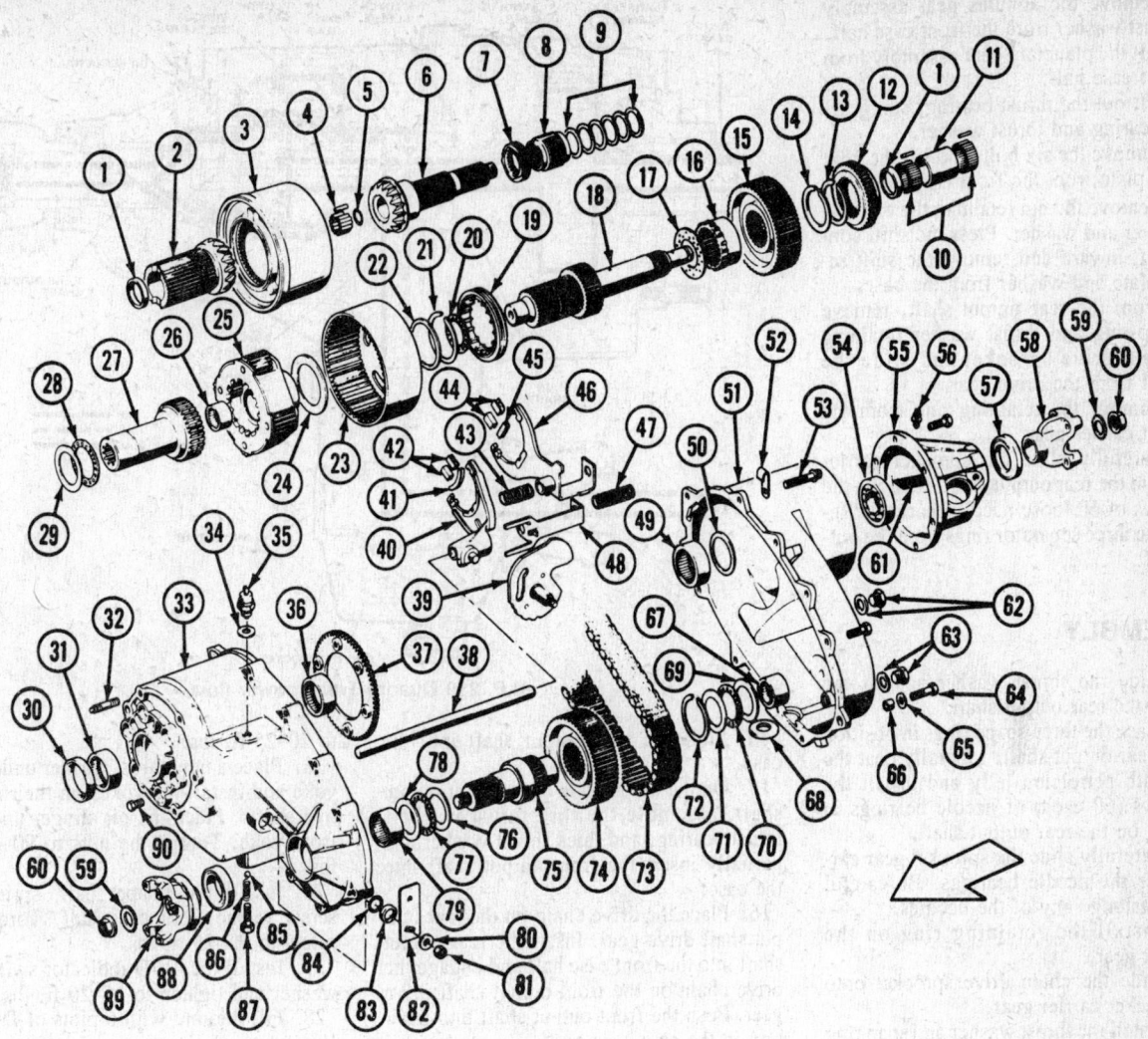

Exploded view of N.P. 219

1 Mainshaft rear bearing spacer—short (2)
2 Side gear
3 Viscous coupling and differential assembly
4 Mainshaft rear pilot roller bearings (15)
5 Mainshaft O-ring
6 Rear output shaft
7 Oil pump
8 Speedometer gear
9 Differential end play shims (selective)
10 Mainshaft needle bearings (82)
11 Mainshaft rear bearing spacer
12 Clutch gear
13 Clutch gear locating ring
14 Drive sprocket locating ring
15 Drive sprocket
16 Side gear clutch
17 Mainshaft thrust washer
18 Mainshaft
19 Clutch sleeve
20 Mainshaft thrust bearing
21 Annulus gear retaining ring
22 Annulus gear thrust washer
23 Annulus gear
24 Planetary thrust washer
25 Planetary assembly
26 Mainshaft front pilot bearing
27 Input gear
28 Input gear thrust bearing
29 Input gear thrust bearing race
30 Input gear oil seal

31 Input gear front bearing
32 Front case mounting stud (6)
33 Front case
34 Lock mode indicator switch gasket
35 Lock mode indicator switch
36 Input gear rear bearing
37 Low range lockplate
38 Shift rail
39 Range sector
40 Range fork
41 Range fork insert
42 Range fork pads
43 Mode fork spring
44 Mode fork pads
45 Mode fork insert
46 Mode fork
47 Shift rail spring
48 Mode fork bracket
49 Rear output shaft bearing
50 Rear output shaft bearing seal
51 Rear case
52 Wiring clip
53 Spline bolt
54 Rear output bearing
55 Rear retainer
56 Vent
57 Output shaft oil seal
58 Rear Yoke
59 Yoke seal washer
60 Yoke locknut
61 Vent chamber seal
62 Fill plug and gasket

63 Drain plug and gasket
64 Rear case bolt
65 Washer (2)
66 Case alignment dowel
67 Front output shaft rear bearing
68 Magnet
69 Front output shaft rear thrust bearing race (thick)
70 Front output shaft rear thrust bearing
71 Front output shaft rear thrust bearing race (thin)
72 Driven sprocket retaining snap ring
73 Drive chain
74 Driven sprocket
75 Front output shaft
76 Front output shaft front thrust bearing race (thin)
77 Front output shaft front thrust bearing
78 Front output shaft front thrust bearing race (thick)
79 Front output shaft front bearing
80 Washer
81 Locknut
82 Operating lever
83 Range sector shaft seal retainer
84 Range sector shaft seal
85 Detent ball
86 Detent spring
87 Detent retaining bolt
88 Front output shaft seal
89 Front yoke
90 Lockplate bolts

15. Remove the annulus gear assembly and thrust washer from the front case half.

16. Lift the planetary gear assembly from the front case half.

17. Lift out the thrust bearing, sun gear, thrust bearing and thrust washer.

18. Remove the six bolts and lift the gear locking plate from the front case half.

19. Remove the nut retaining the external shift lever and washer. Press the shift control shaft inward and remove the shift selector plate and washer from the case.

20. From the rear output shaft, remove the snap-ring and thrust washer retaining the chain drive sprocket and slide the sprocket from the drive gear.

21. Remove the retaining ring from the sprocket carrier gear.

22. Carefully slide the sprocket carrier gear from the rear output shaft. Remove the two rows of 60 loose needle bearings. Remove the three separator rings from the output shaft.

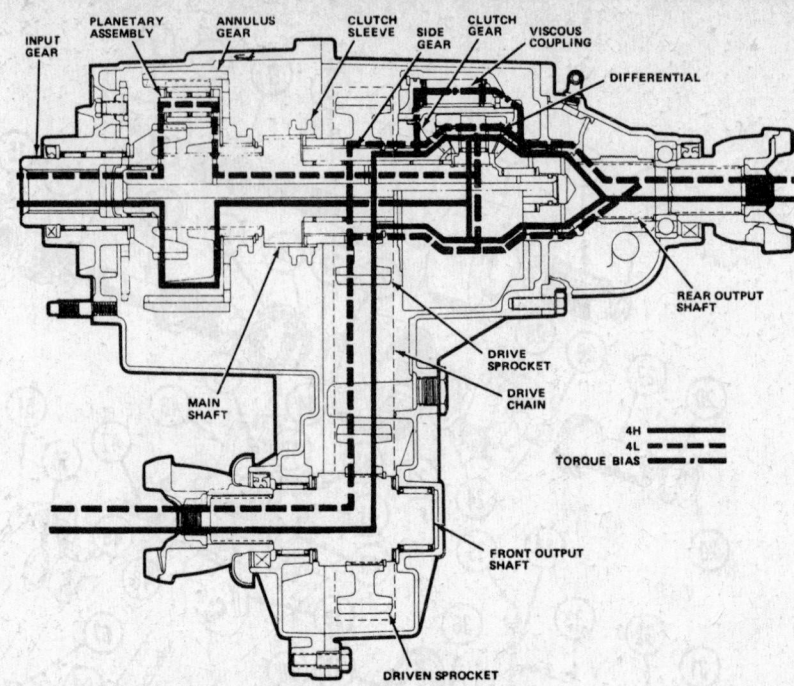

N.P. 219 Quadra-Trac® power flow

ASSEMBLY

1. Slide the thrust washer against the gear on the rear output shaft.

2. Place the three space rings in position on the rear output shaft. Liberally coat the shaft with petroleum jelly and install the two rows (60 each) of needle bearings in position on the rear output shaft.

3. Carefully slide the sprocket gear carrier over the needle bearings. Be careful not to dislodge any of the needles.

4. Install the retaining ring on the sprocket gear.

5. Slide the chain drive sprocket onto the sprocket carrier gear.

6. Install the thrust washer and snap ring on the rear output shaft.

7. Install the shift selector plate and washer through the front of the case.

8. Place the shift lever assembly on the shift control shaft and torque the nut to 14–20 ft.

9. Install the locking plate in the front case half and torque the bolts to 25–35 ft. lbs.

10. Place the thrust bearing and washer over the input shaft of the sun gear. Insert the input shaft through the front case half from the inside and insert the thrust bearing.

11. Install the planetary gear assembly so the fixed plate and planetary gears engage the sun gear.

12. Slide the annulus gear and clutch assembly with the shift fork assembly engaged, over the hub of the planetary gear assembly. The shift fork pin must engage the slot in the shift selector plate. Install the thrust washer and snap ring.

13. Position the shift rail through the shift fork hub in the front case. Tap lightly with a soft hammer to seat the rail in the hole.

14. Position the sliding clutch shift fork on the shift rail and place the sliding clutch and clutch shift spring into the front case

half. Slide the rear output shaft into the case.

15. On the output side of the front output shaft, assemble the thin thrust washer, thrust bearing, and thick thrust washer and partially insert the front output shaft into the case.

16. Place the drive chain on the rear output shaft drive gear. Insert the rear output shaft into the front case half and engage the drive chain on the front output shaft drive gear. Push the front output shaft into position in the case.

17. Assemble the thin thrust washer, thrust bearing and thick thrust washer on the inside of the front output shaft drive gear.

18. Position the magnetic chip collector into position in the front case half.

19. Place a bead of RTV sealant completely around the face of the front case half and assemble the case halves being careful that the shift rail and forward output shafts are properly retained.

20. Alternately tighten the bolts to 20–25 ft. lbs.

21. Slide the oil pump gear over the input shaft and slide the spacer collar into position.

22. Engage the speedometer drive gear onto the rear output shaft and slide the retaining ring into position.

23. Use petroleum jelly to hold the nylon oil pump housing in position at the rear bearing retainer. Apply a bead of RTV sealant around the mounting surface of the retainer and carefully position the retainer assembly over the output shaft and onto the rear case half. The retainer must be installed so that the vent hole is vertical when the case is installed.

24. Torque the retainer bolts alternately

to 20–25 ft. lbs.

25. Place a new thrust washer under each yoke and install the yokes on their respective shafts. Place the oil slinger under the front yoke. Torque the nuts to 90–130 ft. lbs.

26. Install the poppet ball, spring and screw in the front case half. Torque the screw to 20–25 ft. lbs.

27. Install the 4WD indicator switch and washer and tighten to 15–20 ft. lbs.

28. Fill the unit with 6 pints of Dexron® II.

New Process Model 219

Introduced in the 1980 model year of Jeep® vehicles as the Quadra-Trac®, this is a full-time unit. The 4WD mode is fully differentiated in 4H only. The 4L and Lock ranges are undifferentiated. The 4H differentiation is accomplished by a torque biasing viscous coupling and an open differential connected to the coupling. Two drive sprockets and an interconnecting drive chain are used to distribute input torque.

DISASSEMBLY

1. Drain the lubricant from the case.

2. Remove the front and rear output shaft yokes and discard the yoke seal washers and yoke nuts.

3. Mark the rear retainer and rear case for an alignment reference.

4. Unbolt and remove the rear retainer. If necessary, use a soft mallet to loosen the retainer. Under no circumstances should the retainer be pried off.

5. Remove the differential shims and

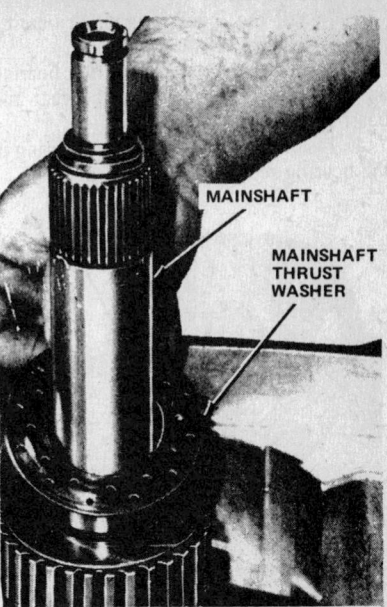

speedometer drive gear from the rear output shaft. Mark the shims for reference.

6. Remove the rear output bearing snap-ring and remove the bearing from the retainer using a soft mallet.

NOTE: The rear output bearing has one side shielded. Note this for reassembly.

7. Remove the rear output shaft seal from the retainer using a screwdriver or punch.

8. Position the front case assembly on wood blocks. The blocks should have V cuts made in them for more positive support of the case.

9. Remove the case halve bolts. The case halves may be pried apart using a screwdriver in the notches provided at the case ends.

NOTE: The two case end bolts have flat washers and alignment dowels. Note their location for assembly.

10. Remove the rear output shaft and viscous coupling as an assembly. Tap the shaft with a plastic mallet if necessary.

11. Remove the O-ring seal and pilot roller bearings from the mainshaft.

12. Remove the rear output shaft from the viscous coupling.

13. Remove the shift rail spring from the rail.

14. Remove the plastic oil pump from the shaft bore in the rear case. Note the pump position for assembly reference. The end with the recess must face the shaft bore when installed.

15. Remove the rear output shaft bearing seal from the case. A screwdriver may be used to pry it out.

16. Remove the front output shaft thrust bearing assembly. Remove the thick washer, bearing and thin washer.

17. Remove the driven sprocket retaining snap-ring.

18. Remove the drive sprocket, drive chain, driven sprocket, side gear clutch and clutch gear as an assembly. Place the assembly on a workbench and mark the components for assembly.

19. Remove the needle bearings and spacers from the mainshaft and side gear bore. A total of 82 bearings and three spacers is used.

20. Remove the side gear/clutch gear assembly from the drive sprocket. Remove two snap-rings and remove the clutch gear from the side gear.

21. Remove the side gear clutch, mainshaft thrust washer and remaining mainshaft needle bearing spacer.

22. Remove the front output shaft and shaft thrust bearing assembly. Note the installation sequence of the bearing assembly.

23. Remove the front output shaft seal from the front case using a screwdriver or punch.

24. Remove the shift rail spring from the shift rail.

25. Remove the clutch sleeve, mode fork and spring as an assembly.

26. Remove the mainshaft thrust washer and mainshaft. Grasp the shaft and pull it straight up and out.

27. Move the range operating lever downward to the last detent position.

28. Disengage the range fork lug from the range sector slot.

29. Remove the annulus gear retaining snap-ring and thrust washer.

30. Remove the annulus gear and range fork.

31. Remove the planetary thrust washer from the hub.

32. Remove the planetary assembly.

33. Remove the mainshaft thrust bearing from the input gear.

34. Remove the input gear and remove the input gear thrust bearing and race.

35. Remove the range selector detent ball and spring retaining bolt and remove the detent ball and spring.

36. Remove the range selector and operating lever attaching nut and lockwasher, and remove the lever.

37. Remove the range selector.

38. Remove the range selector O-ring and retainer.

39. Remove the input gear oil seal from the front case with a screwdriver.

ASSEMBLY

Lubricate all parts before assembly with 10W–30 motor oil. Petroleum jelly will be indicated for some assemblies. Do not use chassis lube or other heavy lubricants.

1. Install new input gear and rear output shaft bearing oil seals. Seat the seals flush with the edge of the seal bore or with the seal groove in the case. Coat the seal lips with petroleum jelly after installation.

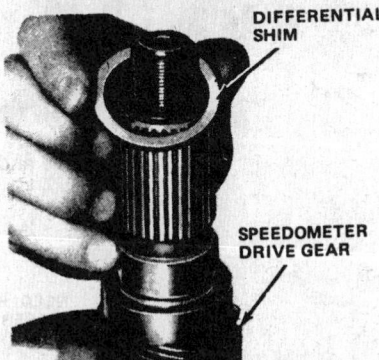

Differential shim, speedometer gear and oil pump

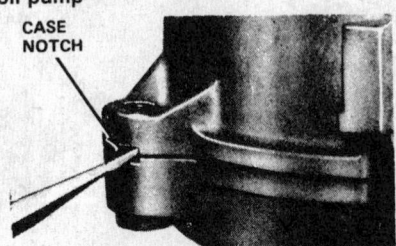

Rear case half removal

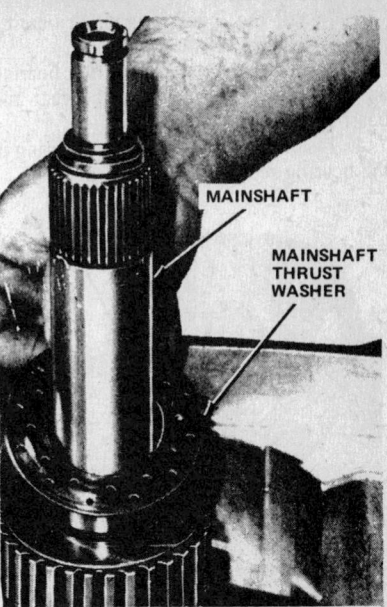

Mainshaft and thrust washer

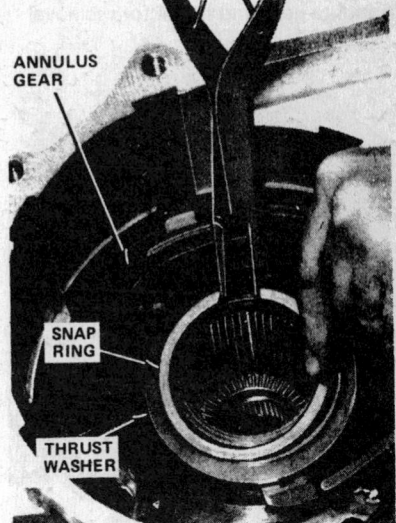

Annulus gear snap-ring and thrust washer

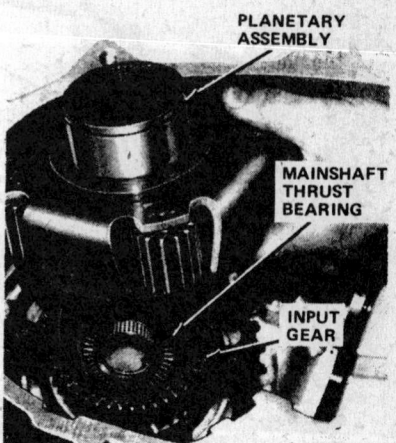

Planetary assembly removal

2. Install the input gear thrust bearing race in the case counterbore.

3. Install the input gear thrust bearing on the input gear and install the gear and bearing in the case.

4. Install the mainshaft thrust bearing in the bearing recess in the input gear.

Annulus gear and range fork removal

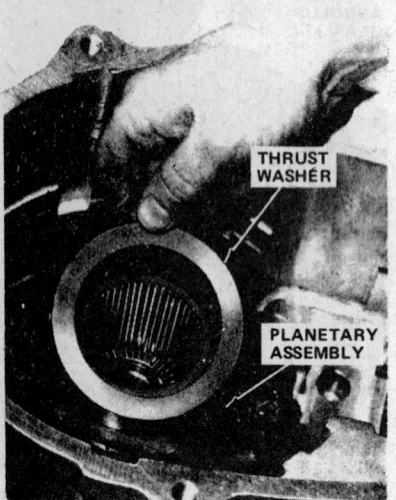

Planetary thrust washer removal

5. Install the planetary assembly on the input gear. Make sure that the planetary pinion teeth mesh fully with the input gear.

6. Install the planetary thrust washer on the planetary hub.

7. Install a new sector shaft O-ring and retainer in the shaft bore in the case.

8. Install the range selector in the front case. Install the operating lever on the sector shaft and install the lever attaching washer and locknut on the shaft. Tighten the locknut to 17 ft. lbs.

9. Install the detent spring, ball and retaining bolt in the front case detent bore. Tighten the bolt to 22 ft. lbs.

10. Move the range selector to the last detent position.

11. Assemble the annulus gear and range fork. Install the assembled fork and gear over the planetary assembly. Be sure that the annulus gear is fully meshed with the planetary pinions.

12. Insert the range fork lug in the range detent slot.

13. Install the annulus thrust washer and retaining ring on the annulus gear hub.

14. Align the mainshaft thrust washer in the input gear, if necessary.

15. Install the mainshaft. Be sure the shaft is fully seated in the input gear.

16. Install the mainshaft thrust washer on the mainshaft.

17. Install the short mainshaft needle bearing spacer on the shaft.

18. Apply a liberal coating of petroleum jelly to the mainshaft needle bearing surface and install 41 of the 82 needle bearings on the shaft. Be sure the bearings seat on the short spacer.

19. Install the long needle bearing spacer on the shaft. Lower the spacer onto the previously installed needle bearings carefully to avoid displacing them.

20. Align the shift rail bore in the case with the bore in the range fork and install the shift rail.

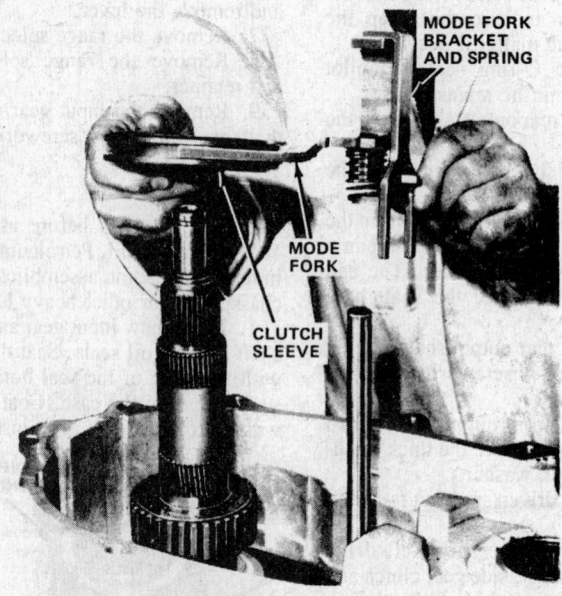

Clutch sleeve and mode fork removal

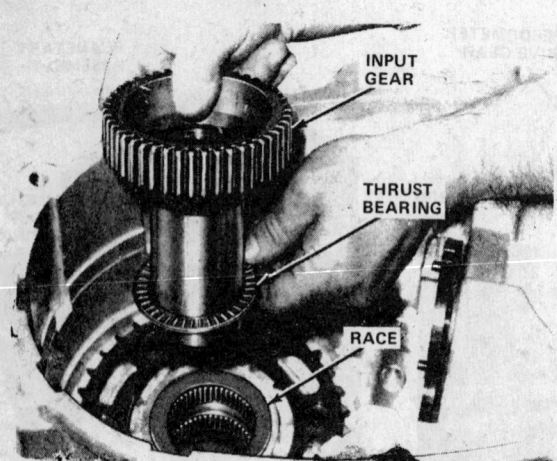

Input gear and thrust bearing removal and installation

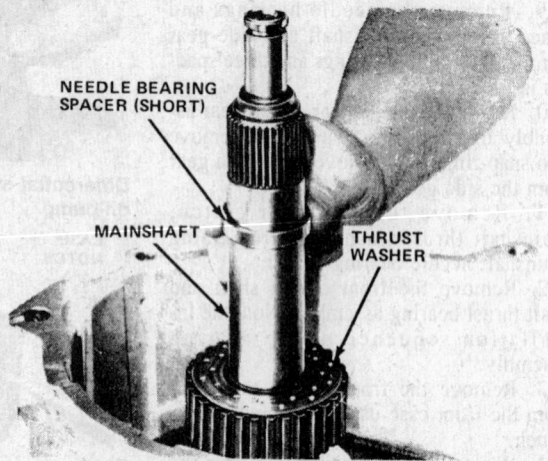

Mainshaft and thrust washer installation

NOTE: Remove all traces of oil from the case shift rail bore before installing the rail. Oil in the case bore may prevent the rail from seating completely and prevent rear case installation.

21. Assemble the mode fork, mode fork spring and mode fork bracket.

22. Install the clutch sleeve in the mode fork. Be sure the sleeve is positioned so that the ID numbers on the sleeve face upward when the sleeve is installed.

23. Align the clutch sleeve and mode fork assembly with the shift rail and install the assembly on the shift rail and mainshaft. Be sure that the clutch sleeve is meshed with the mainshaft gear.

24. Lubricate the remaining 41 needle bearings and place them on the mainshaft.

25. Install the side gear clutch on the mainshaft with the teeth facing downward. Be sure the gear teeth mesh with the clutch sleeve.

26. Install the remaining short mainshaft needle bearing spacer. Install the spacer carefully to avoid displacing previously installed bearings.

27. Install the front output shaft front thrust bearing in the front case. Correct sequence is thick race, bearing, thin race.

28. Install the front output shaft in the front case.

29. Install the clutch gear on the side gear. The tapered side of the clutch gear teeth must face the side gear teeth.

30. Install the clutch gear and drive sprocket locating snap-rings on the side gear. Install the snap-rings so that they face each other.

31. Position the drive and driven sprockets in the drive chain and install the assembled side and clutch gears in the drive sprocket.

32. Install the assembled drive chain, sprockets and side gear on the mainshaft and front output shaft. Align the sprockets with the shaft, keeping the assembly level and carefully lower the assembly onto both shafts simultaneously. Do not displace any of the needle bearings.

33. Install the driven sprocket retaining snap-ring.

34. Install the front output shaft rear thrust bearing assembly on the front output shaft. Correct installation sequence is thin race, thrust bearing, thick race.

35. Install the shift rail spring on the shift rail.

36. Install a new O-ring on the mainshaft pilot bearing hub.

37. Coat the mainshaft pilot roller bearing hub and bearings with a liberal amount of petroleum jelly and install the rollers on the shaft.

38. Install the rear output shaft in the viscous coupling. Be sure it is fully seated.

39. Install the assembled viscous coupling and rear output shaft on the mainshaft. Align the mainshaft pilot hub with the pilot bearing bore in the rear output shaft and carefully lower the assembly onto the mainshaft. Take care to avoid displacing the roller bearings.

40. Align the clutch gear teeth with the viscous coupling teeth and seat the coupling fully onto the clutch gear.

NOTE: When correctly installed, the clutch gear teeth will not be visible or extend out of the coupling.

41. Install the magnet in the front case, if removed.

42. Clean the mating surfaces of the case halves thoroughly.

43. Apply Loctite® 515 or equivalent to the mating surfaces and all attaching bolts.

44. Join the case halves, aligning the dowels and install the bolts. Torque the bolts to 22 ft. lbs.

NOTE: The two end bolts require flat washers.

45. Install the oil pump on the rear output shaft and seat it in the case. The side with the recess should face the inside of the case.

46. Install the speedometer drive gear and differential shift, on the output shaft.

47. Install the vent chamber seal in the rear retainer.

48. Align and install the rear retainer on the case. Make the retainer finger tight only.

49. Install the yoke on the rear output shaft. Make the yoke finger tight only.

50. Mount a dial indicator on the rear retainer. Position the indicator stylus so that it contacts the top of the yoke nut.

51. Install the yoke on the front output shaft and rotate the shaft ten complete revolutions.

52. Rotate the front output shaft again and note the play indicated on the dial. End play should be .002–.010 inch. If the end play must be adjusted, remove the rear retainer and add or subtract shims as required.

53. Remove both output shaft yokes and discard the nuts.

54. Install the front and rear yoke seals.

55. Remove the rear retainer bolts, apply Loctite® 515 or equivalent to the mating surface of the retainer and to the bolts and install the bolts. Torque them to 22 ft. lbs.

56. Install new yoke seal washers on the output shafts, install yokes on the shafts and install new yoke nuts. Tighten the nuts to 110 ft. lbs.

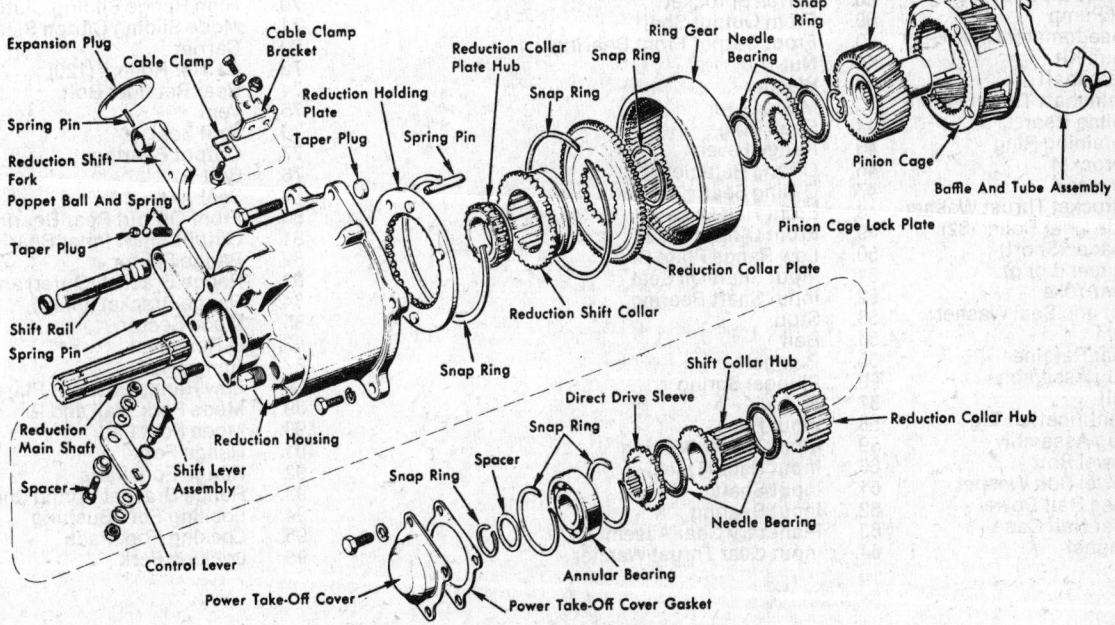

Exploded view of the optional reduction unit (© Jeep Corp.)

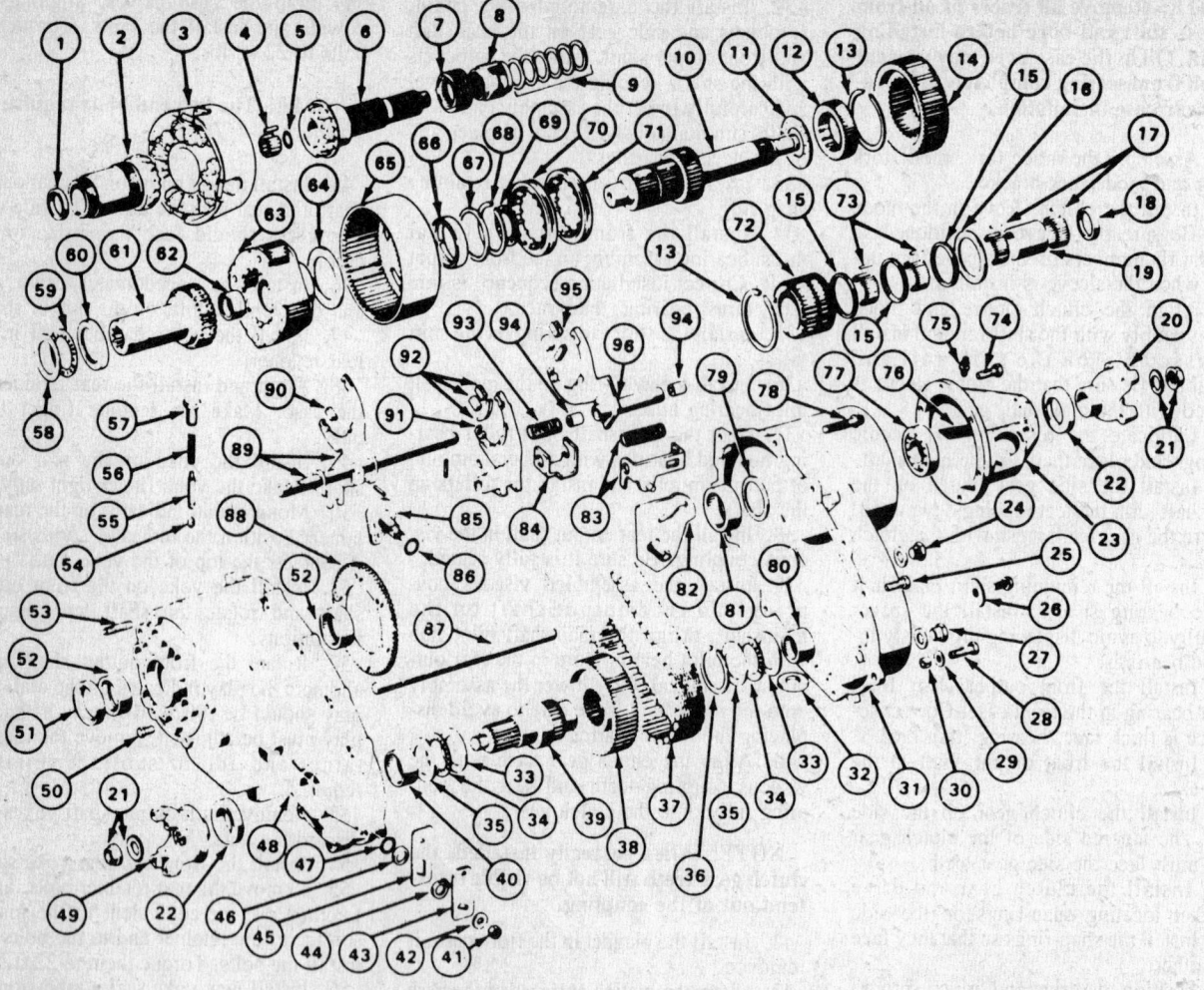

1. Spacer	33. Front Output Shaft Bearing Assembly Race (Thick)	65. Annulus Gear Assembly
2. Side Gear	34. Front Output Shaft Bearing Assembly Thrust	66. Annulus Bushing
3. Differential	35. Front Output Shaft Bearing Assembly Race (Thin)	67. Thrust Washer
4. Pilot Bearing Rollers (15)	36. Retaining Ring	68. Retaining Ring
5. O-Ring Seal	37. Chain	69. Thrust Bearing
6. Rear Output Shaft	38. Driven Sprocket	70. High Range Sliding Clutch Sleeve
7. Oil Pump	39. Front Output Shaft	71. Mode Sliding Clutch Sleeve
8. Speedometer Drive Gear	40. Front Output Front Bearing	72. Carrier
9. Shim Kit	41. Nut	73. Carrier Rollers (120)
10. Mainshaft	42. Washer	74. Rear Retainer Bolt
11. Mainshaft Thrust Washer	43. Mode Lever	75. Vent
12. Spline Gear	44. Snap Ring	76. Vent Seal
13. Retaining Ring	45. Range Lever	77. Output Bearing
14. Sprocket	46. O-Ring Retainer	78. Bolt
15. Spacer	47. O-Ring Seal	79. Seal
16. Sprocket Thrust Washer	48. Front Half Case	80. Front Output Rear Bearing
17. Side Gear Roller (82)	49. Front Output Yoke	81. Output Shaft Inner Bearing
18. Spacer (Short)	50. Low Range Plate Bolt	82. Range Sector
19. Spacer (Long)	51. Input Shaft Oil Seal	83. Range Bracket (Outer) and Spring
20. Rear Yoke	52. Input Shaft Bearing	84. Range Bracket (Inner)
21. Nut and Seal Washer	53. Stud	85. Mode Sector
22. Seal	54. Ball	86. O-Ring Seal
23. Rear Retainer	55. Plunger	87. Range Rail
24. Plug Assembly	56. Plunger Spring	88. Low Range Lockout Plate
25. Bolt	57. Screw	89. Mode Fork, Rail and Pin
26. Identification Tag	58. Input Race	90. Mode Fork Pad
27. Plug Assembly	59. Input Thrust Bearing	91. Range Fork
28. Dowel Bolt	60. Input Race (Thick)	92. Range Fork Pads
29. Dowel Bolt Washer	61. Input Shaft	93. Range Bracket Spring (Inner)
30. Case Half Dowel	62. Input Bearing	94. Locking Fork Bushing
31. Rear Half Case	63. Planetary Gear Assembly	95. Locking Fork Pads
32. Magnet	64. Input Gear Thrust Washer	96. Locking Fork

Model 228 components

57. Install the drain plug and tighten to 18 ft. lbs.

58. Pour 4 pints of 10W-30 motor oil into the case and install the fill plug. Tighten it to 18 ft. lbs.

Selec–Trac Model 228

The Model 228 is a 3 position, full time/part time unit with integral low range and a neutral position. Dexron®II fluid is used.

Range Control Linkage

ADJUSTMENT

1. Place the range control lever in the high position.

2. Insert a $\frac{1}{8}$ inch spacer between the gate and the lever.

3. Hold the lever in this position. Place the transfer case lever in the high range position. Adjust the link to provide a free pin at the transfer case outer lever.

DISASSEMBLY

1. Remove the drain plug and drain the lubricant from the transfer case. Remove the front and rear yokes after match marking them for installation alignment reference.

2. Position the transfer case on wooden blocks that will support it firmly. Mark the rear retainer and rear case for assembly reference. Remove the rear retainer bolts and remove the retainer. Prying slots are provided between the case and retainer.

3. Remove the differential shims and the speedometer drive gear from the rear output shaft.

4. Remove the bolts connecting the front and rear transfer case halves. Note that the bolts used on each end of the transfer case are equipped with flat washers, install in like position. Slots are provided at each end of the transfer case to help in case separation, do not drive any tool between the halves or damage to the cases may occur.

5. Remove the rear case half from the front using two suitable prybars in the slots provided.

6. Remove the oil pump from the rear output shaft. Note the position of the pump for reinstallation, the recessed side of the pump faces the case interior.

7. Remove the rear output shaft from the mainshaft. Remove the 15 mainshaft bearing rollers from the shaft or coupling. The rollers may drop off the shaft during removal. Remove the mainshaft O-ring from the end of the shaft. Remove the differential from the mainshaft and side gear.

8. Remove the front output shaft, driven sprocket and drive chain assembly. Lift the front shaft, sprocket and chain upward. Tilt the front shaft toward the mainshaft. Slide the chain off the drive sprocket and remove the assembly.

9. Remove the front output shaft front thrust bearing assembly from the front of

the case or from the shaft. Remove the drive chain from the front output shaft and sprocket.

10. Remove the snap-ring that retains the driven sprocket on the front output shaft. Mark the sprocket and shaft for assembly reference and remove the sprocket from the shaft.

11. Remove the mainshaft, side gear, drive sprocket and spline gear as an assembly. Place the assembly aside to be serviced later after the front case is disassembled.

12. Remove the mode fork, shift rail and mode sliding clutch sleeve as an assembly. Mark the sleeve and fork for assembly reference and remove the sleeve from the fork. The mode fork and rail are pinned together, remove the pin if separation is necessary.

13. Remove the locking fork, high range sliding clutch sleeve, fork brackets and fork springs as an assembly. Take note of the position of the components for assembly and disassemble for cleaning and inspection.

14. Remove the range selector detent screw and remove the detent spring, plunger and ball. Move the range operating lever downward to the last detent position. Disengage the low range fork lug from the range selector slot.

15. Remove the retaining snap ring from the annulus gear and remove the thrust washer. Remove the annulus gear, range fork and rail as an assembly. Separate the components for cleaning and inspection.

16. Remove the planetary thrust washer from the planetary assembly hub. Grasp the planetary hub and lift the assembly upward and remove it. Remove the mainshaft thrust bearing from the input shaft. Remove the input shaft, input shaft thrust bearing and race.

17. Remove the range sector and operating lever attaching nut and lockwasher. Remove the lever. Remove the range sector and shaft from the front case, then remove the range sector O-ring and retainer.

Mainshaft and Gear

1. Grasp the drive sprocket and lift the side gear upward and off of the mainshaft. Remove the mainshaft needle bearings and two bearing spacers from the mainshaft. 82 needle bearings are used. Note the position of the spacers for reinstallation. Remove the spline gear and thrust washer from the mainshaft.

2. Remove the side gear and thrust washer from the sprocket carrier and drive sprocket. Remove the thrust washer from the side gear. Remove one sprocket carrier snap ring and remove the drive sprocket from the carrier after marking the components for reinstallation reference.

3. Take note that the sprocket carrier and mainshaft needle bearings are different sizes, take care not to mix them. Remove the three bearing spacers and all of the sprocket carrier needle bearings from the

carrier. A total of 120 needle bearings are used.

4. Remove the rear output bearing and rear yoke seal from the rear retainer. The bearing is shielded on one side, take note of shield position for bearing reinstallation.

5. Remove the input gear and front yoke seals from the front case.

Cleaning & Inspection

Wash all components throughly in clean solvent. Apply compressed air to each oil supply port and channel in both case halves to remove dirt. Inspect all gear teeth for excessive wear or damage. Inspect all snap rings and thrust washers for excessive wear, distortion or damage. Replace components as required.

Bearings & Bushings

All bearings and bushings used in the transfer case must be correctly positioned, when replaced, to avoid blocking oil supply holes.

ASSEMBLY

1. During assembly, lubricate all internal parts with Dexron®II fluid or with petroleum jelly. Do not use chassis lube or similar thick lubricant.

2. Install a new input and output shaft oil seal. Seat the seals flush with the edge of the seal bore or in the seal groove in the transfer case. Coat the sealing lips with petroleum jelly.

3. Install the input shaft thrust bearing race in the transfer case counterbore Install the input gear thrust bearing in the input shaft and install the shaft and bearing in the transfer case. Install the mainshaft thrust bearing into the recess on the input shaft.

4. Install the planetary assembly on the input shaft. Ensure that the planetary pinion teeth mesh fully with the input shaft. Install the planetary thrust washer on the planetary hub.

5. Install a new sector shaft O-ring and install the retainer in the transfer case sector shaft bore.

6. Install an O-ring on the mode sector shaft and insert the mode sector through the range sector. Install the range sector in the front case half. Install the operating lever and the snap ring on the range sector shaft. Install the lever, attaching washer and locknut on the mode sector shaft. Tighten the locknut to 17 ft. lbs.

7. Assemble the annulus gear, range fork and rail. Install the assembled fork on and over the planetary assembly. Make sure the annulus gear is fully meshed with the planetary pinions.

8. Engage the range sector lug into the range sector. Install the annulus thrust washer and the retaining ring on the annulus gear hub. Install the detent ball, plunger, spring and retaining bolt in the detent bore. Tighten to 22 ft. lbs.

9. Assemble and install the locking fork, fork bracket, fork springs and high range clutch sleeves. Ensure that the locking lug is full seated in the range sector detent slot. The locking mode clutch sleeve and the high range clutch sleeve are not interchangeable. Make sure that the correct sleeve is installed in the corresponding shift fork.

10. Install the range fork lug into the range sector detent notch. Move the range sector to the high range position. Assemble and install the range fork, shift rail and mode clutch sleeve.

11. Install the thrust washer and new O-ring on the mainshaft. Install the needle bearings and bearing spacers on the mainshaft. Coat the bearing surface on the shaft with petroleum jelly. Install the first 41 needle bearings. Install the first spacer, remaining 41 needle bearings and the remaining spacer. Use additional petroleum jelly to hold the bearings in position if necessary. Take care not to dislodge the bearings when the spacers are installed, or when installing the shaft.

12. Install the spline gear on the mainshaft. Install the sprocket carrier in the drive sprocket and install the sprocket carrier snap rings. Make sure that the carrier and sprocket are aligned to the matchmarks previously made. The sprocket carrier teeth are tapered on one side and the drive sprocket has a deep recess on one side. Make sure that the carrier tapered teeth and sprocket recess are on the same side when assembling.

13. Install the sprocket carrier bearings and spacers. Coat the bore and all off the 120 needle bearings with petroleum jelly. Install the center spacer. Install 60 bearings in each end of the carrier and install the remaining two spacers, one at each side of the carrier. Use additional petroleum jelly if necessary.

14. Install the assembled sprocket carrier and drive sprocket on the mainshaft. Use care so that the mainshaft bearings are not displaced during installation. Make sure that the recessed side of the drive sprocket is facing downwards. Install the thrust washer on the mainshaft. Position the washer on the sprocket carrier. Install the side gear on the mainshaft, be sure it is fully seated in the sprocket carrier. Once again, take care not to dislodge any of the needle bearings.

15. Install the mainshaft and gear assembly in the case, be sure it is fully seated on the input gear.

16. Install the driven sprocket and snapring on the front output shaft. Make sure the reference marks made during disassembly are aligned. Install the thrust washer assembly for the front of the output shaft into the front half of the transfer case.

17. Install the thick race in the transfer case, then install the bearing and thin race.

18. Install the drive chain, front output shaft and driven sprocket by installing the chain on the driven sprocket and then raising and tilting the driven sprocket and chain until the chain can be placed on the drive sprocket.

19. Align the output shaft with the shaft bore in the front transfer case half and install the shaft. Make sure the front shaft bearing assembly is fully seated in the transfer case.

20. Install the front output shaft thrust bearing assembly on the shaft. Install the thin race first, then the bearing and finally the thick race.

21. Fully seat the differential on the side gear. Coat the mainshaft pilot bearing surface and the 15 pilot needle bearings with petroleum jelly. Install the bearings on the shaft.

22. Use care not to dislodge the needle bearings and install the rear output shaft on the mainshaft and into the differential. Make sure the shaft is fully seated.

23. Install the oil pump (recessed side facing down) and the rear output shaft.

24. Install a new rear output shaft seal in the transfer case half.

25. Apply a bead of Loctite 515®, or the equivalent to the mating surface of the rear transfer case half. Attach the rear transfer case half to the front transfer case half. Make sure that the dowels are aligned. If the case halves will not mate, check for: oil in the range fork rail bore, front output shaft rear thrust bearing alignment to rear case, mainshaft not fully seated or oil pump to case half alignment.

26. Tighten the bolts to 23 ft. lbs. Make sure that the flat washers are used on the bolts on the case end where the alignment dowels are located.

27. Install the speedometer drive gear on the rear output shaft.

28. Measure and record the thickness of the rear shim pack.

29. Install a 0.030 in. (approx.) shim on the rear output shaft.

30. Install the rear retainer on the rear case half. Install the bolts snugly (but do not tighten fully). Install the yokes on the shafts. Use the old mounting nuts and tighten them finger tight.

31. Check the differential end play. Set the lever in 4 High position. Position a dial indicator so the probe is above and resting on the top (shaft end) of the yoke nut. Zero the indicator. Pull upward on the rear output yoke. Record the reading. Remove the retainer and add or subtract shims to obtain the correct end play which should be between 0.002–0.010. Recommended is 0.006 in.

32. Remove the yokes. Remove the retainer and apply a bead of Loctite 515® or the equivalent to the retainer mating surface. Install the retainer and tighten the mounting bolts to 23 ft. lbs. Install the yokes using new yoke seal washers and retaining nuts. Tighten the nuts to 120 ft. lbs. while holding the yoke in position with the appropriate tool.

33. Install the detent ball and spring, if not done previously. Apply sealer to the bolt and tighten to 23 ft. lbs.

34. Install the lubricant drain plugs and tighten them to 18 ft. lbs.

35. Install the transfer case. Fill with 7 pts. of Dexron® II fluid and install the fill plug.

Selec–Trac Model 229

The Model 229 is similar in appearance to the Model 219. It is a 3 position, dual range full time/part time unit with integral low range and a neutral position. Dexron® II fluid is used.

ADJUSTMENT

Range Control Linkage

1. Place the range control lever in high range.
2. Insert a 1/8 inch spacer between gate and lever.
3. Hold the lever in this position.
4. Place range control lever in high range position.
5. Adjust as needed.

DISASSEMBLY

1. Remove the drain plug and drain the lubricant from the transfer case.
2. Remove the front and rear yoke nuts and seal washers. Discard the washers.
3. Mark the front and rear yokes for installation alignment reference.
4. Remove the front and rear yokes. Use Tool J-8614-01 or equivalent to remove the yokes if necessary.
5. Place the transfer case on wooden blocks. Cut V-notches in the blocks for clearance for the front case mounting studs.
6. Mark the rear retainer and rear case for assembly reference.
7. Remove the rear retainer bolts and remove the retainer. Use two prybars to pry the retainer off the transfer case. Position the prybars in slots in the retainer and case to pry the retainer loose.
8. Remove the differential shim(s) and speedometer drive gear from the rear output shaft.
9. Remove the bolts attaching the rear transfer case half to the front case half. Note that the bolts used at each end of the transfer case require flat washers.

—— CAUTION ——
Insert two prybars in the slots at each end of the rear transfer case half to loosen it. Do not attempt to wedge the transfer case halves apart or the case mating surfaces will be damaged.

10. Remove the rear transfer case half from the front case half using two prybars.
11. Remove the thrust bearing and races from the front output shaft. Note the position of the bearing and races for assembly reference.

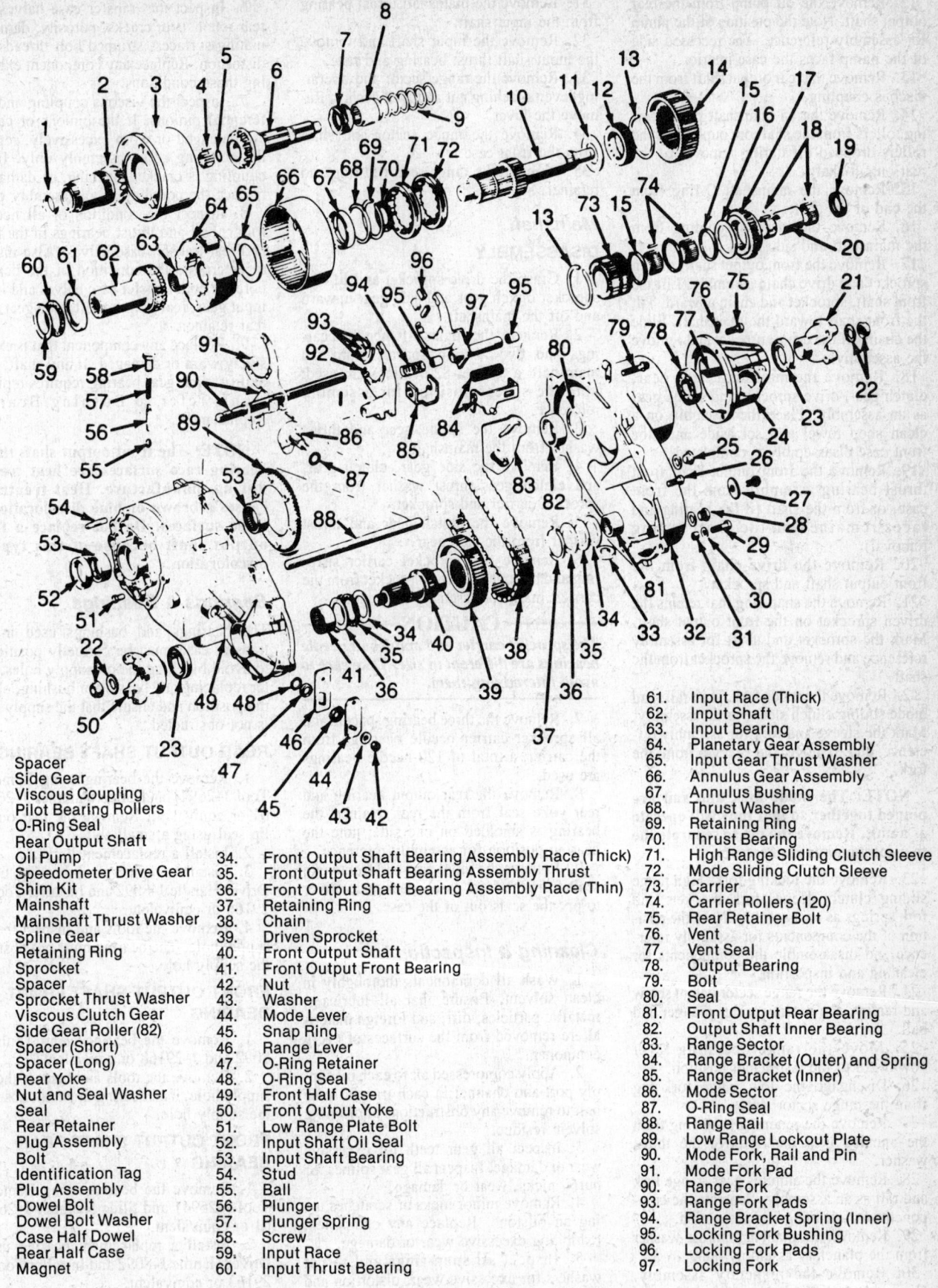

1. Spacer
2. Side Gear
3. Viscous Coupling
4. Pilot Bearing Rollers
5. O-Ring Seal
6. Rear Output Shaft
7. Oil Pump
8. Speedometer Drive Gear
9. Shim Kit
10. Mainshaft
11. Mainshaft Thrust Washer
12. Spline Gear
13. Retaining Ring
14. Sprocket
15. Spacer
16. Sprocket Thrust Washer
17. Viscous Clutch Gear
18. Side Gear Roller (82)
19. Spacer (Short)
20. Spacer (Long)
21. Rear Yoke
22. Nut and Seal Washer
23. Seal
24. Rear Retainer
25. Plug Assembly
26. Bolt
27. Identification Tag
28. Plug Assembly
29. Dowel Bolt
30. Dowel Bolt Washer
31. Case Half Dowel
32. Rear Half Case
33. Magnet

34. Front Output Shaft Bearing Assembly Race (Thick)
35. Front Output Shaft Bearing Assembly Thrust
36. Front Output Shaft Bearing Assembly Race (Thin)
37. Retaining Ring
38. Chain
39. Driven Sprocket
40. Front Output Shaft
41. Front Output Front Bearing
42. Nut
43. Washer
44. Mode Lever
45. Snap Ring
46. Range Lever
47. O-Ring Retainer
48. O-Ring Seal
49. Front Half Case
50. Front Output Yoke
51. Low Range Plate Bolt
52. Input Shaft Oil Seal
53. Input Shaft Bearing
54. Stud
55. Ball
56. Plunger
57. Plunger Spring
58. Screw
59. Input Race
60. Input Thrust Bearing

61. Input Race (Thick)
62. Input Shaft
63. Input Bearing
64. Planetary Gear Assembly
65. Input Gear Thrust Washer
66. Annulus Gear Assembly
67. Annulus Bushing
68. Thrust Washer
69. Retaining Ring
70. Thrust Bearing
71. High Range Sliding Clutch Sleeve
72. Mode Sliding Clutch Sleeve
73. Carrier
74. Carrier Rollers (120)
75. Rear Retainer Bolt
76. Vent
77. Vent Seal
78. Output Bearing
79. Bolt
80. Seal
81. Front Output Rear Bearing
82. Output Shaft Inner Bearing
83. Range Sector
84. Range Bracket (Outer) and Spring
85. Range Bracket (Inner)
86. Mode Sector
87. O-Ring Seal
88. Range Rail
89. Low Range Lockout Plate
90. Mode Fork, Rail and Pin
91. Mode Fork Pad
90. Range Fork
93. Range Fork Pads
94. Range Bracket Spring (Inner)
95. Locking Fork Bushing
96. Locking Fork Pads
97. Locking Fork

Model 229 components

12. Remove the oil pump from the rear output shaft. Note the position of the pump for assembly reference. The recessed side of the pump faces the case interior.

13. Remove the rear output shaft from the viscous coupling.

14. Remove the 15 mainshaft pilot bearing rollers from the shaft or coupling (if the rollers dropped off during removal of the rear output shaft).

15. Remove the mainshaft O-ring from the end of the shaft.

16. Remove the viscous coupling from the mainshaft and side gear.

17. Remove the front output shaft, driven sprocket and drive chain assembly. Lift the front shaft, sprocket and chain upward. Tilt the front shaft toward the mainshaft. Slide the chain off the drive sprocket and remove the assembly.

18. Remove the mainshaft, side gear, clutch gear, drive sprocket and spline gear as an assembly. Place the assembly on a clean shop towel and set aside until the front case disassembly is completed.

19. Remove the front output shaft front thrust bearing assembly from the front case, or from the shaft (if the bearing and races remained on the shaft during removal).

20. Remove the drive chain from the front output shaft and sprocket.

21. Remove the snap-ring that retains the driven sprocket on the front output shaft. Mark the sprocket and shaft for assembly reference and remove the sprocket from the shaft.

22. Remove the mode fork, shift rail, and mode sliding clutch sleeve as an assembly. Mark the sleeve and fork for assembly reference and remove the sleeve from the fork.

NOTE: The mode fork and rail are pinned together so that they will operate as a unit. Remove the pin to separate the two components if necessary.

23. Remove the locking fork, high range sliding clutch sleeve, fork brackets and fork springs as an assembly. Note the position of the components for assembly reference and disassemble the components for cleaning and inspection.

24. Remove the range sector detent screw and remove the detent spring, plunger and ball.

25. Move the range operating lever downward to the last detent position.

26. Disengage the low range fork lug from the range sector slot.

27. Remove the retaining snap-ring from the annulus gear and remove the thrust washer.

28. Remove the annulus gear, range fork and rail as an assembly. Separate the components for cleaning and inspection.

29. Remove the planetary thrust washer from the planetary assembly hub.

30. Remove the planetary assembly. Grasp the planetary hub and lift the assembly upward to remove it.

31. Remove the mainshaft thrust bearing from the input shaft.

32. Remove the input shaft and remove the input shaft thrust bearing and race.

33. Remove the range sector and operating lever attaching nut and lockwasher. Remove the lever.

34. Remove the range sector and shaft from the front case.

35. Remove the range sector O-ring and retainer.

Mainshaft
DISASSEMBLY

1. Grasp the drive sprocket and lift the sprocket clutch gear and side gear upward and off the mainshaft.

2. Remove the mainshaft needle bearings and two bearing spacers from the mainshaft; a total of 82 bearings are used; note the spacer position for assembly reference.

3. Remove the spline gear and thrust washer from the mainshaft.

4. Remove the side gear, clutch gear, and clutch gear thrust washer from the sprocket carrier and sprocket.

5. Remove the clutch gear and thrust washer from the side gear.

6. Remove one sprocket carrier snap-ring and remove the drive sprocket from the carrier; mark for assembly reference.

--- CAUTION ---
The sprocket carrier and mainshaft needle bearings are different in size. Take care to avoid intermixing them.

7. Remove the three bearing spacers and all sprocket carrier needle bearings from the carrier; a total of 120 needle bearings are used.

8. Remove the rear output bearing and rear yoke seal from the rear retainer; the bearing is shielded on one side; note the bearing position for assembly reference.

9. Remove the input gear and front yoke seals from the front case; use a screwdriver to pry the seals out of the case.

Cleaning & Inspection

1. Wash all components thoroughly in clean solvent. Ensure that all lubricant, metallic particles, dirt, and foreign material are removed from the surfaces of every component.

2. Apply compressed air to each oil supply port and channel in each transfer case half to remove any obstructions or cleaning solvent residue.

3. Inspect all gear teeth for excessive wear or damage. Inspect all gear splines for burrs, nicks, wear or damage.

4. Remove minor nicks or scratches using an oilstone. Replace any component exhibiting excessive wear or damage.

5. Inspect all snap-rings and thrust washers for excessive wear, distortion and damage. Replace any component exhibiting these conditions.

6. Inspect the transfer case halves and rear retainer for cracks, porosity, damaged mating surfaces, stripped bolt threads and distortion. Replace any component exhibiting these conditions.

7. Inspect the viscous coupling and differential pinions. If the pinions or carrier are damaged or worn excessively, replace the coupling as an assembly only. If the coupling is cracked, leaking, or damaged, replace the coupling as an assembly only.

8. Inspect the condition of all needle, roller, ball and thrust bearings in the front and rear transfer case halves. Also inspect to determine the condition of the bearing bores in both transfer case halves and in the input gear, rear output shaft, side gear, and rear retainer.

9. Replace any component that is excessively worn or damaged. If any shaft, case half or input gear bearing requires replacement, refer to Bushing/Bearing Replacement.

NOTE: The front output shaft thrust bearing race surfaces are heat treated during manufacture. Heat treatment causes a brown or blue discoloration of these surfaces. Do not replace a front output shaft because of this type of discoloration.

Bearings & Bushings

The bearings and bushings used in the transfer case must be correctly positioned to avoid blocking the oil supply holes. After replacing any bearing or bushing, check the position and ensure that the supply hole is not obstructed.

REAR OUTPUT SHAFT BEARING

1. Remove the bearing using Remover Tool J-26941 and Slide Hammer J-2619-01 or equivalent. Remove the rear output lip seal using a small awl.

2. Install a replacement lip seal.

3. Install a replacement bearing using Driver Handle J-8092 and Installer Tool J-29166 or equivalent.

4. Remove the tools and inspect the oil supply hole. The bearing must not obstruct the supply hole.

FRONT OUTPUT SHAFT FRONT BEARING

1. Remove the bearing using Tools J-8092 and J-29168 or equivalent.

2. Remove the tools and inspect the oil supply hole. The bearing must not obstruct the supply hole.

FRONT OUTPUT SHAFT REAR BEARING

1. Remove the bearing using Remover Tool J-26941 and Slide Hammer J-2619-01 or equivalent.

2. Install a replacement bearing using Driver Handle J-8092 and Installer Tool J-29163 or equivalent.

3. Remove the installer tools and inspect the bearing position to ensure the oil

supply hole is not obstructed. Also ensure that the bearing is seated flush with the edge of the bore in the case to allow clearance for the thrust bearing assembly.

INPUT GEAR FRONT & REAR BEARINGS

1. Remove both bearings simultaneously using Driver Handle J-8092 and Remover Tool J-29170 or equivalent.

2. Install the new bearings one at a time. Install the rear bearing first; then install the front bearing. Use Driver Handle J-8092 and Installer Tool J-29169 or equivalent.

3. Remove the installer tools and inspect the bearing position to ensure the oil supply holes are not obstructed. Also ensure that the bearings are flush with the transfer case bore surfaces.

4. Install a replacement oil seal using seal Installer Tool J-29162 or equivalent.

MAINSHAFT PILOT BUSHING

1. Remove the bushing using Slide Hammer J-2619-01 and Remover Tool J-29369-1 or equivalent.

2. Install a replacement bearing using Driver Handle J-8092 and Installer Tool J-29174 or equivalent.

3. Inspect bushing position to ensure that the oil supply hole is not obstructed.

ANNULUS GEAR BUSHING

1. Remove the bushing using Driver Handle J-8092 and Remover/Installer Tool J-29185 or equivalent.

2. Install a replacement bushing using Tools J-8092 and J-29185-2 or equivalent.

3. Remove any chips generated by the bushing removal and/or installation.

REAR OUTPUT BEARING AND REAR YOKE SEAL

1. Remove the bearing using a brass drift and hammer.

2. Remove the seal from the retainer using a brass drift and hammer.

CAUTION

The rear output bearing is shielded on one side. Ensure that the shielded side faces the transfer case interior after installation.

3. Install a replacement bearing using Driver Handle J-8092 and Installer Tool J-7818 or equivalent.

4. Install a replacement seal in the retainer using Tool J-29162 or equivalent.

ASSEMBLY

NOTE: During assembly, lubricate all transfer case internal components with Dexron II® or petroleum jelly as indicated in the procedure. Do not use chassis lubricant or similar thick lubricants.

1. Install a replacement input shaft and rear output shaft bearing oil seals. Seat the

seals flush with the edge of the seal bore or in the seal groove in the transfer case. Coat the seal lips with petroleum jelly after installation.

2. Install the input shaft thrust bearing race in the transfer case counterbore.

3. Install the input gear thrust bearing on the input shaft and install the shaft and bearing in the transfer case.

4. Install the mainshaft thrust bearing in the bearing recess in the input shaft.

5. Install the planetary assembly on the input shaft. Ensure that the planetary pinion teeth mesh fully with the input shaft.

6. Install the planetary thrust washer on the planetary hub.

7. Install a replacement sector shaft O-ring and install the retainer in the shaft bore in the transfer case.

8. Install the O-ring on the mode sector shaft and insert the mode sector through the range sector.

9. Install the range sector in the front transfer case half. Install the operating lever and the snap-ring on the range sector shaft.

10. Install the lever, attaching washer, and locknut on the mode sector shaft. Tighten the locknut with 17 ft. lbs. torque.

11. Assemble the annulus gear, range fork and rail.

12. Install the assembled fork on and over the planetary assembly.

13. Ensure that the annulus gear is fully meshed with the planetary pinions.

14. Engage the range sector lug into the range sector.

15. Install the annulus thrust washer and the annulus retaining ring on the annulus gear hub.

16. Install the detent ball, plunger, spring and retaining screw in the front transfer case half detent bore.

17. Tighten the bolt with 22 ft. lbs. torque.

CAUTION

The locking mode clutch sleeve and the high range clutch sleeve are not interchangeable. The sleeve splines are different. Ensure that the correct sleeve is installed in the corresponding shift fork.

18. Assemble and install the locking fork, fork bracket, fork springs, and high range clutch sleeves.

19. Ensure that the lug on the fork is seated in the range sector detent slot.

20. Install the range fork lug in the range sector detent notch.

21. Move the range sector to the high range position.

22. Assemble and install the range fork, shift rail and mode clutch sleeve.

23. Install the thrust washer and a replacement O-ring on the mainshaft.

24. Install the needle bearings and bearing spacers on the mainshaft.

25. Coat the shaft bearing surface and all needle bearings with petroleum jelly.

26. Install the first 41 needle bearings.

27. Install the long bearing spacer, the remaining 41 needle bearings and the remaining short spacer.

28. Be careful to avoid displacing the bearings when the spacers are installed.

29. Use additional petroleum jelly to hold the bearings in place if necessary.

30. Install the spline gear on the mainshaft.

31. Take care to avoid displacing the bearings while installing the gear.

32. Install the sprocket carrier in the drive sprocket and install the sprocket carrier snap-rings.

33. Ensure that the carrier and sprocket are aligned according to the reference marks made during disassembly.

NOTE: The sprocket carrier teeth are tapered on one side and the drive sprocket has a deep recess on one side. Ensure that these components are assembled so that the carrier tapered teeth and sprocket recess are on the same side.

34. Install the sprocket carrier bearings and spacers.

35. Coat the carrier bore and all 120 carrier needle bearings with petroleum jelly.

36. Install the center spacer.

37. Install 60 bearings in each end of the carrier and install the remaining two spacers, one at each side of the carrier.

38. Use additional petroleum jelly to hold the bearings in place if necessary.

39. Install the assembled sprocket carrier and drive sprocket on the mainshaft. Do not displace the mainshaft bearings during installation.

40. Ensure that the recessed side of the drive sprocket is facing downward.

41. Install the clutch gear thrust washer in the mainshaft.

42. Position the washer on the sprocket carrier.

43. Install the clutch gear on the side gear.

44. Ensure that the tapered edge of the clutch gear faces the side gear teeth.

45. Install the assembled side gear and clutch gear on the mainshaft. Ensure that the side gear is fully seated in the sprocket carrier.

46. Take care to avoid displacing any of the carrier or mainshaft needle bearings.

47. Install the mainshaft and gear assembly in the case.

48. Ensure that the mainshaft is fully seated in the input gear.

49. Install the driven sprocket on the front output shaft and install the sprocket retaining snap-ring. Ensure that the sprocket is installed according to reference marks made during disassembly.

50. Install the front output shaft front thrust bearing assembly in the transfer case front half.

51. Install the thick race in the transfer case and then install the bearing and the thin race.

52. Install the drive chain, front output shaft and driven sprocket.

53. Install the chain on the driven sprocket.

54. Raise and tilt the driven sprocket and chain and install the opposite end of the chain on the drive sprocket.

55. Align the front output shaft with the shaft bore in the transfer case front half and install the shaft in the transfer case.

56. Ensure that the front shaft thrust bearing assembly is seated in the transfer case.

57. Install the front output shaft rear thrust bearing assembly on the front output shaft.

58. Install the thin race first, then install the bearing and thick race.

59. Install the viscous coupling on the side gear and clutch gear.

60. Ensure that the coupling is fully seated on the clutch gear. The clutch gear should be flush with the coupling and the gear teeth should be visible.

61. Coat the mainshaft pilot bearing surface and all 15 pilot roller bearings with petroleum jelly and install the bearings on the shaft.

62. Use additional petroleum jelly to hold the bearings in place if necessary.

63. Install the rear output shaft on the mainshaft and into the viscous coupling. Ensure that the shaft is completely seated in the coupling.

64. Tap the shaft with a plastic mallet or brass punch to seat it if necessary.

65. Do not displace the pilot bearings during installation of the shaft.

66. Install a replacement rear output shaft bearing seal in the rear transfer case half.

67. Apply a bead of Loctite® 515, or equivalent sealer, to the mating surface of the rear transfer case half.

68. Install the magnet in the case, if removed.

69. Attach the rear transfer case half to the front transfer case half. Ensure that the alignment dowels at the front case half ends are aligned with the bolt holes in the rear case half and mate the rear case half with the front case half.

NOTE: If the rear transfer case half will not mate completely with the front case, inspect for the following: oil in the range fork rail bore, the front output shaft rear thrust bearing assembly is not aligned with the rear case half, the mainshaft is not completely seated, the rear case half is not aligned with the oil pump.

70. Install the rear case half to the front case half bolts. Tighten the bolts with 23 ft. lbs. torque. Ensure that the flat washers are used on the bolts at the case end where the alignment dowels are located.

71. Install the speedometer drive gear on the rear output shaft.

72. Measure the thickness of the shim pack and record.

73. Install a 0.030 in. shim (approximately) on the rear output shaft.

74. Align the rear retainer on the rear transfer case half and install the retainer.

Install the retainer bolts. Tighten the bolts securely but not with the specified torque.

75. Install the front rear output shaft yokes and the original yoke nuts. Tighten the nuts finger-tight only. Check the differential end play.

76. Set the shift lever in the 4–High range position.

77. Position Dial Indicator J–8001 on the rear retainer and position the indicator stylus so it contacts the rear yoke nut.

78. Support the transfer case to prevent the front output yoke from turning.

79. Slowly turn the rear output shaft while maintaining moderate inward pressure on the rear yoke. Turn the rear output shaft at least two full turns to determine the maximum run-out of the shaft.

NOTE: A wrench should be used to turn the yoke to provide the leverage needed to turn the viscous coupling in the transfer case.

80. Set the shaft at its maximum run-out point and zero the dial indicator.

81. Pull upward on the rear output yoke, note the dial indicator pointer position and record it.

82. Remove the retainer. Add or subtract differential shims as necessary to correct the end play. The end play should be between 0.002 to 0.010 in. The recommended end play is 0.006 in.

83. After adjusting the end play, remove the front and rear yokes. Discard the original yoke nuts.

84. Apply a bead of Loctite® 515, or equivalent sealer, to the retainer mating surface and install the retainer.

85. Apply sealer to the retainer bolts and install the bolts. Tighten the bolts with 23 ft. lbs. torque.

86. Position the front and rear yokes. Install replacement yoke seal washers and nuts.

87. Tighten the yoke nuts with 120 ft. lbs. torque. Use Tool J–8614–01 or equivalent to hold the yokes in place while tightening the nuts.

88. Install the detent ball, spring and bolt if these were not installed previously. Apply sealer to the bolt before installing it. Tighten the bolt with 23 ft. lbs. torque.

89. Install the drain plug and washer.

90. Fill the transfer case with 6 pints of Dexron® II and install the fill plug and washer.

91. Tighten the drain and fill plugs with 18 ft. lbs. torque.

92. Install the plug and washer in the front transfer case half, if removed. Tighten the plug with 18 ft. lbs. torque.

93. Reinstall transfer case in vehicle.

Warner Quadra-Trac®

The Quadra-Trac® transfer case provides full-time, four-wheel drive under all driving conditions. The front and rear driveshafts are driven by a limited slip differen-

tial in the transfer case. The limited slip differential is connected to the input shaft by a link-belt type chain. In operation, if the rear axle loses traction, then the engine torque will be transfered through the transfer case differential to the front axle.

The transfer case contains a manually actuated lockout system that locks the front and rear driveshafts together, cancelling the differential action. This feature is used under extreme marginal traction situations.

NOTE: In order to spare the transfer case differential side gears and brake cones from excessive and possibly damaging wear, do not spin the wheels excessively when the vehicle is stuck or bogged down.

An optional gear reduction unit mounted at the rear of the input shaft is available for the Quadra-Trac® unit, making it a two-speed transfer case.

PERFORMANCE CHECKS
TRANSFER CASE
DIFFERENTIAL TORQUE
BIAS CHECK

1. With the lock-out feature NOT engaged and the transmission in Park, raise the vehicle until the front wheels are free of the ground.

2. Disconnect the rear driveshaft from the transfer case.

3. Turn the rear yoke retaining nut with a torque wrench and socket, taking note of how much torque is required to force the cone clutches to slip. They should slip when 110 to 270 ft. lbs. are applied. Slippage below 10 ft. lbs. indicates replacement of the differential is needed. If no slippage occurs at 270 ft. lbs., improper lubrication is indicated. Drain and refill the transfer case and reduction unit, if so equipped, with the proper lubricant mixture.

DRIVE CHAIN TENSION CHECK

1. Drain the lubricant from the transfer case.

2. Remove the chain inspection plug and insert a steel rule into the hole.

3. A new chain will be 1.575 in. from the outer edge of the plug hole. When the slack in the chain reaches 1/2–3/4 in., the chain should be replaced. No adjustment is possible.

4. Reinstall the drain and chain inspection plugs, and refill the unit with the proper lubricant mixture.

REAR CASE COVER

Most Quadra-Trac® components can be serviced without removing the complete unit from the vehicle. To gain access to the rear output shaft, drive sprocket and thrust

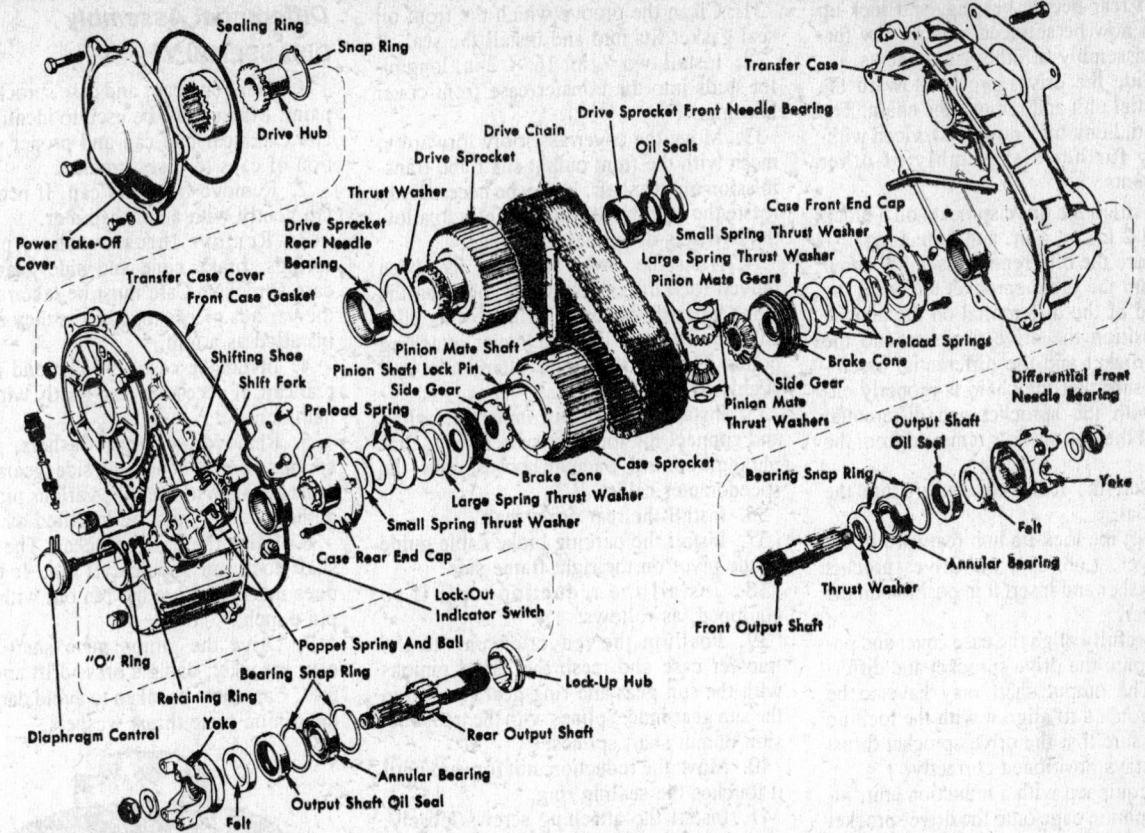

Exploded view of the Warner Quadra-Trac® without the optional reduction unit (© Jeep Corp.)

TORQUE SPECIFICATIONS Model 229

Component	Service Set-To Torque
Detent Retainer Bolt	31 N·m (23 ft-lbs)
Drain and Fill Plugs	24 N·m (18 ft-lbs)
Front/Rear Yoke Nuts	163 N·m (120 ft-lbs)
Operating Lever Locknut	24 N·m (18 ft-lbs)
Rear Case-to-Front Case Bolts (All)	31 N·m (23 ft-lbs)
Rear Retainer Bolts	31 N·m (23 ft-lbs)
Transfer Case-to-Transmission Adapter Nuts	35 N·m (26 ft-lbs)
Universal Joint Strap Bolt-to-Transfer Case	19 N·m (170 in-lbs)

washer, chain, differential and needle bearing, or the diaphragm control system, just the rear cover has to be removed.

1. Lift and support the vehicle.

2. If the vehicle is equipped with a reduction unit, continue on to the next step for the reduction unit removal procedure. If the vehicle is not equipped with a reduction unit, proceed to Step 7.

3. Loosen all the bolts that attach the reduction unit to the transfer case cover.

4. Move the reduction unit backward just enough to allow the oil to drain from the unit.

5. Loosen the cable retaining bolt at the shift control lever. Loosen the cable clamp bolt and remove the control cable from the clamp bracket and control lever.

6. When the oil has drained, remove the bolts which hold the reduction unit to the transfer case cover. Move the reduction unit rearward to clear the transmission output shaft and pinion cage which is attached to the transfer case drive sprocket. The pinion cage will remain with the transfer case assembly.

NOTE: The pinion cage should not be removed if the transfer case cover assembly is to be removed, but may be removed for inspection or replacement if the transfer case cover assembly is to remain in the vehicle. Removal of the pinion cage involves only removing the snap-ring which holds the cage to the sprocket and sliding the cage backward.

7. Remove the transfer case drain plug and allow the unit to drain.

8. Mark the rear output shaft yoke and universal joint to provide an alignment reference during reassembly. Disconnect the rear drive shaft front universal joint from the transfer case rear yoke.

9. Mark the diaphragm control vacuum hoses for identification during reassembly, then disconnect them. Also remove the lock-up indicator switch wire and the speedometer cable. Remove the indicator switch.

10. Disconnect the parking brake cable guide from the pivot on the right frame side.

11. Remove the bolts which attach the case cover assembly to the case (front housing). Carefully slide the cover assembly backward off the front output shaft and the transmission output shaft.

12. To disassemble the unit, remove the rear output shaft yoke.

13. If the unit is NOT equipped with a reduction unit, remove the power take-off cover from the rear of the transfer case cover. Remove the sealing ring from the transfer case cover.

14. Using a piece of wood 2 × 4 in. and 6 in. long, support the cover and drive sprocket.

15. If NOT equipped with a reduction unit, remove the drive hub and sleeve from the drive sprocket rear splines by expanding the internal snap-ring. The ring expanding tabs are accessible through a slot in the outside edge of the drive sleeve.

16. If equipped with a reduction unit, remove the pinion cage snap-ring and carrier.

17. Lift the case cover from the drive sprocket and differential. The cover, rear output shaft, bearings and seal, drive

sprocket rear needle bearing, and lock-up hub can now be serviced without any further disassembly of other components.

18. Slide the drive sprocket toward the differential unit and remove the chain. The differential unit may now be serviced without any further disassembly of other components.

19. Position the drive sprocket on a block of wood 2 in. × 4 in. and 6 in. long.

20. Place the differential assembly about 2 in. from the drive sprocket and with the front end of the differential on the bench.

21. Position the drive chain around the drive sprocket and the differential assembly. Be sure that the chain is properly engaged with the sprocket and differential teeth and that the slack is removed from the chain.

22. Insert the rear output shaft into the differential.

23. Shift the lock-up hub rearward in the case cover. Lubricate the drive sprocket thrust washer and insert it in position on the case cover.

24. Carefully align the case cover and position it onto the drive sprocket and differential. The output shaft may have to be slightly rotated to align it with the lock-up hub. Be sure that the drive sprocket thrust washer stays positioned correctly.

25. If equipped with a reduction unit, install the pinion cage onto the drive sprocket rear splines. Install the snap-ring. Be sure that the snap-ring seats properly in the groove.

26. If the vehicle is NOT equipped with a reduction unit, assemble the drive hub, drive sleeve, and snap-ring, then install them onto the drive sprocket rear splines. Be sure the snap-ring seats properly.

27. Turn the drive sleeve or pinion cage to make sure the drive sprocket thrust washer did not come out of position. No binding should be present.

28. If *not* equipped with a reduction unit, install the power take-off sealing ring and cover and tighten the attaching screws.

29. Install the speedometer gear on the rear output shaft.

30. Install the rear output shaft oil seal and the rear yoke and nut. Tighten the nut to specification.

31. Clean the groove which the front oil seal gasket fits into and install the seal.

32. Install two ³⁄₈ in. 16 × 2 in. long pilot studs into the transfer case front cover housing.

33. Move the cover assembly forward to mesh with the front output shaft and transmission output shaft. It may be necessary to rotate the rear output shaft slightly to allow the two sets of splines to engage.

34. After the cover assembly has been moved forward and is evenly touching the front half of the case, remove the pilot studs and install the rear cover attaching bolts. Tighten the bolts alternately and evenly to specifications.

35. Install the lock out indicator switch and connect the lock out switch wire, diaphragm control vacuum hoses, and the speedometer cable.

36. Install the rear drive shaft.

37. Install the parking brake cable guide to the pivot on the right frame side.

38. Install the reduction unit, if so equipped, as follows:

39. Position the reduction unit to the transfer case and mesh the caged pinions with the sun gear and ring gear, and align the sun gear inner splines with the transmission output shaft splines.

40. Move the reduction unit forward until it touches the sealing ring.

41. Install the attaching screws loosely, then tighten them alternately to specification.

42. Connect the shift control cable and adjust it by first removing the swivel block from the control lever. Move the control lever to the most forward position. Thread the swivel block in or out on the cable end to obtain the correct length to fit the swivel block in the control lever.

43. Install the proper type and amount of lubricant and lower the vehicle.

NOTE: Use 8 oz. of Jeep Lubricant Concentrate Part No. 8123004 or 5356068 or Lubrizol® 762 (there is no substitute) mixed with SAE 30 non-detergent motor oil. 3.5 pints of the mixture is required to fill the transfer case without a reduction unit, 4.5 pints with a reduction unit.

Differential Assembly
DISASSEMBLY

1. Mark end caps and case sprocket with paint. Marks must be used to identify front end cap, rear end cap and proper orientation of caps to case sprocket.

2. Remove front end cap. If necessary, tap gently with a soft hammer.

3. Remove thrust washer, preload springs, brake cone and side gears from case sprocket. Care must be taken to keep the various pieces together as they must be installed as a unit.

4. Invert the case sprocket and remove rear cap. If necessary, tap gently with a soft hammer.

5. Remove the thrust washers, preload springs, brake cone and side gears. Care must be taken to keep the various pieces together as they must be installed as a unit.

6. Raise the case sprocket. The pinion shaft lock pin should fall out. If the pin does not fall, drive the pin out with a ¼ in pin punch.

7. Drive the pinion mate shaft out of case sprocket, using a brass drift and hammer. Care must be taken to avoid damaging the pinion mate thrust washers.

Mark the end caps and case sprocket of the differential assembly before disassembling the differential—Warner Quadra-Trac® (© Jeep Corp.)

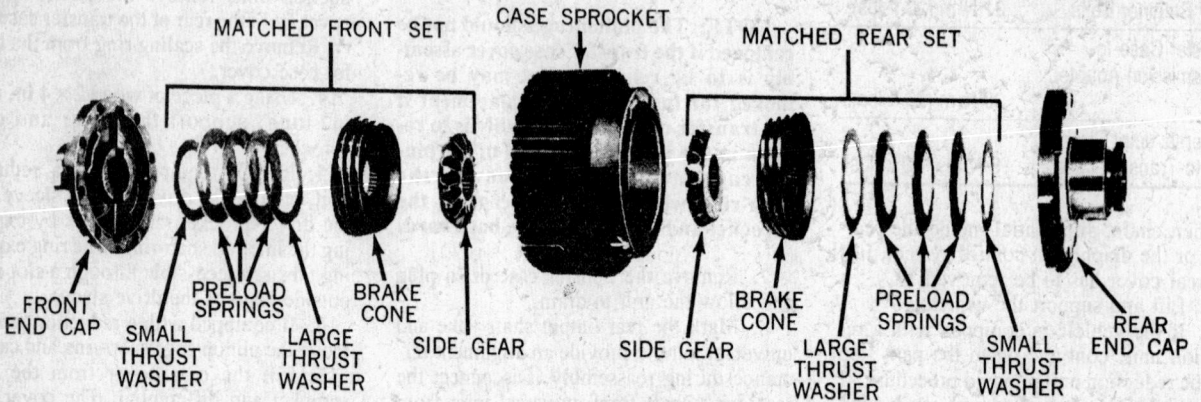

Exploded view of the differential assembly—Warner Quadra-Trac® (© Jeep Corp.)

The Quadra-Trac® transfer case positioned for disassembly (© Jeep Corp.)

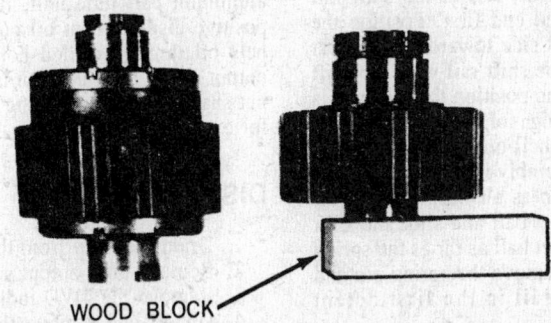

The differential and drive sprocket positioned for installation of the drive chain—Warner Quadra-Trac® (© Jeep Corp.)

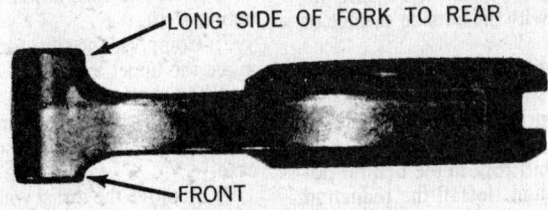

Install the lock-up hub assembly shift fork with the long side of the fork facing toward the rear—Warner Quadra-Trac® (© Jeep Corp.)

8. Thoroughly clean and inspect all component parts. Replace any damaged or worn parts with a complete matched set.

ASSEMBLY

Prelubricate all bearings and thrust surfaces with Jeep Lubricant Concentrate Part No. 8123004 or 5356068 or Lubrizol® 762 (there is no substitute) prior to installation.

1. Slide the pinion mate shaft in the case sprocket three inches.
2. Install the pinion mate thrust washers and gears on shaft in the proper order.
3. Align the pinion mate shaft lockpin hole with hole in case sprocket. Lightly drive the pinion mate shaft into case sprocket until lockpin holes are exactly aligned.

4. Move the pinion mate gears apart until the gears are pressing the washers against the case sprocket.
5. Engage the pinion mate gear with the front side gears. Insert the brake cone over the gear and in case sprocket. Install the large thrust washer and preload springs, concave side of springs facing toward brake cone.
6. Lubricate the small thrust washer and place it on the front end cap. Install the front end cap, secure with attaching screws and alternately tighten to proper torque. Make certain that alignment marks are in order.
7. Invert the case sprocket and end cap and install the pinion shaft lock pin.
8. Mesh remaining side gear with pinion mate gears.

9. Insert the remaining brake cone over the side gear. Install the large thrust washer and preload spring, concave side of springs facing toward brake cone.
10. Lubricate the small thrust washer and place it on the rear end cap. Install the rear cap, tighten attaching screws finger tight.
11. Insert the front and rear output shafts in the differential and rotate until both shafts have aligned with the splines on the brake cones and side gears. Alternately tighten the retaining screws to proper torque.

DIAPHRAGM CONTROL, SHIFT FORK AND LOCK-UP HUB

1. Remove the vent cover and seal ring.
2. Remove the retaining rings positioning the shift fork on diaphragm. Carefully pry the shift fork forward to gain access to the retaining rings. Remove the spring with a magnet.

SPECIFICATIONS

TORQUE (FT. LBS.)

Transfer case breather	6–10
Chain measuring access hole plug	6–14
Drain plug	15–25
Fill plug	15–25
Lock-up cover to transfer case	8–10
Lock-out indicator switch	10–15
Output Shaft Nut	90–150
PTO cover to transfer case bolts: ⅜ in.-16	15–25
⁵⁄₁₆ in.-18	10–20
Speedometer adapter	20–30
Transfer case cover to transfer case	15–25
Transfer case to transmission extension bolt	30–50
Reduction unit cable housing clamp nut	7–12
Fill plug	15–25
Shift lever cable clamp nut	10–20
Shift lever to shift nut	15–25
Reduction PTO cover to case	15–25
Reduction unit to transfer case bolts: ⅜ in.-16	15–25
⁵⁄₁₆ in.-18	8–10

3. Caution must be exercised in removal of the diaphragm control rod as it is retained by a spring loaded detent ball. Insert a magnet into the hole to hold the detent ball. Slip the diaphragm control rod out of case. Remove detent ball and spring.

4. Remove shifting fork, plastic shifting shoes and lock-up hub.

6. Lubricate the shifting shoes and place them in the shift fork.

7. Install the shift fork and lock-up hub assembly in the case cover, long end of shift fork first (toward rear). Make certain that the shift fork does not separate from the lock-up hub by reaching through the needle bearings.

8. Insert the diaphragm control rod in the case and shifting fork, stopping before the detent ball hole is reached.

9. Install the detent ball and spring. Depress the detent ball with a ¼ in. pin punch and slide the diaphragm control rod into place.

10. Install the shift fork retaining pin, (clips) and the diaphragm retaining spring, the spring should be below the surface of the bore.

11. Install the vent cover and seal ring.

Reduction Unit
DISASSEMBLY

1. Remove PTO cover and gasket.

2. Remove snap-ring from reduction main shaft rear end, slide the reduction main shaft and sun gear assembly forward and out. Remove needle bearings.

3. Remove as an assembly, the ring gear, reduction collar plate, pinion cage lock plate, shift collar hub and reduction collar hub. Using a soft hammer, remove the shift collar hub from the pin<chion cage lock plate.

4. Remove the pinion cage lock plate, needle bearing, ring gear, reduction collar plate and shift collar hub. Separate the reduction collar hub and needle bearing from shift collar hub.

5. When necessary, separate the reduction collar plate and ring gear by removing retaining snap-ring.

6. Remove needle bearing and direct drive sleeve from the reduction shift collar.

7. Shift the reduction shift collar to the neutral (center) detent with the control lever. Disengage the shift fork. Place the shift in the direct drive detent position (rear), align the collar outer teeth with the inner teeth in the reduction holding plate. Place the fork and collar in the reduction position (front) detent, and remove the reduction shift collar.

8. Remove the annular bearing rear snap-ring and bearing.

9. Remove the shift fork locating spring pin, large expansion plug, shift rail taper plug, and control lever.

10. Drive the spring pin out of the shift fork and rail with a 3/16 in. pin punch. Slide the shift rail forward out of the shift fork, and remove the shift fork. Remove the

spring shift fork poppet ball. Drive the poppet taper plug into the shift rail bore and remove the plug and spring.

11. Remove the shift lever retaining pin and the shift lever assembly.

12. Remove the reduction holding plate snap-ring and reduction holding plate.

ASSEMBLY

1. Align the shift fork locating spring holes in the reduction holding plate and housing. Install the reduction holding plate. Locating pins should index the plate in the case. Secure the plate with a snap-ring, tabs facing forward.

2. Install the shift lever assembly into the housing, lever towards the rear. Position seal ring on groove in shift lever shaft.

3. Move the shift lever assembly inward and install taper pin.

4. Install the shift rail in the shift rail rear bore, grooved end first. Position the shift rail with flat side towards the poppet spring. Engage the shift rail with the shift lever assembly and position the rail so it is flush with the edge of the poppet bore. Place the poppet ball on the end of spring and insert the assembly in the poppet bore, using a spring pin as an installation tool. Depress the poppet ball and slide the shift rail over the poppet ball as far as the spring pin will allow. Remove the spring pin and place the shift rail in the first detent position.

5. Position the shift rail so the flat side is facing the shift lever assembly and the spring pin bore is aligned with the spring pin bore in the shift fork. Once the spring pin holes are aligned, install the spring pin so that it is flush with the outside surface of the shift fork.

6. Install the shift rail taper plug, poppet bore taper plug, shift rail cover expansion plug, shift fork spring locating pin and the control lever.

7. Place the shift fork in the neutral position (center) detent. Install the reduction shift collar so that the outer teeth engage with the reduction holding plate inner teeth, and the shift collar fork groove forward of the shift fork. Place the shift fork in the direct drive (rear) detent. Move the shift collar away and to the rear of the shift fork until the groove aligns with the shift fork. Engage the collar groove with the shift fork.

8. Place the direct drive sleeve in the reduction shift collar, needle bearing surface facing toward the front. Lubricate and install the needle bearing, against the direct drive sleeve.

9. Assemble the reduction collar plate hub and ring gear. Make certain that snap-rings are seated in their grooves.

10. Install the needle bearing and reduction collar hub on the shift collar hub.

11. Install the ring gear, reduction collar plate and hub on the shift collar hub.

12. Place a needle bearing on the shift collar hub and the reduction collar hub.

13. Tap the pinion cage lock into place

on the shift collar hub with a soft hammer and install assembly in housing. Place needle bearings on the shift collar hub and the pinion cage lock plate.

14. Install the reduction main shaft and sun gear into shift collar hub and through the direct drive sleeve and annular bearing. With a brass drift gently tap the assembly as far to the rear as possible. Place the rear spacer on the main shaft and secure with the selective snap-ring which gives the tightest fit, between .004 in. and .009 in. clearance. Snap rings are available in thicknesses ranging from 0.089 in. to 0.105 in.

15. Install PTO cover and gasket, tighten attaching bolts to proper torque.

Warner Model 13–45

The Warner Model 13–45 is a two piece all aluminum part time unit, lubricated by a positive displacement oil pump that channels oil through drilled holes in the rear output shaft. The pump turns with the output shaft and allows towing of the vehicle for extended distances.

DISASSEMBLY

1. Drain the fluid from the case.

2. Remove both output shaft yokes.

3. Remove the 4WD indicator switch.

4. Unbolt and remove the case cover. The cover may be pried off using a screwdriver in the pry bosses.

5. Remove the magnetic chip collector from the bottom of the case.

6. Slide the shift collar hub off the rear output shaft.

7. Compress the shift fork spring and remove the upper and lower spring retainers from the shaft.

8. Lift the four wheel drive lockup fork and lockup shift collar assembly from the case.

9. Remove the thrust washer being careful not to lose the nylon wear pads on the lockup fork.

10. Remove the snap ring and thrust washer from the front output shaft.

11. Grip the chain and both sprockets and lift them straight up to remove the drive sprocket, driven sprocket and chain from the output shafts.

12. Lift the front output shaft from the case.

13. Remove the four oil pump attaching screws and remove the oil pump rear cover, pickup tube, filter and pump body, two pump pins, pump spring and oil pump front cover from the rear output shaft.

14. Remove the snap ring that holds the bearing retainer inside the case. Lift the rear output shaft while tapping on the bearing retainer with a plastic hammer.

NOTE: Two dowel pins will fall into the case when the retainer is removed.

15. Lift the rear output shaft and bearing retainer from the case. Remove the rear

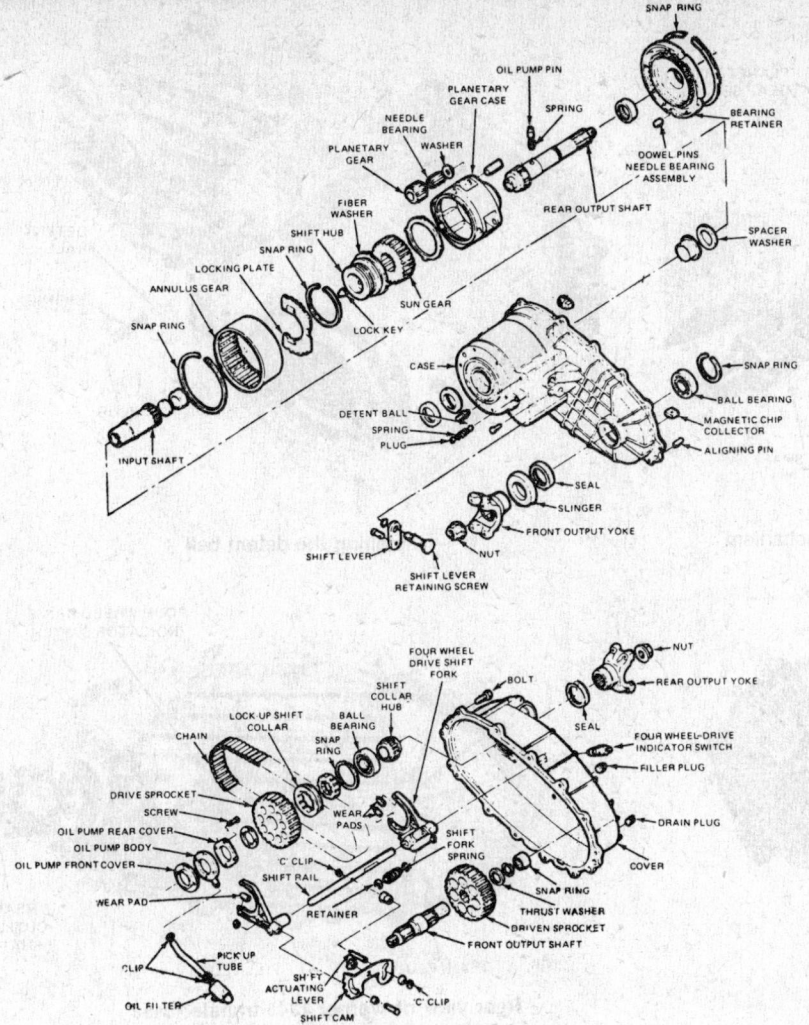

Exploded view of Warner 1345

output shaft from the bearing retainer. If necessary, press the needle bearing assembly out of the retainer.

16. Remove the C-clip that holds the shift cam to the actuating lever inside the case.

17. Remove the retaining screw and lift the shift lever from the case.

NOTE: When removing the lever, the shift cam will disengage from the shift lever shaft and may release the detent ball and spring from the case.

18. Remove the planetary gear set, shift rail, shift cam, input shaft and shift forks, as an assembly, from the case. Be careful not to lose the two nylon wear pads on the shift fork.

19. Remove the spacer washer from the bottom of the case.

20. Drive the plug from the detent spring bore.

ASSEMBLY

Before assembly, lubricate all parts with clean automatic transmission fluid; Dexron® II.

1. Assemble the planetary gear set, shift rail, shift cam, input shaft and shift fork together as a unit. Make sure that the boss on the shift cam is installed toward the case. Install the spacer washer on the input shaft.

2. Place the rear output shaft in the planetary gear set, making sure that the shift cam engages the shift fork actuating pin.

3. Lay the case on its side. Insert the rear output shaft and planetary gear set into the case. Make sure the spacer washer remains on the input shaft.

4. Install the shift rail into the hole in the case. Install the outer roller bushing into the guide in the case.

5. Remove the rear output shaft and position the shift fork in neutral.

6. Place the shift control lever shaft through the cam, and install the clip ring. Make sure that the shift control lever is pointed downward and is parallel to the front face of the case.

7. Check the shift fork and planetary gear engagement.

8. If removed, press a new needle bearing assembly into the bearing retainer.

9. Insert the output shaft through the bearing retainer from the bottom outward.

10. Insert the rear output shaft pilot into the input shaft bushing. Align the dowel holes and the lower bearing.

11. Install the dowel pins. Install the snap ring that retains the bearing retainer in the case.

12. Insert the detent ball and spring in the detent bore in the case. Coat the seal plug with RTV sealant or its equivalent. Drive the plug into the case until the lip of the plug is $1/32$ in. below the surface of the case. Peen the case over the plug in two places.

13. Install the pump front cover over the output shaft with the flanged side down. The word "TOP" must be facing the top of the transfer case.

14. Install the oil pump spring and two pump pins with the flat side outward in the hole in the output shaft. Push both pins in to install the oil pump body, pickup tube and filter.

15. Place the oil pump rear cover on the output shaft with the flanged side outward. The word "TOP" must be positioned toward the top of the case. Apply Loctite® or its equivalent to the oil pump bolts and torque them to 36–40 inch lbs.

16. Install the thrust washer on the rear output shaft nest to the oil pump.

17. Place the drive sprocket on the front output shaft. Install the snap ring and thrust washer.

18. Install the chain on the drive sprocket and driven sprocket. Lower the chain into position in the case. The driven sprocket is installed through the front output shaft bearing and the drive sprocket is installed in the rear output shaft.

19. Engage the 4WD shift fork on the shift collar. Slide the shift fork over the shift shaft and the shift collar over the rear output shaft. Make sure the nylon wear pads are installed on the shift fork tips and the necked-down part of the shift collar is facing downward.

20. Push the 4WD shift spring downward and install the upper spring retainer. Push the spring upward and install the lower retainer.

21. Install the shift collar hub on the rear output shaft.

22. Apply a bead of RTV sealant on the case mounting surface. Lower the cover over the rear output shaft. Align the shift rail with its blind hole in the cover. Make sure the front output shaft is fully seated in its support bearing. Install and tighten the bolts to 40–45 ft. lbs. Allow one hour curing time for the RTV sealant prior to using the case.

23. Install the 4WD indicator switch. Torque to 8–12 ft. lbs.

24. Press the oil slinger on the front yoke. Install the front and rear output shaft yokes. Coat the nuts with Loctite® or equivalent and torque to 100–130 ft. lbs.

25. Fill the unit with 6 pints of Dexron® II. Tighten the fill plug to 18 ft. lbs.

26. Install the unit in the vehicle and start the engine. Remove the level plug. If the

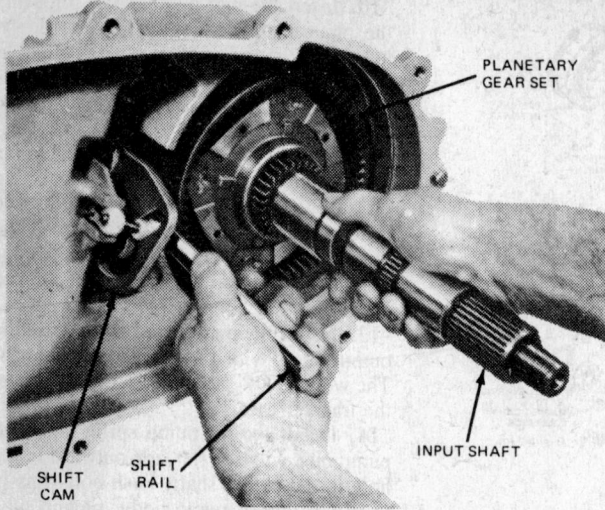

Installing the planetary gear set and shifter mechanism

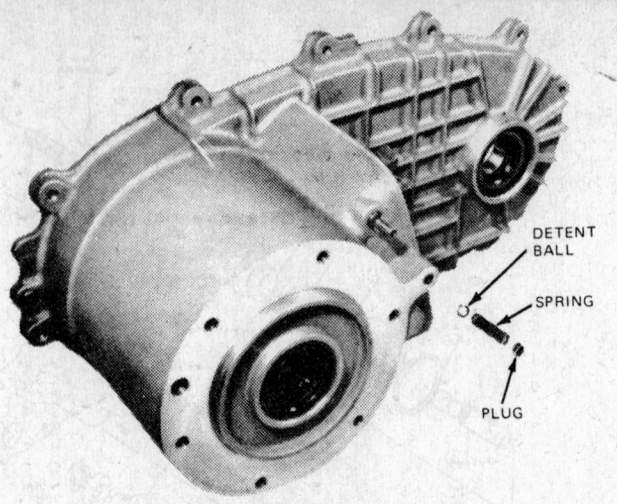

Installing the detent ball

Front view of Warner 1345 transfer case

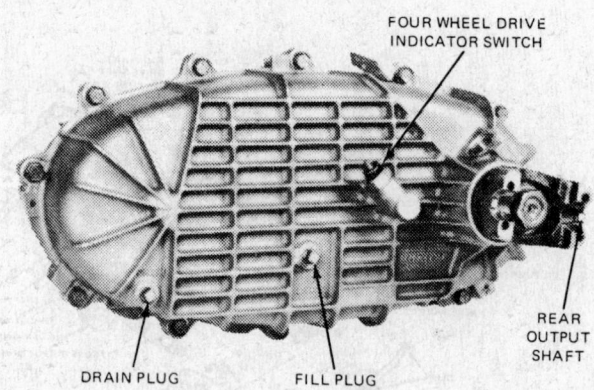

Rear view of Warner 1345 transfer case

fluid is flowing from the hole in a stream, the pump is not operating properly. The fluid should drip slowly from the hole.

Warner Model 13–50

The Warner Model 13–50 is a three-piece aluminum part time transfer case. It transfers power from the transmission to the rear axle and when actuated, also to the front drive axle. The unit is lubricated by a positive displacement oil pump that channels oil flow through drilled holes in the rear output shaft. The pump turns with the rear output shaft and allows towing of the vehicle at maximum legal road speeds for extended distances without disconnecting the front and/or rear driveshaft.

Shift Lever
REMOVAL

NOTE: Remove the shift ball only if the shift ball, boot or lever is to be replaced. If the ball, boot or lever is not being replaced, remove the ball, boot and lever as an assembly.

1. Remove the plastic insert from the shift ball. Warm the ball with a heat gun to 140°–180°F and knock the ball off the lever with a block of wood and a hammer. Be careful not to damage the finish on the shift lever.

2. Remove the rubber boot and floor pan cover.

3. Disconnect the vent hose from the control lever.

4. Unscrew the shift lever from the control lever.

5. Remove the bolts retaining the shifter to the extension housing. Remove the control lever and bushings.

DISASSEMBLY

1. Remove the transfer case from the vehicle.

2. Remove the transfer case drain plug with a ⅜ inch drive ratchet and drain the fluid.

3. Remove the four-wheel drive indicator switch and the breather vent.

4. Remove the rear output shaft yoke by removing the 30mm nut, steel washer and rubber seal from the output shaft.

5. Remove the nine 15mm bolts which retain the front case to the rear cover. Insert a ½ inch drive breaker bar between the three pry bosses and separate the front case from the rear cover. Remove all traces of RTV gasket sealant from the mating surfaces of the front case and rear cover. When removing the RTV sealant, take care not to damage the mating surfaces of the aluminum case.

6. If the speedometer drive gear or ball bearing assembly is to be replaced, first, drive out the output shaft oil seal from either the inside of the rear cover with a brass drift and hammer or from the outside by bending and pulling on the curved-up lip of the oil seal. Remove and discard the oil seal. Remove the speedometer drive gear assembly (gear, clip and spacer). Note that the round end of the speedometer gear clip faces the inside of the rear cover.

7. Remove the internal snap-ring that retains the rear output shaft ball bearing in the bore. From the outside of the case, drive out the ball bearing with Output Shaft Bearing Replacer, T83T-7025-B and Drive Handle, T80T-4000-W or equivalent.

8. If required, remove the front output

shaft caged needle bearing from the rear cover with Puller Collet, D80L–100–S and Impact Slide Hammer, T50T–100–A or equivalent.

9. Remove the 2W–4W shift fork spring from the boss in the rear cover.

10. Remove the shift collar hub from the output shaft. Remove the 2W–4W lock-up assembly and the 2W–4W shift fork together as an assembly. Remove the 2W–4W fork from the 2W–4W lock-up assembly. If required, remove the external clip and remove the roller bushing assembly (bushing, shaft and external clip) from the 2W–4W shift fork.

11. If required to disassemble the 2W–4W lock-up assembly, remove the internal snap-ring and pull the lock-up hub and spring from the lock-up collar.

12. Remove the external snap-ring and thrust washer that retains the driven sprocket to the front output shaft.

13. Remove the chain, driven sprocket and drive sprocket as an assembly.

14. Remove the collector magnet from the notch in the front case bottom.

15. Remove the output shaft and oil pump as an assembly.

16. If required to disassemble the oil pump, remove the four 8mm bolts from the body. Note the position and markings of the front cover, body, pins, spring, rear cover, and pump retainer as removed.

17. Pull out the shift rail.

18. Slip the high-low range shift fork out of the inside track of the shift cam. If required, remove the external clip and remove the roller bushing assembly (bushing, shaft and external clip) from the high-low range shift fork.

19. Remove the high-low shift hub from out of the planetary gearset in the front case.

Removing the input shaft from the planetary gear set

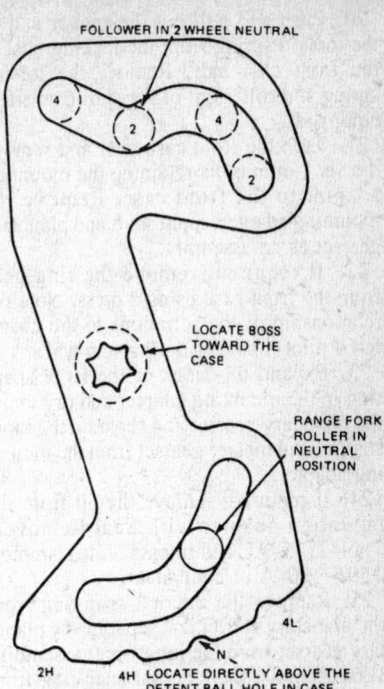

Connect shift cam engagement

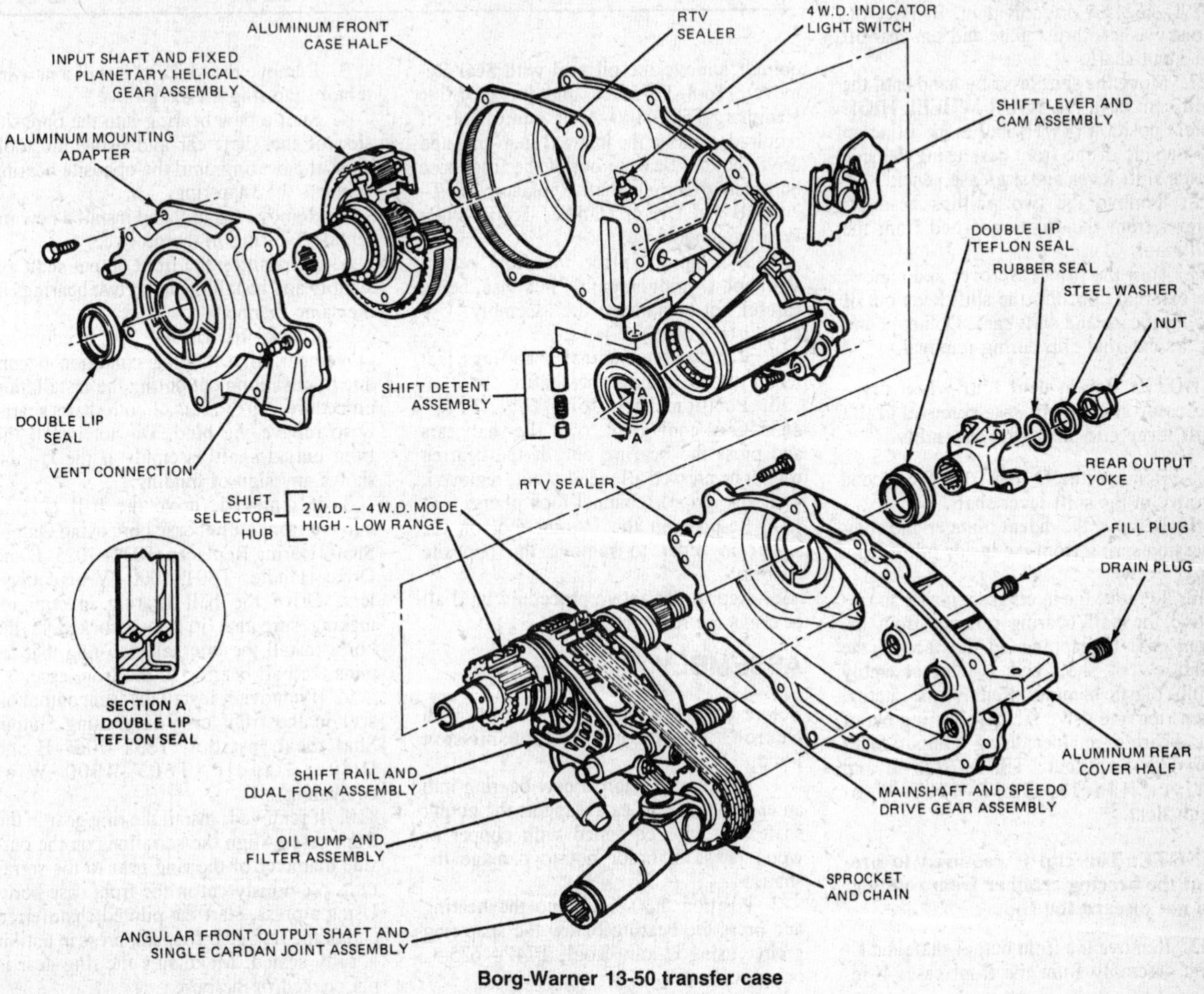

Borg-Warner 13-50 transfer case

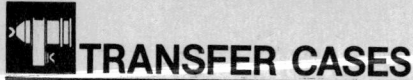

20. Push and pull out the anchor end of the torsion spring from the locking post in the front case half. Remove the torsion spring and roller out of the shift cam (if so equipped).

21. Turn the front case over and remove the six 15mm bolts retaining the mounting adapter to the front case. Remove the mounting adapter, input shaft and planetary gearset as an assembly.

22. If required, remove the ring gear from the front case using a press. Note the relationship of the serrations to the chamfered pilot diameter during removal.

23. Expand the tangs of the large snapring in the mounting adapter and pry under the planetary gearset and separate the input shaft and planetary gearset from the mounting adapter.

24. If required, remove the oil from the mounting adapter with Seal Remover, Tool–1175–AC and Impact Slide Hammer, T50T–100–A or equivalent.

25. Remove the internal snap-ring from the planetary carrier and separate the planetary gearset from the input shaft assembly.

26. Remove the external snap-ring from the input shaft. Place the input shaft assembly in a press and remove the ball bearing from the input shaft using Bearing Splitter, D79L–4621–A or equivalent. Remove the thrust washer, thrust plate and sun gear off the input shaft.

27. Move the shift lever by hand until the shift cam is in the FOUR WHEEL HIGH detent position (4WH) and mark a line on the outside of the front case using the side of the shift lever and a grease pencil.

28. Remove the two phillips head set screws from the front case and from the shift cam.

29. Turn the front case over and remove the external clip. Pry the shift lever out of the front case and shift cam. Do not pound on the external clip during removal.

NOTE: Removal of four-wheel drive indicator switch will ease removal of the shift lever and shift cam assembly.

30. Remove the O-ring from the second groove in the shift lever shaft.

31. Remove the detent plunger and compression spring from the inside of the front case.

32. Remove the internal snap-ring and remove the ball bearing retainer from the front case by tapping on the face of the front output shaft and U-joint assembly with a plastic hammer. Remove the internal snap-ring and drive the ball bearing out of the bearing retainer using Output Shaft Bearing Replacer, T83T–7025–B and Driver Handle, T80T–4000–W or equivalent.

NOTE: The clip is required to prevent the bearing retainer from rotating. Do not discard the clip.

33. Remove the front output shaft and U-joint assembly from the front case. If re-

quired, remove the oil seal with Seal Remover, Tool–1175–AC and Impact Slide Hammer, T50T–100–A or equivalent. If required, remove the internal snap-ring and drive the ball bearing out of the front case bore using Output Shaft Replacer, T83T–7025–B and Driver Handle, T80T–4000–W or equivalent.

34. If required, place the front output shaft and U-joint assembly in a vise, being careful not to damage the assembly. Use copper or wood vise jaws.

35. Remove the internal snap-rings that retain the bearings in the shaft.

36. Position the U-Joint Tool, T74P–4635–C or equivalent, over the shaft ears and press the bearing out. If the bearing cannot be pressed all the way out, remove it with vise grip or channel lock pliers.

37. Re-position the U-joint tool on the spider in order to remove the opposite bearing.

38. Repeat the above procedure until all bearings are removed.

ASSEMBLY

Before assembly, lubricate all parts with Dexron® II, Automatic Transmission Fluid.

1. If removed, start a new bearing into an end of the shaft ear. Support the output shaft in a vise equipped with copper or wood jaws, in order not to damage the shaft.

2. Position the spider into the bearing and press the bearing below the snap-ring groove using U-joint Tool, T74P–4635–C or equivalent.

3. Remove the tool and install a new internal snap-ring on the groove.

4. Start a new bearing into the opposite side of the shaft ear and using the tool, press the bearing until the opposite bearing contacts the snap-ring.

5. Remove the tool and install a new internal snap-ring in the groove.

6. Re-position the front output shaft assembly and install the other two bearings in the same manner.

7. Check the U-joint for freedom of movement. If a binding condition occurs due to misalignment during the installation procedure, tap the ears of both shafts sharply to relieve the bind. Do not install the front output shaft assembly if the U-joint shows any sign of binding.

8. If removed, drive the ball bearing into the front output case bore using Output Shaft Bearing Replacer, T83T–7025–B and Drive Handle, T80T–4000–W or equivalent. Drive the ball bearing in straight, making sure that it is not cocked in the bore. Install the internal snap-ring that retains the ball bearing to the front case.

9. If removed, install the front output oil seal in the front case bore using Output Shaft Seal Installer, T83T–7065–B and Driver Handle, T80T–4000–W or equivalent.

10. If removed, install the ring gear in the front case. Align the serrations on the outside diameter of the ring gear to the serrations previously cut in the front case bore. Using a press, start the piloted chamferred end of the ring gear first and press in until it is fully seated. Make sure the ring gear is not cocked in the bore.

TORQUE SPECIFICATIONS
Borg-Warner 13-50 Transfer Case

Description	Torque N•m	Ft. Lb.
Breather Vent	8–19	6–14
Case to Cover Bolts	31–41	23–30
Drain and Fill Plug	19–30	14–22
Four-Wheel Drive Indicator Switch	34–47	25–35
Front and Rear Driveshaft Bolts	16–20	12–15
Shift Control Bolts—Large	95–122	70–90
Shift Control Bolts—Small	42–57	31–42
Shift Shaft and Shift Cam Set Screw	6.8–9.5	5–7
Skid Plate to Frame Bolt	30–41	22–30
Transfer Case to Transmission Adapter	34–47	25–35
Upper Shift Control Lever and Heat Shield Bolts	37–50	27–37
Yoke Nut	163–203	120–150

Description	N•m	In. Lb.
Oil Pump Bolts	4.0–4.5	36–40
Speedometer Screw	2.3–2.8	20–25

11. If removed, install the ball bearing in the bearing retainer bore. Drive the bearing into the retainer using Output Shaft Bearing Replacer, T83T-7025-B and Driver Handle, T80T-4000-W or equivalent. Make sure the ball bearing is not cocked in the bore. Install the internal snap-ring that retains the ball bearing to the retainer.

12. Install the front output shaft and U-joint assembly through the front case seal. Position the ball bearing and retainer assembly over the front output shaft and install in the front case bore. Make sure the clip on the bearing retainer aligns with the slot in the front case. Tap the bearing retainer into place with a plastic hammer. Install the internal snap-ring that retains the ball bearing and retainer assembly to the front case.

13. Install the compression spring and the detent plunger into the bore from the inside of the front case.

14. Install a new O-ring in the second groove of the shift lever shaft. Coat the shaft and O-ring with Multi-Purpose Long-Life Lubricant.

NOTE: Use a rubber band to fill the first groove so as not to cut the O-ring. Discard the rubber band.

15. Position the shift cam inside the front case with the 4WH detent position over the detent plunger. Holding the shift cam by hand, push the shift lever shaft into the front case to engage the shift cam aligning the side of the shift lever with the mark previously scribed on the front case. Install the external clip on the end of the shift lever shaft.

16. Install the two phillips head set screws in the front case and in the shift cam. Tighten the screws to 5-7 ft. lbs. Make sure the set screw in the front case is in the first groove of the shift lever shaft and not bottomed against the shaft itself. The shift lever should be able to move freely to all detent positions.

17. Slide the sun gear, thrust plate, thrust washer, and press the ball bearing over the input shaft. Install the external snap-ring to the input shaft.

NOTE: The sun gear recessed face and ball bearing snap-ring groove should be toward the rear of the transfer case. The stepped face of the thrust washer should face towards the ball bearing.

18. Install the planetary gear set to the sun gear and input shaft assembly. Install the internal snap-ring to the planetary carrier.

19. Drive the oil seal into the bore of the mounting adapter with Input Shaft Seal Installer, T83T-7065-A and Driver Handle, T80T-4000-W or equivalent.

20. Place the tanged snap-ring in the mounting adapter groove. Position the input shaft and planetary gearset in the mounting adapter and push inward until the

planetary assembly and input shaft assembly are seated in the adapter. When properly seated, the tanged snap-ring will snap into place. Check installation by holding the mounting adapter by hand and tapping the face of the input shaft against a wooden block to ensure that the snap-ring is engaged.

21. Remove all traces of RTV gasket sealant from the mating surfaces of the front case and mounting adapter. Install a bead of RTV gasket sealant on the surface of the front case.

22. Position the mounting adapter on the front case. Install six bolts and tighten to 23-30 ft. lbs.

23. Position the roller on the 90° bent tang of the torsion spring. The larger diameter end of the spring must be installed first.

24. Install the roller into the torsion spring roller track of the shift cam while locating the center of the spring in the pivot groove in the front case. Push the anchor end of the torsion spring behind the locking post adjacent to the ring gear face.

25. Position the High-Low shift hub into the planetary gearset. Slip the High-Low shift fork bushing into the High-Low roller track of the shift cam and the groove of the High-Low shift hub.

NOTE: Make sure the nylon wear pads are installed on the shift fork. Make sure the dot on the pad is installed in the fork hole.

26. Install the shift rail through the high-low fork and make sure the shift rail is seated in the bore in the front case.

27. Place the oil pump cover with the word TOP facing the front of the front case. Install the two pump pins (flats facing upwards) with the spring between the pins and place the assembly in the oil pump bore in the output shaft. Place the oil pump body and pickup tube over the shaft and make sure the pins are riding against the inside of the pump body. Place the oil pump rear cover with the words TOP REAR facing the rear of the front case. The word TOP on the front cover and the rear cover should be on the same side. Install the pump retainer, the four bolts and rotate the output shaft while tightening the bolts to prevent the pump from binding. Tighten the bolts to 36-40 inch lbs.

NOTE: The output shaft must turn freely within the oil pump. If binding occurs, loosen the four bolts and retighten again.

28. Install the output shaft and oil pump assembly in the input shaft. Make sure the external splines of the output shaft engage the internal splines of the high-low shift hub. Make sure the oil pump retainer and oil filter leg are in the groove and notch of the front case.

29. Install the collector magnet in the notch in the front case.

30. Install the chain, drive sprocket and driven sprocket as an assembly over the shafts. Install the thrust washer on the front output shaft and install the external snap-ring over the thrust washer to retain the driven sprocket.

31. If disassembled, assemble the 2W-4W lockup assembly. Install the spring in the lockup collar. Place the lockup hub over the spring and engage the lockup hub in the notches in the lockup collar. Retain the lockup hub to the lockup collar with an internal snap-ring.

32. Install the 2W-4W shift fork to the 2W-4W lockup assembly. If removed, make sure the nylon wear pads are installed on the fork. The dot on the pad must be installed in the hole in the fork. Install the 2W-4W lockup collar and hub assembly over the output shaft and onto the shift rail. If removed, install the shaft, bushing and external clip to the 2W-4W lockup fork.

33. Install the shift collar hub to the output shaft.

34. If removed, drive the caged needle bearing into the rear cover bore with Needle Bearing Replacer, T83T-7127-A and Driver Handle, T80T-4000-W or equivalent.

35. If removed, install the ball bearing in the rear cover bore. Drive the bearing into the rear cover bore with Output Shaft Bearing Replacer, T83T-7025-B and Driver Handle, T80-4000-W or equivalent. Make sure the ball bearing is not cocked in the bore. Install the internal snap-ring that retains the ball bearing to the rear cover.

36. Install the speedometer drive gear assembly into the rear cover bore with the round end of the speedometer gear clip facing towards the inside of the rear cover. Drive the oil seal into the rear cover bore with Output Shaft Seal Installer, T83T-7065-B and Driver Handle, T80T-4000-W or equivalent.

37. Install the 2W-4W shift fork spring on the inside boss of the rear cover.

38. Prior to final assembly of rear cover to front case half, the transfer case shift lever assembly should be shifted into the "4H" detent position to assure positioning of the shift rail to the rear cover.

39. Coat the mating surface of the front case with a bead of Silicone Rubber.

40. Position the rear cover on the front case, making sure that the 2W-4W shift fork spring engages the shift rail and does not fall off the rear cover boss. Install the nine bolts (starting with the bolts on the rear cover) and tighten to 23-30 ft. lbs.

NOTE: If the rear cover assembly does not seat properly, move the rear cover up and down slightly to permit the end of the shift rail to enter the shift rail hole in the rear cover boss.

41. Install the rear yoke on the output shaft. Install the rubber seal, washer and nut. Tighten the nut to 120-150 ft. lbs.

42. Install the four-wheel drive indicator switch and tighten to 25-35 ft. lbs.

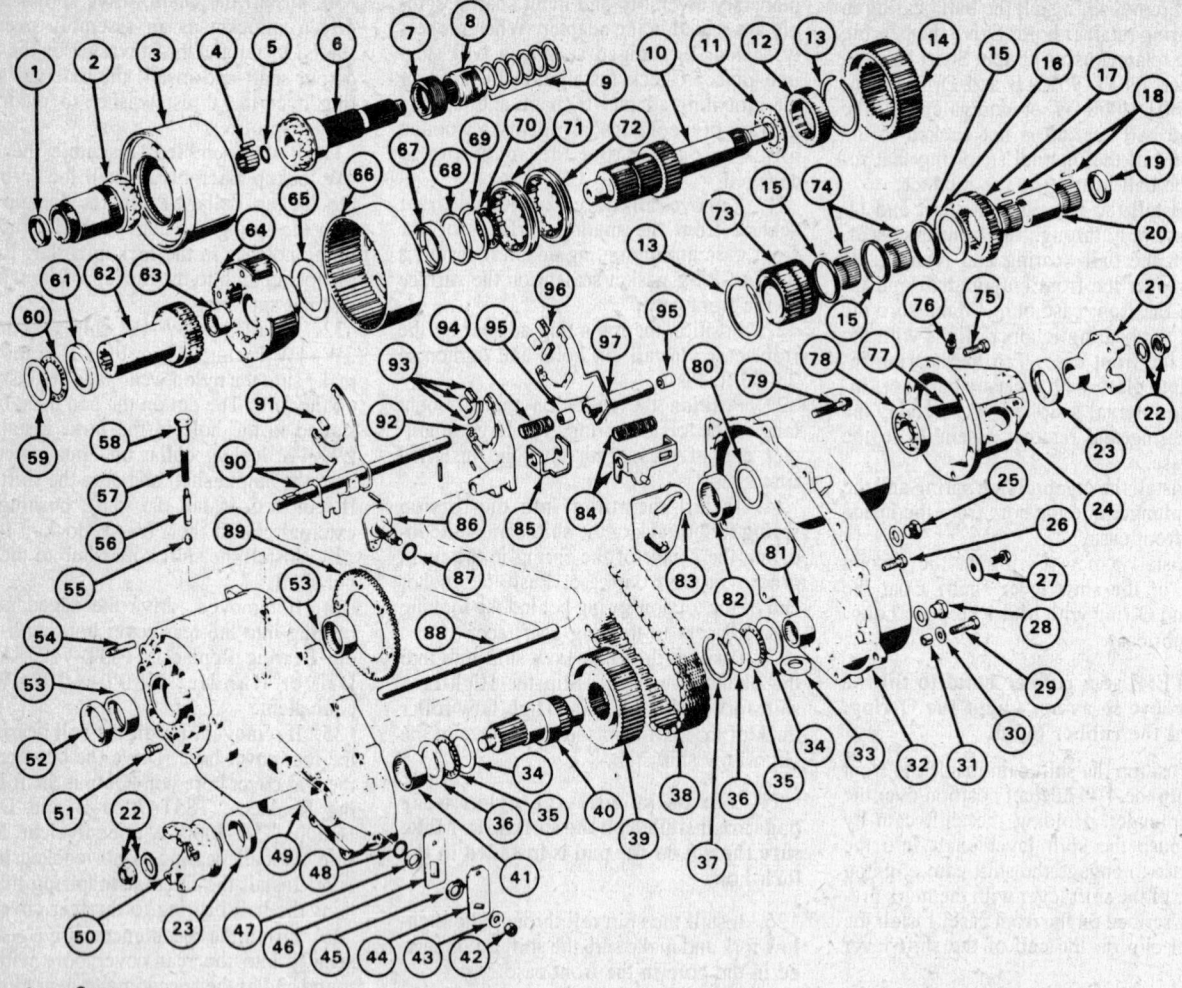

Datsun/Nissan transfer case components

1.	Spacer
2.	Side Gear
3.	Viscous Coupling
4.	Pilot Bearing Rollers
5.	O-Ring Seal
6.	Rear Output Shaft
7.	Oil Pump
8.	Speedometer Drive Gear
9.	Shim Kit
10.	Mainshaft
11.	Mainshaft Thrust Washer
12.	Spline Gear
13.	Retaining Ring
14.	Sprocket
15.	Spacer
16.	Sprocket Thrust Washer
17.	Viscous Clutch Gear
18.	Side Gear Roller (82)
19.	Spacer (Short)
20.	Spacer (Long)
21.	Rear Yoke
22.	Nut and Seal Washer
23.	Seal
24.	Rear Retainer
25.	Plug Assembly
26.	Bolt
27.	Identification Tag
28.	Plug Assembly
29.	Dowel Bolt
30.	Dowel Bolt Washer
31.	Case Half Dowel
32.	Rear Half Case
33.	Magnet

34.	Front Output Shaft Bearing Assembly Race (Thick)
35.	Front Output Shaft Bearing Assembly Thrust
36.	Front Output Shaft Bearing Assembly Race (Thin)
37.	Retaining Ring
38.	Chain
39.	Driven Sprocket
40.	Front Output Shaft
41.	Front Output Front Bearing
42.	Nut
43.	Washer
44.	Mode Lever
45.	Snap Ring
46.	Range Lever
47.	O-Ring Retainer
48.	O-Ring Seal
49.	Front Half Case
50.	Front Output Yoke
51.	Low Range Plate Bolt
52.	Input Shaft Oil Seal
53.	Input Shaft Bearing
54.	Stud
55.	Ball
56.	Plunger
57.	Plunger Spring
58.	Screw
59.	Input Race
60.	Input Thrust Bearing
61.	Input Race (Thick)
62.	Input Shaft
63.	Input Bearing
64.	Planetary Gear Assembly
65.	Input Gear Thrust Washer

66.	Annulus Gear Assembly
67.	Annulus Bushing
68.	Thrust Washer
69.	Retaining Ring
70.	Thrust Bearing
71.	High Range Sliding Clutch Sleeve
72.	Mode Sliding Clutch Sleeve
73.	Carrier
74.	Carrier Rollers (120)
75.	Rear Retainer Bolt
76.	Vent
77.	Vent Seal
78.	Output Bearing
79.	Bolt
80.	Seal
81.	Front Output Rear Bearing
82.	Output Shaft Inner Bearing
83.	Range Sector
84.	Range Bracket (Outer) and Spring
85.	Range Bracket (Inner)
86.	Mode Sector
87.	O-Ring Seal
88.	Range Rail
89.	Low Range Lockout Plate
90.	Mode Fork, Rail and Pin
91.	Mode Fork Pad
92.	Range Fork
93.	Range Fork Pads
94.	Range Bracket Spring (Inner)
95.	Locking Fork Bushing
96.	Locking Fork Pads
97.	Locking Fork

Case components

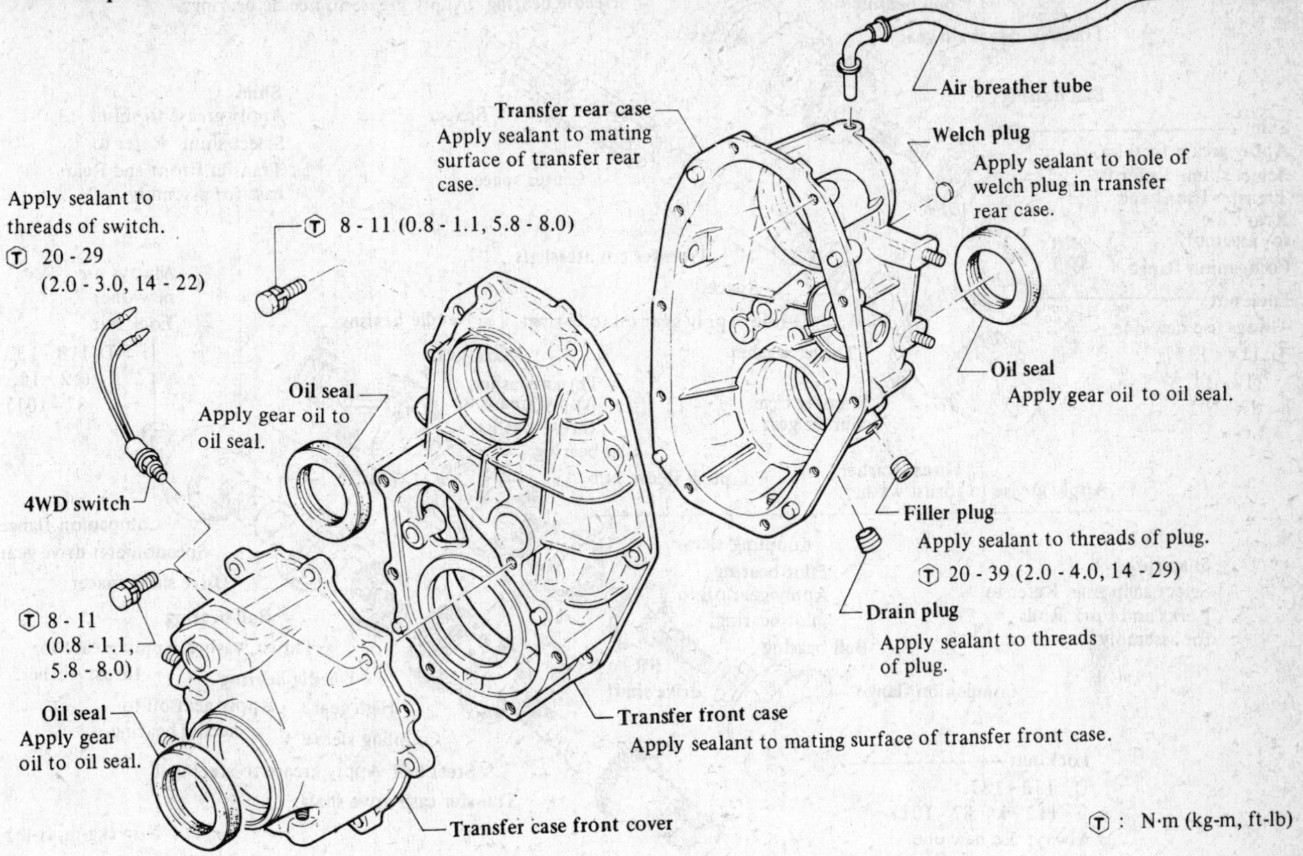

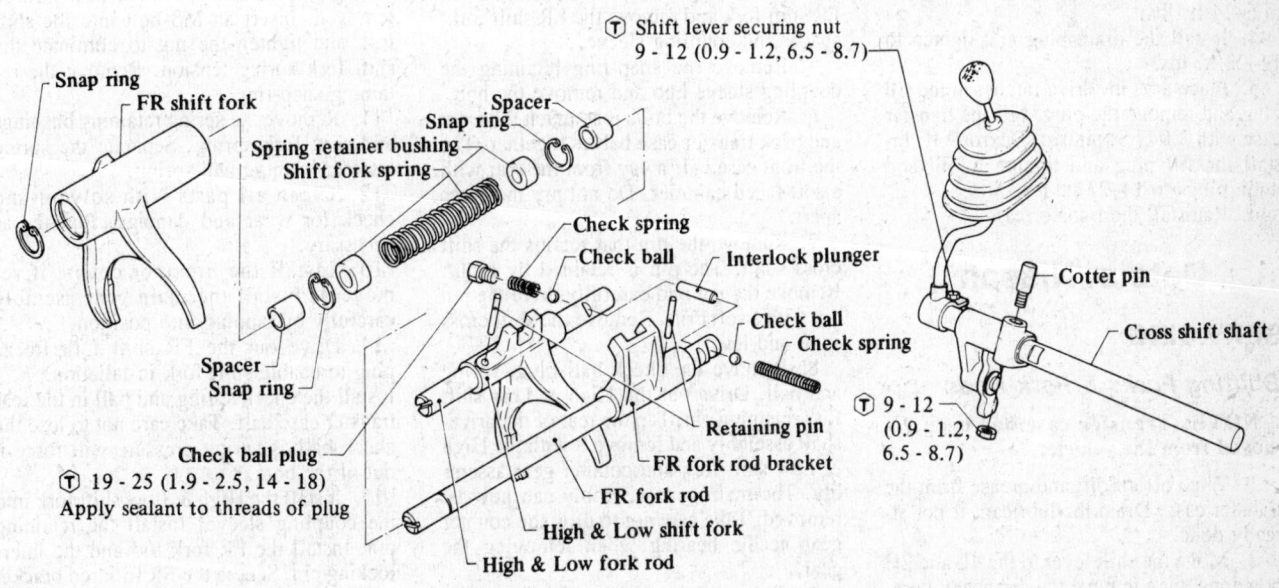

Apply sealant to
threads of switch.
Ⓣ 20 - 29
(2.0 - 3.0, 14 - 22)

Ⓣ 8 - 11 (0.8 - 1.1, 5.8 - 8.0)

Transfer rear case
Apply sealant to mating
surface of transfer rear
case.

Air breather tube

Welch plug
Apply sealant to hole of
welch plug in transfer
rear case.

Oil seal
Apply gear oil to
oil seal.

Oil seal
Apply gear oil to oil seal.

4WD switch

Filler plug
Apply sealant to threads of plug.
Ⓣ 20 - 39 (2.0 - 4.0, 14 - 29)

Ⓣ 8 - 11
(0.8 - 1.1,
5.8 - 8.0)

Drain plug
Apply sealant to threads
of plug.

Oil seal
Apply gear
oil to oil seal.

Transfer front case
Apply sealant to mating surface of transfer front case.

Transfer case front cover

Ⓣ : N·m (kg-m, ft-lb)

Snap ring

FR shift fork

Spring retainer bushing

Shift fork spring

Spacer

Snap ring

Spacer
Snap ring

Check ball plug
Ⓣ 19 - 25 (1.9 - 2.5, 14 - 18)
Apply sealant to threads of plug

Check spring

Check ball

Interlock plunger

Check ball
Check spring

Retaining pin

FR fork rod bracket

FR fork rod

High & Low shift fork

High & Low fork rod

Shift lever securing nut
9 - 12 (0.9 - 1.2, 6.5 - 8.7)

Cotter pin

Cross shift shaft

Ⓣ 9 - 12
(0.9 - 1.2,
6.5 - 8.7)

Ⓣ : N·m (kg-m, ft-lb)

Datsun/Nissan transfer case shift control parts

Gear components

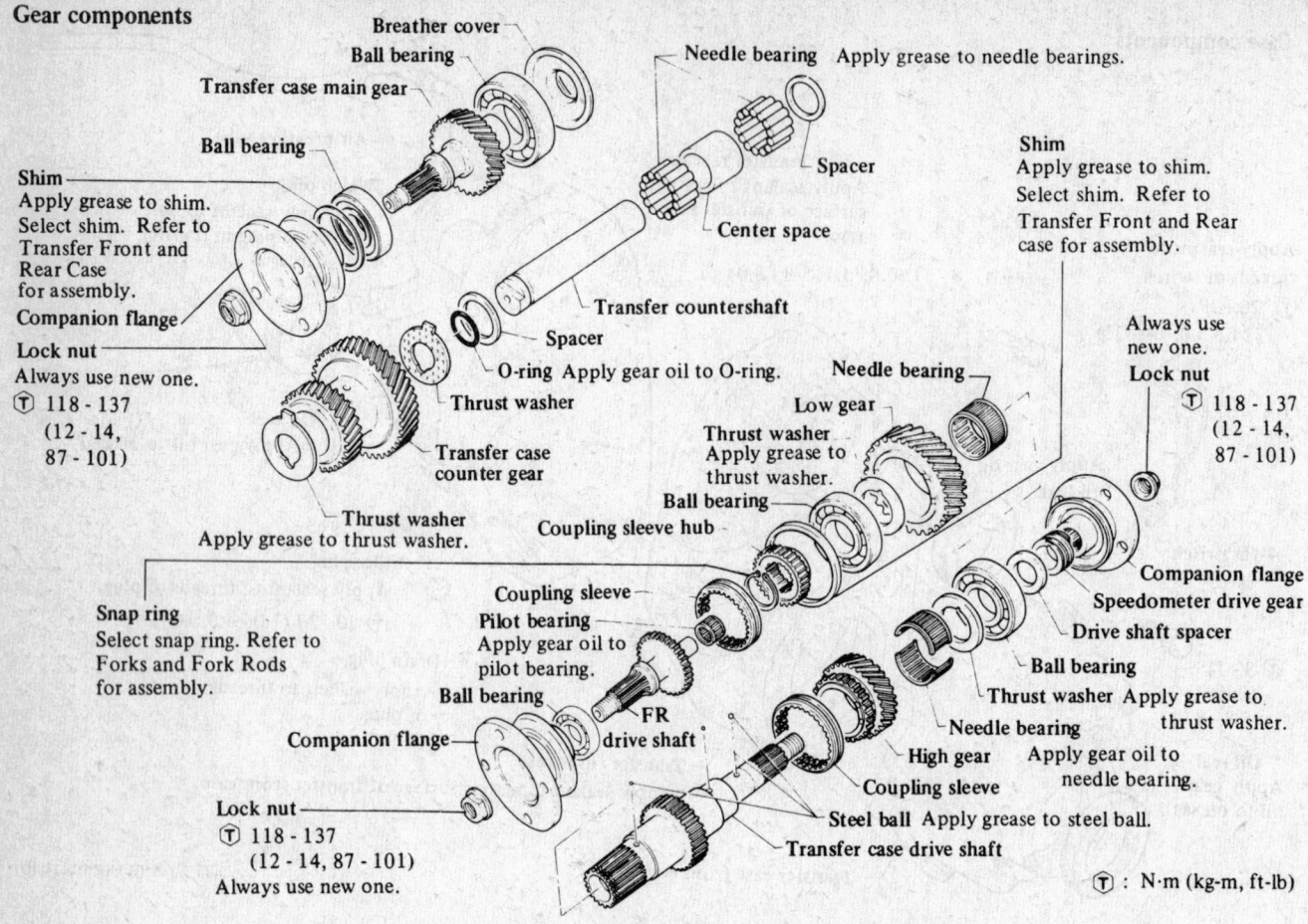

Datsun/Nissan transfer case gear components

43. Install the breather plug and tighten to 6–14 ft. lbs.

44. Install the drain plug and tighten to 14–22 ft. lbs.

45. Place a ³/₈ in. drive ratchet in the fill plug and remove the plug. Fill the transfer case with 3.0 U.S. pints of Dexron® II. Install the fill plug and tighten to fill and drain plugs to 14–22 ft. lbs.

46. Reinstall the transfer case.

Datsun/Nissan

SERVICING

Shifting Forks & Fork Rods

NOTE: Transfer case has been removed from the vehicle.

1. Wipe off all dirt and grease from the transfer case. Drain the lubricant if not already done.

2. Move the shift lever to the 4L and 2H positions, then remove the driveshaft companion flange locknuts. Remove the companion flanges and the 4WD switch.

3. Remove the front cover from the transfer case. Remove the FR driveshaft and needle bearings.

4. Remove the snap-ring retaining the FR shift fork and remove the FR shift fork, spacer and coupling sleeve.

5. Remove the snap-ring retaining the coupling sleeve hub and remove the hub.

6. Remove the bolts that attach the front and back transfer case halves together. Tap the front case half away from the rear with a soft faced hammer. Do not pry the cases apart.

7. Remove the pin that retains the shift cross shaft. The pin is retained by a nut. Remove the nut and carefully drive the pin out with a soft drift. Remove the shift cross shaft and lever.

8. Remove the check ball plug, spring and ball. Drive out the High & Low shift fork retaining pin. Tap the rear of the drive-shaft assembly and remove it with the High & Low shift fork and counter gear assembly. The main gear assembly can now be removed. Take care not to drop the counter gear needle bearing when removing the gear.

9. Remove the front shim from the transfer case. Remove the High & Low and FR shift fork rods, interlock plunger, steel ball and check spring.

10. If servicing is necessary, secure the FR fork rod and carefully drive out the re-taining pin, the fork rod bracket can now be removed. Insert an M8 bolt into the shift fork and tighten the nut to eliminate the shift fork spring tension. Remove the retaining snap-ring.

11. Remove the spring retaining bushings and shift fork spring. Separate the spring retainer bushing and spring.

12. Clean all parts with solvent and check for wear and damage. Replace as necessary.

13. Install the breather cover (if removed). Install the main gear asembly carefully by tapping into position.

14. Drive out the FR shift fork freeze plug to enable shift fork installation. Install the check spring and ball in the rear transfer case half. Take care not to lose the check ball as spring pressure will force it out of the bore.

15. Install the High & Low shift fork into the coupling sleeve. Install the retaining pin. Install the FR fork rod and the interlocking pin. Secure the FR fork rod bracket to the fork rod with the retaining pin.

16. Assemble the snap-ring, spring retainer bushings and shift fork spring to the FR shift fork. Insert an M8 bolt into the bushing and tighten the nut to eliminate spring tension. Install the other retaining

snap-ring and remove the M8 bolt.

17. Lubricate and install a new O-ring to the countershaft and then install the countershaft assembly. Raise the counter gear assembly slightly and install the rear driveshaft assembly making sure the gear teeth mesh.

18. Install the companion flange on the rear end of the driveshaft assembly and tighten the nut finger tight.

19. Install the High & Low fork rod and secure with the retaining pin. Apply sealant to the freeze plug bore edges and install a new freeze plug. Install the check ball and spring. Tighten the check ball retaining bolt to 14–18 ft. lbs. after applying sealer to the threads.

20. Install the coupling sleeve hub and snap-ring. Check end play. The end play should be between 0–0.0079 in. Snap-rings of different thicknesses are available to control the end play.

21. Install the shift lever. Install the shift cross shaft. Apply grease to the main gear front shims and transfer case thrust washer and install them to the front half of the transfer case. Be sure the mating surfaces of the case halves are clean. Apply a bead of sealant to the mating edge and install the front half of the transfer case. Tap with a soft hammer to seat if necessary.

22. Tighten the mounting bolts to 5–8 ft. lbs.

23. Install the spacer, FR shift fork assembly with the coupling sleeve and secure with snap-ring.

24. Lubricate the pilot needle bearing and place in position. Install the FR driveshaft to the transfer case driveshaft. Clean the mating surfaces of the cases. Apply sealant to the extension case surface and install the cover (5–8 ft. lbs.). Install the companion flanges. Tighten the new flange locknuts to 87–101 ft. lbs. Install the 4WD switch to 14–22 ft. lbs.

Gears & Shafts

INSPECTION & ENDPLAY

1. Check all parts for excessive wear, chips or cracks. Replace as necessary.

2. Measure the endplay before and after disassembling the driveshaft. If excessive endplay is present, disassemble shaft and check for worn parts.

3. Standard endplay: High gear & Low gear; 0.0039–0.0079. Coupling sleeve hub; 0–0.0079.

COUNTER GEAR & MAIN GEAR

1. Refer to the previous Shifter Fork procedure and remove the transfer case driveshaft assembly, counter gear assembly, forks and fork rods.

2. Remove the main gear and breather assembly from the rear case half. Remove the main gear front bearing and/or rear bearing with a press, if service is necessary. Remove the needle bearings, center spacer and end spacers from the counter gear for cleaning and any required servicing.

3. Press on the front and rear bearings. Install the breather cover and main gear. Apply grease to the needle bearings and spacers and assemble them in the counter gear. 28 needle bearings are used, 14 at each end.

4. Reassembly in reverse order of removal.

LOW GEAR

1. Press off the front bearing from the driveshaft. Remove the thrust washer and steel ball.

2. Remove the Low gear and needle bearings. Replace parts as required.

3. Lubricate the needle bearings with gear oil and install the needle bearings, coupling sleeve and Low gear.

4. Apply grease to the steel ball and thrust washer and install them on the Low gear. Press on the front bearing while holding the Low gear so that the thrust washer will not drop out of position.

NOTE: If internal parts (i.e. thrust washers, bearings, gears etc.) are replace, check clearances and adjust as necessary using the correct shim(s) to reduce excessive play.

HIGH GEAR

1. Remove the speedometer drive gear. Remove the spacer and steel ball. Press off the rear driveshaft bearing. Remove the thrust washer and steel ball.

2. Remove the high gear, needle bearings and coupling sleeve. Replace parts as required.

3. Lubricate the needle bearings with gear oil and install the bearings and High gear on the shaft.

4. Apply grease to the steel ball and thrust washer and place them on the high gear. Press the bearing onto the shaft. Hold the gear while pressing on the bearing to make sure the thrust washer does not fall.

5. Install the driveshaft spacer. Apply grease to the steel ball and install ball and speedometer drive gear.

NOTE: After servicing components, assemble the unit as described under Shifting Rods and Forks.

Isuzu/LUV

Refer to the Manual Transmission section for servicing procedures.

Mitsubishi

SERVICING

NOTE: Transfer case has been removed from the transmission.

1. Remove the two 4WD switches from the case. Take out the steel balls behind the switches.

2. Remove the speedometer sleeve clamp and the speedometer sleeve assembly.

3. Remove the rear cover, gasket, wave spring and spacer.

4. Take a 3/16 in. punch and drive out the spring pin that retains the H–L shift fork. Remove the two threaded plugs and remove the poppet springs and balls.

5. Pull out the H–L shift rail. Take out the interlock plunger.

6. Remove the rear bearing snap-ring from the rear output shaft. Remove the chain cover, oil guide and side cover.

7. Remove the countershaft locking plate and remove the countershaft. Remove the counter gear assembly through the side cover opening. The gear assembly consists of gear, spacers, needle bearings and thrust washers.

8. Remove the snap-ring, the two spring retainers and the spring from the 2–4WD shift rail.

9. Remove the front output shaft, the rear output shaft and the chain (as an assembly) from the transfer case.

10. Remove the 2–4WD shift rail. Remove the H–L shift fork and clutch sleeve. Remove the needle bearing from the input gear.

11. Remove the snap-ring retaining the input gear assembly and remove the input gear assembly.

12. Remove the snap-ring from the front end of the rear output shaft. Remove the H–L clutch hub, the low speed gear, the thrust bearing and the needle bearing.

13. Raise the detent of the locknut on the rear output shaft and remove the locknut. Remove the rear bearing using a puller or press.

14. Remove the sprocket spacer and balls. Remove the drive sprocket, two needle bearings, sprocket sleeve and steel ball. Remove the 2–4WD clutch sleeve, hub and stop plate. Remove the bearing using a suitable puller or press.

15. Remove the snap-ring retaining the input gear. Press the shaft from the gear. Remove the two bearings from the front output shaft using a suitable puller or press.

16. Clean and inspect all parts. Replace worn parts as necessary. Replace all oil seals. When reassembling the transfer case, coat the new gaskets with sealant. Apply oil to all sliding and rotating parts, pack the oil seal lip spaces with grease. Replace spring pins with new ones.

17. When replacing the control shaft oil seal or input gear oil seal, drive out the spring pin from the transmission control shaft change shifter and remove the adapter from the transfer case. Remove the seals and press fit new ones in position.

18. Assemble the adapter to the transfer case using a new gasket. Before tightening the mounting bolts, install the change shifter over the control shaft. Adapter mounting bolt torque is 22–30 ft. lbs.

19. Press fit the bearing into the input gear. Make sure the bearing turns freely after installation. Install the snap-ring over

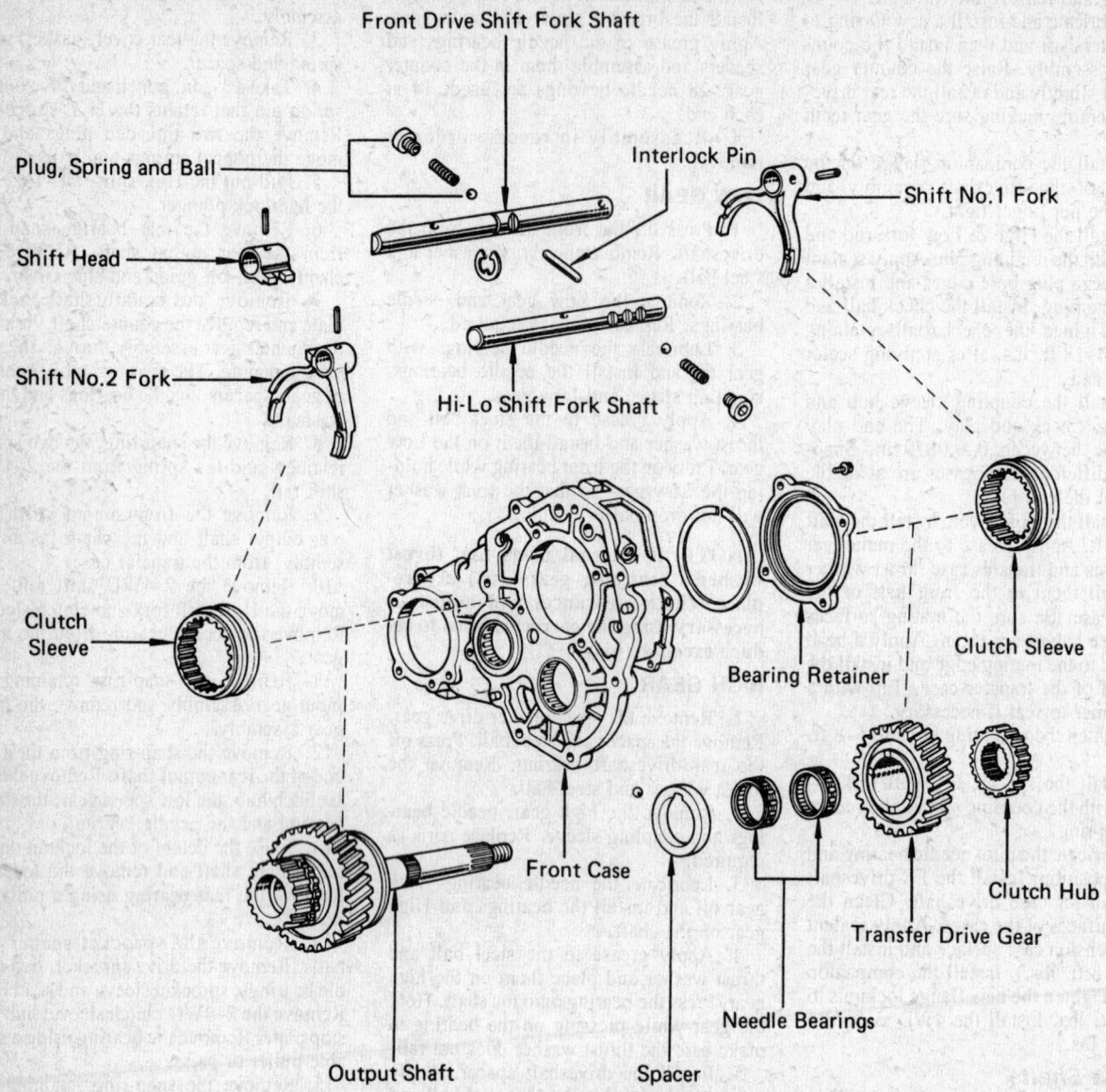

Front Drive Shift Fork Shaft

Plug, Spring and Ball

Interlock Pin

Shift No.1 Fork

Shift Head

Shift No.2 Fork

Hi-Lo Shift Fork Shaft

Clutch Sleeve

Clutch Sleeve

Bearing Retainer

Front Case

Clutch Hub

Transfer Drive Gear

Needle Bearings

Output Shaft

Spacer

Toyota transfer case components

the front end of the input gear. Use the thickest snap-ring that will fit in the groove (five sizes are available). Press fit the two bearings over the front output shaft, make sure they turn freely.

20. Install the bearing on the rear output shaft and make sure the bearing turns freely. Mount the stop plate and install the 2–4WD clutch hub and sleeve. Make sure they are installed facing in the proper direction. Mount the steel ball on the rear output shaft and mount the sprocket sleeve.

21. Mount the two needle berarings on the sprocket sleeve and mouint the drive sprocket. After installing the steel balls and the sprocket spacer, press fit the ball bearing into the inner race, make sure it moves freely.

22. Tighten the mainshaft locknut and lock the tab with a punch. Mount the needle bearing, the thrust washer and the low

speed gear on the rear output shaft from the front end.

23. Mount the H–L clutch hub making sure it is facing in the proper direction. Mount the H–L snapring using the thickest one that will fit in the groove (five sizes are available). Insert the input gear assembly into the transfer case and mount the snapring. Once again using the thickest ring possible.

24. Insert the needle bearing into the input gear. Mount the H–L clutch sleeve and shift fork (sleeve facing in proper direction). Install the 2–4WD shift rail.

25. Securely engage the chain with the front and rear output shaft sprockets. Assemble the 2–4WD clutch sleeve with the 2–4WD shift fork and install the assembly over the shift rail while, at the same time, mount the rear and front input shafts, chain etc.

26. Mount the two spring retainers and spring on the 2–4WD shift rail and secure with the snapring.

27. Install the two needle bearings and spacer into the counter gear. Install the assembly in the transfer case with the thrust washers in place. Insert the counter shaft and locking plate.

28. Install the side cover and gasket. Install the oil guide. Install the chain cover and gasket and make sure that the oilguide end fits in the chain cover opening.

29. Fit the snapring in the groove of the rear output shaft bearing. Insert the interlock plunger. Insert the H–L shift rail and pass it through the shift fork. The shift fork must be to the 4WD side or the shift rail cannot be inserted.

30. Mount the two poppet balls and two springs and mount the seal plug. The smaller end of the spring goes toward the ball.

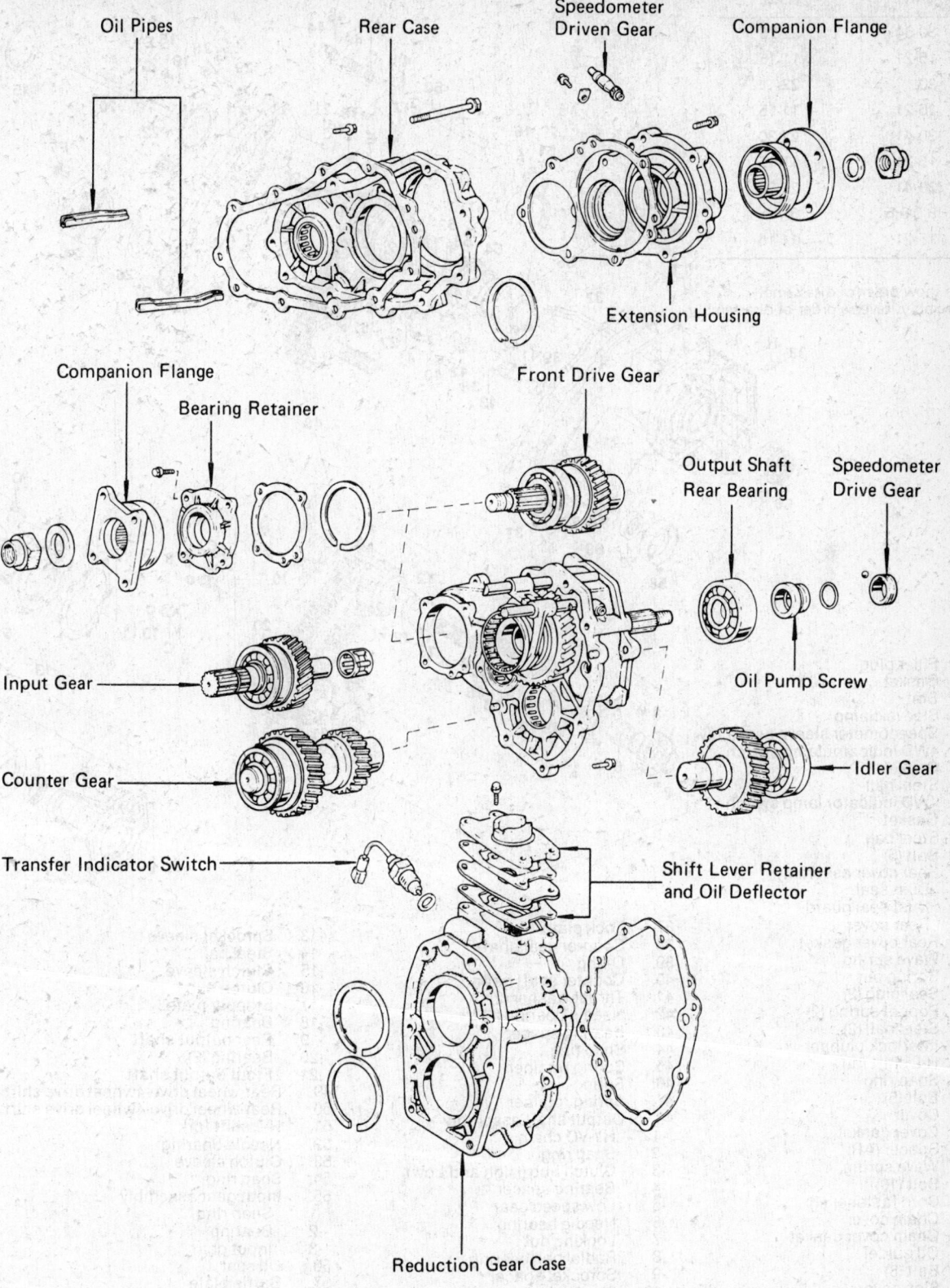

Oil Pipes

Rear Case

Speedometer Driven Gear

Companion Flange

Extension Housing

Companion Flange

Bearing Retainer

Front Drive Gear

Output Shaft Rear Bearing

Speedometer Drive Gear

Input Gear

Oil Pump Screw

Counter Gear

Idler Gear

Transfer Indicator Switch

Shift Lever Retainer and Oil Deflector

Reduction Gear Case

Toyota transfer case components

	Nm	ft.lbs.
A	30-34	22-25
B	15-21	11-15
C	30	22
D	15-21	11-15
E	30-41	22-30
F	15-21	11-15
G	30-41	22-30
H	8.0-9.5	6-7
I	15-21	11-15

NOTE
Numbers show order of disassembly.
For reassembly, reverse order of disassembly.

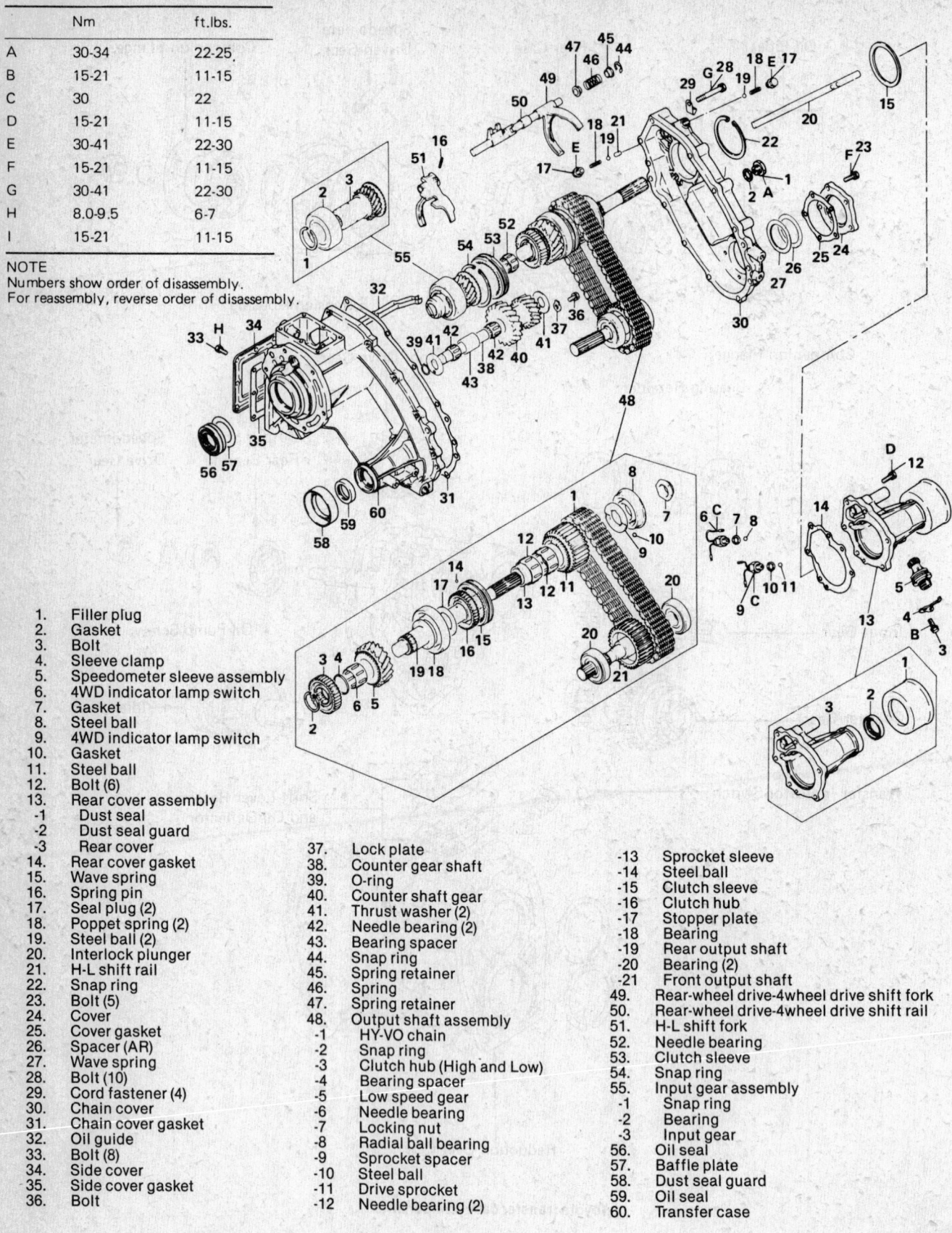

1. Filler plug
2. Gasket
3. Bolt
4. Sleeve clamp
5. Speedometer sleeve assembly
6. 4WD indicator lamp switch
7. Gasket
8. Steel ball
9. 4WD indicator lamp switch
10. Gasket
11. Steel ball
12. Bolt (6)
13. Rear cover assembly
-1 Dust seal
-2 Dust seal guard
-3 Rear cover
14. Rear cover gasket
15. Wave spring
16. Spring pin
17. Seal plug (2)
18. Poppet spring (2)
19. Steel ball (2)
20. Interlock plunger
21. H-L shift rail
22. Snap ring
23. Bolt (5)
24. Cover
25. Cover gasket
26. Spacer (AR)
27. Wave spring
28. Bolt (10)
29. Cord fastener (4)
30. Chain cover
31. Chain cover gasket
32. Oil guide
33. Bolt (8)
34. Side cover
35. Side cover gasket
36. Bolt

37. Lock plate
38. Counter gear shaft
39. O-ring
40. Counter shaft gear
41. Thrust washer (2)
42. Needle bearing (2)
43. Bearing spacer
44. Snap ring
45. Spring retainer
46. Spring
47. Spring retainer
48. Output shaft assembly
-1 HY-VO chain
-2 Snap ring
-3 Clutch hub (High and Low)
-4 Bearing spacer
-5 Low speed gear
-6 Needle bearing
-7 Locking nut
-8 Radial ball bearing
-9 Sprocket spacer
-10 Steel ball
-11 Drive sprocket
-12 Needle bearing (2)

-13 Sprocket sleeve
-14 Steel ball
-15 Clutch sleeve
-16 Clutch hub
-17 Stopper plate
-18 Bearing
-19 Rear output shaft
-20 Bearing (2)
-21 Front output shaft
49. Rear-wheel drive-4wheel drive shift fork
50. Rear-wheel drive-4wheel drive shift rail
51. H-L shift fork
52. Needle bearing
53. Clutch sleeve
54. Snap ring
55. Input gear assembly
-1 Snap ring
-2 Bearing
-3 Input gear
56. Oil seal
57. Baffle plate
58. Dust seal guard
59. Oil seal
60. Transfer case

Mitsubishi transfer case components

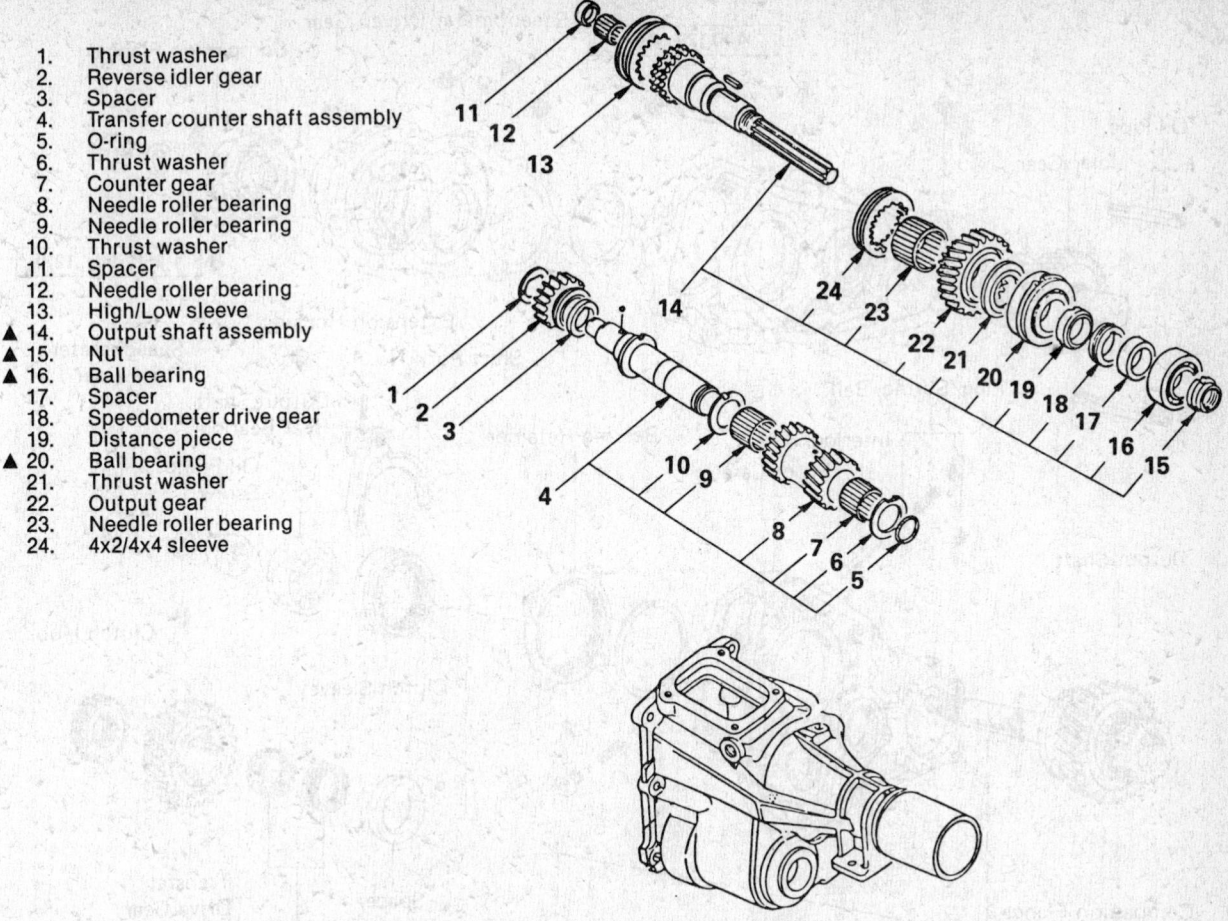

1. Thrust washer
2. Reverse idler gear
3. Spacer
4. Transfer counter shaft assembly
5. O-ring
6. Thrust washer
7. Counter gear
8. Needle roller bearing
9. Needle roller bearing
10. Thrust washer
11. Spacer
12. Needle roller bearing
13. High/Low sleeve
▲ 14. Output shaft assembly
▲ 15. Nut
▲ 16. Ball bearing
17. Spacer
18. Speedometer drive gear
19. Distance piece
▲ 20. Ball bearing
21. Thrust washer
22. Output gear
23. Needle roller bearing
24. 4x2/4x4 sleeve

Isuzu transfer case components

31. Align the H–L shift fork and rail holes and install the spring pin. Install the spring pin with its center slit placed on the center line of the shift rail.

32. Mount the spacer on the rear end of the rear output shaft bearing and install the rear cover and gasket. Check endplay and use a thicker or thinner spacer as needed.

33. Mount the wave spring washer on the rear end of the front output shaft rear bearing and install the cover and gasket. Install the speedometer sleeve assembly and secure. Install the switches and balls. Mount the neutral plungers and springs in the hole on top of the adapter and tighten the seal plug. Install the steel ball, resistance spring and plug.

Toyota

SERVICING

1. Remove the front and rear companion flanges

2. Remove the extension housing. Remove the speedometer drive gear, steel ball, oil pump screw and bearing.

3. Remove the rear case half with the idler gear in position while holding the front half so that the clutch hub and steel ball do not fall out.

4. Remove the idler gear from the rear case.

5. Remove the front bearing retainer and front drive gear.

6. Remove the two oil pipes. Remove shift fork No. 1 and clutch sleeve.

7. Remove the clutch hub and transfer drive gear. Remove the needle roller bearing, No. 2 spacer and steel ball. On models equipped with the 22R engine, remove the transfer case cover and shift lever retainer.

8. Remove the screw cover plugs, springs and locking balls from both sides of the case.

9. Remove the spring pins that retain the shifting forks and remove the front drive fork shaft. Remove the interlocking pin and the high–low fork.

10. Remove the front case, tap lightly with a plastic hammer if necessary.

11. Remove the No. 2 fork, clutch sleeve, and needle roller bearing from the input shaft.

12. Remove the input gear and counter gear from the reduction gear case. Remove the output shaft from the front case.

13. Check the oil and thrust clearance of the transfer low gear: Oil clearance between the gear and shaft with needle bearing installed; 0.0004–0.0022 in. Thrust clearance with spacer and bearing installed; 0.0039–0.0098 in.

14. Check the clearance of the thrust drive gear: Oil clearance between gear and shaft with needle bearing installed: 0.004–0.0020 in. Thrust clearance with clutch hub and spacer installed: 0.0035–0.0106 in.

15. Measure the clearance of the shift forks and clutch sleeves: Max. is 0.039 in.

16. Inspect all parts and bearings, replace as necessary. If bearing replacement is indicated, press the bearing from the shaft. When installing a bearing, use the thickest snap ring that will allow minimum end play.

17. Inspect all oil seals and replace as necessary.

18. Reassembly begins with the installation of the output shaft to the front case. Install the front bearing retainer and torque the mounting bolts to 9 ft. lbs.

19. Install the input gear and counter gear to the reduction gear case. Install the roller bearing on the input shaft.

20. Install the No. 2 hub sleeve and No. 2 shift fork on the input shaft.

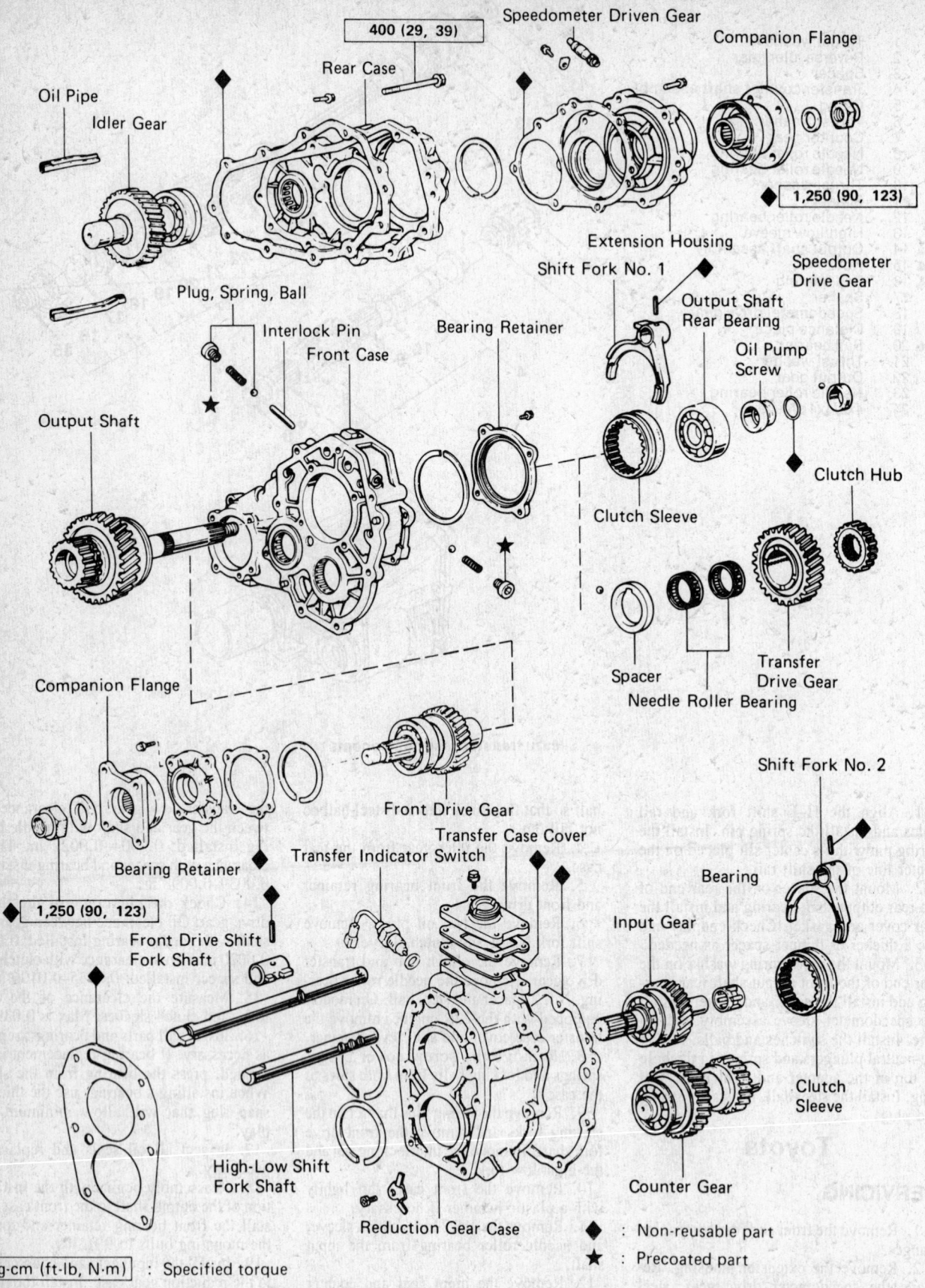

400 (29, 39)

Speedometer Driven Gear

Companion Flange

Oil Pipe

Idler Gear

Rear Case

1,250 (90, 123)

Extension Housing

Shift Fork No. 1

Speedometer Drive Gear

Plug, Spring, Ball

Interlock Pin

Front Case

Bearing Retainer

Output Shaft Rear Bearing

Oil Pump Screw

Output Shaft

Clutch Hub

Clutch Sleeve

Companion Flange

Spacer

Needle Roller Bearing

Transfer Drive Gear

Bearing Retainer

1,250 (90, 123)

Shift Fork No. 2

Front Drive Shift Fork Shaft

Transfer Indicator Switch

Front Drive Gear

Transfer Case Cover

Bearing

Input Gear

High-Low Shift Fork Shaft

Reduction Gear Case

Counter Gear

Clutch Sleeve

◆ : Non-reusable part

★ : Precoated part

kg-cm (ft-lb, N·m) : Specified torque

Toyota transfer case components torque values

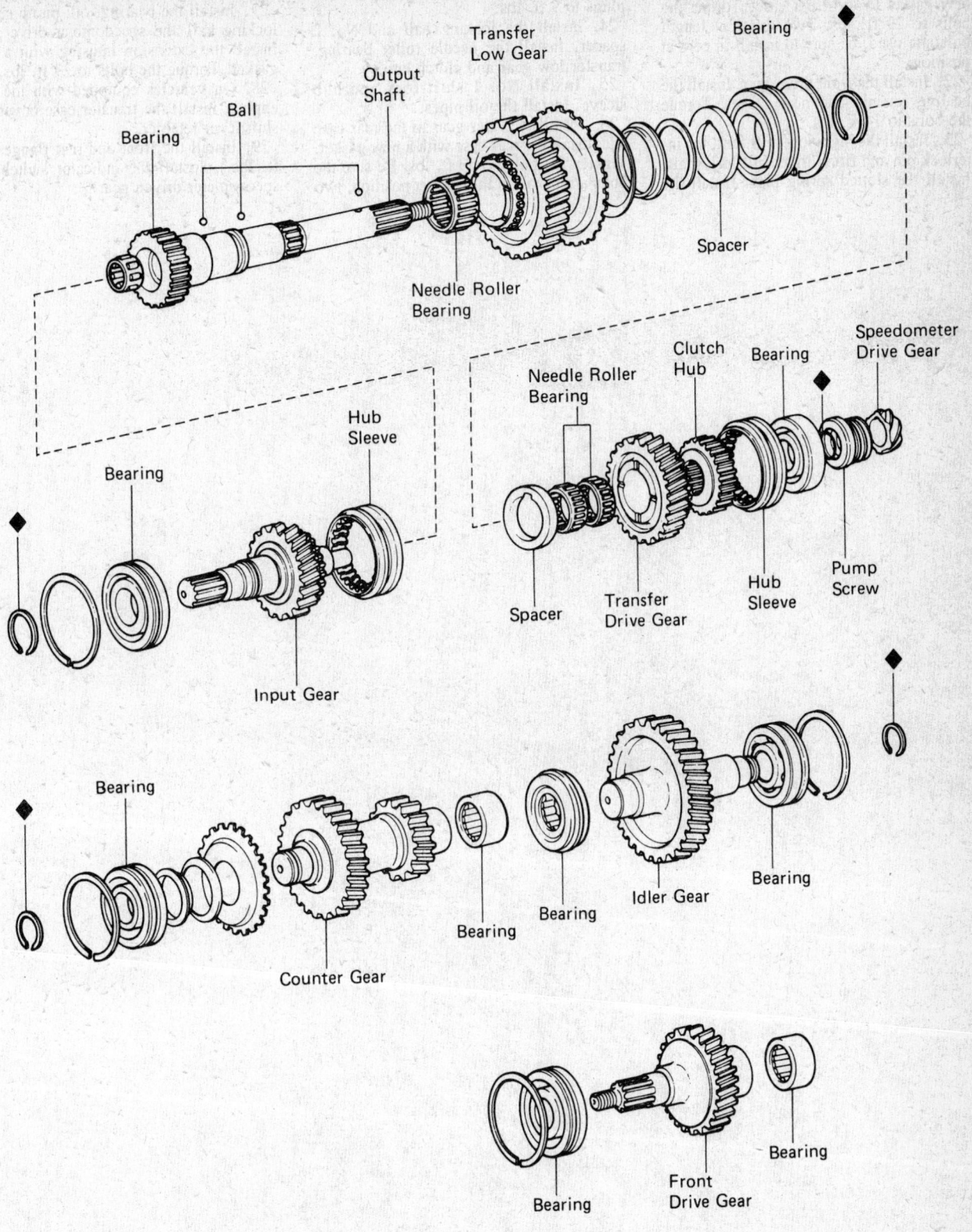

Bearing

Ball

Output Shaft

Transfer Low Gear

Bearing

Needle Roller Bearing

Spacer

Bearing

Hub Sleeve

Input Gear

Needle Roller Bearing

Clutch Hub

Bearing

Speedometer Drive Gear

Spacer

Transfer Drive Gear

Hub Sleeve

Pump Screw

Bearing

Counter Gear

Bearing

Bearing

Idler Gear

Bearing

Bearing

Front Drive Gear

Bearing

◆ : Non-reusable part

Toyota transfer case gear assemblies

21. Install the reduction gear case and new gasket to the front case. Torque the bolts to 29 ft. lbs. Two different length bolts are used, be sure to install in correct position.

22. Install the front drive gear. Install the bearing retainer and new gasket. Torque the bolts to 14 ft. lbs.

23. Install the high–low shift fork, the interlock pin and front drive shift fork shaft. Install the slotted spring pins. Install the balls, springs and side plugs. Tighten the plugs to 9 ft. lbs.

24. Install the locking ball and No. 2 spacer. Install the needle roller bearing, transfer low gear and clutch hub.

25. Install No. 1 shift fork and hub sleeve. Install the oil pipes.

26. Install the idler gear to the rear case and install the rear case with a new gasket. Tighten the bolts to 29 ft. lbs. Be sure the bolts are install in the correct position, two different lengths are used.

27. Install the bearing, oil; pump screw, locking ball and speedometer drive gear. Install the extension housing with a new gasket. Torque the bolts to 29 ft. lbs.

28. On vehicles equipped with the 22R engine, install the transfer case cover and shift lever retainer.

29. Install the front and rear flanges (90 ft. lbs.). Install the indicator switch and speedometer driven gear.

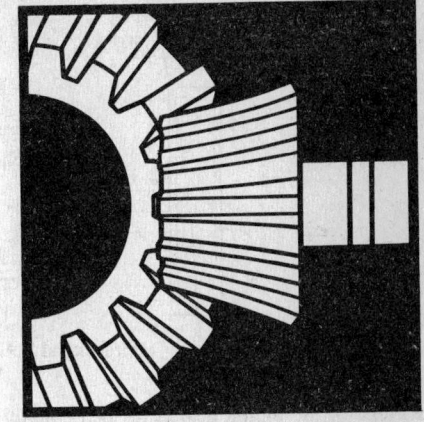

Drive Axles

INDEX

The drive axle must transmit power through 90°. To accomplish this, straight cut bevel gears or spiral bevel gears were used. This type of gear is satisfactory for differential side gears, but since the center-line of the gears must intersect, they rapidly became unsuited for ring and pinion gears. The lowering of the driveshaft brought about a variation of the bevel gear, which is called the hypoid gear. This type of gear does not require a meeting of the gear centerlines and can therefore be underslung, relative to the centerline of the ring gear.

Gear Ratios

The drive axle of a vehicle is said to have a certain axle ratio. This number (usually a whole number and a decimal fraction) is actually a comparison of the number of gear teeth on the ring gear and the pinion gear. For example, a 4.11 rear means that theoretically, there are 4.11 teeth on the ring gear and one tooth on the pinion. Actually, on a 4.11 rear, there are 37 teeth on the ring gear and nine teeth on the pinion gear. By dividing the number of teeth on the pinion gear into the number of teeth on the ring gear, the numerical axle ratio

(4.11) is obtained. This also provides a good method of ascertaining exactly which axle ratio one is dealing with.

Differential Operation

The differential is an arrangement of gears that permits the wheels to turn at different speeds when cornering and divides the torque between the axle shafts. The differential gears are mounted on a pinion shaft and the gears are free to rotate on this shaft.

The pinion shaft is fitted in a bore in the differential case and is at right angles to the axle shafts.

Power flow through the differential is as follows. The drive pinion, which is turned by the driveshaft, turns the ring gear. The ring gear, which is bolted to the differential case, rotates the case. The differential pinion forces the pinion gears against the side gears. In cases where both wheels have equal traction, the pinion gears do not rotate on the pinion shaft, because the input

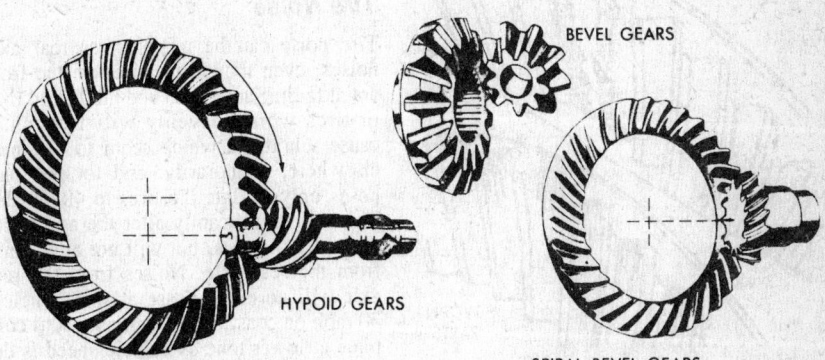

Hypoid gear application **Bevel gear application**

HYPOID GEARS BEVEL GEARS SPIRAL BEVEL GEARS

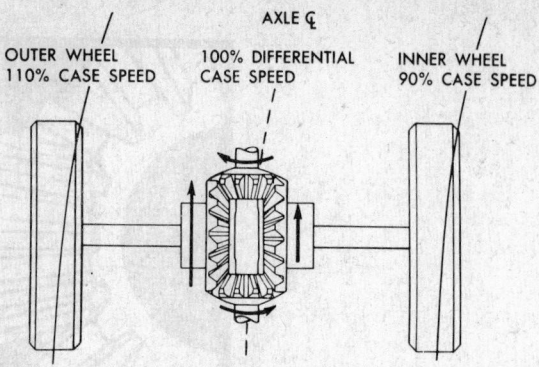

OUTER WHEEL
110% CASE SPEED

AXLE ℄

100% DIFFERENTIAL
CASE SPEED

INNER WHEEL
90% CASE SPEED

Differential action during cornering

force of the pinion gear is divided equally between the two side gears. Consequently the pinion gears revolve with the pinion shaft, although they do not revolve on the pinion shaft itself. The side gears, which are splined to the axle shafts, and meshed with the pinion gears, rotate the axle shafts.

When it becomes necessary to turn a corner, the differential becomes effective and allows the axle shafts to rotate at different speeds. As the inner wheel slows down, the side gear splined to the inner wheel axle shaft also slows down. The pinion gears act as balancing levers by maintaining equal tooth loads to both gears while allowing unequal speeds of rotation at the axle shafts. If the vehicle speed remains constant, and the inner wheel slows down to 90 percent of vehicle speeds, the outer wheel will speed up to 110 percent.

DIFFERENTIAL DIAGNOSIS

The most essential part of rear axle service is proper diagnosis of the problem. Bent or broken axle shafts or broken gears pose little problem, but isolating an axle noise and correctly interpreting the problem can be extremely difficult, even for an experienced mechanic.

Any gear driven unit will produce a certain amount of noise, therefore, a specific diagnosis for each individual unit is the best practice. Acceptable or normal noise can

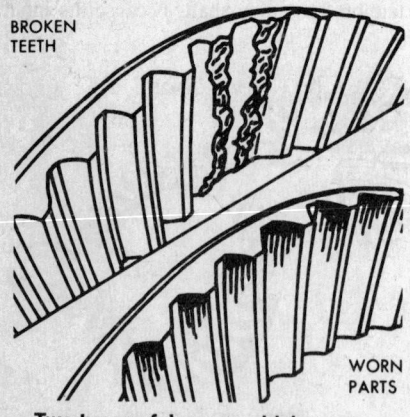

BROKEN
TEETH

WORN
PARTS

Two types of damage which cause gear noise

be classified as a slight noise heard only at certain speeds or under unusual conditions. This noise tends to reach a peak at 40–60 mph depending on the road condition, load, gear ratio and tire size. Frequently, other noises are mistakenly diagnosed as coming from the rear axle. Vehicle noises from tires, transmission, driveshaft, U-joints and front and rear wheel bearings will often be mistaken as emanating from the rear axle. Raising the tire pressure to eliminate tire noise (although this will not silence mud or snow treads), listening for noise at varying speeds and road conditions and listening for noise at drive and coast conditions will aid in diagnosing alleged rear axle noises.

External Noise Elimination

It is advisable to make a thorough road test to determine whether the noise originates in the rear axle or whether it originates from the tires, engine transmission, wheel bearings or road surface. Noise originating from other places cannot be corrected by overhauling the rear axle.

Road Noise

Brick roads or rough surfaced concrete, may cause a noise which can be mistaken as coming from the rear axle. Driving on a different type of road (smooth asphalt or dirt) will determine whether the road is the cause of the noise. Road noise is usually the same on drive or coast conditions.

Tire Noise

Tire noise can be mistaken as rear axle noises, even though the tires on the front are at fault. Snow tread and mud tread tires or tires worn unevenly will frequently cause vibrations which seem to originate elsewhere; temporarily, and for test purposes only, inflate the tires to 40–50 lbs. This will significantly alter the noise produced by the tires, but will not alter noise from the rear axle. Noises from the rear axle will normally cease at speeds below 30 mph on coast, while tire noise will continue at lower tone as vehicle speed is decreased. The rear axle noise will usually change from drive conditions to coast con-

ditions, while tire noise will not. Do not forget to lower the tire pressure to normal after the test is complete.

Engine and Transmission Noise

Engine and transmission noises also seem to originate in the rear axle. Road test the vehicle and determine at which speeds the noise is most pronounced. Stop the vehicle in a quiet place to avoid interfering noises. With the transmission in neutral, run the engine slowly through the engine speeds corresponding to the vehicle speed at which the noise was most noticeable. If a similar noise was produced with the vehicle standing still, the noise is not in the rear axle, but somewhere in the engine or transmission.

Front Wheel Bearing Noise

Front wheel bearing noises, sometimes confused with rear axle noises, will not change when comparing drive and coast conditions. While holding the vehicle speed steady, lightly apply the footbrake. This will often cause wheel bearing noise to lessen, as some of the weight is taken off the bearing. Front wheel bearings are easily checked by jacking up the wheels and spinning the wheels. Shaking the wheels will also determine if the wheel bearings are excessively loose.

Rear Axle Noises

If a logical test of the vehicle shows that the noise is not caused by external items, it can be assumed that the noise originates from the rear axle. The rear axle should be tested on a smooth level road to avoid road noise. It is not advisable to test the axle by jacking up the rear wheels and running the vehicle.

True rear axle noises generally fall into two classes; gear noise and bearing noises, and can be caused by a faulty driveshaft, faulty wheel bearings, worn differential or pinion shaft bearings, U-joint misalignment, worn differential side gears and pinions, or mismatched, improperly adjusted, or scored ring and pinion gears.

Rear Wheel Bearing Noise

A rough rear wheel bearing causes a vibration or growl which will continue with the vehicle coasting or in neutral. A brinelled rear wheel bearing will also cause a knock or click approximately every two revolutions of the rear wheel, due to the fact that the bearing rollers do not travel at the same speed as the rear wheel and axle. Jack up the rear wheels and spin the wheel slowly, listening for signs of a rough or brinelled wheel bearing.

Differential Side Gear & Pinion Noise

Differential side gears and pinions seldom cause noise since their movement is relatively slight on straight ahead driving. Noise produced by these gears will be more noticeable on turns.

Pinion Bearing Noise

Pinion bearing failures can be distinguished by their speed of rotation, which is higher than side bearings or axle bearings. Rough or brinelled pinion bearings cause a continuous low pitch whirring or scraping noise beginning at low speeds.

Side Bearing Noise

Side bearings produce a constant rough noise, which is slower than the pinion bearing noise. Side bearing noise may also fluctuate in the above rear wheel bearing test.

Gear Noise

Two basic types of gear noise exist. First is the type produced by bent or broken gear teeth which have been forcibly damaged. The noise from this type of damage is audible over the entire speed range. Scoring or damage to the hypoid gear teeth generally results from insufficient lubricant, improper lubricant, improper breaking, insufficient gear backlash, improper ring and pinion gear alignment or loss of torque on the drive pinion nut. If not corrected, the scoring will lead to eventual erosion or fracture of the gear teeth. Hypoid gear tooth fracture can also be caused by extended overloading of the gear set (fatigue fracture) or by shock overloading (sudden failure). Differential and side gears rarely give trouble, but common causes of differential failure are shock loading, extended overloading and differential pinion seizure at the cross-shaft, resulting from excessive wheel spin and consequent lubricant breakdown.

The second type of gear noise pertains to the mesh pattern between the ring and pinion gears. This type of abnormal gear noise can be recognized as a cycling pitch or whine audible in either drive, float or coast conditions. Gear noises can be recognized as they tend to peak out in a narrow speed range and remain constant in pitch, whereas bearing noises tend to vary in pitch with vehicle speeds. Noises produced by the ring and pinion gears will generally follow the pattern below.

A.	Drive Noise:	Produced under vehicle acceleration.
B.	Coast Noise:	Produced while the car coasts with a closed throttle.
C.	Float Noise:	Occurs while maintaining constant car speed (just enough to keep speed constant) on a level road.
D.	Drive, Coast and Float Noise:	These noises will vary in tone with speed and be very rough or irregular if the differential or pinion shaft bearings are worn.

BEARING DIAGNOSIS

This section will help in the diagnosis of bearing failure and the causes. Bearing diagnosis can be very helpful in determining the cause of rear axle failure.

When disassembling a rear axle, the general condition of all bearings should be noted and classified where possible. Proper recognition of the cause will help in correcting the problem and avoiding a repetition of the failure. Some of the common causes of bearing failure are:

1. Abuse during assembly or disassembly.
2. Improper assembly methods.
3. Improper or inadequate lubrication.
4. Bearing contact with dirt or water.
5. Wear caused by dirt or metal chips.
6. Corrosion or rust.
7. Seizing due to overloading.
8. Overheating.
9. Frettage of the bearing seats.
10. Brinelling from impact or shock loading.
11. Manufacturing defects.
12. Pitting due to fatigue.

To avoid damage to the bearing from improper handling, it is best to treat a used bearing the same as a new bearing. Always work in a clean area with clean tools. Remove all outside dirt from the housing before exposing a bearing and clean all bearing seats before installing a bearing.

CAUTION

Never spin a bearing, either by hand or with compressed air, as this will lead to almost certain bearing failure.

COMPONENT FAILURE DIAGNOSIS

Scoring & Seizure of Spider and Pinion Gears

The spider arms and pinion gears were badly discolored by heat, caused by the unit operating for a long time after the initial scoring took place. The most probable cause of this type of failure is excessive wheelspin, particularly in off-road or icy road conditions. Other possible causes are inadequate lubrication or overstress. Friction causes the hardened areas to overheat, score, and eventually to seize. The best way to prevent this problem is to avoid wheelspin and overloading under rough terrain or poor traction conditions.

SHOCK FRACTURE

These differential pinion and side gears show a grainy structure which indicates a shock fracture. This type of damage occurs instantaneously. The usual cause is a sudden excessive load, as might be caused by sudden clutch engagement at high engine speed. Another cause is a rapidly spinning wheel suddenly reaching a good traction area. This failure can be prevented by proper clutch operation, and by avoiding wheelspin and overloading under rough terrain or poor traction conditions.

FATIGUE FRACTURE OF THE DIFFERENTIAL SIDE GEARS

This damage occurs in stages. An initial stress caused a crack, and repeated stresses caused complete failure. Some of the gear teeth were broken off in the later stages. The failures can be seen best at points A and B. All differential gears should be checked when this type of failure is found, very often the other gears will be in the initial stages of failure and must be replaced. This is most often caused by abuse such as sudden clutch engagement or incorrect two-speed axle operation, combined with overloading.

SCORED AND SCUFFED GEAR TEETH

This wear pattern is a result of the gear running without enough lubrication between the tooth surfaces. Either poor quality gear lube or low lubrication level can cause this condition. Excessive torque input to the rear can also cause this wear since it will break down even the best of gear lube. Changing gear lube at regular recommended intervals and keeping excessive torque input to a minimum will usually prevent this problem.

FATIGUE FRACTURED PINION GEAR

This type of fracture develops over a period of time. The fracture works through the gear tooth until the tooth is not strong enough to support the load applied. Failure happens and a section of the tooth breaks away. Continued use of pitted gears is the usual cause of this type of gear failure. As the gear pits, the support area is reduced and must carry the entire load of the gear tooth. As this continues the gear tooth fatigues and the final result is failure of the gear. To prevent this problem the ring and pinion must be replaced if there is any pitting on the gears.

MISALIGNMENT FATIGUE FRACTURE

This problem comes from misalignment in the axle shaft. This kind of failure can also happen when the axle shaft breaks. If twisted, bent or sprung axle shaft are are not replaced after they are damaged, this kind of failure to the side gears can occur. Bent axle housing can also cause this to happen. In most cases, this type of failure is not instantaneous. It tends to happen over a period of time. The usual cause of this type of failure is abusive operation of the vehicle and severe overloading.

OVERHEATED GEAR SET

This problem can be caused by one of three, or any combination of the following circumstances. The causes are low gear lubricant level; improper gear lubricant; or infrequent lubricant change. When one or more of these conditions is present in the rear, it causes the lubricant to break down and allows the gear surface to build up heat because of increased friction. In the failure shown, the gears became so hot the pinion bearing fused to the pinion gear and the pinion gear teeth became distorted. To prevent this problem a good quality gear lubricant must be used in the rear to prevent the breakdown of lubricant under a heavy load.

FRACTURE GEAR TEETH

This problem is caused by improper gear adjustment. The picture on the left shows the result of excessive backlash between the ring and pinion gear. Such backlash allows overloading of the heel section of the gear; gear fracture will follow. The picture on the right shows the result of too little backlash thus allowing the toe section of the gear to overload and become fractured. The best way to eliminate this problem is to correctly adjust the ring and piston gears, when necessary, according to specifications.

TWISTED AXLE SHAFT

This problem with the axle comes from abusive and/or extremely severe operation of the vehicle. This is only the first stage of failure where the axle shaft has only twisted, but has not yet started to crack. At this stage the shaft should be replaced. If it is not, the shaft will continue to twist and eventually will break. When this happens it will almost certainly damage other axle parts. To eliminate this problem, the shaft should be replaced if found to be twisted. The driver of the vehicle should be informed to adopt better driving procedures.

PITTED PINION TEETH

This problem is the result of extremely high pressure on the gear teeth due to severe use.

The pitting located at the heel end of the pinion gear teeth happens when overloading of the pinion moves the pinion out of its proper position relative to the ring gear. The result is a concentrated area of contact on the heel part of the gear teeth which will break down the oil film, and thus allow the pinion teeth to pit. Sometimes the ring gear will appear to be undamaged. This is because ring gear damage might not be visible to the naked eye; but the contour of the gear teeth will have changed. The ring and pin<chion gears must be replaced as a pair, or early failure will occur. The best way to eliminate this problem is to use good quality gear lube. The more severely the vehicle is used the better quality the gear lube should be.

SCUFFED GEAR TEETH ON THE COAST SIDE ONLY

This wear can be caused by two different things. The first is worn pinion bearing which allows excessive end play in the pin<chion gear. The result is incorrect contact between the ring and pinion gear teeth on the coast side. This allows excessive pressure to build up on the gear teeth and will break down the oil film, resulting in scoring of the teeth. The second cause is hard, abusive driving in vehicles equipped with a manual transmission. This usually happens when going down a steep grade at high speed and slowing the vehicle by using the clutch to break the speed. The best way to eliminate this problem is to replace the pin<chion bearing if worn and recommend good driving procedures.

Types of Drive Axles

FULL FLOATING AXLES

Support of the vehicle and the payload weight is by the axle housing. The wheels

are driven by splined shafts which "float" within the axle housing.

SEMI-FLOATING AXLE

This axle design provides for the support of the payload and vehicle weight to be carried by the axle shaft through the wheel bearings to the axle housing.

SINGLE REDUCTION AXLE

Final drive ratio is obtained by the use of a single ring gear and pinion set. This type is used for most light and medium duty applications.

AXLE SERVICE & INSPECTION

Cleaning Bearings

Proper bearing cleaning is important. Bearings should always be cleaned separately from other rear axle parts.

1. Soak all bearings in clean kerosene or diesel fuel oil.

— CAUTION —

Ordinary gasoline should not be used. Bearings should not be cleaned in hot solution tank.

2. Slush bearings in cleaning solution until all oil lubricant is loosened. Brush bearings with soft bristled brush until ALL dirt has been removed. Remove loose particles of dirt by striking flat against a wood block.

3. Rinse bearings in clean fluid. While holding races to prevent rotation, blow dry with compressed air.

— CAUTION —

Do not spin bearings while drying.

Front Wheel Drive Halfshaft Troubleshooting

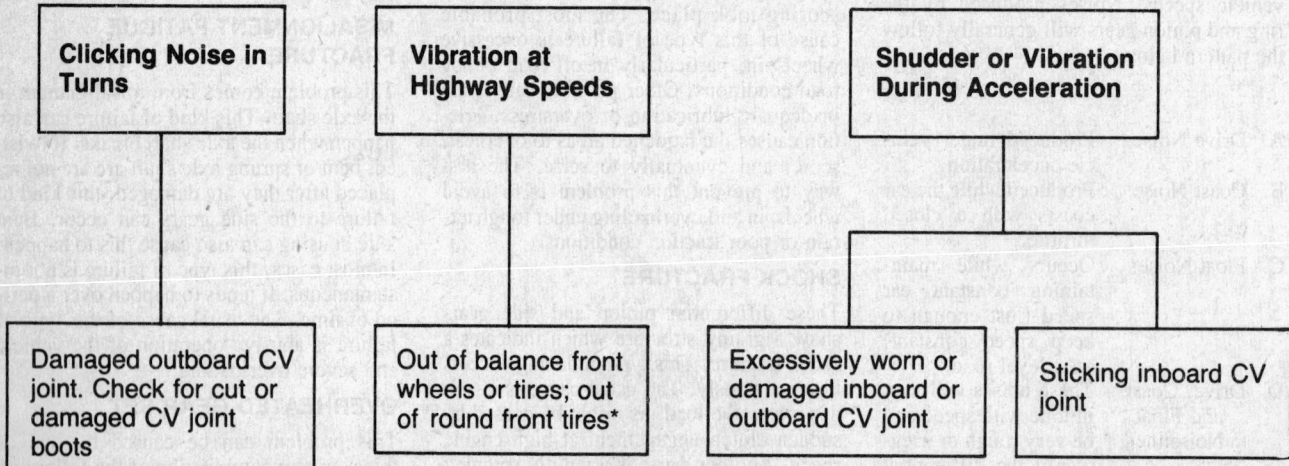

Clicking Noise in Turns	Vibration at Highway Speeds	Shudder or Vibration During Acceleration	
Damaged outboard CV joint. Check for cut or damaged CV joint boots	Out of balance front wheels or tires; out of round front tires*	Excessively worn or damaged inboard or outboard CV joint	Sticking inboard CV joint

*Halfshafts do not usually contribute to rotational vibrations.

4. After bearings have been inspected, lubricate thoroughly with regular axle lubricant; then wrap each bearing in clean cloth until ready to use.

Cleaning Parts

Immerse all parts in suitable cleaning fluid and clean thoroughly. Use a stiff bristle brush as required to remove foreign deposits. Clean all lubricant passages or channels in pinion cage, carrier, caps and retainers. Make certain that interior of housing is thoroughly cleaned. Clean vent plugs and breathers.

Small parts such as cap screws, bolts, studs, nuts etc., should be cleaned thoroughly.

Inspection

Magna Flux all steel parts, except ball and roller bearings, to detect presence of wear and cracks.

Bearings

Rotate each bearing and check to see if the rollers are worn, chipped, rough or in any other way damaged. Check the cage to see if it is in any way damaged. If either the bearing rollers or the cage are damaged the bearing must be replaced.

Gears

Examine drive gear and drive pinion, differential pinions and differential side gears carefully, for damaged teeth, worn spots in surface hardening, distortion and where drive gear is attached to differential case with rivets, inspect rivets for looseness, replace loose rivets. Check radial clearances between differential side gears and differential case. Check fit of differential pinions on spider.

Differential Case

Inspect case for cracks, distortion or damage, if in good condition, thoroughly clean case and cover; then assemble case with bolts and mount in lathe centers of "V" block stand. If lathe is not available, install differential side bearings and mount case in differential carrier. Install dial indicator and check differential case run out.

Differential case with drive gear installed is checked in the same manner, except that dial indicator reading must be taken at gear instead of at case flange.

Whenever run-out exceeds limits, it may be corrected as later described under "Repair" in this section. However, the support case used in the 2 speed axle cannot be repaired and should be replaced with new case.

Axle Shafts

Examine splined end of axle shaft for twisted or cracked splines, twisted shaft, and worn dowel holes in flange. Install new shafts if necessary.

Install axle shaft assembly in lathe centers and check shaft run-out with dial indicator so that indicator shaft end contacts inner surface of flange near outer edge of flange and check flange run-out.

Shims

Carefully inspect shims for uniform thickness. Where various thickness of shims are used in a pack, it is recommended that the thickest shims be used between the thin shims.

Thrust Washers

Replace all thrust washers.

Spider

Carefully inspect spider arms for wear or defects.

Differential Pinion Bushings

Examine bushings (when used) for excessive wear, looseness, or damage. Check fit or gears on spider for excessive clearance.

Axle Housing Sleeves

Sleeves showing damaged threads, wear, or other damage should be replaced if hydraulic press is available, otherwise replace housing.

HOUSING CHECK

Before Removal

A check for bent axle housing can be made with unit in vehicle; however, conventional alignment instruments can be used if available.

1. Raise rear axle with a jack until wheels clear floor. Block up axle under each spring seat.
2. Check wheel bearing adjustment and adjust if necessary, then check wheels for looseness and tighten wheel nuts if necessary.
3. Place a chalk mark on outer side wall of tires at bottom. Measure across tires at chalk marks with a toe-in gauge.
4. Turn wheels half-way around so that chalk marks are positioned at top of wheel. Measure across tires again. If measurement at top is 1/3" or more, smaller than measurement at bottom of wheels, axle housing has sagged and is bent. If measurement at top exceeds bottom dimension by 1/3" or more, axle housing is bent at ends.
5. Turn chalk marks on both wheels so that marks are level with axle and at rear of vehicle. Take measurement with toe-in gauge at chalk marks; then turn both chalk marks to front and level with axle and take another measurement. If measurement at front exceeds rear dimension by 1/3" or more, axle is bent to the rear. If the measurement condition is the reverse, the axle is bent forward.

After Removal

Place two straightedges across the housing flanges and measure the distance between the ends of the straightedges at a point 11 inches from the tube center. Relocate the straightedge 180 degrees and remeasure. If the straightedges are parallel in both measurements within 3/32 inch, the housing is serviceable.

GENERAL REPAIR

Oil Seal Contact Surfaces

Surface of parts, contacted by oil seals must be free of corrosion, pits and grooves. When abrasive cleaning fails to clean up the seal contact surface and restore smooth finish, a new part must be installed.

OIL SEAL

Removal

Oil seals can be removed with a drift pin. When removing a seal, be careful that it does not become cocked and result in damage to the retainer. Clean surface of retainer carefully, so that seal will seat properly in retainer.

Installation

Coat outer surface of seal retainer with a light coat of sealer, to prevent lubricant leaks. Carefully start seal in retainer. Cutting, scratching, or curling of lip of seal seriously impairs its efficiency and usually results in premature replacement. Lip of seal should be coated with a high temperature grease containing zinc oxide to help prevent scoring and damage to parts during installation.

Seals must always be installed so that seal lip is toward the lubricant.

PINION BEARING ADJUSTMENTS (PRE-LOAD)

Pinion bearing must be adjusted for preload before assembly is installed in carrier.

Do not install oil seal until after adjustment is made. Installation of seal would produce false rotating torque.

Cage Type

1. With pinion bearings, and adjusting spacers (or shims) installed in cage, check bearing contact by rotating cage.
2. Using a press, apply pressure (approx. 20,000 lbs.) to outer bearing.
3. Wrap soft wire around cage and pull on horizontal line with spring scale. Rotating (not starting) torque should be within limits recommended by manufacturer.

NOTE: Method of determining inch pounds torque with scale is to determine radius of cage. Multiply radius in inches by pounds pull required to rotate cage to determine inch-pounds torque. Example: An 8-inch diameter divided by 2 equals 4-inch radius. Multiply 4-inch (radius) by 5 pounds (pull) equals 20 inch pounds torque.

4. If press is not available, check pre-load torque by installing propeller shaft yoke, washer, and nut and torque to specifications; then check as previously explained. Remove yoke after correct adjustment is obtained.

BEVEL GEAR SHAFT BEARING ADJUSTMENT

Bevel gear shaft bearings must be adjusted for pre-load before pinion and cage assembly and differential assembly are installed in carrier.

1. Wrap several turns of soft wire around gear teeth on cross shaft, then pull on a horizontal line with spring scale. Rotating (not starting) torque should be used.

NOTE: Method of determining inch-pounds torque with scale is to determine radius. Multiply radius in inches by pounds pull required to rotate shaft to determine inch-pounds torque. Example: An 8-inch diameter divided by 2 equals 4-inch radius times 5 pounds (pull) equals 20 inch pounds torque.

2. Remove or add shims from under cage or cap opposite bevel gear to obtain specified bearing pre-load.

3. When making bevel gear and pinion tooth contact or backlash adjustments it is sometimes necessary to remove or add shims from one side.

NOTE: Always remove or add an equal thickness to the opposite side so to maintain correct pre-load.

GEAR TOOTH CONTACT AND BACKLASH

Pinion Depth Measurement Methods

Methods of adjusting pinions to obtain the proper depths will vary with the axle type and the manufacturers recommendations. Pinion depth settings and gear teeth contact may be determined by the use of pinion setting gauges or by the use of marking dye on the gear teeth.

When using the gauge method, backlash is established after the pinion has been properly set. With the dye method, backlash is obtained first, then the proper pinion tooth contact is established.

The pinion gauge method can be a direct reading micrometer, mounted on or through an arbor bar, set in adapter discs and located in the side carrier bearing cup locations on the differential housing and held in place by the bearing cup caps. The arbor bar coincides and represents the center line of the axle shafts. A reading is taken by the mounted micrometer, from the arbor bar to the head of the pinion to determine the need to add to or remove shims from the shim pack total, to adjust the pinion to the proper nominal assembly dimension or standard pinion depth.

Another method using the arbor bar and discs, is the use of a gauge block with a spring loaded plunger and a thumb screw to lock the plunger upon expansion. A micrometer is used to measure the gauge block after the plunger has been allowed to expand between the arbor bar and the pinion head. As in the mounted micrometer procedure, the shim pack thickness is determined by the reading obtained.

A third method is the use of a gauge block tool, installed in the housing in place of the pinion gear, and a large arbor bar placed in the axle housing differential bearing seats and tightened securely. A measurement is taken between the arbor bar and the pinion tool by either a feeler gauge or the use of individual shims from the shim pack. This measurement represents the shim pack needed for a zero marked pinion.

Setting New Pinion (Without Gauge)

Whenever a pinion setting gauge is not available, the approximate thickness of the pinion shim pack at the rear pinion bearing cup, change the sign of the marking (individual variation distance) on the *new* pin < chion (plus to minus or minus to plus), then add the variation of the old pinion (sign unchanged) which will determine the amount the original shim pack must be changed when installing a new pinion.

On those types of axles where the shims are located between the pinion cage and differential carrier, change the sign of the marking (individual variation distance) on the *old* pinion (plus to minus or minus to plus), then add variation of the new pinion (sign unchanged) which will determine how much the original shim pack must be altered when installing a new pinion.

When the approximate thickness of shim pack has been determined, final check of gear tooth contact must be made using dye method.

Gear Tooth Contact (Dye)

Gear tooth contact cannot be successfully accomplished until pinion and bevel gear bearings are in proper adjustment and gear backlash is within specified limits.

Check for proper tooth contact by painting a few teeth of bevel gear with marking dye. Turn pinion in direction of normal rotation, then check tooth impression on bevel gear.

Gear Backlash

Gears that have been in extended service, form running contacts due to wear of teeth; therefore the original shim pack (between pinion cage and carrier) should be maintained when checking backlash. If backlash exceeds maximum tolerance, reduce backlash only in the amount that will avoid overlap of worn tooth section. Smoothness and roughness can be noted by rotating bevel gear.

If a slight overlap is present at worn tooth section, rotation will be rough.

If new gears are installed, check backlash with dial indicator.

Backlash is increased by moving bevel gear away from pinion, and may be decreased by moving bevel gear toward pin < chion.

When the drive gear is attached to the differential, backlash is accomplished is differential bearing adjusting rings. It should be remembered that when one ring is tightened, the opposite ring must be loosened an equal amount to maintain previously established bearing adjustment.

On axles where the bevel gear is supported by cross shaft, backlash is accomplished by adding or removing shims under bearing cages.

Terms Used

Certain dimensions must be determined when using the pinion setting gauge:

1. *Nominal Assembly Dimension.* (standard pinion depth) This dimension (varying with axle model) is the distance between the center line of the drive gear (or differential carrier bore) and the end of the drive pinion. This dimension may be marked on the pin < chion or listed on the Nominal Assembly Dimension and Adapter Disc chart.

2. *Individual Variation Distance,* (pinion depth variance) This dimension is a plus or minus variation of the *Nominal Assembly Dimension* on each individual pinion which may be caused by manufacturing variations.

3. *Corrected Nominal Dimension* (desired pinion depth) This dimension is the *Nominal Assembly Dimension* plus or minus the *Individual Variation Distance.*

4. *Corrected Micrometer Distance is the Corrected Nominal Dimension* less the thickness of the gauge set step plate (0.400") mounted on end of pinion.

5. *Initial Micrometer Reading* is the dimension taken by micrometer to the gauge step plate.

6. *Shim Pack Correction* is determined by the difference between the *Corrected Micrometer Distance* and the *Initial Micrometer Reading,* and represents the amount of shim pack to be added or removed as later explained.

7. *Measured Pinion Depth.* This measurement is the distance between the axle center line and the top of the pin < chion gear. If a step plate or other type gauge tool is used, this measurement is included in the total.

MARKINGS ON THE PINION AND DRIVE GEARS

Drive gears and pinions are tested at the

time of manufacture to detect machining variances and to obtain desirable tooth contact and quietness. When the correct setting is achieved, the gears are considered matched and a set of numbers, along with other identifying marks are etched on the gear set.

A + (plus) or − (minus) sign is used, followed by a digit to represent the factory setting where the tooth contact and quietness were the best. This is called the *Pinion Depth Variance* or *Individual Variation Distance.*

If the pinion is marked +5 for example, this means the distance from the pinion gear rear face to the axle shaft center line is .005 in. more than the standard setting, and if the pinion gear is marked −5, this means that the distance is .005 in. less than the standard setting. To move the pinion to the standard setting, compensating for the variation, shims must be either added to subtracted from the total shim pack, located under the rear pinion bearing cup, between the pinion cage and the differential carrier, or under the rear pinion bearing, depending upon the differential model being serviced.

The procedures to follow in the adjustment of the pinion and drive gears are outlined in the respective differential model disassembly and assembly chapters.

As a rule of thumb on the addition or removal of shims for the pinion depth adjustment, draw a diagram as shown and determine which way the pinion must be moved to obtain the desired pinion depth.

STANDARD TORQUE SPECIFICATIONS AND CAPSCREW MARKINGS

Because of the varied bolt sizes used in the many models of differentials, the torque specifications are not always available to the technician for a specific bolt. By determining the grade of bolt, size, and thread, the proper torque limit can be found in the following chart.

Limited-Slip Differential Operation

Limited-slip differentials provide driving force to the wheel with the best traction before the other wheel begins to spin. This is accomplished through clutch plates or cones. The clutch plates or cones are located between the side gears and inner wall of the differential case. When they are squeezed together through spring tension and outward force from the side gears, three reactions occur. Resistance on the side gears causes more torque to be exerted on the clutch packs or clutch cones. Rapid one-wheel spin cannot occur, because the side gear is forced to turn at the same speed as the case. Most important, with the side gear and the differential case turning at the same speed, the other wheel is forced to rotate in the same direction and at the same speed as the differential case. Thus driving force is applied to the wheel with the better traction.

LIMITED-SLIP DIFFERENTIAL DIAGNOSIS

Lubrication

The use of proper lubricant is very important in limited-slip type drive axles. The forces applied when cornering tend to apply the clutch pack or clutch cones. The use of the wrong lubricant can cause the clutch surfaces to grab and chatter while turning. Always follow the manufacturer's recommendations regarding drive axle lubrication. When chatter is encountered, the differential lubricant should be drained and refilled with the specified lubricant.

Testing

The clutch operation on all limited-slip type axles can be tested as follows. Refer to the manufacturer in question.

AMERICAN MOTORS "TWIN-GRIP"

1. With the engine off and the transmission in neutral, jack up one rear wheel.
2. Block the other wheel to prevent it from moving.
3. With a socket and torque wrench on the axle shaft nut, turn the raised wheel forward.
4. The torque required to move the wheel should be 70–100 ft lbs for $8\frac{7}{8}$ in. axles or 80–120 ft. lbs. for $7\frac{9}{16}$ in. axles.
5. A breakaway torque which is less than the specified figure, indicates a need for repair or replacement.

CHRYSLER "SURE-GRIP"

1. Raise and support the rear of the vehicle with the engine off and the automatic transmission in Park (manual transmission in low gear).
2. Attempt to rotate the wheel by hand, by gripping the tire.
3. If it is extremely difficult, if not impossible, to rotate either wheel the Sure-Grip differential can be assumed to be performing satisfactorily.
4. If it is relatively easy to continuously turn either rear wheel, the unit should be removed and replaced.

──────── CAUTION ────────
The Sure-Grip differential is serviced as a unit only. Under no circumstances should the unit be disassembled and reinstalled.

FORD "EQUA-LOK"

1. Jack up one rear wheel and remove the wheel cover.
2. Block the other wheel front and rear to prevent the vehicle from moving.
3. Using a 200 ft. lbs. capacity torque wrench on one of the wheel lug nuts, measure the torque required to continuously rotate the wheel. The breakaway torque read-

ing can be disregarded. The minimum torque to continuously rotate the wheel should be as follows: All axles except integral carrier type: 75 ft lbs. Integral carrier type axles: 50 ft lbs.
4. If the minimum torque is not as specified, the differential should be checked for improper assembly.

FORD "TRACTION-LOK"

1. Follow the procedure for the Ford Motor Company Equa-Lok rear. The minimum torque to continuously rotate the wheel (disregarding the breakaway torque) should be at least 40 ft lbs through 1979 and 30 ft lbs for 1980 and later.

GMC "POSITRACTION"

1. Place the transmission in neutral.
2. Raise one rear wheel off the floor and block the other rear wheel (front and rear) to prevent the vehicle from moving.
3. Install a torque wrench and extension on the lug nut and note the torque required to continuously rotate one rear wheel. Disregard the breakaway torque figure, as this may be a great deal higher.
4. The minimum torque to continuously rotate the rear wheel should be at least 35 ft lbs. If it is not, the rear axle is in need of service.

General Diagnosis

Improper operation of a limited-slip type rear axle is generally indicated by clutch slippage or grabbing, which will sometimes produce a whirring or chatter sound. Occasionally, this condition is induced by improper lubrication. Check the unit for the wrong type of lubricant or lubricant which has broken down or become contaminated. Replace the lubricant with the type specified by the manufacturer.

During normal operation, i.e., straight-ahead driving, both wheels are rotating at equal speeds, and the driving force is distributed equally between both wheels. When cornering, the inside wheel delivers extra driving force, causing slippage in both clutch packs. Therefore, if the wheel rotation of both rear wheels is not equal, the unit will constantly be functioning as if the vehicle were cornering. This will cause constant slippage and lead to eventual failure of the unit. It is important that there be no excessive differences in wheel and tire size, wear pattern, or tire pressures between both rear wheels. Swerving on acceleration is an indication of one or more of the above conditions. Before attempting an overhaul or replacement operation, check both rear wheels for identical tire sizes, tire pressure, tire tread depth, and wear pattern.

DRIVE AXLE DISASSEMBLY ANALYSIS

Testing the Gear Tooth Contact Pattern

Once it has been established that the differ-

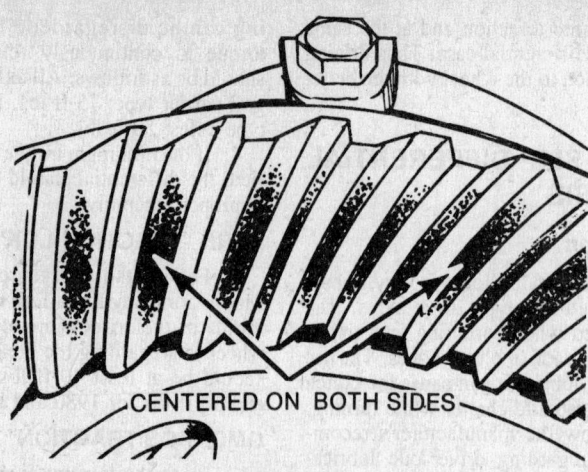

Gear tooth contact pattern showing load centered on gear tooth

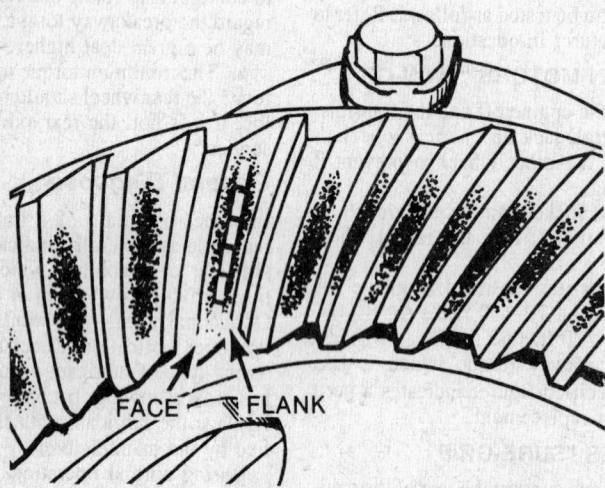

Gear tooth face and flank showing oval gear tooth contact pattern

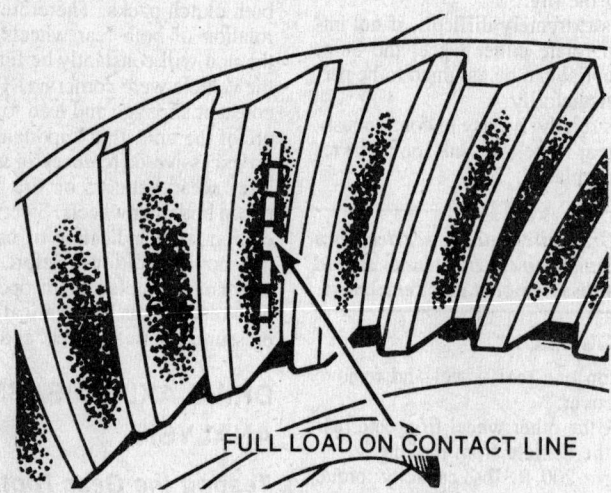

Gear tooth contact pattern showing load centered on gear tooth

ential is indeed in need of service, the worst procedure is to simply plunge ahead and remove the differential and disassemble the parts. Prior to disassembly, a tooth contact pattern test should be made. However, it is worthwhile to first know the nomenclature associated with hypoid gear teeth.

The thick end of the tooth is called the heel and the thin end of the tooth is called the toe. The base half of the tooth is called the flank and the other end of the tooth is known as the face. The imaginary line at the halfway point between the face and flank is known as the pitch line. The space between the meshed pinion and ring gear tooth is known as backlash.

A gear tooth contact pattern can be made with the carrier in or out of the housing depending on the type of carrier. On integral carrier models, the lubricant must be drained and the rear cover removed. The ring gear will now be exposed and the test can be made with the carrier still in the housing. On removable carrier models, drain the lubricant and remove the carrier from the housing. The test can be made on the bench.

Unlike simple spur gears, hypoid gear teeth leave a complex pattern on the ring gear. When hypoid gears turn, the line contact between pinion and ring gear teeth has the same wiping motion as with spur gear teeth. Because of the complicated movement of hypoid gear teeth, the contact area takes an oval shape as opposed to the rectangular shape left by spur gear teeth. Actually, the tooth contact test shows where each gear tooth has been wiped by the movement of the contact line, so that you can tell whether the gears are set correctly. With a properly adjusted ring and pinion (with properly adjusted pinion depth and backlash) the tooth contact will be close to center. In this case, the load is borne by the strongest part of the tooth. If the gear setting is off, the contact line may reach any part of the edge of a tooth, and the metal will be overloaded at that point. When overload occurs, rapid deterioration of the gears will follow.

PREPARING THE TEST

Coat the drive gear teeth with a metallic base artists' oil color such as zinc white or titanium white. The tooth coating material must be smooth and firm enough to spread without running. A consistency somewhat like toothpaste works well. If it is necessary to thicken the material, add a small amount of cup grease.

NOTE: Prussian blue dye does not work well, since the blue tends to smear the pattern.

Thoroughly clean the ring gear and pinion before applying the testing material. Any gear lube left on the teeth will make the pattern quite unreadable. Coat the drive and coast sides of all the ring gear teeth, but leave the pinion gear teeth clean. Do not apply the coating too thickly as the pattern will be smeared.

Because the axle gears are normally easy to rotate, turning resistance must be applied to produce pressure between the pinion and ring gear teeth to make a legible pattern. On a removable carrier type axle, insert a suitable pry bar between the carrier housing and the differential case rim. Apply the load squarely against the case rim while prying out against the upper or lower section of the carrier housing. On integral carrier models, apply the parking brake to a point where it requires approximately 50 ft.lbs. to turn the pinion with a torque wrench. Since the shape and position of the contact pattern will vary, depending on the load, try to use the same load for each test or the results can be misleading. This is especially true when testing after an overhaul.

Once the gears have a load applied, obtain a tooth contact pattern by rotating the ring gear and pinion one complete turn in each direction. This will produce a constant pattern on the coast and drive side of each tooth. Do not rotate the ring gear more than one revolution in each direction as this will tend to obscure the pattern.

NOTE: If the pattern does not look right on the first try, try again.

Making a good gear tooth test takes a little practice; so if it is not right, try again.

INTERPRETING GEAR TOOTH CONTACT PATTERNS

The tooth contact pattern should be the same on every tooth. If the pattern shows heavy and light areas on different teeth, check the ring gear and differential case for excessive run-out.

NOTE: Run-out can be cured in many cases by removing the ring gear from the case, rotating it 90° or 180°, and remounting it.

Since you can only apply test load pressure to the gears, the contact pattern will be less distinct toward the tooth ends. But, when the ring gear and pinion are under operating loads in the vehicle, the tooth contact area spreads out, especially towards the heel end of the tooth. For this reason, do not try to "get by" with a tooth contact pattern that is centered, but favors the heel

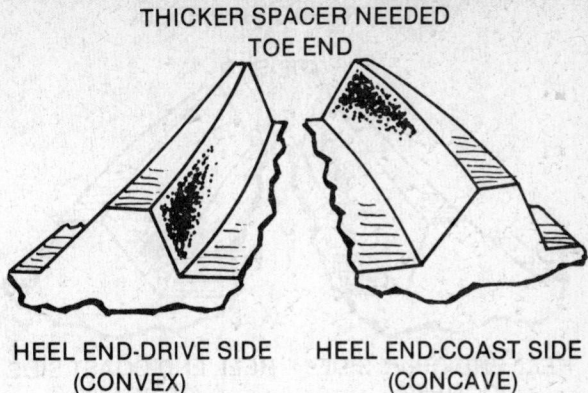

THICKER SPACER NEEDED
TOE END

HEEL END-DRIVE SIDE
(CONVEX)

HEEL END-COAST SIDE
(CONCAVE)

Tooth contact patterns high on the tooth side

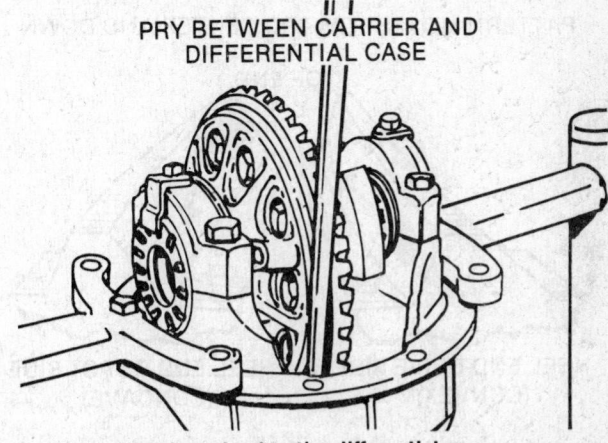

PRY BETWEEN CARRIER AND
DIFFERENTIAL CASE

Applying a load to the differential case

end of the teeth. This will only lead to overloading at the heel ends of the gear teeth. On the other hand, a contact pattern which is reasonably centered, but favors the toe end of the teeth, is acceptable.

Assuming that the tooth contact pattern is even on all teeth, the main problems is to get the most distinct part of the pattern centered on both the teeth. The contact patterns should be nearly opposite each other on both sides of each tooth. In some cases, the pattern will be centered on the drive side and off center on the coast side, or vice versa. The off center pattern can be moved

to a more acceptable position by slightly altering the backlash. This procedure will not seriously affect the other pattern. More often, however, the pattern will be off center on both sides of the teeth. The basic cause of this condition is an improperly adjusted pinion.

ADJUSTING PINION DEPTH

It is necessary to understand that an incorrect pinion depth setting moves the contact pattern away from the center on both sides of the tooth in opposite directions. This means that when you install a thicker or

Drive pinion height adjusting shim

Selecting drive pinion height adjusting shims

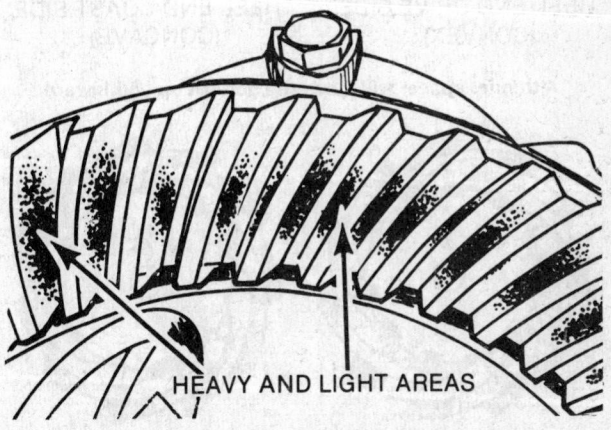

HEAVY AND LIGHT AREAS

Excessive run-out will cause an uneven pattern

THINNER SPACER NEEDED

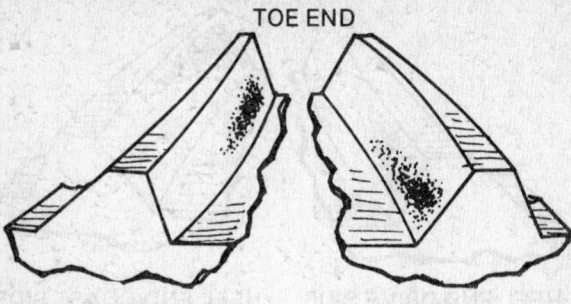

TOE END

HEEL END-DRIVE SIDE
(CONVEX)

HEEL END-COAST SIDE
(CONCAVE)

Gear contact pattern low on tooth side

PATTERN MOVES TOWARD CENTER AND DOWN

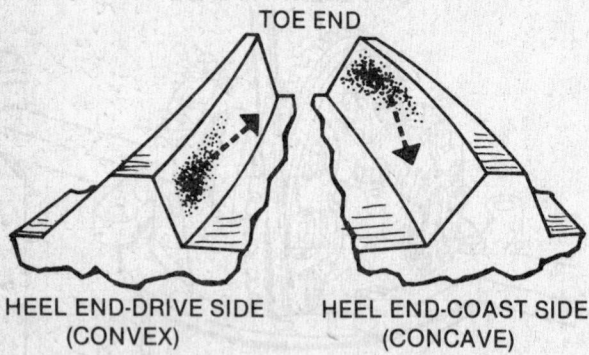

TOE END

HEEL END-DRIVE SIDE
(CONVEX)

HEEL END-COAST SIDE
(CONCAVE)

A thicker spacer moves the pattern in and down

PATTERN MOVES INWARD AND UP

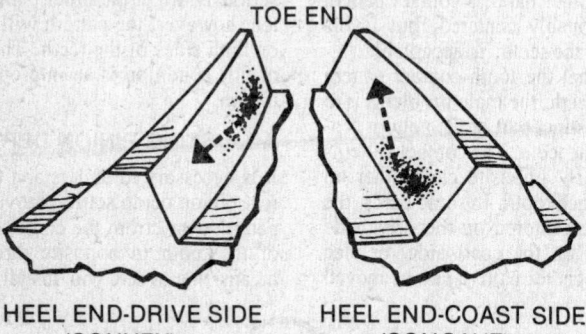

TOE END

HEEL END-DRIVE SIDE
(CONVEX)

HEEL END-COAST SIDE
(CONCAVE)

A thinner spacer will move the pattern up and inward

One example of pinion markings

thinner washer under the pinion head you bring the pattern into the center of the tooth from opposite ends.

When the contact pattern is high on the heel end of the drive side and low on the toe end of the coast side, a thicker washer is needed to bring the pinion in, toward the center of the drive gear. Increasing the thickness of the spacer washer will bring the pattern in, toward the center of the drive gear teeth, and also will move the pattern down from the tooth face. However, this movement is less than the in-or-out movement.

When tooth contact is low on the toe end of the drive side and high on the heel end of the coast side, the pinion must be moved out, by installing a thinner washer under the pinion head. This will move the pattern inward toward the center, and will also result in slight movement of the pattern up from the tooth flank.

A factory service facility will use special tools and gauge blocks to determine the thickness of the spacer under the pinion head. In the absence of such specialized equipment, the following procedure may be used. Bear in mind that with the "hit-or-miss" method, each time you are wrong with the pinion depth, the unit must be disassembled, the spacer thickness changed, and the unit must be completely set up again.

Gather a handful of spacers to cover any thickness and several collapsible pinion spacers (if the unit uses them). Assemble the unit. If the original gear set is being reused, and the tooth contact pattern is reasonably correct, install a new spacer of the same thickness as the old one. This will provide a reasonable starting point. If the gear contact pattern test indicates a need for movement of the pinion, use a new spacer 0.001–0.002 in. thicker or thinner, depending on the direction the pinion must go. If a new gear set is being used, the thickness of the spacer will have to be determined in the following manner. Compare the markings on the old and new pinion. It will usually be marked with a number preceded by a plus (+) or minus (−) sign. This number indicates the production deviation from the nominal pinion, which are known as "zero pinions." In service, zero pinions are rare. Assume that the old pinion is marked with a plus two (+2). Assume that the new pinion is marked with a +3. By comparing the pinion markings, find the numerical difference between the two pinions, in this case +1. With a micrometer, measure the thickness of the original spacer. We will assume that the older spacer is 0.030 in. thick. If the numerical difference between pinions is a positive number (+1) the spacer should be 0.001 in. thinner than the original spacer, or 0.029 in. total. If the numerical difference is a negative number (say, −1) then the spacer should be increased by 0.001 in., to 0.031 in. total. This will only provide a reasonable beginning point.

It is rare that this method works out the

first time. Assemble the pinion, differential, and ring gear with the spacer of calculated thickness. The side bearing preload, backlash, pinion nut torque, and pinion rotating torque must all be set correctly. Obtain a gear tooth pattern on the ring gear teeth and analyze the results. Small deviations from the acceptable pattern can usually be made by varying the backlash within the limits of specifications. If the gear tooth contact pattern is off, the unit must be disassembled and another spacer installed. This spacer must be of suitable thickness to compensate for the contact pattern test.

NOTE: Without special tools, there is absolutely no way of determining exactly how much to increase or decrease the thickness of the pinion shim; it must be estimated.

After estimating the thickness of the new shim, assemble the unit again, setting all preloads and backlash. Check the contact pattern again and act accordingly. If the unit uses a collapsible spacer, be sure a new one is installed each time it is disassembled. Crushed spacers can not be used again. It is well to note that the unit may have to be assembled and disassembled several times before an acceptable contact pattern is obtained.

Adjusting Backlash

The tooth contact pattern can be altered slightly, by varying the backlash adjustment within the limits of the specifications. The backlash adjustment can be used to alter a pattern which is slightly off center on either side of the tooth, but should not be used as a substitute for pinion depth adjustment. This adjustment must always be made after the pinion depth has been adjusted.

AMC/JEEP

All the Jeep, Grand Wagoneer and J–10 models are using a Jeep semi-floating type rear axle with flanges axle shafts. Both standard ($7^9/_{16}$ in.) and heavy duty ($8^7/_8$ in.) are being used. The J–20 models are using the Dana-Model 60 full-floating rear axle. The standard and heavy duty Jeep rear axle housings are made up of a nodular cast iron center section and two steel tubes which are pressed into the center section. The rear drum brake support plates are attached to the mounting flanges at the axle tube outboard ends.

The differential assembly consists of a cast iron case containing two differential side gears, two differential pinion gears and a pinion shaft on which the pinion gears are mounted. The differential side and pinion gears are in constant mesh.

The axle ratio and the ring and pinion gear tooth combinations are stamped on a tag attached to the differential housing cover. On the Jeep rear axles, the axle code letters are stamped on the right side axle housing tube boss.

NOTE: The Trac-Lok limited slip differentials are available as an option.

$7^9/_{16}$ & $8^7/_8$ Inch Ring Gear
REMOVAL

NOTE: It is not necessary to remove the rear axle assembly in order to overhaul the differential.

1. Raise and support the vehicle safely. Remove the axle housing cover and drain the lubricant into a suitable drain pan.
2. Remove the wheels, brakedrums, axle shafts and seals. Keep the left and right-side axle parts separated.
3. Mark the bearing caps with a center punch for assembly reference. Loosen the bearing cap bolts until only several threads are engaged, then pull the bearing caps away from the bearings. This will prevent the differential from falling out and sustaining damage when pried from the axle housing.

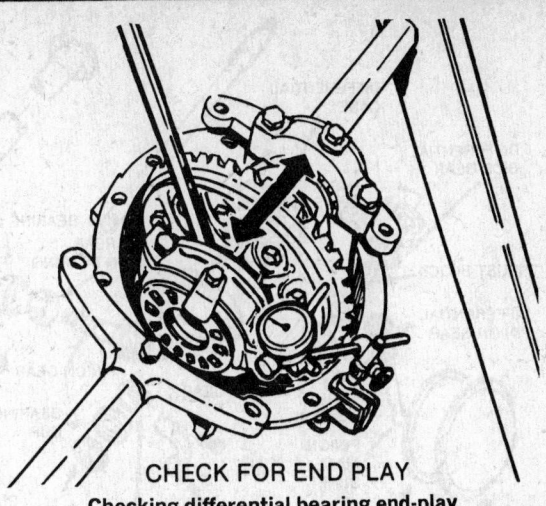

CHECK FOR END PLAY
Checking differential bearing end-play

4. Pry the differential loose in the axle housing. Remove the bearing caps and remove the differential. Tie the differential bearing shims to their respective bearing caps and cups to prevent misplacement.

DISASSEMBLY

1. Use Puller J–29721 and adapters or equivalent to remove the differential bearings. When using this tool, be sure the differential case is secure. When the bearing is removed the differential case can drop if not supported.
2. Remove the ring gear-to-differential case bolts.

NOTE: Do not chisel or wedge the gear from the case.

3. Remove the ring gear from the case. Use a brass drift and hammer to tap the ring gear from the case. Do not nick the ring gear face of the differential case or drop the gear.
4. Remove the pinion mate shaft lockpin using a suitable drift. Remove the pinion mate shaft and remove the thrust block.

TOTAL END PLAY
MEASURED BY
FEELER BLADES

Checking total differential end-play

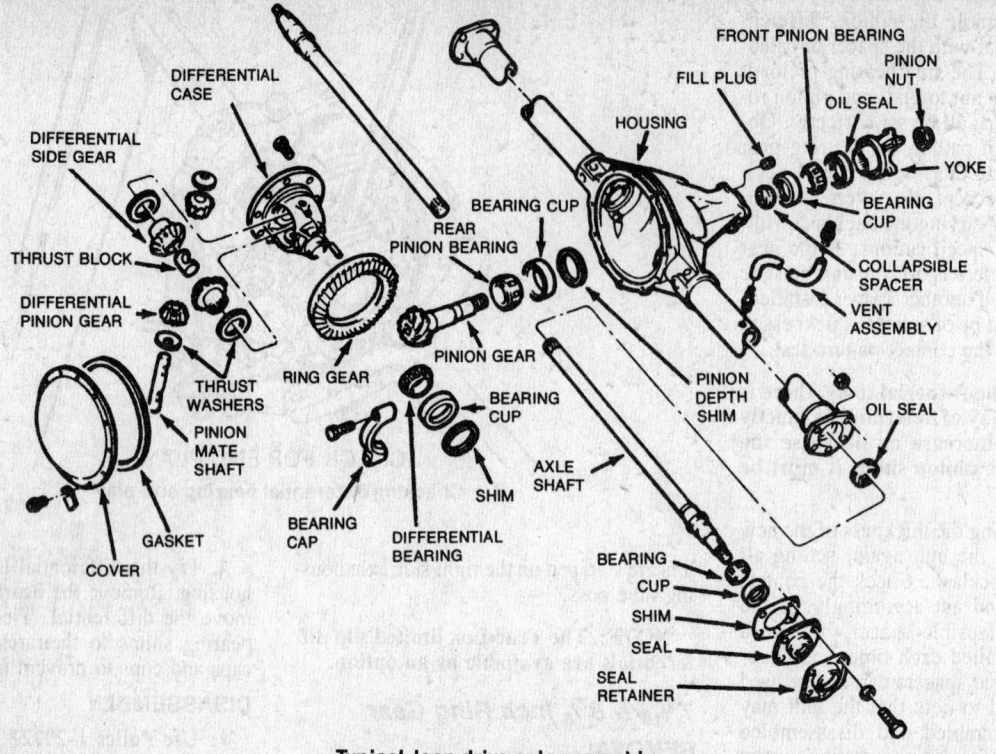

Labels in diagram:
DIFFERENTIAL CASE — DIFFERENTIAL SIDE GEAR — THRUST BLOCK — DIFFERENTIAL PINION GEAR — THRUST WASHERS — PINION MATE SHAFT — GASKET — COVER — RING GEAR — BEARING CAP — DIFFERENTIAL BEARING — SHIM — BEARING CUP — PINION GEAR — REAR PINION BEARING — BEARING CUP — AXLE SHAFT — PINION DEPTH SHIM — FILL PLUG — HOUSING — FRONT PINION BEARING — PINION NUT — OIL SEAL — YOKE — BEARING CUP — COLLAPSIBLE SPACER — VENT ASSEMBLY — OIL SEAL — BEARING CUP — SHIM — SEAL — SEAL RETAINER

Typical Jeep drive axle assembly

5. Rotate the pinion gears on the side gears until the pinion gears are aligned with the case opening. Remove the pinion gears and thrust washers and remove the side gears and thrust washers.

Pinion Gear Removal

6. Remove the pinion nut using Tool J–8614–01 or equivalent. Remove the axle yoke using Tool J–8614–01, –02, –03 or equivalent.

7. Install the axle housing cover to prevent the pinion gear from falling out when the gear is driven out of the bearings and housing. Loosely attach the cover using two bolts.

8. Remove the pinion seal using tool J–9233 or equivalent. Tap the end of the pinion gear with a soft face mallet to drive the pinion gear out of the front bearing. Remove the front bearing and collapsible spacer. Discard the spacer.

NOTE: The collapsible spacer is used to control pinion bearing preload. Discard this spacer after removal, it is not reusable.

9. Remove the axle housing cover and remove the pinion gear and rear bearing from the housing. Remove the rear bearing cup using Tools J–8092 and J–21786 or equivalent

NOTE: The pinion gear depth adjustment shims are located under the rear bearing cup. Tag these shims for assembly reference.

10. Remove the front bearing cup using Tools J–8092 and J–21787 or equivalent.

NOTE: Keep the bearing cup remover tool seated squarely on the cup to prevent damaging the cup bores during removal.

Differential Axle Housing Alignment

1. Place two straightedges across the tube flanges and measure the distance between the flange ends. If the straightedges are parallel within $3/32$ in. at a distance of 11 in. from the tube centerline, the axle housing is serviceable.

2. Perform this inspection with the straightedges placed in horizontal and vertical positions.

Pinion Gear Installation and Depth Adjustment

Ring and pinion gear sets are factory tested to detect machining variances. Tests are started at a standard setting which is then varied to obtain the most desirable tooth contact pattern and quiet operation. When this setting is determined, the ring and pinion gear are etched with identifying numbers. The pinion gear receives two numbers which are separated by a plus (+) or a minus (-) sign.

The second number on the pinion gear indicates pinion position, in relation to the centerline of the axle shafts, where tooth contact was best and gear operation was most quiet. This number represents pinion depth variance and indicates the amount in thousands of an inch that the gear set varied from the standard setting.

The number on the ring gear and first number on the pinion gear identify the gears as a matched set. Do not attempt to use a ring and pinion set having different numbers. The standard setting for AMC/Jeep axles is 2.547 in. If the pinion is marked +2, the gear set varied from standard by + 0.002 in. and will require 0.002 in. less shims than a gear set marked zero (0).

When a gear set is marked plus (+), the distance from the pinion end face to the axle shaft centerline must be more than the standard setting. If the pinion gear is marked -3, the gear set varied from standard by 0.003 in. more shims than a set marked zero (0). When a set is marked minus (-), the distance from the pinion end face to the axle shaft centerline must be less than the standard setting.

NOTE: On some factory installed gear sets, an additional 0.010 or 0.020 in. may have been machined off the pinion gear bottom face. This does not affect the gear operation but does affect the pinion gear marking and depth measurement.

Pinion gears machined in this fashion have different identifying numbers. For example, if the pinion is marked +23, the number 2 indicates that 0.020 in. was removed from the pinion bottom face and the number 3 indicates that variance from the standard setting is + 0.003 in. If the pinion is marked +16, the number 1 indicates that 0.010 in. was removed from the pinion bottom face and the number 6 indicates that variance from the standard setting is + 0.006 in.

Gear sets with additional amounts machined off the pinion bottom face are factory installed items exclusively. All service replacement gear sets will be machined to standard settings only. In addition, replacement gear sets marked + or - 0.009 in. or more, or sets with mismatched identifying numbers must be returned to the parts distributor center. Do not attempt to install these gear sets.

The chart provided in this section will help to determine the approximate "starter shim" thickness needed for the initial pinion depth measurement. However, the chart will not provide the exact shim thickness required for final adjustment and must not be used as a substitute for an actual pinion depth measurement. The chart should be used as follows:

1. Measure the thickness of the original pinion depth shim. Note the pinion depth variance numbers marked on the old and new pinion gears.
2. Now use the chart to determine the starter shim thickness.

Pinion Depth Measurement Adjustment

1. Measure the thickness of the original pinion depth shim. Note the pinion depth variance numbers marked on the old and new pinion gears.
2. Determine the starter shim thickness. With the use of the chart, determine the amount to be added to or subtracted from the original shim thickness for starter shim thickness.

NOTE: The starter shim thickness must not be used as a final shim setting. An actual pinion depth measurement must be performed and the final shim thickness adjusted as necessary.

3. Install the ring bearing on the pinion gear with the large diameter of the bearing cage facing the gear end of the pinion. Press the bearing against the rear face of the gear.
4. Clean the pinion bearing bores in the axle housing thoroughly. This is important in obtaining the correct pinion gear depth adjustment. Install the starter pinion depth shim in the housing rear bearing cup bore. Be sure the shim is centered in the bearing cup bore.

NOTE: If the shim is chamfered, be sure the chamfered side faces the bottom of the bearing cup bore.

5. Install the ring bearing cup using Tools J–8092 and J–8608 or equivalent. Install the front bearing cup using Tools J–8092 and J–8611–01 or equivalent. Install the pinion gear in the rear bearing cup.
6. Install the front bearing, rear universal joint yoke and original pinion nut on the pinion gear. Tighten the pinion nut only enough to remove the bearing end play.

NOTE: Do not install a replacement pinion nut and collapsible spacer at this time as the pinion gear will be removed after depth measurement.

7. Note the pinion depth variance marked on the pinion gear. If the number is preceded by a plus (+) sign, add that amount (in thousandths) to the standard setting for the axle model being overhauled. If the number is preceded by a minus (-) sign, subtract that amount (in thousandths) from the standard setting. The result of this addition or subtraction is the desired pinion depth. Record this figure for future reference.
8. Assemble Arbor tool J–5223–4 and Discs J–5223–23 or equivalent, install the assembled tools in the differential bearing cup bores. Be sure discs are completely seated in bearing cup bores.
9. Install the bearing caps over the discs and install the bearing cap bolts. Tighten the bearing cap bolts securely, but not with the specified torque.
10. Position Gauge Block J–5223–20 or equivalent, on the end face of the pinion gear with the anvil end of the gauge block seated on the gear and the gauge plunger underneath Arbor Tool J–5223–4 or equivalent.
11. Assemble and mount Clamp J–5223–24 and bolt J–5223–29 or equivalent, on the axle housing. Use the axle housing cover bolt to attach the clamp to the housing.
12. Extend the clamp bolt until it presses against the gauge block with enough force to prevent the gauge block from moving. Loosen the gauge block thumbscrew to release the gauge block plunger. When the plunger contacts the arbor tool, tighten the thumbscrew to lock the plunger in position. Do not disturb the plunger position.
13. Remove the clamp and bolt assembly from the axle housing. Remove the gauge block and measure the distance from the end of the anvil to the end of the plunger using a 2–3 in. micrometer. This dimension represents the measured pinion depth. Record this dimension for assembly reference.
14. Remove the bearing caps and remove the arbor tool and discs from the axle housing. Remove the pinion gear, rear bearing cup and pinion depth shim from the axle housing.
15. Measure the thickness or the depth shim. Add this dimension to the measured pinion depth. From this total, subtract the desired pinion depth. The result represents the correct shim thickness required.

NOTE: The desired pinion depth is the standard setting plus or minus the pinion depth variance.

Pinion Gear Bearing Preload Adjustment

1. Install the correct thickness pinion depth shim(s) in the axle housing bearing cup bore. Install the rear bearing cup and pinion gear.

NOTE: The collapsible spacer controls the pinion bearing preload. Do not reuse the old spacer. Use a replacement spacer only.

2. Install the replacement collapsible spacer and front bearing on the pinion gear. Install the pinion oil seal using Tool J–22661 or equivalent.
3. Install the pinion yoke and replacement pinion nut. Tighten the pinion nut finger-tight only. Tighten the pinion nut only enough to remove end play and seat the pinion bearings. Use Tool J–22575 or equivalent to tighten the nut and use Tool J–86141–01 or equivalent to hold the yoke while tightening the nut.
4. Rotate the pinion while tightening the nut to seat the bearings evenly. Remove the tools.

NOTE: Do not exceed the specified preload torque and do not loosen the nut to reduce the preload torque if the specified torque is exceeded.

5. Measure the torque required to turn the pinion gear using an inch-pound torque wrench and Tool J–22575 or equivalent. The correct pinion bearing preload torque is 17–25 in.lbs. torque. Continue tightening the pinion nut until the required preload torque is obtained.
6. If the pinion bearing preload torque is exceeded, remove the pinion gear, replace the collapsible spacer and pinion nut and adjust the preload again.

ASSEMBLY

NOTE: The following items should be done before begining the reassembly of the differential.

Clean each part thoroughly in solvent. Towel dry bearings or allow them to air dry, do not use compressed air to dry bearings as damage might result. Dry all other parts with compressed air or shop towels. If the parts are not to be assembled immediately, cover them to prevent dust or dirt contamination.

Inspect the housing for cracks and sand holes. Replace the housing if it is cracked or porous. Check for burrs and deep scratches or nicks on the gasket and oil seal surfaces. An oil stone or fine tooth file may be used to remove nicks or burrs. The bearing cup bores should be carefully inspected for nicks or burrs that may have been created during bearing cup removal. Inspect and clean the axle tubes. Inspect the vent to be sure that it is not obstructed. Check housing for bent or loose tubes or other physical damage.

Whenever one rear wheel is stationary and the opposite wheel is spinning, the differential pinion shaft is subject to high torque loads. Inspect the shaft for scoring and wear. The shaft should be a press fit of 0.000–0.010 in. in the case. Replace the shaft if worn or scored.

Inspect the side gears for worn, cracked or chipped teeth. The gears should fit snugly on the axle shaft splines. Also inspect the fit of the gears in the differential case bore. With the gears installed, side clearance must not exceed 0.007 in. Excessive side clearance must be corrected to avoid driveline backlash resulting in a "clunk" noise when the transmission is initially engaged in Drive or Reverse (with automatic transmission).

1. Install the differential bearings on the case using Tools J–21784 and J–8092 or equivalent.

2. Install the thrust washers on the differential side gears and install the gears in the differential case. Install the differential pinion gears in the case. Install the thrust washers behind the pinion gears and align the pinion gear bores.

3. Rotate the differential side and pinion gears until the pinion mate shaft bores in the pinion gears are aligned with the shaft bores in the case. Install the thrust block in the case. Insert the block through the side gear bore. Align the bore in the block with the pinion mate shaft bores in the pinion gears and case.

4. Install the pinion mate shaft. Align the lockpin bore in the shaft with the bore in the case and install the shaft lockpin.

DIFFERENTIAL BEARING ADJUSTMENT

1. Place the bearing cup over each differential bearing and install the differential case assembly in the axle housing.

2. Install the shim on each side between the bearing cup and the housing. Use 0.080 in. shims as the starting point.

3. Install the bearing caps and tighten the bolts finger-tight. Mount the Dial Indicator J–8001 or equivalent on the housing. Using an appropriate pry tool, pry between the shims and housing. Pry the assembly to one side and zero the indicator, then pry the assembly to the opposite side and read the indicator.

NOTE: Do not zero or read the indicator while prying.

4. The amount read on the indicator is the shim thickness that should be added to arrive at the zero preload and zero end play. Repeat the procedure to ensure accuracy and adjust if necessary. Shims are available in thicknesses from 0.080 - 0.110 in. in 0.002 in. increments.

5. When sideplay is eliminated, a slight bearing drag will be noticed. Install the bearing caps and tighten the bearing cap bolts with 85 ft. lbs. torque. Attach the dial indicator to the axle housing and check the ring gear mounting face of the differential case for runout. Runout should not exceed 0.002 in.

Remove the case from the housing. Retain the shims used to adjust the sideplay.

RING GEAR INSTALLATION

1. Position the ring gear on the differential case. Install the two ring gear bolts in the opposite holes and tighten the bolts to pull the gear into position.

2. Install the remaining ring gear attaching bolts. Tighten the bolts with 105 ft. lbs. torque.

3. Position the shims previously selected to remove the differential bearing sideplay on the bearing cups and install the differential assembly in the axle housing. Install the bearing cap bolts and tighten the bolts with 85 ft. lbs. torque.

4. Attach the dial indicator to the housing. Position the indicator so the indicator stylus contacts the drive side of a ring gear tooth and at a right angle to the tooth. Move the ring gear back and forth and note the movement registered on the dial indicator. The ring gear backlash should be 0.005–0.009 in., with 0.008 in. desired.

5. Adjust the backlash as follows: to increase the backlash, install the thinner shim on the ring gear side and the thicker shim on the opposite side. To decrease the backlash, reverse the procedure; however, do not change the total thickness of the shims.

NOTE: The following is an example on how to decrease backlash. The sideplay was removed using 0.090 in. shims on each side totaling 0.180 in.. Backlash is checked and found to be 0.011 in. To correct the backlash, add 0.004 in. the the shim on the ring gear side and subtract 0.004 in. from the shim on the opposite side. This will result in 0.094 in. shim on the ring gear side and 0.086 in. shim on the other side. The backlash will be approximately 0.007–0.008 in. The total shim thickness remains 0.180 in.

DIFFERENTIAL INSTALLATION & BEARING PRELOAD ADJUSTMENT

NOTE: The differential bearings must be preloaded to compensate for heat and loads during operation. The differential bearings are preloaded by increasing the shim pack thickness at each side of the differential by 0.004 in. for a total of 0.008 in.

1. Remove the differential assembly from the housing. Be sure to keep the differential bearing shim packs together for the proper assembly. Do not distort the shims in the axle housing bearing bores.

2. Install the differential bearing cups on the differential bearings. The cups should cover the differential bearing rollers completely. Position the differential assembly in the housing so the bearings just start into the housing bearing bores.

NOTE: Slightly tipping the bearing cups will ease starting them into the bores. Also keep the differential assembly square in the housing during installation and push it in as far as possible.

3. Tap the outer edge of the bearing cups until the differential is seated in the housing.

4. Install the differential bearing caps. Position the caps accordingly to the alignment punch marks made at disassembly. Tighten the bearing cap bolts with m 85 ft. lbs. torque. Preloading the differential bearings may change the backlash setting. Check and correct the backlash if necessary.

5. Install the propeller shaft, aligning the index marks made at disassembly. Install the axle shafts, bearings, seals and brake support plates. Fill the rear axle with the specified axle lubricant.

6. Check and adjust the axle shaft end play if necessary. Adjust the end play at the left side of the axle shaft only. Install the hubs, drums and wheels. Lower the vehicle and road test the vehicle to check the rear axle assembly for proper operation.

TRAC-LOK DIFFERENTIAL

Operational Test

If a noisy or rough operation such as a chatter occurs when turning corners, the most probalbe cause of this chatter or noise is incorrect or contaminated lubricant. Before removing the Trac-Lok unit for repair, drain flush and refill the axle with the specified lubricant. A complete lubricant drain and refill with the specified fluid will usally correct the chatter problem. A quick operational test of the Trac-Lok differential can be done easily by performing the following steps.

1. Position one wheel on solid dry pavement and the opposite wheel on ice, mud grease or a similar low traction surface.

2. Gradually increase the engine rpm to obtain the maximum traction prior to a breakaway. The ablity to move the vehicle effectively will demonstrate the proper performance.

NOTE: If the test is performed on extremely slick surfaces suach as ice or grease coated surfaces, some question may exist as to proper performance. Inthese extreme cases, a properly performing Trac-lok will provide greater pulling power by lightly applying the parking brake.

Trac-Lok Differential Dissassembly

1. Remove the differential from the axle housing as previously outlined in this section. Install one axle shaft in the vise with the spline end facing upward and tighten thew vise.

2. Do not allow more than $2^3/4$ in. of the shaft to extend above the top of the vise. This prevents the shaft from fully entering the side gear, causing interfernce with the step plate tool used to remove the differential gears.

3. Mount the differential case on the

axle shaft with the ring gear bolt heads facing upward. Place some shop towels under the ring gear to protect the gear when it is removed from the case.

4. Remove and discard the ring gear bolts. Remove the ring gear from the case using a rawhide hammer. Remove the differential case from the axle shaft and remove the ring gear and remount the differential case on the axle shaft.

5. Use a suitable screwdrivers to disengae the snap rings from the pinion mate shaft. Place a shop towel on the opposite opening of the case to prevent the snap rings from flying out of the case. Remove the pinion mate shaft using a hammer and brass drift.

NOTE: A special gear rotating tool J–23781–3 or equivalent is required to perform the following steps. The tool consists of three parts; the gear rotating tool, forcing screw and step plate.

6. Install step plate tool J–23781–7 or equivalent in the lower differential side gear. Position the pawl end of the gear rotating tool J–23781–3 or equivalent onto the step plate.

7. Insert the forcing screw tool J–8646–2 or equivalent through the top of the case and thread it into the gear rotating tool. Before using the forcing screw tool, apply a small amount of grease to the centering hole in the step plate and oil the threads of the forcing screw.

8. Center the forcing screw in the step plate and tighten the screw to move the differential side gears away from the differential pinion gears. Remove the differential pinion gear thrust washers using a feeler gauge or a shim stock of 0.030 in. thickness. Insert the feeler gauge or shim stock between the washer and the case and withdraw the shim stock with the thrust washer.

9. Tighten the forcing screw until a slight movement of the differential pinion gear is observed. Insert the pawl end of the gear rotating tool between the teeth of one differential side gear.

10. Pull the handle of the tool to rotate the side gears and pinion gears. Remove the pinion gears as they appear in the case opening. It could be necessary to adjust the tension applied on the Belleville springs by the forcing screw before the gears can be rotated in the case.

11. Retain the upper side gear and clutch pack in the case by holding your hand on the bottom of the rotating tool while removing the forcing screw. Remove the rotating tool, upper side gear and clutch pack.

12. Remove the differential case from the axle shaft. Invert the case with the flange or ring gear side up and remove the step plate tool, lower the side gear and clutch pack from the case. Remove ther retainer clips from both the clutch packs to allow separation of the plates and discs.

Trac-Lok Differential Assembly

If any one member of either clutch pack shows evidence of excessive wear or scoring, the complete clutch pack must be replaced on both sides. Clean each part thoroughly in solvent. Towel dry bearings or allow them to air dry, do not use compressed air to dry bearings as damage might result. Dry all other parts with compressed air or shop towels. If the parts are not to be assembled immediately, cover them to prevent dust or dirt contamination.

Inspect the housing for cracks and sand holes. Replace the housing if it is cracked or porous. Check for burrs and deep scratches or nicks on the gasket and oil seal surfaces. An oil stone or fine tooth file may be used to remove nicks or burrs. The bearing cup bores should be carefully inspected for nicks or burrs that may have been created during bearing cup removal. Inspect and clean the axle tubes. Inspect the vent to be sure that it is not obstructed. Check housing for bent or loose tubes or other physical damage.

Inspect the side gears for worn, cracked or chipped teeth. The gears should fit snugly on the axle shaft splines. Also inspect the fit of the gears in the differential case bore.

1. Lubricate all the differential components with the specified gear lubricant. Assemble the clutch packs. Install the plates and discs in the same position as when removed regardless of wheter they are replacement or original parts.

2. Install the clutch retainer clips on the ears of the clutch plates. Be sure the clutch packs are completely assembled and seated on the ears of the plates. Install the clutch packs on the differential side gears and install the assembly in the case.

3. Make sure the clutch pack stays assembled on the side gear splines and that the retainer clips are completely seated in the case pockets. To prevent the pack from falling out of the case, it will be necessary to hold it in place by hand while mounting the case on the axle shaft.

NOTE: When installing the differential case on the axle shaft, make sure that the splines of the side gears are aligned with those of the axle shaft. Make sure the clutch pack is still properly assembled in the case after installing the case on the axle shaft.

4. Mount the case assembly on the axle shaft. Install the step plate tool in the side gear and apply a small amount of grease in the centering hole of the step plate.

5. Install the remaining clutch pack and side gear. Make sure the clutch pack stays assembled on the side gear splines and that the retainer clips are completely seated in the pockets of the case.

6. Position the gear rotating tool in the upper side of the gear. Keep the side gear and rotating tool in position by holding them with your hand. Insert the forcing

screw throught the top of the case and thread it into the rotating tool.

7. Install both of the differential pinion gears in the case. Be sure the bores of the gears are aligned. Hold the gears in place by hand. Tighten the forcing screw to compress the Belleville springs and provide clearance between the teeth of the pinion gears and the side gears.

8. Position the pinion gears in the case and insert the rotating tool pawl between the side gear teeth. Rotate the side gears by pulling on the tool handle and install the pinion gears.

NOTE: If the side gears will not rotate, the Belleville spring load will have to be adjusted. If adjustment is necessary, loosen or tighten the forcing screw slightly until the gears will rotate.

9. Rotate the side gears, using the rotating tool handle, until the shaft bores in both the pinion gears are aligned with the case bore. Lubricate both sides of the pinion gear thrust washers.

10. Tighten or loosen the forcing screw to permit the thrust washer installation. Install the thrust washers and using a suitable tool, guide the washers into position. Make sure the shaft bores in the washers and gears are aligned with the case bores.

11. Remove the forcing screw, rotating tool and step plate. Lubricate the pinion mate shaft and seat the shaft in the case. Be sure the snap ring grooves in the shaft are exposed to allow the snap ring installation.

12. Install the pinion mate shaft snap rings, remove the case from the axle shaft and install the ring gear on the case. Be sure to use replacement ring bolts only. Do not reuse the original bolts.

13. Align the ring gear and case bolt holes and install the ring gear bolts finger tight only. Remove the case on the axle shaft and tighten the bolts down evenly to the proper torque specifications.

14. Install the Trac-Lok differential assembly in the axle housing. and follow the procedures previously outline for the other Jeep axles to complete the differential and axle assembly servicing.

Chevrolet/GMC

The single speed rear axles used on the Chevrolet and GMC light and medium trucks are categorized by the ring gear diameter and are identified as follows: 7½, 8½, 8⅞ inch (corporation)–w/semifloating axles: 9¾, 10½ inch (Dana)–w/full floating axles: 10½ inch (Corporation)–w/full floating axles: 12¼ inch (Corporation)– w/full floating axles.

7½ Inch Ring Gear
CASE REMOVAL

1. Before removing the rear axle case from the housing, ring gear to drive pinion backlash should be checked. This will indi-

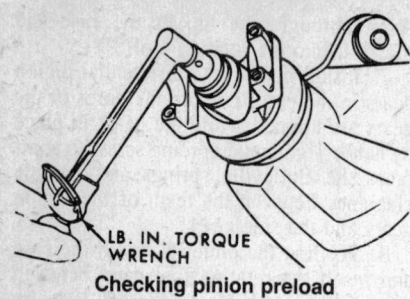

LB. IN. TORQUE WRENCH

Checking pinion preload

cate gear or bearing wear or an error in backlash or preload setting which will help in determining cause of axle noise.

2. Remove rear axle bearing cap bolts. Bearing caps should be marked "R" and "L" to make sure they will be reassembled in their original location.

3. Remove rear axle case. Exercise caution in prying on carrier so that gasket sealing surface is not damaged. Place right and left bearing outer races and shims in sets with marked bearing caps so that they can be reinstalled in their original positions.

DISASSEMBLY

1. If rear axle side bearings are to be replaced, they must be removed using a puller.

2. Remove rear axle pinions, side gears and thrust washers from case. Mark side gears and case so they can be installed in their original location.

3. If ring gear is to be replaced and it is tight on case after removing bolts (L.H. Threads), drive it off using a brass drift and hammer. Do not pry between ring gear and case.

DRIVE PINION, BEARING AND RACES

1. Check drive pinion bearing pre-load. If there is no preload reading, check for looseness of pinion assembly by shaking. Looseness could be caused by defective bearings or worn pinion flange. If rear axle was operated for an extended period with very loose bearings, the ring gear and drive pinion will also require replacement.

2. Remove pinion flange nut and washer.

3. Remove pinion flange.

4. Install drive Pinion Remover J–22536 or its equal and drive out pinion. Apply heavy hand pressure on pinion remover toward rear axle housing to keep front bearing seated to avoid damage to outer race.

<95

BEARING REPLACEMENT

The rear pinion bearing must be removed when it becomes necessary to change the pinion depth adjustment.

1. With drive pinion removed from carrier, press bearing from the pinion gear.

2. Drive pinion oil seal from carrier and remove front pinion bearing. If this bearing is to be replaced, remove outer race from carrier.

3. If rear pinion bearing is to be re-

placed, remove outer race from carrier using a punch in slots provided for this purpose.

CLEANING & INSPECTION

1. Clean all rear axle bearings thoroughly in clean solvent (do not use a brush). Examine bearings visually and by feel. All bearings should feel smooth when oiled and rotated while applying as much hand pressure as possible. Minute scratches and pits that appear on rollers and races at low mileage are due to the initial pre-load, and bearings having these marks should not be rejected.

2. Examine sealing surface of pinion flange for nicks, burrs, or rough tool marks which would cause damage to the seal and result in an oil leak. Replace if damaged.

3. Examine carrier bore and remove any burrs that might cause leaks around the O.D. of the pinion seal.

4. Examine the ring gear and drive pinion teeth for excessive wear and scoring. If any of these conditions exist, replacement of the gear set will be required.

5. Inspect the pinion gear shaft for unusual wear; also check the pinion and side gears and thrust washers.

6. Check the press fit of the side bearing inner race on the rear axle case hub by prying against the shoulder at the puller recess in the case. Side bearings must be a tight press fit on the hub.

7. Diagnosis of a rear axle failure such as: chipped bearings, loose (lapped-in) bearings, chipped gears, etc., is a warning that some foreign material is present; therefore, the axle housing must be cleaned.

DRIVE PINION ASSEMBLY

1. If a new rear pinion bearing is to be installed, install new outer race.

2. If a new front pinion bearing is to be installed, install new outer race.

SETTING PINION DEPTH

Pinion depth is set with Pinion Setting Gauge J–21777–01. The pinion setting gauge provides in effect, a "Normal" or "zero" pinion as a gauging reference. Instructions are included in gauge set.

1. Make certain all of the gauge parts are clean.

2. Lubricate front and rear pinion bearings liberally with rear axle lubricant.

3. While holding pinion bearings in position, install depth setting gauge assembly.

4. Hold stud stationary with a wrench positioned over the flats on the ends of stud and tighten nut to 2.2 N;pdm (20 inch lbs.) torque. Rotate gauge plate assembly several complete revolutions to seat the bearings. Then tighten nut until a torque between 1.6 and 2.2 N;pdm (15 and 25 inch lbs.) is obtained to keep the gauge plate in rotation.

5. Rotate the gauge plate until the gauging areas are parallel with the discs.

6. Make certain rear axle side bearing support bores are clean and free of burrs.

7. Install the correct discs on the gauge shaft.

8. Position the gauge shaft assembly in the carrier so that the dial indicator rod is centered on the gauging area of the gauge block, and the discs seated fully in the side bearing bores. Install side bearing caps and torque bolts to 75 N;pdm (55 ft. lbs.). Use dial indicator J–8001 or an equivalent indicator reading from 0 to 2.5mm (0" to 100").

9. Set dial indicator at ZERO. Then position on mounting post of the gauge shaft with the contact button touching the indicator pad. Push dial indicator downward until the needle rotates approximately ¾ tur clockwise. Tighten the dial indicator in this position and recheck.

10. Rotate gauge shaft slowly back and forth until the dial indicator reads the greatest deflection. At the point of greatest deflection, set the dial indicator to ZERO. Repeat rocking action of gauge shaft to verify the ZERO Setting.

11. After the ZERO setting is obtained, rotate gauge shaft until the dial indicator rod does not touch the gauge block.

12. Record dial reading at pointer position. Example: If pointer moved counterclockwise 1.70mm (.067") to a dial reading of .84mm (.033"), this indicates a shim thickness of .84mm (.033") except as follows: Dial indicator reading should be within the range of .50 to 1.27mm (.020" to .050").

13. Loosen Stud J–21777–43 and remove gauge plate, washer and both bearings from carrier.

14. Position correct shim on drive pinion and install the drive pinion rear bearings.

CASE ASSEMBLY

Before assembling the rear axle case, lubricate all parts with rear axle lubricant.

1. Place side gear thrust washers over side gear hubs and install side gears in case. If same parts are reused, install in original sides.

2. Position one pinion (without washer) between side gears and rotate gears until pinion is directly opposite from loading opening in case. Place other pinion between side gears so that pinion shaft holes are in line; then rotate gears to make sure holes in pinions will line up with holes in case.

3. If holes line up, rotate pinions back toward loading opening just enough to permit sliding in pinion thrust washers.

4. After making certain that mating surfaces of case and ring gear are clean and free of burrs, thread two bolts into opposite sides of ring gear; then install ring gear on case. Install NEW ring gear attaching bolts just snug. NEVER REUSE OLD BOLTS. Torque bolts alternately in progressive stages to 90 ft. lbs.

5. If case side bearings were removed, re-install bearings.

SIDE BEARING PRE-LOAD ADJUSTMENT

The side bearing pre-load adjustment is to be made before installing the pinion. If the pinion is installed, remove ring gear.

Case side bearing pre-load is adjusted by changing the thickness of both the right and left shims by an equal amount. By changing the thickness of both shims equally, the original backlash will be maintained.

Production shims are cast iron and vary in thickness from 0.210–0.272″ in increments of 0.002 in.

Standard service spacers are 0.170″ thick and steel service shims are available from 0.040–0.082″ in increments of 0.002 in.

Do not attempt to reinstall the production shims as they may break when tapped into place. If service shims were previously installed, they can be reused, but (whether using new or old bearings) adhere to the following procedure in all cases.

1. Before installation of the case assembly, make sure that side bearing surfaces in the carrier are clean and free of burrs. If the same bearings are being reused, they must have the original outer races in place.

2. Determine the approximate thickness of shims needed by measuring each production shim or each service spacer and shim pack.

3. In addition to the service spacer, a service shim will be needed. To select a starting point in service shim thickness, use the following chart:

4. Place case with bearing outer races in position in carrier. Slip the service spacer between each bearing race and carrier housing with chamfered edge against housing.

Install the left bearing cap loose so that the case may be moved while checking adjustments. Another bearing cap bolt can be added in the lower right bearing cap hole. This will prevent case from dropping while making shim adjustments.

Select one or two shims totaling the amount shown in the right-hand column and position between the right bearing race and the service spacer. Be sure left bearing race and spacer are against left side of housing.

5. Insert progressively larger feeler gauge sizes .25mm, .30mm, .36mm, etc. (.010″, .012″, .014″, etc.) between the right shim and service spacer until there is noticeable increased drag. Push the feeler gauge downward until the end of the gauge makes contact with the carrier bore so as to obtain a correct reading. The point just before additional drag begins is correct feeler gauge thickness. Rotate case while using feeler gauge to assure an even reading.

The original light drag is caused by weight of the case against the carrier while additional drag is caused by side bearing pre-load. By starting with a thin feeler gauge, a sense of "feel" is obtained so that the beginning of pre-load can be recognized to obtain Zero clearance. It will be

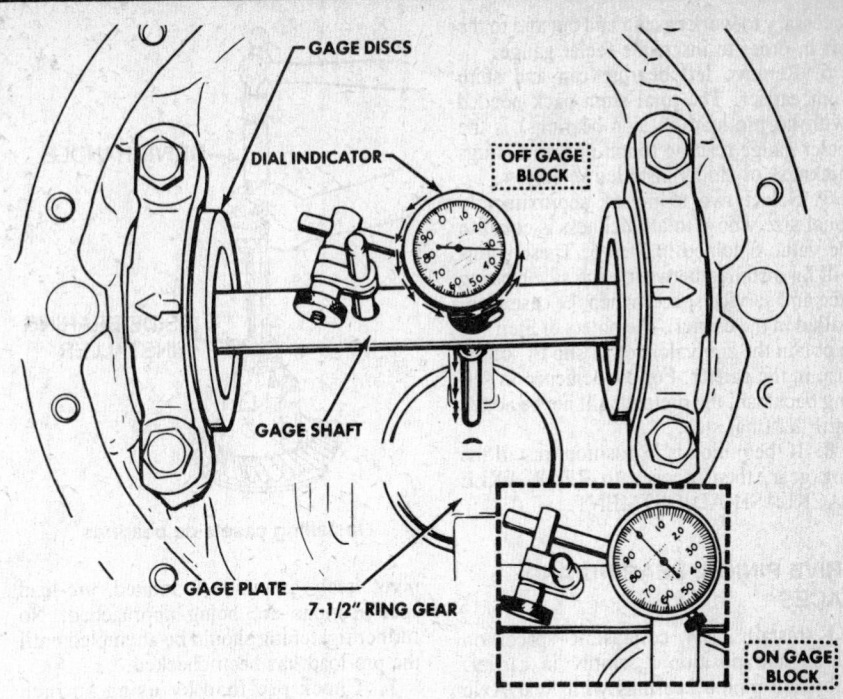

Checking pinion depth

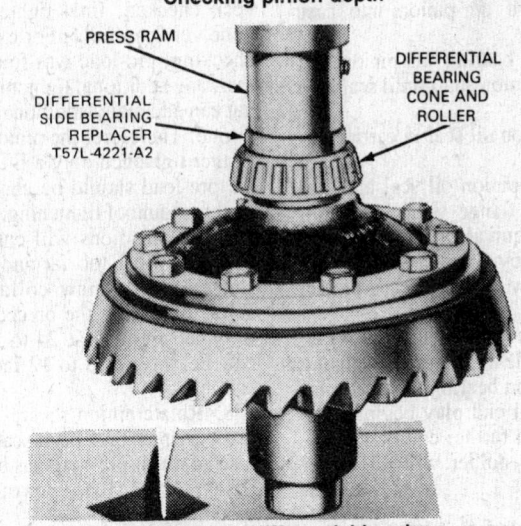

Installing the differential bearing

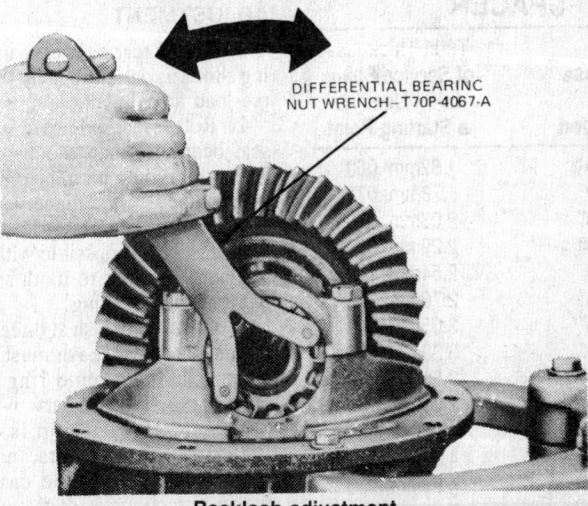

Backlash adjustment

necessary to work case in and out and to the left in order to insert the feeler gauge.

6. Remove left bearing cap and shim from carrier. The total shim pack needed (with no pre-load on side bearings) is the feeler gauge reading found in Step 5 plugs thickness of shims installed in Step 4.

7. Select two shims of approximately equal size whose total thickness is equal to the value obtained in Step 5. These shims will be installed between each side bearing race and service spacer when the case is installed in the carrier. The object of Step 7 is to obtain the equivalent of a "slip fit" of the case in the carrier. For convenience in setting backlash, the preload will not be added until the final step.

8. If the pinion is in position, install the ring gear, then proceed to REAR AXLE BACKLASH ADJUSTMENT.

DRIVE PINION, BEARING AND RACES

1. Install NEW collapsible spacer on pinion and position assembly in carrier. Lubricate pinion bearings with Rear Axle Lubricant before installing pinion.

2. Hold forward on pinion into case assembly.

3. Install front bearing on pinion and drive bearing on pinion shaft until seated in race.

4. Position pinion oil seal in carrier. Install seal.

5. Coat lips of pinion oil seal and seal surface of pinion flange with Lubricant, No. 1050169 or equivalent. Install pinion flange on pinion by tapping with a soft hammer until a few pinion threads project through flange.

6. Install pinion washer and nut. Hold pinion flange. While intermittently rotating pinion to seat pinion bearings, tighten pinion flange nut until end play begins to be taken up. When no further end play is detectable and when holder will no longer

4.32mm (.170″) SERVICE SPACER

Total Thickness of Both Prod. Shims Removed	Total Thickness of Service Shims to be Used as a Starting Point
10.57mm .420″	1.52mm .060″
10.92mm .430″	1.78mm .070″
11.18mm .440″	2.03mm .080″
11.43mm .450″	2.29mm .090″
11.68mm .460″	2.54mm .100″
11.94mm .470″	2.79mm .110″
12.19mm .480″	3.05mm .120″
12.45mm .490″	3.30mm .130″
12.70mm .500″	3.56mm .140″
12.95mm .510″	3.81mm .150″
13.21mm .520″	4.06mm .160″
13.46mm .530″	4.32mm .170″
13.97mm .550″	4.83mm .190″

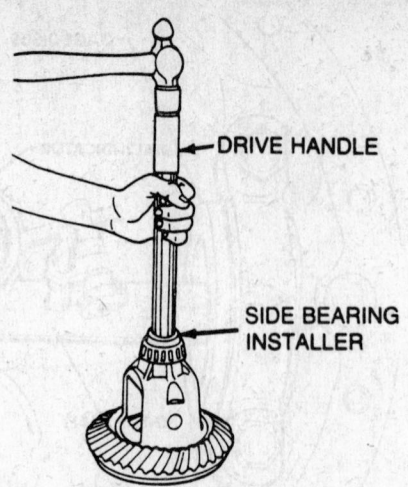

Installing case side bearings

pivot freely as pinion is rotated, pre-load specifications are being approached. No further tightening should be attempted until the pre-load has been checked.

7. Check pre-load by using an inch pound torque wrench. After pre-load has been checked, final tightening should be done very carefully. For example, if when checking, pre-load was found to be 5 inch lbs., any additional tightening of the pinion nut can add many additional inch pounds of torque. Therefore, the pinion nut should be further tightened only a little at a time and the pre-load should be checked after each slight amount of tightening. Exceeding pre-load specifications will compress the collapsible spacer too far and require the installation of a new collapsible spacer. While observing the preceding note, carefully set pre-load at 24 to 32 inch lbs. on new bearings or 8 to 12 inch lbs. on used bearings.

8. Rotate pinion several times to assure that bearings have been seated. Check pre-load again. If pre-load has been reduced by rotating pinion, reset pre-load to specifications.

REAR AXLE BACKLASH ADJUSTMENT

1. Install rear axle case into carrier, using shims as determined by the side bearing pre-load adjustment.

2. Rotate rear axle case several times to seat bearings, then mount dial indicator. Use a small button on the indicator stem so that contact can be made near heel end of tooth. Set dial indicator so that stem is in line as nearly as possible with gear rotation and perpendicular to tooth angle for accurate backlash reading.

3. Check backlash at three or four points around ring gear. Lash must not vary over .05mm (.00″) around ring gear. Pinion must be held stationary when checking backlash. If variation is over .05mm (.002″) check for burrs, uneven bolting conditions or distorted case flange and make corrections as necessary.

4. Backlash at the point of minimum lash should be between .13 and .23mm (.005″ and .009″) for all new gears.

5. If backlash is not within specifications, correct by increasing thickness of one shim and decreasing thickness of other shim the same amount. This will maintain correct rear axle side bearing pre-load. For each .03mm (.001″) change in backlash desired, transfer .05mm (.002″) in shim thickness. To decrease backlash .03mm (.001″), decrease thickness of right shim .05mm (.002″) and increase thickness of left shim 0.5mm (.002″). To increase backlash .05mm (.002″), increase thickness of right shim .10mm (.004″) and decrease thickness of left shim .10mm (.004″).

6. When backlash is correctly adjusted, remove both bearing caps and both shim packs. Keep packs in their respective position, right or left side. Select a shim .10mm (.004″) thicker than one removed from left side, then insert left side shim pack between the spacer and the left bearing race. Loosely install bearing cap.

7. Select a shim .10mm (.004″) thicker than the one removed from right side and insert between the spacer and the right bearing race. It will be necessary to drive the right shim into position.

8. Torque to 55 ft. lbs.

9. Recheck backlash and correct if necessary.

10. Install axles.

11. Install a new cover gasket. Install cover and torque cover bolts to 20 ft. lbs.

12. Fill rear axle to proper level.

8½ & 8⅞ Inch Ring Gear

This axle assembly is the semi-floating type with Hypoid type drive pinion and ring gears. The drive pinion gear is supported by two bearings. The differential case contains two pinion gears. The carrier assembly is not removable since it is part of the axle assembly but the design allows for the differential assembly to be serviced while the axle is still in the vehicle. The ring gear is bolted to a one piece differential case that is supported by two preloaded roller bearings.

CASE REMOVAL

1. Remove the inspection cover from the axle housing and drain the gear lubricant into a pan.

2. Remove the screw or pin that holds the pinion shaft in place and remove the shaft.

3. Push the axle shaft(s) in a little and remove the "C" locks from the ends of the shafts. Remove the axle shafts from the housing.

4. Before going any further, the backlash should be measured and recorded. This will allow the old gears to be reassembled at the same amount of lash to avoid changing the gear tooth pattern. It also helps to indicate if there is gear or bearing wear, and if there is any error in the original backlash setting.

5. Roll the differential pinions and thrust washers out of the case and also remove the side gears and thrust washers. Make sure to mark the pinions and side gears so they can be reassembled in their original position.

6. Mark the bearing caps and housing and loosen the retaining bolts. Tap the caps lightly to loosen them. When the caps are loose, take the bolts all the way out and then reinstall the bolts just a few turns. This will keep the case from falling out of the housing when it is pried loose.

7. With a pry bar, very carefully pry the case assembly loose. Be careful not to damage the gasket surface on the housing when prying. The case assembly may suddenly come free if the bearings were preloaded, so pry very slowly.

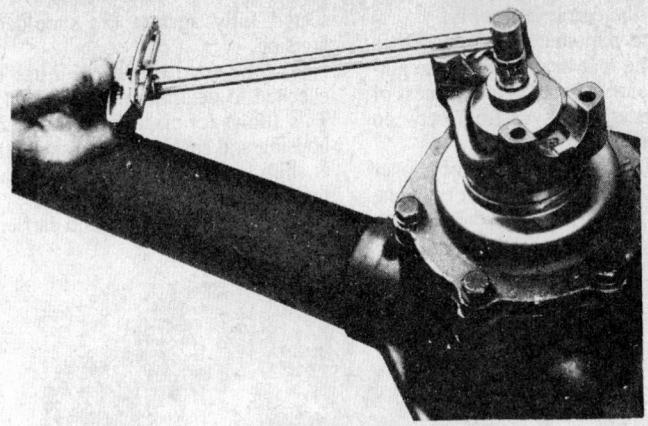
Measuring rotating torque

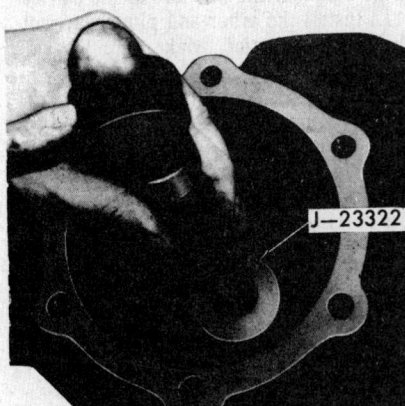
Installing straddle bearing

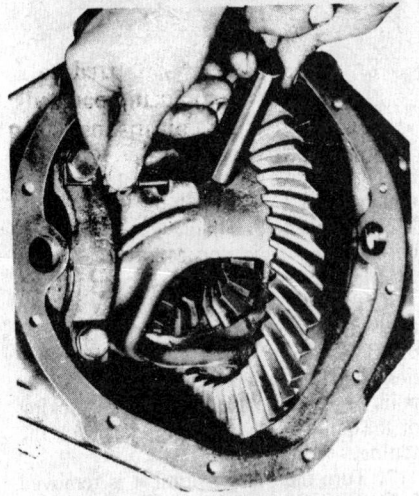
Differential pinion shaft removal

8. When the case assembly is loose, remove the bolts for the bearing caps and remove the caps. Place the caps so they may be reinstalled in the same position. Place any shims that are removed with the cap they were removed from.

DRIVE PINION

1. With the differential removed, check the pinion preload. Do this by checking the amount of torque needed to turn the pinion gear. For a new bearing, it should be 20–25 inch lbs., and for a used bearing it should be 10–15 inch lbs. If there is no preload reading check the pinion for looseness. If there is any looseness the bearing should be replaced.

2. With a holder assembly installed on the flange, use a socket of the proper size and remove the flange nut and washer.

3. Remove the flange by using a puller assembly and drawing the flange off the pinion splines.

4. Thread the pinion nut a few turns onto the pinion shaft. Using a brass drift and hammer, lightly tap the end of the pinion shaft to remove the pinion from the carrier. Be careful not to allow the pinion to fall out of the carrier after it breaks loose.

5. With the pinion removed from the carrier, discard the old seal pinion nut and collapsible spacer and install new ones when reassembling.

CLEANING & INSPECTION

1. Clean all parts in solvent and blow dry.

2. Check all of the parts for any signs of wear, chips, cracks or distortion. Replace any parts that are defective.

3. Check the fit of the differential side gears in the case and the fit of the side gear and axle shaft splines.

DIFFERENTIAL BEARING

1. With a bearing puller attached to the bearing, pull the bearing from the case.

2. Place the new bearing on the case hub with the thick side of the inner race toward the case. Using a bearing driver, drive the bearing onto the case until it seats against the shoulder on the case.

DRIVE PINION BEARING

1. Depending on the bearing that is being replaced, remove the front or rear bear-

Drive pinion flange removal

		CODE NUMBER ON ORIGINAL PINION				
		+2	+1	0	-1	-2
CODE NUMBER ON SERVICE PINION	+2	—	ADD .001	ADD .002	ADD .003	ADD .004
	+1	SUBT. .001	—	ADD .001	ADD .002	ADD .003
	0	SUBT. .002	SUBT. .001	—	ADD .001	ADD .002
	-1	SUBT. .003	SUBT. .002	SUBT. .001	—	ADD .001
	-2	SUBT. .004	SUBT. .003	SUBT. .002	SUBT. .001	—

Pinion depth codes

ing cup from the carrier assembly.

2. With the pinion gear mounted in a press, press the rear bearing from the pinion shaft. Be sure to record the thickness of the shims that are removed from between the bearing and the gear.

3. Using a bearing driver of the proper size, install a new bearing cup for each one that was removed. Make sure the cups are

seated fully against the shoulder in the housing.

4. The pinion depth must now be checked to determine the nominal setting. This allows for machining variations in the housing and enables you to select the proper shim so that the pinion depth can be set for the best bear tooth contact.

5. Clean the housing and carrier assem-

blies to insure accurate measurement of the pinion depth.

6. Lubricate the front and rear pinion bearings with gear lubricant and install them in their races in the carrier assembly.

7. Using a pinion setting gauge, select the proper clover leaf plate, and install it on the preload stud.

8. Insert the stud through the rear bearing, with the proper size pilot on the stud, and through the front bearing using the proper pilot. Install the hex nut and tighten it until it is just snug.

9. Holding the preload stud with a wrench, tighten the hex nut until 20 inch lbs. of torque are required to rotate the bearings.

10. Install the side bearing discs on the ends of the arbor assembly, using the step of the disc that fits the bore of the carrier.

11. Install the arbor and plunger assembly into the carrier. Make sure the side bearing discs fit properly.

12. Install the bearing caps in the carrier assembly finger tight to make sure the discs do not move.

13. Mount a dial indicator on the mounting post of the arbor. Have the contact button resting on the top surface of the plunger.

14. Preload the dial indicator by turning

Pushing the axle shaft inward

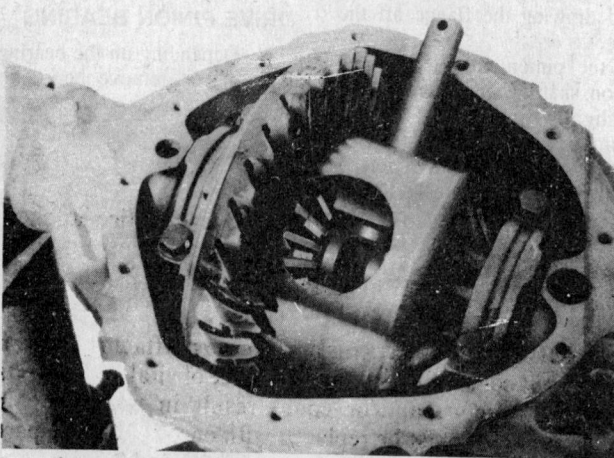
Positioning case for best clearance

Drive pinion nut removal

it one-half revolution and tightening it in this position.

15. Use the button on the gauge plate that corresponds to the ring gear size and turn the plate so the plunger rests on top of it.

16. Rock the plunger rod back and forth across the top of the button until the dial indicator reads the greatest amount of variation. Set the dial indicator to zero at the point of most variation. Repeat the rocking of the plunger several times to check the setting.

17. Turn the plunger until it is removed from the gauging plate button. The dial indicator will now read the pinion shim thickness required to set the nominal pinion depth. Make a note of the reading.

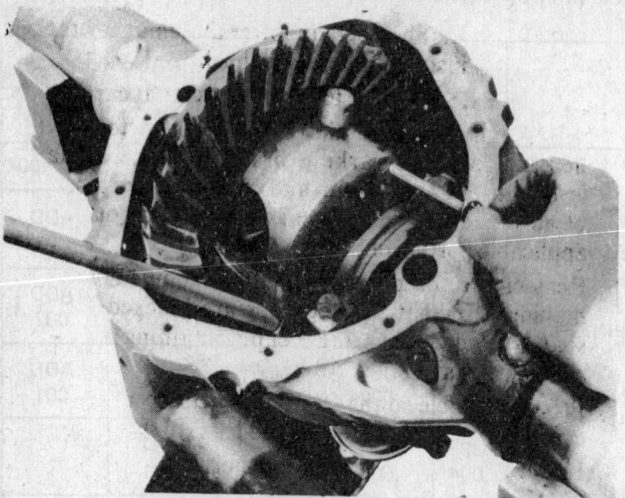
Removing lock screw

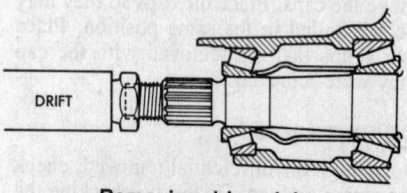
Removing drive pinion

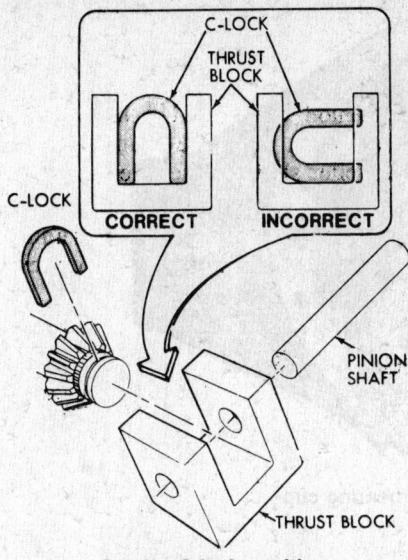

Correct C-lock position

18. Check for the pinion code number on the rear face of the pinion gear being used. This number will indicate the necessary change to the pinion shim thickness. If the pinion is marked with a plus (+) and a number, add that much to the reading you got from the dial indicator. If the pinion has no mark, use the reading from the dial indicator as the correct shim thickness. If the pinion is marked with a minus (−) and a number, subtract that much from the reading on the dial indicator.

19. Remove the depth gauging tools from the carrier assembly and install the proper size shim on the pinion gear.

20. Lubricate the bearing with gear lubricant and using a press, press the bearing into place on the pinion shaft.

PINION GEAR INSTALLATION

1. Lubricate the front bearing with gear lubricant and install it in the front cup.

2. Install the pinion seal in the bore. Using a seal driver and the proper size gauge plate, drive the seal in until the gauge plate is flush with the shoulder of the carrier.

3. Coat the seal lips with gear lubricant and install a new bearing spacer on the pinion gear.

4. Install the pinion gear in the carrier assembly and using a large washer and nut, draw the pinion gear in through the front bearing far enough to get companion flange in place.

5. With the companion flange installed on the pinion shaft, use a holder assembly and tighten the pinion nut until all of the end play is removed from the drive pinion.

6. When there is no more end play the preload should be checked. The preload of the bearing is the amount of torque required to turn the pinion gear. The preload should be 20–25 inch lbs. on new bearings and 10–15 inch lbs. on reused bearings. Tighten the pinion nut until these figures are reached. Do Not over tighten the pinion. This will collapse the spacer too much and make it necessary to replace it.

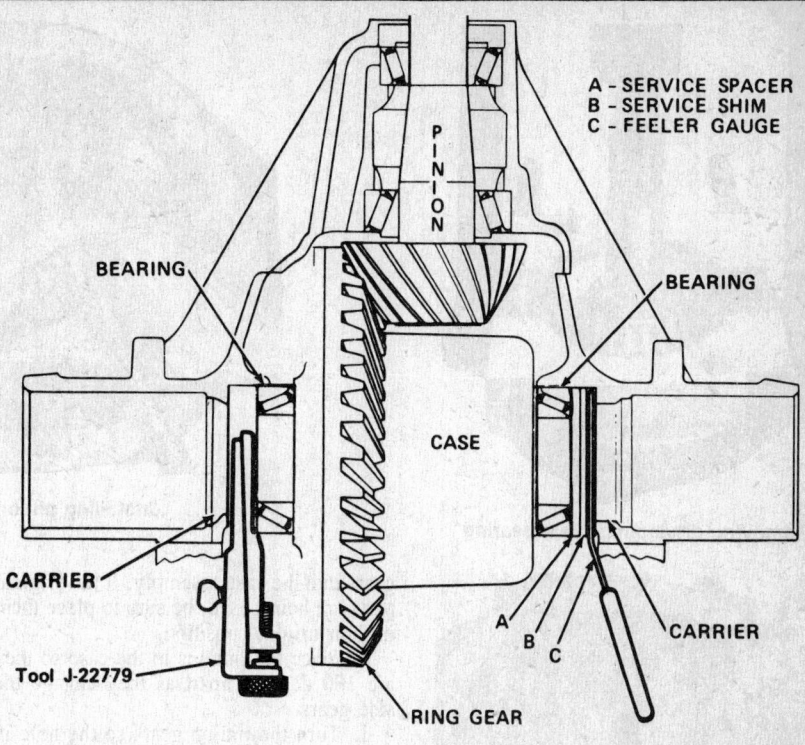

A - SERVICE SPACER
B - SERVICE SHIM
C - FEELER GAUGE

EXAMPLE

	RING GEAR SIDE		OPPOSITE SIDE	
.250"	Thickness of Tool J-22779 required to force ring gear into contact with pinion		Combined total of: Service Spacer (A) Service Shim (B) Feeler Gauge (C)	.265"
− .010" / .240"	TO MAINTAIN PROPER BACKLASH (.005" - .008"), ring gear is moved away from pinion by subtracting .010" shims from ring gear side and adding .010" shims to other side			+ .010" / .275"
+ .004"	TO OBTAIN PROPER PRELOAD on side bearings, add .004" shims to each side.			+ .004"
.244"	Shim dimension required for ring gear side		Shim dimension required for opposite side	.279"

Determining side bearing shim requirements

7. Turn the pinion gear several times to make sure the bearings are seated and recheck the preload.

RING GEAR

1. Remove all of the bolts that hold the ring gear to the differential case and with a soft hammer, tap the ring gear off the case.

NOTE: Do not try to pry the ring gear off the case. This will damage the machined surfaces.

2. Clean all dirt from the case assembly and lubricate the case with gear lube. Align the ring gear bolt holes with the holes in the carrier and lightly press the ring gear onto

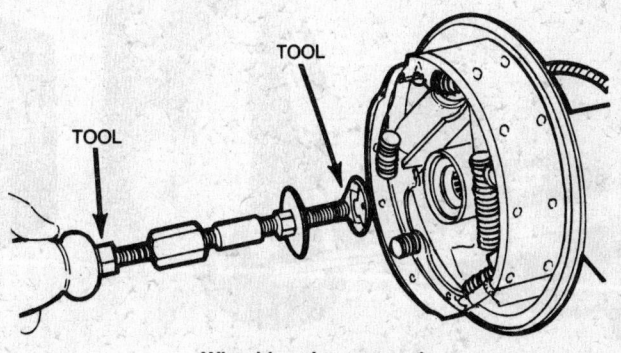

Wheel bearing removal

Removing drive pinion rear bearing

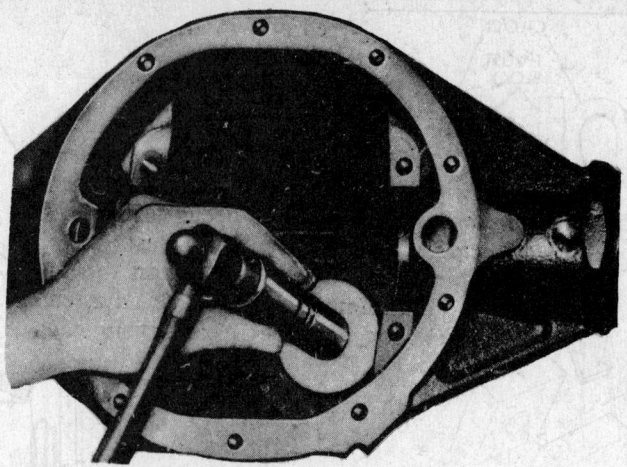

Installing pinion rear bearing cup

Checking ring gear runout

the case assembly. Install all of the bolts and tighten them all evenly, using a criss-cross pattern to avoid cocking the ring gear.

3. When the ring gear is firmly seated against the case, tighten the bolts to 60 ft. lbs.

CASE ASSEMBLY

1. Install the thrust washers and side

gears into the case assembly. If the original parts are being used, be sure to place them in their original position.

2. Place the pinions in the case so they are 180 degrees apart as they engage the side gears.

3. Turn the pinion gears so the hole in the case lines up with the holes in the gears.

When the holes are aligned, install the pinion shaft and lock screw. Do not tighten the lock screw too tightly at this time.

4. Check the bearings, bearing cups, cup seat and carrier caps to make sure they are in good condition.

5. Lubricate the bearings with gear lube. Install the cups on the proper bearings and install the differential assembly in the carrier. Support the carrier assembly to keep it from falling.

6. Install a support strap on the left side bearing and tighten the bearing bolts to an even, snug fit.

7. With the ring gear tight against the pinion gear, insert a gauging tool between the left side bearing cup and the carrier housing.

8. While lightly shaking the tool back and forth, turn the adjusting wheel until a slight drag is felt. Tighten the lock nut.

Differential bearing installation

9. Between the right side bearing and carrier, install a service spacer, 0.170 of an inch thick, a service shim and a feeler gauge. The feeler gauge must be thick enough so a light drag is felt when it is moved between the carrier and the shim.

10. Add the total of the service spacer, service shim and the feeler gauge. Remove

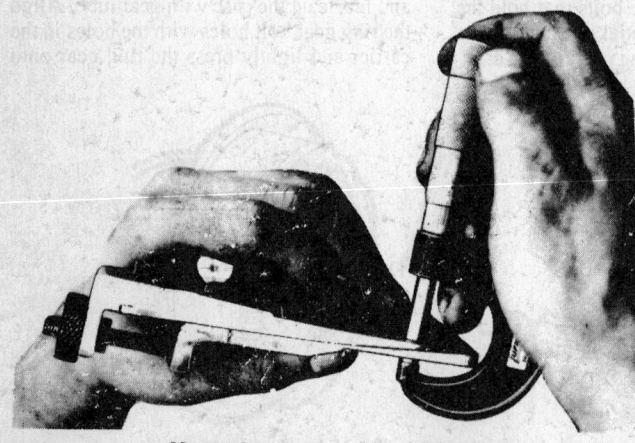

Measuring gauge plate thickness

Checking ring gear backlash

the gauging tool from the left side of the carrier and using a micrometer, measure the thickness in at least three places. Average the readings and record the result.

11. Refer to the chart to determine the proper thickness of the shim packs.

12. Install the left side shim first, then install the right side shim between the bearing cup and spacer. Position the shim so the chamfered side is outward or next to the spacer. If there is not enough chamfer around the outside of the shim, file or grind the chamfer a little to allow for easy installation.

13. If there is difficulty in installing the shim, partially remove the case from the carrier and slide both the shim and case back into place.

14. Install the bearing caps and torque them to 60 ft. lbs. Tighten the pinion shaft lock screw.

NOTE: The differential side bearings are now preloaded. If any adjustments are made in later procedures, make sure not to change the preload. Do Not change the total thickness of the shim packs.

15. Mount a dial indicator on the carrier assembly with the indicator button perpendicular to the tooth angle and in line with the gear rotation.

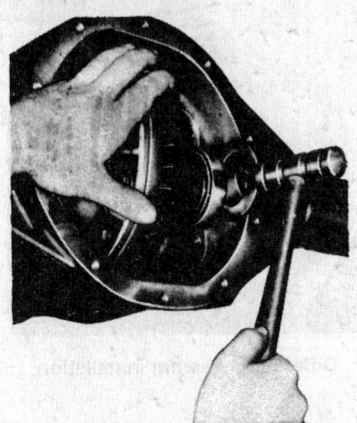

Installing differential shims

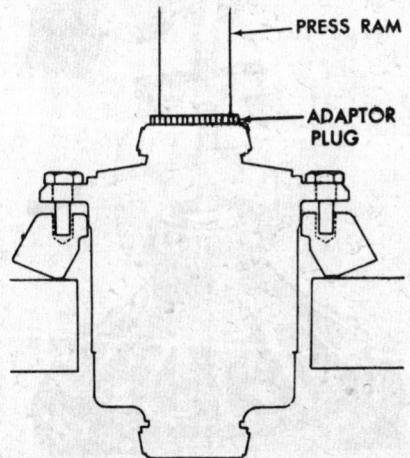

PRESS RAM

ADAPTOR PLUG

Ring gear-to-case installation

Installing rear bearing

16. Measure the amount of backlash between the ring and pinion gears. The backlash should be between 0.005–0.008 of an inch. Take readings at four different spots on the gear. There should not be variations greater than 0.002 of an inch.

17. If there are variations greater than 0.002 of an inch between the readings, check the runout between the case and ring gear. The gear runout should not be greater than 0.003 in. If the runout does exceed 0.003 in. check the case and ring gear for the deformation or dirt between the case and gear.

18. If the gear backlash exceeds 0.008 in., increase the thickness of the shims on the ring gear side and decrease the thick-

Installing pinion front bearing cup

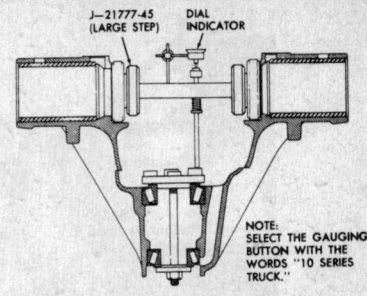

J–21777-45 (LARGE STEP) DIAL INDICATOR

NOTE: SELECT THE GAUGING BUTTON WITH THE WORDS "10 SERIES TRUCK."

Gauging tools installed in carrier

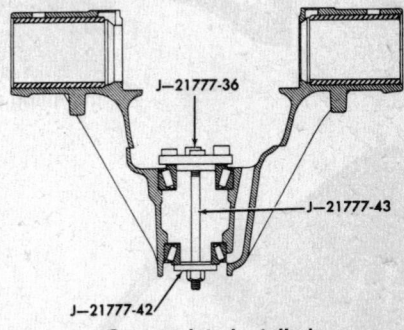

J–21777-36

J–21777-43

J–21777-42

Gauge plate installed

ness of the shims on the opposite side, an equal amount.

19. If the backlash is less than 0.005 in., decrease the shim thickness on the ring gear side and increase the shim thickness on the opposite side an equal amount.

Gear Pattern Check

Before final assembly of the differential, a pattern check of the gear teeth must be made. This determines if the teeth of the ring and pinion gears are meshing properly, for low noise level and long life of the gear teeth. The most important thing to note is if the pattern is located centrally up and down on the face of the ring gear.

1. Wipe any oil out of the carrier and wipe all dirt and oil from the teeth of the ring gear.

REAR PINION BEARING

PINION

PRESS PLATE

Pinion rear bearing removal

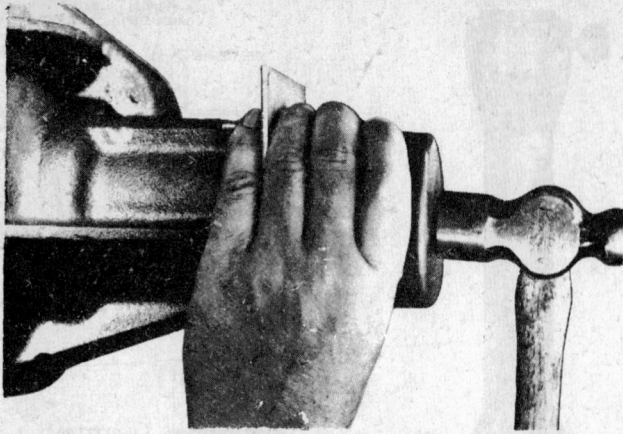

Installing pinion oil seal

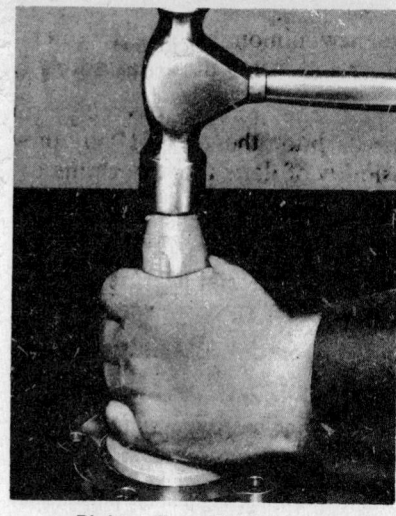

Pinion oil seal installation

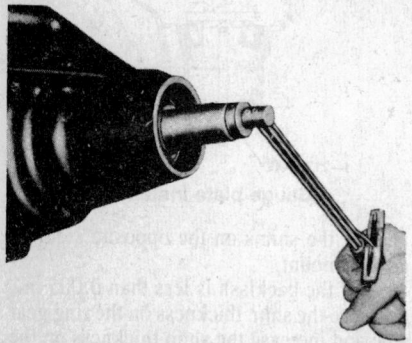

Measuring rotating torque

2. Coat the teeth of the ring gear with a gear marking compound.

3. With the bearing caps torqued to 55 ft. lbs., expand the brake shoes until it takes 20–30 ft. lbs. of torque to turn the pinion gear.

4. Turn the companion flange so the ring gear makes one full rotation in one direction, then turn it one full rotation in the opposite direction.

5. Check the pattern on the teeth and refer to the chart for any adjustments necessary.

6. With the gear tooth pattern checked

and properly adjusted, install the axle housing cover gasket and cover and tighten securely. Fill the axle with gear lube to the correct level.

7. Road test the vehicle to check for any noise and proper operation of the rear.

10½ Inch Ring Gear

This axle is a full floating type that uses special hypoid type drive and pinion gears. The pinion gear is supported by three bearings, two in front of the pinion gear and one behind. The differential assembly has either two or four pinions depending on the application of the axle. This axle assembly must be removed from the vehicle to remove and service the differential.

PINION ASSEMBLY REMOVAL

1. Remove the differential assembly from the axle.

2. Check the pinion bearing for the proper preload. The force required to turn the pinion should be 25–35 inch lbs. for used bearings. If there is no reading, shake the companion flange to check for any looseness in the bearing. If there is any looseness present the bearing should be replaced.

Differential bearing installation

Loosening the adjusting nuts

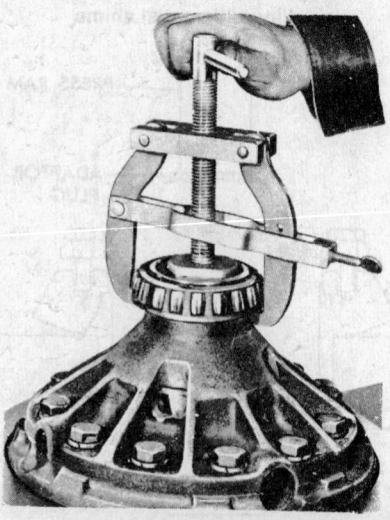

Differential bearing removal

3. Remove the retaining bolts for the pinion bearing from the axle housing.

4. Remove the bearing retainer and pinion assembly from the axle housing. It may be necessary to tap the pilot end of the pinion shaft to help remove the pinion assembly from the carrier.

5. Record the thickness of the shims that are removed from between the carrier assembly and the bearing retainer assembly.

DRIVE PINION

1. With the pinion assembly clamped in a vise, install a holder assembly on the flange.

2. Using the proper size socket, remove the pinion nut and washer from the pinion. When reassembling the pinion use a new nut and washer assembly.

3. With the holder assembly still in place, use a puller to remove the flange from the pinion.

4. With the bearing retainer supported in a press, press the pinion out of the retainer assembly. Be careful not to allow the pinion gear to fall onto the floor because this can damage the gear.

5. Separate the pinion flange, oil seal, front bearing and the bearing retainer. If the oil seal needs to be replaced it may have to be driven from the retainer.

6. Using a drift, drive the front and rear bearing cups from the bearing retainer.

7. Support the pinion assembly in a press, with the bearing supported. Press the bearing from the pinion gear.

8. Using a drift, drive the straddle bearing from the carrier assembly.

CLEANING & INSPECTION

1. Clean off all the parts in solvent and blow dry.

2. Check the pinion gear for signs of wear, chips, cracks or any other imperfections. Check the splines for signs of wear or distortion.

3. Check the bearings for signs of wear or pitting on the rollers and races and check the bearing cage for dents and bends.

Check the bearing retainer for any cracks, pits, grooves or corrosion.

4. Check the pinion flange splines for any signs of wear or distortion.

5. Replace parts that show any of the signs mentioned above.

CASE DISASSEMBLY

1. Scribe a line across the two halves of the differential case so they may be reassembled in the same position, and with the ring gear removed, separate the two halves. To remove the ring gear, remove the ring gear bolts and washers, and using a soft hammer tap the ring gear from the case.

2. Remove the internal parts from the inside of the case and set them aside in order that they may be reassembled in the same position.

CLEANING & INSPECTION

1. Check the differential gears, pinions, thrust washers and spider for any signs of unusual wear, chips, cracks or pitting.

2. Check all mating surfaces for signs of wear.

3. Replace parts that show any of the signs mentioned above.

CASE ASSEMBLY

1. Using a good quality gear lubricant coat all of the parts.

2. Assemble the differential pinions and thrust washers onto the spider and install the assembly into the differential case.

3. Line up the scribe marks on the two halves of the differential case and install the ring gear. Install the ring gear washers and bolts and torque the bolts to approx. 10 ft. lbs.

SIDE BEARING

1. Install a bearing puller on the bearing and remove the bearing assembly from the differential case.

2. Check the bearings for any signs of wear on distortion.

3. Install the new bearing by setting it in place on the differential case and, using a

Removing pinion nut

Removing flange

bearing driver, drive the bearing onto the case assembly until it seats against the shoulder on the case.

DRIVE PINION ASSEMBLY

1. Coat all of the parts with a good quality gear lubricant.

2. With the pinion gear in a press, press the rear bearings onto the pinion assembly.

Removing retainer bolts

Measuring backlash

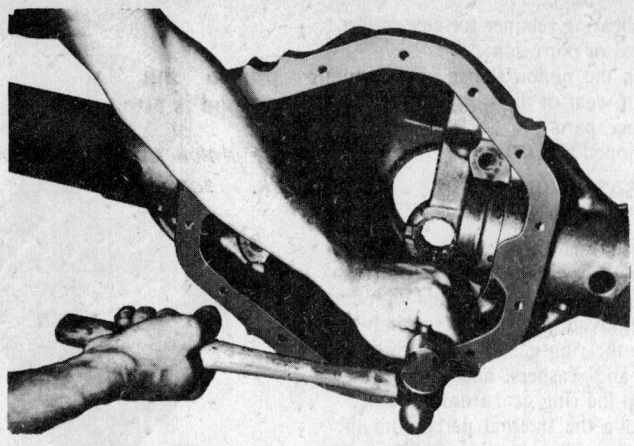

Removing straddle bearing

Pressing drive pinion from bearing retainer

Removing pinion rear bearing

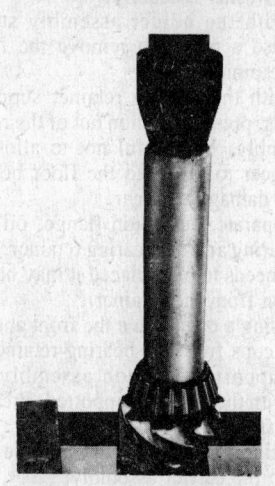

Installing pinion rear bearing

7. Lubricate the oil seal with a good quality high pressure grease and install the seal into the retainer bore. Be sure to press the seal down until it rests against the internal shoulder.

8. Install the pinion flange and oil deflector onto the splines of the pinion gear and install a new lock washer and pinion nut.

9. With the pinion flange clamped in a vise and a holder assembly installed on the flange, tighten the nut to obtain the proper preload. Measure the amount of torque required to turn the pinion gear. For a new bearing the torque required is 25–35 inch lbs. and for an old bearing it is 5–15 inch lbs. To preload the bearing, tighten the pinion nut to approx. 350 ft. lbs. and take a reading of the torque required to turn the pinion. Continue tightening the nut until the proper preload is obtained.

CAUTION

Do not tighten the nut too tightly because it will collapse the spacer too much. This will make replacement necessary.

DRIVE PINION INSTALLATION

1. If installing a new pinion gear, check the top of the new gear for the depth code number.

3. In the bearing retainer, install the front and rear bearing cups using a driver of the proper size.

4. In the axle housing, install the straddle bearing assembly using the proper size driver.

5. Install the bearing retainer with the bearing cups in place on the pinion gear and install a new collapsible spacer.

6. Press the front bearing onto the pinion gear.

Drive pinion front bearing removal

2. Compare the new number with the old number on top of the old pinion and check the pinion depth chart for preliminary setting of the pinion depth.

3. Check the thickness of the original shims removed from the pinion and either add or subtract from the shims according to the chart.

4. Place the shim on the carrier assembly and line the holes up with those in the axle housing. Make sure the surfaces are clean of all dirt and grease.

5. Install the retainer and pinion assembly in the housing making sure the holes line up and install the retaining bolts. Torque the bolts to approx. 45 ft. lbs.

CASE INSTALLATION

1. Place the bearing cups over the side bearings on the differential assembly and place the unit into the carrier in the axle housing.

2. Install the bearing caps making sure the marks are lined up and install the bolts. Tighten the bearing retaining bolts.

3. Loosen the right side nut and tighten the left side nut until the ring gear comes in contact with the pinion gear. Do not force the gears together. This brings the gears to zero lash.

4. Back off the left side adjusting nut

about two slots and install the lock fingers into the nut.

5. In this order tighten the right side adjusting nut firmly to force the case assembly into tight contact with the left side adjusting nut and then loosen the right side nut until it is free from the bearing.

6. Again retighten the right side adjusting nut until it comes in contact with the bearing. Tighten the right adjusting nut about two slots if it is an old bearing or three slots if it is a new bearing.

7. Install the lock retainers into the slots and torque the bearing cap bolts to 100 ft. lbs. This procedure now insures that the bearings are preloaded properly. If more adjustments are made, make sure the preload stays the same. To do this, one adjusting nut must be loosened the same amount the other nuts is tightened.

8. Install a dial indicator on the housing and measure the amount of backlash between the ring and pinion gear. The backlash should measure between 0.003 to 0.012 of an inch with the best figure being between 0.005 to 0.008 of an inch.

9. If the backlash is more than 0.012 of

Pinion rear bearing lock ring installation

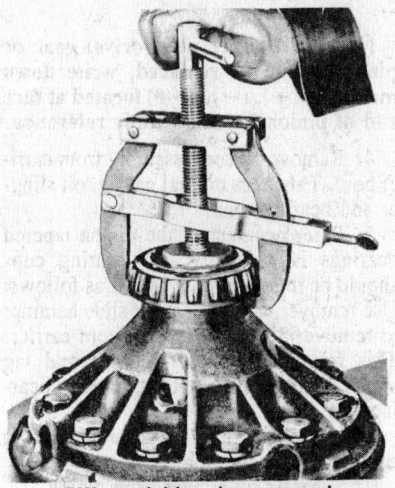

Differential bearing removal

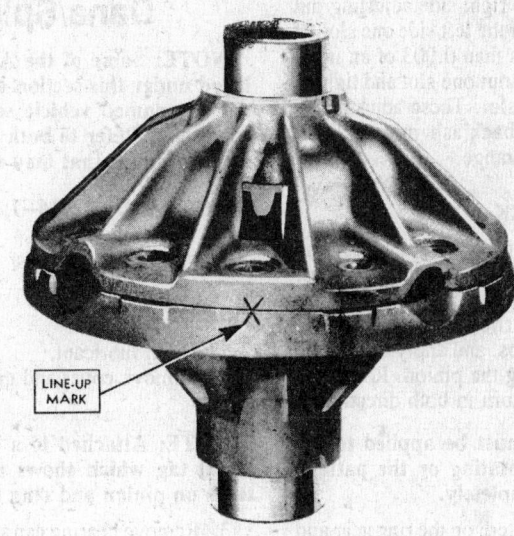

Differential case alignment marks

Differential bearing installation

Checking backlash

Backlash and preload adjustment

an inch, loosen the right side adjusting nut one slot and tighten the left side one slot. If the backlash is less than 0.003 of an inch, loosen the left side nut one slot and tighten the right side one slot. These adjustments should bring the backlash measurement into an acceptable range.

PATTERN CHECK

1. Clean all the oil off the ring gear and using a gear marking compound, coat all of the teeth of the ring gear.

2. Make sure the bearing caps are torqued to 110 ft. lbs. and apply load to the gears while rotating the pinion. Rotate the ring gear one full turn in both directions.

NOTE: Load must be applied to the assembly while rotating or the pattern will not show completely.

3. Check the pattern on the ring gear and following the chart, adjust the assembly to get the contact pattern located centrally on the face of the ring gear teeth.

Dana/Spicer

NOTE: Some of the Axle Assemblies listed under this section also appear under the named vehicle sections by ring gear sizes. Refer to both procedures for any variations that may exist.

Models 30, 44, 44–1, 60–1,2, 70

DIFFERENTIAL

Removal

1. Drain lubricant.
2. Remove cover and gasket.

NOTE: Attached to a cover bolt is a metal tag which shows the number of teeth on pinion and ring (drive) gear.

3. Remove bearing cap screws. Note the matching marks on cap and carrier and make sure caps are reassembled to correct markings.

4. Using a spreader tool, spread carrier a maximum of 0.020 inch and measure amount of spread with a dial indicator.

— CAUTION —

Carrier may be permanently damaged if spread more than 0.020 inch. Do not attempt differential removal without using a spreader.

5. Carefully lift differential assembly out of carrier.

6. Remove the spreader assembly after removing the differential assembly from the housing.

DRIVE PINION

Disassembly

1. Pull flange (yoke) from shaft splines of drive pinion.

2. Using a press or soft hammer, drive pinion and inner bearing cone assembly out of carrier.

3. Remove and tag shim pack from the splined end of pinion.

NOTE: If either ring (drive) gear or pinion are to be replaced, write down markings (+), (−), or (0) located at face end of pinion for reassembly reference.

4. Remove oil seal assembly from carrier bore. This frees oil seal gasket, oil slinger and bearing cone.

5. If replacement of the pinion tapered bearings is necessary, the bearing cups should be removed from carrier as follows: Use remover with a driver or slide hammer to remove inner bearing cup from carrier. This frees shim pack. Remove and tag shims for reassembly. Remove outer bearing cup.

6. Use remover set to separate bearing cone from drive pinion.

7. Separate oil slinger from pinion.

NOTE: This oil slinger is only found on some axle models.

DIFFERENTIAL

Disassembly

1. Remove and label the two bearing cups.

2. Use a suitable type puller to remove the bearing cones. Remove and label adjusting shims.

3. Drive out pinion shaft lock pin.

NOTE: On the Model 70 rear axle, punch-mark the differential case halves (for reassembly reference) and separate. Remove the differential spider, pinion gears, side gears and thrust washers.

4. Separate ring gear from case.

5. Remove pinion shaft, two pinions, two side gears, and four thrust washers from case.

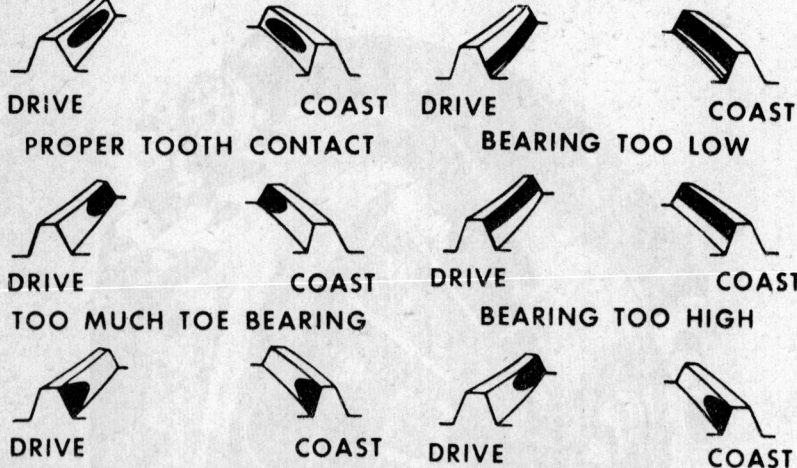

DRIVE COAST	DRIVE COAST
PROPER TOOTH CONTACT	BEARING TOO LOW
DRIVE COAST	DRIVE COAST
TOO MUCH TOE BEARING	BEARING TOO HIGH
DRIVE COAST	DRIVE COAST
TOO MUCH HEEL BEARING	CROSS BEARING

Typical gear tooth contact patterns

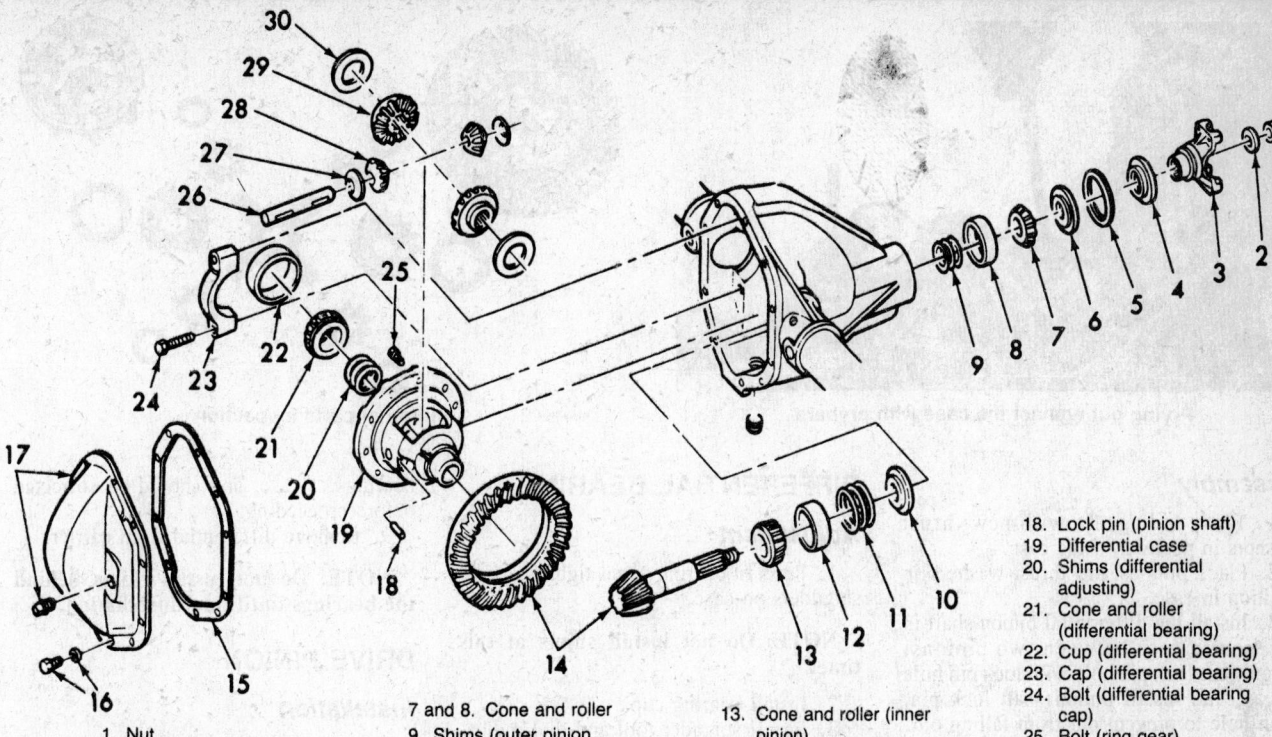

7 and 8. Cone and roller

1. Nut
2. Washer
3. Companion flange
4. Pinion oil seal
5. Gasket
6. Outer pinion oil slinger

9. Shims (outer pinion bearing)
10. Inner pinion oil slinger
11. Shims (inner pinion bearing)
12. Cup (inner pinion bearing)

13. Cone and roller (inner pinion)
14. Ring and pinion
15. Gasket (housing cover)
16. Screw and washer (cover)
17. Cover and plug

18. Lock pin (pinion shaft)
19. Differential case
20. Shims (differential adjusting)
21. Cone and roller (differential bearing)
22. Cup (differential bearing)
23. Cap (differential bearing)
24. Bolt (differential bearing cap)
25. Bolt (ring gear)
26. Pinion shaft
27. Thrust washer (pinion)
28. Pinion
29. Side gear
30. Thrust washer (side gear)

Dana 9¾ inch ring gear axle

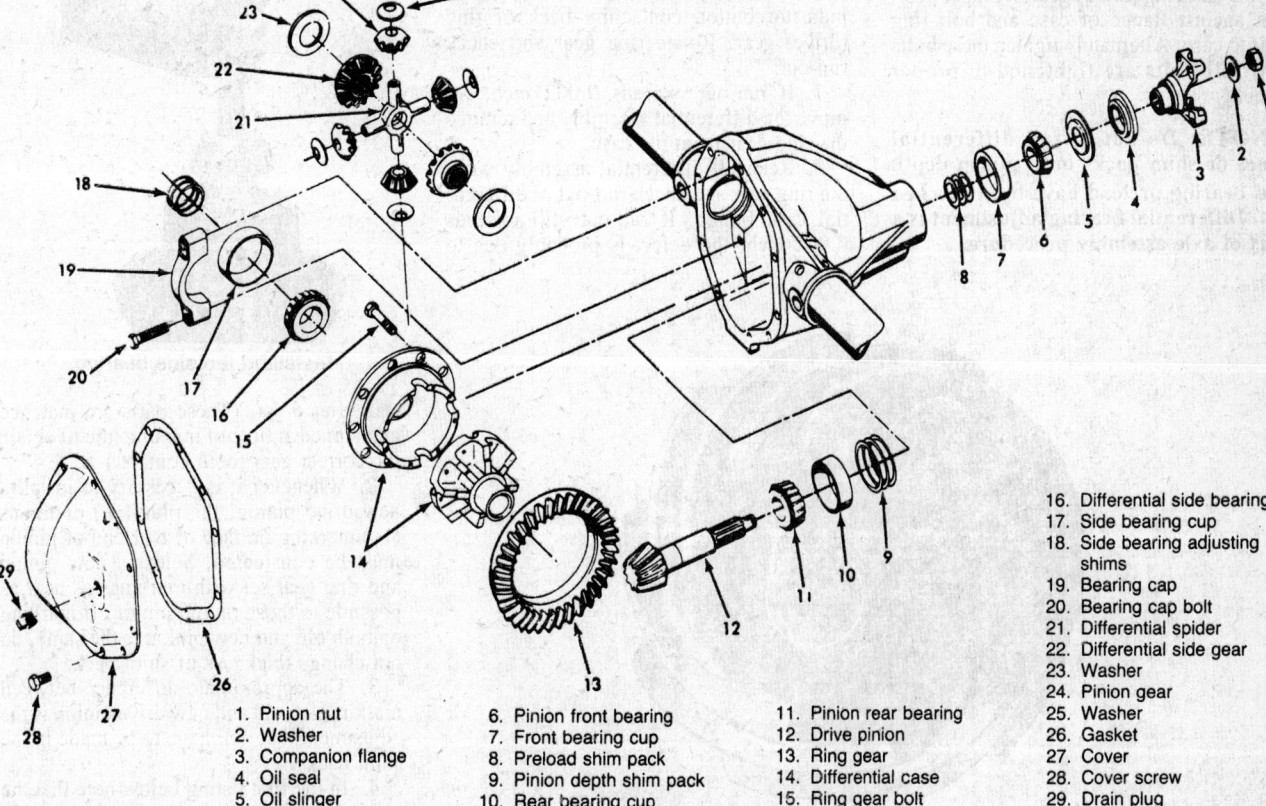

1. Pinion nut
2. Washer
3. Companion flange
4. Oil seal
5. Oil slinger

6. Pinion front bearing
7. Front bearing cup
8. Preload shim pack
9. Pinion depth shim pack
10. Rear bearing cup

11. Pinion rear bearing
12. Drive pinion
13. Ring gear
14. Differential case
15. Ring gear bolt

16. Differential side bearing
17. Side bearing cup
18. Side bearing adjusting shims
19. Bearing cap
20. Bearing cap bolt
21. Differential spider
22. Differential side gear
23. Washer
24. Pinion gear
25. Washer
26. Gasket
27. Cover
28. Cover screw
29. Drain plug

Dana 10½ inch ring gear axle

Prying out against the case with prybars

Internal parts inspection

Assembly

1. Place side gears with new thrust washers in position inside case.

2. Place pinions and thrust washers in position in case.

3. Install the differential pinion shaft in position in case between two pinions. Align shaft lock pin hole with lock pin hole in case and install pinion shaft lock pin. Peen hole to prevent pin from falling out.

NOTE: On the Model 70 rear axle, install the differential spider along with its pinion gears, side gears and thrust washers into the differential case halves. Bolt the two halves together making sure the punch-marks line up.

4. Place ring (drive) gear in proper position against flange of case and bolt ring gear to case. Alternately tighten these bolts until all bolts are tightened to proper torque.

NOTE: Do not install differential cones or shim packs until pinion depth and bearing preload have been checked out. Differential bearing adjustment is a part of axle assembly procedure.

DIFFERENTIAL BEARING

Adjustment

1. Press fit bearing cones tightly against shoulders on case.

NOTE: Do not install shims at this time.

2. Install bearing cups.

3. Install spreader tool and dial indicator, and spread carrier as described in Differential Removal.

4. Place differential assembly into carrier.

5. Install bearing caps using their respective cap screws. Make sure caps are assembled to their correct markings. Hand tighten bearing cap screws.

6. Install dial indicator at carrier with indicator button contacting back of ring (drive) gear. Rotate ring gear and check run-out.

7. If run-out exceeds 0.002 inch, remove the differential assembly and remove the ring gear from the case.

8. Reinstall differential assembly without ring gear and check run-out of differential case flange. If run-out still exceeds 0.002 inch, the defect is probably due to bearings or case, and should be corrected before proceeding.

9. Remove differential from carrier.

NOTE: Do not install shims behind the bearings until final installation.

DRIVE PINION

Installation

1. If either drive pinion or ring (drive) gear must be replaced, they must be in-

Installing the side bearing

stalled as a set. (These parts are matched and lapped at time of manufacture to obtain the correct gear tooth contact.)

2. Whenever it is necessary to install a new drive pinion, the plus (+) or minus (−) marking on face of rear end of pinion must be considered. Select a new pinion and ring gear set with markings as near as possible to those on old pinion. If marking on both old and new pinion is the same, do not change thickness of shim pack.

3. The approximate difference between markings on old and new drive pinion is the adjustment that will have to be made in the shim packs.

4. In the first listing below note that the new pinion is a plus eight (+8) while the old pinion is a plus five (+5). Making a

Pinion code location

difference of plus three (+3). This means that the thickness of each shim pack must be decreased by 0.003 inch.

5. Once proper adjustment in shim packs has been made, place oil slinger, if so equipped, over pinion shaft. Install pinion inner bearing cone over shaft, and use bearing installer and an arbor press to press bearing onto pinion shaft. Bearing must be seated tightly against shoulder or oil slinger.

6. Use pinion front bearing cup installer to install outer bearing cup into carrier bore.

7. Install the selected inner shim pack in carrier. Then use pinion rear bearing cup installer to install inner bearing cup.

8. Insert pinion, oil slinger (when used) and inner bearing cone assembly into carrier and place the selected shim pack into position on outer end of pinion shaft.

9. Place outer bearing cone over pinion shaft, then use installer to seat bearing tight against shim pack.

10. Install pinion flange (yoke), washer and nut. Hold flange while tightening the nut to proper torque.

NOTE: Install oil slinger and oil seal only after pinion depth and pinion bearing preload have been checked out.

Checking Pinion Depth Adjustment

1. A pinion depth gauge and correct adapter, which gives a micrometer reading, should be used to determine pinion depth. The actual pinion depth setting can be determined by adding gauge reading to thickness of step plate and comparing result with the nominal dimension of 2.625 inch (Models 44/60) or 3.125 inch (Model 60), or 3.500 inch (Model 70).

2. If the pinion setting is within minus (−) 0.001-inch to plus (+) 0.003-inch of this nominal dimension, the pinion position can be considered satisfactory.

3. If pinion setting exceeds these limits, it must be corrected by adjusting thickness of shim pack behind the pinion inner bearing cup.

Pinion Bearing Preload Adjustment

1. Use a torque wrench to check pinion bearing preload.

2. Rotating torque of pinion should be from 15 to 30 inch lbs.

3. Add or remove shims from pack just behind outer bearing cone to bring preload within these torque limits.

DIFFERENTIAL

Installation

1. Use dial indicator and spreader tool as described in Differential Removal, to spread carrier a maximum of 0.020 inch.

Old Pinion Marking	New Pinion Marking								
	− 4	− 3	− 2	− 1	0	+ 1	+ 2	+ 3	+ 4
+ 4	+ 0.008	+ 0.007	+ 0.006	+ 0.005	+ 0.004	+ 0.003	+ 0.002	+ 0.001	0
+ 3	+ 0.007	+ 0.006	+ 0.005	+ 0.004	+ 0.003	+ 0.002	+ 0.001	0	− 0.001
+ 2	+ 0.006	+ 0.005	+ 0.004	+ 0.003	+ 0.002	+ 0.001	0	− 0.001	− 0.002
+ 1	+ 0.005	+ 0.004	+ 0.003	+ 0.002	+ 0.001	0	− 0.001	− 0.002	− 0.003
0	+ 0.004	+ 0.003	+ 0.002	+ 0.001	0	− 0.001	− 0.002	− 0.003	− 0.004
− 1	+ 0.003	+ 0.002	+ 0.001	0	− 0.001	− 0.002	− 0.003	− 0.004	− 0.005
− 2	+ 0.002	+ 0.001	0	− 0.001	− 0.002	− 0.003	− 0.004	− 0.005	− 0.006
− 3	+ 0.001	0	− 0.001	− 0.002	− 0.003	− 0.004	− 0.005	− 0.006	− 0.007
− 4	0	− 0.001	− 0.002	− 0.003	− 0.004	− 0.005	− 0.006	− 0.007	− 0.008

Pinion code chart

Bearing cap identification

Spreading rear axle housing

Removing differential pinion shaft lock pin

Measuring differential case drive gear mounting flange runout

2. Install bearing cups and place differential assembly in carrier. Rotate differential and, with a soft hammer, tap ring (drive) gear to assure a proper bearing seating.

3. Reinstall bearing caps in their proper locations as indicated by marks made during the removal procedure. Finger tighten cap screws. Relieve the spreader tool pressure, and tighten cap screws to 70–90 ft. lbs.

4. Move differential assembly tightly against drive pinion.

5. Install dial indicator securely to carrier, then set button at zero and against back of drive gear.

6. Move the differential toward the dial indicator and note the reading. For accuracy, repeat this operation several times.

7. Remove the differential assembly from carrier. Install a shim pack behind differential bearing cone at drive gear side, equal to the dimension indicated by dial indicator.

8. Subtract the indicator reading from the reading previously obtained in paragraph Differential Bearing Adjustment.

9. To the above result should be added 0.015 to 0.020-inch in shims to provide bearing preload.

10. Install the above shim pack behind differential bearing cone at side opposite to drive gear.

11. Spread differential carrier, using spreader tool.

12. Install differential bearing cups then locate differential assembly in carrier.

13. Rotate differential assembly, tapping gear to seat bearings.

14. Install differential bearing caps in their correct location as indicated by marks made upon disassembly. Finger tighten cap screws.

15. Remove differential carrier spreader tool. Tighten differential bearing cap screws to proper torque.

16. Install dial indicator and check drive gear to drive pinion backlash at four equally spaced points around the drive gear. Backlash must be held to 0.003 to 0.006-inch and must not vary more than 0.002-inch between positions checked.

17. Whenever backlash is not within limits, differential bearing shim pack should be corrected.

9¾ & 10½ INCH RING GEAR

The 9¾ and 10½ inch ring gear axle assemblies are basically the same, but with certain exceptions. The differential side bearing shims are located between the side bearing cup assembly and the differential case on the 9¾ inch ring gear axle assembly, while on the 10½ inch ring gear axle assembly, the side bearing shims are located between the side bearing cup and the axle housing. Both axles use inner and outer shims on the pinion gear. The inner shims are used to control the pinion depth in the housing, while the outer shims are used to preload the pinion bearings. The 9¾ inch ring gear axle uses a solid differential carrier with a removable side and pinion gear shaft. The 10½ inch ring gear axle uses a split differential carrier with the side and pinion gears mounted on a cross shaft.

DIFFERENTIAL CASE

Removal

9¾ & 10½ INCH

1. The axle assembly can be overhauled either in or out of the vehicle, depending on the repairman's discretion. Either way, the free-floating axles and wheel assemblies must be removed.

2. Drain the lubricant and remove the rear cover and gasket.

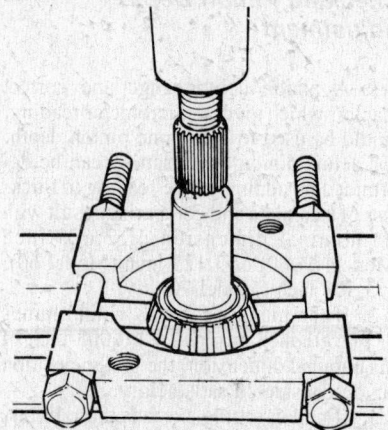

Pinion rear bearing removal

3. Matchmark the bearing caps and the housing for reassembly in the same position. Remove the bearing caps and bolts.

4. Using a spreader tool mounted to the carrier housing, spread the housing a maximum of 0.015 inch.

CAUTION

Do not exceed this measurement. The housing could be permanently damaged. The use of a dial indicator is recommended to prevent over-stretching the housing.

5. Using a pry bar, remove the differential case from the housing. Separate the shims and record the dimensions and location on the 10½ inch ring gear axle. Remove the spreader tool from the housing.

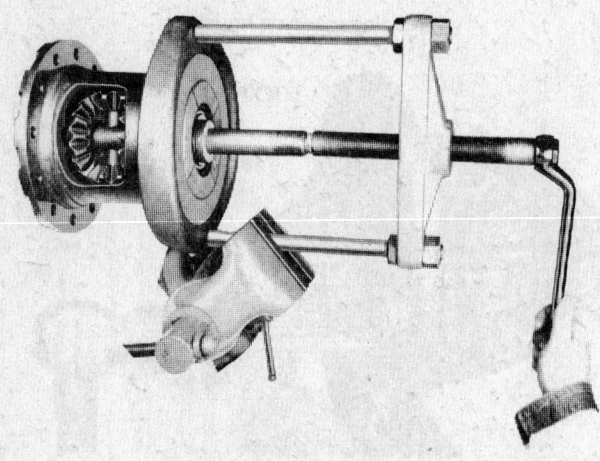

Removing differential bearings

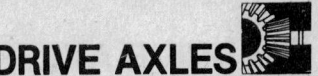
PINION SETTING CHARTS (U.S. AND METRIC)
By utilizing these specifications, proper gear contact should be established.

Old Pinion Marking	New Pinion Marking (U.S. Standards)								
	−4	−3	−2	−1	0	+1	+2	+3	+4
+4	+0.008	+0.007	+0.006	+0.005	+0.004	+0.003	+0.002	+0.001	0
+3	+0.007	+0.006	+0.005	+0.004	+0.003	+0.002	+0.001	0	−0.001
+2	+0.006	+0.005	+0.004	+0.003	+0.002	+0.001	0	−0.001	−0.002
+1	+0.005	+0.004	+0.003	+0.002	+0.001	0	−0.001	−0.002	−0.002
0	+0.004	+0.003	+0.002	+0.001	0	−0.001	−0.002	−0.003	−0.004
−1	+0.003	+0.002	+0.001	0	−0.001	−0.002	−0.003	−0.004	−0.005
−2	+0.002	+0.001	0	−0.001	−0.003	−0.003	−0.004	−0.005	−0.005
−3	+0.001	0	−0.001	−0.002	−0.003	−0.004	−0.005	−0.006	−0.007
−4	0	−0.001	−0.002	−0.003	−0.004	−0.005	−0.006	−0.007	−0.008

Old Pinion Marking	New Pinion Marking (Metric)								
	−10	−8	−5	−3	0	+3	+5	+8	+10
+10	+.20	+.18	+.15	+.13	+.10	+.08	+.05	+.03	0
+8	+.18	+.15	+.13	+.10	+.08	+.05	+.03	0	−.03
+5	+.15	+.13	+.10	+.08	+.05	+.03	0	−.03	−.05
+3	+.13	+.10	+.08	+.05	+.03	0	−.03	−.05	−.08
0	+.10	+.08	+.05	+.03	0	−.03	−.05	−.08	−.10
−3	+.08	+.05	+.03	0	−.03	−.05	−.08	−.10	−.13
−5	+.05	+.03	0	−.03	−.05	−.08	−.10	−.13	−.15
−8	+.03	0	−.03	−.05	−.08	−.10	−.13	−.15	−.18
−10	0	−.03	−.05	−.08	−.10	−.13	−.15	−.18	−.20

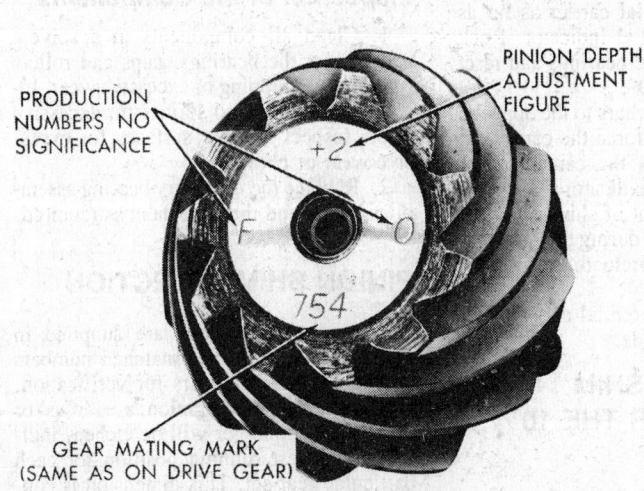

PRODUCTION NUMBERS NO SIGNIFICANCE

PINION DEPTH ADJUSTMENT FIGURE

GEAR MATING MARK (SAME AS ON DRIVE GEAR)

Drive pinion markings

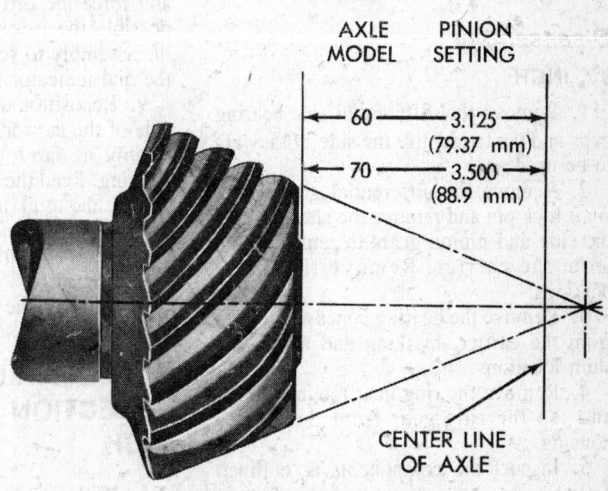

AXLE MODEL	PINION SETTING
60	3.125 (79.37 mm)
70	3.500 (88.9 mm)

CENTER LINE OF AXLE

Pinion setting dimensions

Disassembly

10½ INCH

1. Remove the differential side bearings from the case, using the necessary puller tools.

2. Remove the ring gear bolts and tap the ring gear from the case with a soft-faced hammer.

3. Scribe the case halves for reassembly and remove the retaining bolts.

4. Tap the top half of the case to separate it from the bottom half. Remove the internal gears, washers and cross.

Inspection

9¾ & 10½ INCH

1. Clean the gears, bearings and component parts with solvent and inspect for scoring, chipping or excessive wear.

2. Replace the necessary parts as required.

Assembly

10½ INCH

1. Install new thrust washers to the side gears and lubricate the contact surfaces.

2. Assemble the side gears, pinion bears, washers and cross shaft into the flanged half of the case.

3. Install the top half of the case to the bottom half, making sure the scribe marks are lined up.

4. Install the retaining bolts finger tight. Then tighten the bolts alternately to the proper torque specifications.

5. If a new ring gear is to be installed or the old one used, install it to the differential case and align the bolt holes. Tighten the bolts aternately to the proper torque specifications.

6. Install the side carrier bearings by using the proper installation tools.

7. Cover the assembled unit and set aside until ready for the installation into the housing.

Disassembly

9¾ INCH

1. Remove the differential side bearing cups and tag to identify the side, if they are to be used again.

2. Remove the differential gear pinion shaft lock pin and remove the shaft. Rotate the side and pinion gears to remove them from the carrier. Remove the thrust bearings.

3. Remove the bearing cones and rollers from the carrier, marking and noting the shim locations.

4. Remove the ring gear retaining bolts and tap the ring gear from the carrier housing.

5. Inspect the components as outlined earlier.

Assembly

9¾ INCH

1. Install the differential side gears, the differential pinion gears and new thrust washers into the differential carrier.

2. Align the pinion gear shaft holes and install the pinion shaft into the carrier. Align the lock pin hole in the shaft and carrier. Install the lock pin and peen the hole to avoid having the pin drop from the carrier.

3. Install the differential case side bearings with the proper installation tools. Do not install the shims at this time.

4. Place the carrier assembly into the axle housing with the bearing cups on the bearing cones. Install the bearing caps in their original position and tighten the bearing cap bolts enough to keep the bearing caps in place.

5. Install a dial indicator on the housing so that the indicator button contacts the carrier flange. Press the differential carrier to prevent side play and center the dial indicator. Rotate the carrier and check the flange for run-out. If the run-out is greater than 0.002 inch, the defect is probably due to the bearings or to the carrier and should be corrected.

6. Remove the assembly and install the ring gear. Torque the retaining bolts to specifications and reinstall the assembly into the housing and again install the bearing caps in their original position and tighten the cap bolts enough to keep the bearings caps in place.

7. Again, install the dial indicator and position the indicator button to contact the ring gear back surface. Rotate the assembly and the run-out should be less than 0.002 inch. If over 0.002 inch, remove the assembly and relocate the ring gear 180 degrees. Reinstall the assembly and recheck. If the run-out remains over the 0.002 inch tolerance, the ring gear is defective. If the measurement is within tolerances, continue on with the assembly.

8. Position two pry bars between the bearing cap and the housing on the side opposite the ring gear. Pull on the pry bars and force the differential carrier as far as possible towards the dial indicator. Rock the assembly to seat the bearings and reset the dial indicator to "0".

9. Reposition the prybars to the opposite side of the carrier and force the carrier assembly as far towards the center of the housing. Read the dial indicator scale. This will be the total amount of shims required for setting the backlash during the reassembly, less the bearing preload. Record the measurement.

10. Remove the differential carrier from the housing and set aside.

SIDE BEARING SHIM SELECTION FOR THE 10½ INCH

1. With the pinion gear not in the axle housing, place the bearing cups over the side bearings and install the differential carrier into the axle housing.

2. Place the shim that was originally installed on the ring gear side back into its original position.

3. Install the bearing caps in their proper positions and tighten the bolts enough to keep the bearings in place.

4. Mount a dial indicator on the axle housing with the indicator button contacting the back of the ring gear.

5. Position two prybars between the bearing shim and the housing on the ring gear side of the differential carrier. Force the differential carrier away from the dial indicator and set the indicator to "0".

6. Reposition the prybars to the opposite side of the differential carrier and force the carrier back towards the dial indicator. Repeat several times until the same reading is obtained each time.

7. To the dial indicator reading, add the thickness of the shim and record the results to be used later in the assembly.

DRIVE PINION

Removal

9¾ & 10½ INCH

1. Remove the pinion nut and flange from the pinion gear, using the proper removing tools.

2. Remove the pinion gear assembly from the housing. It may be necessary to tap the pinion from the housing with a soft faced hammer. Catch the pinion so as not to allow it to drop on the floor.

3. With a long drift, remove the inner bearing cup, pinion seal, slinger, gasket, outer pinion bearing and the shim pack. Tag the shim pack for reassembly.

4. Remove the rear pinion bearing cup and shim pack from the housing. Tag the shims for reassembly.

5. Remove the rear pinion bearing from the pinion gear with an arbor press and special plates.

Inspection of the Components

1. Clean all components in a solvent and inspect the bearings, cups and rollers for scoring, chipping or excessive wear. Inspect the flanges and splines for excessive wear. Inspect all gear surfaces for excessive wear or chipping.

2. Replace the necessary bearing assemblies, gears and thrustwashers as required.

PINION SHIM SELECTION

Ring gears and pinions are supplied in matched sets only. The matched numbers are etched on both gears for verification. On the rear face of the pinion, a + (plus) or a − (minus) number will be etched, indicating the best running position for each particular gear set. This dimension is controlled by the shimming behind the inner bearing cup. Whenever baffles or oil sling<chers are used, they become part of the adjusting shim pack. An example: If a pinion is etched +3, this pinion would require 0.003 inch less shims than a pinion etched 0. This means by removing shims, the mounting distance of the pinion is increased by 0.003 inch, which is just what a + (plus) etching indicates. If a pinion is etched −3, it would be necessary to add 0.003 inch more shims than would be required if the pinion was etched 0. By adding the 0.003 inch shims, the mounting distance of the pinion is decreased 0.003 inch, which is just what the − (minus) etching indicates. Pinion adjusting shims are available in thicknesses of 0.003, 0.005 and 0.010 inch. An example: If a new gear set is used and the old pinion reads +2 and the new pinion reads −2, add 0.004 inch shims to the original shim pack.

Assembly

9³/₄ & 10¹/₂ INCH

1. Select the correct pinion depth shims and install in the rear pinion bearing cup bore.

2. Install the rear bearing cup in the axle housing with the proper tool.

3. Add or subtract an equal amount of shim thickness to or from the preload or outer shim pack, as was added or subtracted from the inner shim pack.

4. Install the front pinion bearing cup into its bore in the axle housing.

5. Press the rear pinion bearing onto the pinion gear shaft and install the pinion gear with bearing into the axle housing.

6. Install the preload shims and the front pinion bearing. Do not install the oil seal at this time.

7. Install the flange with the holding bar tool attached, the washer and the nut on the pinion shaft end. Torque the nut to 250 ft. lbs. for the 10¹/₂ inch and 255 ft. lbs. for the 9³/₄ inch.

8. Remove the holding bar from the flange and with an inch pound torque wrench, measure the rotating torque of the pinion gear. The rotating torque should be 10 to 20 inch lbs. with the original bearings and 20 to 40 inch lbs. with new bearings. Disregard the torque reading necessary to start the shaft to turn.

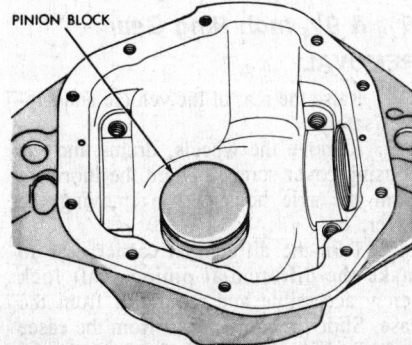

PINION BLOCK

Master pinion block installed in the pinion bore

9. If the preload torque is not in specifications, adjust the shim pack as required: To increase preload, decrease the thickness of the preload shim pack. To decrease preload, increase the thickness of the preload shim pack.

10. When the proper preload is obtained, remove the nut, washer and flange from the pinion shaft.

11. Install a new pinion seal into the housing and reinstall the flange, washer and nut. Using the holder tool, torque the nut to 250 ft. lbs. for the 10¹/₂ inch and 255 ft. lbs. for the 9³/₄ inch.

Assembly of Differential Carrier Into Axle Housing

9³/₄ INCH

1. As outlined in the Differential Carrier Assembly procedure, the amount of shims required for setting the backlash less bearing preload had been selected and the measurement recorded.

2. With the pinion gear installed and properly set, position the differential carrier assembly into the axle housing and install the bearing caps in their proper positions. Tighten the cap bolts just enough to hold the bearing cups in place.

3. Install a dial indicator on the axle housing with the indicator button contacting the back of the ring gear.

4. Position two prybars between the bearing cup and the axle housing on the ring gear side of the case and pry the ring gear into mesh with the pinion gear teeth, as far as possible. Rock the ring gear to allow the teeth to mesh and the bearings to seat. With the pressure still applied by the prybars, set the dial indicator to "0".

5. Reposition the prybars on the opposite side of ring gear and pry the gear as far as it will go. Take the dial indicator reading. Repeat this procedure until the same reading is obtained each time. This reading represents the necessary amount of shims between the differential carrier and the bearing on the ring gear side.

6. Remove the bearing from the differential carrier on the ring gear side and install the proper amount of shims. Reinstall the bearing.

7. Remove the differential carrier bearing from the opposite side of the ring gear. To determine the amount of shims needed, use the following method: Subtract the size of the shim pack just installed on the ring gear side of the carrier from the reading obtained and recorded when measurement was taken without the pinion gear in place during the Differential Carrier Assembly procedure. To this figure, add an additional 0.015 inch to compensate for preload and backlash. An example: If the first reading was 0.085 inch and the shims installed on the ring gear side of the carrier were 0.055 inch, the correct amount of shims whould be 0.085 − 0.055 + 0.015 = 0.045 inch.

8. Install the required shims as determined under Step 7 and install the differen-

C-4171 HANDLE

D-163 PINION SEAL INSTALLER

Installing pinion seal

tial side bearing. The installation of the shims should give the proper preload to the bearings and the proper backlash to the ring and pinion gears.

10¹/₂ INCH

1. Install the differential carrier, with the side bearings and cups installed, in place in the axle housing.

2. Select the smallest of the original shims as a gauging shim and place it between the bearing cup and the housing on the ring gear side.

3. Install the bearing caps and tighten the bolts enough to hold the cups in place.

4. Mount a dial indicator on the ring gear side of the axle housing and position the indicator button on the rear side of the ring gear.

5. Position two prybars between the bearing cup and the housing on the side opposite the ring gear. With the prybars, force the differential carrier towards the dial indicator and set the indicator dial to "0".

6. Reposition the prybars on the ring gear side of the carrier and force the ring gear into mesh with the pinion gear while observing the dial indicator reading. Repeat this operation until the same reading is obtained each time.

7. Add this indicator reading to the gauging shim thickness to determine the correct shim dimension for installation on the ring gear side of the differential carrier.

8. An example: If the gauging shim was 0.115 inch and the indicator reading was 0.017 inch, the correct shim would be 0.115 + 0.017 = 0.172 inch.

9. Remove the gauging shim and install the correct shim into position between the bearing cup and the axle housing on the ring gear side of the housing.

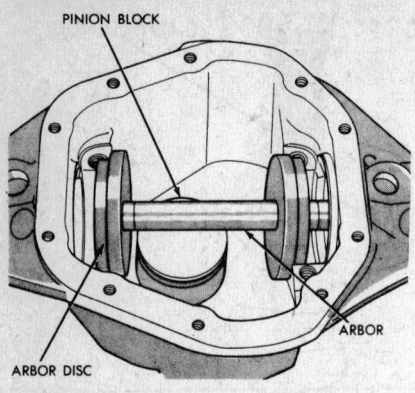

Arbor discs and arbor in position

10. To determine the correct dimension for the remaining shim, refer to the Side Bearing Shim Selection for the 10½ inch and obtain the recorded shim size. From that figure, subtract the size of the shim installed in Step 9 and then add 0.006 inch for the bearing preload and backlash.

11. An example: If the reading of the shim just installed on the ring gear side of the carrier was 0.172 inch and the reading obtained during the checking of clearance without the pinion installed was 0.329, the correct shim dimension would be as follows: 0.329 − 0.172 = 0.157 + 0.006 = 0.163 inch.

Installation of Differential Carrier Into Axle Housing

9¾ INCH

1. Spread the axle housing with the spreader tool no more than 0.015 inch. In-

stall the differential bearing outer cups in their correct locations and install the cups in their respective locations.

2. Install the bolts and tighten finger-tight. Rotate the differential carrier and ring gear and tap with a soft-faced hammer to insure proper seating of the assembly in the axle housing.

3. Remove the spreader tool and torque the cap bolts to specifications.

4. Install a dial indicator and check the ring gear backlash at four equally spaced points of the ring gear circle. The backlash must be within a range of 0.004 to 0.009 inch and must not vary more than 0.002 inch between the points checked.

5. If the backlash is not within specifications, the shim packs must be corrected to bring the backlash within limits.

6. Check the tooth contact pattern and verify.

7. Complete the assembly, fill to proper level with lubricant and operate to verify proper assembly.

10½ INCH

1. Spread the axle housing with a spreader tool, no more than 0.015 inch. The carrier assembly is in place in the housing.

2. Assemble the shim, as determined previously, into place between the bearing cup and the housing. Remove the spreader tool.

3. Install the bearing caps in their marked positions and torque the bolts to specifications.

4. Install a dial indicator and check the ring gear backlash at four equally spaced points around the ring gear.

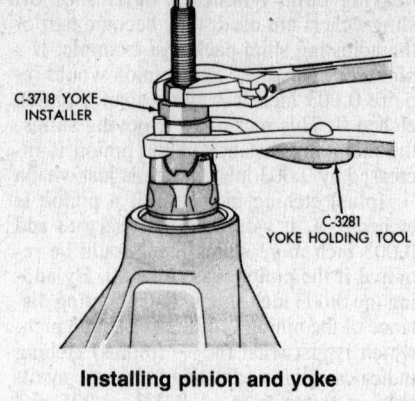

Installing pinion and yoke

NOMINAL ASSEMBLY DIMENSION AND ADAPTOR DISC CHART

Axle Model	Nominal Assembly Dimension	Adapter Disc Tool Number
44	2.625″	SE-1065-9-SS
60	3.125″	SE-1065-9-Y
70①	3.500″	SE-1065-9-Y

① Model 70—Use a 0.375 shim under dial pointer

5. The backlash must be within 0.004 to 0.009 inch and must not vary more than 0.002 inch between the positions checked.

6. Whenever the backlash is not within the allowable limits, it must be corrected. Changing of the shim packs is required: Low backlash is corrected by decreasing the shim on the ring gear side and increasing the opposite side shim an equal amount. High backlash is corrected by increasing the shim on the ring gear side and decreasing the opposite side shim an equal amount.

7. Check the tooth contact pattern and correct as required.

8. Complete the assembly, fill to the correct level and operate to verify correct repairs.

Dodge/Plymouth

8⅜ & 9¼ Inch Ring Gear

REMOVAL

1. Raise the rear of the vehicle and support safely.

2. Remove the wheels, drums and the housing cover screws. Drain the lubricant from the axle housing by removing the cover.

3. Turn the differential carrier case to make the differential pinion shaft lock screw accessible and remove it from the case. Slide the pinion shaft from the case.

4. Push both axle shafts towards the center of the axle assembly and remove the

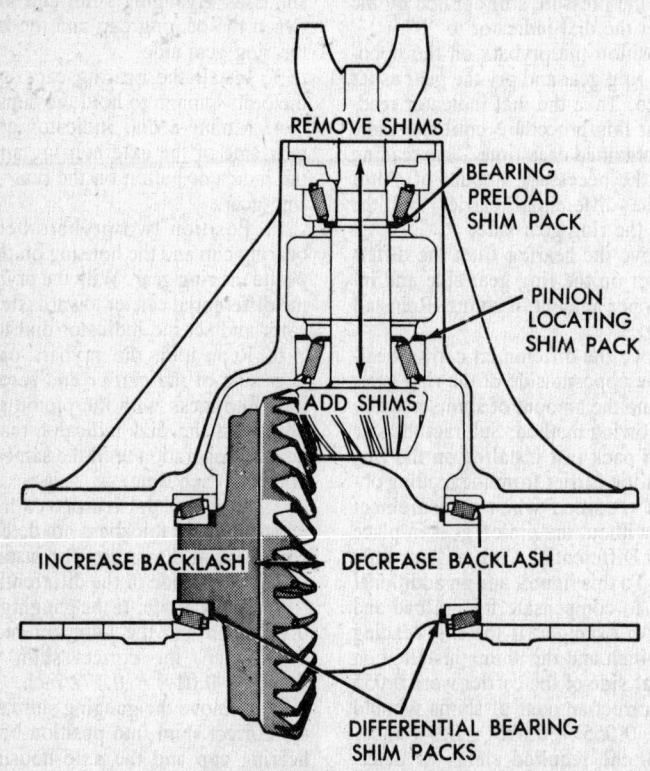

Using shims to adjust gear tooth contact

C-washer clips from the recessed grooves of the axle shafts. Withdraw the axle shafts carefully to avoid damaging the axle shaft bearings in the axle tubes.

5. Clean the inside of the differential case with solvent and blow dry with compressed air.

6. Check for differential side-play by inserting a prybar between the left side of the axle housing and the differential case flange. Using a prying motion, determine whether side-play exists. There should be no side-play.

7. Paint the ring gear teeth and make a gear tooth contact pattern. Determine if proper depth of mesh can be obtained.

8. If side-play was found in Step six, proceed to Step nine. If no side-play was found in Step six, check the drive gear run-out. Mount a dial indicator and index the indicator stem at right angles in the rear face of the ring gear. Rotate the ring gear and mark the ring gear and case at the point of greatest run-out. Total indicator reading should not exceed 0.005 in. If it does, the possibility exists that the case must be replaced.

9. Measure and record the pinion bearing preload. Use an inch lb. torque wrench to measure the preload.

10. Remove the pinion nut, washer and pinion flange.

11. Remove and discard the pinion oil seal.

12. Match-mark the axle housing and the differential bearing caps.

13. Remove the threaded adjusters and the differential bearing caps. There is a special wrench to do this through the axle tube.

14. Remove the differential case from the housing. The differential bearing cups and threaded adjuster must be kept together so they can be installed in their original position.

Disassembly

1. To remove the drive pinion or front bearing cone, drive the pinion rearward out of the bearing. This will result in damage to the bearing and cup. The bearing cone and cup must be replaced with new parts. Discard the collapsible spacer.

2. Drive the front and rear bearing cups from the housing with a brass drift. Remove the shim from behind the rear bearing cup and record the thickness.

3. Remove the rear bearing cone from the pinion stem with a puller.

4. Clamp the differential case and ring gear in a vise with soft jaws.

5. Remove the ring gear bolts (left-hand thread). Tap the ring gear loose with a soft-faced mallet.

6. If the ring gear run-out exceeded 0.005 in., recheck the case as follows. Install the differential case, cups, caps, and adjusters in the housing. Turn the adjusters to eliminate all side-play and tighten the differential cap bolts snugly. Measure the run-out at the ring gear flange face. Total indicator reading should not exceed 0.003

Removing differential pinion shaft lock pin

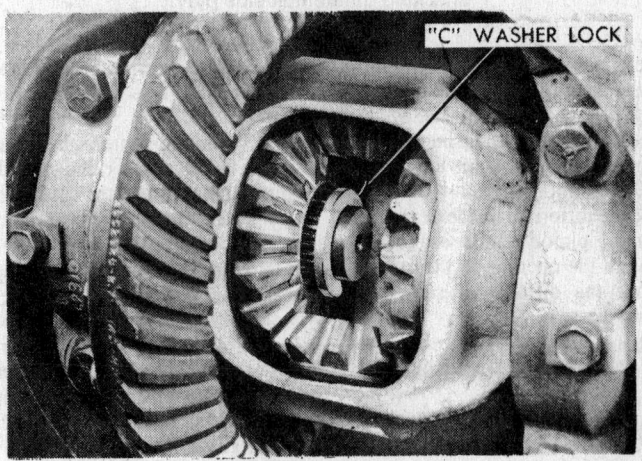

Removing axle shaft C-locks

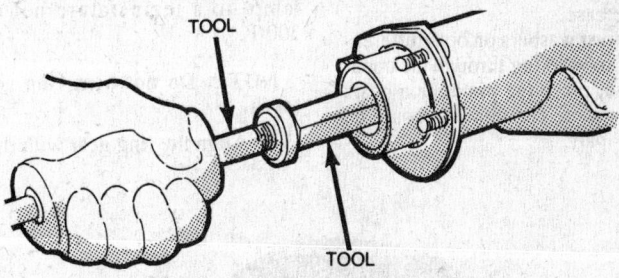

Removing axle shaft bearing and seal

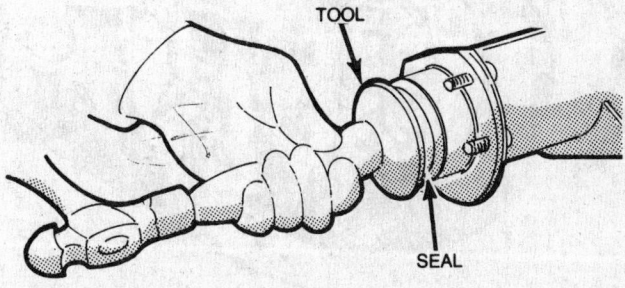

Installing axle shaft oil seal

PINCH BAR

Checking differential side play

in. It is often possible to reduce run-out by removing the ring gear and remounting 180° from its original position. Remove the differential case from the housing.

7. Remove the pinion shaft lock-screw and remove the pinion shaft.

8. Rotate the differential side gears until the differential pinion shafts can be removed through the opening in the case.

9. Remove the differential side gears and thrust washers.

10. Using a puller or a press and press plates, remove the differential side bearings.

Assembly

1. Lubricate all parts, before assembly, with rear axle lubricant.

2. Install the thrust washers on the differential side gears and install the side gears into the case.

3. Place thrust washers on both differential pinions and, working through the opening in the case, mesh the pinion gears with the side gears. The pinions should be exactly 180° apart.

4. Rotate the side gears 90° to align the pinions and thrust washers with the pinion shaft holes.

5. From the pinion shaft lockpin hole side of the case, insert the slotted end of the pinion shaft through the case, conical thrust washer and just through one of the pinion gears.

6. Install a thrust block through the side gear hub, so that the slot is centered between the side gears.

7. Hold all these parts in alignment, and align the lockpin holes in the pinion shaft and case. Install the lockpin from the pinion shaft side of the ring gear flange, temporarily.

8. With a stone, relieve the edge of the chamfer on the inside diameter of the ring gear.

9. Heat the ring gear (fluid bath or heat lamp) to a temperature not exceeding 300°F.

NOTE: Do not heat ring gear with a torch.

10. Align the ring gear with the case. Insert the ring gear screws through the case flange and into the ring gear.

11. Alternately tighten each cap screw to 70 ft. lbs.

12. Position each differential bearing cone on the hub of the differential case (taper away from ring gear) and install the bearing cones. An arbor press may be helpful.

Pinion Depth of Mesh

1. The proper pinion setting (relative to the ring gear) is determined by a shim which has been selected before the pinion is to be installed in the carrier. Pinion bearing shims are available in 0.001 in. increments.

2. The head of the pinion is marked with a "plus" (+) or a "minus" (−) mark that is followed by a number ranging from zero to four. If the old and new pinions have the same marking and the old bearing is being installed, use a shim of the original thickness. If the old pinion is marked zero (0), however, and the new pinion is marked plus two (+ 2), try a shim that is 0.002 in. thinner. If the new pinion is marked axle housing cup bore and install minus two (− 2), try a shim that is 0.002 in. thicker.

3. Position the selected shim in the bore of the rear bearing cup. Install the cup.

NOTE: Special pinion depth measuring tools are available for both the 8³⁄₈ and 9¹⁄₄ inch axles. When using the special tools, follow the manufacturer's recommended procedures. Without the special tools, complete the following procedure and check the pinion depth by examining the pinion to ring gear tooth contact pattern. Correct as required by adding or subtracting shims controlling the pinion depth.

4. Place the rear pinion bearing cone on the pinion stem (small side away from pinion head).

5. Lubricate the front and rear bearing cones and install the rear pinion bearing cone onto the pinion stem with an arbor press.

6. Insert the pinion bearing and collapsible spacer assembly through the carrier and install the front bearing cone. Install the companion flange.

NOTE: During installation of the pinion bearing do not collapse the spacer.

7. Install the drive pinion oil seal into the carrier. Be sure to properly seat the seal.

8. Support the pinion in the carrier.

9. Install the Belleville washer (convex side up) and pinion nut.

10. Hold the companion flange and tighten the pinion nut to remove all end-play, while rotating the pinion to ensure proper bearing seating. Remove the tools and rotate the pinion several revolutions.

11. Torque the pinion nut to 210 ft. lbs.

Measuring drive gear runout

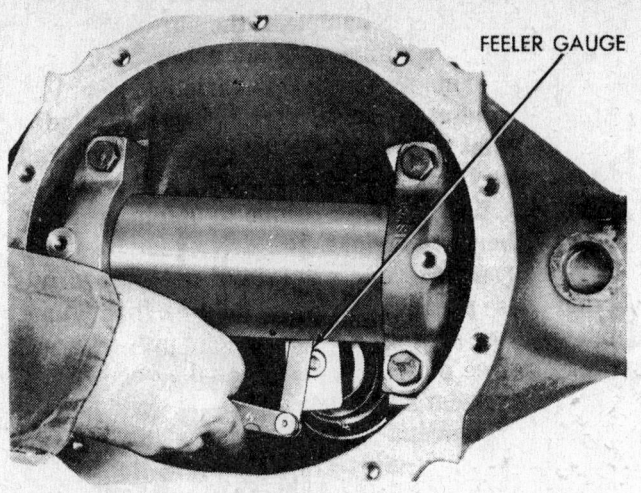

Determining drive pinion depth mesh shim pack thickness

Putting identification marks on bearing caps and housing

With an inch lbs. torque wrench, measure the pinion bearing preload, which would be 20–35 inch lbs. for new bearings or 10 inch lbs. over the original figure if the old pinion bearing is used.

NOTE: The correct preload reading can only be obtained with the carrier nose upright. The final assembly is incorrect if the final pinion nut torque is below 210 ft. lbs. or if the pinion bearing preload is not within specifications. Under no circumstances should the pinion nut be backed off to reduce the pinion bearing preload; if this is done, a new collapsible spacer will have to be installed and the unit adjusted again until proper preload is obtained.

DIFFERENTIAL BEARING PRELOAD AND RING GEAR-TO-PINION BACKLASH

The threaded adjuster uses a hex drive hole, and requires special tool C–4164 to adjust the side bearing preload through the axle tube. An adjuster lock with two pointed teeth which engage in the exposed adjuster thread when the lock is tightened is provided. The shims will range from 0.020–+0.038 in. and will be equipped with internal centering tabs. The shims, marked with a number which represents its thickness in thousandths of an inch, can be installed with either side against the pinion head.

1. Index the gears so that the same gear teeth are in contact throughout the adjustment.

2. The differential bearing cups will not always move with the adjusters. It is important to seat the bearings by rotating them 5–10 times in each direction, each time the adjusters are moved.

3. With the pinion bearings installed and the preload set, install the differential with adjusters, caps and bearings. Lubricate the bearings and adjuster threads. Check to be sure that there are no crossed threads. Tighten the top cap screws on the right and left to 10 ft. lbs. Tighten the bottom cap screws fingertight until the head is just seated on the bearing cap.

4. Using the tool, check to be sure that the adjuster rotates freely. Turn both adjusters in until bearing play is eliminated with some drive gear backlash (0.010 in.). Seat the bearing rollers.

5. Install and register a dial indicator against the drive side of a gear tooth. Check the backlash at four positions to find the point of minimum backlash. Rotate the gear to the position of least backlash and

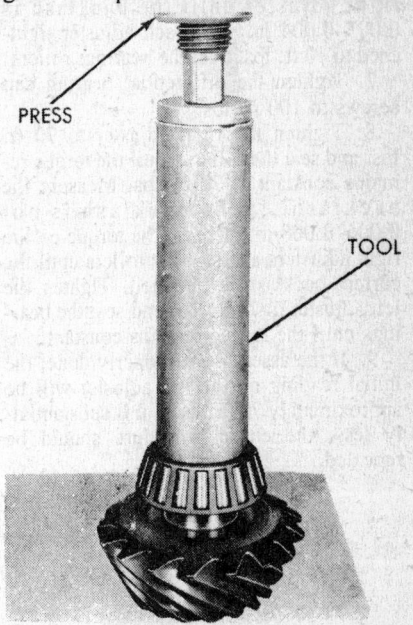

Installing drive pinion rear bearing cone

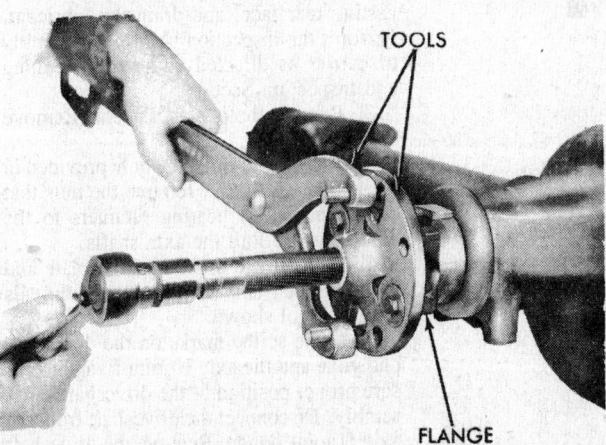

Loosening or tightening hex adjuster

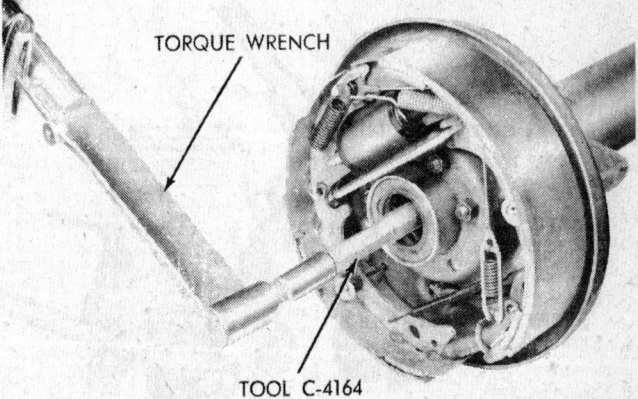

Removing drive pinion companion flange

Checking drive gear mounting flange face

mark the tooth so that all readings will be taken at the same point.

6. Loosen the right adjuster and turn the right adjuster until the backlash is 0.003–0.004 in. with each adjuster tightened to 10 ft. lbs. Seat the bearings rollers.

7. Tighten the differential bearing cap screws to 100 ft. lbs.

8. Tighten the right adjuster to 70 ft. lbs. and seat the rollers, until the torque remains constant at 70 ft. lbs. Measure the backlash. If the backlash is not 0.006–0.008 in. increase the torque on the right adjusters and seat the rollers until the correct backlash is obtained. Tighten the left adjuster to 70 ft. lbs. and seat the bearings until the torque remains constant.

9. If the assembly is properly done, the initial reading on the left adjuster will be approximately 70 ft. lbs. If it is substantially less, the entire procedure should be repeated.

10. After adjustments are complete, install the adjuster locks. Be sure the teeth are engaged in the adjuster threads. Torque the lockscrews to 90 inch lbs.

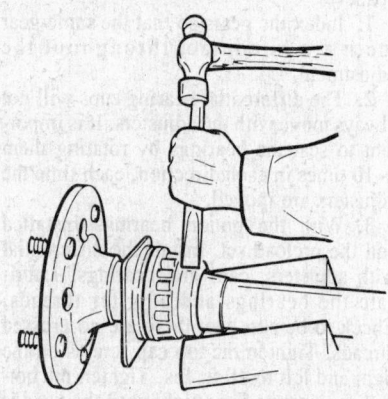

Removing rear wheel bearing retaining ring

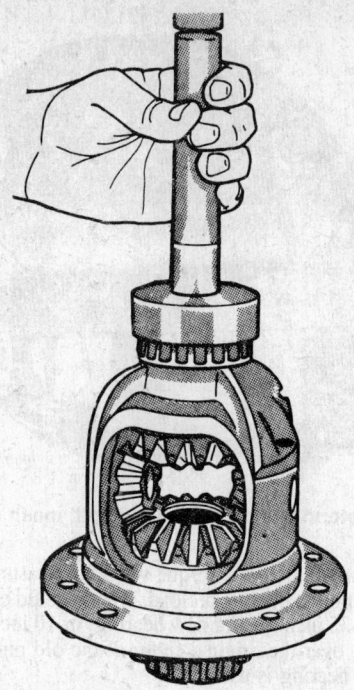

Installing differential bearing cone

Final Assembly

1. Install the axle shafts, C-clips, reinstall the pinion shaft and lock screw and tighten securely.

2. Install the cover on the differential housing, using a new gasket.

3. Refill the rear axle housing with lubricant.

Ford

Bronco II/Ranger 6¾ Inch Ring Gear

DISASSEMBLY

1. Raise the vehicle and support it on the underbody, so that the rear axle drops down as far as the springs and shock absorbers permit.

2. Remove the cover from the carrier casting rear face, and drain the lubricant. Perform the inspection before disassembly of carrier as directed under the Cleaning and Inspection Section.

3. Remove both rear wheels. Remove the brake drums.

4. Working through the hole provided in the axle shaft flange, remove the nuts that attach the wheel bearing retainers to the axle housing. Pull the axle shafts.

5. Remove the brake backing plate and wire it to the frame rail. Remove both seals with the tool shown.

6. Make scribe marks on the driveshaft end yoke and the axle U-joint flange to ensure proper position of the driveshaft at assembly. Disconnect the driveshaft from the axle U-joint flange. Remove the driveshaft from the transmission extension housing.

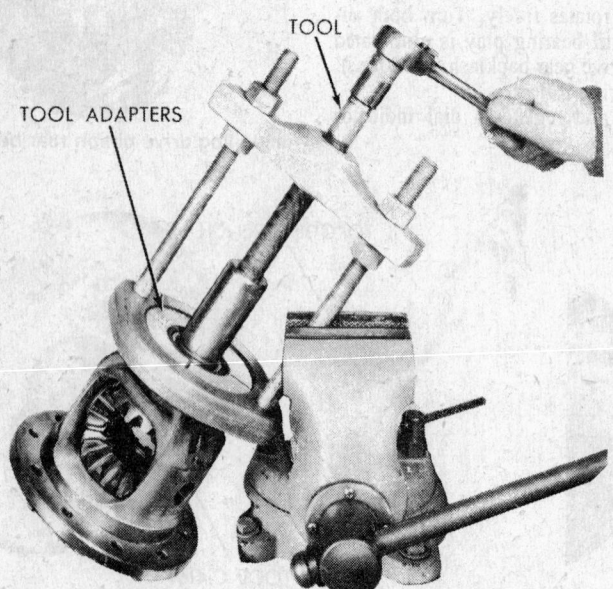

TOOL

TOOL ADAPTERS

Removing differential bearing cone

Install an oil seal replacer tool in the transmission extension housing to prevent transmission leakage.

7. Check and record the ring gear runout and the ring gear backlash.

8. Mark one differential bearing cap and the case to help ensure proper position of the parts during assembly.

9. Loosen the differential bearing cap bolts and bearing caps. Pry the differential case, bearing cups and shims out until they are loose in the bearing caps. Remove the bearing caps and the differential assembly from the carrier.

NOTE: The direction of arrows on bearing caps must be noted. When re-assembled, the arrows must be pointing outward.

10. Mark the companion flange in relation to the pinion shaft. Hold the rear axle companion flange with the proper tool and remove the pinion nut. If a new gear set is being installed, the companion flange need not be marked.

11. Remove the companion flange.

12. With a soft-faced hammer, drive the pinion out of the front bearing cone and remove it through the rear of the carrier casting.

13. Remove the pinion seal and front bearing cone from the front of the carrier casting.

DRIVE PINION

Disassembly

1. Remove the drive pinion, front bearing cone, spacer, and seal.

2. Remove the pinion rear bearing cone.

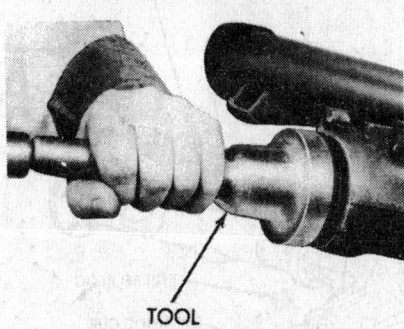

TOOL

Installing drive pinion oil seal

Measure the shim found under the bearing cone with a micrometer. Record the thickness of the shim.

NOTE: Before assembling the rear bearing cone to the pinion, it will be necessary to adjust Pinion Depth.

DRIVE PINION DEPTH ADJUSTMENT

Individual differences in machining the carrier casting and the gear set and variation in bearing widths require a shim between the pinion rear bearing and pinion

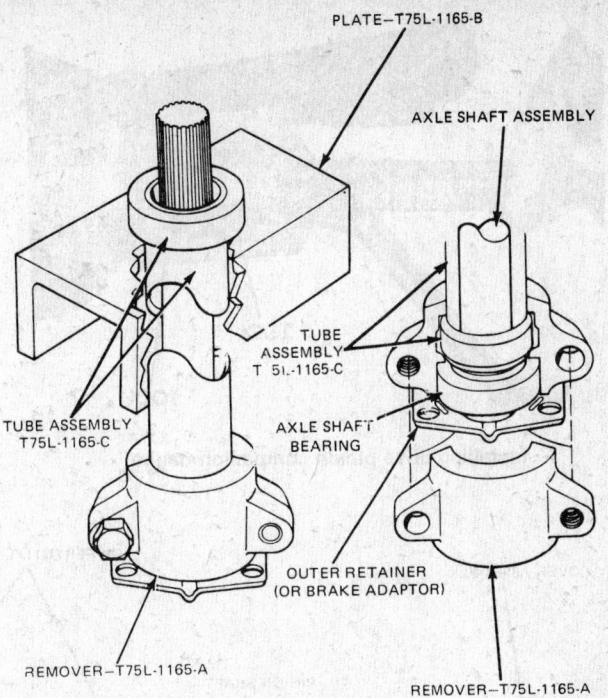

PLATE—T75L-1165-B

AXLE SHAFT ASSEMBLY

TUBE ASSEMBLY T 5L-1165-C

AXLE SHAFT BEARING

TUBE ASSEMBLY T75L-1165-C

OUTER RETAINER (OR BRAKE ADAPTOR)

REMOVER—T75L-1165-A

REMOVER—T75L-1165-A

Bearing remover installed

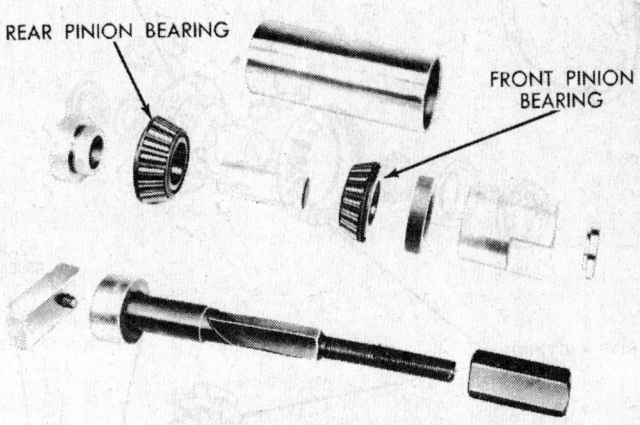

REAR PINION BEARING

FRONT PINION BEARING

8⅜ inch rear axle setting gauge tool

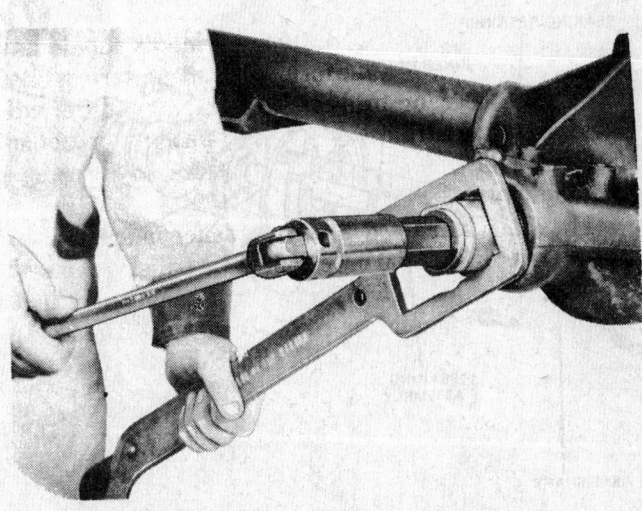

Seating bearing caps in axle housing

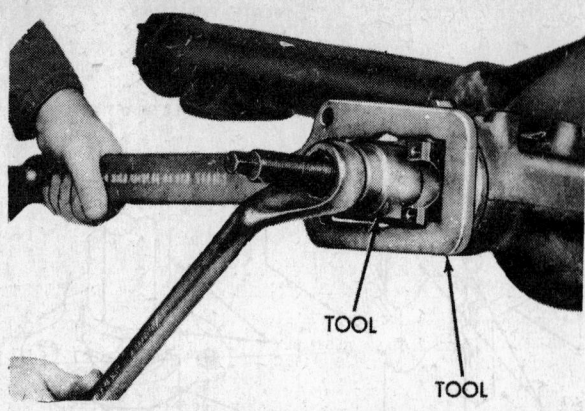

Installing drive pinion companion flange

head, in order to locate the pinion for correct tooth contact with the ring gear. When replacing a ring and pinion gear, the correct shim thickness for the new gear set to be installed is determined by following procedure using Tool T79P-4020-A or equivalent.

1. Place the rear pinion bearing (new or used if in good condition) over the aligning disc and insert it into the pinion bearing cup of the carrier. Place the front bearing into the front bearing cup and assemble the tool handle into the screw and tighten to 20 ft. lbs.

NOTE: The gauge block must be offset to obtain an accurate reading.

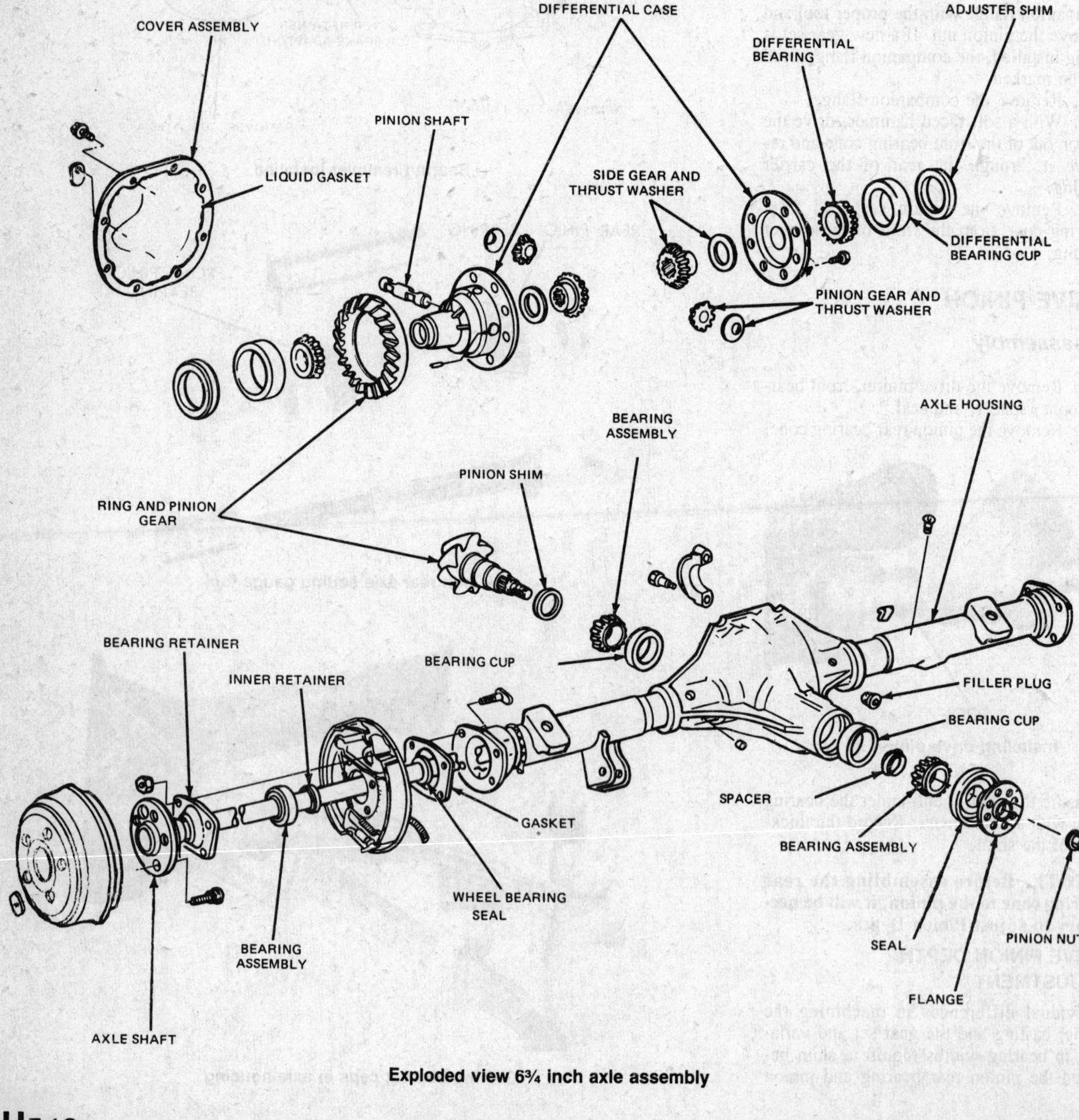

Exploded view 6¾ inch axle assembly

- COVER ASSEMBLY
- LIQUID GASKET
- PINION SHAFT
- DIFFERENTIAL CASE
- SIDE GEAR AND THRUST WASHER
- DIFFERENTIAL BEARING
- ADJUSTER SHIM
- DIFFERENTIAL BEARING CUP
- PINION GEAR AND THRUST WASHER
- RING AND PINION GEAR
- PINION SHIM
- BEARING ASSEMBLY
- AXLE HOUSING
- BEARING CUP
- FILLER PLUG
- BEARING CUP
- BEARING RETAINER
- INNER RETAINER
- SPACER
- BEARING ASSEMBLY
- GASKET
- WHEEL BEARING SEAL
- BEARING ASSEMBLY
- AXLE SHAFT
- SEAL
- PINION NUT
- FLANGE

2. Center the gauge tube into the differential bearing bore. Install the bearing caps and torque the bolts to specification. (Caps to be installed with the arrows pointing outboard). Shims must be flat. Do not use dirty, bent, nicked or mutilated shims as a gauge. If the pinion has a plus (+) marking, subtract this amount from the feeler gauge measurement. If the pinion has a minus (−) marking, add this amount to the feeler gauge measurement.

DRIVE PINION REAR BEARING

NOTE: The same rear pinion bearing used in this procedure must be used in final assembly of the axle.

1. Place the selected shim(s) on the pinion shaft and press the pinion bearing until firmly seated on the shaft.

PINION BEARING CUPS

Do not remove the pinion bearing cups from the carrier casting unless the cups are to be replaced; drive them out of the carrier casting with a drift. Install the new cups with tool T71P–4616–A or equivalent. Make sure the cups are properly seated in their bores. If a 0.0015 inch feeler gauge can be inserted between a cup and the bottom of its bore at any point around the cup, the cup is not properly seated. Whenever the cups are replaced, the cone and roller assemblies should also be replaced.

DIFFERENTIAL CASE

Disassembly

1. If the differential bearings are to be removed, use tool T77F–4220–B or equivalent. Mark the differential case, cover and ring gear for assembly in the original position.
2. Remove the bolts that attach the ring gear to the differential case and discard them. Press the ring gear from the case or tap it off with a soft-faced hammer.
3. Remove the left side of the differential case.
4. With a drift, drive out the differential pinion shaft lock pin.
5. Drive out the pinion shaft with a brass drift. Remove the gears and thrust washers. Clean and inspect all the parts. Repair or replace all parts as indicated by the inspection.

Assembly

1. Install the differential side gears and thrust washers in their bores.
2. Install the pinion shaft aligning the pinion gears and thrust washers.
3. Install the left case on the right and tap them together.
4. Install the pinion shaft lockpin.
5. Install the differential bearing.

ASSEMBLY

The drive pinion should be set to the correct depth before final assembly setting is prop-

BACKLASH CHANGE REQUIRED (INCH)	THICKNESS CHANGE REQUIRED (INCH)	BACKLASH CHANGE REQUIRED (INCH)	THICKNESS CHANGE REQUIRED (INCH)
.001	.002	.009	.012
.002	.002	.010	.014
.003	.004	.011	.014
.004	.006	.012	.016
.005	.006	.013	.018
.006	.008	.014	.018
.007	.010	.015	.020
.008	.010		

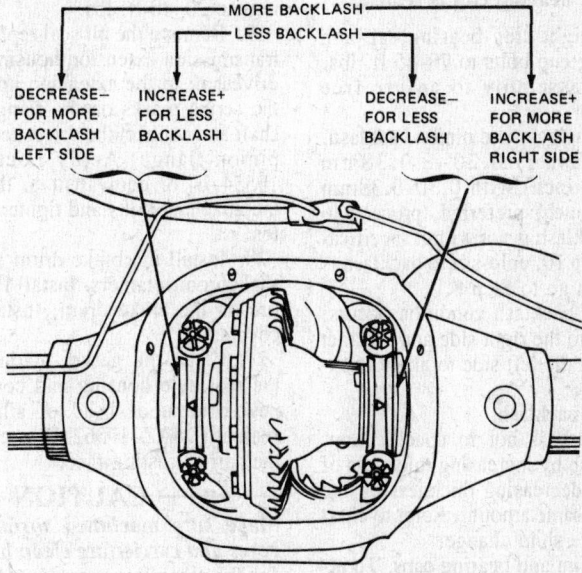

MORE BACKLASH
LESS BACKLASH

DECREASE− FOR MORE BACKLASH LEFT SIDE
INCREASE+ FOR LESS BACKLASH
DECREASE− FOR LESS BACKLASH
INCREASE+ FOR MORE BACKLASH RIGHT SIDE

Shim changes for ring gear and pinion backlash

er. Drive pinion preload requires that a new spacer be used when the pinion is removed. Drive pinion bearing preload is set with the drive pinion assembly installed and the pinion nut tightened to specification. Correct preload is indicated when the rotational torque is 8–14 inch lbs. with original bearings or 16–29 inch lbs. for new bearings.

Differential Bearing Preload and Ring Gear Backlash is adjusted with the drive pinion and differential case installed. Adjustment is performed by the installation of shims between the differential bearing cup and axle housing.

DRIVE PINION AND DRIVE PINION BEARING PRELOAD ADJUSTMENT

1. Install the pinion front bearing.
2. Install the pinion seal.
3. Insert the companion flange into the seal and hold it firmly against the pinion front bearing cone. From the rear of the carrier casting, insert the pinion shaft with a new spacer into the flange.
4. Start a new pinion nut. Hold the flange and tighten the pinion shaft nut. As the nut is tightened, the pinion shaft is pulled into the front bearing cone and into the flange. As the pinion shaft is pulled into the front bearing cone, pinion shaft end play is reduced. While there is still end play in the pinion shaft, the flange and bearing cone will be felt to bottom on the

collapsible spacer. From this point, a much greater torque must be applied to turn the pinion nut, since the spacer must be collapsed. Very slowly, tighten the nut, but check the pinion shaft end play often, to see that the pinion bearing preload does not exceed the limits. If the pinion nut is tightened to the point that pinion bearing preload exceeds the limits, the pinion shaft must be removed and a new collapsible spacer installed. **Do not decrease the preload by loosening the pinion nut.** This will remove the compression between the pinion front and rear bearing cones and the collapsible spacer and may permit the front bearing cone to turn on the pinion shaft.

5. As soon as there is a preload on the bearings, **turn the pinion shaft in both directions several times to set the bearing rollers.**
6. Adjust the bearing preload to specification. Measure the preload with a torque wrench.

DIFFERENTIAL ASSEMBLY

For shim selection after a complete replacement of the rear axle housing, the differential assembly or differential side bearings, use the following instructions. For a ring and pinion replacement only or a backlash adjustment, follow Steps 9 through 13 and Step 15, using the side bearing shims that were originally in the axle.

1. With pinion depth set and pinion installed, place differential case and gear assembly with bearings and cups in carrier.

2. Install a 6.73mm (0.265 inch) shim on left side.

3. Install left bearing cap and tighten bolts finger tight.

4. Install progressively larger shims on the right side until the largest shim selected can be assembled with a slight drag feel.

NOTE: Apply pressure towards left side to ensure bearing cup is seated.

5. Install right side bearing cap and tighten bearing cup bolts to 70–85 ft. lbs.

6. Rotate assembly to ensure free rotation.

7. Check ring gear and pinion backlash. If the backlash is 0.20–+0.38mm (0.008–0.015 inch) with 0.30–0.38mm (0.012–0.015 inch) preferred, proceed to Step 14. If backlash is not within specifications, go to Step 10, unless zero backlash is measured, then go to Step 8.

8. If a zero backlash condition occurs, add .020 inch to the right side and subtract .020 inch from the left side to allow backlash indication.

9. Recheck backlash.

10. If backlash is not to specification, correct backlash by increasing thickness of one shim and decreasing thickness on the other shim the same amount. Refer to chart for approximate shim change.

11. Install shim and bearing caps. Tighten cap bolts to 70–85 ft. lbs.

12. Rotate assembly several times.

13. Recheck backlash. If backlash is within specification, go to Step 14. If backlash is not within specification, repeat Step 10. Backlash specification is 0.20–0.38mm (0.008–0.015 inch). Preferred range is 0.30–0.38mm (0.012–0.015 inch.

14. Increase both left and right shim sizes by .006 and install for correct differential bearing preload. Make sure shims are fully seated and assembly turns freely.

15. Utilize white marking compound to obtain a tooth mesh contact pattern in your assembly. Reincorporation of pattern inspection is intended to allow technicians the ability to detect gross errors in set up prior to complete reassembly. Pattern contact should be within the primary area of the ring gear tooth surface, avoiding any "narrow" or "hard" contact with outer perimeter of tooth (top to root, toe to heel). Pattern inspection should be on the drive (pull) side. Correct assembly of drive pattern will result in satisfactory coast performance. If gross pattern error is detected, with preferred backlash 0.30–0.38mm (0.012–0.015 inch), recheck pinion shim selection.

16. Install bearing caps and tighten cap bolts to 70–85 ft. lbs.

17. Inspect the machined surfaces of the axle shaft and the axle housing for rough spots or other irregularities which would affect the sealing action of the oil seal. Check the axle shaft splines for burrs, wear

or damage. Carefully remove any burrs or rough spots. Replace worn or damaged parts. Install a new gasket on the housing flange and install the brake backing plate.

18. Carefully slide the axle shaft into the housing so that the rough forging of the shaft will not damage the oil seal. Start the axle splines into the side gear, and push the shaft in until the bearing bottoms in the housing. Install the bearing retainer plate on the mounting bolts at the axle housing, and install the attaching nuts. Tighten the nuts to 20–40 ft. lbs.

19. Remove the oil seal replacer from the transmission extension housing. Install the driveshaft in the extension housing. Align the scribe marks on the flange and driveshaft and connect the driveshaft at the drive pinion flange. Apply Loctite (E0AZ-19554–B) or equivalent to the threads of the attaching bolts and tighten to 70–95 ft. lbs.

20. Install the brake drum and attaching shakeproof retainers. Install the wheel and tire on the brake drum. Install the wheel covers.

21. Clean the gasket mating surface of the rear axle housing and cover. Apply a new continuous bead of silicone rubber sealant (D6AZ-19562–B or equivalent) to the carrier casting face.

--- CAUTION ---
Make sure machined surfaces on both cover and carrier are clean before installing the new silicone sealant. Inside of axle must be covered when cleaning the machined surface to prevent axle contamination.

22. Install cover and tighten cover bolts to 25–35 ft. lbs., except the ratio tag bolt, which is tightened to 15–25 ft. lbs.

NOTE: Cover assembly must be installed within 15 minutes of application of the silicone or new sealant must be applied.

23. Add E0AZ-19580–A (ESP-M2C154–A) lubricant or equivalent through the filler hole until the lubricant level reaches the bottom of the filler hole with the axle in the running position. Install filler plug and tighten to 15–30 ft. lbs.

24. Lower vehicle and road test.

Bronco II/Ranger 7½ Inch Ring Gear

NOTE: The Aerostar is equipped with a 7½ in. ring gear and the service procedures are the same as outlined below, except for a few minor changes. There are now two new special tools used in the rear axle service procedures. Tool T85L-4067–AH is a driver which is designed to install shims as needed to adjust the ring gear and pinion backlash. The backlash specification is 0.001 to 0.015 in., but 0.012 to 0.015 in. is preferred. Tool T85L-1225–AH is is used to remove the wheel bearing seal from the

axle. Both of these tools or their equivalent are needed to complete the rear axle overhaul on the Aerostar.

DISASSEMBLY

All service operations on the differential case assembly and the drive pinion can be performed with the axle housing installed in the vehicle.

1. Raise the vehicle and place jack stands under the rear frame crossmember. Lower the hoist so that the axle drops down far enough for working case.

2. Remove the cover from the carrier casting rear face and drain the lubricant. Inspect the case assembly and drive pinion before removal.

3. Remove the rear wheels and brake drums.

4. Remove the axle shafts.

5. Make scribe marks on the driveshaft and yoke and the rear axle companion flange to ensure proper alignment at assembly. Disconnect the driveshaft from the rear axle companion flange. Remove the driveshaft assembly from the vehicle. Insert an oil seal replacement tool in the transmission extension housing to prevent leakage.

6. Check and record the ring gear runout. Check and record the ring gear backlash.

7. Mark on differential bearing cap to help position the caps properly during assembly.

8. Loosen the differential bearing cap bolts and bearing caps.

NOTE: The direction of arrows on bearing caps must be noted. When re-assembled, the arrows must be pointing in the same direction as before removal.

9. Pry the differential case, bearing cups and shims out until they are loose in the bearing caps. Remove the bearing caps and remove the differential assembly out of the carrier. On conventional differentials, if the ring is removed, discard the bolts. Install new bolts, coated with Loctite or equivalent. Tighten to 70–85 ft. lbs.

10. Mark the companion flange in relation to the pinion shaft. Hold the rear axle companion flange with the proper tool and remove the pinion nut. If a new gear set is being installed, the companion flange need not be marked.

11. Remove the companion flange. With a soft-faced hammer, drive the pinion out of the front bearing cone and remove it through the rear of the carrier casting.

12. Remove the drive pinion oil seal with Tool-1125–AC and T50T-100–A or their equivalent. Remove the front pinion bearing cone and roller and slinger from carrier casting.

DRIVE PINION

Disassembly

1. Remove the drive pinion, front bearing cone, spacer, and seal.

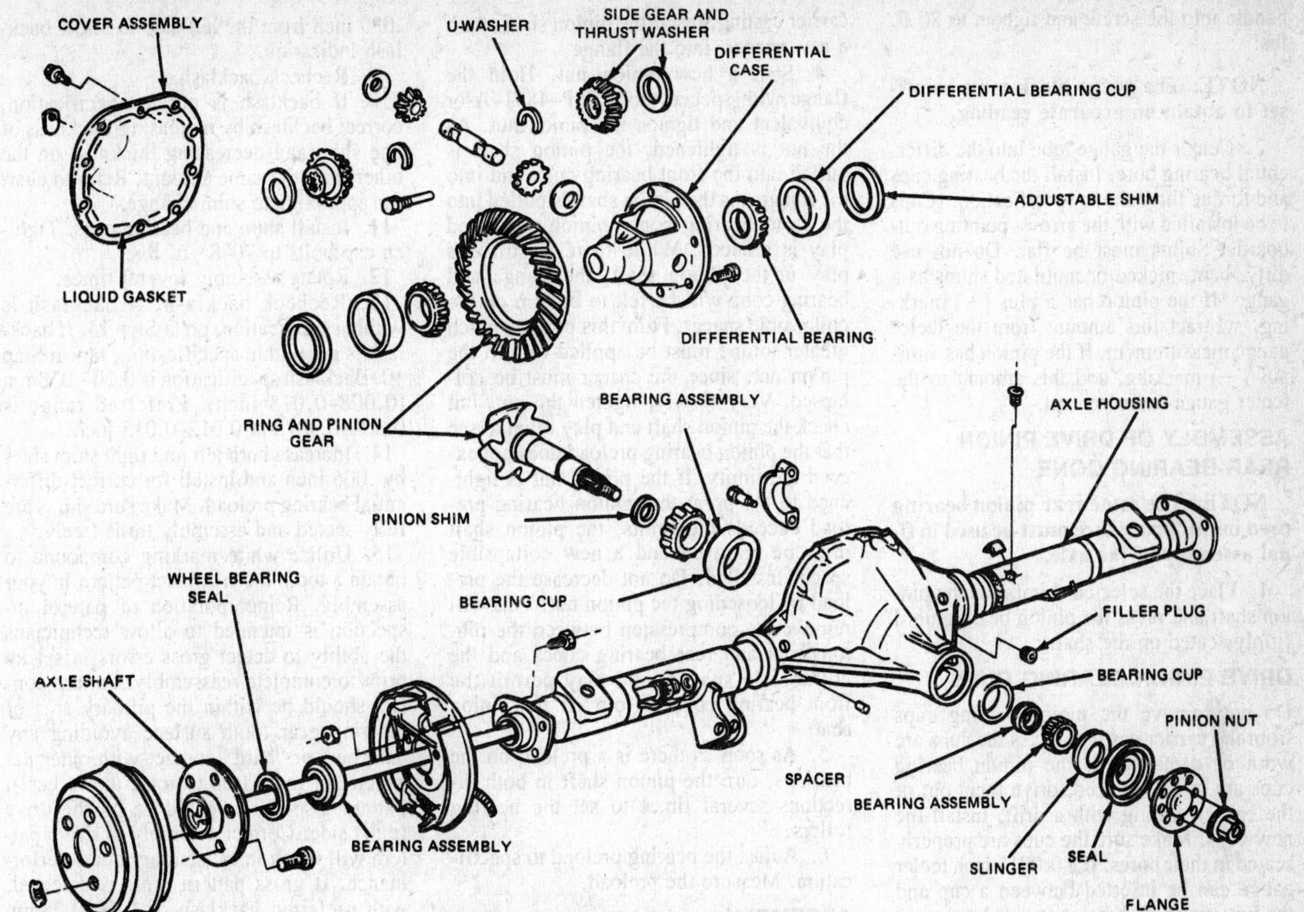

Exploded view diagram labels:

COVER ASSEMBLY

U-WASHER

SIDE GEAR AND THRUST WASHER

DIFFERENTIAL CASE

DIFFERENTIAL BEARING CUP

ADJUSTABLE SHIM

LIQUID GASKET

AXLE HOUSING

DIFFERENTIAL BEARING

RING AND PINION GEAR

BEARING ASSEMBLY

PINION SHIM

BEARING CUP

WHEEL BEARING SEAL

BEARING CUP

FILLER PLUG

BEARING CUP

PINION NUT

AXLE SHAFT

BEARING ASSEMBLY

SPACER

BEARING ASSEMBLY

SLINGER

SEAL

FLANGE

Exploded view 7½ inch axle assembly

2. To remove the pinion rear bearing cone, use tool T71P–4621–B or equivalent. Measure the shim found under the bearing cone with a micrometer. Record the thickness of the shim.

NOTE: Before assembling the rear bearing cone to the pinion, it will be necessary to adjust pinion depth.

DRIVE PINION DEPTH ADJUSTMENT

Individual differences in machining the carrier casting and the gear set and variation in bearing widths require a shim between the pinion rear bearing and pinion head, in order to locate the pinion for correct tooth contact with the ring gear. When replacing a ring and pinion gear, the correct shim thickness for the new gear set to be installed, is determined by following procedure using tool T79P–4020–A or equivalent.

PINION DEPTH TOOL SET

1. Place the rear pinion bearing (new or used if in good condition) over the aligning disc and insert it into the pinion bearing cup of the carrier. Place the front bearing into the front bearing cup and assemble the tool

SHIM CODE CHART

NUMBER OF STRIPES AND COLOR CODE	DIM. A
2 — C-COAL	.3070–.3075
1 — C-COAL	.3050–.3055
5 — BLU	.3030–.3035
4 — BLU	.3010–.3015
3 — BLU	.2990–.2995
2 — BLU	.2970–.2975
5 — PINK	.2930–.2935
4 — PINK	.2910–.2915
3 — PINK	.2890–.2895
2 — PINK	.2870–.2875
1 — PINK	.2850–.2855
5 — GRN	.2830–.2835
4 — GRN	.2810–.2815
3 — GRN	.2790–.2795
2 — GRN	.2770–.2775
1 — GRN	.2750–.2755
5 — WH	.2730–.2735
4 — WH	.2710–.2715
3 — WH	.2690–.2695
2 — WH	.2670–.2675
1 — WH	.2650–.2655
5 — YEL	.2630–.2635
4 — YEL	.2610–.2615
3 — YEL	.2590–.2595
2 — YEL	.2570–.2575
1 — YEL	.2550–.2555
5 — ORNG	.2530–.2535
4 — ORNG	.2510–.2515
3 — ORNG	.2490–.2495
2 — ORNG	.2470–.2475
1 — ORNG	.2450–.2455
2 — RED	.2430–.2435
1 — RED	.2410–.2415

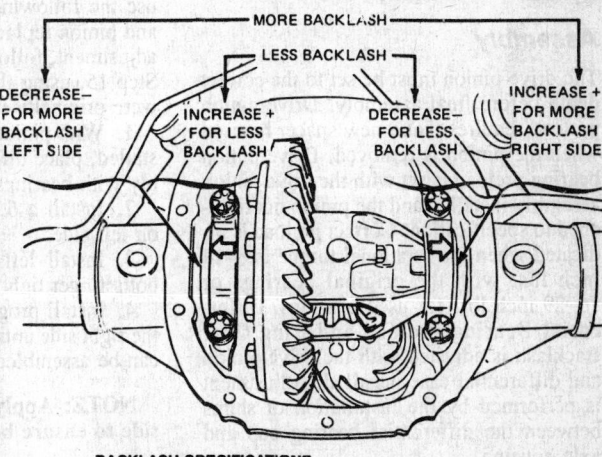

BACKLASH ADJUSTMENTS

MORE BACKLASH

LESS BACKLASH

DECREASE— FOR MORE BACKLASH LEFT SIDE

INCREASE + FOR LESS BACKLASH

DECREASE— FOR LESS BACKLASH

INCREASE + FOR MORE BACKLASH RIGHT SIDE

BACKLASH SPECIFICATIONS

BACKLASH CHANGE REQUIRED (INCH)	THICKNESS CHANGE REQUIRED (INCH)	BACKLASH CHANGE REQUIRED (INCH)	THICKNESS CHANGE REQUIRED (INCH)
.001	.002	.009	.012
.002	.002	.010	.014
.003	.004	.011	.014
.004	.006	.012	.016
.005	.006	.013	.018
.006	.008	.014	.018
.007	.010	.015	.020
.008	.010		

Shim changes for ring gear and pinion backlash

handle into the screw and tighten to 20 ft. lbs.

NOTE: The gauge block must be offset to obtain an accurate reading.

2. Center the gauge tube into the differential bearing bore. Install the bearing caps and torque the bolts to specification. (Caps to be installed with the arrows pointing outboard.) Shims must be flat. Do not use dirty, bent, nicked or mutilated shims as a gauge. If the pinion has a plus (+) marking, subtract this amount from the feeler gauge measurement. If the pinion has a minus (−) marking, add this amount to the feeler gauge measurement.

ASSEMBLY OF DRIVE PINION REAR BEARING CONE

NOTE: The same rear pinion bearing used in this procedure must be used in final assembly of the axle.

1. Place the selected shim(s) on the pinion shaft and press the pinion bearing until firmly seated on the shaft.

DRIVE PINION BEARING CUPS

Do not remove the pinion bearing cups from the carrier casting unless the cups are worn or damaged. If the pinion bearing cups are to be replaced, drive them out of the carrier casting with a drift. Install the new cups. Make sure the cups are properly seated in their bores. If a 0.0015 inch feeler gauge can be inserted between a cup and the bottom of its bore at any point around the cup, the cup is not properly seated. Whenever the cups are replaced, the cone and roller assemblies should also be replaced.

Assembly

The drive pinion must be set to the correct depth before final assembly. Drive pinion preload requires that a new spacer be used when the pinion is removed. Drive pinion bearing preload is set with the drive pinion assembly installed and the pinion nut tightened to specification. Correct preload is indicated when the rotational torque is 8–14 inch lbs. with the original bearings or 16–29 inch lbs. for used bearings. Differential Bearing Preload and Ring Gear Backlash is adjusted with the drive pinion and differential case installed. Adjustment is performed by the installation of shims between the differential bearing cup and axle housing.

DRIVE PINION AND DRIVE PINION BEARING PRELOAD ADJUSTMENT

1. Install the pinion front bearing and slinger.

2. Apply grease, C1AZ–19590–B or equivalent between the lips of the pinion seal and install the pinion seal.

3. Insert the companion flange into the seal and hold it firmly against the pinion front bearing cone. From the rear of the carrier casting, insert the pinion shaft, with a new spacer, into the flange.

4. Start a new pinion nut. Hold the flange with special tool T78P–4851–A or equivalent and tighten the pinion nut. As the nut is tightened, the pinion shaft is pulled into the front bearing cone and into the flange. As the pinion shaft is pulled into the front bearing cone, pinion shaft end play is reduced. While there is still end play in the pinion shaft, the flange and bearing cone will be felt to bottom on the collapsible spacer. From this point, a much greater torque must be applied to turn the pinion nut, since the spacer must be collapsed. Very slowly, tighten the nut, but check the pinion shaft end play often to see that the pinion bearing preload does not exceed the limits. If the pinion nut is tightened to the point that pinion bearing preload exceeds the limits, the pinion shaft must be removed and a new collapsible spacer installed. Do not decrease the preload by loosening the pinion nut. This will remove the compression between the pinion front and rear bearing cones and the collapsible spacer, and may permit the front bearing cone to turn on the pinion shaft.

5. As soon as there is a preload on the bearings, turn the pinion shaft in both directions several times to set the bearing rollers.

6. Adjust the bearing preload to specification. Measure the preload.

ASSEMBLY

For shim selection after a complete replacement of the rear axle housing, the differential assembly or differential side bearings use the following instructions. For a ring and pinion replacement only or a backlash adjustment, follow Steps 9 through 13 and Step 15, using the side bearing shims that were originally in the axle.

1. With pinion depth set and pinion installed, place differential case gear assembly with bearings and cups in carrier.

2. Install a 6.73mm (0.265 inch) shim on left side.

3. Install left bearing cap and tighten bolts finger tight.

4. Install progressively larger shims on the right side until the largest shim selected can be assembled with a slight drag feel.

NOTE: Apply pressure towards left side to ensure bearing cup is seated.

5. Install right side bearing cap and tighten bearing cup bolts to 70–85 ft. lbs.

6. Rotate assembly to ensure free rotation.

7. Check ring gear and pinion backlash. If the backlash is 0.20–0.38mm (0.008–0.015 inch) with 0.304–0.381mm (0.012–0.015 inch) preferred, proceed to Step 14. If backlash is not within specifications, go to Step 10, unless zero backlash is measured, then go to Step 8.

8. If a zero backlash condition occurs, add .020 inch to the right side and subtract .020 inch from the left side to allow backlash indication.

9. Recheck backlash.

10. If backlash is not to specification, correct backlash by increasing thickness of one shim and decreasing thickness on the other shim the same amount. Refer to chart for approximate shim change.

11. Install shim and bearing caps. Tighten cap bolts to 70–85 ft. lbs.

12. Rotate assembly several times.

13. Recheck backlash. If backlash is within specification, go to Step 14. If backlash is not within specification, repeat Step 10. Backlash specification is 0.20–.038mm (0.008–0.015 inch). Preferred range is 0.304–0.381mm 0.012–0.015 inch.

14. Increase both left and right shim sizes by .006 inch and install for correct differential bearing preload. Make sure shims are fully seated and assembly turns freely.

15. Utilize white marking compound to obtain a tooth mesh contact pattern in your assembly. Reincorporation of pattern inspection is intended to allow technicians the ability to detect gross errors in set up prior to complete reassembly. Pattern contact should be within the primary area of the ring gear tooth surface avoiding any "narrow" or "hard" contact with outer perimeter of tooth (top to root, toe to heel). Pattern inspection should be on the drive (pull) side. Correct assembly of drive pattern will result in satisfactory coast performance. If gross pattern error is detected, with preferred backlash of 0.30–0.38mm (0.012–0.015 inch), recheck pinion shim selection.

16. Install bearing caps and tighten cap bolts to 70–85 ft. lbs.

17. Install the axle shafts.

18. Remove the oil seal replacer from the transmission extension housing. Install the driveshaft in the extension housing. Align the scribe marks on the flange and driveshaft and connect the driveshaft at the drive pinion flange. Apply Loctite (E0AZ–19554–B) or equivalent to the threads of the attaching bolts. Tighten attaching bolts to 70–95 ft. lbs.

19. Install the brake drum and attaching shakeproof retainers. Install the wheel and tire on the brake drum. Install the wheel covers.

20. Clean the gasket mating surface of the rear axle housing and cover. Apply a new continuous bead of silicone rubber sealant (D6AZ–19562–B or equivalent) to the carrier casting face.

--- CAUTION ---
MAKE SURE MACHINED SURFACES ON BOTH COVER AND CARRIER ARE CLEAN BEFORE INSTALLING THE NEW SILICONE SEALANT. INSIDE OF AXLE MUST BE COVERED WHEN CLEANING THE MACHINED SURFACE TO PREVENT AXLE CONTAMINATION.

21. Install cover and tighten cover bolts to 25–35 ft. lbs., except the ratio tag bolt,

which is tightened to 15–25 ft. lbs. Cover assembly must be installed within 15 minutes of application of the silicone or new sealant must be reapplied.

22. Add E0AZ–19580–A (ESP–M2C154–A) or equivalent through the filler hole until the lubricant level is 9.5mm (⅜ inch) below the filler hole with the axle in the running position.

23. Lower vehicle and road test.

Removable Carrier Type Ring Gear

This is a conventional type axle used on light duty Ford trucks. The axle design uses a removable carrier with the assembly bolted to the axle housing. The axle uses hypoid type gears and has the pinion gear mounted below the center line on the ring gear. The pinion gear is supported by two bearings in front of the gear and one behind. It is important to refer to the tag showing the axle and model number which is secured to the housing to obtain proper replacement parts.

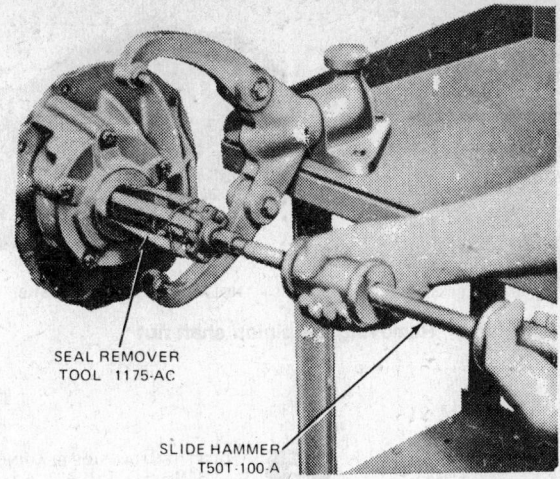

Removing the pinion seal

Installing the pinion seal

CARRIER ASSEMBLY

1. With the vehicle raised on a lift, remove the axle shafts from the housing.

2. Remove the drive shaft from the carrier assembly.

3. With a drain pan under the axle, remove the retaining bolts from the carrier and drain the gear lube.

4. Remove the carrier assembly from the axle.

5. Clean the surfaces of the carrier and the axle housing. Install a new gasket.

6. Position the carrier assembly on the studs in the housing and install the retaining nuts. Torque the nuts to 30–40 ft. lbs.

7. Install the drive shaft and torque the bolts to 13–17 ft. lbs.

8. Install the axles in the housing and secure.

9. Fill the axle housing to the proper level with gear lube and road test for proper operation.

Removing the companion flange

Removing the differential pinion shaft lock pin

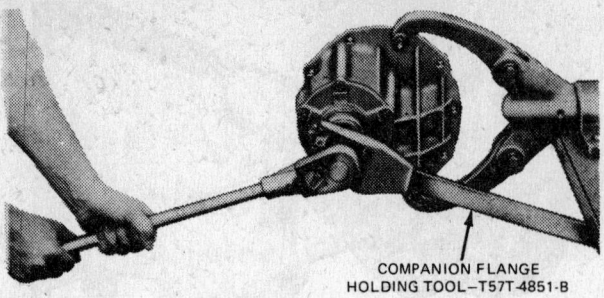

Removing the pinion shaft nut

COMPANION FLANGE
HOLDING TOOL—T57T-4851-B

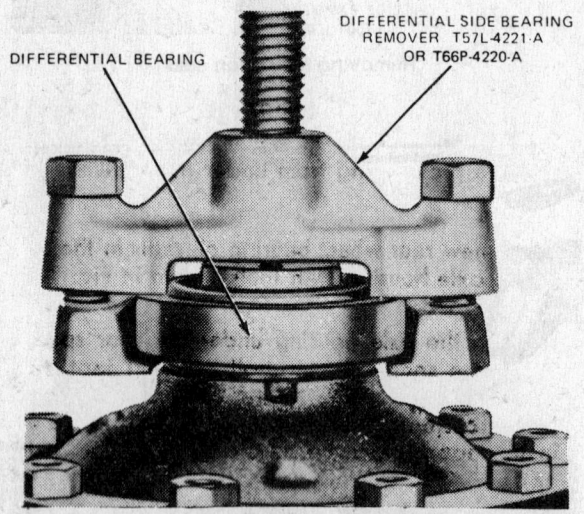

DIFFERENTIAL SIDE BEARING
REMOVER T57L-4221-A
OR T66P-4220-A

DIFFERENTIAL BEARING

Removing the differential bearing

PAINT MARKING INDICATES POSITION IN WHICH GEARS WERE LAPPED

Pinion and ring mesh timing

Installing the pinion front bearing cup

PRESS RAM

Installing the pinion rear bearing cup

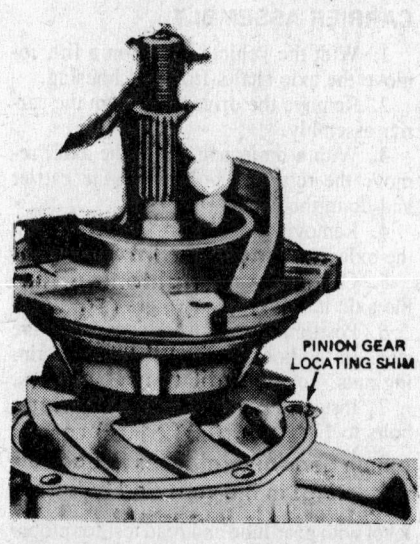

PINION GEAR
LOCATING SHIM

Removing the pinion retainer assembly

DISASSEMBLY

1. Remove the carrier assembly from the axle housing and mount the carrier in a holding fixture.

2. Mark the bearing caps and adjusting nuts so they may be installed in their original positions when assembling.

3. Remove the adjusting nut locks, bearing caps and adjusting nuts.

4. Lift the differential assembly out of the carrier. Using a bearing puller, remove the side bearings from the differential case.

5. Mark the side of the case, the ring gear and the cover so they can be installed in their original positions.

6. Remove the bolts that retain the ring gear to the case and using a soft hammer, tap the ring gear from the case.

7. Using a drift, drive the lock pin from the pinion shaft and separate the halves of the differential case.

8. Drive the pinion shaft out of the case using a brass drift and remove the thrust washers and gears.

DRIVE PINION & BEARING RETAINER

1. With a holding fixture installed on the flange, remove the pinion nut and washer. Leave the holding fixture on the flange and using a puller, remove the flange from the pinion shaft.

2. Using a seal puller, remove the pinion seal from the retainer assembly.

3. Remove the bolts from the retainer assembly and lift the retainer from the carrier. Measure the thickness of the shim that was between the retainer and the carrier assembly. Record the result.

4. Install a piece of hose on the pinion pilot bearing surface in front of the pinion gear. Mount the retainer assembly in a press and press the pinion gear out of the retainer.

5. Mount the pinion shaft in a press and press the rear bearing from the pinion shaft.

PINION BEARING CUP

1. With the retainer assembly mounted in a press, using the proper tool, press the front and rear bearing cups from the assembly.

2. Check the inside surfaces of the retainer for any nicks, dirt or distortion.

3. Install the new cups by pressing them into place with the proper tool. When the cups are installed, make sure they are seated in the retainer by trying to fit a .0015 in. feeler gauge, between the cup and the bottom of the bore.

PILOT BEARING

1. Using a bearing driver, drive the bearing and retainer out of the carrier assembly.

2. Using the same tool, drive the new bearing into place until the driver bottoms against the case.

3. Drive a new retainer into place with the concave side up.

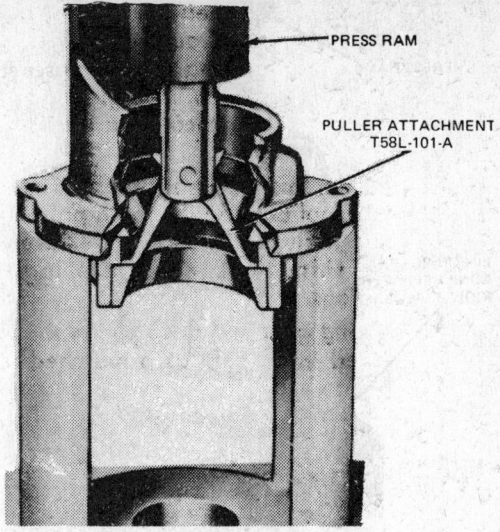

Removing the pinion rear bearing cup

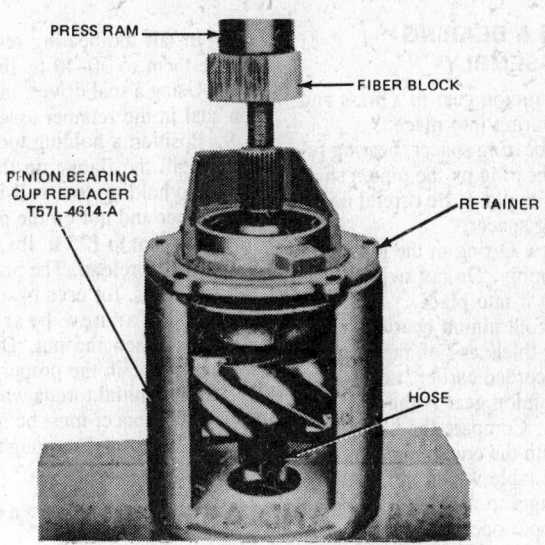

Removing the pinion front bearing cone

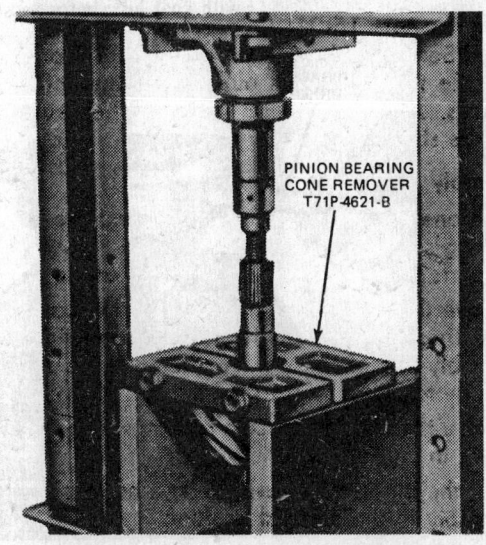

Removing the pinion rear bearing cone

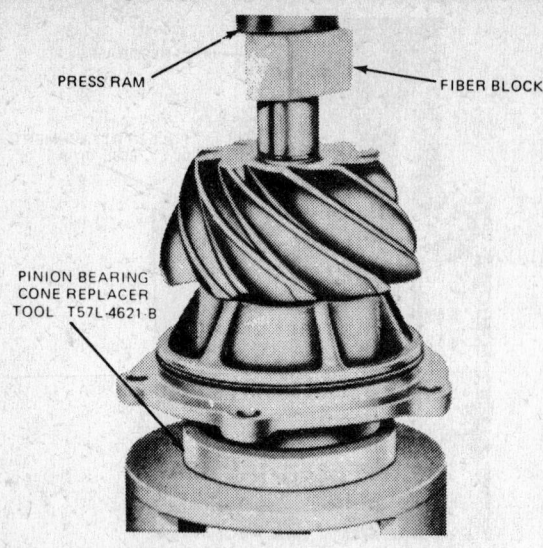

PRESS RAM

FIBER BLOCK

PINION BEARING
CONE REPLACER
TOOL T57L-4621-B

Installing the pinion front bearing

DRIVE PINION & BEARING RETAINER ASSEMBLY

1. Mount the pinion gear in a press and press the rear bearing into place.

2. Install the bearing spacer, bearing retainer and front bearing on the pinion shaft and press them into place. Be careful not to crush the bearing spacer.

3. Install a new O-ring in the groove in the retainer assembly. Do not twist the O-ring when fitting it into place.

4. Lubricate both pinion bearings.

5. Check the thickness of the original shim that was recorded earlier. Located on the head of the pinion gear is the shim adjustment number. Compare the number on the old pinion with the one on the new pinion. Refer to the table which indicates the amount of change to the original shim thickness for proper operation.

6. Install the new shim on the housing and install the pinion and retainer assembly, being careful not to damage the O-ring.

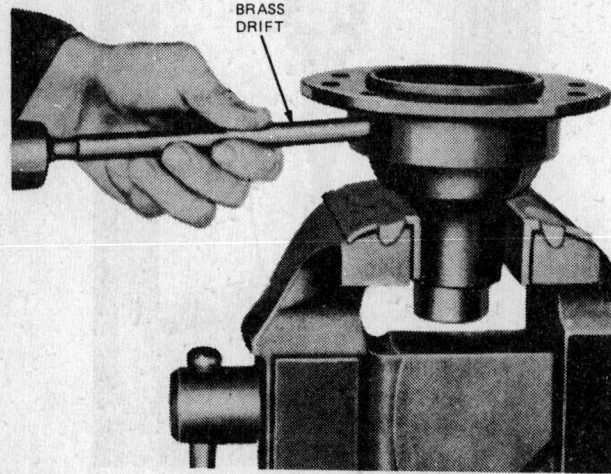

BRASS DRIFT

Driving out the differential pinion shaft

7. Install the bearing retainer bolts and torque them to 30–40 ft. lbs.

8. Using a seal driver, install a new pinion seal in the retainer assembly.

9. Position a holding tool on the flange and install the flange on the pinion shaft. With the holding tool still in place, install the washer and nut on the pinion shaft and torque the nut to 175 ft. lbs. Check the pinion bearing preload. The preload should be 8–14 inch lbs. for used bearings and 22–32 inch lbs. for new bearings. Do not overtight<chen the nut. Do not back off the nut to obtain the proper preload. If the 175 ft. lbs. initial torque was too much, the collapsible spacer must be replaced. Tighten the pinion only enough to obtain the right preload torque.

DIFFERENTIAL CASE

Assembly & Installation

1. Lubricate all of the differential parts with gear lube before assembling.

2. Install a side gear and thrust washer in the case bore. Using a soft hammer, drive the pinion shaft into the case far enough to hold a pinion thrust washer and gear. Place the second pinion thrust washer and gear in position and carefully tap the pinion shaft into place. Be sure to line up the holes for the lock pin in the pinion shaft.

3. With the second side gear and thrust washer in place, install the cover on the differential case. Drive the pinion lock pin into place. Insert an axle shaft spline into the side gear and check for free rotation of the gears.

4. Install two, two inch long $^7/_{16}$ (N.F.) bolts through the differential case and thread them a little way into the ring gear. These will act as a guide when installing the ring gear on the case. Tap the ring gear into place.

5. Remove the guide pins and install the ring gear bolts. Tighten the bolts evenly to 65–85 ft. lbs.

6. If the differential bearings were removed, install the assembly in a press and press the new bearings into place.

7. Coat the bearing bores in the carrier with gear lube and install the bearing cups on the bearings. Place the differential assembly in the carrier.

8. Slide the differential case in the carrier bore until there is a slight amount of backlash between the gears.

9. Install the adjusting nuts in the carrier so that they just contact the bearing cups. The nuts should be engaged about the same number of threads on each side.

10. Position the bearing caps in the carrier. Be careful to line up the marks. Install the cap bolts and torque them to 70–80 ft. lbs. Make sure the adjusting nuts turn freely as the bolts are being tightened.

11. Adjust the backlash and bearing preload as follows: Loosen the bearing cap bolts then retighten them to 35 ft. lbs. Loosen the adjusting nut on the pinion side so that it is away from the bearing cup. Tighten the nut on the opposite side so that the ring gear is forced into the pinion with no backlash. Recheck the nut on the pinion side to make sure it is still loose. Now tighten this nut until it contacts the bearing cup. After it contacts the cup, turn it two more notches. Rotate the ring gear several times in each direction. This helps to seat the bearings in the cups. Again loosen the nut on the pinion side. If there is any backlash between the gears, tighten the nut on the ring gear side until the backlash is removed. Install a dial indicator on the carrier assembly. Tighten the nut on the pinion side until it just contacts the cup. With the dial indicator set at zero, tighten the pinion side nut until the case is spread 0.008–0.012 in. with new bearings and 0.005–0.008 in. with old bearings. As this preload is applied the ring gear is forced away from the pinion and usually results in the correct backlash. Mount the dial indicator on the ring gear and check the gear for

backlash. Make sure the bearing caps are torqued to 75–85 ft. lbs.

12. The backlash should be between 0.008–+0.012 in. If the backlash is not correct, loosen one nut and tighten the other an equal amount to move the ring gear in or out to correct the measurement. When making final adjustments, always move the adjusting nuts in a tightening direction. To do this, if a nut had to be loosened one notch, loosen it two notches and tighten it one. This makes certain the nut is in contact with the cup and will not shift when the vehicle is in operation. Coat the ring gear teeth with a marking compound and check the tooth pattern. If the pattern is not correct make the necessary changes to bring it into adjustment.

13. Install the carrier assembly in the vehicle and road test for proper operation.

Rockwell

12 Inch Ring Gear

This rear axle is a full-floating type with a hypoid drive gear and pinion, and uses a four-pinion differential assembly. The straddle mounted pinion has two tapered roller bearings in front of the pinion teeth and a straight roller bearing behind the pinion teeth. The differential carrier assembly can be removed while the axle remains in the truck.

CARRIER ASSEMBLY

Removal

1. Remove axle shafts.
2. Drain lubricant. Disconnect propeller shaft at pinion shaft yoke.
3. Remove carrier from axle housing and clean thoroughly.

Disassembly

1. Punch mark carrier leg, bearing cap, and bearing adjusting nut to assist in reassembly.
2. Remove screws, adjusting nut locks, bearing caps and adjusting nuts.

Checking backlash

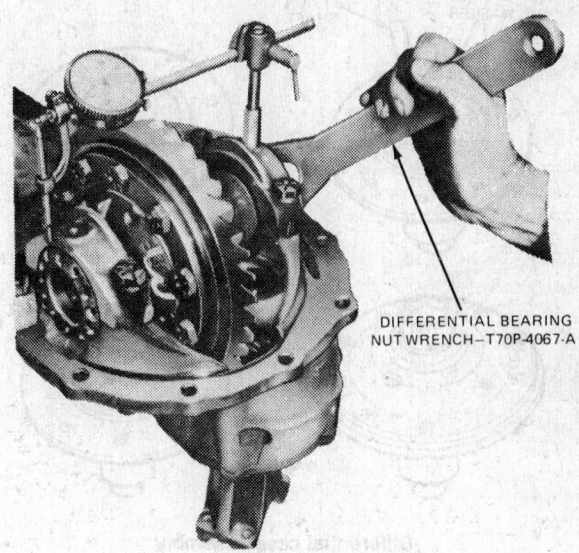
DIFFERENTIAL BEARING NUT WRENCH—T70P-4067-A
Side bearing preload adjustment

Carrier mounted in the repair stand

Punch-marking the carrier leg

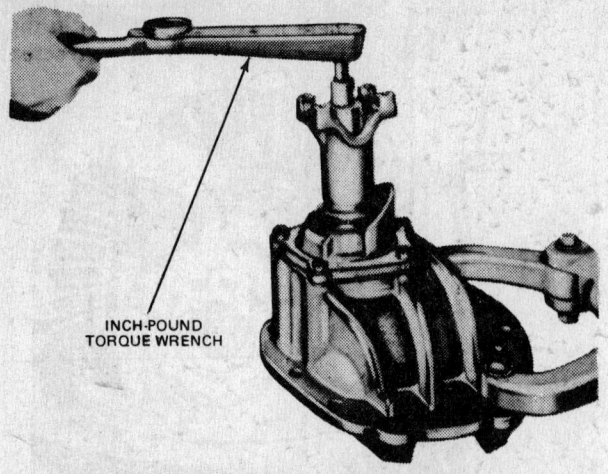

Pinion bearing preload check

INCH-POUND
TORQUE WRENCH

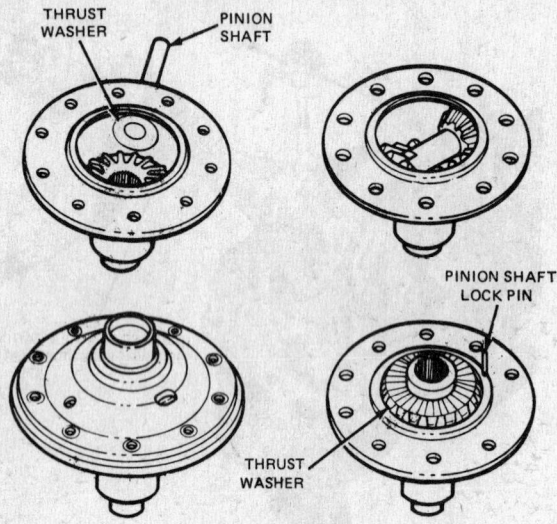

THRUST WASHER

PINION SHAFT

PINION SHAFT LOCK PIN

THRUST WASHER

Differential case assembly

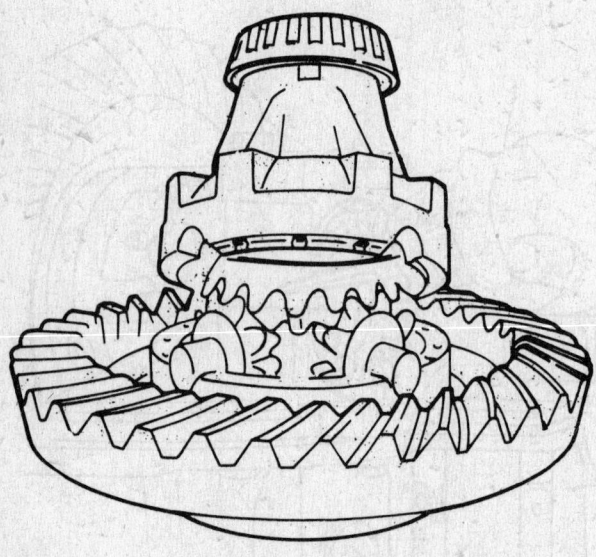

Aligning the mating marks

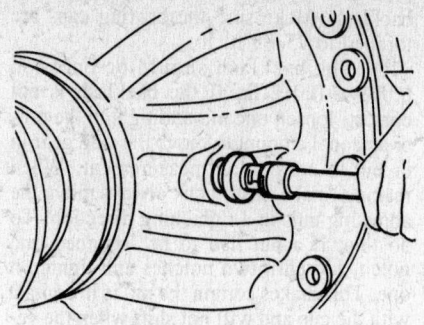

Thrust adjusting screw

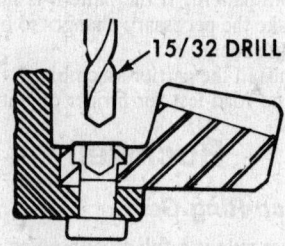

15/32 DRILL

RIGHT

WRONG

Removing the rivets

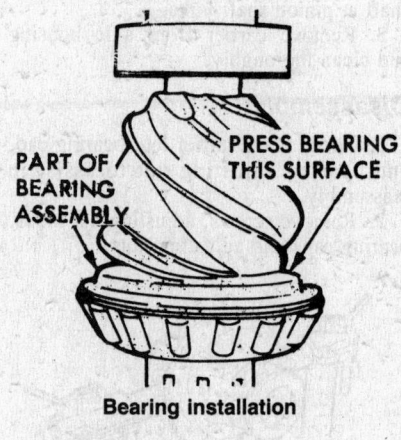

PART OF BEARING ASSEMBLY

PRESS BEARING THIS SURFACE

Bearing installation

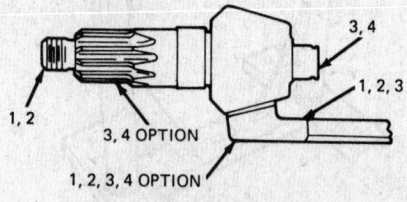

3, 4

1, 2, 3

1, 2

3, 4 OPTION

1, 2, 3, 4 OPTION

STRADDLE MOUNTED PINION
(Shown with parallel sided splines)

Gear set identification

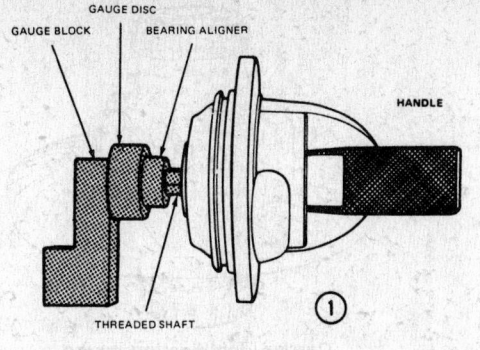
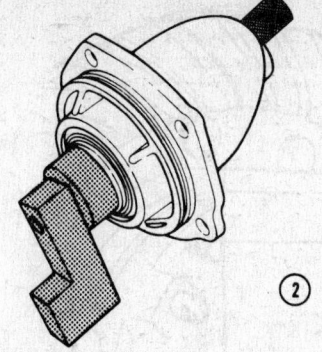
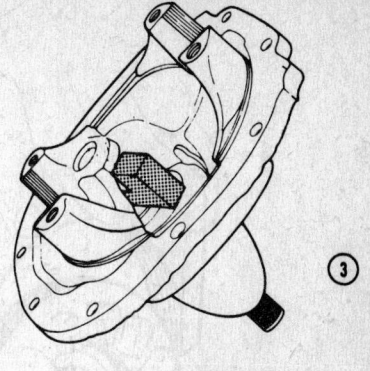
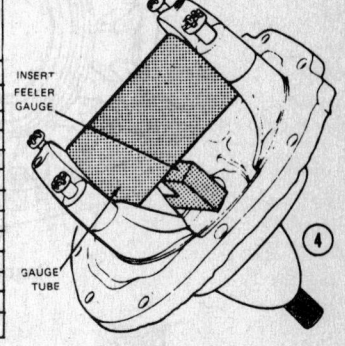

Feeler Gauge Reading		Shim Required		Feeler Gauge Reading		Shim Required		Feeler Gauge Reading		Shim Required	
mm	(in)	(mm)	(in)	(mm)	(in)	(mm)	(in)	(mm)	(in)	(mm)	(in)
.889	.035	.127	.005	.584	.023	.432	.017	.279	.011	.737	.029
.864	.034	.152	.006	.559	.022	.451	.018	.254	.010	.0762	.030
.838	.033	.178	.007	.533	.021	.483	.019	.229	.009	.787	.031
.813	.032	.203	.008	.508	.020	.508	.020	.203	.008	.813	.032
.787	.031	.229	.009	.483	.019	.533	.021	.178	.007	.838	.033
.762	.030	.254	.010	.457	.018	.559	.022	.152	.006	.864	.034
.737	.029	.279	.011	.432	.017	.584	.023	.127	.005	.889	.035
.711	.028	.305	.012	.406	.016	.610	.024	.102	.004	.914	.036
.686	.027	.330	.013	.381	.015	.635	.025	.076	.003	.940	.037
.660	.026	.356	.014	.356	.014	.660	.026	.051	.002	.965	.038
.635	.025	.381	.015	.330	.013	.686	.027				
.610	.024	.406	.016	.305	.012	.711	.028				

Using the pinion depth gauge tool

3. Loosen lock nut and back off drive gear thrust block adjusting screw.

4. Lift differential out of carrier and remove thrust block from end of adjusting screw inside of carrier.

5. Punch mark differential case halves for correct reassembly alignment and separate case halves.

6. Remove pinion shaft, pinions, side gears, thrust washers and differential bearing cones.

7. To remove drive gear, carefully center punch each rivet in center of rivet head. Use a drill $\frac{1}{32}$ inch smaller than the body of rivet to drill through the rivet head. Press out rivets.

8. Remove pinion shaft nut, washer and yoke. Driving yoke off will cause runout.

9. Remove pinion bearing cover and oil seal assembly and, using puller screws, remove bearing cage. Using a pinch bar to remove cage will damage shims. Driving pinion from inner end with a drift will damage bearing lock ring groove.

10. Wire bearing cage shim pack together to facilitate adjustment when reassembling.

11. Tap pinion shaft out of cage with soft mallet or press shaft from cage. Remove bearing from cage.

12. Remove spacers and inner bearing from shaft.

13. Remove pinion shaft rear pilot bearing lock ring, and then bearing.

14. Remove oil seal assembly from bearing cover.

15. Clean all parts thoroughly in a suitable solvent and blow dry with compressed air. Do not spin bearings with air pressure

as they might score due to absence of any lubrication. Inspect all parts for wear or roughness and replace if necessary. Inspect all machined surfaces for nicks, burrs or scratches.

Assembly

1. Press drive pinion inner bearing cone firmly against pinion shoulder.

2. Press rear pilot bearings firmly against pinion shoulder and install lock ring into pinion shaft groove.

3. Press bearing cups firmly against bearing cage shoulders.

4. Lubricate pinion bearings with SAE 90 oil and insert pinion and bearing assembly into pinion cage.

5. Place original spacers on pinion shaft, and install front bearing and press

firmly against spacers. Rotate cage several revolutions to assure normal bearing contact. If a press is not available, install pinion yoke and nut and torque to specifications.

6. While the assembly is in the press under pressure or pinion nut torques to specifications, check pinion bearing pre-load. The correct pressures and torque for checking pre-load are:

7. Wrap a soft wire around cage and pull on horizontal line with pound scale when determining pinion bearing pre-load, first measure diameter of the pinion cage. Assuming cage diameter is 6 inches, the pulling radius would be 3 inches; therefore, 5 pounds pull on the scale would equal 15 inch lbs. pre-load.

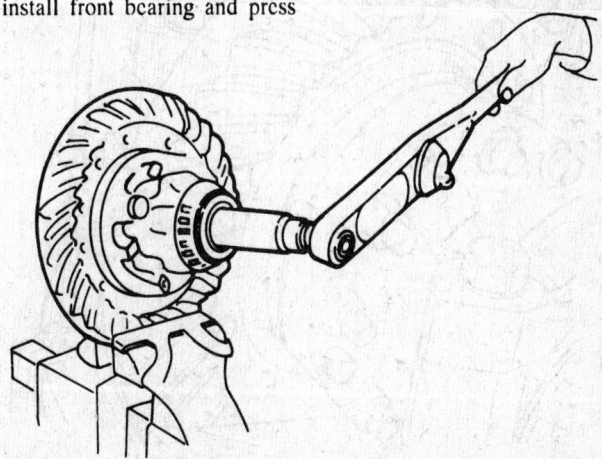

Checking the rotating torque

Installing the bearing caps

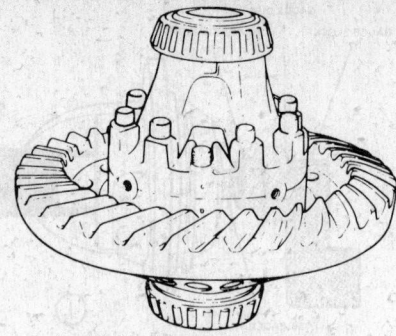

Carrier identification marks

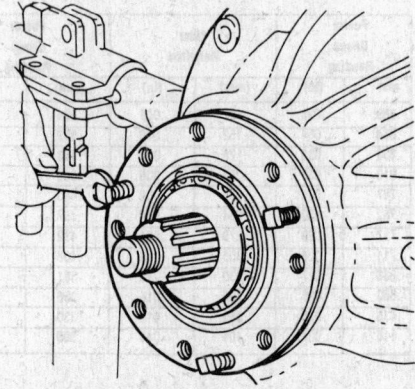

Removing the bearing cage

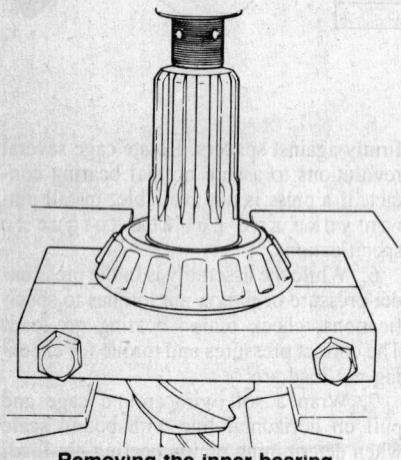

Removing the inner bearing

8. Use rotating torque, not starting torque. If rotating torque is not within 5 to 15 inch lbs., use thinner spacer to increase pre-load or thicker spacer to decrease pre-load. Torque must be near low limit inch-pounds with original pinion bearings and near high limit when using new bearings. Remove yoke and install new oil seal.

9. Lubricate pinion shaft oil seal and lightly coat outer edge of seal body with non-hardening sealing compound. Press seal against cover shoulder. Install new gasket and bearing cover.

10. Install pinion yoke, washer and nut. Place pinion and cage assembly over carrier studs, hold hoke and tighten nut to specified torque. Install cotter key without backing off nut to align cotter key holes.

11. Place original shim pack on carrier studs with thin shims on both sides to create maximum sealing. Position pinion and cage assembly over studs and tap into position with soft mallet. Install lock washers and nuts. Tighten nuts to specified torque. If a new pinion is being installed, consult the General Axle Service section for correct procedure.

12. Rivet drive gear to differential case using new rivets. Rivets should not be heated, but always upset cold. When correct rivet is used, head being formed will be at least $\frac{1}{8}$ inch larger in diameter than rivet

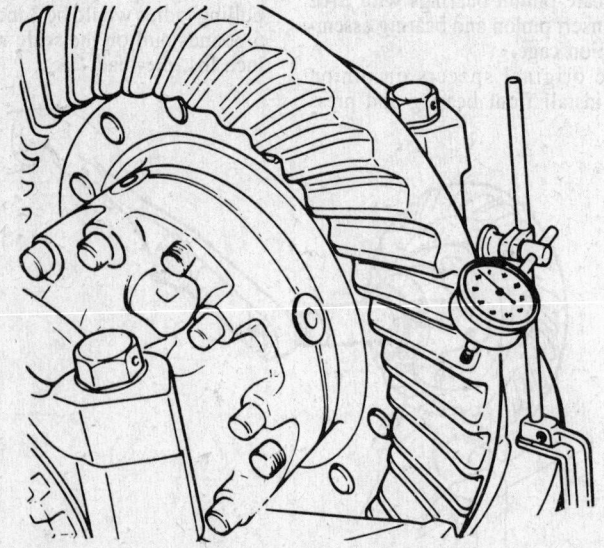

Checking backlash

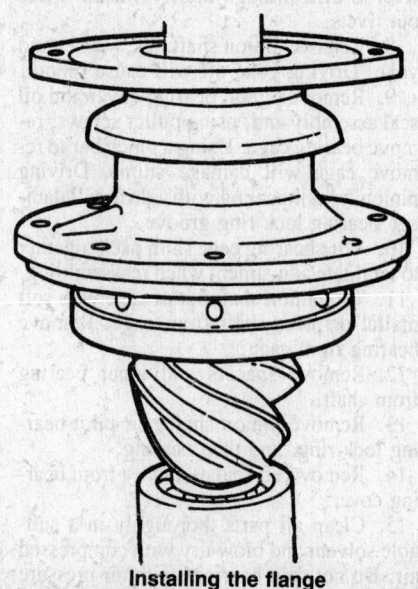

Installing the flange

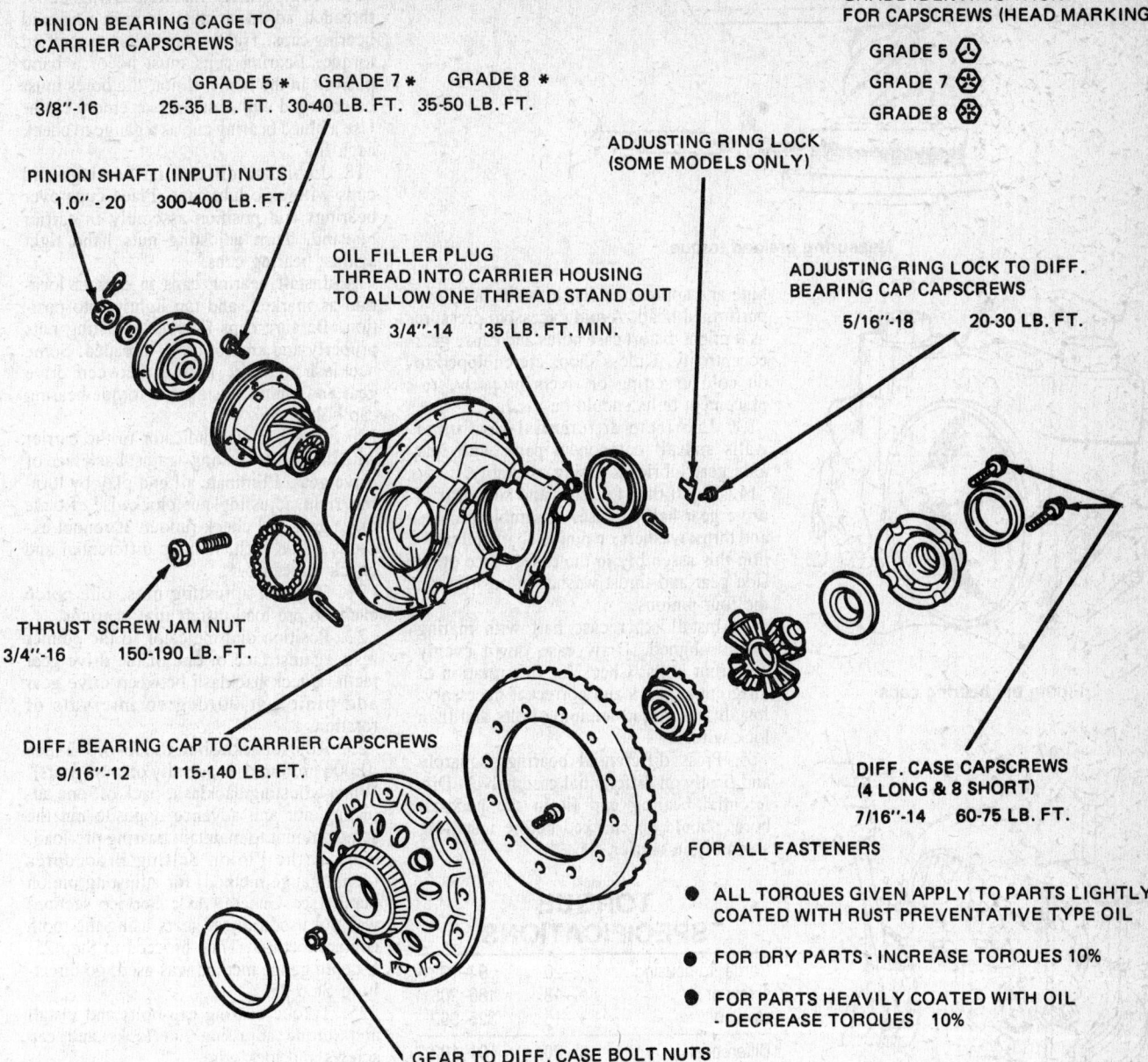

PINION BEARING CAGE TO
CARRIER CAPSCREWS

GRADE 5 *   GRADE 7 *   GRADE 8 *

3/8"-16   25-35 LB. FT.   30-40 LB. FT.   35-50 LB. FT.

PINION SHAFT (INPUT) NUTS

1.0" - 20   300-400 LB. FT.

* GRADE IDENTIFICATION
   FOR CAPSCREWS (HEAD MARKINGS)

GRADE 5
GRADE 7
GRADE 8

ADJUSTING RING LOCK
(SOME MODELS ONLY)

OIL FILLER PLUG
THREAD INTO CARRIER HOUSING
TO ALLOW ONE THREAD STAND OUT

3/4"-14   35 LB. FT. MIN.

ADJUSTING RING LOCK TO DIFF.
BEARING CAP CAPSCREWS

5/16"-18   20-30 LB. FT.

THRUST SCREW JAM NUT

3/4"-16   150-190 LB. FT.

DIFF. BEARING CAP TO CARRIER CAPSCREWS

9/16"-12   115-140 LB. FT.

DIFF. CASE CAPSCREWS
(4 LONG & 8 SHORT)

7/16"-14   60-75 LB. FT.

FOR ALL FASTENERS

● ALL TORQUES GIVEN APPLY TO PARTS LIGHTLY
   COATED WITH RUST PREVENTATIVE TYPE OIL

● FOR DRY PARTS - INCREASE TORQUES 10%

● FOR PARTS HEAVILY COATED WITH OIL
   - DECREASE TORQUES 10%

GEAR TO DIFF. CASE BOLT NUTS

1/2"-20   85-115 LB. FT.

Fastener torques

ROCKWELL–STANDARD BEARING PRELOAD

Axle Model	Pinion Shaft Nut–Thread Size and Torque Limits (ft. lbs.)		Press Ram Pressure for Preload Check (Tons)	Pinion and Cross Shaft Bearing Preload (in. lbs.)	Backlash Limits (in.)	Differential Bearing Preload Adjusting Nut Notches Tighten from Zero End Play (each Adjusting Nut)
Single–speed	1–20	300–400	6	5–15	0.005–0.015	1
Single reduction	1¼ × 18	700–900	11	5–15	0.005–0.015	1
	1¼ × 18	800–1100	14	5–15	0.005–0.015	1
	1½ × 12	800–1100	14	5–15	0.005–0.015	1
	1½ × 18	800–1100	14	5–15	0.005–0.015	1

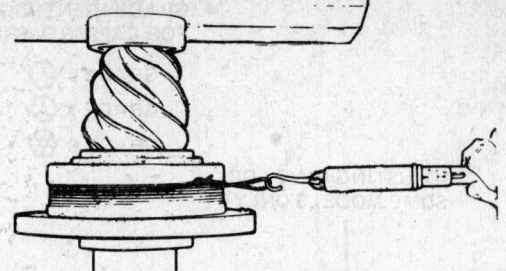

Measuring preload torque

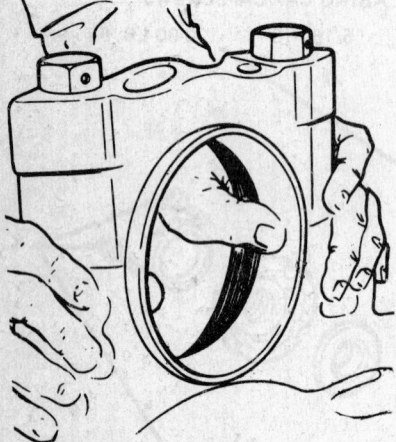

Fitting the bearing caps

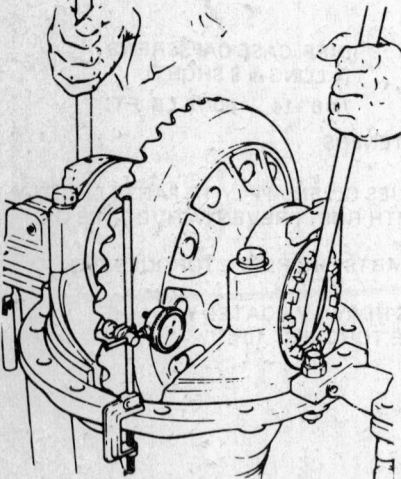

Adjusting the bearing preload

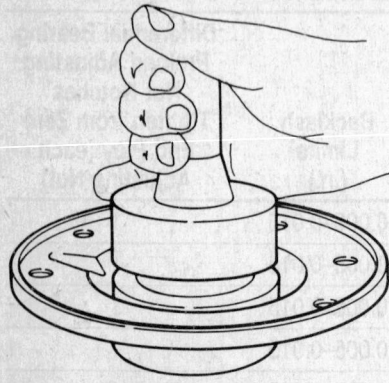

Installing the seal

hole and approximately the same height as performed head. Avoid excessive pressure as it might distort case holes and cause gear eccentricity. Unless shops are equipped to do cold upsetting of rivets properly, replacement bolts should be used.

13. Lubricate differential case inner walls and all component parts with rear axle gear lubricant during assembly.

14. Install thrust washer and side gear in drive gear half of case. Assemble pinions and thrust washers on pinion shaft and position this assembly in the case. Place other side gear and thrust washer in position on the four pinions.

15. Install other case half with mating marks aligned. Draw case down evenly with four bolts. Check for free rotation of differential gears and correct if necessary. Install and torque remining bolts and then lock wire.

16. Press differential bearings squarely and firmly on differential case halves. Differential bearing cup fit in the pedestal bores should be checked before installing assembly in carrier.

TORQUE SPECIFICATIONS

Carrier to housing	½–20	94–102
screw or	⅝–18	186–205
stud nut	¾–16	325–360
Differential	½–20	94–102
case bolt	9⁄16–18	132–145
	⅝–18	186–205
	¾–16	325–360
Differential	⅝–11–18	127–140
bearing cap	¾–10–16	230–250
bolt	⅞–9	345–370
	⅞–14	375–415
	1–12	555–615
Differential	5⁄16–18	15–17
bearing	½–13	85–91
adjuster	9⁄16–12	120–129
lock bolt	⅝–11	168–180
Pinion shaft	1–20	300–400
yoke nut	1–¼–18	700–900
	1–½–18	800–1100
	1–½–12	800–1100
	1–¾–12	800–1100
Shaft flange	7⁄16–20	52–58
stud		

17. Temporarily install bearing cups, threaded adjusting nuts or split ring and bearing caps. Tighten cap bolts to specified torque. Bearing cups must be of a hand push fit in the bores; if not, the bores must be enlarged with a scraper or emery cloth. Use a blued bearing cup as a gauge to check each fit.

18. Lubricate differential bearings and cups with axle lubricant. Place cups over bearings and position assembly in carrier housing. Turn adjusting nuts hand tight against bearing cups.

19. Install bearing caps in correct location as marked, and tap lightly into position. Be sure caps fit over adjusting nuts properly and are not cross-threaded. Some backlash must be present between drive gear and pinion. Install and torque bearing cap bolts.

20. Attach a dial indicator to the carrier with the pointer resting against back face of drive gear. Eliminate all end play by turning right adjusting nut clockwise. Rotate drive gear and check runout. If runout exceeds 0.008 inch, remove differential and check the cause.

21. Tighten adjusting nuts, one notch each, to pre-load differential bearings.

22. Position dial indicator so the pointer rests against face of one of the drive gear teeth. Check backlash between drive gear and pinion at 90 degree intervals of rotation.

23. Adjust backlash to 0.006–0.012 inch (0.006 preferred, especially on new gears). When adjusting backlash, back off one adjusting nut and advance opposite nut the same amount to maintain bearing pre-load.

24. If the Pinion Setting Procedures (depth gauge method) for adjusting pinion depth (see General Axle Service section) was not used, adjust gears using the tooth contact method. Then proceed to Step 25. If depth gauge method was used, go directly to Step 25.

25. Torque bearing cap bolts and install and torque adjusting nut locks and cap screws and lock wire.

26. Hold drive gear thrust block on rear face of gear with heavy grease, rotate gear until hole in thrust block aligns with adjusting screw hole in carrier. Install adjusting screw and lock nut, tighten screw enough to locate thrust block firmly against back face of gear. Back off adjusting screw ¼ turn to create 0.010–0.015 inch clearance and lock securely with nut. Recheck to assure minimum clearance of 0.010 inch during full rotation of drive gear.

Installation

1. Install a new gasket on axle housing flange. Start carrier into clean housing and hold in place with four equally spaced washers and nut. Tighten nuts alternately to draw carrier evenly into housing. Install and torque carrier flange lock washers and nuts.

2. Install axle drive shafts and connect universal joint at pinion flange.

3. Fill axle housing to proper level and road test vehicle.

IMPORTED VEHICLES

Datsun/Nissan

Model H190

PRE-DISASSEMBLY INSPECTION

1. Check backlash of ring gear with a dial indicator at several points. If it is not within specification, .0059–.0079 in., adjust as needed.

2. Check runout of ring gear with a dial indicator. If it is over specification (.0031 in.), hypoid gear set or differential case should be replaced.

NOTE: When backlash varies excessively in different places, the variance may have resulted from foreign matter caught between ring gear and differential case.

3. Check tooth contact.

4. Check backlash of side gear. Using a thickness gauge, measure clearance between side gear and differential case. If it is not within specification, adjust it by selecting side gear thrust washer .0305–.0364 in.

TOOTH CONTACT

Gear tooth contact pattern check is necessary to verify correct relationship between ring gear and drive pinion.

Hypoid gear set which are not positioned properly may be noisy, or have short life or both. With a pattern check, the most desirable contact for low noise level and long life can be assured.

1. Thoroughly clean ring gear and drive pinion teeth.

2. Sparingly apply a mixture of powdered ferric oxide and oil or equivalent to 3 or 4 teeth of ring gear drive side.

3. Hold companion flange steady by hand and rotate the ring gear in both directions.

DISASSEMBLY

1. Put match marks on one side of side bearing cap with paint or punch to ensure that it is replaced in proper position during reassembly.

NOTE: Bearing caps are line-bored during manufacture and should be put back in their original places.

2. Remove side bearing caps.

3. Using a pry bar, remove differential case assembly.

NOTE: Be careful to keep the side bearing outer races together with inner race, don't mix them up.

4. Remove drive pinion nut with special tool ST31530000 or equivalent.

5. Remove companion flange with puller.

6. Remove drive pinion with soft hammer.

7. Remove oil seal by prying up with a large awl, and remove front pinion bearing inner race.

NOTE: Do this carefully so as not to scratch seal bore with awl. Cover end of awl with a rag.

8. Remove pinion bearing outer race using a brass drift.

9. Remove collapsible spacer and washer from drive pinion.

10. Pull out rear bearing inner race with a press and special tool ST30031000 or equivalent.

NOTE: Care should be taken when setting Tool in press to make sure that parting line of Tool is a right angle to support fixture of press. This is to prevent bending Tool.

Case Disassembly

1. Remove side bearing inner race with a puller.

NOTE: To prevent damage to bearing, engage puller paws with groove. Be careful not to confuse left and right hand parts.

2. Remove ring gear by spreading out lock straps and loosening ring gear bolts in a criss-cross fashion.

3. Tap ring gear off gear case using a soft hammer.

NOTE: Tap evenly all around to keep ring gear from binding.

4. Drive out pinion mate shaft lock pin, with drift pin from ring gear side.

5. Draw out pinion mate shaft and thrust block, and rotate pinion mate gears out of the case and remove side gears and thrust washers.

NOTE: Put marks on gears and thrust washers so that they can be reinstalled in their original positions from which they were removed.

INSPECTION

1. Clean disassembled parts completely. Repair or replace any damaged or faulty parts.

NOTE; When replacing drive pinion or ring gear, replace with a new hypoid gear set.

2. The following parts should be replaced by new ones each time they are removed. Gasket. Front oil seal. Collapsible spacer. Lock strap.

Assembly

Assembly should be done in the reverse order of disassembly, while making any necessary inspections and adjustments. Arrange shims and washers to install them correctly. Thoroughly clean the surfaces on which shims, washers bearings and bearing caps are installed. Apply gear oil when installing bearings. Pack recommended multi-purpose grease into cavity between lips when fitting oil seal.

Differential Case

1. Install pinion mate gears, side gears and thrust washers into differential case.

2. Fit pinion mate shaft and thrust block.

3. Adjust clearance between rear face of side gear and thrust washer by selecting side gear thrust washer. Clearance limits 0.0039–0.0079 in.

4. Install pinion mate shaft lock pin using a punch.

NOTE: Make sure lock pin is flush with case.

5. Place ring gear on differential case and install new lock straps and bolts. Tighten bolts in a criss-cross fashion, lightly tapping bolt head with a hammer. Then bend up lock straps to lock the bolts in place.

6. Select side bearing adjusting shims, 0.0020–0.0197 in.

7. Install the shims behind each bearing and press on the bearings, using a press.

Differential Carrier

1. Press fit front and rear bearing outer races using special tools ST3061100, ST30613000, ST30621000 or equivalent.

2. Select pinion height adjusting washer, 0.1016–0.1252 in.

3. Install pinion height adjusting washer in drive pinion, bevel side toward gear, and press fit rear bearing inner race in it, using press and special tool ST30901000 or equivalent.

4. Lubricate front bearing with gear oil and place it in gear carrier.

5. Carefully fit a new oil seal into carrier.

NOTE: Make sure oil seal is flush with end of carrier and apply multi-purpose grease into cavity between lips.

6. Place a washer and a new collapsible spacer on drive pinion and lubricate rear bearing with gear oil, and insert it in gear carrier.

7. Install companion flange and hold it firmly. Insert drive pinion into companion flange by tapping its head with a soft hammer.

8. Hold companion flange and temporarily tighten pinion nut, until there is no axial play.

NOTE: Ascertain that threaded portion of drive pinion and pinion nut are free from oil or grease.

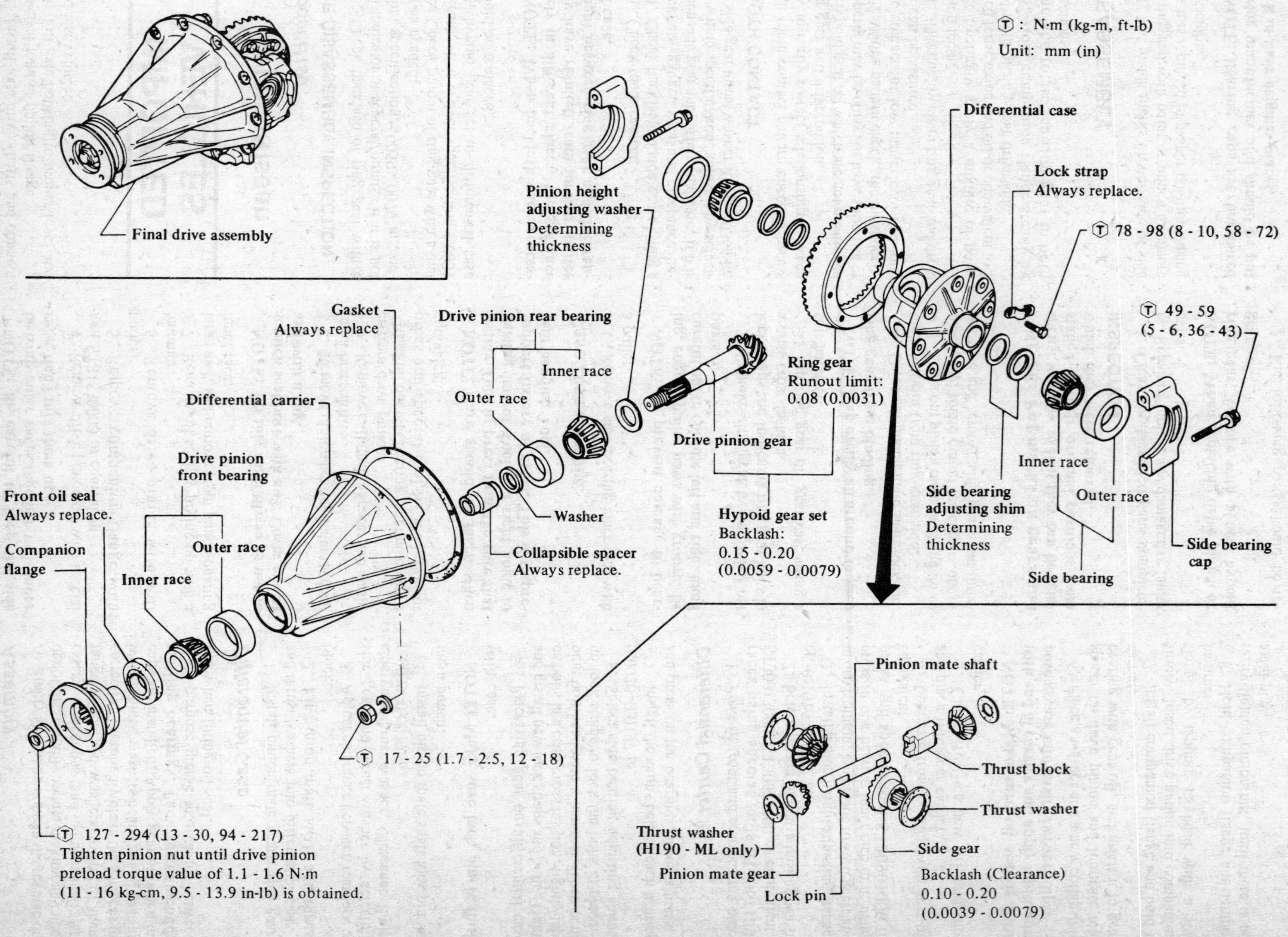

Final drive assembly

ⓉⓉ 127 - 294 (13 - 30, 94 - 217)
Tighten pinion nut until drive pinion
preload torque value of 1.1 - 1.6 N·m
(11 - 16 kg-cm, 9.5 - 13.9 in-lb) is obtained.

Companion
flange

Front oil seal
Always replace.

Inner race

Outer race

Drive pinion
front bearing

Differential
carrier

Gasket
Always replace

Ⓣ 17 - 25 (1.7 - 2.5, 12 - 18)

Outer race

Inner race

Drive pinion
rear bearing

Washer

Collapsible spacer
Always replace.

Pinion height
adjusting washer
Determining
thickness

Drive pinion gear

Hypoid gear set
Backlash:
0.15 - 0.20
(0.0059 - 0.0079)

Ring gear
Runout limit:
0.08 (0.0031)

Thrust washer
(H190 - ML only)

Pinion mate gear

Lock pin

Side gear

Pinion mate shaft

Thrust washer

Thrust block

Backlash (Clearance)
0.10 - 0.20
(0.0039 - 0.0079)

Differential case

Side bearing
adjusting shim
Determining
thickness

Inner race

Outer race

Side bearing

Lock strap
Always replace.

Ⓣ 78 - 98 (8 - 10, 58 - 72)

Ⓣ 49 - 59
(5 - 6, 36 - 43)

Side bearing
cap

Ⓣ : N·m (kg-m, ft-lb)
Unit: mm (in)

Differential carrier—Model H190-ML, H190A

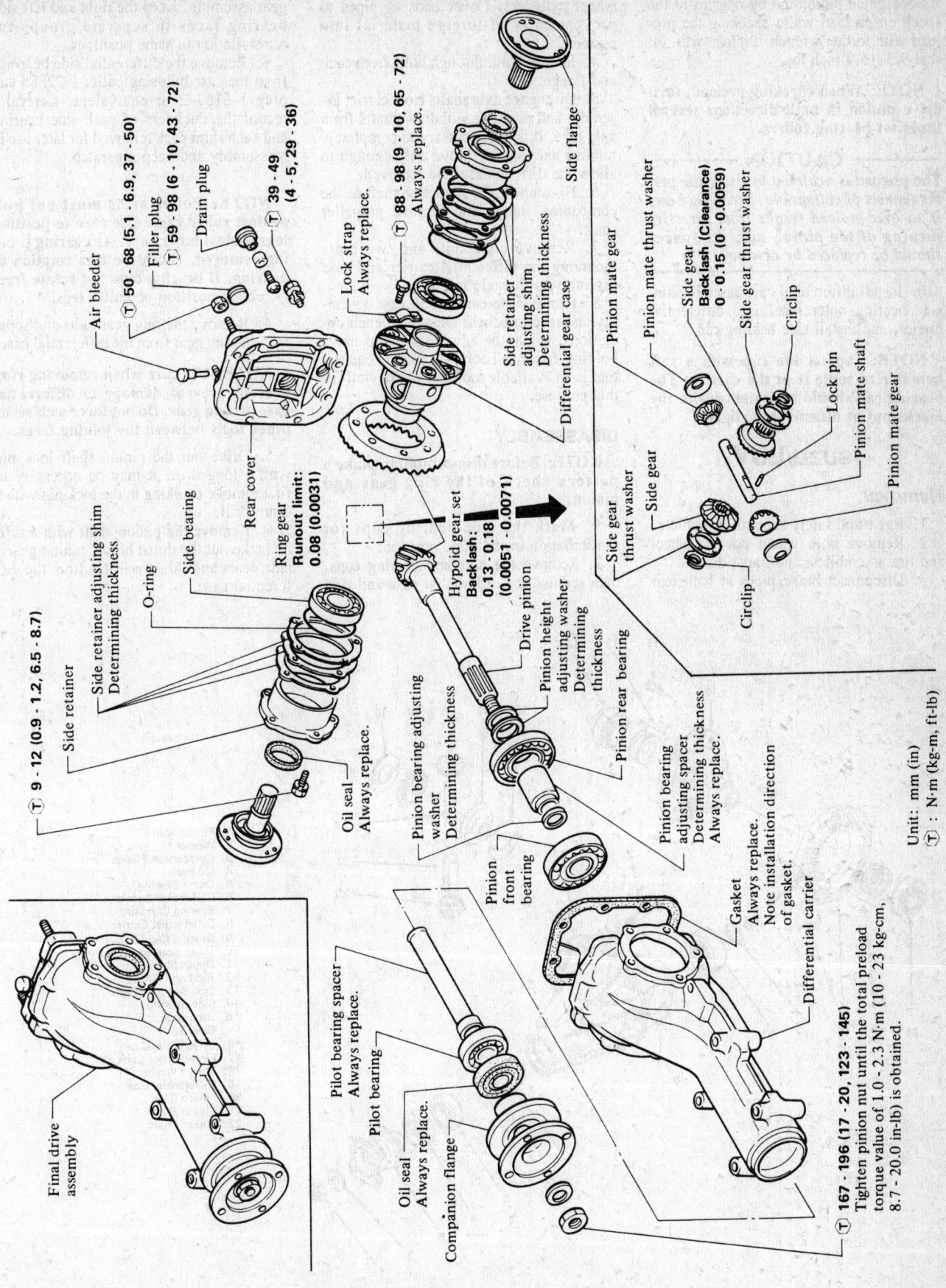

Air bleeder

Ⓣ 50 - 68 (5.1 - 6.9, 37 - 50)

Filler plug

Ⓣ 59 - 98 (6 - 10, 43 - 72)

Drain plug

Ⓣ 39 - 49 (4 - 5, 29 - 36)

Lock strap
Always replace.

Ⓣ 88 - 98 (9 - 10, 65 - 72)

Side flange

Side retainer
adjusting shim
Determining thickness

Differential gear case

Pinion mate gear

Pinion mate thrust washer

Side gear
Backlash (Clearance)
0 - 0.15 (0 - 0.0059)

Side gear thrust washer

Circlip

Lock pin

Pinion mate shaft

Pinion mate gear

Side gear
thrust washer

Side gear

Circlip

Rear cover

Ring gear
Runout limit:
0.08 (0.0031)

Hypoid gear set
Backlash:
0.13 - 0.18
(0.0051 - 0.0071)

Drive pinion

Pinion height
adjusting washer
Determining
thickness

Pinion rear bearing

Side retainer adjusting shim
Determining thickness

O-ring

Side bearing

Ⓣ 9 - 12 (0.9 - 1.2, 6.5 - 8.7)

Side retainer

Oil seal
Always replace.

Pinion bearing adjusting
washer
Determining thickness

Pinion bearing
adjusting spacer
Determining thickness
Always replace.

Gasket
Always replace.
Note installation direction
of gasket.

Differential carrier

Pinion
front
bearing

Final drive
assembly

Pilot bearing spacer
Always replace.

Pilot bearing

Oil seal
Always replace.

Companion flange

Ⓣ 167 - 196 (17 - 20, 123 - 145)
Tighten pinion nut until the total preload
torque value of 1.0 - 2.3 N·m (10 - 23 kg-cm,
8.7 - 20.0 in-lb) is obtained.

Unit: mm (in)
Ⓣ : N·m (kg-m, ft-lb)

Exploded view of a Datsun/Nissan R180 drive axle

9. Tighten pinion nut by degrees to the specified preload while checking the preload with torque wrench. Preload with oil seal 9.5–13.9 inch lbs.

NOTE: When checking preload, turn drive pinion in both directions several times set bearing rollers.

——— CAUTION ———
The preload is achieved by using the permanent set of collapsible spacer. So here, if an over-preload results from excessive turning of the pinion nut, the spacer should be replaced by new one.

10. Install differential case assembly and side bearing outer races into differential carrier, and install side bearing cap.

NOTE: Tap on the cap with a soft hammer to settle it in the carrier. The bearing cap should be installed with the marks put at disassembly aligned.

Isuzu/LUV

Removal

1. Raise and safely support the vehicle.
2. Remove both wheel covers, wheel and tire assemblies, and brake drums.
3. Disconnect brake pipes at both rear wheel cylinders. Cover ends of pipes to prevent entry of foreign material into system.

4. Remove four through bolts from each end-flange.

5. Disengage axle shafts from carrier assembly and partially withdraw shafts from axle tube. It is not necessary to completely remove axle shafts. Move only enough to allow the differential to be removed.

6. Disconnect the propeller shaft at the companion flange and remove propeller shaft.

7. Remove eight nuts and two bolts mounting the differential carrier and case assembly to the axle housing.

8. Remove the carrier and case assembly and take to bench. On some bench operations, it may be advantageous to use a holding fixture. Tool J–21533 or equivalent is an available tool that works well for this purpose.

DISASSEMBLY

NOTE: Before disassembling, make a pattern check of the ring gear and pinion.

1. Mark the side bearing caps for reinstallation in the same position.
2. Remove the nuts and bearing caps, then remove the differential case and ring gear assembly. Keep the right and left side bearing races in separate groups for reinstallation in same positions.

3. Remove the differential side bearings from the case by using puller J–22888 and plug J–8107–2 or equivalent. Carefully record the thickness of each side bearing and each shim pack removed for later use in reassembly and keep separated.

NOTE: Puller arms must not pull against roller cage. Use care to position legs against inner race. As bearing is being removed, check for free rotation of bearing. If bearing does not rotate freely, check position of puller legs.

4. Remove the ring gear bolts and separate the ring gear from the differential case.

NOTE: Use care when removing ring gear to prevent damage to differential case or ring gear. Do not force a chisel or other tools between the joining faces.

5. Drive out the pinion shaft lock pin with a long drift. It may be necessary to first remove caulking in the lock pin with a 5mm drill.

6. Remove the pinion shaft with a drift and take out the thrust block, pinion gears, side gears and thrust washers from the differential case.

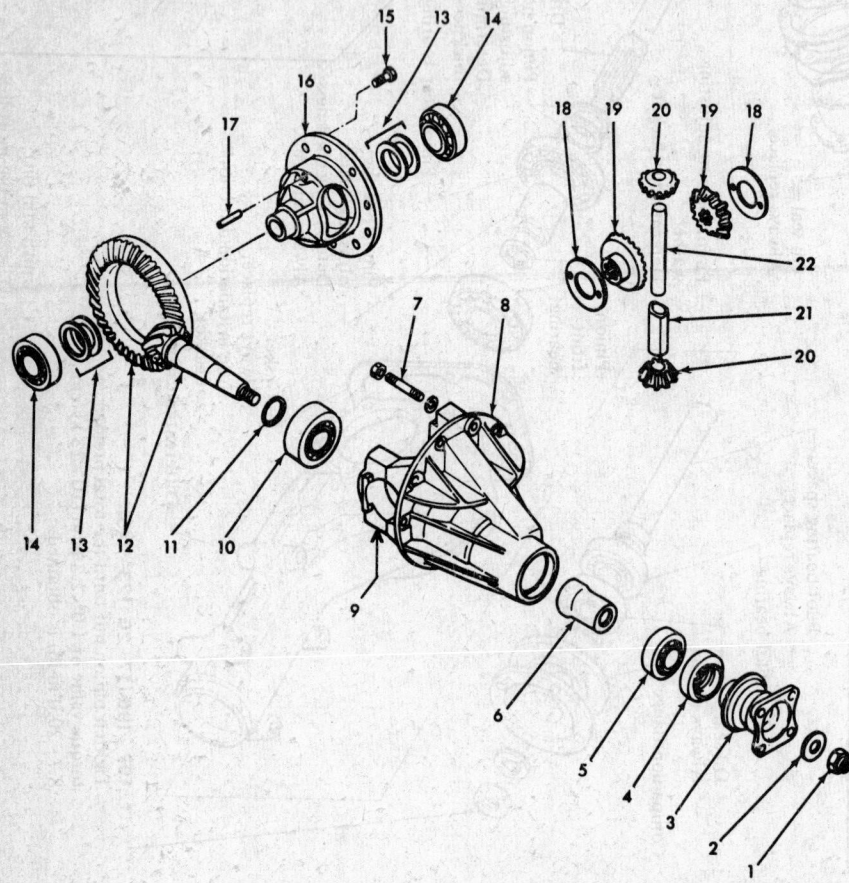

1. Pinion Nut
2. Washer
3. Companion Flange
4. Oil Seal
5. Outer Bearing
6. Collapsible Spacer
7. Bearing Cap Stud
8. Differential Carrier
9. Bearing Cap
10. Inner Bearing
11. Depth Shim
12. Ring and Pinion
13. Side Bearing Shims
14. Side Bearing
15. Ring Gear to Case Bolt
16. Differential Case
17. Pinion Shaft Lock Pin
18. Thrust Washer
19. Differential Gear
20. Pinion Gear
21. Thrust Block
22. Pinion Shaft

LUV/Isuzu rear axle assembly

7. Remove the pinion nut by use of Holder, J-8614-11 or equivalent. Save the pinion nut for pinion bearing preload adjustment procedure.

8. Remove companion flange by use of Nut J-8614-2, and Screw J-8614-3 or equivalent. Install nut onto screw; position nut into J-8614-11 or equivalent and rotate 45 degrees to locking position. Turn screw to draw companion flange off of the drive pinion splines.

9. Hold a soft metal rod against the end of the drive pinion. Drive the pinion from the carrier by using a hammer against the metal rod. The outer (front) bearing will fall loose in the carrier, while the inner (rear) bearing will remain pressed on the drive pinion. Both races will remain in the carrier bores.

10. Remove a rear bearing from the drive pinion by use of a press and tool J-22912-01 or equivalent.

11. Remove the oil seal and then drive the two outer races from the carrier by use of a drift.

Inspection

1. Wash the bearings in a suitable solvent. Then examine bearings carefully for wear, separation, cracks, seizure and other abnormal conditions. Replace bearings as necessary.

2. Check the ring gear and drive pinion teeth for wear, chipping, cracks, pitting, and abnormal contact. Check the drive pinion splines for cracks, distortion and step wear. Replace parts if needed. Ring gears and pinions come only in matched sets. If either item is defective, both parts must be replaced.

3. Check the pinion gears and side gears for wear, chipped teeth and separation, and replace if needed.

4. Check and replace the thrust blocks and thrust washers, if worn.

5. Check the lock pin for bending, dents and other abnormal conditions. Replace if necessary.

6. Check the side gear-to-axle shaft fit. Also check the fit of the pinion shaft to pinion gears.

7. Examine the contact surfaces between side gears and differential case, and between ring gear and case.

NOTE: It is important to clean and assemble parts with care and to follow adjustment procedures. Axles which are contaminated with dirt or other foreign material, or which are incorrectly adjusted may be noisy and have short life. Be sure to use new seals, gaskets and flange nuts when reassembling axle.

Setting Pinion Depth

1. Install the drive pinion front and rear bearing outer races into carrier bores. Use Drive Handle J-8092 with J-24256 for front bearing race and J-24252 or equivalent for rear bearing race.

2. Lubricate and position the front and rear bearings to be used for final assembly into their respective races.

3. Install gauging plate J-23597-7 and preload stud and pilot J-23597-9 or their equivalent through front and rear bearings and tighten nut snugly.

4. Rotate the bearings to insure proper seating and tighten lock nut until 20 inch lbs. of torque are required to rotate new bearings; 8-10 inch lbs. for used bearings.

5. Place discs J-23597-8 onto arbor J-23597-1 or equivalent, and place tool into position in side bearing bores.

6. Install bearing caps snugly.

7. Mount dial indicator J-8001 or equivalent on arbor post, and preload dial one-half revolution. Tighten indicator in this position.

8. Position the indicator plunger on the gauge plate, and slowly swing across until the highest reading is obtained. "Zero" the indicator on the highest reading of the gauge plate.

9. Carefully swing the plunger off the gauge plate. Note the indicator reading. Recheck to verify the reading. The reading on the dial is the correct dimension for the rear pinion depth shim, on a nominal pinion. Shims are available in sizes ranging from 2.18–2.56mm (0.086–0.101 in.). An indicator reading of .000 or 0.03mm (0.001") requires shims of 2.54 and 2.57mm (0.100" and 0.101"), respectively.

10. Examine the head of the drive pinion. The pinion depth code is stamped by chemical ink and is in the lower position of the three numbers. The number indicates a necessary change in the pinion mounting distance. A plus number indicates the need for a greater mounting distance (which can be achieved by decreasing the shim thickness). A minus number indicates the need for a smaller mounting distance (which can be achieved by increasing the shim thickness). If examination reveals no pinion depth code, the pinion is "nominal". Use the chart to determine the proper shim variation to compensate for plus or minus markings.

11. Place the shim on the drive pinion, then install the rear bearing, using J-6133-01 or equivalent.

NOTE: Do not press on roller cage. Press only on inner race.

PINION BEARING PRELOAD

1. Place the drive pinion and spacer into the carrier.

2. Lubricate, then position the front bearing to be used in final assembly into the carrier. Install new oil seal.

3. Mount companion flange to drive pinion. Apply hypoid lubricant to pinion threads. Install new pinion nut and torque to 115 N;pdm (85 lb. ft.), using J-8614-11 or equivalent to hold companion flange.

4. Remove J-8614-11 and rotate drive pinion to insure that bearings are seated.

5. Wind a small amount of string (approximately 4 + + - + +6 windings) around the pinion flange. Connect scale to string. Note the scale reading required to rotate the flange.

6. Continue tightening pinion nut in small increments, using J-8614-11 or equivalent, until the pull required to rotate the flange becomes 17 lbs. for new bearings, 7-9 lbs. for used bearings.

NOTE: The pinion nut should be tightened only in small increments, and the pull scale should be used after each small amount of tightening. Exceeding preload specifications may compress the collapsible spacer too far and require its replacement.

ASSEMBLY

1. Install the side gears and thrust washers in the differential case.

2. Position the pinion gears 180 degrees apart. Roll gears into position, making sure they are in alignment, to allow installation of the pinion shaft.

3. Place the thrust block between the pinion gears, and drive the pinion shaft into position. Make sure that the lock pin hole in cross shaft aligns with the hole in the case.

4. Measure the amount of backlash between the differential gears and the pinion gears. If the backlash is greater than 0.8mm (.003"), make the necessary adjustment with the thrust washers, available in thicknesses of 1.04, 1.14, 1.24 and 1.35mm (.041, .045, .049 and .053"). Increasing washer thickness will decrease backlash. Decreasing washer thickness will increase backlash.

5. Install lock pin into cross shaft and caulk its end to prevent loosening.

6. Clean and apply primer to the bolts.

Pinion Depth Code Number	Alter the Shim Thickness as Determined by Dial Indicator Reading, as Follows:	
+ 10	Subtract	0.13 mm (.005")
+ 8	Subtract	0.10 mm (.004")
+ 6	Subtract	0.08 mm (.003")
+ 4	Subtract	0.05 mm (.002")
+ 2	Subtract	0.03 mm (.001")
0	Use Indicator Reading	
− 2	Add	0.03 mm (.001")
− 4	Add	0.05 mm (.002")
− 6	Add	0.08 mm (.003")
− 8	Add	0.10 mm (.004")
− 10	Add	0.13 mm (.005")

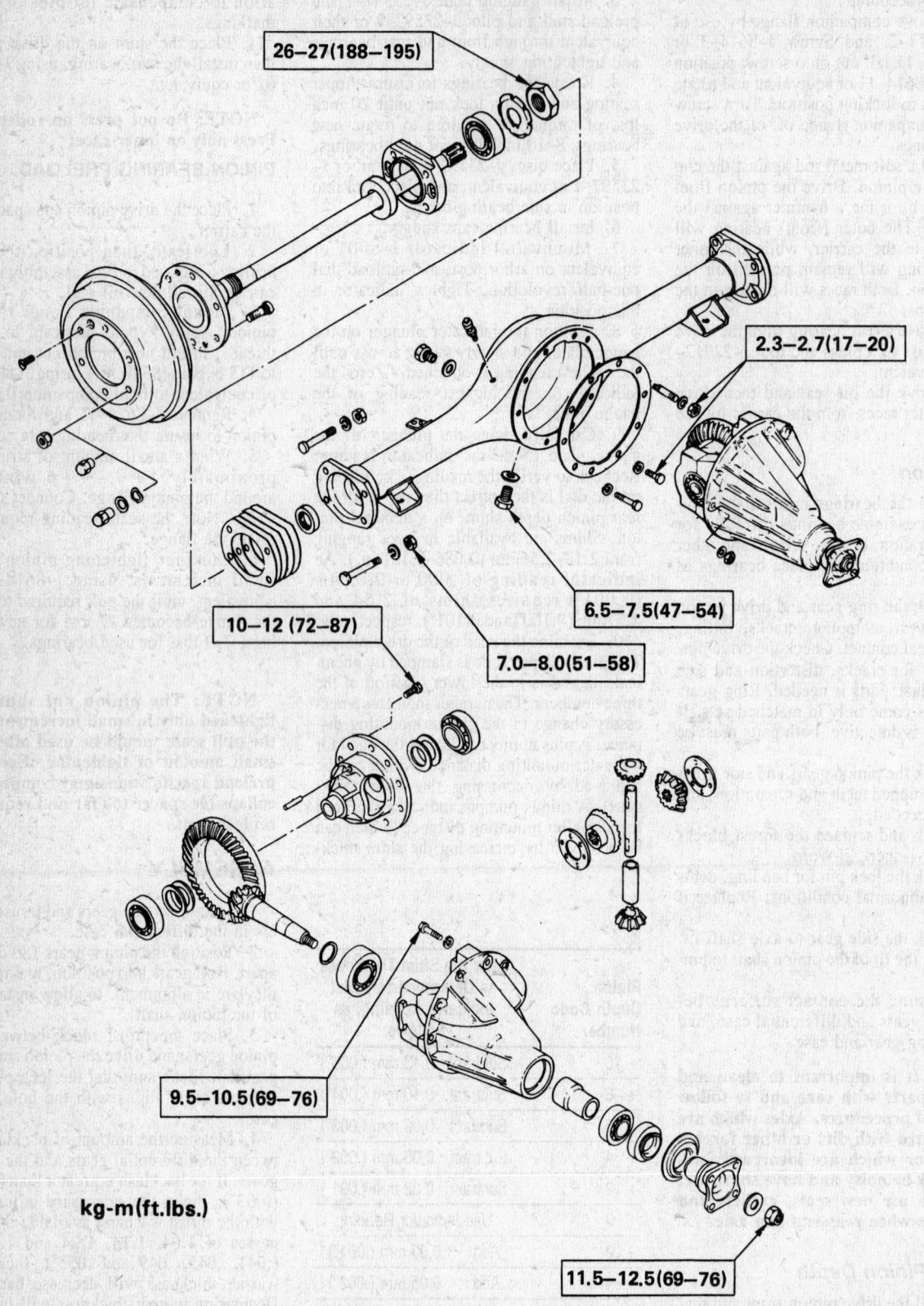

26—27(188—195)

2.3—2.7(17—20)

6.5—7.5(47—54)

10—12 (72—87)

7.0—8.0(51—58)

9.5—10.5(69—76)

11.5—12.5(69—76)

kg-m(ft.lbs.)

Isuzu/LUV drive axle torque specifications

Apply Loctite or equivalent to the threaded portion of the bolts. Install the ring gear in position on the differential case. Tighten the bolts in diagonal sequence to 80–87 ft. lbs.

SIDE BEARING PRELOAD/INITIAL BACKLASH SETTING

If the original side bearings, differential case, ring and pinion, and differential carrier are being reused, and if the pattern check taken before removal showed proper tooth contact, the original shims (or new shims of the same dimension) may be reinstalled in their respective positions.

If new side bearings only are being installed, and if the pattern check taken before removal showed proper tooth contact, this alternate procedure of shim selection may be used: Measure the new bearing with a micrometer, and compare its thickness with the original bearing. If the new bearing is thicker, subtract the numerical difference between new and old bearing from the original shim pack. If the new bearing is thinner, add the numerical difference between old and new bearing to the original shim pack.

If new bearings and/or differential case, ring and pinion, or differential carrier are being installed, new shims will have to be selected for installation behind side bearings.

1. Install the side bearings to be used in final assembly onto the differential case. Do not install shims at this time. Use J–24244 or equivalent for the first bearing installation.

2. Support case on plug J–8107–2 or equivalent for opposite bearing installation.

3. Mount the case into carrier bores.

4. Move the ring gear tightly against the carrier on the ring gear side, away from the drive pinion, and hold in this position. Using a feeler gauge just thick enough to produce a slight drag, carefully measure the clearance between the bearing and differential carrier on the side opposite the ring gear. Record the result.

5. To determine the proper shims for installation, use the following procedure: A predetermined dimension of 0.05mm (.002″) is always needed to establish proper preload. Therefore, add 0.05mm (.002″) (for proper preload) to the clearance measured in Step 4. This will give the necessary combined total thickness of both shim packs. For example, if the measure clearance was 1.12mm (.044″), the combined shims packs would be 1.12 + 0.05 = 1.17mm (.044 + .002 = .046″). Divide the total dimension into two shim packs, so that the numerical difference between the packs equals the numerical difference between the original shim packs. For example, if the original packs measured 1.02 and 0.41mm (.040″ and .016″), the difference between them was 1.02 − 0.41 = 0.61mm (.040 − .016 = .024″). If the new combined shim packs are to total 1.17mm (.046″), the individual shim packs would measure 0.89 and 0.28mm (.035″ and .011″). The difference between 0.89 and 0.28mm (.035″ and .011″), is 0.61mm (.024″); the same numerical difference as between the original packs.

6. Remove the case from the carrier. Carefully remove both side bearings. Install shims as determined in Step 5 behind each bearing, install bearings onto case.

7. Install case onto carrier, tapping carefully into place. Install side bearing caps in original position. Tighten to 75 ft. lbs.

8. Measure the run-out of the ring gear. If the run-out exceeds 0.05mm (.002″), correct by cleaning or replacement of parts.

9. Mount a dial indicator against the ring gear teeth. Make sure indicator button is perpendicular to tooth travel. Measure backlash in three locations. Backlash should be 0.13–0.18mm (.005–.007″).

10. If backlash is not within limits, the shims behind each side bearing will have to be adjusted. In order to maintain proper preload, whenever the backlash is adjusted, the total thickness of both shim packs must not be changed. Therefore, if it is necessary to increase one shim pack, the opposite shim pack must be decreased the same amount.

11. To increase backlash, the right side bearing shim must be increased, and the left side decreased.

12. To decrease backlash, the right side shim must be decreased, while the left side is increased. Backlash changes approximately 0.05mm (.002″) for each 0.08mm (.003″) shim change.

Gear Tooth Pattern Check

Prior to final assembly of the differential, a Gear Tooth Contact Pattern Check is necessary to verify the correct relationship between ring gear and drive pinion. Gear sets which are not positioned properly may be noisy, or have short life, or both. With a pattern check, the most desirable contact between ring gear and drive pinion for low noise level and long life can be assured.

The side of the ring gear tooth which curves outward, or is convex, is referred to as the "drive" side. The concave side is the "coast" side. The end of the tooth nearest center of ring gear is referred to as the "toe" end. The end of the tooth farthest away from center is the "heel" end. Toe end of tooth is smaller than heel end.

Test

1. Wipe oil out of carrier and carefully clean each tooth of ring gear.

2. Use gear marking compound and apply this mixture sparingly to all ring gear teeth using a medium stiff brush. When properly used, the area of pinion tooth contact will be visible when hand load is applied.

3. Hand load the companion flange and rotate the ring gear one revolution in each direction. Excessive turning of ring gear is not recommended.

4. Observe the pattern made on ring gear teeth. The contact pattern should be centrally located up and down on the face of the ring gear teeth.

Adjustments

Two adjustments can be made which will affect tooth contact pattern: (1) backlash, and (2) position of drive pinion in differential carrier. The effects of bearing preloads are not readily apparent on (hand loaded) tooth pattern tests; however, these adjustments should be within specifications before proceeding with backlash and drive pinion adjustments. It may be necessary to adjust both pinion depth and backlash to obtain the correct pattern.

The position of the drive pinion is adjusted by increasing or decreasing the shim thickness between the pinion head and inner race of rear bearing. The shim is used in the differential to compensate for manufacturing tolerances. Increasing shim thickness will move the pinion closer to centerline of the ring gear. Decreasing shim thickness will move pinion farther away from centerline of the ring gear.

Backlash is adjusted by means of the side bearing adjusting shims which moves the entire case and ring gear assembly closer to, or farther from the drive pinion. (The adjusting shims are also used to set side bearing preload.) To increase backlash, increase right shim and decrease left shim an equal amount. To decrease backlash, decrease right shim and increase left shim an equal amount.

Installation

1. Clean the faces of the rear axle case and differential carrier and apply a thin coat of liquid gasket and install the gasket.

2. Mount the differential case and carrier assembly to the rear axle case and tighten the nuts to 24 N;pdm (18 lb. ft.).

3. Install the axle shaft assemblies.

4. Connect the companion flange with the propeller shaft and tighten the bolts to 18 ft. lbs.

NOTE: This propeller shaft to pinion flange fastener is an important attaching part in that it could affect the perform < chance of vital components and systems, and/or could result in major repair expense. It must be replaced with one of the same part number or with an equivalent part if replacement becomes necessary. Do not use a replacement part of lesser quality or substitute design. Torque values must be used as specified during reassembly to assure proper retention of this part.

5. Fill the rear axle case with hypoid gear lubricant, to just below the filler hole.

Mazda/Courier

INSPECTION

An inspection of the adjustments and parts as the carrier is disassembled can assist in learning the cause of the trouble and in determining what corrections are needed.

Mount the carrier in a holding fixture. Wipe the lubricant from the internal working parts and visually inspect them for damage. Rotate the gears to see if there is any roughness which could indicate worn or damaged bearings or chipped gears. Look carefully at the surface of the gear teeth for any scoring, flaking, or signs of abnormal wear.

Mark the ring gear at four points at approx. 90° intervals and mount a dial indicator to the carrier flange. Check to see that the backlash at one of the marked points is 0–.008 in. or for 1986 models 0–.004 in. Also check the backlashes at other three marked points and make sure that the difference between the maximum and minimum backlashes is less than .0028 in.

If no obvious misadjustment or damage is noted, inspect the gear tooth contact. Coat the gear teeth with special compound available, or red lead. Too fluid a mixture will run and smear. Too dry a mixture cannot be squeezed out from between the teeth. Rotate the ring gear back and forth (use a box wrench on the ring gear attaching bolts for a lever), until a clean tooth contact pattern is obtained.

Certain types of gear tooth contact patterns on the ring gear indicate incorrect adjustment. A noise condition caused by incorrect adjustment can often be corrected by readjusting the gears. Gear tooth runout can be detected by an erratic pattern on the teeth. If ring gear runout is suspected, mount a dial indicator to measure the runout of the back face of the ring gear.

SERVICING

1. Remove the carrier from the differential.
2. Mount the carrier in a holding fixture.

3. Apply identification punch marks on the carrier, the differential bearing cap and the adjusters for accurate reassembly.
4. Remove the adjuster lock plates.
5. Loosen the bolts securing the bearing cap and slowly back off the adjuster slightly to relieve the preload.
6. Remove the nuts, bearing caps and adjusters. Keep each bearing cap with its own adjuster.
7. Lift out the differential assembly and keep each bearing outer race with its own bearing.
8. If the differential bearings are to be replaced, remove them using a puller.
9. Remove the bolts and washers retaining the ring gear to the case.
10. Remove the ring gear.
11. Position the assembly in a vise and remove the lock pin with a suitable punch.
12. Remove the pinion shaft and the thrust block.
13. Rotate the differential pinion gears 90 degrees and remove.
14. Remove the differential side gears and thrust washers.
15. Using a holding tool, steady the companion flange and remove the nut.
16. Remove the companion flange.
17. Remove the drive pinion and rear bearing from the carrier, which may require tapping with a plastic or rubber mallet. Guide the pinion to avoid damage to the teeth.
18. Remove the oil seal and the front bearing.
19. The pinion bearing outer races (cups) can be removed from the carrier by using a drift in the slots provided for the purpose.
20. Remove the bearing from the pinion using suitable equipment.
21. With the carrier completely apart, check the drive pinion for damaged or worn teeth, damaged bearing journals or splines. Inspect the ring gear again for worn or chipped teeth. If any of the above conditions are found, replace both drive pinion and ring gear as these are only available in sets.
22. Before reassembly of the carrier, inspect bearing cones and cups and replace any showing wear, flaking or damage. Replace only in sets.

NOTE: Do not use an old cup with a new bearing or an old bearing with a new cup. If this is done, damage will result.

23. Check the companion flange carefully for cracks or worn splines. If either exist, the part should be replaced. Check for rough or scratched oil seal contact surface. If only slight scratches appear, it may be possible to repair with crocus cloth. Otherwise, replace it. Be sure to use a new oil seal when reassembling the carrier. Reassemble the components in the reverse order.

1. Lock nut/washer	8. Lock plate
2. Companion flange	9. Adjuster
3. Oil seal	10. Side bearing
4. Front bearing	11. Thrust washer
5. Collapsible spacer	12. Side gear
6. Carrier	13. Rear bearing
7. Bearing cap	14. Spacer

15. Drive pinion	
16. Ring gear	
17. Pinion gear	
18. Pinion shaft	
19. Pinion shaft pin	
20. Thrust block	
21. Gear case	

Mazda rear axle assembly

Pinion Depth

If you use the same pinion and ring gear, the shim combination found on the pinion may prove satisfactory provided other things are equal.

Individual differences in machining the carrier casting and the gear set require a shim or shims between the pinion rear bearing and the pinion gear to locate the pinion for correct tooth contact with the ring gear.

NOTE: Special tools are required to check and adjust the pinion depth, these include a drive pinion model (49 8531 565), a pinion height adjustment gauge body (49 0727 570) and a gauge block (40 0305 555) or their equivalents. The original factory installed spacers are of the correct thickness to adjust for individual variations in both the carrier casting dimension and in the original gear set dimension. To select the correct spacer thickness when installing a new gear set, follow these steps:

1. Fit the spacer, rear bearing and Collar B (49 8531 568) onto the drive pinion model. Secure the collar with an O-ring and install in the assembly in the carrier.
2. Attach the front bearing, Collar A (49 8351 567), companion flange, washer and nut to the drive pinion model. Use the same spacer and nut which were removed at disassembly. Be careful to install Collars A and B in their correct position facing in the correct direction.
3. Tighten the nut until the drive pinion model can be turned by hand without any apparant play.
4. Install a dial indicator on the pinion height adjustment gauge body. Place the gauge block on top of the drive pinion model and then set the pinion height adjustment gauge body on top of the gauge block.
5. Place the measuring probe of the dial indicator so that it contacts the location where the side bearing is installed in the carrier. Zero and set up the indicator to measure the lowest point. Measure both the left and right sides.
6. Add the two values (right and left side readings) and divide the total by 2. From this result, subtract the result obtained by dividing the number inscribed on the end surface of the drive pinion by 100. (If there is no figure inscribed, use 0). The resulting figure is the pinion height adjustment value.
7. For example, if the measured results obtained are 0.06mm and 0.04mm and the figure inscribed on the end of the drive pinion is -2 the following formula would be used. Thus a spacer which is 0.07mm thicker than the one now used is required. Select the spacer thickness that is closest to requirement.
8. Install the selected spacer on the pinion shaft (facing in the proper direction) and press the bearing on the pinion shaft.
9. Install the drive pinion, spacer, front bearing, collapsible spoacer and compan-

ion flange in the carrier (remove the measuring tools first) and temporarily tighten the locknut. DO NOT install the pinion seal at this time.
10. Adjust the pinion bearing preload at this time.
11. After preload measurements are taken install the pinion seal. Coat the drive flange and seal surface with "moly" grease. Install the flangeand a NEW locknut. Tighten the locknut to torque measurement taken when establishing the required preload.

Pinion Bearing Preload

1. Install the correctly "shimmed" pinion shaft in the carrier with the spacer, flange and nut but NO pinion seal.
2. Turn the flange by hand to seat the bearing.
3. Use a torque wrench to tighten the locknut. Tighten slowly until the required preload drag of 7.8–12.2 inch lbs. is reached with a locknut torque of 94–130 ft. lbs.
4. If the specified preload cannot be maintained within the locknut tightening range, install a new collapsible spacer and repeat the process.
5. Remove the locknut and flange. Install the pinion seal, flange and NEW locknut. Tighten the locknut to the ft. lb. torque giving the correct preload.

Differential Assembly

1. Install the thrust washer on each differential side gear and install these in the case.
2. Through the openings of the case, insert each of two pinion gears exactly 180 degrees opposite each other.
3. Rotate the gears 90 degrees so the pinion shaft holes of the case align with the holes in the two pinion gears.
4. Insert the pinion shaft through the case and pinion gears.
5. Check the backlash of the side gear and pinion gear. The backlash should not exceed 0.008 or 0.004 on 1986 models. If it exceeds this, adjustment can be made with the side gear thrust washers.
6. After adjustment, remove the pinion shaft and install the thrust block so the hole is centered between the differential pinion gears. Reinstall the pinion shaft into the case until the lock pin hole in the pinion shaft is in exact alignment with the hole in the case.
7. Install the lock pin to secure the pinion shaft. Stake the lock pin into position with a punch to prevent it from working out.
8. Install the ring gear on the case and tighten the bolts to the specification listed at the end of this Part.
9. Install each differential bearing to the case using T72J-4221 or equivalent by press or with a hammer. Install the outer races to their respective bearing.

10. Place the differential gear assembly in the carrier.
11. Note the identification marks on the adjusters and install each to its respective side.
12. Install the bearing caps making sure that the identification marks on the caps correspond with those on the carrier and install the bolts.
13. Turn the adjusters with the spanner tool T72J–4067 or equivalent until the bearings are properly positioned in their respective outer race and the end play is eliminated with some amount of backlash existing between the ring gear and drive pinion.
14. Slightly tighten one of the bearing cap bolts on each side and adjust the backlash.
15. Mark the ring gear at four points at approx. 90° intervals and mount a dial indicator to the carrier flange so that the feeler comes in contact at right angles with one of the ring gear teeth. Turn both bearing adjusters equally until the backlash becomes 0.0075–0.0083 in. Check the backlashes at the other three marked points and make sure that the difference between the maximum and minimum backlashes is less than 0.07mm (0.0028 in).
16. Adjust the preload taking care not to disturb the backlash. Use a dial indicator and carefully set the preload at 0.0045.
17. The bearing cap bolts should be tightened to 45 ft. lbs.
18. Install the adjuster lock plates.

Mitsubishi

OVERHAUL

Differential overhaul requires many special tools and access to a range of preload shims and other dealer equipment. If you have never overhauled a rear axle assembly before, it would be wise to let a specialist perform this operation.

1. Remove the lock bolts and plates holding the side bearing nut in place.
2. Remove the side bearing nuts with the special adjusting spanner no. MB990201.
3. Remove the carrier caps and pry out the differential.
4. Pull off the differential side bearings.

NOTE: Be sure to keep the right and left bearings and shims separated.

5. Loosen the ring gear mounting bolts in diagonal sequence. Remove the ring gear.
6. Drive the pinion shaft lock pin out from the rear of the ring gear using a punch, and remove the pinion shaft.
7. Remove the side gears with their spacers. Keep left and right side gears and spacers separate.
8. Hold the end yoke and remove the pinion lock nut.
9. Remove the end yoke.
10. Tap the end of the drive pinion shaft

with a plastic hammer and force out the drive pinion along with its adjusting shim, the rear inner race, the drive pinion spacer and the preload adjusting shim. The rear bearing inner race can be pressed off the pinion shaft.

11. Remove the front and rear pinion bearing outer races. The front race should be removed with its oil seal.

NOTE: Do not reuse the old oil seal. If the unit is to be assembled using no replacement parts except oil seals, the same spacers and shims can generally be used. If either pinion bearing or ring gear and drive pinion are being replaced, new shims should be used. Only replace the drive pinion and ring gear in matched sets.

12. Assemble the side gears in the differential case. Install the spacers in the same positions they were in when removed.

13. With the washers, insert both differential gears at the same time to mesh with the side gears. Insert the pinion shaft.

14. Measure the backlash of the differential pinion gears and the side gears. Backlash should be within 0.002–0.005 in. If not, replace the side gear spacers with the appropriate ones listed below.

15. Align the differential drive pinion shaft with the lock pin hole in the differential case and drive the pin in from the rear of the case. Stake the pin with a small pointed punch to secure it.

16. Remove the old adhesive from the ring gear mounting bolts and apply new adhesive. Snug up all bolts then tighten them on a crisscross pattern of 58–65 ft.lb.

NOTE: To allow the adhesive to set on the bolt threads, keep the unit stationary for about an hour.

17. Press the front and rear bearing outer races into the gear carrier.

————— **CAUTION** —————
Make sure that the races do not tilt and that they sit fully in the case.

18. Look at the top face of the drive pinion (gear side). If there is an etched number, such as −0, −1, −2, +1, +2, etc., complete step 17. If not, skip step 19 and go on to step 20.

19. Insert a shim between the drive pinion and rear bearing. If the original gear set is being replaced, the original shims may be used. If a new gear set is being installed, calculate the shim dimension in the following manner. Assuming the pinion height before disassembly is correct, subtract the new pinion variation marking (on the pinion head) from the old pinion variation marking. If the answer is positive, add shims in the corresponding amount. If the answer is negative, subract shims in the corresponding amount. This will produce a reasonable starting point for assembly. If the shim choice is proved incorrect, the entire pinion must be disassembled, and the

shim changed accordingly. The etched marking on the face of the pinion represents a positive or negative variation from the standard in millimeters.

NOTE: If the original gear set is being reused in the differential case, the original shims may be used.

20. If the drive pinion has no marking on its gear side face, it will be necessary to obtain two dealer special tools: MB990819 and MB990552. Install parts marked 1,6,2,7,3,4, and 5 in the illustration labeled "Measuring pinion height (clearance)" with special tool MB990819 into the carrier case. Gradually tighten the nut to produce 6–9 in. lbs. without the oil seal. Fit special tool MB990552 in the differential caps and replace the caps on the case. Measure the clearance between the two special tools (see illustration) and select a shim of an equivalent thickness to the clearance to make the pinion height within tolerance of ±0.0012 in.

NOTE: If the pinion height has to be adjusted by more than 0.0650 in. use two shims including one 0.0118 in. thick.

21. Install the selected shim between the drive pinion and the rear bearing. Press the bearing onto the drive pinion shaft.

22. Assemble the drive pinion in the case and torque the pinion nut gradually to 137–180 ft.lb. Check the pinion preload. With oil seal, it should be between 9–11 in. lbs. Without the oil seal, it should be 6–9 in. lbs. The preload shim selection ranges from 0.0118 to 0.0917 in.

23. If you have not already done so, apply a thin coat of grease to the drive pinion oil seal and insert it in the case. Refit the yoke and tighten to 137–180 ft.lb.

24. Press the side bearings into the differential case and fit the case into the carrier.

25. Install the carrier caps with their mating marks in line with the marks on the carriers and finger tighten the four set bolts.

26. Install the side bearing nuts, and tighten the carrier cap bolts to 40–47 ft.lb.

27. Screw in the side bearing nuts to adjust the standard backlash value. Each nut should be tightened to 11 lbs. Repeatedly loosen and tighten the bearing nuts to insure smooth operation, then tighten them until they become hard to turn.

28. Attach a dial indicator to the ring gear teeth and make certain the backlash is between 0.005–0.007 in.

NOTE: If the backlash is less than the limit, loosen the bearing nut on the back side of the ring gear and tighten the bearing nut on the teeth side by the same amount.

29. After adjusting backlash, tighten the bearing nuts ¹/₂ pitch.

NOTE: One pitch is the space between two adjacent holes on the side of the bearing nut.

30. Again measure the backlash and install a one or two pronged lock plate whichever lines up with the bearing nut holes. Tighten the lock plate bolts to 11–16 ft.lb.

31. Measure the ring gear runout in four or more spots. Runout should be 0.002 in. or less.

32. Make a ring gear tooth pattern check.

33. Apply gear oil to all moving parts and use sealant when assembling. Install the packing with the embossed portion at about 3 o'clock position on the axle housing.

34. Fit the differential and tighten the mounting nuts to 18–22 ft.lb.

35. Be sure to fill the rear axle with about 3 pints of gear oil before testing.

D50 Arrow

Case Disassembly

1. Remove the lock plate, and then remove the side bearing nut using special tool (MB990201) or equivalent.

2. Remove the carrier cap. Then remove the differential case assembly using the wooden handle of a hammer or similar object so that gears and other parts will not be damaged.

3. Using a bearing puller pull off the side bearing.

NOTE: Keep the side bearings and side bearing nuts separate, so that they do not become mixed at the time of reassembly.

4. Make the mating marks to the differential case and the drive gear.

5. Loosen the drive gear mounting bolts in diagonal sequence, and then remove.

6. Remove the lock pin from the drive gear back side using a long punch. Then pull out the pinion shaft and the pinion gears. The side gears with the side gear thrust spacers can then be removed.

NOTE: The removed side gears and side gear thrust spacers, left and right, should be retained for reassembly.

DRIVE PINION

1. Hold the end yoke with special tool (MB990850) or equivalent and after removing the self-locking nut, remove the end yoke.

2. Make the mating marks to the drive pinion and end yoke.

3. Tap the drive pinion end with a plastic hammer or a wheel puller and force out the drive pinion with the drive pinion height adjusting shim, drive pinion bearing (rear) inner race, drive pinion spacer and the drive pinion preload adjusting shim still installed on the drive pinion.

4. Using special tools (MB990339 and MB990648 or their equal) remove the drive pinion bearing (rear) inner race from the drive pinion. Remove the drive pinion height adjusting shim at the same time.

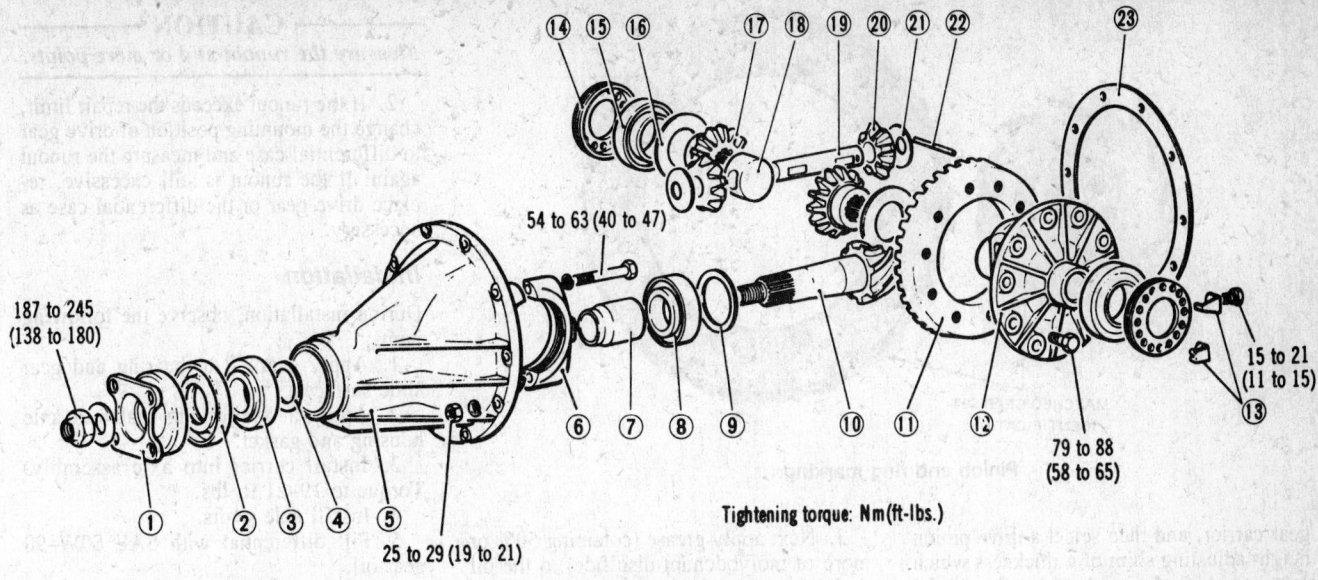

54 to 63 (40 to 47)

187 to 245
(138 to 180)

25 to 29 (19 to 21)

15 to 21
(11 to 15)

79 to 88
(58 to 65)

Tightening torque: Nm(ft-lbs.)

(1) End yoke (companion flange)
(2) Oil seal
(3) Drive pinion bearing, front
(4) Drive pinion preload adjusting shim
(5) Gear carrier
(6) Carrier cap
(7) Drive pinion spacer
(8) Drive pinion bearing, rear

(9) Drive pinion height adjusting shim
(10) Drive pinion
(11) Drive gear
(12) Differential case
(13) Lock plate
(14) Side bearing nut
(15) Side bearing
(16) Side gear thrust spacer

(17) Side gear
(18) Center block
(19) Pinion shaft
(20) Pinion gear
(21) Pinion washer
(22) Lock pin
(23) Packing

Exploded view Mitsubishi/Arrow/D-50 rear axle assembly

NOTE: The drive pinion height adjusting shim should be retained for reassembly.

5. Remove the drive pinion bearing (front and rear) outer race. When removing drive pinion bearing (front) outer race, remove oil seal and drive pinion bearing (front) inner race at the same time.

NOTE: The removed oil seal should not be reused.

INSPECTION

1. Check differential gear tooth contact. Replace any gear that is worn or damaged.
2. Check the bearing race curvature for discoloration caused by seizure, and rough surface. Replace any bearing that is defective.
3. Install the side gear onto the splined end of the axle shaft. Check for looseness of the axle shaft spline using a dial indicator on the side gear.

Description	Service Limit mm (in.)
Axle shaft spline looseness	0.6 (.024)

4. Check the differential pinion and pinion shaft for wear or seizure.

ASSEMBLY & ADJUSTMENT DIFFERENTIAL CASE

1. Install the side gear thrust spacers in their original positions behind the side gears. Then assemble the side gears (left and right) in the differential case.
2. With pinion washers attached to the pinion gears, insert the both pinion gears at the same time and mesh with the side gears by rotating the pinion gears.
3. Complete the temporary assembly of the differential gears by inserting the pinion shaft.
4. Check the pinion gear and side gear backlash. If the backlash exceeds the repair limit, adjust it by selecting a side gear thrust spacer of proper thickness. The backlash, left and right, should be adjusted to an equal value.

Description	Standard Value mm (in.)
Pinion gear and side gear backlash	0.051 to 0.127 (.002 to .005)

TYPES OF SIDE GEAR THRUST SPACER

Part No.	Thickness of Spacer mm (in.)	
MB092034	0.8−0.08 −0.17	(.0315−.0031 −.0067)
MB092035	0.8−0.18 −0.27	(.0315−.0071 −.0106)
MB092036	0.8−0 −0.07	(.0315−0 −.0028)

5. Align the pinion shaft with the lock pin hole in the differential case, and drive the lock pin into the hole from the back side of the drive gear. Securely stake the lock pin at two places with a punch to prevent it from moving.
6. Remove old adhesive from the drive gear mounting bolts using a wire brush and from the internal thread using hand tap. Apply "LOCTITE 271" or equivalent anaerobic adhesive. Temporarily tighten each bolt evenly, and then tighten to the specified torque in a criss-cross fashion.

NOTE: Keep the differential stationary to harden the anaerobic adhesive for half an hour to one hour.

Adjusting Drive Pinion Height

1. Press the drive pinion bearing (front and rear) outer races into the gear carrier using special tools (MB990934, MB990936 and MB990938) or their equal. Use great care so that the outer race does not tilt and be sure the race bottoms in its bore in the gear carrier.
2. Install the drive pinion bearings and special tool (MB990819) or equivalent. Gradually tighten with the nut until the drive pinion preload becomes within 6 to 9 inch lbs.

— **CAUTION** —
When installing the washer, apply a thin coat of grease on the washer.

3. Mount the special tool (MB990552) or equivalent in the side bearing seat of the

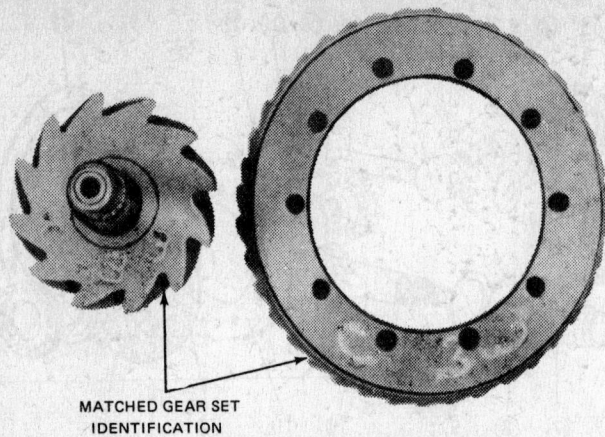

MATCHED GEAR SET
IDENTIFICATION

Pinion and ring markings

gear carrier, and then select a drive pinion height adjusting shim of a thickness which corresponds to the gap between the special tools (MB990552 and MB990819) or their equal.

NOTE: Be sure to clean the side bearing seat thoroughly. When mounting the special tool (MB990552) or equivalent, be sure that the cut-out sections are in the position shown in the illustration, and also confirm that the tool is in close contact with the side bearing seat.

4. Install the selected drive pinion height adjusting shim between the drive pinion and the drive pinion bearing (rear). Using special tool (MB990802) or equivalent, press the drive pinion bearing (rear) onto the drive pinion.

5. If the gear set is to be replaced, install new shims of the same thickness as the shims previously used on the drive pinion.

———— **CAUTION** ————

In determining the thickness of the shim pack, the amount of compression of the shim pack and wear of the bearing (when the old bearing is reused) should be taken into account.

Adjusting Drive Pinion Preload

1. Insert drive pinion preload adjusting shim between drive pinion spacer and drive pinion bearing (front). Tighten end yoke to specified torque, to obtain standard preload 9–11 inch lbs. with seal, 6–9 inch lbs. no seal. If preload is wrong, install a preload adjusting shim of a different thickness until preload is in the standard value.

NOTE: Beside the drive pinion preload adjusting shims, the drive pinion spacers may be used for adjustment.

2. After completion of the drive pinion preload adjustment, remove the end yoke. Apply a thin coat of grease to the periphery of the oil seal, and drive it into the gear carrier using special tool (MB990031) or equivalent.

3. Next apply grease (cotaining 50% or more of molybdenum disulfide) to the oil seal lip contact surface of the end yoke shaft, insert the end yoke, and tighten the self-locking nut to the specified torque.

Side Bearing

1. Press in the side bearing inner race to the differential case using special tool (MB990802) or equivalent.

2. Place differential case assembly into gear carrier. Line up the mating mark on gear carrier with that on carrier cap, and tighten set bolts for tightening bearing cap to gear carrier with fingers. Then install side bearing nut, and tighten it on carrier cap to the specified torque 40–47 ft. lbs.

3. Screw in the side bearing nut on each side of drive gear using the special tool (MB990201) or equivalent to adjust the standard backlash value. Make sure the side bearing nuts are tightened to 11 lbs. at end of special tool. Turn the bearing nuts in and out several times until they operate smoothly, then tighten them to the proper torque.

4. Apply a dial indicator to the drive gear tooth, and make certain that backlash is within the standard value.

NOTE: If the backlash is smaller than the standard value, loosen the side bearing nut on the back side of the drive gear and tighten the side bearing nut on the teeth side by the same amount.

5. After adjustment, tighten side bearing nut on both sides by a half pitch to give preload on side bearings. One pitch means space between two adjacent holes on the side of the side bearing nuts.

6. Measure the backlash again to ensure that it is within the standard value. Choose a lock plate of proper type, and tighten it to the specified torque 11–15 ft. lbs.

Drive Gear

1. Apply a dial indicator to the back of the drive gear and measure the amount of runout. Maximum Runout .002 in.

———— **CAUTION** ————

Measure the runout at 4 or more points.

2. If the runout exceeds the repair limit, change the mounting position of drive gear to differential case and measure the runout again. If the runout is still excessive, replace drive gear or the differential case as necessary.

Installation

During installation, observe the following items:

1. Apply gear oil to bearing and gear slide surfaces.

2. Apply a semi-drying sealer to axle housing and gasket.

3. Install carrier into axle assembly. Torque to 19–21 ft. lbs.

4. Install axle shafts.

5. Fill differential with SAE 80W–90 gear oil.

Toyota

NOTE: If the differential is noisy, perform the following pre-inspection before disassembly to determine the cause of the noise.

1. Check ring gear runout. If the runout is greater than maximum, install a new ring gear. Maximum runout: $\frac{1}{2}$ ton and $\frac{3}{4}$ ton; 0.0028 in.: C&C and 4×4; 0.0039 in.

2. Check ring gear backlash. If the backlash is not within specification, adjust the side bearing preload or repair as necessary. Backlash: 0.0051–0.0071 in.

3. Check the tooth contact.

4. Using a torque meter, measure the total preload. Total preload (Starting): Drive pinion preload plus 3.5–5.2 inch lb.

Disassembly

1. Remove differential case and ring gear. Put alignment marks on the bearing cap and differential carrier. Remove two adjusting nut locks.

2. Remove two bearing caps and two adjusting nuts. Remove the bearing outer races. Remove the differential case from the carrier.

3. Remove companion flange, oil seal and front bearing.

4. Remove drive pinion from differential carrier.

5. Remove drive pinion rear bearing using a universal puller.

6. Remove front and rear drive pinion bearing outer race, using a hammer and punch.

7. Remove ring gear from differential case.

8. Remove the ring gear set bolts and lock plates.

9. Using a brass bar and hammer, tap on the ring gear to separate it from the differential case.

———— **CAUTION** ————

Be careful not to damage the side bearing.

10. Remove the side bearings using a universal puller.

11. Disassemble differential case using a hammer and punch. Drive out the straight pin. Remove the pinion shaft, two pinion gears, two side gears and two thrust washers.

12. Inspect differential case parts.

13. Replace parts that are damaged or worn.

Inspection

1. Clean all parts with solvent.

2. Inspect drive pinion bearings and outer races. If the bearing or outer race are damaged or worn, replace them as a set.

3. Inspect ring gear and drive pinion. If the ring gear or drive pinion are damaged or worn, replace them as a set.

4. Inspect pinion and side gears. If the pinion or side gears are damaged or worn, replace the gears.

5. Inspect side bearings and outer races. If the side bearings or outer races are damaged or worn, replace the bearing and race.

6. Check side gear backlash. Measure the side gear backlash while holding one pinion gear toward the case. Standard backlash: 0.0020–0.0079 in. If the backlash is out of specification, install the correct thrust washers.

Assembly

½ & ¾ TON

1. Adjust drive pinion protrusion. Install the bearings and adjusting gauge (tool 09530–30012 and 09536–30030 or equivalent) in the differential carrier in the following order: Rear bearing. Base rod. Drive pinion front bearing. Collar. Flange. Nut. Base rod head. Bolt.

NOTE: Tighten the bolt only to the point where the drive pinion gear has no play.

2. Place the master gauge on the differential carrier.

3. Align the marks and install the bearing caps. Torque the bearing cap bolts 51–65 ft. lbs.

4. Select a washer that can just be inserted into the clearance between the master gauge and the base rod. Washer thickness 0.0878–1.075 in. Remove the adjusting gauge.

5. Install rear bearing and washer on drive pinion using a press.

NOTE: The chamfered end of the washer should face toward the gear.

6. Coat the bearings with gear oil and install the drive pinion into the carrier.

7. Install new bearing spacer, front bearing and oil slinger.

8. Install new oil seal. Apply multipurpose grease to the oil seal.

9. Install companion seal.

10. Tighten drive pinion nut and adjust preload. Coat threads of a new nut with

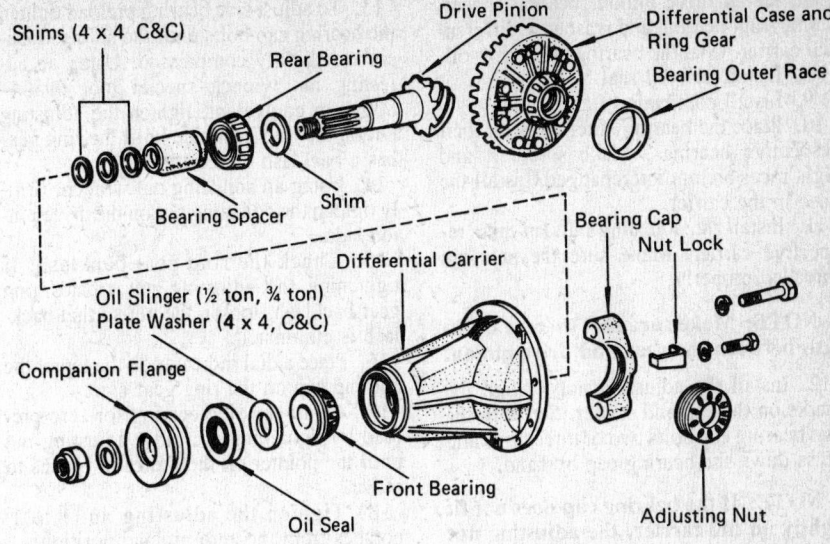

Toyota differential—exploded view

multipurpose grease. Torque nut to 80 ft. lbs. Turn the companion flange several times to snug down the bearing.

11. Using a torque meter, measure the preload of the backlash between the drive pinion and ring gear. Preload: If preload is greater than specification, replace the bearing spacer. If preload is less than specification, retighten the nut 5–10° at a time until the specified preload is reached. If the maximum torque is exceeded while retightening the nut, replace the bearing spacer and repeat the preload procedure. Do not back off pinion nut to reduce the preload. Maximum torque: 173 ft. lb.

12. Using a dial indicator, measure the longitudinal and latitudinal deviation of the companion flange.

CAB CHASSIS AND 4×4

1. Adjust drive pinion preload. Install the bearings, spacer, shim and adjusting gauge etc. into the differential carrier.

NOTE: Do not install the oil seal. Do not install the shim for drive pinion height.

2. Using a wrench to hold the collar, tighten the nut. Torque the nut 123–151 ft. lbs.

3. Using a torque meter, measure the preload.

4. Adjust drive pinion protrusion. Place the master gauge on the differential carrier. Align the marks and install the bearing caps. Torque the bearing cap bolts, 51–65 ft. lbs.

5. Select a washer than can just be inserted into the clearance between the master gauge and the base rod. Remove the adjusting gauge.

6. Using a press and special tool 09506–30011 or equivalent, press the washer and rear bearing on the drive pinion.

NOTE: The chamfered end of the washer should face the gear.

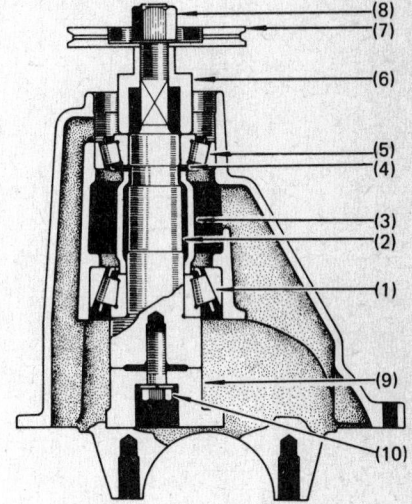

Toyota—4 × 4 and cab/chassis

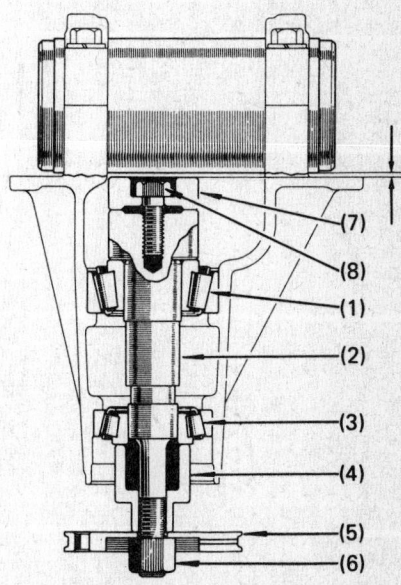

Toyota—½ and ¾ ton

7. Install drive pinion, bearing spacer, shim, front bearing and washer in differential carrier. Coat the bearings with gear oil.

8. Install new oil seal.

9. Install companion flange.

10. Place the bearing outer races on their respective bearings. Make sure left and right races are not interchanged. Install the case in the carrier.

11. Install the adjusting nuts on their respective carrier. Make sure the nuts are threaded properly.

NOTE: Make sure that there is backlash between the gear and drive pinion.

12. Install the adjusting nuts. Align the marks on the cap and carrier. Screw in the two bearing cap bolts two or three turns and press down the bearing cap by hand.

NOTE: If the bearing cap does not fit tightly on the carrier, the adjusting nut threads are not threaded properly. Reinstall adjusting nuts if necessary.

13. To adjust side bearing preload tighten the bearing cap bolts until the spring washers are slightly compressed. Using an adjusting nut wrench special tool 09504–00010 or equivalent, tighten the adjusting nut on the ring gear side until the ring gear has a backlash of about 0.008 in.

14. Using an adjusting nut wrench, firmly tighten the adjusting nut on the drive pinion side.

15. Check the ring gear backlash. If tightening the adjusting nut creates ring gear backlash, loosen the nut so that backlash is eliminated.

16. Place a dial indicator on the top of the bearing cap on the ring gear side.

17. Adjust the side bearing for zero preload by tightening the other adjusting nut until the pointer on the indicator begins to move.

18. Tighten the adjusting nut 1 to 1½ notches from the zero preload position.

19. Using a dial indicator and adjusting nut wrench, adjust the ring gear backlash until the backlash is within specification. Backlash: 0.0051–0.0071 in.

NOTE: The backlash is adjusted by turning the left and right adjusting nuts equal amounts. For example, loosen the nut on the left side one notch and tighten the nut on the right side one notch.

20. Torque the bearing cap bolts 51–65 ft. lbs.

21. Recheck the ring gear backlash.

22. Measure the total preload. Total preload (Starting): Drive pinion preload plus 3.5–5.2 inch lb.

24. Inspect tooth contact between ring gear and drive pinion. Adjust if needed.

25. Stake the drive pinion nut.

26. Install the adjusting nut locks.

27. Install differential carrier assembly in the axle housing.

28. Install axle shafts.

29. Connect the driveshaft.

30. Fill differential with SAE 80W–90 gear oil.

U-Joint/CV-Joint Overhaul

UNIVERSAL JOINTS

U-joint is mechanic's jargon for universal joint. U-joints should not be confused with U-bolts, which are U-shaped bolts used to connect U-joints to the differential pinion flange.

Universal joints provide flexibility between the driveshaft and axle housing to accommodate changes in the angle between them (changes of length are accommodated by the sliding splined yoke between the driveshaft and transmission). The engine and transmission are mounted rigidly on the car frame, while the driving wheels are free to move up and down in relation to the frame. The angles between the transmission, driveshaft and axle change constantly as the car responds to various road conditions.

To give flexibility and still transmit power as smoothly as possible, several types of universal joints are used.

The most common type of universal joint is the cross and yoke type. Yokes are used on the ends of the driveshaft with the yoke arms opposite each other. Another yoke is used opposite the driveshaft and when placed together, both yokes engage a center member, or cross, with four arms spaced 90° apart (the U-joint cross is alternately referred to as a spider, and the arms are called trunnions). A bearing cup (or cap) is used on each arm of the cross to accommodate movement as the driveshaft rotates. The bearings used are needle bearings.

A conventional universal joint will cause the driveshaft to speed up and slow down through each revolution and cause a corresponding change in the velocity of the driven shaft. This change in speed causes

natural vibrations to occur through the driveline, necessitating a third type of universal joint: the constant velocity joint. A rolling ball moves in a curved groove, located between two yoke-and-cross universal joints, connected to each other by a coupling yoke. The result is a uniform motion as the driveshaft rotates, avoiding the fluctuations in driveshaft speed. This type of joint is found in cars with sharp driveline angles, or where the extra measure of isolation is desirable.

CROSS AND YOKE U-JOINT OVERHAUL

There are two types of cross and yoke U-joints. One type retains the cross within the yoke with C-shaped snap rings. The second type of joint is held together by injection molded plastic retainer rings. The second type cannot be reassembled with the same parts, once disassembled. However, repair kits are available.

Snap-Ring Type

1. Remove the driveshaft. For the correct procedure, see the car section for the model you are working on.

2. If the front yoke is to be disassembled, matchmark the driveshaft and sliding splined yoke (transmission yoke) so that driveline balance is preserved upon reassembly. Remove the snap rings which retain the bearing caps.

3. Select two sockets, one small enough

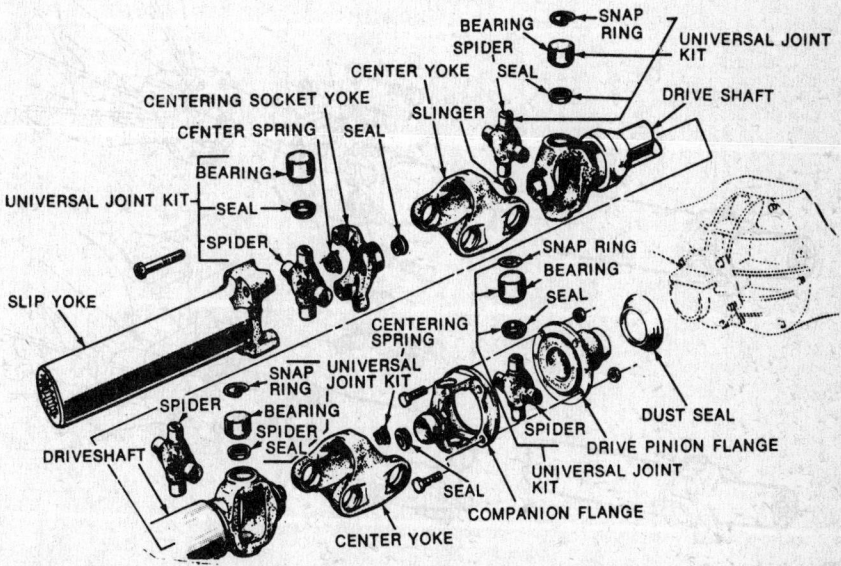

Typical driveshaft with cardan type U-joints

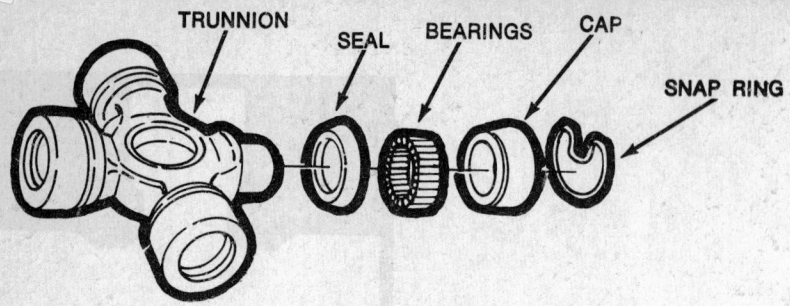

TRUNNION SEAL BEARINGS CAP SNAP RING

Snap ring type universal joint

to pass through the yoke holes for the bearing caps, the other large enough to receive the bearing cap.

4. Using a vise or a press, position the small and large sockets on either side of the U-joint. Press in on the smaller socket so that it presses the opposite bearing cap out of the yoke and into the larger socket. If the cap does not come all the way out, grasp it with a pair of pliers and work it out.

5. Reverse the position of the sockets so that the smaller socket presses on the cross. Press the other bearing cap out of the yoke.

6. Repeat the procedure on the other bearings.

7. To install, grease the bearing caps and needles thoroughly if they are not pre-greased. Start a new bearing cap into one side of the yoke. Position the cross in the yoke.

8. Select two sockets small enough to pass through the yoke holes. Put the sockets against the cross and the cap, and press the bearing cap ¼ inch below the surface of the yoke. If there is a sudden increase in the force needed to press the cap into place,

or if the cross starts to bind, the bearings are cocked. They must be removed and re-started in the yoke. Failure to do so will greatly reduce the life of the bearing.

9. Install a new snap ring.

10. Start a new bearing into the opposite side. Place a socket on it and press in until the opposite bearing contacts the snap ring.

11. Install a new snap ring. It may be necessary to grind the facing surface of the snap ring slightly to permit easier installation.

12. Install the other bearings in the same manner.

13. Check the joint for free movement. If binding exists, smack the yoke ears with a brass or plastic faced hammer to seat the bearing needles. Do not strike the bearings, and support the shaft firmly. Do not install the driveshaft until free movement exists at all joints.

Plastic Retainer Type

Remove and install the bearing caps and trunnion (cross) as described for the snap-ring type universal joints. On an original universal joint, however, the bearing caps will be secured in the yokes with injected

plastic. The plastic will shear when the bearing caps are pressed. Service snap-rings are installed in the groove on the inside (of yoke) of the installed caps.

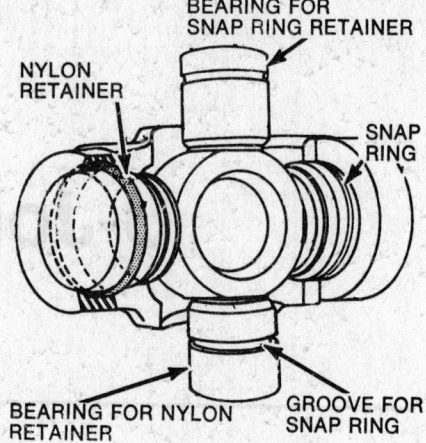

BEARING FOR SNAP RING RETAINER
NYLON RETAINER
SNAP RING
BEARING FOR NYLON RETAINER
GROOVE FOR SNAP RING

U-joint locking methods

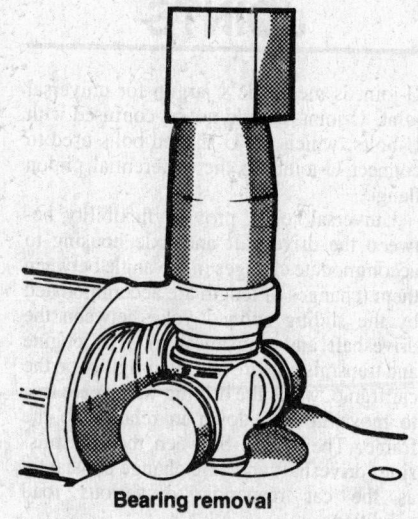

Bearing removal

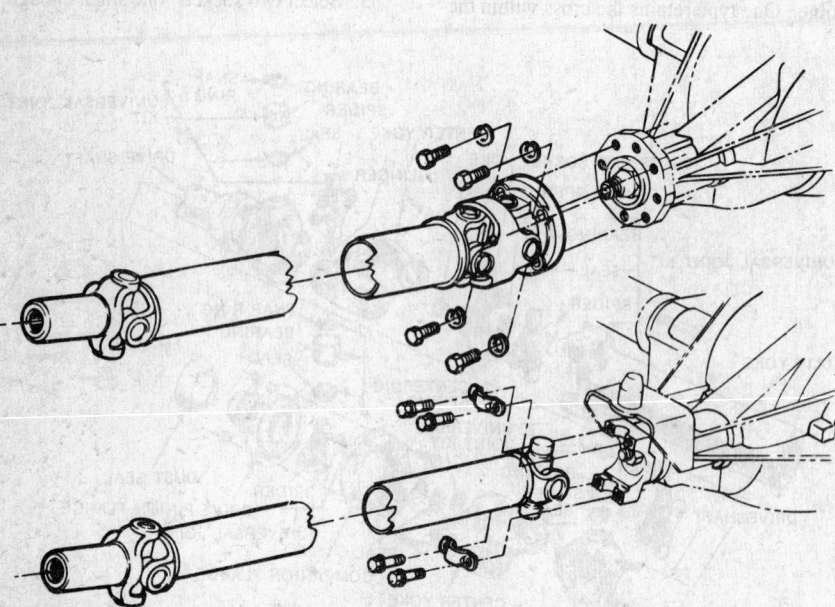

The driveshaft may be retained to the differential pinion by a flange (top) or by U-bolts or straps (bottom)

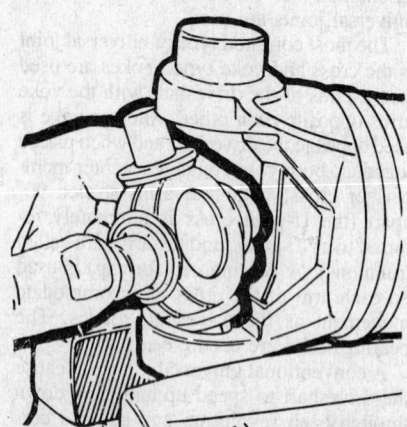

Press a bearing cap into the yoke, then install the cross

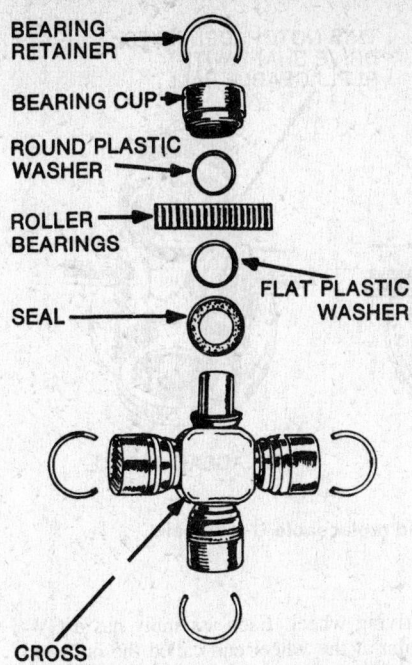

BEARING RETAINER

BEARING CUP

ROUND PLASTIC WASHER

ROLLER BEARINGS

FLAT PLASTIC WASHER

SEAL

CROSS

Plastic retainer U-joint repair kit components

NOTE: The plastic which retains the bearing will be sheared when the bearing cup is pressed out. Be sure to remove the remains of the plastic retainer from the ears of the yoke. It is easier to remove the remains if a small pin or punch is first driven through the injection holes in the yoke. Failure to remove all of the plastic remains may prevent the bearing cups from being pressed into place and the bearing retainers from being properly seated.

CARDAN TYPE U-JOINT OVERHAUL

Some with Cardan type U-joints use snap rings to retain the bearing cups in the yokes. Other cars have plastic retainers. Be sure to obtain the correct rebuilding kit.

1. Use a punch to mark the coupling yoke and the adjoining yokes before disassembly, to ensure proper reassembly and driveline balance.

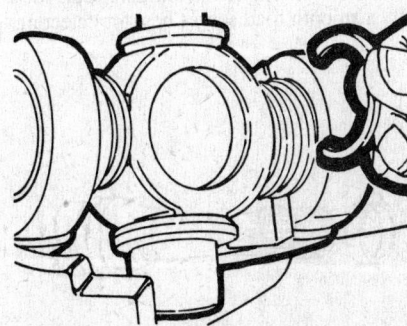

Service snap rings are installed inside the yoke

2. It is easiest to remove the bearings from the coupling yoke first. Follow the order indicated in the illustration.

3. Support the driveshaft horizontally on a press stand, or on the workbench if a vise is being used.

4. If snap rings are used to retain the bearing cups, remove them. Place the rear ear of the coupling yoke over a socket large enough to receive the cup. Place a smaller socket, or a cross press made for the purpose, over the opposite cup. Press the bearing cup out of the coupling yoke ear. If the cup is not completely removed, insert a spacer and complete the operation, or grasp the cup with a pair of slip joint pliers and work it out. If the cups are retained by plastic, this will shear the retainers. Remove any bits of plastic.

5. Rotate the driveshaft and repeat the operation on the opposite cup.

6. Disengage the trunnions of the spider, still attached to the flanged yoke, from the coupling yoke, and pull the flanged yoke and spider from the center ball on the ball support tube yoke.

NOTE: The joint between the shaft and coupling yoke can be serviced without disassembly of the joint between the coupling yoke and flanged yoke.

7. Pry the seal from the ball cavity, remove the washers, spring and three seats. Examine the ball stud seat and the ball stud for scores or wear. Worn parts can be replaced with a kit. Clean the ball seat cavity and fill it with grease. Install the spring, washer, ball seats, and spacer (washer) over the ball.

8. To assemble, insert one bearing cup

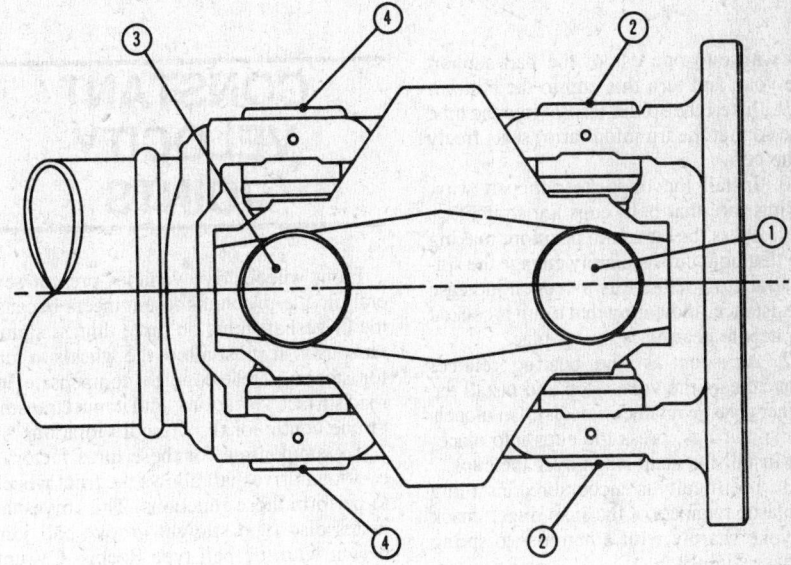

Cardan joint disassembly sequence

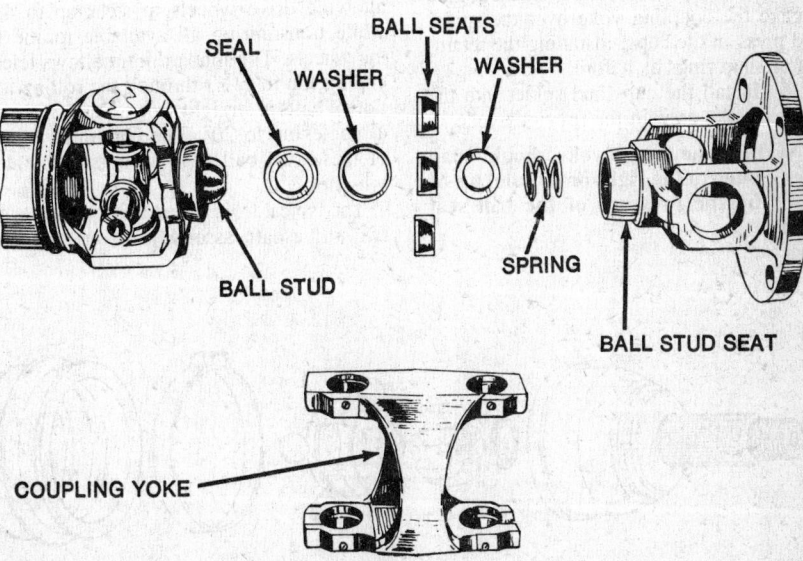

SEAL

BALL SEATS

WASHER

WASHER

BALL STUD

SPRING

BALL STUD SEAT

COUPLING YOKE

Cardan type joint

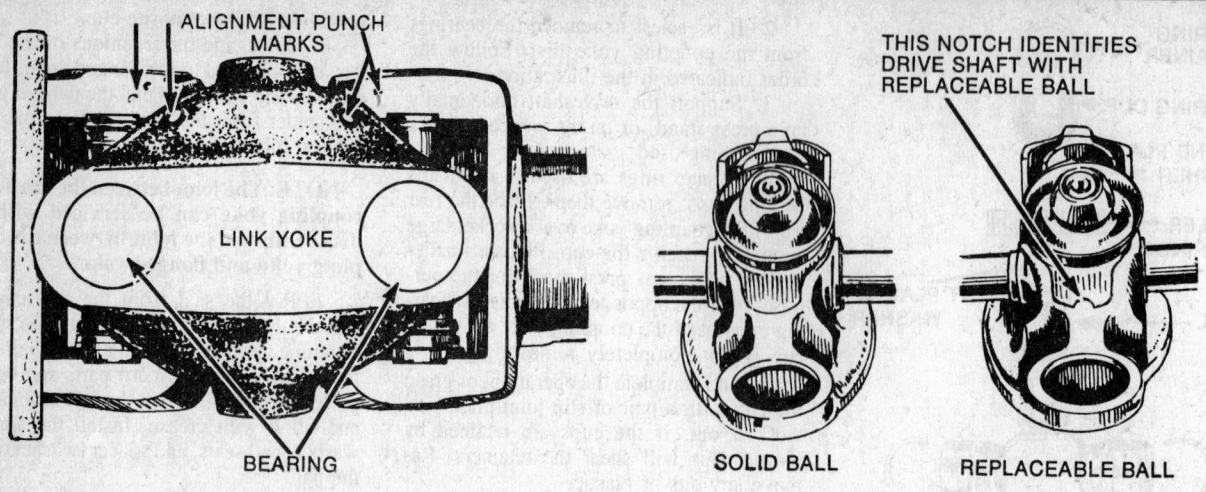

Match marks for double cardan joint

Solid and replaceable U-joint balls

part way into one ear of the ball support tube yoke and turn this cup to the bottom.

9. Insert the spider (cross) into the tube yoke so that the trunnion (arm) seats freely in the cup.

10. Install the opposite cup part way, making sure that both cups are straight.

11. Press the cups into position, making sure that both cups squarely engage the spider. Back off if there is a sudden increase in resistance, indicating that a cup is cocked or a needle bearing is out of place.

12. As soon as one bearing retainer groove clears the yoke, stop and install the retainer (plastic retainer models). On models with snap rings, press the cups into place, then install the snap rings over the cups.

13. If difficulty is encountered installing the plastic retainers or the snap rings, smack the yoke sharply with a hammer to spring the ears slightly.

14. Install one bearing cup part way into the ear of the coupling yoke. Make sure that the alignment marks are matched, then engage the coupling yoke over the spider and press in the cups, installing the retainers or snap rings as before.

15. Install the cups and spider into the flanged yoke as with the previous yoke.

NOTE: The flange yoke should snap over center to the right or left and up or down by the pressure of the ball seat spring.

CONSTANT VELOCITY JOINTS

Front wheel drive vehicles present several unique problems to engineers because the driveshaft must do three things, simultaneously. It must allow the wheels to turn for steering, telescope to compensate for road surface vibrations, and it must transmit torque continuously without vibration.

To compensate for these three factors a two-joint driveshaft allows the front wheels to perform these functions. This driveshaft mates disc type straight groove ball joint design with the bell type Rzeppa CV universal joint.

The Rzeppa joint on the outboard end of each driveshaft provides steering ability by allowing drive wheels to steer up to 43° while transmitting all available torque to the wheels. The inboard joint allows telescoping (up to 1½") through the rolling action of balls in straight grooves and operates at angles up to 20°. The combined action of these two ball type u-joints eliminates vibration.

The typical front wheel drive vehicle uses two driveshaft assemblies—one to each

driving wheel. Each assembly has a CV-joint at the wheel end called the outboard joint. A second joint on each shaft located at the transaxle end is called the inboard joint. This joint may be either the ball or tripode type. It allows the slip motion required when the driveshaft must shorten or lengthen in response to suspension action when traveling over an irregular surface.

Constant velocity joints are precision machined parts that have difficult jobs to perform in a hostile environment. They are exposed to heat, shock, torque, and many thousands of miles of service. For this reason, the lubricants used are specially formulated to be compatible with the rubber boot and give proper lubrication. Most CV-joint repair kits have this special lubricant included.

NOTE: Wear pattern in a used ball or tripode CV-joint are impossible to match during reassembly. If there are any signs of wear, abnormal operating noise, corrosion, heat discoloration, the joint must be replaced.

TROUBLESHOOTING

Noises from the engine, drive axles, suspension and steering in the front drive cars can be misleading to the untrained ear. Ideally a smooth road serves best for detecting

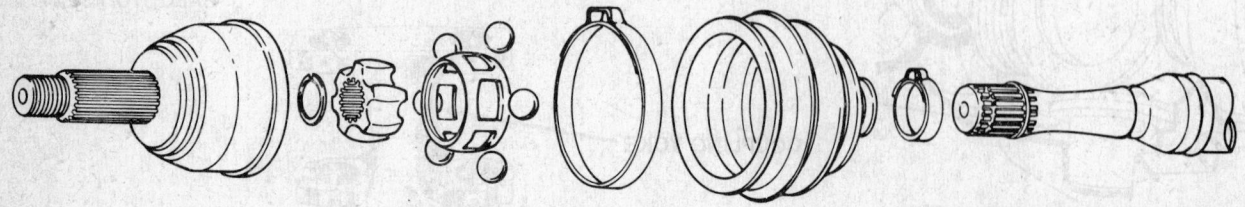

Fixed CV joint

SHAFT REMOVAL

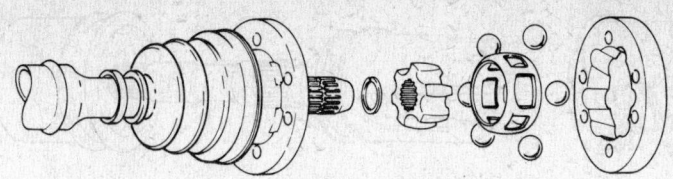

Ball style (Rzeppa) plunging CV joint

operating condition(s) that cause noise.

• A humming noise could indicate that early stage of insufficient or incorrect lubricant.

• Worn driveshaft joints will cause a continuous knock at low speeds.

• A popping or clicking sound on sharp turns indicates trouble in the outer or wheel end joint.

• The clunk noise at acceleration from coasting or deceleration from a load pull indicates two possibilities—damaged inner or transaxle joint or differential problem(s).

• An inner joint will create a vibration during acceleration due to plunging action hanging up and releasing repeatedly. Probable cause would be foreign particles or lack of lubrication, or improper assembly.

• Remember that tires, suspension, engine, and exhaust system are all up front to add their noises.

• Make a check with front wheels elevated off ground. Spin the wheels by hand to determine if wheel bearing could be noisy or if out of round tires are causing vibration. Many wheel bearings are prelubed and sealed at the factory.

─── **CAUTION** ───

Personal injury can occur from spinning wheels by engine power. Spinning a wheel at excess speed may cause damage to CV-joints that could be operating at angles too steep when wheels are allowed to hang. Over speeding might also cause damage to tires and the differential.

1. Remove the hub nut and discard it.
2. Drain the lubricant from the transaxle.

─── **CAUTION** ───
The lubricant may be hot.

3. The speedometer pinion gear assembly must be removed before the right drive shaft can be removed. (Automatic transaxles only).

4. Rotate the driveshaft to view the circlip.

5. Compress the circlip tangs with needle nose pliers as you pry into the side gear. This compresses the circlip in position for shaft removal later. Keep an awl between the differential pinion shaft and the end face of the shaft to prevent circlip reentry to the groove.

NOTE: This applies to Chrysler only.

6. Remove the ball joint clamp bolt. Drop the lower arm too allow clearance. This will permit the front wheel to swing free.

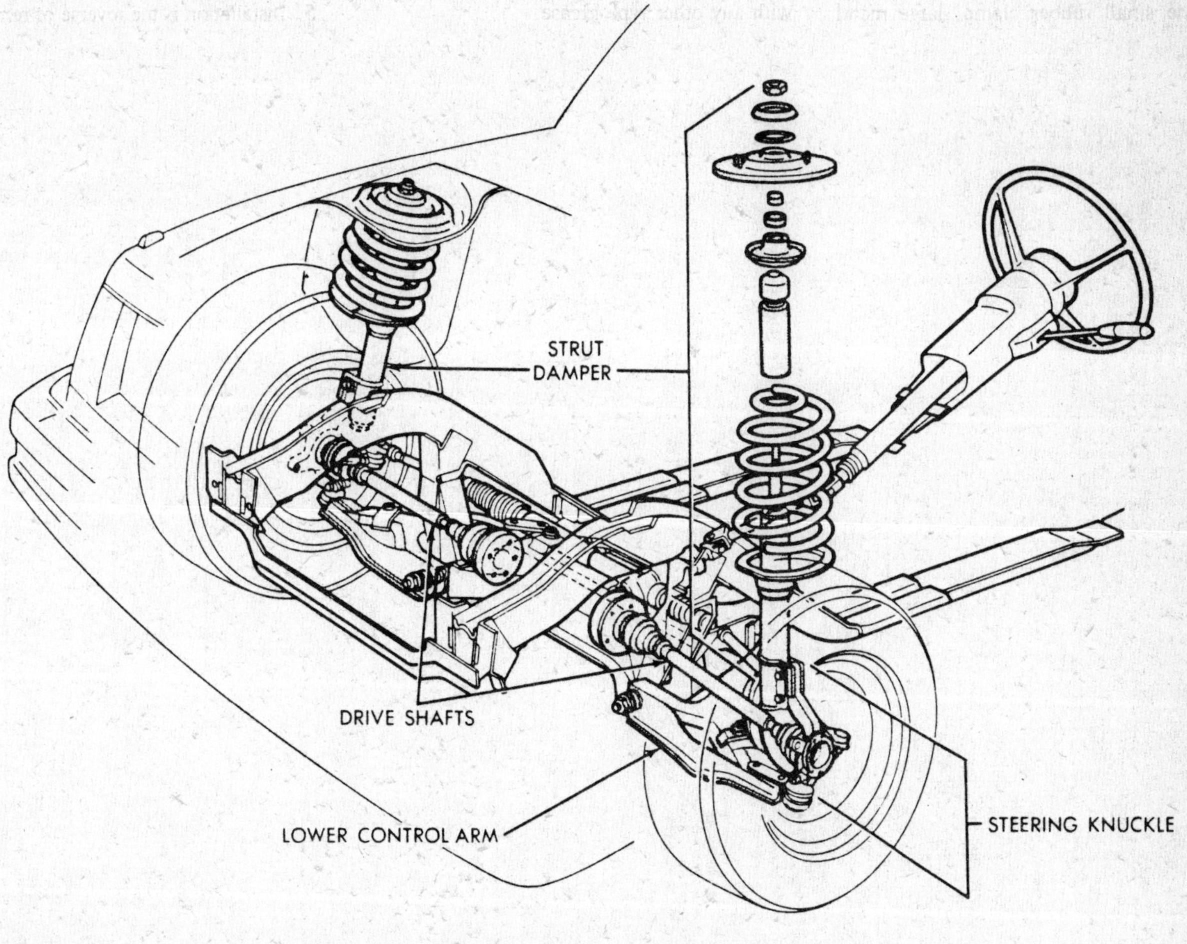

STRUT
DAMPER

DRIVE SHAFTS

LOWER CONTROL ARM

STEERING KNUCKLE

Typical CV driveshaft assembly

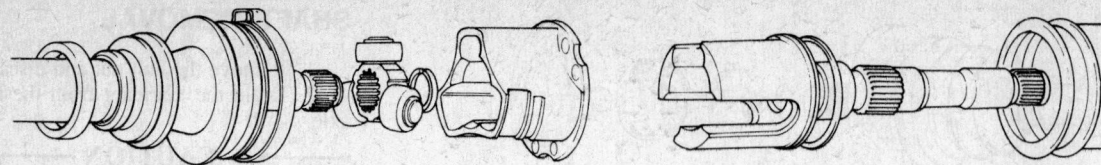

Closed tulip plunging CV joint   **Open tulip plunging CV joint**

7. Pull the outer splined shaft from the wheel hub, when swinging wheel hub away. Do not pull on the shaft. Grasp the joint housing.

8. Remove the inner joint by pulling outward on the inner joint housing. Do not pull the shaft.

NOTE: Do not allow the assembly to hang at either end. This can jam the CV-joint and cause vibration during operation. If necessary, support the shaft at either end by rope or wire.

INNER JOINT/BOOT

9. Place the assembly in a vise. Care must be taken not the crush the tubular shafts. Some shafts are solid steel.

10. If the inner joint needs replacement, cut the small rubber clamp, large metal clamp, and remove the rubber boot. These items must be discarded.

11. Inspect for internal wear and/or damage.

12. Clean the grease by hand from inside the joint housing and around the 3 ball trunnion assembly to inspect. Mark the tripod and housing for proper reassembly, if it is to be reinstalled.

13. To replace the boot, CV-joint, or both, remove the snap ring from the groove and tap the trunnion lightly with a brass drift pin. Leave the tripode bearings on the trunnion. Care must be taken to support the bearings as they may fall off.

14. Installation is the reverse of removal with the following recommendations:
When reinstalling the tripode on the shaft place the chamfer face toward the retainer groove. The grease provided with the repair kit must be used. It can not be substituted with any other type grease.

OUTER JOINT/BOOT

1. Place the shaft in a vise. Be careful not to over tighten the vise thereby damaging the shaft.

2. Remove the boot and clamps. Discard these parts.

3. Using a soft hammer rap sharply on the housing. This forces the inner race over the internal circlip. Never remove the slinger from the housing.

4. Remove and discard the circlip. A new one is included with the boot kit. Leave the lock ring in place.

NOTE: Never disassemble the cage and balls from the housing. Reuse the joint assembly with a new boot kit, unless the grease is contaminated and prior diagnosis indicated trouble. In that case replace the joint and boot.

5. Installation is the reverse of removal.

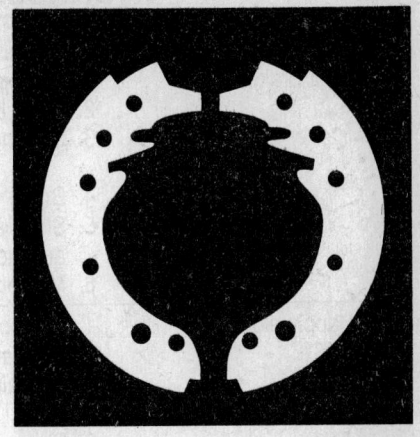

Brakes

INDEX
General Information

HYDRAULIC BRAKE SYSTEM TROUBLE DIAGNOSIS

Condition	Possible Cause	Correction
Insufficient brakes	1. Improper brake adjustment. 2. Worn lining. 3. Sticking brakes. 4. Brake valve pressure low. 5. Slack adjuster to diaphragm rod not adjusted properly. 6. Master cylinder low on brake fluid.	1. Adjust brakes. 2. Replace brake lining and adjust brakes. 3. Lubricate brake pivots and support platforms. 4. Inspect for leaks and obstructed brake lines. 5. Adjust slack adjuster. 6. Fill master cylinder and inspect for leaks.
Brakes apply slowly	1. Improper brake adjustment or lack of lubrication. 2. Low air pressure. 3. Brake valve delivery pressure low. 4. Excessive leakage with brakes applied. 5. Restriction in brake line or hose.	1. Adjust brakes and lubricate linkage. 2. Check belt tension and compressor for output. Adjust as necessary. 3. Check valve pressure and clean or replace as necessary. 4. Inspect all fittings and lines for leaks and repair as necessary. 5. Clean or replace brake line or hose.
Spongy pedal	1. Air in hydraulic system. 2. Swollen rubber parts due to contaminated brake fluid. 3. Improper brake shoe adjustment. 4. Brake fluid with low boiling point. 5. Brake drums ground excessively.	1. Fill and bleed hydraulic system. 2. Clean hydraulic system and recondition wheel cylinders and master cylinder. 3. Adjust brakes. 4. Flush hydraulic system and refill with proper brake fluid. 5. Replace brake drums.
Erratic brakes	1. Linings soaked with grease or brake fluid. 2. Primary and secondary shoes mounted in wrong position.	1. Correct the leak and replace brake lining. 2. Match the primary and secondary shoes and mount in proper position.
Chattering brakes	1. Improper adjustment of brake shoes. 2. Loose front wheel bearings. 3. Hard spots in brake drums. 4. Out-of-round brake drums. 5. Grease or brake fluid on lining.	1. Adjust brakes. 2. Clean, pack and adjust wheel bearings. 3. Grind or replace brake drums. 4. Grind or replace brake drums. 5. Correct leak and replace brake lining.
Squealing brakes	1. Incorrect lining. 2. Distorted brakedrum. 3. Bent brake support plate. 4. Bent brake shoes. 5. Foreign material embedded in brake lining. 6. Dust or dirt in brake drum. 7. Shoes dragging on support plate. 8. Loose support plate. 9. Loose anchor bolts. 10. Loose lining on brake shoes or improperly ground lining.	1. Install correct lining. 2. Grind or replace brake drum. 3. Replace brake support plate. 4. Replace brake shoes. 5. Replace brake shoes. 6. Use compressed air and blow out drums and support plate and shoes. 7. Sand support plate platforms and lubricate. 8. Tighten support plate attaching nuts. 9. Tighten anchor bolts. 10. Replace brake shoes and cam-grind lining.
Brakes fading	1. Improper brake adjustment. 2. Improper brake lining. 3. Improper type of brake fluid. 4. Brake drums ground excessively.	1. Adjust brakes correctly. 2. Replace brake lining. 3. Drain, flush and refill hydraulic system. 4. Replace brake drums.
Dragging brakes	1. Improper brake adjustment. 2. Distorted cylinder cups. 3. Brake shoe seized on anchor bolt. 4. Broken brake shoe return spring. 5. Loose anchor bolt. 6. Distorted brake shoe. 7. Loose wheel bearings.	1. Correct adjust brakes. 2. Recondition or replace cylinder. 3. Clean and lubricate anchor bolt. 4. Replace brake shoe return spring. 5. Adjust and tighten anchor bolt. 6. Replace defective brake shoes. 7. Lubricate and adjust wheel bearings.

HYDRAULIC BRAKE SYSTEM TROUBLE DIAGNOSIS

Condition	Possible Cause	Correction
	8. Obstruction in brake line.	8. Clean or replace brake line.
	9. Swollen cups in wheel cylinder or master cylinder.	9. Recondition wheel or master cylinder.
	10. Master cylinder linkage improperly adjusted.	10. Correctly adjust master cylinder linkage.
Hard pedal	1. Incorrect brake lining.	1. Install matched brake lining.
	2. Incorrect brake adjustment.	2. Adjust brakes and check fluid.
	3. Frozen brake pedal linkage.	3. Free up and lubricate brake linkage.
	4. Restricted brake line or hose.	4. Clean out or replace brake line hose.
Wheel locks	1. Loose or torn brake lining.	1. Replace brake lining.
	2. Incorrect wheel bearing adjustment.	2. Clean, pack and adjust wheel bearings.
	3. Wheel cylinder cups sticking.	3. Recondition or replace the wheel cylinder.
	4. Saturated brake lining.	4. Reline front, rear or all four brakes.
Brakes fade (high speed)	1. Improper brake adjustment.	1. Adjust brakes and check fluid.
	2. Distorted or out of round brake drums.	2. Grind or replace the drums.
	3. Overheated brake drums.	3. Inspect for dragging brakes.
	4. Incorrect brake fluid (low boiling temperature).	4. Drain flush and refill and bleed the hydraulic brake system.
	5. Saturated brake lining.	5. Reline brakes as necessary.

HYDRAULIC BRAKES

Basic Hydraulic System

The hydraulic system controls the braking operation and consists of a master cylinder, hydraulic lines and hoses, control valves and calipers and/or wheel cylinders. When the brake pedal is depressed, the master cylinder forces brake fluid to the calipers and/or cylinders, via lines and hoses. Sliding rubber seals contain the fluid and prevent leakage.

Return springs in the master cylinder help the brake pedal return to the original unapplied position. Check valves (in most cases) regulate the return flow of the fluid to the master cylinder. Other valves, such as the metering valve, proportioning valve, or combination valve, regulate the flow of fluid to the caliper/wheel cylinder, to achieve efficient braking.

Dual Braking Systems

The "dual" system is essentially two master cylinders (usually) formed by aligning two separate pistons and fluid reservoirs into one cylinder bore. Dual brake lines "split" the calipers and/or wheel cylinders into two groups, each actuated by a separate master cylinder piston. In event of failure of one of the "dual" systems, the other should provide enough braking power to safely stop the vehicle. The dual system usually includes a red warning light on the instrument panel which is activated by a pressure differential valve. The valve is sensitive to any loss of hydraulic pressure that might result from a braking failure on either side of the system.

Light trucks are equipped with either a front/rear wheel "split" or a diagonally "split" system. On front/rear systems, the front wheels are connected to one circuit while the rear wheels are connected to the other circuit. Diagonally split systems have diagonally opposite wheels connected to each circuit. Medium and heavy trucks may use the front/rear split or, if equipped with two wheel cylinders per wheel, each circuit will operate one cylinder per wheel.

Service Information

Servicing the hydraulic brake system is chiefly a matter of adjustments, replacement of worn or damaged parts and correcting the damage caused by grit, dirt or contaminated brake fluid. Always make sure the brake system is clean and tightly sealed when a brake job is completed and that only approved heavy duty brake fluid is used.

The approved heavy duty type brake fluid retains the correct consistency throughout the widest range of temperature variation, will not affect rubber cups, helps protect the metal parts of the brake system against failure and assures long trouble-free brake operation.

Never use brake fluid from a container that has been used for any other liquid. Mineral oil, alcohol, anti-freeze, or cleaning solvents, even in very small quantities, will contaminate brake fluid. Contaminated brake fluid will cause piston cups and the valve(s) in the master cylinder to swell or deteriorate.

Brake adjustment is required after installation of new or relined brake shoes. Adjustment is also necessary whenever excessive travel of pedal is needed to start braking action.

LOW PEDAL

Normal brake lining wear reduces pedal reserve. Low pedal reserve may also be caused by the lack of brake fluid in the master cylinder. The wear condition may be compensated for by a minor brake adjustment. Check fluid level in master cylinder and add as required.

FLUID LOSS

If the master cylinder requires constant addition of hydraulic fluid, fluid may be leaking past the piston cups in the master cylinder or brake cylinders, the hydraulic lines; hoses or connections may be loose or broken. Loose connections should be tightened, or other necessary repairs or parts replacement made and the hydraulic brake system bled.

FLUID CONTAMINATION

To determine if contamination exists in the brake fluid, as indicated by swollen, deteriorated rubber cups, the following tests can be made.

Place a small amount of the drained brake fluid into a small clear glass bottle. Separation of the fluid into distinct layers will indicate mineral oil content. Be safe

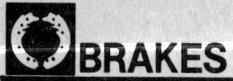

and discard old brake fluid that has been bled from the system. Fluid drained from the bleeding operation may contain dirt particles or other contamination and should not be reused.

BRAKE ADJUSTMENT

Self adjusting brakes usually do not require manual adjustment but in the event of a brake reline it may be advisable to make the initial adjustment manually to speed up adjusting time.

AUTOMATIC ADJUSTER CHECK

Raise and safely support the vehicle, have a helper in the driver's seat to apply brakes. Remove the plug from the adjustment slot to observe adjuster star wheel. Then, to exclude possibility of maximum adjustment; that is, the adjuster refuses to operate because the closest possible adjustment has been reached; the star wheel should be backed off approximately 30 notches. It will be necessary to hold adjuster lever away from star wheel to allow backing off of the adjustment.

Spin the wheel and brake drum in reverse direction and apply brakes vigorously. This will provide the necessary inertia to cause the secondary brake shoe to leave the anchor. The wrap up effect will move the secondary shoe, and a cable or link will pull the adjuster lever away from the starwheel teeth. Upon release of brake pedal, the lever should snap back in position, turning star wheel. Thus, a definite rotation of adjuster star wheel can be observed if automatic adjuster is working properly. If by the described procedure one or more automatic adjusters do not function properly, the respective drum must be removed for adjuster servicing.

HYDRAULIC LINE REPAIR

Steel tubing is used in the hydraulic lines between the master cylinder and the front brake tube connector, and between the rear brake tube connector and the rear brake cylinders. Flexible hoses connect the brake tube to the front brake cylinders or calipers and to the rear brake tube connector.

When replacing hydraulic brake tubing, hoses, or connectors, tighten all connections securely. After replacement, bleed the brake system at the wheel cylinders or calipers and at the booster, if equipped with a bleeder screw.

BRAKE TUBE

If a section of the brake tube becomes damaged, the entire section should be replaced with tubing of the same type, size, shape, and length. Copper tubing should not be used in the hydraulic system. When bending brake tubing to fit the frame or rear axle

contours, be careful not to kink or crack the tube.

All brake tubing should be double flared to provide good leak-proof connections. Always clean the inside of a new brake tube with clean isopropyl alcohol.

BRAKE HOSE

A flexible brake hose should be replaced if it shows signs of softening, cracking, or other damage.

When installing a new brake hose, position the hose to avoid contact with other truck parts.

Hydraulic Control Valves

PRESSURE DIFFERENTIAL VALVE

Also known as a warning valve, dash-lamp valve or system effectiveness indicator. The valve activates a panel warning lamp in event of pressure loss failure. As pressure fails in one "split" system, the other system's normal pressure causes a piston in

the switch to compress a spring and move until an electrical circuit is completed lighting the dash lamp. On some vehicles the spring balanced piston automatically re-centers when the brake pedal is released, thus flashing the warning lamp only during brake application. On other vehicles the lamp will stay on until the cause of pressure loss is corrected.

Valves (pressure differential, metering or proportioning) may be located separately, but are usually part of a combination valve. On some brake systems the valve and switch are part of the master cylinder.

RESETTING VALVES

The pressure differential valve on many vehicles (equipped with a combination valve) will re-center automatically upon brake application after repairs to the system are completed. Just make sure the system is properly bled and that master cylinder reservoirs are kept full of clean, approved brake fluid, and then depress the brake pedal slowly with the ignition switch on (engine off). Other systems require manual resetting. Repair system as required, open a bleeder screw in the half of the system that

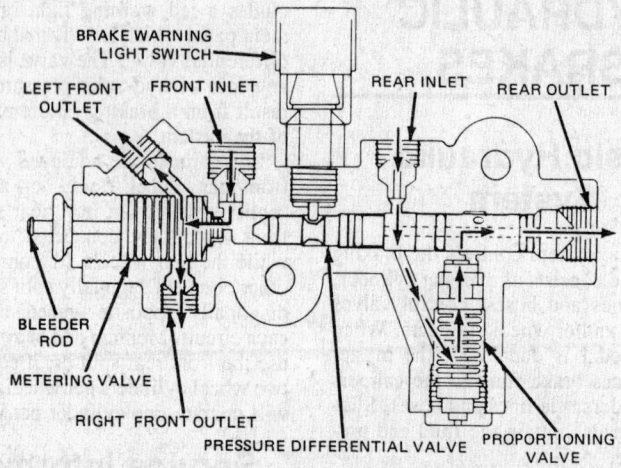

Typical light duty pressure differential valve

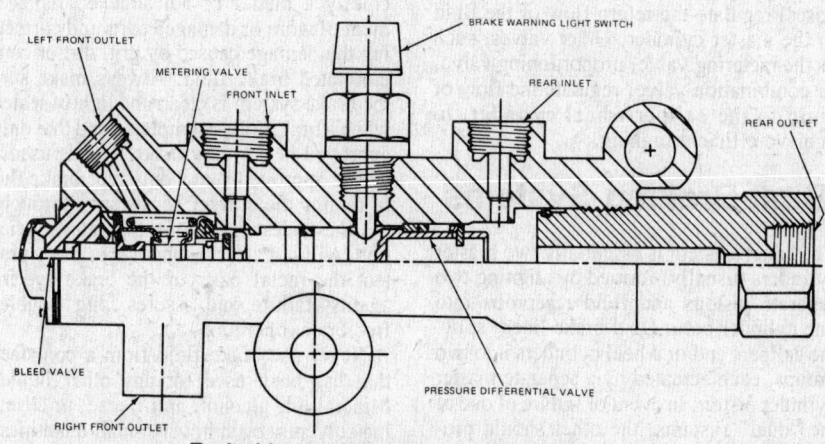

Typical heavy duty pressure differential valve

did not fail (with reservoirs full). Turn on the ignition to light the warning lamp and slowly depress the brake pedal until the lamp goes out. If too much pressure is applied the piston will go to the other side and the procedure will have to be reversed by opening a bleeder screw in the opposite half of the system.

METERING VALVE

Often used on vehicles equipped with front disc and rear drum brakes, the metering valve improves braking balance during light brake applications by preventing application of the front disc brakes until pressure is built-up in the hydraulic system. The built-up hydraulic pressure overcomes the tension of the rear brake shoe return springs. Thus, when the front brake pads contact the rotor the rear brakes shoes move outward to contact the brake drum at the same time.

The metering valve should be inspected whenever the brakes are serviced. A slight amount of moisture inside the boot does not

indicate a defective valve, however a great deal of fluid indicates a worn valve and replacement is indicated. Make sure to install the brake lines in the correct ports when installing a new valve, crossed lines will cause the rear brakes to drag.

If a pressure bleeder is used to bleed a hydraulic system that includes a metering valve, the valve stem (inside the boot on some valves) must either be pushed in or pulled out, depending upon the type of valve. Never apply excessive pressure that might damage the valve. Never use a solid block or clamp to force the valve open. If the valve must be blocked, rig the stem with a yieldable spring load and take care not to exert more than normal pressure.

If the brakes are to be bled manually using the brake pedal, the pressure developed is sufficient to overcome the metering valve and the stem need not be pushed in or pulled out.

PROPORTIONING VALVE

Used on vehicles equipped with front disc and rear drum brakes, the proportioning valve is installed in the line(s) to the rear drum brakes, and in a split system, below

the pressure differential valve. By reducing pressure to the rear drum brakes, the valve helps to prevent premature lock-up during severe brake application and provides better braking balance.

Whenever the brakes are serviced, the valve should be inspected. To check valve operation, install hydraulic gauges ahead and behind the valve and determine that it has an operative transition point above which rear brake pressure is proportioned. If the valve is leaking replacement is required. Make sure the valve port marked "R" is connected to the rear brake line(s).

COMBINATION VALVE

A valve combining two or three functions (metering, proportioning, and/or brake warning) may be used. The combination valve is usually mounted under the hood close to the master cylinder, where the brake lines can be easily routed to the front and rear wheels. The combination valve is a non-serviceable unit, and if found to be malfunctioning, must be replaced as a unit.

Master Cylinder Service

NOTE: This section contains service procedures common to many standard-type designs of master cylinder. Some recent designs of master cylinder employ many special features; these each require specialized servicing procedures. Service procedures for these special types are found after this section.

REMOVAL & DISASSEMBLY

Clean the area around the master cylinder to prevent dirt and grease from contaminating the cylinder or the hydraulic lines. Disconnect the tubes, and cap the openings in the lines. Disconnect the fluid level sensor coupling on those models so-equipped. If the master cylinder uses a remotely mounted reservoir, unclip, disconnect, and then plug the lines. Remove nuts or bolts that secure the master cylinder to fire wall or power brake, and remove the master cylinder.

On vehicles with manual brakes, the push rod must be disconnected from the brake pedal before removing the master cylinder.

1. Remove the reservoir cover, and drain the brake fluid from the reservoir. On many models equipped with a removable reservoir, it is best to remove it. Gently rock the reservoir from side to side to free it and then remove it. Remove sealing grommets, if they are present. Then remove the piston stop bolt, if present, from the master cylinder. Remove boot and snap ring, then slide the primary piston assembly out of the master cylinder. Next, if dual system, remove the secondary piston assembly by tapping the master cylinder, or by using

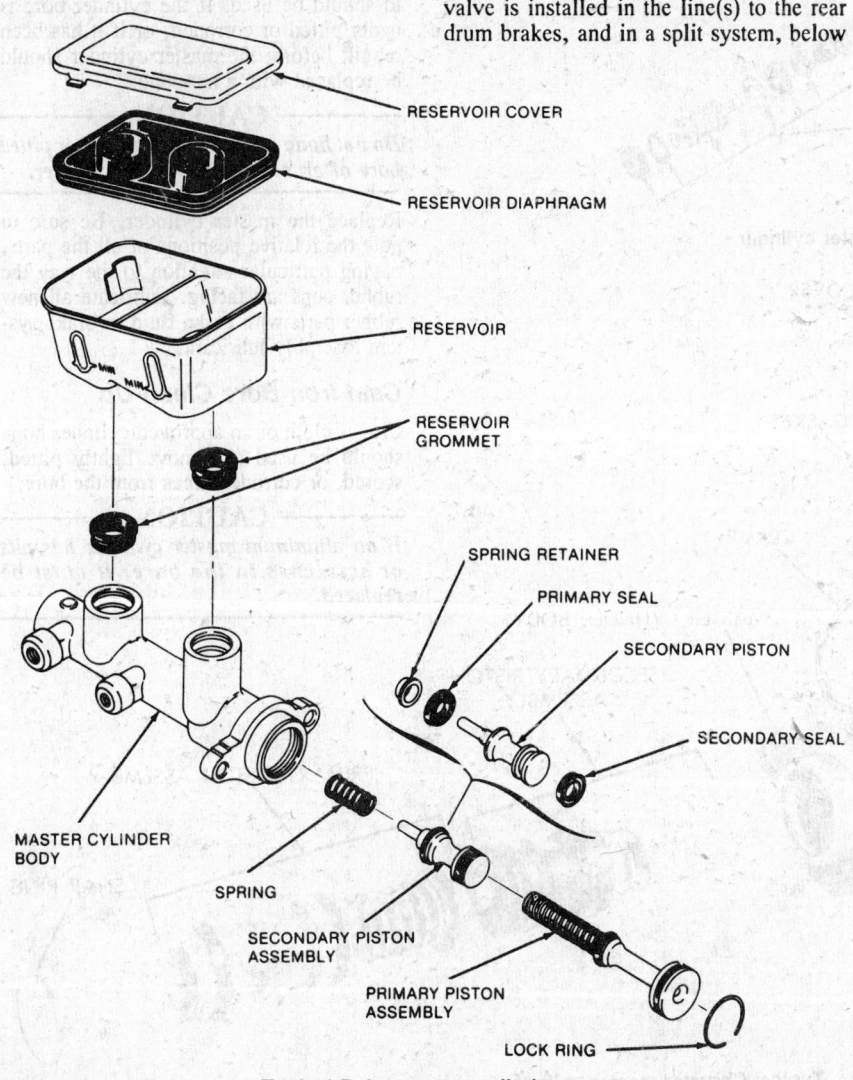

Typical Delco master cylinder

needle nose pliers to pull from bore, or by carefully using compressed air. Disassemble the secondary piston assembly (unless a replacement is provided in the rebuilding kit).

2. Clamp the master cylinder in a vise with outlet ports facing up. Test for presence of check valve by probing with wire through hole in tube seats. Replace tube seat(s) and check valve(s) only if check valve is present and supplied in rebuild kit. Remove the tube seat inserts, if required, by partially threading a self-tapping screw into each tube seat and using two small pry

bars to pry each seat out of the master cylinder. Remove the residual check valve and the spring from the outlet(s) (if present).

Plastic Reservoir Ieaning And Removal

Plastic reservoirs need to be removed only for the following reasons:

If the reservoir is damaged or the rubber grommet(s) between the reservoir and bore is leaking.

For removal of stop pin from Chrysler style plastic reservoir master cylinder to al-

low removal of pistons. Pin is located underneath front reservoir nipple.

To service "Quick Take-up" valve on GM.

The reservoir should be removed by first clamping the flange in a vise. Next remove the reservoir for the Chrysler style. Grasp the reservoir base on one end and pull away from the body. GM reservoirs must be removed by prying between the reservoir and casting with a pry bar. Grommets can be reused if they are in good condition. Whether or not the reservoir is removed, it and the cover or caps should be thoroughly cleaned.

CLEANING AND INSPECTION

Thoroughly clean the master cylinder and any other parts to be reused in clean alcohol. DO NOT USE PETROLEUM PRODUCTS FOR CLEANING. If the bore is not badly scored, rusted or corroded, it is possible to rebuild the master cylinder in some cases. A slight bit of honing is permissible to clean up and smooth out the bore. A master cylinder rebuilding kit and fresh fluid should be used. If the cylinder bore is badly pitted or corroded, or if it has been rebuilt before, the master cylinder should be replaced with a new one.

--- **CAUTION** ---

Do not hone or repair a scratched or pitted bore of an aluminum master cylinder.

Replace the master cylinder. Be sure to note the relative positions of all the parts, paying particular attention to the way the rubber cups are facing. Lubricate all new rubber parts with brake fluid or brake system assembly lubricant.

Cast Iron Bore Clean-Up

Crocus cloth or an approved cylinder hone should be used to remove lightly pitted, scored, or corroded areas from the bore.

--- **CAUTION** ---

If an aluminum master cylinder has pits or scratches in the bore, it must be replaced.

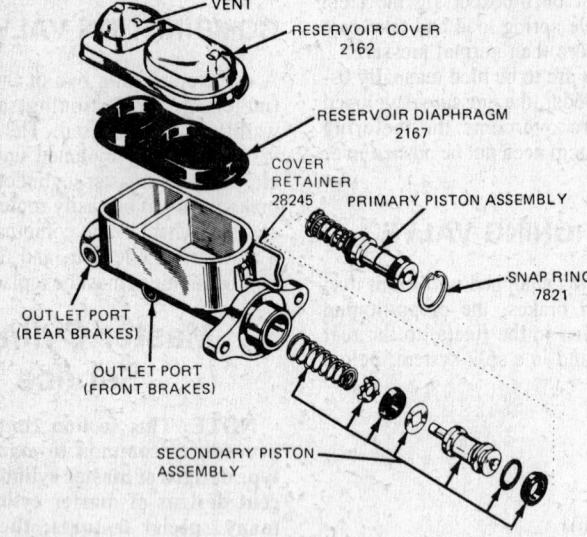

Typical Ford master cylinder

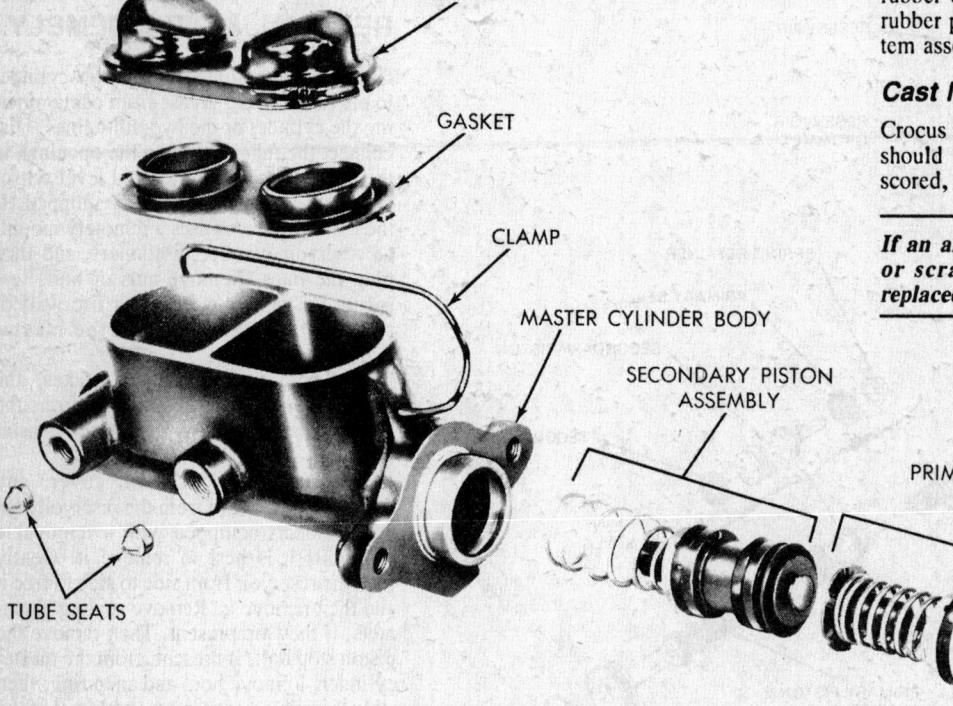

Typical Chrysler master cylinder

Brake fluid can be used as a lubricant while honing lightly. The master cylinder should be replaced if it cannot be cleaned up readily. After using the crocus cloth or a hone, the master cylinder should be thoroughly washed in clean alcohol or brake fluid to remove all dust and grit. If alcohol is used, dry parts thoroughly before reinstalling.

CAUTION

Other solvents should not be used.

Then the clearance between the bore wall and the piston (primary piston of a dual system master cylinder) should be checked. If a narrow (1/8 in. (3.2mm) to 1/4 in. (6.4 mm) wide) 0.006 in. (0.15mm) feeler gauge can be inserted between the wall and a new piston, the clearance is excessive, and the master cylinder should be replaced. The maximum clearance allowed for units containing pistons without replenishing holes is 0.009 in. (0.23mm).

Aluminum Bore Clean-Up

Inspect the bore for scoring, corrosion and pitting. If the bore is scored or badly pitted and corroded the assembly should be replaced. Under no conditions should the bore be cleaned with an abrasive material. This will remove the wear and corrosion resistant anodized surface. Clean the bore with a clean piece of cloth around a wooden dowel and wash thoroughly with alcohol. Do not confuse bore discoloration or staining with corrosion.

REASSEMBLY AND INSTALLATION

1. Carefully install the new cups or seals in the same positions and in reverse order of removal.
2. Use brake fluid or assembly fluid very generously to keep from damaging the seals.
3. Placing the small end of the pressure spring into the secondary piston retainer, slide the assembly into the cylinder bore, taking care not to nick or gouge any rubber part.
4. Place the spring retainer of the primary piston assembly over the secondary piston shoulder and push both assemblies into the bore.
5. Install and tighten the piston retaining screw and gasket, while holding the pistons in their seated positions. At the same time, reinstall any piston snap rings.
6. Install the residual check valve and spring in the proper master cylinder outlet (or both outlets, if originally present). If the tube seat inserts were removed, install new seats in both fluid outlets making sure that they are securely seated.

BLEEDING & CHECKING

NOTE: Bleeding procedures are given in the following section.

1. Bleed the hydraulic system.

NOTE: Be sure to bench bleed a rebuilt or new master cylinder before installation.

2. Check master cylinder vent port clearance by watching for a spurt of brake fluid in both reservoir vent holes when brake pedal is slightly depressed, indicating proper port clearance.

GMC–Quick Take-Up Type

SERVICING

1. Depress the primary piston and remove the snap-ring.
2. Remove the primary and secondary pistons and return springs from the cylinder bore.
3. Disassemble the secondary piston.
4. Inspect the master cylinder bore. If it is corroded, replace the master cylinder. Never use abrasives on the bore.

NOTE: Always lubricate parts with clean, fresh brake fluid before assembly.

5. Install new seals on the secondary piston.
6. Install the spring and secondary piston into the cylinder.
7. Install the primary piston, depress and install the snap-ring.

Bendix Mini-Master

SERVICING

1. Remove the reservoir cover and diaphragm, and drain the fluid from the reservoir.
2. Remove the four bolts that secure the body to the reservoir using Chevrolet special socket number J–25085 or equivalent.

NOTE: Do not remove the two small filters from the inside of the reservoir unless they are damaged and are to be replaced.

3. Remove the small O-ring and the two compensating valve seals from the recessed areas on the bottom side of the reservoir.
4. Depress the primary piston using a tool with a smooth, round end. Then remove the compensating valve poppets and the compensating valve springs from the compensating valve ports in the master cylinder body.
5. Remove the snap-ring at the end of the master cylinder bore. Then release the piston and remove the primary and secondary piston assemblies from the cylinder bore. It may be necessary to plug the front compensating valve port to remove the secondary piston assembly.
6. Lubricate the secondary piston assembly and the master cylinder bore with clean brake fluid.

7. Assemble the secondary spring (shorter of the two springs) in the open end of the secondary piston actuator, and assemble the piston return spring (longer spring) on the projection at the rear of the secondary piston.
8. Insert the secondary piston assembly, actuator end first, into the master cylinder bore and press the assembly to the bottom of the bore.
9. Lubricate the primary piston assembly with clean brake fluid. Insert the primary piston assembly, actuator end first, into the bore.
10. Place the snap-ring over a smooth, round-ended tool and depress the pistons in the bore.
11. Assemble the retaining ring in the groove in the cylinder bore.
12. Assemble the compensating valve seals and the small O-ring seal in the recesses on the bottom of the reservoir. Be sure that all seals are fully seated.
13. While holding the pistons depressed, assemble the compensating valve springs and the compensating valve poppets in the compensating valve ports.
14. Holding the pistons compressed, position the reservoir on the master cylinder body and secure it with the four mounting bolts. Torque the bolts to 12–15 ft. lbs. (16–20 Nm).

Bleeding Brakes

BENCH BLEEDING PROCEDURES

Bench bleeding master cylinders before installation saves time and reduces the possibility of getting air into the lines. In order to expel *all* air trapped in the cylinder, tandem master cylinders must be bench bled before they are installed on the vehicle. Follow this simple procedure for bench bleeding:

1. Route two shortened brake lines from the outlet connection(s) into the fluid reservoir(s), below the normal fluid level.
2. Fill the reservoir(s) with fresh brake fluid and pump the cylinder until air bubbles no longer appear in the reservoir. If the cylinder does not have a check valve at the outlet port, use a *clean* piece of rubber or plastic, or the end of your finger to close off the end of the tubing during the back stroke. Otherwise, the fluid will merely pump back and forth in the tubing.
3. When all air has been purged from the master cylinder, bend the tubes up out of the fluid, and remove them. Refill the cylinder and securely install the master cylinder cap.
4. Install the master cylinder on the vehicle. Attach the lines, but do not tighten the tube connection.
5. Force out any air that might have been trapped in the connection by slowly depressing the pedal several times. Tighten the nut slightly before releasing pedal, and

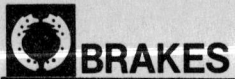
loosen before depressing each time. Catch the fluid in a rag to avoid damaging the car finish. DO NOT BOTTOM THE PISTON. Tighten the connections when air bubbles are no longer present in the fluid. Make sure the master cylinder is adequately filled with brake fluid.

MANUAL BLEEDING

NOTE: See below for GM "Quick Take-Up" master cylinder bleeding sequence.

Bleed the line that is longest first, proceeding from longest to next longest until the shortest line has been bled. This will vary with the individual system. The sequence would start with the front/rear split or diagonally-front wheel, opposite side rear wheel split on most split systems. If a single system, the right rear is usually the longest. During the complete bleeding operation, do not allow the reservoir to run dry. Keep the master cylinder reservoirs filled with the specified brake fluid. Never use brake fluid that has been drained from the hydraulic system.

1. Bleed the master cylinder at the outlet port side of the system being serviced.

NOTE: On a master cylinder without bleed screws, loosen the master cylinder to hydraulic line nut. Operate the brake pedal slowly until the brake fluid at the outlet connection is free of bubbles, then tighten the tube nut to the specified torque. Do not use the secondary piston stop screw located on the bottom of the master cylinder to bleed the brake system. Loosening or removing this screw could result in damage to the secondary piston or stop screw. When bleeding the Bendix-type dual master cylinder it is necessary to solidly cap one reservoir section while bleeding the other to prevent pressure loss through the cap vent hole. Operate the brake pedal slowly until the brake fluid at the outlet connection is free of air bubbles, then tighten the bleed screw.

2. Bleed tha pressure differential valve second, if it has a bleed screw.

3. Clean the bleed screw and the area around it to keep dirt out of the system. Position a suitable size (usually ⅜ in. box wrench) on the bleeder fitting on the cylinder or caliper to be bled. Attach a rubber drain tube to the bleeder fitting. The end of the tube should fit snugly around the bleeder fitting.

3. Submerge the free end of the tube in a container partially filled with clean brake fluid, and loosen the bleeder fitting approximately ⅜ turn.

4. Push the brake pedal down slowly thru its full travel. Close the bleeder fitting, then return the pedal to the full-released position. Repeat this operation until air bubbles cease to appear at the submerged end of the bleeder tube.

5. When the fluid is completely free of air bubbles, close the bleeder fitting and remove the bleeder tube.

6. Repeat this procedure at the brake cylinder or caliper on the other side of the "split" system. Refill the master cylinder reservoir after each cylinder or caliper is bled. When the bleeding is complete, the master cylinder fluid level should be filled to within ¼ in. of the top of the reservoirs.

7. Centralize the pressure differential valve as described above.

PRESSURE BLEEDING DISC BRAKES.

Pressure bleeding equipment should be diaphragm type; placing a diaphragm between the pressurized air supply and the brake fluid. This prevents moisture and other contaminants from entering the hydraulic system.

NOTE: Some front disc/rear drum equipped vehicles use a metering valve which closes off pressure to the front brakes under certain conditions. These systems contain manual release actuators, which must be engaged to pressure bleed the front brakes.

1. Connect the tank hydraulic hose and adapter to the master cylinder.
2. Close hydraulic valve on the bleeder equipment.
3. Apply air pressure to the bleeder equipment.

4. Open the valve to bleed air out of the pressure hose to the master cylinder.

NOTE: Never bleed this system using the secondary piston stopscrew on the bottom of many master cylinders.

5. Open the hydraulic valve and bleed each wheel cylinder and caliper. Bleed rear brake system first when bleeding both front and rear systems.

GM QUICK TAKE UP MASTER CYLINDER

General Motors specifies that these diagonally split dual systems be bled in a specified sequence. Bleed the master cylinder first. To do this, disconnect the left front brake line at the master cylinder, and fill the master cylinder until fluid flows from the port.

Connect the line and tighten fitting.

Depress the brake pedal, one time slowly and hold. Loosen same brake line fitting to purge air from the system. Retighten the fitting and release the brake pedal slowly. Wait 15 seconds. Then repeat the sequence, including the 15 second wait until all air is removed.

Next bleed the right front connection in the same way as the left front.

Bleed the wheel cylinders and calipers only after you are sure that all the air has been removed from the master cylinder. Bleed the wheel calipers and cylinders in this order: Right Rear, Left Front, Left Rear, Right Front. Follow the specified sequence and depress the brake pedal slowly one time before opening bleeder screw to release air. Tighten screw, slowly release pedal, and wait 15 seconds. Repeat all steps, including the 15 second delay until all air has been removed from the system.

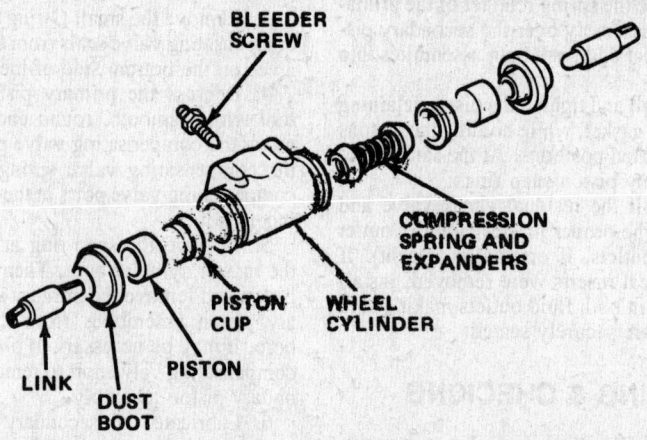

Typical wheel cylinder

SURGE BLEEDING

This method includes both manual and pressure bleeding, and deliberately creates a churning (higher pressure) turbulence in wheel cylinders so that any remaining air can be drawn off in the form of aerated fluid. It is important to remove all possible air before surge bleeding; thus this method is never used unless the routine manual or pressure bleeding method proves inadequate.

The following simple procedure is recommended for surge bleeding:

1. Bleed the brakes at all wheels in a usual manner.
2. At each wheel cylinder, in turn, open the bleeder screw and press the brake pedal down sharply several times. Close the bleeder screw. The action creates a turbulence in each cylinder, forcing out practically all of the remaining trapped air.

BLEEDING THE POWER BRAKE UNIT

On power booster equipped vehicles, the engine should be turned off and the power system purged of vacuum or compressed air by depressing the brake pedal several times. After bleeding the master cylinder, bleed the power brake unit (if equipped with a bleeder screw).

Pressure-multiplying type power units often have bleeder screws to remove the air trapped within the unit. If the unit has more that one bleeder screw, bleed the one at the pressure (main) cylinder first; the control valve second. When bleeding, manually close the bleeder screw before the pedal is allowed to back-stroke each time.

Wheel Cylinders and Calipers

WHEEL CYLINDER OPERATION

The space between the cups in the cylinder bore must remain filled with fluid at all times. After depressing the brake pedal, additional brake fluid is forced into the cylinder bore. As a result of this, cups and pistons move outward in the cylinder bore pushing the shoe links and the brake shoes outward to contact the drum and apply the brakes. On some designs, the end of the shoe web bears directly against the pistons and therefore, shoe links are not used.

SERVICE PROCEDURES

Wheel cylinders may need reconditioning or replacement whenever the brake shoes are replaced or when required to correct a leak condition. On many designs, the wheel cylinders can be disassembled without removing them from the backing plate. On some designs, however, the cylinder is mounted in an indention in the backing plate or a cylinder piston stop is welded to the backing plate. When servicing brakes of this type, the cylinder must be removed from the backing plate before being disassembled.

Diagnostic Inspection and Cleaning

Leaks which coat the boot and the cylinder with fluid, or result in a dropped reservoir fluid level, or dampen and stain the brake linings are dangerous. Such leaks can cause the brakes to "grab" or fail and should be immediately corrected. A leakage, not immediately apparent, can be detected by pulling back the cylinder boot. A small amount of fluid seepage dampening the interior of the boot is normal; a dripping boot is not. Unless other conditions causing a brake to pull, grab, or drag becomes obvious, the wheel cylinder is a suspect and should be included in general reconditioning.

Cylinder binding may be caused by rust, deposits, grime, or swollen cups due to fluid contamination, or by a cup wedged into an excessive piston clearance. If the clearance between the pistons and the bore wall exceeds allowable values, a condition called "heel drag" may exist. It can result in rapid cup wear and can cause the pistons to retract very slowly when the brakes are released.

A ring of a hard, crystal-like substance is sometimes noticed in the cylinder bore where the piston stops after the brakes are released.

Some front wheel cylinders have a baffle located between the opposed pistons. The baffle contains a small hole which causes the cylinder to act as a fluid shock absorber damping servo brake shoes as they become energized. These cylinders cannot be honed and should be replaced if the bore is pitted or corroded.

RECONDITIONING DRUM BRAKE WHEEL CYLINDERS

It is a common practice to recondition a drum brake wheel cylinder without dismounting it, however some brakes are equipped with external piston stops which prevent disassembly unless the cylinder is removed. In order to dismount, remove the shoe springs and spread the shoes apart, disconnect the brake line, remove the mounting bolts or retaining clips, and pull the cylinder free.

Most wheel cylinders are attached to the backing plate with bolts and are easily removed for service or replacement. In recent years, some GM vehicles use a retaining clip for this purpose. To remove this type cylinder, use a special service tool, or this alternate method: Insert 1/8 in. diameter or less awl or pin punch into the slots between wheel cylinder pilot and retainer locking tabs. Bend both tabs away at the same time until tabs spring over the shoulder, releasing cylinder. DISCARD the old retainer.

To replace cylinder, use a new retainer and the following procedure:

1. Hold wheel cylinder against backing plate by inserting a block between the wheel cylinder and axle shaft flange.
2. Position wheel cylinder retainer clip so the tabs will be away from and in horizontal position with the backing plate when installing.
3. Press new retaining clip over wheel cylinder abutment and into position using 1 1/8 in. 12 point socket. Retainer is in place when the tabs are snapped under the retainer abutment. Examine closely to be sure both retainer tabs are properly engaged. Another variation of retainer clip is used on some imported vehicles. The retainer usually consists of two or three separate pieces which when slid together will lock themselves and the wheel cylinder in place. The retainers can be carefully removed without incurring damage which allows them to be reused. If they are damaged or corroded, however, they must be replaced.

Pull the protective dust boots off the cylinder. Internal parts should slide out, or be picked out easily. Parts can be driven out with a wooden dowel, or blown out at low pressure by applying compressed air to the fluid inlet port. Parts which cannot be removed easily indicate they are damaged beyond repair and the cylinder should be replaced. Clean the cylinder and the parts in alcohol and/or brake fluid. (Do NOT use gasoline or other petroleum based products). Use only lint-free wiping cloths. Crocus cloth can be used to clean minute scratches, signs of rust, corrosion or discoloration from the cylinder bore and pistons. Slide the cloth in a circular rather than a lengthwise motion. A clean-up hone may be used. After a cylinder has been honed, inspect it for excessive piston clearance and remove any burrs formed on the edge of fluid intake or bleeder screw ports.

Assemble the cylinder with the internal parts, making sure that the cylinder wall is wet with brake fluid. Insert the cups and pistons from each end of a double-end cylinder; do not slide them through the cylinder. Cup lips should always face inward.

Disc Brake Caliper

An integral part of the caliper, the caliper bore(s) contains the piston(s) that direct thrust against the brake pads supported within the caliper. Since all braking forces (pad application force) are applied on each side of the rotor with no self energization, the cylinder and piston are large in comparison to a drum brake wheel cylinder.

FIXED CALIPER TYPE

A fixed type caliper is mounted solidly to the spindle bracket.

Pistons are located on both sides of the rotor, in inboard and outboard caliper halves. Fluid passes between caliper halves through an external crossover tube or through internal passages. A bleeder screw is located in the inboard caliper half. A dust boot protecting each cylinder fits in a circumferential groove on the piston.

FLOATING/SLIDING CALIPER TYPE

Floating or sliding calipers are free to move in a fixed bracket or support.

The piston is located only on the inboard side of the caliper housing, which straddles the rotor. The cylinder piston applies the inboard brake shoe directly, and simultaneously hydraulic pressure slides the caliper in a clamping action which forces the caliper to apply the outboard brake shoe.

The actual applying movement is small. The unit merely grips during application, relaxes upon release, and the shoes do not retract an appreciable distance from the rotor. The fluid inlet port and the bleeder screw are located on the inboard side of the caliper. A dust boot is fitted into a circumferential groove on the piston and into a recess at or near the outer end of the cylinder bore.

HYDRAULIC SEAL ARRANGEMENTS

Seal arrangements at the caliper pistons vary depending upon the brake manufacturer. Three makes of fixed caliper brakes, Bendix, Budd, and Delco-Moraine, use a ring seal which fits in a circumferential groove on the piston.

A fixed seal is now commonly used in brake calipers. During the very small applying movement of the piston, the elastic-ity of the fixed seal permits some deflection in the cylinder groove. The seal deflects as the brakes are applied and relaxes as the brakes are released, retracting the piston a small amount. Some GM types have a rolling seal that retracts the piston slightly further to reduce pad rubbing friction.

A scratched piston, nicked seal, or a sludge or varnish deposit which lifts the sealing edge away from the piston will cause a fluid leak. A serious leak could develop if calipers are not reconditioned when new pads are installed. Then dust and road grime, gradually accumulating behind the dust boot, could be carried into the seal when the piston is shoved inward to accommodate new thick linings. Old seals may have taken a "set," thus preventing proper seating in the retainer groove and on the piston. Therefore, when reconditioning calipers, new seals should be installed.

SERVICING

Before servicing, syphon or syringe about $2/3$ of the fluid from the master cylinder reservoir; do not, however, lower the fluid level below the cylinder intake port. To prevent a gravity loss of fluid, plug the brake line after disconnecting from the caliper. To recondition, remove the caliper from the vehicle, allow the unit to drain, and remove the brake pads. For benchwork, clamp the caliper housing in a soft jaw vice. On fixed-caliper types, remove the bridge bolts and separate the caliper into halves. Remove the sealing O-rings at cross-over points, if the unit has internal fluid passages across the halves.

Whenever required, use special tools to remove pistons, dust boots, and seals. If compressed air is used, apply it gradually, gently ease the pistons from the cylinders, and trap them in a clean cloth; do not allow them to pop out. Take care to avoid pinching hands or fingers.

While removing stroking type seals and boots, work lip of boot from the groove in the caliper. After the boot is free, pull the piston, and strip the seal and boot from the piston.

While removing fixed position (rectangular ring) seals and boots, pull the piston through the boot. Do not use a metal tool which would scratch the piston. Use a small pointed wooden or plastic tool to lift the boots and seals from the grooves in the cylinder bore.

CLEANING, INSPECTION & INSTALLATION

NOTE: Use only alcohol and/or brake fluid and a lint free wiping cloth to clean the caliper and parts. Other solvents should not be used. Blow out passages with compressed air. Always wear eye protection when using compressed air or cleaning calipers.

To correct minor imperfections in the cylinder bore, polish with a fine grade of crocus cloth working in a circular rather than a lengthwise motion. Do not use any form of abrasive on a plated piston. Discard a piston which is pitted or has signs of plating wear.

Inspect the new seal. It should lie flat and be round. If it has suffered a distorted "set" during its shelf life, do not use it. Lubricate the cylinder wall and parts with brake fluid.

While installing stroking type seals and boots, stretch the boot and the seal over the piston and seat them in position.

Use special alignment tools for inserting lip cup seals. Be sure the seal does not twist or roll.

Where the boot lip is retained inside the cylinder bore the following method works well:

1. Lubricate bottom inside edge of piston and brake seal in caliper with brake fluid.
2. Pull boot over bottom end of piston so that boot is positioned on bottom of piston with lip about $1/4$ in. up from bottom end.
3. Hold piston suspended over bore.
4. Insert back boot lip into groove in caliper.
5. Then tuck the sides of boot into groove and work forward until only one bulge remains.
6. Tuck the final bulge into the front of the groove.
7. Then push the piston carefully through the seal and boot to the bottom of the bore. The inside of the boot should slide on the piston and come to rest in the boot groove.

If the boot lip is retained outside the cylinder bore, first stretch boot over the piston and seat it in its groove, then press the piston through the seal. Fully depress the piston. You'll need 50 to 100 lbs. force to fasten the boot lip in place. On some designs, it is necessary to use a wooden drift or a special tool to seat the metal boot in the caliper counterbore below the face of the caliper.

INSTALLING FIXED CALIPER BRIDGE BOLTS

If the caliper contains internal fluid crossover passages, be sure to install the new O-ring seals at joints. Install high tensile strength bridge bolts on the mated caliper halves. Never replace the bridge bolts with ordinary standard hardware bolts.

ROTOR RUNOUT

Manufacturers differ widely on permissible runout, but too much can sometimes be felt as a pulsation at the brake pedal. A wobble pump effect is created when a rotor is not perfectly smooth and the pad hits the high spots forcing fluid back into the master cylinder. This alternating pressure causes a

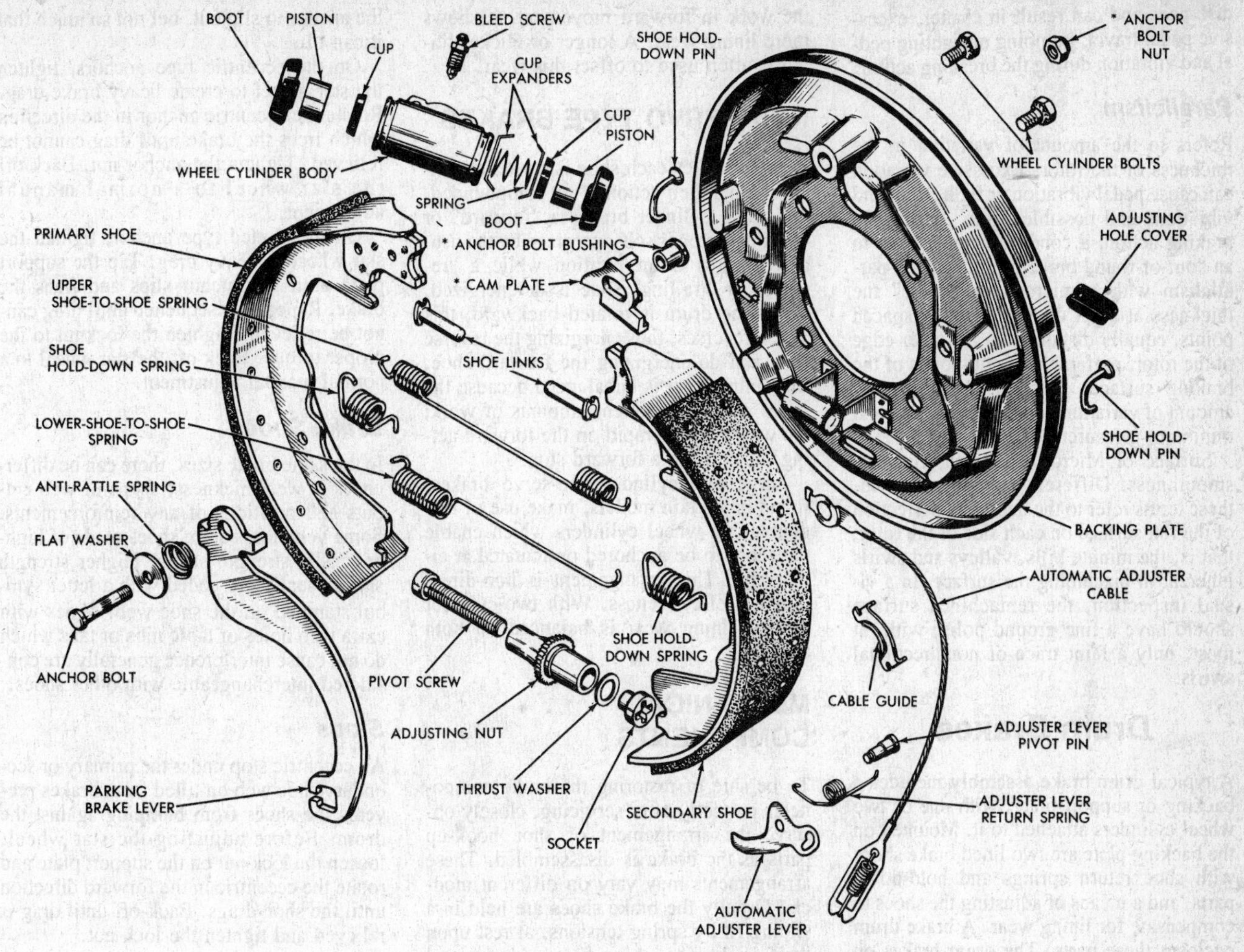

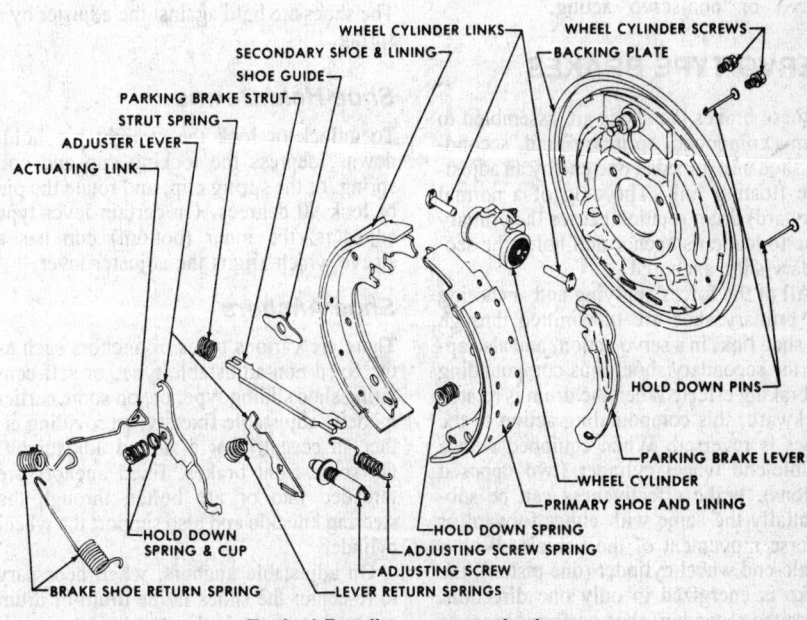

Typical Bendix Duo-Servo brake

- BOOT
- PISTON
- CUP
- BLEED SCREW
- CUP EXPANDERS
- SHOE HOLD-DOWN PIN
- CUP PISTON
- WHEEL CYLINDER BODY
- SPRING
- BOOT
- * ANCHOR BOLT BUSHING
- * CAM PLATE
- PRIMARY SHOE
- UPPER SHOE-TO-SHOE SPRING
- SHOE HOLD-DOWN SPRING
- SHOE LINKS
- LOWER-SHOE-TO-SHOE SPRING
- ANTI-RATTLE SPRING
- FLAT WASHER
- ANCHOR BOLT
- PIVOT SCREW
- ADJUSTING NUT
- SHOE HOLD-DOWN SPRING
- PARKING BRAKE LEVER
- THRUST WASHER
- SOCKET
- SECONDARY SHOE
- AUTOMATIC ADJUSTER LEVER
- * ANCHOR BOLT NUT
- WHEEL CYLINDER BOLTS
- ADJUSTING HOLE COVER
- SHOE HOLD-DOWN PIN
- BACKING PLATE
- AUTOMATIC ADJUSTER CABLE
- CABLE GUIDE
- ADJUSTER LEVER PIVOT PIN
- ADJUSTER LEVER RETURN SPRING

Typical Bendix non-servo brake

- WHEEL CYLINDER LINKS
- SECONDARY SHOE & LINING
- SHOE GUIDE
- PARKING BRAKE STRUT
- STRUT SPRING
- ADJUSTER LEVER
- ACTUATING LINK
- WHEEL CYLINDER SCREWS
- BACKING PLATE
- HOLD DOWN PINS
- PARKING BRAKE LEVER
- WHEEL CYLINDER
- PRIMARY SHOE AND LINING
- HOLD DOWN SPRING
- ADJUSTING SCREW SPRING
- ADJUSTING SCREW
- LEVER RETURN SPRINGS
- HOLD DOWN SPRING & CUP
- BRAKE SHOE RETURN SPRING

pulsating feeling which can be felt at the pedal when the brakes are applied. This excessive runout also causes the brakes to be out of adjustment because disc brakes are self-adjusting; they are designed so that the pads drag on the rotor at all times and therefore automatically compensate for wear. To check the actual runout of the rotor, first tighten the wheel spindle nut to a snug bearing adjustment, end-play removed. Fasten a dial indicator on the suspension at a convenient place so that the indicator stylus contacts the rotor face approximately one inch from its outer edge. Set the dial at zero. Check the total indictor reading while turning the rotor one full revolution. If the rotor is warped beyond the runout specification, it is likely that it can be successfully remachined.

Lateral Runout

A wobbly movement of the rotor from side to side as it rotates. Excessive lateral runout causes the rotor faces to knock back the

disc pads and can result in chatter, excessive pedal travel, pumping or fighting pedal and vibration during the braking action.

Parallelism

Refers to the amount of variation in the thickness of the rotor. Excessive variation can cause pedal vibration or fight, front end vibrations and possible "grab" during the braking action; a condition comparable to an "out-of-round brake drum." Check parallelism with a micrometer, "mike" the thickness at eight or more equally spaced points, equally distant from the outer edge of the rotor, preferably at mid-points of the braking surface. Parallelism then is the amount of variation between maximum and minimum measurements.

Surface of Micro-inch finish, flatness, smoothness: Different from parallelism, these terms refer to the degree of perfection of the flat surface on each side of the rotor; that is, the minute hills, valleys and swirls inherent in machining the surface. In a visual inspection, the remachined surface should have a fine ground polish with, at most, only a faint trace of nondirectional swirls.

Drum Brakes

A typical drum brake assembly includes a backing or support plate, with one or two wheel cylinders attached to it. Mounted on the backing plate are two lined brake shoes with shoe return springs and hold-down parts, and a means of adjusting the shoes to compensate for lining wear. A brake drum encloses these parts. The drum brakes on the rear of most vehicles also normally include the parts required for parking brakes. All of the drum brakes used on modern vehicles have these components but there is a variety of configurations for each.

Drum brakes are designed to be either "servo" or "non-servo" acting.

SERVO TYPE BRAKES

In these brakes the shoes are assembled to form a compound, "primary" and "secondary" shoe unit joined at one end by an adjustable floating link. The drag of a normal (forward) drum rotation causes the primary shoe to leave its anchor and holds the secondary shoe anchored.

All of the forces applying and anchoring the primary shoe are transmitted through the shoe link, in a servo action, and also apply the secondary shoe, thus compounding its braking effect. When the drum is rotated backward, this compounding action of the shoes is reversed. When equipped with a double-end wheel cylinder (two opposed pistons), brake effectiveness can be substantially the same with either forward or reverse movement of the vehicle. With a single-end wheel cylinder (one piston), the brake is energized in only one direction. Since the secondary shoe performs more of

the work in forward movement, it shows more lining wear. A longer or thicker lining is often used to offset this wear.

NON-SERVO TYPE BRAKES

In these brakes each shoe is separately anchored and their action is not compounded. On single cylinder brakes a "forward" or "leading" shoe is self-energized by the usual (forward) drum rotation while a "reverse" or "trailing" shoe is de-energized. When the drum is rotated backward, this action reverses, thus energizing the reverse shoe and de-energizing the forward shoe. The lining wear is unbalanced because the shoes perform different amounts of work; the wear is more rapid on the forward acting shoe during a forward stop.

Large two cylinder non-servo brakes, found on certain models, make use of two double-end wheel cylinders which enable the shoes to be anchored or actuated at either end. This arrangement is non-directional in effectiveness. With two-cylinder brakes, lining wear is balanced on both shoes.

MECHANICAL COMPONENTS

To be sure of restoring the brake components correctly after servicing, closely observe the arrangement of shoe hook-up parts as the brake is disassembled. These arrangements may vary on different models. Usually the brake shoes are held in a sliding fit by spring tensions, at rest upon their anchor by the return springs, and against support pads by spring or clip type hold-downs. Opposite the anchor, a star wheel adjuster links the shoe webs and provides a threaded adjustment which permits the shoes to be expanded or contracted. Some rear brakes have adjustable links. The shoes are held against the adjuster by a spring.

Shoe Hold-Downs

To unlock or lock the straight pin hold-downs, depress the locking cup and coil spring, or the spring clip, and rotate the pin or lock 90 degrees. On certain lever type adjusters, the inner (bottom) cup has a sleeve which aligns the adjuster lever.

Shoe Anchors

There are various types of anchors such as the fixed non-adjustable type, or self-centering shoe sliding type, or, on some earlier models, adjustable fixed type providing either an eccentric or a slotted adjustment. On some front brakes, fixed anchors are threaded into or are bolted through the steering knuckle and also support the wheel cylinder.

On adjustable anchors, when necessary to re-center the shoes in the drum or drum gauge, loosen the locknut enough to permit

the anchor to slip out, but not so much that it can tilt.

On the eccentric type anchors, tighten the star wheel to create heavy brake drag. Rotate the eccentric anchor in the direction which frees the brake until drag cannot be relieved. Tighten the anchor nut. Back off the star wheel to a normal manual adjustment.

On the slotted type anchor, tighten the star wheel to heavy drag. Tap the support plate until the anchor slips and frees the brake. Repeat this sequence until drag cannot be relieved. Tighten the locknut to the proper torque. Back off the star wheel to a normal manual adjustment.

Brake Shoes

In the same brake sizes, there can be differences in web thickness, shape of web cutouts and positions of any reinforcements. Some vehicles require shoes made of higher tensile strength steels. Higher strength shoes usually are coded with a letter symbol stamped on the shoe web. Shoes with extra web holes or table nibs or tabs which do not cause interference generally are considered interchangeable with other shoes.

Stops

An eccentric stop under the primary or secondary shoe web on tilted front brakes prevents the shoes from bumping against the drum. Before adjusting the star wheel, loosen the lock nut on the support plate and rotate the eccentric in the forward direction until the shoe drags. Back-off until drag is relieved and tighten the lock nut.

Piston Stops

If the brake is equipped with piston stops, the wheel cylinder must be dismounted for reconditioning.

BASIC SERVICE

———— CAUTION ————

Do not blow the brake dust out of the drums with compressed air or lung power; always use a damp cloth and wipe it out. Brake linings contain asbestos, a known cancer causing substance. Dispose of the cloth after use. Never work on a vehicle supported only on a jack. Use a hydraulic lift or jack stands to support the vehicle while working.

Raising both front or rear wheels at once and supporting them on jackstands also allows comparison of the brake being serviced to the brake on the opposite side.

Check for Leaks

Press the brake pedal to ensure that there are no leaks in the hydraulic system. If the pedal does not remain hard, and drops to the floor, it is an indication of a leak in the master cylinder, hoses, wheel cylinders, or disc brake calipers. When performing this

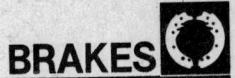
test, the engine should be running if equipped with power brakes. With power brakes it is normal for the pedal to drop slightly when the engine starts. If it continues to drop, start looking for a leak.

Drum Removal

Safely support the vehicle and release the parking brake if working on the rear axle. Remove the lug nuts, the wheel/tire assembly and then pull off the drums. If the brake shoes have expanded too tightly against the drum, or have cut into the friction surface of the brake drum, the drums may be too tight for removal. In such a case, adjust the shoes inward before the brake drum is removed. On vehicles with self-adjusting mechanisms, reach through the adjusting slot with a very small prybar (or similar tool) and carefully push the self-adjusting lever away from the star wheel by a maximum of $\frac{1}{16}$ in. (1.5 mm). While holding the lever back, insert a brake adjusting tool into the slot and turn the star wheel in the proper direction until the brake drum can be removed. On vehicles with manual adjusting mechanisms, try lightly tapping the drum with a rubber mallet. If this does not work, simply reverse the manual adjustment procedures given later in this section until the drum can be removed.

Drum Inspection

Check the drums for any cracks, scores, grooves, or an out-of-round condition. Replace if cracked. Slight scores can be removed with fine emery cloth while extensive scoring requires turning the drum on a lathe.

If the friction surface of the brake drum appears scored or otherwise damaged beyond repair, it will require reconditioning. After machining, the drum diameter must not exceed the diameter cast on the drum or 0.060 in. (1.5 mm) over the original nominal diameter. Consult both the published specifications and any state inspection regulations that may apply and go by the more conservative requirement. Carefully look for signs of grease or oil at the center of the assembly. If any leak is noticed, the seal should be replaced.

Rebuild the Cylinders

It is always a good idea to rebuild or replace the wheel cylinders when relining the brakes. This will help assure a properly operating brake system.

Remove Brake Shoes

It is convenient to disassemble one wheel at a time so the opposite side serves as a reference. Carefully note the colors and locations of different springs and parts. This is necessary to distinguish different springs that appear to be the same but have different tensions. If there are extra unused holes close to the ones in which the springs are located, use a dab of paint or other marking

on the new shoes to identify the holes to be used. Replace any discolored springs and other parts found corroded or distorted. Use special tools whenever necessary. Examine the springs for signs of stretching or other defects and replace if their condition is at all questionable. Examine the flexible brake hoses and replace any that show signs of cracking or other damage.

Clean and Lubricate

With all the brake parts off, clean the backing plate with a damp cloth to avoid raising any asbestos dust, and dispose of the rag after use. Clean any rust with a wire brush. File smooth any ridges or rough edges on the contact points on the backing plate, and lubricate with approved brake lubricant. Clean and lightly lubricate the adjuster threads, and screw the adjuster all the way together to facilitate reassembly later on. Wash the wheel bearings with solvent and repack them with the proper grease. Check backing plate bolts to make sure they are tight.

Reassemble and Install Brake Shoes

Reassemble the brakes in the reverse order of disassembly. Make sure all parts are in their proper locations and that both brake shoes are properly positioned in either end of the adjuster. Also, both brake shoes should correctly engage the wheel cylinder push rods and parking brake links, and should be centered on the backing plate. Parking brake links and levers should be in place on the rear brakes. With all the parts in place, try the fit of the brake drum over the new shoes. If not slightly snug, pull it off and turn the star wheel until a slight drag is felt when sliding the drum on. The use of a brake preset gauge will make this job easy. This makes final brake adjustment simpler. Then install the brake drum, wheel bearings, spindle nuts, cotter pins, dust caps, and wheel/tire assemblies, and make final brake adjustments as specified. Torque the spindle and lug nuts to specifications.

Bleed and Road-Test

Bleed the brakes to make sure of a high, hard brake pedal, and road-test the car. Most self-adjusting mechanisms are activated only during the rearward motion of a car. So, whenever servicing self-adjusting brakes, make sure that the road test includes enough stops, traveling in reverse, to allow the self-adjusters to perform the proper match-up of all wheels. Or, operate the parking brake several times if that activates the automatic adjuster.

CHILTON TIPS

• The primary brake shoe is the one toward the front of the vehicle, and its lining is usually shorter than that on the secondary (rearward) shoe.

• Self-adjusting mechanisms are usually mounted on the secondary shoe.
• The star wheel part of an adjuster usually (but not always) goes toward the rear of the vehicle.
• Different color springs belong in different locations.
• Self-adjusters and related parts are not interchangeable from one side to the other since the direction of adjuster rotation varies from one side to the other. Adjusters on one side usually have right hand threads, and the adjusters on the other side left hand threads.
• Never press the brake pedal when one or more brake drums are off, or wheel cylinders associated with missing drums will pop apart.

DOMESTIC TRUCKS—DISC BRAKE SERVICE

Floating/Sliding Caliper (Single Piston)

The caliper has a single piston mounted in the inboard side. The piston is made of steel or phenolic (plastic) material designed to resist wear and corrosion. The piston is usually equipped with a square cut ring which provides a seal between the piston and the caliper cylinder wall and keeps the piston and brake pads in position as the lining material wears. A rubber dust boot located in a groove in the cylinder helps keep contamination from the piston and cylinder wall.

The caliper is usually mounted on an adapter which is mounted on the steering knuckle.

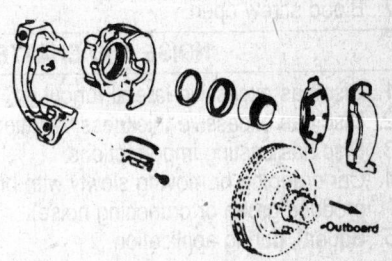

Sliding caliper, typical of AMC/Jeep, Ford

ADJUSTMENT

No adjustment is required on this unit other than applying the pedal several times after the unit has been worked on. This is to seat the pads and after this is done the hydraulic pressure maintains the proper clearance between the brake pads and rotor.

DISC BRAKES—TROUBLE · DIAGNOSIS

Cause	Correction
1. Power brake malfunctioning.	1. Check and repair power unit.
2. Linings soiled with brake fluid, oil or grease.	2. Replace shoes and linings.
3. Lines, hoses or connections dented, kinked, collapsed, clogged or disconnected.	3. Repair or replace defective parts.
4. Master cylinder cups swollen.	4. Drain hydraulic system, flush system with brake fluid and replace combination valve and all cups and seals in complete brake system.
5. Master cylinder bore corroded or rough.	5. Repair or replace master cylinder.
6. Caliper pistons frozen or seized.	6. Disassemble caliper and free pistons (replace if necessary).
7. Caliper cylinder bores corroded or rough.	7. Disassemble caliper and remove corrosion or roughness, or replace caliper.
8. Pedal push rod and linkage binding.	8. Free and lubricate.
9. Metering valve not working.	9. Replace combination valve.

GRABBING OR PULLING (Severe Reaction To Pedal Pressure and Out of Line Stops)

Cause	Correction
1. Linings soiled with brake fluid, oil or grease.	1. Replace shoes and linings.
2. Caliper loose.	2. Tighten caliper mounting bolts to specified torque.
3. Lines, hoses or connection dented, kinked, collapsed or clogged.	3. Repair or replace defective parts.
4. Master cylinder bore corroded or rough.	4. Repair or replace master cylinder.
5. Caliper pistons frozen or seized.	5. Disassemble caliper and free pistons (replace if necessary).
6. Caliper cylinder seals soft or swollen.	6. Drain hydraulic system, flush system with brake fluid and replace all cups and seals in complete brake system.
7. Caliper cylinder bores corroded or rough.	7. Disassemble caliper and remove corrosion or roughness, or replace caliper.
8. Pedal linkage binding (and suddenly releasing).	8. Free and lubricate linkage.
9. Metering valve not functioning properly.	9. Replace combination valve.

FADING PEDAL (Pedal Falling Away Under Steady Pressure)

Cause	Correction
1. Poor quality brake fluid (low boiling point) in system.	1. Drain hydraulic system and fill with approved fluid.
2. Hydraulic connections loose; lines or hoses ruptured (causing leakage).	2. Tighten or replace defective parts.
3. Master cylinder cup worn or damaged. (primary, secondary or both).	3. Repair master cylinder.
4. Master cylinder bore corroded, worn or scored.	4. Repair or replace master cylinder.
5. Caliper cylinder seals worn or damaged.	5. Replace seals.
6. Caliper cylinder bores corroded, worn or scored.	6. Disassemble caliper and remove corrosion or scoring, or replace caliper.
7. Bleed screw open.	7. Close bleed screw and bleed hydraulic system.

NOISE AND CHATTER (May Be Accompanied By Brake Roughness and Pedal Pumping)

Cause	Correction
1. Disc has excessive lateral runout.	1. Replace or machine disc.
2. Disc has excessive thickness variations (out of parallel).	2. Replace or machine disc.
3. Disc has casting imperfections.	3. Replace disc.
4. Car creeping or moving slowly with brakes applied (may produce groan or crunching noise).	4. Increase or decrease pedal effort slightly.
5. Squeal, during application.	5. A small amount of high-pitched squeal is inherent in disc brake design and must be considered normal. Some relief may be obtained with service package backing.

DRAGGING BRAKES (Slow or Incomplete Release of Brakes)

Cause	Correction
1. Lines, hoses or connections dented, kinked, collapsed or clogged.	1. Repair or replace defective parts.
2. Master cylinder compensating port restricted by swollen primary cup.	2. Drain hydraulic system, flush system with brake fluid and replace combination valve and all cups and seals in complete brake system.
3. Residual pressure check valve in lines to front wheels.	3. Remove check valve.
4. Caliper pistons frozen or seized.	4. Disassemble caliper and free pistons (replace if necessary).

DISC BRAKES—TROUBLE DIAGNOSIS

Cause	Correction
1. Master cylinder fluid level low.	1. Fill to proper level with approved fluid. (Fluid level drops as disc brake linings wear.)
2. Poor quality brake fluid (low boiling point) in system.	2. Drain hydraulic system and fill with approved.
3. Air in hydraulic system.	3. Bleed hydraulic system and refill with approved fluid.
4. Hoses soft or weak (expanding under pressure).	4. Replace defective hoses. Combination valve and all cups and seals in complete brakes.

DRAGGING BRAKES (Slow or Incomplete Release of Brakes)

5. Caliper cylinder seals swollen.	5. Drain hydraulic system, flush system with clean brake fluid and replace combination valve and all cups and seals in complete brake system.
6. Caliper cylinder bores corroded or rough.	6. Disassemble caliper and remove corrosion or roughness, or replace caliper.
7. Hydraulic push rod on power brake out of adjustment or binding (causing primary cup to restrict master cylinder compensating port).	7. Adjust or free and lubricate.

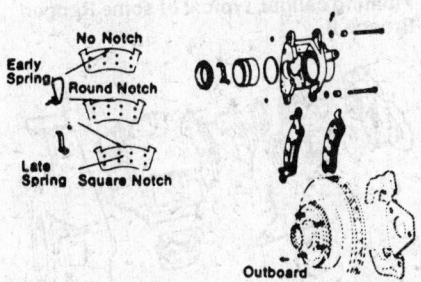

Floating caliper, typical of Chevy/GMC

Sliding caliper, typical of Chrysler Corp.

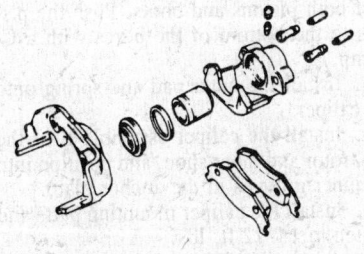

Floating caliper, typical of Chrysler Corp.

BRAKE PADS

Replace the brake pads when the linings are worn to within $\frac{1}{16}$ in. of the shoe or the rivets.

Removal

1. Remove the master cylinder cover and remove the fluid necessary to make the cylinder only $\frac{1}{3}$ full. This is done to prevent any overflow from the cylinder when the piston is pushed into the bore of the caliper.

2. Raise and safely support the vehicle on jackstands. Remove the front wheels.

3. Compress the piston back into the bore by using a large C-clamp and compressing the unit until the piston bottoms in the bore.

4. Depending on design: Remove the two retaining bolts that hold the caliper into the support. Or if the caliper has retaining clips remove the retaining clips and anti-rattle springs. Or if the caliper has (a) key type retainer, or retainers, remove the key retaining screw(s), and using a hammer and drift, punch drive the key(s) out of the mounting groove. Pay attention to the way the pressure spring behind the wedge is mounted (for reinstallation location).

5. Slide the caliper off the rotor disc. Be careful not to damage the dust boot on the piston when removing the caliper.

NOTE: Do not let the caliper hang with the brake hose supporting the weight. This can cause damage to the hose which could result in a loss of brakes. Set the caliper on the front suspension arm or tie rod.

6. Remove the outer pad from the caliper or adaptor. It may be necessary to tap the shoe to loosen it. Depending on design:

Remove the inner pad from the caliper or spindle adaptor assembly depending on how and where the inner pad is mounted. On some models it will be necessary to remove the rotor assembly. Remove the support spring from the caliper piston if so equipped.

Cleaning and Inspection

Clean the sliding surfaces of the caliper and mounting adaptor. Clean any dirt from the mounting bolts, clips or keys.

Inspect the boot on the piston for signs of cracks, cuts or other damage. Check to see if there are signs of fluid leaking around the seal on the piston. This will show up in the boot. If there is an indication of a fluid leak, the entire caliper has to be dis-assembled and the seal replaced.

Installation

1. Make sure that the piston is fully bottomed in the cylinder bore and install the inboard pad on the adapter, or install the pads in the caliper etc., depending on design. If the pad is equipped with a mounting/positioning clip, be sure it is firmly mounted in the caliper piston.

NOTE: On shoes with anti-rattle springs be sure to install the spring before installing the shoe in the caliper.

2. Depending on design: Install the brake rotor and position the outer pad on the caliper/adaptor and press it into place with finger pressure.

3. Position the caliper over the rotor and carefully slide it down into position.

4. Depending on design: Install the caliper mounting bolts and torque them to 35 ft. lbs. On models with retaining clips install the anti-rattle springs and the retaining clips and torque the retaining screws to 200

inch lbs. On models with key type retainers press down the caliper and install the key in its slot and drive it in place with a hammer and drift. Install the retaining screw and torque to 12–18 ft. lbs.

5. Install the wheels and lower the vehicle. Check the master cylinder fluid level and add any fluid necessary to bring it up to the proper level.

6. Pump the brake pedal several times until a firm brake pedal is established, bleed the brake system if necessary. Road test the vehicle to check for proper operation.

DISC CALIPER

Removal

1. Remove the cover on the master cylinder and remove the necessary amount to bring the level down to ⅓ full. This step is necessary to avoid overflow from the master cylinder when the piston is compressed into the cylinder bore.

2. Raise the vehicle and remove the wheel.

3. Compress the piston into the caliper bore and remove the brake hose from the steel line connection and caliper. Tape the end of the hose to prevent dirt from entering the line.

4. Remove the caliper retaining bolts, clips or wedges and remove the caliper from the vehicle.

Disassembly

1. Clean the outside of the caliper with clean brake fluid and drain any fluid from the caliper.

2. Remove the piston from the caliper by connecting the hydraulic line to the caliper and gently stroking the brake pedal. This will push the piston from the caliper bore.

3. With care remove the boot from the caliper piston bore.

4. Remove the piston seal from the caliper bore using a piece of wood or plastic.

NOTE: DO NOT use a metal tool to remove the seal. This can damage the bore or burr the edges of the seal groove.

5. Remove the bleeder valve.

Cleaning and Inspection

1. Clean all the parts with clean brake fluid and blow out all the passages in the caliper.

NOTE: Whenever the caliper is disassembled, discard the boot and piston seal. These parts must not be reused.

2. Inspect the outside of the piston for signs of wear, corrosion, scores or any other defects. If any defects are detected replace the piston.

3. Check the caliper bore for the same defects as the piston. However, the bore can be cleaned up to a point with crocus cloth. If there are any marks that will not clean up with the cloth the caliper must be replaced.

Assembly and Installation

1. Lube the caliper bore and the piston with clean brake fluid and position the seal for the piston in the cylinder bore groove.

2. Install the dust boot into the groove in the piston with the fold faces toward the open end of the piston.

3. Install the piston in the bore being careful not to unseat the piston seal in the bore.

4. With the piston bottomed in the cylinder position the boot in the groove in the caliper. Make sure that the retaining ring in the seal is pressed down evenly around the cylinder.

5. Install the bleeder screw in the caliper and install the caliper back on the vehicle.

6. Connect the brake hoses and bleed the calipers of air. When bleeding is done pump the pedal several times to develop a firm brake pedal.

Sliding Caliper (Double Piston)

BRAKE PADS & CALIPER

Removal

1. Drain about ⅔ of the total brake fluid from the reservoir.

2. Raise and safely support the front of the vehicle. Remove the front wheels.

3. Remove the retaining bolt(s) and drive the wedge and spring from the mounting groove.

4. Lift the caliper off the hub and rotor. If the caliper is to be removed for servicing, disconnect the hydraulic line; if not, lay the caliper on the suspension or support with a length of wire.

5. Remove the brake pads.

Disassembly

1. Drain the brake fluid from the caliper and clean the exterior with clean brake fluid.

2. Place a small block of wood under the caliper pistons and place a protective pad over the exterior. Remove the pistons by directing compressed air into the caliper fluid outlet.

3. Remove and discard piston boots.

4. Remove the piston seals from the groove in the caliper bore.

Assembly

1. Clean all parts in clean brake fluid and blow dry.

2. Dip the new piston seal in clean brake fluid and install it into the cylinder groove.

NOTE: Be sure that the seal is not rolled or twisted in the groove.

3. Install the dust boot in the cylinder groove.

4. Coat the outside diameter of the piston with clean brake fluid. Use something plastic or wood and gradually work the dust boot around the piston.

5. Press the pistons straight into the caliper bores until they bottom. Position the boot in the piston groove.

Installation

1. Install the inner pad and anti-rattle spring on the anchor plate.

2. Push the pistons to the bottom of the piston bore. Place a small block of wood

Floating caliper, typical of some Ranger/ Bronco II

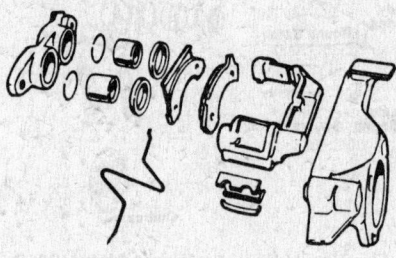

Sliding caliper, typical of Ford

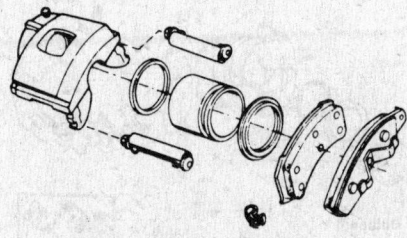

Floating caliper, typical of some Bronco II/Ranger

over both pistons and boots. Push the pistons to the bottom of the bores with a C-clamp.

3. Install the outer pad and spring onto the caliper.

4. Install the caliper assembly over the hub, rotor and inner shoe, and position into the inner grooves in the anchor plate.

5. Install the caliper mounting parts and tighten to 14–22 ft. lbs.

6. Add brake fluid to bring the level to ¼ in. from the top of the reservoir.

7. Bleed the system and add fluid as necessary.

Ford Ranger And Bronco II

INSPECTION

Replace the front pads when the pad thickness is at the minimum thickness recommended by Ford Motor Co. ($\frac{1}{32}$ in.), or at the minimum allowed by the applicable state or local motor vehicle inspection code. Pad thickness may be checked by removing the wheel and looking through the inspection port in the caliper assembly.

CALIPER & BRAKE PADS

Removal & Installation

NOTE: Always replace all disc pad assemblies on an axle. Never service one wheel only.

1. To avoid fluid overflow when the caliper piston is pressed into the caliper cylinder bores, siphon or dip part of the brake fluid out of the larger master cylinder reservoir (connected to the front disc brakes). Discard the removed fluid.
2. Raise the vehicle and install jack stands. Remove a front wheel and tire assembly.
3. Place an eight-inch C-clamp on the caliper and tighten the clamp to bottom the caliper piston in the cylinder bore. Remove the clamp.

NOTE: Do not use a screwdriver or similar tool to pry piston away from the rotor.

4. There are three types of caliper pins used: a single tang type, a double tang type and a split-shell type. The pin removal process is dependent upon how the pin is installed (bolt head direction). Remove the upper caliper pin first.

NOTE: On some applications, the pin may be retained by a nut and torx-head bolt (except the split-shell type).

If the bolt head is on the outside of the caliper, use the following procedure:

5. From the inner side of the caliper, tap the bolt within the caliper pin until the bolt head on the outer side of the caliper shows a separation between the bolt head and the caliper pin.
6. Using a hacksaw or bolt cutter, remove the bolt head from the bolt.
7. Depress the tab on the bolt head end of the upper caliper pin with a screwdriver, while tapping on the pin with a hammer. Continue tapping until the tab is depressed by the v-slot.
8. Place one end of a punch ($\frac{1}{2}$ in. or smaller) against the end of the caliper pin and drive the caliper pin out of the caliper toward the inside of the vehicle. Do not use a screwdriver or other edged tool to help drive out the caliper pin as the v-grooves may be damaged.

CAUTION

Never reuse caliper pins. Always install new pins whenever a caliper is removed.

If the nut end of the bolt is on the outside of the caliper, use the following procedure:

9. Remove the nut from the bolt.
10. Depress the lead tang on the end of the upper caliper pin with a screwdriver while tapping on the pin with a hammer. Continue tapping until the lead tang is depressed by the v-slot.
11. Place one end of a punch ($\frac{1}{2}$ in. or smaller) against the end of the caliper pin and drive the caliper pin out of the caliper toward the inside of the vehicle. Do not use a screwdriver or other sharp-edged tool to help drive out the caliper pin as the v-grooves may be damaged.
12. Repeat the procedure in Step 4 for the lower caliper pin.
13. Remove the caliper from the rotor. If the caliper is to be removed for service, remove the brake hose from the caliper.
14. Remove the outer pad. Remove the anti-rattle clips and remove the inner pad.
15. Place a new anti-rattle clip on the lower end of the inner pad. Be sure the tabs on the clip are positioned properly and the clip is fully seated.
16. Position the inner pad and anti-rattle clip in the pad abutment with the anti-rattle clip tab against the pad abutment and the loop-type spring away from the rotor. Compress the anti-rattle clip and slide the upper end of the pad in position.
17. Install the outer pad, making sure the torque buttons on the pad spring clip are seated solidly in the matching holes in the caliper.
18. Install the caliper on the spindle, making sure the mounting surfaces are free of dirt and lubricate the caliper grooves with Disc Brake Caliper Grease. Install new caliper pins, making sure the pins are installed with the tang in the correct position.
19. The pin must be installed with the lead tang in first, the bolt head facing outward (if equipped) and the pin positioned as shown. Position the lead tang in the v-slot mounting surface and drive in the caliper until the drive tang is flush with the caliper assembly. Install the nut (if equipped) and tighten to 32–47 inch lbs.

CAUTION

Never reuse caliper pins. Always install new pins whenever a caliper is removed.

---

20. If removed, install the brake hose to the caliper.
21. Bleed the brakes as described earlier in this chapter.
22. Install the wheel and tire assembly. Torque the lug nuts to 85–115 ft. lbs.
23. Remove the jack stands and lower the vehicle. Check the brake fluid level and fill as necessary. Check the brakes for proper operation.

DOMESTIC TRUCKS—DRUM BRAKE SERVICE

Non-Servo Type

This brake is a non-servo, floating shoe type brake. Upper ends of shoes extend through wheel cylinder boots and contact inserts in wheel cylinder pistons. Shoe ends are held firmly against pistons by the brake shoe return spring. Lower ends of shoes are held against a fixed anchor plate by the anchor spring. Hold-down spring at center of each shoe holds shoes in alignment. Lining-to-drum clearance adjustment is made through eccentric cam type adjusting studs.

BRAKE SHOE

Removal

1. Back off brake adjustment, then remove brake drum.
2. Remove brake shoe return spring. Spread upper end of shoes until they are clear of wheel cylinders and hold-down springs, then disengage shoes from anchor plate at bottom. Remove anchor spring from shoes.
3. Do not depress brake pedal while shoes are removed.

Cleaning and Inspection

1. Clean all dirt out of brake drum. Inspect drum for roughness, scoring, or out-of-round. Replace or recondition drum as necessary.
2. Carefully pull lower edge of each wheel cylinder boot away from cylinder and note whether interior is excessively wet with brake fluid. Excessive fluid indicates leakage past piston cups, requiring overhaul of wheel cylinder.

NOTE: A slight amount of fluid is nearly always present and acts as a lubricant for pistons.

3. Check backing plate attaching bolts to make sure they are tight. Clean all rust and dirt from ledges on backing plate where shoe rims make contact using fine emery cloth.
4. Inspect the shoe return and anchor springs and hold-down springs. If broken, cracked, or weakened by rust or corrosion, replace springs.
5. If brake linings are worn to the extent that replacement is necessary, replace linings.

Installation

1. Inspect brake shoe lining assemblies and make sure there are no nicks or burrs on edges of shoes which contact backing plate.

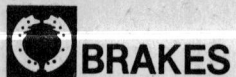

2. Apply a light film of grease at the following points: where shoe webs contact hold-down springs; where anchor ends of shoe webs contact anchor plate; and at six places where shoe rims contact ledges on backing plate.

3. Install hold-down springs on backing plate. Hook anchor spring into slot at bottom of each shoe. Swing upper ends of shoes apart and position at backing plate, with lower ends of shoe webs engaging anchor plate and with anchor spring behind extension on anchor plate.

NOTE: The shoe with the shorter lining must be to the rear of the vehicle.

4. Swing shoes up into position with center of shoe webs engaging hold-down springs, and with upper ends inserted through wheel cylinder boots.

5. Install brake shoe return spring, being sure short end is hooked into slotted hole in rear shoe and long end in round hole in front shoe.

6. Install brake drum and wheel. Adjust brakes. To Adjust the Lining Clearance: On the forward shoe, rotate the drum and the adjuster cam in forward direction until the brake drags. Back-off until the drag is relieved. On reverse shoe, rotate the drum and the adjuster cam in reverse direction until the brake drags. Back-off until the drag is relieved.

Duo Servo Type

REMOVAL & INSTALLATION

Light Duty Type

1. With the vehicle elevated on a hoist, jack or suitable stand, loosen parking brake equalizer nut, remove rear wheel, and drum retaining clips. Back-off on the self-adjuster if necessary. Remove drum.

2. Remove brake shoe return springs. On GMC vehicles, lift upward on the secondary shoe spring to disengage it from the adjuster link.

3. Remove brake shoe hold-down retainers, springs and nails and on GMC vehicles, the adjuster lever.

4. On cable adjuster vehicles loosen the starwheel adjuster, slide eye of automatic adjuster cable off anchor and then unhook from lever. Remove cable, cable guide and anchor plate.

5. Disconnect the adjuster lever spring from lever and disengage from shoe web. Remove spring and lever.

6. Spread the brake shoes apart and remove parking brake strut and spring.

7. Disengage parking brake cable from parking brake lever and remove brake assembly.

8. Remove the primary and secondary brake shoe assemblies and adjusting star wheel from support. Install wheel cylinder clamp to hold pistons in cylinders.

9. Inspect the platforms of support for

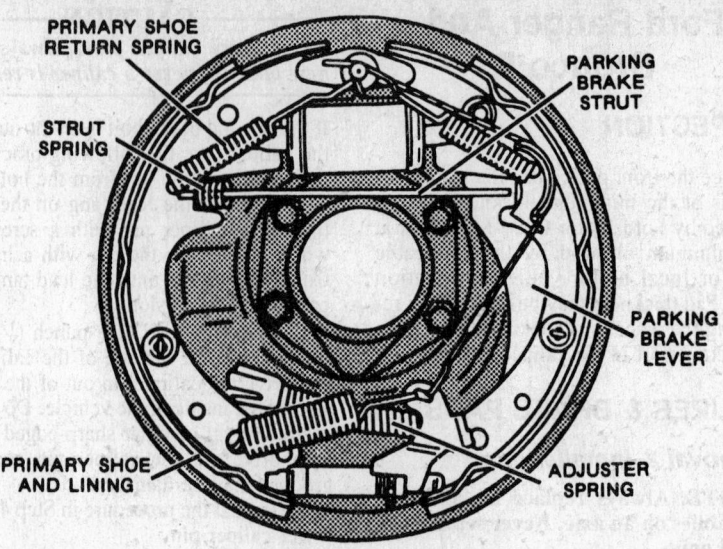

Typical light duty drum brakes

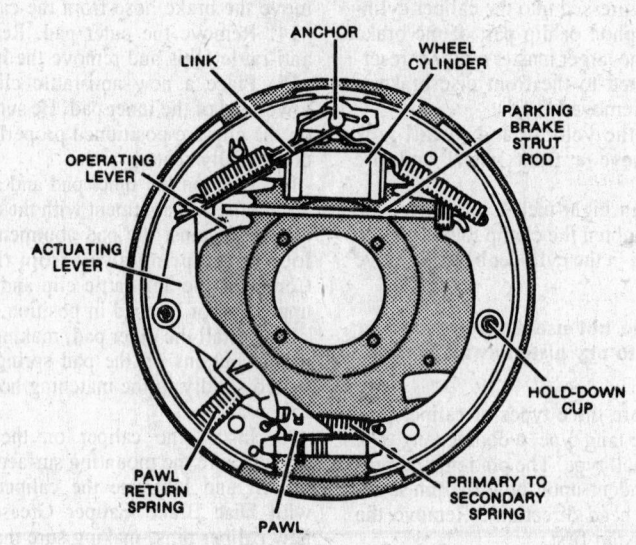

Typical light duty drum brakes

nicks or burrs. Apply a thin coat of lubricant to support platforms.

10. Attach parking brake lever to the back side of the secondary shoe.

11. Place the secondary and primary shoe in their relative position on a work bench.

12. Lubricate threads of adjusting screw and install it between the primary and secondary shoes with star wheel next to secondary shoe. The star adjusting wheels are stamped "R" (right side) and "L" (left side), and indicate their location on vehicle.

13. Overlap anchor ends of the primary and secondary brake shoes and install adjusting spring and lever.

14. Hold the brake shoes in their relative position and engage parking brake cable into parking brake lever.

15. Install parking brake strut and spring between the parking brake lever and primary shoe.

16. Place brake shoes on the support and

install retainer nails, springs and retainers. On GMC vehicles, install the adjuster lever and lower tension spring.

17. Install anchor pin plate.

18. Install "eye" of adjusting cable over anchor pin and install return spring between primary shoe and anchor pin.

19. Install cable guide in secondary shoe then install secondary return spring. (Be sure secondary spring overlaps primary.)

20. Place adjusting cable in groove of cable guide and engage hook of cable into adjusting lever.

21. Install brake drum and retaining clips.

22. Adjust brakes.

Heavy Duty Type

1. Raise the truck until the wheel clears the floor.

2. Remove the wheel, hub and drum assembly, back-off the adjuster if necessary.

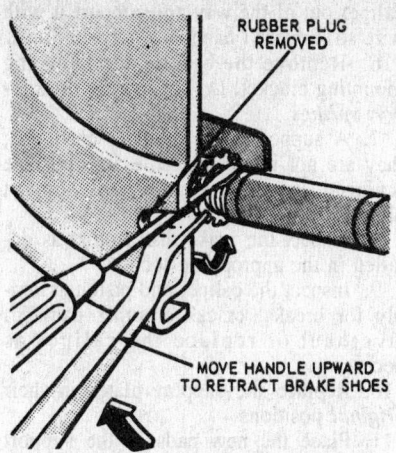

Typical brake adjustment procedure

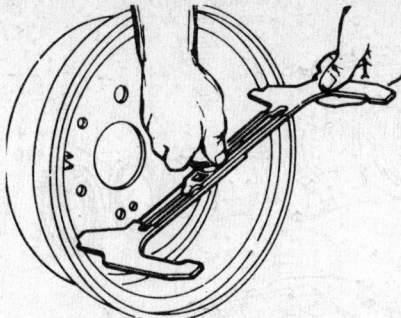

Measuring brake drum diameter

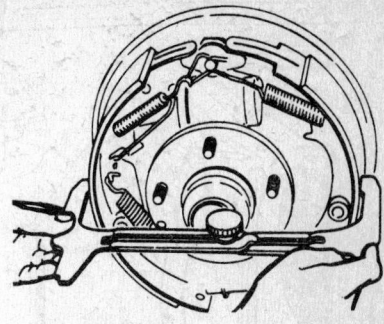

Measuring shoe mounted width

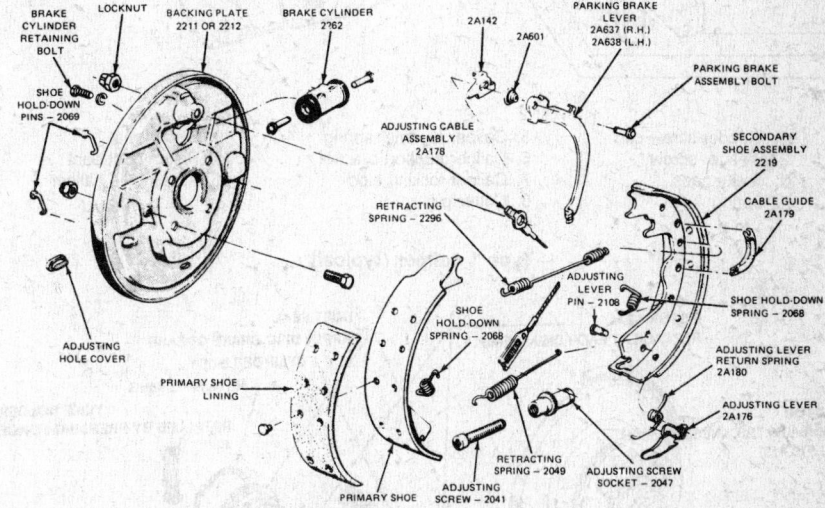

Typical heavy duty drum brakes

3. Remove the parking brake lever assembly retaining nut from behind the backing plate and remove the lever assembly.

4. Remove the adjusting cable assembly from the anchor pin, cable guide and adjusting lever.

5. Remove the brake shoe retracting springs. Remove the mounting hold-down spring from each shoe and remove the brake shoes and adjusting screw assembly.

6. Disassemble the adjuster screw assembly for cleaning and inspection.

7. Clean the ledges on the backing plate. Clean the adjuster assembly and all springs.

8. Apply lithium-based grease to the backing plate ledges, spring contact points and to the adjuster screw and socket.

9. Install the upper retracting spring on the two shoes and place the shoes in position on the backing plate with the cylinder pushrods in place on the shoes.

10. Install the shoe hold-down springs. Install the adjuster assembly with the slot in the head of the adjusting screw toward the primary shoe (Adjusters are marked L and R, do not change sides).

11. Install the lower retracting spring, adjusting lever spring, adjusting lever assembly and connect the adjusting cable to the lever. Position the cable in the cable guide and install the cable anchor fitting on the anchor pin.

12. Install the parking brake lever assembly. Adjust the brakes until the drum just fits over the shoes and complete the assembly in the reverse order of removal.

13. Finish adjusting the brakes, check for firm pedal and road test the vehicle.

Parking Brake

Before attempting parking brake adjustment, make sure that the rear brakes are fully adjusted by making several stops in reverse.

1. Raise and support the rear axle. Release the parking brake.

2. Apply the brake pedal or handle one to four clicks.

3. Adjust the cable equalizer nut under the truck until a moderate drag can be felt when the rear wheels are turned forward.

4. Release the parking brake and check that there is no drag when the wheels are turned forward.

NOTE: If the parking brake cable is replaced, prestretch it by applying the parking brake hard about three times before attempting adjustment.

IMPORTED TRUCKS—DISC BRAKE SERVICE

Type 1

AKEBONO, GIRLING, ETC. SLIDING CALIPER

This unit is a single piston, one-piece caliper that slides on a mounting bracket or frame which bolts to the steering knuckle. The caliper is retained in the mounting bracket by caliper guides (retaining keys) and support springs. Narrow support plates

under each brake pad are utilized to eliminate rattle. One or two caliper guides may be used — — it is imperative that they are replaced as originally found.

Pad Replacement

NOTE: This procedure applies to Mazda trucks through 1984 only. On 1985–86 models, a Type 4 caliper is used. See the pad replacement procedure for the Type 4 caliper for these models.

1. Raise and support the front of the vehicle on jackstands. Remove the front wheel.

2. Siphon a sufficient quantity of brake fluid from the master cylinder reservoir to prevent the brake fluid from overflowing the master cylinder when removing or installing pads. This is necessary as the piston must be forced into the cylinder bore to provide sufficient clearance to remove the pads.

3. Remove the clips or pins that hold the caliper guides in position.

4. Lightly tap out the guides — —there may only be one, so remember the correct positioning.

5. Lift the caliper off of the mounting bracket. It may be necessary to rock it back and forth a bit in order to seat the piston so it will clear the brake pads. Position the

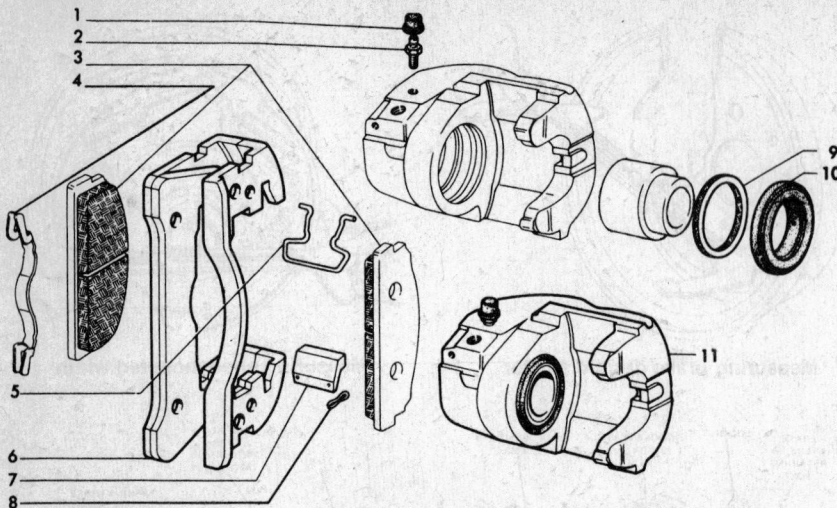

1. Bleeder screw cap
2. Bleeder screw
3. Brake pads
4. Spring
5. Caliper fastener spring
6. Caliper support bracket
7. Caliper locking block
8. Cotter pin
9. Piston seal
10. Piston dust boot
11. Assembled caliper

Type 1 caliper (typical)

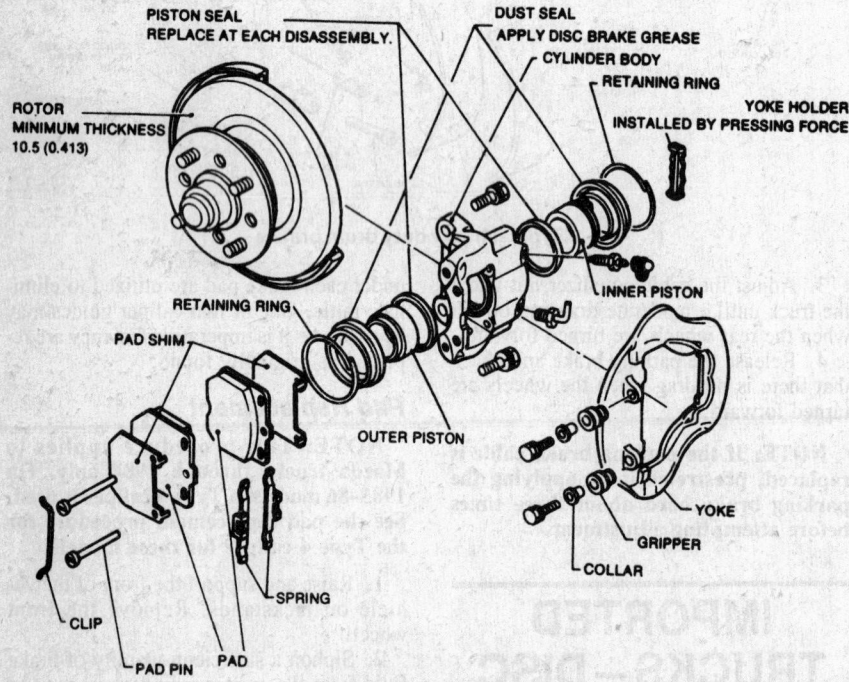

PISTON SEAL
REPLACE AT EACH DISASSEMBLY.

DUST SEAL
APPLY DISC BRAKE GREASE
CYLINDER BODY
RETAINING RING

ROTOR
MINIMUM THICKNESS
10.5 (0.413)

YOKE HOLDER
INSTALLED BY PRESSING FORCE

RETAINING RING

PAD SHIM

INNER PISTON

OUTER PISTON

YOKE

GRIPPER

COLLAR

SPRING

CLIP

PAD PIN   PAD

Type 2 caliper (typical)

INDEX TO CALIPER TYPES

Model	Types
Courier/Mazda	Type 1, 4
Datsun	Type 2
D50/Arrow Mitsubishi	Type 3
LUV/Isuzu	Type 1, 4
Toyota	Type 1, 4, 5
VW	Type 2, 3

caliper out of the way and support it with wire so it doesn't hang by the brake lines.

6. Remove the brake pads from the mounting bracket. *Do not remove the support springs.*

7. A support plate is under each pad; they are not interchangeable and must be replaced correctly. Remove the support plates.

8. Inspect the brake disc (rotor) as detailed in the appropriate section.

9. Inspect the caliper and piston assembly for breaks, cracks or other damage. Overhaul or replace the caliper as necessary.

10. Replace the support plates in their *original* positions.

11. Place the new pads in the support bracket over the support springs.

12. Push the piston all the way back into its bore (a C-clamp may be necessary for this operation).

13. Position the caliper over the pads and onto the mounting bracket.

14. Install the caliper guides (retaining keys) and then install the guide retaining pins or clips.

15. Refill the master cylinder with fresh brake fluid.

16. Install the tire and wheel assembly and then pump the brake pedal several times to bring the pads into adjustment. Road test the vehicle.

NOTE: If a firm pedal cannot be obtained, bleed the system as detailed in "Bleeding the Brakes".

Type 2

GIRLING/ANNETTE SLIDING YOKE CALIPER

This unit is a double piston, one-piece caliper. The cylinder body contains two pistons, back-to-back, in a thru-bore. The cylinder body is bolted to the steering knuckle, with both pistons inboard of the rotor. A yoke, which slides on the cylinder body, is installed over the rotor and the caliper.

When the brakes are applied, hydraulic pressure forces the pistons apart in the double ended bore. The piston closest to the rotor applies force directly to the inboard pad. The other piston applies force to the yoke, which transmits the force to the outer pad, creating a friction force on each side of the rotor.

One variation has a yoke that floats on guide pins screwed into the cylinder body.

Some designs incorporate parking brake mechanisms which are actuated by a lever and cam working between the piston and the yoke. The yokes do not have to be removed to replace the brake pads.

Pad Replacement

1. Raise and support the front of the vehicle on jackstands. Remove the wheel.

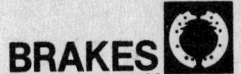
2. Siphon a sufficient quantity of brake fluid from the master cylinder reservoir to prevent the brake fluid from overflowing the master cylinder when removing or installing new pads. This is necessary as the piston must be forced into the cylinder bore to provide sufficient clearance to remove the pads.

3. Disconnect the brake pad lining wear indicator if so equipped.

4. Remove the dust cover and/or anti-rattle (damper) clip if so equipped.

5. Lift off the wire clip(s) which hold the guide pins or retaining pin in place.

6. Remove the upper guide pin and the two hanger springs. Carefully tap out the lower guide pin.

— **CAUTION** —

The lower guide pin usually contains an anti-rattle coil spring, be careful not to lose this spring. If a retaining pin is used, pull the pin out and remove the two hanger springs.

7. Slide the yoke outward and remove the outer brake pad and the anti-noise shim (if so equipped).

8. Slide the yoke inward and repeat Step 7.

9. Check the rotor as detailed in the appropriate section.

10. Inspect the caliper and piston assembly for breaks, cracks or other damage. Overhaul or replace the caliper as necessary.

11. Push the piston next to the rotor back into the cylinder bore until the end of the piston is flush with the boot retaining ring.

— **CAUTION** —

If the piston is pushed further than this, the seal will be damaged and the caliper assembly will have to be overhauled.

12. Retract the piston farthest from the rotor by pulling the yoke toward the outside of the vehicle.

13. Install the outboard pad. Anti-noise shims (if so equipped) must be located on the plate side of the pad with the triangular cutout pointing toward the top of the caliper.

14. Install the inboard pad with the shims (if so equipped) in the correct position.

15. Replace the lower guide pin and the anti-rattle coil spring.

16. Hook the hanger springs under the pin and over the brake pads.

17. Install the upper guide pin over the ends of the hanger springs.

NOTE: If a single two-sided retaining pin is used, install the pin and then install the hanger springs as in Steps 16–17.

18. Insert the wire clip locks into the holes in the guide pins or retaining pin.

19. Refill the master cylinder with fresh brake fluid.

20. Install the tire and wheel assembly. Pump the brake pedal several times to bring

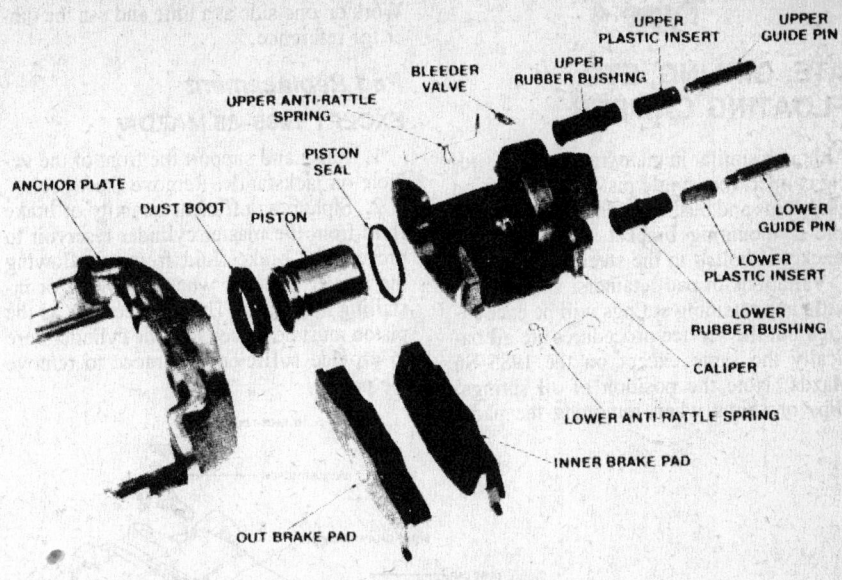

Type 3 caliper (typical)

the pads into adjustment. Road test the vehicle. If a firm pedal cannot be obtained, refer to "Bleeding the Brakes".

Type 3

KELSEY-HAYES FLOATING CALIPER

This unit is a single piston, one-piece caliper which floats on two guide pins screwed into the adapter (anchor plate). The adaptor, in turn, is held to the steering knuckle with two bolts. As the brake pads wear, the caliper floats along the adapter and guide pins during braking.

Pad Replacement

1. Raise the front of the vehicle and support it with jackstands. Remove the wheel.

2. Siphon some brake fluid from the master cylinder reservoir to prevent its overflowing when the piston is retracted into the cylinder bore.

3. Disconnect the brake pad warning indicator if so equipped.

4. Using a pair of needlenose pliers or the like, remove the anti-rattle springs.

5. Using an Allen wrench, back out the two guide pins that attach the caliper to the anchor plate.

NOTE: When replacing pads only, it is not necessary to remove the guide pins completely from the rubber bushings, as they may be difficult to reinstall.

6. Lift off the caliper and position it out of the way with some wire——you need not remove the brake lines.

— **CAUTION** —

Never allow the caliper to hang by its brake lines.

7. Slide the outer pad out of the anchor plate and then remove the inner pad. Check the rotor as detailed in the appropriate section. Check the caliper for fluid leaks or cracked boots. If any damage is found, the caliper will require overhauling or replacement.

8. Carefully clean the anchor plate with a wire brush or some other abrasive material. Install the new brake pads into position on the anchor plate. The inner pad usually has chamfered edges.

NOTE: When replacing brake pads, always replace both pads on both sides of the vehicle. Mixed pads will cause uneven braking.

9. Slowly and carefully push the piston into its bore until it's bottomed and then position the caliper onto the anchor plate. Install the guide pins and tighten them to 25–30 ft. lbs.

NOTE: The upper guide pin is usually longer than the lower one.

— **CAUTION** —

Use extreme care so as not to cross-thread the guide pins when tightening.

10. Install the anti-rattle springs between the anchor plate and brake pads ears. The loops on the springs should be positioned inboard.

11. Fill the reservoir with brake fluid and pump the brake pedal several times to set the piston. It should not be necessary to bleed the system; however, if a firm pedal cannot be obtained, the system must be bled.

12. Install the wheel and lower the vehicle.

Type 4

ATE, GIRLING, ETC. FLOATING CALIPER

Although similar in many respects to a sliding caliper, this single piston unit floats on guide pins and bushings which are threaded into a mounting bracket. The mounting bracket is bolted to the steering knuckle.

Variations in pad retainers, shims, anti-rattle and retaining springs will be encountered but the service procedures are all basically the same except on the 1985–86 Mazda. Note the position of all springs, clips or shims when removing the pads.

Work on one side at a time and use the other for reference.

Pad Replacement
EXCEPT 1985–86 MAZDA

1. Raise and support the front of the vehicle on jackstands. Remove the wheel.
2. Siphon a sufficient quantity of brake fluid from the master cylinder reservoir to prevent the brake fluid from overflowing the master cylinder when removing or installing new pads. This is necessary as the piston must be forced into the cylinder bore to provide sufficient clearance to remove the pads.

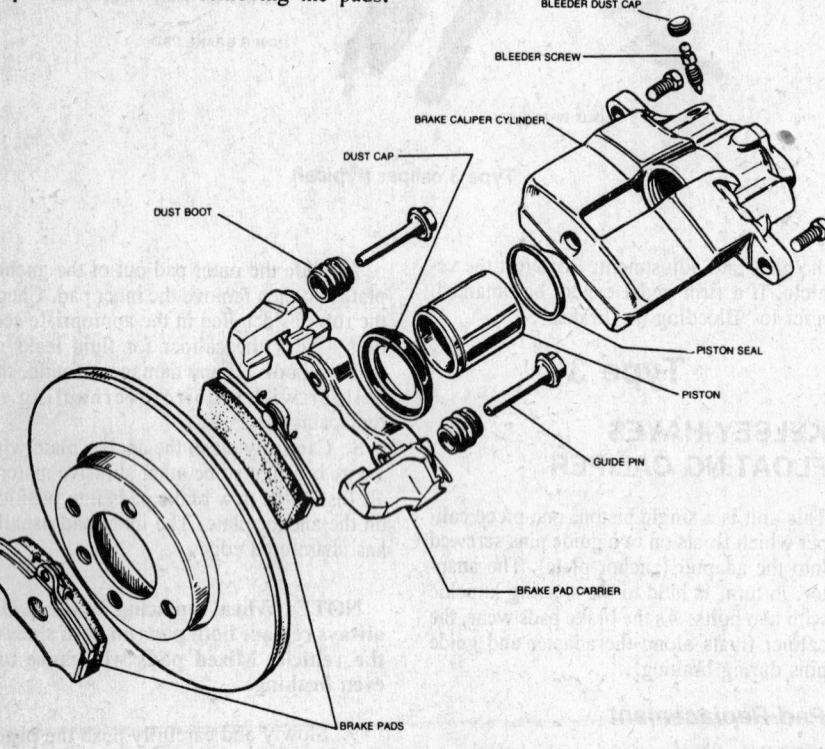

Type 4 caliper (typical)

NOTE: Make sure you perform the next step on each caliper with the opposite caliper fully assembled. If you try to do both sides simultaneously, you may force once piston out of its caliper as you depress the other into the bore.

3. Grasp the caliper from behind and pull it toward you. This will push the piston back into the cylinder bore. If it is too difficult to depress the piston into the bore this way, use a flat, soft object such as a hammer handle to depress the piston directly after you have removed the pads (Step 8).
4. Disconnect the brake pad lining wear indicator if so equipped. Remove any anti-rattle springs or clips if so equipped.

NOTE: Depending on the model and year of the particular caliper, you may not have to remove it entirely to get at the brake pads. If the caliper is the "swing" type, it will have sufficient clearance and brake hose length to permit you to pivot it upward on the upper guide bolt. On these calipers, remove the lower guide bolt, pivot the caliper on the upper bolt and swing it upward exposing the brake pads. If this method is employed, skip to Step 7.

5. Remove the caliper guide pins.
6. Remove the caliper from the rotor by slowly sliding it out and away from the rotor. Position the caliper out of the way and support it with wire so that it doesn't hang by the brake line.
7. On models equipped with anti-rattle springs, remove them. Slide the outboard pad out of the adapter.
8. Remove the inboard pad. Remove any shims or shields behind the pads and note their positions. Remove any support plates (clips) that may be present. If the piston was not depressed back into the caliper in Step 3, do it now.
9. Install the anti-rattle hardware and then the pads (in their proper positions!). On Toyotas with PD60 type calipers, in-

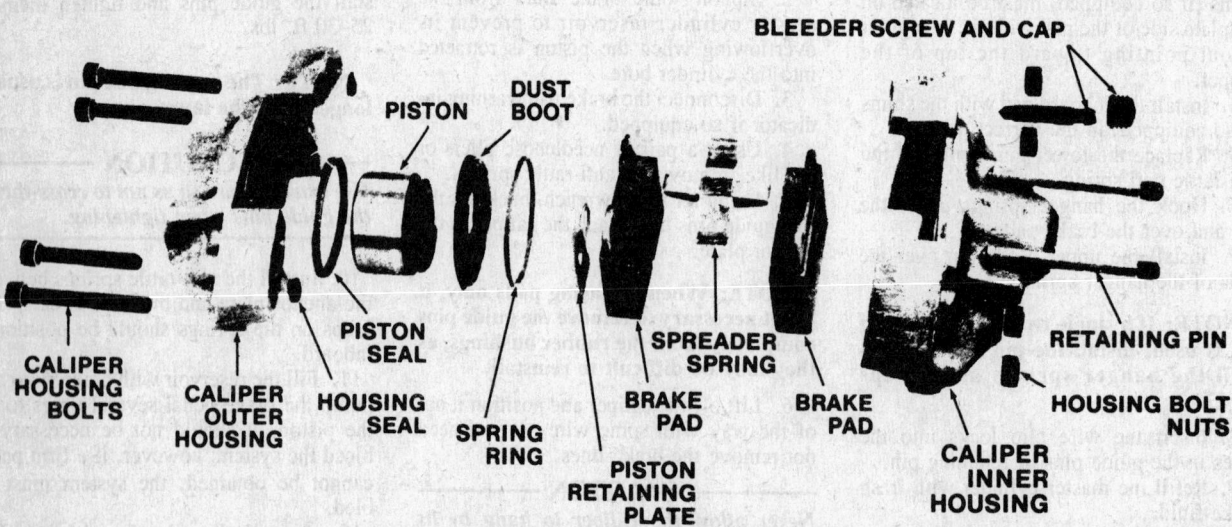

BLEEDER SCREW AND CAP

Type 5 caliper (two-piston, typical)

stall the anti squeal shim onto the surface of the piston. On Toyotas with the FS17 type calipers, replace the single, outboard anti-squeal shim with a new one, facing it toward the rear of the outboard pad. Toyota recommends that the pad support plates or clips used on their disc brakes be replaced with the pads. Note that on FS17 type calipers, anti-rattle springs are used and must be installed last.

10. Install any pad shims or heat shields.
11. Reposition the caliper and install the

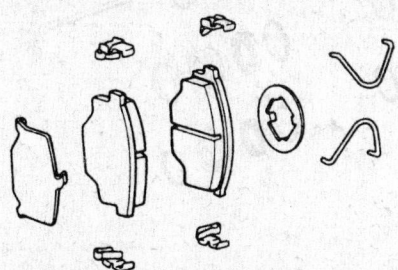

Exploded view of the components that must be removed to replace brake pads—Toyota PD60 caliper, used on later model two wheel drive pickups

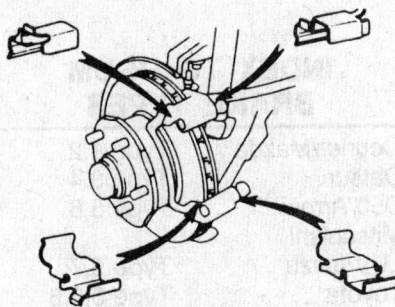

Exploded view of components to be removed/replaced when replacing brake pads on Toyota FS17 type calipers

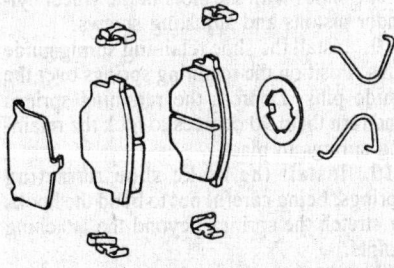

Installing the anti-squeal shim onto the caliper piston of Toyota PD60 type disc brakes

guide pin(s) carefully so as to avoid damaging the rubber boots. On Toyota PD60 calipers, torque the guide pins to 29 ft. lbs. On Toyota FS17 type calipers, torque the pin to 65 ft. lbs.

NOTE: If the caliper is the "swing" type, you need only pivot it back into position and install the lower guide pin.

12. Check to make sure rubber boots are

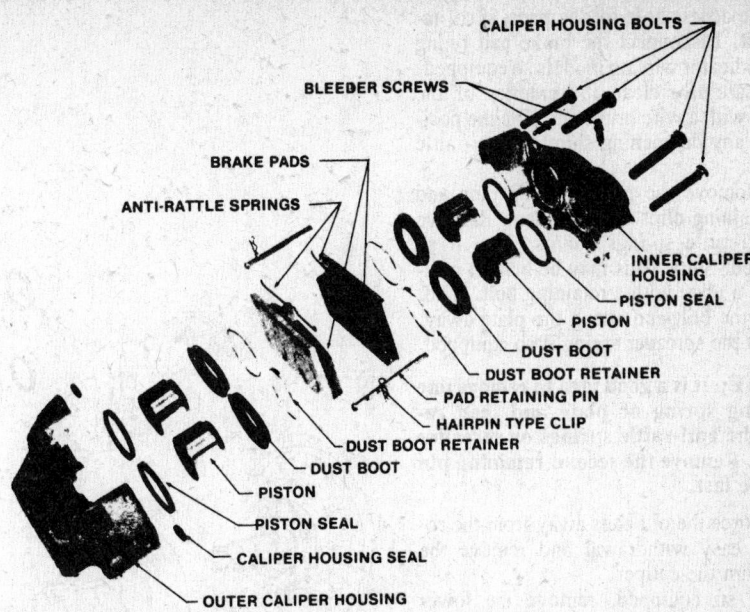

Type 5 caliper (four-piston, typical)

securely seated in their grooves. Make sure they are not bulged out with entrapped air and that they are nowhere pinched. If necessary, relieve air pressure by gently breaking the seal on one end-pulling the boot slightly out of its groove. Where hole plugs are used opposite caliper mounting pins, make sure they are in proper position and do not retain air under pressure, also.

13. Refill the master cylinder with fresh brake fluid.

14. Install the tire and wheel assembly and then pump the brake pedal several times to bring the pads into adjustment. Road test the vehicle.

NOTE: If a firm pedal cannot be obtained, bleed the system.

1985-86 MAZDA

1. Raise and support the front end on jackstands.
2. Remove the wheels.
3. Remove the caliper lockpin bolts.
4. Lift off the caliper and remove the brake pads.
5. Remove about $\frac{1}{2}$ of the fluid from the front brake reservoir of the master cylinder.
6. Position a large C-clamp on the caliper and force the piston back into its bore.
7. Install new pads in the caliper. Shims are used behind the pads on these trucks from the factory. These shims should be discarded and replaced with new ones at each pad change. Some aftermarket pads are too thick to use these shims. In that case, don't try to force new shims in place. Do without them.
8. Position the caliper on the mounting support, install the lockpins and tighten them to 30 ft. lbs.
9. Install the wheels, lower the truck to the ground and refill the master cylinder.

Pump the brake pedal a few times to restore pressure.

NOTE: If a firm pedal cannot be obtained, bleed the system as detailed in "Bleeding the Brakes".

Type 5

ATE, GIRLING, SUMITOMO FIXED CALIPER

These units are either two or four piston, two-piece calipers that are fixed directly to the steering knuckle or spindle.

Brake pads may be changed without removing the caliper on all of these models. There may be some differences in retainers or anti-rattle springs from the illustrations, but all versions are basically the same. Before removing any parts, carefully note the position of any springs, retainers or clips. Change pads on one wheel at a time and use the other as a reference.

All pads on all models are held in position by either retaining pins or retainer plates. The retainer plates are bolted to the caliper housing and need only be loosened and rotated out of the way for pad removal.

Pad Replacement

1. Raise the front (or rear) of the vehicle and support it with jackstands. Remove the wheel.
2. Siphon a sufficient quantity of brake fluid from the master cylinder reservoir to prevent the brake fluid from overflowing the master cylinder when removing or installing new pads. This is necessary as the pistons must be forced into the cylinder bore to provide sufficient clearance to remove the pads.
3. Some models may use a cover plate

over the access hole for the pads, if so, remove it. Disconnect the brake pad lining wear indicator wire on models so equipped.

4. Carefully clean the exterior of the caliper with a wire brush and note the position of any dampening shims or anti-rattle springs.

5. Remove the pad retaining pins and any retaining clips holding them. Remove the anti-rattle springs and/or clips, if so equipped. Some pads may be held in position by a plate with a retaining bolt. If so, loosen the bolt and swing the plate away. Lift out the spreader spring if so equipped.

NOTE: It is a good idea to remove one retaining spring or plate and then remove the anti-rattle springs or spreader spring. Remove the second retaining pin or plate last.

6. Force the old pads away from the rotor for easy withdrawal and remove the pads from the caliper.

7. If so equipped, remove the lower anti-rattle springs and dampening shims using needlenose pliers.

8. Check the brake disc (rotor) as detailed in the appropriate section.

9. Examine the dust boot for cracks or damage and push the pistons back into the cylinder bores. If the pistons are frozen or if the caliper is leaking hydraulic fluid, it must be overhauled.

10. Install the anti-rattle spring or damping shims and slip the new pads into the caliper. If damping shims are used, be sure that the directional arrow on the shims face the forward rotation of the rotor.

11. Install one pad retaining pin and hairpin clip. Position the anti-rattle springs and/or spreader spring and then install the other pad retaining pin and clip.

12. Refill the master cylinder to the correct level with the proper brake fluid.

13. Replace the wheel and lower the vehicle. Pump the brake pedal several times to bring the pads into correct adjustment. Road test the vehicle. If a firm pedal cannot be obtained, the system will require bleeding.

IMPORTED TRUCKS—DRUM BRAKE SERVICE

Type 1

DUAL CYLINDERS—DUAL PISTONS—MANUAL ADJUSTERS

Two dual piston wheel cylinders are used in each rear wheel brake assembly. The dual pistons act together to expand both the

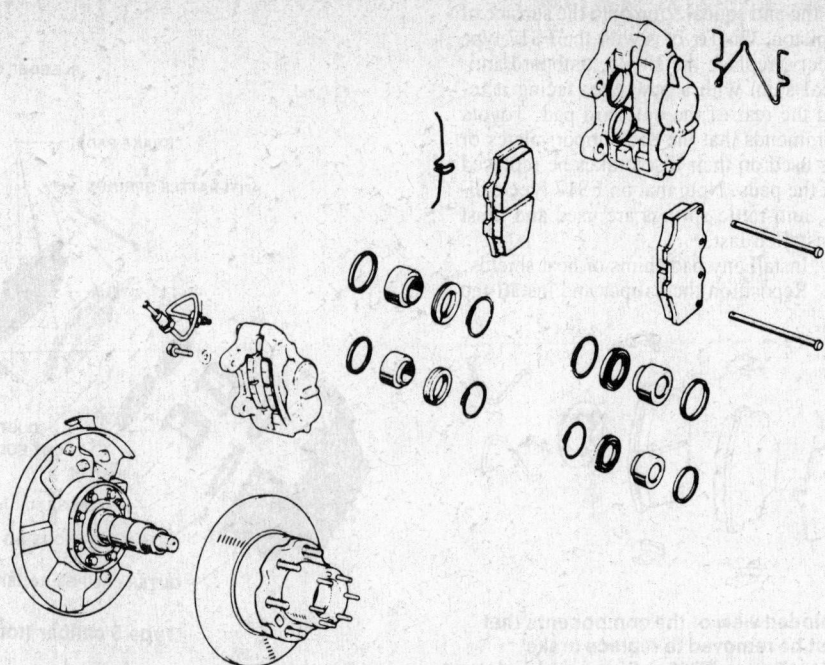

Exploded view of Toyota S12 + 8 type disc brake caliper, used on late model four wheel drive trucks

shoes equally against the brake drum when hydraulic fluid pressure is applied through the brake pedal and master cylinder.

The shoes and linings are interchangeable as are the brake shoe retracting springs.

Removal & Installation

1. Raise the vehicle. Remove the wheel and tire assembly. Remove the brake drum. Remove the brake drum attaching screws and install them into the tapped holes in the brake drum. Turn these screws in evenly to force the brake drum away from the wheel hub and remove the brake drum. Back off adjustment, if necessary, for drum removal.

2. Remove the brake shoe retracting springs.

3. Remove the shoe retaining spring guide pins and retaining spring by holding the guide pin to the backing plate and compressing and turning the retaining spring 90 degrees to release it from the guide pin.

4. Remove the parking brake link.

5. Disengage the parking brake cable from the parking brake lever.

6. Lubricate the threads of the adjusting screws, mating surfaces of the shoe webs and the brake backing plate ledges with a small amount of Lubriplate.

7. Position the parking brake lever on the rear shoe and install its retaining clip. Hold the rear brake shoe assembly near the brake backing plate and install the eye of the parking brake cable on the parking brake operating lever.

8. Position both brake shoes to the backing plate, install the parking brake link be-

INDEX TO DRUM BRAKE TYPES

Courier/Mazda	Type 1,2
Datsun	Type 3,4
D50/Arrow/ Mitsubishi	Type 5,6
LUV/Isuzu	Type 5,7
Toyota	Type 3,5,6
VW	Type 6,8

tween the two shoes and then engage the brake shoes with the slots in the wheel cylinder pistons and adjusting screws.

9. Install the shoe retaining spring guide pins. Position the retaining springs over the guide pins. Depress the retaining springs and turn them 90 degrees to lock the retaining springs in place.

10. Install the brake shoe retracting springs, being careful not to bend the hooks or stretch the springs beyond the attaching points.

11. Install the brake drum.

12. Install the wheel and tire. Torque the wheel stud nuts to 58–65 ft-lbs.

13. Adjust the brakes.

14. Lower the vehicle and check the brakes for proper operation.

Adjustment

The brake drums should be at normal room temperature when the brake shoes are adjusted. If the shoes are adjusted when the drums are hot and expanded, the shoes may drag as the drums cool and contract.

A brake adjustment re-establishes the brake lining-to-drum clearance and com-

pensates for normal lining wear.

The two-cylinder brake assembly brake shoes are adjusted by turning adjusting wheels reached through slots in the backing plate.

The brake adjustment is made with the vehicle raised. Check the brake drag by rotating the drum in the direction of forward rotation as the adjustment is made. Be sure that the parking brake is fully released by disconnecting the equalizer clevis pin.

EXCEPT MAZDA

1. Remove the adjusting slot covers from the backing plate.
2. Turn the lower wheel cylinder adjusting wheel inside the hole to expand the brake shoe until it locks against the brake drum.
3. Back off the adjusting screw (5 notches) so that the drum rotates freely without drag.
4. Repeat the above procedure on the upper wheel cylinder. Connect the parking brake equalizer clevis pin and recheck parking brake adjustment.
5. Replace the adjusting hole covers.
6. After the brake shoes on each wheel have been adjusted, test drive the vehicle to check for equal brake action. Readjust, if necessary.

MAZDA

1. If the shoe retaining spring has been removed, first retract the pushrod fully (drum removed).
2. Raise and support the rear of the truck. The wheels must be free to turn.
3. Make sure the parking brake is fully released.
4. Remove the two adjusting hole plugs from the brake backing plate.
5. An arrow stamped on the backing plate indicates the direction to turn the adjuster starwheel to expand the shoes. Insert a brake spoon through the adjuster hole and

turn the starwheel until the brakes are locked.

6. Insert a drift through the other adjuster hole. Use the drift to hold the pole lever of the self-adjuster firmly. Back off the starwheel three or four notches; the wheel should rotate freely (no drag).
7. Repeat the adjustment on the other wheel. Make sure the adjustment is exactly the same. Road test for equal brake action and readjust as necessary.

Type 2

SINGLE CYLINDER—DUAL PISTONS—AUTOMATIC ADJUSTER

A dual piston wheel cylinder is used in each rear wheel brake assembly. The dual pistons act together to expand both the shoes equally against the brake drum when

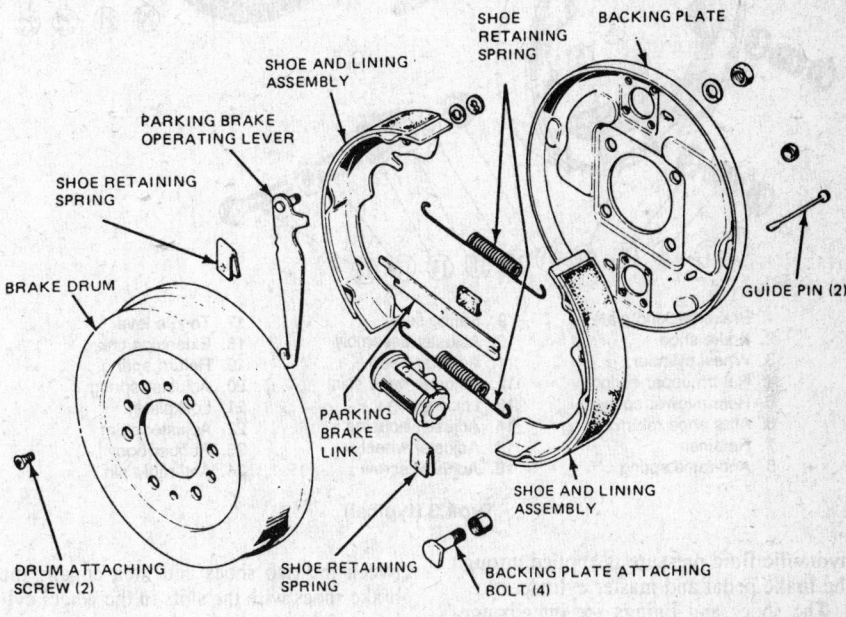

Type 1 (typical)

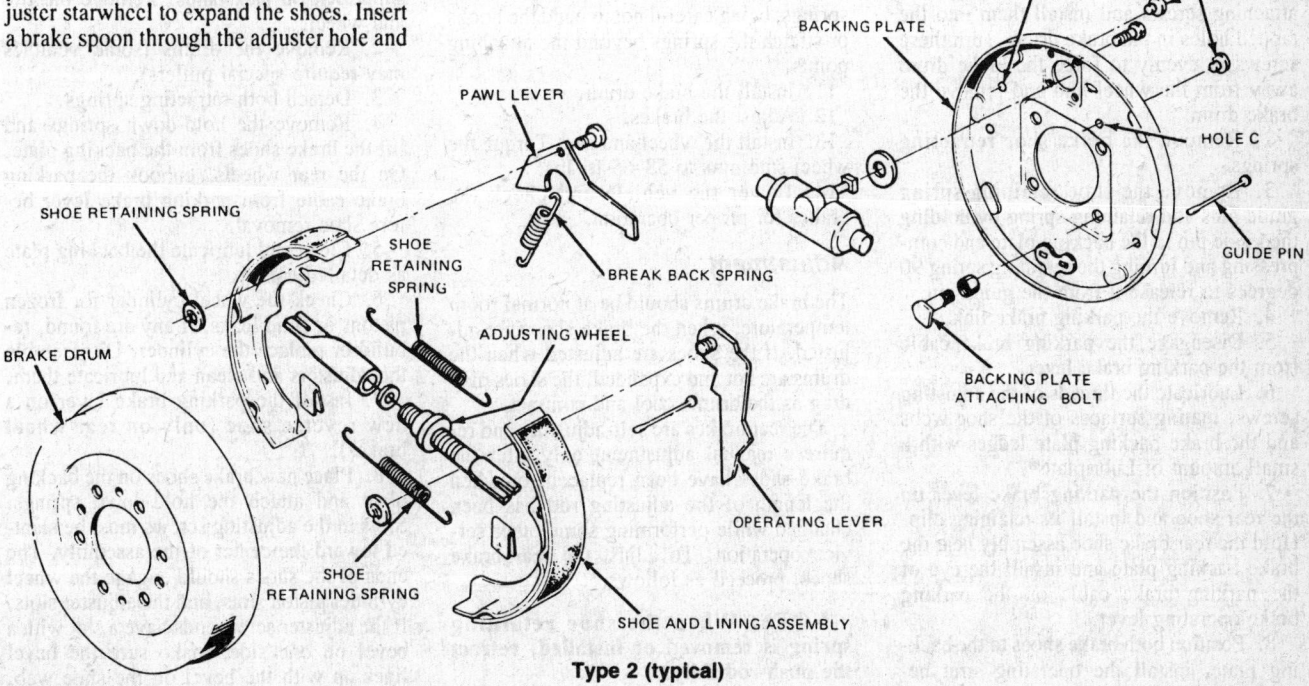

Type 2 (typical)

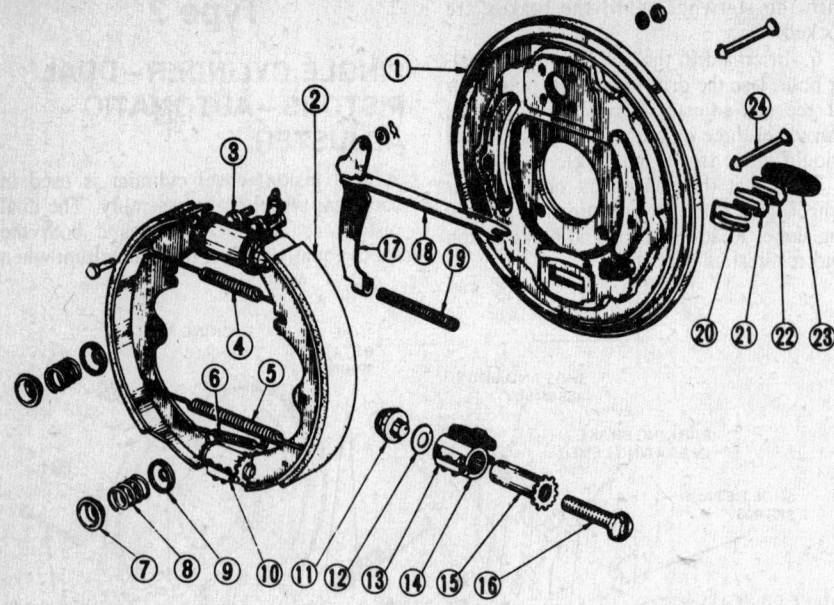

1. Brake backing plate	9. Spring seat	17. Toggle lever
2. Brake shoe	10. Adjuster assembly	18. Extension link
3. Wheel cylinder	11. Adjuster head	19. Return spring
4. Return upper spring	12. Adjuster head shim	20. Adjuster spring
5. Return lower spring	13. Lock-spring	21. Lockplate
6. After shoe return spring	14. Adjuster housing	22. Adjuster shim
7. Retainer	15. Adjuster wheel	23. Rubber boot
8. Anti-rattle spring	16. Adjuster screw	24. Anti-rattle pin

Type 3 (typical)

hydraulic fluid pressure is applied through the brake pedal and master cylinder.

The shoes and linings are interchangeable as are the brake shoe retracting springs.

Removal & Installation

1. Raise the vehicle. Remove the wheel and tire assembly. Remove the brake drum attaching screws and install them into the tapped holes in the brake drum. Turn these screws in evenly to force the brake drum away from the wheel hub and remove the brake drum.

2. Remove the brake shoe retracting springs.

3. Remove the shoe retaining spring guide pins and retaining spring by holding the guide pin to the backing plate and compressing and turning the retaining spring 90 degrees to release it from the guide pin.

4. Remove the parking brake link.

5. Disengage the parking brake cable from the parking brake lever.

6. Lubricate the threads of the adjusting screws, mating surfaces of the shoe webs and the brake backing plate ledges with a small amount of Lubriplate®.

7. Position the parking brake lever on the rear shoe and install its retaining clip. Hold the rear brake shoe assembly near the brake backing plate and install the eye of the parking brake cable on the parking brake operating lever.

8. Position both brake shoes to the backing plate, install the operating strut between the two shoes and then engage the brake shoes with the slots in the wheel cylinder pistons and adjusting screws.

9. Install the shoe retaining spring guide pins. Position the retaining springs over the guide pins. Depress the retaining springs and turn them 90 degrees to lock the retaining springs in place.

10. Install the brake shoe retracting springs, being careful not to bend the hooks or stretch the springs beyond the attaching points.

11. Install the brake drum.

12. Adjust the brakes.

13. Install the wheel and tire. Torque the wheel stud nuts to 58–65 ft. lbs.

14. Lower the vehicle and check the brakes for proper operation.

Adjustment

The brake drums should be at normal room temperature, when the brake shoes are adjusted. If the shoes are adjusted when the drums are hot and expanded, the shoes may drag as the drums cool and contract.

The rear brakes are self-adjusting and require a manual adjustment only after the brake shoes have been replaced, or when the length of the adjusting rod has been changed while performing some other service operation. To adjust the rear brake shoes, proceed as follows:

NOTE: When the shoe retaining spring is removed or installed, retract the push rod fully.

1. Jack up the rear end of the vehicle until the wheels are free to turn. Then, support with stands.

2. Make sure that the parking brake is fully released.

3. Remove the two shoe adjusting hole plugs from the back of the backing plate.

4. Insert a screwdriver into the star wheel of the adjuster through the hole and turn the star wheel toward the arrow direction marked on the backing plate until the wheel is locked.

5. Through the hole, hold the pole lever of the self-adjuster with a suitable drift and back off the adjusting wheel about 3 or 4 notches so that the drum rotates freely without drag.

6. Repeat the above adjustment on the other rear wheel. The adjustments must be the same on both rear wheels.

7. After the brake shoes on each wheel have been adjusted, test drive the vehicle to check for equal brake action. Readjust, if necessary.

Type 3

NON-SERVO–MANUAL ADJUSTER

This brake consists of non-servo forward and reverse shoes with a double-end type wheel cylinder. The shoes anchor upon the slotted adjusting screws which permit them a sliding self centering action. Brakes are mounted with cylinder and adjuster horizontally, or with the cylinder at the top and the adjuster at the bottom.

Removal & Installation

1. Raise the front/rear of the vehicle and support it on jackstands. Remove the tire and wheel.

2. Remove the drums (some vehicles may require special pullers).

3. Detach both retracting springs.

4. Remove the hold-down springs and lift the brake shoes from the backing plate. On the rear wheels, unhook the parking brake cable from parking brake lever before shoe removal.

5. Clean and lubricate the backing plate as detailed earlier.

6. Check the wheel cylinder for frozen pistons or fluid leaks. If any are found, rebuild or replace the cylinder. Disassemble the adjusters and clean and lubricate them.

7. Install the parking brake lever on a new reverse shoe (only on rear wheel brakes).

8. Place new brake shoes on the backing plate and attach the hold-down springs. Slots in the adjusting screws must be slanted toward the center of the assembly. The ends of the shoes should engage the wheel cylinder piston slots, and the adjuster slots. If the adjuster screw ends have a slot with a bevel on one side, make sure the bevel lines up with the bevel on the shoe web.

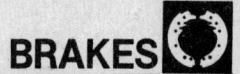

The end of the shoe with a slot for the parking strut should be installed near the wheel cylinder.

9. Hook the parking brake lever on the parking brake cable and then install the parking brake strut.

10. Install the heavier retracting spring between the toe or cylinder ends of the brake shoes.

11. Attach the lighter retracting spring to the heel or anchor ends of the shoes.

12. Replace the drums, bleed and adjust the assembly and road test the car.

Adjustment

Insert an adjusting spoon or a small screwdriver through the adjusting hole in the backing plate and expand the shoe assembly by revolving the notched adjusting wheel in a clockwise direction when facing the end of the wheel cylinder. Adjust the shoe until a heavy drag is felt when turning the wheel and drum; then, back off the adjustment until the wheel spins freely. Adjust one shoe at a time and repeat this procedure at all brake shoes.

Type 4

SERVO – AUTOMATIC ADJUSTER

Removal & Installation

1. Raise and safely support the vehicle. Remove the brake drum.

2. Place the hollow end of a brake spring service tool (available at auto parts stores) on the brake shoe anchor pin and twist it to disengage one of the brake retracting springs. Repeat this operation to remove the other spring.

CAUTION

Be careful the springs do not slip off the tool during removal, as they could cause personal injury.

---

3. Reach behind the brake backing plate and place a finger on the end of one of the brake hold-down spring mounting pins. Using a pair of pliers, grasp the washer on the top of the hold-down spring which corresponds to the pin that you are holding. Push down on the pliers and turn them 90° to align the slot in the washer with the head on the spring mounting pin. Remove the spring and washer and repeat this operation on the hold-down spring on the other brake shoe.

4. Place the tip of a prybar on the top of the brake adjusting screw and move the brake adjusting lever. When there is enough slack in the automatic adjuster cable, disconnect the loop on the top of the cable from the anchor. Grasp the top of each brake shoe and move it outward to disengage it from the wheel cylinder (and parking brake link on rear wheels). When the brake shoes are clear, lift them from the backing plate. Twist the shoes slightly and

the automatic adjuster assembly will disassemble itself.

5. If you are working on rear brakes, grasp the end of the brake cable spring with a pair of pliers and, using the brake lever as a fulcrum, pull the end of the spring away from the lever. Disengage the cable from the brake lever.

6. Transfer the parking brake lever from the old secondary shoe to the new one. This is accomplished by spreading the bottom of the horseshoe clip and disengaging the lever. Position the lever on the new secondary shoe and install the spring washer and the horseshoe clip. Close the bottom of the clip after installing it. Grasp the metal tip of the parking brake cable with a pair of pliers. Position a pair of side cutter pliers on the end of the cable coil spring and, using the pliers as a fulcrum, pull the coil spring back with the side cutters. Position the cable in the parking brake lever.

7. Apply a light coating of high-temperature grease to the brake shoe contact points on the backing plate. Position the primary brake shoe on the front of the backing plate and install the hold-down spring and washer over the mounting pin. Install the secondary shoe on the rear of the backing plate.

8. Install the parking brake link between the notch in the primary brake shoe and the notch in the parking brake lever.

9. Install the automatic adjuster cable loop end on the anchor pin. Make sure that the crimped side of the loop faces the backing plate.

10. Install the return spring in the primary brake shoe and, using the tapered end of a brake spring service tool, slide the top of the spring onto the anchor pin.

CAUTION

Be careful to make sure that the spring does not slip off the tool during installation, as it could cause injury.

---

11. Install the automatic adjuster cable guide in the secondary brake shoe, making sure that the flared hole in the cable guide is inside the hole in the brake shoe. Fit the cable into the groove in the top of the cable guide.

12. Install the secondary shoe return spring through the hole in the cable guide and the brake shoe. Using the brake spring tool, slide the top of the spring onto the anchor pin.

13. Clean the threads on the adjusting screw and apply a light coating of high-temperature grease to the threads. Screw the adjuster closed, then open it one-half turn.

14. Install the adjusting screw between the brake shoes with the starwheel nearest to the secondary shoe. Make sure that the starwheel is in a position that is accessible from the adjusting slot in the backing plate.

15. Install the short hooked end of the automatic adjuster spring in the proper hole in the primary brake shoe.

16. Connect the hooked end of the automatic adjuster cable and the free end of the automatic adjuster spring in the slot in the top of the automatic adjuster lever.

17. Pull the automatic adjuster lever (the lever will pull the cable and spring with it) downward and to the left and engage the pivot hook of the lever in the hole in the secondary brake shoe.

18. Check the entire brake assembly to make sure that everything is installed properly. Make sure that the shoes engage the wheel cylinder properly and are flush on the anchor pin. Make sure that the automatic adjuster cable is flush on the anchor pin and in the slot on the back of the cable guide. Make sure that the adjusting lever rests on the adjusting screw starwheel. Pull upward on the adjusting cable until the adjusting lever is free of the starwheel, then release the cable. The adjusting lever should snap back into place on the adjust-

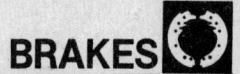

Type 4 (typical)

Labels on diagram:
PARKING BRAKE LINK
PARKING BRAKE LINK SPRING
REAR BRAKE ASSEMBLY
PARKING LEVER PIVOT POINT
SECONDARY SHOE AND LINING ASSEMBLY
FRONT OF TRUCK
PRIMARY SHOE AND LINING ASSEMBLY
AUTOMATIC ADJUSTER SPRING
PARKING BRAKE LEVER
PIVOT HOOK
PARKING BRAKE CABLE

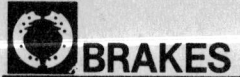

ing screw starwheel and turn the wheel one tooth.

19. Expand the brake adjusting screw until the brake drum will just fit over the brake shoes.

20. Install the wheel and drum and adjust the brakes.

Adjustment

1. Raise the vehicle and support it with safety stands.

2. Remove the rubber plug from the adjusting slot on the backing plate.

3. Insert a brake adjusting spoon into the slot and engage the lowest possible tooth on the starwheel. Move the end of the brake spoon downward to move the starwheel upward and expand the adjusting screw. Repeat this operation until the brakes lock the wheel.

4. Insert a small prybar or piece of firm wire (coat-hanger wire) into the adjusting slot and push the automatic adjuster lever out and free of the starwheel on the adjusting screw.

5. Holding the adjusting lever out of the way, engage the topmost tooth possible on the starwheel with a brake adjusting spoon. Move the end of the adjusting spoon upward to move the adjusting screw starwheel downward and contract the adjusting screw. Back off the adjusting screw starwheel until the wheel spins freely with the minimum of drag. Keep track of the number of turns the starwheel is backed off.

6. Repeat this operation for the other side. When backing off the brakes on the other side, the adjusting lever must be backed off the same number of turns to prevent side-to-side brake pull.

7. Repeat this operation on the other set of brakes.

8. When the brakes are adjusted, make several stops, while backing up the vehicle, to equalize all of the wheels.

9. Road-test the vehicle.

Type 5

SERVO – AUTOMATIC ADJUSTER

Removal & Installation

1. Remove the wheel and the brake drum. If the drum cannot be removed easily, insert a flat-bladed screwdriver through the hole in the backing plate and hold the adjuster lever away from the adjuster. Then, use another screwdriver to turn the lower side of the adjusting bolt toward you and loosen the adjustment.

2. Using a standard brake return spring tool, remove the return spring or springs. On '85–'86 Toyotas with 2WD, this refers the the two springs at the tops of the shoes only.

Loosening the brake adjustment on Toyota duo servo rear drum brakes

3. On '85–'86 Toyotas with 2WD, first push up on the adjusting lever and then remove the cable, shoe guide plate, and cable guide. Then, on all models, remove the adjusting spring and the adjusting lever. Now, on '85–'86 Toyotas with 2WD, remove the return springs located at the lower ends of the shoes.

4. Many models use clips to retain the shoes. Turn these clips 90° and remove them. Remove the brake shoes and, if necessary, remaining parts of the adjusting assembly. Then remove the cable from the parking brake lever.

5. Installation is the reverse of removal with the following notes: Grease those ares of the backing plate where the brake shoes will ride as brakes are applied and released.

6. On most models, after the primary shoes have been installed, install the parking brake cable. On '85–'86 Toyotas with 2WD, first install the brake cable and connect it to the parking brake lever; then, install the rear shoe. Make sure to secure the hold-down springs and clips on those models using them. Set the adjuster assembly, then secure the secondary shoes on all models but '85–'86 Toyota 2WD.

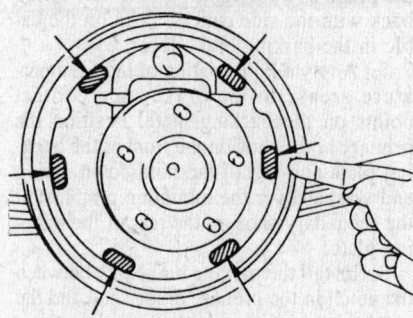

Grease the backing plate for Toyota duo-servo 2WD rear brakes as shown

NOTE: When setting the adjuster, grease the threaded area on the adjuster assembly and make sure it turns smoothly.

7. Except on the '85–'86 Toyota 2WD, install the primary shoe return springs, adjusting cable and the secondary shoe return springs in the given order. On those Toyota models, install the two adjustment tension springs located at the lower ends of the

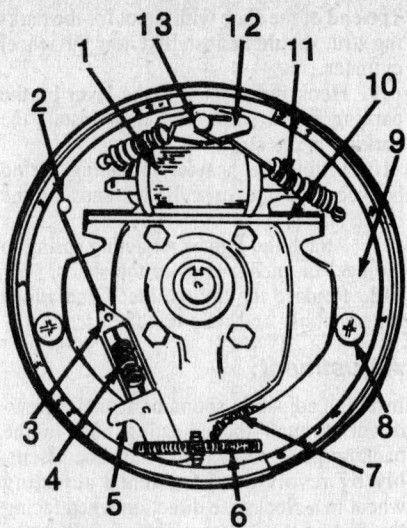

1. Wheel cylinder	8. Shoe hold down
2. Cable guide	spring
3. Cable assembly	9. Primary shoe
4. Lever hold down	10. Parking brake strut
spring	11. Shoe pull—back
5. Adjuster lever	spring
6. Self adjuster	12. Guide plate
7. Shoe return spring	13. Anchor pin

Type 5 (typical)

shoes (one runs between the shoes and the other from one shoe to the backing plate).

NOTE: The springs for the primary shoe and secondary shoe are different colors. Do not mix them, as they are different lengths.

8. On other models, skip to the Step 10. On '85–'86 Toyota with 2WD, grease the threads of the adjuster and the ends with high temperature grease. Then, spread the shoes apart with a screwdriver and install the adjuster. Now, install the shoe guide plate, cable guide, and adjusting cable. Finally, install the front and rear return springs between the tops of the two shoes and the post between the two shoes at the top.

9. Install the tension spring to the rear shoe. Then, hook the adjusting lever to the cable and install the adjusting lever. Hook the tensioning spring to the adjusting lever.

10. To check the adjuster assembly operation: pull the adjuster cable toward you to see if the adjuster lever goes into mesh with the next tooth on the adjuster wheel. Make sure that when the cable is released, the adjuster lever returns to its original position after the adjuster wheel has moved a tooth ahead. On the '85–'86 Toyota with 2WD, loosen the adjustment all the way to permit easy installation of the brake drum. Install the drum and wheel. Then, complete the brake adjustment by backing the vehicle up and applying the brakes hard several times.

11. On all models, bleed the system as necessary.

Adjustment

1. During reassembly of the rear brake

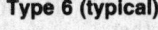

shoes on the remaining models, manually adjust the automatic adjuster so that shoe to drum clearance is no more than 2.5 mm (0.1 in.). An excessive clearance will cause malfunction of the adjuster.

2. After installation is complete, operate the vehicle in reverse and apply the brakes so that the adjuster will equalize.

Type 6

LEADING-TRAILING— AUTOMATIC ADJUSTER

Removal & Installation

1. Raise and safely support the vehicle.
2. Remove the rear wheel and brake drum.
3. Remove the shoe return spring(s) from the two brake shoes. Remove the two brake shoe hold-down springs.
4. Remove the shoes and adjuster as an assembly. Disconnect the parking brake cable from the lever.
5. The wheel cylinder may be removed for servicing if necessary.
6. Transfer the parking brake lever and adjuster lever to the new trailing shoe.
7. Clean and lubricate the adjuster. Assemble the shoes and adjuster as in removal.
8. Lubricate the backing plate shoe contact points.
9. Connect the parking brake cable. Install the brake shoes and adjuster to the backing plate with the holddown pins and springs. Install the return spring(s).
10. Turn adjuster until the brake drum can be installed and turned with a slight drag. Remove drum and back off adjuster until the drum can be turned with no drag.
11. Install wheel and bleed brakes, if necessary.

Adjustment

The brakes are self-adjusting. Operate the vehicle in reverse while applying brakes, stop the vehicle and apply parking brake. Repeat the procedure several times until the adjusters equalize.

Type 7

NON-SERVO—AUTOMATIC ADJUSTER

Removal & Installation

1. Raise and safely support the vehicle. Remove the brake drums.
2. Unhook the brake drum return springs from the anchor pin using a brake tool and remove the springs.
3. Remove the brake shoe hold-down springs using pliers. Depress the spring retainer while rotating it 90° to align the slot in the retainer with the flanged end of the pin.

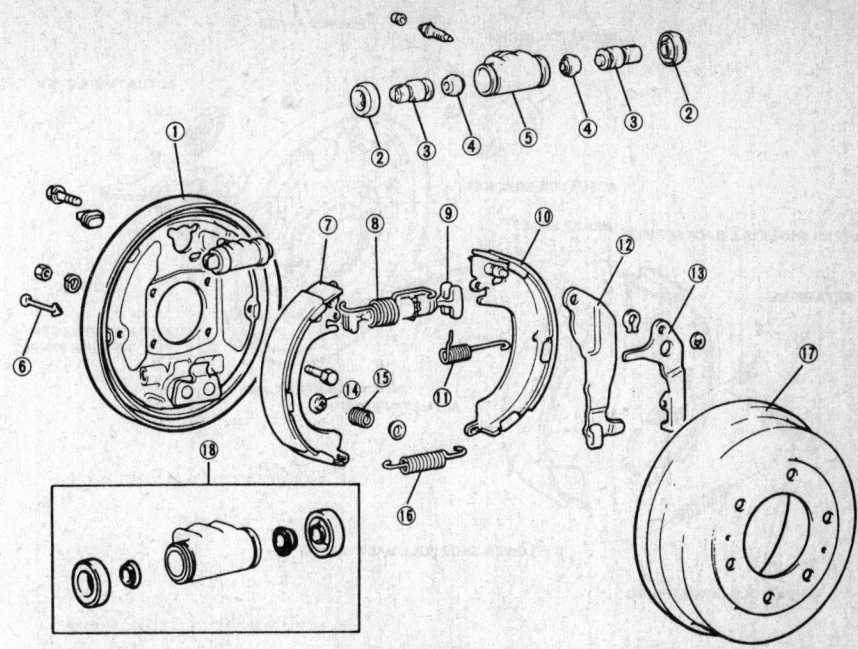

1. Backing plate
2. Wheel cylinder boot
3. Wheel cylinder piston
4. Wheel cylinder piston cup
5. Wheel cylinder body
6. Shoe hold-down pin
7. Shoe and lining assembly
8. Shoe return spring
9. Brake shoe adjuster
10. Shoe and lever assembly
11. Adjusting spring
12. Parking brake lever
13. Autoadjuster lever
14. Shoe hold-down cup
15. Shoe hold-down spring
16. Shoe retaining spring
17. Brake drum
18. Wheel cylinder repair kit

Type 6 (typical)

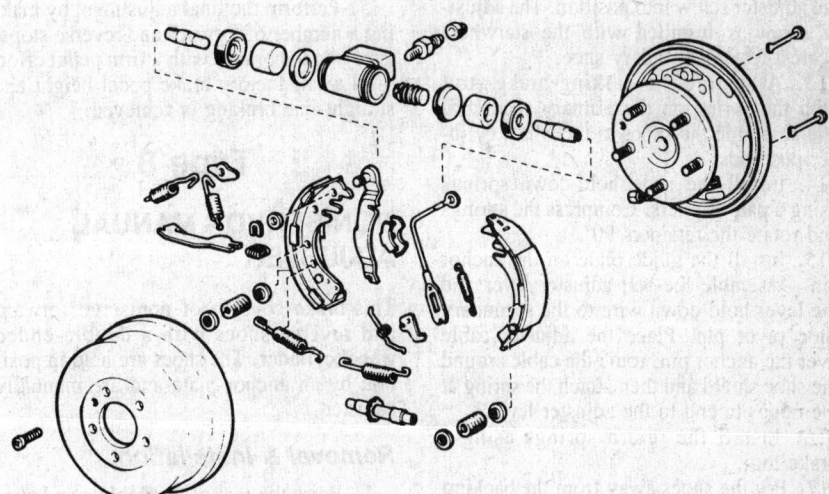

Exploded view of '85–'86 Toyota 2WD rear brake system

4. Remove the self-adjuster cable assembly by disconnecting the spring at the adjuster lever and removing the cable end from the anchor pin. Remove the guide plate from the anchor pin.
5. Remove the adjuster lever and the lever hold-down wire from the shoe pivot.
6. Separate the shoes from the wheel cylinder pushrods.
7. Separate the primary and secondary brake shoes, adjuster, return spring, and parking brake strut assemblies.

NOTE: If the brake shoes are to be reinstalled, be sure to identify them so that they can be reinstalled in their original positions.

8. Separate the parking brake lever and the rear cable. Remove the clip and washer and remove the parking brake lever from the secondary shoe.
9. Lubricate the parking brake cable with Lubriplate®.
10. Assemble the parking brake lever to

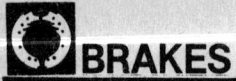

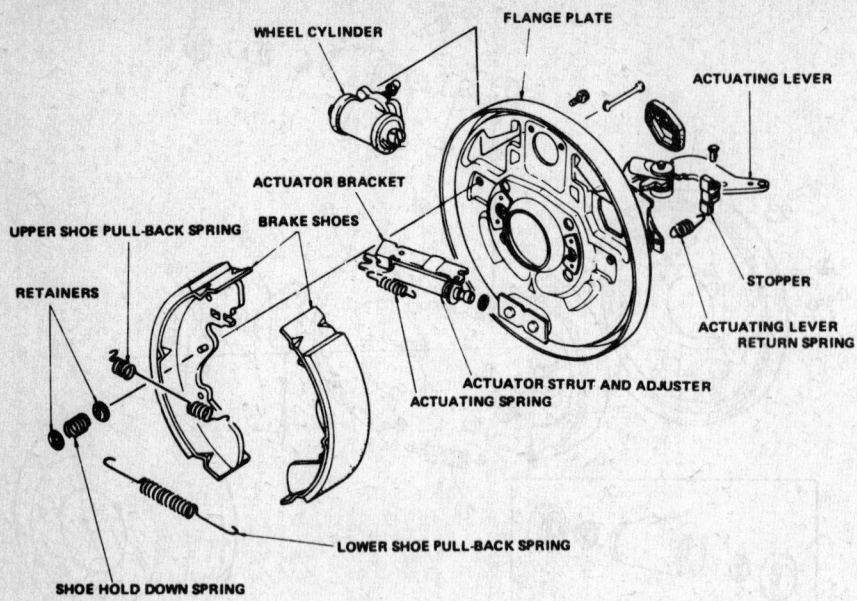

Type 7 (typical)

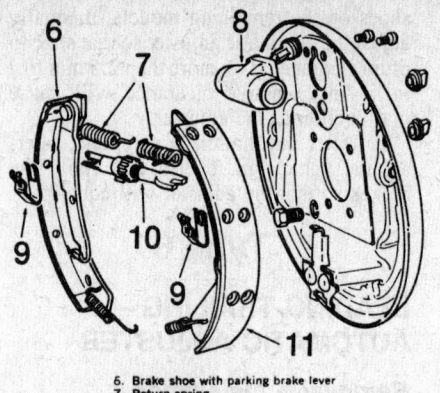

6. Brake shoe with parking brake lever
7. Return spring
8. Wheel cylinder
9. Hold-down spring
10. Adjuster
11. Brake shoe

Type 8 (typical)

the secondary shoe and then assemble the parking brake cable to the lever.

11. Before installation, make sure that the adjusting screw is clean, lubricated and operable.

12. Connect the brake shoes together with bottom return spring and then place the adjuster screw into position. The adjuster screw is installed with the starwheel nearest to the secondary shoe.

13. Assemble the parking brake strut with the spring on the primary shoe end, and assemble the shoes to the wheel cylinder pushrods.

14. Install the shoe hold-down springs using a pair of pliers. Compress the springs and rotate the retainers 90°.

15. Install the guide plate on the anchor pin. Assemble the self-adjuster lever and the lever hold-down wire to the secondary shoe pivot pin. Place the adjuster cable over the anchor pin, route the cable around the shoe shield and then attach the spring at the opposite end to the adjuster lever.

16. Install the return springs using a brake tool.

17. Pry the shoes away from the backing plate and lubricate the shoe contact areas with a thin coat of Lubriplate®.

18. Check the operation of the parking brake. Do not step on the brake pedal.

19. Using a piece of fine (400 grit) sandpaper, evenly rough the surface of the brake linings before installing the brake drums. Do this for the linings on both wheels.

20. Install the brake drum and adjust the brake shoes.

Adjustment

1. With the brake drum removed, remove actuator from the starwheel on the rear brakes.

2. Turn the starwheel until the brake drum slides over the brake shoes with a slight drag.

3. Turn the starwheel on the rear brakes 1¼ turns to retract the shoes.

4. Install the brake drums and wheels and lower the vehicle.

5. Perform the final adjustment by making a number of forward and reverse stops, applying the brakes with a firm pedal effort until a satisfactory brake pedal height and straight-line braking is achieved.

Type 8

NON-SERVO – MANUAL ADJUSTER

This brake consists of non-servo forward and reverse shoes with a double-ended wheel cylinder. The shoes are held in position by an anchor plate and are manually adjusted.

Removal & Installation

1. Raise the rear of the vehicle and support it with jackstands. Remove the tire and wheel assembly.

2. Remove the plug from the brake adjusting hole. Using a small prybar or other suitable tool, release the brake shoes by rotating the shoe adjuster downward on the right side of the vehicle and upward on the left side of the vehicle.

3. Remove the brake drum.

4. Remove the parking brake cable from the parking brake lever by compressing the cable return spring.

5. Remove the shoe-to-anchor springs located at the bottom.

6. Remove the brake shoe hold-down clips and pins.

7. Remove the adjuster screw assembly by spreading the shoes apart making sure that the adjuster screw is fully backed off.

8. Pull the reverse shoe away from the anchor plate to release the tension on the upper return spring. Disengage the shoe and remove the spring. To facilitate the re-assembly operation, note how the upper return spring is positioned on the shoe and how it is connected to the hole in the anchor plate.

9. Remove the forward shoe in the same manner as above.

10. Inspect the wheel cylinder and recondition or replace if necessary.

11. Clean and inspect the adjuster screw assembly. Apply a thin coat of lubricant to the adjuster threads.

12. Inspect the old springs. If old springs are damaged or have been overheated, they should be replaced. Indications of overheated springs are paint discoloration or distortion.

13. Lubricate the bosses on the anchor plate which make contact with the brake shoe tabs.

14. Remove the parking brake lever and attach the parking brake lever to the web of a new reverse shoe.

15. Position the upper return spring on the forward shoe and hook the other end of the spring into the hole in the backing plate.

16. Rotate the shoe outward with the upper part of the shoe against the wheel cylinder piston and insert the bottom part of the shoe under the anchor plate.

17. Repeat the above procedure for the reverse shoe.

18. With the adjuster screw fully retracted, position the straight forked end of the adjuster screw assembly on the parking brake lever. Make sure that the spring lock on the adjuster screw is on the outside and away from the adjusting hole.

19. Rotate the bottom of the forward shoe off the anchor plate and insert the curved fork end of the adjuster screw assembly into the web on the forward shoe.

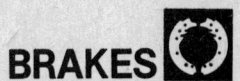

NOTE: Make sure that the curved portion of the forked end is facing downward and that the spring lock is on the outside and away from the adjusting hole.

20. Insert the pins for the hold-down clips through the backing plate and web of the shoes. Install the hold-down clips.
21. Install the shoe-to-anchor springs.
22. Compress the brake cable return spring and attach the cable to the bottom of the parking brake lever.
23. Install the brake drum.
24. Install the wheel and tire assembly.
25. Adjust the brakes.
26. Bleed the system and road test the car.

Adjustment

Adjust the brakes through the adjusting hole located in the backing plate. Adjustment is made manually by spreading the adjuster screw assembly which is located directly under the wheel cylinder. Insert a small prybar or other suitable tool through the hole in the backing plate and rotate the adjuster wheel clockwise until the brakes drag as you turn the wheel in a forward direction. Turn the adjuster in the opposite direction until you just pass the point of drag. Repeat the procedure on the other wheel.

POWER BOOSTER TROUBLE DIAGNOSIS

Condition	Possible Cause	Correction
Loss of fluid	1. Fluid leaking past cup in master cylinder.	1. Recondition master cylinder or replace.
	2. Brake wheel cylinders leaking.	2. Recondition or replace wheel cylinders.
	3. Loose hydraulic hose connectors.	3. Inspect and tighten all hydraulic connections.
	4. Leaking stop light switch.	4. Replace stop light switch.
Presence of brake fluid on hy-power vacuum cylinder	1. Piston cup or push rod seal leaking.	1. Recondition master cylinder.
Pedal kicks back against foot when brakes are applied	1. Vacuum leakage.	1. Inspect and correct vacuum leak.
	2. Dirt under control valve or damaged seat.	2. Clean and recondition booster assembly.
	3. Weak or broken spring.	3. Replace spring.
Brakes are slow to release ①	1. Incorrect pedal linkage adjustment.	1. Adjust and lubricate pedal linkage.
	2. Compensating port of master cylinder plugged.	2. Clean master cylinder with compressed air.
	3. Brake shoes sticking.	3. Free up and lubricate brake shoes.
	4. Weak brake shoe return spring.	4. Replace brake shoe return spring.
	5. Booster control valve piston sticking.	5. Clean booster control valve piston and lubricate.
	6. Booster air filter clogged.	6. Clean air filter in mineral spirits.
	7. Control valve diaphragm return spring missing.	7. Install new control valve return spring.
	8. Defective check valve in slave cylinder piston.	8. Recondition slave cylinder pistons.
	9. Dirt under atmospheric valve disc.	9. Clean atmospheric valve.
Engine runs unevenly at idle with brakes released	1. Vacuum leakage.	1. Inspect and tighten all vacuum fittings.
	2. Dirt under control valve disc or damaged seat.	2. Clean control valve or replace.
	3. Defective spring.	3. Replace defective spring.
Engine runs evenly and pedal is hard with brakes applied	1. Control valve piston assembly not seating on vacuum disc.	1. Clean or replace control valve piston assembly.
	2. Defective control valve plate and diaphragm.	2. Replace control valve plate and diaphragm.
	3. Defective pressure plate and diaphragm.	3. Replace pressure plate and diaphragm.
Brake pedal is hard at different intervals	1. Defective manifold check valve.	1. Clean or replace manifold check valve.
	2. Slave cylinder piston sticking due to dirt or inferior brake fluid.	2. Clean and recondition slave cylinder.
	3. Brake booster air cleaner clogged.	3. Clean air cleaner in mineral spirits and blow dry with compressed air.

① Jack up truck and determine whether or not the brakes are dragging before further testing is done.

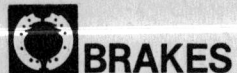

POWER BOOSTER TROUBLE DIAGNOSIS

Condition	Possible Cause	Correction
Vacuum leak (booster in released position)	1. End plate, center plate or control valve body gaskets leak. 2. Distortion of end plate. 3. Misalignment of control valve poppet. 4. Loose vacuum cylinder bolts. 5. Loose control valve body screws. 6. Large control valve poppet spring not centered in spring retainer.	1. Recondition booster unit. 2. Replace end plate. 3. Disassemble, clean and correctly reassemble. 4. Coat vacuum cylinder bolts lightly with a suitable sealing compound and tighten to specified torque. 5. Tighten control valve body screws to specified torque. 6. Disassemble unit and correctly reassemble.
Vacuum leak (booster in applied position)	1. Leak at control valve poppet and seat. 2. Dry or faulty piston leather packing. 3. Faulty control valve disphragm assembly.	1. Clean and inspect poppet and seat for damage and repair as necessary. 2. Clean and lubricate piston leather or replace. 3. Replace faulty parts.
External hydraulic leaks	1. Gasket (O-ring) leaking at hydraulic end plate joint. 2. Fluid leaking at copper gasket under hydraulic cylinder end cap.	1. Disassemble clean and replace (O-ring) gasket and reassemble. 2. Remove end cap and inspect copper gasket and seat install new copper gasket.
Internal hydraulic leak at low pressures	1. Control valve hydraulic piston cup failure. 2. Faulty push rod seal.	1. Recondition control valve unit. 2. Replace push rod seal.
Internal leaks at high pressure	1. Fluid passing copper gasket under hydraulic fitting in control valve. 2. Inspect cups and seals of master cylinder for cuts and scores. 3. Inspect cups of the control valve hydraulic piston.	1. Clean and inspect gasket and fitting, replace if faulty. 2. Hone master cylinder and replace cups and seals. 3. Replace faulty cups.
Hydraulic pressure buildup (without added input)	1. Check hydraulic piston check valve and slot for foreign material under valve.	1. Clean or replace valve and seats as condition indicates.
Failure to release	1. Weak vacuum cylinder piston return spring. 2. Dry vacuum piston leather packing. 3. Swollen rubber cups due to inferior or contaminated brake fluid. 4. Damaged or dented vacuum cylinder shell. 5. Dirty or sticky control valve piston.	1. Replace vacuum cylinder piston return spring. 2. Lubricate vacuum piston leather packing. 3. Flush hydraulic system and recondition or replace all cylinders. 4. Replace vacuum cylinder shell. 5. Recondition control valve assembly.
Failure of booster to operate within specified pressures	1. Rusty, dirty or distorted vacuum cylinder shell. 2. Dry or worn vacuum cylinder leather packing. 3. Swollen rubber cups due to inferior brake fluid. 4. Worn or scored hydraulic cups. 5. Dirt, rust or foreign matter in any component of the system.	1. Clean or replace vacuum cylinder shell. 2. Recondition and lubricate the vacuum booster. 3. Recondition the master cylinder. Replace brake fluid. 4. Recondition the master cylinder. 5. Recondition and lubricate the brake booster assembly.

Strut Overhaul

STRUT SERVICE AND REPAIR

MacPherson struts are appearing on the front (and rear) wheels of more and more cars. The strut design takes up less room in the engine compartment, compared to a conventional upper and lower arm with shock absorber arrangement. The trend toward smaller, lighter and more efficient packaging mandates the use of a strut suspension to permit more room for engine accessories and front wheel drive components.

Strut Suspension Design

In a conventional front suspension, the wheel is attached to a spindle, which is in turn, connected to upper and lower control arms through upper and lower ball joints. A coil spring between the control arms (sometimes on top of the upper arm) supports the weight of the vehicle and a shock absorber controls rebound and dampens oscillations.

In a strut type suspension, the strut performs a shock dampening function, like a shock absorber, but unlike a conventional shock absorber, the strut is a structural part of the vehicle's suspension.

The strut assembly usually contains a spring seat to retain the coil spring that supports the vehicle's weight. The shock absorber is built into the body of the strut housing. The strut is normally attached at the bottom to the lower control arm and at the top to the car body. The upper mount usually features a bearing that permits the coil spring to rotate as the wheels turn for smoother steering. The entire design eliminates the need for the upper control arm,

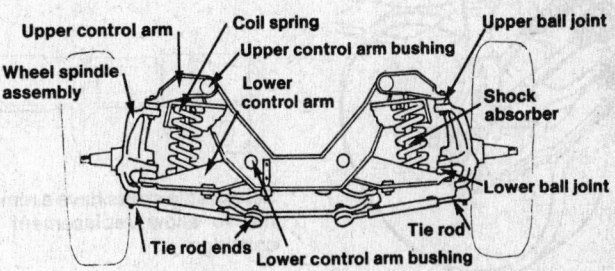

Conventional upper and lower arm suspension

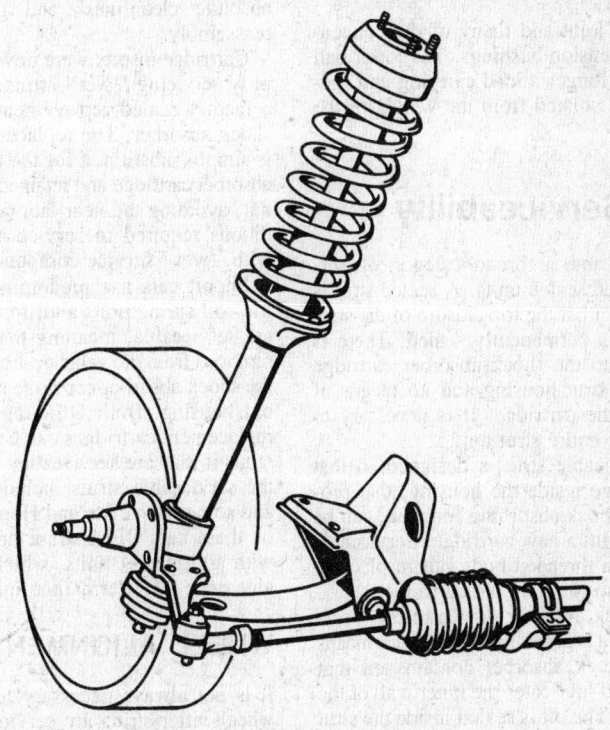

Strut with concentric coil spring (rear wheel drive)

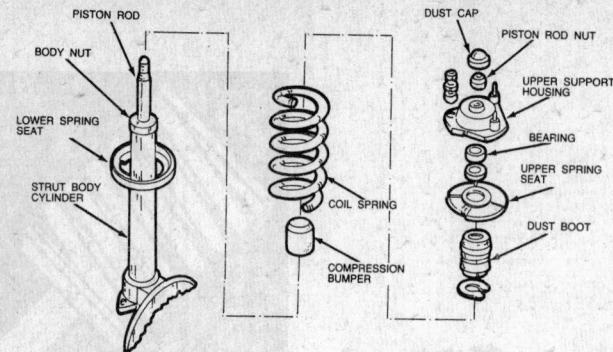

Exploded view of a typical strut

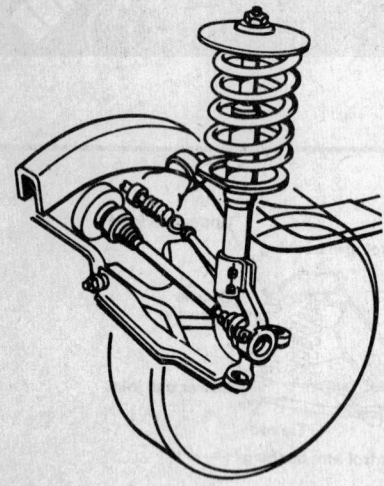

Strut with concentric coil spring (front wheel drive)

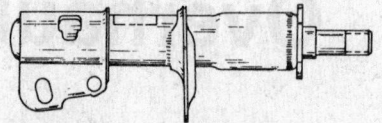

A sealed strut has no body nut and is serviceable by replacement

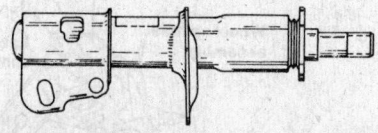

Serviceable struts have a removeable body nut to allow replacement of the strut cartridge

upper ball joint and many of the conventional suspension bushings. The lower ball joint is no longer a load carrying unit, because it is isolated from the weight of the vehicle.

Serviceability

Struts fall into 2 broad categories—serviceable and sealed units. A sealed strut is designed so that the top closure of the strut assembly is permanently sealed. There is no access to the shock absorber cartridge inside the strut housing and no means of replacing the cartridge. It is necessary to replace the entire strut unit.

A serviceable strut is designed so that the cartridge inside the housing, that provides the shock absorbing function, can be replaced with a new cartridge. Serviceable struts use a threaded body nut in place of a sealed cap to retain the cartridge.

The shock absorber device inside a serviceable strut is generally "wet". This means that the shock absorber contains oil that contacts and lubricates the inner wall of the strut body. The oil is sealed inside the strut by the body nut, O-ring and piston rod seal.

Servicing a "wet" strut with the equiv-

alent components involves a thorough cleaning of the inside of the strut body, absolute cleanliness and great care in reassembly.

Cartridge inserts were developed to simplify servicing "wet" struts. The insert is a factory sealed replacement for the strut shock absorber. The replacement cartridge is simply substituted for the original shock absorber cartridge and retained with the body nut, avoiding the near laboratory-like conditions required to service a "wet" strut with "wet" service components.

Import cars use predominantly concentric coil spring units and, for the most part are serviceable, meaning that they can be removed from the vehicle, disassembled and the shock absorber cartridge replaced in the old housing. Both OEM and aftermarket replacement cartridges can be used in these struts if they are serviceable. Exceptions to the serviceable struts include Ford Fiesta and some later Fiats and Hondas, but even on these cars OEM struts can be replaced with aftermarket units, which are serviceable once the aftermarket unit is installed.

WHEEL ALIGNMENT

It is not always necessary to re-align the wheels after struts are serviced. If care is taken matchmarking affected components and in reassembling, alignment may be un-

affected. However, if wheels were not in proper alignment prior to service, or if the entire strut assembly was replaced, a wheel alignment check should be made. Generally, only camber is adjustable, and then only within a narrow range.

Do not attempt to bend components to correct wheel alignment.

On most serviceable import struts, the position of the upper bearing plate or lower mount can be matchmarked and wheel alignment will be maintained during reassembly.

Tools

Without the right tools, a strut job will take longer than necessary and can be dangerous.

A normal selection of hand tools such as open end and box wrenches, sockets, pliers, screwdrivers and hammers are necessary to work on struts. Extensions and universal joints will help reach tight spots. Be sure to have both metric and inch-sized wrenches on hand. Two big time-savers are "crows-feet" and ratcheting box wrenches in assorted sizes. Torx fasteners are also showing up more and more in chassis fasteners.

In addition to the normal handtools, some sort of spanner is necessary to remove the body nut on serviceable struts. Sometimes a pipe wrench can be used successfully.

Strut and cartridge replacement requires a spring compressor.

Makeshift tools for compressing coil springs—threaded rod, chains, wire or other methods—should never be used. The coil spring is under tremendous compression and can fly off causing personal injury and damage to equipment. Use only a good quality spring compressor such as described below.

Economy, or manual, spring compressors are the least expensive but more time consuming to use. Angle hooks grasp the spring coils and must be compressed with a wrench. For those who service struts infrequently, this is probably the wisest investment for purchase.

Other manual spring compressors (jaws type) are faster to operate, have a more positive gripping action and can be used on or off the car. These types are probably not cost effective for the do-it-yourselfer, but can be rented from auto supply stores for single-time use.

For volume work, compressors that are pneumatically or hydraulically operated are best. Air operated compressors are suitable for all types of struts (through use of adaptors), are lightweight and can be used on or off the vehicle. Bench mounted hydraulically operated units are probably the safest, but are also the most expensive and require that the strut be removed from the vehicle, which means separating brake lines and other connections which can be time consuming.

There are also universal kits that fit all struts in either the manual or air operated types.

MAINTAINING WHEEL ALIGNMENT

The location and method of adjusting wheel alignment determines the components that must be match-marked to maintain wheel alignment. There are 4 basic methods of adjusting wheel alignment. Almost all cars use one of these or a slight variation.

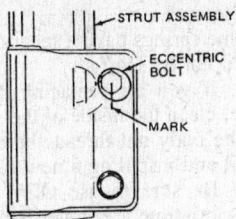

Mark the eccentric (camber adjusting bolt) relative to the clevis mounting bracket.

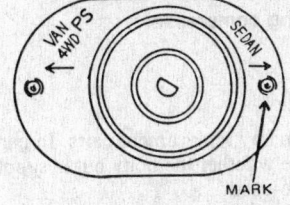

Mark the mounting stud that faces the front of the vehicle. This type of bracket is reversible for varying applications.

Mark the upper support housing relative to the inner fender before removing the strut from the upper mount.

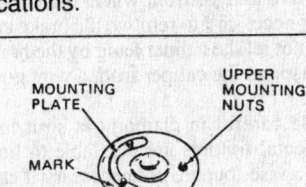

Mark the location of the mounting plate relative to the location on the inner fender.

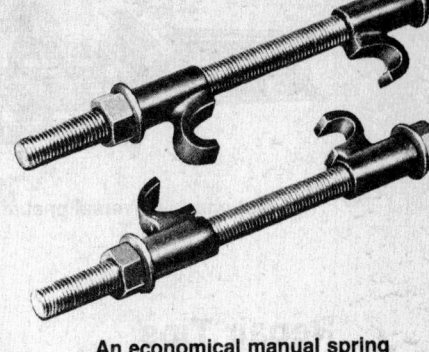

An economical manual spring compressor

"Jaws" type spring compressor

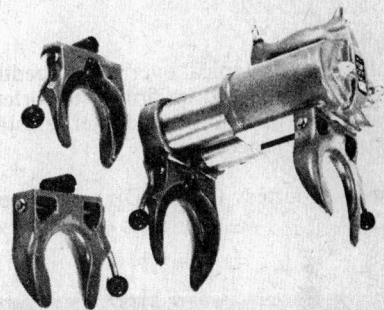

Lightweight, air operated, portable spring compressor can be used on or off the vehicle. Extra shoes are available to handle all strut applications

A simple spanner wrench designed for use with body nuts equipped with recessed lugs. A pipe wrench is a frequent substitute

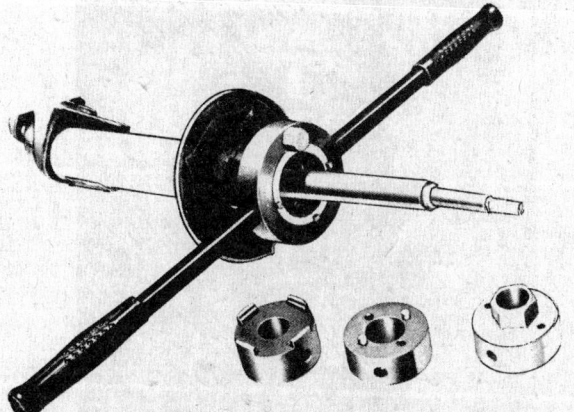

Spanner wrench with adaptor inserts for various applications of body nuts. This type of spanner can be used with a torque wrench for retorqueing the body nut

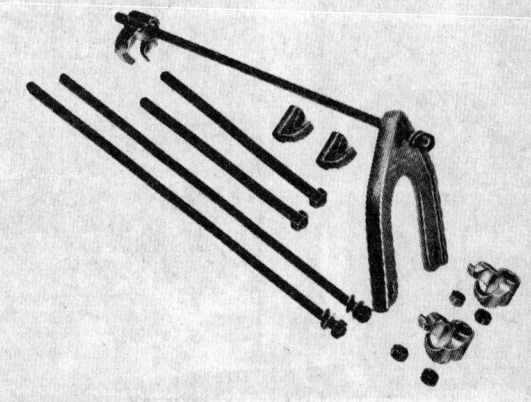

A manual spring compressor with plates or hooks for servicing virtually any strut

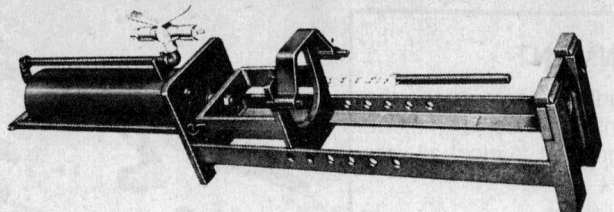

Stationary, universal pneumatic spring compressor

Repair Tips

1. Make sure you have all the tools you'll need. NEVER IMPROVISE A SPRING COMPRESSOR.

2. Normally both front struts should be repaired or replaced at the same time.

3. The easiest way to work on most struts is to remove the entire unit from the vehicle, unless you have access to an air operated spring compressor. Some struts, however, can, and should, be repaired while installed on the vehicle.

4. Always read the instructions packaged with any replacement parts. In particular, note whether the body nut is supplied new or re-used.

5. Mark the position(s) of any bearing plate nuts or cam bolts to assure proper alignment after installation.

6. Be sure to protect the rubber boot on the drive axle of front wheel drive cars.

7. If necessary to remove the brake caliper, do not let the caliper hang by the brake hose. Suspend the caliper from a wire hook or rope.

8. Be careful in clamping a strut in a vise. Special fixtures are available to hold struts in a vise, but are not necessary if care is used to be sure the housing is not crushed or dented. A block of soft wood on either side of the housing will prevent most damage.

9. Use a spring compressor to relieve tension from the spring. Be sure to clean and lubricate the screw threads, particularly on hand operated (manual) spring compressors.

Some springs have a special coating that should not be scuffed.

10. If you are replacing the strut cartridge, clean the inside of the strut housing and the body nut threads before replacing the oil and installing a new cartridge.

11. Be sure to use OEM quality fasteners any time a fastener is replaced.

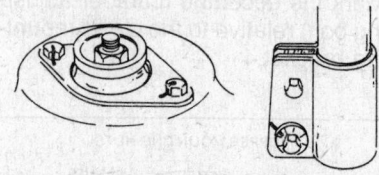

Mark the position of the attachments that control wheel alignment. See Maintaining Wheel Alignment earlier in this section

STRUT OVERHAUL (OFF-CAR)

Following is a typical overhaul procedure of a serviceable MacPherson strut, after having removed the strut from the vehicle. The vehicle should be firmly supported. If it is necessary, to separate the brake line from the strut for strut removal, the brakes will have to be bled after reinstallation. See the manufacturer's car section for specific MacPherson strut removal and installation procedures.

Photos Courtesy Gabriel Div., Maremont Corp.

Step 1. **Examine the strut assembly for damage, dented strut body, spring seat, broken or missing strut mounting parts. Any of these will require replacement of the complete assembly. Also inspect other suspension components for wear or damage**

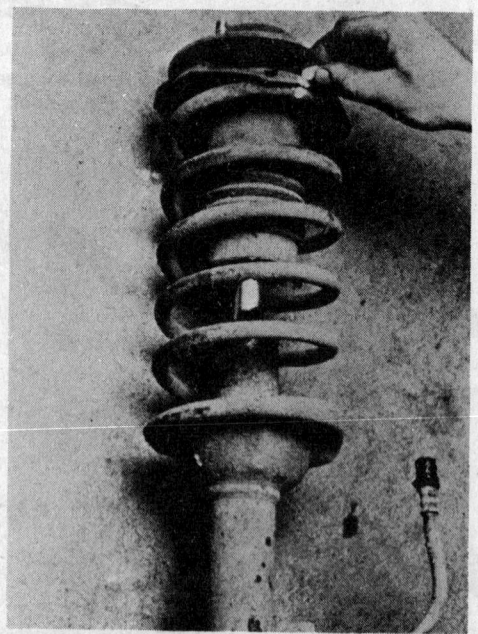

Step 2. **Matchmark the upper end of the coil spring and bearing plate to avoid confusion during reassembly**

Step 3. To make servicing easier, clamp the strut in a strut vise. The strut vise is designed to clamp the strut tight without damage to strut cylinder. It is very handy for strut work and can be used in your shop vise or mounted to any bench

Step 4. Before using the manual spring compressor, lubricate both sides of the thrust washers and the threads with a light coat of grease

Step 5. Install the compressor hooks on opposite sides of the coil spring with the hooks attached to the upper-most and lower-most spring coils. To avoid possible slippage, use tape or small hose clamps on either side of the compressor hooks

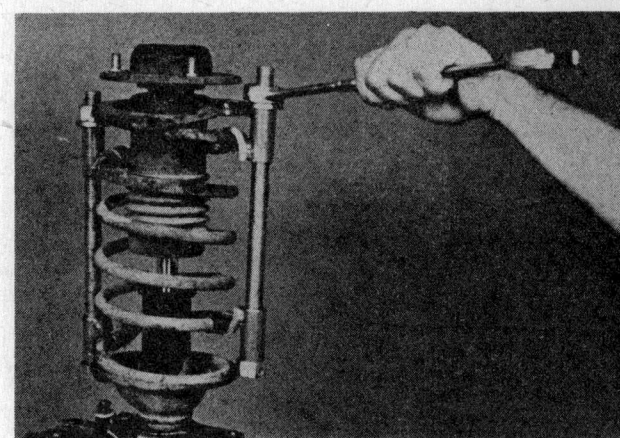

Step 6. Alternately tighten the bolts a few turns at a time until all tension is removed from the spring seat

Step 7. Remove the piston rod nut and disassemble the upper mounting parts, keeping them in order for reassembly. Remove the coil spring. There is no need to remove the compressor from the coil spring

Step 8. An alternative to the manual compressor is the "jaws" type. Turn the load screw to open or close the compressor until the maximum number of spring coils can be engaged

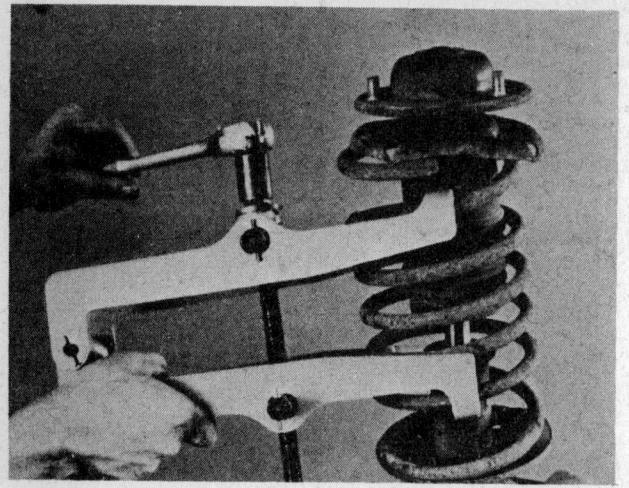

Step 9. Tighten the load screw until the coil spring is loose from the spring seats. There is no need to compress the spring any further

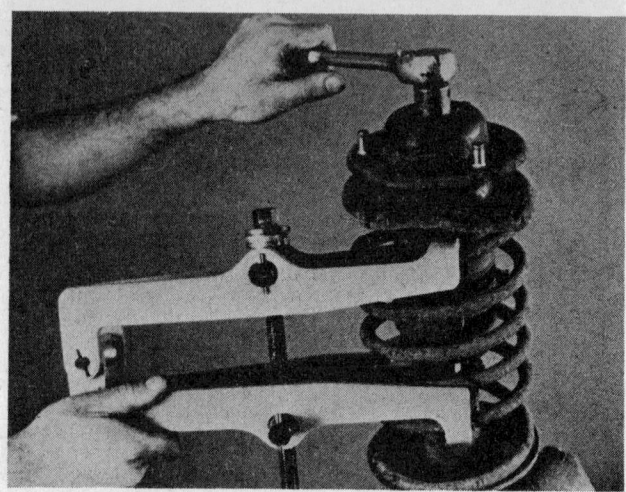

Step 10. Remove the piston rod nut and disassemble the upper mounting parts

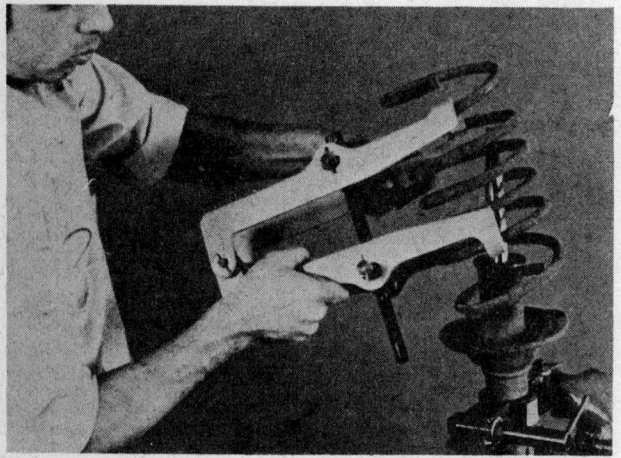

Step 11. Like the manual compressor, there is no need to remove the compressor from the coil spring. Remove the coil spring and compressor

Step 12. Keep the upper mounting parts in order of their removal. They'll be re-assembled in reverse order

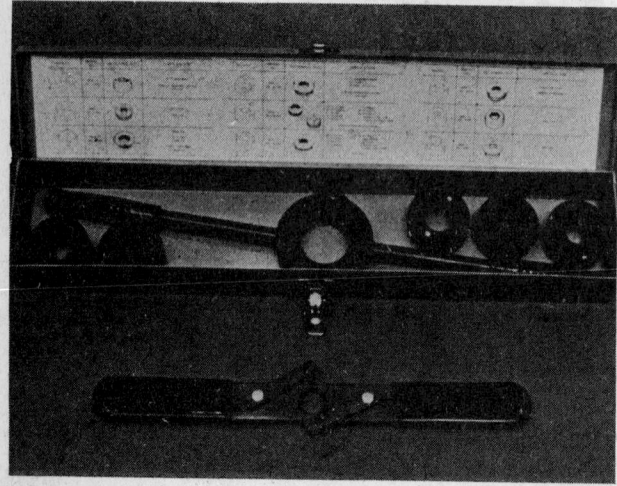

Step 13. A spanner wrench is necessary to remove body nuts, although a pipe wrench will do the job

Step 14. Use the spanner wrench or pipe wrench to loosen the body nut

Step 15. **Remove the body nut and discard if a new body nut came with the replacement cartridge. If not, save the body nut**

Step 16. **Use a scribe or suitable tool to remove the O-ring from the top of the housing**

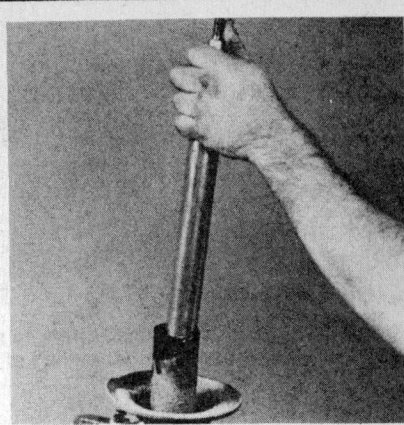

Step 17. **Grasp the piston rod and pull cartridge out of the housing. Remove it slowly to avoid splashing oil. Be sure all pieces come out of the housing**

Step 18. **Pour all of the strut fluid into a suitable container, clean the inside of the strut cylinder, and inspect the cylinder for dents and to insure that all loose parts have been removed from inside of strut body**

Step 19. **Refill the cylinder with one ounce (a shot glass) of the original oil or fresh oil. The oil helps dissipate internal cartridge heat during operation and results in a cooler running, longer lasting unit. Do not put too much oil in—otherwise the oil may leak at the body nut after it expands when heated**

Step 20. **Insert the new replacement cartridge into the strut body**

Step 21. **Push the piston road *all* the way down, to avoid damage to the piston rod if the spanner wrench slips, and start the body nut by hand. Be sure it is not cross-threaded**

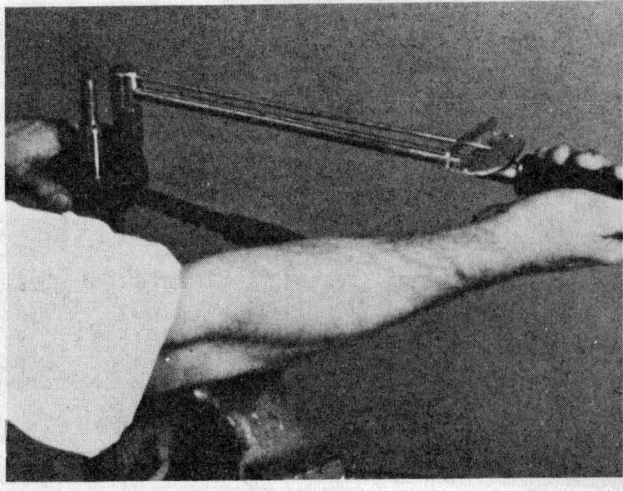

Step 22. **Tighten the body nut securely**

Step 23. **Inspect the loose parts prior to re-assembly. Note the chalk mark location for proper seating of the upper spring seat**

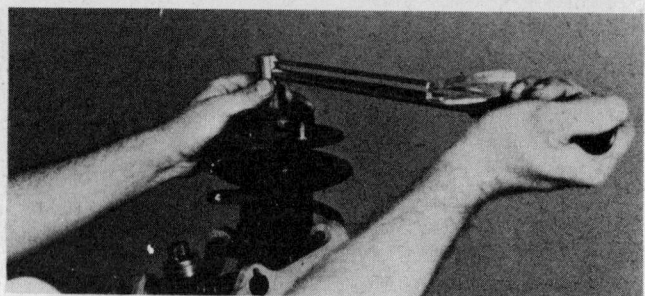

Step 24. **Re-assemble the coil spring and upper mounting parts in reverse order. Tighten the piston rod nut and remove the spring compressor. Install the dust cap. Install the strut in the vehicle. See the car section for details**

STRUT OVERHAUL (ON-CAR)

On some cars, it is best not to completely remove the strut for service. Removal or reassembly of some parts may be difficult without special equipment. Datsun Z-cars are among these. The service methods shown in the following procedure are typical of on-car strut overhaul, although parts or assembly sequences may vary. Refer to the individual manufacturer's car section for details of installations.

Photos Courtesy Gabriel Div., Maremont Corp.

Step 1. **Raise the car to the desired working height and place jack stands under the crossmember, not under the lower control arm. Mark wheel rim and mating lug to assure that wheel balance is not disturbed**

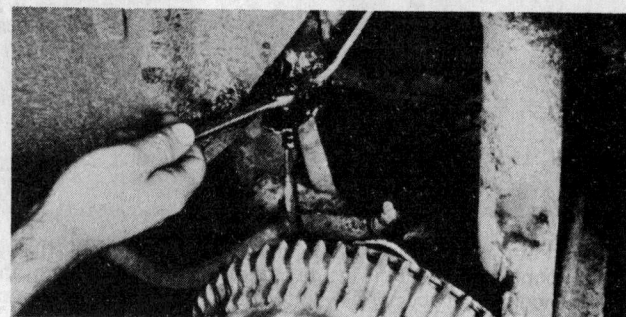

Step 2. **Disconnect brake line at frame bracket**

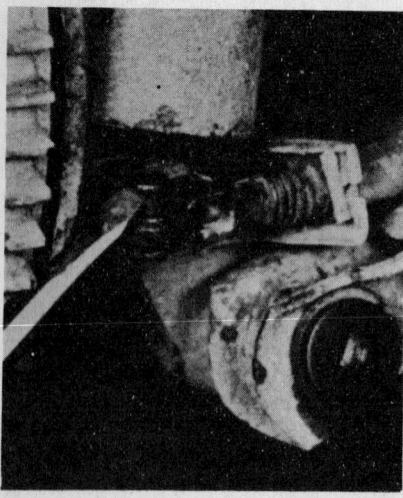

Step 3. **Disconnect the emergency brake cable from actuator arm (Datsun uses a cotter key)**

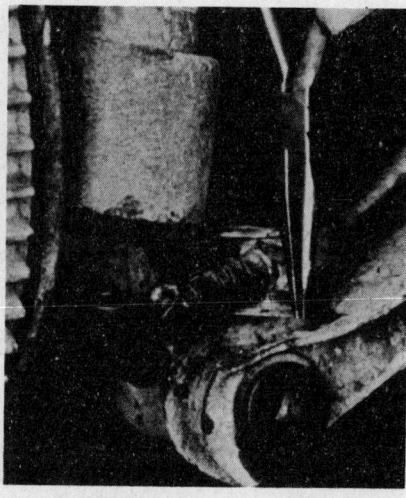

Step 4. **Remove emergency brake cable from bracket by removing the spring clip**

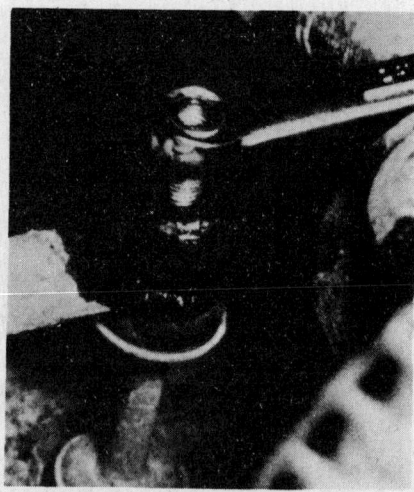

Step 5. **Remove nut, retainer and bushing securing sway bar to its vertical link**

Step 6. Mark the driveshaft flange and hub to assure proper reassembly. Remove the nuts, bolts, and lock washers securing the driveshaft to the hub

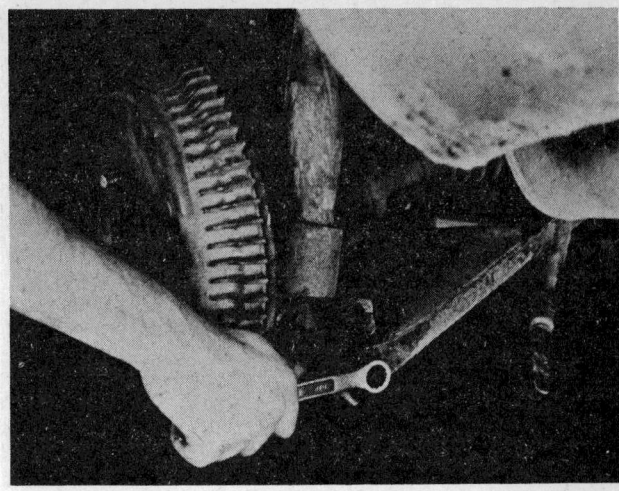

Step 7. Loosen, but do not remove the nut securing the bottom of the strut to the lower control arm

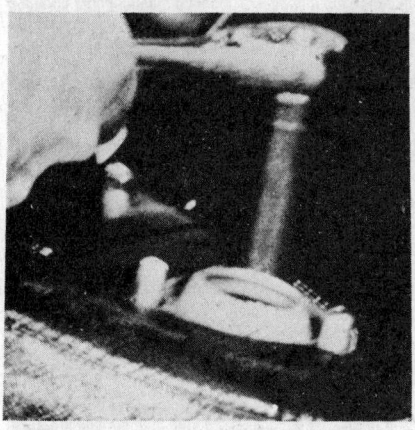

Step 8. From inside the vehicle remove the nuts and lockwashers securing the upper support housing to vehicle fender

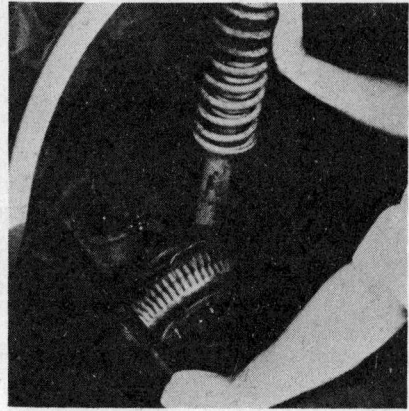

Step 9. Press down on lower control arm with pry bar and swing strut from under fender. (Note: *A slit hose can be used to protect wheel well fender edge*)

Step 10. Either a manual or pneumatic spring compressor can be used to compress the spring. A pneumatic compressor is illustrated. Compress the spring and remove upper mounting parts. Remove the spring from the strut

Step 11. Remove body nut. For convenience, the strut can be propped up while removing body nut

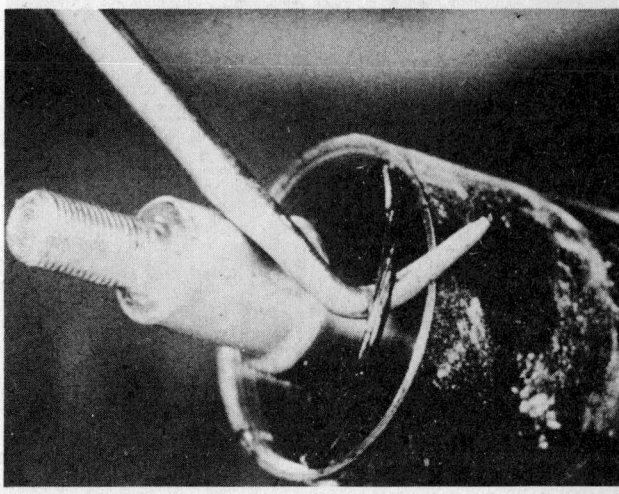

Step 12. Remove O-ring using a small pick or an awl and remove the cartridge

Step 13. Tilt strut down to allow all the oil to drain into a suitable container. Inspect cylinder for dents and to insure all loose parts have been removed from inside of strut body

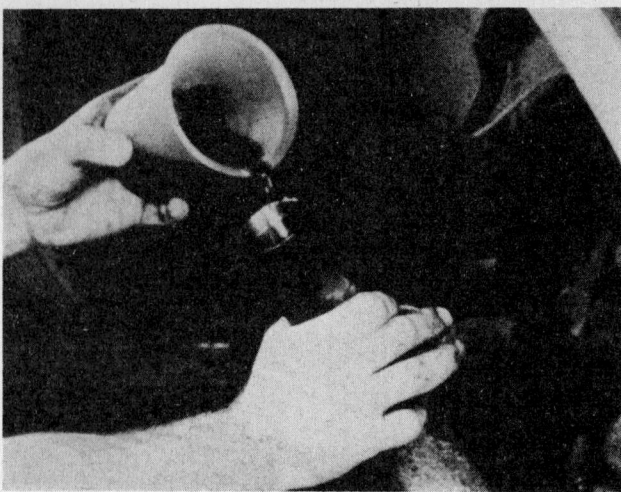

Step 14. Pour about one ounce (a shot glass) of oil back into the strut body for heat transfer. Avoid over-filling or the oil may leak past the body nut

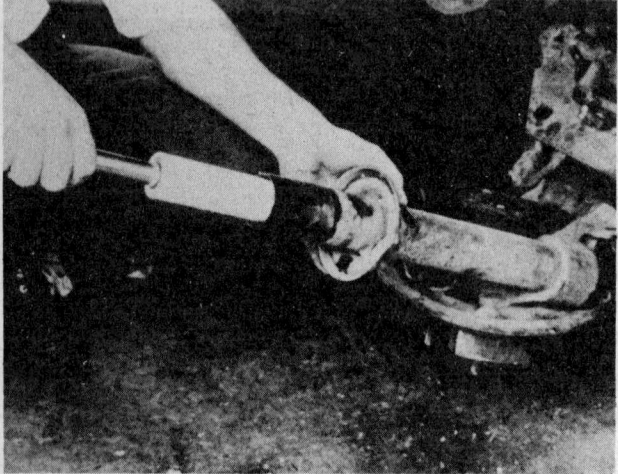

Step 15. Insert the new replacement cartridge into the strut body

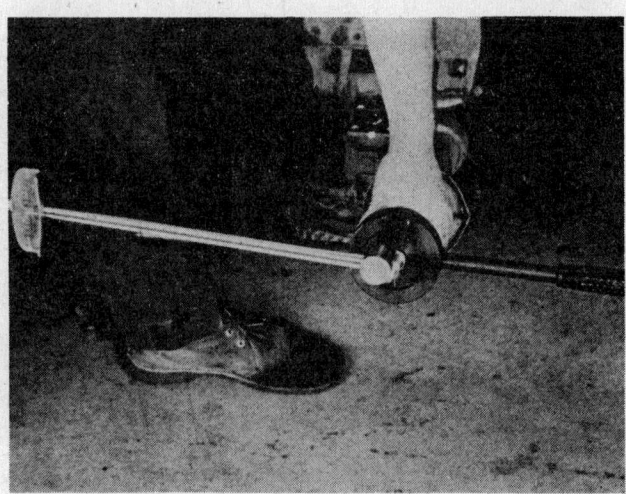

Step 16. Install the body nut and tighten securely

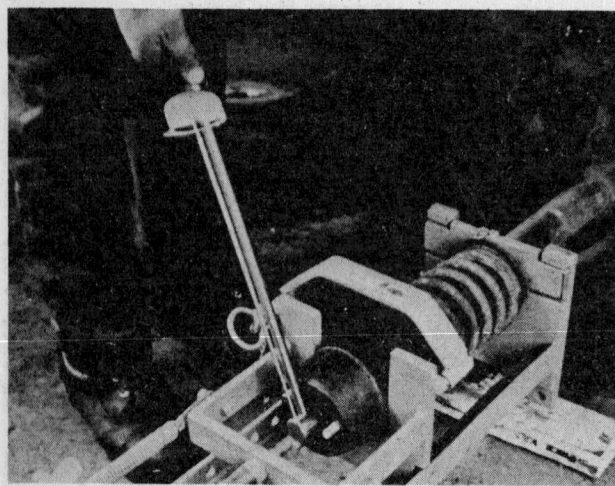

Step 17. Re-install compression bumper, coil spring, spring seat and upper support housing. Be sure to position the support housing studs relative to the holes in the vehicle's inner fender. Install the piston rod nut. Tighten securely. Relieve the spring tension

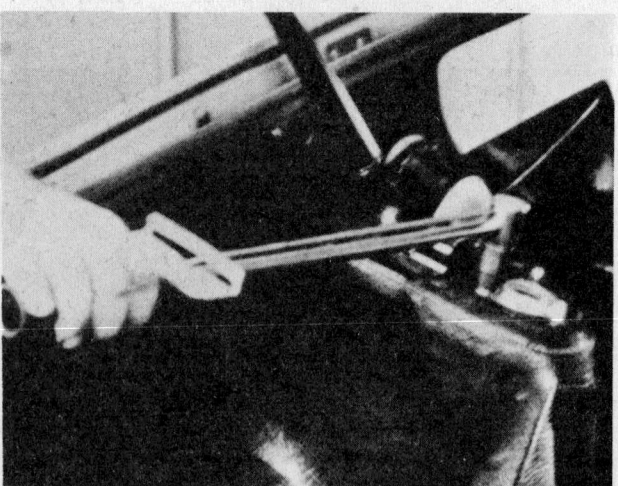

Step 18. Push the strut back under fender and install the nuts and lockwasher securing the upper support housing to the inner fender

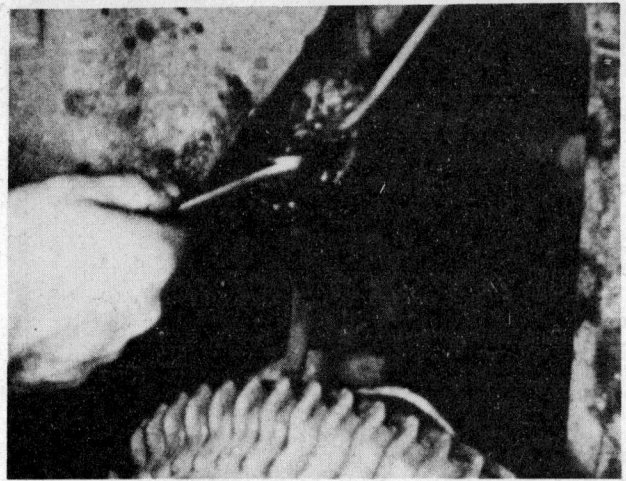

Step 19. Reconnect the brake line

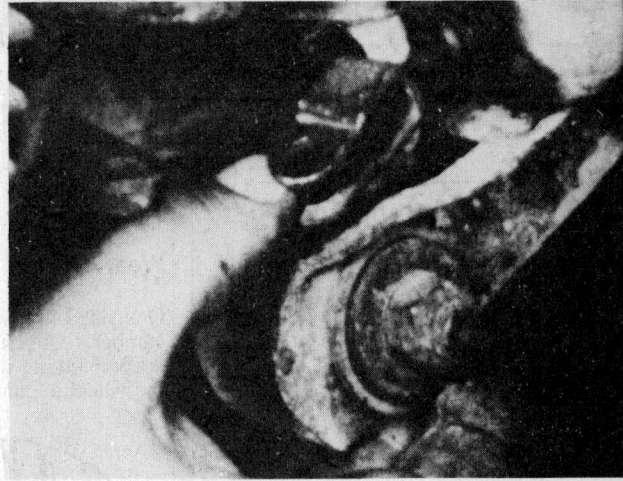

Step 20. Re–assemble the emergency brake cable in the bracket

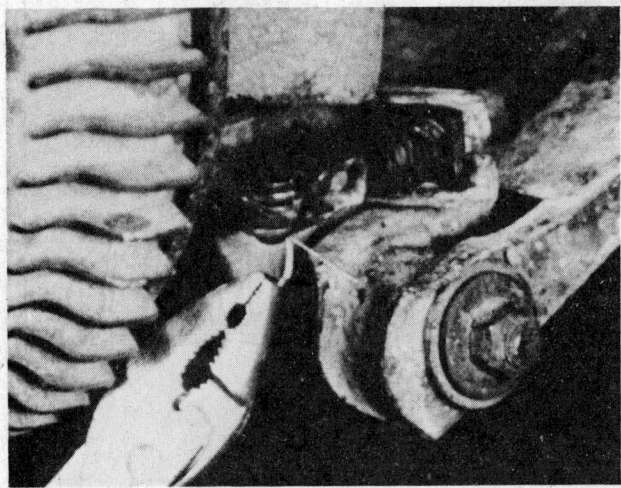

Step 21. Re–assemble brake cable to actuator arm and install cotter pin

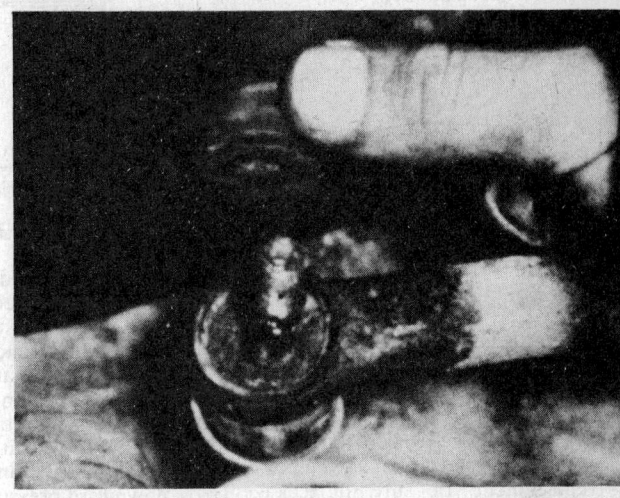

Step 22. Re–install the bushing retainer and nut on sway bar link

Step 23. Align marks made earlier, then install and tighten the four nuts and bolts that secure the driveshaft to the hub

Step 24. Bleed the brakes and install the tire and lower the car to the ground. Tighten the two nuts securing the strut to the lower control arm. (Note: *Torqueing the nuts while the car is not on the ground can cause a hard, choppy ride due to incorrect bushing preload*)

MACPHERSON STRUT PROBLEM DIAGNOSIS

Problems with MacPherson struts generally fall into 3 main categories: suspension, tire wear and steering. In general, the symptoms encountered are not significantly different from those encountered on conventional suspensions.

Suspension

Sag

Vehicle "sag" is a visible tilt of the car from one side to the other or one end to the other while parked on a level surface.

Weak or damaged strut springs could cause this condition and should be repaired immediately.

Sag will also cause steering and tire wear problems to be more pronounced and vehicle instability on rough roads. Front wheel alignment will not solve the problem.

Weak strut springs increase vehicle sag. See "Tire Cupping".

Cartridge Leaks

Strut cartridge leaks (not seepage) indicate the need for cartridge or strut replacement. Be sure the leakage is coming from the strut, and not from elsewhere on the vehicle.

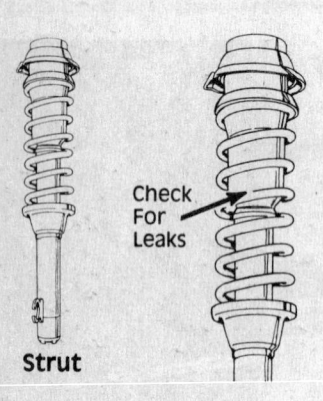

Check For Leaks

Strut

Abnormal Tire Wear

Wear on One Side

One sided tire wear indicates incorrect camber. Check the causes in the accompanying illustration and be sure the wheel alignment is correct.

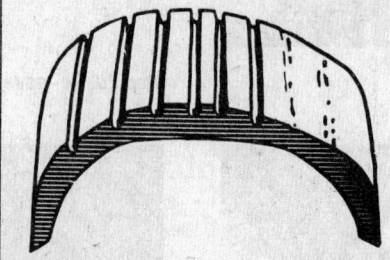

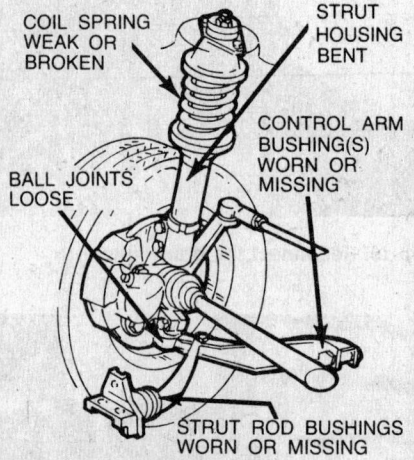

COIL SPRING WEAK OR BROKEN

STRUT HOUSING BENT

CONTROL ARM BUSHING(S) WORN OR MISSING

BALL JOINTS LOOSE

STRUT ROD BUSHINGS WORN OR MISSING

Tire "Cupping"

Cupped tires indicate any or all of the following problems.

1. A weak strut cartridge can be verified by bouncing each corner of the car vigorously and letting go. The car should not bounce more than once, if the shock absorber cartridges are good.

2. Weak strut springs allow sag to increase with only a slight amount of downward pressure. A visual inspection will reveal any broken springs or shiny spots.

3. Check for loose or worn wheel bearings with the weight of the car off of the wheel.

4. Check the wheel balance.

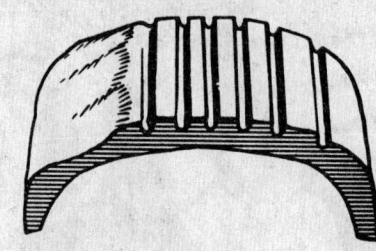

Tread Edge Wear

Wear along tread edges (feathering) indicates a suspension or steering system problem.

1. Strut rod bushings are worn or missing.

2. Tie rod end wear can be determined by grabbing the tie rod end firmly and forcing it up, down or sideways to check for lost motion.

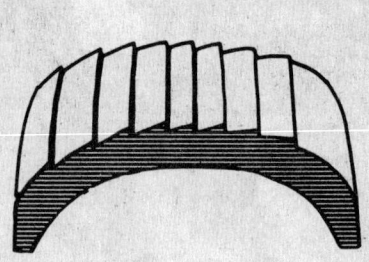

MACPHERSON STRUT PROBLEM DIAGNOSIS

Problems with MacPherson struts generally fall into 3 main categories: suspension, tire wear and steering. In general, the symptoms encountered are not significantly different from those encountered on conventional suspensions.

Tires

Both front tires should match and both rear tires should match. Be sure air pressure is correct.

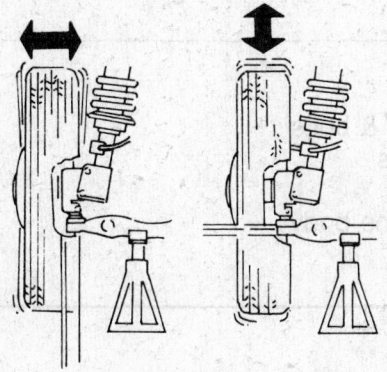

Steering

Ball Joints

Support the car under the frame or crossmember so that the jack does not interfere with the control arm. Rock the tire in and out and up and down. Excessive movement means that both ball joints should be replaced.

Struts with lower weight-carrying ball joints should be supported at the outer edge of the lower control arm. These vehicles usually have wear indicating ball joints that can be checked visually.

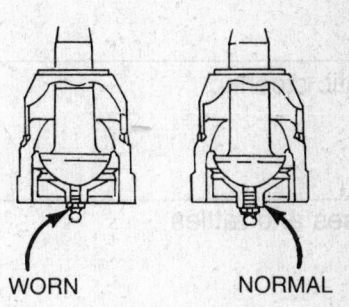

WORN   NORMAL

Stabilizer Bar Bushings

Check for worn bushings or lost motion with the vehicle level and the weight evenly distributed on all wheels.

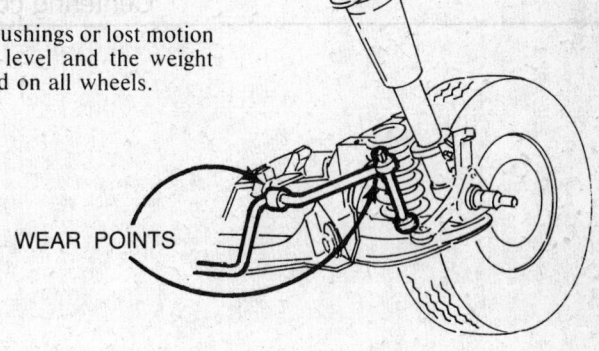

WEAR POINTS

Strut Rod Bushings

Grasp the strut rod and shake it. Any noticeable play indicates excessive wear and need for parts replacement.

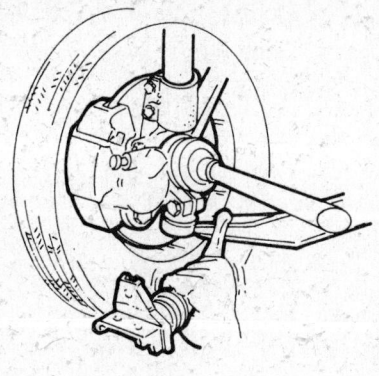

Control Arm Bushings

Support the car under the frame or body and remove the weight from the wheel and control arm. Check for free-play in the bushings at the pivot point, using a pry bar.

NOTE: Some control arm bushings are serviceable only by replacing the entire arm.

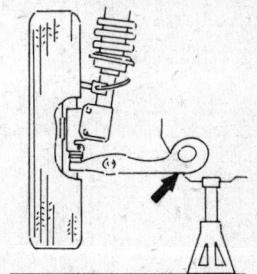

Strut Assembly

Check the strut assembly for cracks or dents in the housing. Look for worn, bent or loose piston rods or dents that will inhibit piston rod movement.

Steering Gear

Check for worn steering gear or loose or worn mounting bolts and bushings.

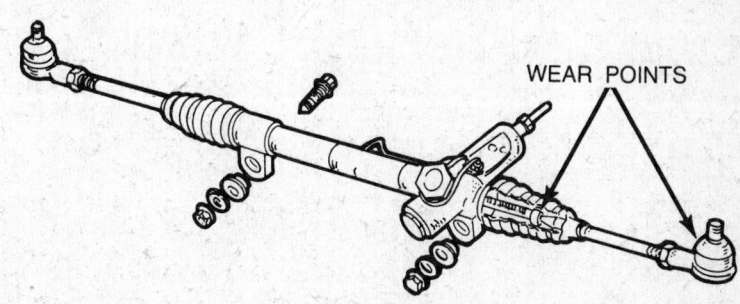

WEAR POINTS

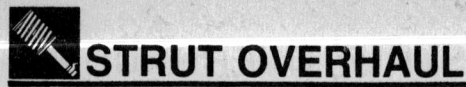

ROAD TEST TROUBLESHOOTING

Following are possible solutions to common potential problems which might be noticed during the road test after strut service is completed. Many are not exclusively strut service related.

Problem	Correction

Brake pedal low or soft	Bleed brakes
	Check for leaks
	    Brake lines
	    Wheel cylinder
	    Caliper piston seal

Erratic steering	Check upper support housing components for proper assembly Check spring assembly right side up Check for spring helix riding correctly on spring seat Check wheel alignment

Noises and rattles	Check torques
	    Piston rod nut
	    Upper support housing nuts & bolts
	    Lower mounting nuts & bolts
	    Body nut
	Check cartridge assembly in the body
	    Spacer used
	    Centering collar used

Diesel Service

NOTE: Most procedures associated with diesel engined cars are similar to gas engined cars, although many parts of the diesel engine are unique compared to their gas engine counterparts. Standard maintenance and service procedures are given here while component removal, installation and adjustment procedures unique to diesel engines can be found in the appropriate section.

HOW THE DIESEL ENGINE WORKS

Four-stroke diesels require four piston strokes for the complete cycle of actions, exactly like a gasoline engine. The difference lies in how the fuel mixture is ignited. A diesel engine does not rely on a conventional spark ignition to ignite the fuel mixture for the power stroke. Instead, a diesel relies on the heat produced by compressing air in the combustion chamber to ignite the fuel and produce a power stroke. This is known as a compression-ignition engine. No fuel enters the cylinder on the intake stroke, only air. At the end of the compression stroke, fuel is sprayed into the precombustion chamber (prechamber). The mixture ignites and spreads out into the main combustion chamber, forcing the piston downward (power stroke). The fuel/air mixture ignites because of the very high combustion chamber temperatures generated by the extraordinarily high compression ratios used in diesel engines. Typically, the compression ratios used in automotive diesels run anywhere from 16:1 to 23:1 A typical spark-ignition engine has a ratio of about 8:1. This is why a spark-ignition engine which continues to run after you have shut off the engine is said to be "dieseling". It is running on combustion chamber heat alone.

Designing an engine to ignite on its own combustion chamber heat poses certain problems. For instance, although a diesel engine has no need for a coil, spark plugs, or a distributor, it does need what are known as "glow plugs". These superficially resemble spark plugs, but are only used to warm the combustion chambers when the engine is cold. Without these plugs, cold starting would be impossible, due to the enormously high compression ratios and the characteristics of the diesel fuel itself.

All diesel engines use fuel injection, be-

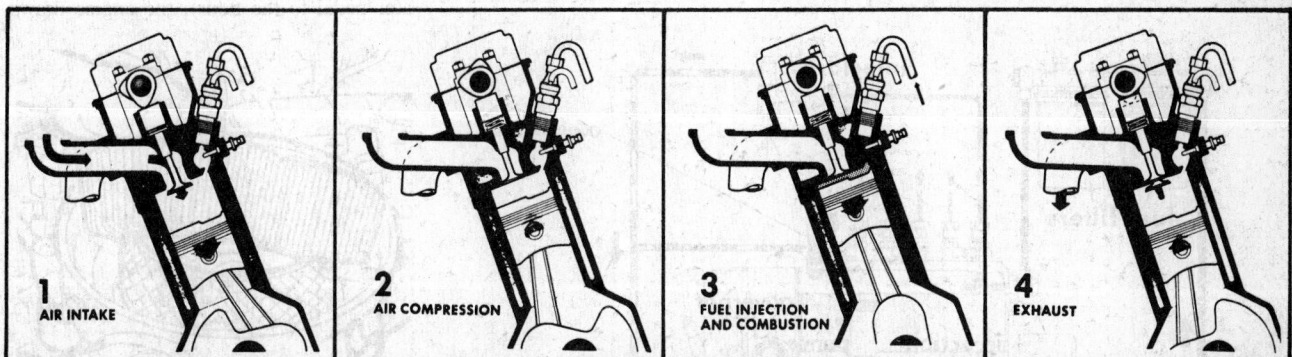

4-stroke diesel engine cycle. At *air intake* (1), rotation of the crankshaft drives a toothed belt that turns the camshaft, opening the intake valve. As the piston moves down, a vacuum is created, sucking fresh air into the cylinder, past the open intake valve. *Air compression* (2): As the piston moves up, both valves are closed, and the air is compressed about 23 times smaller than its original volume. The compressed air reaches a temperature of about 1,650°F., far above the temperature needed to ignite diesel fuel. *Fuel injection and compression* (3): As the piston reaches the top of the stroke, the air temperature is at its maximum. A fine mist of fuel is sprayed into the prechamber, where it ignites, and the flame front spreads rapidly into the combustion chamber. The piston is forced downward by the pressure (about 500 psi) of expanding gases. *Exhaust* (4): As the energy of combustion is spent and the piston begins to move upward again, the exhaust valve opens, and burnt gases are forced out past the open valve. As the piston starts down, the exhaust valve closes, the intake valve opens, and the air intake stroke begins again.

Increasingly, modern diesel engines are being equipped with turbochargers, exhaust gas–driven devices that force more air into the engine to increase power output

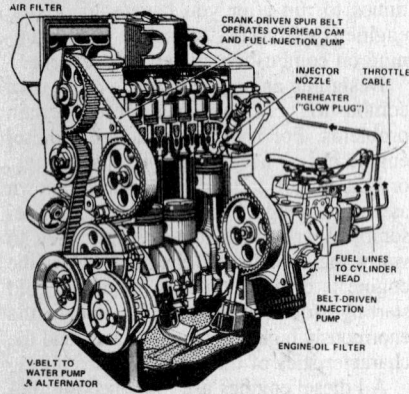

Cutaway view of typical 4-cylinder diesel engine.

Maintenance and Service Procedures

Maintenance procedures for the diesel engine generally fall into three categories:

1. Fuel system
2. Starting system
3. Engine mechanical systems

Of these, the fuel system is usually the most likely source of engine troubles, and should be high on the list for regular maintenance attention.

FUEL SYSTEM

The typical diesel engine fuel system consists of fuel tank, fuel feed and return lines, mechanical fuel injection pump, fuel injectors and lines, and a large capacity fuel filter. On some models, the engine may also be equipped with a small, low pressure fuel pump which feeds the injection pump.

In addition to these, the air intake system (air cleaner, inlet manifold) should be checked over regularly to insure unrestricted air flow into the cylinders.

In operation, fuel is sucked out of the fuel tank by the injection pump (or its feed pump) and fed by the injection pump to the injectors in the cylinder head at a very high pressure. Before the fuel is allowed to enter the main injection pump, it passes through a specially built fuel filter which traps solid particles (and water on some models) in the fuel. Fuel that is not used is pumped back to the fuel tank through the fuel return lines. This recirculated fuel helps cool the injection pump.

Air Cleaner

On a gasoline engine, the volume of air taken in by the engine is controlled by throttle valves. When the throttle valves are closed (engine idling), air intake is restricted. When the throttle valves are wide open (accelerator pedal to the floor), the engine draws

cause unlike spark-ignited engines, the fuel cannot be drawn through the intake tract and into the cylinders. The introduction of fuel into a diesel engine must be precisely timed so that each cylinder "fires" at the proper moment. Also, the fuel injection pressure (at the cylinder) must be great enough to overcome the high compression pressures, and properly atomize the fuel without the aid of a moving air mass (as in a carbureted gas engine). It is not uncommon for diesel engine fuel injection pressures to be set at 1500–1700 psi.

Diesel engines share many of their basic mechanical components with gasoline engines, though the cylinder block, head(s), crankshaft, connecting rods, pistons, etc., are manufactured to be much stronger for use in diesel engines. The additional strength of the components is necessary due to the very high cylinder pressure generated within the diesel engine.

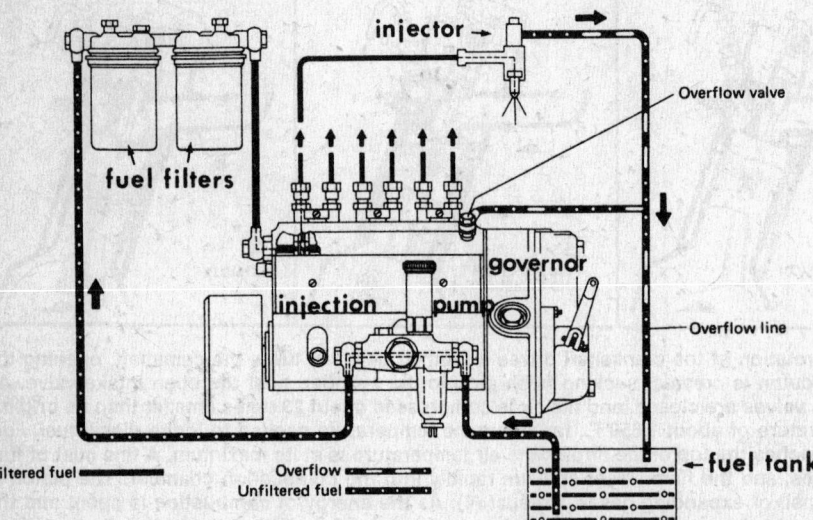

Typical diesel engine fuel system schematic

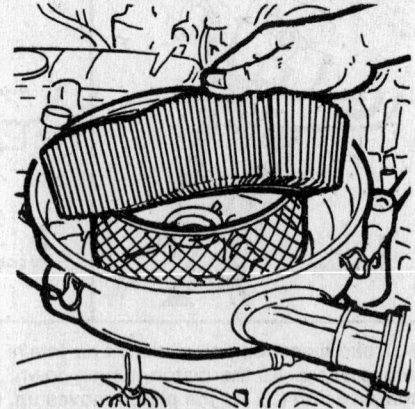

Because a greater quantity of air passes through the diesel engine, air filter maintenance is particularly important. Most diesel air filters on passenger cars are similar to their counterparts on gasoline engines.

in the maximum amount of air it possibly can. This applies to both carbureted and fuel injected gasoline engines.

The speed (rpm) of a diesel engine is controlled by the quantity of fuel which is injected into the engine; no air metering restrictions (throttle valves) are used. Because of this, diesel engines ingest as much air as they possibly can under all conditions. A much greater volume of air passes through the air cleaner of a diesel per mile, therefore, diesel air filters must either be larger or the filter replacement intervals more frequent than those of a similarly sized gasoline engine.

One word of caution: never remove the air cleaner on a diesel with the engine running, and never run the engine with the air cleaner removed. The volume of air drawn through the inlet manifold is very great, and, because the inlet manifold is unobstructed, anything drawn into the inlet manifold (air cleaner wing nut, etc.) goes straight to the combustion chambers, where it can cause major engine damage.

Fuel Filter

The diesel engine fuel filter is usually larger than the filter used on gasoline engines. The extra capacity is needed to trap the suspended particles in diesel fuel, which is generally "dirtier" than gasoline.

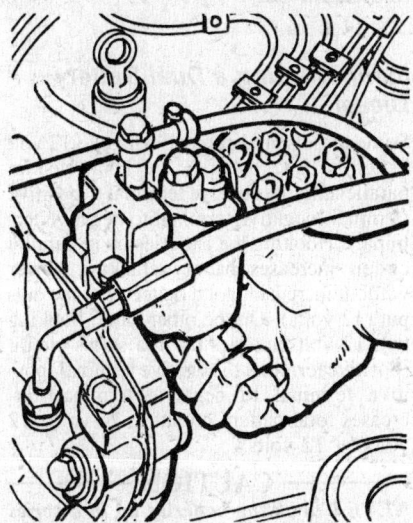

Many diesel engines use a spin-on type primary fuel filter.

On some engines, the fuel filter looks like a second engine oil filter, and is removed and installed in the same manner as the canister-type oil filter.

The fuel filter must be changed according to the manufacturer's suggested interval. See the owner's manual for information.

After installing the fuel filter start the engine and check for leaks. Run the engine for about two minutes, then stop the engine for the same amount of time to allow any air trapped in the injection system to bleed off.

Many diesels also have a small, in-tank filter which is usually maintenance-free.

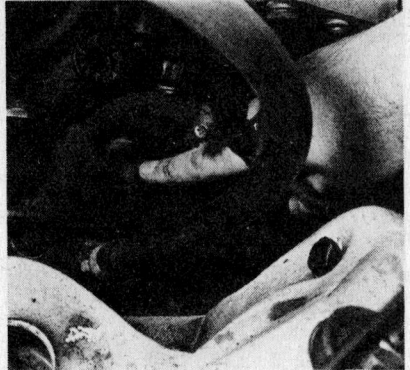

A smaller, in-line secondary filter is used on many engines.

Check the tightness of the clamps securing the injector lines. Note that the injector lines are all the same length.

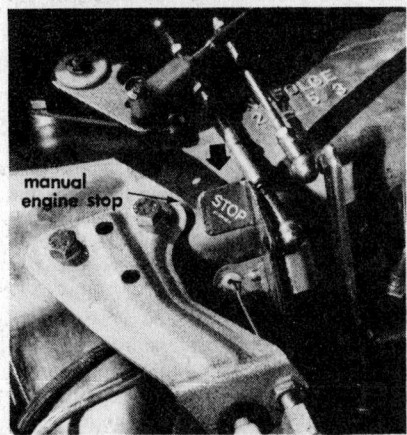

Mercedes-Benz diesel engines use this stop switch, which shuts off fuel delivery

Water In Fuel

Diesel fuel is a hydrophilic fluid, that is, it naturally attracts water. Since diesel fuel and water do not mix, the water remains floating beneath the fuel at the bottom of the tank. This water must be removed every now and then, or it will be sucked into the fuel circuit and pass through the injection system, causing corrosion and possible component failure (injection pumps can cost up to $1,000). Water in the fuel system will also cause the engine to run poorly, if at all.

Most diesel fuel tanks are equipped with a separator which can isolate from 1 to 3 gallons of water from the fuel.

Many diesels are also equipped with "Water in Fuel" lights in the dashboard which warn of the presence of water in the fuel tank. These warning systems can be installed on models not so equipped.

On some diesels, there is a water catcher in the bottom of the fuel filter which can easily be bled off. In addition, there are several bolt-on water filters on the market which attach to the fuel line under the hood and separate water from the fuel. Depending on which kind you buy, draining water from the system is simply a matter of opening the petcock at the bottom of the filter and letting the water drain out, or, if money is no object, a separator is available on which water is drained from the filter simply by activating a switch on the dashboard.

Removing Water from the Fuel Tank

Treat diesel fuel with the same respect you would gasoline, and after the procedure, properly dispose of the fuel.

1. Remove the fuel tank cap.
2. Connect a pump or siphon hose to the ¼ in. fuel return hose (smaller of the two fuel hoses) above the rear axle, or under the hood near the fuel pump (on the passenger's side of the engine, near the front).
3. Siphon until all water is removed from the tank. Do not use your mouth to create siphon vacuum, EVER! The best method is to siphon the water into a large capacity see-through container. The water will collect at the bottom of the container.
4. When all water has been removed from the tank, be sure to reinstall the fuel return hose and fuel cap.

NOTE: If the entire fuel system (not just the tank) is contaminated by water, the vehicle must be stopped immediately and the fuel system must be purged. This includes draining and removing the fuel tank, blowing low pressure compressed air backwards through the fuel feed and return lines, and bleeding the water out of all injection components. This job should be referred to a qualified technician.

Cold Weather Fuel System Maintenance

—— CAUTION ——
NEVER use "starting aids" (e.g.—ether) to help start a diesel engine—serious engine damage will result.

As will be explained later under "Fuel Recommendations", diesel fuel tends to become "cloudy", or thicker, as the temperature drops. The thicker the diesel fuel becomes, the slower it flows through the fuel system, until finally it stops flowing altogether somewhere near the bottom of the thermometer.

One way to fight sluggish fuel flow is to use winterized blends of diesel fuel, straight No. 1 diesel fuel or add cold weather additives to the fuel to improve flow in cold weather.

NOTE: Consult your owners manual for recommendations and be sure to use a fuel conditioner compatible with water separators.

Another way is to install an aftermarket fuel system pre-heater. These are generally canisters which connect into the fuel line and use coolant from the engine cooling system to heat the fuel before it reaches the injection pump. The one drawback with this system is the engine must be started before the pre-heater begins to work. Also available are electric fuel warmers. These preheat the fuel going into the filter and can be used in conjunction with the coolant-type fuel heater.

Cold weather additives and fuel conditioners can help improve cold weather flow of diesel fuel.

Some manufacturers offer an optional electric diesel fuel heater and engine block heaters. The fuel heater is thermostatically controlled to heat the fuel before it enters the fuel filter when fuel temperature is 20°F or lower. The fuel heater works only when the ignition key is in the RUN position. On these models, the fuel tank filter has a by-pass valve which allows fuel to flow to the heater when the tank filter is covered with fuel wax. The engine block heater is equipped with an electrical cord wrapped up in the engine compartment. The cord

Some diesel engines come equipped with a built-in heating system to keep the engine warm in cold temperatures.
Most OEM heaters work from 110-volt house current.

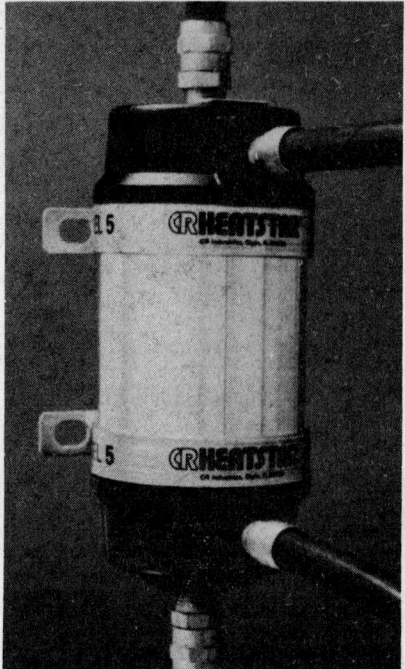

Some aftermarket diesel fuel warmers are thermostatically controlled heat exchangers that use engine coolant to keep diesel fuel above its "cloud point," the temperature at which it gels and forms wax that can clog a fuel system.

plugs into regular 110 volt household current. The block heater can be used, according to the type of oil in the crankcase, up to eight hours or overnight to warm up the block.

STARTING SYSTEM

The diesel starting system includes one (sometimes two) heavy duty batteries, the starter, and the glow plug circuit. In addition to the heavy duty battery(ies), the majority of diesel engines also have starters and battery cables designed specifically as heavy duty items for diesel usage only. Because of the high compression of any diesel, the torque required to turn the engine is much greater than a gasoline engine. The starter must be powerful enough to handle the increased load; the battery cables must be thick enough to withstand the heat generated by the starter load.

For battery maintenance, see the regular "Maintenance" section. Jump starting procedures for a dual battery car are given below. Starter maintenance is included in the appropriate car section.

The glow plug circuit is used on the diesel to initially start the engine. When the ignition switch is turned to the ON position, a light will come on in the instrument panel signalling that the glow plugs are preheating the combustion chambers. After a certain interval (depending on how cold the engine is), the light will go off. This signals that the starter may be engaged and the engine started. If the glow plug circuit mal-

functions, especially in cold weather, the engine will be almost impossible to start.

CAUTION
NEVER use "starting aids" (e.g.—ether) to help start a diesel engine—serious engine damage will result.

Glow Plug Testing

To test each individual glow plug, disconnect the busbar and/or wire connector from the glow plug and connect a test light between the glow plug terminal and the positive battery terminal. If the test light lights, the glow plug is working. Replace individual glow plugs which do not work.

NOTE: Some diesel engines are equipped with either "slow glow" or "fast glow" glow plugs. Do not attempt to interchange any parts of these two glow plug systems.

To test the glow plug circuit, connect a test light to the terminal of one of the glow plugs (glow plug wiring still attached) and turn the ignition to the heating position. The test light should light for a short while. If not, the glow plug circuit is malfunctioning and must be diagnosed and repaired.

NOTE: Perform this operation on a cold engine only.

Jump-Starting a Dual Battery Diesel

Some diesels are equipped with two 12 volt batteries. The batteries are connected in parallel circuit (positive terminal to positive terminal, negative terminal to negative terminal). Hooking the batteries up in parallel circuit increases battery cranking power without increasing total battery voltage output (12 volts). On the other hand, hooking two 12 volt batteries up in a series circuit (positive terminal to negative terminal, positive terminal to negative terminal) increases total battery output to 24 volts (12 volts + 12 volts).

CAUTION
NEVER hook the batteries up in a series circuit; SEVERE electrical system damage will result.

In the event that a dual battery diesel must be jumped started, use the following procedure.

1. Open the hood and locate the batteries.

2. Position the donor car so that the jumper cables will reach from its battery (must be 12 volt, negative ground) to the appropriate battery in the diesel. Do not allow the cars to touch.

3. Shut off all electrical equipment on both vehicles. Turn off the engine of the donor car, set the parking brakes on both vehicles and block the wheels. Also, make sure both vehicles are in Neutral (manual

transmission models) or Park (automatic transmission models).

4. Using the jumper cables, connect the positive (+) terminal of the donor car battery to the positive terminal of one (not both) of the diesel batteries.

5. Using the second jumper cable, connect the negative (−) terminal of the donor battery to a solid, stationary, metallic point on the diesel (alternator bracket, engine block, etc.). Be very careful to keep the jumper cables away from moving parts (cooling fan, alternator belt, etc.) on both vehicles.

6. Start the engine of the donor car and run it at moderate speed.

7. Start the engine of the diesel.

8. When the diesel starts, disconnect the battery cables in the reverse order of attachment.

ENGINE MECHANICAL SYSTEMS

Included are engine lubrication and engine compression.

Although diesel engines are very low in carbon monoxide (CO) and hydrocarbon (HC) emissions, "particulate" emission output is very high from diesel engines. This is evident from the black smoke emitted by diesels, which is most noticeable during hard acceleration or high engine loads. The particulates are made up of mostly soot (carbon) and sulpher particles. The majority of these particulates are released into the atmosphere. However, some of the particulate matter, because it is produced within the engines cylinders, is left inside the engine and gradually contaminates the engine oil. This contamination makes the oil corrosive, due to the sulpher, and abrasive, due to the carbon. Serious engine damage will result if these contaminants continue to accumulate in the oil. Engine oil and filters of diesel engines must be changed more frequently than those of gasoline engines, due to the increased rate at which the contaminants form in the diesel. Consult the "Maintenance" section for oil and filter change procedures. The manufacturer's recommended oil change interval will be given in the owner's manual. An explanation of diesel engine oils is given at the end of this section.

As explained earlier, very high cylinder compression is the key to the operation of the diesel engine. The normal compression of most gasoline engines will rarely exceed 180 psi; whereas with diesel engines, compression pressures of 350–400 psi are commonplace.

─────── **CAUTION** ───────
DO NOT attempt to check the compression of a diesel engine with a standard compression gauge—personal injury could result. A special, high pressure compression gauge is needed to safely check the compression of any diesel.

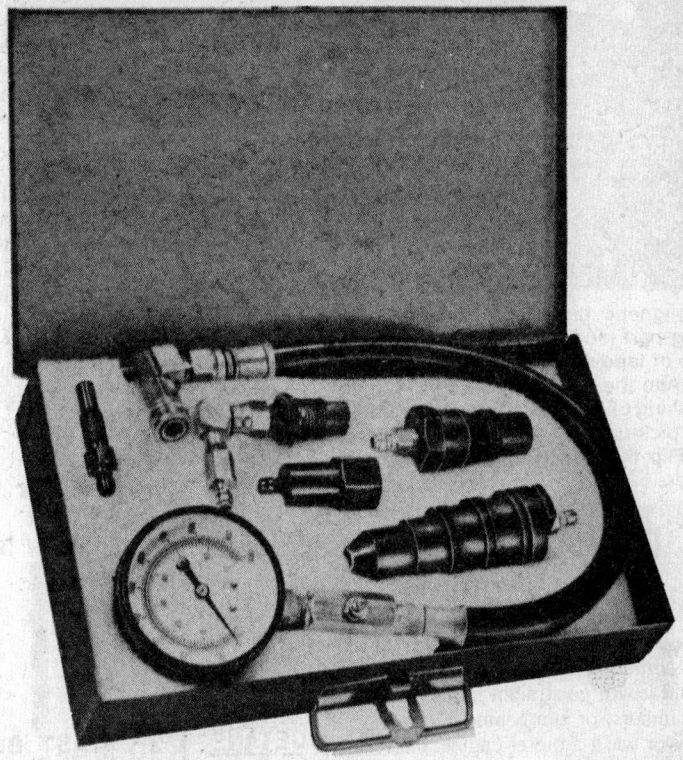

A diesel compression tester kit with adaptors (Courtesy S & G Tools).

Compression Test

1. Remove the air cleaner.
2. Disconnect the wire from the fuel shutoff solenoid terminal of the injection pump.
3. Disconnect the wires from the glow plugs and remove all glow plugs.
4. Screw compression gauge into the glow plug hole in the cylinder being checked.
5. Crank the engine, allowing six "puffs" for each cylinder.

The lowest reading cylinder should not be less than 70% of the highest, and no cylinder should be less than 275 pounds.

Idle Speed Adjustments

Idle speed adjustment procedures for individual diesel engines are given in the car section. Consult the following section for procedures to measure idle speed

Connecting a Tachometer to a Diesel Engine

As mentioned earlier, the diesel engine does not require an electrical ignition system. Because of this, problems arise when attempts are made to connect a tachometer to the engine for the purpose of idle adjustments, etc. The average gasoline engine tachometer senses the ignition spark pulses and converts them into a readable engine rpm signal. This type of tachometer is use-

less on the diesel engine, because of the diesel's compression ignition system.

There are several magnetic and photoelectric tachometers available from various tool manufacturers which were designed specifically for use with the diesel engine. These units can run into a little more money than the average do-it-yourselfer may be willing to spend, in which case any adjustments requiring the monitoring of engine rpm should be performed by a competent service technician.

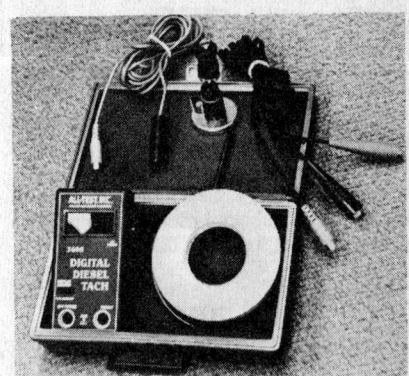

The newest equipment for measuring idle speed on a diesel engine includes (clockwise from lower left) a digital diesel tach display, photomagnetic pick-up with display input, magnetic swivel base (holder), DC power source for the display unit and a roll of magnetic tape.

The magnetic tape is attached to any moving part (such as the balancer). The pieces of tape must be at least 6 inches apart. Aim the photomagnetic pick-up at the moving object and adjust the position of the pick-up until the "on-target" light is lit. Flip the switch to TACH and read the rpm.

Diesel Engine Precautions

- Never run the engine with the air cleaner removed: if anything is sucked into the inlet manifold it will go straight to the combustion chambers, or jam behind a valve.
- Never wash a diesel engine: the reaction of a warm fuel injection pump to cold (or even warm) water can ruin the pump.
- Never operate a diesel engine with one or more fuel injectors removed unless fully familiar with injector testing procedures: some diesel injection pumps spray fuel at up to 1400 psi—enough pressure to allow the fuel to penetrate your skin.
- Do not skip engine oil and filter changes.
- Strictly follow the manufacturer's oil and fuel recommendations as given in the owner's manual.
- Do not use home heating oil as fuel for your diesel.
- Do not use "starting aids" (e.g.—ether) in the automotive diesel engine, as these "aids" can cause severe internal engine damage.
- Do not run a diesel engine with the "Water in Fuel" warning light on in the dashboard.
- If removing water from the fuel tank yourself, use the same caution you would use when working around gasoline engine fuel components.
- Do not allow diesel fuel to come in contact with rubber hoses or components on the engine, as it can damage them.

Fuel and Oil Recommendations

FUEL

Fuel makers produce two grades of diesel fuel, No. 1 and No. 2, for use in automotive diesel engines. Generally speaking, No. 2 fuel is recommended over No. 1 for driving

in temperatures above 20°F. In fact, in many areas, No. 2 diesel is the only fuel available. By comparison, No. 2 diesel fuel is less volatile than No. 1 fuel, and gives better fuel economy. No. 2 fuel is also a better injection pump lubricant.

Two important characteristics of diesel fuel are its cetane number and its viscosity.

The cetane number of a diesel fuel refers to the ease with which a diesel fuel ignites. High cetane numbers mean that the fuel will ignite with relative ease or that it ignites well at low temperatures. Naturally, the lower the cetane number, the higher the temperature must be to ignite the fuel. Most commercial fuels have cetane numbers that range from 35 to 65. No. 1 diesel fuel generally has a higher cetane rating than No. 2 fuel.

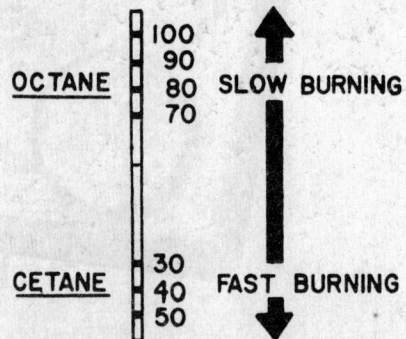

Cetane (diesel engine) versus octane (gasoline engine) ratings. The higher the cetane number, the faster the fuel burns

Viscosity is the ability of a liquid, in this case diesel fuel, to flow. Using straight No. 2 diesel fuel below 20°F can cause problems, because this fuel tends to become cloudy, meaning wax crystals begin forming in the fuel. In extreme cold weather, No. 2 fuel can stop flowing altogether. In either case, fuel flow is restricted, which can result in a "no start" condition or poor engine performance. Fuel manufacturers often "winterize" No. 2 diesel fuel by using various fuel additives and blends (No. 1 diesel fuel, kerosene, etc.) to lower its winter-time viscosity. Generally speaking, though, No. 1 diesel fuel is more satisfactory in extremely cold weather.

NOTE: No. 1 and No. 2 diesel fuels will mix and burn with no ill effects, although the engine manufacturer will undoubtedly recommend one or the other. Consult the owner's manual for information.

Depending on local climate, most fuel manufacturers make winterized No. 2 fuel available seasonally.

Many automobile manufacturers publish pamphlets giving the locations of diesel fuel stations nationwide. Contact the local dealer for information.

Do not substitute home heating oil for automotive diesel fuel. While in some cases, home heating oil refinement levels equal those of diesel fuel, many times they are far below diesel engine requirements. The result of using "dirty" home heating oil will be a clogged fuel system, in which case the entire system may have to be dismantled and cleaned.

One more word on diesel fuels. Don't thin diesel fuel with gasoline in cold weather. The lighter gasoline, which is more explosive, will cause rough running at the very least, and may cause extensive engine damage if enough is used.

OIL

Diesel engines require different engine oil from those used in gasoline engines. Besides doing the things gasoline engine oil does, diesel oil must also deal with increased engine heat and the diesel blow-by gases, which create sulphuric acid, a high corrosive.

Under the American Petroleum Institute (API) classifications, gasoline engine oil codes begin with an "S", and diesel engine oil codes begin with a "C". This first letter designation is followed by a second letter code which explains what type of service (heavy, moderate, light) the oil is meant for. For example, the top of a typical oil can will include: "API SERVICES SC, SD, SE, CA, CB, CC". This means the oil in the can is a good, moderate duty engine oil when used in a diesel engine.

It should be noted here that the further

COMPARISON OF #1 AND #2 DIESEL FUEL

Requirement	1-D	2-D
Flash Point, °F minimum	100	125
Cetane Number, minimum	40	40
Viscosity at 100°F, Centistokes		
Minimum	1.4	2.0
Maximum	2.5	4.3
Water and Sediment, % by volume maximum	Trace	0.05
Sulfur, % by weight maximum	0.5	0.5
Ash, % by weight maximum	0.01	0.01

Flash Point: The temperature at which diesel fuel ignites when exposed to a flame *in the open air.*

Cetane Number: See text

down the alphabet the second letter of the API classification is, the greater the oil's protective qualities are (CD is the severest duty diesel engine oil, CA is the lightest duty oil, etc.). The same is true for gasoline engine oil classifications (SF is the severest duty gasoline engine oil, SA is the lightest duty oil, etc.).

Many diesel manufacturers recommend an oil with both gasoline and diesel engine API classifications. Consult the owner's manual for specifications.

The top of the oil can will also contain an SAE (Society of Automotive Engineers) designation, which gives the oil's viscosity. A typical designation will be: SAE 10W-30, which means the oil is a "winter" viscosity oil, meaning it will flow and give protection at low temperatures.

On the diesel engine, oil viscosity is critical, because the diesel is much harder to start (due to its higher compression) than a gasoline engine. Obviously, if you fill the crankcase with a very heavy oil during winter (SAE 20W-50, for example), the starter is going to require a lot of current from the battery to turn the engine. And, since batteries don't function well in cold weather in the first place, you may find yourself stranded some morning. Consult the owner's manual for recommended oil specifications for the climate you live in.

LUBE OIL ANALYSIS

From an oil sample a laboratory can diagnose many potential engine problems—from piston wear to impending bearing failure. What's more, the laboratory can spot them quicker, and with greater accuracy. Just as easily, the lab can give the diesel a clean bill of health, saving the car owner unnecessary servicing and other routine preventive maintenance, costly in time and money.

There's nothing new about engine lube oil analysis. Thousands of the nation's trucks and buses regularly have their engine's lube oil analyzed by laboratories specializing in this type of work. What is new is the availability of lube oil analysis to individual vehicle owners rather than, as before, almost exclusively to companies operating fleets of diesel equipment.

Lube oil analysis can be a valuable indicator of internal engine condition.

Here's how lube oil analysis works. You write one of the several laboratories that offer individual diesel vehicle owners lube analysis service. By return mail you'll receive an oil sampling kit. It will probably contain a two-ounce plastic oil sampling container with a screw-on plastic top. Instructions tell you how to take the sample. Usually, a lab-bound sample of diesel lube oil may be taken in any of three ways, but always right after the engine has been shut off, so that the sampled oil is as close as possible to normal engine operating temperature. That's important to assure that the lab's test will be accurate. Oil samples can be taken during normal oil changes, when lube oil is drained anyway. Between oil changes, a sample can be drawn from the engine through the dipstick tube (where you normally check the oil's level). In drawing an oil sample from the dipstick tube, a small suction bulb fitted with a length of disposable tubing is used. The tubing is merely inserted into the dipstick tube, the suction bulb depressed, and the oil sample drawn. The third method of sampling is by loosening the drain plug on the engine's by-pass oil filter (if your diesel has one). A little oil is caught in the lube sampling container. In all cases, extreme cleanliness is a must, so as not to contaminate the sample with dirt, grease, or other substances not actually found inside the engine. For example, using a rag that contains solvents, metal filings, or other impurities can contaminate the oil sample, leading to false and even alarming lab reports. A bit of technique is required: In taking a sample of lube oil during a routine oil drain, about half of the crankcase's lube oil should be allowed to drain out before the sample is taken. The sample taken, the date, make and model of the engine, its mileage, mileage since last oil change, and sometimes oil type are noted on the container's label, and the container is mailed to the laboratory.

Shortly, you'll receive the lab's report, which, based on a number of tests, including spectrochemical analysis (using a spectrometer, which can detect the presence of virtually all basic elements and contaminants), tells what's in the oil in what quantities and analyzes both the probable source of what was found and whether it indicates trouble. For one example, the finding of more than trace amounts of copper in an oil sample may strongly point to excessive bearing wear in a particular diesel whose bearings contain copper. Some analyses report on as many as eighteen basic elements that may be found in a diesel's lube oil sample, and in the report's "recommendation" may pinpoint their probable source—as, "indicates piston ring wear." Also indicated is the presence of such contaminants as water, solids (the products of oxidation and engine blow-by), and fuel dilution. Noted, too, is the lubricity of the sample—whether, or not, in the lab's opinion, it is still doing its internal engine lubricating job.

NOTE: Never use lube analysis and a lab's report of "good oil" to extend, beyond the manufacturer's recommendation, the mileage period between oil changes. Follow the manufacturers recommendations.

The more frequently an engine is lube-sampled, the more accurate and meaningful the lab's reports. Infrequent samplings, although they can spot sudden, unusual changes in internal engine condition, may fail to show the gradual deterioration of engine parts. Ideally, you should have the laboratory analyze a lube sample every other oil change. For most automobile diesels, that's every 6,000 miles. Analysis costs from $7 to $11 per sample. Drive an average 18,000 miles a year and you'd change your diesel's oil three times. In that time, you'd submit three samples to the lab at an annual lube analysis cost of $21 to $33.

Aftermarket Fuel System Accessories

Due to reasons described previously, most diesel engine problems can be attributed to either fuel contamination or cold weather fuel performance characteristics. Diesel-engined vehicle manufacturers have designed and installed various systems to combat these problems, but ultimately, their best efforts are limited by cost.

Inconvenience is a major concern to diesel owners. If water accumulates (in substantial quantities) in the diesel fuel system, the fuel and water must be siphoned from the fuel tank and purged from the remainder of the fuel system. It goes without saying that this operation is a messy, time-consuming process. Even if the vehicle is equipped with a water/fuel separator having a drain valve, the owner must manually open the valve from either under the hood or beneath the vehicle.

Although the fuel filter installed by the manufacturer offers adequate performance when maintained properly, the addition of another, separate diesel fuel filter is a wise improvement.

If you live in an extremely cold climate, you've probably experienced cold starting problems due to fuel "waxing", plugged filters, "gelled" fuel, etc. If your vehicle is not factory-equipped with the optional fuel line or cylinder block heaters, these

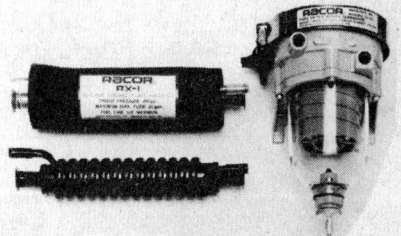

Aftermarket fuel filter/water separator and fuel line heater

heaters can be purchased from the after-market (retail auto parts manufacturers). The installation of either of these items can improve cold-starting dramatically.

WATER/FUEL SEPARATORS

Centrifugal Action

Sometimes referred to as a "cyclonic" water/fuel separator, this device uses baffles which spin the fuel as it comes through the separator inlet. Since water is heavier than diesel fuel, the water will spin away from the fuel, sink to the bottom of the separator, and collect in the sediment bowl.

This type of separator is most efficient in dealing with large water droplets. If the water is in emulsion with the fuel, that is, if the water is equally dispersed through the fuel in very small droplets, some of the water will remain with the fuel to travel through the fuel system.

Coalescing Action

In this type of separator, the fuel must pass through a coalescent filtering media before proceeding through the fuel system. The idea behind the coalescent media is to trap even the smallest droplets of water on the media. As the small droplets combine into larger, heavier droplets, gravity acts on the droplets to pull them downward, off of the media and into the sediment bowl.

FUEL FILTER/SEPARATOR COMBINATION UNITS

Most separators of either the centrifugal or coalescent types are available with disposeable fuel filtering elements which are built into the separator unit. If your car already has a large, disposeable filter, it would probably be more cost-effective to stay with a separator only, and to change the factory-equipped filter at the recommended intervals. Should your vehicle have a fairly small filter, and/or an inconveniently located water drain (or none at all), choose the filter/separator combination. The filter/separator offers both increased fuel filtering ability and efficient water separation.

Convenience Add-Ons

Available with many separators and filter/separators are items such as dash-mounted water-in-fuel indicator lamps, audible water-in-fuel alarms, and dash-controlled water ejection systems. A properly chosen system would warn you of water in the fuel, and allow you to eject the water by simply "flipping" a dash-mounted switch.

Installing a Separator

Clear installation instructions and the necessary installation parts will be provided with the separator kit. Follow those instructions exactly. A general list of suggestions follows:

1. Fuel additives should not be used unless approved by the separator manufacturer.
2. Do not install a separator within 4″ of any exhaust system component.
3. If plastic fittings are supplied with the kit, do not replace them with metal fittings. Also, use extreme caution when tightening the fittings, especially those made of plastic.
4. Use a fuel-proof sealer on all fitting threads, only if the threads are not factory-coated with sealer.
5. Use only fuel-proof hoses for the installation.
6. Do not eliminate the original equipment fuel filter, even if a filter/separator is installed.
7. For new car warranty purposes, a filter/separator should be located BEFORE the original equipment filter. The fuel must pass through the original filter last, before entering the fuel injection pump.
8. If any type of fuel line heater is installed, it is best to position the heater between the fuel tank and the separator inlet.
9. To ease the job of the separator, the separator should be installed between the fuel transfer pump and the tank (unless the separator manufacturer specifies otherwise). Fuel and water which have been churned through the fuel transfer pump will be more difficult to separate.
10. Be sure that any wiring (for warning lamps, water ejection, etc.) is routed and connected properly. If the wiring must pass through a drilled hole, be sure to use a rubber grommet between the drilled component(s) and the wire to prevent damage to the wire.

FUEL LINE HEATERS

Two popular types of fuel line heaters are available for diesel passenger cars. Both types raise the temperature of the fuel to prevent "waxing" and "gelling" of the fuel in the lines during cold weather operation. One type uses engine coolant as a heating source. In order for this type to heat the fuel, the engine must first be started and allowed to run until the coolant temperature increases. Though this type of heater will usually increase fuel mileage, it offers no aid in starting ability.

The other type of heater uses a 12V DC electric heating element. This type is recommended, due to its ability to warm the fuel BEFORE the engine is started. This type of heater will also usually increase the overall fuel mileage.

Installation

Follow the manufacturer's instructions exactly. Also, see suggestions 5, 8, and 9 under "Separator Installation".

CYLINDER BLOCK HEATERS

A cylinder block heater electrically (usually 110V house current) heats the engine coolant, which in turn warms the cylinder block, heads, and engine oil. In this case, the warmth is not used to alter the characteristics of the fuel. Block heaters offer two main advantages when starting a diesel in cold weather:

1. The reduced viscosity (thinning) of the engine oil from the warmth allows the engine to be "turned over" easier (and faster) by the starter. Less strain is imposed on the starting system.
2. Because the diesel relies on the heat of compression to ignite the fuel, the increase in the base combustion chamber temperature results in a higher tempearture during compression. This allows the fuel to ignite easier than if just the glow plugs were used.

Installation

Most cylinder block heaters replace one of the existing freeze (or expansion) plugs of the cylinder block. Follow the manufacturers installation instructions exactly. Also, refer to the manufacturers recommendations for usage.

CATERPILLAR DIESEL ENGINES
3208, 3306, 3406 and 3408 SERIES

GENERAL ENGINE SPECIFICATIONS

Engine Series	No. of Cylinders	Cu. In. Disp.	Bore × Stroke (in.)	Max. Horsepower	Torque SAE @ rpm	Firing Order Right Hand Rotation	Low Idle rpm	Governor Limit rpm	Compression Ratio
3208	8	636	4.5 × 5.0	160 @ 2800	365 @ 1400	1-2-7-3-4-5-6-8	650	3030	16.4:1
3208②	8	636	4.5 × 5.0	165 @ 2600	398 @ 1300	1-2-7-3-4-5-6-8	650	3030	16.4:1
3208③	8	636	4.5 × 5.0	175 @ 2800	425 @ 1400	1-2-7-3-4-5-6-8	650	3030	16.4:1
3208	8	636	4.5 × 5.0	175 @ 2800	400 @ 1400	1-2-7-3-4-5-6-8	650	3030	16.4:1
3208②	8	636	4.5 × 5.0	185 @ 2600	452 @ 1400	1-2-7-3-4-5-6-8	650	3030	16.4:1
3208③	8	636	4.5 × 5.0	200 @ 2800	490 @ 1400	1-2-7-3-4-5-6-8	650	3070	16.4:1
3208	8	636	4.5 × 5.0	210 @ 2800	485 @ 1400	1-2-7-3-4-5-6-8	650	3040	16.4:1
3208②	8	636	4.5 × 5.0	225 @ 2600	560 @ 1400	1-2-7-3-4-5-6-8	650	2800	16.4:1
3208	8	636	4.5 × 5.0	250 @ 2600	610 @ 1400	1-2-7-3-4-5-6-8	650	2825	16.4:1
3208	8	636	4.5 × 5.0	250 @ 2800	610 @ 1400	1-2-7-3-4-5-6-8	650	3030	16.4:1
3306	6	638	4.75 × 6.0	245 @ 2100	860 @ 1350	1-5-3-6-2-4	600	2100	15.0:1
3306③	6	638	4.75 × 6.0	245 @ 2200	820 @ 1300	1-5-3-6-2-4	650	2200	17.5:1
3306	6	638	4.75 × 6.0	250 @ 1800	860 @ 1350	1-5-3-6-2-4	600	1800	15.0:1
3306	6	638	4.75 × 6.0	260 @ 1900	860 @ 1350	1-5-3-6-2-4	600	1900	15.0:1
3306	6	638	4.75 × 6.0	270 @ 2000	860 @ 1350	1-5-3-6-2-4	600	2000	15.0:1
3306③	6	638	4.75 × 6.0	270 @ 2200	775 @ 1400	1-5-3-6-2-4	650	2200	17.5:1
3406	6	893	5.4 × 6.5	350 @ 1800	1200 @ 1200	1-5-3-6-2-4	750	1800	14.5:1
3406③	6	893	5.4 × 6.5	350 @ 1800	1225 @ 1200	1-5-3-6-2-4	750	1800	14.5:1
3406	6	893	5.4 × 6.5	350 @ 2100	1200 @ 1200	1-5-3-6-2-4	750	2100	14.5:1
3406	6	893	5.4 × 6.5	380 @ 2100	1245 @ 1200	1-5-3-6-2-4	750	2100	14.5:1
3406	6	893	5.4 × 6.5	400 @ 2100	1265 @ 1300	1-5-3-6-2-4	750	2100	14.5:1
3406	6	893	5.4 × 6.5	280 @ 2100	1015 @ 1200	1-5-3-6-2-4	750	2100	14.5:1
3406	6	893	5.4 × 6.5	300 @ 2100	1054 @ 1200	1-5-3-6-2-4	750	2100	14.5:1
3406	6	893	5.4 × 6.5	325 @ 2100	1050 @ 1200	1-5-3-6-2-4	750	2100	14.5:1
3406③	6	893	5.4 × 6.5	380 @ 2100	1285 @ 1200	1-5-3-6-2-4	750	2100	14.5:1
3406	6	893	5.4 × 6.5	290 @ 1800	1000 @ 1200	1-5-3-6-2-4	750	1800	14.5:1
3406③④	6	893	5.4 × 6.5	300 @ 1900	1054 @ 1200	1-5-3-6-2-4	750	1900	14.5:1
3406④	6	893	5.4 × 6.5	305 @ 1900	1050 @ 1200	1-5-3-6-2-4	750	①	14.5:1
3406④	6	893	5.4 × 6.5	350 @ 1900	1165 @ 1200	1-5-3-6-2-4	750	1900	14.5:1
3406	6	893	5.4 × 6.5	375 @ 2100	1145 @ 1400	1-5-3-6-2-4	750	①	14.5:1
3406⑤	6	893	5.4 × 6.5	250 @ 1600	1000 @ 1200	1-5-3-6-2-4	750	1600	14.5:1
3406⑤	6	893	5.4 × 6.5	285 @ 1600	1090 @ 1200	1-5-3-6-2-4	750	1600	14.5:1
3406⑤	6	893	5.4 × 6.5	300 @ 1600	1200 @ 1200	1-5-3-6-2-4	750	1600	14.5:1
3406⑤	6	893	5.4 × 6.5	330 @ 1600	1320 @ 1200	1-5-3-6-2-4	750	1600	14.5:1
3408③	V-8	1099	5.4 × 6.0	450 @ 2100	1350 @ 1500	1-8-4-3-6-5-7-2	600	2100	15.3:1
3408	V-8	1099	5.4 × 6.0	450 @ 2100	1460 @ 1200	1-8-4-3-6-5-7-2	700	2100	14.5:1
3408④	V-8	1099	5.4 × 6.0	420 @ 1900	1460 @ 1200	1-8-4-3-6-5-7-2	650	3030	15.5:1

① 280PC—2260   325PC—2285   ③ California use only
  280DI—2300   360PC—2300   ④ Economy (1900 rpm)
② On-highway ratings   ⑤ Economy (1600 rpm)

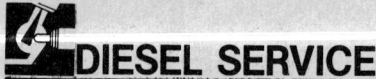

TUNE-UP SPECIFICATIONS

Engine Series	Valve Clearance		Injection Pump Timing (deg.)	Injection Nozzle Pressure (psi)	Idle Speed (rpm)	Maximum No-Load Speed (rpm)
	Intake (in.)	Exhaust (in.)				
3208	.015	.025	16B	2750–2900	650	①
3306	.015	.025	17–19B	1200–2350	600–650	①
3406	.015	.030	②	③	750	①
3408	.015	.030	④	③	600–700	①

① See General Engine Specifications Chart
② Precombustion Chamber Type-All—9–11°BTDC
Direct Injection Type
Pump No. 9–49—28–30° BTDC
 9–798—27–29° BTDC
 9–2626—27–29° BTDC
 9–512 thru 9–215—27–29° BTDC
 9–3384—27–29° BTDC
 9–3702—27–29° BTDC
 9–3746—27–29° BTDC
 9–3839—27–29° BTDC
 9–3939—27–29° BTDC
 9–5435—27–29° BTDC
 9–3702—25.5–27.5° BTDC
 9–3839—21.5–23.5° BTDC

③ Precombustion Chamber Type: 400–750 psi
Direct Injection Type: 2400–3100 psi
④ Precombustion Chamber Type: 10–12° BTDC
Direct Injection Type: 27–29° BTDC

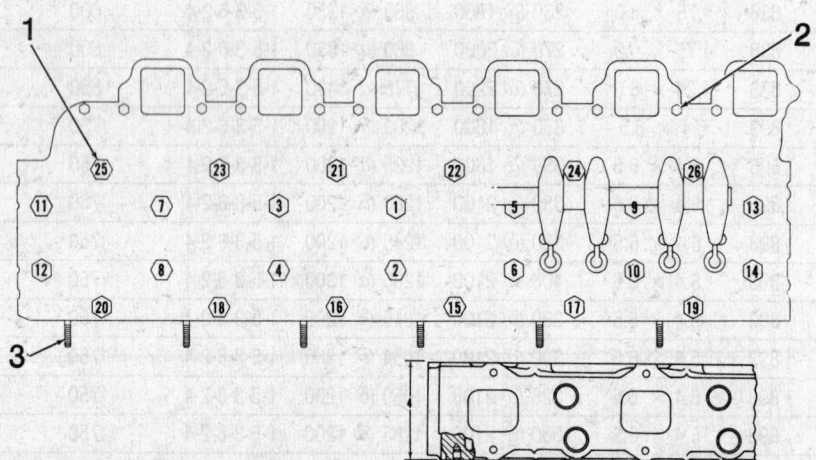

① Large bolts (3/4 inch). Put engine oil on all bolt threads and tighten the bolts in the following sequence:
Step 1: Tighten bolts from 1 to 20 in number sequence to 200 ± 20 ft. lbs. (270 ± 25 N·m)
Step 2: Tighten bolts from 1 to 20 in number sequence to 330 ± 15 ft. lbs. (450 ± 20 N·m)
Step 3: Tighten bolts from 1 to 20 in number sequence again to 330 ± 15 ft. lbs. (450 ± 20 N·m)
Step 4: Install the rocker arm shafts (1) for the engine valves.
Step 5: Tighten bolts from 21 to 26 in number sequence to 200 ± 20 ft. lbs. (270 ± 25 N·m)
Step 6: Tighten bolts from 21 to 26 in number sequence to 330 ± 15 ft. lbs. (450 ± 20 N·m)
Step 7: Tighten bolts from 21 to 26 in number sequence again to 330 ± 15 ft. lbs. (450 ± 20 N·m)
Step 8: Tighten the twelve small bolts (2) to 32 ± 5 ft. lbs. (43 ± 7 N·m)
② Small bolts (3/8 inch). See step 8.
③ Torque for twelve studs in cylinder head 20 ± 3 ft. lbs. (25 ± 4 N·m)
④ Height of cylinder head (new) 4.440 ± .010 in. (112.78 ± 0.25 mm)

Head bolt torque sequence—3406 series

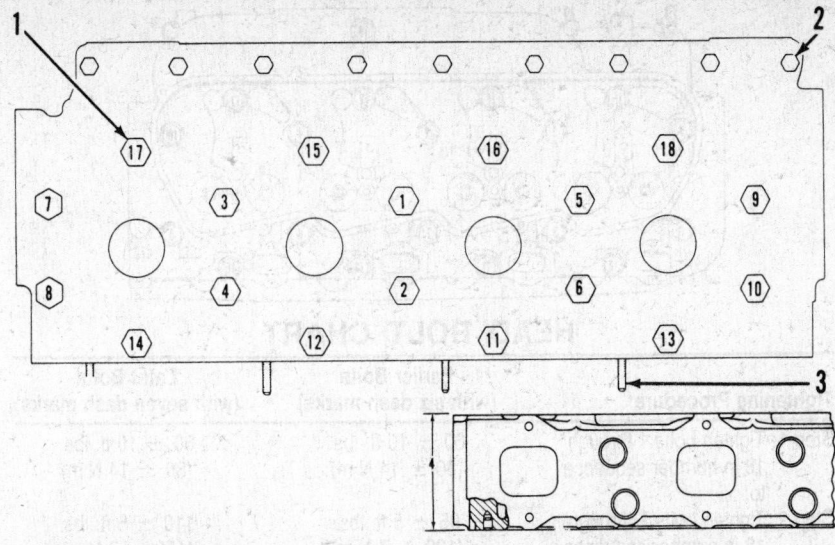

① Large bolts (3/4 inch). Put engine oil on all bolt threads and tighten the bolts in the following step sequence:

Step 1: Tighten bolts from 1 to 14 in number sequence to 200 ± 20 ft. lbs. (270 ± 25 N·m)

Step 2: Tighten bolts from 1 to 14 in number sequence to 330 ± 15 ft. lbs. (450 ± 20 N·m)

Step 3: Tighten bolts from 1 to 14 in number sequence again to 330 ± 15 ft. lbs. (450 ± 20 N·m)

Step 4: Install the rocker arms for the engine valves.

Step 5: Tighten bolts from 15 to 18 in number sequence to 200 ± 20 ft. lbs. (270 ± 25 N·m)

Step 6: Tighten bolts from 15 to 18 in number sequence to 330 ± 15 ft. lbs. (450 ± 20 N·m)

Step 7: Tighten bolts from 15 to 18 in number sequence again to 330 ± 15 ft. lbs. (450 ± 20 N·m)

Step 8: Tighten the nine small bolts (2) to 32 ± 5 ft. lbs. (43 ± 7 N·m)

② Small bolts (3/8 inch). See step 8.

③ Torque for eight studs in each cylinder head 20 ± 3 ft. lbs. (25 ± 4 N·m)

④ Height of cylinder head (new) 4.440 ± .010 in. (112.78 ± 0.25 mm)

Head bolt torque sequence—3408 series

ENGINE TORQUE SPECIFICATIONS

(All measurements in ft. lbs.)

Engine Series	Cylinder Head	Intake Manifold	Exhaust Manifold	Conn. Rod	Main Bearing Caps	Damper to Crankshaft Bolt	Flywheel to Crankshaft	Oil Pump Mounting Bolts	Oil Pan Bolts
3208	①	—	32	30 ②	30 ④	460	55	18	17
3306	⑤	—	32–38	27–33 ③	27–33 ③	210–250	130–170	—	—
3406	315–345	—	33–43	54–66 ④	180–200 ④	—	180–220	—	—
3408	315–345	—	33–43	54–66 ④	180–200 ④	—	180–220	—	—

NOTE: On 3208 engines the early head bolts are identified by six marks; later bolts have seven marks.

① Large bolts (early): 60 then 95 ft. lbs.
  Large bolts (late): 60 then 110 ft. lbs.
  Small bolts: 32 ft. lbs.
② Plus 60°
③ Plus 90°
④ Plus 120°
⑤ Numbered bolts: 115 then 172–198
  Lettered bolts: 27–37

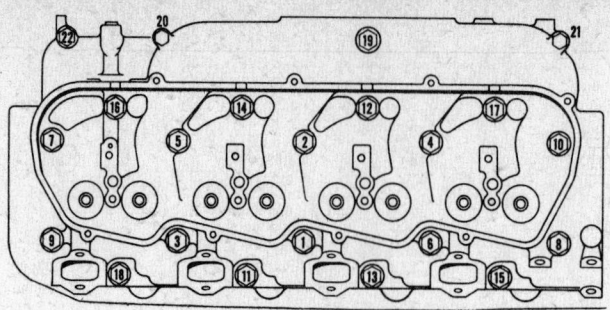

HEAD BOLT CHART

Tightening Procedure	Earlier Bolts (with six dash marks)	Later Bolts (with seven dash marks)
Step 1: Tighten bolts 1 through 18 in number sequence to:	60 ± 10 ft. lbs. (80 ± 14 N·m)	60 ± 10 ft. lbs. (80 ± 14 N·m)
Step 2: Tighten bolts 1 through 18 in number sequence to:	95 ± 5 ft. lbs. (130 ± 7 N·m)	110 ± 5 ft. lbs. (150 ± 7 N·m)
Step 3: Again tighten bolts 1 through 18 in number sequence to:	95 ± 5 ft. lbs. (130 ± 7 N·m)	110 ± 5 ft. lbs. (150 ± 7 N·m)
Torque for head bolts 19 through 22 (tighten in number sequence to)	32 ± 5 ft. lbs. 43 ± 7 N·m	

Cylinder head bolt torque sequence— 3208 series

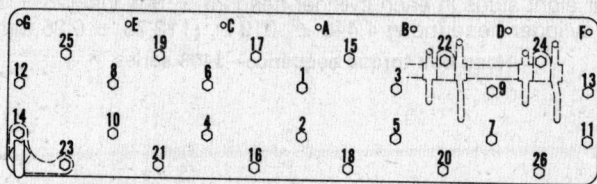

Step 1. Tighten all bolts in number sequence to 115 ft. lbs. (155 Nm)

Step 2. Tighten all bolts in number sequence to 172 ft. lbs. (233 Nm)

Step 3. Tighten all bolts in number sequence to 198 ft. lbs. (267 Nm)

Step 4. Tighten all bolts in letter sequence to 37 ft. lbs. (50 Nm)

Cylinder head bolt torque sequence—3306 series

CHILTON'S THREE "C's" DIESEL ENGINE DIAGNOSIS PROCEDURE

Condition	Cause	Correction
Hard Starting—Engine Turns Freely	1. Cold outside temperatures. 2. No fuel getting to engine-clogged filters, empty tank, plugged or kinked lines. 3. Fuel shut-off solenoid is not energized on 1100 engines, is energized on 1674 engines. 4. Fuel shut-off solenoid is stuck. 5. Fuel transfer pump delivering less than 10 psi. ② 6. Injection timing incorrect. 7. Fuel injection pump drive slipping (on 1100 series only). 8. Incorrect valve adjustment. 9. Defective fuel nozzle. 10. Restriction in exhaust system. 11. Air or water in fuel system.	1. Install block heater. 2. Replace fuel filter, refill fuel tank or clear fuel lines as necessary. 3. Check solenoid operation and repair or replace as necessary. 4. Replace fuel shut-off solenoid. 5. Check fuel pressure at transfer pump. 6. Reset injection timing. 7. Replace injection pump drive. 8. Check and adjust valve clearance. 9. Replace fuel injector. 10. Repair as necessary. 11. Drain fuel system and flush or bleed.
Hard Starting—Engine Will Not Turn or Turns Slowly	1. Battery voltage too low. 2. Defective cable or connection from battery to starter. 3. Defective starter or solenoid. 4. Interior engine condition preventing engine from turning. 5. Transmission or power take-off (PTO) problem, preventing engine from turning.	1. Recharge or replace battery. 2. Repair connection or replace cable. 3. Replace starter or solenoid. 4. Check for seized engine components or hydrostatic lock. 5. Check and repair as necessary.
Engine Misfires ①	1. Fuel injection lines installed in improper firing order. 2. Defective injection nozzles or pump. 3. Incorrect valve lash. 4. Sticking valves. 5. Incorrect injection timing. 6. Fuel transfer pump delivering less than 10 psi. 7. Faulty high pressure fuel line. 8. Poor compression. 9. Air or water in fuel system. 10. Malfunctioning automatic timing advance. 11. Defective head gasket. 12. Fuel leakage from line nut or line adapter. 13. Clogged fuel filter caused by wax build-up in fuel.	1. Check injection line routing to injection pump. 2. Replace injection nozzles or pump assembly. 3. Adjust valve clearance. 4. Overhaul as necessary. 5. Reset injection timing. 6. Check transfer pump pressure. 7. Replace fuel line. 8. Check compression pressures. 9. Bleed or drain as necessary. 10. Replace automatic timer or refer pump to specialist for calibration. 11. Replace head gasket. 12. Tighten fuel connection or replace. 13. Replace fuel filter and install fuel line heater to prevent fuel waxing.
Excessive Black or Gray Smoke (ENGINE RUNS SMOOTH) ②	1. Insufficient combustion air, clogged air cleaner or manifold, malfunctioning turbocharger. 2. Defective fuel nozzle. 3. Incorrect injection timing. 4. Incorrect fuel ratio control setting or injector rack setting. 5. Restriction in exhaust system. 6. Valve leakage or incorrect adjustment.	1. Replace air filter element. Check turbocharger operation. 2. Clean or replace fuel nozzle. 3. Reset injection timing. 4. Remove injection pump and calibrate on test stand. 5. Repair as necessary. 6. Adjust valve lash or overhaul as required.

CHILTON'S THREE "C's" DIESEL ENGINE DIAGNOSIS PROCEDURE

Condition	Cause	Correction
(ENGINE RUNS ROUGH)	1. Engine misfire. 2. Air in fuel system. 3. Malfunctioning automatic timing advance. 4. Incorrect injection timing.	1. Follow procedure outlined above. 2. Bleed air from fuel system. 3. Replace automatic timer assembly. 4. Reset injection timing.
Excessive White or Blue Smoke	1. Cold outside temperatures. 2. Extended idling period. 3. High crankcase oil level. 4. Worn piston rings or cylinder liners. 5. Engine misfire. 6. Incorrect injection timing. 7. Incorrect valve adjustment. 8. Malfunctioning automatic timing advance 9. Defective fuel nozzle.	1. Reset fuel ratio on injector rack. 2. Increase engine rpm periodically. 3. Correct oil level. 4. Replace piston rings or cylinder liners. 5. Follow procedure outlined above. 6. Reset injection timing. 7. Adjust valve lash. 8. Replace automatic timer. 9. Clean or replace fuel nozzle.
Excessive Oil Consumption	1. Worn valve guides. 2. Scored rings or liners. 3. High crankcase oil level. 4. Engine misfire.	1. Replace valve guides. 2. Replace rings or cylinder liner. 3. Correct crankcase oil level. 4. Follow procedure outlined above.
Lack of Power	1. Improperly adjusted accelerator linkage. 2. Failed fuel nozzle. ① 3. Improper grade of fuel. 4. Turbocharger clogged or dragging. 5. Air induction leaks. ③ 6. Improper injection timing. 7. Excessive valve lash. 8. Fuel supply pressure less than 10 psi. ② 9. Faulty timing advance.	1. Adjust accelerator linkage. 2. Clean or replace fuel nozzle. 3. Drain and refill fuel tank. 4. Check turbocharger operation. 5. Correct as necessary. 6. Reset injection timing. 7. Adjust valves. 8. Check fuel supply pump pressure. 9. Check automatic timer operation.
Low Oil Pressure	1. Lubricating oil diluted with fuel. 2. Excessive clearances in bearings in crankcase or timing gears. 3. Defective oil pump or relief valve. 4. Crankcase oil level excessive. 5. Oil temperature too high—faulty cooler or cooler relief valve.	1. Drain and refill crankcase. 2. Replace bearings, check gear backlash. 3. Replace oil pump or relief valve. 4. Correct oil level. 5. Check oil cooler and lines.
Coolant Temperature Too High	1. Combustion gases leaking into coolant. 2. Low coolant level. 3. Faulty water pump. 4. Faulty thermostat. 5. Injection timing incorrect. 6. Pinched shunt line (1100 series).	1. Replace head gasket and check for cracks in block or head. 2. Top up cooling system. 3. Replace water pump. 4. Replace thermostat. 5. Reset injection timing. 6. Replace shunt line.

① When trouble shooting for misfiring, operate the engine at the speed where misfire is most noticeable. Loosen each injector pump fuel nut, one at a time. The cylinder or cylinders which effect performance the least should be tested for a defective fuel nozzle or pump first.

② Fuel transfer pump output should be 10–20 psi at cranking speed. At full load, the rating is 25 psi, and at high idle, 30 psi.

③ In testing for poor performance or smoke, remember that they can result from either fuel system defects or engine mechanical defects. An air intake restriction across the air cleaner of 30 in. of water, or an exhaust system restriction of more than 15 in. of water can cause smoke and poor performance because of lack of oxygen and compression. If fuel system inspection reveals no problem on cylinders that misfire, a compression test and measurement of intake and exhaust manifold pressures should be the next step.

TUNE-UP AND ADJUSTMENTS

Every engine tune-up should include the following checks and/or adjustments.
1. Compression test (if applicable).
2. Valve clearance adjustment.
3. Fuel injection pump timing.
4. Fuel rack settings (if necessary).
5. Engine idle speed setting.
6. Maximum no-load speed setting (governor adjustment).
7. Glow plug system check (if equipped).

NOTE: Efficiency of emission controls and engine performance depends on adherence to proper operation and maintenance recommendations and use of recommended fuels and lubrication oils.

Locating Top Dead Center/Compression Position for No. 1 Piston

3208 SERIES

1. Remove the plug from timing hole in the front cover. Put timing bolt in timing hole.
2. Turn the crankshaft clockwise (as seen from front of engine) until bolt will go into the hole in the drive gear for the camshaft.
3. Remove the valve cover on the right side of the engine (as seen from rear of engine). The two valves at the right front of the engine are the intake and exhaust valves for No. 1 cylinder.
4. The intake and exhaust valves for No. 1 cylinder must now be closed and the timing pointer will be in alignment with the TDC-1 on the damper assembly. The No. 1 piston is now at top center on the compression stroke.

3306 SERIES

NOTE: No. 1 piston at top dead center (TDC) on its compression stroke is the starting point for all timing procedures.

1. Remove the starter motor and install the engine turning tool group or equivalent.
2. Remove the valve cover.
3. Rotate the engine clockwise (as viewed from the flywheel end) 30 degrees to remove the free play from the timing gears.
4. Remove the plug from the timing hole in the flywheel housing. Rotate the crankshaft until a ⅜ in.-16NC bolt, 2.0 in. (50.8mm) long can be threaded into the flywheel through the timing hole in the flywheel housing. No. 1 piston is now at

TDC. If the crankshaft is turned beyond TDC, repeat Steps 3 and 4 again.
5. The intake and exhaust valves for No. 1 cylinder should be closed. Check by moving the rocker arms by hand slightly up and down.
6. If the No. 1 piston is not on its compression stroke, remove the timing bolt, rotate the crankshaft counterclockwise 360° and install the timing bolt as previously described.

3406 SERIES

No. 1 piston at top dead center on the compression stroke is the starting point of all timing procedures. The timing bolt is kept in storage on the engine and can be installed in either the right or left side of the engine.

NOTE: There are two threaded holes in the flywheel which are in alignment with removal plugged holes in the left and right front of the flywheel housing. The two holes in the flywheel are at different distances from the center of the flywheel so that the timing bolt cannot be put in the wrong hole.

1. Remove the bolts and covers from the flywheel housing.
2. Install the engine turning tool or equivalent, into the flywheel housing, until the shoulder of the tool is against the housing. Attach a turning handle to the flywheel rotator and turn the flywheel while holding the timing bolt in either hole in the housing. Stop the rotation when the timing bolt can be installed in the threaded hole of the flywheel.
3. To be assured that the No. 1 piston is at TDC on its compression stroke, remove the valve rocker arm cover. The No. 1 cylinder valves should be closed and the rocker arms should be movable up and down, with hand pressure.
4. If the No. 1 cylinder is not on its compression stroke, rotate the crankshaft 360° and install the timing bolt.
5. If the flywheel was not turned in the direction of normal engine rotation, or was turned past the timing hole, turn the flywheel clockwise (opposite the direction of normal engine rotation) approximately 30 degrees. This operation is to be sure that play is removed from the timing gears when the engine is put on top dead center.

NOTE: The engine is observed from the flywheel end when the direction of rotation is given.

6. Turn the flywheel counterclockwise until the hole in the flywheel is aligned with the timing bolt. When the timing bolt can be turned freely in the threaded hole in the flywheel, the No. 1 piston is on top dead center.
7. If the hole in the flywheel is turned beyond the hole in the flywheel housing, turn the flywheel back (clockwise) a minimum of 30 degrees beyond the hole in the flywheel Then turn the flywheel counterclockwise towards the housing hole again.

3408 SERIES

The No. 1 piston at top dead center on the compression stroke is the starting point for all timing procedures. By referring to the 3406 procedure to locate the TDC position of No. 1 cylinder piston and removing the left rocker arm cover from the engine to observe the valve action, the position of the No. 1 piston on its compression stroke can be ascertained.

Compression Test

PRECOMBUSTION CHAMBER ENGINES ONLY

1. Remove the fuel injection nozzle of the cylinder to be checked, leaving the precombustion chamber in place.
2. Rotate the crankshaft until the piston of the cylinder to be tested is at TDC on the compression stroke.
3. Adapt an air-pressure hose to the precombustion chamber, using a threaded fitting or rubber adapter.
4. Apply approximately 100 psi pressure and listen for air leakage at the air cleaner inlet, exhaust outlet and crankcase breather. On turbocharged engines, it may be necessary to remove the air inlet and exhaust outlet connections to detect leakage. Air through the air intake indicates a leaky intake valve, air through the exhaust indicates a leaky exhaust valve and air through the crankcase breather indicates problems with pistons, rings, or liners. If there is valve leakage, check the valve clearance.

Valve Adjustment

3208 SERIES

1. Remove valve covers.
2. Rotate the crankshaft in a clockwise direction until the piston of No. 1 cylinder is at TDC on the compression stroke. The TDC-1 mark on the damper or pulley will align with the timing pointer.
3. Adjust the lash for intake and exhaust valves of cylinders 1 and 2. Adjust the inlet valve for 0.015 in. lash and the exhaust for 0.025 in. lash. Lash adjustment is accomplished by loosening the locknut, adjusting the screw to the dimension of the feeler gauge and then holding the screw while retightening the locknut to 19–25 ft. lbs. Recheck the adjustment to ensure it was not disturbed during the tightening of the locknut.
4. Rotate the crankshaft 180° clockwise, so the VS mark will align with the pointer. Adjust lash for all valves of cylinders 7 and 3.
5. Rotate the crankshaft 180° clockwise again, until the TDC mark on the damper or pulley aligns with the pointer. Adjust the lash of all valves on cylinders 4 and 5.
6. Again rotate the crankshaft 180° clockwise (the VS mark will again align on engines so equipped). Adjust the lash for cylinders 6 and 8.
7. Replace the valve covers.

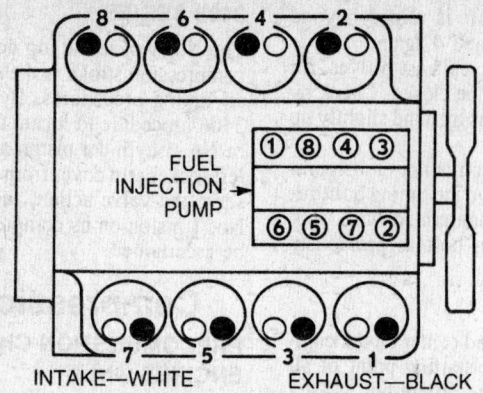

Valve and injector pump location—3208 series

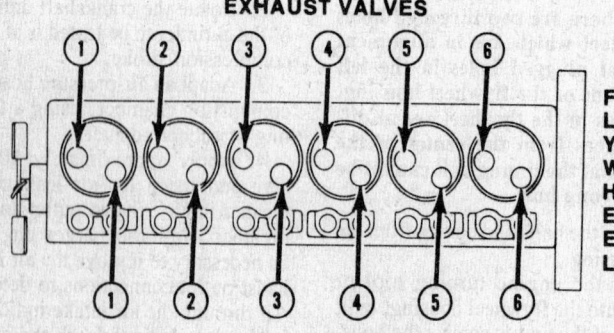

Cylinder and valve location—3306 series

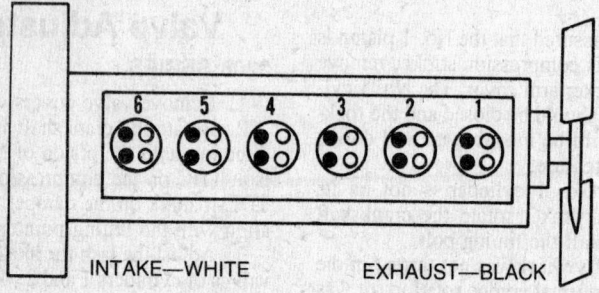

Cylinder and valve location—3406 series

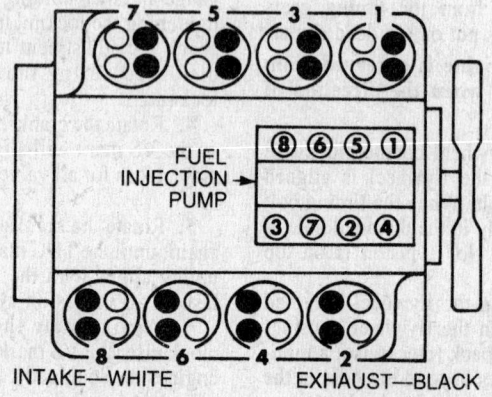

Valve and injector pump location—3408 series

3306 SERIES

Valve clearance is measured between the rocker arm and the valve stem. When the valve clearance is checked, adjustment is not necessary if the measurement is within the range specified in the tune-up chart.

1. Set the No. 1 cylinder at TDC on its compression stroke.
2. Remove the valve cover.
3. Check the valve clearance on the intake valves for cylinders 1, 2 and 4 and adjust if necessary. Check the clearance on the exhaust valves for cylinders 1, 3 and 5 and adjust if necessary.
4. Rotate the crankshaft (or flywheel) 360° (one revolution) in the normal direction of engine rotation.
5. Check the valve clearance on the intake valves for cylinders 3, 5 and 6 and adjust if necessary. Check the clearance on the exhaust valves for cylinders 2, 4 and 6 and adjust if necessary.
6. After all adjustments are complete, tighten the adjustment locknuts to 19–25 ft. lbs. (24–32 Nm) and install the valve cover using a new gasket.

3406 AND 3408 SERIES ENGINES

1. Set No. 1 piston at TDC on the compression stroke.
2. Adjust valve clearance on intake valves (0.015 in.) for cylinders 1, 2 and 4 (cyl. 1, 2 and 7 on 3408 engines). Adjust clearance on exhaust valves (0.030 in.) for cylinders 1, 3 and 5 (1, 3, 4 and 8 on 3408 engines).
3. Following each adjustment, tighten the valve adjustment screw nut to 22 ft. lbs. Recheck adjustment.
4. Remove the timing bolt and rotate the flywheel 360° in the direction of engine rotation. This will align No. 6 piston at TDC on the compression stroke. Replace the timing bolt in the flywheel.
5. Adjust valve clearance on intake valves (0.015 in.) for cylinders 3, 5 and 6 (3, 4, 6 and 8 on 3408 engines). Adjust valve clearance on exhaust valves (0.030 in.) for cylinders 2, 4 and 6 (2, 5, 6 and 7 on 3408 engines).
6. Repeat Step 3.
7. Remove the timing bolt when adjustments are complete.

Bridge Adjustment
3406 AND 3408 SERIES

NOTE: Valves must be fully closed for this procedure.

1. Loosen the adjustment screw locknut and turn adjustment screw out several turns.
2. Keep the bridge in contact with the valve stem opposite the adjustment screw by hand during adjustment.
3. Turn the adjustment screw clockwise until contact with the valve stem is made. Turn the screw another 30 degrees.
4. Tighten the locknut to 22 ft. lbs. (28 Nm).

Injection Pump Timing

3208 SERIES

The timing of the fuel injection pump can be changed to make compensation for movement in the taper sleeve drive or worn timing gears. The timing can be checked and, if necessary, changed using the following method.

1. Remove bolt from the timing pin hole.

2. Turn the crankshaft clockwise (as seen from front of engine) until timing pin goes into the notch in the camshaft for the fuel injection pumps.

3. Remove the plug from timing hole in the front cover. Put timing bolt through the front cover and into the hole with threads in the timing gear.

4. If the timing pin is in the notch in the camshaft for the fuel injection pumps and bolt goes into the hole in the timing gear through timing hole the timing of the fuel injection pump is correct.

5. If bolt does not go in the hole in the timing gear with timing pin in the notch in the camshaft, use the following procedure.

   a. Remove nuts and the cover for the tachometer drive assembly.

   b. Remove the tachometer drive shaft and washer from the camshaft for the fuel injection pumps. Tachometer drive shaft and washer are removed as an assembly.

   c. Put puller on the camshaft for the fuel injection pumps. Tighten bolts until the drive gear on the camshaft for the fuel injection pumps comes loose.

   d. Remove the puller.

   e. Turn the crankshaft clockwise (as seen from front of engine) until timing bolt goes into the hole in the timing gear. With timing pin in the notch in the camshaft for the fuel injection pumps and bolt in the hole in the timing gear, the timing for the engine is correct.

   f. Install washer and tachometer drive shaft. Tighten tachometer drive shaft to:

     Earlier: 80 ± 5 lb. ft. (110 ± 7 N;pdm)

     Later: 110 ± 10 lb. ft. (150 ± 14 N;pdm)

Then remove timing pin.

   g. Turn the crankshaft two complete revolutions clockwise (as seen from front of engine) and put timing pin and bolt in again. If timing pin and bolt can not be installed do Steps a through f again.

   h. Remove bolt from the timing gear and install in holding hole. Install the plug in timing hole. Remove timing pin and install bolt. Install cover for the tachometer drive assembly.

3306 SERIES

1. Set No. 1 piston at TDC on its compression stroke.

2. Remove the timing bolt and rotate the crankshaft 30° clockwise.

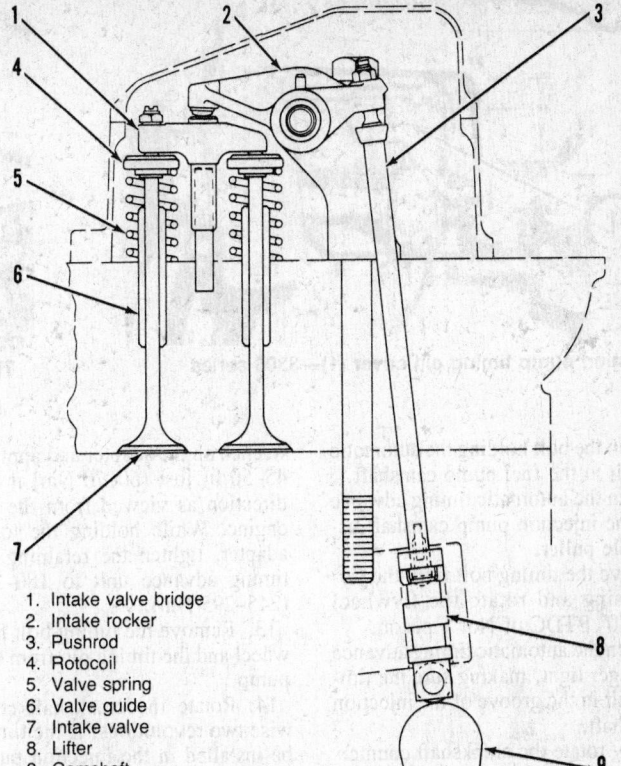

1. Intake valve bridge
2. Intake rocker
3. Pushrod
4. Rotocoil
5. Valve spring
6. Valve guide
7. Intake valve
8. Lifter
9. Camshaft

Typical bridged valve mechanism—intake valve illustrated, exhaust valve similar—3406 and 3408 series

3. Remove the timing pin cover from the side of the fuel injection pump housing.

4. Install timing pin 7N1048 or equivalent into injection pump housing, then slowly rotate the crankshaft counterclockwise until the timing pin slips into the groove in the injection pump camshaft.

5. Place the timing bolt into the timing hole in the flywheel housing. If the bolt can be hand-threaded into the timing hole in the flywheel, the injection timing is correct. If the bolt cannot be easily installed, the injection timing will have to be reset as follows:

6. Remove the mounting nuts and cover from the timing gear housing.

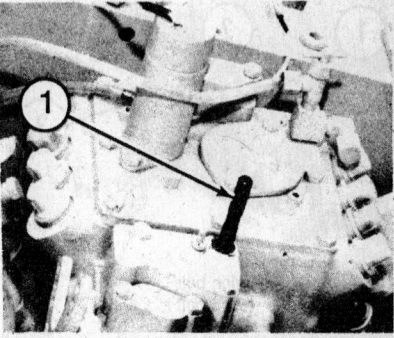

1. Timing pin

Timing pin installed in injector housing —3208 series

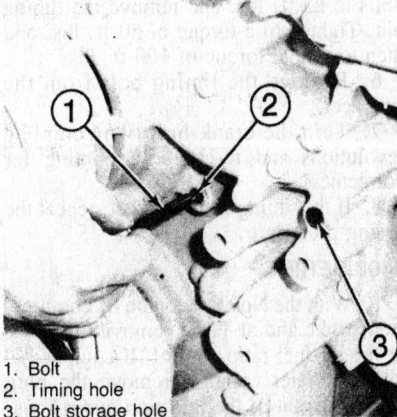

1. Bolt
2. Timing hole
3. Bolt storage hole

Installing timing bolt—3208 series

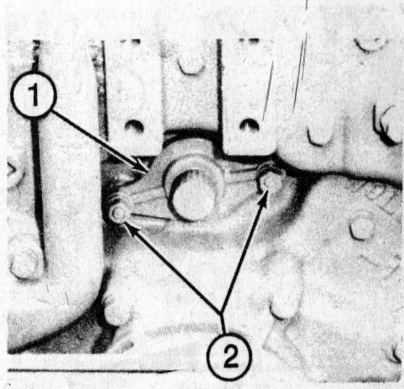

1. Cover   2. Nuts

Location of tachometer drive assembly

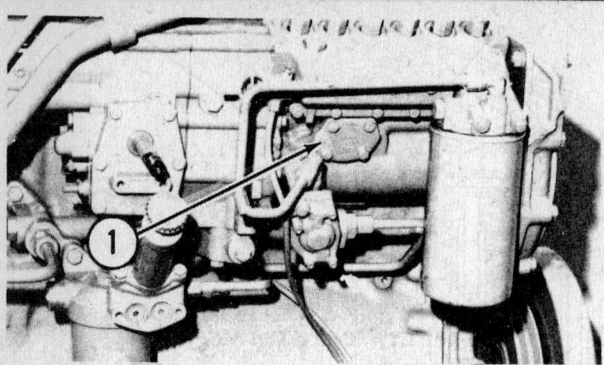

Injection pump timing pin cover (1)—3306 series

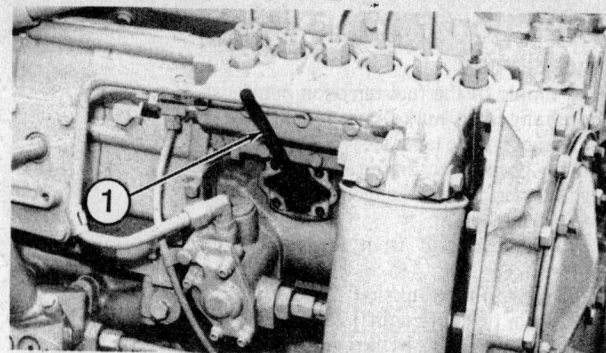

Timing pin (1) installed in injection pump—3306 series

7. Loosen the bolt holding the automatic advance unit to the fuel pump camshaft.

8. Loosen the automatic timing advance unit from the injection pump camshaft using a suitable puller.

9. Remove the timing bolt from the flywheel housing and rotate the flywheel clockwise 60° BTDC of No. 1 piston.

10. Tighten the automatic timing advance unit bolt finger tight, making sure the timing pin is still in the groove of the injection pump camshaft.

11. Slowly rotate the crankshaft counterclockwise until the timing bolt can be installed in the flywheel.

12. Install adapter FT1560 or equivalent on the timing advance unit. Install a torque

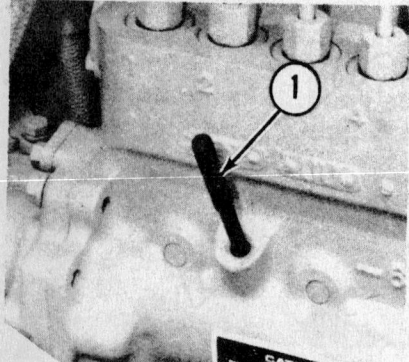

1. Timing bolt
2. Timing bolt location
3. Bolt storage location

Locating TDC for number one piston on its compression stroke—3406 series

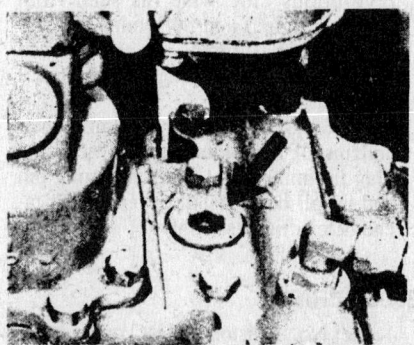

1. Timing pin

Timing pin installed in the injector housing—3406 series

wrench on the adapter and apply a torque of 45–50 ft. lbs. (60–70 Nm) in a clockwise direction as viewed from the front of the engine. While holding the torque on the adapter, tighten the retaining bolt on the timing advance unit to 180–220 ft. lbs. (245–295 Nm).

13. Remove the timing bolt from the flywheel and the timing pin from the injection pump.

14. Rotate the crankshaft counterclockwise two revolutions. If the timing pin can be installed in the injection pump and the timing bolt fits into the flywheel, the injection timing is correct. If not, repeat Steps 6 through 13.

3406 SERIES

1. Install the timing pin through the hole in the pump housing and into the notch in the camshaft.

2. Loosen four bolts (one bolt on earlier engines) holding the automatic timing advance unit to the drive shaft for the fuel injection pump.

3. Tap the automatic timing advance unit with a soft faced hammer to loosen it from the end of the drive shaft for the fuel injection pump.

4. Place the No. 1 piston at TDC, if not previously done.

5. On earlier engines, tighten the bolt to 15 ft. lbs. torque. Remove the timing pin and tighten the bolt to a final torque of 110 ft. lbs. On later engines, tighten the four bolts to 25 ft. lbs. and remove the timing pin. Tighten to a torque of 50 ft. lbs. and then to a final torque of 100 ft. lbs.

6. Remove the timing bolt from the flywheel.

7. Turn the crankshaft two complete revolutions and re-check the timing for correctness.

8. If the timing is not correct, repeat the timing procedure.

3408 SERIES

1. With the No. 1 piston on its compression stroke and at TDC, remove the cover or the air-fuel ratio control (if equipped).

2. On later engines, remove the plug from the front of the injection pump housing. Install the timing pin (tapered end first) through the hole in the pump housing

and into the timing notch in the fuel pump camshaft.

3. If the timing is correct, the timing pin will go into the notch of the camshaft and the timing bolt will turn into the threaded hole in the flywheel. If the timing is not correct, it will have to be changed. If the timing requires adjustment, continue on to the next step.

4. Remove the timing pin, the access cover and loosen the four retaining bolts of the automatic timing advance unit.

NOTE: Be sure the timing pin is removed before the four bolts are loosened.

5. Tighten the four bolts finger-tight to hold the unit against the timing gears when the gears are turned to prevent play or backlash.

6. Remove the timing bolt and turn the flywheel until the timing pin will go into the groove in the injection pump camshaft.

7. Turn the flywheel clockwise (opposite the direction of engine rotation) a minimum of 30° to remove any play in the timing gears at top dead center.

8. Turn the engine in the direction of normal rotation until the No. 1 piston is at the TDC of its compression stroke. Turn the threaded timing bolt into the hole in the flywheel.

9. Tighten the four automatic timing advance unit retaining bolts to 20 ft. lbs. Remove the timing pin from the injection pump housing.

Location of timing hole plug—3408 series (later engines) (© Caterpillar Tractor Co.)

10. Tighten the four advance unit retaining bolts to 100 ft. lbs. and remove the timing bolt from the flywheel.

11. Rotate the crankshaft two complete revolutions and re-check the timing to be sure the timing pin will go into the injector housing camshaft and the timing bolt will go into the flywheel. If the timing is incorrect, complete the procedure again.

12. If the timing is correct, be sure to remove the timing pin and the timing bolt.

Idle Speed Adjustment

3208 SERIES

NOTE: This procedure covers the basic idle and maximum speed adjustment only. If the governor requires overhaul, the fuel injection pump should be removed and referred to a properly equipped specialist for calibration.

1. Check the engine idle speed with an accurate mechanical tachometer and connecting cable.

2. If the low idle requires adjustment, loosen the locknut and turn the low idle adjustment bolt until the correct idle speed is obtained. Tighten the locknut and recheck idle speed.

3. Move the governor lever to the full load position and check the high idle rpm. If the high idle requires adjustment, remove the small cover at the top rear of the injection pump to expose adjusting screw.

4. Loosen the locknut and turn the adjustment screw to obtain the correct high idle specification. Return the governor lever to the low idle position and then back to full load and recheck adjustment.

5. After adjustment is complete, tighten the locknut and install the adjustment screw cover on the injection pump. Whenever the high idle speed is adjusted, the balance point should also be checked.

Checking Balance Point

NOTE: Checking the balance point of the engine is a method of making an engine performance diagnosis. If the balance point and high idle rpm are correct, the fuel system of the engine is operating properly.

1. Connect a mechanical tachometer to the engine by using the tachometer drive takeoff and cable adaptor.

2. Connect a continuity light to the brass terminal screw on the cover for the load stop. Connect the other end to a good ground on the fuel system.

3. Start the engine and allow it to reach operating temperature.

4. Run the engine at high idle and record the engine speed. Add load on the engine slowly until the continuity light just comes on. This is the engine balance point. Record the engine speed at the balance point, repeating the check procedure several times to make sure the readings are accurate.

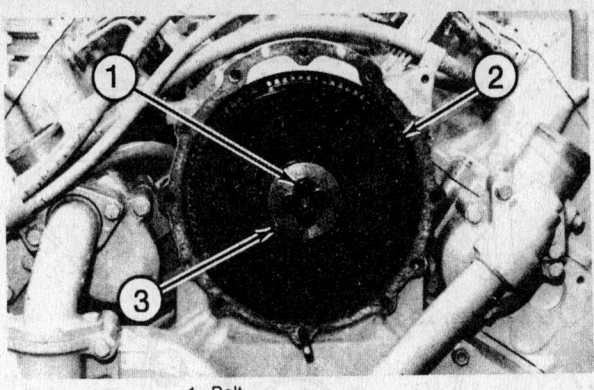

1. Bolt
2. Automatic timing advance unit
3. Retainer

Automatic timing advance unit—3408 series

Timing pin installed in the injector housing—3408 series (early engines)

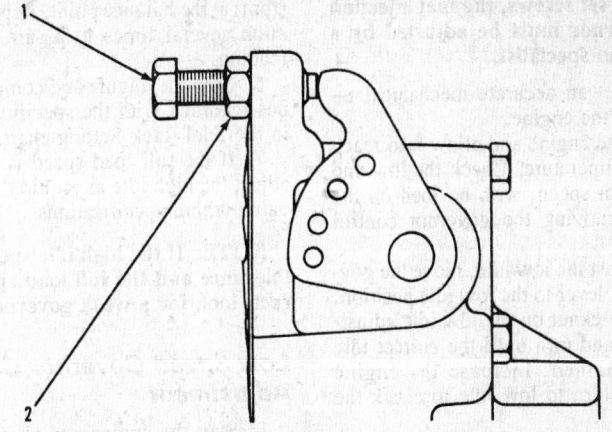

Loosen locknut (2) and turn adjusting bolt (1) to adjust low idle—3208 series

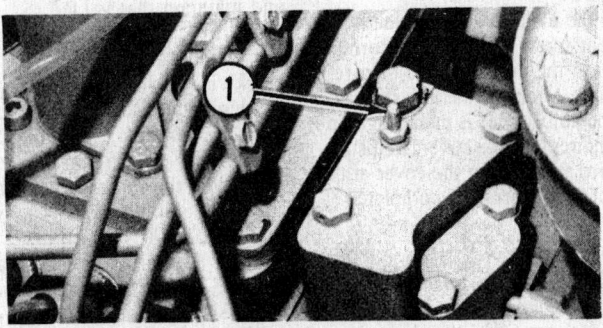

Location of brass terminal (1)—3208 series

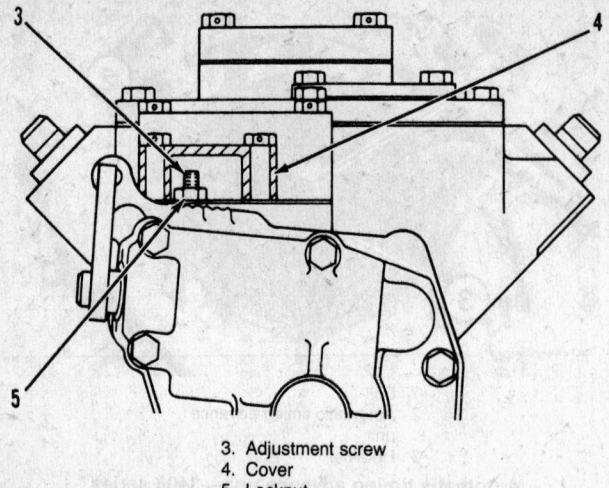

3. Adjustment screw
4. Cover
5. Locknut

High idle adjustment—3208 series

5. Stop the engine and compare the rpm values taken during testing with the Rack Setting Information chart. If the balance point is correct, the fuel system is working properly. If the balance point is incorrect, adjust the high idle rpm adjusting bolt until the correct value is obtained.

Idle Speed Adjustment

3306 SERIES

NOTE: This procedure covers basic adjustment only. If the idle and high rpm values are not obtainable by adjustment of the set screws, the fuel injection pump governor must be adjusted by a fuel injection specialist.

1. Connect an accurate mechanical tachometer to the engine.
2. Start the engine and allow it to reach operating temperature. Check the low and high idle rpm speed, with no load on the engine, by moving the governor control lever.
3. To adjust the low idle, move the governor control lever to the low idle position. Loosen the locknut on the low idle adjustment screw and turn until the correct idle speed is obtained. Increase the engine speed and return to low idle to check the adjustment.
4. To adjust high idle, cut the lockwire and remove the cover over the high idle adjustment screw.
5. Loosen the locknut and turn the high idle adjustment screw to obtain the correct value. Decrease the engine speed then return to high idle to check the adjustment.
6. Once the adjustment is made, tighten the locknut and replace the cover. Reinstall a new lockwire and seal on the cover after all adjustment procedures are complete.

Checking Balance Point

NOTE: The balance point for the 3306 engine is 20 rpm higher than full load speed.

1. Connect a mechanical tachometer to the engine using the tachometer drive.
2. Connect a continuity light to the brass terminal screw on the injection pump, located at the rear of the governor housing. Connect the other end of the light to a good ground on the fuel system.
3. Start the engine and allow it to reach operating temperature.
4. Run the engine at high idle and record the rpm reading.
5. Add load slowly to the engine until the continuity light just comes on. This is the balance point. Record the engine speed (rpm) at the balance point. Repeat the operation several times to assure an accurate reading.
6. Stop the engine and compare the values recorded with the specifications listed in the Fuel Rack Setting chart.
7. If the full load speed is not correct, adjust the high idle as required to bring the value within specifications.

NOTE: If the high idle speed is out of tolerance and the full load speed is correct, look for a weak governor spring.

Low Speed Governor Control Adjustment

1. Start the engine and allow it to idle.
2. engage the governor control. The engine should speed up to 1600 ± 50 rpm and maintain that value.
3. If the rpm needs adjustment, loosen the locknut and turn the adjusting screw clockwise to lower the rpm and counterclockwise to raise the rpm. Once the adjustment is complete, tighten the locknut to 6–12 ft. lbs. (8–16 Nm).
4. With the engine running at 1600 rpm, apply a maximum load. The engine speed should be 1500 ± 45 rpm.
5. If adjustment is necessary, repeat Step 3. Torque the locknut when adjustment is complete.
6. Remove the load and when engine speed is stable at no-load rpm, disengage the governor control to allow the engine to return to low idle.

Idle Speed Adjustment

3406 SERIES

NOTE: This procedure covers basic adjustment only. If the idle speed specifications cannot be obtained with these adjustments, the injection pump will have to be recalibrated by a qualified specialist.

1. Attach an accurate mechanical tachometer to the engine using the 1P7448 tachometer cable or equivalent.
2. Start the engine and allow it to reach operating temperature. Check the low idle rpm reading and compare it to specifications.
3. Before adjusting the idle speed, make sure the governor linkage is in the low idle position. Remove the adjustment bolt cover and turn the low idle adjustment bolt as necessary to either increase or decrease the idle speed.
4. Increase the engine speed and return the governor linkage to the low idle position, then recheck the idle speed setting. Repeat this procedure whenever any adjustment is made until the rpm setting is correct.
5. Move the governor linkage to the high idle position and measure the rpm value with the tachometer.
6. If adjustment is necessary, turn the high idle adjustment bolt as necessary to adjust the high idle engine speed.
7. After adjustment, return the linkage to the low idle position, then back to high idle and record the rpm reading. Repeat this operation until the high idle is correct.
8. Once all adjustment procedures are complete, replace the cover over the adjusting bolts. When the cover is installed on the governor, the idle adjustment screws fit into holes in the cover. The shape of the holes will not let the idle adjustment screws turn once the cover is installed. Install a new wire and seal to the bolt cover.

Checking Balance Point

NOTE: The balance point check of the engine is a method to diagnose engine performance. If the balance point and high idle speed are correct, the fuel system is functioning properly.

1. Connect a mechanical tachometer to the engine tachometer drive hookup.
2. Connect a continuity tester to the brass terminal screw on the governor housing. Connect the other end of the tester to a good ground on the fuel system.
3. Start the engine and allow it to reach normal operating temperature.
4. Raise the engine speed to the high idle position by moving the governor lever. Record the rpm value as measured by the tachometer.

5. Add load on the engine slowly until the continuity light just comes on. This is the engine balance point. Record the engine speed (rpm) at the balance point.

6. Stop the engine and compare the readings taken with the specifications in the Rack Setting Information chart.

7. If the balance point is correct, the governor setting is adjusted correctly. If the balance point is not correct, adjust the high idle rpm until the specifications are obtained.

—————— CAUTION ——————

Do not adjust the rpm above the range given for high idle specifications as damage to the engine can result.

Idle Speed Adjustment

3408 SERIES

NOTE: This procedure is for minor adjustment only. If the idle speed cannot be brought into specifications by these adjustments, the fuel injection pump will have to be recalibrated by a specialist.

1. Attach an accurate mechanical tachometer to the tachometer drive coupling on the engine using the 1P7448 tachometer cable or equivalent.

2. Start the engine and allow it to reach operating temperature.

3. Measure the idle speed and compare it with the specifications listed in the Tune-Up chart.

4. If the low idle needs adjustment, move the governor lever to the low idle position and turn the adjusting screw on the injection pump until the correct low idle rpm is obtained.

5. Raise the engine rpm, then return the governor lever to the low idle position and check the adjustment. Tighten the locknut when all adjustments are complete.

6. Move the governor lever to the high idle position and check the rpm value on the tachometer. Compare high idle speed to the specifications listed in the Tune-Up chart.

7. If the high idle needs adjustment, remove the cover at the top rear of the governor and turn the high idle adjustment bolt as necessary to either increase or decrease the high idle speed.

8. Return the linkage to the low idle position and then back to high idle and check the rpm value again. Repeat this operation several times to assure an accurate reading.

9. Once the correct high idle setting is obtained, replace the cover over the high idle adjustment bolt. When the cover is installed on the governor, the adjustment screw fits into a hole shaped to prevent the idle adjustment from moving after the cover is installed. Install a new wire and seal when all adjustments are complete.

Checking Balance Point

NOTE: The balance point check of the engine is a method of diagnosing en-gine performance. If the balance point and high idle speed are correct, the fuel system is operating properly.

1. Connect a mechanical tachometer to the engine.

2. Connect one end of a continuity tester to the brass terminal screw on the governor housing. Connect the other end to a good ground on the fuel system.

NOTE: On earlier engines, the brass terminal is located on top of the governor cover.

3. Start the engine and allow it to reach normal operating temperature.

4. Move the governor lever to the high idle position and record the engine speed (rpm).

5. Add load to the engine slowly until the circuit tester light just comes on. This is the balance point. Record the engine speed (rpm) when the test light comes on, then repeat the procedure several times to make sure the readings are correct.

6. Stop the engine and compare the readings with the information given in the Rack Setting Information chart.

7. If the balance point is not correct, adjust the high idle rpm until the proper specification is obtained.

—————— CAUTION ——————

Do not adjust the high idle above the range given in the specifications chart. Damage to the engine can result.

Glow Plugs

Removal and Installation
ALL MODELS

1. Remove the valve cover.

2. Disconnect the glow plug clip from the fuel injection line, then remove the wire from the glow plug connector.

3. Remove the glow plug from the precombustion chamber with socket 5P127 or equivalent. Inspect the glow plug for damage, carbon buildup or a melted heating element. Perform a continuity test to verify glow plug operation before installing.

4. Installation is the reverse of removal. Coat the threads with anti-seize compound and install using removal socket tool. Torque the glow plugs to 8–12 ft. lbs. (11–17 Nm).

5. Install wire connector and glow plug clip.

NOTE: The glow plug wire should be on the same side of the fuel line as the glow plug.

Fuel Injection Nozzles

Removal and Installation
3208 SERIES

1. Remove the rocker arm shafts.

2. Disconnect the fuel injection line and nozzle from the adapter. Cap all fuel line fittings immediately to prevent contamination of the fuel system with dirt.

3. Remove the clamp and the spacer holding the fuel injection nozzle in place.

4. Using a twisting motion, carefully pull the fuel injector out of the head. Remove the adapter from the head and remove the injection nozzle.

5. Check the injection nozzle for leakage, opening pressure and spray pattern with a suitable test bench. Remove the carbon seal dam and compression seal and discard.

6. Installation is the reverse of removal. Install a new carbon seal dam and compression seal and make sure the bore in the cylinder head and fuel inlet fittings are clean.

3306 SERIES

1. Remove the fuel injection lines. Do not let the tops of the injection nozzles turn when the fuel lines are loosened or the nozzles will be damaged.

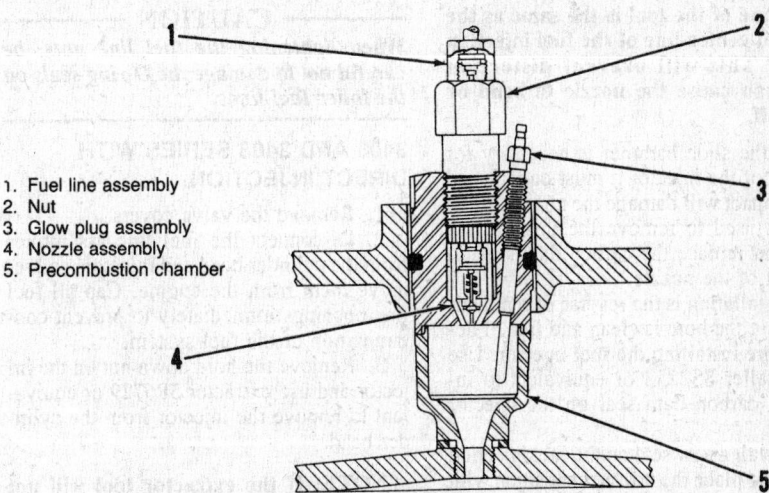

1. Fuel line assembly
2. Nut
3. Glow plug assembly
4. Nozzle assembly
5. Precombustion chamber

Injection valve and precombustion chamber—typical

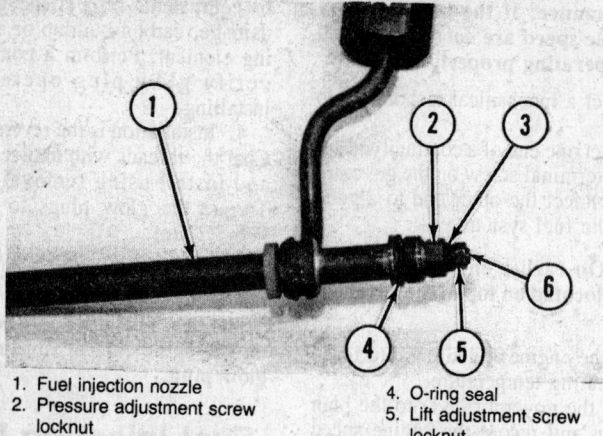

1. Fuel injection nozzle
2. Pressure adjustment screw locknut
3. Pressure adjusting screw
4. O-ring seal
5. Lift adjustment screw locknut
6. Lift adjustment screw

Fuel injection nozzle components

2. Remove the hold down clamp and install nozzle puller 6V3129 or equivalent.

NOTE: Cap all open fuel lines and fittings to prevent contamination of the fuel system during service.

3. Move the screw out of the puller enough to engage the inside lip of the puller on the lower stepped diameter of the nozzle.

4. With the nozzle bleed screw aligned with the clearance hole in the puller, turn the screw in until the tip of the button goes into the thread in the hole for the nozzle clamp bolt.

— **CAUTION** —

Do not exceed 12 ft. lbs. (17 Nm) of torque on the screw in the removal tool to loosen the nozzle in the bore. Over torquing can cause the stem of the nozzle to bend or break off.

5. If the nozzle cannot be removed with the puller, a slide hammer must be used to loosen the nozzle in the bore.

NOTE: Hold the slide hammer so the center line of the tool is the same as the extended center line of the fuel injection nozzle. This will prevent distortion which can cause the nozzle to bend or break off.

6. If the slide hammer is necessary for removal of the injector it must be replaced as the impact will damage the nozzle. If the puller is used to remove the injector, remove and replace the carbon dam seal on the front of the nozzle.

7. Installation is the reverse of removal. Make sure the bore is clean and free of debris before installing the fuel injector. Use seal installer 8S2252 or equivalent to install the carbon dam seal on the injector nozzle.

8. Install a new seal on the fuel injection nozzle and place the nozzle in position with the bleed screw away from the rocker arm cover. The nozzle should slide into the bore

smoothly with a slight twisting motion and moderate pressure. Do not force the injector into the bore. Once the nozzle is installed, install the hold-down clamp and fuel injection lines.

— **CAUTION** —

Do not let the tops of the injection nozzles turn when the fuel lines are tightened. The nozzles will be damaged if the top of the nozzle turns in the body.

3406 AND 3408 SERIES WITH PRECOMBUSTION CHAMBERS

1. Remove the valve cover.
2. Disconnect the clips from the fuel lines and remove the lines from the engine. Cap all fuel line openings immediately to prevent contamination of the fuel system.
3. Remove the hold down nut from the injector and carefully remove the injector from the precombustion chamber.
4. Installation is the reverse of removal. Torque the hold down nut to 50–60 ft. lbs. (68–82 Nm).
5. Install the fuel lines and torque the line nuts to 25–35 ft. lbs. (37–47 Nm).

— **CAUTION** —

When tightening the fuel line nuts, be careful not to damage the O-ring seals on the inner fuel lines.

3406 AND 3408 SERIES WITH DIRECT INJECTION

1. Remove the valve covers.
2. Disconnect the fuel line assemblies from the cylinder head and injectors and remove them from the engine. Cap all fuel line openings immediately to prevent contamination of the fuel system.
3. Remove the hold down nut on the injector and use extractor 5P6729 or equivalent to remove the injector from the cylinder head.

NOTE: If the extractor tool will not loosen the injector, a special puller tool 1P3075 will be necessary for removal.

4. Once the injector is removed from the engine, the nozzle assembly and seal may be removed from the injector body.

5. Installation is the reverse of removal. Coat the injector seal with clean diesel fuel before installing injector into the head. Torque the injector hold down nut to 50–60 ft. lbs. (68–82 Nm).

6. Install the fuel lines and torque the line nuts to 25–35 ft. lbs. (33–47 Nm).

Injection Pump and Governor
Removal and Installation
3208 SERIES

1. Remove the air cleaner, injection lines and intake manifold.

2. Remove the plug from the cover in the pump housing. Install the injection timing pin and turn the engine clockwise until the timing pin drops into the notch in the pump camshaft.

3. Remove the tachometer drive housing and disconnect the tachometer drive shaft.

4. Pull the drive gear free of the shaft using 5P2371 puller plate or equivalent. Tighten the puller bolts evenly until the drive gear is free of the shaft.

5. Remove the plug from the timing hole in the front cover and install a $5/16$ in.–18NC bolt, $2\frac{1}{2}$ in. long. Turn the crankshaft until the bolt can be installed in the timing gear and is in the center of the timing hole. The camshaft for the injection pump is now in correct time to the engine.

6. Remove the fuel line and mounting bolts from the injection pump and carefully lift the injection pump and governor assembly clear as a unit.

7. Installation is the reverse of removal. Align the fuel injection pump housing and governor assembly in position on the engine. Install the mounting bolts and fuel lines.

8. Install the tachometer drive shaft and torque early designs (recognized by the beveled end) to 75–85 ft. lbs. (101–115 Nm). Torque later tachometer drive to 100–120 ft. lbs. (135–163 Nm).

9. Check the injection timing. To check the timing, remove the timing pin and the bolt, then turn the crankshaft two revolutions clockwise and reinstall the timing pin and bolt back into place as before. If the timing pin or bolt cannot be installed, the fuel injection pump camshaft must be put into time again before proceeding on to the next step. See Injection Timing.

10. If the injection timing is correct, remove the bolt from the timing gear and install the plug into the timing hole.

11. Remove the timing pin from the timing slot in the injection pump camshaft and install the tachometer drive cover.

12. Install the air intake manifold, fuel injection lines and air cleaner. Torque all fuel injection line nuts to 25–35 ft. lbs. (33–47 Nm).

Removal and Installation
3306 SERIES

1. Remove the fuel ratio control, shutoff solenoid and fuel injection lines.

NOTE: Cap all fuel lines and connections to prevent dirt contamination of the fuel system.

2. Remove the cover from the timing gear housing.
3. Loosen the timing gear center bolt, leaving a gap of 0.125 in. (3.2mm) between the washer and weight assembly.
4. Use a suitable puller to remove the timing gear assembly from the fuel injection pump camshaft.
5. Remove the tube assembly located just behind the fuel filter housing.
6. Remove the mounting bolts that hold the fuel injection pump and governor assembly to support.
7. Remove the bolts holding the heat shield in position.
8. Remove the turbocharger oil supply line and oil return line.
9. Remove the mounting nuts holding the fuel injection pump and governor in place.
10. Carefully lift the fuel injection pump and governor off the engine as an assembly.

NOTE: Since the fuel injection pump and governor assembly weigh 62 lbs., a nylon strap and shop hoist is recommended for removal and installation of the injection pump.

11. Remove the fuel filter and base, tube assembly and drain lines from the fuel injection pump assembly.
12. Installation is basically the reverse of removal. Set the engine No. 1 piston at TDC on its compression stroke as previously described.
13. Install the timing tool 6V4186 or equivalent in the fuel injection pump housing. Push on the tool lightly while turning the fuel injection pump camshaft until the tool engages the slot in the camshaft. When the timing tool engages the slot, the fuel injection pump is in the No. 1 piston TDC position.
14. Make sure the O-ring seals are in position on the injection pump housing and governor. Coat the O-rings lightly with clean engine oil.
15. Place the fuel injection pump housing and governor assembly in position on the timing gear plate and the oil manifold. Install the mounting nuts that hold the injection pump housing to the timing gear plate and the governor housing to the oil manifold.

━━━ CAUTION ━━━
After the fuel injection pump housing and governor are installed on the timing gear plate, make sure the rack is free to move. The O-ring seal on the drive end of the injection pump can hold the rack and pre-

vent free movement. If the rack does not move freely, the engine can overspeed and be damaged.

16. Install the bolt and washer that hold the weight assembly to the injection pump camshaft. Install the washer with the large outside diameter toward the bolt head, then tighten the bolt finger tight only.
17. Install adapter FT1560 or equivalent on the timing gear. Using a torque wrench, apply a counter-force of 50 ft. lbs. (68 Nm) on the adapter while tightening the timing gear center bolt to 180–220 ft. lbs. (245–295 Nm).
18. Check the injection timing by removing the timing pin from the injection pump and the timing bolt from the flywheel. Turn the crankshaft clockwise (as viewed from the front) approximately ½ turn, then reinstall the timing pin. While the crankshaft is turned slowly clockwise, lightly push on the timing pin until it engages the slot in the injection pump camshaft. Install the timing bolt in the flywheel.

NOTE: The timing is correct when the pin is in the groove on the injection pump camshaft and the timing bolt can be installed into the flywheel. Any results other than these will require the injection timing to be reset. See "Injection Timing—3306 Series."

19. If the injection timing is correct, install the cover on the timing gear housing.
20. Install the heat shield.
21. Install the turbocharger oil supply line and oil return line, along with the clip that holds the lines.
22. Remove the timing pin from the injection pump and install the cover and gasket.
23. Install the tube assembly that runs behind the fuel filter.
24. Remove the timing bolt and install the plug in the flywheel housing. Remove the crankshaft turning tool and reinstall the starter assembly.
25. Install the breather assembly on the rocker arm cover, if removed, install the fuel injection lines, shutoff solenoid and fuel ratio control. Tighten the fuel injection line nuts to 25–35 ft. lbs. (33–47 Nm).

━━━ CAUTION ━━━
Do not let the tops of the fuel nozzles turn when torquing the fuel line nuts to specifications. The nozzles will be damaged if the top of the nozzle turns in the body.

Injection Pump and Governor

Removal
3406 SERIES (EARLY GOVERNOR)

1. Remove the fuel shutoff solenoid and fuel ratio control from the injection pump assembly.

2. Set the engine at TDC, remove the plug from the side of the fuel injection pump housing and install timing pin 8S2291 or equivalent through the hole and into the notch in the injection pump camshaft. If the tool will not fit into the notch, follow the procedures under Injection Pump Timing before proceeding with removal.
3. Disconnect fuel lines from the fuel injection pump housing, then disconnect the fuel injection lines from the pump housing. Cap all open fuel connections to prevent dirt from contaminating the fuel system.
4. Remove the bolts which hold the bracket to the governor housing.
5. Fasten a nylon strap to the injection pump housing and connect to a shop hoist. Remove the bolts holding the injection pump assembly to the cylinder block.
6. Remove the bolts holding the fuel injection pump housing to the fuel injection pump and governor housing, then carefully lift the injection pump and governor clear as an assembly.

NOTE: The use of a shop hoist is recommended since the injection pump and governor assembly weigh 70 lbs.

7. To install, see the procedures under "Installation (Early and Late Governors)."

Removal
3406 SERIES (LATE GOVERNOR)

1. Remove the fuel shutoff solenoid and fuel ratio control from the injection pump assembly.
2. Set the No. 1 piston at TDC on its compression stroke.
3. Install timing bolt 8F8804 or equivalent into the flywheel. Remove the plug from the side of the injection pump housing and install timing pin 8S2291 or equivalent through the hole and into the notch in the injection pump camshaft.

NOTE: If the timing pin cannot be installed into the notch on the injection pump camshaft, follow the instructions under Fuel Injection Timing before proceeding.

4. Disconnect the fuel lines from the injection pump. Disconnect the fuel injection lines from the injection pump housing and cap all open fuel connections to prevent dirt from contaminating the fuel system.
5. Remove the bolts holding the bracket to the governor housing.
6. Fasten a nylon strap and hoist to the injection pump housing and remove the bolts holding the injection pump housing to the cylinder block. Remove the bolts holding the injection pump housing to the injection pump and governor drive housing, then carefully lift the injection pump and governor as an assembly.

NOTE: The use of a shop hoist is necessary since the injection pump assembly weighs 70 lbs.

Installation

3406 SERIES (EARLY AND LATE GOVERNOR)

1. Using a nylon strap and shop hoist, carefully place the fuel injection pump housing and governor in position on the fuel injection pump and governor drive housing.

2. Install the bolts that hold the injection pump housing to the injection pump and governor drive. Install the bolts that hold the injection pump housing to the cylinder block.

3. Install the bracket to the governor housing.

4. Remove the fuel line caps and install the injection lines and fuel lines. Torque all fuel line nuts to 25–35 ft. lbs. (33–47 Nm). Do not let the tops of the fuel nozzles turn when the fuel line nuts are tightened or the nozzles will be damaged.

5. Remove the timing pin from the fuel injection pump and the timing bolt from the flywheel. Turn the crankshaft two complete revolutions clockwise (as seen from the front) and check the fuel injection timing. If the timing is not correct, follow the procedure outlined under Injection Timing.

6. If the injection timing is correct, install the plug into the injection pump housing and flywheel cover.

7. Remove the engine turning tool, if used and install the cover. Install the fuel ratio control and shutoff solenoid.

NOTE: Further disassembly and adjustment of the fuel injection pump and governor assembly should be referred to a qualified specialist.

CAUTION

Whenever starting a diesel engine after any injection pump removal or service, uncover the intake manifold (or turbocharger inlet) and have a flat piece of steel on hand to cover the inlet and shut down the engine should overspeeding occur. Keep fingers away from the portion of the steel plate that covers the air inlet since the suction force is considerable.

Injection Pump and Governor

Removal and Installation

3408 SERIES

1. Remove the fuel injection lines and cap all openings to prevent contamination of the fuel system.

2. Set the No. 1 piston on TDC of its compression stroke and remove the bolts holding the automatic timing advance device cover.

3. Remove the bolts, retainer and automatic timing advance device.

4. Disconnect the fuel lines from the transfer pump.

5. Disconnect the lines from the fuel injection pump housing and air-fuel ratio control. Cap all open lines.

6. Remove the bolt from the governor linkage.

7. Remove the bolts from the bracket holding the fuel injection pump housing to the cylinder block and the bolts holding the housing to the block.

8. Fasten a nylon strap and hoist to the injection pump and governor assembly, then slide the pump and governor assembly to the rear and carefully lift it clear of the engine. Further disassembly of the governor should be referred to a qualified specialist.

NOTE: Since the injection pump and governor assembly weigh 110 lbs., the use of a hoist is necessary during removal and installation procedures.

9. Once the injection pump is clear, remove the three O-ring seals and discard if worn or damaged.

10. Installation is the reverse of removal. Make sure the outer surface of the fuel system and the surfaces that come in contact with it are clean.

11. Install the O-rings on the injection pump housing.

12. Hoist the injection pump and governor assembly as a unit and lower it into position on the engine. Install the bolts that hold the pump housing to the cylinder block.

13. Install the bolts and bracket holding the injection pump housing to the block.

14. Connect the governor linkage and the line to the air-fuel ratio control.

15. Connect the fuel lines to the transfer pump and fuel injection lines to the pump housing. Torque the injection line caps to 25–35 ft. lbs. (33–47 Nm).

16. Place automatic timing advance device in position in the housing. Install the retainer and bolts.

17. Check injection pump timing by inserting timing pin into the groove on the pump camshaft. Remove the timing plug in the pump housing and insert timing pin 5P4185 or equivalent. If the timing pin cannot be inserted into the camshaft groove, the injection timing will have to be reset. See the procedures outlined under "Injection Timing."

18. If the injection timing is correct, tighten the mounting bolts on the timing device to 20 ft. lbs. (25 Nm) and remove the timing pin.

19. Retorque the mounting bolts to 95–105 ft. lbs. (128–142 Nm), then turn the crankshaft two complete revolutions and recheck injection timing by repeating Step 16.

20. If the timing is correct, install the plug into the injection pump housing and reinstall the timing advance device cover.

CAUTION

Whenever starting a diesel engine after any injection pump removal or service, uncover the intake manifold (or turbo-

---

charger inlet) and have a flat piece of steel on hand to cover the inlet and shut down the engine, should overspeeding occur. Keep fingers away from the portion of the steel plate that covers the air inlet since the suction force is considerable.

CUMMINS

TUNE-UP AND ADJUSTMENTS

COMPRESSION TEST

The only accepted means of checking engine compression is by a Blow-By Test which requires the use of a special measuring tools, gauges and engine dynamometer. This procedure is usually accomplished with the engine removed and set up in a test cell. No other compression test procedure is recommended by the manufacturer.

VALVE SET MARK ALIGNMENT (VS)

378, 504, 555 SERIES ENGINES

1. Turn the crankshaft in the proper direction of rotation until the no. 1 "VS" mark on the vibration damper or the crankshaft pulley is aligned with the pointer.

2. In this position, both the intake and exhaust valves must be closed for cylinder no. 1. If this is not the case, advance the crankshaft one revolution. Refer to engine firing order, if required.

NOTE: Do not use the fan blades to rotate the engine.

3. Adjust the injector plunger and then the crossheads and valves of the first cylinder. Turn the crankshaft in the direction of rotation to the next "VS" mark.

NOTE: Two complete revolutions of the crankshaft are needed to set the injectors and valves. Injectors and valves can be adjusted for only one cylinder at any on "VS" setting.

4. Continue turning the crankshaft and making the proper adjustments until all injectors and valves have been serviced.

378, 504, 555 and 855 SERIES ENGINES (TORQUE METHOD)

1. If used, pull the compression release lever to allow crankshaft rotation without engine compression.

2. Loosen the injector rocker lever adjusting nut on all cylinders to distinguish between cylinders adjusted and those not adjusted during the procedure.

3. Turn the engine in normal direction

CUMMINS DIESEL ENGINES
GENERAL ENGINE SPECIFICATIONS

Engine Model	Bore & Stroke	Piston Disp. Cu. In.	Maximum Horsepower @ R.P.M.	Torque @ R.P.M.	Firing Order		Lube Oil Pressure @ Governed Speed
					Right Hand Rotation	Left Hand Rotation	
V-378-155	4⅝ × 3¾	378	155 @ 3300	280 @ 1900	1-4-2-5-3-6	—	—
VT-378-155	4⅝ × 3¾	378	155 @ 3300	280 @ 1900	1-4-2-5-3-6	—	—
V-504-210	4⅝ × 3¾	504	210 @ 3300	375 @ 1900	1-5-4-8-6-3-7-2	1-2-7-3-6-8-4-5	—
VT-504-210	4⅝ × 3¾	504	210 @ 3300	375 @ 1900	1-5-4-8-6-3-7-2	1-2-7-3-6-8-4-5	—
V-555-210	4⅝ × 4⅛	555	210 @ 3300	445 @ 1900	1-5-4-8-6-3-7-2	1-2-7-3-6-8-4-5	40 @ 3300
V-555-225	4⅝ × 4⅛	555	225 @ 3000	425 @ 1800	1-5-4-8-6-3-7-2	1-2-7-3-6-8-4-5	40 @ 3300
V-555-240	4⅝ × 4⅛	555	240 @ 3300	445 @ 1900	1-5-4-8-6-3-7-2	1-2-7-3-6-8-4-5	40 @ 3300
VT-555-225	4⅝ × 4⅛	555	225 @ 3000	425 @ 1800	1-5-4-8-6-3-7-2	1-2-7-3-6-8-4-5	40 @ 3300
VT-555-240	4⅝ × 4⅛	555	240 @ 3300	445 @ 1900	1-5-4-8-6-3-7-2	1-2-7-3-6-8-4-5	40 @ 3300
LTA-10-240	4.9 × 5.4	611	240 @ 2100	750 @ 1300	1-5-3-6-2-4	—	35/45 @ 2460
LTA-10-270	4.9 × 5.4	611	270 @ 2100	840 @ 1300	1-5-3-6-2-4	—	35/45 @ 2460
Super-250	5½ × 6½	927	250 @ 2100	710 @ 1575	1-5-3-6-2-4	1-4-2-6-3-5	50 @ 2100
NHE-225	5½ × 6	855	225 @ 2100	—	1-5-3-6-2-4	1-4-2-6-3-5	—
NTC-230	5½ × 6	855	230 @ 2100	805 @ 1600	1-5-3-6-2-4	1-4-2-6-3-5	50 @ 2100
Formula 230	5½ × 6	855	230 @ 1900	805 @ 1300	1-5-3-6-2-4	1-4-2-6-3-5	40 @ 1900
NH-230	5½ × 6	855	220 @ 2100	—	1-5-3-6-2-4	1-4-2-6-3-5	—
NTC-230-S	5½ × 6	855	—	—	1-5-3-6-2-4	1-4-2-6-3-5	—
NHF-240	5½ × 6	855	240 @ 1800	—	1-5-3-6-2-4	1-4-2-6-3-5	50 @ 2100
Power Torque-240	5½ × 6	855	240 @ 2100	900 @ 1300	1-5-3-6-2-4	1-4-2-6-3-5	50 @ 2100
Formula-240	5½ × 6	855	240 @ 1800	900 @ 1300	1-5-3-6-2-4	1-4-2-6-3-5	—
NTC-250-S	5½ × 6	855	—	—	1-5-3-6-2-4	1-4-2-6-3-5	—
NHH-250	5½ × 6	855	250 @ 2100	—	1-5-3-6-2-4	1-4-2-6-3-5	—
NTC-250	5½ × 6	855	250 @ 2100	855 @ 1575	1-5-3-6-2-4	1-4-2-6-3-5	40 @ 2100
NHC-250	5½ × 6	855	250 @ 2100	850 @ 1300	1-5-3-6-2-4	1-4-2-6-3-5	40 @ 2100
Formula 250	5½ × 6	855	250 @ 1900	850 @ 1300	1-5-3-6-2-4	1-4-2-6-3-5	40 @ 2000
NTE-235	5½ × 6	855	235 @ 2100	650 @ 1575	1-5-3-6-2-4	1-4-2-6-3-5	50 @ 2100
NTF-255	5½ × 6	855	—	—	1-5-3-6-2-4	1-4-2-6-3-5	—
NHF-265	5½ × 6	855	—	—	1-5-3-6-2-4	1-4-2-6-3-5	—
Power Torque-270	5½ × 6	855	270 @ 2100	1000 @ 1300	1-5-3-6-2-4	1-4-2-6-3-5	35 @ 2100 (no load)
Formula-270	5½ × 6	855	270 @ 1800	1000 @ 1300	1-5-3-6-2-4	1-4-2-6-3-5	35 @ 2100 (no load)
NHCT-270	5½ × 6	855	270 @ 2100	740 @ 1575	1-5-3-6-2-4	1-4-2-6-3-5	40 @ 2100
Power Torque-270	5½ × 6	855	270 @ 1950	825 @ 1575	1-5-3-6-2-4	1-4-2-6-3-5	40 @ 2100
NTC-290	5½ × 6	855	290 @ 2100	930 @ 1300	1-5-3-6-2-4	1-4-2-6-3-5	40 @ 2100
NTCC-290	5½ × 6	855	290 @ 2100	835 @ 1575	1-5-3-6-2-4	1-4-2-6-3-5	40 @ 2100
Formula 290	5½ × 6	855	290 @ 1900	930 @ 1300	1-5-3-6-2-4	1-4-2-6-3-5	40 @ 1900
NTCE-290	5½ × 6	855	—	—	1-5-3-6-2-4	1-4-2-6-3-5	—
NTF-295	5½ × 6	855	—	—	1-5-3-6-2-4	1-4-2-6-3-5	—
NTFE-295	5½ × 6	855	—	—	1-5-3-6-2-4	1-4-2-6-3-5	—
Power Torque-300-D	5½ × 6	855	—	—	1-5-3-6-2-4	1-4-2-6-3-5	—
NTC-300	5½ × 6	855	300 @ 2100	1000 @ 1300	1-5-3-6-2-4	1-4-2-6-3-5	35 @ 2100 (no load)

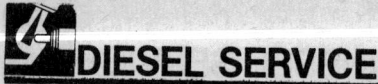

GENERAL ENGINE SPECIFICATIONS

Engine Model	Bore & Stroke	Piston Disp. Cu. In.	Maximum Horsepower @ R.P.M.	Torque @ R.P.M.	Firing Order		Lube Oil Pressure @ Governed Speed
					Right Hand Rotation	Left Hand Rotation	
Formula-300	5½ × 6	855	300 @ 1800	1000 @ 1300	1-5-3-6-2-4	1-4-2-6-3-5	35 @ 2100 (no load)
Power Torque-300	5½ × 6	855	300 @ 2300	875 @ 1600	1-5-3-6-2-4	1-4-2-6-3-5	40 @ 2100
Power Torque-330	5½ × 6	855	330 @ 2300	930 @ 1500	1-5-3-6-2-4	1-4-2-6-3-5	40 @ 2100
NTC-335	5½ × 6	855	335 @ 2100	930 @ 1575	1-5-3-6-2-4	1-4-2-6-3-5	50 @ 2100
NTCC-335	5½ × 6	855	335 @ 2100	930 @ 1500	1-5-3-6-2-4	1-4-2-6-3-5	50 @ 2100
NHHTC-335	5½ × 6	855	—	—	1-5-3-6-2-4	1-4-2-6-3-5	
NTCE-350	5½ × 6	855	350 @ 2100	1120 @ 1300	1-5-3-6-2-4	1-4-2-6-3-5	
NTC-350	5½ × 6	855	350 @ 2100	1120 @ 1300	1-5-3-6-2-4	1-4-2-6-3-5	40 @ 2100
NTCC-350	5½ × 6	855	350 @ 2100	1065 @ 1400	1-5-3-6-2-4	1-4-2-6-3-5	40 @ 2100
Formula 350	5½ × 6	855	350 @ 1900	1120 @ 1300	1-5-3-6-2-4	1-4-2-6-3-5	40 @ 1900
NTF-365	5½ × 6	855	—	—	1-5-3-6-2-4	1-4-2-6-3-5	—
NTA-370	5½ × 6	855	370 @ 2100	1015 @ 1575	1-5-3-6-2-4	1-4-2-6-3-5	50 @ 2100
NT-380	5½ × 6	855	380 @ 2300	855 @ 1600	1-5-3-6-2-4	1-4-2-6-3-5	40 @ 2300
NTA-380	5½ × 6	855	380 @ 2300	855 @ 1500	1-5-3-6-2-4	1-4-2-6-3-5	40 @ 2300
NTA-400	5½ × 6	855	400 @ 2100	1000 @ 1575	1-5-3-6-2-4	1-4-2-6-3-5	40 @ 2100
NTC-400	5½ × 6	855	400 @ 2100	1150 @ 1500	1-5-3-6-2-4	1-4-2-6-3-5	40 @ 2100
NTA-420	5½ × 6	855	420 @ 2300	1045 @ 1725	1-5-3-6-2-4	1-4-2-6-3-5	40 @ 2300
Twin Turbo-475	5½ × 6	855	475 @ 2100	1430 @ 1400	1-5-3-6-2-4	1-4-2-6-3-5	—
V-903	5½ × 4¾	903	280 @ 2600	700 @ 1575	1-5-4-8-6-3-7-2	—	50 @ 2600
VT-903	5½ × 4¾	903	300 @ 2400	795 @ 1600	1-5-4-8-6-3-7-2	—	50 @ 2400
Formula 903	5½ × 4¾	903	290 @ 2200	795 @ 1600	1-5-4-8-6-3-7-2	—	50 @ 2200
V-903	5½ × 4¾	903	280 @ 2600	—	1-5-4-8-6-3-7-2	1-2-7-3-6-8-4-5	—
V-903	5½ × 4¾	903	295 @ 2600	—	1-5-4-8-6-3-7-2	1-2-7-3-6-8-4-5	—
VT-903	5½ × 4¾	903	320 @ 2600	—	1-5-4-8-6-3-7-2	1-2-7-3-6-8-4-5	—
VT-903	5½ × 4¾	903	400 @ 2600	—	1-5-4-8-6-3-7-2	1-2-7-3-6-8-4-5	—
Formula VT-300	5½ × 4¾	903	300 @ 2200	860 @ 1400	1-5-4-8-6-3-7-2	1-2-7-3-6-8-4-5	—
VT-400	5½ × 4¾	903	400 @ 2600	—	1-5-4-8-6-3-7-2	1-2-7-3-6-8-4-5	—
VTA-903-T	5½ × 4¾	903	450 @ 2600	—	1-5-4-8-6-3-7-2	1-2-7-3-6-8-4-5	—
VTB-903	5½ × 4¾	903	275 @ 2100	—	1-5-4-8-6-3-7-2	1-2-7-3-6-8-4-5	—
KT-450	6¼ × 6¼	1150	450 @ 2100	1350 @ 1500	1-5-3-6-2-4	1-4-2-6-3-5	50 @ 2100
KTA-525	6¼ × 6¼	1150	525 @ 2100	1650 @ 1300	1-5-3-6-2-4	1-4-2-6-3-5	50 @ 2100
KTA-600	6¼ × 6¼	1150	600 @ 2100	1650 @ 1600	1-5-3-6-2-4	1-4-2-6-3-5	50 @ 2100

FIRING ORDERS

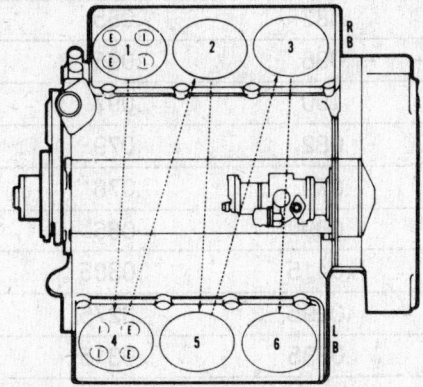

Firing order and cylinder location—378 series

FIRING ORDER 1-5-3-6-2-4

Firing order and cylinder location—611, 855 and KT1150 series engines

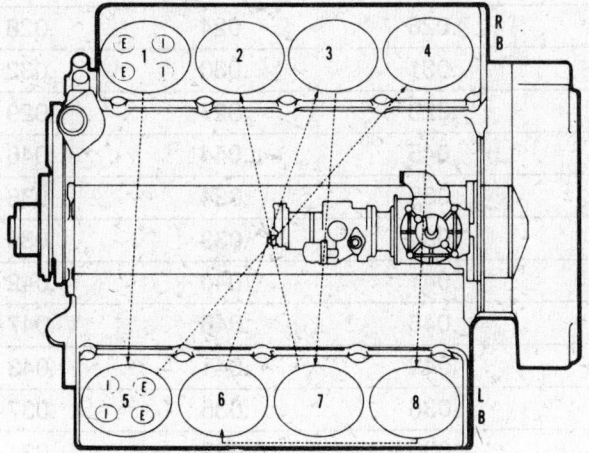

Firing order and cylinder location—504 and 555 series

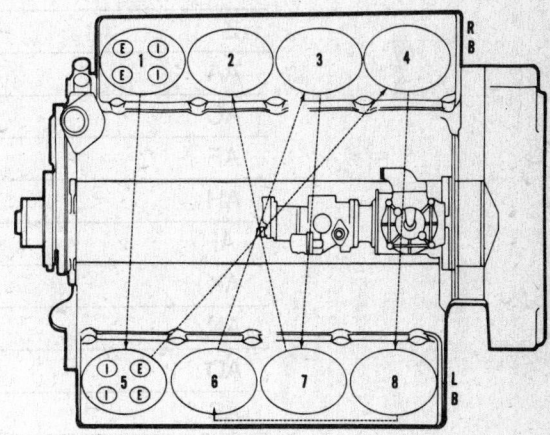

Firing order and cylinder location—903 series

INJECTION TIMING SPECIFICATIONS

Engine Series	Timing Code	Piston Travel (Inches)	Push Rod Travel (Inches)		
			Nominal	Fast	Slow
378, 504, 555	DA	.2032	.050	.048	.052
	DC	.2032	.056	.054	.058
	DD	.2032	.062	.060	.064
	DB	.2032	.061	.059	.063
	DE	.2032	.078	.076	.080
	DG	.2032	.048	.046	.050
	DF	.2032	.065	.063	.067
	DH	.2032	.071	.0695	.0725
	DJ	.2032	.074	.072	.076
	DK	.2032	.102	.100	.104
	DL	.2032	.095	.093	.097

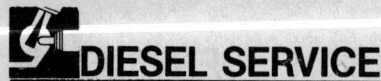

DIESEL SERVICE

INJECTION TIMING SPECIFICATIONS

Engine Series	Timing Code	Piston Travel (Inches)	Push Rod Travel (Inches)		
			Nominal	Fast	Slow
378, 504, 555	DM	.2032	.087	.085	.089
	DN	.2032	.098	.095	.101
	DP	.2032	.100	.097	.103
	DR	.2032	.082	.079	.085
	DS	.2032	.079	.076	.082
	DT	.2032	.069	.066	.072
855, 927	A	.2032	.0415	.0395	.0435
	B	.2032	.0295	.0275	.0315
	C	.2032	.0335	.0315	.0355
	D	.2032	.036	.034	.038
	E	.2032	.029	.028	.030
	Y	.2032	.039	.037	.041
	Z	.2032	.026	.024	.028
	AA	.2032	.031	.030	.032
	AC	.2032	.028	.027	.029
	AF	.2032	.045	.044	.046
	AH	.2032	.035	.034	.036
	AI	.2032	.034	.033	.035
	AK	.2032	.041	.040	.042
	AN	.2032	.046	.045	.047
	AQ	.2023	.042	.041	.043
	AS	.2032	.036	.035	.037
	AT	.2032	.030	.029	.031
	AU	.2032	.049	.048	.050
	AV	.2032	.050	.049	.051
	AW	.2032	.060	.059	.061
	AX	.2032	.055	.054	.056
	AY	.2032	.040	.039	.041
	AZ	.2032	.059	.058	.060
	BA	.2032	.028	.027	.029
	BB,CM	.2032	.100	.099	.101
	BC	.2032	.024	.023	.025
	BD	.2032	.095	.094	.096
	BH,CH	.2032	.052	.051	.053
	BI,CB,CN	.2032	.105	.104	.106
	BM	.2032	.053	.052	.054
	BS	.2032	.072	.071	.073
	BT	.2032	.081	.080	.082
	BU	.2032	.065	.064	.066

INJECTION TIMING SPECIFICATIONS

Engine Series	Timing Code	Piston Travel (Inches)	Push Rod Travel (Inches)		
			Nominal	Fast	Slow
855, 927	BV	.2032	.062	.061	.063
	BW	.2032	.067	.066	.068
	BY	.2032	.070	.069	.071
	CC	.2032	.115	.114	.115
	CD	.2032	.074	.073	.075
	CE	.2032	.026	.025	.027
	CF	.2032	.038	.037	.039
	CO	.2032	.0635	.0625	.0645
	CP	.2032	.120	.119	.121
	CR	.2032	.110	.109	.111
903	J	.2032	.049	.046	.052
	AG	.2032	.066	.063	.069
	BE	.2032	.059	.056	.062
	BK	.2032	.069	.066	.072
	BF	.2032	.079	.076	.082
	Q	.2032	.054	.051	.057
	AB	.2032	.078	.075	.081
	BG	.2032	.094	.091	.097
	BX	.2032	.083	.080	.086
	BZ	.2032	.089	.086	.092
	CG	.2032	.092	.089	.095
	BN	.2032	.084	.081	.087
	CG	.2032	.094	.091	.097
	CS	.2032	.111	.108	.114
1150	AE	.2032	.108	.106	.110
	AM	.2032	.118	.116	.120
611	CX	.2032	.080	.078	.082
	CY	.2032	.075	.073	.077

VALVE AND INJECTOR ADJUSTMENT SPECIFICATIONS—903 AND 611 SERIES
Adjustment Limits Using Indicator Method of Adjustment

Engine Series	Injector Plunger Travel	Valve Clearance Inch (mm)	
		Intake	Exhaust
903	**1.2 TO 1 ROCKER LEVER RATIO—INJECTOR LEVER P/N 196565**		
	0.180 ± 0.001 (4.57 ± 0.003)	0.012 (0.30)	0.025 (0.64)
	1 TO 1 ROCKER LEVER RATIO—INJECTOR LEVER P/N 211399		
	0.187 ± 0.001 (4.75 ± 0.03)	0.012 (0.30)	0.025 (0.64)
611	0.1975–0.1985 (5.01–5.04)	0.014 (0.036)	0.027 (0.069)

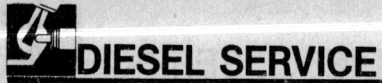

TORQUE SPECIFICATIONS

(For each step, tighten following the specified torque pattern to the torque figure indicated below. All readings are in foot pounds unless otherwise indicated. Degrees of rotation refers to the rotation of the nut or capscrew.)

Operation	903 Series Engines	855 Series Engines	378, 504, 555 Series Engines	1150 Series Engines	611 Series Engines
Cylinder Head Bolts①					
Step 1	50–80	20–25	80–90	40–60	75
Step 2	115–135	80–100	110–115	110–130	125
Step 3	175–185	265–305		180–190	175
Step 4	200–240			250–260	loosen fully
Step 5	280–300				75
Step 6					125
Step 7					175
Main Bearing Capscrew⑨					
Step 1	50	80–90⑩	55–65	190–200	50
Step 2	140–170	160–170	115–125	440–450	105
Step 3	300–320	250–260	165–175	loosen fully	155
Step 4	loosen fully	loosen fully	loosen fully	190–200	loosen fully
Step 5	50	80–90	55–65	440–450	50
Step 6	140–170	160–170	115–125		105
Step 7	300–320	250–260	165–175		155
Side Capscrews					
Step 1	25		35–40		
Step 2	70–75		65–70		
Step 3	140–150		100–110		
Connecting Rod Bolts②					
Step 1	55–60	70–75	25–30	70–80	50
Step 2	90–100	140–150	50–55	140–150	105
Step 3	loosen fully	loosen fully	loosen fully	210–220	155
Step 4	30–40	25–30	25	loosen fully	loosen fully
Step 5	60–70	70–75	60 degrees	70–80	50
Step 6	95–110	140–150		140–150	105
Step 7				210–220	155
Oil Pan Bolts③					
Step 1	30–35	35–40	15–17 (5⁄16″ bolts)	30–35 (3⁄8″ bolts)	35
Step 2				40–45 (7⁄16″ bolts)	
Step 3				60–70 (9⁄16″ bolts)	
Flywheel Housing	50–55	150	70–75	140–160	145
Flywheel④					
Step 1	180–200	200–220	55–60	100–120	50
Step 2			135–145	200–220	95
Step 3					145
Crankshaft Adapter Capscrews	330–350		135–140		
Valve Crosshead Adjustment Locknut⑤	25–30	25–30	25–28	25–30	30
Intake Manifold⑥					
Step 1	30–35	20–25	10	30–35	
Step 2			20		
Step 3			34–37		
Exhaust Manifold⑦	40–45		30–32	40–45	50
Crankshaft Vibration Damper					
Step 1	200–205		30–35	65–75	105
Step 2			110–120		
Crankshaft Drive Pulley					
Step 1			20		105
Step 2			90–100		

TORQUE SPECIFICATIONS

(For each step, tighten following the specified torque pattern to the torque figure indicated below. All readings are in foot pounds unless otherwise indicated. Degrees of rotation refers to the rotation of the nut or capscrew.)

Operation	903 Series Engines	855 Series Engines	378, 504, 555 Series Engines	1150 Series Engines	611 Series Engines
Crankshaft Pulley/Vibration Damper⑧					
Step 1		85	30–35	160–180	105
Step 2			110–120	320–340	

① 555 Series engines; a) 25–30, b) 80–90, c) 135–140
1150 Series engines; torque readings are for Cadium Plated head gasket, if a Lubricated head gasket is used torque gasket as follows: a) 40–60, b) 140–160, c) 240–260, d) 350–370
② 555 Series engines; a) 40–45, b) 75–80, c) loosen fully, d) 55, e) 60 degrees
③ 555 Series engines; 28–31 when ⅜" bolts are used
855 Series engines; torque the capscrews that hold the oil pan to the rear cover plate to 15–20 ft. lbs. Tighten the capscrews that hold the oil pan to the flywheel housing to 70–80 ft. lbs.
903 Series engines; torque the smaller

diameter bolts at the rear of the oil pan to 16–19 foot pounds
④ 855 Series engines; 190–200 when capscrews with safety wire are used
⑤ 903 Series engines; 30–35 when using a torque wrench adapter
855, 378, 504, 555, and 1150 Series engines; 22–26 when using a torque wrench adapter
1150 Series engines; when using a torque wrench adapter on a Jacobs Brake Crosshead
⑥ 855 Series engine; 22–27 when using a steel cover
⑦ 855 Series engine; 40 when washers or lockplates are not used
⑧ 855 Series engines; if ½" grade 8

capscrews are used, torque to 115–125, ⅝" grade 8 capscrews are used, torque to 180–200, ⅝" grade 5 capscrews are used torque to 150–170
555 Series engines; tighten to a final torque of 135–140 ft. lbs.
⑨ 855 Series engines; torque the 1 in. capscrews to a) 100–110, b) 200–210, c) 300–310 d) loosen completely and repeat steps a thru c

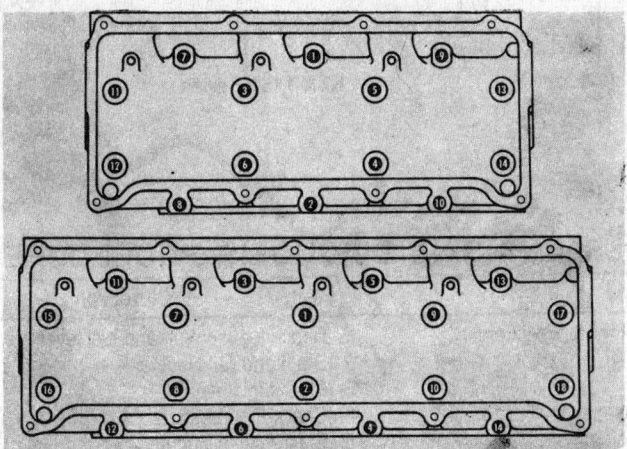

Cylinder head torque sequence—378, 504 and 555 series engines

Cylinder head torque sequence—903 series engines

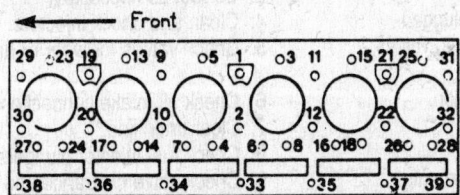

Cylinder head torque sequence—611 series engines

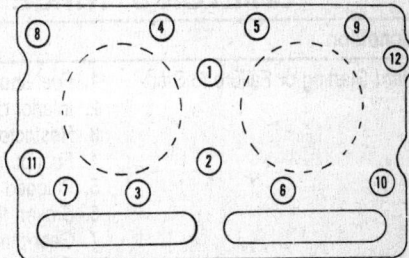

Cylinder head torquing sequence—855 series engines

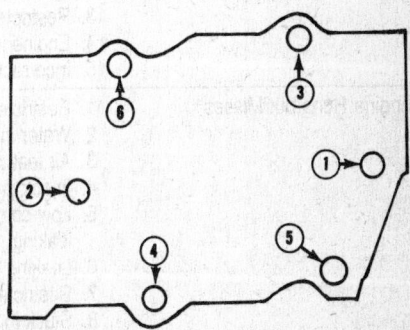

Cylinder head torque sequence—1150 series engine

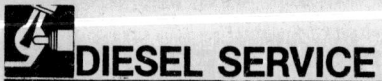

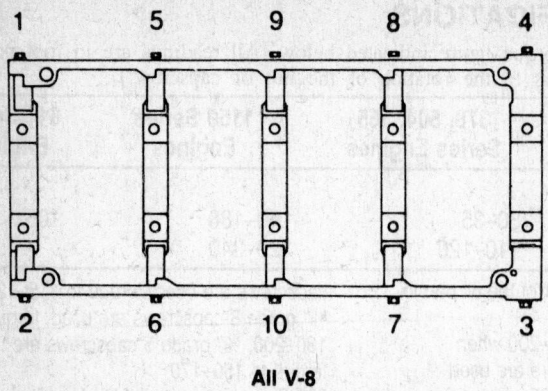

All V-8

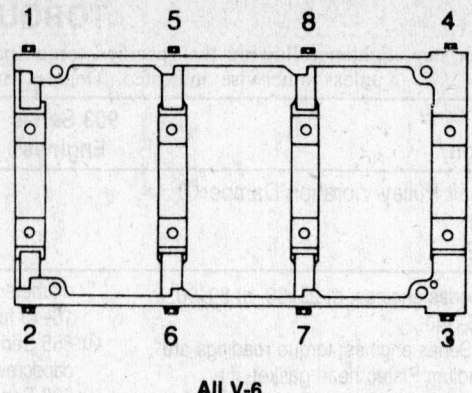

All V-6

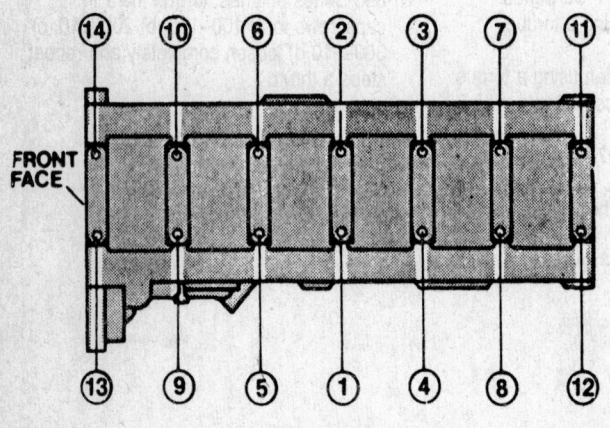

FRONT FACE

L-10 Series

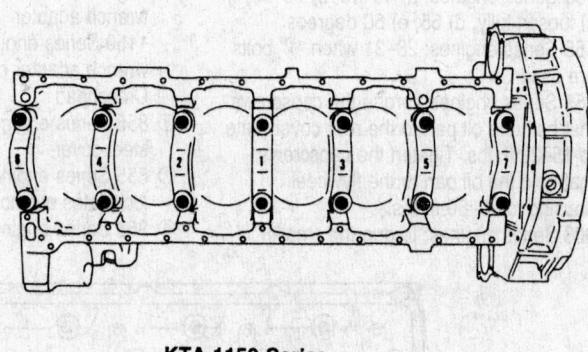

KTA-1150 Series

CHILTON'S THREE "C's" DIESEL ENGINE DIAGNOSIS PROCEDURE

Condition	Cause	Correction
Hard Starting or Failure to Start	1. Fuel shut off valve closed or fuel tank empty.	1. Check fuel shut-off and fuel level.
	2. Inferior quality fuel.	2. Drain and refill fuel tank.
	3. Restricted fuel lies.	3. Clear fuel lines.
	4. Fuel pump pressure regulation faulty.	4. Check fuel pressure.
	5. Plugged injector spray holes.	5. Clean fuel injectors.
	6. Broken fuel pump drive shaft.	6. Repair fuel pump.
	7. Gear pump gears scored or worn.	7. Replace fuel pump.
	8. Injector inlet or drain connections loose.	8. Check connections.
	9. Water in fuel.	9. Drain and refill fuel tank.
	10. Air leaks in fuel suction line.	10. Correct as necessary.
	11. Incorrect injector timing.	11. Reset injection timing.
	12. Valve leakage.	12. Check valves.
	13. Restricted air intake.	13. Clear air intake.
	14. Engine in need of overhaul.	14. Correct as necessary.
	15. Incorrect valve timing.	15. Reset valve timing.
Engine Runs but Misses	1. Restricted fuel lines.	1. Clear fuel lines.
	2. Water in fuel or poor quality fuel.	2. Drain and refill fuel tank.
	3. Air leaks in fuel suction line.	3. Correct as necessary.
	4. Injectors improperly adjusted or plugged.	4. Clean and check injectors.
	5. Low compression, intake or exhaust valves leaking.	5. Check valves for wear or damage.
	6. Leaking supercharger air connection.	6. Check all intake connections.
	7. Restricted drain line.	7. Clear drain line.
	8. Stuck injector plunger.	8. Check fuel injector plungers.
	9. Improper valve and injector adjustments.	9. Check valve clearance and injector opening pressure.

CHILTON'S THREE "C's" DIESEL ENGINE DIAGNOSIS PROCEDURE

Condition	Cause	Correction
Excessive Smoke	1. Restricted fuel system drain lines.	1. Clear fuel lines.
	2. Plugged injector spray holes.	2. Clean fuel injectors.
	3. Inferior quality fuel.	3. Drain and refill fuel system.
	4. Engine fuel rate too high.	4. Reset maximum fuel delivery.
	5. Injectors improperly adjusted.	5. Check injector operation.
	6. Intake manifold or cylinder head gasket leak.	6. Correct as necessary.
	7. Restricted air intake.	7. Replace filter element.
	8. High exhaust back pressure.	8. Check for exhaust restriction.
	9. Broken or worn piston rings.	9. Repair as necessary.
	10. Engine in need of overhaul.	10. Repair as necessary.
	11. Incorrect valve timing.	11. Reset valve timing.
	12. Worn or scored cylinder liners or pistons.	12. Repair as necessary.
Low Power or Loss of Power	1. Inferior quality fuel.	1. Drain and refill fuel system.
	2. Water in fuel.	2. Drain water separator and tank.
	3. Fuel suction line leaking.	3. Repair as necessary.
	4. Restricted fuel lines.	4. Clear fuel lines.
	5. Low fuel pressure.	5. Check delivery pressure.
	6. Plugged injector spray holes.	6. Clean fuel injectors.
	7. Dirty fuel filters or screens.	7. Replace filters and clean screens.
	8. Improper valve and injector adjustments.	8. Check valve clearance and injector operation.
	9. Improperly adjusted throttle linkage.	9. Check throttle linkage.
	10. High speed governor set too low.	10. Adjust governor setting.
	11. Air in system.	11. Bleed air from fuel system.
	12. Sticking stop control (fuel shut off).	12. Replace stop control.
	13. Restricted air intake.	13. Replace filter element.
	14. High exhaust back pressure.	14. Check for exhaust restriction.
	15. Intake manifold or cylinder head gasket leakage.	15. Repair as necessary.
	16. Low compression—intake or exhaust valve leakage.	16. Overhaul as necessary.
	17. Broken or worn piston rings.	17. Replace piston rings.
	18. Incorrect bearing clearances.	18. Replace worn bearings.
	19. Worn or scored cylinder liners or pistons.	19. Replace cylinder liners.
	20. Engine in need of overhaul.	20. Repair as necessary.
	21. Incorrect valve timing.	21. Reset valve timing.
	22. Dirty air cleaner.	22. Replace filter element.
	23. Overheating engine.	23. Check cooling system operation.
Excessive Fuel Consumption	1. Inferior quality fuel.	1. Drain and refill fuel system.
	2. Restricted fuel system drain lines.	2. Clear fuel lines.
	3. Fuel rate set too high.	3. Reset maximum fuel delivery.
	4. Fuel leaks—external or internal.	4. Correct as necessary.
	5. Plugged injector spray holes.	5. Clean fuel injectors.
	6. Injectors not adjusted properly.	6. Check injector opening pressure.
	7. Cracked injector body or cup.	7. Replace injector.
	8. Restricted air intake.	8. Replace air cleaner element.
	9. High exhaust back pressure.	9. Check exhaust restriction.
	10. Engine overloaded.	10. Correct as necessary.
	11. Incorrect bearing clearances.	11. Replace worn bearings.
	12. Engine in need of overhaul.	12. Correct as necessary.
Excessive Oil Consumption	1. External or internal oil leaks.	1. Correct as necessary.
	2. Cylinder oil control not working.	2. Repair as necessary.
	3. Wrong grade oil for climatic conditions.	3. Drain and refill crankcase.
	4. Broken or worn piston rings.	4. Repair as necessary.
	5. Engine in need of overhaul.	5. Overhaul as required.
	6. Worn or scored cylinder liners or pistons.	6. Replace cylinder liners or pistons.
Low Lubrication Oil Pressure	1. Oil suction line restricted.	1. Clear oil suction line.
	2. Oil pump or pressure regulator valve not working properly.	2. Check oil pump and regulator valve.
	3. Crankcase oil level too low.	3. Top up crankcase oil level.
	4. Wrong grade of oil for conditions.	4. Drain and refill crankcase.
	5. Insufficient coolant.	5. Top up cooling system.
	6. Worn water pump.	6. Replace water pump.
	7. Coolant thermostat not working.	7. Replace thermostat.

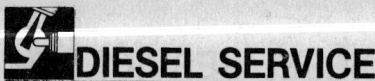
CHILTON'S THREE "C's" DIESEL ENGINE DIAGNOSIS PROCEDURE

Condition	Cause	Correction
Low Lubrication Oil Pressure	8. Loose fan belts.	8. Tighten fan belts.
	9. Clogged coolant passages.	9. Flush cooling system.
	10. Clogged oil cooler.	10. Replace oil cooler.
	11. Radiator core openings restricted.	11. Repair as necessary.
	12. Air in cooling system.	12. Bleed cooling system.
	13. Insufficient radiator capacity.	13. Repair as necessary.
	14. Leaking coolant hoses, connections or gaskets.	14. Correct as necessary.
	15. Incorrect bearing clearances.	15. Replace worn bearings.
	16. Engine in need of overhaul.	16. Repair as necessary.
	17. Engine overloaded.	17. Correct as necessary.

of rotation until the appropriate valve set mark for the cylinder being adjusted aligns with the pointer on the gear case cover.

4. Check the rocker levers on the two cylinders indicated on the crankshaft pulley, only one pair the levers should be loose. This is the cylinder to be adjusted after first setting the injector plunger.

— CAUTION —

If equipped with Top Stop Injectors, the plunger travel can only be adjusted with the injectors removed from the engine, using a special adjusting tool. Bent push rods can result, if not properly adjusted. Refer to Top Stop Injector Adjustment Procedure.

5. The injector plungers are adjusted with a torque wrench and screw driver adapter to specific torque setting. Turn the adjusting screw down until the plunger contacts the cup and advance an additional 15 degrees to squeeze the oil from the cup.

6. Loosen the adjusting screw one turn, then tighten the adjusting screw in two or three passes with a torque wrench and adapter to 60 ft. lbs. (hot or cold engine) on the 378, 504 and 555 series engines and 72 ft. lbs. (hot or cold engine) on the 855 series engine.

NOTE: If the rocker housing is cast iron on the 855 series engines, the cold engine setting is 48 inch lbs. and the hot engine setting is 72 inch lbs.

7. After adjusting the injector plunger travel, set the crossheads and valve clearances to specifications.

NOTE: On the 378, 504 and 555 series engines, after all the injectors and valves have been adjusted and the engine warmed up to 140 degrees F. (oil temperature), reset the injectors.

1150 SERIES ENGINES

NOTE: See the proper valve set mark alignment procedure before making any injector plunger adjustments.

1. Install tool 3375007 (indicator support) or equivalent with its extension on the injector plunger top at number 4 cylinder.

NOTE: Make sure that the indicator extension is secure in the indicator stem and not against the rocker lever.

2. Using tool 3375010 (rocker lever actuator) or equivalent, depress the injector plunger until the plunger is bottomed in the cup. This is necessary to squeeze the oil from the cup.

3. Allow the injector plunger to rise, bottom again and then set the dial indicator at zero with the injector plunger bottomed. At this point, check the extension contact with the plunger top.

4. Allow the plunger to rise and then bottom again to check the zero dial indicator setting.

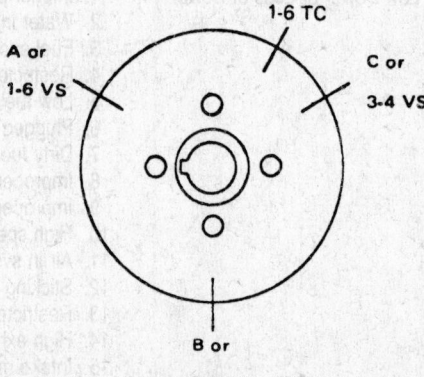

Valve set (VS) marking on pulley—855 series

5. Remove the rocker lever actuator and turn the adjusting screw until 0.304 in. is obtained on the dial indicator, which is the proper adjustment value.

6. Using the rocker lever actuator, bottom out the plunger again, release the lever, the dial indicator must show the injector plunger travel to be within specifications.

7. Torque the adjusting nut 40–45 ft. lbs. (30–35 ft. lbs. when a torque wrench adapter is used).

8. Actuate the injector plunger several times to check the adjustment. Correct if required.

Valve Adjustment

378, 504 and 555 SERIES ENGINES

NOTE: See the proper valve set mark alignment procedure before making any adjustment of the valves.

1. The same crankshaft position used for adjusting the injectors is used for adjusting the valves.

2. While adjusting the valves, be sure that the compression release, if equipped, is in the released position.

3. Loosen the locknut and back off the adjusting screw. Insert the feeler gauge between the rocker lever and the top of the crosshead.

4. Adjust the valve to the proper specification. Turn the screw down until the lever

VALVE AND INJECTOR ADJUSTMENT SPECIFICATIONS—1150 SERIES
Uniform Plunger Travel Adjustment Limits

Oil Temp.	Injector Plunger Travel Inch (mm)		Valve Clearance Inch (mm)	
	Adj. Value	Recheck Limit	Intake	Exhaust
Cold	0.304 (7.72)	0.3035 to 0.3045 (7.709 to 7.734)	0.014 (0.36)	0.027 (0.69)
Hot	0.303 (7.70)	0.3025 to 0.3035 (7.684 to 7.709)	0.014 (0.36)	0.027 (0.69)

just touches the gauge. Lock the adjusting screw in this position with the locknut.

5. Torque the locknut 40–45 ft. lbs. (30–35 ft. lbs. if a torque wrench adapter is used).

NOTE: Make a final valve adjustment after the injectors are adjusted and the engine is at proper operating temperature.

1150 SERIES ENGINES

NOTE: See the proper valve set mark alignment procedure before making any adjustment to the valves.

1. Before adjusting the valves on number 2 cylinder, be sure that the crossheads are properly adjusted.

2. Insert the correct thickness feeler gauge between the rocker lever and the crosshead. Adjust the valve to specifications.

3. Turn the adjusting screw down until the rocker lever just touches the feeler gauge. Torque the adjusting locknut 40–45 ft. lbs.(30–35 ft. lbs. if a torque wrench adapter is used).

4. Continue this procedure until all the valves are properly adjusted.

Injector and Valve Adjustment

903 SERIES ENGINES

1. Rotate the crankshaft in the proper direction of rotation until a "VS" mark on the flange aligns with the pointer on the front cover.

2. There are two cylinder numbers at each "VS" mark. The injector and the valves on one of these two cylinders are now ready to be adjusted.

NOTE: Adjust the valves and the injector on the cylinder that has the intake and exhaust valves closed.

3. After making the proper adjustment, rotate the crankshaft to align the next "VS" mark with the pointer. Adjust the valves and injectors in the same sequence as the engine firing order.

4. Install tool ST-1170 (indicator stand assembly) and tool 3375006 (dial indicator assembly) in place. Install the indicator extension on top of the injector plunger.

NOTE: Be sure the indicator extension is tight in the indicator stem and is not touching the rocker lever.

5. Turn the adjusting screw for the injector lever clockwise until the injector plunger touches the cup.

6. Turn the adjusting screw counterclockwise one-half turn. Turn the adjusting screw clockwise again until the injector plunger touches the injector cup. Set the dial on the indicator to zero.

7. Turn the adjusting screw counterclockwise until the dial indicator tool shows a total reading of .187 in. for a 1 to 1

VALVE AND INJECTOR ADJUSTMENT SPECIFICATIONS—1150 AND 611 SERIES
Injector and Valve Set Position①

Bar in Direction	Pulley Position	Set Cylinder	
		Injector	Valve
Start	A	3	5
Adv. To	B	6	3
Adv. To	C	2	6
Adv. To	A	4	2
Adv. To	B	1	4
Adv. To	C	5	1

① Firing order: 1-5-3-6-2-4

rocker lever ratio-injector lever, p/n–211399 or .180 in. for 1.2 to 1 rocker lever ratio-injector lever p/n–196565.

611 AND 1150 SERIES ENGINES

1. Rotate the crankshaft in the normal direction of engine rotation and align the "A" valve set mark on the accessory drive pulley with the pointer on the gear cover. Make sure the intake and exhaust valves on number 5 cylinder are closed.

2. Install the injector travel adjustment kit or equivalent to the injector to be adjusted. Make sure the stem of the dial indicator is correctly installed and does not touch the rocker lever.

— CAUTION —
Do not tighten the injector adjusting screw more than the amount outlined in the following procedure. The injector can be damaged if the screw is too tight.

3. Turn the adjusting screw clockwise until the injector plunger is at the bottom of its travel. Turn the adjusting screw an additional ¼ turn clockwise to remove the fuel from the injector cup. Turn the adjusting

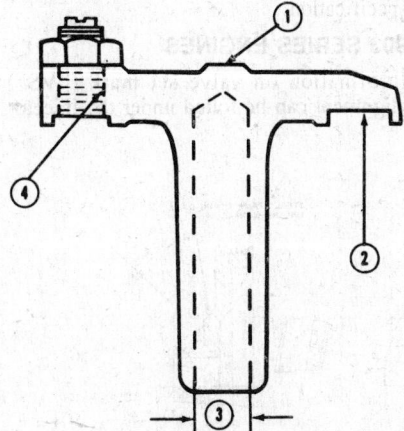

1. Rocker lever contact area
2. Valve stem contact area
3. Crosshead bore
4. Adjusting screw threads

Normal crosshead wear points—typical

screw ½ turn clockwise to the bottom of the injector plunger travel.

4. Adjust the dial indicator to zero.

5. Turn the adjusting screw counterclockwise until the dial indicator shows the value given in the adjustment chart.

6. Once the adjustment is complete, tighten the adjusting screw locknut to 40–45 ft. lbs. Recheck the adjustment.

7. Once the injector adjustment is complete, adjust the intake and exhaust valve clearance to specifications. Refer to the chart for adjustment sequence.

Top Stop Injector Adjustment

855 SERIES ENGINES

1. A cold set zero clearance setting is made at the same injector adjustment as with the dial indicator method.

— CAUTION —
Top Stop injector plunger travel can only be adjusted when the injectors are removed from the engine, using adjusting tool 3375160.

2. When the engine crankshaft has been set in the proper position for the injector to be adjusted, tighten the adjusting screw until all lash is removed from the injector train. Then tighten the adjusting screw one additional turn to properly seat the links and to squeeze oil from the socket surfaces.

3. Back the adjusting screw off until the spring washer contacts stop. Adjust the zero clearance. Using torque wrench, tighten the screw to 5–6 inch lbs. torque.

NOTE: Zero clearance is defined as the condition where the link is slightly loaded. If a torque wrench is not available, zero clearance can be set at the point where the link is slightly loaded, but just free enough to be rotated by hand.

4. Hold the adjusting screw with a screwdriver and tighten the locknut to 40–45 ft. lbs. (30–35 ft. lbs. when using a torque wrench adapter).

Crosshead Adjustment

NOTE: See the proper valve set mark alignment procedure before making any adjustments to the crossheads.

1. On 1150 series engines, position the crossheads over the guides with the adjusting screws toward the water passage in the rocker housing.
2. Loosen valve crosshead adjusting screw locknut and back off screw one turn.
3. Use light finger pressure at the rocker lever contact surface to hold crosshead in contact with valve stem nearest the push rod.
4. Turn adjusting screw down until it contacts its mating valve stem.
5. On some engines it may be necessary to advance the set screw an additional twenty to thirty degrees to straighten the stem on its guide.
6. Hold the adjusting screw in this position and tighten the locknut to the proper specification.
7. Check clearance between crosshead and valve spring retainer with wire gauge. There must be a minimum of 0.025 in. clearance at this point.

Injector Plunger Adjustment

378, 504, 555 AND 855 SERIES ENGINES

NOTE: See the proper valve set mark alignment procedure before making any injector plunger adjustments.

1. On 378, 504 and 555 series engines, tighten the injector holddown capscrew to 30–35 ft. lbs.
2. Turn the adjusting screw until the plunger contacts the cup. Advance the screw an additional 15 degrees to squeeze the oil from the cup.

NOTE: This adjustment will aid in distinguishing between cylinder adjusted and not adjusted.

3. Turn the engine in the direction of rotation until a valve set mark ("VS") aligns with the boss on the gear case cover.

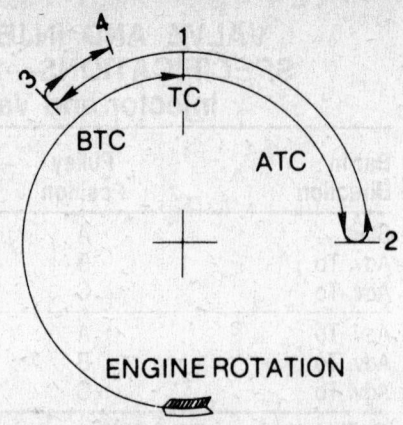

ENGINE ROTATION

Engine rotation method during injection timing procedure—611, 855, 903 and 1150 series engines

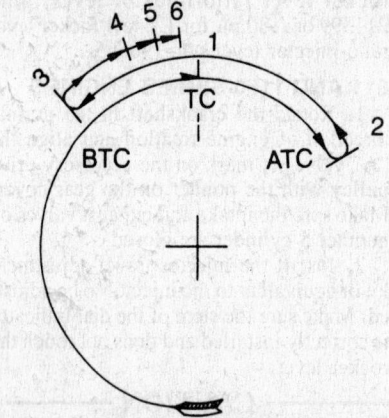

ENGINE ROTATION

Engine rotation method during injection timing procedure—378, 504 and 555 series

4. Check the valve rocker levers on the two cylinders aligned as indicated on the pulley. On one cylinder of the pair, both rocker levers will be free and both valves will be closed; this is the cylinder to adjust.
5. Adjust the injector plunger first, then the crossheads and valves to the proper specification.

903 SERIES ENGINES

Information on valve set mark ("VS") alignment can be found under the Injector

and Valve Adjustment procedure located in this section.

1150 SERIES ENGINES

1. Turn the engine in the direction of rotation until "A" valve set mark on the pulley is aligned with the pointer on the gear housing cover.
2. In this position the injector plunger for no. 3 or no. 4 cylinder will be at the top of its travel and the rocker levers for no. 5 or no. 2 cylinder will be closed.
3. The injector and valves for any one cylinder cannot be set at the same time.

NOTE: CPL (control parts list)—480 engines (California) require injector setting of .184 in..

4. Torque the adjusting screw locknuts 40–45 ft. lbs. (30–35 ft. lbs. when using a torque wrench adapter).
5. Actuate the injector plunger several times to check the adjustment. Valves must be within specification.

NOTE: Use the same engine position used to set the injectors to set the valves for the same cylinder.

6. Adjust the valve clearance by putting the correct feeler gauge between the rocker lever and the crosshead contact pads. Adjust the valves to specification by turning the adjusting screw down until the rocker lever touches the feeler gauge.

NOTE: Make sure that the valve tappet rollers are against the lobe on the camshaft before adjusting the valves.

7. Continue through the firing order until all components requiring adjustments have been adjusted.

NOTE: Be sure to turn the crankshaft to the next "VS" mark each time the valves or injector is adjusted on a particular cylinder.

Injector Timing

378, 504 AND 555 SERIES ENGINES

1. Install the injector tappet without the spring clip on the cylinder used for the timing check.
2. Install the injector push rod in no.2 and no.6 cylinder on V8 engines and in no.5 cylinder on V6 engines.
3. Install the injector timing tool in the injector bore. Tighten the tool in place in the injector mounting capscrew hole.

NOTE: The indicator extension must rest in the socket of the injector push rod.

4. Turn the engine in the direction of rotation to the top center firing position. At the point of maximum piston rise, "zero in" the dial indicator above the piston, indicating piston travel from top center.

NOTE: Both dial indicators will move in a clockwise direction when on the correct stroke for timing.

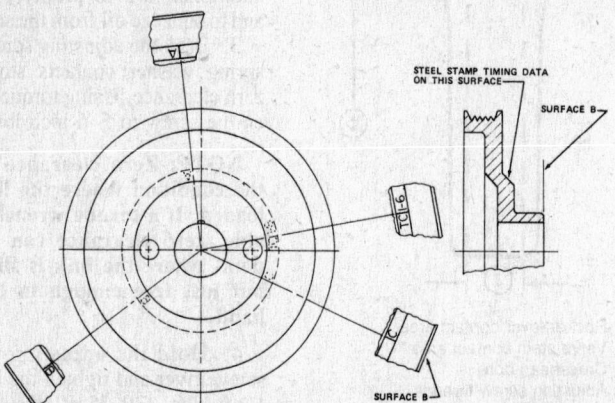

STEEL STAMP TIMING DATA ON THIS SURFACE

SURFACE B

SURFACE B

Valve set (VS) marking location on pulley—KT 1150 series

5. Turn the engine in the direction of rotation to sixty degrees after top center (2). At this point the top of the moving plunger should be in line with the sixty degree groove on the fixed scale of the indicator gauge.

6. Set the indicator above the push rod to 0.020 in. of its fully compressed position and "zero in" the dial indicator.

7. Turn the engine in the opposite direction of rotation to sixty degrees before top center (3), or until the sixty degree mark on the plunger in the injector bore is in line with the groove on the retainer.

NOTE: This alignment mark is the same alignment mark as indicated in Step number 5.

8. Be sure that the piston travel dial indicator has 0.0250 in. travel.

9. Turn the engine in the direction of rotation until the dial indicator above the piston shows that the piston has traveled to the location shown at the first check point.

10. Compare the reading to the piston travel specification reading which is located in the injector timing specification chart.

11. Read the push rod travel on the dial indicator above the push rod and check that reading against the specification in the injector timing specification chart.

12. If the push rod travel is greater than the specification, the timing is slow. If the push rod travel is less than the specification, the timing is fast. A new camshaft key must be installed to bring the reading within specification. Select and install the next advance or retard camshaft timing key to correct the problem.

NOTE: Maximum push rod travel variation between the cylinders should not exceed 0.003 in.. Each 0.006 in. offset of the camshaft key is equal to 0.0025 in. push rod indicator travel at the 0.2032 piston travel check point.

13. Recheck engine injection timing as required.

611 SERIES ENGINE

NOTE: The injection timing of the LTA-10 engine can be adjusted by changing the camshaft key which controls the position of the camshaft lobes during the operating cycles of the engine. If an offset camshaft key is installed so that the arrow marked on the top of the key is toward the engine, the timing will be retarded. If the arrow facing away from the engine, the timing will be advanced.

1. Install the injection timing tool 3375522 or equivalent in the injector bore of the No. 1 cylinder. The open end of the mounting foot must be toward the camshaft.

2. Use the alignment tool to align the push rod plunger rod, then tighten the clamp handle.

3. Install the injector push rod between the injector camshaft follower and the plunger rod.

4. Turn the accessory drive shaft in the direction of engine rotation (clockwise) until both plunger rods of the timing fixture move together in an up direction.

NOTE: The engine is on the compression stroke when both plunger rods move up at the same time.

5. Turn the accessory drive shaft until the piston plunger rod reaches its full upward travel position. The piston is now at TDC.

6. Move the piston travel gauge so the contact tip is in the center of the plunger rod. Lower the gauge so the contact tip is fully compressed and then raise the gauge 0.025 in. (0.063mm) and lock it in this position.

7. Turn the accessory drive shaft in the normal direction of engine rotation (clockwise) to 90° ATDC. Loosen the set screw for the push rod travel dial gauge so the contact tip is in the center of the plunger rod. Lower the gauge so the contact tip is fully compressed, then raise the gauge 0.025 in. (0.063mm) and lock it in this position. Zero the dial gauge.

8. Turn the accessory drive shaft in the opposite direction of engine rotation (counterclockwise) to TDC, then continue turning until the crankshaft is at 45° BTDC. This will take up the gear lash in the engine.

9. Turn the accessory drive shaft slowly clockwise until the piston travel gauge is at 0.2032 in. (5.16mm) BTDC.

——— CAUTION ———
If the crankshaft is turned beyond the 0.2032 in. (5.16 mm) BTDC position, the crankshaft must be turned counterclockwise back to the 45° BTDC position and a new approach started

10. Read the push rod travel gauge. This figure represents the injection timing value. Compare the reading to the specifications listed in the Injection Timing Chart and adjust if necessary by means of an offset camshaft key.

855 SERIES ENGINES
1. Install the injector timing tool in the injector sleeve. engage the rod of the push rod tool into the injector push tube socket. Secure the tool in place by tightening the holddown bolts by hand.

OFFSET CAMSHAFT KEY CHART

Key Part Number	Degree of Offset (To The Camshaft)	Change in Push Rod Travel	
		mm	inch
3030893	0.25	0.051	0.002
3009948	0.50	0.102	0.004
3030894	0.75	0.152	0.006
3009949	1.00	0.203	0.008
3030895	1.25	0.254	0.010
3009950	1.50	0.305	0.012
3030896	1.75	0.356	0.014
3009951	2.00	0.406	0.016
3030897	2.25	0.457	0.018
3030898	2.50	0.508	0.020

VALVE AND INJECTOR ADJUSTMENT SPECIFICATIONS—855 SERIES
Injector and Valve Set Position①

Bar in Direction	Pulley Position	Set Cylinder	
		Injector	Valve
Start	A or 1-6 VS	3	5
Adv. To	B or 2-5 VS	6	3
Adv. To	C or 3-4 VS	2	6
Adv. To	A or 1-6 VS	4	2
Adv. To	B or 2-5 VS	1	4
Adv. To	C or 3-4 VS	5	1

① Engine firing order: right hand: 1-5-3-6-2-4

VALVE AND INJECTOR ADJUSTMENT SPECIFICATIONS—855 SERIES
Uniform Plunger Travel Adjustment

Oil Temp.	Injector Plunger Travel Inch (mm)		Valve Clearance Inch (mm)	
	Adj. Value	Recheck Limit	Intake	Exhaust
ALUMINUM ROCKER HOUSING				
Cold	0.170 (4.32)	0.169 to 0.171 (4.29 to 4.34)	0.011 (0.28)	0.023 (0.58)
Hot	0.170 (4.32)	0.169 to 0.171 (4.29 to 4.34)	0.008 (0.20)	0.023 (0.58)
CAST IRON ROCKER HOUSING				
Cold	0.175 (4.45)	0.174 to 0.176 (4.42 to 4.47)	0.011 (0.28)	0.023 (0.58)
Hot	0.175 (4.45)	0.174 to 0.176 (4.42 to 4.47)	0.008 (0.20)	0.023 (0.58)

NOTE: The two dial indicator gauges used for this procedure must have a total travel of 0.250 in..

2. Loosen both of the indicator supports. Rotate the crankshaft in the direction of rotation to top dead center. In this position the piston travel plunger will be near the full upward position. Adjust both indicators to their fully compressed position.

NOTE: To prevent damage to the indicator dials, raise them approximately 0.020 in. and lock them in place with the setscrew.

3. To assure that the piston is at top dead center on the compression stroke rotate the crankshaft back and forth. Top dead center is indicated by the maximum clockwise position of the piston travel indicator pointer. Align the piston travel indicator face to "zero" with the pointer and lock the tool in place.

NOTE: Both dial indicators move in the same direction when the cylinder is on the proper stroke. If this is not the case, turn the crankshaft one complete revolution to place the cylinder on the compression stroke and then repeat Step number three.

4. Rotate the crankshaft in the direction of rotation to ninety degrees after top dead center. Turn the push tube travel indicator and align "zero" with the pointer. Lock the tool in place.

5. Turn the crankshaft in the opposite direction of rotation to forty-five degrees before top dead center.

6. Turn the crankshaft in the direction of rotation until a reading of 0.2032 in. before top dead center is reached on the piston travel indicator.

7. The push rod travel indicator should read in line with the specification outlined in the injection timing specification chart.

8. If the push rod travel is greater than the specification given, engine timing is slow. If the push rod travel is less than the specification given, engine timing is fast. Correct as required by adding or removing cam follower gaskets.

9. Remove cam follower gaskets on right hand rotation engines to retard timing and add cam follower gaskets on left hand engines to retard engine timing.

NOTE: Adjustments to the engine timing are made by changing the thickness of the cam follower housing gaskets. Before making any adjustments to the cam follower gasket check to see that the cam follower capscrews are torqued 30–35 ft. lbs. Also make sure that the indicator tool is functioning properly and not bottoming or binding.

903 SERIES ENGINES

1. Install the injector timing fixture in the injector bore of cylinder no. 2 or no. 6. The indicator extension must be in the socket of the injector push rod. Using the capscrew hole for mounting the injector, tighten the tool in position.

NOTE: The spring clip must not be on the tappet when checking injection timing. Be sure to install the clip after the injection timing has been set.

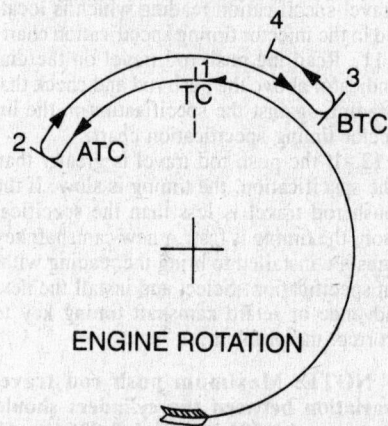

Injection timing procedure—903 VTB

2. Turn the crankshaft in the direction of rotation to the top center of the compression stroke. At the point of maximum piston rise, put the dial indicator above the piston and set it at the "zero" position.

NOTE: Both of the dial indicators will move in a clockwise direction when on the correct stroke for timing.

3. Turn the crankshaft sixty degrees after top center (2). At this point the top of the moving plunger must be in line with the sixty degree groove on the scale.

4. Set the dial indicator above the push rod to the "zero" position.

5. Turn the crankshaft in the direction opposite of rotation (3) until the sixty degree mark on the plunger aligns with the groove on the retainer (same index mark as indicated in Step 3).

6. On VTB engines, turn the crankshaft in the direction opposite rotation (3) until the indicator gauge above the piston reads 0.250–0.300 in. before top center.

7. Turn the crankshaft in the direction of rotation (4) until the dial indicator gauge

VALVE AND INJECTOR ADJUSTMENT SPECIFICATIONS—855 SERIES
Injector and Valve Set Position①

Bar in Direction	Pulley Position	Set Cylinder	
		Injector	Valve
Start	1-6 VS	2	4
Adv. To	3-4 VS	6	2
Adv. To	2-5 VS	3	6
Adv. To	1-6 VS	5	3
Adv. To	3-4 VS	1	5
Adv. To	2-5 VS	4	1

① Engine firing order: left hand 1-4-2-6-3-5

shows that the piston has moved to 0.2032 in. before top center.

8. The travel on the dial indicator gauge above the push rod should be in line with the specification in the injection timing specification chart.

9. If the push rod travel is greater than the specification the timing is slow. If the push rod travel is less than the specification the timing is fast. A new camshaft key must be installed to bring the reading within specification.

10. Select and install the next advance or retard timing key that will bring the reading within specification.

NOTE: The maximum push rod travel difference between cylinders is not to exceed 0.003 in. Each 0.007 in. offset of the key is equal to 0.0025 in. travel of the push rod indicator.

11. Recheck injection timing as required.

1150 SERIES ENGINES

1. Position timing fixture in injector well. Engage push rod indicator in injector push tube socket. Hand tighten hold-downs evenly.

2. If adaptor block is used, attach block to housing then secure rod indicator to block. Mount tools straight in cylinder and over push rod tube. Loosen indicators in their supports to prevent damage when turning the engine.

3. Turn crankshaft in direction of rotation to bring piston to be checked to TDC firing position.

4. Position piston indicator to compress stem within 0.010 in. of inner travel stop. Secure indicator.

5. Check for exact "zero" (TDC) with dial indicator.

6. Turn crankshaft in direction of rotation to 90 degrees ATDC. Position pushrod indicator on follower to 0.020 in. from inner travel stop. Secure indicator.

7. Turn crankshaft back to a position 45 degrees BTDC.

8. Slowly turn engine in direction of rotation until piston indicator is positioned at 0.0032 in. before "zero" on the dial indicator. This is equivalent to 0.2032 in. BTC.

9. The travel on the dial indicator gauge above the push rod should be in line with the specification in the injection timing specification chart.

10. If the push rod travel is greater than the specification the timing is slow. If the push rod travel is less than the specification the timing is fast. A new camshaft key must be installed to bring the reading within specification.

11. Select and install the next advance or retard timing key that will bring the reading within specification.

12. Recheck the injection timing as required.

Governor Adjustments

ALL ENGINES

The accuracy of the following adjustments changes with the condition of the engine, engine loads and the reliability of the test gauges used in performing the operations. The following adjustments can not be performed on a cold engine, engine oil temperature must be at least 165 degrees Fahrenheit. Also, the valves and the injectors must be correctly adjusted.

Idle Speed with "Type R" Fuel Pump

NOTE: Before the engine idle speed is adjusted the engine must be warmed up completely (at least 165 degrees Fahrenheit oil temperature). Idle speed adjustments must never be made on a cold engine.

1. Remove the pipe plug from the spring assembly cover.

2. Install the idle adjusting tool in the spring assembly cover.

NOTE: Be sure to operate the engine long enough to expel all of the air from the fuel system after the idle adjusting tool has been installed in the spring assembly cover.

3. The idle adjusting screw is held in place by a spring clip. Turn the idle adjustment screw in to increase the engine speed and out to decrease the engine speed.

4. 580–620 rpm is the factory recommended idle speed which is intended as a reference point. Changes can be made to this speed. Care must be taken so component cyclic vibrations are not created by extreme variations in the idle speed.

5. Remove the idle adjusting tool from the spring cover assembly. Replace the pipe plug.

NOTE: On the mechanical variable speed (MVS) governor fuel pump, the maximum and minimum idle speed adjusting screws are located on the governor cover.

6. To adjust the idle, loosen the rear idle speed adjustment screw locknut. Turn the adjustment screw in or out until the proper idle has been achieved.

7. Tighten the adjustment screw locknut immediately to prevent air from entering the system.

Idle Speed with "Type G" Fuel Pump

NOTE: Before the engine idle speed is adjusted the engine must be warmed up completely (at least 165 degrees Fahrenheit oil temperature). Idle speed adjustments must never be made on a cold engine.

1. Remove the pipe plug from the spring pack cover.

2. The idle adjustment screw is held in position by a spring clip. Using the idle adjusting tool, adjust the idle by turning the screw in to increase the engine speed or out to decrease the engine speed.

NOTE: The idle adjusting tool will not let the spring pack cover leak while the idle is being adjusted.

3. The factory recommended idle speed, which is intended as a reference point, is 580–620 rpm for 855 and 1150 series engines, 625–675 rpm for 903 series engines and 600–650 rpm for 378, 504 and 555 series engines. Changes can be made from these engine speeds. Extreme care must be exerted so component cyclic vibrations are not created by extreme variations in the idle speed.

NOTE: The engine idle speed specification is located on the engine dataplate. It is under the emission control information on the dataplate. The engine idle speed on some models will exceed the specification given in Step three.

FORWARD THROTTLE SCREW   REAR THROTTLE SCREW

FUEL PUMP LEVER   SPRING PACK COVER   *Screwdriver*
Fuel pump adjustments

4. Replace the pipe plug after removing the idle speed adjusting tool.

NOTE: On the variable speed (VS), mechanical variable speed (MVS) and the special variable speed (SVS) governors the maximum and minimum idle speed adjusting screws are located on the governor cover.

5. To adjust the idle speed, loosen the rear idle adjustment screw locknut. Turn the adjustment screw in or out until the correct idle speed is achieved.

NOTE: Do not set the special variable speed (SVS) governor idle in power take-off applications to less than 1100 rpm.

6. Tighten the idle speed adjusting screw locknut immediately to prevent air from entering the system.

Cut-Off Setting

1. At full throttle increase the engine load until the engine speed is pulled down at least 100 rpm below the rated engine speed.
2. Decrease the engine load slowly, while watching the fuel manifold pressure gauge.

NOTE: The pressure gauge will increase with the decreasing engine load. When the governor begins restricting fuel, the manifold pressure will begin decreasing with the decreasing engine load.

3. Continue decreasing the engine load until the fuel manifold pressure reaches its peak and then decreases one to two psi. This is the governor cut-off point.
4. The governor cut-off speed should be 20–50 rpm higher than the rated engine speed. This is to make sure that the governor is not restricting fuel before the rated engine speed.
5. If the governor cut-off point is higher

or lower than the specification given, remove or add adjusting shims behind the governor high speed spring as required.

6. Recheck the governor cut-off point as required.

Maximum No-Load Speed

NOTE: Before performing this test be sure that the engine is at the proper operating temperature and that all of the air is removed from the fuel system.

1. With the transmission in neutral, or the clutch fully disengaged, open the throttle and hold it fully open.
2. Note the maximum engine speed. This speed should be ten to twelve percent greater than the governor cut-off speed, depending upon the engine loads (fans, pumps etc.).

NOTE: This test must not be used to check or make governor speed adjustments. This test is of secondary importance and should not be considered unless the no-load speed reading is significantly greater than the specification.

3. If the no-load speed is significantly greater than the specification given the governor assembly must be examined for a potential problem.

Throttle Leakage

ALL ENGINES

1. Operate engine to purge all air from system.
2. Adjust throttle control linkage so that pump throttle just contacts the front throttle stop screw when throttle is closed.
3. Place transmission in neutral, open throttle fully and let engine run at high idle — no load.
4. Using a stop watch, check the time required for throttle release to 1000 rpm movement. Repeat several times.

5. If engine begins to stall upon deceleration, increase throttle leakage.
6. Note position of leakage adjusting screw. Turn screw in while checking engine operation until deceleration time is increased 1 to 2 seconds.
7. If engine decelerates too slowly, it may be necessary to decrease leakage.
8. Note position of adjusting screw. Back out screw as engine decelerates until engine tends to stall. Turn screw in until deceleration time is increased 1 to 2 seconds.

Glow Plugs

1. Remove the engine preheater adapter and the glow plug from the intake manifold.
2. Remove the nozzle and the clamping washer from the adapter.
3. Clean the adapter and the nozzle with carburetor cleaner (or equivalent).

NOTE: Be sure that the nozzle screen and the spray holes are open and clean. Check the O-ring for damage.

4. Check the glow plug on a six or twelve volt source, as applicable. Replace as required.

NOTE: Six and twelve volt glow plugs are not interchangeable.

5. Assemble the clamp washer and the nozzle to the adapter.
6. Torque the nozzle to 15–20 ft. lbs. and bend the washer over one of the hexagonal sides of the nozzle.
7. Reinstall the assembled adapter into the intake manifold.

Fuel Pump

The PT (pressure-time) fuel pump is driven by the air compressor at engine speed. The engine speed governor is incorporated in the fuel pump. The tachometer is coupled to the fuel pump main shaft. The fuel pump supplies fuel under high pressure to the injectors; one for each cylinder.

The fuel injectors meter and inject fuel into the cylinders. The injectors are operated and timed by push rods actuated by the camshaft and cam followers. Engine timing can be varied by changing the injector lifter adjustments.

The variable-speed, mechanical governor has two purposes: it maintains sufficient fuel delivery to the injectors for idling with the throttle control in idling position and it cuts off fuel to the injectors above maximum rate rpm.

Removal and Installation
855, 611 and 1150 SERIES
ENGINES

1. Remove, as required, all of the necessary components and linkages in order to properly remove the unit from the vehicle.

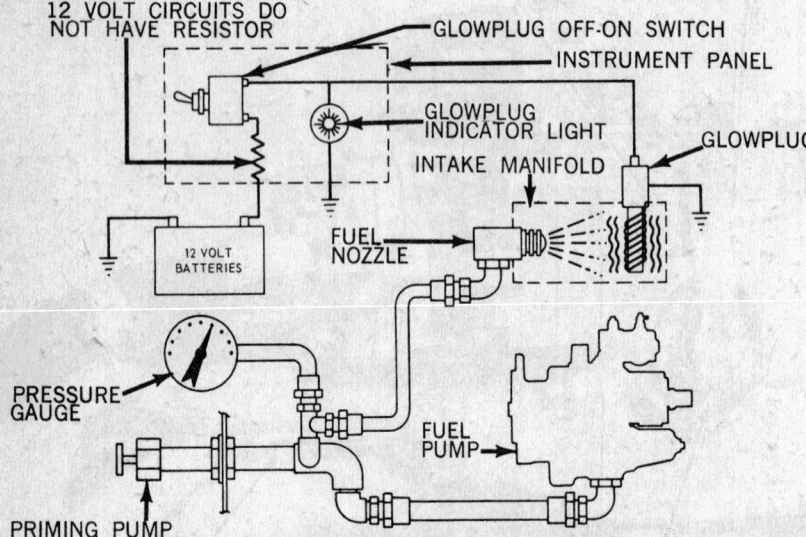

Preheater operation

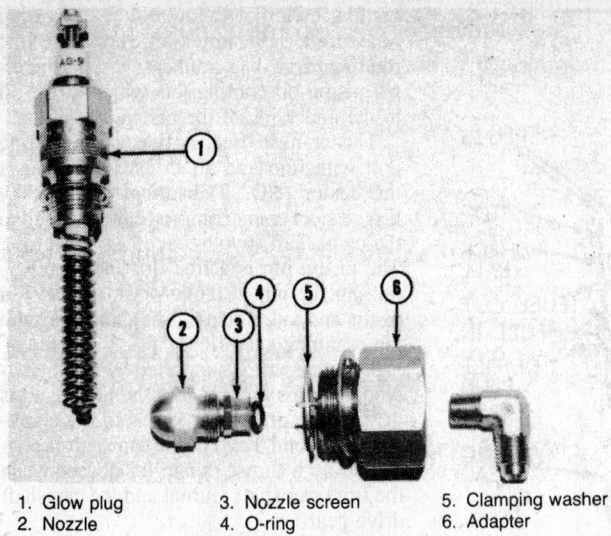

1. Glow plug
2. Nozzle
3. Nozzle screen
4. O-ring
5. Clamping washer
6. Adapter

Exploded view of glow plug

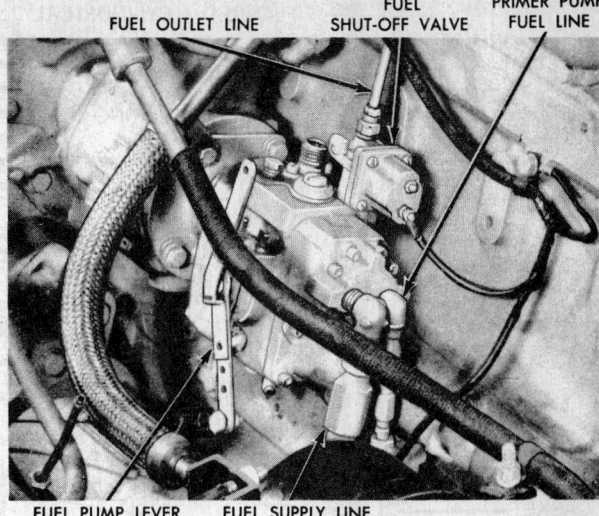

FUEL OUTLET LINE FUEL SHUT-OFF VALVE PRIMER PUMP FUEL LINE

FUEL PUMP LEVER FUEL SUPPLY LINE

Typical fuel pump installation—NH series

2. Remove the fuel supply line and the drain line from the fuel pump and the cylinder head (fuel manifold—1150 series engines).

NOTE: Some engines may have front mounted fuel supply and drain tubing.

3. If equipped, remove the throw-away fuel filter.

4. Remove the fuel pump from the governor drive or the air compressor assembly. Lift out the drive buffer or the splined coupling.

5. Installation is the reverse of removal. Torque the fuel pump mounting bolts 30–35 ft. lbs. Be sure to use new gaskets when making connections that require the use of gaskets.

378, 504, 555 AND 903 SERIES ENGINES

NOTE: On naturally aspirated 378, 504 and 555 series engines using a variable speed (VS) fuel pump and air compressor, it is not necessary to remove the air intake manifold.

1. Remove, as required, all the necessary components and linkages in order to properly remove the unit from the vehicle.

2. Remove the intake manifold retaining bolts. Remove the intake manifold and the intake crossover from the engine.

NOTE: On naturally aspirated engines it will be necessary to remove the air crossover assembly first.

3. Remove the fuel supply line and the drain line from the fuel pump and the cylinder head.

4. Remove the retaining bolts securing the fuel pump to the air compressor. Remove the unit from the vehicle.

5. Remove the drive buffer or spider from the air compressor shaft.

6. Installation is the reverse of removal. Torque the fuel pump mounting bolts

30–35 ft. lbs. Be sure to use new gaskets when making connections that require the use of gaskets.

Injectors

Removal and Installation

1. Remove, as required, all the necessary linkages and components in order to gain access to the engine valve cover.

2. Remove the valve cover retaining bolts and remove the cover.

3. Remove the injector holddown plate.

4. Using the proper injector removal tool, remove the injector from the cylinder head.

NOTE: Do not use a screwdriver or a pry bar to remove the injectors as damage may occur.

5. Before installing the injector assembly, lubricate the O-rings with the proper grade and type lubricating oil.

6. On 378, 504 and 555 series engines, install the injector into the cylinder head. Align the injector so that the bottom screen is toward the center of the engines.

NOTE: To seat Type "D" injectors in the cylinder head, fabricate a T-handle from 5/8 in. bar stock. Drill a 5/16 in. hole, 1/2 in. deep from the bottom and braze in a Type "D" injector link. Place this assembly on top of the injector in place. Install the injector links (if removed) with the part number to the top.

7. On 611 and 855 series engines, start the injector into its bore from the intake side of the engine. The injectors are to be installed with the filter screen at the twelve o'clock position (Type "D" injectors may be turned to any position). Place a valve spring compressor or equivalent on top of the injector plunger coupling and apply force to secure the injector in place.

8. On 903 and 115 series engines, in-

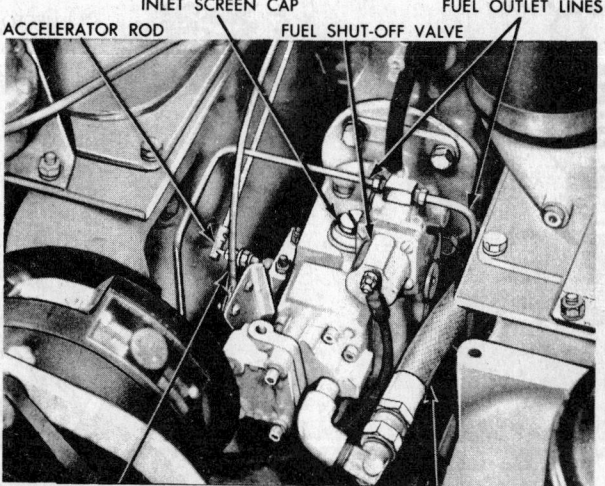

INLET SCREEN CAP FUEL OUTLET LINES
ACCELERATOR ROD FUEL SHUT-OFF VALVE

HAND THROTTLE CONTROL FUEL SUPPLY LINE

Typical fuel pump installation—V6 engines

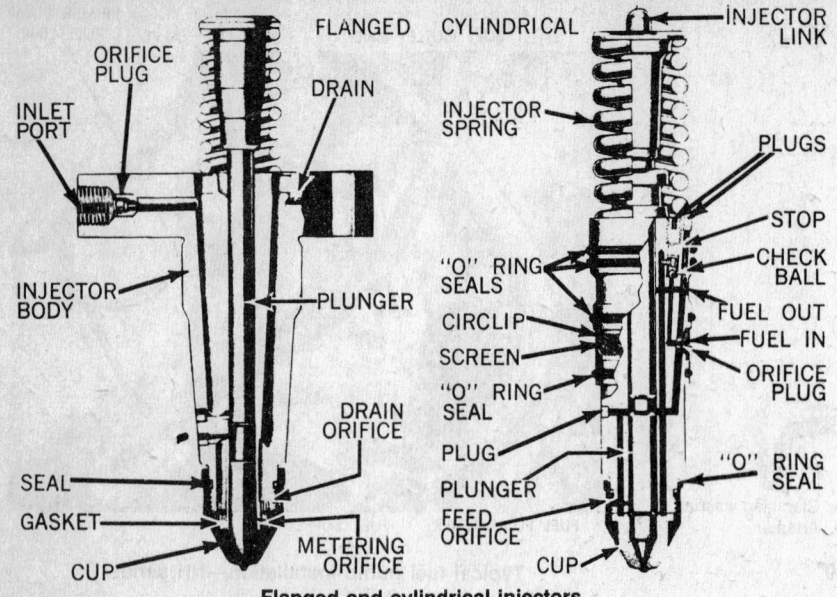

Flanged and cylindrical injectors

stall the injector into the cylinder head. Use the injector removal tool to properly seat the injectors.

NOTE: On 903 series engines, align the injector so that the screen is toward the center of the engine. Install the retaining rings over the injector.

9. On 378, 504 and 555 series engines, install the holddown clamps lockwashers and capscrews. Torque the capscrews 30–35 ft. lbs.

10. On 611, 855 and 1150 series engines, place the holddown plate over the injector body and install the holddown screws. Do not tighten at this time. Carefully insert the injector plunger link. Torque the holddown screws 11–13 ft. lbs.

11. Check the injector plunger for freedom after tightening the retaining clamps on all series engines.

12. On 903 series engines, install the injector links into the center of the injector plunger.

13. Continue the installation procedure in the reverse order of the removal.

FORD V8 6.9L DIESEL

General Description

The 6.9L diesel engine is a four cycle naturally aspirated V–8 with overhead valves. The right bank of cylinders are numbered 1, 3, 5, 7, with number 1 being at the front. The firing order is 1–2–7–3–4–5–6–8.

The crankcase has been especially designed to withstand the loads of diesel operation and utilizes a four bolt main bearing to assure a rigid, inflexible support for the rotating parts. The crankcase also has internal piston oil cooling jets which direct oil to the underside of the piston.

The crankshaft is a five main bearing unit with fore and aft thrust controlled at the center (NO. 3) bearing. Heavy-duty forged steel connecting rods are attached to the crankshaft, two to each bearing throw. The piston pin is a free floating type permitting the pin to move or float freely in piston and rod. The pin is held in place with pin retaining snap-rings.

The camshaft is supported by five insert-type bearings pressed into the block and is driven by a drive gear keyed to the crankshaft. The end thrust of the camshaft is controlled by a thrust flange located between the front camshaft journal and the camshaft drive gear.

The aluminum-alloy pistons are fitted with two compression rings and one oil ring.

The hydraulic valve tappets minimize engine noise and maintain zero valve lash or tappet clearance. This eliminates the need for periodic adjustment. The hydraulic valve tappets also incorporate camshaft roller followers for improved camshaft wear characteristics.

The cylinder head assemblies feature pre-combustion chambers which provide superior combustion characteristics. The cylinder heads used on the engine are equipped with positive valve-rotating mechanisms located at the bottom of the intake and exhaust valve springs.

The engine is equipped with a fully closed crankcase ventilation system. The crankcase depression regulator (CDR) valve is mounted on the intake manifold and provides a connection between the valley pan and the intake manifold, to regulate crankcase pressure.

The rotary-type injection pump is located between the cylinder heads in a recess in the front of the engine. The engine governor is integral with the fuel injection pump. Operating principles and service instructions for the fuel system components are also provided in this Section.

The 6.9L engine is made for Ford Motor Company by International Harvester Company.

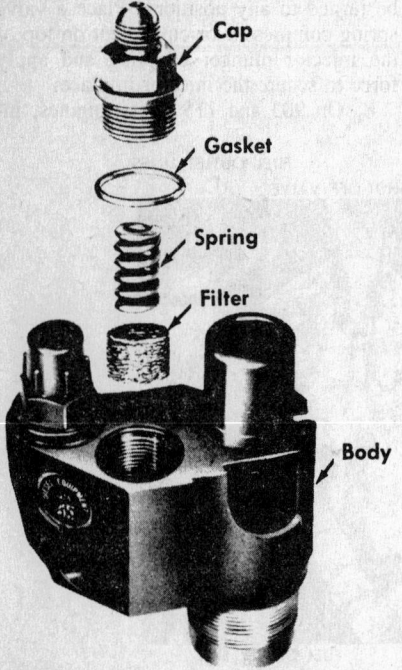

Injector body, filter, spring and cap

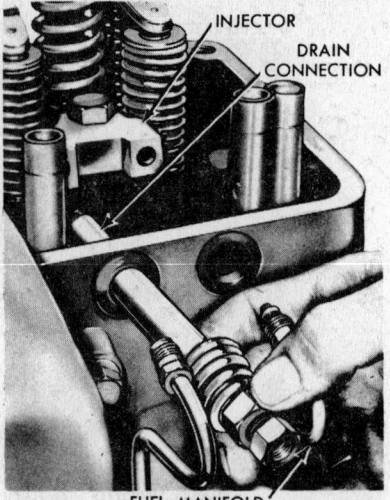

Injector fuel inlet and drain connections

TUNE-UP AND ADJUSTMENTS

Compression Test

1. Be sure that the battery is properly charged. Operate the engine until the engine is at normal operating temperature. Turn the ignition switch off. Remove the air cleaner and disconnect injection pump

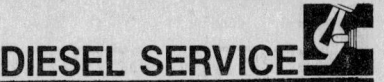
GENERAL ENGINE SPECIFICATIONS

Year	Engine No. Cyl. Displacement (liters)	Fuel Distribution	Horsepower @ rpm	Torque @ rpm (ft. lbs.)	Bore × Stroke (in.)	Compression Ratio	Oil Pressure @ 2000 rpm
'83–	V8 (6.9)	Rotary Injection	170 @ 3,300	307 @ 1,800	4.00 × 4.18	20.7:1	40–60

DIESEL TUNE-UP SPECIFICATIONS

Year	Engine No. Cyl. Displacement (liters)	Static Injection Timing	Fuel Injection Order	Valve Clearance	Injection Nozzle Opening Pressure (psi)	Intake Valve Opens (deg)	Idle Speed ① (rpm) Man.	Idle Speed ① (rpm) Auto.
'83–	V8 (6.9)	①	1-2-7-3-4-5-6-8	②	1850	—	600–700	600–700

Note: The underhood specifications sticker often reflects changes made in production. Sticker figures must be used if they disagree with those in the above chart.
① See underhood sticker for fast idle speed.
② Hydraulic lifters used (not adjustable).

TORQUE SPECIFICATIONS

All readings in ft. lbs.

Year	Engine No. Cyl. Displacement (liters)	Cylinder Head Bolts	Rod Bearing Bolts	Main Bearing Bolts	Crankshaft Pulley Bolt	Flywheel to Crankshaft Bolts	Manifold Intake	Manifold Exhaust
'83–	V8 (6.9)	①	46–51 ②	95 ③	90	44–50 ④	24	30

① First pull 40 ft. lbs., second pull 65 ft. lbs., final 75 ft. lbs.
② First tighten to 38 ft. lbs., then to 46–51 ft. lbs.
③ First tighten to 75 ft. lbs., then to 95 ft. lbs.
④ Apply thread locking sealant before installation.

FIRING ORDERS

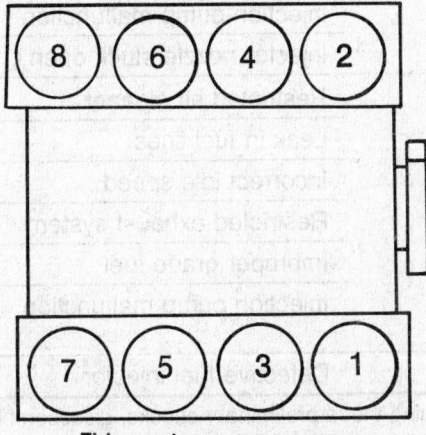

Firing order: 1-2-7-3-4-5-6-8

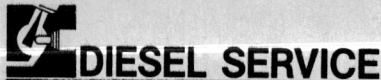

DYNAMIC TIMING SPECIFICATIONS

Fuel Cetane Value	Altitude	
	0-3000 Ft ①	Above 3000 Ft ①
38–42	6° ATDC	7° ATDC
43–46	5° ATDC	6° ATDC
47–50	4° ATDC	5° ATDC

① Installation or resetting tolerance for dynamic timing is ± 1°. Service limit is ± 2°.

CHILTON'S THREE "C's" DIESEL ENGINE DIAGNOSIS PROCEDURE

Condition	Cause	Correction
Rough Idle	Improper adjustment	Adjust idle
	Accelerator control cable binding	Repair or lubricate
	Air or water in the fuel system	Clear air or water from fuel system
	Injection nozzle clogged	Check and clean injector nozzles
	Improper valve clearance	Check valve adjustment
	Injection pump malfunction	Check injection pump
Poor Performance	Air cleaner clogged	Check element
	Accelerator control cable binding	Check control cable for free movement
	Restricted fuel flow (water or air)	Check lines and filter
	Incorrect injection timing	Check injection timing
	Injection pump malfunction	Replace injection pump
Excessive Exhaust Smoke	Restricted air cleaner	Check element
	Air or water in fuel filter	Remove air or water from fuel system
	Improper grade fuel	Check fuel in tank
	Incorrect injection timing	Check injection timing
	Injection pump malfunction	Replace injection pump
	Injector nozzle stuck open	Check injector nozzles
Excessive Fuel Consumption	Restricted air cleaner	Check element
	Leak in fuel lines	Check for leaks
	Incorrect idle speed	Check idle
	Restricted exhaust system	Check exhaust
	Improper grade fuel	Check fuel in tank
	Injection pump malfunction	Check injection pump operation
Loud Knocking In Engine	Defective fuel injector	Replace fuel injector

Note: If the problem persists after performing these preliminary checks, disassembly and inspection of internal engine components may be necessary for further diagnosis.

solenoid leads from injection pump to prevent accidental engine starting. Then remove all the glow plugs.

2. Install a compression gauge Rotunda® 19–0001 or equivalent in No. 1 cylinder glow plug hole.

3. Crank the engine (with the ignition switch off) at least five pumping strokes and record the highest reading indicated. Note and record the approximate number of compression strokes required to obtain the highest reading.

4. Repeat the check on each cylinder, cranking the engine approximately the same number of compression strokes. Record all readings.

CAUTION
Do not add oil to cylinder. This could cause hydrostatic lock.

Test Conclusion

The indicated compression pressures are considered normal if the lowest reading cylinder is at least 75 percent of the highest. Variations lower than 75 percent imply an improperly seated valve or worn or broken piston rings.

Hydraulic Roller Cam Follower

The cam followers are the hydraulic type and are not adjustable. If a hydraulic tappet noise is present, any of the following could be the cause:

1. Excessive collapsed tappet gap.
2. Sticking tappet plunger.
3. Tappet check valve not functioning properly.
4. Air in lubrication system.
5. Leakdown rate too rapid.
6. Excessive valve guide wear.

Excessive collapsed tappet gap may be caused by loose rocker arm fulcrum bolts, or wear of tappet roller, pushrod, rocker arm, rocker arm fulcrum or valve tip. With tappet collapsed, using tool T83T–6500–A, bleed-down wrench or equivalent, check gap between valve tip and rocker arm to determine if any other valve train parts are damaged, worn, or out of adjustment.

A sticking tappet plunger may be caused by dirt, chips, or varnish inside the tappet. The sticking can be corrected by disassembling the tappet and removing the dirt, chips or varnish that is causing the condition.

A tappet check valve that is not functional may be caused by an obstruction such as dirt or chips preventing it from closing when the cam lobe is lifting the tappet, or it may be caused by a broken check valve spring.

Air bubbles in the lubrication system will prevent the tappet from supporting the valve spring load and may be caused by too high or too low an oil level in the oil pan, or by air being drawn into the system through

a hole, crack, or leaking gasket on the oil pump pickup tube.

If the leakdown time is below the specified time for used tappets, noisy operation may result. If no other cause for noisy tappets can be found, the leakdown rate should be checked and any outside the specification should be replaced.

Leakdown Test

CAUTION
Tappets cannot be checked with engine oil in them. Only the testing fluid can be used.

Assembled tappets can be tested with tool 6500–E or equivalent to check the leakdown rate. The leakdown rate specification is the time in seconds for the plunger to move a specified distance of its travel while under a 50 lb. (22.68 kg) load. Test the tappets as follows:

1. Disassemble and clean the tappet to remove all traces of engine oil.

NOTE: Do not mix parts from different tappets. Parts are select-fitted and are not interchangeable.

2. Place the tappet in the tester, with the plunger facing upward. Pour hydraulic tappet tester fluid into the cup to a level that will cover the tappet assembly. The fluid can be purchased from the manufacturer of the tester.

NOTE: Using kerosene or any other fluid will not provide an accurate test.

3. Place the $\frac{5}{16}$ in. steel ball provided with the tester in the plunger cap.

4. Adjust the length of the ram so that the pointer is $\frac{1}{16}$ in. (1.59mm) below the starting mark when the ram contacts the tappet plunger, to facilitate timing as the pointer passes the start timing mark. Use the center mark on the pointer scale as the stop timing point instead of the original stop timing mark at the top of the scale.

5. Work the tappet plunger up and down until the tappet fills with fluid and all traces of air bubbles have disappeared.

6. Allow the ram and weight to force the tappet plunger downward. Measure the exact time it takes for the pointer to travel from the start timing to the stop timing marks of the tester.

7. A tappet that is satisfactory must have a leakdown rate (time in seconds) within

the minimum and maximum limits specifications.

8. If tappet is not within specifications, replace it with a new tappet. It is not necessary to disassemble and clean new tappets before testing, because the oil contained in new tappets is test fluid.

9. Remove fluid from cup and bleed fluid from tappet by working plunger up and down. This step will aid in depressing the tappet plungers when checking the valve clearance.

INJECTOR TIMING

STATIC TIMING

1. Loosen the injection pump to mounting nuts.
2. Rotate the injection pump to bring the mark on the pump into alignment with the mark on pump mounting adapter.
3. Visually recheck the alignment of the timing marks and tighten injection pump mounting nuts.

DYNAMIC TIMING

1. Bring the engine up to normal operating temperature.
2. Stop the engine and install a dynamic timing meter, Rotunda® 78–0100 or equivalent, by placing the magnetic probe pickup into the probe hole.
3. Remove the No. 1 glow plug wire and remove the glow plug, install luminosity probe and tighten to 12 ft. lbs. (16 Nm). Install the photocell over the probe.
4. Connect a dynamic timing meter to the battery and adjust the offset of the meter.
5. Set the transmission in neutral and raise the rear wheels off the ground. Using Rotunda® 14–0302, throttle control, set the engine speed to 1400 rpm with no accessory load. Observe the injection timing on the dynamic timing meter.

NOTE: Obtain a fuel sample from the vehicle and check the cetane value using the tester supplied with the Ford special tools 78–0100 or equivalent. Refer to the dynamic timing chart to find the correct timing in degrees.

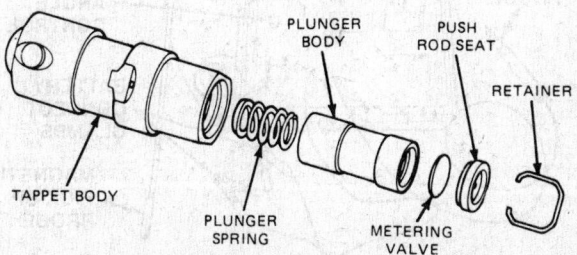

Exploded view of hydraulic lifter assembly

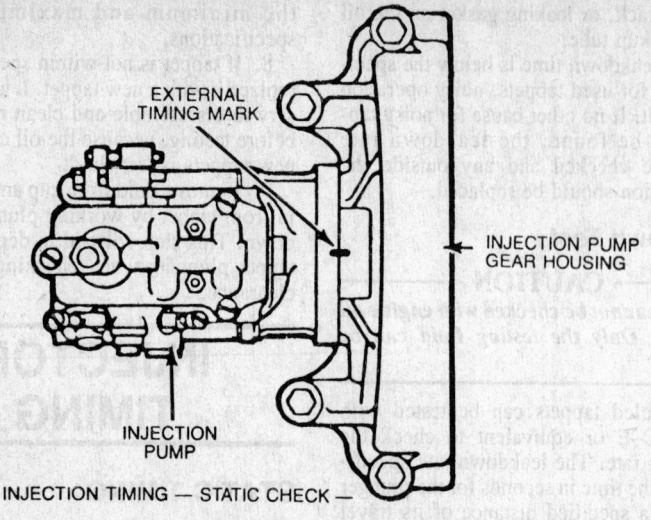

INJECTION TIMING — STATIC CHECK —

Injection pump timing marks

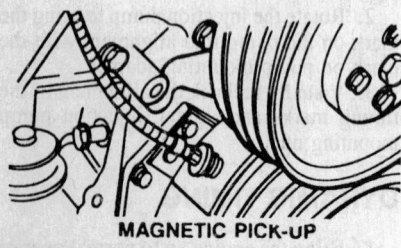

MAGNETIC PICK-UP

Location of magnetic pick-up mounting hole

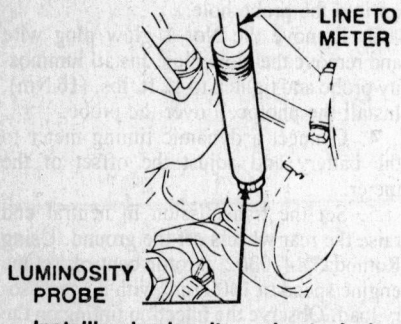

Installing luminosity probe–typical

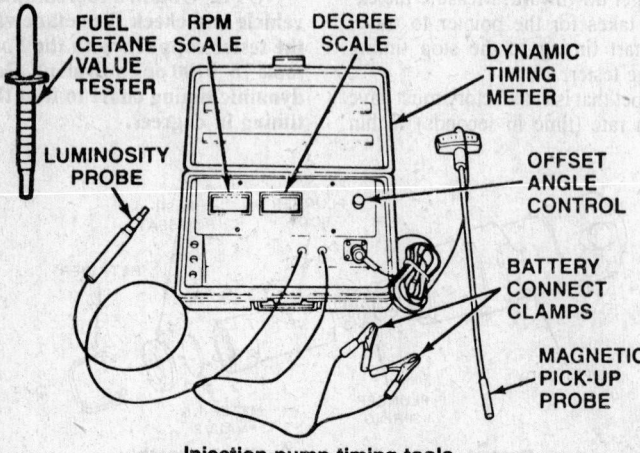

Injection pump timing tools

6. If dynamic timing is not within ± 2° of specification, then injection pump timing will require adjustment.

7. Turn the engine off. Note the timing mark alignment. Loosen the injection pump-to-adapter nuts.

8. Rotate the injection pump clockwise (when viewed from the front of engine) to retard and counterclockwise to advance timing. Two degrees of dynamic timing is approximately 0.030 in. (.75mm) of timing mark movement.

9. Start the engine and recheck the timing. If the timing is not within ± 1° of specification, repeat Steps 7 through 9.

10. Turn off the engine. Remove the dynamic timing equipment. Lightly coat the glow plug threads with anti-seize compound, install the glow plug and tighten to 12 ft. lbs. (16 Nm). Connect the glow plug wires.

Idle Speed

Curb Idle Adjustment

1. Place the transmission in neutral or park.

2. Bring the engine up to normal operating temperature.

3. Idle speed is measured with manual transmission in neutral and automatic transmission in drive with the wheels blocked.

4. Check the curb idle speed, using Rotunda® 99–0001 or equivalent magnetic pick-up tachometer. Adjust the idle speed to 600–700 rpm.

NOTE: Always check the underhood emissions control information label, for the latest idle and adjustment specifications.

5. Place the transmission in neutral or park and momentarily speed up the engine. Allow the rpm to drop to idle and recheck the idle speed. Readjust if necessary.

Fast Idle Adjustment

1. Place the transmission in neutral or park.

2. Start the engine and bring up to normal operating temperature.

3. Disconnect the wire from the fast idle solenoid.

4. Apply battery voltage to activate the solenoid plunger.

5. Speed up the engine momentarily to set the plunger.

6. The fast idle should be between 850–900 rpm. Adjust the fast idle by turning the solenoid plunger in or out.

7. Speed up the engine momentarily and recheck the fast idle. Readjust as necessary.

8. Remove the battery voltage from the solenoid and install the wire to the solenoid.

Glow Plug System

The 6.9L diesel engine utilizes an electric glow plug system to aid in the start of the engine. The function of this system is to pre-heat the combustion chamber to aid ignition of the fuel.

The system consists of eight glow plugs (one for each cylinder), control switch, power relay, after glow relay, wait lamp latching relay, wait lamp and the eight fusible links located between the harness and the glow plug terminal.

On initial start with a cold engine, the glow plug system operates as follows: The glow plug control switch energizes the power relay (which is a magnetic switch) and the power relay contacts close. Battery current energizes the glow plugs. Current to the glow plugs and a wait lamp will be shut off when the glow plugs are hot enough. This takes from 2 to 10 seconds after the key is first turned on. When the wait lamp goes off, the engine is ready to start. After the engine is started the glow plugs begin an on-off cycle for about 40 to 90 seconds. This cycle helps to clear start-up smoke. The control switch (the brain of the operation) is threaded into the left cylinder head coolant jacket. The control unit senses engine coolant temperature. Since the control unit senses temperature and glow plug operation the glow plug system will not be activated unless needed. On a restart (warm engine) the glow plug system will not be activated unless the coolant temperature drops below 165°F (91°C).

The fast start system utilizes 6 volt glow plugs in a 12 volt system to achieve rapid heating of the glow plug, a cycling device is required in the circuit.

CAUTION

Never bypass the power relay of the glow plug system. Constant battery current (12 volts) to glow plugs will cause them to overheat and fail, possibly resulting in severe engine damage.

FUEL SYSTEM

Fuel Supply Pump

Removal

1. Loosen the threaded connections with the proper size wrench (flare nut wrench preferred) and retighten snugly. Do not remove lines at this time.
2. Loosen the mounting bolts one to two turns. Apply force with hand to loosen fuel pump if gasket is stuck. Rotate engine, by nudging starter, until fuel pump cam lobe is at low position. At this position, spring tension against fuel pump bolts will be greatly reduced.
3. Disconnect fuel supply pump inlet, outlet and fuel return line.

CAUTION

Use care to prevent combustion of spilled fuel.

4. Remove fuel pump attaching bolts and remove pump and gasket. Discard old gasket.

Installation

1. Remove all fuel pump gasket material from engine and from fuel supply pump if, reinstalling used pump.
2. Install attaching bolts into fuel supply pump and install a new gasket on bolts. Position fuel supply pump to mounting pad. Turn attaching bolts alternately and evenly and tighten to specification.

NOTE: Cam must be at its low position before attempting to install fuel supply pump. If it is difficult to start the mounting bolts, remove the pump and reinstall with lever on bottom side of cam.

3. Install fuel outlet line. Start fitting by hand to avoid crossthreading.
4. Install inlet line and fuel return line.
5. Start engine and observe all connections for fuel leaks for two minutes.

6. Stop engine and check all fuel supply pump fuel line connections. Check for oil leaks at pump mounting pad.

Fuel Filter

Removal

1. Disconnect battery ground cables from both batteries.
2. Unscrew fuel filter from adapter.

Installation

1. Clean gasket surface of fuel filter adaptor to prevent contamination.
2. Lightly coat filter sealing gasket with clean diesel fuel.
3. Screw new fuel filter onto filter adapter until seal contacts flange.
4. Tighten filter another $\frac{1}{2}$ to $\frac{3}{4}$ turn.

5. Clean up any spilled fuel from top of engine.
6. Connect battery ground cables to both batteries.
7. Run engine and check for fuel leaks.

Fuel/Water Separator

The 6.9L diesel engine is equipped with fuel/water separator in the fuel supply line. A "Water in Fuel" indicator light is provided on the instrument panel to alert the operator. The light should glow when the ignition switch is in the START position to indicate proper light and water sensor function. If the light glows continuously while the engine is running, the water must be drained from the separator as soon as practical to prevent damage to the fuel injection system.

Schematic of glow plug system

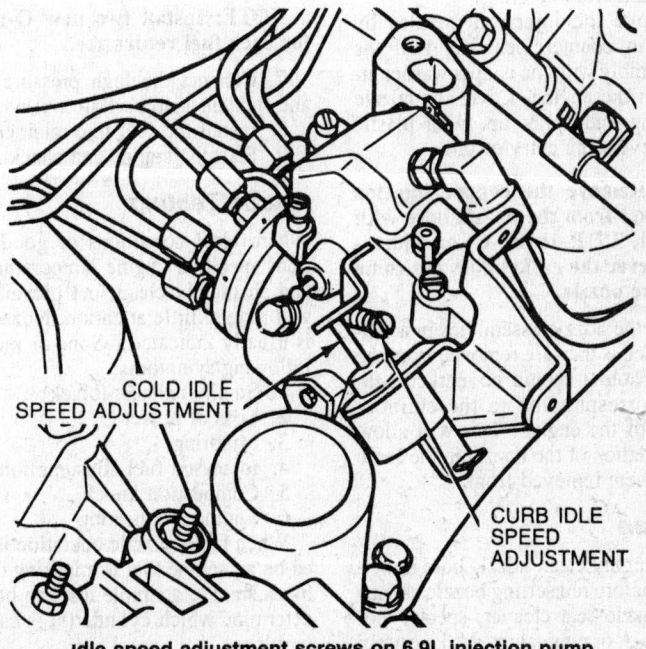

COLD IDLE SPEED ADJUSTMENT

CURB IDLE SPEED ADJUSTMENT

Idle speed adjustment screws on 6.9L injection pump

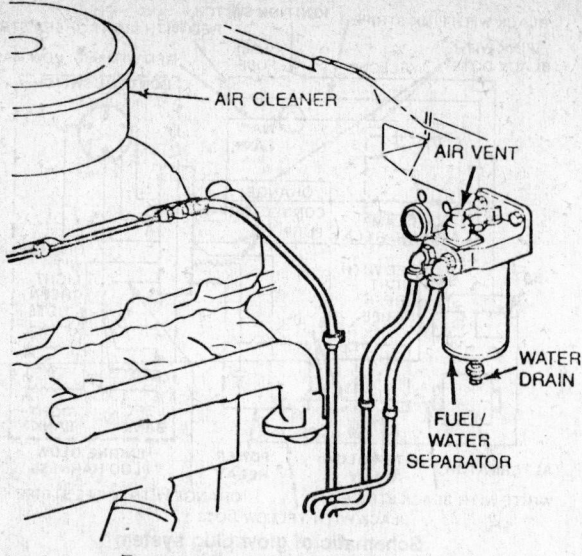

Draining water/fuel separator

Injection Nozzles

Removal

NOTE: Before removing nozzle assemblies, clean exterior of each nozzle assembly and the surrounding area with clean fuel oil or solvent to prevent entry of dirt into the engine when nozzle assemblies are removed. Also, clean fuel inlet and fuel leak-off piping connections. Blow dry with compressed air.

1. Remove the fuel line retaining clamp(s) from the nozzle lines that are to be removed.

2. Disconnect the nozzle fuel inlet (high pressure) and fuel leak-off tees from each nozzle assembly and position out of the way. Cover the open ends of the fuel inlet and outlet or nozzles with protective caps, to prevent dirt from entering.

3. Remove the injection nozzles by turning them counterclockwise. Pull the nozzle assembly with the copper washer attached from the engine. Cover the nozzle fuel opening and spray tip, with plastic caps, to prevent the entry of dirt.

NOTE: Remove the copper injector nozzle gasket from the nozzle bore with special tool, T71P-19703-C, or equivalent, whenever the gasket does not come out with the nozzle.

4. Place the nozzle assemblies in a fabricated holder as they are removed from the heads. The holder should be marked with numbers corresponding to the cylinder numbering of the engine. This will allow for re-installation of the nozzle in the same ports they were removed from.

Installation

1. Thoroughly clean nozzle bore in cylinder head before reinserting nozzle assembly with nozzle seat cleaner, special tool T83T-9527-A or equivalent. Make certain that no small particles of metal or carbon remain on the seating surface. Blow out the particles with compressed air.

2. Remove the protective cap and install a new copper gasket on the nozzle assembly, with a small dab of grease.

NOTE: Anti-seize compound or equivalent should be used on nozzle threads to aid in installation and future removal.

3. Install the nozzle assembly into the cylinder head nozzle bore.

4. Tighten the nozzle assembly to 33 ft. lbs. (45 Nm).

5. Remove the protective caps from nozzle assemblies and fuel lines.

6. Install the leak-off tees to the nozzle assemblies.

NOTE: Install two new O-ring seals for each fuel return tee.

7. Connect the high pressure fuel line and tighten, using a flare nut wrench.

8. Install the fuel line retainer clamps.

9. Start the engine and check for leaks.

Nozzle Testing

Where ideal conditions of good combustion, specified engine temperature control and absolutely clean fuel prevail, nozzles will require little attention. Nozzle trouble is usually indicated by one or more of the following symptoms:

1. Smoky exhaust (black)
2. Loss of power
3. Misfiring
4. Increased fuel consumption
5. Combustion knock
6. Engine overheating

When faulty nozzle operation is suspected on an engine that is misfiring or puffing black smoke, a simple test can be made to determine which cylinder(s) is causing the problem.

1. Run the engine at the rpm, which makes the misfire most pronounced.

2. Momentarily loosen the high pressure fuel inlet line connection on one nozzle assembly one half turn. Then re-tighten the connection.

3. Check each cylinder in the same manner. If one nozzle is found where loosening makes no difference in the misfiring, or puffing of black smoke stops, that nozzle should be tested.

NOTE: It is advisable to test all nozzles before cleaning them.

1. Remove the nozzle(s) to be tested.

2. Prepare Rotunda® #14-6300 (test stand) or equivalent for making the tests. Fill the reservoir with clean calibration fluid. Open the tester valve slightly and operate the tester handle to expel air from the tester and outlet pipe. Operate the tester until solid fluid (without air bubbles) flows from the end outlet pipe. Close the tester valve.

3. Connect the injection nozzle to the test stand.

4. Bleed the air from the nozzles. Open the stand valve and operate the tester handle for several quick strokes to expel air from the injection nozzle. Fluid should flow from the spray holes in the nozzle tip.

CAUTION

Always wear approved safety glasses when operating the nozzle tester. Avoid contacting the spray with any open flame or sparks. Do not smoke during testing.

NOTE: Use only approved SAE No. 208629 calibration fluid in the tester or (SAE J968D or ISO 4113).

CAUTION

During testing keep hands and skin away from the spray, since the liquid leaves the nozzle tip with sufficient force to penetrate the skin and cause serious injury. If available, enclose the nozzle tip in a transparent receptacle.

5. Check the nozzle opening pressure. Open the gauge valve and pump the handle slowly until the nozzle sprays. Observe the gauge to determine opening pressure. Normal operating pressure of the nozzle is 2100-2150 psi (14,480-14,824 kPa) and the service minimum is 1850 psi (12,756 kPa).

NOTE: Chatter will vary from nozzle to nozzle as will the sound. While nozzle chatter is acceptable, a lack of chatter is not a reason to condemn a nozzle.

6. Check the tip leakage by operating the test pump to maintain a constant pressure at 200 psi (1379 kPa) below opening pressure. The nozzle tip should remain dry without an accumulation of fuel drops at the tip spray holes. A slight wetting after about 5 seconds is permissible if no droplets are formed.

NOTE: Wiping the nozzle tip with the fingers will draw fuel from the nozzle giving a false test result. Use a clean, dry and lint free cloth to wipe the tip dry before testing.

7. Check return port fuel leak-off, operate the test pump and observe the return port during spray. A leak-off rate of one or two drops per tester stroke is acceptable. If fuel squirts from the return port, the nozzle is faulty and must be replaced.

8. Operate the tester with smooth even strokes and observe the spray pattern. Concentrate on the first three in.es of spray from the end of the nozzle. The spray should be finely atomized in an even straight pattern. The pattern should be concentric (not lopsided). Fuel should not come out in droplets or a solid stream. If a nozzle fails the spray test, clean the nozzle and repeat the spray test.

9. Soak the nozzles in a cold decarbonizing solution for one hour, or use a sonic nozzle cleaner, if available.

10. Install the nozzle(s), run the engine and check for leaks.

Injection Pump

NOTE: Before removing any fuel lines, clean exterior with clean fuel oil or solvent to prevent entry of dirt into engine when fuel lines are removed.

CAUTION

Do not wash or steam clean engine while engine is running. Serious damage to injection pump could occur.

Removal

1. Disconnect battery ground cables from both batteries.
2. Remove engine oil filler neck.
3. Remove bolts attaching injection pump to drive gear.
4. Disconnect electrical connectors to injection pump.
5. Disconnect accelerator cable and speed control cable from throttle lever, if so equipped.
6. Remove air cleaner and install intake opening cover, Tool T83T-9424-A or equivalent.
7. Remove accelerator cable bracket, with cables attached, from intake manifold and position out of the way.

NOTE: All fuel lines and fittings must be capped using Fuel System Protective Cap Set T83T-9395-A, or equivalent, to prevent fuel contamination.

8. Remove fuel filter-to-injection pump fuel line and cap fittings.
9. Remove and cap injection pump inlet elbow.
10. Remove and cap injection pump fitting adapter.
11. Remove fuel return line on injection pump, rotate out of the way and cap all fittings.

NOTE: It is not necessary to remove injection lines from injection pump to remove injection pump. If lines are to be removed, loosen injection line fittings at injection pump before removing it from engine.

12. Remove fuel injection lines from nozzles and cap lines and nozzles.
13. Remove three nuts attaching injection pump to injection pump adapter.
14. If injection pump is to be replaced, loosen injection line retaining clips and injection nozzle fuel lines and cap all fittings at this time with protective cap set T83T-9395-A or equivalent. Do not install injection nozzle fuel lines until new pump is installed in engine.
15. Lift injection pump, with nozzle lines attached, up and out of engine compartment.

CAUTION

Do not carry injection pump by injection nozzle fuel lines as this could cause lines to bend or crimp.

Installation

1. Install new O-ring on drive gear end of injection pump.
2. Move injection pump down and into position.
3. Position alignment dowel on injection pump into alignment hole on drive gear.
4. Install bolts attaching injection pump to drive gear and tighten to specification.
5. Install nuts attaching injection pump to adapter. Align scribe lines on injection pump flange and injection pump adapter and tighten to 14 ft. lbs. (19 Nm).
6. If injection nozzle fuel lines were removed from injection pump install at this time.
7. Remove caps from nozzles and fuel lines and install fuel line nuts on nozzles and tighten to 22 ft. lbs. (30 Nm).
8. Connect fuel return line to injection pump.

9. Install injection pump fitting adapter with a new O-ring.
10. Clean old sealant from injection pump elbow threads, using clean solvent and dry thoroughly. Apply a light coating of pipe sealant on elbow threads.
11. Install elbow in injection pump adapter and tighten to a minimum of 6 ft. lbs. (8 Nm). Then tighten further, if necessary, to align elbow with injection pump fuel inlet line, but do not exceed 360° of rotation or 10 ft. lbs. (13 Nm).
12. Remove caps and connect fuel filter-to-injection pump fuel line.
13. Install accelerator cable bracket to intake manifold.
14. Remove intake manifold cover and install air cleaner.
15. Connect accelerator and speed control cable, if so equipped, to throttle lever.
16. Install electrical connectors on injection pump.
17. Clean injection pump adapter and oil filler neck sealing surfaces.
18. Apply a 1/8 in. bead of RTV Sealant on adapter housing.
19. Install oil filler neck and tighten to specifications.
20. Connect battery ground cables to both batteries.
21. Run engine and check for fuel leaks.
22. If necessary, purge high pressure fuel lines of air by loosening connector one half to one turn and cranking engine until solid fuel, free from bubbles, flows from connection.
23. Check and adjust injection pump timing.

CAUTION

Keep eyes and hands away from nozzle spray. Fuel spraying from the nozzle under high pressure can penetrate the skin and cause infection. Medical attention should be provided immediately in the event of skin penetration.

Bleeding Fuel System

If necessary, purge high pressure fuel lines of air by loosening connector one half to

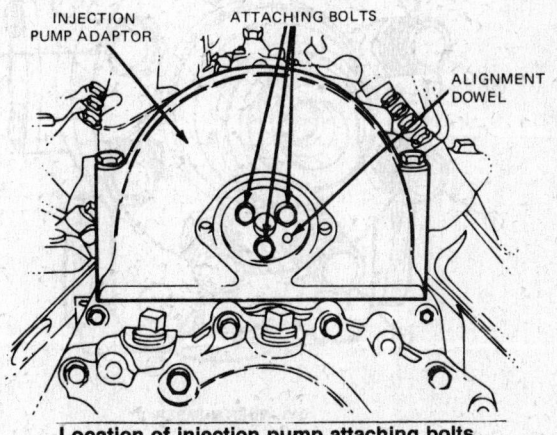

Location of injection pump attaching bolts

one turn and cranking engine until solid fuel, free from bubbles, flows from connection. Tighten the fuel line fitting and start the engine to check for fuel leaks.

Fuel Control

On-off fuel control is provided by an electric solenoid located in the diesel injection pump housing cover. Current is supplied to the solenoid when the ignition switch is turned on. If no fuel is supplied with the ignition switch in the on position, check for current at the solenoid terminal before condemning the solenoid.

Fuel Lines

Removal

NOTE: Before removing any fuel lines, clean exterior with clean fuel oil, or solvent to prevent entry of dirt into fuel system when fuel lines are removed. Blow dry with compressed air.

1. Disconnect battery ground cables from both batteries.
2. Remove air cleaner and cap intake manifold opening with Ford Tool T83T-9424-A or equivalent.
3. Disconnect accelerator cable and speed control cable, if so equipped, from injection pump.
4. Remove accelerator cable bracket from intake manifold and position out of the way with cable(s) attached.

NOTE: To prevent fuel system contamination, cap all fuel lines and fittings with protective cap set.

5. Disconnect fuel line from fuel filter to injection pump and cap all fittings.
6. Disconnect and cap nozzle fuel lines at nozzles.
7. Remove fuel line clamps from fuel lines to be removed.
8. Remove and cap injection pump inlet elbow.

9. Remove and cap inlet fitting adapter.
10. Remove injection nozzle lines, one at a time, from injection pump.

NOTE: Fuel lines must be removed following this sequence: 5–6–4–8–3–1–7–2. Install caps on each end of each fuel line and pump fitting as it is removed and identify each fuel line accordingly.

Installation

1. Install fuel lines on injection pump, one at a time and tighten to 22 ft. lbs. (30 Nm).

NOTE: Fuel lines must be installed in the sequence: 2–7–3–1–8–4–6–5.

2. Clean old sealant from injection pump elbow, using clean solvent and dry thoroughly.
3. Apply a light coating of pipe sealant on elbow threads.
4. Install elbow in injection pump adapter and tighten to a minimum of 6 ft. lbs. (8 Nm) then tighten further, if necessary, to align elbow with injection pump fuel inlet line, but do not exceed 360° of rotation or 10 ft. lbs. (13 Nm).
5. Remove caps from fuel lines and connect lines to nozzles and tighten to 22 ft. lbs. (30 Nm).
6. Uncap and connect fuel line from fuel filter to injection pump and tighten.
7. Install fuel line retaining clamps and tighten.
8. Install accelerator cable bracket on intake manifold.
9. Connect accelerator and speed control cable, if so equipped, to injection pump throttle lever.
10. Remove intake manifold cover and install air cleaner.
11. Connect battery ground cables to both batteries.
12. Run engine and check for fuel leaks.
13. If necessary, purge high pressure fuel lines of air by loosening connector one half to one turn and cranking engine until solid

fuel, free from bubbles, flows from connection.

— CAUTION —
Keep eyes and hands away from nozzle spray. Fuel spraying from the nozzle under high pressure can penetrate the skin.

G.M. DIESEL TUNE-UP AND ADJUSTMENTS

Compression Test

When checking the compression, always make sure that the batteries are at or near full charge. The total reading for any given cylinder is not as important as the difference between all cylinders. The cylinder with the lowest reading should not be less than 70% of the one with the highest reading and no cylinder should be less than 275 psi.

1. Remove the air cleaner and cover the air crossover.
2. Disconnect the wire from the fuel solenoid terminal on the injection pump.
3. Tag and disconnect all glow plug wiring and then remove the glow plugs.
4. Screw a compression gauge into the hole of the cylinder that is being checked.
5. Crank the engine. Six "puffs" per cylinder should be enough for an accurate reading. Normal compression will build up quickly and evenly if the cylinder is OK.

NOTE: Never add oil to any cylinder during a compression test, as extensive damage may result.

6. Installation is in the reverse order.

Valve Adjustment

This engine uses hydraulic valve lifters; no adjustment is necessary or possible.

Injection Timing

Adjustment
1980–81 MODELS

For the engine to be properly timed, the marks on the top of the injection pump adapter and the flange of the injection pump must be in alignment. This is done with the engine turned off.

1. Loosen the three pump retaining nuts with the proper tool.
2. Use a one in. open end wrench on the boss at the front of the injection pump and rotate the pump until the two timing marks align.
3. Tighten the retaining nuts to 35 ft. lbs. and then adjust the throttle rod.

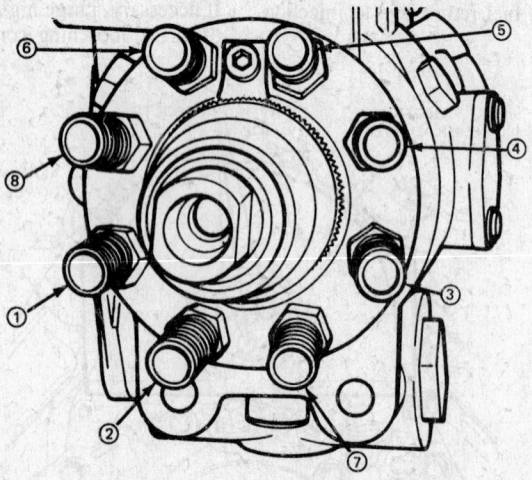

CYLINDER NUMBER
Cylinder location on injection pump delivery lines

G.M. DIESEL ENGINE 5.7 LITER, 350 CU. IN. V8

GENERAL ENGINE SPECIFICATIONS

Year	Engine No. Cyl. Displacement (cu. in.)	Carburetor Type	Horsepower @ rpm	Torque @ rpm (ft. lbs.)	Bore × Stroke (in.)	Compression Ratio	Oil Pressure @ 2000 rpm
1978	8–350	Diesel	120 @ 3600	220 @ 1600	4.057 × 3.385	22.5:1	40
1979–	8–350	Diesel	125 @ 3600	225 @ 1600	4.057 × 3.385	22.5:1	40

TUNE-UP SPECIFICATIONS

Year	Engine No. Cyl. Displacement (cu. in.)	Timing (deg) Man Trans.	Timing (deg) Auto. Trans.	Minimum Compression (lbs)	Valves Intake Opens (deg)	Fuel Pump Pressure (psi)②	Idle Speed (rpm) Man. Trans.	Idle Speed (rpm) Auto. Trans.
1978	8–350	—	①	275	16	5.5–6.5	—	575
1979	8–350	—	①	275	16	5.5–6.5	—	575
1980	8–350	—	①	275	16	5.5–6.5	—	600
1981	8–350	—	①	275	16	5.5–6.5	—	575/600
1982–'84	8–350	—	4 at DC	275	16	5.5–6.5	—	600

Note: The underhood specifications sticker often reflects tune-up specification changes made in production. Sticker figures must be used if they disagree with those in this chart.

①Align the timing marks on the injection pump and drive flange.

②Fuel transfer pressure given—Injection pump = 8-12 PSI @ 1000 rpm (take the reading at the injection pump pressure tap—injector opening can be as high as 1225 psi).

TORQUE SPECIFICATIONS
All readings in ft. lbs.

Year	Engine	Cylinder Head Bolts	Rod Bearing Bolts	Main Bearing Bolts	Crankshaft Bolt	Flywheel to Crankshaft Bolts	Mainfold Intake	Mainfold Exhaust
1978–'84	350	130①	42	120	200–310	60	40①	25

①Clean and dip entire bolt in engine oil before tightening to obtain a correct torque reading.

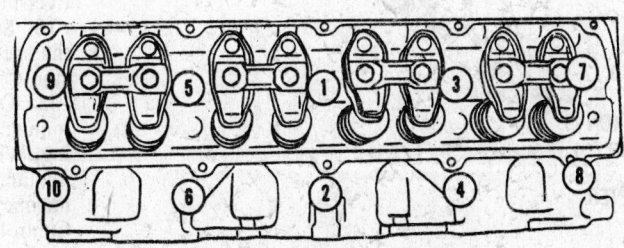

Cylinder nead torque sequence

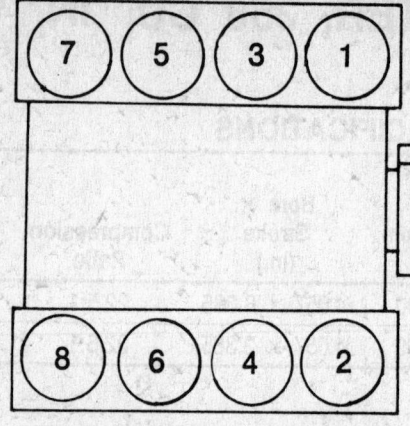

Engine firing order: 1-8-4-3-6-5-7-2

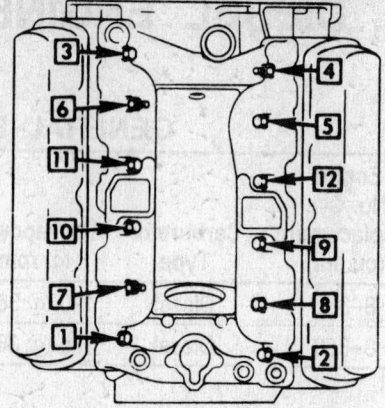

Intake manifold torque sequence

Establishing A New Timing Mark

When a new injection pump adapter has been installed you will need to make a new timing mark also.

1. File off the original mark on the adapter. DO NOT file off the mark on the pump flange.

2. Position the no. 1 cylinder at TDC of the compression stroke.

3. Align the mark on the vibration balancer with the zero mark on the indicator. The position of the injection pump driven gear should be offset to the right when the No. 1 cylinder is at TDC.

4. Install a special timing tool into the pump adapter. Torque the tool toward the no. 1 cylinder to 50 ft. lbs.

5. Mark the pump adapter, remove the special tool and install the injection pump.

Checking Timing

1982 AND LATER

The timing meter J–33075 or equivalent picks up the engine speed and crankshaft position from the crankshaft balancer. It uses a luminosity signal through a glow plug probe to determine combustion timing. Certain engine malfunctions may cause incorrect timing readings. Engine malfunctions should be corrected before a timing adjustment is made. The marks on the pump and adapter flange will normally be aligned within 0.050 in. (1.27mm).

NOTE: Alignment of timing marks may be used in emergency situations (i.e. timing meter not available). However for optimum engine operation, the timing should be adjusted with the timing meter as soon as possible.

1. Place transmission selector lever in park, apply parking brake and block drive wheels.

2. Start the engine and let it run at idle until fully warmed up. Then shut off the engine.

NOTE: Failure to have the engine fully warmed up will result in incorrect timing reading and adjustments.

3. Remove air cleaner assembly and install cover J–26996–1 or equivalent. The EGR valve hose must be disconnected.

4. Clean any dirt from the engine probe holder (RPM counter) and crankshaft balancer rim.

5. Clean the lens on both ends of the glow plug probe and clean the lens in the photo-electric pick-up. Use a dulled tooth pick to scrape the carbon from the combustion chamber side of the glow plug probe. Look through the probe to be sure its clean.

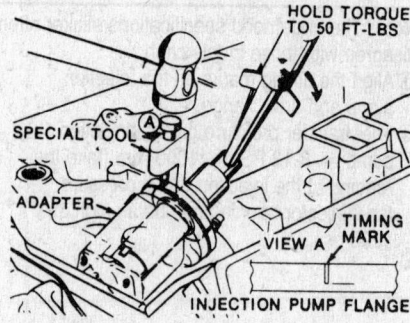

Marking the injection pump adapter

Retarted readings will result if the probe is not clean.

6. Install the rpm probe into the crankshaft rpm counter (probe holder).

7. Remove the glow plug from No. 1 cylinder. Install the glow plug probe in the glow plug opening. Torque the probe to 9 ft. lbs. (12 Nm).

8. Set the timing meter offset selector to the 8 cylinder setting.

9. Connect the battery leads; red to positive, black to negative.

10. Start the engine and adjust the rpm to the speed specified on the "Vehicle Emission Control Information Label".

11. Observe the timing reading then at 2 minute intervals, again observe the reading. When the readings stabilize over the 2 minute interval, compare that reading to the one specified on the "Vehicle Emission Control Information Label". The timing reading, when set to specification will be "Negative" (after top dead center).

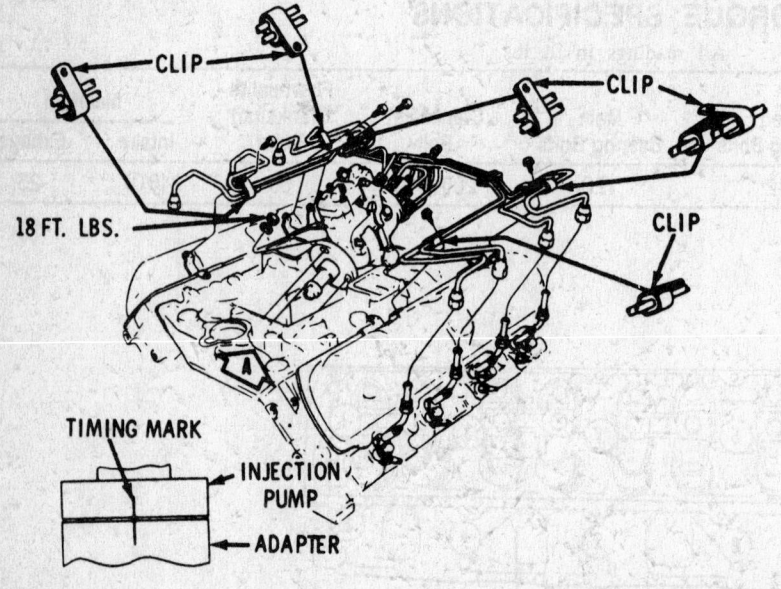

Injection pump timing marks

12. Disconnect the timing meter.

13. Lubricate only the threads of the removed glow plug with lubricant 9985462 or equivalent.

NOTE: Failure to apply the correct lubricant can cause engine damage.

14. Install the removed glow plug. Torque the glow plug to 15 ft. lbs. (21 Nm).

15. Install the air cleaner being certain to reconnect the EGR valve hose.

Adjusting Timing

1982 AND LATER

1. Shut off the engine.

2. Note the relative position of the marks on the pump flange and pump intermediate adapter.

3. Loosen the bolts holding the pump to the adapter to a point where the pump can be rotated. Use a 1″ open end wrench. (Tool J–25304 has the proper offset on the handle to clear the fuel return line).

4. Rotate the pump to the left to advance the timing and to the right to retard the timing. The width of the mark on the intermediate adapter is about 2/3 degree. Move the pump the amount that is needed and tighten the pump retaining bolts to 35 ft. lbs. (47 Nm).

5. Start the engine and recheck the timing reading as outlined previously. Reset and recheck the timing if needed.

6. Reset the fast and curb idle speeds. Both procedures are in this section.

   a. Sooty or dirty probes will result in retarded readings.

   b. The luminosity probe will soot up very fast when used in a cold engine.

   c. Wild needle fluctuations on the timing meter indicate a cylinder not firing properly. Correction of this condition must be made prior to adjusting the timing.

Mechanical Governor

The governor serves the purpose of maintaining the desired engine speed within the operating range under varying load conditions. The limits of throttle travel are set by throttle linkage screws for proper slow idle and maximum high idle. The governor operates automatically and is not adjustable. The maximum high idle is factory set and should not be adjusted at any time. The slow and A/C fast idle settings are adjustable.

IDLE SPEED

Slow Idle Adjustment

1. Run the engine until it reaches normal operating temperature.

2. Insert the probe of a magnetic pickup tachometer into the timing indicator hole.

3. Set the parking brake and block the drive wheels.

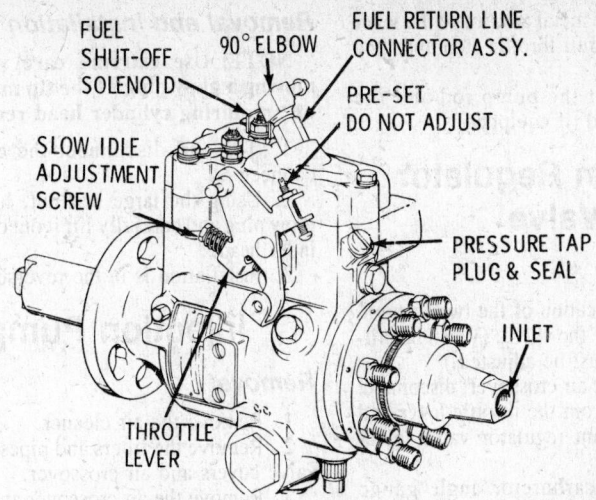

Injection pump slow idle adjustment and fuel solenoid

4. Place the transmission in Drive and turn the A/C off (if so equipped).

5. Turn the slow idle adjustment screw on the injection pump to obtain the idle speed specified on the emission control label.

Fast Idle Solenoid Adjustment

1. Set the parking brake and block the drive wheels.

2. Run the engine until it reaches normal operating temperature.

3. Place the transmission in Drive and disconnect the compressor clutch wire.

4. Turn the A/C on. On cars without A/C, disconnect the solenoid wire and then connect jumper wires to the solenoid terminals. Ground one wire and connect the other to the battery, this will activate the solenoid.

5. Adjust the fast idle solenoid plunger to obtain the specified rpm.

THROTTLE ROD

Adjustment

1. Check timing.

2. Remove the clip from the cruise control rod (if so equipped) and disconnect the rod from the throttle lever assembly.

3. Disconnect the detent cable from the throttle assembly.

4. Loosen the lock nut on the pump rod and shorten it several turns.

5. Rotate the lever assembly to the full throttle position and hold it there.

6. Lengthen the pump rod until the injection pump lever just contacts the full throttle stop.

7. Release the lever assembly and tighten the pump rod lock nut.

8. Remove the pump rod from the lever assembly and reconnect the detent cable.

DETENT CABLE

Adjustment

NOTE: The throttle rod must be adjusted before adjusting the detent cable.

1. Depress and hold the metal lock tab on the cable upper end.

2. Move the slider through the fitting, away from the lever assembly, until it stops against the metal fitting.

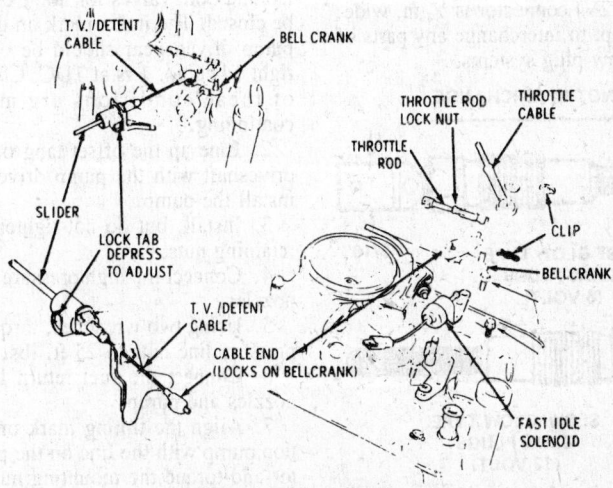

Detent cable adjustment

3. Release the metal tab, rotate the lever assembly to the full throttle stop and then release it.

4. Reconnect the pump rod and the cruise control rod (if equipped).

Vacuum Regulator Valve

Adjustment

1. Note the location of the two vacuum hoses and remove the valve. (When installing, the valve must be adjusted.)

2. Remove the air crossover, disconnect the throttle rod from the throttle lever and loosen the vacuum regulator valve injection pump bolts.

3. Place the carburetor angle gauge adapter on the injection pump throttle lever. Place the angle gauge on the adapter.

4. Rotate the throttle lever to wide open throttle and set the angle gauge to zero degrees. Center the bubble in the level and reset the angle gauge to 50 degrees.

5. Rotate the throttle lever to ceter the bubble. Apply an outside vacuum source of 18–22 in. to the inboard port of the vacuum valve. Rotate the valve clockwise to obtain 7–8 in. of vacuum.

Glow Plugs

Eight glow plugs are used to heat the prechamber to aid in starting. They are essentially small heaters that turn on when the ignition switch is turned to the "RUN" position prior to starting the engine. They remain on for a short time after starting and then automatically shut off.

There are two types of glow plugs used on G.M. diesels; the "fast-glow" type and the "slow-glow" type. The fast-glow type use pulsing current applied to 6 volt glow plugs, while the slow-glow type use a continuous current applied to 12 volt glow plugs.

An easy way to tell the plugs apart is that the fast-glow (6V) plugs have a $^5/_{16}$ in. wide electrical connector plug, while the slow glow (12V) connector is $^1/_4$ in. wide. Do not attempt to interchange any parts of these two glow plug systems.

DO NOT INTERCHANGE

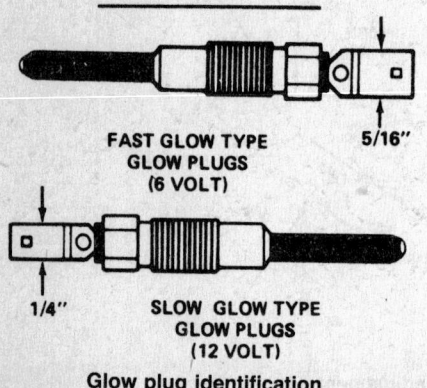

FAST GLOW TYPE GLOW PLUGS (6 VOLT)   5/16"

1/4"   SLOW GLOW TYPE GLOW PLUGS (12 VOLT)

Glow plug identification

Removal and Installation

NOTE: Use extreme care when removing a glow plug as the tip may break off; requiring cylinder head removal.

1. Tag and disconnect the electrical connectors.

2. Using the large hex nut, loosen the glow plug and carefully lift it out of the cylinder head.

3. Installation is in the reverse order.

Injection Pump

Removal

1. Remove the air cleaner.

2. Remove the filters and pipes from the valve covers and air crossover.

3. Remove the air crossover and cap the intake manifold with screened covers or tape.

4. Disconnect the throttle rod and return spring.

5. Remove the bellcrank.

6. Remove the throttle and detent cables from the intake manifold brackets.

7. Disconnect the fuel lines from the filter and remove the filter.

8. Disconnect the fuel inlet line at the pump.

9. Remove the rear A/C compressor (if so equipped) and remove the fuel line.

10. Disconnect the fuel return line from the injection pump.

11. Remove the clamps and pull the fuel return lines from each injection nozzle.

12. Using two wrenches, disconnect the high pressure lines at the nozzles.

13. Remove the three injection pump retaining nuts.

14. Remove the pump and cap all lines and nozzles.

Installation

1. Remove the protective caps from all lines and nozzles. Place the engine on TDC for the no. 1 cylinder. The mark on the harmonic balancer on the crankshaft will be aligned with the zero mark on the timing tab and both valves for no. 1 cylinder will be closed. The index mark on the injection pump driven gear should be offset to the right when no. 1 is at TDC. Check that all of these conditions are met before continuing.

2. Line up the offset tang on the pump driveshaft with the pump driven gear and install the pump.

3. Install, but do not tighten the pump retaining nuts.

4. Connect the high pressure lines at the nozzles.

5. Using two wrenches, torque the high pressure line nuts to 25 ft. lbs.

6. Connect the fuel return lines to the nozzles and pump.

7. Align the timing mark on the injection pump with the line on the pump adaptor and torque the mounting nuts to 35 ft. lbs.

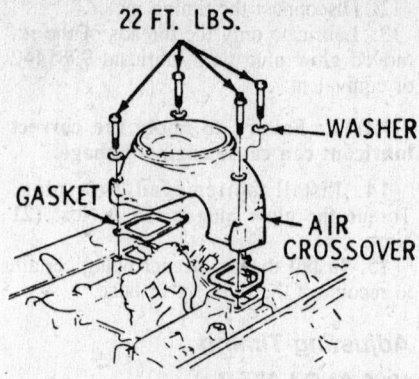

22 FT. LBS.

WASHER

GASKET

AIR CROSSOVER

Typical air crossover

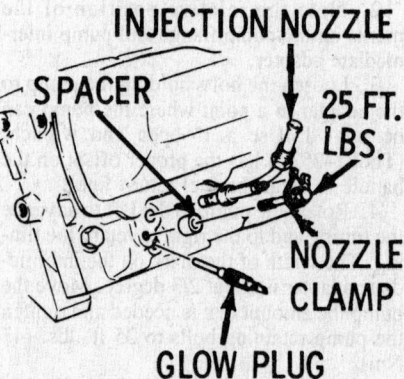

INJECTION NOZZLE

SPACER

25 FT. LBS.

NOZZLE CLAMP

GLOW PLUG

Injector nozzle installation—1978–79

NOTE: A one in. open end wrench on the boss at the front of the injection pump will aid in rotating the pump to align the marks.

8. Adjust the throttle rod.

9. Install the fuel inlet line between the transfer pump and the filter.

10. Install the rear A/C compressor brace (if so equipped).

11. Install the bellcrank and clip.

12. Connect the throttle rod and return spring.

13. Adjust the transmission cable.

14. Start the engine and check for fuel leaks.

15. Remove the screened covers or tape and install the air crossover.

16. Install the tubes in the airflow control valve in the air crossover and install the ventilation filters in the valve covers.

17. Install the air cleaner.

18. Start the engine and allow it to run for two minutes. Stop the engine, let it stand for two minutes, then restart. This permits the air to bleed off within the pump.

Injectors

Removal and Installation

1980–84

The injectors on these engines are simply unscrewed from the cylinder head, after the fuel lines have been removed, much like a spark plug. Be careful not to damage the in-

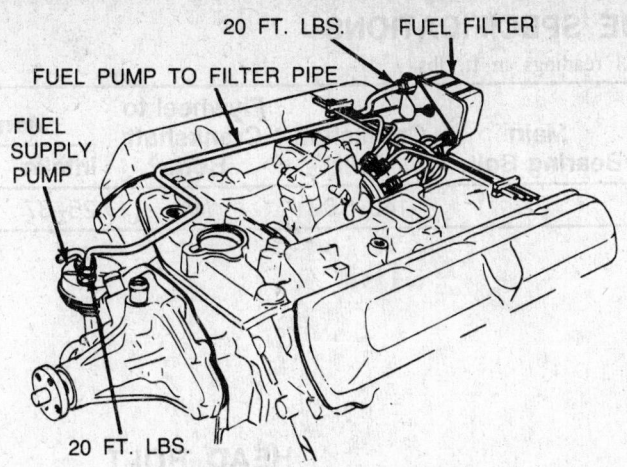

20 FT. LBS. — FUEL FILTER
FUEL PUMP TO FILTER PIPE
FUEL SUPPLY PUMP
20 FT. LBS.

Fuel supply pump, filter and lines

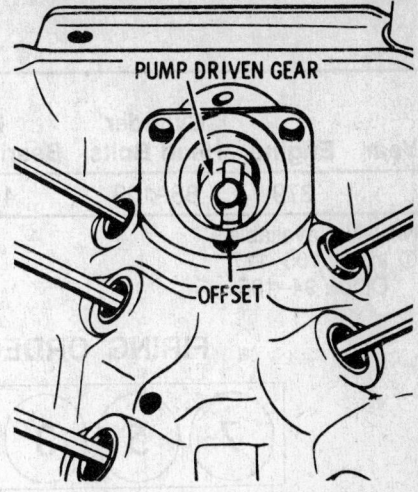

PUMP DRIVEN GEAR
OFFSET

Offset on the pump driven gear

jector tip and make sure that the copper gasket is removed from the cylinder head if it does not come off with the injector.

Clean the carbon build-up from the tip of the injector. Installation is in the reverse.

NOTE: 1981 and later engines use two types of injectors; CAV Lucas and Diesel Equipment. When installing the inlet fittings, tighten to 45 ft. lbs. on the Diesel Equipment injector and to 25 ft. lbs. on the CAV injector.

Injection Pump Fuel Lines

When any fuel lines are to be removed, clean all the fittings before loosening. Immediately cap all lines, nozzles and fittings to maintain system cleanliness.

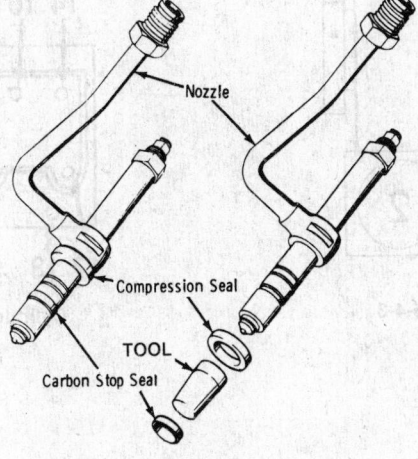

Nozzle
Compression Seal
TOOL
Carbon Stop Seal

Nozzle seal installation—1978–79

G.M. DIESEL ENGINE 6.2 LITER, 379 CU. IN. V8 DESCRIPTION

The 6.2 liter, 379 cu. in. V8, 4 cycle diesel is developed and produced by Chevrolet. It is a totally new engine designed specifically for truck application with heavy duty usage in mind.

The base of the engine (short block) is very similar in design to a V8 gasoline en-

G.M. DIESEL ENGINE 6.2 LITER, 379 CU. IN. V8

GENERAL ENGINE SPECIFICATIONS

Year	Engine No. Cyl. Displacement (cu. in.)	Carburetor Type	Horsepower @ rpm	Torque @ rpm (ft lbs)	Bore × Stroke (in.)	Compression Ratio	Oil Pressure @ 2000 rpm
	8–379	Diesel	130 @ 3600	240 @ 2000	3.98 × 3.80	21.5:1	NA

NA—Not available

TUNE-UP SPECIFICATIONS

Year	Engine No. Cyl. Displacement (cu. in.)	Ignition Timing (deg)① Man Trans.	Auto. Trans.	Compression (lbs)	Valves Intake Opens (deg)	Fuel Pump Pressure (psi)	Idle Speed (rpm)① Slow	Fast
	8–379			275	NA	5.5–6.5	575/550	700

NOTE: The underhood specifications sticker often reflects tune-up specification changes made in the production run. Sticker figures must be used if they differ with those in this chart.

①Where two figures are separated by a slash, the first is for manual trans, the second is for auto trans.

NA—Not available

TORQUE SPECIFICATIONS

All readings in ft. lbs.

Year	Engine	Cylinder Head Bolts	Rod Bearing Bolts	Main Bearing Bolts	Crankshaft Bolt	Flywheel to Crankshaft Bolts	Manifold	
							Intake	Exhaust
	379	88–103	44–52	①	140–162	NA	25–37	18–25

NA—Not available
① Inner: 105–117
   Outer: 94–105

FIRING ORDER

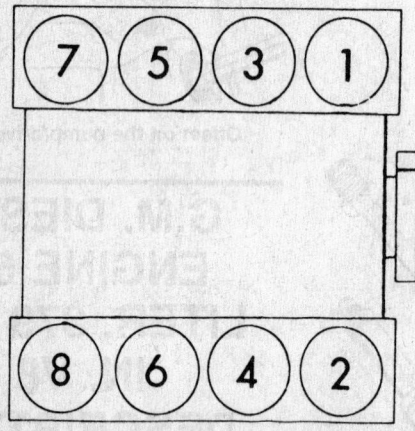

Engine firing order: 1-8-7-2-6-5-4-3

HEAD BOLT TORQUE SEQUENCE

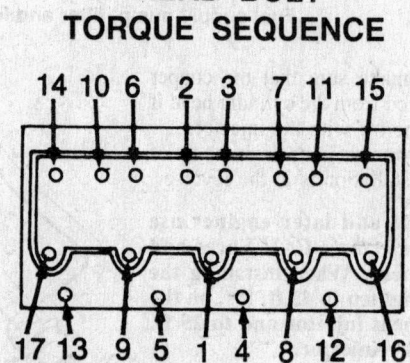

Cylinder head torque sequence

gine; the major difference being the cylinder heads, combustion chamber, fuel distribution system, air intake manifold and the method of ignition. The cylinder block, crankshaft, main bearings, connecting rods and pistons look much the same as their gasoline engine counterparts, although they are of much heavier construction due to the higher compression ratio required to ignite diesel fuel. The intake and exhaust manifolds are of special design and construction.

The cylinder head incorporates a 17 bolt head design which locates 5 bolts around each cylinder. This helps gasket durability. It also includes a high swirl pre-combustion chamber which mixes fuel and air to provide an efficient fuel burn and low emissions. A special cavity in the piston top further assists in mixing the combustion products for complete burning.

Main bearing caps all use 4 bolts instead of the normal 2 to provide rigid support for the crankshaft and minimize stress. The rolled fillet nodular iron crankshaft utilizes a torsional damper, tuned to reduce vibrations.

The engine also uses roller hydraulic lifters running on a forged steel camshaft.

TUNE-UP AND ADJUSTMENT

Compression Test

When checking the compression, always make sure that the batteries are at or near full charge. The total reading for any given cylinder is not as important as the difference between all cylinders. The cylinder with lowest reading should not be less than 70% of the one with the highest reading and no cylinder should be less than 275 psi.

1. Remove the air cleaner and cover the air crossover.
2. Disconnect the wire from the fuel solenoid terminal on the injection pump.
3. Tag and disconnect all glow plug wiring and then remove the glow plugs.
4. Screw a compression gauge into the hole of the cylinder that is being checked.
5. Crank the engine. Six "puffs" per cylinder should be enough for an accurate reading. Normal compression will build up quickly and evenly if the cylinder is OK.

NOTE: Never add oil to any cyinder during a compression test, as extensive damage may result.

6. Installation is in the reverse order.

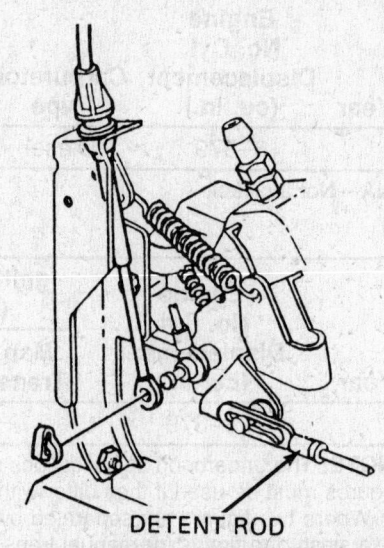

DETENT ROD

Accelerator linkage

Valve Adjustment

This engine uses roller hydraulic valve lifters; no adjustment is necessary or possible.

Injection Timing

Adjustment

For the engine to be properly timed, the marks on top of the engine front cover and the injection pump flange must be aligned. This is done with the engine turned off.

1. Loosen the three pump retaining nuts.
2. Use the proper tool and rotate the pump until the two timing marks are in alignment.
3. Tighten the retaining nuts to 30 ft. lbs. and then adjust the throttle rod.

Establishing a New Timing Mark

When a new front cover has been installed, a new timing mark will also be required.

1. Remove the injection pump and then position the No. 1 cylinder at TDC of the compression stroke.
2. Install a special timing tool into the injection pump. Do not use a gasket.
3. The slot on the injection pump gear should be in the vertical 6 o'clock position and the timing marks on the gears will be aligned. If not, remove the tool and rotate the engine 360°.
4. Fasten the gear to the fixture and tighten.
5. Install a 10mm nut to the upper housing stud to hold the fixture flange nut finger tight.
6. Torque the large bolt (18mm) counterclockwise (toward left bank) to 50 ft. lbs. Tighten the 10mm nut.

7. Make sure that the crankshaft has not rotated and the fixture did not bind on the 10mm nut.
8. Strike a scriber with a mallet to mark the TDC position on the front cover.
9. Remove the tool, install the injection pump and attach the gear to the pump hub.
10. Adjust injection timing.

Throttle Position Switch Adjustment

1. Loose assemble throttle position switch to the injection pump with the throttle lever in the closed position.
2. Attach an ohmmeter across the IGN (pink) and EGR (yellow) terminals or wires.
3. Insert the proper "switch-closed" gauge block between the gauge boss on the injection pump and the wide open stop screw on the throttle shaft.
4. Rotate and hold the throttle lever against the gauge block.
5. Rotate the throttle switch clockwise (facing throttle switch) until continuity just occurs (high meter reading) across the IGN and EGR terminals or wires. Hold the switch body in this position and tighten the mounting screws.

NOTE: The switch point must only be set while rotating the switch body in the clockwise direction.

6. Release the throttle lever and allow it to return to the idle position. Remove the "switch-closed" gauge bar and insert a "switch-open" gauge bar.
7. Rotate the throttle lever against the "switch-open" gauge bar. There should be no continuity across the IGN and EGR terminals or wires.

8. If no continuity exists, the switch is set properly. If there is continuity, the switch must be reset by repeating the entire procedure again.

Transmission Vacuum Regulator Valve

Adjustment

1. Attach the vacuum regulator valve snugly, but loosely, to the injection pump. The switch body must be free to rotate on the pump.
2. Apply approximately 9–10 psi of vacuum to the inboard nipple. Attach a vacuum gauge to the outboard nipple.
3. Insert a vacuum regulator valve gauge bar between the gauge boss on the injection pump and the wide open stop screw on the throttle lever.
4. Rotate and hold the throttle shaft against the gauge bar.
5. Slowly rotate the vacuum regulator valve body clockwise (facing the valve) until the vacuum gauge reads 5.6 psi. Hold the valve at this position and tighten the mounting screws.

NOTE: The valve must only be set while rotating in the clockwise direction.

6. Check by releasing the throttle shaft and allowing it to return to the idle stop position. Rotate the throttle shaft back against the gauge bar and check that the vacuum gauge still reads 5.6 psi. If not, the valve must reset again.

Mechanical Governor

The governor serves the purpose of maintaining the desired engine speed within the

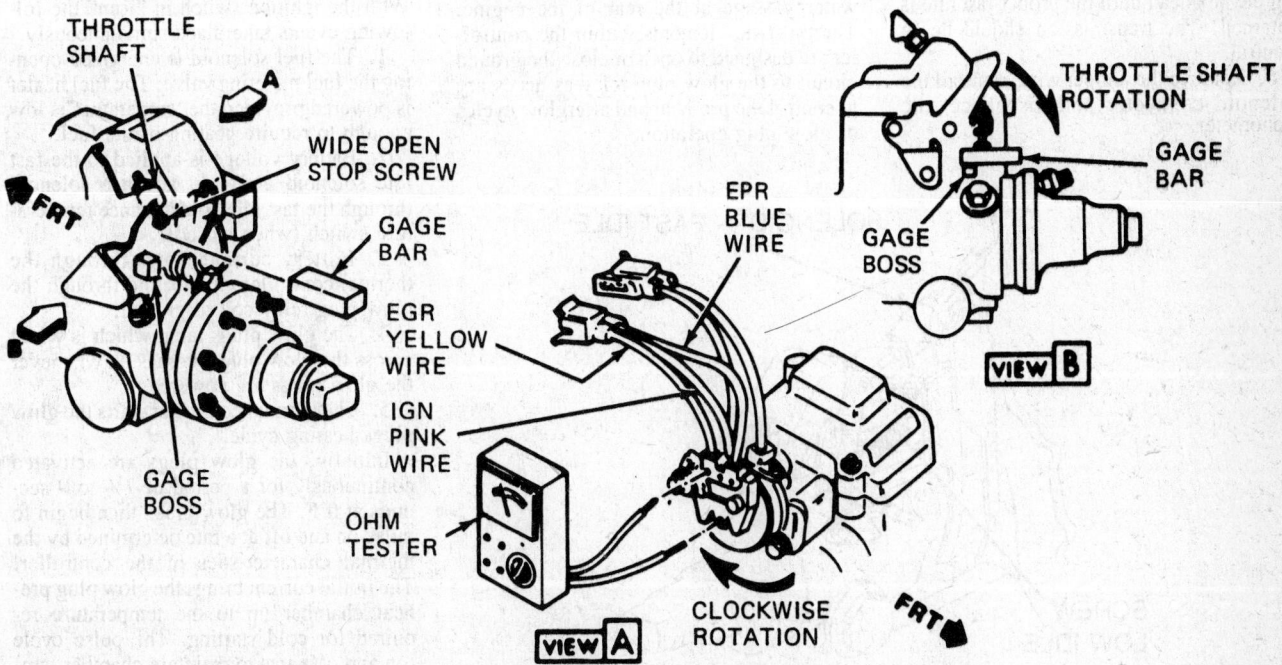

Throttle position switch adjustment

operating range under varying load conditions. The limits of throttle travel are set by throttle linkage screws for proper slow idle and maximum high idle. The governor operates automatically and is not adjustable. The maximum high idle is factory set and should not be adjusted at any time. The slow and A/C fast idle settings are adjustable.

Idle Speed

Slow Idle Adjustment

1. Run the engine until it reaches normal operating temperature.
2. Set the parking brake and block the drive wheels.
3. Remove the air cleaner and turn all accessories off.
4. Install a diesel tachometer
5. Turn the low idle speed screw on the injection pump until the proper idle is obtained. Automatic transmissions should be in Drive and manual transmissions should be in Neutral.
6. Disconnect the tachometer and install the air cleaner.

Fast Idle Speed Adjustment

1. Run the engine until it reaches normal operating temperature.
2. Set the parking brake and block the drive wheels.
3. Disconnect the connector from the fast idle solenoid. Connect an insulated jumper wire between the positive battery terminal and the solenoid terminal. This will energize the terminal.
4. Open the throttle momentarily to ensure that the fast idle solenoid plunger is energized and fully extended.
5. Adjust the extended plunger by turning the hex head until the proper fast idle is obtained. The transmission should be in Neutral.
6. Remove the jumper wire, reinstall the solenoid connector and disconnect the tachometer.

Glow Plugs

Eight glow plugs are used to preheat the chamber as an aid to starting. They are essentially small 12 volt heaters that turn on when the ignition switch is turned to the "Run" position prior to starting the engine. They remain on for a short time after starting and them automatically shut off.

Removal and Installation

NOTE: Use extreme care when removing the glow plugs as the tip may break off; requiring cylinder head removal to retrieve it.

1. Tag and disconnect the electrical connectors.
2. Using the large hex nut, loosen the plug and carefully pull it out of the cylinder head.
3. Installation is in the reverse order.

System Operation

The 6.2 liter diesel glow plug control system consists of a thermal controller, glow plug relay, 6 volt glow plugs and a "Glow Plugs" lamp. Other components which have no function in controlling glow plug operation but are part of the electrical system start and run operations are: fuel solenoid, fast idle and cold advance solenoids, cold advance temperature switch and the TCC, ECR and EPR solenoids.

They are 6 volt glow plugs (operated at 12 volts) that turn on when the ignition key is turned to the run position. They remain pulsing a short time after starting, then automatically turn off.

CONTROLLER

The thermal controller is mounted in the water passage at the rear of the engine. Thermostatic elements within the controller are designed to open or close the ground circuit to the glow plug relay as necessary to control the pre-heat and afterglow cycles of glow plug operation.

GLOW PLUG RELAY

The glow plug relay located on the left inner fender panel provides current to the glow plugs. The relay is pulsed on and off by the thermal controller.

--- CAUTION ---

This relay is automatically controlled. Any attempt to bypass relay with jumper wire or rewire for manual control may result in glow plug failure.

GLOW PLUGS

The glow plugs used in this system are 6 volt plugs which are operated at electrical system voltage (12 volts). They are not designed to burn continuously and are pulsed on and off as needed, by the thermal controller.

GLOW PLUGS LAMP

The glow plugs lamp is mounted in the instrument cluster. The lamp is wired across the glow plugs and is illuminated whenever the glow plugs are heating.

FUEL SOLENOID

The fuel solenoid is activated whenever the ignition switch is on. The solenoid is located in the fuel injection pump housing cover.

COLD ADVANCE SOLENOID

The cold advance solenoid, also located in the injection pump cover, is controlled by a cold advance temperature switch which activates this solenoid and the fast idle solenoid at a specified minimum temperature. The switch should be closed below 90°F and open above 122°F.

CIRCUIT OPERATION—COLD START

With the ignition switch in "Run" the following events take place simultaneously.

1. The fuel solenoid is energized opening the fuel metering valve. The fuel heater is powered provided the temperature is low enough to require heating of the fuel.
2. Battery voltage is applied to the fast idle solenoid and cold advance solenoid through the fast idle/cold advance temperature switch (when closed).
3. Battery current flows through the thermal controller circuits and through the glow plug relay coil to ground.
4. The glow plugs lamp which is wired across the glow plugs, comes on whenever the glow plugs are powered.
5. The thermal controller starts the glow plugs heating cycle.

Initially, the glow plugs are activated continuously for a period of 7½ to 9 seconds at 0°F. The glow plugs then begin to pulse on and off at a rate determined by the thermal characteristics of the controller. The initial current brings the glow plug preheat chamber up to the temperature required for cold starting. The pulse cycle (on and off) acts to maintain chamber temperature to provide stable engine warm up.

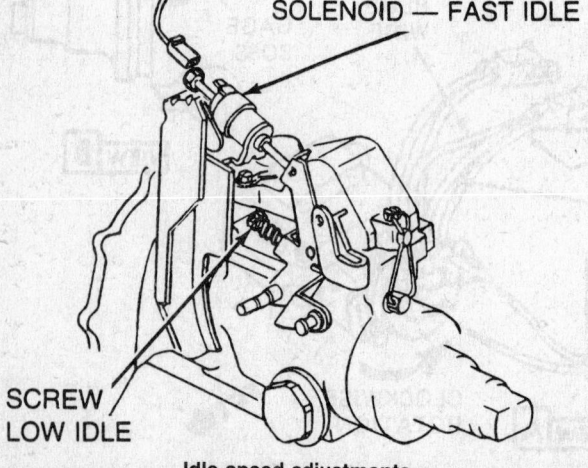

SOLENOID — FAST IDLE

SCREW
LOW IDLE

Idle speed adjustments

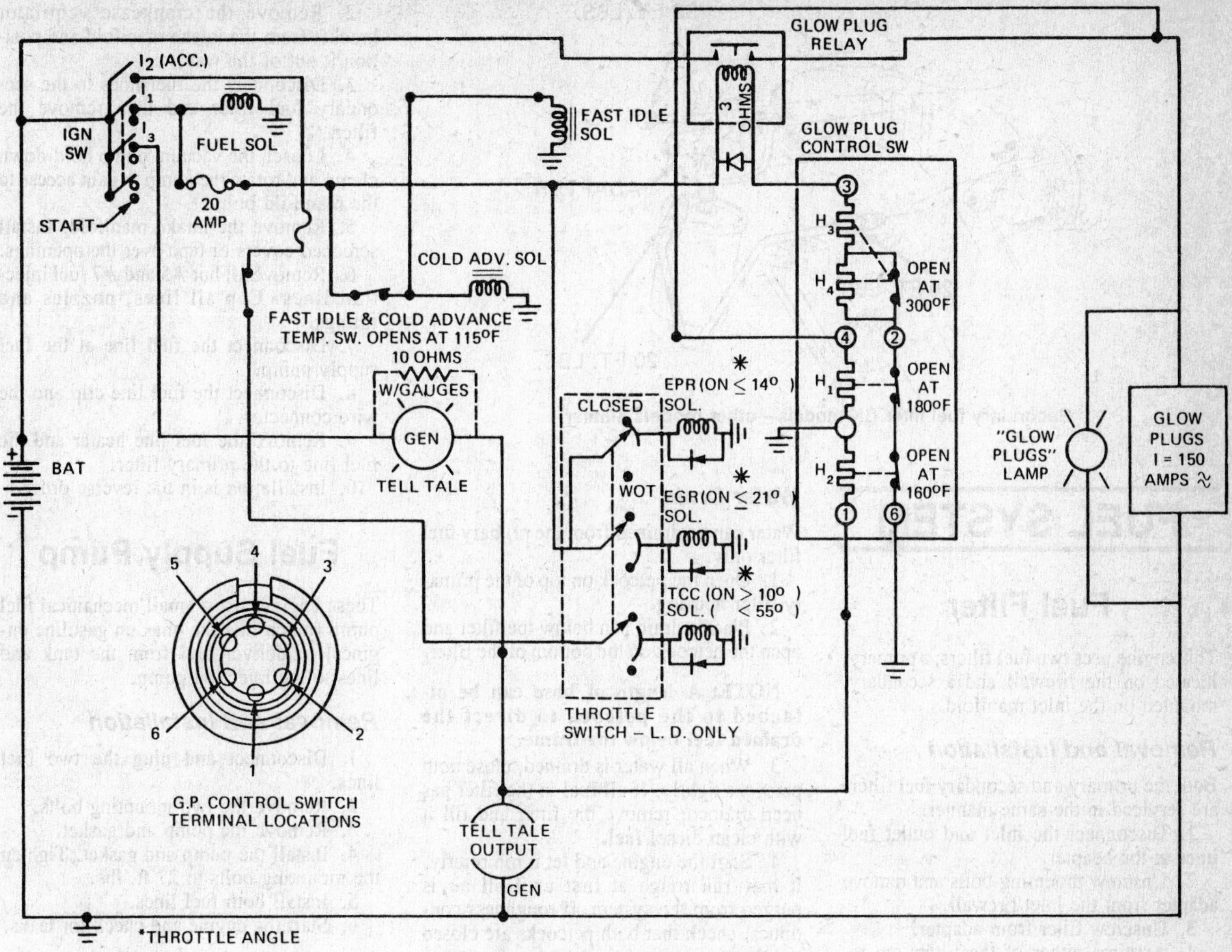

6.2L diesel glow plug wiring diagram

Diagram labels:
- I2 (ACC.)
- IGN SW
- I1, I3
- START
- FUEL SOL
- 20 AMP
- FAST IDLE SOL
- COLD ADV. SOL
- FAST IDLE & COLD ADVANCE TEMP. SW. OPENS AT 115°F
- 10 OHMS W/GAUGES
- GEN TELL TALE
- BAT
- GLOW PLUG RELAY
- 3 OHMS
- GLOW PLUG CONTROL SW
- H3, H4, H1, H2
- OPEN AT 300°F
- OPEN AT 180°F
- OPEN AT 160°F
- "GLOW PLUGS" LAMP
- GLOW PLUGS I = 150 AMPS ≈
- CLOSED
- WOT
- EPR (ON ≤ 14°) SOL.
- EGR (ON ≤ 21°) SOL.
- TCC (ON > 10° SOL. (< 55°)
- THROTTLE SWITCH – L. D. ONLY
- TELL TALE OUTPUT GEN
- G.P. CONTROL SWITCH TERMINAL LOCATIONS (5, 4, 3, 6, 2, 1)
- *THROTTLE ANGLE

As the engine warms up, the thermal controller turns off all current to the relay de-energizing the glow plugs completely. The controller is capable of varying glow plug operation as required (up to one minute) when the engine is started warm and little or no heating is necessary.

Controller failure as in the case of prolonged preheat (more than 9 seconds) would cause a circuit breaker in the controller to open, cutting off glow plug operation completely.

GOVERNORS

The governor is located under the injection pump cover.

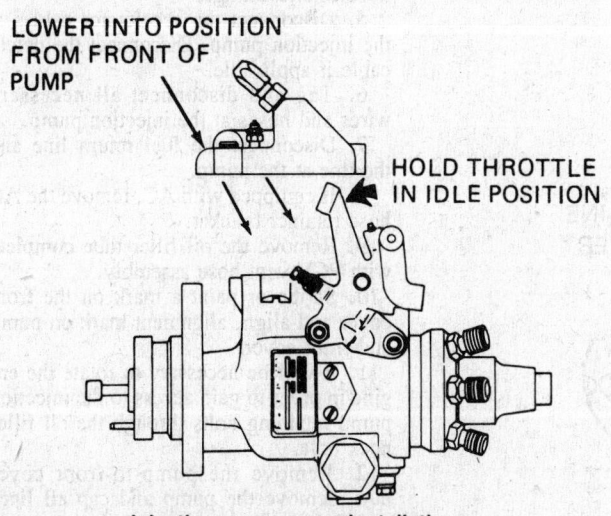

LOWER INTO POSITION FROM FRONT OF PUMP

HOLD THROTTLE IN IDLE POSITION

Injection pump cover installation

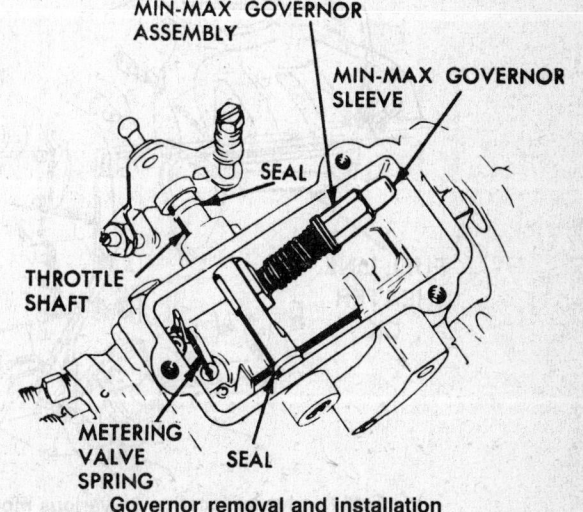

MIN-MAX GOVERNOR ASSEMBLY

MIN-MAX GOVERNOR SLEEVE

SEAL

THROTTLE SHAFT

METERING VALVE SPRING

SEAL

Governor removal and installation

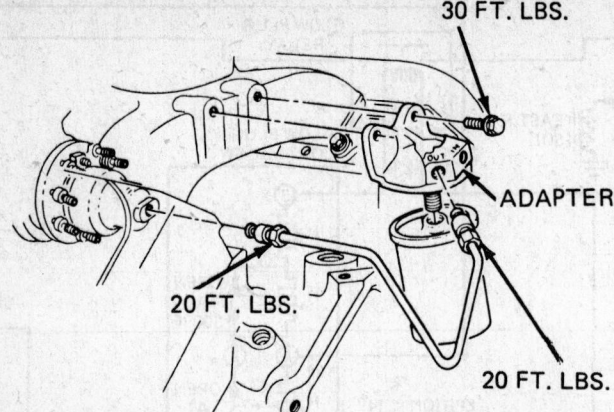

30 FT. LBS.

ADAPTER

20 FT. LBS.

20 FT. LBS.

Secondary fuel filter CK models—other models similar

FUEL SYSTEM

Fuel Filter

This engine uses two fuel filters; a primary, located on the firewall and a secondary, mounted on the inlet manifold.

Removal and Installation

Both the primary and secondary fuel filters are serviced in the same manner.
1. Disconnect the inlet and outlet fuel lines at the adapter.
2. Unscrew mounting bolts and remove adapter from the inlet/firewall.
3. Unscrew filter from adapter.
4. Anytime either of the filters are removed or replaced, refill with clean diesel fuel to prevent stalling after start up and to avoid long engine cranking time.
5. Screw the filter onto the adapter.
6. Remount the adapter and install the fuel lines.
7. Run engine and check for leaks.

Water Drain

Water can be drained from the primary fuel filter only.
1. Open the petcock on top of the primary filter housing.
2. Place a drain pan below the filter and open the petcock on the bottom of the filter.

NOTE: A length of hose can be attached to the petcock to direct the drained fuel below the frame.

3. When all water is drained, close both petcocks tightly. If all fuel in the filter has been drained, remove the filter and fill it with clean diesel fuel.
4. Start the engine and let it run briefly. It may run rough at first until all air is purged from the system. If roughness continues, check that both petcocks are closed tightly.

Fuel Line Heater

Removal and Installation

1. Disconnect the batteries and remove the air cleaner.

FUEL LINE HEATER G

FUEL LINE HEATER CK

Fuel line heater location on various models

2. Remove the crankcase ventilator bracket from the intake manifold and position it out of the way.
3. Disconnect the fuel lines to the secondary fuel filter and then remove the filter.
4. Loosen the vacuum pump hold-down clamp and rotate the pump to gain access to the manifold bolts.
5. Remove the intake manifold. Install screened covers or tape over the openings.
6. Remove all but #5 and #7 fuel injection lines. Cap all lines, nozzles and fittings.
7. Disconnect the fuel line at the fuel supply pump.
8. Disconnect the fuel line clip and the wire connector.
9. Remove the fuel line heater and the fuel line to the primary filter.
10. Installation is in the reverse order.

Fuel Supply Pump

These engines use a small mechanical fuel pump (much like the ones on gasoline engines) to deliver fuel from the tank and lines to the injection pump.

Removal and Installation

1. Disconnect and plug the two fuel lines.
2. Remove the two mounting bolts.
3. Remove the pump and gasket.
4. Install the pump and gasket. Tighten the mounting bolts to 27 ft. lbs.
5. Install both fuel lines.
6. Start the engine and check for leaks.

Injection Pump

Removal

1. Disconnect the batteries.
2. Remove the fan and the fan shroud.
3. Remove the intake manifold.
4. Remove all fuel lines. Cap all lines, nozzles and fittings.
5. Disconnect the accelerator cables at the injection pump. Disconnect the detent cable if applicable.
6. Tag and disconnect all necessary wires and hoses at the injection pump.
7. Disconnect the fuel return line and the line at the pump.
8. If equipped with AC, remove the AC hose retainer bracket.
9. Remove the oil filler tube complete with PCV vent hose assembly.
10. Scribe or paint a mark on the front cover and align, alignment mark on pump and front cover.
11. It will be necessary to rotate the engine in order to gain access to the injection pump retaining bolts through the oil filler neck hole.
12. Remove the pump-to-front cover nuts, remove the pump and cap all lines and fittings.

Testing

1. Drain all fuel from the pump.
2. Connect an air line to the pump inlet connection. Make sure that the air supply is clean and dry.
3. Seal off the return line fitting and completely immerse the pump in a bath of clean test oil.
4. Raise the air pressure in the pump to 20 psi. Leave the pump immersed in the oil for 10 min. to allow any trapped air to escape.
5. Watch for leaks after the 10 min. period. If the pump is not leaking, reduce the pressure to 2 psi for 30 sec. If there is still no leak, increase the pressure to 20 psi again. If still no leaks are seen, the pump is OK.

Installation

1. Replace the gasket.
2. Align the locating pin on the pump hub with the slot in the injection pump gear. At the same time, align the timing marks.
3. Attach the pump to the front cover and tighten the mounting nuts to 30 ft. lbs.
4. Attach pump-to-drive gear and tighten the bolts to 20 ft. lbs.
5. Install the oil filler tube along with the PCV vent hose assembly.
6. Install the AC hose retainer bracket if removed.
7. Install the fuel line at the pump and tighten to 20 ft. lbs. Install the fuel return line.
8. Connect all wires and hoses. Connect the accelerator cable.
9. Connect the injection lines.
10. Install the intake manifold.
11. Install the fan shroud, the fan and connect the batteries.

Injection Nozzle

Removal

1. Disconnect the batteries.
2. Disconnect the fuel line clip and remove the fuel return hose.

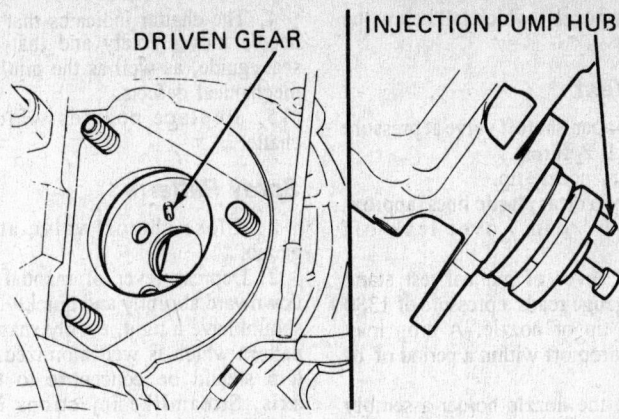

Injection pump locating pin

3. Remove the fuel injection line.
4. Remove the injection nozzle using the special tool if possible. If not, use a 30mm open end wrench. Be sure to remove the nozzle using the large 30mm hex nut. Failure to do this will result in damage to the injection nozzle. Always cap the nozzle and lines to prevent damage and contamination.

Testing

If all of the following tests are satisfied, the nozzle holder can be installed in the engine without any changes. If any one of the tests is not satisfied, the complete nozzle holder assembly must be replaced.

Preparation

1. Connect the nozzle holder assembly to the test line.
2. Close the shutoff valve to the pressure gauge.
3. Fill and flush the nozzle holder assembly with test oil by activating the lever repeatedly and briskly. This will apply test oil to all functionally important areas of the nozzle and purge it of air.

Obtaining Pressure Check

1. Open shutoff valve at pressure gauge ¼ turn.

2. Depress lever of tester slowly. Note at what pressure the needle of the pressure gauge stopped, indicating an increase in pressure (nozzle does not chatter) or at which pressure the pressure dropped substantially (nozzle chatters). The maximum observed pressure is the opening pressure.

3. The opening pressure should not fall below the lower limit of 1600 psi.

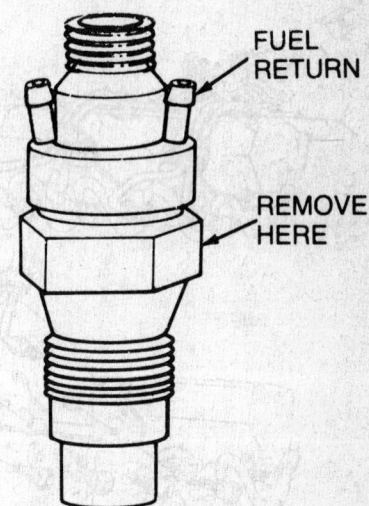

FUEL RETURN

REMOVE HERE

Injection nozzle

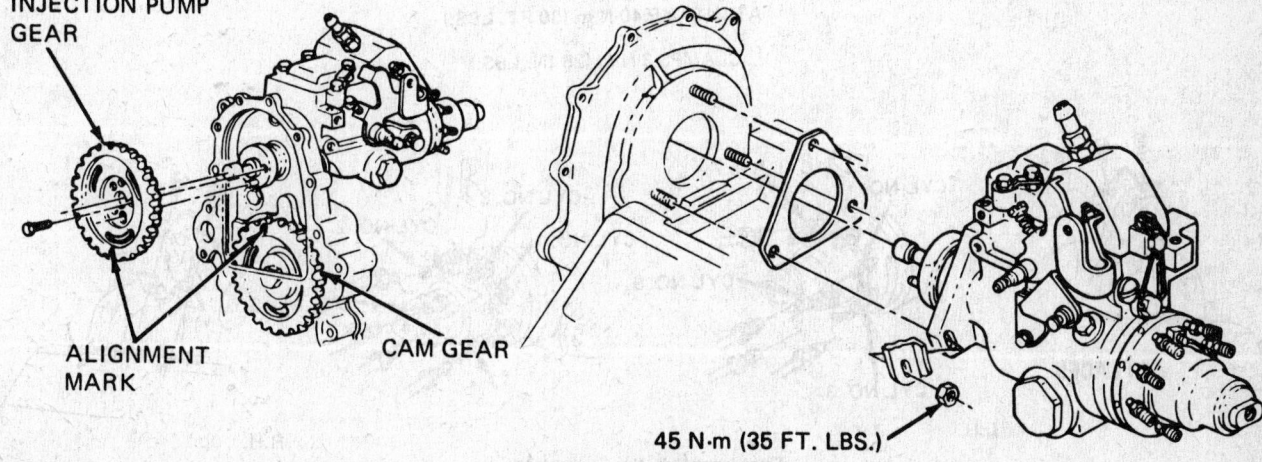

INJECTION PUMP GEAR

ALIGNMENT MARK

CAM GEAR

45 N·m (35 FT. LBS.)

Injection pump installation

4. Replace nozzles which fall below the lower limit.

Leakage Test

1. Further open shutoff valve at pressure gauge (¹⁄₂ to 1 ¹⁄₂ turns).
2. Blow-dry nozzle tip.
3. Install two clear plastic lines (approximately 1–1 ¹⁄₂ in.) over leak-off connections.
4. Depress lever of manual test stand slowly until gauge reads a pressure of 1380 psi. Observe tip of nozzle. A drop may form but not drop off within a period of 10 seconds.
5. Replace the nozzle holder assembly if a droplet drops off the nozzle bottom within the 10 seconds.

Chatter Test

1. Close shutoff lever at pressure gauge.
2. Depress lever of manual test stand slowly noting whether chatter noises can be heard.
3. If no chatter is heard, increase the speed of lever movement until it reaches a point where the nozzle chatters.

4. The chatter indicates that the nozzle needle moves freely and that the nozzle seat, guide, as well as the pintle, have no mechanical defects.
5. Replace nozzles which do not chatter.

Spray Pattern

1. Close shutoff valve at pressure gauge.
2. Depress lever of manual test stand downward abruptly and quickly. The spray should have a tight, evenly shaped conical pattern which is well atomized. This pattern should be concentric to the nozzle axis. Streamlike injections indicate a defect.

Installation

1. Remove protective caps from the nozzle.
2. Install nozzle and torque to 50 ft. lbs.
3. Connect fuel injection line, torque nut to 20 ft. lbs.
4. Install fuel return hose.
5. Install fuel line clip.
6. Connect battery.

INTERNATIONAL HARVESTER TUNE-UP AND ADJUSTMENTS

Compression Test

Refer to the Ford V8 6.9L Diesel Engine section for service information on IH 6.9L Diesel engine.

Valve Adjustment

9.0L, DV–550B, DV–462B, D–150, D–170, D–190 ENGINES

1. Remove valve covers. Allow engine to cool down until all parts are at uniform temperature. See the specifications chart for valve clearance.
2. Rotate the engine in the normal direction of rotation until the no. 1 intake valve

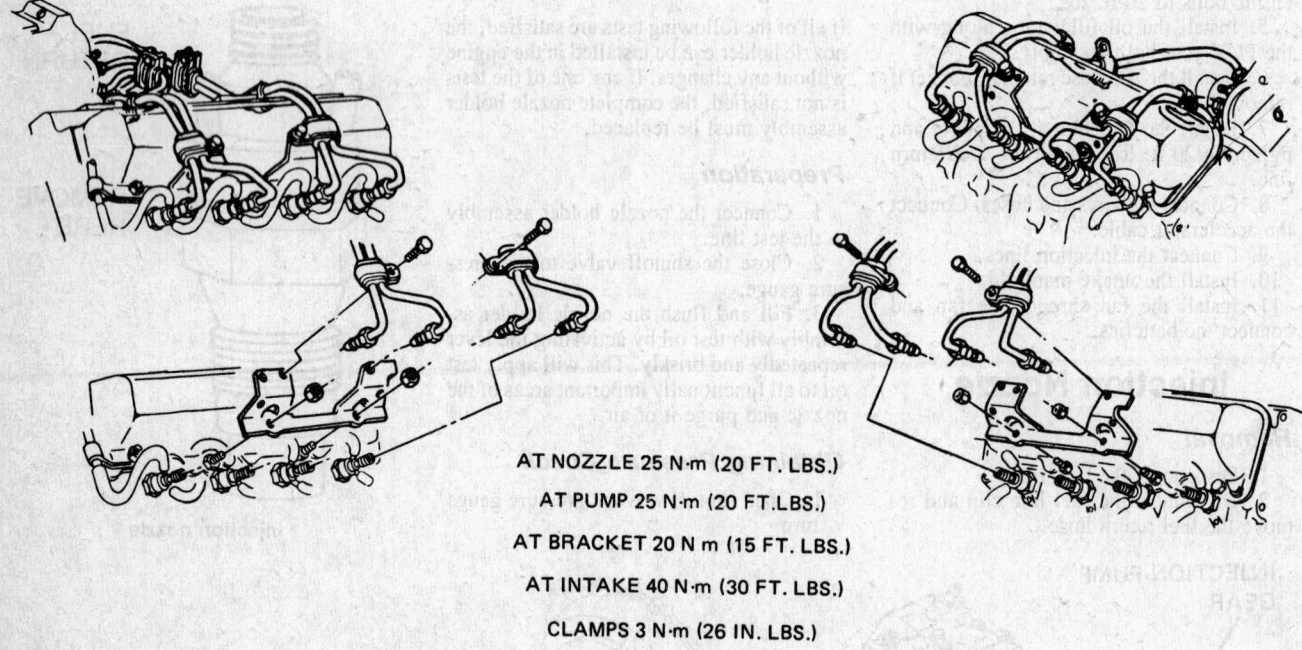

AT NOZZLE 25 N·m (20 FT. LBS.)

AT PUMP 25 N·m (20 FT.LBS.)

AT BRACKET 20 N·m (15 FT. LBS.)

AT INTAKE 40 N·m (30 FT. LBS.)

CLAMPS 3 N·m (26 IN. LBS.)

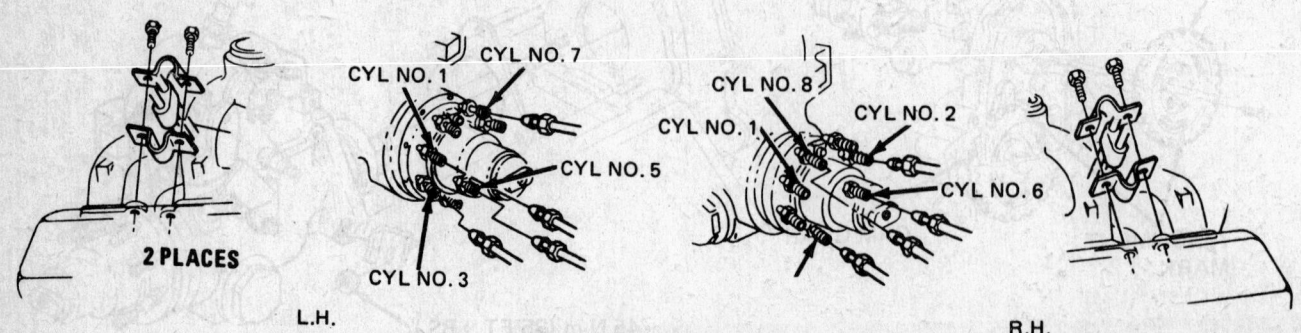

Fuel injection line installations

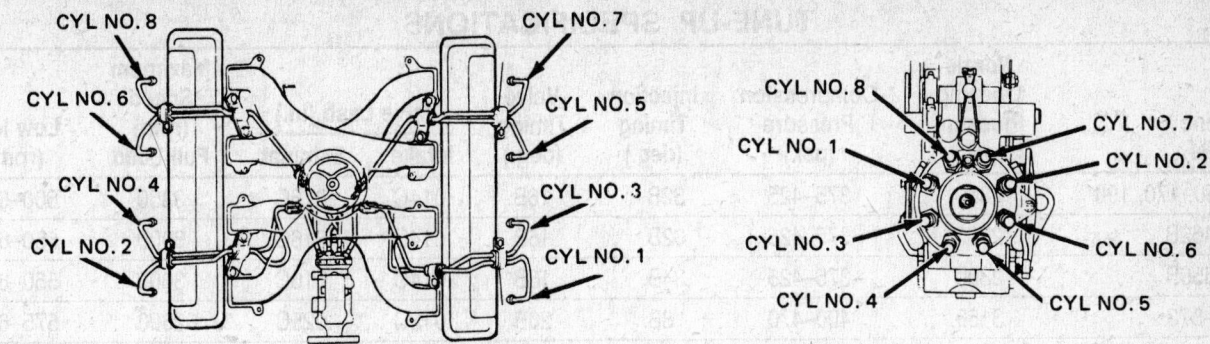

Fuel line routing from injection pump to cylinders

INTERNATIONAL HARVESTER

GENERAL SPECIFICATIONS

Engine Model	Bore & Stroke	Displ. Cu. In.	Horsepower @ rpm	Torque @ rpm	Firing Order	Compression Ratio	Oil Capacity w/Filters (qts)
D-150	4.5 × 4.3125	549	150 @ 3000	320 @ 2000	18736542	16.6:1	14
D-170	4.5 × 4.3125	549	170 @ 3000	340 @ 2000	18736542	16.6:1	14
D-190	4.5 × 4.3125	549	190 @ 3000	360 @ 2000	18736542	16.6:1	14
DV-462B	4.125 × 4.3125	461	160 @ 3000	①	18736542	17.0:1	16
DV-550B	4.5 × 4.3125	549	180 @ 3000	365 @ 2000	18736542	17.0:1	16
DV-550B	4.5 × 4.3125	549	200 @ 3000	389 @ 2000	18736542	17.0:1	16
DT-466	4.30 × 5.35	466	165 @ 2400	②	153624	16.3:1	20
DT-466	4.30 × 5.35	466	180 @ 2400	③	153624	16.3:1	20
DT-466	4.30 × 5.35	466	210 @ 2600	④	153624	16.3:1	20
V-800	5.3125 × 4.5	798	280 @ 2600	812 @ 1600	18736542	16.0:1	38
V-800	5.3125 × 4.5	798	300 @ 2600	725 @ 1800	18736542	16.0:1	38
V-800	5.3125 × 4.5	798	350 @ 2600	820 @ 1800	18736542	16.0:1	38
9.0L	4.510 × 4.312	551	165 @ 2800	366 @ 1200	18736542	19.0:1	14
9.0L ⑤	4.510 × 4.312	551	175 @ 2800	371 @ 1600	18736542	19.0:1	14
9.0L	4.510 × 4.312	551	180 @ 2800	401 @ 1200	18736542	19.0:1	14
6.9L	4 × 4.18	420	170 @ 3300	307 @ 1800	12734568	20.7:1	11

① DV-462B: available in 307 or 341 ft.lbs.
versions @ 2000 rpm
② DT-466: Federal-420 ft.lbs. @ 1600 rpm
Calif.-418 ft.lbs. @ 1600 rpm
③ DT-466: Federal-464 ft.lbs. @ 1600 rpm
Calif.-469 ft.lbs. @ 1600 rpm
④ DT-466: Federal-508 ft.lbs. @ 1800 rpm
Calif.-518 ft.lbs. @ 1800 rpm
⑤ California model

TUNE-UP SPECIFICATIONS

Engine Model	Nozzle Opening Pressure (psi.)	Compression Pressure (psi.)	Injection Timing (deg.)	Valve Timing (deg.)	Valve Lash (in.)		Maximum Speed (rpm) Full Load	Low Idle (rpm)
					Intake	Exhaust		
D-150, 170, 190	2800	375–425	32B	16B	.014C	.016C	3350	600–650
DV-462B	2300	375–425	32B	16B	.014C	.016C	3000	550–600
DV-550B	2300	375–425	34B	16B	.014C	.016C	3000	550–600
DVT-573	3150	400–470	8B	20B	.013C	.025C	2600	575–625
V-800	3100–3200	400–470	22B	30B	.013C	.025C	2600	625–675
DTI-466B	3600–3750	350–400	15B①	24B	.020C	.025C	2600	625–675
DT-466/466B	3600–3750	375–425	17B②	24B	.020C	.025C	2600	625–675
9.0L	3075–3225	450–525	16B	16B	.012	.016	2800	625–675
6.9L	2100–2150	—	③	—	④	④	3300	600–700

① Figure is for engine off. Timing @ 700 rpm is 17B

② Engine off or 700 rpm for 210 hp models
13B w/engine off for 190 hp models
15B @ 700 rpm for 190 hp models

③ Align timing marks and set dynamic timing (see text)

④ Hydraulic followers used (clearance is not adjustable)

ENGINE TORQUE SPECIFICATIONS
(ft. lbs.)

Engine Model	Cyl. Head	Main Brg.	Conn. Rod	Nozzle Clamps	Camshaft Flange	Flywheel	Crankshaft Pulley/ Damper	Camshaft Gear Nut
D-150, 170, 190	105–110	110–115①	55②	14–16	40–50	110–115	260–290	200–225
DV-462B, 550B	110	130①	55	15	40	110	325	200
466 Series	165	115	130	20	20	110–125	125	85
V-800	220	390③	130	35	30	235	425④	30
9.0L	⑤	125–135⑥	55②	14–16	40–45	110–115	260–290	200–225
6.9L	⑦	95⑧	⑨	—	—	38	90	12–18

① Tie bolts: 50 ft. lbs.

② Plus 1/6 turn more

③ Cross bolts: 160 ft. lbs. to be torqued after cap bolts

④ Gear nut

⑤ Step 1-torque to 50 ft. lbs. Step 2-90 ft. lbs. Step 3-110 ft. lbs.

⑥ Tie Bolts: 40–45 ft. lbs.

⑦ Step 1-40 ft. lbs., Step 2-65 ft. lbs., Step 3-75 ft. lbs.

⑧ Step 1-75 ft. lbs., Step 2-95 ft lbs.

⑨ Step 1-38 ft. lbs., Step 2-46–51 ft. lbs.

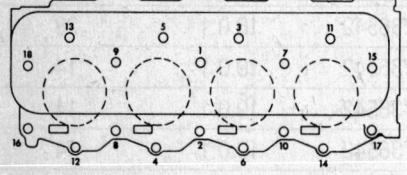

DV-462 and 550 head bolt tightening sequence

HEAD BOLT TORQUE SEQUENCE

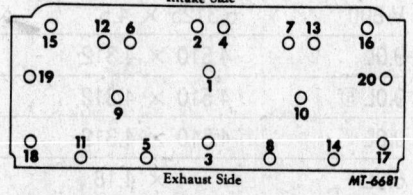

V-800 head bolt tightening sequence

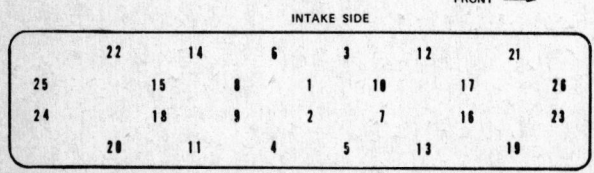

466 series head bolt tightening sequence

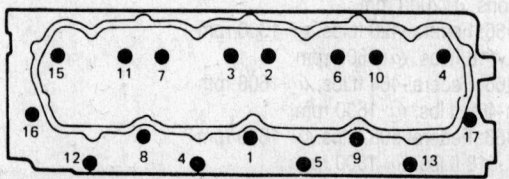

Head bolt torque sequence—6.9L engine

Engine Model	Comp. PSI	Cranking RPM ②
DV–462B, DV–550B	375–425	①
D–150, D–170, D–190	375–425	①
V–800	400–470	235
9.0 Liter	450–525	200
D–466	375–425	①

① Measured at cranking motor speed
② At sea level

Maximum PSI	Minimum PSI
260	195
280	210
300	225
320	240
340	255
360	270
380	285
400	300
420	315
440	330

CHILTON'S THREE "C's" DIESEL ENGINE DIAGNOSIS PROCEDURE

Condition	Cause	Correction
Engine fails to start	1. Tank empty, tank valve closed. 2. Plugged filter or fuel lines. 3. Defective damper valve. 4. Defective transfer pump. 5. Plugged injector line. 6. Defective pump plunger.	1. Refill fuel tank; open valve. 2. Clean or replace as necessary 3. Replace damper valve. 4. Replace transfer pump. 5. Clear fuel line. 6. Replace pump plunger.
Engine hard to start	1. Cranking speed too slow (below 250 rpm). 2. Swirl destroyer stuck open 3. Accelerator fails to reach full fuel position. 4. Improper fuel. 5. Water in fuel. 6. Improper injection timing. 7. Poor compression.	1. Check starter. 2. Check operation. 3. Check linkage operation. 4. Drain and refill fuel tank. 5. Drain and refill fuel tank and lines. 6. Reset injection timing. 7. Check compression.
Erratic engine operation	1. Improper fuel. 2. Inadequate transfer pump pressure. 3. Injection lines leaking. 4. Incorrect injector timing. 5. Faulty injector nozzle. 6. Poor compression.	1. Drain and refill fuel tank. 2. Check fuel pressure. 3. Tighten or replace fuel line(s). 4. Reset injection timing. 5. Replace faulty injector. 6. Check compression.
Low power without smoke	1. Accelerator linkage travel restricted. 2. Governor high idle adjustment incorrect. 3. Low transfer pump pressure. 4. Low fuel supply pressure. 5. Improper maximum fuel setting. 6. Injector plungers worn. 7. Exhaust system restricted. 8. Swirl destroyer in "on" position 9. Air cleaner slightly restricted. 10. Faulty injector nozzles. 11. Improper injection timing.	1. Check linkage operation. 2. Adjust governor. 3. Check fuel pressure. 4. Check fuel pressure. 5. Reset maximum fuel. 6. Replace injectors. 7. Check exhaust flow. 8. Check operation. 9. Replace air cleaner element. 10. Replace injection nozzles. 11. Reset injection timing.
Engine smokes, but with no loss in power	1. Faulty Nozzles. 2. Faulty maximum fuel setting.	1. Replace injection nozzles. 2. Reset maximum fuel delivery.
Engine smokes and lacks power	1. Swirl destroyer on 2. Air cleaner restricted. 3. Faulty nozzles. 4. Injector pump out of time. 5. Loss of compression in one cylinder. 6. Maximum fuel setting substantially too high.	1. Check operation. 2. Replace air cleaner element 3. Replace injection nozzles. 4. Reset injection timing. 5. Check compression. 6. Reset maximum fuel delivery.
Lube oil diluted with fuel	1. Faulty nozzles. 2. Incorrect delivery valve torque. 3. Faulty pump plunger. 4. Damaged pump barrel seat. 5. Cracked pump housing.	1. Replace injection nozzles. 2. Retorque delivery valve. 3. Replace pump plunger. 4. Replace seal. 5. Replace injection pump.

just starts to open. Adjust both valves on no. 6 cylinder.

3. Continue rotating the engine until no. 8 intake valve is just opening and adjust valves on no. 5 cylinder.

4. Rotate the engine until no. 7 intake valve is just opening and adjust the valves on no. 4 cylinder.

5. Rotate the engine until no. 3 intake valve is just opening on no. 2 cylinder.

6. Rotate the engine until no. 6 intake is just opening and adjust the valves on no. 1 cylinder.

7. Rotate the engine until no. 5 intake is just opening and adjust the valves on no. 8 cylinder.

8. Rotate the engine until no. 4 intake is just opening and adjust the valves on no. 7 cylinder.

9. Rotate the engine until no. 2 intake is just opening and adjust the valves on no. 3 cylinder.

V–800 ENGINE

1. Bring the number one piston to its compression stroke position with the timing indicator on the TDC mark. Rotate both the intake and exhaust valve push rods to be sure that the cam is on its off-lift position and the valves are closed on the number one cylinder.

2. With the number one cylinder piston in its compression position at TDC, adjust the intake valves for cylinders 1, 2, 4 and 5. Adjust the exhaust valves for cylinders 1, 3, 7 and 8.

3. Turn the crankshaft one complete revolution until the number six cylinder piston is at its compression position at

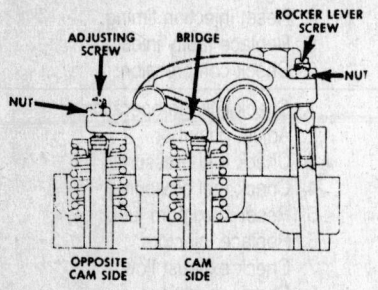

Valve lash adjustment

TDC. Adjust the intake valves for cylinders 3, 6, 7 and 8. Adjust exhaust valves for cylinders 2, 4, 5 and 6.

─────── **CAUTION** ───────

Do not attempt to adjust the valves with the engine running, as severe internal engine damage could result.

─────────────────────

4. The procedure to adjust the bridge valve lash on the V–800 engine is as follows.

a. Loosen the bridge adjusting screw nut and back out the screw. Press down on the rocker arm at the point of contact with the bridge.

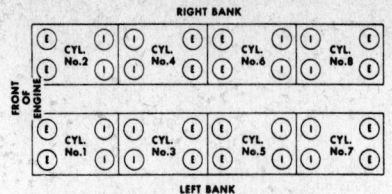

Valve arrangement

b. Turn the adjusting screw down until it contacts the valve stem, then turn the screw an additional 30 degrees. Hold the screw and tighten the locknut to 20–25 ft. lbs. (27–34 Nm).

c. To set the lash, loosen the rocker lever screw and nut. Insert a feeler gauge between the rocker arm and the bridge. Turn the screw until the proper specifications are obtained. Intake 0.013 in. (0.33mm) and exhaust 0.025 in. (0.64mm).

d. Tighten the locknut to 30–35 ft. lbs. (41–47 Nm).

DT–466, DT AND DTI–466 ENGINES

1. With the valve cover removed, turn the crankshaft until the number one piston is on the compression stroke and the timing pointer is on line with the TDC mark on the vibration damper.

2. Six valves are adjusted when the number one piston is at TDC and the remaining six valves are adjusted when the crankshaft is rotated one complete revolution, placing the number six piston at TDC position on the compression stroke.

3. With the number one piston at TDC, adjust the intake valves of cylinders 1, 2 and 4. Adjust the exhaust valves of cylinders 1, 3 and 5.

4. After rotating the crankshaft one full revolution and placing the number six piston at TDC, adjust the intake valves of cylinders 3, 5 and 6. Adjust exhaust valves of cylinders 2, 4 and 6.

Injection Pump Timing

D–150,170 AND 190

1. Position the shut-off control valve in the SHUT-OFF position.

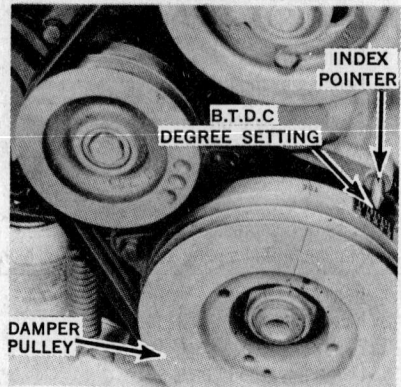

Timing mark and pointer

2. Rotate the engine in the direction of normal rotating until No. 1 cylinder is on the compression stroke. Continue rotating until the 32B mark reaches the pointer.

NOTE: The engine should be turned manually; if the timing mark is passed, back up at least ¼ revolution past the mark and approach it again.

3. Release the shut-off control.

4. Remove the delivery valve, spring and fill piece from the No. 1 pumping element and install a drip spout. The drip spout can be made from a length of injection pipe and a connector nut.

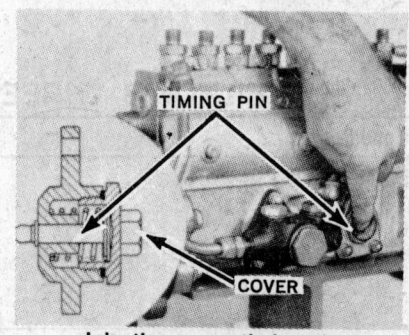

Injection pump timing pin

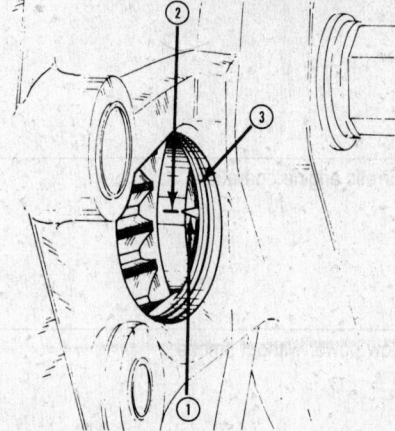

1. Timing pointer
2. Timing mark
3. Pipe plug opening

Injection pump timing marks—late models

5. Position the control rack at the load position as follows:

a. Hold the accelerator lever in the full forward position.

b. Slowly move the shut-off lever rearward. A distinct click will be heard as the rack moves from the excess fuel to the full load position. On pumps with a torque capsule (D–150, 170), a spring clip is used to prevent spring collapse in the torque capsule.

6. Supply fuel to the pump gallery. Fuel should flow from the drip spout at the rate of one drop every three to five seconds. If this rate is not observed, the pump must be removed from the engine and the injection pump drive flange must be repositioned on the engine.

DV–462B AND DV550B ENGINES

1. Put engine shutoff control in "shut-off" position.

2. Remove the cap from the timing pin on the left side of the injector pump.

3. Loosen No. 1 injector nozzle in its bore. Rotate the engine. When the nozzle pops up, start watching the timing mark and pointer on the front of the engine.

NOTE: Nozzle hold-down bolts must be threaded in at least two turns.

4. Stop rotating the engine when the pointer indicates 34 degrees BTDC on 550B engines, or 32 degrees on 462B engines. If damper is turned beyond the proper mark, turn the engine backwards until it has passed at least $1/4$ of a turn beyond the mark.

5. Attempt to insert the timing pin into the slot in the pump camshaft. If the pin cannot be inserted, repeat Steps 3–5. If the pin can be inserted, adjust timing as described below.

6. Remove upper and lower halves of the air cleaner and, where necessary, the manifold crossover adapter. Cover manifold openings.

7. Disconnect the accelerator rod and control cable at the governor.

8. Clean the pump and connections with diesel fuel. Disconnect the low pressure and the injector lines and the pump lube oil line. Cap all openings.

9. Remove the pump stabilizing brackets. Remove the adapter mounting bolts which hold the adapter and housing to the front cover.

10. Pull the pump rearward, freeing the drive flange tangs from the middle disc and remove it.

11. Install the coupling onto the tangs of the drive shaft, with the blind holes in the center of the coupling facing the pump.

12. Position the drive flange of the pump so that its tangs are horizontal and the timing pin can be engaged with the pump camshaft.

13. Locate the pump on the engine, carefully engaging the drive flange tangs with the coupling. Secure the adapter and pump to the rear of the engine front cover with the mounting bolts.

14. Install the stabilizing brackets.

466 SERIES

1. Remove the pump from the engine.

2. Remove the pump drive gear and install the pump and mounting adapter plate on the timing case cover.

3. Secure the pump and adapter plate with two bolts.

4. Set the engine timing mark at the specified static setting. The line on the pump must be in line with the timing pointer.

5. Install the drive gear and mesh it with the idler gear. Do not tighten the bolt at this time.

6. Hold the pump shaft and rotate the

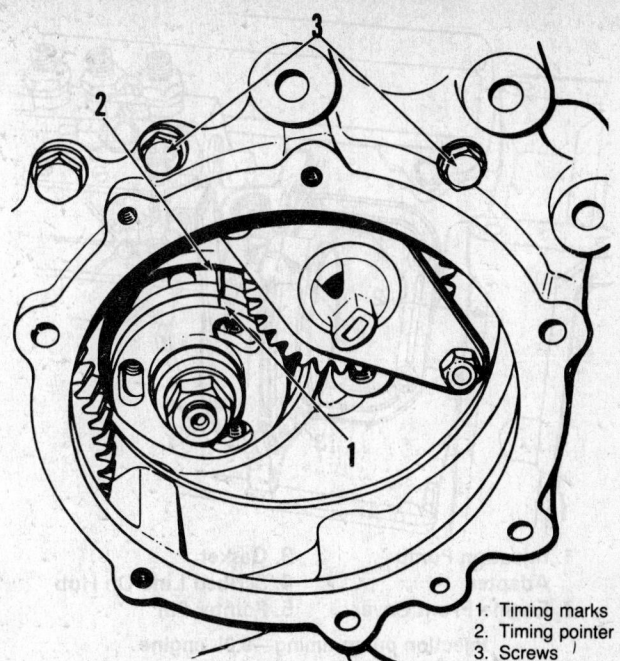

1. Timing marks
2. Timing pointer
3. Screws

Injection pump timing marks—early models

drive gear counterclockwise to remove backlash. Tighten the bolts to 30 ft. lbs.

NOTE: Disregard any timing marks on the idler gear.

7. Connect the fuel lines and oil supply line.

8. Install the injection lines and tighten the fittings to 30 ft. lbs. Secure the line clamps.

9. Connect and adjust the governor control linkage.

10. If the pump is equipped with a port closure lock plug, remove it and install the regular plug.

11. If the engine is not timed correctly, reposition the crankshaft to the specified timing mark. Remove the three bolts from the pump drive gear and rotate the pump driveshaft EXACTLY one revolution. The pump is now timed and the engine will start.

NOTE: It is impossible to align the marks inside the pump and the timing marks on the drive hub at the same time.

12. Install the pump access cover.

V–800

1. Remove pump from engine.

2. Remove the cover from the rear of the mounting adapter.

3. Reach through the opening with a deep socket and loosen the pump drive capscrews. Rotate the pump camshaft to bring each of the six capscrews around into view.

4. Remove the timing pin cover.

5. Rotate the pump camshaft while holding in on the notch in the camshaft. Install the timing pin holding tool to keep the pin depressed.

6. Scribe timing marks on gear and align with scribe mark on advance unit.

7. Install and lubricate a new O-ring on the flange of the pump mounting adapter.

8. Push pump into place on engine, engaging camshaft gear and compressor idle gear.

9. Visually check, through the adapter opening, that the injection pump gear bolt is near the center of the slotted hole in the advance unit. If not, the pump is out of time one tooth in either direction.

10. Push pump into engagement with the camshaft gear and air compressor idler gear. Before O-ring enters bore, align the mounting holes in the pump flange with those in the front plate.

NOTE: When the pump is installed, the gear should have rotated far enough clockwise to allow the gear bolts to be positioned in the middle of the slots in the advance unit. If alignment is not correct, remove the pump, rotate the gear one tooth and reinstall.

11. Install all pump mounting bolts.

12. Using a deep socket, reach in through the access hole and tighten the first gear bolt to 50 ft. lbs.

13. Remove the timing pin holding tool. Make sure that the timing pin is completely released.

14. Torque the five remaining bolts to 50 ft. lbs. each, by rotating the crankshaft through two revolutions, stopping each 120° to tighten the next bolt.

15. Recheck static timing at the vibration damper. It should be 22° BTDC at # 1 compression.

16. Install the access cover and gasket and the timing pin cap.

9.0L ENGINE

1. Block the fuel shut-off control in the OFF position to prevent accidental starting of the engine.

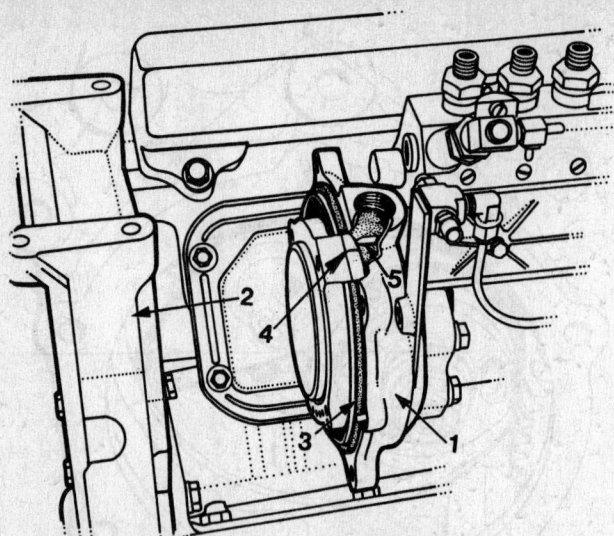

1. Injection Pump Adapter
2. Engine Front Cover
3. Gasket
4. Scribed Line On Hub
5. Pointer Pin

Injection pump timing—9.0L engine

2. Rotate the engine in the normal direction of rotation (clockwise as viewed from the front of the engine) until the No. 1 cylinder is at TDC on the compression stroke. Continue to rotate the engine until the 16 degree mark on the crankshaft pulley is aligned with the engine timing pointer. In this position, the pumping element for the No. 1 cylinder is at the start of injection.

NOTE: If the engine is turned beyond the specified timing mark, back up at least ¼ turn from the mark and begin the approach again in the normal direction of rotation. Do not back engine to align timing marks.

3. To be sure the No. 1 cylinder is on its compression stroke, loosen the injection nozzle hold down bolt from the No. 1 nozzle slightly. Watch for the nozzle to pop up while slowly rotating the engine. To prevent the nozzle from coming out of the head, always make sure three threads on the hold down bolt are engaged in the mounting hole.

4. Remove the sight plug on top of the injection pump adapter housing. The scribed line on the pump must be aligned with the pointer pin. If the marks do not align, it will be necessary to reset the pump to engine timing.

5. Before the timing adjustment can be made, it will be necessary to drain the coolant, remove the fan drive, fan, hub assembly, water outlet, water return tube and drive gear cover.

6. Loosen the injection pump drive gear mounting bolts and rotate the injection pump hub to align the hub with the pointer pin. Once aligned, tighten the mounting bolts and recheck alignment.

7. Once the timing is set, reinstall all removed components and refill cooling system.

NOTE: To minimize timing inaccuracies during pointer and scribed line alignment viewing, threads must be seen around the circumference of the viewing hold.

Idle Speed Adjustment
ALL ENGINES

1. Bring engine to operating temperature. Note tachometer reading. Lo-idle speed is given in the "Tune-Up Specifications" chart.

2. If necessary to adjust the lo-idle, loosen the locknut and turn clockwise to reduce rpm, or counterclockwise to increase it. Tighten the locknut.

3. Operate the engine up to hi-idle three times and allow it to return to lo-idle. If the adjustment is not correct, repair binding in the linkage.

Maximum No-Load Speed Adjustment

ALL ENGINES

NOTE: Maximum no-load governed speed (hi-idle) is the point of engine speed where governor prevents engine rpm from going higher. Hi-idle adjustment is located at accelerator lever stop screw and is initially made on pump calibrating stand. Minor adjustment can be made on engine as follows:

1. With engine running and at normal operating temperature, push control lever to its full forward position-lever against hi-idle stop.

2. Read tachometer and note highest engine rpm. (Make sure tachometer is accurate.)

3. If governed speed does not reach specified limit, check throttle linkage to make sure lever is not restricted.

4. If governed speed is not within specified limits, remove seal on locknut of hi-idle stop screw and reset hi-idle adjustment. Turn stop screw out (counterclockwise) to reduce hi-idle or turn in (clockwise) to increase hi-idle. Tighten locknut and recheck hi-idle speed. When adjustment is correct, reseal locknut.

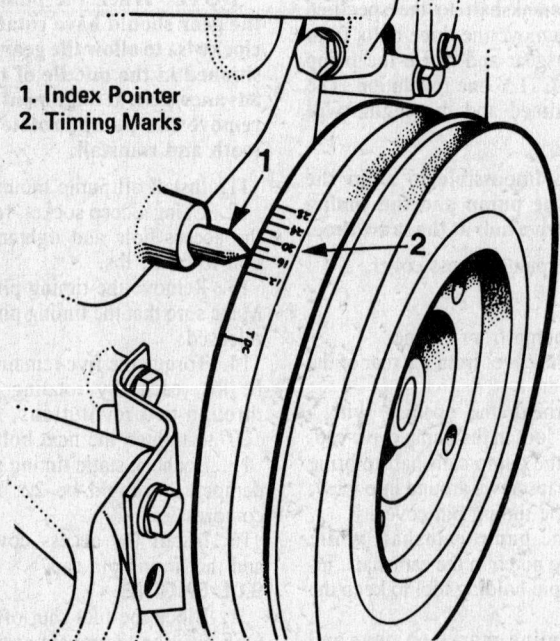

1. Index Pointer
2. Timing Marks

Engine timing marks and pointer location—typical

Fuel Injection Nozzles

Removal and Installation

ALL ENGINES

1. Before removing nozzle assemblies, clean exterior of each nozzle assembly and the surrounding area with clean fuel oil or solvent to prevent entry of dirt into engine when nozzle assemblies are removed. Also, clean fuel inlet and fuel leak-off piping connections.

2. Disconnect fuel inlet (high pressure) and fuel leak-off pipes from each nozzle assembly. Cover open ends of pipes to prevent entry of dirt.

3. Remove nozzle mounting bolt and hold down clamp. Pull nozzle assembly with washer and seal from engine. If assembly seems stuck, rotate it slightly to break it loose from carbon deposits in the cylinder head recess. Be careful not to strike the nozzle tip against any hard surface during removal.

4. Installation is the reverse of removal.

NOTE: Cover nozzle assembly fuel inlet and leak-off openings with plastic caps to prevent entry of dirt. Also protect nozzle tip.

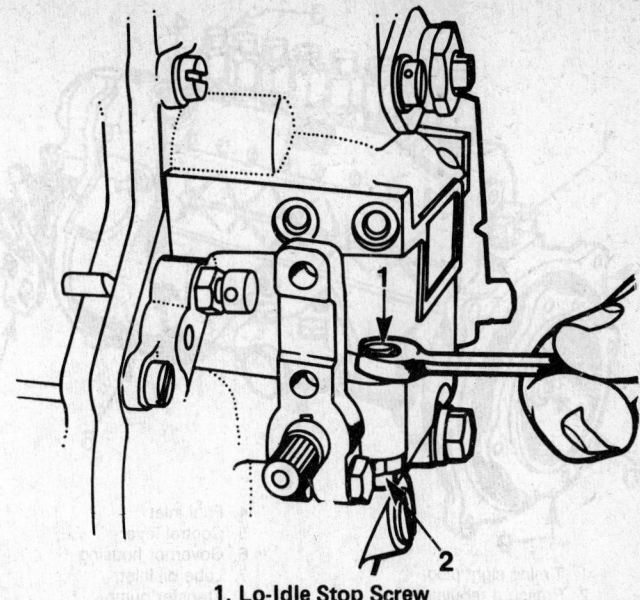

1. Lo-Idle Stop Screw
2. Lock Nut

Idle speed adjustment—typical

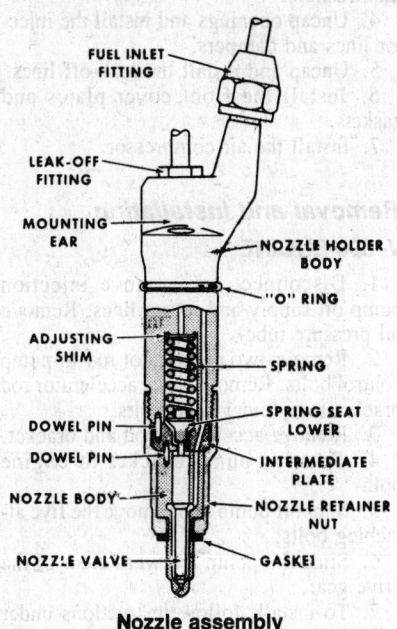

FUEL INLET FITTING
LEAK-OFF FITTING
MOUNTING EAR
NOZZLE HOLDER BODY
"O" RING
ADJUSTING SHIM
SPRING
SPRING SEAT LOWER
DOWEL PIN
DOWEL PIN
INTERMEDIATE PLATE
NOZZLE BODY
NOZZLE RETAINER NUT
NOZZLE VALVE
GASKET

Nozzle assembly

Injection Pump

Removal

DV–462B, DV–550B, D–150, 170 AND 190 ENGINES

1. Remove the upper and lower halves of the air cleaner and, where necessary, the manifold crossover adapter. Cover manifold openings.

2. Disconnect the accelerator rod and

control cable at the governor.

3. Clean the pump and connections with diesel fuel. Disconnect low pressure and injector lines and the pump lube oil line. Cap all openings.

4. Remove the pump stabilizing brackets. Remove the adapter mounting bolts which hold the adapter and housing to the front cover.

5. Pull the pump rearward, freeing the drive flange tangs from the middle disc and remove it.

Inspection

1. Inspect pump mounting flange and the bosses of the rear mounting bracket.

2. Inspect the drive flange for damage, wear, or loose mounting.

3. Torque the securing nut to 75 ft. lbs.

Installation

1. Install the coupling onto the tangs of the drive shaft, with the blind hole in the center of the coupling facing the pump.

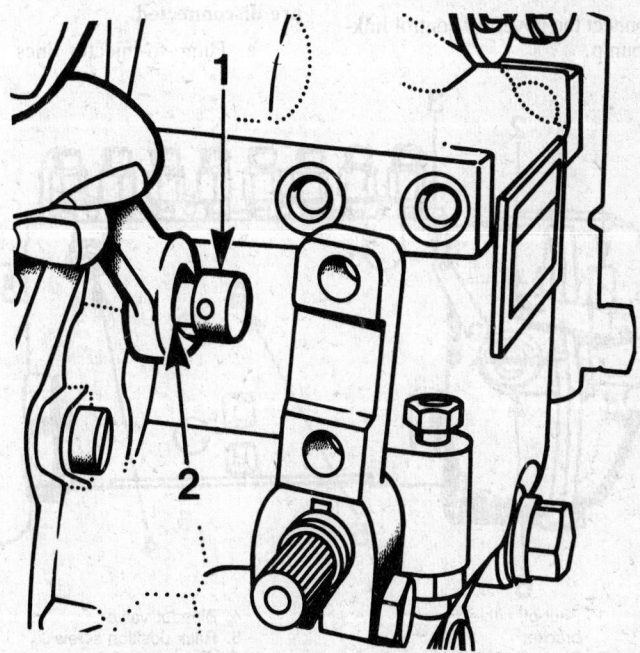

1. Hi-Idle Stop Screw
2. Lock Nut

Maximum no-load adjustment—typical

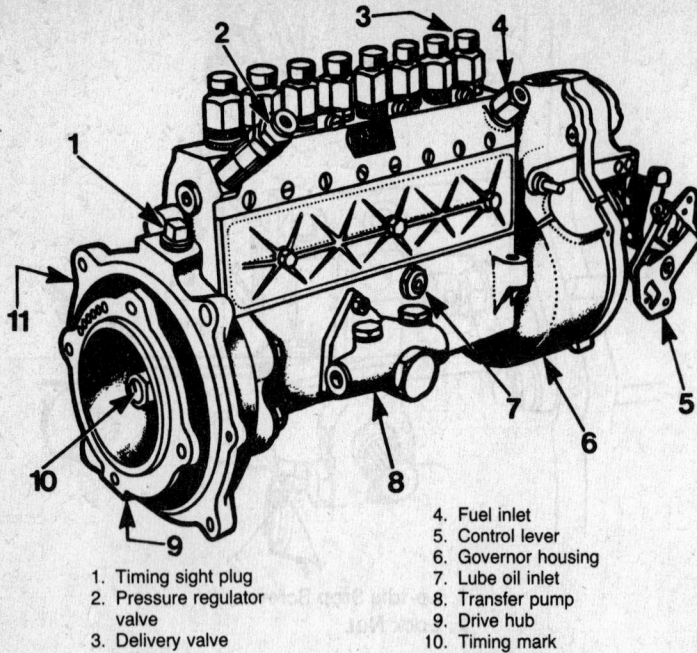

1. Timing sight plug
2. Pressure regulator valve
3. Delivery valve holder
4. Fuel inlet
5. Control lever
6. Governor housing
7. Lube oil inlet
8. Transfer pump
9. Drive hub
10. Timing mark
11. Adapter housing

Typical left side view of injection pump

2. Position the drive flange of the pump so that its tangs are horizontal and the timing pin can be engaged with the pump camshaft.

3. Locate the pump on the engine, carefully engaging the drive flange tangs with the coupling. Secure the adapter and pump to the rear of the engine front cover with the mounting bolts.

4. Install the stabilizer brackets.

466 SERIES ENGINES

Removal and Installation

1. Disconnect the governor control linkage at the pump.

2. On units equipped with a Bowden shut-off wire, disconnect the control linkage from the shut-off lever.

3. On units equipped with an electric shut-off, disconnect the wire.

4. Disconnect the following lines from the pump:
  a. Lube oil supply line
  b. Primary filter-to-supply pump tube
  c. Supply pump-to-final filter tube
  d. Final filter-to-injection pump hose

NOTE: Cap all lines as soon as they are disconnected.

  e. Pump-to-injector lines

  f. Fuel return line

5. Remove the pump access cover.

6. Remove the two bolts securing the pump mounting plate adapter to the front cover. Remove the pump and adapter plate.

7. To install, see the section under Injection Pump Timing.

Removal

DVT–573 ENGINE

1. Remove air compressor.

2. Remove the front cover plates and gaskets from the heads.

3. Remove all leak-off lines at the injector nozzles. Cap all openings.

4. Remove the injector lines and dampers assembled. Cap all openings.

5. Remove the mounting caps and lockwashers.

6. Lift the injection pump and shaft off the engine. Pull the gasket off the drive housing.

Installation

1. Install a new gasket onto the drive housing.

2. Put the pump in place on the engine.

3. Install the mounting screws and lockwashers.

4. Uncap openings and install the injector lines and dampers.

5. Uncap and install the leak-off lines.

6. Install the front cover plates and gaskets.

7. Install the air compressor.

Removal and Installation

V–800 ENGINE

1. Disconnect and remove injection pump oil supply and return lines. Remove oil pressure tube.

2. Remove two accelerator rod-to-pump control bolts. Remove three accelerator rod bracket-to-cylinder head bolts.

3. Remove accelerator rod and bracket.

4. Remove pump bracket-to-engine bolt.

5. Support pump and remove the five attaching bolts.

6. Slide the pump rearward clear of the drive gear.

7. To install, follow instructions under Injector Pump Timing.

Removal

9.0L ENGINE

1. Drain engine coolant from radiator and engine block.

2. Disconnect accelerator and shut off control cables from governor.

3. Before disconnecting fuel lines, clean pump and connections with clean diesel fuel.

4. Disconnect injection lines, low pressure lines and lube oil line from pump. Remove fuel line brackets as necessary. Use plastic caps or plugs to protect openings from dirt.

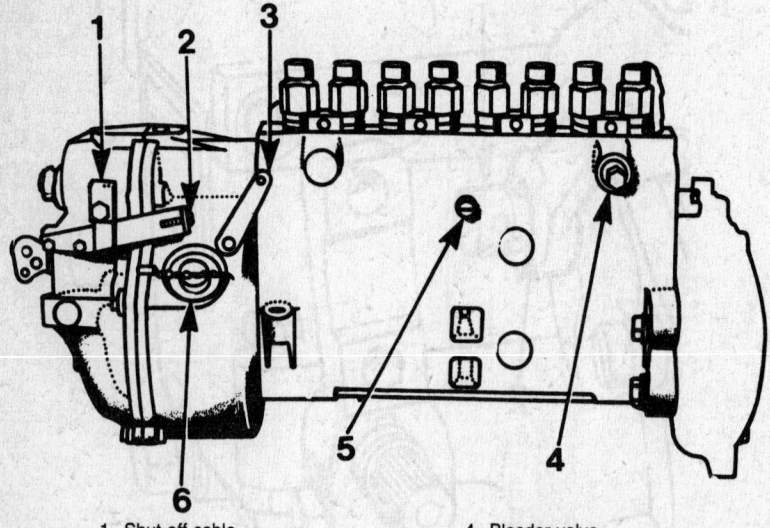

1. Shut-off cable bracket
2. Shut-off lever stop
3. Shut-off lever
4. Bleeder valve
5. Rack position screw
6. Governor adjusting plug

Typical right side view of injection pump

5. Remove fan drive, water outlet and return tube, injection pump drive gear cover and drive gear.

6. Remove four injection pump adapter to engine front cover mounting bolts. Remove two support bracket bolts on right side of pump and two mounting bracket nuts underneath injection pump, remove pump.

7. Installation is the reverse of removal. Inspect pump mounting flange at rear of engine front cover. Also inspect injection pump stabilizer brackets and drive gear for damage or wear.

8. Secure mounting adapter and injection pump to rear of engine front cover with four mounting bolts.

9. Install stabilizing brackets to side at rear of pump housing.

10. Connect low pressure fuel inlet and return lines and lube oil line to injection pump.

11. Connect high pressure injection lines to injection pump and secure fuel line brackets.

12. Connect accelerator rod and stop control cable to governor levers.

13. Install air cleaner or manifold crossover adapter.

14. Unlock primer pump by turning knob counter clockwise. Primer pump is located on top of final filter base.

15. Open bleeder valve located on right front of injection pump. Operate priming pump and allow fuel to flow until all air bubbles are expelled and fuel flows in a solid stream.

16. Close vent and lock primer pump. (Push knob forward and turn clockwise)

17. Start the engine and check the fuel system for leaks.

MACK DIESELS TUNE-UP AND ADJUSTMENTS

A good tune-up and adjustment procedure for diesel engines should include a compression check, valve lash adjustment, injection timing, fuel control adjustment, idle speed adjustment and maximum no-load governor speed adjustment.

Compression

If referral to the chart indicates low compression could be the cause of the malfunction, test the compression of each cylinder as described:

1. Operate the engine at fast idle until it reaches operating temperature. Stop the engine.

2. Disconnect the fuel line at the nozzle of the cylinder to be checked. Place a container near the open end of the line to catch the fuel.

3. Remove the injection nozzle and holder assembly. Clean carbon from the nozzle tip with a special wire brush.

4. Remove carbon from the cylinder head nozzle hole with a special reamer and wire brush. Use an air gun to remove loose carbon. Crank the engine a few times with the starter to remove any remaining loose material.

GENERAL ENGINE AND TUNE-UP SPECIFICATIONS

Engine Model	Bore × Stroke (Inches)	Piston Displ. Cu. In.	Horsepower @ rpm	Torque @ rpm	Firing Order	Governor Speed No Load	Idle Speed rpm	Injection Pump Timing (deg.)
END475	4.53 × 4.92	475	155 @ 2400	385 @ 1500	1-5-3-6-2-4	2600	450	30°B
ENDT475	4.53 × 4.92	475	190 @ 2400	470 @ 1500	1-5-3-5-2-4	2600	500	24°B
ET477	4.53 × 4.92	475	210 @ 2400	510 @ 1500	1-5-3-6-2-4	2600	450–500	18°B
EDT477	4.53 × 4.92	475	157 @ 2400	510 @ 1500	1-5-3-6-2-4	2600	450–500	18°B
ENDT(B)673	4.875 × 6.0	672	225 @ 2100	653 @ 1600	1-5-3-6-2-4	2280	525–575	29°B
ENDT(B)673C	4.875 × 6.0	672	250 @ 2100	700 @ 1500	1-5-3-6-2-4	2280	525–575	28°B
END(B)673E	4.875 × 6.0	672	180 @ 2100	540 @ 1400	1-5-3-6-2-4	2280	500–575	30°B
ET(B)673	4.875 × 6.0	672	260 @ 2100	775 @ 1500	1-5-3-6-2-4	2350	600–650	26°B
ET(B)673E	4.875 × 6.0	672	200 @ 2100	600 @ 1500	1-5-3-6-2-4	2310	600–650	28°B
ETY(B)673E	4.875 × 6.0	672	200 @ 2100	600 @ 1500	1-5-3-6-2-4	2310	600–650	18°B
ETAY(B)673A	4.875 × 6.0	672	315 @ 1900	1050 @ 1450	1-5-3-6-2-4	2225	600–650	21°B
ETAZ(B)673	4.875 × 6.0	672	320 @ 2100	1000 @ 1500	1-5-3-6-2-4	2310	525–575	24°B
ETAZ(B)673A	4.875 × 6.0	672	315 @ 1900	1050 @ 1450	1-5-3-6-2-4	2225	600–650	24°B
ETAZ(B)673C	4.875 × 6.0	672	295 @ 1800	985 @ 1400	1-5-3-6-2-4	2185	600–650	24°B
ENDT(B)675	4.875 × 6.0	672	235 @ 2100	906 @ 1200	1-5-3-6-2-4	2310	525–575	26–27°B
ETY(B)675	4.875 × 6.0	672	235 @ 2100	906 @ 1200	1-5-3-6-2-4	2350	600–650	16°B
ENDT(B)676	4.875 × 6.0	672	285 @ 1800	1080 @ 1200	1-5-3-6-2-4	2310	525–575	22°B
ETA(B)676B	4.875 × 6.0	672	302 @ 2100	1150 @ 1200	1-5-3-6-2-4	2370	600–650	25°B
ETAY(B)676	4.875 × 6.0	672	283 @ 2100	1080 @ 1200	1-5-3-6-2-4	2310	525–575	18°B
ETAY(B)676D	4.875 × 6.0	672	285 @ 1800	1080 @ 1200	1-5-3-6-2-4	2100	525–575	18°B
END707	5.00 × 6.0	707	200 @ 2100	557 @ 1600	1-5-3-6-2-4	2280	500–575	30°B
END711	5.00 × 6.0	707	211 @ 2100	602 @ 1600	1-5-3-6-2-4	2280	500–575	30°B

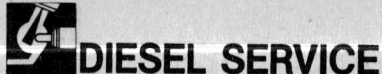

GENERAL ENGINE AND TUNE-UP SPECIFICATIONS

Engine Model	Bore × Stroke (Inches)	Piston Displ. Cu. In.	Horsepower @ rpm	Torque @ rpm	Firing Order	Governor Speed No Load	Idle Speed rpm	Injection Pump Timing (deg.)
E6-200	4.875 × 6.0	672	200 @ 2100	600 @ 1500	1-5-3-6-2-4	2350	600–650	21°B
EC6-235	4.875 × 6.0	672	235 @ 2100	700 @ 1500	1-5-3-6-2-4	2280	525–575	13°B
EM6-237	4.875 × 6.0	672	237 @ 1800	906 @ 1200	1-5-3-6-2-4	2350	①	③
EM6-237R	4.875 × 6.0	672	237 @ 1800	906 @ 1200	1-5-3-6-2-4	2100	525–575	21°B
E6-250	4.875 × 6.0	672	250 @ 2100	750 @ 1500	1-5-3-6-2-4	2350	600–650	19°B
EMC6-250	4.875 × 6.0	672	250 @ 1800	950 @ 1200	1-5-3-6-2-4	2290	525–575	17°B
E6-260	4.875 × 6.0	672	260 @ 2100	775 @ 1500	1-5-3-6-2-4	2350	525–575	20°B
EM6-285	4.875 × 6.0	672	285 @ 1800	1065 @ 1400	1-5-3-6-2-4	2150	525–575	23°B
EM6-285R	4.875 × 6.0	672	285 @ 1700	1065 @ 1400	1-5-3-6-2-4	2050	525–575	22°B
E6-315	4.875 × 6.0	672	315 @ 1900	1050 @ 1450	1-5-3-6-2-4	2225②	①	25°B
EC6-330	4.875 × 6.0	672	330 @ 1950	1065 @ 1400	1-5-3-6-2-4	2150	525–575	16°B
E6-350	4.875 × 6.0	672	350 @ 1950	1132 @ 1400	1-5-3-6-2-4	2150	525–575	22°–23°B
ENDT(B)865	5.25 × 5.0	866	322 @ 2400	1100 @ 1350	1-5-4-8-6-3-7-2	2650	600–650	25°B
ENDT(B)866	5.25 × 5.0	866	375 @ 2200	1040 @ 1600	1-5-4-8-6-3-7-2	2500	600–650	25°B
EM9-400	5.375 × 5.50	998	392 @ 2100	1520 @ 1230	1-5-4-8-6-3-7-2	2420	600–650	17°B
EM9-400R	5.375 × 5.50	998	400 @ 1700	1520 @ 1230	1-5-4-8-6-3-7-2	1970	600–650	17°B
EMC9-400	5.375 × 5.50	998	392 @ 2100	1520 @ 1230	1-5-4-8-6-3-7-2	2420	600–650	13°B
EMC9-400R	5.375 × 5.50	998	400 @ 1700	1520 @ 1230	1-5-4-8-6-3-7-2	1970	600–650	13°B
EM6-225	4.875 × 6.0	672	225 @ 2100	844 @ 1260	1-5-3-6-2-4	2325	525–575	25°B
EM6-250R	4.875 × 6.0	672	250 @ 1700	940 @ 1260	1-5-3-6-2-4	1950	600–650	④
EM6-275	4.875 × 6.0	672	275 @ 2100	1038 @ 1260	1-5-3-6-2-4	2315	525–575	25°B
EM6-275L	4.875 × 6.0	672	275 @ 1700	1275 @ 1020	1-5-3-6-2-4	1900	525–575	22°B
EM6-275R	4.875 × 6.0	672	275 @ 1600	1038 @ 1260	1-5-3-6-2-4	1770	525–575	23°B
EMC6-285	4.875 × 6.0	672	283 @ 2100	1080 @ 1200	1-5-3-6-2-4	2290	525–575	14°B
EMC6-285R	4.875 × 6.0	672	285 @ 2100	1080 @ 1200	1-5-3-6-2-4	2050	525–575	14°B
EM6-300	4.875 × 6.0	672	300 @ 2100	1125 @ 1260	1-5-3-6-2-4	2330	525–575	24°B
EM6-300R	4.875 × 6.0	672	300 @ 1700	1125 @ 1260	1-5-3-6-2-4	2150	525–575	24°B
ER-315R	4.875 × 6.0	672	315 @ 1800	1050 @ 1405	1-5-3-6-2-4	2225	600–650	25°B
ER-325	4.875 × 6.0	672	325 @ 1950	1050 @ 1400	1-5-3-6-2-4	2150	525–575	24°B
ER-325R	4.875 × 6.0	672	325 @ 1950	1050 @ 1400	1-5-3-6-2-4	2030	525–575	24°B
E6-350R	4.875 × 6.0	672	350 @ 1800	1131 @ 1400	1-5-3-6-2-4	2050	525–575	22°–23°B
EC6-350	4.875 × 6.0	672	350 @ 1950	1131 @ 1400	1-5-3-6-2-4	2200	525–575	17°B
EM9-400R	5.375 × 5.50	998	400 @ 1900	1520 @ 1530	1-5-4-8-6-3-7-2	2175	600–650	—
EMC9-400R	5.375 × 5.50	998	400 @ 1900	1520 @ 1230	1-5-4-8-6-3-7-2	2175	600–650	—
E9-440	5.375 × 5.50	998	440 @ 1800	1495 @ 1350	1-5-4-8-6-3-7-2	2100	600–650	—

① American Bosch pump-525-575 rpm
  Robert Bosch pump-600-650 rpm
② American Bosch pump-2100 rpm
③ United Technologies Diesel systems pump-
  21°B
  Robert Bosch pump-20°B
④ United Technologies Diesel systems pump-
  22°B
  Robert Bosch-25°B

ENGINE TORQUE SPECIFICATIONS

Engine Series	Cylinder Head (Ft. Lbs.)	Main Bearing Bolts (Ft. Lbs.)	Connecting Rod Bolts (Ft. Lbs.)	Rocker Bracket Caps (Ft. Lbs.)	Flywheel Mounting Nuts (Ft. Lbs.)
END475/ ENDT475	140	Nut-150 Stud-155	80	40	149④
ET477	140	155	81	35	140④
Current Production 6 Cylinder	⑤⑥	11⁄16"-200 5⁄8"-175	150	35	190①
ENDT 865,866	225	350①②	150③	35	190①
998⑦	⑧	350①②	175①	55	180

① Oiled

② Buttress screw 100 ft. lbs.

③ 170–180 ft. lbs. on part #367GCA3178A (connecting rod only).

④ Use lock plates.

⑤ Oil all cylinder head capscrew bosses, capscrew threads and washers with SAE #30 engine oil prior to assembly. Do not oil the threads in the cylinder block.
Tighten capscrews individually on any one head in the proper sequence. Also:
1. Tighten all to 50 ft. lbs.
2. Repeat, in sequence, to 125 ft. lbs.
3. Finally tighten to 200 ft. lbs.

After run-in procedure, in sequence, back off each capscrew individually until free, then retorque same capscrew to 220 ft. lbs.

⑥ Oil all cylinder head stud nut bosses, stud nut threads, and washers with SAE #30 engine oil prior to assembly. Do not oil threads in cylinder block. Tighten stud nuts individually on any one head in the proper sequence. Also:
1. Tighten all to 50 ft. lbs.
2. Repeat, in sequence, to 125 ft. lbs.
3. Finally tighten to 175 ft. lbs.
After run-in procedure, in sequence, back off each stud nut individually until free, then retorque same nut to 175 ft. lbs.

⑦ 998 cubic inches

⑧ Oil all cylinder head capscrew threads and washers with SAE #30 engine oil prior to assembly. Tighten capscrews individually on any one head in the proper sequence. Also:
1. Tighten all to 50 ft. lbs.
2. Repeat, in sequence, to 175 ft. lbs.
3. Finally tighten to 220 ft. lbs.
After run-in procedure, in sequence, back off each capscrew individually until free, then retorque same capscrew to 220 ft. lbs.

FIRING ORDERS

FRONT OF ENGINE

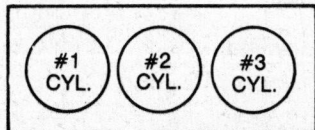 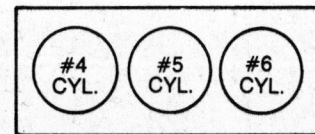

All 6 cylinder engines: 1-5-3-6-2-4

FRONT OF ENGINE

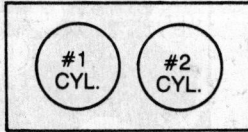 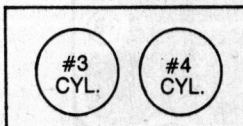

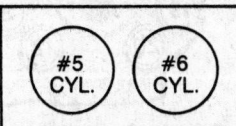 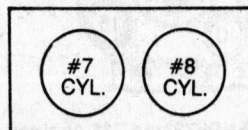

All V8 engines: 1-5-4-8-6-3-7-2

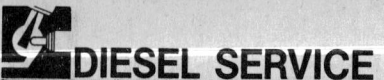
TORQUE SEQUENCES

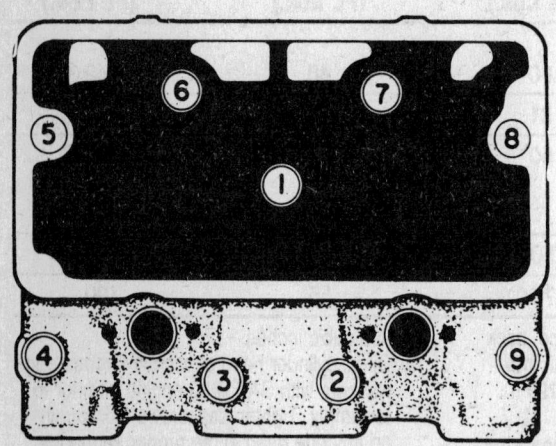

Cylinder head nut torquing sequence, END475 engine series

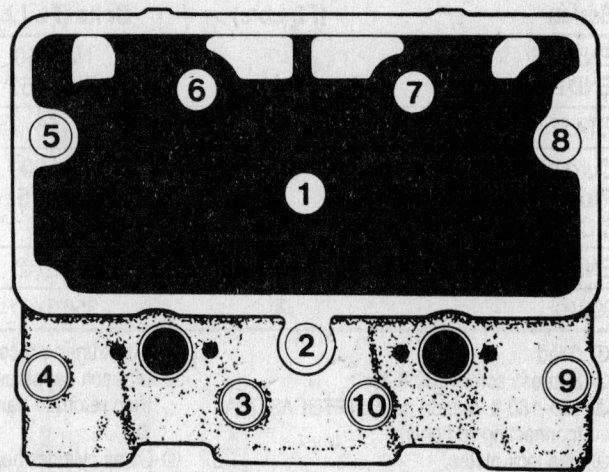

Cylinder head bolt torquing sequence, 475 series 6 cylinder

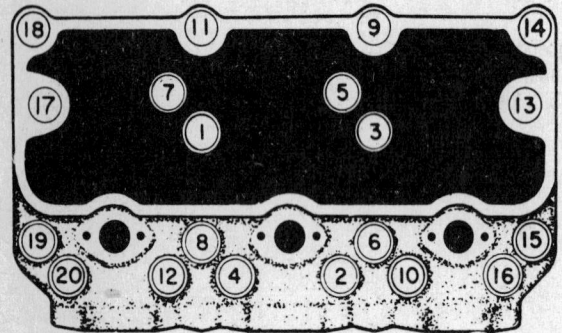

Cylinder head nut torquing sequence, END, T673 and 711 engine series (⅝ in. stud)

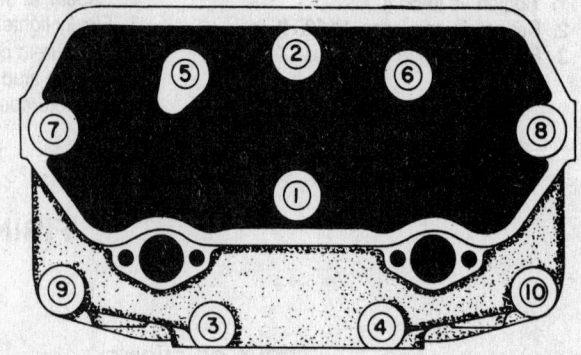

Cylinder head nut torquing sequence, END, T and 864 engine series

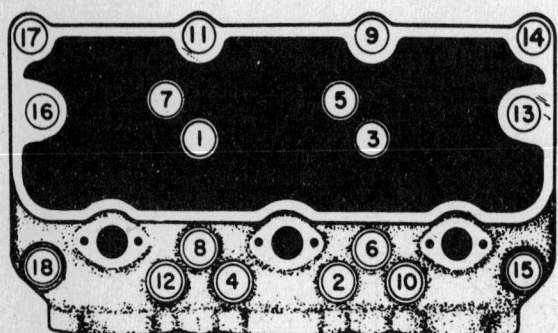

Cylinder head nut torquing sequence, END673 and 711 engine series (¾ in. stud at 15 and 18 locations)

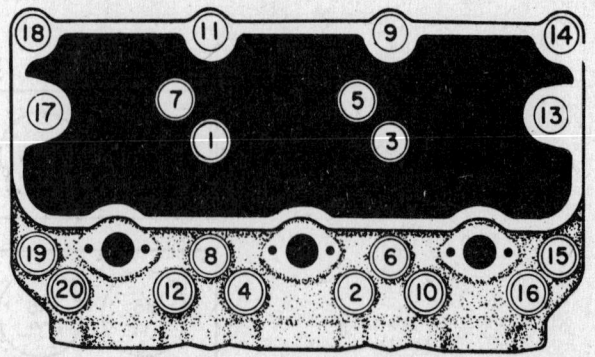

Cylinder head bolt torquing sequence, 6 cylinder current production (⅝ in. studs)

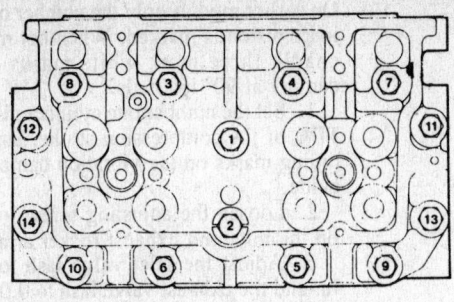

Cylinder nead bolt torquing sequence, EMC9-400R, EMC9-400, EM9-400R and EM9-400 engines

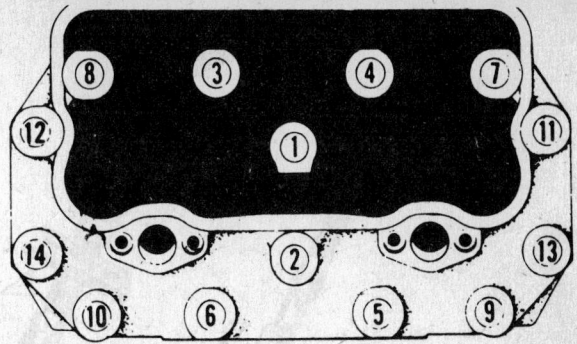

Cylinder head bolt torquing sequence, ENDT and ENDDT V8 series

5. Install a special adapter onto the gauge, place a copper gasket on the adapter and insert the assembly into the nozzle hole. Secure the assembly with the injector hold-down nuts.

6. Start the engine and set the throttle for 1,000 rpm. Compare the reading with the figures on the chart below.

7. Remove the gauge. Reinstall the injection nozzle and holder with a new gasket washer and holder dust seal. Make sure only one gasket washer is used.

8. Repeat the test for the remaining cylinders.

9. Compare readings with each other. They should be within 50 psi.

10. Check all nozzle holder and fuel line fittings for proper torque. Start the engine and make sure there are no fuel leaks.

Valve Adjustment

— CAUTION —

To avoid damage to pistons, always set valve lash under cold static conditions (coolant temperature below 100°F and engine turned to the off position).

6 CYLINDER

Adjust inlet and exhaust valve clearances at TDC of the compression stroke. Follow the firing order when making adjustments. The firing order for all Mack 6 cylinder diesel engines is 1–5–3–6–2–4.

Mark the vibration damper at three 120° increments. Start with the TDC mark for number one cylinder (already marked on damper) and place a white mark 12%/32 in. in both directions from the number one cylinder mark. Use these marks to rotate the engine to TDC for each successive piston in the firing order.

1. Set number one cylinder piston on TDC of the compression stroke, using the mark on the vibration damper.

2. Loosen the adjusting screw locknut on the inlet and exhaust rocker arms.

3. Adjust the inlet and exhaust valve lash to specified clearance.

4. Tighten the adjusting screw locknuts.

5. Recheck the valve lash.

6. Turn the crankshaft, in the direction of normal engine rotation placing number five piston at TDC of the compression stroke. Adjust the valve clearances as previously outlined.

7. Continue rotating the crankshaft through the firing order, stopping at each

NORMAL COMPRESSION SPECIFICATIONS

Engine Series	Normal Compression (psi)
END475	540
ENDT475	470
ET477	470
END(B)673E	530
ENDT(B)673	575
ENDT(B)673C	460
ET(B)673	460
ET(B)673E	460
ETY(B)673E	635
ETAY(B)673A	460
ETAZ(B)673	460
ETAZ(B)673A	460
ETAZ(B)673C	460
ENDT(B)675	460
ETY(B)675	585
ENDT(B)676	460
ETA(B)676B	460
ETAY(B)676	460
ETAY(B)676D	460
END707	530
END711	530

MINIMUM COMPRESSION PRESSURE PSI

Engine Series	Altitude Feet							
	0	2,000	4,000	6,000	8,000	10,000	12,000	14,000
Non-turbocharged six cylinder	530	500	460	430	390	360	340	310
Turbocharged/charged air cooling six cylinder	475	445	415	385	355	325	295	275
V8—866 cu. in. 14.95:1 ①	485	455	425	385	355	335	305	285
V8—866 cu. in. 15.7:1 ①	540	410	480	440	410	380	350	330
V8—400 series ②	585	555	525	495	465	435	405	375

① Compression ratio
② 998 cubic inches

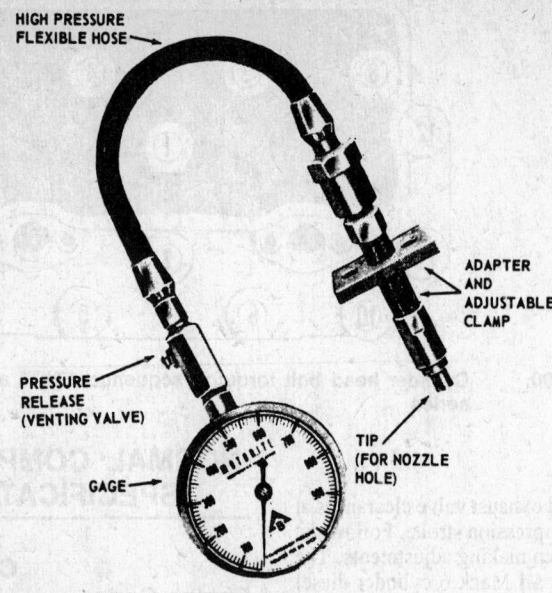

HIGH PRESSURE
FLEXIBLE HOSE

ADAPTER
AND
ADJUSTABLE
CLAMP

PRESSURE
RELEASE
(VENTING VALVE)

GAGE

TIP
(FOR NOZZLE
HOLE)

Compressing tester

On earlier models only the number one cylinder piston is marked. When not marked, chaulk three more white marks on the damper at 90° intervals.

1. Set the number one cylinder piston on TDC of the compression stroke, using the timing marks on the vibration damper as a guide.

2. Loosen the adjusting screw locknuts on the inlet and exhaust rocker arms.

3. Adjust the inlet valve lash to 0.016 in. and the exhaust valve lash to 0.026 in..

NOTE: When adjusting exhaust valve clearances on engines equipped with Mack Dynatard® engine brake use Mack special tool MVT–36–6 or equivalent.

4. Tighten the adjusting screw locknuts and recheck the valve lash.

5. Adjust all valves in the firing order in same manner as outlined for number one cylinder.

Injector Timing

The high pressure port closure stand method must be used to check and adjust Mack injection pump to engine timing.

NOTE: Port closing is defined as the point at which the fuel coming from an injection pump delivery valve changes from a steady stream to a few drops.

When the Mack diesel engine is timed correctly the vibration damper/flywheel timing mark will be at the specified degrees, at the same time as the port closure of the number 1 cylinder fuel delivery valve.

NOTE: Injection timing specifications are found on the injection pump name plate or EPA emission plate.

Adjust the timing by following the same procedure for all Mack 6 or 8 cylinder engines equipped with American Bosch (APE) or Robert Bosch (PES) injection pumps.

VALVE CLEARANCE SPECIFICATIONS

(Valve stem to rocker arm)

Engine Series	Cold Static	
	Inlet	Exhaust
END475	.014	.108
ENDT475	.014	.028
ET477	.014	.028
6 Cyl①	.016	.024
All V8	.016	.026

① Includes all Mack 6 cylinder engines in production as of August 31, 1980.

cylinder TDC compression stroke, to adjust the valve clearance.

NOTE: When adjusting exhaust valve clearances on engines equipped with Mack Dynatard® engine brake, use Mack special tool MVT–36–6 or equivalent.

V8 ENGINES

Set the V8 valve lash in firing order with the piston at TDC of the compression stroke. The firing order of all Mack V8 diesel engines is 1–5–4–8–6–3–7–2. Present production Mack V8 vibration dampers have four timing marks spaced 90° apart, indicating TDC for each cylinder piston.

Special Tools

The following Mach special tools (or equivalent) will be required in order to complete the high pressure port closure timing procedure:

1. Portable high pressure—port closing timer (Bacharach part o. 72–7010) or high pressure hand supply pump (Robert Bosch no. 1–687–222–039).

2. Timing plug gauge for American Bosch injection pump (J24345–1).

3. Timing plug gauge for Robert Bosch injection pump (J24345–2).

High Pressure Port Closing Stand Timing Procedure

1. Cap or connect injection lines on all except number 1 delivery valve outlet.

2. Remove all return fuel lines at the

TOP CENTER
1 & 6

12-29/32 IN.
(120° REF.)

3 & 4

12-29/32 IN.
(120° REF.)

2 & 5

Valve timing mark locations on vibration damper

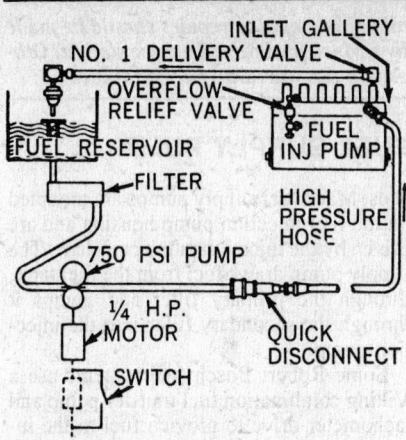

High pressure port closing system diagram

END475 timing marks aligned

Adjusting injection pump coupling on END864 engine

overflow relief valve union fittings. Cap the relief valve port connectors.

3. Connect the high pressure line from the portable closure stand to the fuel inlet of the injection pump gallery.

4. Connect the stand return line from the number 1 cylinder delivery valve holder to the portable port closure stand.

5. Check the timing.

a. Secure the injection pump stop lever in run position.

b. On pumps equipped with a retard start device, remove the control rack cap plug and insert the correct timing plug gauge. (American Bosch J24345-1 and Robert Bosch J24345-2).

— CAUTION —

On Mack Scania engines equipped with a Robert Bosch injection pump the damper cylinder must be removed from the pump to prevent damage to the damper cylinder.

c. When equipped with a Mack Puff Limiter, apply 30 PSI minimum air pressure to the puff limiter air cylinder. The air pressure must be applied before the high pressure fuel is delivered to the pump or mistiming will occur.

d. Activate the throttle lever several times and secure it in the full load position.

e. Introduce fuel pressure to the pump gallery.

NOTE: Fuel pressure applied prior to securing the throttle lever may prevent proper port closing.

f. Slowly rotate the engine, bringing up the number one cylinder piston on its compression stroke. Stop rotating the engine immediately when the fuel stream from number cylinder delivery valve changes to fuel drops (port closing).

g. Check the timing mark on the damper/flywheel. The mark should be at the degrees specified on the pump name plate.

6. If the pump to engine timing differs from specification, then adjust it.

a. Shut off the port closure stand.

b. Loosen the injection pump drive coupling capscrews.

c. Move the injection pump drive coupling in the reverse of normal rotation until the coupling is at the end of its adjusting slots. Lightly tighten the capscrews.

d. Rotate the engine until number one cylinder piston comes up on its compression stroke. Stop rotating the engine, when the specified degrees show on the vibration damper or flywheel.

e. Repeat Step 5. Adjust the drive coupling as needed to correct the timing.

7. When the timing is correct turn off the port closure stand, torque the drive coupling capscrews, remove the timing plug gauge and install the control rack cap plug.

8. Remove the caps from the overflow relief valve union port connections and install all fuel return lines.

Control Rack and Governor

Mack injection pump rack control is accomplished by a mechanical governor mounted on the end of the pump. The governor provides a coupling between the accelerator linkage and the injection pump rack, thereby, regulating fuel delivery in response to pedal position, load and engine speed.

NOTE: Mack fuel injection pumps are factory sealed. No adjustments should be made by other than Mack authorized repair stations, or the engine warranty may be voided.

Adjust Idle Speed

The low idle setting is the only adjustment that can be made on Mack injection pumps without breaking a protective lead seal. All other adjustments require the breaking of seals, which if broken by unauthorized personnel voids the pump and engine warranty.

Special Tools

The Mack special tool J28559 Digistrobe or equivalent magnetic pickup type ta-

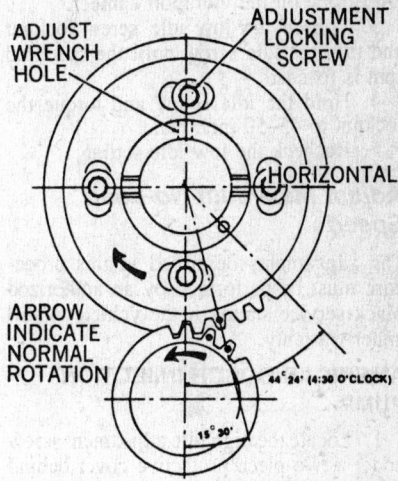

Position of drive coupling tangs and timing marks just prior to pump installation

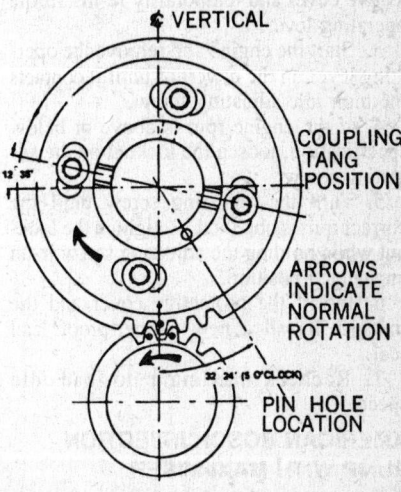

Position of drive coupling tangs and timing marks at No. 1 piston TDC

chometer is needed to check and adjust engine idle speed.

Checking Low Idle Speed

1. Locate and mark the TDC line on the vibration damper.
2. Connect the J28559 Digistrobe to the required power source (Check equipment instructions).
3. Set the Digistrobe at the recommended low idle speed.
4. Start the engine and direct the flash of the Digistrobe at the vibration damper. The mark on the vibration damper should appear to be stopped and the digital read out should equal the specified low idle speed. If the low idle speed is out of specification, adjust it.

Adjusting Low Idle Speed

The low idle adjusting screw is located on the top of the throttle housing cover.
1. Connect the J28559 Digistrobe or equivalent tachometer and set it to the specified low idle rpm.
2. Start the engine and flash the Digistrobe on the vibration damper.
3. Loosen the low idle screw locknut and turn the idle screw until the specified rpm is reached.
4. Hold the idle screw and torque the locknut to 45–50 inch lbs.
5. Recheck the low idle setting.

Adjust Maximum No-Load Speed

The adjustments described in this procedure must be performed by an authorized Mack service station if the vehicle is still under warranty.

AMERICAN BOSCH INJECTION PUMP

1. Locate the high idle adjustment screw under a two-piece protective cover behind the throttle linkage operating lever.
2. Remove the throttle linkage operating lever, break the seal, remove the protective cover and temporarily re-install the operating lever.
3. Start the engine and remove the operating lever on the governor until it contacts the high idle adjusting screw.
4. If the engine rpm is above or below specification, loosen the locknut on the adjusting screw.
5. Turn the adjusting screw until the correct rpm is obtained. Retighten the locknut while holding the adjusting screw in an unchanged position.
6. Install the protective cover and the linkage. Install a new tamper-proof lead seal.
7. Recheck maximum no-load idle speed.

AMERICAN BOSCH INJECTION PUMP WITH MAXI-MISER

1. Apply the truck's spring brakes and place the transmission in neutral.

2. Disconnect the accelerator linkage springs from the injection pump, remove the tamper-proof seals and loosen the air line retaining clip on the valve lifter cover.
3. Remove the air line from the rear of the governor and slide it out of the way.
4. Remove the air inlet fitting from the dual speed governor.
5. Loosen the air inlet cap.
6. Install a master tachometer so that it can be seen as work is performed on the pump. Start the engine, warm it up and secure the accelerator lever in the high idle position.
7. Use an allen wrench to turn the adjusting screw until the correct rpm is obtained.
8. When the rpm is adjusted to specification, tighten the air inlet cap without disturbing the adjusting screw.
9. Install the air inlet fitting and connect the air line to the dual speed governor.
10. Remove the master tachometer, reconnect the accelerator linkage and install new tamper-proof seals.

NOTE: If the vehicle equipped with a Maxi-Miser is to be Dyno tested, the Maxi-Miser must be temporarily deactivated in order to check the power at governed speed. Apply shop air to the governor to deactivate it.

ROBERT BOSCH INJECTION PUMP

1. Locate the high idle adjustment screw on the side of the injection pump.
2. Identify the lead seal and protective cover arrangement which must be removed to gain access to the adjustment screw.
3. Pry out the tamper-proof lead seal.
4. Remove the screw and protective cover. Loosen the high idle locknut.
5. With the throttle lever in the full throttle position, turn the adjusting screw in until there is clearance between the throttle lever and the adjusting screw.
6. Turn the adjusting screw out until it just contacts the throttle lever. Give the screw an additional $\frac{1}{4}$ turn out to prevent the governor's internal linkage from binding.
7. If the engine rpm is still below the governed speed no-load specification, the pump must be removed and recalibrated.
8. If the engine rpm is too high, continue turning the adjusting screw out until the correct rpm is reached. Carefully tighten the locknut, while holding the proper rpm setting.
9. Install the protective cover and a new tamper-proof seal.

FUEL SYSTEM

CAUTION

Mack fuel injection pumps are factory sealed including the pump drive timing gear cover. If the truck is under warranty,

all adjustments or repairs should be made by a Mack authorized service station. Otherwise the warranty may be voided.

FUEL SUPPLY PUMP

Most Mack fuel supply pumps are mounted on the fuel injection pump housing and are driven by the injection pump camshaft. The supply pump draws fuel from the fuel tank, through the primary filter and pumps it through the secondary filter into the injection pump.

Some Robert Bosch fuel systems use a Viking combination fuel transfer pump and tachometer drive to provide fuel to the injection pump and drive the tachometer. The unit is driven off from the auxiliary shaft.

FUEL INJECTION PUMP

All current production Mack engines use either the American Bosch (APE) or the Robert Bosch (PES) injection pumps.

Removal—6 Cylinder

1. Loosen the inner support bracket bolts.
2. Remove the upper and lower support brackets.
3. Remove the lower pump to cylinder block bracket.
4. Remove all fuel line brackets and fuel lines from the pump. Cap all fuel lines to prevent dirt from entering and fuel from spilling.
5. Remove the drive coupling bolts and drive coupling.
6. Remove the pump from the engine.

Installation—6 Cylinder

1. Rotate the engine in the normal direction until number one piston comes up on the compression stroke. At the same time bring the mark on the vibration damper flywheel to the number of degrees specified on the valve cover. The injection pump driveshaft flange lugs should now be in the horizontal position and the indexing pin hole at the 4 o'clock position.
2. Grease the front (non-counterbored) face of the coupling and mount it on the pump driveshaft flange. Center the coupling ring side to side.
3. Remove the adapter inspection hole cover.
4. Assemble the injection pump upper support bracket (if used) to the injection pump and tighten the bolts.
5. Rotate the pump coupling drive flange until the lugs are in a vertical plane and the indexing hole is at a 7 o'clock position.
6. Mount the pump assembly on the engine. As the pump moves into position the indexing pin can be observed through the adaptor inspection hole.
7. Install the adaptor to cylinder block bolts and tighten them.

8. Assemble the injection pump lower support bracket to the cylinder block and tighten the bolts lightly. Install and tighten the support bracket to cylinder block bolts.

9. Check the coupling ring for approximately $3/32$ in. end float.

10. Set the timing by the high pressure (port closing) stand method.

Removal – V8 Engine

1. Remove the hardware and pump bolts.

2. Remove the drive coupling bolts and drive coupling.

3. Remove all fuel line brackets and fuel lines from the injection pump. Cap all fuel lines to prevent dirt from entering and fuel from spilling.

4. Remove the pump from the engine.

Installation – V8 Engine

1. Rotate the engine in the normal direction until the number one piston is at TDC of the compression stroke. The timing mark on the drive coupling should be centered between the two timing marks on the auxiliary drive shaft gear. The pin hole location should be at the five o'clock position and the coupling tangs at $12\frac{1}{2}$ degrees off from horizontal.

2. Rotate the engine backwards approximately 40 degrees.

3. Rotate the engine in the normal direction again until the vibration damper timing reads at the degrees specified on the pump name plate. In this position, the drive coupling tangs will be approximately horizontal and the pin hole will be between the 4 and 5 o'clock position. The gear timing marks will be approximately $15\frac{1}{2}$ degrees to the right.

4. Install the pin and ring assembly on the injection pump drive coupling.

5. Install the injection pump. Connect all fuel lines and fuel line brackets.

6. Recheck the port closing timing.

Fuel Filters

The primary filter is located between the fuel tank and the fuel supply pump. The primary filter is color coded red. A secondary fuel filter is located between the fuel supply pump and the fuel injection pump. The secondary fuel filter is color coded green. For all current production models the filters are the spin on type.

Removal

1. Clean the area around the filter and the adapter with solvent. Dry the area with compressed air.

2. Break the filters loose with a filter wrench. Both filters have right hand threads.

3. Wipe the sealing surface clean on the adapter before installing a new filter.

Installation

NOTE: Mack service filters contain detailed installation procedures. If package procedures conflict with this procedure, the package instructions should be followed.

1. Apply a thin film of engine oil to the filter sealing gasket.

2. Pre-prime the filters by filling them with filtered fuel. Prime the filter through the small outer holes on the top of the filter. Do not add the fuel through the center core.

3. Apply a thin coat of clean engine oil to the sealing gasket. Tighten the filter one full turn by hand after the gasket contacts the adapter.

Fuel Injector Service

The fuel injectors in Mack diesel engines receive metered fuel at injection pressure from the injector pump. The nozzles are used to provide effective atomization of the fuel through use of a nozzle valve and discharge nozzle, which insure sharp start-up and cutoff of fuel flow and production of four high velocity streams which will effectively penetrate to all parts of all combustion chamber.

The nozzle body at the lower end of the injector consists of the hole type nozzle, the seat for the nozzle valve and the fuel passage. The nozzle valve spindle extends upward to the top of the nozzle holder, where there is an adjustable spring for maintenance of precise injector pop pressure. Controlled clearance between the nozzle body and the nozzle valve lubricates the mechanism. A leak-off fitting at the top of the injector returns excess fuel to the fuel tank.

Injector nozzles should be removed and checked after 50,000 to 75,000 miles, or if troubleshooting reveals poor performance. Extreme care must be taken to ensure absolute cleanliness during nozzle service. It must also be remembered that oil leaving an injector nozzle is moving at extreme speed. Contact with the skin will usually result in penetration. Therefore, testing must be carried out in a manner that will protect the skin from nozzle discharge.

NOTE: If an injector is suspected of malfunction, it can be located as follows: run the engine at idle, loosen the high pressure fitting to the suspected injector at the injection pump ($\frac{1}{2}$ turn). This will cut off the fuel to the nozzle. Retighten the fitting. If the cylinder is not restored to firing, then the nozzle should be removed and checked completely.

Removal

1. Clean the cylinder head around the nozzles and the tubing connections with solvent. Blow dry with compressed air.

2. Remove the leak-off lines, carefully recovering the copper gaskets.

3. Remove the high pressure fuel lines and install protector plugs in their open ends.

4. Remove the nuts from the hold down studs. Use a small pry bar to remove nozzles, gripping them under the hold-down flanges at a point near to the nozzle body. Use penetrating oil to aid removal if the nozzle is especially tight.

5. Place nozzles in a rack in order of removal so they may be installed in the engine in original order. Plug the nozzle ports in the cylinder head(s).

Testing

1. Clean carbon from the nozzle with a special wire brush.

2. Mount the nozzle holder assembly in a tester. Make sure the tester is filled with clean fuel.

3. Close the pressure gauge valve (to protect the gauge). Operate the actuating lever at about 25 strokes per minute to expel air and settle the spring and nozzle loading column.

4. Open the pressure gauge valve one half turn and slowly operate the actuating lever to raise the pressure to the point where the nozzle opens. Carefully watch the gauge and note the exact pressure at which the nozzle opens. Also check the characteristics of the flow to ensure that no leakage or dripping course occurs after the end of injection.

5. Compare the opening pressure with the figure specified in the nozzle opening pressure chart and adjust the nozzle to specifications if necessary.

6. Wipe the nozzle tip dry. Operate the actuating lever slowly to bring pressure to within 100 psi of opening pressure (20 psi with END 475 nozzles) and maintain the pressure for five seconds. If drops of fuel form or if the nozzle sprays slightly, reject it.

7. Close the gauge valve (to protect the gauge) and operate the lever at about 15 strokes per minute. The spray pattern formed should be sharp, solid and with uniform quantities and angles between orifices. Use short, rapid strokes on END 475 nozzles to produce a good spray pattern. Reject a nozzle with a poor spray pattern.

8. On all but 475 series nozzles, test the nozzle to make sure it makes a chattering sound. Operate the actuating handle so the stroke takes about two seconds. Close the pressure gauge valve. A distinct and regular chattering sound must be produced, although an occasional variation is acceptable. Reject a nozzle which does not pass the chatter test.

Disassembly

NOTE: The American Bosch ($2^1/_{17}$mm) injectors do not have an opening pressure adjusting screw; therefore, the opening spring pressure cannot be released for disassembly. In order to disassemble this type of injector, soak the assembly in carbon solvent, then place it in the Mack tool disassembly fixture TSE–77108 (or equivalent). The disassembly fixture must be used in order to avoid damage to the index dowel pins, holder and nozzle. With the fixture in a soft jawed vise, loosen the upper cap and then remove the cap by hand turning. Remove the nozzle assembly from the fixture to complete the disassembly.

1. Loosen the opening pressure adjusting screw all the way to relieve all downward pressure.

2. Position the nozzle assembly on a suitable block in a vise. Remove the upper cap nut.

3. Loosen the locknut, if the nozzle is equipped with one and loosen the opening pressure adjusting screw all the way to relieve all downward pressure.

4. Use a special wrench to remove the spring retaining capnut. Remove the pressure adjusting spring and spindle assembly.

5. Invert the holder assembly in a softjawed vise and remove the nozzle capnut. Remove the nozzle body and valve assembly. Keep these two parts together as they are a mated assembly. Reinstall nozzle cap nut loosely.

Inspection

1. Wash all parts in a safe solvent and inspect for wear as described below:

a. Check the spring for corrosion and pitting and replace if they are evident.

b. Check spindle for straightness within 0.010 in. T.I.R.

c. Check the lapped surfaces of the nozzle body and holder for cracks and scratches. Replace parts that are cracked and lap parts which are slightly scratched. If the holder lapped surface is spalled more than .003″ deep in the needle valve area, it must be replaced rather than resurfaced.

d. Remove the locating dowels from the holder very carefully to prevent damaging the lapped surface. Check the spring retaining cap nut to make sure the bleed hole is open.

e. Clean the pressure chamber of the nozzle body with a special scraper, as shown in the illustration.

f. Using a special needle and vise, clean the nozzle body holes. Be careful to avoid breaking the needle inside the discharge holes, as fragments may prove to be impossible to remove.

g. Lift the valve about ⅓ of its length out of the nozzle body, hold the body at 45 degrees from the vertical and release the valve. It should slide back freely. If the valve is not free, work the valve in the nozzle body using a special polishing tallow. Clean the nozzle valve and body in solvent and blow dry.

h. Check the nozzle valve lift using a straightedge and dial indicator. See the injection Nozzle Opening Pressure Chart for specifications. Place the nozzle in a fixture and run a straightedge across the top of the nozzle body. Mount a dial indicator right above the needle valve and zero it. While holding the straightedge, raise the needle valve with a pair of tweezers just until the lower portion of the valve contacts the straightedge. Hold the valve in position while reading the dial indicator. If the valve lift is not to specification, the assembly must be replaced.

i. If the injector uses a filter in the fuel inlet connection, clean it by reverse flushing it with compressed air.

j. Check the ends of the high pressure fuel tubes to make sure they are open. Ream ENDT 864 lines to .085″ and all others to .078″. Flush out the chips.

Assembly

1. Place the injection nozzle holder in a soft jawed vise.

2. Position the nozzle body and valve onto the holder, aligning dowel pins and holes carefully.

3. Install the nozzle cap nut, using a special centering sleeve during the initial tightening. Remove the centering sleeve and torque the cap nut to specification, using a special adapter.

4. Invert the holder and install the spindle and pressure adjusting spring. Tighten the pressure adjusting screw and lock it in position with the locknut.

5. Install the injector in a test stand and adjust opening pressure as described in the section on service.

6. Install the nozzle holder with a new gasket and install and tighten upper capnuts.

Installation

1. Remove copper nozzle tip gasket from the hole in the cylinder head. Clean the nozzle cavity with a special reamer and wire brush. Check the gasket seat for trueness and cleanliness. Crank the engine over by hand to blow loose carbon from the cavity.

2. Apply an anti-seize compound to the outside diameter of nozzle and holder and position the assembly in the head with a new nozzle tip gasket and dust seal O-ring. The nozzle tip gasket may be held in place with a small amount of grease.

3. Seat the nozzle squarely and then install and tighten hold down nuts evenly.

4. Install the high pressure fuel lines and torque nuts carefully so that tubing will not be distorted.

21 MM INJECTION NOZZLE SPECIFICATIONS

Engine Series	Nozzle Opening Pressure—PSI	Valve Lift-In	Injection Pump	Holes	Hole Diameter Inch (mm)
END475	1985 to 2035	0.015	PES	4	0.011 (0.280)
ENDT475	2840 to 2890	0.015	PES	5	0.010 (0.252)
ENDT673	3000 to 3050	0.014	APE/PES	4	0.0126 (0.320)
ENDT673C	3000 to 3050	0.014	APE/PES	5	0.0126 (0.320)
ET673②	3000 to 3050	0.014	APE/PES	5	0.0126 (0.320)
END673E	3000 to 3050	0.019	PES	5	0.0126 (0.320)
END673E	3000 to 3050	0.014	APE	5	0.0126 (0.320)
ENDT675①	3000 to 3050	0.014	APE	5	0.0126 (0.320)
ENDT675	3000 to 3050	0.019	PES	5	0.0126 (0.320)
END707	3000 to 3050	0.019	PES	5	0.0126 (0.320)
END711	3000 to 3050	0.019	PES	5	0.0126 (0.320)

① To November, 1977 on APE injection pumps only
② To January, 1979

21/17MM INJECTION NOZZLE SPECIFICATIONS

Engine Series	Nozzle Opening Pressure—PSI	Valve Lift-In	Injection Pump	Holes	Hole Diameter Inch (mm)
AMERICAN BOSCH INJECTION NOZZLE AND HOLDER					
ENDT676	4200 to 4350	0.014	APE	5	0.0126 (0.320)
ENDT675C	3300 to 3450	0.014	APE	6	0.0126 (0.320)
ETAZ673	4200 to 4350	0.014	APE	5	0.0126 (0.320)
ENDT675①	3800 to 3950	0.014	APE	5	0.0126 (0.320)
ETAY673	4200 to 4350	0.014	APE	5	0.0126 (0.320)
ETAY676	4200 to 4350	0.014	APE	5	0.0126 (0.320)
ETZ675	3800 to 3950	0.014	APE	5	0.0126 (0.320)
ETY675	3800 to 3950	0.014	APE	5	0.0126 (0.320)
ROBERT BOSCH INJECTION NOZZLE AND HOLDER					
ENDT676	4200 to 4400	0.019	PES	5	0.0126 (0.320)
ENDT675C	3800 to 3950	0.014	PES	5	0.0126 (0.320)
ETAZ673	4200 to 4400	0.019	PES	5	0.0126 (0.320)
ET673②	3800 to 3950	0.014	PES	5	0.0126 (0.320)
ETAY676	4200 to 4400	0.019	PES	5	0.0126 (0.320)
ETY675	3800 to 3950	0.014	PES	5	0.0126 (0.320)
ETZ675	3800 to 3950	0.014	PES	5	0.0126 (0.320)
ETAY673	4200 to 4400	0.019	PES	5	0.0126 (0.320)
ETY673E	3800 to 3950	0.014	PES	5	0.0126 (0.320)
MACK SCANIA ONLY					
ET477	2990 to 3060	0.011	PES	5	0.010 (0.252)

① Begining November 1, 1977
② Beginning January, 1979 production

6 CYL.① INJECTION NOZZLE SPECIFICATIONS

Engine Model	Nozzle Opening Pressure—PSI	Valve Lift-In	Injection Pump	Holes	Hole Diameter Inch (mm)
E6-200	3000–3150	—	PES	5	.0126(0.320)
EM6-225	4250–4400	—	PES	5	.0126(0.320)
EC6-235	3800–3950	—	APE	5	.0126(0.320)
EM6-237	②	—	③	5	④
EM6-237R	3800–3950	—	APE	5	.0126(0.320)
E6-250	4250–4400	—	PES	5	.0126(0.320)
EM6-250	4200–4400	—	③	5	④
EM6-250R	4200–4400	—	③	5	⑤
EMC6-250	4200–4350	—	APE	5	.0126(0.320)
E6-260	3800–3950	—	APE	5	.0126(0.320)
EM6-275	4200–4350	—	APE	5	.0126(0.320)
EM6-275L	4200–4350	—	PLM	5	.0126(0.320)
EM6-275R	4200–4350	—	APE	5	.0126(0.320)
EM6-285	4200–4400	—	③	5	.0126(0.320)
EM6-285R	4200–4350	—	APE	5	.0126(0.320)
EMC6-285	4200–4350	—	APE	5	.0126(0.320)
EMC6-285R	4200–4350	—	APE	5	.0126(0.320)
EM6-300	4200–4350	—	APE	5	.0126(0.320)
EM6-300R	4200–4350	—	APE	5	.0126(0.320)
E6-315	4250–4400	—	PES	5	.0126(0.320)

6 CYL.① INJECTION NOZZLE SPECIFICATIONS

Engine Model	Nozzle Opening Pressure—PSI	Valve Lift-In	Injection Pump	Holes	Hole Diameter Inch (mm)
E6-325	4200–4350	—	PLM	5	.0126(0.320)
EC6-330	4200–4350	—	PLM	5	.0126(0.320)
E6-350	4300–4450	—	PLM	5	.0126(0.320)
E6-350R	4200–4350	—	PLM	5	.0126(0.320)
EC6-350	4200–4350	—	PLM	5	.0126(0.320)

①Includes all mack 6 cylinder engines in production as of August 31, 1980.

②United Technologies Diesel systems pump: 3800–3950
Robert Bosch Pump: 4250–4400

③United Technologies Diesel systems pump: APE
Robert Bosch pump: PES

④United Technologies Diesel systems pump: .0126(0.320)
Robert Bosch pump: .0120(0.305)

⑤United Technologies Diesel systems pump: .0118(0.300)
Robert Bosch pump: .0126(0.320)

V-8 INJECTION NOZZLE SPECIFICATIONS

Engine Series	Nozzle Opening Pressure—PSI	Valve Lift-In	Injection Pump	Holes	Hole Diameter Inch (mm)
ENDT(B) 865,866	3800–3950	—	APE	6	0.011(0.28)
400 Series①	4200–4350	—	APE	6	0.011(0.28)

①998 cubic inches

5. Install the leak-off lines with their copper gaskets.

6. Operate the engine and check for leaks. Retighten fittings as necessary.

Bleeding Fuel System

Whenever the injection pump, supply pump, fuel lines or fuel filter(s) have been removed, bleed out all air before attempting to start the engine. Air bubbles in the system will enter the injection pump and cause hard starting and erratic engine performance.

Past production models equipped with a hand priming pump on the injection pump, can be bled (primed) by the following procedure.

1. Disconnect the transfer pump (supply pump) outlet connection.

2. Prime the system with the hand primer, until a solid stream of fuel runs out of the outlet connection.

3. Retighten the transfer pump outlet.

4. Disconnect or loosen the fuel inlet connection at the pump gallery.

5. Prime the system with the hand primer until a solid stream of fuel is obtained.

NOTE: If fuel cannot be obtained by hand priming, check the fuel lines for leaks. Check the lines from the fuel supply tank to the fuel transfer (supply) pump.

6. Connect and tighten the gallery inlet fitting.

7. Disconnect the overflow line on the injection pump and continue priming, until a solid stream of fuel runs out.

8. Connect and tighten the overflow line.

9. Secure the hand priming handle and start the engine.

NOTE: Check the Robert Bosch system for an open bleed line on the secondary filter. If equipped, use the bleed line to prime the system. Start the engine before opening the bleed line. Close the bleed line as soon as a solid stream of fuel is obtained.

10. When the engine is operating normally, check the filters and lines for leaks.

When current production models, not equipped with a hand priming pump, need to be bled, use one of two alternate procedures:

First Alternate Bleeding Procedure

1. Attach a manual or electric low pressure pump (15–20 psi) to the inlet side of the primary fuel filter.

2. Disconnect or loosen the overflow valve fitting.

3. Using clean fuel, pump the system out until all of the air is expelled from the overflow valve.

4. Reconnect and tighten the overflow valve.

Second Alternate Bleeding Method

1. Completely remove the overflow valve from the injection pump gallery, to provide resistance free fuel flow.

2. Press a shop air nozzle wrapped with a shop towel onto the fuel tank filler spout. Pressurize the fuel tank. Any air escaping past the shop towel will prevent excessive pressure in fuel tank.

3. Continue the pressure until a solid stream of fuel is obtained at the overflow port.

4. Install the overflow valve.

Fuel Cut Off Controls

INJECTION PUMP SWITCH

V8 ENGINES WITH AMERICAN BOSCH V-TYPE PUMPS

The switch is located on the rack cap end of the injection pump. It is a micro-type switch. The switch is engaged by the rack extension and activates the system at zero fuel only. The switch is pre-adjusted and under normal operation needs no further adjustment. If adjustment proves necessary:

1. Measure the height of each of the three electrical terminals on the top cover of the Dynatard switch. Each must extend at least $\frac{1}{2}$ in. above the cover. If not, remove the cover and adjust the terminal by loosening the locknuts.

2. Cut and remove the seal wire from the switch side cover, remove the side cover, rectifier and gasket. Position the engine brake toggle switch on the dash to the ON position activate the electrical circuit to the engine brake switch.

3. Depress the microswitch contact button with a screwdriver and check for continuity with a circuit tester.

4. If continuity is good, check the adjustment of the engine brake adjusting screw. Connect the circuit tester from the engine terminals to ground. Push the control rack extension in to the stop position and the light should go on. If not, turn the adjusting screw until it does.

5. Loosen the cap screw on the end of the rack extension just enough to remove the adjusting screw and bracket. Place a straight edge along the forked end of the adjusting screw bracket and measure the clearance between the straight edge and the screw. Loosen the locknut and adjust the screw to obtain a clearance of 0.138–0.142 in. and tighten the locknut.

6. Slide the adjusting screw bracket between the control rack extension and capscrew and tighten the capscrew. Replace the switch side cover and rectifier. Install a new gasket and secure with a pump seal.

SIX AND V8 ENGINE WITH AMERICAN BOSCH INJECTION PUMP, EXCEPT V-TYPE PUMP

This switch is a small contact type and is an integral part of the governor housing cover. The switch assembly is activated by the cam nose and fulcrum lever assembly and activates the system at zero fuel only. If adjustment is necessary follow either of the two following procedures:

WITH INTERNAL PUFF LIMITER

A 0.070 in. feeler gauge and a gauge plug, part #J24659 are necessary for this adjustment.

1. Measure the adjusting screw extension from the first locknut. The screw should extend $\frac{1}{8}$ in. If not, loosen both locknuts and adjust the screw. Tighten the locknuts.

2. Place the pump stop lever in the RUN position.

3. Remove the rack cap plug.

NOTE: On 6–cylinder models, the rack cap may have to be held to prevent its turning when removing the plug or installing the gauge plug.

4. Position the lockring flush against the head of the gauge to allow maximum exposure of the plug gauge screw threads.

5. Screw the special gauge plug into the rack cap until it contacts the rack. Continue to screw it in until it bottoms internally in the injection pump, placing the rack in the full OFF position.

6. Position and lock the lockring by turning it clockwise flush against the rack cap.

--- CAUTION ---
To keep the lockring fixed in this position, screw it in on the set screws placed 180° apart.

7. Turn the special gauge plug counterclockwise until the 0.070 in. feeler gauge can be inserted between the lockring and the cap.

8. Attach the electrical continuity tester to the contact terminal and ground.

9. Loosen the two adjusting screw locknuts on the engine brake switch adjusting screw.

10. Turn the adjusting screw slowly, counterclockwise until continuity is broken.

11. Turn the screw slowly clockwise until continuity is just achieved. Tighten the inner locknut without disturbing the setting.

12. Place the contact terminal in the upright position and tighten the outer locknut.

WITH ANEROID PUFF LIMITER—ROBERT BOSCH PUMP

Special gauge plug #J24660 is necessary for this adjustment.

1. Place the injection pump stop lever in the RUN position.

2. Remove the rack cap plug.

3. Screw the plug gauge into the rack cap until it contacts the rack. Continue to screw it in until it bottoms internally in the full OFF position.

4. Move the throttle from idle to wide open several times. Return and securely retain it against the idle stop by using the throttle return spring or its equivalent.

5. Connect the continuity tester between the contact terminal and ground.

6. Loosen the two adjusting screw locknuts on the engine brake switch adjusting screw.

7. Turn the adjusting screw counterclockwise until continuity occurs at the tester.

8. Turn the adjusting screw slowly clockwise until the point at which continuity breaks. Then turn it one full turn more, clockwise.

9. Tighten the inner locknut without disturbing the setting.

10. Return the terminal to the upright position and tighten the outer locknut.

11. Remove the plug gauge and install the rack cap securely.

PUMPS WITH MACK PUFF LIMITER

1. Run the engine to normal operating temperature and establish a smooth idle of 550 rpm.

2. Loosen both nuts on the pump brake switch.

3. On American Bosch Pumps, turn the adjusting screw clockwise; on Robert Bosch Pumps, turn the adjusting screw counterclockwise until the Dynatard brake comes on.

4. Turn the screw in the opposite direction one full turn.

5. Tighten the jam nut and wire nut. Check the setting by noting that the brake drops out between 1000 and 700 rpm.

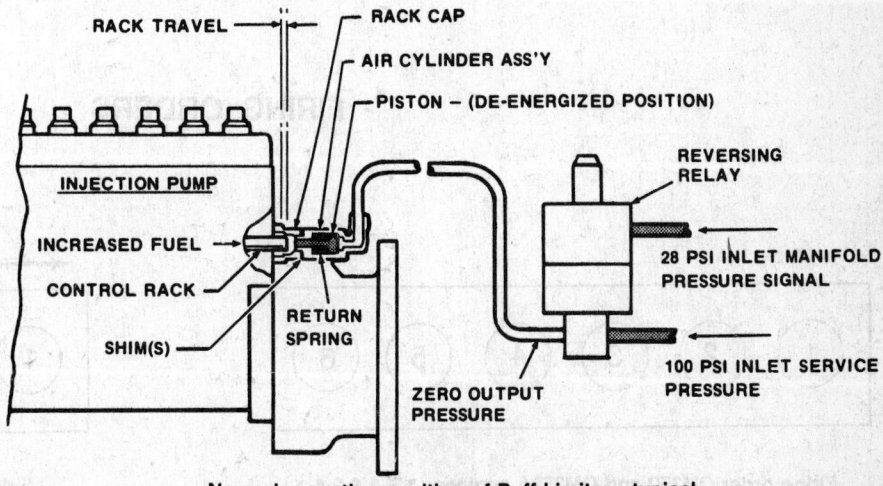

Normal operating position of Puff Limiter—typical

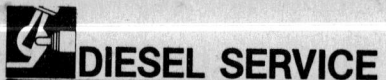

MERCEDES-BENZ OM352, OM352A AND OM355/5 ENGINES

GENERAL ENGINE SPECIFICATIONS

Engine Model	No. Cyl. Displacement (cu. in.)	Horsepower @ rpm (SAE net)	Torque @ rpm (SAE net)	Firing Order	Bore × Stroke	Governed Speed	Oil Pressure @ rpm (psi)
OM352	6-346	130 @ 2800	260 @ 1800	1-5-3-6-2-4	3.82 × 5.04	2800	36.3 @ 2800
OM352A (turbo)	6-346	156 @ 2800	310 @ 2100	1-5-3-6-2-4	3.82 × 5.04	2800	36.3 @ 2800
OM355/5	5-589	181 @ 2200	455 @ 1600	1-2-4-5-3	5.04 × 5.90	2200	36.3 @ 2200

TUNE-UP SPECIFICATIONS

Engine Model	Compression Ratio	Minimum Compression Pressure (psi)	Idle Speed (rpm)	Injection Type	Injector Opening Pressure (psi) New	Injector Opening Pressure (psi) Used	Start of Injection (deg. BTDC)	Valve Clearance (cold—inches) Intake	Valve Clearance (cold—inches) Exhaust
OM352	17:1	284.5	600	Direct	2986	2559	①	0.008	0.012
OM352A (turbo)	16:1	284.5	600	Direct	2986	2559	②	0.008③	0.012③
OM355/5	16:1	284.5	500	Direct	2631	2346	15	0.010	0.016

① Engine production codes:
  344.912–344.937—18°BTDC
  344.942–344.945—17°BTDC
② Engine production codes:
  344.912–344.937—21°BTDC
  344.942–344.945—18°BTDC
③ 1979 and later: intake–0.010 exhaust 0.016

FIRING ORDERS

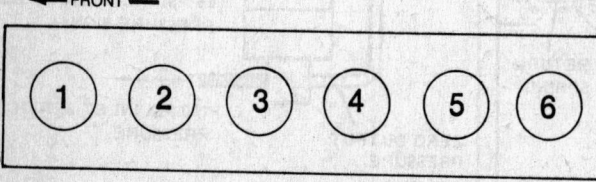

← FRONT →

Firing order OM352 and OM352A engines: 1-5-3-6-2-4

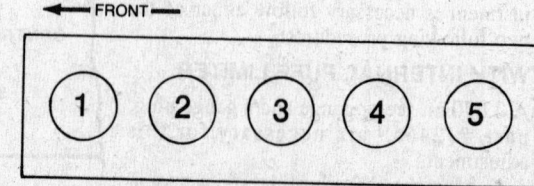

← FRONT →

Firing order OM355/5 engines: 1-2-4-5-3

TORQUE SPECIFICATIONS
(All readings in foot-pounds)

Engine Model	Cylinder Head	Rocker Support	Exhaust Manifold	Connecting Rod	Main Bearing Cap	Damper-to-Crankshaft (center bolt)	Flywheel	Nozzle Holder	Oil Pan	Valve Cover
OM352 & OM352A	①	80	36	79.5 + 90°	42 + 90°	398	29 + 90°	50	③	18
OM355/5	②	72	44	58	167	543	51 + 90°	51	21	18

① Three steps: 1st—43.5, 2nd—65, 3rd—80
② Three steps: 1st—29.0, 2nd—56, 3rd—87
③ M6 Grade bolt: 36
   M8 Grade bolt: 54

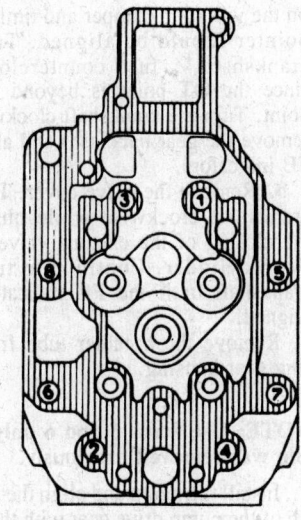

Cylinder head bolt torque sequence—OM355/5 engines

TORQUE SEQUENCES

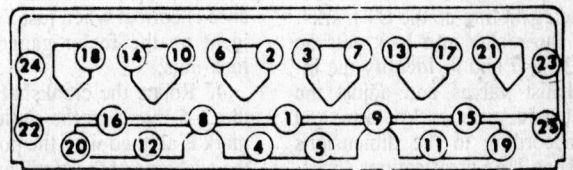

Cylinder head nut torque sequence—OM352 and OM352A engines

TUNE-UP AND ADJUSTMENTS

Every engine tune-up should include the following:
1. Compression test.
2. Valve adjustment.
3. Fuel injection pump timing.
4. Fuel rack settings (if necessary).
5. Engine idle speed setting.
6. Maximum no-load speed setting.

Cylinder Compression

Testing

1. Adjust the valve clearances as outlined later in the tune-up section.
2. Operate the engine until it reaches normal operating temperature.

3. Remove the valve cover. Be careful not to allow foreign matter to enter the engine.
4. Move the accelerator lever of the fuel injection pump to the "fuel shut-off" position and fasten it in this position with wire.
5. Disconnect the fuel injection lines at the injection nozzles.
6. Remove the fuel leak-off line from the cylinder head sidewall.
7. Remove the No. 1 injection nozzle hold-down nut. Using a special nozzle holder puller, remove the No.1 nozzle holder.
8. Clean the nozzle holder seat in the cylinder head and crank the engine to remove any surrounding dirt or carbon flakes.
9. Install a compression test adaptor and secure it to the head with the nozzle holder nut. Tighten the nut to 50 ft. lbs.
10. Connect a compression gauge to the adaptor and crank the engine 5–8 revolutions until the highest reading is obtained.

11. Record the highest compression pressure which was attained.
12. Repeat Steps 7–11 for each cylinder. If any cylinder seems low on compression, first squirt 1 fl.oz. oil into the cylinder and repeat Step 10. If the compression increases, the compression pressure is bypassing the piston rings, which will require engine disassembly for an accurate diagnosis. If the compression does not increase with the extra oil present in the cylinder, the valves, valve guides and/or the cylinder head gasket could be at fault.

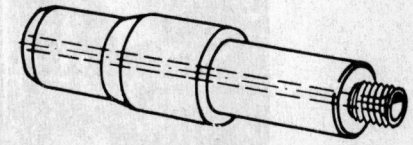

Compression tester adaptor

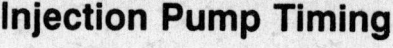

CAUTION

Do not exceed 1 fl.oz. of oil or engine damage may occur.

---

13. If all compression pressure are within limits, reinstall the injection nozzles and the leak-off line. Reconnect the injection lines at the injectors and reinstall the valve cover.

14. Unfasten the accelerator lever of the injection pump and return it to the normal position.

Valve Adjustment

OM352 AND OM352A ENGINES

1. Remove the valve cover. Be careful not to allow foreign matter to enter the engine.

2. Rotate the crankshaft until the OT marks on the vibration damper pulley and the timing case cover are aligned. This position should be top dead center/compression for the No. 1 cylinder. To verify this, hand-spin the No. 1, 2, 3, 5, 7 and 9 push rods. If the pushrods will not spin by hand, turn the crankshaft one more revolution and again line-up the OT marks.

3. Loosen the rocker arm lock nuts on valves 1, 2, 3, 5, 7 and 9. Identify the intake and exhaust valves and adjust the clearance between each rocker arm and valve stem according to the dimensions listed in the Tune-Up Specifications Chart. The clearance is correct when just a slight drag is evident on the feeler gauge.

4. Retighten the rocker arm lock nuts as each valve is adjusted.

5. Rotate the crankshaft one complete revolution and align the OT marks on the vibration damper pulley and the timing case cover. The No. 4, 6, 8, 10, 11 and 12 push rods should be free to rotate.

6. Loosen the rocker arm lock nuts for the No. 4, 6, 8, 10, 11 and 12 valves. Adjust the valve clearances in the same manner as in Step 3.

7. After the adjustments have been completed, turn the engine several revolutions and recheck all of the valve clearances.

8. Reinstall the valve cover using a new valve cover gasket.

NOTE: The valve cover gasket MUST be replaced any time the cover is removed. This gasket is also used as an air intake gasket. Failure to replace the gasket could result in excessive oil consumption.

OM355/5 ENGINES

1. Remove the valve covers. Mark the covers so that they may be reinstalled in their original locations. Be careful not to allow foreign matter to enter the engine.

2. Rotate the engine to position the No. 1 piston at top dead center compression; the OT marks on the crankshaft pulley and the timing case cover must be aligned. To verify this, hand spin the pushrods of the No. 1 cylinder. If the pushrods will not spin by hand, turn the crankshaft one more revolution and again line-up the OT marks.

3. Loosen the rocker arm lock nuts of the No. 1 cylinder. Identify the intake and exhaust valves and adjust the clearance between each rocker arm and valve stem according to the dimensions listed in the Tune-Up Specifications Chart. The clearance is correct when just a slight drag is evident on the feeler gauge. Retighten the lock nuts.

4. Rotate the crankshaft clockwise past the 3–5 mark on the pulley until the 2–4 mark is aligned with the pointer. Hand spin the pushrods of the No. 2 and the No. 4 cylinders to verify the correct crankshaft position.

5. Adjust the valve clearances on cylinders No. 2 and No. 4 in the same manner as in Step 3.

6. Turn the crankshaft clockwise past the OT marks until the 3–5 mark is aligned with the pointer. Adjust the valve clearances on cylinders No. 3 and No. 5 in the same manner as in Step 3.

7. After the adjustments have been completed, turn the engine several revolutions and recheck all of the valve clearances.

8. Install the valve covers in their original locations using new valve cover gaskets.

Injection Pump Timing

OM352 AND OM352A ENGINES

1. Thoroughly clean the injection pump and the surrounding area.

2. Remove the clutch housing dust shield.

3. Bring the No. 1 piston to top dead center compression, aligning the FB (fuel beginning) marks of the vibration damper pulley and the timing case pointer using either of the following two methods:

   a. Remove the valve cover. Turn the crankshaft clockwise while observing the rocker arm movement of the No.6 cylinder. As the No.6 exhaust valve is closing and the No.6 intake valve is opening, the No.1 cylinder should be on its compression stroke and the OT marks on the vibration damper and timing case pointer should be aligned. Turn the crankshaft $1/4$ turn counterclockwise since the OT point is beyond the FB point. Turn the crankshaft clockwise (to remove the gear backlash) and align the FB indicators.

   b. Remove the valve cover. Turn the crankshaft clockwise while observing the closing of the exhaust valve of the No.6 cylinder. Continue to turn the crankshaft until the FB indicators are aligned.

4. Remove the breather tube from the timing gear housing.

NOTE: Use Steps 5 and 6 only if the pump was removed previously.

5. Install the pump and align the marked tooth of the pump drive gear with the housing pointer.

6. Install the five pump support bolts through the support plate and torque the bolts to 36 ft. lbs. Loosen the four screws holding the pump to the support plate.

7. If the pump had not been removed previously, disconnect the No. 1 injection line from the injection pump.

8. Remove the half moon clamps from the No.1 and 2 delivery valve holders.

Fuel reservoir installed for injection timing procedure—OM352 and OM352A engines (optional—see text)

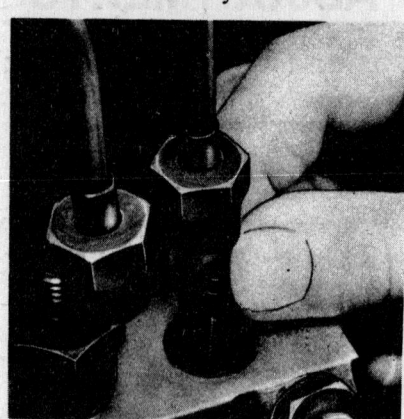

Removing the check valve needle of the delivery valve—OM352 and OM352A engines

9. Remove the No.1 delivery valve holder, filler piece, compression spring and check valve needle.

10. Reinstall only the No. 1 delivery valve holder.

11. Attach a drip tube to the delivery valve holder.

NOTE: Do not use an old high pressure fuel line as a drip tube as this will cause inaccurate results.

12. If the pump had been removed previously, connect the fuel supply lines to the fuel transfer pump and the fuel filter(s).

13. Connect the fuel supply line from the fuel filter to the injection pump.

NOTE: The existing fuel supply of the vehicle or a fuel reservoir attached to the fuel inlet port of the injection pump can be used to supply fuel to the injection pump.

14. Open the bleeder at the front of the injection pump low pressure fuel gallery. Operate the fuel feed hand pump until fuel from the bleeder valve is completely free of air bubbles. Close the bleeder valve.

NOTE: If the vehicle's own fuel supply is used to supply fuel to the pump (as opposed to an external reservoir), actuate the hand pump to maintain a positive pressure.

15. Attach a spring to the injection pump throttle lever to keep the lever in the full forward (full load) position.

16. Push the top of the injection pump towards the engine (retard). A steady stream of fuel should flow from the drip tube attached to the No.1 delivery valve holder.

17. Slowly pull the top of the pump away

Aligning the FB (fuel beginning) marks prior to injection pump timing—OM355/5 engines

from the engine (advance) until the fuel flow is reduced to one drop per 15–20 seconds.

18. Tighten one of the injection pump bolts to lock the injection pump.

19. Recheck the timing as follows:

  a. Turn the engine ¼ turn counterclockwise. A steady stream of fuel should again flow from the drip tube.

  b. Turn the engine clockwise until the fuel flow slows to one drop per 15–20 seconds. The FB indicators should now be aligned.

20. Tighten the remaining injection pump mounting bolts. Torque all of the mounting bolts to 36 ft. lbs.

21. Remove the auxiliary fuel reservoir, if used.

22. Detach the drip tube from the delivery valve holder and remove the delivery valve holder.

23. Reassemble the delivery valve according to the accompanying illustration.

24. Install the delivery valve holder and torque to 33 ft. lbs.

25. Replace the half moon clamping segments and torque the segment screw to 5 ft. lbs.

26. Install the injection line(s) and the fuel supply line (if a fuel reservoir was used).

27. Install the oil line and bleed the fuel system as described later in this section.

28. Install the valve cover (using a new gasket), breather tube and clutch housing cover.

OM355/5 ENGINE

1. Rotate the engine crankshaft clockwise and align the FB mark on the vibration damper with the timing case pointer.

NOTE: If the FB mark was moved past the pointer, turn the crankshaft counterclockwise a minimum of ⅙ of a turn to account for gear backlash.

2. Check that the advance fly weights are in the "rest" position by looking through the inspection hole of the advance mechanism.

3. Check that the FB marks on the injection pump flywheel and the pump pointer are aligned.

4. If the FB marks are not aligned, loosen both nuts on the slotted flange segment and turn the pump flywheel to align the FB marks. Tighten the segment in this position.

5. Install the inspection hole plug into the advance mechanism.

Aligning the FB (fuel beginning) prior to injection pump timing—OM355/5 engines

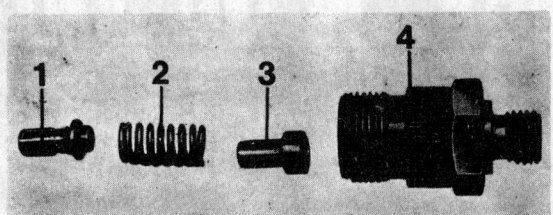

1. Check valve needle 3. Filler piece
2. Spring 4. Delivery valve holder

Delivery valve components

Removing the injection pump drive gear with a puller

Injection Pump

Removal and Installation

OM352 AND OM352A ENGINES

1. Rotate the engine crankshaft until the FB (fuel beginning) mark on the vibration damper is aligned with the FB mark of the pointer on the timing case. Also, check that the marked tooth of the injection pump drive gear is aligned with the pointer on the pump housing.

2. Disconnect the fuel injection, fuel return and oil lines from the injection pump.

3. Disconnect the accelerator rod from the injection pump. Also, on turbocharged engines, disconnect the ancroid line

4. Remove the five injection pump-to-support housing bolts. Remove the injection pump.

NOTE: If the injection pump is to be replaced, remove the drive gear and timing advance unit from the old pump, using a puller. Install the drive gear and advance mechanism on the new pump.

5. Replace the injection pump-to-drive housing gasket. Hold the new gasket in place by applying grease to the flange plate.

6. Install the injection pump with the marked tooth of the drive gear aligned with the pump housing pointer.

7. Connect the lines and accelerator rod to the injection pump.

8. Install the hollow screw with the single small hole on the lube oil filter end of the oil supply line.

9. If a new or remanufactured pump is installed, remove the pump side cover and fill the pump with ½ quart of engine oil.

10. Refer to the injection pump timing procedure to properly time the injection pump.

OM355/5 ENGINE

1. Remove the air filter and the air filter bracket.

2. Drain the coolant from the engine.

3. Remove the power steering pump flange screws. Remove the pump and wire it out of the way.

4. Disconnect the air, water and oil lines from the auxiliary air compressor.

5. Remove the snapring, nut and lock washer from the air compressor pivot bolt.

6. Remove the clamp screw from the compressor mounting bracket and loosen the adjustment screw.

7. Remove the V-belts and remove the air compressor.

1. Marked tooth of the drive gear
2. Pointer on the injection pump housing

Aligning the injection pump drive gear with the pump housing pointer—OM352 and OM352A engines

1, 2. Injection pump-to-flywheel marks
3. Fine adjustment marks
4. Flange-to-advance housing mark

Marks which must be aligned prior to injection pump removal—OM355/5 engines

Advance mechanism flyweight alignment as viewed through the inspection hole—OM355/5 engines

8. Rotate the engine crankshaft clockwise until the FB (fuel beginning) mark on the vibration damper is aligned with the pointer on the timing case cover.

9. Remove the inspection hold plug

Disconnecting the injection pump mounting flange—OM355/5 engines

from the advance mechanism and check the position of the advance flyweights through the inspection hole. The slash mark on each flyweight should be aligned with the mark on the corresponding flyweight.

10. Align the following marks:

   a. injection pump-to-flywheel marks
   b. fine adjustment marks
   c. flange to advance housing marks

NOTE: If the marks from Steps 9 and 10 are not visible, rotate the crankshaft 360° clockwise and again align the FB mark with the timing case pointer. Repeat Steps 9 and 10 if necessary.

11. Disconnect the fuel return and oil lines from the injection pump.

12. Disconnect the accelerator linkage from the injection pump.

13. Disconnect the injection lines from the injection pump as previously outlined.

14. Remove the screws from the injection pump flywheel flange.

15. Remove the injection pump mounting screws and remove the injection pump.

NOTE: If a new or remanufactured pump is to be installed, fill the injection pump with ½ quart of engine oil through the oil return port.

16. Installation is the reverse of the previous steps. Refer to the injection pump timing procedure to properly time the injection pump.

NOTE: With the engine running, a minimum flexing of the pump flex plates should be evident. If the flexing seems excessive, slide the segment flange on the shaft as necessary to minimize the flexing.

Fuel Injection Nozzles

Removal and Installation

OM352 AND OM352A ENGINES

1. Drain the coolant from the engine.
2. Remove the air intake hose and the valve cover.
3. Remove the fuel return lines as previously outlined.
4. Disconnect the injection lines from the nozzle holders.
5. Remove the nozzle holder nut(s) from the cylinder head.
6. Using a special nozzle holder puller remove the nozzle holder(s) from the cylinder head along with the copper sealing washers.

NOTE: If the copper washers must be replaced, use new washers of the same thickness as the original washers.

7. Remove the nozzle holder protective sleeve from the cylinder head using the appropriate special tool.
8. Remove the O-ring from the cylinder head groove. Apply silicone grease to the I.D. of this O-ring during installation.
9. Installation is the reverse of the previous steps. Note the following during installation:

   a. Apply Permatex type sealant to the nozzle holder protective sleeve threads before installing the sleeve.

   b. Torque the protective sleeve to 43 ft. lbs.

Removing the injection nozzle holder with a puller

   c. Make sure the alignment lug of the nozzle holder is positioned in the cylinder head recess.

   d. Torque the nozzle holder nut to 43–51 ft. lbs.

OM355/5 ENGINES

1. Remove the air filter and the air filter bracket.
2. Drain the coolant from the engine.
3. Remove the valve covers. Mark the covers so that they may be reinstalled in their original locations. Be careful not to allow foreign matter to enter the engine.
4. Remove the rocker arm assemblies. Mark each assembly so that it may be reinstalled in its original location.
5. Disconnect the injection lines from the nozzle holder.
6. Remove the fuel filter bowls and the breather from the push rod cover.
7. Loosen the injection line flare and lock nuts at the cylinder head junctions. Turn the lines away from the nozzle holders.
8. Remove the nozzle holder ring nut from each nozzle holder.
9. Remove the nozzle holders using a nozzle holder pulling tool.

1. Nozzle holder
2. Adjusting shim
3. Adjusting shim
4. Compression spring
5. Thrust pin
6. Intermediate spacer with locating pins
7. Nozzle needle
8. Nozzle body
9. Cap nut
10. Copper washer

Exploded view of the injection nozzle

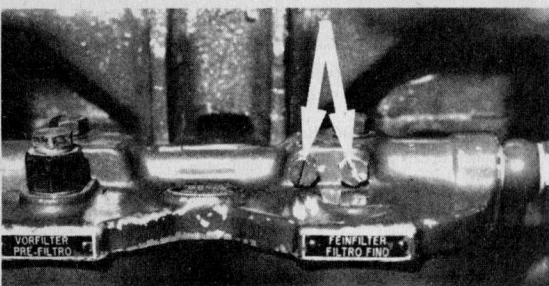

Fuel bleeder valve at the low pressure fuel gallery of the injection pump

Fuel bleeder valves on the fuel filter housing

Fuel System Bleeding

To insure proper engine operation, the fuel system must be completely free of air. Normally, the system is bled continuously through the return line to the fuel filter housing when the engine is running. It is recommended to manually bleed the fuel system after the fuel tank has been run dry or after major fuel system servicing.

1. Open the bleeder valve on the fuel filter housing.
2. Operate the fuel feed hand pump until fuel from the bleeder valve is completely free of air bubbles.
3. Close the fuel filter housing bleeder valve.
4. Open the bleeder valve at the front of the injection pump low pressure fuel gallery.
5. Operate the fuel feed hand pump until fuel from the bleeder valve is completely free of air.
6. Close the low pressure gallery bleeder valve.
7. Tighten the hand pump.

NOTE: Any air remaining in the system will be forced out after a few minutes of engine operation.

10. Remove the copper nozzle holder washers.

NOTE: If the copper washer must be replaced, use new washers of the same thickness as the original washers.

11. Remove the protective sleeves with an appropriate special tool.
12. Remove the O-rings from the cylinder head.
13. Installation is the reverse of the previous steps. Refer to Step 9 of the OM352 and OM352A procedure. It is recommended to bleed the fuel system prior to starting the engine.

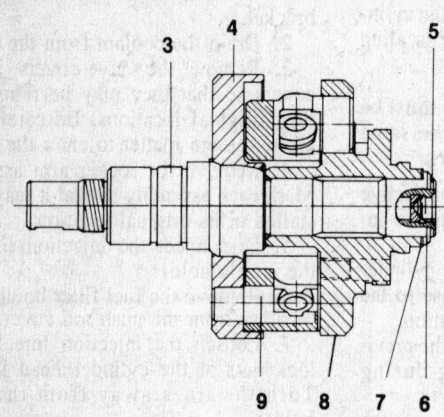

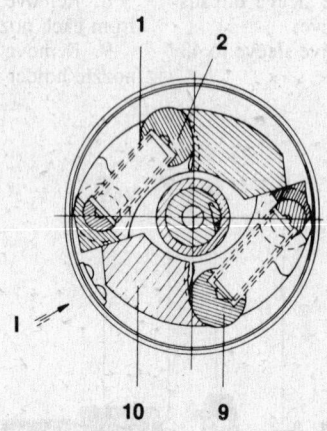

1 Compression Spring
2 Guide Pin—Driven Plate
3 Drive Shaft
4 Drive Plate
5 Snap Ring
6 Washer
7 Adjusting Shim
8 Driven Plate
9 Pivot Pin—Drive Plate
10 Governor Weights

Cross section of injection timing advance mechanism, OM 355/5 engine models

Direction of rotation

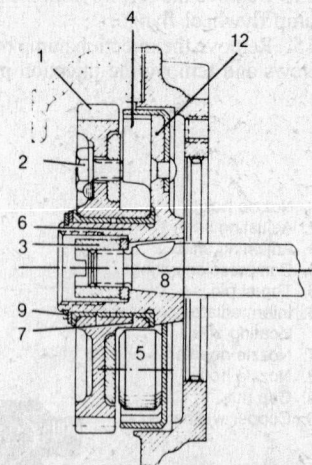

1 Drive Gear
2 Mounting Bolts for Segment Plate (Drive) *)
3 Round Nut
4 Segment Plate-Drive-Injection Timing Device
5 Governor Weights
6 Segment Plate-Driven Injection Timing Device
7 Sleeve Bearing
8 Drive Shaft
9 Retaining Ring
10 Compression Springs
11 Stop Pins
  a) Travel of Governor Weights
12 Driven Assembly

Cross section of injection timing mechanism, OM 352, OM 352A engine models

MITSUBISHI 4 CYL 143.2 Cu. In. (2.3 LITER)
6 CYL 243.5 CU. IN. (4.0 LITER)

GENERAL ENGINE SPECIFICATIONS

Engine Model	Engine Displacement	Bore × Stroke (in.)	Horsepower @ rpm	Torque ft. lbs. @ rpm	Compression Ratio	Compression Pressure (psi)	Oil Pressure @ idle (psi)	Firing Order
2.3L ① (turbo)	2346 cc (143.2 cu. in.)	3.59 × 3.54	80 @ 4200	125 @ 2100	21:1	384 @ 250 rpm	28	1-3-4-2
4.0L ②	3998 cc (243.5 cu. in.)	3.62 × 3.94	100 @ 3700	163 @ 2200	20:1	425	③	1-5-3-6-2-4

① 4 cylinder
② 6 cylinder
③ 42-71 psi @ 2000 rpm

TUNE-UP SPECIFICATIONS

Engine Model	Injection Pressure (psi)	Idle Speed (rpm)	Injection Timing (deg.)	Valve Clearance (in.) Intake	Valve Clearance (in.) Exhaust	Intake Valve Opens (deg.)	Injection Pump Type
2.3L 4 cyl	1707–1849	750	2 ATDC ①	.010 H	.010 H	20 BTDC	Bosch VE
4.0L 6 cyl	1707–1849	750	18 BTDC	.012C	.012C	32 BTDC	Bosch PE56A

ATDC—After Top Dead Center
BTDC—Before Top Dead Center
① TDC for high altitude

TORQUE SPECIFICATIONS
(All measurements in foot pounds)

Engine Model	Cylinder Head	Main Bearing Caps	Connecting Rod Caps	Crankshaft Pulley	Flywheel	Injection Nozzle	Injection Pump②
2.3L	90①	79①	58①	289	65①	44–58	18–25
4.0L	76–83③	55–61	33–34	123–137	94–101	44–50④	15–19

① Oiled
② Mounting bolts
③ 84-90 on hot engine
④ 17-26 on fuel lin cap 33-39 on nozzle body pieces

TORQUE SEQUENCES

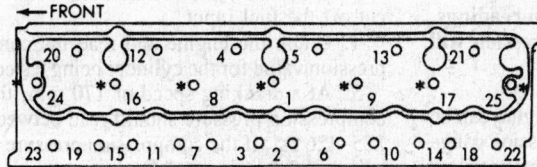

*BOLTS TO BE TIGHTENED TOGETHER WITH THE ROCKER SHAFT BRACKETS.

Head bolt torque sequence—4.0L engine

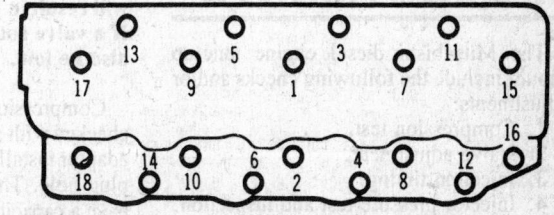

Head bolt torque sequence—2.3L engine

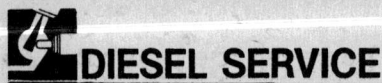

CHILTON'S THREE "C's" DIESEL ENGINE DIAGNOSIS PROCEDURE

Condition	Cause	Correction
Rough Idle	Improper adjustment	Adjust idle
	Accelerator control cable binding	Repair or lubricate
	Air or water in the fuel system	Clear air or water from fuel system
	Injection nozzle clogged	Check and clean injector nozzles
	Improper valve clearance	Check valve adjustment
	Injection pump malfunction	Check injection pump
Poor Performance	Air cleaner clogged	Check element
	Accelerator control cable binding	Check control cable for free movement
	Restricted fuel flow (water or air)	Check lines and filter
	Incorrect injection timing	Check injection timing
	Injection pump malfunction	Replace injection pump
Excessive Exhaust Smoke	Restricted air cleaner	Check element
	Air or water in fuel filter	Remove air or water from fuel system
	Improper grade fuel	Check fuel in tank
	Incorrect injection timing	Check injection timing
	Injection pump malfunction	Replace injection pump
	Injector nozzle stuck open	Check injector nozzles
Excessive Fuel Consumption	Restricted air cleaner	Check element
	Leak in fuel lines	Check fo leaks
	Incorrect idle speed	Check idle
	Restricted exhaust system	Check exhaust
	Improper grade of fuel	Check fuel in tank
	Injection pump malfunction	Check injection pump operation
Loud Knocking In Engine	Defective fuel injector	Replace fuel injector

NOTE: If the problem persists after performing these preliminary checks, disassembly and inspection of internal engine components may be necessary for further diagnosis.

MITSUBISHI TUNE-UP AND ADJUSTMENTS

The Mitsubishi diesel engine tune-up should include the following checks and/or adjustments:
1. Compression test.
2. Valve adjustment.
3. Injection timing.
4. Injector pressure test and inspection.
5. Idle speed adjustment.
6. Glow plug operation check.

Compression Test

2.3L 4 CYLINDER

NOTE: Valve clearances set too close will result in poor compression readings. If a valve rotator fails, compression will also be low.

Compression on the diesel engine can be checked with a screw in compression gauge adaptor installed in the fuel injector or glow plug hole. The compression gauge should have a capacity of at least 500 psi. Individual cylinder pressures should not vary more than 10%.

4.0L 6 CYLINDER

1. Remove the glow plugs.
2. Install cylinder compression pressure adaptor into glow plug mounting hole and attach compression gauge to adaptor.
3. Disconnect the fuel control motor to cut off the fuel input.
4. Crank the engine and read the compression value for the cylinder being tested.
5. At a cranking speed of 170 rpm, the compression pressure should read between 285–426 psi. If the compression pressure is below 285 psi, the need for engine repair is indicated.
6. Test all cylinders and verify that the compression values are within 10% of each other.

Valve Adjustment

2.3L 4 CYLINDER

NOTE: The valves should be adjusted with the engine at normal operating temperature.

1. Set the No. 1 piston at TDC on its compression stroke. Remove the valve cover.

2. Loosen the locknut and adjust the No. 1 and No. 2 intake and the No. 1 and No. 3 exhaust valves to specifications. Tighten the locknut and recheck clearance.

3. Rotate the crankshaft one revolution (No.4 cylinder at TDC), then loosen the locknut and adjust the No.3 and No.4 intake and the No.2 and No.4 exhaust valves to specifications. Tighten the locknut and recheck clearance. Reinstall the valve cover.

4. Check and adjust the idle speed if necessary.

4.0L 6 CYLINDER

1. Stop the engine and remove the valve cover.

2. Set the No. 1 cylinder at TDC on its compression stroke.

3. Insert a feeler gauge of specified thickness into the clearance between the valve stem end and the rocker arm. Adjust as required by loosening the locknut and turning the adjusting screw.

4. Repeat the procedure for each cylinder to adjust the remaining valves. Always use a new gasket when installing the valve cover.

Injection Pump Timing

2.3L 4 CYLINDER

NOTE: This procedure requires the use of a special prestroke measuring adaptor and dial gauge.

1. Set the No. 1 piston at TDC on its compression stroke. Make sure the timing marks on the camshaft sprocket and injection pump sprocket are aligned with the timing marks.

2. Loosen the nuts securing the injection lines to the fuel injection pump at the pump delivery valves. Do not allow the delivery valves on the fuel injection pump to loosen when loosening fuel line connections.

3. Loosen the injection pump mounting bolts slightly to allow pump movement for adjustment.

4. Check that the measuring adaptor push rod protrudes 10mm (0.4 in.). Protrusion may be adjusted by the inner nut on the measuring adaptor.

5. Remove the timing plug from the injection pump and install the measuring adaptor with dial indicator.

6. Turn the crankshaft counterclockwise until the notch on the pulley is 30° BTDC on the compression stroke of No. 1 piston. Zero the dial indicator, then turn the crankshaft slightly from side to side to make sure

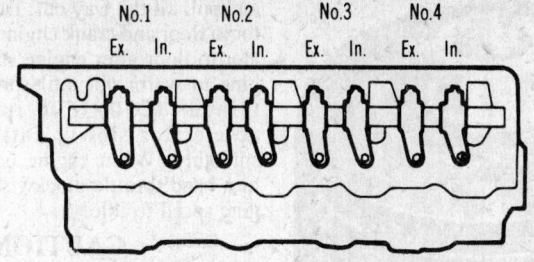

Adjusting valve clearance on 2.3L engine

the dial indicator does not move from the zero position.

7. Turn the crankshaft in the normal direction of rotation (clockwise) to 2° ATDC and check that the dial indicator reads 1 ± 0.03mm (.0394 ± .0011 in.).

8. If the dial indicator does not read the specified value, tilt the injection pump body right or left until the correct reading is obtained. Tighten the mounting bolts.

9. Repeat Steps 7 and 8 to check the adjustment.

10. Remove the timing gauge and measuring adaptor.

11. Install the copper gasket and timing plug in injection pump. Torque fuel line connections to specifications.

4.0L 6 CYLINDER

1. Disconnect the battery ground cable.

2. Disconnect the fuel shutoff rod at the injection pump lever by snapping the rod over the ball stud.

3. Clean all grease and dirt away from the No.1 delivery valve, pipe and pump area.

4. Turn the engine in the normal direction of rotation (clockwise) until the No.1 piston is at TDC on its compression stroke.

5. Continue turning the engine 1¾ turns more.

6. Disconnect the No. 1 injection pipe from the delivery valve.

CAUTION

When disconnecting the injection pipe(s) at the delivery valve(s), hold the delivery valve holder(s) stationary and loosen the injection pipe fitting. Do not turn the delivery valve holder as this will disturb the delivery valve calibration.

7. Turn the engine in the normal direction of rotation very slowly and stop when fuel begins to emerge from the delivery valve holder. This is the point when injection begins.

8. Check the injection timing point on the scale on the back of the crankshaft damper. If the timing is correct, the mark should be at the standard valve shown on the Vehicle Emission Control Information

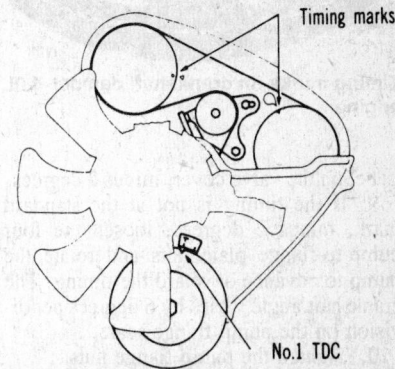

Correct alignment of timing marks at TDC on 2.3L engine (© Mitsubishi Motors Corp.)

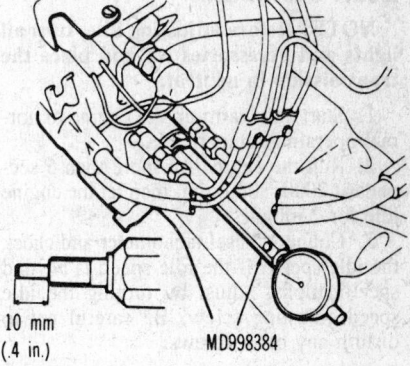

Prestroke measuring adapter installed in injection pump—2.3L engine

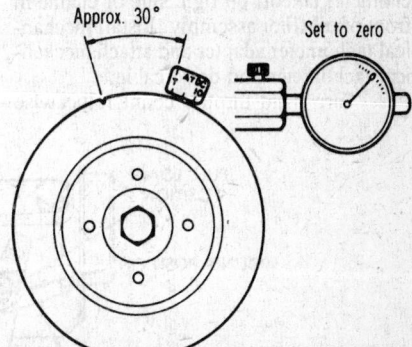

Turn the crankshaft clockwise to zero the dial indicator—2.3L engine

ADJUST VALVES AT TOP OF COMPRESSION STROKE

I = INTAKE VALVE
E = EXHAUST VALVE

CYLINDER NO.	1		2		3		4		5		6	
VALVE ARRANGEMENT	E	I	E	I	E	I	E	I	E	I	E	I

Intake and exhaust valve arrangement—4.0L engine

Timing marks on crankshaft damper-4.0L engine

Label on the valve cover, minus 2 degrees.

9. If the timing is not at the standard mark, minus 2 degrees, loosen the four pump-to-flange plate nuts and rotate the pump to advance or retard the timing. The crankshaft angle varies by 6 degrees per division on the pump flange scale.

10. Tighten the pump flange nuts.

Idle Speed Adjustment

2.3L 4 CYLINDER

NOTE: Before adjusting idle, turn all lights and accessories off and place the transmission in neutral.

1. Start and warm up the engine to normal operating temperature.

2. Run the engine for more than 5 seconds at 2000–3000 rpm, then let the engine idle for 2 minutes.

3. Connect diesel tachometer and check the idle speed. If the idle speed is beyond specifications, adjust by turning the idle speed adjusting screw. Be careful not to disturb any other screws.

4.0L 6 CYLINDER

1. Remove cover and gasket from tachometer takeoff on right side of engine in front of oil filter assembly. Install mechanical tachometer adapter and attach mechanical tachometer and drive cable.

2. Turn hand throttle counterclockwise

and pull all the way out. Depress accelerator to floor and crank engine. Hold accelerator to floor after engine starts. Allow engine to warm up until some speed is attained (1250–1500 rpm). Release accelerator slowly until engine runs smoothly. When engine begins to warm, turn hand throttle clockwise to reduce engine speed to idle.

CAUTION

If a new injection pump has been installed, do not allow engine speed to rise above 1300 rpm. If engine overspeeds, it may run away and damage or destroy itself.

3. Be sure that governor control lever is at idling position before attempting to adjust idle speed.

4. Check tachometer. If idle speed is not between 600 and 70 rpm, adjust idle speed.

5. Loosen idle adjusting screw locknut. Adjust screw as necessary to set idle to specifications. Turn adjusting screw IN to increase idle speed: OUT to decrease idle speed.

6. When idle speed is as specified, tighten idle adjusting screw locknut. Recheck idle speed to be sure it has not shifted.

Glow Plug Systems

2.3L 4 CYLINDER

The Quick Glow System has two main circuits to maintain the glow plug at constant temperature and to shorten the preheating time substantially. One circuit applies battery voltage (12 volts) to the glow plug. The other is the heat stabilization circuit which decreases voltage applied to the glow plug by changing the power source circuit to the dropping resistor when the glow plug reaches the design temperature.

The Quick Glow System operates when the coolant temperature is below 86°F. When the ignition key is turned to the ON position, the indicator light on the dash is lit and No.1 and No.2 glow plug relays are on. Actual current flow to the glow plugs is made by the No.1 glow plug relay only.

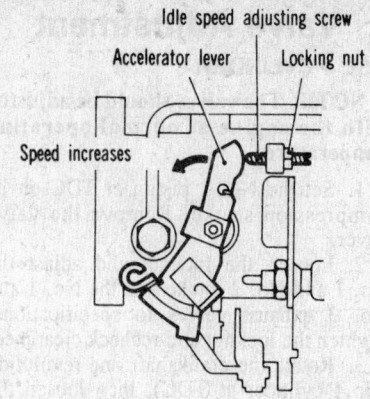

Idle speed adjustment on 2.3L engine

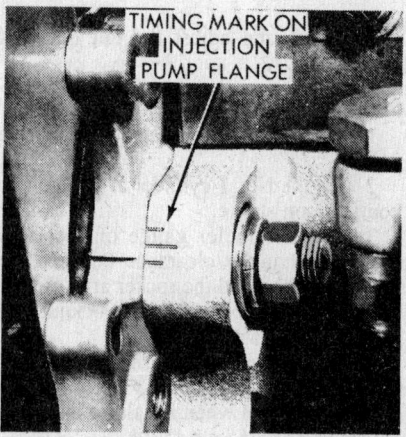

Injection pump timing scale-4.0L engine

When the START light is on and the key is turned to the START position, current flow to the glow plugs is continued by the No.2 relay. After the engine is started, afterglow is continued by the No.2 glow plug relay.

Testing Glow Plug

1. Remove the glow plug and check for damage or deformation of the pin.

2. Check the resistance between the terminal and body, then check for open or short circuit in the plug itself.

NOTE: Standard resistance is 0.1 ohm at 68°F.

Testing Glow Plug Relay

The glow plug relay is functioning if there is continuity between the B terminals when 6 volts is applied between the glow plug relay and coil terminal.

Testing Dropping Resistor

Check the terminals' resistance, then check for a short circuit. Normal resistance value is 130 milliohms. If no shorts are found and the resistance is as specified, the dropping resistor is good.

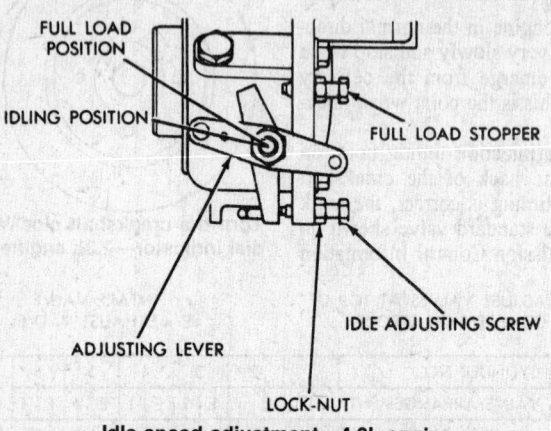

Idle speed adjustment—4.0L engine

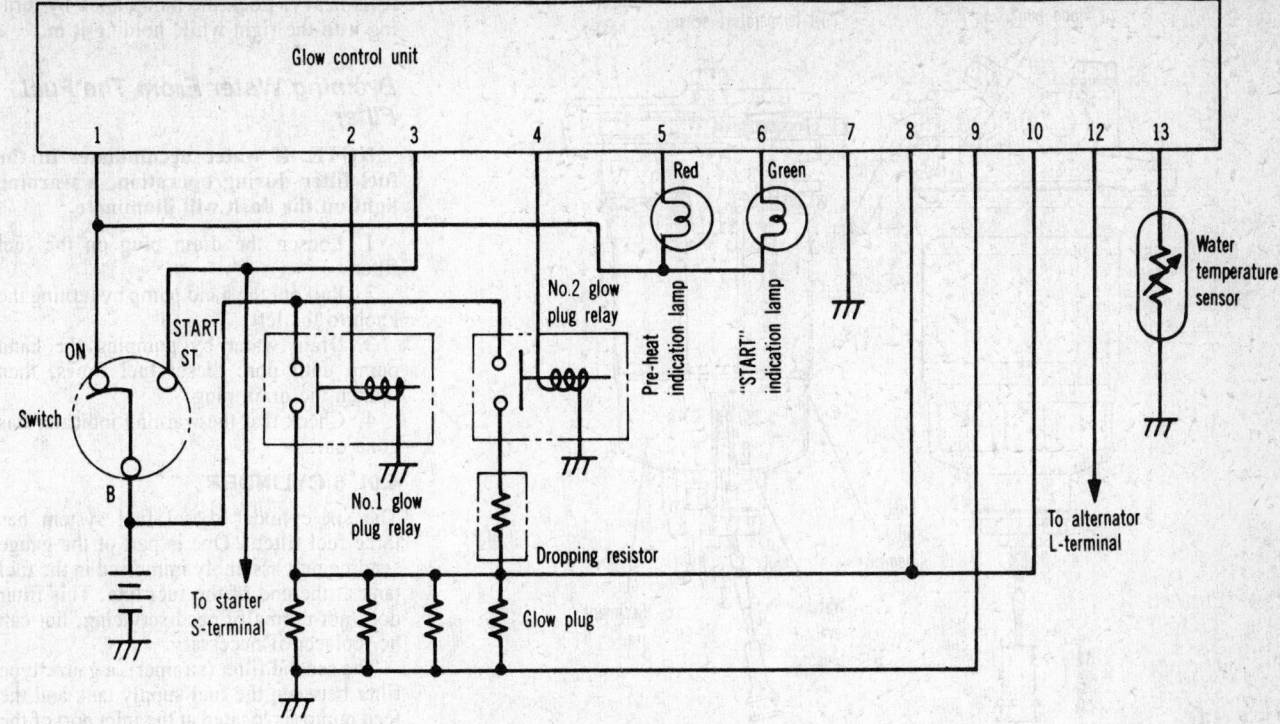

Schematic of quick glow system

4.0L 6 CYLINDER

Electrically operated glow plugs are used to heat the combustion chambers prior to cold weather and initial engine starts. The glow plugs resemble the spark plugs of the conventional gasoline powered engines. A relay controls the length of time the glow plugs are in use. During cold weather, the relay may have to be reactivated through a second glow plug cycle in order to start the engine.

Testing

1. Tag and disconnect the electrical connectors.
2. Using the large hex nut, loosen the glow plug and carefully lift it out of the cylinder head.
3. Check continuity between glow plug end and glow tip with an ohm meter.
4. Installation is in the reverse order.

NOTE: Use extreme care when removing a glow plug as the tip may break off; requiring cylinder head removal. Carefully clean carbon from glow plug tip before re-installing.

Fuel Filter

2.3L 4 CYLINDER

A fuel filter is provided to protect the injection pump from dirt and water in and out of the fuel tank. The fuel filter has a fuel heater built in to prevent the interruption of fuel flow due to parrafin flakes at low ambient temperatures. A hand pump is incorporated by the inlet valve on the fuel filter mount.

The bottom of the fuel filter forms a sediment trap and the accumulation of water will trigger a warning light on the dash.

NOTE: The fuel heater is activated when the fuel temperature is below 38°F as determined by the fuel temperature sensor. The fuel heater light on the dash will illuminate when the fuel heater is functioning.

Removal and Installation

1. Disconnect the water level sensor connector, fuel heater connector and fuel temperature sensor connector.
2. Loosen filter cartridge by hand or by using a band wrench. If the pump body is to be removed, disconnect the fuel lines and loosen the mounting bolts.
3. Remove the water level sensor from the cartridge by lightly clamping the sensor in a vise and turning the cartridge.
4. Transfer the sensor to the new cartridge and lubricate the sealing gasket with diesel fuel.
5. Install in reverse order of removal. Bleed the fuel system.

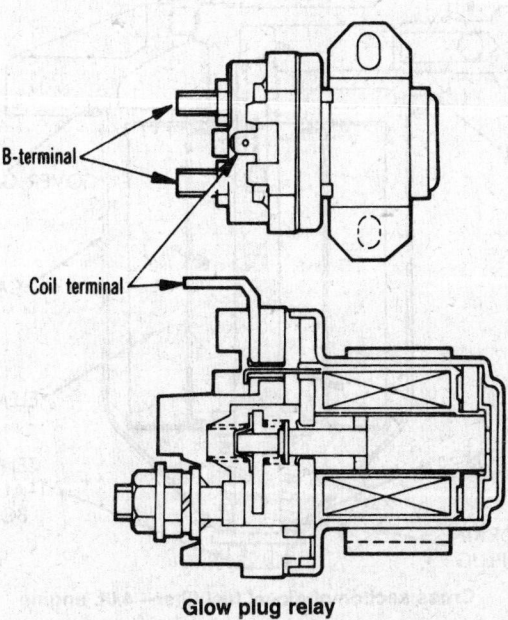

Glow plug relay

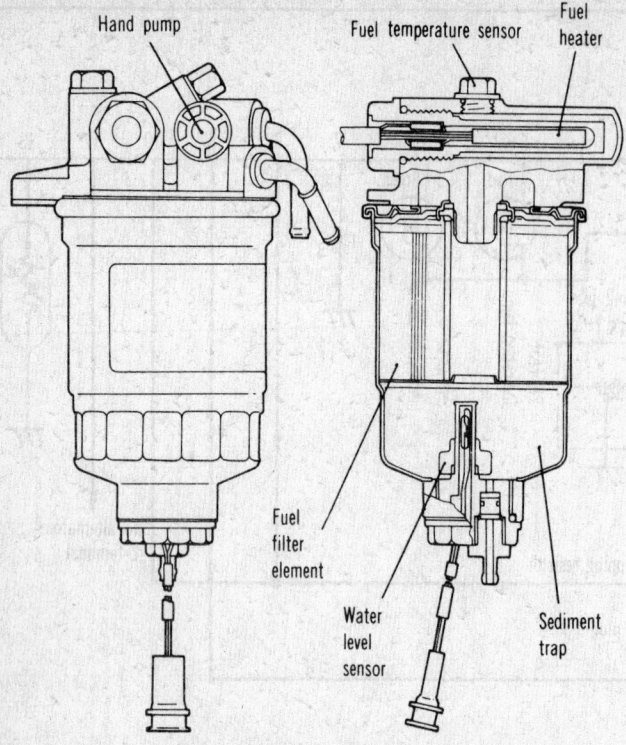

Diesel fuel filter on 2.3L engine

Bleeding the Fuel System

NOTE: The fuel system should be bled whenever the filter is changed or any lines have been disconnected.

1. Loosen the air plug on the fuel filter assembly.

2. Pull out the hand pump knob by turning it to the left.

3. Pump the hand pump until the fuel coming out of the air plug hole has no air bubbles in it.

4. Tighten the air plug, then continue to pump until the operation of the hand pump

feels heavy. Lock the pump knob by turning it to the right while holding it in.

Draining Water From The Fuel Filter

NOTE: If water accumulates in the fuel filter during operation, a warning light on the dash will illuminate.

1. Loosen the drain plug on the fuel filter.

2. Pull out the hand pump by turning the knob to the left.

3. Drain water by pumping the hand pump until pure diesel fuel flows, then tighten the drain plug.

4. Check that the warning indicator has gone out.

4.0L 6 CYLINDER

The six cylinder diesel fuel system has three fuel filters. One is part of the gauge sending unit assembly immersed in the fuel tank at the end of the fuel line. This filter does not normally need servicing, but can be replaced if necessary.

The second filter (strainer), a gauze-type filter between the fuel supply tank and the feed pump, is located at the inlet port of the feed pump. It operates under suction and removes large-size particles of dirt and foreign matter. This filter should be removed and cleaned every 12,000 miles. Clean the wire gauze in a suitable solvent to remove entrapped foreign matter. After cleaning, reinstall filter and fuel line connector and bleed air from the fuel system.

The third filter is a paper element throw away type installed at the back of the intake manifold and installed between the feed pump and the injection pump. A clogged filter will fail to supply a sufficient quantity of fuel to the engine, causing poor performance or erratic operation. Inspect the fuel filter element every 12,000 miles.

Removal and Installation

1. Loosen the air plug at top of fuel filter. Remove the drain plug or open the petcock at the bottom and allow fuel to drain.

2. Remove the center bolt and separate the case from the cover.

3. Inspect the paper filter element for excessive sediment buildup. If the element appears clogged, replace at this time.

4. Clean inside of case thoroughly before installing element.

5. Install new cover gasket and O-ring and reassemble case to cover. Reinstall drain plug or close petcock.

6. Bleed air from fuel system before placing vehicle in operation. Replace the fuel filter element every 24,000 miles.

Bleeding the Fuel System

Air trapped in the fuel system can cause inadequate fuel injection, poor operation and hard starting. Whenever the fuel system is serviced, it should be bled of trapped air in the proper sequence.

Cross section of diesel fuel filter—4.0L engine

1. Loosen the fuel filter petcock or valve and operate the priming pump on the feed pump. If the filter is filled with fuel, fuel containing air bubbles will be discharged from the petcock or valve. Continue pumping until the discharged fuel contains no more air bubbles. Then tighten the fuel valve or petcock securely.

2. Loosen the air bleeder screw at the top of the injection pump and operate the priming pump. Continue pumping until all air is bled from the fuel in the pump reservoir. Then close the air bleeder screw securely.

Injection Nozzles

Removal and Installation

2.3L 4 CYLINDER

1. Remove the fuel delivery lines from the injectors and injection pump. Remove the lines as an assembly and cap all open fuel fittings on the injection pump immediately to prevent contamination with dirt or grease.

2. Remove the fuel return pipe nuts, then remove the fuel return line.

3. Remove the injection nozzle assembly from the cylinder head with special tool MD998387 or equivalent.

4. Installation is the reverse of removal. Torque fuel injector to 44–50 ft. lbs. Replace the heat shields.

NOTE: Exercise care when handling the fuel injector. It is a high precision part that is easily damaged by dirt or dropping.

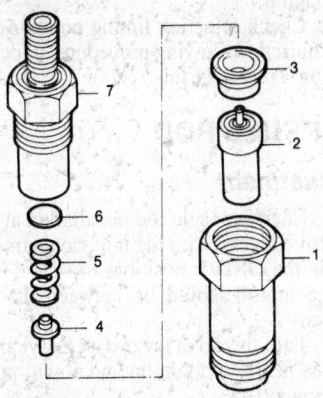

Exploded view of diesel fuel injector on 2.3L engine

4.0L 6 CYLINDER

1. Clean the injection nozzle connection before disassembly.

2. Remove the injection line from the nozzle assembly at the cylinder. Cap the fuel line opening to prevent contamination of the fuel system.

3. Remove the fuel injector from the head using a suitable socket tool.

4. Installation is the reverse of removal. Torque the fuel injector to 44–50 ft. lbs. Torque the fuel line nut to 17–26 ft. lbs.

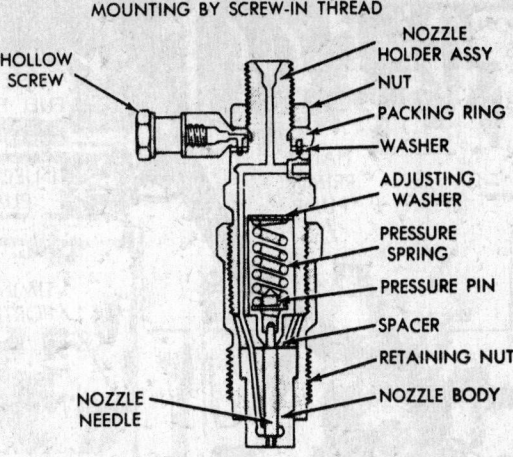

KC TYPE NOZZLE HOLDER MOUNTING BY SCREW-IN THREAD

HOLLOW SCREW
NOZZLE HOLDER ASSY
NUT
PACKING RING
WASHER
ADJUSTING WASHER
PRESSURE SPRING
PRESSURE PIN
SPACER
RETAINING NUT
NOZZLE BODY
NOZZLE NEEDLE

Cross section of fuel injector assembly—4.0L engine

NOTE: Exercise care when handling the fuel injector. It is a high-precision assembly that is easily damaged by dirt or dropping.

Injection Pump

Removal and Installation

2.3L 4 CYLINDER

1. Remove the timing belt upper cover.

2. Remove the nut and washer securing the injection pump sprocket.

NOTE: Be careful not to drop the nut and washer into lower cover.

3. Turn the crankshaft to bring No. 1 piston to TDC on its compression stroke.

4. Use a suitable gear puller to loosen the sprocket from the taper section of the drive shaft. Do not remove the sprocket; carefully set it in the timing belt lower cover with the belt engaged.

5. Remove the two water hoses from the wax element. Keep the end of the removed water hose higher than the cylinder head to prevent coolant drainage.

6. Disconnect the boost compensator hose at the injection pump.

7. Remove the fuel injection lines from the injection pump. Make sure the delivery valves do not turn when loosening the pipe connections at the pump.

8. Remove the injection pump support bracket bolts.

9. Remove the injection pump mounting nuts and remove the injection pump from the engine.

——— CAUTION ———
Do not turn the crankshaft with the injection pump removed

10. Installation is the reverse of removal. Make sure the timing marks on the camshaft sprocket and crankshaft pulley are aligned with their respective timing marks.

11. After mounting injection pump, carefully install injection pump sprocket and belt. Make sure that the injection pump drive shaft key is not misplaced or dropped.

12. Adjust injection timing and bleed the fuel system.

NOTE: If found to be defective, the injection pump must be replaced with a new or rebuilt unit. Any injection pump overhaul should be referred to an authorized Bosch diesel injection specialist.

Injection Pump

Removal

4.0L 6 CYLINDER

1. Disconnect battery negative cable at battery.

2. Disconnect fuel shutoff rod at stop lever. Rod end snaps over stop lever ball stud.

3. Remove steering pump and mounting bracket assembly from engine and set aside.

4. Clean dirt, paint and any other foreign material from fuel line, hose fittings and injection pipes at injection pump.

5. Drain engine oil. Remove dipstick and dipstick tube.

6. Disconnect throttle cable and linkage from injection pump control lever.

7. Remove throttle control bracket assembly from block, injection pump and control motor bracket. Set to one side.

8. Disconnect fuel supply line to fuel feed pump, loosening anchor clamps as necessary.

9. Disconnect fuel filter hoses from fuel feed pump and injection pump. Replace hollow bolts with seals into pumps to prevent dirt entry.

10. Turn engine crankshaft until No. 1 piston is positioned between 7 degrees BTDC and TDC on the compression stroke. Check pointer. It should be about midway between TDC and the 14 degree line on the crankshaft damper.

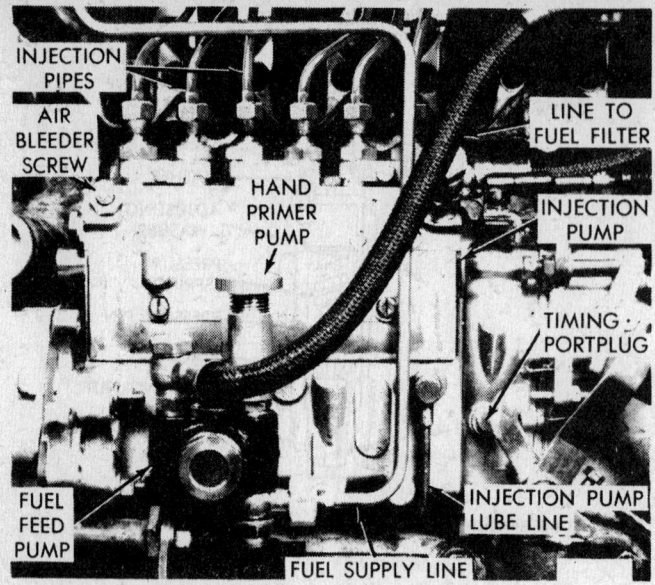

Injection pump assembly—4.0L engine

11. Disconnect injection pipes from delivery valves and move away from block.

CAUTION

When disconnecting the injection pipe(s) at the delivery valve(s), hold the delivery valve holder(s) stationary and loosen the injection pipe fitting. Do not turn the delivery valve holder as this will disturb the delivery valve calibration.

12. Cap open delivery valves to prevent dirt from entering. Disconnect injection pump lube line at block fitting near starter motor forward end.

13. Injection pump assembly is attached to engine by five screws and one bolt. Front screws extend through timing case and engine front plate into pump flange plate. Rear bolt fastens flange plate to engine front plate. Remove these six fasteners.

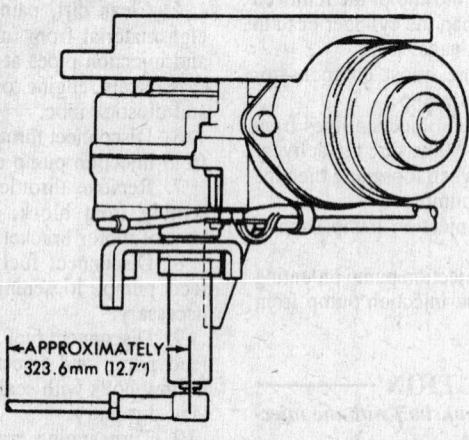

←APPROXIMATELY→
323.6mm (12.7")

Capsule rod adjustment—4.0L engine

14. Pull injection pump rearward to disengage from engine front plate and timing gear case. Twist pump toward block and continue pulling rearward until automatic timer is free of case.

Installation

1. Loosen four nuts attaching injection pump to mounting flange plate. Align center timing mark on pump flange with pointer on plate.

2. Be sure O-ring seal is in place on forward face of pump mounting flange.

3. Remove threaded timing port plug on governor housing behind control lever to expose pump camshaft bushing timing mark. Turn pump drive gear to align timing mark on camshaft bushing with pointer on governor. Guide plate notch on drive gear will be at approximately 8 o'clock point as viewed from front.

4. Be sure that crankshaft is still positioned between TDC and 7 degrees BTDC. See Removal, Step 10, with No. 1 piston on compression stroke.

5. Insert automatic timer into timing gear case. Turn injection pump in until against block. Then turn pump drive gear clockwise or counterclockwise to mesh drive and idler gears. Push pump forward into timing gear case and turn away from block to align attachment holes.

CAUTION

Correct gear mesh is assured by drive gear guide plate. If pump cannot be pushed forward manually until flange plate seal diameter contacts engine front plate, gear mesh is incorrect. DO NOT ATTEMPT TO FORCE PUMP INTO POSITION. Retract pump and turn drive gear as needed to achieve correct gear mesh.

6. Attach the pump to the timing gear case. Turn the crankshaft opposite rotation direction until it reaches the specified timing mark. The governor pointer and pump camshaft bushing timing marks should now be aligned. If not, the pump must be removed and installed again.

7. When the timing marks are aligned, install the governor housing timing port plug and continue with the pump installation by reversing the removal procedure. However, do not yet connect No. 1 injection pipe or battery cable.

8. Refill crankcase with specified engine oil.

9. Bleed air from fuel filter and injection pump.

10. Check injection timing point. Adjust as required, following procedures under Injection Timing in this section.

CAPSULE ROD

Adjustment

1. Check capsule rod installation at motor drive lever. With the injection pump lever in the DRIVE position, the drive lever at the motor should be between the two marks.

2. Turn the drive lever with a prybar until it is nearly straight up and down, pointing upward.

3. Disconnect the capsule rod. Be sure that the injection pump stop lever is in the DRIVE position.

4. Loosen the capsule rod lock nut and adjust rod length until the drive lever is properly positioned between the match marks.

5. Tighten the lock nut and install the capsule rod to the drive lever. The nominal length of the capsule rod is 322.6mm (12.7 in.).

DATSUN/NISSAN 4 CYLINDER 2.2L (SD22) DIESEL ENGINE

GENERAL ENGINE SPECIFICATIONS

Engine Type	Engine Displacement-cc (cu. in.)	Fuel Delivery	Advertised Horsepower @ rpm	Advertised Torque @ rpm (ft. lbs.)	Bore and Stroke (in.)	Advertised Compression Ratio	Oil Pressure (psi/idle)
SD22 2.2L 4 cyl	2164 (132)	Diesel Injection	61 @ 4000	102 @ 1800	3.27 × 3.94	21.6:1	60

DIESEL ENGINE TUNE-UP SPECIFICATIONS

Engine Type	Injector Opening Pressure (psi)	Low Idle (rpm)	Dashpot Speed (rpm)	Valve Clearance (in.) Intake	Valve Clearance (in.) Exhaust	Intake Valve Opens (deg.)	Injection Timing rpm	Firing Order
SD22	1422-1493	550-700	1280-1350	.014	.014	28B	20 BTDC	1-3-4-2

TORQUE SPECIFICATIONS

(All readings in ft. lbs. unless noted)

Engine Type	Engine Displacement cc (cu. in.)	Cylinder Head Bolts	Rod Bearing Bolts	Main Bearing Bolts	Crankshaft Pulley Bolts	Flywheel to Crankshaft Bolts	Manifolds Intake	Manifolds Exhaust
SD22 2.2L 4cyl	2164 (132)	94 large 40 small	36–40	123–127	217–239	33–36	11–13	11–13

TORQUE SEQUENCES

Cylinder head bolt torque sequence

DIESEL SERVICE

CHILTON'S THREE "C's" DIESEL ENGINE DIAGNOSIS PROCEDURE

Condition	CAUSE	CORRECTION
Rough Idle	Improper adjustment	Adjust idle
	Accelerator control cable binding	Repair or lubricate
	Air or water in the fuel system	Clear air or water from fuel system
	Injection nozzle clogged	Check and clean injector nozzles
	Improper valve clearance	Check valve adjustment
	Injection pump malfunction	Check injection pump
Poor Performance	Air cleaner clogged	Check element
	Accelerator control cable binding	Check control cable for free movement
	Restricted fuel flow (water or air)	Check lines and filter
	Incorrect injection timing	Check injection timing
	Injection pump malfunction	Replace injection pump
Excessive Exhaust Smoke	Restricted air cleaner	Check element
	Air or water in fuel filter	Remove air or water from fuel system
	Improper grade fuel	Check fuel in tank
	Incorrect injection timing	Check injection timing
	Injection pump malfunction	Replace injection pump
	Injector nozzle stuck open	Check injector nozzles
Excessive Fuel Consumption	Restricted air cleaner	Check element
	Leak in fuel lines	Check for leaks
	Incorrect idle speed	Check idle
	Restricted exhaust system	Check exhaust
	Improper grade of fuel	Check fuel in tank
	Injection pump malfunction	Check injection pump operation
Loud Knocking In Engine	Defective fuel injector	Replace fuel injector

Note: If the problem persists after performing these preliminary checks, disassembly and inspection of internal engine components may be necessary for further diagnosis.

TUNE-UP AND ADJUSTMENTS

Compression Test

1. Warm up the engine.
2. Remove the injection nozzle.
3. Install compression test adapter into nozzle mounting hole and attach compression gauge.
4. Measure the compression pressure by cranking the engine. Use the same number of compression strokes to measure each cylinder. The standard compression is 427 psi and the limit is 356 psi. There should be no more than 43 psi difference between each cylinder.

NOTE: The engine compression measurement should be made as quickly as possible.

Valve Adjustment

1. Warm up the engine to normal operating temperature.
2. Remove the air cleaner and valve rocker cover.
3. Turn the crankshaft so that the No. 1 cylinder is set at TDC on the compression stroke.
4. Adjust the clearances of the No. 1,2,3 and 5 valves, if necessary, by loosening the rocker arm lock nut and turning the adjusting screw.
5. Rotate the crankshaft so that the No. 4 cylinder is at TDC/compression and adjust the No. 4,6,7 and 8 valves, if necessary.

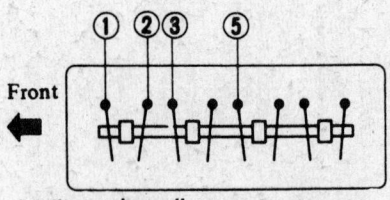

2.2 liter valve adjustment—step one

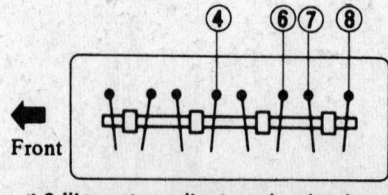

2.2 liter valve adjustment—step two

NOTE: When checking the clearances, the feeler gauge should move with a very slight drag.

6. Install the cylinder head cover.

Injection Timing

1. Check the timing marks on the pump and engine front plate and align if necessary.

2. Turn the crankshaft pulley in the direction of rotation and align the timing marks.

3. Remove all injection tubes and governor hoses.

4. Remove the No. 1 lock plate and delivery valve holder and pull out the delivery valve stopper and spring.

5. Install the delivery valve holder without the stopper and spring.

6. Connect the fuel hose so that fuel can be supplied by the priming pump.

7. Connect test tube to the No. 1 delivery valve holder.

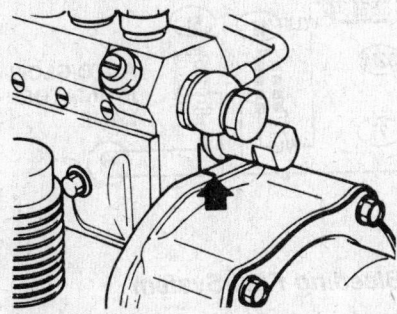

Timing marks on pump and engine front plate

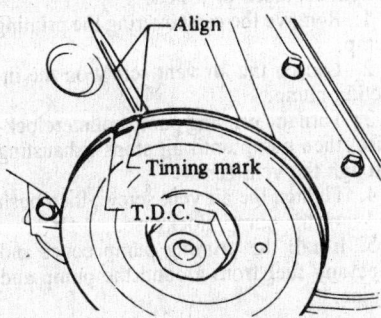

Crankshaft pulley alignment marks

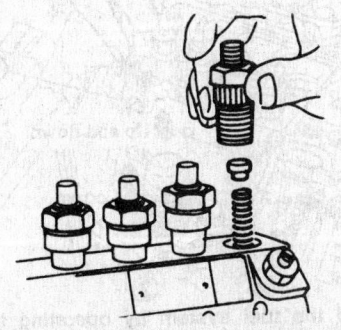

Delivery valve stopper and spring

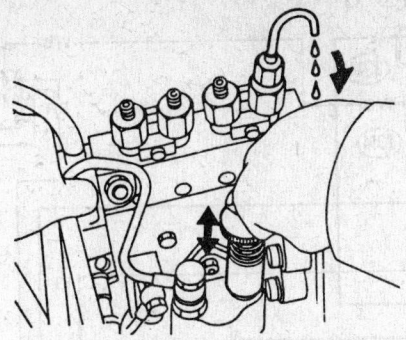

Operating priming pump to check fuel flow

8. Push the injection pump assembly fully down toward the engine side.

9. While feeding fuel by operating the priming pump, slowly move the injection pump until the fuel flow from the No. 1 injection tube stops.

10. Tighten the pump in this position and check that the timing marks on the pump and front plate are aligned. If not, stamp a new mark on the front plate.

11. Remove the No.1 injection tube and delivery valve holder and reinstall spring and stopper, then reinstall delivery valve and torque to 22–25 ft. lbs. (29–34 Nm).

12. Connect governor and fuel hoses and bleed the fuel system.

Idle Speed Adjustment

1. Start the engine and allow it to reach operating temperature. Stop the engine.

2. Attach diesel tachometer to No. 1 injection line. Remove the clamp on No. 1 line to obtain a more accurate reading.

3. Start the engine and check the idle speed with the tachometer. Make sure the accelerator linkage is not binding.

4. If the idle speed is not as specified in the Tune-Up chart, make sure the throttle control knob is pushed all the way in before adjusting.

5. If adjustment is necessary, loosen the idle adjusting screw locknut and turn the idle adjusting screw until the proper idle speed is obtained. Race the engine and allow it to return to idle, then recheck the adjustment.

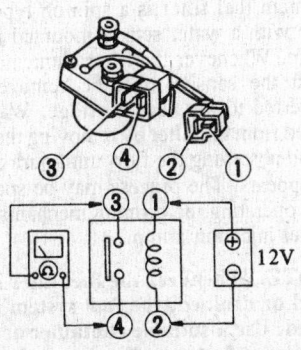

Terminal location on glow plug relay

6. After idle adjustment is complete, tighten the locknut.

NOTE: After every idle speed adjustment, adjust the dashpot (if equipped) by maintaining an engine speed of 1280–1350 rpm, loosening the dash pot locknut, then operating the dash pot and adjusting the tip until it contacts the control lever.

Glow Plug System

The auto-glow system used on the 2.2 liter engine consists of the following components:

1. Glow plug mounted in the engine block.

2. Glow plug relay mounted on the right side of the engine compartment.

3. Water temperature sensor mounted near the thermostat housing.

4. After-glow timer mounted under the glove box.

5. Indcator light mounted on the dash board.

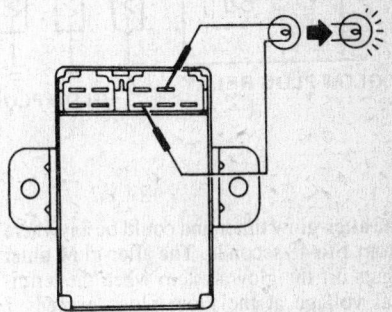

Test light connections on after–glow relay

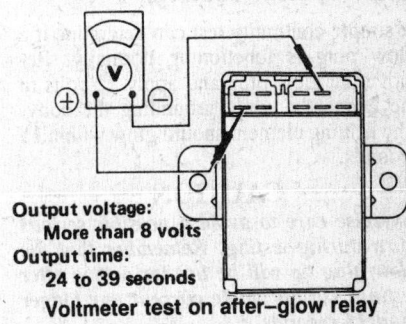

Output voltage:
  More than 8 volts
Output time:
  24 to 39 seconds

Voltmeter test on after–glow relay

GLOW SYSTEM OPERATION

The pre-glow system operates for about six seconds when the engine is cold to warm the combustion chambers and assure sufficient temperatures to fire the fuel on initial starting. The length of time for the pre-glow system operation is controlled by the terminal voltage of the glow plug. After the engine starts, the pre-glow system shuts off and the after-glow system operates for a short length of time. The length of time the after-glow system operates is controlled by

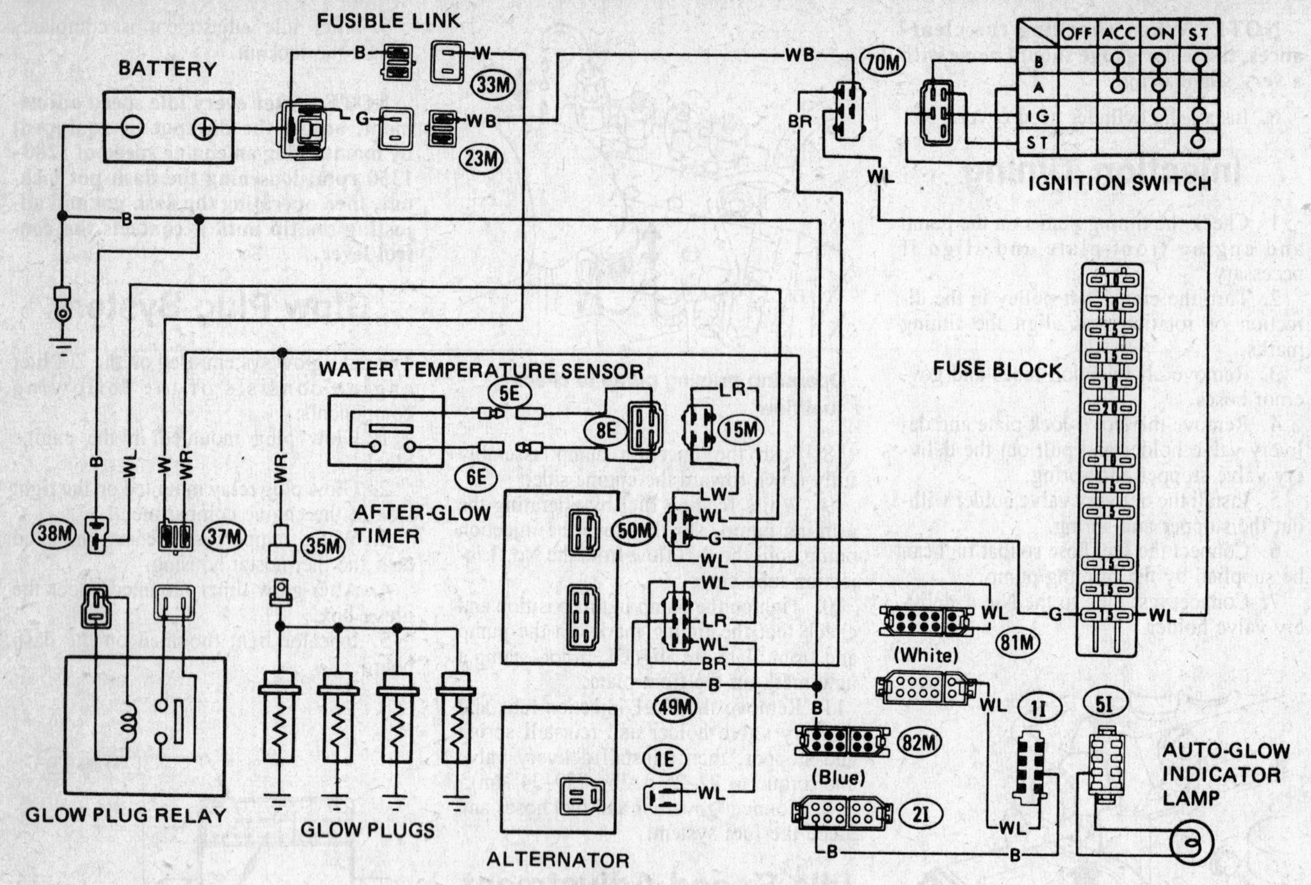

FUSIBLE LINK

BATTERY

	OFF	ACC	ON	ST
B		○	○	○
A		○		
IG			○	○
ST				○

IGNITION SWITCH

WATER TEMPERATURE SENSOR

FUSE BLOCK

AFTER-GLOW TIMER

GLOW PLUG RELAY

GLOW PLUGS

ALTERNATOR

(White)

(Blue)

AUTO-GLOW INDICATOR LAMP

Schematic of auto–glow system, early type

the after-glow timer and could be anywhere from 1 to 48 seconds. The after-glow timer shuts off the glow system when the terminal voltage at the glow plugs exceeds 7 volts.

Glow System Testing

A simple continuity test can determine if a glow plug is functioning normally. Remove the glow plug and apply 12 volts to the connector while grounding the body. The heating element should glow within 15 seconds.

———— CAUTION ————

Exercise care to avoid a possibly serious burn during testing. Remember that the glow plug tip will be hot for a time after testing. Do not apply current any longer than 15 seconds.

The water temperature sensor is checked by measuring the resistance (ohms) across the sensor connector terminals while heating the sensor body in water. The resistance should be about 12 ohms at 14°F and drop to about 1 ohm at 122°F.

The glow plug relay is tested by removing it and making sure continuity exists between terminals 1 and 2. Apply 12 volts between terminals 1 and 2 and check for continuity between terminals 3 and 4. Continuity between terminals 3 and 4 should not exist when battery voltage is removed.

The after-glow timer unit may be checked by connecting a test lamp or voltmeter as illustrated. Turn the ignition ON and measure the length of time the test light stays on. At an engine temperature of around 68°F the test light should glow for approximately 8–12 seconds. Output voltage should be at least 8 volts.

NOTE: Disconnect the lead wire from the S terminal of the starter before testing.

If the glow time and output voltage are as described, the glow system may be considered as functioning normally.

Water/Fuel Separator

The main fuel filter is a spin-on type cartridge with a water sensor mounted in the bottom. Whenever the filter element is replaced the sensor must be remove and transferred to the new cartridge. Water is drained from the filter by removing the sensor and observing the flow until pure diesel fuel appears. The process may be speeded up by operating the priming mechanism on the fuel injection pump.

NOTE: Whenever the fuel filter is replaced or drained, the fuel system must be bled. Use a suitable container or rags to catch any fuel runoff and exercise caution to avoid the risk of fire.

Bleeding Fuel System

Air should be bled out of the fuel system whenever the injection pump is removed or any component in the fuel system is repaired, replaced or tested.

1. Remove the cap covering the priming pump.
2. Loosen the air vent screw on the injection pump.
3. Turn the priming pump counterclockwise, then pump until air stops exhausting through the vent screw.
4. Tighten the air vent screw, then push and turn the pump clockwise.
5. Install the priming pump cover and wipe any fuel from around the pump and engine.

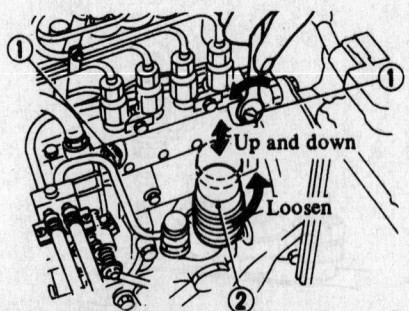

Bleed the fuel system by operating the priming pump (2)

Fuel Injectors

Diagnosis

Diesel fuel injectors can be easily tested for proper operation by using a test bench which measures opening pressure and allows visual inspection of the spray pattern. A defective injector can cause a variety of problems such as hard starting, rough idle, lack of power, excessive smoke or fuel consumption and audible knocks in the engine. An abnormal engine knock is one of the most obvious symptoms of a defective injector. Injector nozzle opening pressure can be adjusted by changing the shim inside the injector body. A shim thickness change of 0.04mm (0.0016 in.) will change opening pressure by 68 psi.

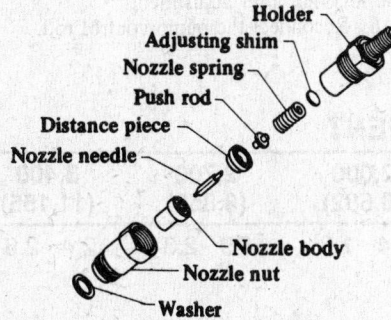

Exploded view of 2.2 liter engine fuel injector

NOTE: All fuel injection systems are sensitive to dirt. When working on the injection system, everything must be kept clean. Make it a practice to wipe every pipe connector before removing it.

Removal and Installation

1. Remove the injection tube assembly.
2. Remove the spill tube assembly. To prevent the spill tube from breaking, remove it by gripping the nozzle holder.
3. Remove the injection nozzle assembly with suitable tool.
4. Installation is the reverse of removal. Torque injection nozzle assembly to 43–51 ft. lbs. (59–69 Nm).
5. Bleed the fuel system.

Injection Pump

GENERAL INFORMATION

The SD22 engine uses a Kiki-Bosch PE type inline injection pump. The pump is driven by a gear with a timing device that automatically advances the start of fuel delivery (timing) in response to increases in engine rpm. A fuel supply pump with a hand primer is mounted on the side and driven by an eccentric located between two cams of the injection pump.

A control mechanism consisting of mechanical linkage and an electronic control unit (D.P.C. module) controls the START, STOP and DRIVE operation of the injection pump. On some models, a high altitude compensator is mounted on a bracket at the rear of the injection pump housing. The oil sump is connected to the engine lubrication system, providing the injection pump with filtered oil, but the pump diaphragm must be oiled at regular intervals as part of normal injection pump maintenance.

NOTE: Datsun/Nissan recommends that all internal injection pump overhaul procedures be performed at an authorized Kiki-Bosch service facility because of the specialized equipment needed for proper calibration.

INJECTION PUMP CONTROL MECHANISM

The injection pump control system is wired through the ignition switch in order to start, operate or stop the diesel injection pump. The control mechanism is controlled by the injection pump control unit (D.P.C. module) to regulate the amount of fuel injection by operating the injection pump lever. When the ignition is in the START position, the injection pump lever is set to increase the amount of fuel delivered. After the engine starts (key in the ON position), the lever moves to the normal driving position. When the key is turned OFF, the pump lever moves to the STOP position and cuts off all fuel to the engine.

NOTE: The engine can be stopped manually by turning the key OFF and moving the control lever away from the injection pump to the STOP position by hand.

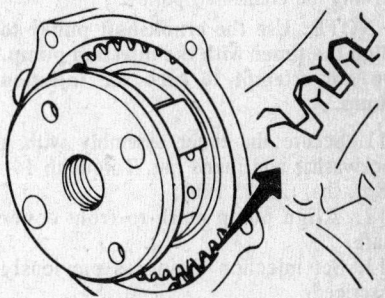

Correct alignment of timing marks on injection pump gears

Removal and Installation

1. Remove Tem-coupling with fan and radiator.
2. Remove the fuel lines as a unit from the pump and injectors. Cap the open ends of the injectors and delivery nozzles to prevent dirt from entering the fuel system.
3. Disconnect the governor hoses, fuel hoses and oil feed pipe bolt.
4. Disconnect pump controller connecting rod.
5. Remove the timing gear cover on injection pump.
6. Remove the timer round nut.
7. Remove the timer assembly using special tool ST19530000 or equivalent.
8. Loosen the mounting bolts and remove the injection pump assembly.
9. To install the pump, first set the en-

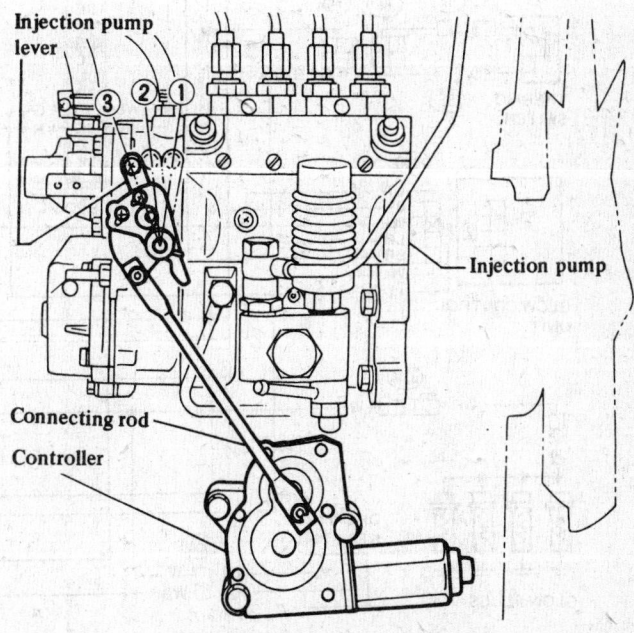

Injection pump control mechanism

gine No. 1 cylinder at TDC on the compression stroke.

10. Mesh the injection pump drive gear with the camshaft idler gear at the "Y" marks, then align the timer gear to the keyway of the injection pump camshaft by turning the crankshaft pulley.

NOTE: Use the crankshaft pulley to align the timer with the injection pump. Do not attempt to turn the injection pump.

11. Secure the timer assembly with a lockwasher and round nut. Torque to 14–18 ft. lbs. (20–25 Nm).
12. Align pump mark to front cover mark.
13. Set injection timing as previously described.
14. Install all disconnected hoses and linkage in the reverse order of removal.
15. Install radiator and fan with Temcoupling. Refill the cooling system.

16. Install the fuel lines and bleed the fuel system.

ALTITUDE COMPENSATOR

Adjustment

1. Check for loose connections before attempting adjustment. Connect hand vacuum pump and make sure compensator control rod is free and moves when vacuum is applied.
2. Disconnect the diesel pump controller rod.
3. Loosen lock nut and cap nut of the compensator and turn the cap nut until it just touches the control rod. Temporarily tighten the lock nut.
4. Determine the altitude (feet above sea level) in the area where the vehicle is to be operated.

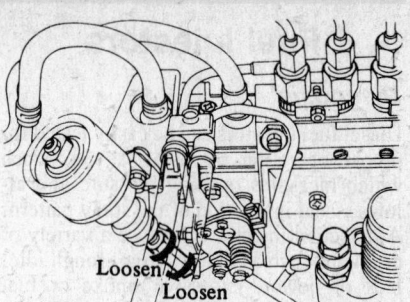

Loosen Loosen

Altitude compensator adjustment

NOTE: Any area that is above 4000 ft. in elevation is considered a High Altitude Area.

5. Refer to the chart to determine the number of cap nut revolutions necessary to adjust the compensator push rod. Tighten the lock nut after adjustment.
6. Reconnect the pump control rod.

ALTITUDE COMPENSATOR ADJUSTMENT

Approximate altitude m (ft)	0 (0)	120 (394)	700 (2,297)	1,300 (4,265)	2,000 (6,562)	2,700 (8,859)	3,400 (11,155)
Amount of loosening of cap nut (No. of revolutions of cap nut)	0	0.1~0.3	0.4~0.8	0.9~1.3	1.4~1.8	1.9~2.3	2.4~2.6

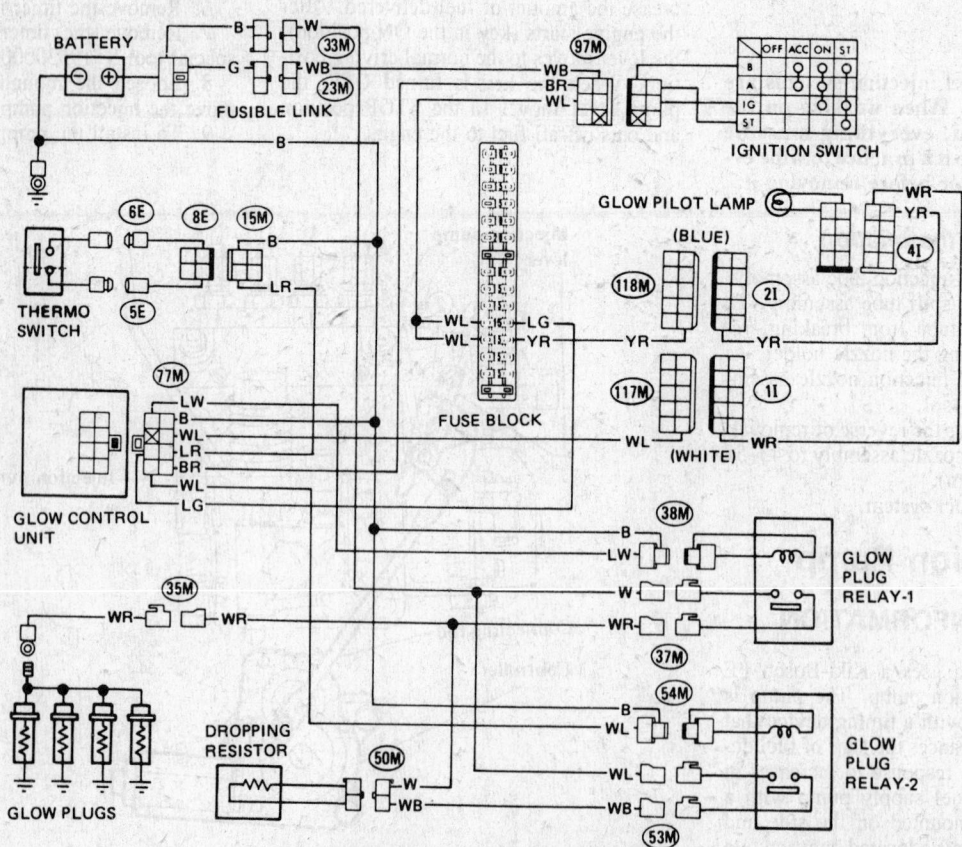

Schematic of auto-glow system later type

TOYOTA DIESEL ENGINES

GENERAL ENGINE SPECIFICATIONS

Year	Engine Type	Engine Displacement Cu. In. (cc)	Fuel Delivery	SAE Net Horsepower (@ rpm)	SAE Net Torque @ rpm (ft lbs)	Bore x Stroke (in.)	Compression Ratio	Oil Pressure @ rpm (psi)
1981-	L	133.5 (2188)	Fuel Injection	62 @ 4200	93 @ 2400	3.54 x 3.38	21.5:1	11.4 @ 700

DIESEL ENGINE TUNE-UP SPECIFICATIONS

Injector Opening Pressure (psi)	Idle Speed (rpm)	Valve Clearance (in.) Intake	Valve Clearance (in.) Exhaust	Cranking Compression Pressure @ 250 rpm	Maximum Compression Variance [3]	Firing Order
1636-1778 [1] 1492-1777 [2]	700	.010	.014	427 psi maximum 284 psi minimum	71 psi	1-3-4-2

[1] New   [3] Between highest and lowest
[2] Used   cylinder readings

FIRING ORDER

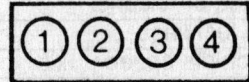

Firing order: 1-3-4-2

TORQUE SPECIFICATIONS

(All readings in ft. lbs.)

Year	Engine Type	Engine Displacement Cu. In. (cc)	Cylinder Head Bolts	Rod Bearing Bolts	Main Bearing Bolts	Crankshaft Pulley Bolt	Flywheel-to-Crankshaft Bolts	Manifolds Intake	Manifolds Exhaust
1981-	L	133.5 (2188)	84-90	37-43	71-81	69-75	84-90	8-11	11-15

TORQUE SEQUENCES

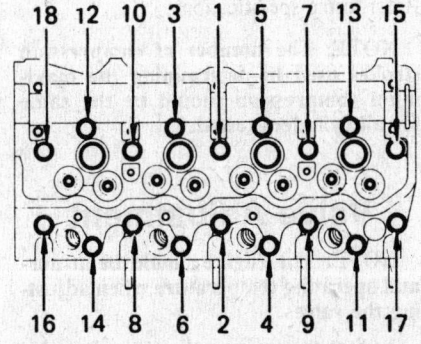

Cylinder head bolt tightening sequence

CHILTON'S THREE "C's" DIESEL ENGINE DIAGNOSIS PROCEDURE

Condition	CAUSE	CORRECTION
Rough Idle	Improper adjustment	Adjust idle
	Accelerator control cable binding	Repair or lubricate
	Air or water in the fuel system	Clear air or water from fuel system
	Injection nozzle clogged	Check and clean injector nozzles
	Improper valve clearance	Check valve adjustment
	Injection pump malfunction	Check injection pump
Poor Performance	Air cleaner clogged	Check element
	Accelerator control cable binding	Check control cable for free movement
	Restricted fuel flow (water or air)	Check lines and filter
	Incorrect injection timing	Check injection timing
	Injection pump malfunction	Replace injection pump
Excessive Exhaust Smoke	Restricted air cleaner	Check element
	Air or water in fuel filter	Remove air or water from fuel system
	Improper grade fuel	Check fuel in tank
	Incorrect injection timing	Check injection timing
	Injection pump malfunction	Replace injection pump
	Injector nozzle stuck open	Check injector nozzles
Excessive Fuel Consumption	Restricted air cleaner	Check element
	Leak in fuel lines	Check for leaks
	Incorrect idle speed	Check idle
	Restricted exhaust system	Check exhaust
	Improper grade of fuel	Check fuel in tank
	Injection pump malfunction	Check injection pump operation
Loud Knocking In Engine	Defective fuel injector	Replace fuel injector

Note: If the problem persists after performing these preliminary checks, disassembly and inspection of internal engine components may be necessary for further diagnosis.

TUNE-UP AND ADJUSTMENTS

The Toyota type L diesel engine tune-up should include the following inspections and/or repairs;

1. Compression test.
2. Valve adjustment.
3. Injection timing check and/or adjustment.
4. Injector pressure test and inspection.
5. Idle speed adjustment.
6. Glow system operation check.

Compression Test

1. Bring the engine to normal operation temperature.
2. Remove all glow plugs. Make sure the load wire is not grounded.

3. Install the compression gauge adapter into the glow plug mounting hole and connect the compression gauge.
4. Disconnect the fuel cut solenoid wire connector.
5. Measure the compression pressure while cranking the engine with the starter. Refer to the specifications.

NOTE: The number of compression strokes used in determining the maximum compression should be the same for all cylinders tested.

Valve Adjustment

NOTE: The engine should be at normal operating temperature when adjusting the valves.

1. Stop the engine and remove the valve cover.

2. Set the No.1 cylinder to TDC on the compression stroke.
3. Adjust the valves indicated by the arrows on the illustration. Valve clearance is measured between the valve stem and the rocker arm adjusting screw.
4. Rotate the crankshaft 360 degrees and adjust the remaining valves.

NOTE: Do not start the engine with the valve cover removed.

Injection Timing

1. Remove the injection pump head bolt.
2. Install the plunger stroke measuring tool and dial indicator to the injection pump head plug.
3. Rotate the crankshaft clockwise and

set either the No. 1 or No. 4 cylinder 45 degrees BTDC on the compression stroke.

4. Zero the dial indicator.

5. Slowly rotate the crankshaft pulley until the No. 1 or No. 4 cylinder is at TDC/compression.

6. Measure the piston plunger stroke on the dial indicator. It should be 1.0mm (0.0394 in.).

7. If the stroke is not correct, loosen the injection pump retaining bolts, the union nuts of the injection pipes on the pump side and the union bolt of the fuel inlet pipe on the pump side.

8. Adjust the piston plunger stroke by slightly tilting the injection pump body. If the stroke is less than specifications, adjust the pump towards the engine, If the stroke exceeds specifications, adjust the pump away from the engine.

9. Tighten the injection pump bolts to 11–15 ft. lbs., making sure the stroke does not change during tightening.

10. Tighten the union nuts to 17–22 ft. lbs. and tighten the fuel inlet fitting to 15–18 ft. lbs.

11. Remove the measuring tool and dial gauge, then torque the distributor head bolt to 8–9 ft. lbs. Replace the washer when installing the head bolt.

NOTE: Bleed air from the fuel pipes by loosening the pipes at the injectors and cranking the engine.

Idle and Maximum Speed Adjustment

1. Warm the engine to normal operating temperature and allow it to idle.

2. Turn the idle adjustor knob counterclockwise; the knob should return to its locked position.

3. Turn the engine off and remove the accelerator connection rod.

4. Connect a tachometer to the engine according to the manufacturers recommendations.

5. Start the engine and check the engine rpm at idle. The idle engine speed should be 700 rpm.

6. If adjustment is necessary, turn the idle adjusting screw on the fuel injection pump as required to obtain the 700 rpm idle speed.

7. Fully depress the injection pump lever, note the maximum engine speed and release the accelerator pedal immediately. The maximum engine speed should be 4900 rpm.

8. If adjustment is necessary, remove the wire seal of the maximum speed adjusting screw, if so equipped.

9. Using Toyota special service tool #092785-54020 or its equivalent, loosen the locknut of the maximum speed adjusting screw.

10. Turn the maximum speed adjusting screw until the proper maximum engine speed is obtained.

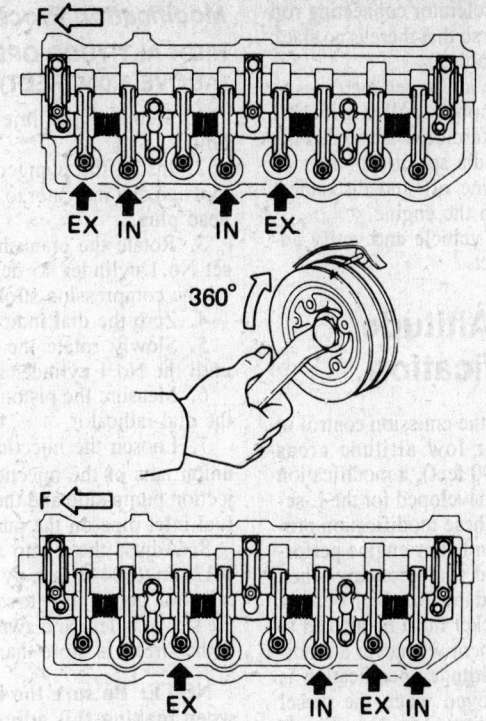

Valve clearance adjustment sequence

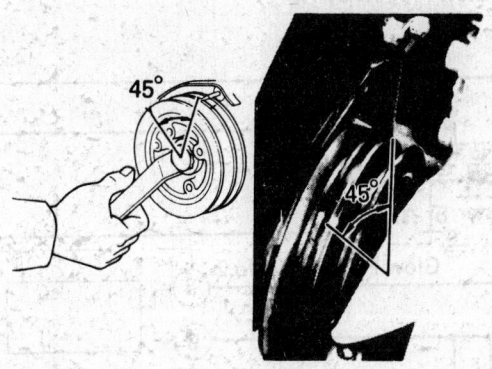

Correct placement of crankshaft to measure injection pump stroke

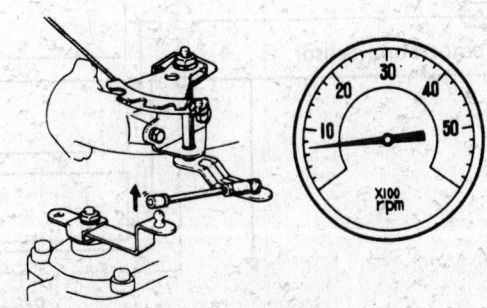

Remove accelerator connecting rod to check idle

11. Install the accelerator connecting rod and adjust its length so that there is no slack in the accelerator cable.

12. Check that the idle speed increases as the idle adjuster knob is pulled outward. Turn the knob counterclockwise so that the rpm returns to the idle specification.

13. Turn the engine off and disconnect the tachometer from the engine.

14. Road test the vehicle and verify adjustments are correct.

High Altitude Modification

In order to improve the emission control in designated high or low altitude areas (above or below 4000 feet), a modification procedure has been developed for the L-series diesel engine. These modification procedures will result in better engine performance and improved fuel economy when the engine is operated in either high or slow altitude areas. A sticker must be affixed to the engine compartment whenever a modification to a high altitude specification is performed and removed when the diesel engine has been re-adjusted for low altitude operation. The emission control labels are available from the manufacturer.

Modification Procedure
HIGH ALTITUDE OPERATION (ABOVE 4,000 FEET)

1. Remove the injection pump head bolt.

2. Install the plunger stroke measuring tool and dial indicator to the injection pump head plug.

3. Rotate the crankshaft clockwise and set No.1 cylinder 45 degrees before TDC on the compression stroke.

4. Zero the dial indicator.

5. Slowly rotate the crankshaft pulley until the No.1 cylinder is at TDC.

6. Measure the piston plunger stroke on the dial indicator.

7. Loosen the injection pump bolts, all union nuts of the injection pipe on the injection pump side and the union bolt of the fuel inlet pipe on the pump side.

8. Adjust the pump plunger stroke to 1.12mm (0.0441 in.), by slightly tilting the injection pump body towards the engine, if the stroke is less and away from the engine if the stroke is more than specifications.

NOTE: Be sure the engine is at TDC when making this adjustment.

9. After adjusting the pump stroke, tighten the injection pump bolts to 11–15 ft. lbs., making sure the stroke does not change during the tightening.

10. Tighten the union nuts the 17–22 ft. lbs. and the fuel inlet bolt to 15–18 ft. lbs.

11. Remove the measuring tool and the dial indicator. Install the pump head bolt with a new washer and torque to 8–9 ft. lbs.

NOTE: Bleed the air from the fuel lines by loosening the pipes at the injectors and cranking the engine.

12. Follow the instructions under "Idle Speed Adjustment" and set the engine idle speed to 700 rpm. Affix a new High Altitude Adjustment label too the underside of the hood, next to the existing Vehicle Emission Control Information label.

Low Altitude Operation (Below 4000 Feet)

To adjust a high altitude vehicle to low altitude specifications, follow the procedure outlined under "Injection Timing." Be sure to remove the high altitude emission label after the adjustment.

Glow Plug System
GENERAL INFORMATION

When the coolant temperature is below 104 degrees F. (40 degrees C.), the glow plug

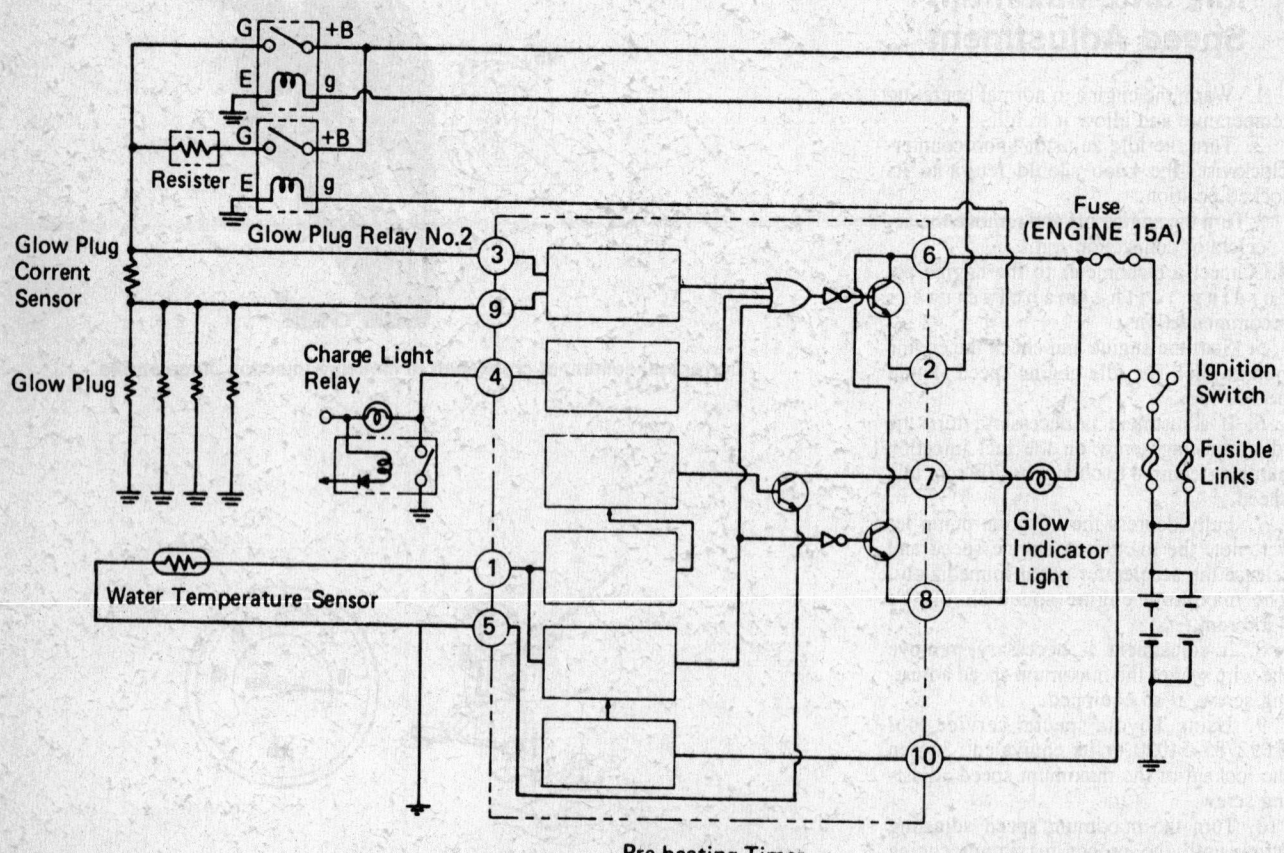

Glow Plug Relay No.1

Schematic of glow plug system

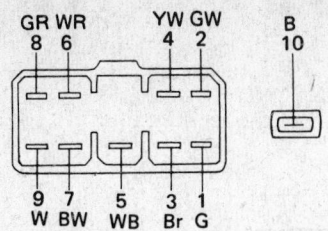

Pre-heating timer connector locations

system is designed to preheat the combustion chambers in order to provide sufficient temperature to fire the fuel on initial starting of the engine. When functioning properly, the glow plug indicator should light for 4.5 seconds when the temperature is below the preset standard and 0.5 seconds when the temperature is above the preset standard. Components of the glow plug system are not serviceable and is found to be defective, must be replaced as a unit.

NOTE: If a glow plug is found to be defective, it is a good policy to replace the complete set.

The glow plug system consists of the following components:

1. Preheating timer located behind the left kick panel.

2. Glow plug relay No.1 located on the left front fender, in the engine compartment.

3. Glow plug relay No.2 located under the right fender, in the engine compartment.

4. Glow plugs mounted on the left side of the engine.

5. Glow plug current sensor located above the No.3 glow plug.

6. Resistor located on the back surface of the intake manifold.

7. Water temperature sensor located on the right side of the engine block.

Glow Plug Diagnosis

Glow System Testing

NOTE: Before beginning any test procedures, make sure the battery voltage is 12 volts with the engine switch off and that the fusible link in the glow system is intact.

1. Test glow plug relay No.1 by checking for continuity between terminal g and E. There should be no continuity between terminals +B and G. Apply 12 volts between terminals g and E and again test for continuity between +B and G. Continuity should, now be present. Any result other than these, replace glow plug relay No.1.

2. Check glow plug relay No.2 by repeating the same procedure used for testing relay No.1. Again, if the test results differ from those described, replace glow plug relay No.2.

3. Measure the glow plug resistance with the glow plug installed in the cylinder head. Attach an ohmmeter leads to the input connector pin of the glow plug and the engine block. The ohmmeter should read 0.14 ohms @ 68 degrees F. If the resistance is not correct, replace the glow plug or set.

4. Check the glow plug current sensor by confirming that there is continuity between the current sensor terminals If no continuity exists, replace the glow plug current sensor.

5. Check the resistor located on the back

Check if indicator lamp lights up with ignition switch ON.
Coolant temp. below 40°C (104°F) : 4.5 secs.
Coolant temp. above 40°C (104°F) : 0.5 secs.

YES → Ignition switch OFF.

Check for battery voltage to No. 2 terminal (GW) of pre-heating timer with itnition switch ON.

No Voltage → Check that there is 1V voltage at No. 3 terminal (Br) and No. 9 terminal (W). If faulty, repair glow plug current sensor. If okay, timer is faulty and should be replaced.

Voltage → Check if voltage to No. 2 terminal (GW) of pre-heating timer is terminated after engine is started.

Yes → Ignition switch OFF.

CONTINUED

No → Check charge lamp relay. Disconnect charge lamp relay and make above check again. If faulty, repair charging system as necessary. If okay, timer is faulty and should be replaced.

NO → Check 15 amp engine fuse.

Fuse Blown → Check for short circuit and repair if necessary.

Fuse OK → Check indicator lamp bulb.

Bulb No Good → Replace bulb.

Bulb OK → Check for battery voltage to No. 7 terminal (BW) of pre-heating timer connector (on wire harness side).
If okay, pre-heating timer is faulty and should be repaired.

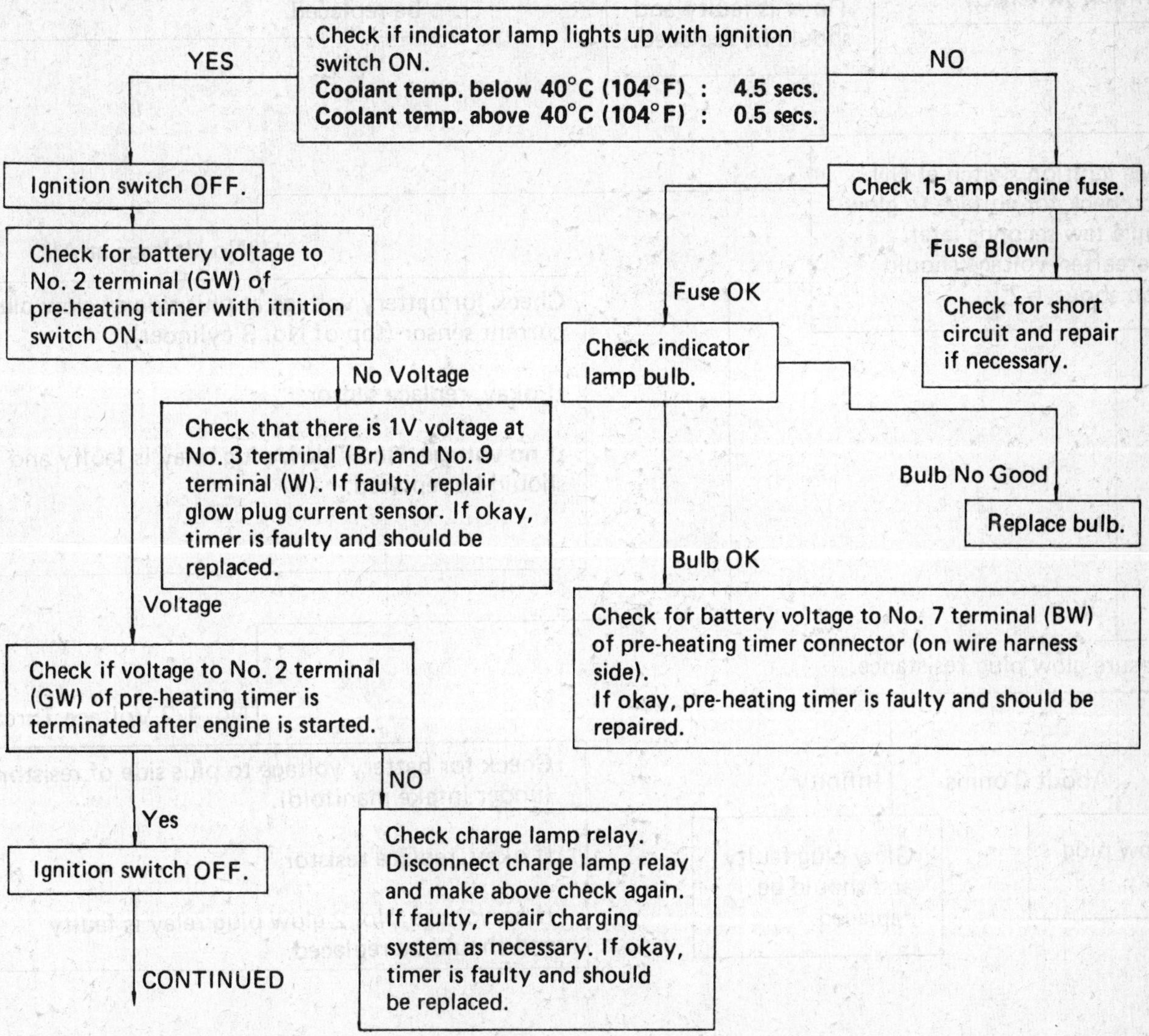

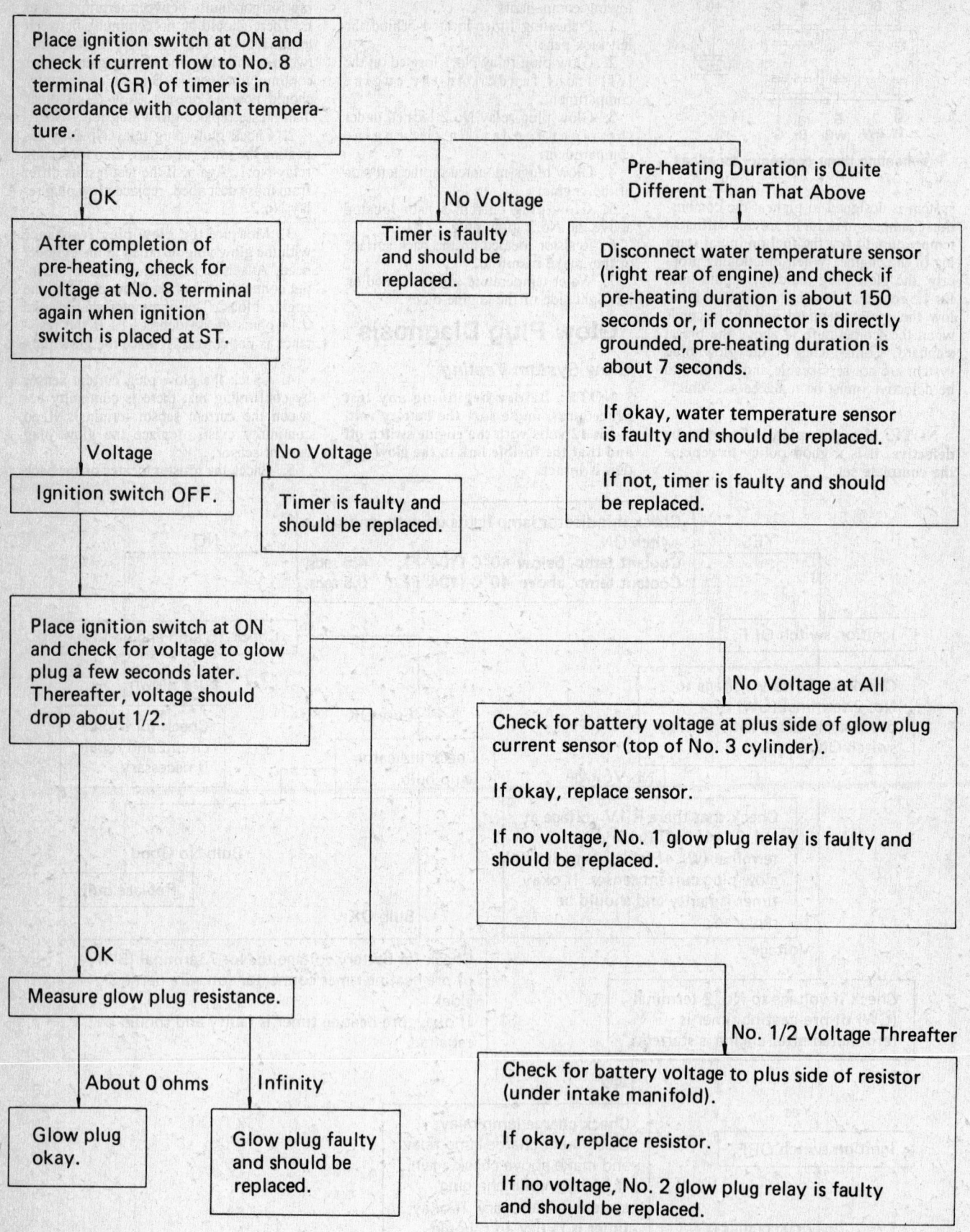

Place ignition switch at ON and check if current flow to No. 8 terminal (GR) of timer is in accordance with coolant temperature.

OK

After completion of pre-heating, check for voltage at No. 8 terminal again when ignition switch is placed at ST.

No Voltage

Timer is faulty and should be replaced.

Pre-heating Duration is Quite Different Than That Above

Disconnect water temperature sensor (right rear of engine) and check if pre-heating duration is about 150 seconds or, if connector is directly grounded, pre-heating duration is about 7 seconds.

If okay, water temperature sensor is faulty and should be replaced.

If not, timer is faulty and should be replaced.

Voltage

Ignition switch OFF.

No Voltage

Timer is faulty and should be replaced.

Place ignition switch at ON and check for voltage to glow plug a few seconds later. Thereafter, voltage should drop about 1/2.

No Voltage at All

Check for battery voltage at plus side of glow plug current sensor (top of No. 3 cylinder).

If okay, replace sensor.

If no voltage, No. 1 glow plug relay is faulty and should be replaced.

OK

Measure glow plug resistance.

No. 1/2 Voltage Threafter

Check for battery voltage to plus side of resistor (under intake manifold).

If okay, replace resistor.

If no voltage, No. 2 glow plug relay is faulty and should be replaced.

About 0 ohms

Glow plug okay.

Infinity

Glow plug faulty and should be replaced.

surface of the intake manifold by measuring the resistance with an ohmmeter between the input pin connection and the resistor body. It should read approximately zero ohms. If the resistance is incorrect, replace the resistor.

6. Check the coolant temperature sensor by removing the connector and measuring the resistance between the water temperature sensor terminals. Refer to the chart for resistance values at various temperatures. Replace the water temperature sensor if the resistance values are incorrect.

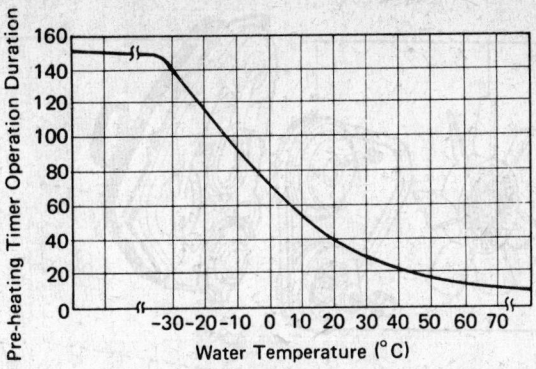

Time–temperature chart for timer test

FUEL SYSTEM

Fuel Filter/Sedimentor (Water Separator)

The Toyota 2.2L, type L diesel engine uses an inline, cartridge type spin-on fuel filter element that is replaced just like an oil filter. The sedimentor function is to separate water and particulates from the fuel before it reaches the injection pump. When the

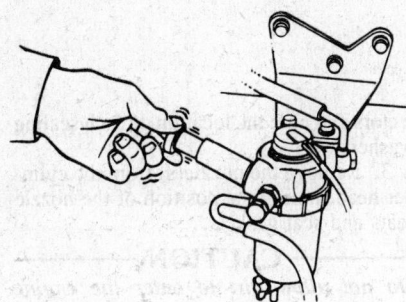

Priming handle on fuel sedimenter

water in the sedimentor reaches a dangerous level, a warning light will illuminate on the dash to indicate that the sedimentor must be drained. To drain the sedimentor, open the drain cock and turn the priming handle counterclockwise to free it. Pump the priming handle until pure diesel fuel appears, then close the drain cock and tighten the priming handle.

Injection Pump

Removal and Installation

1. Disconnect the cables which are positioned above the valve cover and move the cables aside. Remove the valve cover.

2. Disconnect the cables from both batteries.

3. Rotate the engine (clockwise only), until the TDC mark on the pulley is aligned with the pointer. Check that the valves of the No. 1 cylinder are closed (rocker arms

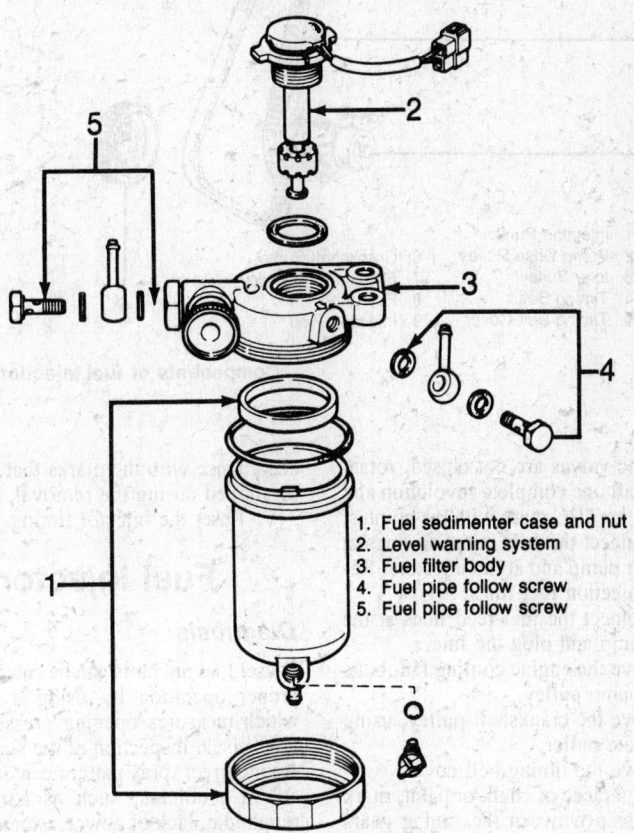

1. Fuel sedimenter case and nut
2. Level warning system
3. Fuel filter body
4. Fuel pipe follow screw
5. Fuel pipe follow screw

Exploded view of fuel sedimenter—typical

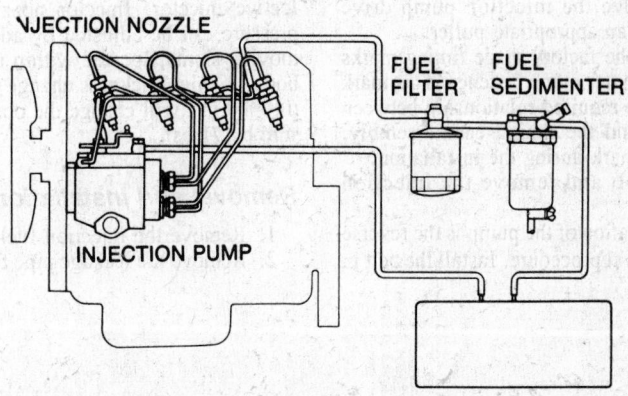

Fuel system components for the 2.2 liter diesel engine

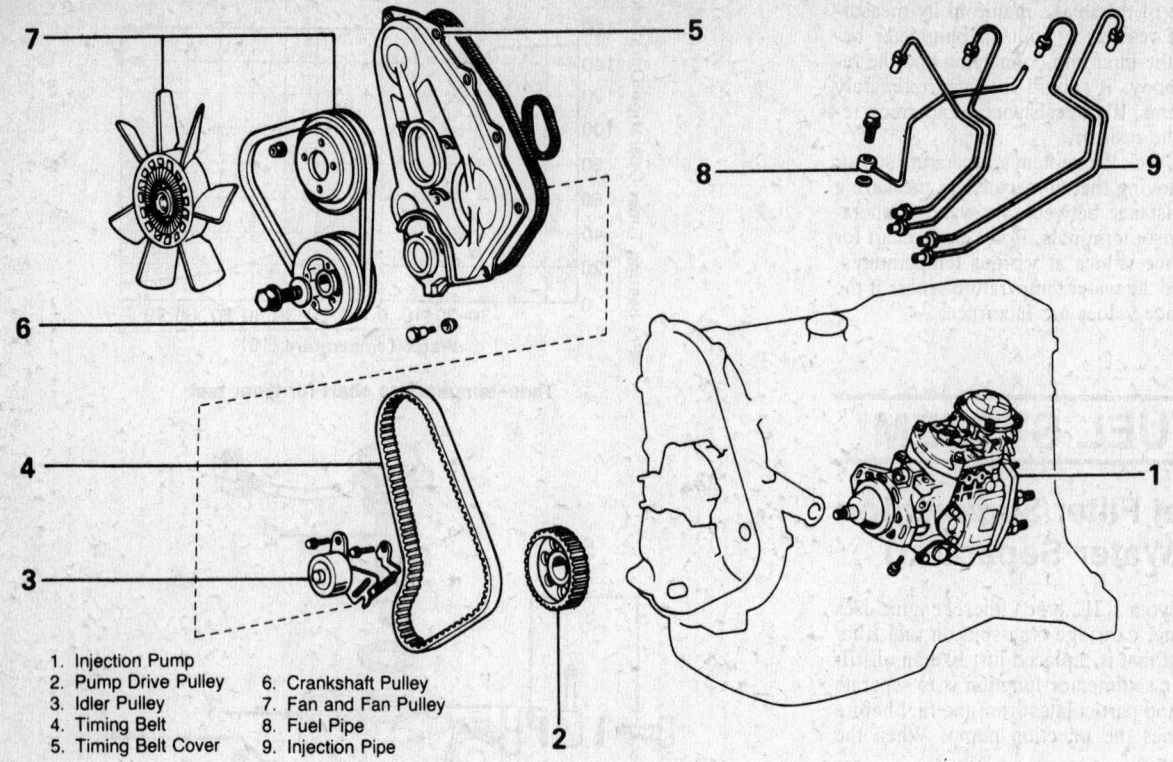

1. Injection Pump
2. Pump Drive Pulley
3. Idler Pulley
4. Timing Belt
5. Timing Belt Cover
6. Crankshaft Pulley
7. Fan and Fan Pulley
8. Fuel Pipe
9. Injection Pipe

Components of fuel injection system

loose). If the valves are not closed, rotate the crankshaft one complete revolution and again align the TDC mark with the pointer.

4. Disconnect the fuel injection lines at the injection pump and at the injectors. Remove the injection fuel lines.

5. Disconnect the fuel feed lines at the injection pump and plug the line.

6. Remove the engine cooling fan, belts and water pump pulley.

7. Remove the crankshaft pulley, using an appropriate puller.

8. Remove the timing belt cover.

9. Using a piece of chalk or paint, mark the relationship between the timing gears and the timing belt.

10. Remove the timing belt idler pulley and the timing belt.

11. Remove the injection pump drive gear, using an appropriate puller.

12. Note the factory made timing marks next to the outer pump fastener. This mark signifies the required relationship between the pump and the timing case assembly. Align this mark during the installation.

13. Unbolt and remove the injection pump.

14. Installation of the pump is the reverse of the removal procedure. Install the belt in accordance with the marks that were made or aligned during the removal.

15. Reset the injector timing.

Fuel Injectors

Diagnosis

Diesel fuel injectors can be easily tested for proper operation by using a test bench which measures opening pressure and allows visual inspection of the spray pattern. An incorrect spray pattern can cause a variety of problems, such as hard starting, rough idle, lack of power, excessive smoke or fuel consumption and audible knocks in the engine. An abnormal engine knock is one of the most obvious symptoms of a defective injector. Injector nozzle opening pressure can be adjusted by adding or removing shims, located within the injector body. A shim thickness change of 0.05mm (0.0020 in.) will change the opening pressure by 71 psi.

Removal and Installation

1. Remove the injection fuel lines.
2. Remove the leakage pipe from the injectors and note the location of each sealing washer.

3. Remove the nozzle(s) from the cylinder head, noting the position of the nozzle seats and seat gaskets.

——————— CAUTION ———————
Do not allow dirt to enter the engine through the nozzle holes.

NOTE: Remove accumulations of carbon from the nozzle holes.

4. Keep the injectors in order so that they may be installed in their original positions.

5. Install the injector assembly, noting that;

   a. The nozzle seat is installed between the injector and the seat gasket.

   b. The nozzle seat must be positioned with the concave side of the seat towards the injector.

6. Position a wrench on the hex of the nozzle body (NOT the nozzle retaining nut) and torque to 44–57 ft. lbs.

NOTE: After any service is performed on the diesel fuel system, pump the priming handle on the fuel sedimentor assembly 30–40 times to purge air from the fuel system.

VOLVO DIESEL ENGINES
D60A, TD60A, TD70D, TD70E, TD70F

GENERAL ENGINE SPECIFICATIONS

Engine Series①	Displacement (cu. in.)	Horsepower @ rpm (net)	Torque @ rpm (ft. lb.)	Bore & Stroke (in.)	Compression Ratio	Oil Pressure (psi)②	Firing Order
D 60 A	334	120 @ 2800	260 @ 1500	3.875 × 4.724	17.0:1	43–71	1-5-3-6-2-4
TD 60 A	334	180 @ 2800	376 @ 1900	3.875 × 4.724	16.0:1	43–71	1-5-3-6-2-4
TD 70 D	409	165 @ 2400	413 @ 1400	4.125 × 5.118	16.0:1	42–71	1-5-3-6-2-4
TD 70 E	409	205 @ 2400	492 @ 1400	4.125 × 5.118	14.5:1	40–70	1-5-3-6-2-4
TD 70 F	409	230 @ 2400	605 @ 1400	4.125 × 5.118	14.5:1	40–65	1-5-3-6-2-4

① All engines are in-line 6 cylinder
② 7 psi @ idle

TUNE-UP SPECIFICATIONS

Engine Series	Injection Timing (deg.)	Injector Opening Pressure (psi)	Low Idle (rpm)	High Idle (rpm)	Maximum Full Load Speed (rpm)	Compression Pressure (psi @ rpm)	Valve Lash (in.) Intake	Valve Lash (in.) Exhaust
D 60 A	22–23B	2844①	600–650	3000–3100	2800	355 @ 200	0.016C	0.018C
TD 60 A	21–22B	2844①	600–650	3000–3100	2800	327 @ 200	0.016C	0.018C
TD 70 D	18–19B	2488②	475–525	2550–2650	2400	340 @ 180	0.016C	0.018C
TD 70 E	20–21B	2844③	475–525	2550–2650	2400	327 @ 180	0.016C	0.022C
TD 70 F	18–19B	3840④	475–550	2650–2750	2400	325 @ 180	0.016C	0.022C

B—Before top dead center
① New nozzle—2958 psi
② New nozzle—2560–2673 psi
③ New nozzle—2915–3029 psi
④ New nozzle—3900–4020 psi

FIRING ORDER

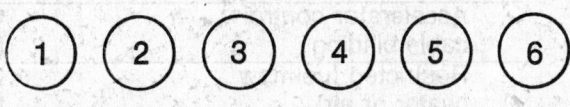

FIRING ORDER 1-5-3-6-2-4

Firing order—all engines

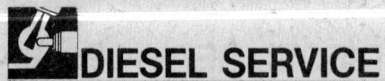

TORQUE SPECIFICATIONS

(All measurements in foot pounds)

Engine Series	Cylinder Head Bolts	Conn. Rod Brg. Bolts	Main Brg. Bolts	Crankshaft Damper Bolt	Flywheel-Crankshaft	Injection Pump Flange	Injectors
60 (All)	123	116	101	188	116–130	14–18①	14
70 (All)	137 long 100 short③	115	100	45②	115–130	50	15

① Element fastener
② Center bolt: 188 ft. lbs.
③ TD70F engine:
    First pass—36 ft. lbs.
    Second pass—118 ft. lbs.
    Final torque—60° turn in sequence

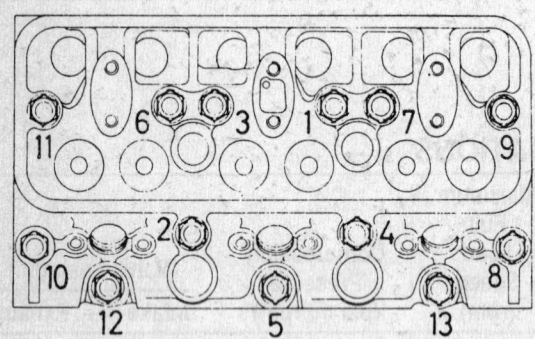

Head bolt torque sequence—series 60

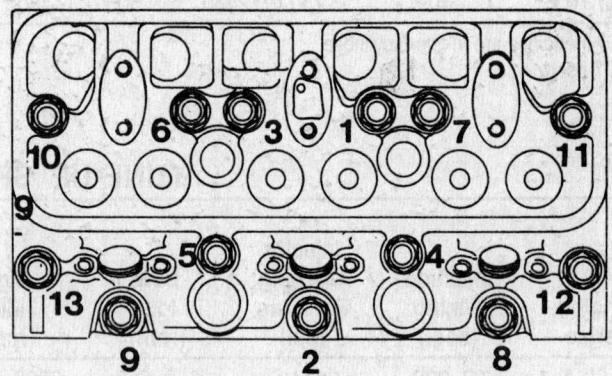

Head bolt torque sequence—series 70

CHILTON'S THREE "C's" DIESEL ENGINE DIAGNOSIS PROCEDURE

Condition	Cause	Correction
Rough Idle	Improper adjustment	Adjust idle
	Accelerator control cable binding	Repair or lubricate
	Air or water in the fuel system	Clear air or water from fuel system
	Injection nozzle clogged	Check and clean injector nozzles
	Improper valve clearance	Check valve adjustment
	Injection pump malfunction	Check injection pump
Poor Performance	Air cleaner clogged	Check element
	Accelerator control cable binding	Check control cable for free movement
	Restricted fuel flow (water or air)	Check lines and filter
	Incorrect injection timing	Check injection timing
	Injection pump malfunction	Replace injection pump
Excessive Exhaust Smoke	Restricted air cleaner	Check element
	Air or water in fuel filter	Remove air or water from fuel system
	Improper grade fuel	Check fuel in tank
	Incorrect injection timing	Check injection timing
	Injection pump malfunction	Replace injection pump
	Injector nozzle stuck open	Check injector nozzles

CHILTON'S THREE "C's" DIESEL ENGINE DIAGNOSIS PROCEDURE

Condition	Cause	Correction
Excessive Fuel Consumption	Restricted air cleaner	Check element
	Leak in fuel lines	Check for leaks
	Incorrect idle speed	Check idle
	Restricted exhaust system	Check exhaust
	Improper grade of fuel	Check fuel in tank
	Injection pump malfunction	Check injection pump operation
Loud Knocking In Engine	Defective fuel injector	Replace fuel injector

Note: If the problem persists after performing these preliminary checks, disassembly and inspection of internal engine components may be necessary for further diagnosis.

TUNE-UP AND ADJUSTMENTS

Valve Adjustment

ALL ENGINES

NOTE: The engine must be cold to adjust the valve clearance. Do not attempt to adjust the valves with the engine running

1. Make sure the stop control is pulled out.
2. Remove the inspection cover beneath the flywheel casing. Remove the rocker covers.
3. Rotate the engine in the normal direction of rotation until the No.1 piston is at top dead center on the compression stroke. (0° on the flywheel).
4. Adjust valves 1,2,4,5,7 and 9 to specifications.
5. Rotate the engine one turn so that the No.6 piston is at top dead center on the compression stroke (0° on the flywheel).
6. Adjust valves 3, 6, 8, 10, 11 and 12 to specifications.
7. Once all adjustments are complete, push in the stop control and install the valve covers.

Injection Pump Timing

60 SERIES ENGINES

1. Set the No.1 piston at top dead center on the compression stroke.
2. Remove the inspection plate from the flywheel housing and check alignment of timing marks. Turn the engine in the normal direction of rotation until the timing gradations show 22–23° (D60) or 21–22° (TD60) opposite the pointer.
3. Install a Wilbar tube or equivalent to the delivery pipe for No.1 cylinder.

4. Bleed the fuel system. Remove the delivery pipe for No. 1 cylinder and bleed the discharge valve and Wilbar tube by turning the injection pump shaft back and forth a few times.
5. Move the throttle control lever to the full throttle position and fasten it there with a spring or similar device.
6. Pull the stop arm back as far as possible and then return it to the operating position. This will set the control rod to the full load position.

NOTE: If the stop arm is not pulled all the way back, then returned to the operating position, the control rod will stop at the cold start position, giving an inaccurate adjustment.

7. Turn the pump shaft in the opposite direction of normal rotation and check that the fuel level in the Wilbar tube moves. Open the valve on the Wilbar tube and allow the level to drop to the middle of the sight glass.
8. Turn the pump shaft in the normal direction of rotation in small increments until the fuel level in the tube starts to rise. The point at which the fuel level just starts to rise is the start of injection for the No. 1 cylinder.

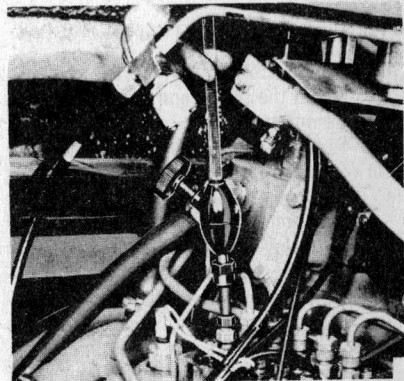

Injection pump adjustment with a Wilbär tube

9. Check that the flywheel marks coincide with the pump adjustments in Steps 7 and 8. If the pump requires adjustment, loosen the pump drive bolts and turn the pump shaft as necessary after setting the timing marks on the flywheel as outlined in Step 2.
10. Once all adjustments are complete, tighten the drive bolts.
11. Remove the Wilbar tube and reinstall the No. 1 delivery pipe. Bleed the fuel systems.

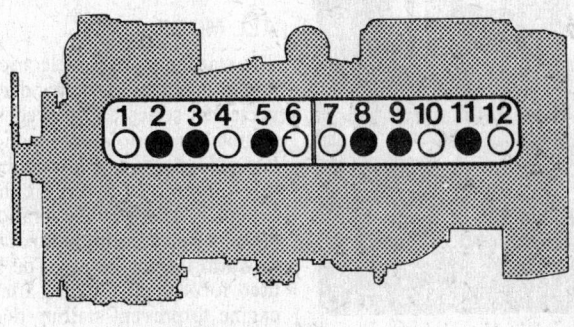

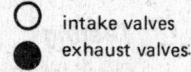

○ intake valves
● exhaust valves

Valve location—all engines

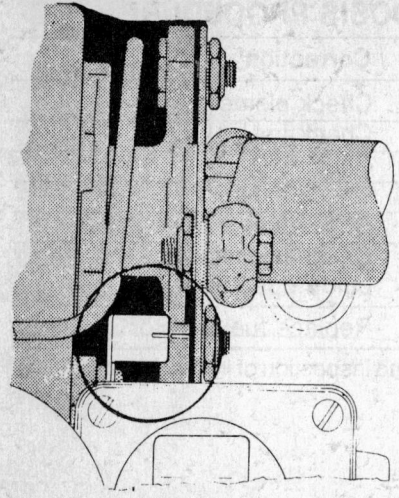

Setting the pump coupling—series 70

12. Reinstall flywheel cover.

70 SERIES ENGINES

1. Clean the pump and components before beginning.

2. Rotate the engine in the normal direction of rotation until the No.1 piston is at top dead center on the compression stroke.

3. Continue to turn the engine slowly until the timing marks on the flywheel align with the proper marks as given in the Tune-Up Specifications chart for injection timing. Make sure the sighting line from the mark on the flywheel over the pointer to the eye is at right angles from the flywheel. If viewed from the side, it is possible to err by a few degrees.

4. Make sure the pump coupling rear flange alignment mark is opposite the mark on the pump end adjusting plate. If the marks do not line up, loosen the coupling bolts and align the marks.

5. Check the adjustment by turning the engine back ½ turn, then foreward in the normal direction of rotation until the firing position for No. 1 cylinder is indicated by the alignment marks on the flywheel.

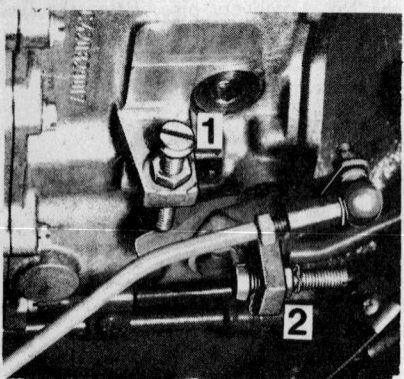

1. Low speed throttle arm stop screw
2. Sealed high speed adjustment stop

Speed setting—series 60

6. Once all adjustments are complete, tighten the flange bolts.

Idle Speed Adjustment
ALL MODELS

1. Check that the accelerator linkage is operating properly and that there is no play.

2. Run the engine to normal operating temperature.

3. Turn the throttle stop screw to obtain 475–525 rpm for the 70 series and 600–650 rpm for the 60 series.

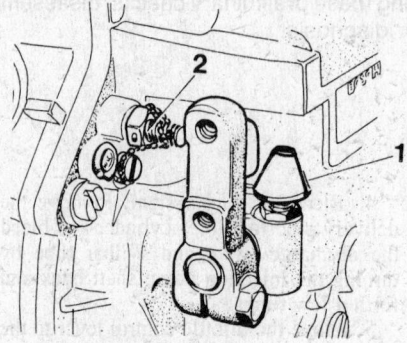

1. Low speed throttle arm stop screw
2. Sealed high speed adjustment stop

Speed setting—series 70

Maximum No-Load Speed Adjustment
ALL MODELS

1. Run the engine to normal operating temperature.

2. Break the lead seal on the speed stop.

3. Run the engine at maximum speed and check that the speed arm on the injection pump touches the maximum speed stop.

4. Adjust the stop to obtain 3000–3100 rpm for the 70 series and 2550–2650 rpm for the 60 series. Replace the lead seal.

Starting Heater
ALL MODELS

The starting heater is electrically operated and consists of three band elements connected in series and placed in-line before the intake manifold. Its function is to facilitate cold starting at low ambient temperatures, thereby reducing exhaust smoke. When energized by the ignition switch, the element heats up to approximately 1292°F to warm the intake air. The heater can be used for several minutes after starting the engine to prevent stalling due to the cold intake air.

Testing

Check the starting heater with a voltmeter by turning the ignition switch to the GLOW

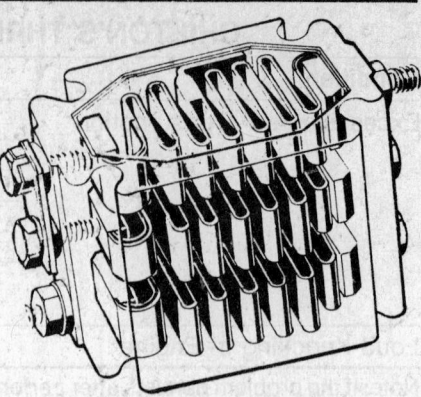

Typical starting heater

position and checking the voltage drop across each outer band element. It should be approximately 8 volts. If no voltage is indicated, check the battery, all cable connections, the ignition switch and the wiring past the relay. If after these check the starting heater is still not the correct voltage, replace the heater.

— **CAUTION** —
Because of the nature of operation, any type of starting spray can be ignited by the starting heater and cause an explosion in the intake manifold. For this reason, do not attempt to use any type of starting fluid on any Volvo diesel engine.

Fuel Injector

Removal and Installation
ALL MODELS

1. Remove the rocker arm covers.

2. Remove the fuel delivery lines and cap them.

3. Remove the injector holddown bolts and pull out the injectors. If injectors are difficult to remove, use tool Volvo #2683 and 2991 or equivalent. Clean the copper sleeve contact surface.

4. Installation is the reverse of removal. Torque bolts to 14 ft. lbs.

Fuel Injector Sleeve

Removal and Installation
ALL MODELS

NOTE: The cab member or gear lever carrier may have to be removed, depending on which sleeve is to be pulled.

1. Remove the injector as described above.

2. Using extractor, Volvo #2128 or its equivalent, pull the sleeve from the head.

3. Remove the O-ring from the head.

4. Clean the O-ring groove and the sealing surface between the head and sleeve. Install a new O-ring.

5. Manually turn the engine until the piston corresponding to the sleeve being worked on, is at bottom dead center. This

Injector removal

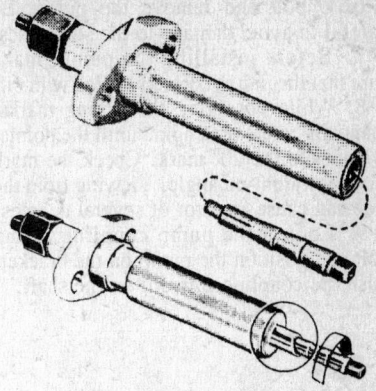

Injector sleeve installation tool

can be determined by removing the inspection cover on the flywheel.

6. Install the sleeve with Volvo tool #6008, or its equivalent, as follows:

   a. Unscrew the tool widening pin and place the sleeve on the tool.

   b. Back off the tool spindle nut.

   c. Screw in the widening pin.

   d. Coat the outside of the sleeve with clean engine oil and push the tool and sleeve into the head. Check that the index mark (recess) for the sleeve points straight upwards.

7. Install the injector holddown nuts and force the widening tool downward until the sleeve bottoms in the head.

8. Hold the widening tool securely and tighten the large nut. The widening pin is pressed through the lower end of the sleeve.

9. Tighten the nut until the tool spindle is free of the sleeve. Pull up on the spindle and remove the rest of the tool from the sleeve.

10. Install the injector and cab member or lever carrier.

BLEEDING THE FUEL SYSTEM

The system is bled at the bleeder screw located on the fuel filter carrier. Open the bleeder screw and prime the system with the hand primer until a clean stream of fuel flows from the nipple. Close the screw while fuel is still flowing.

If the injection pump must be bled, disconnect the bypass valve and prime until bubbles disappear from the stream. Close the connection while fuel is still flowing. Do not bleed at the pressure equalizer.

Fuel Injection Pump

Removal and Installation
60 SERIES ENGINES

1. Clean all related parts.

2. Disconnect the fuel delivery pipes at the injectors and pump. Cap all openings.

3. Disconnect all remaining lines, pipes

and controls from the pump. Cap all openings.

4. Remove the inspection plate from the flywheel housing and manually turn the engine to #1 TDC, on the compression stroke.

5. Remove the bolts securing the flange and pump drive. Remove the intermediate section and bolt.

6. Remove the speed sensor.

7. Remove the pump retaining bolts from the timing gear case and lift off the pump.

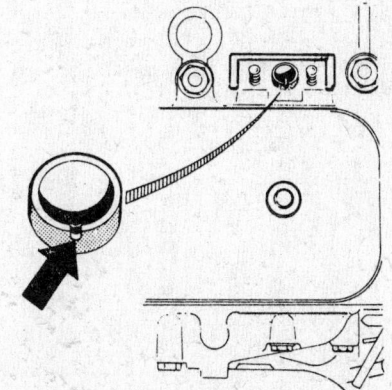

Position of the injector sleeve recess during installation

8. Before installing the pump, make sure that there is 1 pt. of oil in the unit.

9. Turn the engine manually until the timing gradations show 22–23° (D60) or 21–22° (TD60) opposite the pointer.

10. Adjust the pump camshaft until the mark on the end of the shaft inclines about 20° obliquely towards the cylinder block.

11. Apply chassis grease to the sealing ring at the front of the pump.

12. Install the pump on the timing gear case and tighten the bolts.

NOTE: The pump must be positioned so that the stud bolts are opposite the oval holes in the pump.

13. Install the speed sensor.

14. Connect all pipes, except the delivery pipes, to the pump.

15. Install the gear wheel clamp, lock washer and bolts on the pump drive at front of pump. The bolts must be $\frac{5}{16}$ UNC × 45mm. Tighten the bolts snugly. DO NOT OVERTIGHTEN.

16. Install a Wilbar tube, or its equivalent, to the delivery pipe for #1 cylinder. Bleed the fuel system.

17. Remove the delivery pipe for #1 cylinder and bleed the discharge valve and Wilbar tube by turning the pump shaft back and forth a few times.

18. Move the throttle control lever to the full throttle position and hold it there with a spring or similar device.

19. Pull the stop arm back as far as possible and then return it to the operating position. This will set the control rod to the full load position.

NOTE: If the stop arm is not pulled all the way back, then returned to the operating position, the control rod will stop at the cold-start position, giving a faulty adjustment.

20. Turn the pump shaft in the opposite direction of normal rotation and check that the fuel level in the Wilb;auar tube moves. Open the valve on the Wilbar tube and allow the level to drop to the middle of the sight glass.

21. Turn the pump shaft in the normal direction in small increments until the fuel in the level tube starts to rise. The point at which the fuel just starts to rise is the injection point for #1 cylinder.

22. Tighten the bolts at the front of the pump between the flange and the pump gear.

23. Manually turn the engine to check that the flywheel markings coincide with the pump adjustments in Steps 20 and 21.

24. When all adjustments are correct, tighten the pump drive bolts and install the flywheel housing inspection plate.

25. Remove the Wilbar tube, connect the delivery pipes and install the controls. Install the cover on the timing gear case.

26. Start the engine and check for leaks.

Removal and Installation
70 SERIES ENGINES

1. Clean all related parts.

2. Disconnect all pipes, lines and controls from the pump. Cap all openings.

3. Manually turn the engine to #1 cylinder TDC compression. Check the rocker arms and timing marks.

4. Mark the position of the pump coupling nuts for exact reassembly.

5. Remove the bolts from the pump coupling. Separate the rear flange from the intermediate section of the pump coupling.

NOTE: The position of the nuts must not be changed during removal and installation.

6. Unbolt and remove the pump. Be careful to avoid damage to the steel discs.

7. Before installing the pump, make sure that the unit is correctly filled with oil.

8. While observing the timing marks, manually rotate the engine until the pointer is opposite the 20° mark. Check the mark from a straight-on angle. Viewing from the side can cause an error of several degrees.

9. Loosen the pump coupling clamp bolt and position the pump on the bracket. Push the coupling forward on the shaft.

10. Install the pump coupling rear flange on the shaft. Turn it until the index line on the flange is opposite the index line on the setting plate.

11. Install the intermediate section of the pump coupling on the flange by sliding the coupling on the shaft from the auxiliary drive gear end. Make certain the domed washers are located between the rear flange of the coupling and the steel discs. Tighten the bolts. Make certain that the nuts are in the previously marked positions.

12. Tighten the pump coupling clamp bolt and check that the steel disc are not distorted.

13. Check that the timing marks on the pump and coupling coincide with the flywheel indexed at the 20° mark. This can be accomplished by rotating the engine ½ turn opposite normal rotation, then back to the #1 firing position.

14. Install all pipes, lines and controls, then bleed the fuel system. Start the engine and check for leaks.

Troubleshooting and Diagnosis

ENGINE

Gasoline Engine Troubleshooting

See applicable Car or Unit Repair section for specific service procedures

INDEX TO PROBLEMS

Problem Symptom	Begin at Specific Diagnosis, Number
Engine Won't Start	
Starter doesn't turn	1.1, 2.1
Starter turns, engine doesn't	2.1
Starter turns engine very slowly	1.1, 2.4
Starter turns engine normally	3.1, 4.1
Starter turns engine very quickly	6.1
Engine fires intermittently	4.1
Engine fires consistently	5.1, 6.1
Engine Runs Poorly	
Hard starting	3.1, 4.1, 5.1, 8.1
Rough idle	4.1, 5.1, 8.1
Stalling	3.1, 4.1, 5.1, 8.1
Engine dies at high speeds	4.1, 5.1
Hesitation (on acceleration from standing stop)	5.1, 8.1
Poor pickup	4.1, 5.1, 8.1
Lack of power	3.1, 4.1, 5.1, 8.1
Backfire through the carburetor	4.1, 8.1, 9.1
Backfire through the exhaust	4.1, 8.1, 9.1
Blue exhaust gases	6.1, 7.1
Black exhaust gases	5.1
Running on (after the ignition is shut off)	3.1, 8.1
Susceptible to moisture	4.1
Engine misfires under load	4.1, 7.1, 8.4, 9.1
Engine misfires at speed	4.1, 8.4
Engine misfires at idle	3.1, 4.1, 5.1, 7.1, 8.4

SAMPLE SECTION

Test and Procedure	Results and Indications	Proceed to
4.1 Check for spark: Hold each spark plug wire approximately $\frac{1}{4}''$ from ground with gloves or a heavy, dry rag. Crank the engine and observe the spark.	If no spark is evident	4.2
	If spark is good in some cases	4.3
	If spark is good in all cases	4.6

SPECIFIC DIAGNOSIS

This section is arranged so that following each test, instructions are given to proceed to another, until a problem is diagnosed.

SECTION 1—BATTERY

Test and Procedure	Results and Indications	Proceed to
1.1 Inspect the battery visually for case condition (corrosion, cracks) and water level.	If case is cracked, replace battery.	1.4
	If the case is intact, remove corrosion with a solution of baking soda and water. **(CAUTION: Do not get the solution into the battery).** Fill with water.	1.2

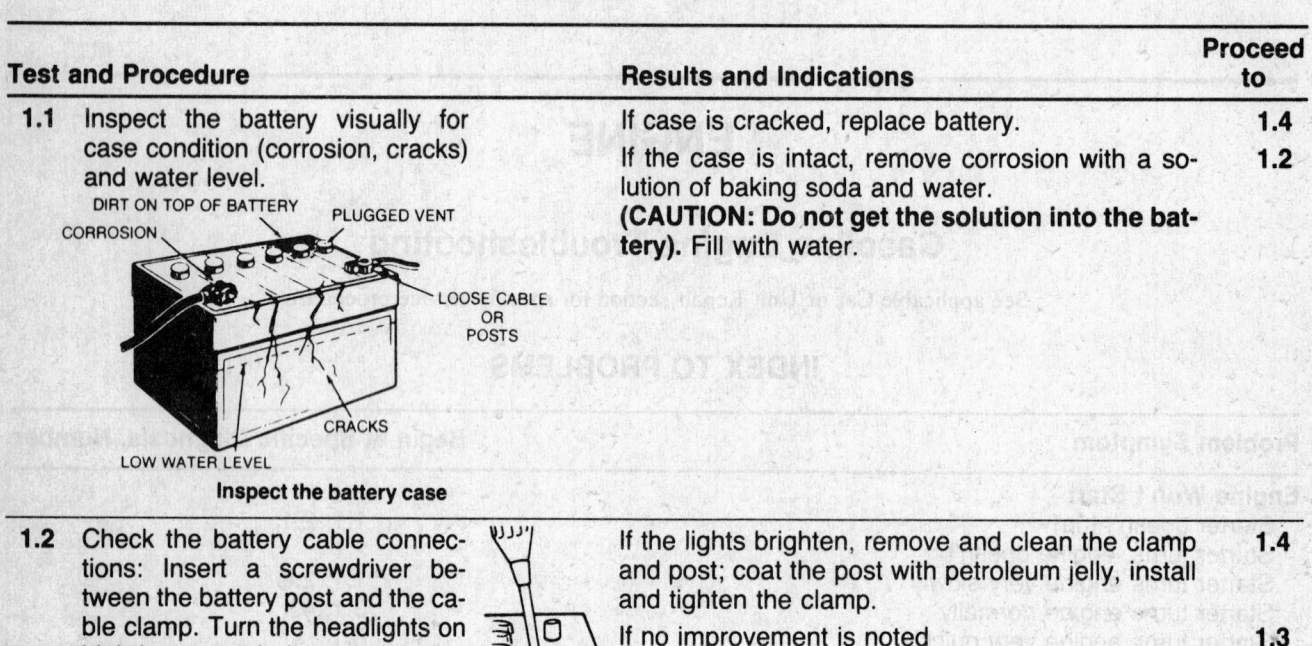

DIRT ON TOP OF BATTERY
PLUGGED VENT
CORROSION
LOOSE CABLE OR POSTS
CRACKS
LOW WATER LEVEL

Inspect the battery case

Test and Procedure	Results and Indications	Proceed to
1.2 Check the battery cable connections: Insert a screwdriver between the battery post and the cable clamp. Turn the headlights on high beam, and observe them as the screwdriver is gently twisted to ensure good metal to metal contact.	If the lights brighten, remove and clean the clamp and post; coat the post with petroleum jelly, install and tighten the clamp.	1.4
	If no improvement is noted	1.3

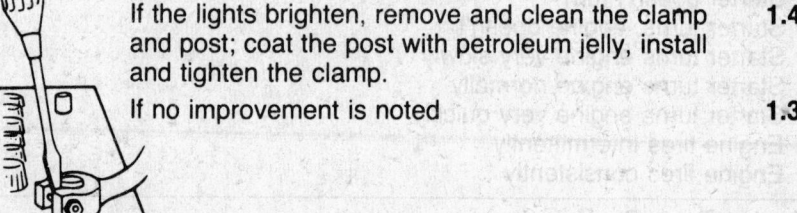

TESTING BATTERY CABLE CONNECTIONS USING A SCREWDRIVER

Test and Procedure	Results and Indications	Proceed to
1.3 Test the state of charge of the battery using an individual cell tester or hydrometer.	If indicated, charge the battery. **NOTE: If no obvious reason exists for the low state of charge (i.e., battery age, prolonged storage), proceed to:**	1.4

Specific Gravity (@ 80° F.)

Minimum	Battery Charge
1.260	100% Charged
1.230	75% Charged
1.200	50% Charged
1.170	25% Charged
1.140	Very Little Power Left
1.110	Completely Discharged

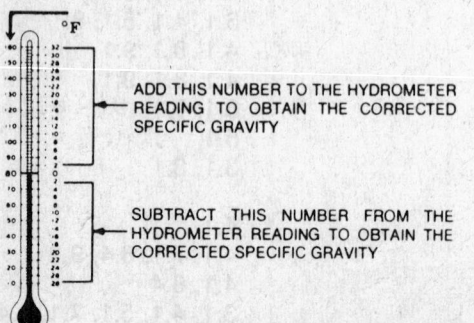

°F

ADD THIS NUMBER TO THE HYDROMETER READING TO OBTAIN THE CORRECTED SPECIFIC GRAVITY

SUBTRACT THIS NUMBER FROM THE HYDROMETER READING TO OBTAIN THE CORRECTED SPECIFIC GRAVITY

The effects of temperature on battery specific gravity (left) and amount of battery charge in relation to specific gravity (right)

Test and Procedure	Results and Indications	Proceed To
1.4 Visually inspect battery cables for cracking, bad connection to ground, or bad connection to starter.	If necessary, tighten connections or replace the cables.	2.1

SECTION 2—STARTING SYSTEM

Test and Procedure	Results and Indications	Proceed to

Note: Tests in Group 2 are performed with coil high tension lead disconnected to prevent accidental starting.

Test and Procedure	Results and Indications	Proceed to
2.1 Test the starter motor and solenoid: Connect a jumper from the battery post of the solenoid (or relay) to the starter post of the solenoid (or relay).	If starter turns the engine normally	2.2
	If the starter buzzes, or turns the engine very slowly	2.4
	If no response, replace the solenoid (or relay).	3.1
	If the starter turns, but the engine doesn't, ensure that the flywheel ring gear is intact. If the gear is undamaged, replace the starter drive.	3.1
2.2 Determine whether ignition override switches are functioning properly (clutch start switch, neutral safety switch), by connecting a jumper across the switch(es), and turning the ignition switch to "start".	If starter operates, adjust or replace switch.	3.1
	If the starter doesn't operate	2.3
2.3 Check the ignition switch "start" position: Connect a 12V test lamp or voltmeter between the starter post of the solenoid (or relay) and ground. Turn the ignition switch to the "start" position, and jiggle the key.	If the lamp doesn't light or the meter needle doesn't move when the switch is turned, check the ignition switch for loose connections, cracked insulation, or broken wires. Repair or replace as necessary.	3.1
	If the lamp flickers or needle moves when the key is jiggled, replace the ignition switch.	3.3

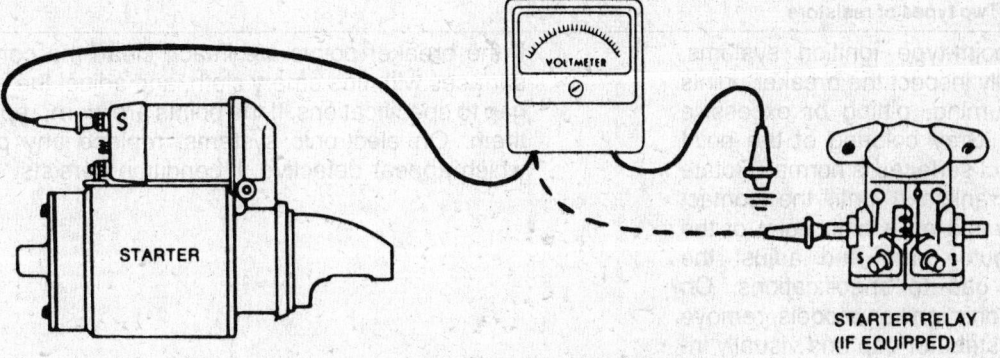

Checking the ignition switch "start" position

Test and Procedure	Results and Indications	Proceed to
2.4 Remove and bench test the starter, according to specifications in the car section.	If the starter does not meet specifications, repair or replace as needed	3.1
	If the starter is operating properly	2.5

Test and Procedure	Results and Indications	Proceed To
2.5 Determine whether the engine can turn freely: Remove the spark plugs, and check for water in the cylinders. Check for water on the dipstick, or oil in the radiator. Attempt to turn the engine using an 18″ flex drive and socket on the crankshaft pulley nut or bolt.	If the engine will turn freely only with the spark plugs out, and hydrostatic lock (water in the cylinders) is ruled out, check valve timing.	9.2
	If engine will not turn freely, and it is known that the clutch and transmission are free, the engine must be disassembled for further evaluation.	See Car Section

SECTION 3—PRIMARY ELECTRICAL SYSTEM

Test and Procedure	Results and Indications	Proceed to
3.1 Check the ignition switch "on" position: Connect a jumper wire between the distributor side of the coil and ground, and a 12V test lamp between the switch side of the coil and ground. Remove the high tension lead from the coil. Turn the ignition switch on and jiggle the key.	If the lamp lights	3.2
	If the lamp flickers when the key is jiggled, replace the ignition switch.	3.3
	If the lamp doesn't light, check for loose or open connections. If none are found, remove the ignition switch and check for continuity. If the switch is faulty, replace it.	3.3

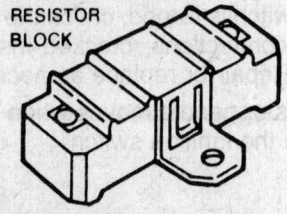

Checking the ignition switch "on" position

3.2 Check the ballast resistor or resistance wire for an open circuit, using an ohmmeter.	Replace the resistor or resistance wire if the resistance is zero. **NOTE: Some ignition systems have no ballast resistor.**	3.3

RESISTOR BLOCK

CALIBRATED RESISTANCE LEAD

Two types of resistors

3.3 On point-type ignition systems, visually inspect the breaker points for burning, pitting or excessive wear. Gray coloring of the point contact surfaces is normal. Rotate the crankshaft until the contact heel rests on a high point of the distributor cam and adjust the point gap to specifications. On electronic ignition models, remove the distributor cap and visually inspect the armature. Ensure that the armature pin is in place, and that the armature is on tight and rotates when the engine is cranked. Make sure there are no cracks, chips or rounded edges on the armature.	If the breaker points are intact, clean the contact surfaces with fine emery cloth, and adjust the point gap to specifications. If the points are worn, replace them. On electronic systems, replace any parts which appear defective. If condition persists	3.4

Test and Procedure	Results and Indications	Proceed To

3.4 On point-type ignition systems, connect a dwell-meter between the distributor primary lead and ground. Crank the engine and observe the point dwell angle. On electronic ignition systems, conduct a stator (magnetic pickup assembly) test. See Electronic Ignition Unit Repair Section.

On point-type systems, adjust the dwell angle if necessary. **3.6**
NOTE: Increasing the point gap decreases the dwell angle and vice-versa.
If the dwell meter shows little or no reading **3.5**
On electronic ignition systems, if the stator is bad, replace the stator. If the stator is good, proceed to the other tests in The Electronic Ignition Unit Repair Section.

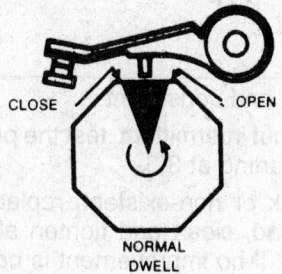 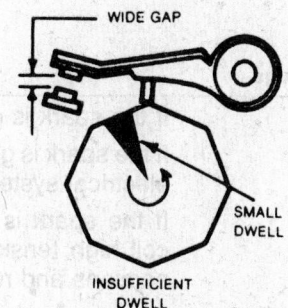 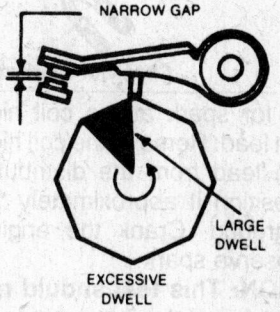

Dwell is a function of point gap

3.5 On the point-type ignition systems, check the condenser for short: connect an ohmeter across the condenser body and the pigtail lead.

If any reading other than infinite is noted, replace the condenser **3.6**

Checking the condenser for short

3.6 Test the coil primary resistance: On point-type ignition systems, connect an ohmmeter across the coil primary terminals, and read the resistance on the low scale. Note whether an external ballast resistor or resistance wire is used. On electronic ignition systems, test the coil primary resistance.

Point-type ignition coils utilizing ballast resistors or resistance wires should have approximately 1.0 ohms resistance. Coils with internal resistors should have approximately 4.0 ohms resistance. If values far from the above are noted, replace the coil. **4.1**

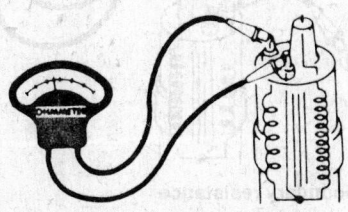

Checking the coil primary resistance

SECTION 4—SECONDARY ELECTRICAL SYSTEM

Test and Procedure	Results and Indications	Proceed to
4.1 Check for spark: Hold each spark plug wire approximately ¼″ from ground with gloves or heavy, dry rag. Crank the engine, and observe the spark.	If no spark is evident	**4.2**
	If spark is good in some cylinders	**4.3**
	If spark is good in all cylinders	**4.6**

Check for spark at the plugs

4.2 Check for spark at the coil high tension lead: Remove the coil high tension lead from the distributor and position it approximately ¼″ from ground. Crank the engine and observe spark. **CAUTION: This test should not be performed on engines equipped with electronic ignition.**	If the spark is good and consistent	**4.3**
	If the spark is good but intermittent, test the primary electrical system starting at 3.3.	**3.3**
	If the spark is weak or non-existent, replace the coil high tension lead, clean and tighten all connections and retest. If no improvement is noted	**4.4**

4.3 Visually inspect the distributor cap and rotor for burned or corroded contacts, cracks, carbon tracks, or moisture. Also check the fit of the rotor on the distributor shaft (where applicable).	If moisture is present, dry thoroughly, and retest per 4.1.	**4.1**
	If burned or excessively corroded contacts, cracks, or carbon tracks are noted, replace the defective part(s) and retest per 4.1.	**4.1**
	If the rotor and cap appear intact, or are only slightly corroded, clean the contacts thoroughly (including the cap towers and spark plug wire ends) and retest per 4.1.	
	If the spark is good in all cases	**4.6**
	If the spark is poor in all cases	**4.5**

CORRODED OR LOOSE WIRE
HIGH RESISTANCE CARBON
EXCESSIVE WEAR OF BUTTON
ROTOR TIP BURNED AWAY

Inspect the distributor cap and rotor

4.4 Check the coil secondary resistance: On point-type systems connect an ohmmeter across the distributor side of the coil and the coil tower. Read the resistance on the high scale of the ohmmeter. On electronic ignition systems, see The Electronic Ignition Unit Repair Section for specific tests.	The resistance of a satisfactory coil should be between 4,000 and 10,000 ohms. If resistance is considerably higher (i.e., 40,000 ohms) replace the coil and retest per 4.1. **NOTE: This does not apply to high performance coils.**

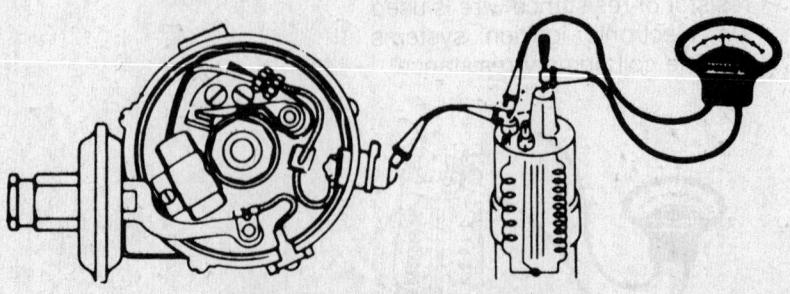

Testing the coil secondary resistance

Spark Plug Analysis

Normal

APPEARANCE
This plug is typical of one operating normally. The insulator nose varies from a light tan to grayish color with slight electrode wear. The presence of slight deposits is normal on used plugs and will have no adverse effect on engine performance. The spark plug heat range is correct for the engine and the engine is running normally.

CAUSE
Properly running engine

RECOMMENDATION
Before reinstalling this plug, the electrodes should be cleaned and filed square. Set the gap to specifications. If the plug has been in service for more than 10–12,000 miles, the entire set should probably be replaced with a fresh set of the same heat range.

Incorrect Heat Range

APPEARANCE
The effects of high temperature on a spark plug are indicated by clean white, often blistered insulator. This can also be accompanied by excessive wear of the electrode, and the absence of deposits.

CAUSE
Check for the correct spark plug heat range. A plug which is too hot for the engine can result in overheating. A car operated mostly at high speeds may require a colder plug. Also check ignition timing, cooling system level, fuel mixture and leaking intake manifold.

RECOMMENDATION
If all ignition and engine adjustments are known to be correct, and no other malfunction exists, install spark plugs one heat range colder.

Oil Deposits

APPEARANCE
The firing end of the plug is covered with a wet, oily coating.

CAUSE
The problem is poor oil control. On high mileage engines, oil is leaking past the rings or valve guides into the combustion chamber. A common cause is also a plugged PCV valve, and a ruptured fuel pump diaphragm can also cause this condition. Oil fouled plugs such as these are often found in new or recently overhauled engines, before normal oil control is achieved, and can be cleaned and reinstalled.

RECOMMENDATION
A hotter spark plug may temporarily relieve the problem, but the engine is probably in need of engine work.

Carbon Deposits

APPEARANCE
Carbon fouling is easily identified by the presence of dry, soft, black, sooty deposits.

CAUSE
Changing the heat range can often lead to carbon fouling, as can prolonged slow, stop-and-start driving. If the heat range is correct, carbon fouling can be attributed to a rich fuel mixture, sticking choke, clogged air cleaner, worn breaker points, retarded timing or low compression. If only one or two plugs are carbon fouled, check for corroded or cracked wires on the affected plugs. Also look for cracks in the distributor cap between the towers of affected cylinders.

RECOMMENDATION
After the problem is corrected, these plugs can be cleaned and reinstalled if not worn severely.

Ash Deposits

APPEARANCE

Ash deposits are characterized by light brown or white colored deposits crusted on the side or center electrodes. In some cases it may give the plug a rusty appearance.

CAUSE

Ash deposits are normally derived from oil or fuel additives burned during normal combustion. Normally they are harmless, though excessive amounts can cause misfiring. If deposits are excessive in short mileage, the valve guides may be worn. Reddish or rusty deposits are caused by manganese, an anti-knock compound replacing lead in unleaded gas. No engine malfunction is indicated.

RECOMMENDATION

Ash-fouled plugs can be cleaned, gapped and reinstalled.

Splash Deposits

APPEARANCE

Splash deposits occur in varying degrees as spotty deposits on the insulator.

CAUSE

These usually occur after a long delayed tune-up. By-products of combustion have accumulated on pistons and valves because of a delayed tune-up. Following tune-up or during hard acceleration, the deposits loosen and are thrown against the hot surface of the plug. If the deposits accumulate sufficiently, misfiring can occur.

RECOMMENDATION

These plugs can be cleaned, gapped and reinstalled.

High Speed Glazing

APPEARANCE

Glazing appears as shiny coating on the plug, either yellow or tan in color.

CAUSE

During hard, fast acceleration, plug temperatures rise suddenly. Deposits from normal combustion have no chance to fluff-off; instead, they melt on the insulator forming an electrically conductive coating which causes misfiring.

RECOMMENDATION

Glazed plugs are not easily cleaned. They should be replaced with a fresh set of plugs of the correct heat range. If the condition recurs, using plugs with a heat range one step colder may cure the problem.

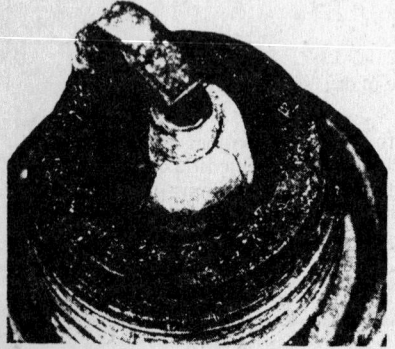

Detonation

APPEARANCE

Detonation is usually characterized by a broken plug insulator.

CAUSE

A portion of the fuel charge will begin to burn spontaneously, from the increased heat following ignition. The explosion that results applies extreme pressure to engine components, frequently damaging spark plugs and pistons.

Detonation can result by over-advanced ignition timing, inferior gasoline (low octane) lean air fuel mixture, poor carburetion, engine lugging or an increase in compression ratio due to combustion chamber deposits or engine modification.

RECOMMENDATION

Replace the plugs after correcting the problem.

Test and Procedure	Results and Indications	Proceed To

4.5 Visually inspect the spark plug wires for cracking or brittleness. Ensure that no two wires are positioned so as to cause induction firing (adjacent and parallel). Remove each wire, one by one, and check resistance with an ohmmeter.

Replace any cracked or brittle wires. If any of the wires are defective, replace the entire set. Replace any wires with excessive resistance (over 8000 Ω per foot for suppression wire), and separate any wires that might cause induction firing.

4.6

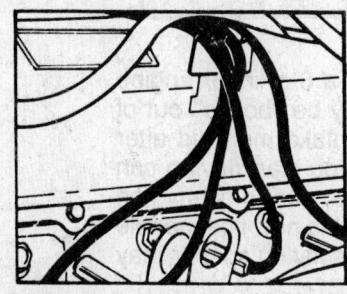

Misfiring can be the result of spark plug leads to adjacent, consecutively firing cylinders running parallel and too close together

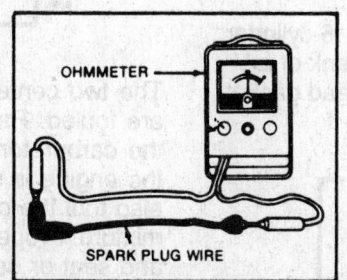

On point-type ignition systems, check the spark plug wires as shown. On electronic ignitions, do not remove the wire from the distributor cap terminal; instead, test through the cap

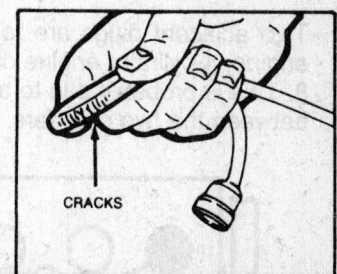

Spark plugs wires can be checked visually by bending them in a loop over your finger. This will reveal any cracks, burned or broken insulation. Any wire with cracked insulation should be replaced

---

4.6 Remove the spark plugs, noting the cylinders from which they were removed, and evaluate according to the chart in this section.

See chart.

See Chart

4.7 Reinstall the spark plugs.
NOTE: Modern electronic ignition systems generate extremely high voltages and high heats. The spark plug boots can soften and actually fuse to the ceramic insulator of the spark plugs after long exposures to high temperature and voltage. If this happens, the boot (and possibly the wire) must be replaced.

To help alleviate this condition, many manufacturers are recommending new silicone compounds to slow the deterioration. The compounds are generally nonconductive, protective lubricants that will not dry out, harden, or melt away. They form a weather-tight seal between rubber or plastic and metal and are found in several typical locations: Inside the insulating boots of spark plug wires, inside primary ignition circuit cable connectors, on distributor and rotor cap electrodes, and under the GM HEI control module.

4.8

Application Point	Silicone Compound
GENERAL MOTORS: Under HEI module	Supplied with new module, or use GE-642 or DC-340
FORD MOTOR COMPANY: Inside spark plug boots, on end of cable when installing new boot, and on rotor and cap electrodes	Ford part number D7AZ-19A331-A or use GE-627 or DC-111
CHRYSLER CORPORATION: ¼" deep within spark control computer connector cavity coating rotor electrode	Use Mopar part number 2932524 or NLGI Grade 2 EP (not a silicone) supplied with new rotor, or use GE-628 or DC-111
AMERICAN MOTORS (Prestolite system): Distributor primary connector—coat male terminal, fill female ¼ full	AMC part number 8127445 or GE-623

GE: General Electric
DC: Dow Corning

Test and Procedure	Results and Indications	Proceed To

4.8 Examine the location of all the plugs.

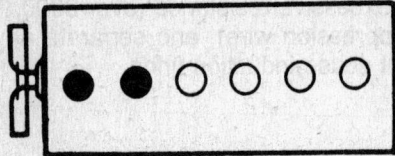

Two adjacent plugs are fouled in a 6-cylinder engine, 4-cylinder engine or either bank of a V-8. This is probably due to a blown head gasket between the two cylinders.

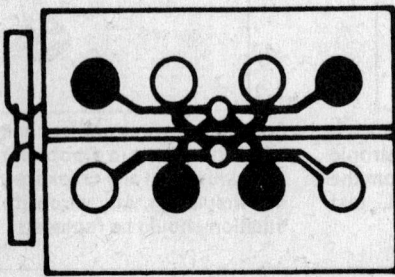

An unbalanced carburetor is indicated. Following the fuel flow on this particular design shows that the cylinders fed by the right-hand barrel are fouled from overly rich mixture, while the cylinders fed by the left-hand barrel are normal.

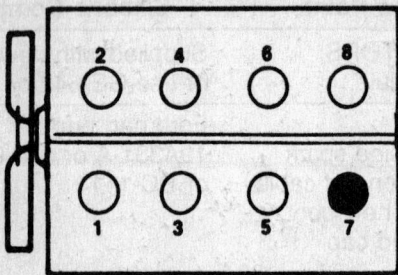

Finding one plug overheated may indicate an intake manifold leak near the affected cylinder. If the overheated plug is the second of two adjacent, consecutively firing plugs, it could be the result of ignition cross-firing. Separating the leads to these two plugs will eliminate cross-fire.

The following diagrams illustrate some of the conditions that the location of plugs will reveal.

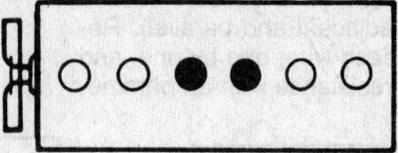

The two center plugs in a 6-cylinder engine are fouled. Raw fuel may be "boiled" out of the carburetor into the intake manifold after the engine is shut-off. Stop-start driving can also foul the center plugs, due to overly rich mixture. Proper float level, a new float needle and seat or use of an insulating spacer may help this problem.

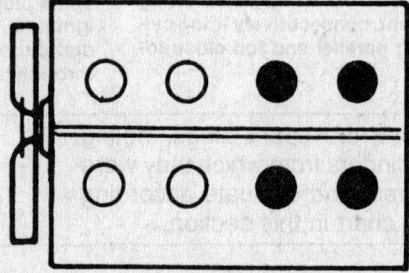

If the four rear plugs are overheated, a cooling system problem is suggested. A thorough cleaning of the cooling system may restore coolant circulation and cure the problem.

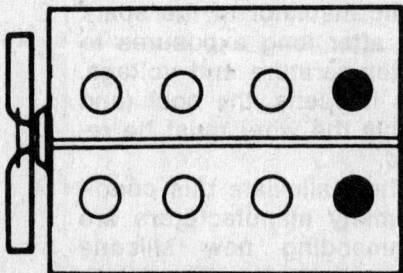

Occasionally, the two rear plugs in large, lightly used V-8's will become oil fouled. High oil consumption and smoky exhaust may also be noticed. It is probably due to plugged oil drain holes in the rear of the cylinder head, causing oil to be sucked in around the valve stems. This usually occurs in the rear cylinders first, because the engine slants that way.

4.9

Test and Procedure	Results and Indications	Proceed To
4.9 Determine the static ignition timing. Using the crankshaft pulley timing marks as a guide, locate top dead center on the compression stroke of the number one cylinder.	The rotor should be pointing toward the No. 1 tower in the distributor cap, and, on electronic ignitions, the armature spoke for that cylinder should be lined up with the stator.	4.10
4.10 Check coil polarity: Connect a voltmeter negative lead to the coil high tension lead, and the positive lead to ground. **NOTE: Reverse the hook-up for positive ground systems.** Crank the engine momentarily.	If the voltmeter reads up-scale, the polarity is correct.	5.1
	If the voltmeter reads down-scale, reverse the coil polarity (switch the primary leads).	5.1

Checking coil polarity

SECTION 5—FUEL SYSTEM

Test and Procedure	Results and Indications	Proceed to
5.1 Determine that the air filter is functioning efficiently: Hold paper elements up to a strong light, and attempt to see light through the filter.	Clean permanent air filters in solvent (or manufacturer's recommendation), and allow to dry. Replace paper elements through which light cannot be seen.	5.2
5.2 Determine whether a flooding condition exists: Flooding is identified by a strong gasoline odor, and excessive gasoline present in the throttle bore(s) of the carburetor.	If flooding is not evident	5.3
	If flooding is evident, permit the gasoline to dry for a few moments and restart.	
	If flooding doesn't recur	5.7
	If flooding is persistent	5.5

If the engine floods repeatedly, check the choke butterfly flap

5.3 Check that fuel is reaching the carburetor: Detach the fuel line at the carburetor inlet. Hold the end of the line in a cup (not styrofoam), and crank the engine.	If fuel flows smoothly	5.7
	If fuel doesn't flow	5.4
	If fuel flows erratically.	
	NOTE: Make sure that there is fuel in the tank	

Check the fuel pump by disconnecting the output line (fuel pump-to-carburetor) at the carburetor and operating the starter briefly

Test and Procedure	Results and Indications	Proceed To
5.4 Test the fuel pump: Disconnect all fuel lines from the fuel pump. Hold a finger over the input fitting, crank the engine (with electric pump, turn the ignition or pump on); and feel for suction.	If suction is evident, blow out the fuel line to the tank with low pressure compressed air until bubbling is heard from the fuel filler neck. Also blow out the carburetor fuel line (both ends disconnected).	5.7
	If no suction is evident, replace or repair the fuel pump. **NOTE: Repeated oil fouling of the spark plugs, or a no-start condition, could be the result of a ruptured vacuum booster pump diaphragm, through which oil or gasoline is being drawn into the intake manifold (where applicable).**	5.7
5.5 Occasionally, small specks of dirt will clog the small jets and orifices in the carburetor. With the engine cold, hold a flat piece of wood or similar material over the carburetor, where possible, and crank the engine.	If the engine starts, but runs roughly the engine is probably not run enough. If the engine won't start.	5.9
5.6 Check the needle and seat: Tap the carburetor in the area of the needle and seat.	If flooding stops, a gasoline additive (e.g., Gumout) will often cure the problem.	5.7
	If flooding continues, check the fuel pump for excessive pressure at the carburetor (according to specifications). If the pressure is normal, the needle and seat must be removed and checked, and/or the float level adjusted.	5.7
5.7 Test the accelerator pump by looking into the throttle bores while operating the throttle.	If the accelerator pump appears to be operating normally	5.8
	If the accelerator pump is not operating, the pump must be reconditioned. Where possible, service the pump with the carburetor(s) installed on the engine. If necessary, remove the carburetor. Prior to removal	5.8

Check for gas at the carburetor by looking down the carburetor throat while someone moves the accelerator

Test and Procedure	Results and Indications	Proceed To
5.8 Determine whether the carburetor main fuel system is functioning: Spray a commercial starting fluid into the carburetor while attempting to start the engine.	If the engine starts, runs for a few seconds, and dies	5.9
	If the engine doesn't start	6.1
5.9 Uncommon fuel system malfunctions: See below:	If the problem is solved	6.1
	If the problem remains, remove and recondition the carburetor.	

Condition	Indication	Test	Prevailing Weather Conditions	Remedy
Vapor lock	Engine will not re-start shortly after running.	Cool the components of the fuel system until the engine starts. Vapor lock can be cured faster by draping a wet cloth over a mechanical fuel pump.	Hot to very hot	Ensure that the exhaust manifold heat control valve is operating. Check with the vehicle manufacturer for the recommended solution to vapor lock on the model in question.
Carburetor icing	Engine will not idle, stalls at low speeds.	Visually inspect the throttle plate area of the throttle bores for frost.	High humidity, 32–40° F.	Ensure that the exhaust manifold heat control valve is operating, and that the intake manifold heat riser is not blocked.
Water in the fuel	Engine sputters and stalls; may not start.	Pump a small amount of fuel into a glass jar. Allow to stand, and inspect for droplets of a layer of water.	High humidity, extreme temperature changes.	For droplets, use one or two cans of commercial gas line anti-freeze. For a layer of water, the tank must be drained, and the fuel lines blown out with compressed air.

SECTION 6—ENGINE COMPRESSION

Test and Procedure	Results and Indications	Proceed to
6.1 Test engine compression: Remove all spark plugs. Block the throttle wide open. Insert a compression gauge into a spark plug port, crank the engine to obtain the maximum reading, and record.	If compression is within limits on all cylinders	7.1
	If gauge reading is extremely low on all cylinders	6.2
	If gauge reading is low on one or two cylinders: (If gauge readings are identical and low on two or more adjacent cylinders, the head gasket must be replaced.)	6.2

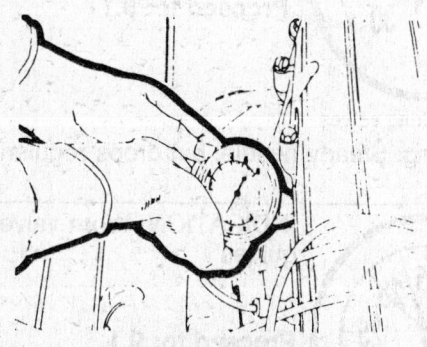

Checking compression

6.2 Test engine compression (wet): Squirt approximately 30 cc. of engine oil into each cylinder, and retest per 6.1.	If the readings improve, worn or cracked rings or broken pistons are indicated:	**See Car Section**
	If the readings do not improve, burned or excessively carboned valves or a jumped timing chain are indicated.	**7.1**
	NOTE: A jumped timing chain is often indicated by difficult cranking.	

SECTION 7—ENGINE VACUUM

Test and Procedure	Results and Indications	Proceed to
7.1 Attach a vacuum gauge to the intake manifold beyond the throttle plate. Start the engine, and observe the action of the needle over the range of engine speeds.	See below.	See below

INDICATION: Normal engine in good condition

Proceed to: 8.1

Normal engine

Gauge reading: Steady, from 17–22 in./Hg.

INDICATION: Sticking valves or ignition miss

Proceed to: 9.1, 8.3

Sticking valves

Gauge reading: Intermittent fluctuation at idle

INDICATION: Late ignition or valve timing, low compression, stuck throttle valve, leaking carburetor or manifold gasket

Proceed to: 6.1

Incorrect valve timing

Gauge reading: Low (10–15 in./Hg) but steady

INDICATION: Improper carburetor adjustment or minor intake leak.

Proceed to: 7.2

Carburetor requires adjustment

Gauge reading: Drifting needle

INDICATION: Ignition miss, blown cylinder head gasket, leaking valve or weak valve spring

Proceed to: 8.3, 6.1

Blown head gasket

Gauge reading: Needle fluctuates as engine speed increases

INDICATION: Burnt valve or faulty valve clearance: Needle will fall when defective valve operates

Proceed to: 9.1

Burnt or leaking valves

Gauge reading: Steady needle, but drops regularly

INDICATION: Choked muffler, excessive back pressure in system

Proceed to: 10.1

Clogged exhaust system

Gauge reading: Gradual drop in reading at idle

INDICATION: Worn valve guides

Proceed to: 9.1

Worn valve guides

Gauge reading: Needle vibrates excessively at idle, but steadies as engine speed increases

White pointer = steady gauge hand   Black pointer = fluctuating gauge hand

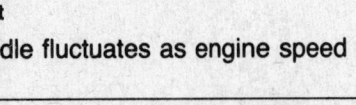

Test and Procedure	Results and Indications	Proceed To
7.2 Attach a vacuum gauge per 7.1, and test for an intake manifold leak. Squirt a small amount of oil around the intake manifold gaskets, carburetor gaskets, plugs and fittings. Observe the action of the vacuum gauge.	If the reading improves, replace the indicated gasket, or seal the indicated fitting or plug:	**8.1**
	If the reading remains low:	**7.3**
7.3 Test all vacuum hoses and accessories for leaks as described in 7.2. Also check the carburetor body (dashpots, automatic choke mechanism, throttle shafts) for leaks in the same manner.	If the reading improves, service or replace the offending part(s):	**8.1**
	If the reading remains low:	**6.1**

SECTION 8—SECONDARY ELECTRICAL SYSTEM

Test and Procedure	Results and Indications	Proceed to
8.1 Remove the distributor cap and check to make sure that the rotor turns when the engine is cranked. Visually inspect the distributor components.	Clean, tighten or replace any components which appear defective.	**8.2**
8.2 Connect a timing light (per manufacturer's recommendation) and check the dynamic ignition timing. Disconnect and plug the vacuum hose(s) to the distributor if specified, start the engine, and observe the timing marks at the specified engine speed.	If the timing is not correct, adjust to specifications by rotating the distributor in the engine: (Advance timing by rotating distributor opposite normal direction of rotor rotation, retard timing by rotating distributor in same direction as rotor rotation.)	**8.3**
8.3 Check the operation of the distributor advance mechanism(s): To test the mechanical advance, disconnect the vacuum lines from the distributor advance unit and observe the timing marks with a timing light as the engine speed is increased from idle. If the mark moves smoothly, without hesitation, it may be assumed that the mechanical advance is functioning properly. To test vacuum advance and or retard systems, alternately crimp and release the vacuum line, and observe the timing mark for movement. If movement is noted, the system is operating.	If the systems are functioning	**8.4**
	If the systems are not functioning, remove the distributor, and test on a distributor tester.	**8.4**
8.4 Locate an ignition miss: With the engine running, remove each spark plug wire, one at a time, until one is found that doesn't cause the engine to roughen and slow down. **CAUTION: Do not pull on the wire to remove the boot from the plug. Be sure your hand is insulated from the wire.**	When the missing cylinder is identified	**4.1**

SECTION 9—VALVE TRAIN

Test and Procedure	Results and Indications	Proceed to
9.1 Evaluate the valve train: Remove the valve cover, and ensure that the valves are adjusted to specifications. A mechanic's stethoscope may be used to aid in the diagnosis of the valve train. By pushing the probe on or near push rods or rockers, valve noise often can be isolated. A timing light also may be used to diagnose valve problems. Connect the light according to manufacturer's recommendations, and start the engine. Vary the firing moment of the light by increasing the engine speed (and therefore the ignition advance), and moving the trigger from cylinder to cylinder. Observe the movement of each valve.	Sticking valves or erratic valve train motion can be observed with the timing light. The cylinder head must be disassembled for repairs.	See Car Section
9.2 Check the valve timing: Locate top dead center of the No. 1 piston, and install a degree wheel or tape on the crankshaft pulley or damper with zero corresponding to an index mark on the engine. Rotate the crankshaft in its direction of rotation, and observe the opening of the No. 1 cylinder intake valve. The opening should correspond with the correct mark on the degree wheel according to specifications.	If the timing is not correct, the timing cover must be removed for further investigation.	See Car Section

SECTION 10—EXHAUST SYSTEM

Test and Procedure	Results and Indications	Proceed to
10.1 Determine whether the exhaust manifold heat control valve is operating: Operate the valve by hand to determine whether it is free to move. If the valve is free, run the engine to operating temperature and observe the action of the valve, to ensure that it is opening.	If the valve sticks, spray it with a suitable solvent, open and close the valve to free it, and retest. If the valve functions properly	10.2
	If the valve does not free, or does not operate, replace the valve.	10.2
10.2 Ensure that there are no exhaust restrictions: Visually inspect the exhaust system for kinks, dents, or crushing. Also note that gases are flowing freely from the tailpipe at all engine speeds, indicating no restriction in the muffler or resonator.	Replace any damaged portion of the system.	11.1

SECTION 11—COOLING SYSTEM

Test and Procedure	Results and Indications	Proceed to
11.1 Visually inspect the fan belt for glazing, cracks, and fraying, and replace if necessary. Tighten the belt so that the longest span has approximately ½″ play at its mid-point under thumb pressure (see Maintenance Section).	Replace or tighten the fan belt as necessary.	11.2

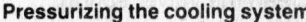

Checking belt tension

Test and Procedure	Results and Indications	Proceed to
11.2 Check the fluid level of the cooling system.	If full or slightly low, fill as necessary.	11.5
	If extremely low	11.3
11.3 Visually inspect the external portions of the cooling system (radiator, radiator hoses, thermostat elbow, water pump seals, heater hoses, etc.) for leaks. If none are found, pressurize the cooling system to 14–15 psi.	If cooling system holds the pressure	11.5
	If cooling system loses pressure rapidly, reinspect external parts of the system for leaks under pressure. If none are found, check dipstick for coolant in crankcase. If no coolant is present, but pressure loss continues	11.4
	If coolant is evident in crankcase, remove cylinder head(s), and check gasket(s). If gaskets are intact, block and cylinder head(s) should be checked for cracks or holes.	
	If the gasket(s) is blown, replace, and purge the crankcase of coolant.	12.6
	NOTE: Occasionally, due to atmospheric and driving conditions, condensation of water can occur in the crankcase. This causes the oil to appear milky white. To remedy, run the engine until hot, and change the oil and oil filter.	
11.4 Check for combustion leaks into the cooling system: Pressurize the cooling system as above. Start the engine, and observe the pressure gauge. If the needle fluctuates, remove each spark plug wire, one at a time, noting which cylinder(s) reduce or eliminate the fluctuation.	Cylinders which reduce or eliminate the fluctuation, when the spark plug wire is removed, are leaking into the cooling system. Replace the head gasket on the affected cylinder bank(s).	**See Car Section**

Pressurizing the cooling system

Test and Procedure	Results and Indications	Proceed To
11.5 Check the radiator pressure cap: Attach a radiator pressure tester to the radiator cap (wet the seal prior to installation). Quickly pump up the pressure, noting the point at which the cap releases.	If the cap releases within ±1 psi of the specified rating, it is operating properly.	11.6
	If the cap releases at more than ±1 psi of the specified rating, it should be replaced.	11.6

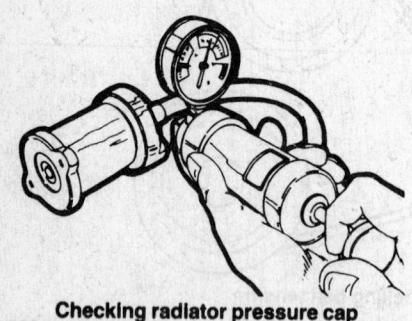

Checking radiator pressure cap

Test and Procedure	Results and Indications	Proceed To
11.6 Test the thermostat: Start the engine cold, remove the radiator cap, and insert a thermometer into the radiator. Allow the engine to idle. After a short while, there will be a sudden, rapid increase in coolant temperature. The temperature at which this sharp rise stops is the thermostat opening temperature.	If the thermostat opens at or about the specified temperature	11.7
	If the temperature doesn't increase (If the temperature increases slowly and gradually, replace the thermostat.)	11.7
11.7 Check the water pump: Remove the thermostat elbow and the thermostat, disconnect the coil high tension lead (to prevent starting), and crank the engine momentarily.	If coolant flows, replace the thermostat and retest per 11.6.	11.6
	If coolant doesn't flow, reverse flush the cooling system to alleviate any blockage that might exist. If system is not blocked, and coolant will not flow, replace the water pump.	See Car Section

SECTION 12—LUBRICATION

Test and Procedure	Results and Indications	Proceed to
12.1 Check the oil pressure gauge or warning light: If the gauge shows low pressure, or the light is on for no obvious reason, remove the oil pressure sender. Install an accurate oil pressure gauge and run the engine momentarily.	If oil pressure builds normally, run engine for a few moments to determine that it is functioning normally, and replace the sender.	—
	If the pressure remains low	12.2
	If the pressure surges	12.3
	If the oil pressure is zero	12.3
12.2 Visually inspect the oil: If the oil is watery or very thin, milky, or foamy, replace the oil and oil filter.	If the oil is normal	12.3
	If after replacing oil the pressure remains low	12.3
	If after replacing oil the pressure becomes normal	—
12.3 Inspect the oil pressure relief valve and spring, to ensure that it is not sticking or stuck. Remove and thoroughly clean the valve, spring, and the valve body.	If the oil pressure improves	—
	If no improvement is noted	12.4

Test and Procedure	Results and Indications	Proceed To
12.4 Check to ensure that the oil pump is not cavitating (sucking air instead of oil): See that the crankcase is neither over nor underfull, and that the pickup in the sump is in the proper position and free from sludge.	Fill or drain the crankcase to the proper capacity, and clean the pickup screen in solvent if necessary. If no improvement is noted	**12.5**
12.5 Inspect the oil pump drive and the oil pump:	If the pump drive or the oil pump appear to be defective, service as necessary and retest per 12.1. If the pump drive and pump appear to be operating normally, the engine should be disassembled to determine where blockage exists.	**12.1**
12.6 Purge the engine of ethylene glycol coolant: Competely drain the crankcase and the oil filter. Obtain a commercial butyl cellosolve base solvent, designated for this purpose, and follow the instructions precisely. Following this, install a new oil filter and refill the crankcase with the proper weight oil. The next oil and filter change should follow shortly thereafter (1000 miles).		

Diesel Engine Troubleshooting

NOTE: The following troubleshooting procedures cover problems usually associated with diesel engines. Those problems common to both gasoline and diesel engines are covered in the gasoline engine troubleshooting procedures.

INDEX TO PROBLEMS

Problem/Symptom	Begin at Specific Diagnosis, Number
Fuel System	Section 1
Engine Starting Difficulty:	
Feed pump does not feed fuel	1.1
Injection pump does not feed fuel	1.2
Incorrect injection timing	1.3
Defective injection nozzles	1.4
Engine Operating Instability:	
Engine shuts off immediately after starting	1.5
Uneven idling	1.6
Engine will not reach maximum rated speed	1.7
Engine exceeds maximum rated speed	1.8
Loss of power	1.9
Engine Knock:	
Associated with exhaust gas problems	1.10
Not associated with exhaust gas problems	1.11
Engine Mechanical	Section 2
Engine Starting Difficulty	2.1
Unusual Noises	2.2
Engine Operating Instability	2.3
Loss of Power	2.4
Exhaust gas Problem	2.5
Engine Shut-Off	2.6
Loss of Oil Pressure	2.7
Oil Leakage	2.8
Compression Pressure Leakage	2.9

SECTION 1—Fuel System

Test and Procedure	Results and Indication	Proceed To
1.1a Check for pressure at the outlet of the feed pump	If pressure exists, there is a clog in the supply line. Clean or replace it. If there is little or no pressure at the outlet, the filter is clogged. Clean or replace the filter. If the filter is clear, the feed pump piston is inoperative. Relace it.	1.1b
1.1b Check the feed pump valves	If the inlet and outlet valves do not operate, the check valve or spring is broken. Replace it.	1.2a
1.2a Check for fuel leakage at the overflow or return line	A clogged filter can result in high pressure causing leakage. Replace the filter.	1.2b
1.2b Check for fuel in the filter leaking at the overflow valve	If leakage is found, the overflow valve is damaged. Replace it.	1.2c
1.2c Check for leakage at the injection pump overflow valve	If leakage is found, it is caused by: damaged overflow valve, sticking plunger, or sticking delivery valve. Replace the defective part(s).	1.2d
1.2d Check the injection pump plunger feed pressures.	If pressure at the plungers is low, replace the plunger(s).	1.2e

	Test and Procedure	Results and Indication	Proceed To
1.2e	Check to make sure the injection pump is operating	An inoperative pump is caused by: a damaged or missing shaft key, or a damaged drive gear train.	1.3a
1.3a	Check that the pump timing marks are correctly aligned in the gear train	Incorrect timing marks alignment must be corrected.	1.3b
1.3b	Check that the injection pump is properly mounted	Remove and install the pump correctly	1.4a
1.4a	Install an injection nozzle on a tester and make sure that fuel is continuously ejected	A broken or intermittent stream is caused by a damaged spring or a sticking nozzle needle	1.4b
1.4b	With the nozzle on the tester as in 1.4a, check that shutoff is clean with no dribble or afterdrip	Dribble is caused by a defective nozzle valve seat. Replace the nozzle.	1.4c
1.4c	Using a tester, check injection pressure	Low pressure is a result of a weak spring. Replace the spring or adjust the initial injection pressure.	1.5a
1.5a	See 1.2a	Proceed as in 1.2a	1.5b
1.5b	Check for water in the fuel	Drain and clean the tank	1.5c
1.5c	Check for air in the fuel lines	Air can be introduced through a damaged fuel inlet line, a loose inlet line connector or a damaged gasket	1.5d
1.5d	Check for insufficient fuel feed	Insufficient fuel feed is caused by: a damaged feed pump, a clogged tank vent, or a clogged filter. Replace or repair as necessary.	1.6a
1.6a	Check the control rack action for smooth operation	Uneven control rack operation is caused by: a sticking plunger, improper meshing of the rack and pinion, poor seating of the plunger spring, insufficient clearance between the plunger and lower spring seat, or an overly tight delivery valve holder. Replace or adjust as necessary.	1.6b
1.6b	Check that the injection pump discharge is uniform	If the output is uneven, adjust as necessary	1.6c
1.6c	Check that the injection pump discharge volume is adequate	An inadequate discharge volume is caused by a worn plunger or a broken spring	1.6d
1.6d	Check for even low speed engine performance	If the engine performs unevenly or erratically at low speed only, a worn feed pump piston or defective feed pump valve is the cause.	1.6e
1.6e	Check for smooth engine operation throughout the operating range	This problem is usually caused by mechanical governor defects such as: a defective low speed spring, defective damper spring, or excessive friction among moving parts. Replace the defective parts.	1.6f
1.6f	Check the injectors on a tester	Improper nozzle operation should be corrected accordingly	1.7a
1.7a	Check the operating governor	A broken or weak spring in the governor will prevent full speed operation.	1.7b
1.7b	Check the injectors on a tester for a drop in injector output	A drop in output is caused by a sticking needle or a dirty nozzle. Replace or clean as necessary.	1.8a

TROUBLESHOOTING AND DIAGNOSIS

Test and Procedure		Results and Indications	Proceed To
1.8a	Check the injection pump for proper rack and pinion action	A catching or dirty rack and pinion will cause overspeeding.	1.8b
1.8b	Check the governor adjustment	An improperly adjusted governor will cause overspeeding. Adjust.	1.9a
1.9a	Check the injection pump output	Low output can be caused by: Incorrect adjustment—Adjust Loose delivery valve—Tighten Broken delivery valve seal—Replace Poor valve seat contact—Replace Broken/weak delivery valve spring—Replace	1.9b
1.9b	Check for unusual noise at the injection pump	A noisy pump is an indication of a broken plunger spring	1.9c
1.9c	Check plunger operation	A sticking injection pump plunger will cause power loss. Replace.	1.9d
1.9d	Check the injection timer	A lag in injection timing is caused by large clearances in the timer due to wear. Replace.	1.9e
1.9e	Check for air or water in the fuel	Bleed the air or drain the fuel and clean the tank and lines	1.9f
1.9f	Check the injection timing	Readjust timing if necessary	1.10a
1.10a	Check the initial injection timing	Adjust if necessary	1.10b
1.10b	Check the injection pressure	High pressure will cause knock. Adjust as necessary	1.10c
1.10c	Check the injector nozzle	A clogged nozzle causes knock. Clean or replace the nozzle.	1.11a
1.11a	Check the injection pump output and timing	Excessive output, coupled with incorrect timing causes knock. Adjust as necessary	1.11b
1.11b	Check the delivery valve seat	Replace a defective seat	1.11c
1.11c	Check the pump plungers	Replace badly worn plungers	1.11d
1.11d	Check injector opening pressure on a tester	Adjust as necessary	1.11e
1.11e	Check the injector	Replace a broken nozzle spring or sticking needle.	

SECTION 2—ENGINE MECHANICAL

Test and Procedure		Results and Indications	Proceed To
2.1a	Check for piston seizing	Seized pistons are caused by low oil pressure, oil breakdown, or overheating. Replace the pistons and liners.	2.1b
2.1b	Check for a damaged flywheel ring gear	A damaged ring gear will cause poor meshing with the starter. Replace the ring gear.	2.1c
2.1c	Make a compression check	Low compression can be caused by: sticking rings, worn rings, worn liners. Replace the rings or liners.	2.2a
2.2a	A knocking noise at idle or during acceleration can be caused by a variety of wear problems.	Use a stethoscope or similar listening device to try to pinpoint the source of the noise. Among other reasons for knocking are: piston pins, rod bearings, loose rod caps, crankshaft journals and/or bearings, crankshaft thrust washer. Replace any worn parts.	2.2b
2.2b	An infrequently encountered noise is a continuous growl during acceleration	This problem is usually caused by problems in the engine timing gears. Poor contact, excessive backlash or loose gears are usually at fault.	2.2c
2.2c	Intermittent noises are the hardest to find. They are usually caused by broken moving parts.	Check the gear train for a chipped or cracked gear; the oil pan for broken parts or foreign objects or the cylinder head for a broken valve or valve spring.	2.3
2.3	Check for oil in the combustion chambers	Oil entering the combustion chambers will cause the engine to overspeed if the amount of oil is too great, or run unevenly. Check for broken or sticking rings, bad head gasket(s) or worn valve guides.	2.4
2.4	Check the compression	Low compression is the main cause of power loss. The main causes for low compression are: worn rings or liners, cracked valves, warped head or block, and bad head gasket.	2.5
2.5	A large amount of black exhaust is caused by low compression	See 2.4 above	2.6
2.6	If the engine stops suddenly during operation, the cause is usually sudden damage	Check the pistons, main bearings or rod bearings for lack of lubrication. A seized camshaft is also a result of low or no lubrication. Check the timing gears for damage.	2.7
2.7	Check for excessive clearance between the bearings and journals on both the mains and rod bearings. Check the oil pressure.	Replace as necessary. Replace the pump as necessary.	2.8
2.8	Aside from the usual leaking gasket problems, check the condition of the combustion chamber O-rings.	Replace as necessary	2.9
2.9	Compression leakage is usually caused by a seal defect between the head and the block	Check the head gasket; check for loose head bolts; check for head or block warpage. Replace or repair as necessary.	

Engine Overheating Troubleshooting

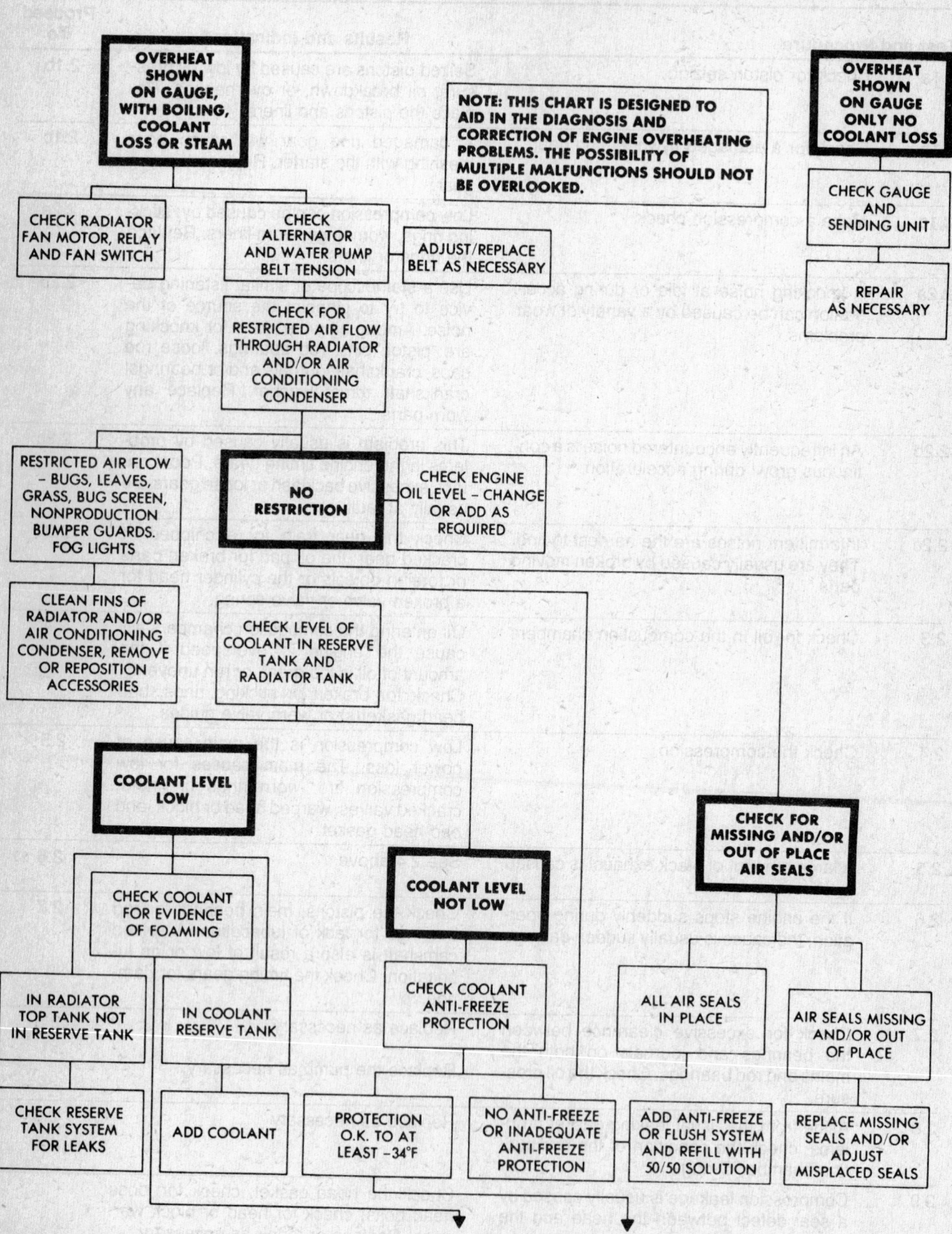

Engine Overheating Troubleshooting

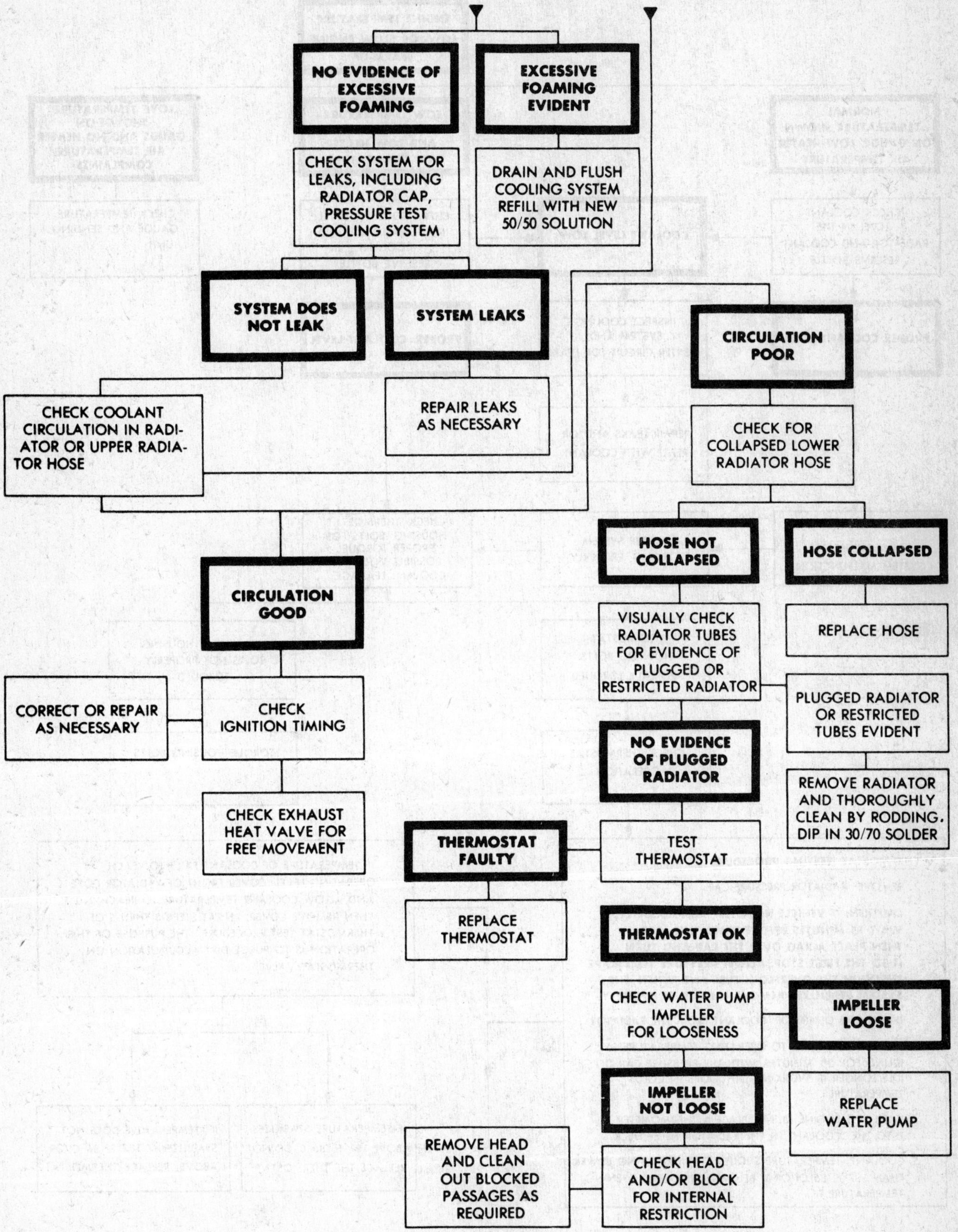

NO EVIDENCE OF EXCESSIVE FOAMING

CHECK SYSTEM FOR LEAKS, INCLUDING RADIATOR CAP, PRESSURE TEST COOLING SYSTEM

EXCESSIVE FOAMING EVIDENT

DRAIN AND FLUSH COOLING SYSTEM REFILL WITH NEW 50/50 SOLUTION

SYSTEM DOES NOT LEAK

CHECK COOLANT CIRCULATION IN RADIATOR OR UPPER RADIATOR HOSE

SYSTEM LEAKS

REPAIR LEAKS AS NECESSARY

CIRCULATION POOR

CHECK FOR COLLAPSED LOWER RADIATOR HOSE

CIRCULATION GOOD

CORRECT OR REPAIR AS NECESSARY

CHECK IGNITION TIMING

CHECK EXHAUST HEAT VALVE FOR FREE MOVEMENT

HOSE NOT COLLAPSED

VISUALLY CHECK RADIATOR TUBES FOR EVIDENCE OF PLUGGED OR RESTRICTED RADIATOR

HOSE COLLAPSED

REPLACE HOSE

PLUGGED RADIATOR OR RESTRICTED TUBES EVIDENT

REMOVE RADIATOR AND THOROUGHLY CLEAN BY RODDING. DIP IN 30/70 SOLDER

NO EVIDENCE OF PLUGGED RADIATOR

THERMOSTAT FAULTY

TEST THERMOSTAT

REPLACE THERMOSTAT

THERMOSTAT OK

CHECK WATER PUMP IMPELLER FOR LOOSENESS

IMPELLER LOOSE

REPLACE WATER PUMP

IMPELLER NOT LOOSE

REMOVE HEAD AND CLEAN OUT BLOCKED PASSAGES AS REQUIRED

CHECK HEAD AND/OR BLOCK FOR INTERNAL RESTRICTION

Low Engine Temperature Troubleshooting

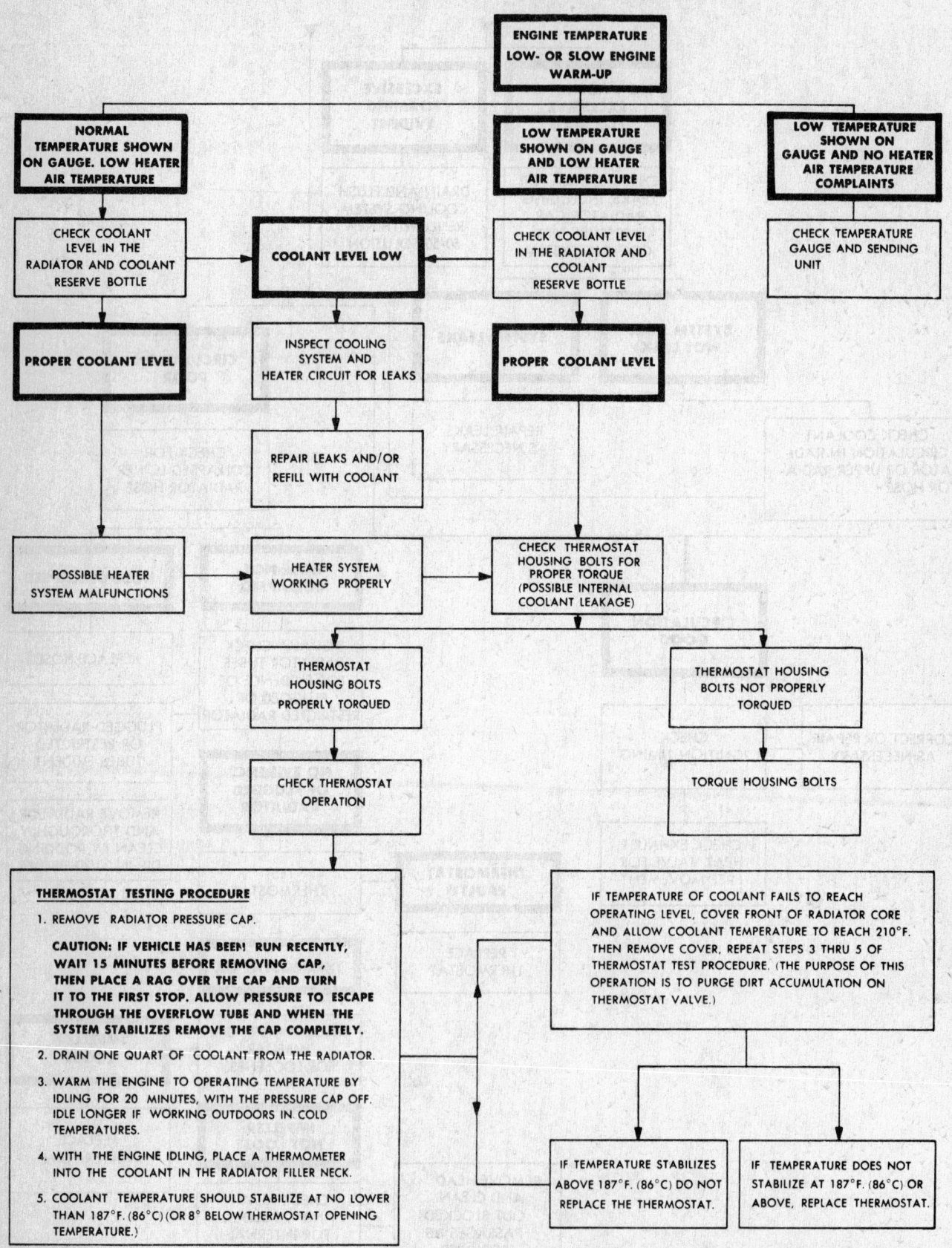

```
                              ENGINE TEMPERATURE
                              LOW- OR SLOW ENGINE
                                   WARM-UP
```

NORMAL TEMPERATURE SHOWN ON GAUGE. LOW HEATER AIR TEMPERATURE

LOW TEMPERATURE SHOWN ON GAUGE AND LOW HEATER AIR TEMPERATURE

LOW TEMPERATURE SHOWN ON GAUGE AND NO HEATER AIR TEMPERATURE COMPLAINTS

CHECK COOLANT LEVEL IN THE RADIATOR AND COOLANT RESERVE BOTTLE

COOLANT LEVEL LOW

CHECK COOLANT LEVEL IN THE RADIATOR AND COOLANT RESERVE BOTTLE

CHECK TEMPERATURE GAUGE AND SENDING UNIT

PROPER COOLANT LEVEL

INSPECT COOLING SYSTEM AND HEATER CIRCUIT FOR LEAKS

PROPER COOLANT LEVEL

REPAIR LEAKS AND/OR REFILL WITH COOLANT

POSSIBLE HEATER SYSTEM MALFUNCTIONS

HEATER SYSTEM WORKING PROPERLY

CHECK THERMOSTAT HOUSING BOLTS FOR PROPER TORQUE (POSSIBLE INTERNAL COOLANT LEAKAGE)

THERMOSTAT HOUSING BOLTS PROPERLY TORQUED

THERMOSTAT HOUSING BOLTS NOT PROPERLY TORQUED

CHECK THERMOSTAT OPERATION

TORQUE HOUSING BOLTS

THERMOSTAT TESTING PROCEDURE

1. REMOVE RADIATOR PRESSURE CAP.

   CAUTION: IF VEHICLE HAS BEEN RUN RECENTLY, WAIT 15 MINUTES BEFORE REMOVING CAP, THEN PLACE A RAG OVER THE CAP AND TURN IT TO THE FIRST STOP. ALLOW PRESSURE TO ESCAPE THROUGH THE OVERFLOW TUBE AND WHEN THE SYSTEM STABILIZES REMOVE THE CAP COMPLETELY.

2. DRAIN ONE QUART OF COOLANT FROM THE RADIATOR.

3. WARM THE ENGINE TO OPERATING TEMPERATURE BY IDLING FOR 20 MINUTES, WITH THE PRESSURE CAP OFF. IDLE LONGER IF WORKING OUTDOORS IN COLD TEMPERATURES.

4. WITH THE ENGINE IDLING, PLACE A THERMOMETER INTO THE COOLANT IN THE RADIATOR FILLER NECK.

5. COOLANT TEMPERATURE SHOULD STABILIZE AT NO LOWER THAN 187°F. (86°C) (OR 8° BELOW THERMOSTAT OPENING TEMPERATURE.)

IF TEMPERATURE OF COOLANT FAILS TO REACH OPERATING LEVEL, COVER FRONT OF RADIATOR CORE AND ALLOW COOLANT TEMPERATURE TO REACH 210°F. THEN REMOVE COVER, REPEAT STEPS 3 THRU 5 OF THERMOSTAT TEST PROCEDURE. (THE PURPOSE OF THIS OPERATION IS TO PURGE DIRT ACCUMULATION ON THERMOSTAT VALVE.)

IF TEMPERATURE STABILIZES ABOVE 187°F. (86°C) DO NOT REPLACE THE THERMOSTAT.

IF TEMPERATURE DOES NOT STABILIZE AT 187°F. (86°C) OR ABOVE, REPLACE THERMOSTAT.

DRIVELINE
Clutch System Troubleshooting

Condition	Possible Cause	Corrective Action
Clutch chatter	1. Grease on driven plate (disc) facing. 2. Binding clutch linkage. 3. Loose, damaged facings on driven plate (disc). 4. Engine mounts loose. 5. Incorrect height adjustment of pressure plate release levers. 6. Clutch housing or housing to transmission adapter misalignment. 7. Loose driven plate hub.	1. Replace plate. 2. Check for worn, bent, broken parts. Replace as required. Lube linkage. 3. Replace driven plate. 4. Tighten mounts. Replace if damaged. 5. Adjust release lever height. 6. Check bore and face run out. Correct as required. 7. Replace driven plate.
Clutch grabbing	1. Oil, grease on driven plate (disc) facing. 2. Broken pressure plate. 3. Warped or binding driven plate. Driven plate binding on clutch shaft.	1. Replace driven plate. 2. Replace pressure plate. 3. Replace warped driven plate. Replace clutch shaft if defective, scored, worn.
Clutch slips	1. Lack of lubrication in clutch linkage (linkage binds, causes incomplete engagement. 2. Incorrect pedal, or linkage adjustment. 3. Broken pressure plate springs. 4. Weak pressure plate springs. 5. Grease on driven plate facings (disc).	1. Lubricate linkage. 2. Adjust as required. 3. Replace pressure plate. 4. Replace pressure plate. 5. Replace driven plate.
Incomplete clutch release	1. Incorrect pedal or linkage adjustment or linkage binding. 2. Incorrect height adjustment on pressure plate release levers. 3. Loose, broken facings on driven plate (disc). 4. Bent, dished, warped driven plate caused by overheating.	1. Adjust as required. Lubricate linkage. 2. Adjust release lever height. 3. Replace driven plate. 4. Replace driven plate.
Grinding, whirring grating noise when pedal is depressed	1. Worn or defective throwout bearing. 2. Starter drive teeth contacting flywheel ring gear teeth.	1. Replace throwout bearing. 2. Look for milled or polished teeth on ring gear. Align clutch housing, replace starter drive or drive spring as required.
Squeal, howl, trumpeting noise when pedal is being released (occurs during first inch to inch and one-half of pedal travel)	1. Pilot bushing worn or lack of lubricant.	1. Replace worn bushing. If bushing appears OK, polish bushing with emery, soak lube wick in oil, lube bushing with oil, apply film of chassis grease to clutch shaft pilot hub, reassemble. **NOTE:** Bushing wear may be due to misalignment of clutch housing or housing to transmission adapter.
Vibration or clutch pedal pulsation with clutch disengaged (pedal fully depressed)	1. Worn or defective engine transmission mounts. 2. Flywheel run out, or damaged or defective clutch components.	1. Inspect and replace as required. 2. Replace components as required. (Flywheel run out at face not to exceed 0.005″).

Manual Transmission Troubleshooting

Condition	Probable Cause
Jumping out of high gear	1. Misalignment of transmission case or clutch housing. 2. Worn pilot bearing in crankshaft. 3. Bent transmission shaft. 4. Worn high speed sliding gear. 5. Worn teeth in clutch shaft. 6. Insufficient spring tension on shifter rail plunger. 7. Bent or loose shifter fork. 8. End-play in clutch shaft. 9. Gears not engaging completely. 10. Loose or worn bearings on clutch shaft or mainshaft.
Sticking in high gear	1. Clutch not releasing fully. 2. Burred or battered teeth on clutch shaft. 3. Burred or battered transmission mainshaft. 4. Frozen synchronizing clutch. 5. Stuck shifter rail plunger. 6. Gearshift lever twisting and binding shifter rail. 7. Battered teeth on high speed sliding gear or on sleeve. 8. Lack of lubrication. 9. Improper lubrication. 10. Corroded transmission parts. 11. Defective mainshaft pilot bearing.
Jumping out of second gear	1. Insufficient spring tension on shifter rail plunger. 2. Bent or loose shifter fork. 3. Gears not engaging completely. 4. End-play in transmission mainshaft. 5. Loose transmission gear bearing. 6. Defective mainshaft pilot bearing. 7. Bent transmission shaft. 8. Worn teeth on second speed sliding gear or sleeve. 9. Loose or worn bearings on transmission mainshaft. 10. End-play in countershaft.
Sticking in second gear	1. Clutch not releasing fully. 2. Burred or battered teeth on sliding sleeve. 3. Burred or battered transmission mainshaft. 4. Frozen synchronizing clutch. 5. Stuck shifter rail plunger. 6. Gearshift lever twisting and binding shifter rail. 7. Lack of lubrication. 8. Second speed transmission gear bearings locked will give same effect as gears stuck in second. 9. Improper lubrication. 10. Corroded transmission parts.
Jumping out of low gear	1. Gears not engaging completely. 2. Bent or loose shifter fork. 3. End-play in transmission mainshaft. 4. End-play in countershaft. 5. Loose or worn bearings on transmission mainshaft. 6. Loose or worn bearings in countershaft. 7. Defective mainshaft pilot bearing.
Sticking in low gear	1. Clutch not releasing fully. 2. Burred or battered transmission mainshaft. 3. Stuck shifter rail plunger. 4. Gearshift lever twisting and binding shifter rail. 5. Lack of lubrication. 6. Improper lubrication. 7. Corroded transmission parts.

Condition	Probable Cause
Jumping out of reverse gear	1. Insufficient spring tension on shifter rail plunger. 2. Bent or loose shifter fork. 3. Badly worn gear teeth. 4. Gears not engaging completely. 5. End-play in transmission mainshaft. 6. Idler gear bushings loose or worn. 7. Loose or worn bearings on transmission mainshaft. 8. Defective mainshaft pilot bearing.
Sticking in reverse gear	1. Clutch not releasing fully. 2. Burred or battered transmission mainshaft. 3. Stuck shifter rail plunger. 4. Gearshift lever twisting and binding shifter rail. 5. Lack of lubrication. 6. Improper lubrication. 7. Corroded transmission parts.
Failure of gears to synchronize	1. Binding pilot bearing on mainshaft, will synchronize in high gear only. 2. Clutch not releasing fully. 3. Detent spring weak or broken. 4. Weak or broken springs under balls in sliding gear sleeve. 5. Binding bearing on clutch shaft. 6. Binding countershaft. 7. Binding pilot bearing in crankshaft 8. Badly worn gear teeth. 9. Scored or worn cones. 10. Improper lubrication. 11. Constant mesh gear not turning freely on transmission mainshaft. Will synchronize in that gear only.
Gears spinning when shifting into gear from neutral	1. Clutch not releasing fully. 2. In some cases an extremely light lubricant in transmission will cause gears to continue to spin for a short time after clutch is released. 3. Binding pilot bearing in crankshaft.
Noisy in all gears	1. Insufficient lubricant. 2. Worn countergear bearings. 3. Worn or damaged main drive gear or countergear. 4. Damaged main drive gear or mainshaft bearings. 5. Worn or damaged countergear anti-lash plate.
Noisy in high gear	1. Damaged main drive gear bearing. 2. Damaged mainshaft bearing. 3. Damaged high speed gear synchronizer.
Noisy in neutral	1. Damaged main drive gear bearing. 2. Damaged or loose mainshaft pilot bearing. 3. Worn or damaged countergear anti-lash plate. 4. Worn countergear bearings.
Noisy in all reduction gears	1. Insufficient lubricant. 2. Worn or damaged drive gear or countergear.
Noisy in second only	1. Damaged or worn second gear constant mesh gears. 2. Worn or damaged countergear rear bearings. 3. Damaged or worn second gear synchronizer.
Noisy in second only	1. Damaged or worn second gear constant mesh gears. 2. Worn or damaged countergear rear bearings. 3. Damaged or worn second gear synchronizer.
Noisy in third only (four speed)	1. Damaged or worn third gear constant mesh gears. 2. Worn or damaged countergear bearings.

Condition	Probable Cause
Noisy in reverse only	1. Worn or damaged reverse idler gear or idler bushing. 2. Worn or damaged mainshaft reverse gear. 3. Worn or damaged reverse countergear. 4. Damaged shift mechanism.
Excessive backlash in all reduction gears	1. Worn countergear bearings. 2. Excessive end–play in countergear.

Automatic Transmission Troubleshooting

Keeping alert to changes in the operating characteristics of the transmission (changing shift points, noises, etc.) can prevent small problems from becoming large ones. If the problem cannot be traced to loose bolts, fluid level, misadjusted linkage, clogged filters or similar problems, you should probably seek professional service.

TRANSMISSION FLUID INDICATIONS

The appearance and odor of the transmission fluid can give valuable clues to the overall condition of the transmission. Always note the appearance of the fluid when you check the fluid level or change the fluid. Rub a small amount of fluid between your fingers to feel for grit and smell the fluid on the dipstick.

If The Fluid Appears	It Indicates
Clear and red colored	Normal operation
Discolored (extremely dark red or brownish) or smells burned	Band or clutch pack failure, usually caused by an overheated transmission. Hauling very heavy loads with insufficient power or failure to change the fluid often results in overheating. Do not confuse this appearance with newer fluids that have a darker red color and a strong odor (though not a burned odor).
Foamy or aerated (light in color and full of bubbles)	The level is too high (gear train is churning oil) An internal air leak (air is mixing with the fluid). Have the transmission checked professionally.
Solid residue in the fluid	Defective bands, clutch pack or bearings. Bits of band material or metal abrasives are clinging to the dipstick. Have the transmission checked professionally.
Varnish coating on the dipstick	The transmission fluid is overheating

Problem	Possible Cause	Correction
Slow initial engagement	1. Improper fluid level. 2. Damaged or improperly adjusted linkage. 3. Contaminated fluid. 4. Faulty clutch and band application, or oil control pressure system.	1. Add fluid as required. 2. Repair or adjust linkage. 3. Perform fluid level check. 4. Perform control pressure test.
Rough initial engagement in either forward or reverse	1. Improper fluid level. 2. High engine idle. 3. Looseness in the driveshaft, U-joints or engine mounts. 4. Incorrect linkage adjustment. 5. Faulty clutch or band application, or oil control pressure system. 6. Sticking or dirty valve body.	1. Perform fluid level check. 2. Adjust idle to specifications. 3. Repair as required. 4. Repair or adjust linkage. 5. Perform control pressure test. 6. Clean, repair or replace valve body.

Problem	Possible Cause	Correction
No drive, slips or chatters in first gear in D. All other gears normal.	1. Faulty one-way clutch.	1. Repair or replace one-way clutch.
No drive, slips or chatters in second gear.	1. Improper fluid level. 2. Damaged or improperly adjusted linkage. 3. Intermediate band out of adjustment. 4. Faulty band or clutch application, or oil pressure control system. 5. Faulty servo and/or internal leaks. 6. Dirty or sticking valve body. 7. Polished, glazed intermediate band or drum.	1. Perform fluid level check. 2. Repair or adjust linkage. 3. Adjust intermediate band. 4. Perform control pressure test. 5. Perform air pressure test. 6. Clean, repair or replace valve body. 7. Replace or repair as required.
No drive in any gear.	1. Improper fluid level. 2. Damaged or improperly adjusted linkage. 3. Faulty clutch or band application, or oil control pressure system. 4. Internal leakage. 5. Valve body loose. 6. Faulty clutches. 7. Sticking or dirty valve body.	1. Perform fluid level check. 2. Repair or adjust linkage. 3. Perform control pressure test. 4. Check and repair as required. 5. Tighten to specification. 6. Perform air pressure test. 7. Clean, repair or replace valve body.
No drive forward—reverse OK.	1. Improper fluid level 2. Damaged or improperly adjusted linkage. 3. Faulty clutch or band application, or oil pressure control system. 4. Faulty forward clutch or governor. 5. Valve body loose 6. Dirty or sticking valve body.	1. Perform fluid level check. 2. Repair or adjust linkage. 3. Perform control pressure test. 4. Perform air pressure test. 5. Tighten to specification. 6. Clean, repair or replace valve body.
No drive, slips or chatters in reverse—forward OK.	1. Improper fluid level 2. Damaged or improperly adjusted linkage. 3. Looseness in the drivehsaft, U-joints or engine mounts. 4. Bands or clutches out of adjustment. 5. Faulty oil pressure control system. 6. Faulty reverse clutch or servo. 7. Valve body loose. 8. Dirty or sticking valve body.	1. Perform fluid level check. 2. Repair or adjust linkage. 3. Repair as required. 4. Adjust as necessary. 5. Perform control pressure test. 6. Perform air pressure test. 7. Tighten to specifications. 8. Clean, repair or replace valve body.
Starts in high—in D drag or lockup at 1–2 shift point or in 2 or 1.	1. Improper fluid level. 2. Damaged or improperly adjusted linkage. 3. Faulty governor. 4. Faulty clutches and/or internal leaks. 5. Valve body loose. 6. Dirty, sticking valve body. 7. Poor mating of valve body to case mounting surfaces.	1. Perform fluid level check. 2. Repair or adjust linkage. 3. Repair or replace governor, clean screen. 4. Perform air pressure test. 5. Tighten to specifications. 6. Clean, repair or replace valve body. 7. Replace valve body or case.

Problem	Possible Cause	Correction
Starts up in 2nd or 3rd but no lockup at 1-2 shift points.	1. Improper fluid level. 2. Damaged or improperly adjusted linkage. 3. Improper band and/or clutch application, or oil pressure control system. 4. Faulty governor. 5. Valve body loose. 6. Dirty or sticking valve body. 7. Cross leaks between valve body and case mating surface.	1. Perform fluid level check. 2. Repair or adjust linkage. 3. Perform control pressure test. 4. Perform governor check. Replace or repair governor, clean screen. 5. Tighten to specification. 6. Clean, repair or replace valve body. 7. Replace valve body and/or case as required.
Shift points incorrect.	1. Improper fluid level. 2. Improper vacuum hose routing or leaks. 3. Improper operation of EGR system. 4. Linkage out of adjustment. 5. Improper speedometer gear installed. 6. Improper clutch or band application, or oil pressure control system. 7. Faulty governor. 8. Dirty or sticking valve body.	1. Perform fluid level check. 2. Correct hose routing. 3. Repair or replace as required. 4. Repair or adjust linkage. 5. Replace gear. 6. Perform shift test and control pressure test. 7. Repair or replace governor—clean screen. 8. Clean, repair or replace valve body.
No upshift at any speed in D.	1. Improper fluid level. 2. Vacuum leak to diaphragm unit. 3. Linkage out of adjustment. 4. Improper band or clutch application, or oil pressure control system. 5. Faulty governor. 6. Dirty or sticking valve bdy.	1. Perform fluid level check. 2. Repair vacuum line or hose. 3. Repair or adjust linkage. 4. Perform control pressure test. 5. Repair or replace governor, clean screen. 6. Clean, repair or replace valve body.
Shifts 1-3 in D.	1. Improper fluid level. 2. Intermediate band out of adjustment. 3. Faulty front servo and/or internal leaks. 4. Polished, glazed band or drum. 5. Improper band or clutch application, or oil pressure control system. 6. Dirty or sticking valve body.	1. Perform fluid level check. 2. Adjust band. 3. Perform air pressure test. Repair front servo and/or internal leaks. 4. Repair or replace band or drum. 5. Perform control pressure test. 6. Clean, repair or replace valve body.
Engine over-speeds on 2-3 shift.	1. Improper fluid level. 2. Linkage out of adjustment. 3. Improper band or clutch application, or oil pressure control system. 4. Faulty high clutch and/or intermediate servo. 5. Dirty or sticking valve body.	1. Perform fluid level check. 2. Repair or adjust linkage. 3. Perform control pressure test. 4. Perform air pressure test. Repair as required. 5. Clean repair or replace valve body.
Mushy 1-2 shift.	1. Improper fluid level 2. Incorrect engine idle and/or performance. 3. Improper linkage adjustment. 4. Intermediate band out of adjustment.	1. Perform fluid level check. 2. Tune, adjust engine idle as required. 3. Repair or adjust linkage. 4. Adjust intermediate band. 5. Perform control pressure test.

Problem	Possible Cause	Correction
Mushy 1-2 shift.	5. Improper band or clutch application, or oil pressure control system. 6. Faulty high clutch and/or intermediate servo release. 7. Polished, glazed band or drum. 8. Dirty or sticking valve body.	6. Perform air pressure test. Repair as required. 7. Repair or replace as required. 8. Clean, repair or replace valve body.
Rough 1-2 shift.	1. Improper fluid level. 2. Incorrect engine idle or performance. 3. Intermediate band out of adjustment. 4. Improper band or clutch application, or oil pressure control system. 5. Faulty intermediate servo. 6. Dirty or sticking valve body.	1. Perform fluid level check. 2. Tune, and adjust engine idle. 3. Adjust intermediate band. 4. Perform control pressure test. 5. Air pressure check intermediate servo. 6. Clean, repair or replace valve body.
Rough 2-3 shift	1. Improper fluid level. 2. Incorrect engine idle or performance. 3. Improper band or clutch application, or oil control pressure system. 4. Faulty intermediate servo apply and release and high clutch piston check ball. 5. Dirty or sticking valve body.	1. Perform fluid level check. 2. Tune and adjust engine idle. 3. Perform control pressure test. 4. Air pressure test the intermediate servo apply and release and the high clutch piston check ball. Repair as required. 5. Clean, repair or replace valve body.
Rough 3-1 shift at closed throttle in D.	1. Improper fluid level. 2. Incorrect engine idle or performance. 3. Improper linkage adjustment. 4. Improper clutch or band application or oil pressure control system. 5. Faulty governor operation. 6. Dirty or sticking valve body.	1. Perform fluid level check. 2. Tune, and adjust engine idle. 3. Repair or adjust linkage. 4. Perform control pressure test. 5. Perform governor test. Repair as required. 6. Clean, repair or replace valve body.
No forced downshifts.	1. Improper fluid level. 2. Linkage out of adjustment. 3. Improper clutch or band application, or oil pressure control system. 4. Faulty internal kickdown linkage. 5. Dirty or sticking valve body.	1. Perform fluid level check. 2. Repair or adjust linkage. 3. Perform control pressure test. 4. Repair internal kickdown linkage. 5. Clean, repair or replace valve body.
No 3-1 shift in D.	1. Improper fluid level. 2. Incorrect engine idle, or performance. 3. Faulty governor. 4. Dirty or sticking valve body.	1. Perform fluid level check. 2. Tune, and adjust engine idle. 3. Perform govenor check. Repair as required. 4. Clean, repair or replace valve body.
Runaway engine on 3-2 downshift.	1. Improper fluid level. 2. Linkage out of adjustment. 3. Intermediate band out of adjustment. 4. Improper band or clutch application, or oil pressure control system.	1. Perform fluid level check. 2. Repair or adjust linkage. 3. Adjust intermediate band. 4. Perform control pressure test. 5. Air pressure test check the intermediate servo. Repair servo and/or seals.

Problem	Possible Cause	Correction
Runaway engine on 3-2 downshift.	5. Faulty intermediate servo. 6. Polished, glazed band or drum. 7. Dirty or sticking valve body.	6. Repair or replace as required. 7. Clean, repair or replace valve body.
No engine braking in manual first gear.	1. Improper fluid level. 2. Linkage out of adjustment. 3. Bands or clutches out of adjustment. 4. Faulty oil pressure control system. 5. Faulty reverse servo. 6. Polished, glazed band or drum.	1. Perform fluid level check. 2. Repair or adjust linkage. 3. Adjust as necessary. 4. Perform control pressure test. 5. Perform air pressure test of reverse servo. Repair reverse clutch or rear servo as required. 6. Repair or replace as required.
No engine braking in manual second gear.	1. Improper fluid level. 2. Linkage out of adjustment. 3. Intermediate band out of adjustment. 4. Improper band or clutch application, or oil pressure control system. 5. Intermediate servo leaking. 6. Polished or glazed band or drum.	1. Perform fluid level check. 2. Repair or adjust linkage. 3. Adjust intermediate band. 4. Perform control pressure test. 5. Perform air pressure test of intermediate servo for leakage. Repair as required. 6. Repair or replace as required.
Transmission noisy—valve resonance.	1. Improper fluid level. 2. Linkage out of adjustment. 3. Improper band or clutch application, or oil pressure control system. 4. Cooler lines grounding. 5. Dirty sticking valve body. 6. Internal leakage or pump cavitation.	1. Perform fluid level check. 2. Repair or adjust linkage. 3. Perform control pressure test. 4. Free up cooler lines. 5. Clean, repair or replace valve body. 6. Repair as required.
Transmission overheats.	1. Improper fluid level. 2. Incorrect engine idle, or performance. 3. Improper clutch or band application, or oil pressure control system. 4. Restriction in cooler or lines. 5. Seized one-way clutch. 6. Dirty or sticking valve body.	1. Perform fluid level check. 2. Tune, or adjust engine idle. 3. Perform control pressure test. 4. Repair restriction. 5. Replace one-way clutch. 6. Clean, repair or replace valve body.
Transmission fluid leaks.	1. Improper fluid level. 2. Leakage at gasket, seals, etc. 3. Vacuum diaphragm unit leaking.	1. Perform fluid level check. 2. Remove all traces of lube on exposed surfaces of transmission. Check the vent for free breathing. Operate transmission at normal temperatures and inspect for leakage. Repair as required. 3. Replace diaphragm.

Automatic Transmission Troubleshooting

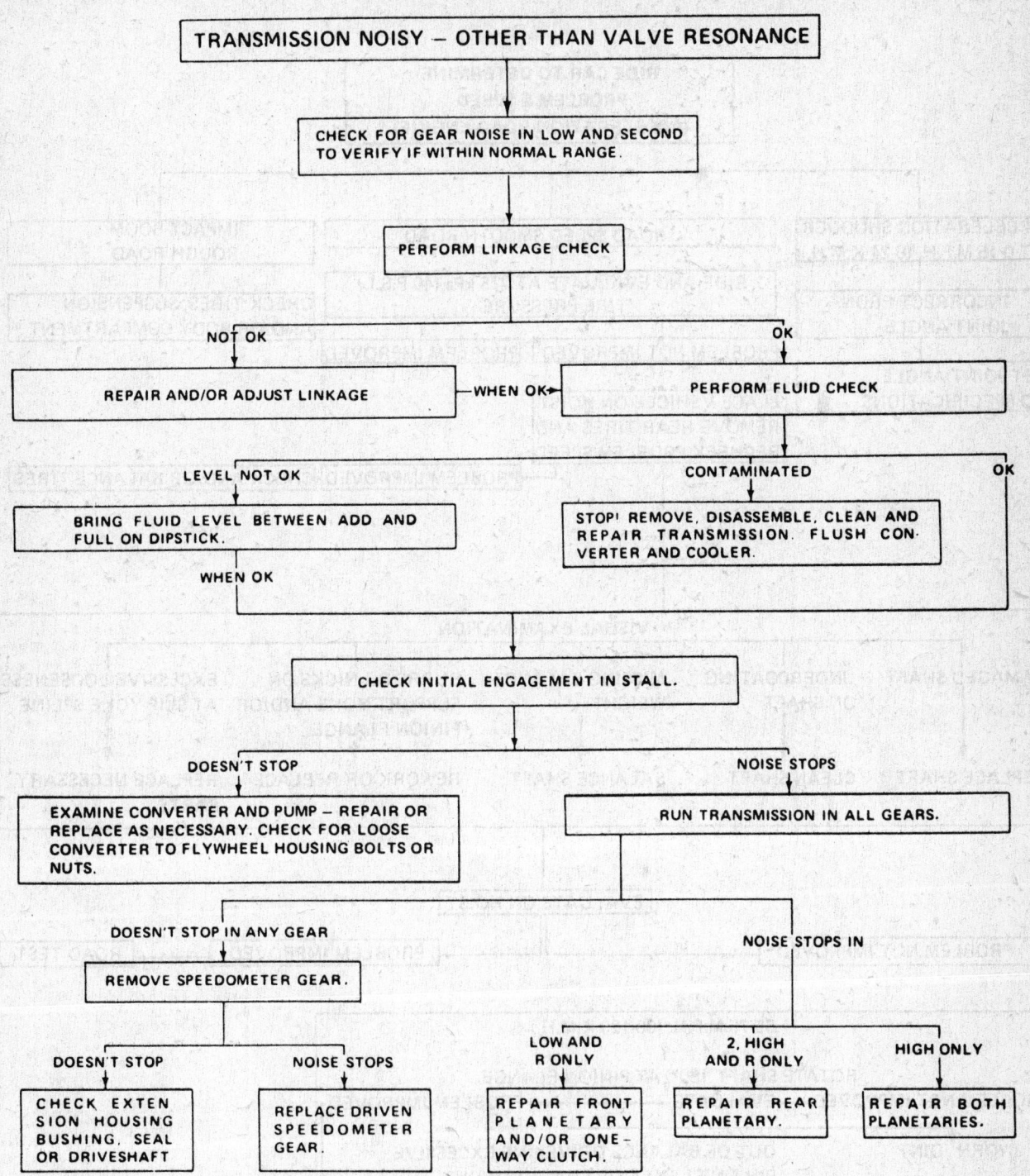

TRANSMISSION NOISY — OTHER THAN VALVE RESONANCE

↓

CHECK FOR GEAR NOISE IN LOW AND SECOND TO VERIFY IF WITHIN NORMAL RANGE.

↓

PERFORM LINKAGE CHECK

NOT OK → REPAIR AND/OR ADJUST LINKAGE — **WHEN OK** →

OK → PERFORM FLUID CHECK

LEVEL NOT OK → BRING FLUID LEVEL BETWEEN ADD AND FULL ON DIPSTICK. — **WHEN OK**

CONTAMINATED → STOP! REMOVE, DISASSEMBLE, CLEAN AND REPAIR TRANSMISSION. FLUSH CONVERTER AND COOLER.

OK

CHECK INITIAL ENGAGEMENT IN STALL.

DOESN'T STOP → EXAMINE CONVERTER AND PUMP — REPAIR OR REPLACE AS NECESSARY. CHECK FOR LOOSE CONVERTER TO FLYWHEEL HOUSING BOLTS OR NUTS.

NOISE STOPS → RUN TRANSMISSION IN ALL GEARS.

DOESN'T STOP IN ANY GEAR → REMOVE SPEEDOMETER GEAR.

DOESN'T STOP → CHECK EXTENSION HOUSING BUSHING, SEAL OR DRIVESHAFT

NOISE STOPS → REPLACE DRIVEN SPEEDOMETER GEAR.

NOISE STOPS IN

LOW AND R ONLY → REPAIR FRONT PLANETARY AND/OR ONE-WAY CLUTCH.

2, HIGH AND R ONLY → REPAIR REAR PLANETARY.

HIGH ONLY → REPAIR BOTH PLANETARIES.

Driveshaft Troubleshooting
Vibration, Roughness, Rumble and/or Boom

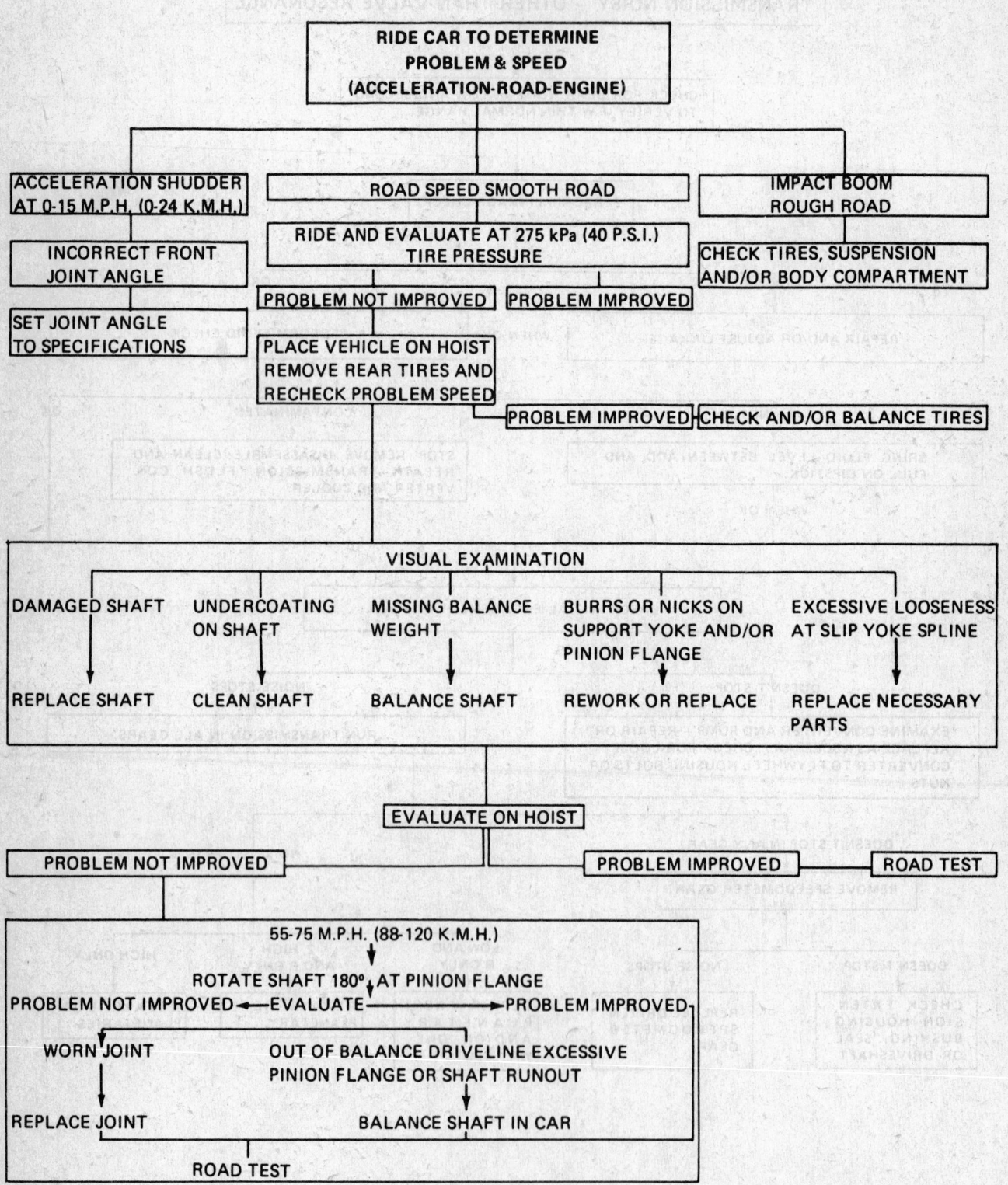

Universal Joint Troubleshooting

Problem	Possible Cause	Correction
Leak at front slip yoke. **NOTE:** An occasional drop of lubricant leaking from splined yoke is normal and requires no attention.	1. Rough outside surface on splined yoke. 2. Defective transmission rear oil seal.	1. Replace seal if cut by burrs on yoke. Minor burrs can be smoothed by careful use of crocus cloth or honing with a fine stone. Replace yoke if outside surface is rough or burred badly. 2. Replace transmission rear oil seal. 3. Bring transmission oil up to proper level after correction.
Knock in drive line, clunking noise when car is operated under floating condition at 10 mph in high gear or neutral.	1. Worn or damaged universal joints. 2. Side gear hub counterbore in differential worn oversize.	1. Disassemble universal joints, inspect and replace worn or damaged parts. 2. Replace differential case and/or side gears as required.
Ping, snap or click in drive line. **NOTE:** Usually occurs on initial load application after transmission has been put into gear, either forward or reverse.	1. Loose upper or lower control arm bushing bolts. 2. Loose companion flange.	1. Tighten bolts to specified torque. 2. Remove companion flange, turn 180° from its original position, apply white lead to splines and reinstall. Tighten pinion nut to specified torque.

Front Wheel Drive Halfshaft Troubleshooting

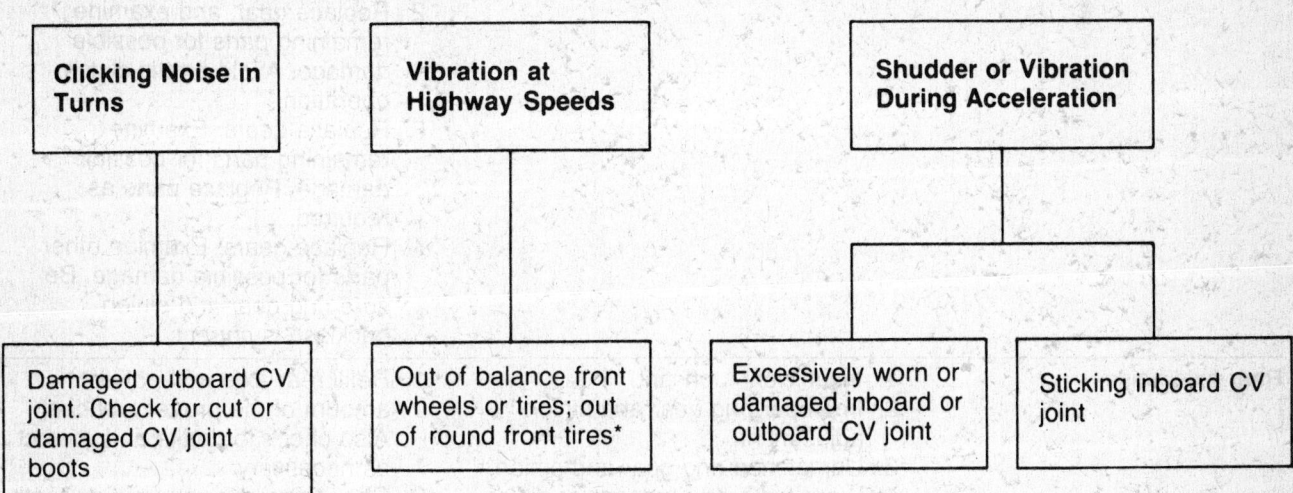

*Halfshafts do not usually contribute to rotational vibrations.

Drive Axle Troubleshooting

Condition	Possible Cause	Correction
Rear wheel noise	1. Loose wheel. 2. Spalled wheel bearing cup or cone. 3. Defective or brinelled wheel bearing. 4. Excessive axle shaft endplay. 5. Bent or sprng axle shaft flange.	1. Tighten loose wheel nuts. 2. Check rear wheel bearings. If spalled or worn, replace. 3. Defective or brinelled bearings must be replaced. Check rear axle shaft end play. 4. Readjust axle shaft end play. 5. Replace bent or sprung axle shaft.
Scoring of differential gears and pinions	1. Insufficient lubrication. 2. Improper grade of lubricant. 3. Excessive spinning of one wheel.	1. Replace scored gears. Scoring marks on the pressure face of gear teeth or in the bore are caused by instantaneous fusing of the mating surfaces. Scored gears should be replaced. Fill rear axle to required capacity with proper lubricant. 2. Replace scored gears. Inspect all gears and bearings for possible damage. Clean and refill axle to required capacity with proper lubricant. 3. Replace scored gears. Inspect all gears, pinion bores and shaft for scoring, or bearings for possible damage.
Tooth breakage (ring gear and pinion)	1. Overloading. 2. Erratic clutch operation. 3. Ice-spotted pavements. 4. Improper adjustments.	1. Replace gear. Examine other gears and bearings for possible damage. Avoid future overloading. 2. Replace gear, and examine remaining parts for possible damage. Avoid erratic clutch operation. 3. Replace gears. Examine remaining parts for possible damage. Replace parts as required. 4. Replace gears. Examine other parts for possible damage. Be sure ring gear and pinion backlash is correct.
Rear axle noise	1. Insufficient lubricant. 2. Improper ring gear and pinion adjustment. 3. Unmatched ring gear and pinion. 4. Worn teeth on ring gear or pinion. 5. End-play in drive pinion bearings. 6. Side play in differential bearings. 7. Incorrect drive gearlash. 8. Limited-slip differential—moan and chatter.	1. Refill rear axle with correct amount of the proper lubricant. Also check for leaks and correct as necessary. 2. Check ring gear and pinion tooth contact. 3. Remove unmatched ring gear and pinion. Replace with a new matched gear and pinion set. 4. Check teeth on ring gear and pinion for contact. If necessary, replace with new matched set. 5. Adjust drive pinion bearing preload.

Problem	Possible Cause	Correction
Rear axle noise		6. Adjust differential bearing preload. 7. Correct drive gear lash. 8. Drain and flush lubricant. Refill with proper lubricant.
Loss of lubricant	1. Lubricant level too high. 2. Worn axle shaft oil seals. 3. Cracked rear axle housing. 4. Worn drive pinion oil seal. 5. Scored and worn companion flange. 6. Clogged vent. 7. Loose carrier housing bolts or housing cover screws.	1. Drain excess lubricant. 2. Replace worn oil seals with new ones. Prepare new seals before replacement. 3. Repair or replace housing as required. 4. Replace worn drive pinion oil seal with a new one. 5. Replace worn or scored companion flange and oil seal. 6. Remove obstructions. 7. Tighten bolts or cover screws to specifications and fill to correct level with proper lubricant.
Overheating of unit	1. Lubricant level too low. 2. Incorrect grade of lubricant. 3. Bearing adjusted too tightly. 4. Excessive wear in gears. 5. Insufficient ring gear-to-pinion clearance.	1. Refill rear axle. 2. Drain, flush and refill rear axle with correct amount of the proper lubricant. 3. Readjust bearings. 4. Check gears for excessive wear or scoring. Replace as necessary. 5. Readjust ring gear and pinion backlash and check gears for possible scoring.

CHASSIS

Shock Absorber and Rear Spring Troubleshooting

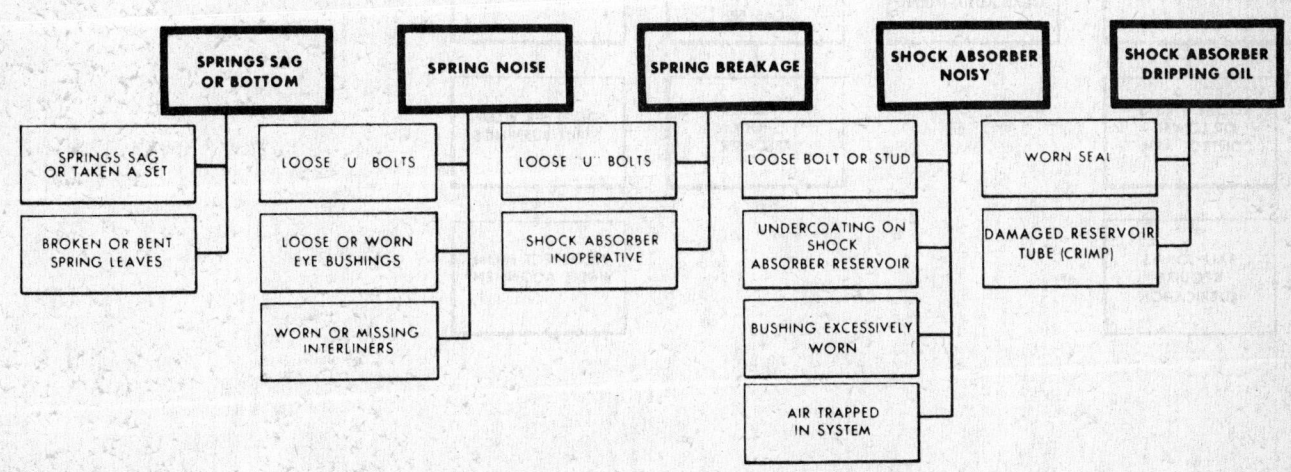

Front Suspension and Steering Linkage
Troubleshooting—Rear Wheel Drive

FRONT END NOISE	EXCESSIVE PLAY IN STEERING	FRONT WHEEL SHIMMY	INSTABILITY	HARD STEERING	CAR PULLS TO ONE SIDE
LOOSE OR WORN FRONT WHEEL BEARINGS	LOOSE OR WORN FRONT WHEEL BEARINGS	LOOSE OR WORN WHEEL BEARINGS	LOW OR UNEVEN TIRE PRESSURE	LOW OR UNEVEN TIRE PRESSURE	BROKEN REAR SPRING
LOOSE OR WORN SHOCK ABSORBER MOUNTING OR SHOCK ABSORBER	LOOSE OR WORN STEERING SHAFT COUPLING	TIRE, WHEEL OUT OF BALANCE	LOOSE WHEEL BEARINGS	LACK OF ASSIST OF POWER STEERING SYSTEM	POWER STEERING CONTROL VALVE OUT OF ADJUSTMENT
LOOSE STEERING GEAR TO FRAME MOUNTING BOLTS	LOOSE STEERING GEAR TO FRAME MOUNTING BOLTS	UNEVEN TIRE WEAR, OR EXCESSIVELY WORN TIRES	BROKEN REAR SPRING	STEERING GEAR NOT ADJUSTED	LOOSE OR WORN STRUT BUSHINGS
STEERING KNUCKLE ARM CONTACTING THE LOWER CONTROL ARM WHEEL STOP	WORN TIE ROD ENDS	WORN TIE ROD ENDS	SHOCK ABSORBER INOPERATIVE	INCORRECT FRONT WHEEL ALIGNMENT (PARTICULARLY CASTER)	INCORRECT FRONT WHEEL ALIGNMENT (PARTICULARLY CASTER)
WORN UPPER CONTROL ARM BUSHINGS	WORN IDLER ARM BUSHING	LOOSE OR WORN STRUT BUSHINGS	IMPROPER STEERING CROSS SHAFT ADJUSTMENT		
WORN LOWER CONTROL ARM SHAFT BUSHINGS	WORN STEERING GEAR PARTS	LOOSE OR WORN UPPER CONTROL ARM BALL JOINTS	STEERING GEAR NOT CENTERED		
LOOSE OR WORN STRUT BUSHINGS	INCORRECT STEERING GEAR ADJUSTMENT	INCORRECT FRONT WHEEL ALIGNMENT (PARTICULARLY CASTER)	WORN IDLER ARM BUSHING		
LOOSE STRUTS OR LOWER CONTROL ARM		WORN SHOCK ABSORBER	LOOSE OR WORN STRUT BUSHINGS		
BALL JOINTS REQUIRE LUBRICATION			INCORRECT FRONT WHEEL ALIGNMENT		

Suspension and Steering Linkage
Troubleshooting— Front Wheel Drive

NOISE	INSTABILITY	EXCESSIVE PLAY IN STEERING	HARD STEERING	CAR PULLS TO ONE SIDE
(DRIVE OR COAST) ROAD/TIRE NOISE	LOW OR UNEVEN TIRE PRESSURE	LOOSE OR WORN HUB BEARINGS	LOW OR UNEVEN TIRE PRESSURE	LOW OR UNEVEN TIRE PRESSURE
(PRONOUNCED ON TURNS) FRONT HUB BEARINGS	LOOSE OR WORN HUB BEARINGS	LOOSE OR WORN STEERING SHAFT COUPLING	LACK OF ASSIST OF POWER STEERING SYSTEM	WHILE BRAKING BRAKE SERVICE
(ON ACCELERATION OR DECELERATION) FRONT WHEEL BEARINGS TRANSAXLE GEARS	BROKEN SPRING OR BENT REAR SUSPENSION	LOOSE STEERING GEAR MOUNTING BOLTS	STEERING GEAR LOW ON LUBRICANT	BROKEN FRONT OR REAR SPRING OR BENT REAR SUSPENSION
(CLUNK-ON ACCELERATION OR DECELERATION) TRANSAXLE BEARINGS OR GEARS	INOPERATIVE SHOCK ABSORBING (STRUTS)	WORN TIE ROD ENDS	INCORRECT WHEEL ALIGNMENT	LOOSE LOWER CONTROL ARM
(CLICKING NOISE ON TURNS) EXCESSIVE WEAR OR BROKEN C.V. JOINT	IMPROPER STEERING GEAR ADJUSTMENT			INCORRECT WHEEL ALIGNMENT
	LOOSE OR WORN STRUT			UNBALANCED STEERING GEAR VALVE (POWER)
	INCORRECT WHEEL ALIGNMENT FRONT OR REAR			

TROUBLESHOOTING AND DIAGNOSIS

Tapered Wheel Bearing Troubleshooting

CONSIDER THE FOLLOWING FACTORS WHEN DIAGNOSING BEARING CONDITION:

1. GENERAL CONDITION OF ALL PARTS DURING DISASSEMBLY AND INSPECTION.

2. CLASSIFY THE FAILURE WITH THE AID OF THE ILLUSTRATIONS.

3. DETERMINE THE CAUSE.

4. MAKE ALL REPAIRS FOLLOWING RECOMMENDED PROCEDURES.

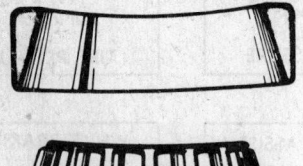

GOOD BEARING

BENT CAGE

CAGE DAMAGE DUE TO IMPROPER HANDLING OR TOOL USAGE.

REPLACE BEARING.

BENT CAGE

CAGE DAMAGE DUE TO IMPROPER HANDLING OR TOOL USAGE.

REPLACE BEARING.

GALLING

METAL SMEARS ON ROLLER ENDS DUE TO OVERHEAT, LUBRICANT FAILURE OR OVERLOAD.

REPLACE BEARING — CHECK SEALS AND CHECK FOR PROPER LUBRICATION.

ABRASIVE STEP WEAR

PATTERN ON ROLLER ENDS CAUSED BY FINE ABRASIVES.

CLEAN ALL PARTS AND HOUSINGS, CHECK SEALS AND BEARINGS AND REPLACE IF LEAKING, ROUGH OR NOISY.

ETCHING

BEARING SURFACES APPEAR GRAY OR GRAYISH BLACK IN COLOR WITH RELATED ETCHING AWAY OF MATERIAL USUALLY AT ROLLER SPACING.

REPLACE BEARINGS — CHECK SEALS AND CHECK FOR PROPER LUBRICATION.

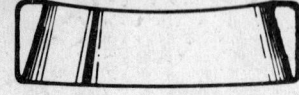

MISALIGNMENT

OUTER RACE MISALIGNMENT DUE TO FOREIGN OBJECT.

CLEAN RELATED PARTS AND REPLACE BEARING. MAKE SURE RACES ARE PROPERLY SEATED.

INDENTATIONS

SURFACE DEPRESSIONS ON RACE AND ROLLERS CAUSED BY HARD PARTICLES OF FOREIGN MATERIAL.

CLEAN ALL PARTS AND HOUSINGS, CHECK SEALS AND REPLACE BEARINGS IF ROUGH OR NOISY.

FATIGUE SPALLING

FLAKING OF SURFACE METAL RESULTING FROM FATIGUE.

REPLACE BEARING — CLEAN ALL RELATED PARTS.

Tapered Wheel Bearing Troubleshooting

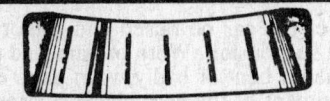

BRINELLING

SURFACE INDENTATIONS IN RACEWAY CAUSED BY ROLLERS EITHER UNDER IMPACT LOADING OR VIBRATION WHILE THE BEARING IS NOT ROTATING.

REPLACE BEARING IF ROUGH OR NOISY.

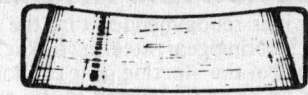

CAGE WEAR

WEAR AROUND OUTSIDE DIAMETER OF CAGE AND ROLLER POCKETS CAUSED BY ABRASIVE MATERIAL AND INEFFICIENT LUBRICATION. CHECK SEALS AND REPLACE BEARINGS.

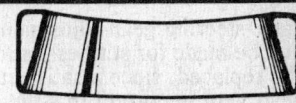

ABRASIVE ROLLER WEAR

PATTERN ON RACES AND ROLLERS CAUSED BY FINE ABRASIVES.

CLEAN ALL PARTS AND HOUSINGS, CHECK SEALS AND BEARINGS AND REPLACE IF LEAKING, ROUGH OR NOISY.

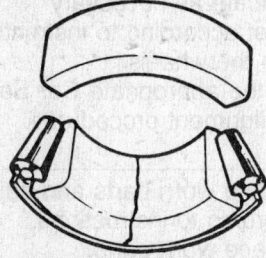

CRACKED INNER RACE

RACE CRACKED DUE TO IMPROPER FIT, COCKING, OR POOR BEARING SEATS.

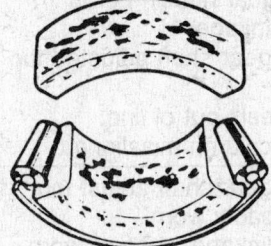

SMEARS

SMEARING OF METAL DUE TO SLIPPAGE. SLIPPAGE CAN BE CAUSED BY POOR FITS, LUBRICATION, OVERHEATING, OVERLOADS OR HANDLING DAMAGE.

REPLACE BEARINGS, CLEAN RELATED PARTS AND CHECK FOR PROPER FIT AND LUBRICATION.

REPLACE SHAFT IF DAMAGED.

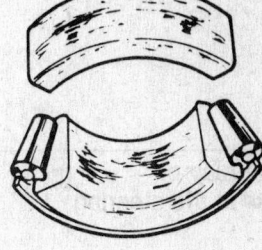

FRETTAGE

CORROSION SET UP BY SMALL RELATIVE MOVEMENT OF PARTS WITH NO LUBRICATION.

REPLACE BEARING. CLEAN RELATED PARTS. CHECK SEALS AND CHECK FOR PROPER LUBRICATION.

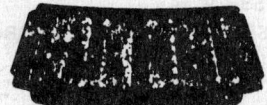

HEAT DISCOLORATION

HEAT DISCOLORATION CAN RANGE FROM FAINT YELLOW TO DARK BLUE RESULTING FROM OVERLOAD OR INCORRECT LUBRICANT.

EXCESSIVE HEAT CAN CAUSE SOFTENING OF RACES OR ROLLERS.

TO CHECK FOR LOSS OF TEMPER ON RACES OR ROLLERS A SIMPLE FILE TEST MAY BE MADE. A FILE DRAWN OVER A TEMPERED PART WILL GRAB AND CUT META, WHEREAS, A FILE DRAWN OVER A HARD PART WILL GLIDE READILY WITH NO METAL CUTTING.

REPLACE BEARINGS IF OVER HEATING DAMAGE IS INDICATED. CHECK SEALS AND OTHER PARTS.

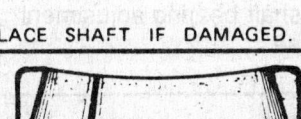

STAIN DISCOLORATION

DISCOLORATION CAN RANGE FROM LIGHT BROWN TO BLACK CAUSED BY INCORRECT LUBRICANT OR MOISTURE.

RE-USE BEARINGS IF STAINS CAN BE REMOVED BY LIGHT POLISHING OR IF NO EVIDENCE OF OVERHEATING IS OBSERVED.

CHECK SEALS AND RELATED PARTS FOR DAMAGE.

Manual Steering Troubleshooting

INSPECTION AND ALIGNMENT

Before any steering gear adjustments are made, it is recommended that the front end of the car be raised and a thorough inspection be made for stiffness or lost motion in steering gear, steering linkage and front suspension. Worn or damaged parts should be replaced, since a satisfactory adjustment of the steering gear cannot be obtained if bent or badly worn parts exist.

It is also very important that the steering gear be properly aligned in the car. Misalignment of the gear places a stress on the steering worm shaft, therefore a proper adjustment is impossible. To align the steering gear, loosen the mounting bolts to permit the gear to align itself. Check the steering gear mounting seat, and if there is a gap at any of the mounting bolts, proper alignment may be obtained by placing shims where excessive gap appears. Tighten the steering gear bolts. Alignment of the gear in the car is very important and should be done carefully so that a satisfactory, trouble-free gear adjustment may be obtained.

Condition	Possible Cause	Corrective Action
Hard steering	1. Low or uneven tire pressure. 2. Insufficient lubricant in the steering gear housing or in steering linkage. 3. Steering gear shaft adjusted too tight. 4. Front wheels out of line. 5. Steering column misaligned.	1. Inflate tires to recommended pressures. 2. Lubricate as necessary. 3. Adjust according to instructions. 4. Align the wheels. 5. See the appropriate Car Section for alignment procedures.
Excessive play or looseness in the steering wheel	1. Steering gear shaft adjust too loose or badly worn. 2. Steering linkage loose or worn. 3. Front wheel bearings improperly adjusted. 4. Steering arm loose on steering gear shaft. 5. Steering gear housing attaching bolts loose. 6. Steering arms loose at steering knuckles. 7. Worn ball joints. 8. Worm shaft bearing adjustment too loose.	1. Replace worn parts and adjust according to instructions. 2. Replace worn parts. 3. Adjust according to instructions. 4. Inspect for damage to the gear shaft and steering arm, replace parts as necessary. 5. Tighten attaching bolts to specifications. 6. Tighten according to specifications. 7. Replace the ball joints as necessary. 8. Adjust worm bearing preload according to instructions.

Power Steering Systems Troubleshooting

Condition	Possible Cause	Corrective Action
Hard steering	1. Improper tire pressure. 2. Loose pump drive belt. 3. Low or incorrect fluid. 4. Loose, bent or poorly lubricated front end parts. 5. Improper front end alignment. 6. Bind in steering column or linkage.	1. Inflate tires to recommended pressures. 2. Tighten or replace belt. 3. Refill reservoir with proper fluid; check for leaks; 4. Tighten or replace parts; lubricate at all fittings. 5. Align front end.

Condition	Possible Cause	Correction Action
Hard steering	7. Air in hydraulic system. 8. Low pump output or leaks in system. 9. Obstruction in lines. 10. Pump valves sticking or out of adjustment.	6. Disassemble and inspect component parts. Repair or replace as necessary. 7. Bleed system, refill and check for leaks. 8. Disassemble pump, check for worn or damaged parts. Check for leaks in the system. 9. Clean or replace lines. 10. Replace or adjust valves.
Loose steering	1. Loose wheel bearings 2. Faulty shocks. 3. Worn linkage components. 4. Loose steering gear mounting or linkage points. 5. Steering mechanism worn or improperly adjusted. 6. Valve spool improperly adjusted	1. Adjust wheel bearings. 2. Relace shocks. 3. Replace worn components. 4. Tighten mountings or linkage. 5. Replace and/or adjust mechanism. 6. Adjust valve spool.
Veer or wander	1. Improper tire pressure. 2. Improper front end alignment. 3. Dragging brakes. 4. Bent frame. 5. Improper rear end alignment. 6. Faulty shocks or springs. 7. Loose or bent front end components. 8. Play in Pitman arm. 9. Loose wheel bearings. 10. Binding Pitman arm. 11. Spool valve sticking or improperly adjusted.	1. Inflate tires to recommended pressures. 2. Align front end. 3. Inspect, replace and/or adjust brakes. 4. Straighten frame. 5. Inspect shocks and control arm torque. Replace and/or adjust as necessary. 6. Replace as necessary. 7. Replace as necessary. 8. Inspect bushings and arm. Replace as necessary. 9. Adjust to specifications. 10. Replace arm. 11. Adjust or replace as necessary.
Wheel oscillation	1. Improper tire pressure. 2. Loose wheel bearings. 3. Improper front end alignment. 4. Bent spindle. 5. Worn, bent or broken front end components. 6. Tires out of round or out of balance. 7. Excessive lateral runout in disc brake rotor.	1. Inflate tires to recommended pressures. 2. Adjust to specifications. 3. Align front end. 4. Replace spindle. 5. Inspect, repair or replace as necessary. 6. Replace or balance tires. 7. Reface or replace rotor.
Noises	1. Loose belts. 2. Low fluid, air in system. 3. Foreign matter in system. 4. Improper lubrication. 5. Interference or chafing in linkage. 6. Steering gear mountings loose. 7. Incorrect adjustment or wear in gear box. 8. Faulty valves or wear in pump.	1. Replace and/or adjust belts. 2. Refill and check for leaks. 3. Disassemble and clean system. 4. Lubricate all fittings. 5. Disassemble, inspect, replace or adjust components. 6. Tighten mountings. 7. Disassemble, inspect, repair, replace and/or adjust parts. 8. Replace parts as necessary.

How To Read Tire Wear

The way your tires wear is a good indicator of other parts of the suspension. Abnormal wear patterns are often caused by the need for simple tire maintenance, or for front end alignment.

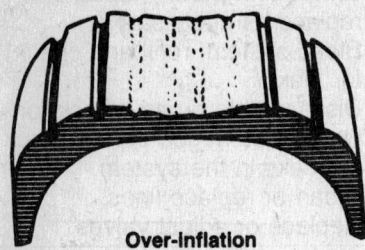

Over-inflation

Excessive wear at the center of the tread indicates that the air pressure in the tire is consistently too high. The tire is riding on the center of the tread and wearing it prematurely. Occasionally, this wear pattern can result from outrageously wide tires on narrow rims. The cure for this is to replace either the tires or the wheels.

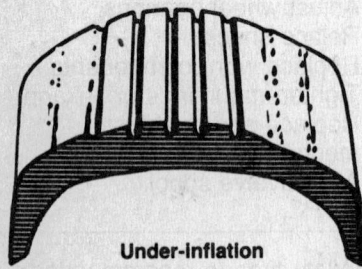

Under-inflation

This type of wear usually results from consistent under-inflation. When a tire is under-inflated, there is too much contact with the road by the outer treads, which wear prematurely. When this type of wear occurs, and the tire pressure is known to be consistently correct, a bent or worn steering component or the need for wheel alignment could be indicated.

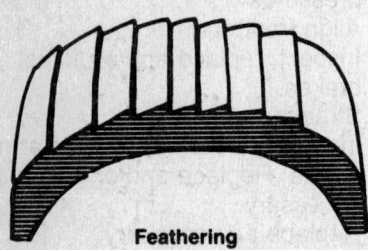

Feathering

Feathering is a condition when the edge of each tread rib develops a slightly rounded edge on one side and a sharp edge on the other. By running your hand over the tire, you can usually feel the sharper edges before you'll be able to see them. The most common causes of feathering are incorrect toe-in setting or deteriorated bushings in the front suspension.

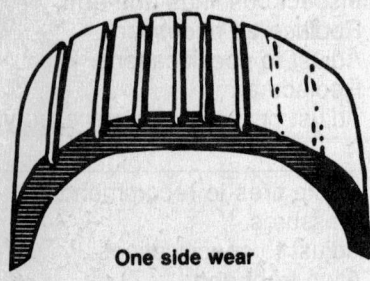

One side wear

When an inner or outer rib wears faster than the rest of the tire, the need for wheel alignment is indicated. There is excessive camber in the front suspension, causing the wheel to lean too much putting excessive load on one side of the tire. Misalignment could also be due to sagging springs, worn ball joints, or worn control arm bushings. Be sure the vehicle is loaded the way it's normally driven when you have the wheels aligned.

Cupping

Cups or scalloped dips appearing around the edge of the tread almost always indicate worn (sometimes bent) suspension parts. Adjustment of wheel alignment alone will seldom cure the problem. Any worn component that connects the wheel to the suspension can cause this type of wear. Occasionally, wheels that are out of balance will wear like this, but wheel imbalance usually shows up as bald spots between the outside edges and center of the tread.

Second-rib wear

Second-rib wear is usually found only in radial tires, and appears where the steel belts end in relation to the tread. It can be kept to a minimum by paying careful attention to tire pressure and frequently rotating the tires. This is often considered normal wear but excessive amounts indicate that the tires are too wide for the wheels.

Drum Brake Troubleshooting

Condition	Possible Cause	Correction Action
Pedal goes to floor	1. Fluid low in reservoir. 2. Air in hydraulic brake system. 3. Improperly adjusted brake. 4. Leaking wheel cylinders. 5. Loose or broken brake lines. 6. Leaking or worn master cylinder. 7. Excessively worn brake lining.	1. Fill and bleed master cylinder. 2. Fill and bleed hydraulic brake system. 3. Repair or replace self-adjuster as required. 4. Recondition or replace wheel cylinder and replace both brake shoes. 5. Tighten all brake fittings or replace brake line. 6. Recondition or replace master cylinder and bleed hydraulic system. 7. Reline and adjust brakes.
Spongy brake pedal	1. Air in hydraulic system. 2. Improper brake fluid (low boiling point). 3. Excessively worn or cracked brake drums. 4. Broken pedal pivot bushing.	1. Fill master cylinder and bleed hydraulic system. 2. Drain, flush and refill with brake fluid. 3. Replace all faulty brake drums. 4. Replace nylon pivot bushing.
Brakes pulling	1. Contaminated lining. 2. Front end out of alignment. 3. Incorrect brake adjustment. 4. Unmatched brake lining. 5. Brake drums out of round. 6. Brake shoes distorted. 7. Restricted brake hose or line. 8. Broken rear spring.	1. Replace contaminated brake lining. 2. Align front end. 3. Adjust brakes and check fluid. 4. Match primary, secondary with same type of lining on all wheels. 5. Grind or replace brake drums. 6. Replace faulty brake shoes. 7. Replace plugged hose or brake line. 8. Replace broken spring.
Squealing brakes	1. Glazed brake lining. 2. Saturated brake lining. 3. Weak or broken brake shoe retaining spring. 4. Broken or weak brake shoe return spring. 5. Incorrect brake lining. 6. Distorted brake shoes. 7. Bent support plate. 8. Dust in brakes or scored brake drums.	1. Cam grind or replace brake lining. 2. Replace saturated lining. 3. Replace retaining spring. 4. Replace return spring. 5. Install matched brake lining. 6. Replace brake shoes. 7. Replace support plate. 8. Blow out brake assembly with compressed air and grind brake drums.
Chirping brakes	1. Out of round drum or eccentric axle flange pilot.	1. Repair as necessary, and lubricate support plate contact areas (6 places).
Dragging brakes	1. Incorrect wheel or parking brake adjustment. 2. Parking brakes engaged. 3. Weak or broken brake shoe return spring. 4. Brake pedal binding. 5. Master cylinder cup sticking. 6. Obstructed master cylinder relief port. 7. Saturated brake lining. 8. Bent or out of round brake drum.	1. Adjust brake and check fluid. 2. Release parking brakes. 3. Replace brake shoe return spring. 4. Free up and lubricate brake pedal and linkage. 5. Recondition master cylinder. 6. Use compressed air and blow out relief port. 7. Replace brake lining. 8. Grind or replace faulty brake drum.

Condition	Possible Cause	Corrective Action
Hard pedal	1. Brake booster inoperative. 2. Incorrect brake lining. 3. Restricted brake line or hose. 4. Frozen brake pedal linkage.	1. Replace brake booster. 2. Install matched brake lining. 3. Clean out or replace brake line or hose. 4. Free up and lubricate brake linkage.
Wheel locks	1. Contaminated brake lining. 2. Loose or torn brake lining. 3. Wheel cylinder cups sticking. 4. Incorrect wheel bearing adjustment.	1. Reline both front or rear of all four brakes. 2. Replace brake lining. 3. Recondition or replace wheel cylinder. 4. Clean, pack and adjust wheel bearings.
Brakes fade (high speed)	1. Incorrect lining. 2. Overheated brake drums. 3. Incorrect brake fluid (low boiling temperature) 4. Saturated brake lining.	1. Replace lining. 2. Inspect for dragging brakes. 3. Drain, flush, refill and bleed hydraulic brake system. 4. Reline both front or rear or all four brakes.
Pedal pulsates	1. Bent or out of round brake drum.	1. Grind or replace brake drums.
Brake chatter and shoe knock	1. Out of round brake drum. 2. Loose support plate. 3. Bent support plate. 4. Distorted brake shoes. 5. Machine grooves in contact face of brake drum. (Shoe Knock). 6. Contaminated brake lining.	1. Grind or replace brake drums. 2. Tighten support plate bolts to proper specifications. 3. Replace support plate. 4. Replace brake shoes. 5. Grind or replace brake drum. 6. Replace either front or rear or all four linings.
Brakes do not self adjust	1. Adjuster screw frozen in thread. 2. Adjuster screw corroded at thrust washer. 3. Adjuster level does not engage star wheel. 4. Adjuster installed on wrong wheel.	1. Clean and free-up all thread areas. 2. Clean threads and replace thrust washer if necessary. 3. Repair, free up or replace adjusters as required. 4. Install correct adjuster parts.

Disc Brake Troubleshooting

Condition	Possible Cause	Correction Action
Noise—Groan—Brake noise emanating when slowly releasing brakes (creep-groan).	1. Not detrimental to function of disc brakes—no corrective action required. (Indicate to operator this noise may be eliminated by slightly increasing or decreasing brake pedal efforts.)	
Rattle—Brake noise or rattle emanating at low speeds on rough roads, (front wheels only).	1. Shoe anti-rattle spring missing or not properly positioned. 2. Excessive clearance between shoe and caliper.	1. Install new anti-rattle spring or position properly. 2. Install new shoe and lining assemblies.

Condition	Possible Cause	Corrective Action
Scraping	1. Mounting bolts too long. 2. Loose wheel bearings.	1. Install mounting bolts of correct length. 2. Readjust wheel bearings to correct specifications.
Front brakes heat up during driving and fail to release	1. Operator riding brake pedal. 2. Stop light switch improperly adjusted. 3. Sticking pedal linkage. 4. Frozen or seized piston. 5. Residual pressure valve in master cylinder. 6. Power brake malfunction.	1. Instruct owner how to drive with disc brakes. 2. Adjust stop light to allow full return of pedal. 3. Free up sticking pedal linkage. 4. Disassemble caliper and free up piston. 5. Remove valve. 6. Replace.
Leaky wheel cylinder	1. Damaged or worn caliper piston seal. 2. Scores or corrosion on surface of cylinder bore.	1. Disassemble caliper and install new seal. 2. Disassemble caliper and hone cylinder bore. Install new seal.
Grabbing or uneven brake action	1. Causes listed under "Pull." 2. Power brake malfunction.	1. Corrections listed under "Pull". 2. Replace.
Brake pedal can be depressed without braking effect	1. Air in hydraulic system or improper bleeding procedure. 2. Leak past primary cup in master cylinder. 3. Leak in system. 4. Rear brakes out of adjustment. 5. Bleeder screw open.	1. Bleed system. 2. Recondition master cylinder. 3. Check for leak and repair as required. 4. Adjust rear brakes. 5. Close bleeder screw and bleed entire system.
Excessive pedal travel	1. Air, leak, or insufficient fluid in system or caliper. 2. Warped or excessively tapered shoe and lining assembly. 3. Excessive disc runout. 4. Rear brake adjustment required. 5. Loose wheel bearing adjustment. 6. Damaged caliper piston seal. 7. Improper brake fluid (boil). 8. Power brake malfunction.	1. Check system for leaks and bleed. 2. Install new shoe and linings. 3. Check disc for runout with dial indicator. Install new or refinished disc. 4. Check and adjust rear brakes. 5. Readjust wheel bearing to specified torque. 6. Install new piston seal. 7. Drain and install correct fluid. 8. Replace.
Brake roughness or chatter (pedal pumping)	1. Excessive thickness variation of braking disc. 2. Excessive lateral runout of braking disc. 3. Rear brake drums out-of-round. 4. Excessive front bearing clearance.	1. Check disc for thickness variation using a micrometer. 2. Check disc for lateral runout with dial indicator. Install new or refinished disc. 3. Reface rear drums and check for out-of-round. 4. Readjust wheel bearings to specified torque.
Excessive pedal effort	1. Brake fluid, oil or grease on linings. 2. Incorrect lining. 3. Frozen or seized pistons. 4. Power brake malfunction.	1. Install new shoe linings as required. 2. Remove lining and install correct lining. 3. Disassemble caliper and free up pistons. 4. Replace.

Condition	Possible Cause	Corrective Action
Pull	1. Brake fluid, oil or grease on linings. 2. Unmatched linings. 3. Distorted brake shoes. 4. Frozen or seized pistons. 5. Incorrect tire pressure. 6. Front end out of alignment. 7. Broken rear spring. 8. Rear brake pistons sticking. 9. Restricted hose or line. 10. Caliper not in proper alignment to braking disc.	1. Install new shoe and linings. 2. Install correct lining. 3. Install new brake shoes. 4. Disassemble caliper and free up pistons. 5. Inflate tires to recommended pressures. 6. Align front end and check. 7. Install new rear spring. 8. Free up rear brake pistons. 9. Check hoses and lines and correct as necessary. 10. Remove caliper and reinstall. Check alignment.

ELECTRICAL

Turn Signal and Flasher Troubleshooting

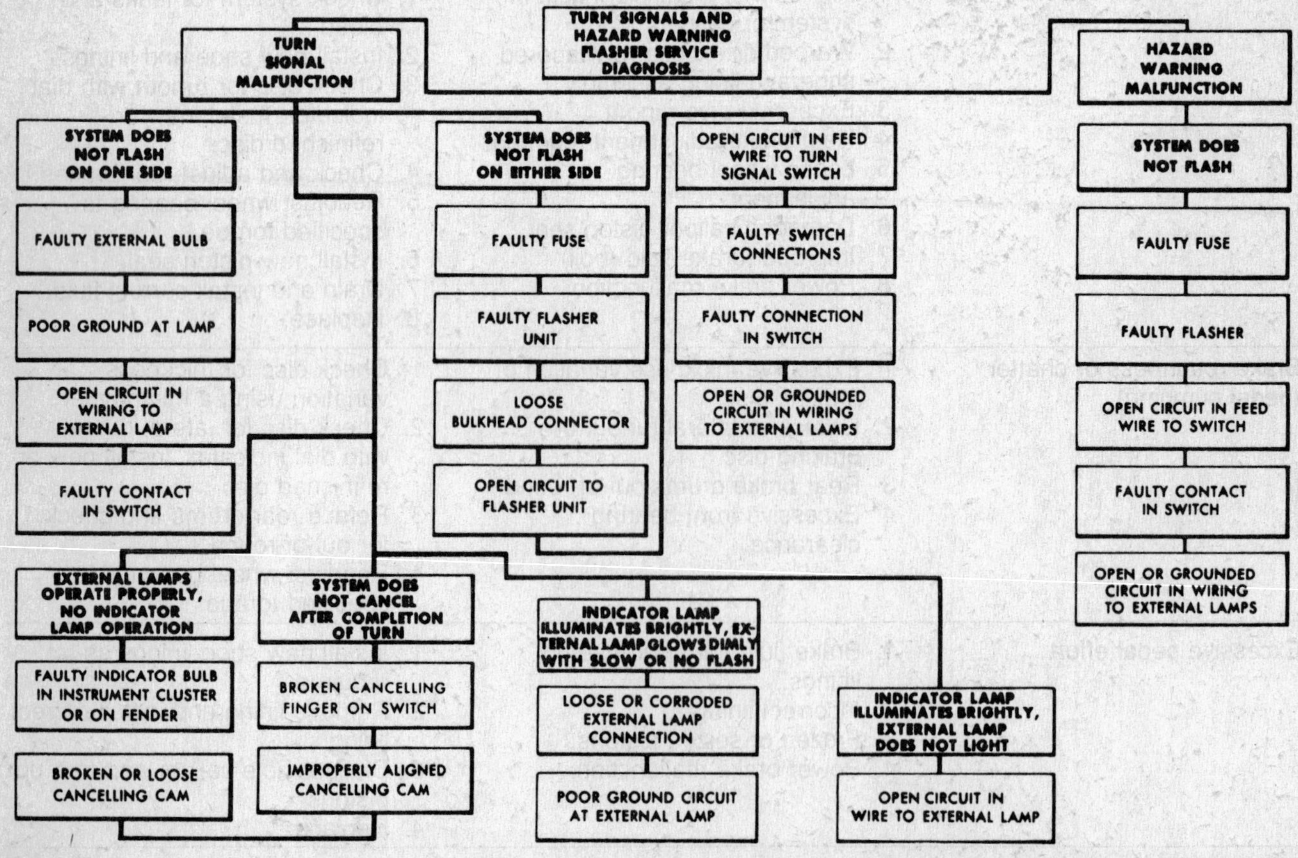

TURN SIGNALS AND HAZARD WARNING FLASHER SERVICE DIAGNOSIS

TURN SIGNAL MALFUNCTION

- **SYSTEM DOES NOT FLASH ON ONE SIDE**
  - FAULTY EXTERNAL BULB
  - POOR GROUND AT LAMP
  - OPEN CIRCUIT IN WIRING TO EXTERNAL LAMP
  - FAULTY CONTACT IN SWITCH

- **SYSTEM DOES NOT FLASH ON EITHER SIDE**
  - FAULTY FUSE
  - FAULTY FLASHER UNIT
  - LOOSE BULKHEAD CONNECTOR
  - OPEN CIRCUIT TO FLASHER UNIT

- **OPEN CIRCUIT IN FEED WIRE TO TURN SIGNAL SWITCH**
  - FAULTY SWITCH CONNECTIONS
  - FAULTY CONNECTION IN SWITCH
  - OPEN OR GROUNDED CIRCUIT IN WIRING TO EXTERNAL LAMPS

HAZARD WARNING MALFUNCTION

- **SYSTEM DOES NOT FLASH**
  - FAULTY FUSE
  - FAULTY FLASHER
  - OPEN CIRCUIT IN FEED WIRE TO SWITCH
  - FAULTY CONTACT IN SWITCH
  - OPEN OR GROUNDED CIRCUIT IN WIRING TO EXTERNAL LAMPS

- **EXTERNAL LAMPS OPERATE PROPERLY, NO INDICATOR LAMP OPERATION**
  - FAULTY INDICATOR BULB IN INSTRUMENT CLUSTER OR ON FENDER
  - BROKEN OR LOOSE CANCELLING CAM

- **SYSTEM DOES NOT CANCEL AFTER COMPLETION OF TURN**
  - BROKEN CANCELLING FINGER ON SWITCH
  - IMPROPERLY ALIGNED CANCELLING CAM

- **INDICATOR LAMP ILLUMINATES BRIGHTLY, EXTERNAL LAMP GLOWS DIMLY WITH SLOW OR NO FLASH**
  - LOOSE OR CORRODED EXTERNAL LAMP CONNECTION
  - POOR GROUND CIRCUIT AT EXTERNAL LAMP

- **INDICATOR LAMP ILLUMINATES BRIGHTLY, EXTERNAL LAMP DOES NOT LIGHT**
  - OPEN CIRCUIT IN WIRE TO EXTERNAL LAMP

Windshield Wiper Troubleshooting

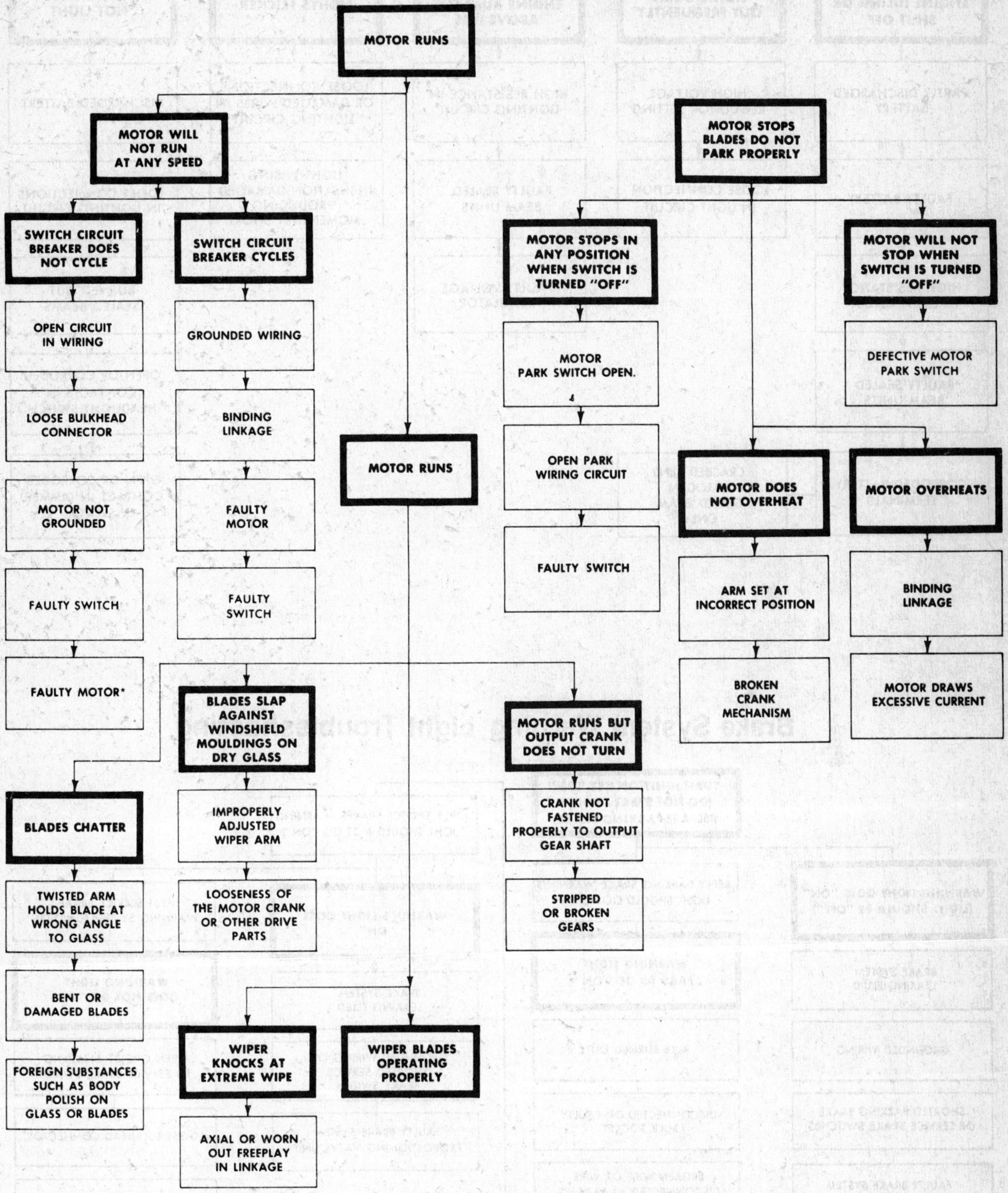
MOTOR RUNS
MOTOR WILL NOT RUN AT ANY SPEED
SWITCH CIRCUIT BREAKER DOES NOT CYCLE
OPEN CIRCUIT IN WIRING
LOOSE BULKHEAD CONNECTOR
MOTOR NOT GROUNDED
FAULTY SWITCH
FAULTY MOTOR*
SWITCH CIRCUIT BREAKER CYCLES
GROUNDED WIRING
BINDING LINKAGE
FAULTY MOTOR
FAULTY SWITCH
MOTOR STOPS BLADES DO NOT PARK PROPERLY
MOTOR STOPS IN ANY POSITION WHEN SWITCH IS TURNED "OFF"
MOTOR PARK SWITCH OPEN.
OPEN PARK WIRING CIRCUIT
FAULTY SWITCH
MOTOR WILL NOT STOP WHEN SWITCH IS TURNED "OFF"
DEFECTIVE MOTOR PARK SWITCH
MOTOR DOES NOT OVERHEAT
ARM SET AT INCORRECT POSITION
BROKEN CRANK MECHANISM
MOTOR OVERHEATS
BINDING LINKAGE
MOTOR DRAWS EXCESSIVE CURRENT
MOTOR RUNS
BLADES CHATTER
TWISTED ARM HOLDS BLADE AT WRONG ANGLE TO GLASS
BENT OR DAMAGED BLADES
FOREIGN SUBSTANCES SUCH AS BODY POLISH ON GLASS OR BLADES
BLADES SLAP AGAINST WINDSHIELD MOULDINGS ON DRY GLASS
IMPROPERLY ADJUSTED WIPER ARM
LOOSENESS OF THE MOTOR CRANK OR OTHER DRIVE PARTS
MOTOR RUNS BUT OUTPUT CRANK DOES NOT TURN
CRANK NOT FASTENED PROPERLY TO OUTPUT GEAR SHAFT
STRIPPED OR BROKEN GEARS
WIPER KNOCKS AT EXTREME WIPE
AXIAL OR WORN OUT FREEPLAY IN LINKAGE
WIPER BLADES OPERATING PROPERLY

Headlamp Troubleshooting

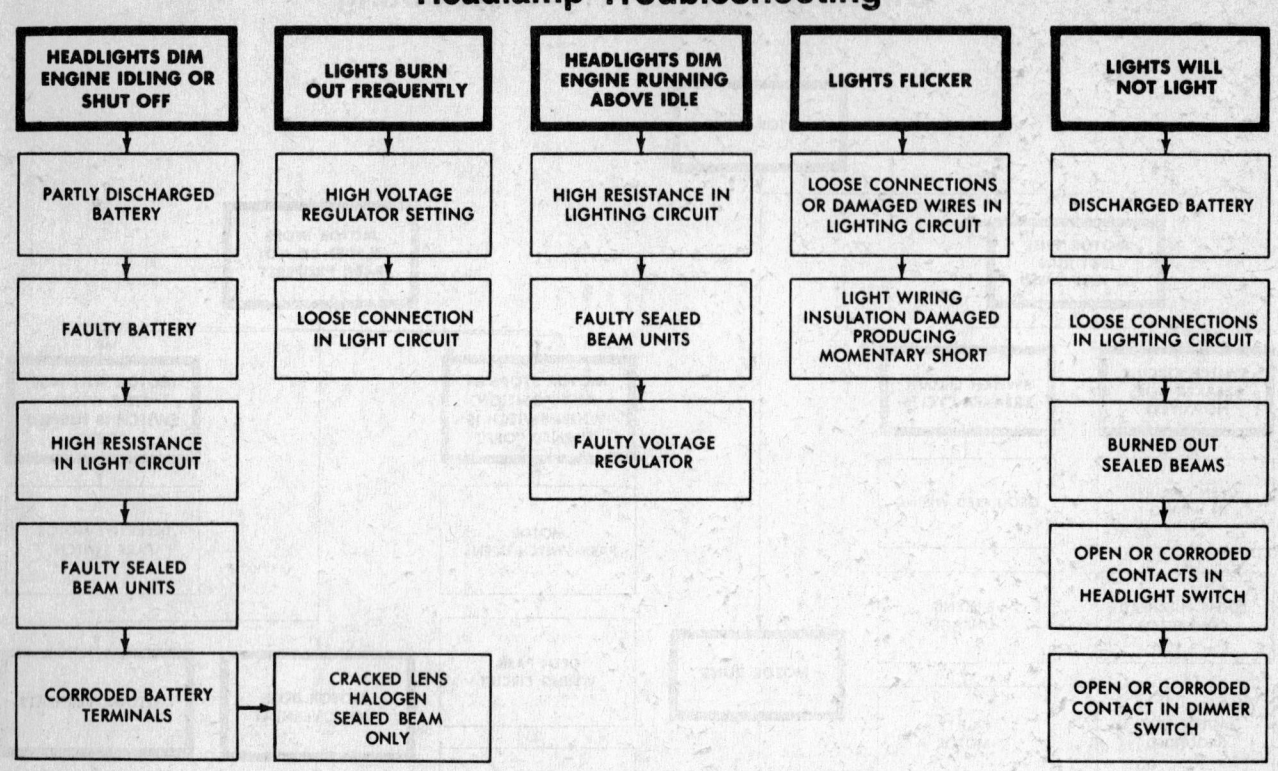

Brake System Warning Light Troubleshooting

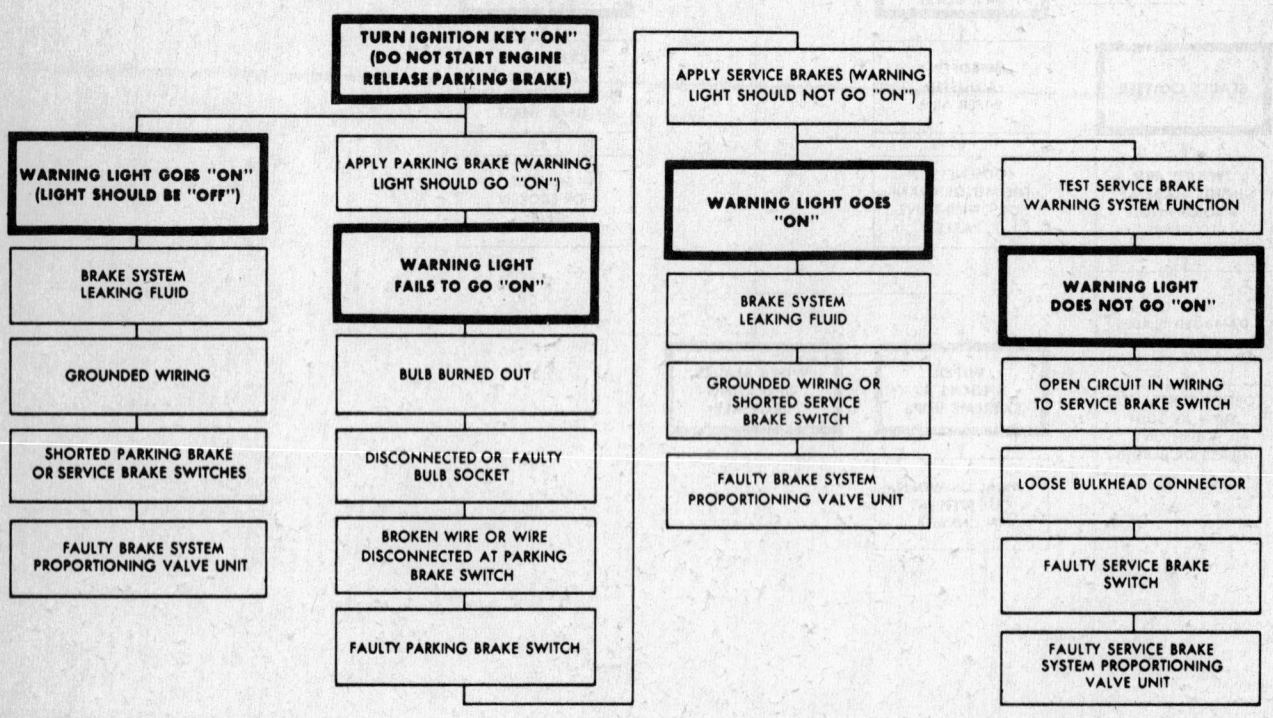

Fuel Gauge System Troubleshooting

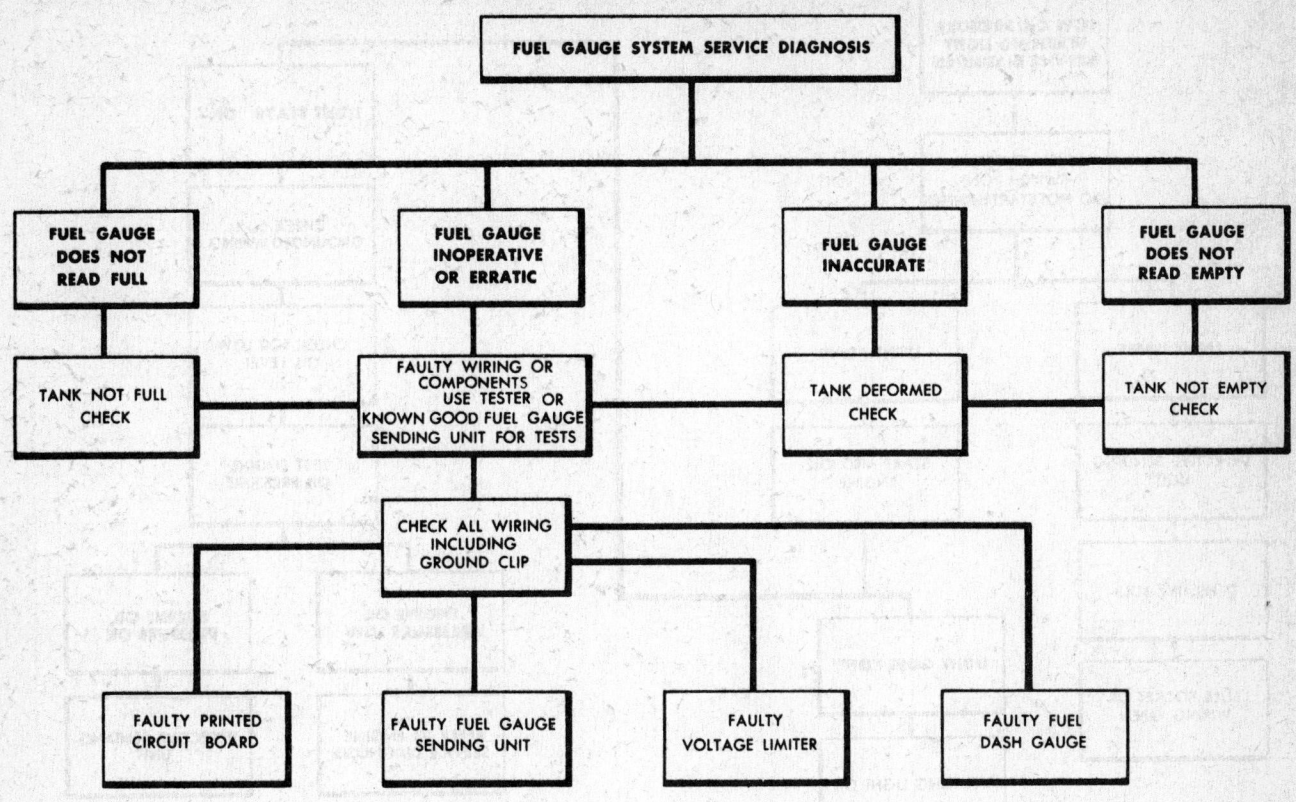

Voltage Limiter Troubleshooting

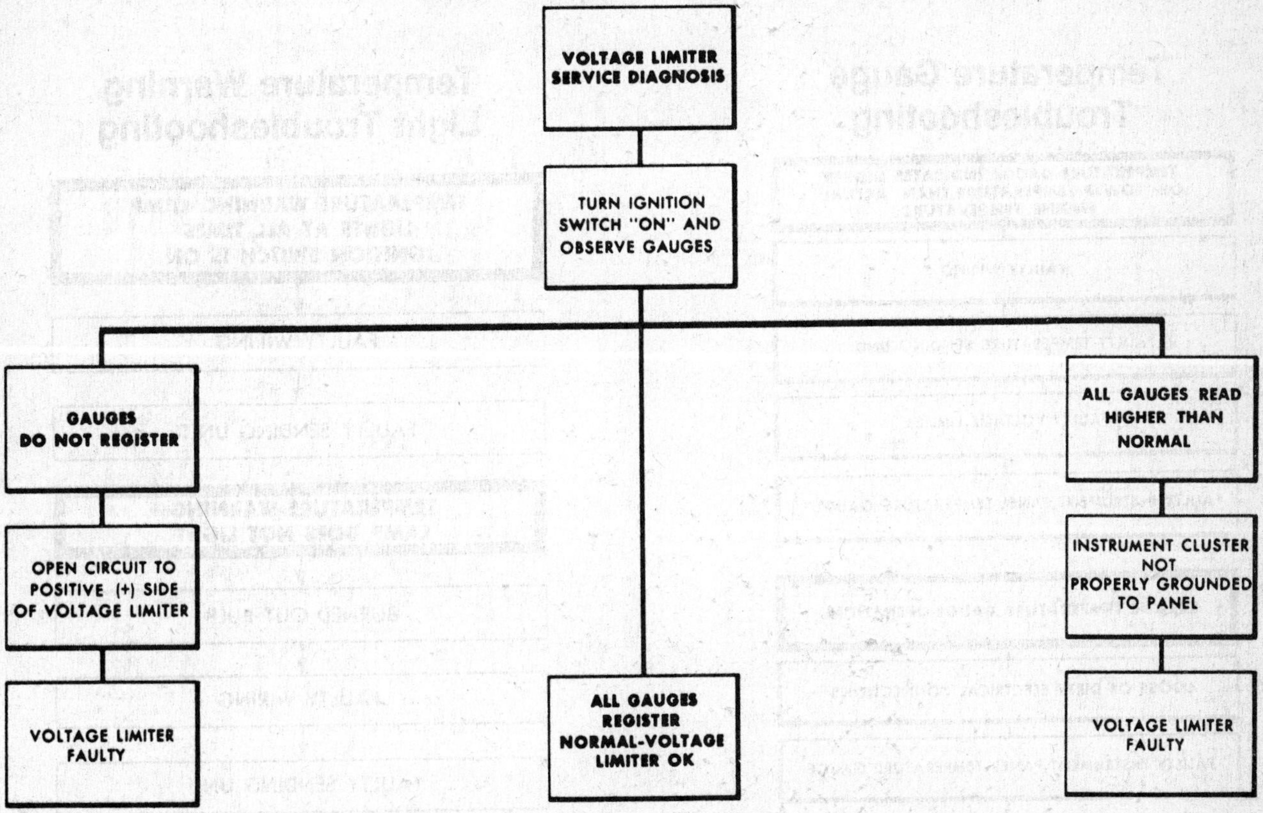

Low Oil Pressure Warning Light Troubleshooting

LOW OIL PRESSURE WARNING LIGHT SERVICE DIAGNOSIS

TURN IGNITION SWITCH "ON" (DO NOT START ENGINE)

LIGHT "OFF"

DEFECTIVE SENDING UNIT

DEFECTIVE BULB

BULB SOCKET OR WIRING OPEN

LIGHT "ON"

START AND IDLE ENGINE

LIGHT GOES "OFF"

WARNING LIGHT OK

LIGHT STAYS "ON"

CHECK FOR GROUNDED WIRING

CHECK FOR LOW OIL LEVEL

TEST ENGINE OIL PRESSURE

ENGINE OIL PRESSURE LOW

REFER TO ENGINE SERVICE DIAGNOSIS

ENGINE OIL PRESSURE OK

DEFECTIVE SENDING UNIT

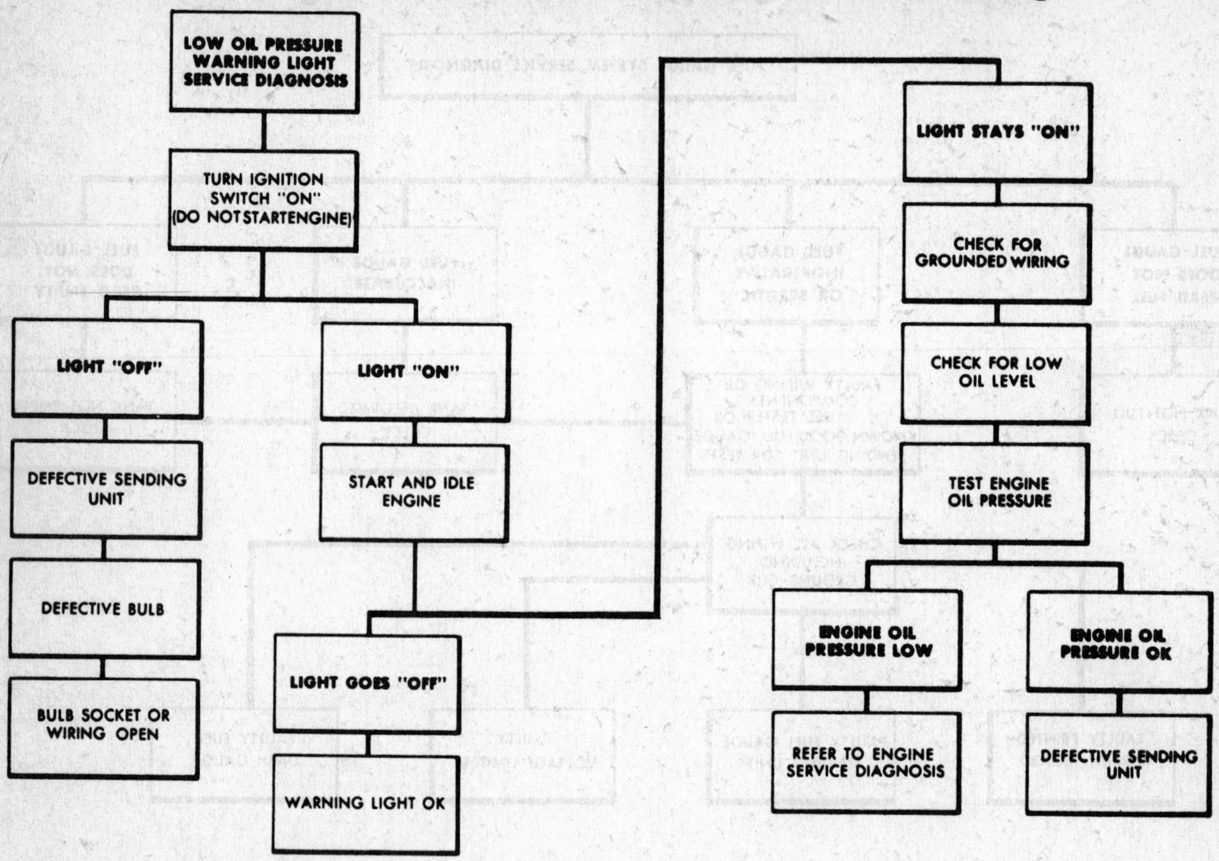

Temperature Gauge Troubleshooting

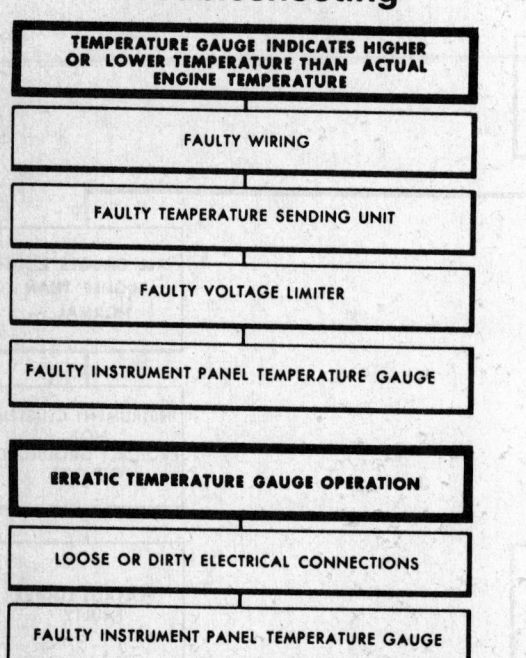

TEMPERATURE GAUGE INDICATES HIGHER OR LOWER TEMPERATURE THAN ACTUAL ENGINE TEMPERATURE

FAULTY WIRING

FAULTY TEMPERATURE SENDING UNIT

FAULTY VOLTAGE LIMITER

FAULTY INSTRUMENT PANEL TEMPERATURE GAUGE

ERRATIC TEMPERATURE GAUGE OPERATION

LOOSE OR DIRTY ELECTRICAL CONNECTIONS

FAULTY INSTRUMENT PANEL TEMPERATURE GAUGE

Temperature Warning Light Troubleshooting

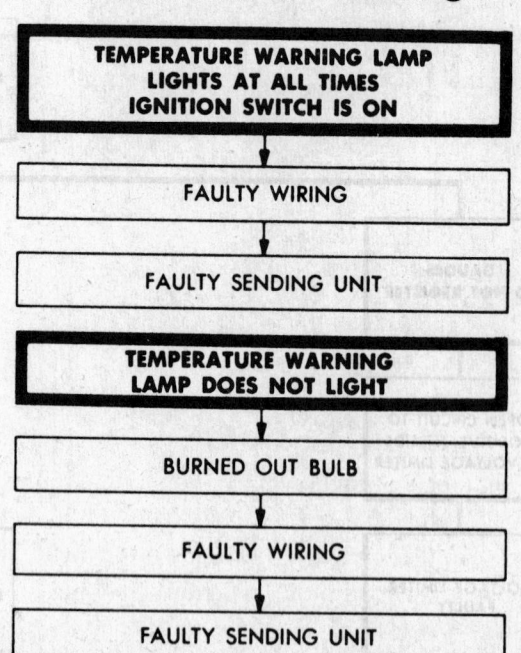

TEMPERATURE WARNING LAMP LIGHTS AT ALL TIMES IGNITION SWITCH IS ON

FAULTY WIRING

FAULTY SENDING UNIT

TEMPERATURE WARNING LAMP DOES NOT LIGHT

BURNED OUT BULB

FAULTY WIRING

FAULTY SENDING UNIT

Mechanics' Data

SI METRIC TABLES

The following tables are given in SI (International System) metric units. SI units replace both customary (English) and the older gavimetric units. The use of SI units as a new worldwide standard was set by the International Committee of Weights and Measures in 1960. SI has since been adopted by most countries as their national standard.

These tables are general conversion tables which will allow you to convert customary units, which appear in the text, into SI units.

The following are a list of SI units and the customary units, used in this book, which they replace:

To measure:	Use SI units:	Which replace (customary units):
mass	kilograms (kg)	pounds (lbs)
temperature	Celsius (°C)	Fahrenheit (°F)
length	millimeters (mm)	inches (in.)
force	newtons (N)	pounds force (lbs)
capacities	liters (l)	pints/quarts/gallons (pts/qts/gals)
torque	newton-meters (N·m)	foot pounds (ft lbs)
pressure	kilopascals (kPa)	pounds per square inch (psi)
volume	cubic centimeters (cm³)	cubic inches (cu in.)
power	kilowatts (kW)	horsepower (hp)

If you have had any prior experience with the metric system, you may have noticed units in this chart which are not familiar to you. This is because, in some cases, SI units differ from the older gravimetric units which they replace. For example, newtons (N) replace kilograms (kg) as a force unit, kilopascals (kPa) replace atmospheres or bars as a unit of pressure, and, although the units are the same, the name Celsius replaces centigrade for temperature measurement.

If you are not using the SI tables, have a look at them anyway; you will be seeing a lot more of them in the future.

DRILL SIZES IN DECIMAL EQUIVALENTS

Inch	Decimal	Wire	mm
1/64	.0156		.39
	.0157		.4
	.0160	78	
	.0165		.42
	.0173		.44
	.0177		.45
	.0180	77	
	.0181		.46
	.0189		.48
	.0197		.5
	.0200	76	
	.0210	75	
	.0217		.55
	.0225	74	
	.0236		.6
	.0240	73	
	.0250	72	
	.0256		.65
	.0260	71	
	.0276		.7
	.0280	70	
	.0292	69	
	.0295		.75
	.0310	68	
1/32	.0312		.79
	.0315		.8
	.0320	67	
	.0330	66	
	.0335		.85
	.0350	65	
	.0354		.9
	.0360	64	
	.0370	63	
	.0374		.95
	.0380	62	
	.0390	61	
	.0394		1.0
	.0400	60	
	.0410	59	
	.0413		1.05
	.0420	58	
	.0430	57	
	.0433		1.1
	.0453		1.15
	.0465	56	
3/64	.0469		1.19
	.0472		1.2
	.0492		1.25
	.0512		1.3
	.0520	55	
	.0531		1.35
	.0550	54	
	.0551		1.4
	.0571		1.45
	.0591		1.5
	.0595	53	

Inch	Decimal	Wire	mm
	.0610		1.55
1/16	.0625		1.59
	.0630		1.6
	.0635	52	
	.0650		1.65
	.0669		1.7
	.0670	51	
	.0689		1.75
	.0700	50	
	.0709		1.8
	.0728		1.85
	.0730	49	
	.0748		1.9
	.0760	48	
	.0768		1.95
5/64	.0781		1.98
	.0785	47	
	.0787		2.0
	.0807		2.05
	.0810	46	
	.0820	45	
	.0827		2.1
	.0846		2.15
	.0860	44	
	.0866		2.2
	.0886		2.25
	.0890	43	
	.0906		2.3
	.0925		2.35
	.0935	42	
3/32	.0938		2.38
	.0945		2.4
	.0960	41	
	.0965		2.45
	.0980	40	
	.0981		2.5
	.0995	39	
	.1015	38	
	.1024		2.6
	.1040	37	
	.1063		2.7
	.1065	36	
	.1083		2.75
7/64	.1094		2.77
	.1100	35	
	.1102		2.8
	.1110	34	
	.1130	33	
	.1142		2.9
	.1160	32	
	.1181		3.0
	.1200	31	
	.1220		3.1
1/8	.1250		3.17
	.1260		3.2
	.1280		3.25

Inch	Decimal	Wire	mm
	.1285	30	
	.1299		3.3
	.1339		3.4
	.1360	29	
	.1378		3.5
	.1405	28	
9/64	.1406		3.57
	.1417		3.6
	.1440	27	
	.1457		3.7
	.1470	26	
	.1476		3.75
	.1495	25	
	.1496		3.8
	.1520	24	
	.1535		3.9
	.1540	23	
5/32	.1562		3.96
	.1570	22	
	.1575		4.0
	.1590	21	
	.1610	20	
	.1614		4.1
	.1654		4.2
	.1660	19	
	.1673		4.25
	.1693		4.3
	.1695	18	
11/64	.1719		4.36
	.1730	17	
	.1732		4.4
	.1770	16	
	.1772		4.5
	.1800	15	
	.1811		4.6
	.1820	14	
	.1850	13	
	.1850		4.7
	.1870		4.75
3/16	.1875		4.76
	.1890		4.8
	.1890	12	
	.1910	11	
	.1929		4.9
	.1935	10	
	.1960	9	
	.1969		5.0
	.1990	8	
	.2008		5.1
	.2010	7	
13/64	.2031		5.16
	.2040	6	
	.2047		5.2
	.2055	5	
	.2067		5.25
	.2087		5.3

Inch	Decimal	Wire & Letter	mm
	.2090	4	
	.2126		5.4
	.2130	3	
	.2165		5.5
7/32	.2188		5.55
	.2205		5.6
	.2210	2	
	.2244		5.7
	.2264		5.75
	.2280	1	
	.2283		5.8
	.2323		5.9
	.2340	A	
15/64	.2344		5.95
	.2362		6.0
	.2380	B	
	.2402		6.1
	.2420	C	
	.2441		6.2
	.2460	D	
	.2461		6.25
	.2480		6.3
1/4	.2500	E	6.35
	.2520		6.4
	.2559		6.5
	.2570	F	
	.2598		6.6
	.2610	G	
	.2638		6.7
17/64	.2656		6.74
	.2657		6.75
	.2660	H	
	.2677		6.8
	.2717		6.9
	.2720	I	
	.2756		7.0
	.2770	J	
	.2795		7.1
9/32	.2810	K	
	.2812		7.14
	.2835		7.2
	.2854		7.25
	.2874		7.3
	.2900	L	
	.2913		7.4
	.2950	M	
	.2953		7.5
19/64	.2969		7.54
	.2992		7.6
	.3020	N	
	.3031		7.7
	.3051		7.75
	.3071		7.8
	.3110		7.9
5/16	.3125		7.93
	.3150		8.0

Inch	Decimal	Letter	mm
	.3160	O	
	.3189		8.1
	.3228		8.2
	.3230	P	
	.3248		8.25
	.3268		8.3
21/64	.3281		8.33
	.3307		8.4
	.3320	Q	
	.3346		8.5
	.3386		8.6
	.3390	R	
	.3425		8.7
11/32	.3438		8.73
	.3445		8.75
	.3465		8.8
	.3480	S	
	.3504		8.9
	.3543		9.0
	.3580	T	
	.3583		9.1
23/64	.3594		9.12
	.3622		9.2
	.3642		9.25
	.3661		9.3
	.3680	U	
	.3701		9.4
	.3740		9.5
3/8	.3750		9.52
	.3770	V	
	.3780		9.6
	.3819		9.7
	.3839		9.75
	.3858		9.8
	.3860	W	
	.3898		9.9
25/64	.3906		9.92
	.3937		10.0
	.3970	X	
	.4040	Y	
13/32	.4062		10.31
	.4130	Z	
	.4134		10.5
27/64	.4219		10.71
	.4331		11.0
7/16	.4375		11.11
	.4528		11.5
29/64	.4531		11.51
15/32	.4688		11.90
	.4724		12.0
31/64	.4844		12.30
	.4921		12.5
1/2	.5000		12.70
33/64	.5156		13.09
17/32	.5312		13.49

Inch	Decimal	mm
	.5315	13.5
35/64	.5469	13.89
	.5512	14.0
9/16	.5625	14.28
	.5709	14.5
37/64	.5781	14.68
	.5906	15.0
19/32	.5938	15.08
39/64	.6094	15.47
	.6102	15.5
5/8	.6250	15.87
	.6299	16.0
41/64	.6406	16.27
	.6496	16.5
21/32	.6562	16.66
	.6693	17.0
43/64	.6719	17.06
11/16	.6875	17.46
	.6890	17.5
45/64	.7031	17.85
	.7087	18.0
23/32	.7188	18.25
	.7283	18.5
47/64	.7344	18.65
	.7480	19.0
3/4	.7500	19.05
49/64	.7656	19.44
	.7677	19.5
25/32	.7812	19.84
	.7874	20.0
51/64	.7969	20.24
	.8071	20.5
13/16	.8125	20.63
	.8268	21.0
53/64	.8281	21.03
27/32	.8438	21.43
	.8465	21.5
55/64	.8594	21.82
	.8661	22.0
7/8	.8750	22.22
	.8858	22.5
57/64	.8906	22.62
	.9055	23.0
29/32	.9062	23.01
	.9219	23.41
59/64	.9252	23.5
15/16	.9375	23.81
	.9449	24.0
61/64	.9531	24.2
	.9646	24.5
31/32	.9688	24.6
	.9843	25.0
63/64	.9844	25.0
1	1.0000	25.4

GENERAL CONVERSION TABLE

Multiply By	To Convert	To	—
Length			—
2.54	Inches	Centimeters	.3937
25.4	Inches	Millimeters	.03937
30.48	Feet	Centimeters	.0328
.304	Feet	Meters	3.28
.914	Yards	Meters	1.094
1.609	Miles	Kilometers	.621
Volume			
.473	Pints	Liters	2.11
.946	Quarts	Liters	1.06
3.785	Gallons	Liters	.264
.016	Cubic inches	Liters	61.02
16.39	Cubic inches	Cubic cms.	.061
28.3	Cubic feet	Liters	.0353
Mass (Weight)			
28.35	Ounces	Grams	.035
.4536	Pounds	Kilograms	2.20
Area			
.645	Square inches	Square cms.	.155
.836	Square yds.	Square meters	1.196
Force			
4.448	Pounds	Newtons	.225
.138	Ft./lbs.	Kilogram/meters	7.23
1.36	Ft./lbs.	Newton-meters	.737
.112	In./lbs.	Newton-meters	8.844
Pressure			
.068	Psi	Atmospheres	14.7
6.89	Psi	Kilopascals	.145
Other			
1.104	Horsepower (DIN)	Horsepower (SAE)	.9861
.746	Horsepower (SAE)	Kilowatts (KW)	1.34
1.60	Mph	Km/h	.625
.425	Mpg	Km/1	2.35
—	To obtain	From	Multiply by

TAP DRILL SIZES

NATIONAL COARSE OR U.S.S.			NATIONAL FINE OR S.A.E.		
Screw & Tap Size / Threads Per Inch / Use Drill Number	Screw & Tap Size / Threads Per Inch / Use Drill Number		Screw & Tap Size / Threads Per Inch / Use Drill Number	Screw & Tap Size / Threads Per Inch / Use Drill Number	
No. 5 ....40....39	1/2 ...13...27/64		No. 5 ....44....37	1/2 ....20....29/64	
No. 6 ....32....36	9/16 ..12...31/64		No. 6 ....40....33	9/16 ...18....33/64	
No. 8 ....32....29	5/8 ...11...17/32		No. 8 ....36....29	5/8 ....18....37/64	
No. 10 ...24....25	3/4 ...10...21/32		No. 10 ...32....21	3/4 ....16....11/16	
No. 12 ...24....17	7/8 ....9...49/64		No. 12 ...28....15	7/8 ....14....13/16	
1/4 ...20.... 8	1 .....8...7/8		1/4 ...28.... 3	1 1/8 ..12....1 3/64	
5/16 ..18.... F	1 1/8 ..7...63/64		5/16 ..24.... 1	1 1/4 ..12....1 11/64	
3/8 ...16....5/16	1 1/4 ..7...1 7/64		3/8 ...24.... Q	1 1/2 ..12....1 27/64	
7/16 ..14.... U	1 1/2 ..6...1 11/32		7/16 ..20.... W		

ENGLISH TO METRIC CONVERSION: MASS (WEIGHT)

Current **mass** measurement is expressed in pounds and ounces (lbs. & ozs.). The metric unit of mass (or weight) is the kilogram (kg). Even although this table does not show conversion of masses (weights) larger than 15 lbs, it is easy to calculate larger units by following the data immediately below.

To convert ounces (oz.) to grams (g): multiply th number of ozs. by 28
To convert grams (g) to ounces (oz.): multiply the number of grams by .035

To convert pounds (lbs.) to kilograms (kg): multiply the number of lbs. by .45
To convert kilograms (kg) to pounds (lbs.): multiply the number of kilograms by 2.2

lbs	kg	lbs	kg	oz	kg	oz	kg
0.1	0.04	0.9	0.41	0.1	0.003	0.9	0.024
0.2	0.09	1	0.4	0.2	0.005	1	0.03
0.3	0.14	2	0.9	0.3	0.008	2	0.06
0.4	0.18	3	1.4	0.4	0.011	3	0.08
0.5	0.23	4	1.8	0.5	0.014	4	0.11
0.6	0.27	5	2.3	0.6	0.017	5	0.14
0.7	0.32	10	4.5	0.7	0.020	10	0.28
0.8	0.36	15	6.8	0.8	0.023	15	0.42

ENGLISH TO METRIC CONVERSION: TEMPERATURE

To convert Fahrenheit (°F) to Celsius (°C): take number of °F and subtract 32; multiply result by 5; divide result by 9

To convert Celsius (°C) to Fahrenheit (°F): take number of °C and multiply by 9; divide result by 5; add 32 to total

Fahrenheit (F)		Celsius (C)		Fahrenheit (F)		Celsius (C)		Fahrenheit (F)		Celsius (C)	
°F	°C	°C	°F	°F	°C	°C	°F	°F	°C	°C	°F
−40	−40	−38	−36.4	80	26.7	18	64.4	215	101.7	80	176
−35	−37.2	−36	−32.8	85	29.4	20	68	220	104.4	85	185
−30	−34.4	−34	−29.2	90	32.2	22	71.6	225	107.2	90	194
−25	−31.7	−32	−25.6	95	35.0	24	75.2	230	110.0	95	202
−20	−28.9	−30	−22	100	37.8	26	78.8	235	112.8	100	212
−15	−26.1	−28	−18.4	105	40.6	28	82.4	240	115.6	105	221
−10	−23.3	−26	−14.8	110	43.3	30	86	245	118.3	110	230
−5	−20.6	−24	−11.2	115	46.1	32	89.6	250	121.1	115	239
0	−17.8	−22	−7.6	120	48.9	34	93.2	255	123.9	120	248
1	−17.2	−20	−4	125	51.7	36	96.8	260	126.6	125	257
2	−16.7	−18	−0.4	130	54.4	38	100.4	265	129.4	130	266
3	−16.1	−16	3.2	135	57.2	40	104	270	132.2	135	275
4	−15.6	−14	6.8	140	60.0	42	107.6	275	135.0	140	284
5	−15.0	−12	10.4	145	62.8	44	112.2	280	137.8	145	293
10	−12.2	−10	14	150	65.6	46	114.8	285	140.6	150	302
15	−9.4	−8	17.6	155	68.3	48	118.4	290	143.3	155	311
20	−6.7	−6	21.2	160	71.1	50	122	295	146.1	160	320
25	−3.9	−4	24.8	165	73.9	52	125.6	300	148.9	165	329
30	−1.1	−2	28.4	170	76.7	54	129.2	305	151.7	170	338
35	1.7	0	32	175	79.4	56	132.8	310	154.4	175	347
40	4.4	2	35.6	180	82.2	58	136.4	315	157.2	180	356
45	7.2	4	39.2	185	85.0	60	140	320	160.0	185	365
50	10.0	6	42.8	190	87.8	62	143.6	325	162.8	190	374
55	12.8	8	46.4	195	90.6	64	147.2	330	165.6	195	383
60	15.6	10	50	200	93.3	66	150.8	335	168.3	200	392
65	18.3	12	53.6	205	96.1	68	154.4	340	171.1	205	401
70	21.1	14	57.2	210	98.9	70	158	345	173.9	210	410
75	23.9	16	60.8	212	100.0	75	167	350	176.7	215	414

ENGLISH TO METRIC CONVERSION: LENGTH

To convert inches (ins.) to millimeters (mm): multiply number of inches by 25.4

To convert millimeters (mm) to inches (ins.): multiply number of millimeters by .04

Inches		Decimals	Milli-meters	Inches to millimeters inches	mm	Inches		Decimals	Milli-meters	Inches to millimeters inches	mm
	1/64	0.051625	0.3969	0.0001	0.00254		33/64	0.515625	13.0969	0.6	15.24
1/32		0.03125	0.7937	0.0002	0.00508	17/32		0.53125	13.4937	0.7	17.78
	3/64	0.046875	1.1906	0.0003	0.00762		35/64	0.546875	13.8906	0.8	20.32
1/16		0.0625	1.5875	0.0004	0.01016	9/16		0.5625	14.2875	0.9	22.86
	5/64	0.078125	1.9844	0.0005	0.01270		37/64	0.578125	14.6844	1	25.4
3/32		0.09375	2.3812	0.0006	0.01524	19/32		0.59375	15.0812	2	50.8
	7/64	0.109375	2.7781	0.0007	0.01778		39/64	0.609375	15.4781	3	76.2
1/8		0.125	3.1750	0.0008	0.02032	5/8		0.625	15.8750	4	101.6
	9/64	0.140625	3.5719	0.0009	0.02286		41/64	0.640625	16.2719	5	127.0
5/32		0.15625	3.9687	0.001	0.0254	21/32		0.65625	16.6687	6	152.4
	11/64	0.171875	4.3656	0.002	0.0508		43/64	0.671875	17.0656	7	177.8
3/16		0.1875	4.7625	0.003	0.0762	11/16		0.6875	17.4625	8	203.2
	13/64	0.203125	5.1594	0.004	0.1016		45/64	0.703125	17.8594	9	228.6
7/32		0.21875	5.5562	0.005	0.1270	23/32		0.71875	18.2562	10	254.0
	15/64	0.234375	5.9531	0.006	0.1524		47/64	0.734375	18.6531	11	279.4
1/4		0.25	6.3500	0.007	0.1778	3/4		0.75	19.0500	12	304.8
	17/64	0.265625	6.7469	0.008	0.2032		49/64	0.765625	19.4469	13	330.2
9/32		0.28125	7.1437	0.009	0.2286	25/32		0.78125	19.8437	14	355.6
	19/64	0.296875	7.5406	0.01	0.254		51/64	0.796875	20.2406	15	381.0
5/16		0.3125	7.9375	0.02	0.508	13/16		0.8125	20.6375	16	406.4
	21/64	0.328125	8.3344	0.03	0.762		53/64	0.828125	21.0344	17	431.8
11/32		0.34375	8.7312	0.04	1.016	27/32		0.84375	21.4312	18	457.2
	23/64	0.359375	9.1281	0.05	1.270		55/64	0.859375	21.8281	19	482.6
3/8		0.375	9.5250	0.06	1.524	7/8		0.875	22.2250	20	508.0
	25/64	0.390625	9.9219	0.07	1.778		57/64	0.890625	22.6219	21	533.4
13/32		0.40625	10.3187	0.08	2.032	29/32		0.90625	23.0187	22	558.8
	27/64	0.421875	10.7156	0.09	2.286		59/64	0.921875	23.4156	23	584.2
7/16		0.4375	11.1125	0.1	2.54	15/16		0.9375	23.8125	24	609.6
	29/64	0.453125	11.5094	0.2	5.08		61/64	0.953125	24.2094	25	635.0
15/32		0.46875	11.9062	0.3	7.62	31/32		0.96875	24.6062	26	660.4
	31/64	0.484375	12.3031	0.4	10.16		63/64	0.984375	25.0031	27	690.6
1/2		0.5	12.7000	0.5	12.70						

ENGLISH TO METRIC CONVERSION: TORQUE

To convert foot-pounds (ft. lbs.) to Newton-meters: multiply the number of ft. lbs. by 1.3

To convert inch-pounds (in. lbs.) to Newton-meters: multiply the number of in. lbs. by .11

in lbs	N-m	in lbs	N-m	in lbs	N-m	in lbs	N-m	in lbs	N-m
0.1	0.01	1	0.11	10	1.13	19	2.15	28	3.16
0.2	0.02	2	0.23	11	1.24	20	2.26	29	3.28
0.3	0.03	3	0.34	12	1.36	21	2.37	30	3.39
0.4	0.04	4	0.45	13	1.47	22	2.49	31	3.50
0.5	0.06	5	0.56	14	1.58	23	2.60	32	3.62
0.6	0.07	6	0.68	15	1.70	24	2.71	33	3.73
0.7	0.08	7	0.78	16	1.81	25	2.82	34	3.84
0.8	0.09	8	0.90	17	1.92	26	2.94	35	3.95
0.9	0.10	9	1.02	18	2.03	27	3.05	36	4.0/

ENGLISH TO METRIC CONVERSION: TORQUE

Torque is now expressed as either foot-pounds (ft./lbs.) or inch-pounds (in./lbs.). The metric measurement unit for torque is the Newton-meter (Nm). This unit—the Nm—will be used for all SI metric torque references, both the present ft./lbs. and in./lbs.

ft lbs	N-m	ft lbs	N-m	ft lbs	N-m	ft lbs	N-m
0.1	0.1	33	44.7	74	100.3	115	155.9
0.2	0.3	34	46.1	75	101.7	116	157.3
0.3	0.4	35	47.4	76	103.0	117	158.6
0.4	0.5	36	48.8	77	104.4	118	160.0
0.5	0.7	37	50.7	78	105.8	119	161.3
0.6	0.8	38	51.5	79	107.1	120	162.7
0.7	1.0	39	52.9	80	108.5	121	164.0
0.8	1.1	40	54.2	81	109.8	122	165.4
0.9	1.2	41	55.6	82	111.2	123	166.8
1	1.3	42	56.9	83	112.5	124	168.1
2	2.7	43	58.3	84	113.9	125	169.5
3	4.1	44	59.7	85	115.2	126	170.8
4	5.4	45	61.0	86	116.6	127	172.2
5	6.8	46	62.4	87	118.0	128	173.5
6	8.1	47	63.7	88	119.3	129	174.9
7	9.5	48	65.1	89	120.7	130	176.2
8	10.8	49	66.4	90	122.0	131	177.6
9	12.2	50	67.8	91	123.4	132	179.0
10	13.6	51	69.2	92	124.7	133	180.3
11	14.9	52	70.5	93	126.1	134	181.7
12	16.3	53	71.9	94	127.4	135	183.0
13	17.6	54	73.2	95	128.8	136	184.4
14	18.9	55	74.6	96	130.2	137	185.7
15	20.3	56	75.9	97	131.5	138	187.1
16	21.7	57	77.3	98	132.9	139	188.5
17	23.0	58	78.6	99	134.2	140	189.8
18	24.4	59	80.0	100	135.6	141	191.2
19	25.8	60	81.4	101	136.9	142	192.5
20	27.1	61	82.7	102	138.3	143	193.9
21	28.5	62	84.1	103	139.6	144	195.2
22	29.8	63	85.4	104	141.0	145	196.6
23	31.2	64	86.8	105	142.4	146	198.0
24	32.5	65	88.1	106	143.7	147	199.3
25	33.9	66	89.5	107	145.1	148	200.7
26	35.2	67	90.8	108	146.4	149	202.0
27	36.6	68	92.2	109	147.8	150	203.4
28	38.0	69	93.6	110	149.1	151	204.7
29	39.3	70	94.9	111	150.5	152	206.1
30	40.7	71	96.3	112	151.8	153	207.4
31	42.0	72	97.6	113	153.2	154	208.8
32	43.4	73	99.0	114	154.6	155	210.2